KIRK-OTHMER

CONCISE ENCYCLOPEDIA OF CHEMICAL TECHNOLOGY

4th EDITION

KIRK-OTHMER

CONCISE ENCYCLOPEDIA OF CHEMICAL TECHNOLOGY

4th EDITION

A WILEY-INTERSCIENCE PUBLICATION

John Wiley & Sons, Inc.

NEW YORK • CHICHESTER • WEINHEIM • BRISBANE • SINGAPORE • TORONTO

Library of Congress Cataloging-in-Publication Data

Kirk-Othmer concise encyclopedia of chemical technology.—4th ed.
p. cm.
Executive editor, J. Kroschwitz.
"A Wiley-Interscience publication."
Abridged version of the 26 volume Kirk-Othmer Encyclopedia of chemical technology. 4th ed. New York: Wiley, c1991–c1998.
Includes index.
ISBN 0-471-29698-8 (alk. paper)
1. Chemistry, Technical—Encyclopedias. I. Kroschwitz, Jacqueline I. II. Encyclopedia of chemical technology. III. Title: Encyclopedia of chemical technology. IV. Title: Concise encyclopedia of chemical technology.

TP9.K54 1999
660′.03–dc21 98-50686
CIP

Printed in the United States of America

PREFACE

This *Concise Encyclopedia of Chemical Technology* is a condensation of the Fourth Edition of the world-renowned *Kirk-Othmer Encyclopedia of Chemical Technology*, published in 27 volumes (25 A–Z volumes plus one supplement volume and one index volume).

The hallmark of the full *Kirk-Othmer Encyclopedia of Chemical Technology* (known to many in the field simply as Kirk-Othmer), and now in its fourth edition, is its excellent coverage of the entire field of chemical technology, in a format of A–Z articles, written by distinguished subject experts under the direction and management of an experienced in-house staff.

In order to produce the *Concise Encyclopedia of Chemical Technology*, a staff of experienced science/chemistry editors reads, reviews, analyzes, and condenses the articles from Kirk-Othmer. The goal of the condensation is to retain basic, fundamental information, as well as key portions of more advanced material—and to provide this material in a form that is even more convenient, more accessible, less voluminous, and less expensive than Kirk-Othmer.

The Fourth Edition of the full Kirk-Othmer was published volume-by-volume, beginning in 1991 and concluding in early 1998. It updates and revises articles from the previous edition, and also has over 150 new articles, including AERATION (BIOTECHNOLOGY); AERATION (WATER TREATMENT); ANTIAGING AGENTS; ANTIVIRAL AGENTS; BIOSENSORS; BLOOD, ARTIFICIAL; CELL CULTURE TECHNOLOGY; CHEMOMETRICS; CAD/CAM; DATABASES; NONLINEAR OPTICAL MATERIALS; RENEWABLE ENERGY RESOURCES; SOL-GEL TECHNOLOGY; and SPACE PROCESSING. The condensation faithfully incorporates this updated, revised, and new material.

The present concise edition is the successor to the previous concise edition, which was published in 1985 and was based on the Kirk-Othmer Third Edition. As with the earlier Concise, all the important chemical industries are covered. The condensed articles retain not only text but also illustrations, tables, graphs, and key references. New topics that appear in the Supplement Volume of Kirk-Othmer are incorporated into the Concise in their appropriate alphabetical order. A representative cross-section of these new late-breaking articles in the Fourth Edition includes the following titles: AEOGELS; BIOREMEDIATION; LITHOGRAPHIC RESISTS; MOLECULAR MODELING; MOLECULAR RECOGNITION; NANOTECHNOLOGY; SMART MATERIALS; SONOCHEMISTRY; and WATER-SOLUABLE POLYMERS.

Feedback since 1985 has indicated that the first Concise was used (and is still used) by many as a first source for chemical information, serving as a collection of abstracts or an index, providing basic information, and directing the user to the full work or to references cited for additional information. We hope that this new Concise will be many things to many people and that it will serve you, the reader, as a useful chemical reference for many years to come and earn a place on your bookshelf close at hand, where it can serve as a guidepost as you traverse the world of chemical information.

THE EDITORS

EDITORIAL STAFF

Executive Editor: **Jacqueline I. Kroschwitz**
Editor: **Mary Howe-Grant**
Editors: **Paula M. Siegel** **Betty Richman** **Elizabeth Zayatz** **Lawrence Altieri**
Managing Editors: **Michalina Bickford** **Lindy Humphreys**

CONTRIBUTORS

Claudio Abaecherli, *Lonza, Inc., Visp, Switzerland,* Ketenes, ketene dimers, and related substances
Michael M. Abbott, *Rensselaer Polytechnic Institute, Troy, New York,* Thermodynamics
Terry E. Acree, *Cornell University, Geneva, New York,* Flavor characterization
Jack H. Adams, *Eagle-Picher Industries, Inc., Quapaw, Oklahoma,* Germanium and germanium compounds
Valery Addes, *Inland Steel Company, East Chicago, Illinois,* Carbonization (under Coal conversion processes)
Rick L. Adkins, *Bayer Corporation, New Martinsville, West Virginia,* Nitrobenzene and nitrotoluenes
Rakesh Agrawal, *Air Products and Chemicals, Inc., Allentown, Pennsylvania,* Cryogenics
Elizabeth R. Aguinaldo, *FMC Corporation, Princeton, New Jersey,* Barium compounds
Jeffrey A. Ahlgren, *U.S. Department of Agriculture, Peoria, Illinois,* Microbial polysaccharides
Gerpreet S. Ahluwalia, *Gillette Research Institute, Gaithersburg, Maryland,* Therapeutic (under Enzyme applications)
Satinder Ahuja, *Ahuja Consulting, Monsey, New York,* Trace and residue analysis
Dorothy Aiken, *Colin A. Houston & Associates, Inc., Mamaroneck, New York,* Market and marketing research
Fazlul Alam, *U.S. Borax Research Corp., Anaheim, California,* Boron halides (under Boron compounds)
Jean-François Alary, *Université de Sherbrooke, Sherbrooke, Quebec, Canada,* Asbestos
Lyle F. Albright, *Purdue University, West Lafayette, Indiana,* Nitration
R. D. Allen, *IBM Research Division, San Jose, California,* Lithographic resists
Tejraj M. Aminabhavi, *Southwest Texas State University, San Marcos, Texas,* Heat-resistant polymers
Bijan Amini, *E. I. du Pont de Nemours & Co., Inc., Deepwater, New Jersey,* Aniline and its derivatives (under Amines, aromatic)
Lowell Ray Anderson, *International Specialty Products, Wayne, New Jersey,* Pyrrole and pyrrole derivatives
Brian Anthony, *The Brewer Company,* Roofing materials
Morton Antler, *Contact Consultants Inc., Columbus, Ohio,* Electrical connectors
Anthony Anton, *E. I. du Pont de Nemours & Co., Inc., Wilmington, Delaware,* Fibers (under Polyamides)
D. H. Antonsen, *International Nickel, Inc., Wycoff, New Jersey,* Nickel compounds
Joseph M. Antonucci, *National Institute of Standards and Technology, Gaithersburg, Maryland,* Dental materials
Frederick J. Antosz, *The Upjohn Company, Kalamazoo, Michigan,* Ansamacrolides (under Antibiotics)
Kazumi Araki, *University of East Asia, Shimonoseki, Japan,* Survey (under Amino acids)
Hazel Aranha-Creado, *Pall Corporation, Port Washington, New York,* Microbial and viral filtration
G. F. Archer, *ASARCO, Inc., Denver, Colorado,* Antimony and antimony alloys
Ronald D. Archer, *University of Massachusetts, Amherst, Massachusetts,* Coordination compounds
Steven Arcidaicono, *U.S. Army Natick Research, Development, & Engineering Center, Natick, Massachusetts,* Silk
Barry Arkles, *Gelest, Inc., Tullytown, Pennsylvania,* Silanes; Silicon esters (under Silicon compounds)
Alton F. Armington, *Consultant, Lexington, Massachusetts,* Synthetic quartz crystals (under Silica)
Mohammad Aslam, *Hoechst-Celanese Corporation, Corpus Christi, Texas,* Esterification
Darryl C. Aubrey, *Chem Systems, Inc., Tarrytown, New York,* Petrochemicals (under Feedstocks)
Thomas A. Augurt, *Propper Manufacturing Company, Inc., Long Island City, New York,* Sterilization techniques
Carmen Avendaño, *Universidad Complutense, Madrid, Spain,* Hydantoin and its derivatives
Mitchell A. Avery, *University of Mississippi, University,* Hormones, adrenal-cortical
Amos A. Avidan, *Mobil Research and Development Corporation, Paulsboro, New Jersey,* Fluidization
Darlene M. Back, *Union Carbide Corporation, Danbury, Connecticut,* Ethylene oxide polymers (under Polyethers)

Marvin O. Bagby, *USDA National Center for Agricultural Utilization Research, Peoria, Illinois,* Survey (under Carboxylic acids)

Webb I. Bailey, *Air Products and Chemicals, Inc., Allentown, Pennsylvania,* Halogens (under Fluorine compounds, organic)

John K. Baird, *Kelco, Division of Merck & Company, Inc., San Diego, California,* Gums

Malcolm H. I. Baird, *McMaster University, Hamilton, Ontario, Canada,* Extraction, liquid–liquid

William X. Bajzer, *Dow Corning Corporation, Midland, Michigan,* Introduction; Poly(fluorosilicones) (both under Fluorine compounds, organic)

C. S. L. Baker, *Malaysian Rubber Producers' Research Association, Brickendonbury, U.K.,* Rubber, natural

Frederick S. Baker, *Westvaco Corp., North Charleston, South Carolina,* Activated carbon (under Carbon)

Richard W. Baker, *Membrane Technology & Research, Inc., Menlo Park, California,* Membrane technology

Robert Bakish, *Fairleigh Dickinson University, Englewood, New Jersey,* Supply and desalination (under Water)

Robert M. Baldwin, *Colorado School of Mines, Golden, Colorado,* Liquefaction (under Coal conversion processes)

Douglas A. Balentine, *Lipton, Englewood Cliffs, New Jersey,* Tea

Lawrence E. Ball, *BP Research, Cleveland, Ohio,* Survey and SAN (under Acrylonitrile polymers)

Bryan Ballantyne, *Union Carbide Corporation,* Toxicology

Pamela Banks-Lee, *North Carolina State University, Raleigh,* Testing (under Textiles)

Brian Bannister, *The Upjohn Company, Kalamazoo, Michigan,* Lincosaminides (under Antibiotics)

Kris Baranwal, *ARDL,* Rubber compounding

Cynthia S. Barcelon-Yang, *E.I. du Pont de Nemours & Co., Inc., Wilmington, Delaware,* Information retrieval

Bruce Barden, *GenCorp Polymer Products, Columbus, Mississippi,* Coated fabrics

Jay A. Bardole, *Vincennes University, Vincennes, Indiana,* Paint and finish removers (under Paint)

Joseph Barendt, *Callery Chemical Company, Pittsburgh, Pennsylvania,* Boron hydrides, commercial aspects (under Boron compounds)

Colin Barker, *University of Tulsa, Oklahoma,* Nomenclature in the petroleum industry; Origin of petroleum (under Petroleum)

J. W. Barlow, *University of Texas, Austin,* Polymer blends

Scott D. Barnicki, *Eastman Chemical Company, Kingsport, Tennessee,* Separations process synthesis

Michael G. Barrett, *Consultant, Columbia, Maryland,* Laboratory information management systems

Robert G. Bartolo, *The Procter & Gamble Company, Cincinnati, Ohio,* Soap

Robert W. Bassemir, *Sun Chemical Corporation, Carlstadt, New Jersey,* Inks

Andrew L. Bastone, *ISORCA, Inc., Granville, Ohio,* Reinforced plastics

Joseph J. Batelka, *Consultant, Savannah, Georgia,* Converting (under Packaging)

Roger G. Bates, *University of Florida, Gainesville,* Hydrogen-ion activity

Brett O. Bateup, *CSIRO, Belmont, Geelong, Australia,* Wool

George L. Batten, Jr., *Georgia-Pacific Corporation, Decatur, Georgia,* Papermaking additives

O. A. Battista, *Research Services Corporation, Fort Worth, Texas,* Cellulose

Ronald S. Bauer, *Shell Development Company, Houston, Texas,* Thermosets (under Composite materials, polymer-matrix)

William Bauer, Jr., *Rohm and Haas Company, Spring House, Pennsylvania,* Acrylic acid and derivatives

Ulrich Baurmeister, *Enka AG, Ohder, Germany,* Dialysis

William F. Beach, *Alpha Metals, Bridgewater, New Jersey,* Xylylene polymers

A. Bean, *Sun Chemical Corporation, Carlstadt, New Jersey,* Inks

Mark C. Bean, *D. D. Bean & Sons Company, Jaffrey, New Hampshire,* Matches

S. L. Bean, *General Chemical Corporation, Claymont, Delaware,* Thiosulfates

John Becher, *CDC Drug Service, Atlanta, Georgia,* Anthelmintics (under Antiparasitic agents)

Donald A. Becker, *National Institute for Standards and Technology, Gaithersburg, Maryland,* Oil (under Recycling)

Gerald W. Becker, *Lilly Research Laboratories, Indianapolis, Indiana,* Human growth hormone (under Hormones)

Gary Beebe, *Rohm and Haas Company, Bristol, Pennsylvania,* Colorants for plastics

Donald H. Beermann, *Cornell University, Ithaca, New York,* Animal (under Growth regulators)

William R. Bell, Jr., *The Johns Hopkins University Hospital, Baltimore, Maryland,* Blood, coagulants and anticoagulants

James Bellows, *Westinghouse Electric Corporation, Orlando, Florida,* Steam

James N. BeMiller, *Whistler Center for Carbohydrate Research, Purdue University, West Lafayette, Indiana,* Carbohydrates

Laszlo Beres, *DuPont-NEN Products, Boston, Massachusetts,* Radioactive tracers

Ronald L. Berglund, *M. W. Kellogg Company, Houston, Texas,* Exhaust control, industrial

Oswald R. Bergmann, *E. I. du Pont de Nemours & Co., Inc., Deepwater, New Jersey,* Explosively clad metals (under Metallic coatings)

Dwight H. Bergquist, *Henningsen Foods, Inc., Omaha, Nebraska,* Eggs

Eric M. Bergtraun, *National Semiconductor, Santa Clara, California,* Maintenance

Bret Berner, *CIBA-GEIGY Corporation, Ardsley, New York,* Drug delivery systems

Peter R. Bernstein, *ICI Americas Corporation, Wilmington, Delaware,* Antiasthmatic agents

Yvette Berry, *Reckitt & Colman Inc., Wayne, New Jersey,* Odor modification

Noelie R. Bertoniere, *USDA, Southern Regional Research Center, New Orleans, Louisiana,* Cellulose; Cotton

Bruce M. Bertram, *Salt Institute, Alexandria, Virginia,* Sodium halides–sodium chloride (under Sodium compounds)

Kenneth E. Bett, *University of London,* High pressure technology

W. D. Betts, *Tar Industries Services, Chesterfield, U.K.,* Tar and pitch

Roop S. Bhakuni, *The Goodyear Tire & Rubber Company, Akron, Ohio,* Tire cord

D. Bhattacharyya, *University of Kentucky, Lexington,* Reverse osmosis

Gabriel Bialy, *National Institute of Health, Bethesda, Maryland,* Contraceptives

Robert M. Biefeld, *Sandia National Laboratories, Albuquerque, New Mexico,* Compound semiconductors (under Semiconductors)

Walter B. Bienert, *Dynatherm Corporation, Lancaster, Pennsylvania,* Heat pipes (under Heat-exchange technology)

Joseph F. Bieron, *Occidental Chemical Corporation, Grand Island, New York,* Benzyl chloride, benzal chloride, and benzotrichloride (under Chlorocarbons and chlorohydrocarbons)

Ernst Billig, *Union Carbide Corporation, South Charleston, West Virginia,* Butyl alcohols; Butyraldehydes; Oxo process

Roger E. Billings, *International Academy of Science, Independence, Missouri,* Hydrogen energy

Gary T. Blair, *Haarmann & Reimer Corporation, Elkhart, Indiana; Springfield, New Jersey,* Citric acid; Hydroxy dicarboxylic acids

James O. Bledsoe, Jr., *Bush Boake Allen, Inc., Jacksonville, Florida,* Terpenoids

Alan Bleier, *Oak Ridge National Laboratory, Oak Ridge, Tennessee,* Colloids

Seymour S. Block, *University of Florida, Gainesville,* Disinfectants and antiseptics

George Blomgren, *Eveready Battery Co., Inc., West Lake, Ohio,* Primary cells (under Batteries)

Bob Blumenthal, *Noah Technologies Corporation, San Antonio, Texas,* Thallium and thallium compounds

Christopher Boerner, *Washington University, St. Louis, Missouri,* Introduction (under Recycling)

Claudio Boffito, *Saes Getters, SpA, Milan, Italy,* Barium

Paul M. Boisde, *Rhône-Poulenc, Inc., Princeton, New Jersey,* Coumarin

Jerry Boisvert, *Magnetrol International, Inc., Downers Grove, Illinois,* Liquid level measurement

John E. Boliek, *E. I. du Pont de Nemours & Co., Inc., Wilmington, Delaware,* Elastomeric (under Fibers)

Allen F. Bollmeier, Jr., *ANGUS Chemical Company, Northbrook, Illinois,* Alkanolamines from nitro alcohols; Nitro alcohols; Nitroparaffins

Tilak V. Bommaraju, *Occidental Chemical Corporation, Niagara Falls, New York; Grand Island, New York,* Chlorine and sodium hydroxide (under Alkali and chlorine products); Hydrogen chloride

Daniel P. Bonner, *Bristol-Myers Squibb Pharmaceutical Research Institute, Princeton, New Jersey,* Monobactams (under Antibiotics, β-lactams)

Jeffrey T. Books, *Ethyl Corporation, Baton Rouge, Louisiana,* Phosphazenes (under Elastomers, synthetic)

E. R. Booser, *Consulting Engineer, Scotia, New York,* Bearing materials; Lubrication and lubricants

Karyn A. Booth, *Donelan, Cleary, Wood & Maser P.C., Washington, D.C.,* Transportation

John K. Borchardt, *Shell Development Company, Houston, Texas,* Enhanced oil recovery (under Petroleum); Paper (under Recycling); Recycling, plastics

Donald Borders, *American Cyanamid Company, Pearl River, New York,* Survey (under Antibiotics)

Sebastian R. Borrello, *Texas Instruments, Inc., Dallas, Texas,* Photodetectors; Thermography

Walter H. Bortle, *General Chemical Corporation, Claymont, Delaware,* Sodium nitrite (under Sodium compounds)

Barbara H. Bory, *Lever Company, Edgewater, New Jersey,* Surfactants

John C. Bost, *The Dow Chemical Company, Midland, Michigan,* Halogenated derivatives (under Acetic acid and derivatives)

Max M. Boudakian, Chemical Consultant, *Pittsford, New York,* Fluorinated aromatic compounds (under Fluorine compounds, organic)

Randy J. Bowers, *Edison Welding Institute, Columbus, Ohio,* Welding

David C. Boyd, *Corning Inc., Corning, New York,* Glass

Carl R. Bozzuto, *ABB Kreisinger Development Laboratory, Windsor, Connecticut,* Furnaces, fuel-fired

Judith M. Bradow, *U.S. Department of Agriculture, New Orleans, Louisiana,* Herbicides

Robert F. Brady, Jr., *U.S. Naval Research Laboratory, Washington, D.C.,* Coatings, marine

John Braithwaite, *Union Carbide Chemicals and Plastics Company, Inc., Texas City, Texas,* Ketones

Donald E. Bray, *Texas A&M University, College Station,* Nondestructive evaluation

James F. Brazdil, *BP Research, Cleveland, Ohio,* Acrylonitrile

John F. Breedis, *Olin Corporation, New Haven, Connecticut,* Wrought copper and wrought copper alloys (under Copper alloys)

Joseph Breen, *U.S. Environmental Protection Agency, Washington, D.C.,* Industrial toxicology (under Lead compounds)

A. B. Brennan, *University of Florida, Gainesville,* Glasses, organic–inorganic hybrids

Thomas E. Breuer, *Humko Chemical, Memphis, Tennessee,* Dimer acids

James A. Brient, *Merichem Company, Houston, Texas,* Naphthenic acids

Dennis W. Brinkman, *Safety-Kleen Corporation, Elk Grove Village, Illinois,* Oil (under Recycling)

Angelo Brisimitzakis, *GE Plastics, Pittsfield, Massachusetts,* ABS resins (under Acrylonitrile polymers)

H. Britt, *Aspen Technology, Inc., Cambridge, Massachusetts,* Computer-aided engineering (CAE)

Ralph Brodd, *Gould, Inc., East Lake, Ohio,* Introduction (under Batteries)

Aaron L. Brody, *Rubbright-Brody, Inc., Devon, Pennsylvania,* Food packaging

Irena Bronstein, *Tropix, Inc., Bedford, Massachusetts,* Chemiluminescence (under Luminescent materials)

Robert T. Brooker, *Olin Chemicals, Charleston, Tennessee,* Perchloric acid and perchlorates

Charlie R. Brooks, *University of Tennessee, Knoxville,* Case hardening (under Metal surface treatments)

Edward Brown, *Kalama Chemical Incorporated, Kalama, Washington,* Benzaldehyde; Benzoic acid

Donald E. Brownlee, *University of Washington, Seattle,* Extraterrestrial materials

Evelyn Brownlee, *E.I. du Pont de Nemours & Co., Inc., Wilmington, Delaware,* Information retrieval

J. R. Brummer, *FMC Corporation, Princeton, New Jersey,* Phosphorus

Daniel J. Brunelle, *General Electric, Schenectady, New York,* Polycarbonates

P. F. Bryan, *Chevron Research and Technology Company, Richmond, California,* BTX processing

David R. Bryant, *Union Carbide Corporation, South Charleston, West Virginia,* Oxo process

James G. Bryant, *Standard Chlorine of Delaware, Inc., Delaware City,* Chlorinated benzenes (under Chlorocarbons and chlorohydrocarbons)

R. C. Buchanan, *University of Illinois at Urbana-Champaign, Urbana,* Ceramics as electrical materials

John E. Bujake, *Brown-Forman Corporation, Louisville, Kentucky,* Beverage spirits, distilled

Daniel B. Bullen, *Iowa State University, Ames,* Nuclear fuel reserves (under Nuclear reactors)

Kathryn R. Bullock, *AT&T Bell Labs, Mesquite, Texas,* Lead–acid (under Batteries, secondary cells)

David Burdick, *Hoffmann-La Roche Inc., Nutley, New Jersey,* Pyridoxine; Thiamine (both under Vitamins)

Thomas H. Burgess, *Fischer & Porter Company, Warminster, Pennsylvania,* Flow measurement

Joseph C. Burnett, *Huntsman Specialty Chemical Corporation, St. Louis, Missouri,* Maleic anhydride, maleic acid, and fumaric acid

Richard Burt, *Tantalum Mining Corporation, Lac Du Bonnet, Manitoba, Canada,* Cesium and cesium compounds

David R. Bush, *PPG Industries, Inc., New Martinsville, West Virginia,* Sodium sulfides (under Sodium compounds)

E. P. Butler, *National Institute of Standards and Technology, Gaithersburg, Maryland,* Ceramic-matrix (under Composite materials)

M. A. Butler, *Sandia National Laboratories, Albuquerque, New Mexico,* Sensors

David Butts, *Great Salt Lake Minerals and Chemical Corporation, Ogden, Utah,* Chemicals from brine; Sodium sulfates (under Sodium compounds)

Christian Butzke, *University of California, Davis,* Wine

David F. Cadogan, *European Council for Plasticizers and Intermediates, Brussels, Belgium,* Plasticizers

Elton J. Cairns, *University of California, Berkeley,* Fuel cells

Timothy A. Calamari, *U.S. Departmentof Agriculture, New Orleans, Louisiana,* Finishing (under Textiles)

Narasimhan Calamur, *Amoco Chemical Company, Naperville, Illinois,* Butylenes; Propylene

Emmett D. Calhoun, *E.I. du Pont de Nemours & Co., Inc., Wilmington, Delaware,* Information retrieval

Timothy A. Calimari, Jr., *U.S. Department of Agriculture, New Orleans, Louisiana,* Flame retardants for textiles

William L. Cameron, *Cyanamid Canada, Inc., Niagara Falls, Ontario, Canada,* Calcium carbide (under Carbides)

William J. Cannella, *Chevron Research & Technology Company, Richmond, California,* Xylenes and ethylbenzene

Robert R. Cantrell, *Union Camp Corporation, Savannah, Georgia, Union College, Jackson, Tennessee,* Branched-chain acids; Fatty acids from tall oil (both under Carboxylic acids); Dicarboxylic acids

W. J. Cantwell, *École Polytechnique Fèdèrale de Lausanne, Écublens, Lausanne, Switzerland,* Survey (under Composite materials)

C. Edward Capes, *National Research Council of Canada, Ottawa,* Size enlargement

Michael Capone, *Exxon Engineering, Florham Park, New Jersey,* Sulfur removal and recovery

C. Robert Cappel, *Eastman Kodak Company, Rochester, New York,* Silver compounds

Bruno A. Caputo, *E.I. du Pont de Nemours & Co., Inc., Wilmington, Delaware,* Information retrieval

S. C. Carapella, *Consultant, Tuckahoe, New York,* Antimony and antimony alloys; Arsenic and arsenic alloys; Tellurium and tellurium compounds

D. A. Carlson, *University of Florida, Gainesville, Florida,* Repellents

Ronald N. Caron, *Olin Corporation, New Haven, Connecticut,* Wrought copper and wrought copper alloys (under Copper alloys)

Dodd S. Carr, *International Lead Zinc Research Organization, Inc., Research Triangle Park, North Carolina; Consultant, Pittsburgh, Pennsylvania,* Cadmium and cadmium alloys; Lead salts (under Lead compounds)

F. Patrick Carr, *OMYA, Inc., Proctor, Vermont ,* Calcium carbonate (under Calcium compounds)

Martin E. Carrera, *Amoco Chemical Company, Naperville, Illinois,* Butylenes; Propylene

Boyce Carsella, Jr., *Magnetrol International, Inc., Downers Grove, Illinois,* Liquid level measurement

John W. Carson, *Jenike and Johanson, Inc., Westford, Massachusetts,* Bulk powders (under Powders, handling)

Robert Casani, *Hoffmann-La Roche Inc., Nutley, New Jersey,* Vitamin E (under Vitamins)

Jeremiah Casey, *Air Products and Chemicals, Inc., Allentown, Pennsylvania,* Cycloaliphatic amines

Patrick E. Cassidy, *Southwest Texas State University, San Marcos, Texas,* Heat-resistant polymers

Geert Cauwenbergh, *Janssen Research Foundation, Beerse, Belgium,* Antimycotics (under Antiparasitic agents)

B. Cavalleri, *Merrell-Dow Research Institute, Gerenzano, Italy,* Glycopeptides (under Antibiotics)

David W. Cawlfield, *Olin Chemicals, Charleston, Tennessee,* Chlorous acid, chlorites, and chlorine dioxide (under Chlorine oxygen acids and salts)

James Cella, *GE Corporate Research and Development, Schenectady, New York,* Silicones (under Silicon compounds)

Peter Cervoni, *Lederle Laboratories, American Cyanamid Company, Pearl River, New York,* Cardiovascular agents; Diuretic agents

Mark Chagnon, *Metalspecialties, Fairfield, Connecticut,* Bismuth and bismuth alloys

Peter S. Chan, *Lederle Laboratories, American Cyanamid Company, Pearl River, New York,* Cardiovascular agents; Diuretic agents

Clarence D. Chang, *Mobil Research & Development Corporation, Princeton, New Jersey,* Liquid fuels (under Fuels, synthetic)

A. R. Chapman, *Rochester Institute of Technology, Rochester, New York,* Containers for industrial materials (under Packaging)

K. Chawla, *New Mexico Institute of Mining and Technology, Socorro,* Metal-matrix composites

Surendra K. Chawla, *The Goodyear Tire & Rubber Company, Akron, Ohio,* Tire cord

Shiou-Shan Chen, *Raytheon Engineers & Constructors, East Weymouth, Massachusetts,* Styrene

Wai-kai Chen, *University of Illinois, Chicago,* Dimensional analysis

X. K. Chen, *University of Edinburgh, Scotland,* Machining methods, electrochemical

K. Y. Chia, *The Carborundum Company, Niagara Falls, New York,* Silicon carbide (under Carbides)

Shiao-Hung Chiang, *University of Pittsburgh, Pennsylvania,* Cleaning and desulfurization (under Coal conversion processes)

Andrew D. Child, *University of Florida, Gainesville,* Electrically conductive polymers

Steven R. Childers, *Wake Forest University, Winston-Salem, North Carolina,* Brain oligopeptides (under Hormones); Opioids, endogenous

Kenneth Chilton, *Washington University, St. Louis, Missouri,* Introduction (under Recycling)

Kuen-Wai Chiu, *Callery Chemical Company, Pittsburgh, Pennsylvania,* Potassium

S. M. Cho, *Foster Wheeler Energy Corporation, Clinton, New Jersey,* Heat transfer (under Heat-exchange technology)

Young I. Cho, *Drexel University, Philadelphia, Pennsylvania,* Heat transfer (under Heat-exchange technology)

Michael S. Cholod, *Rohm and Haas Company, Bristol, Pennsylvania,* Cyanohydrins

Ian Chopra, *American Cyanamid Company, Pearl River, New York,* Tetracyclines (under Antibiotics)

Robert Chorvat, *E. I. du Pont de Nemours & Co., Inc., Wilmington, Delaware,* Memory-enhancing drugs

Lawrence C. Chow, *National Institute of Standards and Technology, Gaithersburg, Maryland,* Dental materials

Kenneth A. Christensen, *University of Michigan, Ann Arbor,* Raman spectroscopy (under Infrared technology and Raman spectroscopy)

John R. Christoe, *CSIRO, Belmont, Geelong, Australia,* Wool

Kevin Chronley, *Hammond Lead Products, Inc., Pittsburgh, Pennsylvania,* Lead salts (under Lead compounds)

Rasik J. Chudgar, *BASF Corporation, Rensselaer, New York,* Azo dyes

Frank H. Y. Chung, *Rhône-Poulenc, Inc., Cranbury, New Jersey,* Chemical leavening agents (under Bakery processes and leavening agents)

Donald Clagett, *Organisation for the Prohibition of Chemical Weapons, the Haag, the Netherlands,* Engineering plastics

David L. Clark, *Glenn T. Seaborg Institute for Transactinium Science, Los Alamos National Laboratory, Los Alamos, New Mexico,* Thorium and thorium compounds; Uranium and uranium compounds

Earl Clark, *Phillips Research Center, Bartlesville, Oklahoma,* Sulfolane and sulfones

Elke M. Clark, *Union Carbide Corporation, Danbury, Connecticut,* Ethylene oxide polymers (under Polyethers)

J. Peter Clark, *Epstein and Sons International, Inc., Chicago, Illinois,* Chemurgy

R. K. Clark, *Shell Development Company, Houston, Texas,* Drilling fluids (under Petroleum)

Margaret A. Clarke, *Sugar Processing Research Institute, Inc., New Orleans, Louisiana,* Cane sugar (under Sugar)

Stephen I. Clarke, *Air Products and Chemicals, Inc., Allentown, Pennsylvania,* Nitric acid

Jay B. Class, *Hercules Incorporated, Wilmington, Delaware,* Resins, natural

Ken G. Claus, *The Dow Chemical Company, Freeport, Texas,* Magnesium and magnesium alloys

Michael Cleary, *Imperial Holly Corporation, Colorado Springs, Colorado,* Beet sugar (under Sugar)

L. D. Clements, *University of Nebraska, Lincoln, Nebraska,* Computer-aided design and manufacturing (CAM/CAM)

Antoine P. Cobb, *Donelan, Cleary, Wood & Maser P.C., Washington, D.C.,* Transportation

James T. Cobb, Jr., *University of Pittsburgh, Pennsylvania,* Cleaning and desulfurization (under Coal conversion processes)

K. J. Coeling, *Nordson Corporation, Amherst, Ohio,* Spray coating (under Coating processes)

A. J. Cofrancesco, *Consultant, Windward Meadows, Delanson, New York,* Anthraquinone; Dyes, natural

Alan P. Cohen, *UOP, Tarrytown, New York,* Desiccants

Edward Cohen, *Du Pont Imaging Systems, Wilmington, Delaware,* Electronics, coatings

Edward D. Cohen, *E. I. du Pont de Nemours & Co., Inc., Wilmington, Delaware,* Survey (under Coating processes)

J. G. Cohn, *Engelhard Corporation, Iselin, New Jersey,* Gold and gold compounds

G. R. "Buck" Coleman, *Consultant, Houston, Texas,* Urea

J. W. Collette, *DuPont Central Research and Development, Wilmington, Delaware,* Research/technology management

Robert Collier, *Monsanto Company, St. Louis, Missouri,* Animals (under Genetic engineering)

Peter J. Collings, *Swarthmore College, Swarthmore, Pennsylvania,* Liquid crystalline materials

Paul W. Collins, *GD Searle & Company, Skokie, Illinois,* Prostaglandins

Steven Collins, *Empire State Electric Energy Research Corporation, New York, New York,* Power generation; Power generation (under Regulatory agencies)

Ward Collins, *Dow Corning Corporation, Midland, Michigan,* Silicon halides (under Silicon compounds)

William J. Colonna, *American Crystal Sugar Company, Moorhead, Minnesota,* Properties of sucrose (under Sugar)

Ann R. Comfort, *CIBA-GEIGY Corporation, Ardsley, New York,* Drug delivery systems

Jim Compton, *Globe Building Materials Company,* Roofing materials

Alan E. Comyns, *Solvay Interox, Chester, England,* Inorganic peroxides (under Peroxides and peroxide compounds)

John A. Conkling, *American Pyrotechnic Association, Chestertown, Maryland,* Pyrotechnics

Gregory S. Conn, *Kalama Chemical, Inc., Kalama, Washington,* Benzaldehyde; Benzoic acid

David A. Cooney, *National Cancer Institute, Bethesda, Maryland,* Therapeutic (under Enzyme applications)

Kenneth W. Cooper, *Poolpak, Inc., Seven Valleys, Pennsylvania,* Air Conditioning

Robert A. Copeland, *DuPont-Merck Pharmaceutical Company, Wilmington, Delaware,* Protein engineering

James Corbin, *University of Illinois, Urbana,* Pet foods (under Feeds and feed additives)

Ray L. Corbin, *Schuller International, Inc., Littleton, Colorado,* Roofing materials

Cajetan F. Cordeiro, *Air Products and Chemicals, Inc., Allentown, Pennsylvania,* Vinyl acetate polymers (under Vinyl polymers)

Michel Costantini, *Rhône-Poulenc Recherches, France,* Hydroquinone, resorcinol, and catechol

Christine A. Costello, *Exxon Research and Engineering Company, Annandale, New Jersey,* Copolymers

J. Kevin Cotchen, *Man GHH Corporation, Pittsburgh, Pennsylvania,* Arc furnaces (under Furnaces, electric)

Gregory L. Cote, *U.S. Department of Agriculture, Peoria, Illinois,* Microbial polysaccharides

Howard Cottam, *University of California, San Diego, California,* Antiaging agents

W. P. Cottom, *Baker Petrolite Corporation, Tulsa, Oklahoma,* Waxes

Charles C. Coutant, *Oak Ridge National Laboratory, Oak Ridge, Tennessee,* Thermal pollution

Joseph A. Cowfer, *The Geon Company, Avon Lake, Ohio,* Vinyl chloride

Charles R. Craig, *West Virginia University, Morgantown,* Stimulants

David L. Crandall, *Lederle Laboratories, Pearl River, New York,* Cardiovascular agents

Louise W. Crandall, *Lilly Research Laboratories, Indianapolis, Indiana,* Polyethers (under Antibiotics)

Paul D. Crane, *DuPont-Merck Pharmaceutical Company, North Billerica, Massachusetts,* Radiopharmaceuticals

John C. Crano, *PPG Industries, Inc., Monroeville, Pennsylvania,* Photochromic (under Chromogenic materials)

Mary H. Crawford, *Sandia National Laboratories, Albuquerque, New Mexico,* Compound semiconductors (under Semiconductors)

Lamberto Crescentini, *Allied-Signal, Inc., Petersburg, Virginia,* Caprolactam

J. M. Criscione, *UCAR Carbon Company, Inc., Cleveland, Ohio,* Carbon and artificial graphite (under Carbon)

Burton B. Crocker, *Consultant, Chesterfield, Missouri,* Air pollution control methods

Ronald D. Crooks, *Consultant, Branderis Associates, Inc., Hockessin, Delaware,* Hardness

Barry A. Crouch, *E. I. du Pont de Nemours & Co., Inc., Wilmington, Delaware,* Fracture mechanics

Daniel A. Crowl, *Michigan Technological University, Houghton,* Hazard analysis and risk assessment

Michael J. Cruickshank, *University of Hawaii at Manoa, Honolulu,* Ocean raw materials

John A. Cuculo, *North Carolina State University, Raleigh, North Carolina,* Cellulose

Benedict S. Curatolo, *BP Research, Cleveland, Ohio,* Survey and SAN (under Acrylonitrile polymers)

L. Calvert Curlin, *OxyTech Systems, Inc., Chardon, Ohio,* Chlorine and sodium hydroxide (under Alkali and chlorine products)

Horace G. Cutler, *USDA/ARS, Athens, Georgia,* Plant (under Growth regulators)

Tom A. Czuppon, *M. W. Kellogg Company, Houston, Texas,* Ammonia; Hydrogen

David R. Dalton, *Temple University, Philadelphia, Pennsylvania,* Alkaloids

Larry R. Dalton, *University of Southern California, Los Angeles,* Nonlinear optical materials

Paul D'Ambra, *AT&T Bell Labs, North Hanover, Massachusetts,* Embedding

Ture Damhus, *Enzyme Technology, Bagsvaerd, Denmark,* Industrial (under Enzyme applications)

Suresh B. Damle, *PPG Industries, Inc., Monroeville, Pennsylvania,* Carbonic and carbonochloridic esters

James R. Daniel, *Purdue University, Lafayette, Indiana,* Starch

R. W. Daniels, *Union Camp Corporation, Savannah, Georgia,* Manufacture (under Carboxylic acids)

Paul S. Danielson, *Corning, Inc., Corning, New York,* Glass

Joseph P. Daniszewski, *E.I. du Pont de Nemours & Co., Inc., Wilmington, Delaware,* Information retrieval

Eli M. Dannenberg, *Consultant, Bellerica, Massachusetts,* Carbon black (under Carbon)

K. Darcovich, *National Research Council of Canada, Ottawa,* Size enlargement

K. V. Darragh, *Rhône-Poulenc, Inc., Cranbury, New Jersey,* Aluminum sulfate and alums (under Aluminum compounds)

Rathin Datta, *Consultant, Chicago, Illinois,* Hydroxycarboxylic acids

Reg Davies, *E. I. du Pont de Nemours & Co., Inc., Wilmington, Delaware,* Sampling

Simon Davies, *Davy Process Technology, London, England,* Methanol

Darwin D. Davis, *E. I. du Pont de Nemours & Co., Inc., Victoria, Texas,* Adipic acid

June P. Davis, *DuPont-Merck Pharmaceutical Company, Wilmington, Delaware,* Protein engineering

William J. Dawson, *Chemical Materials International, Dublin, Ohio,* Hydrothermal processing

Allan G. DeBoos, *CSIRO, Belmont, Geelong, Australia,* Wool

Jeffrey J. DeFraties, *Haarmann & Reimer Corporation, Springfield, New Jersey,* Hydroxy dicarboxylic acids

Ernesto de Guzman, *Sybron Chemicals Inc., Wellford, South Carolina,* Dye carriers

Eckehard Volker Dehmlow, *Universität Bielefeld, Germany,* Catalysis, phase-transfer

Lile H. Deinard, *White & Case, New York,* Trademarks (under Copyrights and trademarks)

Phillip DeLassus, *The Dow Chemical Company, Midland, Michigan,* Barrier polymers

Phillip T. DeLassus, *The Dow Chemical Company, Freeport, Texas,* Vinylidene chloride monomer and polymers

Thomas W. Del Pesco, *E. I. du Pont de Nemours & Co., Inc., Deepwater, New Jersey,* Organic (under Titanium compounds)

Ron J. Denning, *CSIRO, Belmont, Geelong, Australia,* Wool

Alan D. Denniston, *Unocal Corporation, San Ramon, California,* Hydraulic fluids

Maurice Dery, *Akzo Nobel Chemicals Inc., Dobbs Ferry, New York,* Quaternary ammonium compounds

Jean-Roger Desmurs, *Rhône-Poulenc Recherches, Saint-Fons, France,* Chlorophenols

J. P. Dever, *Union Carbide Corporation, South Charleston, West Virginia,* Ethylene oxide

Stephen C. DeVito, *U.S. Environmental Protection Agency, Washington, D.C.,* Industrial toxicology (under Lead compounds); Mercury; Nitriles

Robert DeVries, *Burut Hills, New York,* Diamond, natural (under Carbon)

Martin Dexter, *CIBA-GEIGY Corporation, Ardsley, New York,* Antioxidants

Patrick M. DiBello, *FMC Corporation, Princeton, New Jersey,* Barium compounds

Charles Dickert, *Consultant, Yardley, Pennsylvania,* Ion exchange

Boerge Diderichsen, *Enzyme Technology, Bagsvaerd, Denmark,* Industrial (under Enzyme applications)

R. Bertrum Diemer, Jr., *E.I. du Pont de Nemours & Co., Inc., Wilmington, Delaware,* Incinerators

Steven M. Dinh, *CIBA-GEIGY Corporation, Ardsley, New York,* Drug delivery systems

Christopher P. Dionigi, *U.S. Department of Agriculture, New Orleans, Louisiana,* Herbicides

R. Divakar, *The Carborundum Company, Niagara Falls, New York,* Silicon carbide (under Carbides)

David J. Dixon, *South Dakota School of Mines and Technology, Rapid City,* Supercritical fluids

Lisa A. Dixon, *R. W. Johnson Pharmaceutical Research Institute, Raritan, New Jersey,* Sex hormones (under Hormones)

G. O. Doak, *North Carolina State University, Raleigh, North Carolina,* Antimony compounds; Arsenic compounds; Bismuth compounds

Douglas V. Doane, *Consultant, Ann Arbor, Michigan,* Molybdenum and molybdenum alloys

Robert H. Dobberstein, *Sandoz Pharmaceuticals Corporation, Lincoln, Nebraska,* Expectorants, antitussives, and related agents

Theresa M. Dobel, *Du Pont Chemical Company, Stow, Ohio,* Ethylene–acrylic elastomers (under Elastomers, synthetic)

C. J. Dobratz, *Kalama Chemical, Inc., Kalama, Washington,* Benzaldehyde; Benzoic acid

John C. Dobson, *Rohm and Haas Company, Spring House, Pennsylvania,* Methacrylic acid and derivatives

Edward L. Docks, *U.S. Borax Research Corp., Anaheim, California,* Boric acid esters (under Boron compounds)

Julius E. Dohany, *Consultant, Berwyn, Pennsylvania,* Poly(vinylidene fluoride) (under Fluorine compounds, organic)

Michael Doherty, *University of Massachusetts, Amherst,* Distillation, azeotropic and extractive

T. Dombrowski, *Englehard Corporation, Iselin, New Jersey,* Survey (under Clays)

James A. Doncheck, *Bio-Technical Resources L.P., Manitowoc, Wisconsin,* Malts and malting

Vishu Dosaj, *Dow Corning Corporation, Midland, Michigan,* Chemical and metallurgical (under Silicon and siliconalloys)

Ronald L. Dotson, *Olin Chemicals, Charleston, Tennessee,* Perchloric acid and perchlorates

Ross Dowbenko, *PPG Industries, Allison Park, Pennsylvania,* Allyl monomers and polymers

James Downing, *Consultant, Ellicottville, New York,* Manganese and manganese alloys

Mary Noon Doyle, *Shepherd Chemical Company, Cincinnati, Ohio,* Naphthenic acids

Terrence Doyle, *Bristol-Myers Squibb Company, Wallingford, Connecticut,* Chemotherapeutics, anticancer

Barry A. Dreikorn, *Dow Elanco, Indianapolis, Indiana,* Fungicides, agricultural

Lawrence J. Drew, *U.S. Geological Survey, Reston, Virginia,* Petroleum resources (under Petroleum)

M. P. Dreyfuss, *Consultant, Midland, Michigan,* Tetrahydrofuran and oxetane polymers (under Polyethers)

P. Dreyfuss, *Consultant, Midland, Michigan,* Tetrahydrofuran and oxetane polymers (under Polyethers)

H. G. Drickamer, *University of Illinois, Urbana,* Piezochromic (under Chromogenic materials)

Richard W. Drisko, *U.S. Naval Civil Engineering Laboratory, Port Hueneme, California,* Coatings, marine

Timothy J. Drummond, *Sandia National Laboratories, Albuquerque, New Mexico,* Compound semiconductors (under Semiconductors)

David J. Drury, *BP Chemicals, Ltd., Middlesex, U.K.,* Formic acid (under Formic acid and derivatives)

Beth Dryden, *Callery Chemical Company, Pittsburgh, Pennsylvania,* Boron hydrides, commercial aspects (under Boron compounds)

Paul Duby, *Columbia University, New York, New York,* Extractive metallurgy (under Metallurgy)

David Duchane, *Los Alamos National Laboratory, New Mexico,* Geothermal energy

William G. Dukek, *Consultant, Summit, New Jersey,* Aviation and other gas turbine fuels

Margaret A. Dulany, *Georgia-Pacific Corporation, Decatur, Georgia,* Papermaking additives

Budd L. Duncan, *Olin Chemicals, Charleston, Tennessee,* Chloric acid and chlorates (under Chlorine oxygen acids and salts)

Michael P. Duncan, *Ferro Corporation, Hammond, Indiana,* Sulfurization and sulfur chlorination

Wayne P. Duncan, *Hewlett-Packard Company, Palo Alto, California,* Hyphenated instruments (under Analytical methods)

Frances Duneczky, *Atlas Refinery Inc., Newark, New Jersey,* Castor oil

Kenneth L. Dunlap, *Bayer Corporation, New Martinsville, West Virginia,* Phosgene

Irene Durbak, *USDA Forest Service, Madison, Wisconsin,* Wood

Douglas J. Durian, *University of California, Los Angeles,* Foams

Richard A. Durst, *Cornell University, Geneva, New York,* Hydrogen-ion activity

Seán G. Dwyer, *S. C. Johnson & Sons, Inc., Racine, Wisconsin,* Polishes

Cecil Dybowski, *University of Delaware, Newark,* Chromatography

A. J. Dzermejko, *UCAR Carbon Company, Inc., Cleveland, Ohio,* Carbon and artificial graphite (under Carbon)

H. W. Earhart, *Consultant, Wichita, Kansas,* Polymethylbenzenes
Matthew R. Earlam, *The Dow Chemical Company, Freeport, Texas,* Magnesium and magnesium alloys
Anthony J. East, *Hoechst Celanese Corporation, Summit, New Jersey,* Polyesters, thermoplastic
Alan D. Eastman, *Phillips Petroleum, Bartlesville, Oklahoma,* C_1–C_6; Survey (both under Hydrocarbons)
G. Yale Eastman, *DTX Corporation, Lancaster, Pennsylvania,* Heat pipes (under Heat-exchange technology)
Chris Eberspacher, *UNISUN, Newbury Park, California,* Photovoltaic cells
Frank H. Ebetino, *Norwich Eaton Pharmaceuticals Corporation, Norwich, New York,* Nitrofurans (under Antibacterial agents)
S. Ebnesajjad, *E. I. du Pont de Nemours & Co., Inc., Wilmington, Delaware,* Poly(vinyl fluoride) (under Fluorine compounds, organic)
W. Wesley Eckenfelder, Jr., *Eckenfelder Inc., Nashville, Tennessee,* Wastes, industrial; Pollution (under Water)
J. Eckert, *H. C. Starck Inc., Newton, Massachusetts,* Tantalum and tantalum compounds
Martha R. Edens, *The Dow Chemical Company, Freeport, Texas,* Alkanolamines from olefin oxides and ammonia
David Edgren, *ALZA Corporation, Palo Alto, California,* Pharmaceuticals (under Controlled release technology)
D. Scott Edwards, *DuPont-Merck Pharmaceutical Company, North Billerica, Massachusetts,* Radio pharmaceuticals
Janice W. Edwards, *Monsanto Company, Chesterfield, Missouri,* Plants (under Genetic engineering)
Terry A. Egerton, *Tioxide Group Services Limited, Billingham, U.K.,* Inorganic (under Titanium compounds)
Fred C. Eichmiller, *National Institute of Standards and Technology, Gaithersburg, Maryland,* Dental materials
Peter Eigtved, *Enzyme Technology, Bagsvaerd, Denmark,* Industrial (under Enzyme applications)
Robert G. Eilerman, *Givaudan-Roure Corporation, Clifton, New Jersey,* Cinnamic acid, cinnamaldehyde, and cinnamyl alcohol
Varadaraj Elango, *Hoechst-Celanese Corporation, Corpus Christi, Texas,* Esters, organic
M. Jamal El-Hibri, *Amoco Polymers Inc., Alpharetta, Georgia,* Polysulfones (under Polymers containing sulfur)
George A. Ellestad, *American Cyanamid Company, Pearl River, New York,* Tetracyclines (under Antibiotics)
Arthur J. Elliott, *Halocarbon Products Corporation, North Augusta, South Carolina,* Bromotrifluoroethylene; Fluorinated acetic acids; Fluoroethanols (all under Fluorine compounds, organic)
Thomas D. Ellis, *E.I. du Pont de Nemours & Co., Inc., Wilmington, Delaware,* Incinerators
William R. Ellis, *Raytheon Engineers and Constructors, New York, New York,* Fusion energy
Mark N. Emly, *Brush Wellman Inc., Elmore, Ohio,* Beryllium compounds
William C. Engeland, *University of Minnesota, Minneapolis,* Survey (under Hormones)
Kenneth R. Engh, *CELITE Corporation, Lompac, California,* Diatomite
Alan English, *John Brown E & C, Houston, Texas,* Methanol
Mary G. Enig, *Enig Associates, Inc., Silver Springs, Maryland,* Mineral nutrients
Matthew Ennis, *Stanford University, Stanford, California,* Nitrogen
Royce Ennis, *Du Pont-Beaumont Works, Beaumont, Texas,* Chlorosulfonated polyethylene (under Elastomers, synthetic)
Richard A. Eppler, *Eppler Associates, Cheshire, Connecticut,* Colorants for ceramics
Donald M. Ernst, *Thermacore Inc., Lancaster, Pennsylvania,* Heat pipes (under Heat-exchange technology)
Carolyn A. Ertell, *Rhône-Poulenc, Inc., Cranbury, New Jersey,* Aluminum sulfate and alums (under Aluminum compounds)
Lawrence J. Esposito, *Rhône-Poulenc, Cranbury, New Jersey,* Vanillin
Samuel F. Etris, *The Silver Institute, Wayne, Pennsylvania,* Silver and silver alloys
W. G. Etzkorn, *Union Carbide Chemicals & Plastics Company, Inc., South Charleston, West Virginia,* Acrolein and derivatives
David J. Evans, *CSIRO, Belmont, Geelong, Australia,* Wool
Francis E. Evans, *Consultant, Hamburg, New York,* Boron, boron trifluoride; Sulfur (both under Fluorine compounds, inorganic)
Larry R. Evans, *J. M. Huber Corporation, Havre de Grace, Maryland,* Amorphous silica (under Silica)
Kevin Ewsuk, *Sandia National Laboratories, Albuquerque, New Mexico,* Processing (under Ceramics)
Edward F. Ezell, *The BOC Group, Inc., Murray Hill, New Jersey,* High purity gases
Walter Fabisiak, *Sherwood-Davis and Geck, Connecticut,* Sutures
George C. Fahey, Jr., *University of Illinois, Urbana,* Ruminant feeds (under Feeds and feed additives)
James R. Fair, *The University of Texas at Austin, Austin,* Distillation
James S. Falcone, Jr., *West Chester University, Pennsylvania,* Fillers; Synthetic inorganic silicates (under Silicon compounds)
Daniel F. Farkas, *Oregon State University, Corvallis,* Food processing
Charles E. Farley, *Georgia-Pacific Corporation, Decatur, Georgia,* Papermaking additives
James P. Farr, *The Clorox Company, Pleasanton, California,* Survey (under Bleaching agents)
Michael Fath, *Consultant, Akron, Ohio,* Rubber compounding
Rudolf Faust, *University of Massachusetts, Lowell,* Cationic initiators (under Initiators)
William D. Faust, *Ferro Corporation, Cleveland, Ohio,* Enamels, porcelain or vitreous
Michael Favaloro, *Textron Defense Systems, Gloucester, Massachusetts,* Ablative materials
Darrell C. Fee, *Monsanto Company, St. Louis, Missouri,* Phosphorus compounds
Ray Russel Fell, *Consultant, Akron, Ohio,* Rubber compounding
Timothy R. Felthouse, *Huntsman Specialty Chemicals Corporation, St. Louis, Missouri,* Maleic anhydride, maleic acid, and fumaric acid
Richard Fengl, *Tennessee Eastman, Kingsport, Tennessee,* Cellulose esters, inorganic; Cellulose esters, organic
Richard E. Fernandez, *E. I. du Pont de Nemours & Co., Inc., Wilmington, Delaware,* Fluorinated aliphatic compounds (under Fluorine compounds, organic)
R. Fikentscher, *BASF AG Ludwigshafen, Germany,* Imines, cyclic
K. Thomas Finley, *State University of New York, Brockport,* Quinolines and isoquinolines; Quinones
James A. Finnegan, *Arthur D. Little, Inc., Cambridge, Massachusetts,* Building materials, plastic
Frank Fischetti, Jr., *Craftmaster Flavor Technology Inc., Amityville, New York,* Flavors (under Flavors and spices)
Barry A. J. Fisher, *Scientific Services Bureau, Los Angeles, California,* Forensic chemistry
F. F. Fisher, *UCAR Carbon Company, Inc., Cleveland, Ohio,* Carbon and artificial graphite (under Carbon)
Michael H. Fisher, *Merck, Sharp, and Dohme Research Laboratories, Rahway, New Jersey,* Avermectins (under Antiparasitic agents)
William B. Fisher, *Allied-Signal, Inc., Petersburg, Virginia,* Caprolactam; Cyclohexanol and cyclohexanone
Richard M. Flynn, *3M Company, St. Paul, Minnesota,* Fluoroethers and fluoroamines (under Fluorine compounds, organic)
William George Fong, *Florida Department of Agriculture and Consumer Services, Tallahassee,* Toxicology
M. W. Forkner, *Union Carbide Corporation, South Charleston, West Virginia,* Ethylene glycol and oligomers (under Glycols)
K. Formanek, *Rhône-Poulenc, Cranbury, New Jersey,* Vanillin
Richard D. Fortin, (deceased), *Donelan, Cleary, Wood & Maser P.C., Washington, D.C.,* Transportation
Stephen Fossey, *U.S. Army Natick Research, Development, & Engineering Center, Natick, Massachusetts,* Silk
Peter R. Foster, *Scottish National Blood Transfusion Service, Edinburgh, U.K.,* Plasma fractionation (under Fractionation, blood)
William O. Foye, *Massachusetts College of Pharmacy and Allied Health Sciences, Boston, Massachusetts,* Sulfonamides (under Antibacterial agents)
Anna C. Fraker, *National Institute of Standards and Technology, Gaithersburg, Maryland,* Dental materials
Otto Frank, *Air Products & Chemicals, Inc., Allentown, Pennsylvania,* Solvent recovery, condensation
David K. Frederick, *OMYA, Inc., Proctor, Vermont,* Calcium carbonate (under Calcium compounds)
Leon D. Freedman, *North Carolina State University, Raleigh, North Carolina,* Antimony compounds; Arsenic compounds; Bismuth compounds
Mark B. Freilich, *Memphis State University, Memphis, Tennessee,* Survey (under Calcium compounds); Potassium compounds
Alfred D. French, *Southern Regional Research Center, New Orleans, Louisiana,* Cellulose
Stig E. Friberg, *Clarkson University, Potsdam, New York,* Emulsions
Leslie J. Friedman, *Arthur D. Little, Inc., Cambridge, Massachusetts,* Food additives
Leroy W. Fritch, Jr., *GE Plastics, Pittsfield, Massachusetts,* ABS resins (under Acrylonitrile polymers)

William Fruscella, *Unocal Corporation, Brea, California,* Benzene

C. F. Fulgenzi, *UCAR Carbon Company, Inc., Cleveland, Ohio,* Carbon and artificial graphite (under Carbon)

E. R. Fuller, Jr., *National Institute of Standards and Technology, Gaithersburg, Maryland,* Ceramic-matrix (under Composite materials)

Lance S. Fuller, *Synthetic Chemicals Limited, Wolverhampton, U.K.,* Thiophene and thiophene derivatives

Thomas F. Fuller, *University of California, Berkeley,* Introduction (under Electrochemical processing)

W. S. Fulton, *Malaysian Rubber Producers' Research Association, Brickendonbury, U.K.,* Rubber, natural

Carlo Fumagalli, *LONZA SpA, Scanzorosciate, Italy,* Succinic acid and succinic anhydride

Barbara J. Furches, *The Dow Chemical Company, Midland, Michigan,* Plastics testing

Steven D. Gagnon, *BASF Corporation, Geismar, Louisiana,* Propylene oxide polymers (under Polyethers)

Subhash V. Gangal, *E. I. du Pont de Nemours & Co., Inc., Wilmington, Delaware,* Perfluorinated ethylene–propylene copolymers; Polytetrafluoroethylene; Tetrafluoroethylene–ethylene copolymers; Tetrafluoroethylene–perfluorovinyl ether copolymers (all under Fluorine compounds,organic)

Ashit K. Ganguly, *Schering-Plough Corporation, Bloomfield, New Jersey,* Oligosaccharides (under Antibiotics)

Richard G. Gann, *National Institute of Standards and Technology, Gaithersburg, Maryland,* Overview (under Flame retardants)

John Gannon, *Consultant, Danbury, Connecticut,* Epoxy resins

Richard E. Gannon, *Textron Defense Systems, Everett, Massachusetts,* Acetylene (under Hydrocarbons)

Roque Edward Garcia, *Technicon Instruments Corporation, Tarrytown, New York,* Hematology (under Automated instrumentation)

Rubin G. Garcia, *Schuller International, Inc., Littleton, Colorado,* Roofing materials

David R. Gard, *Monsanto Company, St. Louis, Missouri,* Phosphoric acid and phosphates; Phosphorus compounds

Bruce C. Gates, *University of Delaware Center for Catalytic Science and Technology, Newark, Delaware,* Catalysis

Charles C. Gaver, Jr., *Consultant, Mt. Laural, New Jersey,* Tin and tin alloys

Charles F. Gay, *National Renewable Energy Laboratory, Golden, Colorado,* Photovoltaic cells

Steven Gedon, *Tennessee Eastman, Kingsport, Tennessee,* Cellulose esters, organic

Jon F. Geibel, *Phillips Petroleum Company, Bartlesville, Oklahoma,* Poly(phenylene sulfide) (under Polymers containing sulfur)

Chester H. Gelbert, *E. I. du Pont de Nemours & Co., Inc., Louisville, Kentucky,* Latex technology

Stanley A. Gembicki, *UOP, Des Plaines, Illinois,* Adsorption, liquid separation

Joseph M. Genco, *University of Maine, Orono,* Pulp

David B. George, *Kennecott Corporation, Salt Lake City, Utah,* Copper

Kathleen F. George, *Union Carbide Chemicals and Plastics Company, Inc., South Charleston, West Virginia,* Antifreezes and deicing fluids; Ethylene oxide

Laurie A. George, *National Institute of Standards and Technology, Gaithersburg, Maryland,* Dental materials

H. Robert Gerberich, *Hoechst-Celanese, Corpus Christi, Texas,* Formaldehyde

Brian P. Gersh, *Arthur D. Little, Inc., Cambridge, Massachusetts,* Building materials, plastic

Paul Gherson, *Miles, Inc., Tarrytown, New York,* Clinical chemistry (under Automated instrumentation)

Christen M. Giandomenico, *Johnson Matthew, West Chester, Pennsylvania,* Platinum-group metals, compounds

D. S. Gibbs, *The Dow Chemical Company, Midland, Michigan,* Vinylidene chloride monomer and polymers

Curtis E. Gidding, *DuCoa, Highland, Illinois,* Choline

Melinda B. Gieselman, *3M Specialty Adhesives and Chemicals, St. Paul, Minnesota,* Electrically conductive polymers

Paul Gifford, *Ovonic Battery Company, Troy, Michigan,* Other (under Batteries, secondary cells)

Richard J. Gillenwater, *Carlisle Syntec Systems,* Roofing materials

V. M. Girijavallabhan, *Schering-Plough Corporation, Bloomfield, New Jersey,* Oligosaccharides (under Antibiotics)

Richard S. Givens, *University of Kansas, Lawrence,* Chemiluminescence (under Luminescent materials)

Kenneth H. Glaspey, *E.I. du Pont de Nemours & Co., Inc., Wilmington, Delaware,* Information retrieval

J. Edward Glass, *North Dakota State University, Fargo,* Water-soluble polymers

Jill Glass, *Sandia National Laboratories, Albuquerque, New Mexico,* Mechanical properties and behavior (under Ceramics)

Brian Glover, *Zeneca Colours, Manchester, U.K.,* Dyes, application and evaluation

T. Godel, *F. Hoffmann-La Roche Ltd., Basel, Switzerland,* Psychopharmacological agents

W. L. Godfrey, *BE Inc., Barnwell, South Carolina,* Chemical reprocessing (under Nuclear reactors)

Mary An Godshall, *Sugar Processing Research Institute, Inc., New Orleans, Louisiana,* Sugar analysis (under Sugar)

Harvey M. Goertz, *The O. M. Scott and Sons Company, Marysville, Ohio,* Agricultural (under Controlled release technology)

Michael Golden, *Hoechst Celanese Corporation, Summit, New Jersey,* Polyesters, thermoplastic

Frank E. Goodwin, *International Lead and Zinc Research Organization, Inc., Research Triangle Park, North Carolina,* Zinc and zinc alloys; Zinc compounds

Maximilian B. Gorensek, *The Geon Company, Avon Lake, Ohio,* Vinyl chloride

Myron Gottlieb, *Gas Research Institute, Chicago, Illinois,* Gas, natural

Michael C. Grady, *E. I. du Pont de Nemours & Company, Inc., Philadelphia, Pennsylvania,* Latex technology

Karl F. Graff, *Edison Welding Institute, Columbus, Ohio,* Welding

Gary Grams, *Witco Corporation, Oakland, New Jersey,* Aluminum halides and aluminum nitrate (under Aluminum compounds)

James L. Grant, *DIALOG Information Services, Inc., Palo Alto, California,* Databases

Michael Grant, *Inland Steel Company, East Chicago, Illinois,* Carbonization (under Coal conversion processes)

Derek G. Gray, *Pulp and Paper Research Institute of Canada, Montreal, Canada,* Cellulose

David W. Green, *USDA Forest Service, Madison, Wisconsin,* Wood

Charles B. Greenberg, *PPG Industries, Inc., Pittsburgh, Pennsylvania,* Electrochromic; Thermochromic (both under Chromogenic materials)

N. R. Greening, *Portland Cement Association, Skokie, Illinois,* Cement

C. Gail Greenwald, *Arthur D. Little, Inc., Cambridge, Massachusetts,* Food additives

Peter Gregory, *Zeneca Specialties, Manchester, U.K.,* Dyes and dye intermediates

John J. Gresens, *Merchant and Gould, Minneapolis, Minnesota,* Patents and trade secrets

T. S. Griffen, *U.S. Borax Research Corporation, Anaheim, California,* Organic boron–nitrogen compounds (under Boron compounds)

Werner M. Grootaert, *3M Company, St. Paul, Minnesota,* Fluorocarbon elastomers (under Elastomers, synthetic)

Andrew W. Gross, *Rohm and Haas Company, Spring House, Pennsylvania,* Methacrylic acid and derivatives

Morris P. Grotheer, *Kerr-McGee Corporation, Edmond, Oklahoma,* Inorganic (under Electrochemical processing)

R. A. Guest, *James Robinson, Ltd., Yorkshire, U.K.,* Sulfur dyes

G. Bruce Guise, *CSIRO, Belmont, Geelong, Australia,* Wool

Joseph W. Gunnet, *R. W. Johnson Pharmaceutical Research Institute, Raritan, New Jersey,* Sex hormones (under Hormones)

J. Z. Guo, *R. W. Johnson Pharmaceutical Research Institute, Raritan, New Jersey,* Estrogens and antiestrogens (under Hormones)

Edgar B. Gutoff, *Consultant, Brookline, Massachusetts,* Survey (under Coating processes)

Kenneth A. Gutschick, *National Lime Association, Kensington, Maryland,* Lime and limestone

Harry Gwinnel, *Cabot Corporation, Billerica, Massachusetts,* Carbon black (under Carbon)

C. E. Habermann, *Dow Chemical, USA, Midland, Michigan,* Acrylamide

Kenneth Hacias, *Parker Amchem, Madison Heights, Michigan,* Pickling (under Metal surface treatments)

H. J. Hagemeyer, *Texas Eastman Company, Longview, Texas,* Acetaldehyde

D. W. Hahn, *R. W. Johnson Pharmaceutical Research Institute, Raritan, New Jersey,* Contraceptives; Estrogens and antiestrogens (under Hormones)

Gerald J. Hahn, *General Electric Company, Schenectady, New York,* Design of experiments

Adel F. Halasa, *Goodyear Tire & Rubber Company, Akron, Ohio,* Polybutadiene (under Elastomers, synthetic)

Carl W. Hall, *Engineering Information Services, Arlington, Virginia,* Milk and milk products

Charles M. Hall, *The Upjohn Company, Kalamazoo, Michigan,* Analgesics, antipyretics, and antiinflammatory agents

J. C. Hall, *BE Inc., Barnwell, South Carolina,* Chemical reprocessing (under Nuclear reactors)

Cal Hallada, *Climax Molybdenum Company, Ypsilanti, Michigan,* Molybdenum and molybdenum alloys
Robert L. Hamill, *Lilly Research Laboratories, Indianapolis, Indiana,* Polyethers (under Antibiotics)
H. U. Hammershaimb, *UOP, Des Plaines, Illinois,* Alkylation
Scott Han, *Mobil Research & Development Corporation, Princeton, New Jersey,* Liquid fuels (under Fuels, synthetic)
Stanley E. Handman, *Consultant, Plainview, New York,* Piping systems
William M. Hann, *Rohm & Haas Company, Spring House, Pennsylvania,* Dispersants
James G. Hansel, *Air Products and Chemicals, Inc., Allentown, Pennsylvania,* Oxygen
S. M. Hansen, *E. I. du Pont de Nemours & Co., Inc., Kinston, North Carolina,* Polyester (under Fibers)
Constance B. Hansson, *Occidental Chemical Corporation, Dallas, Texas,* Chlorine and sodium hydroxide (under Alkali and chlorine products)
Thomas L. Hardenburger, *Air Liquide America Corporation, Tualatin, Oregon,* Nitrogen
John Harkness, *Brush Wellman, Inc., Elmore, Ohio,* Beryllium and beryllium alloys
Robert J. Harper, Jr., *U.S. Department of Agriculture, New Orleans, Louisiana,* Flame retardants for textiles; Finishing (under Textiles)
W. James Harper, *Ohio State University, Galena, Ohio,* Dairy substitutes
B. L. Harris, *Consultant, Glenarm, Maryland,* Chemicals in war
Guy H. Harris, *University of California, Berkeley,* Xanthates
Barry V. Harrowfield, *CSIRO, Belmont, Geelong, Australia,* Wool
Robert Hart, *Parker Amchem, Madison Heights, Michigan,* Chemical and electrochemical conversion treatments (under Metal surface treatments)
James A. Harvey, *Hewlett-Packard Company; Oregon Graduate Institute of Science & Technology, Corvallis,* Smart materials
G. L. Hasenhuettl, *Kraft General Foods, Glenview, Illinois,* Fats and fatty oils
Makoto Hattori, *Sumitomo Chemical Company, Tokyo, Japan,* Dyes, anthraquinone
Warren Haupin, *Aluminum Company of America, Alcoa, Pennsylvania,* Aluminum and aluminum alloys
Ronald E. Hebeda, *CPC International Inc., Summit-Argo, Illinois,* Syrups
J. B. Hedge, *UCAR Carbon Company, Inc., Cleveland, Ohio,* Carbon and artificial graphite (under Carbon)
Robert N. Heistand, *Consultant, Englewood, Colorado,* Oil shale
Howard I. Heitner, *Cytec Industries, Stamford, Connecticut,* Flocculating agents
Richard G. Helmer, *Idaho National Engineering Laboratories, Idaho Falls,* Radioisotopes
Richard Helmuth, *Portland Cement Association, Skokie, Illinois,* Cement
J. D. Hem, *U.S. Geological Survey, Menlo Park, California,* Sources and quality issues (under Water)
Ramesh R. Hemrajani, *Exxon Research and Engineering Company, Florham Park, New Jersey,* Mixing and blending
Larry L. Hench, *University of Florida, Gainesville,* Sol–gel technology
Philip B. Henderson, *Air Products and Chemicals, Inc., Allentown, Pennsylvania,* Nitrogen; Tungsten (both under Fluorine compounds, inorganic)
Norman Herron, *E. I. du Pont de Nemours & Co., Inc., Wilmington, Delaware,* Cadmium compounds
Raymond K. Hertz, *Brush Wellman, Inc., Elmore, Ohio,* Beryllium and beryllium alloys
Wayne T. Hess, *E. I. du Pont de Nemours & Co., Inc., Memphis, Tennessee,* Hydrogen peroxide
Stephen G. Hibbins, *Timminco Metals, Timminco Ltd., Haly, Ontario, Canada,* Calcium and calcium alloys; Strontium and strontium compounds
J. C. Hickman, *The Dow Chemical Company, Midland, Michigan,* Tetrachloroethylene (under Chlorocarbons and chlorohydrocarbons)
Jack H. Hicks, *Consultant, Lynchburg, Virginia,* Water chemistry of lightwater reactors (under Nuclear reactors)
Terry L. Highley, *USDA Forest Service, Madison, Wisconsin,* Wood
Jun-ichi Hikasa, *Kuraray Company, Ltd., Osaka, Japan,* Poly(vinyl alcohol) (under Fibers)
James E. Hillis, *The Dow Chemical Company, Freeport, Texas,* Magnesium and magnesium alloys
W. Hillis, *Sun Chemical Corproration, Carlstadt, New Jersey,* Inks
William D. Hinsberg, *IBM Research Division, San Jose, California,* Lithographic resists
Katsumi Hioki, *Kuraray Company, Ltd., Okayama, Japan,* Leather-like materials
Arnold L. Hirsch, *Alpharma Inc., Chicago, Illinois,* Vitamin D (under Vitamins)
Parker W. Hirtle, *Acentech Inc., Cambridge, Massachusetts,* Insulation, acoustic
Mohamed W. M. Hisham, *Occidental Chemical Corporation, Grand Island, New York,* Hydrogen chloride
Brent Hiskey, *University of Arizona, Tucson,* Survey (under Metallurgy)
Joseph J. Hlavka, *American Cyanamid Company, Pearl River, New York,* Tetracyclines (under Antibiotics)
Charles C. Hobbs, *Consultant, Corpus Christi, Texas,* Hydrocarbon oxidation
Albert M. Hochhauser, *Exxon Research and Engineering Company, Linden, New Jersey,* Gasoline and other motor fuels
F. Galen Hodge, *Haynes International, Inc., Kokomo, Indiana,* Cobalt and cobalt alloys
James E. Hoffman, *Jan Reimers and Associates USA Inc., Salt Lake City, Utah,* Selenium and selenium compounds; Tellurium and tellurium compounds
Stanley Hoffman, *Consultant, Danbury, Connecticut,* Transportation
W. C. Hoffman, *Union Carbide Corporation, South Charleston, West Virginia,* Ethylene oxide
George Hoffmeister, *Consultant, Florence, Alabama,* Fertilizers
Eivind Hognestad, *Consultant, Glenview, Illinois,* Cement
Edward Hohmann, *California State Polytechnic University, Pomona,* Network synthesis (under Heat-exchange technology)
A. Höhn, *BASF AG, Lugwigshafen, Germany,* Formamide (under Formic acid and derivatives)
Michael T. Holbrook, *Dow Chemical USA, Plaquemine, Louisiana,* Methyl chloride; Methylene chloride; Chloroform; Carbon tetrachloride (all under Chlorocarbons and chlorohydrocarbons)
Geoffrey Holden, *Holden Polymer Consulting, Inc., Prescott, Arizona,* Thermoplastic elastomers (under Elastomers, synthetic)
Mark W. Holladay, *Abbott Laboratories, Abbott Park, Illinois,* Hypnotics, sedatives, anticonvulsants, and anxiolytics; Neuroregulators
John H. Hong, *Rockwell International Science Center, Thousand Oaks, California,* Holography
Allen T. Hopper, *University of Wisconsin, Madison,* Pharmaceuticals, chiral
Michael Horn, *Hüls A.G., Rheinfelden, Germany,* Alkoxides, metal
Joseph P. Hornak, *Rochester Institute of Technology, Rochester, New York,* Medical imaging technology
Jack Horner, *Allied-Kelite, New Hudson, Michigan,* Electroplating
U. Horns, *Hüls America, Inc.,* Alkoxides, metal
Eugene V. Hort, *GAF Corporation, Wayne, New Jersey,* Acetylene-derived chemicals
Fred H. Hoskins, *Washington State University, Pullman,* Food toxicants, naturally occurring
Kelvin L. Houghton, *ICI Chemicals & Polymers Ltd., Runcorn, Cheshire, United Kingdom,* Chlorinated paraffins (under Chlorocarbons and chlorohydrocarbons)
James L. Howard, *USDA Forest Service, Madison, Wisconsin,* Wood
William L. Howard, *Consultant, The Dow Chemical Company, Freeport, Texas,* Acetone; Chelating agents
Robert A. Howell, *Central Michigan University,* Vinylidene chloride monomer and polymers
Richard D. Howells, *3M, St. Paul, Minnesota,* Waterproofing and water/oil repellency (under Water)
Christopher J. Howick, *European Vinyls Corporation, Cheshire, U.K.,* Plasticizers
Timothy E. Howson, *Wyman-Gordon, Worcester, Massachusetts,* Nickel and nickel alloys
Chia-Lung Hsieh, *Wyeth-Lederle Vaccine and Pediatrics, Pearl River, New York,* Vaccine technology
Chang Samuel Hsu, *Exxon Research and Engineering Company, Annandale, New Jersey,* Composition (under Petroleum)
R. C. Hughes, *Sandia National Laboratories, Albuquerque, New Mexico,* Sensors
Derk T. A. Huibers, *Union Camp Corporation, Princeton, New Jersey,* Tall oil
Bill Humphries, *CSIRO, Belmont, Geelong, Australia,* Wool
W. Hunkeler, *F. Hoffmann-La Roche Ltd., Basel, Switzerland,* Psychopharmacological agents
James Hunter, *Eveready Battery Co., Inc., West Lake, Ohio,* Primary cells (under Batteries)
Jim Hunter, *Gardner Asphalt Corporation,* Roofing materials
William N. Hunter, *Celanese Canada Inc., Edmonton, Alberta, Canada,* Alcohols, polyhydric
T. R. Hupp, *UCAR Carbon Company, Inc., Cleveland, Ohio,* Carbon and artificial graphite (under Carbon)
Mickey G. Huson, *CSIRO, Belmont, Geelong, Australia,* Wool
Shuen-Cheng Hwang, *The BOC Group, Inc., Murray Hill, New Jersey,* Gases (under Helium group); High purity gases
Deborah Illman, *University of Washington, Seattle,* Chemometrics
T. Imai, *UOP, Des Plaines, Illinois,* Alkylation

Philip F. Jackisch, *Ethyl Technical Center, Baton Rouge, Louisiana,* Bromine; Bromine compounds

Laurence A. Jackman, *Teledyne Allvac, Monroe, North Carolina,* Metal treatments

M. G. Jacko, *Allied-Signal Inc., Troy, Michigan,* Brake linings and clutch facings

L. C. Jackson, *Martin Marietta Magnesia Specialties, Inc., Baltimore, Maryland,* Magnesium compounds

Michael D. Jackson, *Acurex Corporation, Mountain View, California,* Alcohol fuels

Henry I. Jacoby, *Discovery Research Consultants, Rydal, Pennsylvania,* Gastrointestinal agents

E. E. Jaffe, *Ciba-Geigy Corporation, Newport, Delaware,* Organic (under Pigments)

Norman C. Jamieson, *Mallinckrodt Specialty Chemicals Company, St. Louis, Missouri,* Standards (under Fine chemicals)

Sei-Joo Jang, *Pennsylvania State University, University Park,* Ferroelectrics

Linda H. Jansen, *Callery Chemical Company, Pittsburgh, Pennsylvania,* Boron, elemental

Monique M.-L. Janssens, *Janssen Research Foundation, Beerse, Belgium,* Histamine and histamine antagonists

Hiremagalur N. Jayaram, *Indiana University School of Medicine, Indianapolis,* Therapeutic (under Enzyme applications)

F. Jenck, *F. Hoffmann-La Roche Ltd., Basel, Switzerland,* Psychopharmacological agents

Frank Jenkins, *Akron Consulting Company, Akron, Ohio,* Rubber compounding

Arnold W. Jensen, *E. I. du Pont de Nemours & Co., Inc., Wilmington, Delaware,* Elastomeric (under Fibers)

Thomas C. Johns, *E.I. du Pont de Nemours & Co., Inc., Wilmington, Delaware,* Information retrieval

James A. Johnson, *UOP, Des Plaines, Illinois,* Adsorption, liquid separation

Peter K. Johnson, *APMI International, Princeton, New Jersey,* Powder metallurgy (under Metallurgy)

Richard M. Johnson, *U.S. Department of Agriculture, New Orleans, Louisiana,* Herbicides

Robert W. Johnson, *Union Camp Corporation, Savannah, Georgia,* Branched-chain acids, Fatty acids from tall oil, Manufacture (all under Carboxylic acids); Dicarboxylic acids

Thomas A. Johnson, *Air Products and Chemicals, Inc., Allentown, Pennsylvania,* Aliphatic amines

Keith P. Johnston, *University of Texas, Austin,* Supercritical fluids

Carmel R. Jolicoeur, *Université de Sherbrooke, Sherbrooke, Quebec, Canada,* Asbestos

Weldon B. Jolley, *ICN Pharmaceuticals Corporation, Costa Mesa, California,* Antiaging agents

Francis W. Jones, *CSIRO, Belmont, Geelong, Australia,* Wool

Les N. Jones, *CSIRO, Belmont, Geelong, Australia,* Wool

Roger W. Jones, *Ames Laboratory, USDOE, Ames, Iowa,* Infrared technology (under Infrared technology and Raman spectroscopy)

Steven Jones, *Clarkson University, Potsdam, New York,* Emulsions

August H. Jorgensen, *Zeon Chemicals, USA, Louisville, Kentucky,* Nitrile rubber (under Elastomers, synthetic)

Stuart M. Kaback, *Exxon Research and Engineering Company, Linden, New Jersey,* Patents, literature

Alexy D. Kachkovski, *National Academy of Sciences of the Ukraine, Kiev,* Polymethine dyes

Jerry J. Kaczur, *Olin Chemicals, Charleston, Tennessee,* Chlorous acid, chlorites, and chlorine dioxide (under Chlorine oxygen acids and salts)

Dennis Kaegi, *Inland Steel Company, East Chicago, Illinois,* Carbonization (under Coal conversion processes)

Guenther Kaempf, *University of Aachen, Germany,* Optical (under Information storage materials)

Herbert D. Kaesz, *University of California, Los Angeles,* Carbonyls

Mary A. Kaiser, *E. I. du Pont de Nemours & Co., Inc., Newark, Delaware,* Chromatography

Steven W. Kaiser, *Union Carbide Chemicals and Plastics Company, Inc., South Charleston, West Virginia,* Diamines and higher amines, aliphatic

Norbert W. Kaleta, *National Gypsum, Gold Bond Research Center, Buffalo, New York,* Calcium sulfate (under Calcium compounds)

Peter Kamarchik, Jr., *PPG Industries, Inc., Allison, Pennsylvania,* Rheological measurements

Conrad W. Kamienski, *Consultant, Gastonia, North Carolina,* Lithium and lithium compounds

Jon Kapecki, *Eastman Kodak Company, Rochester, New York,* Color photography

David L. Kaplan, *U.S. Army Natick Research, Development, & Engineering Center, Natick, Massachusetts,* Silk

Lawrence Karas, *Arco Chemical Company, Newton Square, Pennsylvania,* Ethers

Subhash Karkare, *Amgen, Inc., Thousand Oaks, California,* Cell culture technology

Curtis R. Kates, *Advanced Aromatics, Baytown, Texas,* Naphthalene derivatives

Joseph J. Katz, *Argonne National Laboratories, Argonne, Illinois,* Deuterium and tritium

Steven G. Katz, *ISORCA, Inc., Granville, Ohio,* Reinforced plastics

Tetsuya Kawakita, *Ajinomoto Company, Saga, Japan,* Monosodium glutamate (MSG) (under Amino acids)

Brian Kaye, *Laurentian University, Ontario, Canada,* Size measurement of particles

Philip C. Kearney, *University of Maryland, College Park,* Soil chemistry of pesticides

Ronald J. Keeling, *Formica Corporation, Cincinnati, Ohio,* Laminated materials, plastic

J. A. Keely, *FMC Corporation, Princeton, New Jersey,* Phosphorus

Michael J. Keenan, *Exxon Chemical Company, Baton Rouge, Louisiana,* Trialkylacetic acids (under Carboxylic acids); Cyclopentadiene and dicyclopentadiene

Thomas R. Keenan, *Kind & Knox Gelatine, Inc., Sioux City, Iowa,* Gelatin

Robert J. Keller, *Vinings Industries, Inc., Atlanta, Georgia,* Aluminates (under Aluminum compounds)

James F. Kelly, *Pacific Northwest National Laboratory, Richland, Washington,* Spectroscopy, optical

J. Robert Kelly, *National Institute of Standards and Technology, Gaithersburg, Maryland,* Dental materials

Donald R. Kemp, *E. I. du Pont de Nemours & Co., Inc., Victoria, Texas,* Adipic acid

George L. Kenyon, *University of California, San Francisco,* Enzyme inhibitors

D. Webster Keogh, *Glenn T. Seaborg Institute for Transactinium Science, Los Alamos National Laboratory, Los Alamos, New Mexico,* Thorium and thorium compounds; Uranium and uranium compounds

James F. Kerwin, Jr., *Abbott Laboratories, Abbott Park, Illinois,* Neuroregulators

Henno Keskkula, *University of Texas, Austin,* Polymer blends

Robert Kessler, *Textron Defense Systems, Everett, Massachusetts,* Magnetohydrodynamics

Riaz Khan, *Polytech, Trieste, Italy,* Sugar derivatives (under Sugar)

Richard Kieffer, *Technical University, Vienna, Austria,* Survey (under Carbides)

David J. Kiemle, *SUNY-Syracuse, Syracuse, New York,* Magnetic spin resonance

G. Kientz, *Rhône-Poulenc, Cranbury, New Jersey,* Vanillin

Barry Kilbourn, *Molycorp Inc., White Plains, New York,* Cerium and cerium compounds

William Kilmartin, *Arco Chemical Company, Newton Square, Pennsylvania,* Ethers

D. K. Kim, *The Goodyear Tire & Rubber Company, Akron, Ohio,* Tire cord

Yung K. Kim, *Dow Corning Corporation, Midland, Michigan,* Introduction; Poly(fluorosilicones) (both under Fluorine compounds, organic)

Glenn E. Kinard, *Air Products and Chemicals, Inc., Allentown, Pennsylvania,* Cryogenics

Alicia P. King, *E.I. du Pont de Nemours & Co., Inc., Wilmington, Delaware,* Information retrieval

Chris J. H. King, *Monsanto Chemical Group, Pensacola, Florida,* Organic (under Electrochemical processing)

Desmond F. King, *Chevron Research and Technology Company, Richmond, California,* Fluidization

M. G. King, *Selenium–Tellurium Development Association, Salt Lake City, Utah,* Selenium and selenium compounds

Michael King, *ASARCO, Inc., Salt Lake City, Utah,* Lead

Larry W. Kingston, *National Gypsum, Gold Bond Research Center, Buffalo, New York,* Calcium sulfate (under Calcium compounds)

Kimio Kinoshita, *University of California, Berkeley,* Fuel cells

Herbert Kirst, *Eli Lilly and Company, Indianapolis, Indiana,* Macrolides (under Antibiotics)

Fred Kish, *Hewlett-Packard Company, San Jose, California,* Light-emitting diodes (under Light generation)

Ganesh M. Kishore, *Monsanto Company, Chesterfield, Missouri,* Plants (under Genetic engineering)

Yury V. Kissin, *Mobil Chemical Company, Edison, New Jersey,* Introduction, High density polyethylene, Linear low density polyethylene, Polymers of higher olefins (under Olefin polymers)

Walter Klamp, *Consultant, Akron, Ohio,* Rubber compounding

Donald L. Klass, *Entech International, Inc., Barryington, Illinois,* Fuels from biomass

Daniel L. Klayman, *Walter Reed Army Institute of Research, Washington, D.C.,* Antiprotozoals (under Antiparasitic agents)

Martin Klein, *Rutgers University, Piscataway, New Jersey,* Alkaline (under Batteries, secondary cells)

D. Kline, *Sun Chemical Corporation, Carlstadt, New Jersey,* Inks

William Klingensmith, *Akron Consulting Company, Akron, Ohio,* Rubber compounding

Jerome M. Klosowski, *Dow Corning Corporation, Midland, Michigan,* Sealants

Edward A. Knaggs, *Consultant, Deerfield, Illinois,* Sulfonation and sulfation

Jerome W. Knapczyk, *Monsanto Company, Indian Orchard, Massachusetts,* Vinyl acetal polymers (under Vinyl polymers)

Jeffrey P. Knapp, *E. I. du Pont de Nemours & Co., Inc.,* Distillation, azeotropic and extractive

Chris Kneupper, *The Dow Chemical Company, Freeport, Texas,* Allyl chloride (under Chlorocarbons and chlorohydrocarbons)

Stan A. Knez, *M. W. Kellogg Company, Houston, Texas,* Ammonia; Hydrogen

Raymond S. Knorr, *Monsanto Company, Pensacola, Florida,* Acrylic (under Fibers)

Ted M. Knowlton, *Institute of Gas Technology, Chicago, Illinois,* Fluidization

Edmond I. Ko, *Carnegie Mellon University, Pittsburgh, Pennsylvania,* Aerogels

Dianna S. Kocurek, *Tischler/Kocurek, Round Rock, Texas,* Waste treatment, hazardous waste

I. Fred Koenigsberg, *White & Case, New York, New York,* Copyrights (under Copyrights and trademarks)

Ranga Komanduri, *Oklahoma State University, Stillwater,* Tool materials

Andrew P. Komin, *Koch Chemical Company, Wichita, Kansas,* Polymethylbenzenes

Paul A. Konowicz, *Polytech, Trieste, Italy,* Sugar derivatives (under Sugar)

F. X. N. M. Kools, *Philips Components, Evreux, France,* Ferrites

Peter W. Kopf, *Arthur D. Little, Cambridge, Massachusetts,* Phenolic resins

Gabe I. Kornis, *Pharmacia & Upjohn Inc., Kalamazoo, Michigan,* Pyrazoles, pyrazolines, and pyrazolones

Sandra Kosinski, *AT&T Bell Laboratories, Murray Hill, New Jersey,* Fiber optics

William C. Koskinen, *USDA-Agricultural Research Service, Beltsville, Maryland,* Soil chemistry of pesticides

Jack L. Kosmala, *3M Company, St. Paul, Minnesota,* Polychlorotrifluoroethylene (under Fluorine compounds, organic)

Steven H. Kosmatka, *Portland Cement Association, Skokie, Illinois,* Cement

William H. Koster, *Bristol-Myers Squibb Pharmaceutical Research Institute, Princeton, New Jersey,* Monobactams (under Antibiotics)

Roger H. Kottke, *Great Lakes Chemical Corporation, West Lafayette, Indiana,* Furan derivatives

Joseph Kozakiewicz, *American Cyanamid Company, Stamford, Connecticut,* Acrylamide polymers

David M. Krentz, *E.I. du Pont de Nemours & Co., Inc., Wilmington, Delaware,* Information retrieval

Charles T. Kresge, *Mobil Research and Development Corporation, Paulsboro, New Jersey,* Molecular sieves

Edward Kresge, *Exxon Chemical Company, Linden, New Jersey,* Butyl rubber (under Elastomers, synthetic)

M. A. Krevalis, *Exxon Chemical Company, Baton Rouge, Louisiana,* Trialkylacetic acids (under Carboxylic acids)

A. B. Krewinghaus, *Shell Development Company, Houston, Texas,* Gasification (under Coal conversion processes)

Larry J. Kricka, *University of Pennsylvania, Philadelphia,* Chemiluminescence (under Luminescent materials)

Pieter Krijgsman, *CEC Company, Switzerland,* Hydrothermal processing

Ramesh Krishnamurti, *Occidental Chemical Corporation, Grand Island, New York,* Ring-chlorinated toluenes (under Chlorocarbons and chlorohydrocarbons)

Gerald A. Krulik, *Applied Electroless Concepts, Inc., El Toro, California,* Electroless plating; Survey (under Metallic coatings)

Léon Krumenacker, *Rhône-Poulenc Recherches, France,* Hydroquinone, resorcinol, and catechol

Katharine Ku, *Stanford University, Palo Alto, California,* Licensing

Volker Kuellmer, *Hoffmann-La Roche Inc., Nutley, New Jersey,* Ascorbic acid (under Vitamins)

Günter H. Kühl, *Consultant, Cherry Hill, New Jersey,* Molecular sieves

Donald M. Kulich, *GE Plastics, Pittsfield, Massachusetts,* ABS resins (under Acrylonitrile polymers)

S. M. Kunz, *The Carborundum Company, Niagara Falls, New York,* Silicon carbide (under Carbides)

Ralf Kuriyel, *Millipore Corporation, Bedford, Massachusetts,* Ultrafiltration

J. J. Kurland, *Union Carbide Chemicals & Plastics Company, Inc., South Charleston, West Virginia,* Acrolein and derivatives

Roger Kust, *Tetra Chemicals, Houston, Texas; Tetra Technologies, Inc., The Woodlands, Texas,* Calcium chloride (under Calcium compounds); Sodium halides–sodium bromide (under Sodium compounds)

Florence H. Kvalnes, *E.I. du Pont de Nemours & Co., Inc., Wilmington, Delaware,* Information retrieval

Vernon L. Kyllingstad, *Zeon Chemicals, Louisville, Kentucky,* Polyethers (under Elastomers, synthetic)

Y. Labat, *Elf Atochem, Artix, France,* Thioglycolic acid

Michael R. Ladisch, *Purdue University, West Lafayette, Indiana,* Bioseparations

Richard J. Lagow, *University of Texas at Austin,* Direct fluorination (under Fluorine compounds, organic)

Yu-Chin Lai, *Bausch & Lomb, Rochester, New York,* Contact lenses

Leonard Lamberson, *Western Michigan University, Kalamazoo, Michigan,* Materials reliability

Gregory Lambeth, *Schenectady Chemicals, Inc., Schenectady, New York,* Alkylphenols

Marvin Landau, *Huls America, Inc., Piscataway, New Jersey,* Driers and metallic soaps

L. M. Landoll, *Hercules Inc., Wilmington, Delaware,* Olefin (under Fibers)

David Lang, *Portland Cement Association, Skokie, Illinois,* Cement

Julie B. Lang, *Arthur D. Little, Inc., Cambridge, Massachusetts,* Building materials, plastic

Howard Lanza, *Miles, Inc., Tarrytown, New York,* Clinical chemistry (under Automated instrumentation)

George R. Lappin, *Ethyl Corporation, Albemarle Corporation, Baton Rouge, Louisiana,* Synthetic processes (under Alcohols, higher aliphatic); Olefins, higher

P. A. Larson, *E. I. du Pont de Nemours & Co., Inc., Wilmington, Delaware,* Dimethylacetamide (under Acetic acid and derivatives)

Richard R. Lattime, *The Goodyear Tire & Rubber Company, Akron, Ohio,* Styrene–butadiene rubber

Robert P. Lattimer, *The BF Goodrich Company, Brecksville, Ohio,* Antiozonants

S. K. Lau, *The Carborundum Company, Niagara Falls, New York,* Silicon carbide (under Carbides)

Armin Lauterbach, *SQM Iodine Corporation, Chile,* Iodine and iodine compounds

Hubert Lauvray, *Rhône-Poulenc, France,* Gallium and gallium compounds

B. C. Lawes, *E. I. Du Pont de Nemours & Co., Inc., Wilmington, Delaware,* Sulfuric and sulfurous esters

Mary E. Lawson, *SPI Polyols, Inc., New Castle, Delaware,* Sugar alcohols

Robert W. Layer, *The BF Goodrich Research Center, Brecksville, Ohio,* Antiozonants; Phenylenediamines, Diarylamines (both under Amines, aromatic)

Joel L. Lazewatsky, *DuPont-Merck Pharmaceutical Company, North Billerica, Massachusetts,* Radiopharmaceuticals

Tom Leahy, *Mettler-Toledo, Inc., Inman, South Carolina,* Weighing and proportioning

M. E. Leaphart II, *University of South Carolina, Columbia,* Nitrides

Stuart E. Lebo, Jr., *LignoTech USA, Inc., Rothchild, Wisconsin,* Lignin

Thomas D. Lee, *Kraft Foods, Tarrytown, New York,* Sweeteners

Harold Leeper, *ALZA Corporation, Palo Alto, California,* Pharmaceuticals (under Controlled release technology)

William F. Lehmann, *Consultant, Federal Way, Washington,* Wood-based composites and laminates

H. David Leigh, III, *Clemson University, Clemson, South Carolina,* Refractories

John Leland, *Phillips 66, Bartlesville, Oklahoma,* Poly(phenylene sulfide) (under Polymers containing sulfur)

Arthur A. Leman, *Rohm and Haas Company, Spring House, Pennsylvania,* Architectural (under Paint)

Charles H. Lemke, *E. I. du Pont de Nemours & Company, Inc., Niagara Falls, New York and the University of Delaware, Newark,* Sodium and sodium alloys

Joseph J. Len, *Vinings Industries, Inc., Atlanta, Georgia,* Aluminates (under Aluminum compounds)

Marguerite L. Leng, *Leng Associates, Midland, Michigan,* Pesticides

George R. Lenz, *The BOC Group Technical Center, Murray Hill, New Jersey,* Anesthetics

J. A. Lepinski, *P.T. Perkasa Indobaja, Indonesia,* Iron; Iron by direct reduction

George Y. Lesher, *Sterling-Winthrop Research Institute, Rensselaer, New York,* Quinolones (under Antibacterial agents)

Patricia M. Lesko, *Rohm and Haas Company, Spring House, Pennsylvania,* Methacrylic polymers

Gerd Leston, *Consultant, Pittsburgh, Pennsylvania,* (Polyhydroxy)benzenes

Alan Letki, *Alfa Laval Separation Inc., Warminster, Pennsylvania,* Centrifugal separation (under Separation)

Robert Leurs, *Leiden/Amsterdam Center for Drug Research,* Histamine and histamine antagonists

S. P. Levings, *Martin Marietta Magnesia Specialties, Inc., Baltimore, Maryland,* Magnesium compounds

Cynthia L. Levinson, *University of California, San Francisco,* Enzyme inhibitors
Ronald Levy, *Arthur D. Little, Inc., Cambridge, Massachusetts,* Building materials, plastic
B. A. Lewis, *Cornell University, Ithaca, New York,* Dietary fiber
I. C. Lewis, *UCAR Carbon Company, Inc., Cleveland, Ohio,* Carbon and artificial graphite (under Carbon)
Larry Lewis, *GE Corporate Research and Development, Schenectady, New York,* Silicones (under Silicon compounds)
O. Griffin Lewis, *Consultant, Norwalk, Connecticut,* Sutures
T. Li, *ASARCO, Inc., Denver, Colorado,* Antimony and antimony alloys
Jason Liang, *Kalama Chemical, Inc., Kalama, Washington,* Benzaldehyde; Benzoic acid
Richard B. Lieberman, *Montell Polyolefins, Elkton, Maryland,* Polypropylene (under Olefin polymers)
Thomas A. Liederbach, *Electrode Corporation, Chardon, Ohio,* Metal anodes
JoAnn S. Lighty, *University of Utah, Salt Lake City,* Incinerators
Henry C. Lin, *Occidental Chemical Corporation, Grand Island, New York,* Benzyl chloride, benzal chloride, and benzotrichloride; Ring-chlorinated toluenes (both under Chlorocarbons and chlorohydrocarbons)
K. F. Lin, *Hercules Incorporated, Wilmington, Delaware,* Alkyd resins
Stephen Y. Lin, *LignoTech USA, Inc., Rothchild, Wisconsin,* Lignin
Youlin Lin, *Mallinckrodt Medical, Inc., St. Louis, Missouri,* Radiopaques
Charles B. Lindahl, *Elf Atochem North America, Inc., Tulsa, Oklahoma,* Introduction; Antimony; Arsenic; Barium; Calcium; Germanium; Phosphorus; Tantalum; Tin; Zinc (all under Fluorine compounds, inorganic)
Klaus R. Linder, *Bristol-Myers Squibb Pharmaceutical Research Institute, Princeton, New Jersey,* Monobactams (under Antibiotics)
Victor Lindner, *Armament Research, Development, and Engineering Agency, Dover, New Jersey,* Explosives; Propellants (both under Explosives and propellants)
Noam Lior, *University of Pennsylvania, Philadelphia,* Supply and desalination (under Water)
David Lipp, *American Cyanamid Company, Stamford, Connecticut,* Acrylamide polymers
John H. Litchfield, *Battelle Memorial Institute, Columbus, Ohio,* Foods, nonconventional
Kou-Chang Liu, *International Specialty Products, Wayne, New Jersey,* Pyrrole and pyrrole derivatives
Teh C. Lo, *T. C. Lo & Associates, Wayne, New Jersey,* Extraction, liquid–liquid
Gary W. Loar, *McGean-Rohco, Inc., Cleveland, Ohio,* Chromium compounds
J. Fred Lochary, *The Dow Chemical Company, Freeport, Texas,* Alkanolamines from olefin oxides and ammonia
David J. Locker, *Kodak Research Laboratories, Rochester, New York,* Photography
J. C. Lodder, *University of Twente, the Netherlands,* Magnetic (under Information storage materials)
Gerd Loebbert, *BASF Corporation, Holland, Michigan,* Phthalocyanine compounds; Pigment dispersions
Richard Loehman, *Sandia National Laboratories, Albuquerque, New Mexico,* Ceramic processing; Electronic properties and material structure; Glass structure and properties; Mechanical properties and behavior; Nonlinear optical and electrooptic ceramics; Overview (all under Ceramics)
Robert B. Login, *Sybron Chemicals Inc., Wellford, South Carolina,* Vinyl ether monomers and polymers; *N*-Vinylamide polymers (both under Vinyl polymers)
John E. Logsdon, *Union Carbide Corporation, Texas City,* Ethanol; Isopropyl alcohol (under Propyl alcohols)
Richard A. Loke, *Union Carbide Corporation, Texas City, Texas,* Isopropyl alcohol (under Propyl alcohols)
D. A. Lomas, *UOP Research Center, Des Plaines, Illinois,* Regeneration, FCC (under Catalysts)
G. Gilbert Long, *North Carolina State University, Raleigh, North Carolina,* Antimony compounds; Arsenic compounds; Bismuth compounds
George M. Long, *Institute of Gas Technology, Chicago, Illinois,* Pipelines
J. C. Long, *UCAR Carbon Company, Inc., Cleveland, Ohio,* Carbon and artificial graphite (under Carbon)
Gabriel P. López, *University of New Mexico, Albuquerque,* Nanotechnology
José A. López, *Baylor College of Medicine, Houston, Texas,* Nanotechnology
John F. Lorenc, *Schenectady Chemicals, Inc., Schenectady, New York,* Alkylphenols
Steven Lowenkron, *The Dow Chemical Company, LaPorte, Texas,* Methylenedianiline (under Aromatic amines)
R. Derric Lowery, *Exxon Chemical Company, Baton Rouge, Louisiana,* Hydrocarbon resins
Peter Luckie, *Pennsylvania State University, University Park, Pennsylvania,* Size separation (under Separation)
Matthew Lucy, *University of Missouri, Columbia,* Animals (under Genetic engineering)
R. P. Lukens, *American Society for Testing and Materials, Woodbury, New Jersey,* Conversion Factors, Abbreviations, and Unit Symbols
Hugh M. Lybarger, *The Goodyear Tire and Rubber Company, Akron, Ohio,* Isoprene
Jeremiah Lynch, *Exxon Chemical Company, East Millstone, New Jersey,* Industrial hygiene
Matthew L. Lynch, *The Procter & Gamble Company, Cincinnati, Ohio,* Soap
M. Bruce Lyne, *International Paper, Tuxedo, New York,* Paper
Jesse L. Lynn, Jr., *Lever Brothers Company, Edgewater, New Jersey,* Detergency; Surfactants
Michael J. Lysaght, *Cyto Therapeutics, Inc., Providence, Rhode Island,* Dialysis
John B. MacChesney, *AT&T Bell Laboratories, Murray Hill, New Jersey,* Fiber optics
Kenneth O. MacFadden, *W.R. Grace & Company, Columbia, Maryland,* Laboratory information management systems
Warren C. MacKellar, *Lilly Research Laboratories, Indianapolis, Indiana,* Human growth hormone (under Hormones)
K. J. Mackenzie, *Consultant, Greenville, South Carolina,* Film and sheeting materials
Donald Mackey, *Zeon Chemicals, USA, Louisville, Kentucky,* Nitrile rubber (under Elastomers, synthetic)
George MacZura, *Aluminum Company of America, Pittsburgh, Pennsylvania,* Alumina, calcined and tabular (under Aluminum compounds)
Arun Madan, *MVSystems, Inc., Golden, Colorado,* Amorphous semiconductors (under Semiconductors)
Hubert Maehr, *Hoffmann-LaRoche, Inc., Nutley, New Jersey,* Elfamycins (under Antibiotics)
Patrick Maestro, *Rhône-Poulenc, France,* Lanthanides
Douglas Magde, *University of California at San Diego,* Kinetic measurements
Uday Mahagaokar, *Shell Development Company, Houston, Texas,* Gasification (under Coal conversion processes)
Tariq Mahmood, *Elf Atochem North America, Inc., Tulsa, Oklahoma,* Introduction; Antimony; Arsenic; Calcium; Germanium; Phosphorus; Tantalum; Tin; Zinc (all under Fluorine compounds, inorganic)
Bernard Maisonneuve, *Akzo Chemicals, Inc., Dobbs Ferry, New York,* Amine oxides
Thomas G. Majewicz, *Aqualon Company, Wilmington, Delaware,* Cellulose ethers
Harry V. Makar, *Consultant, Ellicott City, Maryland,* Ferrous metals (under Recycling)
Subhash Makhija, *Consultant, Summit, New Jersey,* Polyesters, thermoplastic
Christian Maliverney, *Rhône-Poulenc Recherches, France,* Hydroxybenzaldehydes
Henrik Malmos, *Enzyme Technology, Bagsvaerd, Denmark,* Industrial (under Enzyme applications)
Daniel Maloney, *Universal Foods Technical Center, Milwaukee, Wisconsin,* Yeasts
Nenad V. Mandich, *HBM Engineering Company, Lansing, Illinois,* Survey (under Metallic coatings)
James L. Manganaro, *FMC Corporation, Princeton, New Jersey,* Barium compounds
W. C. Mangum, *University of Kentucky, Lexington,* Reverse osmosis
Ganpat Mani, *Allied-Signal Inc., Morristown, New Jersey,* Boron, boron trifluoride; Sulfur (both under Fluorine compounds, inorganic)
M. L. Maniocha, *Martin Marietta Magnesia Specialties, Inc., Baltimore, Maryland,* Magnesium compounds
Mark J. Manning, *U.S. Borax Research Corporation, Anaheim, California,* Organic boron–nitrogen compounds (under Boron compounds)
Robert M. Manyik, *Union Carbide Corporation, South Charleston, West Virginia,* Acetylene (under Hydrocarbons)
Chien-Pei Mao, *Delavan Inc, West Des Moines, Iowa,* Sprays
Joseph Marinelli, *Peabody Solids Flow, Charlotte, North Carolina,* Bulk powders (under Powders, handling)
Vernon H. Markant, *E. I. du Pont de Nemours & Company, Inc., Niagara Falls, New York,* Sodium and sodium alloys
Eric J. Markel, *University of South Carolina, Columbia,* Nitrides
Kathleen Markey, *Synergen, Inc., Boulder, Colorado,* Electrophoresis (under Electroseparations)
Daniel Marmion, *Allied-Signal, Buffalo, New York,* Colorants for foods, drugs, cosmetics, and medical devices
John A. Marsella, *Air Products and Chemicals, Inc., Allentown, Pennsylvania,* Dimethylformamide (under Formic acid and derivatives)
F. Lennart Marten, *Air Products and Chemicals, Inc., Allentown, Pennsylvania,* Vinyl alcohol polymers (under Vinyl polymers)
Alton E. Martin, *The Dow Chemical Company, Freeport, Texas,* Propylene glycols (under Glycols)
James R. Martin, *F. Hoffmann-La Roche Ltd., Basel, Switzerland,* Psychopharmacological agents

S. J. Martin, *Sandia National Laboratories, Albuquerque, New Mexico,* Sensors

Robert T. Mason, *Koppers Industries, Inc., Pittsburgh, Pennsylvania,* Naphthalene

J. M. Massie, *Goodyear Tire & Rubber Company, Akron, Ohio,* Polybutadiene (under Elastomers, synthetic)

Louis R. Matricardi, *Consultant, Tonawanda, New York,* Manganese and manganese alloys

Robert L. Matteri, *U.S. Department of Agriculture, Columbia, Missouri,* Anterior pituitary hormones; Anterior pituitary-like hormones (both under Hormones)

Donald M. Mattox, *Management Plus, Inc., Albuquerque, New Mexico,* Film formation techniques (under Thin films)

Ignacio Maturana, *SQM Nitratos SA, Santiago, Chile,* Sodium nitrate (under Sodium compounds)

F. Mauger, *Rhône-Poulenc, Cranbury, New Jersey,* Vanillin

V. Maureaux, *Rhône-Poulenc, Cranbury, New Jersey,* Vanillin

J. Wilson Mausteller, *Consultant, Evans City, Pennsylvania,* Oxygen-generation systems

Ivo Mavrovic, *Consultant, New York, New York,* Urea

William J. Mazzafro, *Air Products and Chemicals, Inc., Allentown, Pennsylvania,* Nitric acid

William H. McBride, *UCLA Medical Center, Los Angeles, California,* Radioprotective agents

Robert B. McBroom, *U.S. Borax Research Corporation, Anaheim, California,* Boron oxides, boric acid, and borates (under Boron compounds)

Edward F. McCarthy, *Luzenac America, Englewood, Colorado,* Talc

W. B. McCormack, *E. I. Du Pont de Nemours & Co., Inc., Wilmington, Delaware,* Sulfuric and sulfurous esters

Paul H. McCormick, *Drying Unincorporated, Newark, Delaware,* Drying

Walter C. McCrone, *McCrone Research Institute, Chicago, Illinois,* Microscopy

C. E. McDonald, *E. I. du Pont de Nemours & Co., Inc., Deepwater, New Jersey,* Chlorosulfuric acid

Daniel P. McDonald, *Consultant, Belmont, North Carolina,* Lithium and lithium compounds

Joseph A. McDonough, *Hoechst-Celanese Corporation, Corpus Christi, Texas,* Esters, organic

Thomas McDonough, *Institute of Paper Science and Technology, Atlanta, Georgia,* Pulp and paper (under Bleaching agents)

Robin S. McDowell, *Pacific Northwest National Laboratory, Richland, Washington,* Spectroscopy, optical

Harold J. McElhone, Jr., *Ciba-Geigy Corporation, Greensboro, North Carolina,* Fluorescent whitening agents

Thomas McEntee, *Morton International, Inc., Beverly, Massachusetts,* Industrial antimicrobial agents

J. A. McGeough, *University of Edinburgh, Scotland,* Machining methods, electrochemical

Vincent D. McGinniss, *Battelle Columbus Laboratory, Columbus, Ohio,* Radiation curing

Donald McGregor, *Bristol-Myers Squibb Company, Wallingford, Connecticut,* Aminoglycosides (under Antibiotics)

J. L. McGuire, *R. W. Johnson Pharmaceutical Research Institute, Raritan, New Jersey,* Contraceptives

David B. McKeever, *USDA Forest Service, Madison, Wisconsin,* Wood

Arthur McKenna, *Witco Corporation, Chicago, Illinois,* Aluminum carboxylates (under Aluminum compounds)

Ronald J. McKinney, *E. I. du Pont de Nemours & Co., Inc., Wilmington, Delaware,* Nitriles

Wayne A. McRae, *Consultant, Mannedork, Switzerland,* Electrodialysis (under Electroseparations)

David E. Mears, *Unocal Corporation, Brea, California,* C_1–C_6; Survey (both under Hydrocarbons)

Roland E. Meissner III, *The Ralph M. Parsons Company, Pasadena, California,* Plant layout; Plant location

Charlene Mello, *U.S. Army Natick Research, Development, & Engineering Center, Natick, Massachusetts,* Silk

Sudhir K. Mendiratta, *Olin Chemicals, Charleston, Tennessee,* Chloric acid and chlorates (under Chlorine oxygen acids and salts); Perchloric acid and perchlorates

J. Carlos Menéndez, *Universidad Complutense, Madrid, Spain,* Hydantoin and its derivatives

Dieter Mergel, *University of Essen, Germany,* Optical (under Information storage materials)

Philip H. Merrell, *Mallinckrodt, Inc., St Louis, Missouri,* Sodium halides–sodium iodide (under Sodium compounds)

James A. Mertens, *Dow Chemical USA, Midland, Michigan,* Dichloroethylene; Trichloroethylene (both under Chlorocarbons and chlorohydrocarbons)

Stanley H. Mervis, *Polaroid Corporation, Cambridge, Massachusetts,* Color photography, instant

Keith Mesch, *Morton International, Inc., Cincinnati, Ohio,* Heat stabilizers

Dayal T. Meshri, *Advance Research Chemicals, Inc., Catoosa, Oklahoma,* Aluminum; Cobalt; Copper; Iron; Lead; Mercury; Molybdenum; Nickel; Rhenium; Silver; Titanium; Zirconium (all under Fluorine compounds, inorganic)

Robert L. Metcalf, *University of Illinois at Urbana-Champaign,* Insect control technology

Walter C. Meuly, *Rhône-Poulenc, Inc., Princeton, New Jersey,* Coumarin

Theodore Michelsen, *Schuller International, Littleton, Colorado,* Survey (under Building materials)

Jana B. Milford, *University of Connecticut, Storrs, Connecticut,* Atmospheric modeling

Carol J. Miller, *Wayne State University, Detroit, Michigan,* Groundwater monitoring

Charles E. Miller, *Westvaco Corp., North Charleston, South Carolina,* Activated carbon (under Carbon)

David J. Miller, *Union Carbide Chemicals & Plastics Corporation, South Charleston, West Virginia,* Aldehydes

F. M. Miller, *Portland Cement Association, Skokie, Illinois,* Cement

George H. Miller, *Schering-Plough Research, Bloomfield, New Jersey,* Chloramphenicol and analogues (under Antibiotics)

J. A. Miller, *DuPont Central Research and Development, Wilmington, Delaware,* Research/technology management

Matt C. Miller, *Dow Chemical USA, Freeport, Texas,* Ethyl chloride (under Chlorocarbons and chlorohydrocarbons)

Raimund J. Miller, *Lonza, Inc., Fair Lawn, New Jersey,* Ketenes, ketene dimers, and related substances

Regis B. Miller, *USDA Forest Service, Madison, Wisconsin,* Wood

T. M. Miller, *University of Florida, Gainesville,* Glasses, organic–inorganic hybrids

Norman Milleron, *SEN Vac Services, Berkeley, California,* Vacuum technology

George H. Millet, *3M Company, St. Paul, Minnesota,* Fluorocarbon elastomers (under Elastomers, synthetic); Polychlorotrifluoroethylene (under Fluorine compounds, organic)

Keith R. Millington, *CSIRO, Belmont, Geelong, Australia,* Wool

Luray M. Minkiewicz, *E.I. du Pont de Nemours & Co., Inc., Wilmington, Delaware,* Information retrieval

C. A. Mintmier, *Martin Marietta Magnesia Specialties, Inc., Baltimore, Maryland,* Magnesium compounds

Paul E. Minton, *Union Carbide Corporation, South Charleston, West Virginia,* Heat-transfer media other than water (under Heat-exchange technology)

Chanakya Misra, *Aluminum Company of America, Alcoa Center, Pennsylvania,* Alumina, hydrated (under Aluminum compounds)

Scott F. Mitchell, *Huntsman Specialty Chemicals Corporation, St. Louis, Missouri,* Maleic anhydride, maleic acid, and fumaric acid

Stephen Mitchell, *University of London, Birmingham, England,* Aminophenols

Irving Moch, Jr., *E. I. du Pont de Nemours & Co., Inc., Wilmington, Delaware,* Hollow-fiber membranes

R. T. Moll, *Alfa Laval Separation Inc., Warminster, Pennsylvania,* Centrifugal separation (under Separation)

Robert C. Monroe, *Hudson Products Corporation, Houston, Texas,* Fans and blowers

Victor Monroy, *General Tire, Inc., Akron, Ohio,* Anionic initiators (under Initiators)

Anthony J. Montana, *Warner-Lambert Company, Morris Plains, New Jersey,* Trends (under Analytical methods)

Braja D. Mookherjee, *International Flavor & Fragrances, Union Beach, New Jersey,* Benzyl alcohol and β-phenethyl alcohol; Oils, essential

John J. Mooney, *Engelhard Corporation, Iselin, New Jersey,* Exhaust control, automotive

Colin Moore, *Consultant, Woodbury, Connecticut,* Mass spectrometry

Patrick J. Moran, *United States Naval Academy, Annapolis, Maryland,* Corrosion and corrosion control

J-L. Moreau, *F. Hoffmann-La Roche Ltd., Basel, Switzerland,* Psychopharmacological agents

Jeffrey O. Moreno, *Donelan, Cleary, Wood & Maser P.C., Washington, D.C.,* Transportation

Booker Morey, *SRI International, Menlo Park, California,* Dewatering

Bradley P. Morgan, *Pfizer, Inc., Groton, Connecticut,* Steroids

Don Morgan, *O.S. Walker Company, Milwaukee, Wisconsin,* Magnetic separation (under Separation)

James J. Morgan, *California Institute of Technology, Pasadena,* Properties (under Water)

Earl D. Morris, *The Dow Chemical Company, Midland, Michigan,* Halogenated derivatives (under Acetic acid and derivatives)

Michael D. Morris, *University of Michigan, Ann Arbor,* Raman spectroscopy (under Infrared technology and Raman spectroscopy)

Lowen Morrison, *Procter & Gamble, Cincinnati, Ohio,* Glycerol

R. Steve Morrow, *Coca-Cola USA, Atlanta, Georgia,* Carbonated beverages

Roy E. Morse, *Consultant, Clemmons, North Carolina,* Fat replacers

Lester R. Morss, *Argonne National Laboratory, Argonne, Illinois,* Plutonium and plutonium compounds

Joseph Mort, *Xerox Corporation, Webster, New York,* Electrophotography

J. Morton, *Virginia Polytechnic Institute and State University, Blacksburg, Virginia,* Survey (under Composite materials)

Maurice Morton, *Consultant, Beachwood, Ohio,* Survey (under Elastomers, synthetic)

Armand Moscovici, *The Kerite Company, Woodbridge, Connecticut,* Insulation, electric

Carl B. Moyer, *Acurex Corporation, Mountain View, California,* Alcohol fuels

Melinda S. Moynihan, *Pfizer, Inc., Groton, Connecticut,* Steroids

Helmut Mrozik, *Merck, Sharp, and Dohme Research Laboratories, Rahway, New Jersey,* Avermectins (under Antiparasitic agents)

Werner H. Mueller, *Hoechst-Celanese Corporation, Charlotte, North Carolina,* Sodium (under Fluorine compounds, inorganic)

Michel Mulhauser, *Rhône-Poulenc Recherches, France,* Hydroxybenzaldehydes

Thomas L. Muller, *DuPont Chambers Works, Deepwater, New Jersey,* Sulfuric acid and sulfur trioxide

Michael E. Mullins, *Michigan Technological University, Hougton,* Engineering, chemical data correlation

Michael Mummey, *Huntsman Specialty Chemicals Corporation, St. Louis, Missouri,* Maleic anhydride, maleic acid, and fumaric acid

Ted Munday, *FMC Corporation, Princeton, New Jersey,* Phosphorus

Toru Murakami, *UBE Industries Limited, Tokyo, Japan,* Oxalic acid

Donald P. Murphy, *Parker Amchem, Madison Heights, Michigan,* Cleaning (under Metal surface treatments)

Frank H. Murphy, *The Dow Chemical Company, Freeport, Texas,* Propylene glycols (under Glycols)

David Murray, *University of Alberta, Cross Cancer Institute, Edmonton, Canada,* Radioprotective agents

Raymond L. Murray, *Consultant, Raleigh, North Carolina,* Introduction, Reactor types, Waste management (under Nuclear reactors)

Ramiah Murugan, *Reilly Industries, Inc., Indianapolis, Indiana,* Pyridine and pyridine derivatives

Angelika Muscate, *University of California, San Francisco,* Enzyme inhibitors

Jeffrey C. Myers, *Midrex Direct Reduction Corporation, Charlotte, North Carolina,* Iron

Philip Myers, *Chevron Research and Technical Company, Orinda, California,* Tanks and pressure vessels

Terry N. Myers, *Elf Atochem North America, Inc., Buffalo, New York,* Free-radical initiators (under Initiators); Organic peroxides (under Peroxides and peroxide compounds)

Tattanahalli Nagabhushan, *Schering-Plough Research, Bloomfield, New Jersey,* Chloramphenicols and analogues (under Antibiotics)

D. R. Nagaraj, *CYTEC Industries, Stamford, Connecticut,* Minerals recovery and processing

Vasantha Nagarajan, *E. I. du Pont de Nemours & Co., Inc., Wilmington, Delaware,* Microbes (under Genetic engineering)

Nobuyuki Nagato, *Showa Denko K.K., Tokyo, Japan,* Allyl alcohol and its derivatives

Raman Nambudripad, *Beth Israel Hospital, Boston, Massachusetts,* Proteins

Raymond Narr, *General Electric Company, Pittsfield, Massachusetts,* Engineering plastics

Kurt Nassau, *Nassau Consultants, Lebanon, New Jersey,* Color; Gemstone materials, Gemstone treatment (both under Gemstones)

Michael Nastasi, *Los Alamos National Laboratory, Los Alamos, New Mexico,* Ion implantation

Samuel Natansohn, *GTE Laboratories, Inc., Waltham, Massachusetts,* Structural ceramics (under Advanced ceramics)

Paul M. Natishan, *Naval Research Laboratory, Washington, D.C.,* Corrosion and corrosion control

Patrick T. Naughtin, *CSIRO, Belmont, Geelong, Australia,* Wool

Frank C. Naughton, *CasChem/CAMREX, Bayonne, New Jersey,* Castor oil

Robert J. Naumann, *University of Alabama in Huntsville,* Space processing

Behrooz Nazer, *E.I. du Pont de Nemours & Co., Inc., Wilmington, Delaware,* Information retrieval

Jeffrey T. Neil, *GTE Laboratories, Inc., Waltham, Massachusetts,* Structural ceramics (under Advanced ceramics)

W. D. Neilsen, *Union Carbide Chemicals & Plastics Company, Inc., South Charleston, West Virginia,* Acrolein and derivatives

Lev Nelik, *Roper Pumps Company, Commerce, Georgia,* Pumps

Ralph D. Nelson, Jr., *E. I. du Pont de Nemours & Company, Inc., Wilmington, Delaware,* Dispersion of powders in liquids (under Powders, handling)

L. H. Nemec, *Albemarle Corporation, Baton Rouge, Louisiana,* Olefins, higher

Marshall J. Nepras, *Stepan Company, Northfield, Illinois,* Sulfonation and sulfation

Mary P. Neu, *Glenn T. Seaborg Institute for Transactinium Science, Los Alamos National Laboratory, Los Alamos, New Mexico,* Thorium and thorium compounds; Uranium and uranium compounds

John Newman, *University of California, Berkeley,* Introduction (under Electrochemical processing)

Robert E. Newnham, *Pennsylvania State University, University Park, Pennsylvania,* Electronic ceramics (under Advanced ceramics)

D. S. Newsome, *M. W. Kellogg Company, Houston, Texas,* Hydrogen

William E. Newton, *Virginia Polytechnic Institute and State University, Blacksburg,* Nitrogen fixation

Kirstin Nichols, *ALZA Corporation, Palo Alto, California,* Pharmaceuticals (under Controlled release technology)

R. Terrell Nichols, *Ford Motor Company, Dearborn, Michigan,* Laminated materials, glass

Henrik Kim Nielsen, *Enzyme Technology, Bagsvaerd, Denmark,* Industrial (under Enzyme applications)

Peder Holk Nielsen, *Enzyme Technology, Bagsvaerd, Denmark,* Industrial (under Enzyme applications)

Ralph H. Nielsen, *Teledyne Wah Chang Corporation, Albany, Oregon,* Hafnium and hafnium compounds

Robert Nielsen, *Consultant, Montgomery, Texas,* Supported (under Catalysts)

Tage Kjaer Nielsen, *Enzyme Technology, Bagsvaerd, Denmark,* Industrial (under Enzyme applications)

Alvin W. Nienow, *University of Birmingham, Birmingham, United Kingdom,* Biotechnology (under Aeration)

J. F. Nissen, *Brygerifoneningen (The Danish Brewer's Association), Kobenbarn K, Denmark,* Beer

Jacobus W. N. Noordermeer, *DSM Elastomers Europe BV, Geleen, the Netherlands,* Ethylene–propylene–diene rubber (under Elastomers, synthetic)

P. M. Norling, *DuPont Central Research and Development, Wilmington, Delaware,* Research/technology management

Ronald W. Novak, *Rohm and Haas Company, Spring House, Pennsylvania,* Survey (under Acrylic ester polymers); Methacrylic polymers

Mirek Novotny, *Cerdec Corporation, Washington, Pennsylvania,* Inorganic (under Pigments)

Milton Nowak, *Troy Chemical Corporation, Newark, New Jersey,* Mercury compounds

Richard A. Nugent, *The Upjohn Company, Kalamazoo, Michigan,* Analgesics, antipyretics, and antiinflammatory agents

Gustavo Ober, *SQM Iodine Corporation, Chile,* Iodine and iodine compounds

B. E. Obi, *The Dow Chemical Company, Midland, Michigan,* Vinylidene chloride monomer and polymers

J. T. O'Connor, *Loctite Corporation, Newington, Connecticut,* 2-Cyanoacrylic ester polymers (under Acrylic ester polymers)

Anders Oestergaard, *Enzyme Technology, Bagsvaerd, Denmark,* Industrial (under Enzyme applications)

Julia I. O'Farrelly, *Johnson Matthey Technology Center, Reading, U.K.,* Platinum-group metals

Robert F. Ohm, *R. T. Vanderbilt Company, Inc., Norwalk, Connecticut,* Rubber chemicals; Rubber compounding

George A. Olah, *University of Southern California, Los Angeles,* Friedel-Crafts reactions

J. E. Oldfield, *Oregon State University, Corvallis,* Tellurium and tellurium compounds

Hans Sejr Olsen, *Enzyme Technology, Bagsvaerd, Denmark,* Industrial (under Enzyme applications)

D. L. Olsson, *Rochester Institute of Technology, Rochester, New York,* Containers for industrial materials (under Packaging)

E. F. Olszewski, *ABB Lummus Crest, Inc., Bloomfield, New Jersey,* Ethylene

Suzan Onel, *McKenna & Cuneo, LLP, Washington, D.C.,* Pharmaceuticals and cosmetics (under Regulatory agencies)

Jarl L. Opgrande, *Kalama Chemical, Inc., Kalama, Washington,* Benzaldehyde; Benzoic acid

Ross Opsahl, *Akzo Chemicals Corporation, McCook, Illinois,* Amides, fatty acid
Rodrigo Orefice, *University of Florida, Gainesville,* Sol–gel technology
T. A. Orofino, *AstroTurf Industries, Dalton, Georgia,* Recreational surfaces
Anil R. Oroskar, *UOP, Des Plaines, Illinois,* Adsorption, liquid separation
Neal F. Osborne, *SmithKline Beecham Pharmaceuticals, Surrey, United Kingdom,* Carbapenems and penems (under Antibiotics, β-lactams)
John Osepchuk, *Raytheon Company, Lexington, Massachusetts,* Microwave technology
K. Oshima, *Pall Corporation, Port Washington, New York,* Microbial and viral filtration
Robert A. Outten, *Hoffmann-La Roche, Inc., Nutley, New Jersey,* Biotin (under Vitamins)
Michael J. Owen, *Dow Corning Corporation, Midland, Michigan,* Defoamers; Release agents
W. John Owen, *Dow Elanco Europe, Wantage, Oxon, U.K.,* Fungicides, agricultural
S. Ted Oyama, *Clarkson University, Potsdam, New York,* Survey (under Carbides)
Toshitsugu Ozeki, *Kyowa Hakko Kogyo Company, Yokkaichi, Japan,* Survey (under Amino acids)
E. Dickson Ozokwelu, *Amoco Chemical Company, Naperville, Illinois,* Toluene
John E. Pace, *GE Plastics, Pittsfield, Massachusetts,* ABS resins (under Acrylonitrile polymers)
Morton Pader, *Consumer Products Development Resources, Inc., Teaneck, New Jersey,* Dentifrices
Billie J. Page, *McGean-Rohco, Inc., Cleveland, Ohio,* Chromium compounds
D. J. Page, *UCAR Carbon Company, Inc., Cleveland, Ohio,* Carbon and artificial graphite (under Carbon)
Richard P. Palluzi, *Exxon Research and Engineering Company, Florham Park, New Jersey,* Pilot plants
Richard A. Palmer, *Dow Corning Corporation, Midland, Michigan,* Sealants
Robert J. Palmer, *Du Pont de Nemours International SA, Geneva Switzerland,* Plastics (under Polyamides)
Y. C. Pao, *University of Nebraska, Lincoln, Nebraska,* Computer-aided design and manufacturing (CAD/CAM)
Anthony J. Papa, *Union Carbide Corporation, South Charleston, West Virginia,* Amyl alcohols
John R. Papcun, *Atotech, Cleveland, Ohio,* Ammonium; Boron, fluoroboric acid and fluoroborates; Lithium; Magnesium; Potassium (all under Fluorine compounds, inorganic)
Peter G. Pape, *Dow Corning Corporation, Midland, Michigan,* Silylating agents (under Silicon compounds)
Lyn Paquin, *Cabot Corporation, Billerica, Massachusetts,* Carbon black (under Carbon)
F. Parenti, *Merrell-Dow Research Institute, Gerenzano, Italy,* Glycopeptides (under Antibiotics)
Chang-Man Park, *Amoco Chemical Company, Naperville, Illinois,* Phthalic acids and other benzenepolycarboxylic acids
Edward E. Parry, *National Institute of Standards and Technology, Gaithersburg, Maryland,* Dental materials
Angela K. G. Parsons, *E.I. du Pont de Nemours & Co., Inc., Wilmington, Delaware,* Information retrieval
T. E. Parsons, *Eastman Chemical Company, Kingsport, Tennessee,* Other glycols (under Glycols)
Carol E. Parssinen, *Acentech Inc., Cambridge, Massachusetts,* Insulation, acoustic
Marina R. Pascucci, *GTE Laboratories, Inc., Waltham, Massachusetts,* Structural ceramics (under Advanced ceramics)
Nancy R. Passow, *Lonza Inc., Fair Lawn, New Jersey,* Survey; Chemical process industry (both under Regulatory agencies)
Baldev K. Patel, *Cyanamid Canada Inc., Niagara Falls, Ontario,* Cyanamides
Cliff Patenaude, *Bird, Inc.,* Roofing materials
Robert E. Patterson, *The PQ Corporation, Conshohocken, Pennsylvania,* Introduction (under Silica)
David B. Paul, *Sandoz Pharmaceuticals Corporation, Lincoln, Nebraska,* Expectorants, antitussives, and related agents
Donald R. Paul, *University of Texas, Austin,* Polymer blends
John P. Paul, *Carter & Burgess, Inc., Fort Worth, Texas,* Rubber (under Recycling)
Harry Paxton, *Carnegie Mellon University, Pittsburgh, Pennsylvania,* Steel
Joel D. Payne, *Kansas State University, Manhattan, Kansas,* Yeast-raised products (under Bakery processes and leavening agents)
Alan Pearson, *Aluminum Company of America, Alcoa Center, Pennsylvania,* Alumina, activated (under Aluminum compounds)
Dennis Pearson, *Hoechst Celanese Corporation, Corpus Christi, Texas,* *n*-Propyl alcohol (under Propyl alcohols)
Lloyd W. Pebsworth, *Polyethylene Technology, Morris, Illinois,* Low density polyethylene (under Olefin polymers)
Michael Pecht, *University of Maryland, College Park,* Electronic materials (under Packaging)
Michael C. Peck, *Georgia-Pacific Corporation, Decatur, Georgia,* Papermaking additives
Jan Pegram, *North Carolina State University, Raleigh,* Testing (under Textiles)
Milton Pelavin, *Miles, Inc., Tarrytown, New York,* Clinical chemistry (under Automated instrumentation)
Mel Pell, *E. I. du Pont de Nemours & Co., Inc., Wilmington, Delaware,* Fluidization
Jeanne E. Pemberton, *University of Arizona, Tucson,* Surface and interface analysis
Thomas W. Penrice, *Consultant, Mt. Juliet, Tennessee,* Tungsten and tungsten alloys; Tungsten compounds
Carol R. Perrotto, *E.I. du Pont de Nemours & Co., Inc., Wilmington, Delaware,* Information retrieval
David W. Pershing, *University of Utah, Salt Lake City,* Incinerators
Lawrence D. Pesce, *E. I. du Pont de Nemours & Co., Inc., Memphis, Tennessee,* Cyanides
Elizabeth M. Peters, *Mallinckrodt, Inc., St. Louis, Missouri,* Sodium halides–sodium iodide (under Sodium compounds)
Richard A. Peters, *The Procter & Gamble Company, Cincinnati, Ohio,* Survey and natural alcohols manufacture (under Alcohols, higher aliphatic)
Donald J. Petersen, *National Gypsum, Gold Bond Research Center, Buffalo, New York,* Calcium sulfate (under Calcium compounds)
Richard L. Petersen, *Memphis State University, Memphis, Tennessee,* Survey (under Calcium compounds); Potassium compounds
Michael Petschel, *Parker Amchem, Madison Heights, Michigan,* Chemical and electrochemical conversion treatments (under Metal surface treatments)
Roger C. Petterson, *USDA Forest Service, Madison, Wisconsin,* Wood
Alex Pettigrew, *Ethyl Technical Center, Baton Rouge, Louisiana,* Halogenated flame retardants (under Flame retardants)
Carol Phifer, *Sandia National Laboratories, Albuquerque, New Mexico,* Glass structure and properties (under Ceramics)
David G. Phillips, *CSIRO, Belmont, Geelong, Australia,* Wool
Mark Phillips, *Sandia National Laboratories, Albuquerque, New Mexico,* Nonlinear optical and electrooptic ceramics (under Ceramics)
W. J. Piel, *Arco Chemical Company, Newton Square, Pennsylvania,* Ethers
Ronald Pierantozzi, *Air Products and Chemicals, Inc., Allentown, Pennsylvania,* Carbon dioxide; Carbon monoxide
Tony J. Pierlot, *CSIRO, Belmont, Geelong, Australia,* Wool
John R. Pierson, *Johnson Controls, Inc., Milwaukee, Wisconsin,* Lead–acid (under Batteries, secondary cells)
William Pietro, *York University, North York, Ontario, Canada,* Biosensors
Linda R. Pinckney, *Corning, Inc., Corning, New York,* Glass-ceramics
Edwin M. Piper, *Piper Designs LLC, Littleton,* Colorado, Oil shale
Kenneth Pisarczyk, *Carus Chemical Company, LaSalle, Illinois,* Manganese compounds
Sarma V. Pisupati, *Pennsylvania State University, University Park,* Combustion science and technology
Alphonsus V. Pocius, *The 3M Company, St. Paul, Minnesota,* Adhesives
Thomas J. Podlas, *Aqualon Company, Wilmington, Delaware,* Cellulose ethers
Stanley Pohl, *Clairol Inc., Stamford, Connecticut,* Hair preparations
Ludwik Pokorny, *SQM Nitratos SA, Santiago, Chile,* Sodium nitrate (under Sodium compounds)
Malcolm Polk, *Georgia Institute of Technology, Atlanta,* High performance fibers
Peter Pollak, *Lonza Ltd., Basel, Switzerland,* Production (under Fine chemicals); Malonic acid and derivatives
Charles M. Pollock, *Union Camp Corporation, Savannah, Georgia,* Dicarboxylic acids
R. J. Ponsford, *SmithKline Beecham Pharmaceuticals, Surrey, United Kingdom,* Penicillins and others (under Antibiotics, β-lactams)
P. Pontal, *Rhône-Poulenc Recherches, France,* Hydroquinone, resorcinol, and catechol
Joseph G. Ponte, Jr., *Kansas State University, Manhattan, Kansas,* Yeast-raised products (under Bakery processes and leavening agents)
Louise C. Potter, *Johnson Matthey Technology Center, Reading, U.K.,* Platinum-group metals
Asohk Prabhu, *Nitto Denko America Inc., San Jose, California,* Electronic materials (under Packaging)
G. K. Surya Prakash, *University of Southern California, Los Angeles,* Friedel-Crafts reactions

R. David Prengaman, *RSR Corporation, Dallas, Texas,* Lead alloys

Duane B. Priddy, *The Dow Chemical Company, Midland, Michigan,* Styrene plastics

Ralph Priester, Jr., *Dow Chemical USA, Freeport, Texas,* Isocyanates, organic

Roger C. Prince, *Exxon Research & Engineering Company, Annandale, New Jersey,* Bioremediation

David E. Prinzing, *Enserch, Sacramento, California,* Fuel resources

Michael Prior, *Hosokawa Micron Ltd., Runcorn, Cheshire, U.K.,* Size reduction

Salvatore Profeta, Jr., *Monsanto, St. Louis, Missouri,* Molecular modeling

Gerfried Pruckmayr, *Du Pont Specialty Chemicals, Wilmington, Delaware,* Tetrahydrofuran and oxetane polymers (under Polyethers)

Richard W. Prugh, *Process Safety Engineering, Inc., Wilmington, Delaware,* Plant safety

Thomas A. Pugsley, *Warner-Lambert, Parke-David Pharmaceutical, AnnArbor, Michigan,* Epinephrine and norepinephrine

R. D. Putnam, *Putnam Environmental Services, Research Triangle Park, North Carolina,* Tellurium and tellurium compounds

Donald E. Putzig, *E. I. du Pont de Nemours & Co., Inc., Deepwater, New Jersey,* Organic (under Titanium compounds)

Elizabeth Quadros, *CIBA-GEIGY Corporation, Ardsley, New York,* Drug delivery systems

Christine M. Quinn, *Coca-Cola USA, Atlanta, Georgia,* Carbonated beverages

Roderic P. Quirk, *University of Akron, Ohio,* Anionic initiators (under Initiators)

Anatol Rabinkin, *Allied-Signal Inc., Parsippany, New Jersey,* Solders and brazing filler metals

Peter P. Radecki, *Michigan Technological University, Houghton,* Engineering, chemical data correlation

S. Raharjo, *Colorado State University, Fort Collins,* Meat products

Richard B. Rajendaen, *Aeromix Systems, Inc., Minneapolis, Minnesota,* Water treatment (under Aeration)

Philip E. Rakita, *Elf Atochem Japan, Chuo-ku, Tokyo,* Grignard reactions

Ramesh Ramachandran, *Union Carbide Corporation, Danbury, Connecticut,* Ethylene oxide polymers (under Polyethers)

Venkoba Ramachandran, *ASARCO, Inc., Salt Lake City, Utah,* Lead

T. S. Ramesh, *Mobil Research and Development Corporation, Princeton, New Jersey,* Expert systems

Francis J. Randall, *S. C. Johnson & Sons, Inc., Racine, Wisconsin,* Polishes

Mary E. Rasenberger, *White & Case, New York, New York,* Trademarks (under Copyrights and trademarks)

Serge Ratton, *Rhône-Poulenc Specialties Chimiques, Saint-Fons, France,* Chlorophenols

Frank Rauh, *FMC Corporation, Princeton, New Jersey,* Sodium carbonate (under Alkali and chlorine products)

Thimma R. Rawalpally, *Hoffmann-La Roche, Inc., Nutley, New Jersey,* Folic acid; Pantothenic acid (both under Vitamins)

John F. Ready, *Honeywell Systems and Research Center, Edina, Minnesota,* Lasers

Ludwig Rebenfeld, *TRI/Princeton, Princeton, New Jersey,* Survey (under Fibers); Survey (under Textiles)

R. L. Reddy, *UCAR Carbon Company, Inc., Cleveland, Ohio,* Carbon and artificial graphite (under Carbon)

V. Prakash Reddy, *University of Southern California, Los Angeles,* Friedel-Crafts reactions

V. Sreenivasulu Reddy, *Southwest Texas State University, San Marcos,* Heat-resistant polymers

Daniel J. Reed, *Dow Chemical USA, Freeport, Texas,* Survey (under Chlorocarbons and chlorohydrocarbons)

Richard W. Rees, *E.I. du Pont de Nemours & Co., Inc., Wilmington, Delaware,* Ionomers

Warren Rehman, *Kraft General Foods Corporation, White Plains, New York,* Coffee

Jill Rehmann, *Fordham University, Brooklyn, New York,* Nucleic acids

Nancy L. Reichenback, *Temple University School of Medicine, Philadelphia, Pennsylvania,* Nucleosides (under Antibiotics)

Austin H. Reid, Jr., *E. I. Du Pont de Nemours & Co., New Johnsonville, Tennessee,* Technical service

Kenneth I. G. Reid, *Tetra Chemicals, Houston, Texas,* Calcium chloride (under Calcium compounds)

Abraham Reife, *CIBA-GEIGY Corporation, Tom's River, New Jersey,* Dyes, environmental chemistry

Albert J. Repik, *Westvaco Corporation, North Charleston, South Carolina,* Activated carbon (under Carbon)

Paul R. Resnick, *E. I. du Pont de Nemours & Co., Inc., Wilmington, Delaware,* Perfluoroepoxides (under Fluorine compounds, organic)

Michael Reuman, *Sterling-Winthrop Research Institute, Rensselaer, New York,* Quinolones (under Antibacterial agents)

Ganapathi R. Revankar, *Triplex Pharmaceutical Corporation, The Woodlands, Texas,* Antiviral agents

A. H. Reyes, *Martin Marietta Magnesia Specialties, Inc., Baltimore, Maryland,* Magnesium compounds

John R. Reynolds, *University of Florida, Gainesville,* Electrically conductive polymers

C. K. Rhee, *The BF Goodrich Co., Brecksville, Ohio,* Antiozonants

S. K. Rhee, *Allied-Signal Inc., Troy, Michigan,* Brake linings and clutch facings

A. J. Ricco, *Sandia National Laboratories, Albuquerque, New Mexico,* Sensors

Jonathan Rich, *GE Corporate Research and Development, Schenectady, New York,* Silicones (under Silicon compounds)

H. Wayne Richardson, *CP Chemicals, Inc., Phibro-Tech, Inc., Sumter, South Carolina,* Cobalt compounds; Copper compounds; Nonferrous metals (under Recycling)

Douglas S. Richart, *Mortow International, Inc., Reading, Pennsylvania,* Powder technology (under Coating processes)

W. Frank Richey, *The Dow Chemical Company, Freeport, Texas,* Chlorohydrins

Reinhard H. Richter, *Consultant, Freeport, Texas,* Isocyanates, organic

William A. Rickelton, *Cytec Canada Inc., Niagara Falls, Canada,* Phosphine and its derivatives

Martin M. Rieger, *M & A Rieger Associates, Morris Plains, New Jersey,* Cosmetics

Ralph M. Riggin, *Lilly Research Laboratories, Indianapolis, Indiana,* Human growth hormone (under Hormones)

John A. Rippon, *CSIRO, Belmont, Geelong, Australia,* Wool

Mary B. Ritchey, *Wyeth-Lederle Vaccine and Pediatrics, Pearl River, New York,* Vaccine technology

G. Robert, *Rhône-Poulenc, Cranbury, New Jersey,* Vanillin

John Roberts, *Hoffmann-La Roche,Inc., Nutley, New Jersey,* Cephalosporins (under Antibiotics, β-lactams)

John S. Roberts, *Phillips Petroleum Company/CH&A Corporation, Kingwood, Texas,* Thiols

Winston K. Robbins, *Exxon Research and Engineering Company, Annandale, New Jersey,* Composition (under Petroleum)

Roland K. Robins, *ICN Nucleic Acid Research Institute, Costa Mesa, California,* Antiviral agents

D. W. Robinson, *UOP Research Center, Des Plaines, Illinois,* Regeneration, noble metals (under Catalysts)

J. Robinson, *Betz Dearborn, Inc., Trevose, Pennsylvania,* Industrial water treatment (under Water)

John H. Robson, *Union Carbide Corporation, South Charleston, West Virginia,* Ethylene glycol and oligomers (under Glycols)

Brendon Rodgers, *Consultant, Akron, Ohio,* Rubber compounding

James Rodgers, *Eastman Kodak Company, Rochester, New York,* Color photography

James A. Rogers, *Consultant, Ramsey, New Jersey,* Spices (under Flavors and spices)

Tony N. Rogers, *Michigan Technological University, Houghton,* Engineering, chemical data correlation

Gérard Romeder, *Lonza Ltd., Basel, Switzerland,* Malonic acid and derivatives

John P. N. Rosazza, *University of Iowa, Iowa City,* Microbial transformations

Kenneth Rose, *Rensselaer Polytechnic Institute, Troy, New York,* Silicon-based semiconductors (under Semiconductors)

R. D. Roseman, *University of Illinois at Urbana-Champaign, Urbana,* Ceramics as electrical materials

Stephen L. Rosen, *University of Missouri, Rolla,* Polymers

Jack L. Rosette, *Forensic Packaging Concepts, Inc., Fort Mill, South Carolina,* Cosmetics and pharmaceuticals (under Packaging)

C. Philip Ross, *Creative Opportunities, Inc., Laguna Niguel, California,* Glass (under Recycling)

Ronald W. Rousseau, *Georgia Institute of Technology, Atlanta, Georgia,* Crystallization

Rob A. Rottenbury, *CSIRO, Belmont, Geelong, Australia,* Wool

Jerry M. Rovner, *M. W. Kellogg Company, Houston, Texas; John Brown E & C, Houston, Texas,* Ammonia; Methanol

Roger M. Rowell, *USDA Forest Service, Madison, Wisconsin,* Wood

Howard C. Rowles, *Air Products and Chemicals, Inc., Allentown, Pennsylvania,* Cryogenics

Aroop K. Roy, *Dow Corning Corporation, Midland, Michigan,* Inorganic high polymers

Slawomir Rubinsztajn, *GE Silicones, Waterford, New York,* Silicones (under Silicon compounds)

Scott Rudge, *Synergen, Inc., Boulder, Colorado,* Electrophoresis (under Electroseparations)

Charles V. Rue, *Norton Company, Worcester, Massachusetts,* Abrasives

Carlos N. Ruiz, *Teledyne Allvac, Monroe, North Carolina,* Metal treatments

Wolfgang Runde, *Glenn T. Seaborg Institute for Transactinium Science, Los Alamos National Laboratory, Los Alamos, New Mexico,* Thorium and thorium compounds; Uranium and uranium compounds

W. R. Runyon, *Texas Instruments Incorporated, Dallas, Texas,* Pure silicon (under Silicon and silicon alloys)

Nelson W. Rupp, *National Institute of Standards and Technology, Gaithersburg, Maryland,* Dental materials

Armistead G. Russell, *Carnegie Mellon University, Pittsburgh, Pennsylvania,* Atmospheric modeling

Ian M. Russell, *CSIRO, Belmont, Geelong, Australia,* Wool

W. E. Rusterholz, *Sun Chemical Corporation, Carlstadt, New Jersey,* Inks

Douglas M. Ruthven, *University of New Brunswick, Fredricton, New Brunswick, Canada,* Adsorption

J. L. Ryans, *Eastman Chemical Company, Kingsport, Tennessee,* Pressure measurement

Lester Saathoff, *The Dow Chemical Company, Freeport, Texas,* Allyl chloride (under Chlorocarbons and chlorohydrocarbons)

Jean Louis Sabot, *Rhône-Poulenc, Shelton, Connecticut,* Gallium and gallium compounds; Lanthanides

Stephen H. Safe, *Texas A&M University, College Station,* Toxic aromatics (under Chlorocarbons andchlorohydrocarbons)

M. R. V. Sahyun, *3M Center, St. Paul, Minnesota,* Survey (under Photochemical technology)

Alvin J. Salkind, *Rutgers University, Piscataway, New Jersey,* Alkaline (under Batteries, secondary cells)

Upasiri Samaraweera, *American Crystal Sugar Company, Moorhead, Minnesota,* Properties of sucrose (under Sugar)

José Sanchez, *Elf Atochem North America, Inc., Buffalo, New York,* Free-radical initiators (under Initiators); Organic peroxides (under Peroxides and peroxide compounds)

James R. Sandifer, *Eastman Kodak Company, Rochester, New York,* Electroanalytical techniques

Stanley R. Sandler, *Elf Atochem North America, Inc., King of Prussia, Pennsylvania,* Sulfur compounds

John Paul San Giovanni, *Jockey Hollow Technologies, Boston, Massachusetts,* Process control

A. T. Santhanam, *Kennametal Inc., LaTrobe, Pennsylvania,* Cemented carbides; Industrial hard carbides (both under Carbides)

P. B. Sargeant, *E. I. du Pont de Nemours & Co., Inc., Kinston, North Carolina,* Polyester (under Fibers)

E. T. Sauer, *The Procter & Gamble Company, Cincinnati, Ohio,* Economic aspects (under Carboxylic acids)

J. D. Sauer, *Albemarle Corporation, Baton Rouge, Louisiana,* Olefins, higher

Patricia Savu, *3M Company, St. Paul, Minnesota,* Fluorinated higher carboxylic acids; Perfluoroalkanesulfonic acids (both under Fluorine compounds, organic)

Hiroyuki Sawada, *UBE Industries Limited, Tokyo, Japan,* Oxalic acid

Irene Sawchyn, *Consultant, Whippany, New Jersey,* Integrated circuits

Alan W. Scaroni, *Pennsylvania State University, University Park,* Combustion science and technology

P. E. Scheerer, *Martin Marietta Magnesia Specialties, Inc., Baltimore, Maryland,* Magnesium compounds

William Scheffer, *Schenectady Chemicals, Inc., Schenectady, New York,* Alkylphenols

Paul M. Schermerhorn, *Corning, Inc., Corning, New York,* Vitreous silica (under Silica)

G. Scherr, *BASF AG, Ludwigshafen, Germany,* Imines, cyclic

Michael Scherrer, *Morton International, Inc., Woodstock, Illinois,* Polysulfides (under Polymers containing sulfur)

L. McDonald Schetky, *Memry Corporation, Brookfield, Connecticut,* Shape-memory alloys

Henry W. Schiessl, *Olin Corporation, Cheshire, Connecticut,* Hydrazine and its derivatives

Hans Erik Schiff, *Enzyme Technology, Bagsvaerd, Denmark,* Industrial (under Enzyme applications)

Steven L. Schilling, *Mobay Corporation, New Martinsville, West Virginia,* Amines by reduction

James H. Schlewitz, *Teledyne Wah Chang, Albany, Oregon,* Niobium and niobium compounds

Francis J. Schmidt, *University of Missouri, Columbia,* Procedures (under Genetic engineering)

Glenn R. Schmidt, *Colorado State University, Fort Collins,* Meat products

Rosalie A. Schnick, *U.S. Fish and Wildlife Service, La Crosse, Wisconsin,* Aquaculture chemicals

Hollis G. Schoepke, *Anaquest, Murray Hill, New Jersey,* Anesthetics

Clifford K. Schoff, *PPG Industries, Inc., Allison, Pennsylvania,* Rheological measurements

William L. Schreiber, *International Flavors and Fragrances, Union Beach, New Jersey,* Perfumes

Gary J. Schrobilgen, *McMaster University, Ontario, Canada,* Compounds (under Helium group)

David M. Schubert, *U.S. Borax Research Corporation, Anaheim, California,* Boron hydrides (under Boron compounds)

Donald N. Schulz, *Exxon Research and Engineering Company, Annandale, New Jersey,* Copolymers

R. C. Schulz, *UOP, Des Plaines, Illinois,* Cumene

Thomas J. Schwan, *Norwich Eaton Pharmaceuticals Corporation, Norwich, New York,* Nitrofurans (under Antibacterial agents)

Robert Schwartz, *Sandia National Laboratories, Albuquerque, New Mexico,* Electronic properties and material structure (under Ceramics)

John J. Sciarra, *Sciarra Aeromed Development Corporation, Brooklyn, New York,* Aerosols

John W. Scott, *Hoffmann-La Roche Inc., Nutley, New Jersey,* Survey; Vitamin B_{12} (under Vitamins)

Eric F. V. Scriven, *Reilly Industries, Inc., Indianapolis, Indiana,* Pyridine and pyridine derivatives

Glenn T. Seaborg, *University of California, Berkeley, Berkeley, California,* Actinides and transactinides

Stan R. Seagle, *Consultant, Warren, Ohio,* Titanium and titanium alloys

George C. Seaman, *Hoechst-Celanese, Corpus Christi, Texas,* Formaldehyde

Oldrich K. Sebek, *Consultant, Kalamazoo, Michigan,* Microbial transformations

Dana E. Selley, *Wake Forest University, Winston-Salem, North Carolina,* Opioids, endogenous

Jeffrey Selley, *Consultant, Durham, North Carolina,* Polyesters, unsaturated

Daniel R. Sempolinski, *Corning, Inc., Corning, New York,* Vitreous silica (under Silica)

Surjit S. Sengha, *Sterling Winthrop, Inc., West Chester, Pennsylvania,* Fermentation

J. Senior, *James Robinson, Ltd., Yorkshire, U.K.,* Sulfur dyes

Paul Sennett, *Engelhard Corporation, Iselin, New Jersey,* Uses (under Clays)

J. Sentenac, *Rhône-Poulenc Recherches, France,* Hydroquinone, resorcinol, and catechol

Michael Senyek, *The Goodyear Tire & Rubber Company, Akron, Ohio,* Polyisoprene (under Elastomers, synthetic)

George A. Serad, *Hoechst-Celanese Corporation, Charlotte, North Carolina,* Cellulose esters (under Fibers)

Nick Serpone, *Concordia University, Montreal, Canada,* Photocatalysis (under Photochemical technology)

Richard J. Seymour, *Johnson Matthey Technology Center, Reading, U.K.,* Platinum-group metals

J. Shacter, *Martin Marietta Energy Systems, Oak Ridge, Tennessee,* Diffusion separation methods

Wayne Shannon, *Magnetrol International, Inc., Downers Grove, Illinois,* Liquid level measurement

Leonard Shapiro, *Consultant, Warminster, Pennsylvania,* Centrifugal separation (under Separation)

Anand S. G. Sharangpani, *BASF Corporation, Holland, Michigan,* Pigment dispersions

Reza Sharifi, *Pennsylvania State University, University Park,* Combustion science and technology

Philip E. Shaw, *U.S. Department of Agriculture, Winter Haven, Florida,* Fruit juices

Richard J. Sheehan, *Amoco Chemical Company, Naperville, Illinois,* Phthalic acids and other benzenepolycarboxylic acids

Daniel R. Shelton, *USDA-Agricultural Research Service, Beltsville, Maryland,* Soil chemistry of pesticides

Frederick A. Shelton, *Chatterton Petrochemical Corporation, Delta, British Columbia, Canada,* Benzaldehyde; Benzoic acid

John D. Sherman, *UOP, Tarrytown, New York,* Adsorption, gas separation

Marc E. Sherwin, *Sandia National Laboratories, Albuquerque, New Mexico,* Compound semiconductors (under Semiconductors)

George Shia, *Allied-Signal, Buffalo, New York,* Fluorine

Gary J. Shiflet, *University of Virginia, Charlottesville,* Glassy metals

A. Ray Shirley, Jr., *Applied Chemical Technology, Muscle Shoals, Alabama,* Urea

Steven E. Shoelson, *Joslin Diabetes Center, Harvard Medical School, Boston, Massachusetts,* Insulin and other antidiabetic agents

M. M. Shreehan, *ABB Lummus Crest, Inc., Bloomfield, New Jersey,* Ethylene

Jean'ne M. Shreeve, *University of Idaho, Moscow,* Oxygen (under Fluorine compounds, inorganic)

Thomas R. Shrout, *Pennsylvania State University, University Park, Pennsylvania,* Electronic ceramics (under Advanced ceramics)

D. Shuttleworth, *The Goodyear Tire & Rubber Company, Akron, Ohio,* Tire cord

Theodore Hein Smit Sibinga, *Haemonetics, Braintree, Massachusetts,* Cell separation (under Fractionation, blood)

Howard W. Sibley, *Carrier Corporation, Syracuse, New York,* Refrigeration

Scott Sibley, *Goucher College, Baltimore, Maryland,* Semiconductors, organic

Kristine S. Siefert, *Nalco Chemical Company, Naperville, Illinois,* Polyaluminum chlorides (under Aluminum compounds)

Jeffrey J. Siirola, *Eastman Chemical Company, Kingsport, Tennessee,* Separations process synthesis

Geoffrey D. Silcox, *University of Utah, Salt Lake City,* Incinerators

Gary S. Silverman, *Atochem North America, King of Prussia, Pennsylvania,* Grignard reactions

Irwin Silverstein, *International Specialty Products, Wayne, New Jersey,* Quality assurance

Laura J. Sim, *Wake Forest University, Winston-Salem, North Carolina,* Opioids, endogenous

Edlyn S. Simmons, *Hoechst Marion Roussel, Inc., Cincinnati, Ohio,* Patents, literature

Howard E. Simmons III, *DuPont Imaging Systems, Wilmington, Delaware,* Electronics, coatings

Merete Simonsen, *Enzyme Technology, Bagsvaerd, Denmark,* Industrial (under Enzyme applications)

William T. Simpson, *USDA Forest Service, Madison, Wisconsin,* Wood

Norman Singer, *Nutrasweet, Mount Prospect, Illinois,* Fat replacers

William Singer, *Troy Chemical Corporation, Newark, New Jersey,* Mercury compounds

Navjot Singh, *GE Corporate Research and Development, Schenectady, New York,* Silicones (under Silicon compounds)

Vernon L. Singleton, *University of California, Davis,* Wine

J. E. Singley, *Environmental Science & Engineering, Inc., Gainesville, Florida,* Municipal water treatment (under Water)

Kenneth E. Skog, *USDA Forest Service, Madison, Wisconsin,* Wood

E. A. Skrabek, *Orbital Sciences Corporation, Germantown, Maryland,* Thermoelectric energy conversion

James A. Slattery, *Indium Corporation of America, Utica, New York,* Indium and indium compounds

A. J. Sleight, *F. Hoffmann-La Roche Ltd., Basel, Switzerland,* Psychopharmacological agents

William C. Sleppy, *Aluminum Company of America, Alcoa Center, Pennsylvania,* Introduction (under Aluminum compounds)

Bruce E. Smart, *E. I. du Pont de Nemours & Co., Inc., Wilmington, Delaware,* Fluorinated aliphatic compounds (under Fluorine compounds, organic)

David E. Smith, *FMC Corporation, Princeton, New Jersey,* Carbon disulfide

David M. Smith, *Martin Marietta Magnesia Specialties, Inc., Baltimore, Maryland,* Magnesium compounds

Duane H. Smith, *Technical Solutions, Morgantown, West Virginia,* Microemulsions

Mark D. Smith, *Allied-Signal Aerospace Company, Kansas City, Missouri,* Plasma technology

Peter A. S. Smith, *University of Michigan, Ann Arbor,* Nomenclature

Robert A. Smith, *U.S. Borax Research Corporation, Anaheim, California; Allied-Signal, Morristown, New Jersey,* Boron oxides, boric acid, and borates (under Boron compounds); Hydrogen (under Fluorine compounds, inorganic)

Roy E. Smith, *CIBA-GEIGY Corporation, Greensboro, North Carolina,* Azine dyes; Dyes, reactive; Stilbene dyes

Temple F. Smith, *Boston University, Massachusetts,* Proteins

William L. Smith, *The Clorox Company, Pleasanton, California,* Survey (under Bleaching agents)

Ronald Smorada, *BBA Nonwovens, Old Hickory, Tennessee,* Fabrics, spunbonded (under Nonwoven fabrics)

J. F. Smullen, *Hershey Foods Corporation, Hershey, Pennsylvania,* Chocolate and cocoa

Gayle Snedecor, *The Dow Chemical Company, Freeport, Texas,* Other chloroethanes (under Chlorocarbons and chlorohydrocarbons)

W. M. Snellings, *Union Carbide Corporation, South Charleston, West Virginia,* Ethylene glycol and oligomers (under Glycols)

L. G. Snow, *E. I. du Pont de Nemours & Co., Inc., Wilmington, Delaware,* Poly(vinyl fluoride) (under Fluorine compounds, organic)

Asta Sokov, *Université de Sherbrooke, Sherbrooke, Quebec, Canada,* Asbestos

Z. Solc, *University of Pardúbice, Czech Republic,* Inorganic (under Pigments)

I. J. Solomon, *IIT Research Institute, Chicago, Illinois,* Oxygen (under Fluorine compounds, inorganic)

Richard A. Sommer, *Consultant, Warren, Ohio,* Introduction; Induction furnaces (both under Furnaces, electric)

J. S. Son, Shell Development Company, Houston, Texas, Fluid mechanics

H. Soo, *Union Carbide Corporation, South Charleston, West Virginia,* Ethylene oxide

Henry E. Sostmann, *Consultant, Albuquerque, New Mexico,* Temperature measurement

Thomas F. Soules, *General Electric Company, Cleveland, Ohio,* Phosphors (under Luminescent materials)

Robert Southgate, *SmithKline Beecham Pharmaceuticals, Surrey, United Kingdom,* Carbapenems and penems (under Antibiotics, β-lactams)

Robert M. Sowers, *Ford Motor Company, Dearborn, Michigan,* Laminated materials, glass

William C. Spangenberg, *Hammond Lead Products, Inc., Pittsburgh, Pennsylvania,* Lead salts (under Lead compounds)

Robert A. Sparks, *Siemens Analytical X-Rays Systems, Inc., Madison, Wisconsin,* X-Ray technology

Arno F. Spatola, *University of Louisville, Kentucky,* Posterior pituitary hormones (under Hormones)

Theodore C. Spaulding, *Anaquest, Murray Hill, New Jersey,* Anesthetics

James G. Speight, *Western Research Institute, Laramie, Wyoming,* Asphalt; Gaseous fuels (under Fuels, synthetic); Refinery processes, survey (under Petroleum); Tar sands

A. L. Spelta, *EniChem Elastomer, Milano, Italy,* Acrylic elastomers (under Elastomers, synthetic)

William Spiegelberg, *Brush Wellman, Inc., Elmore, Ohio,* Beryllium and beryllium alloys

Alok M. Srivastava, *General Electric Company, Cleveland, Ohio,* Phosphors (under Luminescent materials)

Philip Staal, *Haarmann & Reimer Corporation, Elkhart, Indiana,* Citric acid

Howard D. Stahl, *Kraft General Foods Corporation, White Plains, New York,* Coffee

Apryll M. Stalcup, *University of Cincinnati, Ohio,* Chiral separations

James T. Staley, *Aluminum Company of America, Alcoa, Pennsylvania,* Aluminum and aluminum alloys

J. G. Stam, *Pfizer Central Research, Groton, Connecticut,* β-Lactamase inhibitors (under Antibiotics, β-lactams)

Ferris C. Standiford, *Consultant, Bellevue, Washington,* Evaporation

Jeffrey W. Stansbury, *National Institute of Standards and Technology, Gaithersburg, Maryland,* Dental materials

David M. Stark, *Monsanto Company, Chesterfield, Missouri,* Plants (under Genetic engineering)

Marshall W. Stark, *FMC Corporation, Bessemer City, North Carolina,* Lithium and lithium compounds

John B. Starr, *Hoechst Celanese Corporation, Summit, New Jersey,* Acetal resins

Nicholas J. Stathis, *White & Case, New York, New York,* Trademarks (under Copyrights and trademarks)

I. R. Steel, *Sun Chemical Corporation, Carlstadt, New Jersey,* Inks

Barbara A. Stefl, *Union Carbide Chemicals and Plastics Company, Inc., South Charleston, West Virginia,* Antifreezes and deicing fluids

Dale S. Steichen, *The Clorox Company, Pleasanton, California,* Survey (under Bleaching agents)

Judith Stein, *GE Corporate Research and Development, Schenectady, New York,* Silicones (under Silicon compounds)

Dan Steinmeyer, *Monsanto Company, St. Louis, Missouri,* Process energy conservation

Eric W. Stern, *Engelhard Corporation, Iselin, New Jersey,* Gold and gold compounds

Horst D. Sterzenberger, *Technochemie GmbH-Verfahrenstechnik, Gutenbergstrasse, Germany,* Thermosets (under Composite materials, polymer-matrix)

U. Steuerle, *BASF AG, Ludwigshafen, Germany,* Imines, cyclic

Clare A. Stewart, Jr., *Consultant, Wilmington, Delaware,* Chloroprene (under Chlorocarbons and chlorohydrocarbons)

Steven L. Stewart, *Shell Development Company, Houston, Texas,* Thermosets (under Composite materials, polymer-matrix)

Robert R. Stickney, *Texas A&M University, Bryan, Texas,* Aquaculture

Edward I. Stiefel, *Exxon Research and Engineering Company, Annandale, New Jersey,* Molybdenum compounds

Robert A. Stokes, *Stokes Associates, Golden, Colorado,* Solar energy

Milan Stolka, *Xerox Corporation, Webster, New York,* Electrophotography

W. M. Stoll, *Kennametal Inc., LaTrobe, Pennsylvania,* Industrial hard carbides (under Carbides)

Norman S. Stoloff, *Rensselaer Polytechnic Institute,* High temperature alloys

Alan M. Stolzenberg, *West Virginia University, Morgantown,* Iron compounds

A. James Stonehouse, *Brush Wellman Inc., Elmore, Ohio,* Beryllium and beryllium alloys; Beryllium compounds

D. Stoppels, *Philips Components, Eindhoven, the Netherlands,* Ferrites

Steven G. Streitel, *Day-Glo Color Corporation, Cleveland, Ohio,* Fluorescent pigments (daylight) (under Luminescent materials)

Werner Stumm, *Swiss Federal Institute of Technology, Duebendorf, Switzerland,* Properties (under Water)

David M. Sturmer, *Eastman Kodak Company, Rochester, New York,* Cyanine dyes; Dyes, sensitizing

C. Su, *Sun Chemical Corporation, Carlstadt, New Jersey,* Inks

Kyung W. Suh, *The Dow Chemical Company, Granville, Ohio,* Foamed plastics

Robert J. Suhadolnik, *Temple University School of Medicine, Philadelphia, Pennsylvania,* Nucleosides (under Antibiotics)

L. David Suits, *New York State Department of Transportation, Albany,* Geotextiles

Don A. Sullivan, *Shell Chemical Company, Houston, Texas,* Solvents, industrial

Edward A. Sullivan, *Morton International, Danvers, Massachusetts,* Hydrides

The Sulphur Institute, *Washington, D.C.,* Sulfur

James W. Summers, *The Geon Company, Avon Lake, Ohio,* Vinyl chloride polymers (under Vinyl polymers)

H. N. Sun, *Exxon Chemical Company, Baytown, Texas,* Butadiene

K. M. Sundaram, *ABB Lummus Crest, Inc., Bloomfield, New Jersey,* Ethylene

Richard J. Sundberg, *University of Virginia, Charlottesville,* Indole

Gregory D. Sunvold, *University of Illinois, Urbana,* Ruminant feeds (under Feeds and feed additives)

Kenneth S. Suslick, *University of Illinois, Urbana,* Sonochemistry

Boyce Sutton, Jr., *Sybron Chemicals, Inc., Wellford, South Carolina,* Dye carriers

Michio Suzuki, *Nissan Chemical Industries, Ltd., Tokyo, Japan,* Sulfamic acid and sulfamates

Ladislav Svarovsky, *Consultant Engineers and Fine Particle Software, Heaton/Bradford/West Yorkshire, U.K.,* Filtration; Sedimentation

W. A. Sweeney, *Chevron Research and Technology Company, Richmond, California,* BTX processing

H. G. Sweenie, *AstroTurf Industries, Dalton, Georgia,* Recreational surfaces

Graham Swift, *Rohm and Haas Research Laboratories, Spring House, Pennsylvania,* Polymers, environmentally degradable

Michael Szycher, *PolyMedica Industries, Inc., Woburn, Massachusetts,* Prosthetic and biomedical devices

Arthur J. Taggi, *DuPont Printing and Publishing, Boothwyn, Pennsylvania,* Printing processes

Tohru Takekoshi, *General Electric, Schenectady, New York,* Polyimides

Mannan Talukder, *Advanced Aromatics, Baytown, Texas,* Naphthalene derivatives

James T. Tanner, *North Carolina State University, Asherville,* Mica

Barry L. Tarmy, *TBD Technology, Berkeley Heights, New Jersey,* Reactor technology

Edward Tarnell, *Colin A. Houston & Associates, Inc., Mamaroneck, New York,* Market and marketing research

Roger Tate, *Delavan, Inc, West Des Moines, Iowa,* Sprays

Kwoliang D. Tau, *Hoechst-Celanese Corporation, Corpus Christi, Texas,* Esters, organic

Harold A. Taylor, Jr., *Bureau of Mines, U.S. Department of the Interior, Washington, D.C.,* Natural graphite (under Carbon)

J. J. Taylor, *Electric Power Research Institute, Palo Alto, California,* Safety in nuclear power facilities (under Nuclear reactors)

Jean A. Taylor, *Consultant, Columbia, Maryland,* Laboratory information management systems

Paul Taylor, *GAF Corporation, Wayne, New Jersey,* Acetylene-derived chemicals

R. Ray Taylor, *Phillips Petroleum Company, Bartlesville, Oklahoma,* Liquefied petroleum gas

Richard F. Taylor, *Arthur D. Little, Inc., Cambridge, Massachusetts,* Immunoassay

M. T. Tayyabkhan, *Tayyabkhan Consultants Inc., Princeton, New Jersey,* Computer-aided engineering (CAE)

Henryk Temkin, *Colorado State University, Fort Collins,* Semiconductor lasers (under Light generation)

John A. Tesk, *National Institute of Standards and Technology, Gaithersburg, Maryland,* Dental materials

Dean Thetford, *Zeneca Specialties, Manchester, U.K.,* Triphenylmethane and related dyes

Curt Thies, *Washington University, St. Louis, Missouri,* Microencapsulation

Dennis Thomas, *Eagle-Picher Industries, Inc., Quapaw, Oklahoma,* Germanium and germanium compounds

Mary R. Thomas, *The Dow Chemical Company, Midland, Michigan,* Salicylic acid and related compounds

David A. Thompson, *Corning, Inc., Corning, New York,* Glass

G. J. Thompson, *UOP, Des Plaines, Illinois,* Alkylation

Norman S. Thompson, *Consultant, Appleton, Wisconsin,* Hemicellulose

Quentin E. Thompson, *Monsanto Company, Millstadt, Illinois,* Biphenyl and terphenyls

F. M. Thomson, *Consultant, Wilmington, Delaware,* Conveying

Merle Thorpe, *Thorpe Thermal Technologies, Inc., Stuart, Florida,* Refractory coatings

Thomas C. Thorstensen, *TSG, North Cholmsford, Massachusetts,* Leather

John K. Tien, *Columbia University, New York, New York,* Nickel and nickel alloys

Jefferson W. Tilley, *Hoffmann-LaRoche Inc., Nutley, New Jersey,* Antiobesity drugs

David A. Tillman, *Enserch, Sacramento, California,* Fuel resources; Fuels from waste

Hendrik Timmerman, *Leiden/Amsterdam Center for Drug Research,* Histamine and histamine antagonists

Robert W. Timmerman, *FMC Corporation, Princeton, New Jersey,* Carbon disulfide

George A. Timmons, *Consultant, Ann Arbor, Michigan,* Molybdenum and molybdenum alloys

David B. Todd, *Stevens Institute of Technology, Hoboken, New Jersey,* Plastics processing

Elizabeth A. Todd, *University of Michigan, Ann Arbor,* Raman spectroscopy (under Infrared technology and Raman spectroscopy)

E. Donald Tolles, *Westvaco Corp., North Charleston, South Carolina,* Activated carbon (under Carbon)

Joseph E. Toomey Jr., *Reilly Industries, Inc., Indianapolis, Indiana,* Pyridine and pyridine derivatives

G. Paull Torrence, *Hoechst-Celanese Corporation, Corpus Christi, Texas,* Esterification

Edward A. Torrero, *Empire State Electric Energy Research Corporation, New York, New York,* Renewable energy resources

Bernard Toseland, *Air Products and Chemicals, Inc., Allentown, Pennsylvania,* Diaminotoluenes (under Amines, aromatic)

Laszlo Toth, *The Western Sugar Company, Denver, Colorado,* Sugar economics (under Sugar)

Irving Touval, *Touval Associates, Sparta, New Jersey,* Antimony and other inorganic flame retardants (under Flame retardants)

G. A. Townes, *BE Inc., Barnwell, South Carolina,* Chemical reprocessing (under Nuclear reactors)

Paul M. Treichel, Jr., *University of Wisconsin, Madison,* Rhenium and rhenium compounds

David Trent, *The Dow Chemical Company, Freeport, Texas,* Propyleneoxide

David J. Triggle, *State University of New York, Buffalo,* Pharmacodynamics

Terrance B. Tripp, *H. C. Starck Inc., Newton, Massachusetts,* Tantalum and tantalum compounds

M. Trojan, *University of Pardúbice, Czech Republic,* Inorganic (under Pigments)

Remi Trottier, *Aluminum Company of America, Alcoa Center, Pennsylvania,* Size measurement of particles

F. Truchet, *Rhône-Poulenc, Cranbury, New Jersey,* Vanillin

Ron W. Tucker, *Lilly Industries, Inc., High Point, North Carolina,* Stains, industrial

Paul S. Tully, *Stepan Company, Northfield, Illinois,* Sulfonic acids

John H. Tundermann, *Inco Alloys International Inc., Huntington, West Virginia,* Nickel and nickel alloys

Albin F. Turbak, *Consultant, Sandy Springs, Georgia,* High performance fibers

Michael G. Turcotte, *Air Products and Chemicals, Inc., Allentown, Pennsylvania,* Aliphatic amines

D. L. Turk, *UCAR Carbon Company, Inc., Cleveland, Ohio,* Carbon and artificial graphite (under Carbon)

John Turrell, *Technicon Instruments Corporation, Tarrytown, New York,* Hematology (under Automated instrumentation)

Samuel M. Tuthill, *Mallinckrodt Specialty Chemicals Company, St. Louis, Missouri,* Standards (under Fine chemicals)

Daniel L. Twarog, *American Foundrymen's Society, Des Plaines, Illinois,* Cast copper alloys (under Copper alloys)

Aaron Twerski, *Brooklyn Law School, New York,* Product liability

Ronald P. Tye, *Consultant, Surrey, England,* Insulation, thermal

Abraham Ulman, *Polytechnic University, Brooklyn, New York,* Monomolecular layers (under Thin films)

Henri Ulrich, *Consultant, Guilford, Connecticut,* Urethane polymers

Jerry D. Unruh, *Hoechst Celanese Corporation, Corpus Christi, Texas,* *n*-Propyl alcohol (under Propyl alcohols)

L. L. Upson, *UOP Research Center, Des Plaines, Illinois,* Regeneration, FCC (under Catalysts)

Arthur Usmani, *Firestone, Carmel, Indiana,* Medical diagnostic reagents

Rajesh Vaidya, *University of New Mexico, Albuquerque,* Nanotechnology

Donald Valentine, Jr., *Cytec Industries Inc., Stamford, Connecticut,* Soil stabilization

Hardarshan Valia, *Inland Steel Company, East Chicago, Illinois,* Carbonization (under Coal conversion processes)

Susan D. Van Arnum, *Hoffmann-La Roche, Inc., Nutley, New Jersey,* Niacin, nicotinamide, and nicotinic acid; Vitamin A; Vitamin K (all under Vitamins)

J. A. A. M. van Asten, *National Institute for Public Health and the Environment, Bilthoven, the Netherlands,* Sterilization techniques

Hendrick C. Van Ness, *Rensselaer Polytechnic Institute, Troy, New York,* Thermodynamics

P. J. Van Opdorp, *UOP, Des Plaines, Illinois,* Cumene

Jan F. VanPeppen, *Allied-Signal Inc., Petersburg, Virginia,* Cyclohexanol and cyclohexanone

Lambertus van Zelst, *Conservation Analytical Laboratory, Smithsonian Institution, Washington, D.C.,* Fine art examination and conservation

Joseph Varco, *Clairol Inc., Stamford, Connecticut,* Hair preparations

Kanwal J. Varma, *Schering-Plough Research, Bloomfield, New Jersey,* Chloramphenicols and analogues (under Antibiotics)

E. A. Vaughn, *Clemson University, Clemson, South Carolina,* Staple fibers (under Nonwoven fabrics)

Joseph G. Venner, *BASF Structural Materials, Inc., Charlotte, North Carolina,* Carbon and graphite fibers

David Vietti, *Morton International, Inc., Woodstock, Illinois,* Polysulfides (under Polymers containing sulfur)

Tyrone L. Vigo, *U.S. Department of Agriculture, New Orleans, Louisiana,* High performance fibers

Kenneth Visek, *Akzo Chemicals, Inc., McCook, Illinois,* Fatty amines

Dan Vlastelica, *Miles Inc., Tarrytown, New York,* Clinical chemistry (under Automated instrumentation)

E. Von Halle, *Consultant, Martin Marietta Energy Systems, Oak Ridge, Tennessee,* Diffusion separation methods; Isotope separation (under Nuclear reactors)

Urs von Stockar, *École Polytechnique Fédérale, Lausanne, Switzerland,* Absorption

B. V. Vora, *UOP, Des Plaines, Illinois,* Alkylation

Karl S. Vorres, *Argonne National Laboratory, Argonne, Illinois,* Coal; Lignite and brown coal

Dolatrai Vyas, *Bristol-Myers Squibb Company, Wallingford, Connecticut,* Chemotherapeutics, anticancer

M. P. Wachter, *R. W. Johnson Pharmaceutical Research Institute, Raritan, New Jersey,* Estrogens and antiestrogens (under Hormones)

Walter H. Waddell, *Exxon Chemical Company, Baytown, Texas,* Amorphous silica (under Silica)

Mark V. Wadsworth, *Texas Instruments, Inc., Dallas, Texas,* Photodetectors

Frank S. Wagner, *Strem Chemicals Inc., Newbury Port, Massachusetts,* Carbonyls; Rubidium and rubidium compounds

Frank S. Wagner, Jr., *Nandina Corporation, Corpus Christi, Texas,* Acetamide; Acetic acid; Acetic anhydride; Acetyl chloride (all under Acetic acid and derivatives)

John D. Wagner, *Ethyl Corporation, Albermarle Corporation, Baton Rouge, Louisiana,* Synthetic processes (under Alcohols, higher aliphatic); Olefins, higher

M. T. Wajer, *Martin Marietta Magnesia Specialties, Inc., Baltimore, Maryland,* Magnesium compounds

Richard J. Wakeman, *University of Exeter, Devon, U.K.,* Extraction, liquid–solid

Park Waldroup, *University of Arkansas, Fayetteville,* Nonruminant feeds (under Feeds and feed additives)

Peter Walker, *DuPont Printing and Publishing, Boothwyn, Pennsylvania,* Printing processes

Jim Wallace, *M. W. Kellogg Company, Houston, Texas,* Phenol

Paul Wallace, *Clairol Inc., Stamford, Connecticut,* Hair preparations

G. M. Wallraff, *IBM Research Division, San Jose, California,* Lithographic resists

Leonard E. Walp, *Humko Chemical, Division of the Witco Corporation, Memphis, Tennessee,* Antistatic agents

Kevin C. Walter, *Los Alamos National Laboratory, Los Alamos, New Mexico,* Ion implantation

M. D. Walter, *Martin Marietta Magnesia Specialties, Inc., Baltimore, Maryland,* Magnesium compounds

Robert R. Walton, *Wellman Furnaces, Inc., Shelbyville, Tennessee,* Resistance furnaces (under Furnaces, electric)

Vivian K. Walworth, *Polaroid Corporation, Concord, Massachusetts,* Color photography, instant

H. C. Wang, *Exxon Chemical Company, Linden, New Jersey,* Butyl rubber (under Elastomers, synthetic)

Ying Wang, *Du Pont Company, Wilmington, Delaware,* Photoconductive polymers

D. J. Ward, *Consultant, Des Plaines, Illinois,* Cumene

Thomas J. Ward, *Clarkson University, Potsdam, New York,* Economic evaluation

Timothy Ward, *Millsap College, Jackson, Mississippi,* Analytical techniques (under Biopolymers)

Rosemary Waring, *University of Birmingham, Birmingham, England,* Aminophenols

Jeffrey Warshauer, *Enserch, Sacramento, California,* Fuel resources

O. Wasilewski, *Sun Chemical Corporation, Carlstadt, New Jersey,* Inks

Gerald Wasserman, *Kraft General Foods Corporation, White Plains, New York,* Coffee

Richard W. Waterstrat, *National Institute of Standards and Technology, Gaithersburg, Maryland,* Dental materials

Ian Watson, *Consultant, Rochester, New York,* Computer technology

J. C. Watts, *E. I. du Pont de Nemours & Co., Inc., Wilmington, Delaware,* Dimethylacetamide (under Acetic acid and derivatives)

Richard A. Waugh, *Thermal Ceramics, Augusta, Georgia,* Refractory fibers

A. Dinsmoor Webb, *University of California, Davis,* Vinegar

Edwin Weber, *Technische Universität Bergakademie Freiberg, Institut Für organische Chemie, Germany,* Inclusion compounds; Molecular recognition

Joseph N. Weber, *Du Pont Nylon, Wilmington, Delaware,* General (under Polyamides)

Amie H. Webster, *E.I. du Pont de Nemours & Co., Inc., Wilmington, Delaware,* Information retrieval

Owen W. Webster, *E. I. du Pont de Nemours & Co., Inc., Wilmington, Delaware,* Cyanocarbons

Edward D. Weil, *Polytechnic University, Brooklyn, New York,* Phosphorus flame retardants (under Flame retardants); Sulfur compounds

David A. Weitz, *Exxon Research & Engineering Company, Annandale, New Jersey,* Foams

William R. Weltmer, Jr., *The BOC Group, Inc., Murray Hill, New Jersey,* Gases (under Helium group)

Armin Wendel, *Rhône-Poulenc Rorer, Cologne, Germany,* Lecithin

Jeff Wengrovius, *GE Silicones, Waterford, New York,* Silicones (under Silicon compounds)

Robert H. Wentorf, Jr., *Greenwich, New York,* Diamond, synthetic (under Carbons); Refractory boron compounds (under Boron compounds)

Leonard A. Wenzel, *Lehigh University, Bethlehem, Pennsylvania,* Simultaneous heat and mass transfer

Jack Wernick, *Murray Hill, New Jersey,* Bulk; Thin films and particles (both under Magnetic materials)

R. A. Wessling, *The Dow Chemical Company, Midland, Michigan,* Vinylidene chloride monomer and polymers

Peter J. Wessner, *Merichem Company, Houston, Texas,* Naphthenic acids

Jack H. Westbrook, *Sci-Tech Knowledge Systems, Brookline Technologies, Ballston Spa, New York,* Chromium and chromium alloys; Materials standards and specifications

Charles W. Weston, *Freeport Research and Engineering Company, Bell Chase, Louisiana,* Ammonium compounds

J. Marc Whalen, *McMaster University, Ontario, Canada,* Compounds (under Helium group)

Thomas P. Whaley, *Consultant, Sun City, Arizona,* Pipelines

Roy L. Whistler, *Purdue University, Lafayette, Indiana,* Starch

David L. White, *Kwick Kleen Industrial Solvents, Inc., Vincennes, Indiana,* Paint and finish removers (under Paint)

Dwain M. White, *General Electric, Schenectady, New York,* Aromatic (under Polyethers)

J. S. White, *White Technical Research Group, Argenta, Illinois,* Special sugars (under Sugar)

Robert H. White, *USDA Forest Service, Madison, Wisconsin,* Wood

Walter H. Whitlock, *The BOC Group, Inc., Murray Hill, New Jersey,* High purity gases

Peter Whitman, *Kraft General Foods Corporation, White Plains, New York,* Coffee

Zeno W. Wicks, Jr., *Consultant, Las Cruces, New Mexico,* Coatings; Drying oils

U. Widmer, *F. Hoffmann-La Roche Ltd., Basel, Switzerland,* Psychopharmacological agents

Paul Wight, *Zeneca Specialties, Manchester, U.K.,* Xanthene dyes

William R. Wilcox, *Clarkson College of Technology, Potsdam, New York,* Zone refining

Charles R. Wilke, *University of California, Berkeley, Berkeley, California,* Absorption

Rodney Willer, *Gaylord Chemical Corporation, Slidell, Louisiana,* Sulfoxides

Laurence L. Williams, *American Cyanamid Company, Stamford, Connecticut,* Amino resins and plastics

M. E. Williams, *EET Corporation, Knoxville, Tennessee,* Reverse osmosis

Martha E. Williams, *University of Illinois, Urbana, Illinois,* Databases

Michael Williams, *Abbott Laboratories, Abbott Park, Illinois,* Hypnotics, sedatives, anticonvulsants, and anxiolytics

Ted Williams, *Magnetrol International, Inc., Downers Grove, Illinois,* Liquid level measurement

R. A. Wilsak, *Amoco Chemical Company, Naperville, Illinois,* Butylenes

Alan C. Wilson, *Bausch & Lomb, Rochester, New York,* Contact lenses

Clifford B. Wilson, *The Dow Chemical Company, Freeport, Texas,* Magnesium and magnesium alloys

David A. Wilson, *Designed Chemicals R&D, Freeport, Texas,* Chelating agents

Richard Wilson, *International Flavor & Fragrances, Union Beach, New Jersey,* Benzyl alcohol and β-phenethyl alcohol; Oils, essential

Jerrold E. Winandy, *USDA Forest Service, Madison, Wisconsin,* Wood

Robert M. Winslow, *University of California, San Diego,* Blood, artificial

William T. Winter, *SUNY-Syracuse, Syracuse, New York,* Magnetic spin resonance

Edmund M. Wise Jr., *The Wellcome Research Laboratories, Research Triangle Park, North Carolina,* Peptides (under Antibiotics)

John S. Wishnok, *Massachusetts Institute of Technology, Cambridge,* *N*-Nitrosamines

Jan With, *Chatterton Petrochemical Corp., Delta, British Columbia, Canada,* Benzaldehyde; Benzoic acid

Donald T. Witiak, *University of Wisconsin, Madison,* Pharmaceuticals, chiral

J. T. Witkawski, *Martin Marietta Magnesia Specialties, Inc., Baltimore, Maryland,* Magnesium compounds

W. K. Witsiepe, *Consultant, Louisville, Kentucky,* Polychloroprene (under Elastomers, synthetic)

Suhad Wojkowski, *U.S. Department of Agriculture, New Orleans, Louisiana,* Herbicides

John A. Wojtowicz, *Consultant, Olin Corporation, Cheshire, Connecticut,* Dichlorine monoxide, hypochlorous acid, and hypochlorites (under Chlorine oxygen acids and salts); Chloramines and bromamines; Cyanuric and isocyanuric acids; Ozone; Treatment of swimming pools, spas, and hot tubs (under Water)

Clarence J. Wolf, *Michigan Molecular Institute, Midland, Michigan,* Thermoplastics (under Composite materials, polymer-matrix)

Walter J. Wolf, *U.S. Department of Agriculture, Peoria, Illinois,* Soybeans and other oilseeds

George T. Wolff, *General Motors, Warren, Michigan,* Air pollution

Leszek J. Wolfram, *Clairol Inc., Stamford, Connecticut,* Hair preparations

C. P. Wong, *AT&T Bell Labs, Princeton, New Jersey,* Embedding

Stewart Wong, *Jing Xing Health and Safety Resources, Inc., Annandale, Virginia,* Immunotherapeutic agents

W. E. Wood, *James Robinson, Ltd., Yorkshire, U.K.,* Sulfur dyes

Calvin R. Woodings, *Courtaulds, Coventry, U.K.,* Regenerated cellulosics (under Fibers)

Kermit E. Woodcock, *Consultant, Gowanda, New York,* Gas, natural

Gayle Woodside, *IBM Corporation, Austin, Texas,* Waste treatment, hazardous waste

Mike Woolery, *U.S. Vanadium, Hot Springs, Arkansas,* Vanadium compounds

John R. Woolfrey, *University of Mississippi, University,* Hormones, adrenal-cortical

Allen R. Worm, *3M Company, St. Paul, Minnesota,* Fluorocarbon elastomers (under Elastomers, synthetic)

Paul R. Worsham, *Eastman Chemical Company, Kingsport, Tennessee,* Coal chemicals (under Feedstocks)

Andrew J. Woytek, *Air Products and Chemicals, Inc., Allentown, Pennsylvania,* Halogens; Nitrogen; Tungsten (all under Fluorine compounds, inorganic)

Jeremy Wright, *ALZA Corporation, Palo Alto, California,* Pharmaceuticals (under Controlled release technology)

J. P. Wristers, *Exxon Chemical Company, Baytown, Texas,* Butadiene

Victor J. Wroblewski, *Lilly Research Laboratories, Indianapolis, Indiana,* Human growth hormone (under Hormones)

C. J. Wust, Jr., *Hercules Inc., Wilmington, Delaware,* Olefin (under Fibers)

Marino Xanthos, *Stevens Institute of Technology, Hoboken, New Jersey,* Plastics processing

Eli Yablonovitch, *University of California, Los Angeles,* Electronic materials

Chen-Hsyong Yang, *Monsanto Company, St. Louis, Missouri,* Phosphorus compounds

Baki Yarar, *Colorado School of Mines, Golden, Colorado,* Flotation

Gary L. Yingling, *McKenna & Cuneo, LLP, Washington, D.C.,* Pharmaceuticals and cosmetics (under Regulatory agencies)

Carmen M. Yon, *UOP, Tarrytown, New York,* Adsorption, gas separation

Fumio Yoneda, *Fujimoto Pharmaceutical Corporation, Osaka, Japan,* Riboflavin (under Vitamins)

Katsumasa Yoshikubo, *Nissan Chemical Industries, Ltd., Tokyo, Japan,* Sulfamic acid and sulfamates

Clyde T. Young, *North Carolina State University, Raleigh,* Nuts

Raymond A. Young, *University of Wisconsin, Madison,* Vegetable (under Fibers)

Robert Zaczek, *E. I. du Pont de Nemours & Co., Inc., Wilmington, Delaware,* Memory-enhancing drugs

Marek Zaidlewicz, *Nicolaus Copernicus University, Poland,* Hydroboration

Aleksey Zaks, *Schering-Plough Research Institute, Union, New Jersey,* Enzymes in organic synthesis

Paul Zanowiak, *Temple University, Philadelphia, Pennsylvania,* Pharmaceuticals

Steve G. Zantos, *Bausch & Lomb, Rochester, New York,* Contact lenses

David Zelmanovic, *Technicon Instruments Corporation, Tarrytown, New York,* Hematology (under Automated instrumentation)

John I. Zerbe, *USDA Forest Service, Madison, Wisconsin,* Wood

Edmund G. Zey, *Hoechst-Celanese Corporation, Corpus Christi, Texas,* Esterification

J. Richard Zietz, *Ethyl Corporation, Baton Rouge, Louisiana,* Synthetic processes (under Alcohols, higher aliphatic)

B. L. Zoumas, *Hershey Foods Corporation, Hershey, Pennsylvania,* Chocolate and cocoa

CONVERSION FACTORS, ABBREVIATIONS AND UNIT SYMBOLS

SI Units (Adopted 1960)

The International System of Units (abbreviated SI), is being implemented throughout the world. This measurement system is a modernized version of the MKSA (meter, kilogram, second, ampere) system, and its details are published and controlled by an international treaty organization (The International Bureau of Weights and Measures) (1).

SI units are divided into three classes:

BASE UNITS

length	meter† (m)
mass	kilogram (kg)
time	second (s)
electric current	ampere (A)
thermodynamic temperature‡	kelvin (K)
amount of substance	mole (mol)
luminous intensity	candela (cd)

SUPPLEMENTARY UNITS

plane angle	radian (rad)
solid angle	steradian (sr)

DERIVED UNITS AND OTHER ACCEPTABLE UNITS

These units are formed by combining base units, supplementary units, and other derived units (2–4). Those derived units having special names and symbols are marked with an asterisk in the list below.

Quantity	Unit	Symbol	Acceptable equivalent
*absorbed dose	gray	Gy	J/kg
acceleration	meter per second squared	m/s^2	
*activity (of a radionuclide)	becquerel	Bq	1/s
area	square kilometer	km^2	
	square hectometer	hm^2	ha (hectare)
	square meter	m^2	
concentration (of amount of substance)	mole per cubic meter	mol/m^3	
current density	ampere per square meter	A/m^2	
density, mass density	kilogram per cubic meter	kg/m^3	g/L; mg/cm^3
dipole moment (quantity)	coulomb meter	C·m	

†The spellings "metre" and "litre" are preferred by ASTM; however, "-er" is used in the *Encyclopedia*.

‡Wide use is made of Celsius temperature (t) defined by

$$t = T - T_0$$

where T is the thermodynamic temperature, expressed in kelvin, and $T_0 = 273.15$ K by definition. A temperature interval may be expressed in degrees Celsius as well as in kelvin.

Quantity	Unit	Symbol	Acceptable equivalent
*dose equivalent	sievert	Sv	J/kg
*electric capacitance	farad	F	C/V
*electric charge, quantity of electricity	coulomb	C	A·s
electric charge density	coulomb per cubic meter	C/m^3	
*electric conductance	siemens	S	A/V
electric field strength	volt per meter	V/m	
electric flux density	coulomb per square meter	C/m^2	
*electric potential, potential difference, electromotive force	volt	V	W/A
*electric resistance	ohm	Ω	V/A
*energy, work, quantity of heat	megajoule	MJ	
	kilojoule	kJ	
	joule	J	N·m
	electronvolt†	eV†	
	kilowatt-hour†	kW·h†	
energy density	joule per cubic meter	J/m^3	
*force	kilonewton	kN	
	newton	N	$kg{\cdot}m/s^2$
*frequency	megahertz	MHz	
	hertz	Hz	1/s
heat capacity, entropy	joule per kelvin	J/K	
heat capacity (specific), specific entropy	joule per kilogram kelvin	J/(kg·K)	
heat-transfer coefficient	watt per square meter kelvin	$W/(m^2{\cdot}K)$	
*illuminance	lux	lx	lm/m^2
*inductance	henry	H	Wb/A
linear density	kilogram per meter	kg/m	
luminance	candela per square meter	cd/m^2	
*luminous flux	lumen	lm	cd·sr
magnetic field strength	ampere per meter	A/m	
*magnetic flux	weber	Wb	V·s
*magnetic flux density	tesla	T	Wb/m^2
molar energy	joule per mole	J/mol	
molar entropy, molar heat capacity	joule per mole kelvin	J/(mol·K)	
moment of force, torque	newton meter	N·m	
momentum	kilogram meter per second	kg·m/s	
permeability	henry per meter	H/m	
permittivity	farad per meter	F/m	
*power, heat flow rate, radiant flux	kilowatt	kW	
	watt	W	J/s
power density, heat flux density, irradiance	watt per square meter	W/m^2	
*pressure, stress	megapascal	MPa	
	kilopascal	kPa	
	pascal	Pa	N/m^2
sound level	decibel	dB	
specific energy	joule per kilogram	J/kg	
specific volume	cubic meter per kilogram	m^3/kg	
surface tension	newton per meter	N/m	
thermal conductivity	watt per meter kelvin	W/(m·K)	
velocity	meter per second	m/s	
	kilometer per hour	km/h	
viscosity, dynamic	pascal second	Pa·s	
	millipascal second	mPa·s	
viscosity, kinematic	square meter per second	m^2/s	
	square millimeter per second	mm^2/s	
volume	cubic meter	m^3	
	cubic diameter	dm^3	L (liter) (5)
	cubic centimeter	cm^3	mL
wave number	1 per meter	m^{-1}	
	1 per centimeter	cm^{-1}	

†This non-SI unit is recognized by the CIPM as having to be retained because of practical importance or use in specialized fields (1).

In addition, there are 16 prefixes used to indicate order of magnitude, as follows:

Multiplication factor	Prefix	Symbol	Note
10^{18}	exa	E	
10^{15}	peta	P	
10^{12}	tera	T	
10^{9}	giga	G	
10^{6}	mega	M	
10^{3}	kilo	k	
10^{2}	hecto	h[a]	[a]Although hecto, deka, deci, and centi are SI prefixes, their use should be avoided except for SI unit-multiples for area and volume and nontechnical use of centimeter, as for body and clothing measurement.
10	deka	da[a]	
10^{-1}	deci	d[a]	
10^{-2}	centi	c[a]	
10^{-3}	milli	m	
10^{-6}	micro	μ	
10^{-9}	nano	n	
10^{-12}	pico	p	
10^{-15}	femto	f	
10^{-18}	atto	a	

For a complete description of SI and its use the reader is referred to ASTM E380 (4) and the article UNITS AND CONVERSION FACTORS.

A representative list of conversion factors from non-SI to SI units is presented herewith. Factors are given to four significant figures. Exact relationships are followed by a dagger. A more complete list is given in the latest editions of ASTM E380 (4) and ANSI Z210.1 (6).

Conversion Factors to SI Units

To convert from	To	Multiply by
acre	square meter (m^2)	4.047×10^{3}
angstrom	meter (m)	1.0×10^{-10}†
are	square meter (m^2)	1.0×10^{2}†
astronomical unit	meter (m)	1.496×10^{11}
atmosphere, standard	pascal (Pa)	1.013×10^{5}
bar	pascal (Pa)	1.0×10^{5}†
barn	square meter (m^2)	1.0×10^{-28}†
barrel (42 U.S. liquid gallons)	cubic meter (m^3)	0.1590
Bohr magneton (μ_B)	J/T	9.274×10^{-24}
Btu (International Table)	joule (J)	1.055×10^{3}
Btu (mean)	joule (J)	1.056×10^{3}
Btu (thermochemical)	joule (J)	1.054×10^{3}
bushel	cubic meter (m^3)	3.524×10^{-2}
calorie (International Table)	joule (J)	4.187
calorie (mean)	joule (J)	4.190
calorie (thermochemical)	joule (J)	4.184†
centipoise	pascal second (Pa·s)	1.0×10^{-3}†
centistokes	square millimeter per second (mm^2/s)	1.0†
cfm (cubic foot per minute)	cubic meter per second (m^3/s)	4.72×10^{-4}
cubic inch	cubic meter (m^3)	1.639×10^{-5}
cubic foot	cubic meter (m^3)	2.832×10^{-2}
cubic yard	cubic meter (m^3)	0.7646
curie	becquerel (Bq)	3.70×10^{10}†
debye	coulomb meter (C·m)	3.336×10^{-30}
degree (angle)	radian (rad)	1.745×10^{-2}
denier (international)	kilogram per meter (kg/m)	1.111×10^{-7}
	tex‡	0.1111
dram (apothecaries')	kilogram (kg)	3.888×10^{-3}
dram (avoirdupois)	kilogram (kg)	1.772×10^{-3}
dram (U.S. fluid)	cubic meter (m^3)	3.697×10^{-6}
dyne	newton (N)	1.0×10^{-5}†
dyne/cm	newton per meter (N/m)	1.0×10^{-3}†
electronvolt	joule (J)	1.602×10^{-19}
erg	joule (J)	1.0×10^{-7}†

†Exact.

‡See footnote on p. xxviii.

To convert from	To	Multiply by
fathom	meter (m)	1.829
fluid ounce (U.S.)	cubic meter (m^3)	2.957×10^{-5}
foot	meter (m)	0.3048†
footcandle	lux (lx)	10.76
furlong	meter (m)	2.012×10^{-2}
gal	meter per second squared (m/s^2)	1.0×10^{-2}†
gallon (U.S. dry)	cubic meter (m^3)	4.405×10^{-3}
gallon (U.S. liquid)	cubic meter (m^3)	3.785×10^{-3}
gallon per minute (gpm)	cubic meter per second (m^3/s)	6.309×10^{-5}
	cubic meter per hour (m^3/h)	0.2271
gauss	tesla (T)	1.0×10^{-4}
gilbert	ampere (A)	0.7958
gill (U.S.)	cubic meter (m^3)	1.183×10^{-4}
grade	radian	1.571×10^{-2}
grain	kilogram (kg)	6.480×10^{-5}
gram force per denier	newton per tex (N/tex)	8.826×10^{-2}
hectare	square meter (m^2)	1.0×10^{4}†
horsepower (550 ft·lbf/s)	watt (W)	7.457×10^{2}
horsepower (boiler)	watt (W)	9.810×10^{3}
horsepower (electric)	watt (W)	7.46×10^{2}†
hundredweight (long)	kilogram (kg)	50.80
hundredweight (short)	kilogram (kg)	45.36
inch	meter (m)	2.54×10^{-2}†
inch of mercury (32°F)	pascal (Pa)	3.386×10^{3}
inch of water (39.2°F)	pascal (Pa)	2.491×10^{2}
kilogram-force	newton (N)	9.807
kilowatt hour	megajoule (MJ)	3.6†
kip	newton (N)	4.448×10^{3}
knot (international)	meter per second (m/S)	0.5144
lambert	candela per square meter (cd/m^3)	3.183×10^{3}
league (British nautical)	meter (m)	5.559×10^{3}
league (statute)	meter (m)	4.828×10^{3}
light year	meter (m)	9.461×10^{15}
liter (for fluids only)	cubic meter (m^3)	1.0×10^{-3}†
maxwell	weber (Wb)	1.0×10^{-8}†
micron	meter (m)	1.0×10^{-6}†
mil	meter (m)	2.54×10^{-5}†
mile (statute)	meter (m)	1.609×10^{3}
mile (U.S. nautical)	meter (m)	1.852×10^{3}†
mile per hour	meter per second (m/s)	0.4470
millibar	pascal (Pa)	1.0×10^{2}
millimeter of mercury (0°C)	pascal (Pa)	1.333×10^{2}†
minute (angular)	radian	2.909×10^{-4}
myriagram	kilogram (kg)	10
myriameter	kilometer (km)	10
oersted	ampere per meter (A/m)	79.58
ounce (avoirdupois)	kilogram (kg)	2.835×10^{-2}
ounce (troy)	kilogram (kg)	3.110×10^{-2}
ounce (U.S. fluid)	cubic meter (m^3)	2.957×10^{-5}
ounce-force	newton (N)	0.2780
peck (U.S.)	cubic meter (m^3)	8.810×10^{-3}
pennyweight	kilogram (kg)	1.555×10^{-3}
pint (U.S. dry)	cubic meter (m^3)	5.506×10^{-4}
pint (U.S. liquid)	cubic meter (m^3)	4.732×10^{-4}
poise (absolute viscosity)	pascal second (Pa·s)	0.10†
pound (avoirdupois)	kilogram (kg)	0.4536
pound (troy)	kilogram (kg)	0.3732
poundal	newton (N)	0.1383
pound-force	newton (N)	4.448
pound force per square inch (psi)	pascal (Pa)	6.895×10^{3}
quart (U.S. dry)	cubic meter (m^3)	1.101×10^{-3}
quart (U.S. liquid)	cubic meter (m^3)	9.464×10^{-4}

†Exact.

To convert from	To	Multiply by
quintal	kilogram (kg)	$1.0 \times 10^{2\dagger}$
rad	gray (Gy)	$1.0 \times 10^{-2\dagger}$
rod	meter (m)	5.029
roentgen	coulomb per kilogram (C/kg)	2.58×10^{-4}
second (angle)	radian (rad)	$4.848 \times 10^{-6\dagger}$
section	square meter (m^2)	2.590×10^{6}
slug	kilogram (kg)	14.59
spherical candle power	lumen (lm)	12.57
square inch	square meter (m^2)	6.452×10^{-4}
square foot	square meter (m^2)	9.290×10^{-2}
square mile	square meter (m^2)	2.590×10^{6}
square yard	square meter (m^2)	0.8361
stere	cubic meter (m^3)	$1.0^{\dagger}$
stokes (kinematic viscosity)	square meter per second (m^2/s)	$1.0 \times 10^{-4\dagger}$
tex	kilogram per meter (kg/m)	$1.0 \times 10^{-6\dagger}$
ton (long, 2240 pounds)	kilogram (kg)	1.016×10^{3}
ton (metric) (tonne)	kilogram (kg)	$1.0 \times 10^{3\dagger}$
ton (short, 2000 pounds)	kilogram (kg)	9.072×10^{2}
torr	pascal (Pa)	1.333×10^{2}
unit pole	weber (Wb)	1.257×10^{-7}
yard	meter (m)	$0.9144^{\dagger}$

†Exact.

Abbreviations and Unit Symbols

Following is a list of common abbreviations and unit symbols used in the *Encyclopedia*. In general they agree with those listed in *American National Standard Abbreviations for Use on Drawings and in Text* (*ANSI Y1.1*) (6) and *American National Standard Letter Symbols for Units in Science and Technology* (*ANSI Y10*) (6). Also included is a list of acronyms for a number of private and government organizations as well as common industrial solvents, polymers, and other chemicals.

Rules for Writing Unit Symbols (4):

1. Unit symbols are printed in upright letters (roman) regardless of the type style used in the surrounding text.
2. Unit symbols are unaltered in the plural.
3. Unit symbols are not followed by a period except when used at the end of a sentence.
4. Letter unit symbols are generally printed lower-case (for example, cd for candela) unless the unit name has been derived from a proper name, in which case the first letter of the symbol is capitalized (W, Pa). Prefixes and unit symbols retain their prescribed form regardless of the surrounding typography.
5. In the complete expression for a quantity, a space should be left between the numerical value and the unit symbol. For example, write 2.37 lm, *not* 2.37lm, and 35 mm, *not* 35mm. When the quantity is used in an adjectival sense, a hyphen is often used, for example, 35-mm film. *Exception:* No space is left between the numerical value and the symbols of degree, minute, and second of plane angle, degree Celsius, and the percent sign.
6. No space is used between the prefix and unit symbol (for example, kg).
7. Symbols, not abbreviations, should be used for units. For example, use “A,” not “amp,” for ampere.
8. When multiplying unit symbols, use a raised dot:

   $$\mathrm{N{\cdot}m} \quad \text{for} \quad \text{newton meter}$$

   In the case of W·h, the dot may be omitted, thus:

   $$\mathrm{Wh}$$

   An exception to this practice is made for computer printouts, automatic typewriter work, etc, where the raised dot is not possible, and a dot on the line may be used.
9. When dividing unit symbols, use one of the following forms:

   $$\mathrm{m/s} \quad or \quad \mathrm{m{\cdot}s^{-1}} \quad or \quad \frac{\mathrm{m}}{\mathrm{s}}$$

   In no case should more than one slash be used in the same expression unless parentheses are inserted to avoid ambiguity. For example, write:

   $$\mathrm{J/(mol{\cdot}K)} \quad or \quad \mathrm{J{\cdot}mol^{-1}{\cdot}K^{-1}} \quad or \quad \mathrm{(J/mol)/K}$$

but *not*

J/mol/K

10. Do not mix symbols and unit names in the same expression. Write:

joules per kilogram *or* J/kg *or* $J{\cdot}kg^{-1}$

but *not*

joules/kilogram *nor* joules/kg *nor* $joules{\cdot}kg^{-1}$

ABBREVIATIONS AND UNITS

A	ampere
A	anion (eg, HA)
A	mass number
a	atto (prefix for 10^{-18})
AATCC	American Association of Textile Chemists and Colorists
ABS	acrylonitrile–butadiene–styrene
abs	absolute
ac	alternating current, *n.*
a-c	alternating current, *adj.*
ac-	alicyclic
acac	acetylacetonate
ACGIH	American Conference of Governmental Industrial Hygienists
ACS	American Chemical Society
AGA	American Gas Association
Ah	ampere hour
AIChE	American Institute of Chemical Engineers
AIME	American Institute of Mining, Metallurgical, and Petroleum Engineers
AIP	American Institute of Physics
AISI	American Iron and Steel Institute
alc	alcohol(ic)
Alk	alkyl
alk	alkaline (not alkali)
amt	amount
amu	atomic mass unit
ANSI	American National Standards Institute
AO	atomic orbital
AOAC	Association of Official Analytical Chemists
AOCS	American Oil Chemists' Society
APHA	American Public Health Association
API	American Petroleum Institute
aq	aqueous
Ar	aryl
ar-	aromatic
as-	asymmetric(al)
ASHRAE	American Society of Heating, Refrigerating, and Air Conditioning Engineers
ASM	American Society for Metals
ASME	American Society of Mechanical Engineers
ASTM	American Society for Testing and Materials
at no.	atomic number
at wt	atomic weight
av(g)	average
AWS	American Welding Society
b	bonding orbital
bbl	barrel
bcc	body-centered cubic
BCT	body-centered tetragonal
Bé	Baumé
BET	Brunauer-Emmett-Teller (adsorption equation)
bid	twice daily
Boc	*t*-butyloxycarbonyl
BOD	biochemical (biological) oxygen demand
bp	boiling point
Bq	becquerel
C	coulomb
°C	degree Celsius
C-	denoting attachment to carbon
c	centi (prefix for 10^{-2})
c	critical
ca	circa (approximately)
cd	candela; current density; circular dichroism
CFR	Code of Federal Regulations
cgs	centimeter-gram-second
CI	Color Index
cis-	isomer in which substituted groups are on same side of double bond between C atoms
cl	carload
cm	centimeter
cmil	circular mil
cmpd	compound
CNS	central nervous system
CoA	coenzyme A
COD	chemical oxygen demand
coml	commercial(ly)
cp	chemically pure
cph	close-packed hexagonal
CPSC	Consumer Product Safety Commission
cryst	crystalline
cub	cubic
D	debye
D-	denoting configurational relationship
d	differential operator
d	day; deci (prefix for 10^{-1})
d	density
d-	*dextro-*, dextrorotatory
da	deka (prefix for 10^1)
dB	decibel
dc	direct current, *n.*
d-c	direct current, *adj.*
dec	decompose
detd	determined
detn	determination
Di	didymium, a mixture of all lanthanons
dia	diameter
dil	dilute
DIN	Deutsche Industrie Normen
dl-; DL-	racemic
DMA	dimethylacetamide
DMF	dimethylformamide

DMG	dimethyl glyoxime
DMSO	dimethyl sulfoxide
DOD	Department of Defense
DOE	Department of Energy
DOT	Department of Transportation
DP	degree of polymerization
dp	dew point
DPH	diamond pyramid hardness
dstl(d)	distill(ed)
dta	differential thermal analysis
(*E*)-	entgegen; opposed
ϵ	dielectric constant (unitless number)
e	electron
ECU	electrochemical unit
ed.	edited, edition, editor
ED	effective dose
EDTA	ethylenediaminetetraacetic acid
emf	electromotive force
emu	electromagnetic unit
en	ethylene diamine
eng	engineering
EPA	Environmental Protection Agency
epr	electron paramagnetic resonance
eq.	equation
esca	electron spectroscopy for chemical analysis
esp	especially
esr	electron-spin resonance
est(d)	estimate(d)
estn	estimation
esu	electrostatic unit
exp	experiment, experimental
ext(d)	extract(ed)
F	farad (capacitance)
F	faraday (96,487 C)
f	femto (prefix for 10^{-15})
FAO	Food and Agriculture Organization (United Nations)
fcc	face-centered cubic
FDA	Food and Drug Administration
FEA	Federal Energy Administration
FHSA	Federal Hazardous Substances Act
fob	free on board
fp	freezing point
FPC	Federal Power Commission
FRB	Federal Reserve Board
frz	freezing
G	giga (prefix for 10^9)
G	gravitational constant $= 6.67 \times 10^{11}\ \mathrm{N{\cdot}m^2/kg^2}$
g	gram
(g)	gas, only as in $H_2O(g)$
g	gravitational acceleration
gc	gas chromatography
gem-	geminal
glc	gas–liquid chromatography
g-mol wt; gmw	gram-molecular weight
GNP	gross national product
gpc	gel-permeation chromatography
GRAS	Generally Recognized as Safe
grd	ground
Gy	gray
H	henry
h	hour; hecto (prefix for 10^2)
ha	hectare
HB	Brinell hardness number
Hb	hemoglobin
hcp	hexagonal close-packed
hex	hexagonal
HK	Knoop hardness number
hplc	high performance liquid chromatography
HRC	Rockwell hardness (C scale)
HV	Vickers hardness number
hyd	hydrated, hydrous
hyg	hygroscopic
Hz	hertz
^{i} (eg, Pr^i)	iso (eg, isopropyl)
i-	inactive (eg, *i*-methionine)
IACS	International Annealed Copper Standard
ibp	initial boiling point
IC	integrated circuit
ICC	Interstate Commerce Commission
ICT	International Critical Table
ID	inside diameter; infective dose
ip	intraperitoneal
IPS	iron pipe size
ir	infrared
IRLG	Interagency Regulatory Liaison Group
ISO	International Organization Standardization
ITS-90	International Temperature Scale (NIST)
IU	International Unit
IUPAC	International Union of Pure and Applied Chemistry
IV	iodine value
iv	intravenous
J	joule
K	kelvin
k	kilo (prefix for 10^3)
kg	kilogram
L	denoting configurational relationship
L	liter (for fluids only) (5)
l-	*levo*-, levorotatory
(l)	liquid, only as in $NH_3(l)$
LC_{50}	conc lethal to 50% of the animals tested
LCAO	linear combination of atomic orbitals
lc	liquid chromatography
LCD	liquid crystal display
lcl	less than carload lots
LD_{50}	dose lethal to 50% of the animals tested
LED	light-emitting diode
liq	liquid
lm	lumen
ln	logarithm (natural)
LNG	liquefied natural gas
log	logarithm (common)
LOI	limiting oxygen index
LPG	liquefied petroleum gas
ltl	less than truckload lots
lx	lux
M	mega (prefix for 10^6); metal (as in MA)
M	molar; actual mass
$\overline{M}_w$	weight-average mol wt
$\overline{M}_n$	number-average mol wt
m	meter; milli (prefix for 10^{-3})
m	molal
m-	meta
max	maximum
MCA	Chemical Manufacturers' Association (was Manufacturing Chemists Association)
MEK	methyl ethyl ketone
meq	milliequivalent
mfd	manufactured

mfg	manufacturing
mfr	manufacturer
MIBC	methyl isobutyl carbinol
MIBK	methyl isobutyl ketone
MIC	minimum inhibiting concentration
min	minute; minimum
mL	milliliter
MLD	minimum lethal dose
MO	molecular orbital
mo	month
mol	mole
mol wt	molecular weight
mp	melting point
MR	molar refraction
ms	mass spectrometry
MSDS	material safety data sheet
mxt	mixture
μ	micro (prefix for 10^{-6})
N	newton (force)
N	normal (concentration); neutron number
N-	denoting attachment to nitrogen
n (as n_D^{20})	index of refraction (for 20°C and sodium light)
^{n} (as Bu^n), *n*-	normal (straight-chain structure)
n	neutron
n	nano (prefix for 10^9)
na	not available
NAS	National Academy of Sciences
NASA	National Aeronautics and Space Administration
nat	natural
ndt	nondestructive testing
neg	negative
NF	*National Formulary*
NIH	National Institutes of Health
NIOSH	National Institute of Occupational Safety and Health
NIST	National Institute of Standards and Technology (formerly National Bureau of Standards)
nmr	nuclear magnetic resonance
NND	New and Nonofficial Drugs (AMA)
no.	number
NOI-(BN)	not otherwise indexed (by name)
NOS	not otherwise specified
nqr	nuclear quadruple resonance
NRC	Nuclear Regulatory Commission; National Research Council
NRI	New Ring Index
NSF	National Science Foundation
NTA	nitrilotriacetic acid
NTP	normal temperature and pressure (25°C and 101.3 kPa or 1 atm)
NTSB	National Transportation Safety Board
O-	denoting attachment to oxygen
o-	ortho
OD	outside diameter
OPEC	Organization of Petroleum Exporting Countries
o-phen	*o*-phenanthridine
OSHA	Occupational Safety and Health Administration
owf	on weight of fiber
Ω	ohm
P	peta (prefix for 10^{15})
p	pico (prefix for 10^{-12})
p-	para
p	proton
p.	page
Pa	pascal (pressure)
PEL	personal exposure limit based on an 8-h exposure
pd	potential difference
pH	negative logarithm of the effective hydrogen ion concentration
phr	parts per hundred of resin (rubber)
p-i-n	positive-intrinsic-negative
pmr	proton magnetic resonance
p-n	positive-negative
po	per os (oral)
POP	polyoxypropylene
pos	positive
pp.	pages
ppb	parts per billion (10^9)
ppm	parts per million (10^6)
ppmv	parts per million by volume
ppmwt	parts per million by weight
PPO	poly(phenyl oxide)
ppt(d)	precipitate(d)
pptn	precipitation
Pr (no.)	foreign prototype (number)
pt	point; part
PVC	poly(vinyl chloride)
pwd	powder
py	pyridine
qv	quod vide (which see)
R	univalent hydrocarbon radical
(*R*)-	rectus (clockwise configuration)
r	precision of data
rad	radian; radius
RCRA	Resource Conservation and Recovery Act
rds	rate-determining step
ref.	reference
rf	radio frequency, *n.*
r-f	radio frequency, *adj.*
rh	relative humidity
RI	Ring Index
rms	root-mean square
rpm	rotations per minute
rps	revolutions per second
RT	room temperature
RTECS	Registry of Toxic Effects of Chemical Substances
^{s} (eg, Bu^s); *sec*-	secondary (eg, secondary butyl)
S	siemens
(*S*)-	sinister (counterclockwise configuration)
S-	denoting attachment to sulfur
s-	symmetric(al)
s	second
(s)	solid, only as in $H_2O(s)$
SAE	Society of Automotive Engineers
SAN	styrene–acrylonitrile
sat(d)	saturate(d)
satn	saturation
SBS	styrene–butadiene–styrene
sc	subcutaneous
SCF	self-consistent field; standard cubic feet
Sch	Schultz number
sem	scanning electron microscope(y)
SFs	Saybolt Furol seconds
sl sol	slightly soluble
sol	soluble
soln	solution
soly	solubility
sp	specific; species
sp gr	specific gravity
sr	steradian

std	standard
STP	standard temperature and pressure (0°C and 101.3 kPa)
sub	sublime(s)
SUs	Saybolt Universal seconds
syn	synthetic
^{t} (eg, Bu^t), *t*-, *tert*-	tertiary (eg, tertiary butyl)
T	tera (prefix for 10^{12}); tesla (magnetic flux density)
t	metric ton (tonne)
t	temperature
TAPPI	Technical Association of the Pulp and Paper Industry
TCC	Tagliabue closed cup
tex	tex (linear density)
T_g	glass-transition temperature
tga	thermogravimetric analysis
THF	tetrahydrofuran
tlc	thin layer chromatography
TLV	threshold limit value
trans-	isomer in which substituted groups are on opposite sides of double bond between C atoms
TSCA	Toxic Substances Control Act
TWA	time-weighted average
Twad	Twaddell
UL	Underwriters' Laboratory
USDA	United States Department of Agriculture
USP	*United States Pharmacopeia*
uv	ultraviolet
V	volt (emf)
var	variable
vic-	vicinal
vol	volume (not volatile)
vs	versus
v sol	very soluble
W	watt
Wb	weber
Wh	watt hour
WHO	World Health Organization (United Nations)
wk	week
yr	year
(*Z*)-	zusammen; together; atomic number

Non-SI (Unacceptable and Obsolete) Units		Use
Å	angstrom	nm
at	atmosphere, technical	Pa
atm	atmosphere, standard	Pa
b	barn	cm^2
bar[†]	bar	Pa
bbl	barrel	m^3
bhp	brake horsepower	W
Btu	British thermal unit	J
bu	bushel	m^3; L
cal	calorie	J
cfm	cubic foot per minute	m^3/s
Ci	curie	Bq
cSt	centistokes	mm^2/s
c/s	cycle per second	Hz
cu	cubic	exponential form
D	debye	C·m
den	denier	tex
dr	dram	kg
dyn	dyne	N
dyn/cm	dyne per centimeter	mN/m
erg	erg	J
eu	entropy unit	J/K
°F	degree Fahrenheit	°C; K
fc	footcandle	lx
fl	footlambert	lx
fl oz	fluid ounce	m^3; L
ft	foot	m
ft·lbf	foot pound-force	J
gf den	gram-force per denier	N/tex
G	gauss	T
Gal	gal	m/s^2
gal	gallon	m^3; L
Gb	gilbert	A
gpm	gallon per minute	(m^3/s); (m^3/h)
gr	grain	kg
hp	horsepower	W
ihp	indicated horsepower	W
in.	inch	m
in. Hg	inch of mercury	Pa
in. H_2O	inch of water	Pa
in.-lbf	inch pound-force	J
kcal	kilo-calorie	J
kgf	kilogram-force	N
kilo	for kilogram	kg
L	lambert	lx
lb	pound	kg
lbf	pound-force	N
mho	mho	S
mi	mile	m
MM	million	M
mm Hg	millimeter of mercury	Pa
mμ	millimicron	nm
mph	miles per hour	km/h
μ	micron	μm
Oe	oersted	A/m
oz	ounce	kg
ozf	ounce-force	N
η	poise	Pa·s
P	poise	Pa·s
ph	phot	lx
psi	pounds-force per square inch	Pa
psia	pounds-force per square inch absolute	Pa
psig	pounds-force per square inch gage	Pa
qt	quart	m^3; L
°R	degree Rankine	K
rd	rad	Gy
sb	stilb	lx
SCF	standard cubic foot	m^3
sq	square	exponential form
thm	therm	J
yd	yard	m

[†]Do not use bar (10^5 Pa) or millibar (10^2 Pa) because they are not SI units, and are accepted internationally only for a limited time in special fields because of existing usage.

BIBLIOGRAPHY

1. The International Bureau of Weights and Measures, BIPM (Parc de Saint-Cloud, France) is described in Appendix X2 of Ref. 4. This bureau operates under the exclusive supervision of the International Committee for Weights and Measures (CIPM).
2. *Metric Editorial Guide* (*ANMC-78-1*), latest ed., American National Metric Council, 5410 Grosvenor Lane, Bethesda, Md. 20814, 1981.
3. *SI Units and Recommendations for the Use of Their Multiples and of Certain Other Units* (*ISO 1000-1981*), American National Standards Institute, 1430 Broadway, New York, 10018, 1981.
4. Based on *ASTM E380-89a* (*Standard Practice for Use of the International System of Units* (*SI*)), American Society for Testing and Materials, 1916 Race Street, Philadelphia, Pa. 19103, 1989.
5. *Fed. Reg.*, Dec. 10, 1976 (41 FR 36414).
6. For ANSI address, see Ref. 3.

R. P. Lukens
ASTM Committee E-43 on SI Practice

A

ABLATIVE MATERIALS

Ablative materials are used to protect vehicles from atmospheric reentry, to protect rocket nozzles and ship hulls from propellant gas erosion, as protection from laser beams, and to protect land-based structures from high heat environments.

The Ablation Environment

The functional requirements of the ablative heatshield must be well understood before selection of the proper material can occur. Ablative heatshield materials not only protect a vehicle from excessive heating, they can also act as an aerodynamic body and a structural component. Intensity and duration of heating, thermostructural requirements and shape stability, potential for particle erosion, weight limitations and reusability are some of the factors which must be considered in selection of an ablative material.

A variety of test methods and facilities have been developed to address the process of ablation. These utilize lasers, chemical flames, plasma arcs, electric arc heaters and other heat sources and sometimes include high velocity wind tunnel facilities that introduce particles to simulate high speed erosion.

The Ablation Process

Thermophysically, the ablation process can be described as the elimination of a large amount of thermal energy by sacrifice of surface material. Principles operating during this highly efficient and orderly heat and mass transfer process are (*1*) phase changes such as melting, vaporization and sublimation, (*2*) conduction and storage of heat in the material substrate, (*3*) absorption of heat by gases as they are forced to the surface, (*4*) heat convection in a liquid layer, (*5*) transpiration of gases and liquids, and subsequent heat absorption from the surface into the boundary layer, (*6*) exothermic and endothermic chemical reactions, and (*7*) radiation on the surface and in bulk.

Ablative Materials

Ablative materials are characterized according to dominant ablation mechanism. There are three groups: subliming or melting ablators, charring ablators, and intumescent ablators. Figure 1 shows the physical zones of each. Because of the basic thermal and physical differences, the classes of ablative materials are used in different types of applications.

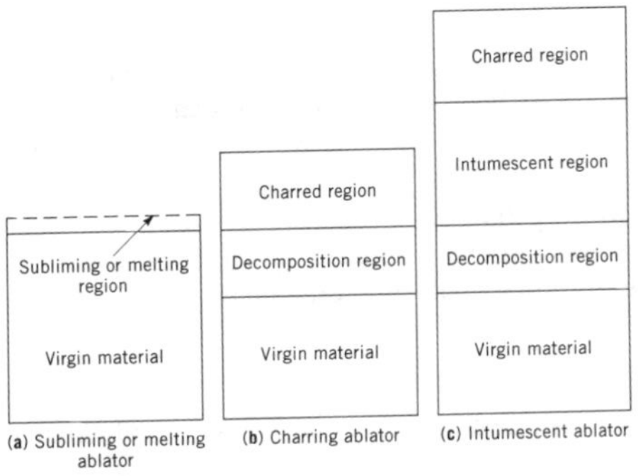

Figure 1. Physical zones of ablators. Typical time-integrated heat flux, J/m^2, (**a**) >500 (**b**) 5000, (**c**) <50; maximum instantaneous heat flux, MW/m^2, (**a**) 0.5, (**b**) >1, (**c**) 0.1. To convert J to cal, divide by 4.184.

Subliming and Melting Ablators. Subliming ablators act as heat sinks to the incident heat flux until the temperature on the surface reaches the sublimation or melting temperature, also known as the reaction temperature. At this time, the sublimation or melting action removes heat from the insulation material. In the sublimation process, the convective transfer of heat from the boundary layer to the material surface is also blocked by the gas evolving from the ablative material, concurrently thickening the boundary layer. This blocking action can reduce the net heating of the ablative material by more than 50%.

One of the first subliming ablative materials to be identified was polytetrafluoroethylene (Teflon), which offers light weight, good insulation properties as a result of its decomposition temperature (ca 500°C), and a high endothermic value for the depolymerization or ablative heat of reaction. Teflon has been combined with high strength, high temperature fibers such as quartz to reduce the necessary wall thickness and improve overall thermostructural performance, as well as to retain dielectric performance during reentry. Carbon, another subliming ablator, has been identified as having the highest heat of ablation. Consequently, it is used in high heating environments where a minimum of shape change is tolerable, such as in missile nosetips and rocket nozzle throats. Since pure forms of carbon, such as pyrolitic graphite, are subject to thermal fracture, composites with carbon matrix and carbon fiber reinforcement (carbon-carbon composites) are typically used where high thermal stresses are anticipated. Ceramic ablators are sometimes considered when dielectric performance is important. These include boron nitride, fused silica, nitroxyceram, and AS3DX, which is a silica composite material.

Charring Ablators. Charring ablators are used in a greater variety of thermal environments than either subliming or intumescent ablators because of their ability to withstand a much higher heat flux. In the charring ablator, the surface material acts as a heat sink, absorbing all of the incident heat flux and causing the surface temperature to increase quickly. At reaction temperature, endothermic chemical decomposition occurs: the matrix pyrolizes into charred material and gaseous products. The passage of heat-absorbing gases through the charred surface provides further insulative performance and thickens the boundary layer, reducing the convective heat transfer. The charring is a continuous process; as the charred surface is eroded, more char forms to take its place.

Charing ablators are commonly used with subliming or melting reinforcement materials, such as silica or carbon. Carbon fiber reinforced phenolic composites are high density charring ablators commonly used as heatshields for high load reentry vehicles. The reinforcement improves thermostructural performance, can minimize thermal conductivity, and can improve shear resistance. Low density charring ablators are used for low shear environments. An example is the Apollo mission, where epoxy novolac resin with phenolic microballoons and silica fiber were used to fill a fiber glass–phenolic honeycomb (which improved shear resistance). Cork is mixed with silicone or phenolic resin for an effective low cost charring ablator.

Intumescent Ablators. Additives in intumescent ablators form a foamlike region on exposure to heat, resulting in improved insulation performance. There are basic differences between intumescent and charring ablators. An intumescent reaction results in a net decrease in thermal conductivity, and a net increase in specific heat as the material temperature increases. Conversely, a charring reaction produces a net increase in thermal conductivity and a decrease in specific heat as the material temperature increases. Thus, as the net material temperature increases, the thermal diffusivity of an intumescent ablator decreases and that of a charring ablator increases.

Several forms of intumescent ablators are available. CHARTEK is a sprayable or trowelable epoxy-based material from Textron than can be used to coat load-carrying beams on oil rigs, allowing for increased performance of the beam in the event of a fire. INTERAM by 3M is a rubber wrap material that can be used to cover explosive material to prevent ignition during a fire.

Michael R. Favaloro
Textron Defense Systems

G. F. D'Alelio and J. A. Parker, eds. *Ablative Plastics,* Marcel Dekker, Inc., New York, 1971.

H. Hurwicz and J. E. Rogan, "Ablation," in W. M. Rohsenow and J. P. Hartlett, eds. *Handbook of Heat Transfer,* McGraw-Hill Book Co., Inc., New York, 1973.

ABRASIVES

An abrasive is a substance used to abrade, smooth, or polish an object. If the object is soft, such as wood, then relatively soft abrasive materials may be used. Usually, however, abrasive connotes very hard substances ranging from naturally occurring sands to the hardest material known, diamond.

There are three basic forms of abrasives: grit (loose, granular, or powdered particles); bonded materials (particles are bonded into wheels, segments, or stick shapes); and coated materials (particles are bonded to paper, plastic, cloth, or metal).

Properties of Abrasive Materials

Hardness. Table 1 lists the various scales of hardness (qv) used for abrasives.

Toughness. An abrasive's toughness is often measured and expressed as the degree of friability, the ability of an abrasive grit to withstand impact without cracking, spalling, or shattering.

Refractoriness (Melting Temperature). Instantaneous grinding temperatures may exceed 3500°C at the interface between an abrasive and the workpiece being ground. Hence melting temperature is an important property.

Chemical Reactivity. Any chemical interaction between abrasive grains and the material being abraded affects the abrasion process.

Thermal Conductivity. Abrasive materials may transfer heat from the cutting tip of the grain to the bond posts, retaining the heat in a bonded wheel or coated belt. The cooler the cutting point, the harder it is.

Fracture. Fracture characteristics of abrasive materials are important, as well as the resulting grain shapes. Equiaxed grains are generally preferred for bonded abrasive products and sharp, acicular grains are preferred for coated ones. How the grains fracture in the grinding process determines the wear resistance and self-sharpening characteristics of the wheel or belt.

Microstructure. Crystal size, porosity, and impurity phases play a major role in fixing the fracture characteristics and toughness of an abrasive grain.

Table 1. Scales of Hardness

Material	Mohs' scale	Ridgeway's scale	Woodell's scale	Knoop hardness[a], kN/m^{2b}
talc	1			
calcite	3			
apatite	5			
vitreous silica		7		
topaz	8	9		13
corundum	9		9	20
fused ZrO_2/Al_2O_3[c]				16
SiC		13	14	24
cubic boron nitride				46
diamond	10	15	42.5	78

[a] At a 100-g load (K-100) average.
[b] To convert kN/m^2 to kgf/mm^2 divide by 0.00981.
[c] 39% ZrO_2 (NZ Alundum).

Natural Abrasives

Naturally occuring abrasives are still an important item of commerce, although synthetic abrasives now fill many of their former uses. They include diamonds, corundum, emery, garnet, silica, sandstone, tripoli, pumice, and pumicite.

Manufactured Abrasives

Manufactured abrasives include silicon carbide, fused aluminum oxide, sintered aluminum oxide, sol–gel sintered aluminum oxide, fused zirconia–alumina, synthetic diamond, cubic boron nitride, boron carbide, slags, steel shot, and grit.

Sizing, Shaping, and Testing of Abrasive Grains

Sizing. Manufactured abrasives are produced in a variety of sizes that range from a pea-sized grit of 4 (5.2 mm) to submicrometer diameters.

Shaping. Desired shapes are obtained by controlling the method of crushing and by impacting or mulling. In general, cubical particles are preferred for grinding wheels, whereas high aspect-ratio acicular particles are preferred for coated abrasive belts and disks.

Testing. Chemical analyses are done on all manufactured abrasives, as well as physical tests such as sieve analyses, specific gravity, impact strength, and loose poured density (a rough measure of particle shape). Special abrasives such as sintered sol–gel aluminas require more sophisticated tests such as electron microscope measurement of α-alumina crystal size, and indentation microhardness.

Coated Abrasives

Coated abrasives consist of a flexible backing on which films of adhesive hold a coating of abrasive grains. The backing may be paper, cloth, open-mesh cloth, vulcanized fiber (a specially treated cotton rag base paper), or any combination of these materials. The abrasives most generally used are fused aluminum oxide, sol–gel alumina, alumina–zirconia, silicon carbide, garnet, emery, and flint.

A new form of coated abrasive has been developed that consists of tiny aggregates of abrasive material in the form of hollow spheres. As these spheres break down in use, fresh cutting grains are exposed; this maintains cut-rate and keeps power low.

Bonded Abrasives

Grinding wheels are by far the most important bonded abrasive product both in production volume and utility. They are produced in grit sizes ranging from 4, for steel mill snagging wheels, to 1200, for polishing the surface of rotogravure rolls.

Marking System. Grinding wheels and other bonded abrasive products are specified by a standard marking system which is used throughout most of the world. This system allows the user to recognize the type of abrasive, the size and shaping of the abrasive grit, and the relative amount and type of bonding material.

Bond Type. Most bonded abrasive products are produced with either a vitreous (glass or ceramic) or a resinoid (usually phenolic resin) bond.

Special Forms of Bonded Abrasives. Special forms of bonded abrasives include honing and superfinishing stones, pulpstone wheels, crush-form grinding wheels, and creep feed wheels.

Superabrasive Wheels

Superabrasive wheels include diamond wheels and cubic boron nitride (CBN) wheels.

Grinding Fluids

Grinding fluids or coolants are fluids employed in grinding to cool the work being ground, to act as a lubricant, and to act as a grinding aid. Soluble oil coolants in which petroleum oils are emulsified in water have been developed to impart some lubricity along with rust-preventive properties.

Loose Abrasives

In addition to their use in bonded and coated products, both natural and manufactured abrasive grains are used loose in such operations as polishing, buffing, lapping, pressure blasting, and barrel finishing.

Jet Cutting

High pressure jet cutting with abrasive grit can be used on metals to produce burn-free cuts with no thermal or mechanical distortion.

Health and Safety

Except for silica and natural abrasives containing free silica, the abrasive materials used today are classified by NIOSH as nuisance dust materials and have relatively high permissible dust levels.

CHARLES V. RUE
Norton Company

L. Coes, Jr., *Abrasives,* Springer-Verlag, New York, Vienna, 1971.

M. C. Shaw, ed., *New Developments in Grinding, Proceedings of the International Grinding Conference 1972,* Carnegie Press, Carnegie-Mellon University, Pittsburgh, Pa., 1972.

T. Ishikawa, *1986 Proceedings of the 24th Abrasive Engineering Society Conference,* Abrasive Engineering Society, Pittsburgh, Pa., 1986, pp. 32–51.

C. A. Sluhan, *Lub. Eng.,* 352–374 (Oct. 1970).

ABSORPTION

Absorption, or gas absorption, is a unit operation used in the chemical industry to separate gases by washing or scrubbing a gas mixture with a suitable liquid. One or more of the constituents of the gas mixture dissolves or is absorbed in the liquid and can thus be removed from the mixture. In some systems, this gaseous constituent forms a physical solution with the liquid or the solvent, and in other cases, it reacts with the liquid chemically.

The purpose of such scrubbing operations may be any of the following: gas purification (eg, removal of air pollutants from exhaust gases or contaminants from gases that will be further processed), product recovery, or production of solutions of gases for various purposes. Several examples of applied absorption processes are shown in Table 1.

Gas absorption is usually carried out in vertical countercurrent columns as shown in Figure 1. The solvent is fed in at the top of the absorber, whereas the gas mixture enters from the bottom. The absorbed substance is washed out by the solvent and leaves the absorber at the bottom as a liquid solution. The solvent is often recovered in a subsequent *stripping* or desorption operation. This second step is essentially the reverse of absorption and involves countercurrent contacting of the liquid loaded with solute using an inert gas or water vapor. The absorber may be a packed column, plate tower, or simple spray column, or a bubble column.

The fundamental physical principles underlying the process of gas absorption are the solubility of the absorbed gas and the rate of mass transfer. Information on both must be available when sizing equipment for a given application. In addition to the fundamental design concepts based on solubility and mass transfer, many practical details have to be considered during actual plant design and construction which may affect the performance of the absorber significantly (see also DISTILLATION; HEAT-EXCHANGE TECHNOLOGY.)

Design of Packed Absorption Columns

Standard Absorber Design Methods. *Operating Line.* As a gas mixture travels up through a gas absorption tower the solute is transferred to the liquid phase and thus gradually removed from the gas. The liquid accumulates solute on its way down through the column so x increases from the top to the bottom of the column. The steady-state concentrations y and x at any given point in the column are interrelated through a mass balance around either the upper or lower part of the column (eq. 2), whereas the four concentrations in the streams entering and leaving the system are interrelated by the overall material balance.

Since the total gas and liquid flow rates per unit cross-sectional area vary throughout the tower, rigorous material balances should be based on the constant inert gas and solvent flow rates G'_M and L'_M, respectively, and expressed in terms of mole ratios Y' and X'. A balance around the upper part of the tower yields

$$G'_M Y' + L'_M X'_2 = G'_M Y'_2 + L'_M X' \tag{1}$$

which may be rearranged to give

$$Y' = \frac{L'_M}{G'_M}\left(X' - X'_{A,2}\right) + Y'_{A,2} \tag{2}$$

where G'_M, L'_M are in kg · mol/(h · m^2)[lb · mol/(h · ft^2)] and $Y' \equiv y/(1 - y)$ and $X' \equiv x/(1 - x)$. The overall material balance is obtained by substituting $Y' = Y'_1$ and $X' = X'_1$. For dilute gases the total molar gas and liquid flows may be assumed to be constant, and a similar mass balance yields

$$y = \frac{L_M}{G_M}(x - x_2) + y_2 \tag{3}$$

A plot of either equation 2 or 3 is called the operating line of the process, as shown in Figure 2.

Design Procedure. The packed height of the tower required to reduce the concentration of the solute in the gas stream from $y_{A,1}$ to an acceptable residual level of $y_{A,2}$ may be calculated by combining point values of the mass-transfer rate and a differential material balance for the absorbed component.

Multicomponent Mass-Transfer Effects. *Equimolar Counterdiffusion.* Just as unidirectional diffusion through stagnant films represents the situation in an ideally simple gas absorption process, equimolar counterdiffusion prevails as another special case in ideal distillation columns. In this case, the total molar flows L_M and G_M are constant, and the mass balance is given by equation 3.

Nonisothermal Gas Absorption. *Nonvolatile Solvents.* In practice, some gases tend to liberate such large amounts of heat when they are absorbed into a solvent that the operation cannot be assumed to be isothermal, as has been done thus far. The resulting temperature variations over the tower will displace the equilibrium line on a y–x diagram considerably because the solubility usually depends strongly on temperature. Thus nonisothermal operation affects column performance drastically.

Axial Dispersion Effects. *Effect of Axial Dispersion on Column Performance.* Another assumption underlying standard design methods is that the gas and the liquid phases move in plug-flow fashion through the column. In reality, considerable departure from this ideal flow assumption exists and different fluid particles travel through the packing at varying velocities.

This effect, called axial dispersion, counteracts the countercurrent contacting scheme for which the column is designed and this lowers the driving forces throughout the packed bed. Neglect of axial dispersion results in an overestimation of the driving forces and in an underestimation of the number of transfer units needed. It may therefore lead to an unsafe design.

Capacity Limitations. Thus far the discussion has been confined to factors affecting the tower height required to perform a specific absorption job. The necessary tower diameter, on the other hand, depends primarily on the total amount of gas and liquid that must be handled. At a given set of flow rates, the diameter of the packing can only be decreased at the expense of a large pressure drop, which in turn generates higher operating costs because more power is needed to blow the gas through the packing. The reason for this is the fact that handling a given total gas flow rate in a smaller tower diameter

Table 1. Typical Commercial Gas Absorption Processes

Treated gas	Absorbed gas, solute	Solvent	Function
coke oven gas	ammonia	water	by-product recovery
coke oven gas	benzene and toluene	straw oil	by-product recovery
reactor gases in manufacture of formaldehyde from methanol	formaldehyde	water	product recovery
drying gases in cellulose acetate fiber production	acetone	water	solvent recovery
refinery gases	hydrogen sulfide	alkaline solutions	pollutant removal
natural and refinery gases	hydrogen sulfide	solution of sodium 2,6- (and 2,7-)anthraquinonedisulfonate	pollutant removal
products of combustion	sulfur dioxide	water	pollutant removal
	carbon dioxide	ethanolamines	by-product recovery
wet well gas	propane and butane	kerosene	gas separation
ammonia synthesis gas	carbon monoxide	ammoniacal cuprous chloride solution	contaminant removal
roast gases	sulfur dioxide	water	production of calcium sulfite solution for pulping

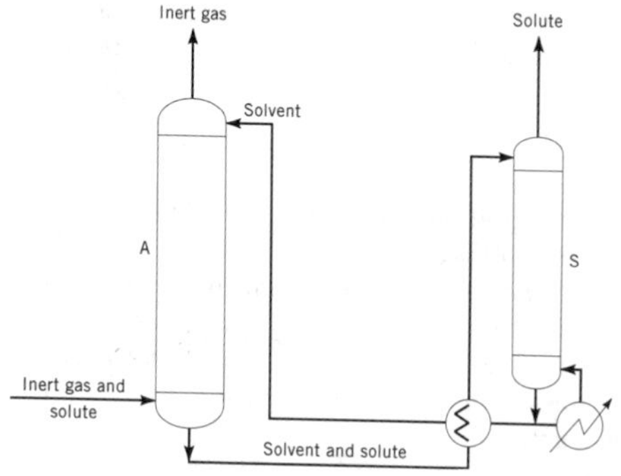

Figure 1. Absorption column arrangement with a gas absorber A and a stripper S to recover solvent.

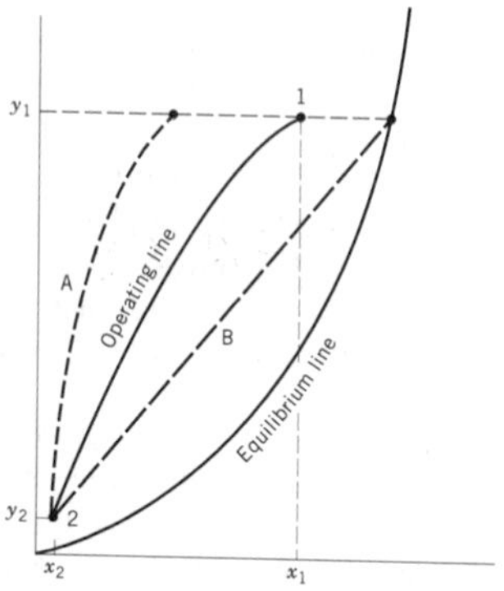

Figure 2. Operating lines for an absorption system: line A, high L_M/G_M ratio; solid line, medium L_M ratio; line B, L_M/G_M ratio at theoretical minimum necessary for the removal of the specified quantity of solute. Subscript 1 represents the bottom of tower, 2, the top of tower.

increases the superficial velocity at which the gas has to be pushed through the packing.

Bubble Tray Absorption Columns

General Design Procedure. Bubble tray absorbers may be designed graphically based on a so-called McCabe-Thiele diagram. An operating line and an equilibrium line are plotted in y–x, Y'–X', or Y^0–X^0 coordinates using the principles for packed adsorbers outlined above (Fig. 3). The minimum number of plates required for a specified recovery

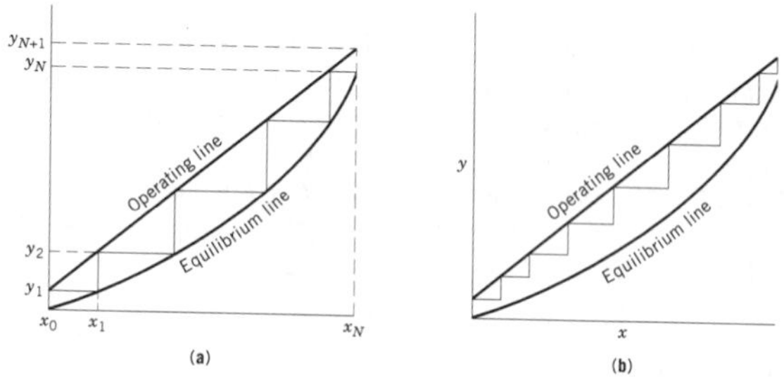

Figure 3. McCabe-Thiele diagram: (**a**) number of theoretical plates, 5; (**b**) number of actual plates, 8.

may be computed by assuming that equilibrium is reached between the two phases on each bubble tray. Thus the gas and the liquid leaving a tray are at equilibrium and a hypothetical tray capable of equilibrating the phase streams is termed a theoretical plate. Starting the calculation at the bottom of the tower, where the concentrations are y_{N+1} and x_N, the concentration leaving the lowest theoretical plate y_N may be found on the design diagram (Fig. 3**a**) by moving from the operating line vertically to the equilibrium line, because y_N is at equilibrium with x_N. Since the concentrations between two plates are always related by the operating line, x_{N-1} may be found from y_N by moving horizontally to the operating line. By repeating this sequence of steps until the desired residual gas concentration y_1 is reached, the number of theoretical plates can be counted.

The required number of actual plates, N_P, is larger than the number of theoretical plates, N_{TP}, because it would take an infinite contacting time at each stage to establish equilibrium. The ratio N_{TP}:N_P is called the overall column efficiency. This parameter is difficult to predict from theoretical considerations, however, or to correct for new systems and operating conditions. It is therefore customary to characterize the single place by the so-called Murphree vapor plate efficiency, E_{MV}:

$$E_{MV} \equiv \frac{y_n - y_{n+1}}{y_n^* - y_{n+1}} \tag{4}$$

which indicates the fractional approach to equilibrium achieved by the plate. An efficiency of 80% means that the reduction in solute gas concentration effected by the plate is 80% of the reduction obtained from a theoretical plate. Corresponding actual plates may therefore be stepped off by moving from the operating line vertically only 80% of the distance between operating and equilibrium line (Fig. 3**b**). In some special cases having negligible resistance in the gas phase, E_{MV} values may become unreasonably small. It is then more logical to define a Murphree liquid plate efficiency, E_{ML}, simply by reversing the role of liquid and gas and by focusing on the change in liquid composition across the plate with respect to an equilibrium given by the leaving vapor.

Nonisothermal Gas Absorption. The computation of nonisothermal gas absorption processes is difficult because of all the interactions involved as described for packed columns. A computer is normally required for the enormous number of plate calculations necessary to establish the correct concentration and temperature profiles through the tower. Suitable algorithms have been developed, and nonisothermal gas absorption in plate columns has been studied experimentally and the measured profiles compared to the calculated results.

Capacity Limitations. The fluid flow capacity of a bubble tray may be limited by any of three principal factors: flooding, entrainment, and large hydraulic gradient at high liquid flow.

URS VON STOCKAR
École Polytechnique Fédérale, Lausanne

CHARLES R. WILKE
University of California, Berkeley

T. K. Sherwood, R. L. Pigford, and C. R. Wilke, *Mass Transfer,* McGraw-Hill Book Co., Inc., New York, 1975.

R. H. Perry and D. Green, eds., *Perry's Chemical Engineer's Handbook,* 6th ed., McGraw-Hill Book Co., Inc., New York, 1984.

R. C. Reid, J. M. Prausnitz, and B. E. Poling, *The Properties of Gases and Liquids,* 4th ed., McGraw-Hill Book Co., New York, 1988.

J. R. Hufton, J. L. Bravo, and J. R. Fair, *Ind. Eng. Chem. Res.* **27,** 2096 (1988).

ACETALDEHYDE

Acetaldehyde is a product of most hydrocarbon oxidations. It is an intermediate product in the respiration of higher plants and occurs in trace amounts in all ripe fruits that have a tart taste before ripening. The aldehyde content of volatiles has been suggested as a chemical index of ripening during cold storage of apples. Acetaldehyde is also an intermediate product of fermentation (qv), but it is reduced almost immediately to ethanol. It may form in wine (qv) and other alcoholic beverages after exposure to air, imparting an unpleasant taste; the aldehyde reacts to form diethyl acetal and ethyl acetate. Acetaldehyde is an intermediate product in the decomposition of sugars in the body and hence occurs in trace quantities in blood.

Physical Properties

Acetaldehyde is a colorless, mobile liquid having a pungent, suffocating odor that is somewhat fruity and quite pleasant in dilute concentrations. Its physical properties are given in Table 1.

Acetaldehyde is miscible in all proportions with water and most common organic solvents, eg, acetone, benzene, ethyl alcohol, ethyl ether, gasoline, paraldehyde, toluene, xylenes, turpentine, and acetic acid.

Table 1. Physical Properties of Acetaldehyde

Properties	Values
formula weight	44.053
melting point, °C	−123.5
boiling point at 101.3 kPa[a] (1 atm), °C	20.16
density, g/mL	
$d^0{}_4$	0.8045
$d^{15}{}_4$	0.7846
coefficient of expansion per °C (0–30°C)	0.00169
refractive index, $n^{20}{}_D$	1.33113
vapor density (air = 1)	1.52
surface tension at 20°C, mN/m (= dyn/cm)	21.2
absolute viscosity at 15°C, mPa·s(= cP))	0.02456
specific heat at 0°C, J/(g·K)[b]	
15°C	2.18
25°C	1.41
$\alpha = C_p/C_v$ at 30°C and 101.3 kPa[a] (1 atm)	1.145

[a] To convert kPa to psi, multiply by 0.14503.
[b] To convert J to cal, divide by 4.187.

Chemical Properties

The limits and products of the various combustion zones for acetaldehyde–oxygen and acetaldehyde–air have been defined.

Acetaldehyde is a highly reactive compound exhibiting the general reactivity of aldehydes (qv). Acetaldehyde undergoes numerous condensation, addition, and polymerization reactions; under suitable conditions, the oxygen or any of the hydrogens can be replaced.

Decomposition. Acetaldehyde decomposes at temperatures above 400°C, forming principally methane and carbon monoxide.

Manufacture

Since 1960, the liquid-phase oxidation of ethylene has been the process of choice for the manufacture of acetaldehyde. There is, however, still some commercial production by the partial oxidation of ethyl alcohol and hydration of acetylene. The economics of the various processes are strongly dependent on the prices of the feedstocks. Acetaldehyde is also formed as a coproduct in the high temperature oxidation of butane. A more recently developed rhodium catalyzed process produces acetaldehyde from synthesis gas as a coproduct with ethyl alcohol and acetic acid.

Health and Safety Factors

Acetaldehyde appears to paralyze respiratory muscles, causing panic. It has a general narcotic action which prevents coughing, causes irritation of the eyes and mucous membranes, and accelerates heart action. When breathed in high concentration, it causes headache and sore throat. The threshold limit value (TLV) of acetaldehyde in air is 100 ppm.

Economic Aspects and Uses

Acetaldehyde production is linked with the demand for acetic acid, acetic anhydride, cellulose acetate, vinyl acetate resins, acetate esters, pentaerythritol, synthetic pyridine derivatives, terephthalic acid, and peracetic acid. In 1976 acetic acid production represented 60% of the acetaldehyde demand. That demand has diminished as a result of the rising cost of ethylene as feedstock and methanol carbonylation as the preferred route to acetic acid (qv).

H. J. HAGEMEYER
Texas Eastman Company

"Acetaldehyde, 601.500H," *Chemical Economics Handbook,* SRI International, Menlo Park, Calif., 1989.

ACETAL RESINS

The term "acetal resins" commonly denotes the family of homopolymers and copolymers whose main chains are completely or essentially composed of repeating oxymethylene units $(\text{—CH}_2\text{—O—})_n$. The polymers are derived chiefly from formaldehyde or methanal, either directly or through its cyclic trimer, trioxane or 1,3,5-trioxacyclohexane.

Today there are 10 producers of acetal resins, mostly copolymers. Throughout the remainder of this article the term homopolymer refers to Delrin (a registered trademark) acetal resin manufactured and sold

Table 1. Mechanical Properties of Acetal Resin

Property	ASTM test method	Homopolymer	Copolymer
tensile strength, yield, MPa[a], 23°C	D638	68.9	60.6
elongation, break, %	D638	25–75	40–75
tensile modulus, MPa[a], 23°C	D638	3100	2825
flexural strength, MPa[a],23°C	D790	97.1	89.6
flexural modulus, MPa[a], 23°C	D790	2830	2584
compressive stress, MPa[a], 23°C	D695	35.8	31
1% deflection		35.8	31
shear strength, MPa[a], 23°C	D732	65	53
Izod impact strength, notched, 3.175 mm, J/m[b], 23°C		69–122	53–80
specific gravity	D792	1.42	1.41

[a] To convert MPa to psi, multiply by 145.
[b] To convert J/m to ft · lb/in., divide by 53.39.

by Du Pont; the term copolymer refers to Celcon acetal copolymer resins (registered trademark of Hoechst Celanese Corporation).

Structure and Properties

The many commercially attractive properties of acetal resins are due in large part to the inherent high crystallinity of the base polymers. Values reported for percentage crystallinity (x-ray, density) range from 60 to 77%. The lower values are typical of copolymer. Polyoxymethylene most commonly crystallizes in a hexagonal unit cell with the polymer chains in a 9/5 helix.

The high crystallinity of acetal resins contributes significantly to their excellent resistance to most chemicals, including many organic solvents.

Mechanical Properties. Stiffness, resistance to deformation under constant applied load (creep resistance), resistance to damage by cyclical loading (fatigue resistance), and excellent lubricity are mechanical properties for which acetal resins are perhaps best known and which have contributed significantly to their excellent commercial success.

Typical values of important properties of general purpose acetal resins (homopolymer and copolymer) are collected in Table 1.

Electrical Properties. The dielectric constant is constant over the temperature range of most interest (−40 to 50°C).

Chemical Structure and Properties. Homopolymer consists exclusively of repeating oxymethylene units. The copolymer contains alkylidene units (eg, ethylidene—CH_2—CH_2—) randomly distributed along the chain. The number-average molecular weight of most commercially available acetal resins is between 20,000 and 90,000.

The ionic polymerizations of formaldehyde and trioxane are equilibrium reactions. Unless suitable measures are taken, polymer will begin to revert to monomeric formaldehyde at processing temperatures by depolymerization (called unzipping) which begins at chain ends.

Acetal resins are generally stable in mildly alkaline environments. However, bases can catalyze hydrolysis of ester end groups, resulting in less thermally stable polymer. Properly end-capped acetal resins, substantially free of ionic impurities, are relatively thermally stable. Like most other engineering thermoplastics, acetal resins are susceptible to photooxidation by oxidative radical chain reactions.

Manufacturing

Homopolymer. Formaldehyde polymerizes by both anionic and cationic mechanisms. Strong acids are needed to initiate cationic polymerization. Anionic polymerization can be initiated by relatively weak bases (eg, pyridine). Homopolymer is typically treated to replace hemiformal endgroups with more stable endgroups (eg, ester) in a process known as end-capping.

Copolymer. Copolymerization of trioxane with cyclic ethers or formals is accomplished with cationic initiators. Raw copolymer is typically treated in melt, suspension, or solution to depolymerize unstable fractions.

Product from melt or suspension treatment is obtained directly as crumb or powder. Polymer recovered from solution treatment is obtained by precipitative cooling or spray drying. Polymer with stable end groups may be washed and dried to remove impurities, especially acids or their precursors, prior to finishing operations.

Processing and Fabrication

Finishing. All acetal resins contain various stabilizers introduced by the supplier in a finishing extrusion (compounding) step.

Fabrication. Acetal resins are most commonly fabricated by injection molding.

Scrap and Recycle. Acetal resins can be processed with very little waste. Sprues, runners, and out-of-tolerance parts can, in general, be ground and the resins reused.

Resin Grades. Nonfilled and unmodified (except for stabilizers) grades of acetal resin are generally differentiated on the basis of melt index.

Health and Safety

When processed and used according to manufacturer's recommendations, acetal resins present no extraordinary health risks.

Uses

Acetal resins are used in conveying devices, gears, plumbing and irrigation applications, automotive parts, and many household appliances.

JOHN B. STARR
Hoechst Celanese Corporation

O. Vogl, ed., *Polyaldehydes,* Marcel Dekker, New York, 1967.

R. N. MacDonald, *Macromolecular Synthesis,* Vol. 3, John Wiley & Sons, Inc., New York, 1968.

A. Serle in J. M. Margolis, ed., *Engineering Thermoplastics,* Marcel Dekker, New York, 1985.

ACETATE AND TRIACETATE FIBERS. See FIBERS, CELLULOSE ESTERS.

ACETIC ACID AND DERIVATIVES

ACETIC ACID

Acetic acid, CH_3COOH, is a corrosive organic acid having a sharp odor, burning taste, and pernicious blistering properties. It is found in ocean water, oilfield brines, rain, and at trace concentrations in many plant and animal liquids. It is central to all biological energy pathways. Fermentation of fruit and vegetable juices yields 2–12% acetic acid solutions, usually called vinegar (qv). Any sugar-containing sap or juice can be transformed by bacterial or fungal processes to dilute acetic acid.

Most of the acetic acid is produced in the United States, Germany, Great Britain, Japan, France, Canada, and Mexico. Total annual production in these countries is close to four million tons. Uses include the manufacture of vinyl acetate and acetic anhydride. Vinyl acetate is used to make latex emulsion resins for paints, adhesives, paper coatings, and textile finishing agents. Acetic anhydride is used in making cellulose acetate fibers, cigarette filter tow, and cellulosic plastics.

Physical Properties

Acetic acid, fp 16.635°C, bp 117.87°C at 101.3 kPa, is a clear, colorless liquid. Water is the chief impurity in acetic acid. Traces of acetaldehyde, acetic anhydride, formic acid, biacetyl, methyl acetate, ethyl acetoacetate, iron, and mercury are sometimes found.

A summary of the physical properties of glacial acetic acid is given in Table 1.

Chemical Properties

Decomposition Reactions. Minute traces of acetic anhydride are formed when very dry acetic acid is distilled.

Acid–Base Chemistry. Acetic acid dissociates in water, $pK_a = 4.76$ at 25°C. It is a mild acid which can be used for analysis of bases too weak to detect in water. It readily neutralizes the ordinary hydroxides of the alkali metals and the alkaline earths to form the corresponding acetates.

Acetylation Reactions. Alcohols may be acetylated without catalysts by using a large excess of acetic acid.

$$CH_3COOH + ROH \rightarrow CH_3COOR + H_2O$$

Nearly all commercial acetylations are realized using acid catalysts.

Economic Aspects

Acetic acid has a place in organic processes comparable to sulfuric acid in the mineral chemical industries and its movements mirror the industry. Growth of synthetic acetic acid production in the United States was greatly affected by the dislocations in fuel resources of the 1970s. The growth rate for 1988 was 1.5%.

About half of the world production comes from methanol carbonylation and about one-third from acetaldehyde oxidation. Another tenth of the world capacity can be attributed to butane–naphtha liquid-phase oxidation. Appreciable quantities of acetic acid are recovered from reactions involving peracetic acid.

Health and Safety

Acetic acid has a sharp odor and the glacial acid has a fiery taste and will penetrate unbroken skin to make blisters. Prolonged exposure to air containing 5–10 mg/m^3 does not seem to be seriously harmful, but there are pronounced, undesirable effects from constant exposure to as high as 26 mg/m^3 over a 10-day period.

Glacial acetic acid is dangerous, but its precise toxic dose is not known for humans. The LD_{50} for rats is said to be 3310 mg/kg, and for rabbits 1200 mg/kg. Ingestion of 80–90 g must be considered extraordinarily dangerous for humans. Vinegar, on the other hand, which is dilute acetic acid, has been used in foods and beverages since the most ancient of times.

Table 1. Properties of Glacial Acetic Acid

Property	Value
freezing point, °C	16.635
boiling point, °C	117.87
density, g/mL at 20°	1.0495
refractive index, n_D^{25}	1.36965
heat of vaporization ΔH_v, J/g^a at bp	394.5
specific heat (vapor), J/(g·K)^a at 124°C	5.029
flammability limits, vol % in air	4.0 to 16.0
autoignition temperature, °C	465

[a] To convert J to cal, divide by 4.184.

FRANK S. WAGNER, JR.
Nandina Corporation

D. Ambrose and N. B. Ghiassee, *J. Chem. Thermodyn.* **19,** 505–519 (1987)

J. F. Roth, *Catal. Today* **13**(1), 1–12 (1992); J. R. Zoeller and co-workers, *ibid.*, 73–91 (1992).

A. Popoff in J. J. Lagowski, ed., *Chemistry of Nonaqueous Solvents,* Vol. 3, Academic Press, New York, 1970.

ACETAMIDE

Acetamide, C_2H_5NO, mol wt 59.07, is a white, odorless, hygroscopic solid derived from acetic acid and ammonia. The melt is a solvent for organic substances; it is used in electrochemistry and organic synthesis. Pure acetamide has a bitter taste. It is found in coal mine waste dumps.

Physical and Chemical Properties

Table 1 lists many of acetamide's important physical properties. Acetamide, CH_3CONH_2, dissolves easily in water, exhibiting amphoteric behavior. It is slow to hydrolyze unless an acid or base is present. It combines with acids, eg, HBr, HCl, HNO_3, to form solid complexes.

Preparation and Manufacture

Most commercial routes for the production of acetamide involve dehydration of ammonium acetate:

$$NH_4OOCCH_3 \rightarrow H_2O + CH_3CONH_2$$

Health and Safety

Acetamide has been used experimentally as a source of nonprotein nitrogen for sheep and dairy cattle. It does not appear to be toxic in amounts of about 2–3% of ration.

Economic Aspects

Heico Chemicals is the only producer of acetamide in the United States. Acetamide appears to have a wide spectrum of applications. It suppresses acid buildup in printing inks, lacquers, explosives, and perfumes. It is a mild moisturizer and is used as a softener for leather, textiles, paper, and certain plastics. It finds some applications in the synthesis of pharmaceuticals, pesticides, and antioxidants for plastics.

Table 1. Physical Properties of Acetamide

Property	Value
melting point (trigonal), °C	80.0–80.1
triple point, K	353.33
heat of melting, ΔH_m, kJ/kg^a	264
dielectric constant	59

[a] To convert kJ to kcal, divide by 4.184.

Frank S. Wagner, Jr.
Nandina Corporation

H. G. M. De Wit and co-workers, *J. Chem. Thermodyn.* **15,** 651–663, 891–902 (1983).

D. H. Kerridge, *Chem. Soc. Rev.* **17,** 181–227 (1988).

M. Ravindranatha, N. Kalyanam, and S. Sivaram, *J. Org. Chem.* **47,** 4812–4813 (1982).

ACETIC ANHYDRIDE

Acetic anhydride $(CH_3CO)_2O$, is a mobile, colorless liquid that has an acrid odor and is a more piercing lacrimator than acetic acid. It is the largest commercially produced carboxylic acid anhydride: U.S. production capacity is over 900,000 t yearly. Its chief industrial application is for acetylation reactions; it is also used in many other applications in organic synthesis, and it has some utility as a solvent in chemical analysis.

Physical and Chemical Properties

No dimerization of acetic anhydride has been observed in either the liquid or solid state. Decomposition, accelerated by heat and catalysts such as mineral acids, leads slowly to acetic acid. Acetic anhydride is soluble in many common solvents, including cold water. Its common physical properties are given in Table 1.

Table 1. Physical Properties of Acetic Anhydride

Property	Value
freezing point, °C	−73.13
boiling point, °C at 101.3 kPa[a]	139.5
density, $d^{20}{}_4$, g/cm^3	1.0820
specific heat, J/kg[b] at 20°C	1817
viscosity, mPa·s(= cP) at 15°C	0.971
electric conductivity, S/cm	2.3×10^{-8}

[a] To convert kPa to mm Hg, multiply by 7.5.
[b] To convert J to cal, divide by 4.184.

Manufacture

Manufacturing processes include the acetic acid process, the acetaldehyde oxidation process, and methyl acetate carbonylation. Methyl acetate–dimethyl ether carbonylation seems to be the leading new route to acetic anhydride production.

Health and Safety

Acetic anhydride penetrates the skin quickly and painfully, forming burns and blisters that are slow to heal.

Frank S. Wagner, Jr.
Nandina Corporation

"Acetic Anhydride Design Problem" in J. J. McKetta and W. A. Cunningham, eds., *Encyclopedia of Chemical Processing and Design,* Vol. 1, Marcel Dekker, Inc., New York, p. 271.

V. H. Agreda, D. M. Pond, and J. F. Zoeller, *Chemtech,* 172–181 (March 1992).

V. H. Agreda, L. H. Partin, and W. H. Heise, *Chem. Eng. Prog.* **86**(2), 40 (1990).

J. F. Roth, *Catal. Today* **13**(1), 1–12 (1992); J. R. Zoeller and co-workers, *ibid.,* 73–91 (1992).

ACETYL CHLORIDE

Acetyl chloride, C_2H_3OCl, mol wt 78.50, is a colorless, corrosive, irritating liquid that fumes in air. It has a stifling odor and reacts very rapidly with water, readily hydrolyzing to acetic acid and hydrochloric acid. As little as 0.5 parts per million activate the flow of tears, and often provoke a burning sensation in the eyes, nose, and throat. Acetyl chloride is toxic. Its high reactivity with hydroxyl, sulfhydryl, and amine groups leads to modifications that block the action of many important enzymes needed by living tissue.

Table 1. Physical Properties of Acetyl Chloride

Property	Value
freezing point, °C	−112.0
boiling point, °C at 101.3 kPa[a]	50.2
density, g/mL at 20°C	1.1051
heat of vaporization at bp, ΔH_v, kJ/g[b]	0.36459

[a] To convert kPa to atm, divide by 101.3.
[b] To convert kJ to kcal, divide by 4.184.

Physical and Chemical Properties

The common physical properties of acetyl chloride are given in Table 1.

Reactions of acetyl chloride that are formally analogous to hydrolysis occur with alcohols, mercaptans, and amines: primary or secondary compounds form corresponding acetates or amides; tertiary alcohols generally yield the tertiary alkyl chlorides.

Manufacture

Acetyl chloride is manufactured commercially in Europe and the Far East. Some acetyl chloride is produced in the United States for captive applications such as acetylation of pharmaceuticals.

Acetyl chloride is manufactured by the reaction of sodium acetate or acetic acid and phosphorus trichloride.

Uses

A small amount of acetyl chloride is consumed in the start-up of acetic acid chlorination to monochloroacetic acid. Acetyl chloride is a powerful acetylating agent. It is used in the manufacture of aspirin, acetaminophen, acetanilide, and acetophenone. Liquid crystal compositions for optical display and memory devices frequently require acetyl chloride.

Frank S. Wagner, Jr.
Nandina Corporation

D. R. Stull, E. F. Westrum, Jr., and G. C. Sinke, *The Chemical Thermodynamics of Organic Compounds,* John Wiley & Sons, Inc., New York, 1969, pp. 536–537.

DIMETHYLACETAMIDE

Dimethylacetamide, DMAC, mol wt 87.12, $CH_3CON\ (CH_3)_2$, is a colorless, high boiling polar solvent. DMAC is a good solvent for a wide range of organic and inorganic compounds and it is miscible with water, ethers, esters, ketones, and aromatic compounds. Unsaturated aliphatics are highly soluble, but saturated aliphatics have limited solubility in DMAC. The polar nature of DMAC enables it to act as a combined solvent and reaction catalyst, in many instances producing high yields and pure product in short time periods.

Physical Properties

Selected physical properties of DMAC are boiling point, 166.1°C; melting point, −20°C; vapor pressure at 25°C, 0.27 kPa (2 mm Hg); density at 15.6°C, 0.945 g/mL; viscosity at 25°C, 0.92 mPa·s (= cP); surface tension at 30°C, 32.43 mN/m (= dyn/cm); refractive index n_D^{25}, 1.4356; heat of vaporization at 166°C, 43.1 kJ/mol (10.3 kcal/mol);

heat of combustion, 2544 kJ/mol (608 kcal/mol); thermal conductivity at 22.2°C, 0.1835 W/(m·K) (0.1579 kcal/(mh·°C)); flash point (Tag closed up), 63°C; and ignition temperature, 490°C.

Chemical Properties

The chemical reactions of DMAC are typical of those of disubstituted amides.

Manufacturing Processes

Dimethylacetamide can be produced by the reaction of acetic acid and dimethylamine:

$$CH_3COOH + (CH_3)_2NH \rightarrow CH_3CON(CH_3)_2 + H_2O$$

DMAC can also be made by the reaction of acetic anhydride and dimethylamine:

$$(CH_3CO)_2O + 2\ (CH_3)_2NH \rightarrow 2\ CH_3CON(CH_3)_2 + H_2O$$

Health and Safety Factors

DMAC is capable of producing systemic injury when repeatedly inhaled or absorbed through the skin. The principal effect is cumulative damage to the liver and kidney. DMAC has a low order of acute toxicity when swallowed or upon brief contact of the liquid vapor with the eyes or skin.

The U.S. Department of Labor (OSHA) has ruled that an employee's exposure to dimethylacetamide in any 8-h work shift of a 40-h work week shall not exceed a time-weighted average of 10 ppm DMAC vapor in air by volume or 35 mg/m^3 in air by weight.

Uses

The uses of dimethylacetamide are very similar to those for dimethylformamide. DMAC is employed most often where higher temperatures are needed for solution of resins or activation of chemical reactions.

J. C. WATTS
P. A. LARSON
E. I. du Pont de Nemours & Co., Inc.

G. L. Kennedy, *CRC Crit. Rev. Toxicol.* **17,** 129 (1986).

Dimethylacetamide, Material Safety Data Sheet, E. I. du Pont de Nemours & Co., Inc., Wilmington, Del., 1988.

HALOGENATED DERIVATIVES

The most important of the halogenated derivatives of acetic acid is chloroacetic acid. Fluorine, chlorine, bromine, and iodine derivatives are all known, as are mixed halogenated acids. For a discussion of the fluorine derivatives see FLUORINE COMPOUNDS, ORGANIC.

Chloroacetic Acid

Physical Properties. Pure chloroacetic acid ($ClCH_2COOH$), mol wt 94.50, $C_2H_3ClO_2$, is a colorless, white deliquescent solid. It has been isolated in three crystal modifications: α, mp 63°C, β, mp 56.2°C, and γ, mp 52.5°C. Commercial chloroacetic acid consists of the α form. Physical properties are given in Table 1.

Chemical Properties and Industrial Uses. Chloroacetic acid has wide applications as an industrial chemical intermediate. It is bifunctional since both the carboxylic acid group and the α-chlorine are very reactive. Major industrial uses for chloroacetic acid are in the manufacture of cellulose ethers (mainly carboxymethylcellulose, CMC), herbicides, and thioglycolic acid. Other industrial uses include manufacture of glycine, amphoteric surfactants, and cyanoacetic acid.

Manufacture. Most chloroacetic acid is produced by the chlorination of acetic acid using a suitable catalyst such as acetic anhydride. The remainder is produced by the hydrolysis of trichloroethylene with sulfuric acid or by reaction of chloroacetyl chloride with water.

Table 1. Physical Properties of Chloroacetic Acid

Property	Value
boiling point, °C	189.1
density, at 25°C, g/mL	1.4043
dielectric constant at 60°C	12.3
dissociation constant K_a	1.4×10^{-3}

A major disadvantage of the chlorination process is residual acetic acid and overchlorination to dichloroacetic acid. Chloroacetic acid is usually purified by crystalization. High purity 99% chloroacetic acid will contain less than 0.5% of either acetic acid or dichloroacetic acid.

Toxicity and Handling. Chloroacetic acid is extremely corrosive and will cause serious chemical burns. It also is readily absorbed through the skin in toxic amounts. Contamination of 5–10% of the skin is usually fatal.

Sodium Chloroacetate

Sodium chloroacetate, mol wt 116.5, $C_2H_2ClO_2Na$, is produced by reaction of chloroacetic acid with sodium hydroxide or sodium carbonate. In many applications chloroacetic acid or the sodium salt can be used interchangeably.

Dichloroacetic Acid

Dichloroacetic acid ($Cl_2CHCOOH$), mol wt 128.94, $C_2H_2Cl_2O_2$, is a reactive intermediate in organic synthesis. Physical properties are mp 13.9°C, bp 194°C, density 1.5634 g/mL, and refractive index 1.4658, both at 20°C. It has been manufactured by the chlorination of acetic and chloroacetic acids, reduction of trichloroacetic acid, hydrolysis of pentachloroethane, and hydrolysis of dichloroacetyl chloride.

Trichloroacetic Acid

Trichloroacetic acid (Cl_3CCOOH), mol wt 163.39, $C_2HCl_3O_2$, forms white deliquescent crystals and has a characteristic odor. Physical properties are given in Table 2.

Trichloroacetic acid is manufactured in the United States by the exhaustive chlorination of acetic acid.

Sodium trichloroacetate, $C_2Cl_3O_2Na$, is used as a herbicide for various grasses and cattails.

Chloroacetyl Chloride

Chloroacetyl chloride ($ClCH_2COCl$) is the corresponding acid chloride of chloroacetic acid. Physical properties include mol wt 112.94, $C_2H_2Cl_2O$, mp −21.8°C, bp 106°C, vapor pressure 3.3 kPa (25 mm Hg) at 25°C, 12 kPa (90 mm Hg) at 50°C, and density 1.4202 g/mL and refractive index 1.4530, both at 20°C.

Since chloroacetyl chloride can react with water in the skin or eyes to form chloroacetic acid, its toxicity parallels that of the parent acid.

Chloroacetyl chloride is manufactured by reaction of chloroacetic acid with chlorinating agents such as phosphorus oxychloride, phosphorus trichloride, sulfuryl chloride, or phosgene.

Table 2. Physical Properties of Trichloroacetic Acid

Property	Value
melting point, °C	59
boiling point, °C	197.5
density, at 64°C, g/mL	1.6218
refractive index at 61°C	1.4603
heat of combustion, kJ/g[a]	3.05

[a] To convert kJ/g to kcal/g, divide by 4.184.

Table 3. Physical Properties of Chloroacetate Esters

Property	Methyl ester	Ethyl ester
molecular weight	108.52	122.55
melting point, °C	−32.1	−26
boiling point, °C	129.8	143.3
density, at 20°C, g/mL	1.2337	1.159
flash point, °C	57	66

Much of the chloroacetyl chloride produced is used captively as a reactive intermediate. It is useful in many acylation reactions and in the production of adrenalin, diazepam, chloroacetophenone, chloroacetate esters, and chloroacetic anhydride.

Chloroacetate Esters

Two chloroacetate esters of industrial importance are methyl chloroacetate, $C_3H_5ClO_2$, and ethyl chloroacetate, $C_4H_7ClO_2$. Their properties are given in Table 3.

Chloroacetate esters are usually made by removing water from a mixture of chloroacetic acid and the corresponding alcohol. Both methyl and ethyl chloroacetate are used as agricultural and pharmaceutical intermediates, specialty solvents, flavors, and fragrances.

Other Derivatives

Bromoacetic acid ($BrCH_2COOH$), mol wt 138.96, $C_2H_3BrO_2$, occurs as hexagonal or rhomboidal hygroscopic crystals, mp 49°C, bp 208°C, d^{50} 1.9335, n^{50}_D 1.4804. It is soluble in water, methanol, and ethyl ether. Bromoacetic acid can be prepared by the bromination of acetic acid in the presence of acetic anhydride and a trace of pyridine, by the Hell-Volhard-Zelinsky bromination catalyzed by phosphorus, and by direct bromination of acetic acid at high temperatures or with hydrogen chloride as catalyst. Dibromoacetic acid ($Br_2CHCOOH$), mol wt 217.8, $C_2H_2Br_2O_2$, mp 48°C, bp 232–234°C (decomposition), is soluble in water and ethyl alcohol. It is prepared by adding bromine to boiling acetic acid, or by oxidizing tribromoethene with peracetic acid. Tribromoacetic acid (Br_3CCOOH), mol wt 296.74, $C_2HBr_3O_2$, mp 135°C, bp 245°C (decomposition), is soluble in water, ethyl alcohol, and diethyl ether. Tribromoacetic acid can be prepared by the oxidation of bromal or perbromoethene with fuming nitric acid and by treating an aqueous solution of malonic acid with bromine.

Iodoacetic acid (ICH_2COOH), mol wt 185.95, $C_2H_3IO_2$, is commercially available. Iodoacetic acid has been prepared by iodination of acetic anhydride in the presence of sulfuric or nitric acid. Diiodoacetic acid ($I_2CHCOOH$), mol wt 311.85, $C_2H_2I_2O_2$, mp 110°C, occurs as white needles and is soluble in water, ethyl alcohol, and benzene. It has been prepared by heating diiodomaleic acid with water and by treating malonic acid with iodic acid in a boiling water solution. Triiodoacetic acid (I_3CCOOH), mol wt 437.74, $C_2HO_2I_3$, mp 150°C (decomposition), is soluble in water, ethyl alcohol, and ethyl ether. It has been prepared by heating iodic acid and malonic acid in boiling water.

EARL D. MORRIS
JOHN C. BOST
The Dow Chemical Company

G. Koenig, E. Lohmar, and N. Rupprich, "Chloroacetic Acids," in *Ullmann's Encyclopedia of Industrial Chemistry,* Vol. A6, VCH Publishers, New York, 1986.

Chemical Economics Handbook, SRI International, Menlo Park, Calif., Dec. 1988, p. 676.1000D.

ACETONE

Acetone (2-propanone, dimethyl ketone, CH_3COCH_3), molecular weight 58.08 (C_3H_6O), is the simplest and most important of the ketones. It is a colorless, mobile, flammable liquid with a mildly pungent, somewhat aromatic odor, and is miscible in all proportions with water and most organic solvents. Acetone is an excellent solvent for a wide range of gums, waxes, resins, fats, greases, oils, dyestuffs, and cellulosics. It is used as a carrier for acetylene, in the manufacture of a variety of coatings and plastics, and as a raw material for the chemical synthesis of a wide range of products such as ketene, methyl methacrylate, bisphenol A, diacetone alcohol, methyl isobutyl ketone, hexylene glycol (2-methyl-2,4-pentanediol), and isophorone.

Most of the world's manufactured acetone is obtained as a coproduct in the process for phenol from cumene and most of the remainder from the dehydrogenation of isopropyl alcohol. U.S. manufacturers include Allied Signal Corporation (Frankford, Pa.), Aristech Chemical Corporation (Haverhill, Ohio), BTL Specialty Resins Corporation (Blue Island, Ill.), Dow Chemical USA (Oyster Creek, Tex.), General Electric Company (Mount Vernon, Ind.), Georgia Gulf Corporation (Plaquemine, La.), Shell Oil Company (Deer Park, Tex.), Texaco Corporation (El Dorado, Kans.), Union Carbide Corporation (Institute, W. Va.), Eastman Kodak Company (Kingsport, Tenn.), and The Goodyear Tire and Rubber Company, Chemical Division (Bayport, Tex.).

Numerous natural sources of acetone make it a normal constituent of the environment. It is readily biodegradable.

Table 1. Physical Properties

Property	Value
melting point, °C	−94.6
boiling point at 101.3 kPa[a], °C	56.29
refractive index, n_D	
at 20°C	1.3588
at 25°C	1.35596
electrical conductivity at 298.15 K, S/cm	5.5×10^{-8}

[a] To convert kPa to mm Hg, multiply by 7.501.

Physical Properties

Selected physical properties are given in Table 1.

Chemical Properties

The closed cup flash point of acetone is −18°C and open cup −9°C. The auto ignition temperature is 538°C, and the flammability limits are 2.6 to 12.8 vol % in air at 25°C.

Acetone shows the typical reactions of saturated aliphatic ketones.

Health and Safety

Acetone is among the solvents of comparatively low acute and chronic toxicity. Acetone can be handled safely if common-sense precautions are taken. It should be used in a well-ventilated area, and because of its low flash point, ignition sources should be absent.

WILLIAM L. HOWARD
The Dow Chemical Company

American Institute of Chemical Engineers, *Design Institute for Physical Property Data,* (DIPPR File), University Park, Pa., 1989. For other listings of properties, see *Beilsteins Handbuch der Organischen Chemie,* Springer-Verlag, Berlin, Vol. 1 and supplement; and J. A. Riddick, W. B. Bunger, and T. K. Sakano, "Organic Solvents, Physical Properties, and Methods of Purification," in *Techniques of Organic Chemistry,* Vol. 2, John Wiley & Sons, Inc., New York, 1986.

Acetone, Form No. 115-598-84, product bulletin of The Dow Chemical Co., Midland, Mich., 1984.

C. S. Read with T. Gibson and Z. Sedaghat-Pour, "Acetone" in *Chemical Economics Handbook,* SRI International, Menlo Park, Calif., 1989. An excellent information source.

ACETONITRILE. See NITRILES.

ACETYLENE. See HYDROCARBONS, ACETYLENE.

ACETYLENE-DERIVED CHEMICALS

Acetylene, C_2H_2, is an extremely reactive hydrocarbon, principally used as a chemical intermediate. (see HYDROCARBONS, ACETYLENE). Because of its thermodynamic instability, it cannot easily or economically be transported for long distances.

Because of its relatively high price, there have been continuing efforts to replace acetylene in its principal applications with cheaper raw materials. Such efforts have been successful, particularly in the United States, where ethylene has displaced acetylene as raw material for acetaldehyde, acetic acid, vinyl acetate, and chlorinated solvents. Only a small percentage of U.S. vinyl chloride production is still based on acetylene. Propylene has replaced acetylene as feed for acrylates and acrylonitrile. Even some recent production of traditional Reppe acetylene chemicals, such as butanediol and butyrolactone, is based on new raw materials.

Ethynylation Reaction Products

The name ethynylation was coined by Reppe to describe the addition of acetylene to carbonyl compounds.

$$HC{\equiv}CH + RCOR' \rightarrow HC{\equiv}CC(OH)RR'$$

Principal Reppe acetylene chemicals include propargyl alcohol, butynediol, butenediol, butanediol, butyrolactone, methylpyrrolidone, and vinylpyrrolidone. Some of their physical properties are given in Table 1.

Other Alcohols and Diols. Secondary acetylenic alcohols and glycols are prepared by ethynylation of aldehydes higher than formaldehyde, and tertiary acetylenic alcohols and glycols by ethynylation of ketones.

Table 2 and Table 3 list the physical properties of acetylenic alcohols and glycols, respectively.

Uses of Reppe Chemicals. Propargyl alcohol is used in oil-well acidizing compositions, inhibiting the attack of mineral acids on steel. It is also used as an intermediate for manufacture of a miticide, Omite, and of sulfadiazine.

Butynediol is principally consumed in manufacture of butenediol and butanediol.

Butenediol is used as an intermediate for manufacture of the insecticide Endosulfan and of pyridoxine (vitamin B_6).

Butanediol is consumed in manufacture of butyrolactone and tetrahydrofuran. Large amounts are also used as a monomer for polyesters and polyurethanes.

Butyrolactone is principally used as an intermediate for manufacture of pyrrolidones. Substantial amounts are used as a solvent for agricultural chemicals and polymers.

Health and Safety. Secondary and tertiary acetylenic alcohols are stable under normal conditions. They are toxic orally, through skin absorption, and through inhalation. The glycols are relatively low in toxicity.

Propargyl alcohol is stable when pure, but violent reactions can occur when it is heated with contaminants such as bases or strong acids. It is a primary skin irritant and severe eye irritant, highly toxic by all means of ingestion.

Butynediol is similar to propargyl alcohol, but somewhat more stable. Because of its high boiling point, its vapors do not ordinarily present a problem.

Butenediol, butanediol, and butyrolactone are stable and only moderately toxic.

Vinylation Reaction Products

Unlike ethynylation, in which acetylene adds across a carbonyl group and the triple bond is retained, in vinylation a labile hydrogen compound adds to acetylene, forming a double bond. Catalytic vinylation has been applied to a wide range of alcohols, phenols, thiols, carboxylic acids, and certain amines and amides.

Vinyl Ethers. The principal commercial vinyl ethers are methyl vinyl ether (methoxyethene, C_3H_6O); ethyl vinyl ether (ethoxyethene, C_4H_8O); and butyl vinyl ether (1-ethenyloxybutane $C_6H_{12}O$) (see Table 4 for physical properties).

Table 1. Physical Properties of Reppe Chemicals

Property	Propargyl alcohol	Butynediol	Butenediol	Butanediol	Butyrolactone
mp, °C	−52	58	11.8	20.2	−44
bp, °C, at 101.3 kPa[a]	114	248	234	228	204
specific gravity, d_4^{20}	0.948	1.114	1.070[b]	1.017	1.129
flash point, open cup, °C	36	152	128	121	98

[a] To convert kPa to mm Hg, multiply by 7.5
[b] d_{15}^{25}

Table 2. Physical Properties of Acetylenic Alcohols

Property	Hexynol	Ethyloctynol	Methylbutynol	Methylpentynol
molecular weight	98	154	84	98
freezing point, °C	−80	−45	2.6	−30.6
boiling point, °C	142	197.2	103.6	121.4
specific gravity, d_{20}^{20}	0.882	0.873	0.8672	0.8721

Table 3. Physical Properties of Acetylenic Glycols

Property	Dimethylhexynediol	Dimethyloctynediol	Tetramethyldecynediol
molecular weight	142	170	226
melting point, °C	96–97	49–51	37–38
boiling point, °C	206	222	260

Table 4. Physical Properties of Vinyl Ethers[a]

Property	Methyl	Ethyl	Butyl
molecular weight	58	72	100
freezing point, °C	−122.8	−115.4	−91.9
boiling point, °C	5.5	35.7	93.5

[a] Lower vinyl ethers are miscible with nearly all organic solvents.

Uses. Vinyl ethers are used in the manufacture of glutaraldehyde and as monomers.

Health and Safety. Lower vinyl ethers represent a severe fire hazard. Inhalation should be avoided, although oral toxicity is low.

EUGENE V. HORT
PAUL TAYLOR
GAF Corporation

S. A. Miller, *Acetylene, Its Properties, Manufacture and Uses,* Vol. 1, Academic Press, Inc., New York, 1965, pp. 24–28, 42–44.

M. J. Haley with T. Ball and S. Yoshikawa, "Acetylene," CEH Product Review, *Chemical Economics Handbook*, SRI International, Menlo Park, Calif., Oct. 1988.

S. A. Miller, *Acetylene, Its Properties, Manufacture and Uses,* Vol. 2, Academic Press, Inc., New York, 1965.

J. W. Copenhaver and M. H. Bigelow, *Acetylene and Carbon Monoxide Chemistry,* Reinhold Publishing Corp., New York, 1949.

ACID RAIN. See AIR POLLUTION; ATMOSPHERIC MODELING; ENVIRONMENTAL IMPACT.

ACROLEIN AND DERIVATIVES

Acrolein (2-propenal), C_3H_4O, is the simplest unsaturated aldehyde ($CH_2{=}CHCHO$). The primary characteristic of acrolein is its high reactivity due to conjugation of the carbonyl group with a vinyl group. More than 80% of the refined acrolein that is produced today goes into the synthesis of methionine. Much larger quantities of crude acrolein are produced as an intermediate in the production of acrylic acid. More than 85% of the acrylic acid produced worldwide is by the captive oxidation of acrolein.

Acrolein is a highly toxic material with extreme lacrimatory properties. At room temperature acrolein is a liquid with volatility and flammability somewhat similar to acetone; but unlike acetone, its solubility in water is limited. Commercially, acrolein is always stored with hydroquinone and acetic acid as inhibitors. Special care in handling is required because of the flammability, reactivity, and toxicity of acrolein.

The physical and chemical properties of acrolein are given in Table 1.

Economic Aspects

Presently, worldwide refined acrolein nameplate capacity is about 113,000 t/yr. Degussa has announced a capacity expansion in the United States by building a 36,000 t/yr acrolein plant in Theodore, Alabama to support their methionine business. The key producers of refined acrolein are Union Carbide (United States), Degussa (Germany), Atochem (France), and Daicel (Japan).

Reactions and Derivatives

Acrolein is a highly reactive compound because both the double bond and aldehydic moieties participate in a variety of reactions, including oxidation, reduction, reactions with alcohols yielding alkoxy propionaldehydes, acrolein acetals, and alkoxypropionaldehyde acetals, addition of mercaptans yielding 3-methylmercaptopropionaldehyde, reaction with ammonia yielding β-picoline and pyridine, Diels-Alder reactions, and polymerization.

Table 1. Properties of Acrolein

Property	Value
Physical properties	
molecular formula	C_3H_4O
molecular weight	56.06
specific gravity at 20/20°C	0.8427
boiling point, °C at 101.3 kPa[a]	52.69
Chemical properties	
autoignition temperature in air, °C	234
heat of combustion at 25 °C, kJ/kg[b]	5383

[a] To convert kPa to mm Hg, multiply by 7.5.
[b] To convert kJ to kcal, divide by 4.184.

Direct Uses of Acrolein

Because of its antimicrobial activity, acrolein has found use as an agent to control the growth of microbes in process feed lines, thereby controlling the rates of plugging and corrosion (see WASTES, INDUSTRIAL).

Acrolein at a concentration of <500 ppm is also used to protect liquid fuels against microorganisms.

W. G. ETZKORN
J. J. KURLAND
W. D. NEILSEN
Union Carbide Chemicals & Plastics Company Inc.

R. C. Schulz in J. I. Kroschwitz, ed., *Encyclopedia of Polymer Science and Engineering,* 2nd ed., Vol. 1, Wiley-Interscience, New York, 1985, pp. 160–169.

T. Ohara, T. Sato, N. Shimizu, G. Prescher, H. Schwind, and O. Weiberg, *Ullman's Encyclopedia of Industrial Chemistry,* 5th ed., Vol. A1, 1985, pp. 149–160.

C. W. Smith, ed., *Acrolein,* John Wiley & Sons, Inc., New York, 1962.

ACRYLAMIDE

Acrylamide (NIOSH No. A533250) has been commercially available since the mid-1950s and has shown steady growth since that time, but is still considered a small-volume commodity. Its formula, $H_2{=}CHCONH_2$ (2-propeneamide), indicates a simple chemical, but it is by far the most important member of the series of acrylic and methacrylic amides. Water-soluble polyacrylamides have the most important applications, including potentially large uses in enhanced oil recovery as mobility-control agents in water flooding, additives for oilwell drilling fluids, and aids in fracturing, acidifying, and other operations. Other uses include flocculants for waste-water treatment, the mining industry, and various other process industries, soil stabilization, papermaking aids, and thickeners. Smaller but nonetheless important uses include dye acceptors; polymers for promoting adhesion; additives for textiles, paints, and cement; increasing the softening point and solvent resistance of resins; components of photopolymerizable systems; and cross-linking agents in vinyl polymers.

Physical Properties

The physical properties of solid acrylamide monomer are summarized in Table 1. Typical physical properties of 50% solution in water appear in Table 2.

Table 1. Physical Properties of Solid Acrylamide Monomer

Property	Value
molecular weight	71.08
melting point, °C	84.5 ± 0.3
boiling point, °C at 0.67 kPa[a]	103

[a] To convert kPa to mm Hg, multiply by 7.5.

Table 2. Physical Properties of 50% Aqueous Acrylamide Solution

Property	Value
pH	5.0–6.5
refractive index range, 25°C (48–52%)	1.4085–1.4148
viscosity, mPa (= cP) at 25°C	2.71
specific gravity, at 25°C	1.0412
boiling point at 101.3 kPa[a], °C	99–104

[a] To convert kPa to mm Hg, multiply by 7.5.

Chemical Properties

Acrylamide, C_3H_5NO, is an interesting difunctional monomer containing a reactive electron-deficient double bond and an amide group, and it undergoes reactions typical of those two functionalities. It exhibits both weak acidic and basic properties.

Manufacture

The current routes to acrylamide are based on the hydration of inexpensive and readily available acrylonitrile (C_3H_3N, 2-propenenitrile, vinyl cyanide, VCN, or cyanoethene) (see ACRYLONITRILE).

Health and Safety Considerations

Contact with acrylamide can be hazardous and should be avoided. The most serious toxicological effect of exposure to acrylamide monomer is as a neurotoxin. In contrast, polymers of acrylamide exhibit very low toxicity.

Economic Aspects

The largest production of acrylamide is in Japan; the United States and Europe also have large production facilities. The principal producers in North America are The Dow Chemical Company, American Cyanamid Company, and Nalco Chemical Company (internal use).

C. E. HABERMANN
Dow Chemical, USA

D. C. MacWilliams, in R. H. Yocum and E. B. Nyquist, eds., *Functional Monomers,* Vol. 1, Marcel Dekker, Inc., New York, 1973, pp. 1–197.

Chemistry of Acrylamide, Bulletin PRC 109, Process Chemicals Department, American Cyanamid Co., Wayne, N.J., 1969.

U.S. Pat. 3,597,481 (Aug. 3, 1971), B. A. Tefertiller and C. E. Habermann (to The Dow Chemical Co.); U.S. Pat. 3,631,104 (Dec. 28, 1971), C. E. Habermann and B. A. Tefertiller (to The Dow Chemical Co.); U.S. Pat. 3,642,894 (Feb. 15, 1972), C. E. Habermann, R. E. Friedrich, and B. A. Tefertiller (to The Dow Chemical Co.); U.S. Pat. 3,642,643 (Feb. 15, 1972), C. E. Habermann (to The Dow Chemical Co.); U.S. Pat. 3,642,913 (Mar. 7, 1972), C. E. Habermann (to The Dow Chemical Co.); U.S. Pat. 3,696,152 (Oct. 3, 1972), C. E. Habermann and M. R. Thomas (to The Dow Chemical Co.); U.S. Pat. 3,758,578 (Sept. 11, 1973), C. E. Habermann and B. A. Tefertiller (to The Dow Chemical Co.); U.S. Pat. 3,767,706 (Oct. 23, 1972), C. E. Habermann and B. A. Tefertiller (to The Dow Chemical Co.).

Environmental and Health Aspects of Acrylamide, A Comprehensive Bibliography of Published Literature 1930 to April 1980, EPA Report No. 560/7-81-006, 1981.

ACRYLAMIDE POLYMERS

Acrylamide,

$$(CH_2{=}CH\overset{\overset{O}{\|}}{C}{-}NH_2)$$

polymerizes in the presence of free-radical initiators to form polyacrylamide chains with the following structure:

$$(-CH_2-\underset{\underset{CONH_2}{|}}{CH}-)_n$$

In this article the term *acrylamide polymer* refers to all polymers which contain acrylamide as a major constituent. Consequently, acrylamide polymers include functionalized polymers prepared from polyacrylamide by postreaction and copolymers prepared by polymerizing acrylamide (2-propenamide, C_3N_5NO) with one or more comonomers.

Manufacturing processes have been improved by use of on-line computer control and statistical process control leading to more uniform final products. Production methods now include inverse (water-in-oil) suspension polymerization or polymerization in water on moving belts. Conventional azo, peroxy, redox, and gamma-ray initiators are used in batch and continuous processes.

Physical Properties

Solid Polymer. Completely dry polyacrylamide is a brittle white solid.

The physical properties of nonionic polyacrylamide are listed in Table 1.

Polymers in Solution. Polyacrylamide is soluble in water at all concentrations, temperatures, and pH values.

In general nonionic polyacrylamides do not interact strongly with neutral inorganic salts.

Flow Properties. In water, high molecular weight polyacrylamide forms viscous homogeneous solutions.

Chemical Properties

The preparation of polyacrylamides and postpolymerization reactions on polyacrylamides are usually conducted in water. Reactions on the amide groups of polyacrylamides are often more complicated than reactions of simple amides because of neighboring groups' effects.

Post-reactions of polyacrylamide to introduce anionic, cationic, or other functional groups are often attractive from a cost standpoint. This approach can suffer, however, from side reactions resulting in cross-linking or the introduction of unwanted functionality. Reactions include hydrolysis, sulfomethylation, methylol formation, reaction with other aldehydes, transamidation, Hoffman degradation, and reaction with chlorine.

Uses

Polyacrylamides are classified according to weight-average molecular weight ($\overline{m}_w$) as follows:

Table 1. Physical Properties of Solid Polyacrylamide

Property	Value
density, g/cm^3	1.302
glass-transition temp, °C	188
chain structure	mainly heterotactic linear or branched, some head-to-head addition
crystallinity	amorphous (high mol wt)

Table 2. Suppliers of Polyacrylamide

Region	Companies
United States	Allied Colloids, Inc.
	American Cyanamid Co.
	Aqua Ben Corp.
	Betz Laboratories, Inc.
	Calgon Corp. (Merck & Co.)
	Chemtall, Inc. (SNF Floerger)
	Dearborn Chemical Co. (W. R. Grace & Co.)
	The Dow Chemical Company
	Drew Chemical Corporation (Ashland Chemical, Inc.)
	Exxon Chemical Co.
	Hercules, Inc.
	Nalco Chemical Co.
	Polypure, Inc.
	Secodyne, Inc.
	Stockhausen, Inc.
Europe	Allied Colloids, Ltd.
	American Cyanamid Co.
	BASF AG
	Chemische Fabrik Stockhausen & Cie
	The Dow Chemical Company
	SNF Floerger (France)
	Kemira Oy (Finland)
	Rohm GmbH
	Rhône-Poulenc Specialties Chimiques (France)
Japan	Dai-Ichi Kogyo Seiyaku Co., Ltd.
	Kurita Water Industries, Ltd.
	Kyoritsu Yuki Co., Ltd.
	Mitsubishi Chemical Industries, Ltd.
	Mitsui-Cyanamid, Ltd.
	Sankyo Kasei Co., Ltd.
	Sanyo Chemical Industries, Ltd.
	Takenaka Komuten Co., Ltd.
	Toa Gosei Chemical Industry Co., Ltd.

high 15×10^6
low 2×10^5
very low 2×10^3

Most uses for high molecular weight polyacrylamides in water treating, mineral processing, and paper manufacture are based on the ability of these polymers to flocculate small suspended particles by charge neutralization and bridging. Low molecular weight polymers are employed as dispersants, crystal growth modifiers, or selective mineral depressants. In oil recovery, polyacrylamides adjust the rheology of injected water so that the polymer solution moves uniformly through the rock pores, sweeping the oil ahead of it. Other applications such as superabsorbents and soil modification rely on the very hydrophilic character of polyacrylamides.

Safety and Health

Dry nonionic and cationic material caused no skin and minimal eye irritation during primary irritation studies with rabbits. Dry anionic polyacrylamide did not produce any eye or skin irritation in laboratory animals. Emulsion nonionic polyacrylamide produced severe eye irritation in rabbits, while anionic and cationic material produced minimal eye irritation in rabbits. Polyacrylamides are used safely for numerous indirect food packaging applications, potable water, and direct food applications. Suppliers of polyacrylamide are listed in Table 2.

JOSEPH KOZAKIEWICZ
DAVID LIPP
American Cyanamid Company

W. M. Kulicke, R. Kniewske, and J. Klein, *Progr. Polym. Sci.* **8,** 373–468 (1982).

V. F. Kurenkov and L. I. Abramova, *Polym.-Plast. Technol. Eng.* **31**(7&8), 659–704 (1992).

V. F. Kurenkov and V. A. Myagchenkov, *Polym.-Plast. Technol. Eng.* **30**(4), 367–404 (1991).

V. A. Myagchenkov and V. F. Kurenkov, *Polym.-Plast. Technol. Eng.* **30**(2&3), 109–135 (1991).

ACRYLIC ACID AND DERIVATIVES

Acrylic acid (propenoic acid) was first prepared in 1847 by air oxidation of acrolein. Interestingly, after use of several other routes over the past half century, it is this route, using acrolein from the catalytic oxidation of propylene, that is currently the most favored industrial process.

Acrylates are primarily used to prepare emulsion and solution polymers. The emulsion polymerization process provides high yields of polymers in a form suitable for a variety of applications. Acrylate emulsions are used in the preparation of both interior and exterior paints, floor polishes, and adhesives. Solution polymers of acrylates, frequently with minor concentrations of other monomers, are employed in the preparation of industrial coatings. Polymers of acrylic acid can be used as superabsorbents in disposable diapers, as well as in formulation of superior, reduced-phosphate-level detergents.

The polymeric products can be made to vary widely in physical properties through controlled variation in the ratios of monomers employed in their preparation, cross-linking, and control of molecular weight. They share common qualities of high resistance to chemical and environmental attack, excellent clarity, and attractive strength properties (see ACRYLIC ESTER POLYMERS).

Physical Properties

Physical properties of acrylic acid and representative derivatives appear in Table 1.

Reactions

Acrylic acid and its esters may be viewed as derivatives of ethylene, in which one of the hydrogen atoms has been replaced by a carboxyl or carboalkoxyl group. This functional group may display electron-withdrawing ability through inductive effects of the electron-deficient carbonyl carbon atom, and electron-releasing effects by resonance involving the electrons of the carbon–oxygen double bond. Therefore, these compounds react readily with electrophilic, free-radical, and nucleophilic agents.

Specialty Acrylic Esters

Higher alkyl acrylates and alkyl-functional esters are important in copolymer products, in conventional emulsion applications for coatings and adhesives, and as reactants in radiation-cured coatings and inks. In general, they are produced in direct or transesterification batch processes because of their relatively low volume.

Health and Safety Factors

The toxicity of common acrylic monomers has been characterized in animal studies using a variety of exposure routes. Toxicity varies with level, frequency, duration, and route of exposure. The simple higher esters of acrylic acid are usually less absorbed and less toxic than lower esters. In general, acrylates are more toxic than methacrylates.

Table 1. Physical Properties of Acrylic Acid Derivatives

Property	Acrylic acid	Acrolein	Acrylic anhydride	Acryloyl chloride	Acrylamide
molecular formula	$C_3H_4O_2$	C_3H_4O	$C_6H_6O_3$	C_3H_3OCl	C_3H_5ON
melting point, °C	13.5	−88			84.5
boiling point[a], °C	141	52.5	38[b]	75	125[c]
refractive index[d], n_D	1.4185[e]	1.4017	1.4487	1.4337	

[a] At 101.3 kPa = 1 atm unless otherwise noted.
[b] At 0.27 kPa.
[c] At 16.6 kPa.
[d] At 20°C, unless otherwise noted.
[e] At 25°C.

Current TLV/TWA values are provided in Material Safety Data Sheets provided by manufacturers on request.

WILLIAM BAUER, JR.
Rohm and Haas Company

Acrylic and Methacrylic Monomers—Specifications and Typical Properties, Bulletin 84C2, Rohm and Haas Co., Philadelphia, Pa., 1986.

Hydrocarbon Process. **60**(11), 124 (1981).

T. P. Snyder and C. G. Hill, Jr., *Catal. Rev. Sci. Eng.* **31,** 43–95 (1989).

ACRYLIC AND MODACRYLIC FIBERS. See FIBERS, ACRYLIC.

ACRYLIC ESTER POLYMERS

SURVEY

Acrylic esters are represented by the generic formula

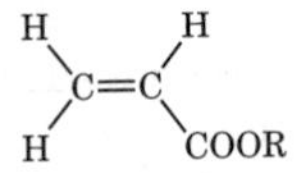

The nature of the R group determines the properties of each ester and the polymers it forms. Polymers of this class are amorphous and are distinguished by their water-clear color and their stability on aging. Acrylic monomers are extremely versatile building blocks. They are relatively moderate to high boiling liquids that readily polymerize or copolymerize with a variety of other monomers. Copolymers with methacrylates, vinyl acetate, styrene, and acrylonitrile are commercially significant. Polymers designed to fit specific application requirements ranging from soft, tacky adhesives to hard plastics can be tailored from these versatile monomers. Although the acrylics have been higher in cost than many other common monomers, they find use in high quality products where their unique characteristics and efficiency offset the higher cost.

Physical Properties

To a large extent, the properties of acrylic ester polymers (Table 1) depend on the nature of the alcohol radical and the molecular weight of the polymer. As is typical of polymeric systems, the mechanical properties of acrylic polymers improve as molecular weight is increased; however, beyond a critical molecular weight, which often is about 100,000 to 200,000 for amorphous polymers, the improvement is slight and levels off asymptotically.

Chemical Properties

Under conditions of extreme acidity or alkalinity, acrylic ester polymers can be made to hydrolyze to poly(acrylic acid) or an acid salt and the corresponding alcohol. However, acrylic polymers and copolymers have a greater resistance to both acidic and alkaline hydrolysis than competitive poly(vinyl acetate) and vinyl acetate copolymers.

Acrylic polymers are fairly insensitive to normal uv degradation since the primary uv absorption of acrylics occurs below the solar spectrum.

Table 1. Physical Properties of Acrylic Polymers

Polymer[a]	Monomer molecular formula	T_g, °C
methyl acrylate	$C_4H_6O_2$	6
ethyl acrylate	$C_5H_8O_2$	−24
propyl acrylate	$C_6H_{10}O_2$	−45
isopropyl acrylate	$C_6H_{10}O_2$	−3
n-butyl acrylate	$C_7H_{12}O_2$	−50
sec-butyl acrylate	$C_7H_{12}O_2$	−20
isobutyl acrylate	$C_7H_{12}O_2$	−43
tert-butyl acrylate	$C_7H_{12}O_2$	43
hexyl acrylate	$C_9H_{16}O_2$	−57
heptyl acrylate	$C_{10}H_{18}O_2$	−60
2-heptyl acrylate	$C_{10}H_{18}O_2$	−38
2-ethylhexyl acrylate	$C_{11}H_{20}O_2$	−65
2-ethylbutyl acrylate	$C_9H_{16}O_2$	−50
dodecyl acrylate	$C_{15}H_{28}O_2$	−30
hexadecyl acrylate	$C_{19}H_{36}O_2$	35
2-ethoxyethyl acrylate	$C_7H_{12}O_3$	−50
isobornyl acrylate	$C_{13}H_{20}O_2$	94
cyclohexyl acrylate	$C_9H_{14}O_2$	16

[a] Density (g/cm^3) and refractive index (n_D) for methyl acrylate, ethyl acrylate, and *n*-butyl acrylate: 1.22, 1.479; 1.12, 1.464; 1.08, 1.474. Density for isopropyl acrylate = 1.08 g/cm^3.

Acrylic Ester Monomers

Some of the physical properties of the principal commercial acrylic esters are given in Table 2.

There are currently two principal processes used for the manufacture of monomeric acrylic esters: the semicatalytic Reppe process

Table 2. Physical Properties of Acrylic Monomers

Acrylate	Molecular weight	bp, °C[a]	d^{25}, g/cm^3
methyl	86	79–81	0.950
ethyl	100	99–100	0.917
n-butyl	128	144–149	0.894
isobutyl	128	61–63[b]	0.884
t-butyl	128	120	0.879
2-ethylhexyl	184	214–220	0.880

[a] At 101.3 kPa unless otherwise noted.
[b] At 6.7 kPa = 50 mm Hg.

and the propylene oxidation process. The newer propylene oxidation process is preferred because of economy and safety.

The toxicities of acrylic monomers range from moderate to slight. In general, they can be handled safely and without difficulty by trained personnel following established safety practices.

Radical Polymerization

Usually, free-radical initiators such as azo compounds or peroxides are used to initiate the polymerization of acrylic monomers. Photochemical and radiation-initiated polymerizations are also well known. Methods of radical polymerization include bulk, solution, emulsion, suspension, graft copolymerization, radiation-induced, and ionic with emulsion being the most important.

The free-radical polymerization of acrylic monomers follows a classical chain mechanism in which the chain-propagation step entails the head-to-tail growth of the polymeric free radical by attack on the double bond of the monomer.

The vast majority of all commercially prepared acrylic polymers are copolymers of an acrylic ester monomer with one or more different monomers. Copolymerization greatly increases the range of available polymer properties and has led to the development of many different resins suitable for a broad variety of applications.

In general, acrylic ester monomers copolymerize readily with each other or with most other types of vinyl monomers by free-radical processes.

Health and Safety Factors

Acrylic polymers are considered to be nontoxic. In fact, the FDA allows certain acrylate polymers to be used in the packaging and handling of food.

Potential health and safety problems of acrylic polymers occur in their manufacture. During manufacture, considerable care is exercised to reduce the potential for violent polymerizations and to reduce exposure to flammable and potentially toxic monomers and solvents.

Uses

Acrylic ester polymers are used primarily in coatings, textiles, adhesives, and paper.

RONALD W. NOVAK
Rohm and Haas Company

E. H. Riddle, *Monomeric Acrylic Esters,* Reinhold Publishing Corp., New York, 1954.

F. W. Billmeyer, Jr., *Textbook of Polymer Chemistry,* Interscience Publishers, New York, 1957.

2-CYANOACRYLIC ESTER POLYMERS

The polymers of the 2-cyanoacrylic esters, more commonly known as the alkyl 2-cyanoacrylates, are hard glassy resins that exhibit excellent adhesion to a wide variety of materials. The polymers are spontaneously formed when their liquid precursors or monomers are placed between two closely fitting surfaces. The spontaneous polymerization of these very reactive liquids and the excellent adhesion properties of the cured resins combine to make these compounds a unique class of single-component, ambient-temperature-curing adhesives of great versatility (Table 1). The materials that can be bonded run the gamut from metals, plastics, most elastomers, fabrics, and woods to many ceramics.

The utility of these adhesives arises from the electron-withdrawing character of the groups adjacent to the polymerizable double bond, which accounts for both the extremely high reactivity or cure rate and their polar nature, which enables the polymers to adhere tenaciously to many diverse substrates.

At present, a number of manufacturers in the United States, Europe, Japan, and elsewhere market extended lines of these adhesives all over the world. Some of the major producers and their trademarks include Loctite (Prism and Superbonder), Toagosei (Aron Alpha, Krazy Glue), Henkel (Sicomet), National Starch (Permabond), Sumitomo (Cyanobond), Three Bond (Super Three), and Alpha Giken (Alpha Ace, Alpha Techno).

Manufacture and Processing

The cyanoacrylic esters are prepared via the Knoevenagel condensation reaction, in which the corresponding alkyl cyanoacetate reacts with formaldehyde in the presence of a basic catalyst to form a low molecular weight polymer. The polymer slurry is acidified and the water is removed. Subsequently, the polymer is cracked and redistilled at a high temperature onto a suitable stabilizer combination to prevent premature repolymerization. Strong protonic or Lewis acids are normally used in combination with small amounts of a free-radical stabilizer.

Adhesives formulated from the 2-cyanoacrylic esters typically contain stabilizers and thickeners, and may also contain tougheners, colorants, and other special property-enhancing additives.

Economic Aspects

Production of the 2-cyanoacrylic ester adhesives on a worldwide basis is estimated to be approximately 2400 metric tons. This amounts to only 0.02% of the total volume of adhesives produced but about 3% of the dollar volume.

Because of the high costs of raw materials and the relatively complex synthesis, the 2-cyanoacrylic esters are moderately expensive materials when considered in bulk quantities. In typical bonding applications, where single drops are adequate for bonding, the adhesives are very economical to use.

Health and Safety Factors

The 2-cyanoacrylic esters have sharp, pungent odors and are lacrimators, even at very low concentrations. The TLV for methyl 2-cyanoacrylate is 2 ppm and the short-term exposure limit is 4 ppm. Good ventilation when using the adhesives is essential.

Eye and skin contact should be avoided because of the adhesive's rapid tissue-bonding capabilities.

Both the liquid and cured 2-cyanoacrylic esters support combustion.

Uses

Some of the market segments served by these versatile materials include automotive, electronic, sporting goods, toys, hardware, morticians, law enforcement, cosmetics, jewelry, and medical devices. Although they are not approved for such use in the United States, their strong tissue bonding characteristics have led to their use as chemical sutures and hemostatic agents in other countries around the world.

Table 1. Adhesive Bond Properties of 2-Cyanoacrylic Esters with Metals and Various Polymeric Materials

Property	Ester type					
	Methyl	Ethyl[a]	Butyl	Isobutyl	Methoxyethyl	Ethoxyethyl
			Set time, s			
steel	20	10	30	20	15	5
nitrile rubber	5	3	5	5	5	3
ABS	20	10	20	20	5	3
polycarbonate	20	10	20	20	60	20
PVC	5	3	2	5		10
phenolic resin	5	3	30	5	25	5
			Bond strength, kPa[b]			
steel	206	172	151	96	206	165
ABS	48	48	96	48	48	48
polycarbonate	69	69	90	69		41
PVC	96	96	62	83	55	69
phenolic resin	69	76	90	62	62	55

[a] Set times for allyl esters are similar to those for ethyl esters, as are bond strengths to steel, ABS, and PC.
[b] To convert kPa to psi multiply by 0.145.

J. T. O'CONNOR
Loctite Corporation

H. W. Coover, D. W. Dreifus, and J. T. O'Connor, in I. Skeist, ed., *Handbook of Adhesives,* 3rd ed., Van Nostrand Reinhold Co., Inc., New York, 1990, Chapt. 27.

H. Lee, ed., *Cyanoacrylate Resins—The Instant Adhesives Monograph,* Pasadena Technology Press, Calif., 1986.

ACRYLONITRILE

Today over 90% of the approximately 4,000,000 metric tons of acrylonitrile (also called acrylic acid nitrile, propylene nitrile, vinyl cyanide, and propenoic acid nitrile) produced worldwide each year use the Sohio-developed ammoxidation process. Acrylonitrile is among the top 50 chemicals produced in the United States as a result of the tremendous growth in its use as a starting material for a wide range of chemical and polymer products. Acrylic fibers remain the largest use of acrylonitrile; other significant uses are in resins and nitrile elastomers and as an intermediate in the production of adiponitrile and acrylamide.

Physical Properties

Acrylonitrile (C_3H_3N, mol wt = 53.064) is an unsaturated molecule having a carbon–carbon double bond conjugated with a nitrile group. It is a polar molecule because of the presence of the nitrogen heteroatom. Tables 1 and 2 list some physical properties and thermodynamic information, respectively, for acrylonitrile.

Acrylonitrile is miscible in a wide range of organic solvents, including acetone, benzene, carbon tetrachloride, diethyl ether, ethyl acetate, ethylene cyanohydrin, petroleum ether, toluene, some kerosenes, and methanol.

Acrylonitrile has been characterized using infrared, Raman, and ultraviolet spectroscopies, electron diffraction, and mass spectroscopy.

Table 1. Physical Properties of Acrylonitrile

Property	Value
appearance/odor	clear, colorless liquid with faintly pungent odor
boiling point, °C	77.3
freezing point, °C	−83.5
density, 20°C, g/cm^3	0.806
vapor density (air = 1)	1.8
pH (5% aqueous solution)	6.0–7.5
viscosity, 25°C, mPa · s(= cP)	0.34

Table 2. Thermodynamic Data[a]

Property	Value
flash point, °C	0
autoignition temperature, °C	481
heat of combustion, liquid, 25°C, kJ/mol	1761.5
heat of vaporization, 25°C, kJ/mol	32.65

[a] To convert kJ to kcal, divide by 4.184.

Chemical Properties

Acrylonitrile undergoes a wide range of reactions at its two chemically active sites, the nitrile group and the carbon–carbon double bond. Detailed descriptions of specific reactions have been given.

Manufacturing and Processing

Acrylonitrile is produced in commercial quantities almost exclusively by the vapor-phase catalytic propylene ammoxidation process developed by Sohio.

$$C_3H_6 + NH_3 + \tfrac{3}{2}\,O_2 \xrightarrow{\text{catalyst}} C_3H_3N + 3\,H_2O$$

Economic Aspects

More than half of the worldwide acrylonitrile production is situated in Western Europe and the United States. In the United States, production is dominated by BP Chemicals, with more than a third of the domestic capacity. The export market has been an increasingly important outlet for U.S. production, exports growing from around 10% in the mid-1970s to 53% in 1987 and 43% in 1988.

Storage and Transport

Acrylonitrile is transported by rail car, barge, and pipeline. Department of Transportation (DOT) regulations require labeling acrylonitrile as a flammable liquid and poison.

Health and Safety Factors

Acrylonitrile is highly toxic if ingested, with an acute LDL_0 value for laboratory rats of 113 mg/kg. It is moderately toxic if inhaled

Table 3. Worldwide Acrylonitrile Uses and Consumption, 10^3 t

Use	1988	1985	1980	1976
acrylic fibers	2,520	2,410	2,040	1,760
ABS resins	550	435	300	270
adiponitrile	310	235	160	90
other (including nitrile rubber, SAN resin, acrylamide, and barrier resins)	460	390	240	420

(rat LCL_0 = 500ppm/4h), and it is extremely irritating and corrosive to skin and eyes. Acrylonitrile is categorized as a cancer hazard by OSHA.

Acrylonitrile will polymerize violently in the absence of oxygen if initiated by heat, light, pressure, peroxide, or strong acids and bases. It is combustible and ignites readily, producing toxic combustion products such as hydrogen cyanide, nitrogen oxides, and carbon monoxide.

Federal regulations (40 CFR 261) classify acrylonitrile as a hazardous waste, and it is listed as Hazardous Waste Number U009. Disposal must be in accordance with federal (40 CFR 262, 263, 264), state, and local regulations only at properly permitted facilities.

Uses

The trend in consumption is shown in Table 3 for the principal uses of acrylonitrile: acrylic fiber, acrylonitrile–butadiene–styrene (ABS) resins, adiponitrile, nitrile rubbers, elastomers, and styrene–acrylonitrile resins (SAN).

JAMES F. BRAZDIL
BP Research

M. A. Dalin, I. K. Kolchin, and B. R. Serebryakov, *Acrylonitrile,* Technomic, Westport, Conn., 1971.

The Chemistry of Acrylonitrile, 1st ed., American Cyanamid Co., New York, 1951.

U.S. Pat. 3,193,480 (July 6, 1965), M. M. Baizer, C. R. Campbell, R. H. Fariss, and R. Johnson (to Monsanto Chemical Co.).

Chem. Mark. Rep. **235,** 50 (1989).

ACRYLONITRILE POLYMERS

SURVEY AND SAN

Acrylonitrile has found its way into a great variety of polymeric compositions based on its polar nature and reactivity. Some of these areas include adhesives and binders, antioxidants, medicines, dyes, electrical insulations, emulsifying agents, graphic arts materials, insecticides, leather, paper, plasticizers, soil-modifying agents, solvents, surface coatings, textile treatments, viscosity modifiers, azeotropic distillations, artificial organs, lubricants, asphalt additives, water-soluble polymers, hollow spheres, cross-linking agents, and catalyst treatments.

SAN Physical Properties and Test Methods

Styrene–acrylonitrile (SAN) resins possess many physical properties desired for thermoplastic applications. They are characteristically hard, rigid, and dimensionally stable with load bearing capabilities. They are also transparent, have high heat distortion temperatures, possess excellent gloss and chemical resistance, and adapt easily to conventional thermoplastic fabrication techniques.

SAN polymers are random linear amorphous copolymers. Physical properties are dependent on molecular weight and the percentage of acrylonitrile. An increase of either generally improves physical properties, but may cause a loss of processibility or an increase in yellowness. Various processing aids and modifiers can be used to achieve a

Table 1. Properties of Injection-Molded Commercial SAN Resins[a]

Property	Monsanto Lustran-35	Dow Tyril-880
tensile strength, MPa[b]	79.4	82.1
ultimate elongation, %	3.0	3.0
Izod impact strength, J/m[c]	24.0	26.7
hardness—Rockwell M	83	80
coefficient of linear thermal expansion, cm/(cm·°C)	6.8×10^{-5}	6.6×10^{-5}
flammability, cm/min		2.0
specific heat, J/(g·K)[d]		1.3
dielectric constant, kHz (MHz)		3.18 (3.02)
water absorption, % in 24 h	0.25	0.35
specific gravity	1.07	1.08

[a] Data taken from Monsanto and Dow product data sheets.
[b] To convert MPa to psi multiply by 145.
[c] To convert J/m to ftlb/in. divide by 53.39.
[d] To convert J to cal divide by 4.184.

specific set of properties. Modifiers may include mold release agents, uv stabilizers, antistatic aids, elastomers, flow and processing aids, and reinforcing agents such as fillers and fibers. Some typical physical properties are listed in Table 1.

SAN Chemical Properties and Analytical Methods

SAN resins show considerable resistance to solvents and are insoluble in carbon tetrachloride, ethyl alcohol, gasoline, and hydrocarbon solvents. They are swelled by solvents such as benzene, ether, and toluene. Polar solvents such as acetone, chloroform, dioxane, methyl ethyl ketone, and pyridine will dissolve SAN.

The properties of SAN are significantly altered by water absorption. The equilibrium water content increases with temperature while the time required decreases. A large decrease in T_g can result. Strong aqueous bases can degrade SAN by hydrolysis of the nitrile groups.

SAN Manufacture

Commercially, SAN is manufactured by three processes: emulsion, suspension, and continuous bulk.

Processing. SAN copolymers may be processed using the conventional fabrication methods of extrusion, blow molding, injection molding, thermoforming, and casting.

Other Copolymers

Acrylonitrile copolymerizes readily with many electron-donor monomers other than styrene. Hundreds of acrylonitrile copolymers have been reported, and a comprehensive listing of reactivity ratios for acrylonitrile copolymerizations is readily available.

Copolymers of acrylonitrile and methyl acrylate and terpolymers of acrylonitrile, styrene, and methyl methacrylate are used as barrier polymers. Acrylonitrile copolymers and multipolymers containing butyl acrylate, ethyl acrylate, 2-ethylhexyl acrylate, hydroxyethyl acrylate, methyl methacrylate, vinyl acetate, vinyl ethers, and vinylidene chloride are also used in barrier films, laminates, and coatings. Environmentally degradable polymers useful in packaging are prepared from polymerization of acrylonitrile with styrene and methyl vinyl ketone.

Economic Aspects

Since its introduction in the 1950s, SAN has shown steady growth. The combined properties of SAN copolymers such as optical clarity, rigidity, chemical and heat resistance, high tensile strength, and flexible molding characteristics, along with reasonable price have secured their market position. The largest portion of SAN (80%) is incorporated into ABS resins, and their markets are inexorably joined.

There are two major producers of SAN resin in the United States, Monsanto Chemical Company and The Dow Chemical Company, which market these materials under the names of Lustran and Tyril, respectively.

Health and Toxicology

SAN resins themselves appear to pose few health problems in that they have been approved by the FDA for food packaging use. The main concern is that of toxic residuals, eg, acrylonitrile, styrene, or other polymerization components such as emulsifiers, stabilizers, or solvents. Each component must be treated individually for toxic effects and safe exposure level.

LAWRENCE E. BALL
BENEDICT S. CURATOLO
BP Research

The Chemistry of Acrylonitrile, 2nd ed., American Cyanamid Co., Petrochemical Division, New York, 1959.

F. M. Peng, in J. I. Kroschwitz, ed. *Encyclopedia of Polymer Science and Engineering,* 2nd ed., Vol. 1, Wiley-Interscience, New York, 1985, p. 463.

F. L. Reithel, in R. Juran, ed., *Modern Plastics Encyclopedia 1989,* **65**(11), McGraw-Hill Book Co., Inc., New York, p. 105.

N. W. Johnston, *J. Macromol. Sci. Rev. Macromol. Chem.* **C14**, 215 (1973).

ABS RESINS

Acrylonitrile–butadiene–styrene (ABS) polymers are composed of elastomer dispersed as a grafted particulate phase in a thermoplastic matrix of styrene and acrylonitrile copolymer (SAN). The presence of SAN grafted onto the elastomeric component, usually polybutadiene or a butadiene copolymer, compatabilizes the rubber with the SAN component. Property advantages provided by this graft terpolymer include excellent toughness, good dimensional stability, good processibility, and chemical resistance. Property balances are controlled and optimized by adjusting elastomer particle size, morphology, microstructure, graft structure, and SAN composition and molecular weight. Therefore, although the polymer is a relatively low cost engineering thermoplastic the system is structurally complex. This complexity is advantageous in that altering these structural and compositional parameters allows considerable versatility in the tailoring of properties to meet specific product requirements. This versatility may be even further enhanced by adding various monomers to raise the heat deflection temperature, impart transparency, confer flame retardancy, and, through alloying with other polymers, obtain special product features. Consequently, research and development in ABS systems is active and continues to offer promise for achieving new product opportunities.

Physical Properties

The range of properties typically available for general purpose ABS is illustrated in Table 1. Numerous grades of ABS are available including new alloys and specialty grades for high heat, plating, flaming-retardant, or static dissipative product requirements.

Chemical Properties

The behavior of ABS may be inferred from consideration of the functional groups present within the polymer.

Chemical Resistance. The polar character of the nitrile group reduces interration of the polymer with hydrocarbon solvents, mineral and vegetable oils, waxes, and related household and commercial materials. Good chemical resistance provided by the presence of acrylonitrile as a comonomer combined with the relatively low water absorptivity (<1%) results in high resistance to the staining agents typically encountered in household applications.

Processing Stability. As with elastomers or other rubber modified polymers, the presence of double bonds in the elastomeric phase increases sensitivity to thermal oxidation either during processing or end use. Antioxidants are generally added at the compounding step to ensure retention of physical properties. Physical effects can also have marked effects on mechanical properties due to orientation, molded-in stress, and the agglomeration of dispersed rubber particles under very severe conditions. Proper drying conditions are essential to prevent moisture-induced splay. Discoloration can be minimized by reducing stock temperature during molding or extrusion.

Thermal Oxidative Stability. ABS undergoes autoxidation with the polybutadiene component more sensitive to thermal oxidation than the styrene–acrylonitrile component. Antioxidants substantially improve oxidative stability. Studies on the oven aging of molded parts have shown that oxidation is limited to the outer surface (<0.2 mm), ie, the oxidation process is diffusion limited.

Photoxidative Stability. Unsaturation present as a structural feature in the polybutadiene component of ABS (also in high impact polystyrene, rubber-modified PVC, and butadiene-containing elastomers) also increases lability with regard to photooxidative degradation. Such degradation also only occurs in the outermost layer, and impact loss upon irradiation can be attributed to embrittlement of the rubber and possibly to scission of the grafted styrene–acrylonitrile copolymer. Applications involving extended outdoor exposure, especially in direct sunlight, require protective measures such as the use of stabilizing additives, pigments, and protective coatings and film.

Flammability. The general-purpose grades are usually recognized as 94 HB according to the requirements of Underwriters' Laboratories UL94 and also meet the requirements, dependent on thickness, of the Motor Vehicle Safety Standard 302.

Polymerization

In all manufacturing processes, grafting is achieved by the free-radical copolymerization of styrene and acrylonitrile monomers in the presence of an elastomer. Ungrafted styrene–acrylonitrile copolymer is formed during graft polymerization or added afterward.

Manufacturing

There are three commercial processes for manufacturing ABS: emulsion, mass, and mass-suspension. ABS is sold as an unpigmented product for on-line coloring using color concentrates during molding, or as precolored pellets matched to exacting requirements.

Analysis

Analytical investigations may be undertaken to identify the presence of an ABS polymer, characterize the polymer, or identify nonpolymeric ingredients. Fourier transform infrared (ftir) spectroscopy is the method of choice to identify the presence of an ABS polymer and determine the acrylonitrile–butadiene–styrene ratio of the composite polymer. Confirmation of the presence of rubber domains is achieved by electron microscopy. Comparison with available physical property data serves to increase confidence in the identification or indicate the presence of unexpected structural features. Phase-separation techniques can be used to provide detailed compositional analyses.

Processing

Good thermal stability plus shear thinning allow wide flexibility in viscosity control for a variety of processing methods. ABS exhibits non-Newtonian viscosity behavior. ABS can be processed by all the techniques used for other thermoplastics: compression and injection molding, extrusion, calendering, and blow-molding. Clean, undergraded regrind can be reprocessed in most applications (plating excepted), usually at 20% with virgin ABS. Post-processing operations include cold forming; thermoforming; metal plating; painting;

Table 1. Material Properties of General Purpose and Heat Distortion Resistant ABS

Properties	High impact	Medium impact	Heat resistant
notched Izod impact at RT, J/m[a]	347–534	134–320	107–347
tensile strength, MPa[b]	33–43	30–52	41–52
tensile modulus, GPa[c]	1.7–2.3	2.1–2.8	2.1–2.6
flexural modulus, GPa[c]	1.7–2.4	2.2–3.0	2.1–2.8
Rockwell hardness	80–105	105–112	100–111
heat deflection[d], °C at 455 kPa[e]	99–107	102–107	110–118
coefficient of linear thermal expansion, $\times 10^5$ cm/cm·°C	9.5–11.0	7.0–8.8	6.5–9.2
dielectric strength, kV/mm	16–31	16–31	14–35
dielectric constant, $\times 10^6$ Hz	2.4–3.8	2.4–3.8	2.4–3.8

[a] To convert J/m to ft · lb/in. divide by 53.4.
[b] To convert MPa to psi multiply by 145.
[c] To convert GPa to psi multiply by 145,000.
[d] Annealed.
[e] To convert kPa to psi multiply by 0.145.

Table 2. Markets for ABS Plastics by Region in 1988, 10^3 t

	United States and Canada	Western Europe	Japan	Total	%
transportation	139	120	96	355	*25*
appliances	95	97	117	309	*22*
business machines	124	60	99	283	*20*
pipe and fittings	91	23		114	*8*
other	98	144	102	344	*25*
Total	*547*	*444*	*414*	*1405*	*100*

hot stamping; ultrasonic, spin, and vibrational welding; and adhesive bonding.

Applications

Its broad property balance and wide processing window have allowed ABS to become the largest selling engineering thermoplastic. ABS enjoys a unique position as a "bridge" polymer between commodity plastics and other higher performance engineering thermoplastics. Table 2 summarizes estimates for 1988 regional consumption of ABS resins by major use.

DONALD M. KULICH
JOHN E. PACE
LEROY W. FRITCH, JR.
ANGELO BRISIMITZAKIS
GE Plastics

C. T. Pillichody and P. D. Kelley in I. I. Rubin, ed., *Handbook of Plastic Materials and Technology*, John Wiley & Sons, Inc., New York, 1990, Chapt. 3.

D. M. Kulich, P. D. Kelley, and J. E. Pace in J. I. Kroschwitz, ed., *Encyclopedia of Polymer Science and Engineering*, 2nd ed., Vol. 1, Wiley-Interscience, New York, 1985, p. 396.

Cycolac Brand ABS Resin Design Guide, Technical Publication CYC-350, GE Plastics, Pittsfield, Mass., 1990.

Chemical Economics Handbook, SRI International, Menlo Park, Calif., 1989, 580.0180D.

ACTINIDES AND TRANSACTINIDES

ACTINIDES

The actinide elements are a group of chemically similar elements with atomic numbers 89 through 103, and their names, symbols, and atomic numbers are given in Table 1 (see RADIOACTIVE TRACERS; THORIUM AND THORIUM COMPOUNDS; URANIUM AND URANIUM COMPOUNDS; PLUTONIUM AND PLUTONIUM COMPOUNDS; NUCLEAR REACTORS; and RADIOISOTOPES). Each of the elements has a number of isotopes, all radioactive and some of which can be obtained in isotopically pure form.

Table 1. The Actinide Elements

Atomic number	Element	Symbol	Atomic weight[a]
89	actinium	Ac	227
90	thorium	Th	232
91	protactinium	Pa	231
92	uranium	U	238
93	neptunium	Np	237
94	plutonium	Pu	242
95	americium	Am	243
96	curium	Cm	248
97	berkelium	Bk	249
98	californium	Cf	249
99	einsteinium	Es	254
100	fermium	Fm	257
101	mendelevium	Md	258
102	nobelium	No	259
103	lawrencium	Lr	260

[a] Mass number of longest-lived or most available isotope.

Thorium, uranium, and plutonium are well known for their role as the basic fuels (or sources of fuel) for the release of nuclear energy. The importance of the remainder of the actinide group lies at present, for the most part, in the realm of pure research, but a number of practical applications are also known. The actinides present a storage-life problem in nuclear waste disposal and consideration is being given to separation methods for their recovery prior to disposal (see NUCLEAR REACTORS, WASTE MANAGEMENT).

Source

Only the members of the actinide group through Pu have been found to occur in nature. Thorium and uranium occur widely in the earth's

Table 2. The Oxidation States of the Actinide Elements

Atomic number and element														
89 Ac	90 Th	91 Pa	92 U	93 Np	94 Pu	95 Am	96 Cm	97 Bk	98 Cf	99 Es	100 Fm	101 Md	102 No	103 Lr
						(2)			(2)	(2)	2	2	**2**	
3	(3)	(3)	3	3	3	**3**	**3**	**3**	**3**	**3**	**3**	**3**	3	**3**
	4	4	4	4	**4**	4	4	4	(4)					
		5	5	**5**	5	5								
			6	6	6	6								
				7	(7)									

Table 3. Properties of Actinide Metals

Element	Melting point, °C	Heat of vaporization, ΔH_v, kJ/mol (kcal/mol)	Boiling point, °C
actinium	1100 ± 50	293 (70)	
thorium	1750	564 (130)	3850
protactinium	1575		
uranium	1132	446.4 (106.7)	3818
neptunium	637 ± 2	418 (100)	3900
plutonium	646	333.5 (79.7)	3235
americium	1173	230 (55)	2011
curium	1345	386 (92.2)	3110
berkelium	1050		
californium	900 ± 30		
einsteinium	860 ± 30		

Table 4. Properties and Crystal Structure Data for Important Actinide Binary Compounds

Compound	Color	Melting point, °C	Symmetry	Space group or structure type	Density, g/mL
AcH_2	black		cubic	fluorite ($Fm3m$)	8.35
ThH_2	black		tetragonal	$F4/mmm$	9.50
Th_4H_{15}	black		cubic	$I\bar{4}3d$	8.25
α-PaH_3	gray		cubic	$Pm3n$	10.87
β-PaH_3	black		cubic	β-W	10.58
α-UH_3	?		cubic	$Pm3n$	11.12
β-UH_3	black		cubic	β-W ($Pm3n$)	10.92
NpH_2	black		cubic	fluorite	10.41
NpH_3	black		trigonal	$P\bar{3}c1$	9.64
PuH_2	black		cubic	fluorite	10.40
PuH_3	black		trigonal	$P\bar{3}c1$	9.61
AmH_2	black		cubic	fluorite	10.64
AmH_3	black		trigonal	$P\bar{3}c1$	9.76
CmH_2	black		cubic	fluorite	10.84
CmH_3	black		trigonal	$P\bar{3}c1$	10.06
BkH_2	black		cubic	fluorite	11.57
BkH_3	black		trigonal	$P\bar{3}c1$	10.44
Ac_2O_3	white		hexagonal	La_2O_3 ($P\bar{3}m1$)	9.19
Pu_2O_3	?		cubic	$Ia3$ (Mn_2O_3)	10.20
Pu_2O_3	black	2085	hexagonal	La_2O_3	11.47
Am_2O_3	tan		hexagonal	La_2O_3	11.77
Am_2O_3	reddish brown		cubic	$Ia3$	10.57
Cm_2O_3	white to faint tan	2260	hexagonal	La_2O_3	12.17
Cm_2O_3			monoclinic	$C2/m$ (Sm_2O_3)	11.90
Cm_2O_3	white		cubic	$Ia3$	10.80
Bk_2O_3	light green		hexagonal	La_2O_3	12.47
Bk_2O_3	yellow-green		monoclinic	$C2/m$	12.20
Bk_2O_3	yellowish brown		cubic	$Ia3$	11.66
Cf_2O_3	pale green		hexagonal	La_2O_3	12.69
Cf_2O_3	lime green		monoclinic	$C2/m$	12.37
Cf_2O_3	pale green		cubic	$Ia3$	11.39
Es_2O_3	white		hexagonal	La_2O_3	12.7
Es_2O_3	white		monoclinic $C2/m$	$C2/m$	12.4

Table 4. Properties and Crystal Structure Data for Important Actinide Binary Compounds *(continued)*

Compound	Color	Melting point, °C	Symmetry	Space group or structure type	Density, g/mL
Es_2O_3	white		cubic	*Ia*3	11.79
ThO_2	white	ca 3050	cubic	fluorite	10.00
PaO_2	black		cubic	fluorite	10.45
UO_2	brown to black	2875	cubic	fluorite	10.95
NpO_2	apple green		cubic	fluorite	11.14
PuO_2	yellow-green to brown	2400	cubic	fluorite	11.46
AmO_2	black		cubic	fluorite	11.68
CmO_2	black		cubic	fluorite	11.92
BkO_2	yellowish-brown		cubic	fluorite	12.31
CfO_2	black		cubic	fluorite	12.46
Pa_2O_5	white		cubic	fluorite-related	11.14
Np_2O_5	dark brown		monoclinic	$P2_1/c$	8.18
α-U_3O_8	black-green	1150 (dec)	orthorhombic	$C2mm$	8.39
β-U_3O_8	black-green		orthorhombic	$Cmcm$	8.32
γ-UO_3	orange	650 (dec)	orthorhombic	$Fddd$	7.80
$AmCl_2$	black		orthorhombic	$Pbnm$ ($PbCl_2$)	6.78
$CfCl_2$	red-amber		?		
$AmBr_2$	black		tetragonal	$SrBr_2$ ($P4/n$)	7.00
$CfBr_2$	amber		tetragonal	$SrBr_2$	7.22
ThI_2	gold		hexagonal	$P6_3/mmc$	7.45
AmI_2	black	ca 700	monoclinic	EuI_2 ($P2_1/c$)	6.60
CfI_2	violet		hexagonal	CdI_2 ($P\bar{3}m1$)	6.63
CfI_2	violet		rhombohedral	$CdCl_2$ ($R\bar{3}m$)	6.58
AcF_3	white		trigonal	LaF_3 ($P\bar{3}c1$)	7.88
UF_3	black	>1140(*dec*)	trigonal	LaF_3	8.95
NpF_3	purple		trigonal	LaF_3	9.12
PuF_3	purple	1425	trigonal	LaF_3	9.33
AmF_3	pink	1393	trigonal	LaF_3	9.53
CmF_3	white	1406	trigonal	LaF_3	9.85
BkF_3	yellow-green		orthorhombic	YF_3 ($Pnma$)	9.70
BkF_3	yellow-green		trigonal	LaF_3	10.15
CfF_3	light green		orthorhombic	YF_3	9.88
CfF_3	light green		trigonal	LaF_3	10.28
$AcCl_3$	white		hexagonal	UCl_3 ($P6_3/m$)	4.81
UCl_3	green	835	hexagonal	$P6_3/m$	5.50
$NpCl_3$	green	ca 800	hexagonal	UCl_3	5.60
$PuCl_3$	emerald green	760	hexagonal	UCl_3	5.71
$AmCl_3$	pink or yellow	715	hexagonal	UCl_3	5.87
$CmCl_3$	white	695	hexagonal	UCl_3	5.95
$BkCl_3$	green	603	hexagonal	UCl_3	6.02
α-$CfCl_3$	green	545	orthorhombic	$TbCl_3$ ($Cmcm$)	6.07
β-$CfCl_3$	green		hexagonal	UCl_3	6.12
$EsCl_3$	white to orange		hexagonal	UCl_3	6.20
$AcBr_3$	white		hexagonal	UBr_3($P6_3/m$)	5.85
UBr_3	red	730	hexagonal	$P6_3/m$	6.55
$NpBr_3$	green		hexagonal	UBr_3	6.65
$NpBr_3$	green		orthorhombic	$TbCl_3$($Cmcm$)	6.67
$PuBr_3$	green	681	orthorhombic	$TbCl_3$	6.72
$AmBr_3$	white to pale yellow		orthorhombic	$TbCl_3$	6.85
$CmBr_3$	pale yellow-green	625 ± 5	orthorhombic	$TbCl_3$	6.85
$BkBr_3$	light green		monoclinic	$AlCl_3$($C2/m$)	5.604
$BkBr_3$	light green		orthorhombic	$TbCl_3$	6.95
$BkBr_3$	yellow green		rhombohedral	$FeCl_3$ ($R\bar{3}$)	5.54
$CfBr_3$	green		monoclinic	$AlCl_3$	5.673
$CfBr_3$	green		rhombohedral	$FeCl_3$	5.77
$EsBr_3$	straw		monoclinic	$AlCl_3$	5.62
PaI_3	black		orthorhombic	$TbCl_3$ ($Cmcm$)	6.69
UI_3	black		orthorhombic	$TbCl_3$	6.76
NpI_3	brown		orthorhombic	$TbCl_3$	6.82
PuI_3	green		orthorhombic	$TbCl_3$	6.92
AmI_3	pale yellow	ca 950	hexagonal	BiI_3 ($R\bar{3}$)	6.35
AmI_3	yellow		orthorhombic	$PuBr_3$	6.95

Table 4. *(continued)*

Compound	Color	Melting point, °C	Symmetry	Space group or structure type	Density, g/mL
CmI_3	white		hexagonal	BiI_3	6.40
BkI_3	yellow		hexagonal	BiI_3	6.02
CfI_3	red-orange		hexagonal	BiI_3	6.05
EsI_3	amber to light yellow		hexagonal	BiI_3	6.18
ThF_4	white	1068	monoclinic	$UF_4(C2/c)$	6.20
PaF_4	reddish-brown		monoclinic	UF_4	6.38
UF_4	green	960	monoclinic	$C2/c$	6.73
NpF_4	green		monoclinic	UF_4	6.86
PuF_4	brown	1037	monoclinic	UF_4	7.05
AmF_4	tan		monoclinic	UF_4	7.23
CmF_4	light gray-green		monoclinic	UF_4	7.36
BkF_4	pale yellow-green		monoclinic	UF_4	7.55
CfF_4	light green		monoclinic	UF_4	7.57
α-$ThCl_4$	white		orthorhombic		4.12
β-$ThCl_4$	white	770	tetragonal	$UCl_4(I4_1/amd)$	4.60
$PaCl_4$	greenish-yellow		tetragonal	UCl_4	4.72
UCl_4	green	590	tetragonal	$I4_1/amd$	4.89
$NpCl_4$	red-brown	518	tetragonal	UCl_4	4.96
α-$ThBr_4$	white		tetragonal	$I4_1/a$	5.94
β-$ThBr_4$	white		tetragonal	UCl_4	5.77
$PaBr_4$	orange-red		tetragonal	UCl_4	5.90
UBr_4	brown	519	monoclinic	2/c-/-	
$NpBr_4$	dark red	464	monoclinic	2/c-/-	
ThI_4	yellow	556	monoclinic	$P2_1/n$	6.00
PaI_4	black				
UI_4	black				
PaF_5	white		tetragonal	$I\bar{4}2d$	
α-UF_5	grayish white		tetragonal	$I4/m$	5.81
β-UF_5	pale yellow		tetragonal	$I\bar{4}2d$	6.47
NpF_5			tetragonal	$I4/m$	
$PaCl_5$	yellow	306	monoclinic	$C2/c$	
α-UCl_5	brown		monoclinic	$P2_1/n$	3.81
β-UCl_5	red-brown		triclinic	$P\bar{1}$	
α-$PaBr_5$			monoclinic	$P2_1/c$	
β-$PaBr_5$	orange-brown		monoclinic	$P2_1/n$	
UBr_5	brown		monoclinic	$P2_1/n$	
PaI_5	black		orthorhombic		
UF_6	white	64.02[a]	orthorhombic	$Pnma$	5.060
NpF_6	orange	55	orthorhombic	$Pnma$	5.026
PuF_6	reddish-brown	52	orthorhombic	$Pnma$	4.86
UCl_6	dark green	178	hexagonal	$P\bar{3}m1$	3.62

[a] At 151.6 kPa; to convert kPa to atm, divide by 101.3.

crust in combination with other elements, and, in the case of uranium, in significant concentrations in the oceans. With the exceptions of uranium and thorium, the actinide elements are synthetic in origin for practical purposes, ie, they are products of nuclear reactions. High neutron fluxes are available in modern nuclear reactors, and the most feasible method for preparing actinium, protactinium, and most of the actinide elements is through the neutron irradiation of elements of high atomic number.

Experimental Methods of Investigation

All of the actinide elements are radioactive and, except for thorium and uranium, special equipment and shielded facilities are usually necessary for their manipulation.

The study of the chemical behavior of concentrated preparations of short-lived isotopes is complicated by the rapid production of hydrogen peroxide in aqueous solutions and the destruction of crystal lattices in solid compounds. These effects are brought about by heavy recoils of high energy alpha particles released in the decay process.

Special techniques for experimentation with the actinide elements other than Th and U have been devised because of the potential health hazard to the experimenter and the small amounts available. In addition, investigations are frequently carried out with the substance present in very low concentration as a radioactive tracer. Tracer studies offer a method for obtaining knowledge of oxidation states, formation of complex ions, and the solubility of various compounds. These techniques are not applicable to crystallography, metallurgy, and spectroscopic studies. Microchemical or ultramicrochemical techniques are used extensively in chemical studies of actinide elements.

Electronic Structure

Measurements of paramagnetic susceptibility, paramagnetic resonance, light absorption, fluorescence, and crystal structure, in addition to a consideration of chemical and other properties, have provided a great deal of information about the electronic configuration of the aqueous actinide ions in which the electrons are in the 5*f* shell. There are exceptions, such as U_2S_3, and subnormal compounds, such as Th_2S_3, where 6*d* electrons are present.

Properties

The close chemical resemblance among many of the actinide elements permits their chemistry to be described for the most part in a correlative way.

Oxidation States. The oxidation states of the actinide elements are summarized in Table 2. The most stable states are designated by boldface type and those which are very unstable are indicated by parentheses.

The actinide elements exhibit uniformity in ionic types. Corresponding ionic types are similar in chemical behavior, although the oxidation–reduction relationships and therefore the relative stabilities differ from element to element.

Hydrolysis and Complex Ion Formation. Of the actinide ions, the small, highly charged M^{4+} ions exhibit the greatest degree of hydrolysis and complex ion formation.

The degree of hydrolysis or complex ion formation decreases in the order $M^{4+} > MO_2^{2+} > M^{3+} > MO_2^{+}$. Presumably the relatively high tendency toward hydrolysis and complex ion formation of MO ions is related to the high concentration of charge on the metal atom. On the basis of increasing charge and decreasing ionic size, it could be expected that the degree of hydrolysis for each ionic type would increase with increasing atomic number.

Metallic State. The actinide metals, like the lanthanide metals, are highly electropositive. They can be prepared by the electrolysis of molten salts or by the reduction of a halide with an electropositive metal, such as calcium or barium. Their physical properties are summarized in Table 3.

Solid Compounds. Thousands of compounds of the actinide elements have been prepared, and the properties of some of the important binary compounds are summarized in Table 4.

Crystal Structure and Ionic Radii. Crystal structure data have provided the basis for the ionic radii (coordination number = CN = 6). For both M^{3+} and M^4 ions there is an actinide contraction, analogous to the lanthanide contraction, with increasing positive charge on the nucleus. As a consequence of the ionic character of most actinide compounds and of the similarity of the ionic radii for a given oxidation state, analogous compounds are generally isostructural.

Absorption and Fluorescence Spectra. The absorption spectra of actinide and lanthanide ions in aqueous solution and in crystalline form contain narrow bands in the visible, near-ultraviolet, and near-infrared regions of the spectrum.

TRANSACTINIDES

The elements beyond the actinides in the periodic table can be termed the transactinides. These begin with the element having atomic number 104 and extend, in principle, indefinitely. Although only six such elements, numbers 104–109, were definitely known in 1991, there are good prospects for the discovery of a number of additional elements just beyond number 109 or in the region of larger atomic numbers. They are synthesized by the bombardment of heavy nuclides with heavy ions.

On the basis of the simplest projections it is expected that the half-lives of the elements beyond element 109 will become shorter as the atomic number is increased, and this is true even for the isotopes with the longest half-life for each element.

Turning to consideration of electronic structure, upon which chemical properties must be based, modern high speed computers have made possible the calculation of such structures. The calculations show that elements 104 through 112 are formed by filling the 6*d* electron subshell, which makes them, as expected, homologous in chemical properties with the elements hafnium ($Z = 72$) through mercury ($Z = 80$). Elements 113 through 118 result from the filling of the 7*p* subshell and are expected to be similar to the elements thallium ($Z = 81$) through radon ($Z = 86$).

It can be seen that elements in and near the island of stability based on element 114 can be predicted to have chemical properties as follows: element 114 should be a homologue of lead, that is, should be eka-lead; and element 112 should be eka-mercury, element 110 should be eka-platinum, etc. If there is an island of stability at element 126, this element and its neighbors should have chemical properties like those of the actinide and lanthanide elements.

GLENN T. SEABORG
University of California, Berkeley

G. T. Seaborg, *The Transuranium Elements,* Yale University Press, New Haven, Conn., 1958.

J. J. Katz, G. T. Seaborg, and L. R. Morss, eds., *The Chemistry of the Actinide Elements,* 2nd ed., Chapman and Hall, London, 1986.

G. T. Seaborg, *Ann. Rev. Nucl. Sci.* **18,** 53 (1968); O. L. Keller, Jr., and G. T. Seaborg, *Ann. Rev. Nucl. Sci.* **27,** 139 (1977).

G. Hermann, *Superheavy Elements, International Review of Science, Inorganic Chemistry,* Series 2, Vol. 8, Butterworths, London, and University Park Press, Baltimore, Md., 1975; G. T. Seaborg and W. Loveland, *Contemp. Physics* **28,** 233 (1987).

ACTIVATION ANALYSIS. See FINE ARTS EXAMINATION AND CONSERVATION; NONDESTRUCTIVE EVALUATION.

ADHESIVES

An *adhesive* is a material capable of holding together solid materials by means of surface attachment. *Adhesion* is the physical attraction of the surface of one material for the surface of another. An *adherend* is the solid material to which the adhesive adheres and the *adhesive bond* or *adhesive joint* is the assembly made by joining adherends together by means of an adhesive. *Practical adhesion* is the physical strength of an adhesive bond. It primarily depends on the forces of adhesion, but its magnitude is determined by the physical properties of the adhesive and the adherend, as well as the engineering of the adhesive bond.

The *interphase* is the volume of material in which the properties of one substance gradually change into the properties of another. The interphase is useful for describing the properties of an adhesive bond. The *interface,* contained within the interphase, is the plane of contact between the surface of one material and the surface of another. Except in certain special cases, the interface is imaginary. It is useful in describing surface energetics.

Theories of Adhesion

There is no unifying theory of adhesion describing the relationship between practical adhesion and the basic intermolecular and interatomic interactions which take place between the adhesive and the adherend either at the interface or within the interphase. The existing adhesion theories are, for the most part, rationalizations of observed phenomena, although in some cases, predictions regarding the relative ranking of practical adhesion can actually be made.

Diffusion Theory. The diffusion theory of adhesion is mostly applied to polymers. It assumes mutual solubility of the adherend and adhesive to form an interphase.

Electrostatic Theory. The basis of the electrostatic theory of adhesion is the differences in the electronegativities of adhering materials which leads to a transfer of charge between the materials in contact. The attraction of the charges is considered the source of adhesion.

Surface Energetics and Wettability Theory. The surface energetics and wettability theory of adhesion is concerned with the effect of intermolecular and interatomic forces on the surface energies of the adhesive and the adherend and the interfacial energy between the two.

Mechanical Interlocking Theory. A practical adhesion can be enhanced if the adhesive is applied to a surface which is microscopically rough.

Guidelines for Good Adhesion. The various adhesion theories can be used to formulate guidelines for good adhesion:

1. An adhesive should possess a liquid surface tension that is less than the critical wetting tension of the adherend's surface.
2. The adherend should be mechanically rough enough so that the asperities on the surface are on the order of, or less than, one micrometer in size.
3. The adhesive's viscosity and application conditions should be such that the asperities on the adherend's surface are completely wetted.
4. If an adverse environment is expected, covalent bonding capabilities at the interface should be provided.

For good adhesion, the adhesive and the adherend should, if possible, display mutual solubility to the extent that both diffuse into one another, providing an interphasal zone.

Advantages and Disadvantages of Using Adhesives

Adhesive Advantages. In comparison to other methods of joining, adhesives provide several advantages. First, a properly applied adhesive provides a joint having a more uniform stress distribution under load than a mechanical fastener which requires a hole in the adherend. Second, adhesives provide the ability to bond dissimilar materials such as metals without problems such as galvanic corrosion. Third, using an adhesive to make an assembly increases fatigue resistance. Fourth, adhesive joints can be made of heat- or shock-sensitive materials. Fifth, adhesive joining can bond and seal simultaneously. Sixth, use of an adhesive to form an assembly usually results in a weight reduction in comparison to mechanical fasteners since adhesives, for the most part, have densities which are substantially less than that of metals.

Adhesive Disadvantages. There are some limitations in using adhesives to form assemblies. The major limitation is that the adhesive joint is formed by means of surface attachment and is, therefore, sensitive to the substrate surface condition. Another limitation of adhesive bonding is the lack of a nondestructive quality control procedure. Finally, adhesive joining is still somewhat limited because most designers of assemblies are simply not familiar with the engineering characteristics of adhesives.

Mechanical Tests of Adhesive Bonds

The three principal forces to which adhesive bonds are subjected are a shear force in which one adherend is forced past the other, peeling in which at least one of the adherends is flexible enough to be bent away from the adhesive bond, and cleavage force. The cleavage force is very similar to the peeling force, but the former applies when the adherends are nondeformable and the latter when the adherends are deformable. Appropriate mechanical testing of these forces are used. Fracture mechanics tests are also typically used for structural adhesives.

Table 1 provides the approximate load-bearing capabilities of various adhesive types. Because the load-bearing capabilities of an adhesive are dependent upon the adherend material, the loading rate, temperature, and design of the adhesive joint, wide ranges of performance are listed.

Table 1. Load-Bearing Capabilities of Adhesives[a]

Adhesive type	Shear load, MPa[b]	Peel load, N/m[c]
pressure sensitive	0.005–0.02[d]	300–600
rubber based	0.3–7	1000–7000
emulsion	10–14	
hot melt	1–15	1000–5000
natural product (structural)	10–14	
polyurethane	6–17	2000–10,000
acrylic	6–20	900–6000
epoxy	14–50	700–18,000
phenolic	14–35	700–9000
polyimide	13–17	350–1760

[a] Load bearing capabilities are dependent upon the adherend, joint design, rate of loading, and temperature. Values given represent the type of adherends normally used at room temperature. Lap shear values approximate those obtainable from an overlap of 3.2 cm^2.
[b] To convert from MPa to psi, multiply by 145.
[c] To convert from N/m to ppi, divide by 175.
[d] Pressure-sensitive adhesives normally are rated in terms of shear holding power, ie, time to fail in minutes under a constant load.

Chemistry and Uses of Adhesives

Structural Adhesives. A structural adhesive is a resin system, usually a thermoset, that is used to bond high strength materials in such a way that the bonded joint is able to bear a load in excess of 6.9 MPa (1000 psi) at room temperature. Structural adhesives are the strongest form of adhesive and are meant to hold loads permanently. They exist in a number of forms. The most common form is the two-part adhesive, widely available as a consumer product. The next most familiar is that which is obtained as a room temperature curing liquid. Less common are primer–liquid adhesive combinations which cure at room temperature. Structural adhesive pastes which cure at 120°C are widely available in the industrial market.

Structural adhesives are formulated from epoxy resins, phenolic resins, acrylic monomers and resins, high temperature-resistant resins (eg, polyimides), and urethanes. Structural adhesive resins are often modified by elastomers.

Natural-product-based structural adhesives include protein-based adhesives, starch-based adhesives, and cellulosics.

Pressure-Sensitive Adhesives. A pressure-sensitive adhesive, a material which adheres with no more than applied finger pressure, is aggressively and permanently tacky. It requires no activation other than the finger pressure, exerts a strong holding force, and should be removable from a smooth surface without leaving a residue.

Applications and Formulation. Pressure-sensitive adhesives are most widely used in the form of adhesive tapes. The general formula for a pressure-sensitive adhesive includes an elastomeric polymer, a tackifying resin, any necessary fillers, various antioxidants and stabilizers, if needed, and cross-linking agents.

Hot-Melt Adhesives. Hot-melt adhesives are 100% nonvolatile thermoplastic materials that can be heated to a melt and then applied as a liquid to an adherend. The bond is formed when the adhesive resolidifies. The oldest example of a hot-melt adhesive is sealing wax.

Solvent- and Emulsion-Based Adhesives. *Solvent-Based Adhesives.* Solvent-based adhesives, as the name implies, are materials that are formed by solution of a high molecular weight polymer in an appropriate solvent. Solvent-based adhesives are usually elastomer-based and formulated in a manner similar to pressure-sensitive adhesives.

Emulsion Adhesives. The most widely used emulsion-based adhesive is that based upon poly(vinyl acetate)–poly(vinyl alcohol) copolymers formed by free-radical polymerization in an emulsion system. Poly(vinyl alcohol) is typically formed by hydrolysis of the poly(vinyl acetate). This is also known as "white glue."

Economic Aspects

Although the manufacture and sale of adhesives is a worldwide enterprise, the adhesives business can be characterized as a fragmented industry. The 1987 Census of Manufacturers obtained reports from 712 companies in the United States, each of which considers itself to be in the adhesives or sealants business; only 275 of these companies had more than 20 employees. Phenolics, poly(vinyl acetate) adhesives, rubber cements, and hot-melt adhesives are the leading products in terms of monetary value. These products are used primarily in the wood, paper, and packaging industries. The annual growth rate of the adhesives market is 2.3%, and individual segments of the market are expected to grow faster than this rate.

Health and Safety

Health and safety information is available from the manufacturer of every adhesive sold in the United States.

ALPHONSUS V. POCIUS
The 3M Company

I. M. Skeist, ed., *Handbook of Adhesives,* 3rd ed., Van Nostrand Reinhold Co., Inc. New York, 1990. A basic resource for practitioners of this technology.

D. Satas, ed., *Handbook of Pressure Sensitive Adhesive Technology,* Van Nostrand Reinhold Co., Inc., New York, 1989.

S. R. Hartshorn, ed., *Structural Adhesives: Chemistry and Technology,* Plenum, New York, 1986.

S. Wu, *Polymer Interface and Adhesion,* Marcel Dekker, Inc., New York, 1982. A basic textbook covering surface effects on polymer adhesion.

ADIPIC ACID

Adipic acid, hexanedioic acid, 1,4-butanedicarboxylic acid, mol wt 146.14, $HOOCCH_2CH_2CH_2CH_2COOH$, is a white crystalline solid with a melting point of about 152°C. Little of this dicarboxylic acid occurs naturally, but it is produced on a very large scale at several locations around the world. The majority of this material is used in the manufacture of nylon-6,6 polyamide, which is prepared by reaction with 1,6-hexanediamine.

Chemical and Physical Properties

Adipic acid is a colorless, odorless, sour-tasting crystalline solid. Its fundamental chemical and physical properties are listed in Table 1.

Chemical Reactions

Adipic acid undergoes the usual reactions of carboxylic acids, including esterification, amidation, reduction, halogenation, salt formation, and dehydration. Because of its bifunctional nature, it also undergoes several industrially significant polymerization reactions.

Table 1. Physical and Chemical Properties of Adipic Acid

Property	Value
molecular formula	$C_6H_{10}O_4$
molecular weight	146.14
melting point, °C	152.1 ± 0.3
specific gravity	1.344 at 18°C (sol)
	1.07 at 170°C (liq)
vapor pressure, Pa[a]	
solid at °C	
18.5	9.7
47.0	38.0
liquid at °C	
205.5	1300
244.5	6700
specific heat, kJ/kg·K[b]	1.590 (solid state)
	2.253 (liquid state)
	1.680 (vapor, 300°C)
heat of fusion, kJ/kg[b]	115
melt viscosity, mPa·s (= cP)	4.54 at 160°C
heat of combustion, kJ/mol[b]	2800

[a] To convert Pa to mm Hg, divide by 133.3.
[b] To convert J to cal, divide by 4.184.

Manufacture and Processing

Adipic acid historically has been manufactured predominantly from cyclohexane and, to a lesser extent, phenol. During the 1970s and 1980s, however, much research has been directed to alternative feedstocks, especially butadiene and cyclohexene, as dictated by shifts in hydrocarbon markets. All current industrial processes use nitric acid in the final oxidation stage. Growing concern with air quality may exert further pressure for alternative routes as manufacturers seek to avoid NO_x abatement costs, a necessary part of processes that use nitric acid.

Since adipic acid has been produced in commercial quantities for almost 50 years, it is not surprising that many variations and improvements have been made to the basic cyclohexane process. In general, however, the commercially important processes still employ two major reaction stages. The first reaction stage is the production of the intermediates cyclohexanone and cyclohexanol, usually abbreviated as KA, KA oil, ol-one, or anone-anol. The KA (ketone, alcohol), after separation from unreacted cyclohexane (which is recycled) and reaction byproducts, is then converted to adipic acid by oxidation with nitric acid. An important alternative to this use of KA is its use as an intermediate in the manufacture of caprolactam, the monomer for production of nylon-6. The latter use of KA predominates by a substantial margin on a worldwide basis, but not in the United States.

Storage, Handling, and Shipping

When dispersed as a dust, adipic acid is subject to normal dust explosion hazards. The material is an irritant, especially upon contact with the mucous membranes. Thus protective goggles or face shields should be worn when handling the material.

The material should be stored in corrosion-resistant containers, away from alkaline or strong oxidizing materials.

Economic Aspects

Adipic acid is a very large-volume organic chemical. It is one of the top 50 chemicals produced in the United States in terms of volume. Demand is highly cyclic, reflecting the automotive and housing markets especially. Prices usually follow the variability in crude oil prices. Adipic acid for nylon takes about 60% of U.S. cyclohexane production; the remainder goes to caprolactam for nylon-6, export, and miscellaneous uses.

Toxicity, Safety, and Industrial Hygiene

Adipic acid is relatively nontoxic; no OSHA PEL or NIOSH REL have been established for the material.

DARWIN D. DAVIS
DONALD R. KEMP
E.I. du Pont de Nemours & Co., Inc.

V. D. Luedeke, "Adipic Acid", in *Encyclopedia of Chemical Processing and Design,* J. McKetta and W. Cunningham, eds., Vol. 2, Marcel Dekker, Inc., New York, 1977, pp. 128–146.

Y. C. Yen and S. Y. Wu, *Nylon-6,6, Report No. 54B, Process Economics Program,* SRI International, Menlo Park, Calif., 1987, pp. 1–148.

A. Castellan, J. C. J. Bart, and S. Cavallaro, *Catalysis Today* **9,** 237–322 (1991).

A. K. Suresh, T. Sridhar, and O. E. Potter, *AIChE J.* **34**(1), 55–93 (1988).

ADSORPTION

Adsorption is the term used to describe the tendency of molecules from an ambient fluid phase to adhere to the surface of a solid. This is a fundamental property of matter, having its origin in the attractive forces between molecules. The force field creates a region of low potential energy near the solid surface and, as a result, the molecular density close to the surface is generally greater than in the bulk gas. Furthermore, and perhaps more importantly, in a multicomponent system the composition of this surface layer generally differs from that of the bulk gas since the surface adsorbs the various components with different affinities. Adsorption may also occur from the liquid phase and is accompanied by a similar change in composition, although, in this case, there is generally little difference in molecular density between the adsorbed and fluid phases.

The enhanced concentration at the surface accounts, in part, for the catalytic activity shown by many solid surfaces, and it is also the basis of the application of adsorbents for low pressure storage of permanent gases such as methane. However, most of the important applications of adsorption depend on the selectivity, ie, the difference in the affinity of the surface for different components. As a result of this selectivity, adsorption offers, at least in principle, a relatively straightforward means of purification (removal of an undesirable trace component from a fluid mixture) and a potentially useful means of bulk separation.

Fundamental Principles

Forces of Adsorption. Adsorption may be classified as chemisorption or physical adsorption, depending on the nature of the surface forces. In physical adsorption the forces are relatively weak, involving mainly van der Waals (induced dipole–induced dipole) interactions, supplemented in many cases by electrostatic contributions from field–dipole or field–gradient–quadrupole interactions. By contrast, in chemisorption there is significant electron transfer, equivalent to the formation of a chemical bond between the sorbate and the solid surface. Such interactions are both stronger and more specific than the forces of physical adsorption and are obviously limited to monolayer coverage.

Selectivity. Selectivity in a physical adsorption system may depend on differences in either equilibrium or kinetics, but the great majority of adsorption separation processes depend on equilibrium-based selectivity. Significant kinetic selectivity is, in general, restricted to molecular sieve adsorbents—carbon molecular sieves, zeolites, or zeolite analogues.

Hydrophilic and Hydrophobic Surfaces. Polar adsorbents such as most zeolites, silica gel, or activated alumina adsorb water (a small polar molecule) more strongly than they adsorb organic species, and, as a result, such adsorbents are commonly called hydrophilic. In contrast, on a nonpolar surface where there is no electrostatic interaction, water is held only very weakly and is easily displaced by organics. Such adsorbents, which are the only practical choice for adsorption of organics from aqueous solutions, are termed hydrophobic.

Capillary Condensation. In a porous adsorbent the region of multilayer physical adsorption merges gradually with the capillary condensation regime, leading to upward curvature of the equilibrium isotherm at higher relative pressure. In the capillary condensation region the intrinsic selectivity of the adsorbent is lost.

Practical Adsorbents

To achieve a significant adsorptive capacity an adsorbent must have a high specific area, which implies a highly porous structure with very small micropores. Such microporous solids can be produced in several different ways. Adsorbents such as silica gel and activated alumina are made by precipitation of colloidal particles, followed by dehydration (see ALUMINUM COMPOUNDS–ALUMINUM OXIDE; SILICA–AMORPHOUS SILICA). Carbon adsorbents are prepared by controlled burn-out of carbonaceous materials such as coal, lignite, and coconut shells (see CARBON–ACTIVATED CARBON). The crystalline adsorbents (zeolite and zeolite analogues) are different in that the dimensions of the micropores are determined by the crystal structure and there is therefore virtually no distribution of micropore size (see MOLECULAR SIEVES). Although structurally very different from the crystalline adsorbents, carbon molecular sieves also have a very narrow distribution of pore size. The adsorptive properties depend on the pore size and the pore size distribution as well as on the nature of the solid surface.

Adsorption Equilibrium

Henry's Law. Like any other phase equilibrium, the distribution of a sorbate between fluid and adsorbed phases is governed by the principles of thermodynamics. Equilibrium data are commonly reported in the form of an isotherm, which is a diagram showing the variation of the equilibrium adsorbed-phase concentration or loading with the fluid-phase concentration or partial pressure at a fixed temperature. In general, for physical adsorption on a homogeneous surface at sufficiently low concentrations, the isotherm should approach a linear form, and the limiting slope in the low concentration region is commonly known as the Henry's law constant. The Henry constant is a thermodynamic equilibrium constant and the temperature dependence therefore follows the usual van't Hoff equation:

$$\lim p \rightarrow 0\left(\frac{\delta q}{\delta p}\right)_T \equiv K' = K_0' e^{-\Delta H_0/RT} \tag{1}$$

in which $-\Delta H_0$ is the limiting heat of adsorption at zero coverage. Since adsorption, particularly from the vapor phase, is usually exothermic, $-\Delta H_0$ is a positive quantity and K' therefore decreases with increasing temperature.

Henry's law corresponds physically to the situation in which the adsorbed phase is so dilute that there is neither competition for surface sites nor any significant interaction between adsorbed molecules. At higher concentrations both of these effects become important and the form of the isotherm becomes more complex. The isotherms have been classified into five different types (Fig. 1). Isotherms for a microporous adsorbent are generally of type I; the more complex forms are associated with multilayer adsorption and capillary condensation.

Langmuir Isotherm. Type I isotherms are commonly represented by the ideal Langmuir model:

$$\frac{q}{q_s} = \frac{bp}{1 + bp} \tag{2}$$

where q_s is the saturation limit and b is an equilibrium constant which is directly related to the Henry constant ($K' = bq_s$).

Freundlich Isotherm. The isotherms for some systems, notably hydrocarbons on activated carbon, conform more closely to the Freundlich equation:

$$q = bp^{1/n} (n > 1.0) \tag{3}$$

Adsorption of Mixtures. The Langmuir model can be easily extended to binary or multicomponent systems:

$$\frac{q_1}{q_{s1}} = \frac{b_1 p_1}{1 + b_1 p_1 + b_2 p_2 + \cdots}; \frac{q_2}{q_{s2}} = \frac{b_2 p_2}{1 + b_1 p_1 + b_2 p_2; + \cdots} \tag{4}$$

Thermodynamic consistency requires $q_{s1} = q_{s2}$, but this requirement can cause difficulties when attempts are made to correlate data for sorbates of very different molecular size. For such systems it is common practice to ignore this requirement, thereby introducing an additional model parameter. This facilitates data fitting but it must be recognized that the equations are then being used purely as a convenient empirical form with no theoretical foundation.

Ideal Adsorbed Solution Theory. Perhaps the most successful general approach to the prediction of multicomponent equilibria from single-component isotherm data is ideal adsorbed solution theory. In essence, the theory is based on the assumption that the adsorbed

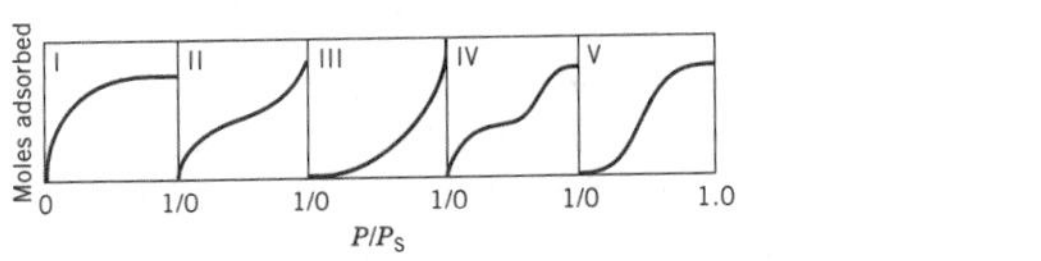

Figure 1. The Brunaner classification of isotherms (I–V).

phase is thermodynamically ideal in the sense that the equilibrium pressure for each component is simply the product of its mole fraction in the adsorbed phase and the equilibrium pressure for the pure component *at the same spreading pressure.* The theoretical basis for this assumption and the details of the calculations required to predict the mixture isotherm are given in standard texts on adsorption. Whereas the theory has been shown to work well for several systems, notably for mixtures of hydrocarbons on carbon adsorbents, there are a number of systems which do not obey this model. Azeotrope formation and selectivity reversal, which are observed quite commonly in real systems, are not consistent with an ideal adsorbed phase and there is no way of knowing *a priori* whether or not a given system will show ideal behavior.

Adsorption Kinetics

Intrinsic Kinetics. Chemisorption may be regarded as a chemical reaction between the sorbate and the solid surface, and, as such, it is an activated process for which the rate constant (k) follows the familiar Arrhenius rate law:

$$k = k_0 e^{-E/RT} \tag{5}$$

Depending on the temperature and the activation energy (E), the rate constant may vary over many orders of magnitude.

In practice the kinetics are usually more complex than might be expected on this basis, since the activation energy generally varies with surface coverage as a result of energetic heterogeneity and/or sorbate–sorbate interaction. As a result, the adsorption rate is commonly given by the Elovich equation:

$$q = \frac{1}{k'} \ln(1 + k''t) \tag{6}$$

where k' and k'' are temperature-dependent constants.

In contrast, physical adsorption is a very rapid process, so the rate is always controlled by mass transfer resistance rather than by the intrinsic adsorption kinetics. However, under certain conditions the combination of a diffusion-controlled process with an adsorption equilibrium constant that varies according to equation 1 can give the appearance of activated adsorption.

A porous adsorbent in contact with a fluid phase offers at least two and often three distinct resistances to mass transfer: external film resistance and intraparticle diffusional resistance. When the pore size distribution has a well-defined bimodal form, the latter may be divided into macropore and micropore diffusional resistances. Depending on the particular system and the conditions, any one of these resistances may be dominant, or the overall rate of mass transfer may be determined by the combined effects of more than one resistance. The magnitude of the intraparticle diffusional resistances, or any surface resistance to mass transfer, can be conveniently determined by measuring the adsorption or desorption rate, under controlled conditions, in a batch system.

Adsorption Column Dynamics

In most adsorption processes the adsorbent is contacted with fluid in a packed bed. An understanding of the dynamic behavior of such systems is therefore needed for rational process design and optimization. What is required is a mathematical model which allows the effluent concentration to be predicted for any defined change in the feed concentration or flow rate to the bed. The flow pattern can generally be represented adequately by the axial dispersed plug-flow model, according to which a mass balance for an element of the column yields, for the basic differential equation governing the dynamic behavior,

$$-D_L \frac{\delta^2 c_i}{\delta z^2} + \frac{\delta}{\delta z}(vc_i) + \frac{\delta c_i}{\delta t} + \left(\frac{1-\epsilon}{\epsilon}\right)\frac{\delta \bar{q}_i}{\delta t} = 0 \tag{7}$$

The term $\delta\bar{q}_i/\delta t$ represents the overall rate of mass transfer for component i (at time t and distance z) averaged over a particle. This is governed by a mass transfer rate expression which may be thought of as a general functional relationship of the form

$$\frac{\delta \bar{q}}{\delta t} = f(c_i, c_j, \cdots q_i, q_j, \cdots) \tag{8}$$

This rate equation must satisfy the boundary conditions imposed by the equilibrium isotherm and it must be thermodynamically consistent so that the mass transfer rate falls to zero at equilibrium.

Equilibrium Theory. The general features of the dynamic behavior may be understood without recourse to detailed calculations since the overall pattern of the response is governed by the form of the equilibrium relationship rather than by kinetics. If the equilibrium isotherm is of "favorable" form (ie, slope decreasing with increasing concentration as in Figure 1,I) the concentration front, for adsorption, will assume the form of a travelling shock wave, whereas for desorption the front will assume the form of a simple wave which spreads as it propagates through the column.

Constant Pattern Behavior. In a real system the finite resistance to mass transfer and axial mixing in the column lead to departures from the idealized response predicted by equilibrium theory. In the case of a favorable isotherm the shock wave solution is replaced by a constant pattern solution. The concentration profile spreads in the initial region until a stable situation is reached in which the mass transfer rate is the same at all points along the wave front and exactly matches the shock velocity. In this situation the fluid-phase and adsorbed-phase profiles become coincident. This represents a stable situation and the profile propagates without further change in shape—hence the term constant pattern.

Length of Unused Bed. The constant pattern approximation provides the basis for a very useful and widely used design method based on the concept of the length of unused bed (LUB). In the design of a typical adsorption process the basic problem is to estimate the size of the adsorber bed needed to remove a certain quantity of the adsorbable species from the feed stream, subject to a specified limit (c') on the effluent concentration. The length of unused bed, which measures the capacity of the adsorber which is lost as a result of the spread of the concentration profile, is defined by

$$\text{LUB} = (1 - q'/q_o)L = (1 - t'/\bar{t})L \tag{9}$$

where q' is the capacity at the break time t' and $\bar{t}$ is the stoichiometric time (see Fig. 2). The values of t', $\bar{t}$, and hence the LUB are easily determined from an experimental breakthrough curve since, by overall

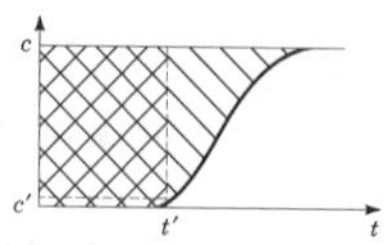

Figure 2. Sketch of breakthrough curve showing break time t' and the method of calculation of the stoichiometric time $\bar{t}$ and LUB. ⧅ = the integral of equation 10; ⊠ = integral of equation 11.

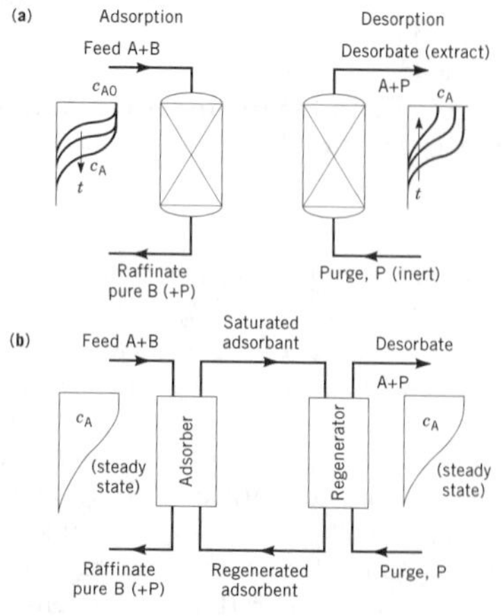

Figure 3. The two basic modes of operation for an adsorption process: (**a**) cyclic batch system; (**b**) continuous countercurrent system with adsorbent recirculation.

Table 1. Factors Governing Choice of Regeneration Method

Method	Advantages	Disadvantages
thermal swing	good for strongly adsorbed species; small change in T gives large change in q^*	thermal aging of adsorbent
	desorbate may be recovered at high concentration	heat loss means inefficiency in energy usage unsuitable for rapid cycling, so adsorbent cannot be used with maximum efficiency
	gases and liquids	in liquid systems the latent heat of the interstitial liquid must be added
pressure swing	good where weakly adsorbed species is required at high purity	very low P may be required mechanical energy more expensive than heat
	rapid cycling—efficient use of adsorbent	desorbate recovered at low purity
displacement desorption	good for strongly held species	product separation and recovery needed (choice of desorbent is crucial)
	avoids risk of cracking reactions during regeneration avoids thermal aging of adsorbent	

mass balance:

$$\bar{t} = \frac{L}{v}\left[1 + \left(\frac{1-\epsilon}{\epsilon}\right)\left(\frac{q_0}{c_0}\right)\right] = \int_{\infty}^{0}\left(1 - \frac{c}{c_0}\right)dt \quad (10)$$

$$t' = \frac{L}{v}\left[1 + \left(\frac{1-\epsilon}{\epsilon}\right)\left(\frac{q'}{c_0}\right)\right] = \int_{L'}^{0}\left(1 - \frac{c}{c_0}\right)dt \quad (11)$$

The length of column needed for a particular duty can then be found simply by adding the LUB to the length calculated from equilibrium considerations, assuming a shock concentration front.

Proportionate Pattern Behavior. If the isotherm is unfavorable (as in Fig. 1,III), the stable dynamic situation leading to constant pattern behavior can never be achieved. The equilibrium adsorbed-phase concentration then lies above rather than below the actual adsorbed-phase profile. As the mass transfer zone progresses through the column it broadens, but the limiting situation, which is approached in a long column, is simply local equilibrium at all points ($c = c^*$) and the profile therefore continues to spread in proportion to the distance traveled. This difference in behavior is important since the LUB approach to design is clearly inapplicable under these conditions.

Adsorption Chromatography. In a linear multicomponent system (several sorbates at low concentration in an inert carrier) the wave velocity for each component depends on its adsorption equilibrium constant. Thus, if a pulse of the mixed sorbate is injected at the column inlet, the different species separate into bands which travel through the column at their characteristic velocities, and at the outlet of the column a sequence of peaks corresponding to the different species is detected. Measurement of the retention time ($\bar{t}$) under known flow conditions thus provides a simple means of determining the equilibrium constant (Henry constant).

In an ideal system with no axial mixing or mass-transfer resistance, the peaks for the various components propagate without spreading. However, in any real system the peak broadens as it propagates and the extent of this broadening is directly related to the mass transfer and axial dispersion characteristics of the column. Measurement of the peak broadening therefore provides a convenient way of measuring mass-transfer coefficients and intraparticle diffusivities.

APPLICATIONS

The applications of adsorbents are many and varied. They may be classified as "regenerative" and "nonregenerative". Most process applications, in which the adsorbent is used as a means of purifying or separating the components of a gas or liquid mixture are regenerative. The process operates in a cyclic manner so that the adsorbent is alternately saturated and regenerated. Nonregenerative applications include the use of adsorbents in cigarette filters, in some water purification systems, as deodorants in health care products and as desiccants in storage, packaging and dual-pane windows.

Adsorption Separation and Purification Processes. Adsorption processes can be classified according to the flow system (cyclic batch or continuous countercurrent) and the method by which the adsorbent is regenerated. The two basic flow schemes are illustrated in Figure 3. The cyclic batch scheme is simpler but less efficient. It is generally used where selectivity is relatively high. Countercurrent or simulated countercurrent schemes are more expensive in initial cost and are generally used only for difficult separations in which selectivity is limited or mass-transfer resistance is high.

The three common methods of regeneration are thermal swing, pressure swing, and displacement. The main factors governing this choice are summarized in Table 1.

NOTATION

b = Langmuir equilibrium constant
c = sorbate concentration in fluid phase
c_0 = initial value of c
D_L = axial dispersion coefficient
E = activation energy
$-\Delta H_0$ = limiting heat of adsorption
K' = Henry's law constant
K'_0 = preexponential factor
k = rate constant
k_0 = preexponential factor
k', k'' = constants in Elovich equation
L = bed length

DOUGLAS M. RUTHVEN
University of New Brunswick, Canada

D. M. Ruthven, *Principles of Adsorption and Adsorption Processes,* Wiley-Interscience, New York, 1984.

M. Suzuki, *Adsorption Engineering,* Kodansha-Elsevier, Tokyo, 1990.

R. T. Yang, *Gas Separation by Adsorption Processes,* Butterworths, Stoneham, Mass., 1987.

P. Wankat, *Large Scale Adsorption and Chromatography,* CRC Press, Boca Raton, Fla., 1986.

ADSORPTION, GAS SEPARATION

Gas-phase adsorption is widely employed for the large-scale purification or bulk separation of air, natural gas, chemicals, and petrochemicals (Table 1). In these uses it is often a preferred alternative to the older unit operations of distillation and absorption.

An adsorbent attracts molecules from the gas, and the molecules become concentrated on the surface of the adsorbent and are removed from the gas phase. Many process concepts have been developed to allow the efficient contact of feed gas mixtures with adsorbents to carry out desired separations and to allow efficient regeneration of the adsorbent for subsequent reuse. In nonregenerative applications, the adsorbent is used only once and is not regenerated.

Most commercial adsorbents for gas-phase applications are employed in the form of pellets, beads, or other granular shapes, typically about 1.5 to 3.2 mm in diameter. Most commonly, these adsorbents are packed into fixed beds through which the gaseous feed mixtures are passed. Normally, the process is conducted in a cyclic manner. When the capacity of the bed is exhausted, the feed flow is stopped to terminate the loading step of the process, the bed is treated to remove the adsorbed molecules in a separate regeneration step, and the cycle is then repeated.

The growth in both variety and scale of gas-phase adsorption separation processes, particularly since 1970, is due in part to continuing discoveries of new porous, high surface-area adsorbent materials (particularly molecular sieve zeolites) and, especially, to improvements in the design and modification of adsorbents. These advances have encouraged parallel inventions of new process concepts. Increasingly, the development of new applications requires close cooperation in adsorbent design and process cycle development and optimization.

Adsorption Principles

The design and manufacture of adsorbents for specific applications involves manipulation of the structure and chemistry of the adsorbent to provide greater attractive forces for one molecule compared to another, or, by adjusting the size of the pores, to control access to the adsorbent surface on the basis of molecular size. Adsorbent manufacturers have developed many technologies for these manipulations, but they are considered proprietary and are not openly communicated. Nevertheless, the broad principles are well known.

Adsorption Forces. Coulomb's law allows calculations of the electrostatic potential resulting from a charge distribution, and of the potential energy of interaction between different charge distributions. Various elaborate computations are possible to calculate the potential energy of interaction between point charges, distributed charges, etc.

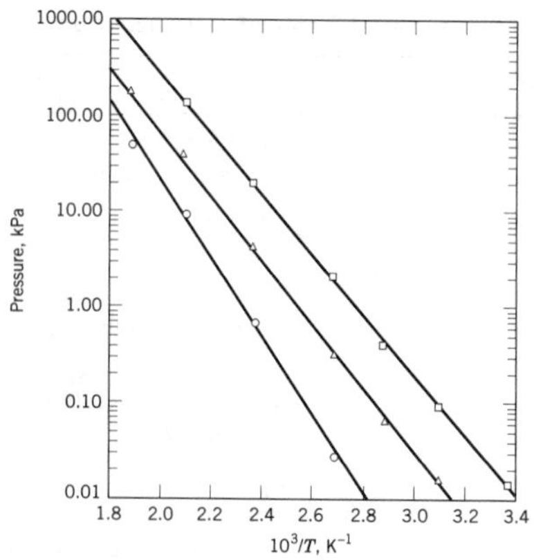

Figure 1. Adsorption isosteres, water vapor on 4A (NaA) zeolite pellets. H_2O loading: □, 15 kg/100 kg zeolite; △, 10 kg/100 kg, ○, 5 kg/100 kg. To convert kPa to mm Hg, multiply by 7.5. Courtesy of Union Carbide.

Adsorption Selectivities. For a given adsorbent, the relative strength of adsorption of different adsorbate molecules depends on the relative magnitudes of the polarizability α, dipole moment μ, and quadrupole moment Q of each. Often, just the consideration of the values of α, μ, and Q allows accurate qualitative predictions to be made of the relative strengths of adsorption of given molecules on an adsorbent or of the best adsorbent type (polar or nonpolar) for a particular separation.

Heats of Adsorption. The integral heat of adsorption is the total heat released when the adsorbate loading is increased from zero to some final value at isothermal conditions. The differential heat of adsorption δH_{iso} is the incremental change in heat of adsorption with a differential change in adsorbate loading. This heat of adsorption δH_{iso} may be determined from the slopes of adsorption isosteres (lines of constant adsorbate loading) on graphs of ln P vs $1/T$ (Fig. 1) through the Clausius-Clapeyron relationship:

$$\frac{d \ln P}{d(1/T)} = -\frac{\delta H_{iso}}{R}$$

where R is the gas constant, P the adsorbate absolute pressure, and T the absolute temperature.

Table 1. Commercial Adsorption Separations

Separation	Adsorbent
Gas bulk separations	
normal paraffins, isoparaffins, aromatics	zeolite
N_2/O_2	zeolite
O_2/N_2	carbon molecular sieve
CO, CH_4, CO_2, N_2, Ar, NH_3/H_2	zeolite, activated carbon
acetone/vent streams	activated carbon
C_2H_4/vent streams	activated carbon
H_2O/ethanol	zeolite
Gas purifications	
H_2O/olefin-containing cracked gas, natural gas, air, synthesis gas, etc	silica, alumina, zeolite
CO_2/C_2H_4, natural gas, etc	zeolite
organics/vent streams	activated carbon, others
sulfur compounds/natural gas, hydrogen, liquified petroleum gas (LPG), etc	zeolite
solvents/air	activated carbon
odors/air	activated carbon
NO_x/N_2	zeolite
SO_2/vent streams	zeolite
Hg/chlor–alkali cell gas effluent	zeolite

Isotherm Models. *Thermodynamically Consistent Isotherm Models.* These models include both the statistical thermodynamic models and the models that can be derived from an assumed equation of state for the adsorbed phase plus the thermodynamics of the adsorbed phase.

Statistical Thermodynamic Isotherm Models. These approaches were pioneered by Fowler and Guggenheim and Hill and this approach has been applied to modeling of adsorption in microporous adsorbents.

Semiempirical Isotherm Models. Some of these models have been shown to have some thermodynamic inconsistencies and should be used with due care. They include models based on the Polanyi adsorption potential (Dubinin-Radushkevich, Dubinin-Astakhov, Radke-Prausnitz, Toth, UNILAN, and BET).

Isotherm Models for Adsorption of Mixtures. Of the following models, all but the ideal adsorbed solution theory (IAST) and the related heterogeneous ideal adsorbed solution theory (HIAST) have been shown to contain some thermodynamic inconsistencies. They include Markham and Benton, the Leavitt loading ratio correlation (LRC) method, the ideal adsorbed solution (IAS) model, the heterogeneous ideal adsorbed solution theory (HIAST), and the vacancy solution model (VSM).

Adsorption Dynamics. An outline of approaches that have been taken to model mass-transfer rates in adsorbents has been given. Extensive literature exists on the interrelated topics of modeling of mass-transfer rate processes in fixed-bed adsorbers, bed concentration profiles, and breakthrough curves and the related simple design concepts of WES, WUB, and LUB for constant-pattern adsorption.

Reactions on Adsorbents. To permit the recovery of pure products and to extend the adsorbent's useful life, adsorbents should generally be inert and not react with or catalyze reactions of adsorbate molecules. These considerations often affect adsorbent selection or require limits be placed upon the severity of operating conditions to minimize reactions of the adsorbate molecules or damage to the adsorbents.

Adsorbent Principles

Principal Adsorbent Types. Commercially useful adsorbents can be classified by the nature of their structure (amorphous or crystalline), by the sizes of their pores (micropores, mesopores, and macropores), by the nature of their surfaces (polar, nonpolar, or intermediate), or by their chemical composition. All of these characteristics are important in the selection of the best adsorbent for any particular application. However, the size of the pores is the most important initial consideration because if a molecule is to be adsorbed, it must not be larger than the pores of the adsorbent.

Adsorption Properties. Not only do the more highly polar molecular sieve zeolites adsorb more water at lower pressures than do the moderately polar silica gels and alumina gels, but they also hold onto the water more strongly at higher temperatures. For the same reason, temperatures required for thermal regeneration of water-loaded zeolites are higher than for less highly polar adsorbents.

Physical Properties. Physical properties of importance include particle size, density, volume fraction of intraparticle and extraparticle voids when packed into adsorbent beds, strength, attrition resistance, and dustiness. These properties can be varied intentionally to tailor adsorbents to specific applications (see ADSORPTION, LIQUID SEPARATION; ALUMINUM COMPOUNDS–ALUMINUM OXIDE; CARBON–ACTIVATED CARBON; ION EXCHANGE; MOLECULAR SIEVES; and SILICON COMPOUNDS–SYNTHETIC INORGANIC SILICATES).

Deactivation. Gradual adsorbent degradation by chemical attack or physical damage commonly occurs in many uses, accompanied by declining separation performance. Allowance for this must be taken into account in design of the process and in scheduling the replacement of spent adsorbents.

Adsorption Processes

Adsorption processes are often identified by their method of regeneration. Temperature-swing adsorption (TSA) and pressure-swing (PSA) are the most frequently applied process cycles for gas separation. Purge-swing cycles and nonregenerative approaches are also applied to the separation of gases. Special applications exist in the nuclear industry. Others take advantage of reactive sorption. Most adsorption processes use fixed beds, but some use moving or fluidized beds.

Design Methods

Design techniques for gas-phase adsorption range from empirical to theoretical. Methods have been developed for equilibrium, for mass transfer, and for combined dynamic performance. Approaches are available for the regeneration methods of heating, purging, steaming, and pressure swing. Several broad reviews have been published on analytical equations describing adsorption, on experimental adsorption processes, and on adsorption design considerations.

Future Directions

Advances in fundamental knowledge of adsorption equilibrium and mass transfer will enable further optimization of the performance of existing adsorbent types. Continuing discoveries of new molecular sieve materials will also provide adsorbents with new combinations of useful properties. New adsorbents and adsorption process will be developed to provide needed improvements in pollution control, energy conservation, and the separation of high value chemicals. New process cycles and new hybrid processes linking adsorption with other unit operations will continue to be developed.

JOHN D. SHERMAN
CARMEN M. YON
UOP

G. E. Keller, II, R. A. Anderson, and C. M. Yon, in R. W. Rousseau, ed., *Handbook of Separation Process Technology,* John Wiley & Sons, Inc., New York, 1987, pp. 644–696.

R. M. Barrer, *Zeolites and Clay Minerals as Adsorbents and Catalysts,* Academic Press, London, 1978, pp. 164, 174, and 185.

D. W. Breck, *Zeolite Molecular Sieves—Structure, Chemistry, and Use,* John Wiley & Sons, Inc., New York, 1974.

R. N. Macnair and G. N. Arons in P. N. Cheremisinoff and F. Eleerbusch, eds., *Carbon Adsorption Handbook,* Ann Arbor Science, Ann Arbor, Mich., 1978, pp. 819–859.

ADSORPTION, LIQUID SEPARATION

Liquid-phase adsorption has long been used for the removal of contaminants present at low concentrations in process streams. In most cases, the objective is to remove a specific feed component; alternatively, the contaminants are not well defined, and the objective is the improvement of feed quality defined by color, taste, odor, and storage stability (see WASTES, INDUSTRIAL; WATER, INDUSTRIAL WATER TREATMENT).

In contrast to trace impurity removal, the use of adsorption for bulk separation in the liquid phase on a commercial scale is a relatively recent development. This article is devoted mainly to the theory and operation of these liquid-phase bulk adsorptive separation processes.

Adsorbate–Adsorbent Interactions

An adsorbent can be visualized as a porous solid having certain characteristics. When the solid is immersed in a liquid mixture, the pores fill with liquid, which at equilibrium differs in composition from that of the liquid surrounding the particles. These compositions can then be related to each other by enrichment factors that are analogous to relative volatility in distillation. The adsorbent is selective for the component that is more concentrated in the pores than in the surrounding liquid.

Table 1. Molecular Sieve Pore Structures

Common name	Ring size, number of atoms	Free aperture, nm	Pore structure	Formula
faujasite	12	0.74	3-D	$(Ca, Mg, Na_2, K_2)_{29.5}[(AlO_2)_{59}(SiO_2)_{133}]\cdot 235H_2O$
mordenite	8	0.29 × 0.57	1-D	$Na_8[(AlO_2)_8(SiO_2)_{40}]\cdot 24H_2O$
	12	0.67 × 0.7	1-D	
L	12	0.71	1-D	$K_9[(AlO_2)_9(SiO_2)_{27}]\cdot 22H_2O$
ZSM-5	10	0.54 × 0.56	1-D	$(Na, TPA^a)_3$ $[(AlO_2)_3(SiO_2)_{93}\cdot 16H_2O]$
	10	0.51 × 0.56	1-D	
Erionite	8	0.36 × 0.52	2-D	$(Ca, Mg, Na_2, K_2)_{4.5}[(AlO_2)_9(SiO_2)_{27}]\cdot 27H_2O$
A	8	0.42	3-D	Na_{12} $[(AlO_2)_{12}(SiO_2)_{12}]\cdot 27H_2O$

[a] TPA = tetrapropylammonium.

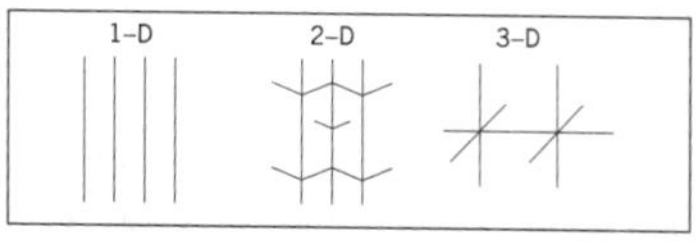

Figure 1. Schematic diagram of molecular sieve pore structure. See Table 1.

A significant advantage of adsorbents over other separative agents lies in the fact that favorable equilibrium-phase relations can be developed for particular separations; adsorbents can be produced that are much more selective in their affinity for various substances than are any known solvents. This selectivity is particularly true of the synthetic crystalline zeolites containing exchangeable cations.

An example of unique selectivity is provided by the use of 5A molecular sieves for the separation of linear hydrocarbons from branched and cyclic types. In this system only the linear molecules can enter the pores; others are completely excluded because of their larger cross section. Thus the selectivity for linear molecules with respect to other types is infinite. In the more usual case, all the feed components access the selective pores, but some components of the mixture are adsorbed more strongly than others. A selectivity between the different components that can be used to accomplish separation is thus established.

Practical Adsorbents

The search for a suitable adsorbent is generally the first step in the development of an adsorption process. A practical adsorbent has four primary requirements: selectivity, capacity, mass transfer rate, and long-term stability. The requirement for adequate adsorptive capacity restricts the choice of adsorbents to microporous solids with pore diameters ranging from a few tenths to a few tens of nanometers.

Traditional adsorbents such as silica, SiO_2; activated alumina, Al_2O_3; and activated carbon, C, exhibit large surface areas and micropore volumes. The surface chemical properties of these adsorbents make them potentially useful for separations by molecular class. However, the micropore size distribution is fairly broad for these materials. This characteristic makes them unsuitable for use in separations in which steric hindrance can potentially be exploited (see ALUMINUM COMPOUNDS, ALUMINA; SILICON COMPOUNDS, SYNTHETIC INORGANIC SILICATES).

In contrast to these adsorbents, zeolites offer increased possibilities for exploiting molecular-level differences among adsorbates. Zeolites are crystalline aluminosilicates containing an assemblage of SiO_4 and AlO_4 tetrahedra joined together by oxygen atoms to form a microporous solid, which has a precise pore structure. Nearly 40 distinct framework structures have been identified to date. Table 1 and Figure 1 summarizes some of those structures that have been widely used in the chemical industry. The versatility of zeolites lies in the fact that widely different adsorptive properties may be realized by the appropriate control of the framework structure, the silica-to-alumina ratio (Si/Al), and the cation form.

Commercial Processes

Industrial-scale adsorption processes can be classified as batch or continuous. In a batch process, the adsorbent bed is saturated and regenerated in a cyclic operation. In a continuous process, a countercurrent staged contact between the adsorbent and the feed and desorbent is established by either a true or a simulated recirculation of the adsorbent. The efficiency of an adsorption process is significantly higher in a continuous mode of operation than in a cyclic batch mode. For difficult separations, batch operation may require 25 times more adsorbent inventory and twice the desorbent circulation rate than does a continuous operation. In addition, in a batch mode, the four functions of adsorption, purification, desorption, and displacement of the desorbent from the adsorbent are inflexibly linked, whereas a continuous mode allows more degrees of freedom with respect to these functions, and thus a better overall operation.

Continuous Countercurrent Processes

The need for a continuous countercurrent process arises because the selectivity of available adsorbents in a number of commercially important separations is not high. In the *p*-xylene system, for instance, if the liquid around the adsorbent particles contains 1% *p*-xylene, the liquid in the pores contains about 2% *p*-xylene at equilibrium. Therefore, one stage of contacting cannot provide a good separation, and multistage contacting must be provided in the same way that multiple trays are required in fractionating materials with relatively low volatilities.

Since the 1960s the commercial development of continuous countercurrent processes has been almost entirely accomplished by using a flow scheme that simulates the continuous countercurrent flow of adsorbent and process liquid without the actual movement of the adsorbent. The idea of a simulated moving bed (SMB) can be traced back to the Shanks system for leaching soda ash.

Such a concept was originally used in a process developed and licensed by UOP under the name UOP Sorbex. The extent of commercial of Sorbex processes is shown in Table 2. Other versions of the SMB system are also used commercially. Toray Industries built the Aromax process for the production of *p*-xylene. Illinois Water Treatment and

Table 2. UOP Sorbex Processes for Commodity Chemicals

UOP Processes	Separation	Licensed units
Parex	*p*-xylene from C_8 aromatics	53
Molex	*n*-paraffins from branched and cyclic hydrocarbons	33
Olex	olefins from paraffins	6
Cymex	*p*- or *m*-cymene from cymene isomers	1
Cresex	*p*- or *m*-cresol from cresol isomers	1
Sarex	fructose from dextrose plus polysaccharides	5
Total		*99*

Mitsubishi have commercialized SMB processes for the separation of fructose from dextrose.

Cyclic-Batch Processes

Continuous processes have wide application in different areas of the chemical industry. The separation efficiency of a continuous process is generally higher than that of a batch or cyclic-batch process. However, in some applications the cyclic-batch process may be preferred because of the complexity of design and the difficulty of controlling the continuous processes. Examples of commercial cyclic-batch adsorption processes operating in liquid phase include the UOP methanol recovery (UOP MRU) and oxygenate removal (UOP ORU) processes, which separate oxygenates from C_4 hydrocarbons; the UOP Cyclesorb process, which separates fructose from glucose; and ion-exclusion processes for recovering sucrose from molasses.

Liquid Chromatography

Conventional liquid chromatography has not attained great commercial significance in the area of large-scale bulk separations from the liquid phase. In analytical chromatography, the primary objective is to maximize the resolution between two components subject to some restrictions on the maximum time of elution. As a result, the feed pulse loading is minimized, and the number of theoretical plates is maximized. In preparative chromatography, the objective is to maximize production rate as well as reduce capital and operating costs at a given separation efficiency. The adsorption column is therefore commonly run under overload conditions with a finite feed pulse width. The choice of operating conditions for preparative chromatography has been discussed. In production chromatography, the optimal pulse sequence occurs when the successive pulses of feed are introduced at intervals such that the feed components are just resolved both within a given sample and between adjacent samples.

Outlook

Liquid adsorption processes hold a prominent position in several applications for the production of high purity chemicals on a commodity scale. Many of these processes were attractive when they were first introduced to the industry and continue to increase in value as improvements in adsorbents, desorbents, and process designs are made. The UOP Parex process alone has seen four generations of adsorbent and four generations of desorbent.

The value of many chemical products from pesticides to pharmaceuticals to high performance polymers, is based on unique properties of a particular isomer from which the product is ultimately derived. Often the purity requirement for the desired product includes an upper limit on the content of one or more of the other isomers. This separation problem is a complicated one, but one in which adsorptive separation processes offer the greatest chances for success.

STANLEY A. GEMBICKI
ANIL R. OROSKAR
JAMES A. JOHNSON
UOP

C. L. Mantell, *Adsorption,* 2nd ed., McGraw-Hill, Inc., New York, 1951.

D. B. Broughton, *Chem. Eng. Prog.* **64,** 60 (1968).

D. W. Breck, *Zeolite Molecular Sieves,* John Wiley & Sons, Inc., New York, 1974.

D. M. Ruthven, *Principles of Adsorption and Adsorption Processes,* John Wiley & Sons, Inc., New York, 1984.

ADVANCED CERAMICS

ELECTRONIC CERAMICS

Electronic ceramics is a generic term describing a class of inorganic, nonmetallic materials utilized in the electronics industry. Although the term electronic ceramics, or electroceramics, includes amorphous glasses and single crystals, it generally pertains to polycrystalline inorganic solids comprised of randomly oriented crystallites (grains) intimately bonded together. This random orientation of small, micrometer-size crystals results in an isotropic ceramic possessing equivalent properties in all directions. The isotropic character can be modified during the sintering operation at high temperatures or upon cooling to room temperature by processing techniques such as hot pressing or poling in an electric or magnetic field (see CERAMICS AS ELECTRICAL MATERIALS).

The properties of electroceramics are related to their ceramic microstructure, ie, the grain size and shape, grain–grain orientation, and grain boundaries, as well as to the crystal structure, domain configuration, and electronic and defect structures. Electronic ceramics are often combined with metals and polymers to meet the requirements of a broad spectrum of high technology applications, computers, telecommunications, sensors (qv), and actuators. Roughly speaking, the multibillion dollar electronic ceramics market can be divided into six equal parts as shown in Figure 1. In addition to SiO_2-based optical fibers and displays, electronic ceramics encompass a wide range of materials and crystal structure families used as insulators, capacitors, piezoelectrics, magnetics, semiconductor sensors, conductors, and the recently discovered high temperature superconductors. Currently the growth of the electronic ceramic industry is driven by the need for large-scale integrated circuitry giving rise to new developments in materials and processes. The development of multilayer packages for the microelectronics industry, composed of multifunctional three-dimensional ceramic arrays called monolithic ceramics (MMC), continues the miniaturization process begun several decades ago to provide a new generation of robust, inexpensive products.

Electroceramic Processing

Fabrication technologies for all electronic ceramic materials have the same basic process steps, regardless of the application: powder preparation, powder processing, green forming, and densification.

Processing of Multilayer Ceramics

Rapid advances in integrated circuit technology have led to improved processing and manufacturing of multilayer ceramics (MLC) especially for capacitors and microelectronic packages. The increased reliability has been the result of an enormous amount of research aimed at understanding the various microstructural–property relationships involved in the overall MLC manufacturing process. This includes powder processing, thin sheet formation, metallurgical interactions, and testing.

Presently, multilayer capacitors and packaging make up more than half the electronic ceramics market. For multilayer capacitors, more than 20 billion units are manufactured a year, outnumbering by far any other electronic ceramic component. Multilayer ceramics and hybrid packages consist of alternating layers of dielectric and metal electrodes, as shown in Figure 2.

The driving force for these compact configurations is miniaturization. For capacitors, the capacitance (C) measured in units of farads,

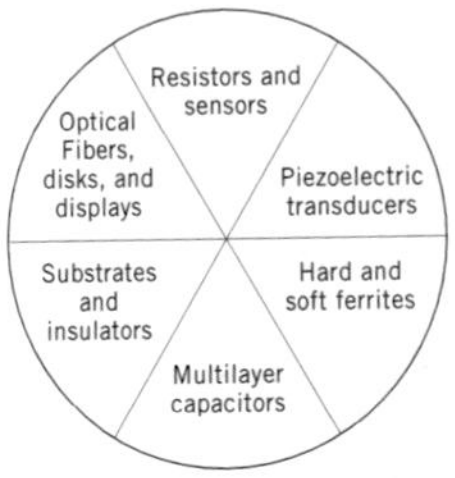

Figure 1. Electronic ceramics market.

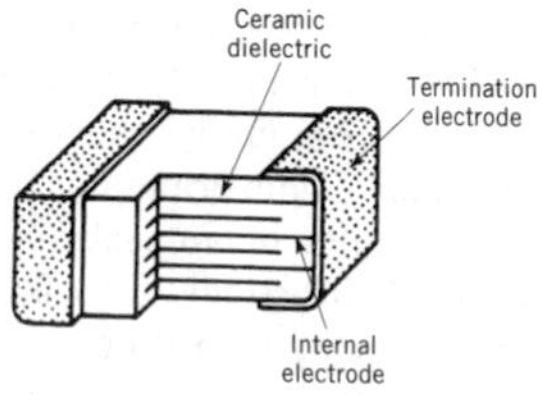

Figure 2. Schematic cross-section of a conventional MLC capacitor.

F, is

$$C = \frac{\epsilon_0 AK}{t}$$

where K is the dielectric constant (unitless); ϵ_0 the permittivity of free space = 8.85×10^{-12} F/m; A the electrode area, m^2; and t the thickness of dielectric layer, m. Thus C increases with increasing area and number of layers and decreasing thickness.

A number of processing steps, shown in Figure 3, are used to obtain the multilayer configuration(s) for the ceramic–metal composites.

Thick-Film Technology

Equally important as tape casting in the fabrication of multilayer ceramics is thick-film processing. Thick-film technology is widely used in microelectronics for resistor networks, hybrid integrated circuitry, and discrete components, such as capacitors and inductors along with metallization of MLC capacitors and packages as mentioned above.

In principle, the process is equivalent to the silk-screening technique whereby the printable components, paste or inks, are forced through a screen with a rubber or plastic squeegee (see Fig. 3). Generally, stainless steel or nylon filament screens are masked using a polymeric material forming the desired printed pattern in which the composition is forced through to the underlying substrate.

Thick film compositions possess three parts: (*1*) functional phase, (*2*) binder, and (*3*) vehicle. The functional phase includes various metal powders for conductors, electronic ceramics for resistors, and dielectrics for both capacitors and insulation. The binder phase, usually a low ($<1000°C$) melting glass adheres the fired film to the substrate whereas the fluid vehicle serves to temporarily hold the unfired film together and provide proper rheological behavior during screen printing. Thick film processing for hybrid integrated circuits typically takes place below 1000°C, providing flexible circuit designs.

Current and Future Developments in Multilayer Electronic Ceramics

Advances in the field of electronic ceramics are being made in new materials, novel powder synthesis methods, and in ceramic integration. Monolithic multicomponent components (MMC) take advantage of three existing technologies: (*1*) thick film methods and materials, (*2*) MLC capacitor processes, and (*3*) the concept of cofired packages.

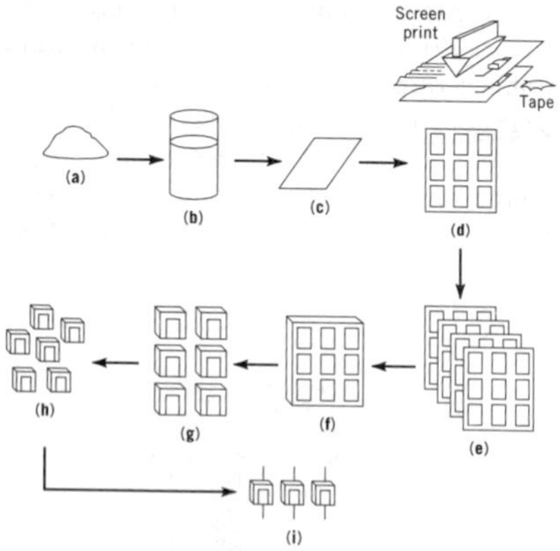

Figure 3. Fabrication process for MLC capacitors. Steps are (**a**) powder; (**b**) slurry preparation; (**c**) tape preparation; (**d**) electroding; (**e**) stacking; (**f**) lamination; (**g**) dicing; (**h**) burnout and firing; and (**i**) termination and lead attachment.

New materials for packaging include aluminum nitride, AlN, silicon carbide, SiC, and low thermal expansion glass-ceramics, replacing present-day alumina packaging technology. Properties include higher thermal conductivity, lower dielectric constant, cofire compatibility, and related packaging characteristics such as thermal expansion matching of silicon and high mechanical strength as compared to Al_2O_3.

Greater dimensional control and thinner tapes in multilayer ceramics are the driving forces for techniques to prepare finer particles. Metal organic decomposition and hydrothermal processing are two synthesis methods that have the potential to produce submicrometer powders having low levels of agglomeration to meet the demand for more precise tape fabrication.

The continuing miniaturization of electronic packaging should see the replacement of components and processes using such thin-film technologies developed for semiconductors as sputtering, chemical vapor deposition, and sol–gel.

ROBERT E. NEWNHAM
THOMAS R. SHROUT
Pennsylvania State University

J. M. Herbert, *Ceramic Dielectrics and Capacitors,* Gordon and Breach Science Publishers, New York, 1985.

R. R. Tummala and E. J. Rymaszewski, *Microelectronics Packaging Handbook,* Van Nostrand Reinhold Co., Inc., New York, 1989.

R. C. Buchanan, ed., *Ceramic Materials for Electronics,* Marcel Dekker, Inc., New York, 1986.

L. M. Levinson, ed., *Electronic Ceramics,* Marcel Dekker, Inc., New York, 1988.

STRUCTURAL CERAMICS

Advanced structural ceramics are those ceramics intended for use as load-bearing members. They are materials that combine the properties and advantages of traditional ceramics (qv), such as chemical inertness, high temperature capability, and hardness, with the ability to carry a significant mechanical stress. Like all ceramics, they are inorganic and nonmetallic; in addition, they are often multicomponent or multiphased materials having complex crystal structures. These materials are usually intended to be fully dense and to have tight dimensional tolerances. In addition to being designed to withstand substantially higher levels of mechanical and thermal stress, there are other important features which make advanced structural ceramics different from traditional ones. Starting powders, compositions, processing, and resulting microstructure must be carefully controlled to provide required levels of performance. Consequently, advanced structural ceramics are more expensive than traditional ceramics.

Most of the advanced structural ceramics under development today are based on silicon nitride, Si_3N_4, (Table 1); silicon carbide, SiC (Table 2); zirconia, ZrO_2 (Table 3); or alumina, Al_2O_3. In addition, materials such as titanium diboride, TiB_2; aluminum nitride, AlN; silicon aluminum oxynitride, SiAlON; and some other ceramic carbides and nitrides are often classified as advanced or high-tech ceramics because of processing methods or applications. Ceramic matrix composites are also receiving increasing attention as advanced structural ceramics (see COMPOSITE MATERIALS, CERAMIC-MATRIX).

Physical Properties

Advanced structural ceramics typically possess some combination of high temperature capabilities, high strength, toughness or flaw tolerance, high hardness, mechanical strength retention at high temperatures, wear resistance, corrosion resistance, thermal shock resistance, creep resistance, and long-term durability.

Table 1. Properties of Silicon Nitride Ceramics

Property	Material densification mode			
	Reaction-bonded	Sintered[a]	Hot-pressed isostatically[b]	Hot-pressed[c]
density, kg/m^3	2.5	3.26	3.23	3.2
elastic modulus, GPa[d]	180	300	310	310
hardness, kg/mm^2	1350	1370	1620	1800
flexure strength, MPa[e] at ambient temperature	340	700	900	700
1000°C		600		610
1200°C		480		570
1370°C		210	580	310
fracture toughness, MPa$\sqrt{m}$	3–4	4.6	4.7–5.5	4.9
thermal expansion coeff., 25–1000°C, 10^{-6}/°C	3	3.9	3.9	3.5
thermal conductivity, W/(m·K) at 25°C	12	32	38	32

[a] GTE Laboratories AY-6, 6 wt % yttria + 2 wt % alumina.
[b] Norton NT154, 4 wt % yttria.
[c] Norton NC 132, 1 wt % magnesia.
[d] To convert GPa to psi, multiply by 145 × 10^3.
[e] To convert MPa to psi, multiply by 145.

Table 2. Properties of Silicon Carbide Ceramics

Property	Material densification mode				
	Reaction-bonded	Sintered alpha[a]	Sintered beta	Hot-pressed (Al_2O_3)[b]	Sintered (Y_2O_3)[c]
density, kg/m^3	3.1	3.1	3.0	3.3	3.2
hardness, kg/mm^2	1620	2800		2400	
flexure strength, MPa[d] at 25°C	245	460	490	702	917
Young's modulus, GPa[e]	383	410	372	446	
Poisson's ratio, GPa[a]	0.24	0.14	0.16	0.17	
thermal expansion coeff. ×10^{-6}/°C	4.8	4.02	4.4	4.6	
thermal conductivity, W/(m·K) at 25°C	135	126	71	80	

[a] Material subjected to a post-sintering HIP treatment; sinter/HIP alpha SiC has a density of 3.2 g/mL and a flexural strength of 530 MPa at 25°C.
[b] Contains about 2% alumina.
[c] Contains some yttria; see text.
[d] To convert MPa to psi, multiply by 145.
[e] To convert GPa to psi, multiply by 145,000.

Table 3. Properties of Zirconia Ceramics

Property	PSZ[a]	TZP[b]	ZTA[c]
density, kg/m^3	5.7	6.0	4.2
hardness, kg/mm^2	1000	1300	1600
flexural strength, MPa[d]	300–700	1000–2500	400–900
fracture toughness, MPa$\sqrt{m}$	4–8	5–15	5–10
elastic modulus, GPa[e]	200	200	340
thermal expansion coeff. ×10^{-6}/K	9–10	10–11	8–9
thermal conductivity,W/(m·K)	2.0–2.5	2.7	7–10
maximum service temperature, °C	950	500	1700

[a] Properties of PSZ depend on whether CaO, MgO, or Y_2O_3 is used as the stabilizing agent.
[b] Properties of TZP depend on whether CeO_2 or Y_2O_3 is used as the stabilizing agent.
[c] Properties of zirconia toughened alumina, ZTA, depend on the specific microstructure and the proportions of zirconia and alumina.
[d] To convert MPa to psi, multiply by 145.
[e] To convert GPa to psi, multiply by 145 × 10^3.

Processing and Fabrication Technology

The relationship between processing and properties is especially critical for advanced structural ceramics because subsequent successful operation in severe environments often requires carefully controlled compositions and microstructures. Fabrication generally takes place in four steps: powder processing, consolidation/forming, densification, and finishing.

Applications

Throughout the development of structural ceramics the focus has been on applications for gas turbine, diesel, and spark-ignited engines. Advanced structural ceramics are also under investigation for use in numerous other high performance applications including antifriction roller and ball bearings, metal-cutting and shaping tools, hot extrusion and hot forging dies, industrial wear parts (eg, sand-blast nozzles, pump seals, thread guides, chute liners), and various military applications (eg, armor, radomes, ir domes, gun barrel liners). More comprehensive works on the processing, properties, and applications of advanced ceramics are available.

Environmental Aspects

Exposure limits for silicon carbide and powders of zirconium compounds (including zirconium dioxide) have been established by ACGIH. TLV–TWAs are 10 mg/m^3 and 5 mg/m^3, respectively. OSHA guidelines for zirconium compounds call for a PEL of 5 mg/m^3. There are no exposure limits for silicon nitride powder, but prudent practice suggests a TLV–TWA of 0.1 mg/m^3. The solid ceramics present no apparent health hazard.

Economic Aspects

It is projected that in the year 2000 the market for these ceramic components will be in the $2.7–4.2 billion range. Structural ceramics producers are given in Table 4.

Table 4. Producers of Structural Ceramics

Location, company	Ceramic type
Australia	
Nilcra	ZrO_2
Europe	
ASEA	Si_3N_4
ESK	SiC, Si_3N_4
Feldmühle[a]	ZrO_2
Japan	
Kyocera Corp.[a]	SiC, Si_3N_4, ZrO_2
NGK Insulators	SiC, Si_3N_4
NTK Technical Ceramics	Si_3N_4
United States	
Carborundum Co.	SiC
Ceramatec	ZrO_2
Coors[b]	ZrO_2
Corning	ZrO_2
Dow Chemical	ZrO_2
WR Grace	ZrO_2
GTE	Si_3N_4
Norton Co.	SiC, Si_3N_4, ZrO_2
Zircoa[b]	ZrO_2

[a] Kycocera Corp. licenses Feldmühle and Max Planck Institute zirconia technology.
[b] Largest U.S. suppliers of ZrO_2 ceramics.

MARINA R. PASCUCCI
JEFFREY T. NEIL
SAMUEL NATANSOHN
GTE Laboratories Incorporated

W. Bunk and H. Hausner, eds., *Proc. of 2nd Int. Symp. on Ceramic Materials and Components for Engines,* Verlag Deutsche Keramische Gesellschaft, Bad Honeff, Germany, 1986.

P. F. Becher, M. V. Swain, and S. Somiya, eds., *Advanced Structural Ceramics, Materials Research Society Symp. Proc.,* Vol. 78, Materials Research Society, Pittsburgh, Pa., 1987.

J. B. Wachtman, Jr., ed., *Structural Ceramics: Treatise on Materials Science and Technology,* Vol. 29, Academic Press, New York, 1989.

G. L. Leatherman and R. N. Katz, *Superalloys, Supercomposites and Superceramics,* Academic Press, New York, 1989.

AERATION

BIOTECHNOLOGY

The supply of oxygen to a growing biological species, aeration, in aerobic bioreactors is one of the most critical requirements in biotechnology. It was one of the biggest hurdles that had to be overcome in designing bioreactors (fermenters) capable of turning penicillin from a scientific curiosity to the first major antibiotic. Aeration is usually accomplished by transferring oxygen from the air into the fluid surrounding the biological species, from where it is in turn transferred to the biological species itself. The rate at which oxygen is demanded by the biological species in a bioreactor depends very significantly on the species, on its concentration, and on the concentration of the other nutrients in the surrounding fluid (see CELL CULTURE TECHNOLOGY). There is no unique set of units used to define this rate requirement, but some typical figures are given in Table 1. The very wide range is noteworthy; during the course of a batch bioreaction, oxygen demand often passes through a marked maximum when the species is most biologically active.

The main reason for the importance of aeration lies in the limited solubility of oxygen in water, a value which decreases in the presence of electrolytes and other solutes and as temperature increases.

Table 1. Oxygen Demands of Biological Species

Biological species	kg $O_2/(m^3 \cdot h)$
bacteria/yeasts	1 to 7
plant cells	0.03 to 0.3
seed priming[a]	1 to 8 $\times 10^{-2}$
mammalian cells[b]	2 to 10 $\times 10^{-3}$

[a] Based on a seed density of 100 kg/m^3.
[b] Based on a cell density of 10^{12} cells/m^3.

In addition to each bioreaction demanding oxygen at a different rate, there is a unique relationship for each between the rate of reaction and the level of dissolved oxygen. A typical generalized relationship is shown in Figure 1 for a particular species, eg, *Penicillium chrysogenum* or yeast. The shape of the curve is such that a critical oxygen concentration, C_{crit}, can be defined above which the rate of the bioreaction is independent of oxygen concentration.

Principles of Oxygen Transfer

The Basic Mass Transfer Steps. The steps through which oxygen must pass in moving from air (or oxygen-enriched air) to the reaction site in a biological species consist of transport through the gas film inside the bubble, across the bubble–liquid interface, through the liquid film around the bubble, across the well-mixed bulk liquid (broth), through the liquid film around the biological species, and finally transport within the species (eg, cell, seed, microbial floc) to the bioreaction site.

The Basic Air–Broth Mass-Transfer Relationship. The basic principles which underlie oxygenation (aeration) are exactly the same as those which determine the rate of transfer of any sparingly soluble gas (oxygen) from the gas stream (air) to the unsaturated liquid (broth). The rate at which this transfer takes place is dependent on four principal parameters. The first is the area of contact between the gas and the liquid. The other three mass-transfer rate parameters are the driving force available (ie, the difference in concentration of oxygen in the two phases); the two-phase fluid dynamics (including the effect of viscosity); and the chemical composition of the liquid.

Aeration in Bioreactors

A huge variety of bioreactors has been developed and a thorough review is available. A useful subdivision has been made into three generic types involving the way in which air is dispersed to give the desired specific surface area. These are bioreactors driven by rotating agitators (stirred tanks), bioreactors driven by gas compression (bubble columns/loop fermenters), and bioreactors driven by circulating liquid (jet loop reactors). The first two are the most important.

Applications to Different Biological Species

Aeration is used in mycelial fermentation, xanthan gum fermentation, high oxygen demanding fermentation, mammalian cell culture, plant cell culture, seed priming bioreactors, single-cell protein, and biological aerobic wastewater treatment.

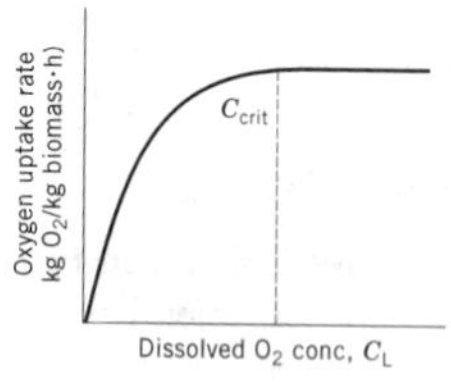

Figure 1. The relationship between rate of oxygen uptake and dissolved oxygen, concentration where C_{crit} is the critical oxygen concentration.

ALVIN W. NIENOW
University of Birmingham (U.K.)

J. E. Bailey and D. F. Ollis, *Biochemical Engineering Fundamentals,* 2nd ed., McGraw-Hill Book Co., New York, 1986.

A. W. Nienow, *Trends Biotechnol.* **8,** 224 (1990).

M. Y. Chisti, *Airlift Bioreactors,* Elsevier Applied Science, New York, 1989.

N. Kioukia, A. W. Nienow, A. N. Emery, and M. Al-Rubeai, *Food Bioproduct Process. (Trans. Inst. Chem. Eng., Pt. C)* **70,** 143 (1992).

WATER TREATMENT

Aeration for water treatment, the transfer of oxygen, O_2, from air to water, is well-studied. The basic purpose of aeration, which is used primarily for the treatment of wastewater, is to improve water quality for subsequent usage (see WATER). Aeration can bring about the physical removal of taste- and odor-producing substances such as hydrogen sulfide, H_2S, and other volatiles as well as the chemical removal of metals (iron, manganese), gases (hydrogen sulfide), and other compounds (organics and inorganics) through oxidation and settling. Additionally, aeration is used extensively for the biological oxidation of both domestic and industrial organic wastes.

The function of aeration in a wastewater treatment system is to maintain an aerobic condition.

Oxygen Solubility

The solubility of a gas in water is affected by temperature, total pressure, the presence of other dissolved materials, and the molecular nature of the gas. Oxygen solubility is inversely proportional to the water temperature and, at a given temperature, directly proportional to the partial pressure of the oxygen in contact with the water.

Diffusion

The driving force in diffusion involves differences in the concentration of the diffusing substance. The molecular diffusion of a gas into a liquid is dependent on the characteristics of the gas and the liquid, the temperature of the liquid, the concentration deficit, the gas to liquid contact area, and the period of contact.

One important factor affecting oxygen transfer from air to water is the resistance generated by the film at the air–water interface (Fig. 1). The interface exhibits properties that are strikingly different from those of the main body of water or air. The degree of resistance depends on the presence of surface active materials at the interface. Minute amounts of materials such as soap, detergent, or organic acids are capable of causing considerable additional resistance to gas–liquid transfer.

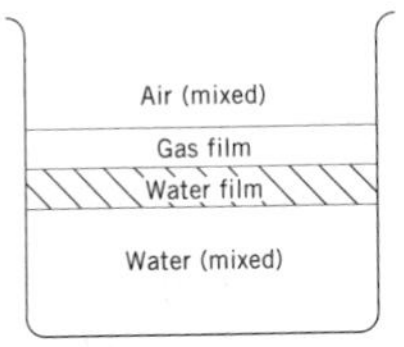

Figure 1. Schematic representation of the two films at the air–water interface.

Methods for Determining Oxygen Demand

Several methods have been developed to estimate the oxygen demand in wastewater treatment systems. Commonly used laboratory methods are biochemical oxygen demand (BOD), chemical oxygen demand (COD), total oxygen demand (TOD), total organic carbon (TOC), and theoretical oxygen demand (ThOD).

Aerating Systems

Aerators designed to facilitate the transfer of oxygen from air to water increase interfacial area by producing liquid turbulence and circulation. There are four basic types of aerators, summarized in Table 1.

Oxygen Requirements

The oxygen requirements of wastewater depend on the treatment system utilized.

Oxygen Transfer Rate

Oxygen transfer rate (OTR) is estimated by the following standard procedure, where the rate of mass transfer per unit volume of liquid is taken to be directly proportional to the driving force of the system

$$\text{OTR} = K_{\text{L}}aV(C_{\infty}^{*} - C)$$

where OTR is the oxygen transfer occuring in the liquid volume V; $K_{\text{L}}a$ is the overall volumetric transfer coefficient in clean water based on the liquid film resistance; C_{∞}^{*} is the dissolved oxygen concentration at saturation, approached at infinite aeration time; and C is the dissolved oxygen concentration (DO).

Table 1. Aerator Specifications

Aerator type	Materials of construction	Oxygen transfer efficiency, OTE, %	Oxygen transfer rate, OTR, g/(W·h)
diffused aerator			
coarse bubble	noncorrosive metal or plastic	4–20	0.73–1.09
fine bubble	ceramic, cast iron, or plastic	10–30	1.09–1.52
static tube	noncorrosive metal or plastic	7–20	1.09–1.58
jet aerator	noncorrosive metal	10–25	1.52–2.13
mechanical aerator			
low speed surface	large diameter turbine having gear reducers (fixed bridge or platform mounted)		1.22–2.74
high speed surface	noncorrosive metal small diameter propeller having direct drive		1.22–1.52
rotary brush and disk	noncorrosive metal gear reducers or direct drive		1.52–2.11
horizontally mixing aspirating aerator	noncorrosive metal having hollow or solid shaft, small marine propeller at high speed, direct drive		0.61–1.82

The oxygen transfer rate for aerators is normally reported at standard conditions. Thus, in order to make meaningful comparisons, the ORT under working or field conditions should be adjusted to standard conditions oxygen requirement for treatment (SORT) by means of

$$\text{SORT} = \text{ORT} \div \alpha\left[\frac{\beta C_{\text{walt}} - C_{\text{L}}}{9.09}\right]\theta^{(T-20)}$$

where β is the salinity–surface tension correction factor = (wastewater C^*_∞)/(clean water C^*); C_{walt} is the oxygen saturation concentration for wastewater at a given temperature and altitude (mg/L); C_{L} is the residual oxygen concentration (mg/L); T is the operating temperature of the wastewater (C); α is the oxygen transfer correction factor = (wastewater $K_{\text{L}}a$)/(clean water $K_{\text{L}}a$).

RICHARD RAJENDEN
Aeromix Systems, Inc.

W. K. Lewis and W. G. Whitman, *Ind. Eng. Chem.* **16,** 1215–1220 (1924).

W. F. Langelier, *J. Amer. Water Works Ass.* **24,** 62–72 (1932).

W. W. Eckenfelder, Jr., and E. L. Barnhart, *AIChE J.* **7,** 631–634 (1961).

L. C. Brown and C. R. Baillod, *M. ASCE Modeling and Interpreting Oxygen Transfer Data* **108,** EE4, 17240, 607–627 (1982).

AEROGELS

Aerogels are solid materials that are so porous that they contain mostly air. Almost all applications of aerogels are based on the unique properties associated with a highly porous network. Envision an aerogel as a sponge consisting of many interconnecting particles which are so small and so loosely connected that the void space in the sponge, the pores, can make up over 90% of its volume. The ability to prepare materials of such low density, and perhaps more importantly, to vary the density in a controlled manner, is indeed what make aerogels attractive in many applications.

Sol–Gel Chemistry

Inorganic Materials. Sol–gel chemistry involves first the formation of a sol, which is a suspension of solid particles in a liquid, then of a gel, which is a diphasic material with a solid encapsulating a solvent. A detailed description of the fundamental chemistry is available in the literature. The chemistry involving the most commonly used precursors, the alkoxides ($M(OR)_m$), can be described in terms of two classes of reactions:

Hydrolysis $\quad -M-OR + H_2O \rightarrow -M-OH + ROH$

Condensation $\quad -M-OH + XO-M- \rightarrow -M-O-M + XOH$

where X can either be H or R, an alkyl group

The important feature is that a three-dimensional gel network comes from the condensation of partially hydrolyzed species. Thus, the microstructure of a gel is governed by the rate of particle (cluster) growth and their extent of crosslinking or, more specifically, by the *relative* rates of hydrolysis and condensation.

Acid- and base-catalyzed gels yield micro- (pore width less than 2 nm) and meso-porous (2–50 nm) materials, respectively, upon heating. An acid-catalyzed gel which is weakly branched and contains surface functionalities that promote further condensation collapses to give micropores. This example highlights a crucial point: *the initial microstructure and surface functionality of a gel dictates the properties of the heat-treated product.*

Besides pH, other preparative variables that can affect the microstructure of a gel, and consequently, the properties of the dried and heat-treated product include water content, solvent, precursor type and concentration, and temperature.

In the preparation of a two-component system, the minor component can either be a network modifier or a network former. In the latter case, the distribution of the two components, or mixing, at a molecular level is governed by the *relative* precursor reactivity. Qualitatively good mixing is achieved when two precursors have similar reactivities. When two precursors have dissimilar reactivities, the sol–gel technique offers several strategies to prepare well-mixed two-component gels. Two such strategies are prehydrolysis, which involves prereacting a less reactive precursor and chemical modification, which involves slowing down a more reactive precursor. The ability to control microstructure *and* component mixing is what sets sol–gel apart from other methods in preparing multicomponent solids.

Organic Materials. The sol–gel chemistry of organic materials is similar to that of inorganic materials. The first organic aerogel was prepared by the aqueous polycondensation of resorcinol with formaldehyde using sodium carbonate as a base catalyst.

Resorcinol–formaldehyde gels are dark red in color and do not transmit light. The preparation of melamine–formaldehyde gels, which are colorless and transparent, is also aqueous-based. Since water is deleterious to a gel's structure at high temperatures and immiscible with carbon dioxide (a commonly used supercritical drying agent), these gels cannot be supercritically dried without a tedious solvent-exchange step. In order to circumvent this problem, an alternative synthetic route of organic gels that is based upon a phenolic–furfural reaction using an acid catalyst has been developed. The solvent-exchange step is eliminated by using alcohol as a solvent. The phenolic–furfural gels are dark brown in color.

Carbon aerogels can be prepared from the organic gels mentioned above by supercritical drying with carbon dioxide and a subsequent heat-treating step in an inert atmosphere.

Despite these changes, the carbon aerogels are similar in morphology to their organic precursors, underscoring again the importance of structural control in the gelation step. Furthermore, changing the sol–gel conditions can lead to aerogels that have a wide range of physical properties.

Inorganic–Organic Hybrids. One of the fastest growing areas in sol–gel processing is the preparation of materials containing both inorganic and organic components, because many applications demand special properties that only a combination of inorganic and organic materials can provide. In this regard, sol–gel chemistry offers a real advantage because its mild preparation conditions do not degrade organic polymers, as would the high temperatures that are associated with conventional ceramic processing techniques. The voluminous literature on the sol–gel preparation of inorganic–organic hybrids can be found in several recent reviews and the references therein.

Preparation and Manufacturing

Supercritical Drying. The development of aerogel technology from the original work of Kistler to about late 1980s has been reviewed. Over this period, supercritical drying was the dominant method in preparing aerogels. Several advances, summarized in Table 1, have made possible the relatively safe supercritical drying of aerogels in a matter of hours. In recent years, the challenge has been to produce aerogel-like materials without using supercritical drying at all in an attempt to deliver economically competitive products.

Supercritical drying should be considered as part of the aging process, during which events such as condensation, dissolution, and reprecipitation can occur. The extent to which a gel undergoes aging during supercritical drying depends on the structure of the initial gel network. A higher drying temperature changes the particle structure of base-catalyzed silica aerogels but not that of acid-catalyzed ones. Gels that have uniform-sized pores can withstand the capillary forces during drying better because of a more uniform stress distribution. Such gels can be prepared by a careful manipulation of sol–gel parameters such as pH and solvent or by the use of so-called drying control chemical additives (DCCA).

Table 1. Important Developments in the Preparation of Aerogels

Decade	Developments
1930	Using inorganic salts as precursors, alcohol as the supercritical drying agent, and a batch process; a solvent-exchange step was necessary to remove water from the gel.
1960	Using alkoxides as precursors, alcohol as the supercritical drying agent, and a batch process; the solvent exchange step was eliminated.
1980	Using alkoxides as precursors, carbon dioxide as the drying agent, and a semicontinuous process; the drying procedure became safer and faster. Introduction of organic aerogels.
1990	Producing aerogel-like materials without supercritical drying at all; preparation of inorganic–organic hybrid materials.

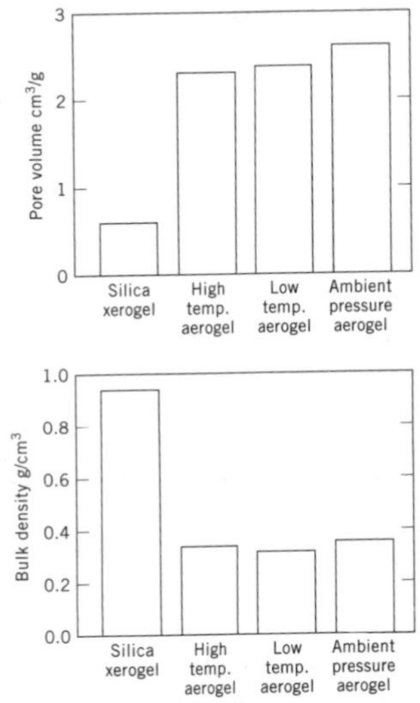

Figure 1. Comparison of physical properties of silica xerogels and aerogels. Note the similar properties of the aerogels prepared with and without supercritical drying. Reproduced from C. J. Brinker and co-workers, *Mat. Res. Soc. Symp. Proc.* **271,** 567 (1992). Courtesy of the Materials Research Society.

Carbon dioxide is the drying agent of choice if the goal is to stabilize kinetically constrained structure, and materials prepared by this low-temperature route are referred to by some people as *carbogels.* In general, carbogels are also different from aerogels in surface functionality, in particular hydrophilicity.

However, even with carbon dioxide as a drying agent, the supercritical drying conditions can affect the properties of a product. Other important drying variables include the path to the critical point, composition of the drying medium, and depressurization.

For some applications it is desirable to prepare aerogels as thin films that are either self-supporting or supported on another substrate. All common coating methods such as dip coating, spin coating, and spray coating can be used to prepare gel films.

In all the processes discussed above, the gelation and supercritical drying steps are done sequentially. Recently a process that involves the direct injection of the precursor into a strong mold body followed by rapid heating for gelation and supercritical drying to take place was reported. By eliminating the need of forming a gel first, this entire process can be done in less than three hours per cycle. Besides saving time, gel containment minimizes some stresses and makes it possible to produce near net-shape aerogels and precision surfaces. The optical and thermal properties of silica aerogels thus prepared are comparable to those prepared with conventional methods.

Ambient Preparations. Economic and safety considerations have provided a strong motivation for the development of techniques that can produce aerogel-like materials at ambient conditions, ie, without supercritical drying. The strategy is to minimize the deleterious effect of capillary pressure which is given by:

$$P = 2\sigma \cos(\theta)/r$$

where P is capillary pressure, σ is surface tension, θ is the contact angle between liquid and solid, and r is pore radius.

The equation above suggests that one approach would be to use a pore liquid that has a low surface tension. In fact, with a pore liquid that has a sufficiently small surface tension, ambient pressure acid catalyzed aerogels with comparable pore volume and with bulk density to those prepared with supercritical drying (see Fig. 1) have been produced.

For base-catalyzed silica gels, it has been shown that modifying the surface functionality is an effective way to minimize drying shrinkage. In particular, surface hydroxyl groups, the condensation of which leads to pore collapse, can be "capped off" via reactions with organic groups such as tetraethoxysilane and trimethylchlorosilane. This surface modification approach (also referred to as surface derivatization), initially developed for bulk specimens, has recently been applied to the preparation of thin films.

In changing surface hydroxyls into organosilicon groups, surface modification has an additional advantage of producing hydrophobic gels. This feature, namely the immiscibility of surface-modified gel with water, has led to the development of a rapid extractive drying process shown in Figure 2. This ambient pressure process offers improved heat transfer rates and, in turn, greater energy efficiency without compromising desirable aerogel properties.

Another approach to produce aerogels without supercritical drying is freeze drying, in which the liquid–vapor interface is eliminated by freezing a wet gel into a solid and then subliming the solvent to form what is known as a *cryogel.* The limited data available on freeze drying suggest that it might not be as attractive as the above ambient approaches in producing aerogels on a commercial scale.

Properties

Table 2 summarizes the key physical properties of silica aerogels. A range of values is given for each property because the exact value is dependent on the preparative conditions and, in particular, on density.

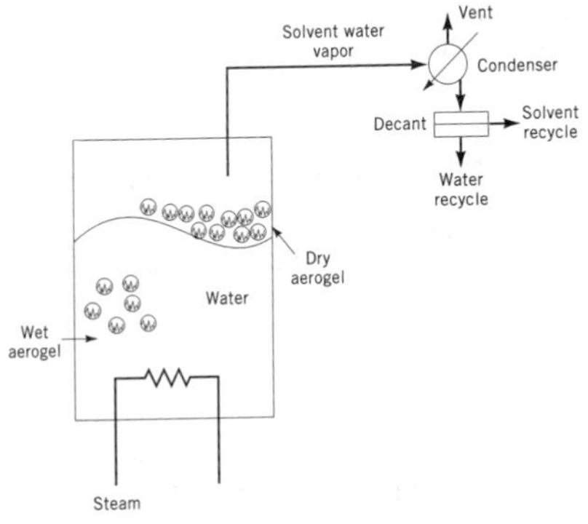

Figure 2. Schematic diagram of an extractive drying process that produces aerogels at ambient pressure. Reproduced from D. M. Smith and co-workers, *Mat. Res. Soc. Symp. Proc.* **431,** 291 (1996). Courtesy of the Materials Research Society.

Table 2. Typical Values of Physical Properties of Silica Aerogels

Property	Values
density, kg/m^3	3–500
surface area, m^2/g	800–1000
pore sizes, nm	1–100
pore volume, cm^3/g	3–9
porosity, %	75–99.9
thermal conductivity, W/(m·K)	0.01–0.02
longitudinal sound velocity, m/s	100–300
acoustic impedance, $kg/(m^2 \cdot s)$	10^3–10^6
dielectric constant	1–2
Young's modulus, N/m^2	10^6–10^7

Applications

Aerogels are used in thermal insulation, catalysis, detection of high energy particles, piezoceramic, ultrasound transducers, integrated circuits, and as dehydrating agents.

Summary

Aerogels have the potential of being marketable both as a commodity chemical (eg, in thermal insulation) and as a specialty chemical (eg, in electronic applications) because of their unique and tailorable properties. The next few years will be critical in assessing whether aerogels can penetrate and grow in either end of the market, as the field is changing rapidly with the development of cost-competitive technologies and novel applications.

EDMUND I. KO
Carnegie Mellon University

C. J. Brinker and G. W. Scherer, *Sol-Gel Science: The Physics and Chemistry of Sol-Gel Processing,* Academic Press, New York, 1990.

J. Livage, M. Henry, and C. Sanchez, *Prog. Solid State Chem.* **18,** 259 (1988).

M. Schneider and A. Baiker, *Catal. Rev. - Sci. Eng.* **37**(4), 515 (1995).

J. Fricke, *Sci. Amer.* **256**(5), 92 (1988).

AEROSOLS

The aerosol container has enjoyed commercial success in a wide variety of product categories. Insecticide aerosols were introduced in the late 1940s. Additional commodities, including shave foams, hair sprays, antiperspirants, deodorants, paints, spray starch, colognes, perfumes, whipped cream, and automotive products, followed in the 1950s. Medicinal metered-dose aerosol products have also been developed for use in the treatment of asthma, migraine headaches, and angina.

Personal products are the fastest growing segment of the aerosol industry and represent the largest of the categories.

Advantages of Aerosol Packaging

Aerosol products are hermetically sealed, ensuring that the contents cannot leak, spill, or be contaminated. The packages can be considered to be tamper-proof. They deliver the product in an efficient manner, generating little waste, often to sites of difficult access. By control

Table 1. Properties of Hydrofluorocarbon and Hydrochlorofluorocarbon Propellants

Property	Propellant 22	Propellant 142b	Propellant 152a
chemical name	chlorodifluoromethane	1-chloro-1,1-difluoroethane	1,1-difluoroethane
formula	$CHClF_2$	$C_2H_3ClF_2$	$C_2H_4F_2$
molecular weight	86.5	100.5	66.1
boiling point, °C	−40.8	−9.44	−23.0
vapor pressure, kPa[a]			
21°C	834	200	434
54°C	2048	669	1220
density, g/mL at 21°C	1.21	1.12	0.91
solubility in water, wt % at 21°C	3.0	0.5	1.7
Kauri-butanol value	25	20	11
flammability limit, vol % in air	nonflammable	6.3–14.8	3.9–16.9
flash point, °C	none	none	< -50

[a] To convert kPa to psi, multiply by 0.145.

Table 2. Properties of the Compressed Gas Propellants

Property	Carbon dioxide	Nitrous oxide	Nitrogen
formula	CO_2	N_2O	N_2
molecular weight	44.0	44.0	28.0
boiling point, °C	−78	−88	−196
critical temperature, °C	31	37	−111
vapor pressure, kPa[a] at 21°C	5772	4966	
solubility in water, vol gas/vol liq at 21°C and 101.3 kPa[a]	0.82	0.6	0.016
flammability limits, vol % in air	nonflammable	nonflammable	nonflammable
flash point, °C	none	none	none

[a] To convert kPa to psi, multiply by 0.145.

Table 3. Physical Properties of Chlorofluorocarbon and Hydrocarbon Propellants

Property	Propellant					
	11	12	114	A-108	A-31	A-17
chemical name	trichloromonofluoromethane	dichlorodifluoromethane	1,2-dichloro-1,1,2,2-tetrafluoroethane	propane	isobutane	*n*-butane
formula	CCl_3F	CCl_2F_2	$CClF_2CClF_2$	C_3H_8	$HC(CH_3)_3$	C_4H_{10}
molecular weight	137.4	120.9	170.9	44.1	58.1	58.1
boiling point, °C	23.8	−29.8	3.6	−42.2	−10.2	−0.6
vapor pressure, kPa[a]						
21°C	194	585	190	846	315	214
54°C	269	1349	507	1893	763	556
liquid density, g/mL, 21°C	1.485	1.325	1.468	0.5005[b]	0.5788[b]	0.5571[b]
flammability limit, vol % in air	nonflammable	nonflammable	nonflammable	2.3–7.3	1.8–8.4	1.6–6.5

[a] To convert kPa to psi, multiply by 0.145.
[b] At 68°C.

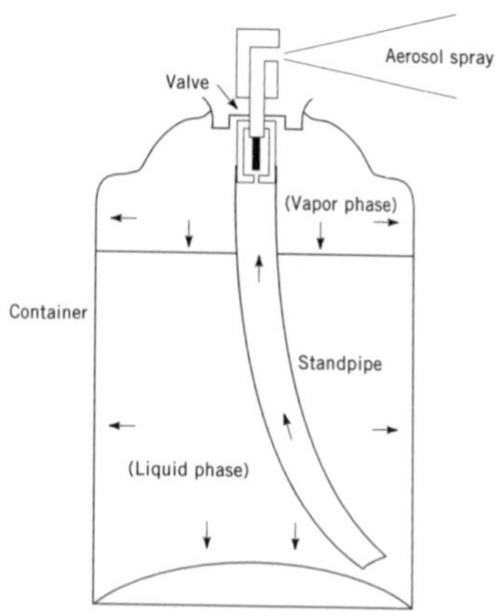

Figure 1. Solution-type aerosol system in which internal pressure is typically 240 kPa at 21°C. To convert kPa to psi, multiply by 0.145.

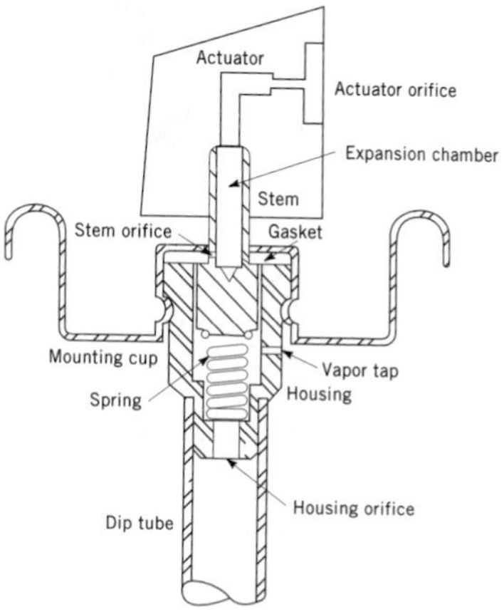

Figure 2. Aerosol valve components.

of particle size, spray pattern, and volume delivered per second, the product can be applied directly without contact by the user.

Formulation of Aerosols

All aerosols consist of product concentrate, propellant, container, and valve (including an actuator and dip tube). A typical aerosol system in shown in Figure 1.

Product Concentrate. An aerosol's product concentrate contains the active ingredient and any solvent, co-solvent, or dispersing agent necessary. Various propellent and valve systems, which must consider the solvency, density, and viscosity of the concentrate–propellent blend, may be used to deliver the product from the aerosol container. Systems can be formulated as solutions, emulsions, dispersions, creams, or ointments.

Propellants. The propellant, said to be the heart of an aerosol system, maintains a suitable pressure within the container and expels the product once the valve is opened. Propellants may be either a liquefied halocarbon, hydrocarbon, or halocarbon–hydrocarbon blend (Table 1), or a compressed gas such as carbon dioxide (qv), nitrogen (qv), or nitrous oxide (Table 2).

Worldwide, use of CFC propellants, designated as Propellants 11, 12, and 114, is strictly limited to specialized medicinal aerosol products such as metered-dose inhalers. The physical properties and chemical names of these propellents are given in Table 3. For all other products, compressed gas propellants, hydrocarbon propellants, hydrofluorocarbon, and hydrochlorofluorocarbon propellants are used.

Containers. Aerosol containers are manufactured from tin-plated steel, aluminum, and glass. The most popular aerosol container is the three-piece, tin-plated steel container.

Valves. The dispensing valve and actuator serve to close the opening through which the product and frequently the propellant enter the container, to retain the pressure within the container, and to dispense the product in the precise form and dosage intended by the manufacturer and expected by the consumer. An aerosol valve, shown in Figure 2, consists of seven components. Many variations exist both for special purposes and to avoid infringing existing patents.

Filling of Aerosols

All aerosols are produced by either a cold-filling or pressure-filling process.

Barrier Packs. Barrier packs utilize a plastic bag or piston to separate the product from the propellant.

JOHN J. SCIARRA
Sciarra Aeromed Development Corporation

Handbook of Pharmaceutical Excipients, American Pharmaceutical Association/The Pharmaceutical Society of Great Britain, Washington, D.C. and London, 1986, pp. 19, 99, 101, 145, 240, 333.

M. A. Johnsen, *The Aerosol Handbook,* 2nd ed., Wayne Dorland Company, Mendham, N.J., 1982.

P. A. Sanders, *Handbook of Aerosol Technology,* 2nd ed., Van Nostrand Reinhold Co., Inc., New York, 1979.

J. J. Sciarra and L. Stoller, *The Science and Technology of Aerosol Packaging,* John Wiley & Sons, Inc., New York, 1974.

AGRICULTURAL CHEMICALS.

See FERTILIZERS; FUNGICIDES; HERBICIDES; INSECT CONTROL TECHNOLOGY; SOIL CHEMISTRY OF PESTICIDES.

AIR CONDITIONING

Basic Principles

Thermodynamic principles govern all air conditioning processes (see HEAT EXCHANGE TECHNOLOGY, HEAT TRANSFER). Of particular importance are specific thermodynamic applications both to equipment performance which influences the energy consumption of a system and to the properties of moist air which determine air conditioning capacity. The concentration of moist air defines a system's load.

Thermodynamics. Many definitions and formulations exist for the laws of thermodynamics, a detailed treatment of which may be found in standard engineering texts. Definitions that apply best to air conditioning are as follows:

First Law. This is the law of conservation of energy which states that the flow of energy into a system must equal the flow of energy out of the same system minus the energy that remains inside the system boundary. For an open system in which the energy flows are not time dependent and in which there is no accumulation of energy in the system, the first law may be written as

$$\sum_{in} \dot{Q} + \sum_{in} \dot{m}_i \left(h_i + Z_i + \frac{V_i^2}{2g} \right) = \sum_{out} \dot{W} + \sum_{out} \dot{m}_j \left(h_j + z_j + \frac{V_j^2}{2g} \right)$$

where $\dot{Q}$ = rate of heat transfer to the system, $\dot{m}$ = mass flow rate, h = enthalpy of substance, Z = elevation of boundary above a horizontal reference, V = velocity of substance,

Table 1. Typical Industrial Inside-Design Conditions[a]

Industry	Process	Dry-bulb temperature, °C	Rh, %
bakery	storage:		
	dried ingredients	21	55–65
	fresh ingredients	0–7	80–85
ceramics	refractory	43–65	50–90
	molding room	27	60–70
cereal	packaging	24–27	45–50
plastic	manufacturing:		
	thermosetting compounds	27	25–30
pharmaceutical	powder storage:		
	before manufacturing	21–27	30–35
	milling room	27	35
	tablet compressing	21–27	40
	tablet coating	27	35
	hypodermic tablet	24–27	30
	colloids	21	30–50
	ampule manufacturing	27	35
	gelatin capsule	78	40–50
	capsule storage	24	35–40
	biological manufacturing	27	35
textiles	rayon:		
	spinning	27–32	50–60
	throwing	27	55–60

[a] Listed conditions are typical; final design conditions are established by customer requirements.

g = gravitational acceleration, and $\dot{W}$ = work done by the system. Open steady-flow systems, which include almost all air conditioning processes, follow this law. Examples include the energy flows in a cooling and dehumidifying coil or an evaporative cooling system.

Second Law. This law defines the maximum theoretical performance for air conditioning equipment and provides a means of identifying energy losses in a system. It states that no heat engine operating in a closed cycle may produce work when communicating with a single temperature source. Air conditioning is the result of a heat engine operating in reverse. This means that work is added to the system and there must always be at least two temperatures, a low temperature source from which heat is received and a high temperature sink to which heat is rejected. The Carnot cycle is formulated directly from the second law of thermodynamics. It is a perfectly reversible, adiabatic cycle consisting of two constant entropy processes and two constant temperature processes. It defines the ultimate efficiency for any process operating between two temperatures. The coefficient of performance (COP) of the reverse Carnot cycle (refrigerator) is expressed as

$$\text{COP} = \frac{\text{useful effect}}{\text{work input}} = \frac{T_1}{T_2 - T_1}$$

where T_1 = absolute temperature of the cold source and T_2 = absolute temperature of the hot sink. Real processes always involve losses and irreversibilities and thus deviate from theory. Typical inefficiencies arise from temperature differences between the air stream and the heat exchange fluid, friction between moving parts, fluid pressure drops through heat exchangers and ducts, and pressure drops through pipes and valves.

Psychrometrics. Psychrometrics is the branch of thermodynamics that deals specifically with moist air, a binary mixture of dry air and water vapor. The properties of most air are frequently presented on psychrometric charts.

The quantities found on a psychrometric chart are

1. *Dry-bulb temperature* (DB).
2. *Dew-point temperature (DPT)* is the temperature at which the condensation of water vapor in a space begins for a given state of humidity and pressure as the temperature is reduced.
3. *Enthalpy,* is the total heat or heat content.
4. *Humidity ratio (Ordinate),* is the weight of water vapor in the air per unit weight of dry air.
5. *Relative humidity (rh),* is the ratio of the mole fraction of water vapor present in the air to the mole fraction of water vapor present in saturated air at the same temperature and barometric pressure.
6. *Saturation temperature* is the temperature at which the water vapor in moist air is in equilibrium with liquid water.
7. *Sensible heat factor,* is the ratio of the change in sensible (constant moisture content) cooling enthalpy to the change in total cooling enthalpy.
8. *Specific volume* is the volume of air per unit mass.
9. *Wet-bulb temperature* the equilibrium temperature which air attains if adiabatically saturated by water from a condensed phase.

The psychrometric chart may be used to determine the change in properties of air required for a condition or a process.

Human Comfort. ASHRAE has extensively researched the effect of air conditioning on human comfort. Thermal comfort may be defined as "that condition of mind in which satisfaction is expressed with the thermal environment". It is thus defined by a statistically valid sample of people under very specific and controlled conditions. No single environment is satisfactory for everybody, even if all wear identical clothing and perform the same activity.

There are no significant age or gender-related differences in thermal environment preference when all other factors such as weight of clothing and activity level are the same. Whereas people often accept thermal environments outside of their comfort range, there is no evidence that they adapt to these other conditions. Their environmental preference does not change. Similarly there is no evidence that there

Table 2. Owning and Operating Cost Data and Summary

	Owning costs
Initial cost of system (amortized)	*Amortization factors*
equipment	Annual fixed charges
control systems (complete)	taxes
wiring and piping costs attributable to system	insurance
installation costs	rent
	Operating costs
Annual energy and fuel costs	ventillation
electric energy costs	domestic water heating
chiller or compressor	cooking and food service equipment
pumps	air conditioning
fans	water
boiler auxiliaries	*Wages of engineers and operators*
heat pump	*Annual maintenance allowances*
domestic water heating	replacement or servicing of oil, air, or water filters
lighting	contracted maintenance service
cooking and food service	lubricating oil and grease
miscellaneous	general housekeeping costs
gas, oil, coal, or purchased steam costs	replacement of worn parts
treating	refrigerant
	Summary
equivalent uniform annual cost	
total annual fixed charges	
total annual fuel and energy costs	
wages of engineers and operators	
total annual maintenance allowance	
Total Annual Owning and Operating Costs	

is any seasonal or circadian rhythm influence on a person's thermal preference.

Local areas of thermal discomfort, ie, one part of the body warmer or cooler than preferred, may cause a person to be uncomfortable when the overall temperature and humidity would normally produce a sensation of thermal comfort. Some causes of this are nonuniform thermal radiation, such as hot or cold windows, walls, panels, floors, and ceilings.

Design Conditions

Fundamental to the design of any air conditioning system is the determination of the operating conditions of temperature and humidity. Worker comfort must also be considered.

Typical inside dry-bulb temperatures and relative humidities used for preparing, processing, manufacturing and storing are listed in Table 1. In some instances, the conditions have been compromised for the sake of worker comfort and do not represent the optimum for the product. In others, the conditions listed have no effect on the product or process other than to increase worker efficiency.

Equipment Size Requirements. Determining the proper size of air conditioning and heating equipment requires detailed study and calculation. A comprehensive statement of requirements and allowable variations must be supplied so that the best comfort conditioning system choice can be made. With such information it is possible to estimate not only equipment size, but also yearly energy requirements using any of several comprehensive computer or manual methods. Different systems can be designed and compared to ensure that the owner has a cost-effective yet energy-efficient system.

An analysis of the building structure must be conducted to determine the effects of heat gain from the sun. This analysis includes building orientation, type of construction, surrounding vegetation and structures, and reflective surfaces. People, lighting, machinery loads, and heat gains from chemical processes are evaluated as well as hooded (ventilated) processes, and lengths of operation. Outdoor air required both for ventilation to remove odors and contaminants, and for replacement of air exhausted through hoods must be conditioned.

Air Conditioning and Humidification Systems

Air conditioning may involve heating or cooling air, humidifying or drying it, and the control of chemical impurities to maintain the desired space conditions. Proper controls and energy conserving practices are also important to air conditioning and humidification.

Typical Air Conditioning Systems. Two broad categories of air conditioning systems exist, unitary and applied. Unitary systems are self-contained units that are "off the shelf." They use electricity for cooling, and may use electricity, natural gas, fuel oil, or propane for heating. Multiple unitary systems may be employed to provide greater overall reliability and to permit individual control of various sections of a plant. A typical unit for rooftop mounting contains means for heating, ventilating, and cooling.

More flexibility is obtained with applied equipment which is normally used to condition a relatively large area of a plant. This is usually part of a "field erected" system. Most systems are electrically powered; however, steam or gas turbines are used occasionally. Absorption chillers are frequently used when a suitable supply of "waste" heat is available. Low pressure steam, hot water, and process streams may provide the motive force.

Air Conditioning Control. When sized to meet design conditions, a heating or cooling system normally operates over a wide range of temperatures and loads; thus proper control becomes important. The system must be adjusted and maintained in order to provide operation for many years. The simplest control system which will produce the necessary results is best. The design professional should have the information for all anticipated operating conditions. For energy efficiency, the specifications should contain wide tolerances.

Uses of Air Conditioning in Industry

Many industrial processes require accurate environmental control. Examples include chemical reactions and processes that are affected

Table 3. Initial Costs

energy and fuel service costs
  electrical service entrance and distribution equipment
heat producing equipment
  boilers and furnaces
  steam-water converters
  heat pumps or resistance heaters
  make-up heaters
  heat producing equipment auxiliaries
  refrigeration equipment
  compressors, chillers, or absorption
  cooling towers
  refrigeration equipment auxiliaries
heat distribution equipment
  pumps, reducing valves, piping
  terminal units or devices
(cooling distribution equipment)
  pumps, piping, condensate drains,
  terminal units, mixing boxes
  air treatment and distribution equipment
  air heaters, humidifiers, dehumidifiers
  fans, ducts, duct insulation, dampers
  exhaust and return systems
  system and controls automation
  terminal or zone controls
  system program control
  alarms and indicator system
  building construction and alteration
  mechanical and electric space
  building insulation
  solar radiation controls
  acoustical and vibration treatment
  distribution shafts, machinery
  foundations, furring

by atmospheric conditions; biochemical reactions; quality, uniformity, and standardization of certain products; factors such as rate of crystallization and size of crystals; product moisture content or regain; deliquescence, lumping, and caking of hygroscopic materials; expansion and contraction of machines and products; physical, chemical, and biological cleanliness; effects of static electricity; odors and fumes; conditions in storage and packaging; quality of painted and lacquered finishes; simulation of stratosphere or space conditions; and productivity and comfort of workers. Controlled atmospheric conditions are especially important to the textile, pharmaceutical, food processing, explosives, and photographic materials industries. Analytical laboratories, clean rooms, and computer control rooms also require air conditioned environments.

Economic Evaluation of Air Conditioning Systems

The total economic picture, including life cycle costs and energy expenditures, must be considered in selecting an efficient air conditioning system. For many systems, the ratio of annual energy usage to first cost ranges from 0.25 to 2. Thus, over its useful life, a system consumes many times its initial costs.

Comparing two or more complex alternatives is more difficult than examining equipment capacity or first cost. Characteristics of alternatives should be weighted for relative importance and measured on a common scale to allow proper evaluation. Many characteristics such as first cost, capacity, space requirement, and annual energy use can be measured objectively and used for system comparisons. Experience has shown that items such as maintenance expense, component life, and downtime can also be reliably estimated. Other factors, eg, system maintainability, flexibility, and comfort, are more arbitrary.

Life cycle cost analysis is the proper tool for evaluation of alternative systems. The total cost of a system, including energy cost, maintenance cost, interest, cash flow, equipment replacement and/or salvage value, taxes, inflation, and energy cost escalation, can be estimated over the useful life of each alternative system.

A list of partial life cycle cost items which may be considered for each system is presented in Tables 2 and 3.

KENNETH W. COOPER
Poolpak, Inc.

ASHRAE Handbooks, American Society of Heating, Refrigeration, and Air Conditioning Engineers, Inc., Publications Department, Atlanta, Ga. four vols.: *Fundamentals, Equipment, Systems and Applications,* and *Refrigeration.* One volume is revised each year.

F. C. McQuiston and J. D. Parker, *Heating, Ventilating and Air Conditioning,* 3rd ed., John Wiley & Sons, Inc., New York, 1988.

ASHRAE Standards, American Society of Heating, Refrigerating, and Air Conditioning Engineers, Inc., Publications Department, Atlanta, Ga., Standard 55, 1981. Standards are upgraded on a regular basis.

R. T. Ruegg and S. R. Petersen, *Comprehensive Guide for Least-cost Energy Decisions,* NBS Special Publication 709, January 1987.

ASHRAE Journal (monthly) and *ASHRAE Transactions* (semiannually) American Society of Heating, Refrigerating and Air Conditioning Engineers, Atlanta, Ga.

R. W. Haines, *HVAC System Design Handbook,* TAB Books, Blue Ridge Summit, Pa., 1988.

G. W. Gupton Jr., *HVAC Controls, Operation and Maintenance,* Fairmont Press, Lilburn, Ga., 1987.

AIR, LIQUID. See CRYOGENICS.

AIR POLLUTION

Air pollution is the presence of any substance in the atmosphere at a concentration high enough to produce an objectionable effect on humans, animals, vegetation, or materials, or to significantly alter the natural balance of any ecosystem. Substances can be solids, liquids, or gases, and can be produced by anthropogenic activities or natural sources. In this article only nonbiological material is considered and the discussion of airborne radioactive contaminants is limited to radon, (see HELIUM-GROUP GASES), which is discussed in the context of indoor air pollution.

The original six criteria pollutants, so named because the Environmental Protection Agency is required to summarize published information on each and the summaries are called criteria documents, were sulfur dioxide, SO_2; carbon monoxide (qv), CO; nitrogen dioxide, NO_2; ozone (qv), O_3; suspended particulates; and nonmethane hydrocarbons, NMHC. The NMHC are now referred to as volatile organic compounds (VOC). The NMHC were dropped from the list shortly after the criteria pollutants were so designated. In the late 1970s, lead (qv) Pb, was added to the list and in 1987, so was particulate matter having an aerodynamic diameter of less than or equal to 10 μm, PM_{10}.

There have been several developments since the establishment of the criteria pollutants. In the mid-1970s it was shown that high concentrations of O_3 and sulfate haze could be transported hundreds of

Table 1. National Ambient Air Quality Standards for Criteria Pollutants[a]

	Primary		Secondary		
Pollutant	μg/m^3	ppm	μg/m^3	ppm	Averaging time
PM_{10}	50		50		annual arithmetic mean
	150		150		24-h[b]
SO_2	80	(0.03)			annual arithmetic mean
365	(0.14)			24-h[b]	
			1300	(0.50)	3-h[b]
CO	(10)	9			8-h[b]
	(40)	35			1-h[b]
NO_2	(100)	0.053	(100)	0.053	annual arithmetic mean
Pb	1.5		1.5		maximum quarterly average
O_3	(235)	0.12	(235)	0.12	maximum daily[c] 1-h average

[a] Parenthetical value is an approximately equivalent concentration.
[b] Not to be exceeded more than once per year.
[c] Not to be exceeded on more than three days in three years.

miles, and acid deposition studies in the 1980s clearly illustrated the international and global aspects of this transport. Then stratospheric O_3 depletion and global warming became issues and air pollution was finally viewed in a global context. At the same time that the geographic scale of air pollution was expanding, the number of pollutants of concern also increased and detection capabilities improved, leading to the establishment of a hazardous air pollutant category which includes any potentially toxic substance in the air that is not a criteria pollutant.

Air Pollution Components

Air pollution can be considered to have three components: sources, transport and transformations in the atmosphere, and receptors. The source emits airborne substances that, when released, are transported through the atmosphere. Some of the substances interact with sunlight or chemical species in the atmosphere and are transformed. Pollutants that are emitted directly to the atmosphere are called primary pollutants; pollutants that are formed in the atmosphere as a result of transformations are secondary pollutants. The reactants that undergo transformation are referred to as precursors. An example of a secondary pollutant is O_3, and its precursors are VOC and nitrogen oxides, NO_x, a combination of nitric oxide, NO, and NO_2. The receptor is the person, animal, plant, material, or ecosystem affected by the emissions.

Air Quality Management

In the United States, the framework for air quality management is the Clean Air Act (CAA), which defines two categories of pollutants; criteria and hazardous. For the criteria pollutants, the CAA requires that EPA establish National Ambient Air Quality Standards (NAAQS) and emissions standards for stationary sources and for motor vehicles, and gives the primary responsibility for designing and implementing air quality improvement programs to the states. For the hazardous air pollutants, only emissions standards for some sources are required, but the number is growing rapidly. The NAAQS apply uniformly across the United States whereas emission standards for criteria pollutants depend on the severity of the local air pollution problem and whether an affected source already exists or is proposed. In addition, individual states have the right to set their own ambient air quality and emissions standards (which must be at least as stringent as the federal standards) for all pollutants and all sources except motor vehicles. With respect to motor vehicles, the CAA allows the states to choose between two sets of emissions standards: the federal standards or the more stringent California ones.

The two levels of NAAQS, primary and secondary, are listed in Table 1. Primary standards were set to protect public health within an adequate margin of safety; secondary standards, where applicable, were chosen to protect public welfare, including vegetation. According to the CAA, the scientific bases for the NAAQS are to be reviewed every five years so that the NAAQS levels reflect current knowledge. In practice, however, the review cycle takes considerably longer.

Air Pollution Issues

Current air pollution concerns include photochemical smog, volatile organic compounds (VOC), nitrogen oxides (NO_x), sulfur oxides (SO_x), carbon monoxide (CO), particulate matter, lead, air toxics, odors, visibility, acid deposition, global warming (the greenhouse effect), stratospheric O_3 depletion, and indoor air pollution.

GEORGE T. WOLFF
General Motors Research and Environmental Staff

P. Warneck, *Chemistry of Natural Atmospheres,* Academic Press, New York, 1988.

Intergovernmental Panel on Climate Change, *Scientific Assessment of Climate Change,* United Nations, New York, 1990.

J. H. Seinfeld, *Atmospheric Chemistry and Physics of Air Pollution,* John Wiley & Sons, Inc., New York, 1986.

AIR POLLUTION CONTROL METHODS

Air pollution (qv), recognized in the National Ambient Air Quality Standards (NAAQS) as being characterized by a time–dosage relationship, may be rendered less harmful by reducing the concentration of contaminants, the exposure time, or both.

Selection of pollution control methods is generally based on (*1*) the need to control ambient air quality in order to achieve compliance with standards for criteria pollutants, (*2*) the need to reduce emission to the atmosphere of a hazardous air pollutant, or (*3*), in the case of nonregulated contaminants, to protect human health and vegetation. There are three elements to a pollution problem: a source, a receptor affected by the pollutants, and the transport of pollutants from source to receptor. Modification or elimination of any one of these elements can change the nature of a pollution problem.

There are three main classes of pollutants: gases, particulates (which may be either liquid or solid or a combination), and odors (which may originate as gases or particulates). Although odors are controlled similarly to other pollutants, they are often discussed separately because of the different methods used for sensing and measurement (see ODOR MODIFICATION). To achieve air pollution control, reliable measurements are needed to quantify both the pollutant concentration and the contribution of individual sources. These data are necessary for designing control equipment, for monitoring emissions, and for maintaining acceptable ambient air quality.

Measurement of Air Pollution

Measurement techniques are divided into two categories: ambient and source measurement. Ambient air samples often require detection and measurement in the ppmv to ppbv (parts by volume) range, whereas source concentrations can range from tenths of a volume percent to a few hundred ppmv. Federal regulations (CFR 40, parts 50–99) require periodic ambient air monitoring at strategic locations in a designated air quality control region.

Ambient sampling may fulfill one or more of the following objectives: (*1*) establishing and operating a pollution alert network, (*2*) monitoring the effect of an emission source, (*3*) predicting the effect of a proposed installation, (*4*) establishing seasonal or yearly trends, (*5*) locating the source of an undesirable pollutant, (*6*) obtaining permanent sampling records for legal action or for modifying regulations, and (*7*) correlating pollutant dispersion with meteorological, climatological, or topographic data, and with changes in societal activities.

The problems of source sampling are distinct from those of ambient sampling. Source gas may have a high temperature or contain high concentrations of water vapor or entrained mist, dust, or other interfering substances. In addition, particulates or gases may be deposited on or absorbed into the grain structure of the gas-extractive sampling probes and thus be lost from the sample obtained. Depending on the objective or regulations, source sampling may be infrequent, occasional, intermittent, or continuous. Typical objectives are: (*1*) demonstrating compliance with regulations, (*2*) obtaining emission data, (*3*) measuring product loss or optimizing process operating variables, (*4*) obtaining data for engineering design, (*5*) determining collector efficiency or acceptance testing of purchased equipment, and (*6*) determining the need for maintenance of process or control equipment.

Status of Air Pollution and Control Regulations

There has been considerable improvement, especially in industrial areas, in U.S. air quality since the adoption of the Clean Air Act of 1972. Appreciable reductions in particulate emissions and in SO_2 levels are especially evident.

The U.S. Congress adopted a new Clean Air Act in 1990 which has three areas of emphasis: acid rain reduction in the northeastern United States; severe limitation on atmospheric emissions of 189 chemicals on the hazardous or Toxic Substance list; and tightened regulations on vehicular exhaust, reformulated vehicular fuels, and vehicles capable of using alternative fuels (ozone compliance and smog reduction).

Minimizing Pollution Control Cost

Although the first impulse for emission reduction is often to add a control device, this may not be the environmentally best or least costly approach. Process examination may reveal changes or alternatives that can eliminate or reduce pollutants, decrease the gas quantity to be treated, or render pollutants more amenable to collection. Following are principles to consider for controlling pollutants without the addition of specific treatment devices, ie, the fundamental means of reducing or eliminating pollutant emissions to the atmosphere:

- Eliminate the source of the pollutant.
  - Seal the system to prevent interchanges between system and atmosphere.
  - Use pressure vessels.
  - Interconnect vents on receiving and discharging containers.
  - Provide seals on rotating shafts and other necessary openings.
  - Change raw materials, fuels, etc., to eliminate the pollutant from the process.
  - Change the manner of process operation to prevent or reduce formation of, or air entrainment of, a pollutant.
  - Change the type of process step to eliminate the pollutant.
  - Use a recycle gas or recycle the pollutants rather than using fresh air or venting.
- Reduce the quantity of pollutant released or the quantity of carrier gas to be treated.
  - Minimize entrainment of pollutants into a gas stream.
  - Reduce number of points in system in which materials can become airborne.
  - Recycle a portion of process gas.
  - Design hoods to exhaust the minimum quantity of air necessary to ensure pollutant capture.
- Use equipment for dual purposes, such as a fuel combustion furnace to serve as a pollutant incinerator.

Steps such as the substitution of low sulfur fuels or nonvolatile solvents, change of raw materials, lowering of operation temperatures to reduce NO_x formation or volatilization of process material, and installion of well-designed hoods at emission points to effectively reduce the air quantity needed for pollutant capture are illustrations of the above principles.

Selection of Control Equipment

Engineering approaches for the design and selection of pollution control equipment include knowledge of the properties of pollutants: chemical species, physical state, particle size, concentration, quantity of conveying gas, and of effects of pollutant on surrounding environment. The design must consider likely future collection requirements. Advantages of alternative collection techniques must be determined, eg

1. Collection efficiency.
2. Ease of reuse or disposal of recovered material.
3. Ability of collector to handle variations in gas flow and loads at required collection efficiencies.
4. Equipment reliability and freedom from operational and maintenance attention.
5. Initial investment and operating cost.
6. Possibility of recovery or conversion of contaminant into a saleable product. Known engineering principles should be applied even in areas of extremely dilute concentration.

With such information, a list of possible treatment methods can be made (Table 1).

Control of Gaseous Emissions

Five methods are available for controlling gaseous emissions: absorption, adsorption, condensation, chemical reaction, and incineration. Atmospheric dispersion from a tall stack, considered as an alternative in the past, is now less viable. Absorption is particularly attractive for pollutants in appreciable concentration; it is also applicable to dilute concentrations of gases having high solvent solubility. Adsorption is desireable for contaminant removal down to extremely low levels (less than 1 ppmv) and for handling large gas volumes that have quite dilute contaminant levels. Condensation is best for substances having rather high vapor pressures. Elimination of noncondensible diluents is desirable to permit use of smaller condensers, especially when the final step requires refrigeration. Incineration, suitable only for combustibles, is used to remove organic pollutants and small quantities of H_2S, CO, and NH_3. Specific problem gases such as sulfur and nitrogen oxides require combinations of methods and are discussed below.

Specific Problem Gases. Sulfur dioxide, nitrogen oxides, and vehicular exhaust gases are widespread gas pollutants that present specific problems. (AIR POLLUTION; ATMOSPHERIC MODELING). The U.S. Clean Air Reauthorization Act of 1990 requires greater control of the emissions

Table 1. Checklist of Applicable Devices for Control of Pollutants

Equipment type	Pollutant classification			
			Particulate	
	Gas	Odor	Liquid	Solid
absorption	•	•		
aqueous solution				
nonaqueous				
adsorption	•	•		
throw-away canisters				
regenerable stationary beds				
regenerable traveling beds				
chromatographic adsorption				
air dispersion (stacks)	•	•	•	•
condensation	•	•		
centrifugal separation (dry)			•	•
chemical reaction	•	•		
coagulation and particle growth			•	•
filtration				
fabric and felt bags				•
granular beds			•	•
fine fibers			•	•
gravitational settling				•
impingement (dry)				•
incineration	•	•	•	•
precipitation, electrical				
dry			•	•
wet	•	•	•	•
precipitation, thermal			•	•
wet collection[a]	•	•	•	•

[a] Includes cyclonic, dynamic, filtration, inertial impaction (wetted targets, packed towers, turbulent targets), spray chambers, and venturi.

of these gases. Germany and Denmark had previously adopted acid rain regulations on sulfur and nitrogen oxide emissions.

Major sources of sulfur dioxide are the combustion of sulfur-containing fossil fuels, the manufacture of sulfuric acid and sludge acid purification, sulfur recovery from petroleum processing, non-ferrous smelting, and pulp and paper manufacture. Combustion emissions are controlled by substituting a low sulfur fuel source, by fuel desulfurization and refining (see COAL; PETROLEUM; FUELS, SYNTHETIC), and by sulfur removal either in the combustion process or from the flue gas (see SULFUR REMOVAL AND RECOVERY). Many methods of sulfur removal from flue gas have been developed and voluminous literature is available. Flue gas desulfurization (FGD) systems can be classified as (*1*) throwaway vs regenerative and (*2*) wet vs dry.

Major sources of nitrogen oxide emission are nitrogen fixation during high temperature combustion, nitric acid manufacture and concentration (see NITRIC ACID), organic nitrations (see NITRATION), and vehicular emissions. NO formation in combustion may be reduced by maintaining low excess air (0.5% O_2 or less in flue gas), employing two-stage combustion where the first stage is fuel-rich and reducing (high temperature) and the second stage is oxidizing (1000–1100°C), flue gas recirculation, burner design, combustion chamber modifications, and burner placement. These combustion modification methods are valid for reducing NO emission levels only to 200–300 ppmv NO_x. Other process methods developed for NO_x destruction land capable of achieving NO_x emission levels of 80–100 ppmv are being developed for combustion NO_x emissions are selective catalytic reduction (SCR), Thermal Denox, and urea reduction.

Control of NO_x emission from nitric acid and nitration operations is usually achieved by NO_2 reduction to N_2 and water using natural gas in a catalytic decomposer (see EXHAUST CONTROL, INDUSTRIAL). NO_x from nitric acid–nitration operations is also controlled by absorption in water to regenerate nitric acid.

Control of Particulate Emissions

The removal of particles (liquids, solids, or mixtures) from a gas stream requires deposition and attachment to a surface. The surface may be continuous, such as the wall and cone of a cyclone or the collecting plates of an electrostatic precipitator, or it may discontinuous such as spray droplets in a scrubbing tower. Once deposited on a surface, the collected particles must be removed at intervals without appreciable reentrainment in the gas stream. Gravity settling is efficient only for large ($D_p < 40–50$ μm) particles; flow-line interception and inertial impaction are effective for particles down to 2–3 μm; diffusional deposition and thermal precipitation, increasingly efficient with a decrease in particle size, are highly efficient for particles ≤ 0.5 μm; and electrostatic forces are the strongest forces available to act on fine particles, which are loosely defined as $\leq 2–3$ μm. There is a gap in collectability between 0.2 and 2.0 μm. Particles in this range are the most difficult to charge electrically as well as collect by any other method.

In some instances, a few other principles such as diffusiophoresis and methods of particle growth and agglomeration have also been used. The magnitude of the force required to move a particle toward a collecting surface is influenced markedly by the size and shape of the particle.

Particle Filtration. Devices for particle collection by filtration can be divided into three categories: cloth filtration using either woven or felted fabrics in a bag or envelope, paper and mat filters, and in-depth aggregate bed filtration. The first type is used for dry particle removal from gases, but cannot be employed when liquid particulates are present or condensation is imminent. Bag filters, most commonly used for dry, fine dust control, are available for bag cleaning by shaking, reverse flow, and pulse-jet cleaning. Other filter types for handling occasional light dust loads are cloth pocket filters, canister filters, and filter media in frames. Choice of bag fabric depends upon chemical compatibility with the dust to be collected (Table 2) and temperature resistance (Table 3) as well as fabric cost. The pocket filter has low dust-handling capacity, and when pressure drop becomes too high, the element must be removed and either discarded or manually cleaned. It is used primarily for very low dust loads, occasional emissions, or as a safeguard against broken bags after a normal baghouse filter. Likewise, the multiple canister filters are difficult to clean and must be replaced once high pressure drop occurs. Filters in the form of fiber pads or pleated paper in frames are used for preparing clean air for process use or ventilation. They have limited dust-holding capacity and are seldom used for air pollution control.

Several types of aggregate-bed filters are available which provide in-depth filtration. Both gravel and particle-bed filters have been developed for removal of dry particulates but have not been used extensively. Filters have also been developed using a porous ceramic or porous metal filter surface. Mesh beds of knitted wire mesh, plastic, or glass fibers are used for the removal of liquid particulates and mist.

Table 2. Chemical Compatibility of Fibers in Dust Collector Bags

Resistance	In acid media	In alkaline media
excellent	polypropylene	polypropylene
	polyethylene	polyethylene
	Saran	Dynel
	Teflon	nylon-6,6
	Orlon	Teflon
good	Dacron	cotton
	Dynel	nylon-6
	glass	Nomex nylon
	wool	Saran
unsuitable	cotton	wool
	nylon-6,6	glass
	nylon-6	
	Nomex nylon	

Table 3. Maximum Desirable Operating Temperature for Filter Bags[a]

Material	Temperature, °C	Material	Temperature, °C
polyethylene	70	Arnel	120
Saran	70	Microtain	125
cotton	80	Kodel	135
Dynel	85	Dacron	140
polypropylene	90	Darvan	150
wool	100	Nomex nylon	230
nylon-6,6 and -6	105	Teflon	260
Orlon	120	glass	290
Acrilan	120		

[a] Longer bag life is obtained if bags are not operated at their maximum temperature.

They will also remove solid particles, but plug rapidly unless irrigated or flushed with a particle-dissolving solvent.

Wet Scrubbing. Scrubbers can be highly effective for both particulate collection and gas absorption. Capital costs can be quite reasonable for the required efficiency, but the addition of water treatment for recycle or for waste disposal (depending on treatment complications) may make the total cost as great as any other collection method. In addition, efficient collection of particles finer than 3–4 μm, and especially submicrometer particles, can require very high pressure drop and energy consumption. Although scrubbers automatically provide cooling of hot gases, the water-saturated effluent may produce offensive plume condensation in cold weather. Many moist effluents become more corrosive than dry ones. Solid accumulation may occur at wet–dry interfaces and icing problems may occur around the stack in winter.

Scrubbers make use of a combination of particulate collection mechanisms. It is difficult to classify scrubbers predominantly by any one mechanism, but for some systems, inertial impaction and direct interception predominate.

Wet-scrubber collection efficiency may be unexpectedly enhanced by particle growth. Vapor condensation, high turbulence, and thermal forces acting within the confines of narrow passages can all lead to particle growth or agglomeration. Of these, vapor condensation produced by cooling is the most common.

Developing Particulate Control Technology. Present control methods for particulates are least efficient in the size range from 0.2 to 2.5 μm; this range is the most costly to collect and very energy intensive. Health studies indicate that particles in this size range are also those which penetrate most deeply into, and often become deposited in, the human respiratory system. This is the main reason for the U.S. EPA change in the ambient air quality standard from total suspended particulate (TSP) concentration to a PM_{10} standard (ambient air particles equal to or smaller than 10-μm aerodynamic diameter). Attractive means for improved collection would be the use of a separating force which is independent of gas velocity or the growth of particles to a size which can be more readily collected. Particle growth can be accomplished through coagulation (agglomeration), chemical reaction, condensation, and electrostatic attraction. Promising separation forces are the "flux forces" involving diffusiophoresis, thermophoresis, electrophoresis, and Stefan flow. Although particle growth techniques and flux-force collection theoretically can be considered independently, both phenomena are applied in many practical devices.

Odor Control

Odor control involves any process that gives a more acceptable perception of smell, whether as a result of dilution, removal of the offending substance, or counteraction or masking (see ODOR MODIFICATION; PERFUMES).

Odor Measurement. Both static and dynamic measurement techniques exist for odor. The objective is to measure odor intensity by determining the dilution necessary so that the odor is imperceptible or doubtful to a human test panel. The odor threshold is the lowest concentration at which an odor stimulus may be detected. The recognition threshold is a higher value at which the chemical entity is recognized.

Static and Dynamic Dilution Methods. In the static dilution method, a known volume of odorous sample is diluted with a known amount of nonodorous air, mixed, and presented statically or quiescently to the test panel. In the dynamic dilution method, odor dilution is achieved by continuous flow. Advantages are more accurate results, simplicity, reproducibility, and speed. Devices known as dynamic olfactometers control the flow of both odorous and pure diluent air, provide for ratio adjustment to give desired dilutions, and present multiple, continuous samples for test panel observers at ports beneath odor hoods.

Odor Control Methods. Absorption, adsorption, and incineration are all typical control methods for gaseous odors; odorous particulates are controlled by the usual particulate control methods. However, carrier gas odorized by particulates may require gaseous odor control treatment even after the particulates have been removed. For oxidizable odors, treatment with oxidants such as hydrogen peroxide (qv), ozone (qv), and $KMnO_4$ may sometimes be practiced; catalytic oxidation has also been employed (see EXHAUST CONTROL, INDUSTRIAL).

BURTON B. CROCKER
Consultant

S. Calvert and H. M. Englund, eds., *Handbook of Air Pollution Technology,* John Wiley & Sons, Inc., New York, 1984.

A. J. Buonicore and W. T. Davis, eds., *Air Pollution Engineering Manual,* Van Nostrand Reinhold, New York, 1992.

L. K. Wang and N. C. Pereira, eds., *Air and Noise Pollution Control, Handbook of Environmental Engineering,* Vol. 1, The Humana Press, Clifton, N.J., 1979.

K. Wark and C. F. Warner, *Air Pollution: Its Origin and Control,* 2nd ed., Harper & Row, New York, 1981.

Code of Federal Regulations 40 (CFR 40), *Fed. Reg.,* C-50–99.

AIR SEPARATION. See CRYOGENICS; MEMBRANE TECHNOLOGY; MOLECULAR SIEVES; NITROGEN.

ALCOHOL FUELS

A wide variety of countries have, at various times, explored the use of either methanol or ethanol alcohol fuels as alternatives to diesel and gasoline fuels. Recently the United States has demonstrated light and heavy-duty vehicles using alcohols (mostly methanol (qv)), but otherwise has not yet passed beyond the use of limited amounts of alcohols, and of ethers produced from alcohols, as gasoline components. The potential benefits of alcohol fuels include increased energy diversification in the transportation sector, accompanied by some energy security and balance of payments benefits, and air quality improvements as a result of the reduced emissions of photochemically reactive products (see AIR POLLUTION). The Clean Air Act of 1990, emission standards set out by the State of California, and the 1992 Comprehension National Energy Policy Act (see below), may serve to encourage the substantial use of alcohol fuels, unless gasoline and diesel technologies can be developed that offer comparable advantages.

Properties of Alcohol Fuels

Table 1 summarizes key properties of ethanol and methanol as compared to other fuels. Both alcohols make excellent motor fuels, although the high latent heats of vaporization and the low volatilities can make cold-starting difficult in vehicles having carburetors or fuel injectors in the intake manifold where the fuel must be vaporized prior to being introduced into the combustion chamber. This is not the case for direct injection diesel-type engines where the high heat of vaporization and low volatility of methanol (or ethanol) are overcome

Table 1. Properties of Fuels

Properties	Methanol	Ethanol	Propane	Methane	Isoctane	Unleaded gasoline	Diesel fuel #2
constituents	CH_3OH	CH_3CH_2OH	C_3H_8	CH_4	C_8H_{18}	$C_nH_{1.87n}$ (C_4 to C_{12})	$C_nH_{1.8n}$ (C_8 to C_{20})
molecular weight	32.04	46.07	44.10	16.04	114.23	110	170
element composition, wt %							
C	37.49	52.14	81.71	74.87	84.12	86.44	86.88
H	12.58	13.13	18.29	25.13	15.88	13.56	13.12
O	49.93	34.73	0	0	0	0	0
density at 16°C and 101.3 kPa[a], kg/m^3	794.6	789.8	505.9	0.6776	684.5	721–785	833–881
boiling point at 101.3 kPa[a], °C	64.5	78.3	6.5	−161.5	99.2	38–204	163–399
freezing point, °C	−97.7	−114.1	−188.7	−182.5	−107.4		<−7
vapor pressure at 38°C, kPa[a]	31.9	16.0	1.297	0.5094	11.8	48–108	negligible
heat of vaporization, ΔH_v, MJ/kg[b]	1.075	0.8399	0.4253	0.5094	0.2712	0.3044	0.270
autoignition temperature, °C	464	363	450	537	418	260–460	257
flame temperature, °C	1,871	1,916	1,988	1,949	1,982	2,027	1,993
flash point, °C	11	13			4	−43to − 39	52–96
octane ratings							
research	106	107	112	120	100	92–98	
motor	92	89	97	120	100	80–90	
cetane rating	0–5						>40
flammability limits, vol % in air	6.72–36.5	3.28–18.95	2.1–9.5	5.0–15.0	1.0–6.0	1.4–7.6	1.0–5.0
stoichiometric air–fuel mass ratio	6.46	8.98	15.65	17.21	15.10	14.6	14.5
stoichiometric air–fuel volumetric ratio	7.15	14.29	23.82	9.53	59.55	55	85
sulfur content, wt %	0	0	0	0	0	<0.06	<0.5

[a] To convert kPa to psi, multiply by 0.145.
[b] To convert MJ/kg to Btu/lb, multiply by 430.3.

by injecting near the top of the compression stroke where gases are at high temperatures and pressures. Both methanol and ethanol have high octane values and allow high compression ratios providing increased efficiency and improved power output per cylinder. Both have wider combustion envelopes than gasoline and can be run at lean air–fuel ratios with better energy efficiency. However, ethanol and methanol have very low cetane ratings and cannot directly be used in compression ignition, diesel-type engines unless the in-cylinder gas temperatures are high enough at the time of injection or a cetane improver is used. Manufacturers of heavy-duty engines have developed several types of systems to assist in alcohol fuel ignition. These include glow plugs or spark plugs, reduced engine cooling, and increased amounts of exhaust gas recirculation or, in the case of two-stroke engines, reduced scavenging. Additives to improve cetane number have also been effective for assisting ignition in compression ignition engines. Dual fuel approaches, in which a small amount of diesel fuel is used to ignite the alcohol, has also been demonstrated.

The properties of methanol and ethanol also result in different vehicle fuel storage considerations.

There are particular alcohol fuel safety considerations. Unlike gasoline or diesel fuel, the vapor of methanol or ethanol above the liquid fuel in a fuel tank is usually combustible at ambient temperatures. This poses the risk of an explosion should a spark or flame find its way to the tank such as during refueling. Additionally, a neat methanol fire has very little luminosity and, consequently, fire fighting efforts can be difficult in daylight. However, low luminosity also implies low radiative fluxes from the fire. This, combined with the high latent heat of vaporization, means that the heat release of a methanol fire is low relative to one of gasoline or diesel fuel. Burn severity resulting from methanol flames in accidents will, therefore, be less than for comparable gasoline and diesel burn exposures, because the destructive radius of the flames is shorter and because longer direct flame contact is needed to cause comparable tissue damage. Because methanol or ethanol are both water-soluble, fires can be successfully controlled by dilution with large amounts of water, a tactic that simply spreads gasoline fires. Nevertheless, fire-extinguishing foams (see FLAME RETARDANTS) are the preferred alcohol fire-fighting method.

Some potential problems of alcohol fuels have been addressed by adding small amounts of gasoline or specific hydrocarbons to the fuel, reducing the flammability envelope and providing luminosity in case of fire. The 85% methanol–15% gasoline fuel in use in California and elsewhere is commonly called M85. Flame luminosity purchased through the addition of gasoline to the fuel compromises the emissions performance of both light- and heavy-duty vehicles. M85 also increases the evaporative emissions throughout the fuel storage and distribution system.

Energy Diversification and Energy Security

In the 1990s world events precipitated renewed interest in energy diversification strategies for the U.S. transportation sector. However, few measures are in place to encourage fuel alternatives outside of the exemption from the Federal excise tax on motor fuel granted to ethanol blends. The Alternative Motor Fuels Act of 1988 did extend credits to automobile manufacturers in the calculation of corporate average fuel economy (CAFE) for vehicles that use methanol or ethanol or natural gas; electric vehicles had previously been granted such a credit. Under the Act's provisions, neither ethanol nor methanol is counted as fuel consumed in the calculation of fuel economy. Thus vehicles that use alcohol have very high fuel economy ratings, reflecting the value of these vehicles in reducing oil imports. The credits took effect in model-year 1993. The fuel economy calculation assumes that methanol and ethanol fuels in commerce contain 15% by volume gasoline. Vehicles that can use either alcohol fuels or gasoline receive a reduced CAFE credit. The maximum credit that can be earned by a manufacturer for selling vehicles capable of using petroleum fuels is capped because alcohol fuels usage by these vehicles is not assured. In 1992 the Comprehensive National Energy Policy Act was approved. This legislation requires that federal and state fleets of 50 or more vehicles, capable of central fueling, and operated within population areas of 250,000 or more, use alternative fuels. Alternative fuels are defined as methanol, ethanol, natural gas, liquid petroleum gas, hydrogen, electricity, and other fuels derived form coal and biomass feedstocks. The goal of this legislation is to reduce the United States' imports of petroleum fuels by 10% by the year 2000 and 30% by the year 2030. Additionally, at least 50% of the alternative fuels used to replace petroleum fuels should come from domestic feedstocks.

Alcohol Availability

Methanol. If methanol is to compete with conventional gasoline and diesel fuel it must be readily available and inexpensively produced. Thus methanol production from a low-cost feedstock such as natural gas or coal is essential (see FEEDSTOCKS). There is an abundance of natural gas (see GAS, NATURAL) worldwide and reserves of coal are even greater than those of natural gas.

Ethanol's largest fuel use is as an additive to gasoline for conventional automobiles. Methanol passenger vehicles primarily use M85. The M85 retail network is still in its infancy. To overcome the fuel availability problem, many automobile manufacturers have developed what are called flexible-fuel or variable-fuel vehicles. These passenger cars can use any combination of M85 and conventional gasoline, allowing drivers to refuel with conventional gasoline where M85 is not available. Flexible fuel vehicles are the dominant alcohol burning light-duty vehicles in the United States, but in heavy-duty use M100 is dominant.

Ethanol. Ethanol is primarily produced from a variety of crops and crop by-products, generically called renewable biomass. Currently, corn is the principal feedstock for predominantly small-scale production plants. Research is focusing on genetic enhancement of the bacteria responsible for fermentation and other ways to lower crop costs and production techniques. The wide range of potential ethanol feedstocks and production techniques make it problematic to establish upper boundary levels for possible production. Estimates range from 412 million gasoline-equivalent barrels per year to over 8 billion gasoline-equivalent barrels per year. The latter figure assumes predominant use of wood feedstocks. By contrast, 1990 consumption of gasoline in the United States was 2.6 billion barrels.

Economic Aspects

Alcohol Production. Studies to assess the costs of alcohol fuels and to compare the costs to those of conventional fuels contain significant uncertainties. In general, the low cost estimates indicate that methanol produced on a large scale from low cost natural gas could compete with gasoline when oil prices are around 14¢/L ($27/bbl). This comparison does not give methanol any credits for environmental or energy diversification benefits. Ethanol does not become competitive until petroleum prices are much higher.

Vehicle Technology and Vehicle Emissions

One of the reasons that U.S. automobile manufacturers showed more interest in alcohols as alternative fuels in the late 1970s and early 1980s is because alcohol's energy density is closer to gasoline and diesel than other alternatives such as compressed natural gas. They reasoned that consumers would be more comfortable with liquid fuels, envisioning little change in the fuel distribution of alcohols. Most of the research in the 1970s focused on converting light-duty vehicles, to alcohol fuels. Toward the late 1970s, researchers also began to turn their attention to heavy-duty applications. In heavy-duty engines the emissions benefits of alcohols are far greater than in light-duty vehicles. However, it is also much harder to design heavy-duty engines to use the low cetane number alcohols.

It was not until the early 1980s that the potential air quality benefits of alcohol fuels started to be investigated. It was about five years later that proponents argued that alcohols could provide significant air quality benefits in addition to energy security benefits. Low level blends of ethanol and gasoline were argued to provide lower carbon monoxide emissions. The exhaust from light-duty methanol vehicles was thought to be less reactive in the formation of ozone. Uses of alcohol fuels in heavy-duty engines showed substantially reduced mass emissions of nitrogen oxides (NO_x) and particulates compared to diesel-fueled engines. By contrast, light-duty experience showed about the same mass emissions of hydrocarbons and NO_x but a reduced reactivity of the exhaust hydrocarbons.

Air Quality Benefits of Alcohol Fuels

The most comprehensive air quality study, supported by the California ARB and the South Coast Air Quality Management District, showed that if gasoline and methanol cars emitted the same amounts of carbon, an assumption that seemed reasonable based on emissions test data taken throughout the 1980s, and if methanol cars had formaldehyde emissions controlled to 9.3 mg/km (equal to the current California formaldehyde emissions standard for methanol automobiles), then substituting M85 for gasoline would produce a 9% reduction in the peak summer-day afternoon ozone level and a 19% reduction in exposure to ozone levels above the Federal standard of 0.12 ppm. These reductions constituted a substantial fraction of the reductions that would be obtained by eliminating all the emissions from vehicles. Additional assumptions were that exhaust carbon emissions were at the level of 0.15 g/km, equal to the planned certification standard for new vehicles in California beginning in 1993 (but not really expected to be characteristic of in-use vehicles), and that the distribution of hydrocarbon species in the exhaust resembled that of cars tested in the 1980s.

The results of the modeling study, conducted at Carnegie Mellon University, are now generally accepted as one of the best available guides for the smog-reducing benefits of a methanol substitution strategy, at least for the conditions prevailing in the Los Angeles basin. Overall, replacement of a conventional gasoline vehicle by an equivalent M85 vehicle should provide about a 30% reduction in smog-forming potential. A 100% result would be earned by eliminating the vehicle entirely. Vehicles using M100 would provide double the ozone reduction benefits of M85 vehicles.

Methanol lacks the hundreds of known and probable carcinogens found in gasoline and diesel fuels, such as benzene, 1,3-butadiene, diesel particulates, and polynuclear aromatic hydrocarbons. Exhaust levels of formaldehyde are higher in methanol-powered light-duty vehicles than in their gasoline equivalents, but gasoline exhaust can cause secondary atmospheric formation of formaldehyde that leads to higher ambient levels of this chemical. Methanol is expected to reduce overall public exposure to air toxic contaminants that result from the refining, distribution, and use of petroleum fuels.

Public Safety Issues

Several investigators have assessed the comparative safety of methanol and conventional hydrocarbon fuels. The ingestion toxicity of methanol has been of some concern because of the number of gasoline ingestions associated with siphoning and in-home accidental ingestions. Skin contact with methanol may present a greater health threat than skin contact with gasoline and diesel fuel and is being evaluated.

The risk of fires in methanol vehicles appears to be substantially smaller than for gasoline vehicles, but greater than the fire hazard of diesel vehicles.

In reviewing the full range of health and safety issues associated with all alternative fuels, the California Advisory Board determined that there were no roadblocks that would prevent the near-term deployment of either methanol or ethanol, assuming that adequate safety practices were followed appropriate to the specific nature of each fuel.

The Future of Alcohol Fuels

The City Air Resources Board adopted a regulatory package for light-duty vehicles in 1990 that modifies the historically uniform approach to vehicle emissions, in which each and every vehicle in a regulated class must meet the same emissions standard. The new approach adopts emissions standards that apply on the average to the entire sales mix of vehicles sold by each manufacturer in each of several broad weight classes of vehicles. Thus vehicles that use fuels such as methanol and ethanol having air quality benefits in the form of lower levels of photochemical reactivity have the emissions adjusted to reflect the lower smog-forming tendency of these fuels. This regulatory approach provides a powerful incentive for vehicle manufacturers to certify at least some of the sales mix of vehicles on fuels such as methanol and ethanol.

The future market response to the new form of emissions regulation is unknown. For the purpose of meeting new vehicle emissions standards, however, it is still not clear whether some combination of new emissions control approaches and reformulated gasolines can provide benefits equal to those of methanol and ethanol. It is possible that the new emissions standards will simply result in improved gasoline technologies, and that, despite the prospective air quality advantages of the alcohol fuels, the market result of the new standards will simply be cleaner gasolines. However, in 1990 the U.S. Alternative Fuels Council agreed on a goal of a 25% share of nonpetroleum transportation fuels by 2005. This was the first official statement of a specific goal to substitute for the use of petroleum in U.S. transportation. Alcohol fuels could capture a large part of this 25% share of nonpetroleum fuels, although vehicles powered by natural gas and electric energy will no doubt win some acceptance. For the ordinary passenger car, alcohol fuels may offer the most gasolinelike alternative in terms of range, comparable costs, and compatibility with the current gasoline/diesel storage and distribution infrastructure.

MICHAEL D. JACKSON
CARL B. MOYER
Acurex Environmental Corporation

Acurex Corporation, *California's Methanol Program: Evaluation Report,* Vols. 1 (Executive Summary) and 2 (Technical Analyses), Pub. #P500-86-012 and #P500-86-012A, California Energy Commission, Sacramento, Calif., 1987.

California Advisory Board on Air Quality and Fuels, Vol. 1, *Executive Summary;* Vol. 2, *Energy Security Report;* Vol. 3, *Environmental Health and Safety Report;* Vol. 4, *Economics Report;* Vol. 5, *Mandates and Incentives Report,* San Francisco, Calif., June 13, 1990.

Russell, J. Harris, J. Milford, and D. St. Pierre, *Quantitative Estimate of Air Quality Effects of Methanol Fuel Use,* prepared for the California Air Resources Board and the South Coast Air Quality Management District, ARB No. A6-048-32, Carnegie Mellon University, Pittsburgh, Pa., 1989.

D. Sperling, *New Transportation Fuels,* University of California Press, Berkeley, Calif., 1988.

ALCOHOLS, HIGHER ALIPHATIC

SURVEY AND NATURAL ALCOHOLS MANUFACTURE

The monohydric aliphatic alcohols of six or more carbon atoms are generally referred to as higher alcohols. Historically, the higher alcohols, particularly those of 12 or more carbon atoms, were derived from natural fats, oils, and waxes and were called fatty alcohols (see FATS AND FATTY OILS); but now similar alcohols are widely available from synthetic processes using petrochemical feedstocks (qv). Although the natural and synthetic alcohols are used interchangeably for many applications, for some applications the distinction still remains. The higher alcohols can be separated into the plasticizer range alcohols, generally 6–11 carbon atoms, and the detergent range alcohols, 12 or more carbon atoms. There is, however, considerable overlap in use. Production of higher alcohols in North America, Europe, and Japan in 1985 was about 2,600,000 tons and United States production was 35% of that total. About three-fourths of the U.S. output was plasticizer range alcohols, which are used primarily as ester derivatives in plasticizers (qv) and lubricants (see LUBRICATION AND LUBRICANTS). The detergent range alcohols are used mainly as sulfate, ethoxy, and ethoxysulfate derivatives in a wide variety of detergent and surfactant applications (see DETERGENCY; SURFACTANTS).

Most higher alcohols of commercial importance are primary alcohols; secondary alcohols have more limited specialty uses. Detergent range alcohols are apt to be straight chain materials and are made either from natural fats and oils or by petrochemical processes. The plasticizer range alcohols are more likely to be branched chain materials and are made primarily by petrochemical processes. Whereas alcohols made from natural fats and oils are always linear, some petrochemical processes produce linear alcohols and others do not.

Physical Properties

The homologous series of primary normal alcohols exhibits definite trends in physical properties: for each additional CH_2 unit the normal boiling point increases by about 20°C, the specific gravity increases by about 0.003 units, and the melting point increases by about 10°C in the lower end of the range and about 4°C in the upper end. The water solubility decreases with increasing molecular weight and the oil solubility increases. In general, the higher alcohols are soluble in lower alcohols such as ethanol and methanol and in diethyl ether and petroleum ether. The solubility of water in 1-hexanol and 1-octanol is appreciable, but drops off rapidly as alcohol molecular weight increases. Enough solubility remains, however, to make even 1-octadecanol slightly hygroscopic. Mixtures of alcohols, such as 1-octadecanol and 1-hexadecanol, are considerably more hygroscopic. Below C_{12} the normal alcohols are colorless, oily liquids with light, rather fruity odors. At room temperature pure 1-dodecanol solidifies to soft, crystalline platelets and the physical form of higher molecular weight alcohols progresses from these soft platelets to crystalline waxes. Although 1-dodecanol has a slight odor, the higher homologues are essentially odorless. The secondary and branched primary alcohols are oily liquids at room temperature and have light, fruity odors. They are soluble in alcohol solvents and diethyl ether, and also show less affinity for water as molecular weights increase. The members of this group do not have well-defined freezing points; they set to a glass at very low temperatures. Physical properties are often ill-defined because of difficulties in obtaining pure samples.

Chemical Properties

The higher alcohols undergo the same chemical reactions as other primary or secondary alcohols. Similar to other chemicals having long carbon chains, however, reactivity decreases as molecular weight or chain branching increase. This lower reactivity and concommitant decreased solubility in water and in other solvents means that more rigorous reaction conditions, or even use of different reaction schemes as compared to shorter chain alcohols, are generally required. Typical reactions of the higher alcohols include esterification, sulfation, etherification, halogenation, dehydration, oxidation, and amination.

Economic Aspects

In the United States, the lion's share of detergent range alcohol production is by synthetic processes; Shell Chemical is the largest producer in a plant having a 250,000-t capacity (see Table 1). Linear synthetic alcohols can be used interchangeably with natural alcohols except where the presence of minor amounts of chain branching or secondary alcohols precludes use of the synthetics. The more highly branched alcohols are used where branching is not a problem, is desired, or the alcohols are ethoxylated. Ethoxylation reduces the physical and chemical effects of chain branching.

Domestic manufacturers of representative plasticizer range alcohols are given in Table 2. There has been a reduction in the number of manufacturers of 2-ethyl hexanol and other branched chain alcohols. The volume of most branched alcohols has been static, however, and 2-ethylhexanol volume has doubled; the volume of linear alcohols has also grown. A substantial portion of these materials is used in plasticizers for poly(vinyl chloride) (PVC), so plasticizer range alcohol fortunes are tied to variations in the PVC industry. The plasticizers are

Table 1. U.S. Manufacturers of Detergent Range Alcohols

Manufacturer	Process	Feedstock	Products
Procter & Gamble	catalytic hydrogenolysis	coconut and palm kernel oils, tallow, palm oil	C_6–C_{18}
Sherex	catalytic hydrogenolysis	tallow	C_{16}, C_{18}, oleyl
Shell Chemical	modified oxo	ethylene/olefins	C_9–C_{15}
Vista	Ziegler	ethylene	C_6–C_{22}
Ethyl	modified Ziegler	ethylene	C_6–C_{22}
Exxon	modified oxo	olefins	C_{13}, C_{15}

Table 2. Prices and Manufacturers of Plasticizer Range Alcohols

Material	Price, U.S.$/kg[a]	Manufacturer
hexanol	1.74	Ethyl
		Vista
4-methyl-2-pentanol	1.32	Union Carbide
octanol	2.01	Ethyl
		Vista
octanol, perfumer's grade	3.09	
isooctyl alcohol	0.97	Exxon
2-ethylhexanol	0.93	BASF
		Eastman
		Shell Chemical
		Tenn-USS
		Union Carbide
decanol	1.34	Ethyl
		Vista
decanol, perfumer's grade	1.65	

[a] Delivered price May 1989. The listed price is not necessarily the price listed by the indicated manufacturer.

mainly diesters of the alcohols and phthalic acid; di(2-ethylhexyl) phthalate is the highest volume product.

Most manufacturers sell a portion of their alcohol product on the merchant market, retaining a portion for internal use, typically for the manufacture of plasticizers.

Properties

The sales brochures of the manufacturers describe the plasticizer range alcohols available on the merchant market. Typical properties of several commercial plasticizer range alcohols are presented in Table 3. Because in most cases these are mixtures of isomers or alcohols with several carbon chains, the properties of a particular material can vary somewhat from manufacturer to manufacturer. Both odd and even carbon chain alcohols are available, in both linear and highly branched versions.

Toxicological Properties

The higher alcohols are among the less toxic of commonly used chemicals and, in general, their toxic effects are reduced as the number of carbon atoms is increased.

Primary human skin irritation of tetradecanol, hexadecanol, and octadecanol is nil; they have been used for many years in cosmetic creams and ointments. Inhalation hazard, further mitigated by the low vapor pressure of these alcohols, is slight.

Manufacture from Fats and Oils

Fats and oils from a number of animal and vegetable sources are the feedstocks for the manufacture of natural higher alcohols. These materials consist of triglycerides: glycerol esterified with three moles of a fatty acid. The alcohol is manufactured by reduction of the fatty acid functional group.

Hydrogenolysis Process. Fatty alcohols are produced by hydrogenolysis of methyl esters or fatty acids in the presence of a heterogeneous catalyst at 20,700–31,000 kPa (3000–4500 psi) and 250–300°C in conversions of 90–98%. A higher conversion can be achieved using more rigorous reaction conditions, but it is accompanied by a significant amount of hydrocarbon production.

$$RCOOCH_3 + 2\ H_2 \xrightarrow{\text{catalyst}} RCH_2OH + CH_3OH$$

$$RCH_2CH_2OH + H_2 \xrightarrow{\text{high pressure}} RCH_2CH_3 + H_2O$$

Fatty esters (wax esters), formed by ester interchange of the product alcohol and the starting material in the hydrogenolysis reactors, are later separated from the product by distillation. Unreacted methyl esters are also converted to fatty esters in the distillation step

$$RCOOCH_3 + R'OH \rightarrow RCOOR' + CH_3OH$$

so that they too can be separated from the product. Fatty esters are recycled to the hydrogenolysis reactors since they can undergo hydrogenation in a manner similar to methyl esters, in this case yielding two moles of fatty alcohol per mole of ester. Fatty acids can also be used for the higher alcohol production.

High Pressure Hydrogenolysis. There are three principal hydrogenolysis processes in worldwide use: the methyl ester, slurry catalyst process operated by Procter & Gamble, Henkel, and Kao; the methyl ester, fixed-bed catalyst process operated by Henkel and Oleofina; and the fatty acid, slurry catalyst process developed by Lurgi and operated by several licensees. Each process typically uses a copper chromite or copper-zinc catalyst that is modified to meet the needs of the individual producer. Copper chromite when prepared is nominally a complex mixture of primarily copper(II) oxide and copper(II) chromite. But in use it is believed to be reduced to a mixture of metallic copper, copper(II) oxide, and copper(II) chromite, the metallic copper playing an important, but as yet undefined, role in the catalysis of the reaction. Barium, manganese, or other metal ions are sometimes added to improve stability, and silica or other binders may be put in to make a physically strong, fixed-bed catalyst pellet. Hydrogen is usually generated on site from methane or propane. The hydrogen should be of high purity to avoid catalyst poisons, such as sulfur and carbon dioxide, and to prevent buildup of inert gases in the system; pressure swing adsorption (PSA) is often used to remove gaseous impurities.

Uses of Detergent Range Alcohols

The detergent range alcohols and their derivatives have a wide variety of uses in consumer and industrial products either because of surface-active properties, or as a means of introducing a long-chain moiety into a chemical compound. The major use is as surfactants (qv) in detergents and cleaning products. They also are used extensively in cosmetics and pharmaceuticals. Only a small amount of the alcohol is used as is; rather, most is used as derivatives such as the poly(oxyethylene) ethers and the sulfated ethers, the alkyl sulfates, and the esters of other acids, eg, phosphoric acid and monocarboxylic and dicarboxylic acids.

Table 3. Typical Properties of Commercial Plasticizer Range Alcohols

Name	Molecular formula	Hydroxyl value	Acidity, % as acetic	Carbonyl, wt % O	Boiling range, °C	Color, APHA	Moisture, %	Flash point,[a] °C
hexyl	($C_6H_{14}O$)		0.001	<0.003	152–160	5	0.05	63
2-ethylhexanol	$C_8H_{18}O$)	431	<0.007	<0.02	182–186	<10	<0.10	84[b]
isooctyl	($C_8H_{18}O$)		0.001	<0.003	184–190	5	0.05	84
isononyl	($C_9H_{20}O$)		0.001	<0.003	202–213	5	0.05	91
hexyl decyl		408	<0.004	0.003	168–203	5	0.01	81[c]
octanol	($C_8H_{18}O$)	431	<0.005	0.003	184–195	5	0.03	88[c]
decanol	($C_{10}H_{22}O$)	355	<0.01	0.003	226–230	5	0.03	113
tridecyl	($C_{13}H_{28}O$)	283	0.001	<0.003	254–263	5	<0.05	127

[a] Pensky-Martens closed cup unless otherwise noted.
[b] Cleveland open cup.
[c] Tag closed cup.

Table 4. Uses of Plasticizer Range Alcohols

Industry	Use as alcohol	Use as derivative
plastics	emulsion polymerization	plasticizer, flame retardant, oxidation and uv stabilizer, heat stabilizer, polymerization initiator
petroleum and lubrication	defoamer	lubricant, grease, lubricant additive, hydraulic fluid, diesel fuel additive
agriculture	stabilizer, tobacco sucker control, herbicide, fungicide	surfactant, insecticide, herbicide
mineral processing	solvent, extractant, antifoam	extractant, surfactant
textile	leveling agent, defoamer	surfactant
coatings	solvent, smoothing agent	surfactant, drying agent, solvent
metal working	solvent, lubricant, protective coating	lubricant, surfactant
chemical processing	antifoam, solvent	solvent
food		flavoring agent
cosmetics	perfume ingredient	

Uses of Plasticizer Range Alcohols

The plasticizer range alcohols are utilized primarily in plasticizers, but they also have a wide range of uses in other industrial and consumer products, as shown in Table 4. As in the case of the detergent range alcohols, the plasticizer range materials are little used as is, but rather are employed as the ester derivatives of acids such as phthalic, adipic, and trimellitic.

RICHARD A. PETERS
The Procter & Gamble Company

U. R. Kreutzer, *J. Am. Oil Chem. Soc.* **61,** 343–348 (1984).

Fatty Alcohols, Raw Materials, Methods, Uses, Henkel K.-G.a.A., Düsseldorf, 1982. Also published in German as *Fettalkohole.*

SYNTHETIC PROCESSES

Higher aliphatic alcohols (C_6–C_{18}) are produced in a number of important industrial processes using petroleum-based raw materials. These processes are summarized in Table 1, as are the principal synthetic products and most important feedstocks (qv).

Table 1. Synthetic Industrial Processes for Higher Aliphatic Alcohols

Process	Feedstock(s)	Principal products	Worldwide capacity, millions of tons
Ziegler (organoaluminum)	ethylene, triethylaluminum	primary C_6–C_{18} linear alcohols	0.3
oxo (hydroformylation)	olefins based on ethylene, propylene, butylene, or paraffins	primary alcohols	4.2
aldol	*n*-butyraldehyde	2-ethylhexanol	[a]
paraffin oxidation	paraffin hydrocarbons	secondary alcohols	0.2
Guerbet	lower primary alcohols	branched primary alcohols	[b]
Total			*4.7*

[a] Included in oxo process total.
[b] Less than 0.05.

By far the largest volume synthetic alcohol is 2-ethylhexanol, $C_8H_{18}O$, used mainly in production of the poly(vinyl chloride) plasticizer bis(2-ethylhexyl) phthalate, $C_{24}H_{38}O_4$, commonly called dioctyl phthalate or DOP (see PLASTICIZERS). A number of other plasticizer primary alcohols in the C_6–C_{11} range are produced, as are large volumes of C_{10}–C_{18} synthetic, mainly primary, alcohols used as intermediates to surfactants (qv) for detergents. Other lower volume synthetic alcohol application areas include solvents and specialty esters.

JOHN D. WAGNER
GEORGE R. LAPPIN
J. RICHARD ZIETZ
Ethyl Corporation

E. R. Freitas and C. R. Gum, *Chem. Eng. Prog.*, 73 (Jan. 1979).

K. Noweck and H. Ridder, *Ullmann's Encyclopedia of Industrial Chemistry,* 5th ed., VCH Verlagsgesellschaft mbh, Weinheim, Germany, 1987.

J. A. Monick, *Alcohols, Their Chemistry, Properties, and Manufacture,* Reinhold, New York (1968).

E. J. Wickson and H. P. Dengler, *Hydrocarbon Process.* **51**(11), 69 (1972).

ALCOHOLS, POLYHYDRIC

Polyhydric alcohols or polyols contain three or more CH_2OH functional groups. The monomeric compounds have the general formula $R(CH_2OH)_n$, where $n = 3$ and R is an alkyl group or $C.CH_2OH$; the dimers and trimers are also commercially significant.

Each polyhydric alcohol is a white solid, ranging from the crystalline pentaerythritols to the waxy trimethylol alkyls. The trihydric alcohols are very soluble in water, as is ditrimethylolpropane. Pentaerythritol is moderately soluble and dipentaerythritol and tripentaerythritol are less soluble. Table 1 lists the physical properties of these alcohols. Pentaerythritol and trimethylolpropane have no known toxic or irritating effects. Finely powdered pentaerythritol, however, may form explosive dust clouds at concentrations above 30 g/m^3 in air. The minimum ignition temperature is 450°C.

Reactions

Direct acetylation of pentaerythritol using acetic acid in aqueous solution or in toluene produces a mixture of acetates which can be fairly readily separated by chromatographic methods or distillation.

Long-chain esters of pentaerythritol have been prepared by a variety of methods.

Polyhydric alcohol mercaptoalkanoate esters are prepared by reaction of the appropriate alcohols and thioester using *p*-toluenesulfonic acid catalyst.

Pentaerythritol can be oxidized to 2,2-bis(hydroxymethyl) hydracrylic acid, $C_5H_{10}O_5$. Bromohydrins can be prepared directly from polyhydric alcohols. Borolane products of mixed composition can be synthesized by direct addition of boric acid to pentaerythritol.

Reaction between pentaerythritol and phosphorus trichloride yields the spirophosphite, 3,9-dichloro-2,4,8,10-tetraoxa-3,9-diphosphaspiro[5,5]-undecane, $C_5H_8Cl_2O_4P_2$.

The commercially important explosive pentaerythritol tetranitrate (PETN), $C_5H_8N_4O_{12}$, is produced by direct reaction of pentaerythritol in nitric or nitric–sulfuric acid media.

Aminoalkoxy pentaerythritols are obtained by reduction of the cyanoethoxy species obtained from the reaction between acrylonitrile, pentaerythritol, and lithium hydroxide in aqueous solution.

Tosylates of pentaerythritol and the higher homologues can be converted to their corresponding tetra, hexa, or octaazides by direct reaction of sodium azide, and azidobenzoates of trimethylolpropane and dipentaerythritol are prepared by reaction of azidobenzoyl chloride and the alcohols in pyridine medium.

Pentaerythritol can be converted to the biscyclic formal, 2,4,8,10-tetra-oxaspiro[5,5]undecane, $C_7H_{12}O_4$, by heating in the presence of formaldehyde or paraformaldehyde and an acid catalyst.

Simple alkyl and alkenyl ethers of pentaerythritol are produced on direct reaction of the polyol and the required alkyl or alkenyl chloride in the presence of quaternary alkylamine bromide.

Manufacture

Pentaerythritol is produced by reaction of formaldehyde and acetaldehyde in the presence of a basic catalyst, generally an alkali or alkaline-earth hydroxide.

Dipentaerythritol and tripentaerythritol are obtained as by-products of the pentaerythritol process and may be further purified by fractional crystallization or extraction. Trimethylolethane and trimethylolpropane may be prepared using the appropriate aldehyde in place of acetaldehyde. Ditrimethylolpropane is obtained as a by-product of the trimethylolpropane synthesis.

Economic Aspects

Production of pentaerythritol in the United States has been erratic. Demand decreased in 1975 because of an economic recession and grew only moderately to 1980.

The world's largest producers are Perstorp AB (Sweden, the United States, and Italy), Hoechst Celanese Corporation (the United States and Canada), Degussa (Germany), and Hercules (the United States).

The world's largest producers of trimethylolpropane are Perstorp AB, Hoechst Celanese, and Bayer. Dipentaerythritol is sold by Perstorp AB and by Hercules (the United States), and ditrimethylolpropane by Perstorp AB, both in relatively pure form. Tripentaerythritol is also available; however, the purity is limited. Trimethylolethane is produced commercially by Alcolac (the United States) and Mitsubishi Gas Chemicals (Japan).

Health and Safety Factors

Pentaerythritol and trimethylolpropane are classified as nuisance particulate and dust, respectively. They are both nontoxic to animals by ingestion or inhalation and are essentially nonirritating to the skin or eyes.

Uses

The most important industrial use of pentaerythritol is in a wide variety of paints, coatings, and varnishes, where the cross-linking capability of the four hydroxy groups is critical.

The explosives and rocket fuels formed by nitration of pentaerythritol to the tetranitrate using concentrated nitric acid are generally used as a filling in detonator fuses.

Pentaerythritol is used in self-extinguishing, nondripping, flame-retardant compositions with a variety of polymers, including olefins, vinyl acetate and alcohols, methyl methacrylate, and urethanes. Polymer compositions containing pentaerythritol are also used as secondary heat-, light-, and weather-resistant stabilizers with calcium, zinc, or barium salts, usually as the stearate, as the prime stabilizer. The polymers may be in plastic or fiber form.

Pentaerythritol in rosin ester form is used in hot-melt adhesive formulations, especially ethylene–vinyl acetate (EVA) copolymers, as a tackifier.

Table 1. Physical Properties of Polyhydric Alcohols

Property	Pentaerythritol	Dipentaerythritol	Tripentaerythritol	Trimethylolethane	Trimethylolpropane	Ditrimethylolpropane[a]
molecular formula	$C_5H_{12}O_4$	$C_{10}H_{22}O_7$	$C_{15}H_{32}O_{10}$	$C_5H_{12}O_3$	$C_6H_{14}O_3$	$C_{12}H_{26}O_5$
melting point, °C	261–262	221–222.5	248–250	202	58.8	112–114
boiling point, °C	276 (4 kPa)			283	289	210 (0.12 kPa)
solubility, g/100 g water at 25°C	7.23	0.28[b]	0.018[b]	soluble	soluble	2.6
density, g/mL	1.396	1.369	1.30		1.09	1.18

[a] Data supplied by Perstorp AB.
[b] Estimated value.

E. Berlow, R. H. Barth, and E. J. Snow, *The Pentaerythritols*, Reinhold Publishing Corp., New York, 1958.

WILLIAM N. HUNTER
Celanese Canada Inc.

ALDEHYDE RESINS. See ACETAL RESINS.

ALDEHYDES

Aldehydes are carbonyl-containing organic compounds in which the carbonyl function is at a terminal carbon. These compounds are extremely reactive. The carbonyl group is susceptible to both oxidation and reduction, yielding acids and alcohols, respectively. Additionally, the carbonyl group is susceptible to nucleophilic addition, providing a means by which to form new chemical bonds. Furthermore, the presence of the carbonyl activates the hydrogens bound to the alpha carbon and thus provides an additional site of reactivity. Ketones are a similar class of compounds where the carbonyl group is nonterminal (see KETONES).

Nomenclature

The common method of naming aldehydes corresponds very closely to that of the related acids (see CARBOXYLIC ACIDS), in that the term *aldehyde* is added to the base name of the acid.

Physical Properties

The C-1 and C-2 carbon aliphatic aldehydes, formaldehyde and acetaldehyde, are gases at ambient conditions whereas the C-3 (propanal) through C-11 (undecanal) aldehydes are liquids, and higher aldehydes are solids at room temperature, as can be seen in Table 1. Aldehydes are usually soluble in common organic solvents and, except for the C-1 to C-5 aldehydes, are only sparingly soluble in water. The lower, C-1 to C-8 aldehydes have pungent, penetrating, unpleasant odors, some of which may be attributed to the presence of the corresponding acids that form by air oxidation. Above C_8, aldehydes have more pleasant odors and some higher aldehydes are used in the perfume and flavoring industries (see PERFUMES). Interestingly, the C_9 aldehyde, nonanal, is reported possibly to be a human sex pheromone. Aldehydes must be kept from contact with air (oxygen) to retain purity.

Chemical Properties

Aldehydes are very reactive compounds. Reactions generally fall into two classes: those directly affecting the carbonyl group and those involving the adjacent carbon atom.

Manufacture

Recent advances in technology involving a new class of highly reactive phosphite-promoted catalysts permit the manufacture of higher (C-7 to C-15) aldehydes and the hydroformylation of internal olefins at low temperatures and pressures. Fatty aldehydes, generally produced by dehydrogenation of corresponding alcohols in the presence of a suitable catalyst, can now be manufactured using oxo technology.

The direct oxidation of ethylene is used to produce acetaldehyde (qv) in the Wacker-Hoechst process. Another commercial aldehyde synthesis is the catalytic dehydrogenation of primary alcohols at high temperature in the presence of a copper or a copper–chromite catalyst.

Production and Economic Aspects

U.S. aldehyde producers include Aristech Chemical Corporation, BASF Corporation, Borden Chemical, E.I. du Pont de Nemours, Eastman, Georgia-Pacific Corporation, Givauden Corporation, Hoechst Celanese, Koch Industries, Penta Manufacturing, Reichhold Limited, Rhone-Polenc Inc., and Union Carbide.

Characterization

Aldehydes can be characterized qualitatively through the use of Tollens' or Fehling's reagents as well as by spectroscopic means.

Toxicology

Interest in the toxicity of aldehydes has focused primarily on specific compounds, particularly formaldehyde, acetaldehyde, and acrolein. Little evidence exists to suggest that occupational levels of exposure to aldehydes would result in mutations, although some aldehydes are clearly mutagenic in some test systems. There are however, acute effects of aldehydes: eye, skin, and respiratory tract irritation and sensitization, anesthesia, and organ pathology (particularly respiratory tract and pulmonary edema).

Table 1. Properties of Aldehydes

Aldehyde	Molecular formula	Molecular weight	Melting point, °C	Boiling point, °C
formaldehyde	CH_2O	30.03	−92	−21
acetaldehyde	C_2H_4O	40.05	−121	20
propionaldehyde	C_3H_6O	58.08	−81	49
butanal (*n*-butyraldehyde)	C_4H_8O	72.1	−99	75
2-methylpropanal (isobutyraldehyde)	C_4H_8O	72.1	−66	64
pentanal (*n*-valeraldehyde)	$C_5H_{10}O$	86.13	−91	103
3-methylbutanal (isovaleraldehyde)	$C_5H_{10}O$	86.13	−51	93
hexanal (caproaldehyde)	$C_6H_{12}O$	100.16	−56	131
benzaldehyde	C_7H_6O	106.13	−26	179
heptanal (heptaldehyde)	$C_7H_{14}O$	114.19	−42	155
octanal (caprylaldehyde)	$C_8H_{16}O$	128.21		171
phenylacetaldehyde	C_8H_8O	120.16	33	194
o-tolualdehyde	C_8H_8O	120.14		202
m-tolualdehyde	C_8H_8O	120.14		199
p-tolualdehyde	C_8H_8O	120.14		205
salicylaldehyde (*o*-hydroxybenzaldehyde)	$C_7H_6O_2$	122.12	2	197
p-hydroxybenzaldehyde (4-formylphenol)	$C_7H_6O_2$	122.12	116	
p-anisaldehyde (*p*-methoxybenzaldehyde)	$C_8H_8O_2$	136.14	0	248

Uses

Aldehydes find the most widespread use as chemical intermediates. A large proportion of aldehydes are converted to alcohols (eg, butyraldehyde and butanol) for use as solvents and also in the manufacture of plasticizers (see PLASTICIZERS). Fatty aldehydes C-8–C-13 are used in nearly all perfume types and aromas. (see PERFUMES). Polymers and copolymers of aldehydes exist and are of commercial significance.

DAVID J. MILLER
Union Carbide Chemicals and Plastics Corporation

J. Buckingham, *Dictionary of Organic Compounds,* 5th ed., Chapman and Hall, New York, 1982.

S. Patai, *The Chemistry of the Carbonyl Group,* Wiley-Interscience, New York, 1966 and 1970.

C. D. Gutche, *The Chemistry of Carbonyl Compounds,* Prentice-Hall, Inc., Englewood Cliffs, N.J., 1967.

ALKALI AND CHLORINE PRODUCTS

CHLORINE AND SODIUM HYDROXIDE

Production and Manufacture

Alkali and chlorine products are a group of commodity chemicals which include chlorine, Cl_2; sodium hydroxide (caustic soda), NaOH; sodium carbonate (soda ash), Na_2CO_3; potassium hydroxide (caustic potash), KOH; and hydrochloric acid (muriatic acid or anhydrous), HCl. Chlorine and caustic soda are the two most important products in this group, ranking among the top ten chemicals in the United States. The applications for chlorine and the alkalies are so varied that there is hardly a consumer product which is not dependent on one or both of them at some manufacturing stage.

Electrolytic Decomposition of Sodium Chloride. Chlorine and caustic soda are coproducts of the electrolysis of aqueous solutions of sodium chloride, NaCl (commonly called brine), following the overall chemical reaction

$$2\ NaCl + 2\ H_2O \overset{energy}{\rightarrow} 2\ NaOH + Cl_2 + H_2$$

This reaction has a positive free energy of 422.2 kJ (100.9 kcal) at 25°C and hence energy has to be supplied in the form of d-c electricity to

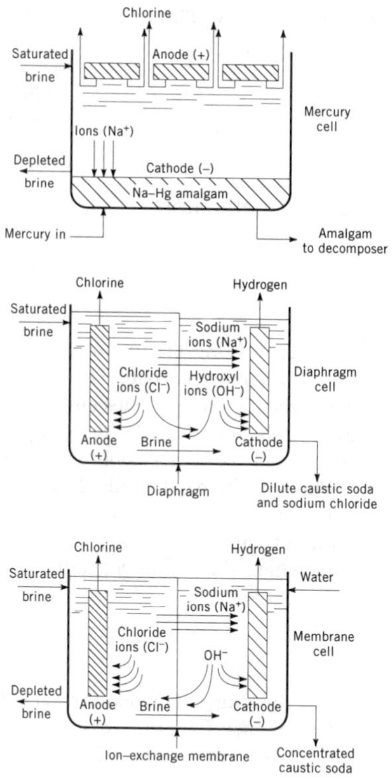

Figure 1. Chlorine electrolysis cells. Courtesy of McGraw-Hill, Inc.

drive the reaction in a net forward direction. The amount of electrical energy required for the reaction depends on electrolytic cell parameters such as current density, voltage, anode and cathode material, and the cell design.

Conversion of aqueous NaCl to Cl_2 and NaOH is achieved in three types of electrolytic cells: the diaphragm cell, the membrane cell, and the mercury cell. The distinguishing feature of these cells is the manner by which the electrolysis products are prevented from mixing with each other, thus ensuring generation of products having the proper purity.

Efficiency and Energy Consumption. The electrical energy consumed during the electrolysis of brine to produce chlorine gas and sodium amalgam is greater than that used to generate Cl_2 and H_2 in the diaphragm or membrane cell. However, the latter processes also use

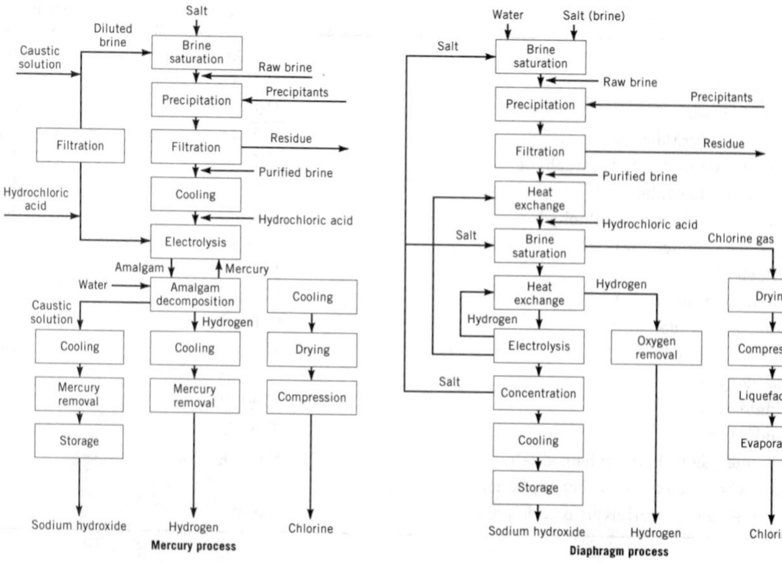

Figure 2. Flow diagrams of the Mercury and Diaphragm chlor–alkali processes.

Table 1. Components of Diaphragm, Membrane, and Mercury Cells

Component	Mercury cell	Diaphragm cell	Membrane cell
anode	RuO_2 + TiO_2 coating on Ti substrate	RuO_2-based coating on Ti substrate	RuO_2-based coating on Ti substrate
cathode	mercury on steel	steel or steel coated with activated nickel	steel- or Ni-based catalytic coating on nickel
diaphragm	none	asbestos, polymer-modified asbestos, or Polyramix (nonasbestos)	ion-exchange membrane
cathode product	sodium amalgam	10−12% NaOH + 15−17% NaCl and H_2	30−33% NaOH+ >0.01% NaCl and H_2
decomposer product	50% NaOH and H_2	none	none
evaporator product	none	50% NaOH with ~1.1% salt and solid salt	50% NaOH with ~0.01% salt
steam consumption	none	1500−2300 kg/t NaOH	450−550 kg/t NaOH
cell voltage, V	4−5	3−4	2.8−3.3
current density, kA/m^2	7−10	0.5−3	2−5

Table 2. Summary of Current Membrane Cell Technologies

	Bipolar cells			Monopolar cells					
	Asahi Chemical Acilyzer	De Nora bipolar	Uhde	Asahi Glass AZEC-M	CEC DMC-404 × 2	De Nora DD	Lurgi	ICI FM-215P	OxyTech MGC
effective membrane area, m^2	2.88−5.4	0.9−5.12	1−3	0.2	3.03	0.9−1.7	0.8	0.21	1.5
cells per electrolyzer	50−100	10−32	2−120	30−540	4	10−32		1−120	2−30
current load, kA	5.8−21.6	2.7−20.5	2−18	18−340	24−48	27−218		1−100	6−225
current density, kA/m^2	2−4	2−4	2−6	3−4	2−4	3−4	1−4	1.5−4.1	2−5

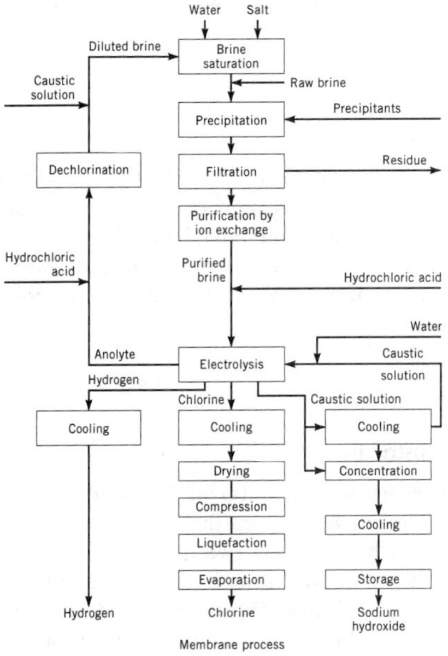

Figure 3. Flow diagram of the membrane chlor–alkali process.

energy in the form of steam for evaporation of the cell liquor. Table 1 summarizes the major differences in the three technologies. The minimum energy required to convert salt to Cl_2, H_2, and 50% NaOH (6.06 GJ/t of Cl_2) is, of course, the same in all of these processes.

Manufacturing Processes. Solution mining of salt and the availability of asbestos resulted in the dominance of the diaphragm process in North America, whereas solid salt and mercury availability led to the dominance of the mercury process in Europe. Japan imported its salt in solid form and, until the development of the membrane process, also favored the mercury cell for production.

Anodes. All but a very few installations have now converted to the exclusive use of the RuO_2 + TiO_2-coated titanium anodes. These are supplied under the trade name DSA (for dimensionally stable anode). (DSA is a trademark of Eltech Systems Corp.).

Electrolytic Cell Operating Characteristics. Currently the greatest volume of chlorine production is by the diaphragm cell process, followed in order by those of the mercury cell and the membrane cell. However, because of the ecological and economic advantages of the membrane process over the other systems, membrane cells are currently favored for new production facilities. The basic characteristics of the three cell processes are shown in Figure 1. A summary of the current membrane cell technologies is provided in Table 2.

Catalytic Cathodes. The cathode material generally found in diaphragm cells is low carbon steel in either mesh or perforated form, whereas nickel or stainless steel is used in membrane cell electrolyzers. Energy savings by reducing the Cathodic overpotential by as much as 200 to 250 mV are realizable, in principle, by using high surface-area catalytic coatings on the cathode substrates. Materials generally chosen for cathode coatings are nickel or noble metal based.

Chlorine Plant Auxiliaries. Flow diagrams for the three electrolytic chlor–alkali processes are give in Figures 2 and 3. Although they differ somewhat in operation, auxiliary processes such as brine purification and chlorine recovery are common to each.

Sodium Hydroxide Processing. Sodium hydroxide is usually produced as a 50% water-based solution, although 73% and anhydrous sodium hydroxide are also marketed. High purity sodium hydroxide is available directly from the mercury and membrane cell processes.

Hydrogen Processing. The hydrogen produced in all electrolytic chlor–alkali processes is relatively (>99.9%) pure and requires only cooling to remove water along with entrained salt and caustic. The heat is often recovered into the brine system. Hydrogen from the mercury process must also be scrubbed to remove mercury.

Table 3. Physical Constants of Chlorine

Property	Value
atomic number	17
atomic weight	35.453
stable isotope abundance, %	
^{35}Cl	75.53
^{37}Cl	24.47
electronic configuration in the ground state	$[Ne]3s^2 3p^5$
melting point, °C	−100.98
boiling point, °C, at 101.3 kPa[a]	−34.05
density relative to air	2.48
critical density, kg/m^3	565.00
critical pressure, MPa[a]	7.71083
critical volume, m^3/kg	0.001745
critical temperature, K	417.15
density, kg/m^3 at 0°C and 101.3 kPa[a]	3.213
viscosity (gas), Pa·s[b] at 20°C	14.0
viscosity (liquid), Pa·s[b] at 20°C	340
latent heat of vaporization, J/g[c]	287.4
enthalpy of fusion ΔH_f, kJ/kg[c]	90.33
enthalpy of vaporization ΔH_v, kJ/kg[c]	287.1
standard electrode potential, V	1.359
enthalpy of dissociation, kJ/mol[c]	2.3944
electron affinity, eV	3.77
enthalpy of hydration of Cl^-, kJ/mol[c]	405.7
ionization energies, eV	13.01, 23.80, 39.9, 53.3, 67.8, 96.6, and 114.2
specific heatC_p, kJ/kg·K[c]	0.481
specific heat C_v, kJ/kg·K[c]	0.357
specific magnetic susceptibility, m^3/kg at 20°C	-7.4×10^{-9}
electrical conductivity of liquid Cl_2, $(\Omega\ cm)^{-1}$ at −70 °C	10^{-16}
dielectric constant for wavelengths 10 m at 0°C	1.97

[a] To convert kPa to mm Hg, multiply by 7.5.
[b] To convert Pa·s to P, multiply by 10.
[c] To convert J to cal, divide by 4.184.

The hydrogen can be used for organic hydrogenation, catalytic reductions, and ammonia synthesis. It can also be burned with chlorine to produce high quality HCl and used to provide a reducing atmosphere in some applications. In many cases, however, it is used as a fuel.

Other Chlorine Production Processes. Although electrolytic production of Cl_2 and NaOH from NaCl accounts for most of the chlorine produced, other commercial processes for chlorine are also in operation, eg, chlorine from potassium hydroxide manufacture, chlorine from HCl, chlorine from the magnesium process, chlorine from the titanium process, and chemical oxidation of HCl.

CHLORINE

Physical Properties

Chlorine, a member of the halogen family, is a greenish-yellow gas having a pungent odor at ambient temperatures and pressures and a density 2.5 times that of air. In liquid form it is clear amber; Solid chlorine forms pale yellow crystals. The principal properties of chlorine are presented in Table 3.

Chlorine is soluble in water and in salt solutions, the solubility decreasing with salt strength and temperature. It is partially hydrolyzed in aqueous solution as

$$Cl_2 + H_2O \rightleftharpoons HCl + HOCl$$

Chemical Properties

The chemical properties of chlorine have been discussed. Chlorine generally exhibits a valence of −1 in compounds, but it also exists in the formal positive oxidation states of +1 (NaClO), +3 ($NaClO_2$), +5 ($NaClO_3$), and +7 ($NaClO_4$). Molecular chlorine is a strong oxidizer and a chlorinating agent, adding to double bonds in aliphatic compounds or undergoing substitution reactions with both aliphatic and aromatic ones. Significant industrial reaction products are presented in Tables 4 and 5.

Materials of Construction

The choice of construction material for handling chlorine depends on equipment design and operating conditions. Dry chlorine, with less than 40 ppm by weight of water, can be handled safely below 120°C in equipment made from iron, steel, stainless steels, Monel metal, nickel, copper, brass, bronze, and lead. Silicone materials, titanium, and high surface-area materials such as steel wool should be avoided.

Liquid chlorine is generally stored in vessels made from nonalloyed carbon steel or cast steel.

Gaskets in both dry gas and liquid chlorine systems are made of rubberized compressed asbestos.

Storage and Transportation

Surveys of existing national and international regulations for the handling and transportation of chlorine are available from the Chlorine Institute. Chlorine is liquefied and stored at ambient or low

Table 4. Chlorine Derivatives[a]

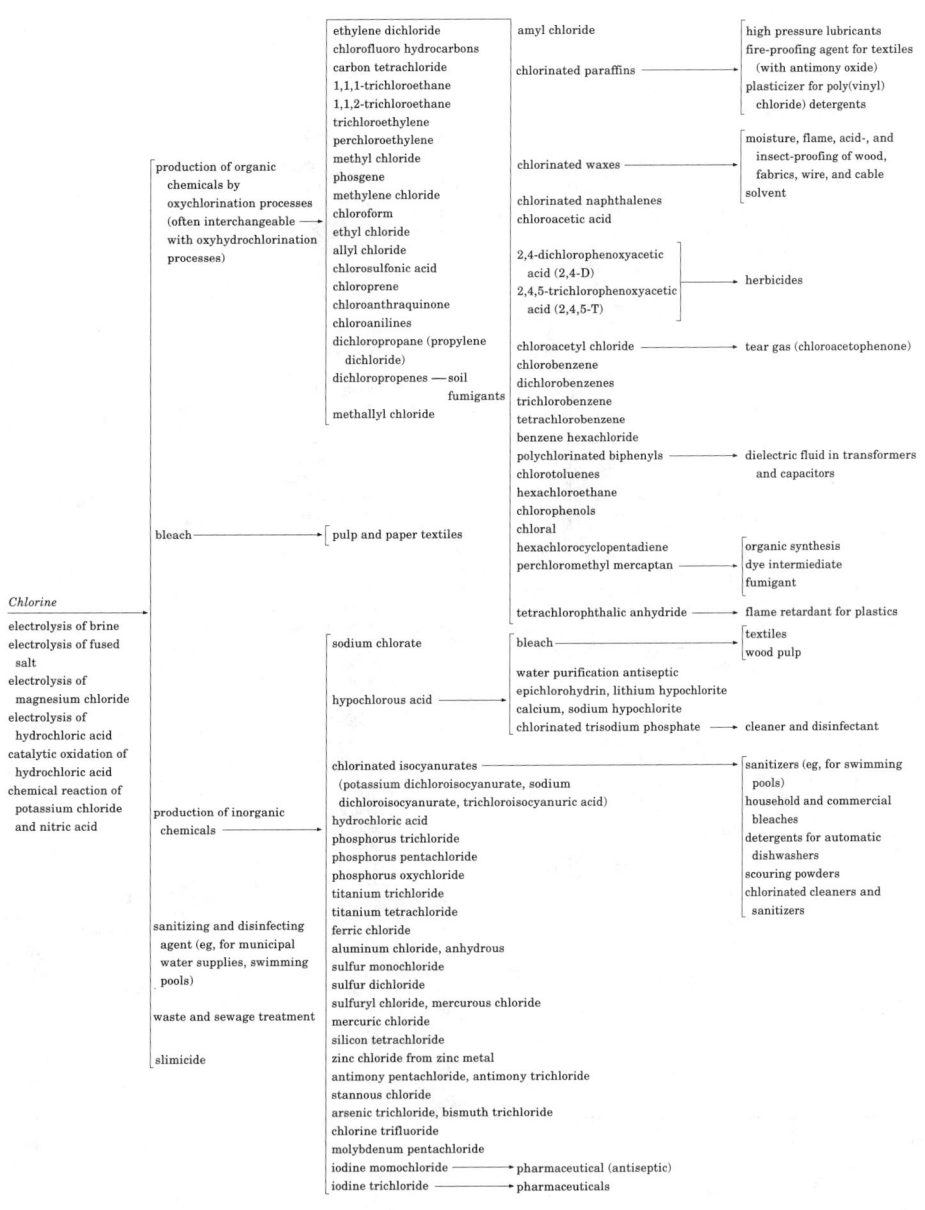

Chlorine →
- electrolysis of brine
- electrolysis of fused salt
- electrolysis of magnesium chloride
- electrolysis of hydrochloric acid
- catalytic oxidation of hydrochloric acid
- chemical reaction of potassium chloride and nitric acid

Uses:

- production of organic chemicals by oxychlorination processes (often interchangeable with oxyhydrochlorination processes) →
  - ethylene dichloride
  - chlorofluoro hydrocarbons
  - carbon tetrachloride
  - 1,1,1-trichloroethane
  - 1,1,2-trichloroethane
  - trichloroethylene
  - perchloroethylene
  - methyl chloride
  - phosgene
  - methylene chloride
  - chloroform
  - ethyl chloride
  - allyl chloride
  - chlorosulfonic acid
  - chloroprene
  - chloroanthraquinone
  - chloroanilines
  - dichloropropane (propylene dichloride)
  - dichloropropenes — soil fumigants
  - methallyl chloride
  - amyl chloride
  - chlorinated paraffins → high pressure lubricants; fire-proofing agent for textiles (with antimony oxide); plasticizer for poly(vinyl) chloride) detergents
  - chlorinated waxes → moisture, flame, acid-, and insect-proofing of wood, fabrics, wire, and cable; solvent
  - chlorinated naphthalenes
  - chloroacetic acid
  - 2,4-dichlorophenoxyacetic acid (2,4-D); 2,4,5-trichlorophenoxyacetic acid (2,4,5-T) → herbicides
  - chloroacetyl chloride → tear gas (chloroacetophenone)
  - chlorobenzene
  - dichlorobenzenes
  - trichlorobenzene
  - tetrachlorobenzene
  - benzene hexachloride
  - polychlorinated biphenyls → dielectric fluid in transformers and capacitors
  - chlorotoluenes
  - hexachloroethane
  - chlorophenols
  - chloral
  - hexachlorocyclopentadiene
  - perchloromethyl mercaptan → organic synthesis; dye intermiediate; fumigant
  - tetrachlorophthalic anhydride → flame retardant for plastics
- bleach → pulp and paper textiles
- production of inorganic chemicals →
  - sodium chlorate
  - hypochlorous acid → bleach → textiles; wood pulp; water purification antiseptic; epichlorohydrin, lithium hypochlorite; calcium, sodium hypochlorite; chlorinated trisodium phosphate → cleaner and disinfectant
  - chlorinated isocyanurates (potassium dichloroisocyanurate, sodium dichloroisocyanurate, trichloroisocyanuric acid) → sanitizers (eg, for swimming pools); household and commercial bleaches; detergents for automatic dishwashers; scouring powders; chlorinated cleaners and sanitizers
  - hydrochloric acid
  - phosphorus trichloride
  - phosphorus pentachloride
  - phosphorus oxychloride
  - titanium trichloride
  - titanium tetrachloride
  - ferric chloride
  - aluminum chloride, anhydrous
  - sulfur monochloride
  - sulfur dichloride
  - sulfuryl chloride, mercurous chloride
  - mercuric chloride
  - silicon tetrachloride
  - zinc chloride from zinc metal
  - antimony pentachloride, antimony trichloride
  - stannous chloride
  - arsenic trichloride, bismuth trichloride
  - chlorine trifluoride
  - molybdenum pentachloride
  - iodine momochloride → pharmaceutical (antiseptic)
  - iodine trichloride → pharmaceuticals
- sanitizing and disinfecting agent (eg, for municipal water supplies, swimming pools)
- waste and sewage treatment
- slimicide

[a] Courtesy of SRI International.

temperature, ensuring that the pressure in the storage system corresponds to the vapor pressure of liquefied chlorine at the temperature in the tank.

Chlorine is classified by the U.S. Department of Transportation (DOT) as a nonflammable compressed gas requiring a green label. Chlorine must be packaged in containers complying with DOT or Coast Guard regulations related to construction, loading, and labeling, and state, local, and insurance regulations. Transportation of about 70% of liquid chlorine is by rail, 20% by pipe lines, 7% by barges, and the remainder in cylinders.

Safety

No attempt should be made to handle chlorine for any purpose without a thorough understanding of its properties and the hazards involved. This information is available from the Chlorine Institute and from various handbooks.

Chlorine gas is a respiratory irritant and is readily detectable at concentrations of <1 ppm in air because of its penetrating odor. Chlorine gas, after several hours of exposure, causes mild irritation of the eyes and of the mucous membrane of the respiratory tract. At high concentrations and in extreme situations, increased difficulty in breathing can result in death through suffocation.

SODIUM HYDROXIDE (CAUSTIC SODA)

Physical Properties

Sodium hydroxide, NaOH, mol wt 39.998, is a brittle, white, translucent crystalline solid. Because of its corrosive action on all human

Table 5. Hydrochloric Acid Derivatives[a]

Hydrochloric acid
- by-product
- from salt and sulfuric acid
- from chlorine and hydrogen

→
- steel pickling → ferric chloride (by-product)
- modifier in ore benefication (flotation) processes
- refining of tin and tantalum
- chemical milling baths (metal working)
- gaseous flux for babbitting operation
- chemical reagent and intermediate →
  - production of chlorinated hydrcarbons via oxyhydrochlorination processes (often interchangeable with oxychlorination processes) →
    - vinyl chloride from acetylene
    - ethyl chloride
    - other chlorinated hydrocarbons
  - production of alkyl chlorides via substitution reactions →
    - methyl chloride
    - ethyl chloride
    - 1,4-dichlorobutane (tetramethylene dichloride) → adiponitrile; other organic synthesis products
    - higher alkyl chlorides
    - Grignard reagents: organometallics, alkychlorosilanes → silicone resins and elastomers
  - production of metallic chlorides and other inorganic chemicals →
    - calcium chloride (waste by-product)
    - aluminum chloride, anhydrous
    - barium chloride
    - cadmium chloride
    - cobaltous chloride
    - cupric chloride
    - magnesium chloride from brines
    - nickel chloride
    - stannic chloride
    - titanium tetrachloride
    - chlorosulfonic acid
    - zinc chloride
    - antimony trichloride
    - arsenic trioxide → arsenic trichloride
    - bismuth trichloride
    - lithium chloride
    - zirconium tetrachloride ↓
      - zirconium → refractories; alloys; photoflash bulbs; pyrotechnics
      - water repellant agent
      - pigments
    - chlorine
    - silica gel
    - silica, hydrated
    - other inorganic chemicals
    - aniline hydrochloride ↓
      - diphenylamine → rubber accelerator; plastics stabilizer; rocket propellants; pesticides
      - dyes
  - production of hydrochlorides →
    - amine hydrochlorides → pharmaceuticals
    - hydrzine dihydrochloride → chlorine scavenger for hydrogen chloride chloride gas streams
    - other hydrochlorides
- oil well acidizing
- food processing (eg, in sugar refining, brewing industry, and for manufacture of monosodium glutamate and gelatin)
- catalyst for production of aniline by nitrobenzene reduction
- latex coagulating agent
- deliming agent for hides
- desulfurization agent for petroleum
- pH control
- scale removal
- veterinary medicine

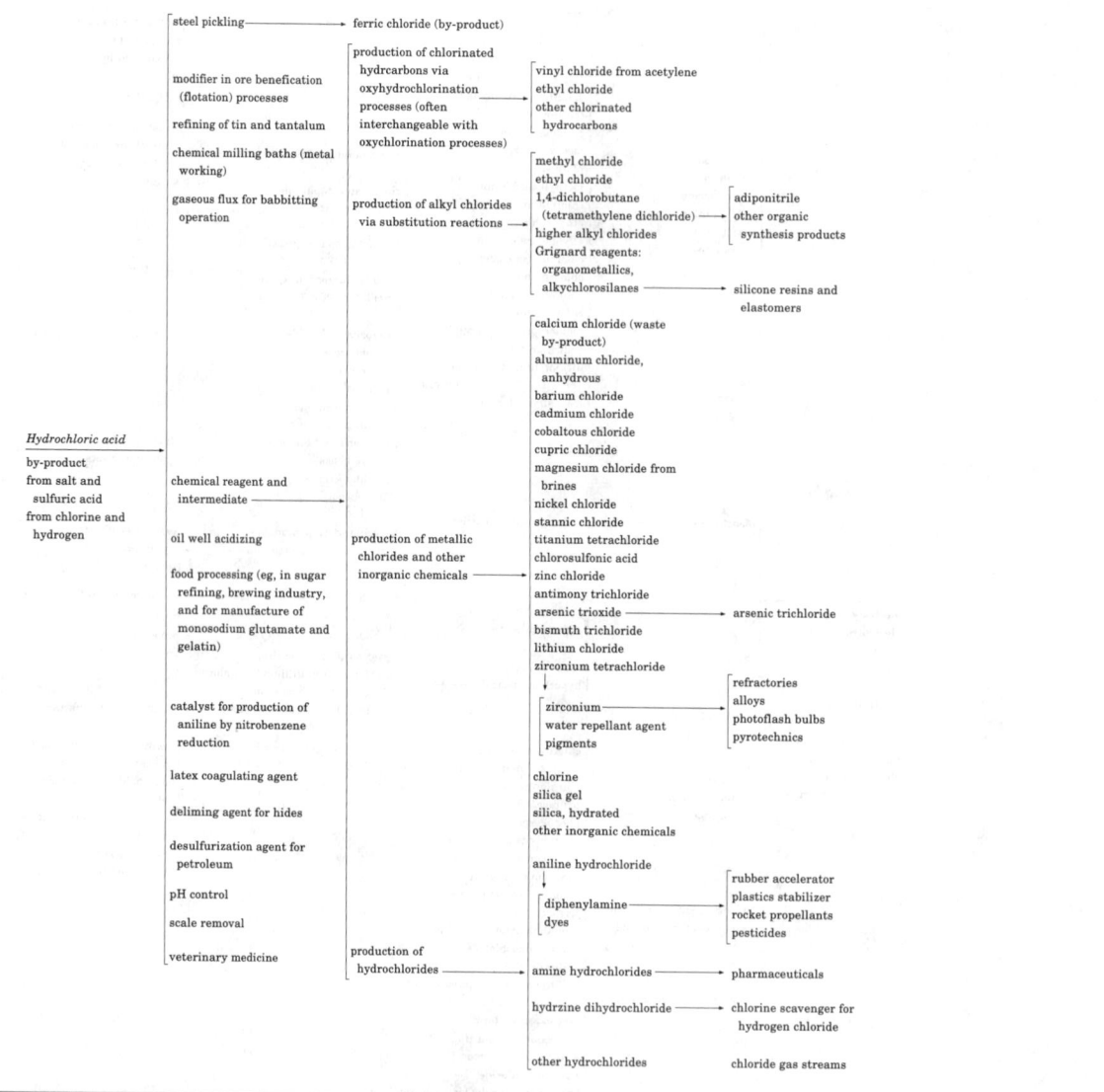

[a] Courtesy of SRI International.

body tissue, it is also known as caustic soda. Physical properties of the pure material are noted in Table 6. Sodium hydroxide is produced and shipped in the anhydrous state in the form of solid cakes, flakes, or beads, but is used in solution. Properties of aqueous sodium hydroxide solutions relevant to industrial operations are available from manufacturers and in handbooks. Table 7 lists the significant uses.

Chemical Properties

Aqueous solutions of caustic soda are highly alkaline. Hence caustic soda is primarily used in neutralization reactions to form sodium salts.

Reactions of NaOH with natural products are complex. They include solubilization of cotton in rubber reclaiming, starch dextrination, cotton scouring, refining of vegetable oils, and removal of lignin and hemicellulose in the Kraft pulping process.

Other Processes for NaOH Production

The only caustic soda production process besides electrolysis is the soda–lime process involving the reaction of lime with soda ash:

$$Ca(OH)_2 + Na_2CO_3 \rightarrow CaCO_3 + 2\ NaOH$$

The lime–soda process is practiced mainly in isolated areas in some process operations, in the Kraft recovery process, and in the production of alumina. It is not as efficient a route as electrolytic production.

Special Grades of Caustic Soda

Three forms of caustic soda are produced to meet customer needs: purified diaphragm caustic (50% Rayon grade), 73% caustic, and anhydrous caustic. Regular 50% caustic from the diaphragm cell process is suitable for most applications and accounts for about 85% of the NaOH

Table 6. Physical Constants of Pure Sodium Hydroxide

Property	Value
molecular weight	39.998
specific gravity at 20°C	2.130
melting point, °C	318
boiling point, °C at 101.3 kPa[a]	1388
specific heat, J/g·°C[b] at 20°C	1.48
refractive index at 589.4 nm	
320°C	1.433
420°C	1.421
latent heat of fusion, J/g[b,c]	167.4
lattice energy, kJ/mol[d]	737.2
entropy, J/(mol·K)[b] at 25°C and 101.3 kPa[a]	64.45
heat of formation ΔH_f, kJ/mol[a]	
α form	422.46
β form	426.60
heat of transition from α to β form, J/g[b]	103.3
transition temperature, °C	299.6
free energy of formation ΔG_f, kJ/mol[d] at 25°C and 101.3 kPa[a]	−379.5

[a] To convert kPa to mm Hg, multiply by 7.5.
[b] To convert J to cal, divide by 4.184.
[c] To convert J/g to Btu/lb, multiply by 0.4302.
[d] To convert kJ to kcal, divide by 4.184.

consumed in the United States. However, it cannot be used in operations such as the manufacture of rayon, the synthesis of alkyl aryl sulfonates, or the production of anhydrous caustic because of the presence of salt, sodium chlorate, and heavy metals. Membrane and mercury cell caustic, on the other hand, is of superior quality and meets the high purity market requirements.

Materials of Construction

Steel is an acceptable material of construction for handling solutions of up to 50% NaOH below 40°C. Nickel is the ideal material for handling caustic at all concentrations and temperatures, including molten anhydrous caustic up to 480°C.

Storage and Transportation

Caustic soda is classified as a corrosive material by the DOT, and DOT regulations and specifications must be followed for handling, labeling, and transportation in containers.

Safety

Caustic soda in liquid or solid form has a marked corrosive action on all body tissue, so that even dilute solutions have a deleterious effect on tissue after prolonged contact. Inhalation of the dust or mist can cause damage to the upper respiratory tract; ingestion causes damage to the mucous membrane of the exposed tissue. It is therefore important that all the properties of caustic and the safety precautions be reviewed before handling. During handling, all persons should wear proper protective clothing, safety goggles (sometimes a full face shield), rubber gloves, boots, and a caustic-resistant apron or suit.

Disposal of waste or spilled caustic soda must meet all federal, state, and local regulations and be carried out by properly trained personnel.

Table 8. Energy Consumption of Operating Cells, Kilowatt-Hours per Ton of Chlorine

Energy	Cell type		
	Diaphragm	Mercury	Membrane
electricity for electrolysis	2800–3000	3200–3600	2600–2800
steam requirements[a]	600–800	0	200–300
Total	*3400–3800*	*3200–3600*	*2800–3100*

[a] One ton of steam is assumed to be ~400 kW·h.

Chlorine Future Growth

Chlorine Consumption. In the latter part of the 1980s U.S. chlorine consumption grew an average of 3.2% annually from 9.7×10^6 t in 1986 to 11×10^6 t in 1990. Increased exports of chlorine derivatives, aided by the weaker dollar, have added to chlorine's recovery. The high growth rates of the late 1980s are unlikely to continue.

U.S. Caustic Soda Market and Future Growth

Increases in U.S. demand for caustic soda have been unpredictably high recently. However, the caustic soda market is mature and new areas of significant growth have not surfaced in recent years. The unexpected demand was generally related to two factors: the pick-up in the U.S. economy after the slump of 1986 and pulp mills operating at full capacity, leading to less efficient caustic use.

Economic Aspects

The choice of technology, the associated capital, and operating costs for a chlor–alkali plant are strongly dependent on local factors. Especially important are local energy and transportation costs, as are environmental constraints. The primary differences in operating costs among diaphragm, mercury, and membrane cell plants result from variations in electricity requirements for the three processes (Table 8) so that local energy and steam costs are most important.

Table 7. Sodium Hydroxide Derivatives

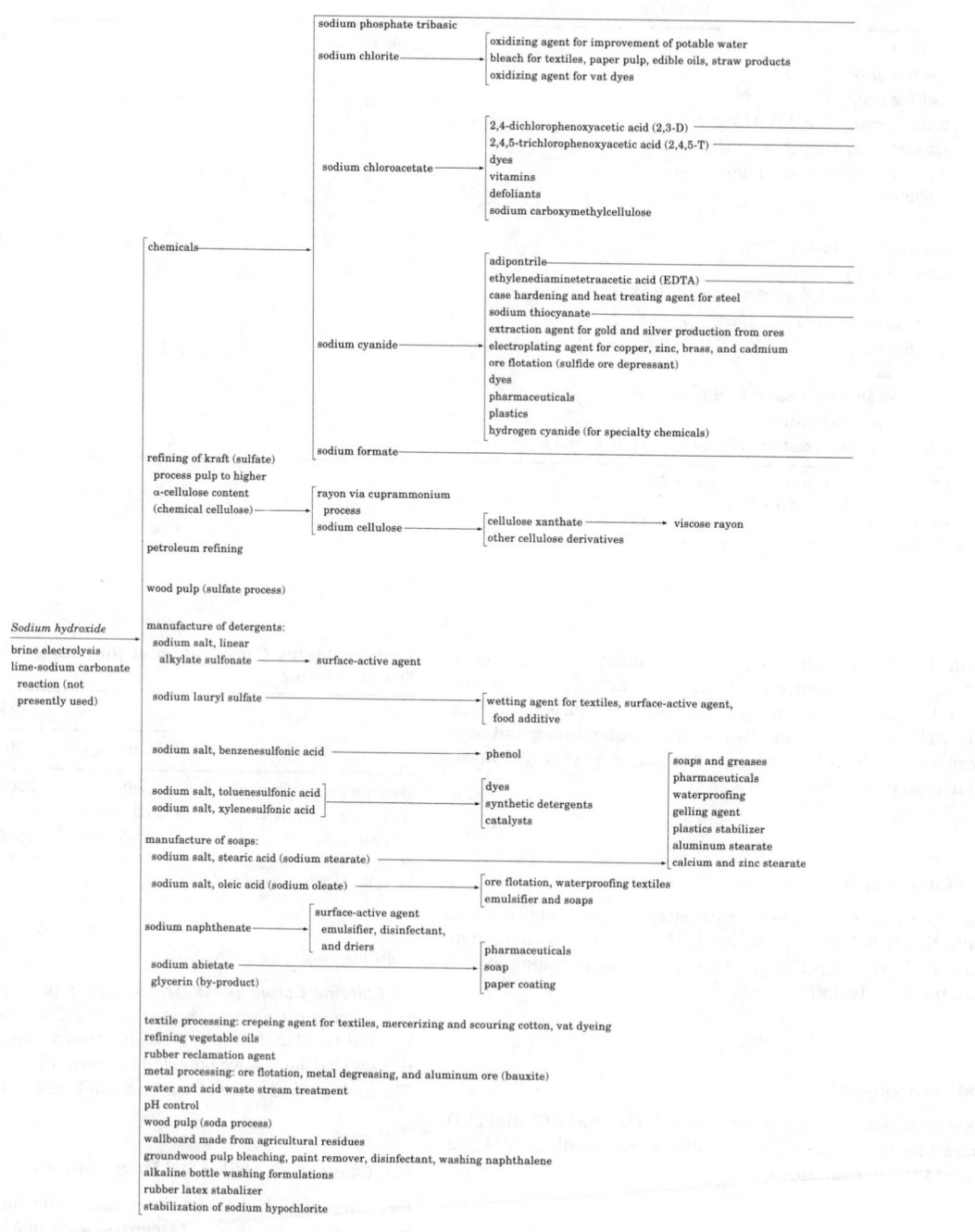

Sodium hydroxide
brine electrolysis
lime-sodium carbonate reaction (not presently used)

- chemicals →
  - sodium phosphate tribasic
  - sodium chlorite →
    - oxidizing agent for improvement of potable water
    - bleach for textiles, paper pulp, edible oils, straw products
    - oxidizing agent for vat dyes
  - sodium chloroacetate →
    - 2,4-dichlorophenoxyacetic acid (2,3-D)
    - 2,4,5-trichlorophenoxyacetic acid (2,4,5-T)
    - dyes
    - vitamins
    - defoliants
    - sodium carboxymethylcellulose
  - sodium cyanide →
    - adipontrile
    - ethylenediaminetetraacetic acid (EDTA)
    - case hardening and heat treating agent for steel
    - sodium thiocyanate
    - extraction agent for gold and silver production from ores
    - electroplating agent for copper, zinc, brass, and cadmium
    - ore flotation (sulfide ore depressant)
    - dyes
    - pharmaceuticals
    - plastics
    - hydrogen cyanide (for specialty chemicals)
  - sodium formate
- refining of kraft (sulfate) process pulp to higher α-cellulose content (chemical cellulose) →
  - rayon via cuprammonium process
  - sodium cellulose →
    - cellulose xanthate → viscose rayon
    - other cellulose derivatives
- petroleum refining
- wood pulp (sulfate process)
- manufacture of detergents:
  - sodium salt, linear alkylate sulfonate → surface-active agent
  - sodium lauryl sulfate → wetting agent for textiles, surface-active agent, food additive
  - sodium salt, benzenesulfonic acid → phenol
  - sodium salt, toluenesulfonic acid; sodium salt, xylenesulfonic acid →
    - dyes
    - synthetic detergents
    - catalysts
- manufacture of soaps:
  - sodium salt, stearic acid (sodium stearate) →
    - soaps and greases
    - pharmaceuticals
    - waterproofing
    - gelling agent
    - plastics stabilizer
    - aluminum stearate
    - calcium and zinc stearate
  - sodium salt, oleic acid (sodium oleate) →
    - ore flotation, waterproofing textiles
    - emulsifier and soaps
- sodium naphthenate →
  - surface-active agent
  - emulsifier, disinfectant, and driers
- sodium abietate; glycerin (by-product) →
  - pharmaceuticals
  - soap
  - paper coating
- textile processing: crepeing agent for textiles, mercerizing and scouring cotton, vat dyeing
- refining vegetable oils
- rubber reclamation agent
- metal processing: ore flotation, metal degreasing, and aluminum ore (bauxite)
- water and acid waste stream treatment
- pH control
- wood pulp (soda process)
- wallboard made from agricultural residues
- groundwood pulp bleaching, paint remover, disinfectant, washing naphthalene
- alkaline bottle washing formulations
- rubber latex stabalizer
- stabilization of sodium hypochlorite

Table 7. *(Continued)*

- → herbicide
- → herbicide
- → hexamethylenediamine →
  - nylon-6,6 → films; fibers; resins
  - nylon-6,10 → monofilaments
- → chelating agent
- dyeing
- → pharmaceuticals
- aritficial mustard oil
- rubber treatment
- black nickel plating
- polyacrylate solvent
- → reducing agent
- pharmaceuticals
- formic acid
- oxalic acid
- mordant
- nickel formate
- leather tanning agent
- wallpaper printing
- plating
- catalyst

- sodium propionate → pesticides; mold preventative for food
- sodium phenolate → antiseptic; salicylic acid
- sodium metasilicate
- sodium orthosilicate
- sodium picramate → dye intermediate
- sodium stannate → tin, metal (electrolytic process); blueprint papers; dye mordant; ceramics and glass; tin electroplating; textile fireproofing; stabilizer for hydrogen peroxide; alkaline electroplating of tin; immersion tinning of aluminum alloys
- sodium polysulfides → sulfur dyes; insecticides; synthetic rubber; petroleum additives; electroplating
- sodium sulfite (from sodium benzene sulfonate)
- sodium arsenite
- sodium bromite → textile desizer
- sodium dimethyldithiocarbamate → zinc dimethyldithiocarbamate → fungicide; activator for rubber vulcanization
- sodium dinitro-o-cresylate → herbicide; control agent for fruit setting; fungicide
- sodium diuranate
- sodium tungstate
- sodium fluoroacetate → rodenticide
- sodium formaldehyde sulfoxylate → textile stripping; bleaching agent for molasses and soap
- monosodium glutamate → flavor enhancer
- sodium molybdate
- cupric hydroxide
- manganous hydroxide
- nickel hydroxide
- beryllium hydroxide
- barium hydroxide
- cadmium hydroxide
- cobalt hydroxide
- lead hydroxide
- amyl alcohol
- ethylene oxide
- hydrazine (Raschig process)
- 1-naphthol
- 2-naphthol
- phenol (chlorobenzene process)
- resorcinol
- vanillin (from waste sulfite pulping liquor) → flavors; perfumes
- vinyl chloride (from ethylene dichloride)
- sodium salt, oil-soluble petroleum sulfonate → lubricating oil additives
- rosin size → paper
- cryolite, synthetic (sodium aluminum fluoride)

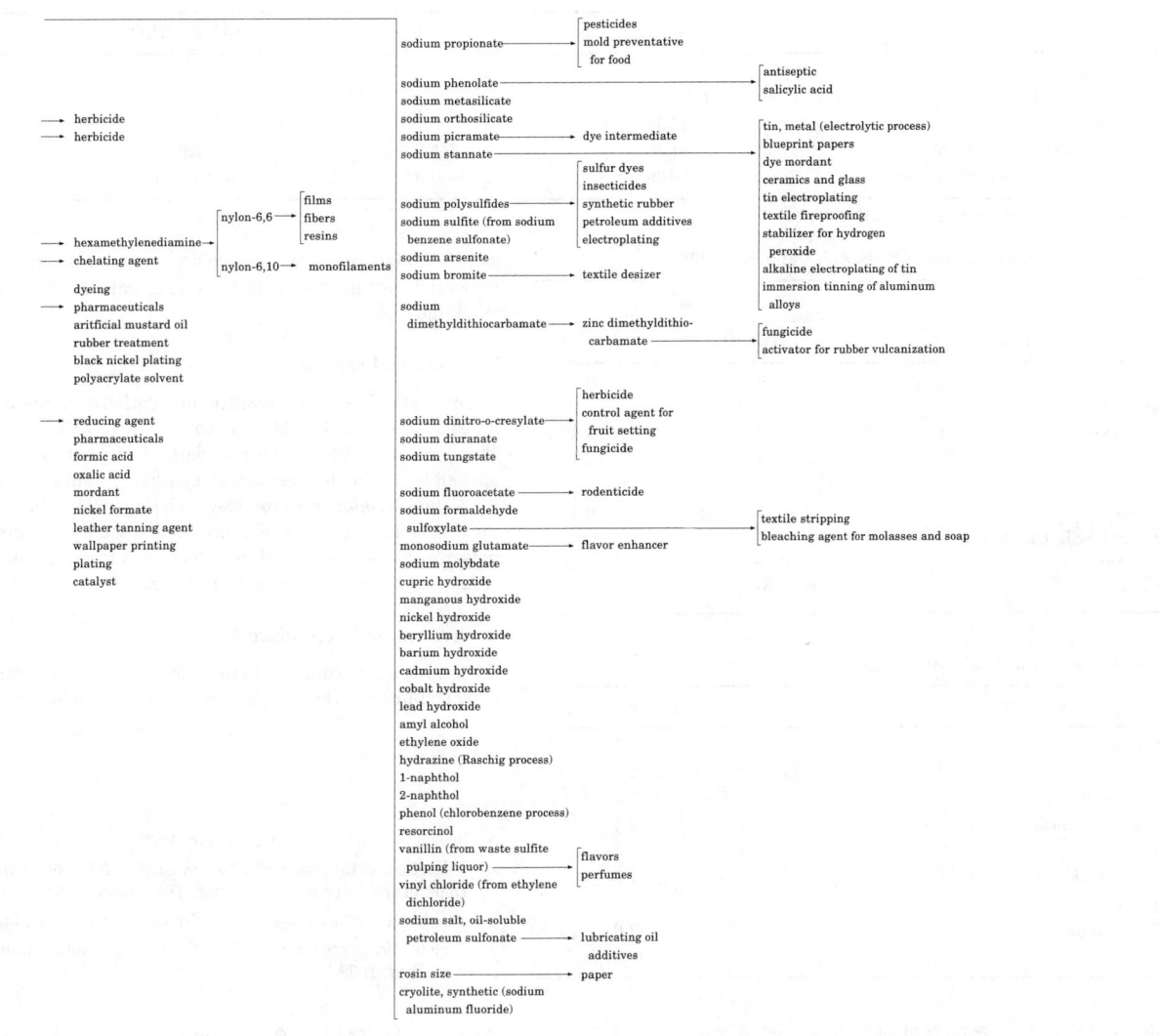

L. Calvert Curlin
OxyTech Systems, Inc.
Tilak V. Bommaraju
Constance B. Hansson
Occidental Chemical Corporation

Chlorine and Its Derivatives: A World Survey of Supply, Demand, and Trade to 1992, Tecnon Consulting Group, London, 1988.

Caustic Soda in the 1990's: A World Survey of Supply, Demand, and Trade, Tecnon Consulting Group, London, 1989.

P. Schmittinger and co-workers, in *Ullmann's Encyclopedia of Industrial Chemistry,* 5th ed., Vol. A6, 1986, p. 399.

J. E. Currey and G. G. Pumplin in J. J. McKetta and W. A. Cunningham, eds., *Encyclopedia of Chemical Engineering and Design,* Vol. 7, Marcel Dekker, Inc., New York, 1978, p. 305.

Y. C. Chen, *Process Economics Program Report,* Stanford Research Institute, Menlo Park, Calif., No. 61, 1970; No. 61A, 1974; No. 61B, 1978; No. 61C, 1982; No. 61D, 1992.

SODIUM CARBONATE

Sodium carbonate, Na_2CO_3, formula wt 105.99, is a white crystalline solid known as soda ash and, less commonly, ash, soda, or calcined soda. It is readily soluble in water and in strongly alkaline. It is the eleventh largest world commodity chemical. About 75% of world production is synthetic ash made from sodium chloride and limestone via the Solvay or similar processes; the remaining 25% is produced from natural sodium carbonate-bearing deposits. Over half of the world's production is consumed in the glass industry, and another 22% is used in the production of sodium-based chemicals. Sodium carbonate is also used in detergents, pulp (qv) and paper (qv), and environmental control (water treatment and flue gas desulfurization). The normal article of commerce is highly purified (>99%). Differences in bulk density are the only major distinction between the various grades. Minor amounts of sodium carbonate monohydrate, $Na_2CO_3 \cdot H_2O$, and sodium carbonate decahydrate, $Na_2CO_3 \cdot 10H_2O$, are sold and used in specialty applications. Aqueous solutions are alkaline. At 25°C the pH of 1, 5, and 10 wt % Na_2CO_3 solutions is 11.37, 11.58, and 11.70, respectively. Physical properties and solubility data are given in Table 1.

Table 1. Physical Properties of Sodium Carbonate

Property	Value			
	Na_2CO_3	$Na_2CO_3 \cdot H_2O$	$Na_2CO_3 \cdot 7H_2O$	$Na_2CO_3 \cdot 10H_2O$
melting point	825			
bulk density, g/mL	0.59–1.04			
specific gravity	2.533			
heat of formation, ΔH_f, kJ/mol[a] at 0°C	−1131	−1459	−3201	−4082
temperature, °C, stable solid phase	>109	35.4–109.0	32.0–35.4	0–32.0

[a] To convert kJ/mol to kcal/mol, divide by 4.184.

Table 2. World Distribution of Soda Ash Capacity, 1989

Region	Soda ash, 10^6 t/yr	
	Natural	Synthetic
North American	9.5	0.4
Western Europe		6.7
Latin America	0.2	0.8
Africa	0.3	0.1
Asia		4.3
Oceania		0.4
(former) USSR, China, Eastern Europe	0.5	14.5
Total	*10.5*	*27.2*

Table 3. Soda Ash Producers, 1989

Company	Capacity, 10^6 t/yr
Solvay & Cie	4.3
FMC	2.6
Rhône-Poulenc	2.4
General Chemical	2.4
ICI	1.3
North American Chemical Co.	1.0
Tenneco	1.0
Elf Acquitaine	1.0
Total	*16.0*

Table 4. Typical Properties of Commercial Soda Ash

Property	Synthetic ash		Natural ash (U.S.)	
	Light	Dense	Intermediate	Dense
bulk density, g/mL	0.59	1.04	0.80	1.04
NaCl, ppm	8000–1000	5000–300	300	300
Na_2SO_4, ppm	200	200	300	300
Fe_2O_3, ppm	80–20	80–20	5	3

Economic Aspects

Natural and synthetic soda ash capacity is shown in Table 2. As indicated in Table 3, eight companies represent about 75% of the Western world's soda ash capacity.

Although each production process yields ash that is essentially chemically equivalent, the various products differ in physical properties and in contaminants as shown in Table 4. Hopper cars, pneumatic trucks, supersacks, and multiwall kraft bags with polyethylene liners are the usual shipping containers.

Safety and Handling

Under National Fire Production Association (NFPA) Designation 704, soda ash is classified as a moderate health hazard. Exposure to soda ash dust may cause severe eye and slight nose and throat irritation. Repeated contact may affect the skin causing redness, dryness, and cracking.

Environmental Aspects

Synthetic Processes. Traditional synthetic plants produce large volumes of aqueous, chloride-containing waste which must be discharged. This fact, in addition to a noncompetitive cost position, is largely responsible for the demise of U.S. synthetic plants.

Natural Production Processes. The natural soda ash processes produce no large volumes of associated wastes. The major waste products are the tailings, insoluble shale, and minerals associated with the trona and removed during processing.

By-Products and Coproducts

By-products and coproducts include calcium chloride, ammonium chloride, sodium bicarbonate, and sodium hydroxide.

FRANCIS RAUH
FMC Corporation

A. Russell, *Ind. Min. London,* 19 (Jan. 1990).

D. S. Kostick, *Soda Ash and Sodium Sulfate Minerals Yearbook—1988,* U.S. Dept. of the Interior, Washington, D.C., 1988.

The Economics of Soda Ash, Roskill Information Services Ltd., London, 1989.

T. P. Hou, *Manufacture of Soda* (ACS Monogr. Ser.), Hafner Publishing Co., New York, 1969.

ALKALI METALS. See CESIUM; LITHIUM; POTASSIUM; RUBIDIUM; SODIUM.

ALKALINE-EARTH METALS. See BARIUM; CALCIUM; MAGNESIUM; STRONTIUM.

ALKALOIDS

Alkaloids are not simply defined. However, most scientists working in the field of alkaloid chemistry would agree that most alkaloids, in addition to being products of secondary metabolism, are organic nitrogen-containing bases of complex structure; many have physiological activity.

Occurrence, Detection, and Isolation

Given the massive amount of material available, the following discussion is necessarily incomplete and the interested reader is directed to the references listed at the end of this article for more detailed information.

Current compendia of alkaloids list compounds that, for the most part, occur in flowering plants and it is probably true that the highest concentrations of these substances are to be found there. However, as detection methods improve it is almost certain that some concentration of alkaloids will be found almost everywhere. For example, in addition to flowering plants, alkaloids have also been found in birds, butterflies and beetles; in millipedes and in algae and fungi. They are found in toads, eg, bufotenine, an established hallucinogen (in humans); and in the musk deer, eg, muscopyridine, a presumed attractant (in humans). Even in humans, it appears trace quantities of alkaloids are naturally occurring components of cerebrospinal fluid.

The concentration of alkaloids, as well as the specific area of occurrence or localization of the material within a plant or animal, varies. Usually, they are minor plant constituents and they are present in only miniscule quantities in animals.

Initially, the search for alkaloids in plant material depended on reports of specific plant use for definite purposes (folklore) or observations of the effect specific plants had on indigenous animals among native populations. Newer methods of analysis (eg, mass spectroscopy) are now employed and they have the potential to screen plants and animals for alkaloid constituents rapidly.

Until separation techniques such as chromatography and countercurrent extraction had advanced sufficiently to be of widespread use, only the major alkaloids could be isolated (and purified) from plant extracts. The minor constituents were either discarded or remained uninvestigated. With the improvement in isolation techniques, materials of physiological significance, even when present in very low concentrations, can be obtained in commercial quantities.

Properties

Since most alkaloids are basic they are generally separated from accompanying neutrals and acids by extraction with dilute mineral acid. The purified bases tend to be colorless and crystalline, with definite melting points. Most are chiral; only one enantiomer being isolated. However, among nearly ten thousand individual compounds, these descriptions are overgeneralizations. Some alkaloids are not basic, some are liquid, some brightly colored, some achiral, and in a few cases both enantiomers have been isolated in equal amounts from the same plant.

Organization

Early investigators grouped alkaloids (*1*) according to the plant families in which they are found, or (*2*) structural types based on their carbon framework or their principal heterocyclic systems. However, as it became clear that the alkaloids, as secondary metabolites, were derived from compounds of primary metabolism (eg, amino acids, carbohydrates, etc), biogenetic hypotheses evolved to link the more elaborate skeletons of alkaloids with their simpler proposed progenitors. These hypotheses continue to serve as valuable organizational tools.

The building blocks of primary metabolism from which biosynthetic studies have shown the large majority of alkaloids to be built are few. They include the common amino acids ornithine, lysine, phenylalanine, tyrosine, and tryptophan. Other small fragments such as nicotinic acid, anthranilic acid, histidine, and the nonnitrogenous acetate derived fragments mevalonic acid and loganin are also common.

Ornithine-Derived Alkaloids. The simple pyrrolidine alkaloids as well as the widely found pyrrolizidine alkaloids and the bicyclic tropane bases such as atropine, scopolamine, and cocaine (**1**) are all derived from the same, common amino acid precursor ornithine (**2**).

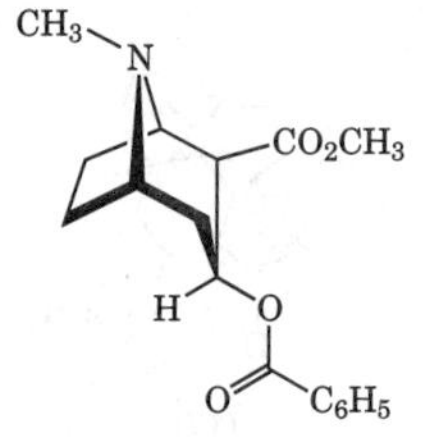

(**1**)

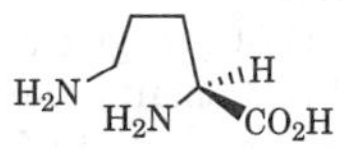

(**2**)

Lysine-Derived Alkaloids. The simple monocyclic bases and their more elaborate polycyclic relatives found in pomegranate and various members of the common genera *Sedum* and *Lycopodium,* are derived from the amino acid lysine (**3**). These alkaloids include, for the first, pelletierine. Sedamine and annotinine, respectively, are found in the last two.

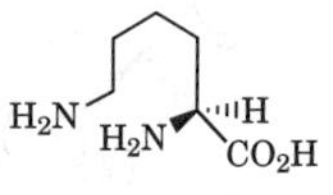

(**3**)

Tobacco Alkaloids. Nicotine (**4**), containing both nicotinic acid (**5**) and (ornithine derived) pyrrolidine rings, as well as its less common relatives, such as anabasine, which has nicotinic acid and (lysine-derived) piperidine fragments arranged as in nicotine, are commonly found only in various tobaccos.

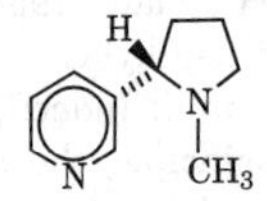

(**4**)

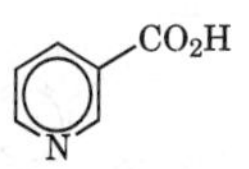

(**5**)

Phenylalanine- and Tyrosine-Derived Alkaloids. There is a relatively large number of alkaloids which may be considered as derived from phenethylamine, $C_6H_5CH_2CH_2NH_2$, or tyramine, *p*-$HOC_6H_4CH_2CH_2NH_2$. These bases are, in turn, considered derived from phenylalanine (**6**, R = H; for the former) and tyrosine (**6**, R = OH; for the latter). Interestingly, although the two amino acids (**6**, R = H and R = OH) are clearly related to each other (both coming from shikimic acid), they appear to be separately used (ie, not interconvertable). The alkaloids derived from them include the relatively simple monocyclic bases such as mescaline (**7**) as well as its somewhat more complicated relatives, ipecoside and emetine.

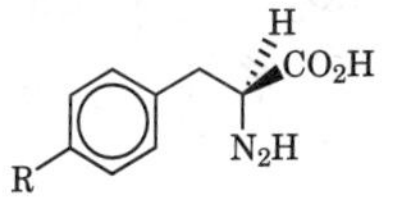

(**6**)

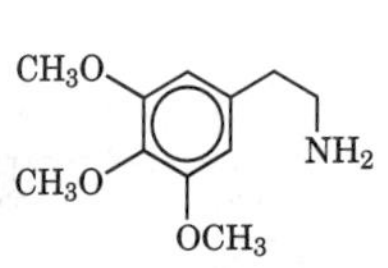

(**7**)

Interestingly, it appears that rather elaborate systems of many alkaloids incorporating these two amino acids are formed from the rather simple generalization that tyrosine is used as a six carbon + two carbon framework ($C_6 + C_2$) and phenylalanine is used as a six carbon + one, two, or three carbon framework ($C_6 + C_1$ or $C_6 + C_2$ or $C_6 + C_3$), depending upon the plant, etc.

There are only two groups of alkaloids that appear to be derived from tyrosine utilized as a C_6–C_2 fragment and the C_6–C_1 unit from phenylalanine. The first is that small group found only in the Orchidaceae, exemplified by cryptostylene I (**8**). The second, is a very large group of compounds collectively called the alkaloids of the Amaryllidaceae. This cosmopolitan family of related compounds includes over one hundred isolated and characterized members of known structure. In every case examined, the C_6–C_2 unit is derived from tyrosine (never from phenylalanine) and the C_6–C_1 unit from phenylalanine (never from tyrosine).

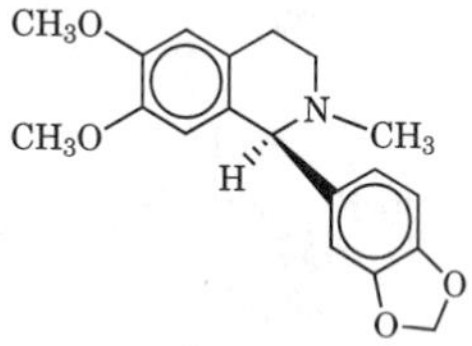

(**8**)

For this large number of compounds, it is now believed that a single progenitor, ie, norbelladine (**9**), derived from the original coupling of the $C_6)$–C_2 unit and the C_6–C_1 unit, accounts for them all. It is further believed that norbelladine, or its enzyme bound equivalent, undergoes a variety of enzyme-catalyzed free-radical intra-molecular cyclization reactions, followed by late-stage oxidation, elimination, rearrangement, and/or *N*- and *O*-alkylations to produce the vast number of daughter molecules found. Indeed, working from this generalization as an organizing principle, the majority of known Amaryllidaceae alkaloids can be divided into only a handfull of structural classes.

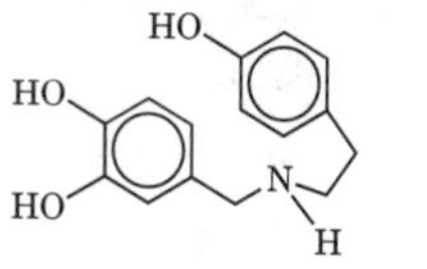

(**9**)

Just as for norbelladine (or its enzyme bound equivalent) serving as the precursor for the C_6-C_2 + C_6-C_1 family of alkaloids, so it is that norlaudanosoline (**10**, or its enzyme bound equivalent) is considered as the progenitor for the very large number of C_6-C_2 + C_6-C_2 bases common, among others, in the opium poppy (*Papaver Somniferum*) and including codeine (**11**, R = CH_3) and morphine (**11**, R = H).

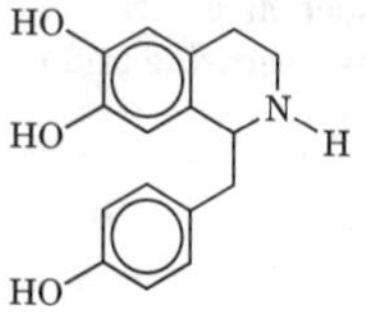

(**10**)

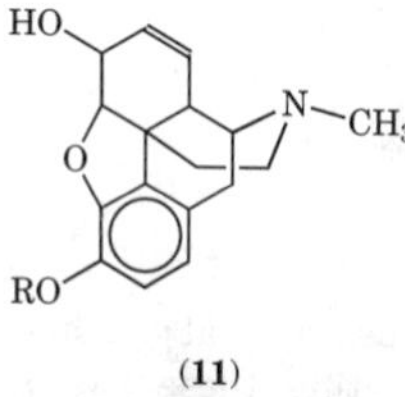

(**11**)

The last possibility is to use tyrosine as the C_6–C_2 fragment and phenylalanine as a C_6–C_3 fragment. This somewhat smaller group of alkaloids includes materials such as colchicine (**12**).

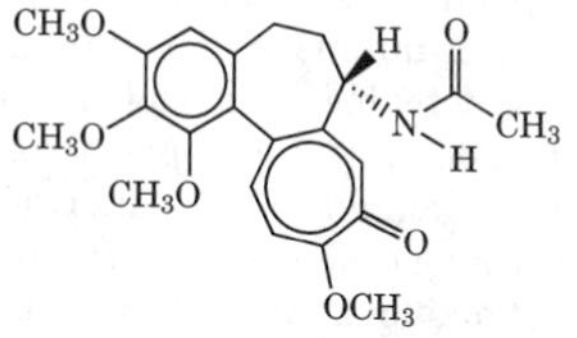

(**12**)

Tryptophan-Derived Alkaloids. There are volumes written exclusively on alkaloids known to be derived from the amino acid tryptophan (**13**) with and without additional carbon framework. The simpler alkaloids include seratonin (**14**). However, more complex bases can be easily built up with only a few building blocks. Thus, the addition of a five carbon unit (from mevalonic acid) yields lysergic acid (**15**), a ten carbon unit (from loganin, itself derived from two units of mevalonic acid) yields the related materials catharanthine (**16**), tabersonine (**17**), strychnine (**18**), and quinine (**19**) and these materials themselves can dimerize yielding more complex bases such as vincaleukoblastine (**20**).

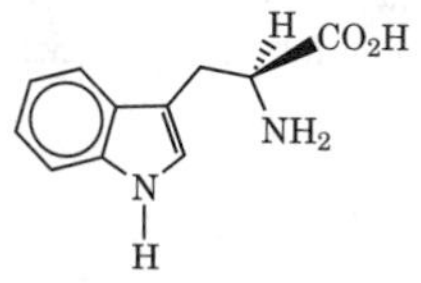

(**13**)

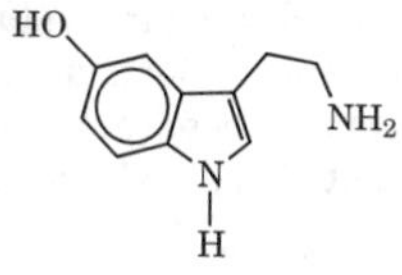

(**14**)

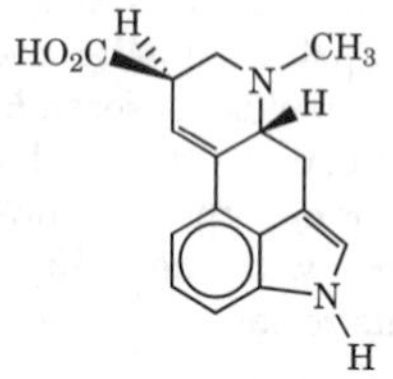

(**15**)

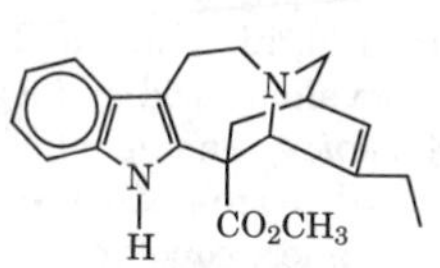

(**16**)

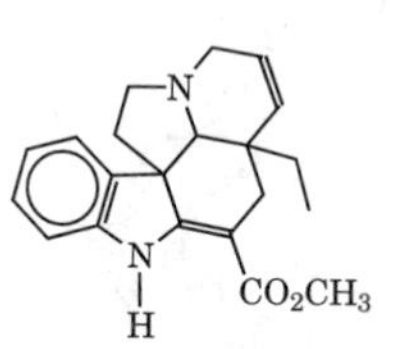

(**17**)

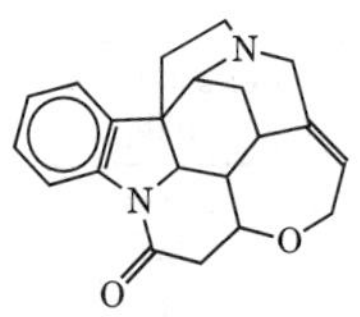

(18)

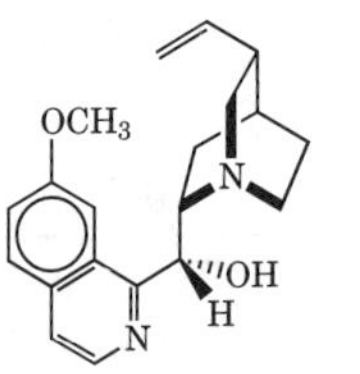

(19)

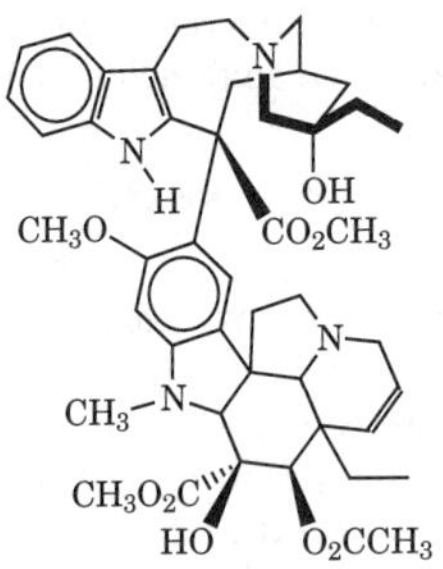

(20)

Purine Alkaloids. The purine skeleton (**21**) does not appear to be derived from the amino acid histidine, as might be initially imagined. Indeed, as is now widely appreciated, the nucleus common to xanthine, caffeine (**22**), theophylline, and theobromine is created from much smaller fragments early attached to a ribosyl unit.

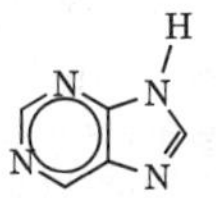

(21)

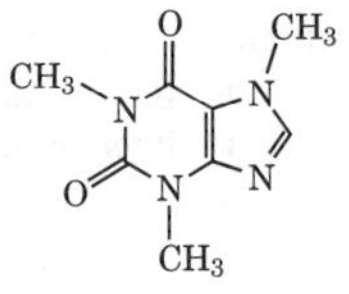

(22)

Economic Aspects

As the twentieth century draws to a close, many alkaloids such as atropine and reserpine, that have served humanity since early history as pallatives are being replaced by synthetic materials. Others, such as the *Vinca* bases, eg, vincristine, remain as powerful medical tools. Replacement of naturally occurring alkaloids may be desirable in order to maintain and augment favorable properties while eliminating those deleterious.

There are, currently, four broad classes of alkaloids and/or materials whose alkaloid content is important enough to be of sizeable economic value. As a consequence, markets are made, treaties and alliences formed, etc, regarding (*1*) the opiates such as morphine (**11**, R = H) and codeine (**11**, R = CH_3); (*2*) cocaine (**1**), licit and illicit; (*3*) caffeine (**22**) and related bases in coffee and tea, and; (*4*) the tobacco alkaloids such as nicotine (**4**).

DAVID R. DALTON
Temple University

I. W. Southon and J. Buckingham, eds., *Dictionary of Alkaloids,* Chapman and Hall, New York, 1989.

R. H. F. Manske and H. L. Holmes, eds. *The Alkaloids: Chemistry and Physiology,* Vol. 1, Academic Press, Inc., New York, 1950. This series gives a detailed exposition of the chemistry and pharmacology of the alkaloids, by structural class. Vol. 36, A Brossi, ed., was published in 1989.

ALKANOLAMINES

ALKANOLAMINES FROM OLEFIN OXIDES AND AMMONIA

Ethylene oxide, propylene oxide, or butylene oxide react with ammonia to produce alkanolamines. Ethanolamines, $NH_{3\text{-}n}(C_2H_4OH)_n(C_2H_{OH})_n$ ($n = 1, 2, 3$, mono-, di-, and tri-), are derived from the reaction of ammonia with ethylene oxide. Isopropanolamines, $NH_{3\text{-}n}$ $(CH_2CHOHCH_3)_n$ (mono-, di-, and tri-), result from the reaction of ammonia with propylene oxide. Secondary butanolamines, NH_{3-n} $(CH_2CHOHCH_2CH_3)_n$ (mono-, di-, and tri-), are the result of the reaction of ammonia with butylene oxide. Mixed alkanolamines can be produced from a mixture of oxides reacting with ammonia.

Ethanolamines have been commercially available for over 50 years and isopropanolamines, for over 40 years. *sec*-Butanolamines have been prepared in research quantities, but are not available commercially. Primary butanolamines, eg, 2-amino-1-butanol are made by a different chemical route.

A variety of substituted alkanolamines, shown in Table 1, are also available commercially, but have not reached the volume popularity of the ethanolamines and isopropanolamines.

Physical Properties

The freezing points of alkanolamines are moderately high, as shown in Table 2. The ethanolamines, monoisopropanolamine and mono-*sec*-butanolamine, are colorless liquids at or near room temperature.

Table 1. Some Physical Properties of Substituted Alkanolamines

Common name	Freezing point, °C	Boiling point[a], °C
dimethylethanolamine	-59	135
diethylethanolamine		162
aminoethylethanolamine (AEEA)	-38[b]	244
methylethanolamine	-4.5	160
butylethanolamine	-2	199
N-acetylethanolamine	16	decompn
phenylethanolamine	11	285
dibutylethanolamine	-75[c]	229
diisopropylethanolamine	-39	191
phenylethylethanolamine	37[d]	decompn
methyldiethanolamine	-21	247
ethyldiethanolamine	<-44[c]	253
phenyldiethanolamine	57[b]	
dimethylisopropanolamine	-85[c]	126
N-(2-hydroxypropyl)ethylenediamine	-50[c]	155[e]

[a] At 101.3 kPa = 1atm unless otherwise noted.
[b] Pour point.
[c] Sets to a glass-like solid below this temperature.
[d] Melting point.
[e] At 8 kPa (60 mm Hg).

Table 2. Physical Properties of Alkanolamines Prepared from Ammonia and Olefin Oxides

Common name	Molecular formula	Freezing point, °C	Boiling point[a], °C	Water solubility[b], g/100 g	Viscosity[b], mPa·s(= cP)
monoethanolamine (MEA)	C_2H_7NO	10	171	∞	19
diethanolamine (DEA)	$C_4H_{11}NO_2$	28	268	∞	54 (60°C)
triethanolamine (TEA)	$C_6H_{15}NO_3$	21	340	∞	600
monoisopropanolamine (MIPA)	C_3H_9NO	3[c]	159	∞	23
diisopropanolamine (DIPA)	$C_6H_{15}NO_2$	44[c]	249	1200	86 (54°C)
triisopropanolamine (TIPA)	$C_9H_{21}NO_3$	44[c]	306	> 500	100 (60°C)
mono-*sec*-butanolamine	$C_4H_{11}NO$	3	169	∞	29
di-*sec*-butanolamine	$C_8H_{19}NO_2$	68–70	256	∞	890
tri-*sec*-butanolamine	$C_{12}H_{27}NO_3$	41–47	310	ca 7	ca 6000

[a] At 101.3 kPa = 1 atm.
[b] Approximate, at 25°C unless otherwise noted.
[c] Supercools; freezing points may show variation.

Di- and triisopropanolamine and di- and tri-*sec*-butanolamine are white solids at room temperature.

All the ethanolamines and isopropanolamines except monoisopropanolamine are available in low freezing grades, to provide liquid handling at room temperature.

Alkanolamines have a mild ammoniacal odor and are extremely hygroscopic. The mono- and dialkanolamines have a basicity similar to aqueous ammonia; the trialkanolamines are slightly weaker bases.

Chemical Reactions

Alkanolamines are bifunctional molecules because of the alcohol and the amine functional groups in the same compound. This allows them to react in a wide variety of ways, with similarities to primary, secondary, and tertiary amines, and primary and secondary alcohols.

Economic Aspects

Consumption of ethanolamines in the United States has changed dramatically since the 1960s. Consumption in gas conditioning applications has peaked and chemical processing intermediates (captive use for ethyleneamine and surfactant applications) have increased significantly.

Analytical Test Methods

Generally, alkanolamines are analyzed by gas chromatography or wet test methods.

Storage and Handling

Stainless steel, 315L and 304L, is the preferred material of construction for shipment and storage of alkanolamines, if product quality is of importance.

Storage tanks, lines, and pumps should be heat traced and insulated to enable product handling. Temperature control is required to prevent product degradation because of color.

Health and Safety

Oral Toxicity. Alkanolamines generally have low acute oral toxicity, but swallowing substantial quantities could have serious toxic effects, including injury to mouth, throat, and digestive tract.

Vapor Toxicity. Laboratory exposure data indicate that vapor inhalation of alkanolamines presents low hazards at ordinary temperatures (generally, alkanolamines have low vapor pressures). Heated material may cause generation of sufficient vapors to cause adverse effects, including eye and nose irritation.

Eye Irritation. Exposure of the eye to undiluted alkanolamines can cause serious injury.

Skin Irritation. Monoethanolamine and monoisopropanolamine, being strongly alkaline, are skin irritants, capable of producing serious injury in concentrations of 10% or higher upon repeated or prolonged contact.

Special Precautions. Use of sodium nitrite or other nitrosating agents in formulations containing alkanolamines could lead to formation of suspected cancer-causing nitrosamines.

Strong oxidizers and strong acids are incompatible with alkanolamines.

Uses

Alkanolamines and their derivatives are used in a wide variety of household and industrial applications. Nonionic surfactants (alkanolamides) can be formed by the reaction of alkanolamines with fatty acids, at elevated temperatures. The amides can be liquid, water-soluble materials as produced from a 2:1 ratio, or solid, poorly water-soluble materials, or "super" amides as produced from a 1:1 ratio of reactants. These products are useful as foam stabilizers and aid cleaning in laundry detergents, dishwashing liquids, shampoos, and cosmetics. They are also used as antistatic agents, glass coatings, fuel gelling agents, drilling mud stabilizers, demulsifiers, and in mining flotation. Reaction of alkanolamines with a fatty acid at room temperature produces neutral alkanolamine soaps. Alkanolamine soaps are found in cosmetics, polishes, metalworking fluids, textile applications, agricultural products, household cleaners, and pharmaceuticals.

Alkanolamine salts are anionic surfactants formed from the reaction of alkanolamines and the acids of synthetic detergents, such as alkylarylsulfonates, alcohol sulfates, and alcohol ether sulfates. These add to the surfactants line used in detergents, cosmetics, textiles, polishes, agricultural sprays, household cleaners, pharmaceutical ointments, and metalworking compounds. Salts of alkanolamines and inorganic acids are useful chemical intermediates, and are also used in corrosion inhibitors, antistatic agents, glass coatings, electroplating, high octane fuels, inks, metalworking, dust control in mining, and in textiles.

MARTHA R. EDENS
J. FRED LOCHARY
The Dow Chemical Company

The Alkanolamines Handbook, The Dow Chemical Company, Midland, Mich., 1988.

Expert Panel of the Cosmetic Ingredient Review, *Final Report on the Safety Assessment for Diisopropanolamine, Triisopropanolamine, Isopropanolamine, Mixed Isopropanolamines,* Sept. 26, 1986; *Final Report for the Safety Assessment for Triethanolamine, Diethanolamine, Monoethanolamine,* May 19, 1983.

ALKANOLAMINES FROM NITRO ALCOHOLS

The nitro alcohols obtained by the condensation of nitroparaffins with formaldehyde may be reduced to a unique series of alkanolamines (β-amino alcohols):

Table 1. Properties of Alkanolamines

Name	Common designation	Molecular weight	Boiling point, °C	Melting point, °C	Specific gravity
2-amino-2-methyl-1-propanol	AMP	89.14	165[a]	30–31	0.934[b]
2-amino-2-ethyl-1,3-propanediol	AEPD	119.16	152–153[c]	37.5–38.5	1.099[b]
2-dimethylamino-2-methyl-1-propanol	DMAMP	117.19	160[a]	19	0.90[d]
2-amino-2-(hydroxymethyl)-1,3-propanediol[e]	TRIS AMINO	121.14	219–220[c]	171–172	
2-amino-2-methyl-1,3-propanediol	AMPD	105.16	151–152[c]	109–111	
2-amino-1-butanol	AB	89.14		−2	0.944[b]

[a] At 101.3 kPa (1 atm).
[b] At 20/20°C.
[c] At 1.3 kPa (10 mm Hg).
[d] At 25/25°C.
[e] Common name is tris(hydroxymethyl)aminomethane.

$$RCH_2NO_2 + CH_2O \longrightarrow \underset{NO_2}{RCHCH_2OH} \xrightarrow{[H]} \underset{NH_2}{RCHCH_2OH}$$

Only the five primary amino alcohols discussed in the following are manufactured on a commercially significant scale. *N*-Substituted derivatives of these compounds also have been prepared, but only 2-dimethylamino-2-methyl-1-propanol has been available in commercial quantities.

Physical Properties

Physical properties of the six commercial alkanolamines are given in Table 1. These compounds are highly soluble in water. They are generally very soluble in alcohols, slightly soluble in aromatic hydrocarbons, and nearly insoluble in aliphatic hydrocarbons; tris(hydroxymethyl) aminomethane is appreciably soluble only in water (80 g/100 mL at 20°C) and methanol.

According to current DOT regulations, AMP, AMP-95, DMAMP, DMAMP-80, AEPD, and AB are all classified as combustible liquids.

All alkanolamines have slight amine odors in the liquid state; the solids are nearly odorless.

Chemical Properties

The alkanolamines discussed here exhibit the chemical reactivity of both amines and alcohols, as is the case with other alkanolamines. Typically, they attack copper, brass, and aluminum, but not steel or iron. Alkanolamines are useful as amination agents; however, the reactivity of both the amino and alcohol group must be considered in attempting any specific synthetic scheme.

Manufacture and Economic Aspects

The reduction of nitro alcohols to alkanolamines is readily accomplished by hydrogenation in the presence of Raney nickel catalyst.

Production statistics on alkanolamines are not available, but they are sold in thousand-ton quantities and are available for bulk shipment except for DMAMP, AB, TRIS AMINO, and AMPD (the latter two being crystalline solids).

ANGUS Chemical Company is the only basic manufacturer of technical-grade TRIS AMINO (Tromethamine). However, ANGUS and numerous processors offer recrystallized, high purity grades of this alkanolamine for biomedical applications.

Health and Safety Factors

Alkanolamines are only slightly toxic by ingestion (acute oral LD_{50} in rodents = 1.0−5.5 g/kg.

Undiluted DMAMP, AMP-95, and AB cause eye burns and permanent damage, if not washed out immediately. They are also severely irritating to the skin, causing burns by prolonged or repeated contact.

The 40% aqueous solution of TRIS AMINO is nonirritating to the eyes and skin. In general, the toxicology of the alkanolamines is typical of alkaline materials, ie, the greater the base strength, the greater the effect.

Uses

Because they are similar, the alkanolamines often can be used interchangeably. However, cost–performance considerations generally dictate a best choice for specific applications. They are used in emulsions, pigment dispersion, resin solubilizers, catalysts, boiler water treatment, formaldehyde scavenging, applications in oil and gas production, biomedical applications, and synthetic applications.

ALLEN F. BOLLMEIER, JR.
ANGUS Chemical Company

H. B. Haas and E. F. Riley, *Chem. Rev.* **32,** 373 (1943).

G. G. Nahas and co-workers, *Ann. NY Acad. Sci.* **92,** 333 (1961).

C. Fernandez Palomero and J. Perez Pallares, *Rev. Soc. Nucl. Esp.* **1**(56), 27 (1987).

J. A. Frump, *Chem. Rev.* **71,** 483 (1971).

ALKOXIDES, METAL

Metal alkoxides are compounds in which a metal is attached to one or more alkyl groups by an oxygen atom. Alkoxides are derived from alcohols by the replacement of the hydroxyl hydrogen by metal.

Physical Properties

The metal alkoxides exhibit great differences in physical properties (Tables 1 and 2), depending primarily on the position of the metal in the periodic table, and secondarily on the alkyl group. Many alkoxides are strongly associated by intermolecular forces which depend on the size and shape of the alkyl groups. This explains the fact that many metal methoxides are solid, nondistillable compounds, because the small methyl group has little screening effect on the metal atom. With a larger number of methyl groups and a smaller atomic radius of the metal, methoxides become sublimable and even distillable.

Many metal alkoxides are soluble in the corresponding alcohols, but magnesium alkoxides are practically insoluble.

Chemical Properties

The most outstanding property of the metal alkoxides is the ease of hydrolysis. This is used for sol–gel applications.

Metal alkoxides and phenol usually form phenolates smoothly.

Enols and alkoxides give chelates with elimination of alcohol.

Metal alkoxides catalyze the Tishchenko condensation of aldehydes, the transesterification of carboxylic esters, the Meerwein-Ponndorf reaction, and other enolization and condensation reactions.

Commercial Alkoxides

Commercial alkoxides include alkali metal alkoxides (sodium methylate, sodium ethylate, sodium *tert*-butylate, sodium *tert*-amylate,

Table 1. Physical Properties of Metal Ethoxides

Alkoxide	mp, °C	bp, °C/Pa[a]
$LiOC_2H_5$	20–24	nd[b]
$NaOC_2H_5$	260 dec	nd
KOC_2H_5	250 dec	nd
$TlOC_2H_5$	9.5	nd
$Mg(OC_2H_5)_2$	270 dec	nd
$Ca(OC_2H_5)_2$	270 dec	nd
$Sr(OC_2H_5)_2$	300 dec	nd
$Ba(OC_2H_5)_2$	270 dec	nd
$Zn(OC_2H_5)_2$		
$Sn(OC_2H_5)_2$	200 dec	
$Mn(OC_2H_5)_2$		
$Al(OC_2H_5)_3$	140	189/400
$Ga(OC_2H_5)_3$	144.5	180–190/50[c]
$Cr(OC_2H_5)_3$		
$Fe(OC_2H_5)_3$	120	155/10
$Sb(OC_2H_5)_3$		37–38/5
$VO(OC_2H_5)_3$	ca 8	91/1100
$Ti(OC_2H_5)_4$	ca 40	124/160
$Zr(OC_2H_5)_4$	172	180/13
$Hf(OC_2H_5)_4$	ca 180	180–200/13
$Th(OC_2H_5)_4$	300 dec	
$Ce(OC_2H_5)_4$	200 dec	
$V(OC_2H_5)_4$		100/5
$Ge(OC_2H_5)_4$	−72	54.5/670
$Sn(OC_2H_5)_4$	unmeltable	nd
$U(OC_2H_5)_4$	dec	
$Nb(OC_2H_5)_5$	6	156/6.6
$Ta(OC_2H_5)_5$	22	137/6.6
$W(OC_2H_5)_5$		120/6.6
$U(OC_2H_5)_5$	180 dec	160/6.6
$Sb(OC_2H_5)_5$	46	135–145/20
$U(OC_2H_5)_6$		72/0.13

[a] To convert Pa to mm Hg, divide by 133.3.
[b] nd = not distillable.
[c] Sublimes.

potassium methylate, potassium ethylate, and potassium *tert*-butylate), alkaline-earth metal alkoxides (magnesium methylate, magnesium ethylate, and calcium methylate and ethylate), aluminum alkoxides (aluminum isopropylate and aluminum *sec*-butylate), transition-metal alkoxides (titanium alkoxides, zirconium alkoxides, and vanadium alkoxides), and antimony trialkoxides.

Table 2. Physical Properties of Double Alkoxides

Alkoxide[a]	mp, °C	bp, °C/Pa[b]
$K[Li(OC_3H_7)_2]$		
$Na[Zr_2(OC_3H_7)_9]$[c]	168–180	260/1
$K[Zr_2(OC_3H_7)_6]$[c]		subl 200/26.6
$Li[Zr_2(OC_3H_7)_9]$		260/26.6
$Tl[Zr_2(OC_3H_7)_6]$		subl 220/66.5
$Mg[Al(OC_2H_5)_4]_2$	129	220–228/53.5
$Ca[Al(OC_3H_7)_4]_2$	124	230–240/400
$K[Al(OC_4H_9)_4]$	164–165	
$Mg[Al(OC_3H_7)_4]_2$	20	130–142/260
$Co[Al(OC_2H_5)_4]_2$[c,d]		
$Na[Sn_2(OC_2H_5)_9]$[c]	260 dec	
$Ca[U(OC_2H_5)_6]_2$[e]		subl 200/0.13
$Al[U(OC_2H_5)_6]_3$[c,f]		111–115/0.16
$U[Al(OC_3H_7)_4]_4$[c,g]		95–97/0.13
$Na[U(OC_2H_5)_6]$[e]	dec	

[a] White solids unless otherwise noted.
[b] To convert Pa to mm Hg, divide by 133.3.
[c] Soluble in organic solvents.
[d] Violet solid.
[e] Green solid.
[f] Green liquid.
[g] Green oil.

Handling, Shipment, and Toxicology

Metal alkoxides are strongly caustic and are decomposed by the humidity of the air or moisture of the skin, requiring the use of protective glasses and gloves.

The health hazard presented by metal alkoxides reflects the toxicity of the metals they contain and the metallic hydroxides and alcohols they form on hydrolysis.

Applications

Metal alkoxides are mainly used as catalysts (in Ziegler-Natta polymerization, transesterifications, and condensations), with partial or complete hydrolysis, alcoholysis, transesterification in coatings for plastics, textiles, glass, and metals, and in additives for adhesives and paints, for sol–gel applications, for synthesis of minerals capable of safely enclosing radioactive nuclear waste and for the cross-linking or hardening of natural and synthetic materials. Alkali metal alkoxides find their principal use in organic synthesis where they act as strong bases.

M. HORN
Hüls AG
U. HORNS
Hüls America, Inc.

D. C. Bradley, in W. L. Jolly, ed., *Preparative Inorganic Reactions,* Vol. 2, Interscience Publishers, New York, 1965, pp. 169–186.

D. C. Bradley, in F. A. Cotton, ed., *Progress in Inorganic Chemistry,* Vol. 2, Interscience Publishers, New York, 1960, pp. 303–361.

Beilstein, *Handbuch der organischen, Chemie,* Vol. 1 and 1–4, Ergänzungswerk, Germany, 1918–1973.

J. H. Harwood, *Industrial Application of the Organometallic Compounds,* Reinhold Publishing Corp., New York, 1963, pp. 199–329.

ALKYD RESINS

In spite of challenges from many new coating resins developed over the years, alkyd resins as a family have maintained a prominent position for two principal reasons, their high versatility and low cost.

Fundamental Reactions and Resin Structure

The main reactions involved in alkyd resin synthesis are polycondensation by esterification and ester interchange. Figure 1 uses the following symbols to represent the basic components of an alkyd resin.

As Figure 1 implies, there is usually some residual acidity as well as free hydroxyl groups left in the resin molecules.

Classification of Alkyd Resins

Alkyd resins are usually referred to by a brief description based on certain classification schemes. From the classification the general

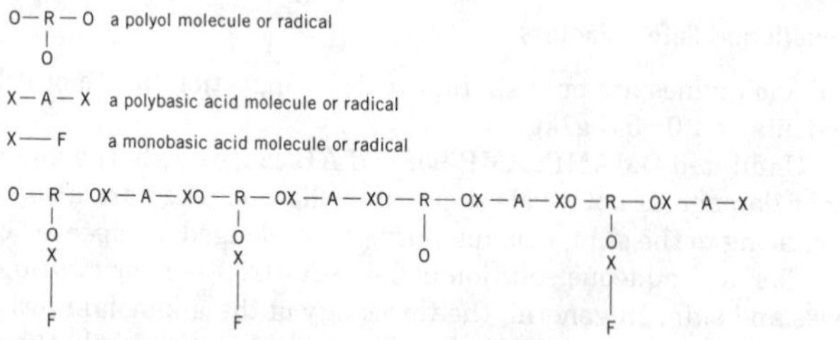

Figure 1. Schematic representation of an alkyd resin molecule.

Table 1. Property Changes With Oil Length of Alkyd Resins[a]

Property	Oil length		
	Long	Medium	Short
requirement of aromatic/polar solvents	———	———	——→
compatibility with other film-formers	———	———	——→
viscosity	———	———	——→
ease of brushing	←——	———	———
air dry time, set-to-touch	←——	———	———
through-dry	←——	—— ——	——→
film hardness	———	———	——→
gloss	———	———	——→
gloss retention	———	———	——→
color retention	———	———	——→
exterior durability	←——	—— ——	——→

[a] Primarily drying-type alkyds.

properties of the resin become immediately apparent. Classification is based on the nature of the fatty acid and oil length.

Oil Length–Resin Property Relationship

The oil length of an alkyd resin has profound effects on the properties of the resin (Table 1).

Alkyd Ingredients

For each of the three principal components of alkyd resins, the polybasic acids, the polyols, and the monobasic acids, there is a large variety to be chosen from. The selection of each of these ingredients affects the properties of the resin and may affect the choice of manufacturing processes. Thus, to both the resin manufacturers and the users, the selection of the proper ingredients is a significant decision.

Polybasic Acids and Anhydrides. The principal polybasic acids used in alkyd preparation include phthalic anhydride (mol wt 148, eq wt 74), isophthalic acid (mol wt 166, eq wt 83), maleic anhydride (mol wt 98, eq wt 49), fumaric acid (mol wt 116, eq wt 58), adipic acid (mol wt 146, eq wt 73), azelaic acid (mol wt 160, eq wt 80), sebacic acid (mol wt 174, eq wt 87), chlorendic anhydride (mol wt 371, eq wt 185.5), and trimellitic anhydride (mol wt 192, eq wt 64).

Polyhydric Alcohols. The principal types of polyol used in alkyd synthesis are shown in Table 2.

Monobasic Acids. The overwhelming majority of monobasic acids used in alkyd resins are long-chain fatty acids of natural occurrence. They may be used in the form of oil or free fatty acid. Free fatty acids are usually available and classified by their origin, *viz,* soya fatty acids, linseed fatty acids, coconut fatty acids, etc. Fats and oils commonly used in alkyd resins include castor oil, coconut oil, cottonseed oil, linseed oil, oiticica oil, peanut oil, rapeseed oil, safflower oil, soyabean oil, sunflowerseed oil, and tung oil.

The drying property of fats and oils is related to their degree of unsaturation, and hence, to iodine values.

Table 2. Polyols for Alkyd Synthesis

Type	Mol wt	Eq wt
pentaerythritol	136	34
glycerol	92	31[a]
trimethylolpropane	134	44.7
trimethylolethane	120	40
ethylene glycol	62	31
neopentyl glycol	104	52

[a] Because glycerol is usually supplied at 99% purity (1% moisture), its eq wt is commonly assumed to be 31 in recipe calculations.

Linolenic acid is responsible for the high yellowing tendency of alkyds based on linseed oil fatty acids. Alkyds made with nondrying oils or their fatty acids have excellent color and gloss stability. They are frequently the choice for white industrial baking enamels and lacquers.

The Concept of Functionality and Gelation

The concept of functionality and its relationship to polymer formation was greatly expanded the theoretical consideration and mathematical treatment of polycondensation systems. Thus if a dibasic acid and a diol react to form a polyester, assuming there is no possibility of other side reactions to complicate the issue, only linear polymer molecules are formed. When the reactants are present in stoichiometric amounts, the average degree of polymerization, $\bar{x}_n$ follows the equation:

$$\bar{x}_n = 1/(1 - p) \tag{1}$$

where p is the fractional extent of reaction. Thus when the reaction is driven to completion, theoretically, the molecular weight approaches infinity and the whole mass forms one giant polymer molecule. Although the material should theoretically still be soluble and fusible, the molecular weight would be so high that it would not be processible by any of the existing methods. For all practical purposes it is a gel; this is the sole example of difunctional monomers being polymerized to gelation.

The functionality of the system, f, is the sum of all of the functional groups, ie, equivalents, divided by the total number of moles of the reactants present in the system. Thus, in the above equimolar reaction system,

$$f = (1 \times 2 + 1 \times 2)/(1 + 1) = 2 \tag{2}$$

Microgel Formation and Molecular Weight Distribution

The behavior of alkyd resin reactions often deviates from that predicted by the theory of Flory. To explain this, a mechanism of microgel formation by some of the alkyd molecules at a relatively early stage of the reaction was proposed. The microgel particles are dispersed and stabilized by smaller molecules in the remaining reaction mixture. As polyesterification proceeds, more microgel particles are formed, until finally a point is reached where they can no longer be kept separated. The microgel particles then coalesce or flocculate, phase inversion occurs, and the entire reaction mass gels. The drying capability of an alkyd resin comes primarily from the microgel fraction. For example, when the highest molecular weight fraction representing about 20% of the total was removed through fractionation, a residual linoleic alkyd lost all ability to air dry to a hard film.

Principles for the Designing of Alkyd Resins

The process of alkyd resin designing should begin with the question, "What are the intended applications of the resin?" The application dictates property requirements, such as solubility, viscosity, drying characteristics, compatibility, film hardness, film flexibility, acid value, water resistance, chemical resistance, and environmental endurance. With the targets in mind, a selection of oil length and a preliminary list of alternative choices of ingredients can then be made. For commercial production, the raw material list is screened based on considerations of material cost, availability, yield, impact on processing cost, and potential hazard to health, safety, and the environment. The list may be further narrowed by limitations imposed by the production equipment or other considerations. Once the oil length and ingredients are chosen, the first draft of a detailed formulation for the resin can be made.

A simple molecular approach is favored by some alkyd chemists for deriving a starting formulation. The basic premise of this approach is that when the total number of moles of the polyols is equal to or

Table 3. Excess Hydroxyl Content Required in Alkyd Formulations

Oil length, fatty acid, %[a]	Excess OH based on diacid equivalents, %
62 or more	0
59–62	5
57–59	10
53–57	18
48–53	25
38–48	30
29–38	32

[a] Based on C-18 fatty acids with average eq wt of 280. If the average eq wt of the monobasic acids is significantly different, adjustment is necessary.

slightly larger than that of the dibasic acids, and the hydroxyl groups are present in an empirically prescribed excess amount, the probability of gelation is very small. Table 3 lists the empirical requirements for excess hydroxyl groups at various oil (fatty acid) lengths of the alkyd.

Chemical Procedures for Alkyd Resin Synthesis

Different chemical procedures may be used for the synthesis of alkyd resins. The choice is usually dictated by the selection of the starting ingredients. Procedures include the alcoholysis process, the fatty acid process, the fatty acid–oil process, and the acidolysis process.

Alkyd Resin Production Processes

Depending on the requirements of the chemical procedures, the processing method may be varied with different mechanical arrangements to remove the by-product, water, in order to drive the esterification reaction toward completion. Methods include the fusion process and the solvent process.

Process Control. The progress of the alkyd reaction is usually monitored by periodic determinations of the acid number and the solution viscosity of samples taken from the reactor. The frequency of sampling is commonly every half-hour.

Safety and Environmental Precautions

The manufacturing of alkyd resins involves a wide variety of organic ingredients. Whereas most of them are relatively mild and of low toxicity, some, such as phthalic anhydride, maleic anhydride, solvents, and many of the vinyl (especially acrylic) monomers, are known irritants or skin sensitizers and are poisonous to humans. The hazard potential of the chemicals should be determined by consulting the Material Safety Data Sheets provided by the suppliers, and recommended safety precautions in handling the materials should be practiced.

With the ever-increasing awareness of the need of environment protection, the emission of solvent vapors and organic fumes into the atmosphere should be prevented by treating the exhaust through a proper scrubber. The solvent used for cleaning the reactor is usually consumed as part of the thinning solvent. Aqueous effluent should be properly treated before discharge.

Modification of Alkyd Resins by Blending With Other Polymers

One of the important attributes of alkyds is their good compatibility with a wide variety of other coating polymers. This good compatibility comes from the relatively low molecular weight of the alkyds, and the fact that the resin structure contains, on the one hand, a relatively polar and aromatic backbone, and, on the other hand, many aliphatic side chains with low polarity. An alkyd resin in a blend with another coating polymer may serve as a modifier for the other film-former, or it may be the principal film-former and the other polymer may serve as the modifier for the alkyd to enhance certain properties. Examples of compatible blends follow.

Nitrocellulose-based lacquers often contain short or medium oil alkyds to improve flexibility and adhesion. The principal applications are furniture coatings, top lacquer for printed paper, and automotive refinishing primers.

Amino resins are probably the most important modifiers for alkyd resins. Many industrial baking enamels, such as those for appliances, coil coatings, and automotive finishes (especially refinishing enamels), are based on alkyd–amino resin blends. Some of the so-called catalyzed lacquers for finishing wood substrate require very low bake or no bake at all.

Chlorinated rubber is often used in combination with medium oil drying-type alkyds. The principal applications are highway traffic paint, concrete floor, and swimming pool paints.

Vinyl resins, ie, copolymers of vinyl chloride and vinyl acetate which contain hydroxyl groups from the partial hydrolysis of vinyl acetate or carboxyl groups, eg, from copolymerized maleic anhydride, may be formulated with alkyd resins to improve their application properties and adhesion. The blends are primarily used in making marine top-coat paints.

Synthetic latex house paints sometimes contain emulsified long oil or very long oil drying alkyds to improve adhesion to chalky painted surfaces.

Silicone resins with high phenyl contents may be used with medium or short oil alkyds as blends in air-dried or baked coatings to improve heat or weather resistance; the alkyd component contributes to adhesion and flexibility. Applications include insulation varnishes, heat-resistant paints, and marine coatings.

Chemically Modified Alkyd Resins

Although blending with other coating resins provides a variety of ways to improve the performance of alkyds, or of the other resins, chemically combining the desired modifier into the alkyd structure eliminates compatibility problems and gives a more uniform product. Several such chemical modifications of the alkyd resins have gained commercial importance. They include vinylated alkyds, silicone alkyds, urethane alkyds, phenolic alkyds, and polyamide alkyds.

High Solids Alkyds

There has been a strong trend in recent years to increase the solids content of all coating materials, including alkyds, to reduce solvent vapor emission. In order to raise solids and still maintain a manageable viscosity, the molecular weight of the resin must be reduced. Consequently, film integrity must be developed through further chain extension or cross-linking of the resin molecules during the "drying" step. A high cross-linking density necessitated by the lower molecular weight of the resin builds high stress in the film and causes it to be prone to cracking. Therefore, adequate flexibility should be designed into the resin structure. Chain extension and cross-linking of high solids alkyd resins are typically achieved by the use of polyisocyanato oligomers or amino resins.

Water-Reducible Alkyds

Replacing solvent-borne coatings with water-borne coatings not only reduces solvent vapor emission, but also improves the safety against the fire and health hazards of organic solvents. Alkyd resins may be made water-reducible either by converting the resin into an emulsion form or by incorporating "water-soluble" groups in the molecules.

Economic Aspects

Alkyd resins, as a family, have remained the workhorse of the coatings industry for decades. The top alkyd resin manufacturers in the United States are Cargill, Reichhold, a subsidiary of Dainippon Ink & Chemicals, Inc., and Spencer Kellog, now a part of NL Industries, Inc.

Future Prospects. Because of the efforts of the coatings industry to reduce solvent emission, there has been a clear gradual decline in the market share of alkyds as a group relative to all synthetic coating resins. However, their versatility and low cost will undoubtedly maintain them as significant players in the coatings arena. Alkyds are much more amenable to development of higher solids compositions than most other coating resins. Great strides in the development of water-borne types have also been made in recent years. Another good reason to remain optimistic about alkyds for the future is that a significant portion of their raw material, fatty acids, is renewable.

K. F. LIN
Hercules Incorporated

R. G. Mraz and R. P. Silver, in N. M. Bikales, ed., *Encyclopedia of Polymer Science and Technology,* Vol. 1, John Wiley & Sons, Inc., 1964, pp. 663–734.

H. J. Lanson in J. I. Kroschwitz, ed., *Encyclopedia of Polymer Science and Engineering,* 2nd ed., Vol. 1, John Wiley & Sons, Inc., 1985, pp. 644–679.

T. C. Patton, *Alkyd Resin Technology, Formulating Techniques and Allied Calculations,* Wiley-Interscience, New York-London, 1962.

ALKYLATION

The alkylation described in this article is the substitution of a hydrogen atom bonded to the carbon atom of a paraffin or aromatic ring by an alkyl group. The alkylations of nitrogen, oxygen, and sulfur are described in separate articles.

Significant technological development has been made in the area of alkylation in recent years. Environmental concerns associated with mineral acid catalysts have encouraged process changes and the development of solid-bed alkylation processes. The application of heterogeneous catalysts, especially zeolite catalysts, has led to new alkylation technologies. Research efforts to develop environmentally acceptable, economical technologies by applying new materials as alkylation catalysts will continue, and more new technologies are expected to be commercialized in the 1990s.

Alkylation of Paraffinic Hydrocarbons

Paraffin alkylation as discussed here refers to the addition reaction of an isoparaffin and an olefin. The desired product is a higher molecular weight paraffin that exhibits a greater degree of branching than either of the reactants.

The principal industrial application of paraffin alkylation is in the production of premium quality fuels for spark-ignition engines. Future gasoline specifications will continue to favor the clean-burning characteristics and the low emissions typical of alkylate. Alkylate production capacity was more than 50 million tons per year in 1989 and is expected to grow as worldwide gasoline specifications become more stringent.

Catalysts and Reactions

Although the alkylation of paraffins can be carried out thermally, catalytic alkylation is the basis of all processes in commercial use. The reaction steps include the formation of a light tertiary cation, the addition of the cation to an olefin to form a heavier cation, and the production of a heavier paraffin (alkylate) by a hybride transfer from a light isoparaffin. This last step generates another light tertiary cation to continue the chain.

The catalysts used in the industrial alkylation processes are strong liquid acids, either sulfuric acid (H_2SO_4) or hydrofluoric acid (HF).

Feedstock and Products

Isobutane. Although other isoparaffins can be alkylated, isobutane is the only paraffin commonly used as a commercial feedstock.

Butylenes. Butylenes are the primary olefin feedstock of alkylation and produce a product high in trimethylpentanes. The research octane number, which is typically in the range of 94–98, depends on isomer distribution, catalyst, and operating conditions.

Propylene. Propylene alkylation produces a product that is rich in dimethylpentane and has a research octane typically in the range of 89–92.

Amylenes. Amylenes (C-5 monoolefins) produce alkylates with a research octane in the range of 90–93.

Alkylation of Aromatic Hydrocarbons

Most of the industrially important alkyl aromatics used for petrochemical intermediates are produced by alkylating benzene with monoolefins. The most important monolefins for the production of ethylbenzene, cumene, and detergent alkylate are ethylene, propylene, and olefins with 10–18 carbons, respectively.

Acid Catalysts and Reaction Mechanism

Acid catalysts promote the addition of alkyl groups to aromatic rings. Olefins, alcohols, ethers, halides, and other olefin-producing compounds can be used as alkylating reagents. In addition to traditional protonic acid catalysts (H_2SO_4, HF, phosphoric acid) and Friedel-Crafts type catalysts ($AlCl_3$, boron fluoride), any solid acid catalyst having a comparable acid strength is effective for aromatic alkylation. Typical solid acid catalysts are amorphous and crystalline alumino-silicates, clays, ion-exchange resins, mixed oxides, and supported acids. Among these solid acid catalysts, ZSM-5 and Y-type zeolites have become the new commercial catalysts for aromatic alkylation.

Base Catalysts and Reaction Mechanism

Alkali metals and their derivatives can catalyze the alkylation of aromatics with olefins. In contrast to acid-catalyzed alkylation, in which the aromatic ring is alkylated, an olefin is added to the alkyl group of aromatics over a base catalyst through a carbanion intermediate. The carbanion intermediate is produced from an aromatic compound by the abstraction of benzylic hydrogen as a proton by a base. The carbanion reacts with an olefin to grow the side chain of the aromatic compound.

Industrial Application

Ethylbenzene. This alkylbenzene is almost exclusively used as an intermediate for the manufacture of styrene monomer. A small amount (<1%) is used as a solvent and as an intermediate in dye manufacture.

Ethylbenzene is primarily produced by the alkylation of benzene with ethylene, although a small percentage of the world's ethylbenzene capacity is based on the superfractionation of ethylbenzene from mixed xylenes streams.

Cumene. The demand for cumene has risen at an average rate of 2–3%/yr since 1970, and this trend is expected to continue throughout the 1990s.

Currently (1993), almost all cumene is produced commercially by two processes: *(1)* a fixed-bed, kieselguhr-supported phosphoric acid catalyst system developed by UOP and *(2)* a homogeneous $AlCl_3$ and hydrogen chloride catalyst system developed by Monsanto.

Two new processes using zeolite-based catalyst systems were developed in the late 1980s. Unocal's technology is based on a conventional fixed-bed system. CR&L has developed a catalytic distillation system based on an extension of the CR&L MTBE technology.

Cymene. Methylisopropylbenzene can be produced over a number of different acid catalysts by alkylation of toluene with propylene.

Detergent Alkylate. The synthetic detergent industry has become one of the largest chemical process industries (see DETERGENCY).

Industrial Processes. Two catalysts, HF and $AlCl_3$, are used for the alkylation of benzene with higher olefins (C-10–C-15 detergent-range olefins). In some earlier units, H_2SO_4 was used. In addition to the alkylation activity, all of these acid catalysts possess, in varying degrees, activity to shift the olefinic double bond along the chain. Thus, regardless of the position of the double bond in the olefin feed, the position of the phenyl group in the final product is specific to the catalyst system used. Regardless of the catalyst system used, minor side reactions, like the formation of dialkylbenzene, take place.

Future Developments. The most recent advance in detergent alkylation is the development of a solid catalyst system. UOP and Compania Española de Petroleos SA (CEPSA) have disclosed the joint development of a fixed-bed heterogeneous aromatic alkylation catalyst system for the production of LAB.

Xylenes. The main application of xylene isomers, primarily *p*- and *o*-xylenes, is in the manufacture of plasticizers and polyester fibers and resins. Demands for xylene isomers and other aromatics such as benzene have steadily been increasing over the last two decades. The major source of xylenes is the catalytic reforming of naphtha and the pyrolysis of naphtha and gas oils.

Polynuclear Aromatics. The alkylation of polynuclear aromatics with olefins and olefin-producing reagents is effected by acid catalysts. The alkylated products are more complicated than those produced by the alkylation of benzene because polynuclear aromatics have more than one position for substitution.

Future Technology Trends

Over the years, improvements in aromatic alkylation technology have come in the form of both improved catalysts and improved processes. This trend is expected to continue into the future.

Catalysts. Nearly all of the industrially significant aromatic alkylation processes of the past have been carried out in the liquid phase with unsupported acid catalysts. Since 1976, these forms of acids have become a significant environmental concern from both a physical handling and disposal perspective. This concern has fueled much development work toward solid acid catalysts, including zeolites, silica–aluminas, and clays.

Process. As solid acid catalysts have replaced liquid acid catalysts, they have typically been placed on conventional fixed-bed reactors. An extension of fixed-bed reactor technology is the concept of catalytic distillation being offered by CR&L. In catalytic distillation, the catalytic reaction and separation of products occur in the same vessel. The concept has been applied commercially for the production of MTBE and is also being offered for the production of ethylbenzene and cumene.

A new alkylation process called Alkymax was introduced by UOP in 1990. In addition to lowering the benzene content, the alkylate formed has a high octane value and can typically boost the octane of the gasoline pool by 0.5 RON.

Other Alkylations

Alkylation of Phenol. The hydroxyl group activates the alkylation of the benzene ring because it is a strong electron-donating group; therefore, the alkylation of phenol can be achieved with olefins and olefin-producing reagents under milder conditions than the alkylation of aromatic hydrocarbons.

Alkylated phenol derivatives are used as raw materials for the production of resins, novolaks (alcohol-soluble resins of the phenol–formaldehyde type), herbicides, insecticides, antioxidants, and other chemicals.

Alkylation of Aromatic Amines and Pyridines. Commercially important aromatic amines are aniline, toluidine, phenylenediamines, and toluenediamines (see AMINES, AROMATIC). The ortho alkylation of these aromatic amines with olefins, alcohols, and dienes to produce more valuable derivatives can be achieved with solid acid catalysts.

The alkylation of pyridine takes place through nucleophilic or homolytic substitution because the π-electron-deficient pyridine nucleus does not allow electrophilic substitution, eg, Friedel-Crafts alkylation.

Health and Safety Factors

In industrial applications, specialized procedures are required to ensure the safe handling of these materials. Replacing these materials with solid acid catalysts will become more important in the future. The solid acid catalysts themselves present a disposal problem that favors the development of regenerable catalysts or the implementation of recycling procedures.

H. U. HAMMERSHAIMB
T. IMAI
G. J. THOMPSON
B. V. VORA
UOP

G. Stefanidakis and J. E. Gwyn, in J. J. McKetta and W. A. Cunningham, eds., *Encyclopedia of Chemical Processing and Design,* Vol. 2, Marcel Dekker, New York, 1977, p. 357.

W. Keim and M. Roper, in W. Gerhartz, ed., *Ullmann's Encyclopedia of Industrial Chemistry,* Vol. A1, VCH Verlagsgesellschaft, Weinheim, Germany, 1985, p. 185.

T. Hutson and G. E. Hays, in L. F. Albright and A. R. Goldsby, eds., *Industrial and Laboratory Alkylations* (ACS Symposium Series) American Chemical Society, Washington, D.C., 1977, pp. 27–56.

ALKYLPHENOLS

The production of alkylphenols (1993) exceeds 450,000 t/yr on a worldwide basis. Alkylphenols of greatest commercial importance have alkyl groups ranging in size from one to twelve carbons. The direct use of alkylphenols is limited to a few minor applications such as epoxy-curing catalysts and biocides. The vast majority of alkylphenols are used to synthesize derivatives which have applications ranging from surfactants to pharmaceuticals. The four principal markets are nonionic surfactants, phenolic resins, polymer additives, and agrochemicals.

Nonionic surfactants and phenolic resins based on alkylphenols are mature markets and only moderate growth in these derivatives is expected. Concerns over the biodegradability and toxicity of these alkylphenol derivatives to aquatic species may limit their use in the future. The use of alkylphenols in the production of both polymer additives and monomers for engineering plastics is expected to show above-average growth as plastics continue to replace traditional building materials.

Alkylphenols containing 3–12-carbon alkyl groups are produced from the corresponding alkenes under acid catalysis. Alkylphenols containing the methyl group are produced by the alkylation of phenol with methanol.

Nomenclature

An alkylphenol is a phenol derivative wherein one or more of the ring hydrogens has been replaced by an alkyl group(s). Phenol is a heading parent in the CAS indexing system. Appropriate names of alkylphenols for abstract citations can be derived by using the appropriate aids. The names generated in this manner are unambiguous and refer to a specific compound, but are lengthy and cumbersome to use. Common names are used on a daily basis and are especially prevalent for alkylphenols that have gained commercial importance.

Physical Properties

The physical properties of alkylphenols are comparable to phenol. The properties are strongly influenced by the type of alkyl substituent and its position on the ring. Alkylphenols, like phenol, are typically solids at 25°C. Their form is affected by the size and configuration of the alkyl group, its position on the ring, and purity. They appear colorless, or white, to a pale yellow when pure (Table 1).

The solubility of alkylphenols in water falls off precipitously as the number of carbons attached to the ring increases. They are generally soluble in common organic solvents: acetone, alcohols, hydrocarbons, toluene. Solubility in alcohols or heptane follows the generalization that "like dissolves like."

Table 1. Commercially Important Alkylphenols

Name	Molecular formula	Molecular weight	Boiling point, °C[a]	Freezing point, °C	Density[b], g/mL	Flash point, °C
4-*tert*-amylphenol	$C_{11}H_{16}O$	164.0	249	90.0	0.915^{107}	121
4-*tert*-butylphenol	$C_{10}H_{14}O$	150.2	237	97.5	0.890^{107}	117
2-*sec*-butylphenol	$C_{10}H_{14}O$	150.2	224	20.0	0.938^{43}	>93
4-cumylphenol	$C_{15}H_{16}O$	212.0	335	70.0	1.029^{93}	188
4-dodecylphenol	$C_{18}H_{30}O$	262.0	334	<20.0	0.914^{20}	>100
4-nonylphenol	$C_{15}H_{24}O$	220.3	310	<20.0	0.933^{43}	146
4-*tert*-octylphenol	$C_{14}H_{22}O$	220.3	290	81.0	0.940^{25}	132
2,4-di-*tert*-amylphenol	$C_{16}H_{26}O$	234.4	275	23.0	0.900^{49}	104
2,4-di-*tert*-butylphenol	$C_{14}H_{22}O$	206.3	263	52.0	0.867^{82}	115
2,6-di-*tert*-butylphenol	$C_{14}H_{22}O$	206.3	253	36.0	0.898^{43}	<99
di-*sec*-butylphenol	$C_{14}H_{22}O$	206.3		<20.0	0.902^{66}	127
2,4-dicumylphenol	$C_{24}H_{26}O$	330.0		65.0	1.030^{66}	462
2-methylphenol	C_7H_8O	108.1	191	30.0	$1.049^{15.5}$	81
3-methylphenol	C_7H_8O	108.1	202	10.0	$1.042^{15.5}$	86
4-methylphenol	C_7H_8O	108.1	202	34.0	1.022^{25}	86
2,6-dimethylphenol	$C_8H_{10}O$	122.1	203	48.0	1.020^{25}	88

[a] At 101.3 kPa = 1 atm.
[b] At the temperature indicated by the superscript, °C.

Synthesis of Alkylphenols

Alkylphenols can be synthesized by several approaches, including alkylation of a phenol, hydroxylation of an alkylbenzene, dehydrogenation of an alkyl-cyclohexanol, or ring closure of an appropriately substituted acyclic compound. The choice of approach depends on the target alkylphenol, availability of the starting materials, and cost of processing.

Reactions of Alkylphenols

Alkylphenols undergo a variety of chemical transformations, involving the hydroxyl group or the aromatic nucleus that convert them to value-added products.

The Hydroxyl Group. The unshared pairs of electrons on hydroxyl oxygens seek electron-deficient centers. Alkylphenols tend to be less nucleophilic than aliphatic alcohols as a direct result of the attraction of the electron density by the aromatic nucleus. The reactivity of the hydroxyl group can be enhanced in spite of the attraction of the ring current by use of a basic catalyst which removes the acidic proton from the hydroxyl group, leaving the more nucleophilic alkylphenoxide. Reactions include esterification and etherification.

Reactions Involving the Ring. The aromatic nucleus of alkylphenols can undergo a variety of aromatic electrophilic substitutions. Electron density from the hydroxyl group is fed into the ring. Besides activating the aromatic nucleus, the hydroxyl group controls the orientation of the incoming electrophile.

Manufacture and Processing

Alkylphenols of commercial importance are generally manufactured by the reaction of an alkene with phenol in the presence of an acid catalyst. The alkenes used vary from single species, such as isobutylene, to complicated mixtures, such as propylene tetramer (dodecene). The alkene reacts with phenol to produce monoalkylphenols, dialkylphenols, and trialkylphenols. The monoalkylphenols comprise ~ 85% of all alkylphenol production.

The choice of catalyst is based primarily on economic effects and product purity requirements. More recently, the handling of waste associated with the choice of catalyst has become an important factor in the economic evaluation. Catalysts that produce less waste and more easily handled waste by-products are strongly preferred by alkylphenol producers. Some commonly used catalysts are sulfuric acid, boron trifluoride, aluminum phenoxide, methanesulfonic acid, toluene–xylene sulfonic acid, cationic-exchange resin, acidic clays, and modified zeolites.

Reactors. Reactors used to produce alkylphenols are simple batch reactors, complex batch reactors, and continuous reactors.

Purification. The method used to recover the desired alkylphenol product from the reactor output is highly dependent on the downstream use of the product and the physical properties of the alkylphenol.

Some alkylphenol applications can tolerate "as is" reactor products, most significantly in the production of alkylphenol–formaldehyde resins. However, most alkylphenols sold today require refinement. Distillation is by far the most common separation route.

Production and Shipment

Most commercially important alkylphenol production is of three types, unrefined alkylphenols, monoalkylphenols, and dialkylphenols. Together, these processes comprise over 95% of all alkylphenol production in the United States.

Economic Aspects

Among the key variables in strategic alkylphenol planning are feedstock quality and availability, equipment capability, environmental needs, and product quality. In the past decade, environmental needs have grown enormously in their effect on economic decisions. The manufacturing cost of alkylphenols includes raw-material cost, nonraw-material variable cost, fixed cost, and depreciation.

Raw-material costs are the largest cost items over the lifetime of a plant and typically make up between 40 and 90% of the total manufacturing cost. The placement of plants near production facilities making alkenes or phenol is important to producers of alkylphenols.

Health and Safety Factors

The toxicity of alkylphenols as a class of compounds ranges from moderately toxic (oral rat LD_{50} 50–500 mg/kg) to practically nontoxic (oral rat LD_{50} 5,000–15,000 mg/kg) and most are irritants or corrosive toward skin.

Commercial Uses and Derivatives of Alkylphenols

4-tert-Amylphenol. 4-*tert*-Amylphenol is used in phenolic resins (novolaks and resoles).

4-tert-Butylphenol. Phenolic resin applications account for 60–70% of all 4-*tert*-butylphenol consumed worldwide.

2-sec-Butylphenol. A significant volume of 2-*sec*-butyl-4,6-dinitrophenol is used worldwide as a polymerization inhibitor in the production of styrene where it is added to the reboiler of the styrene distillation tower to prevent the formation of polystyrene.

Because of environmental concerns about 2-*sec*-butylphenol-based derivatives, the market growth is expected to be negative in the future, with the exception of possible significant growth in the use of the carbamate insecticide.

4-Cumylphenol. The major use of 4-cumylphenol is as a chain terminator for polycarbonates.

4-Dodecylphenol. The major use of technical grade 4-dodecylphenol is in lube oil additives. High purity 4-dodecylphenol is used to produce specialty surfactants by its reaction with ethylene oxide.

2-Methylphenol. The majority of 2-methylphenol is used in the production of novolak phenolic resins.

3-Methylphenol. A major use of 3-methylphenol is in the production of phenolic-based antioxidants, which are particularly good at stabilizing polymers in contact with copper against thermal oxidative degradation. Another significant use of 3-methylphenol is in the production of herbicides and insecticides.

4-Methylphenol. The bulk of 4-methylphenol is used in the production of phenolic antioxidants.

4-Nonylphenol. The major use for 4-nonylphenol is in the production of nonionic surfactants.

Another significant use of 4-nonylphenol is in the production of tris(4-nonylphenyl) phosphite (TNPP), a secondary antioxidant which protects organic materials against oxidative degradation by decomposing hydroperoxides.

4-tert-Octylphenol. 4-*tert*-Octylphenol reacts with ethylene oxide under base catalysis and the resulting ethoxylates are used in many of the same applications as the 4-nonylphenol-based surfactants. Another important application for 4-*tert*-octylphenol is in the production of phenolic resins.

Other applications for 4-*tert*-octylphenol are in chain termination of polycarbonates and the production of uv stabilizers.

Dialkylated Phenols. A significant use for 2,4-di-*tert*-amylphenol is in the production of uv stabilizers; the principal one is a benzotriazole-based uv absorber, 2-(2′-hydroxy-3′,5′-di-*tert*-amylphenyl)-5-chlorobenzotriazole, which is widely used in polyolefin films, outdoor furniture, and clearcoat automotive finishes. Another significant use for 2,4-di-*tert*-amylphenol is in the photographic industry.

The primary use for 2,4-di-*tert*-butylphenol is in the production of substituted triaryl phosphites.

The principal use for 2,6-di-*tert*-butylphenol is in the production of hindered phenolic antioxidants, and this application accounts for 80–90% of all of this compound produced.

The only significant use for di-*sec*-butylphenol is a specialty nonionic surfactant produced by reaction with ethylene oxide under base catalysis. This surfactant is registered with EPA for use in emulsifying agrochemicals.

The largest use for 2,4-dicumylphenol is in the production of a uv stabilizer of the benzotriazole class, 2-(2′-hydroxy-3′,5′-dicumylphenyl)benzotriazole, which is used in engineering thermoplastics where high molding temperatures are encountered.

The oxidative coupling of 2,6-dimethylphenol to yield poly(phenylene oxide) represents 90–95% of the consumption of 2,6-dimethylphenol.

JOHN F. LORENC
GREGORY LAMBETH
WILLIAM SCHEFFER
Schenectady Chemicals, Inc.

S. H. Patinkin and B. S. Friedman in G. A. Olah, ed., *Friedel-Crafts and Related Reactions*, Vol. 2, Wiley-Interscience, New York, 1964.

A. S. Lindsey and H. Jeskey, *Chem. Rev.* **57** 583–620 (1957).

P. R. Dean, *Index of Commercial Antioxidants and Antiozonants*, Technical Bulletin, Goodyear Chemicals, Akron, Ohio, 1983.

C. B. Campbell and A. Onopchenko, *Ind. Eng. Chem. Res.* **31**, 2278–2281 (1992).

ALLYL ALCOHOL AND MONOALLYL DERIVATIVES

The technology of introducing a new functional group to the double bond of allyl alcohol was developed in the mid-1980s. Allyl alcohol is accordingly used as an intermediate compound for synthesizing raw materials such as epichlorohydrin and 1,4-butanediol, and this development is bringing about expansion of the range of uses of allyl alcohol.

Physical Properties

Allyl alcohol is a colorless liquid having a pungent odor; its vapor may cause severe irritation and injury to eyes, nose, throat, and lungs. It is also corrosive. Allyl alcohol is freely miscible with water and miscible with many polar organic solvents and aromatic hydrocarbons, but is not miscible with n-hexane. It forms an azeotropic mixture with water and a ternary azeotropic mixture with water and organic solvents. Allyl alcohol has both bacterial and fungicidal effects. Properties of allyl alcohol are shown in Table 1.

Chemical Properties

Addition Reactions. The C═C double bond of allyl alcohol undergoes addition reactions typical of olefinic double bonds.

Hydroformylation. Hydroformylation of allyl alcohol is a synthetic route for producing 1,4-butanediol, a raw material for poly(butylene terephthalate), an engineering plastic.

Substitution of Hydroxyl Group. The substitution activity of the hydroxyl group of allyl alcohol is lower than that of the chloride group of allyl chloride and the acetate group of allyl acetate. However, allyl alcohol undergoes substitution reactions under conditions in which saturated alcohols do not react. Reactions proceed in catalytic systems in which a π-allyl complex is considered as an intermediate.

Oxidation. The C═C double bond of allyl alcohol undergoes epoxidation by peroxide, yielding glycidol. This epoxidation reaction is applied in manufacturing glycidol as an intermediate for industrial production of glycerol.

Industrial Manufacturing Processes for Allyl Alcohol

There are four processes for industrial production of allyl alcohol. One is alkaline hydrolysis of allyl chloride. A second process has two steps. The first step is oxidation of propylene to acrolein and the second step is reduction of acrolein to allyl alcohol by a hydrogen transfer reaction, using isopropyl alcohol. At present, neither of these two processes is being used industrially. Another process is isomerization of propylene oxide. Until 1984, all allyl alcohol manufacturers were using this process. Since 1985 Showa Denko K.K. has produced allyl alcohol industrially by a new process which they developed. This process, which

Table 1. Properties of Allyl Alcohol

Property	Value
molecular formula	C_3H_6O
molecular weight	58.08
boiling point, °C	96.90
freezing point, °C	−129.00
density, d^{20}_{4}	0.8520
refractive index, n^{20}_{D}	1.413
viscosity at 20°C, mPa·s(= cP)	1.37
flash point[a] °C	25
solubility in water at 20°C, wt %	infinity

[a] Closed cup.

Table 2. Properties of Important Allyl Compounds

Property	Allyl chloride	Allyl acetate	Allyl methacrylate	AGE[a]	Allyl amine	DMAA[b]
molecular formula	C_3H_5Cl	$C_5H_8O_2$	$C_7H_{10}O_2$	$C_6H_{10}O_2$	C_3H_7N	$C_5H_{11}N$
molecular weight	76.53	100.12	126.16	114.14	57.10	85.15
boiling point, °C	44.69	104	150	153.9	52.9	64.5
freezing point, °C	−134.5	−96	−60	−100	−88.2	
density, $d^{20}{}_4$	0.9382	0.9276	0.934	0.9698	0.7627	0.72
viscosity at 20°C, mPa · s(= cP)	3.36	0.52	13	1.20		0.44
flash point, °C	−31.7	6	33	57.2	−29	−23

[a] Allyl glycidyl ether.
[b] Dimethylallylamine.

was developed partly for the purpose of producing epichlorohydrin via allyl alcohol as the intermediate, has the potential to be the main process for production of allyl alcohol. The reaction scheme is as follows:

$$CH_2{=}CHCH_3 + CH_3COOH + 1/2\ O_2 \xrightarrow{Pd} CH_2{=}CHCH_2O\overset{O}{\overset{\|}{C}}CH_3$$

$$CH_2{=}CHCH_2O\overset{O}{\overset{\|}{C}}CH_3 + H_2O \xrightleftharpoons{H^+} CH_2{=}CHCH_2OH + CH_3COOH$$

The world's manufacturers of allyl alcohol are ARCO Chemical Company, Showa Denko K.K., Daicel Chemical Industries, and Rhône-Poulenc Chimie; total production is approximately 70,000 tons per year.

Monoallyl Derivatives

In this article, mainly monoallyl compounds are described. Diallyl and triallyl compounds used as monomers are covered elsewhere.

Reactivity of Allyl Compounds

Hydrosilylation. The addition reaction of silane

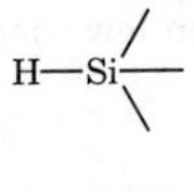

to the C=C double bond of allyl compounds is applied in the industrial synthesis of silane coupling agents.

π-Allyl Complex Formation. Allyl halide, allyl ester, and other allyl compounds undergo oxidative addition reactions with low atomic valent metal complexes to form π-allyl complexes.

Physical Properties of Derivatives

The physical properties of some important monoallyl compounds are summarized in Table 2.

Allyl Chloride

This derivative, abbreviated AC, is a transparent, mobile, and irritative liquid. It can be easily synthesized from allyl alcohol and hydrogen chloride. However, it is industrially produced by chlorination of propylene at high temperature.

Uses. Allyl chloride is industrially the most important allyl compound among all the allyl compounds. It is used mostly as an intermediate compound for producing epichlorohydrin, which is consumed as a raw material for epoxy resins.

Allyl Esters

Allyl Acetate. Allyl acetate is produced mostly for manufacturing allyl alcohol.

Allyl Methacrylate. At present, allyl methacrylate, AMA, is used mostly as a raw material for silane coupling agents.

Allyl Ethers

The C—H bond of the allyl position easily undergoes radical fission, especially in the case of allyl ethers, reacting with the oxygen in the air to form peroxide compounds.

Therefore, in order to keep allyl ether for a long time, it must be stored in an air-tight container under nitrogen.

Allyl Glycidyl Ether. This ether is used mainly as a raw material for silane coupling agents and epichlorohydrin rubber.

Allyl Amines

Allylamine. This amine can be synthesized by reaction of allyl chloride with ammonia at the comparatively high temperature of 50–100°C, or at lower temperatures using $CuCl_2$ or CuCl as the catalyst.

Dimethylallylamine. Dimethylallylamine is used in the production of insecticides and pesticides.

Safety and Handling

Most allyl compounds are toxic and many are irritants. Those with a low boiling point are lachrymators. Precautions should be taken at all times to ensure safe handling.

Nobuyuki Nagato
Showa Denko K.K.

Allyl Alcohol, Technical Publication SC: 46-32, Shell Chemical Corp., San Francisco, Calif., Nov. 1, 1946.

C. E. Schildknecht, *Allylic Compounds and Their Polymers,* John Wiley & Sons, Inc., New York, 1973.

H. Raech, Jr., *Allylic Resins and Monomers,* Reinhold Publishing Corporation, New York, 1965.

ALLYL MONOMERS AND POLYMERS

Allyl compounds comprise a large group of ethylenic compounds having unique reactivities and uses often contrasting with those of typical vinyl-type compounds (styrenes, acrylics, vinyl esters and ethers, and related compounds). In allyl compounds the double bond is not substituted by a strong activating group to promote polymerization but is attached to a carbon which generally bears one or more reactive hydrogen atoms. Unlike monovinyl compounds, monoallyl compounds do not form homopolymers of high molecular weight by free-radical or conventional ionic mechanisms; in general, only viscous liquid homopolymers of limited use have been obtained. This is explained by the low reactivity of the ethylenic double bond together with the high reactivity of hydrogen atoms on the allylic carbon in reducing the molecular weight by degradative chain transfer.

In contrast, many allyl compounds containing two or more reactive double bonds yield solid, high molecular weight polymers by initiation with suitable free-radical catalysts. A number of polyfunctional allyl esters have achieved importance in polymerization and copolymerization, especially to obtain heat-resistant cast sheets and thermoset moldings.

Diallyl Carbonate Cast Plastics

From a number of diallyl esters investigated, diallyl diglycol carbonate or diethylene glycol bis(allyl carbonate), DADC, was developed to produce by bulk polymerization cast sheets, lenses, and other shapes of outstanding scratch resistance and optical and mechanical properties.

DADC Monomers. Reaction of allyl alcohol in the presence of alkali with diethylene glycol bis(chloroformate), obtained from the glycol and phosgene, gives the DADC monomer.

DADC monomer is a colorless liquid of mild odor and a viscosity of 9 mPa·s(= cP) at 25°C. It is low in toxicity, but can produce skin irritation. It is fairly resistant to saponification by dilute alkali. Contact with strong alkali at higher temperature produces the more toxic allyl alcohol. Properties are given in Table 1 and the trade literature. DADC is soluble in common organic solvents and in methyl methacrylate, styrene, and vinyl acetate. It is partially soluble in amyl alcohol, gasoline, and ligroin. It is insoluble in ethylene glycol, glycerol, and water.

The DADC monomer is available from several manufacturers, such as PPG Industries (CR-39), Akzo (Nouryset 200), Enichem (RAV), Rhône-Poulenc (XR-80), Tokuyama Soda (TS-16), and Mitsui Toatsu (MR-3).

DADC Homopolymerization. Bulk polymerization of CR-39 monomer gives clear, colorless, abrasion-resistant polymer castings that offer advantages over glass and acrylic plastics in optical applications. Free-radical initiators are required for thermal or photochemical polymerization.

Casting of DADC. Sheets, rods, and lens preforms are cast from CR-39 or prepolymer syrup by methods similar to those used for methacrylate ester syrups.

Usage of DADC polymers in impact-resistant, lightweight eyewear lenses has grown rapidly and is now the principal application.

Coatings. In recent years methods have been developed to improve abrasion resistance of DADC polymer surfaces in optical devices and glazings by means of special coatings. Hard or glass-like coatings may be applied by near-vacuum vapor deposition of quartz (silica) or by hydrolysis of alkoxysilanes.

Modified Polymers and Copolymers. DADC pure monomer and mixtures with small amounts of comonomers or other additions are commercially available for casting. Monomer formulations are available including agents for protecting the eyes against uv light. Another grade is designed to absorb infrared radiation, and several modified monomers give copolymers of increased heat resistance and hardness.

Polymeric Nuclear-Track Detectors. DADC polymer is used in solid-state track detectors (SSTD) of nuclear particles, including alpha-particles, fast neutrons, cosmic rays, and ions of elements of atomic number 10 and above.

Diallyl Phthalates and Their Polymerization

The three isomeric diallyl phthalates are colorless liquids of mild odor, low volatility, and relatively slow polymerization in the early stages. At ca 25% conversion, the viscous liquid undergoes gelation and polymerization accelerates; however, the last monomer disappears at a slow rate.

The monomers are prepared by conventional esterification. Properties of two diallyl phthalate monomers, $C_{14}H_{14}O_4$, are given in Table 2.

DAP Copolymerization. The diallyl phthalates copolymerize readily with monomers bearing strong electron-attracting groups attached to the ethylenic group. These include maleic anhydride, maleate and fumarate esters, and unsaturated polyesters.

Diallyl Isophthalate. DAIP polymerizes faster than DAP, undergoes less cyclization, and yields cured polymers of better heat resistance, eg, up to ca 200°C. Besides application as heat-resistant molding powders for electronic and other applications, DAIP copolymers have been proposed for optical applications.

Telomerization. Polymerization of DAP is accelerated by telogens such as CBr_4, which are more effective chain-transfer agents than the monomer itself; gelation is delayed.

Applications. The largest use of diallyl phthalate thermoset polymers is in moldings and coatings for electronic devices requiring high reliability under long-term adverse environmental conditions.

Other Diallyl Esters

Tables 3 and 4 give properties of some diallyl esters. Dimethallyl phthalate has been copolymerized with vinyl acetate and benzoyl peroxide, and reactivity ratios have been reported.

Allyl–Vinyl Compounds

Monomers such as allyl methacrylate and diallyl maleate have applications as cross-linking and branching agents selected especially for the different reactivities of their double bonds; some physical properties are given in Table 4. These esters are colorless liquids soluble in most organic liquids but little soluble in water; DAM and DAF have pungent odors and are skin irritants.

Allyl Methacrylate (AMA). Of the compounds containing both allyl and vinyl-type double bonds, allyl methacrylate is the most important. AMA is used as cross-linking agent with methacrylate esters in contact lenses. AMA is also used in low concentrations in curable acrylic coatings.

Table 1. Typical Properties of Commercial DADC Monomer

Property	Value
specific gravity	$1.15^{20}{}_4$
refractive index, $n^{20}{}_D$	1.452
boiling point at 266 Pa[a], °C	166
melting point (supercooled), °C	−4 to 0

[a] To convert Pa to mm Hg, multiply by 0.0075.

Table 2. Properties of Commercial Diallyl Esters[a]

Property	DAP[b]	DAIP[c]
boiling point, °C at 0.53 kPa[d]	161	181
density, g/mL	1.117^{25}	1.124^{20}
refractive index, $n^{25}{}_D$	1.518	1.5212
viscosity at 20°C, mPa·s(= cP)	12	17
freezing point, °C	below −70	−3

[a] Sources: Osaka Soda Co., Hardwick Chemical Co., and FMC Corp.
[b] Diallyl phthalate.
[c] Diallyl isophthalate.
[d] To convert kPa to mm Hg, multiply by 7.5.

Table 3. Properties of Some Diallyl Esters

Diallyl ester	Bp_{Pa}[a]	n_D^{20}	d_4^{20}
oxalate	$107_{1.9}$	1.4460	1.0081
malonate	$119_{1.9}$	1.4489	1.060
succinate	$94_{0.13}$	1.4507	1.056
adipate	$115_{0.13}$	1.4542	1.023
sebacate	$164_{0.26}$	1.4550	0.978
tartrate	$171_{1.3}$	1.187	

[a] To convert Pa to mm Hg, multiply by 0.0075.

Table 4. Properties of Allyl–Vinyl Monomers

Property	Allyl methacrylate, AMA	Diallyl maleate, DAM	Diallyl fumarate, DAF
boiling point, °C$_{kPa}$[a]	55_4	$112_{0.53}$	$140_{0.40}$
density at 25°C, g/cm^3	0.930	1.070	1.0516
refractive index, n^{25}_D	1.453	1.4664[b]	1.4669
viscosity, mPa·s(= cP)	13	4.3	3.0

[a] To convert kPa to mm Hg, multiply by 7.5.
[b] At 20°C.

Table 5. Properties of TAC and TAIC

Property	TAC	TAIC
melting point, °C	31	24
boiling point, °C	$140_{67\ Pa}$[a]	$126_{40\ Pa}$[a]
density at 30°C, g/cm^3	1.1133	1.1720
refractive index, n^{25}_D	1.5049	1.5115

[a] To convert Pa to mm Hg, multiply by 0.0075.

Polyfunctional Allyl Nitrogen Monomers

Triallyl Cyanurate as Cross-Linking Agent. Triallyl cyanurate (TAC), 2,4,6-tris(allyloxy)-*s*-triazine, and its isomer triallyl isocyanurate (TAIC) are used as cross-linking agents with comonomers and for aftercuring preformed polymers such as olefin copolymers in electrical insulations. TAC monomer melts at 20–25°C. It is prepared by gradual addition of cyanuric chloride to an excess of allyl alcohol in the presence of aqueous alkali. Properties of TAC and TAIC are given in Table 5.

Diallyl Ammonium Polymers. *N,N*-Diallyldimethyl(DADM)ammonium salts are used for the preparation of polyelectrolytes. Copolymers of diallyldimethylammonium chloride with acrylamide have been used in electroconductive coatings. Molded polyamide surfaces can be hardened by grafting with *N,N*-diallylacrylamide monomer under exposure to electron beam. *N,N*-Diallyltartardiamide is a cross-linking agent for acrylamide reversible gels in electrophoresis.

R. DOWBENKO
PPG Industries

C. E. Schildknecht, *Allyl Compounds and Their Polymers,* Wiley-Interscience, New York, 1973.

CR-39 Allyl Diglycol Carbonate Monomer Bulletin 45A and *Casting and Material Safety Data Sheet,* PPG Industries, Inc., Pittsburgh, Pa., 1984.

ALUMINUM AND ALUMINUM ALLOYS

Aluminum, Al, is a silver-white metallic element in Group III of the Periodic Table having an electronic configuration of $1s^22s^22p^63s^23p^1$. Aluminum exhibits a valence of +3 in all compounds except for a few high temperature gaseous species in which the aluminum may be monovalent or divalent. Aluminum is the most abundant metallic element on the surfaces of the earth and moon, comprising 8.8% by weight (6.6 atomic %) of the earth's crust. However, it is rarely found free in nature. Nearly all rocks, particularly igneous rocks, contain aluminum as aluminosilicate minerals.

Aluminum reflects radiant energy throughout the spectrum. It is odorless, tasteless, nontoxic, and nonmagnetic. Because of its many desirable physical, chemical, and metallurgical properties, aluminum is the most widely used nonferrous metal. The utility of the metal is enhanced by the formation of a stable adherent oxide surface that resists corrosion. Because of high electrical conductivity and lightness, aluminum is used extensively in electrical transmission lines. High purity aluminum is soft and lacks strength but its alloys, containing small amounts of other elements, have high strength-to-weight ratios. Alloys of aluminum are readily formable by many metalworking processes; they can be joined, cast, or machined and accept a wide variety of finishes. Aluminum, having a density about one-third that of ferrous alloys, is used in transportation and structural applications where weight-saving is important.

Physical Properties

The properties of aluminum vary significantly according to purity and alloying. Physical properties for aluminum of a minimum of 99.99% purity are summarized in Table 1.

Chemical Properties

Reactions with Elements and Inorganic Compounds. Aluminum reacts with oxygen, O_2, having a heat of reaction of −1675.7 kJ/mol(−400.5 kcal/mol) Al_2O_3 produced.

Aluminum does not combine directly with hydrogen, but it does react with nitrogen, N_2, sulfur, and carbon in oxygen-free atmospheres at high temperatures.

Very high purity aluminum, resistant to attack by most acids, is used in the storage of nitric acid, concentrated sulfuric acid, organic acids, and other chemical reagents. Aluminum is, however, dissolved by aqua regia.

Aluminum is attacked by salts of more noble metals. In particular, aluminum and its alloys should not be used in contact with mercury or mercury compounds.

Reaction with Organic Compounds. Aluminum is not attacked by saturated or unsaturated, aliphatic or aromatic hydrocarbons. Halogenated derivatives of hydrocarbons do not generally react with aluminum except in the presence of water, which leads to the formation of halogen acids. The chemical stability of aluminum in the presence of alcohols is very good and stability is excellent in the presence of aldehydes, ketones, and quinones.

Manufacture and Processing

Raw Materials. Aluminum, the third most abundant element in the earth's crust, is usually combined with silicon and oxygen in rock. When aluminum silicate minerals are subjected to tropical weathering, aluminum hydroxide may be formed. Rock that contains high concentrations of aluminum hydroxide minerals is called bauxite. Although bauxite is, with rare exception, the starting material for the production of aluminum, the industry generally refers to metallurgical grade alumina, Al_2O_3, extracted from bauxite by the Bayer process, as the ore. Aluminum is obtained by electrolysis of this purified ore (Fig. 1).

Energy Considerations. Table 2 gives a breakdown of the energy required to produce aluminum. Note that smelting consumes about 65% of the required energy. In the United States most of this energy comes from fossil fuels. Hydropower, which supplies 33% for smelting and

Table 1. Physical Properties of Aluminum

Property	Value
density at 25°C, kg/m^3	2698
melting point, °C	660.2
boiling point, °C	2494
thermal conductivity at 25°C, W/(m·K)	234.3
latent heat of fusion, J/g[a]	395
latent heat of vaporization at bp, ΔH_v, kJ/g[a]	10,777
electrical conductivity	65% IACS[b]
electrical resistivity at 20°C, Ω·m	2.6548×10^{-8}
tensile strength, MPa[c]	50

[a] To convert J to cal, divide by 4.184.
[b] International Annealed Copper Standard.
[c] To convert MPa to psi, multiply by 145.

Table 2. Energy Consumption Per Metric Ton of Aluminum Produced[a]

Operation	Thermal, MJ[b]	Electric, kW·h	Total energy	
			Fossil and hydro, MJ[b]	If all fossil, MJ[b]
mining and refining	30,000	480	35,200	35,200
smelting	19,000[c]	15,000	146,400	182,500
mill processing	19,000	1,830	33,700	38,900
Total	*68,000*	*17,310*	*215,300*	*256,600*

[a] Values are approximate. Actual energy consumption depends upon the particular plant, alloy produced, and product formed.
[b] To convert MJ to Mcal, divide by 4.184.
[c] Includes forming, baking, and fuel value of anodes.

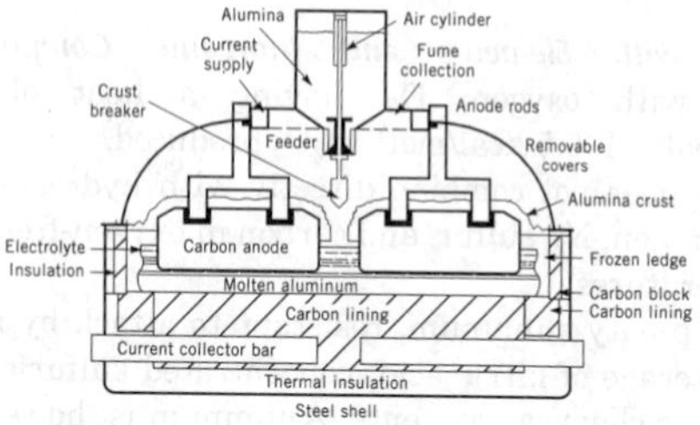

Figure 1. Aluminum electrolyzing cell with prebaked anode.

17% for fabricating, is not used for mining and refining; little nuclear power is used.

Energy Conservation. The U.S. Department of Energy (DOE) is sponsoring research on inert anodes and refractory hard metal (RHM) composite cathodes. Success in these developments could significantly lower the energy required to reduce alumina. Inert anodes would eliminate the consumption of anode carbon and allow improved sealing of the cell for reduced heat loss and reduced fluoride emission. RHM cathodes would be wetted with a thin film of aluminum which would drain to a sump and provide stable cathodic surface rather than the present aluminum pool that sloshes about owing to electrohydrodynamic effects. The stable cathode should improve current efficiency and also allow closer interelectrode spacing for reduced power consumption. The most promising route to reduced energy requirement, however, is through recycling of scrap aluminum. Recycling scrap requires less than 5% of the energy required to produce new metal. About 60% of aluminum beverage can scrap was recycled in 1990. This should go higher and extend to other aluminum scrap.

High Purity Aluminum. The Hall-Héroult process cannot ensure aluminum purity higher than 99.9%. Techniques such as electrolytic refining and fractional crystallization are required to produce metal of higher purity.

Production

The tremendous growth of the aluminum industry as compared to that of other nonferrous metals is shown in Table 3. The principal markets for aluminum in the United States are containers, packaging, building and construction, and transportation.

Economic Aspects

Aluminum prices have historically been more stable than those of other nonferrous metals. Beginning in the 1970s, however, aluminum prices have fluctuated. These fluctuations reflect increased energy costs as well as increased costs of raw materials. Improvements in production processes as well as a rebalancing of demand and supply are expected to stabilize aluminum prices in the future.

Worldwide primary aluminum capacity continues to grow but mostly in countries where there is low cost electric power. The United States and other developed countries are expected to concentrate more on converting raw aluminum into high value-added products.

Table 3. Annual World Production of Aluminum, Copper, Magnesium, Lead, and Zinc, 10^3 t/yr

Year	Al	Cu	Mg	Pb	Zn
1900	5.7	449	0.01	877	479
1950	1516	2791	21.2	1748	1958
1960	4732	4400	93.1	2630	3070
1970	9780	6376	222.8	4003	5097
1980	16043	6083	317.4	5399	6153
1988	17304	5959	338.1	3371	7224

Environmental Considerations

Fluoride emission from aluminum smelting cells has long been an area of great concern. Treatment consists of highly (over 99%) efficient dry scrubbers that catch particulates and sorb HF on alumina that is subsequently fed to the cells. Hence, nearly all the fluoride evolved is fed back into the cell.

Hydrocarbon fumes evolved during anode baking are generally disposed of by burning. Handling of alumina and coke presents dusting problems. Hoods and exhaust systems collect the dust, which is then separated from the exhaust air either by cyclones, electrostatic precipitators, filter bags, or a combination of these methods, and recycled to the process. Fumeless fluxing procedures remove hydrogen and undesirable metallic impurities.

The linings of aluminum reduction cells must be replaced periodically. These spent linings represent the largest volume of waste associated with the smelting process. Because they contain fluorides and cyanide, they must be either stored under roof or buried in landfills lined with impervious materials to prevent leaching and contamination of the environment.

Aluminum Alloys

Many of the properties of aluminum alloy products depend on metallurgical structure which is controlled both by the chemical composition and by processing. In addition to features such as voids, inclusions, grains, subgrains, dislocations, and vacancies which are present in virtually all metallic products, the structure of aluminum alloys is characterized by three types of intermetallic particles. Aluminum metallurgists refer to these as constituent particles, dispersoid particles, and precipitate particles. Constituent particles are formed during solidification, generally as a byproduct of a divorced eutectic reaction, and range in size in the final product from about 1–20 micrometers. These particles, present in all aluminum alloys, play a role in grain-size control and negatively affect toughness of high strength alloy products. Dispersoids and precipitates both form by a solid-state reaction. The particles known as dispersoids characteristically form during thermal treatment of an ingot by precipitation of a solid solution which exceeds maximum solid solubility because of nonequilibrium conditions during ingot solidification. Dispersoids, about 10–200 nm in the largest dimension, are present in most aluminum alloy products. Their primary function is to control grain size, grain orientation (texture), and degree of recrystallization. Particles classified as precipitates form during heat treatment of the final mill product by precipitation from a supersaturated solid solution that does not exceed the maximum equilibrium solid solubility. In the final product, their size may range from disks a few atoms thick by a few nm in diameter up to needlelike or platelike particles which may exceed 1 micrometer in the largest dimension. Precipitates may confer high strength. The nature of the constituent, dispersoid, and precipitate particles depends strongly on the phase diagrams of the particular alloy.

Binary Alloys. Aluminum-rich binary phase diagrams show three types of reaction between liquid alloy, aluminum solid solution, and other phases: eutectic, peritectic, and monotectic.

Al–Fe. The Al–Fe system is important because virtually all commercial aluminum alloys contain some iron, Fe. The system has a eutectic at 1.9% Fe, but solid solubility of only 0.05% Fe.

Al–Mn. The Al–Mn system, the basis for the oldest yet most widely used aluminum alloys, is characterized by a eutectic at 1.95% Mn and 658°C. Maximum solid solubility is 1.76% manganese, Mn, and the intermetallic phase in aluminum-rich alloys is Al_6Mn.

Al–Cu. Many structural aluminum alloys contain significant amounts of copper, Cu. There is a eutectic in the Al–Cu system at 33.2% Cu and 548°C, but the important feature is the maximum solubility of 5.7% Cu at 548°C which decreases drastically at lower temperatures. This decreasing solubility with decreasing temperature is necessary for the phenomenon known as age hardening or precipitation strengthening.

Al–Mg. Almost every commercial precipitation-strengthened aluminum alloy contains magnesium as an alloying element. The Al–Mg system has a eutectic at 35% magnesium Mg, and 451°C. Maximum solid solubility is 14.9% Mg, and solubility decreases to about 0.8% Mg at room temperature. Despite this decreased solubility, precipitation strengthening by the metastable β'-phase precursor to the equilibrium β-phase precursor to the equilibrium β-phase Al_3Mg_2 precipitates is observed only at very high magnesium levels.

Al–Si. Al–Si alloys possess high fluidity and castability and are consequently used for weld wire, brazing, and as casting alloys. This system has a eutectic at 12.6% Si and 577°C; maximum solid solubility of Si is 1.65%.

Al–Li. Alloys containing about two to three percent lithium, Li, received much attention in the 1980s because of their low density and high elastic modulus. Each weight percent of lithium in aluminum alloys decreases density by about 3% and increases elastic modulus by about 6%. The system is characterized by a eutectic reaction at 8.1% Li at 579°C. The maximum solid solubility is 4.7% Li.

Al–Cr. Although no commercial alloys are based on this system, chromium, Cr, is an ingredient of several complex and commercially significant alloys. The Cr is added for control of grain structure. The Al–Cr system has a peritectic portion at 661°C where solid solubility is 0.7% Cr and liquid solubility is 0.4% Cr.

Al–Pb. Both lead, Pb, and bismuth, Bi, which form similar systems, are added to aluminum alloys to promote machinability by providing particles to act as chip breakers.

Al–Zn. Aluminum-rich binary alloys are not age hardenable to any commercial significance, and zinc, Zn, additions do not significantly increase the ability of aluminum to strain harden. Al–Zn alloys find commercial use as sacrificial claddings on high strength Al–Cu–Mg–Zn aircraft alloy sheet. The eutectoid composition near 78% Zn has found use as a superplastic sheet alloy.

Al–Zr. This system has a peritectic reaction at 660.8°C at which solubility is 0.28% zirconium, Zr, solid and 0.11% Zr, liquid.

Ternary Alloys. Almost all commercial alloys are of ternary or higher complexity. Alloy type is defined by the nature of the principal alloying additions, and phase reactions in several classes of alloys can be described by reference to ternary phase diagrams. Minor alloying additions may have a powerful influence on properties of the product because of the influence on the morphology and distribution of constituents, dispersoids, and precipitates.

Al–Fe–Si. Iron and silicon, present in primary aluminum, may also be added to produce enriched alloys for specific purposes.

Al–Mg–Si. An important class of commercial alloys is based on the Al–Mg–Si system because of its precipitation-hardening capabilities and good corrosion resistance.

Al–Mg–Mn. The basis for the alloys used as bodies, ends, and tabs of the cans used for beer and carbonated beverages is the Al–Mg–Mn alloy system. It is also used in other applications that exploit the excellent weldability and corrosion resistance. These alloys have the unique ability to be highly strain-hardened and yet retain a high degree of ductility.

Al–Cu–Mg. The first precipitation-hardenable alloy was an Al–Cu–Mg alloy. There is a ternary eutectic at 508°C, and there are nine binary and five ternary intermetallic phases. For aluminum-rich alloys, only four phases are encountered, in addition to the aluminum solid solution. Several commercial alloys are based on the age-hardening characteristics of the metastable precursors of θ, Al_2Cu, or S-phase, Al_2CuMg, principally θ' or S′.

Al–Mg–Zn. Although neither aluminum-rich binary Al–Zn nor Al–Mg alloys are precipitation hardenable, the ternary system is a source of alloys strengthened in this manner. Alloy compositions are selected for precipitation of an M- or η-phase, $MgZn_2$, precursor because T-phase, $Al_2Mg_3Zn_3$, is less effective as a strengthener. Commercially important alloys always contain more zinc than magnesium to provide attractive combinations of strength, extrudability, and weldability.

Al–Cu–Li. Although the addition of Cu to Al–Li alloys increases density, the boost in strength more than offsets the density increase so that Al–Cu–Li alloy products develop higher specific strengths (strength/density) than do binary Al–Li alloy products. Furthermore, the fracture toughness and corrosion resistance of products manufactured from Al–Cu–Li alloys are higher than these properties in binary Al–Li alloy products. These alloys are strengthened by δ', Al_7Li, and T_1, Al_2CuLi.

Al–Li–Mg. In aluminum-rich alloys the ternary phase, T, sometimes designated as Al_2LiMg, is encountered in addition to AlLi (δ), Al_3Mg_2 (β), and $Al_{12}Mg_{17}$ (γ). Assessment of the composition of the T-phase indicates that it contains 15.5 atomic % Mg and 32 atomic % Li.

Quaternary and Higher Alloys. Further additions to commercial aluminum alloys usually are made either to modify the metastable strengthening precipitates or to produce dispersoids.

Modifications to Precipitates. Silicon is sometimes added to Al–Cu–Mg alloys to help nucleate S′ precipitates without the need for cold work prior to the elevated temperature aging treatments. Additions of elements such as tin, Sn, cadmium, Cd, and indium, In, to Al–Cu alloys serve a similar purpose for θ' precipitates. Copper is often added to Al–Mg–Si alloys in the range of about 0.25% to 1.0% Cu to modify the metastable precursor to Mg_2Si. The copper additions provide a substantial strength increase. When the copper addition is high, the quaternary $Al_4CuMg_5Si_4$ Q-phase must be considered and dissolved during solution heat treatment.

The highest strength aluminum alloy products are based on the Al–Cu–Mg–Zn system and all are strengthened by precursors to the η-phase.

When combined with magnesium, silver, Ag, has found commercial use as an alloying element in several aluminum alloys for specialized applications. It was added to an Al–Cu–Mg–Zn forging alloy to increase the resistance to stress-corrosion cracking and to an Al–Cu–Mg casting alloy to increase strength. It is an essential component with the new Weldalite series of alloys.

Dispersoid Formers. The three elements commonly added to precipitation hardenable alloys to form dispersoids are manganese, chromium, and zirconium. The amounts customarily used (0.5% Mn, 0.2% Cr, and 0.1% Zr) remain in supersaturated solid solution during ingot casting and precipitate as dispersoids during thermal treatment

Table 4. Mechanical Properties of Aluminum Foundry Alloys

Alloy and temper	Tensile strength, MPa[a]	Yield strength, MPa[a]	Elongation, %	Brinell hardness
		Sand castings		
208.0–F	145	95	2.5	55
242.0–T77	205	160	2	75
295.0–T6	250	165	5	75
319.0–T6	250	165	2	80
356.0–T6	225	165	3	70
514.0–F	170	85	9	50
A712.0–T5	240	170	5	75
		Permanent mold castings		
242.0–T61	325	290	0.5	110
319.0–F	235	130	2.5	85
F332.0–T5	250	195	1	105
356.0–T6	260	185	5	80
		Die castings		
360.0	305	170	3	75
A413.0	290	130	3	80

[a] To convert MPa to psi, multiply by 145.

of the ingot. These dispersoids serve to minimize recrystallization during solution heat treatment of products such as plate, forgings, and extrusions which are hot-worked, and to maintain a fine recrystallized grain size in sheet and tubing which is cold-worked.

Foundry Alloys and Their Characteristics. Unalloyed aluminum does not have either mechanical properties or casting characteristics suitable for general foundry use yet both can be greatly improved by the addition of other elements. The most common addition is silicon which enhances fluidity, increases resistance to hot-cracking, and improves pressure tightness. Because binary Al–Si alloys have relatively low strengths and ductility, other elements such as copper and magnesium are added to obtain higher strengths through heat treatment. Typical mechanical properties are presented in Table 4.

Wrought Alloys and Their Characteristics. Alloys for the production of wrought products are selected for fabricability as well as their physical, chemical, and mechanical properties. Usually, these alloys are less highly alloyed than those for foundry use and contain less iron and silicon. A series of alloys based on the eutectic Al–Fe–Si composition, however, has been developed to provide good combinations of strength and formability in thin sheet products.

Thermal Treatments

Aluminum alloys are subjected during manufacture to a variety of thermal treatments that range from heating to assist fabrication, to heating for control of final properties. Although the natural oxide film on aluminum provides good protection against surface oxidation and deterioration during such treatments, controlled atmospheres are sometimes employed for products requiring minimum surface oxide such as foil and sheet for reflectors.

Homogenization. Ingots are usually preheated prior to rolling, forging, or extrusion to increase workability. The process is commonly referred to as homogenization because chemical segregation of the major alloying elements that are completely soluble in the solid state is reduced.

Annealing. The resistance to further deformation of aluminum alloy products at elevated temperatures reaches a steady value after a modest strain when the rate of formation of fresh dislocations is balanced by the rate of annihilation. This process is known as dynamic recovery. In-process annealing is employed to decrease the dislocation density, thereby increasing the plasticity of the hot-rolled metal prior to cold rolling. This thermal process also modifies the crystallographic texture, a very important consideration in producing products requiring control of anisotropy.

Annealing is also employed as a final mill operation to produce a material having high formability for subsequent customer shaping or forming operations.

Solution Heat Treatment. Solution heat treatment is the first stage of a series of operations to achieve precipitation hardening.

Quenching. After solution treatment, the product is generally cooled to room temperature at a rate necessary for it to retain essentially all of the solute in solution.

Precipitation Heat Treatment. The supersaturated solution produced by the quench from the solution temperature is unstable, and the alloys tend to approach equilibrium by precipitation of solute. Because the activation energies required to form equilibrium precipitate phases are higher than those to form metastable phases, the solid solution decomposes to form G-P zones at room temperature (natural aging). Metastable precursors to the equilibrium phases are formed at the temperatures employed for commercial precipitation heat treatments (artificial aging).

Shaping and Fabricating

Aluminum alloys are commercially available in a wide variety of cast forms and in wrought mill products produced by rolling, extrusion, drawing, or forging. The mill products may be further shaped by a variety of metal-working and forming processes and assembled by conventional joining procedures into more complex components and structures.

Corrosion

Aluminum and aluminum alloys are employed in many applications because of the ability to resist corrosion. Corrosion resistance is attributable to the tightly adherent, protective oxide film present on the surface of the products. This film is 5–10 nm thick when formed in air; if disrupted it begins to form immediately in most environments. The loss in strength as a result of atmospheric weathering and corrosion is small, and the rate decreases with time. The amount of corrosion that occurs is a function of the alloy as well as the severity of the corrosive environment. Wrought alloys of the Al, Al–Mn, Al–Mg, and Al–Mg–Si types have excellent corrosion resistance in most weathering exposures, including industrial and seacoast atmospheres. Alloys based on additions of copper, or copper, magnesium, and zinc, have significantly lower resistance to corrosion.

Uses

Packaging has replaced the building and construction industry as the largest consumer of aluminum in the United States because aluminum is impermeable to gas, resistant to corrosion, and recyclable. The most prominent has been the use of aluminum for beer and carbonated beverages.

The largest market worldwide for aluminum products is in the building and construction industry.

Because of aluminum's low density, the field of transportation is another large market for aluminum alloys.

Aluminum is used in the home as household foil (0.18 mm thick), cooking utensils (the first commercial use of aluminum), refrigerators, air conditioners, appliances, insect screening, and hardware. It is also used for toys, sporting equipment, lawn furniture, lawn mowers, and portable tools.

Aluminum is an excellent conductor of electricity, having a volume conductivity 62% of that of copper. It also has many applications in the chemical and petrochemical industries such as for piping and tanks in alloys 1100, 3003, 6061, 6063, and the Al–Mg alloys.

JAMES T. STALEY
WARREN HAUPIN
Aluminum Company of America

J. Hatch, ed., *Aluminum: Properties and Physical Metallurgy*, American Society for Metals, Metals Park, Ohio, 1984.

A. K. Vasudévan and R. D. Doherty, eds., *Aluminum Alloys: Contemporary Research and Applications*, Academic Press, San Diego, Calif., 1989.

T. Khan and G. Effenburg, eds., *Advanced Aluminum and Magnesium Alloys*, Proceedings of an International Conference on Light Metals, ASM International, Metal Park, Ohio, 1990.

ALUMINUM COMPOUNDS

INTRODUCTION

The CAS registry lists 5037 aluminum-containing compounds exclusive of alloys and intermetallics. Some of these are listed in Table 1.

Table 1. Aluminum Compounds Referred to in Text

Compounds	Molecular formula
alum	$KAl(SO_4)_2{\cdot}12H_2O$
alumina	Al_2O_3
aluminum bromide	$AlBr_3$
aluminum chlorhydroxide (ACH)	$Al_2Cl(OH)_5$
aluminum(I) chloride	$AlCl$
aluminum(III) chloride	$AlCl_3$
aluminumchloride hexahydrate	$AlCl_3{\cdot}6H_2O$
aluminum(I) fluoride	AlF
aluminum(III) fluoride	AlF_3
aluminum hydroxide	$Al(OH)_3$
aluminum iodide	AlI_3
aluminum(II) oxide	AlO
aluminum silicate	$Al_2(SiO_3)_3$
aluminum sulfate	$Al_2(SO_4)_3$
aluminum sulfate octadecahydrate	$Al_2(SO_4)_3{\cdot}18H_2O$
alunite	$K_2Al_6(SO_4)_4(OH)_{12}$
anorthite	$CaO{\cdot}Al_2O_3{\cdot}2SiO_2$
bauxite	
boehmite	$AlO(OH)$
calcium aluminate	$Al_2O_3{\cdot}3CaO$
corundum	α-Al_2O_3
diaspore	α-$AlO(OH)$
gibbsite	α-$Al(OH)_3$
halloysite	$Al_2Si_2O_5(OH)_4{\cdot}2H_2O$
kaolin	$H_2Al_2Si_2O_8{\cdot}H_2O$
kaolinite	$Al_2O_3{\cdot}2SiO_2{\cdot}2H_2O$
kyanite	$H_6O_5Si{\cdot}2Al$
montmorillonite	
nepheline	$NaAl(OH)SiO_3$
nepheline	$NaAl_2(OH)_2(SiO_3)_2{\cdot}H_2O$
nepheline	AlH_4O_4Si
sapphire	Al_2O_3
sodium aluminate	$NaAlO_2$
triethylaluminum	$(C_2H_5)_3Al$
triisobutylaluminum	$(C_4H_9)_3Al$
zeolite A	$Na_{12}[(Al_{12}Si_{12})O_{48}]{\cdot}27H_2O$
zeolite Y	$Na_{56}[(AlO_2)_{56}(SiO_2)_{136}]{\cdot}250H_2O$
zeolite X	$Na_{86}[(AlO_2)_{86}(SiO_2)_{106}]{\cdot}264H_2O$

Commercially Significant Compounds

The aluminum containing compound having the largest worldwide market is metal-grade alumina. Second is aluminum hydroxide. The split between additive and feedstock applications for $Al(OH)_3$ is roughly 50:50. Additive applications include those as flame retardants in products such as carpets, and to enhance the properties of paper, plastic, polymer, and rubber products. Significant quantities are also used in pharmaceuticals, cosmetics, adhesives, polishes, dentifrices, and glass.

Feedstock applications of $Al(OH)_3$ for production of other chemicals include almost all of the 5000 plus compounds listed in the CAS registry.

Aluminum Sulfate (Alum). Aluminum sulfate, $Al_2(SO_4)_3{\cdot}18H_2O$, also known as alum cake, is industrially produced by reaction of $Al(OH)_3$ and sulfuric acid, H_2SO_4, in agitated pressure vessels at about 170°C. Aluminum sulfate has largely replaced alums for the major applications as a sizing agent in the paper industry and as a coagulant to clarify municipal and industrial water supplies. In terms of worldwide production, it ranks third behind alumina and aluminum hydroxide, with markets in excess of $3 \times 10_6$ t/yr.

Aluminum Halides. All the halogens form covalent aluminum compounds having the formula AlX_3. The most important commercially are the anhydrous chloride and fluoride, and aluminum chloride hexahydrate.

Anhydrous aluminum chloride, $AlCl_3$, is manufactured primarily by reaction of chlorine vapor with molten aluminum and used mainly as a catalyst in organic chemistry.

Aluminum chloride hexahydrate, $AlCl_3{\cdot}6H_2O$, manufactured from aluminum hydroxide and hydrochloric acid, HCl, is used in pharmaceuticals and cosmetics as a flocculant and for impregnating textiles. Conversion of solutions of hydrated aluminum chloride with aluminum to the aluminum chlorohydroxy complexes serve as the basis of the most widely used antiperspirant ingredients.

Organoaluminum Compounds. The alkyls and aryls, R_3Al (in monomer form), are colorless liquids or low melting solids easily oxidized and hydrolyzed when exposed to the atmosphere. Triethylaluminum (TEA), one of the most commercially important members of this family of chemicals, is so reactive it bursts into flame on contact with air, ie, it is pyrophoric, and it reacts violently with water. This behavior is typical and special techniques are necessary for the safe handling and use of organoaluminum compounds.

The alkylaluminum halides, R_nAlX_{3-n}, where X is Cl, Br, I, and R is methyl, ethyl, propyl, iso-butyl, etc, in monomer form, and n = 1 or 2, are less easily oxidized and hydrolyzed than the trialkyls. Organoaluminum hydrides such as diisobutylaluminum hydride, $(iso$-$C_4H_9)_2AlH$, are also available. Organoaluminum compounds are used commercially in multimillion kg/yr quantities as catalysts or starting materials for the manufacture of organic compounds such as plastics, elastomers biodegradable detergents, and organometallics containing zinc, phosphorus, or tin.

Sodium Aluminate. Sodium aluminate is manufactured by dissolving high purity $Al(OH)_3$ in 50% sodium hydroxide solution.

Sodium aluminate is used in water purification, in the paper industry, for the after-treatment of TiO_2 pigment, and in the manufacture of aluminum-containing catalysts and zeolite. Worldwide markets are in the range of 125,000 t/yr.

Zeolites A large and growing industrial use of aluminum hydroxide and sodium aluminate is the manufacture of synthetic zeolites. Zeolites are aluminosilicates with Si/Al ratios between 1 and infinity. There are 40 natural, and over 100 synthetic, zeolites. All the synthetic structures are made by relatively low (100–150°C) temperature, high pH hydrothermal synthesis.

Zeolite-based materials are extremely versatile: uses include detergent manufacture, ion-exchange resins (ie, water softeners), catalytic applications in the petroleum industry, separation processes (ie, molecular sieves), and as an adsorbent for water, carbon dioxide, mercaptans, and hydrogen sulfide.

WILLIAM C. SLEPPY
Aluminum Company of America

C. Misra, *Industrial Alumina Chemicals, ACS Monogr. Ser. No. 184,* American Chemical Society, Washington, D.C., 1986.

K. Wefers and C. Misra, *Oxides and Hydroxides of Aluminum, Alcoa Technical Paper No. 19,* revised, Alcoa Laboratories, Alcoa Division, Aluminum Company of America, Pittsburgh, Pa., 1987.

L. D. Hart, ed., *Aluminum Chemicals: Science and Technology Handbook,* American Ceramics Society, Columbus, Ohio, 1990.

W. Buchner, R. Schliebs, G. Winter, and K. H. Buchel, *Industrial Inorganic Chemistry,* VCH Publishers, New York, 1989, pp. 247–255.

ALUMINATES

Among industrial users of sodium aluminate are producers of paper, paint pigments, silica, alumina or alumina-based catalysts, dishwasher detergents, molecular sieves, concrete, antacids, and others. Sodium aluminate is used in removal of phosphates from municipal and industrial waste waters and for clarification of industrial process and potable water. Commercial sodium aluminate products are available as liquids, and to a lesser degree, in solid form. The formula of anhydrous sodium aluminate is variously given as $NaAlO_2$ (aluminum sodium oxide), $Na_2O{\cdot}Al_2O_3$, or $Na_2Al_2O_4$. Commercial sodium aluminates are not accurately represented by these formulas because the products contain more than the stoichiometric amount of sodium oxide, Na_2O. The amount of excess caustic in commercial products is indicated by ratios of Na_2O/Al_2O_3 that are typically between 1.05 and 1.15 for dry products, and 1.26 and 1.5 for liquids.

Physical and Chemical Properties

Commercial grades of sodium aluminate contain both waters of hydration and excess sodium hydroxide. In solution, a high pH retards the reversion of sodium aluminate to insoluble aluminum hydroxide. The chemical identity of the soluble species in sodium aluminate solutions has been the focus of much work. Solutions of sodium aluminate appear to be totally ionic. The aluminate ion is monovalent and the predominant species present is determined by the Na_2O concentration.

Manufacture

Small amounts of sodium aluminate are prepared in the lab by fusion of equimolar quantities of sodium carbonate and aluminum acetate, $Al(C_2H_3O_2)_3$, at 800°C. Other methods involve reaction of sodium hydroxide with amorphous alumina or aluminum metal. Commercial quantities of sodium aluminate are made from hydrated alumina, in the form of aluminum hydroxy oxide, AlO(OH), or aluminum hydroxide, $Al(OH)_3$, a product of the Bayer process which is used to refine bauxite, the principal aluminum ore.

Commercial grades of sodium aluminate are obtained by digestion of aluminum trihydroxide in aqueous caustic at atmospheric pressure and near the boiling temperature.

Uses

Sodium aluminate is used in the treatment of industrial and municipal water supplies. It is also an effective precipitant for soluble phosphate in sewage and is especially useful in wastewater having low alkalinity.

Large quantities of sodium aluminate are used in papermaking where it improves sizing, filler retention, and pitch deposition. The addition of sodium aluminate to titanium dioxide paint pigment improves the nonchalking performance of outdoor paints.

Sodium aluminate is widely used in the preparation of alumina-based catalysts. Aluminosilicate can be prepared by impregnating silica gel with alumina obtained from sodium aluminate and aluminum sulfate. Reaction of sodium aluminate with silica or silicates has produced porous crystalline aluminosilicates which are useful as adsorbents and catalyst support materials, ie, molecular sieves.

ROBERT J. KELLER
JOSEPH J. LEN
Vinings Industries Inc.

J. R. Glastonbury, *Chem. Ind.* (London), 121 (Feb. 1969).

U.S. Pat. 382,505 (May 8, 1888), K. J. Bayer.

U.S. Pat. 515,859 (Mar. 6, 1894), K. J. Bayer.

ALUMINUM CARBOXYLATES

Aluminum salts of carboxylic acids, aluminum carboxylates, may occur as aluminum tricarboxylates (normal aluminum carboxylates), $Al(OOCR)_3$; monohydroxy (monobasic) aluminum dicarboxylates, $(RCOO)_2Al(OH)$; and dihydroxy (dibasic) aluminum monocarboxylates, $RCOOAl(OH)_2$. Aluminum carboxylates are used in three general areas: textiles, gelling, and pharmaceuticals. Derivatives of low molecular weight carboxylic acids have been mainly associated with textile applications; those of fatty carboxylic acids are associated with gelling salts; and more complex carboxylates find applications in pharmaceuticals.

Commercial products can be combinations of mono-, di-, and tricarboxylates (the latter is the least stable to hydrolysis) and not just the individual species.

The aluminum carboxylates of most commercial interest are listed in Table 1.

Table 1. Commercially Important Aluminum Carboxylates

Name	Empirical formula	Mol wt
monobasic aluminum formate	$C_2H_3AlO_5$	134.02
normal aluminum formate	$C_3H_3AlO_6$	162.03
monobasic aluminum acetate[a]	$C_4H_7AlO_5$	162.08
normal aluminum acetate	$C_6H_9AlO_6$	204.09
dihydroxyaluminum aminoacetate	$C_2H_6AlNO_4$	135.06
dihydroxyaluminum aminoacetate hydrate	$C_2H_6AlNO_4{\cdot}xH_2O$	
aluminum octanoate	$C_{16}H_{31}AlO_5$	474.4
aluminum 2-ethylhexanoate	$C_{16}H_{31}AlO_5$	474.4
aluminum monostearate[b]	$C_{18}H_{37}AlO_4$	344.5
aluminum distearate[c]	$C_{36}H_{71}AlO_5$	610.9
aluminum tristearate[d]	$C_{54}H_{108}AlO_6$	877.4

[a] Mp = 54°C.
[b] Mp = 155°C.
[c] Mp = 145°C.
[d] Mp = 115°C.

ARTHUR L. MCKENNA
Witco Corporation

United States International Trade Commission, *Synthetic Organic Chemicals, United States Production and Sales, 1987;* U.S. Government Printing Office, Washington, D.C., 1987.

United States Pharmacopeia: National Formulary, 22nd ed., Mack Publishing, Easton, Pa., 1990, p. 50.

ALUMINUM HALIDES AND ALUMINUM NITRATE

The aluminum halides and aluminum nitrates have similar properties with the exception of the family of aluminum fluoride compounds which are discussed elsewhere. Of the remaining members in this aluminum halide family, chloride derivatives are the most commercially important; aluminum bromide, $AlBr_3$, aluminum iodide, AlI_3, and aluminum nitrate, $Al(NO_3)_3$ are of only minor commercial interest.

ALUMINUM CHLORIDE

The chemistry of aluminum chloride is influenced significantly by hydration. Aluminum chloride hexahydrate, $AlCl_3{\cdot}6H_2O$, is a crystalline solid that dissolves easily in water, forming ionic species. Heating the hydrate results in the loss of hydrogen chloride, HCl, and formation of aluminum oxide, Al_2O_3. On the other hand, anhydrous aluminum chloride reacts violently with water, evolving heat, a gas consisting of hydrogen chloride and steam, and aluminum oxide particulates. Anhydrous aluminum chloride sublimes at 180°C, leaving no residue. The uses of anhydrous aluminum chloride and the hydrated form are also very different. The anhydrous material is a Lewis acid used as an alkylation catalyst. The hydrate is used principally as a flocculating aid.

Commercially, aluminum chloride is available as the anhydrous $AlCl_3$, as the hexahydrate, $AlCl_3{\cdot}6H_2O$, or as a 28% aqueous solution designated 32°Be′. Polyaluminum chloride, or poly(aluminum hydroxy) chloride is a member of the family of basic aluminum chlorides. These are partially neutralized hydrates having the formula $Al_2Cl_{6-x}(OH)_x{\cdot}6H_2O$ where $x = 1–5$.

Anhydrous Aluminum Chloride

Properties. Anhydrous aluminum chloride is a hygroscopic, white solid that reacts with moisture in air. Properties are shown in Table 1.

Manufacture. In the United States anhydrous aluminum chloride is manufactured by the exothermic reaction of chlorine, Cl_2, vapor with molten aluminum.

Economic Aspects. The North America demand for anhydrous aluminum chloride, estimated to be from 25,000 to 30,000 metric tons per year, is divided among the applications shown in Figure 1

Table 1. Physical Properties of Anhydrous Aluminum Chloride

Property	Value
molecular weight	133.34054
density at 25°C, g/mL	2.46
sublimation temperature, °C	180.2
triple point, °C, 233 kPa[a]	192.5 ± 0.2
heat of formation, 25°C, kJ/mol[b]	−705.63 ± 0.84
heat of sublimation of dimer, 25°C, kJ/mol[b]	115.52 ± 2.3
heat of solution, 20°C, kJ/mol[b]	−329.1
heat of fusion, kJ/mol[b]	35.35 ± 0.84
entropy, 25°C, J/(K·mol)[b]	109.29 ± 0.42
heat capacity, 25°C, J/(K·mol)[b]	91.128

[a] To convert kPa to psi, multiply by 0.145.
[b] To convert J to cal, divide by 4.184.

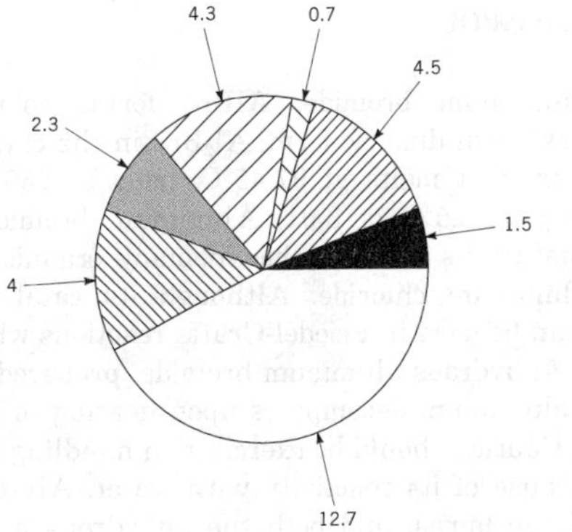

Figure 1. Annual anhydrous aluminum chloride market, 1984, 10^3 t/yr. ■ Dyestuffs; ▨, ethylbenzene; ⧅, ethyl chloride; ▨, hydrocarbon resin; ▦, pharmaceuticals; ▧, titanium dioxide; and □, other applications.

Safety and Handling. In storage, some reaction with moisture may occur and over time can lead to a pressure build-up from HCl in the container. Containers should be carefully vented before being opened. Safety goggles or face shields, rubber gloves, rubber shoes, and coveralls made of acid-resistant material should be used in handling. A NIOSH/OSHA-certified respirator is also required to prevent breathing fumes and dust. Aluminum chloride reacts with moisture in the skin, in the eyes, ears, nose, and throat.

Environmental Protection. Fumes resulting from exposure of anhydrous aluminum chloride to moisture are corrosive and acidic. Collection systems should be provided to conduct aluminum chloride dusts or gases to a scrubbing device. The choice of equipment, usually one of economics, ranges from simple packed-tower scrubbers to sophisticated high energy devices such as those of a Venturi design.

Spills should be picked up before flushing thoroughly with water and neutralizing with soda ash or lime.

Aluminum Chloride Hexahydrate

The hexahydrate of aluminum chloride is a deliquescent, crystalline solid soluble in water and alcohol and usually made by dissolving aluminum hydroxide, $Al(OH)_3$, in concentrated hydrochloric acid.

Roofing granules and mineral aggregate for bituminous products are treated with aluminum chloride solution to improve adhesion of the asphalt. Pigmented coatings, containing sodium silicate, Na_2SiO_3, and used to color roofing granules, are insolubilized by spraying with aluminum chloride solution and then heating. Aluminum chloride hydrates are the alumina sources used in the manufacture of special forms of alumina and alumina–silica refractories.

Aluminum chloride hydrate is used in textile finishing to impart crease recovery and nonyellowing properties to cotton fabrics, antistatic characteristics to polyester, polyamide, and acrylic fabrics, and to improve the flammability rating of nylon. Dye-bleeding of printed textile may be blocked by treatment with aluminum chloride and zinc acetate, $Zn(O_2CCH_3)_2$, followed by solubilizing with ethylenediamine tetraacetic acid, and washing from the fabric.

Basic Aluminum Chlorides

The class of compounds identified as basic aluminum chlorides is used primarily in deodorant, antiperspirant, and fungicidal preparations. They have the formula $Al_2(OH)_{6-x}Cl_x$, where $x = 1–5$, and are prepared by the reaction of an excess of aluminum with 5–15% hydrochloric acid at a temperature of 67–97°C.

Hydrates of aluminum chloride and basic aluminum chlorides are also effective in a number of difficult water treatment problems.

ALUMINUM BROMIDE

Anhydrous aluminum bromide, $AlBr_3$, forms colorless trigonal crystals and exists in dimeric form, Al_2Br_6, in the crystal and liquid phases. This product melts at 97.45°C, boils at 256°C, and has a specific gravity at 25°C of 3.01. Aluminum bromide, because of its covalent nature, is more soluble in many organic solvents than anhydrous aluminum chloride. Although its catalytic activity is moderate, it can be used in Friedel-Crafts reactions where selectivity is important. Anhydrous aluminum bromide, prepared from bromine and metallic aluminum, decomposes upon heating in air to bromine and alumina. Caution should be exercised in handling this hazardous compound because of its reactivity with water. Aluminum bromide may cause tissue burns, and both the anhydrous and the hydrate forms may be toxic upon ingestion.

Aluminum bromide hexahydrate, $AlBr_3{\cdot}6H_2O$, may be made by dissolving aluminum or aluminum hydroxide in hydrobromic acid, HBr. This white, crystalline solid is precipitated from aqueous solution.

ALUMINUM IODIDE

Aluminum iodide, AlI_3, is a crystalline solid with a melting point of 191°C. Aluminum iodide hexahydrate, $AlI_3{\cdot}6H_2O$, and aluminum iodide pentadecahydrate, $AlI_3{\cdot}15H_2O$, are precipitated from aqueous solution. They may be prepared by the reaction of hydroiodic acid, HI, with aluminum or aluminum hydroxide.

ALUMINUM NITRATE

Aluminum nitrate is available commercially as aluminum nitrate nonahydrate, $Al(NO_3)_3{\cdot}9H_2O$. It is a white, crystalline material with a melting point of 73.5°C, that is soluble in cold water, alcohols, and acetone.

Aluminum nitrate nonahydrate is prepared by dissolving aluminum or aluminum hydroxide in dilute nitric acid, and crystallizing the product from the resulting aqueous solution. It is made commercially from aluminous materials such as bauxite.

Anhydrous aluminum nitrate is covalent in character, easily volatilized, and decomposes on heating. Hydrated aluminum nitrate is used in the preparation of insulating papers, on transformer core laminates, and in cathode-ray tube heating elements.

G. W. GRAMS
Witco Corporation

JANAF Thermochemical Tables, 3rd ed., American Chemical Society, Washington, D.C., and American Institute of Physics, New York, for the National Bureau of Standards, 1985.

C. A. Thomas, *Anhydrous Aluminum Chloride in Organic Chemistry,* ACS Monogr. Ser. #87, Reinhold Publishing Corp., New York, 1941. Out of print, but available in facsimile by the University Microfilms, a Xerox Company, Ann Arbor, Mich.

G. A. Olah, *Friedel-Crafts Chemistry,* John Wiley & Sons, Inc., New York, 1973.

E. W. Post and J. C. Kotz, in M. F. Lappert, ed., *Int. Rev. Sci., Inorganic Chemistry, Series 2,* Vol. 1, Butterworth, London, Chapt. 7, p. 219.

ALUMINUM OXIDE (ALUMINA)

ACTIVATED

The activated aluminas comprise a series of nonequilibrium forms of partially hydroxylated aluminum oxide, Al_2O_3. The chemical composition can be represented by $Al_2O_{(3-x)}\ (OH)_{2x}$ where x ranges from about 0 to 0.8. They are porous solids made by thermal treatment of aluminum hydroxide precursors and find application mainly as adsorbents, catalysts, and catalyst supports. Activated alumina, for purposes of this discussion, refers to thermal decomposition products (excluding α-alumina of aluminum trihydroxides, oxide hydroxides, and nonstoichiometric gelatinous hydroxides. The term "activation" is used in this article to indicate a change in properties resulting from heating (calcining).

Physical and Chemical Properties

In general, as a hydrous alumina precursor is heated, hydroxyl groups are driven off, leaving a porous solid structure of activated alumina. The transformation is topotactic and little change in size or shape of the material is observed at low magnifications. At magnifications higher than about 10,000, changes in texture resulting from recrystallization can be seen. The physical properties of the material are set by the choice of precursor, the forming process, and the activation conditions.

Decomposition of Boehmite. Boehmite, AlO(OH), can be synthesized having surface areas ranging from about 1 to over 800 m^2/g depending upon the method of preparation. The properties of activated boehmite products are strongly influenced by the crystallite size of the precursor material.

Activation Products of Aluminum Hydroxide. As gibbsite, α-$Al(OH)_3$, is heated, surface area reaches a maximum of 300 m^2/g or more at about 650 K. As temperature is increased further, surface area decreases and the skeletal structure becomes more dense, reflecting increased ordering of the crystalline structure during the progression from chi to kappa to alpha. At about 1450 K conversion to alpha alumina occurs, with a major rearrangement of crystal structure and corresponding decrease in surface area to about 5 m^2/g. These trends vary somewhat according to precursor crystal size, purity, and the atmosphere of heating.

Economic Aspects

In 1990 U.S. production of activated alumina was about 50,000 t/yr. The least expensive products are those derived directly from Bayer-process gibbsite, and powders are generally less expensive than formed products. The soda content (0.2–0.3% Na_2O) of Bayer gibbsite makes it unattractive for many catalytic applications. Gel-based products are normally used where low soda level is required.

Shaped products used for adsorbent purposes are generally less sophisticated and therefore less expensive than catalytic products. North American producers of Bayer process-based activated aluminas include Alcoa, La Roche (formerly Kaiser Chemicals), Discovery, and Alcan. Gel-based activated aluminas are produced by La Roche, Vista, and several of the major catalyst manufacturers. In Europe, principal sources of supply are Rhône-Poulenc and Condea.

Safety and Handling

Activated alumina is a relatively innocuous material from a health and safety standpoint. It is nonflammable and nontoxic. Fine dusts can cause eye irritation and there is some record of lung damage because of inhalation of activated alumina dust mixed with silica and iron oxide. Normal precautions associated with handling of nuisance dusts should be taken. Activated alumina is normally shipped in moisture-proof containers (bags, drums, sling bins) because of its strong desiccating action.

ALAN PEARSON
Aluminum Company of America

C. Misra, *Industrial Alumina Chemicals, ACS Monogr. 184,* American Chemical Society, Washington, D.C., 1986, p. 78.

K. Wefers and C. Misra, *Oxides and Hydroxides of Aluminum, Alcoa Technical Paper 19,* revised, Alcoa Laboratories, Aluminum Company of America, Pittsburgh, Pa., 1987, p. 52.

R. K. Oberlander in B. E. Leach, ed., *Aluminas for Catalysis: Their Preparation and Properties,* Vol. 3, *Applied Industrial Catalysis,* Academic Press, New York, 1984, pp. 98–102.

L. D. Hart, ed., *Alumina Chemicals Handbook,* American Ceramic Society, Westerville, Ohio, 1990.

CALCINED, TABULAR, AND ALUMINATE CEMENTS

CALCINED ALUMINA

Calcined aluminas are generally obtained from Bayer process gibbsite α-Al(OH)$_3$, thermal decomposition of which follows the transition through the generic gamma alumina phases to α-alumina (corundum), α-Al_2O_3. Nonmineralized metal-grade or smelter-grade alumina (SGA) for aluminum production is calcined at lower temperatures and usually contains about 20 to 50% α-Al_2O_3. The remainder consists of higher temperature transition aluminas, usually theta, kappa, delta, and gamma, depending upon the consolidation of the original gibbsite structure, impurities, heating rate, and furnace atmosphere.

Preparation

Calcination of gibbsite has been done in rotary kilns for many years. Specialty calcined aluminas can also be prepared in stationary or fluid bed calciners similar to those used for producing SGA.

Ground, Calcined, and Reactive Aluminas. Most ceramic grade aluminas are supplied dry ground to about 95% −325 mesh (44 μm) using 85–90% Al_2O_3 ceramic ball, attrition, vibro-energy, or fluid-energy milling. Particles larger than 44 μm can be removed by air classification during continuous milling to produce 99 + % − 325 mesh product. More fully ground, or superground, calcined aluminas having particle size distributions that approximate the natural or ultimate crystal size of the Bayer grain as calcined are often desired.

Thermally reactive aluminas contain submicrometer crystals. These must be separated from the Bayer agglomerate during grinding to permit dense compaction upon ceramic forming, and thus, enhance densification upon sintering at lower temperatures. Such superground, thermally reactive aluminas exhibit higher densification rates when compacted and sintered into ceramic products, and complete densification is obtained about 200°C lower than when using the coarser, continuously ground aluminas.

Specialty Aluminas. Process control techniques permit production of calcined specialty aluminas having controlled median particle sizes differentiated by about 0.5 μm. This broad selection enables closer shrinkage control of high tech ceramic parts. Production of pure 99.99% Al_2O_3 powder from alkoxide precursors, apparently in spherical form, offers the potential of satisfying the most advanced applications for calcined aluminas requiring tolerances of ±0.1% shrinkage.

Uses

Calcined alumina markets consume slightly less than 50% of the specialty alumina chemicals production. Worldwide usage is estimated to be about 50% for refractories, 20% for abrasives, and 25% for ceramics. Calcined aluminas are also used in the manufacture of tabular alumina and calcium aluminate cements (CAC). Quantities are estimated to be over 200,000 and 100,000 t, respectively.

TABULAR ALUMINA

Tabular alumina is a high density, high strength form of α-Al_2O_3 made by sintering an agglomerated shape of ground, calcined alumina. It is available in the form of smooth balls having diameters from 3 to 25 mm and imperfect 19 mm diameter spheres, which are crushed, screened, and ground to obtain a wide variety of graded, granular, and powdered products having various particle size distributions ranging from a top size of 12.7 mm to −325 mesh (44 μm).

Uses

The large α-Al_2O_3 crystals containing closed round pores make tabular alumina an excellent refractory raw material. Tabular alumina is the ideal base material for high alumina brick and monolith liners in the metal, ceramic, and petrochemical industries. Tabular alumina also offers advantages over other materials as an aggregate in castables made from calcium aluminate cement as the binder and in phosphate-bonded monolithic furnace linings in all thermal processing industries. Other applications include their use in electrical insulators, electronic components, and kiln furniture.

ALUMINATE CEMENT

Refined calcined alumina is commonly used in combination with high purity limestone to produce high purity calcium aluminate cement (CAC). The manufacture, properties, and applications of CAC from bauxite limestone, as well as high purity CAC, has been described. High purity CAC sinters readily in gas-fired rotary kiln calcinations at 1600–1700 K. CAC reactions are considered practically complete when content of free CaO is less than 0.15% and loss on ignition is less than 0.5% at 1373 K.

Uses

High purity CA cements are primarily used as binders for high strength refractory castables to form linings up to about 1.0 m thick, as, for example, in iron blast furnaces.

The high purity CAC finds extensive use as an efficient binder for other aggregates such as fire clays, kaolin, andalusite, kyanite, pyrophyllite, sillimanite, mullite, and refractory-grade bauxite, having the added advantage of increasing the refractoriness of some of these aggregates. The many applications cited for tabular alumina in refractories are also common for high purity CAC.

High purity CAC is also used as a steel slag conditioner during ladle refining of steel.

GEORGE MACZURA
Aluminum Company of America

L. D. Hart, ed., *Alumina Chemicals: Science and Technology Handbook,* The American Ceramic Society, Columbus, Ohio, 1990, pp. 99–184.

W. H. Gitzen, ed., *Alumina as a Ceramic Material,* The American Ceramic Society, Columbus, Ohio, 1970, pp. 1–253.

K. Wefers and G. M. Bell, *Oxides and Hydroxides of Aluminum, Technical Paper No. 19,* Aluminum Company of America, Pittsburgh, Pa., 1972, pp. 1–51.

L. Hudson, C. Misra, and K. Wefers, "Aluminum Oxide," in W. Gerhartz, ed., *Ullmann's Encyclopedia of Industrial Chemistry,* 5th ed., Vol. A1, VCH Publishers, Deerfield Beach, Fla., 1985.

X. Zhong, J. Lu, X. Yan, and M. Li, editors, *Proceedings of the Second International Symposium on Refractories—Refractory Raw Materials in High Performance Refractory Products,* The Chinese Silicate Society and The Chinese Society for Metals, International Academic Publishers, Beijing, China, 1992, pp. 64–78.

HYDRATED

The term alumina hydrates or hydrated aluminas is used in industry and commerce to designate aluminum hydroxides. These compounds are true hydroxides and do not contain water of hydration. The most well-defined crystalline forms are the trihydroxides, Al(OH)$_3$: gibbsite, bayerite, and nordstrandite. In addition, two aluminum oxide–hydroxides, AlO(OH), boehmite and diaspore, have been clearly defined.

Table 1. Mineralogical Properties of Aluminum Hydroxides

	Index of refraction[a]				
Material	α	β	γ	Brittleness	Mohs' hardness
gibbsite	1.568	1.568	1.587	tough	2 1/2–3 1/2
boehmite	1.649	1.659	1.665		3 1/2–4
diaspore	1.702	1.722	1.750	brittle	6 1/2–7

[a] The average index of refraction for bayerite is 1.583.

The terms gelatinous alumina or alumina gel cover a range of products in which colloidal hydrated alumina is the predominant solid phase. Structural order varies from x-ray indifferent (amorphous) to some degree of crystallinity. The latter product has been named pseudoboehmite or gelatinous boehmite. Its x-ray diffraction pattern shows broad bands that coincide with the strong reflections of the well-crystallized boehmite.

Crystalline Alumina Hydrates

The mineralogical, structural, and thermodynamic properties of the various crystalline alumina hydrates are listed in Tables 1, 2, and 3 respectively. X-ray diffraction methods are commonly used to differentiate between materials.

Gelatinous Aluminum Hydroxides

Apart from the crystalline forms, aluminum hydroxide often forms a gel. Fresh gels are usually amorphous, but crystallize on aging, and gel composition and properties depend largely on the method of preparation. Gel products have considerable technical use.

Phase Relations in the Al_2O_3–H_2O System

Under equilibrium vapor pressure of water, the crystalline trihydroxides, $Al(OH)_3$ convert to oxide–hydroxides at above 100°C. Below 280–300°C, boehmite is the prevailing phase, unless diaspore seed is present. Although spontaneous nucleation of diaspore requires temperatures in excess of 300°C and 20 MPa (200 bar) pressure, growth on seed crystals occurs at temperatures as low as 180°C. For this reason it has been suggested that boehmite is the metastable phase although its formation is kinetically favored at lower temperatures and pressures. The ultimate conversion of the hydroxides to corundum Al_2O_3, the final oxide form, occurs above 360°C and 20 MPa.

Production

Aluminum hydroxides are technically the most widely used members of the alumina chemicals family. The most important source of aluminum hydroxides is the bauxite refining plant (Bayer process) for alumina production. A small amount of somewhat purer aluminum hydroxide is produced by the Sinter process.

Several commercial grades of aluminum hydroxide are produced. The properties of some grades are given in Table 4.

Hydroxide grades can be surface-treated to modify dispersion behavior and rheological properties. The most widely used surface coating agents are stearic acid and stearates. Additionally, compounds from the silane group have been used as coupling agents to give improved adhesion to polymers when the hydroxide is used as a filler.

Shipping and Analysis

Shipping of aluminum hydroxide powders is usually in paper bags of 10- to 25-kg size. Bulk shipment by road or rail wagons is also common. Aluminum hydroxides are not hygroscopic but could be dusty, and precautions against dust inhalation should be taken during handling.

Economic Aspects

About 90% of world production in 1988 came from the Bayer process; the remaining came from Sinter, Ziegler, and gel processes. Although many alumina plants possess the capability to make the normal Bayer-grade hydroxide, specialty grade aluminum hydroxides for the chemical industry are generally produced in alumina refining plants dedicated to nonmetallurgical alumina products.

Bayer aluminum hydroxides in most grades are sold by all major U.S. alumina producers. Other firms offering aluminum hydroxide fillers probably operate reprocessing facilities to grind or otherwise treat hydroxide obtained from the primary producers. Countries exporting small amounts to the United States are Japan, Germany, Canada, and the U.K.

Health and Safety Information

Aluminum hydroxides are minimally absorbed by the body and LD_{50} values for ingestion are unavailable. Death upon ingestion occurs from intestinal blockage rather than systemic aluminum toxicity. In recognition of the possible adverse effects of long-term exposure to alumina

Table 3. Thermodynamic Data for Crystalline Aluminum Hydroxides at 298.15 K and 0.1 MPa[a]

Substance	Molecular weight	Molar vol, cm^3/mol	ΔH°_f, kJ/mol[b]
gibbsite	78.004	31.956	−1293.2
bayerite	78.004		−1288.2
boehmite	59.989	19.55	−990.4
diaspore	59.989	17.76	−999.8

[a] To convert MPa to psi, multiply by 145.
[b] To convert J to cal, divide by 4.184.

Table 2. Structural Properties of Aluminum Hydroxides

			Unit axis length, nm				
Material	Crystal system[a]	Space group	a	b	c	Angle	Density, g/cm^3
			Al(OH)₃				
gibbsite	monoclinic[b]	C^5_{2h}	0.8684	0.5078	0.9136	94°34′	2.42
bayerite	monoclinic	C^5_{2h}	0.5062	0.8671	0.4713	90°27′	2.53
nordstrandite	triclinic	C^1_1	0.5114	0.5082	0.5127	70°16′	
						74° 0′	
						58°28′	
			AlO(OH)				
boehmite	orthorhombic	D^{17}_{2h}	0.2868	0.1223	0.3692		3.01
diaspore	orthorhombic	D^{16}_{2h}	0.4396	0.9426	0.2844		3.44

[a] Unit cell contains two molecules unless otherwise indicated.
[b] Unit cell contains four molecules.

Table 4. Properties of Commercial-Grade Aluminum Hydroxides

Property	Normal coarse grade[a]	Normal white grade[b]	Ground[c]	Fine precipitated[d]
Al_2O_3, wt %	65.0	65.0	65.0	64.7
SiO_2, wt %	0.012	0.01	0.02	0.04
Fe_2O_3, wt %	0.015	0.004	0.03	0.01
Na_2O (total), wt %	0.40	0.15	0.30	0.45
Na_2O (soluble), wt %	0.05	0.05	0.05	0.1–0.25
LOI at 1200°C, wt %[e]	34.5	34.5	34.5	34.5
moisture at 100°C, wt %	0.1	0.1	0.4	0.3–1.0
specific gravity	2.42	2.42	2.42	2.42
bulk density (loose), g/cm^3	1.2–1.4	1.0–1.1	0.7–1.25	0.13–0.22
surface area, m^2/g	0.1	0.15	2–4	6–8
color	off-white	white	off-white	white
refractive index	1.57	1.57	1.57	1.57
Mohs' hardness	2.5–3.5	2.5–3.5	2.5–3.5	2.5–3.5

[a] Alcoa C-30.
[b] Alcoa C-31.
[c] Alcoa C-330.
[d] Alcoa Aydral 710.
[e] Loss on ignition.

dusts, threshold limit values have been established by the ACGIH as follows: 10 mg/m^3 TLV–TWA and 20 mg/m^3 TLV–STEL. Aluminum hydroxide and aluminum hydroxide oxide are reported in EPA TSCA inventory.

CHANAKYA MISRA
Aluminum Company of America

C. Misra, *Industrial Alumina Chemicals, ACS Monogr. 184,* American Chemical Society, Washington, D.C., 1986.

K. Wefers and C. Misra, *Oxides and Hydroxides of Aluminum, Technical Paper No. 19,* Revised, Aluminum Company of America, Pittsburgh, Pa., 1987.

H. Ginsberg and K. Wefers, *Aluminum and Magnesium,* Vol. 15, Die Metallischen Rohstoffe, Enke Verlag, Stuttgart, Germany, 1971.

L. D. Hart, ed., *Alumina Chemicals Science and Technology Handbook,* The American Ceramic Society, Westerville, Ohio, 1990.

ALUMINUM SULFATE AND ALUMS

Aluminum sulfate octadecahydrate, $Al_2(SO_4)_3 \cdot 18H_2O$, and its aqueous solutions are used primarily in the paper industry for sizing and as a flocculating agent in water and wastewater treatment. This material is often called papermakers' alum or alum. Because this salt is precipitated from aqueous solution, aluminum sulfate hydrate, $Al_2(SO_4)_3 \cdot nH_2O$, can have variable composition and is sometimes referred to as cake alum or patent alum. The solid commercial hydrate, generally written as the 18-hydrate, is typically dehydrated to correspond to from 17.0–17.5% Al_2O_3 where $n = 13-14$. This dehydrated form is called dry alum, ground or lump. Aluminum sulfate solutions are typically 7.5–8.5% Al_2O_3 and are known as liquid alum.

Anhydrous aluminum sulfate, $Al_2(SO_4)_3$, is a specialty item used in food applications.

Properties

Over 50 acidic, basic, and neutral aluminum sulfate hydrates have been reported. Only a few of these are well-characterized because the exact compositions depend on conditions of precipitation from solution. Variables such as supersaturation, nucleation and crystal growth rates, occlusion, nonequilibrium conditions, and hydrolysis can each play a role in the final composition. Commercial dry alum is likely not a single crystalline hydrate, but rather it contains significant amounts of amorphous material.

Manufacture

In the United States, aluminum sulfate is usually produced by the reaction of bauxite or clay with sulfuric acid.

Other Alums

The word alum is derived from the Latin *alumen,* which was applied to several astringent substances, most of which contained aluminum sulfate. Unfortunately, the term alum is now used for several different materials. Papermakers' alum or simply alum refers to commercial aluminum sulfate. Common alum or ordinary alum usually refers to potash alum which can be written in the form $K_2SO_4 \cdot Al_2(SO_4)_3 \cdot 24H_2O$, or it can refer to ammonium alum, ammonium aluminum sulfate. The term is also applied to a whole series of crystallized double sulfates $[M(I)M(III)(SO_4)_2 \cdot 12H_2O]$ having the same crystal structure as the common alums, in which sodium and other univalent metals may replace the potassium or ammonium, and other metals may replace the aluminum. Even the sulfate radical may be replaced, by selenate, for example. Some examples of alums are cesium alum, $CsAl(SO_4)_2 \cdot 12H_2O$; iron alum, $KFe(SO_4)_2 \cdot 12H_2O$; chrome alum, $KCr(SO_4)_2 \cdot 12H_2O$; and chromoselenic alum $KCr(SeO_4)_2 \cdot 12H_2O$.

Pseudoalums are a series of double sulfates, such as iron(II) aluminum sulfate, $FeSO_4 \cdot Al_2(SO_4)_3 \cdot 24H_2O$, containing a bivalent metal ion in place of the univalent element of ordinary alums. These pseudoalums have different crystal structures from those of the ordinary alums.

In industrial practice it is generally the aluminum content of alums that is important. Because aluminum sulfate is widely available, other alums are more in the nature of specialty items and are no longer produced in quantities comparable to those of aluminum sulfate.

K. V. DARRAGH
C. A. GREEN
Rhône-Poulenc, Inc.

J. W. Mellor, *A Comprehensive Treatise on Inorganic and Theoretical Chemistry,* Vol. 5, Longmans, Green and Co. Ltd., London, U.K., pp. 332–357.

Chemical Economics Handbook, Stanford Research Institute, Menlo Park, Calif., Sept. 1991, parts 702.1000H, 702.1003U–1004P.

W. Gerhartz, *Ullmann's Encyclopedia of Industrial Chemistry,* 5th ed., Vol. A1, VCH, Deerfield Beach, Fla., 1985, pp. 527–534.

POLYALUMINUM CHLORIDES

Aluminum chloride hydroxide, $AlCl(OH)_2$, $AlCl_2(OH)$, products, commonly known as polyaluminum chlorides (PAC), are used for a wide variety of industrial applications. Other names for PAC are basic aluminum chloride, polybasic aluminum chloride, aluminum hydroxychloride, aluminum oxychloride, and aluminum chlorohydrate. The presence of polymeric, aluminum-containing cations, the distribution of which can differ greatly, typifies PAC products. Although the formation of polynuclear aluminum species in solution has been studied for over a century, there is still much controversy concerning aluminum polymerization reactions and the resulting product compositions.

Commercially, PAC has been used in water and wastewater treatment in Japan, the former USSR, and Europe since about 1970, and in the United States since the early 1980s. Aluminum chlorohydrate, $Al_2Cl(OH)_5$, a specialty PAC product, has been utilized as an antiperspirant for over 50 years. Other PAC uses include preparation of pillared clay catalysts, stabilization of formation clays, paper sizing in papermaking, catalysts in durable-press finishing of fabrics, and preparation of alumina-coated silica sols.

Properties

Physical and chemical properties of the numerous PAC products can vary considerably. PAC products are usually aqueous solutions, although solid products are also sold. Solutions range from colorless to amber and from clear to hazy in appearance; specific gravities at 25°C vary from about 1.2 to 1.35. Product viscosities, as measured by a Brookfield viscometer at 25°C, are generally about 10–50 mPa·s(= cP), but can be much greater than 10, 000 mPa·s(= cP) for certain aged compositions.

Manufacture

A generic manufacturing process for PAC involves the addition of base to aluminum chloride solution:

$$n\ AlCl_3 + m\ OH^- \rightarrow Al_n(OH)_mCl_{3n-m} + m\ Cl^-$$

Typical values for m/n are 0.5 to 2.5. Commercially used bases include sodium hydroxide, potassium hydroxide, calcium hydroxide (lime), magnesium hydroxide, sodium carbonate, sodium aluminate, calcium carbonate, or various mixtures. For certain applications, PAC can be made from waste grades of aluminum chloride such as spent catalyst solutions from Friedel-Crafts synthesis.

Aluminum Chlorohydrate. A common process for the manufacture of aluminum chlorohydrate involves the addition of metallic Al to aluminum chloride.

Shipping and Handling

Liquid polyaluminum chloride is acidic and corrosive to common metals. Suitable materials for construction of storage and handling facilities include synthetic rubber-lined steel, corrosion-resistant fiber glass-reinforced plastics (FRP), ceramics, tetrafluoroethylene polymer (PTFE), poly(vinylidene fluoride) (PVDF), polyethylene, polypropylene, and poly(vinyl chloride) (PVC).

Specifications and Safety

Polyaluminum chloride products used in the treatment of potable (drinking) water must be approved by the National Sanitation Foundation (NSF). NSF certification has superseded EPA approval. Aluminum chlorohydrate for topical use as an antiperspirant is regulated by the FDA.

Contact with polyaluminum chloride products may cause burns or irritation to the eyes or skin; thus, protective clothing and eye protections are recommended.

Kristine S. Siefert
Nalco Chemical Company

B. A. Demsey, H. Sheu, T. M. Tanzeer Ahmed, and J. Mentink, *J. Amer. Water Works Assoc.* **77**(3), 74–80 (1985).

T. J. Pinnavaia, M. Tzou, S. D. Landau, and R. H. Raythatha, *J. Mol. Catal.* **27,** 195–212 (1984).

M. G. Reed, *J. Pet. Technol.* **24,** 860 (1972).

J. C. Ginocchio, *Sulzer Tech. Rev.* **65**(3), 12 (1983).

AMIDES, FATTY ACID

Fatty acid amides are of the general formula

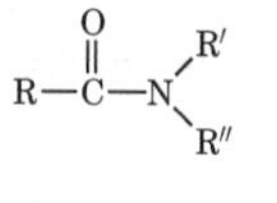

in which R may be a saturated or unsaturated alkyl chain derived from a fatty acid. They can be divided into three categories. The first is primary monoamides in which R is a fatty alkyl or alkenyl chain of C_5–C_{23} and $R' = R = H$. The second is substituted monoamides, including secondary, tertiary, and alkanolamides in which R is a fatty alkyl or alkenyl chain of C_5–C_{23}; R′ and R″ may be a hydrogen, fatty alkyl, aryl, or alkylene oxide condensation groups with at least one alkyl, aryl, or alkylene oxide group. The third category is bisamides of the general formula:

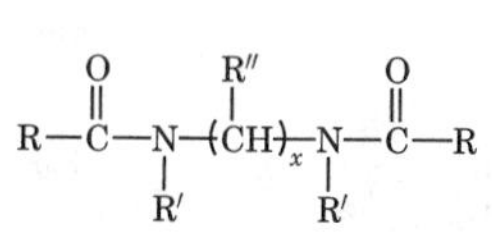

where R groups are fatty alkyl or alkenyl chains. R′ and R″ may be hydrogen, fatty alkyl, aryl, or alkylene oxide condensation groups. Other amides include halogenated amides and multifunctional amides such as amidoamines and polyamides.

Physical Properties

Many of the physical properties of fatty acid amides have been explained on the basis of the tautomeric structures:

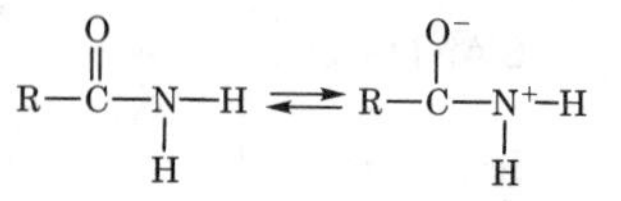

Primary and secondary amides show strong hydrogen bonding which accounts for their high melting points and low solubilities in most solvents. With tertiary amides (disubstituted amides), hydrogen bonding is not possible, as exhibited by their increased solubility and lower melting points. Many fatty acid amides are essentially insoluble in water. Polar solvent solubilities decrease with longer alkyl chain lengths. Amides with alkyl lengths greater than C-12, with the exception of alkanolamides, have low solubility in all solvents. In nonpolar solvents, solubility is low and varies irregularly with chain length.

Amides have a strong tendency to reduce friction on various surfaces by forming a layer on surfaces. This coating action may be attributed to their hydrophobic character and strong hydrogen bonding. Primary, secondary, and bisamides are widely used as lubricating or slip agents and alkanolamides are commonly used as surfactants. Detailed descriptions of their properties can be found in the literature.

Chemical Properties

Amides in general are stable to elevated processing temperatures, air oxidation, and dilute acids and bases. Stability is reduced in amides

Table 1. Primary Fatty Acid Amides

Common name	Trade name	Manufacturer	Physical form
coco fatty amide	Armid C	Akzo	solid, flake
	Adogen 60	Sherex	solid
stearamide	Armid 18	Akzo	flake
	Kemamide S	Humko	beads
	Adogen 42	Sherex	beads
hydrogenated-tallow fatty amide	Armid HT	Akzo	flake, powder
	Adogen 40	Sherex	beads
docosanamide	Adogen I	Sherex	beads
	Kemamide B	Humko	powder
octadecenamide	Adogen 72	Sherex	solid, flake
	Armid O	Akzo	solid, flake
	Armoslip CP	Akzo	pellet, flake
	Kemamide U	Humko	powder
13-docosenamide	Kemamide E	Humko	powder, flake
	Armid E	Akzo	flake, pellet
	Armoslip E	Akzo	pellet, powder

Table 2. Substituted Fatty Acid Amides

Chemical identity	Trade name	Manufacturer	Form
N,N-bis(2-hydroxyethyl) dodecanamide	Ninol AA62 Extra	Stepan	wax
N,N-bis(2-hydroxyethyl)coco fatty acid amide	Ninol 2021 Extra	Stepan	flakes
N-(2-hydroxypropyl)-dodecanamide	Ninol AD31	Stepan	flakes
octadecenylpeptide, sodium salt	Maypon K	Maywood	44% soln
N-methyl-*N*-(1-oxododecyl)-glycine	Sarkosyl L	CIBA GEIGY	95% powder
N-methyl-*N*-(1-oxooctadecyl)-glycine	Sarkosyl S	CIBA GEIGY	powder
N-methyl-*N*-(1-oxooctadecenyl)-glycine	Sarkosyl O	CIBA GEIGY	paste
N,N-dimethylhexanamide	Hallcomid M6	Hall	liquid
N,N-dimethyldodecanamide	Hallcomid M12	Hall	liquid
N,N-dimethyloctadecanamide	Hallcomid M18	Hall	solid
N,N-dimethyloctadecenamide	Hallcomid M18-OL	Hall	liquid
1,2-ethanediylbis-octadecenamide	Acrawax C	Glyco	powder/beads
	Kemamide W40	Humko	powder
	Armowax	Akzo	powder
1,2-ethanediylbis-octadecenamide	Kemamide W20	Humko	powder
Sodium N-acyl-N-methylaminoethanesulfonates			
from coco fatty acid	Igepon TC-42	GAF	22% soln
from tallow fatty acid	Igepon TE-42	GAF	24% soln
from octadecenoic acid	Igepon T-77	GAF	67% flake
from hexadecanoic acid	Igepon TN-71	GAF	14.5% soln

containing unsaturated alkyl chains; unsaturation offers reactive sites for many reactions.

Hydrolysis of primary amides catalyzed by acids or bases is very slow. Even more difficult is the hydrolysis of substituted amides. The dehydration of amides which produces nitriles is of great commercial value.

Synthesis and Manufacture

Unsubstituted Amides. The most widely used synthetic route for primary amides is the reaction of fatty acid with anhydrous ammonia.

Substituted Amides. Most monosubstituted and disubstituted amides can be synthesized with or without solvents from fatty acids and alkylamines.

Bisamides. Most bisamides are prepared by the reaction of the primary fatty amide and formaldehyde in the presence of an acid catalyst, or by the reaction of ethylene diamine with fatty acid.

Commercial Aspects

Many primary fatty amides which are available from various manufacturers are listed in Table 1.

Table 2 lists many of the commercially available substituted fatty acid amides. The *N,N*-dimethylamides, ethoxylated amides, and other specialty substituted amides are available in commercial quantities, but are of lower commercial volume.

ROSS OPSAHL
Akzo Chemicals Corporation

A. L. McKenna, *Fatty Amides,* Witco Chemical Corporation, Tenn., 1982.

S. H. Shapiro in E. S. Pattison, ed., *Fatty Acids and Their Industrial Application,* Vol. 5, Marcel Dekker, New York, 1968, p. 77.

U.S. Pat. 2,546,521 (Mar. 27, 1951), R. H. Potts (to Akzo Chemicals).

L. W. Burnette in M. Schick, ed., *Nonionic Surfactants,* Marcel Dekker, New York, 1967, pp. 395–418.

AMINE OXIDES

Amine oxides, known as *N*-oxides of tertiary amines, are classified as aromatic or aliphatic, depending on whether the nitrogen is part of

Table 1. Commercial Amine Oxides

Name	Molecular formula	Structural formula
dimethyldodecylamine oxide	$C_{14}H_{31}NO$	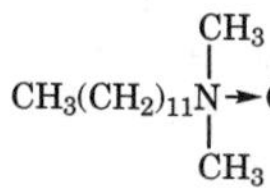
dihydroxyethyldodecylamine oxide	$C_{16}H_{35}NO_3$	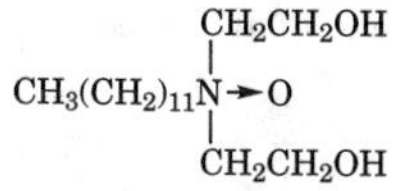
dimethyltetradecylamidopropyl amine oxide	$C_{20}H_{40}NO_2$	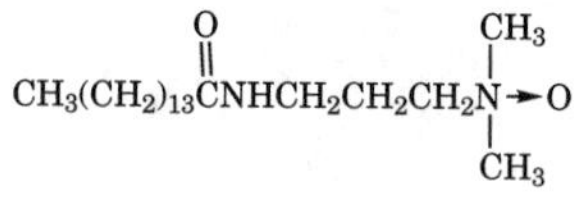
N-dodecylmorpholine *N*-oxide	$C_{16}H_{33}NO_2$	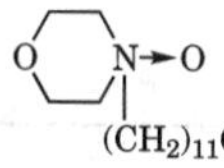
1-hydroxyethyl-2-octadecyl imidazoline oxide	$C_{23}H_{46}N_2O_2$	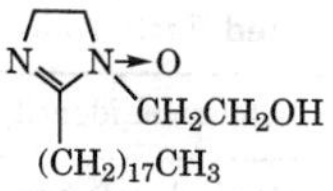
N,N′,N′-hydroxyethyl-*N*-octadecyl-1,3-propylenediamine oxide	$C_{27}H_{58}N_2O_5$	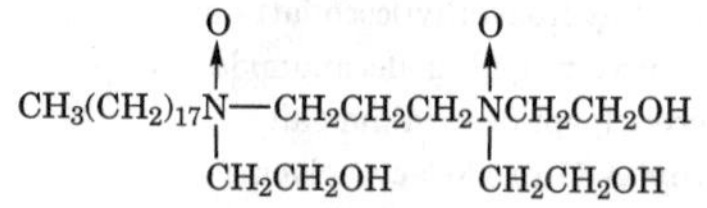

an aromatic ring system or not. This structural difference accounts for the difference in chemical and physical properties between the two types.

The higher aliphatic amine oxides are commercially important because of their surfactant properties and are used extensively in detergents. Amine oxides that have surface-acting properties can be further categorized as nonionic surfactants; however, because under acidic conditions they become protonated and show cationic properties, they have also been called cationic surfactants. Typical commercial amine oxides include the types shown in Table 1.

Aromatic amine oxides, produced on a much smaller scale and having some pharmaceutical importance, do not demonstrate the surface-acting properties that the aliphatic amine oxides do.

Reactions

Decomposition. Most amine oxides undergo thermal decomposition between 90 and 200°C. Aromatic amine oxides generally decompose at higher temperatures than aliphatic amine oxides and yield the parent amine.

Reduction. Just as aromatic amine oxides are resistant to decomposition reactions, they are more resistant than aliphatic amine oxides to reduction.

Alkylation. Alkylating agents such as dialkyl sulfates and alkyl halides react with aliphatic amine oxides to form trialkylalkoxyammonium quaternaries.

Acylation. Aliphatic amine oxides react with acylating agents such as acetic anhydride and acetyl chloride to form either *N,N*-dialkylamides and aldehyde, the Polonovski reaction, or an ester, depending on the polarity of the solvent used.

Manufacturing and Processing

Linear alpha-olefins are the source of the largest volume of aliphatic amine oxides. The olefin reacts with hydrogen bromide in the presence of peroxide catalyst, to yield primary alkyl bromide, which then reacts with dimethylamine to yield the corresponding alkyldimethylamine. Fatty alcohols and fatty acids are also used to produce amine oxides.

Amine oxides used in industry are prepared by oxidation of tertiary amines with hydrogen peroxide solution using either water or water and alcohol solution as a solvent.

Uses

Amine oxides are used in the detergent, organic synthesis, textile, and pharmaceutical industries.

Economic Aspects

The major producers of fatty amine oxides are Jordon Chemical Company, Procter and Gamble, Lonza, Stepan, Sherex Chemicals, and Akzo Chemicals, Inc.

Health and Safety Factors

Aliphatic amine oxides such as alkyldimethylamine oxides and alkylbis(2-hydroxyethyl)amine oxides range from practically nontoxic to slightly toxic.

Among the aromatics, it was found that 4-nitroquinoline *N*-oxide is a powerful carcinogen, producing malignant tumors when painted on the skin of mice. It was further established that the 2-methyl, 2-ethyl, and 6-chloro derivatives of 4-nitroquinoline oxide are also carcinogens.

B. MAISONNEUVE
Akzo Chemicals, Inc.

P. A. S. Smith, *The Chemistry of Open-Chain Organic Nitrogen Compounds,* Vol. II, W. A. Benjamin, Inc., New York, 1966, pp. 21–28.

L. W. Burnette, in M. J. Shick, ed., *Nonionic Surfactants,* Vol. I, Marcel Dekker, Inc., New York, 1967, pp. 403–410.

J. D. Sauer, in J. M. Richmond, ed., *Surfactant Science Series,* Vol. 34, Marcel Dekker, New York, 1990, pp. 275–295.

AMINES

LOWER ALIPHATIC AMINES

Lower aliphatic amines are derivatives of ammonia with one, two, or all three of the hydrogen atoms replaced by alkyl groups of five carbons or less. Amines with higher alkyl groups are known as fatty amines. Amines are toxic, colorless gases or liquids, highly flammable, and have strong odors. Lower mol wt amines are water soluble and are sold as aqueous solutions and in pure form. Amines react with water and acids to form alkylammonium compounds analogous to ammonia. The base strengths in water of the primary, secondary, and tertiary amines and ammonia are essentially the same. Primary and secondary amines can also act as very weak acids ($K_a \sim 10^{-33}$). They react with acyl halides, anhydrides, and esters with rates depending on the size of the alkyl group(s). The lower aliphatic amines are widely used as intermediates in the manufacture of medicinal, agricultural, textile, rubber, and plastic chemicals.

Physical Properties

Table 1 lists the physical properties of the commercially important alkylamines.

Chemical Properties

The formation of salts with acids is the most characteristic reaction of amines.

Allylamines are somewhat unique in that both amine and olefin functionalities are available.

Table 1. Alkylamine Physical Properties

Alkylamine	Mp, °C	Bp, °C	Vapor pressure at 20°C, kPa[a]	Density[b]
methylamine	−93.0	−6.3	288	0.67
dimethylamine	−93.0	6.9	170	0.656
trimethylamine	−117.0	2.9	191	0.633
ethylamine	−81.0	16.6	116	0.683[c]
diethylamine	−50.0	55.9	25.9	0.7062[c]
triethylamine	−114.7	88.8	7.2	0.729[c]
n-propylamine	−83.0	47.8	33.9	0.718[c]
di-*n*-propylamine	−40.0	109.3	2.8	0.7401[c]
tri-*n*-propylamine	−93.5	151.0	0.3	0.7567
isopropylamine	−95.2	32.4	63.7	0.689[c]
diisopropylamine	−61.0	83.9	8	0.7178[c]
allylamine	−88.2	52.9		0.7627
diallylamine	−88.4	110.4		0.7874
triallylamine	−70.0	149.5		0.80
n-butylamine	−50.0	77.8	9.6	0.74
di-*n*-butylamine	−61.9	159.6	0.3	0.76
tri-*n*-butylamine	<−70	214.0		0.78[c]
isobutylamine	−85.5	68.5	13.3	0.736[c]
diisobutylamine	−70.0	139.5	1.3	0.745
triisobutylamine	−21.8	191.5		0.7684
sec-butylamine	−104.5	63.0	20.0	0.7246
t-butylamine	−72.7	44.5		0.69[d]
ethyl-*n*-butylamine	−78.0	111.0	2.4	0.7398
dimethyl-*n*-butylamine		95.0		0.7206
n-amylamine	−55.0	104.5		0.7547
di-*n*-amylamine		202–203		0.7771
tri-*n*-amylamine		240–245		0.7907

[a] To convert kPa to mm Hg, multiply by 7.5.
[b] d_4^{20} unless otherwise noted.
[c] d_{20}^{20}
[d] d_{25}^{25}

Alkylamines are corrosive to copper, copper-containing alloys (brass), aluminum, zinc, zinc alloy, and galvanized surfaces.

Manufacture

Lower aliphatic amines can be prepared by a variety of methods, using many different types of raw materials. By far the largest commercial applications involve the reaction of alcohol with ammonia to form the corresponding amines. Other methods are employed depending on the particular amine desired, raw material availability, plant economics, and the ability to sell co-products. The following manufacturing methods are used commercially to produce the lower alkylamines.

Method 1. Alcohol amination–acid catalyzed: high temperature amination of an alcohol over a solid acid catalyst.

Method 2. Alcohol amination–metal catalyzed: amination of an alcohol over a metal catalyst under reducing conditions.

Method 3. Reductive alkylation: reaction of an amine or ammonia and hydrogen with an aldehyde or ketone over a hydrogenation catalyst.

Method 4. Ritter reaction: reaction of hydrogen cyanide with an olefin in an acidic medium to produce a primary amine.

Method 5. Nitrile reduction: reaction of a nitrile with hydrogen over a hydrogenation catalyst.

Method 6. Olefin amination: reaction of an olefin with ammonia.

Method 7. Alkyl halide amination: reaction of ammonia or alkylamine with an alkyl halide.

Shipment and Handling

The U. S. Department of Transportation requires labeling of all shipments of amines commensurate with the associated hazards.

Health and Safety, Toxicology

Alkylamines are toxic. Both the liquids and vapors can cause severe irritation to mucous membranes, eyes, and skin. Protective butyl rubber gloves, aprons, chemical face shields, and self-contained breathing apparatus should be used by all personnel handling alkylamines.

MICHAEL G. TURCOTTE
THOMAS A. JOHNSON
Air Products and Chemicals, Inc.

R. T. Morrison and R. N. Boyd, *Organic Chemistry,* 3rd ed., Allyl and Bacon, Boston, Mass., 1973, pp. 729–730.

D. R. Stull, E. F. Westrum, and G. C. Sinke, *The Chemical Thermodynamics of Organic Compounds,* John Wiley & Sons, Inc., New York, 1969, pp. 457–467.

M. Deeba, M. E. Ford, and T. A. Johnson, in D. W. Blackburn, ed., *Catalysis of Organic Reactions,* Marcel Dekker, New York, 1990, pp. 241–260.

Threshold Limit Values for Chemical Substances in Workroom Air Adopted by ACGIH for 1989–1990, American Conference of Governmental Industrial Hygienists, Cincinnati, Ohio, 1976.

CYCLOALIPHATIC AMINES

Cycloaliphatic amines are comprised of a cyclic hydrocarbon structural component and an amine functional group external to that ring. Included in an extended cycloaliphatic amine definition are aminomethyl cycloaliphatics. Although some cycloaliphatic amine and diamine products have direct end use applications, their major function is as low cost organic intermediates sold as moderate volume specification products.

Table 1. Properties of Primary Aminocycloalkanes

Cycloaliphatic amine	Molecular formula	Boiling point, °C	Flash point, °C	Specific gravity, g/mL	Refractive index, n_D
cyclopropylamine	C_3H_7N	49	-26	0.824	1.4210
cyclobutylamine	C_4H_9N	82	-4	0.833	1.4363
cyclopentylamine	$C_5H_{11}N$	108	17	0.863	1.4478
cyclohexylamine	$C_6H_{13}N$	134	32	0.868	1.4565
cycloheptylamine	$C_7H_{15}N$	169	42		1.4724
cyclooctylamine	$C_8H_{17}N$	190	80	0.928	1.4804
cyclododecylamine	$C_{12}H_{25}N$	280[a]	121		

[a] Melting point, 27°C.

Table 2. Properties of Cycloaliphatic Diamines

Diamine	Molecular formula	Boiling point[a],°C	Flash point, °C
cis,trans-1,2-cyclohexanediamine	$C_6H_{14}N_2$	183	75
cis-1,2-cyclohexanediamine	$C_6H_{14}N_2$	182	72
(±)*trans*-1,2-cyclohexanediamine	$C_6H_{14}N_2$		
(+)*trans*-1,2-cyclohexanediamine	$C_6H_{14}N_2$		
(-)*trans*-1,2-cyclohexanediamine	$C_6H_{14}N_2$		
cis, trans-1,3-cyclohexanediamine	$C_6H_{14}N_2$		91
cis-1,3-cyclohexanediamine	$C_6H_{14}N_2$	198	
trans-1,3-cyclohexanediamine	$C_6H_{14}N_2$	203	
methylcyclohexanediamine	$C_7H_{16}N_2$	99(1.66)	83
cis, trans-1,3-cyclohexanediamine,2-methyl			
cis, trans-1,3-cyclohexanediamine,4-methyl			
cis, trans-1,4-cyclohexanediamine	$C_6H_{14}N_2$	181	80
cis-1,4-cyclohexanediamine	$C_6H_{14}N_2$		
trans-1,4-cyclohexanediamine	$C_6H_{14}N_2$	197	71
cis,trans-1,8-menthanediamine	$C_{10}H_{22}N_2$	210	102
cis,trans-1,3-di(aminomethyl)cyclohexane	$C_8H_{18}N_2$		106
cis-1,3-di(aminomethyl)cyclohexane		114 (1.07)	
trans-1,3-di(aminomethyl)cyclohexane		117 (1.33)	
cis, trans-1,4-di(aminomethyl)cyclohexane	$C_8H_{18}N_2$	245	107
cis-1,4-di(aminomethyl)cyclohexane	$C_8H_{18}N_2$		
trans-1,4-di(aminomethyl)cyclohexane			
cis,trans-isophoronediamine	$C_{10}H_{22}N_2$	252	112
methylenedi(cyclohexylamine)	$C_{13}H_{26}N_2$	162 (2.40)	127
isopropylidenedi(cyclohexylamine)	$C_{15}H_{30}N_2$	182(1.32)	178
3,3′-dimethylmethylene-di(cyclohexylamine)	$C_{15}H_{30}N_2$	160 (0.27)	174
cis,trans-tricyclodecanediamine[b]	$C_{12}H_{22}N_2$	~ 314	165

[a] At 101.3 kPa unless otherwise indicated by the value (in kPa) in parentheses. To convert kPa to mm Hg, multiply by 7.5.
[b] (4,7-Methano-1*H*-indene-dimethaneamine, octahydro).

Physical Properties

For simple primary amines directly bonded to a cycloalkane by a single C—N bond to a secondary carbon the homologous series is given in Table 1. Up through C_8 each is a colorless liquid at room temperature. The ammoniacal or fishy odor and high degree of water solubility decrease with increased molecular weight and boiling point for these corrosive, hygroscopic mobile fluids.

Table 2 lists cycloaliphatic diamines.

Chemical Properties

Cycloaliphatic amines are strong bases with chemistry similar to that of simpler primary, secondary, or tertiary amines. Upon reaction with nitrous acid, primary amines evolve nitrogen and generate alcohols; secondary amines form mutagenic nitrosamines.

Salt formation with Brønsted and Lewis acids and exhaustive alkylation to form quaternary ammonium cations are part of the rich derivatization chemistry of these amines.

Primary cycloaliphatic amines react with phosgene to form isocyanates.

Cycloaliphatic diamines react with dicarboxylic acids or their chlorides, dianhydrides, diisocyanates and di- (or poly-) epoxides as comonomers to form high molecular weight polyamides, polyimides, polyureas, and epoxies.

Manufacture and Processing

Cycloaliphatic amine synthesis routes may be described as distinct synthetic methods, though practice often combines, or hybridizes, the steps that occur: amination of cycloalkanols, reductive amination of cyclic ketones, ring reduction of cycloalkenylamines, nitrile addition to alicyclic carbocations, reduction of cyanocycloalkanes to aminomethylcycloalkanes, and reduction of nitrocycloalkanes or cyclic ketoximes.

Production and Shipment

Larger volume cycloaliphatic amines and diamines and their worldwide major manufacturers are shown in Table 3. Shipment of these liquid products is by nitrogen-blanketed tank truck or tank car. Drum shipments are usually in carbon steel, DOT-17E.

Table 3. Commercial Cycloaliphatic Amines

Amine	Volume, 10^3 t/yr	Price,[a] $/kg	Manufacturers
cyclohexylamine	20	2.90	Air Products, BASF, Bayer, Hoechst, ICI, Kanto Denka, New Japan
isophoronediamine	18	5.40	Hüls
methylenedi(cyclohexylamine)	10	6.80	Air Products, BASF, Hüls, New Japan
3,3′-dimethylmethylene/di(cyclohexylamine)	3	8.90	BASF
dicyclohexylamine	2	3.75	Air Products, BASF, Hoechst
dimethylcyclohexylamine	2	4.85	Air Products, BASF, ICI
1,2-cyclohexanediamine	1	3.65	Du Pont/Milliken

[a] Jan. 1990; approximate.

Health and Safety, Toxicology

Cycloaliphatic amines and diamines are extreme lung, skin, and eye irritatants. MSD sheets universally carry severe personal protective equipment use warnings because of the risk of irreversible eye damage.

Derivatives

Before a 1/1/70 FDA ban (rescission proposed in early 1990), cyclamate noncaloric sweeteners were the major derivatives driving cyclohexylamine production.

Cyclohexylamine condensed with mercaptobenzothiazole produces the large volume moderate rubber accelerator *N*-cyclohexyl-2-benzothiazolesulfenamide.

1,3-Dicyclohexylcarbodiimide is an important peptide-condensing agent and analytical reagent.

trans-1,2-Cyclohexanediamine is derivatized by Mannich reaction of formaldehyde and HCN, then hydrolyzed to the tetraacetate and sold as a chelating agent by Eastman Kodak and Takeda.

Methylenedi(cyclohexylisocyanate) (MDCHI, Desmodur W) is the dominant derivative of MDCHA and is used in light-stable urethanes.

Isophoronediisocyanate made by phosgenation of IPD competes effectively in this same polyurethane market, predominantly coatings, and is the major commercial application of isophoronediamine.

1,4-Cyclohexanediamine from hydrogenation of *p*-phenylenediamine may be easily phosgenated, unlike the corresponding 1,2- and 1,3- isomers, to produce a useful diisocyanate for performance polyurethanes efficiently, particularly *trans*-1,4-cyclohexanediisocyanate (CHDI).

A representative agrochemical application of cycloaliphatic amines is the reaction of the commercial 30/70 *cis/trans* isomer mixture of 2-methylcyclohexylamine with phenylisocyanate to give the crabgrass and weed control agent Siduron (1-(2-methylcyclohexyl)-3-phenylurea. The preplant herbicide Cycloate, used for sugar beets, vegetable beets, and spinach, (*S*-ethyl-*N*-ethyl-*N*-cyclohexylthiocarbamate, incorporates *N*-ethylcyclohexylamine. The herbicide Hexazinone (3-cyclohexyl-6-dimethylamino-1-methyl-1,3,5-triazine-2,4-dione is prepared from cyclohexylisocyanate.

JEREMIAH P. CASEY
Air Products and Chemicals

R. L. Augustine, *Catalytic Hydrogenation,* Marcel Dekker, New York, 1965.

M. Freifelder, in *Catalytic Hydrogenation in Organic Synthesis: Procedures and Commentary,* John Wiley & Sons, Inc., New York, 1978, Chaps. 5, 7.

P. N. Rylander, *Hydrogenation Methods,* Academic Press, London, 1985.

FATTY AMINES

Fatty amines are nitrogen derivatives of fatty acids, olefins, or alcohols prepared from natural sources, fats and oils, or petrochemical raw materials. Commercially available fatty amines consists of either a mixture of carbon chains or a specific chain length from C-8–C-22. The amines are classified as primary, secondary, or tertiary depending on the number of hydrogen atoms of an ammonia molecule replaced by fatty alkyl or methyl groups (Fig. 1). The amino nitrogen is most frequently found on a primary carbon atom, but secondary and tertiary carbon substitution derivatives have been made and are commercially available. Fatty amines are cationic surface-active compounds, which strongly adhere to surfaces by either physical or chemical bonding, thus modifying surface properties. Important commercial products are prepared using fatty amines as reactive intermediates.

Commercially available fatty amines are most frequently prepared from naturally occurring materials (see FATS AND FATTY OILS) by hydrogenation of a fatty nitrile intermediate using a variety of catalysts.

Fatty amines derived from fats and oils, containing several carbon-chain-length moieties, are designated as such by common names which describe these mixtures: tallowalkyl-, cocoalkyl-, and soyaalkylamines, for example. High purity fatty amines are also commercially available. These amines are prepared by distillation of either the precursor fatty acid or amine product mixture. Trade names are commonly used for commercial products.

Fatty alcohols, prepared from fatty acids or via petrochemical processes, aldol or hydroformylation reactions, or the Ziegler process, react with ammonia or a primary or secondary amine in the presence of a catalyst to form amines.

Physical Properties

Data on physical properties of fatty amines have been well documented and summarized in many reference works on fatty acids and nitrogen derivatives. It is evident that (*1*) melting points within a homologous series of single-chain-length fatty amines increase with molecular weight; (*2*) symmetrical secondary amines have a higher melting point than the primary amine of the same alkyl group, but are lower melting than a primary amine with the same number of carbon atoms (hydrogen bonding); (*3*) symmetrical tertiary amines are lower melting than

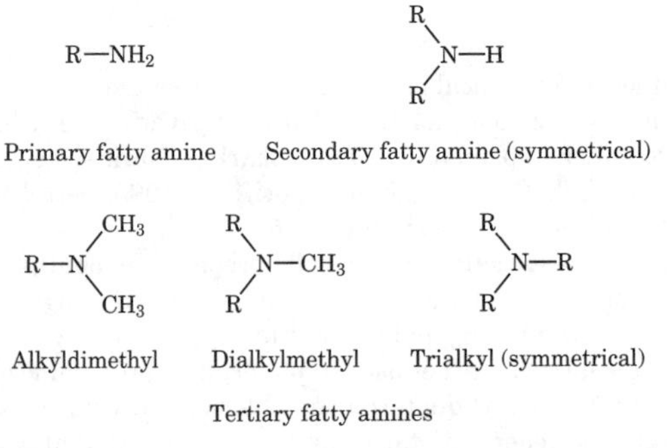

Figure 1. Types of commercially available fatty amines. R = C-8–C-22.

a symmetrical secondary amine of the same alkyl group; (*4*) symmetrical tertiary amines are lower melting than a primary or secondary amine containing the same number of carbon atoms; and (*5*) unsaturation lowers the melting point of the fatty amine, eg, oleylamine versus 1-octadecylamine and ditallowalkylamines versus dihydrogenated tallowalkylamines.

Boiling points of fatty amines have been reported. A direct correlation between molecular weight and boiling point is observed. Mixtures of primary fatty amines prepared from fats and oils can be separated into component amines by fractional distillation; an approximately 10°C increment in boiling point per carbon in the chain length is maintained throughout the series.

Fatty amines are insoluble in water, but soluble in organic solvents to varying degrees. Water, however, is soluble in the amines, and hydrates are formed.

Chemical Reactions

General amine chemistry is applicable to fatty amines. Many chemical reactions using fatty amines as reactive intermediates are run on an industrial scale to produce a wide range of important products. Important industrial reactions are salt formation, methylation of primary and secondary fatty amines, quaternization, ethoxylation and propoxylation, oxidation by hydrogen peroxide, and cyanoethylation.

Commercial Availability

U.S. consumption of fatty nitrogen derivatives was estimated to be 280,000 t in 1990. It was estimated in 1986 that plant capacities of the major U.S. producers of fatty amines and derivatives were 207,000 t. Thus, plant expansion was needed to keep up with the increasing usage of fatty amines and nitrogen derivatives. Plant expansions have occurred at Akzo Chemicals Inc., Ethyl Corporation, Humko Chemical (Witco Corporation), Jetco Chemicals, Inc. (The Procter and Gamble Company), Tomah Products, Inc. (Exxon Chemical Company), and Sherex Chemical Co. (now Witco).

Fatty amine products are normally shipped in 55-gal (208-L), lined and unlined steel drums, or in tank cars or tank trucks for bulk shipments. High melting amines can be flaked and shipped in cardboard cartons or paper bags.

Health and Safety Factors

Skin and Eye Irritation. Fatty alkylamines are generally considered to be irritating to both the skin and eyes.

Oral Toxicity. Depending on the chemical class, most fatty amines range from moderately toxic to practically nontoxic by acute oral ingestion.

Dermal Toxicity. Fatty alkylamines are not considered especially toxic with regard to skin penetration and systemic absorption into the body; certain polyamines may be absorbed through the skin to a much greater degree.

Inhalation. Long-chain amines are not considered an inhalation hazard at ambient conditions because of their relatively low volatility.

Uses

Fatty amines and chemical products derived from the amines are used in many industries. Uses for the nitrogen derivatives may be broken down as follows as a percentage of total market: fabric softeners (46%), oil field chemicals (15%), asphalt emulsifiers (10%), petroleum additives (10%), mining (4%), and others (15%).

Amine salts, especially acetate salts prepared by neutralization of a fatty amine with acetic acid, are useful as flotation agents (collectors), corrosion inhibitors, and lubricants.

The single largest market use for quaternary fatty amines is in fabric softeners. Monoalkyl quaternaries (chloride) have been used in liquid detergent softener antistat formulations (LDSA), dialkyldimethyl quaternaries (chloride) in the rinse cycle, and dialkyldimethyl quaternaries (sulfate) as dryer softeners.

Another significant use for dialkyldimethyl quaternary ammonium salts and alkylbenzyldimethylammonium salts is in preparing organoclays for use as drilling muds, paint thickeners, and lubricants.

Betaines, or specialty quaternaries, are used in the personal care industry in shampoos, conditioners, foaming, and wetting agents.

A major use for ethoxylated and propoxylated amines is as an antistatic agent in the textile and plastics industry. Ethoxylates are also used in the agricultural area as adjuvants.

Examples of uses for amine oxides include detergent and personal care areas as a foam booster and stabilizer, as a dispersant for glass fibers, and as a foaming component in gas recovery systems.

Important uses for the diamines include: corrosion inhibitors, gasoline and fuel oil additives, flotation agents, pigment wetting agents, epoxy curing agents, herbicides, and asphalt emulsifiers. Fatty amines and derivatives are widely used in the oil field, as corrosion inhibitors, surfactants, emulsifying/deemulsifying and gelling agents. In the mining industry, amines and diamines are used in the recovery and purification of minerals, flotation, and benefication. A major use of fatty diamines is as asphalt emulsifiers for preparing asphalt emulsions. Diamines have also been used as epoxy curing agents, corrosion inhibitors, gasoline and fuel oil additives, and pigment wetting agents. Oleyl amine is a petroleum additive useful as a detergent in gasoline. In addition, derivatives of the amines, amphoterics, and long-chain alkylamines are used as anionic and cationic surfactants in the personal care industry.

K. VISEK
Akzo Chemicals Inc.

R. A. Reck, in R. W. Johnson and E. Fritz, ed., *Fatty Acids in Industry,* Marcel Dekker, Inc., New York, 1989, pp. 177–199, 201–215.

Oleochemicals: Fatty Acids, Fatty Alcohols, Fatty Amines, Course sponsored by the Education Committee of the American Oil Chemists' Society, Kings Island, Ohio, Sept. 13–16, 1987.

N. O. V. Sonntag, "Nitrogen Derivatives," in K. S. Markley, ed., *Fatty Acids,* Part 3, John Wiley & Sons, Inc., New York, 1964, pp. 1551–1715.

S. H. Shapiro, "Commercial Nitrogen Derivatives of Fatty Acids," in E. Pattison, ed., *Fatty Acids and Their Industrial Applications,* Marcel Dekker, Inc., New York, 1968, pp. 77–154.

AMINES, AROMATIC

ANILINE AND ITS DERIVATIVES

Aniline (benzenamine) is the simplest of the primary aromatic amines.

Aromatic amines can be produced by reduction of the corresponding nitro compound, the ammonolysis of an aromatic halide or phenol, and by direct amination of the aromatic ring. At present, the catalytic reduction of nitrobenzene is the predominant process for manufacture of aniline. To a smaller extent aniline is also produced by ammonolysis of phenol.

Important analogs of aniline include the toluidines, xylidines, anisidines, phenetidines, and its chloro-, nitro-, *N*-acetyl, *N*-alkyl, *N*-aryl, *N*-acyl, and sulfonic acid derivatives.

Physical Properties

Pure, freshly distilled aniline is a colorless, oily liquid that darkens on exposure to light and air. It has a characteristic sweet, amine-like aromatic odor. Aniline is miscible with acetone, ethanol, diethyl ether, and benzene, and is soluble in most organic solvents.

The physical properties of aniline are given in Table 1.

Chemical Properties

Aromatic amines are usually weaker bases than aliphatic amines.

Table 1. Physical Properties of Aniline

Property	Value
molecular formula	C_6H_7N
molecular weight	93.129
boiling point, °C at 101.3 kPa[a]	184.4
freezing point, °C	−6.03
density, liquid, g/mL, 20/4°C	1.02173
viscosity, mPa · s(= cP) at 20°C	4.35
ignition temperature, °C	615
lower flammable limit, vol %	1.3

[a] To convert kPa to mm Hg, multiply by 7.5.

Aromatic amines form addition compounds and complexes with many inorganic substances, such as zinc chloride, copper chloride, uranium tetrachloride, or boron trifluoride. Various metals react with the amino group to form metal anilides; and hydrochloric, sulfuric, or phosphoric acid salts of aniline are important intermediates in the dye industry. Important reactions include *n*-alkylation, ring alkylation, acylation, condensation, cyclization, reaction with nitrous acid, oxidation, halogenation, sulfonation, nitration, and reduction.

Manufacturing and Processing

The predominant process for manufacture of aniline is the catalytic reduction of nitrobenzene with hydrogen.

Du Pont uses a liquid-phase hydrogenation process that employs a palladium–platinum-on-carbon catalyst. The process uses a plug-flow reactor that achieves essentially quantitative yields, and the product exiting the reactor is virtually free of nitrobenzene.

Economic Aspects

Table 2 lists the manufacturers of aniline in the United States at the end of 1989.

In the 1980s manufacturing capacity for aniline underwent some major changes. It is estimated that aniline capacity utilization was about 50% of nameplate capacity when Aristech's new 91,000 t/yr plant came on stream. That same year American Cyanamid closed its 23,000-t plant at Willow Island, W. Va., and withdrew from the aniline business. Mobay shut down its larger plant (45,000-t) at New Martinsville, W. Va. in 1983; and Du Pont idled its 77,000-t facility in 1984.

These reductions in capacity, coupled with the growth in aniline demand, led to shortages in aniline supply in 1987 and 1988. The shortage is expected to persist into the early 1990s until new capacity comes on stream. Mobay and BASF have announced plans to build new plants (113,000 and 54,000 t, respectively), which were to start in 1992.

Storage and Handling

The flash point of aniline (70°C) is well above its normal storage temperature, but aniline should be stored and used in areas with minimum fire hazard.

Aniline is slightly corrosive to some metals. It attacks copper, brass, and other copper alloys, and use of these metals should be avoided in equipment that is used to handle aniline. For applications in which color retention is critical, the use of 400-series stainless steels is recommended.

Aniline is shipped in tank truck and tank car quantities and is classified by the U.S. Department of Transportation (DOT) as a Class B poison (UN 1547), and must carry a poison label.

Wastes contaminated with aniline may be listed as RCRA Hazardous Waste, and if disposal is necessary, the waste disposal methods used must comply with U.S. federal, state, and local water pollution regulations.

Table 2. U.S. Manufacturers of Aniline

Company	Location	Capacity, 10^3 t
Aristech Chemical	Haverhill, Ohio	91
Du Pont	Beaumont, Tex.	113
First Chemical	Pascagoula, Miss.	136
Mobay	New Martinsville, W. Va.	18
Rubicon	Geismar, La.	172

Health Hazards and Safety Precautions

Aniline is highly toxic and may be fatal if swallowed, inhaled, or absorbed through the skin. Aniline vapor is mildly irritating to the eye, and in liquid form it can be a severe eye irritant and cause corneal damage.

The U.S. Department of Labor (OSHA) has ruled that an employee's exposure to aniline in an 8-h work shift of a 40-h work week shall not exceed an 8-h time-weighted average (TWA) of 5 ppm vapor in air. The American Conference of Governmental Industrial Hygienists (ACGHI) recommends a threshold limit value (TLV) of 2 ppm aniline vapor in air, TWA for an 8-h work day.

Based on tests with laboratory animals, aniline may cause cancer.

In view of the above, aniline should be handled in areas with adequate ventilation and skin exposure should be avoided by the wearing of proper safety equipment.

Uses

The major uses of aniline are in the manufacture of polymers, rubber, agricultural chemicals, dyes and pigments, pharmaceuticals, and photographic chemicals.

Derivatives

Most derivatives of aniline are not obtained from aniline itself, but are prepared by hydrogenation of their nitroaromatic precursors. The exceptions, eg, *N*-alkylanilines, *N*-arylanilines, sulfonated anilines, or the *N*-acyl derivatives, can be prepared from aniline. Nitroanilines are usually prepared by ammonolysis of the corresponding chloronitrobenzene. Special isolation methods may be required for some derivatives if the boiling points are close and separation by distillation is not feasible.

BIJAN AMINI
E. I. du Pont de Nemours & Co., Inc.

P. H. Groggins, *Aniline and Derivatives,* Van Nostrand Rheinhold Co., New York, 1924, p. 177.

Du Pont Aniline Properties, Uses, Storage, and Handling Bulletin, E. I. du Pont de Nemours & Co., Inc., 1983.

N. Irving Sax and R. J. Lewis, Jr., *Dangerous Properties of Industrial Materials,* Vol. II, 7th ed., Van Nostrand Reinhold Co., New York, 1989, p. 262.

DIAMINOTOLUENES

Toluenediamine (diaminotoluene, *m*-TDA) is an important industrial chemical intermediate; it is produced in the largest volume of any arylamine and is the lowest priced diamine. TDA producers in the United States include Air Products and Chemicals Corporation (merchant only), Mobay Corporation (Bayer USA), BASF Corporation, Olin Corporation, Dow Chemical USA, and Rubicon Corporation (ICI). The principal use for TDA is in the manufacture of toluenediisocyanate (TDI), the predominant diisocyanate in the flexible foams and elastomers industries.

Table 1. Physical Properties of the Toluendiamine Isomers

Diaminotoluene isomers	Melting point, °C	Boiling point, °C	Vapor pressure, kPa[a] (at °C)
2,3	63–64	255	1.20 (150)
			1.87 (160)
			2.67 (180)
2,4	99	292	1.47 (150)
			2.27 (160)
			4.80 (180)
2,5	64	273–274	
2,6	106		2.13 (150)
			3.33 (160)
			7.60 (180)
3,4	89–90	265 (subl)	
3,5	<0	283–285	

[a] To convert kPa to mm Hg, multiply by 7.5.

Physical Properties

Although all six possible toluenediamine isomers are made in the commercial synthesis, only two products are available commercially. The properties of the individual isomers are summarized in Table 1.

Reactions

The aromatic toluenediamines undergo typical amine reactions. The general chemistry is similar to that of the phenylenediamines or the cyclohexanediamines, except that the aromatic diamine is a weaker base than its cycloaliphatic counterpart. Reactions of industrial importance involving commercial toluenediamines include phosgenation, ring alkylation, alkoxylation, and diazotization.

Manufacture

Dinitrotoluenes can be catalytically hydrogenated to toluenediamines under a wide variety of temperatures, pressures, and solvents; the catalyst can be supported noble metal, eg, Pd/C, or nickel, either supported or Raney type. The reduction requires six moles of hydrogen per mole of DNT and produces four moles of water.

Toxicity

Toluenediamine is classed as toxic. TDA is readily absorbed through the skin and this is the major route of human exposure.

BERNARD A. TOSELAND
MARK S. SIMPSON
Air Products and Chemicals Corporation

U.S. Pat. 3,499,034 (Mar. 3, 1970), R. A. Gonzalez (to E. I. du Pont de Nemours & Co., Inc.).

U.S. Pat. 3,761,521 (Sept. 25, 1973), L. Alheritiere and co-workers (to Melle-Bezons).

U.S. Pat. 3,420,752 (Jan. 7, 1969), V. Kriss and co-workers (to Allied Chemical Company).

U.S. Pat. 4,218,543 (Aug. 19, 1980), C. Weber and H. Schafer (to Bayer AG).

DIARYLAMINES

Diarylamines are compounds that have two aromatic groups and one hydrogen atom attached to nitrogen. Diphenylamine (DPA) or

Table 1. Physical Properties of Diarylamines

Diarylamine	Formula	Mp, °C	Bp, °C[a]
diphenylamine	$C_{12}H_{11}N$	53	302
2-methyldiphenylamine	$C_{13}H_{13}N$	41	175 (2.93)
3-methyldiphenylamine	$C_{13}H_{13}N$	27.5	
4-methyldiphenylamine	$C_{13}H_{13}N$	88	
4-(1,1-dimethylethyl)diphenylamine	$C_{16}H_{19}N$	67	139 (0.05)
4-octyldiphenylamine	$C_{20}H_{27}N$	48	152 (0.03)
4,4′-bis(1,1-dimethylethyl)diphenylamine	$C_{20}H_{27}N$	110	
4,4′-bis(1-phenylethyl)diphenylamine	$C_{28}H_{27}N$		
4,4′-bis(1-methyl-1-phenylethyl)diphenylamine	$C_{30}H_{31}N$	101	
4,4′-dioctyldiphenylamine	$C_{28}H_{43}N$	102	
2,2′-diethyldiphenylamine	$C_{16}H_{19}N$		
2,2′-bis(1-methylethyl)diphenylamine	$C_{18}H_{23}N$		
2,4,4′-tris(1-methyl-1-phenylethyl)diphenylamine	$C_{39}H_{41}N$	122	
4-hydroxydiphenylamine	$C_{12}H_{11}NO$	73	200 (1.33)
4,4′-dimethoxydiphenylamine	$C_{14}H_{15}NO_2$	103	
N-phenyl-1-naphthylamine	$C_{16}H_{13}N$	62	244 (16)
N-[4-(1-methyl-1-phenylethyl)phenyl]-1-naphthylamine	$C_{25}H_{23}N$	92	
N-phenyl-2-naphthylamine	$C_{16}H_{13}N$	108	395
N-[4-(1-methyl-1-phenylethyl)phenyl]-1-(1-methyl-1-phenylethyl)-2-naphthylamine	$C_{34}H_{33}N$	122	
di-6-chrysenylamine	$C_{36}H_{23}N$	314	
N-nitrosodiphenylamine	$C_{12}H_{10}N_2O$	68	
N,N′-diphenyl-*p*-phenylenediamine	$C_{18}H_{16}N_2$	152	200 (0.07)
N,N′-di-2-naphthyl-*p*-phenylenediamine	$C_{26}H_{20}N_2$	234	
9*H*-carbazole	$C_{12}H_9N$	245	
9,10-dihydro-9,9-dimethyl acridine	$C_{15}H_{15}N$	137	
10*H*-phenothiazine	$C_{12}H_9NS$	193	
8-octyl-10*H*-phenothiazine	$C_{20}H_{26}NS$	118	

[a] At 101.3 kPa unless otherwise indicated in kPa in parentheses. To convert kPa to mm Hg, multiply by 7.5.

N-phenylbenzenamine, is the most commercially significant diarylamine. Today, it is manufactured by heating aniline by itself, or with phenol, and with an acid catalyst at high temperatures. It is used as a stabilizer for nitrocellulose propellants. When alkylated, diphenylamine finds its largest application as an antioxidant and stabilizer for oils, greases, polymers, and elastomers. Diarylamines that contain other functional groups, such as hydroxyl or amino, are useful dye intermediates, antiozonants, hair dyes, and color photography intermediates.

Physical Properties

Selected physical properties for representative diarylamines are given in Table 1.

Reactions

Important reactions include C-alkylation and oxidation.

Manufacture

Diarylamines are manufactured by the self-condensation of a primary aromatic amine in the presence of an acid, or the reaction of an arylamine with a phenol at high temperatures.

Laboratory Procedures

Specific diarylamines not easily obtained by the above methods can be prepared by the Ullmann and Chapman reactions.

Toxicity

Industrial poisoning by diphenylamine has been encountered and appears clinically as bladder symptoms, tachycardia, hypertension, and skin problems. There is no federal standard for permissible exposure limits in air, but the American Conference of Government Industrial Hygienists (1983/1984) has adopted a time-weighted average value of 10 mg/m^3 and set a short-term exposure limit of 20 mg/m^3. The alkylated diphenylamines, used as antioxidants, have much higher molecular weights and are relatively nonvolatile.

N-Nitrosodiphenylamine can act as a nitrosating agent for other amines with all the consequences thereof (see *N*-NITROSAMINES).

As with the parent aromatic hydrocarbons, diarylamines based on polycyclic aromatic amines also tend to be more harmful.

ROBERT W. LAYER
The BF Goodrich Company

J. Pospisil in G. Scott, ed., *Developments in Polymer Stabilization,* Vol. 7, Elsevier Applied Science Publishers, New York, 1984, Chapt. 1.

E. G. Rozantsev, *Free Nitroxyl Radicals,* Plenum Press, New York, 1970.

W. L. Fath, D. B. Keyes, and R. L. Clark, *Industrial Chemicals,* 2nd ed., John Wiley & Sons, Inc., New York, 1959, pp. 330–333.

METHYLENEDIANILINE

Commercial production of 4,4′-methylenedianiline (4,4′-MDA) is carried out by the acid-catalyzed reaction of formaldehyde with aniline. All processes produce polymeric MDA (PMDA), which consists of mixtures of isomers and oligomers of MDA. More than 99% of the manufactured PMDA products are used in reactions with phosgene to produce the corresponding isocyanates for use in polyurethanes. The resultant polymeric isocyanates (PMDI) are either sold commercially or are purified to isolate 4,4′-methylenediphenyldiisocyanate (MDI). Only 15–20% of the total PMDI manufactured in the United States is consumed in the monomeric form. MDI is an important intermediate in the manufacture of spandex fibers, thermoplastic resins, and coatings, and it is used in reaction injection molding (RIM) for automotive applications. The primary use of PMDI products is in rigid foam insulation, but they are also used in semiflexible foams, foundry core binders, and particle board manufacture. Nonisocyanate uses for MDA or PMDA include epoxy curing agents, filament wound pipe, wire coatings, and military applications.

Table 1. Physical and Chemical Properties of MDA and PMDA

Property	4,4′-MDA	PMDA[a]
molecular formula	$C_{15}H_{14}N_2$	
molecular weight	198.3	
melting point, °C	93	60–80
specific heat, J/(g·°C)[b]	2.1	2.1
heat of fusion, kJ/mol[b]	19.6	~19.6
boiling point, °C	238 at 1.33 kPa[c]	398 at 101.3 kPa[c]
viscosity, mPa·s(= cP)	8.3 at 100°C	80 at 70°C
approximate solubility, g/100 mL solvent at 25°C		
acetone	273.0	
carbon tetrachloride	0.7	
methanol	143.0	
water	0.1	

[a] For PMDA containing approximately 70% MDA.
[b] To convert kJ to kcal, or J to cal, divide by 4.184.
[c] To convert kPa to mm Hg, multiply by 7.5

Table 2. 1989 Global Production Capacity for MDI and Polymeric MDI

Manufacturer	Location	Capacity[a], 10^{3}t
United States		
BASF	Geismar, La.	68
Dow	La Porte, Tex.	150
Mobay	Baytown, Tex.	100
	New Martinsville, W. Va.	50
Rubicon	Geismar, La.	141
Total United States		*509*
Western Europe		
BASF	Antwerp, Belgium	60
Bayer	Krefeld, Germany	126
Bayer	Leverkusen, Germany	20
Bayer	Tarragona, Spain	6
Bayer-Shell	Antwerp, Belgium	26
ICI	Fleetwood, U.K.	45
ICI	Rosenberg, Holland	40
Montedison	Brindisi, Italy	70
Dow	Isopor, Portugal	50
Dow	Delfzijl, Holland	70
Total Western Europe		*513*
Asia		
Dow-MDK	Yokkaichi, Japan	36
Mitsui Toatsu	Omuta, Japan	35
Nippon Polyurethane	Nanyo, Japan	50
Sumitomo Bayer	Niihama, Japan	36
Total Asia		*157*
other (Eastern Europe and South America)		ca 61
Global total		*1,240*

[a] The conversion factor for PMDA to PMDI is 1.25–1.35; eg, 80 kg of MDA produces 100 kg of MDI.

Physical Properties

The physical and chemical properties of 4,4′-MDA and a typical PMDA are listed in Table 1.

Chemical Properties

MDA reacts similarly to other aromatic amines under the proper conditions. For example, nitration, bromination, acetylation, and diazotization all give the expected products. Much of the chemistry carried out on MDA takes advantage of the difunctionality of the molecule in reacting with multifunctional substrates to produce low and high molecular weight polymers.

Economic Aspects

The data in Table 2 represent the total MDI (MDI and PMDI) was produced on a global basis, compiled in 1989.

Health and Safety, Toxicology

All of the toxicity data on MDA have been collected using either 4,4′-MDA or the corresponding hydrochloride salt. The information discussed in this section can also be used for commercial products containing MDA or PMDA. Because MDA is a potentially hazardous chemical, worker exposure should be kept to a minimum.

The major exposure route in workers who experience MDA poisoning is by skin contact.

MDA is a suspected human carcinogen, although there are no reports of cancer in humans as a result of MDA exposure.

STEVEN LOWENKRON
The Dow Chemical Company

E. C. Wagner, *J. Org. Chem.* **19**, 1862 (1954).

H. J. Twitchett, *Chem. Soc. Rev.* **3**, 209 (1974).

C. J. Elskamp, *OSHA Analytical Method #57—4,4′-Methylenedianiline (MDA)*, OSHA Analytical Laboratory, Salt Lake City, Utah, Jan. 1986.

N. M. Bernholc, S. C. Morris, III, and J. E. Brower, *Biological Effects Summary Report 4,4′-Methylenedianiline* (BNL-51903), Brookhaven National Laboratory, Upton, N.Y., 1985.

PHENYLENEDIAMINES

Phenylenediamines are aromatic amines with two amino groups attached to benzene. They are called benzenediamines by *Chemical Abstracts.* There are three phenylenediamines; ortho, meta, and para; or 1,2-; 1,3-; and 1,4-benzenediamines, respectively. Since they are difunctional, they are easily converted to polymers. The *m*-phenylenediamines are the least expensive isomers and are used in the largest volumes for the manufacture of polyurethanes. *p*-Phenylenediamine is used for the manufacture of polyamides with exceptional tensile strength. The use of *o*-phenylenediamines for polymer formation is limited, but polybenzimidazoles and polyquinoxalines have been made. However, it is the unique chemistry of the *p*-phenylenediamine derivatives that makes them useful in the photographic and dye industries, and widely used as antioxidants and antiozonants for rubbers and plastics.

Physical Properties

The three phenylenediamines are all white solids when pure, but darken after standing in air. They are all very soluble in hot water. See also Table 1.

Table 1. Physical Properties of Phenylenediamines

Phenylenediamine	Molecular formula	Appearance	Mp, °C	Bp[a], °C	Synthetic route
ortho[b]	$C_6H_8N_2$	white solid	102	256–258	ammonol, reduction of *o*-chloronitrobenzene
meta[b]	$C_6H_8N_2$	white solid	63	287	catalytic reduction of *m*-dinitrobenzene
para[b]	$C_6H_8N_2$	white solid	140	267	ammonol, reduction of *p*-chloronitrobenzene
toluene-2,4-diamine	$C_7H_{10}N_2$		99	280	reduction of 2,4-dinitrotoluene
toluene-2,6-diamine	$C_7H_{10}N_2$		105		reduction of 2,6-dinitrotoluene
2,3,5,6-tetramethyl-*p*-	$C_{10}H_{16}N_2$	tan colored	151–152		nitration, reduction of durene
N,N-dimethyl-*p*-	$C_8H_{12}N_2$		41	262	nitrosation, reduction of *N,N*-dimethylaniline
N,N-diethyl-*p*-	$C_{10}H_{16}N_2$	pale yellow oil		260–262	reduction of *p*-nitro-*N,N*-diethylaniline
N,N′-bis(1-methylpropyl)-*p*-	$C_{14}H_{24}N_2$	pale yellow oil		98 (26.6 Pa[c])	reductive alkylation of *p*-phenylenediamine
N,N′-bis(1-methylheptyl)-*p*-	$C_{22}H_{40}N_2$	low melting white solid		190 (40 Pa[c])	reductive alkylation of *p*-phenylenediamine
N,N′-bis(1-methylpropyl)-*N,N′*-dimethyl-*p*-	$C_{16}H_{28}N_2$	oil		115 (20 Pa[c])	Leuckart-Wallach reaction
N-phenyl-*p*-	$C_{12}H_{12}N_2$		75	354	nitrosation, reduction of diphenylamine
N,N′-diphenyl-*p*-	$C_{18}H_{16}N_2$	white solid	146		from hydroquinone and aniline, acid cat
N,N′-di-2-naphthalenyl-*p*-	$C_{26}H_{20}N_2$		236		from 2-naphthol and *p*-phenylenediamine, heat
N-1-methylethyl-*N′*-phenyl-*p*-	$C_{15}H_{18}N_2$		78		reductive alkylation of *N*-phenyl-*p*-phenylenediamine
N-(1,3-dimethylbutyl)-*N′*-phenyl-*p*-	$C_{18}H_{24}N_2$		48–50		reductive alkylation of *p*-aminodiphenylamine with methyl isobutyl ketone
N-cyclohexyl-*N′*-phenyl-*p*-	$C_{18}H_{22}N_2$				reductive alkylation of *p*-aminodiphenylamine with cyclohexanone

[a] At 101.3 kPa = 1 atm unless otherwise noted.
[b] The K_bs of *o*-, *m*-, and *p*-phenylenediamine are 3.2×10^{-10}, 7.6×10^{-10}, and 1.1×10^{-8}, respectively.
[c] To convert Pa to mm Hg, multiply by 0.0075.

Chemical Properties

The phenylenediamines are aromatic amines which undergo typical amine reactions, such as *N*-acylation and *N*-alkylation. However, unlike most aromatic amines, they cannot easily be alkylated on the aromatic ring using an olefin and an acid catalyst, since protonation of the second amino group deactivates the ring. Consequently, most of the chemistry of the phenylenediamines involves reactions on the nitrogen. Commercially the largest-volume usage of phenylenediamines is in the manufacture of diisocyanates; large volumes of *N,N′*-alkyl/aryl-*p*-phenylenediamines are also manufactured.

Production

The *m*-phenylenediamines are easily obtained by dinitrating, followed by catalytically hydrogenating, an aromatic hydrocarbon.

Health and Safety

Like aniline, phenylenediamines may be toxic. Appropriate substitution of *p*-phenylenediamine, generally on the nitrogen with phenyl or larger alkyl groups, significantly reduces its toxicity.

ROBERT W. LAYER
The BF Goodrich Company

N. V. Sidgwick, *The Organic Chemistry of Nitrogen,* The Clarendon Press, Oxford, U.K., 1937.

P. Wiseman, *An Introduction to Industrial Organic Chemistry,* John Wiley & Sons, New York, 1976.

R. W. Layer and R. P. Lattimer, *Rubber Chem. Technol.* **63,** 426 (1990).

AMINES BY REDUCTION

Amines are derivatives of ammonia in which one or more of the hydrogens is replaced with an alkyl, aryl, cycloalkyl or heterocyclic group. When more than one hydrogen has been replaced, the substituents can either be the same or different. Amines are classified as primary, secondary, or tertiary depending on the number of hydrogens which have been replaced. General structures for ammonia as well as primary, secondary, and tertiary amines are shown below.

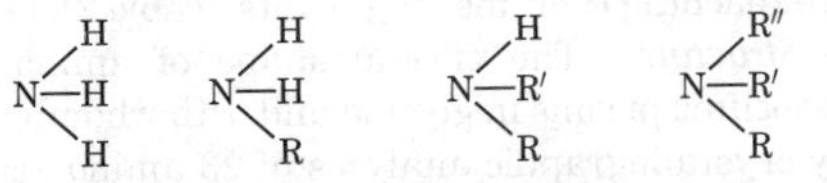

ammonia primary amine secondary amine tertiary amine

In reductive methods of making amines, the nitrogen is already incorporated in the molecule, and the amine is formed by reducing the oxidation state of the compound with the addition of hydrogen. In theory, many different types of nitrogen-containing compounds can be reduced to amines. In practice, however, nitriles or nitro compounds are usually used because they are the most easily obtained starting materials.

There are several commercial processes for reducing nitro or nitrile groups to amines. Most large volume aromatic and aliphatic amines are made by continuous high-pressure catalytic hydrogenation. Nitro compounds can also be reduced in good yields with iron and hydrochloric acid in the Bèchamp process. Other more specialized methods used for making amines by reduction are also used on occasion.

Catalytic Hydrogenation

In catalytic hydrogenation, a compound is reduced with molecular hydrogen in the presence of a catalyst.

Nitro $R—NO_2 + 3\,H_2 \rightarrow R—NH_2 + H_2O$

Nitrile $R—CN + 2\,H_2 \rightarrow R—CH_2NH_2$

Catalytic hydrogenation is the most efficient method for the large-scale manufacture of many primary aromatic and aliphatic amines. Aromatic amines are usually made by hydrogenating the corresponding nitro compound, whereas the aliphatic amines generally start with the corresponding nitrile. Certain aliphatic amines can be prepared by reduction of corresponding aromatic amines using catalytic hydrogenation.

Bèchamp Process

In the Bèchamp process, nitro compounds are reduced to amines in the presence of iron and an acid. This is the oldest commercial process for preparing amines, and is still used in the dyestuff industry for the production of small volume amines and for the manufacture of iron oxide pigments where aniline is produced as a by-product. The overall reaction in the Bèchamp process is as follows.

$$4\,RNO_2 + 9\,Fe + 4\,H_2O \xrightarrow{FeCl_2} 4\,RNH_2 + 3\,Fe_3O_4$$

Miscellaneous Reductions

The method of reducing aromatic nitro compounds with divalent sulfur is known as the *Zinin reduction.* This reaction is carried out in a basic media using sulfides, polysulfides, or hydrosulfides as the reducing agent. Sodium bisulfite ($NaHSO_3$) is occasionally used to perform simultaneous reduction of a nitro group to an amine and the addition of a sulfonic acid group. Both nitro compounds and nitriles can be reduced electrochemically. One advantage of an electrochemical reduction is the cleanness of the operation, which results in a minimum of by-products. Metal hydrides and amalgams are sometimes the preferred method of reducing various functional groups in the laboratory, especially when the necessary equipment for catalytic hydrogenations is unavailable. However, these reagents are usually too expensive for use on a large commercial scale.

Environmental and Safety Aspects

Amines, nitro compounds, nitriles, and the various solvents and reagents used in the preparation of amines by reduction vary widely in the hazards they may pose. Some of these materials are acutely toxic by ingestion, inhalation, or absorption through the skin. Others are skin irritants or sensitizers. Still others, by chronic exposure, may cause damage to organs such as the liver or may be carcinogenic. Since amines vary so widely in their potential danger, no general rules can govern their safe use in all cases. Regulations governing the safe handling and shipping of amines are given in U.S. Department of Transportation publications. Specific information on the safe handling and hazards associated with a particular amine can be found in the Material Safety Data Sheet for that material.

STEVEN L. SCHILLING
Mobay Corporation

P. H. Groggins, ed., *Unit Processes in Organic Synthesis,* 5th ed., McGraw-Hill Book Co., Inc., New York, 1958, pp. 170–171.

P. N. Rylander, *Hydrogenation Methods,* Academic Press, London, 1985.

PB 77729 (also issued as BIOS Report No. 1144), Report of the British Intelligence Objectives Subcommittee on (1) The Manufacture of Nitration Products of Benzene, Toluene, and Chlorobenzene at Griesheim and Leverkusen, (2) The Manufacture of Aniline and Iron Oxide Pigments at Uerdigen, BIOS Trip Report No. 2526, Sept.–Oct. 1946, pp. 25–32.

Toxic and Hazardous Industrial Chemicals Safety Manual, The International Technical Information Institute, Tokyo, Japan, 1975.

AMINOACETIC ACID. See AMINO ACIDS (SURVEY).

AMINO ACIDS

SURVEY

Amino acids are the main components of proteins. Approximately 20 amino acids are common constituents of proteins and are called protein amino acids, or primary protein amino acids because they are found in proteins as they emerge from the ribosome in the translation process of protein synthesis, or natural amino acids.

Hydroxylated amino acids (eg, 4-hydroxyproline, 5-hydroxylysine) and *N*-methylated amino acids (eg, *N*-methylhistidine) are obtained by the acid hydrolysis of proteins. γ-Carboxyglutamic acid occurs as a component of some sections of protein molecules; it decarboxylates spontaneously to L-glutamate at low pH. These examples are formed upon the nontranslational modification of protein and are often called secondary protein amino acids.

The presence of many nonprotein amino acids has been reported in various living metabolites, such as in antibiotics, some other microbial products, and in nonproteinaceous substances of animals and plants. Plant amino acids and seleno amino acids have been reviewed.

The general formula of an α-amino acid may be written:

$$\mathrm{R{-}{*}CH{-}COOH} \quad (\text{with } \mathrm{NH_2} \text{ on the } {*}\mathrm{C})$$

The asterisk signifies an asymmetric carbon. All of the amino acids, except glycine, have two optically active isomers designated D- or L-. Isoleucine and threonine also have centers of asymmetry at their β-carbon atoms. Protein amino acids are of the L-α-form as illustrated in Table 1.

Amino acids are important components of the elementary nutrients of living organisms. For humans, ten amino acids are essential for existence and must be ingested in food. The nutritional value of proteins is governed by the quantitative and qualitative balance of individual essential amino acids.

The nutritional value of a protein can be improved by the addition of amino acids of low abundance in that protein. Thus the fortification of plant proteins such as wheat, corn, and soybean with L-lysine, DL-methionine, or other essential amino acids (L-tryptophan and L-threonine) is expected to alleviate some food problems. Such fortification has been widespread in the feedstuff of domestic animals.

Proteins are metabolized continuously by all living organisms, and are in dynamic equilibrium in living cells. The role of amino acids in protein biosynthesis has been described. Most of the amino acids absorbed through the digestion of proteins are used to replace body proteins. The remaining portion is metabolized into various bioactive substances such as hormones and purine and pyrimidine nucleotides (the precursors of DNA and RNA), or is consumed as an energy source.

All of the protein amino acids are currently available commercially and their uses are growing. Amino acids and their analogues have their own characteristic effects in flavoring, nutrition, and pharmacology.

In the food industries a number of amino acids have been widely used as flavor enhancers and flavor modifiers (see FLAVORS AND SPICES). For example, monosodium L-glutamate is well-known as a meat flavor-enhancer and an enormous quantity of it is now used in various food applications (see AMINO ACIDS–L-MONODOSIUM GLUTAMATE (MSG)). Protein, hydrolyzed by acid or enzyme to be palatable, has been used for a long time in flavoring agents. The addition of L-glutamate, L-aspartate, glycine, DL-alanine, and other palatable amino acids can improve flavoring by these protein hydrolyzates. In addition, some nucleotides, such as 5′-inosinic acid and 5′-guanylic acid, have a synergistic effect on the meat flavor enhancing effects of L-glutamate and L-aspartate. Tricholomic acid (**1**) and ibotenic acid, (**2**), nonprotein amino acids found in mushrooms, have 4 to 25 times stronger umami

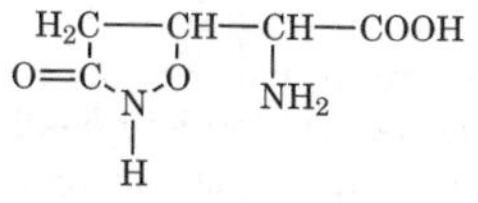

(**1**)

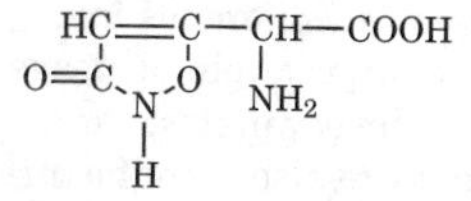

(**2**)

taste than L-glutamic acid. Umami taste, which is typically represented by L-glutamic acid salt (and some 5′-nucleotide salts), makes food more palatable and is recognized as a basic taste, independent of the four other classical basic tastes of sweet, sour, salty, and bitter.

Some peptides have special tastes. L-Aspartyl phenylalanine methyl ester is very sweet and is used as an artificial sweetener (see SWEETENERS). In contrast, some oligopeptides (such as L-ornithinyltaurine·HCl and L-ornithinyl- β-alanine·HCl), and glycine methyl or ethyl ester·HCl have been found to have a very salty taste.

Amino acids are also used in medicine. Amino acid infusions prepared from crystalline amino acids are used as nutritional supplements for patients before and after surgery. Some amino acids and their analogues are used for treatment of major diseases. L-DOPA, L-3-(3,4-dihydroxyphenyl)alanine, is an important drug used in the treatment of Parkinson's disease, and L-glutamine and its derivatives are used for treatment of stomach ulcers. α-Methyl-DOPA is an effective antidepressant (see PSYCHOPHARMACOLOGICAL AGENTS). Some peptides, eg, oxytocin, angiotensin, gastrin, and cerulein, have hormonal effects which have medical utility. The physiological effect of glutathione (L-glutamyl-L-cysteinyl glycine) has been reviewed.

Amino acid polymers like poly(γ-methyl-L-glutamate) have been developed as raw materials for artificial leathers (see LEATHER-LIKE MATERIALS). Derivatives of amino acids are now finding new applications in industry and agriculture.

Physical Properties

Melting Point. Amino acids are solids, even the lower carbon-number amino acids such as glycine and alanine. The melting points of amino acids generally lie between 200 and 300°C. Frequently amino acids decompose before reaching their melting points (Table 2).

Crystalline Structures. The crystal shape of amino acids varies widely, eg, monoclinic prisms in glycine and orthorhombic needles in L-alanine. X-ray crystallographic analyses of 23 amino acids have been described.

Dipole. Every amino acid molecule has two equal electric charges of opposite sign caused by the amino and carboxyl groups on the α-carbon atom.

$$\mathrm{R{-}CH{-}COO^-} \quad (\text{with } \mathrm{NH_3^+} \text{ on CH})$$

The dielectric constants of amino acid solutions are very high. Their ionic dipolar structures confer special vibrational spectra (Raman, ir), as well as characteristic properties (specific volumes, specific heats, electrostriction).

Optical Configuration. With the exception of glycine, all α-amino acids contain at least one asymmetric carbon atom and may be characterized by their ability to rotate light to the right (+) or to the left (−), depending on the solvent and the degree of ionization. Specific rotations are given in Table 2. They are also characterized by the stereochemical configuration of the asymmetric carbon based on the configuration of glyceraldehyde; D,L-notation is popular for amino acids, but *R,S*-notation is a more precise designation of chirality.

Solubility. In all instances there are at least two polar groups acting synergistically on the solubility in water. The solubility of

Table 1. α-Amino Acids

Common name	Abbreviation	Systematic name	Formula	Molecular weight
		Monocarboxylic		
Aliphatic				
glycine	Gly	aminoacetic acid	H_2NCH_2COOH	75.07
alanine	Ala	2-aminopropanoic acid	$CH_3\underset{\substack{\vert\\ NH_2}}{C}HCOOH$	89.09
valine[a]	Val	2-amino-3-methylbutanoic acid	$(CH_3)_2CH\underset{\substack{\vert\\ NH_2}}{C}HCOOH$	117.15
leucine[a]	Leu	2-amino-4-methylpentanoic acid	$(CH_3)_2CHCH_2\underset{\substack{\vert\\ NH_2}}{C}HCOOH$	131.17
isoleucine[a]	Ileu	2-amino-3-methylpentanoic acid	$CH_3CH_2\underset{\substack{\vert\\ CH_3}}{C}H{—}\underset{\substack{\vert\\ NH_2}}{C}HCOOH$	131.17
Aliphatic containing —OH, —S—, —NH—group				
serine	Ser	2-amino-3-hydroxypropanoic acid	$HOCH_2\underset{\substack{\vert\\ NH_2}}{C}HCOOH$	105.09
threonine[a]	Thr	2-amino-3-hydroxybutanoic acid	$CH_3\underset{\substack{\vert\\ OH}}{C}H{—}\underset{\substack{\vert\\ NH_2}}{C}HCOOH$	119.12
cysteine	Cys	2-amino-3-mercaptopropanoic acid	$HSCH_2\underset{\substack{\vert\\ NH_2}}{C}HCOOH$	121.16
cysteine	$(Cys)_2$	3,3′-dithio-bis-(2-aminopropanoic acid)	$\begin{array}{l}SCH_2\underset{\substack{\vert\\ NH_2}}{C}HCOOH\\ \vert\\ SCH_2\underset{\substack{\vert\\ NH_2}}{C}HCOOH\end{array}$	240.30
methionine[a]	Met	2-amino-4-methylthiobutanoic acid	$CH_3SCH_2CH_2\underset{\substack{\vert\\ NH_2}}{C}HCOOH$	149.21
lysine[a]	Lys	2,6-diaminohexanoic acid	$H_2N(CH_2)_4\underset{\substack{\vert\\ NH_2}}{C}HCOOH$	146.19
arginine[b]	Arg	2,-amino-5-guanidopentanoic acid	$HN{=}\underset{\substack{\vert\\ H_2N}}{C}NH(CH_2)_3\underset{\substack{\vert\\ NH_2}}{C}HCOOH$	174.20
Aromatic				
phenylalanine[a]	Phe	2-amino-3-phenylpropanoic acid	$C_6H_5CH_2\underset{\substack{\vert\\ NH_2}}{C}HCOOH$	165.19
tyrosine	Tyr	2-amino-3-(4-hydroxyphenyl)-propanoic acid	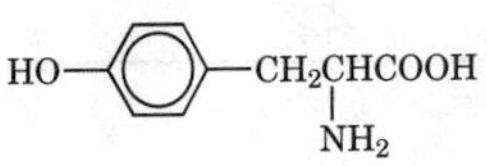	181.19
Heterocyclic				
proline	Pro	2-pyrrolidine-carboxylic acid	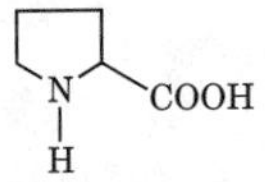	115.13
hydroxyproline	Hypro	4-hydroxy-2-pyrrolidine-carboxylic acid	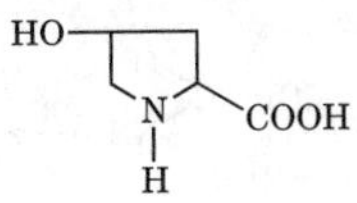	131.13
histidine[b]	His	2-amino-3-imidazole-propanoic acid	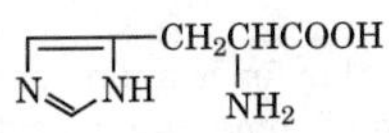	155.16

Table 1. α-Amino Acids[ab] *(continued)*

Common name	Abbreviation	Systematic name	Formula	Molecular weight
		Dicarboxylic		
tryptophan[a]	Trp	2-amino-3-indoyl-propanoic acid	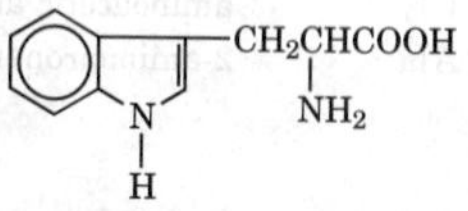	204.22
aspartic acid	Asp	2-aminobutanedioic acid	$HOOCCH_2CHCOOH$ $\vert$ NH_2	133.10
glutamic acid	Glu	2-aminopentanedioic acid	$HOOCCH_2CH_2CHCOOH$ $\vert$ NH_2	147.13
asparagine	Asn	2-amino-3-carbamoyl-propanoic acid	$H_2NCOCH_2CHCOOH$ $\vert$ NH_2	132.12
glutamine	Gln	2-amino-4-carbamoyl-butanoic acid	$H_2NCOCH_2CH_2CHCOOH$ $\vert$ NH_2	146.15

[a] Essential amino acid.
[b] Arginine and histidine are also essential for children.

amino acids having additional polar groups, eg, —OH, —SH, —COOH, $—NH_2$, is even more enhanced.

Dissociation. In aqueous solution, amino acids undergo a pH-dependent dissociation:

$$\underset{\text{cationic form}}{\overset{\text{at pH = 1}}{H_3N^+—\underset{R}{\underset{|}{CH}}—COOH}} \underset{+H^+}{\overset{-H^+}{\rightleftharpoons}}\ (\text{reaction 1}, K_1)\ \underset{\text{ampholyte}}{\overset{\text{at pH = 6}}{H_3N^+—\underset{R}{\underset{|}{CH}}—COO^-}} \underset{+H^+}{\overset{-H^+}{\rightleftharpoons}}\ (\text{reaction 2}, K_2)\ \underset{\text{anionic form}}{\overset{\text{at pH = 11}}{H_2N—\underset{R}{\underset{|}{CH}}—COO^-}}$$

where

$$K_1 = \frac{[H^+][H_3N^+CH(R)COO^-]}{[H_3N^+CH(R)COOH]}$$

$$K_2 = \frac{[H^+][H_2NCH(R)COO^-]}{[H_3N^+CH(R)COO^-]}$$

Chemical Properties

Synthesis of α-Amino Acids

Many methods for chemical synthesis of α-amino acids have been established. Because excellent reviews have been published, well-known reactions are introduced here only by their names and synthetic pathways.

Strecker Synthesis

$$RCHO \xrightarrow[NH_4Cl]{NH_3\text{ or}} \underset{NH_2}{\underset{|}{RCHOH}} \xrightarrow[NaCN]{HCN\text{ or}} \underset{NH_2}{\underset{|}{RCHCN}} \xrightarrow[OH^-]{H^+\text{ or}} \underset{NH_2}{\underset{|}{RCHCOOH}}$$

Bucherer Synthesis

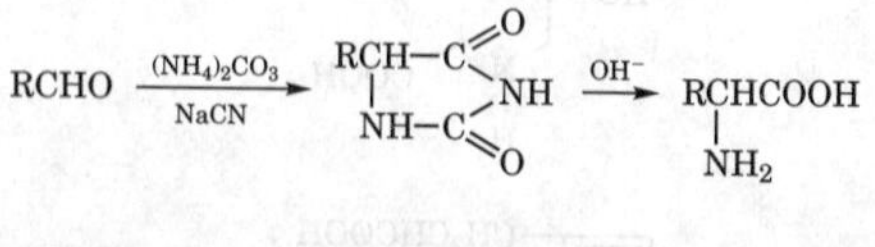

These two methods are popular for α-amino acid synthesis, and used in the industrial production of some amino acids since raw materials are readily available.

Amination of α-Halogeno Carboxylic Acids. *Original Method*

$$R—\underset{X}{\underset{|}{CH}}—COOH \xrightarrow{NH_3} R—\underset{NH_2}{\underset{|}{CH}}—COOH$$

Gabriel's Modification

Table 2. Physical Constants of Amino Acids

Amino acid	Melting point, °C	Density, d_{t1}^{t2}	Specific rotation $[\alpha]_D$	t, °C	c, %	Solvent
Ala L-	297 dec	1.401	+2.8	25	6	H_2O
L-·HCl	314 dec	1.432^{23}	+2.8	25	6	H_2O
	204 dec		+8.5	26	9.3	
D-	314 dec		−13.6	25	1	6 *N* HCl
DL-	264 dec	1.424				
	295 dec	1.424				
Arg L-	244 dec		+12.5	20	3.5	H_2O
L-·HCl	235 dec		+12.0	20	4	
DL-	217–218					
Asn L-·H_2O	234–235	$1.543^{15}{}_4$	−5.42	20	1.3	
D-·H_2O	215		+5.41	20	1.3	
	234.5	$1.543^{15}{}_4$	+5.41	20	1.3	
DL-·H_2O	182–183	$1.4540^{15}{}_4$				
Asp L-	270–271	$1.661^{12.5}$	+25.0	20	1.97	6 *N* HCl
	324 dec	$1.6613^{13}{}_{13}$	+24.6	24	2	6 *N* HCl
D-			−23.0	27	2.30	6 *N* HCl
	269–271	$1.6613^{13}{}_{13}$	−25.5	20		HCl
DL-	338–339	$1.6632^{13}{}_{13}$				
Cys L-			+6.5	25		5 *N* HCl
	240 dec		+9.8	30	1.3	H_2O
L-·HCl	175–178		+5.0	25		5 *N* HCl
(Cys)$_2$ L-	260–261 dec	1.677	−223.4	20	1	1 *N* HCl
D-			+223	20		1 *N* HCl
	247–249		+224	20	1	1 *N* HCl
DL-	260					
Glu L-	247–249 dec	$1.538^{20}{}_4$	+31.4	22.4		6 *N* HCl
	224–225 dec	$1.538^{20}{}_4$	+31.4	22	1	6 *N* HCl
L-·HCl	214 dec		+24.4	22	6	
D-			−30.5	20	1.0	6 *N* HCl
	213 dec	$1.538^{20}{}_4$	−31.7	25		1.7 *N* HCl
DL-	225–227 dec	1.4601^{20}				
	199 dec	$1.4601^{20}{}_4$				
Gly	233 dec	1.1607				
	262 dec	0.828^{17}				
His L-	287 dec		−39.74	20	1.13	
	287 dec		−39.7	20	1.13	H_2O
L-HCl·H_2O	259 dec		+8.0	26	2	3 *N* HCl
D-	287 dec		+40.2	20		H_2O
DL-	285 dec					
Ileu L-	284 dec		+11.29	20	3	
			+40.61	20	4.6	6.1 *N* HCl
	285–286 dec		+12.2	25	3.2	H_2O
			+36.7		4	1 *N* HCl
D-	283–284 dec		−12.2	20	3.2	H_2O
			−40.7		1	5 *N* HCl
DL-	280 dec					
Leu L-	293–295 dec		−10.8	25	2.2	
	293–295	$1.293^{18}{}_4$	−10.42	25	22	H_2O
D-	293		+10.34	20		
DL-	332 dec					
	293–295	$1.293^{18}{}_4$				
Lys L-	224.5 dec		+25.9	23	2	6 *N* HCl
	224–245 dec		+14.6	20	6	H_2O
L-·HCl	263–264		+14.6	25	2	0.6 *N* HCl
L-·2HCl	193		+15.3	20	2	
	201–202		+15.29	20		H_2O
DL-·HCl	260–263					
DL-·2HCl	187–189					
Met L-	280–282 dec		−8.2	25		
	283 dec		−8.2	25	1	H_2O
DL-	281 dec	1.340				
Phe L-	283 dec		−35.1	20	1.94	

Table 2. Physical Constants of Amino Acids *(continued)*

Amino acid	Melting point, °C	Density, d^{t2}_{t1}	Specific rotation $[\alpha]_D$	t, °C	c, %	Solvent
D-	285 dec		+35.0	20	2.04	
DL-	271–273 dec					
Pro L-	220–222 dec		−52.6	20	0.58	0.5 N HCl
	220–222 dec		−80.9	20	1	H_2O
DL-	205 dec					
Hyp L-	274		−76.5		2.5	H_2O
Ser L-	228 dec		−6.83	20	10	H_2O
D-	228 dec		+6.87	20	10	H_2O
DL-	246 dec	1.537				
	246 dec	$1.603^{22.5}$				
Thr L-	255–257 dec		−28.3	26	1.1	
DL-	229–230 dec					
Trp L-	289 dec		−31.5	23	1	
			+2.4	20		0.5 N HCl
			+0.15	20	2.43	0.5 N NaOH
	290–292 dec		−31.5	20	0.5	H_2O
			+6.1	20	11	1 N NaOH
D-	281–282		+33	20		H_2O
DL-	282					
Tyr L-	342–344 dec	1.456	−10.6	22	4	1 N HCl
			−13.2	18	4	3 N NaOH
D-	310–314 dec		+10.3	25	4	1 N HCl
DL-	316 dec					
	340 dec					
Val L-	315	1.230	+22.9	23	0.8	20% HCl
	93–96(?)	1.230	+22.9	23	0.8	20% alc
D-	156–157.5		−29.4	20		20% alc
DL-	298 dec	1.310				

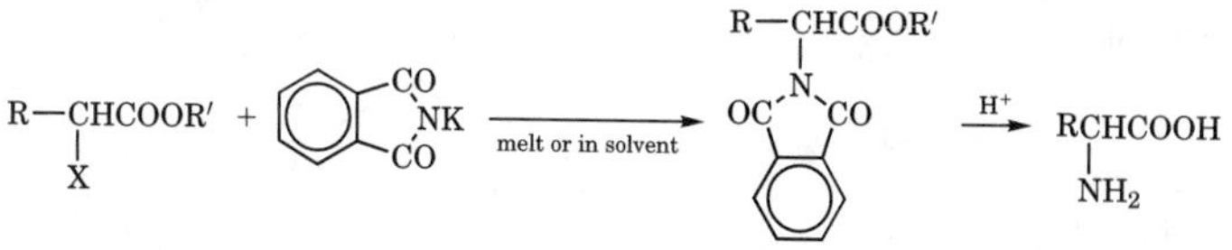

Alkylation of Active Methylene Compounds. *Erlenmeyer Synthesis and Others.* Hydantoin azlactone, diketopiperazine, etc, are readily available, so that these methods are often utilized.

Amination of α-Keto Acids. α-Keto acids are catalytically reduced in the presence of ammonia.

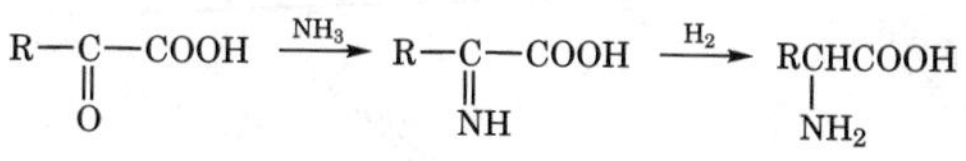

α-Keto acids are readily prepared by hydrolysis of substituted hydantoins, or double carbonylation of benzyl halide in the case of phenylpyruvic acid. Enzymatic amination of α-keto acids has been developed by many research groups.

Reduction of α-Ketoxime

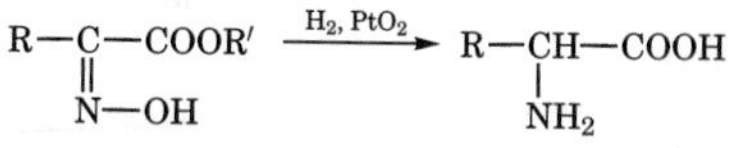

Reduction of α-Nitro Carboxylic Acid

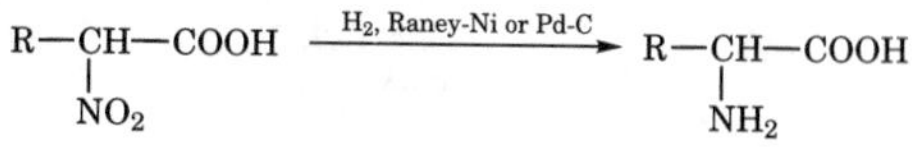

Hofmann Degradation

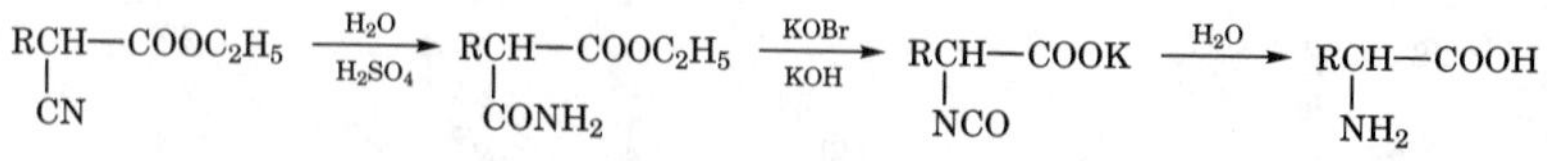

Schmidt Reaction

$$RCH\!-\!COOC_2H_5 + HN_3 \xrightarrow{H_2SO_4} \underset{NHCOCH_3}{RCHCOOC_2H_5} \xrightarrow{H_2O} \underset{NH_2}{RCHCOOH}$$
$$\underset{COCH_3}{|}$$

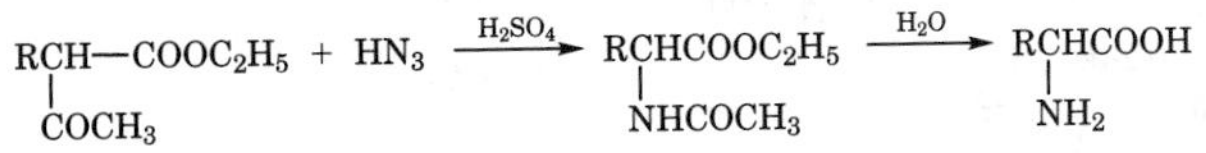

$$RCH(COOH)_2 + HN_3 \xrightarrow{H_2SO_4} \underset{NH_2}{RCHCOOH}$$

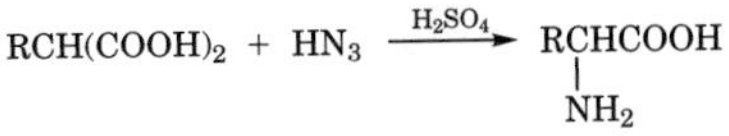

Curtius Degradation

$$\underset{COOC_2H_5}{RCH\!-\!COOK} \xrightarrow{NH_2NH_2} \underset{CONHNH_2}{RCH\!-\!COOK} \xrightarrow{HNO_2} \underset{CON_3}{RCH\!-\!COOK} \xrightarrow[H_2O]{H^+} \underset{NH_2}{RCHCOOH}$$

$$\underset{COOC_2H_5}{RCHCN} \xrightarrow{NH_2NH_2} \underset{CONHNH_2}{RCHCN} \xrightarrow{HNO_2} \underset{CON_3}{RCHCN} \xrightarrow[H_2O]{H^+} \underset{NH_2}{RCHCOOH}$$

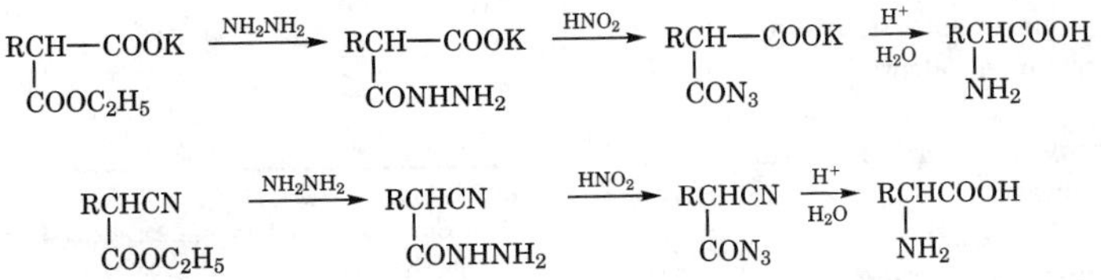

Amine Addition to Double Bond. Production of D,L-aspartic acid from maleic acid ester or fumaric acid ester is a typical example.

$$C_2H_5OOCCH\!=\!CHCOOC_2H_5 \xrightarrow{NH_3} H_2NCOCH_2CH\underset{NH-CO}{\overset{CO-NH}{<}}CHCH_2CONH_2 \xrightarrow{H^+} \underset{NH_2}{HOOCCH_2CHCOOH}$$

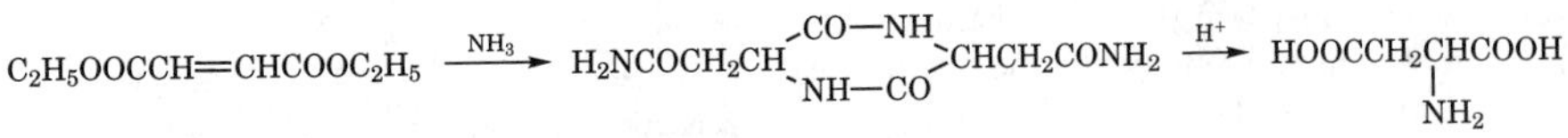

Carbonylation of Aldehyde. This method is noteworthy as an efficient one-step synthesis.

Wakamatsu Reaction

$$R\!-\!CHO + R'CONH_2 + CO \xrightarrow[Co_2(CO)_8]{H_2} \underset{NHCOR'}{R\!-\!CH\!-\!COOH} \longrightarrow \underset{NH_2}{R\!-\!CH\!-\!COOH}$$

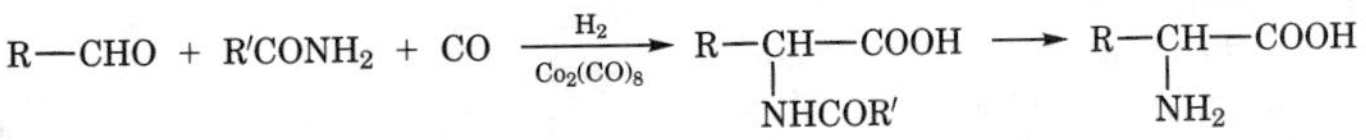

Modified Method

$$\underset{O}{C_6H_5CH\!-\!CH_2} + CH_3CONH_2 + CO \xrightarrow[Co_2(CO)_8,\ Ti(O\text{-}iC_3H_7)_4]{H_2} \underset{NHCOCH_3}{C_6H_5CH_2CHCOOH}$$

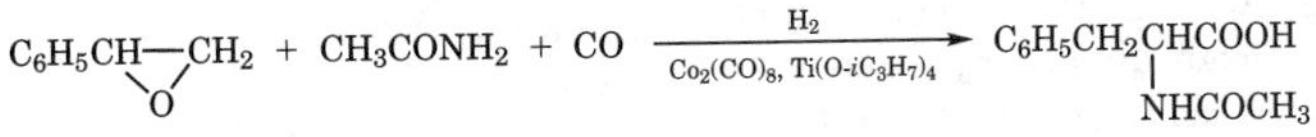

Optical Resolution

In many cases only the racemic mixtures of α-amino acids can be obtained through chemical synthesis. Therefore, optical resolution is indispensable to get the optically active L- or D-forms in the production of expensive or uncommon amino acids. The optical resolution of amino acids can be done in two general ways: physical or chemical methods which apply the stereospecific properties of amino acids, and biological or enzymatic methods which are based on the characteristic behavior of amino acids in living cells or in the presence of enzymes.

Asymmetric Synthesis

Asymmetric synthesis is a method for direct synthesis of optically active amino acids, and finding efficient catalysts is a great target for researchers. Many excellent reviews have been published. Asymmetric syntheses are classified as either enantioselective or diastereoselective reactions. Asymmetric hydrogenation has been applied for practical manufacturing of L-DOPA and L-phenylalanine, but conventional methods have not been exceeded because of the short life of catalysts.

Reactions of α-Amino Acids

α-Amino acids are ampholytic compounds. The chemical reactions of amino acids can be classified according to their carboxyl, amino, and side-chain groups. Most of the reactions have been well known for a long time; the details of these reactions have been reviewed.

Reactions of the amino group include n-acylation, reaction with phosgene, formation of Schiff-bases, the Maillard reaction (nonenzymatic glycation), and substitution reactions.

Reactions of the carboxyl group include esterification, amidation, acid chloride formation, reduction to amino alcohols, and anhydride formation.

Reactions depending on both amino and carboxyl groups include formation of diketopiperazines, formation of hydantoin, Strecker degradation (oxidative deamination), formation of n-carboxy-α-amino acid anhydride (NCA), and ninhydrin-color reaction.

Other reactions include salt formation and metal chelation, synthesis of peptide, and induction of asymmetry by amino acids.

Biology of Amino Acids

Nutrition. Protein amino acids, which are not synthesized by the body and should be supplied as nutrients to maintain life, are called essential amino acids. For humans, L-arginine, L-histidine, L-isoleucine, L-leucine, L-lysine, L-methionine, L-phenylalanine, L-valine, L-threonine, and L-tryptophan are essential amino acids. However, in adults, L-arginine and L-histidine are somewhat synthesized in cells. For histidine, there is evidence that it is dietetically essential for the maintenance of nitrogen balance. On the other hand, those amino acids which are synthesized in apparently adequate amounts are nonessential amino acids: L-alanine, L-asparagine, L-aspartic acid, L-cysteine, L-glutamic acid, L-glutamine, glycine, L-proline, L-serine, and L-tyrosine. Of these, L-tyrosine and L-cysteine are essential for children. Recent advances in nutritional studies of amino acids have led to development of amino acid transfusion.

Biosynthesis of Protein. The human body is maintained by a continuous equilibrium between the biosynthesis of proteins and their degradative metabolism, where the nitrogen lost as urea (about 85% of total excreted nitrogen) and other nitrogen compounds is about 12 g/d under ordinary conditions. The details of protein biosynthesis in living cells have been described (see also PROTEINS).

Toxicity of α-Amino Acid. LD_{50} values of α-amino acids are listed in Table 3. L-Lysine and L-arginine are mutually antagonistic.

Metabolism of Amino Acids. The amino acids are metabolized, principally in the liver, to a variety of physiologically important metabolites, eg, creatine (creatinine), purines, pyrimidines, hormones, lipids, amino sugars, urea, ammonia, carbon dioxide, and energy sources.

As Neurotransmitters. Several amino acids serve as specialized neurotransmitters in both vertebrate and invertebrate nervous systems. These amino acids can be classified as inhibitory transmitters, eg, γ-aminobutyric acid (GABA) and glycine, and excitatory amino acids, eg, L-glutamic acid and L-aspartic acid.

Modification of Amino Acid in Protein Molecules. Protein kinases, whose activities are regulated by secondary messengers, such as cyclic nucleotide and Ca^{2+}, modify physiologically important proteins by phosphorylating the hydroxy moiety of serine, threonine, and tyrosine in protein molecules. Consequently, various cellular functions, cell growth, and cell differentiation are seriously affected.

Table 3. Toxicity[a] of Amino Acids

Amino acid	LD_{50}[b]	
	Oral	Intraperitoneal
L-ArgHCl	12 g	
L-Cys	5580 mg	1620 mg
L-CysHCl		1250 mg[c]
L-(Cys)$_2$	25 g	
L-His	7930 mg	
L-Ileu		6822 mg
L-Leu		5379 mg
D-Leu		6429 mg
L-Lys·HCl	10 g	4019 mg
L-Met	36 g	4328 mg
L-Phe		5287 mg
D-Phe		5452 mg
DL-Thr		3098 mg
L-Trp		1634 mg
D-Trp		4289 mg
L-Val		5390 mg
D-Val		6093 mg

[a] Rat, unless otherwise noted.
[b] N. I. Sax and R. J. Lewis, *Dangerous Properties of Industrial Materials,* 7th ed., Van Nostrand Reinhold, New York, 1989.
[c] Mouse

Table 4. Production[a] of Amino Acids

Amino acid	Method[a]	Production, t/yr	
		Japan	World
DL-alanine	C	1,500	
L-alanine	Enz	150	
L-arginine	Ext, F	700	1,000
L-aspartic acid	Enz	2,000	4,000
L-asparagine	Ext	30	
L-citrulline	Enz	50	
L-cysteine and cystine	Ext, Enz	300	1,000
glycine	C	3,500	6,000
L-glutamic acid	F	80,000	340,000
L-glutamine	F	850	
L-histidine	F	250	
L-isoleucine	F	200	
L-leucine	Ext, F	200	
L-lysine	F	30,000	70,000
DL-methionine	C	30,000	250,000
L-methionine	Enz	150	
L-ornithine	F	70	
L-phenylalanine	C, F, Enz	1,500	3,000
L-proline	F	150	
L-serine	F, Enz	60	
DL-serine	C		
L-threonine	C, F	200	
L-tryptophan	C, F, Enz	250	
L-tyrosine	Ext, F	60	
L-valine	C, F	200	

[a] S. Kinoshita, *Proc. 4th Eur. Cong. on Biotechnol.* Amsterdam, the Netherlands, 1987. C, chemical synthesis; Enz, enzymatic synthesis; Ext, extraction; F, fermentation.

Analysis. Methods have been developed for analysis or determination of free amino acids in blood, food, and feedstocks. In proteins, the first step is hydrolysis, then separation if necessary, and finally, analysis of the amino acid mixture.

Manufacture and Processing

Since the discovery of amino acids in animal and plant proteins in the nineteenth century, most amino acids have been produced by extraction from protein hydrolyzates. However, there are many problems in the efficient isolation of the desired amino acid in the pure form.

DL-Alanine is the first amino acid which was synthesized chemically. Glycine and DL-methionine have also been supplied by this method. However, amino acids formed by the chemical method are racemic, and it is necessary to resolve the mixture to get the L- or D-form amino acid which is usually demanded.

In the 1950s, a group of coryneform bacteria which accumulate a large amount of L-glutamic acid in the culture medium were isolated. The use of mutant derivatives of these bacteria offered a new fermentation process for the production of many other kinds of amino acids. The amino acids which are produced by this method are mostly of the L- form, and the desired amino acid is singly accumulated. Therefore, it is very easy to isolate it from the culture broth. Rapid development of fermentative production and enzymatic production have contributed to the lower costs of many protein amino acids and to their availability in many fields as economical raw materials.

Economic Aspects

An estimation made in 1987 of the amount of amino acid production and the production methods used are shown in Table 4.

Uses

Amino acids are used in feeds, food, parenteral and enteral nutrition, medicine, cosmetics, and raw materials for the chemical industry.

KAZUMI ARAKI
University of East Asia
TOSHITSUGU OZEKI
Kyowa Hakko Kogyo Company

T. Kaneko and co-eds., *Synthetic Production and Utilization of Amino Acids,* John Wiley & Sons, Inc., New York 1974.

G. C. Barrett, ed., *Chemistry and Biochemistry of the Amino Acids,* Chapman and Hall, London, 1985.

K. Aida and co-eds, *Biotechnology of Amino Acid Production,* Elsevier, Amsterdam, the Netherlands, 1986.

A. Yoshida and co-eds., *Nutrition: Proteins and Amino Acids,* Japan Science Society Press, Tokyo, 1990.

L-MONOSODIUM GLUTAMATE (MSG)

Monosodium glutamate (MSG), more specifically monosodium L-glutamate, is used in large quantities as a flavor enhancer throughout the world. The world capacity for production of MSG continues to increase. The demand for MSG is expected to increase in developing countries as its use in commercially prepared packaged foods, ready-made soups, and as a table seasoning increase in Western and Asian countries.

Properties

Monosodium L-glutamate, $C_5H_8NO_4Na \cdot H_2O$ (mol wt 187.13), crystallizes from aqueous solution at room temperature as rhombic prisms. Its structure, as determined by x-ray crystallography, indicates that the sodium ions are coordinated octahedrally by four (3α and 1γ) carboxyl oxygen atoms and two water molecules as follows:

crystal system: orthorhombic
space group: $P2_12_12_1$
$z = 8$(2 formula units in the asymmetric unit)
$a = 1.5267(9)$ nm, $b = 1.7937(9)$ nm, $c = 0.5562(4)$ nm
$V = 1.520\ \text{nm}^3, D_x = 1.635\ \text{g/cm}^3, D_m = 1.63\ \text{g/cm}^3$

Commercially preferred crystals for use as flavor enhancement are obtained by crystallization in the presence of amino acids such as alanine.

The solubility of MSG may be expressed by the following equation:

$$S = 35.30 + 0.098t + 0.0012t^2$$

where S is % of anhydrous monosodium glutamate in a saturated solution at t°C and the remainder at equilibrium is the monohydrate.

Crystalline MSG is less hygroscopic than sodium chloride.

The α-carbon of glutamic acid is chiral. A convenient and effective means to determine the chemical purity of MSG is measurement of its specific rotation.

L-Glutamic acid does not racemize in neutral solution, even at 100°C. Deviation of pH from neutral to greater than 8.5 results in thermal racemization with loss of taste characteristics.

L-Glutamic acid is split into α-ketoglutaric acid (2-oxo-pentanedioic acid) and ammonia by glutamate dehydrogenase. By the reverse reaction, L-glutamic acid is synthesized from α-ketoglutaric acid, a component of the tricarboxylic acid (TCA) cycle of glycolysis.

Production

Fermentation Process. Glutamic acid-producing microorganisms are distributed widely throughout the natural environment. They are classified taxonomically as the *Micrococcus*, *Brevibacterium*, *Corynebacterium*, and *Aerobacter* sp. and *Microbacterium* genera. Most of these microorganisms are gram-positive, nonspore-forming, nonmotile, and require biotin for growth. They all have intense glutamate dehydrogenase activity and oxidative degradability, to both L-glutamic acid and α-ketoglutaric acid.

The carbon sources for biosynthesis of glutamic acid include acetic acid and the commonly used carbohydrates. For industrial production, molasses and starch hydrolyzate are generally used at present.

Uses

Monosodium L-glutamate elicits a unique taste, known as umami, which is different from the four basic tastes of sweet, salty, sour, and bitter.

L-Glutamic acid is used as a neutralizing agent for basic compounds, eg, arginine glutamate is employed as a pharmaceutical and raw material for cosmetics. Glutamic acid hydrochloride is proposed as an acidifying agent in the stomach. As an amphoteric electrolyte it may be applied as a chelating agent or a builder for detergents. Sodium pyroglutamate, obtained by dehydration of L-glutamic acid with heating and then neutralization by sodium oxide, is used as a component of a natural moisturizing factor for human skin because of its hygroscopic property.

Safety

The acute toxicity of MSG is low.

The main use of MSG is as a food ingredient, and so its safety when used in the diet is the most important aspect of its safety for use. Both short-term and chronic toxicity studies on MSG in the diet of several species at doses of up to 4% (approximately 6–8 g/kg body weight per day) in the diet showed no specific toxic effects, and no evidence for carcinogenicity or mutagenicity.

TETSUYA KAWAKITA
Ajinomoto Co., Inc.

C. Sano, N. Nagashima, T. Kawakita, and Y. Iitaka, *Anal. Sci.* **5,** 121 (1989).

S. Kinoshita, S. Umada, and M. Shimono, *J. Gen. Appl. Microbiol.* **3,** 193 (1957).

Y. Kawamura and M. R. Kare, *Umami: A Basic Taste,* Marcel Dekker, New York, 1987.

Commission of the EC Reports of the Scientific Committee for Food **25**, 16 (1990).

AMINOLYSIS. See AMMONIA.

AMINOPHENOLS

Aminophenols and their derivatives are of commercial importance, both in their own right and as intermediates in the photographic, pharmaceutical, and chemical dye industries. They are amphoteric and can behave either as weak acids or weak bases, but the basic character usually predominates. 3-Aminophenol (**2**) is fairly stable in air unlike 2-aminophenol (**1**) and 4-aminophenol (**3**) which easily undergo oxidation to colored products. The former are generally converted to their acid salts, whereas 4-aminophenol is usually formulated with low concentrations of antioxidants which act as inhibitors against undesired oxidation.

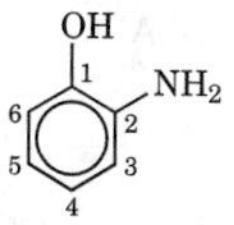

(**1**)

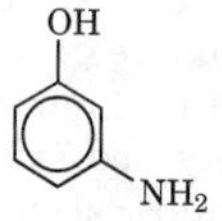

(**2**)

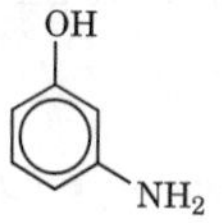

(**2**)

Physical Properties

The simple aminophenols exist in three isomeric forms depending on the relative positions of the amino and hydroxyl groups around the benzene ring. At room temperature they are solid crystalline compounds. In the past the commercial-grade materials were usually impure and colored because of contamination with oxidation products, but now virtually colorless, high purity commercial grades are available. General properties are listed in Table 1.

2-Aminophenol. This compound forms white orthorhombic, bipyramidal needles when crystallized from water or benzene, which readily become yellow-brown on exposure to air and light.

3-Aminophenol. This is the most stable of the isomers under atmospheric conditions. It forms white prisms when crystallized from water or toluene.

4-Aminophenol. This compound forms white plates when crystallized from water. The base is difficult to maintain in the free state and deteriorates rapidly under the influence of air to pink-purple oxidation products.

Chemical Properties

The chemical properties and reactions of the aminophenols and their derivatives are to be found in detail in many standard chemical texts. The acidity of the hydroxyl function is depressed by the presence of an amino group on the benzene ring; this phenomenon is most pronounced with 4-aminophenol. The amino group behaves as a weak base, giving salts with both mineral and organic acids. The aminophenols are true ampholytes, with no zwitterion structure; hence they exist either as neutral molecules (**4**), or as ammonium cations (**5**), or phenolate ions (**6**), depending on the pH value of the solution.

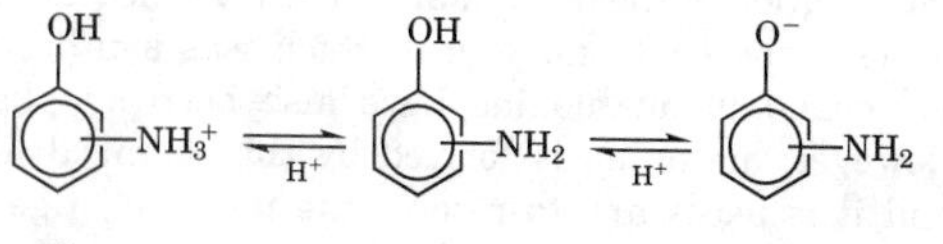

(**5**) (**4**) (**6**)

The aminophenols are chemically reactive, undergoing reactions involving both the aromatic amino group and the phenolic hydroxyl moiety, as well as substitution on the benzene ring. Oxidation leads to the formation of highly colored polymeric quinoid structures. 2-Aminophenol undergoes a variety of cyclization reactions. Important reactions include alkylation, acylation, diazonium salt formation, cyclization reactions, condensation reactions, and reactions of the benzene ring.

Manufacture and Processing

Aminophenols are either made by reduction of nitrophenols or by substitution. Reduction is accomplished with iron or hydrogen in the presence of a catalyst. Catalytic reduction is the method of choice for the production of 2- and 3-aminophenol (see AMINES BY REDUCTION). Electrolytic reduction is also under industrial consideration and substitution reactions provide the major source of 3-aminophenol.

Purification

Contaminants and by-products which are usually present in 2- and 4-aminophenol made by catalytic reduction can be reduced or even removed completely by a variety of procedures. These include treatment with 2-propanol, with aliphatic, cycloaliphatic, or aromatic ketones, with aromatic amines, with toluene or low mass alkyl acetates, or with phosphoric acid, hydroxyacetic acid, hydroxypropionic acid, or citric acid. In addition, purity may be enhanced by extraction with methylene chloride, chloroform, or nitrobenzene. Another method employed is the treatment of aqueous solutions of aminophenols with activated carbon.

Economic Aspects

Production figures for the aminophenols are scarce, the compounds usually being classified along with many other aniline derivatives. Most production of the technical grade materials (95% purity) occurs on-site, as they are chiefly used as intermediate reactants in continuous chemical syntheses. World production of the fine chemicals (99% purity) is probably no more than a few hundred metric tons yearly.

Storage

Under atmospheric conditions, 3-aminophenol is the most stable of the three isomers. Both 2- and 4-aminophenol are unstable; they darken on exposure to air and light and should be stored in brown glass containers, preferably in an atmosphere of nitrogen. The use of activated iron oxide in a separate cellophane bag inside the storage container, or the addition of stannous chloride or sodium bisulfite inhibits the discoloration of aminophenols. The salts, especially the hydrochlorides, are more resistant to oxidation and should be used where possible.

Health and Safety Factors

In general, aminophenols are irritants. Their toxic hazard rating is slight to moderate and their acute oral toxicities in the rat (LD_{50}) are quoted as 1.3 g/kg, 1.0 g/kg, and 0.375 g/kg body weight for the 2-, 3-, and 4-isomer, respectively.

4-Aminophenol is a selective nephrotoxic agent and interrupts proximal tubular function.

Table 1. General Properties of Aminophenols

Property	2-Aminophenol	3-Aminophenol	4-Aminophenol
alternative names	2-hydroxyaniline 2-amino-1-hydroxybenzene	3-hydroxyaniline 3-amino-1-hydroxybenzene	4-hydroxyaniline 4-hydroxy-1-aminobenzene
molecular formula	C_6H_7NO	C_6H_7NO	C_6H_7NO
molecular weight	109.13	109.13	109.13
melting point, °C	174	122–123	189–190[a]
boiling point, °C			
1.47 kPa	153[b]	164	174
101.3 kPa			284
ΔH_f, kJ/mol[c]	−191.0 ± 0.9	−194.1 ± 1.0	−190.6 ± 0.9[d]

[a] Decomposes.
[b] Sublimes. To convert kPa to mm Hg, multiply by 7.5.
[c] In the crystalline state. To convert kJ to kcal, divide by 4.184.
[d] −179.1 is also noted.

Teratogenic effects have been noted with 2- and 4-aminophenol in the hamster, but 3-aminophenol was without effect in the hamster and rat.

Obviously, care should be taken in handling these compounds, with the wearing of chemical-resistant gloves and safety goggles; prolonged exposure should be avoided.

The addition of slaked lime and the initiation of polymerization reactions with H_2O_2 and ferric or stannous salts are techniques employed to remove aminophenols from wastewaters.

Uses

The aminophenols are versatile intermediates and their principal use is as synthesis precursors; their products are represented among virtually every class of stain and dye.

Derivatives

The derivatives of the aminophenols have important uses in both the photographic and pharmaceutical industries. They are also extensively employed as precursors and intermediates in the synthesis of more complicated molecules, especially those used in the staining and dye industry. All of the major classes of dyes have representatives that incorporate substituted aminophenols; those compounds produced commercially as dye intermediates have been reviewed. Details of the more commonly encountered derivatives of the aminophenols can be found in standard organic chemistry texts.

STEPHEN MITCHELL
University of London
ROSEMARY WARING
University of Bermingham, England

A. R. Forester and J. L. Wardell in S. Coffey, ed., *Rodd's Chemistry of Carbon Compounds,* 2nd ed., Vol. 3A, Elsevier Publishing Co., Amsterdam, The Netherlands, 1971, Chapt. 4, pp. 352–363.

Colour Index, 3rd ed., The Society of Dyers and Colorists, Bradford, England, Vol. 4, 1971, pp. 4001–4863 and Vol. 6, 1975, pp. 6391–6410.

Beilstein's Handbuch der Organischen Chemie, Julius Springer, Berlin, 1918, Section 13, pp. 354–549 and Section 13(2), pp. 164–308.

S. Coffey, ed., *Rodd's Chemistry of Carbon Compounds,* 2nd ed., Vol. 3A, Elsevier Publishing Co., Amsterdam, The Netherlands, 1971, pp. 352–363.

AMINO RESINS AND PLASTICS

Amino resins are thermosetting polymers made by combining an aldehyde with a compound containing an amino (—NH_2) group. Urea–formaldehyde (U/F) accounts for over 80% of amino resins; melamine–formaldehyde accounts for most of the rest. Other aldehydes and other amino compounds are used to a very minor extent.

The principal attractions of amino resins and plastics are water solubility before curing, which allows easy application to and with many other materials; colorlessness, which allows unlimited colorability with dyes and pigments; excellent solvent resistance in the cured state; outstanding hardness and abrasion resistance; and good heat resistance. Limitations of these materials include release of formaldehyde during cure and, in some cases, such as in foamed insulation, after cure, and poor outdoor weatherability for urea moldings. Repeated cycling of wet and dry conditions causes surface cracks. Melamine moldings have relatively good outdoor weatherability.

Amino resins are manufactured throughout the industrialized world to provide a wide variety of useful products. Adhesives (qv), representing the largest single market, are used to make plywood, chipboard, and sawdust board. Other types are used to make laminated wood beams, parquet flooring, and for furniture assembly (see WOOD-BASED COMPOSITES AND LAMINATES). Some amino resins are used as additives to modify the properties of other materials. Amino resins are also often used for the cure of other resins such as alkyds and reactive acrylic polymers.

Aminoplasts and other thermosetting plastics are molded by an automatic injection-molding process similar to that used for thermoplastics, but with an important difference. Instead of being plasticized in a hot cylinder and then injected into a much cooler mold cavity, the thermosets are plasticized in a warm cylinder and then injected into a hot mold cavity where the chemical reaction of cure sets the resin to the solid state. The process is best applied to relatively small moldings. Melamine plastic dinnerware is still molded by standard compression-molding techniques. The great advantage of injection molding is that it reduces costs by eliminating manual labor, thereby placing the amino resins in a better position to compete with thermoplastics (see POLYMER BLENDS).

The future for amino resins and plastics seems secure because they can provide qualities that are not easily obtained in other ways. New developments will probably be in the areas of more highly specialized materials for treating textiles, paper, etc, and for use with other resins in the formulation of surface coatings, where a small amount of an amino resin can significantly increase the value of a more basic material. Additionally, since amino resins contain a large proportion of nitrogen, a widely abundant element, they may be in a better position to compete with other plastics as raw materials based on carbon compounds become more costly.

Raw Materials

Most amino resins are based on the reaction of formaldehyde with urea or melamine.

Chemistry of Resin Formation

The first step in the formation of resins and plastics from formaldehyde and amino compounds is the addition of formaldehyde to introduce the hydroxymethyl group, known as methylolation or hydroxymethylation:

$$R{-}NH_2 + HCHO \rightarrow R{-}NH{-}CH_2OH$$

The second Step is a condensation reaction that involves the linking together of monomer units with the liberation of water to form a dimer, a polymer chain, or a vast network. This is usually referred to as methylene bridge formation, polymerization, resinification, or simply cure, and is illustrated in the following equation:

$$RNH{-}CH_2OH + H_2NR \rightarrow RNH{-}CH_2{-}NHR + H_2O$$

Success in making and using amino resins largely depends on the precise control of these two chemical reactions. Consequently, these reactions have been much studied.

Manufacture

Precise control of the course, speed, and extent of the reaction is essential for successful manufacture. Important factors are mole ratio of reactants, catalyst (pH of reaction mixture), and reaction time and temperature. Amino resins are usually made by a batch process.

Regulatory Concerns

Both urea– and melamine–formaldehyde resins are of low toxicity.

Melamine–formaldehyde resins may be used in paper which contacts aqueous and fatty foods according of 21 CFR 121.181.30. However, because a lower PEL has been established by OSHA, some mills are looking for alternatives.

Economic Aspects

Japan produces more amino resin than any other country; the United States is next, with the former USSR, France, the United Kingdom, and Germany following.

LAURENCE L. WILLIAMS
American Cyanamid Company

J. F. Walker, *Formaldehyde, American Chemical Society Monograph, No. 159*, 3rd ed., Reinhold Publishing Corp., New York, 1964. U.F. Concentrate-85, Technical Bulletin, Allied Chemical Corp., New York, 1985.

W. Lindlaw, *The Preparation of Butylated Urea–Formaldehyde and Butylated Melamine Formaldehyde Resins Using Celanese Formcel and Celanese Paraformaldehyde,* Technical Bulletin, Celanese Chemical Co., New York, Table XIIA. *Technical Bulletin S-23-8,* 1967, *Supplement to Technical Bulletin S-23-8,* 1968, Celanese Chemical Co., Example VIII.

AMMINES. See COORDINATION COMPOUNDS.

AMMONIA

Ammonia, NH_3, a colorless alkaline gas, is lighter than air and possesses a unique, penetrating odor.

The synthesis of ammonia directly from hydrogen (qv) and nitrogen (qv) on a commercial scale was pioneered by Haber and Bosch in 1913, for which they were awarded Nobel prizes. Further developments in economical, large-scale ammonia production for fertilizers have made a significant impact on increases in the world's food supply.

Physical Properties

Table 1 lists the important physical properties of ammonia. The flammable limits of ammonia in air are 16 to 25% by volume; in oxygen the range is 15 to 79%. Such mixtures can explode, although ammonia–air mixtures are quite difficult to ignite. The ignition temperature is about 650°C.

Table 1. Physical Properties of Anhydrous Ammonia

Property	Value
molecular weight	17.03
boiling point, °C	−33.35
freezing point, °C	−77.7
critical temp, °C	133.0
critical pressure, kPa[a]	11,425
specific heat, J/(kg·K)[b]	
0°C	2097.2
100°C	2226.2
200°C	2105.6
heat of formation of gas, ΔH_f, kJ/mol[b]	
0 K	−39, 222
298 K	−46, 222
solubility in water, wt %	
0°C	42.8
40°C	23.4
specific gravity	
0°C	0.639
40°C	0.580

[a] To convert kPa to psi, multiply by 0.145.
[b] To convert J to cal, divide by 4.184.

Ammonia is readily absorbed in water to make ammonia liquor. Additional thermodynamic properties may be found in the literature. Considerable heat is evolved during the solution of ammonia in water: approximately 2180 kJ (520 kcal) of heat is evolved upon the dissolution of 1 kg of ammonia gas.

Ammonia is an excellent solvent for salts, and has an exceptional capacity to ionize electrolytes. Many organic compounds such as amines (qv), nitro compounds, and aromatic sulfonic acids, also dissolve in liquid ammonia. Ammonia is superior to water in solvating organic compounds such as benzene (qv), carbon tetrachloride, and hexane.

Chemical Properties

Ammonia is comparatively stable at ordinary temperatures, but decomposes into hydrogen and nitrogen at elevated temperatures. The rate of decomposition is greatly affected by the nature of the surfaces with which the gas comes into contact: glass is very inactive; porcelain and pumice have a distinct accelerating effect; and metals such as iron, nickel, osmium, zinc, and uranium have even more of an effect.

Ammonia reacts readily with a large variety of substances (see AMMONIUM COMPOUNDS; AMINES BY REDUCTION; AMINES). Oxidation at a high temperature is one of the more important reactions, giving nitrogen and water.

Of major industrial importance is the reaction of ammonia and carbon dioxide, giving ammonium carbamate, $CH_6N_2O_2$,

$$2\ NH_3 + CO_2 \rightarrow NH_2CO_2NH_4$$

which then decomposes to urea (qv) and water:

$$NH_2CO_2NH_4 \rightarrow NH_2CONH_2 + H_2O$$

Manufacture

Thermodynamics and Kinetics. Ammonia is synthesized by the reversible reaction of hydrogen and nitrogen.

$$N_2 + 3\ H_2 \leftrightarrows 2\ NH_3$$

Essential for synthesis considerations is the ability to determine the amount of ammonia present in an equilibrium mixture at various temperatures and pressures. Reliable data on equilibrium mixtures for pressures ranging from 1,000 to 101,000 kPa (10–1,000 atm) were developed early on and resulted in the determination of the reaction equilibrium constant. Experimental data indicate that K_p is dependent not only on temperature and pressure, but also on the ratio of hydrogen and nitrogen present.

Future Sources of Synthesis Gas. The energy-intensive nature of ammonia production and the worldwide energy crisis in the 1970s led to the proposal of new concepts for synthesis gas generation that do not require hydrocarbon feedstock. For example, the use of waste heat from nuclear installations has been proposed for replacing the fuel required for synthesis gas generation, as has the use of thermal energy from the ocean (see THERMAL POLLUTION).

In the 1980s, however, the prices of oil and natural gas reversed their upward trends. Natural gas discoveries, both on-shore and off-shore, have considerably increased the world's energy supply, and oil discoveries, many with associated gas, contributed more feedstock potential for ammonia production.

Based on these developments, the foreseeable future sources of ammonia synthesis gas are expected to be mainly from steam-reforming of natural gas, supplemented by associated gas from oil production and hydrogen-rich off-gases (especially from methanol plants). At the start of the 1990s, almost 70% of the world's ammonia production was based on steam-reforming of natural gas.

Environmental Considerations

Ammonia production *per se* is relatively clean compared to other chemical process industries, and presents no unique environmental problems. Synthesis gas generation is the principal area requiring environmental controls and the nature of the controls depends on the feedstock and method of processing.

Coal feedstocks present the most serious environmental problems. Particulate emissions from the coal handling and processing facilities must be controlled (see AIR POLLUTION; EXHAUST CONTROL, INDUSTRIAL).

Partial oxidation of heavy-liquid hydrocarbons requires somewhat simpler environmental controls, relative to those required for coal. The principal source of particulates is carbon, or soot, formed by the high temperature of the oxidation step.

Reforming of natural gas or naphtha, respectively, constitutes the cleanest synthesis gas generation operations. Most natural gases contain sufficiently low levels of sulfur that it is possible to remove it in a simple fixed-bed adsorption system. Similarly, organic sulfur compounds in naphtha are hydrotreated over a cobalt–molybdenum catalyst and stripped as hydrogen sulfide. Process condensate from reforming operations is commonly treated by steam stripping.

All fired equipment, whether it be a process furnace or a utility boiler, is also subject to regulation, usually in the form of sulfur and nitrous oxide limitations.

Storage, Shipping, and Handling

Storage. Anhydrous ammonia is ordinarily stored in refrigerated tanks at the plant site at −33.3 °C and atmospheric pressure.

Shipping. Distribution of anhydrous ammonia in the United States is facilitated by pipeline, where three companies serve 11 states having lines almost 4800 km in total length; by water, where over 4800 km of river-barge transport capability exists; by rail, where an extensive network in the continental United States has tie-ins to Canada and Mexico; and by truck, used mainly for interstate or local delivery.

Handling. Gasket material most suitable for anhydrous ammonia services is hard finished, rubber-frictioned asbestos (qv) sheet. Magnetic or rotary gauges are preferred to gauge glasses. Pressure-reducing valves should be of steel construction, designed for minimum and maximum operation conditions.

Economic Aspects

Capacity, Production, and Consumption. Ammonia production has worldwide significance: about 85% of the ammonia produced is used for nitrogen fertilizers. As the primary source of fertilizer nitrogen, it is key to solving world food production requirements. The remaining 15% goes into various industrial products such as fibers, animal feeds, explosives, refrigerant, etc.

Health and Safety

Ammonia is a strong local irritant which also has a corrosive effect on the eyes and the membranes of the pulmonary system. Current OSHA standards specify the threshold limit value (TLV) 8-h exposure to ammonia as 50 ppm (35 mg/m^3). However, the ACGIH recommends a TLV of 25 ppm. Respiratory protection should be provided for workers exposed to ammonia. Protective clothing such as rubber aprons, boots, gloves, and goggles should be worn when handling ammonia.

By-Product Ammonia

The recovery process depends on the scrubbing medium employed, sulfuric acid, milk of lime, and phosphoric acid, respectively. Ammonium sulfate recovery by the so-called semidirect process is the most widely employed.

Chevron's WWT (wastewater treatment) process treats refinery sour water for reuse, producing ammonia and hydrogen sulfide as by-products.

Nonconventional Sources of Ammonia

The search for a high yield alternative energy route to ammonia, in an effort to meet fertilizer demands and conserve natural gas reserve, is a continuing one. Energy sources such as photochemical, microwave, magnetohydrodynamic, and chemonuclear are being explored in the laboratory for fixing nitrogen as ammonia.

T. A. CZUPPON
S. A. KNEZ
J. M. ROVNER
M. W. Kellogg Company

A. V. Slack and G. R. James, eds., *Ammonia,* Vols. 1–4, Marcel Dekker, New York, 1973.

M. V. Twigg, ed., *Catalyst Handbook,* Imperial Chemical Industries, Ltd., Wolfe Publishing, Frome, U.K., 1989.

AMMONIUM COMPOUNDS

There are a considerable number of stable crystalline salts of the ammonium ion; NH_4^+. Several are of commercial importance because of large-scale consumption in fertilizer and industrial markets. The ammonium ion is about the same size as the potassium and rubidium ions, so these salts are often isomorphous and have similar solubility in water. Compounds in which the ammonium ion is combined with a large, uninegative anion are usually the most stable. Ammonium salts containing a small, highly charged anion generally dissociate easily into ammonia (qv) and the free acid.

Ammonium Acetates

Both normal or neutral ammonium acetate, $NH_4C_2H_3O_2$, and the acid salt are known. The normal salt results from exact neutralization of acetic acid using ammonia; the acid salt is composed of the neutral salt and acetic acid.

The normal salt, CH_3COONH_4, is a white, deliquescent, crystalline solid, formula wt 77.08, having a specific gravity of 1.073. It is quite soluble in water or ethanol: 148 g dissolve in 100 g of water at 4°C.

Ammonium acetate has limited commercial use. It serves as an analytical reagent, and in the production of foam rubber and vinyl plastics; it is also used as a diaphoretic and diuretic in pharmaceutical applications.

Ammonium Carbonates

Ammonium Bicarbonate. Ammonium bicarbonate, also known as ammonium hydrogen carbonate or ammonium acid carbonate, is easily formed. However, it decomposes below its melting point, dissociating into ammonia, carbon dioxide, and water.

Ammonium bicarbonate is produced as both food and standard grade, and the available products are normally very pure.

Ammonium Carbonate. Normal ammonium carbonate, mp 43°C, formula wt 96.09, is a crystalline solid. The commercial product may be produced by passing carbon dioxide into an absorption column containing aqueous ammonia solution and causing distillation. Ammonium carbonate is the principal ingredient of smelling salts because of its characteristic strong ammonia odor. It is also used for other medicinal purposes and as a leavening agent.

Ammonium Citrate

Diammonium citrate, $(NH_4)_2C_6H_6O_7$, mol wt 226.19, is soluble in an equal weight of water, but is only slightly soluble in ethanol. The pH of a 0.1 *M* solution is 4.3. It is made by neutralization of citric acid with ammonia; the crystalline or granular product is used as a chemical reagent and pharmaceutically as a diuretic.

Ammonium Halides

Ammonium chloride, NH_4Cl, ammonium bromide, NH_4Br, and ammonium iodide, NH_4I, are crystalline, ionic compounds of formula wts 53.49, 97.94, and 144.94, respectively. Their densities $d^{20}{}_4$ systematically follow the increase in formula weight: 1.53, 2.40, and 2.52. All three exist in two crystal modifications: the chloride, bromide, and iodide have the CsCl structure below temperatures of 184.5, 137.8, and −17.6°C, respectively; each reversibly transforms to the NaCl structure at higher temperatures.

The solubility of the ammonium halides in water also increases with increasing formula weight.

All ammonium halides exhibit high vapor pressures at elevated temperatures, and thus, sublime readily. The vapor formed on sublimation consists not of discrete ammonium halide molecules, but is composed primarily of equal volumes of ammonia and hydrogen halide.

Aqueous solutions of ammonium halides, like the other ammonium salts of strong acids, are acidic.

Ammonium Chloride. *Manufacture.* Production by direct reaction of ammonia and hydrochloric acid is simple but usually economically unattractive; a process based on metathesis or double decomposition is generally preferred.

Several commercial grades are available: fine crystals of 99 to 100% purity, large crystals, pressed lumps, rods, and granular material.

Economic Aspects and Uses. Ammonium chloride is used as a nitrogen source for fertilization of rice, wheat, and other crops in Japan, China, India, and Southeast Asia. Ammonium chloride has a number of industrial uses, most importantly in the manufacture of dry-cell batteries, where it serves as an electrolyte. It is also used to make quarrying explosives, as a hardener for formaldehyde-based adhesives, as a flame suppressant, and in etching solutions in the manufacture of printed circuit boards.

Ammonium Bromide and Iodide. *Manufacture.* Ammonium bromide and iodide are manufactured either by the reaction of ammonia with the corresponding hydrohalic acid or, more economically, by the reaction of ammonia with elemental bromine or iodine.

Economic Aspects and Uses. Ammonium bromide is available as a dry technical grade or as 38 to 45% solutions. It is used to manufacture chemical intermediates, and in photographic chemicals; it also has some flame-retardant applications.

Ammonium Nitrate

Ammonium nitrate, NH_4NO_3, formula wt 80.04, is the most commercially important ammonium compound both in terms of production volume and usage. It is the principal component of most industrial explosives and nonmilitary blasting compositions; however, it is used primarily as a nitrogen fertilizer.

One general disadvantage of nitrogen fertilizers, and ammonium nitrate in particular, is that the nitrate ion is more prone to leach through the soil profile and enter the groundwater. The presence of nitrate in groundwater became an important environmental issue in the 1980s (see GROUNDWATER MONITORING).

Physical and Chemical Properties. Ammonium nitrate is a white, crystalline salt, $d_4^{20} = 1.725$, that is highly soluble in water. Although it is very hygroscopic, it does not form hydrates.

Solid ammonium nitrate occurs in five different crystalline forms (Table 1) detectable by time temperature cooling curves.The specific heat of solid β-phase ammonium nitrate is 1.70 J/g (0.406 cal/g) between 1 and 31°C.

Ammonium nitrate has a negative heat of solution in water, and can therefore be used to prepare freezing mixtures.

Decomposition and Detonation Hazard. Ammonium nitrate is considered a very stable salt. When the salt is heated to temperatures from 200 to 230°C, exothermic decomposition occurs. The reaction is rapid, but it can be controlled, and it is the basis for the commercial preparation of nitrous oxide. Above 230°C, exothermic elimination of N_2 and NO_2 begin. The final violent exothermic reaction occurs with great rapidity when ammonium nitrate detonates.

When used in blasting, ammonium nitrate is mixed with fuel oil and sometimes sensitizers such as powdered aluminum.

Manufacture Modern commercial processes rely almost exclusively on the neutralization of nitric acid (qv), produced from ammonia through catalyzed oxidation, with ammonia.

Safety Considerations Ammonium nitrate can be considered a safety material if treated and handled properly. Potential hazards include those associated with fire, decomposition accompanied by generation of toxic fumes, and explosion.

Ammonium Nitrate Limestone

Many plants outside of North America prill or granulate a mixture of ammonium nitrate and calcium carbonate. Production of this mixture, often called calcium ammonium nitrate, essentially removes any explosion hazard.

Ammonium Nitrite

Ammonium nitrite, NH_4NO_2, a compound of questionable stability, can be prepared by reaction of barium nitrite and aqueous ammonium sulfate.

Table 1. Crystalline Forms of Ammonium Nitrate

Designation	Temperature range, °C	Crystal system
α	<−18	tetragonal
β	−18−32.1	rhombic
γ	32.1−84.2	rhombic
δ	84.2−125.2	tetragonal
ε	125.2−169.6	cubic

Ammonium Sulfate

Ammonium sulfate, $(NH_4)_2SO_4$, is a white, soluble, crystalline salt having a formula wt of 132.14. The crystals have a rhombic structure; d_4^{20} is 1.769.

The solubility of ammonium sulfate in 100 g of water is 70.6 g at 0°C and 103.8 g at 100°C. It is insoluble in ethanol and acetone, does not form hydrates, and deliquesces at only about 80% relative humidity.

Manufacture. Ammonium sulfate is produced from the direct neutralization of sulfuric acid with ammonia.

Economic Aspects and Uses. Almost all ammonium sulfate is used as a fertilizer; for this purpose it is valued both for its nitrogen content and for its readily available sulfur content.

Ammonium Sulfides

Ammonia combines with hydrogen sulfide, sulfur, or both, to form various ammonium sulfides and polysulfides. Generally these materials are somewhat unstable, tending to change in composition on standing. Ammonium sulfides are used by the textile industry. They include ammonium sulfide and ammonium hydrosulfide.

CHARLES W. WESTON
Freeport Research and Engineering Company

W. L. Jolly, *The Inorganic Chemistry of Nitrogen,* W. A. Benjamin, Inc., 1964, pp. 40–42.

A. F. Wells, *Structural Inorganic Chemistry,* 5th ed., Clarendon Press, Oxford, U.K., 1984, pp. 362–363.

V. Sauchelli, ed., *Fertilizer Nitrogen, Its Chemistry and Technology,* Reinhold Publishing Corp., New York, 1964, pp. 237–241.

C. Keleti, ed., *Nitric Acid and Fertilizer Nitrates,* Marcel Dekker, Inc., New York, 1985, pp. 208–222, 251–259.

AMORPHOUS MAGNETIC MATERIALS. See MAGNETIC MATERIALS.

AMYL ALCOHOLS

Amyl alcohol describes any saturated aliphatic alcohol containing five carbon atoms. This class consists of three pentanols, four substituted butanols, and a disubstituted propanol, ie, eight structural isomers $C_5H_{12}O$: four primary, three secondary, and one tertiary alcohol. In addition, 2-pentanol, 2-methyl-1-butanol, and 3-methyl-2-butanol have chiral centers and hence two enantiomeric forms.

The odd-carbon structure and the extent of branching provide amyl alcohols with unique physical and solubility properties and often offer ideal properties for solvent, surfactant, extraction, gasoline additive, and fragrance applications. Amyl alcohols have been produced by various commercial processes in past years. Today the most important industrial process is low pressure rhodium-catalyzed hydroformylation (oxo process) of butenes.

Mixtures of isomeric amyl alcohols (1-pentanol and 2-methyl-1-butanol) are often preferred because they are less expensive to produce commercially; also, the different degree of branching of the mixture imparts a more desirable combination of properties. One such mixture is a commercial product sold under the name Primary Amyl Alcohol by Union Carbide Chemicals and Plastics Company Inc.

Physical Properties

With the exception of neopentyl alcohol (mp 53°C), the amyl alcohols are clear, colorless liquids under atmospheric conditions, with characteristic, slightly pungent and penetrating odors. They have relatively higher boiling points than ketonic or hydrocarbon counterparts and are considered intermediate boiling solvents for coating systems (Table 1).

Commercial primary amyl alcohol is a mixture of 1-pentanol and 2-methyl-1-butanol, in a ratio of ca 65 to 35 (available from Union Carbide Chemicals and Plastics Company Inc. in other ratios upon request). Typical physical properties of this amyl alcohol mixture are listed in Table 2.

Like the lower alcohols, amyl alcohols are completely miscible with numerous organic solvents and are excellent solvents for nitrocellulose, resin lacquers, higher esters, and various natural and synthetic gums and resins. However, in contrast to the lower alcohols, they are only slightly soluble in water. Only 2-methyl-2-butanol exhibits significant water solubility.

Chemical Properties

The amyl alcohols undergo the typical reactions of alcohols which are characterized by cleavage at either the oxygen–hydrogen or carbon–oxygen bonds. Important reactions include dehydration, esterification, oxidation, amination, etherification, and condensation.

Manufacture

Three significant commercial processes for the production of amyl alcohols include separation from fusel oils, chlorination of C-5 alkanes with subsequent hydrolysis to produce a mixture of seven of the eight isomers (Pennsalt), and a low pressure oxo process, or hydroformylation, of C-4 olefins followed by hydrogenation of the resultant C-5 aldehydes.

The oxo process is the principal one in practice today; only minor quantities, mainly in Europe, are obtained from separation from fusel oil. *tert*-Amyl alcohol is produced on a commercial scale in lower volume by hydration of amylenes (Dow, BASF).

Health and Safety Factors

The main effects of prolonged exposure to amyl alcohols are irritation to mucous membranes and upper respiratory tract, significant depression of the central nervous system, and narcotic effects from vapor inhalation or oral absorption. All the alcohols are harmful if inhaled or swallowed, appreciably irritating to the eyes and somewhat irritating to uncovered skin on repeated exposure. Prolonged exposure causes nausea, coughing, diarrhea, vertigo, drowsiness, headache, and vomiting. 3-Methyl-1-butanol has demonstrated carcinogenic activity in animal studies.

All of the amyl alcohols are TSCA and EINECS (European Inventory of Existing Commercial Chemical Substances) registered.

The amyl alcohols are readily flammable substances; *tert*-amyl alcohol is the most flammable (closed cup flash point, 19°C). Their vapors can form explosive mixtures with air.

Shipping and Storage

Amyl alcohols are best stored or shipped in either aluminum, lined steel, or stainless steel tanks.

Economic Aspects

All eight amyl alcohol isomers are available from fine chemical supply firms in the United States. Five of them, 1-pentanol, 2-pentanol, 2-methyl-1-butanol, 3-methyl-1-butanol, and 2-methyl-2-butanol (*tert*-amyl alcohols) are available in bulk in the United States; in Europe all but neopentyl alcohol are produced.

Table 1. The Amyl Alcohols and Some of Their Physical Properties

Properties	1-Pentanol	2-Pentanol	3-Pentanol	2-Methyl-1-butanol	3-Methyl-1-butanol	2-Methyl-2-butanol	3-Methyl-2-butanol	2,2-Dimethyl-1-propanol
common name	*n*-amyl alcohol	*sec*-amyl alcohol			isoamyl alcohol	*tert*-amyl alcohol		neopentyl alcohol
critical temperature, °C	315.35	287.25	286.45	291.85	306.3	272.0	300.85	276.85
critical pressure, kPa[a]	3868	3710	3880	3880	3880	3880	3960	3880
critical specific volume, mL/mol	326.5	328.9	325.3	327	327	327	327	327
critical compressibility	0.25810	0.26188	0.27128	0.27009	0.26335	0.27992	0.27133	0.27745
boiling point at pressure, °C								
101.3 kPa[a]	137.8	119.3	115.3	128.7	130.5	102.0	111.5	113.1
40 kPa	111.5	93.8	90.9	103.5	105.6	78.3	87.2	89.0
1.33 kPa	44.6	32.0	27.7	40.2	43.0	21.0	26.0	25.0
vapor pressure[b], kPa[a]	0.218	0.547	0.761	0.274	0.200	1.215	0.810	0.929
melting point, °C	−77.6	−73.2	−69.0	<−70	−117.2	−8.8	forms glass	54.0
heat of vaporization at normal boiling point, kJ/mol[c]	44.83	43.41	42.33	44.75	43.84	40.11	41.10	41.35
ideal gas heat of formation[d], kJ/mol[c]	−298.74	−313.80	−316.73	−302.08	−302.08	−329.70	−314.22	−319.07
liquid density[b], kg/m^3	815.1	809.4	820.3	819.1	810.4	809.6	818.4	851.5[e]
liquid viscosity[b], mPa·s(= cP)	4.06	4.29	6.67	5.11	4.37	4.38	3.51[d]	2.5[e]
surface tension[b], mN/m(= dyn/cm)	25.5	24.2	24.6	25.1[d]	24.12	22.7	23.0[d]	14.87[e]
refractive index[d]	1.4080	1.4044	1.4079	1.4086	1.4052	1.4024	1.4075	1.3915
solubility parameter[d], $(MJ/m^3)^{0.5}$[f]	22.576	21.670	21.150	22.274	22.322	20.758	21.607	19.265[e]
solubility in water[b], wt %	1.88	4.84	5.61	3.18	2.69	12.15	6.07	3.74
solubility of water in[b], wt %	9.33	11.68	8.19	8.95	9.45	24.26	11.88	8.23

[a] To convert kPa to mm Hg, multiply by 7.5.
[b] At 20°C unless otherwise noted.
[c] To convert kJ/mol to cal/mol, multiply by 239.
[d] At 25°C.
[e] At the melting point.
[f] To convert (MJ/m^3) to $(cal)^{0.5}$, divide by 2.045.

Table 2. Physical Properties of Primary Amyl Alcohol, Mixed Isomers

Property	Value[a]
molecular weight	88.15
boiling point at 101.13 kPa[b], °C	133.2
freezing point, °C	−90[c]
specific gravity 20/20 °C	0.8155
absolute viscosity at 20°C, mPa·s(= cP)	4.3
vapor pressure at 20°C, kPa[b]	0.27
flash point (closed cup), °C	45
solubility at 20°C, by wt %	
in water	1.7
water in	9.2

[a] 65/35 blend, ie, a mixture of 1-pentanol and 2-methyl-1-butanol, 65/35 wt %, respectively.
[b] To convert kPa to mm Hg, multiply by 7.5.
[c] Sets to glass below this temperature.

ANTHONY J. PAPA
Union Carbide Chemicals and Plastics Company Inc.

Kirkpatrick Chemical Engineering Achievement Award, *Chem. Eng.* **84,** 110 (Dec. 5, 1977).

H. Bieber, *Encycl. Chem. Process Des.* **3,** 278 (1977).

ANALEPTICS. See STIMULANTS.

ANALGESICS, ANTIPYRETICS, AND ANTIINFLAMMATORY AGENTS

Pain, pyresis, and inflammation are distinct physiological responses which can occur independently; they are often associated as the body mounts a response to an injury or insult. Each is an important signal from injured tissue and the signal's continued presence can guide a physician in diagnosing and treating the condition which led to the occurrence.

The twentieth century has seen considerable progress in the understanding of pain and inflammation and the relationship of one to the other. It is increasingly apparent that the central and peripheral nervous systems are capable of causing the production of mediators which can attract and activate inflammatory cells, thereby initiating or amplifying an inflammatory response. At the same time, inflammatory cells are also capable of releasing substances which not only respond to the original insult, but also stimulate the nervous systems.

A number of interdependent physiological mechanisms, each providing new targets for therapeutic intervention, and allowing for the development of new treatments for pain, pyresis, and inflammation, have been discovered. At least eight distinct types of opioid receptors have been identified and the corresponding individual functions are beginning to be understood. Moreover, a great deal has been learned about the role of lipid mediators in the inflammatory response and how antiinflammatory agents control their production.

Opioid Agonists

Opium is the dried, powdered sap of the unripe seed pod of *Papaver somniferum*, a poppy plant indigenous to Asia Minor. More than 20 different alkaloids (qv) of two different classes comprise 25% of the weight of dry opium. The benzylisoquinolines, characterized by papaverine (1.0%), a smooth muscle relaxant, and noscapine (6.0%), an antitussive agent, do not have any analgesic effects. The phenanthrenes, the second group, are the more common and include 10% morphine (**1,** R′ = R = H), 0.5% codeine, $C_{18}H_{21}NO_3$, (**1,** R′ = H, R = CH_3), and 0.2% thebaine, $C_{19}H_{21}NO_3$, (**2**).

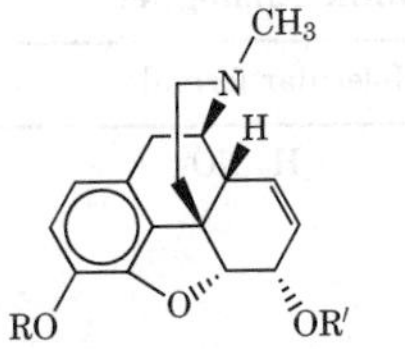

(1)

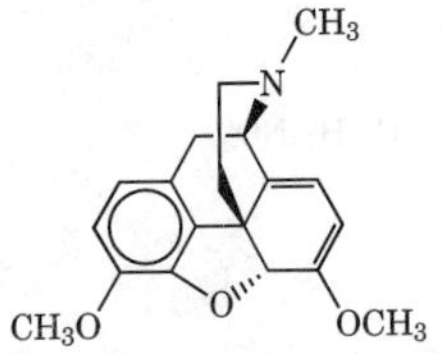

(2)

Morphine, mol wt 285.3, effectively relieves pain and increases an individual's ability to tolerate a painful experience. It also produces a remarkably broad range of other effects, including drowsiness, mood changes, respiratory depression, nausea, decreased gastrointestinal motility, and vomiting. Morphine behaves as a receptor agonist, acting preferentially at the μ receptor, but also exhibiting appreciable affinity for other opioid receptors. A standard therapeutic dose is 10–15 mg, usually administered subcutaneously. Analgesia peaks in about one hour and lasts for four to five hours. An important feature of morphine and related drugs is the development of physical dependence on, and tolerance to, some of the effects.

Codein, mol wt 299.3, is a significantly less potent analgesic than morphine, requiring 60 mg (0.20 mmol) to equal the effectiveness of 10 mg (0.04 mmol) of morphine. However, codeine is orally effective, and it is less addictive and associated with less nausea than morphine.

Introduced in 1898, heroin (**1,** R = R′ = $COCH_3$) was heralded as a nonaddictive alternative to morphine. Subsequent clinical experience showed it to be highly addictive and preferred by addicts over morphine.

Synthetic and Semisynthetic Agonists

In attempts to discover drugs demonstrating fewer undesirable side effects than morphine, many synthetic analogues have been prepared. Some of these are shown in Table 1.

Several common structural features necessary for opioid, analgesic activity have been identified from the action of the analogues. Systematic simplification demonstrated that much of the morphine ring structure could be modified or even eliminated. The piperidine ring, in a chair conformation, and in particular the nitrogen atom of that ring, appears essential to pharmacological activity. The nitrogen is believed to attach to an anionic center in the receptor. Also crucial is the presence of the phenyl ring which, through van der Waals forces, binds to the hydrophobic portion of the receptor.

Opioid Antagonists and Partial Agonists

The replacement of the *N*-methyl group on the nitrogen atom of the piperidine ring of morphine and analogues, by allyl, isopropyl, or methyl cyclopropyl, an isopropyl isostere, results in compounds which antagonize opioid responses, especially respiratory depression. Naloxone (**7**), $C_{19}H_{21}NO_4$, and naltrexone (**8**), $C_{20}H_{23}NO_4$, are both derived from oxymorphone (Table 1) and exhibit agonist activity only at doses that are of little clinical significance. In the absence of opioid drugs, naloxone does not cause analgesia, respiratory depression, or sedation. However, when administered with an opioid analgesic, the effects produced by the opioid agonist are promptly reversed. The ability to antagonize opioids at all of the different opioid receptors makes naloxone useful for the treatment of opioid overdose. Naltrexone has a similar profile, but it is orally active and has a significantly longer half-life.

Table 1. Synthetic Morphine Analogues

Name	Molecular formula	Structure
levorphanol	$C_{17}H_{23}NO$	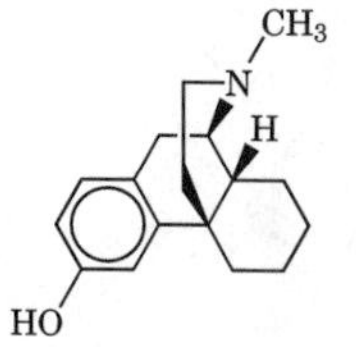 **(3)**
hydromorphone	$C_{17}H_{19}NO_3$	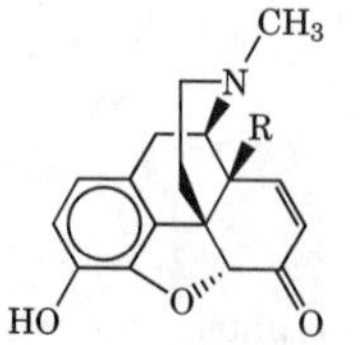 **(4, R = H) (4, R = OH)**
oxymorphone	$C_{17}H_{19}NO_4$	
phenazocine	$C_{22}H_{27}NO$	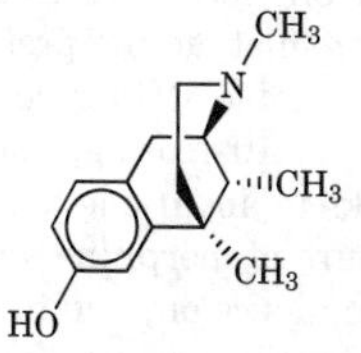 **(5)**
fentanyl	$C_{22}H_{29}NO_2$	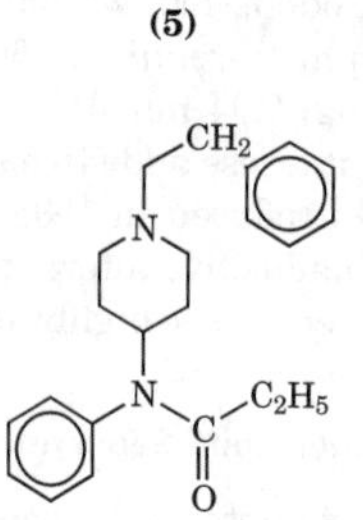 **(6)**

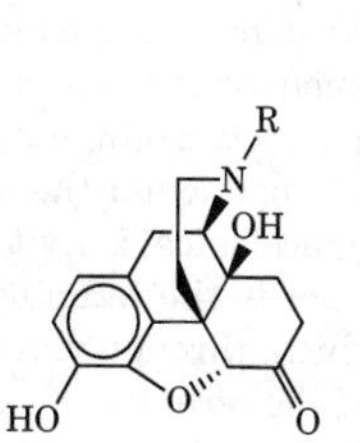

(7)

R = CH_2CHCH_2

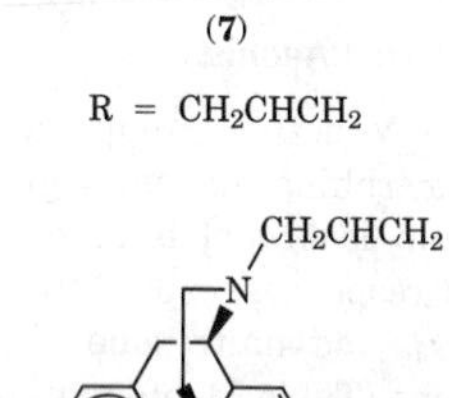

(8)

R = CH_2– 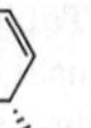

The quest for compounds that combined the analgesic properties of morphine, were nonaddictive and lacked side effects, led to the development of the drugs that have both agonist and antagonist activities. Nalbuphine and buprenorphine are semisynthetic materials derived from oxymorphone and thebaine respectively, whereas pentazocine and butorphanol are benzomorphan and morphan derivatives. Although structurally similar, they display different receptor affinities: pentazocine is a weak μ-antagonist, but a strong agonist of the κ-receptor; nalbuphine is a competitive antagonist for the μ-receptor, blocking the effects of the morphinelike drugs, but is a partial agonist for the κ and ς receptors; and buprenorphine is a partial agonist for the μ-receptor.

Pharmacologically, the effects of these drugs resemble those of opioid agonists. All four have analgesic potency equal to or greater than morphine and, like morphine, they cause respiratory depression. A ceiling effect is reached, however, above which increased doses do not increase respiratory depression or do not produce proportionally greater depression.

Other Analgesic Agents

Most analgesic agents rely on agonism of the μ receptor for their activity, however the ability of the κ and δ receptors to induce analgesia is also well documented. Some nonmorphine analgesics which may preferentially bind to κ and δ receptors are found in Table 2.

Antiinflammatories and Antipyretics

Most of the time, the powerful analgesia supplied by morphine and the other opioid analgesics is not needed. Rather, a mild analgesic, such as aspirin, the most commonly employed analgesic agent, can be used for the treatment of simple pain associated with headaches, minor muscle pain, mild trauma, arthritis, cold and flu symptoms, and fever.

Aspirin, the oldest of the nonsteroidal antiinflammatory drugs (NSAIDs), is a member of the salicylate group.

Aspirin's pain relief results, not through direct action on the central nervous system, but rather through peripheral action. It has been proposed that aspirin and aspirinlike drugs inhibit the enzymatic production of prostaglandins, a group of endogenous agents which are well known to cause erythema, edema, pain, and fever. Aspirin does not act as a prostaglandin receptor antagonist; rather it blocks the cyclooxygenase enzyme-catalyzed conversion of arachidonic acid, $C_{20}H_{32}O_2$, to cyclic endoperoxides, a prostaglandin precursor.

Table 2. Nonmorphine Analgesics

Name	Molecular formula	Structure
acetorphan	$C_{21}H_{23}NO_4S$	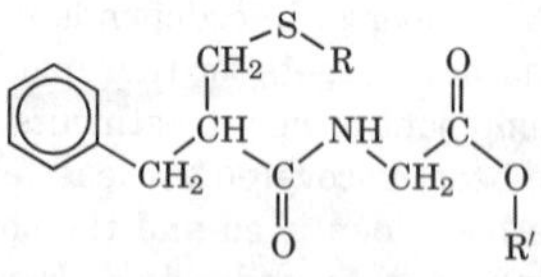 **(9, R = COCH$_3$, R′ = CH$_2$C$_6$H$_5$) (9, R = H, R′ = H)**
thiorphan	$C_{12}H_{15}NO_3S$	
spiradoline	$C_{22}H_{30}Cl_2N_2O_2$	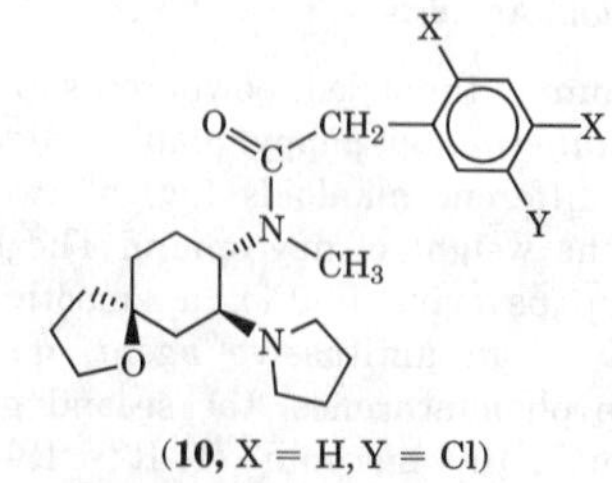 **(10, X = H, Y = Cl)**

The action of endogenous pyrogens on the hypothalmus produces fever, because of a readjustment in the central set point controlling the body's internal temperature. Salicylates and other NSAIDs achieve their antipyretic effect by controlling the prostaglandin-induced release of pyrogens.

A second class of NSAIDs, the so-called coal-tar analgesics, are derived from acetanilide. Although it is no longer used therapeutically, its analogues, phenacetin and the active metabolite, acetaminophen, are effective alternatives to aspirin. They have analgesic and antipyretic effects that do not differ significantly from aspirin, but they do not cause the gastric irritation, erosion, and bleeding that may occur after salicylate treatment. In contrast to aspirin, however, they are not cyclooxygenase inhibitors and have no antiinflammatory properties. Clinically, acetaminophen is preferred over phenacetin, because it has less overall toxicity.

A more recently introduced, nonprescription analgesic is the aryl propionic acid, ibuprofen, which offers significant advantages over aspirin. Ibuprofen, a cyclooxygenase inhibitor, displays good antiinflammatory activity. It is more potent than aspirin and has a lower incidence of gastrointestinal irritation, although at high doses or chronic exposure, gastric irritation, as well as some renal toxicity, has been observed. Ibuprofen is more effective than propoxyphene in relieving episiotomy pain, pain following dental extractions, and menstrual pain.

The adrenal cortex produces steroidal hormones that are associated with carbohydrate, fat, and protein metabolism, electrolyte balance, and gonadal functions. One of these, cortisone, $C_{21}H_{28}O_5$ (**11**), demonstrated a remarkable ability to relieve the symptoms of inflammatory conditions. Other glucocorticoid steroids, such as dexamethasone, $C_{22}H_{29}FO_5$ (**12**, R = F, R′ = CH_3), and prednisolone, $C_{21}H_{28}O_5$ (**12**, R = R′ = H), also have antiinflammatory properties.

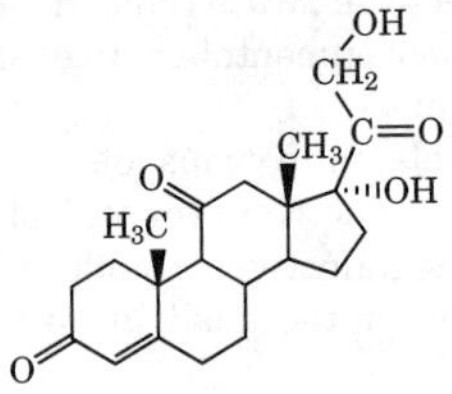

(11)

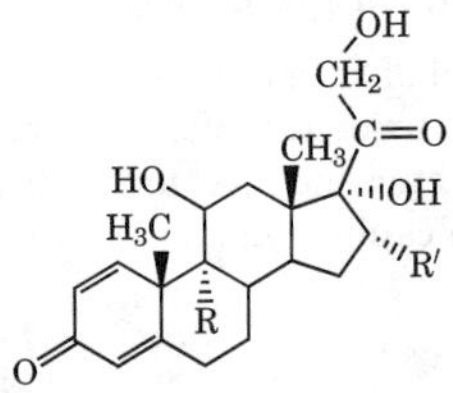

(12)

These steroids are capable of preventing or suppressing the development of the swelling, redness, local heat, and tenderness which characterize inflammation.

Unfortunately steroids merely suppress the inflammation while the underlying cause of the disease remains. Another serious concern about steroids is that of toxicity. The abrupt withdrawal of glucocorticoid steroids results in acute adrenal insufficiency. Long-term use may induce osteoporosis, peptidic ulcers, the retention of fluid, or an increased susceptibility to infections. Because of these problems, steroids are rarely the first line of treatment for any inflammatory condition, and their use in rheumatoid arthritis begins after more conservative therapies have failed.

Economic Aspects

Analgesics and antiarthritics represent significant worldwide pharmaceutical markets. Table 3 lists trade names and producers of some of the more commercially important agents.

Table 3. Trade Names and Producers of Analgesics, Antipyretics, and Antiinflammatory Drugs

Compound name	Trade name	Producer
levorphanol	Dromoran	Hoffmann-LaRoche
phenazocine	Prinadol	Smith Kline & Beecham
meperidine	Demerol	Winthrop
propoxyphene	Darvon	Eli Lilly
naloxone	Narcan	Du Pont
naltrexone	Trexan	Du Pont
nalbuphine	Nubain	Du Pont
buprenorphine	Buprenex	Norwich Eaton
pentazocine	Talwin	Winthrop
butorphanol	Stadol	Bristol
acetaminophen	Tylenol	McNeil
ibuprofen	Motrin	Upjohn
naproxen	Naprosyn	Syntex
ketoprofen	Orudis	Wyeth
flurbiprofen	Ansaid	Upjohn
indomethacin	Indocin	Merck Sharpe & Dohme
sulindac	Clinoril	Merck Sharpe & Dohme
diclofenac	Voltaren	CIBA-GEIGY
meclofenamate	Meclomen	Parke-Davis
piroxicam	Feldene	Pfizer
diflunisal	Dolobid	Merck Sharpe & Dohme

RICHARD A. NUGENT
CHARLES M. HALL
Upjohn Company

A. Herz, ed., *Opioids I*, in *Handbook of Experimental Pharmacology*, Vol. 104, Springer-Verlag, New York, 1993.

A. Herz, ed., *Opioids II*, in *Handbook of Experimental Pharmacology*, Vol. 104, Springer-Verlag, New York, 1993.

J. G. Lombardino, ed., *Nonsteroidal Antiinflammatory Drugs*, John Wiley & Sons, Inc., New York, 1985.

J. I. Gallin, I. M. Goldstein, and R. Snyderman, eds., *Inflammation: Basic Principles and Clinical Correlates*, Raven Press, New York, 1992.

ANALYTICAL METHODS

SURVEY

Analytical methods are utilized by all branches of the chemical industry. Sometimes the goal is the qualitative determination of elemental and molecular constituents of a selected specimen of matter; at other times the goal is the quantitative measurement of the fractional distribution of those constituents; and sometimes it is to monitor a process stream or a static system. Information concerning the various individual analytical methods may be found in separate articles dispersed alphabetically throughout the *Encyclopedia*. The articles are introductions to topics each of which is the subject of numerous books and other publications.

TRENDS

Analytical chemistry, the science of the measurement and characterization of systems, is being revolutionized by the rapid advances in computer technology (qv), microelectronics, and materials science. New techniques in chemical analysis, as well as significant developments in more established instrumentation and methodology, have resulted in the production of more analytical information of higher quality in less time than ever before. These developments are being driven by five separate factors: (*1*) the quest for truly automated or intelligent instruments (see EXPERT SYSTEMS); (*2*) the continued development of multidimensional hyphenated instruments, which involve the marriage of two instruments having complementary capabilities by way of a sampling interface and a common computer, thus facilitating the analysis of complex samples (see ANALYTICAL METHODS, HYPHENATED INSTRUMENTS); (*3*) continuous improvements in analyte detection limits, bringing single atom analysis closer to reality; (*4*) increased capabilities in data manipulation, miniaturization of hardware, and remote *in vivo* sensing, leading to new instrumentation capable of both separation and specification; and (*5*) the development of nondestructible, stable, field-operable instrumentation utilizing fiber optic technology to serve as in-field sampling devices and process control (qv) tools (see FIBER OPTICS).

Analytical Instrumentation for Process Control

It is becoming more and more desirable for the analytical chemist to move away from the laboratory and into the field via in-field instruments and remote, point-of-use measurements. As a result, process analytical chemistry has undergone an offensive thrust in regard to problem-solving capability. *In situ* analysis enables the study of key process parameters for purposes of definition and subsequent optimization. On-line analysis capability has already been extended to gc, lc, ms, and ftir techniques as well as to icp-emission spectroscopy, flow injection analysis, and near-infrared spectrophotometry. In addition, new technology developments, such as on-line Raman spectroscopy, x-ray diffraction and gc/lc are under commercialization for the 1990s.

Analytical Capabilities

The increased demand for analyses in applications, such as industrial processes, the environment, and health, pushed worldwide sales of analytical instruments from $300,000 to over $3,000,000 during the 1980s. U.S. instrument manufacturers have also proliferated.

Refinement of existing techniques and instrumentation, rather than introduction of a whole new host of methods and instruments, is expected for the 1990s. One technology challenge is clearly the migration of analytical techniques out of the area of research and development and into the quality control and process control environments. Another challenge involves continued advances in smaller sample size, higher sensitivity and specificity,and lowered costs per sample analysis. Instruments should also become more user-friendly and more integrated within the total laboratory environment. Expert systems are expected to expand to the point where usage for both identification and quantification is routine in the analytical laboratory. Essential to this usage is the proper application of sampling (qv) and the understanding of chemistry.

ANTHONY J. MONTANA
Warner-Lambert Company

M. B. Denton, *Analyst* **112**, 347 (1987).

T. Hirschfeld, *Science* **230**, 286 (1985).

K. Eckschlager and V. Stepanek, *Information Theory as Applied to Chemical Analysis,* Wiley-Interscience, New York, 1984.

L. A. Currie, ed., *Detection in Analytical Chemistry,* American Chemical Society, Washington, D.C., 1988.

HYPHENATED INSTRUMENTS

In 1990, Chemical Abstracts Service listed over 10 million substances in their Registry. Moreover, the growth of new compounds is exponential, leading to a doubling of known chemicals every eleven years. Thus there is an ever-increasing need to efficiently identify substances and quantitate material with a high degree of confidence. Hyphenated instruments, combinations of accepted instrumental techniques where the sample is passed from one instrument directly into another, were developed to aid in solving this problem.

Hyphenated analytical methods provide more complementary information in a shorter time period leading to faster and more reliable results, than data obtained from traditional instrumental methods. The number of types of analytical instruments that can be joined is very large, depending only on the nondestruction of samples after the initial analytical procedure and the ability of the manufacturer to interface the instrumental techniques. Combinations include separation–separation, separation–identification, and identification–identification techniques (see ANALYTICAL METHODS, SURVEY).

Table 1 lists the most significant of the hyphenated instruments that were commercially available as of 1990. The instruments and methods discussed in detail herein all have both separation and identification capability.

The combination of gas chromatography and mass spectrometry offers several advantages over the use of these techniques individually. Capillary gas chromatography provides very high separation efficiency for complex organic mixtures such as those found in environmental, foods and flavors, and general industrial applications. The mass spectrometer adds to this separation capability the unique ability to provide molecular structural information leading to both identification and quantitation of the separated components. Thus the gc/ms is a significant labor- and time-saving device, accomplishing separation, component transfer, identification, and quantitation, often accompanied by a printed final report, all in a single instrument.

Applications. One of the most frequently used tools for systems requiring both qualitative and quantitative chemical analysis is gc/ms. It has been particularly valuable in the environmental field, where laboratories responsible for the monitoring of hazardous compounds depend heavily on gc/ms to separate components of a usually complex mixture present in an environmental sample, and then to identify and quantitate each component.

A technique that is related to gc/ms and is starting to grow significantly in use is that of inductively coupled plasma/mass spectrometry (icp/ms). In icp/ms the inductively coupled plasma serves as a good interface and ion source for the quadrupole mass spectrometer (see PLASMA TECHNOLOGY).

Gas Chromatography/Infrared Spectroscopy

Gc/ir instruments are all of the gc/ftir variety that utilize an interferometer and require a Fourier transform of the signal, as opposed to employing a monochrometer and mechanical scanning. At least 3 scan/s are needed across the capillary gc peaks which are typically 3–6 s wide.

Gas Chromatography/Mass Spectrometry/Infrared Spectroscopy

Gas chromatography/fourier transform infrared spectroscopy (gc/ftir), itself a very useful analytical technique, is especially powerful when combined with mass spectrometry. One of the factors involved in making the ternary gc/ir/ms system a viable analytical tool was the development of faster, more powerful computers to acquire, analyze, and report the large quantities of data generated by the mass spectrometer and the ftir.

Applications. The number of capabilities of a gc/ir/ms in separating and identifying components in complex mixtures is very high for a broad spectrum of analytical problems. One area where ir information particularly complements ms data is in the differentiation of isomeric compounds.

Another analysis handled effectively by use of gc/ir/ms is essential oil characterization, which is of interest to the foods, flavors, and fragrances industries (see OILS, ESSENTIAL).

The confidence level in the identification capability for gc/ir/ms is enhanced by the ability to perform computer library searching of large spectral databases.

Table 1. Commercially Available Hyphenated Instruments

Instrument[a]	Manufacturers	Price range, $ × 10^3	Estimated sales 1990[b], $ × 10^6
gas chromatograph/gas chromatograph (gc/gc)	Siemens AG, Carlo Erba	25	
gas chromatograph/liquid chromatograph (gc/lc)	Carlo Erba	35	
gas chromatograph/mass spectrometer (gc/ms)	Hewlett-Packard, Finnigan, VG Instruments, Varian	50–500	175
gas chromatograph/infrared spectrometer (gc/ir)	Hewlett-Packard, Nicolet, Digilab, Perkin-Elmer	70–250	26
gas chromatograph/atomic emission spectrometer (gc/ae)	Hewlett-Packard	90	
liquid chromatograph/mass spectrometer (lc/ms)	Hewlett-Packard, Finnigan, VG Instruments	130–300	25
mass spectrometer/mass spectrometer (ms/ms)	Finnigan, VG Instruments, Kratos, Extrel	250–500	258[c]
gas chromatograph/mass spectrometer/infrared spectrometer (gc/ms/ir)	Hewlett-Packard	115	
super critical fluid chromatograph/ infrared spectrometer (scfr/ir)	Nicolet, Digilab	100–200	
inductively-coupled plasma/mass spectrometer (icp/ms)	VG Instruments, Perkin Elmer	150–600	23

[a] Whereas analytical techniques are often abbreviated using capital letters, *Encyclopedia* style is to use the lower case.
[b] Strategic Directions International, Inc., Los Angeles, Calif.
[c] Includes all quadruple and magnetic sector ms (no gc/ms) of which ms/ms is a subset.

Liquid Chromatography/Mass Spectrometry

The replacement of a gas chromatograph with a liquid chromatograph (lc) substantially enhances the separation of compounds which might decompose at the operating temperatures of the gc or which are not readily volatilized. However, lc/ms coupling presents a challenge in terms of removing the large quantities of lc solvent.

Applications. The primary advantage of an lc/ms is in combining the separation of large, potentially thermally labile compounds with the qualitative and quantitative features of the mass spectrometer. As the need to analyze mixtures containing larger biological molecules increases in the fields of pharmaceuticals (qv), agricultural chemistry, and biotechnology, the demand for lc/ms should also increase.

Mass Spectrometry/Mass Spectrometry

Tandem mass spectrometry or ms/ms was first introduced in the 1970s and gained rapid acceptance in the analytical community. The technique has been used for structure elucidation of unknowns and has the ability to provide sensitive and selective analysis of complex mixtures with minimal sample clean-up. Developments in the mid-1980s advancing the popularity of ms/ms included the availability of powerful data systems capable of controlling the ms/ms experiment and the viability of soft ionization techniques which essentially yield only molecular ion species. Ms/ms is based on the characterization of selected ions in the mass spectrum of a sample through further fragmentation and analysis.

Applications. Ms/ms has found application in areas such as trace analysis of biological tissue, complex hazardous waste site samples, and human blood serum, as well as in drug testing.

WAYNE P. DUNCAN
Hewlett-Packard Company

T. Hirschfeld, *Anal. Chem.* **52,** 297A (1980).

R. D. Smith and co-workers, *Anal. Chem.* **62,** 882 (1990).

F. W. McLafferty, ed., *Tandem Mass Spectrometry,* John Wiley & Sons, Inc., New York, 1983.

W. M. Muck and J. D. Henion, *Biomed. Environ. Mass Spectrom.* **19,** 37 (1990).

ANESTHETICS

The term anesthesia comes from the Greek *anaisthai* or insensibility and constitutes a state in which perception of noxious events such as surgical procedures is imperceptible. This state may or may not be accompanied by loss of consciousness. A complete or general anesthetic given by the inhalation or intravenous routes produces hypnosis (profound sleep), analgesia, muscle relaxation, and protection against the increase in blood pressure and heart rate resulting from surgical stress (maintains homeostasis). An anesthetic which blocks the neural transmission of painful stimuli through afferent nerves and does not affect the level of consciousness can be classified as a local anesthetic.

The induction of general anesthesia produces a progressive deepening of the anesthetic state and represents, in the anatomical sense, a descending desensitization of the central nervous system (CNS). A progression of clinical signs are useful for estimating the depth of anesthesia. These signs vary somewhat for each anesthetic, but in general four stages can be defined: *(1)* state of altered consciousness and analgesia indicative of action on cerebral canticle areas begin; *(2)* loss of consciousness, often accompanied by delirium and excitement, occurs. Irregular respiration and motor movement result from depression of higher motor inhibitory centers with the release of lower motor mechanisms; *(3)* stage of surgical anesthesia is reached in which spinal cord and spinal reflexes are abolished, providing relaxation of skeletal musculature. Within the four planes in this state are loss of corneal, conjunctival, pharyngeal, and laryngeal reflexes as the depth of anesthesia progresses; *(4)* onset of respiratory paralysis occurs resulting from significant depression of the medullary respiratory center. Subsequent cardiovascular collapse ensues.

Most signs which require a skeletal muscle reflex would not be apparent after treatment with a neuromuscular blocking drug or would be altered significantly by preoperative drugs. In these cases central nervous system monitoring via an electroencephalogram (EEG), and hemodynamic and blood gas monitoring, help assess the depth of anesthesia. The potency of inhaled agents is expressed as the minimum alveolar concentration (MAC) that is required to prevent spontaneous

Table 1. Properties and Partition Coefficients of Inhalation Anesthetics

Agent	Molecular formula	Partition coefficient		Boiling point, °C	Year introduced
		Oil/gas	Blood/gas		
ethyl ether	$C_4H_{10}O$	65	12.1	35	1842
chloroform	$CHCl_3$	394	8.4	61	1847
trichloroethylene	C_2HCl_3	200–250	9.0	87	1934
fluoroxene	$C_4H_5F_3O$	47	1.37	43.1	1960
halothane	$C_2HBrClF_3$	224	2.3	50.2	1956
isoflurane	$C_3H_2ClF_5O$	90.8	1.4	48.5	1980
enflurane	$C_3H_2ClF_5O$	96.5	1.9	56.5	1972
methoxyflurane	$C_3H_4Cl_2F_2O$	970	12.0	105	1962
sevoflurane	$C_4H_3F_7O$	42	0.6	58	1989
desflurane	$C_3H_2F_6O$	18.7	0.42	23.5	
nitrous oxide	N_2O	1.4	0.46	−89.5	1850s
cyclopropane	C_3H_6			−33	1934

Table 2. Clinically Used Local Anesthetic Agents

Agent	Method of application	Comment
	Amino esters	
procaine	infiltration, spinal	slow onset, short duration
chloroprocaine	peripheral nerve and obstetric extradural	fast onset, short duration, low systemic toxicity
tetracaine	spinal	slow onset, long duration, high systemic toxicity
	Amino amides	
lidocaine	infiltration, iv regional, peripheral nerve block, extradural block, spinal and topical	fast onset, moderate duration, low systemic toxicity, most versatile agent
mepivacaine	infiltration, peripheral nerve block, extradural	similar to lidocaine
prilocaine	similar to lidocaine	safest amino amide, methemoglobinemia at high doses
bupivacaine	infiltration, peripheral nerve block, extradural	moderate onset, long duration, potential for cardiovascular side effects, sensory/motor separation
etidocaine	infiltration, peripheral nerve block, extradural	fast onset, long duration, profound motor block

movement in response to a surgical or equivalent stimulus in 50% of patients.

The onset of action is fast (within 60 s) for the intravenous anesthetic agents and somewhat slower for inhalation and local anesthetics. The induction time for inhalation agents is a function of the equilibrium established between the alveolar concentration relative to the inspired concentration of the gas. Onset of anesthesia can be enhanced by increasing the inspired concentration to approximately twice the desired alveolar concentration, then reducing the concentration once induction is achieved. The onset of local anesthetic action is influenced by the site, route, dosage (volume and concentration), and pH at the injection site.

Theories of General Anesthesia

Although the modern practice of anesthesia is exceedingly sophisticated, identification of the basic molecular mechanism underlying it is still lacking. Even whether the target sites of the inhalational anesthetics are lipid or protein remains unknown, although it is likely that proteins are intimately involved. The inhalational anesthetics are very diverse chemically and extrapolation from the effect of a particular agent to the physiological state of anesthesia is problematic. However, anesthesia theories may be roughly divided into two categories: lipid theories and protein theories.

Lipid Theories. Although there are many varieties of the lipid theory all postulate that inhalational anesthetics exert their primary effects by dissolving in the lipid portions of nerves, thereby altering the conductivity. The primary site of action is postulated to be the lipid matrix of cell membranes.

Protein Theories. The direct interaction of inhalation anesthetics and proteins (qv) has been proposed as the cause of anesthesia. An inhalation agent, whether a noble gas or a fluorinated ether, could dissolve asymmetrically in a protein. Resultant conformational changes in the protein, if these changes occur, could then cause changes in biological activity.

Anesthetic Agents

Inhalation Agents. An ideal inhalation anesthetic would exhibit physical, chemical, and pharmacological properties allowing safe usage in a variety of surgical interventions. The agent should be odorless, nonflammable at concentrations which are likely to be used in the operating room, and stable both on storage and to soda lime, which is used as the CO_2 absorber in the anesthetic circuit. Induction of, and recovery from, anesthesia should be rapid, and minimal side effects should be observed on the cardiovascular (depression and epinephrine compatibility) or central nervous systems (EEG activation) (see EPINEPHRINE AND NOREPINEPHRINE; NEUROREGULATORS). The drug should not be metabolized.

A number of inhalation anesthetics have been introduced to clinical practice, some of which are listed in Table 1. All agents introduced after 1950, except ethyl vinyl ether, contain fluorine. Agents such as ether, chloroform, trichloroethylene (Trilene), cyclopropane, and fluoroxene (Fluoromar), which were once used, have been displaced by the newer fluorinated anesthetics.

Intravenous Anesthetic. The intravenous (iv) anesthetic agents are of two types: those which are used to induce, but not maintain, anesthesia, and those which are useful not only for induction, but also for maintenance. The period of induction is perhaps the most crucial part of the anesthesia. The need is for an anesthetic agent having an extremely fast rate of onset and limited side effects. Fast onset minimizes the stress and agitation which could arise during a more lengthy induction. Various classes of compounds have been used. Among these are: barbiturates, opioids, steroids, benzodiazepines, and hindered phenols. But the ideal agent for both induction and maintenance has not yet been found. For this reason, balanced anesthesia is often used: a potent opioid is combined with an inhalation agent.

A major difference between the inhalational agents and the iv anesthetics is that the former probably exert their biological activity through physical effects while the latter, in most cases, function through a biological receptor-mediated pathway.

Opioid Reversal Agents. At the end of a surgery, when the decision is made to reverse opioid-induced activity, the primary concern is to eliminate the respiratory depressant effects inherent in the potent opiates used as anesthetics. This reversal, brought about by opioid reversal agents, can also diminish the analgesic effects of the opiates. There are two types of reversal agents: pure antagonists and mixed agonist–antagonists. The antagonists naloxone and naltrexone are the most commonly used, although the use of the agonist–antagonist nalbuphine is increasing because it maintains a moderate level of analgesia while reversing the side effects. Nalmefene, a long-acting reversal agent, is in clinical trials.

Local Anesthetics. Nerve impulses are initiated by membrane depolarization, effected by the opening of a sodium-ion channel and an influx of sodium ions. Local anesthetics act by inhibiting the channel's opening; they bind to a receptor located in the channel's interior. The degree of blockage on an isolated nerve depends not only on the amount of drug, but also on the rate of nerve stimulation.

Local anesthetic activity is usually demonstrated by compounds which possess both an aromatic and an amine moiety separated by a lipophilic hydrocarbon chain and a polar group. In the clinically useful agents (Table 2) the polar group is an ester or an amide. Activity may be maintained, however, when the polar function is an ether, thioether, ketone, or thioester.

Pharmacological Profile. The profile of the ideal local anesthetic agent depends largely on the type and length of the surgical procedure for which it is applied. Procedures could include neuraxial (spinal and epidural) anesthesia, nerve and plexus blocks, or field blocks (local infiltration). In general, the ideal agent should have a short onset of anesthesia and be useful for multiple indications such as infiltration, nerve blocks, intravenous, extradural, spinal, and topical administration. The therapeutic indexes for systemic CNS and cardiovascular toxicity should be high and the agent should be compatible with the vasoconstrictor epinephrine.

GEORGE R. LENZ
BOC Group Technical Center
HOLLIS G. SCHOEPKE
THEODORE C. SPAULDING
Anaquest

A. G. Gilman, L. S. Goodman, T. W. Rall, and F. Murad, eds., *The Pharmacological Basis of Therapeutics,* Macmillan Publishing Company, New York, 1985.

K. W. Miller and M. E. Wolff, eds., *Burger's Medicinal Chemistry,* 4th ed., Vol. 3, John Wiley & Sons, Inc., New York, 1981, p. 623.

G. R. Lenz, S. M. Evans, D. E. Walters, and A. J. Hopfinger, *Opiates,* Academic Press, Orlando, Fla., 1986.

B. G. Covino and H. G. Vasallo, *Local Anesthetics: Mechanism of Action and Clinical Use,* Grune and Stratton, New York, 1976.

ANETHOLE. See ETHERS.

ANILINE AND ITS DERIVATIVES. See AMINES, AROMATIC AMINES, ANILINE AND ITS DERIVATIVES.

ANIMAL FEEDS. See FEEDS AND FEED ADDITIVES.

ANOREXIANTS. See ANTIOBESITY AGENTS.

ANSAMACROLIDES. See ANTIBIOTICS, ANSAMACROLIDES.

ANTACIDS, GASTRIC. See GASTROINTESTINAL AGENTS.

ANTHRACENE. See HYDROCARBONS.

ANTHRAQUINONE

In 1869, Graebe and Liebermann obtained anthraquinone by oxidizing anthracene with bichromate and sulfuric acid, which later became the first commercial method for the manufacture of anthraquinone from anthracene.

For many years, anthraquinone provided the dyestuff industry with one of the greatest and most prolific building blocks for the manufacture of valuable dyestuffs noted for their outstanding fastness properties. Since 1969, production figures have not been made public. However, in recent times, production has fallen off considerably due to the high cost of manufacturing and the discovery of other classes of dyestuffs with good fastness properties.

Physical Properties

When sublimed, anthraquinone forms a pale yellow, crystalline material, needlelike in shape. Unlike anthracene, it exhibits no fluorescence. It melts at 286°C and boils at 379–381°C. At much higher temperatures, decomposition occurs. Anthraquinone has only a slight solubility in alcohol or benzene and is best recrystallized from glacial acetic acid or high boiling solvents such as nitrobenzene or dichlorobenzene. It is very soluble in concentrated sulfuric acid.

Chemical Properties

In general, anthraquinone is a relatively inert compound exhibiting stability toward oxidation. Only the reduction products involving the keto groups are of any academic or industrial importance.

Economic Aspects

In the dyestuff industry, anthraquinone still ranks high as an intermediate for the production of dyes and pigments having properties unattainable by any other class of dyes or pigments. In the United States and abroad, anthraquinone is manufactured by a few large chemical companies. Although several radically different processes for manufacturing anthraquinone have been attempted, at this writing (1993) only two processes for its production come into consideration: manufacture by the Friedel-Crafts reaction utilizing benzene, phthalic anhydride, and anhydrous aluminum chloride, and by the vapor-phase catalytic oxidation of anthracene; the latter method is preferred.

Health, Safety, and Environmental Factors

Anthraquinone is a comparatively safe compound: LD_{50} (rat) is 3500 mg/kg. It is a mild allergen and, as a fine powder, may cause skin irritation. It presents only a slight fire hazard on exposure to heat.

Of the two processes used for the manufacture of anthraquinone, the one involving the vapor-phase catalytic oxidation of anthracene has the least effect on health and the environment. On the other hand, the use of benzene in the Friedel-Crafts process presents serious safety and health problems.

Uses

Aside from its major use in the manufacture of intermediates for anthraquinone dyes and pigments, anthraquinone is finding increasing interest as a catalyst in the pulping of wood, in the polymerization of various materials for plastics, and in the isomerization of vegetable oils. It has been used to make seeds distasteful to birds and as an accelerator in nickel electroplating.

A. J. COFRANCESCO
Consultant

BIOS, 1148 (1946).

P. H. Groggins, *Ind. Eng. Chem.* **23,** 152–160 (1931).

L. F. Fieser, *Experiments in Organic Chemistry,* 2nd ed., D. C. Heath & Co., Englewood, N.J., 1941, p. 189.

U.S. Pat. 2,652,408 (Sept. 15, 1953), H. Z. Lecher and K. C. Whitehouse (to American Cyanamid Co.); U.S. Pat. 2,938,913 (May 31, 1960), R. G. Weyker, F. H. Megson, T. Hoffman, and G. Wiesner (to American Cyanamid Co.).

ANTIAGING AGENTS

The human aging process is extremely complex and appears to affect all bodily functions equally; there are no known tissues that appear to age significantly faster than others. The tremendous advances in molecular biology over the past decade have made it possible to begin to study the basic mechanisms of aging at the molecular level and relate those findings to specific age-related diseases at the cellular, tissue, and organ level. It is likely that these advances will demonstrate that there are multiple mechanisms underlying the aging process at all levels and that multiple interventions will therefore be necessary to arrest, slow, or reverse the aging process. Factors already known to be intimately involved in the aging process include environmental, dietary, and genetic influences; each appears to be involved in the various age-related disease states to a greater or lesser degree. Consequently, no single agent is likely to be found that will arrest or reverse all aging processes. Indeed, current research efforts in the discovery of antiaging agents focus mostly on modulation of single aging processes and the effects of such modulation on one or two major organ systems of the body. Agents are discussed here relative to the prevention and treatment of the aging process and age-related disorders characteristic of the principal organ systems that are most affected.

Skin

It is essential to distinguish intrinsic aging from age-dependent abnormalities; in this case, the cumulative assault of sunlight on the skin. A brief review of these differences has appeared.

Tretinoin, the all-*trans* isomer of retinoic acid, was shown in the 1960s to be useful for the treatment of disorders associated with abnormal epithelial differentiation.

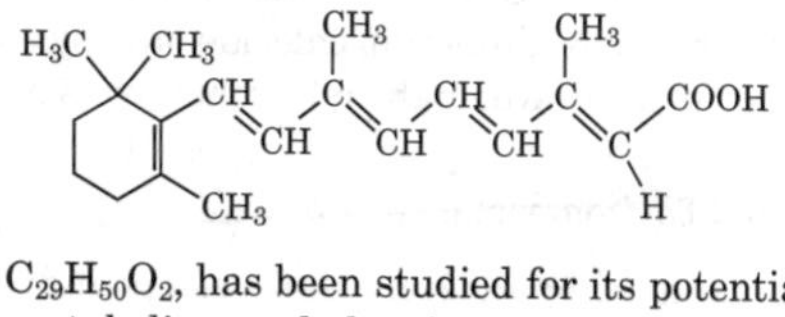

Vitamin E $C_{29}H_{50}O_2$, has been studied for its potential to modulate prostaglandin metabolism and alter immune response in aged mice.

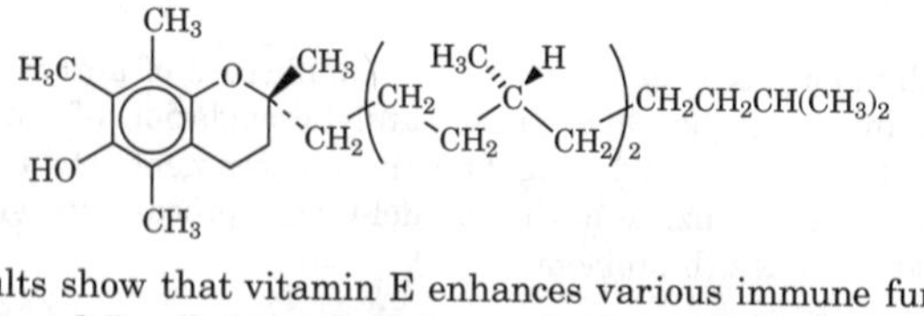

Results show that vitamin E enhances various immune functions (both T- and B-cell dependent) in aged mice and the enhancement appears to be mediated by decreased prostaglandin synthesis.

Sunscreens block the cutaneous absorption of ultraviolet (uv) light radiation (280–315 nm) and prevent sunburning, premature aging, and cancer of the skin (see COSMETICS). Photoaged skin, long believed to be irreversibly damaged, has now been shown to undergo significant repair when uv exposures are stopped.

Central Nervous System

Aging is associated with progressive deterioration in CNS function. A number of factors besides genetic considerations contribute to those degenerative changes, including nutritional elements and environmental toxins. The blood brain barrier (BBB) is a major modulator of nutrient delivery to the CNS. Senescence is associated with significant changes in BBB function including a decrease in BBB choline transport and brain glucose influx.

Studies have confirmed that enhancing the cholinergic system could lead to reversal of age-related memory loss. Direct and indirect evidence indicating an activation of cholinergic mechanisms exist for pyrrolidinone derivatives including piracetam, oxiracetam, aniracetam, pyroglutamic acid, tenilsetam, and pramiracetam, and for miscellaneous chemical structures such as vinpocetine, naloxone, ebiratide (a hexapeptide), and phosphatidylserine. All these drugs prevent scopolamine-induced disruptions of several learning and memory paradigms in animals and humans.

Free radicals have been implicated in the pathogenesis of a number of neuropsychiatric disorders including aging of the CNS. The known nootropic effects of the drug centrophenoxine (an alternative name for meclofenoxate), a form of dimethylaminoethanol, $(CH_3)_2NCH_2CH_2OH$, can be interpreted on the basis of its OH radical scavenger properties and is thus thought to act as a protective agent of neuronal membranes into which it is incorporated in the form of phosphatidyldimethylaminoethanol. At the molecular level, prevention of cross-linking of proteins is viewed as a major beneficial antiaging effect of centrophenoxine.

Cardiovascular System

Cardiovascular disease remains the most common age-related disease of all major disease types and is the most common cause of morbidity and mortality in old age. Besides age, however, risk factors that initiate and aggravate cardiovascular disease include hypertension, personality traits, genetics, diet, smoking, obesity, physical indolence, dyslipidemia, impaired glucose tolerance, and sex. Most risk factors are more prevalent in the elderly than in the young adult. The management of elevated blood lipids represents a most important objective in the prevention and treatment of cardiovascular disease. A concise review of pharmacological agents such as lovastatin for the treatment of hypercholesterolemia has appeared (see CARDIOVASCULAR AGENTS). In addition, antiatherosclerogenic diets low in saturated fat and cholesterol, rich in fiber, and with substitution of polyunsaturated fat and restricted calories tend to normalize serum lipids and to cause lesions to involute.

A review of cardiac failure and the mechanism of the dysfunction of the heart and the vasculature has pointed out that a knowledge of cellular and subcellular mechanisms is critical if alteration and corrective therapies are to effect modulation of the aging process. Whatever the etiology, one important feature of aging is the remodeling of the ventricle and blood vasculature. Various agents such as vasodilators, hydroalazine, and isosorbide dinitrate have been used.

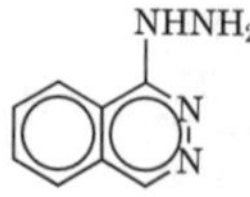

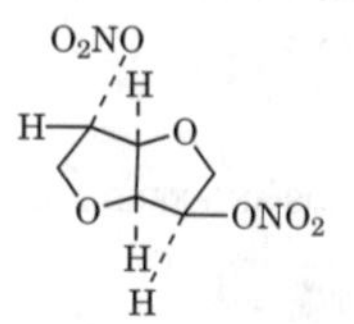

Other possibilities include α-blockers and calcium antagonists. Other unproven agents include the β-adrenoreceptor agonists, phosphodiesterase inhibitors that affect movement of sodium and potassium through cellular channels, and calcium contractile protein sensitizers. Any of these agents and the appropriate cellular mechanisms that prevent or slow morphologic alteration in the heart or blood vessels should counteract the aging process.

Immune System

Normal aging is associated with a gradual decline in the immune response of all species studied, including humans. Because of the important role that the immune system plays with respect to overall health in the aged, a number of approaches have been devised to prevent the loss of immune responsiveness or to restore it once it is declining or has been lost. These immunological interventions have been reviewed.

One component of the age-related decline in immune function is decreased production of the lymphokine that promotes the growth of T-cells, interleukin 2 (IL-2). Recovery of T-regulatory effects on B-cell differentiation has been reported in human cells from elderly patients treated with IL-1 or IL-2.

IL-2, endotoxin, and thymic epithelial cell products, but not interleukin 1, were found to promote functional maturation of immature thymocytes. Two classes of drugs show thymomimetic actions. Levamisole, $C_{11}H_{12}N_2S$, and the sodium salt of diethyl dithiocarbamate (imuthiol), as well as certain other sulfur-containing compounds restore T-cell function via induction of a thymic hormone-like factor. Imuthiol, $C_5H_{10}NS_2Na$, can also generate cytotoxic T-cells in athymic nude mice and in euthymic old animals. The other class consists of hypoxanthine derivatives such as isoprinosine and NPT 15392. These compounds induce T-cell maturation directly and promote T-cell function *in vitro* and *in vivo*.

Endogenous substances have been studied as immunoregulatory agents. Melatonin, the pineal neurohormone *N*-acetyl-5-methoxytryptamine, has been shown to antagonize the immunosuppressive effects of anxiety stress in mice, as measured by antibody production, thymus weight, and by the capacity of stressed- and evening-melatonin-treated mice to react against a lethal virus.

Administration of coenzyme Q_{10} has been reported to have an adjuvant effect on humoral responses in mice, and to protect them from viral- and carcinogen-induced tumor formation.

Natural killer (NK) cells and macrophages play an important role in host defense against a variety of pathogenic challenges including viruses, bacteria, parasites, and cancer cells. A number of drugs have been developed to activate this arm of the immune response and to restore the diminished response in the aged. One such agent, bropirimine, $C_8H_{10}BrN_3O$, is a pyrimidinone that has shown good NK cell activation as measured by their enhanced ability to kill a susceptible tumor target.

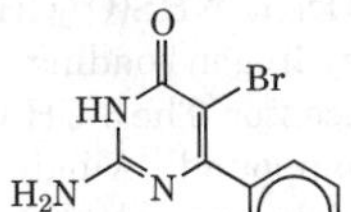

Another class of compounds that has been found to stimulate nonspecific, as well as some specific immune responses, is the 8-substituted guanosines.

The transmembrane signaling component of cellular activation has been an area of intense research interest. A report on the reduced proliferation of T-cells in aged humans indicated that the age-related defect in proliferation was mainly in the CD8+ (T-suppressor) cells and that is was predominantly in the CD4+ (T-helper) subset that decreases in transmembrane signaling were observed. Certain thiol compounds have been shown to enhance the T-cell-dependent immune response of mice in vivo and the proliferation of T-cells *in vitro*.

The interactions between nutrition and immunity have been the subject of much investigation and have been reviewed. It is now recognized that even mild deficiencies involving protein–calorie malnutrition and those involving single nutrients are associated with impaired immune responses. Large doses of "essential" nutrients (eg, zinc) may have a deleterious effect on the immune system. However, corrections of deficiencies of iron, zinc, and vitamins C, E, and B complex are associated with improved immune responses in the elderly (see VITAMINS).

Finally, in another study related to nutrition and the immune response in the aged, old mice were given oral doses of two amino acids (qv), lysine, and arginine. The treated mice showed evidence of recovered mitogenic responsiveness, expression of T-cell markers, and production of thymic serum factor (thymulin).

W. B. JOLLEY
ICN Pharmaceuticals Inc.
H. B. COTTAM
University of California, San Diego

A. Oikarinen, *Photo-Dermatol.* **7,** 3–4 (1990).
M. Pahlavani, *Drugs of Today* **23,** 611–624 (1987).
R. Chandra, *Age Ageing* **19** (Suppl.), 25–31 (1990).
R. G. Cutler, *Am. J. Clin. Nutr.* **53,** 373S–379S (1991).

ANTIALLERGIC AGENTS. See ANTIASTHMATIC AGENTS.

ANTIALLERGY AGENTS. See HISTAMINE AND HISTAMINE ANTAGONISTS.

ANTIANEMIA PREPARATIONS. See IRON COMPOUNDS; PHARMACEUTICALS; VITAMINS.

ANTIANGINAL AGENTS. See CARDIOVASCULAR AGENTS.

ANTIANXIETY AGENTS. See PSYCHOPHARMACOLOGICAL AGENTS.

ANTIARRYTHMIC DRUGS. See CARDIOVASCULAR AGENTS.

ANTIASTHMATIC AGENTS

Asthma is an extremely complex condition characterized by variable and reversible airways obstruction combined with nonspecific bronchial hypersensitivity. Asthma affects 3–5% of the population and is one of the most common chronic illnesses.

Clinically, there are several ways of classifying asthma and treatment varies depending upon the classification. Extrinsic asthma, also called allergic asthma, is experienced by adults and, most commonly, by children. In extrinsic asthma, it is possible to demonstrate specific causal agents, usually antigens, eg, animal danders, foods, drugs, house dust, pollens, or mold spores.

Intrinsic asthma, also called idiopathic asthma, usually develops in adulthood. In intrinsic asthma allergic factors are not demonstrable. Episodes of intrinsic asthma may be triggered by a variety of stimuli, eg, emotional state, exposure to cold air, or inert dusts. Both intrinsic and extrinsic asthmatics can be prone to exercise-induced attacks. Individuals who experience a combination of extrinsic and intrinsic asthmatic reactions have mixed asthma.

Current asthma treatments are not curative and historically have relied on pharmacologic intervention with bronchodilators (Fig. 1) to prevent or relieve the symptoms of asthma. More recently the focal points of both treatment and research efforts have shifted from bronchodilators to agents which reduce the underlying inflammatory state. Management of extrinsic asthma usually includes manipulation

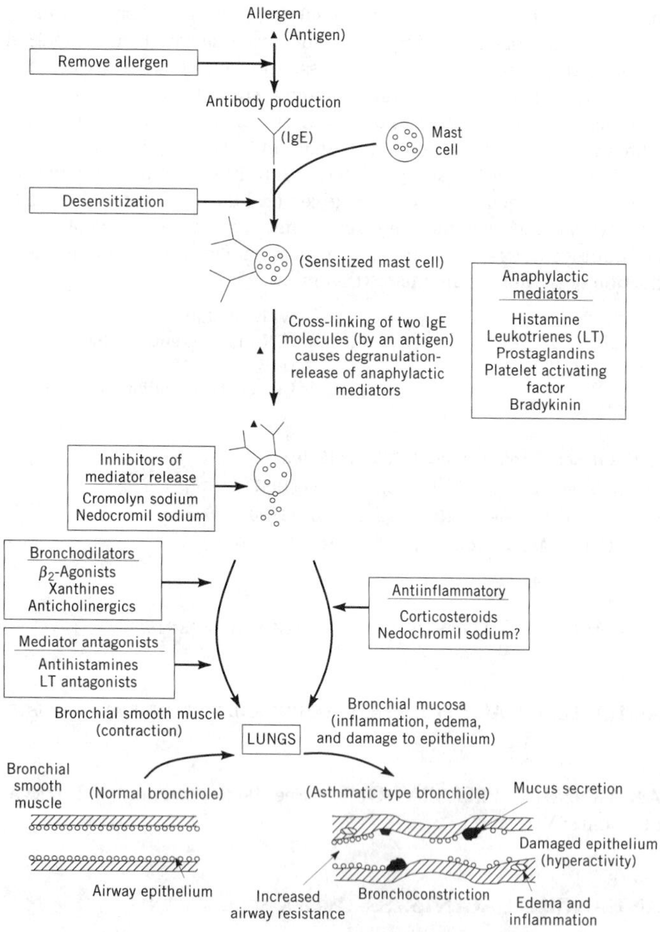

Figure 1. Schematic diagram showing possible sites of action of antiasthmatic drugs.

of the patient's environment to minimize or completely eliminate the causal agent. This can be especially beneficial in occupationally induced asthma. The cascade of events involved in the asthmatic response and potential points for pharmacologic intervention are shown in Figure 1.

The order of preference of drug treatment varies from country to country (Table 1). In part this difference reflects the lack of a "magic bullet" for the treatment of asthma. However, this difference may also be explained by differences in marketing approval. Some agents are not yet available in the United States and the exposure of U.S. physicians to some of the newer inhalation formulations has been limited.

Table 1. Order of Preference for First Choice Antiasthmatic Agent Maintenance Therapy

	Order of preference		
	U.S.	U.K.	1989 U.S. market share, %
oral xanthines	1	4	~27
inhaled β_2-agonist	2	1	~26
oral β_2-agonist	3		~18
inhaled steroid	4	2	~18
inhaled inhibitor of mediator release	5	3	~5

Improvements in asthma treatment include the development of more effective, safer formulations of known drugs. The aerosol administration of β_2-agonists or corticosteroids results in a decrease in side effects. Also, the use of reliable sustained release formulations has revolutionized the use of oral xanthines which have a very narrow therapeutic index (see CONTROLLED RELEASE TECHNOLOGY). For many individuals, asthma symptoms tend to worsen at night and the inhaled bronchodilators do not usually last through an entire night's sleep.

β-Adrenergic Stimulants

β_2-Agonists are widely used in the symptomatic treatment of asthma. Although both oral and aerosol formulations of these bronchodilators have been available for many years, advances have occurred in delivery technology with the development of dry powder aerosols (qv) (see DRUG DELIVERY SYSTEMS). The ease of usage of these breath-activated systems has improved patient compliance and therapeutic response. There are several detailed reviews on β_2-agonist therapy of bronchial asthma, and on the structure–activity relationships of this class of drugs.

Because of the widespread nature of adrenoceptors, nonselective β-agonists can produce many undesirable side effects. Therefore, before adrenergic agonists could become widely used in the treatment of asthma, some selectivity in action was needed. Whereas epinephrine and ephedrine have significant agonist activity at both α and β adrenoceptors, isoproterenol is a selective agonist at the β receptor. However, isoproterenol does not distinguish between the β_1 and β_2 receptors and it is not active orally.

Aerosol administration of isoproterenol produces a prompt (2–5 min) intense bronchodilatation of relatively short (1-h) duration. The lack of β_2-selectivity leads, in many cases, to tachycardia and blood pressure elevation. Also, use of isoproterenol, like all other known β-agonists, results in a down-regulation, or desensitization, of β-adrenergic receptors. This desensitization is only partial, and after time (depending on dose, patient, and agent), a stable, less responsive state is achieved in which β-agonists remain effective. Isoproterenol has been widely used for many years.

A significant advance in β-agonist therapy occurred with the discovery of metaproterenol, $C_{11}H_{17}NO_3$. Replacing the catechol subgroup with a resorcinol unit results in a compound which is no longer susceptible to metabolism by COMT and therefore has a longer (4-h) duration of action.

Changing the *N*-substituent of metaproterenol to a *tert*-butyl group gives rise to terbutaline, $C_{12}H_{19}NO_3$, an orally active agent having a duration of action of from 6–7 h. Terbutaline has about six times the selectivity of isoproterenol and is claimed to produce fewer side effects.

Variation of the substituents on the aromatic ring led to the discovery that the 3-hydroxy could be replaced with a variety of groups, eg, CH_2OH, $NHCH_3$, $NHCONH_2$, $NHS(O)_2CH_3$, that successfully mimic the electron donating and hydrogen bonding properties of the hydroxide moiety at the adrenergic receptor. The 3-CH_2OH analogue has been the most studied in this series of compounds. Albuterol, $C_{13}H_{21}NO_3$, also called salbutamol, is the most widely prescribed β_2-agonist for either aerosol or oral administration both in the United States and worldwide.

More recent research efforts have focused on the development of longer acting β-agonists which could be administered less frequently and be more efficacious in controlling nocturnal asthma. There are several agents currently under clinical evaluation: bitolterol, $C_{28}H_{31}NO_5$, clenbuterol, $C_{12}H_{18}Cl_2N_2O$, and salmeterol, $C_{25}H_{37}NO_4$.

Although β_2-agonists are useful in the treatment of asthma, the profiles would be improved if the bronchodilating effects could be further separated from the side effects. It is not clear how this could be accomplished, but it has been shown that a β_3-receptor exists in humans and other uncharacterized subclasses of receptors have also been postulated. The investigation of these other β-receptors could lead to more selective agents.

Table 2. Xanthine and Xanthine Derivatives Used as Oral Antiasthmatic Agents

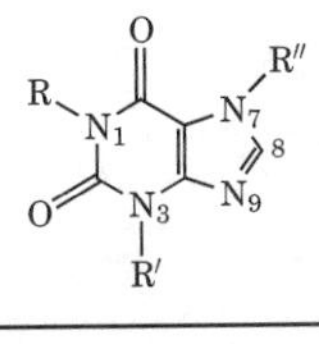

Compound	Molecular formula	R	R′	R″
xanthine	$C_5H_4N_4O_2$	H	H	H
theophylline	$C_7H_8N_4O_2$	CH_3	CH_3	H
theobromine	$C_7H_8N_4O_2$	H	CH_3	CH_3
caffeine	$C_8H_{10}N_4O_2$	CH_3	CH_3	CH_3
enprofylline	$C_8H_{10}N_4O_2$	H	n-C_3H_7	H

Xanthine Derivatives

For many years oral xanthines, shown in Table 2, were the preferred first-line treatment for asthma in the United States. Within this class of compounds theophylene, or one of its various salt forms, such as aminophylline (theophylline: ethylenediamine::2:1), have been the predominant agents. Theophylline, 1,3-dimethylxanthine, is but one member of a class of naturally occurring alkaloids. Two more common alkaloids are theobromine, isomeric with theophylline and the principal alkaloid in cacao beans, and caffeine, 1,3,7-trimethylxanthine, found in coffee and tea.

Common side effects of theophylline therapy include headache, dyspepsia, and nausea. More serious side effects such as lethal seizures or cardiac arrythmias can occur if blood levels are too high. Many derivatives of theophylline have been prepared in an effort to discover an analogue without these limitations. However, the most universal solution has resulted from the development of reliable sustained-release formulations. This technology limits the peaks and valleys in serum blood levels that occur with frequent dosing of immediate release formulations. Controlled release addresses the problems inherent in a drug which is rapidly metabolized but which is toxic at levels (>20 μg/mL that are only slightly higher than the therapeutically efficacious ones (10–20 μg/mL). Furthermore, such once-a-day formulations taken just before bedtime have proven especially beneficial in the control of nocturnal asthma.

The effectiveness of theophylline in the treatment of asthma seems to result from a combination of biological properties which are not clearly understood. Detailed discussions of the possible role of xanthines in asthma may be found.

Theophylline's predominant mode of action appears to be bronchodilation. However, it has also been shown that prophylactic administration of theophylline provides some protection from asthma attacks and suppresses the late-phase response.

An alternative mechanism which has more recently been suggested is antagonism by theophylline of adenosine receptors. This is a well-documented effect which has been blamed for causing many of the theophylline side effects. Adenosine has a bronchoconstrictor effect when given by inhalation to asthmatics but not when given to controls. Also, asthmatics release adenosine into the circulation following antigen-induced bronchoconstriction.

Although the benefit of theophylline treatment in asthma is without question, its use is decreasing as a result of the introduction of other effective drugs which do not have the same potential for serious side effects.

Steroids

Steroids are widely used for the preventative treatment of asthma. Especially useful in the management of severe cases, their usage appears to be increasing. However, they have had a checkered history in asthma treatment. Shortly after the first synthetic corticosteroids became available, po cortisone was tried as an antiasthmatic agent and showed remarkable success. However, within a few years the number and severity of side effects reported from the systemic administration of nonselective corticosteroids was deemed unacceptable.

Resurgent interest in the use of steroids for the treatment of asthma has been prompted by several developments. First came the discovery of steroids such as prednisone, beclomethasone dipropionate, and budesonide. These newer, more selective compounds successfully separate glucocorticoid (antiinflammatory) and mineralocorticoid (electrolyte regulation) activities and do not suppress serum hydrocortisone. Steroid usage has increased because of the identification of asthma as an inflammatory disease. Thus aerosolized antiinflammatory steroids are often used for first-line asthma treatment. Those is use today include cortisone ($C_{12}H_{28}O_5$), prednisone ($C_{21}H_{26}O_5$), beclomethasone dipropionate ($C_{22}H_{29}ClO_5$), budesonide ($C_{25}H_{34}O_6$).

Although the mechanism of glucocorticosteroid action in bronchial asthma is not fully understood, various possibilities have been discussed in depth. The time course of action of steroids is slower than that of β_2-agonists or theophylline and therefore steroids are not considered to be bronchodilators. It is known that steroids bind to cytosolic receptors. The steroid–receptor complex then enters the cell nucleus where the complex acts at specific sites and affects protein synthesis. The effect is to reduce the inflammatory response as well as the concentration of bronchoconstricting mediators. In addition, glucocorticoid treatment is known to reverse β_2-agonist-induced adrenergic subsensitivity and to increase the number of β_2-adrenergic receptors in lung cells. The resultant increase in sensitivity to the natural circulating levels of norepinephrine could help induce bronchorelaxation.

Single-dose or short-term treatment with aerosolized steroids inhibits both the late asthmatic response and allergen-induced bronchial hyperresponsiveness. However it does not affect the early asthmatic response nor does it induce bronchodilation.

Inhibitors of Mediator Release

Whereas disodium chromoglycate (DSCG), $C_{23}H_{14}Na_2O_{11}$, enjoys some modest success as a preventative antiasthmatic agent, it has never achieved the same level of popularity in the United States as it has in other markets, such as in the United Kingdom. Its properties have been covered in detail in several reviews.

DSCG is very poorly absorbed following oral dosing and is therefore administered by aerosol inhalation, four times a day. Metabolism is minimal: DSCG is excreted mostly as unchanged drug. Thus, although it is considered one of the safest available antiasthmatic agents, it is not a cure-all and seems to work best as an adjunct to other therapy, allowing a reduction in the dosage of the other agents. Clinical studies show that DSCG offers no protection if administered following antigen challenge, but if administered prophylactically, it protects both extrinsic and intrinsic asthmatics. DSCG is uniquely effective against both the early- and late-phase asthmatic responses; it also protects against exercise-induced asthma.

With the shift in preference to aerosolized agents, nedocromil sodium has been introduced as a follow-up agent to DSCG. Nedocromil sodium, structurally distinct from DSCG, is significantly more potent in humans, and can be administered twice daily.

Anticholinergic Agents

At this writing anticholinergic agents are not widely used for the symptomatic treatment of asthma, although compounds such as atropine, $C_{17}H_{23}NO_3$, have been used for centuries.

The beneficial effect of anticholinergics in asthma relies upon bronchial smooth muscle exhibiting a cholinergically mediated tone (resting state tension). Although atropine is effective in preventing exercise-induced asthma, it and other anticholinergic agents have no effect on bronchial hyperresponsiveness, the release of other mediators, or on the inflammatory process. Significant dose-related side effects such as blurred vision, dry mouth, and inhibition of gastric motility occur. These side effects result from systematic distribution of the drugs, including penetration into the central nervous system, and their widespread antagonism of other muscarinic receptors.

Ipratropium bromide, $C_{20}H_{30}BrNO_3$, is an example of a newer anticholinergic agent. Ipratropium bromide has gained limited acceptance as an antiasthmatic agent and seems to be more useful in patient populations that show limited response to β_2-adrenergic agonists.

Agents Undergoing Clinical Evaluation

Agents undergoing clinical evaluation include antihistamines [clemastine ($C_{21}H_{26}$ ClNO) chloropheniramine ($C_{16}H_{19}N_2Cl$), terfenadine ($C_{20}H_{41}NO_2$), and astemizole ($C_{28}H_{31}N_4FO$)] and leukotriene antagonists [FPL-55712 ($C_{27}H_{30}O_9$), LY-171883 ($C_{16}H_{22}N_4O_3$), MK-571 (formerly L-660711) ($C_{24}H_{23}N_2O_3ClS_2$), and ICI-204219 ($C_{31}H_{33}N_3O_6S$)].

PETER R. BERNSTEIN
ICI Americas Corporation

D. R. Buckle and H. Smith, eds., *Development of Anti-Asthma Drugs*, Butterworths, London, 1984.

E. B. Weiss, M. S. Segal, and M. Stein, eds., *Bronchial Asthma: Mechanism and Therapeutics*, 2nd ed., Little, Brown and Co., Toronto, Canada, 1985.

P. J. Barnes, ed., *New Drugs for Asthma*, IBC Technical Services, London, 1989.

ANTIBACTERIAL AGENTS, SYNTHETIC

QUINOLONES

Quinolone carboxylic acids are a class of totally synthetic antibacterial agents which have the general structure (**1**).

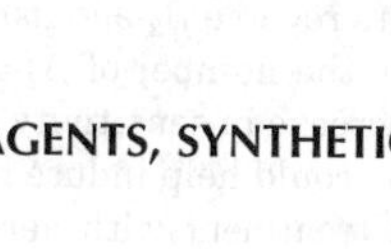

(**1**)

This representation is intended to encompass 4-oxo-3-quinolinecarboxylic acids as well as the corresponding 1,8-naphthyridines, cinnolines, and pyrido [2,3-*d*]-pyrimidines. These classes are illustrated by ciprofloxacin (**2**), nalidixic acid (**3**), cinoxacin (**4**), and piromidic acid (**5**), respectively.

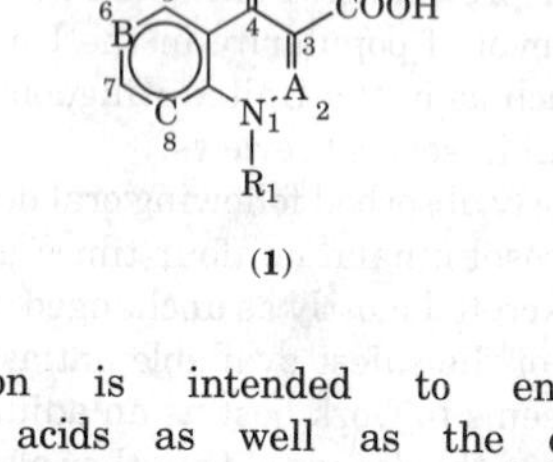

(**2**)

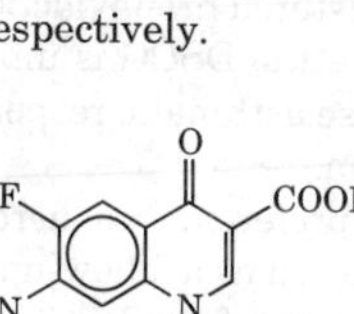

(**3**)

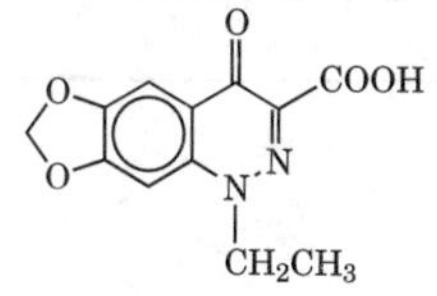

(**4**)

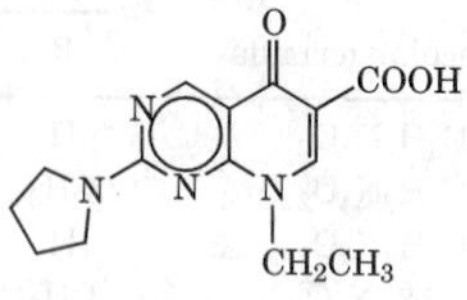

(**5**)

Established quinolone antibacterial agents are ciprofloxacin, ofloxacin, enoxacin, norfloxacin, and pefloxacin.

Structure–Activity Relationships

The most important structural features necessary for meaningful antibacterial activity of quinolones include a carboxylic acid attached to the 3-position of the quinolone nucleus and a small alkyl or aryl group in the 1-position. In addition to these more or less rudimentary requirements for activity, all of the newer generation quinolones have a fluorine attached to the 6-position and a nitrogen heterocycle attached to the 7-position. This heterocycle is often a piperazine or pyrrolidine derivative.

Quinolone–Cephalosporin Codrugs. Quinolones have been covalently linked to cephalosporins in order to generate a codrug containing one molecule of each type of antibacterial agent. An example is the fleroxacin–cefotaxime combination, Ro 23-9424. In essence, the cephalosporin acts as a carrier for the quinolone.

Mechanism of Action

Quinolones exert their antibacterial activity by interfering with the replication of bacterial DNA by inhibition of the enzyme DNA gyrase through the formation of a quinolone-DNA-enzyme complex. The critical reaction, catalyzed by the enzyme, is the negative supercoiling of bacterial DNA, a process involving the breaking and resealing of double-stranded circular DNA.

Preparation of Quinolones

The general method by which most newer fluoro quinolones are prepared involves a ring closure reaction to form the quinolone nucleus by forming one of the highlighted bonds, *a* or *b*, shown in general structure (**6**).

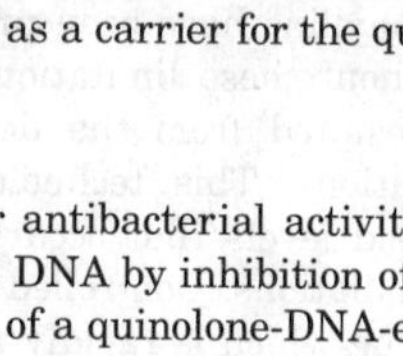

(**6**)

Safety of Quinolone Antibacterial Agents

For the most part quinolones are well tolerated with few reports of adverse reactions, but there are some concerns, one of which is that they cause arthropathy in juvenile animals. This is a condition in which the cartilage of weight-bearing joints is damaged. Because of this concern, quinolones are contraindicated for children and during pregnancy. A second problem is that of adverse side effects including those of the central nervous system (CNS) and also adverse reactions when quinolones are taken with other drugs. Adverse reactions should decrease with more recent quinolones such as lomefloxacin, sparfloxacin, and temafloxacin.

Resistance. Resistance to quinolones, particularly to newer agents, is not a significant problem at this time.

Economic Aspects

The domestic antiinfective market is in excess of \$3 billion and the global market surpasses \$8 billion. Earlier quinolones have had little impact on this market as their utility was limited primarily to urinary tract infections. Newer quinolones, because of their safety, seemingly low propensity toward bacterial resistance, and vastly improved therapeutic utility, now enjoy a much greater share of this market.

M. Reuman
G. Y. Lesher
Sterling-Winthrop Inc.

W. A. Goss, W. H. Dietz, and T. M. Cook, *J. Bacteriol.* **89,** 1068 (1965).

A. Bauernfeind and G. Grummer, *Chemotherapia* **10,** 95 (1965).

G. C. Crumplin and J. T. Smith, *Antimicrob. Agents Chemother.* **8,** 251 (1975).

M. Gellert, K. Mizuuchi, M. H. O'Dea, and H. A. Nash, *Proc. Natl. Acad. Sci. USA* **73,** 3872 (1976).

M. Gellert, K. Mizuuchi, M. H. O'Dea, T. Itoh, and J. Tomizawa, *Proc. Natl. Acad. Sci. USA* **74,** 4772 (1977).

NITROFURANS

The nitrofurans encompass a class of synthetic antibacterials characterized by the 5-nitro-2-furanyl group:

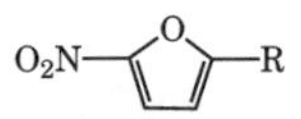

With relatively few exceptions, the R function contains the azomethine (—CH=N—) moiety. The most prominent exceptions contain the olefinic (—C=C—) moiety. Nitrofurans include furium ($C_9H_7N_3O_4S$), furazolidone ($C_{14}H_{12}N_6O_6 \cdot HCl$), *Z*-furan ($C_7H_6N_2O_4$), furylfuramide ($C_{11}H_8N_2O_5$), nitrovin ($C_{14}H_{12}N_6O_6 \cdot HCl$), furalazine ($C_9H_7N_5O_3$), acetylfuratrizine ($C_{11}H_9N_5O_4$), panfuran-S ($C_{11}H_{11}N_5O_5$), nifuroxime ($C_5H_4N_2O_4$), nitrofurazone ($C_6H_6N_4O_4$), nifuraldezone ($C_7H_6N_4O_5$), nihydrazone ($C_7H_7N_3O_4$), nitrofurantoin ($C_8H_6N_4O_5$), nifuratel ($C_{10}H_{11}N_3O_5S$), nitrofurathiazide ($C_{13}H_{12}N_4O_6S$), nifurtoinol ($C_9H_8N_4O_6$), and nifurprinol ($C_{12}H_{10}N_2O_4$).

Several nitrofurans have been marketed either regionally or worldwide in the human and veterinary areas because of their broad spectrum of activity, relatively mild toxicity, and low tendency for resistance development.

Chemical Properties

Nitrofurans are generally stable, water-insoluble crystalline solids, which decompose (darken) upon prolonged exposure to light or alkali. Their strong ultraviolet absorption provides the basis for analytical procedures for these materials.

Biological Mechanism of Action

The mechanism of action of selected members of this class has been investigated. Nitrofurantoin is bactericidal at the concentrations achieved in the urinary tract and highly effective against the corresponding pathogens. Since its introduction for use in uncomplicated urinary tract infections, essentially no resistance development has occurred, unlike nearly all other antimicrobials. This is believed to be because of the ability of the nitrofurans to affect multiple cytoplasmic targets, such as citric acid cycloenzyme, bacterial DNA and RNA, and ribosomes essential for bacterial function.

Mutagenic and Carcinogenic Activity

Although *in vitro* mutagenicity tests suggest that some nitrofurans in general provoke a positive response, use of *in vivo* mammalian systems has produced equivocal or negative results. While it is not possible to generalize the carcinogenicity of this class of compounds, nitrogurazone, furazolidone, and nitrofurantoin have been used therapeutically for over 30 yr with no reports of human neoplasia.

Preparation

Most commercial nitrofurans are derived from 5-nitro-2-furancarboxaldehyde or the corresponding diacetate.

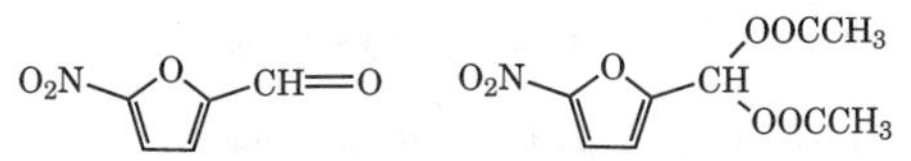

Thomas J. Schwan
Frank H. Ebetino
Procter & Gamble Pharmaceuticals

C. C. McOsker, J. R. Pollack, and J. A. Anderson, in L. H. Harrison, ed., *Management of Urinary Infections,* Royal Society of Medicine International Congress and Symposium Series No. 154, Royal Society of Medicine Service Limited, London, 1989.

W. H. Butler, T. C. Graham, and M. L. Sutton, *Food Chem. Toxicol.* **28,** 49 (1990).

J. R. Ames, M. D. Ryan, and P. Kovacic, *J. Free Radical Bio. Med.* **2**(5–6), 377, 1986.

R. E. Chamberlain, *J. Antimicrob. Chemother.* **2,** 325 (1976).

SULFONAMIDES

Sulfonamides derived from sulfanilamide (*p*-aminobenzenesulfonamide) are commonly referred to as sulfa drugs. Therapeutically active derivatives are usually substituted on the N^1 nitrogen; the N^4 position is generally unsubstituted. These features are illustrated by the structures of sulfanilamide (**1**) and sulfadiazine (**2**).

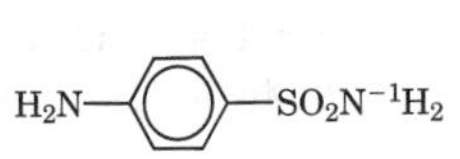

(**1**)

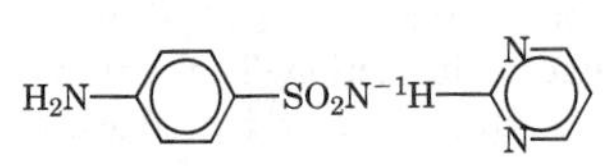

(**2**)

The sulfa drugs are still important as antimicrobials, although they have been replaced in many systemic infections by the natural and semisynthetic antibiotics. They are of great value in Third-World countries where problems of storage and lack of medical personnel make appropriate use of antibiotics difficult. They are especially useful in urinary tract infections, particularly the combination of sulfamethoxazole with trimethoprim. Their effectiveness has been enhanced by co-administration with dihydrofolate reductase inhibitors, and the combination of sulfamethoxazole with trimethoprim is of value in treatment of a number of specific microbial infections. The introduction of this combination (cotrimoxazole) in the late 1960s (1973 in the United States) resulted in increased use of sulfonamides.

The sulfas also remain clinically useful in the treatment of chancroid, lymphogranuloma venereum, trachoma, inclusion conjunctivitis, and the fungus-related nocardiosis. In combination with pyrimethamine, they are recommended for toxoplasmosis and have been used for chloroquine-resistant falciparium malaria. There has also been some use of sulfas for the prophylaxis of rheumatic fever. The sulfone, dapsone, remains an accepted treatment for all forms of leprosy.

The clinical usefulness of the sulfonamides depends not only on antimicrobial effectiveness, but on other factors such as aqueous and liposolubility, protein-binding, half-life, and metabolism. Currently used sulfas vary widely in their absorption, distribution, and excretion patterns. Some of those in clinical practice, past or present, are sulfanilamide, sulfacetamide, sulfadiazine, sulfadimethoxine, sulfaguanidine, sulfisomidine, sulfisoxazole, sulfamethazine, sulfamethizole, sulfamethoxazole, sulfamethoxypyridazine, sulfamoxole, sulfaphenazole, sulfapyridine, sulfapyrazine, sulfathiazole, N^4-phthalysulfathiazole, N^4-succinylsulfathiazole, mafenide, sulfasalazine, sulfamidochrysoidine (Prontosil), dapsone, and acedapsone.

Their antimicrobial activity includes species requiring a folic acid synthetic pathway, comprising many gram-positive and gram-negative cocci and bacilli, mycobacteria, some large viruses, protozoa, and fungi. The action of the sulfa drugs and related sulfones is bacteriostatic rather than bactericidal. It is believed to be due to interference with the assembly of folic acid at the step that adds *p*-aminobenzoic acid (PABA).

Physiochemical Properties

The sulfa drugs are weak acids, the more important ones generally having a pK_a in the range of 5–8. This acidity, due to the sulfonamide function, makes the sulfas soluble in basic aqueous solution. The pK_a is modified by the presence of the N^1-substituent, but the clinically useful sulfas generally have pK_a values which give the compounds good solubility at physiologic pH.

The amino group is readily diazotized in aqueous solution, and this reaction forms a basis for the assay of sulfas. Aldehydes also react to form colored anils, useful in detection.

Lipid solubilities of the sulfonamides vary over a wide range. Oil–water partition coefficients using an aqueous ethylene chloride system have been determined. The differences among individual members unquestionably influence their pharmacokinetics as well as microbial activity.

Numerous studies have been made to find a correlation between the physicochemical properties of the sulfonamides and their bacteriostatic activity. Degree of ionization, lipid–water solubility, electron distribution values, and protein binding have all been observed.

Biological Mechanism of Action

The sulfa drugs act on microorganisms by limiting or halting growth rather than by a bactericidal action. They inhibit growth of bacteria *in vitro* only if the medium is free of inactivating substances, mainly peptones and *p*-aminobenzoic acid.

Development of Resistance. One of the principal disadvantages of sulfonamide therapy is the emergence of drug-resistant strains of bacteria. Resistance develops by several mechanisms: overproduction of PABA; altered permeability of the organisms to sulfonamides; and reduced affinity of dihydropteroate synthetase for sulfonamides while the affinity for PABA is retained. Sulfonamides also show cross-resistance to other sulfonamides but not to other antibacterials. In plasmodia, resistance may occur by means of a bypass mechanism in which the organisms can use preformed folic acid.

Structure–Activity Relationships

The following generalizations arose from a review of more than 5000 sulfonamides.

1. The amino and sulfonyl groups on the benzene ring should be in the 1,4 positions; the amino group should be unsubstituted or converted to a free amino *in vivo*.
2. Replacement of the benzene ring by other ring systems, or the introduction of additional substituents, decreases or abolishes activity.
3. Exchange of the SO_2NH by SOC_6H_4-*p*-NH_2, $CONH_2$, CONHR, or COC_6H_4R generally reduces activity.
4. N^1-Monosubstitution may result in greater activity, and will increase activity with a number of heterocycles; N^1-disubstitution in general leads to inactive compounds.

A later review confirmed these generalizations.

Preparation and Manufacture

The most common method for the preparation of sulfonamides is by the action of *N*-acetylsulfanilyl chloride with the appropriate amine. Excess amine or suitable base is used to neutralize the hydrochloric acid formed.

$$CH_3CONHC_6H_4SO_2Cl + RNH_2 \xrightarrow{\text{base}} CH_3CONHC_6H_4SO_2NHR + HCl$$

The resulting acetyl compound is usually hydrolyzed with aqueous alkali to give the free amine.

Analysis

Sulfonamides having a free *p*-amino group are readily assayed by titration with nitrous acid. The sulfonamide function may also be titrated with base, such as lithium methoxide. The majority of the sulfas listed in the *U.S. Pharmacopeia XXII*, however, are assayed by chromatographic methods, particularly high performance liquid chromatography.

Toxicity

A small percentage of patients treated with sulfa drugs have shown toxic effects, such as drug fever, rashes, mild peripheral neuritis, and mental disturbance. In general, these effects are more prevalent with higher blood levels, and may accompany poor excretion or overdosing. In 1966 the FDA required that two long-acting sulfas, sulfamethoxypyridazine and sulfadimethoxine carry a label warning of the possibility of death due to Stevens-Johnson syndrome, an extremely severe dermatologic reaction.

Crystalluria, due to formation of insoluble *N*-acetyl metabolic products, was one of the more serious toxic effects observed with sulfonamide therapy. This is much less of a problem than with the early sulfas, mainly because of the discovery of agents highly soluble at urinary pH, the development of long-acting sulfas that build up adequate blood levels at doses low enough to avoid crystallization, and the discovery of compounds that are excreted chiefly as highly soluble glucuronides.

Blood dyscrasias are quite uncommon, but if they occur may be serious enough to cause discontinuance of the therapy. Both topical and systemic administration of sulfas can cause hypersensitivity reactions, such as urticaria, exfoliative dermatitis, photosensitization, erythema nodosum, and in its most severe form, erythema multiformexudativum (Stevens-Johnson syndrome). In general, however, use of sulfonamide therapy is considered relatively safe.

WILLIAM O. FOYE
Massachusetts College of Pharmacy
and Allied Health Sciences

E. H. Northey, *The Sulfonamides and Allied Compounds,* Reinhold Publishing Corp., New York, 1948, pp. 517–577.

N. Anand, in W. O. Fove, ed., *Principles of Medicinal Chemistry,* 3rd ed., Lea & Febiger, Philadelphia, 1988, pp. 637–659.

P. G. De Benedetti in B. Testa, ed., *Advances in Drug Research,* Vol. 16, Academic Press, London, 1987, pp. 228–279.

G. L. Mandell and M. E. Sande, in A. G. Gilman, L. S. Goodman, T. W. Rall, and F. Murad, eds., *Goodman and Gilman's The Pharmacological Basis of Therapeutics,* 7th ed., Macmillan Pub. Co., New York, 1985, pp. 1095–1109.

ANTIBIOTICS

SURVEY

Antibiotics are chemical substances produced by microorganisms and other living systems that are capable in low concentrations of inhibiting the growth of bacteria or other microorganisms. This inhibitory effect can be *in vitro* or *in vivo.* Antibiotics having both *in vivo* activity and low mammalian toxicity have been extremely valuable in treating infectious diseases. There are over 10,000 antibiotics produced by microorganisms that have been reported in the scientific literature, approximately 6100 of which have been characterized and have had molecular structures assigned. These substances range from the very simple to extremely complex, but most antibiotics are in the 300–800 mol wt range. Many thousands of semisynthetic variations of the naturally occurring antibiotics have been prepared. Only relatively few (~200) have become commercial products for human and veterinary uses (see also ANTIBACTERIAL AGENTS, SYNTHETIC.)

The mechanism of action of a number of antibiotics with regard to the inhibition of bacteria, fungi, or other organisms has been established. The more common mechanisms include inhibition of bacteria cell wall biosynthesis, inhibition of protein, RNA, or DNA synthesis, and damaging of membranes. Cell wall biosynthesis is a target present in bacteria but not in mammalian cells. Thus the β-lactams, which are very effective against bacteria, are relatively nontoxic to humans. In contrast, antibiotics that damage DNA, like adriamycin, are relatively toxic to both types of cells. However, adriamycin, which was found to be more toxic to rapidly proliferating tumor cells than to most normal cells with slower turnover rates, shows significant selectivity against tumor cells to find clinical application as an antitumor agent.

The number of naturally occurring antibiotics increased from about 30 known in 1945 to 150 in 1949, 450 in 1953, 1,200 in 1960, and to 10,000 by 1990. Table 1 lists the years of historical importance to the development of antibiotics used for treatment in humans. Most of the antibiotics introduced since the 1970s have been derived from synthetic modifications of the β-lactam antibiotics.

Table 1. Year of Discovery or Market Introduction of Some of the More Important Antibiotics

Antibiotic	Year	
	Discovery	Introduction
penicillin	1929	
tyrothricin	1939	
griseofulvin	1939	
streptomycin	1944	
bacitracin	1945	
chloramphenicol		1947
polymyxin	1948	
chlortetracycline		1948
cephalosporin C,N,P	1948	
neomycin	1949	
oxytetracycline		1950
nystatin	1950	
erythromycin		1952
novobiocin	1955	
kanamycin		1957
ampicillin[a]		1962
fusidic acid		1961
cephalothin[a]	1962	
lincomycin		1963
gentamicin		1963
carbenicillin[a]	1964	
cephalexin[a]	1966	
clindamycin[a]		1967
cephaloxidine and cephalothin[a,b]		1969
minocycline[a]		1971
amoxycillin[a]		1972
cefoxitin[a,c]		1978
tricarcillin[a]		1979
mezlocillin[a]		1980
piperacillin[a]		1980
cefotaxime[a]		1980
moxalactam[a]		1981
augmentin[d]		1984
aztreonam[e]		1984
imipenem[a,f]		1985

[a] Semisynthetic products.
[b] First oral cephalosporins.
[c] First commercial cephamycin.
[d] First β-lactamase inhibitor combination.
[e] First monobactam and a synthetic product.
[f] First carbapenem.

Classification of Antibiotics

A chemical classification of some of the commercially more important antibiotic families is given here.

Aminoglycosides. Antibiotics in the aminoglycoside group characteristically contain amino sugars and deoxystreptamine or streptamine. This family of antibiotics has frequently been referred to as aminocyclitol aminoglycosides. Representative members are streptomycin, neomycin, kanamycin, gentamicin, tobramycin, and amikacin. These antibiotics all inhibit protein biosynthesis.

Ansamacrolides. Antibiotics in the ansamacrolide family are also referred to as ansamycins. They are benzenoid or naphthalenoid aromatic compounds in which nonadjacent positions are bridged by an aliphatic chain to form a cyclic structure. One of the aliphatic–aromatic junctions is always an amide bond. Rifampin is a semisynthetically derived member of this family and has clinical importance. It has selective antibacterial activity and inhibits RNA polymerase.

β-Lactams. All β-lactams are chemically characterized by having a β-lactam ring. Substructure groups are the penicillins, cephalosporins, carbapenems, monobactams, nocardicins, and clavulanic acid. Commercially this family is the most important group of antibiotics used to control bacterial infections. The β-lactams act by inhibition of bacterial cell wall biosynthesis.

Chloramphenicol. Only chloramphenicol and a few closely related analogues fall into this group. Chloramphenicol, a nitro benzene derivative of dichloroacetic acid, inhibits protein biosynthesis.

Glycopeptides. Vancomycin, avoparcin, and teicoplanin are examples of glycopeptide antibiotics. This family has cyclic peptide structures and biphenyl containing amino acids. Sugars are attached to the peptide unit resulting in compounds frequently in the molecular weight range of 1400–2000. These antibiotics inhibit bacterial cell wall biosynthesis by binding to D-alanyl-D-alanine units found in the cell walls.

Lincomycin. The lincomycins and celesticetins are a small family of antibiotics that have carbohydrate-type structures. Clindamycin, a chemical modification of lincomycin, is clinically superior. Antibiotics in this family inhibit gram-positive aerobic and anaerobic bacteria by interfering with protein biosynthesis.

Macrolides. Antibiotics in the macrolide group are macrocyclic lactones that can be further classified into two main subgroups: *(1)* polyene macrolides that are antifungal agents and include compounds like nystatin and amphotericin B; and (2) antibacterial antibiotics represented by erythromycin and tylosin. A number of other subfamilies of antibacterial and antifungal antibiotics fall into the broad category of macrolides.

Polyethers. Antibiotics within this family contain a number of cyclic ether and ketal units and have a carboxylic acid group. They form complexes with mono- and divalent cations that are soluble in nonpolar organic solvents. They interact with bacterial cell membranes and allow cations to pass through the membranes causing cell death. Because of this property they have been classified as ionophores. Monensin, lasalocid, and maduramicin are examples of polyethers that are used commercially as anticoccidial agents in poultry and as growth promotants in ruminants.

Tetracyclines. The tetracyclines are a small group of antibiotics characterized as containing a polyhydronaphthacene nucleus. Commercially the tetracyclines are important. They have been used clinically against gram-positive and gram-negative bacteria, spirochete, mycoplasmas, and rickettsiae and have veterinary applications in promoting growth and feed efficiency. The mode of action is inhibition of protein synthesis. Some of the more important members of this family are tetracycline, minocycline, and doxycycline.

Production

Most of the microorganisms used to produce antibiotics were isolated from soil samples. These microorganisms occur as heterogeneous populations and generally inhabit the top few centimeters of soil. Families demonstrated to produce antibiotics include actinomycetes, bacteria, and fungi. Actinomycetes are in numbers the most productive for antibiotics.

To obtain reproducible antibiotic production by fermentation, it is necessary to obtain a pure culture of the producing organism. Pure cultures are isolated from mixed soil sample populations by various streaking and isolation techniques on nutrient media. Once a pure culture has been found that produces a new antibiotic typically on a mg/L scale, improvement in antibiotic yield is accomplished by modification of the fermentation medium or strain selection and mutation of the producing organism. Production of g/L quantities may take years to accomplish.

The vast majority of new antibiotics result from screening soil microorganisms or by semisynthetic modification of naturally occurring antibiotics. Genetic engineering technology has begun to evolve that allows modifications of a microorganism's DNA so that it will produce new antibiotics.

Commercial fermentations are conducted in large bioreactors which are usually referred to as fermentors and are designed for operation in batch, fed-batch, or continuous fermentation modes. The batch and fed-batch procedures are used for most commercial antibiotic fermentations.

Uses

Antibiotics are used as antibacterial agents, anticancer agents, antituberculin agents, antifungal agents, antiviral agents, and in veterinary products and animal feed supplements for growth promotion.

DONALD BORDERS
American Cyanamid Company

A. A. Higton and A. D. Roberts, in B. W. Bycroft, ed., *Dictionary of Antibiotics and Related Substances,* Chapman and Hall, New York, 1988.

J. Bérdy, A. Aszalos, and K. L. McNitt, *CRC Handbook of Antibiotic Compounds,* Vol. 14, CRC Press, Inc., Boca Raton, Fla., 1987.

County NatWest WoodMac, International Pharmaceutical Service, County NatWest Securities Limited, London, 1989; Part 4, 1990.

A. Kucers, N. Mck. Bennett, and R. J. Kemp, *The Use of Antibiotics,* 4th ed., William Heinemann Medical Books, London, 1987, pp. 914, 1418–1528.

AMINOGLYCOSIDES

The term *aminoglycoside* is commonly used to refer to members of the class of antibacterial antibiotics, the structures of which are derived from D-streptamine (**1**, R = OH), D-2-deoxystreptamine (**1**, R = H), or closely related compounds. The term *aminocyclitol* is also sometimes used to identify this group of compounds. A typical member of this class, tobramycin, has the structure (**2**).

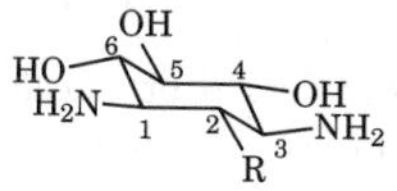

(**1**)

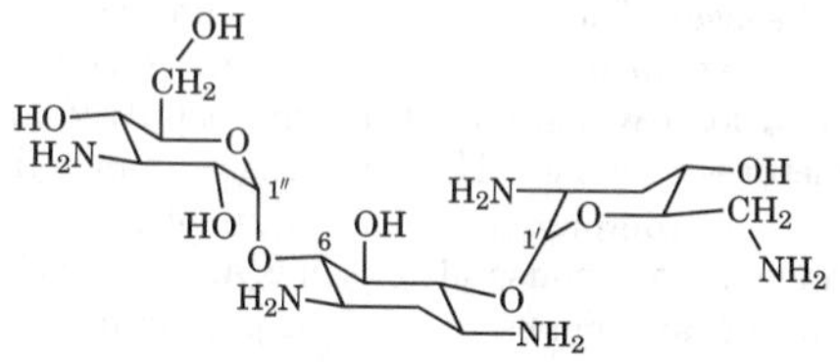

(**2**)

Aminoglycosides in Medical Usage

In 1991, the most widely used aminoglycosides in medical practice were gentamicin, tobramycin, amikacin, and netilmicin. Other aminoglycosides used to a lesser extent include dibekacin, isepamicin, neomycin, astromicin, spectinomycin, kanamicin A, sisomicin, and streptomycin.

Medical and Biological Properties

General Antibacterial Properties. In general, the aminoglycosides are useful for the treatment of serious infections involving aerobic or facultative gram-negative bacilli, especially in the compromised host. Particular advantages of the aminoglycosides include the findings that, in general, the bactericidal concentration is not significantly higher than the growth inhibitory concentration, and that the bactericidal effect is rapid and concentration-dependent. Clinical usage has been extensively reviewed in the medical literature.

Bacterial Resistance Mechanisms. The most common resistance mechanism involves the inactivation of the aminoglycoside by reactions catalyzed by plasmid-borne enzymes. In general, amikacin and isepamicin tend to be least susceptible to inactivation by this mechanism, while netilmicin and dibekacin are intermediate and gentamicin and tobramycin are most susceptible. Less common resistance mechanisms include decreased affinity for the antibiotic by the bacterial ribosome, and decreased rate of transport into the bacterial cytoplasm.

Pharmacokinetics. The aminoglycosides are not reliably absorbed following oral dosing, so they are administered primarily by intravenous infusion or intramuscular injection. Distribution throughout the vascular and interstitial space is fairly rapid, but intracellular, cerebrospinal fluid, and bronchial secretion levels are generally low. Most of the administered dose is eliminated unmetabolized in the urine.

Toxicology. Potential toxicity is a primary limiting factor in the clinical use of aminoglycosides. The most important toxicities are nephrotoxicity, ototoxicity, and to a lesser extent, neuromuscular blockade. Although there is some variation, all the aminoglycosides in medical practice are capable of causing these adverse events. The effect of this potential is to prevent the use of significantly increased doses to cover difficult infections. In addition, serum aminoglycoside concentrations can reach toxic levels after normal dosing due to variations in glomerular filtration efficiency.

Mechanism of Antibacterial Action. The bactericidal mechanisms employed by aminoglycosides are incompletely understood. Initially, the cationic aminoglycoside binds nonspecifically to anionic groups on the bacterial cell surface. Passage through the cell wall is via porins or self-promoted defects. Uptake into the cell cytoplasm appears to be a two-step, energy-requiring process. In the cytoplasm, ribosomal binding leads to misreading of the genetic code and production of abnormal proteins and, at higher concentrations, protein synthesis inhibition. The cause of cell death is uncertain.

Aminoglycoside Biosynthesis. The biosynthesis of aminoglycosides has been extensively studied. Probably the most interesting aspect is the biosynthesis of 2-deoxystreptamine, in which the C-1 and C-6 of a D-glucose molecule become the C-1 and C-2 of 2-deoxysptreptamine by way of the intermediate 2-deoxy-*scyllo*-inosose.

Structure-Activity Relationships Among Aminoglycoside Derivatives

The aminoglycosides possess properties which make them valuable for the control of bacterial infectious disease, but they also have distinct limitations, especially in the areas of toxicity and susceptibility to bacterial resistance mechanisms. Thus, a large amount of research has been conducted aimed at reducing the limitations while maintaining the advantages by modification of the aminoglycoside molecular structure. The principal approaches to novel structural variations have been (*1*) a search for new microorganisms which produce novel aminoglycosides directly (eg, tobramycin); (*2*) chemical modification of available aminoglycosides (semisynthesis) (eg, amikacin); and (*3*) generation of microorganism mutants which require a modifiable exogenous substrate for the biosynthesis of the aminoglycoside. Overall, structural modification has been successful in reducing susceptibility to bacterial inactivation mechanisms. It has not been possible, however, to substantially dissociate the toxicity potential from the antibacterial activity.

DONALD MCGREGOR
Bristol-Myers Squibb Company

A. Schatz, E. Bugie, and S. A. Waksman, *Proc. Soc. Exp. Biol. Med.* **55,** 66 (1944).

F. A. Keuhl, R. L. Peck, C. E. Hoffhine, Jr., and K. Folkers, *J. Am. Chem. Soc.* **70,** 2325 (1948).

S. Neidle, D. Rogers, and M. B. Hursthouse, *Tetrahedron Lett.* 4725 (1968).

S. Umezawa, Y. Takahasi, T. Usui, and T. Tsuchiya, *J. Antibiot.* **27,** 997 (1974).

ANSAMACROLIDES

The ansamacrolides or ansamycins are a family of antibiotics characterized by an aliphatic ansa-bridge that connects two nonadjacent positions of an the aromatic nucleus. Ansamacrolides can be divided into two groups based on the nature of the aromatic nucleus. One group contains a naphthoquinoid nucleus and includes the streptovaricins, the rifamycins, tolypomycin, the halomycins, the naphthomycins, actamycin, the diastovaricins, kanglemycin, awamycin, and ansathiazin. The other group contains a benzoquinoid nucleus and includes geldanamycin, the maytansinoids, the herbimycins, the macbecins, the mycotrienins, the trienomycins, the ansatrienins, and the ansamitocins. Table 1 summarizes the biological activity of these antibiotics.

Table 1. Biological Activity of the Ansamacrolides

Ansamacrolide	Biological activity
streptovaricins	antibacterial (gram-positive and mycobacteria), antiviral, inhibitors of reverse transcriptase
rifamycins	antibacterial (gram-positive, gram-negative, and mycobacteria), antiviral, inhibitors of reverse transcriptase
tolypomycins	antibacterial (gram-positive)
halomycins	antibacterial (gram-positive)
naphthomycins	antibacterial (gram-positive), vitamin K antagonist
actamycins	inhibitors of fatty acid synthesis
diastovaricins	antileukemic
kanglemycins	antibacterial (gram-positive)
awamycins	antibacterial (gram-positive), antitumor
ansathiazins	antibacterial (gram-positive)
geldanamycins	antiprotozoal, herbicidal, inhibitors of reverse transcriptase
herbimycins	herbicidal, antitumor, antiviral, inhibitors of tyrosine kinase
macbecins	antibacterial (gram-positive), antiprotozoal, antifungal, antitumor
mycotrienins	antifungal
trienomycins	antitumor
ansatrienins	antifungal
maytansinoids	antileukemic, antitumor
ansamitocins	antiprotozoal, antifungal, antitumor

Naphthoquinoids

Streptovaricins. The streptovaricins are produced by *Streptomyces Spectablis* n. sp. and are isolated as a crude complex.

Chemical Properties and Derivatives. All of the streptovaricins except streptovaricin D react with one mole of sodium periodate to yield the corresponding streptovals. The streptovaricins undergo thermal isomerization to the corresponding atropisostreptovaricins. In the natural streptovaricins the ansa-bridge lies above the aromatic nucleus but in the atropisostreptovaricins this bridge lies below the aromatic nucleus. Most spectral properties of the isomers are nearly identical, but the optical rotations, although of approximately equal magnitude, are of opposite sign.

Assay Methods. The primary assay for the streptovaricins is the microbiological assay using the agar diffusion method or a turbidimetric procedure. The streptovaricins can also be identified by paper or thin-layer chromatography.

Rifamycins. The rifamycins were first isolated from a broth of *Nocardia mediterranei* (the producing organism was originally identified as *Streptomyces mediterranei*). The rifamycins were originally designated as rifomycins. Only rifamycin B, which accounts for 10–15% of the crude complex, can be isolated easily as a stable crystalline compound.

The structures of the rifamycins were arrived at by chemical degradation studies and confirmed by x-ray crystallography. The absolute configuration of the ansa-bridge is 6(*S*), 7 (*S*), 8 (*R*), 9 (*R*), 10 (*R*), 11 (*S*), 12 (*R*), and 13 (*S*). Studies of ^{13}C nmr and ir have been reported and mass spectra of the rifamycins have been obtained.

Chemical Properties and Derivatives. There have been thousands of rifamycin derivatives prepared in an attempt to obtain a broader-spectrum antibiotic having good oral absorption. Rifamycins B, O, and S have served as starting materials for the preparation of numerous classes of derivatives. Several of the semisynthetic derivatives are more active, have a broader spectrum of biological activity, and are therapeutically more effective than the parent antibiotics.

Manufacture and Processing. Although fermentation procedures have not been reported, assumptions concerning fermentation media and optimal conditions have been made. The transformation of the biologically inactive rifamycin B to the biologically active rifamycin S is usually accomplished chemically. Several rifamycin B oxidases have been isolated that can enzymatically transform rifamycin B to rifamycin O, which is hydrolyzed in the fermentation medium to rifamycin S. The enzymes from *Monocillium spp.* ATC 20621 and *Humicold spp.* ATCC 20620 are intracellular, whereas the enzyme from *Curvularia lunata* var. *aeri* is extracellular. The use of a fluidized bed reactor containing immobilized whole cells of *Humicola* for the transformation of rifamycin B to rifamycin S has been described. Rifamycin SV-producing strains have been isolated, but it is not known if these strains are used commercially.

Assay Methods. A large number of assays exist for the determination of the various rifamycins. Rifamycin SV and rifampicin can be determined by a microbiological assay using *Sarcina lutea* ATTC 9341 as the test organism, and rifampicin can be determined using *S. aureus* 560. Rifamycins B, S, and SV can be separated by electrophoresis on agar gel and determined microbiologically using *B. subtilis* or *S. lutea.* Spectrophotometric assays exist for the rifamycins and for rifampicin. Rifamycins B, O, S, and SV can be determined via polarography or by amperometric titration. Rifamycins B, O, S, and SV can be separated by thin-layer chromatography on silica gel or by paper chromatography. Fluorimetric assays exist for rifamycin B and rifampicin. High performance liquid chromatographic (hplc) procedures exist for rifamycins B, O, S, and SV, for rifampicin in formulations, in body fluids, in mixtures of antibiotics, and for rifapentine in plasma.

Tolypomycins. The addition of small amounts of iron salts to the fermentation medium increases the production of tolypomycin Y, the structure of which was arrived at by chemical degradation and confirmed by x-ray crystallographic analysis.

Tolypomycin Y shows strong antibacterial activity against gram-positive bacteria and *Neisseria gonorrheae.*

A differential bioassay was developed to distinguish tolypomycin Y from rifamycin B.

Halomycins. The halomycins are a group of four antibiotics produced by *Micromonospora halophytica* and separated by partition chromatography on Chromosorb W coated with formamide. Further purification was accomplished using preparative tlc.

The halomycins are active against gram-positive bacteria. The halomycin complex exhibited high activity against bacterial strains resistant to penicillin G.

Naphthomycins, Naphthoquinomycins, Actamycin, and Diastovaricins. The naphthomycins are a group of closely related antibiotics differing in the substituent at C-2 and C-30, and in the geometry about the C-4 and C-6 double bonds. The naphthoquinomycins, diastovaricins, and actamycin are all closely related to the naphthomycins. Naphthomycin A is isolated from a fermentation beer of *Streptomyces collinus* (Tü 105).

Naphthomycin B is produced by *Streptomyces galbus* (Tü 353) whereas naphthomycin C is produced by *Streptomyces diastatochromogenes* (Tü 1892).

Naphthoquinomycins A and B are isolated from *Streptomyces S*-1998 and their structures are assigned on the basis of spectral data. Actamycin is obtained from *Streptomyces* sp. E/784, and its structure arrived at on the basis of spectral data and degradation studies.

Diastovaricins I and II are produced by *Streptomyces diastochromogenes.* Diastovaricins I and II are active against Friend mouse leukemia cells. Spectral data are used to determine the structures.

Kanglemycin. Kanglemycin is isolated from the fermentation broth filtrate of *Nocardia mediterranei* var *kanglensis* and its structure determined by x-ray crystallographic studies. The antibiotic is active against gram-positive bacteria.

Awamycin and Ansathiazin. Awaymycin and ansathiazin are produced by *Streptomyces* sp. No. 80-217. The structures for awamycin and ansathiazin were assigned on the basis of spectral data. Both antibiotics are active against gram-positive bacteria, and awamycin is reported to have antitumor activity.

Benzoquinoids

Geldanamycin. Geldanamycin is isolated from the filtered beer of *Streptomyces hygroscopicus* var. *geldanus* var. *nova.* This organism also produces nigericin nocardamine, and a libanamycinlike activity depending on the composition of the fermentation medium. The structure of geldanamycin was assigned in great part on the basis of nmr studies. Unlike the naphthoquinoid ansamacrolides, geldanamycin has little antibacterial activity, being primarily active against protozoa and fungi, especially *Tetrahymena pyriformis* and *Crithidia fasciculata* (see ANTIPARASITIC AGENTS). Geldanamycin also has herbicidal activity (see HERBICIDES).

Herbimycins. Herbimycins A, B, and C along with some derivatives, are isolated from the fermentation broth of *Streptomyces hygroscopicus* AM-3672. The structure of herbimycin A was assigned on the basis of spectral data and confirmed by x-ray crystallographic studies. The structures for herbimycins B and C were derived by comparing spectral data to those for herbimycin A. The herbimycins possess strong herbicidal activity and exhibit some antitumor and antiviral activity.

Several derivatives of herbimycin A have been prepared that possess greater antitumor activity than the parent.

Macbecins. Macbecin I and II are isolated from the fermentation broth of *Nocardia sp.* C-14919. The structures were assigned on the basis of ^{1}H nmr studies on the intact antibiotics as well as on several degradation products. The assigned structures were confirmed by x-ray crystallographic studies. The macbecins are active against gram-positive bacteria, fungi, and protozoa and exhibit *in vitro* antitumor activity against murine leukemia P 388 and melanoma B 16 (see CHEMOTHERAPEUTICS, ANTICANCER).

Mycotrienins, Mycotrienols, Trienomycins, and Ansatrienins. The mycotrienins are produced by *Streptomyces rishiriensis* T-23. The structures for mycotrienins I and II were assigned primarily on the basis of nmr spectral analysis.

Streptomyces rishiriensis T-23 also produces mycotrienols I and II, the structures of which were based on spectral analysis. The mycotrienins possess no antibacterial activity but are active against fungi and yeasts (qv), and exhibit weak antitumor activity. The mycotrienols are of an order of magnitude less active than the mycotrienins, suggesting that the cyclohexanecarbonylalanine group is important for biological activity.

The trienomycins are isolated from *Streptomyces* sp. 83-16. The assigned structures were based on spectral data. The trienomycins have no antimicrobial activity but have good antitumor activity. Trienomycin A is the most active, exhibiting good *in vivo* antitumor activity against sarcoma 180 and P 388 leukemia in mice.

The ansatrienins are produced by *Streptomyces collinus* Tü 1982. The structures were assigned on the basis of spectral data of the intact antibiotics as well as several derivatives. The ansatrienins are active against fungi.

Maytansinoids and Maytansides (Ansamitocins). *Isolation and Structure Proof.* The maytansinoids were the first ansamacrolides to be found in plants. The term maytansinoids refers to those ansamacrolides related to maytansine, whereas the term maytansides refers to maytansinoids lacking the ester side chain at C-3 as well as the corresponding elimination products. Maytansine was first isolated from the alcoholic extract of *Maytenus ovatus* Loes. Several other maytansinoids and maytansides have been isolated from this species. The structure of maytansine was established by x-ray crystallographic analysis, and the structures of the other maytansinoids and maytansides were arrived at by comparative nmr studies using maytansine. The absolute configuration of maytansine is *3(S), 4(S), 5(S), 6(R), 7(S), 10(R)*, and *2′(S).*

Colubrinol and colubrinol acetate are isolated from *Colubrina texensis* Gray (Rhamnaceae) along with maytanbutine. Colubrinol is also isolated from *Trewia nudiflora* (Euphorbiaceae). The structures for colubrinol and colubrinol acetate were established by high resolution ms and the comparison of their nmr spectra with that of the known maytanbutine.

Normaytansine is isolated from *Maytenus buchananii*, and the maytansinoids trewiasine, dehydrotrewiasine, and demethyltrewiasine are isolated from the ethanolic extract of the seed from *Trewia nudiflora* L. (Euphorbiaceae). Also isolated from *Trewia nudiflora* are the maytansinoids treflorine, trenudine, and *N*-methyltrenudone, all of which contain an additional macrocyclic ring linking C-3 to the aromatic amide nitrogen.

Another large group of maytansinoids are produced by the microorganism *Nocardia* sp. C-25003 (N-1) and are designated ansamitocins.

The structures of the ansamitocins were determined by spectral analysis. By comparison of reported physical data, it was concluded that ansamitocins P-0, P-1, and P-2 were identical to maytansinol, maytanacine, and maytansinol propionate, respectively.

Chemical Properties and Derivatives. Procedures for the total synthesis of several of the maytansinoids have been thoroughly reviewed. A variety of bacteria, actinomycetes, yeasts, and fungi were screened for their ability to modify the ansamitocins.

Several semisynthetic maytansinoids have been prepared by acylating the C-3 hydroxyl group of maytansinol. Some of these derivatives have antiprotozoal and antitumor activity similar to maytansine and ansamitocin P-3. 3-Epimaytansinoids have been synthesized and were not biologically active.

Biological Activity. The maytansinoids possess antitumor activity, particularly against P 388 lymphocytic leukemia, B 16 melanocarcinoma, and Lewis lung carcinoma. A number of semisynthetic esters of maytansinol have been prepared and exhibit good antileukemic activity. The maytansides lack antitumor activity, indicating that the ester at C-3 is a requirement for activity. The carbinolamide also appears to be necessary for antitumor activity. The maytansinoids do not inhibit bacterial RNA polymerase as do the other ansamacrolides. Besides having antitumor activity, the ansamitocins have antiprotozoal and antifungal activity. Maytansine has undergone Phase I and II clinical studies and does not appear to be effective.

Mode of Action and Biosynthesis

The mode of action of the naphthoquinoid ansamacrolides was established through studies using the rifamycins and streptovaricins. The ansamacrolides inhibit bacterial growth by inhibiting RNA synthesis. This is accomplished by forming a tight complex with DNA-dependent RNA polymerase.

The ansamacrolides form no such complex with mammalian RNA polymerase and thus have low mammalian toxicity.

The antiviral activity of the ansamacrolides does not result from inhibition of RNA polymerase but rather from the inhibition of the assembly of the virus particles.

The antitumor activity of geldanamycin and its derivatives appears to result from inhibition of DNA synthesis, whereas RNA synthesis is not affected. The antitumor activity of the maytansinoids also appears to result from the inhibition of DNA synthesis.

The ansa-chain of the ansamycins streptovaricins, rifamycins, geldanamycin, and herbimycin has been shown to be polyketide in origin, being made up of propionate and acetate units with the *O*-methyl groups coming from methionine. The remaining aromatic C_7N portion of the ansamacrolides is derived from 3-amino-5-hydroxybenzoic acid, which is formed via shikimate precursors. Based on the precursors of the rifamycins and streptovaricins isolated from mutant bacteria strains, a detailed scheme for the biosynthesis of most of the ansamacrolides has been proposed.

Commercially Available Ansamacrolides

Rifampicin, the only commercially available ansamacrolide, is manufactured by Merrell Dow under the trade name Rifadin, and by CIBA under the trade name Rimactane. Rifampicin is also supplied in combination with isoniazid or pyrazinamide. The rifampicin–isoniazid combination is known as Rifamate (Merrell Dow), Rifinah (Merrell Dow), and Rimactazid (CIBA); the rifampicin–pyrazinamide as Rifater (Merrell Dow). Several other rifamycin derivatives including rifabutin and rifapentine are undergoing clinical studies.

FREDERICK J. ANTOSZ
The Upjohn Company

K. L. Rinehart, Jr. and L. S. Shield, *Fortschr. Chem. Org. Naturst.* **33**, 231 (1976).

O. Ghisalba, J. A. L. Auden, T. Schupp, and J. Nüsch, in E. J. VanDamme, ed., *Biotechnology of Industrial Antibiotics,* Marcel Dekker, New York, 1984, Chapt. 9.

G. Lancini, in H. Pape and H.-J. Rehm, eds., *Biotechnology: Microbial Products II,* Vol. 4, VCH, New York, 1986, Chapt. 14.

C. R. Smith, Jr. and R. G. Powell, in S. W. Pelletier, ed., *Alkaloids: Chemical and Biological Perspectives,* John Wiley & Sons, Inc., New York, 1984, Chapt. 4.

CHLORAMPHENICOL AND ANALOGUES

Chloramphenicol (**1**, R = NO_2), $C_{11}H_{12}Cl_2N_2O_5$, is a commercially significant antibacterial agent and its status in clinical practice has been reviewed. Although widespread use of this antibiotic declined in the United States in the 1960s because of reports of serious toxic effects, this situation changed a decade later when ampicillin-resistant *Hemophilus influenzae* emerged on the clinical scene. The appearance of *Bacteroides* species and of *Streptococcus pneumoniae* resistant to β-lactam antibiotics contributed further to the resurgence. In the 1970s, chloramphenicol also became important in the treatment of serious *Salmonella* invasive gastroenteritis in infants less than three months of age. Because chloramphenicol crosses the blood brain barrier, it is indicated in infections of the central nervous system caused by susceptible organisms.

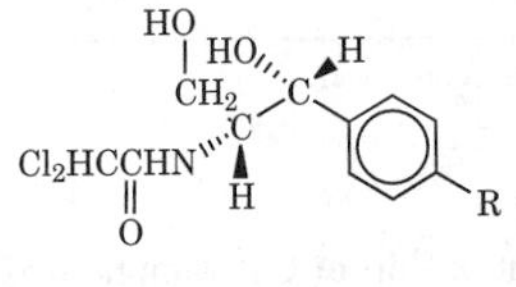

(**1**)

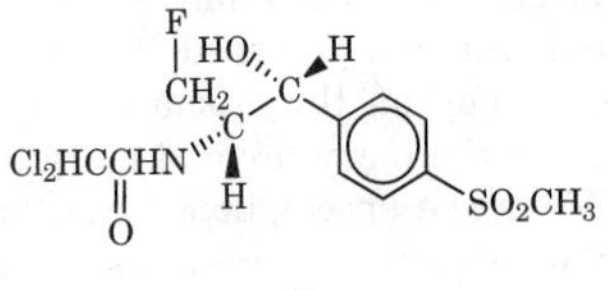

(**2**)

The emergence of quinolones (see ANTIBACTERIAL AGENTS, SYNTHETIC) and other antibiotics is expected to curtail the use of chloramphenicol in the future, but this drug is relatively inexpensive, orally active, and the toxicity, except for the rare idiosyncratic aplastic anemia, can be managed through monitoring of blood levels by sensitive modern analytical procedures. However, clinical use is being further curtailed by the emergence of chloramphenicol-resistant organisms. In Table 1, the median *in vitro* susceptibilities of chloramphenicol, thiamphenicol, and a fluoroanalogue, florfenicol (**2**), $C_{12}H_{14}Cl_2FNO_4S$, against a host of chloramphenicol-sensitive and -resistant organisms are given. Bacteria resistant to chloramphenicol are also resistant to thiamphenicol.

Both chloramphenicol and thiamphenicol cause reversible bone marrow suppression. The irreversible, often fatal, aplastic anemia, however, is only seen for chloramphenicol. This rare (1 in 10,000–45,000) chloramphenicol toxicity has been linked to the nitroaromatic function. Thiamphenicol, which is less toxic than chloramphenicol in regard to aplastic anemia, lacks potency and has never found much usage in the United States. An analogue of thiamphenicol having antimicrobial potencies equivalent to chloramphenicol was sought. Florfenicol (**2**) was selected for further development from a number of closely related structures.

Bacterial Resistance of Amphenicol

Of the many mechanisms of bacterial resistance to chloramphenicol and thiamphenicol, the plasmid-mediated transmissible resistance conferred by the presence in resistant bacteria of chloramphenicol-acetyltransferases (CAT) is the most important.

Table 1. *In Vitro* Susceptibilities of Amphenicols

Organism strain	No. of strains tested	Susceptibility[b]	MIC, μg/mL[a]		
			Chloramphenicol	Thiomphenicol	Florfenicol
Enterobacter	4	S	4	64	4
	14	R	512	1024	8
Citrobacter	3	S	4	32	8
	3	R	512	1024	128
E. coli	9	S	4	64	8
	20	R	256	1024	8
Klebsiella	9	S	4	64	8
	20	R	512	1024	4
Providencia	4	S	16	128	8
	12	R	128	1024	8
Pseudomonas	13	R	128	128	256
Serratia	6	S	16	512	64
	18	R	512	1024	64
Salmonella	15	S	4	32	8
	7	R	256	1024	8
Shigella	9	S	1	2	2
Proteus	23	R	256	512	8
Acinetobacter	4	R	64	512	128
Staphylococcus aureus	9	S	4	8	8
	7	R	64	512	8
Streptococcus pneumonae	3	R	8	64	4

[a] Agar dilution, 24 h, Mueller-Hinton agar.
[b] S = susceptible; R = resistant.

Structure–Activity Relationship of Chloramphenicol

Structure–activity and mechanism of action studies indicate that the requirements for chloramphenicol activity are the D-threo-configuration, the 1,3-propanediol moiety, and a strong electron-withdrawing group on the aromatic ring. The L-threo, the mirror image of (**1**), and the D-erythro and L-erythro isomers are not biologically active. Thus the speculation arose that certain specific intramolecular dipolar attractive interactions must exist in chloramphenicol leading to greater stabilization of one particular conformer over the others where biological activity results from the most stable conformer. The three basic conformational isomers of chloramphenicol are shown in Figure 1.

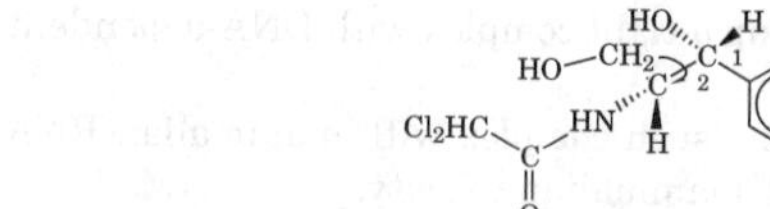

(**1**)

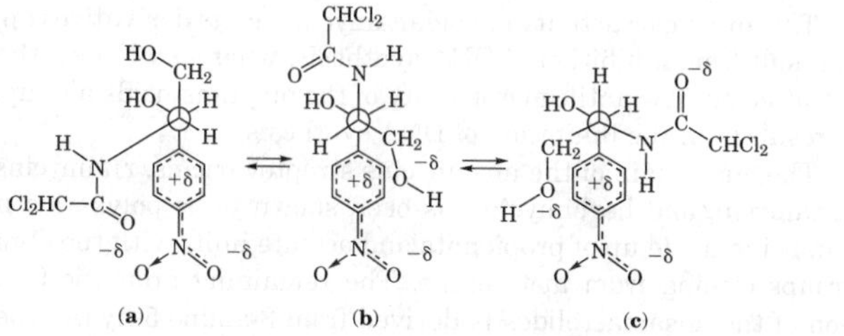

Figure 1. Conformers (**a**), (**b**), and (**c**) of chloramphenicol (**1**, R = NO_2).

Fluoroanalogues

Because the lack of biological activity of 3-substituted chloramphenicols reported previously might result from the inability to exist in the "active" (**a**) type conformation, it was speculated that the size and nature of the C-3 substituent, maintenance of a low barrier to rotation about the C-2–C-3 bond, and the length of the carbon-substituent atom bond at C-3 were highly critical for achieving a conformational preference of the (**a**) type. Thus, on the basis of the van der Waals radii of fluorine and oxygen being the same (0.14 nm) and the average C–O and C–F bond lengths being close (0.131 nm and 0.138 nm, respectively), the C-3-hydroxyl group of chloramphenicol was replaced by a fluorine atom. Optical rotation measurements of 3-fluoro-chloramphenicol, $C_{11}H_{11}Cl_2FN_2O_4$, in ethanol, gave $[\alpha]_D = +24.4°$ and in dimethylformamide gave $-23.4°$. Thus, as in the case of chloramphenicol, the optical rotation changed from a positive to a negative value on going from a protic to a dipolar aprotic solvent. The solid-state conformation was determined by single crystal x-ray structure analysis and the crystals contained two rotameric structures in the asymmetric unit.

Biological Activity. The biological activity of 3-fluorochloramphenicol against chloramphenicol-sensitive and -resistant organisms was determined. Potencies against sensitive strains are similar to those of chloramphenicol. Additionally, this fluoroderivative is highly active against chloramphenicol-resistant organisms having MICs ranging from 1 to 16 μg/L. This result prompted the synthesis and biological evaluation of a number of amphenicols containing a fluorine atom at the 3-position. The most promising florfenicol (**2**), 3-fluorothiamphenicol, is not only active against the chloramphenicol–thiamphenicol-resistant strains, but the potency of florfenicol against sensitive organisms is also superior to any of the other amphenicols.

Veterinary Potential of Florfenicol. The absolute ban on the use of chloramphenicol in food-producing animals in the United States and Canada has accentuated the need for an effective broad-spectrum antibiotic in animal food medicine. Florfenicol and other antibiotics commonly used in veterinary medicine have been evaluated *in vitro* against a variety of important veterinary and aquaculture pathogens. Florfenicol was broadly active, having MICs lower than those of chloramphenicol in each of the genera tested. Florfenicol was also superior to chloramphenicol, thiamphenicol, oxytetracycline, ampicillin, and oxolinic acid against the most commonly isolated bacterial pathogen of fish in Japan.

Structure–Activity Relationships of 3-Fluoro-Amphenicols. A number of analogues of 3-fluorochloramphenicol and florfenicol (**2**) have been synthesized and the biological activities examined. Replacement of the dichloroacetyl group by a difluoroacetyl function in both series led basically to retention of potency and the spectrum of activity of the parent structures. However, changing the difluoroacetyl to a trifluoro- or a chlorodifluoroacetyl group abolished the antimicrobial activity almost completely. Reduced level of potency was also seen when the dichloroacetyl group was changed to a chlorofluoroacetyl group. Other amide functions such as methoxyacetyl or methylsulfonylacetyl did not give any appreciable activity. In the florfenicol structure, changing the methylsulfonyl group to methylsulfoxide greatly reduced the potency and the methylthio analogue was practically inactive.

Mechanism of Action of Florfenicol. The inhibitory activities of chloramphenicol (**1**, R = NO_2), thiamphenicol (**1**, R = SO_2CH_3), and florfenicol (**2**) against a sensitive *E. coli* strain have been studied. In two different liquid media, both chloramphenicol and florfenicol allowed only 20–30% residual growth at a drug concentration of 2 mg/L, whereas a thiamphenicol concentration of 25 mg/L was required to produce a similar effect. Florfenicol was also found to be a selective inhibitor of prokaryotic cells. At concentrations of 1 mg/L chloramphenicol and florfenicol, and at a concentration of 25 mg/L, thiamphenicol, inhibited protein synthesis.

In Vivo Effects of Florfenicol. Florfenicol is similar to thiamphenicol in acute toxicity by oral and subcutaneous (sc) administration, but is comparable to chloramphenicol by intraperitoneal (ip) and intravenous (iv) routes.

The efficacy of florfenicol *in vivo* was determined by measuring the dose required to obtain values for protection from infection in 50% of the animals (PD_{50}) against 10 chloramphenicol-resistant strains and two chloramphenicol-sensitive isolates. Florfenicol, chloramphenicol, and thiamphenicol were evaluated concurrently against each strain. Against sensitive *Enterobacter*, PD_{50} by the subcutaneous and oral routes was similar for florfenicol and chloramphenicol (25 mg/kg sc, 5 mg/kg oral), but higher for thiamphenicol (30 mg/kg sc, 20 mg/kg oral). A dramatic effect was seen for florfenicol against *Shigella* (3 mg/kg sc, 2 mg/kg oral) as compared to chloramphenicol and thiamphenicol (100 mg/kg by both routes). Against resistant strains of *Enterobacter, Klebsiella, Providencia, Serratia, Salmonella, and Staphylococcus*, the PD_{50} values for florfenicol ranged from 5 to 60 mg/kg whereas chloramphenicol and thiamphenicol were practically ineffective.

Pharmacokinetics in Nonrodents

The pharmacokinetic disposition of florfenicol (**2**) was studied in preruminant veal calves after administration of a single 22-mg/kg dose intravenously, orally after a 12-h fast, and orally 5 min postfeeding. The disposition of florfenicol in veal calves following a single iv dose was adequately described by a two-compartment open model where there was no significant effect of the animal's age on the pharmacokinetic parameters. Calves given the oral doses had a complex absorption pattern and delayed absorption. Administering florfenicol with milk delayed the onset of absorption and therefore the time to peak concentration. The disposition of the serum concentration of florfenicol in veal calves given by either oral method could be adequately described by a one-compartment pharmacokinetic model with first-order drug absorption and first-order drug elimination. The bioavailability of florfenicol was significantly less when given with milk replacer than when given on an empty stomach: after a 12-h fast median bioavailability was 88% of the dose; and given 5 min post-feeding, median bioavailability of the drug was 65%.

The elimination half-life of florfenicol after a single iv dose of 22 mg/kg (138–204 min) compares well with the elimination half-life of chloramphenicol (**1**, R = NO_2) reported in cattle, except in very young calves.

Florfenicol concentrations in tissues and body fluids of male veal calves were studied after 11 mg/kg intramuscular doses administered at 12-h intervals. Concentrations of florfenicol in the lungs, heart, skeletal muscle, synovia, spleen, pancreas, large intestine, and small intestine were similar to the corresponding serum concentrations, indicating excellent penetration of florfenicol into these tissues. Because the florfenicol concentration in these tissues decreased over time as did the corresponding serum concentrations, it was deemed that florfenicol equilibrated rapidly between these tissues and the blood. Thus serum concentrations of florfenicol can be used as an indicator of drug concentrations in these tissues.

Florfenicol has a wide tissue distribution, similar to that reported for chloramphenicol in calves and thiamphenicol in humans.

TATTANAHALLI NAGABHUSHAN
GEORGE H. MILLER
KANWAL J. VARMA
Schering-Plough Research

T. W. Schafer, E. L. Moss, T. L. Nagabhushan, and G. H. Miller, *Curr. Chemother. Infect. Dis.* **1**, 444–446 (1980).

T. L. Nagabhushan, D. Kandasamy, H. Tsai, W. N. Turner, and G. H. Miller, *Proc. 11th ICC 19th ICAAC Am. Soc. Microbiol.* **1**, 442–443 (1980); U.S. Pat. 4,235,892 (1980), T. L. Nagabhushan.

T. L. Nagabhushan and A. T. McPhail, unpublished data, 1979.

H. C. Neu, K. P. Fu, and K. Kong, *Curr. Chemother. Infect. Dis.* **1**, 446–447 (1980).

ELFAMYCINS

The elfamycins are so named because they exhibit antimicrobial activity through the inhibition of protein biosynthesis via binding to the *el*ongation *fa*ctor Tu. All of the known elfamycins are listed in Table 1. These antibiotics are distinguished by low mammalian toxicity, narrow-range antimicrobial activity, and positive effects on feed utilization and growth promotion in farm animals. Elfamycins also improve milk production in lactating ruminants.

Elfamycins are natural products. Aurodox and efrotomycin have been synthesized chemically.

Properties

Elfamycins are slightly acidic because of the 4-hydroxy-2-pyridone or the carboxylic acid moiety. They are soluble in most polar organic solvents and the alkali and ammonium salts are water-soluble. The extractability of the free acids from aqueous solution into solvents such as dichloromethane and ethyl acetate is utilized in their isolation from fermentation broths.

Production, Biosynthesis, and Chemistry

The production of elfamycins has been described in the literature. Fermentation yield improvements with aurodox has proved difficult because of feedback inhibition. Aurodox-resistant strains, however, respond positively to conventional mutagenic methods, leading to yield increases from 0.4 to 2.5 g/L. Scale-up of efrotomycin fermentations was found to be particularly sensitive to small changes in sterilization conditions of the oil-containing medium used.

Biological Properties

Mode of Action. Elfamycins block bacterial protein biosynthesis at the level of elongation factor Tu (EF-Tu).

Antimicrobial Activity. The elfamycins' antimicrobial specificity and lack of toxicity in animals can be explained in view of species-dependent specificity of elfamycin binding to EF-Tu. Inefficient cellular uptake or the presence of a nonresponding EF-Tu were cited

Table 1. Elfamycins

Antibiotic	Molecular formula	Producing organism
aurodox	$C_{44}H_{62}N_2O_{12}$	*Streptomyces goldiniensis* var. *goldiniensis*
kirromycin[a]	$C_{43}H_{60}N_2O_{12}$	*S. collinus* Tü 365
azdimycin	unknown	*S. diastatochromogenes* ATCC 31013
efrotomycin	$C_{59}H_{88}N_2O_{20}$	*S. lactamdurans* NRRL 3802
dihydromocimycin	$C_{43}H_{62}N_2O_{12}$	*S. ramocissimus* CBS190.69
heneicomycin	$C_{44}H_{62}N_2O_{11}$	*S. filipinensis* NRRL 11044
kirrothricin	$C_{44}H_{64}N_2O_{10}$	*S. cinnamomeus* Tü 89
factumycin	$C_{44}H_{62}N_2O_{10}$	*S. lavendulae* ATCC 31312
MSD A63A	unknown	*Streptoverticillum hiroshimense*
L681,217	$C_{36}H_{53}N_1O_{10}$	*S. cattleya* ATCC 39203
SB22484, factor 3	$C_{41}H_{56}N_2O_{11}$	*S.* strain NRRL 15496
SB22484, factor 4	$C_{42}H_{58}N_2O_{11}$	*S.* strain NRRL 15496
phenelfamycin A	$C_{51}H_{71}N_1O_{15}$	*S. violaceoniger* str. AB999F-80, AB1047T-33
phenelfamycin B	$C_{61}H_{71}N_1O_{15}$	*S. violaceoniger* str. AB999F-80, AB1047T-33
phenelfamycin C	$C_{58}H_{83}N_1O_{18}$	*S. violaceoniger* str. AB999F-80, AB1047T-33
phenelfamycin D	$C_{58}H_{83}N_1O_{18}$	*S. violaceoniger* str. AB999F-80, AB1047T-33
phenelfamycin E	$C_{65}H_{95}N_1O_{21}$	*S. violaceoniger* str. AB999F-80, AB1047T-33
phenelfamycin F	$C_{65}H_{95}N_1O_{21}$	*S. violaceoniger* str. AB999F-80, AB1047T-33
unphenelfamycin	$C_{43}H_{65}N_1O_{14}$	*S. violaceoniger* str. AB999F-80, AB1047T-33
LL-E19020α	$C_{65}H_{95}N_1O_{21}$	*S. lydicus* sp. *tanzanius* NRRL 18036
LL-E19020β	$C_{65}H_{95}N_1O_{21}$	*S. lydicus* sp. *tanzanius* NRRL 18036
UK-69,753	$C_{58}H_{86}N_2O_{18}$	*Amycolatopsis orientalis* ATCC 53550
N-demethylefrotomycin	$C_{58}H_{86}N_2O_{20}$	*Nocardia* ATCC 53758

[a] Identical to mocimycin which was isolated from *S. ramocissimus* CB5190.69.

as responsible factors for the natural resistance in *Halobacterium cutirubrum*, *Lactobacillus brevis*, and in actinomycetes. The low activity of elfamycins against *S. aureus* was also attributed to an elfamycin-resistant EF-Tu system. However, cross-resistance with other antibacterial agents has not been observed.

Elfamycins have similar *in vitro* antimicrobial spectra and the activity against *Moraxella, Pasteurella, Yersinia, Haemophilus, Streptococcus, Corynebacterium,* and *Neisseria* appears to be common.

Growth Promotion. Elfamycins, in general, enhance the growth of farm animals (see GROWTH REGULATORS, ANIMAL). Growth improvement and feed conversion was studied using aurodox in chicks and turkeys. Efrotomycin is being developed as a growth-promoting agent for swine.

Stable growth promoting feed additives were prepared by granulation of the drug with alginic acid and magnesium hydroxide and adding it to oiled rice hulls or by adsorption of elfamycins onto corn cob grits and coating with 10% tristearin. A molecular efrotomycin dispersion with (2-vinyl-pyridine)-styrene copolymer served as a postrumen effective dosage form for oral administration.

Improvement of Lactation. Qualitative and quantitative improvements in milk production (see MILK AND MILK PRODUCTS) are dependent on changes of ruminal volatile fatty acid (VFA) production and VFA composition. Increased propionate production at the expense of acetate and butyrate is responsible for growth enhancement and is achieved by additives such as polyether antibiotics (see ANTIBIOTICS, POLYETHERS). Relatively high levels of acetate and butyrate, however, are required to maintain adequate fat content in milk. Oral administration of elfamycins at concentrations of 1–10 mg/kg of animal body weight per day has been shown to increase milk volume from 2–15% relative to untreated animals without compromising fat content. The methods of administration to ruminants such as dairy cows and goats were similar to those employed for improving feed utilization and growth promotion.

Economic Aspects

The potential usefulness of elfamycins as growth promotors and feed-conversion enhancers is now generally recognized. Low original fermentation yields and difficulties in yield improvements discouraged early attempts to develop aurodox and mocimycin (kirromycin) commercially. A development program for efrotomycin, however, is ongoing as of this writing. Some of the newer elfamycins, such as the LL-E19020 pair, are considerably more active growth promotors than either aurodox or mocimycin, pointing toward the emergence of a second generation of elfamycins.

H. MAEHR
Hoffmann-LaRoche Inc.

D. L. Miller and H. Weissbach, *Arch. Biochem. Biophys.* **141,** 26 (1970).

A. Parmegggiani and G. W. M. Swart, *Ann. Rev. Microbiol.* **39,** 557 (1985).

A. Parmeggiani and G. Sander, *Top. Antibiot. Chem.* **5,** 159 (1980).

U.S. Pat. 4,808,412 (Feb. 28, 1989), E. P. Smith and S. H. Wu.

U.S. Pat. 4,336,250 (June 22, 1982), C. C. Scheifinger.

GLYCOPEPTIDES (DALBAHEPTIDES)

The vancomycin–ristocetin family of glycopeptides is a subclass of linear sugar-containing peptides composed of seven amino acids cross-linked to generate a specific stereochemical configuration. This configuration forms the basis of a particular mechanism of action, eg, complexation with the D-alanyl-D-alanine terminus of bacterial cell wall components. Because the mechanism of action is the distinguishing feature of these peptides, the term dalbaheptide, from D-al (anyl-D-alanine)b(inding)a(ntibiotics) having hept(apept)ide structure, has been proposed to distinguish them within the larger and diverse groups of glycopeptide antibiotics.

About 40 different naturally produced dalbaheptides have been reported (Table 1). They correspond to a larger number of chemical entities, because many are groups of strictly related factors called complexes. Among them, vancomycin and teicoplanin are used clinically as a result of high activity against gram-positive pathogens such as many coagulase-negative *Staphylococci* (CNS), corynebacteria, *Clostridium difficile*, multiresistant *Staphylococcus aureus*, and highly gentamicin-resistant *Enterococci* which are refractory to established drugs. Eremomycin is under clinical evaluation. Many patents

Table 1. Naturally Occurring Dalbaheptides

Antibiotic	Producing organism	Year[a]	Company or Institute
ristocetin A, B	*Nocardia lurida* NRRL 2430	1953	Abbott
vancomycin	*Streptomyces orientalis* NRRL 2450	1955	Lilly
actinoidin A, B	*Proactinomyces actinoides*	1956	Research Institute for the Discovery of New Antibiotics, Moscow
K-288	*Streptomyces haranomachiensis* n. sp.	1961	Tohoku University, Sendai
ristomycin A, B	*Proactinomyces fructiferi* var. *ristomycini*	1962	Research Institute for the Discovery of New Antibiotics, Moscow
avoparcin (LL-AV290 complex)	*Streptomyces candidus* NRRL 3218	1966	American Cyanamid
AM374	*Streptomyces eburosporeus* NRRL 3582	1969	American Cyanamid
A477	*Actinoplanes* sp. NRRL 3884	1971	Lilly
actaplanin (A4696 complex)	*Actinoplanes missouriensis* ATCC 23342	1971	Lilly
AB-65	*Saccharomonospora viride* T-80	1973	Dainippon Pharmaceuticals
teicoplanin (teichomycin complex)	*Actinoplanes teichomyceticus* ATCC 31121	1975	Lepetit
A35512 complex	*Streptomyces candidus* NRRL 8156	1976	Lilly
OA-7653A, B	*Streptomyces hygroscopicus* subsp. *hiwasaensis* ATCC 31613	1978	Otsuka Pharmaceutical
A51568A (*N*-demethyl-vancomycin), B	*Nocardia orientalis* NRRL 15232	1982	Lilly
A41030 complex	*Streptomyces toyocaensis* NRRL 15156	1982	Lilly
A47934	*Streptomyces toyocaensis* NRRL 15009	1982	Lilly
ardacin (AAD-216 complex, aridicin)	*Kibdelosporangium aridum* ATCC 39323	1983	Smith, Kline and French
chloropolysporin A, B, C	*Faenia interjecta* sp. nov. FERM BP-538	1983	Sankyo
A40926 complex	*Actinomadura* sp. ATCC 39727	1984	Lepetit
M43 complex	*Nocardia orientalis* NRRL 2450	1984	Lilly
kibdelin (AAD-609 complex)	*Kibdelosporangium aridum* subsp. *largum* ATCC 39922	1985	Smith, Kline and French
izupeptin A, B	*Nocardia* sp. FERM P-8656	1986	Kitasato Institute, Tokyo
parvodicin (AAJ-271)	*Actinomadura parvosata* ATCC 53463	1986	Smith, Kline and French
synmonicin A, B, C (CWI-785)	*Synnemomyces mamnoorii* ATCC 53296	1986	Eskayef Ltd.; Smith, Kline and French
actinoidin A_2	*Nocardia* sp. SKF-AAJ-193	1987	Smith, Kline and French
orienticin (PA 42867-A, B, C, D)	*Nocardia orientalis* PA-42867	1987	Shionogi
eremomycin	*Actinomyces* sp. INA-238	1987	Institute of New Antibiotics, Academy of Medical Sciences, Moscow
A42867	*Nocardia* sp. nov. ATCC 53492	1987	Lepetit
A82846 A, B, C	*Amycolatopsis orientalis* NRRL 18098	1987	Lilly
UK-68,597	*Actinoplanes* sp. ATCC 53533	1987	Pfizer
helvecardin A, B	*Pseudonocardia compacta* subsp. *helvetica* SANK 65185	1988	Sankyo
chloroorienticins A, B, C, D, E (PA-45052)	*Amycolatopsis orientalis* PA-45052	1988	Shionogi
A80407 A, B	*Kibdelosporangium philippinensis* NRRL 18198	1989	Lilly
MM 45289, MM 47756	*Amycolatopsis orientalis* NCIB 12531	1989	Beecham
MM 47761, MM 49721	*Amycolatopsis orientalis* NCIB 12608	1989	Beecham
MM 47766, MM 47767, MM 55256, MM 55260	*Amycolatopsis orientalis* NCIB 40011	1989	Beecham
MM 49728, MM 55266, MM 55267, MM 55268	*Amycolatopsis* sp. NCIB 40089	1989	Beecham
decaplanin	*Actinomyces* sp. DSM 4763	1990	Hoechst
UK-72,051	*Amycolatopsis orientalis* n.sp.	1990	Pfizer

[a] Year of patent or first publication.

claim feed-utilization efficiency increase and growth promotion in domestic animals for several dalbaheptides but only avoparcin (Avotan) is commercially used. The platelet aggregation ability of ristocetin A renders it a suitable diagnostic agent for von Willebrand's disease, a hematologic disorder of genetic origin.

Knowledge of the mechanism of action and investigations on the physicochemical characteristics of the therapeutically used dalbaheptides has stimulated the transformation of natural antibiotics into new derivatives using both chemical and biosynthetic modification.

Screening. All the dalbaheptides are produced by strains belonging to the order of *Actinomycetales*. They were at first isolated by conventional screening procedures. More recently specific discovery assays have been used. For example, differential inhibition of a vancomycin-resistant *S. aureus* strain and its susceptible parent, and an assay based on antagonism of the antibacterial activity by *N,N*-diacetyl-L-Lys-D-Ala-D-Ala, a tripeptide analogue of the dalbaheptides receptor.

Fermentation. Dalbaheptides are produced by submerged fermentation (qv) and this topic has been reviewed. In a few cases addition of a biosynthetic precursor of the core aglycones such as tyrosine and *p*-hydroxyphenylglycine increased the yields.

Recovery and Purification. The dalbaheptides are present in both the fermentation broth and the mycelial mass, from which they can be extracted with acetone or methanol, or by raising the pH of the harvested material. A detailed review on the isolation of dalbaheptides has been written. Recovery from aqueous solution is made by ion pair (avoparcin) or butanol (teicoplanin) extraction.

Physical Properties. The molecular weight of dalbaheptides ranges from about 1150 to 2200. Pure dalbaheptides are obtained as colorless or whitish amorphous powders that usually retain water and solvents. Dalbaheptides are generally water-soluble. Teicoplanin can be obtained as an internal salt or as a partial monoalkaline (sodium) salt depending on the pH of the aqueous solution in the final purification step. Other dalbaheptides are obtained as acidic salts, such as hydrochlorides (vancomycin, actaplanin) or sulfates (ristocetin A, avoparcin, eremomycin). The presence of amino, carboxyl, benzylic, and phenolic hydroxyl functions, sugars, and aliphatic chains influences both water solubility and total charge.

Dalbaheptides are levorotatory. The absolute configuration of vancomycin was determined by x-ray analysis of degradation product CDP-I.

Biosynthesis. Biochemical studies on dalbaheptides have been reviewed. Experiments with ^{13}C and ^{2}H have shown that in vancomycin, D-tyrosine is the precursor of D-*p*-hydroxyphenylglycine and β-hydroxy-*m*-chlorotyrosine, and acetate the precursor of the two *m,m'*-dihydroxyphenylglycines. Similar results using either ^{13}C or radioactively labeled material have been reported for avoparcin, ristocetin, ardacin, and A47934.

Kibdelins are converted into the corresponding ardacins by cultures of *Kibdelosporangium aridum*, indicating that the oxidation of carbon-6 of glucosamine to a carboxyl group is the last biosynthetic step.

Biological Properties

Mechanism of Action. The basis of the antibacterial action of dalbaheptides is the ability to form a complex with the terminal D-Ala-D-Ala residues of growing peptidoglycan chains, thus preventing transglycosylation, eg, chain elongation, and transpeptidation, eg, cross-linking. The consequent defective cell wall stops bacterial growth and eventually leads to cell death. The mechanism of action has been reviewed both at the molecular and at the biochemical level.

Antibacterial Activity. *In Vitro Properties.* The antibacterial spectrum of most dalbaheptides is known. Vancomycin or teicoplanin is generally introduced as a reference compound. Most aerobic and anaerobic gram-positive bacteria are sensitive to the action of dalbaheptides. Gram-negative bacteria are generally insensitive with the partial exception of *Neisseria gonorrhoeae*.

In Vivo Properties. The efficacy of dalbaheptides has been assessed in various models of experimental infections. In general there was good correlation between the ED_{50}s in the septicemia model in mice and the MICs on test strains.

Pharmacokinetics. Pharmacokinetic studies in mice via iv administration have been done mainly on vancomycin, eremomycin, and lipodalbaheptides. A systematic investigation of the relationship between pI, lipophilicity, and pharmacokinetic parameters of ardacin and its hydrolysis products in comparison to those of ristocetin, teicoplanin, and vancomycin has been carried out. Ardacin and its pseudoaglycone having pIs $\approx$3.8 yielded high and prolonged serum concentrations and the half-life ranged from 226 to 492 min. In contrast, vancomycin and ristocetin, which have pIs $\approx$8, had $t_{1/2}$ = 20 and 62 min, respectively; teicoplanin and ardacin aglycone, pIs $\approx$5, had intermediate elimination rates, $t_{1/2}$ ranged from 118 to 155 min. For dalbaheptides having similar pIs, clearance decreased and half-life increased as lipophilicity increased.

Mechanism of Resistance. Resistance to high levels of dalbaheptides has been described in enterococci. In some isolates, resistance is inducible. Transfer of resistance, in some cases plasmid-mediated, has been described. More recently, strains highly resistant to vancomycin but sensitive to teicoplanin have been isolated in the United States. Among the resistant enterococci, the VanA strains have been the most intensively studied and a reasonable picture of the resistance mechanism is emerging.

Chemical Modifications

Although it has been known for some time that hydrolysis of ristocetin leads to derivatives having different or increased antimicrobial activity, little chemical modification work was reported in the literature until the mid-1980s. Selective removal of sugar moieties and glycosylation, deacylation, deamination, dechlorination, introduction of bromine or chlorine atoms, esterification or amidation of the terminal carboxyl, acylation or alkylation of the terminal (or sugar) amino groups have all since been described, mainly in patent applications. Some of the aglycones and pseudoaglycones showed a weak activity against selected gram-negative bacteria. Among the most interesting compounds produced were SKF 104662 and MDL 62,873. SKF 104662 was obtained from the B′ epimer of synmonicin. The *in vitro* activity and therapeutic efficacy in experimental infection in mice for SKF 104662 were similar to those for vancomycins or teicoplanins. SKF 104662 was less toxic than vancomycin in mice, and the pharmacokinetic profile in mice, rats, and dogs indicated that the half-life was also longer than that of vancomycin. MDL 62,873 is the $-NH(CH_2)_3N(CH_3)_2$ amide of teicoplanin factor A2-2. The combined effect of a moderate basicity and a slightly increased lipophilicity at neutral pH probably led to a better penetration through the cell wall. MDL 62,873 was consistently more active than teicoplanin against CNS clinical isolates.

Economic Aspects

Only three dalbaheptides are commercialized: vancomycin and teicoplanin for human health, and avoparcin for animal usage.

Clinical Use. Vancomycin and teicoplanin as formulated drugs are lyophilized powders to be reconstituted with sterile water for injection.

B. Cavalleri
F. Parenti
Merrell-Dow Research Institute

A. Cassetta, E. Bingen, and N. Lambert-Zechovsky, Pathol. Boil. *39,* 700 (1991).

D. M. Campoli-Richards, R. N. Brogden, and D. Faulds, *Drugs* **40,** 449 (1990).

G. Cassani in M. E. Bushell and U. Gräfe, eds., *Bioactive Metabolites from Microorganisms, Progress in Industrial Microbiology,* Vol. 27, Elsevier, Amsterdam, the Netherlands, 1989, p. 221.

G. C. Lancini and B. Cavalleri in H. Kleinkauf and H. von Doehren, eds., *Biochemistry of Peptide Antibiotics, Recent Advances in the Biotechnology of β-Lactams and Microbial Bioactive Peptides,* W. de Gruyter and Co., Berlin, 1990, pp. 159–178.

β-LACTAMS

CARBAPENEMS AND PENEMS

In the period up to 1970 most β-lactam research was concerned with the penicillin and cephalosporin group of antibiotics. Since that time, however, a wide variety of new mono- and bicyclic β-lactam structures have been described. The carbapenems, characterized by the presence of the bicyclic ring system (**1**, X = CH_2) originated from natural sources; the penem ring (**1**, X = S) and its derivatives are the products of the chemical synthetic approach to new antibiotics. The chemical names are: 7-oxo-(*R*)-1-azabicyclo[3.2.0]hept-2-ene-2-carboxylic acid, $C_7H_7NO_3$, and 7-oxo-(*R*)-4-thia-1-azabicyclo[3.2.0]hept-2-ene-2-carboxylic acid, $C_6H_5NO_3S$, respectively.

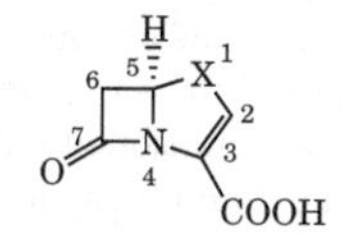

(**1**)

Carbapenems

Carbapenems include thienamycin ($C_{11}H_{16}N_2O_4S$), MM 4550 ($C_{13}H_{16}N_2O_9S_2$), MM 13902 ($C_{13}H_{16}N_2O_8S_2$), MM 17880 ($C_{13}H_{18}N_2O_8S_2$),

PS-5 ($C_{13}H_{18}N_2O_4S$), carpetimycin A ($C_{14}H_{18}N_2O_6S$), asparenomycin A ($C_{14}H_{16}N_2O_6S$), and pluracidomycin A ($C_9H_{11}NO_{10}S_2$).

Occurrence, Fermentation, and Biosynthesis. Although a large number of *Streptomyces* species have been shown to produce carbapenems, only *S. cattleya* and *S. penemfaciens* have been reported to give thienamycin. Generally the antibiotics occur as a mixture of analogues or isomers and are often coproduced with penicillin N and cephamycin C.

Properties. Thienamycin is isolated as a colorless, hygroscopic, zwitterionic solid, although the majority of carbapenems have been obtained as sodium salts and, in the case of the sulfated olivanic acids, as disodium salts. Concentrated aqueous solutions of the carbapenems are generally unstable, particularly at low pH. All the substituted natural products have characteristic uv absorption properties that are often used in assay procedures. The ir frequency of the β-lactam carbonyl is in the range 1760–1790 cm^{-1}.

Structure Determinations. The structural elucidation of the early carbapenems, thienamycin, and the olivanic acids, followed a fairly similar sequence making use of both spectroscopic and degradation studies. Infrared absorption spectra suggested the presence of a β-lactam ring (ν_{max} 1765 cm^{-1}) and in the case of thienamycin a trans-arrangement of β-lactam protons was indicated by the small coupling constant ($J_{5,6} < 3$ Hz) for the β-lactam hydrogens in the nmr. For the sulfated olivanic acids, the coupling constant ($J_{5,6} \approx 6$ Hz) indicated the more familiar *cis*-β-lactam stereochemistry found in the penicillins and cephalosporins.

Reactions. Although carbapenems are extremely sensitive to many reaction conditions, a wide variety of chemical modifications have been carried out. Many derivatives of the amino, hydroxy, and carboxy group of thienamycin have been prepared primarily to study structure–activity relationships.

Synthesis One consequence of the discovery of the carbapenem natural products has been the development of new synthetic methods, the impetus for which was provided by the exceptional antibacterial potential of the compounds coupled with the extremely poor fermentation yields. Only chemical synthesis could provide the quantities of *N*-formimidoyl thienamycin (MK 0787) necessary for clinical trials and commercial production.

A synthetic approach that involves the [3,4] bond formation using a carbene insertion reaction has been highly successful and is illustrated by the enantioselective synthesis of (+-thienamycin) starting from L-aspartic acid, $C_4H_7NO_4$.

A second method makes use of the lactone from acetone dicarboxylate and for which a synthesis form (−)-carvone has been reported.

Biological Properties. Thienamycin, the olivanic acids, and the majority of carbapenems are highly active broad-spectrum antibiotics having good stability to β-lactamases. Of the natural products, thienamycin is the most potent, having a spectrum of activity encompassing both aerobic and anaerobic gram-positive and gram-negative bacteria, including *Pseudomonas* species.

Penems

Historically, the development of penems is contemporary with that of the naturally occurring carbapenems, and the direction of penem research has clearly been influenced by the structures of the closely related natural products. The origins of the two groups of compounds are, however, quite different. Unlike carbapenems, no penems have been found in nature.

Synthesis. *Woodward's Phosphorane Route.* The first penem synthesis utilized an intramolecular Wittig reaction to form the [2, 3] double bond of the thiazoline ring. Reductive acylation of the penicillin-derived disulfide gave the thioester. Ozonolysis of the latter provided the oxalimide which on mild methanolysis gave the azetidinone. Well-established methods were applied to convert to the phosphorane which underwent thermal cyclization to the penem ester. Catalytic hydrogenation gave the penem acid, which was shown to possess antibacterial activity in spite of its rather limited stability. Penems include SCH 29482 ($C_{10}H_{13}NO_4S_2$), SCH 34343 ($C_{11}H_{14}N_2O_6S_2$), FCE 221201 ($C_{10}H_{12}N_2O_6S$), FCE 22891 ($C_{13}H_{16}N_2O_8S$), HRE 664 ($C_{15}H_{14}N_2O_6S$), SUN 5555 ($C_{12}H_{15}NO_5S$), CGP 31608 ($C_9H_{12}N_2O_4S$), CP 65207 ($C_{12}H_{15}NO_5S_3$), and FCE 25199 ($C_{15}H_{17}NO_7S$).

Biological Properties. In marked contrast to the antibacterially inactive penicillanic and cephalosporanic acids, 6-unsubstituted penems exhibit good activity against both gram-positive and gram-negative bacteria.

Economic Aspects

Extensive carbapenem and penem antibiotic research has been ongoing since thienamycin was discovered in 1978. However, only the imipenem–cilastatin combination has become a commercial product. Meropenem was expected to be the second carbapenem on the market by 1998.

ROBERT SOUTHGATE
NEAL F. OSBORNE
SmithKline Beecham Pharmaceuticals (U.K.)

R. W. Ratcliffe and G. Albers-Schönberg, "The Chemistry of Thienamycin and Other Carbapenem Antibiotics" and I. Ernest, "The Penems" in R. B. Morin and M. Gorman, eds., *Chemistry and Biology of β-Lactam Antibiotics*, Academic Press, New York, 1982, Chapts. 4 and 5.

R. Southgate and S. Elson, *Progr. Chem. Org. Nat. Prod.* **47**, 1 (1985).

A. G. Brown, M. J. Pearson, and R. Southgate, *Comprehens. Med. Chem.* **2**, 655 (1990).

CEPHALOSPORINS

The cephalosporins, a subgroup of β-lactam antibiotics, consist of a 4-membered lactam ring fused through the nitrogen and the adjacent tetrahedral carbon atom to a second heterocycle forming a 6-membered dihydrothiazine ring. Other structural features common to all the cephalosporins are a carboxyl group on the dihydrothiazine ring on the carbon next to the ring nitrogen and a functionalized amino group on C-7, the carbon of the β-lactam ring opposite the nitrogen. These features are evidenced in 7-aminocephalosporanic acid (7-ACA), $C_{10}H_{12}N_2O_5S$ (**1**). Cephalosporins, like all β-lactam antibiotics, exert their antibacterial effect by interfering with the synthesis of the bacterial cell wall. These antibiotics tend to be "irreversible" inhibitors of cell wall biosynthesis and they are usually bactericidal at concentrations close to their bacteriostatic levels. Cephalosporins are widely used for treating bacterial infections. They are highly effective antibiotics and have low toxicity.

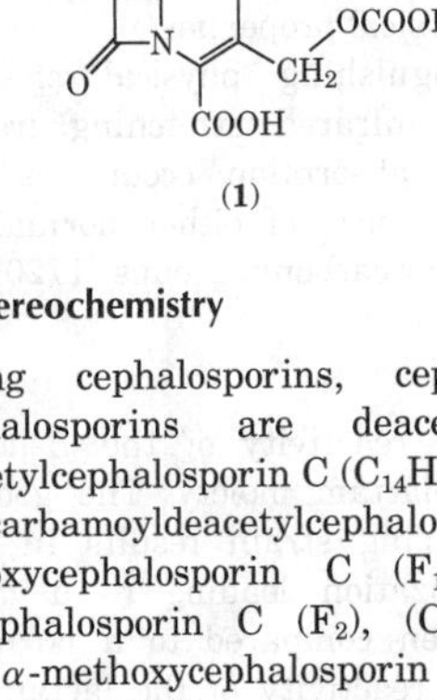

(1)

Nomenclature and Stereochemistry

Naturally occurring cephalosporins, cephamycins, and the 7-formamido cephalosporins are deacetoxycephalosporin C ($C_{14}H_{19}N_3O_8S$), deacetylcephalosporin C ($C_{14}H_{19}N_3O_7S$), cephalosporin C ($C_{16}H_{12}N_2O_5S$), *O*-carbamoyldeacetylcephalosporin C ($C_{15}H_{19}N_4O_8S$), 3′-methylthiodeacetoxycephalosporin C (F_1) ($C_{15}H_{21}N_3O_6S_2$), 3′-sulfothiodeacetoxycephalosporin C (F_2), ($C_{14}H_{19}N_3O_9S_3$), C43-219 ($C_{19}H_{28}N_4O_8S_2$), 7α-methoxycephalosporin C ($C_{17}H_{23}N_3O_9S$), cephamycin C ($C_{16}H_{22}N_4O_9S$), cephamycin A ($C_{25}H_{29}N_3O_{14}S_2$), cephamycin B ($C_{25}H_{29}N_3O_{11}S$), Takeda C2801X ($C_{25}H_{29}N_3O_{12}S$), SF-1623 ($C_{15}H_{21}N_3O_{10}S_3$), SQ 28, 516 ($C_{36}H_{55}N_{11}O_{15}S_1$), SQ 28, 517 ($C_{36}H_{56}N_{12}O_{14}S_1$), cephabacin F_1 or chitinovorin A ($C_{26}H_{41}N_9O_{11}S$), cephabacin F_2 or chitinovorin B ($C_{29}H_{46}N_{10}O_{12}S$), cephabacin F_3

($C_{32}H_{51}N_{11}O_{13}S$), cephabacin F_4 ($C_{26}H_{41}N_9O_{12}S$), cephabacin F_5 ($C_{29}H_{46}N_{10}O_{14}S$), cephabacin F_6 ($C_{32}H_{51}N_{11}O_{15}S$), cephabacin F_7 ($C_{26}H_{41}N_7O_{12}S$), cephabacin F_8 ($C_{29}H_{46}N_8O_{14}S$), cephabacin F_9 ($C_{32}H_{51}N_9O_{15}S$), cephabacin H_1 ($C_{25}H_{40}N_8O_{10}S$), cephabacin H_2 ($C_{28}H_{45}N_9O_{11}S$), cephabacin H_3 ($C_{31}H_{50}N_{10}O_{12}S$), cephabacin H_4 ($C_{25}H_{40}N_8O_{11}S$), cephabacin H_5 ($C_{28}H_{45}N_9O_{13}S$), cephabacin H_6 ($C_{31}H_{50}N_{10}O_{14}S$), cephabacin M_1 ($C_{31}H_{50}N_8O_{13}S$), cephabacin M_2 ($C_{34}H_{55}N_9O_{15}S$), cephabacin M_3 ($C_{37}H_{60}N_{10}O_{16}S$), cephabacin M_4 ($C_{41}H_{69}N_{11}O_{15}S$), cephabacin M_5 ($C_{44}H_{74}N_{12}O_{17}S$), and cephabacin M_6 ($C_{47}H_{79}N_{13}O_{18}S$).

Biogenesis

The biosynthesis of cephalosporins and penicillins both start from the amino acids and proceed via δ-(L-α-aminoadipyl)-L-cysteinyl-D-valine (LLD-ACV), $C_{14}H_{25}N_3O_6S$, often referred to as the Arnstein tripeptide. Because LLD-ACV is not transported into intact cells, a cell-free system was required to determine that this intermediate is the precursor of the penicillins. The cell-free system, obtained from *C. acremonium,* converts LLD-ACV into isopenicillin N, $C_{14}H_{21}N_3O_6S$. Isopenicillin N (IPN) synthetase, the enzyme which catalyzes this conversion, requires oxygen, Fe^{2+}, a reducing agent such as ascorbate, and a thiol group such as that of dithiothreitol for high activity. Isopenicillin N synthetase is present in *P. chrysogenum* and in species of *Streptomyces* as well as in *C. acremonium.* In the *Cephalosporium* species and the *Streptomyces* species, an epimerase is present that converts isopenicillin N into penicillin N, $C_{14}H_{21}N_3O_6S$, which then undergoes a ring expansion to deacetoxycephalosporin C. The ring expansion enzyme (REX) from *C. acremonium* appears to be bifunctional; it also catalyzes the subsequent hydroxylation of deacetoxycephalosporin C to deacetylcephalosporin C, itself a precursor of cephalosporin C. The introduction of the methoxyl group is also a two-step process involving molecular oxygen. Using cephalosporin C or the corresponding carbamoyl derivative as a substrate, another dioxygenase catalyzes the incorporation of a 7α-hydroxy function. The resulting 7α-hydroxycephalosporin is then methylated using *S*-adenosylmethionine to form the corresponding 7α-methoxycephalosporin. The details of these steps are discussed in depth in the literature. Rapid advances in this area have been made possible by the successful cloning and expression of isopenicillin N synthetase (IPNS) and ring expansion-hydroxylase (REX).

Physical Properties

Most cephalosporin antibiotics are white, off-white, tan, or pale yellow solids that are usually amorphous, but can sometimes be obtained crystalline. The cephalosporins do not usually have sharp melting points, but rather decompose upon heating at elevated temperatures. The acid strength, pK_a, of the carboxyl group on the dihydrothiazine ring depends on environment. Representative pK_a values are given in Table 1, as are other physical properties.

One of the distinguishing physical characteristics of the cephalosporins is the infrared stretching frequency of the β-lactam carbonyl. This absorption occurs at higher frequencies ($1770-1815\,cm^{-1}$) than those of either normal secondary amides ($1504-1695\,cm^{-1}$) or ester carbonyl groups ($1720-1780\,cm^{-1}$).

Chemical Properties

Much of the chemical reactivity of the β-lactam antibiotics is associated with the β-lactam moiety. The geometry and the accompanying increased ring strain results in very little, if any, amide-resonance stabilization leading to a marked increase in chemical reactivity when compared to a normal amide. In fact, in many instances the reactivity of the lactam carbonyl is analogous to that of a carboxylic acid anhydride. Fused β-lactam antibiotics are readily attacked by nucleophiles with resultant ring opening and loss of biological activity. The cephalosporins are more resistant to ring opening than the penicillins.

Biological Properties

The clinical effectiveness of the cephalosporins depends on a number of properties. The antibiotic must inhibit, or preferably kill, bacteria at acceptable concentrations of the drug (*in vitro* activity); it must be capable of achieving host serum and tissue levels greater than those required to inhibit the pathogenic organism; and the selective toxicity profile must allow for safe administration to the host.

Classification. As of 1991, there were approximately fifty different cephalosporins in clinical use or at an advanced stage of evaluation and development. These include oral cephalosporins [cephalexin ($C_{16}H_{17}N_3O_4S$), cefaclor ($C_{15}H_{14}N_3O_4S_1Cl$), cephradine ($C_{16}H_{19}N_3O_4S$), cefadroxil ($C_{16}H_{17}N_3O_5S$), cefixime ($C_{16}H_{15}N_5O_7S_2$), ceftibuten ($C_{15}H_{14}N_4O_6S_2$), ceprozil ($C_{18}H_{19}N_3O_5S$), and $C_{15}H_{15}N_5O_6S_2$]; oral cephalosporins-prodrugs [cefuroximeaxetil ($C_{20}H_{22}N_4O_{10}S_1$), cefpodoximeproxetil ($C_{21}H_{27}N_5O_9S_2$), and ($C_{18}H_{19}N_5O_7S_2$)]; cephamycins [cefoxitin ($C_{15}H_{15}N_3O_6$), cefmetazole ($C_{14}H_{15}N_7O_4S_3$), cefminox ($C_{15}H_{19}N_7O_6S_3$), and cefotetan ($C_{16}H_{15}N_7O_7S_4$)]; parenteral cephalosporins [cephalothin ($C_{16}H_{16}N_2O_6S_2$), cephacetrile ($C_{13}H_{13}N_3O_6S$), cephapirin ($C_{17}H_{17}N_3O_6S_2$), cefamandole ($C_{11}H_{18}N_6O_5S_2$), cefonicid ($C_{18}H_{17}N_6O_8S_3Na$), cefazolin ($C_{14}H_{14}N_8O_4S_3$), ceforanide ($C_{20}H_{19}N_7O_6S_2$), cefoperazone ($C_{25}H_{27}N_9O_8S_2$), cefuroxime ($C_{16}H_{16}N_4O_8S$), cefotaxime ($C_{16}H_{17}N_5O_7S_2$), ceftizoxime ($C_{13}H_{13}N_5O_5S_2$), cefmenoxime ($C_{16}H_{17}N_9O_5S_3$), ceftriaxone ($C_{18}H_{17}N_8O_7S_3Na$), ceftazidime ($C_{22}H_{22}N_6O_7S_2$), cefsulodin ($C_{22}H_{20}N_4O_8S_2$), cefpiramide ($C_{25}H_{24}N_8O_7S_2$), cefpirome ($C_{22}H_{22}N_6O_5S_2$), cefpimizole ($C_{28}H_{26}N_6O_{10}S_2$), cefepime ($C_{19}H_{24}N_6O_5S_2$), $C_{28}H_{21}N_9O_{11}S_3Na_2$, and $C_{31}H_{31}N_8O_8S_2F_3$]; and oxadethiacephalosporins [moxalactam latamoxef ($C_{20}H_{20}N_6O_8S$) and flomoxef ($C_{15}H_{18}N_6O_6S_2F_2$)]. Cephalosporins may be classified for convenience by their clinical pharmacology, β-lactamase resistance, chemical structure, or their antibacterial spectrum. The most common classification, which is somewhat arbitrary, divides the cephalosporins into three groups or generations, based primarily on their antibacterial spectrum. First-generation cephalosporins are characterized by good gram-positive activity and modest to weak gram-negative activity. Third-generation cephalosporins have an expanded gram-negative spectrum and are the most active against enteric gram-negative bacilli, including penicillinase-producing strains, as well as *Serratia* and *Citrobacter.*

Some other compounds tentatively labeled as examples of the as yet undefined "fourth generation" have appeared in the literature, but these are probably best thought of as third-generation cephalosporins having slight advantages over earlier examples of this group. This group includes cefpirome, cefepime, and others undergoing clinical trials.

In Vitro Antibacterial Activity and Structure-Activity Relationships. The *in vitro* antibacterial activity of any particular cephalosporin is a combination of the degree and type of activity at the target site, the ease with which it can penetrate to the target, and the ability to resist the attack of destructive enzymes. The nature and complexity of the biochemical target(s) for β-lactam antibiotics is fairly well-established and the mechanisms of penetration are also understood to some extent. However, many factors are involved, including pharmacokinetic and pharmacodynamic properties, and the antibiotic may not perform as predicted.

Structure–activity relationships can be inferred by comparison of the antibacterial properties of the clinical agents and related compounds. Different acyl side chains can result in significant changes in the antibacterial activity, both with respect to potency and to breadth of spectrum. The highest activities are observed when the acylamino side chain at C-7 is a substituted acetic acid. Homologation of the acetic acid moiety lowers activity dramatically as exemplified by the naturally occurring cephalosporins, which all have weak activity.

One of the principal deficiencies of the older cephalosporins was the lack of resistance to β-lactamases. Compounds with improved β-lactamase resistance have one or more of the following characteristics: a second, monovalent substituent on the α-carbon of the C-7 acyl

Table 1. Properties of Cephalosporins

Name	pK_a[b]	β-Lactam ir stretching frequency, cm^{-1}[c]	Uv absorption, λ max, nm (ϵ, $cm^{-1}M^{-1}$)[d]	Nmr absorption, ppm[a] H-7[e]	H-6[f]	$J_{6,7}$, Hz[g]
7-ACA	1.75, 4.63	1806	261(8500)	5.53d 4.83d	5.13d	4.5
cephalosporin C	<2.6, 3.1, 9.8	1780	260(8900)	5.66d	5.15d	4.7
cephalothin	5.0	1760	265(9000)	5.70d	5.14d	4.5
cephalexin	5.2, 7.3	1775	260(7750)	6.10d	5.45d	4.2
cephamycin C	4.2, 5.6, 10.4[h]	1770	264(6900) 242(5700)		5.19s	

[a] In D_2O relative to external standard tetramethyl silane (TMS); s = singlet, d = doubet.
[b] Values for aqueous solutions unless otherwise noted, either by direct determination or by extrapolation from mixed solvents.
[c] Nujol mull.
[d] In aqueous solution.
[e] Range 5.23–6.21.
[f] Range 4.24–5.46.
[g] $J_{6,7}(cis) = 4-5$ Hz, $J_{6,7}(trans) = 1.5-2$ Hz, $J_{7\text{-NH}} = 8-11$ Hz.
[h] In 66% DMF.

group such as is found in cefamandole and cefoperazone; a *syn*-oxime substituent, eg, cefuroxime and ceftriaxone; a methoxy, or formamido, substituent on the β-lactam ring at the 7 α-position such as in cefoxitin. However, these various substituents have different effects, and increased resistance to one enzyme does not indicate resistance to all β-lactamases.

Intrinsic Activity. β-Lactam antibiotics affect sensitive bacteria by inhibiting late stages in the biosynthesis of their cell wall peptidoglycan.

Resistance. Resistance to the cephalosporins may result from the alteration of target penicillin-binding sites (PBPs), decreased permeability of the bacterial cell wall and outer membrane, or by inactivation via enzyme-mediated hydrolysis of the lactam ring. This resistance can be either natural or acquired.

Transport and Cell Penetration. One of the causes of bacterial resistance to the cephalosporins is poor transport of the antibiotic through the outer membrane of gram-negative bacteria. This lipid-bilayer membrane carries receptor proteins for the recognition and transport of essential nutrients, but provides an effective barrier to large molecules. In the case of the cephalosporins there can be a considerable difference between the concentration required to inhibit intact cells and the concentrations required to saturate the target enzymes in broken cell preparations.

Pharmacokinetics. The pharmacokinetic properties of the cephalosporins depend to a large extent on the substituent at C-3. The 3′-acetoxy group is metabolized in the body to the less active 3′-carbamate, most other substituents, including the 3′-carbamate, are metabolically stable. Most cephalosporins are eliminated rapidly, having serum half-lives of 1–2 h.

Manufacture and Chemical Synthesis

At present all of the cephalosporins are manufactured from one of four β-lactams, cephalosporin C, penicillin V, penicillin G, and cephamycin C, which are all produced in commercial quantities by fermentation. The manufacturing process consists of three steps: fermentation, isolation, and chemical modification.

Nuclear Analogues of Cephalosporins

In the search for improved antibacterials not only has the effect produced by the variation of the C-7 amido side chain and the 3′ substituent been studied, but so also has the more synthetically challenging question of the effect of changes in the cephem nucleus. Nuclear analogues have been studied since the early 1970s but only the oxacephem class has reached the marketplace.

Cephalosporins With Special Properties

Chromogenic Cephalosporins. A 3-substituted pyridinium cephalosporin known as PADAC, $C_{27}H_{26}N_6O_4S_2$, is purple in color but on hydrolysis or treatment with β-lactamase releases the 3′-pyridinium group with concomitant loss of the purple color. PADAC is an example of the chromogenic cephalosporins which are useful in studying interactions with β-lactamases.

Uses

The cephalosporins are used for treating infectious diseases of bacterial origin in both humans and animals. First-generation cephalosporins such as cephalothin and cephalexin are the most active against staphylococci and nonenterococcal streptococci and are effective alternatives to the penicillins in patients with endocarditis, osteomyelitis, septic arthritis, and cellulitis. They are especially useful for treating patients who are allergic to the pencillins or who have mixed infections from gram-positive and gram-negative bacteria. Although these drugs have proved useful in treating infections such as bacteremias, urinary tract infections, and pneumonias, caused by gram-negative bacilli, their use as single agents in this regard is not recommended because activity against gram-negative organisms is somewhat weak and unpredictable. The first-generation cephalosporins have been widely used for prophylaxis in cardiovascular, orthopedic, biliary, pelvic, and intraabdominal surgery.

Whereas third-generation cephalosporins do have some coverage against gram-positive infections, they are not the agents of choice. Similarly, most community-acquired infections are better treated with drugs other than the third-generation cephalosporins. The treatment of meningitis is an important exception.

The third-generation cephalosporins are effective in the treatment of bacteremias, pneumonias, urinary tract infections, intraabdominal infections, and skin and soft tissue infections.

Toxicity

The cephalosporins generally cause few side effects. Thrombophlebitis occurs as a result of intravenous administration of all cephalosporins. Hypersensitivity reactions related to the cephalosporins are the most common side effects observed, but these are less common than found with the penicillins.

Although immediate reactions of anaphylaxis, bronchospasm, and urticaria have been reported, most commonly patients exhibiting an adverse reaction develop a maculopapular rash, usually after several days of therapy. They may also develop fever and eosinophilia.

JOHN ROBERTS
Hoffmann-La Roche, Inc.

J. E. Baldwin in P. H. Bentley and R. Southgate, eds., *Recent Advances in the Chemistry of β-Lactam Antibiotics: Proceedings of the Fourth International Symposium,* The Royal Society of Chemistry, London, 1989, p. 1.

G. L. Mandell in G. L. Mandell, R. G. Douglas, Jr., and J. E. Bennett, *Anti-infective Therapy,* John Wiley & Sons, Inc., New York, 1985, p. 76.

E. Squires and R. Cleeland, *Microbiology and Pharmacokinetics of Parenteral Cephalosporins,* Roche Laboratories, Nutley, N.J., 1985.

G. R. Donowitz and G. L. Mandell, *N. Engl. J. Med.* **318,** 490 (1988).

β-LACTAMASE INHIBITORS

The antibacterial effectiveness of penicillins, cephalosporins, and other β-lactam antibiotics depends on selective acylation and consequently, inactivation, of transpeptidases involved in bacterial cell wall synthesis. This acylating ability is a result of the reactivity of the β-lactam ring (**1**). Bacteria that are resistant to β-lactam antibiotics often produce enzymes called β-lactamases that inactivate the antibiotics by catalyzing the hydrolytic opening of the β-lactam ring to give products (**2**) devoid of antibacterial activity.

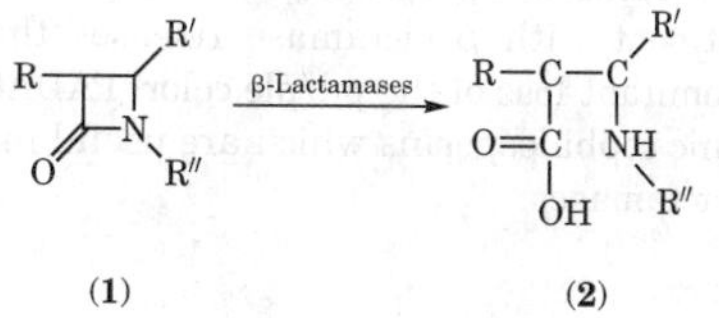

(**1**) active (**2**) inactive

Based on sequence data, it has been suggested that β-lactamases evolved from the enzymes involved in bacterial cell wall synthesis.

One approach to combating antibiotic resistance caused by β-lactamase is to inhibit the enzyme (see ENZYME INHIBITORS). Effective combinations of enzyme inhibitors with β-lactam antibiotics such as penicillins or cephalosporins result in a synergistic response, lowering the minimal inhibitory concentration (MIC) by a factor of four or more for each component. However, inhibition of β-lactamases alone is not sufficient. Pharmacokinetics, stability, ability to penetrate bacteria, cost, and other factors are also important in determining whether an inhibitor is suitable for therapeutic use. Almost any class of β-lactam is capable of producing β-lactamase inhibitors. Several reviews have been published on β-lactamase inhibitors, detection, and properties.

Table 1 shows the clinically most important bacteria that produce β-lactamases, separated into gram-positive and gram-negative organisms. The prevalence of β-lactamase is indicated as is the origin and Richmond-Sykes classification. Based on these data the most important β-lactamases to inhibit clinically are the gram-positive penases, the gram-negative TEM, which are Richmond-Sykes type III, and the gram-negative chromosomal cephalosporinases–cephases which are Richmond-Sykes type I. These enzymes are subsequently referred to as penase, TEM(III), and cephase(I).

Mechanistic Aspects of β-Lactamase Inhibition

The clinically important β-lactamases, eg, the penases, TEM(III), and cephases(I), are serine proteases that form an acyl enzyme intermediate with β-lactam substrates and β-lactam-derived β-lactamase inhibitors. Mechanistic studies using several β-lactamase inhibitors have been extensively reviewed and a general inhibition scheme is illustrated in Figure 1.

Active site-directed β-lactam-derived inhibitors have a competitive component of inhibition, but once in the active site they form an acyl enzyme species which follows one or more of the pathways outlined in Figure 1.

β-Lactam-Based Inhibitors

Penicillins, Cephalosporins, and Monobactams. Early attempts at inhibiting β-lactamases using inorganics or penicillin fragments were not successful. The discovery that cephalosporin C and methicillin inhibited β-lactamases resulted in the screening of numerous antibiotics, and a number of β-lactamase-resistant penicillins and cephalosporins were found to be β-lactamase inhibitors. These compounds act by forming a transiently inhibited acyl enzyme species as a result of conformational change and are inhibitory substrates (Fig. 1, route C). No clinically useful inhibitors have been identified from this class. These efforts have been extensively reviewed.

Modern β-lactamase-resistant cephalosporins have been reported to be inhibitors of type I cephases. β-Lactamase inhibition occurs through the transiently inhibited enzyme species, which requires a good C-3 leaving group.

Several monobactams have been reported to be inhibitory substrates for type I cephases.

Clavulanic Acid Class of β-Lactamase Inhibitors. Clavulanic acid has only weak antibacterial activity, but is a potent irreversible inhibitor for many clinically important β-lactamases, including penases and Richmond-Sykes types II, III, IV, V, VI (*Bacteroides*).

Carbapenem-β-Lactamase Inhibitors. Carbapenems are another class of natural product β-lactamase inhibitors discovered about the same time as clavulanic acid.

Penem β-Lactamase Inhibitors. The synthesis and antibacterial properties of penems have been reviewed. Like the closely related carbapenems, many of the penems are potent antibacterials. Additionally, penems are also susceptible to degradation by renal dipeptidase, but to a lesser extent.

Penam Sulfone β-Lactamase Inhibitors. Natural product discoveries stimulated the rational design of β-lactamase inhibitors based on the readily accessible penicillin nucleus. An early success was penicillanic acid sulfone, (2(*S*)-*cis*)-3,3-dimethyl-7-oxo-4,4-dioxide-4-thia-1-azabicyclo [3.2.0]heptane-2-carboxylic acid (sulbactam) ($R = R^1 = H, R^2 = R^3 = CH_3$), $C_8H_{11}NO_5S$. The synthesis, microbiology, and clinical use of sulbactam have been reviewed. Sulbactam, with minor exceptions, is a weak antibacterial, but is a potent irreversible inactivator of many β-lactamases, including penases and Richmond-Sykes type II, III, IV, V, and VI (*Bacteroides*) β-lactamases. Sulbactam is better than clavulanic acid against type I cephases, and synergy is observed for combinations of many penicillins and cephalosporins. Because sulbactam is not well absorbed orally, prodrug forms have been developed. Numerous other penicillin sulfones have been reported to be β-lactamase inhibitors.

Penam β-Lactamase Inhibitors. Penam is the trivial name of 4-thia-1-azabicyclo[3.2.0]heptane. The report that 6-β-bromopenicillanic acid, [2(*S*)-(2α,5α,6β)]-6-bromo-3,3-dimethyl-7-oxo-4-thia-1-azabicyclo[3.2.0]heptane-2-carboxylic acid, ($R = Br, R^1 = H, R^2 = R^3 = CH_3$) is a potent inhibitor led to intense study both of this compound and analogues. The microbiology profile of 6-β-bromopenicillanic acid has been reported and the compound has progressed to clinical trials. Mechanistic studies have demonstrated that the dihydrothiazine derivative is responsible for inactivation of β-lactamases.

Other Unusual β-Lactam-Based Inhibitors. There are a number of other unusual β-lactams reported to have β-lactamase inhibition activity. In general these compounds are not very potent and are not irreversible inhibitors. Data are also very limited.

Economic Aspects

Although a broad range of β-lactamase inhibitors has been discovered, only clavulanic acid and sulbactam have been commercialized. Clavulanic acid manufactured by SmithKline Beecham is sold as an oral and parenteral product in combination with amoxicillin under the trade name Augmentin. A parenteral product in combination with ticarcillin, $C_{15}H_{16}N_2O_6S$, has the trade name, Timentin.

Table 1. β-Lactamase-Producing Bacteria

Organism	β-Lactamase-producers, %[a]	Enzyme origin, %		Richmond-Sykes classification
		Chromosomal	Plasmid	
Gram-positive				
Staphylococcus aureus	80		penase	
Staphylococcus epidermidis	80		penase	
Gram-negative				
Escherichia coli	25(16–76)	15	TEM, 80; OXA-1, 7.5	I,III,V
Haemophilus influenzae	25–60		TEM, 92; ROB-1, 8	III
Neisseria gonorrhea	1–10		TEM, 100	III
Salmonella	6		TEM, 100	III
Shigella			TEM	III
Klebsiella pneumoniae	60–90		TEM, 24; SHV-1, 76	III,IV
Enterobacter	25–30	73	TEM, 27	I,III
Citrobacter	20–50	77	TEM, 23	I,III
Pseudomones aeruginosa	23	44	TEM, 9; PSE, 10; others	I,III
Bacillus catarrhalis	87	90^+	BRO-1	
Bacillus fragilis	87	~100		VI

[a] Clinically resistant bacteria, virtually 100% of the *Enterobacteriaceae,* produce a low level of chromosomal enzyme that can clinically be selected for higher levels.

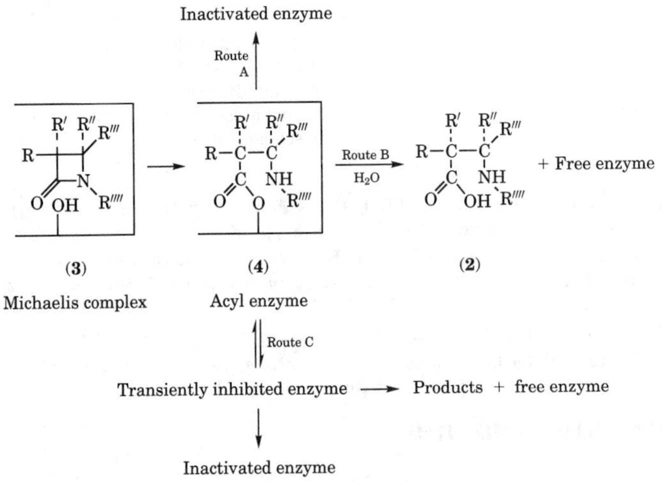

Figure 1. Scheme for the interaction of β-lactamase inhibitors and β-lactamases where the enzyme is represented by [OH].

Sulbactam is produced by Pfizer. The oral version of sulbactam in combination with ampicillin is called Unasyn Oral, which is the mutual prodrug sultamicillin. Two sulbactam parenteral products are sold, a combination product with ampicillin called Unasyn and a combination with cefoperazone called Sulperazon. In addition, sulbactam is sold alone for parenteral use with any β-lactam antibiotic as Betamaze.

J. G. STAM
Pfizer Central Research

J. M. T. Hamilton Miller and J. T. Smith, eds., *Beta-Lactamases,* Academic Press, New York, 1979.

C. Reading and M. Cole, *J. Enzyme Inhib.* **1,** 83 (1986).

J. R. Knowles, *Acc. Chem. Res.* **18,** 97 (1985).

K. Bush and R. B. Sykes, *J. Antimicrob. Chemother.* **11,** 97 (1983).

MONOBACTAMS

β-Lactam antibiotics are one of the best established classes of antimicrobial agents for the treatment of infectious diseases.

By screening strains of bacteria that were specifically responsive to β-lactam antibiotics, monocyclic β-lactams, ie, monobactams, varying in substitution at the C-3 position, were identified. The naturally occurring monobactams include SQ 26,180 ($C_6H_{10}N_2O_6S{\cdot}K$), SQ 26,445 ($C_{12}H_{20}N_4O_9S$), sulfazecin ($C_{12}H_{20}N_4O_9S$), and isosulfazecin ($C_{12}H_{20}N_4O_9S$). More recently, the 4β-methyl analogue of SQ 26,445/sulfazecin was isolated using a differential antibacterial assay.

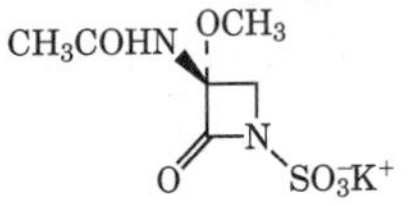

SQ 26, 180

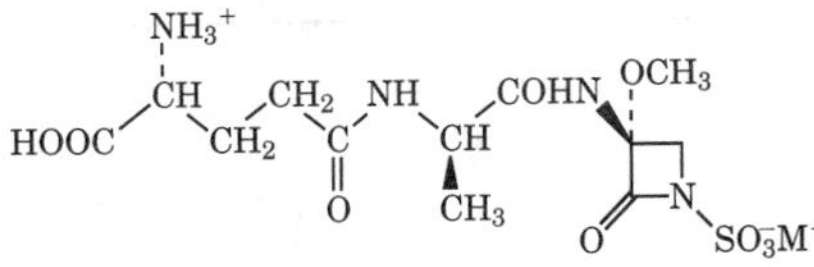

SQ 26, 445/ sulfazecin

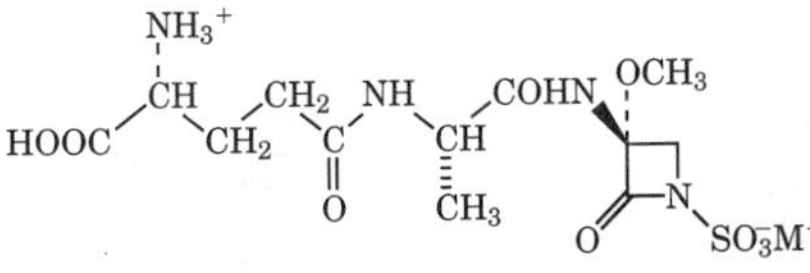

isosulfazecin

Synthesis

Initial syntheses employed the sulfonation of an N-1 unsubstituted azetidinone as the key step. The natural product SQ, 26,180 as well as other methoxylated monobactams were synthesized, starting from either 7-aminocephalosporanic acid (7-ACA) or 6-aminopenicillanic acid (6-APA), via sulfonation of the N-1 unsubstituted degradation products. Subsequently, many more C-3 side-chain analogues were prepared by this method.

A second, conceptually distinct chiral synthesis of monobactams was developed from β-hydroxy amino acids. Cyclization of the acylsulfamate of an amino-protected O-mesylserine derivative leads directly to the monobactam nucleus. This methodology was also applied to the synthesis of 4α- and 4β-methyl monobactams from L-threonine and allothreonine, respectively.

The monobactams, like pencillins and cephalosporins, interfere with the synthesis of bacterial cell walls. β-Lactam antibiotics bind to a series of penicillin-binding proteins (PBPs) on the cytoplasmic membrane and their antibacterial effect is believed to result from inhibition of a subset of these PBPs known as peptidoglycan transpeptidases.

The biosynthetic origin of monobactams has been elucidated by fermentation experiments using radioactively labeled amino acids. The monocyclic ring of the naturally occurring C-4 unsubstituted monobactams is derived from serine. Similar techniques have shown that the methyl moiety of the methoxyl group in 3α-methoxylated monobactams is derived from methionine.

All of the naturally occurring monobactams discovered as of this writing have exhibited poor antibacterial activity. However, as in the case of the penicillins and cephalosporins, alteration of the C-3 amide side chain led to many potent new compounds.

Aztreonam. Aztreonam is a totally synthetic compound having an antibacterial spectrum that is unique among β-lactam antibiotics. It exhibits potent and specific activity against a wide range of both β-lactamase-producing and nonproducing aerobic gram-negative bacteria, including *Pseudomonas aeruginosa,* but displays minimal inhibition against anaerobic and gram-positive aerobic bacteria, eg, staphylococci and streptococci.

Overall, aztreonam appears to be a safe agent having toxicity side effects similar to those of other β-lactams. The safety profile suggests that aztreonam may be useful as a replacement for aminoglycoside therapy. The biological properties of aztreonam have been extensively reviewed.

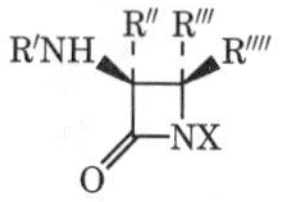

Table 1. Biologically Active Monobactam Subclasses

Subclass	Substituent, X
monobactams	SO_3H
monophosphams	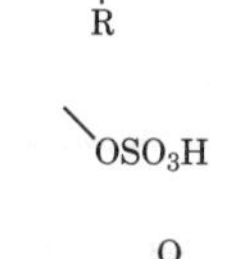
monosulfactams	OSO_3H
monophosphatams	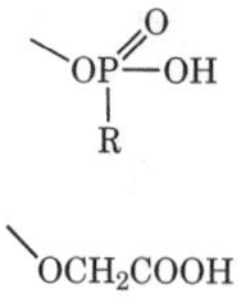
monoxacetams (oxamazins)	OCH_2COOH
monocarbams	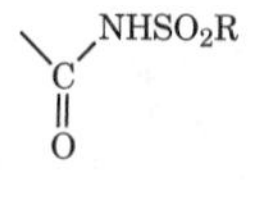
tetrazole-activated monobactams	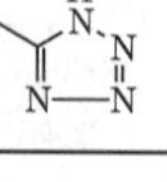

Alternative N-1 Activating Groups

β-Lactam antibiotics exert their antibacterial effects via acylation of a serine residue at the active site of the bacterial transpeptidases. Critical to this mechanism of action is a reactive β-lactam ring having a proximate anionic charge that is necessary for positioning the ring within the substrate binding cleft.

All of the naturally occurring monobactams and aztreonam are characterized by the presence of the N-1 sulfonate group, which serves a dual function. The electron-withdrawing sulfonate moiety renders the β-lactam ring more reactive toward nucleophilic attack and at the same time provides the anionic charge necessary for binding. A variety of monobactam subclasses bearing N-1 activating groups having the necessary physical properties for antibacterial activity are listed in Table 1.

Economic Aspects

Two monobactams were in clinical use as of 1990. Aztreonam, manufactured by Bristol-Myers Squibb, has the worldwide trademark of Azactam. The global experience with aztreonam and its clinical acceptance signifies that the monobactams are recognized as an important new class of antibacterial agents. Carumonam, manufactured by Takeda, received approval for human usage in 1988 in Japan. Carumonam has the trademark Amasulin.

KLAUS R. LINDNER
DANIEL P. BONNER
WILLIAM H. KOSTER
Bristol-Myers Squibb Pharmaceutical
Research Institute

W. L. Parker, J. O'Sullivan, and R. B. Sykes, *Advances in Applied Microbiology,* Vol. 31, Academic Press, New York, 1986, pp. 181–205.

W. H. Koster, C. M. Cimarusti, and R. B. Sykes in R. B. Morin and M. Gorman, eds., *Chemistry and Biology of β-Lactam Antibiotics,* Vol. 3, Academic Press, New York, 1982, pp. 339–375.

J. F. Acar and H. C. Neu, eds., *Rev. Infect. Disease,* **1** (Suppl. 4) (1985).

W. H. Koster and D. P. Bonner in H. Umezawa, ed., *Frontiers of Antibiotic Research,* Academic Press, New York, 1987, pp. 211–226.

PENICILLINS AND OTHERS

The basic structural features of the penicillin nucleus (**1**) include a β-lactam ring fused through nitrogen and the adjacent tetrahedral carbon to a second heterocycle which, in natural penicillin is a 5-membered thiazolidine ring. Biologically active penicillins are generally characterized by a functionalized amino group in the 6β-position of the β-lactam ring and a carboxyl group in the 3-position in the thiazolidine ring. The unsubstituted bicyclic ring system of the penicillins is designated as penam, C_5H_7NOS (**2**) and the penicillins (**1**) are generally 6-acylamino-2,2-dimethylpenam-3-carboxylic acids. A further simplification is the use of the term penicillanic acid, $C_8H_{11}NO_3S$, to designate the penicillin ring system having the substituents indicated in (**3**).

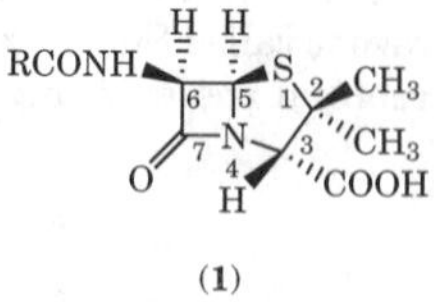

(**1**)

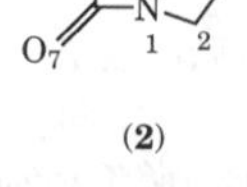

(**2**)

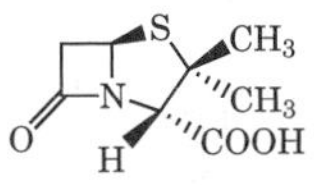

(3)

In general, penicillins exert their biological effect, as do the other β-lactams, by inhibiting the synthesis of essential structural components of the bacterial cell wall. These components are absent in mammalian cells so that inhibition of the synthesis of the bacterial cell wall structure occurs with little or no effect on mammalian cell metabolism. Additionally, penicillins tend to be irreversible inhibitors of bacterial cell-wall synthesis and are generally bactericidal at concentrations close to their bacteriostatic levels. Consequently penicillins have become widely used for the treatment of bacterial infections and are regarded as one of the safest and most efficacious classes of antibiotics. Penicillins in clinical use include limited spectrum penicillins [benzyl penicillin (penicillin G) ($C_{16}H_{18}N_2O_4S$), phenoxymethyl penicillin (penicillin V) ($C_{16}H_{18}N_2O_5S$), and phenethicillin ($C_{17}H_{20}N_2O_5S \cdot K$)], β-lactamase stable penicillins [methicillin ($C_{17}H_{20}N_2O_6S \cdot Na$), oxacillin ($C_{19}H_{19}N_3O_5S$), cloxacillin ($C_{19}H_{18}ClN_3O_5S$), dicloxacillin ($C_{19}H_{17}Cl_2N_3O_5S$), flucloxacillin ($C_{19}H_{17}ClFN_3O_5S$), and nafcillin ($C_{21}H_{22}N_2O_5SNa$)], broad spectrum penicillins [ampicillin ($C_{16}H_{19}N_3O_4S$), hetacillin ($C_{19}H_{23}N_3O_4S$), pivampicillin ($C_{22}H_{29}N_3O_6S$), talampicillin ($C_{24}H_{23}N_3O_6S$), bacampicillin ($C_{21}H_{27}N_3O_7S$), ciclacillin ($C_{15}H_{23}N_3O_4S$), amoxicillin ($C_{16}H_{19}N_3O_5S$), carbenicillin ($C_{17}H_{18}N_2O_6S$), ticarcillin ($C_{15}H_{16}N_2O_6S_2$), sulbenicillin ($C_{16}H_{18}N_2O_7S_2$), azlocillin ($C_{20}H_{23}N_5O_6S$), mezlocillin ($C_{21}H_{25}N_5O_8S_2$), and piperacillin ($C_{23}H_{27}N_5O_7S$)], and directed spectrum β-lactamase stable penicillins [temocillin ($C_{16}H_{18}N_2O_7S$)].

Physical Properties

Penicillins have several properties that are characteristic of β-lactam antibiotics. They are obtained in relatively pure form as off-white, tan, or yellow freeze-dried or spray-dried solids that are usually amorphous. Alternatively they are sometimes obtained as crystalline solids, often as hydrates. Penicillins do not usually have sharp melting points, but decompose upon heating to elevated temperatures. Most natural members have a free carboxyl group and commercial preparations are generally either supplied as salts, most frequently as sodium salts, or in zwitterionic form as hydrates, eg, amoxicillin trihydrate. The acid strength of the carboxyl group in aqueous solution varies from $pK_{a1} = 2.73$ for oxacillin to $pK_{a1} = 3.06$ for carbenicillin.

Spectral Characteristics. The infrared stretching frequency of the penicillin β-lactam carbonyl group normally occurs at relatively high frequencies (1770–1815 cm^{-1}), as compared to the absorptions for the secondary amide (1504–1695 cm^{-1}) and ester (1720–1780 cm^{-1}) carbonyl groups. There is little difference between solution and KBr spectra. The nuclear magnetic resonance spectrum of penicillins invariably provides information about the integrity of the ring, attachments to the ring, and the stereochemistry of those attachments. The disposition of the C-5 and C-6 resonances is also characteristic. More recently nmr spectroscopy has also proved to be a valuable tool for detecting and characterizing penicillin metabolites in biofluids.

Stereochemistry. The absolute stereochemistry of the penicillins is 3(*S*):5(*R*):6(*R*) and the stereochemistry of the substituents attached to the ring is designated by the α and β notations. Thus the β-lactam hydrogens are α, the acylamino group is β, and the penicillin carboxyl is α as in (**1**).

Chemical Properties

Penicillin to Other β-Lactam Conversions. Penicillin G (**1**, R = $C_6H_5CH_2$), penicillin V (**1**, R = $C_6H_5OCH_2$), and 6-APA (**1**, RCO = H) have proven to be cheap and versatile starting materials for a number of conversions to novel β-lactam systems including 7-aminodeacetoxycephalosporanic acid (7-ADCA), the deacetoxycephalosporins, cephalexin and cephradin, moxalactam and related oxacephems, novel penems, and the carbapenem antibiotic thienamycin.

Synthesis

The only penicillins used in their natural form are benzylpenicillin (penicillin G) and phenoxymethylpenicillin (penicillin V). The remainder of penicillins in clinical use are derived from 6-APA and most penicillins having useful biological properties have resulted from acylation of 6-APA using standard procedures.

A variety of coupling methods have been employed including: acid chorides, mixed anhydrides, mixed sulfonic acid anhydrides, *N,N*-dicyclohexylcarbodiimide and similar condensing agents, activated esters with *N*-hydroxysuccinimide, and *N*-hydroxybenzotriazole together with other acylating agents commonly used in peptide synthesis.

The use of protecting groups is common in penicillin chemistry: the amino function is normally protected by a trityl, benzyloxycarbonyl, *p*-nitrobenzyloxycarbonyl, trichloroethyloxycarbonyl, or trimethylsilyl group; and the carboxylic acid is usually protected as a benzyl, *p*-nitrobenzyl, *p*-methoxybenzyl, or trichloroethyl ester. Acylations may thus be carried out in aqueous or nonaqueous media with subsequent removal of the protecting group as required.

Chemical Modification. Chemical modification of most positions in the penicillin nucleus have been carried out. Apart from acylation of 6-APA, few of these modifications have proven profitable in terms of improving the biological properties of the derived penicillins. However, one of the modifications that has led to beneficial properties is substitution at the 6α-position.

Degradation. Penicillins are rapidly hydrolyzed by aqueous alkali to the corresponding penicilloic acids which are stable as salts, but which decarboxylate on acidification to yield penilloic acids. Penicillins are also degraded by aqueous acids via initial reaction of the side-chain carbonyl group with the β-lactam.

Biological Properties

Structure–Activity Relationships. Biological evaluation of penicillins yields information such as *in vitro* and *in vivo* antibacterial activities, minimum inhibitory concentration (MIC), minimum bactericidal concentration (MBC), protective effectiveness in laboratory animals (PD_{50}), and pharmacokinetic characteristics including efficiency of absorption, serum levels, tissue distribution, urinary excretion, recycling, etc. Penicillins are also tested for ability to resist inactivation by β-lactamase produced by both gram-positive and gram-negative bacteria.

Penicillin G remains probably the most active penicillin against gram-positive organisms. However, the majority of *Staphylococcus aureus* strains are resistant to penicillin by virtue of β-lactamase production. The β-lactamase-resistant penicillins such as oxacillin, cloxacillin, flucloxacillin, and nafcillin are active against most penicillin-resistant *Staphylococci* but lack activity against methicillin-resistant *Staphylococci* (MRSA) and the majority of gram-negative organisms.

Ampicillin and its congeners amoxicillin, bacampicillin, and ciclacillin, have largely similar antibacterial spectra, exhibiting activity against both penicillin-sensitive gram-positive and gram-negative microorganisms. The susceptibility of ampicillin or amoxicillin to β-lactamase may be overcome by combination with a β-lactamase inhibitor. Clavulanic acid and sulbactam are two such β-lactamase inhibitors in clinical use. The products Augmentin, a combination of clavulanic acid and amoxycillin for oral use primarily, and Unasyn, a combination of sulbactam and ampicillin for use by injection, are highly active against both gram-positive aerobic and anaerobic organisms in addition to many important gram-negative pathogens. Carbenicillin and ticarcillin possess moderate broad-spectrum activity and were the first penicillins to show activity against *Pseudomonas*

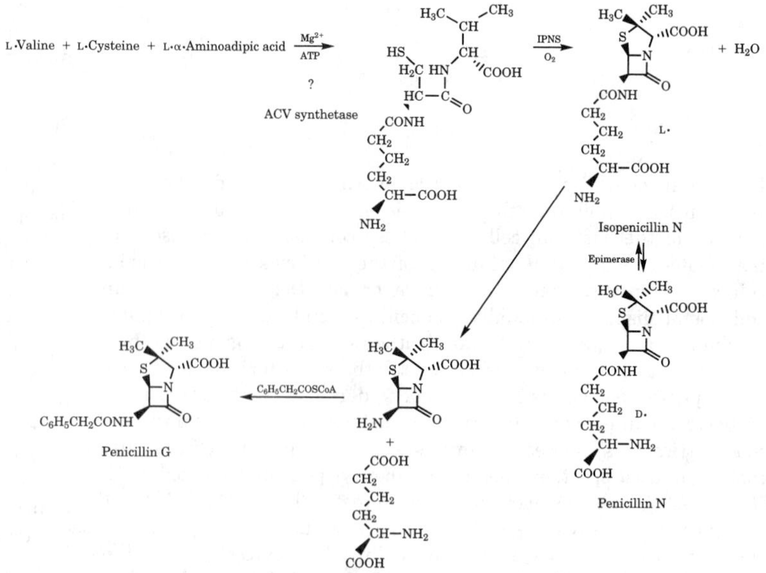

Figure 1. Biosynthesis of penicillins when ACV is aminoadipoly cysteinyl valine and IPNS is isopenicillin N synthase and $C_6H_5CH_2COSCoA$ represents benzyl coenzyme A. ACV synthetase is thought to catalyze the first step of this reaction sequence. Courtesy of J. E. Baldwin, Royal Society of Chemistry.

species. Azlocillin, mezlocillin, and piperacillin possess largely the same characteristics but have slightly greater potency against many gram-negative species. Temocillin is the first penicillin to possess a high level of activity against gram-negative organisms, notably the *Enterobacteriaceae,* as well as excellent β-lactamase stability. The introduction of the 6α-methoxy group and the concomitant β-lactamase stability compromises activity against both gram-positive organisms and *Pseudomonas* species. Similarly BRL 36650, having the 6α-formamido substituent, exhibits high bacterial β-lactamase stability. Some activity is retained against most species of *Streptococci* but the compound is inactive against *Staphylococci.* BRL 36650 is highly potent against most gram-negative organisms including refractory gram-negative species such as *Acinetobacter* and *Pseudomonas.* The level of potency is much greater than either that of ticarcillin or piperacillin.

The pharmacology of penicillins differs markedly from compound to compound. The majority of derivatives, including penicillin G and the antipseudomonal penicillins, are unstable in gastric acid and are not available orally. The isoxazolyl penicillins are relatively acid-stable but not consistently well absorbed by the oral route. Nafcillin and oxacillin are poorly absorbed orally; cloxacillin, dicloxacillin, and flucloxacillin are more reliable. Penicillin V, ampicillin, and particularly amoxicillin are relatively well absorbed orally. Esters of ampicillin such as bacampicillin, pivampicillin, and talampicillin improve the level of oral absorption of ampicillin to that achieved by amoxicillin. Absorption can be diminished by food after oral administration, however, and peak blood levels, usually achieved after 1 to 2 h, are somewhat delayed after ingestion of food.

The penicillins in general are renowned for their lack of toxicity. The most common adverse effect of the use of penicillins is an allergic reaction which can change from a mild rash to fatal anaphylactic shock in rare cases. All penicillins cross the placenta and are excreted in maternal milk. However, the relative freedom from toxicity renders these compounds valuable agents during pregnancy and lactation.

Biosynthesis. The microbial synthesis of penicillins from eukaryotes such as *C. acremonium* and *P. chrysogenum* has been comprehensively reviewed. In essence the biosynthesis of penicillins is described in Figure 1 although certain stages have yet to be fully characterized. Products are isopenicillin N, penicillin G, and penicillin N, $C_{14}H_{21}N_3O_6S$.

Mode of Action. Penicillins exert their antibacterial effect by inhibiting the high molecular weight penicillin-binding proteins (PBPs) that are implicated in the final stages of peptidoglycan synthesis.

Production, Manufacture, and Processing

Most methods used for the production of the commercially important α-amino penicillins, such as ampicillin and amoxicillin, are based on modifications of an enamine process employing the appropriate phenylglycine and methylacetoacetate followed by coupling with 6-APA. Other aspects of the fermentation, strain maintenance, equipment, inoculum development, media, and procedures used in the production of penicillin are well covered in previous editions of the *Encyclopedia.* Developments in these areas have been reviewed.

R. J. PONSFORD
SmithKline Beecham Pharmaceuticals

M. F. Parry, *Med. Clin. North Am., Update on Antibiotics 1, The Penicillins* **71,** 1093 (1987).

W. B. Pratt and R. Fekety, eds., *The Antimicrobial Drugs,* Oxford University Press, London, 1986.

A. A. W. Long, J. H. C. Nayler, H. Smith, T. Taylor, and N. Ward, *J. Chem. Soc.* **(C)**, 1920 (1971).

J. O. Neway, ed., *Fermentation Process Development of Industrial Organisms,* Marcel Dekker Inc., New York, 1989.

LINCOSAMINIDES

Lincomycin (**1**, R = OH, R′ = H), $C_{18}H_{34}N_2O_6S$, the first lincosaminide antibiotic to which a structure was assigned, is defined chemically as methyl 6,8-dideoxy-6-(1-methyl-*trans*-4-propyl-L-pyrrolidin-2-ylcarbonylamino)-1-thio-D-erythro-D-galacto-octopyranoside. Both lincomycin and the semisynthetic clindamycin (**1**, R = H, R′ = Cl), $C_{18}H_{33}ClN_2O_5S$, are widely used in clinical practice. The trivial name of the sugar fragment of this antibiotic, methyl α-thiolincosaminide, has lent itself to the other members of this family, whether produced as secondary metabolites of soil microorganisms or derived semisynthetically by chemical modification. Thus celesticetin (**2**, R = OC– 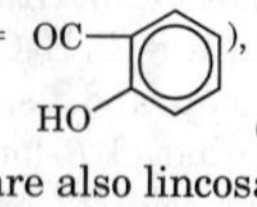), $C_{24}H_{36}N_2O_9S$, and desalicetin (**2**, R = H), $C_{17}H_{32}N_2O_7S$, are also lincosaminides.

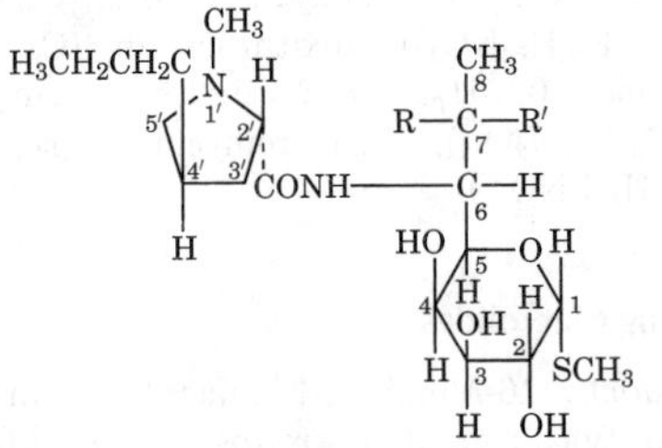

(1)

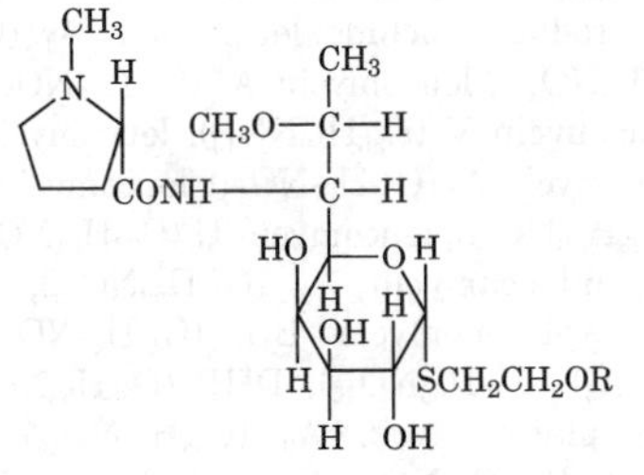

(2)

Lincomycin

The discovery and biological properties of lincomycin (**1**, R = OH, R′ = H) were described in 1962. This antibiotic is active *in vitro* and *in vivo* against most of the common gram-positive pathogens. Resistance by Staphylococci is developed slowly in a stepwise manner, based on *in vitro* serial subculture experiments, and its activity is not influenced by body fluids up to concentrations of 50% in the assay medium.

Lincomycin has been produced by a variety of *Streptomyces* strains and by strain 1146 of *Actinomyces roseolus*. Extraction by standard procedures using *Sarcina lutea* as the assay organism on agar trays leads to a crystalline hydrochloride having molecular formula $C_{18}H_{34}N_2O_6S{\cdot}HCl{\cdot}\frac{1}{2}H_2O$.

Biosynthesis. The terminal *C*-methyl of the propyl side chain, the *S*-methyl, and the *N*-methyl groups are derived from methionine. *trans*-4-Propyl-L-proline was shown to accumulate when *Streptomyces lincolnensis* is grown in media deficient in sulfur, and the addition of L-tyrosine or L-dihydroxyphenylalanine (DOPA) was shown to stimulate this production.

Mechanism of Action. The earliest studies on the mechanism of action of lincomycin showed that lincomycin had the immediate effect on *Staphylococcus aureus* of complete inhibition of protein synthesis. This inhibition results from the blocking of the peptidyltransferase site of the 50S subunit of the bacterial ribosome. Little effect on DNA and RNA synthesis was observed.

Resistance to Lincomycin. Resistance to lincomycin is developed slowly, and is usually caused by modification of 23S ribosomal RNA, which leads to coresistance to macrolide, lincosaminide, and streptogramin B antibiotics.

Pharmacology and Uses. Lincomycin hydrochloride (Lincocin) is available in oral dosage forms and as a sterile solution for injection.

Lincomycin has found use in the treatment of diseases of the ear, throat, nose, respiratory tissue, skin and soft tissue, bone, joint, dental, and septicemic infections caused by staphylococci, pneumonococci, and streptococci (other than enterococci). It has also been used in the treatment of diphtheria and a variety of anaerobic infections, including actinomycosis.

Celesticetin and Other Lincosaminide Metabolites

The production and isolation of the antibiotic celesticetin

(**2**, R = OC 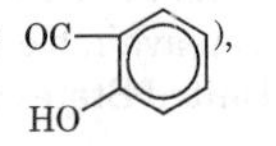),

a salicylate ester of the β-hydroxyethylthio-substituent, was reported as early as the 1950s, although its structure was not determined until 1968.

Antibiotic Bu-2545, 7-*O*-methyl-4′-depropyllincomycin (**1**, R = OCH, R = H′3 but lacking the 4′-propyl group), $C_{16}H_{30}N_2O_6S$, produced by *Streptomyces* strain No. H 230-5, possesses structural features in common with both celesticetin and lincomycin.

Accompanying lincomycin in *S. lincolnensis* fermentations is a small amount of the analogue 4′-depropyl-4′-ethyllincomycin, $C_{17}H_{32}N_2O_6S$, which has considerably lower antibacterial activity than the parent compound. Extension of the normal six-day fermentation to twelve days resulted in the formation of lincomycin sulfoxide, $C_{18}H_{34}N_2O_7S$, and 1-demethylthio-1-hydroxylincomycin (lincomycose), $C_{17}H_{32}N_2O_7$, both of which have greatly reduced antibacterial activity.

Clindamycin

Clindamycin, 7(*S*)-7-chloro-7-deoxylincomycin, (**1**, R = H, R = Cl′), also known as Cleocin, first resulted from the reaction of lincomycin and thionyl chloride; improved synthetic methods involve the reaction of lincomycin and triphenylphosphine dichloride or triphenylphosphine in carbon tetrachloride. Clindamycin is significantly more active than lincomycin against gram-positive bacteria *in vitro*, and is absorbed rapidly following oral administration. Clindamycin 2-palmitate, the 2-palmitate ester of clindamycin, is tasteless and antibacterially inactive. However, following oral administration, esterase cleavage occurs to give good blood levels of clindamycin, and this ester has been developed as a pediatric formulation of the antibiotic Cleocin Pediatric.

Uses. Clindamycin has found use in the treatment of common infections caused by gram-positive cocci. It is also efficacious in the treatment of anaerobic infections, including actinomycosis. Clindamycin has been shown to be active against strains of *Plasmodium* in animals.

Resistance to Clindamycin. Cross-resistance between lincomycin and clindamycin is complete, and co-resistances of lincomycin also apply to clindamycin. However, the inactivation of clindamycin by clinical isolates of *Staphylococcus haemolyticus* and *Staphylococcus aureus* is caused by adenylylation at the 4-position to form clindamycin 4-(5′-adenylate) in contrast to the lincomycin 3-(5′-adenylate) that forms.

Biomodification of Clindamycin. When added to fermentations of *Streptomyces punipalus*, clindamycin is converted into de-*N*-methylclindamycin. However, when clindamycin is incubated with *Streptomyces armentosus*, clindamycin sulfoxide, $C_{18}H_{33}ClN_2O_6S$, which has low antibacterial activity, is formed. Clindamycin 3-phosphate, antibacterially inactive *in vitro*, and the ribonucleotides clindamycin 3-(5′-cytidylate), clindamycin 3-(5′-adenylate), clindamycin 3-(5′-uridylate), and clindamycin 3-(5′-guanylate), all inactive *in vitro*, can be generated. All of these derivatives protect mice infected with *Staphylococcus aureus*, however, presumably because of biotransformation into clindamycin.

Other Changes in the Amino Acid Fragment

The influence of the size of the cyclic amino acid in analogues of clindamycin has been examined. The (2-(*S*)-*cis*)-4-ethylpipecolic acid analogue (6-ring), $C_{17}H_{31}ClN_2O_5S$, was highly active both *in vitro* and *in vivo*, but the *cis-(R)*-isomer had minimal activity. Significant activity was shown for the azepine (7-ring) analogue, $C_{16}H_{29}ClN_2O_5S$, but the azetidine (4-ring) analogue, $C_{13}H_{23}ClN_2O_5S$, showed little activity.

Economic Aspects

The composition of matter patents in the United States issued to The Upjohn Company on clindamycin phosphate and hydrochloride expired at the end of 1986 and in early 1987, respectively. Since then, these compounds have been available generically from more than two dozen companies in the United States alone.

BRIAN BANNISTER
The Upjohn Company

R. J. Fass in E. M. Kagan, ed., *Antimicrobial Therapy,* 2nd ed., W. B. Saunders Co., Philadelphia, Pa., 1974, p. 83.

MACROLIDES

Macrolide antibiotics are well-established antimicrobial agents in both clinical and veterinary medicine. These agents can be administered orally and are generally used to treat infections in the respiratory tract, skin and soft tissues, and genital tract caused by gram-positive organisms, *Mycoplasma* species, and certain susceptible gram-negative and anaerobic bacteria.

The macrolide class is large and structurally diverse. Macrolides are produced by fermentation of soil microorganisms. Additionally, structural modifications using both chemical and microbiological means have yielded biologically active semisynthetic derivatives.

The term macrolide was introduced to denote the class of substances produced by *Streptomyces* species containing a macrocyclic lactone ring. The generalized structure is a highly substituted monocyclic lactone (aglycone) to which is attached one or more saccharides glycosidically linked to hydroxyl groups on either the aglycone or another saccharide. The aglycones are derived via similar polyketide biosynthetic pathways and thus share many structural features in terms of pattern and stereochemistry of substituents. Traditional macrolide antibiotics are divided into three families according to the size of the aglycone, which can be 12-, 14-, or 16-membered.

Because one or more aminosugars are usually present, these compounds are basic and can form acid addition salts. In addition, one or more neutral sugars are often present. A few macrolides possess no aminosugar. The saccharides share some common features: they tend to be highly deoxygenated and *N*- or *O*-methylated; and the amino groups are located at either position 3 or 4.

12-Membered Ring Macrolides

12-Membered ring macrolides include methymycin ($C_{25}H_{43}NO_7$), neomethymycin ($C_{25}H_{43}NO_7$), and YC-17 ($C_{25}H_{43}NO_6$).

14-Membered Ring Macrolides

Natural Products. Naturally occurring 14-membered macrolides include erythromycin A ($C_{37}H_{67}NO_{13}$), erythromycin B ($C_{37}H_{67}NO_{12}$), erythromycin C ($C_{36}H_{65}NO_{13}$), erythromycin D ($C_{36}H_{65}NO_{12}$), erythromycin F ($C_{37}H_{67}NO_{14}$), erythromycin E ($C_{37}H_{65}NO_{14}$), oleandomycin ($C_{35}H_{61}NO_{12}$), oleandomycin Y ($C_{34}H_{59}NO_{12}$), pikromycin ($C_{28}H_{47}NO_8$), narbomycin ($C_{28}H_{47}NO_7$), 5-*O*-mycaminosylnarbonolide ($C_{28}H_{47}NO_8$), 10,11-dihydropikromycin ($C_{28}H_{49}NO_8$), kayamycin ($C_{28}H_{49}NO_8$), 12-deoxykromycin ($C_{20}H_{30}O_4$), kromycin ($C_{20}H_{30}O_5$), megalomicin A ($C_{44}H_{80}N_2O_{15}$), megalomicin B ($C_{46}H_{82}N_2O_{16}$), megalomicin C_1 ($C_{48}H_{84}N_2O_{17}$), megalomicin C_2 ($C_{49}H_{86}N_2O_{17}$), XK-41-B_2 ($C_{47}H_{84}N_2O_{16}$), lankamycin (kujimycin B) ($C_{42}H_{72}O_{16}$), kujimycin A ($C_{40}H_{70}O_{15}$), 15-dehydrolankamycin ($C_{42}H_{70}O_{16}$), 15-*O*-(4-*O*-acetylarcanosyl)-lankamycin ($C_{52}H_{88}O_{20}$), 3″-*O*-demethyl-2″,3″-anhydrolankamycin ($C_{41}H_{68}O_{15}$), and 23672 RP ($C_{48}H_{82}O_{20}$).

Semisynthetic Derivatives. Erythromycin has been the principal subject of modification of 14-membered macrolides; some of the derivatives are being commercially launched. Derivatives of erythromycin and oleandomycin include 2′-*O*-acetylerythromycin ($C_{39}H_{69}NO_{14}$), 2′-*O*-propionylerythromycin ($C_{40}H_{71}NO_{14}$), erythromycin ethyl carbonate ($C_{40}H_{71}NO_{15}$), erythromycin ethyl succinate ($C_{43}H_{75}NO_{16}$), tri-*O*-acetyloleandomycin ($C_{41}H_{67}NO_{15}$), erythromycin-11,12-carbonate ($C_{38}H_{65}NO_{14}$), roxithromycin ($C_{41}H_{76}N_2O_{15}$), 9(*S*)-erythromycylamine ($C_{37}H_{70}N_2O_{12}$), dirithromycin ($C_{42}H_{78}N_2O_{14}$), azithromycin ($C_{38}H_{72}N_2O_{12}$), clarithromycin ($C_{38}H_{69}NO_{13}$), and flurithromycin ($C_{37}H_{66}FNO_{13}$).

16-Membered Ring Macrolides

Natural Products. 16-Membered macrolides are divided into leucomycin- and tylosin-related groups, which differ in the substitution pattern of their aglycones. Multifactor complexes are usually produced and some compounds have been isolated from culture broths of different organisms and then been given different names. Natural products include leucomycin A_1 ($C_{40}H_{67}NO_{14}$), leucomycin A_5 ($C_{39}H_{65}NO_{14}$), leucomycin A_7 ($C_{38}H_{63}NO_{14}$), leucomycin A_9 ($C_{37}H_{61}NO_{14}$), leucomycin V ($C_{35}H_{59}NO_{13}$), leucomycin A_3 (josamycin) ($C_{42}H_{69}NO_{15}$), leucomycin A_4 ($C_{41}H_{67}NO_{15}$), leucomycin A_6 ($C_{40}H_{65}NO_{15}$), leucomycin A_8 ($C_{39}H_{63}NO_{15}$), leucomycin U ($C_{37}H_{61}NO_{14}$), platenomycin A_1 ($C_{43}H_{71}NO_{15}$), midecamycin A_2 ($C_{42}H_{69}NO_{15}$), midecamycin A_1 (espinomycin A_1) (platenomycin B_1), ($C_{41}H_{67}NO_{15}$), platenomycin C_2 (espinomycin A_3) ($C_{40}H_{65}NO_{15}$), DHP ($C_{38}H_{63}NO_{14}$), espinomycin A_2 ($C_{42}H_{69}NO_{15}$), platenomycin A_0 ($C_{44}H_{73}NO_{15}$), maridomycin I (platenomycin C_3) ($C_{43}H_{71}NO_{16}$), maridomycin VII ($C_{42}H_{69}NO_{16}$), maridomycin III (platenomycin C_1) ($C_{41}H_{67}NO_{16}$), maridomycin IV ($C_{40}H_{65}NO_{16}$), maridomycin VI ($C_{39}H_{63}NO_{16}$), carbomycin A (deltamycin A_4) ($C_{42}H_{67}NO_{16}$), deltamycin A_3 ($C_{41}H_{65}NO_{16}$), deltamycin A_2 ($C_{40}H_{63}NO_{16}$), deltamycin A_1 ($C_{39}H_{61}NO_{16}$), deltamycin X (EOA) ($C_{37}H_{59}NO_{15}$), EOP ($C_{38}H_{61}NO_{15}$), carbomycin B ($C_{42}H_{67}NO_{15}$), platenomycin W_1 ($C_{43}H_{69}NO_{15}$), platenomycin W_2 ($C_{44}H_{71}NO_{15}$), niddamycin ($C_{39}H_{63}NO_{14}$), midecamycin A_4 ($C_{42}H_{67}NO_{15}$), midecamycin A_3 ($C_{41}H_{65}NO_{15}$), DOA ($C_{37}H_{59}NO_{14}$), and DOP ($C_{38}H_{61}NO_{14}$).

The spiramycin complex, also discovered as foromacidine, was isolated from culture broths of *S. ambofaciens.* The spiramycins are distinguished by a second aminosugar, forosamine, attached to the 9-hydroxyl group by a β-glycosidic linkage. Spiramycins include spiramycin I (foromacidine A) ($C_{43}H_{74}N_2O_{14}$), spiramycin II (foromacidine B) ($C_{45}H_{76}N_2O_{15}$), spiramycin III (foromacidine C) ($C_{46}H_{78}N_2O_{15}$), neospiramycin I ($C_{36}H_{62}N_2O_{11}$), spiramycin U ($C_{42}H_{71}NO_{16}$), spiramycin S ($C_{42}H_{73}NO_{16}$), and forocidin I ($C_{28}H_{47}NO_{10}$).

A second large group of 16-membered macrolides differs from the leucomycins in the substitution pattern of the aglycone. One difference is a methyl or hydroxymethyl group at C-14. If hydroxymethyl is present, it may be glycosidically substituted by a neutral sugar such as mycinose or an analogue. The most prominent member of this group is tylosin, an important veterinary antibiotic produced by *S. fradiae.* Tylosin and related products include tylosin ($C_{46}H_{77}NO_{17}$), relomycin (20-dihydrotylosin) ($C_{46}H_{79}NO_{17}$), macrocin ($C_{45}H_{75}NO_{17}$), *O*-demethylmacrocin ($C_{44}H_{73}NO_{17}$), desmycosin ($C_{39}H_{65}NO_{14}$), lactenocin ($C_{38}H_{63}NO_{14}$), *O*-demethyllactenocin (DOML) ($C_{37}H_{61}NO_{14}$), 23-*O*-demycinosyltylosin ($C_{38}H_{63}NO_{13}$), 23-(demycinosyloxy)tylosin ($C_{38}H_{63}NO_{12}$), GS-77-1 ($C_{38}H_{65}NO_{11}$), and GS-77-3 ($C_{38}H_{65}NO_{12}$).

Many 16-membered macrolides possess a tylosin-type aglycone with only one saccharide, an aminosugar that is either mycaminose or desosamine. Rosaramicin is an example. These compounds also differ in their degree of oxidation at C-20, C-23, and C-9 to C-13 (dienone or epoxyenone). Hydrolysis of both neutral sugars from tylosin yields 5-*O*-mycaminosyltylonolide (OMT). However, this compound is more conveniently obtained by hydrolysis of mycarose from the fermentation-derived demycinosyltylosin. Analogous hydrolyses of mycarose from DMOT, GS-77-3, and GS-77-1 yield, respectively, 23-deoxy-5-*O*-mycaminosyltylonolide (DOMT), 20-deoxo-5-*O*-mycaminosyltylonolide (GS-77-4), and 5-*O*-mycaminosyltylactone (GS-77-2).

The mycinamicin complex, also found as AR-5 complex, was isolated from culture broths of *M. griseorubida.* These compounds contain a 2,3-double bond and a methyl group rather than a two-carbon substituent at C-6 of the aglycone. A hydroxyl group is present at C-14 in mycinamicins II and V. X-ray crystallography established the stereochemistry and absolute configuration of the aglycone as identical to tylosin.

A few neutral 16-membered macrolides have been isolated. The structure of chalcomycin, produced by *S. bikiniensis,* was established as a 16-membered lactone having two neutral sugars, D-chalcose and D-mycinose. Its aglycone resembles that of the mycinamicins, but differs by a hydroxyl group, the stereochemistry of which is uncertain, at C-8 and a methyl group at C-15.

The aldgamycins, produced by *S. lavendulae,* contain a novel bicyclic sugar, aldgarose.

As is the case for the leucomycin group, several aglycones exist in the tylosin group which differ in the pattern of oxidation and substitution of the lactone. If the aglycone contains an aldehyde, cleavage of the aminosugar yields the hemiketal, such as tylonolide and rosaranolide.

Semisynthetic Derivatives. 3″-*O*-Acyl derivatives have not been found via fermentation, but chemical acylation of the 3″-hydroxyl group yields products having good antibiotic activity and better pharmacokinetics than the parent macrolides. Two such compounds have been developed: 3″-*O*-propionyl-leucomycin A_5 (rokitamycin) $C_{42}H_{69}NO_{15}$, formerly TMS-19-Q,; and 9,3″-di-*O*-acetylmidecamycin (miokamycin) $C_{45}H_{71}NO_{17}$. At least part of the *in vivo* improvement was attributed to slower elimination of active metabolites from serum.

The enhanced activity from acylation of the 3- and 4″-hydroxyl groups of leucomycin prompted analogous studies of tylosin. Bioconversion of tylosin by *S. thermotolerans* yielded 3- or 4″-*O*-acyl derivatives possessing increased activity against certain resistant microorganisms and higher concentrations of antibiotic in serum after oral administration. From this study, 3-*O*-acetyl-4″-*O*-isovaleryltylosin (AIV-tylosin), $C_{53}H_{87}NO_{19}$, was developed as a new veterinary antibiotic.

Tilmicosin, $C_{46}H_{80}N_2O_{13}$, was selected as a therapeutic agent to treat pneumonia in cattle and pigs because of its activity against *Pasturelle* species, oral bioavailability, and prolonged concentrations *in vivo.*

Hybrid Macrolides

Other macrolides have been prepared which represent hybrids of structures within the 14-membered family, within the 16-membered family, or between the two families. These hybrids have been made by chemical, bioconversion, or genetic manipulations.

Aglycones obtained from one macrolide have been bioconverted by microorganisms producing either the same or a different macrolide; such studies have been important for production of new compounds and elucidation of biosynthetic pathways.

The advent of molecular biology has opened new possibilities for producing hybrid macrolides. Insertion of DNA from the oleandomycin-producer into a mutant strain of the erythromycin-producer produced derivatives of erythromycin lacking the 2-methyl group. Genetic manipulations of biosynthetic pathways in macrolide-producing microorganisms complement traditional chemical and microbiological approaches. For example, inactivation of the *Ery F* gene coding for a 6-hydroxylase in *Sa. Erythraea* produced 6-deoxyerythromycin.

Table 1. Commercial Macrolide Products

Macrolide	Trade name[a]	Clinical route	Manufacturer
erythromycin[b]	Ery-tab	oral	Abbott
erythromycin[c]	Erythromycin base filmtab	oral	Abbott
erythromycin[d]	Erythromycin delayed-release capsules	oral	Abbott
erythromycin[e]	E-Mycin	oral	Boots
erythromycin[b]	Ilotycin	oral	Dista
erythromycin[d]	Eryc	oral	Parke-Davis
erythromycin topical solution	Eryderm 2%	topical	Abbott
erythromycin topical solution	ETS-2%	topical	Paddock
erythromycin ophthalmic ointment	Ilotycin ointment	topical	Dista
erythromycin stearate	Erythrocin stearate	oral	Abbott
erythromycin lactobionate	Erythrocin lactobionate-iv	iv	Abbott
	Erythrocin piggyback	iv	Abbott
erythromycin gluceptate	Ilotycin gluceptate	iv	Dista
erythromycin ethyl succinate	E.E.S.	oral	Abbott
	Eryped	oral	Abbott
erythromycin estolate	Ilosone	oral	Dista
erythromycin acistrate	Erasis (Finland)	oral	Orion
erythromycin stinoprate	Erythrocist (Italy)	oral	Refarmed
propionylerythromycin mercaptosuccinate	Zalig (Italy)	oral	Pierrel
erythromycin-11,12-carbonate	Davercin (Poland)	oral	Tarchomin
roxithromycin	Rulid (France)	oral	Roussel-Uclaf
clarithromycin	Biaxin	oral	Abbott
azithromycin	Zithromax	oral	Pliva
triacetyloleandomycin	TAO	oral	Roerig
josamycin	Josamycin (Japan)	oral	Yamanouchi
miokamycin	Miocamycin (Japan)	oral	Meiji Seika
rokitamycin	Ricamycin (Japan)	oral	Toyo Jozo

[a] Country in parenthesis represents launch outside the United States.
[b] Enteric-coated tablets.
[c] Film-coated tablets.
[d] Enteric-coated pellets.
[e] Coated tablets.

Biological Properties

Antimicrobial Properties. Macrolides inhibit growth of gram-positive bacteria, *Mycoplasma* species, and certain gram-negative and anaerobic bacteria. Susceptible gram-positive bacteria include many species of *Staphylococcus* and *Streptococcus;* susceptible gram-negative bacteria include *Bordetella pertussis, Legionella pneumophila, Moraxella catarrhalis* (formerly *Branhamella*), and *Haemophilus ducreyi.* Although erythromycin has some *in vitro* activity against *Haemophilus influenzae,*it does not exhibit a high level of *in vivo* efficacy. An important susceptible anaerobe is *Propionibacterium acnes.* Comparative evaluations have been published, and monographs devoted to particular macrolides and reviews are available.

Azithromycin expanded the traditional *in vitro* spectrum because of increased activity against gram-negative bacteria, including *H. influenzae.*

Macrolides inhibit growth of bacteria by inhibiting protein synthesis on ribosomes. Bacterial resistance to macrolides is often accompanied by cross-resistance to lincosamide and streptogramin B antibiotics (MLS-resistance), which can be either inducible or constitutive. 14-Membered macrolides generally induce resistance to themselves, whereas 16-membered macrolides do not; consequently, one advantage of the latter is their activity against bacteria which are inducibly resistant to erythromycin. Both 14- and 16-membered macrolides lack activity against constitutively resistant strains.

Bacterial resistance to antibiotics usually results from modification of a target site, enzymatic inactivation, or reduced uptake into or increased efflux from bacterial cells.

Pharmacokinetics and Pharmacology. Older macrolides such as erythromycin exhibit relatively low serum concentrations, short *in vivo* half-lives, highly variable oral absorption, and low oral bioavailability. Improvements in these pharmacokinetic parameters have been accomplished for newer derivatives. The principal side effects of macrolides are gastrointestinal problems, such as pain, indigestion, diarrhea, nausea, and vomiting.

Biosynthetic Patterns and Conformational Analysis

Macrolides are obtained by controlled submerged aerobic fermentations of soil microorganisms. Although species of *Streptomyces* have dominated, species of *Saccharopolyspora, Micromonospora,* and *Streptoverticillium* are also well represented. New techniques such as enzyme-linked immunosorbent assay (ELISA) may prove beneficial for discovering new structures.

Economic Aspects

Macrolide antibiotics are used clinically to treat infections resulting from susceptible organisms in the upper and lower respiratory tract, skin and soft tissues, and genital tract. They are generally used orally, although they can be given intravenously. For the latter purpose, a water-soluble salt is used, because macrolides are poorly soluble in water as free bases. Commonly employed acid-addition salts include the lactobionate, gluceptate, tartrate, and phosphate. For oral administration, acid-stable esters, salts, formulations, and coatings are necessary for erythromycin, which is unstable as its unprotected free base under acidic conditions such as those in the stomach. They are not administered by intramuscular injection because of severe pain.

Reported worldwide sales of macrolides are estimated as \$800,000,000 annually; approximately 25% of this figure are U.S. sales. Table 1 provides a partial list of products commercially available for clinical use.

Macrolides are regarded as among the safest of antibiotics.

Relatively few macrolides are used in veterinary medicine. The most important is tylosin (Tylan, Elanco Products), which is used to control chronic respiratory disease caused by *Mycoplasma gallisepticum* in poultry. It is also used to treat and control infections in pigs resulting from gram-positive bacteria and *Mycoplasma,* and as a growth promotant for pigs and poultry. Other macrolides in veterinary use include erythromycin, oleandomycin, and spiramycin. Newer macrolides being developed for veterinary applications are 3-*O*-acetyl-4″-*O*-isovaleryltylosin and tilmicosin. The pharmacokinetics of macrolides in animals have been reviewed, and a book devoted to pharmacokinetics of veterinary antibiotics has been published.

H. A. KIRST
Eli Lilly and Company

A. Bryskier, J. P. Butzler, H. C. Neu, and P. M. Tulkens, eds., *The Macrolides,* Arnette-Blackwell Publishers, Oxford, U.K., 1993.

H. C. Neu, L. S. Young, and S. H. Zimmer, eds., *The New Macrolides, Azalides, and Streptogramins: Pharmacology and Clinical Applications,* Marcel Dekker, Inc. New York, 1993.

H. A. Kirst, *Prog. Med. Chem.* **30** (1993).

H. A. Kirst, *Prog. Med. Chem.* **31** (1994)

NUCLEOSIDES AND NUCLEOTIDES

The naturally occurring nucleoside and nucleotide antibiotics exist as either the *C*- or *N*-glycosides (see NUCLEIC ACIDS). They include ezomycin A_1 ($C_{26}H_{38}N_8O_{15}S$), ezomycin B_1 ($C_{26}H_{39}N_7O_{17}S$), ezomycin C_1 ($C_{26}H_{37}N_7O_{16}S$), ezomycin A_2 ($C_{19}H_{26}N_6O_{12}$), ezomycin B_2 ($C_{19}H_{25}N_5O_{13}$), ezomycin C_2 ($C_{19}H_{25}N_5O_{13}$), showdomycin ($C_9H_{11}NO_6$), isoshowdomycin ($C_9H_{11}NO_6$), maleimycin ($C_7H_7NO_3$), oxazinomycin ($C_9H_{11}NO_7$), pyrazomycin (pyrazofurin) ($C_9H_{13}N_3O_6$), formycin ($C_{10}H_{13}N_5O_4$), formycin B ($C_{10}H_{12}N_4O_5$), oxoformycin B ($C_{10}H_{12}N_4O_6$). These antibiotics contain a variety of purine and pyrimidine rings, including the diazepin, maleimide, indole, imidazole, pyrrolopyrimidine, pyrazolopyrimidine, pyrazole, oxazine, triazene, hydantoin, and the purine ring having a 6-*N*-phosphoramidate substituent. The structures of the carbohydrate moieties also vary. Some nucleosides have additional carbon atoms attached to the ribosyl moiety and in some cases ribose has been replaced by other sugars such as allulose, olefinic sugars, 2-, 3-, 4-, or 5-amino sugars, 4-fluororibose, 4-aminohexoses, aminouronic acids, disaccharides, or a tricyclic dodecose.

The naturally occurring nucleoside/nucleotide antibiotics, which have been isolated from bacteria, fungi, blue-green algae, and marine sponges, have proven to be useful biochemical probes in eucaryotic, procaryotic, viral, fungal, and plant systems. Some excellent reviews on the nucleoside/nucleotide antibiotics are available (see also ANTIPARASITIC AGENTS).

C-Nucleosides

The naturally occurring *C*-nucleosides containing *C*-glycosyl linkages include ezomycins, showdomycin, oxazinomycin, pyrazomycin, and pyrazolopyrimidine nucleosides.

N-Nucleosides

The naturally occurring nucleoside analogues contain the *N*-glycosyl linkage and either purine, pyrimidine, imidazole, diazepin, or indole rings. The purine nucleosides inhibit protein synthesis, RNA and DNA synthesis, and methyltransferases; they have antimycoplasmal, antiviral, hypotensive, antifungal, antimycobacterial, and antitumor activities and induce sporulation. The pyrimidine nucleosides inhibit protein synthesis, virus replication, RNA and DNA synthesis, and cAMP phosphodiesterase. The imidazole nucleosides inhibit nucleic acid synthesis. The diazepin nucleosides inhibit adenosine deaminase (ADA). The indole nucleosides inhibit bacteria, yeast, fungi, and viruses.

Purine N-Nucleosides. The purine *N*-nucleoside antibiotics include 2′-amino-2′-deoxyadenosine ($C_{10}H_{14}N_6O_3$), 2′-amino-2′-deoxyguanosine ($C_{10}H_{14}N_6O_4$), 3′-amino-3′deoxyadenosine ($C_{10}H_{14}N_6O_3$), cordycepin (3′-deoxyadenosine) ($C_{10}H_{13}N_5O_3$), puromycin ($C_{22}H_{29}N_7O_5$), homocitrullylaminoadenosine ($C_{17}H_{27}N_9O_5$), lysylaminoadenosine ($C_{16}H_{26}N_8O_4$), chryscandin ($C_{20}H_{23}N_7O_6$), A201A ($C_{37}H_{50}N_6O_{14}$), psicofuranine ($C_{11}H_{15}N_5O_5$), decoyinine (angustmycin A) ($C_{11}H_{13}N_5O_4$), arabinofuranosyladenine (ara-A, vidarabine) ($C_{10}H_{13}N_5O_4$), sinefungin ($C_{15}H_{23}N_7O_5$), A9145A ($C_{15}H_{23}N_7O_5$), A9145C ($C_{15}H_{21}N_7O_5$), herbicidins A, B, E, F, and G ($C_{23}H_{30}N_5O_{10}$), doridosine ($C_{11}H_{15}N_5O_5$), aristeromycin ($C_{11}H_{15}N_5O_3$), neplanocin A ($C_{11}H_{13}N_5O_3$), neplanocin B ($C_{11}H_{13}N_5O_4$), neplanocin C ($C_{11}H_{13}N_5O_4$), neplanocin D ($C_{11}H_{12}N_4O_4$), neplanocin F ($C_{11}H_{13}N_5O_3$), nucleocidin ($C_{10}H_{13}FN_6O_6S$), ascamycin ($C_{13}H_{18}ClN_7O_7S$), nebularine ($C_{10}H_{12}N_4O_4$), crotonoside ($C_5H_5N_5O$), griseolic acid ($C_{14}H_{13}N_5O_8$), griseolic acid B ($C_{14}H_{13}N_5O_9$), griseolic acid C ($C_{14}H_{15}N_5O_7$), oxetanocin A ($C_{10}H_{13}N_5O_3$), and oxanosine ($C_{10}H_{12}N_4O_6$).

Pyrrolopyrimidine Nucleosides. The pyrrolopyrimidine *N*-nucleside antibiotics include tubercidin ($C_{11}H_{14}N_4O_4$), toyocamycin ($C_{12}H_{13}N_5O_4$), sangivamycin ($C_{12}H_{15}N_5O_5$), cadeguomycin ($C_{12}H_{14}N_4O_7$), kanagawamicin ($C_{13}H_{17}N_5O_6$), mycalisine A ($C_{13}H_{13}N_5O_3$), and mycalisine B ($C_{13}H_{12}N_4O_4$).

Diazepin Nucleosides. Four naturally occurring diazepin nucleosides, coformycin ($C_{11}H_{16}N_4O_5$), 2′-deoxycoformycin ($C_{11}H_{16}N_4O_4$), adechlorin or 2′-chloro-2′-deoxycoformycin ($C_{11}H_{15}ClN_4O_4$), and adecypenol ($C_{12}H_{16}N_4O_4$), have been isolated.

Bredinin, Neosidomycin, and SF-2140. Bredinin ($C_{19}H_{13}N_3O_6$) isolated from the culture filtrates of *Eupenicillium brefeldianum,* inhibits the multiplication of L5178Y, HeLa S3, RK-13, mouse L-cells, and Chinese hamster cells.

Neosidomycin ($C_{17}H_{20}N_2O_6$) and SF-2140 ($C_{18}H_{20}N_2O_6$) are indole *N*-glycosides produced by *S. hygroscopicus* and *Actinomadura,* respectively. Both show activity against gram-positive bacteria, yeast, fungi, and viruses.

Pyrimidine–Nucleoside Antibiotics. *Fatty acyl Nucleosides.* The nucleoside antibiotics with fatty acyl groups containing adenine, uracil, and dihydrouracil aglycons, the tunicamycins, streptovirudins, corynetoxins include tunicamycins [I(A_0) ($C_{36}H_{58}N_4O_{16}$), II (C) ($C_{37}H_{60}N_4O_{16}$), III (A_2) ($C_{37}H_{60}N_4O_{16}$), IV (B_2) ($C_{38}H_{62}N_4O_{16}$), V (A) ($C_{38}H_{62}N_4O_{16}$), VI ($C_{38}H_{62}N_4O_{16}$), VII (B) ($C_{39}H_{64}N_4O_{16}$), VIII (C_2), ($C_{39}H_{64}N_4O_{16}$), IX (D_1) ($C_{40}H_{66}N_4O_{16}$), and X (D) ($C_{40}H_{66}N_4O_{16}$)], streptovirudins (A_1–D_1) [A_1 ($C_{35}H_{58}N_4O_{16}$), B_1 ($C_{36}H_{60}N_4O_{16}$), B_{1a} ($C_{36}H_{60}N_4O_{16}$), C_1 ($C_{37}H_{62}N_4O_{16}$), D_1 ($C_{38}H_{64}N_4O_{16}$)], streptovirudins (A_2–D_2) [A_2 ($C_{35}H_{56}N_4O_{16}$), B_2 ($C_{36}H_{58}N_4O_{16}$), B_{2a} ($C_{36}H_{58}N_4O_{16}$), C_2 ($C_{37}H_{60}N_4O_{16}$), and D_2 ($C_{38}H_{62}N_4O_{16}$)], and corynetoxins [H16*i* ($C_{39}H_{66}N_4O_{17}$), H18*i* ($C_{41}H_{70}N_4O_{17}$), H17*a* ($C_{40}H_{68}N_4O_{17}$), H19*a* ($C_{42}H_{72}N_4O_{17}$), U16*i* ($C_{39}H_{64}N_4O_{16}$), U18*i* ($C_{41}H_{68}N_4O_{16}$), U17*a* ($C_{40}H_{66}N_4O_{16}$), U19*a* ($C_{42}H_{70}N_4O_{16}$), S16*i* ($C_{39}H_{66}N_4O_{16}$), S18*i* ($C_{41}H_{70}N_4O_{16}$), S15*a* ($C_{38}H_{64}N_4O_{16}$), S17*a* ($C_{40}H_{68}N_4O_{16}$), and S19*a* ($C_{42}H_{72}N_4O_{16}$)].

Peptidyl N-Nucleoside Antibiotics. *Polyoxins and Neopolyoxins.* The polyoxins and neopolyoxins are peptidylpyrimidine nucleoside antibiotics that have achieved use as agricultural fungicides (see FUNGICIDES, AGRICULTURAL).

Mureidomycins and Pacidamycins. The mureidomycins and pacidamycins are listed in Table 1. The four peptidylnucleosides, mureidomycins A–D, are produced by *S. flavidovirens.*

4-Aminohexose Nucleosides. The 4-aminohexose nucleosides include gougerotin ($C_{16}H_{25}N_7O_8$), blasticidin S ($C_{17}H_{26}N_8O_5$), amicetin ($C_{29}H_{42}N_6O_9$), bamicetin ($C_{28}H_{40}N_6O_9$), oxamicetin ($C_{29}H_{42}N_6O_{10}$), plicacetin ($C_{25}H_{35}N_5O_7$), norplicacetin ($C_{24}H_{33}N_5O_7$), hikizimycin (anthelmycin) ($C_{21}H_{37}N_5O_{14}$), bagougeramine A ($C_{17}H_{28}N_{10}O_7$), bagougeramine B ($C_{24}H_{44}N_{12}O_7$), arginomycin ($C_{18}H_{28}N_8O_5$), mildiomycin ($C_{19}H_{30}N_8O_9$), and SCH 36605 (rodaplutin) ($C_{24}H_{39}N_9O_7$). A biosynthetic relationship between the 4-aminohexose peptidyl nucleoside antibiotics and the pentopyranines has been proposed. The 4-aminohexose pyrimidine nucleoside antibiotics block peptidyl transferase activity and inhibit transfer of amino acids from aminoacyl-tRNA to polypeptides.

Nikkomycins. The nikkomycins, isolated from *S. tendae,* are nucleoside–peptide antibiotics. They include Z ($C_{20}H_{25}N_5O_{10}$), X ($C_{20}H_{25}N_5O_{10}$), J ($C_{25}H_{32}N_6O_{13}$), I ($C_{25}H_{32}N_6O_{13}$), B_z, B_x ($C_{21}H_{26}N_4O_{10}$), K_z ($C_{19}H_{23}N_5O_9$), K_x ($C_{19}H_{23}N_5O_9$), O_z ($C_{19}H_{23}N_5O_{10}$), O_x ($C_{19}H_{23}N_5O_{10}$), Q_z ($C_{24}H_{32}N_6O_{12}$), Q_x ($C_{24}H_{32}N_6O_{12}$), P_x ($C_{20}H_{25}N_5O_9$), R_z ($C_{25}H_{32}N_6O_{12}$), R_x ($C_{25}H_{32}N_6O_{12}$), W_z ($C_{19}H_{21}N_4O_9$), W_x ($C_{19}H_{21}N_4O_9$), pseudo-Z ($C_{20}H_{25}N_5O_{10}$), and pseudo-J ($C_{25}H_{32}N_6O_{13}$).

Glycosyl Nucleosides. There are five glycosyl antibiotics with either the adenine, uracil, or 7-deazaguanine aglycon. They are dapiramicin A ($C_{21}H_{29}N_5O_{10}$), epidapiramicin A ($C_{21}H_{29}N_5O_{10}$), dapiramicin B ($C_{21}H_{29}N_5O_{11}$), capuramycin ($C_{23}H_{31}N_5O_{12}$), and adenomycin ($C_{25}H_{39}N_7O_{18}S$).

Table 1. The Mureidomycins and the Pacidamycins

Name	Molecular formula
Mureidomycins	
mureidomycin A	$C_{38}H_{48}N_8O_{12}S$
mureidomycin B[a]	$C_{38}H_{50}N_8O_{12}S$
mureidomycin C	$C_{40}H_{51}N_9O_{13}S$
mureidomycin D[a]	$C_{40}H_{53}N_9O_{13}S$
Pacidamycins	
pacidamycin 1	
pacidamycin 2	$C_{39}H_{49}N_9O_{12}$
pacidamycin 3	$C_{39}H_{49}N_9O_{13}$
pacidamycin 4	
pacidamycin 5	$C_{36}H_{44}N_8O_{11}$
pacidamycin 6	
pacidamycin 7	$C_{38}H_{47}N_9O_{12}$

[a] Positions 5, 6 of the uracil moiety are saturated to give dihydrouracil.

Octosyl Acids. Three octosyl uronic acid nucleosides are produced by *S. cacaoi* sub sp. *asoensis.*

Arabinosylpyrimidine Nucleosides. 1-β-D-Arabinofuranosylthymine (ara-T), $C_{10}H_{14}N_2O_6$, and 1-β-D-arabinofuranosyluracil (ara-U), $C_9H_{12}N_2O_6$, also known as spongouridine, were first isolated from the sponge *C. crypta.*

5-Azacytidine. 5-Azacytidine, $C_8H_{12}N_4O_5$, 4-amino-1-β-D-ribofuranosyl-*S*-triazine, 2-(1*H*-one), was chemically synthesized in 1964 and subsequently isolated from culture filtrates of *S. ladakanus.*

5,6-Dihydro-5-azathymidine. 5,6-Dihydro-5-azathymidine, $C_9H_{15}N_3O_5$, contains the *s*-triazine ring and is isolated from the culture filtrates of *S. platensis* var. *clarensis.*

Clitocine. Clitocine, $C_9H_{13}N_5O_6$, isolated from the mushroom, *Clitocybe inversa,* is 5-nitro-4-(β-D-ribofuranosylamino)pyrmidine-6-amine. The crystal structure indicates that the nitro group is hydrogen bonded to the 4-amino hydrogen atom. The base is in the anti conformation and the sugar moiety is disordered. Clitocine shows strong insecticidal activity against *Pectinophoro gossypiella* and is a substrate and inhibitor of adenosine kinase.

Uridine Analogues. 5-Formyloxymethyluridine, $C_{11}H_{14}N_2O_8$, produced by *Serratia plymuthica,* inhibits bacterial growth. 3-Methylpseudouridine, $C_{10}H_{14}N_2O_6$, has been isolated and identified from the culture filtrates of *Nocardia lactamdurans.* 1-Methylpseudouridine was previously isolated from *S. platensis.* 4-Thiouridine, $C_9H_{12}N_2O_5S$, has been isolated from an actinomycete resembling *S. hygroscopicus.*

Hydantocidin. Hydantocidin, $C_7H_{10}N_2O_6$, is elaborated by *S. hygroscopicus.* It is unique in that the anomeric carbon of the ribosyl moiety forms the spiro bond of hydantoin. The ribofuranose moiety which has been reported to be in a C_2-*endo* conformation has been synthesized. Hydantocidin is a herbicidal nucleoside with activity against monocotyledenous and dicotyledenous plants.

N-Nucleotides

The *N*-nucleotide antibiotics include agrocin 84 ($C_{21}H_{34}N_6O_{16}P_2$), thuringiensin ($C_{22}H_{30}N_5O_{19}P$), phosmidosine ($C_{16}H_{24}N_7O_8P$), fosfadecin ($C_{13}H_{19}N_5O_{10}P_2$), and fosfocytocin ($C_{12}H_{20}N_4O_{13}P_2$). Agrocin 84 is an adenine 6-*N*-phosphoramidate nucleotide analogue that contains adenine, phosphate, and D-glucose in a 1:2:1 ratio. It is produced by *Agrobacterium radiobacter* strain K-84.

Thuringiensin, produced by *B. thuringiensis,* is a β-exotoxin that exerts its toxic action on insects and mammals through the inhibition of RNA polymerases.

The structure of the nucleotide antibiotic, phosmidosine, isolated from the culture filtrates of *Streptomyces* sp. RK-16 has been elucidated. Phosmidosine inhibits pore formation of *Botrytis cinerea* and *Aspergillus niger.*

Fosfadecin and fosfocytocin are adenine and cytosine nucleotide antibiotics isolated from culture filtrates of *Pseudomonas* species. Fosfadecin and fosfocytocin inhibit gram-positive and gram-negative bacteria.

ROBERT J. SUHADOLNIK
NANCY L. REICHENBACH
Temple University School of Medicine

K. Isono, *J. Antibiot.* **41,** 1711 (1988).

K. Isono, *Pharmacol. Therapeut.,* **52,** 269 (1991).

R. J. Suhadolnik, *Nucleoside Antibiotics,* John Wiley & Sons, Inc., New York, 1970.

R. J. Suhadolnik, *Nucleosides as Biological Probes,* John Wiley & Sons, Inc., New York, 1979.

OLIGOSACCHARIDES

Oligosaccharide antibiotics represented by the everninomicins [everninomicin B ($C_{66}H_{99}Cl_2NO_{36}$), everninomicin C ($C_{63}H_{93}Cl_2NO_{34}$), everninomicin D ($C_{66}H_{99}Cl_2NO_{35}$), everninomicin 2 ($C_{58}H_{86}Cl_2O_{31}$), everninomicin 3 ($C_{66}H_{98}Cl_2O_{33}$), everninomicin 7 ($C_{66}H_{100}Cl_2O_{34}$), everninomicin-13-384-Component 1 ($C_{70}H_{97}Cl_2NO_{38}$), and everninomicin-13-384-Component 5 ($C_{70}H_{99}Cl_2O_{36}$)], flambamycins [flambamycin ($C_{61}H_{88}Cl_2O_{33} \cdot H_2O$)], avilamycins [avilamycin A ($C_{61}H_{88}Cl_2O_{32}$), avilamycin A_1, avilamycin B ($C_{59}H_{84}Cl_2O_{32}$), avilamycin C ($C_{61}H_{90}Cl_2O_{32}$), avilamycin D_1 ($C_{57}H_{82}Cl_2O_{31}$), avilamycin D_2, avilamycin E, avilamycin F ($C_{60}H_{87}ClO_{32}$), avilamycin G ($C_{62}H_{90}Cl_2O_{32}$), avilamycin H ($C_{61}H_{89}ClO_{32}$), avilamycin I ($C_{60}H_{86}Cl_2O_{32}$), avilamycin J ($C_{60}H_{86}Cl_2O_{32}$), avilamycin K ($C_{61}H_{88}Cl_2O_{33}$), avilamycin L ($C_{60}H_{86}Cl_2O_{32}$), avilamycin M ($C_{60}H_{86}Cl_2O_{32}$), and avilamycin N ($C_{60}H_{86}Cl_2O_{32}$)], and curamycins [curamycin A ($C_{59}H_{84}Cl_2O_{32}$)] have complex and unique structural features. They are sometimes referred to as orthosomycins because characteristically these oligosaccharides possess two acid-sensitive ortho ester linkages, cleavage of which results in the complete loss of antibiotic activity. Another structural feature common to all the oligosaccharide antibiotics is the presence of a substituted phenol ester derived from dichloroisoeverninic acid attached to the sugar ring B at C-13. This acidic phenolic group, which could form salts with organic or inorganic bases, eg, the sodium or *N*-methylglucamine salt, is also essential for the antibiotic activity.

Properties

Physical Properties. Oligosaccharide antibiotics are colorless solids, which are often crystalline and have defined melting points and optical rotations.

Chemical Properties and Structure. Oligosaccharide antibiotics are sensitive to acid pH because of the ortho ester linkages. Yet all of them have an acidic phenolic group which makes the molecule relatively unstable to handling conditions. The ortho ester connecting the C and D rings is comparatively more sensitive to acid pH than the one linking the G and H residues. Other chemically reactive groups in these compounds are the glycosidic linkages, the hydroxyl and carbonyl groups, and any nitro groups present.

Everninomicins

Everninomicin D is the principal component from cultures of *Micromonospora carbonacae*. Chemical modification of the nitro group in everninomicin D has resulted in the formation of amino, mono- and dialkylamino, *N*-acylamino, and *N*-hydroxylamino (and its nitrone derivatives) everninomicin D, all of which possess great antibiotic activity against gram-positive bacteria.

Everninomicin B, though a secondary component, has been extensively investigated because of its improved pharmacokinetic properties over those of everninomicin D.

Flambamycin, Curamycin, and Avilamycins

Flambamycin. Flambamycin, produced by *Streptomyces hygroscopicus* DS 23230, also has undergone substantial chemical degradative experiments during structure elucidation. The structure, chemical reactions, and biological properties have been thoroughly reviewed; it melts at 202–203°C.

Curamycin. Curamycin A is the primary component of the culture *Streptomyces curacoi*. Curamycin A melts at 192–199°C.

Avilamycins. At least 16 avilamycins have been reported. Avilamycins A (mp 181–182°C) and C (mp 188–189°C) are the primary components produced by the strain *Streptomyces viridochromogenes*.

Other Oligosaccharides and Relevant Work

Sporacuracin A, $C_{63}H_{94}Cl_2O_{35}$, mp 145–148°C, $[\alpha]_D$ −26.3, and sporacuracin B, $C_{63}H_{94}Cl_2O_{35}$, mp 162–165°C, $[\alpha]_D$ −18.8, are other probable members of the oligosaccharide antibiotic family.

Biological Properties. The *in vitro* activity is such that oligosaccharides, in general, are highly potent, but are narrow-spectrum antibiotics. Everninomicins are active against a wide variety of gram-positive aerobes and anaerobes; *Neisseria, Mycoplasma,* and some *Mycobacteria.* Comparatively, flambamycin is less potent than everninomicin D.

In Vivo Activity. Potential for oligosaccharide antibiotics in both human and animal health care has been claimed in various patents.

V. M. Girijavallabhan
Ashit K. Ganguly
Schering-Plough Corporation

A. K. Ganguly, in P. Sammes, ed., *Oligosaccharide Antibiotics, Topics in Antibiotic Chemistry,* Vol. 2, Ellis Horwood Publishers, Chichester, U.K. 1978, p. 49.

A. K. Ganguly, O. Z. Sarre, D. Greeves, and J. Morton, *J. Amer. Chem. Soc.* **97,** 1982 (1975).

A. K. Ganguly and co-workers, *Heterocycles* **28,** 83 (1989).

PEPTIDES

Peptide antibiotics are classified according to their overall shape, which can be linear or cyclic, and by the nature of the bonds joining the constituent amino acids (qv) and other carboxylic acids (qv), which can be all amide bonds or amide plus ester bonds. Most peptide antibiotics are cyclic peptides that do not contain disulfide linkages.

Peptide antibiotics differ in many respects from proteins (qv) and from peptides having hormonal or other functions in higher animals and plants (see Hormones). Among the significant differences are low molecular weight, where the vast majority of peptide antibiotics have molecular weights in the 500–1500 range. The average protein has a mol wt of 40,000. Many peptide antibiotics have unusual fatty acids and amino acids, such as D-amino acids, *N*-methyl amino acids, or imino acids, and they usually lack methionine and histidine. Additionally, peptide antibiotics often have other nonamino acid moieties, such as the chromophore of dactinomycin. Ring closure in cyclic peptide antibiotics is by amide or ester bonds, not by disulfide bonds as in normal proteins; or, if the peptide antibiotic has a tail, a diamino acid or a hydroxy amino acid provides the amide or ester bonds for ring closure. Peptide antibiotics are normally resistant to the usual proteases and peptidases. Also they are usually synthesized on multienzyme complexes much as are fatty acids, and not on ribosomes as are proteins. Nisin and related antibiotics are exceptions. Most peptide antibiotics are synthesized, and are often marketed, as groups of closely related structures, usually reflecting a lack of specificity of the biosynthetic enzymes. Even the small fraction of peptide antibiotics that have therapeutic usefulness are quite toxic. For these useful peptide antibiotics the therapeutic index, the ratio of the minimum toxic dose to the maximum effective dose, is smaller than for most nonpeptide antibiotics.

Although historically most useful antibiotics have come from spore forming microorganisms, marine organisms have yielded the candidate antitumor peptide didemnin B and cytostatic peptides such as the patellamides. Many of the marine peptides have little or no antimicrobial activity. Antibacterial peptides called magainins are found in frog skin and antibacterial polypeptides called defensins are found in mammalian white blood cells and other tissues.

Peptide antibiotics are not often the drugs of first choice for systemic therapy of important human disease. However, the World Health Organization, which chooses drugs especially for Third World use based on efficacy, safety, quality, price, and availability, includes as essential such peptide antibiotics as bleomycin, dactinomycin, and bacitracin (as an ointment containing neomycin), plus several β-lactams (see Antibiotics, β-lactams). Systemic use of peptide antibiotics is many times limited by nephrotoxicity and other toxicities. Semisynthesis or complete chemical synthesis of analogues of peptide antibiotics has most often not resulted in improved drugs.

Table 1. Peptide Antibiotics of Special Interest

Name	Related to	Producing microbe	Antibiotic activity[a]	Mol wt (number of amino acids)	Number in family
From bacteria of the genus Bacillus					
bacillomycin F		*B. subtilis*	F	1080 (8)	2
bacilysin		*B. subtilis, B. pumilis*	P, N	270 (2)	
bacitracin (ayfivin)		*B. licheniformis*	P	1410 (11)	5
circulins	polymyxins	*B. circulans*	N	1150 (10)	2
colistin (polymyxin E)	polymyxins, circulins	*B. colistinus*	N	1150 (10)	10
edeine A_1		*B. brevis*	P, N, F	755 (5)	6
gramicidins, linear, Dubos		*B. brevis*	P	1900 (15)	6
gramicidin S		*B. brevis*	P	1141 (10)	3
iturins A	bacillomycin B	*B. subtilis*	F	1050 (8)	9
micrococcin P_1	thiocillins	*B. pumilis*	P	1144 (5)	4
mycobacillin		*B. subtilis*	F	1528 (13)	1
polymyxins	circulins	*B. polymyxa*	N	1200 (10)	10
subtilin	nisin	*B. subtilis*	P	3321 (26)	1
tyrocidines		*B. brevis*	P, N	1300 (10)	5
tyrothricin		*B. brevis*	P, N	mixture	
From higher bacteria of the genus Streptomyces					
albomycin		*S. subtropicus, S. griseus*	P	950 (5)	5
amphomycin (glumamycin)		*S. canis, S. violaceus, S. albo-griseolus*	P	1290 (10)	5
bialaphos (SF 1293)	phosphinothricin	*S. hydroscopicus,* other *S.* species	P, N,[b]	323 (3)	2
bicozamycin (bicyclomycin)		*S. sapporonensis, S. aizunensis*	N	302 (2)	
bleomycins	phleomycin, peplomycin, liblomycin	*S. verticillus*	P, N, M, F, T	1416 (12)	
bottromycin		*S. bottropensis*	P, N	810 (6)	5
capreomycins	viomycin, enviomycin, tuberoactinomycin	*S. capreolus*	P, N, M	670 (5)	4
dactinomycin (actinomycin D)	cactinomycin	*S. parvullus*	P, T	1255 (10)	30
daptomycin (deptomycin, LY 146032, semisynthetic)		from A21978C of *S. roseosporus*	P	1619 (14)	6
distamycin A (stallimycin)	netropsin	*S. distallicus*	F, T, V	482 (4)	
enduracidins (enramycin)		*S. fungicidicus*	P, M	2355 (17)	4
enviomycin (tuberactinomycin N)	capreomycin	*S. griseoverticilatus* var. *tuberacticus*	P, N, M	686 (6)	3
ferrimycin A	sideromycins	*S. griseoflavus,* other *S.* species	P, N	978 (6)	4
globomycin		*S. halstedii, S. cinnamomeus*	N	657 (5)	
ilamycin A (rufomycin A)		*S. islandicus, S. atratus*	P, M	1026 (7)	5
neoviridogrisein II (etamycin B)		*S. griseoviridis*	P	863 (7)	3
netropsin	distamycin	*S. netropsis, S. chromogens,* others	P, N, M, T, V	430 (2)	8
quinomycin C	triostin	*S. aureus*	P, N, T, V	1141 (8)	3
stendomycin		*S. antimycoticus*	P, N, F	1850 (14)	3
streptogramin A (pristinamycin IIA, virginiamycin M1)	virginiamycin M2	*S. virginiae,* others	P	527 (3)	2
streptogramin B (pristinamycin IA)	virginiamycin S_1	*S. virginiae,* others	P	867 (6)	12
streptothricin A (polymycin A)	noureothricins, racemomycins	*S. lavendulae,* others	P, N, M, V	1143 (6)	7
telomycin		*S. canus,* others	P	1272 (11)	1
thiopeptin		*S. tateyamensis*	P	1550 (6)	8
thiostrepton (bryamycin)		*S. aureus*	P	1665 (9)	1
valinomycin		*S. fulvissimus, S. tsusimaensis*	P, N, F, T, V	1111 (6)	
viridogrisein (etamycin A)		*S. lavendulae*	P	879 (7)	2
zinostatin (neocarzinostatin)		*S. carcinostaticus*	P, T	11400 (113)	1
From other bacteria					
nisin	subtilin	*Streptococcus lactis*	P	3354 (34)	
From fungi					
alamethicin		*Trichoderma viridis*	F	1970 (19)	2
cilofungin (semisynthetic)	aculeacin B, sporiofungin	from echinocandin B of *Aspergillus* spp.	F	1029 (4)	

Table 1. Peptide Antibiotics of Special Interest *(continued)*

Name	Related to	Producing microbe	Antibiotic activity[a]	Mol wt (number of amino acids)	Number in family
cyclosporine (cyclosporin A)[c]	other cyclosporins	*Cylindrocarpon lucidium, Tolypocladium inflatum*	F[c]	1203 (11)	25
destruxin B		*Oospora destructor,* others	[d]	584 (5)	
echinocandin B		*Aspergillus rugulosus, A. nidulans* var. *roseus*	F	1060 (4)	8
enniatin A		*Fusarium orthoceras, F. sciroi*	P, M, F	682 (6)	3
tentoxin		*Alternaria tenuis, A. alternata*	[b]	415 (4)	2

[a] Activity against gram-positive bacteria = P; gram-negative bacteria = N; mycobacteria = M; fungi = F; tumors = T; and viruses = V.
[b] Active as a herbicide
[c] Active against the immune system.
[d] Active as an insecticide.

The complex structure of peptide antibiotics adds considerably to the problems of synthesis, but more recent efforts toward improved peptide antibiotics are encouraging (see FERMENTATION). Methods of bioassay and other laboratory use of economic antibiotics are available. Table 1 is a list of important peptide antibiotics. More extensive listings of minor peptide antibiotics have been published.

EDMUND M. WISE, JR.
The Wellcome Research Laboratories

H. Kleinkauf and H. von Döhren, eds., *Biochemistry of Peptide Antibiotics,* W. de Gruyter, Berlin, 1990.

V. Lorian, ed., *Antibiotics in Laboratory Medicine,* 3rd ed., Williams & Wilkins, Baltimore, Md., 1991.

B. W. Bycroft, ed., *Dictionary of Antibiotics and Related Substances,* Chapman and Hall, London, 1988.

A. I. Laskin and H. A. Lechevalier, eds., *CRC Handbook of Microbiology,* 2nd ed., Vol. 9, Parts A and B, CRC Press, Boca Raton, Fla., 1988.

POLYETHERS

The polyether antibiotics were first recognized as a separate class with the publication of the structure of monensin in 1967. Several members of the group have since found commercial application as anticoccidials in poultry farming and in improvement of feed efficiency for ruminants (see FEEDS AND FEED ADDITIVES).

These antibiotics are characterized by multiple tetrahydrofuran and tetrahydropyran rings connected by aliphatic bridges, direct C–C linkage, or spiro linkage. Other features include a free carboxyl function, many lower alkyl groups, and a variety of functional oxygen groups. These structural features enable the molecule to form a cyclic conformation with the oxygen functions at the center and the alkyl groups on the outer surface. This conformation results in lipid solubility, even for the salt forms, enabling transport of cations across lipid membranes.

Individual polyethers exhibit varying specificities for cations. Some polyethers have found application as components in ion-selective electrodes for use in clinical medicine or in laboratory studies involving transport studies or measurement of transmembrane electrical potential.

Properties

The polyether class can be subdivided based on the number of carbon atoms in the backbone. The carbon backbone or skeleton refers to the longest chain of contiguous carbons between the carboxyl group and the terminal carbon. The 30 C skeleton group accounts for about 60% of the polyethers for which structures have been determined. Most of these contain one or two sugar moieties, usually 2,3,6-trideoxy-4-*O*-methyl-D-*erythro* pyranose. Monensin **(1)**, widely used as an anticoccidial and feed efficiency enhancer, is an example of the 26 C backbone class. Maduramicin **(2)**, another commercially important polyether, is shown as an example of the 30 C group with a sugar moeity.

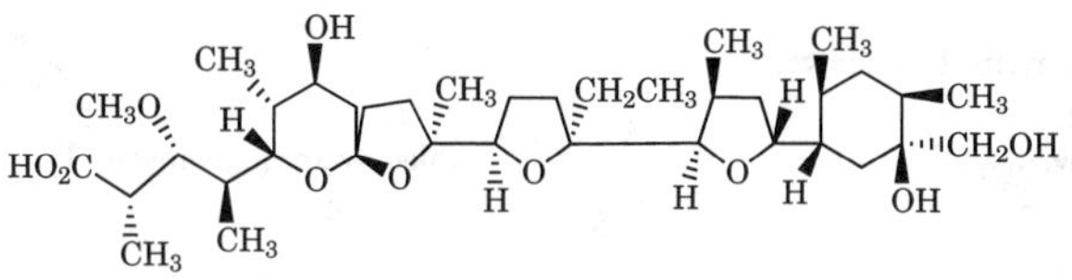

(1) Monensin A

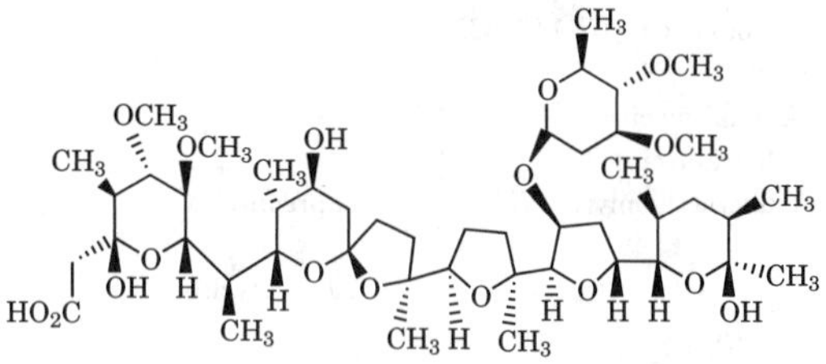

(2) Maduramicin alpha

Purification and Production

Polyethers are usually found in both the filtrate and the mycelial fraction, but in high yielding fermentations they are mostly in the mycelium because of their low water-solubility. The high lipophilicity of both the free acid and the salt forms of the polyether antibiotics lends these compounds to efficient organic solvent extraction and chromatography (qv) on adsorbents such as silica gel and alumina. Many of the production procedures utilize the separation of the mycelium followed by extraction using solvents such as methanol or acetone. A number of the polyethers can be readily crystallized, either as the free acid or as the sodium or potassium salt, after only minimal purification.

Polyethers such as monensin, lasalocid, salinomycin, and narasin are sold in many countries in crystalline or highly purified forms for incorporation into feeds or sustained-release bolus devices (see CONTROLLED-RELEASE TECHNOLOGY). There are also mycelial or biomass products, especially in the United States. The mycelial products are generally prepared by separation of the mycelium and then drying by azeotropic evaporation, fluid-bed driers, continuous tray driers, flash driers, and other types of commercial driers. In countries allowing biomass products, crystalline polyethers may be added to increase the potency of the product.

Biological Activities

The polyether antibiotics exhibit a broad range of biological, antibacterial, antifungal, antiviral, anticoccidial, antiparasitic, and insecticidal activities. They improve feed efficiency and growth performance in ruminant and monogastric animals. Only the anticoccidial activity in poultry and cattle, and the effect on feed efficiency in ruminants such as cattle and sheep are of commercial interest.

The discovery of the activity of the polyethers against *Eimeria* sp. has greatly altered the prevention and control of coccidiosis. It is estimated that the polyether ionophores constitute more than 80% of the total worldwide usage of anticoccidials. The enhancement of feed efficiency in ruminants correlates with increased production of propionic acid in the rumen. A decrease in the production of acetic acid, butyric acid, and methane often accompanies the increased propionic acid production. These volatile fatty acids are produced in the rumen by the degradation of carbohydrates by microorganisms. The polyethers apparently selectively inhibit certain of the microorganisms to achieve greater propionate production. Propionate is thought to be more effectively utilized in energy metabolism in the host animal than acetate or butyrate.

Biosynthesis

Bacteria belonging to the order Actinomycetales are the organisms reported to produce all of the polyethers. Most are secondary metabolites of *Streptomyces* sp. with the species *hygroscopicus* and *albus* accounting for about one-third of the antibiotics. Other genera represented are *Streptoverticillium, Dactylosporangium, Actinomadura, Nocardia,* and *Nocardiopsis.* The taxonomy of these producing organisms has been reviewed.

Economic Aspects

As of 1998 the worldwide usage of polyether antibiotics for controlling coccidiosis is approximately $190 million compared to a total market of $210–220 million. Monensin and salinomycin represent about 65–70% of this market; lasalocid, narasin, and maduramicin make up the remainder. Other compounds for coccidiosis control include nicarbazine, halofunginone, amprolium, and robenidine. Worldwide usage is in excess of 3 million kg of product.

LOUISE W. CRANDALL
ROBERT L. HAMILL
Lilly Research Laboratories

J. W. Westley, ed., *Polyether Antibiotics,* Vols. 1 and 2, Marcel Dekker, Inc., New York, 1982.

J. A. Robinson, "Chemical and Biochemical Aspects of Polyether–Ionophore Antibiotic Biosynthesis" in W. Herz, G. W. Kirby, W. Steglich, Ch. Tamm, eds., *Progress in the Chemistry of Organic Natural Products,* Springer-Verlag, Wien, New York, 1991, pp. 1–82.

O. Yonemitsu and K. Horita, "Total Synthesis of Polyether Antibiotics" G. Lukacs and M. Ohno, eds. in *Recent Progress in the Chemical Synthesis of Antibiotics,* Springer Verlag, Berlin, Heidelberg, 1990, pp. 448–466.

TETRACYCLINES

The tetracyclines are a group of antibiotics having an identical 4-ring carbocyclic structure as a basic skeleton and differing from each other chemically only by substituent variation. Figure 1 shows the principal tetracycline derivatives now used commercially.

The first tetracycline discovered was produced by a soil organism, *Streptomyces aureofaciens,* and is now known as chlortetracycline (2), $C_{22}H_{23}ClN_2O_8$. This compound ushered in a new era in antibacterial chemotherapy because it was effective orally and against a broad range of gram-positive and gram-negative bacteria.

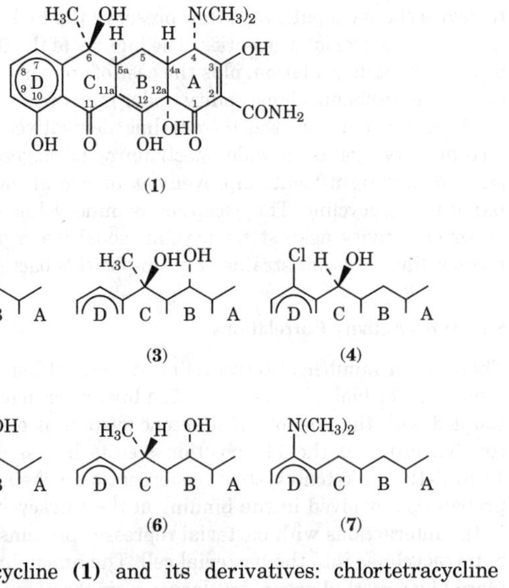

Figure 1. Tetracycline **(1)** and its derivatives: chlortetracycline (7-chlorotetracycline) **(2)**; oxytetracycline (5-hydroxytetracycline) **(3)**; demeclocycline (6-demethyl-7-chlorotetracycline) **(4)**; methacycline (6-demethyl-6-deoxy-5-hydroxy-6-methylenetetracycline) **(5)**; doxycycline (6α-deoxy-5-hydroxytetracycline) **(6)**; and minocycline (6-demethyl-6-deoxy-7-dimethylamino tetracycline) **(7)**. Substituents at positions not specifically shown or mentioned remain as in **(1)**. Courtesy of Blackwell Scientific Publications, Ltd.

The three tetracyclines most recently marketed were made by a semisynthetic pathway. The first of these were methacycline (6-methylene oxytetracycline) **(5)**, $C_{22}H_{22}N_2O_8$, and its reduction product doxycycline **(6)**, $C_{22}H_{24}ClN_2O_8$. The latter compound is a potent antibiotic which is well-absorbed and slowly excreted, thus allowing small and infrequent (once or twice a day) dosage schedules. Finally, the most recent addition to the commercial tetracyclines is minocycline **(7)**, $C_{23}H_{27}N_3O_7$, which is also well-absorbed and slowly excreted.

Physical Properties

In general, the tetracyclines are yellow crystalline compounds that have amphoteric properties (Fig. 2). They are soluble in both aqueous acid and aqueous base.

The tetracyclines are strong chelating agents. This ability to chelate to metals, such as calcium, results in tooth discoloration when tetracycline is administered to children.

Semisynthetic Modifications

The tetracycline molecule (1) presents a special challenge with regard to the study of structure–activity relationships. The difficulty has been

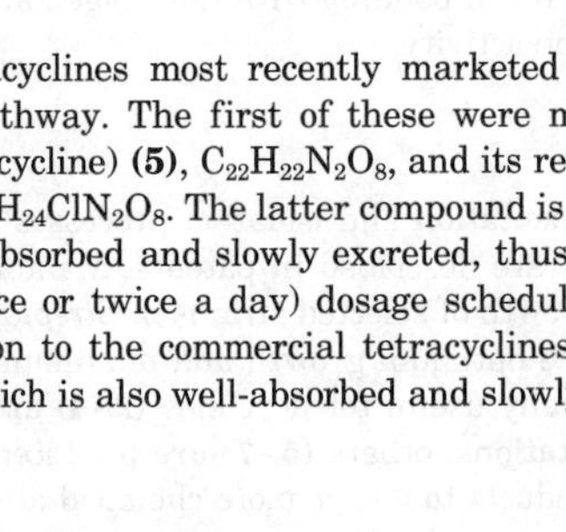

Figure 2. Tetracycline **(1)** indicating the titratable hydrogens and showing **(a)** the BCD-chromophore and **(b)** the A-chromophore.

to devise chemical pathways that preserve the BCD ring chromophore and its antibacterial properties. The lability of the 6-hydroxy group to acid and base degradation, plus the ease of epimerization at position 4, contribute to chemical instability under many reaction conditions.

Although many of the tetracycline derivatives showed useful *in vivo* activity against a wide spectrum of pathogenic organisms, few showed any significant improvement in overall activity when compared to tetracycline. The exception is minocycline **(7)** which exhibits superior activity against tetracycline-sensitive organisms and many tetracycline-resistant strains of gram-positive bacteria.

Structure–Activity Correlations

There are a number of tetracycline structural features that are prerequisites for biological activity. The linear arrangement of the rings, coupled with the phenolic β-diketone system, is essential. Any structure variation at the 11a position results in loss of activity. The C-11 to C-12 β-diketone system has exceptional chelating qualities, and probably is involved in the binding of the tetracyclines to ribosomes, in the interactions with bacterial repressor proteins, and in transport of tetracyclines into the bacterial cell. The amide hydrogen can be replaced by a methyl group, but larger residues, if not rapidly cleaved in water, bring about a reduction in activity.

The configuration at the chiral centers C-4a, C-5a, and C-12a determine the conformation of the molecule. In order to retain optimum *in vitro* and *in vivo* activity, these centers must retain the natural configuration. The hydrophobic part of the molecule from C-5 to C-9 is open to modification in many ways without losing antibacterial activity. However, modification at C-9 may be critical because steric interactions or hydrogen bonding with the oxygen atom at C-10 may be detrimental to the activity.

Manufacture

Most of the fermentation and isolation processes for manufacture of the tetracyclines are described in patents. Manufacture begins with the cultivated growth of selected strains of *Streptomyces* in a medium chosen to produce optimum growth and maximum antibiotic production. Some clinically useful tetracyclines **(2–4)** are produced directly in these fermentations; others **(5–7)** are produced by subjecting the fermentation products to one or more chemical alterations. The purified antibiotic produced by fermentation is used as the starting material for a series of chemical transformations.

Economic Aspects

The development of the semisynthetic β-lactam antibiotics (see ANTIBIOTICS–β-LACTAMS) and emergence of resistance to the tetracyclines has steadily diminished the clinical usefulness of tetracyclines.

In the United States, the manufacturers of fermentation-derived tetracyclines **(1)**, **(2)**, and **(3)** are the Lederle Laboratories, a division of American Cyanamid Company, Charles Pfizer Inc., Bristol Laboratories, and Rachelle Laboratories. There are also several manufacturers abroad. Tetracycline is now sold generically by many companies. Pfizer's doxycycline **(6)** and Lederle's minocycline **(7)**, both semisynthetic tetracyclines, are the only members of the group that have increasing sales. Table 1 lists the commercial tetracyclines and the corresponding trade names.

Biological Aspects

It has been known for some time that tetracyclines are accumulated by bacteria and prevent bacterial protein synthesis. Furthermore, inhibition of protein synthesis is responsible for the bacteriostatic effect. Inhibition of protein synthesis results primarily from disruption of codon–anticodon interaction between tRNA and mRNA so that binding of aminoacyl–tRNA to the ribosomal acceptor (A) site is prevented. The precise mechanism is not understood. However, inhibition is likely to result from interaction of the tetracyclines with the 30S ribosomal subunit because these antibiotics are known to bind strongly to a single site on the 30S subunit.

Table 1. Tetracyclines Used for the Therapy of Infectious Diseases

Generic name	Trade name	Year of discovery
chlortetracyline	Aureomycin	1948
oxytetracycline	Terramycin	1948
tetracycline	Achromycin	1953
demeclocycline	Declomycin	1957
methacycline	Rondomycin	1965
doxycycline	Vibramycin	1967
minocycline	Minocin	1972

Uses

Clinical Uses. The emergence of bacterial resistance to tetracyclines has limited the use of these agents as the drugs of first choice in the treatment of many infections for which they were previously effective. Nevertheless, they are still the treatment of first choice in the following cases: *(1)* for bacterial infections causing brucellosis, cholera, chancroid, granuloma inguinale, and Lyme disease; *(2)* for rickettsial infections; *(3)* for chlamydial infections; *(4)* in the treatment of nonspecific urethritis because of *Chlamydia* or *Ureaplasmas;* and *(5)* in the treatment of acne vulgaris and rosacea.

Tetracyclines are used as alternative drugs in a variety of circumstances when the patient is unable to take the drug of choice, eg, in patients allergic to penicillin.

Veterinary Uses. Tetracyclines are widely used for veterinary therapy. The types of pathogens encountered are frequently different from those for which tetracyclines are used in humans. Tetracyclines are also used in animal husbandry as growth promoters (see GROWTH REGULATORS–ANIMAL).

Resistance to Tetracyclines. *Mechanisms of Resistance.* Three distinct biochemical mechanisms of resistance to tetracyclines have been identified. The energy-dependent efflux of antibiotic mediated by resistance proteins located in the bacterial cytoplasmic membrane is one mechanism. The intracellular tetracycline concentration remains too low for effective binding to ribosomes. Ribosomal protection, whereby tetracyclines no longer bind productively to the bacterial ribosome, is a second mechanism. In a tetracycline-resistant cell, tetracycline accumulation within the cell is similar to that in the sensitive cell, but the ribosome is modified so that tetracycline no longer binds productively to the ribosome. Finally, chemical alteration of the tetracycline molecule by a reaction in the cytoplasm that requires oxygen renders the drug inactive as an inhibitor of protein synthesis. The altered tetracycline then diffuses out of the cell.

JOSEPH J. HLAVKA
GEORGE A. ELLESTAD
IAN CHOPRA
American Cyanamid Company

R. W. Broschard and co-workers, *Science* **109,** 199 (1949); B. M. Duggar, *Ann. N.Y. Acad. Sci.* **51,** 177 (1948).

R. K. Blackwood and co-workers, *J. Amer. Chem. Soc.* **85,** 3943 (1963).

C. R. Stephens and co-workers, *J. Amer. Chem. Soc.* **85,** 2643 (1963).

M. J. Martell, Jr. and J. H. Boothe, *J. Med. Chem.* **10,** 44 (1967); G. S. Redin in G. L. Hobby, ed., *Antimicrobial Agents and Chemotherapy,* Williams & Wilkins, Baltimore, Md., 1966, p. 371.

J. J. Hlavka and J. H. Boothe, eds., *Handbook of Experimental Pharmacology,* Vol. 78, Springer-Verlag, New York, 1985, p. 86.

ANTIFREEZES AND DEICING FLUIDS

An antifreeze is defined as a chemical which, when added to a water-based fluid, reduces the freezing point of the mixture. Antifreezes are used in a wide variety of mechanical equipment during the winter months to prevent freezing of aqueous heat-transfer fluids. Most commonly, antifreeze refers to the freeze-protected fluid which cools automotive engines, although antifreeze liquids are also used in ice skating rinks, refrigeration systems (as a secondary coolant), heating and air conditioning systems, solar energy units, and many other applications. Chemical antifreezes include brines, such as calcium chloride; alcohols, such as methanol, ethanol, and 2-propanol; and glycerol and glycols. Since 1960 ethylene glycol has held the majority of the antifreeze market share because of its availability and superior performance characteristics.

Colligative Properties

Many chemicals when added to water cause a freezing point depression, as shown in Table 1, and thus are termed antifreezes. The antifreeze properties of these chemicals vary widely as a function of their colligative, or concentrative, properties. The reduction in freeze point depends both on the chemical itself and the concentration of the chemical in water. The freeze point depression increases as the antifreeze chemical is added to the water, until a characteristic concentration is achieved. Further addition of the antifreeze chemical to water will either result in insolubility or serve to increase the freezing point of the mixture.

The colligative properties of antifreeze chemicals may also result in boiling point elevation. As the chemical is added to water, the boiling point of the mixture increases. Unlike the freeze depression, the boiling elevation does not experience a maximum; the boiling point versus concentration curve is a smooth curve that achieves its maximum at the 100% antifreeze level. The boiling point elevation can be another important characteristic for antifreeze fluids in certain heat-transfer applications.

Engine Coolants

Besides freeze protection, antifreezes provide many other performance properties that enhance the operation of a heat-transfer system. Because the internal combustion engine is by far the largest antifreeze application, and ethylene glycol is the predominant antifreeze in use, the following focuses on the performance properties of an ethylene glycol-based antifreeze and their relationship to engine cooling.

Heat Transfer. The primary role of the antifreeze liquid in an internal combustion engine is to remove heat and thus cool the engine. If heat is not removed, the engine will be destroyed. The antifreeze fluid removes heat by passing through the engine where it picks up heat, then through the radiator where it is cooled by air. The fluid operates in a closed loop system and thus is continuously reused. To provide efficient cooling, the antifreeze must have a high specific heat and thermal conductivity and low viscosity at operating temperatures.

Table 1. Freeze Point Depression of Antifreeze Chemicals

Component	Molecular formula	Concentration in water, wt %	Freeze point depression, °C
calcium chloride	$CaCl_2$	32[a]	−50
calcium magnesium acetate	$C_2H_4O_2{\cdot}x\text{Ca}{\cdot}x\text{Mg}$	32.5[a]	−28
ethanol	C_2H_6O	50	−38
ethylene glycol	$C_2H_6O_2$	50	−36
glycerol	$C_3H_8O_3$	50	−22
methanol	CH_4O	50	−50
potassium acetate	$C_2H_4O_2{\cdot}K$	50	−50
potassium chloride	KCl	13[a]	−6.5
propylene glycol	$C_3H_8O_2$	50	−32
seawater		100 (6% salt)	−3
sodium chloride	NaCl	23[a]	−21
sucrose	$C_{12}H_{22}O_{11}$	42[a]	−5
urea	CH_4N_2O	44[a]	−18

[a] Saturated solution at this concentration. Becomes insoluble at higher concentrations.

Freeze Point Depression. The slight heat-transfer penalty incurred when an antifreeze is added to the aqueous heat-transfer fluid is necessitated by the need for increased operating temperature range in most internal combustion engines. Because most parts of the world achieve temperatures below freezing during some time of the year, an antifreeze fluid is required to keep equipment operational in these subfreezing temperatures.

The colligative properties of antifreeze fluids dictate the freeze point depression of the antifreeze solution. For ethylene glycol solutions, the maximum freezing point is achieved by using a 68% aqueous solution; this concentration is known as the eutectic concentration.

Boiling Point Elevation. Boiling point elevation may also be beneficial in heat-transfer applications. As engine power increases, the heat given off by the engine also increases. This increase in engine operating temperature increases engine efficiency by making more heat available to power the engine. However, if too much heat remains in the engine, overheating of both the engine and the oil may occur and engine efficiency decreases. Enabling the coolant to remove more heat most often requires no capital changes, and can be achieved by either increasing the flow rate of the coolant, or increasing the system pressure, and thus the boiling point of the coolant, which would allow the coolant to circulate at a higher maximum temperature. Today, most engines operate slightly pressurized to increase the maximum use temperature of the coolant and thus eliminate boiling of the coolant under extreme operating conditions.

Corrosion Inhibition. Another important property of antifreeze solutions is the corrosion protection they provide. Most cooling systems contain varied materials of construction including multiple metals, elastomeric materials, and rigid polymeric materials. The antifreeze chosen must contain corrosion inhibitors that are compatible with all the materials in a system. Additionally, the fluid and its corrosion inhibitor package must be suitable for the operating temperatures and conditions of the system.

Today, all antifreezes contain inhibitors designed to minimize corrosive attack, including both organic and inorganic compounds. The most common corrosion inhibitors for glycol-based antifreezes include borates, molybdates, nitrates, nitrites, phosphates, silicates, amines, triazoles, and thiazoles. Since no single compound provides complete corrosion protection to all metals under all conditions, antifreeze formulations contain a combination of these inhibitors. Additionally, since cooling system design varies from automobile to automobile, and from year to year, the antifreeze formulation must have enough latitude to account for these changes and yet maintain adequate corrosion protection for most systems.

Foaming. Another factor in maintaining heat-transfer efficiency and good system operation, is minimizing the foaming tendency of the fluid. Foaming or air entrainment in the fluid minimizes the amount of fluid that is in contact with the heat-transfer surface, and thus reduces the amount of heat that the fluid can remove from the system.

Although glycol–water solutions are not inherently prone to foaming, mechanical and chemical conditions may cause foam to form in the system. To counteract these effects, most antifreeze formulations contain antifoaming agents, such as silicones, polyglycols, or oils.

Effect on Elastomers. The antifoamers and corrosion inhibitors used in antifreeze formulations are designed not only to function effectively in their assigned role, but also to be compatible with the other materials of construction in the system, such as the elastomers and plastics. Various components in the coolant can swell and soften elastomers or cause them to become brittle and crack. In general, ethylene glycol causes the elastomer to shrink when compared to the effect of pure water. Therefore, once a system is charged with a specific type of coolant, replacement with another chemical may require replacement

Table 2. Compatibility of Ethylene Glycol with Elastomeric Materials[a]

Trade name or common name	Material	Temperature 25°C	80°C	160°C
Adriprene L-100[b]	urethane	good	poor	poor
black rubber 3773	polyisoprene, synthetic	good	poor	poor
buna N	acrylonitrile–butadiene	good	good	
buna S	styrene–butadiene	good	fair	poor
butyl rubber	isobutylene–isoprene	good	good	
EPDM	ethylene–propylene–diene	good	good	good
EPR rubber	ethylene-propylene rubber	good	good	good
Hycar, D-24[b]	butadiene–styrene, butadiene–acrylonitrile, acrylate emulsion	good	fair	
Hypalon[b]	chlorosulfonated polyethylene	good	poor	poor
Kalrez[b]	tetrafluoroethylene, perfluoromethyl vinyl ether cure site monomer	good	good	good
natural rubber gum	polyisoprene, natural	good	poor	poor
neoprene 7797	chloroprene	good	fair	
red rubber #107	polyisoprene, synthetic	good	poor	poor
silicone #65	polydimethylsiloxane	good	good	
Viton A[b]	vinylidene fluoride and hexafluoropropylene	good	good	poor

[a] See ELASTOMERS, SYNTHETIC.
[b] Registered trademarks.

of the gaskets and hoses in the system. Table 2 shows the compatibility of various elastomers with ethylene glycol.

Service Life. The service life offered by a coolant is dependent on many factors, including the initial condition of the coolant and the cooling system, the type of water used for solution, the metals of construction in the system, the type of corrosion inhibitors and SCAs used, the system operating temperature, and the type of operation the coolant is in: light-duty/intermittent, heavy-duty/diesel, or continuous stationary. Before initial charging of a cooling system with fresh coolant, the system should be free of dirt, oil, and other contaminants.

Similarly, the antifreeze itself must be prepared properly to extend the life of the fluid. Because many antifreeze solutions contain inorganic corrosion inhibitors such as phosphates and molybdates, the fluids are sensitive to dilution by hard water. In addition to the hardness ions, poor quality water of dilution can result in the introduction of corrosive ions, chloride and sulfate, to the system.

The service life of an antifreeze can be affected both by the absolute time a coolant is in a system, and by the actual service hours of the engine. Thus, the type of engine operation and the type of corrosion inhibitors in the fluid play a significant role in determining the life of the fluid.

Despite all these safeguards to extend the service life of the antifreeze, fluid replacement is required periodically. Typically, fluids are replaced because of irreversible damage caused by one of four conditions: contamination, gel formation because of glycol–silicate reaction, extensive glycol degradation caused by overheating or excessive oxygen exposure, or inhibitor depletion.

Fluid Maintenance. The frequency of fluid replacement can be minimized and the heat-transfer efficiency and overall condition of the system maximized by proper maintenance of the engine cooling system. The most critical requirement for an antifreeze is that it provides adequate freeze protection. Because of evaporative losses of water and system makeup, it is necessary to monitor the freezing point of the antifreeze to ensure adequate protection. The freezing point of a glycol-based fluid can be determined in the field by specific gravity or refractive index methods.

Fluid Specifications. The performance characteristics of all antifreeze solutions are governed by fluid specifications that have been developed over the years by industry standards committees, such as the American Society for Testing and Materials (ASTM) and the Society of Automotive Engineers (SAE). Additionally, most engine or cooling system manufacturers have their own compositional specifications to which the fluids must conform.

Deicing Fluids

Antifreeze chemicals, because of their characteristic freeze point depression, find additional applications in the deicing of aircraft and runways. Because of the stringent requirements of corrosion protection of aircraft surfaces and the necessity of a low flash point for use around aircraft engines, aircraft deicing fluids use glycols exclusively (over brines and alcohols) to achieve freeze point depression; runway deicing is commonly achieved with either urea or glycol-based fluids that contain urea, although two acetate salts are being studied for runway usage. Calcium magnesium acetate and potassium acetate are being reviewed by the Federal Aviation Administration (FAA) and the Air Transport Association (ATA).

Materials Compatibility. In addition to the runway deicing and aircraft deicing/antiicing fluids' primary roles of removing frozen accumulations and protecting against the refreezing of aircraft and runway surfaces, these fluids are subject to many other performance criteria, as defined by the airlines, the aircraft manufacturers, and the aircraft engine manufacturers. The fluids must contain corrosion inhibitors to minimize the corrosive effects on aircraft metals of construction. Additionally, the fluids must not adversely affect the acrylics and polycarbonates used for aircraft windows by crazing or staining, and should not stain, discolor, or blister painted and unpainted aircraft surfaces.

Deicing and Antiicing Specifications. The performance criteria for deicing fluids have traditionally been governed by specifications of the Aerospace Division of Society of Automotive Engineers (SAE). These specifications are widely used by most airlines in North America.

In 1987, a proposal was made to the International Standards Organization (ISO) to prepare international specifications for deicing and antiicing fluids, equipment, and procedures, in an effort to standardize winter ground operations and set minimum requirements for deicing and antiicing at airports throughout the world, and thus aid the pilot in understanding and defining the protection against refreezing as a function of the fluid used at a given airport. These specifications for worldwide standardization are currently being developed by the ISO.

Toxicity and Environmental Issues

The toxicity of antifreeze and deicing fluids is predominantly a function of the main component, the freezing point depressant. For ethylene glycol-based fluids, the toxicity is well-defined, as the toxicity of ethylene glycol has been studied extensively because of its wide usage in varied applications.

Ethylene glycol is acutely toxic to humans and animals by oral exposure.

Because of these toxicity concerns, propylene glycol is sometimes used instead of ethylene glycol in these applications. Studies on propylene glycol have shown no evidence of teratogenic effects and it has a lower oral toxicity. Propylene glycol has, however, been shown to be slightly more irritating to the skin than ethylene glycol.

Aquatic Toxicity. In formulating deicing and antiicing fluids, the additives play a significant role in defining the aquatic toxicity. In particular, many surfactants or wetting agents are highly toxic to aquatic life, whereas both ethylene and propylene glycol are essentially nontoxic; concentrations exceeding 10,000 mg/L show no effect on aquatic life.

Biodegradation and Oxygen Demand. Another important consideration in the use of all chemicals, including antifreezes and deicing fluids, is the rate and extent of biodegradation, or the length of time the chemical persists in the environment. In the case of ethylene glycol-based fluids, laboratory tests indicate complete biooxidation occurs within 20 days. The rate of biooxidation is steady throughout this 20-day period and thus exerts no undue oxygen demand on waterways (causing fish kill) or waste treatment facilities. Propylene glycol, on the other hand, biodegrades rapidly during the first five days of the test (62%); then the biooxidation rate slows down substantially such that only 79% biooxidation is indicated at the end of the 20-day test. The rapid initial biooxidation imposes a great oxygen demand on the receiving stream or waste treatment facility.

Spent fluids contain varied contaminants that can drastically affect the toxicity and environmental impact of the fluid. Most pronounced is the impact of heavy-metal contaminants in spent antifreeze. Data on spent and recycled antifreeze, compiled by the ASTM Committee on Engine Coolants, show an average lead level 11 ppm, as well as various other metal contaminants (iron, copper, zinc). The presence of these contaminants in a used fluid may require special disposal techniques for the fluids.

Fluid Recycling

When antifreeze becomes unsuitable for use, either because of depletion of inhibitors, presence of corrosion products or corrosive ions, or degradation of the fluid, recycling and reuse of the antifreeze rather than disposal may be considered.

Two methods of recycling spent antifreeze, filtration or distillation, have been reported. The filtration method consists of a filter and a system for reinhibition of the filtered glycol. The other recycling method is fractional distillation.

Because of the problems with some recycling options, the ASTM committee on engine coolants is exploring a specification for recycled coolants.

With the increasing emphasis on recycling in general and environmental awareness in the United States, further growth in the recycling of antifreezes and deicers is anticipated. Emerging technologies may allow portable shop systems to properly clean spent antifreeze to provide a good base to build an acceptable antifreeze product.

BARBARA A. STEFL
KATHLEEN F. GEORGE
Union Carbide Chemicals and Plastics Company, Inc.

Engine Coolants, Cooling System Materials, and Components, SP-960, Society of Automotive Engineers, Warrendale, Pa., 1992.

1992 Annual Book of ASTM Standards, Vol. 15.05, American Society for Testing and Materials, Philadelphia, Pa., 1992.

Hazards Following Ground Deicing and Ground Operations in Conditions Conductive to Aircraft Icing, FAA Advisory Circular AC 20-117A, Federal Aviation Administration, Washington, D.C., Fall, 1993.

Aerospace Material Specifications, 1424 and 1428, Jan. 1993; *Aerospace Recommended Practice 4737,* (Oct. 1992), Society of Automotive Engineers, Warrendale, Pa.

ANTIMONY AND ANTIMONY ALLOYS

Antimony, Sb, belongs to Group 15 (VA) of the Periodic Table which also includes the elements arsenic and bismuth. It is in the second long period of the table between tin and tellurium. Antimony, which may exhibit a valence of +5, +3, 0, or −3 (see ANTIMONY COMPOUNDS), is classified as a nonmetal or metalloid, although it has metallic characteristics in the trivalent state. There are two stable antimony isotopes that are both abundant and have masses of 121 (57.25%) and 123 (42.75%).

History and Occurrence

The word antimony (from the Greek *anti* plus *monos*) means "a metal seldom found alone" and, in fact, native antimony is seldom found in nature because of its high affinity for sulfur and metallic elements such as copper, lead, and silver. Over a hundred naturally occurring minerals of antimony have been identified. Occasionally native metallic antimony is found; however, the most important source of the metal is the mineral stibnite (antimony trisulfide), Sb_2S_3. In areas where stibnite has been exposed to oxidation, it is converted to oxides of antimony. The important oxide minerals stibiconite, $Sb_2O_4{\cdot}H_2O$, cervantite, Sb_2O_4 or $Sb_2O_3{\cdot}Sb_2O_5$, valentinite, orthorhombic Sb_2O_3, and senarmontite, cubic Sb_2O_3; kermesite, $2Sb_2S_3{\cdot}Sb_2O_3$, an oxysulfide ore, is also of commercial importance.

World reserves of antimony are estimated to be from 4–5 million metric tons. Approximately 80% of the world's reserves are located in China, Bolivia, Russia, Thailand, the Republic of South Africa, and Mexico; China has the world's largest reserves. The United States possesses only 2% of the world's reserves. Most of the antimony produced in the United States is from the complex antimony deposits found in Idaho, Nevada, Alaska, and Montana. Ores of the complex deposits are mined primarily for lead, copper, zinc, or precious metals; antimony is a by-product of the treatment of these ores.

Properties

Physical properties of antimony are given in Table 1. Antimony, a silvery white, brittle, crystalline solid, is a poor conductor of electricity and heat. On solidification, pure antimony contracts 0.79 ± 0.14 vol %.

At room temperature, antimony is ordinarily quite stable and not readily attacked by air or moisture. At higher temperatures, antimony reacts with oxygen to form the oxides Sb_2O_3, Sb_2O_4, and Sb_2O_5.

Process Metallurgy

The antimony content of commercial ores ranges from 5 to 60%, and determines the method of treatment, either pyrometallurgical or hydrometallurgical. In general, the lowest grades of sulfide ores, 5–25% antimony, are volatilized as oxides; 25–40% antimony ores are smelted in a blast furnace; and 45–60% antimony ores are liquated or treated by iron precipitation. The blast furnace is generally used for mixed sulfide and oxide ores, and for oxidized ores containing up to

Table 1. Physical Properties of Antimony

Property	Value
at wt	121.75
mp, °C	630.8
bp, °C	1753
density at 25°C, kg/m^3	6684
crystal system	hexagonal (rhombohedral)
hardness, Mohs' scale	3.0–3.5
latent heat of fusion, J/mol[a]	19,874
latent heat of vaporization, J/mol[a]	195,250
coefficient of linear expansion at 20°C, μm/(m·°C)	8–11
electrical resistivity at 0°C, μΩ·cm	39
magnetic susceptibility at 20°C, cgs	-99.0×10^{-6}
specific heat at 25°C, J/(mol·K)[a]	25.2
thermal conductivity at 0°C, W/(m·K)	25.5

[a] To convert J to cal, divide by 4.184.

about 40% antimony; direct reduction is used for rich oxide ores. Some antimony ores are treated by leaching and electrowinning to recover the antimony. The concentrates may be leached directly or converted into a complex matte first. The most successful processes use an alkali hydroxide or sulfide as the solvent for antimony (see METALLURGY, EXTRACTIVE).

By-Product and Secondary Antimony. Antimony is often found associated with lead ores. The smelting and refining of these ores yield antimony-bearing flue, baghouse, and Cottrell dusts, drosses, and slags. These materials may be treated to recover elemental antimony or antimonial lead from which antimony oxide or sodium antimonate may be produced.

Recycling of antimony provides a large proportion of the domestic supply of antimony. Secondary antimony is obtained from the treatment of antimony-bearing lead and tin scrap such as battery plates, type metal, bearing metal, antimonial lead, etc. The scrap are charged into blast furnaces, reverberatory furnaces, or rotary furnaces, and an impure lead bullion or lead alloy is produced. Pure lead or antimony is then added to meet the specifications of the desired lead-antimony alloy.

Refining. The metal produced by a simple pyrometallurgical reduction is normally not pure enough for a commercial product and must be refined. Impurities present are usually lead, arsenic, sulfur, iron, and copper. The iron and copper concentrations may be lowered by treating the metal with stibnite or a mixture of sodium sulfate and charcoal to form an iron-bearing matte which is skimmed from the surface of the molten metal. The metal is then treated with an oxidizing flux consisting of caustic soda or sodium carbonate and niter (sodium nitrate) to remove the arsenic and sulfur. Lead cannot be readily removed from antimony, but material high in lead may be used in the production of antimony-bearing, lead-based alloys. The yield of refined antimony from the matting and fluxing technique is 85–90%.

Impure metal may be refined by electrolysis, although this procedure is not as economical as the pyrometallurgical treatment.

Economic Aspects

The United States is not self-sufficient in its requirements for antimony and is heavily dependent on imports of both ore and metal.

Production. In the 1980s, mine production in the United States accounted for only a small percentage of the annual domestic supply of antimony. An important component of the domestic supply is the recovery of antimony from old scrap; such as that recovered from the recycling of scrapped batteries. Other factors that influence the supply of secondary antimony are the percentage of available batteries being recycled, the demand for batteries, and the prices of lead. The remainder of the U.S. supply of antimony is imported.

Total smelter output in the United States has been growing steadily since 1982 because of the growth in antimony oxide production. Primary antimony metal output has decreased since the 1970s because of the falling demand for antimony metal, and the availability of low cost metal from China.

Imports and Exports. The availability of economical foreign sources of antimony, mainly from China, the Republic of South Africa, Mexico, Bolivia, and Russia, has resulted in an increase in the quantity of antimony imported for consumption. Since 1986, over 50% of the antimony imported by the United States originated in China. The majority of the antimony imported into the United States is now in the form of antimony oxide. Exports of antimony, as antimony oxide, increased during the 1980s.

Industrial Consumption. The total consumption of primary antimony fell during the period from 1970 to 1986 because of the declining demand for antimony in most types of metallic uses. Since 1986, the demand for primary antimony in antimonial lead has increased, probably because of an increase in demand for starting–lighting–ignition (SLI) batteries. Total consumption in nonmetallic uses has remained stable. However, an increasing proportion of this is made up of flame retardant uses. Currently, batteries and flame retardants are the two largest markets for antimony.

Uses and Specifications

Antimony in the unalloyed state is extremely brittle and is not easily fabricated. For this reason the use of the pure metal is restricted to ornamental applications.

Antimony Alloys. Approximately one-half of the total antimony demand is for metal used in antimony alloys. Antimonial lead is a term used to describe lead alloys containing antimony in proportions of up to 25%. Most commercial lead-antimony alloys have antimony contents less than 11%.

The largest application for antimonial lead is its use as a grid metal alloy in the lead acid storage battery (see BATTERIES, SECONDARY, LEAD-ACID). In the manufacture of grid metal, antimony imparts fluidity, increased creep resistance, and fatigue strength, as well as electrochemical stability to the lead which is particularly advantageous for battery plates required in heavy-duty cycling.

Demand for high performance SLI batteries has led to the development of smaller, lighter batteries that require less maintenance. The level of antimony in these batteries is 1.75–2.75% compared with 3–5% in conventional batteries. Lead alloys that contain no antimony have also been introduced. Hybrid batteries use a low antimony–lead alloy in the positive plate and a calcium–lead alloy in the negative plate.

Tin–antimony–copper and lead–antimony–tin white bearing alloys, commonly referred to as babbitt, are used to reduce friction and wear in machinery and help prevent failure by seizure or fatigue. These alloys exhibit good rubbing characteristics even under extreme operating conditions such as high loads, fatigue, or high temperatures. Addition of antimony to babbits increases strength and hardness.

Soldering is a method of joining two metallic surfaces by flowing between them a low melting point alloy. Many different alloys are used for this purpose; however, tin–lead alloys are the most widely used. Other metals are added in small amounts, depending on the desired properties. Antimony is added to increase the hardness of tin–lead alloys. Rising concern over the contamination of drinking water by lead has resulted in the use of lead-free alloys for soldering copper pipes.

Type metal, another tin–antimony–lead alloy, is used primarily in relief or letterpress printing. Antimony is added to increase hardness, minimize shrinkage, permit sharp definition, and reduce the melting point of the alloy. There has been a substantial decrease in the use of type metals as a result of the emergence of less expensive typesetting techniques.

Antimony hardens the lead used in the manufacture of small arms ammunition. Antimony alloyed with lead is also used in cable covering, sheet and pipe, and collapsible tubes. In these applications, antimony is utilized to increase strength and inhibit corrosion.

Semiconductor and Solar Cells. High purity (up to 99.9999%) antimony has a limited but important application in the manufacture of semiconductor devices (see SEMICONDUCTORS) and is utilized in such applications as infrared devices, diodes, and Hall-effect components. More recently, high efficiency solar cells have been produced that are comprised of two layers: one of gallium arsenide and the other of gallium antimonide (see SOLAR ENERGY). Antimony with a purity as low as 99.9 + % is an important alloying ingredient in the bismuth telluride, Bi_2Te_3, class of alloys which are used for thermoelectric cooling.

Antimony Compounds. The greatest use of antimony compounds is in flame retardants (qv) for plastics, paints, textiles, and rubber. Antimony trioxide and sodium antimonate are added to specialty glasses as decolorizing and fining agents, and are used as opacifiers in porcelain enamels. Antimony oxides are used as white pigments in paints, whereas antimony trisulfide and pentasulfide yield black, vermillion, yellow, and orange pigments. Camouflage paints contain antimony trisulfide which reflects infrared radiation. In the production of red rubber, antimony pentasulfide is used as a vulcanizing agent. Antimony compounds are also used in catalysts, pesticides, ammunition, and medicines (see ANTIMONY COMPOUNDS).

Health, Safety, and the Environment

Although metallic antimony may be handled freely without danger, it is recommended that direct skin contact with antimony and its alloys be avoided. Properly designed exhaust ventilation systems and/or approved respirators are required for operations that create dusts or fumes. As with other heavy metals, orderly housekeeping practice and good personal hygiene are necessary to prevent ingestion of (or exposure to) antimony.

Toxicology. Antimony is not known to cause cancer, birth defects, or affect reproduction in humans. However, antimony has been shown to cause lung cancer in laboratory animals that inhaled antimony-containing dusts and prolonged exposure to antimony can cause irritation of the eyes, skin, lungs, and stomach, in the form of vomiting and diarrhea. Antimony generally enters the body through the lungs where it is transported to the blood and then to other organs. The body eliminates antimony over several weeks through urine and feces.

Stibine (SbH_3), a highly toxic gas, can form when nascent hydrogen is present with antimony metal. Adequate safeguards against overexposure to this gas are advised when handling antimony and its alloys.

Antimony in the Environment. Antimony enters the environment through mining and processing of antimony-containing ores; in the production and working of antimony metals, alloys, compounds, and to a lesser extent, by incinerators and coal burning plants. Most of the released antimony, after being suspended in the air, ends up in the soil where it attaches strongly to iron, manganese, and aluminum. Because of the stability of antimony in aqueous systems, the concentration of dissolved antimony in rivers is very small (about 5 ppb antimony). Also, despite the fact that antimony is in the solder used in water pipes, it does not appear to dissolve in the drinking water.

Disposal of Antimony. Antimony and its compounds have been designated as priority pollutants by the EPA. As a result users, transporters, generators, and processors of antimony-containing material must comply with regulations of the Federal Resource Conservative and Recovery Act (RCRA).

T. LI
G. F. ARCHER
S. C. CARAPELLA, JR.
ASARCO Incorporated

C. Y. Wang, *Antimony,* Charles Griffin Co., London, 1952.

P. A. Plunkert, *Mineral Facts and Problems, Preprint of Antimony Chapter in Bulletin 675,* U.S. Dept. of the Interior, U.S. Bureau of Mines, Washington, D.C., 1985.

T. O. Llewellyn, *Annual Report, Antimony,* U.S. Dept. of the Interior, U.S. Bureau of Mines, Washington, D.C., 1991 and 1992.

ANTIMONY COMPOUNDS

Antimony is the fourth member of the nitrogen family and has a valence shell configuration of $5s^25p^3$. The utilization of these orbitals and, in some cases, of one or two $5d$ orbitals permits the existence of compounds in which the antimony atom forms three, four, five, or six covalent bonds.

A recommended exposure limit for antimony compounds of 0.5 mg/m^3 (as Sb) has been given. Disposal may be effected by washing residues down the drain at very high dilution unless prohibited by local regulations.

Inorganic Compounds of Antimony

Stibine. Stibine, SbH_3; mp, −88°C; bp, −18°C; density of the liquid at its bp, 2.204 g/mL; sp gr at 18°C with respect to air as 1.000, 4.344, is a colorless, poisonous gas having a disagreeable odor. It is the only well-characterized binary compound of antimony and hydrogen, although distibine, Sb_2H_4, has been reported.

Stibine is readily oxidized and may be ignited in the presence of air or oxygen to form water and antimony trioxide; at lower temperatures metallic antimony and water are slowly formed. Sulfur and selenium react with stibine at 100°C in the presence of light to form antimony trisulfide, Sb_2S_3, and antimony selenide, Sb_2Se_3, respectively. At elevated temperatures stibine reacts with most metals to give antimonides.

High purity stibine is used as an n-type, gas-phase dopant for Si in semiconductors.

Metallic Antimonides. Numerous binary compounds of antimony with metallic elements are known. The most important of these are indium antimonide, InSb, gallium antimonide, GaSb, and aluminum antimonide, AlSb, which find extensive use as semiconductors.

Antimony Trioxide. Antimony(III) oxide (antimony sesquioxide), Sb_2O_3, is dimorphic, existing in an orthorhombic modification; valentinite is colorless (sp gr 5.67) and exists in a cubic form; and senarmontite, Sb_4O_6, is also colorless (sp gr 5.2).

Common methods of preparation include direct combination of metallic antimony with air or oxygen, roasting of antimony trisulfide, and alkaline hydrolysis of an antimony trihalide and subsequent dehydration of the resulting hydrous oxide; when heated too vigorously in air, some of the Sb(III) is converted to Sb(V).

Antimony trioxide is insoluble in organic solvents and only very slightly soluble in water.

Antimony trioxide has numerous practical applications. Its principal use is as a flame retardant in textiles and plastics (see FLAME RETARDANTS; FLAME RETARDANTS FOR TEXTILES). It is also used as a stabilizer for plastics, as a catalyst and as an opacifier in glass (qv), ceramics (qv), and vitreous enamels (qv).

Antimony Tetroxide. Antimony(III,V) oxide, antimony dioxide, SbO_2 and Sb_2O_4, occurs in two modifications. Orthorhombic antimony tetroxide has long been known as the mineral cervantite, α-b_2O_4, (colorless, sp gr 4.07). More recently a monoclinic modification, β-Sb_2O_4, has been recognized.

Antimony tetroxide finds use as an oxidation catalyst, particularly for the dehydrogenation of olefins.

Antimony Pentoxide Hydrates. Antimonic acid (antimony(V) acid), and antimony(V) oxide, $Sb_2O_5 \cdot nH_2O$, are both hydrates of Sb_2O_5. Commercial antimony pentoxide is either hydrated Sb_2O_5 or at times $\beta-Sb_2O_4$.

Hydrated antimony pentoxide (antimonic acid) is essentially insoluble in nitric acid solutions and only very slightly soluble in water, but dissolves in aqueous KOH. Numerous hydrated antimonate(V) salts have been reported in which the Sb(V) atom is octahedrally surrounded by six OH groups.

Antimonic acid has been used as an ion-exchange material for a number of cations in acidic solution.

Antimony Trifluoride. Antimony(III) fluoride, SbF_3, is a white, crystalline orthorhombic solid; vapor pressure at the mp, 26.34 kPa (0.26 atm); Sb–F bond energy, 437.4 kJ (104.5 kcal). Antimony trifluoride is extremely soluble in water, the solubility being increased by the presence of hydrofluoric acid. It is also very soluble in polar solvents such as methanol, 154 g/100 mL, and acetone.

Antimony(III) fluoride may be prepared by treating antimony trioxide or trichloride with hydrofluoric acid.

Antimony trifluoride is used as a fluorinating agent to replace nonmetal chloride with fluorine.

Antimony Trichloride. Antimony(III) chloride, $SbCl_3$, is a colorless, crystalline solid, readily soluble in hydrochloric acid; water, ca 9% at 25°C, increasing with temperature; $CHCl_3$, 22%; CCl_4 13%; benzene; CS_2; and dioxane.

Antimony trichloride may be prepared by chlorination of antimony metal, Sb_2O_3, or Sb_2S_3, or by reaction of Sb_2O_3 with concentrated HCl.

Antimony trichloride is used as a catalyst or as a component of catalysts to effect polymerization of hydrocarbons and to chlorinate olefins. It is also used in hydrocracking of coal (qv) and heavy hydrocarbons (qv), as an analytic reagent for chloral, aromatic hydrocarbons, and vitamin A, and in the microscopic identification of drugs. Liquid $SbCl_3$ is used as a nonaqueous solvent.

Antimony Tribromide and Triiodide. Antimony(III) bromide, $SbBr_3$, is a colorless, crystalline solid.

Antimony(III) iodide, SbI_3, forms red rhombohedral crystals, intermediate in structure between a molecular and an ionic crystal.

Both antimony tribromide and antimony triiodide are prepared by reaction of the elements. Their chemistry is similar to that of $SbCl_3$ in that they readily hydrolyze, form complex halide ions, and form a wide variety of adducts with ethers, aldehydes, mercaptans, etc.

Antimony Pentafluoride. Antimony(V) fluoride, SbF_5, is a colorless, hygroscopic, viscous liquid that has SbF_6 units with *cis*-fluorines bridging to form polymeric units.

Antimony pentafluoride may be prepared by fluorination of SbF_3 or by treatment of $SbCl_5$ with HF.

Antimony pentafluoride is a strong Lewis acid and a good oxidizing and fluorinating agent.

Antimony Pentachloride. Antimony(V) chloride, $SbCl_5$, is a colorless, hygroscopic, oily liquid that is frequently yellow because of the presence of dissolved chlorine; it cannot be distilled at atmospheric pressure without decomposition, but the extrapolated normal boiling point is 176°C.

Antimony pentachloride is usually prepared by chlorination of molten $SbCl_3$.

Antimony pentachloride is a strong Lewis acid and a useful chlorine carrier.

Antimony Trisulfide.

Antimony(III) sulfide (antimony sesquisulfide), Sb_2S_3, exists as a black crystalline solid, stibnite, and as an amorphous red to yellow-orange powder. Stibnite melts at 550°C and has $\Delta H°_{f,298}$, −175 kJ/mol(−41.8 kcal/mol); $S°_{298}$, 182 J/(mol·K) [43.5 cal/(mol·K)]; for the amorphous solid $\Delta H°_{f,298}$ is −147 kJ/mol(−35.1 kcal/mol).

Amorphous Sb_2S_3 can be prepared by treating an $SbCl_3$ solution with H_2S or with sodium thiosulfate, or by heating metallic antimony or antimony trioxide with sulfur. Antimony trisulfide is almost insoluble in water but dissolves in concentrated hydrochloric acid or in excess caustic.

Antimony trisulfide is used in fireworks, in certain types of matches, as a pigment, and in the manufacture of ruby glass.

Antimony Pentasulfide. Antimony pentasulfide, Sb_2S_5, is a yellow to orange to red amorphous solid of indefinite composition. It is frequently given the formula Sb_2S_5, but actually consists of Sb(III) with a variable quantity of sulfur. The product is prepared commercially by the conversion of Sb_2S_3 to tetrathioantimonate(V) by boiling with sulfur in alkaline solution. The product is commercially known as golden sulfide of antimony, and is used in vulcanization to produce a red variety of rubber. The material is also used as a pigment and in fireworks.

Antimony(III) Salts. Concentrated acids dissolve trivalent antimony compounds. From the resulting solutions it is possible to crystallize normal and basic salts, eg, antimony(III) sulfate, $Sb_2(SO_4)_3$; antimonyl sulfate, $(SbO)_2SO_4$; antimony(III) phosphate, $SbPO_4$; antimony(III) acetate, $Sb(C_2H_3O_2)_3$; antimony(III) nitrate, $Sb(NO_3)_3$; and antimony(III) perchlorate trihydrate, $Sb(ClO_4)\cdot 3\ H_2O$. The normal salts all hydrolyze readily.

Compounds Containing Sb–O–C or Sb–S–C Linkages. A large number of compounds have been prepared in which the antimony atom is linked to carbon through an oxygen or sulfur atom.

By far the largest group of compounds containing the Sb–O–C linkage are those obtained by reaction of an antimony oxide with an α-hydroxy acid, *o*-dihydric phenol, sugar alcohol, or some other polyhydroxy compound containing at least two adjacent hydroxyl groups. The best known compound of this type is antimony potassium tartrate (tartar emetic). Tartar emetic has been used as an antiparasitic agent in medicine, as an insecticide, and as a mordant in the textile and leather industries.

Organoantimony Compounds

A wide variety of compounds containing the Sb–C bond is known. Organoantimony compounds can be broadly divided in Sb(III) and Sb(V) compounds.

Primary and Secondary Stibines Relatively few primary ($RSbH_2$) and secondary (R_2SbH) stibines are known. Methylstibine, CH_5Sb, ethylstibine, C_2H_7Sb, isopropylstibine, C_3H_9Sb, and butylstibine, $C_4H_{11}Sb$, have been prepared by the reduction of the corresponding alkyldichlorostibines using lithium aluminum hydride or sodium borohydride. All of the alkylstibines are thermally unstable, easily oxidizable, colorless liquids with strong alliaceous odors. Diethylstibine, $C_4H_{11}Sb$, di-*tert*-butylstibine, $C_8H_{19}Sb$, and dicyclohexylstibine, $C_{12}H_{23}Sb$, have been obtained by reduction of the corresponding dialkylhalostibines with lithium aluminum hydride. Dimethylstibine, C_2H_7Sb, was first prepared by the interaction of dimethylbromostibine, C_2H_6BrSb, and $LiHB(OCH_3)_2$ at temperatures below −40°C.

Phenylstibine, C_6H_7Sb, has been obtained by the reduction of phenyldiiodostibine, $C_6H_5I_2Sb$, or phenyldichlorostibine, $C_6H_5Cl_2Sb$, with lithium borohydride. Diphenylstibine, $C_{12}H_{11}Sb$, can be prepared by the interaction of diphenylchlorostibine, $C_{12}H_{11}Sb$, with either lithium borohydride or lithium aluminum hydride.

Tertiary Stibines A large number of trialkyl- and triarylstibines are known. They are usually prepared by the interaction of a reactive organometallic compound and an antimony trihalide, a halostibine, or a dihalostibine.

Tertiary stibines have been widely employed as ligands in a variety of transition-metal complexes, and they appear to have numerous uses in synthetic organic chemistry, eg, for the olefination of carbonyl compounds. They have also been used for the formation of semiconductors by the metal–organic chemical vapor deposition process, as catalysts or cocatalysts for a number of polymerization reactions, as ingredients of light-sensitive substances, and for many other industrial purposes.

Halostibines, Dihalostibines, and Related Compounds Alkyldichloro- and alkyldibromostibines are readily prepared by the alkylation of the corresponding antimony trihalide with an organolead reagent.

Dialkylchlorostibines are obtained in good yields by the cleavage of tetraalkyldistibines using sulfuryl chloride. Dialkylbromo- and dialkyliodostibines can similarly be prepared by the cleavage of tetraalkyldistibines using equimolar amounts of bromine or iodine.

Compounds $ArSbX_2$ and Ar_2SbX, in which X is Cl or Br and Ar is an aryl group, have also been obtained by the reduction of the corresponding stibonic or stibinic acids in hydrochloric or hydrobromic acid.

Distibines and Distibenes A considerable number of tetraalkyl- and tetraaryldistibines have been investigated. These are usually obtained by the reduction of a dialkyl- or diarylhalostibine with sodium hypophosphite or magnesium.

Although distibenes, the antimony analogues of azo compounds, have never been isolated as free, monomeric molecules, a tungsten complex, tritungsten pentadecacarbonyl [μ_3-η^2-diphenyldistibene], $C_{27}H_{10}O_{15}Sb_2W_3$, has been prepared by the reductive dehalogenation of phenyldichlorostibine.

Cyclic and Polymeric Substances Containing Antimony–Antimony Bonds A number of organoantimony compounds containing rings of four, five, or six antimony atoms have been prepared. The first such compound to be adequately characterized, tetrakis-1,2,3,4-*tert*-butyltetrastibetane, $C_{16}H_{36}Sb_4$, was obtained by the interaction of a di-(*tert*-butyl)stibide and iodine. More recently, it has been prepared by the dehalogenation of *tert*-butyldichlorostibine, $C_4H_9Cl_2Sb$, with magnesium.

Antimonin and Its Derivatives Antimonin(stibabenzene), C_5H_5Sb, the antimony analogue of pyridine, can be prepared by the dehydrohalogenation of a cyclic chlorostibine using 1,5-diazabicyclo[4.3.0]non-5-ene.

Stibonic and Stibinic Acids The stibonic and stibinic acids are polymeric compounds of unknown structure and are very weak acids.

Methylstibonic acid, the only alkylstibonic acid known with certainty, was not reported until 1990.

Dialkylstibinic acids can be readily prepared. The preferred method is the aqueous hydrolysis of trialkoxydialkylantimony compounds.

Stibine Oxides and Related Compounds Both aliphatic and aromatic stibine oxides, R_3SbO, or their hydrates, $R_3Sb(OH)_2$, are known. Thus both dihydroxotrimethylantimony, $C_3H_{11}O_2Sb$, and trimethylstibine oxide, C_3H_9OSb, have been prepared. The former may be readily obtained by passing an aqueous solution of dichlorotrimethylantimony, $C_3H_9Cl_2Sb$, through an anionic-exchange resin.

Both trialkyl- and triarylstibine sulfides and selenides are known.

Pentacovalent Antimony Halides and Related Compounds Antimony halides of the types $RSbX_4$, R_2SbX_3, R_3SbX_2, and R_4SbX, where X is a halogen, are known, but compounds of the first type have only been isolated and characterized where R is aryl. The aryl compounds are unstable substances which decompose on standing and are hydrolyzed in moist air. The chlorides are readily prepared by the action of hydrochloric acid on the corresponding arylstibonic acids.

Both dialkyl- and diaryltrihaloantimony compounds are known, although only a few dialkyl compounds have been described. The trichlorides have been obtained by the chlorination of either dialkylchlorostibines or tetraalkyldistibines with sulfuryl chloride.

In addition to the trihalo compounds, triacetatodiphenylantimony, $C_{18}H_{19}O_2Sb$, has been prepared. The best known of the halides are the trialkyldihalo- and triaryldihaloantimony compounds. The dichloro, dibromo, and diiodo compounds are generally prepared by direct halogenation of the corresponding tertiary stibines. The difluoro compounds are obtained by metathasis from the dichloro or dibromo compounds and silver fluoride.

Dichlorotriphenylantimony has been suggested as a flame retardant and as a catalyst for the polymerization of ethylene carbonate. Dibromotriphenylantimony has been used as a catalyst for the reaction between carbon dioxide and epoxides to form cyclic carbonates and for the oxidation of α-keto alcohols to diketones.

In addition to the trialkyldihalo- and triaryldihaloantimony compounds, mixed dihalo compounds such as chloroiodotriphenylantimony, $(C_6H_5)_3SbClI$, have been reported. Compounds of the type $R_3Sb(OSO_2R')_2$, and $R_3Sb(O_2CR')_2$, where R is an alkyl or an aryl group, have been prepared from a dihydroxide or oxide and the appropriate acid.

Tetraalkyl and tetraaryl compounds, R_4SbX, are well-known and are often referred to as stibonium salts. There is evidence, however, that most of the tetraaryl compounds contain pentacovalent antimony. The perchlorate $[(C_6H_5)_4Sb](ClO_4)$, however, is ionic. The tetraalkyl halides are readily prepared by quaternization of the corresponding tertiary stibines. The tetraaryl compounds can be prepared by employing anhydrous aluminum chloride. Both the tetraalkyl and tetraaryl compounds can be prepared by cleavage of the corresponding pentaalkyl- or pentaarylantimony by a halogen or a hydrogen halide.

Stibonium Ylids and Related Compounds In contrast to phosphorus and arsenic, only a few antimony ylids have been prepared. A new method, utilizing an organic copper compound as a catalyst, has resulted in the synthesis of a number of new antimony ylids:

$$(X)(Y)C{=}N_2 + (C_6H_5)_3Sb \xrightarrow[C_6H_6,\ 80^{\circ}C]{[(CF_3CO)_2CH]_2Cu} (X)(Y)C{=}Sb(C_6H_5)_3 + N_2$$

Among the ylids prepared by this method are those in which X and Y are $C_6H_5SO_2$, $4\text{-}CH_3C_6H_4SO_2$, or CH_3CO or where X is CH_3CO and Y is C_6H_5CO. These ylids are solids, stable in a dry atmosphere, but readily hydrolyzed by traces of moisture. Closely related to the ylids are imines of the type $R_3Sb{=}NR'$, where R is either an alkyl or an aryl group.

Organoantimony Compounds With Five Sb–C Bonds A number of pentaalkyl- and pentaalkenylantimony compounds have been prepared from tetraalkyl- or tetraalkenylstibonium halides and alkyl or alkenyllithium or Grignard reagents. Rather than using the stibonium halide, a trialkyl- or trialkenyldihaloantimony compound can be used, as in the preparation of pentavinylantimony, $C_{10}H_{15}Sb$. Pentaarylantimony compounds can be readily prepared in a similar way.

Antimony Compounds Used in Medicine

Compounds of antimony have been used as therapeutic agents for thousands of years. There is evidence that the ancient Egyptians used a preparation of naturally occurring stibnite as a cosmetic and as a prophylactic agent against diseases of the eyes. References to the use of antimonials as drugs occur in the Bible and in ancient writings from India, China, and Mexico. During the Middle Ages antimony compounds were widely employed in European medicine. The toxicity of these substances was noted at an early date, and in 1566 their use as drugs was condemned by the physicians of Paris and banned by an act of the Paris Parliament. This prohibition was withdrawn in 1667 when Louis XIV became convinced that his recovery from a serious illness (possibly typhoid) resulted from the administration of an antimony compound. The therapeutic use of antimonials reached a peak during the eighteenth or early nineteenth century. By the beginning of the twentieth century, antimony compounds were virtually obsolete in medical practice and were used only rarely for veterinary purposes.

The introduction of antimony compounds for the treatment of parasitic diseases (eg, American mucocutaneous leishmaniasis, visceral leishmaniasis, and oriental sore) is undoubtedly one of the important milestones in the history of therapeutics (see ANTIPARASITIC AGENTS).

Therapeutic compounds include antimony potassium tartrate (tartar emetic), antimony(V) sodium gluconate, and meglumine antimonate.

LEON D. FREEDMAN
G. O. DOAK
G. GILBERT LONG
North Carolina State University

T. O. Llewellyn, *Mineral Yearbook for 1988,* Vol. 1, *Metals and Minerals,* U.S. Department of the Interior, Bureau of Mines, U.S. Government Printing Office, Washington, D.C., 1990, pp. 125–132.

M. Mirbach and M. Wieber in M. Wieber, *Gmelin Handbook of Inorganic Chemistry, Sb Organoantimony Compounds, Part 5,* 8th ed., Springer-Verlag, Berlin, Germany, 1990.

L. D. Freedman and G. O. Doak, *Chem. Met.-Carbon Bond,* **5**, 397 (1989).

M. Wieber, *Gmelin Handbook of Inorganic Chemistry, Sb Organoantimony Compounds, Parts 1–4,* 8th ed., Springer-Verlag, Berlin, Germany, 1981, 1981, 1982, 1986.

ANTIOBESITY DRUGS

Obesity is a difficult condition to treat. Appetite suppressants have been widely used as an adjunct to dietary restriction and sympathomimetic amines have traditionally been used for this purpose. These agents have not proven particularly useful and frequently cause unacceptable side effects; hence their popularity has been waning for several years. The most promising newer drugs work through a serotoninergic mechanism and hold considerable promise at least for certain obese patients.

The majority of formerly obese patients eventually regain excess weight lost and thus a truly successful program must include behavior modification. Whereas no drugs are available for long-term use in a maintenance program, such as the thermogenic agents are being studied clinically.

Sympathomimetic Agents

A large variety of peripheral organ functions are affected by the sympathetic nervous systems including heart rate, cardiac contractile force, blood pressure, bronchopulmonary tone, and metabolism. In the central nervous system, sympathetic nervous stimulation results in increases in wakefulness, psychomotor activity, and reduction of appetite.

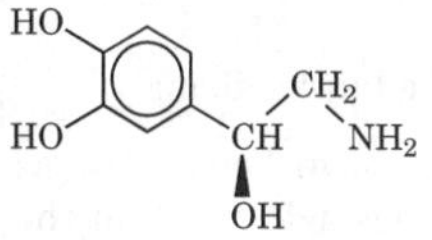

(1)

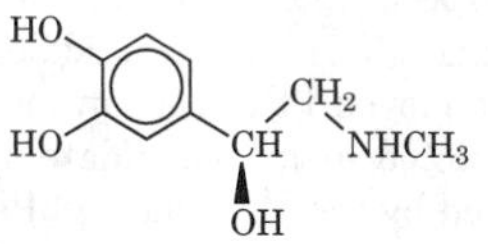

(2)

Compounds structurally related to the endogenous sympathomimetic amines norepinephrine (**1**) and epinephrine (**2**) have classically been employed as appetite suppressants. These compounds act primarily by indirect mechanisms involving displacement of norepinephrine from presynaptic nerve storage vesicles or by prevention of its reuptake, rather than by a direct effect at the receptor level. To a lesser extent, certain agents of this class affect dopaminergic or serotoninergic neurons. The overall pharmacological profile of members of the noncatechol sympathomimetic amine class depends on individual tissue distribution and mechanism of action.

Whereas they vary in degree, all of them share similar liabilities of cardiovascular side effects, the potential for central nervous system (CNS) stimulation, the development of tolerance, and abuse potential. All, with the exception of mazindol, are derivatives of phenethylamine. The introduction of an oxygen atom on the β-carbon of the side chain tends to reduce CNS stimulant properties without decreasing the anorectic activity.

Following the Federal Controlled Drug Act of 1970, compounds available in the United States were classified into one of five schedules according to medical utility and abuse potential. They include Schedule II: [amphetamine sulfate ($C_9H_{14}SO_4N$, Benzedrine), dextroamphetamine sulfate ($C_9H_{14}NO_4S$, Dexedrine), methamphetamine hydrochloride ($C_{10}H_{16}ClN$, Desoxyn), and phenmetrazine hydrochloride ($C_{11}H_{16}ClNO$, Preludin)]; Schedule III: [phendimetrazine tartrate ($C_{16}H_{23}NO_7$, Plegine) and benzphetamine hydrochloride ($C_{17}H_{22}ClN$, Didrex)]; and Schedule IV: [phenteramine hydrochloride ($C_{10}H_{16}NCl$, Fastin, Adipex-P, Ionamin), diethylpropion hydrochloride ($C_{13}H_{20}ClNO$, Tenuate, Tepanil), mazindol ($C_{16}H_{13}ClN_2O$, Sanorex, Mazanor), and phenylpropanolamine hydrochloride ($C_9H_{14}ClNO$, Dexatrim, Acutrim)]. Only drugs in Schedules III and IV are commonly employed for obesity and of them, phenylpropanolamine is available over the counter.

Seratoninergic Agents

Serotoninergic neurons utilize serotonin, $C_{10}H_{12}N_2O$, which is 5-hydroxytryptamine (5-HT), as a neurotransmitter. Central serotoninergic neurons play an important role in determining mood and food intake. Selective serotoninergic agents are receiving increased attention as antidepressants, antiemetics, and treatments for migraine. Fenfluramine ($C_{12}H_{17}ClF_3N$, Pondimim) is a close analogue of amphetamine. However, fenfluramine acts as a serotoninergic agonist primarily by inhibiting serotonin reuptake and to a lesser extent by stimulation of serotonin release. The *o*-isomer of fenfluramine is more selective than the racemate and is being developed clinically in the United States.

Fluoxetine ($C_{17}H_{19}ClF_3NO$, Prozac) is a serotonin reuptake inhibitor. It is more selective than fenfluramine and is being marketed by Eli Lilly as an antidepressant (see PSYCHOPHARMACOLOGICAL AGENTS). Since it was observed in animal studies that fluoxetine decreased meal size and promoted weight loss, these findings have been extended to humans in a number of clinical trials.

Gut Hormones

A number of peptidic hormones are involved in the regulation of food intake and meal termination. Prominent among these is cholecystokinin (CCK) that consists of a family gut and neurohormones of which the 33-amino acid peptide CCK-33 and the equipotent 8 amino acid peptide CCK-8 are the most abundant bioactive forms. Peripheral activities of CCK include stimulation of enzyme secretion of pancreatic acinar cells, inhibition of gastric emptying, and stimulation of vagal afferents terminating in the nucleus of the solitary tract and the area postrema. exogenous administration of CCK to animals or humans induces a sequence of behavior mimicking normal satiety and it has been proposed that CCK released endogenously from the small intestine is a signaling hormone mediating post-prandial satiety. Research is focused on developing selective CCK agonists which may be useful as adjuncts to dietary therapy by limiting meal size.

Lipid Adsorption Inhibitors

In a situation where maintenance or induction of weight loss is desired with a minimal impact on meal composition, an attractive approach to limiting caloric intake is to minimize the absorption of fat by inhibiting pancreatic lipase. Lipstatin is a pancreatic lipase inhibitor isolated from *Streptomyces toxytricini.* The corresponding tetrahydro derivative, (**3**) has been shown to decrease fat absorption in mice, rats, and monkeys and to lower the rate of body weight gain in rats on a high calorie diet. This compound is undergoing clinical trials and potentially represents a new approach to antiobesity therapy.

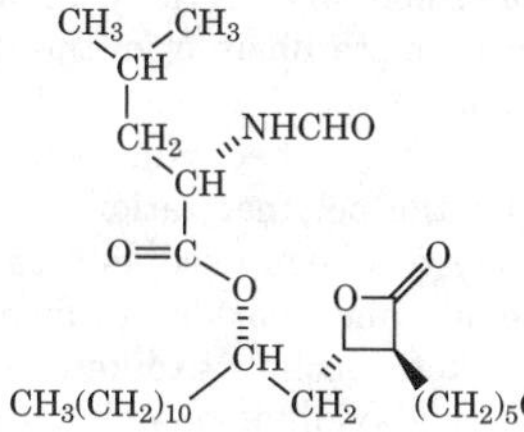

(3)

Thermogenic Agents

Drugs which stimulate or block peripheral adrenergic receptors have been in widespread use as, for example, antihypertensives, antianginal agents, and bronchodilators (see ANTIASTHMATIC AGENTS; CARDIOVASCULAR AGENTS). Detailed investigations with some of these has led to the proposal of a novel type of β-adrenergic receptor, termed the β_3-adrenergic receptor, which mediates catecholamine-induced lipolysis in brown adipose tissue leading to increases in thermogenesis. Compounds such as ICI 201, 651, $C_{19}H_{23}NO_6$ (**4**), and BLR 37, 344, $C_{19}H_{22}ClNO_4$ (**5**), have been identified which selectively stimulate this receptor and lead to increases in metabolic rate and thermogenesis in rodents with little effect on food intake. In humans such compounds could be expected to decrease the efficiency of food utilization and could be particularly helpful to those patients whose metabolic set-points favor an inappropriate body mass.

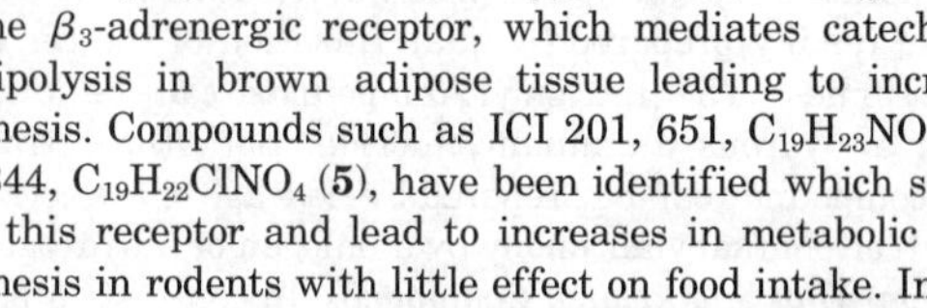

(4)

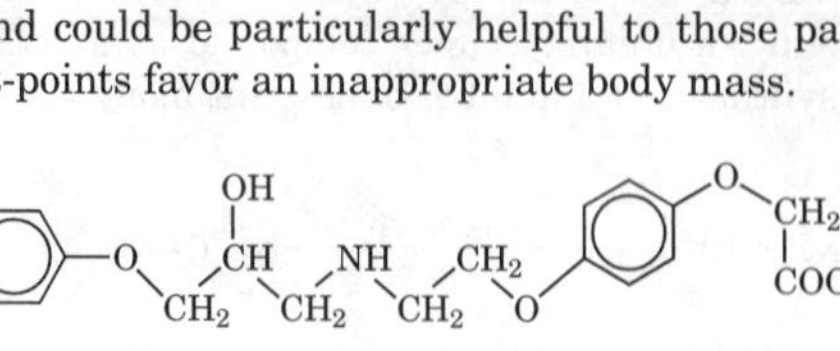

(5)

JEFFERSON W. TILLEY
Hoffmann-La Roche Inc.

I. D. Caterson, *Drugs* **39**(Suppl. 3), 20–32 (1990).

ANTIOXIDANTS

Antioxidants are used to retard the reaction of organic materials with atmospheric oxygen. Such reaction can cause degradation of the mechanical, aesthetic, and electrical properties of polymers; loss of flavor and development of rancidity in foods; and an increase in the viscosity, acidity, and formation of insolubles in lubricants. The need for antioxidants depends on the chemical composition of the substrate and the conditions of exposure. Relatively high concentrations of antioxidants are used to stabilize polymers such as natural rubber and polyunsaturated oils. Saturated polymers have greater oxidative stability and require relatively low concentrations of stabilizers.

Mechanism of Uninhibited Autoxidation

The mechanism by which an organic material (RH) undergoes autoxidation involves a free-radical chain reaction:

Initiation

$$\text{RH} \rightarrow \text{Free radicals, eg, R}\cdot, \text{ROO}\cdot, \text{RO}\cdot, \text{HO}\cdot \quad (1)$$

$$\text{ROOH} \rightarrow \text{RO}\cdot + \text{OH}\cdot \quad (2)$$

$$2\ \text{ROOH} \rightarrow \text{RO}\cdot + \text{ROO}\cdot + \text{H}_2\text{O} \quad (3)$$

$$\text{ROOR} \rightarrow 2\ \text{RO}\cdot \quad (4)$$

Propagation

$$\text{R}\cdot + \text{O}_2 \rightarrow \text{ROO}\cdot \quad (5)$$

$$\text{ROO}\cdot + \text{RH} \rightarrow \text{ROOH} + \text{R}\cdot \quad (6)$$

Termination

$$2\ \text{R}\cdot \rightarrow \text{R—R} \quad (7)$$

$$\text{ROO}\cdot + \text{R}\cdot \rightarrow \text{ROOR} \quad (8)$$

$$2\ \text{ROO}\cdot \rightarrow \text{nonradical products} \quad (9)$$

Radical Scavengers

Hydrogen-donating antioxidants (AH), such as hindered phenols and secondary aromatic amines, inhibit oxidation by competing with the organic substrate (RH) for peroxy radicals. This shortens the kinetic chain length of the propagation reactions.

$$\text{ROO}\cdot + \text{AH} \xrightarrow{k_{10}} \text{ROOH} + \text{A}\cdot \quad (10)$$

$$\text{ROO}\cdot + \text{RH} \xrightarrow{k_6} \text{ROOH} + \text{R}\cdot \quad (11)$$

Because k_1 is much larger than k_6, hydrogen-donating antioxidants generally can be used at low concentrations. The usual concentrations in saturated thermoplastic polymers range from 0.01 to 0.05%, based on the weight of the polymer. Higher concentrations, ie, ca 0.5–2%, are required in substrates that are highly sensitive to oxidation, such as unsaturated elastomers and ABS.

Peroxide Decomposers

Thermally induced homolytic decomposition of peroxides and hydroperoxides to free radicals (eqs. 2–4) increases the rate of oxidation. Decomposition to nonradical species removes hydroperoxides as potential sources of oxidation initiators. Most peroxide decomposers are derived from divalent sulfur and trivalent phosphorus.

Application of Antioxidants in Polymers

Nearly all polymeric materials require the addition of antioxidants to retain physical properties and to ensure an adequate service life. The selection of an antioxidant or system of antioxidants is dependent upon the polymer and the anticipated end use.

Polyolefins. Low concentrations of stabilizers (<0.01%) are added to polyethylene and polypropylene after synthesis and prior to isolation to retard oxidation of the polymers exposed to air.

These polymers are subjected to high temperatures, ca 300°C, during extrusion and injection molding. Processing stabilizers are used to decrease both the change in viscosity of the polymer melt and the development of color. A phosphite, such as tris(2,4-di-*tert*-butylphenyl)phosphite or bis(2,4-di-*tert*-butylphenyl)pentaerythritol diphosphite, in combination with a phenolic antioxidant such as octadecyl 3,5-di-*tert*-butyl-4-hydroxyhydrocinnamate, may be used.

Stabilization of Elastomers. Polyunsaturated elastomers are sensitive to oxidation. Stabilizers are added to the elastomers prior to vulcanization to protect the rubber during drying and storage. Nonstaining antioxidants such as butylated hydroxytoluene, 2,4-bis(octylthiomethyl)-6-methylphenol, 4,4′-bis(α, α-dimethylbenzyl)-diphenylamine, or a phosphite such as tris(nonylphenyl)phosphite may be used in concentrations ranging from 0.01% to 0.5%. Staining antioxidants such as *N*-isopropyl-*N*′-phenyl-*p*-phenylenediamine are preferred for the manufacture of tires (see also AMINES, AROMATIC–PHENYLENEDIAMINES).

Stabilization of Fuels and Lubricants. Gasoline and jet engine fuels contain unsaturated compounds that oxidize on storage, darken, and form gums and deposits. Radical scavengers such as 2,4-dimethyl-6-*tert*-butylphenol, 2,6-di-*tert*-butyl-*p*-cresol, 2,6-di-*tert*-butylphenol, and alkylated paraphenylene diamines are used in concentrations of about 5–10 ppm as stabilizers.

Lubricants for gasoline engines are required to withstand harsh conditions. Relatively high concentrations of primary antioxidants and synergists are used to stabilize lubricating oils. Up to 1% of a mixture of hindered phenols, of the type used for gasoline, and secondary aromatic amines, such as alkylated diphenylamine and alkylated phenyl-α-naphthylamine, are used as the primary antioxidants. About 1% of a synergist, zinc dialkyldithiophosphonate, is added as a peroxide decomposer. Zinc dialkyldithiophosphates are cost-effective, multifunctional additives.

Stabilization of Foods. The oxidation of food containing fats and oils results in a loss of sensory appeal and eventually rancidness. FDA regulations covering the use of direct food additives are stringent and few new materials have been regulated. A number of products in use today, such as citric acid, and α-tocopherol, are in the GRAS list. The most commonly used materials are butylated hydroxytoluene *n*-propyl gallate, α-tocopherol, and butylated hydroxyanisole.

Health and Safety Factors

Safety is assessed by subjecting the antioxidant to a series of animal toxicity tests, eg, oral, inhalation, eye, and skin tests. Mutagenicity tests are also carried out to determine possible or potential carcinogenicity. Stabilizers are being granulated and liquid products are receiving greater acceptance to minimize the inhalation of dust and to improve flow characteristics.

A number of antioxidants have been accepted by the FDA as indirect additives for polymers used in food applications. Acceptance is determined by subchronic or chronic toxicity in more than one animal species and by the concentration expected in the diet, based on the amount of the additive extracted from the polymer by typical foods or solvents that simulate food in their extractive effects.

MARTIN DEXTER
Consultant

G. Scott, *Atmospheric Oxidation and Antioxidants*, Elsevier, Amsterdam, the Netherlands, 1965.

J. R. Shelton in W. L. Hawkins, ed., *Polymer Stabilization*, Wiley-Interscience, New York, 1972, pp. 97–98.

ASTM Bull., 19103 (1990).

M. Dexter in J. I. Kroschwitz, ed., *Encyclopedia of Polymer Science and Engineering*, Vol. 2, Wiley-Interscience, New York, 1985, pp. 73–91.

ANTIOZONANTS

The term antiozonant, in its broadest sense, denotes any additive that protects an elastomer against ozone deterioration. Most frequently, the protective effect results from a reaction with ozone, in which case the term used is chemical antiozonant. Ozone is generated naturally by electrical discharge and also by solar radiation in the stratosphere. Ozone is also produced in urban centers by ultraviolet photolysis. Only a few pphm ozone in air can cause rubber cracking, which may destroy the usefulness of elastomer products. Ozone attacks any elastomer with backbone unsaturation. Degradation results from the reaction of ozone with rubber double bonds. Commercial antiozonants have been available since the early 1950s.

Desirable Properties of Antiozonants

A physical antiozonant must provide an effective barrier against the penetration of ozone at the rubber surface. A chemical antiozonant, on the other hand, must first of all be extremely reactive with ozone. The antiozonant should possess adequate solubility and diffusivity characteristics. Since ozone attack is a surface phenomenon, the antiozonant must migrate to the surface of the rubber to provide protection. The antiozonant should have no adverse effects on the rubber processing characteristics, eg, mixing, fabrication, vulcanization, or physical properties. The antiozonant should be effective under both static and dynamic conditions. It should persist in the rubber over its entire life cycle. For noncarbon-black-filled rubbers, the antiozonant must be nondiscoloring and nonstaining. The antiozonant should have a low toxicity and should be nonmutagenic, and the antiozonant should be acceptable economically, eg, have low manufacturing and end use costs.

Hydrocarbon Waxes

Waxes are one of the two general classes of commercial antiozonants. Waxes are derived from petroleum and are of two common types, paraffin and microcrystalline. Commercial waxes are usually blends of paraffin and microcrystalline waxes.

Chemical Antiozonants

Chemical antiozonants comprise the second general class of commercial antiozonants. Of the many compounds reported to be chemical antiozonants, nearly all contain nitrogen. Compound classes include derivatives of 1,2-dihydro-2,2,4-trimethylquinoline, *N*-substituted ureas or thioureas, substituted pyrroles, and nickel or zinc dithiocarbamate salts (see also ANTIOXIDANTS). The most effective antiozonants, however, are derivatives of *p*-phenylenediamine (*p*-PDA):

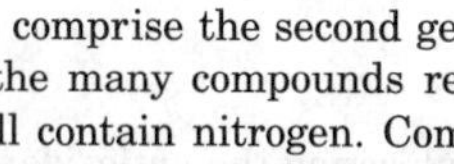

(see AMINES–AROMATIC, PHENYLENEDIAMINES). The commercial materials fall into three classes: *N,N′*-dialkyl-*p*-PDAs, *N*-alkyl-*N′*-aryl-*p*-PDAs, and *N,N′*-diaryl-*p*-PDAs.

The principal objection to *p*-PDA antiozonants is their staining characteristics. The lack of suitable alternative antiozonants for light-colored diene rubber articles is one of the outstanding problems in rubber technology. Few chemical antiozonants outside the class of *p*-PDAs have much commercial importance. One of the few exceptions to this rule is 6-ethoxy-1,2-dihydro-2,2,4-trimethylquinoline, one of the first commercial antiozonants.

Mechanism of Action of Chemical Antiozonants

Several theories have appeared in the literature regarding the mechanism of protection by *p*-PDA antiozonants. The scavenger theory states that the antiozonant diffuses to the surface and preferentially reacts with ozone, with the result that the rubber is not attacked until the antiozonant is exhausted. The protective film theory is similar, except that the ozone–antiozonant reaction products form a film on the surface that prevents attack. The relinking theory states that the antiozonant prevents scission of the ozonized rubber or recombines severed double bonds. A fourth theory states that the antiozonant reacts with the ozonized rubber or carbonyl oxide to give a low molecular weight, inert self-healing film on the surface.

The literature suggests that more than one mechanism may be operative for a given antiozonant, and that different mechanisms may be applicable to different types of antiozonants. All of the evidence, however, indicates that a combined scavenger–protective film mechanism is the most important.

Manufacture and Production

The *N,N′*-dialkyl-*p*-PDAs are manufactured by reductively alkylating *p*-PDA with ketones. Alternatively, these compounds can be prepared from the ketone and *p*-nitroaniline with catalytic hydrogenation.

Alternatives to Antiozonants

Any diene rubber article subjected to flexing, bending, or folding requires protection against ozone. Chemical antiozonants, primarily *p*-PDAs, are used in general-purpose commercial applications (mainly tires, hoses, flat belts, and transmission belts) where staining and discoloration are not serious problems. Waxes act as antiozonants for diene rubbers under conditions requiring little or no flexing, such as tie-down straps. Under these conditions, waxes are more effective than the *p*-PDA antiozonants. However, in applications involving dynamic stress where discoloration cannot be tolerated, alternatives to the traditional antiozonants must be used. Typical applications include white tire sidewalls, gaskets, weather stripping, gloves, and sporting goods. The two principal alternatives to antiozonants are ozone-resistant elastomers and blends of ozone-resistant elastomers and diene rubbers.

Health and Safety Factors

Current higher molecular weight *N,N′*-dialkyl or *N*-alkyl-*N′*-aryl derivatives are not primary skin irritants. A notable exception is *N*-(1-methylethyl)-N′-phenyl-*p*-PDA, which causes dermatitis. However, since some individuals are more sensitive than others, antiozonants should always be handled with care. In case of eye contact, flush well with water. Inhalation of rubber chemicals should be avoided, and respiratory equipment should be used in dusty areas.

Uses and Formulations

Chemical antiozonants are routinely used to protect diene rubbers (NR, IR, BR, SBR, NBR) against atmospheric ozone for extended periods of time. Large volumes are used in tire, belt, and hose applications. The *N*-alkyl-*N′*-aryl-*p*-PDAs have largely displaced the *N,N′*-dialkyl-*p*-PDAs as the materials of choice since they are less scorchy, more persistent, and more easily handled. The *N,N′*-diaryl-*p*-PDAs are more effective antiozonants for chloroprene rubber (CR).

ROBERT P. LATTIMER
ROBERT W. LAYER
The BFGoodrich Company
C. K. RHEE
The Uniroyal-Goodrich Tire Company

P. M. Lewis, *Polym. Degrad. Stab.* **15,** 33 (1986).

F. Jowett, *Rubber World* **188**(2), 24 (1983).

R. W. Layer and R. P. Lattimer, *Rubber Chem. Technol.* **63,** 426 (1990).

ANTIPARASITIC AGENTS

ANTHELMINTICS

Human infections caused by worms (helminths) represent one of the most important public health problems in the world. Helminths form three main categories or phyla: Platyhelminths, flatworms; Aschelminthes, roundworms; and Nemathelminthes, thorny-headed worms.

Over the last several years, substantial progress in the discovery and development of anthelmintic drugs has been made. Effective agents are available for most human gastrointestinal infections; however, drugs that are effective in treating the extraintestinal complications of many helminthic infections are still needed. Agents include praziquantel for the treatment of flukes (*C. sinensis, F. buski, P. westermani, S. haematobium, S. mansoni, S. japonicum*), tapeworms (*H. nana, T. solium, T. saginata, D. latum*), and larval stage (cysticercosis); oxamniquine for the treatment of flukes (*S. mansoni*); metrifonate for the treatment of blood flukes (*Schistosoma haematobium*); bithionol for the treatment of lung flukes (*Paragonimu westermani*), sheep liver flukes (*Fasciola hepatica*); tetrachloroethylene for the treatment of intestinal fluke (*Fasicolopsis buski*); niclosamide for the treatment of tapeworms (*T. saginata, T. solium,* and *D. latum*); quinacrine for the treatment of pork tapeworm (*Taenia solium*); piperazine citrate for the treatment of roundworms (*Ascaris lumbricoides*); diethylcarbamazine citrate for the treatment of lymphatic filariases (*Wuchereria bancrofti, Loa loa,* and other filaria); pyrantel pamoate for the treatment of pinworms (*Enterobius vermiculus*), roundworms (*Ascaris lumbricoides*), hookworms (*Ancyclostoma duodenale, Necator americanus*); mebendazole for the treatment of (whipworms (*Trichuris trichiura*), hookworms (*Ancylostoma duodenale, Nector americanus*), filariasis (*Mansonella perstans*), roundworm (*Ascaris lumbricoides*), Trichinosis (*Trichinella spiralis*); thiabendazole for the treatment of hookworms (*Trichostrongylus* sp., cutaneous larva migrans, *Angiostrongylus costaricensis,* and *Strongyloides stercoralis*); and ivermectin for the treatment of filariasis (*Onchocerca volvulus*).

Frequently, the treatment of helminthic diseases requires adjunct medication. Allergic reactions are commonly seen as a result of tissue invasion by worms or as a consequence of anthelmintic therapy. Antihistamines and corticosteroids may be necessary adjuncts to therapy. Anemia, indigestion, and secondary bacterial infections can also occur and may require concomitant therapy with hematopoietic drugs and appropriate antibiotics.

Treatment of Trematode Infections

Blood Flukes (Schistosomiasis). Three main species of blood flukes cause schistosomiasis in humans: *Schistosoma haematobium, S. mansoni,* and *S. japonicum.* The effectiveness of antischistosomal drugs depends on the reduction or arrest of egg production. Infection only occurs as a result of the penetration of the intact skin by free-swimming cercaria. After they develop to preadult forms in the skin and lung, the parasites migrate through host blood vessels to various tissues. Adult worms that are approximately 2 cm in length remain paired within the portal or intestinal vasculature of the host. The female, however, travels as far as possible toward the capillary beds to lay eggs, a portion of which embolize to the liver and induce granuloma formation. During the early stages of the disease, patients may experience fever, gastrointestinal distress, headache, and fatigue. In later stages of the disease, signs of hepatic fibrosis and ascites are seen. The number of worm pairs found in a patient may vary from a few to over 100. In untreated patients, adult worms are now generally assumed to live less than 5–10 years.

Praziquantel This drug **(1)**, $C_{19}H_{24}N_2O_2$, can be used to treat schistosomiasis caused by any one of the three major species. Praziquantel is an acylated pyrazino–isoquinoline derivative that has replaced the traditional (and more toxic) trivalent antimonial compounds.

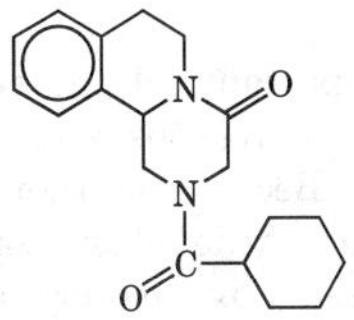

(1)

JOHN BECHER
CDC Drug Service

P. C. Beaver, R. C. Jung, and E. W. Cupp, *Clinical Parasitology,* 9th ed., Lea and Febiger, Philadelphia, Pa., 1984, 825 pp.

W. C. Campbell and R. S. Rew, eds., *Chemotherapy of Parasitic Diseases,* Plenum Press, New York, 1986, 655 pp.

J. F. Reynolds and co-workers, *Martindale, The Extra Pharmacopeia,* 29th ed., The Pharmaceutical Press, London, 1989, pp. 47–69, 505–521.

L. T. Webster in L. S. Goodman, A. Gilman, T. W. Rall, and F. Murad, eds., *The Pharmacological Basis of Therapeutics,* 7th ed., Macmillan, New York, 1985, pp. 1004–1028.

ANTIMYCOTICS

Since the 1960s, there has been such a rapid evolution in the pharmacotherapy of mycoses that three phases of the development of antimycotic agents can be distinguished: (*1*) compounds that existed before griseofulvin (1939); (*2*) compounds presently available on the market, including itraconazole (1987); (*3*) new antimycotics that are being studied now and will be prescribed tomorrow. Both milestones (griseofulvin and itraconazole) are intended for oral administration. Topical treatment of superficial dermatomycoses (trichophytosis, favus, and microsporosis) has been complemented by systemic oral treatment.

The world market for antimycotics in 1980 was valued at a total of $350 million. About 10% of this came from the sale of systemic (intravenous or oral) antifungals. In 1990, the antifungal world market was estimated at $1560 million of which almost $280 million was accounted for by systemic antifungals. This important increase in market value has induced more interest in research activities in the field of antifungals leading to more and more costly new treatment modalities. The five most prescribed antifungals today are miconazole, clotrimazole, ketoconazole, nystatin, and econazole. As this list indicates, among antifungal drugs the azoles are most significant; roughly 70% of all antifungals used today belong to this chemical category.

Topical Antimycotics

Older Compounds. Some antimycotics have been used for a long time; Whitfield's ointment is a typical example. The ointment usually contains 6% benzoic acid, $C_7H_6O_2$, and 3% salicylic acid, $C_7H_6O_3$. The action is attributed to the keratolytic effect of the salicylic acid and the direct effect of benzoic acid on the fungus. The main advantage of this ointment is its low price. This combination has now been replaced by more active modern antimycotics.

Antimycotic Antibiotics. Polyene antimycotics include nystatin, natamycin ($C_{33}H_{47}NO_{13}$), amphotericin B ($C_{47}H_{73}NO_{17}$), candicidin, filipin, homycin, etruscomycin ($C_{36}H_{53}NO_{13}$), and trichomycin.

Nystatin is mainly used to treat vaginal and oral infections and localized skin lesions, including *Candida* intertrigo and *Candida* nappy dermatitis. It may also be used as prophylaxis during treatment with antibiotics.

Natamycin is indicated for skin and nail infections with *C. albicans,* intertrigo and fissures at the corners of the mouth (perleche) caused by *C. albicans, Candida vulvitis,* and vaginitis. Natamycin plays an important role in the treatment of mycotic keratitis. Natamycin also appears to be active against the protozoan *Trichomonas vaginalis.* Side effects are nausea or diarrhea during oral treatment.

Amphotericin B, an important polyene antibiotic, is administered almost exclusively via the intravenous route.

Synthetic Antimycotics. Because of their limited activity, small spectrum, and side effects, the older topical antimycotics have generally been surpassed by newer antimycotic chemotherapeutic agents. These newer antimycotics for topical use include the azole derivatives clotrimazole ($C_{22}H_{17}ClN_2$), miconazole ($C_{18}H_{14}Cl_4N_2O$), econazole ($C_{18}H_{15}C_3N_2O$), isoconazole ($C_{18}H_{14}Cl_4N_2O$), ketoconazole ($C_{26}H_{28}Cl_2N_4O_4$), butoconazole ($C_{19}H_{17}Cl_3N_2S$), oxiconazole ($C_{18}H_{13}Cl_4N_3O$), omoconazole ($C_{20}H_{17}Cl_3N_2O_2$), sulconazole ($C_{18}H_{15}Cl_3N_2S$), itraconazole ($C_{35}H_{38}Cl_2N_8O_4$), fluconazole ($C_{13}H_{12}F_2N_6O$), saperconazole, fenticonazole ($C_{24}H_{20}Cl_2N_2OS$), bifonazole ($C_{22}H_{18}N_2$), terconazole ($C_{26}H_{31}Cl_2N_5O_3$), tioconazole ($C_{16}H_{13}Cl_3N_2OS$), and zinoconazole. The introduction of the azole derivatives represents a milestone in the treatment of mycoses.

Clotrimazole (**1**) or 1-(*o*-chloro-α, α-diphenylbenzyl) imidazole is a water-insoluble antimycotic for topical application, with a broad-spectrum activity against mycoses of the skin and the vagina.

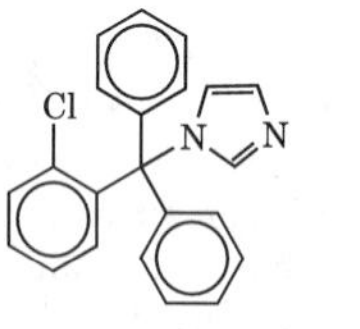

(**1**)

Miconazole is indicated for infections of skin and nails due to dermatophytes or *Candida* species and vulvovaginal infections due to *Candida* species. Miconazole is used orally for prophylactic purposes in patients who have been treated with cytostatics or immunosuppressants, in particular, to prevent candidosis of the mouth and digestive tract. Side effects include possible nausea during oral treatment.

Econazole and isoconazole are structurally very similar to miconazole.

Naftifine (**2**) belongs to the allylamines, a new class of antimycotics. It is used to treat superficial mycoses and is particularly active against dermatophytes.

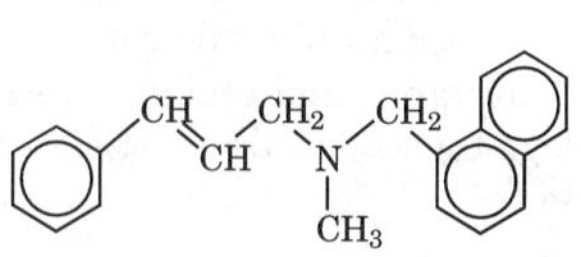

(**2**)

Ciclopiroxolamine is used mainly in the treatment of mycoses of the skin and nails.

Initial observations indicating that oral administration of ketoconazole produced good results in seborrheic eczema and dandruff, led to the development of a 2% cream and a 2% shampoo (scalp gel) of this antimycotic.

Haloprogin is active against dermatophytes.

Systemic Antimycotics

Systemic antimycotics include potassium iodide, griseofulvin, amphotericin B, flucytosin, miconazole, ketoconazole, itraconazole, and fluconazole.

Future Antimycotics for Systemic Treatment. Two new antimycotics for systemic use are a triazole and fluoride analogue of itraconazole, named saperconazole, which is in development. The second molecule is currently marketed as terbinafine, an allylamine, $C_{21}H_{25}N$, that appears to be particularly active against dermatophytes, just like topical naftifine.

GEERT CAUWENBERGH
Janssen Research Foundation

D. Kerridge and L. E. Whelan in A. P. J. Trinci and J. F. Ryley, eds., *Mode of Action of Antifungal Agents,* British Mycological Society Symposium 9, Cambridge University Press, Cambridge, U.K., 1984, pp. 343–375.

E. F. Gale, E. Cundliffe, P. E. Reynolds, M. H. Richmond, and M. J. Waring, *The Molecular Basis of Antibiotic Action,* 2nd ed., John Wiley & Sons, Inc., 1981, pp. 201–219.

G. Medoff and G. A. Kobayashi in D. C. E. Speller, ed., *Antifungal Chemotherapy,* John Wiley & Sons, Inc., New York, 1980, pp. 3–33.

A. Polak and D. M. Dixon in R. A. Fromtling, ed., *Recent Trends in the Discovery, Development and Evaluation of Antifungal Agents,* J. R. Prous Science Publishers, SA, 1987, pp. 555–573.

ANTIPROTOZOALS

Diseases caused by protozoa affect more people than those brought on by any other biological cause. There are over 60,000 species of protozoa, of which some 10,000 are parasitic. In humans, protozoa chiefly infect the gastrointestinal tract, vagina, urethra, blood, and blood-forming organs. Malaria is the most widespread of the protozoan diseases, and is responsible for the greatest number of deaths due to infection. Although protozoan diseases occur throughout the world, they impact most severely on people of tropical areas where there is widespread malnutrition, minimal health education, and poor sanitation. Fatal protozoal infections are occurring to an increased extent in the more developed countries among immunosuppressed individuals, especially those with AIDS.

Amebiasis

Amebiasis, a widespread disease of humans, causes an estimated 500 million cases (without the inclusion of China) annually. It is believed to be the third leading cause of death due to parasites. The disease is caused by pathogenic strains of the protozoan *Entamoeba histolytica,* which exists both in a stable infective cyst form and in a more fragile, potentially invasive, trophozoite form. Amebiasis antiprotozoal agents include metronidazole ($C_6H_9N_3O_3$), iodoquinol ($C_9H_5I_2NO$), oxytetracycline ($C_{22}H_{24}N_2O_9$), emetine (HCl) ($C_{29}H_{40}N_2O_4 \cdot 2\ ClH$), dehydroemetine ($C_{29}H_{38}N_2O_4$), chloroquine (diphosphate) ($C_8H_{26}ClN_3 \cdot 2\ H_3O_4P$), tetracycline ($C_{22}H_{24}N_2O_8$), paromomycin ($C_{23}H_{45}N_5O_{14}$), diloxanide furoate ($C_{14}H_{11}Cl_2NO_4$), niridazole ($C_6H_6N_4O_3S$), tinidazole ($C_8H_{13}N_3O_4S$), gossypol ($C_{30}H_{30}O_8$), bithionol ($C_{12}H_6Cl_4O_2S$), pimaricin ($C_{33}H_{47}NO_{13}$), amphotericin B ($C_{47}H_{73}NO_{17}$), miconazole ($C_{18}H_{14}Cl_4N_2O$), rifampin ($C_{43}H_{58}N_4O_{12}$), and minocycline ($C_{28}H_{27}N_3O_7 \cdot ClH$.

Anaplasmosis

Anaplasmosis is a tick-borne disease of cattle and other ruminants, such as deer in the western United States and elk in the former USSR. It is caused by the protozoan *Anaplasma marginale.* Horseflies are also vectors of the disease. Anaplasmosis antiprotozoal agents include imidocarb ($C_{19}H_{20}N_6O$), chlortetracycline ($C_{22}H_{23}ClN_2O_8$), oxytetracycline ($C_{22}H_{24}N_2O_9$), chloroquine, and amodiaquine ($C_{20}H_{22}ClN_3O$).

Babesiasis

Babesiasis (babesiosis, piroplasmosis), primarily a disease of cattle, is caused by a protozoan *Babesia microti,* related to the malaria parasite. Humans do not appear to be reservoir for the infection and are not commonly affected by the disease. However, they can acquire babesiasis by blood transfusion. Babesiasis antiprotozoal agents include quinine (sulfate), ($C_{20}H_{24}N_2O_2{\cdot}1/2\ H_2O_4S$) plus clindamycin ($C_{18}H_{33}ClN_2O_5S$), chloroquine pentamidine ($C_{19}H_{24}N_4O_2$), diminazene aceturate ($C_{22}H_{29}N_9O_6$), imidocarb, amicarbalide ($C_{15}H_{16}N_6O$), phenamidine isethionate ($C_{18}H_{26}N_4O_9S_2$), and babesan ($C_{19}H_{14}N_4O$).

Balantidiasis

Balantidiasis (balantidiosis, balantidial dysentery), an intestinal disease seen almost worldwide, is caused by the large ciliated protozoan, *Balantidium coli.* The organism is usually found in the lumen of the large intestine of humans and animals. The infection can be cured most readily with tetracycline, oxytetracycline, or chlorotetracycline. Metronidazole and iodoquinol are also effective. Additional effective drugs include paromomycin, tinidazole, sulfadiazine, and carbarsone (*p*-ureidobenzenearsonic acid, $C_7H_9AsN_2O_4$).

Coccidiosis

Coccidiosis is a widespread disease that occurs most often in fowl, such as chickens and turkeys, and other farm animals (cows, sheep, swine, horses, and rabbits). It is caused by a combination of infective *Eimeria* protozoa.

Anticoccidial drugs are administered in the drinking water or feed of poultry and are primarily prophylactic, rather than curative, in nature. The most common ones in use are salinomycin ($C_{42}H_{70}O_{11}$) and the sodium salt of monensin ($C_{36}H_{62}O_{11}$). Other coccidiosis antiprotozoal agents include lasalocid A ($C_{34}H_{54}O_8$), maduramicin ($C_{47}H_{83}NO_{17}$), sulfadimethoxine ($C_{12}H_{14}N_4O_4S$), sulfaquinoxaline ($C_{14}H_{12}N_4O_2S$), roxarsone ($C_6H_6AsNO_6$), nitrofurazone ($C_6H_6N_4O_4$), diclazuril ($C_{17}H_9Cl_3N_4O_2$), toltrazuril ($C_{18}H_{14}F_3N_3O_4S$), nicarbazin ($C_{13}H_{10}N_4O_5{\cdot}C_6H_8N_2O$), pyrimethamine, clopidol ($C_7H_7Cl_2NO$), robenidine ($C_{15}H_{13}Cl_2N_5$), halofuginone ($C_{16}H_{17}BrClN_3O_3$), amprolium ($C_{14}H_{19}N_4Cl$), lincomycin ($C_{18}H_{34}N_2O_6S$). Cryptosporidiosis, caused by a *Cryptosporidium* species is taxonomically related. It occurs in humans as well as animals, and especially in those with a suppressed immune system (eg, AIDS). Spiramycin is an effective treatment.

Giardiasis

Giardiasis is a waterborne enteric disease of protozoan origin that occurs throughout the world. It is the most prevalent protozoal disease found in humans in the United States; the Rocky Mountain region is a particularly highly endemic area. Infection of the small intestine is caused by the flagellated protozoan, *Giardia lamblia.*

Treatment of giardiasis is most successful with quinacrine ($C_{23}H_{30}ClN_3O$) or its hydrochloride. Other giardiasis antiprotozoal agents include metronidazole, furazolidone ($C_8H_7N_3O_5$), paromomycin, tinidazole and acranil ($C_{21}H_{28}Cl_3N_3O_2$).

Hexamitosis

Hexamitosis is a disease of chickens, turkeys, quail, and pheasants in which there is an infectious catarrhal enteritis in the duodenum and small intestine. The disease, caused by the protozoan *Hexamita meleagridis,* occurs in the United States, the United Kingdom, South America, and parts of Europe.

There is no effective treatment for hexamitosis. Penicillin, oxytetracycline, chlorotetracycline, enheptin (2-amino-5-nitrothiazole), and streptomycin sulfate were found to have limited value.

Histomoniasis

Histomoniasis (histomonas, enterohepatitis, blackhead disease), caused by *Histomonas meleagridis,* is primarily an affliction of chickens and turkeys, but it also affects wild populations of peafowl, guinea fowl, pheasant, grouse, quail, and partridge.

A favored treatment for histomoniasis is dimetridazole ($C_5H_7N_3O_2$). Other hexamitosis and histomoniasis antiprotozoal agents include enheptin ($C_3H_3N_3O_2S$), streptomycin ($C_{21}H_{39}N_7O_{12}$), ronidazole ($C_6H_8N_4O_4$), dimetridazole ($C_5H_7N_3O_2$), acetylenheptin ($C_5H_5N_3O_3S$), nithiazide ($C_6H_8N_4O_3S$), and ipronidazole ($C_7H_{11}N_3O_2$).

Leishmaniasis

Leishmaniasis affects some 12 million humans annually in an area where 350 million are at risk. It is a complex of at least two protozoan (species of *Leishmania*) diseases, consisting primarily of cutaneous and visceral forms. A mucocutaneous form is considered by some to be another distinct variety. Leishmaniasis can often be fatal, especially in the visceral form. The vector is the female sand fly of the genus Phlebotomus.

Pentavalent antimony preparations constitute the primary treatments for all forms of leishmaniasis, the most important of which are sodium stibogluconate ($C_{12}H_{20}O_{17}Sb_2{\cdot}9H_2O{\cdot}3\ Na$) and glucantime (*N*-methylglucamine antimonate, $C_7H_{18}NO_8Sb$, meglumine antimonate). Other leishmaniasis antiprotozoal agents include pentamidine, stilbamidine ($C_{16}H_{16}N_4$), amphotericin B, allopurinol ($C_5H_4N_4O$), ketoconazole ($C_{26}H_{28}Cl_2N_4O_4$), and paromomycin.

Malaria

Malaria affects an estimated 270 million people and causes 2–3 million deaths annually, approximately one million of which occur in children under the age of five. While primarily an affliction of the tropics and subtropics, it has occurred as far north as the Arctic Circle. The responsible protozoa are from the genus *Plasmodium.* The species that infect humans are *P. falcipanium, P. vivax, P. malariae,* and *P. ovale.* The vector is the female *Anopheles* mosquito.

Antimalarials Antimalarials can be categorized according to their mode of action.

Tissue Schizonticides. These eradicate the liver stages of the parasite and thereby prevent their entry into the blood. As a class, therefore, they are useful for prophylaxis. Some tissue schizonticides can act on the long-lived tissue forms (hypnozoites) of *P. vivax* and *P. ovale* and thus can cure the latter infections by preventing relapses.

Blood Schizonticides. These destroy the erythrocytic stages of the parasites and are useful for the clinical cure of falciparum malaria or suppression of relapsing infections.

Gametocytocides. These annihilate the sexual forms of the plasmodia (gametocytes) and also destroy the stages of the parasites in the *Anopheles* mosquito.

Sporontocides. These act against the sporozoites and oocysts in the mosquito and thereby prevent the transmission of the disease.

It should be noted that drugs may operate by more than one mechanism, and may possess a specific mode of action against one species of plasmodium but lack efficacy against others. In addition, antimalarial drugs may be classified according to their structural types.

Antimalarials include 4-quinolinemethanols [quinine, quinidine (sulfate) ($C_{20}H_{24}N_2O_2{\cdot}1/2\ H_2O_4S$), and mefloquine ($C_{17}H_{16}F_6N_2O$)], phenanthrene-methanols [halofantrine ($C_{26}H_{30}Cl_2F_3NO$)], 4-aminoquinolines [chloroquine, amodiaquine, and hydroxychloroquine ($C_{18}H_{26}ClN_3O$)], 8-aminoquinolines [pamaquine ($C_{19}H_{29}N_3O$) and primaquine (phosphate) ($C_{15}H_{21}N_3O{\cdot}2\ H_3O_4P$)], antifolates [dapsome ($C_{12}H_{12}N_2O_2S$), sulfadoxine ($C_{12}H_{14}N_4O_4S$), sulfadiazine, sulfalene ($C_{11}H_{12}N_4O_3S$), pyrimethamine, trimethoprim ($C_{14}H_{18}N_4O_3$), chlorguanide ($C_{11}H_{16}ClN_5$), and cycloguanil ($C_{11}H_{14}ClN_5$)]. Other antimalarial drugs include menoctone ($C_{24}H_{32}O_3$), pyronaridine ($C_{29}H_{32}ClN_5O_2$, tetracycline, doxycycline ($C_{22}H_{24}N_2O_8$), minocycline chloramphenicol ($C_{11}H_{12}Cl_2N_2O_5$), clindamycin, febrifugine ($C_{16}H_{19}N_3O_3$), artemisinin ($C_{15}H_{22}O_5$), artemether ($C_{16}H_{26}O_5$), arteether ($C_{17}H_{28}O_5$), the sodium salt of artesunic acid ($C_{19}H_{28}O_8$), and the sodium salt of artelinic acid ($C_{23}H_{30}O_7$).

Pneumocystosis

The organism responsible for pneumocystosis in humans, *Pneumocystis carinii,* having the characteristics of both protozoan and fungi, has defied simple taxonomic classification. Recent evidence suggests that it falls into the fungi category. Acute pneumocystosis rarely strikes healthy individuals, although the organism is harbored by a wide variety of animals and most people without any apparent adverse effect. *P. carinii* becomes active only in those individuals who have a serious impairment of their immunologic systems (eg, AIDS patients).

The drug of choice is the antifolate mixture, trimethoprim and sulfamethoxazole ($C_{10}H_{11}N_3O_3S$) given orally or intravenously. Other pneumocystosis antiprotozoal agents are pyrimethamine plus sulfadiazine or sulfadoxine, pentamidine isethionate, eflornithine ($C_6H_{12}F_2N_2O_2$), and a combination of clindamycin and primaquine.

Theileriasis

Theileriasis (theileriosis) is a tick-transmitted protozoan disease of cattle seen primarily in central and eastern Africa. The disease not only affects domesticated cattle, but also ox, zebu, water buffalo, and African buffalo. The most significant form of the disease is caused by the protozoan *Theileria parva (Piroplasma kochi, Piroplasma parvum, Theileria kochi, Theileria lawrencei)* and is known as African Coast fever, East Coast fever, bovine theileriasis, or Corridor disease.

There are no fully effective therapeutic agents for the treatment of theileriasis. Chlorotetracycline and oxytetracycline have therapeutic activity during the incubation period. Pamaquine was reported to have a specific effect on the erythrocyctic forms. Other theileriasis antiprotozoal agents include imidocarb, halofuginone, menoctone, parvaquone ($C_{16}H_{16}O_3$), buparvaquone ($C_{21}H_{26}O_3$), and methotrexate ($C_{20}H_{22}N_8O_5$).

Toxoplasmosis

Toxoplasmosis, a coccidial disease, affects both humans and animals throughout the world. Approximately one-third of the population has antibodies to it but is asymptomatic. Although rare in the past. Toxoplasmosis is an increasingly important opportunistic disease that afflicts immunocompromised hosts such as post-transplant patients taking immunosuppressive drugs and AIDS patients.

Toxoplasmosis is caused by the protozoan *Toxoplasma gondii.* In humans the disease is generally acquired by eating inadequately cooked meat or by association with infected felines.

The cyst of *T. gondii* is persistant, making long-term therapy a requirement in patients whose immune systems are compromised. Treatment of cerebral or ocular toxoplasmosis is generally accomplished with antifolate compounds consisting of a combination of pyrimethamine and a long-acting sulfonamide such as sulfadiazine. Sometimes the latter is replaced by triple sulfonamides or sulfalene. There is evidence that the related mixture of trimethoprim plus sulfamethoxazole is nearly as effective but less toxic than the pyrimethamine combinations. Spiramycin and Clindamycin have also been used.

Trichomoniasis

Trichomoniasis is a widespread disease of the urogenital tract caused by the protozoan *Trichomonas vaginalis.* This sexually transmitted disease affects humans only and has no known animal reservoirs.

Treatment by the 5-nitroimidazole class of antiprotozoal agents can destroy most strains of *T. vaginalis;* however, some strains have been reported to be resistant to these drugs. The agent of choice is orally administered metronidazole. Other members of this therapeutic group include niridazole, tinidazole, and ornidazole ($C_7H_{10}ClN_3O_3$), but in general, they offer no distinct advantage over metronidazole.

Trypanosomiasis

The disease trypanosomiasis has two geographically delineated forms, the so-called African and America varieties. Because of their distinct differences they will be discussed separately.

African Trypanosomiasis. According to the World Health Organization, African trypansomiasis, caused by *Trypanosoma brucci,* affects some 25,000 people annually in 36 sub-Saharan countries where 50 million inhabitants are at risk. It is the cause of sleeping sickness, a disease transmitted by the bite of either the female or male tsetse fly.

The early stages of African trypanosomiasis, ie, before central nervous system involvement and the sleeping sickness take effect, can be best treated with intravenous suramin sodium ($C_{51}H_{34}N_6Na_6O_{23}S_6$). Suramin plus tryparsamide and pentamidine have also been used.

For late-stage disease, in which the central nervous system is implicated, the compound of choice until recently was melarsoprol ($C_{12}H_{15}AsN_6OS_2$) for the *T. b. gambiense* or *T. b. rhodesiense.*

The first new treatment in 40 years that is showing great promise is eflornithine. It is an irreversible inhibitor of ornithine decarboxylase that blocks putrescine biosynthesis in *T. brucei in vitro* and inhibits the growth and multiplication of the parasite *in vivo.* A mixture of eflornithine and nifurtimox ($C_{10}H_{13}N_3O_5S$) shows good activity against *T. b. gambiense.*

American Trypanosomiasis. American trypanosomiasis, known as Chagas' Disease, is limited to South and Central America, where it affects 16–18 million people annually in an area where 90 million are at risk. The protozoan causative agent, *Trypanosoma cruzi,* is harbored in domesticated animals, such as dogs, cats, and pigs, as well as wild animals, eg, rats, bats, foxes, opossums, and monkeys.

The nitrofuran nifurtimox is the most effective drug against *T. cruzi,* being cidal against the trypomastigote and amastigote forms. Benznidazole ($C_{12}H_{12}N_4O_3$), an alternative drug, can prevent the spread of the parasites from one tissue to another although relapses are common. Primaquine is effective against extracellular trypanosomes in the blood.

DANIEL L. KLAYMAN
Walter Reed Army Institute of Research

G. T. Strickland, ed., *Hunter's Tropical Medicine,* 7th ed., W. B. Saunders, Philadelphia, Pa., 1991, p. 547.

J. H. Leech, M. A. Sande, and R. K. Root, eds., *Parasitic Infections,* Churchill Livingston, New York, 1988.

W. C. Campbell and R. S. Rew, *Chemotherapy of Parasitic Diseases,* Plenum Press, New York, 1986.

G. C. Cook, *Parasitic Disease in Clinical Practice,* Springer-Verlag, London, 1990.

AVERMECTINS

In 1976 scientists at the Merck Corporation discovered a complex of eight closely related natural products, subsequently named avermectins A_{1a} through B_{2b}, in a culture of *Streptomyces avermitilis* MA-4680 (NRRL8165) originating from a soil sample collected at Kawana, Ito City, Shizuoka Prefecture, Japan and isolated by the Kitasato Institute. Their structures are shown in Figure 1. They are among the most potent anthelmintic, insecticidal, and acaricidal compounds known.

Avermectin	Molecular formula
A_{1a}	$C_{49}H_{74}O_{14}$
A_{1b}	$C_{48}H_{72}O_{14}$
A_{2a}	$C_{49}H_{76}O_{15}$
A_{2b}	$C_{48}H_{74}O_{15}$
B_{1a}	$C_{48}H_{72}O_{14}$
B_{1b}	$C_{47}H_{70}O_{14}$
B_{2a}	$C_{48}H_{74}O_{15}$
B_{2b}	$C_{47}H_{72}O_{15}$

Two avermectins, abamectin (the avermectin B_1) and ivermectin, which is saturated at C-22–C-23, have been commercialized to date.

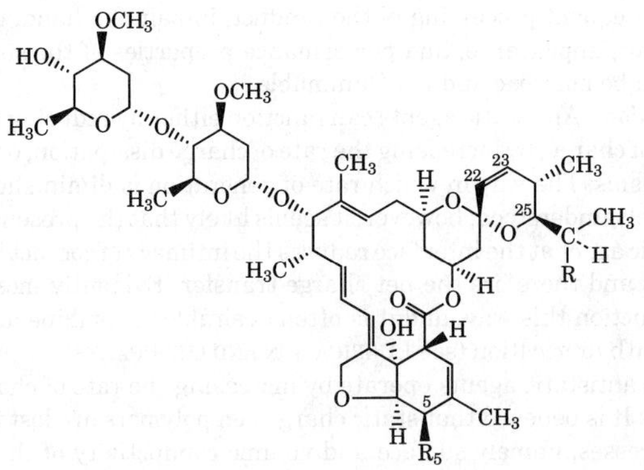

Figure 1. The avermectins show variations at positions 5, 22–23, and 25. In the A series $R_5 = OCH_3$; for the B series $R_5 = OH$. The designation 1 corresponds to $\underset{22}{-CH}\!=\!\underset{23}{CH-}$ as shown. The designation 2 indicates $\underset{22}{-CH_2}-\underset{23}{CHOH}$. In ivermectin, C-22–C-23 is saturated. The avermectins are isolated as mixtures wherein the major component a has a *sec*-butyl group at position 25($R = C_2H^5$), and component b has an isopropyl group ($R = CH_3$) at C-25.

Abamectin

Avermectin B_1 is the most effective of the avermectin family of natural products against agriculturally important insects and mites. Abamectin is also used to control the imported red fire ant *Solenopsis invicta*.

Ivermectin

Ivermectin is active against two significant phyla of animal parasite: the Nemathelminthes or nematodes (roundworms) and the Arthropoda (insects, ticks, and mites). Ivermectin is inactive against platyhelminthes (flukes and tapeworms). Ivermectin has an extremely broad spectrum of antinematodal activity in a variety of domestic animals.

Biosynthesis

The proposed pathway for the biosynthesis of the avermectins (Fig. 2) has been described in a review.

Chemistry

Stability. Avermectins are highly lipophilic substances and dissolve in most organic solvents. Their solubility in water is correspondingly low, only 0.006–0.009 ppm (= mg/L). Avermectins are acid-sensitive.

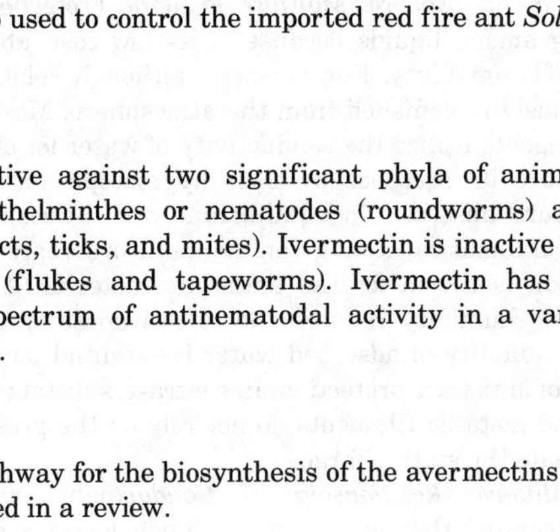

Figure 2. Biosynthetic scheme of avermectins, ■ $\overset{18}{=}$ O, $\overrightarrow{CH_3COOH}$, CH_3CH_2COOH.

Reactivity. Avermectin B_1 and ivermectin both contain two secondary and one tertiary hydroxy group. The two secondary alcohols are readily acetylated with acetic anhydride in pyridine.

Monosaccharides and Aglycones

The macrocyclic lactone of all avermectins has an α-L-oleandrosyl-α-L-oleandrosyloxy substituent at carbon 13, which is a 2-deoxy glycoside, relatively sensitive to acid hydrolysis or alcoholysis. A solution of ivermectin in methanol containing a strong acid such as 1% sulfuric acid readily gives a good yield of the aglycone after 16–24 h at RT. When 2-propanol is substituted for methanol in this reaction, a good yield of the monosaccharide is obtained, as the bulkier 2-propanol preferentially attacks the sterically less hindered 1′-position over the 1′-position, which is attached directly to the C-13 of the macrocycle, which in turn is flanked on either side by the C-12 and C-14 methyl groups, respectively. These procedures readily yield the monosaccharides of ivermectin and avermectin B_1 and the aglycone of ivermectin. The preparation of the aglycone of avermectin B_1, however, is complicated: under the reaction conditions, addition of methanol to the 22,23-double bond occurs, yielding mainly the two epimeric 22,23-dihydro-23-methoxy monosaccharides and/or aglycones. The aglycone of the avermectin B_1 therefore must be prepared with aqueous acid; it can be obtained as a 1:1 mixture of aglycone and monosaccharide using 10% sulfuric acid in aqueous THF solution.

Removal of the 13-hydroxy group from avermectin aglycones gives 13-deoxyavermectin aglycones which are closely related to certain milbemycins.

Chemical Reactions

Oxidations. Avermectin B_1 and ivermectin have two secondary hydroxy groups susceptible to oxidation to the corresponding ketones.

Reductions Including Ivermectin Preparation. Reductions of avermectin B_1 containing five double bonds with Pd or Pt catalysts proceed at almost comparable rates at the two disubstituted 10,11- and 22,23-double bonds to give a mixture of 10,11- and 22,23-dihydro derivatives.

Radiolabeled Derivatives. Since ivermectin (= 22,23–dihydroavermectin B_1) is obtained by catalytic reduction of avermectin B_1, the same procedure using tritium gas conveniently affords tritiated ivermectin (22,23-$^3[H]$-22,23-dihydroavermectin B_1).

Interconversions of Avermectins. Methylation of avermectins B_1 and B_2 leads to the corresponding derivatives of the A series.

Syntheses

Avermectin aglycones, monosaccharides, and the naturally occurring disaccharides themselves have been further modified by attaching various sugars to the different hydroxy groups. Most of these methods have used 1-bromo sugars via the Konigs-Knorr procedure.

Economic Aspects

The worldwide acceptance of ivermectin in livestock production and health care of companion animals has made it the largest selling animal health drug. Abamectin is in commercial use as an agricultural pesticide and its applications are continuing to expand.

MICHAEL H. FISHER
HELMUT MROZIK
Merck, Sharp and Dohme Research Laboratories

R. W. Burg and co-workers, *Antimicrob. Agents Chemother.* **15**, 361 (1979).

G. Albers-Schönberg and co-workers, *J. Am. Chem. Soc.* **103**, 4216 (1981).

M. H. Fisher and H. Mrozik, in S. Omura, ed., *Macrolide Antibiotics,* Academic Press, New York, 1984, pp. 553–606.

T. Blizzard, M. H. Fisher, H. Mrozik, and T. L. Shih, in G. Lukacs and M. Ohno, eds., *Recent Progress in the Chemical Synthesis of Antibiotics,* Springer-Verlag, Berlin, Heidelberg, Germany, 1990, pp. 65–102.

ANTISTATIC AGENTS

Electrification is the process of producing an electric charge on an object. If the charge is confined to the object it is said to be electrostatic. The term static electricity refers to accumulated, immobile electrical charges in contrast to charges in rapid flow, which is the subject of electrodynamics.

Static is a surface phenomenon. When two surfaces are brought into contact, electrons pass across the interface in both directions. This exchange is not symmetrical, and even with two identical bodies one of the two acquires an excess of electrons at the expense of the other. As long as contact is maintained, there is no further effect. However, when the surfaces are separated part of the charge is discharged into air, and the remainder is left on the material. One of the two bodies has an excess of electrons and the other has a shortage of electrons, ie, the objects are electrically charged with the same magnitude of charge but opposite signs. The magnitude of the charge depends on the rate of separation of the two bodies. The electrons in a conductor move freely, thus the excess of electrons is eliminated by backflow. However, the electrons in an insulator are not mobile, and the phenomenon of static electricity becomes noticeable.

Methods of Controlling Static Charges

The amount of electrostatic charge developed on a material represents a balance between the rate of generation and the rate of dissipation. Good conductors disperse a static charge quickly; however, textiles and plastics have a high surface resistivity and charge decay can occur only at a low rate. It is feasible to reduce accumulation of static charges by either reducing the rate of generation or by increasing the rate of dissipation. It is not possible to completely eliminate the generation of a static charge, so measures must be taken to increase the rate of charge dissipation.

Increasing Conductivity. Increasing the electrical conductivity of the material by increasing its electrolytic (ionic) or electronic conductivity is a means of controlling static charges. The electrolytic conductivity can be improved by increasing the moisture content of the surrounding atmosphere (humidification), by application of internal or external antistatic agents, or by chemically modifying the material. Chemical modification is especially used for textile fibers and involves surface modification to increase hygroscopic properties or grafting of functional groups leading to ionic configurations or greater moisture sensitivity. The electrical conductivity of the material can also be improved by increasing its electronic conductivity. This can be achieved by coating fibers with a thin layer of a conductor such as silver or carbon black, by blending fine metal fibers with static-prone fibers in the textile material, or by incorporating carbon black in plastics.

Static Dissipative Materials

Materials that have the ability to dissipate a charge formed by any means including tribocharging and induction on the material are referred to as static dissipative. There is a correlation between static dissipation and surface resistance. EIA-541 currently defines the static dissipative range as 10^5 to 10^{12} Ω.

The term antistatic agent is generally used to describe a substance that is added to a material to make that material static dissipative. The two biggest markets for antistatic agents are textiles and plastics.

Antistatic Agents for Textile Materials. Antistatic finishes are defined as antistatic agents in combination with water, mineral oil, composite finishes applied by producers during fiber manufacture, oils used to facilitate throwing or coning, textile softeners or lubricants, and hand building compounds. The terms antistatic agents and antistatic finishes have often been used interchangeably in the literature.

Effective antistatic agents must act at a relative humidity below 40%, preferably below 15%. The agent must form a film on various surfaces and be applied from a solution or dispersion in water or other inexpensive solvents. The antistatic agent must not interfere with subsequent processing of the product, impair the hand, or affect color, odor, appearance, and performance properties of the substrate. It should be nontoxic and nonflammable.

Function. Antistatic agents can function either by reducing the generation of charge, by increasing the rate of charge dissipation, or by both mechanisms. The way in which rate of generation is diminished is not completely understood; however, it seems likely that the presence of the antistatic agent at the interface reduces the intimacy of contact between surfaces and therefore the net charge transfer. Evidently most lubricants function this way, and it is often desirable to combine antistatic action with lubrication (see LUBRICATION AND LUBRICANTS).

Most antistatic agents operate by increasing the rate of charge dissipation. It is believed that static charges on polymers are lost through two processes, namely surface and volume conductivity of the charge through the substrate, and loss of charge through air by radiation. The radiation process is at least as important as the conduction process in effecting charge dissipation.

Chemical Classification.

Substances of high electrical conductivity are effective antistatic agents. Numerous chemicals have been proposed and reviewed. In general, most antistatic agents belong to one of the following classes: nitrogen compounds such as long-chain amines, amides, and quaternary ammonium salts; esters of fatty acids and their derivatives; sulfonic acids and alkyl aryl sulfonates; polyoxyethylene derivatives; polyglycols and their derivatives; polyhydric alcohols and their derivatives; phosphoric acid derivatives; solutions of electrolytes in liquids with high dielectric constants; molten salts; metals; carbon black; semiconductors; and liquids with high dielectric constant, such as water, which are usually volatile and have a temporary effect only.

Role of Atmospheric Moisture in Static Protection. Water is outstanding among liquids because of its low cost, absence of toxicity, and nonflammability. Furthermore, although volatile, water can be continuously replenished from the atmosphere. Most nonmetallic antistatic agents utilize the conductivity of water for charge dissipation and ensure its presence by their hygroscopic nature. The amount of moisture absorbed in equilibrium with atmospheres of various relative humidities at a given temperature represents a moisture absorption isotherm for the antistatic material. A reduction of atmospheric humidity reduces the effectiveness of antistatic agents. A small quantity of adsorbed water is essential for ionization in the process of antistatic protection. In contrast, substances such as carbon black and metallic filaments do not rely on the presence of moisture to dissipate the static charge.

Quantitative Relationship of Conductivity and Antistatic Action. Assuming that an antistatic finish forms a continuous layer, the conductance it contributes to the fiber is proportional to the volume or weight and specific conductance of the finish. As long as the assumption of continuity is fulfilled it does not matter whether the finish surrounds fine or coarse fibers. Assuming a cylindrical filament of length 1 cm and radius r, denoting the thickness of the finish layer as Δr and the specific conductance of the finish k, the conductance K of the finish layer is given by the equation.

$$K = k\frac{2\pi r \Delta r}{l}$$

The specific conductance of the finish on the filament k is not necessarily the specific conductance it exhibits in its bulk condition. For instance, absorption of ions from the finish by the fiber can reduce the conductivity. The specific conductance greatly depends on the amount of moisture present.

Unless some measures are taken to ensure uniformity, about 0.1% of antistatic agent is needed for antistatic protection of textile fabrics. The exact amount depends on the efficiency of the agent, relative humidity, temperature, diameter of fiber, fabric structure, and degree of antistatic protection desired.

Although it has been generally demonstrated that antistatically treated fabrics exhibit increased surface conductivities, many examples have been found where static behavior is not always related to the conductivity of the fabric or the material. One of the main reasons for this is the fact that static charges decay not only by conduction but also by radiation.

Durability of Finishes. *Nondurable Antistatic Finishes.* Most surface-active agents reduce static properties of materials to which they are applied and are classified as nondurable antistatic agents. They appear to work on the basis of moisture absorption, and their effectiveness decreases as the humidity and temperature in the atmosphere are reduced, reaching a cutoff in effectiveness when ionization is inhibited. Antistatic agents are often combined with lubricating oils to allow emulsification of the oils and to facilitate their removal during scouring and dyeing of textile materials. Surface-active agents of the nonionic or cationic type are commonly used as the antistat, and compatibility with lubricants becomes an added requirement.

The structures that render a substance surface-active are its hydrophobic hydrocarbon residue and a terminal hydrophilic group.

Surface-active agents increase the conductivity of oils quite significantly, and addition of water, probably dissolved at the interface with the surfactant, further increases the conductivity. Nonionic and cationic surface-active agents are preferred to anionic surface-active agents probably because of their higher solubility in oils and higher hygroscopicity. Many anionic surfactants have adequate antistatic efficiency, but they are used less frequently.

The relationship between chemical constitution and antistatic action of an array of surface-active agents has been evaluated. Residual amounts of surface-active detergents in hydrophobic garments after laundering can also be effective in aiding charge dissipation. Some examples of commercial nondurable antistatic agents are given in Table 1. Nondurable antistatic agents are suitable for textile processing aids in such operations as spinning, winding, weaving, and fiber manufacture, but cannot provide the material with antistatic protection in consumer use.

Durable Antistatic Finishes. The difficulty with nondurable finishes, as far as the consumer is concerned, is that they are water-soluble and thus easily removed by washing. An effective antistatic finish must be durable and capable of withstanding repeated laundering and drycleaning cycles. Only a small number of durable antistatic agents are available for textiles.

Some commercial durable antistatic finishes have been listed in Table 1. Essentially, durable antistats are polyelectrolytes, and the majority of useful products involve variations of cross-linked polyamines containing polyethoxy segments.

Application of Antistatic Finishes. Antistatic finishes are commonly applied to textile materials by padding (dipping and squeezing off excess finish), exhausting (absorption from solution due to affinity of the antistat for the textile material), spraying, and coating (see TEXTILES, FINISHING).

Effect of Aging on Antistatic Protection. Aging is known to improve the laundering resistance of durable antistatic finishes that have been incompletely cross-linked by curing, but a gradual reduction of static protection on aging has been observed with some nondurable antistats.

Soiling of Antistatic Finishes. Soiling of fabrics having a tendency for accumulation of charges has been assumed to be an electrostatic phenomenon, and therefore it follows that if static is eliminated, soiling will be reduced. However, most antistatic agents have been developed and used for reasons other than the reduction of soiling.

Inherently Static-Free Textile Fibers and Fabrics. There have been many efforts to develop fibers that are inherently static-free. One method involves incorporation of internal antistatic additives in the bulk of the polymer before spinning, and a second method consists of coating the fiber with conducting metals or carbon black and incorporating fine metal fibers in the textile material. Both methods increase the electrical conductivity of the whole material rather than the surface conductivity alone. However, the first method increases ionic conductivity, whereas the second increases the electronic conductivity.

Consumption of Antistatic Agents in Textile Applications. In textile finishing of synthetic fibers and their blends, durable antistatic agents represent a potential market of a million kg/yr or more; however, volumes have not been disclosed. Many thousand metric tons of nondurable antistatic agents are also used in textile processing, and the volume increases as use of synthetic fibers and production speeds increase.

Antistatic compositions are being sold for application during household laundering, to be added in the rinse cycle or during tumble drying. Most of these products, listed as antistatic agents for textiles, are nondurable and based on cationic and nonionic surface-active agents (see SURFACTANTS).

Table 1. Antistatic Agents for Textiles

Trade name	Manufacturer	Chemical type
	Durable antistats	
Aerotex Antistatic D	American Cyanamid Corp.	
Aston 123	Refinex Onyx	polyamine resin
Cirrasol Z	ICI Americas	
Nonax 1166	Henkel	polyamine resin
Permalose T	ICI Americas	nonionic
Stanax 1166	Standard Chemical Products	polyamine resin
	Cationic antistatic softeners	
Ampitol VAC	Dexter Chemical Co.	quaternary ammonium
Avitex DN	DuPont	cationic
Kemamine Q9702C	Humko Chemical	quaternary ammonium compound
	Unclassified	
Antistatin D	BASF	
Arkostat AC	Hoechst	
Aston MS	Refinex Onyx	quaternized (cationic) fatty amine condensate
Aston AP	Refinex Onyx	amine condensate
Cassastat	Cassella	cationic
Catanac SN	American Cyanamid Corp.	cationic
Elfugin UW	Sandoz	
Igepal CO 430	GAF Corp.	nonylphenol ethylene oxide
Neutro-stat[a]	Simco Co.	
Siligen APE[b]	BASF	quaternary ammonium compound
Statexan HA	Bayer	
Tinorex TC	CIBA-GEIGY	
Zelec DP	DuPont	cationic
Zero C	Lutex Chemical Corp.	nitrogeneous polymer

[a] Spray.
[b] For polyester.

Antistatic Agents for Plastics

Plastics are excellent insulators and have a great tendency to generate and retain static charges.

Antistatic agents may be applied to the surface of the finished article or incorporated in the bulk of the polymer during processing (see POLYMER BLENDS). They function by decreasing the rate of charge generation, by increasing the rate of charge dissipation, or by both mechanisms.

Types of Antistats. Plastics are rendered static dissipative by either adding a conductive material to the plastic in sufficient quantity that there is a three-dimensional conductive pathway through the plastics, or by adding to the plastic a surfactant-type chemical that will attract moisture to the surface of the plastic. Since water is a conductor, the surface layer of moisture then dissipates a static charge. The source of conductive material can be either an inherently conductive polymer, a metal or metal-containing material, or carbon black. The conductive additive can, if desired, be added to the plastic at a concentration sufficient to make it static dissipative, but not completely conductive. Surfactant-type additives are of several different chemical types. They can be applied to the surface of the finished plastic article or incorporated in the bulk of the polymer during processing.

Inherently Conducting Polymers. Conducting polymers are polymers with a pi-electron backbone capable of passing an electrical current. These polymers generally are not sufficiently conductive as neat polymers but require the inclusion of an oxidizing or reducing agent (dopant) to render them conductive.

Common conductive polymers are polyacetylene, polyphenylene, poly(phenylene sulfide), polypyrrole, and polyvinylcarbazole (see ELECTRICALLY CONDUCTIVE POLYMERS).

Metal-Containing Polymers. Metal-containing polymers function simply by adding sufficient quantities of a metal to form a three-dimensional conduction pathway through the plastic. The metal is in the form of a powder, micrometer-sized needles, or a thin coating on glass spheres, carbon fibers, or mica. The metals normally employed are nickel, zinc, stainless steel, copper, and aluminum. At higher levels these materials make the polymer capable of shielding electromagnetic impulses (EMI).

Carbon Blacks. The high electrical conductivity of carbon black is utilized where its color is not objectionable and its reinforcing action is used (see FILLERS; COMPOSITE MATERIALS).

Surfactant-Type Antistats. Inherently conductive antistats have the advantage of not being dependent on atmospheric moisture to function. Their drawbacks include expense, coloration of the plastic, and alteration of the mechanical properties of the plastic. The added stiffness caused by conductive fillers may not be a problem with a rigid container, but it can be a problem for a flexible bag.

Surfactant-type antistats find the widest use because of their low cost and minimal effect on the mechanical properties of the plastic. Ease of use is another favorable aspect to surfactants. They can be mixed with the bulk of the plastic prior to processing or can be applied to the surface of the finished plastic article as the need dictates.

Internal surfactant antistats are physically mixed with the plastic resin prior to processing. Because the antistatic effect only occurs after the surfactant has migrated to the surface, the solubility of the surfactant in the polymer is an important consideration. Surfactants can generally be classified as one of four chemical types: cationic, where the hydrophilic portion has a positive charge; anionic, where the hydrophilic portion has a negative charge; nonionic, where the hydrophile does not have a charge; and amphoteric, where the molecule contains both positive and negative charges.

Cationic surfactants perform well in polar substrates like styrenics and polyurethane. Examples of cationic surfactants are quaternary ammonium chlorides, quaternary ammonium methosulfates, and quaternary ammonium nitrates (see QUATERNARY AMMONIUM COMPOUNDS). Anionic surfactants work well in PVC and styrenics. Examples of anionic surfactants are fatty phosphate ester and alkyl sulfonates. Nonionic surfactants perform well in nonpolar polymers such as polyethylene and polypropylene. Examples of nonionic surfactants are ethoxylated fatty amines, fatty diethanolamides, and mono- and diglycerides (see AMINES, FATTY AMINES; ALKANOLAMINES). Amphoteric surfactants find little use in plastics.

Surface-Applied Surfactants. Antistat agents can be applied directly to the surface of a plastic part. Usually the antistat is diluted in water or in a solvent. The antistat solution is applied by spraying, dipping, or wiping on the surface. In practice, the quaternary ammonium compounds find the most use. They are soluble in water and effective at low concentrations. The antistatic protection provided by surface treatment is excellent while it lasts. However, surface treatments provide only temporary protection. The antistat layer is extremely vulnerable to rubbing and wiping, especially when wet.

Consumption of Antistatic Agents for Use in Plastics Applications. Antistatic agents are widely used in the plastic industry, and their economic significance is greater than in the textile industry. Food and drug packaging accounts for well over half the market, and FDA-acceptable products such as polyethylene, poly(vinyl chloride), and polypropylene film and bottles are predominant. The use of antistatic agents in medical and surgical applications of flexible poly(vinyl chloride) and polyethylene film and sheet has increased, providing higher safety in places where a buildup of static charges can trigger explosions. The use of antistatic agents in plastics used for the electronics industry has grown to the point that virtually all of these plastic products contain some type of antistatic agent (see ELECTRONIC MATERIALS).

LEONARD E. WALP
Humko Chemical
Division of the Witco Corporation

N. Sclater, *Electrostatic Discharge Protection for Electronics,* Tab Books, Blue Ridge, Pa., 1990.

K. Johnson, *Antistatic Compositions for Textiles and Plastics,* Noyes Data Corp., Park Ridge, N.J., 1976.

H. Mark, N. S. Wooding, and S. M. Atlas, eds., *Chemical Aftertreatment of Textiles,* Wiley-Interscience, New York, 1971.

J. Frommer, *The Potential for Applications of Conducting Polymers,* 43rd Annual Technical Conference Proceedings, Society of Plastics Engineers, Brookfield, Conn., 1985.

ANTIVIRAL AGENTS

Viral infections are among the greatest causes of human morbidity, and it is estimated that in developed countries more than 60% of all the episodes of human illness result from viral infections. High virus infection rates also occur among pets, livestock, and plants. The high morbidity and the resulting economic loss caused by these infections have generated tremendous efforts in recent years to develop means to combat viral infections using antiviral agents.

Since viruses propagate only within living cells, the development of antiviral drugs which would disrupt the viral replication without affecting the metabolism of the host cell was initially believed to be difficult, if not impossible. However, dramatic progress in viral molecular biology has now made it possible to identify enzymatic processes which are unique to virus-infected cells. As a consequence, it is becoming feasible to design chemical compounds which identify infected cells, block a specific step in viral replication, and leave uninfected cells unharmed.

The majority of antiviral drugs which are under clinical development today generally interrupt viral nucleic acid synthesis. These compounds often do not affect host cell metabolism and possess considerable selectivity against virus-induced enzymes. This article discusses agents exhibiting significant antiviral activity against viral infections in animal model systems.

Agents Active Against DNA Viruses

One of the simplest molecules found to inhibit the replication of DNA viruses in animals is phosphonoformic acid (PFA), CH_3O_5P **(1)**. Both PFA (as the trisodium salt CNa_3O_5P, foscarnet) and its homologue phosphonoacetic acid (PAA), $C_2H_5O_5P$ **(2)**, were developed by Astra Pharmaceuticals and show selective inhibition of DNA polymerase in various herpes viruses.

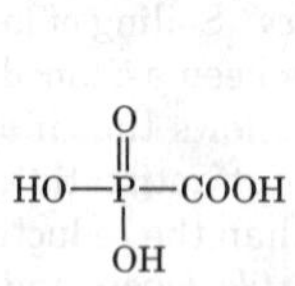

(1)

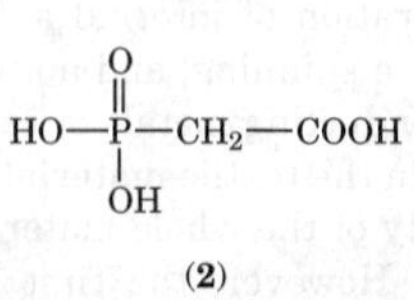

(2)

PFA has recently undergone clinical evaluation in humans for the treatment of recurrent genital herpes, hepatitis B viral infection, and acquired immunodeficiency syndrome (AIDS), as well as cytomegalovirus (CMV) infection of bone marrow and renal transplant patients.

Levamisole (6-phenyl-2,3,5,6-tetrahydroimidazol[2,1-*b*]thiazole), $C_{11}H_{12}N_2S$ **(3)**, was found to be effective against herpes virus infections in humans.

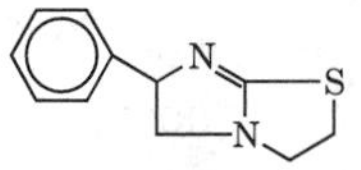

(3)

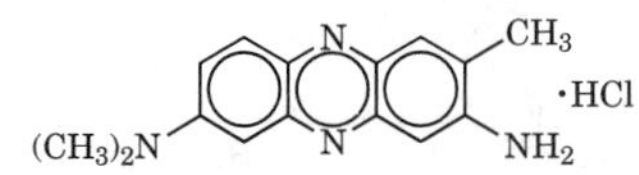

(4)

Certain heterocyclic dyes such as neutral red, $C_{15}H_{17}ClN_4$ **(4)**, acridine orange, and proflavine, which act by binding to viral nucleic acid, absorbing visible light, and inactivating the virus by oxidation, have shown antiviral activity of clinical potential against cutaneous herpes virus infections. However, concern has been raised that photoinactivation of herpes viruses may be clinically dangerous because the inactivated virus may be capable of transforming normal cells to a malignant state. Moreover, these dyes have a strong affinity for the nuclei of normal cells and may damage their genetic makeup.

Several pyrimidine bases have been found to inhibit herpes virus-induced keratitis in rabbits, eg, 2-amino-4,6-dichloropyrimidine, $C_4H_3Cl_2N_3$ **(5)**, and 1-allyl-6-chloro-3,5-diethyluracil, $C_{11}H_{15}ClN_2O_2$ **(6)**.

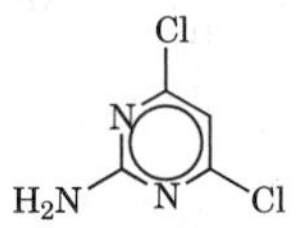

(5)

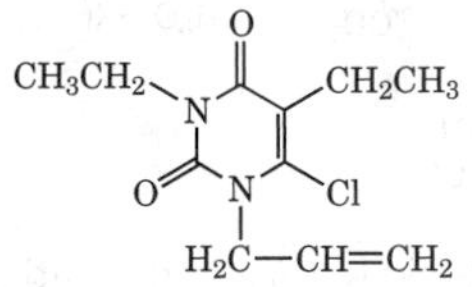

(6)

The most successful clinical antiviral agents belong to the nucleoside category. Nucleoside analogues with potent antiviral activity have been known since idoxuridine and trifluridine were shown to be efficacious against herpes keratitis more than 30 years ago. However, despite their antiviral efficacy the therapeutic usefulness of these early nucleosides was limited because of mutagenic, teratogenic, carcinogenic, cytostatic, or cytotoxic side effects. Recently the specificity of antiviral action of this class of compounds has been significantly improved; potent and highly selective nucleoside antiviral agents have now been developed. These are either antiviral nucleosides which are only activated by virus-infected cells, or antiviral agents which specifically inhibit virus-induced enzymes required for viral replication.

Cytosine Derivatives. 1-β-D-Arabinofuranosylcytosine (ara-C), $C_9H_{13}N_3O_5$ reportedly has had significant therapeutic effects in patients with localized herpes zoster, herpes eye infections, and herpes encephalitis, although several negative results have also been reported. Ara-C, also known as cytarabine, is quite toxic and is only recommended for very severe viral infections. A number of derivatives of ara-C have been prepared in an effort to improve on antiviral activity and to reduce the toxicity.

Purine Nucleoside Derivatives. A number of purine nucleoside analogues are also found to be active against several DNA viruses (Fig. 1). They include ara-A (9-β-D-arabinofuranosyladenine, vidarabine), $C_{10}H_{13}N_5O_4$ **(7)**, ara-HxMP, $C_{10}H_{13}N_4O_8P$ **(8)**, cyclaradine, $C_{11}H_{15}N_5O_3$ **(9)** and the xylofuranosyl analogue of tubercidin, $C_{11}H_{14}N_4O_4$ of tubercidin, **(10)**.

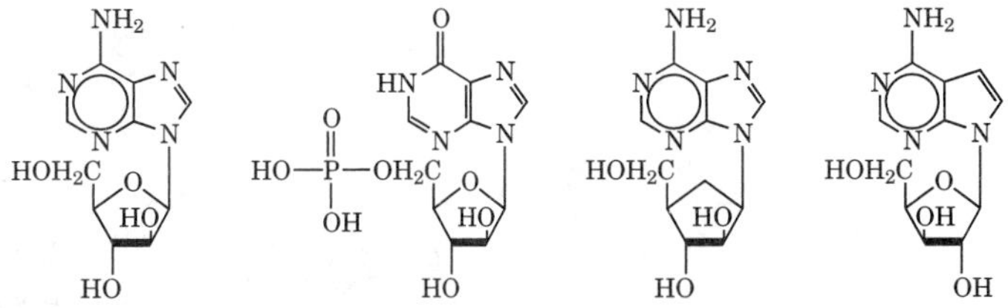

(7) **(8)** **(9)** **(10)**

Figure 1. Purine nucleoside analogues found to be active against DNA viruses.

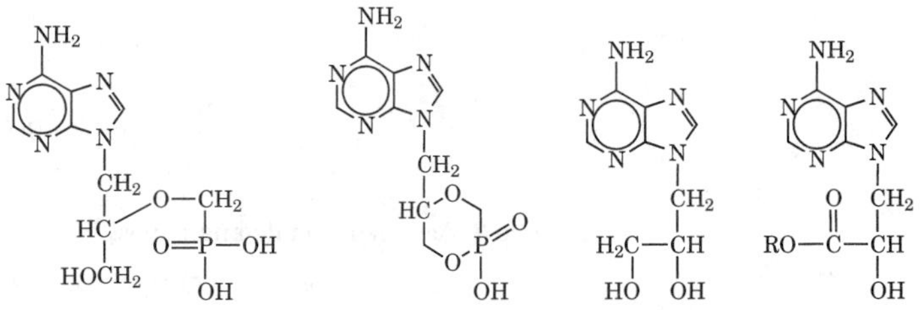

(11) **(12)** **(13)** **(14)**

Figure 2. Aliphatic adenosine analogues with broad-spectrum antiviral activity.

Adenine Derivatives. These include (*S*)-9-(3-hydroxy-2-phosphonylmethoxypropyl) adenine [(*S*)-HPMPA], $C_9H_{14}N_5O_5P$ **(11)**, the cyclic phosphonate of (*S*)-HPMPA, (S)-cHPMPA, $C_9H_{12}N_5O_4P$ **(12)**, (*S*)-DHPA, $C_8H_{11}N_5O_2$ **(13)**, and adeninylhydroxypropanoic acid alkyl esters [(*R,S*)-AHPA esters] **(14)** (Fig. 2).

Acyclic Purine Nucleosides. As a consequence of earlier studies involving adenosine deaminase as a model enzyme for examining structural features required for antiviral activity, a number of acyclic purine nucleosides have emerged as antiviral agents (Fig. 3). They include acyclovir [9-(2-hydroxyethoxymethyl) guanine], $C_8H_{11}N_5O_3$ **(15)**, 6-deoxyacyclovir, $C_8H_{11}N_5O_2$ **(16)**, DHPG [9-(1,3-dihydroxy-2-propoxymethyl)guanine, ganciclovir], $C_9H_{13}N_5O_4$ **(17)** DHPG cyclic phosphate $C_9H_{11}N_5O_6P \cdot Na$ **(18)**, 3,4-dihydroxybutylguanine (DHBG), $C_9H_{13}N_5O_3$ **(19)**, (±)-(1α, 2β, 3α)-9-[2,3-bis(hydroxymethyl) cyclobutyl]guanine [(± − BHCG)], $C_{11}H_{15}N_5O_3$ **(20)**, the ara-carbocyclic analogue of 7-deazaguanosine (±)-2-amino-3,4-dihydro-7-[(1α, 2α, 3β, 4α)-2,3-dihydroxy-4-(hydroxymethyl)-1-cyclopentyl] pyrrolo[2,3-*d*]-pyrimidin-4-one, $C_{12}H_{16}N_4O_4$ **(21)**, and 9-β-D-+xylofuranosylguanine (xylo-G), $C_{10}H_{13}N_5O_5$ **(22)**.

Agents Active Against RNA Viruses

A large number of α-hydroxybenzylbenzimidazole (HBB), $C_{14}H_{12}N_2O$ **(23)**, derivatives have been prepared and extensively studied as selective inhibitors of the RNA-containing enteroviruses. Although none of these derivatives have shown any antiviral activity in animals, 1,2-bis(5-methoxy-2-benzimidazol-2-yl)-1,2-ethanediol, $C_{18}H_{18}N_4O_4$ **(24)**, was found to be active against an experimentally induced rhino virus infection in chimpanzees. However, the *in vivo* antiviral efficacy was accompanied by significant toxicity.

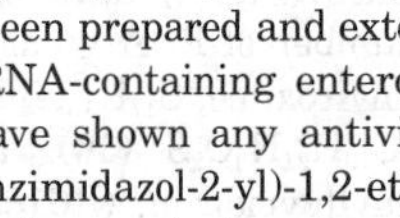

(23)

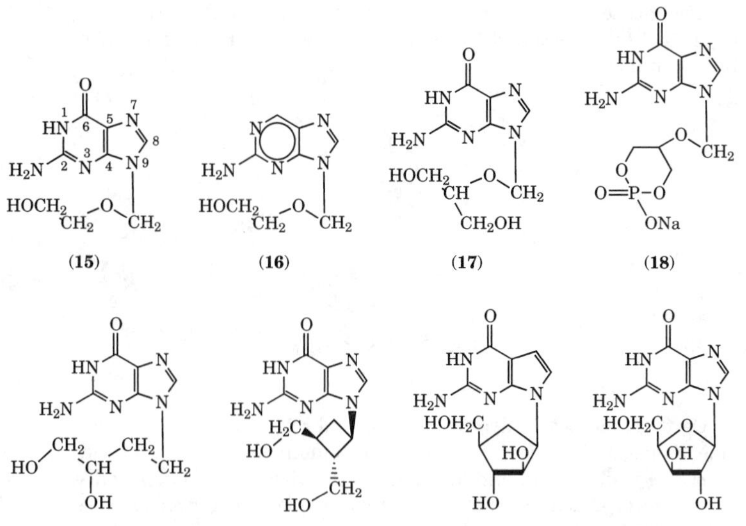

(15) (16) (17) (18)

(19) (20) (21) (22)

Figure 3. Acyclovir and its analogues.

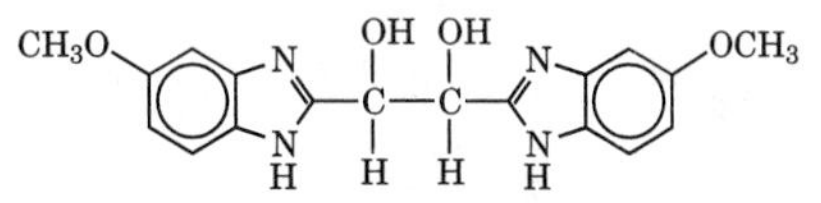

(24)

Amantadine hydrochloride (1-adamantanamine hydrochloride), $C_{10}H_{17}N \cdot HCl$ **(25)**, is a good example of a narrow-spectrum agent active only against influenza A virus. It became the first antiviral drug available for systemic use in the United States when it was approved by the FDA in 1966 for use against Asian influenza. A structurally related drug, rimantadine hydrochloride, (α-methyl-1-adamantane-methylamine hydrochloride), $C_{12}H_{21}N \cdot HCl$ **(26)**, is widely used in Russia to treat influenza A virus.

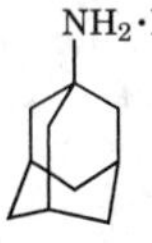

(25)

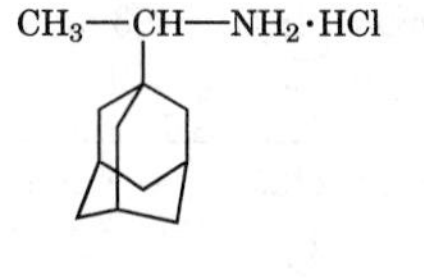

(26)

Lipophilic β-Diketones. Arildone, $C_{20}H_{29}ClO_4$ (**27**), and several other lipophilic β-diketones (Fig. 4) have exhibited significant *in vivo* activity against a number of RNA viruses. These include WIN 51711, $C_{20}H_{26}N_2O_3$ (**28**), enviroxime, $C_{17}H_{18}N_4O_3S$ (**29**), the enviroxime surrogate enviradene, $C_{19}H_{21}N_3O_2S$ (**30**), and 3-methoxy-6-[4-(3-methylphenyl)-1-piperazinyl]pyridazine (R61837), $C_{16}H_{20}N_4O$ (**31**).

Adenosine and Guanosine Analogues. Several adenosine analogues have been found to be active against both RNA and DNA viruses (Fig. 5). They include 3-deazaadenosine $C_{11}H_{14}N_4O_4$ (**32**), the carbocyclic analogue of 3-deazaadenosine, [($\pm$)-3-deazaaristeromycin], $C_{12}H_{16}N_4O_4$ (**33**), neplanocin A, $C_{11}H_{13}N_5O_3$ (**34**), sinefungin $C_{15}H_{23}N_7O_5$ (**35**), 3-deazaguanine, $C_6H_6N_4O$ (**36**), and 3-deazaguanosine, $C_{11}H_{14}N_4O_5$ (**37**).

The naturally occurring nucleoside antibiotic pyrazofurine, $C_9H_{13}N_3O_6$ **(38)**, shows broad-spectrum antiviral activity and

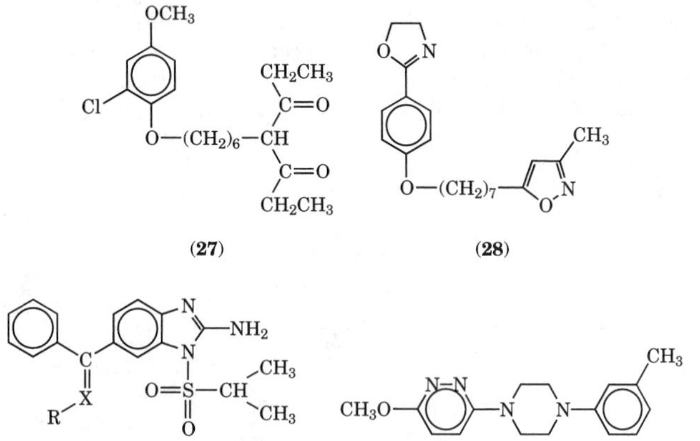

(27) (28)

(29) X = N, R = OH
(30) X = CH, R = CH_3

(31)

Figure 4. Lipophilic β-diketones and derivatives active against RNA viruses, especially rhinovirus.

a broad safety margin against both RNA and DNA viruses *in vitro.*

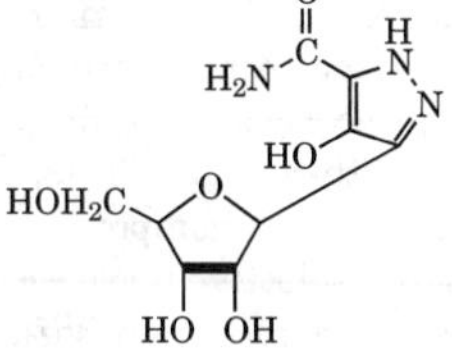

(37)

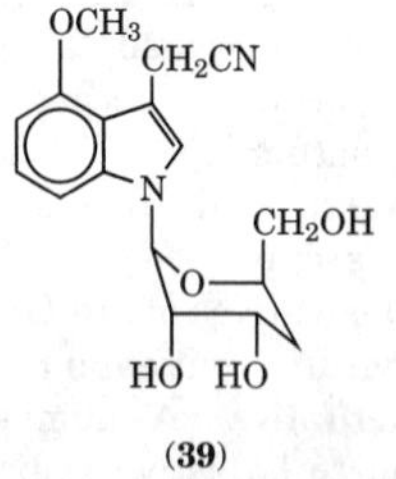

(39)

Another naturally occurring nucleoside antibiotic SF-2140, $C_{17}H_{20}N_2O_5$ **(39)** a 3-cyanomethyl-4-methoxyindole nucleoside, is

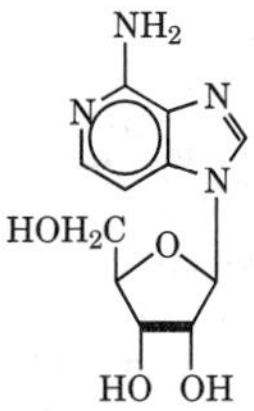

(32)

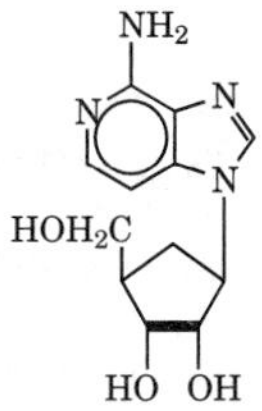

(33)

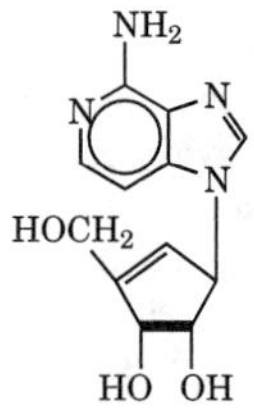

(34)

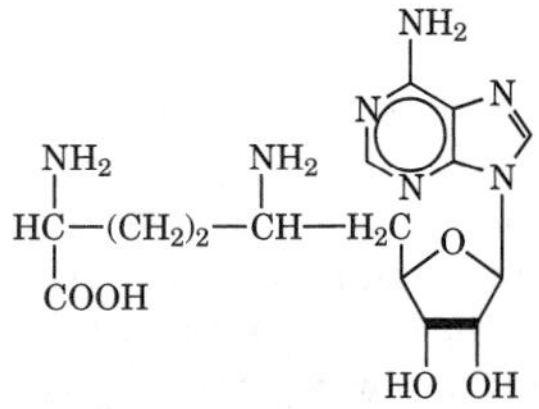

(35)

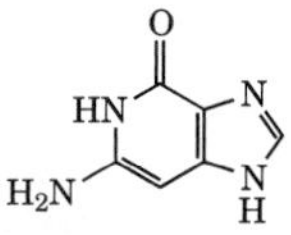

(36)

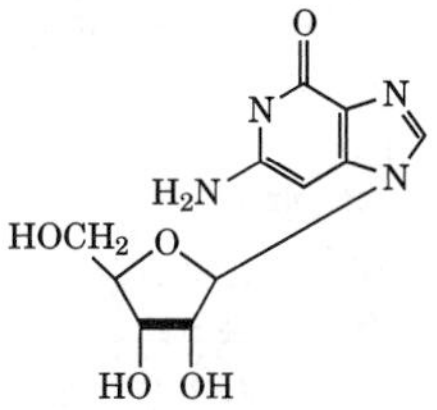

(37)

Figure 5. Adenosine and guanosine analogues with both RNA and DNA antiviral effectiveness.

found to be as active as amantadine in protecting mice against an APR-8 strain of influenza A.

Ribavirin and Structural Analogues. One of the broad-spectrum antiviral agents that emerged from ICN Pharmaceuticals is an azole ribonucleoside, 1-β-D-ribofuranosyl-1,2,4-triazole-3-carboxamide, designated as ribavirin, $C_8H_{12}N_4O_5$ **(40)**. Ribavirin has been studied in more animals and against more viruses than any other antiviral agent known today. It is active in cell culture against approximately 85% of all viruses studied, although it is inactive against polio virus, certain coxsackie viruses, most corona viruses, pseudorabies virus, and HBV.

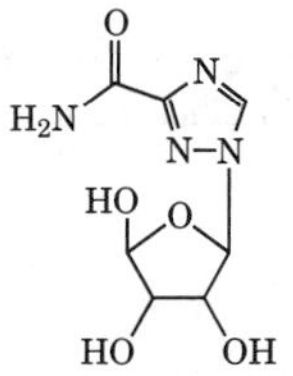

(40)

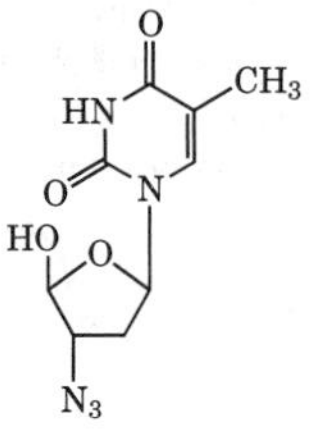

(41)

Unlike idoxuridine, BVdU, and acyclovir, viral strains susceptible to ribavirin have not been found to develop a resistance to the drug. The resistance against ribavirin is less likely because the drug exhibits multiple sites of antiviral action.

Agents Active Against Persistent Viral Infections

Persistent viral infection is a difficult challenge for antiviral chemotherapy. Retroviruses as a class are often found to be responsible for persistent viral infections. Retroviruses are unique RNA viruses characterized by the transcription of their single-stranded RNA into the double-stranded DNA of the host cell using the viral enzyme reverse transcriptase. AIDS is an example of such a persistent and latent human viral infection.

Following the identification of a retrovirus, HIV, as the etiological agent of AIDS, an intense effort has been made to identify drugs for the treatment or prevention of this debilitating, lethal disease. Several 2′,3′-dideoxyribonucleosides of purine and pyrimidine were discovered to be potent inhibitors of HIV replication *in vitro*. Considerable data has also accumulated on *in vitro* antiHIV testing of acyclic and carbocyclic nucleoside analogues. Although more than 100 nucleosides have been shown to have an *in vitro* antiHIV activity commensurate with development to clinical trials, only 3′-azido-3′-deoxythymidine (retrovir, zidovudine, AZT), $C_{10}H_{13}N_5O_4$ **(41)** 2,′3′-dideoxycytidine (DDC), $C_9H_{13}N_3O_3$ **(42)**, and 2′,3′-dideoxyinosine (DDI), $C_{10}H_{12}N_4O_3$ **(43)**, have become widely available for the treatment of AIDS and approved by the FDA for treatment of advanced AIDS cases.

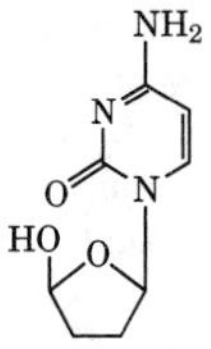

(42)

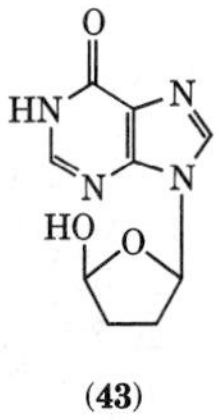

(43)

GANAPATHI R. REVANKAR
Triplex Pharmaceutical Corporation
ROLAND K. ROBINS
ICN Nucleic Acid Research Institute

R. K. Robins, *Chem. Eng. News* **64,** 28 (1986).

G. B. Elion, *Am. J. Med.* **73,** 7 (1982).

R. W. Sidwell, G. R. Revankar, and R. K. Robins, in D. Shugar, ed., *Intl. Enc. Pharmacol. Ther.,* Sec. 116, Vol. 2, Pergamon Press, New York, 1985, pp. 49–108.

E. De Clercq, ed., in *Design of Anti-Aids Drugs,* Elsevier, New York, 1990, pp. 1–24.

AQUACULTURE

One definition of aquaculture is the rearing of aquatic organisms under controlled or semicontrolled conditions. Another, promulgated by the Food and Agriculture Organization of the United Nations, is that aquaculture is, "the farming of aquatic organisms, including fish, molluscs, crustaceans, and aquatic plants". Included within those broad definitions are activities in fresh, brackish, marine, and even hypersaline waters. The term mariculture is often used in conjunction with aquaculture in the marine environment.

Public sector aquaculture involves production of aquatic animals to augment or establish recreational and commercial fisheries. Public sector aquaculture is widely practiced in North America and to a lesser extent in other parts of the world. The FAO definition of aquaculture also indicates that farming implies ownership of the organisms being cultured, which would seem to exclude public sector aquaculture.

In recent years, aquaculture has been increasingly used as a means of aiding in the recovery of threatened and endangered species. Those efforts are currently public sector activities, although there is interest in the private sector to become involved. As global awareness of endangered species issues grows, recovery programs for aquatic threatened and endangered species may arise in many more countries. Going hand in hand with attempts to recover endangered species are enhancement stocking programs aimed at releasing juvenile animals to build back stocks of aquatic animals that have been reduced due to overfishing. Examples of enhancement programs currently in existence include the stocking of cod in Norway, flounders in Japan, and red drum in the United States.

The bulk of global production from aquaculture is utilized directly as human food, with public aquaculture playing a minor role in many nations or being absent. Private aquaculture is not only about human food production, however. There is, in some regions, well-developed private sector aquaculture involved in the production of bait and ornamental fishes and invertebrates. Most of the information available in the literature relates to the production of such aquatic animals as molluscs, crustaceans, and finfish.

Aquatic plants are cultured in many regions of the world. In fact, aquatic plants, primarily seaweeds, account for nearly 25% of the world's aquaculture production. The origins of aquaculture are rooted in China and may go back some 4000 years. Asia dominates the world in aquaculture production. The United States has become one of the leaders in aquaculture research and development, although production, while significant, makes up only a fraction of the world's total. The United States commercial aquaculture industry is dominated by channel catfish, trout, salmon, minnows, oysters, mussels, clams, and crawfish.

Aquaculture production continues to grow annually, but increasing competition for suitable land and water, problems associated with wastewater from aquaculture facilities, disease outbreaks, and potential shortages of animal protein for aquatic animal feeds are having, or may have, negative effects on future growth. New technology, including the application of molecular genetic approaches to improving performance and disease resistance in aquaculture species, along with the development of water reuse (recirculating) systems and the establishment of offshore facilities, may provide the impetus for a resurgence in growth of the industry.

Economics

The production of aquatic animals for recreation, in nations where that type of aquaculture exists, is typically funded through user fees such as fishing licenses that support hatcheries and the personnel to run them. In order for most private aquaculture companies to get started, outside funding is required. Funding may come through banks and other commercial lending sources or from venture capitalists.

A key factor in obtaining funding support for aquaculture is development of a sound business plan. The plan needs to demonstrate that the prospective culturist has identified all costs associated with establishment of the facility and its day-to-day operation. One or more suitable sites should have been identified and the species to be cultured selected before the business plan is submitted. Cost estimates should be verifiable. Having actual bids for a specific task at a specific location; eg, pond construction, well drilling, building construction, and vehicle costs helps strengthen the business plan. The business plan also needs to provide projections of annual production and feed costs. For many aquaculture ventures, between 40 and 50% of the variable costs involved in aquaculture can be attributed to feed.

Each business plan should project income and expenses projected over the term of all loans in order to demonstrate to the lending agency or venture capitalist that there is a high probability the investment will be repaid. The projected income should be based on a realistic estimate of farmgate value of the product and an accurate assessment of anticipated production.

Regulation

The extent to which governments regulate aquaculture varies greatly from one nation to another. In some parts of the world, particularly in developing nations, there has historically been little or no regulation. Unregulated expansion of aquaculture in some countries has led to pollution problems, destruction of valuable habitats such as mangrove swamps, and has enhanced the spread of disease from one farm to another. The need for imposing regulations is now becoming evident around the world.

In developed countries there may or may not be a standardized set of national regulations. The United States is an example of a mixture of local, state, and federal regulations.

In general, it is easier to establish an aquaculture facility on private land than in public waters such as a lake or coastal embayment. Obtaining permits to use public waters is often not simple. In most cases it is necessary to contact a number of state agencies to apply for permits. Public hearings may be required, and the process can take months or even years to complete. The costs can be substantial and the outcome uncertain.

Most states now have an aquaculture coordinator, usually housed in the state department of agriculture, who can assist prospective aquaculturists.

Species Under Cultivation

This article emphasizes aquatic animal production, but many hundreds of thousands of people are involved, worldwide, in aquatic plant production. The quantity of brown seaweeds, red seaweeds, green seaweeds, and other algae alone produced in nineteen ninety-two was estimated at over 4.8 million metric tons (Table 1).

Animal aquaculture is concentrated on finfish, molluscs, and crustaceans (Table 1).

Culture Systems

At one extreme aquaculture can be conducted with a small amount of intervention from humans and the employment of little technology. At the other is total environmental control and the use of computers, molecular genetics, and complex modern technology. Many aquaculturists operate between the extremes. The range of culture approaches can be described as running from extensive to intensive, or even hyper-intensive, with extensive systems being relatively simple and intensive systems being complex to very complex. In general, as the level of culture intensity increases, stocking density, and as a consequence, production per unit area of culture system or volume of water, increases.

The most extensive types of aquaculture involve minimal human intervention to promote increases in natural productivity. One of the most extensive forms of culture involves placing oyster, clam, or other types of shell (cultch) on the bottom in intertidal areas that are known to have good oyster reproduction, but limited natural cultch material. The next step in increasing oyster culture intensity might involve hatchery production and settling of spat (oyster larvae) on cultch. Control of predators such as starfish and oyster drills could easily be a

Table 1. World Aquaculture Production in 1992 for Selected Groups of Aquaculture Species

Species group	Production, 10^3 t
carp and other cyprinids	6,652
tilapia and other cichlids	473
salmon and trout	628
flatfish	120
freshwater crustaceans	624
shrimp	884
oysters	954
mussels	109
scallops	549
clams, cockles, arkshells	765
brown seaweeds	3,640
red seaweeds	1,133
green seaweeds and other algae	17
miscellaneous aquatic plants	600
Total[a]	*19,311*

[a] Includes species categories not listed.

part of culture at all levels. The highest level of intensity with respect to oysters involves hatchery production of spat and the rearing of them suspended from rafts, long-lines, or as cultchless oysters in trays.

The stocking of ponds, lakes, and reservoirs to increase the production of desirable fishes that depend on natural productivity for their food supply and are ultimately captured by recreational fishermen or for subsistence is another example of extensive aquaculture. Most of the aquaculture practiced around the world is conducted in ponds (Fig. 1). Fertilization of ponds to increase productivity is the next level of intensity with respect to fish culture.

With the application of increased technology and control over the culture system, intensity continues to increase. Utilization of specific pathogen-free animals, provision of nutritionally complete feeds, careful monitoring and control of water quality, and the use of animals bred for good performance, can lead to impressive production levels.

Where water is plentiful and inexpensive, raceway culture is an attractive option and one which allows for production levels well in excess of what is possible in ponds. Linear raceways are commonly used by trout and salmon culturists both for commercial production and for hatchery programs conducted by government agencies.

Salmon, steelhead trout, and a variety of marine fishes are currently being reared in net-pens. Net-pen technology has only been widely employed commercially for salmon production since the 1980s when the Norwegian salmon farming industry was developed. Most net-pens are located in protected waters since they are easily damaged or destroyed by storms.

Competition by various user groups for space in protected coastal waters in much of the world has led to strict controls and in some cases

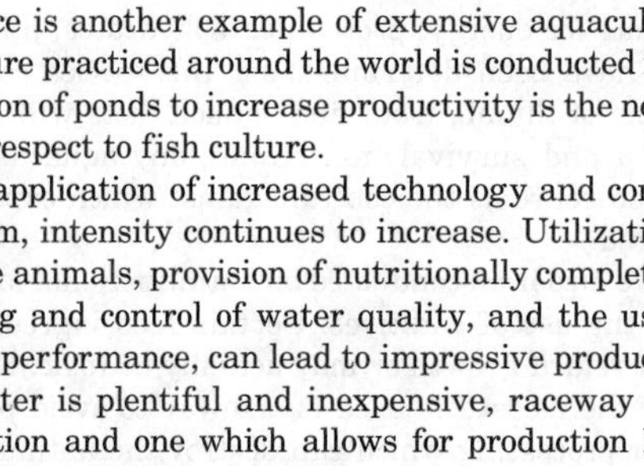

Figure 1. Aquaculture ponds are often rectangular in shape. They should be equipped with plumbing for both inflow and drainage of water.

prohibitions against the establishment of inshore net-pen facilities. As a result, there is growing interest in developing the technology to move offshore.

The highest levels of intensity that can be found in aquaculture systems are associated with totally closed systems, often called recirculating systems. In these systems, all water passing through the chambers in which the finfish or shellfish are held is continuously treated and reused. It is necessary to add some water to such systems to make up for that lost to evaporation, splashout, and in conjunction with solids removal. Many of the recirculating systems in use today are operated in a mode between entirely closed and completely open. In most there is a significant percentage of replacement water added either continuously or intermittently on a daily basis.

The heart of a recirculating water system is the biofilter, a device that contains a medium on which bacteria that help purify the water become established. Fish and aquatic invertebrates produce ammonia as a primary metabolite. If not removed or converted to a less toxic chemical, ammonia can quickly reach lethal levels. Two genera of bacteria are responsible for ammonia removal in biofilters: *Nitrosomonas* and *Nitrobacter*.

Recirculating systems typically also employ one or more settling chambers or mechanical filters to remove solids such as unconsumed feed, feces, and mats of bacteria that slough from the biofilter into the water. Each recirculating system requires a mechanical means of moving water from component to component, eg, mechanical pumping or air-lifts.

Control of circulating bacteria and oxidation of organic matter can be obtained through ozonation of the water. However, ozone (O_3) is highly toxic to aquatic organisms and must be allowed to dissipate prior to exposing the water to the aquaculture animals. With time, and with the assistance of aeration, ozone can be driven off or converted to molecular oxygen.

Ultraviolet (uv) light has also been used to sterilize the water in aquaculture systems. The effectiveness of uv decreases with the thickness of the water column being treated, so the water is usually flowed past uv lights as a thin film (alternatively, the water may flow through a tube a few cm in diameter that is surrounded by uv lights). Uv systems require more routine maintenance than ozone systems.

Recirculating systems often feature other types of apparatus, such as foam strippers and supplemental aeration. The technology for denitrifying nitrate to nitrogen gas has developed to the point that it may find a place in commercial culture systems in the near future. Computerized water-quality monitoring systems that will sound alarms and call emergency telephone numbers to report systems failures to the culturists are finding increased use among culturists using recirculating systems.

The technology involved makes recirculating systems expensive to construct and operate. However, recirculating systems can make aquaculture feasible in locations where conditions would otherwise not be conducive to successful operations.

Water Sources and Quality

Sources of water for aquaculture include municipal supplies, wells, springs, streams, lakes, reservoirs, estuaries, and the ocean. The water may be used directly from the source or it may be treated in some fashion prior to use (see WATER).

Many municipal water sources are chlorinated and contain sufficiently high levels of chlorine so as to be toxic to aquatic life. Chlorine can be removed by passing the water through activated charcoal filters or through the use of sodium thiosulfate metered into the incoming water.

The most commonly used pretreatments of well water (both fresh and saline) include temperature alteration (either heating or cooling); aeration to add oxygen or to remove or oxidize such substances as carbon dioxide, hydrogen sulfide, and iron and increasing salinity (in mariculture systems). Pretreatment may also include adjusting pH, hardness, and alkalinity.

To heat or cool water requires large amounts of energy. A major consideration in locating an aquaculture facility is to have water at or near the optimum temperature for growing the species that has been selected. Many aquaculture facilities that utilize surface waters and wells are required to pump the water into their facilities. Pumping costs can be a major expense, particularly when the facility requires continuous inflow.

Surface water can sometimes be obtained through gravity flow by locating aquaculture facilities at elevations below those of adjacent springs, streams, lakes, or reservoirs. Coastal facilities may be able to obtain water through tidal flow.

The most common treatment of incoming surface water is removal of particulate matter. This can be effected through the use of settling basins or filtration to remove suspended inorganic material such as clay, silt, and sand, and organic material, including living organisms.

For many freshwater species that can be characterized as warmwater (such as channel catfish and tilapia) or coldwater (such as trout), the conditions outlined in Table 2 should provide an acceptable environment. So-called midrange species are those with an optimum temperature for growth of about 25°C (examples are walleye, northern pike, muskellunge, and yellow perch). Typically they do well under the conditions, other than temperature, specified in Table 2 for coldwater species. Recommended water quality conditions for marine fish production systems are presented in Table 3.

The water quality criteria for each species should be determined from the literature or through experimentation when literature information is unavailable. Synergistic effects that occur among water quality variables can have an influence on the tolerance a species has under any given set of circumstances.

Biocides should not be present in water used for aquaculture. Trace metal levels should be low as indicated in Tables 2 and 3.

Most aquaculture facilities release water constantly or periodically into the environment without passing it through a municipal sewage treatment plant. The effects of those effluents on natural systems have become a subject of intense scrutiny. Research is currently underway to develop feeds containing reduced levels of nutrients or to provide nutrients in forms that can better be utilized by the culture animals. The goal in both approaches is to reduce losses of nutrients to the environment through excretion.

Nutrition and Feeding

There are cases in which intentional fertilization is used by aquaculturists in order to produce desirable types of natural food for the species under culture. Provision of live foods is currently necessary for the early stages of many aquaculture species because acceptable prepared feeds have yet to be developed. However, some of the most popular aquaculture species do accept prepared feeds from first feeding. Included are catfish, tilapia, salmon, and trout.

Requirements for energy, protein, carbohydrates, lipids, vitamins and minerals have been determined for the species commonly cultured. As a rule of thumb, trout and salmon diets will, if consumed, support growth and survival in virtually any aquaculture species. Such diets often serve as the control against which experimental diets are compared.

Studies have also been conducted on antinutritional factors in feedstuffs and on the use of additives. Certain feed ingredients contain chemicals that retard growth or may actually be toxic. Restriction on the amount of the feedstuffs used is one way to avoid problems. Another is proper processing which can destroy the antinutritional factor.

When color development is an important consideration whether external or of the flesh, it can be achieved by incorporating ingredients that contain pigments or by adding extracts or synthetic compounds. One class of additives that impart color is the carotenoids.

Odor also may be used to induce ingestion when the pelleted feed is not readily accepted.

Prepared feeds are marketed in various forms from very fine particles through crumbles, flakes, and pellets. Pelleted rations may be hard, semimoist, or moist. The most widely used types of prepared feeds are produced by pressure pelleting or extrusion.

Nearly all aquaculture feeds contain at least some animal protein since the amino acid levels in plant proteins cannot meet the requirements of most aquatic animals. Fish meal is the most commonly used source of animal protein. Most formulations contain a few percent of added fat as well as added vitamins and minerals. Purified amino acids, binders, carotenoids, and antioxidants are other components found in many feeds. Growth hormone and antibiotics are sometimes used. Regulations on the incorporation of hormones along with other chemicals and drugs into aquatic animal feeds are in place in the United States and some other countries (Table 4). Few such regulations have been promulgated in developing nations.

Feeding practices vary from species to species. It is important not to overfeed since waste feed can lead to degradation of water quality. Most species require only three to four percent of body weight in dry feed daily for optimum growth. Very young animals are an exception because they are growing rapidly and consume a greater percentage of body weight daily.

Table 2. General Water Quality Requirements for Trout and Warmwater Aquatic Animals in Fresh Water

Variable	Acceptable level or range	
	Cold water	Warm water
temperature, °C	<20	26–30
alkalinity, mg/L	10–400	50–400
dissolved oxygen, mg/L	>5	≥5
hardness, mg/L	10–400	50–400
pH	6.5–8.5	6.5–8.5
total ammonia, mg/L	<0.1	<1.0
ferrous iron, mg/L	0	0
ferric iron, mg/L	0.5	0–0.5
carbon dioxide, mg/L	0–10	0–15
hydrogen sulfide, mg/L	0	0
cadmium, μg/L	<10	<10
chromium, μg/L	<100	<100
copper, μg/L	<25	<25
lead, μg/L	<100	<100
mercury, μg/L	<0.1	<0.1
zinc, μg/L	<100	<100

Table 3. Suggested Water Quality Conditions for Marine Fish Production Facilities

Variable	Acceptable level or range
temperature, °C	1–40 (depends on species)
salinity, g/kg	1–40 (depends on species)
dissolved oxygen, mg/L	>6
pH	<7.9–8.2
total ammonia, μg/L as NH_3	<10
iron, μg/L	100
carbon dioxide, mg/L	<10
hydrogen sulfide, μg/L	<1
cadmium, μg/L	<3
chromium, μg/L	<25
copper, μg/L	<3
mercury, μg/L	<0.1
nickel μg/L	<5
lead, μg/L	<4
zinc, μg/L	<25

Table 4. Therapeutants and Disinfecting Agents Approved for Use in United States Aquaculture

Name of compound	Use of compound
Therapeutants	
copper	antibacterial for shrimp
formalin	parasiticide for various species
furanace (Nifurpyrinol)	antibiotic for aquarium fishes
oxytetracycline (Terramycin)	antibiotic for fishes and lobsters
sodium chloride	osmoregulatory enhancer for fishes
sulfadimethoxine (Romet)	antibacterial for salmonids and catfish
trichlorofon (Masoten)	parasiticide for baitfish and goldfish
Disinfectants	
calcium hypochlorite (HTH)	used in raceways and on equipment
didecyl dimethyl ammonium chloride (Sanaqua)	used in aquaria and fish-holding equipment
povidone–iodine compounds (Argentyne, Betadine, Wescodyne)	disinfection of fish eggs

Reproduction and Genetics

Selective breeding has been long-practiced as a mean of improving aquaculture stocks. Most of the species that are being reared in significant quantities around the world are produced in hatcheries using either captured or cultured broodstock.

Fish breeders have worked with varying degrees of success to improve growth and disease resistance in a number of species. As genetic engineering techniques are adapted to aquatic animals, dramatic and rapid changes in the genetic makeup of aquaculture species may be possible. However, since it is virtually impossible to prevent escapement of aquacultured animals into the natural environment, potential negative impacts of such organisms on wild populations cannot be ignored.

Diseases and Their Control

Aquatic animals are susceptible to a variety of diseases including those caused by viruses, bacteria, fungi, and parasites. A range of chemicals and vaccines has been developed for treating the known diseases, although some conditions have resisted all control attempts to date and severe restrictions on the use of therapeutants in some nations has impaired that ability of aquaculturists to control disease outbreaks. The United States is a good example of a nation in which the variety of treatment chemicals is limited (see Table 4). Maintenance of conditions in the culture environment that keep stress to a minimum is one of the best methods of avoiding diseases.

Harvesting, Processing, and Marketing

Harvesting techniques vary depending on the type of culture system involved. Seines are often used to capture fish from ponds, or the majority of the animals can be collected by draining the pond through netting. Fish pumps are available that can physically remove aquatic animals directly onto hauling trucks from ponds, raceways, cages and net-pens without causing damage to the animals.

Aquaculturists may harvest, and even process their own crops, although custom harvesting and hauling companies are available. Some processing plants also provide harvesting and live-hauling services.

Centralized processing plants specifically designed to handle regional aquaculture crops are established in areas where production is sufficiently high. In coastal regions, aquacultured animals are often processed in plants that also service capture fisheries.

Marketing can be done by aquaculturists who operate their own processing facilities. Most aquaculture operations depend on a regional processing plant to market the final product.

ROBERT STICKNEY
Texas A&M University

R. R. Stickney, *Principles of Aquaculture,* John Wiley & Sons, Inc., New York, 1994.

FAO, *Agriculture Production Statistics, 1974–1993,* Food and Agriculture Organization of the United Nations, 1995.

FAO, *FAO Fisheries Circular 815, revision 2,* Food and Agriculture Organization of the United Nations, 1991.

Aquaculture Buyer's Guide '95 and Industry Directory, Vol. 8, 1995 Aquaculture Magazine, Asheville, N.C.

AQUACULTURE CHEMICALS

Intensive or extensive culture of aquatic animals requires chemicals that control disease, enhance the growth of cultured species, reduce handling trauma to organisms, improve water quality, disinfect water, and control aquatic vegetation, predaceous insects, or other nuisance organisms. The aquacultural chemical needs for various species have been described for rainbow trout, *Oncorhynchus mykiss;* Atlantic and Pacific salmon, *Salmo salar* and *Oncorhynchus* sp.; channel catfish, *Ictalurus punctatus;* striped bass, *Morone saxatilis;* milkfish, *Chanos chanos;* mollusks; penaeid (*Penaeus* sp.) shrimp; and a variety of other marine species.

Laws and regulations on the use of chemicals in aquaculture vary by country and serve to ensure safe and effective use and protection of humans and the environment. Regulations and therapeutants or other chemicals that are approved or allowed for use in the United States, Canada, Europe, and Japan are presented below.

Regulation of Aquacultural Chemicals in the United States

In the United States, the application of chemicals to organisms or to their environments is regulated by the U.S. Food and Drug Administration (FDA) and the U.S. Environmental Protection Agency (EPA). FDA controls the use of drugs and anesthetics, and EPA controls the application of chemicals and pesticides to the environment. In cases that involve water treatments to control pathogens, the jurisdiction becomes unclear and has been changed over time. Each agency develops appropriate guidelines and policies to implement the laws for its field of responsibility.

Registered Aquacultural Chemicals in the United States

Antibacterials. Few therapeutants are registered in the United States for use on any cultured aquatic species. In the most critical area of antibacterials, only two (Terramycin for Fish and Romet-30) are fully registered and available.

Fungicides. Formalin (Formalin-F, Parasite-S, Paracide-F) is the only registered fungicide and is labeled for use on eggs of trout, salmon, and esocids at 1,000–2,000 mg/L for 15 min.

Parasiticides. Formalin (Formalin-F, Parasite-S, Paracide-F) is the only parasiticide currently registered and available. It is only registered for use on trout, salmon, catfish, largemouth bass (*Micropterus salmoides*), and bluegill (*Lepomis macrochirus*). Compounds considered to be Low Regulatory Priority Substances as parasiticides include acetic acid, calcium oxide, garlic, magnesium sulfate, onion and sodium chloride.

Disinfectants. Povidone–iodine compounds can be used to disinfect the surface of fish eggs because FDA has classified them as low regulatory priority substances. Several chemicals can be used on hard surfaces as disinfectants where no fish are present: calcium hypochlorite, sodium hypochlorite, and didecyl dimethyl ammonium chloride, and other quaternary ammonium compounds, among others.

Water Treatment Compounds. Of particular interest is potassium permanganate which is exempted from registration by EPA when used as an oxidizer or detoxifier. Compounds considered to be Low Regulatory Priority Substances as water treatment compounds include calcium chloride, potassium chloride, calcium carbonate, calcium hydroxide, sodium hydroxide, tris buffer, and sodium sulfite.

Anesthetics. Tricaine is the only currently registered anesthetic and requires a 21-day withdrawal time. Compounds considered to be Low Regulatory Priority Substances as anesthetics include carbon dioxide gas and sodium bicarbonate.

Herbicides. An array of herbicides are registered for use in aquatic sites, but copper sulfate, chelated copper compounds, and diquat dibromide are of additional interest because they also have therapeutic properties.

Piscicides. The two piscicides antimycin and rotenone are both used in ponds to control nuisance fish.

Regulation and Registration of Aquacultural Chemicals Outside the United States

The control of aquacultural drugs varies according to country from no regulation to restrictive regulations. Generally, few requirements are needed for a therapeutant to be licensed or registered in South America, Africa, and most of Asia. Seafood-exporting countries are increasingly concerned because importing countries may no longer accept products without a guarantee that the products contain no chemical residues of concern.

Canada. Except for environmental studies, requirements for registration data in Canada are similar to requirements in the United States. However, Canada has significantly different regulations and approval processes. Canadian aquaculturalists use drugs that are either licensed for other food animals and prescribed by veterinarians or used in an emergency under the direction of the Canadian Bureau of Veterinary Drugs (BVD). Chemicals authorized for use in aquaculture in Canada include oxytetracycline, $C_{22}H_{24}N_2O_9$; sulfadimethoxine (Romet-30), $C_{12}H_{14}N_4O_4S$, and ormetoprim (Romet-B), $C_{14}H_{18}N_4O_2$ erythromycin, $C_{37}H_{67}NO_{13}$; dichlorvos (Nuvan), $C_4H_7Cl_2O_4P$; oxolinic acid, $C_{13}H_{11}NO_5$; penicillin G, $C_{16}H_{18}N_2O_4S$; sulfamerazine $C_{11}H_{12}N_4O_2S$; chloramine-T, $C_7H_7ClNNaO_2S$; formalin, CH_2O; sodium chloride, NaCl; and malachite green, $C_{23}H_{25}ClN_2$.

Europe. Chemicals authorized for use in aquaculture in certain European countries include ampicillin, $C_{16}H_{19}N_3O_4S$; chloramphenicol, $C_{11}H_{12}Cl_2N_2O_5$; chlortetracycline, $C_{22}H_{23}ClN_2O_8$; dibutyltin oxide, $C_8H_{18}SnO$; dimetridozole, $C_5H_7N_3O_2$; estomycine; flumequine, $C_{14}H_{12}FNO_3$; furaltadone, $C_{13}H_{16}N_4O_6$; furazolidone $C_8H_7N_3O_5$; nifurpracine hydrochloride; 3-nitro-2-thiazolylamine; nitrofurazone (Tricofuron), $C_6H_6N_4O_4$; oxolinic acid (Aqualinic Powder, etc), $C_{13}H_{11}NO_5$; oxytetracycline (Terramycin, etc), $C_{22}H_{24}N_2O_9$;paramomycinsulfate; sulfadiazine, $C_{10}H_{10}N_4O_2S$, and trimethoprim (Tribrissen, etc), $C_{14}H_{18}N_4O_3$; sulfamerazine, $C_{11}H_{12}N_4O_2S$; sulfadimethoxine, $C_{12}H_{14}N_4O_4S$, and trimethoprim, $C_{14}H_{18}N_4O_3$; sulfamethazine, $C_{12}H_{14}N_4O_2S$; sulfadoxine, $C_{12}H_{14}N_4O_4S$; chloramine-T, $C_7H_7ClNNaO_2S$; copper sulfate, $CuSO_4$; dichlorvos (Nuvan EC, Aquaguard), $C_4H_7Cl_2O_4P$; formalin, CH_2O; potassium permanganate, $KMnO_4$; povidone–iodine compounds; sodium chloride, NaCl; trichlorfon $C_4H_8C_{13}O_9P$; chlorobutanol, $C_4H_7Cl_3O$; carbon dioxide CO_2; and tricaine, $C_{10}H_{15}NO_5S$.

Japan. In Japan, registration of drugs for aquatic species requires the same data as those required for drugs on other animals. The Ministry of Agriculture, Forests, and Fisheries and the Ministry of Welfare control the use of chemicals and aquaculture in Japan. As of July 1990, more chemicals were registered for aquacultural use in Japan than in any other country. They include amoxicillin, $C_{16}H_{19}N_3O_5S$; ampicillin, $C_{16}H_{19}N_3O_4S$; colistin, $C_{53}H_{100}N_{16}O_{13}$; doxycycline $C_{22}H_{24}N_2O_8H_2O$; erythromycin, $C_{37}H_{67}NO_{13}$; florfenicol, $C_{12}H_{14}Cl_2FNO_4S$; flumequine, $C_{14}H_{12}FNO_3$; josamycin, $C_{42}H_{69}NO_{15}$; lincomycin, $C_{18}H_{34}N_2O_6S$; miloxacin, $C_{12}H_9NO_6$; nalidixic acid, $C_{12}H_{12}N_2O_3$; nifurstylenic acid; novobiocin, $C_{31}H_{36}N_2O_{11}$; oleandomycin, $C_{35}H_{61}NO_{12}$; oxolinic acid, $C_{13}H_{11}NO_5$; oxytetracycline, $C_{22}H_{24}N_2O_9$; piromidic acid, $C_{14}H_{16}N_4O_3$; spiramycin, $C_{43}H_{74}N_2O_{14}$; sulfamonomethoxine, $C_{11}H_{12}N_4O_3S$, and ormetoprim, $C_{14}H_{18}N_4O_2$; sulfisozole; tetracycline, $C_{22}H_{24}N_2O_8$; thiamphenicol, $C_{12}H_{15}Cl_2NO_5S$; eugenol, $C_{10}H_{12}O_2$; $C_9H_8N_2S{\cdot}BrH$; tricaine, $C_{10}H_{15}NO_5S$; hydrogen peroxide, H_2O_2; and trichlorfon, $C_4H_8Cl_3O_4P$.

Promising Chemicals for Registration for Aquaculture

More therapeutants and vaccines may soon be added to the medicine chest of fish farmers. A variety of chemicals have potential for registration and use in aquaculture. The U.S. Fish and Wildlife Service (USFWS) and other agencies and organizations are developing data under investigational new animals drug permits (INAD) for several of these compounds.

Antibacterials. Research has been conducted on three important antibacterial compounds in the United States under INAD's: chloramine-T, sarafloxacin, erythromycin, oxytetracycline (extension), copper sulfate, and potassium permanganate.

Fungicides and Parasiticides. USFWS is working on several fronts to improve the availability of fungicides and parasiticides.

Fungicide testing at the National Fisheries Research Center (La Crosse, Wisco.) has produced several candidates, including compounds that have been used in the past as fungicides: formalin and sodium chloride. USFWS is working on defining the use patterns of these compounds and extending the use of formalin as a fungicide under an INAD to other fish species and the fish themselves. Hydrogen peroxide is another potential candidate fungicide.

USFWS has initiated INADs on copper sulfate, chelated copper compounds, potassium permanganate, and formalin (extension to other species) to develop data for their use as parasiticides.

Disinfectants. Promising disinfectants include ultraviolet (uv) light and ozone, O_3.

Anesthetics. Ethyl aminobenzoate (benzocaine), $C_9H_{11}NO_2$, is the only anesthetic candidate that might allow spawned-out broodstock carcasses to be used for pet or human food. Electronarcosis is an alternative to chemical anesthesia that uses varying electrical frequencies to rapidly anesthetize fishes and allow gentle recovery.

ROSALIE A. SCHNICK
U.S. Fish and Wildlife Service

D. J. Alderman in J. F. Muir and R. J. Roberts, eds., *Recent Advances in Aquaculture,* Vol. 3, Timber Press, Portland, Oreg., 1988, p. 1.

F. P. Meyer and R. A. Schnick, *Rev. Aquat. Sci.* **1,** 693 (1989).

C. Michel and D. J. Alderman, eds., *Chemotherapy in Aquaculture: from Theory to Reality,* Office International des Epizooties, Paris, France, 1992.

Joint Subcommittee on Aquaculture, Working Group on Quality assurance in aquaculture production, Federal regulation of Drugs, biologicals, and chemicals used in aquaculture production, National Agricultural Library, Beltsville, MD., 1992.

ARAMID FIBERS. See HIGH PERFORMANCE FIBERS.

ARGON. See HELIUM-GROUP GASES.

ARSENIC AND ARSENIC ALLOYS

Arsenic, although often referred to as a metal, is classified chemically as a nonmetal or metalloid and belongs to Group 15 (VA) of the Periodic Table (as does antimony). The principal valences of arsenic are +3, +5, and −3. Only one stable isotope of arsenic having mass 75 (100% natural abundance) has been observed.

Properties

Physical properties of α-crystalline metallic arsenic are given in Table 1. The properties of β-arsenic are not completely defined. The density of β-arsenic is 4700 kg/m^3; it transforms from the amorphous to

Table 1. Physical Properties of Arsenic

Property	Value
atomic weight	74.9216
mp at 39.1 MPa[a], °C	816
bp, °C	615[b]
density at 26°C, kg/m^3	5,778
specific heat at 25°C, J/(mol·K)[c]	24.6
linear coefficient of thermal expansion at 20°C, μm/(m·°C)	5.6
electrical resistivity at 0°C, $\mu\Omega$·cm	26
hardness, Mohs' scale	3.5

[a] To convert MPa to psi, multiply by 145.
[b] Sublimes.
[c] To convert to cal/(mol·K), divide by 4.184.

the crystalline form at 280°C; and the electrical resistivity is reported to be 107 Ω·cm.

Metallic arsenic is stable in dry air, but when exposed to humid air the surface oxidizes, giving a superficial golden bronze tarnish that turns black upon further exposure. The amorphous form is more stable to atmospheric oxidation. Upon heating in air, both forms sublime and the vapor oxidizes to arsenic trioxide, As_2O_3.

Elemental arsenic combines with many metals to form arsenides.

Arsenic is widely distributed about the earth and has a terrestrial abundance of approximately 5 g/t. The most important commercial source of arsenic, however, is as a by-product from the treatment of copper, lead, cobalt, and gold ores.

Economic Aspects

The demand for arsenic metal is limited. The 1990 U.S. requirement for metallic arsenic was supplied by the People's Republic of China. Commercial arsenic metal is sold at a typical purity of 99% in fragment or lump (5–7.5-cm) form.

Arsenic metal is also offered in high (ranging from 99.99% to in excess of 99.999%+) purity form for semiconductor applications. The United States is not self-reliant in its requirements for high purity arsenic and depends on imports primarily from Japan, Canada, and the United Kingdom.

Uses

The predominant use of arsenic in the United States is in the manufacture of chemicals. During the 1980s, the market for arsenic chemicals had shifted from cotton farming, where its use is now restricted because of environmental considerations, to wood (qv) preservatives for the protection of lumber and other wood products. Arsenic trioxide is the basic commodity of commerce from which a number of important chemicals are manufactured.

Alloys. Arsenic metal is used primarily in alloys in combination with lead and, to a lesser extent, copper.

Trace quantities of arsenic are added to lead–antimony grid alloys used in lead–acid batteries (see BATTERIES, SECONDARY CELLS–LEAD–ACID).

Minor additions of arsenic (0.02–0.5%) to copper (qv) raise the recrystallization temperature and improve corrosion resistance. In some brass alloys, small amounts of arsenic inhibit dezincification and minimize season cracking.

Phosphorized deoxidized arsenical copper (alloy 142) is used for heat exchangers and condenser tubes.

Health and Safety Factors

The toxicity of arsenic ranges from very low to extremely high depending on the chemical state. Metallic arsenic and arsenious sulfide, As_2S_3, have low toxicity. Arsine is extremely toxic. The toxicity of other organic and inorganic arsenic compounds varies.

Arsenic is classified as a carcinogen by the International Agency for Research on Cancer (IARC). The handling of arsenic in the workplace should be in compliance with the U.S. Occupational Safety and Health Administration (OSHA) regulations. Precautions should be taken to avoid accidental generation of arsine gas. Disposal of arsenical products should be in compliance with Federal and local government environmental regulations.

S. C. CARAPELLA, JR.
Consultant

J. W. Mellor, *Comprehensive Treatise on Inorganic and Theoretical Chemistry,* Vol. 9, Longmans, Green & Co., Inc., New York, 1930, pp. 3–9, 16–19.

S. Wallden and H. Hilmer, *Ulmanns Encyclopadie der Technischen Chemie,* Verlag Chemie, GmbH, Weinheim, Germany, 1974, pp. 53–55.

R. J. Bauer, in W. H. Lederer and R. J. Fensterheim, eds., *Arsenic: Industrial, Biomedical, Environmental Perspectives,* Van Nostrand Reinhold Co., New York, 1983, pp. 45–55.

Code of Federal Regulations, Title 29, Part 1910.1018, U.S. Food and Drug Administration, Washington, D.C., rev. July 11, 1988.

ARSENIC COMPOUNDS

Arsenic is the third member of the nitrogen family of elements and hence possesses an outermost shell having the electron configuration of $4s^24p^3$. The most common oxidation states of arsenic are −3, +3, and +5, although compounds containing the simple As^{3-}, As^{3+}, and As^{5+} ions are unknown. In the majority of arsenic compounds the arsenic atom is in the tetrahedral valence state.

Arsenic compounds have numerous practical applications. Although a variety of inorganic and organic arsenicals are used in commerce, arsenic trioxide, As_2O_3, accounted for 98% of the arsenic consumed in 1988.

The commercial uses of arsenic compounds in 1988, measured in terms of elemental arsenic, are wood (qv) preservatives, 69%; agricultural products (herbicides (qv) and desiccants (qv)), 23%; glass (qv), 4%; nonferrous alloys and electronics, 2%; and animal feed additives and pharmaceuticals (qv), 2% (see FEEDS AND FEED ADDITIVES). Chromated copper arsenate (CCA) is the most widely used arsenic-based wood preservative. The U.S. Environmental Protection Agency has, however, restricted the use of arsenical wood preservatives to certified applicators.

Arsenic compounds must be considered extremely poisonous. In spite of the toxicity of arsenic compounds, there is evidence that arsenic is an essential nutrient for several animal species.

Inorganic Compounds

Arsenic Hydrides. Although there are occasionally reports of other arsenic hydrides, eg, As_2H_4, As_2H_2 (or AsH), and As_4H_2, the only well-characterized binary compound of arsenic and hydrogen is arsine.

Arsine, AsH_3, is a colorless, exceedingly poisonous gas with an unpleasant garlic-like odor; mp, −116.3°C; bp, −62.4°C; density of liquid at −64.3°C, 1.640 g/mL; $\Delta H°f_{298}$, 66.44 kJ/mol (16 kcal/mol); $\Delta S°$222.7 J/(mol·K) (53 cal/(mol·K)). Arsine is soluble to the extent of 2 mL at 101 kPa (1 atm) per 100 g of water at RT.

Arsine is formed when any inorganic arsenic-bearing material is brought in contact with zinc and sulfuric acid.

Arsine is not particularly stable and starts to decompose into its elements well below 300°C. If moisture is present, light effects the decomposition. Arsine is a good reducing agent, capable of reducing many substances.

Arsine is used for the preparation of gallium arsenide, GaAs, and there are numerous patents covering this subject (see ARSENIC AND ARSENIC ALLOYS).

Other arsenic hydrides include diarsine, As_2H_4, arsenic monohydride, As_2H_2 or AsH, and hydrogen diarsenide, As_4H_2 or As_2H.

Arsenic Halides. Arsenic forms a complete series of trihalides, but arsenic pentafluoride is the only well-known simple pentahalide. All of the arsenic halides, the physical properties of which are given in Table 1, are covalent compounds that hydrolyze in the presence of water. The trihalides form pyramidal molecules similar to the trivalent phosphorus analogues and may be prepared by direct combination of the elements.

Arsenic Oxides and Acids. The only arsenic oxides of commercial importance are the trioxide and the pentoxide. These are readily soluble in alkaline solution, forming arsenites and arsenates, respectively.

Arsenic trioxide (arsenic(III) oxide, arsenic sesquioxide, arsenous oxide, white arsenic, arsenic), As_2O_3, is the most important arsenic compound of commerce. The octahedral or cubic modification, arsenolite, $\Delta H°f_{298}$, -1313.9 kJ/mol(-314 kcal/mol); S°_{298}, 214 J/(mol·K) (51 cal/(mol·K))f, is the most common form and has been known from early times.

Arsenic trioxide may be made by burning arsenic in air or by the hydrolysis of an arsenic trihalide. Commercially, it is obtained by roasting arsenopyrite, FeAsS.

Arsenous Acids and the Arsenites. Arsenous acid, As_3O_3, is known to exist only in solution. It is a weak acid with a dissociation constant of 8×10^{-16} at 25°C. Arsenic pentoxide [arsenic oxide, arsenic (V) oxide], As_2O_5, is made up of equal numbers of AsO_6 octahedra and AsO_4 tetrahedra sharing corner oxygens to give cross-linked strands. The compound is thermally unstable and begins to decompose near the melting point, ca 300°C. Arsenic pentoxide is an oxidizing agent capable of liberating chlorine from hydrogen chloride.

Arsenic Acids and the Arsenates. Commercial arsenic acid corresponds to the composition, one mole of arsenic pentoxide to four moles of water, and probably is the arsenic acid hemihydrate, $H_3AsO_4{\cdot}05\ H_2O$.

Arsenates are oxidizing agents and are reduced by concentrated hydrochloric acid or sulfur dioxide. Arsenates of calcium or lead are used as insecticides; sodium arsenate is used in printing inks and as a mordant.

Arsenous arsenate (arsenic dioxide, arsenic tetroxide), As_2O_4, is known and probably corresponds to $As(AsO_4)$.

Arsenic Sulfides. The physical properties of the common arsenic sulfides are given in Table 2. Numerous arsenic sulfides have been reported as well as compounds containing the $As_3S_4^+$ cation.

Other arsenic compounds include arsenic trisulfate, $As_2(SO_4)_3$, arsenyl sulfate, $(AsO)_2SO_4$, and arsenic triacetate, $As(C_2H_3O_2)_3$, mp 82°C, bp 165–170°C.

Organic Compounds

Arsenic combines readily with carbon to form a wide variety of compounds containing one or more As—C bonds. These may be broadly divided into As(III) and As(V) compounds. The As(III) compounds contain from one to four organic groups; the As(V) compounds from one to six organic groups.

Primary and Secondary Arsines. Primary arsines are commonly prepared by the zinc–hydrochloric acid reduction of substances containing one organic group attached to arsenic (such as arsonic acids, dihaloarsines, and compounds with arsenic—arsenic bonds); the zinc is often amalgamated or coated with copper.

Secondary arsines, which can be synthesized by methods analogous to those used for primary arsines, are obtained in good yields by the reduction of arsinic acids or haloarsines with amalgamated zinc and hydrochloric acid. Methylarsine, trifluoromethylarsine, and bis(trifluoromethyl)arsine, C_2HAsF_6, are gases at room temperature; all other primary and secondary arsines are liquids or solids. These compounds are extremely sensitive to oxygen, and in some cases are spontaneously inflammable in air. They readily undergo addition reactions with alkenes, alkynes, aldehydes (qv), ketones (qv), isocyanates, and azo compounds. They also react with diborane and a variety of other Lewis acids. Alkyl halides react with primary and secondary arsines to yield quaternary arsenic compounds.

Tertiary Arsines. An enormous number of trialkyl- and triarylarsines are known. They are usually prepared by the reaction of an organometallic compound with an arsenic trihalide, a haloarsine, or a dihaloarsine.

Trimethylarsine, C_3H_9As, has been identified as the toxic volatile arsenical, once known as "Gosio gas," produced by the reaction of certain molds that grow on wallpaper paste and react with inorganic arsenic compounds present in the paper.

Most trialkylarsines are volatile liquids with intensely disagreeable odors. They react readily with oxygen, and in some cases they ignite spontaneously when exposed to air. Triarylarsines are solids that can usually be handled in air without danger of oxidation. They are, however, easily converted to triarylarsine oxides with suitable oxidizing agents.

Tertiary arsines have been widely employed as ligands in a variety of transition metal complexes, and they appear to be useful in synthetic organic chemistry, eg, for the olefination of aldehydes.

Haloarsines, Dihaloarsines, and Related Compounds. Halo- and dihaloarsines (R_2AsX and $RAsX_2$, where X = Cl or Br) are easily obtained by the reduction of the corresponding arsinic or arsonic acids in hydrochloric or hydrobromic acid.

Dihalo- and haloarsines are very reactive substances.

Diarsines and Diarsenes. A number of diarsines have been obtained by the reduction of arsinic acids with phosphorous or hypophosphorous acid. Diarsines can also be prepared by the treatment of a metal dialkyl- or diarylarsenide with iodine or a 1,2-dihaloethane.

Diarsines are extremely reactive compounds.

Diarsenes, the arsenic analogues of azo compounds, have been obtained in good yield by the base-promoted interaction of a primary arsine and a dichloroarsine, both of which were sterically hindered.

Cyclic and Polymeric Substances Containing Arsenic–Arsenic Bonds. A number of organoarsenic compounds containing rings of four, five, or six arsenic atoms have been prepared (cyclic polyarsines). The first such four-membered ring compound to be adequately characterized, tetrakis(trifluoromethyl)tetrarsetane, was obtained by the interaction of a diiodoarsine and mercury. Other compounds containing rings of four or five arsenic atoms can be prepared by the treatment of dichloroarsines with sodium.

Cyclic polyarsines undergo a number of reactions with transition metal compounds to form complexes containing both As—As and As—metal bonds.

A large number of polymeric substances, $(RAs)_x$ or $(ArAs)_x$, are also known. They are usually prepared by the reduction of arsonic acids with hypophosphorous acid or sodium dithionite.

Arsenin and Its Derivatives. Arsenin (arsabenzene), C_5H_5As, the arsenic analogue of pyridine, can be prepared by the treatment of 1,4-dihydro-1,1-dibutylstannabenzene with arsenic trichloride. A large number of arsenin derivatives have also been studied.

Arsonic and Arsinic Acids. The arsonic acids, compounds of the type $RAsO(OH)_2$, are among the most important organic arsenicals. The aliphatic arsonic acids are generally prepared by the Meyer reaction; ie, heating an alkyl halide with As_4O_6 in alkaline solution. Aromatic arsonic acids are generally prepared by the Bart reaction from an aromatic diazonium salt and sodium arsenite.

Both arsonic and arsinic acids give precipitates with many metal ions, a property which has found considerable use in analytical chemistry. Of particular importance are certain azo dyes (qv) containing both arsonic and sulfonic acid groups which give specific color reactions with a wide variety of transition, lanthanide, and actinide metal ions.

Arsine Oxides. Both aliphatic and aromatic arsine oxides, compounds of the type R_3AsO, are well known. These compounds have been prepared by the oxidation of trialkylarsines using mercury(II) oxide or 30% hydrogen peroxide, with the rigid exclusion of oxygen. The products, which are purified by sublimation, are extremely hygroscopic and can only be handled under dry box conditions.

Table 1. Physical Properties of Arsenic Halides.

Arsenic halide	Color and physical state at 25°C	Mp, °C	Bp, °C	Specific gravity[a]	Heat of formation, ΔH°_{298}, kJ/mol[b]	Entropy, S°_{298}, J/(mol·K)[b]
arsenic trifluoride (AsF_3)	colorless liquid	−6.0	62.8	2.666^{0}	−956.25	
arsenic pentafluoride (AsF_5)	colorless gas	−79.8	2.8	2.33^{-53}		
arsenic trichloride ($AsCl_3$)	colorless liquid	−16.2	130.2	2.205^{0}	−305	208
arsenic tribromide ($AsBr_3$)	yellow solid	31.2	221	3.66^{15}	−197	363.8[c]
arsenic triiodide (AsI_3)	red solid	140.4	ca 400	4.39^{15}	−58.2	213.0

[a] Temperature, °C, of measurement given as a superscript.
[b] To convert J to cal, divide by 4.184.
[c] Gaseous phase.

Table 2. Physical Properties of Arsenic Sulfides

Arsenic sulfides	Molecular formula	Color and physical state at 25°C	Heat of formation, ΔH_{298}, kJ/mol[a]
arsenous sulfide (orpiment)[b]	As_2S_3	yellow solid	−338
arsenic sulfide (realgar)	As_4S_4	red or orange solid	−285
arsenic pentasulfide	As_4S_{10}	yellow solid	
tetraarsenic trisulfide	As_4S_3	orange-yellow	
tetraarsenic pentasulfide	As_4S_5		

[a] To convert J to cal, divide by 4.184.
[b] The entropy, S°_{298}, is 327 J/(mol · K)).

Trialkyl- and triarylarsine sulfides have been prepared by several different methods. The reaction of sulfur with a tertiary arsine, with or without a solvent, gives the sulfides in almost quantitative yields.

Haloarsoranes. Halides of the types $RAsX_4$, R_2AsX_3, R_3AsX_2, and R_4AsX are known. The R_4AsX compounds are ionic in nature and are discussed under arsonium salts. The tetrahalides are unstable compounds which frequently decompose on standing and are readily hydrolyzed in moist air to form arsonic acids.

Trialkyl- and triaryldihaloarsoranes have been studied to a much greater extent than the tri- and tetrahaloarsoranes. The dihalo compounds are stable crystalline species, although they decompose on heating. The usual method of preparing these compounds is by direct halogenation of tertiary arsines.

Arsonium Salts. Arsonium salts are compounds of the type R_4AsY, where R may be either an alkyl or aryl group and Y is a wide variety of negative groups, such as halogen, nitrate, sulfate, and perchlorate.

Arsonium salts have found considerable use in analytical chemistry.

Arsonium Ylides. Arsonium ylides were first prepared by reaction between an arsonium halide and phenyllithium. Arsenic ylides of the types $R_3As{=}CH_2$ and $R_3As{=}CHR'$ are unstable and generally react with carbonyl compounds to give either a tertiary arsine and an epoxide, or a mixture of these two compounds and the normal Wittig reaction products (an alkene and a tertiary arsine oxide). By contrast, arsenic ylides of the type $R_3AsCHC(O)R'$ are stable and react with carbonyl compounds to give the normal Wittig reaction products.

Pentaalkyl- and Pentaarylarsoranes. Compounds of the type R_5As, where R may be aliphatic or aromatic, have assumed considerable importance in arsenic chemistry. Although only a few pentaalkylarsoranes have been described in the chemical literature, a large number of pentaarylarsoranes have been prepared, including a number of spirocyclic compounds. These compounds are of particular interest in studies on the stereochemistry of five-covalent compounds.

Arsenic Compounds Used in Medicine

Since 1943, when penicillin was shown to be effective for the therapy of syphilis (see ANTIBIOTICS, β-LACTAMS–PENICILLINS AND OTHERS), there has been much less work on the use of organoarsenic compounds in medicine. No important new arsenical drug has been introduced. However, arsenicals are still important for the treatment of African trypanosomiasis; they are probably indispensable for the late neurological stage of the disease (see ANTIPARASITIC AGENTS, ANTHELMINTICS). Toxic reactions caused by arsenicals can often be successfully treated using 2,3-dimercapto-1-propanol (dimercaprol, BAL), $C_3H_8OS_2$.

Arsenamide (thiacetarsamide), $C_{11}H_{12}AsNO_5$, is a thioarsenite that is employed in veterinary medicine. Although relatively toxic, it has proved useful for the treatment of *Dirofilaria immitis* (heartworm) infestation in dogs.

G. O. DOAK
G. GILBERT LONG
LEON D. FREEDMAN
North Carolina State University

Minerals Yearbook for 1988, Metals and Minerals, Vol. 1, U.S. Department of the Interior, Washington, D.C., 1990, pp. 1059–1062.

S. Samaan, *Houben-Weyl Methoden der Organischen Chemie. Metallorganische Verbindungen As, Sb, Bi,* Georg Thieme Verlag, Stuttgart, Germany, 1978.

G. O. Doak and L. D. Freedman, *Organometallic Compounds of Arsenic, Antimony, and Bismuth,* John Wiley & Sons, Inc., New York, 1970.

F. G. Mann, *The Heterocyclic Derivatives of Phosphorus, Arsenic, Antimony and Bismuth,* 2nd ed., Wiley-Interscience, New York, 1970.

ASBESTOS

The term asbestos is a generic designation referring usually to six types of naturally occurring mineral fibers which are or have been commercially exploited. These fibers are extracted from certain varieties of hydrated alkaline silicate minerals comprising two families: serpentines and amphiboles. The serpentine group contains a single fibrous variety: chrysotile; five fibrous forms of amphiboles are known: anthophyllite, amosite, crocidolite, tremolite, and actinolite.

These fibrous minerals share several properties which qualify them as asbestiform fibers: (*1*) they are found in large clusters which can be easily separated from the host matrix or cleaved into thinner fibers; (*2*) the fibers exhibit high tensile strengths; (*3*) they show high length:diameter ratios, from a minimum of 20 up to >1000; (*4*) they are sufficiently flexible to be spun; and (*5*) macroscopically, they resemble organic fibers such as cellulose.

Since asbestos fibers are all silicates, they exhibit several other common properties, such as incombustibility, thermal stability, resistance to biodegradation, chemical inertia toward most chemicals, and low electrical conductivity.

The usual definition of asbestos fiber excludes numerous other fibrous minerals which could be qualified as asbestiform following the criteria listed above. However, it appears the term asbestos has traditionally been attributed only to those varieties which are commercially exploited.

The fractional breakdown of the recent world production of the various fiber types shows that the industrial applications of asbestos fibers have now shifted almost exclusively to chrysotile. Two types of amphiboles, commonly designated as amosite and crocidolite are still being used, but their combined production is currently less than 2% of the total world production. The other three amphibole varieties, anthophyllite, actinolite, and tremolite, have no significant industrial applications presently. This statement excludes asbestiform amphiboles which may occur in other industrial minerals.

Geology and Fiber Morphology

The genesis of asbestos fibers as mineral deposits required certain conditions with regard to chemical composition, nucleation, and fiber growth; such conditions must have prevailed over a period sufficiently long and perturbation-free to allow a continuous growth of the silicate chains into fibrous structures. Some of the important geological or mineralogical features of the industrially significant asbestos fibers are summarized in Table 1.

Crystal Structure of Asbestos Fibers

The microscopic and macroscopic properties of asbestos fibers stem from their intrinsic, and sometimes unique, crystalline features. As with all silicate minerals, the basic building blocks of asbestos fibers are the silicate tetrahedra which may occur as double chains $(SiO)^{6-}114$, as in the amphiboles, or in sheets $(SiO)^{4-}104$, as in chrysotile.

Chrysotile. In the case of chrysotile, an octahedral brucite layer having the formula $(MgO_4(OH)_8)^{4-}6$ is intercalated between each silicate tetrahedra sheet.

Amphiboles. The crystalline structure common to amphibole minerals consists of two ribbons of silicate tetrahedra placed back to back.

Properties of Asbestos Fibers

Asbestos fibers used in most industrial applications consist of aggregates of smaller units (fibrils). This is most evident with chrysotile which exhibits an inherent, well-defined unit fiber. Typical diameters of fibers in bulk industrial samples may reach several tens of micrometers; fiber lengths are on the order of one to ten millimeters.

The mechanical processes employed to extract the fibers from the host matrix, or to further separate (defiberize, open) the aggregates, can impart significant morphological alterations to the resulting fibers. Typically, microscopic observations on mechanically opened fibers reveal fiber bends and kinks, partial separation of aggregates, fiber end-splitting, etc. The resulting product thus exhibits a wide variety of morphological features. The consequences of the peculiar morphology of fiber shapes are difficult to assess, but it is quite obvious that a proper dimensional characterization of these fibers requires a shape factor in addition to diameter and length.

The morphological variance appears more important with chrysotile than with amphiboles. The intrinsic structure of chrysotile, its higher flexibility, and interfibril adhesion allow a variety of intermediate shapes when fiber aggregates are subjected to mechanical shear. Amphibole fibers are generally more brittle and accommodate less morphological deformation during mechanical treatment.

Fiber Length Distribution. For industrial applications, the fiber length and length distribution are of primary importance because they are closely related to the performance of the fibers in matrix reinforcement. Representative distributions of fiber lengths and diameters can be obtained through measurement and statistical analysis of microphotographs; fiber length distributions have also been obtained recently from automated optical analyzers.

Physico-Chemical Properties. The industrial applications of chrysotile fibers were developed taking advantage of their particular combination of properties: fibrous morphology, high tensile strength, resistance to heat and corrosion, low electrical conductivity, and high friction coefficient. In many applications, the surface properties of the fibers also play an important role; in such cases, a distinction between chrysotile and amphiboles can be observed because of their differences in chemical composition and surface microstructure. Technologically relevant physical and chemical properties of asbestos fibers are given in Table 2.

Analytical Methods and Identification of Asbestos Fibers

In a general way, the identification of asbestos fibers can be performed through morphological examination, together with specific analytical methods to obtain the mineral composition and/or structure. Morphological characterization in itself usually does not constitute a reliable identification criterion. Hence, microscopic examination methods and other analytical approaches are usually combined.

Production

After a peak near 1980, production leveled off after 1985 at $4.2-4.3 \times 10^6$ t.

Mining and Milling Technologies

The finding and mapping of chrysotile asbestos ore deposits usually relies on magnetometric surveys. Indeed, magnetite is generally associated with asbestos fiber deposits, except in the case of ore bodies located in sedimentary formations. As in other mining operations, core drilling is used for a precise evaluation of the grade and volume of the ore body.

The choice of a particular mining method depends on a number of parameters, typically the physical properties of the host matrix, the fiber content of the ore, the amount of sterile materials, the presence of contaminants, and the extent of potential fiber degradation during the various mining operations. However, most of the asbestos mining operations are of the open pit type, using bench drilling techniques.

The fiber extraction (milling) process must be chosen so as to optimize recovery of the fibers in the ore, while minimizing reduction of fiber length. Since the asbestos fibers have a chemical composition similar to that of the host rock, the separation processes must rely on differences in the physical properties between the fibers and the host rock rather than on differences in their chemical properties. Dry milling operations are the most widely used.

Fiber Classification and Standard Testing Methods

In the production, or industrial applications, of asbestos fibers, several parameters are considered critically important and are used as standard evaluation criteria: length (or length distribution), degree of opening and surface area, performance in cement reinforcement, and dust and granule content. The measurement of fiber length is important since the length determines the product category in which the fibers will be used and, to a large extent, their commercial value.

Dry Classification Method. The most widely accepted method for chrysotile fiber length characterization in the industry is the Quebec Standard test (QS).

Wet Classification Method. A second industrially important fiber-length evaluation technique is the Bauer-McNett (BMN) classification.

Other classification techniques have been developed which provide some insight on fiber lengths, typically the Ro-Tap test, the Suter-Webb Comb, and the Wash test.

Table 1. Geological Occurrence of Asbestos Fibers

	Chrysotile	Amosite	Crocidolite	Tremolite
mineral species	chrysotile	cummingtonite-grunerite	riebeckite	tremolite
structure	as veins in serpentine	lamellar, coarse to fine, fibrous and asbestiform	fibrous in ironstones	long, prismatic, and fibrous aggregates
origin	alteration and metamorphism of basic igneous rocks rich in magnesium silicates	metamorphic	regional metamorphism	metamorphic
essential composition	hydrous silicates of magnesia	hydroxy silicate of Fe and Mg	hydroxy silicate of Na, Mg, and Fe	hydroxy silicate of Ca and Mg

Table 2. Physical and Chemical Properties of Asbestos Fibers

Property	Chrysotile	Amosite	Crocidolite	Tremolite
color	usually white to grayish green; may have tan coloration	yellowish gray to dark brown	cobalt blue to lavender blue	gray-white, green, yellow, blue
luster	silky	vitreous to pearly	silky to dull	silky
hardness, Mohs	2.5–4.0	5.5–6.0	4.0	5.5
specific gravity	2.4–2.6	3.1–3.25	3.2–3.3	2.9–3.2
optical properties	biaxial positive parallel extinction	biaxial positive parallel extinction	biaxial oblique extinction	biaxial negative oblique extinction
refractive index	1.53–1.56	1.63–1.73	1.65–1.72	1.60–1.64
flexibility	high	fair	fair to good	poor, generally brittle
texture	silky, soft to harsh	coarse but somewhat pliable	soft to harsh	generally harsh
spinnability	very good	fair	fair	poor
tensile strength, MPa[a]	1100–4400	1500–2600	1400–4600	< 500
resistance to:				
acids	weak, undergoes fairly rapid attack	fair, slowly attacked	good	good
alkalies	very good	good	good	good
surface charge, mV (zeta potential)	+13.6 to +54[b]	−20 to −40	−32	
decomposition temperature, °C	600–850	600–900	400–900	950–1040
residual products	forsterite, silica, eventually enstatite	Fe and Mg pyroxenes, magnetite, haematite, silica	Na and Fe pyroxenes, haematite, silica	Ca, Mg, and Fe pyroxenes, silica

[a] To convert MPa to psi, multiply by 145.
[b] Chrysotile fibers tend to become negative after weathering and/or leaching.

Industrial Applications

Asbestos fibers have been used in a broad variety of industrial applications. In the peak period of asbestos consumption in industrialized countries, some 3000 applications, or types of products, have been listed. Because of recent restrictions, many of these applications have now been abandoned and others are pursued under strictly regulated conditions.

The main characteristic properties of asbestos fibers that can be exploited in industrial applications are their thermal, electrical, and sound insulation; nonflammability; matrix reinforcement (cement, plastic, and resins); adsorption capacity (filtration, liquid sterilization); wear and friction properties (friction materials); and chemical inertia (except in acids). These properties have led to several main classes of industrial products or applications: fire protection and heat or sound insulation, fabrication of papers and felts for flooring and roofing products, pipeline wrapping, electrical insulation, thermal and electrical insulation, friction products in brake or clutch pads, asbestos–cement products, reinforcement of plastics, fabrication of packings and gaskets, friction materials for brake linings and pads, reinforcing agents, vinyl or asphalt tiles, and asphalt road surfacing.

Alternative Industrial Fibers and Materials

Table 3 lists some of the materials and fibers that have been suggested or used in the development of asbestos-free products.

Health and Safety

The relationship between workplace exposure to airborne asbestos fibers and respiratory diseases is one of the most widely studied subjects of modern epidemiology.

The research efforts resulted in significant consensus in some areas, although strong controversies remain in other areas. Typically, it is widely recognized that the inhalation of long (considered usually as >5 μm), thin, and durable fibers can induce or promote lung cancer. It is also widely accepted that asbestos fibers can be associated with three types of diseases: asbestosis: a lung fibrosis resulting from long-term, high level exposures to airborne fibers; lung cancer: usually resulting from high level exposures and often correlated with asbestosis; mesothelioma: a rare form of cancer of the lining of the thoracic and abdominal cavities (mesothelium).

A further consensus developed within the scientific community regarding the relative carcinogenicity of the different types of asbestos fibers. There is strong evidence that the genotoxic and carcinogenic po-

Table 3. Asbestos Substitutes and Relative Costs[a]

Minerals	Synthetic mineral fibers	Synthetic organic fibers
	<2 $/kg[b]	
attapulgite	mineral wool	
diatomite	glass wool	
mica		
perlite		
sepiolite		
talc		
vermiculite		
wollastonite		
asbestos, grades 3–7		
	2–10 $/kg	
	steel fibers	polypropylene (PP)
	continuous filament glass	poly(vinyl alcohol) (PVA)
	alkali-resistant glass	polyacrylonitrile (PAN)
	aluminosilicates	
	10–20 $/kg	
	continuous filament glass	polytetrafluoroethylene (PTFE)
	<20 $/kg	
	alumina fibers	polybenzimidazole (PBI)
	silica fibers	aramid fibers
		pitch and PAN carbon fibers

[a] In U.S. $, 1989.
[b] The natural organic fiber, cellulose (pulp), also falls in the <$2/kg range.

tentials of asbestos fibers are not identical; in particular, mesothelial cancer is mostly, if not exclusively, associated with amphibole fibers.

Regulation. The identification of health risks associated with asbestos fibers, together with the fact that huge quantities of these minerals were used ($\approx 5 \times 10^6$t/yr) in a variety of applications, has prompted strict regulations to limit the maximum exposure of airborne fibers in workplace environments.

CARMEL R. JOLICOEUR
JEAN-FRANCOIS ALARY
ASTA SOKOV
Université de Sherbrooke, Canada

M. Ross, R. A. Kuntze, and R. A. Clifton, in B. Levadie, ed., *Definition for Asbestos and Other Health Related Silicates,* ASTM STP 834, American Society for Testing and Materials, Philadelphia, Pa., 1984, pp. 139–147; W. J. Campbell and co-workers, *Selected Silicate Minerals and Their Asbestiform Varieties,* IC 8751, U.S. Bureau of Mines, Washington, D.C., 1977, pp. 5–17, 33.

A. A. Hodgson, in L. Michaels and S. S. Chissick, eds., *Asbestos: Properties, Applications and Hazards,* Vol. 1, John Wiley & Sons, Inc., New York, 1979; A. A. Hodgson, *Scientific Advances in Asbestos, 1967 to 1985,* Anjalena Publication, Crowthorne, UK, 1986.

H. C. W. Skinner, M. Ross, and C. Frondel, *Asbestos and Other Fibrous Materials,* Oxford Press, New York, 1988, pp. 21–23, 25, 31, 34, 35.

A. A. Hodgson, ed., *Alternatives to Asbestos, The Pros and Cons,* John Wiley & Sons, Inc., New York, 1989, p. xi.

"International Symposium on Man-Made Mineral Fibers in the Working Environment," Copenhagen, Denmark, Oct. 28–29, 1986, *Ann. Occup. Hyg.* **31,** 4B (1987).

ASPHALT

Asphalt is defined as a dark brown to black cementitious material in which the predominating constituents are bitumens that occur in nature or are obtained in petroleum processing. Asphalts or bituminous materials are further classified as solids, semisolids, or liquids.

Asphalt characteristically contains very high molecular weight molecular polar species, called asphaltenes, which are soluble in carbon disulfide, pyridine, aromatic hydrocarbons, chlorinated hydrocarbons, and tetrahydrofuran.

Naturally Occurring Materials

Naturally occurring materials include native asphalt (bitumen), lake asphalt, and rock asphalt.

Asphalt (bitumen) also occurs in various oil sand (also called tar sand) deposits which occur widely scattered through the world, and the bitumen is available by means of various extraction technologies.

Manufactured Materials

Petroleum asphalts derive their characteristics from the nature of their crude origins with some variation possible by choice of manufacturing process.

Straight Run Asphalt. In crude-oil refining, the crude oil at 340°C is injected into a fractionating column. The lighter fractions are separated as overhead products and the residuum is straight-run (straight-reduced) asphalt.

Propane Asphalt. Asphalt is also a product of the propane deasphalting and fractionation process which involves the precipitation of asphalt from a residuum stock by treatment with liquid propane under controlled conditions. The petroleum charge stock is usually atmospheric-reduced residue from a primary distillation tower.

Air-Blown Asphalt. Air-blowing is an exothermic process in which an asphalt (flux) is converted to a harder product by air contact at 200–275°C. Dehydrogenation and condensation are involved; oxygen is not retained in the asphalt product, except to a very minor extent.

Air-blown asphalts, more resistant to weather and changes in temperature than the types mentioned previously, are produced by batch and continuous methods.

Thermal Asphalt. Thermal or cracked asphalts differ from other asphalts in that they are products of a cracking (thermal decomposition) process. They have relatively high specific gravity, low viscosity, and high temperature susceptibility, and contain coke precursors (carbenes) as indicated by various tests.

Blended Asphalt. Any particular refinery may produce two grades of asphalt, one at each end of the viscosity spectrum of the entire product grade requirements. Intermediate grades are prepared by blending (proportioning) the extreme. The preparation of asphalts in liquid form by blending (cutting back) an asphalt with a petroleum distillate fraction is customary. There are three general types of cutback asphalt which differ mainly in the diluent used (Table 1).

Asphalt Emulsions. In the most common asphalt emulsion, the oil-in-water type, the asphalt is the dispersed (internal) phase, and water is the continuous (external) phase. The acid number of asphalt is an indicator of its ability to form emulsions and reflects the presence of high molecular weight acids.

Emulsion processes vary and the emulsions are made in rapid-, medium-, and slow-setting types for diverse application techniques in the road-building industry.

Table 1. Three Types of Liquid Asphalt Made by Cutting Back with Petroleum Diluents

Type asphalt	Diluent type	Diluent, %	Viscosity range[a], mm^2/s (= cSt)
slow curing (SC)	gas oil	0–50	70–6000
medium curing (MC)	kerosene	15–45	30–6000
rapid curing (RC)	naphtha	15–45	70–6000

[a] The liquid cutback asphalts are prepared in a number of viscosity grades, ranging generally from 70 to 6000. The grade number indicates the viscosity at 60°C.

Composition and Properties

Determination of the components of asphalts has always presented a challenge because of the complexity and high molecular weights of the constituents.

The methods employed can be conveniently arranged into a number of categories: *(1)* fractionation by precipitation; *(2)* fractionation by distillation; *(3)* separation by chromatographic techniques; *(4)* chemical analysis using spectrophotometric techniques (infrared, ultraviolet, nuclear magnetic resonance, x-ray fluorescence, emission, neutron activation), titrimetric and gravimetric techniques, elemental analysis; and *(5)* molecular weight analysis by mass spectrometry, vapor pressure osmometry, and size-exclusion chromatography.

Colloidal State. The principal outcome of many of the composition studies has been the delineation of the asphalt system as a colloidal system at ambient or normal service conditions.

Elemental Analysis. Asphalt is not composed of a single chemical species, but is rather a complex mixture of organic molecules that vary widely in composition from nonpolar, saturated hydrocarbons to highly polar and condensed aromatic ring systems. Although asphalt molecules are composed predominantly of carbon and hydrogen, most molecules contain one or more of the heteroatoms nitrogen, sulfur, and oxygen, together with trace amounts of metals, principally vanadium and nickel. Because the heteroatoms often impart functionality and polarity to the molecules, their presence may make a disproportionately large contribution to the differences in physical properties among asphalts from different sources. Generally, most asphalts are 79–88 wt % C, 7–13 wt % H, trace–8 wt % S, 2–8 wt % O, and trace–3 wt % N.

Molecular Weight. The molecular weights of the individual fractions of asphalt have received more attention, and have been considered to be of greater importance, than the molecular weight of the asphalt itself. Molecular weight data are used as an indication of the intermolecular interactions in the asphalt, which can confer property differences in asphalts.

Acid Number. Asphalt contains a small amount of organic acids and saponifiable material and the acid values of asphalt, 0.1–2.8 mg potassium hydroxide per g of asphalt, are related to emulsification conditions and ease of dispersion. The acid content is largely determined by the percentage of naphthenic (cycloparaffinic) acids of higher molecular weight that are originally present in the crude oil.

Rheology. Asphalt is a viscoelastic material whose rheological properties reflect crude type and, to a lesser extent, processing. The ability of the asphalt to perform under many conditions depends on flow behavior.

Durability. The term "durable" has several meanings, but in the present context it is used to describe an asphalt that possesses the necessary chemical and physical properties required for the specified pavement performance, being resistant to change during the in-service conditions that are prevalent during the life of the pavement.

Asphalts are used as protective films, adhesives, and binders because of their waterproof and weather-resistant properties. One particularly valuable property, the ability of the asphalt to undergo movement without fracture, can occur because of the viscous (sol) nature. In addition, asphalts have long and continuous satisfactory service because of their slow rate of hardening from heat, oxidation, fatigue, and weathering.

Failure of an asphalt film is reflected in the appearance of typical crack patterns because of a decrease in the plasticity of the asphalt through insufficient lower molecular weight constituents remaining in the continuous phase.

Mineral fillers are often added to asphalts to influence their flow properties, reduce costs, and are commonly used as stabilizers in roofing coatings at concentrations up to 60 wt %. Mineral-filled films show improved resistance to flow at elevated temperatures, improved impact resistance, and better flame-spread resistance. Fillers may increase the water absorption of asphalts. Fillers commonly used are ground limestone, slate flours, finely divided silicas, trap rocks, and mica; they are often produced as by-products in rock-crushing operations. Opaque fillers offer protection from weathering. Asbestos (qv) filler has special properties because of its fibrous structure, high resistance to flow, and toughness. It has been used in asphalt paving mixes to increase the resistance to movement under traffic and in roofing materials for fire-retardant purposes.

Specifications and Tests

In 1903 an American Society for Testing and Materials (ASTM) Committee on Road and Paving Materials was formed to develop test methods and specifications for highway materials. Test methods for volatilization, penetration, and bitumen were developed by the Office of Public Roads and were adopted by ASTM in 1911.

Test Procedures. Most tests applied to petroleum asphalts are empirical in nature. The ASTM tests are not the only ones applied to asphalt testing. Private tests of a proprietary nature have been used within companies. Some of these tests are now being made public and mention of them occurs from time to time in literature reports. Such tests may become a part of the ASTM standards.

Uses

Properties, and therefore uses, depend upon the method of manufacture (Table 2). There are a multitude of uses for asphalt, including hydraulics (dam facings, canal linings, pond linings), recreation (substrate for artificial surfaces, tennis courts, running tracks), agriculture (mulches, underground water barriers, stockyard paving), transportation (railroad ballast treatment and roadbeds), and metals (ore leaching pads, ore and coal briquetting, etc).

Straight-Run Asphalt. The largest use of straight-run asphalts is in the paving industry where they serve primarily as binders in paving mixes and as bases in liquid asphalts used as seal coatings, surface treatments, road mixes, and soil stabilizers. The most important technical innovation in asphalt paving has been to utilize asphalt throughout the entire pavement structure (termed total asphalt) to provide more efficient distribution of traffic stresses to the subgrade and provide better protection of the base from intrusion of outside materials such as water and soil.

Air-Blown Asphalts. Air-blown asphalts can be used in prepared roofings as well as for saturating the felts used in asphalt shingles and mineral-surfaced roll roofing.

Thermal Asphalt. Thermal asphalt products are especially useful as binders and as saturants for fiberboards, and as sizings for fiberboard.

Liquid Asphalt. Liquid asphalt products comprise cutback asphalts and emulsions. The lower viscosity products are used for dust-laying purposes and as tack coats, prior to laying asphalt surface courses. The heavier grades are used for mix-in-place road asphalts.

Modified Asphalts

A variety of materials have been proposed to modify the properties of asphaltic binders to enhance the properties of the mix, including fillers and fibers to reinforce the asphalt–aggregate mixture, sulfur to strengthen or harden the binder, polymers, rubber, epoxy–resin composites, antistripping agents, metal complexes, and lime. All of these additives serve to improve the properties of the binder and, ultimately, the properties of the asphalt–aggregate mix. The addition of fossil fuel-derived materials has also been investigated.

Aggregates

Although much of the focus tends to be on the composition and behavioral characteristics of the asphalt (ie, the binder or cement), the properties of the aggregate also play a significant role in determining the ultimate properties of the asphalt–aggregate mix. There are many factors to consider and they include the functionalities in the asphalt, absorption of the asphalt into the pores of the aggregate, the aging properties of the asphalt when it is on the aggregate, as well as methods for coating the asphalt on to the aggregate.

Table 2. Properties of Asphalts

Property	Straight-run, residual	Thermal	Air-blown residual
softening point (ring and ball), °C	46	113	93
penetration of 100 g at 25°C for 5 s, mm/10	90	0	20
ductility at 25°C, 5 cm/min, cm[a]	150+	too hard	3.2
specific gravity, 15.6/15.6°C	1.03	1.12	1.05
mean coefficient of cubical expansion/°C at 15.6–65.6°C	0.00063	0.00058	0.00063
specific heat, J/(kg·K)[b] at 93.3°C	1968	1842	1926
thermal conductivity at 26.7°C, W/(m·K)	0.16	0.16	0.16
permeability constant at 25°C, kg·m/(m·s·N/m)22			
water vapor	$0.62-1.93 \times 10^{-15}$	1.1×10^{-15}	$1.25-2.4 \times 10^{-15}$
oxygen			0.08×10^{-15}
water absorption of 10-mil[c] films on aluminum panels, wt % at 50 wks			1.5–10
surface tension, mN/m (= dyn/cm) at 25°C	34		32
dielectric strength, spherical electrodes, V/m	$11-45 \times 10^6$	36×10^6	$30-35 \times 10^6$
dielectric constant 50 Hz at 20°C	2.7	3.0	2.7

[a] A + sign after a number indicates a minimum value.
[b] To convert J to cal, divide by 4.184.
[c] 10 mil = 6.254 mm.

Health and Safety

The word asphalt has been carelessly used in that it is not adequately differentiated from thermally degraded materials, especially coal tar and derivatives. It is essential to differentiate asphalt from these materials which contain known carcinogens and health hazards. For this reason, the use of cracked asphalts must be treated with caution.

Steps to minimize potential safety hazards in the handling of asphalt include *(1)* sudden pressure increases from hot asphalt in contact with moisture in enclosed tanks or transports; *(2)* exposure to air at 150°C or above; *(3)* local overheating above heating coils; flashing of asphalt volatiles in the presence of an ignition source or possible autoignition; and *(4)* hydrogen sulfide from high temperature operations.

JAMES G. SPEIGHT
Western Research Institute

H. Abraham, *Asphalts and Allied Substances,* 6th ed., Vol. 1, Van Nostrand Co. Inc., Princeton, N.J., 1960, Chapt. 2.

J. G. Speight, *Fuel Science and Technology Handbook,* Marcel Dekker, Inc., New York, 1990.

J. G. Speight, *The Chemistry and Technology of Petroleum,* 2nd ed., Marcel Dekker, Inc., New York, 1991.

J. C. Petersen, *Fuel Sci. Technol. Int.* **6,** 255 (1988).

ATMOSPHERIC MODELING

Mathematical air quality models provide a powerful framework for understanding the dynamics of pollutants in the atmosphere and for assessing the impact emission sources have on pollutant concentrations. Two classes of models are commonly used. Empirical models provide an understanding of source impacts by statistically analyzing historical air quality data. Diagnostic models provide a comprehensive description of the detailed physics and chemistry of compounds in the atmosphere, following the evolution of pollutant emissions to their ultimate fate. At the heart of the model is a system of mathematical routines that integrate the effects of individual processes. The complexity and computational intensity of modern models have necessitated the development of algorithms for fast and accurate mathematical solution techniques. Air quality models are being applied to solving such problems as urban smog, acid deposition, regional ozone, haze in scenic regions, and the destruction of the protective stratospheric ozone layer.

Mathematical models have grown increasingly detailed in descriptions of air pollutant dynamics and are thus key tools for gaining scientific understanding of atmospheric processes. These models also are the most practical and scientifically defensible means of relating pollutant emissions to air quality. They are widely used in regulatory planning and analysis. This article focuses on the models used to understand air pollution dynamics.

Development of a mathematical air quality model proceeds through four stages. In the conceptual stage, a working set of relationships approximating the physical system is derived. Next, these relationships are expressed as mathematical equations, giving a formal description of the idealized system. The third step is computational implementation of the model, including development of the algorithms and computer code needed to solve the equations, given various inputs. The final step is the application of the model, including acquisition and processing of the necessary input data, and evaluation of model results.

Air Quality Models

Models used in air pollution analysis fall into two classes: empirical–statistical and deterministic, as shown in Figure 1.

Empirical–Statistical Models. Empirical–statistical models are based on establishing a relationship between historically observed air quality and the corresponding emissions. The linear rollback model is the simplest, easiest to use, and requires little data, and for these reasons has been widely applied.

A second class of empirical/statistical models of continuing interest is the receptor-oriented model, which has been used extensively for estimating the contributions that distinguishable sources such as automobiles or municipal incinerators make to particulate matter concentrations. Receptor models compare the measured chemical composition of particulate matter at a receptor site with the chemical composition of emissions from the major sources to identify the source contributions at the monitoring location. There are three principal categories of receptor models: chemical mass balance, multivariate, and microscopic. Hybrid analytical and receptor (or combined source/receptor) models have been proposed and used, but further investigation into their capabilities is required. Receptor models are powerful tools for source apportionment of particulates because a vast amount of particulate species characterization data have been collected at many sampling sites worldwide, and because many aerosol species are primary pollutants.

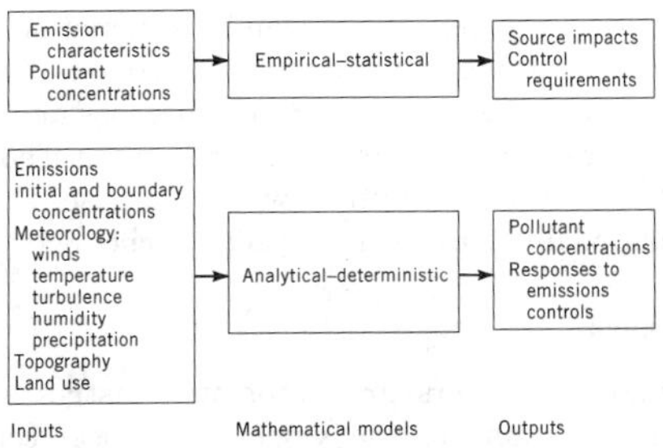

Figure 1. Inputs, types of models, and outputs used in air quality modeling studies.

Deterministic Models. Deterministic air quality models described in a fundamental manner the individual processes that affect the evolution of pollutant concentrations. These models are based on solving the atmospheric diffusion-reaction equation, which is in essence the conservation-of-mass principle for each pollutant species.

There are two distinct reference frames from which to view pollutant dynamics. The most natural is the Eulerian coordinate system which is fixed at the earth's surface. In that case, a succession of different air parcels are viewed as being carried by the wind past an observer who is fixed to the earth's surface. The second is the Lagrangian reference frame in which the frame of reference moves with the flow of air, in effect maintaining the observer in contact with the same air parcel over extended periods of time. Because pollutants are carried by the wind, it is often convenient to follow pollutant evolution in a Lagrangian reference frame, and this perspective forms the basis of Lagrangian trajectory and Lagrangian marked-particle or particle-in-cell models. In a Lagrangian marked-particle model, the center of mass of parcels of emissions are followed, traveling at the local wind velocity, while diffusion about that center of mass is simulated by an additional random translation corresponding to the atmospheric diffusion rate. Lagrangian trajectory models can be viewed as following a column of air as it is advected in the air basin at the local wind velocity. Simultaneously, the model describes the vertical diffusion of pollutants, deposition, and emissions into the air parcel.

One of the most basic and widely used transport models is the Gaussian plume model. This model describes a plume with a Gaussian distribution of pollutant concentrations, with plume dispersion characterized by the standard deviations of the mean concentration in the y and z directions. Gaussian plume models are easy to use and require relatively few input data. Of the Eulerian models, the box model is the easiest to conceptualize. The atmosphere over the modeling region is envisioned as a well-mixed box, and the evolution of pollutants in the box is calculated following conservation-of-mass principles including emissions, deposition, chemical reactions, and a changing mixing height (or inversion-base).

Eulerian "grid" models are the most complex, but potentially the most powerful, air quality models, involving the least-restrictive assumptions. They are also the most computationally intensive. Grid models include temporal and spatial variation of the meteorological parameters, emission sources, and surface characteristics. Grid models divide the modeling region into a large number of cells, horizontally and vertically, which interact with each other by simulating diffusion, advection, and sedimentation (for particles) of pollutant species. Input data requirements for grid models are similar to those for Lagrangian trajectory models, but in addition require data on background concentrations (boundary conditions) at the edges of the grid system used. Eulerian grid models produce pollutant concentration predictions throughout the entire airshed, which can be examined over successive time periods to observe the evolution of pollutant concentrations and how they are affected by transport and chemical reaction.

Temporal and Spatial Resolution. The temporal and spatial resolutions of models can vary from minutes to a year and from meters to hundreds of kilometers. The minimum meaningful resolution of a model is determined by the input data resolution and the structure of the model. Statistical models generally rely on several years' worth of measurements of hourly or daily pollutant concentrations. The resolution of the input data represents the minimum resolution of a statistical model. Resolution of analytical models is limited by the spatial and temporal resolution of the emissions inventory, the meteorological fields, and the grid size chosen for model implementation.

Model Components

A model's ability to correctly predict pollutant dynamics and to apportion source contributions depends on the accuracy of the individual process descriptions and input data, and the fidelity with which the framework reflects the interactions of the processes.

Turbulent Transport and Diffusion. There are two pollutant transport terms: an advection term, in which pollutants are carried along with the time-averaged mean wind flow; and a dispersion term representing transport resulting from local turbulence. The averaging time that determines the mean winds is related to the spatial scale of the system being modeled. Minutes may be appropriate for urban-scale simulations, multihour averages for the regional scale, and daily to weekly averages for determining long-term concentrations of nonreactive pollutants.

Turbulent transport is determined by complex interactions between meteorological conditions and topography. In addition to gross topographical features, surface "roughness" scales have been devised to parameterize surface characteristics according to land-use categories.

Removal Processes. Pollutant removal processes, particularly dry deposition and scavenging by rain and clouds, are a primary factor in determining the dynamics and ultimate fate of pollutants in the atmosphere.

Representation of Atmospheric Chemistry Through Chemical Mechanisms. A complete description of atmospheric chemistry within an air quality model would require tracking the kinetics of many hundreds of compounds through thousands of chemical reactions. Fortunately, in modeling the dynamics of reactive compounds such as peroxyacetyl nitrate (PAN), $C_2H_3NO_5$, O_3, and NO_2, it is not necessary to follow every compound. Instead, a compact representation of the atmospheric chemistry is used.

Three different types of chemical mechanisms have evolved as attempts to simplify organic atmospheric chemistry: surrogate, lumped, and carbon bond. These mechanisms were developed primarily to study the formation of O_3 and NO_2 in photochemical smog, but can be extended to compute the concentrations of other pollutants, such as those leading to acid deposition.

Aerosol Dynamics. Inclusion of a description of aerosol dynamics within air quality models is of primary importance because of the health effects associated with fine particles in the atmosphere, visibility deterioration, and the acid deposition problem. Aerosol dynamics differ markedly from gaseous pollutant dynamics in that particles come in a continuous distribution of sizes and can coagulate, evaporate, grow in size by condensation, be formed by nucleation, or be deposited by sedimentation. Furthermore, the species mass concentration alone does not fully characterize the aerosol. The particle size distribution, which changes as a function of time, and size-dependent composition determine the fate of particulate air pollutants and their environmental and health effects.

Air Quality Model Inputs. Inputs to analytical air quality models can be broadly grouped as those dealing with meteorology, emissions, topography, and atmospheric concentrations. Meteorological inputs generally control the transport rate of pollutants and are used to determine reaction rates and the depositional flux of compounds. Topography influences transport and deposition. Observed compound concentrations are used to specify both initial and boundary conditions for model simulations. Especially for pollution problems involving organic compounds, emissions are a key input subject to considerable uncertainty. Although emissions from primary industrial facilities or utilities may be reasonably well known, emissions from residential or commercial facilities, mobile sources, and natural sources are often roughly estimated and difficult to verify.

Mathematical and Computational Implementation. Solution of the complex systems of partial differential equations governing both the evolution of pollutant concentrations and meteorological variables, eg, winds, requires specialized mathematical techniques.

Historically, the computational intensity of the more complex chemically active models have limited their application. For example, modeling only the gas-phase dynamics over an area like Los Angeles requires solving about 500,000 simultaneous, nonlinear equations. Until the late 1980s, computational power severely limited the use of chemically active models, and particularly inhibited the development and use of regional oxidant and acid deposition models. The rapid increase in computational power is making it possible to address much larger problems, for example, regional and global scale pollution. One aspect of air quality models is that they are highly parallelizable. The development of parallel computers should allow for significantly more detailed modeling of large systems.

Application of Air Quality Models

Both receptor and analytical air quality models have proven to be powerful tools for understanding atmospheric pollutant dynamics and for determining the impact of sources on air quality.

Receptor Models. Receptor models, by their formulation, are effective in determining the contributions of various sources to particulate matter concentrations.

Analytical Modeling Studies. Analytical air quality models have been used the most in modeling the dynamics of pollutants at local and urban scales.

ARMISTEAD G. RUSSELL
Carnegie Mellon University
JANA B. MILFORD
University of Connecticut

J. G. Watson, R. C. Henry, J. A. Cooper, and E. S. Macias, in E. S. Macias and P. K. Hopke, eds., *Atmospheric Aerosol: Source/Air Quality Relationships,* Symp. Ser. No. 167, American Chemical Society, Washington, D.C., 1981, pp. 89–106.

P. K. Hopke, *Receptor Modeling in Environmental Chemistry,* John Wiley & Sons, Inc., New York, 1985.

R. J. Yamartino, in S. C. Dattner and P. K. Hopke, eds., *Receptor Models Applied to Contemporary Pollution Problems,* Air Pollution Control Association, Pittsburgh, Pa., 1983, pp. 285–295.

S. K. Friedlander, *Smoke, Dust and Haze,* Wiley-Interscience, New York, 1977.

ATOMIC ABSORPTION SPECTROSCOPY. See SPECTROSCOPY, OPTICAL.

ATTRACTANTS. See INSECT CONTROL TECHNOLOGY.

AUTOMATED INSTRUMENTION

CLINICAL CHEMISTRY

Clinical chemistry, initiated in the 1940s, involves the biochemical testing of body fluids to provide objective information on which to base clinical diagnosis. The ever-increasing demand for high quality, routine clinical testing stimulated the development of automated techniques, and as early as the 1960s automation in the clinical laboratory was the rule rather than the exception.

Although growth was initially driven by automation, in the 1990s the growth in U.S. laboratory testing may be attributed to several factors. Among them are the discovery of new diseases and the introduction of new therapies, as well as better understanding of body chemistry and the aging of the U.S. population, which increases the risk of contracting age-related illnesses requiring clinical chemistry analyses.

Assay Automation

Clinical chemistry analyzers are automated instruments used for measuring concentrations of the various chemical constituents of blood or other body fluids. For a discussion of the related category of instruments used for the measurement of blood cell parameters, see AUTOMATED INSTRUMENTATION–HEMATOLOGY.

Before the advent of automation, the steps a clinical technologist had to follow in performing a single clinical chemistry assay included sample preparation, identification, centrifugation, and filtering; reagent preparation; manual metering and addition of sample and reagents into a reaction vessel; mixing and incubation in the reaction vessel; optical measurement of the mixture in a separate cuvette; and calculation and recording of the results. With the exception of the sample and reagent preparation procedure, which is sometimes done separately, these steps are all performed by an automated analyzer such as that shown in Figure 1.

Technology

An automated system for clinical analysis consists of the instrument (hardware), the reagents, and the experimental conditions (time, temperature, etc) required for each determination. The reagents plus the experimental conditions are sometimes referred to as the chemistry of the system. The chemistry employed is generally similar to that used in manual assays because most automated assay methods have been adapted from the manual ones. However, automated analyzers rarely afford the flexibility of experimental procedure that is possible in manual analyses. Chemical determinations available on automated analyzers include those for albumin, calcium, chloride, creatinine, iron, total bilirubin, total protein, alanine aminotransferase (ALT), alkaline phosphatase, aspartate transaminase (AST), carbon dioxide, cholesterol, creatine kinase (CK), gamma glutamyl transferase (GGT), glucose, lactate dehydrogenase (LD), triglycerides, and urea nitrogen.

Essential features of an automated method are the specificity, ie, the assay should be free from interference by other serum or urine constituents, and the sensitivity, ie, the detector response for typical sample concentration of the species measured should be large enough compared to the noise level to ensure assay precision. Also important are the speed, ie, the reaction should occur within a convenient time interval (for fast analysis rates), and adequate range; the result for most samples should fall within the allowable range of the assay.

The assay methods for the various biochemical species can be classified according to reaction rate behavior, eg, end point vs kinetic methods, blanking schemes, or reaction principle and type of reagents employed.

Measurement Methods

The majority of the various analyte measurements made in automated clinical chemistry analyzers involve optical techniques such as absorbance, reflectance, luminescence, and turbidimetric and nephelometric detection means.

Classification of Clinical Chemistry Analyzers

Automated clinical analyzers can be divided into discrete and continuous-flow systems. In discrete systems, the sample–reagent mixtures from different specimens are kept in individual reaction cuvettes during incubation and optical reading. In continuous-flow analyzers, the sample–reagent mixtures form liquid segments flowing through a tube. Adjacent segments are separated by air bubbles. Clinical analyzers can also be classified according to their degree of

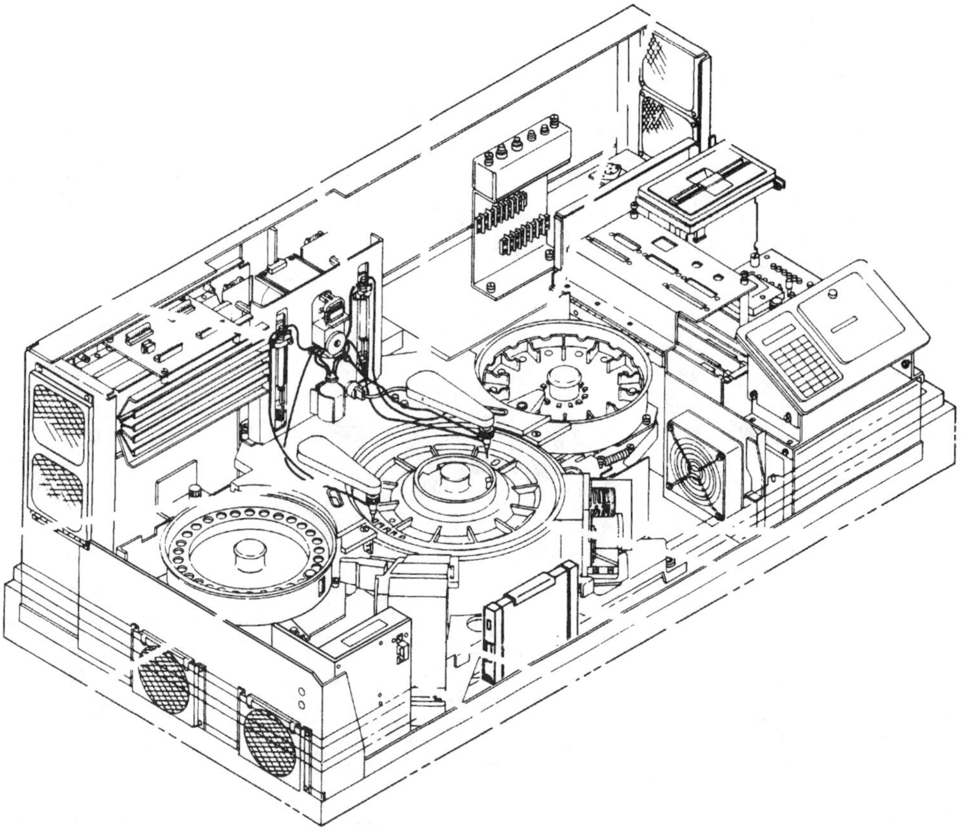

Figure 1. Internal view of a clinical chemistry analyzer (Technicon RA-XT). Courtesy of Miles Inc.

flexibility. Most of the modern systems are random access analyzers, for which the tests on various specimens are performed in any order programmed by the operator.

The throughput range, number of assays performed per hour, of clinical analyzers reflects the diversity of the market. Systems that perform less than 400 tests per hour are usually referred to as small analyzers; medium analyzers cover the range from about 500 tests per hour to 1,500 tests per hour, while high throughput analyzers can process up to 10,000 tests per hour. Table 1 lists a number of representative automated systems and their characteristics.

Functional Elements of Clinical Chemistry Analyzers

Sample Handling System. Venous or capillary blood, urine, and cerebrospinal fluid are specimens routinely used in medical diagnostic testing. Of these biological fluids, the use of venous blood is by far the most prevalent. Collection devices such as syringes and partial vacuum test tubes, eg, Vacutainer, are used to draw 10 mL or less of venous blood. At collection time, the test tubes are carefully labeled for later identification.

The specimen, as drawn, contains cells, platelets, fibrin, and particulates. For most chemistry tests, eg, determination of glucose, cholesterol, etc, it is necessary to first separate the cellular blood fraction which, if present during the assay, would interfere with the determination and adversely affect the accuracy of the measurement. The cellular components are separated by centrifugation.

After centrifugation, the specimens are placed in a holder specifically designed to work with a particular instrument. Once loaded, the sample holder is placed onto the carrier transport mechanism of the system, which is referred to as the sampler. The function of the sampler is to bring each specimen into a position where a sample of it can be taken by the aspiration probe, then to transport it further to the unloading area.

Sample Aspiration and Dispensing. Precisely metered amounts of the sample have to be aspirated rapidly and without allowing intersample contamination, known as sample carry-over. Generally, pipetting is accomplished by motor-driven syringes connected through a fluid line to a thin aspiration probe (see Fig. 1)

In most analyzers, sample carry-over is minimized by one or more of the following techniques: sensing the liquid level in the sample container, and limiting the probe immersion in the liquid to a few millimeters; or rinsing the aspiration probe inside and out after each immersion.

Reagent Handling System. Many analyzers can be programmed to preform a wide variety of assays. However, reagents for only a limited number of tests, usually referred to as resident tests, are available on the instrument at any one time. The main reason for this limitation is that, for infrequently requested tests, the time period until reagent depletion may exceed the chemical stability time limit.

Precise metering of the amount of liquid reagent needed for a test is generally done using a motor-driven syringe.

Two notable technological developments related to the reagent handling system occurred in the late 1980s. One of these was the introduction of the dry slide reagent technology where the reagents needed for a particular assay are deposited in thin layers on a slide.

Another unique development, which significantly reduced the amount of reagent necessary for an assay, is an extension of the continuous-flow analyzer technology called capsule chemistry. The core of the system is a capillary Teflon tube coated on the inside with a thin, flowing film of fluorocarbon oil. Sample and liquid reagents, alternating with air bubbles, are aspirated into the tube, forming a segmented flow.

Table 1. Features of Automated Analyzers

Manufacturer and system	Throughput per hour: Samples	Throughput per hour: Tests	Number of resident methods	Amount of sample needed per test, μL	Incubator temperature, °C	Optical system[a]	Distinctive features
Roche COBAS MIRA	variable	125	30	2–95	37	AMW,P	disposable cuvettes
Technicon RA-1000	variable	240	12	2–30	30, 37	AMW,P	no probe wash
Ames CLINISTAT	up to 80	up to 80		10	37	R	dry chemistry
Abbott SPECTRUM	variable	400–600	23 3	1.25–25	25, 30, 37	AMW,P,T	polychromatic sample blanking
Beckman SYNCHRON CX3	up to 75	600	8	122/8 tests	37	AMW,P	STAT analyzer[b] Beckman liquid calibrators and controls
Kodak EKTACHEM 700	300 max	600	26	10, 11	37	R	dry chemistry slides, routine/STAT[b]
Baxter PARAMAX	240	720	32	2–50	37	AMW,P	closed container sampling option; unit dose dry tablet reagent
BM Hitachi 737	300	1,200	23	3–20	30, 37	AMW,P	
Technicon CHEM 1	variable	1,800	35	1	30, 37	AMW,P	continuous-flow capsule chemistry
BM Hitachi 736-50	300	8,100	27	10	30, 37	AMW,P	
Technicon DAX = 96	300	10,200	34	3–67	37	AMW,P	

[a] AMW = absorption, multiple wavelength per test; N = nephelometry; T = turbidimetry; P = potential; and R = reflectance.
[b] A specimen that has to be processed with priority vs other specimens is known as STAT sample.

Optical Detection System. The optical detection system includes dedicated or time-multiplexed arrangements of the various elements such as the light source, cuvette or flowcell, wavelength isolation elements, and detector. A common layout may have a single light source with light distribution via fiber optics to many cuvettes or flowcell read stations. On the output side of the reaction cuvette or flowcell, fiber optics pipe the light signals to an optical chopper, through dedicated interference filters, and to a common multiplexed photomultiplier. A variation of this arrangement uses a rotating mirror device to multiplex the light output of many cuvettes into the input of a single grating–array photodiode.

Computer System. The brain of the modern clinical chemistry analyzer is its computer system. The part of the computer system that controls the functional aspects of the analyzer is known as the process control computer or analytical processor (AP); the test results are handled by the data management computer, also known as the results processor (RP).

Economic Aspects

It is estimated that the worldwide clinical chemistry diagnostics market is about \$3 billion. This amount includes an estimated \$700 million in instrument sales, and \$2.3 billion in sales of reagents and consumables. Some of the principal instrument manufacturers are Hitachi (Japan), Miles Laboratories (United States), E. I. du Pont de Nemours (United States), Beckman Instruments (United States), Eastman Kodak (United States), Abbott Laboratories (United States), Olympus (Japan), Toshiba (Japan), Hoffmann-La Roche (Switzerland), and Ciba Corning Diagnostics (United States).

PAUL GHERSON
HOWARD LANZA
MILTON PELAVIN
DAN VLASTELICA
Miles Inc.

N. W. Tietz, ed., *Textbook of Clinical Chemistry,* W. B. Saunders Co., Philadelphia, Pa., 1986.

L. J. Kaplan and A. J. Pesce, *Clinical Chemistry, Theory, Analysis, and Correlation,* The C. V. Mosby Co., 1989.

G. Kessler and L. R. Snyder, *Technicon International Congress,* Vol. 1, Mediad Press, Tarrytown, N.Y., 1977, pp. 28–35.

I. M. Kolthoff, P. J. Elving, and E. J. Meehan, *Optical Methods of Analysis,* Vol. 8, *Treatise on Analytical Chemistry,* Part I, John Wiley & Sons, Inc., New York, 1986.

HEMATOLOGY

Hematology analyzers provide information about blood cells and their constituents. The three basic blood cell types are erythrocytes or red blood cells, leukocytes or white blood cells, and thrombocytes or platelets. Hemoglobin is the principal nonaqueous component of red blood cells. Its physiological importance gives it the status of a primary hematological constituent.

ApertureImpedanceInstruments

The aperture impedance principle of blood cell counting and sizing, also called the Coulter principle, exploits the high electrical resistivity of blood cell membranes. Red blood cells, white blood cells, and blood platelets can all be counted. In the aperture impedance method, blood cells are first diluted and suspended in an electrolytic medium, then drawn through a narrow orifice (aperture) separating two electrodes (Fig. 1). In the simplest form of the method, a d-c current flows between the electrodes, which are held at different electrical potentials. The resistive cells reduce the current as the cells pass through the aperture, and the current drop is sensed as a change in the aperture resistance.

Aperture impedance counters are flow cytometers, as are all current automated cell counters. Flow cytometers measure cells as the cells flow hydraulically, one at a time, through sensing zones.

An alternative approach is to model the probability of coincidence as a function of either sample concentration or the fraction of time the

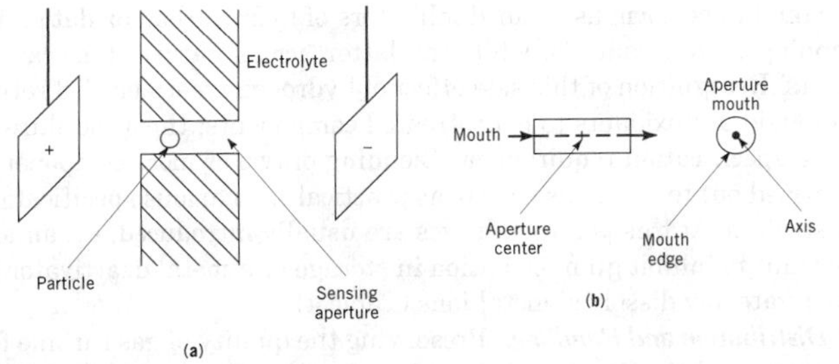

Figure 1. (**a**) Schematic diagram of the electrical impedance method for counting and sizing blood cells. The electrodes are represented as charged rectangles. As particles flow through an aperture, the impedance and, hence, the flow of current are modified; (**b**) schematic diagram of the sensing aperture of a flow cytometer.

sensing zone is occupied by passing cells (dead time). The model produces a coincidence-correction factor which is applied to the count. Still another approach is to reduce the size of the sensing zone, which reduces the probability of coincidence. In practice, commercial counters combine the latter two approaches.

Flow cytometer cell counts are much more precise and more accurate than hemocytometer counts. Hemocytometer cell counts are subject both to distributional and sampling errors.

Aperture impedance counters provide cell volume information as well as cell counts. They also provide mean red cell volume (MCV) and mean platelet volume (MPV) as well as cell volume distribution information because they measure volume on a cell-by-cell basis.

Aperture impedance counters provide cell volume information as well as cell counts.

In summation, aperture impedance counters provide information on WBC, RBC, HCT, MCV, PLT, MPV, RDW, PDW, and three-part Diff, for d-c current only, or four-part Diff using a d-c/r-f combination.

Light-Scattering Instruments

The light-scattering principle of cell counting is based on the observation that microscopic particles, such as blood cells, scatter into small (0–15°) angles, most of the visible light incident upon them. This principle is used to count red blood cells, white blood cells, and platelets. In the basic form of the light-scattering method, a dilute suspension of cells and a sheathing fluid having a matching refractive index flow concentrically through an optical flow cell onto which light is focused. The cells scatter the incident light as they pass through the flow cell, and the light scattered into a small-angle interval, typically 1–3°, is sensed by an optical detector. The purpose of the sheath is to narrow the cell stream and reduce the cell passage to single file while centering the stream in the flow cell. The unscattered component of focused light is much larger than the scattered component and must be blocked from the detector so that it does not swamp the desired small-angle scatter signal.

The basic single-angle interval light-scattering method cannot accurately measure individual red blood cell or platelet volumes, but it can provide MCV and MPV. Red cells are bi-concave disks, and platelets are rod to disk shaped. Scattering intensities depend on the orientation in the flow cell. Because the cells can interrupt the optical path in random orientations, individual scattering intensities are not proportional to cell volume. However, because thousands of cells of each type pass through the flow cell, the effects of orientation can be averaged.

Light-scattering measurements of sphered cells are not subject to orientation effects, and a method for the rapid sphering and fixing of red cells for the purpose of measuring them in a light-scattering flow cytometer has been developed.

Light-scattering measurements of red cell volume are accurate only if the measurements account for the effects of cell refractive index on scattering intensity. Basic single-angle interval light-scattering measurements of red cell volume, even for sphered and fixed red cells, are only accurate to within approximately 10%. Mie Scattering Theory predicts that the angular scattering intensity pattern for sphered red cells depends on cell refractive index as well as cell size. Red cell refractive index is linearly related to cellular hemoglobin concentration. Therefore, two normal red cells of the same volume but of different cellular hemoglobin concentration scatter light differently. A sphered cell's volume and the hemoglobin concentration can both be accurately determined from Mie Scattering Theory calculations by making simultaneous measurements of light-scatter intensity over two suitably chosen angle interval. Two suitable angle intervals for red cells are 2–3°, known as low angle scatter (SL), and 5–15°, high angle scatter (SH). In contrast to the aperture impedance method, this method for determining red cell volume does not suffer from inaccuracies as a result of cellular hemoglobin concentration variability.

The two-angle interval scattering method can provide CBC, RDW, PDW, PCT, and HDW.

The single-angle interval light-scattering intensity counts white cells but cannot provide five-part Diffs. White cell light-scattering intensity depends on cell size, cell refractive index, and internal cell structure. White cell size ranges overlap and the cells differ in refractive index. Scattering measurements can distinguish among cell types based on differences in internal structure alone, but only if the cell types have the same overall size and refractive index.

Light-scattering measurements made over two suitable chosen angle intervals can combine with depolarized light-scattering measurements to provide five-part Diffs.

Cytochemistry

Cytochemical techniques can be combined with light-scattering and absorption measurements to provide five-part Diffs. Cytochemistry concerns the chemical reactions of cell components. The reactions for automated white blood cell differential analysis include those that bind chromophores to the granules of white cell types, based on the presence of various substrates and enzymes in the granules. These reactions yield products suitable for light-scattering and optical absorption measurements. Other reactions exploit the differential resistance of white cell types to cytoplasmic stripping by the lysing action of surfactants. The reaction products are suitable for light-scattering and aperture impedance measurements.

Various method combinations can be used to produce five-part Diffs including d-c/r-f aperture impedance plus light scattering; two-angle interval light-scattering plus depolarized light scattering; light scattering plus cytochemistry; and d-c/r-f aperture impedance plus cytochemistry. Flow cytometric methods produce five-part Diffs that are generally more precise and accurate than manual Diffs. However, manual methods can provide more information.

Photometric Hemoglobinometry

Most automated hemoglobinometry methods are derivatives of the manual ICSH reference method. The manual method follows three steps: (*1*) The blood sample is diluted in a reagent containing Triton X-100, a nonionic surfactant. The surfactant lyses the RBC membrane, releasing hemoglobin into solution. (*2*) Ferricyanide present in the diluent diffuses into the hemoglobin molecule and oxidizes heme from the Fe_{2+} to the Fe_{3+} form, yielding methemoglobin. (*3*) Cyanide diffuses into methemoglobin's interior and reacts with heme to yield cyanmethemoglobin (HiCN), which has a characteristic absorption spectrum that is measured spectrophotometrically.

The automated method differs from the ICSH method chiefly in that oxidation and ligation of heme iron occur after the hemes have been released from globin. Therefore, ferricyanide and cyanide need not diffuse into the hemoglobin and methemoglobin, respectively. Because diffusion is rate-limiting in this reaction sequence, the overall reaction time is reduced from approximately three minutes for the manual method to 3–15 s for the automated method.

DAVID ZELMANOVIC
ROQUE A. GARCIA
JOHN TURRELL
Technicon Instruments Corporation

J. L. Grant and co-workers, *Amer. J. Clin. Path.* **33,** 138–143 (1960).

M. Wales and co-workers, *Rev. Sci. Inst.* **32,** 1132–1136 (1961).

H. C. van de Hulst, *Light Scattering by Small Particles,* John Wiley & Sons, Inc., New York, 1957, Chapt. 8.

M. J. Malin and co-workers, *J. Clin. Path.* **92**(3), 286–294 (1989).

AVIATION AND OTHER GAS TURBINE FUELS

Gas Turbine Products

Composition. Because the jet aircraft is a weight-limited vehicle, a high premium is assigned to hydrocarbon fuels (Table 1) with a maximum gravimetric heat content or hydrogen-to-carbon ratio.

Specifications. The global nature of jet aircraft operations has mandated that aviation fuel quality be closely controlled in every part of the world. Specifications tend to be industry standards issued by a consensus organization, like ASTM, or by a government body rather than by manufacturer's requirements. Table 2 lists some of the requirements for the principal grades of civil and military aviation turbine fuels in use throughout the world.

Liquid fuels for ground-based turbines are best defined today by ASTM Specification D2880. Table 3 lists the detailed requirements for five grades which cover the volatility range from naphtha to residual fuel.

Manufacture. Aviation turbine fuels are primarily blended from straight-run distillates rather than cracked stocks because of specification limitations on olefins and aromatics. However, in refineries where heavy gas oil is hydrocracked, ie, a process that involves catalytic cracking in a hydrogen atmosphere, aviation fuels can include hydrocracked components since they are free of olefins, are low in sulfur, and are stable. Ground turbine fuels are equivalent to their fuel oil counterparts and are manufactured as dual-purpose products.

Most of the crudes available in greatest quantity today are high in sulfur, and yield products that must be desulfurized to meet specifications or environmental standards. The process most widely used is mild catalytic hydrogenation.

Hydrogen treating removes oxygen-containing species, such as phenols and naphthenic acids, that are found in some crudes. The former tend to perform as natural inhibitors of hydrocarbon oxidation by trapping peroxy radicals, while the latter act as natural lubricating agents. Recognition of this side effect of hydrogen treatment led refiners to add antioxidants to hydrotreated components; this procedure is now a specification requirement. Blending of two or more components is carried out to match as closely as practical the various specification constraints. At this point, additives are usually introduced, eg, an antioxidant to inhibit gum formation in storage or a metal deactivator to deactivate any dissolved metal ions (Table 4).

Table 1. Fuels Used in Gas Turbine Applications

Fuel	Aircraft propulsion	Ground power: Aircraft type	Advanced industrial	Heavy duty
hydrogen	[a]	X	X	
methane (natural gas)	[a]	X	X	
synthetic gas		X	X	
LPG		X	X	
light naphtha		X	X	
methanol (ie, nonhydrocarbons)		X	X	
gasoline	[b]			
wide-cut fuel	X	X	X	
kerosene	X	X	X	
No. 2 fuel oil (diesel)		X	X	X
other gas oils		X	X	X
residual fuel			[c]	X
crude oil				X

[a] Future application as cryogenic liquid.
[b] Emergency fuel only.
[c] Limited by metal content of fuels.

Distribution and Handling. Preserving the quality of gas turbine fuels between the refinery and the point of use is an important but difficult task. The difficulty arises because of the complicated distribution systems of multiproduct pipelines which move fuel and sometimes introduce contaminants. It is common practice to install several stages of filter-coalescers between the storage tank and the aircraft delivery point.

The extensive processing and cleanup steps carried out on gas turbine fuels produce a purified liquid dielectric of high resistivity which is capable of retaining electrical charges long enough for buildup of large surface voltages. The result is a possible hazard—a tank filled with charged fuel that under some circumstances can discharge its energy to ground in a spark capable of igniting fuel mists or vapors.

This aspect of fuel handling has received much attention because of a number of accidents that have resulted in tank explosions, most often in filling tank trucks but also in fueling aircraft. Several approaches have been taken to reduce the risk of static discharge. The most common method requires introduction of an additive to increase the electrical conductivity of the fuel, ie, to speed up charge relaxation to a fraction of a second.

Performance

Combustion. The primary reaction carried out in the gas turbine combustion chamber is oxidation of a fuel to release its heat content at constant pressure. The heat content of the fuel is therefore a primary measure of the attainable efficiency of the overall system in terms of fuel consumed per unit of work output. Table 5 lists the net heat content of a number of typical gas turbine fuels. The most desirable gas turbine fuels for use in aircraft, after hydrogen, are hydrocarbons. Fuels that are liquid at normal atmospheric pressure and temperature are the most practical and widely used aircraft fuels; kerosene, with a distillation range from 150 to 300°C, is the best compromise to combine maximum mass–heat content with other desirable properties. For ground turbines, a wide variety of gaseous and heavy fuels are acceptable.

Stability. Aviation fuel is exposed to a wide range of thermal environments, and these greatly influence required fuel properties. On its way to the engine the fuel absorbs heat from air frame and engine components and in fact is used as a coolant for engine lubricant. Fuel stability assumes primary importance since freedom from deposits within the fuel system is essential for both performance and life.

The fuel systems of ground-based turbines are far less critical, since coolants other than fuel can be used and fuel lines can be well insulated. The tendency for deposit formation in fuel is not a concern in ground systems.

Oxidation of hydrocarbons by the air dissolved in fuel is catalyzed by metals and leads to polymer formation, ie, varnish and sludge deposits, by a chain reaction mechanism involving free radicals. Since it is impossible to exclude air dissolved in fuel, oxidation stability is controlled by eliminating species prone to form free radicals and by introducing antioxidants (qv).

It is possible to deactivate a metal ion by adding a compound such as disalicylidene alkyl diamine, which readily forms a chelate with most metal atoms to render them ineffective.

Volatility. The volatility of aircraft gas turbine fuel is controlled primarily by the aircraft itself and by its operating environment. For example, limits of the vapor pressure of aviation gasoline were

Table 2. Selected Specification Properties of Civil and Military Aviation Gas Turbine Fuels

Characteristic	ASTM D1655[a] Jet A kerosene A/L and	ASTM D1655[a] Jet B wide-cut Gen Avn	Mil-T-5624-N[a] JP-4 wide-cut USAF	Mil-T-5624-N[a] JP-5 kerosene USN	Mil-T-83133C[a] JP-8 kerosene USAF
composition					
aromatics, vol % max	25[b]	25[b]	25	25	25
sulfur, wt % max	0.3	0.3	0.4	0.4	0.3
volatility					
dist. temp. max °C, 10% rec'd	205			205	205
dist. temp. max °C, 50% rec'd		190	190		
dist. temp. max °C, end pt	300		270	300	300
flash pt, °C min	38			60	38
vapor pressure at 38°C, kPa max (psi)		21 (3)	14–21 (2–3)		
density at 15°C, kg/m³	775–840	751–802	751–802	788–845	775–840
fluidity					
freezing pt, °C max	−40[c]	−50	−58	−46	−47
viscosity at −20°C, mm²/s(= cSt) max	8.0			8.5	8.0
combustion					
heat content, MJ/kg, min[d]	42.8	42.8	42.8	42.6	42.8
smoke pt, mm, min	18[e]	18[e]	20	19	19
H_2 content, wt % min			13.5	13.4	13.4
stability					
test temp[f], °C min	245	245	260	260	260

[a] Full specification requires other tests.
[b] For aromatics above 20% (22% for Jet A1) users must be notified.
[c] International airlines use Jet A1 with −47°C freeze point.
[d] To convert MJ to kcal, multiply by 239.
[e] Plus naphthalenes 3 vol % max unless smoke point exceeds 25.
[f] Thermal stability test by D3241 to meet 3.3 kPa (25 mm Hg) pressure drop and Code 3 Deposit Rating.

Table 3. Specifications for Ground Gas Turbine Fuels

Property	0-GT[a] Naphtha	1-GT[a] Light distillate	2-GT[a] Medium distillate	3-GT[a] Heavy distillate	4-GT Heavy residual
90% distillation, °C		288 max	282–338		
density, kg/m³, min		850	876		
flash pt, °C max		38	38	55	66
pour pt, °C max		−18	−6		
carbon residue, % max	0.15	0.15	0.35		
ash, wt % max	0.01	0.01	0.01	0.03	
viscosity at 40°C, mm²/s(= cSt)		1.3–2.4	1.9–4.1		
viscosity at 50°C, mm²/s(= cSt)max				638	638
water and sediment, vol % max	0.05	0.05	0.05	1.0	1.0

[a] Trace metal limits at point of delivery are established at 0.5 ppm by wt maximum for vanadium, sodium and potassium, calcium, and lead.

dictated by the vapor and liquid entrainment losses that could occur in a piston aircraft capable of climbing to an altitude of about 6000 m; the vapor pressure of the warm fuel exceeds atmospheric pressure at that altitude (48 kPa or 7 psi). Since early military jet aircraft could climb even higher (12,000 m) and faster, it was necessary to further limit the vapor pressure of military gas turbine fuels to 21 kPa (3 psi). The volatility of kerosene fuel is measured not by its vapor pressure but by its temperature at the point where its vapors just prove to be flammable, ie, the flash point.

Commercial aviation utilizes low volatility kerosene defined by a flash point minimum of 38°C. The flammability temperature has been invoked as a safety factor for handling fuels aboard aircraft carriers.

Ground turbine fuels are not subject to the constraints of an aircraft operating at reduced pressures of altitude. The temperature of fuel in ground tanks varies over a limited range, eg, 10–30°C, and the vapor pressure is defined by a safety-handling factor such as flash point temperature. Volatile fuels such as naphtha (No. 0-GT) are normally stored in a ground tank equipped with a vapor recovery system to minimize losses and meet local air quality codes on hydrocarbons.

A minimum volatility is frequently specified to assure adequate vaporization under low temperature conditions. It can be defined either by a vapor pressure measurement or by initial distillation temperature limits.

Low Temperature Fuel Flow. The decrease in the temperature of fuel in the tank of an aircraft during a long-duration flight produces a number of effects which can influence flight performance. Fuel viscosity increases, necessitating more pumping energy in the tank boost pump and in the engine pump. Fuel becomes saturated with water, and droplets of free water form and settle. Those carried with fuel may form ice on the cold filter which protects the fuel control. For this reason filter heaters are used in civil aircraft to avoid ice blockage and fuel starvation. Military aircraft avoid the complication of a filter heater and depend instead on an antiicing additive in the fuel to depress the freezing point of water. At still lower temperatures, crystals of wax form in fuel. The temperature at which these wax crystals

Table 4. Additives Used in Aviation Gas Turbine Fuels

Class of additive[b]	Chemical types	Purpose	Specification status[a] Civil	Military
antioxidant	alkyl phenylene diamines, hindered alkylphenols	improve oxidation stability–storage and high temperature	O	O
metal deactivator	disalicylidene alkyl diamine	remove metal ions that catalyze oxidation	O	O
antiicing	glycol ether	prevent low temperature filter icing		R
corrosion inhibitor	dimer acid, phosphate ester	reduce development of corrosion products, improve lubrication		R
antistatic	alkyl chromium salt and calcium sulfosuccinate, olefin–sulfur dioxides and polyamines	increase conductivity	OR	R
biocide	boron ester	inhibit organic growth in fuel tanks	A	

[a] O = optional, R = required, A = airline use in maintenance.
[b] See also ANTIOXIDANTS; ANTIFREEZES AND DEICING FLUIDS; ANTISTATIC AGENTS; CORROSION AND CORROSION CONTROL; INDUSTRIAL ANTIMICROBIAL AGENTS.

Table 5. Net Heat Content of Gas Turbine Fuels

Fuel	Heat content, MJ/kg[a]
hydrogen	120
natural gas (methane)	50
light naphtha (0-GT)	44
wide-cut fuel (JP-4)	43
Jet A kerosene (1-GT)	43
No. 2-GT fuel	42
No. 3-GT fuel	41
No. 4-GT fuel	40
methanol	20

[a] To convert MJ to kcal, multiply by 239.

disappear is called the freezing point and distinguishes Jet A1, used by long-range international airlines, from Jet A, used by domestic airlines for relatively short-duration flights.

Water in Fuel. The effects of water in fuel inside a tank, particularly an aircraft tank, are important because of the demonstrated proclivity of free water to form undrainable pools where microorganisms can flourish. In the aircraft the fungi and yeast growth usually takes place on tank surfaces, forming a fungal mat under which metabolic products such as organic acids penetrate polymeric coatings to attack the aluminum skin itself. The growth may also affect the capacitance gauge used to read liquid level.

It is common to curb growth of organisms by biocidal treatment. In storage tanks a water-soluble agent is used. Aircraft tanks are opened periodically for hand cleaning and subsequent treatment with a fuel-soluble boron-containing biocide. The antiicing additive used by the military to lower the freezing point of water, 2-methoxyethanol, happens also to inhibit growth of organisms effectively. Therefore, aircraft tank treatment to remove fungal mats is needed only for commercial transports.

Compatibility and Corrosion. Gas turbine fuels must be compatible with the elastomeric materials and metals used in fuel systems. Elastomers are used for O-rings, seals, and hoses as well as pump parts and tank coatings. Polymers tend to swell and to improve their sealing ability when in contact with aromatics, but the degree of swell is a function of both elastomer-type and aromatic molecular weight. Rubbers can also be attacked by peroxides that form in fuels that are not properly inhibited (see ELASTOMERS, SYNTHETIC; RUBBER, NATURAL).

Boundary lubrication of rubbing surfaces such as those found in high speed pumps and fuel controls has been found to be related to the presence in fuel of species capable of forming a chemisorbed film that reduces friction and wear. Corrosion inhibitors, which tend to form tenacious films on metal surfaces, are generally excellent lubricity agents. Refining processes to reduce sulfur, remove olefins, and control acidity tend to degrade a fuel's lubrication propensity. Corrosion inhibitors are mandatory in military fuels since they also improve the lubricity of fuels by extending the life of pumps and controls.

Economic Aspects

The exacting list of specification requirements for aviation gas turbine fuels and the constraints imposed by delivering clean fuel safely from refinery to aircraft are the factors that affect the economics. Compared with other distillates such as diesel and burner fuels, kerosene jet fuels are narrow-cut specialized products, and usually command a premium price over other distillates. The prices charged for jet fuels tend to escalate with the basic price of crude, a factor which seriously undermined airline profits during the Persian Gulf war as crude prices increased sharply. Availability and cost of jet fuels are also affected by the interrelationships of the market for other petroleum products.

Future Trends

Aircraft Fuels. Demand for aviation gas turbine fuels has been growing more rapidly than demand for other petroleum products since 1960, about 3–5% per year compared with 1% for all oil products. This strong demand reflects a current and predicted growth in worldwide air traffic of 4–7% annually until the end of the century. Total world oil demand will be up by 15% by the year 2000, but aviation fuel demand will increase by 50–125%. However, the fraction of the oil barrel devoted to aviation, now about 8%, will increase only slightly.

Synthetic fuels derived from shale or coal will have to supplement domestic supplies from petroleum in the future, and aircraft gas turbine fuels producible from these sources have been assessed. Production of liquid jet fuel from processing of abundant natural gas is a more promising and cheaper source of high quality product than shale or coal.

Fuels for Advanced Aircraft. The first commercial supersonic transport, the Concorde, operates on Jet A1 kerosene but produces unacceptable noise and exhaust emissions. Moreover, it is limited in capacity to 100 passengers and to about 3000 miles in range.

NASA is considering a more advanced aircraft, eg, a Mach 5 version, to cut Pacific travel time to about 3 h, but in this case kerosene fuel is no longer acceptable, and liquefied natural gas or liquefied hydrogen would be needed to provide the necessary cooling and stability. However, a completely new fueling system would be required at every international airport to handle these cryogenic fluids.

Ground Turbine Fuels. Unlike the outlook for aviation fuels, the demand for ground turbine fuels has stabilized because the ground turbine can be replaced by a more fuel-efficient engine for passenger cars and trucks.

W. G. DUKEK
Consultant

W. M. Bustin and W. G. Dukek, *Electrostatic Hazards in the Petroleum Industry,* John Wiley & Sons, Inc., Chichester, U.K., 1983.

M. Smith, *Aviation Fuels,* G. T. Foulis & Co., Ltd., Henley-on-Thames, Oxfordshire, UK, 1970.

H. C. Barnett and R. R. Hibbard, *Properties of Aircraft Fuels,* NASA TN 3276, NASA, Lewis Research Center, Cleveland, Ohio, Aug. 1956.

Factors in Using Kerosene Jet Fuel of Reduced Flash Point, ASTM STP 688, American Society for Testing and Materials, Philadelphia, Pa., 1979.

AZINE DYES

Azine, oxazine, and thiazine dyes were among the earliest of synthetic dyes. The names are derived from the 6-member heterocyclic ring system present in all dyes of these classes: 1,4-diazine (**1**), 1,4-oxazine (**2**), and 1,4-thiazine (**3**).

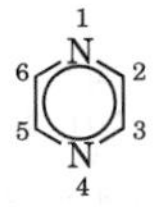

(**1**)

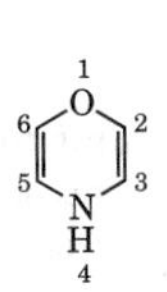

(**2**)

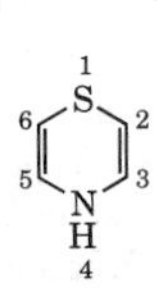

(**3**)

Azine, oxazine, and thiazine dyes were historically more important than they are at present. However, at least one example of each, introduced more than 100 years ago, is still offered commercially today. Azo and anthraquinone dyes have largely displaced them in commercial application. Azo dyes (qv) offer better fastness and broader shade ranges at more economical prices.

All classes of azine dyes are vattable, ie, they are reduced to "colorless" forms, then oxidized back to the dye. They therefore offer good fastness to oxidation. Because of this property, many find uses as redox indicators in titrations.

Dyes in these classes are generally basic dyes, ie, the chromophore is cationic.

Azines

Azine dyes are used extensively as biological stains. Colors are mostly yellow to red.

Synthesis. Methods of synthesis have changed little since the synthesis of Basic Red 2. One method starts with *o*-toluidine, C_7H_9N, which is diazotized and coupled to form amino azotoluene, $C_{14}H_{15}N_3$. Amino azotoluene is reduced to one mole each of *o*-toluidine and 2,5-diaminotoluene, $C_7H_{10}N_2$. Condensation and oxidation with aniline, C_6H_7N, gives the desired dye.

Oxazine Dyes

There have been approximately 80 references to oxazine dyes in the past 15 years. Several are references to their use as laser dyes and titration indicators. Thirty of these references are to oxazine dyes for either acrylic fibers or leather (qv) (see FIBERS, ACRYLIC).

Oxazine dyes containing sulfonic acid groups and claimed to be suitable for dyeing leather in brown shades include $C_{13}H_8N_2O_{11}S_2 \cdot xNa$ (**4**) and $C_{13}H_7ClN_2O_8S$ (**5**).

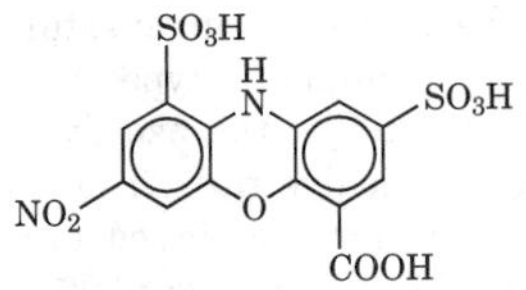

(**4**)

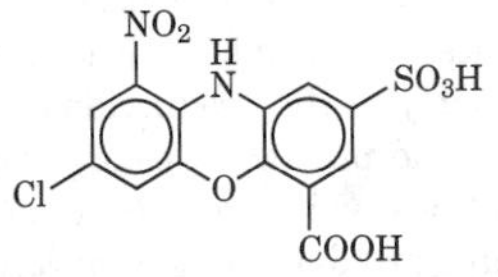

(**5**)

Cationic oxazine dyes suitable for dyeing acrylic fibers have been reported.

Synthesis. Several synthetic methods have been reported for oxazine dyes. The following are representative.

Condensation of phenol, C_6H_6O, with *p*-nitroso-*N,N*-dimethylaniline, $C_8H_{10}N_2O$, in acetic acid, and condensation of an *o*-nitrosophenol with an amine hydrochloride in acetic acid.

Thiazine Dyes

Methylene Blue CI Basic Blue 9(CI 52015), is the classic thiazine dye still in use today. This dye was first reported in 1876.

Thiazine is not important in dyes as such, but it is a part of some reactive dyes. Thiazine is important in sulfur dyes (qv).

Synthesis. A method of synthesis for Methylene Blue is still the stepwise method of choice for thiazine dyes. *N,N*-Dimethyl-*p*-phenylene diamine, $C_8H_{12}N_2$, reacts with sodium thiosulfate to form the thiosulfonic acid which condenses with *N,N*-dimethylaniline, $C_8H_{11}N$, in the presence of sodium dichromate to the indamine, then with copper sulfate and sodium dichromate to Methylene Blue.

ROY E. SMITH
CIBA-GEIGY Corporation

K. Venkataraman, *The Chemistry of Synthetic Dyes and Pigments,* Vol. 2, Academic Press, Inc., New York, 1952, pp. 761–795.

Eur. Pat. Appl. 84,718 (Aug. 3, 1983), A. H. M. Renfrew (to ICI); Ger. Pat. 3,330,547 (Mar. 21, 1985), W. Harms, G. Franke, and K. Wunderlich (to Bayer AG); Eur. Pat. Appl. 135,381 (Mar. 27, 1985), R. D. McClelland and A. Renfrew (to ICI); Ger. Pat. 3,336,362 (Apr. 18, 1985), H. Jaeger (to Bayer AG).

Ger. Pat. 2,518,587 (Nov. 13, 1975), P. Moser and A. N. Nicopoulos (to CIBA-GEIGY); Ger. Pat. 2,458,347 (June 16, 1976), R. Mohr and E. Fleckenstein (to Hoechst AG); Ger. Pat. 2,065,887 (Oct. 21, 1976), E. Mundlos, R. Mohr, and L. Herz (to Hoechst AG); Brit. Pat. 1,488,609 (Oct. 12, 1977) (to Hoechst AG); Ger. Pat. 2,631,040 (Jan. 27, 1977), E. Brunn (to CIBA-GEIGY); Ger. Pat. 2,631,207 (Jan. 19, 1978), K. Lehment, F. Klaus, and U. Trense (to Bayer AG); U.S. Pat. 4,116,622 (Sept. 26, 1978), S. Koller and J. Koller (to CIBA-GEIGY); U.S. Pat. 4,196,286 (Apr. 1, 1980), E. Brunn (to CIBA-GEIGY); Ger. Pat. 3,033,439 (Sept. 5, 1980), U. Mayer (to BASF); U.S. Pat. 4,288,227 (Sept. 8, 1981), B. Gertiser and B. Henz (to Sandoz); Eur. Pat. Appl. 55,223 (June 30, 1982), P. Loew (to CIBA-GEIGY); Switz. Pat. 642,987 (May 15, 1984), P. Moser (to CIBA-GEIGY).

Ger. Pat. 2,518,587 (Nov. 13, 1975), P. Moser and A. Nicopoulos (to CIBA-GEIGY); Ger. Pat. 2,516,920 (Oct. 21, 1976), W. Loehr and co-workers; Ger. Pat. 139,268 (Dec. 19, 1979), G. Mann and co-workers.

AZO DYES

The term azo dyes is applied to those synthetic organic colorants that are characterized by the presence of the chromophoric azo group (—N=N—). This divalent group is attached to sp^2 hybridized carbon atoms: on one side, to an aromatic or heterocyclic nucleus; on the other, it may be linked to an unsaturated molecule of the carbocyclic, heterocyclic, or aliphatic type. No natural dyes contain this chromophore. Commercially, the azo dyes are the largest and most versatile class of organic dyestuffs. There are more than 10,000 Colour Index (CI) generic names assigned to commercial colorants; approximately 4,500 are in use, and over 50% of these belong to the azo class. Synthetic dyes are derived in whole or in part from cyclic intermediates. Approximately two-thirds of the dyes consumed in the United States are used by the textile industry to dye natural and synthetic fiber or fabrics, about one-sixth is used for coloring paper, and the rest is used chiefly in the production of organic pigments and in the dyeing of leather and plastic. Dyes are sold as pastes, powders, and liquids; concentrations vary from 6–100%. The concentration, form, and purity of a dye are determined largely by the use for which it is intended.

Classification and Designations. The most authoritative compilation covering the constitution, properties, preparations, manufacturers, and other coloring data is the publication *Colour Index,* which is edited jointly by the Society of Dyers and Colourists and the American Association of Textile Chemists and Colorists (AATCC).

The Chemistry of Synthesis

Peter Griess synthesized the first azo dye soon after his discovery of the diazotization reaction in 1858. The two reactions which form the basis for azo dye chemistry are diazotization (eq. 1) and coupling (eq. 2):

$$ArNH_2 + 2\ HX + NaNO_2 \rightarrow ArN_2^+X^- + NaX + 2\ H_2O$$
$$\text{where } X = Cl^-, Br^-, NO_3^-, HSO_4^-, BF_4^-, \text{etc} \quad (1)$$

$$ArN_2^+X^- + RH \rightarrow ArN{=}N{-}R + HX \quad (2)$$

where Ar = Aryl and R represents an alkyl or aryl radical whose conjugate acid RH is capable of coupling.

Azo Coupling. The coupling reaction between an aromatic diazo compound and a coupling component is the single most important synthetic route to azo dyes. Of the total dyes manufactured, about 60% are produced by this reaction. Other methods include oxidative coupling, reaction of arylhydrazine with quinones, and oxidation of aromatic amines. These methods, however, have limited industrial applications.

All coupling components used to prepare azo dyes have the common feature of an active hydrogen atom bound to a carbon atom. Compounds of the following types can be used as azo coupling components: *(1)* aromatic hydroxy compounds such as phenols and naphthols; *(2)* aromatic amines; *(3)* compounds that possess enolizable ketone groups of aliphatic character.

Broad principles governing the activity of coupling components may be summarized as follows:

1. Diazo coupling follows the rules of orientation of substituents in aromatic systems in accordance with the mechanism of electrophilic aromatic substitution and the concept of resonance.
2. Generally, phenols (such as the phenolate anion) couple more readily than amines, and members of the naphthalene series more readily than the members of the benzene series.
3. Electron-attracting substituents in the coupling components such as halogen, nitro, sulfo, carboxyl, and carbonyl, are deactivating and tend to retard coupling.
4. A lower alkyl or alkoxy group substituted in the ortho or meta position to an amino group may promote coupling. Good couplers are obtained from dimethylaniline when lower alkyl, lower alkoxy, or both groups are present in the 2- and 5-position.
5. It is possible for diazo compounds to attack both the ortho and para position of hydroxyl and amino coupling components when these positions are not already occupied.

Technologically, the most important examples of such couplers are 1-naphthylamine, 1-naphthol, and sulfonic acid derivatives of 1-naphthol. Of great importance in the dyestuff industry are derivatives of 1-naphthol-3-sulfonic acid, such as H-acid (8-amino-1-naphthol-3,6-disulfonic acid), J-acid (6-amino-1-naphthol-3-sulfonic acid) (**1**), and gamma acid (7-amino-1-naphthol-3-sulfonic acid) (**2**).

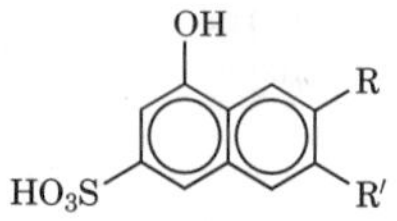

(**1**) R = H, R′ = NH_2
(**2**) R = NH_2, R′ = H

The azo coupling reaction proceeds by the electrophilic aromatic substitution mechanism.

Classification of Azo Dyes

In addition to classification according to the number of azo groups, further subdivision is achieved, first according to whether the compound is water-soluble and, secondly, according to the types of component used. Another system of classification is based on dyeing classes. All colorants are divided to indicate the chief method of application or to indicate principal use. Azo dyes are found in the acid, basic (cationic), direct, disperse, mordant, and reactive dyeing classes.

In the disazo group of azo dyes, primary and secondary types are distinguished. The former covers compounds made from two molecules of a diazo derivative and one molecule of a bifunctional coupling component. In both cases, the monofunctional reagent may consist of two molecules of one compound or one molecule of each of two substances used stepwise, the first alternative yielding symmetrical products.

Secondary Disazo Dyes. These dyes are made by diazotizing an aminoazo compound, the amino group of which derives from the original coupling component, and coupling it to a suitable intermediate.

Miscellaneous Disazo Dyes. Another group of disazo dyes is prepared by condensation of two identical or different aminoazo compounds commonly with phosgene, cyanuric chloride, or fumaryl dichloride, the fragments of which act as blocking groups between chromophores.

Trisazo and Polyazo Dyes. These are mostly direct dyes, the hues are predominantly brown, black, or dark blue or green. Some are leather dyes. Benzidine, which used to be an important bisdiazo component, has been replaced by 4,4′-diaminobenzanilide, 4,4′-diaminodiphenylamine-2-sulfonic acid, etc. Benzidine dyes are almost never produced any longer because of their carcinogenicity.

Dyeing Classes: Structure, Application, Uses

Of all classes of dyestuffs, azo dyes have attained the widest range of usage because variations in chemical structure are readily synthesized and methods of application are generally not complex. There are azo dyes for dyeing all natural substrates such as cotton, paper, silk, leather, and wool; and there are azo dyes for synthetics such as polyamides, polyesters, acrylics, polyolefins, viscose rayon, and cellulose acetate; for the coloring of paints, varnishes, plastics, printing inks, rubber, foods, drugs, and cosmetics; for staining polished and absorbed surfaces; and for use in diazo printing and color photography. The shades of azo dyes cover the whole spectrum.

Table 1. Yellow and Orange Commercial Azo Acid Dyes, U.S. 1988

CI name	CI number	Chemical type
Acid Yellow 34	18890	monoazo
Acid Yellow 36	13065	monoazo
Acid Yellow 49	18640	monoazo
Acid Yellow 59	18690	monoazo, metallized
Acid Yellow 65	14170	monoazo
Acid Yellow 99	13900	monoazo, metallized
Acid Yellow 135	14255	monoazo
Acid Yellow 151	13906	monoazo, metallized
Acid Yellow 200	18930	monoazo
Acid Orange 7	15510	monoazo
Acid Orange 10	16230	monoazo
Acid Orange 24	20170	disazo
Acid Orange 60	18732	monoazo, metallized
Acid Orange 156	26501	disazo

Table 2. Other Commercial Azo Acid Dyes, U.S. 1988

CI name	CI number	Chemical type
Acid Red 4	14710	monoazo
Acid Red 14	14720	monoazo
Acid Red 18	16255	monoazo
Acid Red 73	27290	disazo
Acid Red 85	22245	disazo
Acid Red 88	15620	monoazo
Acid Red 114	23635	disazo
Acid Red 137	17755	monoazo
Acid Red 151	26900	disazo
Acid Red 186	18810	monoazo, metallized
Acid Red 266	17101	monoazo
Acid Violet 3	16580	monoazo
Acid Violet 7	18055	monoazo
Acid Violet 12	18075	monoazo
Acid Blue 92	13390	monoazo
Acid Blue 113	26360	disazo
Acid Blue 118	26410	disazo
Acid Green 20	20495	disazo
Acid Brown 14	20195	disazo
Acid Black 1	20470	disazo
Acid Black 52	15711	monoazo, metallized
Acid Black 60	18165	monoazo, metallized
Acid Black 63	12195	monoazo, metallized

Acid Dyes. Commercial acid dyes contain one or more sulfonate groups, thereby providing solubility in aqueous media. These dyes are applied in the presence of organic or mineral acids (pH 2–6). Such acids protonate any available cationic sites on the fiber, thereby making possible bonding between the fiber and the anionic dye molecule.

There are three general classifications of acid dyes grouped according to their method of application: acid dyes that dye directly from the dyebath, mordant dyes that are capable of forming metallic lakes on the fiber when aftertreated with metallic salts, and premetallized dyes. Yellow and orange azo acid dyes are listed in Table 1. Other azo acid dyes are listed in Table 2.

Metal Complexes of Azo Dyes. Metal complexes of certain *o,o′*-dihydroxyazo, *o*-carboxy-*o′*-hydroxyazo, *o*-amino-*o′*-hydroxyazo, arylazosalicyclic acid, and formazan compounds are used as dyes for wool, nylon, and cotton with generally much improved washfastness and lightfastness properties when compared to their respective unmetallized precursors. Dyes that are chelated with the metal on the substrate during the dyeing process are termed metallizable or mordant dyes. Conversely, those dyes that have been metallized by the dye manufacturer prior to use by the dyer, are classified as premetallized dyes. The two important types of premetallized dyes are the 1:1 and 2:1 complexes, eg, complexes with 1:1 and 2:1 ligand-to-metal ratios, respectively.

Chromium is the principal metal used with mordant dyes for wool, whereas both chromium and cobalt are used extensively in premetallized types for wool and nylon. Copper(II) is employed almost exclusively as the chelating metal ion in both metallizable and premetallized direct dyes for cotton.

Direct Dyes. Direct dyes are defined as anionic dyes substantive to cellulosic fibers (cotton, viscose, etc), when applied from an aqueous bath containing an electrolyte.

Direct dyes are one of the most versatile classes of dyestuff. In worldwide usage for cellulosic textiles, direct dyes are the second largest class of dyestuff. The AATCC *Buyers Guide* (July 1991) lists over 180 different CI categories for direct dyes representing nearly 850 commercially available products. The important direct yellows and oranges of revealed chemical composition are listed in Table 3.

Direct Reds. The principal commercially produced direct reds are of disazo composition, except for Direct Red 80, which is polyazo.

CI name	CI number
Direct Red 2	23500
Direct Red 4	29165
Direct Red 16	27680
Direct Red 23	29160
Direct Red 24	29185
Direct Red 26	29190
Direct Red 72	29200
Direct Red 73	29180
Direct Red 80	35780
Direct Red 81	28160
Direct Red 83	29225

Direct Blues. Table 4 shows some direct blues.

Direct Violets, Greens, Browns, and Blacks. Direct violets and greens are small-volume products.

Two important browns, other than benzidine derivatives, are of azo chemical composition: Direct Brown 30 and Direct Brown 44.

Table 5 lists some direct blacks.

Azoic or Naphthol Dyes. Azoic dyes (known also as ice colors and ingrain colors) are water-insoluble azo pigments, free from solubilizing groups, formed on the fiber by reaction of a diazo component with

Table 3. Yellow and Orange Shade Commercial Direct Dyes, U.S. 1988

CI name	CI number	Chemical type
Direct Yellow 4	24890	disazo
Direct Yellow 6	40001–40006	stilbene
Direct Yellow 11	40000	stilbene
Direct Yellow 28	19555	thiazole
Direct Yellow 34	29060	disazo
Direct Yellow 44	29000	disazo
Direct Yellow 106	40300	stilbene
Direct Yellow 118	29042	disazo
Direct Orange 6	23375	disazo
Direct Orange 8	22130	disazo
Direct Orange 15	40002–40003	stilbene
Direct Orange 26	29150	disazo
Direct Orange 34	40215–40220	stilbene
Direct Orange 39	40215	stilbene
Direct Orange 72	29058	disazo
Direct Orange 102	29156	disazo

Table 4. Blue Shade Commercial Direct Dyes

CI name	CI number	Chemical type
Direct Blue 15	24400	disazo
Direct Blue 22	24280	disazo
Direct Blue 25	23790	disazo
Direct Blue 75	34220	trisazo
Direct Blue 76	24410	disazo
Direct Blue 80	24315	disazo
Direct Blue 86	74180	phthalocyanine
Direct Blue 98	23155	disazo
Direct Blue 108	51320	oxazine
Direct Blue 218	24401	disazo metallized

Table 5. Black Shade Commercial Direct Dyes

CI name	CI number	Chemical type
Direct Black 19	35255	polyazo
Direct Black 22	35435	polyazo
Direct Black 150	32010	trisazo
Direct Black 166	30026	trisazo

a coupling component, a so-called Naphthol AS compound, such as an arylide of 3-hydroxy-2-naphthoic acid. The discovery that 3-hydroxy-2-naphthoic acid arylides have greater substantivity for cotton than 2-naphthol, tremendously increased the range of bright and fast shades, and led to the introduction of Naphthol AS derivatives, fast color salts, rapid fast dyes, and the rapidogen dyes from diazoamino compounds.

Disperse Azo Dyes. Disperse dyes are coloring substances having very low aqueous solubility which are applied to hydrophobic fibers from an aqueous system in which the dye is present in a highly dispersed state. The sharp increase in the importance of disperse dyes in the 1970s and 1980s can be attributed directly to the emergence of polyester and nylon as the principal synthetic fibers.

Azo and anthraquinone compounds comprise the two principal structural types which are used as disperse dyes. Other compounds used to a much lesser extent include methines, cyanostyryls, hydroxyquinophthalones, and nitrodiarylamines.

Disperse dyes are used mainly in the coloring of polyester, cellulose acetate, and triacetate fibers in textile applications. Another important use is the dyeing of nylon, especially carpets and hosiery in which disperse dyes effectively cover barré defects in the fiber caused by processing inconsistencies. Yellow and orange shade commercial disperse azo dyes are listed below.

CI name	CI number
Disperse Yellow 3	11855
Disperse Yellow 4	12770
Disperse Yellow 5	12790
Disperse Yellow 7	26090
Disperse Yellow 8	12690
Disperse Yellow 10	12795
Disperse Yellow 23	26070
Disperse Yellow 60	12712
Disperse Orange 1	11080
Disperse Orange 3	11005
Disperse Orange 5	11100
Disperse Orange 13	26080
Disperse Orange 25	11227
Disperse Orange 29	26077
Disperse Orange 30	11119
Disperse Orange 56	12650
Disperse Orange 62	11239
Disperse Orange 138	11145

Dispersion Technology. Manufacturing procedures for producing dye dispersions are generally not disclosed. The principal dispersants in use include long-chain alkyl sulfates, alkaryl sulfonates, fatty amine–ethylene oxide condensates, fatty alcohol–ethylene oxide condensates, naphthalene–formaldehyde–sulfuric acid condensates, and the lignin sulfonic acids.

All dispersions are thermodynamically unstable, since the interfacial area and hence the surface energy tend to decrease, ie, agglomeration occurs. The primary function of dispersing agents is to stabilize dispersions. Additional commercial disperse dyes are listed below.

CI name	CI number
Disperse Red 1	11100
Disperse Red 5	11215
Disperse Red 7	11150
Disperse Red 13	11115
Disperse Red 17	11210
Disperse Red 19	11130
Disperse Red 31	11250
Disperse Red 32	11190
Disperse Red 58	11135
Disperse Red 65	11228
Disperse Red 72	11114
Disperse Red 73	11116
Disperse Red 90	11117
Disperse Violet 24	11200
Disperse Violet 33	11218
Disperse Blue 11	11260
Disperse Blue 79	11345
Disperse Blue 165	11077
Disperse Blue 183	11078
Disperse Brown 1	11152
Disperse Black 1	11365

Application Techniques, Structural Variations, and Fastness Properties. Since 1950, there has been a steady development of new disperse dyes to meet the demands imposed by the changing application methods and to provide the much needed improvement in fastness properties. Six different methods of applying disperse dyes have been developed since the introduction of polyester fibers: *(a)* dyeing at the boil in the presence of a carrier is used for delicate fabrics, polyester–wool blends, etc.; *(2)* dyeing at 120–135°C in pressurized vessels gives better exhaustion and often improved fastness to light, rubbing, and perspiration; *(3)* thermofixation techniques at 190–220°C are used for the continuous processing of certain types of fabrics; *(4)* transfer printing, generally at 210°C for 30 seconds, is an important development; *(5)* solvent dyeing methods are available, but not popular; *(6)* printing and continuous dyeing processes have been developed for polyester–cotton blends using specialist dyes and application techniques such as the Dybln (DuPont), Cellestren (BASF), and Dispersol (ICI) ranges.

Disperse dyes are classified as high energy or low energy types. The use of higher dyeing temperatures for polyester fibers compared with those used for cellulose acetate has made possible the use of dyes of higher molecular weight, the so-called high energy dyes.

A use for disperse dyes which has undergone rapid growth since 1970 is in inks for the heat-transfer printing of polyester, especially double knit fabrics. This simple method consists of printing the desired design on paper and then transferring the design from the paper to the fabric with heat. In the heat-transfer process the dye volatilizes, is adsorbed onto the fiber surface, and then diffuses into the fiber. Generally low and medium energy types are used for this purpose.

Dye structures have been modified to achieve high resistance to sublimation, good lightfastness, and good build-up of color by either increasing the molecular weight or introducing more polar groups, or both, and by decreasing further their slight solubility in water.

Heterocyclic Disperse Dyes. Diazotizable aminoheterocyclic compounds are also used in the production of disperse dyes.

Oil-Soluble Azo Dyes. The oil-soluble, water-insoluble, azo dyes dissolve in oils, fats, waxes, etc. Generally, yellow, orange, red, and brown oil colors are azo structures and greens, blues, and violets are primarily anthraquinones (see DYES, ANTHRAQUINONE). Blacks are usually nigrosines and indulines of the azine type (see AZINE DYES).

Substitution by chloro, nitro, and similar groups increases the molecular weight and improves sublimation fastness but lowers the oil solubility of this group of dyes.

Spirit-Soluble Azo Dyes. Spirit-soluble azo dyes dissolve in polar solvents, such as alcohol and acetone, and find application in the coloring of lacquers, plastics, printing inks, and ball-point pen inks. Of the two principal types of azo structures used, the most important are the insoluble salts of azo dyes containing sulfo groups and relatively complex organic amines. Mono- and dicyclohexylamine, isoamylamine, and the arylguanidines often serve as the amine, and the anionic component is chosen from the class of acid dyes for wool.

The second type is comprised of 2:1 metal complexes of *o,o'*-dihydroxy azo dyes which generally do not contain sulfo or other strongly hydrated groups as found in the premetallized 2:1 complexes for wool. Thus their solubility in esters, ketones, and alcohols is relatively increased.

Basic (Cationic) Azo Dyes. Basic dyes of the azo class are the simplest and oldest known synthetic dyes. Current cationic dyes are used for modified acrylics, modified nylons, modified polyesters, leather, unbleached papers, and inks. An important application is for conversion into pigments. Principal chemical classes include azo, anthraquinone, triarylmethane, methine, thiazine, oxazine, etc. The dyes are applied in acidic dyebaths to fibers made of negatively charged polymer molecules.

Cationic azo dyes carry a positive charge in the chromophore portion of the molecule. The salt-forming counterion is usually a chloride or acetate. CI basic dyes are ammonium, sulfonium, or oxonium salts. Commercial basic azo dyes for which chemical structures are revealed by U.S. producers are listed below.

CI name	CI number
Basic Yellow 15	11087
Basic Yellow 24	11480
Basic Yellow 25	11450
Basic Yellow 57	12719
Basic Orange 1	11320
Basic Orange 2	11270
Basic Red 18	11085
Basic Red 22	11055
Basic Red 24	11088
Basic Red 29	11460
Basic Red 39	11465
Basic Red 76	12245
Basic Blue 41	11105
Basic Blue 54	11052
Basic Blue 65	11076
Basic Blue 66	11075
Basic Blue 67	11185
Basic Brown 1	21000
Basic Brown 2	21030
Basic Brown 4	21010
Basic Brown 16	12250
Basic Brown 17	12251
Basic Black 2	11825

Azo Pigments. Organic pigments are an important class of organic colorants. Expanding areas of usage include the mass coloration of synthetic fibers and textile printing in the textile field, and in the non-textile area, plastics. A pigment is insoluble in the medium in which it is used. The physical properties of pigments are of great significance since the coloring process does not involve solution of the colorant. Azo pigments can be grouped as metal toners, metal chelates, and metal-free azo pigments.

Metal toners usually contain one sulfonic acid group and often a carboxylic acid group. Metal chelation is also a means of insolubilizing organic molecules.

In the metal-free class of azo pigments it is remarkable that the simplest derivatives of 2-naphthol such as Hansa Red B (CI Pigment Red 3; CI 12120, Toluidine Red) and Para Red, both known since 1905, are still of importance.

Because these pigments are organic in nature they tend to bleed in resins and solvents. Increasing the molecular weight often reduces this tendency.

RASIK J. CHUDGAR
BASF Corporation

Colour Index, 3rd ed., Vol. 8, The Society of Dyers and Colourists, Bradford, Yorkshire, U.K., 1987.

K. Venkataraman, *The Chemistry of Synthetic Dyes,* Vols. 1 and 2, Academic Press, New York, 1952.

H. E. Fierz-David and L. Blangey, *Grundlegende Operationen der Farbenchemie,* 8th ed., Vienna, Austria, 1952.

K. H. Saunders and R. L. M. Allen, *Aromatic Diazo Compounds,* Edward Arnold, London, U.K., 1985.

G. Hallas, in J. Griffiths, ed., *Developments in the Chemistry and Technology of Organic Dyes,* Blackwell, Scientific Publishing, Oxford, U.K., 1984.

B

BAKERY PROCESSES AND LEAVENING AGENTS

YEAST-RAISED PRODUCTS

The U.S. baking industry comprises an important portion of the food industry, the nation's largest business. Sales of various bakery foods by the baking industry currently exceed $41 billion per year. The industry is diverse in both products made and size of production units.

Bakery foods are produced in over 48,000 bakeries throughout the country. Over 45,000 of these facilities are made up of retail and in-store bakeries where products are typically baked and sold directly to the consumer at the same location. Only about 33% of the dollar volume of the baking industry is attributed to these small producing facilities. Of the bakery foods produced and consumed in the United States, about 67% are made in approximately 3,000 wholesale bakeries.

Another source of bakery foods is the food service industry. Over 200,000 establishments, including restaurants, hospitals, and prisons, produce bakery foods with a combined value exceeding $8 billion.

Differentiated from the quickly perishable bakery foods are the dry bakery products such as cookies, crackers, pretzels, and ice cream cones. These latter items possess a much longer shelf life and may be distributed over a wider area from typically very large manufacturing facilities. According to the 1987 Census of Manufacturers, there are 380 establishments producing these dry-type bakery foods, and the value added by such manufacturing facilities amounts to over $4 billion.

Per capita consumption of all bakery products in the United States was about 50 kg in 1990. Bread accounted for about 45% of this consumption, while all yeast-leavened items comprised approximately 74% of the total. The U.S. Department of Commerce's *1991 U.S. Industrial Outlook* projects that per capita consumption of bakery foods would increase 2.2% annually from 1991 to 1996.

Yeast. The role of baker's yeast (*Saccharomyces cerevisiae*) in producing leavened bread depends on two factors: the ability of yeast to generate carbon dioxide and alcohol through the breakdown of simple sugars, and the unique ability of wheat flour proteins to form films in dough that trap evolved gases (see FERMENTATION; YEASTS). Basic bread is made with flour, water, salt, and yeast. Product variety is achieved by incorporating varying amounts of additional ingredients; by altering the breadmaking process; by shaping or cutting or putting toppings on the dough prior to baking; or by the method of baking. Each of these practices may be utilized alone or in combination to produce a virtually limitless number of yeast-raised products. Of the total annual sales of the baking industry, yeast-raised goods constitute about 61%.

Chemical Leavening. Chemical leaveners (primarily baking powders) are utilized in cakes, cookies, some crackers, refrigerated dough, and quick bread manufacturing. These products make up roughly 39% of the value of the baking industry's total production (see BAKERY PROCESSES AND LEAVENING AGENTS, CHEMICAL LEAVENING AGENTS).

Ingredients

A great many ingredients may be utilized in the production of innumerable bakery foods. Key ingredients include flour, yeast, yeast foods, sugar, shortening, surfactants, milk and milk replacers, eggs, salt, water, enzymes, mold inhibitors (antimycotics), flavorings, and enriching ingredients.

Procedures and Equipment

Bulk Ingredient Handling. Mills ship flour in bulk directly to large commercial bakeries via specially designed railroad cars or trucks. Upon receipt, this flour is transferred pneumatically from the cars or trucks to bins in the plant, from which it is conveyed (also pneumatically) to the mixing process as required. Smaller plants may instead use a tote bin system of bulk handling. Flour used in small quantities, or flour used by bakers with no bulk handling facilities, is delivered in multiwalled paper bags and is handled on skids with forklift trucks.

Most of the sugar used by commercial bread bakers is in syrup form; high fructose corn syrup is primarily used. Following shipment in tank cars, it is piped into bakery storage tanks, from which it is metered into mixers upon demand.

Shortening and oil can also be handled in bulk. If the shortening is plastic, it is melted so that it can be pumped through pipes. These pipes must be heated throughout the plant to prevent the shortening from congealing.

Yeast (compressed and cream) and other perishable ingredients are stored under refrigeration; freezers are required for frozen eggs and fruits. Minor ingredients may be weighed directly from their containers or they may be suspended in water slurries which are subsequently metered into mixers.

Dough Processes for Bread Production. Principal process categories used in the manufacture of yeast-raised products include the sponge and dough method, the straight dough method, and highly accelerated short-time methods that include frozen dough processing, continuous mix, and liquid ferment processes. Considerable variation exists among commercial bakeries within each of these categories. Rolls, buns, and sweet yeast-raised products are produced in ways analogous to those described for bread production. Principal differences are that special makeup machinery is required to produce the various shapes, and finishing equipment is required to fill and ice sweet roll and coffee cake units.

Standards of Identity for Bakery Foods

Federal standards of identity exist for white bread, enriched bread, milk bread, raisin bread, and whole wheat bread that move in interstate commerce. Many states have adopted the federal standards or have promulgated nearly identical laws. Most bakery foods sold in the United States are produced by large commercial bakeries engaged in interstate commerce. As a result, nearly all bread and rolls necessarily conform to the federal standards. The FDA has taken the position that if an ingredient or packaging material has crossed state boundaries, then products containing that ingredient or wrapped in that packaging material are covered under regulations pertaining to interstate commerce. Most specialty breads, such as rye, multigrain, pita, and French, are not covered by standards of identity. All ingredients permitted in standardized bakery foods are considered optional ingredients and therefore must be declared in ingredient legends on bread wrappers. Since 1975, the FDA has required that all enriched bread-type products bear nutrition labeling.

JOSEPH G. PONTE, JR.
JOEL D. PAYNE
Kansas State University

E. J. Pyler, *Baking Science and Technology,* 3rd ed., Vols. 1 and 2, Sosland Publishing Co., Merriam, Kans., 1988.

S. A. Matz, *Formulas and Processes for Bakers,* Pan-Tech International, Inc., McAllen, Tex., 1987.

R. C. Hoseney, *Principles of Cereal Science and Technology,* American Association of Cereal Chemists, Inc., St. Paul, Minn., 1986.

B. S. Kamel and C. E. Stauffer, *Advances in Baking Technology,* Blackie Academic & Professional (VCH Publishers, Inc.), 1993.

CHEMICAL LEAVENING AGENTS

Chemical leavening involves the action of an acid on bicarbonate to release carbon dioxide gas for aeration of a dough or batter during mixing and baking. The aeration provides a light, porous cell structure, fine grain, and a texture with desirable appearance along with palatability to baked goods.

Table 1. Baking Acids Commercially Produced for Leavening Systems

Common name [Chemical Abstracts name]	Formula	Neutralizing value	Uses[a]
monocalcium phosphate monohydrate, MCP [phosphoric acid, calcium salt (2:1), monohydrate]	$CaH_4(PO_4)_2 \cdot H_2O$	80	H, L, C
monocalcium phosphate anhydrous, AMCP coated [phosphoric acid, calcium salt (2:1)]	$CaH_4(PO_4)_2$	83	(H)[b], L
sodium acid pyrophosphate, SAPP [diphosphoric acid, disodium salt]	$Na_2H_2P_2O_7$	72	C, L
sodium aluminum phosphate, SALP			4
[phosphoric acid, aluminum sodium salt (8:3:1)]	$NaH_{14}Al_3(PO_4)_8 \cdot 4H_2O$	100	C, L
[phosphoric acid, aluminum sodium salt (8:2:3)]	$Na_3H_{15}Al_2(PO_4)_8$		
[phosphoric acid, aluminum sodium salt]	$Na_xAl_x(PO_4)_x$		
dicalcium phosphate, dihydrate, DPD [phosphoric acid, calcium salt (1:1)]	$CaHPO_4 \cdot 2H_2O$	33	L
sodium aluminum sulfate, SAS (soda alum) [sulfuric acid, aluminum sodium salt (2:1:1)]	$Al_2(SO_4)_3 \cdot Na_2SO_4$	100	H
glucono-delta-lactone, GDL [D-gluconic acid-δ-lactone]	$C_6H_{10}O_6$	50	L
potassium hydrogen tartrate (cream of tartar) [butanedioic acid, 2,3-dihydroxy-, [*R*-(*R**, *R**)]-monopotassium salt]	$KHC_4H_4O_6$	45	(H)[b]

[a] C = commercial baking powders; H = household baking powders; L = leavening agents for preleavened products.
[b] () = some use, but limited.

Composition of Chemical Leavening Systems

There are essentially two components in a chemical leavening system: bicarbonate that supplies carbon dioxide gas and an acid which triggers the liberation of carbon dioxide from bicarbonate upon contact with moisture (see CARBON DIOXIDE).

Sodium bicarbonate (baking soda) is the primary source of carbon dioxide gas in practically all chemical leavening systems. Other bicarbonates of considerable commercial importance are ammonium bicarbonate and potassium bicarbonate.

The prevalent baking acids in modern chemical leavening systems are sodium or calcium salts of ortho, pyro, and complex phosphoric acids in which at least two active hydrogen ions are attached to the molecule (Table 1). Alum, although not a protonic acid, is used in retail baking powder formulations. Some organic acids are also used in refrigerated dough products.

Characteristics of Leavening Agents

The evolution of carbon dioxide essentially follows the stoichiometry of acid–base reactions. Baking soda determines the amount of carbon dioxide evolved, whereas the type of acid controls the speed of liberation.

Leavening Agents for Preleavened Mixes and Doughs

By far, the greatest use of leavening in the home is in preleavened mixes. Pancake mixes, biscuit mixes, and cake mixes have long been established as commercial products.

Sales of frozen or refrigerated doughs and batters have increased rapidly in recent years. This trend is attributable to increased refrigeration and freezer capacity in stores and to the development of new and improved ingredients. The latter include leavening acids, emulsifiers, gums, and starches which provide greater dough and batter stability during storage. For many of these products, leavening is the most critical component. Practically all of the leavening acids used in the mix industry today are of the phosphate type.

Manufacture of Baking Powders

The blending of baking powder is essentially a physical mixing of the various components in a large-scale batch mixer and is often carried out in automated plants. The order in which mixing occurs may influence the stability of the product. Mixing of starch or other inert components with individual reactive components tends to maintain the separation of reactive components so as to prevent premature reaction during storage.

FRANK H. Y. CHUNG
Rhône-Poulenc, Inc.

E. B. Bennion and co-workers, *Cake Making,* Leonard Hill Books, London, U.K., 1966, p. 76.

U.S. Pat. 2,160,232 (May 30, 1939), W. H. Knox and J. R. Schlaeger (to Victor Chemical Works/Stauffer Chemical Co.) (now Rhône-Poulenc Basic Chemicals Co.).

A. D. F. Toy, *Phosphorus Chemistry in Everyday Living,* American Chemical Society, Washington, D.C., 1976, p. 36.

BARIUM

Barium, Ba, is a member of Group 2 (IIA) of the Periodic Table, where it lies between strontium and radium. Along with calcium and strontium, barium is classed as an alkaline-earth metal, and is the densest of the three. Barium metal does not occur free in nature; however, its compounds occur in small but widely distributed amounts in the earth's crust, especially in igneous rocks, sandstone, and shale. The principal barium minerals are barytes (barium sulfate) and witherite (barium carbonate), which is also known as heavy spar. The latter mineral can be readily decomposed via calcination to form barium oxide, BaO, which is the ore used commercially for the preparation of barium metal.

Barium is prepared commercially by the thermal reduction of barium oxide with aluminum. Barium metal is highly reactive, a property which accounts for its principal uses as a getter for removing residual gases from vacuum systems and as a deoxidizer for steel and other metals.

Physical Properties

Pure barium is a silvery-white metal, although contamination with nitrogen produces a yellowish color. The metal is relatively soft and ductile and may be worked readily. It is fairly volatile (though less so than magnesium), and this property is used to advantage in commercial production. Barium has a bcc crystal structure at atmospheric pressure, but undergoes solid-state phase transformations at high pressures. Because of such transformations, barium exhibits pressure-induced superconductivity at sufficiently low temperatures.

Additional physical properties of barium are given in Table 1.

Table 1. Physical Properties of Barium

Property	Value
atomic weight (12C = 12.000)	137.33
stable isotopes	
weight	
130 132 134 135 136 137 138	
natural abundance, %	
0.106 0.101 2.42 6.59 7.85 11.23 71.7	
density or specific gravity at 20°C, kg/m^3	3.51×10^3
melting point, °C	729
boiling point, °C	1640
heat of fusion, ΔH_f, kJ/mol[a]	7.64
heat of vaporization, ΔH_v, kJ/mol[a]	188.2
specific heat at 25°C, J/(g·K)[a]	0.204
coefficient of thermal expansion, m/(m·°C)	1.8×10^{-5}
superconducting temperature at 8.8×10^6 kPa[b], K	3.05

[a] To convert J to cal, divide by 4.184.
[b] Increases with increasing pressure at the rate of 1.3×10^{-7} K/kPa. To convert kPa to atm, divide by 101.3.

Chemical Properties

Barium has a valence electron configuration of $6s^2$ and characteristically forms divalent compounds. It is an extremely reactive metal, and its compounds possess large free energies of formation.

Economic Aspects

Barium is no longer produced in the United States or Canada. In Europe it is produced by Degussa AG, Hanau, Germany, in the form of rods.

Health and Safety Factors

Barium metal and most barium compounds are highly poisonous. A notable exception is barium sulfate, which is nontoxic because of its extreme insolubility in water. Barium ions act as a muscle stimulant and can cause death through ventricular fibrillation of the heart.

Barium also presents a hydrogen explosion hazard if allowed to come into contact with water or atmospheric moisture, and must always be kept dry and preferably sealed in the shipping containers.

CLAUDIO BOFFITO
Saes Getters SpA

Barium Bibliography, Mineral Products Division, Food Machinery and Chemical Corp., New York, 1961 (esp. pp. 95–155 and 199–238).

R. C. Weast, ed., *Handbook of Chemistry and Physics,* 67th ed., CRC Press, Boca Raton, Fla., 1986, pp. B2, B9, B309–311, B216.

K. T. Jacob and Y. Waseda, *J. Less Common Met.* **139,** 269 (1988).

Metals Handbook, 9th ed., Vol. 2, American Society for Metals, Metals Park, Ohio, pp. 716–717.

BARIUM COMPOUNDS

Barite

Barite, natural barium sulfate, $BaSO_4$, commonly known as barytes, and sometimes as heavy spar, till, or cawk, occurs in many geological environments in sedimentary, igneous, and metamorphic rocks. Commercial deposits are of three types: vein and cavity filling deposits; residual deposits; and bedded deposits. Most commercial sources are replacement deposits in limestone, dolomitic sandstone, and shales, or residual deposits caused by differential weathering that result in lumps of barite enclosed in clay. Barite is widely distributed, with minable deposits in many countries.

Mineralogically, barite crystallizes in the dipyramidal class of the orthorhombic system. Barite is most commonly associated with quartz, chert, jasperoid, calcite, dolomite, siderite, rhodochrosite, celestite, gluorite, various sulfide minerals, and their oxidation products.

Barite is a moderately soft crystalline mineral, Mohs' hardness 3–3.5; sp gr 4.3–4.6; n_D 1.64. The ore is white opaque to transparent, but impurities can produce pale shades of yellow, green, blue, brown, red, or gray-black. The most important impurities are Fe_2O_3, Al_2O_3, SiO_2, and $SrSO_4$, all of which are undesirable in chemical-grade barite. When the barite is used for drilling mud, the iron content can be permitted to be much higher than for other uses.

Residual barite is usually mined by open-pit methods. Bedded and vein deposits may be mined by either open-pit or underground methods, depending on the characteristics of each deposit.

Uses. About 80% of the world's barite production is used as a weighting agent for the muds circulated in rotary drilling of oil and gas wells (see PETROLEUM, DRILLING FLUIDS AND OTHER OIL RECOVERY CHEMICALS).

Finely ground barite which may be bleached, usually by sulfuric acid, or unbleached, is used as a filler or extender in paint (qv), especially in automotive undercoats, where its low oil absorption, easy wettability in oils, and good sanding properties are advantageous (see FILLERS). It is also used as a filler in plastics and rubber products.

In the glass (qv) and ceramic industry (see CERAMICS), barite can be used both as a flux, to promote melting at a lower temperature or to increase the production rate, and as an additive to increase the refractive index of glass.

Barium Acetate

Barium acetate, $Ba(C_2H_3O_2)_2$, crystallizes from an aqueous solution of acetic acid and barium carbonate or barium hydroxide. Barium acetate is used in printing fabrics, lubricating grease, and as a catalyst for organic reactions.

Barium Bromide

Barium bromide, $BaBr_2$, mp 854°C, density 4.781 g/mL, also exists as barium bromide dihydrate, $BaBr_2{\cdot}H_2O$, dehydration temperature 120°C, density 3.58 g/mL. Barium bromide is very soluble in methanol, yet almost insoluble in ethanol. Reported uses of barium bromide include fabrication of phosphors; as a crystallization nucleating agent to control supercooling of $CaBr_2$ solutions; and in the production of halide glasses having ir transmission properties.

Barium Carbonate

Most barium compounds are prepared from reactions of barium carbonate, $BaCO_3$, which is commercially manufactured by the "black ash" process from barite and coke in a process identical to that for strontium carbonate production.

Precipitated or synthetic barium carbonate is the most commercially important of all the barium chemicals except for barite. Barium carbonate is an unusually dense material that is almost insoluble in water and only slightly soluble in carbonated water. It does dissolve in dilute hydrochloric, nitric, and acetic acids and is also soluble in ammonium nitrate and ammonium chloride solutions.

Uses. There are several different grades of barium carbonate manufactured to fit the specific needs of a wide variety of applications: very fine, highly reactive grades are made for the chemical industry; coarser and more readily handleable grades are mainly supplied to the glass industry.

In 1989, 30% of the barium carbonate produced was used in glass manufacturing, 30% in brick and clay products, 20% in the manufacture of barium chemicals, 5% in the manufacture of barium ferrites, 5% in the production of photographic papers, and 10% in miscellaneous uses, including titanates.

Barium Chloride

Both anhydrous barium chloride, $BaCl_2$, mol wt 208.25, density 3.856 g/mL, and barium chloride dihydrate, $BaCl_2{\cdot}2H_2O$, mol wt 244.28, density 3.097 g/mL, are produced from a filtered aqueous solution formed by the reaction of hydrochloric acid and $BaCO_3$ or BaS. If BaS is used, the H_2S generated must be appropriately handled.

$BaCl_2$ is used in heat treating baths because of the eutectic mixtures it readily forms with other chlorides. $BaCl_2$ is also used to set up porcelain enamels for sheet steel (see ENAMELS, PORCELAIN OR VITREOUS; STEEL), and it is used to produce blanc fixe.

Barium 2-Ethylhexanoate

Barium 2-ethylhexanoate, $Ba(C_8H_{16}O_2)_2$, also known as barium octanoate or barium octoate, is usually used in synergistic combination with cadmium or zinc organic salts as a thermal stabilizer for PVC. Barium ethylhexanoate is a liquid; barium stearate, $Ba(C_{18}H_{36}O_2)_2$, is a powder that also serves as a thermostabilizer for PVC.

Barium Hydrosulfide

Barium hydrosulfide, $Ba(HS)_2$, is formed by absorption of hydrogen sulfide into barium sulfide solution. On addition of alcohol, barium hydrosulfide tetrahydrate, $Ba(HS)_2{\cdot}4H_2O$, crystallizes as yellow rhombic crystals that decompose at 50°C. Solid barium hydrosulfide is very unstable.

Barium Hydroxide

Barium hydroxide is the strongest base and has the greatest water-solubility of the alkaline-earth elements. Barium hydroxide (barium hydrate, caustic baryta) exists as the octahydrate, $Ba(OH)_2{\cdot}8H_2O$, the monohydrate, $Ba(OH)_2{\cdot}H_2O$, or as the anhydrous material, $Ba(OH)_2$. The octahydrate and monohydrate have sp gr 2.18 and 3.74, respectively. The mp of the octahydrate and anhydrous are 77.9°C and 407°C, respectively.

Barium hydroxide is used in the manufacture of barium greases and plastic stabilizers such as barium 2-ethylhexanoate, in papermaking, in sealing compositions (see SEALANTS), vulcanization accelerators, water purification, pigment dispersion, in a formula for self-extinguishing polyurethane foams, and in the protection of objects made of limestone from deterioration (see FINE ART EXAMINATION AND CONSERVATION; LIME AND LIMESTONE).

Barium Iodide

Barium iodide dihydrate, $BaI_2{\cdot}2H_2O$, crystallizes from hot aqueous solution. BaI_2 is useful in making other iodides. BaI_2 has been cited for producing ir transparent glasses that are useful in power transmission from CO and CO_2 lasers (qv) from the $ZnI_2{-}CsI{-}BaI_2$ system; as a catalyst promoter, for such catalysts as rhodium(III) chloride, in carbonylation reactions; as being useful in chemical vapor deposition as a precursor in forming the superconducting composition $YBa_2Cu_3O_{7-x}$ as a sintering aid for aluminum nitride, and in phosphor formulations for cathode-ray tubes.

Barium Metaborate

Barium metaborate monohydrate, $Ba(BO_2)_2{\cdot}H_2O$, has a sp gr of 3.25–3.35, and can be prepared from the reaction of a solution of BaS and sodium tetraborate.

$Ba(BO_2)_2{\cdot}H_2O$, used in flame retardant plastic formulations as a synergist for phosphorus or halogen compounds and as a partial or complete replacement for antimony oxide (see FLAME RETARDANTS), is excellent as an afterglow suppressant. The low refractive index of $Ba(BO_2)_2$ results in greater transparency and brighter colors in formulated plastics. Barium metaborate has been reported in paint formulations to convey insecticidal properties (see INSECT CONTROL TECHNOLOGY). $Ba(BO_2)_2$ has been reported to be used in antibacterial coatings for aluminum heat exchanger surfaces of air conditioners. $Ba(BO_2)_2$ crystals have been used in nonlinear optics.

Barium Nitrate

Barium nitrate, $Ba(NO_3)_2$, occurs as colorless crystals; mp 592°C; sp gr 3.24. Its solubility in water is

Temperature, °C	0	25	40	60	80	100	135
Solubility, wt % $Ba(NO_3)_2$	4.72	9.27	12.35	16.9	21.4	25.6	32.0

Barium nitrate is prepared by reaction of $BaCO_3$ and nitric acid, filtration and evaporative crystallization. The precipitate is centrifuged, washed, and dried. Barium nitrate is used in pyrotechnic green flares, tracer bullets, primers, and in detonators. These make use of its property of easy decomposition as well as its characteristic green flame. A small amount is used as a source of barium oxide in enamels.

Barium Nitrite

Barium nitrate, $Ba(NO_2)_2$, crystallizes from aqueous solution as barium nitrite monohydrate, $Ba(NO_2)_2{\cdot}H_2O$, which has yellowish hexagonal crystals, and sp gr 3.173. The monohydrate loses its water of crystallization at 116°C. Barium nitrite may be prepared by crystallization from a solution of equivalent quantities of barium chloride and sodium nitrite, by thermal decomposition of barium nitrate in an atmosphere of NO, or by treating barium hydroxide or barium carbonate with the gaseous oxidation products of ammonia. It has been used in diazotization reactions.

Barium Oxide

Barium oxide, BaO, occurs as colorless cubic or hexagonal crystals; mp 1923°C; sublimation ca 2000°C; bp ca 3088°C; sp gr (cubic) 5.72, (hexagonal) 5.32.

Of the alkaline-earth carbonates, $BaCO_3$ requires the greatest amount of heat to undergo decomposition to the oxide. Thus carbon in the form of coke, tar, or carbon black, is added to the carbonate to lower reaction temperature from about 1300°C in the absence of carbon to about 1050°C.

BaO is used to impart improved strength to porcelain, as a solid base catalyst, in specialty cements, and for drying gases.

Barium Peroxide

When heated in air or oxygen to 500°C, barium oxide is converted readily to barium peroxide, BaO_2.

Reported uses of BaO_2 include in the cathodes of fluorescent lamps, formation of $YBa_2Cu_3O_{7-x}$ superconducting phase from CuN_3, BaO_2, and Y_2O_3, and as a drying agent for lithographic inks (qv).

Barium Sodium Niobium Oxide

Barium sodium niobium oxide, $Ba_2NaNb_5O_{15}$, finds application for its dielectric, piezoelectric, nonlinear crystal and electrooptic properties. It has been used in conjunction with lasers for second harmonic generation and frequency doubling. The crystalline material can be grown at high temperature, mp ca 1450°C.

Barium Sulfate

Barium sulfate, $BaSO_4$, occurs as colorless rhombic crystals, mp 1580°C (dec); sp gr 4.50. Solubility 0.000285 g/100 g H_2O at 30°C and 0.00118 at 100°C. Precipitated $BaSO_4$ is known as blanc fixe.

Because of its extreme insolubility $BaSO_4$ is not toxic. In medicine, barium sulfate is widely used as an x-ray contrast medium (see X-RAY TECHNOLOGY). It is also used in photographic papers, filler for plastics, and in concrete as a radiation shield.

Barium Sulfide

Impure barium sulfide with 20–35% contaminants is produced in large volume by the black ash kiln. Pure barium sulfide, BaS, occurs as colorless cubic crystals, sp gr 4.25 and as hexagonal plates of barium sulfide hexahydrate, $BaS{\cdot}6H_2O$. BaS melts at 2227°C.

BaS is used in the manufacture of lithophone, useful as a white pigment in paints. BaS also has been used in the production of thin-film electroluminescent phosphors. Similarly, infrared-triggered phosphors may be fabricated from Ba or Sr sulfide.

Barium Titanate

The basic crystal structure of barium titanate, $BaTiO_3$, the perovskite structure, so-called after the mineral $CaTiO_3$, leads to unique, outstanding dielectric properties. Barium titanate has widespread use in the electronics industry, eg, for miniature capacitors, and also sees application in numerous sonic and ultrasonic devices.

Barium titanate is usually produced by the solid-state reaction of barium carbonate and titanium dioxide.

Yttrium–Barium–Copper Oxide

Yttrium–barium–copper oxide, $YBa_2Cu_3O_{7-x}$, is a newly developed high T_c material which has been found to be fully superconductive at temperatures above 90 K, a temperature that can be maintained during practical operation.

Ultrapure powders of yttrium–barium–copper oxide that are sinterable into single-phase superconducting material at low temperatures are required creating a worldwide interest in high purity barium chemicals.

A number of promising new routes to chemically synthesize ultrapure, ultrahomogeneous particles of controlled particle size and particle size distribution are currently under development. A modified sol–gel method has been developed to produce thermoplastic gels that are compatible with fiber spinning technology (see SOL–GEL TECHNOLOGY), as has a process which avoids the formation of barium carbonate, a troublesome impurity at the grain boundaries which seriously detracts from the critical current.

Health and Safety

The average adult human body contains 22 mg Ba, of which 93% is present in bone. The remainder is widely distributed throughout the soft tissues of the body in very low concentrations.

Environmental Levels and Exposures. Barium constitutes about 0.04% of the earth's crust. Agricultural soils contain Ba^{2+} in the range of several micrograms per gram. The Environmental Protection Agency, under the Safe Drinking Water Act, has set a limit for barium of 1 mg/L for municipal waters in the United States.

Toxicity. The toxicity of barium compounds depends on solubility.

Soluble Compounds. The mechanism of barium toxicity is related to its ability to substitute for calcium in muscle contraction. Toxicity results from stimulation of smooth muscles of the gastrointestinal tract, the cardiac muscle, and the voluntary muscles, resulting in paralysis. Skeletal, arterial, intestinal, and bronchial muscle all seem to be affected by barium.

PATRICK M. DIBELLO
JAMES L. MANGANARO
ELIZABETH R. AGUINALDO
FMC Corporation

F. A. Cotton and G. Wilkinson, *Advanced Inorganic Chemistry,* John Wiley & Sons, Inc., New York, 1980, pp. 271–273.

Industrial Minerals and Rocks, 4th ed., American Institute of Mining, Metallurgical, and Petroleum Engineers, Inc., New York, 1975, pp. 427–442.

U.S. Bureau of Mines Mineral Commodity Summaries, Washington, D.C., 1989.

A. L. Reeves, in L. Friberg, G. F. Nordberg, and V. B. Vouk, eds., *Handbook on the Toxicology of Metals,* 2nd ed., Vol. 2, Elsevier, New York, 1986, pp. 84–94.

BARRIER POLYMERS

Barrier polymers are used for many packaging and protective applications. As barriers they separate a system, such as an article of food or an electronic component, from an environment.

The Permeation Process

Barrier polymers limit movement of substances, hereafter called permeants. The movement can be through the polymer or, in some cases, merely into the polymer. The overall movement of permeants through a polymer is called permeation, which is a multistep process. First, the permeant molecule collides with the polymer. Then, it must adsorb to the polymer surface and dissolve into the polymer bulk. In the polymer, the permeant "hops" or diffuses randomly as its own thermal kinetic energy keeps it moving from vacancy to vacancy while the polymer chains move. The random diffusion yields a net movement from the side of the barrier polymer that is in contact with a high concentration or partial pressure of the permeant to the side that is in contact with a low concentration of permeant. After crossing the barrier polymer, the permeant moves to the polymer surface, desorbs, and moves away.

Permeant movement is a physical process that has both a thermodynamic and a kinetic component. For polymers without special surface treatments, the thermodynamic contribution is in the solution step. The permeant partitions between the environment and the polymer according to thermodynamic rules of solution. The kinetic contribution is in the diffusion. The net rate of movement is dependent on the speed of permeant movement and the availability of new vacancies in the polymer.

Small Molecule Permeation

Permanent Gases. Table 1 lists the permeabilities of oxygen, nitrogen, and carbon dioxide for selected barrier and nonbarrier polymers at 20°C and 75% rh. The effect of temperature and humidity are discussed later. For many polymers the permeabilities of nitrogen, oxygen, and carbon dioxide are in the ratio 1:4:14.

The traditional definition of a barrier polymer required an oxygen permeability less than 2 nmol/(m·s·GPa) (originally, less than (1 cc·mil)/(100 in.^{2}·d·atm)) at room temperature. This definition was based partly on function and partly on conforming to the old commercial unit of permeability. The old commercial unit of permeability was created so that the oxygen permeability of Saran Wrap brand plastic film, a trademark of The Dow Chemical Company, would have a numerical value of 1.

Table 1. Permeabilities of Selected Polymers

Polymer	Gas permeability, nmol/m·s·GPad		
	Oxygen	Nitrogen	Carbon dioxide
vinylidene chloride copolymers	0.02–0.30	0.005–0.07	0.1–1.5
ethylene–vinyl alcohol copolymers, dry	0.014–0.095		
at 100% rh	2.2–1.1		
nylon-MXD6[a]	0.30		
nitrile barrier polymers	1.8–2.0		6–8
nylon-6	4–6		20–24
amorphous nylon (Selar[b] PA 3426)	5–6		
poly(ethylene terephthalate)	6–8	1.4–1.9	30–50
poly(vinyl chloride)	10–40		40–100
high density polyethylene	200–400	80–120	1200–1400
polypropylene	300–500	60–100	1000–1600
low density polyethylene	500–700	200–400	2000–4000
polystyrene	500–800	80–120	1400–3000

[a] Trademark of Mitsubishi Gas Chemical Co.
[b] Trademark of E. I. du Pont de Nemours & Co., Inc.

Table 2. Diffusion and Solubility Coefficients for Oxygen and Carbon Dioxide in Selected Polymers at 23°C, Dry

	Oxygen		Carbon dioxide	
			Polymer	*D*, m^2/s
S, $nmol/(m^3 \cdot GPa)$[a]	*D*, m^2/s	*S*, $nmol/(m^3 \cdot GPa)$[a]		
vinylidene chloride copolymer	1.2×10^{-14}	1.01×10^{13}	1.3×10^{-14}	3.2×10^{13}
ethylene–vinyl alcohol copolymer[b]	7.2×10^{-14}	2.4×10^{12}		
acrylonitrile barrier polymer	1.0×10^{-13}	1.0×10^{13}	9.0×10^{-14}	4.4×10^{13}
poly(ethylene terephthalate)	2.7×10^{-13}	2.8×10^{13}	6.2×10^{-14}	8.1×10^{14}
poly(vinyl chloride)	1.2×10^{-12}	1.2×10^{13}	8.0×10^{-13}	9.7×10^{13}
polypropylene	2.9×10^{-12}	1.1×10^{14}	3.2×10^{-12}	3.4×10^{14}
high density polyethylene	1.6×10^{-11}	7.2×10^{12}	1.1×10^{-11}	4.3×10^{13}
low density polyethylene	4.5×10^{-11}	2.0×10^{13}	3.2×10^{-11}	1.2×10^{14}

[a] Solubility coefficient in $cc(STP)/(cm^3 \cdot atm) \times (4.04 \times 10^{14})$ = solubility coefficient in $nmol/(m^3 \cdot GPa)$.
[b] 42 mol% ethylene.

Poly(ethylene terephthalate) (PET), with an oxygen permeability of 8 nmol/(m·s·GPa), is not considered a barrier polymer by the old definition; however, it is an adequate barrier polymer for holding carbon dioxide in a 2-L bottle for carbonated soft drinks. The solubility coefficients for carbon dioxide are much larger than for oxygen. For the case of the PET soft drink bottle, the principal mechanism for loss of carbon dioxide is by sorption in the bottle walls as 500 kPa (5 atm) of carbon dioxide equilibrates with the polymer. For an average wall thickness of 370 μm (14.5 mil) and a permeability of 40 nmol/(m·s·GPa), many months are required to lose enough carbon dioxide (15% of initial) to be objectionable.

The diffusion and solubility coefficients for oxygen and carbon dioxide in selected polymers have been collected in Table 2.

Polymers With Good Barrier to Permanent Gases. The polymers that are good barriers to permanent gases, especially oxygen, have important commercial significance.

Vinylidene chloride copolymers are available as resins for extrusion, latices for coating, and resins for solvent coating. Comonomer levels range from 5 to 20 wt %. Common comonomers are vinyl chloride, acrylonitrile, and alkyl acrylates. The permeability of the polymer is a function of type and amount of comonomer. As the comonomer fraction of these semicrystalline copolymers is increased, the melting temperature decreases and the permeability increases. The permeability of vinylidene chloride homopolymer has not been measured.

Vinylidene chloride copolymers are marketed under a variety of trade names. Saran is a trademark of The Dow Chemical Company for vinylidene chloride copolymers. Other trade names include Daran (W.R. Grace), Amsco Res (Union Oil), and Serfene (Morton Chemical) in the United States; and Haloflex (Imperial Chemical Industries, Ltd.), Diofan (BASF), Ixan (Solvay and Cie SA), and Polyidene (Scott-Bader) in Europe.

Hydrolyzed ethylene–vinyl acetate copolymers, commonly known as ethylene–vinyl alcohol (EVOH) copolymers, are usually used as extrusion resins, although some may be used in solvent-coating applications.

Copolymers of acrylonitrile are used in extrusion and molding applications. Commercially important comonomers for barrier applications include styrene and methyl acrylate.

Polyamide polymers can provide a good-to-moderate barrier to permeation by permanent gases.

Two often-used polymers have adequate properties for some applications. Poly(ethylene terephthalate) (PET) is used to make films and bottles. Poly(vinyl chloride) (PVC) is a moderate barrier to permanent gases. Plasticized poly(vinyl chloride) is used as a household wrapping film. The plasticizers greatly increase the permeabilities.

Water Vapor Transmission. Table 3 lists water vapor transmission (WVTR) values for selected polymers. Comparison of Tables 1 and 3 shows that often there is a reversal of roles. Those polymers that are good oxygen barriers are often poor water-vapor barriers and vice versa. This can be rationalized as follows. Barrier polymers often rely on dipole–dipole interactions to reduce chain mobility and, hence, diffusional movement of permeants. These dipoles can be good sites for hydrogen bonding. Water molecules are attracted to these sites. Polymer molecules without dipole–dipole interactions, such as polyolefins, dissolve very little water and have low WVTR and permeability values. The low values of S more than compensate for the naturally higher values of D.

Table 3. Water-Vapor Transmission Rates of Selected Polymers[a]

Polymer	WVTR, nmol/(m·s)
vinylidene chloride copolymers	0.005–0.05
high density polyethylene (HDPE)	0.095
polypropylene	0.16
low density polyethylene (LDPE)	0.35
ethylene–vinyl alcohol, 44 mol % ethylene[b]	0.35
poly(ethylene terephthalate) (PET)	0.45
poly(vinyl chloride) (PVC)	0.55
ethylene–vinyl alcohol, 32 mol % ethylene[b]	0.95
nylon-6,6, nylon-11	0.95
nitrile barrier resins	1.5
polystyrene	1.8
nylon-6	2.7
polycarbonate	2.8
nylon-12	15.9

[a] At 38°C and 90% rh unless otherwise noted.
[b] Measured at 40°C.

Large Molecule Permeation

The permeation of flavor, aroma, and solvent molecules in polymers follows the same physics as the permeation of small molecules. However, there are two significant differences. For these larger molecules, the diffusion coefficients are much lower and the solubility coefficients are much higher. This means that steady-state permeation may not be reached during the storage time of some packaging situations. Hence, large molecules from the environment might not enter the contents, or loss of flavor molecules would be limited to sorption into the polymer. However, since the solubility coefficient is large, the loss of flavor could be important solely from sorption in the polymer. Furthermore, the large solubility coefficient can lead to enough sorption of the large molecule that plasticization occurs in the polymer, which can increase the diffusion coefficient.

Table 4 contains some selected permeability data, including diffusion and solubility coefficients for flavors in polymers used in food packaging. Generally, vinylidene chloride copolymers and glassy polymers such as polyamides and EVOH are good barriers to flavor and

Table 4. Examples of Permeation of Flavor and Aroma Compounds in Selected Polymers at 25°C[a], Dry[b]

Flavor/aroma compound	Permeant formula	P, MZU[c]	D, m^2/s	S, $kg/(m \cdot Pa^3)$
Vinylidene chloride copolymer				
ethyl hexanoate	$C_8H_{16}O_2$	570	8.0×10^{-18}	0.71
ethyl 2-methylbutyrate	$C_7H_{14}O_2$	3.2	1.9×10^{-17}	1.7×10^{-3}
hexanol	$C_6H_{14}O$	40	5.2×10^{-17}	7.7×10^{-3}
trans-2-hexenal	$C_6H_{10}O$	240	1.8×10^{-17}	0.14
d-limonene	$C_{16}H_{16}$	32	3.3×10^{-17}	9.7×10^{-3}
3-octanone	$C_8H_{16}O$	52	1.3×10^{-18}	0.40
propyl butyrate	$C_7H_{14}O_2$	42	4.4×10^{-18}	9.4×10^{-2}
dipropyl disulfide	$C_6H_{14}S_2$	270	2.6×10^{-18}	1.0
Ethylene–vinyl alcohol copolymer				
ethyl hexanoate		0.41	3.2×10^{-18}	1.3×10^{-3}
ethyl 2-methylbutyrate		0.30	6.7×10^{-18}	4.7×10^{-4}
hexanol		1.2	2.6×10^{-17}	4.6×10^{-4}
trans-2-hexenal		110	6.4×10^{-17}	1.8×10^{-2}
d-limonene		0.5	1.1×10^{-17}	4.5×10^{-4}
3-octanone		0.2	1.0×10^{-18}	2.0×10^{-3}
propyl butyrate		1.2	2.7×10^{-17}	4.5×10^{-4}
Low density polyethylene				
ethyl hexanoate		4.1×10^6	5.2×10^{-13}	7.8×10^{-2}
ethyl 2-methylbutyrate		4.9×10^5	2.4×10^{-13}	2.3×10^{-2}
hexanol		9.7×10^5	4.6×10^{-13}	2.3×10^{-2}
trans-2-hexenal		8.1×10^5		
d-limonene		4.3×10^6		
3-octanone		6.8×10^6	5.6×10^{-13}	1.2×10^{-1}
propyl butyrate		1.5×10^6	5.0×10^{-13}	3.0×10^{-2}
dipropyl disulfide		6.8×10^6	7.3×10^{-14}	9.3×10^{-1}
High density polyethylene				
d-limonene		3.5×10^6	1.7×10^{-13}	2.5×10^{-1}
menthone	$C_{10}H_{18}O$	5.2×10^6	9.1×10^{-13}	4.7×10^{-1}
methyl salicylate	$C_8H_8O_3$	1.1×10^7	8.7×10^{-14}	1.6
Polypropylene				
2-butanone	C_4H_8O	8.5×10^3	2.1×10^{-15}	4.0×10^{-2}
ethyl butyrate	$C_6H_{12}O_2$	9.5×10^3	1.8×10^{-15}	5.3×10^{-2}
ethyl hexanoate		8.7×10^4	3.1×10^{-15}	2.8×10^{-1}
d-limonene		1.6×10^4	7.4×10^{-16}	2.1×10^{-1}

[a] Values for vinylidene chloride copolymer and ethylene–vinyl alcohol are extrapolated from higher temperatures.
[b] Permeation in the vinylidene chloride copolymer and the polyolefins is not affected by humidity; the permeability and diffusion coefficient in the ethylene–vinyl alcohol copolymer can be as much as 1000 times greater with high humidity.
[c] MZU $= (10^{-20}\ kg \cdot m)/(m \cdot s \cdot Pa)^2$.

aroma permeation, whereas the polyolefins are poor barriers. Comparison to Table 2 shows that the large-molecule diffusion coefficients are 1000 or more times lower than the small-molecule coefficients.

Physical Factors Affecting Permeability

Several physical factors can affect the barrier properties of a polymer. These include temperature, humidity, orientation, and cross-linking.

Temperature The temperature dependence of the permeability arises from the temperature dependencies of the diffusion coefficient and the solubility coefficient. Typically, the permeability increases 5 to 10% for every increase of 1°C.

Humidity. When a polymer equilibrates with a humid environment, it absorbs water. This can plasticize the polymer and increase the permeability.

Orientation. The effect of orientation on the permeability of polymers is difficult to assess because the words orientation and elongation or strain have been used interchangeably in the literature. Diffusion in some polymers is unaffected by orientation; in others, increases or decreases are observed.

Cross-Linking. Cross-linking has been shown in a few cases to decrease the diffusion coefficient.

Barrier Structures

Barrier polymers are often used in combination with other polymers or substances. The combinations may result in a layered structure either by coextrusion, lamination, or coating. The combinations may be blends that are either miscible or immiscible. In each case, the blend seeks to combine the best properties of two or more materials to enhance the value of a final structure.

Predicting Permeabilities

Reasonable prediction can be made of the permeabilities of low molecular weight gases such as oxygen, nitrogen, and carbon dioxide in many polymers. The diffusion coefficients are not complicated by the shape of the permeant, and the solubility coefficients of each of these molecules do not vary much from polymer to polymer. Hence, all that is required is some correlation of the permeant size and the size of holes in the polymer matrix. Reasonable predictions of the permeabilities of larger molecules such as flavors, aromas, and solvents are not easily made. The diffusion coefficients are complicated by the shape of the permeant, and the solubility coefficients for a specific permeant can vary widely from polymer to polymer.

The permachor method is an empirical method for predicting the permeabilities of oxygen, nitrogen, and carbon dioxide in polymers. In

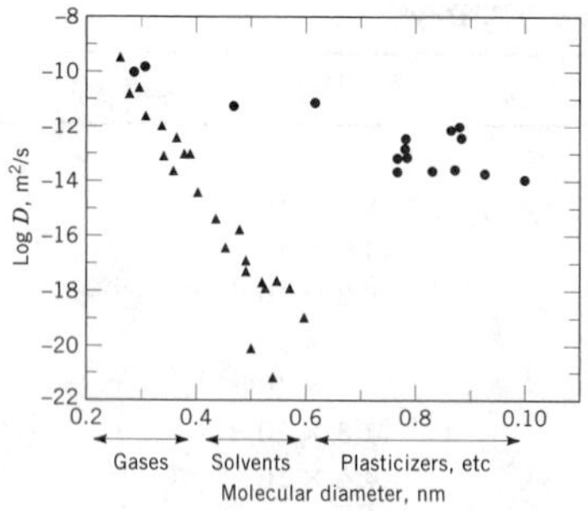

Figure 1. Diffusivities of penetrants in rigid (▲) and plasticized (●) poly(vinyl chloride) versus molecular diameter at 30°C.

this method a numerical value is assigned to each constituent part of the polymer. An average number is derived for the polymer, and a simple equation converts the value into a permeability. This method has been shown to be related to the cohesive energy density and the free volume of the polymer. The model has been modified to liquid permeation with some success.

For larger molecules, independent predictions of the diffusion coefficients and the solubility coefficients are required. Figure 1 shows how the diffusion coefficient varies as a function of permeant size in poly(vinyl chloride) (PVC). The two sets of data represent glassy PVC, which is below its glass-transition temperature and plasticized PVC, which is above its glass-transition temperature. Other glassy polymers show the steep slope, and other rubbery polymers show the shallow slope. The points near the lines represent spherical permeants whereas the points above the lines represent linear permeants. Predicting the diffusion coefficient for a permeant in a polymer requires knowing one other diffusion coefficient in the polymer.

The solubility coefficients are more difficult to predict. Although advances are being made, the best method is probably to use a few known solubility coefficients in the polymer to predict others with a simple plot of S vs $(\delta_{poly} - \delta_{perm})^2$ where δ_{poly} and δ_{perm} are the solubility parameters of the polymer and permeant, respectively. When insufficient data are available, S at 25°C can be estimated with equation 1 where $\kappa = 1$ and the resulting units of cal/cm³ are converted to kJ/mol by dividing by the polymer density and multiplying by the molecular mass of the permeant and by 4.184.

$$S = \kappa(\delta_{poly} - \delta_{perm})^2 \quad (1)$$

The boiling temperature of a permeant can be used to predict the solubility coefficient only when the solubility coefficients of other permeants of the same chemical family are known.

Measuring Barrier Properties

Measuring the barrier properties of polymers is important for several reasons. The effects of formulation or process changes need to be known, new polymers need to be evaluated, data are needed for a new application before a large investment has been made, and fabricated products need to have performance verified.

Oxygen Transport. The most widely used methods for measuring oxygen transport are based upon the Ox-Tran instrument (Modern Controls, Inc.). Several models exist, but they all work on the same principle. The most common application is to measure the permeability of a film sample.

Water Transport. Two methods of measuring water-vapor transmission rates (WVTR) are commonly used. The newer method uses a Permatran-W (Modern Controls, Inc.). The other method is the ASTM cup method.

Carbon Dioxide Transport. Measuring the permeation of carbon dioxide occurs far less often than measuring the permeation of oxygen or water. A variety of methods are used; however, the simplest method uses the Permatran-C instrument (Modern Controls, Inc.).

Flavor and Aroma Transport. Many methods are used to characterize the transport of flavor, aroma, and solvent molecules in polymers. Each has some value, and no one method is suitable for all situations. Any experiment should obtain the permeability, the diffusion coefficient, and the solubility coefficient. Furthermore, experimental variables might include the temperature, the humidity, the flavor concentration, and the effect of competing flavors.

Applications

The primary application for barrier polymers is food and beverage packaging. Barrier polymers are also used for packaging medical products, agricultural products, cosmetics, and electronic components and in moldings, pipe, and tubing.

Safety and Health Factors

The use of safe materials is vital for barrier applications, particularly for food, medical, and cosmetics packaging. Suppliers of specific barrier polymers can provide the necessary details, such as material safety data sheets, to ensure safe processing and use of barrier polymers.

PHILLIP DE LASSUS
The Dow Chemical Company

W. J. Karos, ed., *Barrier Polymers and Structures,* ACS Symposium Series No. 423, American Chemical Society, Washington, D.C., 1990.

S. A. Risch and J. H. Hotchkiss, eds., *Food and Packaging Interactions II,* ACS Symposium Series, No. 473, American Chemical Society, Washington, D.C., 1991.

J. Comyn, ed., *Polymer Permeability,* Elsevier Applied Science Publishers, Ltd., Barking, U.K., 1985.

W. R. Vieth, *Diffusion In and Through Polymers,* Hanser Publishers, Munich, Germany 1991.

BATTERIES

INTRODUCTION

Batteries, storehouses for electrical energy "on demand," range in size from large house-sized batteries for utility storage, cubic meter-sized batteries for automotive starting, lighting, and ignition, down to tablet-sized batteries for hearing aids and paper-thin batteries for memory protection in electronic devices.

In bulk chemical reactions, an oxidizer (electron acceptor) and fuel (electron donor) react to form products resulting in direct electron transfer and the release or absorption of energy as heat. By special arrangements of reactants in devices called batteries, it is possible to control the rate of reaction and to accomplish the direct release of chemical energy in the form of electricity on demand, without intermediate processes.

Figure 1 schematically depicts an electrochemical reactor in which the chemical energy stored in the electrodes is manifested directly as a voltage and current flow. The electrons involved in the chemical reactions are transferred from the active materials undergoing oxidation to the oxidizing agent by means of an external circuit. The passage of electrons through this external circuit generates an electric current, providing a direct means for energy utilization without going through heat as an intermediate step. As a result, electrochemical reactors can be significantly more efficient than Carnot Cycle heat engines.

The three main types of batteries are primary, secondary, and reserve. A primary battery is used or discharged once and discarded. Secondary or rechargeable batteries can be discharged, recharged, and used again. Reserve batteries are normally special constructions of primary battery systems that store the electrolyte apart from the

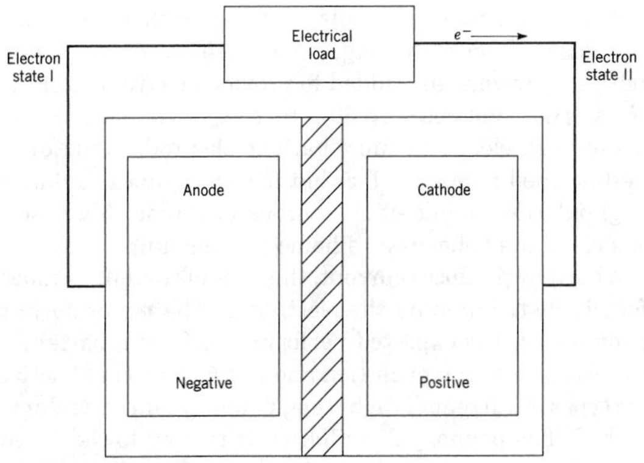

Figure 1. Schematic representation of a battery system also known as an electrochemical transducer where the anode, also known as electron state I, may be comprised of lithium, magnesium, zinc, cadmium, lead, or hydrogen, and the cathode, or electron state II, depending on the composition of the anode, may be lead dioxide, manganese dioxide, nickel oxide, iron disulfide, oxygen, silver oxide, or iodine.

electrodes, until put into use. They are designed for long-term storage before use. Fuel cells (qv) are not discussed herein.

Economic Aspects

The U.S. primary battery market is usually divided according to the chemical system used in the batteries, whereas the secondary battery market is usually divided according to usage. The lead–acid battery accounts for over 85% of the secondary battery market.

Thermodynamics

Batteries are miniature chemical reactors that convert chemical energy into electrical energy on demand. The thermodynamics of battery systems follow directly from that for bulk chemical reactions. For the general reaction

$$aA + bB \rightleftarrows cC + dD \qquad (1)$$

the basic thermodynamic equations for a reversible electrochemical transformation are given as

$$\Delta G = \Delta H - T\Delta S \qquad (2)$$

$$\Delta G^\circ = \Delta H^\circ - T\Delta S^\circ \qquad (3)$$

where ΔG is the Gibbs free energy, or the energy of a reaction available for useful work, ΔH is the enthalpy, or the energy released by the reaction, ΔS is the entropy, or the heat associated with the organization of material, and T is the absolute temperature. The superscript ° is used to indicate that the value of the function is for the material in the standard state at 25°C and unit activity. Although the Helmholtz free energy ΔA is used to describe constant volume situations found in battery systems, the use of the Gibbs free energy ΔG is adequate to describe practical battery systems.

The terms ΔG, ΔH, and ΔS are state functions and depend only on the identity of the materials and the initial and final state of the reaction.

Because ΔG is the net useful energy available from a given reaction, in electrical terms, the net available electrical energy from a reaction is given by

$$-\Delta = nFE \qquad (4)$$

and

$$-\Delta^\circ = nFE^\circ \qquad (5)$$

where n is the number of electrons transferred in the reaction, F is Faraday's constant, E is the voltage or electromotive force (emf) of the cell, and E° is that voltage at 25°C and unit activity. The voltage is unique for each group of reactants comprising the battery system. The amount of electricity produced is determined by the total amount of materials involved in the reaction. The voltage may be thought of as an intensity factor, and the term nF may be considered a capacity factor.

The more negative the value of ΔG, the more energy or useful work can be obtained from the reaction. Reversible processes yield the maximum output. In irreversible processes, a portion of the useful work or energy is used to help carry out the reaction. The cell voltage or emf also has a sign and direction. Spontaneous processes have a positive emf; the reaction, written in a reversible fashion, goes in the forward direction.

Electrolytes

Electrolytes are a key component of electrochemical cells and batteries. Electrolytes are formed by dissolving an ionogen into a solvent. When salts are dissolved in a solvent such as water the salt dissociates into ions through the action of the dielectric, water. Strong electrolytes, ie, salts of strong acids and bases, are completely dissociated in solution into positive and negative ions. The ions are solvated but positive ions tend to interact more strongly with the solvent than do the anions. The ions of the electrolyte provide the path for the conduction of electricity by movement of charged particles through the solution. The electrolyte also provides the physical separation of the positive and negative electrodes needed for electrochemical cell operation.

Transport properties of the electrolye, as well as electrode reactions, have a significant impact on battery operation. The electrode reactions and ionic transference that occur during discharge result in considerable modifications to the solution composition at each electrode compartment. The negative and positive electrode compartments can lose or gain electrolyte and solvent depending on the transference numbers of the ions and the electrode reactions. The composition of the electrolyte in the separator between the two compartments generally remains unchanged.

Battery electrolytes are concentrated solutions of strong electrolytes and the Debye-Hückel theory of dilute solutions is only an approximation. Typical values for the resistivity of battery electrolytes range from about 1 ohm·cm for sulfuric acid, H_2SO_4, in lead–acid batteries and for potassium hydroxide, KOH, in alkaline cells to about 100 ohm·cm for organic electrolytes in lithium Li, batteries.

Each electrolyte is stable only within certain voltage ranges. Exceeding these limits results in decomposition. The stable range depends on the solvent, electrolyte composition, and purity level. In aqueous systems, hydrogen and oxygen form if the voltage limit is exceeded. In the nonaqueous organic solvent-based systems used for lithium batteries, exceeding the voltage limit can result in polymerization or decomposition of the solvent system. It is especially important to remove traces of water from the nonaqueous electrolytes as water can catalyze the electrolytic decomposition of the organic solvent.

In addition to the liquid conductors described above, two types of solid-state ionic conductors have been developed; one involves inorganic compounds and the other is based on polymeric materials. Several inorganic solids have been found to have excellent conductivity resulting wholly from ionic motion in the crystal lattice. One solid electrolyte, lithium iodide, LiI, has found application in heart-pacer batteries even though it has a fairly low conductivity.

A second type of solid ionic conductors based around polyether compounds such as poly(ethylene oxide) (PEO) has been discovered and characterized. The polyethers can complex and stabilize lithium ions in organic media. They also dissolve salts such as $LiClO_4$ to produce conducting solid solutions. The use of these materials in rechargeable lithium-batteries has been proposed.

Electrical Double Layer

When two conducting phases come into contact with each other, a redistribution of charge occurs as a result of any electron energy level difference between the phases. If the two phases are metals, electrons flow from one metal to the other until the electron levels equilibrate. When an electrode, ie, electronic conductor, is immersed in an electrolyte, ie, ionic conductor, an electrical double layer forms at the electrode–solution interface resulting from the unequal tendency for distribution of electrical charges in the two phases. Because overall electrical neutrality must be maintained, this separation of charge between the electrode and solution gives rise to a potential difference between the two phases, equal to that needed to ensure equilibrium.

On the electrode side of the double layer the excess charges are concentrated in the plane of the surface of the electronic conductor. On the electrolyte side of the double layer the charge distribution is quite complex. The potential drop occurs over several atomic dimensions and depends on the specific reactivity and atomic structure of the electrode surface and the electrolyte composition. The electrical double layer strongly influences the rate and pathway of electrode reactions.

Electrically, the electrical double layer may be viewed as a capacitor with the charges separated by a distance of the order of molecular dimensions. The measured capacitance ranges from about two to several hundred microfarads per square centimeter, depending on the structure of the double layer, the potential, and the composition of the electrode materials.

Kinetics and Transport

Activation Processes. To be useful in battery applications reactions must occur at a reasonable rate. The rate or ability of battery electrodes to produce current is determined by the kinetic processes of electrode operations, not by thermodynamics, which describes the characteristics of reactions at equilibrium when the forward and reverse reaction rates are equal. Electrochemical reaction kinetics follow the same general considerations as those of bulk chemical reactions. Two differences are a potential drop that exists between the electrode and the solution because of the electrical double layer at the electrode interface, and the reaction that occurs at a two-dimensional interfaces rather than in three-dimensional space.

Transport Processes. The velocity of electrode reactions is controlled by the charge-transfer rate of the electrode process, or by the velocity of the approach of the reactants, to the reaction site. The movement or transport of reactants to and from the reaction site at the electrode interface is a common feature of all electrode reactions. Transport of reactants and products occurs by diffusion, by migration under a potential field, and by convection. The complete description of transport requires a solution to the transport equations. A full account is given in texts and discussions on hydrodynamic flow (see FLUID MECHANICS). Molecular diffusion in electrolytes is relatively slow. Although the process can be accelerated by stirring, enhanced mass transfer (qv) by stirring or convection is not possible in most battery designs. Natural convection from density changes does occur but does not greatly enhance transport in battery operation. Lead–acid batteries, used for motive power and stationary applications, are given a gassing overcharge on a regular basis. The gas evolution stirs up the electrolyte and equalizes the sulfuric acid concentration in the electrolyte.

Practical Battery Systems

Most battery electrodes are porous structures in which an interconnected matrix of solid particles, consisting of both nonconductive and electronically conductive materials, is filled with electrolyte. When the active mass is nonconducting, conductive materials, usually carbon or metallic powders, are added to provide electronic contact to the active mass. The solids occupy 50% to 70% of the volume of a typical porous battery electrode. Most battery electrode structures do not have a well-defined planar surface but have a complex surface extending throughout the volume of the porous electrode. Macroscopically, the porous electrode behaves as a homogeneous unit.

When a battery produces current, the sites of current production are not uniformly distributed on the electrodes. The nonuniform current distribution lowers the expected performance from a battery system, and causes excessive heat evolution and low utilization of active materials. Two types of current distribution, primary and secondary, can be distinguished. The primary distribution is related to the current production based on the geometric surface area of the battery construction. Secondary current distribution is related to current production sites inside the porous electrode itself. Most practical battery constructions have nonuniform current distribution across the surface of the electrodes. This primary current distribution is governed by geometric factors such as height (or length) of the electrodes, the distance between the electrodes, and the resistance of the anode and cathode structures; by the resistance of the electrolyte; and by the polarization resistance or hinderance of the electrode reaction processes.

Cell geometry, such as tab/terminal positioning and battery configuration, strongly influence primary current distribution. The monopolar construction is most common. Several electrodes of the same polarity may be connected in parallel to increase capacity. The current production concentrates near the tab connections unless special care is exercised in designing the current collector. Bipolar construction, wherein the terminal or collector of one cell serves as the anode and cathode of the next cell in pile formation, leads to greatly improved uniformity of current distribution.

Whereas current-producing reactions occur at the electrode surface, they also occur at considerable depth below the surface in porous electrodes. Porous electrodes offer enhanced performance through increased surface area for the electrode reaction and through increased mass-transfer rates from shorter diffusion path lengths. The key parameters in determining the reaction distribution include the ratio of the volume conductivity of the electrolyte to the volume conductivity of the electrode matrix, the exchange current, the diffusion characteristics of reactants and products, and the total current flow. The porosity, pore size, and tortuosity of the electrode all play a role.

Mathematical formulations of the models of primary and secondary current distribution permit rapid optimization in the design of new battery configurations. Models to describe and predict porous electrode performance in the lead–acid battery system have been developed. The high rate performance of the present starting, lighting, and ignition (SLI) automotive batteries have evolved directly from coupling collector designs with the porous electrode compositions identified from modeling studies.

The positive electrode in a battery system is most often a metal oxide, but it may also be a metal sulfide or halide. Generally, these materials are relatively poor electrical conductors and exhibit extremely high ohmic polarizations (impedances) if not combined with supporting electronic conductors such as graphite, lead, silver, copper, or nickel in the form of powder, rod, mesh, wire, grid, or other configurations. In almost all cases the negative electrode is a metallic element of sufficient conductivity to require only minimal supporting conductive structures. Exceptions are the oxygen (air) positive and hydrogen gas negative electrodes, which require a substantial conductive, catalytically active surface support, which also serves as current collector.

Although there are a multitude of chemical reactions that release energy, only a few reactions have the characteristics requisite for use in commercial batteries. A set of criteria can be established to characterize reactions suitable for battery development. The principal features necessary for battery reactions are as follow. *(1) Mechanical and chemical stability.* The reactants or active masses and cell

Table 1. Commercial Primary Battery Systems

Common name	Cell reaction	Nominal voltage	Energy content[a] W·h/mL	Energy content[a] W·h/kg	Comments	Manufacturers
Leclanché	$Zn + 2MnO_2 + 2NH_4Cl \rightarrow Zn(NH_3)_2Cl_2 + H_2O + Mn_2O_3$	1.5	0.20	80	low cost; general purpose, wide range of sizes	Eveready, Rayovac, Bright Star
zinc chloride	$4Zn + 8MnO_2 + ZnCl_2 + 9H_2O \rightarrow 8MnOOH + ZnCl_2 \cdot 4ZnO \cdot 5H_2O$	1.5	0.20	125	intermediate cost and performance	Eveready, Rayovac
alkaline	$2Zn + 3MnO_2 \rightarrow 2ZnO + Mn_3O_4$	1.5	0.25	130	sets standard for cylindrical cells	Duracell, Eveready, Rayovac
silver	$Zn + Ag_2O \rightarrow 2Ag + ZnO$	1.6	0.5	200	good pulse, higher voltage than mercury or Zn–air cells	Duracell, Eveready, Rayovac
mercury	$Zn + HgO \rightarrow Hg + ZnO$	1.35	0.5	165	sets standard for button cells; environmental problems	Duracell, Rayovac, Alexander
air	$2Zn + O_2(air) \rightarrow 2ZnO$	1.4	1.0	300	twice capacity of mercury and silver cells, limited active stand	Duracell, Rayovac
Li–CuO	$2Li + CuO \rightarrow Li_2O + Cu$	1.5	0.6	300	potential replacement for Leclanché and zinc chloride	Panasonic
Li–FeS	$2Li + FeS \rightarrow Li_2S + Fe$	1.6	0.4	160	replacement for mercury and silver cells, no mercury	Eveready
Li–CF_x	$xLi + CF_x \rightarrow xLiF + C$	2.7	0.5	300	high voltage, long shelf life, wide operating temperature	Rayovac, Eagle Picher
Li–MnO_2	$Li + MnO_2 \rightarrow LiMnO_2$	2.8	0.5	330	high voltage, long shelf life, wide operating temperature	Duracell, Eveready, Power Conversion
Li–I_2	$2Li + I_2 \rightarrow 2LiI$	2.8	0.9	290	solid electrolyte, 10+ year life, welded construction, used in most of heart pacers, iodine charge transfer complex	Medtronic, Catalyst, Wilson Greatbatch
Li–SO_2	$2Li + 2SO_2 \rightarrow Li_2S_2O_4$	2.8	0.4	330	military battery, low temperature, excellent storage	SAFT, Power Conversion
Li–$SOCl_2$	$4Li + 2SOCl_2 \rightarrow 4LiCl + SO_2 + S$	3.6	1.0	530	high voltage, high energy density	Electrochem Industries, Eagle Picher, Power Conversion

[a] Approximate values.

components must be stable over time (5 years or more) in the operating environment and must reform in their original condition on recharge. *(2) Energy content.* The reactants must have sufficient energy content to provide a useful voltage and current level, measured in W·h/L or W·h/kg. *(3) Power density.* The reactants must be capable of reacting at rates sufficient to deliver useful rates of electricity, measured in terms of W/L or W/kg. *(4) Temperature range.* The reactants must be able to maintain energy, power, and stability over a normal operating environment. The military often specifies −50 to 75°C. The average consumer has a less severe range of operating requirements, usually −10 to 50°C. *(5) Safety.* The battery must be safe in the normal operating environment as well as under mild abusive conditions. *(6) Cost.* The reactants and the materials of construction should be inexpensive and in good supply.

Table 1 and 2 contain characteristics of various primary and secondary battery systems, respectively. Table 3 contains performance parameters for promising rechargeable battery systems in various stages of research and commercial development.

Table 2. Commercial Rechargeable Battery Systems

Common name	Cell reaction	Nominal voltage	Energy content[a] W·h/L	Energy content[a] W·h/kg	Comments	Manufacturers
lead–acid	$Pb + PbO_2 + 2H_2SO_4 \rightleftarrows 2PbSO_4 + 2H_2O$	2.10	80	35	lowest cost, largest sales, available sealed	Johnson Controls Delco Exide GNB
nickel–cadmium	$Cd + 2NiOOH + 2H_2O \rightleftarrows Cd(OH)_2 + 2Ni(OH)_2$	1.35	80	38	high rate, available sealed	Gates SAFT
nickel–metal hydride	$H_2(M) + 2NiOOH \rightleftarrows 2Ni(OH)_2 + M$	1.35	160	55	hydrogen absorbing alloy, good cycle life, high self-discharge	Ovonics
nickel–iron	$Fe + 2NiOOH + 2H_2O \rightleftarrows Fe(OH)_2 + 2Ni(OH)_2$	1.25	90	30	limited production, very long cycle life, almost indestructible, old technology	Gates SAB
nickel–hydrogen	$H_2(g) + 2NiOOH \rightleftarrows 2Ni(OH)_2$	1.35	90	45	special space battery, very long cycle life, high self-discharge	Eagle Picher Hughes
silver–zinc	$Zn + 2AgO \rightleftarrows ZnO + Ag_2O$	1.86	200	100	two-step discharge, limited cycle life, high energy density	Eagle Picher Yardney
	$Zn + Ag_2O \rightleftarrows ZnO + 2Ag$	1.60				
lithium–MoS_2	$Li + MoS_2 \rightleftarrows LiMoS_2$	2.3	150	80	under development, small sealed cell	Moli Energy
lithium–MnO_2	$Li + MnO_2 \rightleftarrows LiMnO_2$	3.2	140	55	sealed coin cell	Sanyo
lithium–V_2O_5	$Li(C) + V_2O_5 \rightleftarrows LiV_2O_5 + C$	3.0	75	30	sealed coin cell	Toshiba
lithium–carbon	$Li + C \rightleftarrows Li(C)$ (intercalate)	3.0	6	2.1	sealed coin cell, "supercapacitor"	Panasonic

[a] Approximate values.

Table 3. Rechargeable Battery Systems in Various Stages of Research and Development

Common name	Cell reaction	Nominal voltage	Energy content[a] W·h/L	Energy content[a] W·h/kg	Comments
nickel–zinc	$Zn + 2NiOOH + H_2O \rightleftarrows ZnO + 2Ni(OH)_2$	1.65	95	60	limited cycle life, high rate capability
zinc–bromine	$Zn + Br_2 \rightleftarrows ZnBr_2$	1.85	75	65	bromine complex with quaternary ammonium salt, circulating electrolyte
lithium–FeS	$2LiAl + FeS \rightleftarrows Li_2S + 2Al + Fe$	1.33	90	95	low cost, high temperature fused salt, second step possible
sodium–sulfur	$2Na + 3S \rightleftarrows Na_2S_3$	2.1	120	160	solid β-Al_2O_3 separator, high temperature operation
aluminum–air	$4Al + 3O_2 + 2H_2O \rightleftarrows 2Al_2O_3 \cdot H_2O$	1.6	360	250	circulating electrolyte, low cost, low energy efficiency, replaceable negative electrode
lithium–V_6O_{13}	$8Li + V_6O_{13} \rightleftarrows Li_8V_6O_{13}$ (intercalate)	2.5	200	200	solid polymer electrolyte

[a] Approximate values.

RALPH BRODD
Gould Inc.

D. Linden, ed., *Handbook of Batteries and Fuel Cells,* McGraw-Hill Book Co., Inc., New York, 1984.

C. A. Vincent, B. Scrosati, M. Lazzari, and F. Bonino, *Modern Batteries,* Edward Arnold, Ltd., London, 1984.

J. Koryta and J. Dvorak, *Principles of Electrochemistry,* John Wiley & Sons, Inc., New York, 1987.

A. J. Bard and L. R. Faulkner, *Electrochemical Methods, Fundamentals and Applications,* John Wiley & Sons, Inc., New York, 1980.

PRIMARY CELLS

Primary cells are galvanic cells designed to be discharged only once, and attempts to recharge them can present possible safety hazards. The cells are designed to have the maximum possible energy in each cell size because of the single discharge. Thus, comparison between battery types is usually made on the basis of the energy density in $W{\cdot}h/cm^3$. The specific energy, W·h/kg, is often used as a secondary criterion for primary cells, especially when the application is weight-sensitive, as in space applications. The main categories of primary cells are carbon–zinc, known as heavy-duty and general purpose; al-

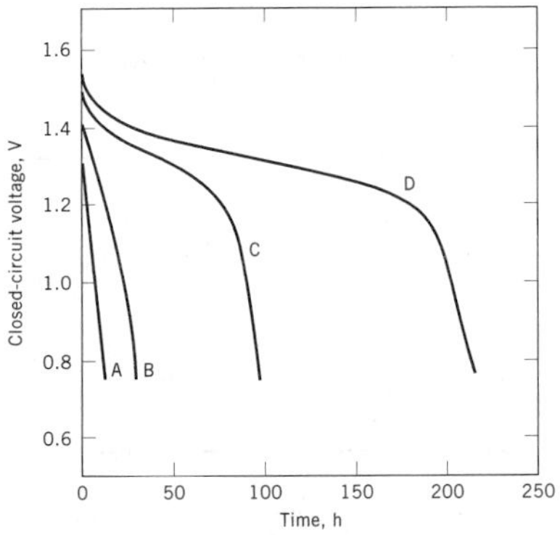

Figure 1. Hours of service on 40-Ω discharge for 4 h/d radio test at 21°C for A, RO3 "AAA"; B, R6 "AA"; C, R14 "C"; and D, R20 "D" paper-lined, heavy-duty zinc chloride cells.

kaline, cylindrical, and miniature; lithium; and reserve or specialty cells.

Carbon–Zinc Cells

Carbon–zinc batteries are the most commonly found primary cells worldwide and are produced in almost every country. Traditionally there are a carbon rod, for cylindrical cells, or a carbon-coated plate, for flat cells, to collect the current at the cathode and a zinc anode. There are two basic versions of carbon–zinc cells: the Leclanché cell and the zinc chloride, $ZnCl_2$, or heavy-duty cell. Both have zinc anodes, manganese dioxide, MnO_2, cathodes, and include zinc chloride in the electrolyte. The Leclanché cell also has an electrolyte saturated with ammonium chloride, NH_4Cl. Additional undissolved ammonium chloride is usually added to the cathode, whereas the zinc chloride cell has at most a small amount of ammonium chloride added to the electrolyte. Both types are dry cells, in the sense that there is no excess liquid electrolyte in the system. The zinc chloride cell is often made using synthetic manganese dioxide and gives higher capacity than the Leclanché cell, which uses inexpensive natural manganese dioxide for the active cathode material. The MnO_2 is only a modest conductor. Thus the cathodes in both types of cell contain 10–30% carbon black in order to distribute the current. Because of the ease of manufacture and the long history of the cell, this battery system can be found in many sizes and shapes.

Performance. Carbon–zinc cells perform best under conditions of intermittent use, and many standardized tests have been devised that are appropriate to such applications as light and heavy flashlight usage, radios, cassettes, and motors (toys). The most frequently used tests are American National Standards Institute (ANSI) tests. The tests are carried out at constant resistance and the results reported in minutes or hours of service. Figure 1 shows typical results under a light load for different size cells.

To compare one battery with another, it is useful to compute the energy density from these data. Because the voltage declines with capacity, the average voltage during the discharge is used to compute an average current, which is then multiplied by the service in hours to give the ampere-hours of capacity. Watt-hours of energy can be obtained by multiplying again by the average voltage.

Cylindrical Alkaline Cells

Primary alkaline cells use sodium hydroxide or potassium hydroxide as the electrolyte. They can be made using a variety of chemistries and physical constructions. The alkaline cells of the 1990s are mostly of the limited electrolyte, dry cell type. Most primary alkaline cells are made using zinc as the anode material; a variety of cathode materials can be used. Primary alkaline cells are commonly divided into two classes, based on type of construction: the larger, cylindrically shaped batteries, and the miniature, button-type cells. Cylindrical alkaline batteries are mainly produced using zinc–manganese dioxide chemistry, although some cylindrical zinc–mercury oxide cells are made.

Cylindrical alkaline cells are zinc–manganese dioxide cells having an alkaline electrolyte, which are constructed in the standard cylindrical sizes, R20 "D", R14 "C", R6 "AA", RO3 "AAA", as well as a few other less common sizes. They can be used in the same types of devices as ordinary Leclanché and zinc chloride cells. Moreover, the high level of performance makes them ideally suited for applications such as toys, audio devices, and cameras.

Performance. Alkaline manganese dioxide batteries have relatively high energy density, as can be seen from Table 1. This results in part from the use of highly pure materials, formed into electrodes of near optimum density. Moreover, the cells are able to function well with a rather small amount of electrolyte. The result is a cell having relatively high capacity at a fairly reasonable cost.

Miniature Alkaline Cells

Miniature alkaline cells are small, button-shaped cells which use alkaline NaOH or KOH electrolyte and generally have zinc anodes, but may have a variety of cathode materials. They are used in watches, calculators, cameras, hearing aids, and other miniature devices.

Cylindrical alkaline cells are made in only a few standard sizes and have only one important chemistry. In contrast, miniature alkaline cells are made in a large number of different sizes, using many different chemical systems. Whereas the cylindrical alkaline batteries are multipurpose batteries, used for a wide variety of devices under a variety of discharge conditions, miniature alkaline batteries are highly specialized, with the cathode material, separator type, and electrolyte all chosen to match the particular application.

Zinc–Mercuric Oxide Batteries. Miniature zinc–mercuric oxide batteries have a zinc anode and a cathode containing mercuric oxide, HgO.

Miniature zinc–mercuric oxide batteries function efficiently over a wide range of temperatures and have good storage life.

Although the zinc–mercuric oxide battery has many excellent qualities, increasing environmental concerns have led to a de-emphasis in the use of this system. The main environmental difficulty is in the disposal of the cell. Both the mercuric oxide in the fresh cell and the mercury reduction product in the used cell have long-term toxic effects.

Zinc–Silver Oxide Batteries. Miniature zinc–silver oxide batteries have a zinc anode, and a cathode containing silver oxide, Ag_2O. Miniature zinc–silver oxide batteries are commonly used in electronic watches and in other applications where high energy density, a flat discharge profile, and a higher operating voltage than that of a mercury cell are needed. These batteries function efficiently over a wide range of temperatures and are comparable to mercury batteries in this respect. Miniature zinc–silver oxide batteries have good storage life.

Divalent Silver Oxide Batteries. It is possible to produce a silver oxide in which the silver has a higher oxidation state, approaching a composition of AgO. This material can provide both higher capacity and higher energy density than Ag_2O. Alternatively, a battery can be made with the same capacity as a monovalent silver cell, but with cost savings. However, some difficulties with regard to material stability and voltage regulation must be addressed.

Zinc–Manganese Dioxide Batteries. The combination of a zinc anode and manganese dioxide cathode, which is the dominant chemistry in large cylindrical alkaline cells, is used in some miniature alkaline cells as well. Overall, this type of cell does not account for a large share of the miniature cell market. It is used in cases where an economical power source is wanted and where the devices can tolerate the sloping discharge curve shown in Figure 2.

Zinc–Air Batteries. Zinc–air batteries offer the possibility of obtaining extremely high energy densities. Instead of having a cathode material placed in the battery when manufactured, oxygen from the atmosphere is used as cathode material, allowing for a much more efficient design. The construction of a miniature air cell is shown in

Table 1. Characteristics of Aqueous Primary Batteries

Parameter	Carbon–zinc (Zn/MnO_2)	Alkaline manganese dioxide ($Zn-MnO_2$)	Mercuric oxide ($Zn-HgO$)	Silver oxide ($Zn-Ag_2O$)	Zinc–air ($Zn-O_2$)
nominal voltage, V	1.5	1.5	1.35	1.5	1.25
working voltage, V	1.2	1.2	1.3	1.55	1.25
specific energy, W·h/kg	40–100	80–95	100	130	230–400
energy density, W·h/mL	0.07–0.17	0.15–0.25	0.40–0.60	0.49–0.52	0.70–0.80
temperature range, °C					
storage	−40−50	−40−50	−40−60	−40−60	−40−50
operating	−5−55	−20−55	−10−55	−10−55	−10−55

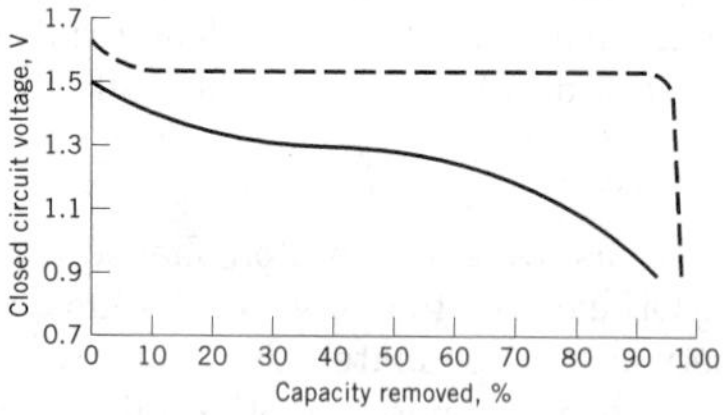

Figure 2. Discharge curves for miniature zinc–silver oxide batteries (----), and zinc–manganese dioxide batteries (—). Courtesy of Eveready Battery Co.

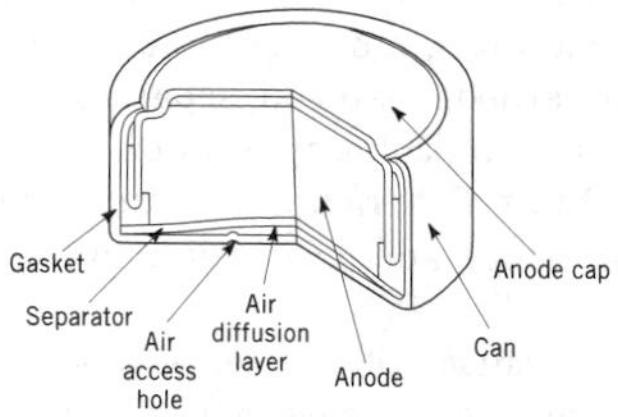

Figure 3. Cutaway view of a miniature air cell battery. Courtesy of Eveready Battery Co.

Figure 3. From the outside, the cell looks like any other miniature cell, except for the air access holes in the can. On the inside, however, the anode occupies much more of the internal volume of the cell. Rather than the thick cathode pellet, there is a thin layer containing the cathode catalyst and air distribution passages. Air enters the cell through the holes in the can and the oxygen reacts at the surface of the cathode catalyst. The air access holes are often covered with a protective tape, which is removed when the cell is placed in service.

The performance level of air cells is exceptional, but these are not general-purpose cells. They must be used in applications where the usage is largely continuous, and where the discharge level is relatively constant and well-defined. The reasons for these limitations lies in the fact that the cell must be open to the atmosphere and the holes that allow oxygen into the cell also allow other gases to enter or leave the cell.

Miniature air cells are mainly used in hearing aids, where they are required to produce a relatively high current for a relatively short time period such as a few weeks. In this application they provide exceptional performance compared to other batteries.

Lithium Cells

Cells having lithium anodes are generally called lithium cells regardless of the cathode. They can be conveniently separated into two types: cells having solid cathodes and cells having liquid cathodes. Cells having liquid cathodes also have liquid electrolytes and in fact, at least one component of the electrolyte solvent and the active cathode material are one and the same. Cells having solid cathodes may have liquid or solid electrolytes but, except for the lithium–iodine system, those having solid electrolytes are not yet commercial.

All of the cells take advantage of the inherently high energy of lithium metal and its unusual film-forming property.

Much analytical study has been required to establish the materials for use as solvents and solutes in lithium batteries. Among the best organic solvents are cyclic esters, such as propylene carbonate (PC), $C_4H_6O_3$, ethylene carbonate (EC), $C_3H_4O_3$, and butyrolactone, $C_4H_6O_2$, and ethers, such as dimethoxyethane (DME), $C_4H_{10}O_2$, the glymes, tetrahydrofuran (THF), and 1,3-dioxolane, $C_3H_6O_2$. Among the most useful electrolyte salts are lithium perchlorate, $LiClO_4$, lithium trifluoromethanesulfonate, $LiCF_3SO_3$, lithium tetrafluoroborate, $LiBF_4$, and lithium hexafluoroarsenate, $LiAsF_6$. A limitation of these organic electrolytes is the relatively low conductivity, compared to aqueous electrolytes. This limitation, combined with the generally slow kinetics of the cathode reactions, has forced the use of certain designs, such as thin electrodes and very thin separators, in all lithium batteries. This usage led to the development of coin cells rather than button cells for miniature batteries and jelly or Swiss roll designs rather than bobbin designs for cylindrical cells.

Many of the cylindrical cells have glass-to-metal hermetic seals, although this is becoming less common because of the high cost associated with this type of seal. Alternatively, cylindrical cells have compression seals carefully designed to minimize the ingress of water and oxygen and the egress of volatile solvent. These construction designs are costly and the high price of the lithium cell has limited its use. However, the energy densities are superior.

Solid Cathode Cells. Solid cathode cells include lithium–manganese dioxide cells, lithium–carbon monofluoride cells, lithium–iron disulfide cells, and lithium–iodine cells.

Liquid Cathode Cells. Liquid cathode cells include lithium–sulfur dioxide cells and lithium–thionyl chloride cells.

Reserve Batteries

Reserve batteries have been developed for applications that require a long inactive shelf period followed by intense discharge during which high energy and power, and sometimes operation at low ambient temperature, are required. These batteries are usually classified by the mechanism of activation which is employed. There are water-activated batteries that utilize fresh or seawater; electrolyte-activated batteries, some using the complete electrolyte, some only the solvent; gas-activated batteries where the gas is used as either an active cathode material or part of the electrolyte; and heat-activated or thermal batteries which use a solid salt electrolyte activated by melting on application of heat.

Activation of these batteries involves adding the missing component which can be done in a simple way, such as pouring water into an opening in the cell, for water-activated cells, or in a more complicated way by using pistons, valves, or heat pellets activated by gravitational or electric signals for the case of the electrolyte- or thermal-activation types. Such batteries may be stored for 10–20 yr while awaiting use. Reserve batteries are usually manufactured under contract for various government agencies such as the U.S. Department of Defense, although occasional industrial or safety uses have been found. Many of the electrochemical systems involved in these batteries are beyond the scope of this article.

The lithium–thionyl chloride, or the lithium–sulfur dioxide, system is often used in a reserve battery configuration in which the electrolyte is stored in a sealed compartment which upon activation may be forced by a piston or inertial forces into the interelectrode space. Most applications for such batteries are in mines and fuse applications in military ordnance.

One variant of the liquid cathode reserve battery is the lithium–water cell in which water serves as both the liquid cathode and the electrolyte. A certain amount of corrosion occurs, but sufficient lithium is provided to compensate. These cells are mostly used in the marine environment where water is available or compatible with the cell reaction product. Common applications are for torpedo propulsion and to power sonobuoys and submersibles.

The last type of reserve cell is the thermally activated cell. The older designs use calcium or magnesium anodes; newer types use lithium alloys as anodes.

The heat pellet used for activation in these batteries is usually a mixture of a reactive metal such as iron or zirconium, and an oxidant such as potassium perchlorate. An electrical or mechanical signal ignites a primer, which then ignites the heat pellet which in turn melts the electrolyte. Sufficient heat is given off by the high current to sustain the necessary temperature during the lifetime of the application. Many millions of these batteries have been manufactured for military ordnance and employed in rockets, bombs, missiles, etc.

GEORGE BLOMGREN
JAMES HUNTER
Eveready Battery Company, Inc.

G. W. Heise and N. C. Cahoon, eds., *The Primary Battery,* Vol. 1 and 2, John Wiley & Sons, Inc., New York, 1971, 1976.

D. Linden, ed., *Handbook of Batteries and Fuel Cells,* McGraw-Hill, New York, 1984.

T. R. Crompton, *Battery Reference Book,* Butterworths, London, 1990.

R. J. Brodd, *Batteries for Cordless Appliances,* John Wiley & Sons, Inc., New York, 1987.

SECONDARY CELLS

ALKALINE

Alkaline electrolyte storage battery systems are more suitable than others in applications where high currents are required, because of the high conductivity of the electrolyte. Additionally, in almost all of these battery systems, the electrolyte which is usually an aqueous solution containing 25–40% potassium hydroxide, KOH, does not enter into the chemical reaction. Thus concentration and cell resistance are invariant with state of discharge and these battery systems give high performance and have long cycle life. The annual production value of alkaline storage batteries is growing: it was over $\$2 \times 10^9$ worldwide in 1990, representing approximately 20% of the value for all secondary batteries.

Positive electrode active materials have been made from the oxides or hydroxides of nickel, silver, manganese, copper, mercury, and from oxygen. Negative electrode active materials have been fabricated from various geometric forms of cadmium, Cd, iron, Fe, and zinc, Zn, and from hydrogen. Two different types of hydrogen electrode designs are common: those used in space, which employ hydrogen as a gas, and those used in consumer batteries, where the hydrogen is used as a metallic hydride. As indicated in Table 1, nine electrode combinations exist in some scale of commercial production. Five system combinations are in the research and development stage as of this writing, and two have been abandoned before or after commercial production for reasons such as short life, high cost, low voltage, low energy density, and excessive maintenance.

The annual production value of small, sealed nickel–cadmium cells is over $\$1.2 \times 10^9$. However, environmental considerations relating to cadmium are necessitating changes in the fabrication techniques, as well as recovery of failed cells. Battery system designers are switching to nickel–metal hydride (MH) cells for some applications, typically in "AA"-size cells, to increase capacity in the same volume and avoid the use of cadmium.

Many of the most recent applications for alkaline storage batteries require higher energy density and lower cost designs than were previously available. Materials such as foam or fiber nickel, Ni, mats as substrates, and new processing techniques including plastic bounded, pasted, or electroplated electrodes, have enabled the alkaline storage battery to meet these new requirements, while reducing environmental problems in the manufacturing plants. In addition, substantial technical efforts have been devoted to the recovery of used batteries. The most recent innovations in materials relate to the development of metal–hydride alloys for the storage and electrochemical utilization of hydrogen. Modifications to the chemical structure or the cell design of manganese dioxide MnO_2, electrodes have resulted in sufficient improvement to allow the reintroduction of the rechargeable MnO_2–zinc cell to the market as a lower cost, albeit lower performance, alternative to nickel–cadmium consumer size cells. Improvements in materials science and electrical circuits have led to better separators, seals, welding techniques, feedthroughs, and charging equipment.

Nickel–Cadmium Cells

Electrodes. A number of different types of nickel oxide electrodes have been used. The term nickel oxide is common usage for the active materials that are actually hydrated hydroxides at nickel oxidation state 2+, in the discharged condition, and nickel oxide hydroxide, $NiO{\cdot}OH$, nickel oxidation state 3+, in the charged condition. Nickelous hydroxide, $Ni(OH)_2$, can be precipitated from acidic solutions of bivalent nickel either by the addition of sodium hydroxide or by cathodic processes to cause an increase in the interfacial pH at the solution–electrode surface (see NICKEL AND NICKEL ALLOYS; NICKEL COMPOUNDS).

The many varieties of practical nickel electrodes can be divided into two main categories. In the first, the active nickelous hydroxide is prepared in a separate chemical reactor and is subsequently blended, admixed, or layered with an electronically conductive material. This active material mixture is afterwards contained in a confining porous metallic structure or pasted onto a metallic mat or grid.

The other type of nickel electrode involves constructions in which the active material is deposited *in situ.* This includes the sintered-type electrode in which nickel hydroxide is chemically or electrochemically deposited in the pores of a 80–90% porous sintered nickel substrate that may also contain a reinforcing grid.

Almost all the methods described for the nickel electrode have been used to fabricate cadmium electrodes. However, because cadmium, cadmium oxide, CdO, and cadmium hydroxide, $Cd(OH)_2$, are more electrically conductive than the nickel hydroxides, it is possible to make simple pressed cadmium electrodes using less substrate (see CADMIUM AND CADIUM ALLOYS; CADMIUM COMPOUNDS). These are commonly used in button cells.

Electrochemistry and Crystal Structure. The solid-state chemistry of the nickel electrode is complex. Nickel hydroxide in the discharged state has a hexagonal layered lattice, where planes of Ni^{2+} ions are sandwiched between planes of OH^-. This structure, similar to that of cadmium iodide, CdI_2, is common to seven metal hydroxides, including those of cadmium and cobalt. There are various hydrated and nonhydrated nickel hydroxides that have slightly different crystal habitats and electrochemical potentials. The most common form of charged material observed in batteries is NiOOH, density = 4.6 g/mL. In comparison, $Ni(OH)_2$ has a density of 4.15 g/mL. Thus the theoretical change in density on charge–discharge is only 9%, and the kinetics involve only a proton transfer.

The chemistry, electrochemistry, and crystal structure of the cadmium electrode is much simpler than that of the nickel electrode. The

Table 1. Rechargeable Alkaline Storage Battery Systems

System[a]	Historical name	Voltage, V	Production[b]
nickel–cadmium	Jungner	1.30	vl
nickel–iron	Edison	1.37	s
nickel–zinc	Drumm	1.70	vs
nickel–hydrogen		1.30	l
silver–cadmium		1.38 and 1.16[c]	vs
silver–iron	Jirsa	1.45 and 1.23[c]	s
silver–zinc	Andre	1.86 and 1.60[c]	s
silver–hydrogen		1.38 and 1.16[c]	vs
manganese–zinc[d]		1.52	vs
mercury–cadmium		0.92	r
air (oxygen)–zinc		1.60	r
air (oxygen)–iron		1.40	r
air (oxygen)–aluminum			r
copper–lead		1.20	r
copper–cadmium	Darrieus	0.45	n
copper–zinc	Waddell-Entz, Edison-LeLande, Lelande-Chaperon	0.85	n

[a] The substance named first represents the positive electrode; the substance named second is the negative electrode. In all cases except for air (oxygen) systems, the active electrode material is the oxide or the hydroxide of the named species.
[b] vl =>100 × 10^6 3A·h/yr product; l =>25 × 10^6 A·h/yr; s => 5 × 10^6 A·h/yr; vs =< 5 × 10^6 A · h/yr; r = research and development phase; and n = no longer in production.
[c] Silver electrodes have two voltages plateaus.
[d] Secondary system.

overall reaction is generally recognized as:

$$Cd + 2OH^- \underset{\text{charge}}{\overset{\text{discharge}}{\rightleftarrows}} Cd(OH)_2 + 2e^-$$

However, there is a strong likelihood of a soluble intermediate in the formation of $Cd(OH)_2$. Cadmium has an appreciable solubility in alkaline solutions: ~2 × 10^{-4} mol/L in 8 *M* potassium hydroxide at room temperature. In general, it is believed that the solution process consists of anodic dissolution of cadmium ions in the form of complex hydroxides (see CADMIUM COMPOUNDS).

Sealed Cells. Most sealed cells are based on the principles appearing in patents of the early 1950s where the virtues of limited electrolyte, a separator that would absorb and retain electrolyte, and providing free passage for the oxygen from the positive to the negative plate were described. Although both pocket and sintered electrodes of the nickel–cadmium type have been used in sealed-cell construction, the preponderant majority of cells in commercial production use sintered positive (nickel) electrodes, and either sintered or pasted negative (cadmium) electrodes.

Cell Fabrication Methods. *Pocket Cells.* The essential steps of positive (nickel) electrode construction are (*1*) cold-rolled steel ribbon is cut to proper width and is perforated using either needles or rolls; (*2*) the perforated steel ribbon is nickel-plated and usually annealed in hydrogen. The ribbon is formed into a trough shape, is filled with active material by either a briquetting or a powder-filling technique; (*3*) a second strip is formed into a lid that covers and locks with the filled trough; (*4*) the filled strips are cut to length and are arranged to form an electrode sheet by interleaving. This operation, carried out by means of rollers in a forming roll, is often combined with the pressing of a pattern into the electrode sheet in order to ensure good contact between ribbon and active material and to add mechanical strength to the construction; and (*5*) the electrode sheet is then cut to pieces of appropriate size, and side bedding and lugs are attached to form a metallic frame. The frame material is usually also cold-rolled steel ribbon.

The pockets are usually arranged horizontally in the electrodes, but in a few cases vertical pockets are used. No significant difference has been observed between the two arrangements.

Pocket-type cadmium electrodes are made by a procedure similar to that described for the positive electrode. Because cadmium active material is more dense than nickel active material, and because cadmium has a 2+ valence, cadmium electrodes, when fabricated to equal thicknesses, have almost twice the working capacity of the nickel electrode.

After the individual pocket electrodes are fabricated, they are assembled into electrode groups. Electrodes of the same polarity are electrically and mechanically connected to each other and to a pole bolt. Plates of opposite polarity are interleaved with separators.

To complete the assembly of a cell, the interleaved electrode groups are bolted to a cover and the cover is sealed to a container. Originally, nickel-plated steel was the predominant material for cell containers, but more recently plastic containers have been used for a considerable proportion of pocket nickel–cadmium cells. Polyethylene, high impact polystyrene, and a copolymer of propylene and ethylene have been the most widely used plastics.

Tubular Cells. Although the tubular nickel electrode invented by Edison is almost always combined with an iron negative electrode, a small quantity of cells is produced in which nickel in the tubular form is used with a pocket cadmium electrode. This type of cell construction is used for low operating temperature environments, where iron electrodes do not perform well or where charging current must be limited.

Sintered Cells. The fabrication of sintered electrode batteries can be divided into five principal operations: preparation of sintering-grade nickel powder; preparation of the sintered nickel plaque; impregnation of the plaque with active material; assembly of the impregnated plaques (often called plates) into electrode groups and into cells; and assembly of cells into batteries.

Other Cells. Other methods to fabricate nickel–cadmium cell electrodes include those for the button cell, used for calculators and other electronic devices. This cell, the construction of which is illustrated in Figure 1, is commonly made using a pressed powder nickel electrode mixed with graphite that is similar to a pocket electrode. The cadmium electrode is made in a similar manner. The active material, graphite blends for the nickel electrode, are almost the same as that used for pocket electrodes, ie, 18% graphite.

Lower cost and lower weight cylindrical cells have been made using plastic bound or plated active material pressed into a metal screen. These cells suffer slightly in utilization at high rates compared to a sintered-plate cylindrical cell, but they may be adequate for most applications.

Applications. Nickel–cadmium cells represent almost 20% of the market for all storage batteries, including lead–acid, manufactured in the world. Uses are divided into three categories: pocket cells are used

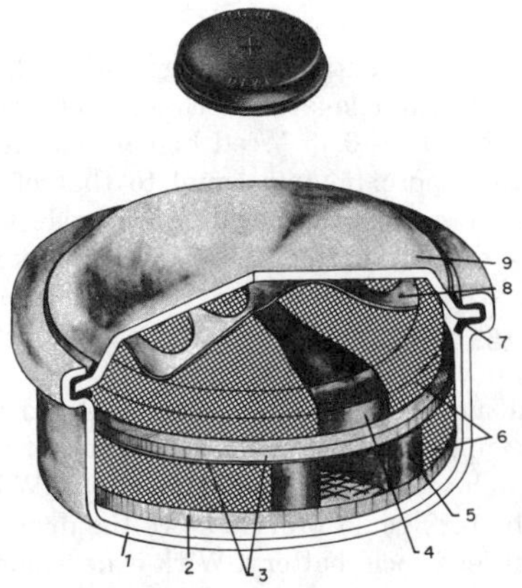

Figure 1. Section of disk-type cell where: 1 is the cell cup; 2, the bottom insert; 3, the separator; 4, the negative electrode; 5, the positive electrode; 6, the nickel wire gauze; 7, the sealing washer; 8, the contact spring; and 9, the cell cover.

in emergency lighting, diesel starting, and stationary and traction applications where the reliability, long life, medium-high rate capability, and low temperature performance characteristics warrant the extra cost over lead–acid storage batteries; sintered, vented cells are used in extremely high rate applications, such as jet engine and large diesel engine starting as well as some electric vehicle designs; and sealed cells, both the sintered and button types, are used in computers, phones, cameras, portable tools, electronic devices, calculators, cordless razors, toothbrushes, carving knives, flashguns, and in space applications, where nickel–cadmium is optimum because it can be recharged a great number of cycles and not suffer from prolonged trickle overcharge. Cells of this category are generally made in sizes comparable to conventional dry cells, such as "D," "C," "AA," etc.

Charger Technology. Alkaline storage batteries are commonly charged from rectified d-c equipment, solar panels, or other d-c sources and have fairly good tolerance to ripple and transient pulses. Advances in electronic control circuitry in chargers and redesign of small sealed nickel–cadmium cells permit the rapid (15-min) recharge of special designs.

Nickel–Iron Cells. There has been renewed interest in the system for electric vehicles (EV). The EV design is based on a high rate, usually sintered, iron electrode as well as high rate nickel electrodes.

Electrochemistry and Kinetics. The electrochemistry of the nickel–iron battery and the crystal structures of the active materials depends on the method of preparation of the material, degree of discharge, the age (life cycle), concentration of electrolyte, and type and degree of additives, particularly the presence of lithium and cobalt. A simplified equation representing the charge–discharge cycle can be given as:

$$2NiOOH^* + \alpha - Fe + 2H_2O \rightleftarrows 2Ni(OH)_2^* + Fe(OH)_2$$

where the asterisks indicate adsorbed water and KOH.

Electrode Structures. The classical iron active material for pocket and pasted iron electrodes was formed by roasting recrystallized ferrous sulfate, $FeSO_4$, in an oxidizing atmosphere to ferric oxide, Fe_2O_3, and then reducing the latter in hydrogen. The α-iron formed was then heated to a mixture of Fe_3O_4 and Fe. As such it was pure enough to be used for pharmaceutical purposes. For battery use, a small amount of sulfur, as FeS, was added, as were other additives which were believed to increase the cycle life by acting as depassivating agents, ie, helping to reduce the tendency of iron to evolve hydrogen upon standing in alkaline electrolyte.

A study of sintered iron electrodes claimed advantages of high rate capability, long life, and low hydrogen evolution.

Sintered nickel electrodes used in nickel iron cells are usually thicker than those used in Ni–Cd cells. These result in high energy density cells, because very high discharge rates are usually not required.

Performance Characteristics. The sintered nickel-sintered iron design battery has outstanding power characteristics at all states of discharge, making them attractive to the design of electric vehicles (EV) which must accelerate with traffic even when almost completely discharged.

Silver–Zinc Cells

The silver–zinc battery has the highest attainable energy density of any rechargeable system in use as of this writing. In addition, it has an extremely high rate capability coupled with a very flat voltage discharge characteristic. Its use, in the early 1990s, is limited almost exclusively to the military for various aerospace applications such as satellites and missiles, submarine and torpedo propulsion applications, and some limited portable communications applications. The main drawback of these cells is the rather limited lifetime of the silver–zinc system. Life is normally limited to less than 200 cycles with a total wet-life of no more than about two years. The silver–zinc system also carries a very high cost and applications are justified only where cost is a minor factor. The high cost of silver battery systems is attributable to the cost of the active silver material used in the positive electrodes.

Cellophane or its derivatives have been used as the basic separator for the silver–zinc cell since the 1940s. The cellophane is the principal limitation to cell life. Oxidation of the cellophane in the cell environment degrades the separator and within a relatively short time short circuits may occur in the cell.

A second lifetime limitation is the zinc anode. In spite of the separator and cell designs, some zinc material is solubilized during the charge–discharge reaction. Over a period of cycling there is a shift of active material, originally distributed evenly over the face of the electrode, to the center and bottom areas of the electrode. This shape change limits the life of the cell as exemplified by a fading of the capacity and a build-up of internal pressure that may eventually lead to a short circuit.

Electrochemistry. Silver–zinc cells have some unusual thermodynamic properties. The equations indicate that the higher valence silver oxide is AgO, silver(II) oxide. However, in the crystallographic unit cell, which is monolithic, there are four silver atoms and four oxygen atoms, and none of the Ag–O bonds conforms to a silver(II) bond length. Instead there are two Ag–O bonds of 0.218 nm corresponding to silver(I) and two Ag–O bonds of 0.203 nm corresponding to silver(III). This structure has also been proposed on the basis of magnetic and semiconductor properties and confirmed using neutron diffraction.

Electrodes. All of the finished silver electrodes have certain common characteristics: the grids or substrates used in the electrodes are made exclusively of silver, although in some particular cases silver-plated copper is used.

There are three methods of silver electrode fabrication: (*1*) the slurry pasting of monovalent or divalent silver oxide to the grid, drying, reducing by exposure to heat, and then sintering to agglomerate the fine particles into an integral, strong structure; (*2*)) the dry processing of fine silver powders by pressing in a mold or by a continuous rolling operation onto a silver grid followed by sintering; and, (*3*) the use of plastic-bonded active material formed by embedding the active material (fine silver powder) in a plastic vehicle such as polyethylene, which can then be milled into flexible sheets. These sheets are cut to size, pressed in a mold on both sides of a conductive grid, and the pressed electrode subjected to sintering where the plastic material is fired off, leaving the metallic silver.

Silver electrodes prepared by any of the three methods are almost always subjected to a sintering operation prior to cell or battery assembly.

Zinc electrodes for secondary silver–zinc batteries are made by one of three general methods: the dry-powder process, the slurry-pasted process, or the electroformed process. The active material used in any of the processes for the manufacture of electrodes is a finely divided zinc oxide powder, USP grade 12.

Electrolyte. The electrolyte in silver–zinc cells is 30–45% KOH. The lower concentrations in this range have higher conductivities and are preferred for high rate cells. Higher concentrations have a less deleterious effect on cellulosic separators and are preferable for extended life characteristics.

Cell Hardware. Cell jars are constructed almost exclusively of injection-molded plastics, which are resistant to the strong alkali electrolyte.

Cell terminal connections are usually brought out by two-threaded terminals that protrude through the cell jar cover. They are usually steel, brass, or copper with a hollow construction.

Performance. *Charging.* Charging of silver–zinc cells can be done by one of several methods. The constant-current method which is most common consists of a single rate of current, usually equivalent to a full input within the 12–16-h period.

Discharge. Silver–zinc cells have one of the flattest voltage curves of any practical battery system known, although there are two voltage steps caused by the two different valence states of silver oxide.

Performance of silver–zinc cells is normally considered to be adequate in the temperature range of 10–38°C. If a wider temperature range is desired silver–zinc cells and batteries may be used in the range 0–71°C without any appreciable derating.

Cell Life. Silver–zinc cells are usually manufactured as either low or high rate cells. Approximately 10–30 cycles can be expected for high rate cells, depending on the temperature of use, the rate of discharge, and methods of charging. Low rate cells have been satisfactorily used for 100–300 cycles under the proper conditions. In general, the overall life of the silver–zinc cells with the separator systems normally in use is approximately 1–2 yr.

Other Silver Positive Electrode Systems

Silver–Cadmium Cells. In satellite applications the nonmagnetic property of the silver–cadmium battery is of utmost importance because magnetometers were used on satellites to measure radiation and the effects of magnetic fields of energetic particles. Satellites have to be constructed of nonmagnetic components in sealed batteries.

Silver–cadmium satellite batteries have been used in cyclic periods of five hours or more with discharge times of 30–60 min. Operational and test programs have shown cycle life periods of 3 yr at low temperatures. At temperatures of 40°C and 50°C, the cycle life is 1 yr and 0.2 yr, respectively. The cycle life at intermediate temperatures is 1.4–2.0 yr.

Another application for silver–cadmium batteries is propulsion power for submarine simulator-target drones.

Silver–Iron Cells. The silver–iron battery system combines the advantages of the high rate capability of the silver electrode and the cycling characteristics of the iron electrode. Commercial development has been undertaken to solve problems associated with deep cycling of high power batteries for ocean systems operations.

Cells consist of porous sintered silver electrodes and high rate iron electrodes. The latter are enclosed with a seven-layered, controlled porosity polypropylene bag which serves as the separator. The electrolyte contains 30% KOH and 1.5% LiOH.

Applications have been found for these batteries in emergency power applications for telecommunications systems in tethered balloons. Unfortunately, the system is expensive because of the high cost of the silver electrode. Applications are, therefore, generally sought where recovery and reclamation of the raw materials can be made.

Applications have been found for these batteries in emergency power applications for telecommunications systems in tethered balloons. Unfortunately, the system is expensive because of the high cost of the silver electrode. Applications are, therefore, generally sought where recovery and reclamation of the raw materials can be made.

Nickel–Zinc Cells

Nickel–zinc cells offer potential advantages over other rechargeable alkaline systems. The single-level discharge voltage, 1.60–1.65 V/cell is approximately 0.35–0.45 V/cell higher than nickel–cadmium or nickel–iron and approximately equal to that of silver–zinc. In addition, the use of zinc as the negative electrode should result in a higher energy density battery than either nickel–cadmium or nickel–iron and a lower cost than silver–zinc. In fact, nickel–zinc cells having energy densities in the range of 40–60 W·h/kg have been successfully demonstrated.

A commercial nickel–zinc battery is considered to be the most likely candidate for electric vehicle development. If the problems of limited life and high installation cost ($100–150/kW·h) are solved, a nickel–zinc EV battery could provide twice the driving range for an equivalent weight lead–acid battery. Work is developmental; there is no commercial production of nickel–zinc batteries.

Cell Construction. Nickel–zinc batteries are housed in molded plastic cell jars of styrene, SAN, or ABS material for maximum weight savings. Nickel electrodes can be of the sintered or pocket type; however, these types are not cost-effective, and several different types of plastic-bonded nickel electrodes have been developed.

Nickel hydrate, usually 5–10% cobalt added, serves as the active material and is mixed with a conductive carbon, eg, graphite. The active mass is mixed with an inert organic binder such as polyethylene or poly(tetrafluoroethylene) (TFE). The resultant mass is rolled into sheets on a compounding mill or pressed into electrodes as a dry powder on a nickel grid.

Negative electrodes are fabricated of zinc oxide by any of the methods (pasting, pressing, etc) described. Binders, usually TFE, are used to reduce the solubility of the electrode in KOH.

Separators are both of the organic and inorganic type.

Performance. The limited life of nickel–zinc batteries is the principal drawback to widespread use.

Nickel–Hydrogen Cells

There are two types of nickel–hydrogen cells: those that employ a gaseous H_2 electrode and those that utilize a metal hydride, MH. The use of MH electrodes in small-scaled. However, the market for consumer-type cells is rapidly growing. The system has also been proposed for electric vehicles.

Other Cell Systems

Silver–Hydrogen Cells. With the development of the nickel–hydrogen system limited attention was directed to the development of a silver–hydrogen cell. The main characteristics of interest were the potential for a higher gravametric energy density based on the lighter weight of the silver electrode vs that of the nickel. The packaging approach utilized for this battery is similar to that of nickel–hydrogen single cylindrical cells as shown in Figure 2. The silver electrode is typically the sintered type used in rechargeable silver–zinc cells. The hydrogen electrode is a Teflon-bonded platinum black gas diffusion electrode.

Because the silver oxide electrode is slightly soluble in the potassium hydroxide electrolyte the separator is of a barrier type to minimize silver diffusion to the opposite electrode.

Zinc–Oxygen Cells. On the basis of reactants the zinc–oxygen or air system is the highest energy density system of all the alkaline rechargeable systems with the exception of the $H_2 \cdot O_2$ one. The reactants are cheap and abundant and therefore a number of attempts have been made to develop a practical rechargeable system.

Iron–Air Cells. The iron–air system is a potentially low cost, high energy system being considered mainly for mobile applications. The iron electrode, similar to that employed in the nickel–iron cell, exhibits long life and therefore this system could be more cost effective than the zinc–air cell.

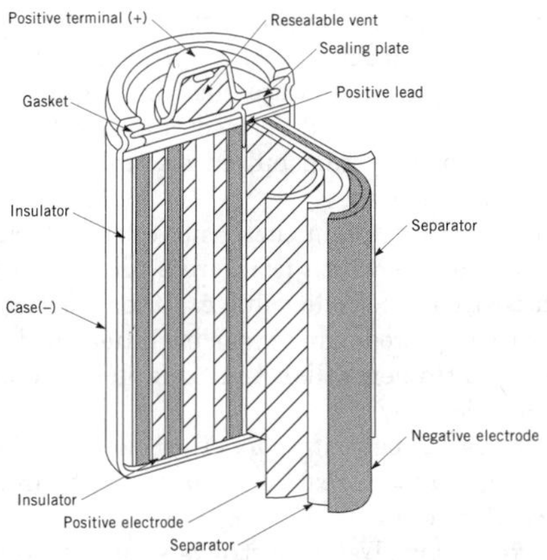

Figure 2. Schematic diagram of a Ni–H (MH) cell.

Hydrogen–Oxygen Cells. The hydrogen–oxygen cell can be adapted to function as a rechargeable battery, although this system is best known as a primary one (see FUEL CELLS).

Mechanically Rechargeable Batteries. To avoid the time required for electric recharge, the problems of *in situ* electric recharge, or to utilize anodes that are not electrically rechargeable in aqueous electrolytes, mechanically rechargeable batteries have been studied. These systems are metal–air couples. The anodes that have received attention are zinc, lithium, and aluminum.

The most significant results with these battery types focus on aluminum as the anode. Figure 3 shows an aluminum–air cell being developed for electric vehicle applications. The aluminum hydroxide reaction product would be returned to the factory to be reprocessed into fresh aluminum anodes. One set of anodes could yield up to 800 km range before replacement. However, the corrosion reaction of the aluminum with the electrolyte is still a problem. Additionally, the system is complex and it is anticipated that replacing the anodes repeatedly will be problematic. The system has poor energy efficiency when consideration is given to the full cycle of electric generation, aluminum production, battery efficiency, and reprocessing of the battery reaction product back to aluminum.

Electrolyte

Potassium hydroxide is the principal electrolyte of choice for the above batteries because of its compatibility with the various electrodes, good conductivity, and low freezing point temperature.

Safety and Disposal

The potassium hydroxide electrolyte used in alkaline batteries is a corrosive, hazardous chemical. It is a poison and if ingested attacks the throat and stomach linings. Immediate medical attention is required. It slowly attacks skin if not rapidly washed away. Extreme care should be taken to avoid eye contact that can result in severe burns and blindness. Protective clothing and face shields or goggles should be worn when filling cells with water or electrolyte and performing other maintenance on vented batteries.

Alkaline batteries generate hydrogen and oxygen gases under various operating conditions. This can occur during charge, overcharge, open circuit stand, and reversal. In vented batteries free ventilation should be provided to avoid hydrogen accumulations surrounding the battery.

Alkaline batteries are capable of high current discharges and accidental short circuits should be avoided. Spontaneous low resistance internal short circuits can develop in silver–zinc and nickel–cadmium batteries.

Because of increasing environmental concerns, the disposal of all batteries is being reviewed. Traditionally silver batteries were reclaimed for the silver metal and all other alkaline batteries were disposed of in landfills or incinerators. Some aircraft and industrial nickel–cadmium batteries are rebuilt to utilize the valuable components.

To reduce or eliminate the scattering of cadmium in the environment, the disposal of nickel–cadmium batteries is under study. Already a large share of industrial batteries are being reclaimed for the value of their materials.

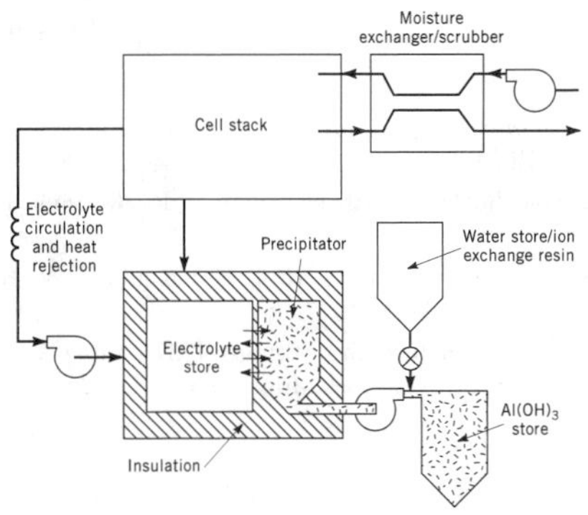

Figure 3. Aluminum–air power cell system. The design provides for forced convection of air and electrolyte, heat rejection, electrolyte concentration control via $Al(OH)_3$ precipitation, and storage for reactants and products.

ALVIN J. SALKIND
MARTIN KLEIN
Rutgers University

L. Ojefors, *Proceedings of the 9th International Society for Electrochemistry,* Marcoussis, France, 1975; B. Andersson and L. Ojefors, *11th International Power Sources Symposium,* Brighton, U.K., 1978; *J. Electrochem. Soc.* **123,** 824 (1976).

R. Swaroop, *Reports on Electric Vehicle Batteries,* Electric Power Research Institute (EPRI), Palo Alto, Calif., 1989–1991.

M. J. Sulkes, "Nickel-Zinc Secondary Batteries," *Proceedings of the 23rd Annual Power Sources Conference,* 1969.

M. Klein, in D. H. Collins, ed., *Power Sources,* Vol. 5, Academic Press, New York, 1974.

LEAD–ACID

The lead–acid battery is one of the most successful electrochemical systems and the most successful storage battery developed. About 80% of the lead (qv), Pb, consumption in the United States was for batteries in that year.

The lead–acid battery consists of a number of cells in a container. These cells contain positive (PbO_2) and negative (Pb) electrodes or plates, separators to keep the plates apart, and sulfuric acid, H_2SO_4, electrolyte. The battery reactions are highly reversible, so that the battery can be discharged and charged repeatedly. The number of charge–discharge cycles that can be obtained depends strongly on the use mode and can vary from several hundred to thousands of cycles.

Each cell has a nominal voltage of 2 V and capacities typically vary from 1 to 2000 ampere-hours. Lead–acid cells can be operated with coulombic efficiencies as high as 95% and with energy efficiencies greater than 80%. The many cell designs available for a wide variety of uses can be divided into three main categories: automotive, industrial, and consumer. Automotive batteries, starting, lighting, and ignition (SLI) for cranking of internal combustion engines accounted for nearly 73% of the lead–acid battery sales in 1988. Industrial batteries are used for heavy-duty application such as motive and standby power. More recently, the use of batteries for utility peak shaving has been increasing. Consumer batteries are used for emergency lighting, security alarm systems, cordless convenience devices and power tools, and small engine starting. This is one of the fastest growing markets for the lead–acid battery.

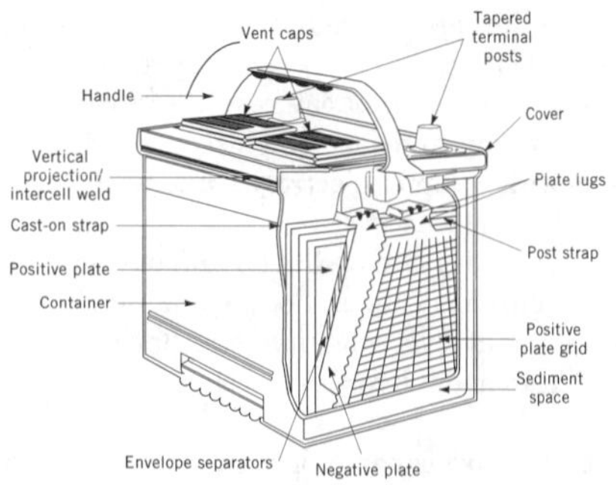

Figure 1. Cutaway view of an automotive SLI lead–acid battery container and cell element. Courtesy of Johnson Controls, Inc.

In Figure 1, the cutaway view of the automotive battery shows the components used in its construction. Automotive and industrial motive power batteries have the standard free electrolyte systems and operate only in the vertical position.

Two types of batteries having immobilized electrolyte systems are also made. They are most common in consumer applications, but their use in industrial and SLI applications is increasing. Both types have low maintenance requirements and usually can be operated in any position. They are sometimes called valve regulated or recombinant batteries because they are equipped with a one-way pressure relief vent and normally operate in a sealed condition with an oxygen recombination cycle to reduce water loss.

In the gelled electrolyte battery, the sulfuric acid electrolyte has been immobilized by a thixotropic gel. This is made by mixing an inorganic powder such as silicon dioxide, SiO_2, with the acid. Other cells use a highly absorbent separator to immobilize the electrolyte.

Cell Thermodynamics

The chemical reaction of the lead–acid battery was explained as early as 1882. The double sulfate theory has been confirmed by a number of methods as the only reaction consistent with the thermodynamics of the system. The thermodynamics of the lead–acid battery has been reviewed in great detail.

Lead sulfate is formed as the battery discharges, and sulfuric acid is regenerated as the battery is charged. The open circuit voltage of the lead–acid battery is the function of the acid concentration and temperature. A review of this subject is available. The Nernst equation may be used to calculate the open circuit cell voltage. The battery voltage is then obtained by multiplying the cell voltage by the number of cells.

Lead Grid Corrosion

The corrosion of the lead grid at the lead dioxide electrode is one of the primary causes of lead–acid battery failure. The mechanisms of lead corrosion in sulfuric acid have been studied and good reviews of the literature are available.

Charge–Discharge Processes

An excellent review covers the charge and discharge processes in detail and ongoing research on lead–acid batteries may be found in two symposia proceedings. Detailed studies of the kinetics and mechanisms of lead–acid battery reactions are published continually.

At high discharge rates, such as those required for starting an engine, the voltage drops sharply primarily because of the resistance of the lead current collectors. This voltage drop increases with the cell height and becomes significant even at moderate discharge rates in large industrial cells. Researchers have measured this effect in industrial cells and have developed a model which has been used to improve grid designs for automotive batteries.

Self-Discharge Processes. The shelf life of the lead–acid battery is limited by self-discharge reactions, which proceed slowly at room temperature. High temperatures reduce shelf life significantly. The reactions which can occur are well defined, and self-discharge rates in lead–acid batteries having immobilized electrolyte and limited acid volumes have been measured.

The lead current collector in the positive lead–dioxide plate corrodes and the compounds which form are a function of the acid concentration and positive electrode voltage. Other reactions which take place at the positive electrode are oxygen evolution, oxidation of organics, sulfation of PbO (in new cells), and oxidation of additives such as antimony, in the grid alloy.

Similar reactions can be written for other metallic additives. At the negative electrode two more reactions can occur, hydrogen evolution and oxygen recombination.

Overcharge Reactions. Water electrolysis during overcharge is an irreversible process. Theoretically, water should decompose at a voltage below the voltage required to recharge a lead–acid battery. However, the rate of water electrolysis is much slower than the rate of the recharge reaction. Thus the lead–acid battery can operate with as little as 5% excess charge to compensate for water electrolysis. Use of lead–antimony alloys for the current collectors in lead–acid batteries increases water loss. Some of these batteries need regular maintenance by addition of water to replace the water lost on overcharge. Many newer designs, however, use either lower concentrations of antimony in the alloy or lead–calcium alloys to reduce water loss (see LEAD ALLOYS). This is the basis for the maintenance-free batteries.

Material Fabrication and Manufacturing Processes

The lead–acid battery is comprised of three primary components: the element, the container, and the electrolyte. The element consists of positive and negative plates connected in parallel and electrically insulating separators between them. The container is the package which holds the electrochemically active ingredients and houses the external connections or terminals of the battery. The electrolyte, which is the liquid active material and ionic conductor, is an aqueous solution of sulfuric acid (see Fig. 1).

Economic Aspects

Whereas automotive batteries have the majority of the market, other types of lead–acid batteries, such as sealed and small maintenance-free varieties, are making inroads into various applications. The automotive battery's operating environment has changed substantially in the last 10 years. Underhood temperature has risen and electrical loads have increased. This trend is expected to continue as car manufacturers reevaluate their design strategies and objectives. Battery design is changing to meet these needs.

KATHRYN R. BULLOCK
JOHN R. PIERSON
Johnson Controls, Inc.

Storage Battery Technical Service Manual, 10th ed., Battery Council International, Chicago, Ill., 1987.

G. W. Vinal, *Storage Batteries,* 4th ed., John Wiley & Sons, Inc., New York, 1955.

H. Bode, *Lead–Acid Batteries* (trans. R. J. Brodd and K. V. Kordesch), John Wiley & Sons, Inc., New York, 1977.

"SAE Standard Test Procedure for Storage J537—June '86," *SAE Recommended Practices,* SAE, New York, June 1986.

OTHER

The proliferation of portable electronic devices has fueled rapid market growth for the rechargeable battery industry. Miniaturization of electronics coupled with consumer demand for lightweight batteries providing ever longer run times continues to spur interest in advanced battery systems. Interest also continues to run strong in electric vehicles (EVs) and the large auto manufacturers continue to develop prototype EVs. It is clear that advances in battery technology are required for a widely acceptable EV. Advanced batteries continue to play a strong role in other applications such as load leveling for the electric utility industry and satellite power systems for aerospace.

Ambient Temperature Lithium Systems

Traditionally, secondary battery systems have been based on aqueous electrolytes. Whereas these systems have excellent performance, the use of water imposes a fundamental limitation on battery voltage because of the electrolysis of water, either to hydrogen at cathodic potentials or to oxygen at anodic potentials. The application of nonaqueous electrolytes affords a significant advantage in terms of achievable battery voltages. By far the most actively researched field in nonaqueous battery systems has been the development of practical rechargeable lithium batteries. These are systems that are based on the use of lithium metal, Li, or a lithium alloy, as the negative electrode (see LITHIUM AND LITHIUM COMPOUNDS).

The use of lithium as a negative electrode for secondary batteries offers a number of advantages. Lithium has the lowest equivalent weight of any metal and affords very negative electrode potentials when in equilibrium with solvated lithium ions, resulting in very high theoretical energy densities for battery couples. These high theoretical energy densities have prompted a wealth of research activity in a wide variety of experimental battery systems. However, realization of the technology to commercialize these systems has been slow.

A key technical problem in developing practical lithium batteries has been poor cycle life attributable to the lithium electrode. The highly reactive nature of freshly plated lithium leads to reactions with electrolyte and impurities to form passivating films that electrically isolate the lithium metal.

The choice of battery electrolyte is of paramount importance in achieving acceptable cycle life because of the high reducing power of the metallic lithium. The formation of surface films on the lithium electrode imparts the apparent stability of the electrolyte to the electrode. It is critical to determining lithium cycling efficiency. In addition to providing a stable film in the presence of lithium, the electrolyte must satisfy additional requirements, including good conductivity, being in the liquid range over the battery operating temperature, and electrochemical stability over a wide voltage range. Solubility of the electrolyte salt in the solvent system is important in achieving good conductivity. In order to satisfy the various electrolyte system requirements, the use of mixed solvent electrolytes has become common in practical cells. Examples are tetrahydrofuran, C_4H_8O, -based electrolytes or ethylene carbonate $C_2H_4O_3$, –propylene carbonate, $C_4H_6O_3$, mixed solvent systems.

A second class of important electrolytes for rechargeable lithium batteries are solid electrolytes. Of particular importance is the class known as solid polymer electrolytes (SPEs). SPEs are polymers capable of forming complexes with lithium salts to yield ionic conductivity. The best known of the SPEs are the lithium salt complexes of poly(ethylene oxide) (PEO), $—(CH_2CH_2Oh)_n—$, and poly(propylene oxide) (PPO).

The lithium or lithium alloy negative electrode systems employing a liquid electrolyte can be categorized as having either a solid positive electrode or a liquid positive electrode. Systems employing a solid electrolyte employ solid positive electrodes to provide a solid-state cell. Another class of lithium batteries are those based on conducting polymer electrodes. Several of these systems have reached advanced stages of development or initial commercialization such as the Seiko Bridgestone lithium polymer coin cell.

The most important rechargeable lithium batteries are those using a solid positive electrode within which the lithium ion is capable of intercalating. These intercalation, or insertion, electrodes function by allowing the interstitial introduction of the Li^+ ion into a host lattice. A large number of inorganic compounds have been investigated for their ability to function as a reversible positive electrode in a lithium battery. Intercalation electrodes have found wide application in systems employing both solid or liquid electrolytes.

High Temperature Systems

Lithium–Aluminum/Metal Sulfide Batteries. The use of high temperature lithium cells for electric vehicle applications has been under development since the 1970s. Advances in the development of lithium alloy–metal sulfide batteries have led to the Li–Al/FeS system, where the following cell reaction occurs.

$$2\ LiAl_x + FeS \leftrightarrows Fe + Li_2S + 2x\ Al$$

The cell voltage is 1.33 V to give a theoretical energy density of 458 W·h/kg. The cell employs a molten salt electrolyte, most commonly a lithium chloride/potassium chloride, LiCl–KCl eutectic mixture. The cell is generally operated at 400–500°C. The negative electrode is composed of lithium–aluminum alloy, which operates at about 300 mV positive of pure lithium. The positive electrode is composed of iron sulfide mixed with a conductive agent such as carbon or graphite. Electrodes are constructed by cold pressing powder onto current collectors.

Development of practical and low cost separators has been an active area of cell development. Cell separators must be compatible with molten lithium, restricting the choice to ceramic materials. Early work employed boron nitride, BN, but a more desirable separator has been developed using magnesium oxide, MgO, or a composite of MgO powder–BN fibers.

Li–Al/FeS cells have demonstrated good performance under EV driving profiles and have delivered a specific energy of 115W·h/kg for advanced cell designs. Cycle life expectancy for these cells is projected to be about 400 deep discharge cycles. This system shows considerable promise for use as a practical EV battery.

A similar system under development employs iron disulfide, FeS_2, as the positive electrode. Whereas this system offers a higher theoretical energy density than does Li–Al/FeS, the FeS_2 cell is at a lower stage of development.

Sodium–Sulfur. The best known of the high temperature batteries is the sodium–sulfur, Na–S, battery. The cell reaction is best represented by the equation:

$$2\ Na + 3\ S \leftrightarrows Na_2S_3$$

occurring at a cell voltage of 1.74 V, to give a specific energy of 760 W·h/kg. The cell is constructed using a solid electrolyte typically consisting of β-alumina, β-Al_2O_3, ceramic, although borate glass fibers have also been used. These materials have high conductivities for the sodium ion. The negative electrode consists of molten sodium metal and the positive electrode of molten sulfur. Because sulfur is not conductive, a current collection network of graphite is required. The cell is operated at about 350°C.

The Na–S battery couple is a strong candidate for applications in both EVs and aerospace. Projected performance for a sodium–sulfur-powered EV van is shown in Table 1 for batteries having three different energies.

The Na–S system is expected to provide significant increases in energy density for satellite battery systems. In-house testing of Na–S cells designed to simulate midaltitude (MAO) and geosynchronous orbits (GEO) demonstrated over 6450 and over 1400 cycles, respectively.

Table 1. Electric Vehicle Battery Performance

	Battery			
Parameter	Lead–acid	Sodium–sulfur		
battery energy, kW·h	40.0	40.0	60.0	85.0
range, km	84.0	113.0	169.0	242.0
max payload, t	0.9	1.7	1.6	1.6
battery weight, kg	1250.0	330.0	424.0	580.0

Difficulties with the Na–S system arise in part from the ceramic nature of the alumina separator: the specific β-alumina is expensive to prepare, and the material is brittle and quite fragile. Separator failure is the leading cause of early cell failure. Cell failure may also be related to performance problems caused by polarization at the sodium/solid electrolyte interface. Lastly, seal leakage can be a determinant of cycle life. In spite of these problems, however, the safety and reliability of the Na–S system has progressed to the point where pilot plant production of these batteries is anticipated for EV and aerospace applications.

A battery system closely related to Na–S is the Na–metal chloride cell. The cell design is similar to Na–S; however, in addition to the β-alumina electrolyte, the cell also employs a sodium chloroaluminate, $NaAlCl_4$, molten salt electrolyte. The positive electrode active material consists of a transition metal chloride such as iron(II) chloride, $FeCl_2$, or nickel chloride, $NiCl_2$, in lieu of molten sulfur. This technology is in a younger state of development than the Na–S.

Miscellaneous Systems

Rechargeable cells employing aluminum, Al, as a negative electrode in room temperature molten salts have been investigated.

Redox flow batteries, under development since the early 1970s, are still of interest primarily for utility load leveling applications. Unlike other batteries, the active materials are not contained within the battery itself but are stored in separate tanks. The reactants each flow into a half-cell separated from the other by a selective membrane. An oxidation and reduction electrochemical reaction occurs in each half-cell to generate current. Examples of this technology include the iron–chromium, Fe–Cr, battery and the vanadium redox cell.

Other flow batteries investigated for both electric vehicle applications and utility load leveling include zinc–chlorine, $Zn–Cl_2$, and zinc–bromine, $Zn–Br_2$, batteries.

Economic Aspects

As of this writing, there is little commercialization of advanced battery systems.

Efforts to develop commercially viable EV versions of advanced battery systems continue. The ultimate goal is to develop battery technology suitable for practical, consumer-acceptable electric vehicles. The United States Advanced Battery Consortium (USABC) has been formed with the express purpose of accelerating development of practical EV batteries.

PAUL R. GIFFORD
Ovonic Battery Company

Y. Matsuda and C. Schlaikjer, eds., *Practical Lithium Batteries*, JEC Press, Inc., 1988.

J. Bockris and co-eds., *Comprehensive Treatise of Electrochemistry*, Vol. 3, Plenum Press, New York, 1981.

J. L. Sudworth and A. R. Tilley, *The Sodium Sulfur Battery*, Chapman and Hall Ltd., U.K., 1985, and references therein.

D. Pletcher, *Industrial Electrochemistry*, Chapman and Hall Ltd., U.K., 1984, pp. 272–274.

BEARING MATERIALS

Economic Aspects

Production trends for bearings and bearing materials closely parallel general industrial activity.

Ball and roller bearings represent the largest business segment, with worldwide production estimated at $14 billion in 1988. Except for rolling-element bearings, only a few bearing types are of general commercial significance. U.S. shipments of plain bearings were $375 million in 1987. Powdered metal bearing production is expected to be about $63 million in 1993, jewel bearings $13 million, and wood $14 million. Production of air bearings is expected to increase 20% annually and reach $76 million in 1993 for use in light-load, high speed applications such as air circulators in aircraft, lasers, and dental drills. Environmental concerns have resulted in diminished use of lead in recent years in lead babbitt, porous metal bearings, and related bearing materials.

Some other bearing materials for which production volume is less well defined also find extensive use. Filled plastics such as nylon, acetal resin, PTFE, and phenolics are formed and molded into bearings in a wide variety of mechanical structures. Tin, lead, and bronze alloys are used for oil-film bearings in heavy industrial and power-generating equipment, frequently in custom bearings manufactured directly as machine components.

Distinctive Property Requirements

Friction, Wear, and Compatibility. Even bearings operating primarily with full oil-film lubrication may rub the shaft during starting and stopping, at initial run-in, under high transient loads, and during interruption of lubricant supply. During this sliding contact, the bearing material must avoid either welding to the shaft or scoring and galling under the localized high surface strains and high temperature at microscopic asperities.

Comprehensive tests for compatibility of metallic elements rubbing against a common low carbon shaft steel gave the results shown in Figure 1. Adding more of the good element, eg, lead in a copper alloy, generally improves score resistance, whereas adding more of the poor metal zinc will degrade score resistance.

Various plastics and other nonmetallics also provide excellent compatibility, low friction, low wear, and good scoring resistance. Their application is usually limited to slow surface speeds, however, where their low thermal conductivity does not lead to overheating.

Ease of Embedding and Conformability. When a shift surface forces dirt, machining chips, or grinding debris against a bearing, the bearing material is required to absorb the foreign particles to minimize scoring and wear. Experimental observations rank materials in the same order as anticipated from Table 1, with low modulus of elasticity giving good embedability. The soft babbitts are unsurpassed for both embedability and conformability.

Strength. An alloy too low in strength is prone to extrude under load, whereas too high strength may be accompanied by brittleness, poor embedding of foreign particles, and inability to conform to misalignment. In general, bearing surface hardness should be no greater than 1/3 to 1/2 the journal hardness to avoid self-propagating shaft scoring.

Corrosion Resistance. Materials containing lead, cadmium, copper, and zinc are susceptible to corrosion by the organic acids and peroxides formed in lubricating oil during its oxidation in service. This difficulty can be minimized by selecting oils with good oxidation inhibitors, by keeping the operating temperature low, and with periodic oil changes.

Thermal Properties. Conducting frictional heat out through a bearing can be a significant requirement, particularly for high speeds. For operation over a range of temperatures, matching thermal expansion coefficients of the bearing and shaft is important to maintain suitable clearance.

Good	Fair	Poor	Very poor	
germanium	carbon	magnesium	beryllium	molybdenum
silver	copper	aluminum	silicon	rhodium
cadmium	selenium	copper	calcium	palladium
indium	cadmium	zinc	titanium	cerium
tin	tellurium	barium	chromium	tantalum
antimony		tungsten	iron	iridium
thallium			cobalt	platinum
lead			nickel	gold
bismuth			zirconium	thorium
			columbium	uranium

Figure 1. Score resistance of elements against 1045 steel.

Table 1. Physical Properties of Sliding Bearing Materials

Material	Hardness[a]	Specific gravity	Tensile strength, MN/m^2[b]	Modulus of elasticity, GN/m^2[b]	Thermal conductivity, W/(m·K)	Coefficient of expansion $10^{-6}/°C$
Metals						
lead babbitt	21	10.1	69	29	24	25
tin babbitt	25	7.4	79	52	55	23
copper lead	25	9.0	55	52	290	20
silver	25	10.5	160	76	410	20
cadmium	35	8.6		55	92	30
aluminum alloy	45	2.9	150	71	210	24
lead bronze	60	8.9	230	97	47	18
tin bronze	70	8.8	310	110	50	18
zinc alloy	95	5.1	320	79	125	
steel	150	7.8	520	210	50	12
cast iron	180	7.2	240	160	52	10
Porous metals						
bronze	40	6.4	120	11	29	16
iron	50	6.1	170		28	10
aluminum	H55[c]	2.3	100		137	23
Plastics						
nylon	M79[c]	1.14	79	2.8	0.24	80
acetal	M94	1.42	69	2.8	0.22	80
PTFE[d]	D60[e]	2.17	21	0.6	0.24	130
phenolic	M100	1.36	69	6.9	0.28	40
polyester	D78[e]	1.45	59	2.3	0.19	95
polyimide	E52[e]	1.43	73	3.2	0.36	50
Other nonmetallics						
carbon graphite	75[f]	1.7	14	14	9	5
wood		0.68	8	12	0.19	5
rubber	65[f]	1.2	10	0.04	0.16	77
tungsten carbide	A91[f]	14.2	900	560	70	6
silicon nitride	1430[g]	3.2		310	17	3
Al_2O_3	2500[h]	3.9	210	340	24	8

[a] Brinell, unless otherwise noted.
[b] To convert MN/m^2 to psi, multiply by 145.
[c] Rockwell.
[d] Polytetrafluoroethylene.
[e] Shore Durometer.
[f] Shore Scleroscope.
[g] Shore A.
[h] Knoop.

Oil-Film Bearing Materials

Lubricant-film bearings primarily employ the white-metal babbitts and a variety of copper and aluminum alloys. Steel and cast iron structural parts are frequently used as oil-film bearing materials. Silver, zinc, and cadmium find limited use. For small bearings and bushings in light-duty and intermittent service, materials with self-lubricating properties are commonly used.

Dry and Semilubricated Bearing Materials

Porous bronze and iron, a variety of plastics, carbon–graphite, wood, and rubber are widely used in dry sliding or under conditions of sparse lubrication. These materials have commonly allowed design simplifications, freedom from regular maintenance, reduced sensitivity to contamination, and good performance at low speeds and with intermittent lubrication. Although these materials are often used dry or with sparse lubrication, performance normally improves the closer the approach to full-film lubrication.

High Temperature Materials

As the temperature limits for lubricating oils (150–250°C) and solid lubricants (350–400°C) are exceeded, bearing materials must accommodate either dry, low speed sliding or operate with very poor lubricants such as gas, pressurized water, or liquid metals. This involves new frontiers as continually higher temperatures are being encountered by bearings in gas turbines, diesel engines, automotive engines, superchargers, nuclear plant equipment, and rocket engines. Prototype testing is commonly required as a final step in bearing material selection.

High temperature strength often leads to selection of alloys of nickel, cobalt, and chromium for use from 500–850°C. Although wear of these materials is often high at ordinary temperatures, a transition temperature is encountered above which wear drops. Above this transition temperature, a smooth surface oxide forms with sufficient rapidity to eliminate significant metal-to-metal contact. Following are approximate transition temperatures for a number of metals:

steel 185°C
molybdenum 460°C
titanium 575°C
chromium 630°C
nickel 630°C

Ceramics (qv) find high temperature use to over 800°C. Advanced ceramics finding interest include alumina, partially stabilized zirconia, silicon nitride, boron nitride, silicon carbide, boron carbide, titanium diboride, titanium carbide, and sialon (Si–Al–O–N) (see ADVANCED CERAMICS, STRUCTURAL).

Desirable bearing material properties offered by ceramics are high compressive strength, fatigue resistance, corrosion resistance, low density, and retention of mechanical properties at elevated temperatures. Drawbacks include low fracture resistance, and difficulty in processing and fabrication. Use of nickel, cobalt, molybdenum, or chromium is often desirable for bonding ceramics in bearing materials to provide increased toughness, ductility, and shock resistance.

Plasma-sprayed, flame-plated, or electrolytically deposited coatings of powders of Al_2O_3, Cr_2O_3, TiN, WC, and TiO_2 can be applied as wear-resistant ceramics on metal substrates with or without Co, Ni, or Cr incorporated to improve mechanical properties. Silver, barium fluoride–calcium fluoride, and other modifying materials have also been found useful in ceramic coatings for improved friction and wear properties. Diamond coatings are also being developed.

Rolling Bearing Materials

Ball bearings are almost exclusively made with through-hardened materials such as the industry standard 52100 and stainless 440C. Case-hardened steels, commonly containing a lower carbon content of about 0.20%, are used for the rollers and races in many roller bearings for automobiles and railroad equipment to obtain better resistance both to shock load and to cracking with heavy interference fits during mounting on shafts.

Ceramic materials, and especially silicon nitride, Si_3N_4, are being applied in a variety of demanding applications. Additives of tungsten carbide, W_2C, and other metal oxides and carbides are useful in reducing ceramic wear.

Ceramic ball bearings are also sometimes effective in operation with water which would result in rapid failure with steel bearings.

Most ball and roller bearings use a retainer, also called a cage or separator, to properly space the balls or rollers between the stationary and rotating rings of the bearing. Since stresses on the retainer are normally low, low carbon strip steel has commonly been selected for simple manufacture at low cost. However, a continuing trend has been to polymeric retainers.

E. R. BOOSER
Consulting Engineer

C. D. Corte and H. E. Sliney, *ASLE Trans.* **30,** 77–83 (1987).

P. K. Bachmann and R. Messier, *Chem. Eng. News,* 24–39 (1987).

P. Eschmann, L. Hasbargen, and W. Weigand, *Ball and Roller Bearings,* John Wiley & Sons, Inc., New York, 1985.

E. R. Booser *Handbook of Lubrication,* Vol. 2, CRC Press, Boca Raton, Fla., 1984.

BEER

Beer is generally defined as an alcoholic beverage made by fermentation of a farinaceous extract that is obtained from a starchy raw material, barley, in the form of malt. Although it is possible to replace some part of the barley with other starchy materials (eg, corn, rice, wheat, oats, or potatoes), it is usually the main constituent of brewing materials. Between the ripe barley and the refreshing glass of beer, there are many production steps involving some of the most difficult disciplines of biochemical and physical science. Despite all the scientific knowledge developed in the last century, some brewing information remains culturally derived.

Various conditions such as tradition, taxation, and other peculiarities have resulted in the beer market of today, one characterized by numerous types varying in strength, color, alcohol content, and bitterness.

Bottom-Fermented Beers

Pilsner. Pilsner is a pale beer with a medium hoppy taste. It contains 3.9–4.7% by vol alcohol and, although traditionally lagered 2–3 mo, is now lagered 1–3 wk. The water for this type of beer is soft and contains a small amount of salt. Some 70–80% of all beer consumed in the world is of this light-larger type.

Dortmund. Dortmund is a pale beer with fewer hops than Pilsner but more body and taste. The alcohol content is 3.9–4.7% by vol and storage time is 3–4 mo. The brewing water is hard and contains large amounts of carbonates, sulfates, and chlorides.

Munich Beers. Munich beers are dark brown with a full-bodied, slightly sweet taste. They are only mildly hopped. The alcohol content is 3.2–6.3% by vol and storage time is 3–5 mo. The brewing water has many carbonates but only small amounts of other salts.

Top-Fermented Beers

Ale. Ale is pale with a pronounced hop taste and aroma, and 5.1–5.7% by vol alcohol. In Burton-on-Trent, England, where the best ales are made, the brewing water is rich in calcium sulfate; elsewhere it is usual to burtonize the water by adding calcium sulfate. There are two kinds of ale: pale (bitter) and mild.

Porter. Porter is a dark brown, full-bodied beer with a heavy foam. It is less hoppy and slightly sweeter in taste than ale; it contains 6.3% vol alcohol and is made with some dark or black malts.

Stout. Stout is a very dark beer with a sweet, slightly burned taste and a strong malt flavor. It is heavier than porter and is strongly hopped. It contains 6.3–8.3% by vol alcohol. Storage time is about six months and fermentation usually occurs in the bottle.

Lambic Beer. Lambic beer is made in Brussels and is one of the few top-fermented beers still brewed in Belgium. It is made from 60% barley malt and 40% unmalted wheat. The beer is strongly hopped and fermentation is spontaneous with wild yeasts, lactic acid bacteria, and bretanomyces. Fermentation and storage take place in the same vats and the beer is kept there for two or more years.

Low Alcohol Beers

The difficulty in making a beer with low alcohol content and a low degree of fermentation is its pronounced wort taste. The following measures can be used in making such beers: interruption of the fermenting process; scalding beer in the copper kettle; vacuum distillation; or reverse osmosis.

An improved beer quality has been obtained during recent years by using the modern and more advanced filtration techniques based on more and more sophisticated membrane-flux characteristics. By using the right membrane, pressure, and time it is possible to reduce the alcohol content of beer to lower levels with minimum flavor loss.

Properties of Beer

The properties of the finished beer vary with the type of beer and place of origin. The figures in Table 1 do not, however, show much about the quality of the beer; this can only partly be expressed in figures based on objective measurements. The quality consists of aroma, taste, appearance (color, clarity), formation, and stability of foam.

Nutritional Value of Beer. Beer is drunk primarily as a source of liquid and for its pleasant and refreshing taste; nonetheless, its nutritional properties are of great importance. The caloric content of beer is significant but not especially high. A 355-mL (12-oz) bottle of average beer (about 10.7°P = 1.045 original gravity) yields approximately 560 kJ (135 kcal). The calories in beer are provided by the unfermented residues and by the alcohol. The metabolic role of the latter is not fully understood but it replaces carbohydrates, fats, and proteins so that there may be a gain in body weight. Besides its caloric value, beer also contributes to the mineral requirements of the body and supplies useful quantities of B-complex vitamins (see VITAMINS).

A special use of beer is for the control of sodium intake in the treatment of diseases such as congestive heart failure, high blood pressure, and certain kidney and liver ailments.

Beer Defects. Among beer defects turbidity or haze is of primary importance. It may be either biological or physicochemical in nature.

Table 1. Properties of Beer

Property	Pilsner Urquell[a]	U.S. lager	Danish pilsner	English ale	English stout	Munich Löwenbrau	Dortmund
original wort extract, °P[b]	12.1	11.5–12.0	10.6	15.0	21.1	13.3	13.6
real extract content, °P[b]	5.3	5.5	3.1	5.0	8.7	6.4	5.5
alcohol content, vol %	4.4	4.3–4.7	4.7	6.6	7.2	4.4	3.8
protein content, wt %		0.28–0.35	0.3	0.6	0.8	0.5	0.6
CO_2 content, wt %		0.53	0.50	0.40	0.41		0.42
color, EBC[c]	10	2.7	5			40	8
air in bottle, mL		1.5	2	8	10		6
pH		4.25–4.50	4				
real degree of attenuation, %		60–75	69	66	59	48	60

[a] Urquell is a trade name belonging to Bürgerliches Brauhaus in Pilsen, Czechoslovakia.
[b] °P = °Plato, wt % extract (sugar).
[c] EBC = European Brewery Convention.

In a modern brewery the importance of oxygen and traces of metals is no longer a problem as far as haze-stability is concerned. On the other hand, oxygen is still of importance in considering taste and taste stability, because oxidized beers have a distinct cardboard flavor correlated to the amount of *trans*-2-nonenal, $C_9H_{16}O$, present in the beer (0.5–1.0 pbb).

The effect of sunlight, ie, the sunstruck flavor in beer, is caused by the formation of mercaptans.

Wild or gushing beer is a defect observed as a rather violent overfoaming from the bottle immediately after opening. This defect, however, does not affect the taste of the beer, but it is the most disastrous and complex problem a brewery might experience.

Brewing Materials

Barley. Barley is the predominant raw material of beer in most countries, except where other cereals are cultivated, eg, rice in China and kafir in Africa. Barley has technical advantages that make it superior for malting and brewing.

During the malting process the raw, hard, flat-tasting barley is changed into a crisp, mellow, sweet-tasting malt. The germ is only a small part of the barley kernel; the rest is a proteinaceous cell tissue filled with starch. However, starch and proteins are not directly soluble in water. The aim of malting is to bring forth enzymes, that will hydrolyze starch and proteins to less complex, water-soluble compounds, ie, amino acids, fermentable sugars, and small peptides. When these compounds are dissolved in water the resulting liquid is known as wort.

The character of the malt plays a large role in the resulting beer. The two extremes in malt character are reflected in Pilsner and Munich beers. For Pilsner, a pale malt with no pronounced aroma must be used; the drying of the malt takes 20–24 h at 80–90°C. Munich beer demands malts with a pronounced aroma; in this case the drying takes up to 48 h at 100–110°C.

New techniques, ie, genetic engineering or genetic manipulation, have been used in the field of barley improvement. The most interesting result has been a barley variety with little or no content of the tannin fraction called proanthocyanogen in the husks. This barley opens new possibilities for controlling the haze-stability problems mentioned earlier.

Adjuncts. Adjuncts in the form of cereals other than barley are often part of the brewing materials when the demand is for a stable, nonsatiating, sparkling beer. Since adjuncts are essentially starch with very little protein, they are a source of additional alcohol but contribute little to the color, taste aroma, or protein content of the beer. Corn, rice, brewing sugars, and soybeans are often used.

Hops. Hops are the blossoms of the female hop plant, wild in North and Middle Europe, North Asia, and North America.

Originally, the principal aim of adding hops was to compensate for the insipid, sweet taste of the unhopped beer with the characteristic bitter taste and aroma of hops. Other assets of adding hops include increasing the biological stability of the beer and improving the head retention and body of the beer. The amount of hops added varies from 0.4–4.0 g/L.

Brewers Yeast. Yeast for brewing is originally propagated from single-cell cultures and is recovered after fermentation for use throughout several generations. There are two types of yeast in brewing: top yeast, which forms spores, ferments vigorously at elevated temperature, and tends to float on top of the beer; and bottom yeast, which does not usually form spores, is well adapted to slow fermentation at low temperatures, and settles to the bottom of the tank at the end of fermentation (see YEASTS). There are two types of bottom-fermenting yeasts, named respectively after the weakly attenuating Saaz and the strongly attenuating Frohberg yeasts.

Perhaps the greatest preoccupation in relation to brewing yeasts is with stability, ie, whether a strain or mixture of strains remains unchanged and uncontaminated with prolonged use, and whether the range of performance of the constant strain or mixture of strains is small enough to provide adequate constancy of beer quality despite inevitable minor variations in wort composition. Various strains of yeast have individual flavor characteristics; thus brewers yeasts are not selected solely on the basis of fermenting power, but decidedly on the flavor they give beer.

Water. The character of the water has a great influence on the character of the beer, and the hardness of water (alkalinity) manifests itself by the extent of its reaction with the weak acids of the mash. Certain ions are harmful to brewing; nitrates slow down fermentation, iron destroys the colloidal stability of beer, and calcium ions give beer a purer flavor than magnesium or sodium ions.

Brewing water plays so large a role that some of the world's best known beer types, such as Pilsner, Munich, Dortmunder, and Burton Pale Ale, are special because of the properties of water used in their production.

Production

The most important phases during brewing are mashing, wort boiling, fermenting, lagering, filtering, and bottling or keging. During mashing, a mixture of finely crushed malt and warm water is exposed to enzymatic activity, thus converting starch into miscellaneous sugars and protein into peptides and amino acids. The dissolved product from mashing is called wort and the insoluble remainder (mostly husks of the malt) is called spent grain, which is sold as cattle fodder. The wort is boiled with hops and, during boiling, the enzymes are destroyed while bitter substances are extracted from the hops. Boiling of the wort also causes a certain amount of unconverted protein to coagulate and flocculate. This flocculation appears in the boiled wort as flocs; the brewer says that the wort has a fine break. After separation from the wort it is called a sludge.

After cooling of the wort to about 10°C, the yeast is added in order to convert the sugar into alcohol and carbon dioxide during fermentation. After fermentation most of the yeast is harvested, followed by a slow after-fermentation and maturing of the green beer in lagering tanks. By the end of the lagering there is a sediment in the tank, called draff, consisting of the remaining yeast together with precipitated proteins and tannins. During this process the temperature decreases from 5 to 1°C, or the beer may be transferred to conditioning tanks while cooling down to −1°C, and this temperature is held from two to six d.

The beer is filtered into pressure tanks and transferred to the bottling or keging machine. It is then pasteurized to avoid biological spoilage. The finished product is completed in five to six weeks for pilsner beer and six to nine weeks for the stronger beers.

Grinding of Malt. To extract the malt substances quickly and efficiently the malt must be crushed before being mixed with water. The degree of crushing depends on the method of separating the wort from the spent grains.

Treatment of Adjunct. Solid adjuncts, eg, corn grits or rice, contain high mol wt polysaccharides that are insoluble in water; they are not pretreated, ie, there is no germination or enzyme activation. Adjuncts must be treated to be accessible to attack by the malt enzymes, usually by boiling.

Mashing. Enzymatic breakdown of polysaccharides and proteins is actually started during malting, and mashing is a continuation of this breakdown, yielding the extract of the wort. It consists of dissolving water-soluble substances, enzymatic breakdown followed by solution of a series of substances important for the type and character of the beer, and separation of the undissolved substances. The breakdown during the mashing is regulated by time, temperature, pH, and concentration of the mash.

Separation of Mash. During mashing, all the valuable substances of malt and adjunct are broken down and dissolved. The mash contains sugars, nitrogenous substances, and husks. The undissolved part of the mash, called spent grains, is separated from the wort. Normally the strong wort is strained off first as completely as possible and the residual sugar is then sparged out using hot water of 75–78°C.

This separation process is time-consuming and consequently much effort has been spent to minimize the problem. Various practical ways have been invented, ie, the lauter tun, the mash filter, and the strainmaster.

Wort Boiling. The purpose of the boiling is to stabilize the wort microbiologically and enzymatically, to extract and isomerize the valuable substances of the hops to give the beer its characteristic taste and aroma, and to evaporate a certain amount of water to give the wort its desired density. During the boiling the wort is sterilized effectively, all enzymes are destroyed so that their breakdown cannot continue during fermentation, and the so-called hot break sludge, colloidal protein, coagulates and precipitates. This process is aided by the action of tannins from malt and hops on proteins.

Treatment of the Wort. The hot wort produced in the brewhouse cannot be transferred directly to the fermenting room. If natural hops are used they must be separated by a hop strainer. During boiling, protein–tannin complexes are precipitated in the form of warm sludge. After this separation the wort is cooled in heat exchangers to the desired temperature for the addition of yeast. During cooling, at ca 50°C, the cold break begins to precipitate and many breweries choose to separate this as well.

Fermentation. During fermentation, sugars are converted into alcohol and carbon dioxide by the yeast.

Fermentation is carried out in two different, very distinct ways: top fermentation and bottom fermentation. The governing principles are the same in both processes; the chief differences are in the type of yeast and temperature employed, and consequently, the method used for collecting the yeast after fermentation is finished. The alcohol content and, to a high degree, the taste and stability of the beer, are directly dependent on the normal progress of the fermentation.

New Developments Since the 1960s the fermentation of beer has undergone dramatic changes and developments, perhaps the most dramatic in the history of brewing so far.

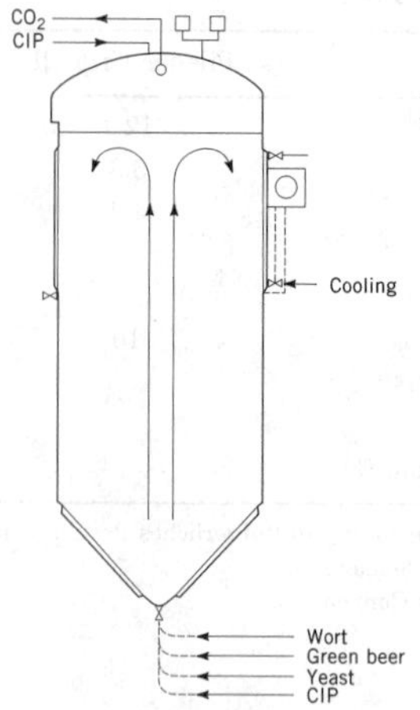

Figure 1. Modern fermenting tank with conical bottom; CIP (cleaning-in-place) = automatic cleaning system.

In the early 1970s open fermentors and the continuous fermenting systems were found to be obsolete. The batch process was going to survive, and many new fermentor constructions appeared. The cylindroconical fermentor seemed to be the preferred solution for both a single- and a combi-vessel fermentation system, ie, fermentation and lagering in the same vessel (Fig. 1).

Genetic manipulation or cloning offers many possibilities and perhaps there will be yeast strains especially designed for special beers, ie, types, which are useful because of low diacetyl formation, high–low ester formation, and insensitivity to pressure or high fermentation temperatures or extracellular enzymatic abilities (β-glucanases).

Beer Contaminants. Besides the culture yeasts, there are many wild yeasts, molds, and bacteria that may contaminate the beer. By constant cleaning, brewers try to keep to a minimum the number of harmful microorganisms. The infecting organisms are nonpathogenic because the low pH of the beer and antiseptic substances from the hops reduce the possibility of growth of pathogenic germs. The infecting organisms spoil the beer more or less according to the degree of infection in various ways, ie, increased acidity and off-flavors.

Foreign organisms capable of growing in beer are most often introduced through contamination of the inoculated yeast.

Lagering. The product of the primary fermentation process known as "green" beer is generally lacking in taste and stability. The taste is harsh and bitter, and has a peculiar yeasty aroma probably resulting from higher alcohols and aldehydes. Both the biological and the physico-chemical stabilities are unsatisfactory. This green beer must undergo a maturing process in which yeast and amorphous substances have sufficient time to settle to the bottom, the beer is saturated with carbon dioxide, taste and aroma are improved, and the chill-haze complex coagulates so that it can later be filtered off; at that point the beer must not have contact with air. Simple sedimentation does not make the beer satisfactory. To reach this goal the beer must be filtered.

The time for lagering varies with different types of beer. For every type of beer there is an optimal lagering time, and longer lagering is detrimental to beer quality.

Filtering. A high quality beer must be clear and totally sterile and have colloidal stability; yeast must be removed to allow the beer to have biological stability. The protein–tannin complexes must also be removed so as not to upset the colloidal stability. These substances are normally removed from the beer by filtration performed in different ways using different techniques, ie, pulp or mass, kieselguhr, sheet filtration, or a combination.

Packaging. The beer in pressure tanks is transferred to bottling, canning, and racking, or in some cases, to road tankers. During this filling operation it is important that the beer not come into contact

Table 2. Properties of Low Alcohol Beers

Property	Belgian beer	Maltibeer	Antimille	Hümmer	Alcohol reduced	Lunch beer
original gravity, °P[a]	6.4	8.1	7.7	6.9	7.3	11.3
alcohol, vol%	0.8	0.65	1.13	1.3	2.2	3.4
extract, real, °P[a]	5.2	7.1	5.9	4.9	3.9	6.3
degree of fermentation, %	23.2	15.2	28.9	36.4	58.0	55.0
pH	4.64	4.77	5.06	5.18		4.70
bitter substances, mg/L	12.2		25.6	36.4		
kJ/L[b]	1075	1370	1280	1140	1175	1850

[a] °P = °Plato, wt % extract (sugar).
[b] To convert kJ to kcal, divide by 4.184.

Table 3. Composition of Effluents

Process	Effluent content	BOD level
mashing	cellulose, sugars, amino acids, cleaning compounds	low
mash filtering	spent grains, sugars, amino acids, cleaning compounds	high
wort boiling	hops, wort, cleaning compounds	low
hop strainer	spent hops, wort, cleaning compounds	high
whirlpool	sludge, wort, cleaning compounds	high
fermentation	yeast, sludge, beer, cleaning compounds	high
lagering	yeast, protein, beer, cleaning compounds	high
beer filtering	kieselguhr, yeast, protein, beer, cleaning compounds	high
filling	beer, glass, crowns, cleaning compounds	high
bottle washing	beer, glass, labels, glue, oil, cleaning compounds	high

with oxygen, lose carbon dioxide, or be contaminated by molds, yeast, or bacteria.

New Developments. Changing the basic methods of production worries brewers because these changes may cause a deterioration in the quality of their product. Much has changed in the technology of beer making; however, the methods of transforming barley into beer have not undergone profound modification, because of the lack of fundamental knowledge.

The separation of hops from the boiled wort has been accelerated by the use of hop pellets or hop extract.

High gravity brewing involves the production of worts with high initial gravity, and the introduction of a substantial portion of water at as late a stage in the conventional brewing process as possible, usually after primary or final filtration. This technique is popular because brewhouse, fermenting, and storage facilities do not have to be enlarged for increased production and it has no negative effect on the quality of the product if the additional water volume is less than about 20%.

A further development in the process is the use of universal refrigerant-cooled tanks, designed to ferment, age, and finish beer in a single tank without the usual transfers.

In recent years, automated regulation of processes has steadily gained a foothold in the brewing industry. The aim has been to produce beer with a better, more even quality and at lower cost.

Since the 1970s an interesting evolution has been the concept of minibreweries or pub breweries, which began in the United States, especially in towns of a reasonable size without their own local brewery. In the large houses affiliated with chains or big independent pubs the installation of the so-called beer drive system, ie, beer supplied by tankers to relatively big cellar tanks combined with sophisticated beer dispensing systems, has been used successfully. The beer drive system might be seen as an alternative to the pub breweries originating prior to the pub brewery period.

Alcohol-Free Beer and Low Alcohol Beer. Many breweries are trying to produce beers with little or no alcohol, which also taste and smell like normal beer. The following methods have been suggested: interruption of fermentation, vacuum distillation, and reverse osmosis. Of these methods reverse osmosis is becoming more and more dominant, as far as reduced alcohol beers are concerned (Table 2).

Cleaning and Disinfection. Cleaning and disinfection of brewery equipment at every stage of the process is of vital importance. In breweries both chemical and bacteriological degrees of cleaning are nearly always obtained.

Service Processes of the Brewery

The brewing process consumes great amounts of energy in the form of electricity and heat. Heat is used either as steam or heated oil. High temperature oil systems are popular in newer constructions because they are easier to regulate and maintain and the investments are lower. The total heat consumption varies considerably from one brewery to another depending on equipment, method of production, and size. As in other industries, precautions are taken to minimize energy consumption. A well-managed brewery ($<1 \times 10^6$hL/yr) uses 216–252 MJ/L (60–70 kWh/hL = 7750–9050 BTU/U.S. gallons) total energy for heating, cooling, and lighting.

Electricity is often bought from the local distributors as high voltage current to 3–20 kV, which is transformed to normal working voltage of 380 V by the brewery's transformer.

Cooling is performed in a closed circuit where the cooling medium, ammonia, changes from liquid to gas. The gas phase is compressed from 100–200 kPa to 1–2 MPa (1–2 atm to 10–12 atm) and is cooled by water, and thus condensed and liquified.

Compressed air is used for various purposes. Air which might come into contact with the product must be free of odor, particles, and oil, have a pressure dew point at less than 2°C, and be sterile.

Carbon dioxide is used to increase the natural CO_2 content of the beer and as counterpressure in tanks and filling machines. It must be free of water and any aroma. The consumption is 1–10 g/L of beer produced.

Environmental Problems

Breweries must consider pollution of their effluent and also the availability of good quality incoming water; the price of the latter is going up rapidly. In the 1970s the beer: water ratio used was 1:10 or more, but today, in well managed breweries, it is 1:5–6, which is nearly impossible to improve further. The biochemical oxygen demand (BOD) level of effluent from breweries varies between 1000–2000 mg/L. Normal household sewage water has a BOD level of 600 mg/L; average BOD level exceeding 600 mg/L is charged extra. Thus every company has an economic interest in producing effluents at as low a BOD level as possible. Table 3 shows the composition of the effluent in various steps of the brewing process, and likewise the most critical points. To minimize the polluting substances in the effluent, and consequently the payment, several things can be done. Traditionally the by-products of the brewery, ie, spent grains, surplus yeast, and carbon dioxide, have been separated and used reasonably for other purposes, but this could be done with more care. By careful handling of all relevant methods, ie, beer transfers, cleaning, cleaning solutions, disinfectants, spent kieselguhr, and sludge, the BOD level can be reduced to below 1000 mg/L, and if the total water consumption has been reduced to 6 L per liter beer produced, this is probably the lower limit by traditional methods. Brewery effluents have favorable C/N and N/P ratios, ie, 8–10 times better than needed. In some cases the payment for pollution is so high that it becomes reasonable to build a plant for prepurifying the brewery effluent, and this has been done in several breweries.

In addition to considering the external environment, recommendations for the internal environment have been set. Every room in which human activity is required has maximum limits for noise (85–90 dB), carbon dioxide, solvents, radiation, temperature, etc.

J. F. NISSEN
The Danish Brewers' Association

H. M. Broderick, *The Practical Brewer,* Masters Brewer Association of the Americas, Madison, Wis., 1977.

J. R. A. Pollack, ed., *Brewing Science,* Academic Press, London, 1981.

D. E. Briggs, J. S. Hough, R. Stevens, and T. W. Young, *Malting and Brewing Science,* 2nd ed., Vols. 1 and 2, Chapman and Hall, Ltd., London, 1982.

Proceedings of the 22nd Congress of the European Brewery Convention, Zurich, IRL Press, Ltd., Oxford, U.K. 1989.

BENZALDEHYDE

Benzaldehyde, C_6H_5CHO, is the simplest and quite possibly the most industrially useful member of the family of aromatic aldehydes. Benzaldehyde exists in nature, primarily in combined forms such as a glycoside in almond, apricot, cherry, and peach seeds. The characteristic benzaldehyde odor of oil of bitter almond occurs because of trace amounts of free benzaldehyde formed by hydrolysis of the glycoside amygdalin.

Physical Properties

Physical properties of benzaldehyde are listed in Table 1.

Table 1. Physical Properties of Benzaldehyde

Property	Value
molecular formula	C_7H_6O
molecular weight	106.12
boiling point, °C at 101.3 kPa[a]	179
melting point, °C	−26
flash point, closed cup, °C	63
autoignition temperature, °C	192
refractive index, n^{20}	1.5455
viscosity, mPa·s(= cP) at 25°C	1.321
density, g/cm^3 at 25°C	1.046
specific heat (liquid) at 25°C, J/g·K[b]	1.615
latent heat of vaporization[c], J/g[b]	362
standard heat of combustion, kJ/g[b]	−31.9
solubility in water at 20°C, wt %	~0.6
solubility of water in at 20°C, wt %	~1.5

[a] To convert kPa to atm, divide by 101.3.
[b] To convert J to cal, divide by 4.184.
[c] At the boiling point (179°C).

Chemical Properties

Benzaldehyde is a versatile intermediate because of its reactive aldehyde hydrogen, its carbonyl group, and the benzene ring.

Manufacture

The only industrially important processes for the manufacturing of synthetic benzaldehyde involve the hydrolysis of benzal chloride and the air oxidation of toluene. The hydrolysis of benzal chloride, which is produced by the side-chain chlorination of toluene, is the older of the two processes. It is no longer utilized in the United States. Other processes, including the oxidation of benzyl alcohol, the reduction of benzoyl chloride, and the reaction of carbon monoxide and benzene, have been utilized in the past, but they no longer have any industrial application.

Economic Aspects

Benzaldehyde is produced in the United States by Kalama Chemical Incorporated, Kalama, Washington.

Specifications and Test Methods

Benzaldehyde is sold as technical grade or as meeting the specifications of the *National Formulary* (NF), the *Food Chemicals Codex* (FCC), or the *British Pharmacopeia* (BP). The test methods used for the analysis of benzaldehyde are standard methods, with the exception of the assay method.

Health Effects

The oral LD_{50} for benzaldehyde is reported as 1300 mg/kg in rats and as 1000 mg/kg in guinea pigs. Based upon these values, benzaldehyde is considered a moderately toxic substance when ingested. Studies of the carcinogenic effects of benzaldehyde are currently in progress. In the industrial setting, exposure to benzaldehyde through eye and skin contact and inhalation is far more prevalent than ingestion incidence. Overexposure to benzaldehyde vapors is irritating to the upper respiratory tract and produces central nervous system depression, with respiratory failure possible.

Safety and Handling

The low autoignition temperature of benzaldehyde (192°C) presents safety problems, since benzaldehyde can be ignited by exposure to low pressure steam piping, for example. Benzaldehyde may also spontaneously ignite when soaked into rags or clothing or adsorbed onto activated carbon.

Bulk storage of benzaldehyde should be made under a nitrogen blanket, since benzaldehyde is easily oxidized to benzoic acid upon exposure to air. All storage tank openings should be easily accessible for cleaning, since they will have a tendency to plug with benzoic acid. Benzaldehyde is stored in noninsulated type 304 stainless steel storage tanks.

Uses

Benzaldehyde is a synthetic flavoring substance, sanctioned by the U.S. Food and Drug Administration (FDA) to be generally recognized as safe (GRAS) for foods (21 CFR 182.60). Both "pure almond extract" and "imitation almond extract" are offered for sale. Each contains 2.0–2.5 wt % benzaldehyde in an aqueous solution containing approximately one-third ethyl alcohol.

Benzaldehyde is also recognized as safe for use as a bee repellant in the harvesting of honey (21 CFR 180.2), and is authorized for use in denatured alcohol (27 CFR 21.151).

Benzaldehyde's most important use is in organic synthesis, where it is the raw material for a large number of products. In this regard, a considerable amount of benzaldehyde is utilized to produce various aldehydes, such as cinnamic, methylcinnamic, amylcinnamic, and hexylcinnamic.

Derivatives

Derivatives include benzoin, 2-hydroxy-2-phenylacetophenone, $C_6H_5CH(OH)COC_6H_5$ (mp, 133–137°C; bp, 343–344°C at 101.3 kPa); benzil, diphenyl-α, β-diketone, $C_6H_5COCOC_6H_5$ (mp, 95°C; bp, 346–348°C at 101.3 kPa); benzyl alcohol, $C_6H_5CH_2OH$ (bp, 205.4°C at 101.3 kPa); benzoyl chloride, C_6H_5COCl (mp, −1 °C; bp, 197.2°C at 101.3 kPa); benzylamine, $C_6H_5CH_2NH_2$ (bp, 184°C at 101.3 kPa); benzylideneacetone, $C_6H_5CH{=}CHCOCH_3$ (bp, 260–262°C at 101.3 kPa; mp, 35–39°C); benzylacetone, $C_6H_5CH_2CH_2COCH_3$ (bp, 233–234°C at 101.3 kPa); dibenzylamine, $C_6H_5CH_2NHCH_2C_6H_5$ (bp, 300°C at 101.3 kPa); cinnamaldehyde, $C_6H_5CH{=}CHCHO$ (bp, 253°C at 101.3 kPa); α-methylcinnamaldehyde, $C_6H_5CH{=}C(CH_3)CHO$; α-amylcinnamaldehyde, $C_6H_5CH{=}C(C_5H_{11})CHO$ (bp, 140°C at 0.7 kPa); and α-hexylcinnamaldehyde, $C_6H_5CH{=}C(C_6H_{13})CHO$ (bp, 174–176°C at 2 kPa).

JARL L. OPGRANDE
C. J. DOBRATZ
EDWARD BROWN
JASON LIANG
GREGORY S. CONN
Kalama Chemical, Inc.

FREDERICK J. SHELTON
JAN WITH
Chatterton Petrochemicals Corporation

W. L. Faith, D. B. Keyes, and R. L. Clark, *Industrial Chemicals,* John Wiley & Sons, Inc., New York, 1965, pp. 120–124.

W. W. Kaeding and co-workers, *Indust. Eng. Chem. Proc. Des. Dev.* **4**(1), 97–101 (1965).

BENZANTHRONE. See DYES, ANTHRAQUINONE.

BENZENE

Benzene, C_6H_6, is a volatile, colorless, and flammable liquid aromatic hydrocarbon possessing a distinct, characteristic odor. Benzene is used as a chemical intermediate for the production of many important industrial compounds, such as styrene (polystyrene and synthetic rubber), phenol (phenolic resins), cyclohexane (nylon), aniline (dyes), alkylbenzenes (detergents), and chlorobenzenes. These intermediates, in turn, supply numerous sectors of the chemical industry producing pharmaceuticals, specialty chemicals, plastics, resins, dyes, and pesticides. In the past benzene has been used in the shoe and garment industry as a solvent for natural rubber. Benzene has also found limited application in medicine for the treatment of certain blood disorders, such as polycythemia and malignant lymphoma, and in veterinary medicine as a disinfectant. Benzene, along with other light high octane aromatic hydrocarbons such as toluene and xylene, is used as a component of motor gasoline. Although this use has been substantially reduced in the United States, benzene is still used extensively in many countries for the production of commercial gasoline. Benzene is no longer used in appreciable quantity as a solvent because of the health hazards associated with it.

Since the 1950s, benzene production from petroleum feedstocks has been very successful and accounts for about 95% of all benzene obtained. Less than 5% of commercial benzene is derived from coke oven light oil.

Benzene is the simplest and most important member of the aromatic hydrocarbons.

Table 1. Physical and Thermodynamic Properties of Benzene[a]

Property	Value
mol wt	78.115
freezing point, °C in air at 101.3 kPa[b]	5.530
boiling point, °C at 101.3 kPa[b]	80.094
density, g/cm^3	
20°C	0.8789
25°C	0.8736
vapor pressure, 25°C, kPa[c]	12.6
refractive index, n_D, 25°C	1.49792
surface tension, 25°C, mN/m (= dyn/cm)	28.20
viscosity, absolute, 25°C in mPa·s(= cP)	0.6010
critical temperature, °C	289.01
critical pressure, kPa[b]	4.898×10^3
critical volume, cm^3/mol	259.0
heat of formation	
g, kJ/mol	82.93
L, kJ/mol	49.08
heat of combustion, kJ/mol[d,e]	3.2676×10^3
heat of fusion, kJ/mol	9.866
heat of vaporization, 25°C, kJ/mol	33.899
solubility in H_2O, 25°C, g/100 g H_2O	0.180

[a] Courtesy of the Thermodynamics Research Center, The Texas A&M University System.
[b] To convert kPa to atm, divide by 101.3.
[c] To convert kPa to mm Hg, multiply by 7.5.
[d] To convert kJ to kcal, divide by 4.184.
[e] At 298.15 K and constant pressure to CO_2 and H_2O.

Physical Properties

The physical and thermodynamic properties of benzene are shown in Table 1. Benzene forms minimum-boiling azeotropes with many alcohols and hydrocarbons. Benzene also forms ternary azeotropes.

Structure. The representation of the benzene molecule has evolved from the Kekule ring formula (**1**) to the more electronically accurate (**2**), which indicates all carbon–carbon bonds are identical.

(**1**)

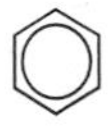

(**2**)

Resonance Stabilization. Benzene has great thermal stability. It has a lower heat of formation from the elements than the corresponding structure (**1**) possessing three fixed, ethylene-type double bonds. Similarly, when benzene is decomposed into carbon and hydrogen, it absorbs more energy than is predicted by the Kekule formula.

Chemical Properties

Benzene undergoes substitution, addition, and cleavage of the ring; substitution reactions are the most important for industrial applications.

Manufacture

Petroleum-derived benzene is commercially produced by reforming and separation, thermal or catalytic dealkylation of toluene, and disproportionation. Benzene is also obtained from pyrolysis gasoline formed in the steam cracking of olefins.

Table 2. U.S. Producers of Benzene from Petroleum and Annual Capacities[a]

Producers	Capacities, 10^3 t[b]
Continental United States	
American Petrofina Company, Inc.	214
Amoco Corp.	368
Aristech Chemical Corp.	150
Ashland Chemical Co.	191
British Petroleum America (incl. Sohio Oil Co.)	601
Champlin Refining and Chemicals, Inc.	183
Chevron Corp.	561
Citgo Petroleum Corp.	83
Coastal Corp.	116
Dow Chemical U.S.A.	835
Exxon Corp.	752
Hoechst-Celanese Corp.	50
Kerr-McGee Corp.	57
Koch Industries, Inc.	317
Lyondell Petrochemical Co.	474
Mobil Corp.	394
Occidental Petroleum Corp.	601
Phillips Petroleum	37
Salomon, Inc.	17
Shell Oil Co.	685
Sun Company, Inc.	301
Texaco Chemical	301
Unocal Corp.	63
USX Corp.	23
Virgin Islands and Puerto Rico	
Arochem International	301
Hess Oil Virgin Islands Corp.	257
Phillips Puerto Rico Core, Inc.	281
Total	*8063*

[a] As of Jan. 1, 1990.
[b] To convert t to gal, multiply by 300.

Economic Aspects

The United States is the largest producer of benzene and accounts for about 30% of world production. U.S. producers of benzene from petroleum and their approximate production capacities are shown in Table 2. These figures are inexact because the size of the market and instability of benzene prices causes frequent changes in capacity.

Specifications, Standards, and Test Methods.

Several different grades of benzene are commercially available. The most common grades are benzene 535, benzene 485 (nitration grade), benzene 545, and thiophene-free benzene. Specifications and the corresponding ASTM test procedures for these various types are shown in Table 3.

Handling and Shipping

Manufacturers of benzene are required by federal law to publish Material Safety Data Sheets (MSDS) that describe in detail the procedures for its safe handling. Benzene is classified as a flammable liquid and should be stored away from any potential source of ignition.

Benzene is shipped by rail tank cars, trucks, barges, and tankers. Because of the flammability, toxicity, and volatility of benzene, transfers from one vessel to another are conducted in closed systems. Benzene should be handled only where adequate ventilation is provided; protective clothing and self-contained respirators are recommended.

Environmental Considerations

Benzene is classified as a hazardous waste by the EPA under subtitle C of the Resource and Recovery Act (RCRA). Effective Sept. 25, 1990, solid wastes containing more than 0.5 mg/mL benzene must be treated in accordance with applicable RCRA regulations. Benzene is also subject to annual reporting of environmental releases as described in Section 313 of the Emergency and Community Right to Know Act of 1986. Benzene emissions and effluent streams from petroleum refineries or benzene processing plants are also subject to strict federal regulations. Federal waste management procedures must be complied with for any industrial process involving manufacture, transport, treatment, or disposal of benzene. A complete description of the new EPA regulations concerning benzene and other hazardous wastes is found in the *Federal Register.* Further information regarding the handling and disposal of toxic or hazardous wastes is in the CFR, Vol. 40.

Table 3. Specifications for Commercial Grades of Benzene

ASTM test	Benzene 535[a]	Benzene 485[b]	Industrial-grade[c]
appearance	clear liquid, free from sediment or haze at 18–20°C	clear liquid, free from sediment or haze at 18–24°C	clear liquid, free from sediment at 18–24°C
relative density, 14.56–15.56°C, D3505	0.8820–0.8860	0.8820–0.8860	0.875–0.886
density, 20°C, g/cm^3, D4052	0.8780–0.8820	0.8780–0.8820	0.871–0.882
color pt-co scale, D1209	20 max	20 max	20 max
total distillation range, 101.3 kPa[d], D850	1.0°C max, including the temperature of 80.1°C	1.0°C max, including the temperature of 80.1°C	2.0°C max, including the temperature of 80.1°C
solidification point, D852	5.35°C (anhydrous)	not lower than 4.85°C (anhydrous)	
acid wash color, D848	1 max	2 max	3 max
acidity, D847	none detected	none detected	none detected
H_2S and SO_2, D853	none detected	none detected	none detected
thiophene, D1685	1 mg/kg max		
copper corrosion, D849	pass	pass	copper strip shall not show iridescence, a gray or black deposit, or discoloration
nonaromatics, D2360	0.15 wt % max		

[a] ASTM D2359-85a.
[b] Nitration-grade, ASTM D835-85.
[c] ASTM D836-84.
[d] To convert kPa to mm Hg, multiply by 7.5.

Table 4. National Exposure Limits for Benzene

Country	Year	Concentration[a] ppm	Status
Australia	1978	10	guideline
Belgium	1978	10	regulation
Czechoslovakia	1976	160	regulation
		256[b]	
Finland	1975	10	regulation
Hungary	1974	64	regulation
Italy	1978	10	guideline
Japan	1978	25[c]	guideline
the Netherlands	1978	10	guideline
Poland	1976	64[c]	regulation
Romania	1976	160[d]	regulation
Sweden	1978	5	guideline
		10[e]	
Switzerland	1978	2	regulation
United States	1987	1	regulation
USSR	1980	16[c]	regulation
Yugoslavia	1971	15[c]	regulation

[a] TWA, unless otherwise noted.
[b] 10 min ceiling.
[c] Ceiling.
[d] Maximum.
[e] 15 min maximum.

Health and Safety

At room temperature and atmospheric pressure benzene is sufficiently vaporized to pose an inhalation hazard. Benzene is a toxic substance which can produce both acute and chronic adverse health effects. It is generally recognized that prolonged or repeated exposure to benzene can result in serious damage to the blood-forming elements.

Inhalation of 3,000 ppm benzene can be tolerated for 0.5–1 h; 7,500 ppm causes toxic effects in 0.5–1 h; and 20,000 ppm is fatal in 5–10 min. The lethal oral dose for an adult is approximately 15 mL. Repeated skin contact is reported to cause drying, defatting, dermatitis, and the risk of secondary infection if fissuring occurs.

In chronic benzene intoxication, mild poisoning produces headache, dizziness, nausea, stomach pain, anorexia, and hypothermia. In severe cases, pale skin, weakness, blurred vision, and dyspnea occur on exertion. Hemorrhagic tendencies include petechia, easy bruising, and bleeding gums. Bone marrow depression produces a decrease in circulating peripheral erythrocytes and leucocytes. Fatalities from chronic exposure show at autopsy severe bone marrow aplasia, and necrosis or fatty degeneration of the heart, liver, and adrenals.

Acute benzene poisoning results in CNS depression and is characterized by an initial euphoria followed by staggered gait, stupor, coma, and convulsions.

Treatment for acute exposure to benzene vapor involves removing the subject from the affected area, followed by artificial respiration with oxygen; intubation and cardiac monitors may be necessary for severe acute exposures. Because of its low surface tension, benzene poses a significant aspiration hazard if the liquid enters the lungs. Emesis is indicated in alert patients if more than 1 mL of benzene per kg of body weight has been ingested and less than two hours have passed between ingestion and treatment.

Treatment for chronic benzene poisoning is supportive and symptomatic, with chemotherapy and bone marrow transplants as therapeutic agents for leukemia and aplastic anemia.

Regulations

Because of the potential hazards associated with benzene, exposure to benzene in the workplace has been heavily regulated in the United States. Benzene is considered one of the approximately 40 known human carcinogens. Fifteen countries have been reported to limit occupational exposure to benzene by regulation or recommended guideline. These occupational exposure limits are shown in Table 4.

WILLIAM FRUSCELLA
Unocal Corporation

E. G. Hancock, *Benzene and Its Industrial Derivatives,* Halsted Press, a division of John Wiley & Sons, Inc., New York, 1975, p. 55.

R. A. Meyers, ed., *Handbook of Petroleum Refining Processes,* McGraw-Hill, New York, 1986.

J. M. Wakim and Z. Sedaghat-Pour, *Chemical Economics Handbook Marketing Research Report BENZENE,* Stanford Research Institute, Menlo Park, Calif., Jan. 1990.

J. D. Graham, L. C. Green, and M. J. Roberts, *In Search of Safety: Chemicals and Cancer Risk,* Harvard University Press, Cambridge, Mass., 1988, Chapt. 5, pp. 115–150.

BENZOIC ACID

Benzoic acid, C_6H_5COOH, is the simplest member of the aromatic carboxylic acid family.

In the United States, virtually all benzoic acid is manufactured by the continuous liquid-phase air oxidation of toluene. Benzoic acid and its salts and esters are very useful, finding application in medicinals, food and industrial preservatives, cosmetics, resins, plasticizers, dyestuffs, and fibers.

Occurrence

Benzoic acid in the free state, or in the form of simple derivatives such as salts, esters, and amides, is widely distributed in nature.

Properties

Selected physical properties of benzoic acid are given in Table 1.

In its chemical behavior benzoic acid shows few exceptional properties; the reactions of the carboxyl group are normal, and ring substitutions take place as would be predicted.

Economic Aspects

The growth of demand for benzoic acid is expected to increase at a rate of between 1 and 2% per year. Glycol dibenzoate plasticizers have been growing at close to 10% annually for several years, in part due to environmental concerns with regard to phthalate plasticizers (qv). The growth of the diet soft drink market has increased the demand for sodium and potassium benzoates.

All of the benzoic acid producers in the United States employ the liquid-phase toluene air oxidation process. If the attractiveness of toluene as an octane booster continues, the cost of producing benzoic acid will most likely increase.

The principal North American producers of benzoic acid and their estimated production capacities as of the early 1990s are Kalama Chemical, Kalama, Wash. 80×10^3t/yr and Velsicol Chemical, Chattanooga, Tenn., 32.5×10^3t/yr.

The bulk of this benzoic acid production capacity is consumed internally by the producers. Kalama converts over half of its production to phenol. A large portion of Velsicol's benzoic acid production is utilized in the manufacture of glycol dibenzoate plasticizer esters.

Specifications, Analysis, Packaging, and Shipment

Benzoic acid is available in industrial and technical grades, and in grades meeting the specifications of the *United States Pharmacopeia,* the *Food Chemicals Codex,* or the *British Pharmacopeia.*

Table 1. Physical Properties of Benzoic Acid

Parameter	Value
molecular formula	$C_7H_6O_2$
mp, °C	122.4
bp, at 101.3 kPa[a] °C	249.2
density	
solid, d_4^{24}	1.316
liquid, d_4^{180}	1.029
refractive index, n_D[b], liquid	1.504
viscosity at 130°C mPa·s(= cP)	1.26
surface tension at 130°C, mN/m (= dyn/cm)	31
specific heat, J/g[c]	
solid	1.1966
liquid	1.774
heat of fusion, J/g[c]	147
heat of combustion, kJ/mol[c]	3227
heat of formation at 26.16°C, kJ/mol[c], solid	−385
heat of vaporization,	
at 140°C, J/g[c]	534
at 249°C, J/g[c]	425
dissociation constant, K_a, at 25°C	6.339×10^{-5}
flash point, °C	121–131
autoignition temperature, °C, in air	573
pH of saturated aqueous solution at 25°C	2.8

[a] To convert kPa to atm, divide by 101.3.
[b] At 131.9°C.
[c] To convert J to cal, divide by 4.184.

Toxicity, Safety, and Handling

Benzoic acid's toxicity is rated as moderate (3 on a scale of 1–6) based upon its LD_{50} (oral–rat) of 2530 mg/kg. Healthy individuals may tolerate small doses (under 0.5 g of benzoates per day) mixed with food without ill effects. Manufacturers' product and information bulletins provide an excellent source for information regarding the safety and handling of benzoic acid.

The principal safety concern in handling molten benzoic acid is its elevated temperature. Thermal burns may result from improper handling of the molten product.

Benzoic Acid Derivatives

Benzoic acid derivatives include benzoyl chloride (C_6H_5COCl, mp, −1°C; bp, 197.2°C at 101.3 kPa); benzoic anhydride ($(C_6H_5CO)_2O$, mp, 42°C; bp, 360°C at 101.3 kPa); benzoic acid salts [ammonium benzoate ($C_6H_5COONH_4$, mp, 198°C), sodium benzoate (C_6H_5COONa), and potassium benzoate (C_6H_5COOK)]; and benzoic acid esters [benzyl benzoate ($C_6H_5COOCH_2C_6H_5$, mp, 21°C; bp, 323–324°C at 101.3 kPa), butyl benzoate ($C_6H_5COOC_4H_9$, mp, −22°C; bp 250°C at atm = 101.3 kPa), ethyl benzoate ($C_6H_5COOC_2H_5$, mp, −35°C; bp, 212°C at 101.3 kPa) *n*-hexyl benzoate ($C_6H_5COOC_6H_{13}$, bp, 272°C at 103.9 kPa), methyl benzoate ($C_6H_5COOCH_3$, bp, 198–200°C at 101.3 kPa), and phenyl benzoate ($C_6H_5COOC_6H_5$, mp, 70–71°C; bp, 314°C at 101.3 kPa)].

JARL L. OPGRANDE
C. J. DOBRATZ
EDWARD E. BROWN
JASON C. LIANG
GREGORY S. CONN
Kalama Chemical, Inc.
JAN WITH
FREDERICK J. SHELTON
Chatterton Petrochemical Corporation

W. W. Kaeding, *Hydrocarbon Process.* **43,** 173 (1964).

W. W. Kaeding and co-workers, *I & EC Process Des. Dev.* **4**(1), 97 (Jan. 1965).

Hydrocarbon Process. **56**(11), 134 (Nov. 1977).

BENZYL ALCOHOL AND β-PHENETHYL ALCOHOL

Benzyl alcohol (**1**) and β-phenethyl alcohol (**2**) (2-phenylethanol) are the simplest of the aromatic alcohols, and, as such, are chemically similar. Their physical properties are given in Table 1.

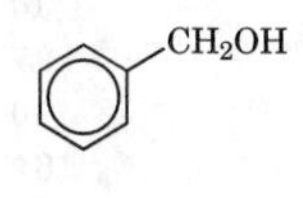

(**1**)

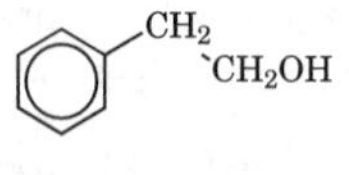

(**2**)

Benzyl Alcohol

Benzyl alcohol (**1**) occurs widely in essential oils both as the free alcohol, and, more importantly from a fragrance standpoint, in the form of various esters.

Although benzyl alcohol itself is rather bland in odor, combined with its much more fragrant esters it is an important part of the odor of jasmine, ylang-ylang, gardenia, some rose varieties, narcissus and peony, as well as castoreum, balsams of peru and tolu, and propolis. Benzyl alcohol occurs primarily in flower oils and tree exudates.

Benzyl alcohol readily undergoes the reactions characteristic of a primary alcohol, such as esterification and etherification, as well as halide formation. In addition, it undergoes ring substitution. Catalytic oxidation over copper oxide yields benzaldehyde; benzoic acid is obtained by oxidation with chromic acid or potassium permanganate.

Manufacture. Today benzyl alcohol is almost universally manufactured from toluene which is first chlorinated to give benzyl chloride. This is then hydrolyzed to benzyl alcohol by treatment with aqueous sodium carbonate.

World Consumption and Uses. In the soap, perfume, and flavor industries benzyl alcohol is primarily used in the form of its aliphatic esters. Benzyl alcohol is commercially available in five grades.

Table 1. Physical Properties of Benzyl Alcohol and β-Phenethyl Alcohol

Property	Benzyl alcohol	β-Phenethyl alcohol
molecular formula	C_7H_8O	$C_8H_{10}O$
mp, °C	−15	−25.8
bp at 101.3 kPa[a]	205.4–205.7	219.5–220
bp at 1.33 kPa[a], °C	89.0–89.5	99–100
d_{25}^{25}	1.0441	1.017
flash point, closed cup, °C	100.4	
open cup, °C	104.4	
autoignition temp, °C	436	
vapor density (air = 1)	3.7	
vapor pressure at 58°C, kPa[a]	0.133	0.133
100°C, kPa[a]	2.02	1.33
surface tension at 20°C, mN/m(= dyn/cm)	39	
80°C, mN/m(= dyn/cm)	33	
viscosity at 25°C, mPa·s(= cP)	5.05	7.58
solubility at 25°C, water	1 g/25 mL	1 g/51 mL
50% ethanol	1 g/1.5 mL	1 g/1.7 mL

[a] To convert kPa to mm Hg, multiply by 7.5., °C

The largest proportion of benzyl alcohol is for use in the photographic and textile industries although the latter use has been declining. The textile grade is used as a dyeing assistant for wool and nylon.

The pharmaceutical industry makes use of benzyl alcohol's local anesthetic, antiseptic, and solvent properties. It is used in nail lacquers and as a color developer in hair dyes by the cosmetics industry, and in acne treatment preparations.

Because of its strong polarity and limited water solubility, the technical grade of benzyl alcohol is used in rug cleaners as a degreasing agent, in leather dyeing, in ballpoint inks, as a cleaner for soldering, and as an extractive distillation solvent for xylenes and cresols. It is used as a stabilizer in insecticidal formulations by the agriculture industry and in treating fruits and vegetables. In addition, benzyl alcohol is used extensively in the polymer industry and in the manufacture of automobile tires.

Health and Safety. This material has a Generally Recognized As Safe (GRAS) status indicated by the Flavor and Extract Manufacturers' Association for use in flavors and by the Council of Europe for use as a flavor. Benzyl alcohol satisfies the most current guidelines published by the International Fragrance Association (IFRA) which governs the use of fragrance materials.

β-Phenethyl Alcohol

Of all the aromatic organic molecules β-phenethyl alcohol (PEA) (**2**) is probably the most prestigious aroma chemical in the world of perfumery. This is because of its exquisite odor of natural rose petals.

Physical properties of PEA are shown in Table 1. The compound undergoes the usual chemical reactions of alcohols or aromatic compounds.

In insect control, PEA has been considered as a mosquito repellant, and its acetate has been used as an ingredient in Japanese beetle bait. The alcohol also has bacteriostatic action and antifungicidal properties, and it has been claimed as a surface-active agent.

Because of factors of low cost, stability, and odor quality, PEA is ideally suited for use in bar soap fragrances, where its use can be up to 30–50% of the fragrance.

Manufacture. Current commercial methods for making PEA include Grignard synthesis, Friedel-Crafts process, and catalytic hydrogenation of styrene oxide.

Future methods of production include catalytic air oxidation and microbiological oxidation.

Purification. Purification problems are primarily solved by two methods: continuous vacuum fractionation and chemical combination to yield a high boiling ester, separation of the noncombining impurities by distillation, and hydrolysis of the ester.

World Consumption. Approximately 85% of the PEA is employed for fragrance use.

Health and Safety. The use of β-phenethyl alcohol generally presents no health problems. This material has Generally Recognized As Safe (GRAS) status as indicated by the Flavor and Extract Manufacturers Association and is approved by the U.S. Food and Drug Administration and the Council of Europe for use in flavors. PEA satisfies the most current guidelines published by the International Fragrance Association (IFRA), which governs the use of fragrance materials.

BRAJA D. MOOKHERJEE
RICHARD A. WILSON
International Flavors & Fragrances

P. Z. Bedoukian, *Perfumery and Flavoring Synthetics,* 2nd revised ed., Elsevier Publishing Co., New York, 1967, p. 55.

A. Bujold, *Chem. Econ. Handbook Program,* Stanford Research Institute, Menlo Park, Calif., 1990.

BERYLLIUM AND BERYLLIUM ALLOYS

Beryllium, Be, specific gravity = 1.848 g/mL, and mp of 1287°C, is the only light metal having a high melting point. The majority of the beryllium commercially produced is used in alloys, principally copper–beryllium alloys (see COPPER ALLOYS). The usage of unalloyed beryllium is based on its nuclear and thermal properties, and its uniquely high specific stiffness, ie, elastic modulus/density values. Beryllium oxide ceramics (qv) are important because of the very high thermal conductivity of the oxide while also serving as an electrical insulator. The only commercial extraction plant operating in the Western world is that of Brush Wellman at Delta, Utah using both beryl and bertrandite ores as input.

Occurrence

The beryllium content of the earth's surface rocks has been estimated at 4–6 ppm. Although ca 45 beryllium-containing minerals have been identified, only beryl and bertrandite are of commercial significance.

Properties

A summary of physical and chemical constants for beryllium is compiled in Table 1. One of the more important characteristics of beryllium is its pronounced anisotropy resulting from the close-packed hexagonal crystal structure. This factor must be considered for any property that is known or suspected to be structure sensitive.

At ambient temperatures beryllium is quite resistant to oxidation; highly polished surfaces retain their brilliance for years.

Beryllium is susceptible to corrosion under aqueous conditions, especially when exposed to solutions containing the chloride ion. It is rapidly attacked by seawater. Protective systems used for beryllium include chromic acid passivation, chromate conversion coatings, chromic acid anodizing, electroless plating (qv), and paints.

Beryllium reacts readily with sulfuric, hydrochloric, and hydrofluoric acids. It reacts with fused alkali halides releasing the alkali metal until an equilibrium is established. It does not react with fused halides of the alkaline-earth metals to release the alkaline-earth metal.

Table 1. Physical and Chemical Properties of Beryllium

Parameter	Value
transformation pt, HCP to BCC, K	1527
mp, °C	1287
bp, °C	2472
density, g/mL	
at 298 K	1.8477
at 1773 K	1.42
heat of fusion, ΔH_{fus}, J/g[a]	1357
heat of sublimation, ΔH_s, kJ/g[a]	35.5–36.6
heat of vaporization, ΔH_v, kJ/g[a]	25.5–34.4
heat of transformation, J/g[a]	837
standard entropy, $S°$, J/(g·K)[a]	1.054
standard enthalpy, $H°$, J/g[a]	216
contraction on solidification, %	3
vapor pressure, MPa[b] at 500 K	5.7×10^{-29}
specific heat, J/(g·K)[a] at 298 K	1.830
thermal conductivity at 298 K, W/(m·K)	220
linear coefficient of thermal expansion, 278–333 K[c]	11.4×10^{-6}
electrical resistivity at 298 K, Ω · m	4.31×10^{-8}
reflectivity, %	
white light	50–55
infrared (10.6 μm)	98
sound velocity, m/s	12,600

[a] To convert J to cal, divide by 4.184.
[b] To convert MPa to psi, multiply by 145.
[c] Value is for unworked, isostatically pressed powder metallurgy metal.

Chemically, beryllium is closely related to aluminum, from which complete separation is difficult.

Ore Processing

Sulfate Extraction of Beryl. The Kjellgren-Sawyer sulfate process is used commercially for the extraction of beryl.

Extraction of Bertrandite. Bertrandite-containing tuff from the Spor Mountain deposits is wet milled to provide a thixotropic, pumpable slurry of below 840 μm ((−20 mesh) particles. This slurry is leached with sulfuric acid at temperatures near the boiling point. The resulting beryllium sulfate solution is separated from unreacted solids by countercurrent decantation thickener operations. Water conservation practices are essential in semiarid Utah, so the wash water introduced in the countercurrent decantation separation of beryllium solutions from solids is utilized in the wet milling operation by magnesium.

Production of Beryllium Metal

Reduction of Beryllium Fluoride With Magnesium. The Schwenzfeier process is used to prepare a purified, anhydrous beryllium fluoride, BeF_2, for reduction to the metal.

Electrolytic Processes. The electrolytic procedures for both electrowinning and electrorefining beryllium have primarily involved electrolysis of the beryllium chloride, $BeCl_2$, in a variety of fused-salt baths. The chloride readily hydrolyzes, making the use of dry methods mandatory for its preparation (see BERYLLIUM COMPOUNDS). For both ecological and economic reasons there is no electrolytically derived beryllium available in the marketplace.

Vacuum Melting and Casting. A vacuum melting operation is required for beryllium regardless of its origin.

Because beryllium is primarily used as a powder metallurgy product or as an alloying agent, casting technology in the conventional metallurgical sense is not commonly utilized with the pure metal.

Fabrication

Most beryllium hardware is produced by powder metallurgy techniques achieving fine-grained microstructure having a nearly random crystallographic orientation, thus providing a strong material with substantial ductility at room temperature. For some specialized applications, sheet and foil have been rolled from cast beryllium ingot. Such material exhibits an average grain size of 50–100 μm as compared to the typical 12 μm or less of the powder metallurgy products.

Beryllium powder is manufactured from vacuum-cast ingot using impact grinding or jet milling.

Hot-isostatic-pressing (HIP) is replacing the vacuum hot-pressing procedure for all but the largest shapes. Cold-isostatic-pressing followed by vacuum sintering or HIP is also used to manufacture smaller intricate shapes.

Beryllium sheet is produced by rolling powder metallurgy billets clad in steel cans at 750–790°C. Beryllium foil down to 12.5 μm (0.0005 in.) in gauge is commercially available.

Safe Handling

Beryllium, beryllium-containing alloys, and beryllium oxide ceramic in solid or massive form present no hazard whatsoever. Solid shapes may be safely handled with bare hands; however, inhalation of fine airborne beryllium may cause chronic beryllium disease, a serious lung disease in certain sensitive individuals. However, the vast majority of people, perhaps as many as 99%, do not react to beryllium exposure at any level. The biomedical and environmental aspects of beryllium have been summarized.

Safe Exposure Levels. The U.S. Occupational Safety and Health Administration (OSHA) has adopted workplace exposure limits designed to keep airborne concentrations well below the levels known to cause health problems.

Recycling. Beryllium is typically recycled; thus, it is not a waste disposal problem.

Uses

Beryllium is used extensively as a radiation window, both in source and detector applications, because of its ability to transmit radiation, particularly low energy x-rays. Although it is considered a moderator material for neutrons in nuclear reactors, its actual usage has been in nuclear weapons and as a neutron reflector in high flux test reactors.

Alloys Containing Beryllium

A small beryllium addition produces strong effects in several base metals. In copper and nickel this alloying element promotes strengthening through precipitation hardening. In aluminum alloys a small addition improves oxidation resistance, castability, and workability. Other advantages are produced in magnesium, gold, and zinc. Many other alloying compositions have been researched, but no alloy with commercial importance approaching these dilute alloys has emerged.

Copper–Beryllium Alloys. Wrought copper–beryllium alloys rank high among copper alloys in attainable strength and, at this high strength, useful levels of electrical and thermal conductivity are retained (see COPPER ALLOYS). Applications include uses in electronic components where their strength-formability-elastic modulus combination leads to use as electronic connector contacts; electrical equipment where fatigue strength, conductivity, and thermal relaxation resistance lead to use as switch and relay blades; control bearings where antigalling features are important; housings for magnetic sensing devices where low magnetic susceptibility is critical; and resistance welding systems where hot-hardness and conductivity are important in structural components.

A variety of copper–beryllium casting alloys exhibit strengths nearly as high as the wrought products with castability advantages. Casting alloys are used in molds and cores for plastic molding, and in undersea, aerospace, sports equipment, and jewelry applications by virtue of their high replication of detail in the investment casting process.

Nickel–Beryllium Alloys. Dilute alloys of beryllium in nickel, like their copper–beryllium counterparts, are age hardenable. Nickel–beryllium alloys are distinguished by very high strength; good bend formability in strip; and high resistance to fatigue, elevated temperature softening, stress relaxation, and corrosion. Wrought nickel–beryllium is available as strip, rod, and wire and is used in mechanical, electrical, and electronic components that must exhibit good spring properties at elevated temperatures. Examples include thermostats, bellows, as well as pressure-sensing diaphragms and burn-in connectors and sockets.

Beryllium in Aluminum Alloys. Small additions of beryllium to aluminum systems are known to improve consistency. When as little as 0.005–0.05 wt % beryllium is added as a master alloy to an aluminum alloy during melting, a protective surface oxide film is formed. Castings have improved surface finish, consistent strength, and higher ductility. Additional benefits cited include reduced tarnishing, improved buffing and polishing response, and consistency of aging response, particularly in alloys containing magnesium or silicon. Applications include aircraft skin panels and aircraft structural castings in alloy A357.

Beryllium and aluminum are virtually insoluble in one another in the solid state. The potential therefore exists for an aluminum–beryllium metal matrix composite with lower density and higher elastic modulus. At least one wrought composite system with nominally 62 wt % Be and 38 wt % Al has seen limited use in aerospace applications (see COMPOSITE MATERIALS).

A. JAMES STONEHOUSE
RAYMOND K. HERTZ
WILLIAM SPIEGELBERG
JOHN HARKNESS
Brush Wellman Inc.

D. Webster and G. J. London, eds., *Beryllium Science and Technology,* Vol. 1, Plenum Press, New York, 1979.

D. R. Floyd and J. N. Lowe, eds., *Beryllium Science and Technology,* Vol. 2, Plenum Press, New York, 1979.

M. D. Rossman, O. P. Preuss, and M. B. Powers, eds., *Beryllium–Biomedical and Environmental Aspects,* Williams & Wilkins, Baltimore, Md., 1991, 319 pp.

J. C. Harkness, W. D. Spiegelberg, and W. R. Cribb, "Beryllium–Copper and Other Beryllium-Containing Alloys," *Metals Handbook,* 10th ed., Vol. 2, *Properties and Selection,* ASM International, Metals Park, Ohio, 1990, 25 pp.

BERYLLIUM COMPOUNDS

Beryllium Carbide

Beryllium carbide, Be_2C, may be prepared by heating a mixture of beryllium oxide and carbon to 1950–2000°C, or heating a blend of beryllium and carbon powders to 900°C under mechanical pressure of 3.5–6.9 MPa (500–1000 psi). The metal–carbon reaction is easier to carry out and is accompanied by a substantial exotherm.

The melting point of beryllium carbide is 2250–2400°C. This compound is not used industrially, but Be_2C is a potential first-wall material for fusion reactors, one on the very limited list of possible candidates (see FUSION ENERGY).

Beryllium Carbonates

Beryllium carbonate tetrahydrate $BeCO_3 \cdot 4H_2O$, has been prepared by passing carbon dioxide through an aqueous suspension of beryllium hydroxide. Beryllium oxide carbonate is precipitated when sodium carbonate is added to a beryllium salt solution.

Soluble beryllium carbonate complexes are produced by dissolving beryllium oxide carbonate or hydroxide in ammonium carbonate. The solid beryllium oxide carbonate intermediates are obtained by a laboratory procedure for preparing pure beryllium salt solutions by reaction with aqueous mineral or organic acids.

Beryllium Carboxylates

The beryllium salts of organic acids can be divided into normal carboxylates, $Be(RCOO)_2$, and beryllium oxide carboxylates, $Be_4O(RCOO)_6$. The latter are prepared by dissolving beryllium oxide, hydroxide, or the oxide carbonate in an organic acid, followed by evaporation to give either a solid or an oily liquid. The oxide carboxylate is extracted using chloroform or petroleum ether and recrystallized from the solvent.

Beryllium Halides

The properties of the fluoride differ sharply from those of the chloride, bromide, and iodide. Beryllium fluoride is essentially an ionic compound, whereas the other three halides are largely covalent. The fluoroberyllate anion is very stable.

Beryllium fluoride, BeF_2, is produced commercially by the thermal decomposition of diammonium tetrafluoroberyllate, $(NH_4)_2BeF_4$.

Beryllium chloride, $BeCl_2$, is prepared by heating a mixture of beryllium oxide and carbon in chlorine at 600–800°C.

Beryllium bromide, $BeBr_2$, and beryllium iodide, BeI_2, are prepared by the reaction of bromine or iodine vapors, respectively, with metallic beryllium at 500–700°C. They cannot be prepared by wet methods. Neither compound is of commercial importance and special uses are unknown.

Beryllium Hydride

Beryllium hydride, BeH_2, is best prepared by the controlled pyrolysis of di-*t*-butyl beryllium, $C_8H_{18}Be$, at 200°C. Interest in beryllium hydride has centered on its potential use as a solid propellant rocket fuel. Theoretically, BeH_2 has the highest specific impulse of any fuel material except solid hydrogen.

Beryllium Hydroxide

Beryllium hydroxide, $Be(OH)_2$, exists in three forms. On addition of alkali to a beryllium salt solution to obtain a slightly basic pH, a slimy, gelatinous beryllium hydroxide is produced. Aging this amorphous product results in a metastable tetragonal crystalline form, which after months of standing transforms into a stable orthorhombic crystalline form. The orthorhombic modification is also precipitated from a sodium beryllate solution containing more than 5 g/L Be by hydrolysis near the boil.

Beryllium Intermetallic Compounds

Beryllium forms intermetallic compounds, referred to as beryllides, with most metals. They are usually prepared by a solid-state reaction of the blended powder constituents at about 1270°C (see METALLURGY, POWDER). The properties exhibited by some beryllides include excellent oxidation resistance, high strength at elevated temperature, good thermal conductivity, and low densities as compared with refractory metals and ceramic materials (see CERAMICS; REFRACTORIES).

The beryllides continue to be of interest for high temperature aerospace applications. The limited strain capacity of the materials, particularly at low temperatures,has thus far prevented actual use.

Beryllium Nitrate

Beryllium nitrate tetrahydrate, $Be(NO_3)_2 \cdot 4H_2O$, is prepared by crystallization from a solution of beryllium hydroxide or beryllium oxide carbonate in a slight excess of dilute nitric acid.

Beryllium Nitride

Beryllium nitride, Be_3N_2, is prepared by the reaction of metallic beryllium and ammonia gas at 1100°C.

Beryllium Oxalate

Beryllium oxalate trihydrate, $BeC_2O_4 \cdot 3H_2O$, is obtained by evaporating a solution of beryllium hydroxide or oxide carbonate in a slight excess of oxalic acid. Beryllium oxalate is important for the preparation of ultrapure beryllium hydroxide by thermal decomposition above 320°C. The latter is frequently used as a standard for spectrographic analysis of beryllium compounds.

Beryllium Oxide

Beryllium oxide, BeO, is the most important high purity commercial beryllium chemical. In the primary industrial process, beryllium hydroxide extracted from ore is dissolved in sulfuric acid. The solution is filtered to remove insoluble oxide and sulfate impurities. The resulting clear filtrate is concentrated by evaporation and upon cooling high purity beryllium sulfate, $BeSO_4 \cdot 4H_2O$, crystallizes. This salt is calcined at carefully controlled temperatures between 1150°C and 1450°C, selected to give tailored properties of the beryllium oxide powders as required by the individual beryllia ceramic fabricators.

High purity beryllium oxide powder is fabricated by classical ceramic-forming processes such as dry pressing, isostatic pressing, extrusion, tape casting, and slip casting. Additives consisting of the oxides of magnesium, aluminum, or silicon, or various combinations are frequently included in the ceramic mixes to improve the reproducibility of sintering and resultant properties.

Beryllia ceramics offer the advantages of a unique combination of high thermal conductivity and heat capacity with high electrical resistivity.

Beryllia ceramic parts are frequently used in electronic and microelectronic applications requiring thermal dissipation (see CERAMICS AS ELECTRICAL MATERIALS).

Economic Aspects

Beryllium is principally consumed in the metallic form, either as an alloy constituent or as the pure metal. Consequently, there is no industry associated with beryllium compounds except for beryllium oxide, BeO, which is commercially important as a ceramic material.

Safe Handling

Care must be taken in the fabrication and processing of beryllium products to avoid inhalation of airborne beryllium particulate matter such as dusts, mists, or fumes in excess of prescribed work place limits. Inhalation of fine airborne beryllium may cause chronic beryllium disease, a serious lung disorder, in certain sensitive individuals. However, most people, perhaps as many as 99%, do not react to beryllium exposure at any level (see BERYLLIUM AND BERYLLIUM ALLOYS).

A. JAMES STONEHOUSE
MARK N. EMLY
Brush Wellman Inc.

W. W. Beaver, in D. W. White and J. E. Burke, eds., *The Metal Beryllium,* American Society for Metals, Novelty, Ohio, 1955, pp. 570–598.

D. A. Everest, *The Chemistry of Beryllium,* Elsevier Publishing Co., Amsterdam/London/New York, 1964.

M. D. Rossman, O. P. Preuss, and M. B. Powers, eds., *Beryllium—Biomedical and Environmental Aspects,* Williams & Wilkins, Baltimore, Md., 1991, 319 pp.

BEVERAGE SPIRITS, DISTILLED

Government Regulations and Taxation

Distilled spirits and the industry they support have always been subject to heavy taxation. Not only has it been an excellent source of revenue for many governments throughout the world, but high taxes can also be rationalized as having an inhibitory effect on consumption.

In January 1991, the U.S. Federal Excise Tax on distilled spirits was raised to $3.56 per liter or $13.50 per proof gallon. In addition, many states have substantially increased the state excise taxes on distilled spirits.

In the United States, the Alcohol Tax Unit came into being with the repeal of Prohibition in 1933, and it became the Alcohol and Tobacco Tax division of the Internal Revenue Service in 1952. The Bureau of Alcohol, Tobacco, and Firearms (ATF), established in 1972, and the Department of the Treasury closely regulate the manufacture of distilled spirits.

Production and Consumption Patterns

Although total consumption peaked in 1981 at 1.7 individual adult consumption peaked in 1971 at 11.6 L per adult (Fig. 1). The decline has continued through 1990 and is projected to continue because of increasing taxes and changing attitudes and lifestyles. The consumption trend is toward lower proof, lighter alcoholic beverages covering an ever-widening spectrum of products, including premixed cocktails, cordials, creams, light whiskeys, and wine coolers. Although overall spirit consumption is declining, the premium segment, which includes whiskey, vodka, and flavored products, has recently shown growth, indicating a more selective, upscale consumption pattern.

Spirit Types

In spite of a decline over the past 50 years, whiskeys are still the most popular distilled alcoholic beverage group in the United States (Table 1). However, vodka consumption has increased significantly to 22% of total distilled spirits in 1990.

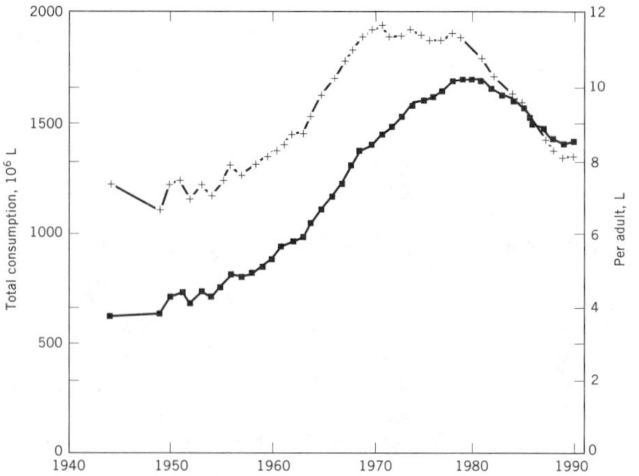

Figure 1. Consumption of distilled spirits from 1940 to 1990; ■ total; +, per adult.

Because of the economic interest in distilled spirits, each country has established standards for their various types of distilled beverages, and countries mutually respect each other's alcoholic beverage standards. U.S. Standards of Identity are given by the Bureau of Alcohol, Tobacco, and Firearms (ATF).

Within each type of distilled spirits, wide variations of flavor can be achieved by the type and amount of starting grains or other fermentable materials, methods of preparation, types of yeasts, fermentation conditions, distillation process, maturation time and temperature, blending, and use of new technologies such as membrane separation.

The flavor and aroma of distilled spirits are derived primarily from minor constituents called congeners that are produced and augmented in the fermentation and maturation processes. The congener profiles for various distilled spirits are shown in Table 2.

Canadian. By government regulation, Canadian whiskeys contain no distilled spirits less than three years old. They are usually blended products and are often up to six years of age. Canadian whisky tends to be light bodied and delicate in flavor. The Canadian government sets no limitations as to mashing formulas, distilling proofs, or types of cooperage used in maturation.

As in the United States, Canadians use corn, rye, and barley malt. Their process is essentially the same as the one used by many distilleries in the United States. Since they have no limitations on distillation proofs, distillers operate their systems for optimum separation and congener concentration. In addition, they are permitted to add blenders or flavoring components up to 9.06% by volume in the final blending after the aging process.

Scotch. In 1988, the Scotch Association Council approved a new, tighter definition for Scotch whisky which is as follows: "Scotch whisky is a potable spirit—

a) which has been produced in Scotland:

(i) from water and malted barley, with or without whole grains of other cereals, wholly processed at the distillery into a mash, converted to a fermentable substrate solely by the indigenous enzyme systems, fermented with the addition of yeast only;

(ii) by distillation of the wash obtained there at an alcoholic strength by volume of less than 94.8% in such a way that the distillate has an aroma and flavor derived from the said materials and process; and

Table 1. United States Liquor Consumption by Type, %

Type	1949	1960	1966	1978	1990	1993[a]
blends	66.2	31.3	24.3	10.4	5.9	5.5
bourbon	8.7	25.4	23.9	14.0	10.6	10.1
bonds	5.4	4.1	2.4	0.7	0.1	0.1
Scotch	4.4	8.1	10.4	12.0	8.5	7.1
Canadian	2.7	5.2	6.9	11.4	13.0	12.1
other	0.4	0.3	0.3	0.1	0.1	0.1
Total whiskey	*87.8*	*74.4*	*68.2*	*48.6*	*38.2*	*35.0*
gin	7.1	9.3	10.5	9.5	8.6	8.4
vodka	0.0	7.8	10.4	20.0	22.7	22.5
cordials	2.2	3.8	4.3	7.9	11.1	10.6
brandy	1.3	2.6	3.2	3.8	4.9	4.6
rum	1.3	1.6	2.2	5.8	8.3	8.0
tequila, cocktails, and other	0.3	0.5	1.2	4.4	6.2	10.9
Total nonwhiskey	*12.2*	*25.6*	*31.8*	*51.4*	*61.8*	*65.0*
Total consumption $\times 10^6$ L[b]	*641.5*	*888.3*	*1169.2*	*1786.1*	*1386.0*	*1349.8*

[a] Projected.
[b] To convert liters to gallons, divide by 3.785.

Table 2. Congeneric Content of Various Distilled Alcoholic Beverages[a]

Component	Canadian	Scotch	Bourbon whiskey	Kentucky whiskey	Cognac brandy	Tequila
fusel oil	53.0	105.0	199.0	195.0	193.0	195.0
total acids[b]	20.0	15.0	69.0	63.0	36.0	
esters[c]	4.6	11.4	23.0	18.0	41.0	12.9
aldehydes[d]	2.0	4.1	3.9	3.2	7.6	5.3
furfural	0.11	0.11	0.45	0.90	0.67	
total solids, g/100 mL	82.0	109.0	102.0	80.0	698.0	
tannins	18.0	8.0	52.0	48.0	25.0	21.0
color at 420 nm	5.4	5.6	9.5	8.0	11.0	

[a] Grams per 100 liters at 100° proof (50%). Determinations were made according to the official methods of analysis of the Association of Official Analytical Chemists, 15th ed., 1990.
[b] As acetic acid.
[c] As ethyl acetate.
[d] As acetaldehyde.

b) which has been matured in Scotland:

(i) in oak cask of a capacity not exceeding 700 liters;

(ii) for a period of not less than three years;

(iii) in excise warehouse, which for the purpose of this definition has the meaning assigned to it by Section One of the Customs, Excise Management Act 1979; and

c) which retains the color, aroma, and taste resulting naturally from the above process; and

d) to which no substances may be added other than:

(i) water, and

(ii) spirit caramel and

e) whose alcoholic strength apart from any natural evaporation losses may be reduced only by the addition of water to a bottling strength of not less than 40% alcohol by volume."

The single malt Scotch or malt Scotch, which has recently become popular in the United States, is made from a mash of only malted barley. Single malts are usually darker with heavier flavor than blended Scotches because of increased aging and the absence of the lighter grain whisky.

Irish Whiskey. Irish whiskeys are blends of grain and malt spirits three or more years of age that are produced in either the Republic of Ireland or Northern Ireland and comply with the respective laws regulating their manufacture. Since no peat is used in the malting process, Irish whiskey lacks the smokey character of Scotch.

United States Spirits. The manufacture of distilled spirits is tightly controlled within narrow limits. ATF regulations require that a detailed statement of the production process be submitted for approval prior to placing any process in operation.

Distilled beverages are classified according to type, materials, composition, distillation and maturation proofs, types of barrels, and maturation time.

Whiskey. Whiskey refers to any alcoholic distillate made from a fermented grain mash at less than 190° proof (95%) in such a manner that it possesses the taste, aroma, and characteristics generally attributed to whiskey. It is matured in new or used charred oak barrels.

Neutral Spirits. Neutral spirits are produced from any fermentable material and are distilled at or above 190° proof and bottled at 80° proof or higher. The substrate must be specified unless it is grain.

Grain Spirits. Grain spirits are neutral spirits from grain that are matured in used oak barrels and bottled at 80° proof or higher. The period of aging in oak may be declared on the bottle.

Vodka. Vodka is a neutral spirit made from any fermentable material and distilled in such a manner that is without any distinctive character, taste, aroma, or color. Charcoal filtration is often used in processing vodka which is bottled at 80° proof or higher. In the United States, the substrate must be specified if it is not grain. Any flavoring, if added, must be stated. The product must be bottled at not less than 70° proof and called a flavored vodka.

Light Whiskey. Light whiskey is distilled at not less than 160° proof and not more than 190° proof. It is matured in used charred-oak barrels or new uncharred barrels.

Bourbon. Bourbon, and also rye, wheat, malt, and rye malt whiskeys, are made from a fermented mash not less than 51% corn, rye, wheat, malt, or rye malt, respectively. They are distilled at not over 160° proof and matured at not more than 125° proof in new charred oak barrels and bottled at not less than 80° proof. If stored for less than four years, it must be declared on the label.

Corn Whiskey. Corn whiskey must be distilled from a fermentable mash that contains at least 80% corn and at not over 160° proof. It is usually matured in new uncharred oak barrels or used oak barrels and bottled at not more than 125° proof.

Straight Whiskey. Straight whiskey is distilled at not over 160° proof and barreled at not more than 125° proof. It must be matured for at least two years in new charred oak barrels and bottled at not less than 80° proof.

Sour mash fermentations must have not less than 20% stillage added back (backset) to the mash and be fermented for not less than 72 h. A lactic culture is used and is permitted to develop for a period of not less than 6 h.

Blended Whiskey. Blended whiskey is made with at least 20% of 100° proof straight whiskey either separately or in combination with whiskey or grain neutral spirits.

Tennessee Whiskey. Tennessee whiskey is a product made by Tennessee distillers and processed in a manner similar to bourbon. However, Tennessee whiskey is filtered through maple charcoal prior to maturing which gives it its distinctive flavor.

U.S. regulations define two types of gin; distilled gin and compounded gin. Distilled gin is produced from the original mash or the redistillation of neutral spirits with juniper berries and other botanicals. Compounded gin is produced by adding extracts of juniper berries and other botanicals to high proof neutral spirits. This gin is perceived to be of a lower quality than distilled gin and not much is produced by this method.

Other Types of Spirits. *Brandy.* Brandy is a distillate from fermented juice, mash, fruit wine, or fruit residues. It is distilled at less than 190° proof in such a manner as to produce the taste, aroma, and characteristics generally attributed to brandy. Fruit brandy is distilled solely from the fermented juice or mash of whole, ripe fruit or from standard grape, citrus, or other fruit wine.

Brandies are distilled using batch or continuous systems. Variations of the pot still are used in France. Elsewhere, both systems are used. The batch system yields a more flavorful product, whereas the continuous still yields a lighter flavor.

The most famous brandy comes from the Cognac region of France. In the United States about 95% of the brandy comes from California.

Rum. Rum is a distillate from the fermented juice of sugar cane, sugar cane syrup, molasses, sugar beets, or other by-products distilled at less than 190° proof in such a manner that it possesses the taste, aroma, and characteristics generally attributed to rum. It is bottled at not less than 80° proof. There are three types of rum: light or amber, full-bodied, and aromatic.

Tequila. Tequila is an alcoholic distillate produced in Mexico from the fermented juice of the heads of the Agave Tequilana Wever (blue variety) cactus. It is cultivated and takes 8–12 yr to mature.

Tequila is usually bottled at 80–86° proof. It is sold unaged as white tequila, or it can be matured in oak barrels. Aging gives Tequila a golden color and a pleasant mellowness without altering its basic taste.

Tequila can only be produced in an area of Mexico known as Tequila in the state of Jalisco, about from Guadalajara. It is called mezcal or maguey when produced outside of Tequila.

Cordials and Liqueurs. Cordials and liqueurs are the same products, with the different names being the American and European designations, respectively. They are produced by blending or redistilling neutral spirits, brandy, or other distilled spirits, with fruits, flowers, plants, juices, or concentrates, and other natural flavoring materials or extracts derived from infusions, percolations, or macerations of such materials. Cordials must contain a minimum of 2.5% (w/w) of sugar or dextrose or a combination of both. If the added sugar and dextrose are less than 10% (w/w), the cordial may be designated as dry. Most cordials contain larger amounts of sugar and other sweeteners. U.S. cordials containing synthetic or artificial flavoring materials must be labeled and are considered a spirits specialty.

Manufacturing Process

Ethyl alcohol, C_2H_6O, is produced by the fermentation of materials containing sugar or substances convertible to sugar, such as starches and fruit processing residues. Cereal grains are usually used in the production of beverage distilled spirits. Beverage alcohol is always ethyl alcohol, CH_3CH_2OH. Higher alcohols may be present in distilled spirits and are referred to as fusel oils or by specific name.

Composition of grains varies considerably and depends on such factors as climate, soil, and hybrid variety. Another variable is the malt; it is generally germinated barley, though rye malt or wheat malt can be used. The malting process develops the active enzymes (amylases) in the grain that convert grain starch into dextrins and then to maltose, a fermentable sugar. The malt and malting technique also can affect the final flavor and aroma of the alcohol, as in the case of Scotch whisky.

Grain Handling and Milling. Distilleries receive grain in either hopper railcars or trucks. It is usually transferred from the unloading pit by a pneumatic conveyor system or auger system. Milling breaks the outer cellulose protective wall around the kernel and exposes the starch to the cooking and conversion processes. Distillers require an even grind with as small a particle size as can be physically handled by the facility.

Milling is usually accomplished by two methods: hammer mills or cage mills.

Mashing. The mashing process consists of cooking the starch (gelatinization) and converting it (saccharification) to grain sugar (maltose). Cooking can be carried out at or above atmospheric pressure in either a batch or continuous system. After cooling, conversion is accomplished in the cooking vessel by adding barley malt or enzymes from other sources to the cooked grain. The converted mash is cooled and pumped to the fermentors.

Conversion. Primary conversion refers to the saccharification taking place during conversion and is in the order of 75–85% of the available starches. The remainder of the conversion to fermentable sugar takes place during the fermentation process and is referred to as secondary conversion.

Fermentation. The saccharified grain mash is cooled to around 20°C prior to setting the fermentor and inoculation with yeast. Selected yeast strains of *Saccharomyces cerevisiae* are used to inoculate the mash. Yeast (qv) metabolizes maltose and glucose sugars via the Embden-Meyerhof pathway to pyruvate, and via acetaldehyde to ethanol. All distillers' yeast strains can be expected to produce 6% (v/v) ethanol from a mash containing 11% (w/v) starch. secondary products (congeners) arise during fermentation and are retained in the distillation of whiskey.

Distillation. Distillation separates and concentrates the alcoholic products of yeast fermentation from the fermented grain mash. In addition to the alcohol and the desirable congeners, the fermented mash contains solid grain particles, yeast cells, water-soluble proteins, mineral salts, lactic acid, fatty acids, and traces of glycerol and succinic acid. Although a great number of different distillation processes are available, the most common systems used in the United States include: the continuous whiskey separating column (beer still), with or without an auxiliary doubler unit for the production of straight whiskeys; the continuous multicolumn system, used for the production of grain neutral spirits; and the batch rectifying column and kettle unit, used primarily in the production of grain neutral spirits that are subsequently stored in barrels for maturation purposes. In the batch and extractive distillation systems, the head and tail fractions are separated from the product resulting from the middle portion of the distillation cycle.

By-Products. After the removal of alcohol, the fermentation residues are processed to produce distillers grains. These residues consist of proteins, fats, minerals, vitamins, and fiber that are concentrated threefold by removal of the starch. Distillers grains are usually divided into one of four groups including distillers dry grains (DDG), distillers dry solubles (DDS), distillers dry grains with solubles (DDG/S), and condensed distillers solubles (CDS).

Maturation. The oak barrels used for aging distilled spirits play a significant role in determining the final aroma and flavor of the beverage. Newly distilled whiskey is colorless, grainy, and harsh. The new whiskey undergoes many types of physical and chemical change in the maturation process that smooth it out and give it character. These changes include extraction of the wood compounds, decomposition and diffusion of the wood macromolecules into the alcohol, reactions of the wood and distillate components with each other, and diffusion through the wood and evaporation of components.

Alcohol Reduction

Pervaporation. Vapor-arbitrated pervaporation is used to remove ethanol from whiskey by selective passage of the alcohol through a membrane. Whiskey flows on one side of a membrane. A water-vapor stream flows on the other side and sweeps away the ethanol that permeates the membrane. Thus, alcohol reduction and selective retention of flavor and aroma components can be achieved using membranes with a particular porosity. The ethanol can be recovered by condensing or scrubbing the vapor stream. Pervaporation systems operate at or slightly above atmospheric pressure.

Pervaporation has been successfully used both to reduce the alcohol content and to concentrate the congeners in bourbon whiskey.

Reverse Osmosis. A reverse osmosis (RO) process has been developed to remove alcohol from distilled spirits without affecting the sensory properties. It consists of passing barrel-strength whiskey through a permeable membrane at high pressure, causing the alcohol to permeate the membrane and concentrating the flavor components in the retentate.

Health and Safety Factors

Ethyl Carbamate. In November 1985, the Canadian Government indicated that it had detected ethyl carbamate (urethane), a suspected carcinogen, in some wines and distilled spirits. Since that time, the U.S. distilled spirits industry has mounted a serious effort to monitor and reduce the amount of ethyl carbamate (EC) in its products. In December 1985, the Canadian Government set limits of 150 ppb in distilled spirits and 400 ppb in fruit brandies, cordials, and liqueurs. The FDA accepted a plan in 1987 from the Distilled Spirits Council of the United States (DISCUS) to reduce ethyl carbamate in whiskey to 125 ppb or less, beginning with all new production in January 1989.

Distillers often employ a process that is in some ways unique to make various products and hence have tailored specific approaches to control and reduce ethyl carbamate that are peculiar to their own particular process. Some of the methods used are the use of copper packing in the rectifying section of stills, increased frequency of cleaning stills and other equipment, and using a cool-down period in the cleaning procedure. Increased rectification also reduces ethyl carbamate. Keeping the system clean is critical to minimizing the amount of ethyl carbamate.

Packaging

Packaging for distilled spirits intended for domestic distribution is regulated by the Federal Bureau of Alcohol, Tobacco, and Firearms. This strict supervision establishes acceptable container size, labeling, and sealing requirements, as well as guidelines for the disclosure of information on the shipping container. Furthermore, local and state distilled spirits' labeling and packaging requirements must also be met.

Flavor Applications

Flavoring beverage alcohol products presents some interesting challenges since ethyl alcohol itself has flavor. In high proof beverages (ie, over 80°), which contain pure alcohol, and not whiskey or rum that have inherent flavor characteristics, flavors must be added to help smooth out the harshness and singularity of the flavor profile. At proofs below 10°, the ethyl alcohol flavor is often below optimum sensory profiles.

Most flavors that are designed for beverage alcohol products use ethanol as the primary solvent for the flavor. Glycerol, propylene glycol, and water are other common solvents in liquid flavors.

Many beverage alcohol products depend heavily on the addition of compounded flavors, distillates, percolates, and extracts to carry the organoleptic profile of the product. Cordials, liquors, and schnapps at various proofs, such as crème de cacao, peppermint schnapps, fruit-flavored cordials and schnapps, and spirit coolers are examples. Beverage alcohol flavor applications require the consideration of many factors that affect the finished product. These include proof, inherent flavors, added flavors, source of ethanol, and overall composition.

Future Developments

The decline in distilled spirits consumption is likely to continue, but will be somewhat ameliorated in the increased consumer interest in high price premium products and the increased activity in the international markets.

The trend toward lower proof beverages will also likely continue because of new consumer preferences, cost reduction, and tax savings opportunities. Pressure to improve production efficiencies and lower costs will increase and new technology must play a greater role in this area.

JOHN E. BUJAKE
Brown-Forman Corporation

J. R. Piggott, R. Sharp, and R. E. B. Duncan, eds., *Science and Technology of Whiskies,* Longman Scientific and Technical/John Wiley & Sons, Inc., New York, 1989.

L. Bluhm, in H. J. Rehm and G. Reed, eds., *Biotechnology,* Vol. 5, Verlag Chemie, Basel, Switzerland, 1983, p. 447.

R. A. Lipinski and K. A. Lipinski, *Alcoholic Beverages,* Van Nostrand Reinhold, New York, 1989.

J. R. Piggott and A. Paterson, eds., *Distilled Beverage Flavour,* Ellis Horwood, Ltd., Chichester, U.K., 1989.

BIOCIDES. See INDUSTRIAL ANTIMICROBIAL AGENTS.

BIODEGRADABLE POLYMERS. See POLYMERS, ENVIRONMENTALLY DEGRADABLE.

BIOPOLYMERS

Biopolymers are the naturally occurring macromolecular materials that are the components of all living systems. There are three principal categories of biopolymers, each of which is the topic of a separate article in the *Encyclopedia:* proteins (qv); nucleic acids (qv); and polysaccharides (see CARBOHYDRATES; MICROBIAL POLYSACCHARIDES). Biopolymers are formed through condensation of monomeric units; ie, the corresponding monomers are amino acids (qv), nucleotides, and monosaccharides for proteins, nucleic acids, and polysaccharides, respectively. The term biopolymers is also used to describe synthetic polymers prepared from the same or similar monomer units as are the natural molecules.

In addition to being necessary for all forms of life, biopolymers, especially enzymes (proteins), have found commercial applications in various analytical techniques (see AUTOMATED INSTRUMENTATION, CLINICAL CHEMISTRY; AUTOMATED INSTRUMENTATION, HEMATOLOGY; BIOSENSORS; IMMUNOASSAY); in synthetic processes (see ENZYME APPLICATIONS, INDUSTRIAL; ENZYMES IN ORGANIC SYNTHESIS); and in prescribed therapies (see ENZYME APPLICATIONS, THERAPEUTIC; IMMUNOTHERAPEUTIC AGENTS; VITAMINS). Other naturally occurring biopolymers having significant commercial importance are the cellulose (qv) derivatives, eg, cotton (qv) and wood (qv), which are complex polysaccharides.

Analytical Techniques

Analytical techniques that utilize biopolymers, ie, natural macromolecules such as proteins, nucleic acids, and polysaccharides that compose living substances, represent a rapidly expanding field. The number of applications is large and thus uses herein are limited to chiral chromatography, immunology, and biosensors.

Biopolymers in Chiral Chromatography. Biopolymers have had a tremendous impact on the separation of nonsuperimposable, mirror-image isomers known as enantiomers. Enantiomers have identical physical and chemical properties in an achiral environment except that they rotate the plane of polarized light in opposite directions. Thus separation of enantiomers by chromatographic techniques presents special problems. Direct chiral resolution by liquid chromatography (lc) involves diastereomeric interactions between the chiral solute and the chiral stationary phase. Because biopolymers are chiral molecules and can form diastereomeric interactions with chiral solutes, they are ideal for use as chiral stationary phases. This property has led to a rapid growth of chromatographic stationary phases utilizing biopolymers to separate chiral molecules. They include cyclodextrin chromatographic phases, α_1-acid glycoprotein chromatographic phases, bovine serum albumin chromatographic phases, and cellulose triacetate and cellulose derivatives.

Biopolymers in Immunology. Biopolymers are employed in many immunological techniques, including the analysis of food, clinical samples, pesticides, and in other areas of analytical chemistry. Immunoassays (qv) are specific, sensitive, relatively easy to perform, and usually inexpensive. For repetitive analyses, immunoassays compare very favorably with many conventional methods in terms of both sensitivity and limits of detection.

Antigens. One condition that must be met for the application of an immunochemical method is that the analyte must be capable of stimulating an immune response leading to the formation of antibodies in the immunized animal. These antibodies can then be isolated and used as highly specific analytical reagents (immunoassays). Analytes that can combine with the corresponding antibodies are called antigens. There are physical and chemical restrictions on the types of analytes that may be used as immunoassay antigens. In general, large, rigid, chemically complex molecules make good antigens.

Antibodies. Antibodies are proteins, found in many body fluids such as tears, saliva, and urine, that are present in highest concentrations in blood serum. Because antibodies are proteins (qv), they may be characterized by such physical properties as solubility, electrostatic charge, isoelectric point, and molecular weight. The particular proteins which exhibit antibody activity are the immunoglobulins (Ig). The principal immunoglobulin in blood serum is immunoglobulin G (IgG), the structure of which is similar to the other immunoglobins.

There are estimated to be approximately 10 million potential combinations of antigen-binding specificities resulting from light-and heavy-chain combinations in the immunoglobulins. The possibility of utilizing all of these combinations as reagents in immunochemical methods is highly interesting though improbable.

Biopolymers as Biosensors. Selectivity is an important consideration in analytical chemistry. Biologically derived polymers can be used as highly selective immobilized reagents in analytical applications.

Immobilized Enzymes. The immobilized enzyme electrode is the most common immobilized biopolymer sensor, consisting of a thin layer of enzyme immobilized on the surface of an electrochemical sensor. The enzyme catalyzes a reaction that converts the target substrate into a product that is detected electrochemically. The advantages of immobilized enzyme electrodes include minimal pretreatment of the sample matrix, small sample volume, and the recovery of the enzyme for repeated use. Several reviews and books have been published on immobilized enzyme electrodes.

Enzyme Immunosensors. Enzyme immunosensors are enzyme immunoassays coupled with electrochemical sensors. These sensors (qv) require multiple steps for analyte determination, and either sandwich assays or competitive binding assays may be used. Both of these assays use antibodies for the analyte of interest attached to a membrane on the surface of an electrochemical sensor.

Economic Aspects

Enantiomeric separations are expected to continue to have a considerable economic impact on the development of new drugs and therapy in the biomedical field.

The importance of immunoassays for food monitoring and in the detection of diseases is expected to continue to grow as techniques and detection limits improve.

The development of biosensors is expected to benefit monitoring therapeutic drug levels, office testing, and implantable devices because of the advantages of cost-saving automation and improved data handling.

TIMOTHY WARD
Millsaps College

D. W. Armstrong and S. M. Han, *CRC Crit. Rev. Analyt. Chem.* **19**(3), 175 (1988).

E. Stinshoff, W. Stein, W. G. Wood, and P. Laska, *Anal. Chem.* **59,** 339R (1987).

E. P. Diamandis and T. K. Christopoulos, *Anal. Chem.* **62,** 1149A (1990).

M. Thompson and U. J. Krull, *Anal. Chem.* **63,** 393A (1991).

BIOREMEDIATION

Bioremediation is the process of judiciously exploiting biological processes to minimize an unwanted environmental impact; usually it is the removal of a contaminant from the biosphere.

Bioremediation is already a commercially viable technology, with estimates of aggregate bioremediation revenues of $2–3 billion for the period 1994–2000. There are significant opportunities to enlarge upon this success. Bioremediation has applications in the gas phase, in water, and in soils and sediments. For water and soils, the process can be carried out *in situ,* or after the contaminated medium has been moved to some sort of contained reactor (*ex situ*). The former is generally rather cheaper, but the latter may result in such a significant increase in rate that the additional cost of manipulating the contaminated material is overshadowed by the time saved. Bioremediation may explicitly exploit bacteria, fungi, algae, or higher plants. Each, in turn, may be part of a complex food-web, and optimizing the local ecosystem may be as important as focusing solely on the primary degraders or accumulators.

Bioremediation usually competes with alternative approaches to achieving an environmental goal. It is typically among the least expensive options, but an additional important consideration is that in many cases bioremediation is a permanent solution to the contamination problem, since the contaminant is completely destroyed or collected. Some of the alternatives technologies, such as thermal desorption and destruction of organics, are also permanent, solutions, but the simplest, removing the contaminant to a dump site, merely moves the problem, and may well not eliminate the potential liability. Furthermore, by its very nature bioremediation addresses the bioavailable part of any contamination. The same cannot necessarily be said for nonbiological technologies, which may leave bioavailable contaminants at low levels.

Bioremediation also has the advantage that it can be relatively nonintrusive, and can sometimes be used in situations where other approaches would be severely disruptive. On the other hand, bioremediation is usually slower than most physical techniques, and may not always be able to meet some very strict cleanup standards.

Table 1. Some Technological Definitions Relevant to Bioremediation

Technology	Description
air sparging; aquifer sparging; biosparging	injection of air to stimulate aerobic degradation; may also stimulate volatilization
air stripping	injection of air to stimulate volatilization
aquifer bioremediation	*in situ* bioremediation in an aquifer, usually by adding nutrients or co-substrates
aquifer sparging	injection of air into a contaminated aquifer to stimulate aerobic degradation, may also stimulate volatilization
batch reactor	a bioreactor loaded with contaminated material, and run until the contaminant has been consumed, then emptied, and the process is repeated
bioactive barrier; bioactive zone; biowall	a zone, usually subsurface, where biodegradation of a contaminant occurs so that no contaminant passes the barrier
bioaugmentation	addition of exogenous bacteria with defined degradation potential (or rarely indigenous bacteria cultivated in a reactor and reapplied)
biofilm reactor	a reactor where bacterial communities are encouraged on a high surface area support, biofilms often have a redox gradient so that the deepest layer is anaerobic while the outside is aerobic
biofiltration	usually an air filter with degrading organisms supported on a high surface area support such as granulated activated carbon
biofluffing	augering soil to increase porosity
bioleaching	extracting metallic contaminants at acid pH
biological fluidized bed; fluidized-bed bioreactor	bioreactor where the fluid phase is moving fast enough to suspend the solid phase as a fluid-like phase
biopile; soil heaping	an engineered pile of excavated contaminated soil, with engineering to optimize air, water, and nutrient control
bioslurping	vacuum extraction of the floating contaminant, water, and vapor from the vadose zone; the air flow stimulates biodegradation
biostimulation	optimizing conditions for the indigenous biota to degrade the contaminant
biotransformation	the biological conversion of a contaminant to some other form, but not to carbon dioxide and water
biotrickling filter	a reactor where a contaminated gas stream passes up a reactor with immobilized microorganisms on a solid support, while nutrient liquor trickles down the reactor
bioventing	vacuum extraction of contaminant vapors from the vadose zone, thereby drawing in air that stimulates the biodegradation of the remainder
borehole bioreactor; in-well bioreactor	the addition of nutrients and electron acceptor to stimulate the biodegradation *in situ* in a contaminated aquifer
closed-loop bioremediation	groundwater recovery, a bioreactor, and low-pressure reinjection to maximize nutrient use, and maintain temperature in cold climates
composting	addition of biodegradable bulking agent to stimulate microbial activity; optimal composting generally involves self-heating to 50–60°C
constructed wetland	artificial marsh for bioremediation of contaminated water
continuous stirred tank reactor (CSTR)	a completely mixed bioreactor
digester	usually an anaerobic bioreactor for digestion of solids and sludges that generates methane
ex-situ bioremediation	usually the bioremediation of excavated contaminated soil in a biopile, compost system or bioreactor
fixed-bed bioreactor	bioreactor with immobilized cells on a packed column matrix
land-farming; land treatment	application of a biodegradable sludge as a thin layer to a soil to encourage biodegradation; the soil is typically tilled regularly
natural attenuation; intrinsic bioremediation	unassisted biodegradation of a contaminant
phytoextraction	the use of plants to remove and accumulate contaminants from soil or water to harvestable biomass
phytofiltration	the use of completely immersed plant seedlings, to remove contaminants from water
phytoremediation	the use of plants to effect bioremediation
phytostabilization	the use of plants to stabilize soil against wind and water erosion
pump and treat	pumping groundwater to the surface, treating, and reinjection or disposal
rhizofiltration	the use of roots to immobilize contaminants from a water stream
rotating biological contactor	bioreactor with rotating device that moves a biofilm through the bulk water phase and the air phase to stimulate aerobic degradation
sequencing batch reactor	periodically aerated solid phase or slurry bioreactor operated in batch mode
soil-vapor extraction	vacuum-assisted vapor extraction

General Technological Aspects

Table 1 lists some of the technologies in use today in bioremediation.

Organic Contaminants

Hydrocarbons. *Constituents.* Hydrocarbons get into the environment from biogenic and fossil sources. Methane is produced by anaerobic bacteria in enormous quantities in soils, sediments, ruminants and termites, and it is consumed by methanotrophic bacteria on a similar scale. Submarine methane seeps support substantial oases of marine life, with a variety of invertebrates possessing symbiotic methanotrophic bacteria. Thus, methanotrophic bacteria are ubiquitous in aerobic environments. Plants generate large amounts of volatile hydrocarbons, including isoprene and a range of terpenes. These compounds provide an abundant substrate for hydrocarbon-degrading organisms.

Crude oil has been part of the biosphere for millennia, leaking from oil seeps on land and in the sea. Crude oils are very complex mixtures, primarily of hydrocarbons although some components do have heteroatoms such as nitrogen (eg, carbazole) or sulfur (eg, dibenzothiophene). Chemically, the principal components of crude oils and refined products can be classified as aliphatics, aromatics, naphthenics, and asphaltic molecules. When crude oils reach the surface environment the lighter molecules evaporate, and are either destroyed by atmospheric photooxidation or are washed out of the atmosphere in rain, and are biodegraded. Some molecules, such as the smaller aromatics (benzene, toluene, etc) have significant solubilities, and can be washed out of floating slicks, whether these are at sea, or on terrestrial water tables. Fortunately the majority of molecules in crude oils, and refined products made from them, are biodegradable, at least under aerobic conditions.

Biodegradation. Methane and the volatile plant terpenes are fully biodegradable by aerobic organisms, and most refined petroleum products are essentially completely biodegradable under aerobic conditions. The least biodegradable material, principally polar molecules and asphaltenes, lacks the "oily" feel and properties that are associated with oil. These are essentially impossible to distinguish from more recent organic material in soils and sediments, such as the humic and fulvic acids, and appear to be biologically inert.

Numerous bacterial and fungal genera have species able to degrade hydrocarbons aerobically and the pathways of degradation of representative aliphatic, naphthenic and aromatic molecules have been well characterized in at least some species. It is a truism that the hallmark of an oil-degrading organism is its ability to insert oxygen atoms into the hydrocarbon, and there are many ways in which this is achieved. Figures 1 and 2 show the most well-studied.

In recent years it has become clear that at least some hydrocarbons are oxidized by bacteria under completely anaerobic conditions, where the oxygen is probably coming from water. Limited hydrocarbon biodegradation has now been shown under sulfate-, nitrate-, carbon dioxide- and ferric iron-reducing conditions (Table 2). Figure 3 shows the intermediates identified in anaerobic toluene degradation in different organisms. It is noteworthy that while organisms capable of aerobic oil biodegradation seem to be ubiquitous, organisms capable of the anaerobic degradation of hydrocarbon have to date only been found in a few places.

Although the majority of molecules in crude oils and refined products are hydrocarbons, the U.S. Clean Air Act amendment of 1990 mandated the addition of oxygenated compounds to gasoline in many parts of the United States. The requirement is usually that 2% (w/w) of the fuel be oxygen, which requires that 5–15% (v/v) of the gasoline be an oxygenated additive (eg, methanol, ethanol, methyl *tert*-butyl ether (MTBE), etc). Although methanol and ethanol are readily degraded under aerobic conditions, the degradability of MTBE remains something of an open question. The compound was previously very rare in the environment, but now it is one of the major chemicals in commerce. At first it seemed that the compound was completely resistant to biodegradation, but complete mineralization has now been reported. Whether biodegradation can be optimized for effective bioremediation remains to be seen.

Bioremediation. Crude oil and refined products are readily biodegradable under aerobic conditions, but they are only incomplete foods since they lack any significant nitrogen, phosphorus, and essential trace elements. Bioremediation strategies for removing large quantities of hydrocarbon must therefore include the addition of fertilizers to provide these elements in a bioavailable form.

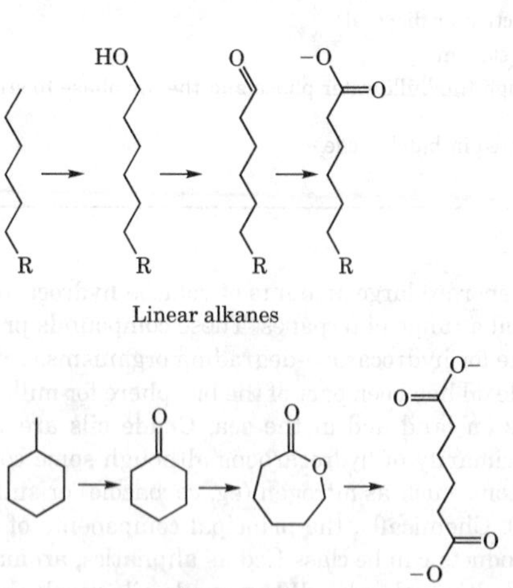

Figure 1. Initial steps in the biodegradation of linear and cyclic alkanes.

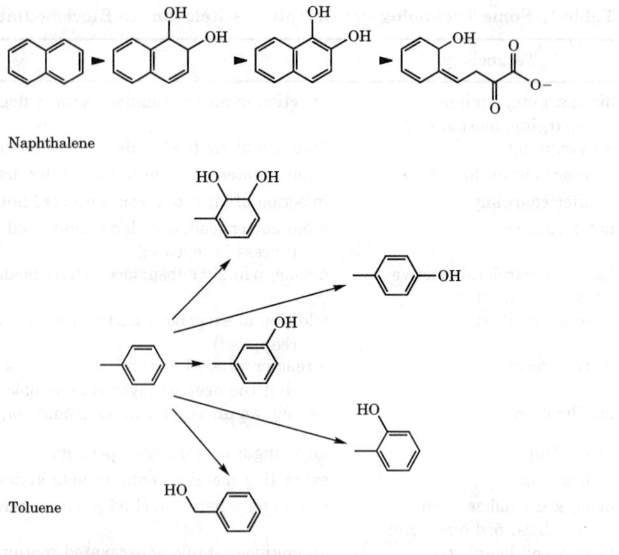

Figure 2. Initial steps in the aerobic degradation of naphthalene, as a representative multiringed aromatic, and toluene. The different initial steps of toluene degradation are examples of the diversity found in different organisms.

Table 2. Hydrocarbons That Have Been Shown to be Biodegraded Under Anaerobic Conditions

Electron acceptor	Substrate
nitrate (to nitrogen)	heptadecene
	toluene, ethylbenzene, xylene
	naphthalene
	terpenes
iron(III) (to iron(II))	toluene
manganese(IV) (to Mn(II))	toluene
sulfate (to sulfide)	hexadecane, alkylbenzenes
	benzene
	naphthalene, phenanthrene
CO_2 (to methane)	toluene, xylene

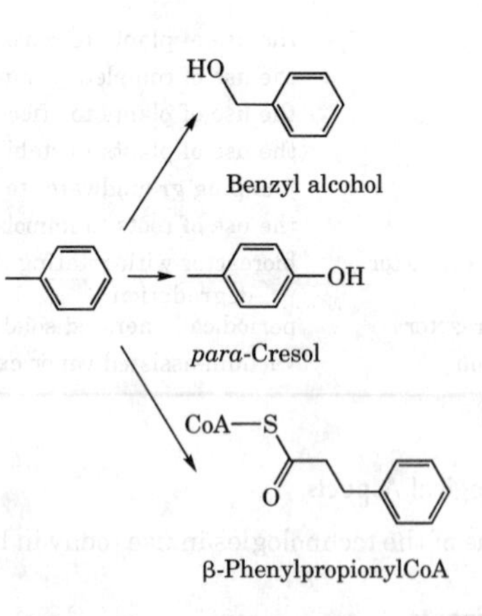

Figure 3. Proposed initial steps in the anaerobic biodegradation of toluene in different organisms.

Air. Hydrocarbon vapors in air are readily treated with biofilters. These are typically rather large devices with a very large surface area provided by bulky material such as a bark or straw compost. The contaminated air, perhaps from a soil vapor-extraction treatment, or from a factory using hydrocarbon solvents, is blown through the filter, and organisms, usually indigenous to the filter material or provided by a soil or commercial inoculum, grow and consume the hydrocarbons.

Sea. Significantly more oil reaches the world's oceans from municipal sewers than widely covered crude oil spills. Physical collection of spilled oil is the preferred remediation option, but if skimming is unable to collect the oil, biodegradation and perhaps combustion or photooxidation are the only routes for eliminating of the spill. One approach to stimulating biodegradation is to disperse the oil with chemical dispersants. Modern dispersants and application protocols stimulate biodegradation by increasing the surface area of the oil available for microbial attachment, and perhaps providing nutrients to stimulate microbial growth.

Shorelines. The successful bioremediation of shorelines affected by the spill from the *Exxon Valdez* in Prince William Sound, Alaska, was perhaps the largest project to date. Bioremediation focused on the addition of nitrogen and phosphorus fertilizers to partially remove the nutrient-limitation on oil degradation. Of course the addition of fertilizers was complicated by the fact that oiled shorelines were washed by tides twice a day. Two fertilizers were used in the full-scale applications; one, an oleophilic product known as Inipol EAP22 (trademark of CECA, Paris, France), was a microemulsion of a concentrated solution of urea in an oil phase of oleic acid and trilaurethphosphate, with butoxyethanol as a cosolvent. This product was designed to adhere to oil, and to release its nutrients to bacteria growing at the oil-water interface. The other fertilizer was a slow-release formulation of inorganic nutrients, primarily ammonium nitrate and ammonium phosphate, in a polymerized vegetable oil skin. This product, known as Customblen (trademark of Grace-Sierra, Milpitas, California), released nutrients with every tide, and these were distributed throughout the oiled zone as the tide fell. Fertilizer application rates were carefully monitored so that the nutrients would cause no harm, and the rate of oil biodegradation was stimulated between two- and five-fold.

Areas where there are currently few remediation options but where bioremediation may provide an option include oiled marshes, mangroves, and coral reefs. Bioremediation also offers options for dealing with oiled material, such as seaweed, that gets stranded on shorelines; composting has been shown to be effective.

Groundwater. Spills of refined petroleum product on land, and leaking underground storage tanks, sometimes contaminate groundwater. Stand-alone bioremediation is an option for these situations, but "pump and treat" is the more usual treatment. Contaminated water is brought to the surface, free product is removed by flotation, and the cleaned water re-injected into the aquifer or discarded. Adding a bioremediation component to the treatment, typically by adding oxygen and low levels of nutrients, is an appealing and cost-effective way of stimulating the degradation of the residual hydrocarbon not extracted by the pumping. This approach is becoming widely used.

Typically only small aromatic molecules, the infamous BTEX (benzene, toluene, ethylbenzene, and xylenes), are soluble enough to contaminate groundwater. With the advent of oxygenated gasolines, it is expected that these oxygenates [ethanol, methanol, MTBE (methyl-*tert*-butyl ether) etc] will also be found in groundwater. These contaminants are biodegradable, and some biodegradation is probably already occurring when the contamination is discovered. The cheapest approach to remediation is, thus, to allow this intrinsic process to continue.

Intrinsic bioremediation is becoming an acceptable option in locations where the contaminated groundwater poses little threat to environmental health. Nevertheless, it may not be the lowest cost option if there are extensive monitoring and documentation costs involved for several years. In such cases it may well be more cost effective to optimize conditions for biodegradation.

One approach is to optimize the levels of electron acceptors. Slow release formulations of inorganic peroxides, such as magnesium peroxide, have recently been used with success. Nitrate may be added, although there are sometimes regulatory limitations on the amount of this material that may be added to groundwater. Ferric iron availability may be manipulated by adding ligands.

If there are significant amounts of both volatile and nonvolatile contaminants, remediation may be achieved by a combination of liquid and vapor extraction of the former, and bioremediation of the latter. This combination has been termed "bioslurping".

The majority of remediation operations include stopping the source of the contamination, but in some cases this is impossible, either because of the location of the spill, or because it is over a large area, and not a point source. In these situations it may be possible to intercept the flow of contaminated groundwater on-site, and ensure that no contamination passes. Approaches include biowall, trench biosparge, funnel and gate, bubble curtain, sparge curtain and engineered trenches and gates. Both aerobic and anaerobic designs have been successfully installed.

Where there are large volumes of contaminated water under a small site, it is sometimes most convenient to treat the contaminant in a biological reactor at the surface.

Of course the presence of a liquid phase of hydrocarbon in a soil gives rise to vapor contamination in the vadose zone above the water table. This can be treated by vacuum extraction, and the passage of the exhaust gases through a biofilter (see above) can be a cheap and effective way of destroying the contaminant permanently.

Soil. Spills from production facilities and pipelines often involve both oil and brine. Successful bioremediation strategies must therefore include remediating the brine. In wet regions the salt is eventually diluted by rainfall, but in arid regions, and to speed the process in wetter regions, gypsum is often added to restore soil porosity.

Many hydrocarbons bind quite tightly to soil components, and are thereby less available to microbial degradation. Intrinsic biodegradation occurs, but it usually only removes the lightest refined products, such as gasoline, diesel and jet fuel. Active intervention is typically required. Usually the least expensive approach is *in situ* remediation, typically with the addition of nutrients, and the attempted optimization of moisture and oxygen by tilling.

Deeper contamination may be remedied with bioventing, where air is injected through some wells, and extracted through others to both strip volatiles and provide oxygen to indigenous organisms. Fertilizer nutrients may also be added. This is usually only a viable option with lighter refined products.

A recent suggestion has been to use plants to stimulate the microbial degradation of the hydrocarbon (hydrocarbon phytoremediation). The plants are proposed to help deliver air to the soil microbes, and to stimulate microbial growth in the rhizosphere by the release of nutrients from the roots. The esthetic appeal of an active phytoremediation project can be very great.

When soil contamination extends to some depth it may be preferable to excavate the contaminated soil and put it into "biopiles" where oxygen, nutrient and moisture levels are more easily controlled. Composting by the addition of readily degradable bulking agents is also a useful option for relatively small volumes of excavated contaminated soil.

Slurry bioreactors offer the most aggressive approach to maximizing contact between the contaminated soil and the degrading organisms. Slurry bioreactors are usually the most expensive bioremediation option because of the large power requirements, but under some conditions this cost is offset by the rapid biodegradation that can occur.

In all these cases it is important to bear in mind that although the majority of hydrocarbons are readily biodegraded, some are very resistant to microbial attack. It is thus important to run laboratory studies to ensure that the contaminant is sufficiently biodegradable that clean-up targets can be met.

Halogenated Organic Solvents. *Constituents.* Halogenated organic solvents are widely used in metal processing, electronics, dry cleaning and paint, paper and textile manufacturing and are fairly widespread contaminants. Unlike the hydrocarbons, the halogenated solvents typically have specific gravities greater than 1, and they generally sink to the bottom of any groundwater, and float on the bedrock. For this reason they are sometimes known as DNAPLs for dense nonaqueous phase liquids.

Biodegradation. Halogenated solvents are degraded under aerobic and anaerobic conditions. The anaerobic process is typically a reductive dechlorination that progressively removes one halide at a time.

The simplest chlorinated alkanes, alkenes, and alcohols (eg, chloromethane, dichloromethane, chloroethane, 1,2-dichloroethane, vinyl chloride, and 2-chloroethanol) serve as substrates for aerobic growth for some bacteria, but the majority of halogenated solvents cannot support growth. Nevertheless these compounds are mineralized under aerobic conditions.

Bioremediation. Air. Biofilters are an effective way of dealing with air from industrial processes that use halogenated solvents that support aerobic growth. Both compost-based dry systems and trickling filter wet systems are in use. Similar filters could be incorporated into pump-and-treat operations.

Groundwater and Soil. Pumping out the liquid phase is an obvious first step if the contaminant is likely to be mobile, but *in situ* bioremediation is a promising option. Thus, the U.S. Department of Energy is investigating the use of anaerobic *in situ* degradation of carbon tetrachloride with nitrate as electron acceptor, and acetate as electron donor.

Trichloroethylene, the most frequent target of remediation, is only metabolized co-metabolically. Remediation operations thus incorporate the addition of co-metabolized substrate.

Plants may have a role to play in enhancing microbial biodegradation of halogenated solvents, for it has recently been shown that mineralization of radiolabelled trichlorothylene is substantially greater in vegetated rather than unvegetated soils, indicating that the rhizosphere provides a favorable environment for microbial degradation of organic compounds.

Methane has been used in aerobic bioreactors that are part of a pump-and-treat operation, and toluene and phenol have been used as co-substrates at the pilot scale. Anaerobic reactors have also been developed for treating trichloroethylene.

Groundwater contaminated with other halogenated solvents can also be treated in aboveground reactors. Aerobic reactors are useful for those compounds that can support growth. Sequential anaerobic and aerobic reactors are capable of mineralizing tetrachloroethylene.

Halogenated Organic Compounds. *Constituents.* Complex halogenated organic compounds have been widely used in commerce in the last fifty years. Representative examples are pentachlorophenol, (2,4-dichlorophenoxy)acetate, DDT and polychlorinated biphenyls (PCBs). They may not seem a good target for bioremediation but some successful applications have been developed.

Biodegradation. An important characteristic of degradation is the cleavage of carbon–chlorine bonds, and the enzymes that catalyze these reactions, the dehalogenases, are being characterized. The reductive dechlorination seen with carbon tetrachloride and tetrachloroethylene seems to be a general phenomenon, and even compounds as persistent as DDT and the polychlorinated biphenyls are reductively dechlorinated under some conditions, particularly under methanogenic conditions. Some compounds, such as pentachlorophenol, can be completely mineralized under anaerobic conditions, but the more recalcitrant ones require aerobic degradation after reductive dehalogenation.

Bioremediation. Soil. Pentachlorophenol has been the target of bioremediation at a number of wood-treatment facilities, and good success has been achieved in several applications. *In situ* degradation has been stimulated by bioventing. Just as with the halogenated solvents, it seems that plants stimulate microbial degradation of pentachlorophenol in the rhizosphere.

The kinetics of such *in situ* degradation are rather slow, however, and more active bioremediation is usually attempted. For example, contaminated soil at the Champion Superfund site in Libby, Montana, was placed into 1-acre land treatment units in 6-in. layers, and irrigated, tilled, and fertilized. Under these conditions, the half-lives of pentachlorophenol, pyrene, and several other polynuclear aromatic hydrocarbons, initially present at around 100–200 ppm, were on the order of 40 days. Composting, and bioremediation focusing on the use of white-rot fungi, has also met with success at the pilot scale. Others have used fed-batch or fluidized-bed bioreactors to stimulate the biodegradation of pentachlorophenol. This allows significant optimization of the process and increases in rates of degradation by tenfold.

A major concern when remediating wood-treatment sites is that pentachlorophenol was often used in combination with metal salts, and these compounds, such as chromated-copper-arsenate, are potent inhibitors of at least some pentachlorophenol degrading organisms. Sites with significant levels of such inorganics may not be suitable candidates for bioremediation.

The phenoxy-herbicide, 2,4-D, has been successfully bioremediated in a soil contaminated with such a high level of the compound (710 ppm) that it was toxic to microorganisms. Success relied on washing a significant fraction of the contaminant off the soil and adding bacteria enriched from a less contaminated site. Success was achieved in remediating both soil washwater and soil in a bioslurry reactor. 2,4-D is also effectively degraded in composting, with about half being completely mineralized, and the other half becoming incorporated in a nonextractable form in the residual soil organic matter.

Cultures are being found that can degrade both polychlorinated biphenyls and petroleum hydrocarbons. There is also interest in the role of rhizosphere organisms in polychlorinated biphenyl degradation, particularly since some plants exude phenolic compounds into the rhizosphere that can stimulate the aerobic degradation of the less chlorinated biphenyls.

Groundwater. A successful groundwater bioremediation of pentachlorophenol is being carried out at the Libby Superfund site described above. A shallow aquifer is present at 5.5 to 21 m below the surface, and a contaminant plume is nearly 1.6 km in length. Nutrients and hydrogen peroxide were added at the source area and approximately half way along the plume, and pentachlorophenol concentrations decreased from 420 ppm to 3 ppm where oxygen concentrations were successfully raised.

Pentachlorophenol is readily degraded in biofilm reactors, so bioremediation is a promising option for the treatment of contaminated groundwater brought to the surface as part of a pump-and-treat operation.

River and Pond Sediments. Much of the work on polychlorinated biphenyls has focused on the remediation of aquatic sediments, particularly from rivers, estuaries, and ponds. Harkness and coworkers have successfully stimulated aerobic biodegradation in large caissons in the Hudson River by adding inorganic nutrients, biphenyl, and hydrogen peroxide, but found that repeated addition of a polychlorinated-biphenyl degrading bacterium (*Alcaligenes eutrophus* H850) had no beneficial effect. Whether this approach can be scaled-up for large-scale use, with a net environmental benefit, remains to be seen.

Nonchlorinated Pesticides and Herbicides. *Constituents.* It is unusual for these compounds to become contaminants where they are applied correctly, but manufacturing facilities, storage depots and rural airfields where crop-dusters are based have had spills that can lead to long lasting contamination.

Biodegradation. The vast majority of pesticides, herbicides, fungicides, and insecticides in use today are biodegradable, although the intrinsic biodegradability of individual compounds is one of the variables used in deciding which compound to use for which task.

Compounds with organophosphate moieties, such as Diazinon, Methyl Parathion, Coumaphos and Glyphosate are usually hydrolyzed at the phosphorus atom. Indeed several *Flavobacterium* isolates are able to grow using parathion and diazinon as sole sources of carbon.

Very few pure cultures of microorganisms are able to degrade triazines such as Atrazine, although some *Pseudomonads* are able to use the compound as sole source of nitrogen in the presence of citrate or other simple carbon substrates. The initial reactions seem to be the removal of the ethyl or isopropyl substituents on the ring, followed by complete mineralization of the triazine ring.

Nitroaromatic compounds, such as Dinoseb, are degraded under aerobic and anaerobic conditions. The nitro group may be cleaved from the molecule as nitrite, or reduced to an amino group under either aerobic or anaerobic conditions. Alternatively, the ring may be the subject of reductive attack. Recent work has isolated a *Clostridium bifermentans* able to anaerobically degrade dinoseb cometabolically in the presence of a fermentable substrate. The dinoseb was degraded to below detectable levels, although only a small fraction was actually mineralized to CO_2.

Carbamates such as Aldicarb undergo degradation under both aerobic and anaerobic conditions.

Bioremediation. Groundwater. Atrazine dominated the world herbicide market in the 1980s, and contamination of groundwater has been reported in several locations in the U.S., Europe, and South Africa. Successful biodegradation has been achieved with indigenous organisms in laboratory mesocosms after a lag phase, and once activity was found, it remained. It is clear that intrinsic remediation is likely to lead to the disappearance of atrazine from groundwaters.

If more active treatment is required, such as pump and treat, it is possible that biological reactors will be a cost-effective replacement for activated carbon filters.

Marsh and Pond Sediments. Herbicides and pesticides are detectable in marsh and pond sediments, but intrinsic biodegradation is usually found to be occurring.

Soil. Herbicides and pesticides are of course metabolized in the soil to which they are applied, and there are many reports of isolating degrading organisms from such sites. Little work has yet been presented where the biodegradation of these compounds has been successfully stimulated by a bioremediation approach.

It is a general observation that herbicide degradation occurs more readily in cultivated than fallow soil, suggesting that rhizosphere organisms are effective herbicide degraders. Whether this can be effectively exploited in a phytoremediation strategy remains to be seen.

Military Chemicals. *Constituents.* The military use a range of chemicals such as explosives and propellants that are sometimes termed "energetic molecules." Generally speaking, modern explosives are cyclic, often heterocyclic, composed of carbon, nitrogen and oxygen, eg; 2,4,6-trinitrotoluene; RDX (Royal Demolition eXplosive; hexahydro-1,3,5-trinitro-1,2,3-triazine); HMX (High Melting eXplosive; octahydro-1,3,5,7-tetranitro-1,3,5,7-tetrazocine); and *N,N*-dimethylhydrazine. These compounds are sometimes present at quite high levels in soils and groundwater on military bases and production sites. Bioremediation is a promising new technology for treating sites contaminated with such compounds.

Bioremediation may also be an appropriate tool for dealing with chemical agents such as the mustards and organophosphate neurotoxins, but little work on actual bioremediation has been published.

Biodegradation. Nitrosubstituted compounds are subject to a variety of degradative processes. Under anaerobic conditions TNT is readily reduced to the corresponding aromatic amines and subsequently deaminated to toluene. As shown in the section on hydrocarbons, the latter can be mineralized under anaerobic conditions, leading to the potentially complete mineralization of TNT in the absence of oxygen.

Under aerobic conditions TNT can be mineralized by a range of bacteria and fungi, often co-metabolically with the degradation of a more degradable substrate. There is even evidence that some plants are able to deaminate TNT reductively.

RDX and HMX are rather more recalcitrant, especially under aerobic conditions, but there are promising indications that biodegradation can occur under some conditions, especially composting.

Little work has been reported on the biodegradation of dimethylhydrazine.

Bioremediation. Groundwater. Nitrotoluenes have been detected in groundwater in some areas, and intrinsic remediation may be occurring at some sites by anaerobic degradation.

A commercial technology, the SABRE process, treats contaminated water and soil in a two-stage process by adding a readily degradable carbon and an inoculum of anaerobic bacteria able to degrade the contaminant. An initial aerobic fermentation removes oxygen so that the subsequent reduction of the contaminant is not accompanied by oxidative polymerization.

Soil. Composting of soils contaminated by high explosives is being carried out at the Umatilla Army Depot near Hermiston, Oregon. If this is successful, there are 30 similar sites on the National Priority List that could be treated in a similar way.

Other Organic Compounds. The majority of organic compounds in commerce are biodegradable, so bioremediation is a potential option for cleaning up after industrial and transportation accidents.

Inorganic Contaminants

Nitrogen Compounds. *Constituents.* Nitrate levels are regulated in groundwater because of concerns for human and animal health. Ammonia is regulated in streams and effluents as a potential fish toxicant, and any nitrogenous contaminant is a potential problem in water because of its stimulatory effect on the growth of algae. Other nitrogenous contaminants include cyanides in mine waters. Fortunately, all are amenable to biological treatment.

Biodegradation. The biological mineralization of fixed nitrogen is well studied; ammonia is oxidized to nitrite, and nitrite to nitrate, by autotrophic bacteria, and nitrate is reduced to nitrogen by anaerobic bacteria. On the other hand, ammonia and nitrate are essential nutrients for plant and bacterial growth, so one option is to use these organisms to take up and use the contaminants.

Bioremediation. Surface Water. One example of exploiting biology to handle excess nitrogen in a surface water is the case of the Venice Lagoon in Italy. About a million tons of sea lettuce (*Ulva*) grows in the lagoon annually because of the high levels of nitrogenous nutrients in this relatively landlocked bay. This material is harvested, composted and sold as a low-cost remediation of this problem.

A more constrained opportunity for nitrate bioremediation arose at the U.S.-DoE Weldon Spring Site near St. Louis, Missouri which had been a uranium and thorium processing facility. Two pits had nitrate levels that required treatment before discharge. Bioremediation by the addition of calcium acetate as a carbon source successfully treated more than 19 million liters of water at a reasonable cost.

Groundwater. One approach to minimizing the environmental impact of excess nitrogen in groundwater migrating into rivers and aquifers is to intercept the water with rapidly growing trees, such as poplars, that will use the contaminant as a fertilizer.

An alternative approach is to add a readily degradable substrate to the contaminated aquifer, in the absence of oxygen, to stimulate bacterial denitrification.

Metals and Metalloids. A wide range of metallic and nonmetallic elements are present as contaminants at industrial and agricultural sites throughout the world, both in ground and surface water, and in soils. They pose a quite different problem from that of organic contaminants, since they cannot be degraded so that they disappear. Some metal and metalloid elements have radically different bioavailabilities and toxicities depending on their redox state, so one option is stabilization by converting them to their least toxic form. This can be a very effective way of minimizing the environmental impact of a contaminant, but if the contaminant is not removed from the environment, there is always the possibility that natural processes, biological or abiological, may reverse the process. Removing the contaminants from water phases is relatively straightforward, and the wastewater treatment industry practices this on an enormous scale. Pump-and-treat systems that mimic wastewater treatments are already being used for several contaminants, and less complex systems involving biological mats are a promising solution for less demanding situations.

Table 3. Will Bioremediation be a Suitable Treatment for a Site Contaminated with Organic, Nitrogenous, or Organic Contaminants?

Organic
Is the contaminant biodegradable? If the contaminant is a complex mixture of components, are the individual chemical species biodegradable? If the contaminant has been at the site for some time, biodegradation of the most readily degradable components may have already occurred. Is the residual contamination biodegradable?
Are degrading organisms present at the site?
What is limiting their growth and activity? Can this be added effectively?
Are the levels of contaminant amenable to bioremediation? Are they toxic to microorganisms? Are they so abundant that even substantial microbial activity will take too long to clean the site?
Are the clean-up standards reasonable? Are biological processes known to degrade substrates down to the levels required?
Nitrogenous
Are appropriate microorganisms present at the site?
What is limiting their growth and activity? Can this be added effectively?
Are the levels of contaminant amenable to bioremediation? Are they toxic to microorganisms? Are they so abundant that even substantial microbial activity will take too long to clean the site?
Can the nitrogenous compound be used by plants?
Are the clean-up standards reasonable? Are biological processes known to degrade substrates down to the levels required?
Inorganic
Can the contaminant be made less hazardous by changing its redox state?
Can the contaminant be brought to a reactor or constructed wetland where biological systems, microbial or plant, can extract and immobilize the contaminant?
Can plants extract the contaminant from the soil matrix?
Are the clean-up standards reasonable? Are biological processes known to accumulate contaminants down to the levels required?

In the past, removing metal and metalloid contaminants from soil has been impossible, and site clean-up has meant excavation and disposal in a secure landfill. An exciting new approach to this problem is phytoextraction, where plants are used to extract contaminants from the soil and harvested.

Bioremediation. Water. Groundwater can be treated in anaerobic bioreactors that encourage the growth of sulfate reducing bacteria, where the metals are reduced to insoluble sulfides, and concentrated in the sludge.

Phytoremediation is not yet being used commercially, but results at several field trial suggest that commercialization is not far away. Perhaps the biggest success to date is the successful rhizofiltration of radionuclides from a Department of Energy site at Ashtabula, Ohio, where uranium concentrations of 350 ppb were reduced to less than 5 ppb, well below groundwater standards, by Sunflower roots.

Mine Drainage. In recent years it has become clear that the environmental impact of acid mine drainage can be minimized by the construction of artificial wetlands that combine geochemistry and biological treatments. These systems are being designed for a range of wastewaters, most of which fall outside the scope of this article.

Soil. The results of the first reported field trial of the use of hyperaccumulating plants to remove metals from a soil contaminated by sludge applications were positive. However, the rates of metal uptake suggest a time scale of decades for complete cleanup. Trials with higher biomass plants, such as *B. juncea,* are underway.

The bacterial reduction of Cr(VI) to Cr(III) is also being used to reduce the hazards of chromium in soils and water.

Conclusions

Bioremediation has many advantages over other technologies, both in cost and in effectively destroying or extracting the pollutant. An important issue is thus when to consider it, and a series of questions may lead to the appropriate answer (Table 3).

If the answers to the questions in Table 3 lead to the selection of bioremediation, it then becomes important to assess the success of the bioremediation strategy in achieving the cleanup criteria. A major disadvantage of bioremediation is that it is typically rather slower than competing technologies such as thermal treatments. How can regulators and responsible parties gain confidence during this time that success will indeed be achieved? The National Research Council has recently addressed this issue, and suggested a three-fold strategy for "proving" bioremediation: (*1*) a documented loss of contaminants from the site; (*2*) laboratory tests showing the potential of endogenous microbes to catalyze the reactions of interest; and (*3*) some evidence that this potential is achieved in the field.

Finally a caveat. Despite its documented success in many situations, bioremediation may not always be able to meet current clean up criteria for a particular site. Some standards are so tight that they are essentially "detection limit" standards, and it is not clear that biological processes will be able to remove contaminants to such low levels. Bioremediation will be more likely to fulfill its promise as an important tool in contaminated site remediation if there is progress towards standards based on bioavailability and net environmental benefit from the clean up, rather than on arbitrary absolute standards.

Roger C. Prince
Exxon Research and Engineering Company

R. L. and D. L. Crawford, *Bioremediation: Principles and Practice,* Cambridge University Press, 1996.

Bioremediation; when does it work?, National Research Council, National Academy Press, 1993.

M. A. Alexander, *Biodegradation and Bioremediation,* Academic Press, New York, 1994.

L. Y. Young and C. E. Cerniglia, eds., Wiley-Liss, *Microbial Transformation and Degradation of Toxic Organic Compounds,* New York, 1995.

BIOSENSORS

The detection of trace and low levels of biologically active substances is among the most significant and challenging analytic technologies (see Trace and residue analysis). Any discrete sensing device that relies on a biologically derived component as an integral part of its detection mechanism is known as a biosensor (see Biopolymers).

Basic Components

All biosensors are composed of two basic parts, the molecular recognition component and the transducer component. The molecular recognition component is typically a complex chemical system usually extracted or derived directly from a biological organism. The function of the molecular recognition component is to interact specifically with a target compound, ie, the chemical to be detected. The molecular recognition component must be capable of discerning the presence of a target in a solution containing possibly hundreds of other compounds, some of which may have molecular structures closely resembling that of the target. The term selectivity is used to quantify the degree to which a molecular recognition system responds specifically to a target while rejecting compounds having related molecular structures. There are three principal classes of molecular recognition components used in biosensors: enzymes or catalytic proteins (qv); immunoglobins, which are biological macromolecules that selectively bind to foreign substances invading an organism; and chemoreceptors, biomolecular units responsible for sensory reception in living organisms (see ENZYME APPLICATIONS). Each class has advantages and disadvantages, and each class comprises a vast number of highly selective possible molecular recognition components.

Once the molecular recognition component interacts with its target, a user-readable signal must be generated. This is the purpose of the transducer component. Ideally, the generated signal should not only report the presence of target but should also relay information concerning the amount or concentration of target compound in the test solution. As of this writing, there were four main types of analytical transducer components employed in biosensors: optical, electrochemical, field effect transistor, and piezoelectric.

The term sensitivity is used to quantify the ability of a biosensor to reliably report target level. There are two separate uses of the term. First, the detection limit of a biosensor is the lowest concentration of target for which the biosensor provides a reliable, reproducible, and unambiguous response. Increasing the sensitivity of a biosensor frequently refers to decreasing the biosensor's detection limit. Second, sensitivity can also refer to the smallest change in concentration from some nominal level of target compound that the biosensor can unambiguously report. Increasing sensitivity in these latter terms is usually a more difficult task than decreasing the detection limit.

Biosensor design involves consideration of many factors. Molecular recognition components are frequently large complex biomolecules or biological macroassemblies and usually perform efficiently within narrow ranges of pH, temperature, and ionic strength. Degradation and the limited lifetimes of biologically derived components can be important. Additionally, immobilization of enzymes sometimes results in decreased target activity. Factors influencing the selection of an appropriate transducer component include compatibility with the molecular recognition component and the environment in which the biosensor is to be used. Development and optimization of biosensor techniques is a rapidly growing research area; as of the early 1990s, publications on fundamental and applied aspects of biosensing were appearing at a rate approaching 400 papers per year.

Factors Affecting Biosensor Detection Limits

Biosensor designers are striving for lower detection limits, higher sensitivity, greater selectivity, and lower occurrences of false-positive signals. The detection limit for a biosensor is related to the nature of both the molecular recognition component and the transducer component. In general, antibodies and chemoreceptors are well-suited for low detection limit biosensors because of the high binding constants for target molecules.

In general, low level detection is masked by the noise level inherent in any measuring device. Electrochemical methods are susceptible to electrical interference from external sources, variations in reference electrode parameters resulting from aging or contamination, and interference from redox contaminants in the test solution.

An ingenious method of decreasing detection limits in biosensors that is frequently employed is enzyme amplification. An example is enzyme multiplied immunoassay technology (emit).

WILLIAM PIETRO
York University

A. Heller and Y. Degani, in G. Dryhurst and K. Niki, eds., *Redox Chemistry and Interfacial Behavior of Biological Molecules,* Plenum Publishing Corp., New York, 1988, pp. 151–170.

Recent Advances in the Development and Analytical Application of Biosensing Probes, Vol. 20, Chemical Rubber Co. Press, Boca Raton, Fla., 1988.

A. P. F. Turner, I. Karube, and G. S. Wilson, eds., *Biosensors: Fundamentals and Applications,* Oxford University Press, New York, 1987.

BIOSEPARATIONS

The large-scale purification of proteins and other bioproducts is the final production step, prior to product packaging, in the manufacture of therapeutic proteins, specialty enzymes, diagnostic products, and value-added products from agriculture. These separation steps, taken to purify biological molecules or compounds obtained from biological sources, are referred to as bioseparations. Large-scale bioseparations combine art and science. Bioseparations often evolve from laboratory-scale techniques, adapted and scaled up to satisfy the need for larger amounts of extremely pure test quantities of the product. Uncompromising standards for product quality, driven by commercial competition, applications, and regulatory oversight, provide many challenges to the scale-up of protein purification. The rigorous quality control embodied in current good manufacturing practices, and the complexity and lability of the macromolecules being processed provide other practical issues to address.

Recovery and purification of new biotechnology products is the fastest growing area of bioseparations.

Manufacturing approaches for selected bioproducts of the new biotechnology impact product recovery and purification. The most prevalent bioseparations method is chromatography (qv). Thus the practical tools used to initiate scaleup of process liquid chromatographic separations starting from a minimum amount of laboratory data are given.

Economic Aspects

The development of biotechnology processes in the biopharmaceutical and bioproduct industries is driven by the precept of being first to market while achieving a defined product purity, and developing a reliable process to meet validation requirements. The economics of bioseparations are important, but are likely to be secondary to the goal of being first to market.

The three main sources of competitive advantage in the manufacture of high value protein products are first to market, high product quality, and low cost. The first company to market a new protein biopharmaceutical, and the first to gain patent protection, enjoys a substantial advantage. In the absence of patent protection, product differentiation becomes very important. Differentiation reflects a product that is purer, more active, or has a greater lot-to-lot consistency.

Biopharmaceuticals and Protein Products. Purification of proteins is a critical and expensive part of the production process, often accounting for ≥50% of total production costs. Hence, bioseparation processes have a significant impact on manufacturing costs. For small-volume, very high value biotherapeutics, however, these costs may be considered secondary to the first to market principle unless a lower cost competitor surfaces.

Bioproduct Separations

The task of quickly specifying, designing, and scaling-up a bioproduct separation is not simple. These separations are carried out in a liquid phase using macromolecules which are labile, and where conformation and heterogeneous chemical structure undergoing even subtle change during purification may result in an unacceptable product. A typical purification scheme for biopharmaceutical proteins involves the harvesting of protein-containing material or cells, concentration of protein using ultrafiltration (qv), initial chromatographic steps, viral clearance steps, additional chromatographic steps, again concentration of protein using ultrafiltration, and finally formulation.

Biosynthetic Human Insulin from E. coli. Human insulin was the first animal protein to be made in bacteria in a sequence identical to the human pancreatic peptide. Expression of separate insulin A and B chains were achieved in *Escherichia coli* K-12 using genes for the insulin A and B chains synthesized and cloned in frame with the β-galactosidase gene of plasmid pBR322.

Recovery and Purification. The production of Eli Lilly's human insulin requires an estimated 31 principal processing steps of which 27 are associated with product recovery and purification. The production process for human insulin is based on a fermentation which yields proinsulin. Whereas the exact sequence has not been published, the principle steps in the purification scheme are outlined in Figure 1.

Ion-exchange chromatography removes most of the impurities that form during processing and is followed by reversed-phase chromatography which separates insulin from structurally similar insulin-like components. Then size-exclusion (gel-permeation) chromatography is introduced to remove salts and other small molecules from the insulin. This sequence follows the principle of orthogonality of separation sequence, ie, each step is based on a different property, in this case charge, solubility, and size, respectively. Near the end of the chromatography sequence, the insulin may be concentrated by precipitation to form insulin zinc crystals.

Yield Losses. The numerous steps incur a built-in yield loss. For example, if only 2% yield loss were to be associated with each step, the overall yield for a purification sequence of 10 steps would be as in equation 1:

$$\eta = 100(1 - L/100)^n = 100(1 - 0.02)^{10} = 81.7\% \quad (1)$$

where η denotes yield, L the percent yield loss at each step, and n the number of steps. Maximizing recovery at each step is important.

The purification of human insulin involves five separate alterations in the molecular structure, and hence, changes in physicochemical properties during its recovery and purification. The final purification steps rely on multiple properties of the insulin, such as size hydrophobicity, ionic charge, and cystallizability. The final purity level is reported to be >99.99%.

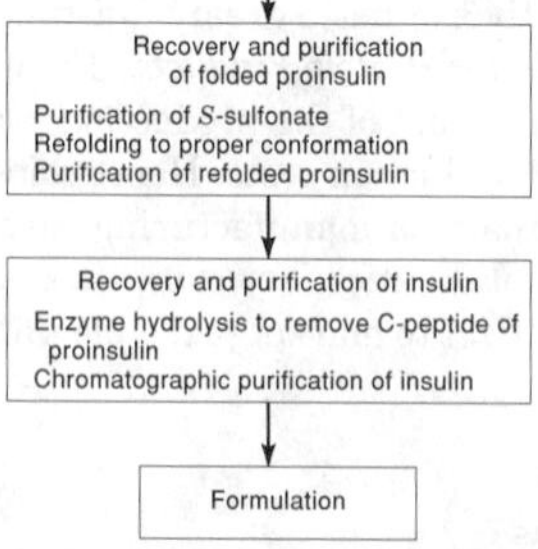

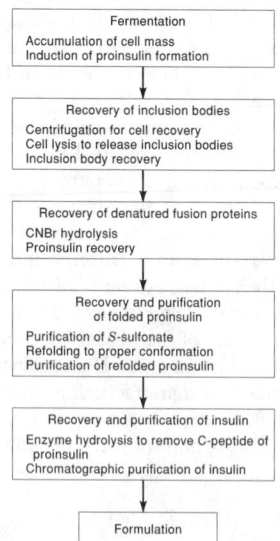

Figure 1. Process flow sheet for human insulin production, recovery, and purification.

Tissue Plasminogen Activator from Mammalian Cell Culture. In 1981 the Bowes melanoma cell line was found to secrete tissue plasminogen activator (known as mt-PA) at 100 × higher concentrations than normal, making possible the isolation and purification of this enzyme in sufficient quantities that antibodies could be generated and assays developed to lead to cloning of the gene for this enzyme and subsequent expression of the enzyme in both *E. coli* and a chinese hamster ovary (CHO) cell line. Comparison of the melanoma and recombinant forms of the enzymes showed the same activity toward dissolution of blood clots.

Characteristics of t-PA. Tissue plasminogen activator, a proteolytic, hydrophobic enzyme, has a molecular weight of 66,000, 12 disulfide bonds, 4 possible glycosylation sites, and a bridge of 6 amino acids connecting the principal protein structures. Only three of the sites (Asn-117–118, −448) are actually glycosylated. When administered to heart attack victims it dissolves clots consisting of platelets in a fibrin protein matrix and acts by clipping plasminogen, an active precursor protein found in the blood, to form plasmin, a potent protease that degrades fibrin. Concentrations of plasminogen activator are low in blood and tissues.

t-PA Production. Recombinant technology provides the only practical means of rt-PA production. The amount of t-PA required per dose is on the order of 100 mg. Cell lines of transformed CHO cells, selected for high levels of rt-PA expression using methotrexate, are grown in large fermentors. The purification steps for rt-PA must therefore separate out cells, virus, and DNA. Recovering a protein derived from mammalian cells involves a number of steps. In one possible scheme, the culture medium is separated from the cell by sterile filtration (see MICROBIAL AND VIRAL FILTRATION). This is followed by additional removal by cross-flow filtration, ultrafiltration, and chromatography to remove DNA and remaining viruses. The product protein then undergoes purification by chromatography.

The separation of cells from the culture media or fermentation broth is the first step in a bioproduct recovery sequence. Whereas centrifugation is common for recombinant bacterial cells (see CENTRIFUGAL SEPARATION), the final removal of CHO cells utilizes sterile-filtration techniques.

The possibility that DNA from recombinant immortal cell lines such as CHO cells could cause oncogenic (gene altering) events resulting in cancer was a concern during development of the rt-PA purification sequence. The goal for rt-PA purification is to reduce the DNA to <10 pg/dose (10^{-11} g/dose). A level of 0.1 pg was achieved. Special assays are required to detect and quantify these very low levels of DNA in the final product.

Ultrafiltration followed by ion-exchange (qv) chromatography (qv) and then a final round of ultrafiltration concentrate the dilute protein while purifying it.

Independent Assays for Proving Virus Removal. Retroviruses and viruses can also be present in culture fluids of mammalian cell lines. Certainly the absence of virus can be difficult to prove.

Viral clearance can be achieved by use of chaotropes such as urea or guanidine, pH extremes, detergents, heat, formaldehyde, proteases or DNA'ses organic solvents such as formaldehyde, or ion-exchange or size-exclusion chromatography. Because only the inactivation or removal which can be measured counts as validation, a sequence of orthoganol removal/fractionation steps must be used.

Manufacture of Biologics and Government Regulation

The definition of biologics versus drugs continues to evolve. Assignment is made on a case by case basis. Section 351 of the Public Health Service Act defines a biologic product as "any virus, therapeutic serum, toxin, antitoxin, vaccine, blood, blood component or derivative, allergenic product, or analogous product...applicable to the prevention, treatment, or cure of diseases or injuries in man." To compensate for the incomplete analytical capability to define biologics, regulatory agencies include parameters of the process used to make biologics in the control and monitoring. Changes in the process may yield a different product from that previously reviewed and approved. A different product requires a new license. Thus substantial barriers exist in terms of effort, money, and time to making significant changes in processes used to produce licensed biologies, although regulations are evolving.

Biologics are subject to licensing provisions that require that both the manufacturing facility and the product be approved. All licensed products are subject to specific requirements for lot release by the FDA.

The design of bioseparation unit operations is influenced by these governmental regulations. The constraints on process development grow as a recovery and purification scheme undergo licensing for commercial manufacture. Given changes in the products, their regulation, process, and analytical technology, regulations are continually evolving as well.

Protein Chromatography

Proteins and nucleotides are macromolecular biomolecules. Mixtures of biomolecules are fractionated based on differences in charge; molecular weight, shape, and size; solubility in organic solvents; surface hydrophobic character; and types of active sites using ion-exchange, size-exclusion (gel-permeation), reversed-phase, hydrophobic interaction (surface hydrophobicity), and affinity chromatographies, respectively. The appropriate separation may be selected from these five basic classes of chromatography. More than 30% of the purification steps for laboratory-scale protein purification procedures use ion-exchange and/or gel filtration, and at least 20% use affinity chromatography. This pattern is likely to be consistent with industrial practice.

Reversed-phase chromatography is widely used as an analytical tool for protein chromatography, but it is not as commonly found on a process scale for protein purification because the solvents which make up the mobile phase, ie, acetonitrile, isopropanol, methanol, and ethanol, reversibly or irreversibly denature proteins. Hydrophobic interaction chromatography appears to be the least common process chromatography tool, possibly owing to the relatively high costs of the salts used to make up the mobile phases.

Liquid Chromatographs. The basic equipment for liquid chromatography is shown in Figure 2. The column is packed with an adsorbent, ie, the stationary phase. The mixture to be separated is pushed through the column by the eluent or mobile phase. Isocratic chromatography, carried out at a constant flow rate, buffer composition, and temperature, is usually associated with size-exclusion separations. Gradient chromatography typically uses a constant flow rate and temperature, but the composition of the element is altered by mixing two or more buffer reservoirs to achieve a steadily changing salt concentration or changes in pH. The gradients formed are reported in

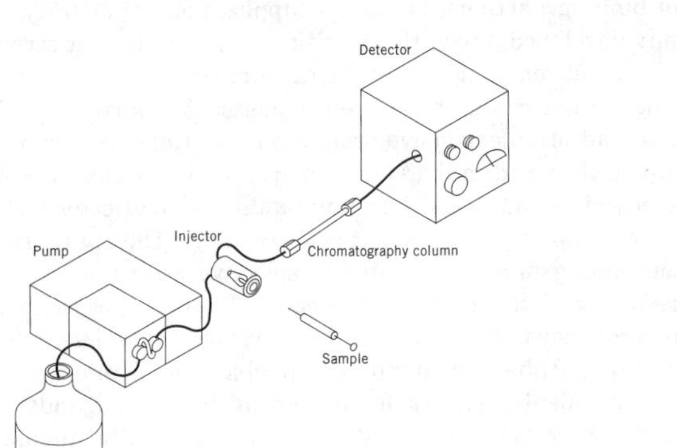

Figure 2. Schematic of a process liquid chromatography system. Courtesy of K. Hanaker.

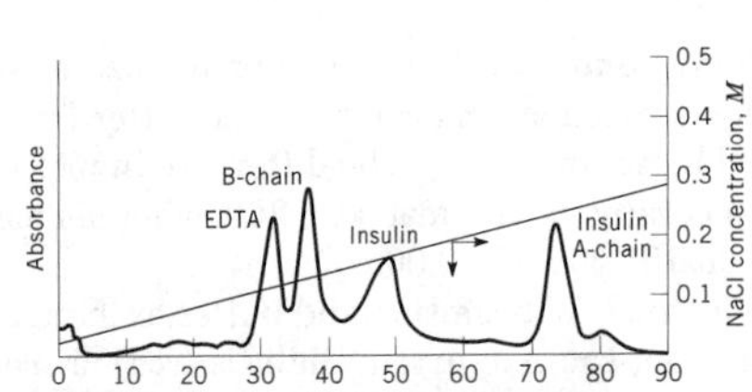

Figure 3. Anion-exchange separation of insulin, and insulin A- and B-chains, over diethylaminoethyl (DEAE) in a 10.9 × 200 mm column having a volume of 18.7 mL. Sample volume is 0.5 mL and protein concentration in 16.7 m*M* Tris buffer at pH 7.3 is 1 mg/mL for each component in the presence of EDTA. Eluent (also 16.7 m*M* Tris buffer, pH 7.3) flow rate is 1.27 mL/min, and protein detection is by uv absorbance at 280 nm. The straight line depicts the salt gradient. Courtesy of the American Chemical Society.

terms of concentration at the inlet of the chromatography column; protein is detected at the column outlet. A chromatogram of the type illustrated in Figure 3 results.

Ion-Exchange Chromatography. Ion-exchange chromatography is initiated by eluting an injected sample through a column using a buffer but no NaCl or other displacing salt. The protein, which has charged sites spread over its surface, displaces anions or cations previously equilibrated on the stationary phase, ie, the protein sites exchange with the salt counterions associated with the ion-exchange stationary phase. A protein having a greater number and/or density of charged sites displaces or exchanges more ions and hence binds more strongly than a protein having a lower charge number or charge density.

After the column is loaded, proteins of similar size and shape are separated by differential desorption from the ion exchanger by using an increasing salt gradient of the mobile phase. The more weakly bound macromolecules elute first; the most tightly bound elute last, at the highest salt concentration. Figure 3 is an example of an anion-exchange separation.

Size Exclusion (Gel-Permeation) Chromatography Size-exclusion chromatography is often referred to as gel-permeation chromatography because the stationary phases are usually made up of soft spherical particles which resemble gels. Separation occurs by a molecular sieving effect (see MOLECULAR SIEVES; SIZE SEPARATION).

The apparatus utilized to carry out size-exclusion (gel-permeation) chromatography is analogous to that used for isocratic operating conditions. The column is packed with a gel-filtration stationary phase, selected according to the molecular weight of the protein of interest.

An example of a size-exclusion chromatogram is given in Figure 4, for a bench-scale (23.5 mL column) separation. The stationary phase is Sepharose CL-6B, a cross-linked agarose with a nominal molecular weight range of $\sim 5000 - 2 \times 10^6$.

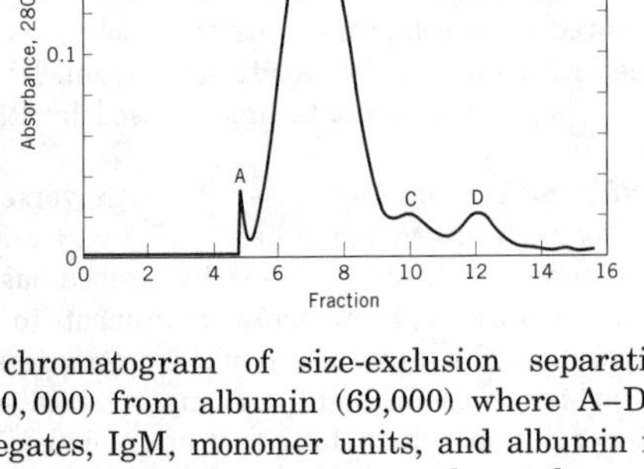

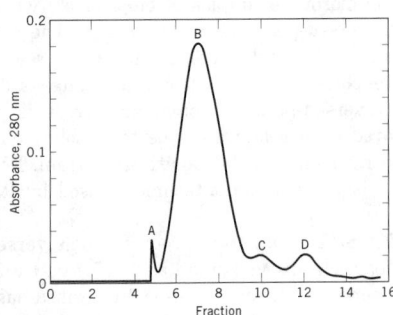

Figure 4. chromatogram of size-exclusion separation of IgM (mol wt = 800,000) from albumin (69,000) where A–D correspond to IgM aggregates, IgM, monomer units, and albumin respectively, using FPLC Superose 6 in a 1 × 30 − cm long column. Courtesy of American Chemical Society.

Buffer Exchange and Desalting. A primary use of size-exclusion chromatography (sec) is for removal of salt or buffer from the protein, ie, desalting and buffer exchange. The difference in molecular weights is large; salts generally have a mol wt <200, whereas mol wts of proteins are between 10,000 and 60,000.

Alternative methods of desalting and buffer exchange include continuous diafiltration, countercurrent dialysis (ccd), a membrane separation technique, and cross-flow filtration, which uses membranes (see MICROBIAL AND VIRAL FILTRATION). Buffer exchange, used to remove denaturing agents in order to induce refolding of proteins, to remove buffers between purification steps, or to remove buffers and other reagents from the final product, is usually carried out at later steps in a recovery sequence.

The use of sec is likely to continue to be a widely practiced technology in industry. Rapid size-exclusion columns for the purpose of buffer exchange have been developed which enable desalting to be achieved at linear velocities of 500–600 cm/h, significantly increasing throughput and reducing operating time and plant floor space. Further, sec using gel-filtration media on cellulosic-based material has a special niche for the partial and controlled separation of denaturing salts from recombinant proteins for purposes of refolding. The development of such rapid desalting techniques is important because of the larger volumes of proteins needing to be processed in industry.

Product Monitoring and Peptide Mapping. One purpose in monitoring a protein product is to detect the presence of a change in which as little as one amino acid has been chemically or biologically altered or replaced during the manufacturing process. Variant amino acid(s) in a protein may not affect protein retention during reversed-phase chromatography if the three-dimensional structure of the polypeptide shields the variant residue from the surface of the reversed-phase support. Reversed-phase chromatography discriminates between different molecules on the basis of hydrophobicity. Because large proteins may contain only small patches of hydrophobic residues, these patches may not correlate to the molecular modifications which a reversed-phase analytical method seeks to detect. The reversed-phase method must therefore be completely validated, and preferably combined with controlled chemical and/or proteolytic hydrolysis followed by chromatography or electrophoresis (see ELECTROSEPARATIONS) of the cleared protein to give a map of the resulting peptide fragments.

A peptide map is generated by cleaving a previously purified protein using chemicals or enzymes. Hydrolytic agents having known specificity are used to perform limited proteolysis followed by resolution and identification of all the peptide fragments formed. Identification of changes, and reconstruction of the protein's primary structure, is then possible. Reagents and enzymes which cleave specific bonds are discussed in the literature.

Reversed-Phase Process Chromatography. Polypeptides, peptides, antibiotics, alkaloids, and other low molecular weight compounds are amenable to process chromatography by reversed-phase methods. There are numerous examples of bioproducts which have been purified using reversed-phase chromatography. The manufacture of salmon calcitonin, a 32-residue peptide used for treatment of postmenopausal osteoporosis, hypercalcemia, and Paget's disease of the bone, includes reversed-phase chromatography. This peptide, commercially prepared on a kilogram scale by a solid-phase synthesis, is then purified by a multimodal purification train. Reversed-phase chromatography is the dominant technique used by Rhône-Poulenc Rorer.

Small Particle Silica Columns. Process-scale reversed-phase supports can have particle sizes as small as 5–25 μm. Unlike polymeric reversed-phase sorbents, these small-particle silica-based reversed-phase supports require high pressure equipment to be properly packed and operated. The introduction of axial compression columns has helped promote the use of high performance silica supports on a process scale. Resolution approaching that of an analytical-scale separation can be achieved using these columns that can also be quickly packed.

Hydrophobic Interaction Chromatography. Hydrophobic interactions of solutes with a stationary phase result in their adsorption on neutral or mildly hydrophobic stationary phases. The solutes are adsorbed at a high salt concentration, and then desorbed in order of increasing surface hydrophobicity, in a decreasing kosmotrope gradient. This characteristic follows the order of the lyotropic series for the anions: phosphates > sulfates > acetates > chlorides > nitrates > thiocyanates. Anions which precipitate proteins less effectively than chloride (nitrates and thiocyanates) are chaotropes or water structure breakers, and have a randomizing effect on water's structure; the anions preceding chlorides, ie, phosphates, sulfates, and acetates, are polar kosmotropes or water structure makers. These promote precipitation of proteins. Kosmotropes also promote adsorption of proteins and other solutes onto a hydrophobic stationary phase. These kosmotropes have other beneficial characteristics which include increasing the thermal stability of enzymes, decreasing enzyme inactivation, protecting against proteolysis, increasing the association of protein subunits, and increasing the refolding rate of denatured proteins. Hence, utilization of hydrophobic interaction chromotography is attractive for purification of proteins where recovery of a purified protein in an active and stable conformation is desired.

Various types of proteins have been purified using hydrophobic interaction chromatography including alkaline phophatase, estrogen receptors, isolectins, strepavidin, calmodulin, epoxide hydrolase, proteoglycans, hemoglobins, and snake venom toxins. In the case of cobra venom toxins, the order of elution of the six cardiotoxins supports the hypothesis that the mechanism of action is related to hydrophobic interactions with the phospholipids in the membrane.

The recovery of recombinant chymosin from a yeast fermentation broth showed that large-scale hydrophobic interaction chromatography could produce an acceptable product in one step.

Affinity Chromatography. The concept of affinity chromatography, credited to the discovery of biospecific adsorption in 1910, was reintroduced as a means to purify enzymes in 1968. Substrates and substrate inhibitors diffuse into the active sites of enzymes irreversibly or reversibly binding there. Conversely, if the substrate or substrate inhibitor is immobilized through a covalent bond to a solid particle of stationary phase having large pores, the enzyme should be able to diffuse into the stationary phase and bind with the substrate or inhibitor. Because the substrate is small (mol wt <500) and the enzyme large (>15,000), the diffusion of the enzyme to its binding partner at a solid surface can be sterically hindered. The placement of the substrate at the end of an alkyl or glycol chain tethered to the stationary phase's surface reduces hindrance and forms the basis of affinity chromatography. This concept has also been applied to ion-exchange chromatography under the names of tentacle or fimbriated stationary phases.

The realization that enzymes could be selectively retained in a chromatography column packed with particles of immobilized substrates or substrate analogues led to experiments with other pairs of binding partners. Numerous applications of affinity chromatography developed, given the specific and reversible yet strong affinity of biological macromolecules for numerous specific ligands or effectors. These interactions have been exploited for purposes of highly selective, but often expensive protein purifications, recovery of messenger ribonucleic acid (mRNA) in some recombinant DNA applications, and study of mechanisms of protein binding with effector molecules.

Minimization of Nonspecific Binding. The purpose of affinity chromatography is the highly selective adsorption and subsequent recovery of the target biomolecule. Loss of specificity occurs when macromolecules, other than the targeted materials, adsorb onto the stationary phase owing to hydrophobic or ionic interactions.

The ideal matrices for anchoring binding ligands are nonionic, hydrophilic, chemically stable, and physically robust. The most popular matrices are polysaccharide based, principally owing to their hydrophilic character and history of use as size-exclusion or gel-permeation gels, although glass beads, polyacrylamide gels, cross-linked dextrans (Sephadex), and agarose synthesized into a bead form have all been used.

Multistep Processes. An excellent synopsis and industrial viewpoint of affinity chromatography and its fit with other bioseparations unit operations is available. Ligands range from the low molecular weight components, eg, arginine and benzamidine, which both bind trypsin-like proteases, triazine dyes, and metal chelates; to high molecular weight ligands, eg, protein A, immunoglobins, and monoclonal antibodies. The blood factor VIII, purified by monoclonal affinity techniques, was approved by the U.S. FDA in 1988. Limitations of affinity chromatography as an industrial separation technique can be due to leaching of bound ligands from the column into the product at ppm levels, nonspecific interactions resulting in contamination of the target molecule, and failure of the affinity ligand to differentiate all variant forms of a protein or polypeptide. Because many antibody preparations cannot differentiate between minor structural changes in proteins, affinity chromatography must be followed by other separation steps, and does not provide a one-step purification.

Receptor Affinity Chromatography. Receptor affinity chromatography is a selective form of immunoaffinity chromatography which is based on antigen-antibody interactions. Protein or polypeptide ligands used in preparing receptor affinity supports are themselves products of fermentation of recombinant microorganisms and are subjected to a separate sequence of purification steps, prior to being reacted with a functionalized stationary phase to form the affinity support. The resulting affinity chromatography columns are expensive when viewed on the basis of cost of support/unit volume of stationary phase. The cost/benefit ratio would still be attractive because process-scale columns can be small (volumes on the order of 1–10 L). Moreover, as with other types of affinity chromatography, purification of dilute but highly active protein is possible.

MICHAEL R. LADISCH
Purdue University

R. C. Willson and M. R. Ladisch, in M. R. Ladisch, R. C. Willson, C-C. Painton, and S. E. Builder, eds., *ACS Symp. Ser.* **427,** 1–13 (1990).

Committee on Bioprocess Engineering, National Research Council, *Putting Biotechnology to Work: Bioprocess Engineering,* National Academy of Sciences, Washington, D.C., 1992, pp. 2–22.

S. M. Wheelwright, *Protein Purification: Design and Scale-up of Downstream Processing,* Hanser Publishers, Munich, Germany, 1991, pp. 1–9, 61, 213–217.

C. A. Bisbee, *GEN* **13**(14), 8–9 (1993).

S. E. Builder, R. van Reis, N. Paoni, and J. Ogez, *Proc. 8th Int. Biotechnol. Symp.,* Paris, July, 1989.

BIOTECHNOLOGY

This article offers an overview of the coverage of biotechnology, ie, the use of biochemical and biological materials and processes, in the *Encyclopedia.* Biotechnology has long had a role in chemical technology, and information on the various processes and materials is well integrated in articles throughout the *Encyclopedia.*

In the early years of the chemical industry, use of biological agents centered on fermentation (qv) techniques for the production of food products, eg, vinegar (qv), cheeses (see MILK AND MILK PRODUCTS), beer (qv), and of simple organic compounds such as acetone (qv), ethanol (qv), and the butyl alcohols (qv). By the middle of the twentieth century, most simple organic chemicals were produced synthetically. Fermentation was used for food products and for more complex substances such as pharmaceuticals (qv) (see also ANTIBIOTICS). Moreover, supports were developed to immobilize enzymes for use in industrial processes such as the hydrolysis of starch (qv) (see ENZYME APPLICATIONS).

Advances in molecular biology and genetic engineering (qv) during the latter part of the twentieth century have widened the scope of possibilities for the use of biotechnological methods and resulted in increased interest on the part of the chemical industry. Microorganisms and mammalian cells are grown on an industrial scale (see AERATION, BIOTECHNOLOGY; CELL CULTURE TECHNOLOGY) to be harvested for their chemical output (see GROWTH REGULATORS; HORMONES-HUMAN GROWTH HORMONE; INSULIN AND OTHER ANTIDIABETIC DRUGS; and VACCINE TECHNOLOGY). Enzymes and microorganisms are utilized industrially to effect chemical modification of materials or to direct the outcome of synthetic reactions (see ENZYMES IN ORGANIC SYNTHESIS; MICROBIAL TRANSFORMATIONS). Customized biological molecules are biologically produced to meet the needs of industry (see ENZYME INHIBITORS; IMMUNOTHERAPEUTIC AGENTS; PHARMACEUTICALS; PROTEIN ENGINEERING). Biopolymers (qv), ie, carbohydrates (qv), enzymes, nucleic acids (qv), and proteins (qv), are used in clinical and chemical analyses both for detection (see AUTOMATED INSTRUMENTATION-CLINICAL CHEMISTRY; BIOSENSORS; IMMUNOASSAY; MEDICAL DIAGNOSTIC REAGENTS) and for separation (see BIOPOLYMERS; CHROMATOGRAPHY) of materials.

G. P. Bailon and D. R. Weber, *Nature* **335**(6193), 839–840 (1995).

BIOTIN. See VITAMINS.

BIPHENYL AND TERPHENYLS

Biphenyl (diphenyl, phenylbenzene) and terphenyl are the lowest members of a family of polyphenyls in which benzene rings are attached one to another in a chainlike manner, $C_6H_5(C_6H_4)_mC_6H_5$. Many higher polyphenyls are known, but only biphenyl and the terphenyls are of commercial significance.

Physical Properties

Pure biphenyl is a white crystalline solid that separates from solvents as plates or monoclinic prismatic crystals. Commercial samples are often slightly yellow or tan in color. Similarly, pure terphenyls are white crystalline solids whereas commercial grades are somewhat yellow or tan. Physical and chemical constants for biphenyl and the three isomeric terphenyls, respectively, are given in Tables 1 and 2.

Chemical Properties

Biphenyl and terphenyls may be regarded as substituted benzenes that undergo acylation, alkylation, halogenation, nitration, sulfonation, and other reactions common to benzene (qv). The points of initial attack on chlorination, nitration, and sulfonation of biphenyl occur at the 2- and 4-positions; the latter group predominates.

Terphenyls, like biphenyl, undergo the usual reactions of aromatic hydrocarbons. The *ortho-* and *para-* isomers nitrate initially at the 4-position.

Manufacture

Dow, Monsanto, and Koch Chemical Company are the principal biphenyl producers, with lesser amounts coming from Sybron Corporation and Chemol, Inc. With the exception of Monsanto, the

Table 1. Physical Properties of Biphenyl

Property	Value
melting point, °C	69.2
freezing point commercial grades, °C	68.5–69.4
boiling point at 101.3 kPa[a] °C	255.2 ± 0.2
critical properties	
temperature, °C	515.7
pressure, MPa[a]	4.05
density, g/mL	0.314
flash point, °C	113.0
fire point, °C	123.0
autogenous ignition temperature, °C	560.0
heat of combustion, kJ/mol[b]	6243.2
heat of fusion, kJ/mol[b]	18.60[c,d]

[a] To convert kPa to mm Hg, multiply by 7.5. [b] To convert MPa to psi, multiply by 145. [c] To convert J to cal, divide by 4.184. [d] To convert W/(cm · K) to (cal·cm)/(s·cm^2·°C), divide by 4.184.

Table 2. Physical Properties of Pure Terphenyl Isomers

Property	*Ortho-*	*Meta-*	*Para-*
melting point, °C	56.2	87.5	212.7
boiling point at 101.3 kPa[a], °C	332.0	365.0	376.0
heat of vaporization at 101.3 kPa[a]	253.0	279.0	272.0
flash point, °C	171.0	206.0	210.0
fire point, °C	193.0	229.0	238.0
auto ignition temperature, °C	530.0	555.0	555.0
vapor pressure, kPa[a]			
93°C	0.01172	0.00165	
315.6°C	64.40	27.3	
density of liquid, g/L			
93°C	1022.0	1039.0	solid
315.6°C	842.0	871.0	879.0
heat capacity of liquid, kJ/kg[b]			
93°C	1.007	0.970	
398.9°C	1.400	1.397	1.116
viscosity of liquid, mPa·s(= cP)			
100°C	4.34	3.87	solid
300°C	0.30	0.40	0.43
thermal conductivity of liquid, W/(m·K)[c]			
100°C	0.1316	0.1347	
210°C	0.1206	0.1356	0.1359
heat of vaporization, J/g at 252°C	280.0	298.0	305.0
heat of fusion, kJ/kg[b]	55.2	73.7	146.5
critical temperature, K	891.0	927.0	926.0
critical pressure, MPa[d]	3.903	3.503	3.330

[a] To convert kPa to mm Hg, multiply by 7.5.
[b] To convert J to cal, divide by 4.184.
[c] To convert W/(m · K) to (cal·cm)/(s·cm^2·°C), divide by 418.4.
[d] To convert MPa to psi, multiply by 145.

above suppliers recover biphenyl from high boiler fractions that accompany the hydrodealkylation of toluene to benzene.

High purity biphenyl is currently produced by Monsanto in the United States and United Kingdom by direct dehydrocondensation of benzene. Terphenyls are also obtained from the higher boiling polyphenyl by-products that accompany the biphenyl.

Shipping

By-product biphenyl is usually sold as a dye carrier in the molten state in tank truck or tank car lots. Grades of higher purity are also sold in the molten state or as flakes in 22.7-kg bags.

Biphenyl is defined as a toxic chemical under, and subject to, reporting requirements of Section 313 of Title III of the Superfund Amendments and Reauthorization Act (SARA) of 1986 and 40 CFR, Part 372, under the name biphenyl. It is identified as a hazardous chemical under criteria of the OSHA Hazard Communication Standard (29 CFR 1910.1200).

The small amount of mixed terphenyls that are sold as such are shipped in the form of flaked solids in 22.7-kg multiwall bags. The U.S. freight classification is Plastics, synthetic other than liquid, NOIBN. Like biphenyl, mixed terphenyls fall under the hazardous chemical criteria of the OSHA Hazard Communication Standard (29 CFR 1910.1200).

Economic Aspects

Reliable estimates of annual production of biphenyl in the United States are difficult to obtain. About 10% of the biphenyl derived from HDA sources is consumed as 93–95% grade in textile dye carrier applications. The remainder is used for alkylation or upgraded to ≥99.9% grades for heat-transfer purposes. Essentially all of the high purity biphenyl produced by dehydrocondensation of benzene is used as alkylation feedstock or is utilized directly in heat-transfer applications.

As in the case of biphenyl, current worldwide production figures for terphenyls are not readily obtainable. Currently, most of the terphenyl produced is converted to a partially hydrogenated form.

Health and Safety Factors

Although biphenyl and the terphenyls fall under the hazardous chemical criteria of the OSHA Hazard Communications Standard, the products themselves are fairly low in toxicity and do not constitute a serious industrial hazard.

Because biphenyl is often transported in the molten state, a moderate fire hazard does exist under these circumstances.

Environmental Considerations

The widespread use of biphenyl and methyl-substituted biphenyls as dye carriers (qv) in the textile industry has given rise to significant environmental concern because of the amount released to the environment in wastewater effluent. Although biphenyl and simple alkylbiphenyls are themselves biodegradable, the prospect of their conversion by chlorination to PCBs in the course of wastewater treatment has been a subject of environmental focus. Despite the fact that the lower chlorinated biphenyls are also fairly biodegradable continued environmental concern has resulted in decreased use of biphenyl as a dye carrier (see DYES, ENVIRONMENTAL CHEMISTRY).

Terphenyls in heat-transfer applications are used in relatively smaller quantities with negligible release to the environment. They are sufficiently biodegradable so as not to constitute an environmental threat.

Derivatives

Short-chain alkylated biphenyls are the principal biphenyl derivatives in commercial use. They are generally produced by liquid-phase

Friedel-Crafts alkylation of biphenyl with ethylene, propylene, or mixed butenes.

Ortho- and *para*-phenylphenols are commercially significant biphenyl derivatives that do not involve biphenyl as a starting material. Both are produced as by-products from the hydrolysis of chlorobenzene with aqueous sodium hydroxide. *o*-Phenylphenol, ie, 1,1-biphenyl-2-ol, particularly as its sodium salt, is widely used as a germicide or fungicide. *para*-Phenylphenol with formaldehyde forms a resin used in surface coatings.

Several functionalized biphenyls either are, or show promise of becoming, commercially significant polymer building blocks.

QUENTIN E. THOMPSON
Monsanto Company

W. M. Meylan and P. H. Howard, *Chemical Market Input/Output Analysis of Selected Chemical Substances to Assess Sources of Environmental Contamination: Task II. Biphenyl and Diphenyl Oxide,* EPA Contract No. 68-1-3224, Syracuse, N.Y., 1976, p. 2.

H. Mandel, *Heavy Water Organic Cooled Reactor, Physical Properties of Some Polyphenyl Coolants,* AEC Report A 1-CE-15, Apr. 15, 1966.

W. E. Taylor, in R. F. Makens, ed., *Organic Coolant Summary Report, AEC Accession No. 15554, Report No. IDO-11401,* Idaho Operations Office, U.S. Atomic Energy Commission, CFSTI, Washington, D.C., 1964, pp. 9–38.

R. Dasgupta and B. Maiti, *Ind. Eng. Chem. Process Des. Dev.* **25**(2), 381 (1986); U.S. Pat. 2,143,509 (1939), C. Conover and A. E. Huff (to Monsanto Co.).

BISMUTH AND BISMUTH ALLOYS

The element bismuth, Bi, found in Group 15 (VA) of the Periodic Table, has at no. 83, at wt 208.98. Its valences are +5 and +3. Bismuth is a silvery metal having a high metallic luster and exhibiting a slightly pink tinge on a cleanly broken surface. The metal itself is brittle in nature and easily broken.

Occurrence

Bismuth is referred to as a minor metal. Bismuth occurs in the earth's crust in a concentration of approximately 0.1 ppm on the average.

Properties

The physical properties of bismuth, summarized in Table 1, are characterized by a low melting point, a high density, and expansion on solidification. The solid metal floats on the liquid metal as ice floating on water. Bismuth is the most diamagnetic of the metals, and it is a poor electrical conductor. The thermal conductivity of bismuth is lower than that of any other metal except mercury.

Table 1. Physical Properties of Bismuth

Property	Value
bp, °C	1.560
crystal structure	rhombohedral
density, kg/m^3	
20°C	9,800
271°C[a]	10,070
electrical resistivity, Ω·cm	
0°C	106×10^{-6}
mp, °C	271.3
vapor pressure, kPa[b]	
400°C	1.013×10^{-4}
880°C	1.013×10^{2}
viscosity, mPa·s(= cP)	
285°C	1.610
365°C	1.460
600°C	0.998

[a] Liquid.
[b] To convert kPa to psi, divide by 6.895×10^3.

Production

Bismuth is mined primarily as a by-product of the processing of ores of other metals, mostly copper and lead. The countries that mine significant quantities of bismuth are Australia, Bolivia, Canada, China, Japan, Mexico, Peru, and the United States.

The principal portion of the bismuth in copper (qv) ores follows the copper into the matte. During the conversion of the matte to blister copper most of the bismuth fumes off. The fumes are caught in the baghouse or Cottrell system along with other elements such as lead (qv), arsenic, and antimony. These dusts are transferred to the lead-smelting operation. The portion of the bismuth remaining with the blister copper is separated during the electrolytic refining in the slimes. The procedure for handling the slimes results in the bismuth being collected in the lead bullion.

The bismuth that is found in the lead ore accompanies the lead through the smelting operation right up to the last refining steps. The removal of bismuth then requires special techniques, the most common being the Betterton-Kroll and the Betts processes.

Recovery of Bismuth from Tin Concentrates. Bismuth is leached from roasted tin concentrates and other bismuth-bearing materials by means of hydrochloric acid.

The Sperry process for making white lead in an electrolytic cell recovers bismuth as a by-product in the anode slimes.

Refining. The alloy of bismuth and lead from the separation procedures is treated with molten caustic soda to remove traces of such acidic elements as arsenic and tellurium. It is then subjected to the Parkes desilverization process to remove the silver and gold present. This process is also used to remove these elements from lead.

The desilverized alloy now contains bismuth as well as lead and zinc. To remove the lead and zinc, advantage is taken of the fact that zinc and lead chlorides are formed before bismuth chloride, $BiCl_3$, when the alloy is treated at 500°C with chlorine gas. Zinc chloride, $ZnCl_2$, forms first, and after its removal lead chloride, $PbCl_2$, forms preferentially.

Fabrication. There are four basic forms of bismuth that are readily available commercially: ingot, needle, pellet, and powder.

Economic Aspects

The United States is highly dependent on bismuth imports because domestic usage greatly outruns domestic production. The supply of bismuth metal is dependent on the supply of the associated metals with which it is mined.

Uses

The three primary categories of uses of bismuth in industry are chemical, metallurgical additive, and fusible alloy. The chemical category can be broken down into pharmaceuticals, cosmetics, catalysts, industrial pigments, and electronics, and the metallurgical additive category into steel, aluminum, and cast-iron additives. The fusible alloy category is divisible into more than a dozen subcategories dependent on a specific application.

Bismuth Alloys

Because bismuth expands on solidification and because it alloys with certain other metals to give low melting point alloys, bismuth is particularly well-suited for a number of uses. Alloys of bismuth can be made that expand, shrink, or remain dimensionally stable on solidification. Bismuth alloys and uses are summarized in Table 2.

Safety

No industrial poisoning from bismuth has been reported. However, precautions should be taken against the careless handling of bismuth

Table 2. Alloy Compositions and Uses

Melting point, °C	Composition, wt %					Uses[a]
	Bismuth	Lead	Tin	Cadmium	Indium	
47	44.7	22.6	8.3	5.3	19.1	LB
58	49.0	18.0	12.0	10.0	21.0	LB
70	50.0	26.37	13.3	10.0	0.0	RS, W, LB
101	39.4	29.8	30.8	0.0	0.0	FSD
124	55.5	44.5	0.0	0.0	0.0	PC
138	58.0	0.0	42.0	0.0	0.0	W, SMF, FC
138/170	40.0	0.0	60.0	0.0	0.0	IC

[a] LB = lens blocking; W = work holding; RS = radiation shielding; FSD = fusible safety device; PC = proof casting; SMF = sheet metal forming; FC = cores; IC = investment casting.

and its compounds; ingestion and inhalation of dusts and fumes should be avoided.

MARK J. CHAGNON
Metalspecialties, Inc.

J. W. Hasler, M. H. Miller, and R. M. Chapman, *United States Mineral Resources,* Geological Survey Professional Paper 820, U.S. Dept. of the Interior, Washington, D.C., 1973, p. 96.

A. R. Powell, in *Papers of the Institute of Mining and Metallurgy Symposium,* Institute of Mining and Metallurgy, London, U.K., 1950, pp. 245–257.

J. O. Betterton and Y. E. Levebeff, *Metallurgy of Lead and Zinc,* American Institute of Mining and Metallurgy Engineers, New York, 1936.

BISMUTH COMPOUNDS

Bismuth is the fifth member of the nitrogen family of elements and, like its congeners, possesses five electrons in its outermost shell, $(6s^2)(6p^3)$. In many compounds, the bismuth atom utilizes only the three 6*p* electrons in bond formation and retains the two 6*s* electrons as an inert pair. Compounds are also known where bismuth is bonded to four, five, or six other atoms. Many bismuth compounds do not have simple molecular structures and exist in the solid state as polymeric chains or sheets.

Technical information concerning bismuth and its compounds is distributed periodically by the Bismuth Institute, a nonprofit organization incorporated in La Paz, Bolivia, that has an information center in Brussels.

Inorganic Compounds of Bismuth

Inorganic compounds of bismuth include bismuthine, bismuthides, bismuth halides, bismuth oxide halides, bismuth oxides and bismuthates, higher oxides of bismuth and related compounds, sulfides and related compounds, and bismuth salts. Properties of important inorganic bismuth compounds are given in Table 1.

Organobismuth Compounds

In a manner similar to phosphorus, arsenic, and antimony, the bismuth atom can be either tri- or pentacovalent. However, organobismuth compounds are less stable thermally than the corresponding phosphorus, arsenic, or antimony compounds, and there are fewer types of organobismuth compounds.

The chemistry of organobismuth compounds has been described in several publications. The use of organobismuth compounds, as well as organoantimony ones, in organic synthesis has been exhaustively reviewed.

Primary and Secondary Bismuthines. Only one primary and one secondary bismuthine are known, namely methylbismuthine, CH_3BiH_2, and dimethylbismuthine, $(CH_3)_2BiH$. They are prepared by the lithium aluminum hydride reduction of methyldichlorobismuthine, CH_3BiCl_2, and dimethylchlorobismuthine, respectively, in a nitrogen atmosphere at $-150°C$.

Tertiary bismuthines appear to have a number of uses in synthetic organic chemistry. They have also been employed for the preparation of (qv) of superconducting bismuth strontium calcium copper oxide, as cocatalysts for the polymerization of alkynes, as inhibitors of the flammability of epoxy resins, and for a number of other industrial purposes.

Halobismuthines, Dihalobismuthines, and Related Compounds. Chloro-, dichloro-, bromo-, and dibromobismuthines are best prepared by the reaction of a tertiary bismuthine and bismuth trichloride or tribromide. Iodo- and diiodobismuthines are easily obtained by the reaction of the corresponding chloro-, dichloro-, bromo-, or dibromobismuthine with sodium or potassium iodide.

Dibismuthines. Only about a dozen tetraalkyl- and tetraaryldibismuthines are known. These compounds can be obtained by the reaction of a sodium dialkyl- or diarylbismuthide and a 1,2-dihaloethane. Several dibismuthines have also been obtained by the addition of the stoichiometric amount of sodium to a solution of a halobismuthine in liquid ammonia. The best method for the synthesis of tetraphenyldibismuthine, $C_{24}H_{20}Bi_2$, involves the reduction of diphenyliodobismuthine, $C_{12}H_{10}BiI$, using bis(cyclopentadienyl)cobalt(II), $C_{10}H_{10}Co$. Dibismuthines tend to be thermally unstable. They also are very sensitive to oxidation.

Bismin and Its Derivatives. Bismin (bismabenzene), C_5H_5Bi, the bismuth analogue of pyridine, has never been isolated, but it can be formed in solution by the dehydrohalogenation of 1-chloro-1,4-dihydrobismin, C_5H_6BiCl, using 1,8-diazabicyclo[5.4.0]undec-7-ene (DBU) at low temperatures.

Diarylbismuthinic Acids and Their Esters. Although organobismuth(V) compounds containing three, four, or five Bi—C bonds are well known, no compounds containing two such bonds had been prepared until a number of methyl diarylbismuthinates (diarylmethoxybismuth oxides) were reported in 1988.

Because organobismuth(V) compounds have found considerable use as oxidizing agents, the oxidizing ability of methyl di-1-naphthylbismuthinate, $C_{21}H_{17}BiO_2$, was investigated.

Trialkyl- and Triarylbismuth Dihalides and Related Compounds. After the tertiary bismuthines, the triarylbismuth dihalides constitute the most important class of organobismuth compounds and are by far the largest class of compounds containing pentacovalent bismuth. However, only two trialkylbismuth dihalides have been prepared. These are *cis*-tripropenylbismuth dibromide, $C_9H_{15}BiBr_2$, and *trans*-tripropenylbismuth dibromide, $C_9H_{15}BiBr_2$, prepared by oxidative bromination of the corresponding bismuthines at $-55°C$.

In addition to use in organic synthesis, triarylbismuth dihalides and related compounds have found limited industrial use. A patent has been issued for the use of such compounds as antifungal agents on plastics or fibrous material. Triarylbismuth dihalides have been used as catalysts for the carbonation of epoxides to form cyclic carbonates.

Table 1. Physical Properties of Bismuth Compounds

Bismuth compound	Formula	Mp, °C	Bp, °C	$\Delta H°_{f,298}$, kJ/mol[a]	$S°_{298}$, J/mol·K[a]	$\Delta H°_{fusion}$, kJ/mol[a]	$\Delta S°_{fusion}$, J/mol·K[a]
bismuth trifluoride	BiF_3	649[b]	900 ± 10	−900 ± 13	123 ± 4	21.6 ± 0.6	23.4 ± 0.8
bismuth pentafluoride	BiF_5	151	230				
bismuth trichloride	$BiCl_3$	233.5	44.7	−379	174 ± 6	23.9	
bismuth monochloride	BiCl			−131	94.5		
bismuth oxychloride	BiOCl			−367	120		
bismuth tribromide	$BiBr_3$	219	441	−276 ± 2	190 ± 1	21.8	
bismuth triiodide	BiI_3	408.6	542[c]	−151 ± 4	224.8 ± 0.8	39.1 ± 0.3	57.3 ± 0.4
bismuth trioxide	Bi_2O_3[d]	824		−574	151.5		
bismuth trisulfide	Bi_2S_3	850		−143	200		
bismuth tritelluride	Bi_2Te_3			−77.4	260.9		

[a] To convert J to cal, divide by 4.184.
[b] The mp frequently cited is 120°C higher than this and is, apparently, for material contaminated with oxyfluoride.
[c] The normal bp has been extrapolated from vapor-pressure data.
[d] Monoclinic.

Quaternary Bismuth Compounds. It was not until 1952 that tetraphenylbismuth bromide, $C_{24}H_{20}BiBr$, was obtained from pentaphenylbismuth, $C_{30}H_{25}Bi$, and one molar equivalent of bromine at $-70°C$. In a similar manner tetraphenylbismuth chloride, $C_{24}H_{20}BiCl$, and tetraphenylbismuthonium tetrafluoroborate, $C_{24}H_{20}BBiF_4$, are obtained from pentaphenylbismuth and hydrogen chloride or hydrogen tetrafluoroborate, respectively. A number of other tetraarylbismuth compounds Ar_4BiY, where Y is a group, such as NO_3^-, ClO_4^-, OCN^-, N_3^-, etc, have been prepared from the chloride by metathesis.

Quaternary bismuth compounds are generally unstable. They have not found extensive use in industry or in organic synthesis.

Bismuthonium Ylides. Prior to 1988 the only bismuthonium ylides known were (**1**) and (**2**).

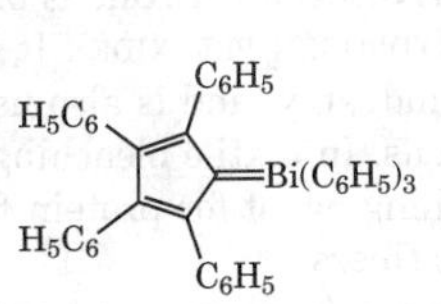

(**1**)

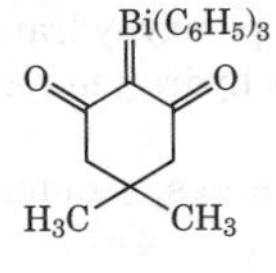

(**2**)

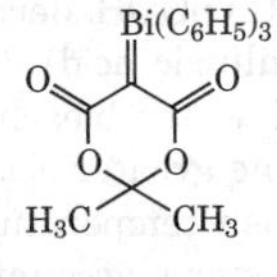

(**3**)

Structure (**1**) is an unstable blue solid which cannot be obtained in a pure state; structure (**2**), however, is stable. Structure (**2**) was obtained from triphenylbismuth carbonate and dimedone. More recently a number of bismuthonium ylides, eg, (**3**), have been prepared and their reactions studied.

Quinquenary Bismuth Compounds. No pentaalkylbismuth compounds have been reported, but a number of pentaarylbismuth compounds are known. Pentaphenylbismuth, $C_{30}H_{25}Bi$, was first prepared by means of the reaction,

$$(C_6H_5)_3BiCl_2 + 2\ C_6H_5Li \longrightarrow (C_6H_5)_5Bi + 2\ LiCl$$

It can also be prepared by the reaction of phenyllithium with tetraphenylbismuth chloride or the *N*-triphenylbismuth derivative of 4-toluenesulfonamide. Pentaphenylbismuth is a violet-colored, crystalline compound that decomposes spontaneously after standing for several days in a dry nitrogen atmosphere.

Pentaphenylbismuth has been studied as a reagent in organic synthesis where it can act either as an oxidizing or an arylating agent.

Bismuth Compounds Used in Medicine

Antibiotics (qv), especially penicillin, have made both arsenic and bismuth compounds completely obsolete for the treatment of syphilis.

Bismuth subsalicylate, Pepto-Bismol, is a basic salt of varying composition, corresponding approximately to *o*-$HOC_6H_4CO_2(BiO)$. Like a number of other insoluble bismuth preparations, it is not currently approved in the United States for the treatment of peptic ulcer disease but is under active investigation for this purpose. It does appear to be effective for the relief of mild diarrhea and for the prevention of travelers' diarrhea.

Bismuth subcarbonate (basic bismuth carbonate) is a white or pale yellow powder that is prepared by interaction of bismuth nitrate and a water-soluble carbonate. It has been widely used as an antacid.

De-Nol, tripotassium dicitratobismuthate (bismuth subcitrate), is a buffered aqueous suspension of a poorly defined, water-insoluble bismuth compound. It is said to be very effective for the treatment of gastric and duodenal ulcers.

Bismuth subnitrate (basic bismuth nitrate) can be prepared by the partial hydrolysis of the normal nitrate with boiling water. It has been used as an antacid and in combination with iodoform as a wound dressing.

Bismuth subgallate (basic bismuth gallate), Dermatol, is a bright yellow powder that can be prepared by the interaction of bismuth nitrate and gallic acid in an acetic acid medium. It has been employed as a dusting powder in some skin disorders and as an ingredient of suppositories for the treatment of hemorrhoids. It has been taken orally for many years by colostomy patients in order to control fecal odors, but the drug may cause serious neurological problems.

G. GILBERT LONG
LEON D. FREEDMAN
G. O. DOAK
North Carolina State University

S. Samaan, *Methoden der Organischen Chemie. Metallorganische Verbindungen As, Sb, Bi,* Band XIII, Teil 8, Georg Thieme Verlag, Stuttgart, Germany, 1978.

M. Wieber, *Gmelin Handbuch der Anorganischen Chemie,* Band 47, Springer-Verlag, Berlin, 1977.

L. D. Freedman and G. O. Doak, *Chem. Rev.* **82,** 15 (1982).

G. O. Doak and L. D. Freedman, *Organometallic Compounds of Arsenic, Antimony, and Bismuth,* Wiley-Interscience, New York, 1970.

BITUMEN. See ASPHALT; ROOFING MATERIALS.

BLEACHING AGENTS

SURVEY

A bleaching agent is a material that lightens or whitens a substrate through chemical reaction. The bleaching reactions usually involve oxidative or reductive processes that degrade color systems. These processes may involve the destruction or modification of chromophoric groups in the substrate as well as the degradation of color bodies into smaller, more soluble units that are more easily removed in the bleaching process. The most common bleaching agents generally fall into two categories: chlorine and its related compounds (such as sodium hypochlorite) and the peroxygen bleaching agents, such as hydrogen peroxide and sodium perborate. Reducing bleaches represent another category. Bleaching agents are used for textile, paper, and pulp bleaching as well as for home laundering.

Chlorine-Containing Bleaching Agents

Chlorine-containing bleaching agents are the most cost-effective bleaching agents known. They are also effective disinfectants, and water disinfection is often the largest use of many chlorine-containing bleaching agents. They may be divided into four classes: chlorine, hypochlorites, *N*-chloro compounds, and chlorine dioxide.

The first three classes are called available chlorine compounds and are related to chlorine by the equilibria in equations 1–4. These equilibria are rapidly established in aqueous solution, but the dissolution of some hypochlorite salts and *N*-chloro compounds can be quite slow.

$$Cl_2(gas) \rightleftharpoons Cl_2(aq) \tag{1}$$

$$Cl_2(aq) + H_2O \rightleftharpoons HOCl + H^+ + Cl^- \tag{2}$$

$$HOCl \rightleftharpoons H^+ + OCl^- \tag{3}$$

$$RR'NCl + H_2O \rightleftharpoons HOCl + RR'NH \tag{4}$$

The total concentration or amount of chlorine-based oxidants is often expressed as available chorine, or less frequently as active chlorine. Available chlorine is the equivalent concentration or amount of Cl_2 needed to make the oxidant according to equations 1–4. Active chlorine is the equivalent concentration or amount of Cl atoms that can accept two electrons. This is a convention, not a description of the reaction mechanism of the oxidant. Because Cl_2 only accepts two electrons as does HOCl and monochloramines, it only has one active Cl atom according to the definition. Thus the active chlorine is always one-half of the available chlorine. The available chlorine is usually measured by iodometric titration. The weight of available chlorine can also be calculated by equation 5, where 70.9 represents the mol wt of Cl_2 and moles of oxidant can be represented wt oxidant/mol wt of oxidant.

$$\text{weight available} = 70.9 \times \text{moles of oxidant} \times \frac{\text{number active Cl atoms}}{\text{molecule}} \tag{5}$$

In solutions, the concentration of available chlorine in the form of hypochlorite or hypochlorous acid is called free-available chlorine.

Commercially important solid available chlorine bleaches are usually more stable than concentrated hypochlorite solutions. They decompose very slowly in sealed containers. But most of them decompose quickly as they absorb moisture from air or from other ingredients in a formulation. This may release hypochlorite that destroys other ingredients as well.

Chlorine. Except to bleach wood pulp and flour, chlorine itself is rarely used as a bleaching agent.

Hypochlorites. The principal form of hypochlorite produced is sodium hypochlorite, NaOCl.

Other hypochlorites include calcium hypochlorite, bleach liquor, bleaching powder and tropical bleach, dibasic magnesium hypochlorite, lithium hypochlorite, chlorinated trisodium phosphate, and hypochlorous acid.

N-Chloro Compounds. *Chlorinated Isocyanurates.* The principal solid chlorine bleaching agents are the chlorinated isocyanurates. The one used most often for bleaching applications is sodium dichloroisocyanurate dihydrate, with 56% available chlorine.

Other *N*-chloro compounds include halogenated hydantoins, sodium *N*-chlorobenzenesulfonamide (chloramine B), sodium *N*-chloro-*p*-toluenesulfonamide (chloramine T), *N*-chlorosuccinimide, and trichloromelamine.

Chlorine Dioxide. Chlorine dioxide, ClO_2, is a gas that is more hazardous than chlorine. Large amounts for pulp bleaching are made by several processes in which sodium chlorate is reduced with chloride, methanol, or sulfur dioxide in highly acidic solutions by complex reactions. For most other purposes chlorine dioxide is made from sodium chlorite.

Peroxygen Compounds

Peroxygen compounds contain the peroxide linkage (—O—O—) in which one of the oxygen atoms is active.

Hydrogen Peroxide. Hydrogen peroxide is one of the most common bleaching agents (see HYDROGEN PEROXIDE). It is the primary bleaching agent in the textile industry, and is also used in pulp, paper, and home laundry applications. In textile bleaching, hydrogen peroxide is the most common bleaching agent for protein fibers, and is also used extensively for cellulosic fibers.

Solid Peroxygen Compounds. Hydrogen peroxide reacts with many compounds, such as borates, carbonates, pyrophosphates, sulfates, silicates, and a variety of organic carboxylic acids, esters, and anhydrides to give peroxy compounds or peroxyhydrates. A number of these compounds are stable solids that hydrolyze readily to give hydrogen peroxide in solution.

Compounds include perborates, sodium carbonate peroxyhydrate, and peroxymonosulfate.

Peracids. Peracids are compounds containing the functional group —OOH derived from an organic or inorganic acid functionality. Typical structures include $CH_3C(O)OOH$ derived from acetic acid and $HOS(O)_2OOH$ (peroxymonosulfuric acid) derived from sulfuric acid. Peracids have superior cold water bleaching capability versus hydrogen peroxide because of the greater electrophilicity of the peracid peroxygen moiety. Lower wash temperatures and phosphate reductions or bans in detergent systems account for the recent utilization and vast literature of peracids in textile bleaching.

Peracids can be introduced into the bleaching system by two methods. They can be manufactured separately and delivered to the bleaching bath with the other performance components or as a separate product. Peracids can also be formed *in situ* utilizing the perhydrolysis reaction shown in equation 6.

$$R{-}\overset{\overset{\displaystyle O}{\|}}{C}{-}Z + {}^-OOH \longrightarrow R\overset{\overset{\displaystyle O}{\|}}{C}OOH + Z^- \tag{6}$$

Peracid Precursor Systems. Compounds that can form peracids by perhydrolysis are almost exclusively amide, imides, esters, or anhydrides. Two compounds were commercially used for laundry bleaching as of 1990. Tetraacetylethylenediamine (TAED) is utilized in over

50% of Western European detergents. Nonanoyloxybenzene sulfonate (NOBS) is used in detergent products in the United States and Japan.

Preformed Peracids. Peracids can be generated at a manufacturing site and directly incorporated into formulations without the need for *in situ* generation. Two primary methods are utilized for peracid manufacture. The first method uses the equilibrium shown in equation 7 to generate the peracid from the parent acid.

$$\mathrm{R\overset{O}{\overset{\|}{C}}OH} + H_2O_2 \rightleftharpoons \mathrm{R\overset{O}{\overset{\|}{C}}OOH} + H_2O \qquad (7)$$

The equilibrium is shifted by removal of the water or removal of the peracid by precipitation. Peracids can also be generated by treatment of an anhydride with hydrogen peroxide to generate the peracid and a carboxylic acid.

$$\mathrm{R\overset{O}{\overset{\|}{C}}O\overset{O}{\overset{\|}{C}}R} + H_2O_2 \longrightarrow \mathrm{R\overset{O}{\overset{\|}{C}}OOH} + \mathrm{R\overset{O}{\overset{\|}{C}}OH} \qquad (8)$$

The latter method (eq. 8) typically requires less severe conditions than the former because of the labile nature of the organic anhydride. Both of these reactions can result in explosions, and significant precautions should be taken prior to any attempted synthesis of a peracid.

Peracid Decomposition. Peracids, whether preformed or formed *in situ* via the perhydrolysis reaction, are susceptible to decomposition in an aqueous bleaching bath. The decomposition is caused by the occurrence of one of four reactions. The peracid can decompose as a result of oxidation of the bleachable material. Transition metals present even at extremely low concentration in the bath from the incoming water can decompose the peracid catalytically. To minimize this effect, metal-sequestering agents have been proposed to prevent the degradation of the peracid in solution. Peracids can also hydrolyze to the parent acid and hydrogen peroxide because of the large excess of water present in the aqueous bleaching bath. This is generally a kinetically slow process. A final decomposition mechanism involves the reaction of two moles of peracid generating two moles of parent acid and a mole of oxygen.

Reducing Bleaches

The reducing agents generally used in bleaching include sulfur dioxide, sulfurous acid, bisulfites, sulfites, hydrosulfite (dithionites), sodium sulfoxylate formaldehyde, and sodium borohydride. These materials are used mainly in pulp and textile bleaching (see SULFUR COMPOUNDS; BORON COMPOUNDS).

The Mechanism of Bleaching

Bleaching is a decolorization or whitening process that can occur in solution or on a surface. The color-producing materials in solution or on fibers are typically organic compounds that possess extended conjugated chains of alternating single and double bonds and often include heteroatoms, carbonyl, and phenyl rings in the conjugated system. The portion of molecule that absorbs a photon of light is referred to as the chromophore (Greek: *color bearer*). For a molecule to produce color the conjugated system must result in sufficiently delocalized electrons such that the energy gap between the ground and excited states is small enough so that photons in the visible portion of the light spectrum are absorbed (see COLOR).

Bleaching and decolorization can occur by destroying one or more of the double bonds in the conjugated chain, by cleaving the conjugated chain, or by oxidation of one of the other moieties in the conjugated chain. The result of any one of the three reactions is an increase in the energy gap between the ground and excited states, so that the molecule then absorbs light in the ultraviolet region, and no color is produced. Bleaching may also increase the water solubility of organic compounds after reaction. Conversion of an olefin to a vicinal diol, for example, dramatically increases the polarity and consequently the water solubility of the compound. A variety of bleaching agents can affect this transformation. The increased solubility allows actual removal of the bleached substance from a surface.

Chlorine bleaches react with more chromophores than oxygen bleaches. They react irreversibly with aldehydes, alcohols, ketones, carbon–carbon double bonds, acidic carbon–hydrogen bonds, nitrogen compounds, sulfur compounds, and aromatic compounds.

The mechanism of bleaching of hydrogen peroxide is not well understood. It is generally believed that the perhydroxyl anion (HOO^-) is the active bleaching species since both the concentration of this anion and the rate of the bleaching process increase with increasing pH. Hydrogen peroxide and other peroxygen compounds can destroy double bonds by epoxidation. This involves addition of an oxygen atom across the double bond usually followed by hydrolysis of the epoxide formed to 1,2-diols under bleaching conditions.

Peracids undergo a variety of reactions which result in bleaching. Peracids can add an oxygen across a double bond to give an epoxide, which can undergo further reactions including hydrolysis to give a vicinal diol. Peracids can oxidize aldehydes to acids, sulfur compounds to sulfoxides and sulfones, and nitrogen compounds to amine oxides, hydroxylamines, and nitro compounds. Peracids can also oxidize α-diketone compounds to anhydrides and ketones to esters.

Reducing agents are thought to work by reduction of the chromophoric carbonyl groups in textiles or pulp.

Applications of Bleaching Compounds

Laundering and Cleaning. *Home and Institutional Laundering.* The most widely used bleach in the United States is liquid chlorine bleach, an alkaline aqueous solution of sodium hypochlorite. This bleach is highly effective at whitening fabrics and also provides germicidal activity at usage concentrations. Liquid chlorine bleach is sold as a 5.25% solution, and 1 cup provides 200 ppm of available chlorine in the wash. Liquid chlorine bleaches are not suitable for use on all fabrics. Dry and liquid bleaches that deliver hydrogen peroxide to the wash are used to enhance cleaning on fabrics. They are less efficacious than chlorine bleaches but are safe to use on more fabrics. The dry bleaches typically contain sodium perborate in an alkaline base whereas the liquid peroxide bleaches contain hydrogen peroxide in an acidic solution. Detergents containing sodium perborate tetrahydrate are also available.

The worldwide decreasing wash temperatures, which decrease the effectiveness of hydrogen peroxide-based bleaches, have stimulated research to identify activators to improve bleaching effectiveness. Tetraacetylethylenediamine is widely used in European detergents to compensate for the trend to use lower wash temperatures. TAED generates peracetic acid in the wash in combination with hydrogen peroxide. TAED has not been utilized in the United States, where one activator nonanoyloxybenzene sulfonate (NOBS) has been commercialized and incorporated into several detergent products. NOBS produces pernonanoic acid when combined with hydrogen peroxide in the washwater and is claimed to provide superior cleaning in contrast to perborate bleaches.

In industrial and institutional bleaching, either liquid or dry chlorine bleaches are used because of their effectiveness, low cost, and germicidal properties. Dry chlorine bleaches, particularly formulated chloroisocyanurates, are used in institutional laundries.

Hard Surface Cleaners and Cleansers. Bleaching agents are used in hard surface cleaners to remove stains caused by mildew, foods, etc, and to disinfect surfaces. Disinfection is especially important for many industrial uses. Alkaline solutions of 1–5% sodium hypochlorite that may contain surfactants and other auxiliaries are most often used for these purposes. These are sometimes thickened to increase contact times with vertical surfaces. A thick, alkaline cleaner with 5% hydrogen peroxide is also sold in Europe. Liquid abrasive cleansers with suspended solid abrasives are also available and contain about 1% sodium hypochlorite. Powdered cleansers often contain 0.1–1% available chlorine and they may contain abrasives. Sodium dichloroisocyanurate is the most common bleach used in powdered cleansers, having largely replaced chlorinated trisodium phosphate. Calcium hypochlorite is also used. Dichloroisocyanurates are also used in effervescent tablets that dissolve quickly to make cleaning solutions. In-tank toilet cleaners use calcium hypochlorite, dichloroisocyanurates, or *N*-chloro compounds to release hypochlorite with each flush.

Automatic Dishwashing and Warewashing. The primary role of bleach in automatic dishwashing and warewashing is to reduce spotting and filming by breaking down and removing the last traces of adsorbed soils. They also remove various food stains such as tea. All automatic dishwashing and warewashing detergents contain alkaline metal salts or hydroxides.

Textile Bleaching. Many textiles are bleached to remove any remaining soil and colored compounds before dyeing and finishing (see TEXTILES). Bleaching is usually preceded by washing in hot alkali to remove most of the impurities in a process called scouring. Bleaching is usually done as part of a continuous process, but batch processes are still used. Bleaching conditions vary widely, depending on the equipment, the bleaching agent, the type of fiber, and the amount of whiteness required for the end use.

Cotton and Cotton–Polyester. Cotton is the principal fiber bleached today, and almost all cotton is bleached. About 80–90% of all cotton and cotton–polyester fabric is bleached with hydrogen peroxide.

Other Cellulosics. Rayon is bleached similarly to cotton but under milder conditions since the fibers are more easily damaged and since there is less colored material to bleach.

Synthetic Fibers. Most synthetic fibers are sufficiently white and do not require bleaching. When needed, synthetic fibers and many of their blends are bleached with sodium chlorite solutions. Solutions of 0.1% peracetic acid are also used.

Wool and Silk. Wool must be bleached carefully, in order to avoid fiber damage. It is usually bleached with 1–5% hydrogen peroxide solutions. Silk is bleached similarly, but at slightly higher temperatures.

Bleaching of Other Materials. *Hair.* Hydrogen peroxide is the most satisfactory bleaching agent for human hair.

Fur. The coloring matter in fur is usually bleached using hydrogen peroxide stabilized with sodium silicate.

Foodstuffs, Oils. Sulfur dioxide is used to preserve grapes, wine (qv), and apples; the process also results in a lighter color. During the refining of sugar (qv), sulfur dioxide is added to remove the last traces of color. Flour can be bleached with a variety of chemicals including chlorine, chlorine dioxide, oxides of nitrogen, and benzoyl peroxide. Bleaching agents such as chlorine dioxide or sodium dichromate are used in the processing of nonedible fats and fatty oils for the oxidation of pigments to colorless forms (see FOOD PROCESSING).

JAMES P. FARR
WILLIAM L. SMITH
DALE S. STEICHEN
The Clorox Company

Household and Industrial Bleach Systems, North America Forecast to 2000, Colin A. Houston and Associates, Mamaroneck, N.Y., 1988, pp. 2,3.

W. H. Sheltmire, in J. S. Sconce, ed., *Chlorine: Its Manufacture, Properties And Uses,* American Chemical Society Monograph Series 154, Reinhold Publishing Co., New York, 1962, pp. 512–542.

D. M. Coons, *J. Amer. Oil Chemist's Soc.* **55,** 104 (1978).

PULP AND PAPER

Worldwide, more than 50×10^6 t of pulp is bleached annually, making the pulp and paper industry one of the largest consumers of bleaching chemicals.

The principal objective in bleaching any type of pulp is usually to increase the whiteness of the pulp, as measured by its brightness, which is defined as reflectance measured at a wavelength of 457 nm.

Chemical Pulp

Virtually all of the color of any pulp resides in its lignin component. It is not possible to remove all of the lignin from the wood during pulping because concurrent cellulose damage would seriously weaken the fibers. Pulping is therefore terminated when 5–8% of the original lignin remains. Bleaching may therefore be regarded as a more selective continuation of the pulping process, but there are important distinctions to be made between these two phases of delignification. Side reactions during pulping cause the residual lignin to be both darker in color and more tightly bound to the fiber than native lignin. From a practical point of view, a more important distinction concerns the fate of the organic material removed. Spent pulping liquors are recycled to a chemical recovery system, where they are concentrated and burned to generate energy. Bleaching effluent, on the other hand, is typically contaminated with chloride ion and chlorine compounds, making it difficult or impossible to recycle it to the recovery system because of potential for corrosion. It is therefore necessary to treat and discharge the bleaching effluent. Associated environmental concerns are presently causing rapid changes in bleaching technology.

Bleaching Sequences. Chemical pulps are invariably bleached in multistage processes, usually with washing between stages to limit chemical consumption. A shorthand notation for bleaching sequences uses single capital letters to designate individual bleaching agents, as follows: C = chlorine, E = caustic extraction, D = chlorine dioxide, O = oxygen, P = hydrogen peroxide, H = hypochlorite, and Z = ozone. Stages employing only one chemical are designated by a single letter, those using combinations by a variety of notations, such as C_D, D/C, (C + *D*), etc. Sequences of from three to six stages are common. The first two or three usually remove the bulk of the residual lignin and, depending on the exact nature of the sequence, may not increase the brightness significantly. The final one to four stages accomplish a large brightness increase with little lignin removal and comprise the brightening partial sequence.

Desirable Attributes of Bleaching Agents. An ideal bleaching agent for chemical pulp would have high selectivity, reactivity, efficiency, and particle bleaching ability, while simultaneously having low equivalent weight and low potential for harm to the environment. With the exception of caustic, all important chemical pulp bleaching agents are oxidants.

Chlorine. Chlorine, Cl_2, is used either in the first stage of the sequence or in the stage immediately following an oxygen predelignification stage. Its chief function is to render most of the residual lignin soluble in alkali, and it is always followed by a caustic extraction stage.

Caustic Extraction. A relatively small amount of lignin dissolves during chlorination; most is removed in the subsequent caustic extraction stage.

Chlorine Dioxide. Chlorine dioxide, ClO_2, offers a unique combination of low equivalent weight (13.5 vs 35.5 for chlorine), extremely high selectivity for lignin, high efficiency, ability to bleach to very high brightness, and good particle bleaching ability. It is versatile, inasmuch as it can be used both to brighten and to delignify, and in the latter role it interacts synergistically with chlorine. Furthermore, it is less likely than chlorine to have significant environmental impact since it forms a much smaller amount of chlorinated organic byproducts for a given bleaching effect.

Hypochlorite. Calcium hypochlorite, $Ca(OCl)_2$, was among the earliest chemicals to be used for bleaching pulp. Both calcium and sodium hypochlorite are cheap and capable of bleaching pulp to high brightness. Hypochlorite is still used, but its use is decreasing because the bleaching reaction generates chloroform, $CHCl_3$, a by-product that is strictly regulated by government agencies.

Hydrogen Peroxide. Hydrogen peroxide, H_2O_2, has a variety of applications in chemical pulp bleaching. These include final brightening; addition, alone or with oxygen, to caustic extraction stages; and in a predelignification stage at the beginning of the sequence.

Oxygen. The use of oxygen, O_2, with alkali to bleach chemical pulps was first practiced on a commercial scale in 1970. Since then it has grown remarkably, as a result of continuing efforts by the industry to improve effluent quality. Oxygen predelignification is now becoming widespread and most new mills and mill expansions include it. In addition, the use of oxygen to enhance the effectiveness of the first caustic extraction stage has become nearly universal.

Ozone. Ozone, O_3, bleaching has been the subject of laboratory and pilot plant studies for many years, but it remained uncommercialized until the announcement of the 1992 start-up of a full-scale plant at Union Camp's Franklin, Virginia mill. The laboratory studies have shown that ozone rapidly and extensively delignifies chemical pulps over wide ranges of consistency and other conditions.

Newer Developments. Research in progress points to the possible commercialization of several other chemical pulp bleaching technologies. These are chemical pretreatments to improve the selectivity of oxygen bleaching, enzymatic pretreatments to facilitate subsequent delignification, and recycling of bleach plant effluents.

Mechanical Pulp

Much new high yield pulping technology has been developed, including production of thermomechanical pulp (TMP) and chemithermomechanical pulp (CTMP). Because little material is lost in these processes, the pulps have a very high lignin content. Consequently, it is not feasible to bleach them by removing all of the lignin, as in chemical pulp bleaching. Instead, the lignin must be decolorized, a process sometimes referred to as brightening, to distinguish it from lignin-removing bleaching methods. Only two brightening agents, hydrogen peroxide and sodium hydrosulfite (sodium dithionite), are of commercial importance.

Health and Safety Aspects

Because pulp bleaching agents are, for the most part, reactive oxidizing agents, appropriate precautions must be taken in their handling and use.

Hazards associated with the use of pulp bleaching agents include their potential for damage resulting from contact with skin or eyes, their ability, as oxidizers, to cause fires or explosions upon contact with some kinds of organic matter or certain metals and inorganic compounds, and the possibility of explosive decomposition.

Additional information on health and safety aspects should be sought by consulting material safety data sheets available from suppliers of the chemical in question. In addition, most suppliers of bleaching chemicals, upon request, provide on-site training sessions by experts on the safe use and handling of their products.

THOMAS MCDONOUGH
Institute of Paper Science and Technology

T. J. McDonough, in *1991 Bleach Plant Operations Short Course Notes,* TAPPI Press, Atlanta, Ga., 1991, pp. 57–65.

T. J. McDonough, *Tappi J.* **69**(6), 46 (1986); L. Tench and S. Harper, *Tappi J.* **70**(11), 55 (1987).

R. W. Allison, *APPITA '91 Conference Proceedings,* Rotorua, New Zealand, 1991, pp. 201–208; N. Liebergott, B. van Lierop, and A. Skothos, *Proceedings of the 1991 TAPPI Pulping Conference,* Book 1, TAPPI Press, Atlanta, Ga., 1991, pp. 1–23.

Bleach Plant Operations Short Course Notes, Technical Association of the Pulp and Paper Industry (TAPPI) Press, Atlanta, Ga., 1983–1991.

BLENDING. See MIXING AND BLENDING.

BLOOD, ANIMAL. See MEAT PRODUCTS.

BLOOD, ARTIFICIAL

Artificial blood is herein defined as consisting of red cell substitutes. Red cell substitutes are solutions intended for use in patients whose red cells are either not available or their use is to be avoided for other reasons. Despite enormous effort, more than 100 years of research have not produced a solution that can be used safely in humans.

Hemoglobin Modifications

Reactivity. Hemoglobin can exist in either of two structural conformations, corresponding to the oxy (*R,* relaxed) or deoxy (*T,* tense) states. The key differences between these two structures are that the constrained *T* state has a much lower oxygen affinity than the *R* state and the *T* state has a lower tendency to dissociate into subunits that can be filtered in the kidneys. Therefore, stabilization of the *T* conformation would be expected to solve both the oxygen affinity and renal excretion problems.

The transition between the *T* and *R* states of hemoglobin is also deeply involved in the Bohr effect and cooperativity. Therefore stabilization of either of the two structures should diminish these effects, which have important physiologic consequences. The clinical consequences of stabilization are not known.

Stabilization of the *T* conformation under normal conditions is illustrated by the reaction of 2,3-diphosphoglycerate, (2,3-DPG) (Fig. 1).

Many of the reactions considered to be useful in the production of hemoglobin-based blood substitutes use chemical modification at one or more of the sites in Figure 1. Table 1 lists the different types of hemoglobin modifications with examples of the most common reactions for each. Differences in the reactions are determined by the dimensions and reactivity of the cross-linking reagents. Because the function of hemoglobin in binding and releasing oxygen is intricately connected to the transition between *T* and *R* conformations, it is not surprising that P_{50} and yield are highly variable. Even small differences among structures of the reagents can yield products having very different properties. In addition, the conditions of the reaction are very important, not only in regard to the state of ligation, ie, oxygen saturation, but also in regard to the presence of agents or molecules that block or compete for certain reactive sites.

A further complication of these reactions is that many nonhemoglobin proteins contain reactive groups and may also be modified to produce new, potentially toxic, contaminants. It has been difficult to produce a pure modified hemoglobin for toxicity studies because most processes start with relatively crude, stroma-free hemoglobin.

Hemoglobin Sources

Purification. Hemoglobin is provided by the red blood cell in highly purified form. However, the red cell contains many enzymes and other

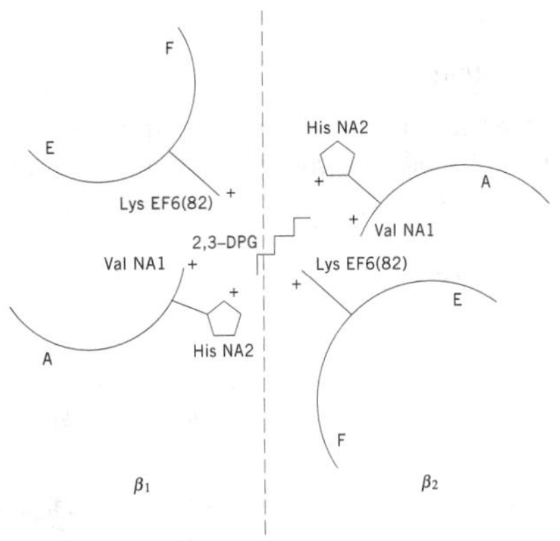

Figure 1. Reaction of diphosphoglycerate (2,3-DPG) and deoxyhemoglobin. The molecule fits into the central cavity of hemoglobin and forms salt bridges with valine NA(1) β, histidines NA2(2)β, H21(143)β, and lysine EF6(82)β. A, E, and F correspond to specific hemoglobin helices, and NA is the sequence from the amino-terminals to segment A.

Table 1. Classes of Hemoglobin Modifications

Class	Examples
amino-terminal modification	carbamylation, carboxymethylation, pyridoxylation, acetaldehyde
lysine EF6(82)β modification	mono(3,5-dibromosalicyl)fumarate
valine NA1(1)β-lysine EF6(82)β cross-link	2-nor-2-formylpyridoxal 5′-phosphate (NFPLP), bis-pyridoxal tetraphosphate (bis-PL)P_4
lysine G6(99) α_1–lysine G6(99) α_2 cross-link	bis(3,5-dibromosalicyl)fumarate
2,3-DPG analogue surface, multisite	pyridoxal 5′-phosphate glutaraldehyde, polyaldehydes, ring-opened dials, diimidate esters
conjugated hemoglobin	dextran–aldehyde, poly(ethylene glycol), polyoxyethylene

proteins, and red cell membranes contain many components that could potentially cause toxicity problems. Furthermore, plasma proteins and other components could cause toxic reactions in recipients of hemoglobin preparations.

Rabiner's method for the filtration purification of hemoglobin was thought to be a significant advance over older centrifugation methods. However, hemoglobin prepared in this way still caused unwanted reactions in human recipients. The crystallization method showed fewer toxic effects in animals, but batch-to-batch reproducibility was uneven. Ultrapurification of hemoglobin using ion-exchange chromatographic technique is possible, but tedious and expensive.

Outdated Human Blood. If clinical efficacy and safety of hemoglobin solutions can be shown, the demand for product would soon outstrip the supply of outdated human blood. About 12×10^6 units of blood (1 unit $\cong$ 480 mL) are used in the United States each year, and only about 5×10^5 outdate.

Bovine Hemoglobin. One solution to the hemoglobin supply problem is to use as a starting material blood from nonhuman sources. For example, bovine hemoglobin is being developed for use. The ultimate success of bovine, or any other hemoglobin, depends on demonstration of safety, not on supply. One problem in using bovine hemoglobin is the fear of bovine spongiform encephalitis (BSE) virus. This virus, related to the Scrapie organism, has been detected in cows in Europe as well as other mammals in North America. Although there are no known human disease cases related to BSE, the FDA is concerned about bovine products of all types. This virus is especially resistant to heat treatment.

Recombinant Hemoglobin. An alternative and novel source of hemoglobin for modification is from microorganisms the genome of which has been modified to contain globin genes for recombinant hemoglobin (rHb) production (see GENETIC ENGINEERING). Significant strides have been made in this approach, and it is possible to express both human α- and β-globin chains in *Escherichia coli.*

Status of Blood Substitutes

Several of the products discussed herein are under intense development. One product, based on recombinant hemoglobin, is in early human trials as of this writing. Other hemoglobin-based solutions are also under review at the FDA. Replacement of red blood cells using massive amounts of protein, free in solution, is an unprecedented therapeutic adventure.

ROBERT M. WINSLOW
University of California, San Diego

R. M. Winslow, *Hemoglobin-Based Red Cell Substitutes,* Johns Hopkins Press, Baltimore, Md., 1992.

S. F. Rabiner, J. R. Helbert, H. Lopas, and L. H. Friedman, *J. Exp. Med.* **126,** 1127–1142 (1967).

S. M. Christensen and co-workers, *J. Biochem. Biophys. Meth.* **17,** 143–154 (1988).

S. J. Hoffman and co-workers, *Proc. Natl. Acad. Sci. USA* **87,** 8521–8525 (1990).

BLOOD, COAGULANTS AND ANTICOAGULANTS

The conversion of inert procoagulant glycoproteins to coagulant glycoproteins via proteolytic processing involves delicately balanced interaction between many different proteins, proteases, phospholipids, and the divalent cation calcium. The coagulation process can be activated and proceed through one of two possible sequential pathways: the intrinsic system path, components present within the circulating blood, and the extrinsic system, components present in the extravascular and intravascular compartment. Cooperative integration of these two systems, along with circulating platelets, maintains vascular integrity and preserves hemostasis.

Biochemically, coagulation of the blood results from proteolytic processing of many different inert glycoproteins that originate in or migrate into the circulating blood. The participants in this complex process have been designated factors, and most have been assigned roman numerals. More recently identified participants in coagulation of the blood are designated by the name of the person who recognized the given factor, eg, von Willebrand protein, or by a name that indicates the composition of the substance, eg, high molecular weight kininogen. All factors listed in Tables 1 and 2 are present in plasma except III, and all are present in serum except for I, II, III, V, and VIII, though XIII is decreased to trace quantities. Factor IIa (thrombin) may be present in serum. Factors I, VIII, and XIII are present in Cohn's fraction I; Factors II, V, VII, IX, X, XI, and XII are present in Cohn's fraction III; Factors II, VII, IX, XI, and XII are present in Cohn's fraction IX. Recently, the following factors, not previously reported, have been detected: Fletcher (prekininogenin [prekallikrein]), Fitzgerald (high molecular weight) kininogen (also designated as Williams or Flaujeac trait), and Passovoy (bleeding diathesis-prolonged partial thromboplastin time, normal known coagulation factors, autosomal dominant). Roman numerals have not yet been assigned to these factors. Factor VI is not listed in the tables because it is obsolete.

Coagulant Factor Replacement Therapy

The optimal technique for the treatment of hemorrhage associated with congenital factor deficiency states, including hemophilia A, hemophilia B, and von Willebrand disease, is intravenous replacement of the missing factor. This can be accomplished by the transfusion of whole blood, fresh-frozen plasma, cryoprecipitate, and factor concentrates. Depending on the circumstances and individual patient involved, one source of the necessary factor can be optimally selected. Because of the frequency of transmission by transfusion of these blood products of certain diseases, such as hepatitis A, hepatitis B, non A–non B hepatitis, hepatitis C, hepatitis delta, HIV, and additional rare viral and bacterial infections, recently there has been a concerted effort to pasteurize these sources of coagulation proteins so that they are free of contaminants. These pasteurization purification techniques utilize methodology including monoclonal antibody-specific selective separation of the factor from all other proteins in the plasma, dry and wet heat, solvent detergent, heat suspension in organic solvents, and molecular biologic recombinant expression of these individual proteins by mammalian cells. These techniques have been designed to produce safer, efficacious, more

Table 1. Physical and Chemical Properties of Plasma Coagulation Factors

Factor (synonym)	Protein type[a]	Molecular weight,[b] daltons	Isoelectric point	PPTA,[c] %
I (fibrinogen)	GP	340,000	5.5	25
II (prothrombin)[d]	GP[e,f]	73,000[g]	4.2	50
III (thromboplastin)	LP	47,000		
IV (calcium)		40		
V (proaccelerin, plasma acglobulin)	GP	300,000		
VII (SPCA, proconvertin)[d,h]	GP[e,f]	48,000–100,000[i]		50
VIII (antihemophilic globulin, AHG, antihemophilic factor, AHF)[h,j]	GP	2,000,000[i]		33
VIII:vWFAg (von Willebrand protein)	GP[f]	30,000,000[i]		33
IX (plasma thromboplastin component, PTC, Christmas factor)[d,h]	GP[e,f]	55,000–70,000[i]	4.50	50
X (Stuart-Prower Factor)[d]	GP[e,f]	100,000[g]	5.5	50
XI (plasma thromboplastin antecedent, PTA)	[f]	200,000[k]	4.50	33
XII (Hageman Factor)[l]	Sialo-GP[f]	100,000[m]		50
XIII (fibrin-stabilizing factor)		350,000[g]		33
Protein C_a[d]	GP	102,000		40
Protein S[d]	GP	78,000		40
Protein Z	GP	62,000		40

[a] GP = glycoprotein; LP = lipoprotein.
[b] Of conjugated protein or of noted globulin.
[c] PPTA = degree of saturation $(NH_4)_2SO_4$ solution necessary for precipitation of factor.
[d] Synthesis is vitamin K-dependent.
[e] Adsorbed by $BaSO_4$ or $Au(OH)_3$ or $Ca_3(PO_4)_2$.
[f] Adsorbed by Celite (infusorial earth) or kaolin or carboxymethylcellulose (CMC).
[g] α-Globulin.
[h] Factor migrates in an electrophoretic field between α- and β-globulins.
[i] α-β-Globulin.
[j] Molecular weight of procoagulant subunit is ~200, 000.
[k] β- or γ-Globulin.
[l] Factor migrates in an electrophoretic field between β- and γ-globulins.
[m] β-γ-globulin.

suitable, and less expensive factor replacement products. Recombinant factor VIII:C is now available.

Table 3 contains products available for Factor VII, Factor VIII:C (hemophilia A), Factor IX, and von Willebrand protein deficiency. Table 4 lists miscellaneous hemostatics and their proposed mechanisms of action.

Anticoagulants

In Vitro Anticoagulants. A number of substances have been identified that prevent coagulation of the blood when it is removed from the vascular compartment of the body. Most of these substances remove a vital constituent of the blood that is essential in the mediation of transformation of liquid blood into a solid.

Ethylenediamine tetraacetic acid (EDTA) (Sequestrene), an anticoagulent at 1 mg of the disodium salt per mL blood, complexes with and removes calcium, Ca^{2+}, from the blood. Oxalate, citrate, and fluoride ions form insoluble salts with Ca^{2+} and chelate calcium from the blood. Salts containing these anticoagulants include lithium oxalate, $Li_2C_2O_4$, 1 mg/mL blood; sodium oxalate $Na_2C_2O_4$, 2 mg/mL blood; potassium oxalate monohydrate, $K_2C_2O_4{\cdot}H_2O$, 2 mg/mL blood; sodium fluoride NaF, 2 mg/mL blood; trisodium citrate $C_6H_5Na_3O_7$, 0.42 mg/mL blood; sodium polyanetholesulfonate (Liquoid), 1–2.5 mg/mL blood; and heparin a micropolysaccharide, 0.1 unit/mL blood.

Plasma Inhibitors, In Vivo Anticoagulants. Fourteen naturally occurring compounds that normally exert an inhibiting effect on the activity of coagulation, platelet function, and fibrinolytic activity and complement systems have been identified within the circulating blood.

α_1-antichymotrypsin	C-1 esterase inhibitor
α_1 antiplasmin	inter α-trypsin inhibitor
α_2-antiplasmin	plasminogen activator inhibitor-1
α_1-antitrypsin	plasminogen activator inhibitor-2
α_2-antiactivator	protein C
α_2-macroglobulin	protein S
antithrombin III	protein Z

Therapeutic Anticoagulants. Therapeutic anticoagulants include heparin and coumarinic acid compounds (Table 5).

New Therapeutic Anticoagulant Agents. There are some patients who cannot receive heparin or warfarin but who need anticoagulation. To deal with this need, appreciable investigation is in progress to identify new anticoagulants. Several such agents are on the horizon, including hirudin, arvin, activated protein C (C_a), argatroban, and brodifacoum.

Anticoagulant Antagonists. Heparin, because of its high polyanionic charge density, is highly negatively charged and can be readily neutralized by polycation substances that are positively charged. A number of polycationic substances can be employed, including protamine sulfate. Vitamin K competitively antagonizes the coumarinic acid anticoagulants.

The Fibrinolytic System

The human fibrinolytic system is a proteolytic enzyme system consisting of several components found in different locations in the body including the blood, vascular endothelium, and several tissues extrinsic to the vascular compartment. The central component of the human fibrinolytic system is plasminogen synthesized by the liver and present in the euglobulin fraction of the blood. This inert precursor or proenzyme can be activated by an agent intrinsic to the blood designated proactivator that is converted to an activator by the combined interaction of prekallikrein, high molecular weight kininogen, kallikrein, and Factor XIIa. The action of the activator on plasminogen yields the potent proteolytic enzyme plasmin. Another path of activation that is intrinsic to the body but extrinsic to the

Table 2. Biological Aspects of Plasma Coagulation Factors

					Deficiency state^{a}		
Factor	Synthesis site	Biological half-life, h	Volume of distribution, MPV^{b}	Hemostasis concentration, %^{c}	Per 10^6 population	Inheritance pattern^{d}	Chromosome
I	liver	95–120	2.5	20	hypofibrinogenemia^{e} or afibinogenemia, <0.5	AR	4q26–q28
II	liver^{f}	72	1.5–2.5	20	hypoprothrombinemia, <0.5	AR	11p11q12
III	all body tissues				not reported	AR	1pter–p12
IV							
V	liver	12–36	2	10–20	factor V deficiency, <0.5	AR	
VII	liver^{f}	4–6	2–4	10	factor VII deficiency, <0.5	AR	13q34
VIII	hepatocytes	6–10^{g} 15–18^{h}	1–1.5	30–40	hemophilia A, 60–80	X-LR	xq28
VIII:vWFAg	vascular, endothelial cells	18–24	1–1.5	60	von Willebrand disease, 80–100	AD	12pter–p12
						AR	
IX	liver^{f}	6–10^{g} 25–30^{h}	2–3	30–40	hemophilia B Christmas disease, 15–20	X-LR	xq26–27
X	liver^{f}	40–50	1	10–20	Stuart-Prower deficiency, <0.5	AR	13q34
XI	liver^{i}	50–80	1	10–20	PTA trait, <1.0	AR^{j}	
XII	liver	50–60			Hageman trait, <1.0	AR	5
XIII	liver^{i}, megakaryocytes	95–120	1–2	1–2	factor XIII deficiency, <0.5	AR	
Protein C_a	liver^{f}	6–8	1–2	10–20	1–5	AR	
Protein S	endothelial cells^{e}	42.5	1–2	10–20	<1	AR	
Protein Z	endothelial cells^{e}, hepatocytes	60	1–2	10–20			

^{a} Condition in which coagulation factor level in a given patient is below the acceptable normal level.
^{b} Multiples of plasma volume, ie, number given multiplied by plasma volume equals total volume in which the factor is located.
^{c} The approximate concentration required to produce hemostasis, in percent of normal concentration.
^{d} AD = autosomal dominant; AR = autosomal recessive; X-LR = sex-linked recessive.
^{e} Some pedigrees reported to have hypofibrinogenemia may be examples of dysfibrinogenemia; this condition is inherited as an autosomal dominant.
^{f} Synthesis is vitamin K-dependent.
^{g} Initial; the shorter initial half-lives for Factors VIII and IX are probably because of distribution into extravascular compartments.
^{h} Secondary.
^{i} Data are suggestive, but conclusive evidence is not yet available.
^{j} Reported to be inherited both as autosomal recessive and as autosomal dominant.

blood originates in the vascular endothelium. The agent released by the vascular endothelium, designated vascular plasminogen activator, is capable of direct activation of plasma plasminogen, independent of the blood activator. Additional agents extrinsic to the vascular compartment, designated tissue plasminogen activators, have been extracted from the uterus, pancreas, lung, kidney, and prostate. The activity of all these endogenous activators is normally modulated by several intrinsic, naturally occurring inhibitors. Activation of the human plasminogen–plasmin proteolytic enzyme system is also achieved by a variety of exogenous agents from several different sources including bacteria, fungi, synthetic substances, and human urine. The fibrinolytic system in humans occupies several vital positions not only in the removal of intravascular fibrin-thrombi but also participates in other biological functions including ovulation, embryo implantation, neoplastic transformation, tissue repair, and macrophage function.

Action of Plasmin on Fibrin and Fibrinogen. Fibrin is the principal physiologic substrate of plasmin. Other elements of the coagulation cascade, including Factor VIII, Factor V, and fibrinogen, are degraded by plasmin. The human fibrinogen molecule has a molecular weight of about 340,000 daltons and consists of three pairs of nonidentical peptide chains represented by the formula $(A\alpha B\beta\gamma)_2$. When the coagulation system is activated, thrombin cleaves the A and B fragments from fibrinogen, yielding the fibrin polymer $(\alpha\beta\gamma)_2$. Activated Factor XIII subsequently cross-links the fibrin polymers to form an insoluble fibrin clot.

The actions of plasmin on both fibrin and fibrinogen have been studied extensively. Plasmin cleaves fibrin and fibrinogen into a family of fragments known as fibrinogen and fibrin (FDP-fdp) degradation products.

The anticoagulant effects of plasmin result not only from the destruction or inactivation of fibrin, fibrinogen, and other procoagulants, but also from the coagulation-inhibitingproperties of the fibrin(ogen) degradation products themselves. Fragments D and Y both inhibit coagulation in the clotting time test. Fragment X retains the ability to induce ADP-dependent platelet aggregation. Fragments Y, D, and E have lost the ability to induce and support platelet aggregation.

Activators of the Fibrinolytic System. There are several different pathways that lead to the activation of the plasminogen–plasmin proteolytic enzyme system. However, therapeutic utilization of this system occurs only by the exogenous administration of agents that initiate the conversion of inert plasminogen to the potent proteolytic enzyme plasmin. Currently, four thrombolytic agents have been approved for worldwide clinical use: streptokinase (SK), urokinase (U.K.), recombinantly produced tissue plasminogen activator (rTPA), and acylated Lys-plasminogen–streptokinase activator complex (APSAC). Some properties of these agents are listed in Table 6.

Inhibitors of Fibrinolysis. Inhibitors of the fibrinolytic system are either endogenous, naturally occurring inhibitors or modulators, or exogenous inhibitors employed primarily for therapeutic reasons.

The naturally occurring endogenous inhibitors/modulators of *in vivo* activity of the plasminogen–plasmin proteolytic enzyme system have been considered and listed previously. Some of these inhibitors increase in concentration during certain physiologic states, ie,

Table 3. Products Available for Treatment of Specific Factor Deficiencies

Product name	Content[a]	Method of viral inactivation[b]	Manufacturer
Factor VII			
Proplex-T	VII, rVII, VII,II,IX,X	DH 153 h at 60°C	Immuno, Novo,[c] Hyland[d]
Factor VIII			
Coagulation Factor VIII-SD	VIII:C	SD	New York Blood Center[e]
Coagulation Factor VIII-SD	VIII:C	SD	American Red Cross
Monoclate-P	VIII:C	MA-DH	Armour
Hemofil M	VIII:C	MA-SD	Hyland
AHF-M	VIII:C	MA-SD	American Red Cross
Humate-P	VIII:C	AHP 10 h at 60°C	Behringwerke[f]
Koate-HS	VIII:C	AHP 10 h at 60°C	Cutter[g]
Koate-HP	VIII:C	SD	Cutter
Profilate-OSD	VIII:C	SHP 20 h at 60°C	Alpha
Koate HT	VIII:C	DH	Cutter
Cryoprecipitate	VIII:C VIII:vWFAg fibrinogen plasminogen factor XIII		American Red Cross
fresh-frozen plasma	all coagulation factors, r VIII:C		American Red Cross
Autoplex-T	IIa, VIIa, IXa, Xa	DH	Hyland, Immuno
FEIBA	VIIa		
Factor IX			
Konyne-HT	IX, VII, X, II	DH 72 h at 68°C	Cutter
Proplex-T	IX, VII, X, II	DH 153 h at 60°C	Hyland
Proplex SX-T	IX, X, II	DH 153 h at 60°C	Hyland
Profilnine	IX, X, II	SHP 20 h at 60°C	Alpha
Alphanine	IX, X, II	MA-SD-AHP	Alpha
von Willebrand Factor products			
Cryoprecipitate	VIII:vWFAg VIII:C, fibrinogen plasminogen factor XIII		American Red Cross

[a] Coagulation factors; r indicates recombinant.
[b] AHP = anhydrous heat pasteurized; DH = dry heat; MA = moist atmosphere; SD = solvent detergent; SHP = solvent heat pasteurized.
[c] Novo = Nordisck AS, Copenhagen, Denmark
[d] Hyland Division = Baxter Healthcare Corp., Glendale, Calif.
[e] New York Blood Center = Melville Biologics, Inc., a division of the New York Blood Center, Melville, N.Y.
[f] Behringwerke is distributed by Armour Pharmaceutical, Blue Bell, Pa.
[g] Cutter = Miles, Inc., West-Haven, Conn.

Table 4. Miscellaneous Topical[a] Hemostatics and Action Mechanisms

Preparation	Trade name	Mechanism	Manufacturer
carbazochrome salicylate	Adrenosem[b]	reduce capillary permeability	SK Beecham Pharmaceutical
oxidized cellulose	Oxycel	support structure clot formation	Parke Davis
gelatin sponge	Surgical	support structure clot formation	Johnson & Johnson
gelatin sponge	Gelfoam	support structure clot formation	Upjohn Co.
microfibrillar collagen	Avitene	induce platelet aggregation	Alcon Laboratories
topical thrombin	Thrombin	induce fibrin polymerization	Parke Davis
fibrin glue	Tisseel	support structure clot formation	Immuno
desmopressin	DDAVP[c]	increase concentration of von Willebrand protein	Rhône-Poulenc Rorer
conjugated estrogens	Premarin[c]	induce hormone balance	Wyeth-Ayerst

[a] Unless otherwise noted.
[b] Oral or intramuscular administration.
[c] Intravenous administration.

pregnancy when PAI-1 increases three to fourfold above normal and decreases the normal degree of fibrinolytic activity. In various pathologic conditions, several of these inhibitors are increased manyfold above normal and exercise a paralytic effect on the fibrinolytic system.

Several synthetic amino acids have been identified that excite inhibition of the fibrinolytic system (Table 7). Some of these agents prevent the conversion of plasminogen to plasmin, whereas others block the degradative action of plasmin. A few of these agents inhibit both the conversion of plasminogen and hence the proteolytic action of plasmin, as well as the activity of the complement system. These agents experience wide use whenever there is a need to reduce the activity of the fibrinolytic system. In addition, these agents prevent excessive blood loss in patients with ulcerative colitis; menometrorrhagia; post-surgical prostatectomy; hereditary angioneurotic edema; oral surgery of any type, but in particular, dental extractions in patients with hemophilia A, hemophilia B, and von Willebrand disease; and immune and nonimmune-mediated thrombocytopenia. These antifibrinolytic agents inhibit complement activity and are often useful in immunologically-mediated phenomena such as asthma, post-organ transplantation, post-transfusion hemolysis, and neoplasia.

Table 5. Properties of Coumarinic Acid Anticoagulants

Name	Molecular formula	Melting point, °C
bis-4-hydroxycoumarin	$C_6H_{12}O_6$	287–293
warfarin	$C_{19}H_{16}O_4$	161
4-hydroxycoumarin	$C_6H_6O_3$	213–214
cyclocumarol	$C_{20}H_{18}O_4$	166
acenocoumarin	$C_{19}H_{15}NO_6$	196–199
coumachlor	$C_{19}H_{15}ClO_4$	164–165
ethyl biscoumacetate	$C_{22}H_{16}O_8$	177–182 154–157[a]

[a] Dimorphous.

Table 6. Properties of Approved Thrombolytic Agents

Category	SK[a]	U.K.[b]	rTPA[c]	APSAC[d]
source	streptococcal culture	heterologous mammalian tissue culture	heterologous mammalian tissue culture	streptococcal culture
mol wt, daltons	47,000	32,000–54,000	70,000	131,000
type of agent	bacterial proactivator	tissue plasminogen activator	tissue plasminogen activator	bacterial proactivator
plasma clearance, min	12–18	15–20	2–6	40–60
fibrinolytic activation	systemic	systemic	systemic	systemic
fibrin specificity	minimal	moderate	moderate	minimal
antigenic	yes	no	no	yes
allergic reactions	yes	no	no	yes
trade names	Streptase	Abbokinase	Alteplase (Activase)	Eminase
producer	Hoechst	Abbott Laboratories	Genentech Inc.	SmithKline Beecham Pharmaceuticals

[a] SK = streptokinase.
[b] UK = urokinase.
[c] rTPA = recombinant tissue plasminogen activator.
[d] APSAC = acylated plasminogen streptokinase activator complex.

Table 7. Synthetic Therapeutic Fibrinolytic Inhibitors

Names	Trade name, company	Dose
ε-aminocaproic acid, 6-aminohexanoic acid	Amicar, Lederle	4–6 g loading dose followed by 1 g/2–4 h iv or oral
transexamic acid, *trans*-4-(aminomethyl)cyclohexanecarboxylic acid	Amstat, Lederle	10 mg/kg 3 times daily iv or 10–20 mg/kg orally
aprotinin[a], trypsin inhibitor	Trasylol, Delbay	initial dose of 100,000 kIU[b] iv followed by 100,000 kIU iv/h
p-aminomethylbenzoic acid (PAMBA)		not applicable for chemical use

[a] Protein, mol wt 6200.
[b] Kallikrein international units.

William R. Bell, Jr.
The Johns Hopkins University School of Medicine

W. G. Bigelow, *Mysterious Heparin,* McGraw-Hill, Ryerson, Toronto, Canada, 1990, pp. 1–205.

W. R. Bell, *Hemostasis Thrombosis,* J. B. Lippincott, Philadelphia, Pa., 1987, pp. 886–900.

W. R. Bell, *Hematology: Basic Principles and Practice,* Churchill Livingstone, New York, 1991, p. 1450.

BLOOD, FRACTIONATION. See Fractionation, blood.

BLOWING AGENTS. See Foamed plastics.

BLUE PRINTING. See Printing processes.

BORDEAUX MIXTURE. See Fungicides, agricultural.

BORON, ELEMENTAL

Boron, B, is unique in that it is the only nonmetal in Group 13 (IIIA) of the Periodic Table. Boron, at wt 10.81, at no. 5, has more similarity to carbon and silicon than to the other elements in Group 13. There are two stable boron isotopes, ^{10}B and ^{11}B, which are naturally present at 19.10–20.31% and 79.69–80.90%, respectively.

There is a very low cosmic abundance of boron, but its occurrence at all is surprising for two reasons. First, boron's isotopes are not involved in a star's normal chain of thermonuclear reactions, and second, boron should not survive a star's extreme thermal condition. The formation of boron has been proposed to arise predominantly from cosmic ray bombardment of interstellar gas in a process called spallation.

Boron is the 51st most common element present in the earth's crust at a concentration of three grams per metric ton.

Properties

Elemental boron has a diverse and complex chemistry, primarily influenced by three circumstances. First, boron has a high ionization energy, 8.296 eV, 23.98 eV, and 37.75 eV for first, second, and third ionization potentials, respectively. Second, boron has a small size. Third, the electronegativities of boron (2.0), carbon (2.5), and hydrogen (2.1) are all very similar resulting in extensive and unusual covalent chemistry.

Boron has electronic structure $(1s)^2(2s)^2(2p)$ and an expected valence of three. Because of the high ionization energies there is no formation of univalent compounds as for the other Group 13 elements.

Boron also has a high affinity for oxygen-forming borates, polyborates, borosilicates, peroxoborates, etc. Boron reacts with water at temperatures above 100°C to form boric acid and other boron compounds (qv).

The physical properties of elemental boron are significantly affected by purity and crystal form. In addition to being an amorphous powder, boron has four crystalline forms: α-rhombohedral, β-rhombohedral, α-tetragonal, and β-tetragonal. The α-rhombohedral form has mp 2180°C, sublimes at approximately 3650°C, and has a density of 2.45 g/mL. Amorphous boron, by comparison, has mp 2300°C, sublimes at approximately 2550°C, and has a density of 2.35 g/mL.

Boron is an extremely hard refractory solid having a hardness of 9.3 on Mohs' scale and a very low (1.5×10^{-6} ohm^{-1} cm^{-1}) room temperature electrical conductivity so that boron is classified as a metalloid or semiconductor. These values are for the α-rhombohedral form.

The electron-deficient character of boron also affects its allotropic forms. The high ionization energies and small size prevent boron from adopting metallic bonding to compensate for its electron deficiency and that of other hypoelectronic elements.

Preparation

Three methods, electrolytic reduction, chemical reduction, and thermal decomposition, are used on a laboratory scale. A high purity (greater than 99%) boron comes from the direct thermal decomposition of boron hydrides such as diborane, B_2H_6.

Production

The Moissan process, the reduction of boric oxide with magnesium, is the most widely used commercial process for producing boron.

Another commercial process yields high purity boron of greater than 99%. In this process boron hydrides, such as diborane, are thermally decomposed.

Applications

Elemental boron is used in very diverse industries ranging from metallurgy (qv) to electronics. Other areas of application include ceramics (qv), propulsion, pyrotechnics, and nuclear chemistry. Boron is nontoxic. Workplace hygienic practices, however, include avoiding the breathing of boron dust or fine powder.

LINDA H. JANSEN
Callery Chemical Company

N. N. Greenwood and A. Earnshaw, *Chemistry of the Elements,* Pergamon Press, Oxford, U.K., 1984, p. 16.

S. H. Bauer, in J. F. Liebman, ed., *Advances in Boron and the Boranes,* Vol. 19, VCH Publishers, Inc., New York, 1988, p. 391.

H. E. Boyer and T. L. Gall, eds., *Metal Handbook, Desk Edition,* American Society for Metals, Metals Park, Ohio, 1985, pp. 4–11.

BORON COMPOUNDS

BORON OXIDES, BORIC ACID, AND BORATES

Borate Minerals

There are about 150 known boron-containing mineral. In nature boron always occurs in chemical combination with oxygen in the form of borates. Borax (tincal), kernite, colemanite, ulexite, probertite, hydroboracite, inderite, datolite, and szaibelyite (ascharite) are the only borate minerals of commercial importance. Borax and colemanite are the most important. Borate production comes mostly from seven countries: the United States, Turkey, Russia, Kazakhstan, Argentina, China, Peru, and Chile. Deposit areas and reserves in these countries are shown in Table 1.

Boron Oxides

Boric Oxide. Boric oxide, B_2O_3, formula wt 69.62, is the only commercially important oxide. It is also known as diboron trioxide, boric anhydride, or anhydrous boric acid. B_2O_3 is normally encountered in the vitreous state. This colorless, glassy solid has a Mohs' hardness of 4 and is usually prepared by dehydration of boric acid at elevated temperatures. Boric oxide is an excellent Lewis acid.

The physical properties of vitreous boric oxide (Table 2) are somewhat dependent on moisture content and thermal history.

The uses of boric oxide relate to its behavior as a flux, an acid catalyst, or a chemical intermediate. The fluxing action of B_2O_3 is important in preparing many types of glass, glazes, frits, ceramic coatings, and porcelain enamels.

Boric oxide is used as a catalyst in many organic reactions. It also serves as an intermediate in the production of boron halides, esters, carbide, nitride, and metallic borides.

Boron Monoxide and Dioxide. High temperature vapor phases of BO, B_2O_3, and BO_2 have been the subject of a number of spectroscopic and mass spectrometric studies aimed at developing theories of bonding, electronic structures, and thermochemical data. Values for the principal thermodynamic functions have been calculated and compiled for these gases.

Lower Oxides. A number of hard, refractory suboxides have been prepared either as by-products of elemental boron production or by the reaction of boron and boric acid at high temperatures and pressures. It appears that the various oxides represented as B_6O, B_7O, $B_{12}O_2$, and $B_{13}O_2$ may all be the same material in varying degrees of purity.

Boric Acid

The name boric acid is usually associated with orthoboric acid, which is the only commercially important form of boric acid and is found in nature as the mineral sassolite. Three crystalline modifications of metaboric acid also exist. All these forms of boric acid can be regarded as hydrates of boric oxide and formulated as $B_2O_3 \cdot 3H_2O$ for orthoboric acid and $B_2O_3 \cdot 3H_2O$ for metaboric acid.

Properties. The standard heats of formation of crystalline orthoboric acid and the three forms of metaboric acid are $\Delta H^\circ = -1094.3$ kJ/mol(-261.54 kcal/mol) for $B(OH)_3$; -804.04 kJ/mol(-192.17 kcal/mol) for HBO_2-I; -794.25 kJ/mol (-189.83kcal/mol) for HBO_2-II; and -788.77 kcal/mol (188.52 kcal/mol) for HBO_2-III. Values for the principal thermodynamic functions of $B(OH)_3$ are given in Table 3.

The solubility of boric acid in water increases rapidly with temperature. The heat of solution is somewhat concentration-dependent. The presence of inorganic salts may enhance or depress the aqueous solubility of boric acid: it is increased by potassium or sodium sulfate but decreased by lithium and sodium chlorides.

Boric acid is quite soluble in many organic solvents.

Manufacture. The majority of boric acid is produced by the reaction of inorganic borates with sulfuric acid in an aqueous medium. Sodium borates are the principal raw material in the United States. European manufacturers have generally used partially refined calcium borates, mainly colemanite from Turkey. Turkey uses both colemanite and tincal to make boric acid.

Uses. Boric acid has a surprising variety of applications in both industrial and consumer products. It serves as a source of B_2O_3 in many fused products, including textile fiber glass, optical and sealing glasses, heat-resistant borosilicate glass, ceramic glazes, and porcelain

Table 1. Distribution of Borate Minerals

Country	Area	Principal minerals	Reserves, 10^6 t of B_2O_3
United States	Boron, Calif.	tincal, kernite brine	41–50
	Searles Lake, Calif.		15
	Death Valley, Calif.	colemanite, ulexite, probertite	several
Turkey	Bigadic	colemanite, priceite, ulexite	
	Emet	colemanite	23
	Kirka	tincal, colemanite, ulexite	122
Kazakhstan	Inder	szaibelyite	54
Russia	Dal'negorsk	datolite	54
Argentina	Tincalayu	tincal, kernite, ulexite	23
China	Liaoning	szaibelyite	27

Table 2. Physical Properties of Vitreous Boric Oxide

Property	Value
vapor pressure[a], 1331–1808 K	$\log P = 5.849 - \frac{16960}{T}$
heat of vaporization, ΔH_{vap}, kJ/mol[b], at 1500 K	390.4
boiling point, extrapolated, °C	2316
viscosity, log η, mPa·s(= cP)	
350°C	10.60
1000°C	4.00
density, g/mL	
0°C	1.8766
18–25°C	1.844
500°C[c]	1.648
index of refraction, 14.4°C	1.463
heat capacity (specific), J/(kg·K)[b]	
298 K	62.969
700 K	132.63
heat of formation[d], ΔH_f, kJ,[b] 298.15 K	−1252.2 ± 1.7

[a] P is in units of kPa; T is in K. To convert kPa to torr, multiply by 7.5.
[b] To convert J to cal, divide by 4.184.
[c] Quenched.
[d] For $2B(s) + 3/2O_2(g) \rightarrow B_2O_3$ (glass).

enamels (see ENAMELS, PORCELAIN AND VITREOUS). It also serves as a component of fluxes for welding and brazing (see SOLDERS AND BRAZING ALLOYS; WELDING).

A number of boron chemicals are prepared directly from boric acid. These include synthetic inorganic borate salts, boron phosphate, fluoborates, boron trihalides, borate esters, boron carbide, and metal alloys such as ferroboron.

Boric acid catalyzes the air oxidation of hydrocarbons and increases the yield of alcohols by forming esters that prevent further oxidation of hydroxyl groups to ketones and carboxylic acids (see HYDROCARBON OXIDATION).

The bacteriostatic and fungicidal properties of boric acid have led to its use as a preservative in natural products such as lumber, rubber latex emulsions, leather, and starch products.

NF-grade boric acid serves as a mild, nonirritating antiseptic in mouthwashes, hair rinse, talcum powder, eyewashes, and protective ointments (see DISINFECTANTS AND ANTISEPTICS). With the addition of an anticaking agent, they have been used to control cockroaches and to protect wood against insect damage (see INSECT CONTROL TECHNOLOGY).

Inorganic boron compounds are generally good fire retardants.

Because boron compounds are good absorbers of thermal neutrons, owing to isotope ^{10}B, the nuclear industry has developed many applications. High purity boric acid is added to the cooling water used in high pressure water reactors (see NUCLEAR REACTORS).

Solutions of Boric Acid and Borates

Polyborates and pH Behavior. Whereas boric acid is essentially monomeric in dilute aqueous solutions, polymeric species may form at concentrations above 0.1 *M*. The conjugate base of boric acid in aqueous systems is the tetrahydroxyborate anion sometimes called the metaborate anion, $B(OH)_4^-$. This species is also the principal anion in solutions of alkali metal (1:1) borates such as sodium metaborate, $Na_2O{\cdot}B_2O_3{\cdot}4H_2O$. Mixtures of $B(OH)_3$ and $B(OH)_4^-$ appear to form classical buffer systems where the solution pH is governed primarily by the acid:salt ratio, ie, $[H^+] = K_a[B(OH)_3]/[B(OH)_4^-]$. This relationship is nearly correct for solutions of sodium or potassium (1:2) borates, eg, borax, where the ratio $B(OH)_3{:}B(OH)_4^- = 1$, and the pH remains near 9 over a wide range of concentrations. However, for solutions that have pH values much greater or less than 9, the pH changes greatly on dilution as shown in Figure 1.

This anomalous pH behavior results from the presence of polyborates, which dissociate into $B(OH)_3$ and $B(OH)_4^-$ as the solutions are diluted.

Solubility Trends. Formation of polyborates greatly enhances the mutual solubilities of boric acid and alkali borates.

Sodium borate solutions near the $Na_2O{:}B_2O_3$ ratio of maximum solubility can be spray-dried to form an amorphous product with the approximate composition $Na_2O{\cdot}4B_2O_3{\cdot}4H_2O$, commonly referred to as sodium octaborate. This material dissolves rapidly in water without any decrease in temperature to form supersaturated solutions.

The Polyborate Species. From a series of very rigorous pH studies, a series of equilibrium constants involving the species $B(OH)_3$, $B(OH)_4^-$, and the polyions $B_3O_3(OH)_5^{2-}$, $B_3O_3(OH)_4^-$, $B_5O_6(OH)_4^-$, and $B_4O_5(OH)_4^{2-}$ have been calculated. The relative populations of these species as functions of pH are shown in Figure 2.

Table 3. Thermodynamic Properties of Crystalline Boric Acid, $B(OH)_3$

Temperature, K	$C°_p$, J/(kg·K)[a]	$S°$, J/K[a]	$H°-H°_{298}$, J/mol[a]
0	0	0	−13393
100	35.92	28.98	−11636
200	58.74	61.13	−6866
298	81.34	88.74	0
400	100.21	115.39	9284

[a] To convert J to cal, divide by 4.184.

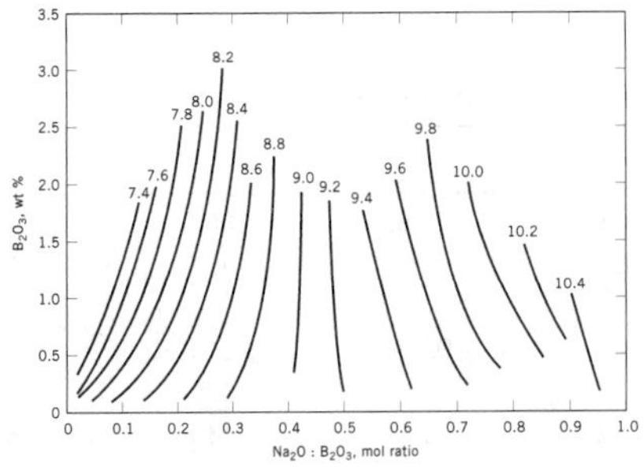

Figure 1. Values of pH in the system $Na_2O—B_2O_3—H_2O$ at 25°C.

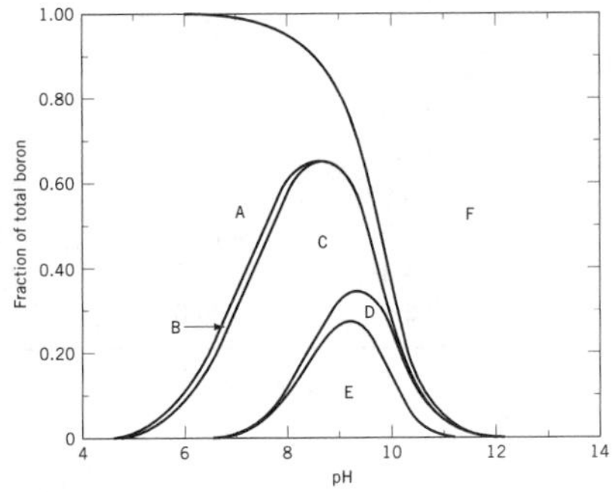

Figure 2. Distribution of boron in A, $B(OH)_3$; B, $B_5O_6(OH)_4^-$; C, $B_3O_3(OH)_4^-$; D, $B_3O_3(OH)_5^{2-}$; E, $B_4O_5(OH)_4^{2-}$; F, $B(OH)_4^-$; where total B_2O_3 concentration is 13.93 g/L. At a given pH, the fraction of the total boron in a given ion is represented by the portion of a vertical line falling within the corresponding range.

Sodium Borates

The solubility–temperature data for the $Na_2O—B_2O_3—H_2O$ system are given in Table 4.

Disodium Tetraborate Decahydrate (Borax Decahydrate). Disodium tetraborate decahydrate, $Na_2B_4O_7{\cdot}10H_2O$ or $Na_2O{\cdot}2B_2O_3{\cdot}10H_2O$, formula wt, 381.36; monoclinic; sp gr, 1.71; specific heat 1.611 kJ/(kg · K) [0.385 kcal/(g°C)] at 25–50°C; heat of formation, −6.2643 MJ/mol(−1497.2 kcal/mol), exists in nature as the mineral borax. Its crystal habit, nucleation, and growth rate are sensitive to inorganic and surface active organic modifiers.

Disodium Tetraborate Pentahydrate (Borax Pentahydrate). Although referred to as borax pentahydrate, well-formed crystals actually contain not five but 4.67 moles of water, $Na_2B_4O_7{\cdot}4.67H_2O$ or $Na_2O{\cdot}2B_2O_3{\cdot}4.67H_2O$. The structural formula is best represented as $Na_2[B_4O_5(OH)_4]{\cdot}2.67H_2O$, formula wt, 285.29; trigonal; rhombohedral crystal shape; sp gr, measured 1.880, crystallographic 1.912; specific heat, 1.32 kJ/(kg·K)[0.316kcal/(g·°C)]; heat of formation, −4.7844 MJ/mol(−1143.5 kcal/mol). It is found in nature as a fine-grained mineral, tincalconite, formed by dehydration of borax. Solubility data in water are given in Table 4.

Disodium Tetraborate Tetrahydrate. Disodium tetraborate tetrahydrate, $Na_2B_4O_7{\cdot}4H_2O$ or $Na_2O{\cdot}2B_2O_3{\cdot}4H_2O$, formula wt, 273.27; monoclinic; sp gr, 1.908; specific heat, ca 1.2 kJ/(kg·K) [0.287 kcal/(g·°C)]; heat of formation, −4.4890 MJ/mol(−1072.9 kcal/mol), exists in nature as the mineral kernite and has a structural formula $Na_2[B_4O_6(OH)_2]{\cdot}3H_2O$. The crystals have two perfect cleavages and when ground, form elongated splinters.

The water solubility of kernite is shown in Table 4.

Disodium Tetraborate (Anhydrous Borax). Disodium tetraborate, $Na_2B_4O_7$ or $Na_2O{\cdot}2B_2O_3$, formula wt, 201.21; sp gr (glass), 2.367, (α-crystalline form), 2.27; heat of formation (glass), −3.2566 MJ/mol(−778.34 kcal/mol), (α-crystalline form), −3.2767 MJ/mol(−783.2 kcal/mol), exists in several crystalline forms as well as a glassy form.

Disodium Octaborate Tetrahydrate. The composition of a commercially available sodium borate hydrate, 66.3 wt % B_2O_3, POLYBOR, corresponds quite closely to that of a hypothetical compound, disodium octaborate tetrahydrate, $Na_2B_8O_{13}{\cdot}4H_2O$ or $Na_2O{\cdot}4B_2O_3{\cdot}4H_2O$. This product dissolves rapidly in water without the temperature decrease, which occurs when the crystalline borates dissolve, and easily forms viscous supersaturated solutions at elevated temperatures. The solution pH decreases as the concentration increases.

Sodium Pentaborate Pentahydrate. Sodium pentaborate pentahydrate, $NaB_5O_8{\cdot}5H_2O$ or $Na_2O{\cdot}5B_2O_3{\cdot}10H_2O$ (formula wt, 295.11; monoclinic; sp gr, 1.713) exists in nature as the mineral sborgite. Heat capacity, entropy, and other thermal measurements have been made at 15–345 K.

Sodium pentaborate can easily be crystallized from a solution having a $Na_2O{:}B_2O_3$ mol ratio of 0.2. Its water solubility (Table 4) exceeds that of borax and boric acid. Its pH decreases with solution concentration.

Sodium Metaborate Tetrahydrate. Sodium metaborate tetrahydrate, $NaBO_2{\cdot}4H_2O$ or $Na_2O{\cdot}B_2O_3{\cdot}8H_2O$(formula wt, 137.86; triclinic; sp gr, 1.743) is easily formed by cooling a solution containing borax and an amount of sodium hydroxide just in excess of the theoretical value. It is the stable phase in contact with its saturated solution between 11.5 and 53.6°C. At temperatures above 53.6°C, the dihydrate, $NaBO_2{\cdot}2H_2O$, becomes the stable phase. The water solubility of sodium metaborate is given in Table 4.

Sodium Metaborate Dihydrate. Sodium metaborate dihydrate, $NaBO_2{\cdot}2H_2O$ or $Na_2O{\cdot}B_2O_3{\cdot}4H_2O$ (formula wt, 101.83; triclinic; sp gr, 1.909) can be prepared by heating a slurry of the tetrahydrate above 54°C, by crystallizing metaborate solutions at 54–80°C, or by dehydrating the tetrahydrate in vacuum. The water solubility for the dihydrate is shown in Table 4.

Sodium Perborate Hydrates. Peroxyborates are commonly known as perborates, written as if the perborate anion were BO_3^-. X-ray crystal structure has shown that they contain the dimeric anion $[(HO)_2B(O_2)_2B(OH)_2]^{2-}$ (**1**). Three sodium perborate hydrates, $NaBO_3{\cdot}xH_2O$ (x = 1, 3, and 4), are known. Only the mono- and tetrahydrate are of commercial importance, primarily as bleaching agents (qv) in laundry products.

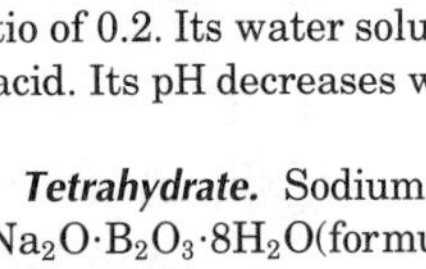

(**1**)

Sodium perborate tetrahydrate, $NaBO_3{\cdot}4H_2O$ or $Na_2B_2(O_2)_2(OH)_4{\cdot}6H_2O$, is triclinic, heat of formation, −2112 kJ/mol (−504.8 kcal/mol) (crystal), −921 kJ/mol (−220.2 kcal/mol) (1 M soln), and contains 10.4 wt% active oxygen. It melts at 63°C by dissolving in its own water of hydration and on heating to 250°C decomposes rapidly and completely to oxygen and sodium metaborate. In water its decomposition, which is important in its use as a bleach, is accelerated by catalysts or elevated temperature.

Sodium perborate trihydrate, $NaBO_3{\cdot}3H_2O$ or $Na_2B_2(O_2)_2(OH)_4{\cdot}4H_2O$, triclinic, contains 11.8 wt% active oxygen (96). It has been claimed to have better thermal stability than the tetrahydrate but has not been used commercially. The trihydrate can be made by dehydration of the tetrahydrate or by crystallization from a sodium metaborate and hydrogen peroxide solution in the present of trihydrate seeds.

Sodium perborate monohydrate, $NaBO_3{\cdot}H_2O$ or $Na_2B_2(O_2)_2(OH)_4$, 16.0 wt% active oxygen, is commercially prepared by dehydration of the tetrahydrate.

Manufacture. *Borax Decahydrate and Pentahydrate.* Borax decahydrate and pentahydrate are produced from sodium borate ores, dry lake brines, colemanite, or magnesium borate ores.

Production from sodium borate ores takes place in the United States, Turkey, and Argentina.

Table 4. Aqueous Solubilities of Alkali Metal and Ammonium Borates at Various Temperatures

	Solubility, wt % anhydrous salt, at °C						
Compound	0	20	30	50	70	90	100
$Li_2O\cdot 5B_2O_3\cdot 10H_2O$[a]				20.88	27.98	36.2	41.2
$Li_2O\cdot B_2O_3\cdot 4H_2O$	2.2–2.5	2.81	3.01	3.50	4.08	4.75	5.17
$Li_2O\cdot B_2O_3\cdot 16H_2O$[b]	0.88	2.51	4.63				
$Li_2O\cdot B_2O_3\cdot 4H_2O$				7.84	9.43	11.8	13.4[c]
						9.7	9.70
$Na_2\cdot 5B_2O_3\cdot 10H_2O$	5.77	10.55	13.72	21.72	32.25	44.3	51.0
$Na_2O\cdot 2B_2O_3\cdot 10H_2O$	1.18	2.58	3.85	9.55			
$Na_2O\cdot 2B_2O_3\cdot 4.67H_2O$[d]					19.49	28.37	34.63
$Na_2O\cdot 2B_2O_3\cdot 4H_2O$[e]					17.12	23.31	28.22
$Na_2O\cdot B_2O_3\cdot 8H_2O$[f]	14.5	20.0	23.6	34.1			
$Na_2O\cdot B_2O_3\cdot 4H_2O$					40.7	47.4	52.4
$K_2O\cdot 5B_2O_3\cdot 8H_2O$	1.56	2.82	3.80	6.88	11.7	18.3	22.3
$K_2O\cdot 2B_2O_3\cdot 4H_2O$		12.1	15.6	24.0	33.3	43.2	48.4
$K_2O\cdot B_2O_3\cdot 2.5H_2O$		43.0	44.0	46.1	48.2	50.3	
$Rb_2O\cdot 5B_2O_3\cdot 8H_2O$	1.58	2.67	3.58	6.52	11.4	18.1	23.75[g]
$Cs_2O\cdot 5B_2O_3\cdot 8H_2O$[h]	1.6	2.5	3.52	6.4	10.5	18.0	23.45[i]
$(NH_4)_2O\cdot 2B_2O_3\cdot 4H_2O$	3.75	7.63	10.8	21.2	34.4	52.7	
$(NH_4)_2O\cdot 5B_2O_3\cdot 8H_2O$	4.00	7.07	9.10	14.4	22.4	30.3	

[a] Incongruent solubility below 37.5 or 40.5°C.
[b] Transition point to tetrahydrate, 36.9 or 40 °C.
[c] At 101.2°C.
[d] Commonly known as the five hydrate, transition point to decahydrate, 60.7°C, 16.6% $Na_2B_4O_7$.
[e] Transition point to decahydrate, 58.2°C, 14.55% $Na_2B_4O_7$.
[f] Transition point to tetrahydrate, 53.6°C, 36.9% $Na_2B_2O_4$.
[g] At 102°C.
[h] Dicesium tetraborate pentahydrate, $Cs_2O\cdot 2B_2O_3\cdot 5H_2O$, and dicesium diborate heptahydrate $Cs_2O\cdot B_2O_3\cdot 7H_2O$, also exist. The former has incongruent solubility; the latter has a solubility of 36.8 wt% anhydrous salt at 18°C.
[i] At 101.65°C.

Anhydrous Borax. Anhydrous borax is produced from its hydrate forms, borax decahydrate or pentahydrate, by fusing in a furnace at elevated temperatures.

Sodium Perborate. The common commercial practice for the manufacture of sodium perborate tetrahydrate involves the reaction of sodium metaborate and hydrogen peroxide and subsequent crystallization of sodium perborate tetrahydrate.

Health and Safety. Cases of industrial intoxication on exposure to inorganic borates have not been reported. There is a large body of literature on the toxicology of boric acid and borax. Studies indicate no evidence of carcinogenic or mutagenic activity. Boric acid and borax are poorly absorbed through healthy skin and do not cause skin irritation. Gloves, goggles, and a simple dust mask should be used when handling sodium metaborate powder.

Boron in the form of borate is an essential micronutrient for the healthy growth of plants and is present in the normal daily human diet at an estimated level of 3–40 mg as boron. It is not a proven essential micronutrient for animals.

Uses. In the United States over 54% of the total B_2O_3 consumption is for glass (qv) manufacture. Approximately 24% is used in fiber glass insulation where the borate is added to increase fiber durability. Approximately 16% is used in textile fiber glass, where the B_2O_3 is added to aid in weathering resistance. Other glass applications also use B_2O_3, eg, the production of borosilicate glasses, where the B_2O_3 imparts a low coefficient of thermal expansion, ie, Pyrex glass.

Borates are used as fluxing agents for porcelain enamels and ceramic glazes (see CERAMICS; ENAMELS, PORCELAIN AND VITREOUS). This market accounts for about 3% of the total usage for the United States. Approximately 8% of total consumption goes into soap and cleaning compositions.

Approximately 5% of the U.S. consumption of B_2O_3 is in agriculture. Boron is a necessary trace nutrient for plants and is added in small quantities to a number of fertilizers.

Other Alkali Metal and Ammonium Borates

Dipotassium Tetraborate Tetrahydrate. Dipotassium tetraborate tetrahydrate, $K_2B_4O_7\cdot 4H_2O$ or $K_2O\cdot 2B_2O_3\cdot 4H_2O$ (formula wt, 305.49; orthorhombic; sp gr, 1.919), is much more soluble than borax in water. Solubility data are given in Table 4.

Potassium Pentaborate Tetrahydrate. Potassium pentaborate tetrahydrate, $KB_5O_8\cdot 4H_2O$ or $K_2O\cdot 5B_2O_3\cdot 8H_2O$, formula wt, 293.20; orthorhombic prisms; sp gr, 1.74; heat capacity, 329.0 J/(mol·K)[78.6 cal/(mol·K)] at 296.6 K, is much less soluble than sodium pentaborate (Table 4).

Diammonium Tetraborate Tetrahydrate. Diammonium tetraborate tetrahydrate, $(NH_4)_2B_4O_7\cdot 4H_2O$ or $(NH_4)_2O\cdot 2B_2O_3\cdot 4H_2O$ (formula wt, 263.37; monoclinic; sp gr, 1.58) is readily soluble in water (Table 4). The pH of solutions of diammonium tetraborate tetrahydrate is 8.8 and independent of concentration.

Ammonium Pentaborate Tetrahydrate. Ammonium pentaborate tetrahyrate, $NH_4B_5O_8\cdot 4H_2O$ or $(NH_4)_2O\cdot 5B_2O_3\cdot 8H_2O$, formula wt 272.13; sp gr, 1.567; heat capacity, 359.4 J/(mol·K)[85.9 cal/(mol·K)] at 301.2 K, exists in two crystal line forms, orthorhombic (α) and monoclinic (β). Solubility data are given in Table 4.

Lithium Borates. Two lithium borates are of minor commercial importance, the tetraborate trihydrate and metaborate hydrates.

Manufacture. Potassium tetraborate tetrahydrate may be prepared from an aqueous solution of KOH and boric acid having a B_2O_3:K_2O ratio of about 2 or by separation from a KCl–borax solution. Potassium pentaborate is prepared in a manner analogous to that used for the tetraborate, but the strong liquor has a B_2O_3:K_2O ratio near 5.

Ammonium tetraborate tetrahydrate is prepared by crystallization from an aqueous solution of boric acid and ammonia having a B_2O_3:$(NH_4)_2O$ ratio of 1.8:2.1. Ammonium pentaborate is similarly produced from an aqueous solution of boric acid and ammonia having a B_2O_3:$(NH_4)_2O$ ratio of 5. Supersaturated solutions are easily formed and the rate of crystallization is proportional to the extent of supersaturation. A process for the production of ammonium pentaborate by precipitation from an aqueous ammonium chloride–borax mixture has been patented.

Health and Safety. Few toxicological data are available on borates other than boric acid and borax. Most water-soluble borates have the same toxicological effects as borax when adjusted to account for differences in B_2O_3 content.

Uses. Dipotassium tetraborate tetrahydrate is used to replace borax in applications where an alkali metal borate is needed but sodium salts cannot be used or where a more soluble form is required. The potassium compound is used as a solvent for casein, as a constituent in welding fluxes, and a component in diazotype developer solutions. Potassium pentaborate tetrahydrate is used in fluxes for welding and brazing of stainless steels for nonferrous metals. Diammonium tetraborate tetrahydrate is used when a highly soluble borate is desired but alkali metals cannot be tolerated. It is used mostly as a neutralizing agent in the manufacture of urea–formaldehyde resins and as an ingredient in flameproofing formulations. Ammonium pentaborate tetrahydrate is used as a component of electrolytes for electrolytic capacitors, as an ingredient in flameproofing formulations, and in paper coatings.

Calcium-Containing Borates

Dicalcium Hexaborate Pentahydrate. Dicalcium hexaborate pentahydrate, $Ca_2B_6O_{11}\cdot 5H_2O$ or $2CaO\cdot 3B_2O_3\cdot 5H_2O$, formula wt, 411.08; monoclinic; sp gr, 2.42; heat of formation, −3.469 kJ/mol (−0.83 kcal/mol), exists in nature as the mineral colemanite. Its solubility in water is about 0.1% at 25°C and 0.38% at 100°C.

Sodium Calcium Pentaborate Octahydrate. Sodium calcium pentaborate octahydrate, $NaCaB_5O_9\cdot 8H_2O$ or $Na_2O\cdot 2CaO\cdot 5B_2O_3\cdot 16H_2O$ (formula wt, 405.23; triclinic; sp gr, 1.95) exists in nature as the mineral ulexite. The solubility in water at 25°C is 0.5% as $NaCaB_5O_9$.

Sodium Calcium Pentaborate Pentahydrate. Sodium calcium pentaborate pentahydrate, $NaCaB_5O_9\cdot 5H_2O$ or $Na_2O\cdot 2CaO\cdot 5B_2O_3\cdot 10H_2O$ (formula wt 351.19; monoclinic; sp gr, 2.14) exists in nature as the mineral probertite.

Manufacture. The alkaline-earth metal borates of primary commercial importance are colemanite and ulexite. Both of these borates are sold as impure ore concentrates from Turkey, the principal world supplier.

Uses. Colemanite, $2CaO\cdot 3B_2O_3\cdot 5H_2O$, is used in the production of boric acid and borax, as well as in several direct applications. It is a highly desirable material for the manufacture of the E-glass used in textile glass fibers and plastic reinforcement (where sodium cannot be tolerated). High As or Fe levels in the ore concentrate can limit its use in this application. Colemanite has seen limited application as a slagging material in steel manufacture. It is also used in some fire retardants and as a precursor to some boron alloys.

Ulexite, $NaCaB_5O_9\cdot 8H_2O$, and probertite, $NaCaB_5O_9\cdot 5H_2O$, have found application in the production of insulation fiber glass and borosilicate glass as well as in the manufacture of other borates.

Other Metal Borates

Borate salts or complexes of virtually every metal have been prepared. For most metals, a series of hydrated and anhydrous compounds may be obtained by varying the starting materials or reaction conditions. Some have achieved commercial importance.

In general, hydrated borates of heavy metals are prepared by mixing aqueous solutions or suspensions of the metal oxides, sulfates, or halides and boric acid or alkali metal borates such as borax. The precipitates formed from basic solutions are often sparingly soluble amorphous solids having variable compositions. Crystalline products are generally obtained from slightly acidic solutions.

Anhydrous metal borates may be prepared by heating the hydrated salts to 300–500°C, or by direct fusion of the metal oxide with boric acid or B_2O_3. Many binary and tertiary anhydrous systems containing B_2O_3 form vitreous phases over certain ranges of composition.

Barium Metaborate. Three hydrates of barium metaborate, $BaO\cdot B_2O_3\cdot xH_2O$, are known. Barium metaborate is used as an additive to impart fire-retardant and mildew-resistant properties to latex paints, plastics, textiles, and paper products.

Copper, Manganese, and Cobalt Borates. Borate salts of copper, manganese, and cobalt are precipitated when borax is added to aqueous solutions of the metal(II) sulfates or chlorides. However, these materials are no longer produced commercially.

Zinc Borates. A series of hydrated zinc borates have been developed for use as fire-retardant additives and anticorrosive pigments in coatings and polymers.

Boron Phosphate

Boron phosphate, BPO_4, is a white, infusible solid that vaporizes slowly above 1450°C, without apparent decomposition. It is normally prepared by dehydrating mixtures of boric acid and phosphoric acid at temperatures up to 1200°C.

The principal application of boron phosphate has been as a heterogeneous acid catalyst.

ROBERT A. SMITH
ROBERT B. MCBROOM
U.S. Borax Research Corporation

N. P. Nies and G. W. Campbell, in R. M. Adams, ed., *Boron, Metallo–Boron Compounds, and Boranes,* Interscience Publishers, New York, 1964.

R. Thompson and A. J. E. Welch, eds., *Mellor's Comprehensive Treatise on Inorganic and Theoretical Chemistry,* Vol. V, *Boron,* Part A, *Boron–Oxygen Compounds,* Longman, New York, London, 1980.

R. Will, Y. Sakuma, and R. Willhalm, "Boron Minerals and Chemicals Report," in *Chemical Economics Handbook,* SRI International, Menlo Park, Calif., Sept. 1990.

N. J. Travis and E. J. Cocks, *The Tincal Trail,* Harrap Ltd., London, 1984.

BORIC ACID ESTERS

The general formula for boric acid esters if $B(OR)_3$. The lower molecular weight esters such as methyl, ethyl, and phenyl are most commonly referred to as methyl borate, ethyl borate, and phenyl borate, respectively. Some of the most common boric acid esters used in industrial applications are listed in Table 1.

Physical Properties

Most reported boric acid esters are trialkoxy or triaryloxy boranes. The esters range from colorless low boiling liquids to solids that possess high melting points. Boric acid esters usually have an odor similar to the hydroxy compound from which they are derived. A more complete description of the physical properties of the compounds, given in Table 1, has been published.

Chemical Properties

Alkyl boric acid esters derived from straight-chain alcohols and aryl boric acid esters are stable to relatively high temperatures. Trialkoxyboranes from branched-chain alcohols are much less stable, and boranes from tertiary alcohols can even decompose at 100°C. Decomposition of branched-chain esters leads to mixtures of olefins, alcohols, and other derivatives.

Boric acid esters are very susceptible to hydrolysis in the presence of water, or in some cases atmospheric moisture.

Preparative Methods

There are a number of methods used for the preparation of borate esters.

Table 1. Properties of Boric Acid Ester Derivatives

Compound name	Mp, °C	Bp, °C[a]	d^t_4	n^b_D
trimethyl borate	−29	68.0–68.5	0.920^{20}	1.3548
trimethyl borate azeotrope		54.3	0.8804^{25}	1.3472
triethyl borate	−84.8	117–119	0.859^{26}	1.3723
triethyl borate azeotrope		76.6		
tri-*n*-propyl borate		176–179	0.356^{24}	1.3933
triisopropyl borate		139–140	0.815^{23}	1.3750
tri-*n*-butyl borate		227	0.856^{25}	1.4077
triphenyl borate	80–90[c]	360–370 224–230[d]		
tricresyl borate[e]		185–200[f]		
trimethoxyboroxine		dec	1.2286^{25}	
triisopropoxyboroxine	52–54	235–239 dec		
2,2′-oxybis[4,4,6-trimethyl-1,3,2-dioxaborinane]		114–115[f]	1.013^{24}	1.4308
2,2′-[1-methyl-1,3-propanediyl]bis(oxy)-bis[4-methyl-1,3,2-dioxaborinane]		207–213[d]	1.071^{25}	1.4464[g]
2,2′-[1,1,3-trimethyltri-methylene-dioxy]bis[4,4,6-trimethyl-1,3,2-dioxaborinane]		274–276	0.932^{21}	1.4381
triethanolamineborate	235–239	170–172[h]		

[a] At 101.3 kPa (760 mm Hg) unless otherwise indicated.
[b] At 25°C unless otherwise indicated.
[c] Triphenyl borate has been reported to melt at temperatures from 38 to 146°C. Most values are in the range of 80–90°C.
[d] At 2.3 kPa (17 mm Hg).
[e] Mixture of the m- and p-isomers.
[f] At 0.27 kPa (2 mm Hg).
[g] At 17°C.
[h] At 2.7 kPa (20 mm Hg).

From Boric Acid. The most common method for the preparation of trialkoxy- and triaryloxyboranes is the esterification of boric acid using three moles of an alcohol or phenol.

From Boric Oxide and Alcohol. To avoid removing water, boric oxide, B_2O_3, can be used in place of boric acid.

Transesterification. Transesterification is another method that does not require the removal of water.

From Boron Halides. Using boron halides is not economically desirable because boron halides are made from boric acid. However, this method does provide a convenient laboratory synthesis of boric acid esters.

Miscellaneous Methods. Other methods for preparing borate esters have also been described. These include alcoholysis of borax or other alkali metal borates in either neutral or acidic media, disproportionation, decomposition or reaction of alkoxyhaloboranes, and disproportionation of trialkoxyboroxines. A simple and convenient method for synthesizing trialkoxyboranes and trialkoxyboroxines using calcium hydride as a drying agent has been published.

Manufacture and Economic Aspects

There are relatively few producers of boric acid esters in the United States. Eight domestic producers of these compounds are Anderson Development Company, Akzo America, Inc., E. I. du Pont de Nemours & Co., Inc., Eagle-Picher Industries, Inc., The Gas Flux Company, Morton International, Callery Chemical Company, and U.S. Borax & Chemical Corporation. In addition, Rhone-Poulenc Chemicals, Manchester, U.K., produces commercial quantities of selected boric acid esters.

Methyl borate is believed to be the boric acid ester produced in the largest quantity, approximately 8600 metric tons per year. Most methyl borate is produced by Morton International and used captively to manufacture sodium borohydride.

A combination of the two dioxaborinanes, is marketed as a fuel microbiocide by Hammonds Fuel Additives, Inc., Houston, Texas, under the trademark BIOBOR JF. Annual U.S. production and consumption is estimated at 140–230 metric tons.

U.S. Borax, Valencia, California, markets the biborate BORESTER 7.

Handling and Shipping

Procedures for shipping boric acid esters depend on the particular compound. Aryl borates produce phenols when in contact with water and are therefore subject to shipping regulations governing such materials and must carry a Corrosive Chemical label. Lower alkyl borates are flammable; flash points of methyl, ethyl, and butyl borates are 0, 32, and 94°C, respectively, and must be stored in approved areas. Other compounds are not hazardous, and may be shipped or stored in any convenient manner. Because borate esters are susceptible to hydrolysis, the more sensitive compounds should be stored and transferred in an inert atmosphere, such as nitrogen.

Health and Safety

The toxicity of a few boric acid esters has been summarized. In general the toxicities are directly related to the toxicity of the alcohol or phenol produced on hydrolysis.

Uses

Boric acid esters are used in the production of sodium borohydride, gas fluxing, as polymer additives, in hydraulic fluids and lubricants, as biocides, and in hydrocarbon oxidation.

EDWARD L. DOCKS
U.S. Borax

H. Steinberg, *Organboron Chemistry, Boron–Oxygen and Boron–Sulfur Compounds,* Vol. 1. Interscience Publishers, New York, 1964.

D. B. Green, *Mellor's Comprehensive Treatise on Inorganic and Theoretical Chemistry (Supplement) 1, Part A,* Vol. 5, Longman, London, 1980, pp. 703–720.

R. J. Brotherton, C. J. Weber, C. R. Guibert, and J. L. Little, *Ullmanns Encyclopedia of Industrial Chemistry,* 5th ed., Vol. A4, VCH Publishers, Deerfield, Fla., 1985, pp. 309–330.

REFRACTORY BORON COMPOUNDS

Borides have metallic characteristics such as high electrical conductivity and positive coefficients of electrical resistivity. Many of them, particularly the borides of metals of Groups 4 (IVB), 5 (VB), and 6 (VIB), the MB_6 compounds of Groups 2(II) and 13(III), and the borides of aluminum and silicon, have high melting points, great hardness, low coefficients of thermal expansion, and good chemical stability.

Borides are inert toward nonoxidizing acids; however, a few, such as Be_2B and MgB_2, react with aqueous acids to form boron hydrides. Most borides dissolve in oxidizing acids such as nitric or hot sulfuric acid and they are also readily attacked by hot alkaline salt melts or fused alkali peroxides; forming the more stable borates. In dry air, where a protective oxide film can be preserved, borides are relatively resistant to oxidation.

The structures of borides range from the isolated boron atoms in the M_2B borides through single chains in MB borides, double chains (M_3B_4), two-dimensional hexagonal nets (MB_2), cross-linked nets (MB_4), and interconnected B_6 octahedra (MB_6), to cages of 24 boron atoms surrounding the central metal atom in the MB_{12} borides.

Preparation. The simplest method of preparation is a combination of the elements at a suitable temperature, usually in the range of 1100–2000°C. On a commercial scale, borides are prepared by the reduction of mixtures of metallic and boron oxides using aluminum, magnesium, carbon, boron, or boron carbide, followed by purification. Borides can also be synthesized by vapor-phase reaction or electrolysis.

Uses. In spite of unique properties, there are few commercial applications for monolithic shapes of borides. They are used for resistance-heated boats (with boron nitride), for aluminum evaporation, and for sliding electrical contacts.

Boron Carbide

Boron and carbon form one compound, boron carbide, B_4C, although excess boron may dissolve in boron carbide, and a small amount of boron may dissolve in graphite.

Properties. Boron carbide has a rhombohedral structure consisting of an array of nearly regular icosahedra, each having twelve boron atoms at the vertices and three carbon atoms in a linear chain outside the icosahedra. Thus a descriptive chemical formula would be $B_{12}C_3$.

Boron carbide is resistant to most acids but is rapidly attacked by molten alkalies.

Preparation. Boron carbide is most commonly produced by the reduction of boric oxide with carbon in an electric furnace between 1400–2300°C.

Uses. Applications for boron carbide relate either to its hardness or its high neutron absorptivity (^{10}B isotope). Hot-pressed boron carbide finds use as wear parts, sandblast nozzles, seals, and ceramic armor plates.

Boron carbide is used in the shielding and control of nuclear reactors (qv).

Boron Nitride

Boron and nitrogen form one compound, boron nitride, BN, which may exist in a hexagonal, graphite-like form, hBN, having a layered structure and planar 6-membered rings of alternating boron and nitrogen atoms, or in two denser forms having tetrahedrally bonded B and N atoms on a cubic (zinc-blend) or a hexagonal (wurtzite) lattice.

Properties. Under nitrogen pressure hexagonal boron nitride melts at about 3000°C but sublimes at about 2500°C at atmospheric pressure. Despite the high melting point, the substance is mechanically weak because of the relatively easy sliding of the sheets of rings past one another. The denser forms are nearly as hard as diamond; the cubic form is the more stable of the two.

Preparation. Hexagonal boron nitride can be prepared by heating boric oxide with ammonia, or by heating boric oxide, boric acid, or its salts with ammonium chloride, alkali cyanides, or calcium cyanamide at atmospheric pressure. The denser forms are prepared by exposing the hexagonal form to very high pressures and high temperatures, usually in the presence of a salt-like nitride catalyst.

Uses. Hot-pressed hBN is useful for high temperature electric or thermal insulation, vessels, etc, especially in inert or reducing atmospheres, and for special materials such as III-V semiconductors (qv).

The greatest use of cubic boron nitride is as an abrasive under the name Borazon, in the form of small crystals, 1–500 μm in size.

The cubic BN crystals may also be bonded into strong bodies that make excellent cutting tools for hard iron and nickel-based alloys. Such tools produce red-hot chips and permit the wider use of tough, high temperature alloys which would otherwise be prohibitively difficult to shape.

ROBERT H. WENTORF, JR.
Consultant

N. N. Greenwood, in J. C. Bailar and co-eds., *Comprehensive Inorganic Chemistry,* Vol. 1, Pergamon Press, New York, 1973, pp. 665–993.

Handbook on Boron Carbide, Elemental Boron, and Other Stable, Boron-Rich Materials, Norton Co., Worcester, Mass., 1955.

F. P. Bundy and R. H. Wentorf, Jr., *J. Chem. Phys.* **38,** 1144 (1963).

R. H. Wentorf, Jr., R. C. DeVries, and F. P. Bundy, *Science* **208,** 873–888 (1980).

Table 1. Physical Properties of the Boron Trihalides

Property	BCl_3	BBr_3	BI_3
mp, °C	-107	-46	-49.9
bp, °C	12.5	91.3	210
density[a], g/mL (liq)	$1.434^{0}{}_{4}$ $1.349^{11}{}_{4}$	$2.643^{18}{}_{4}$	3.35
critical temperature, °C	178.8	300	
critical pressure, kPa[b]	3901.0		
vapor pressure, kPa[b]			
−80 °C	0.53	[c]	
0°C	63.5		
80°C	689		
viscosity, mPa·s(= cP)	[d]	[d]	
ΔH, kJ/mol, gas[e]	−403	−206	+18
C_p, J/(mol·C), for gas at 25°C[e]	62.8	67.78	
ΔH_{hydrol}, kJ/mol, liquid at 25°C[e]	−289	−351	
ΔH_{fus}, J/g at mp[e]	18		

[a] For BCl_3: $\rho = 1.3730 - 2.159 \times 10^{-3}$ °C − 8.377×10^{-7} °C; from −44 to 5°C. Superscript indicates temperature of measurement. For BBr_3: $-\rho = 2.698 - 2.996 \times 10^{-3}$ °C; from −20 to 90°C.

[b] To convert kPa to mm Hg, multiply by 7.50.

[c] For BBr_3: log(pressure) = [6.9792 − 1311/(C + 230)] − 0.8752; from 0–90°C.

[d] For BCl_3: $-\eta = 0.34417/(1 - 6.9662 \times 10^{-3}$ °C − 5.9013×10^{-6} °C); from −40 to 10°C. For BBr_3: $\log \eta = (333/K) - 1.257$; from 0–90°C.

[e] To convert J to cal, divide by 4.184.

BORON HALIDES

The boron trihalides boron trifluoride, BF_3, boron trichloride, BCl_3, and boron tribromide, BBr_3, are important industrial chemicals having increased usage as Lewis acid catalysts and in chemical vapor deposition (CVD) processes (see ELECTRONIC MATERIALS). Boron halides are widely used in the laboratory as catalysts and reagents in numerous types of organic reactions and as starting material for many organoboron and inorganic boron compounds.

BCl_3, BBr_3, and BI_3 undergo exchange reactions to yield mixed boron halides.

Reactions of boron trihalides that are of commercial importance are those of BCl_3 and, to a lesser extent, BBr_3, with gases in chemical vapor deposition (CVD).

Manufacturing and Processing

Boron Trichloride. Boron trichloride is prepared on a large scale by the reaction of Cl_2 and a heated mixture of borax, $Na_2BO_4O_7 \cdot 10H_2O$, and crude oil residue in a rotary kiln heated to 1038°C. On a smaller scale, BCl_3 can be prepared by the reaction of Cl_2 and a mixture of boron oxide, B_2O_3, petroleum coke, and lampblack in a fluidized bed.

BORON TRIHALIDES

Properties

Boron trihalides, BX_3, are trigonal planar molecules which are $(sp)^2$ hybridized. The X–B–X angles are 120°. Important physical and thermochemical data are presented in Table 1.

The boron trihalides are strong Lewis acids; however, the order of relative acid strengths, $BI_3 \geq BBr_3 > BCl_3 > BF_3$, is contrary to that expected based on the electronegativities and atomic sizes of the halogen atoms.

A convenient laboratory method for the preparation of BCl_3 is by the reaction of $AlCl_3$ and BF_3 or BF_4.

Boron Tribromide. Boron tribromide is produced on a large scale by the reaction of Br_2 and granulated B_4C at 850–1000°C or by the reaction of HBr with CaB_6 at high temperatures.

Boron Triiodide. Boron triiodide is not manufactured on a large scale. Small-scale production of BI_3 from boron and iodine is possible in the temperature range 700–900°C.

Uses

Boron Trichloride. Approximately 75–95% of the BCl_3 consumed in the United States is used to prepare boron filaments by CVD. Another important use of BCl_3 is as a Friedel-Crafts catalyst in various polymerization, alkylation, and acylation reactions, and in other organic synthesis (see FRIEDEL-CRAFTS REACTION).

BCl_3 is also used for the production of halosilanes, in the preparation of many boron compounds, in the production of optical wave guides, and for the prevention of solid polymer formation in liquid SO_3; for the removal of SiO_2 from SiC powders, carbochlorination of kaolinitic ores, and removal of impurities from molten Mg.

Boron Bromide. Approximately 30% of BBr_3 produced in the United States is consumed in the manufacture of proprietory pharmaceuticals (qv). BBr_3 is used in the manufacture of isotopically enriched crystalline boron, as a Friedel-Crafts catalyst in various polymerization, alkylation, and acylation reactions, and in semiconductor doping and etching.

BBr_3 is a very useful reagent for cleaving ethers, esters, lactones, acetals, and glycosidic bonds; it is used to deoxygenate sulfoxides and in the preparation of image-providing materials for photography.

Boron Triiodide. There are no large-scale commercial uses of boron triiodide.

BORON SUBHALIDES

Boron subhalides are binary compounds of boron and the halogens, where the atomic ratio of halogen to boron is less than 3. The boron monohalides, BCl, bromoborane(1), BBr, and iodoborane(1), BI, are unstable species that have been observed spectroscopically when the respective trihalides were subjected to a discharge. Boron dihalide radicals have been studied, and structural and thermochemical data for these species ($\cdot BX_2$) have been deduced.

Diboron tetraflouride, B_2F_4, diboron tetrachloride, B_2Cl_4, diboron tetrabromide, B_2Br_4, and dibórontetraiodide, B_2I_4, are well-known but thermally unstable compounds.

FAZLUL ALAM
U.S. Borax Research Corporation

B. R. Gragg, in K. Niedenzu, K. C. Buschbeck, and P. Merlet, eds., *Gmelin Handbook of Inorganic Chemistry, Borverbindungen,* Vol. 19, Springer-Verlag, Berlin, Germany, 1978, pp. 109, 168.

R. Will, *Chemical Economics Handbook,* Stanford Research Institute, Menlo Park, Calif., 1991, p. 510.5000j.

A. Finch and P. J. Gardner, in R. J.-Brotherton and H. Steinberg, eds., *Progress in Boron Chemistry,* Pergamon Press, New York, 1970, p. 177.

N. N. Greenwood and B. S. Thomas, in A. F. Trotman-Dickerson, ed., *Comprehensive Inorganic Chemistry,* Pergamon Press, Oxford, U.K., 1973, p. 956.

BORON HYDRIDES, HETEROBORANES, AND THEIR METALLA DERIVATIVES

The boron hydrides, including the polyhedral boranes, heteroboranes, and their metalla derivatives, encompass an amazingly diverse area of chemistry. This class contains the most extensive array of structurally characterized cluster compounds known. Included here are many novel clusters possessing idealized molecular geometries ranging over every point group symmetry from identity (C_1) to icosahedral (I_h). Because boron hydride clusters may be considered in some respects to be progenitorial models of metal clusters, their development has provided a framework for the development of cluster chemistry in general as well as for chemical bonding theory.

Structural Systematics

Because the polyhedral boron hydrides are cage molecules, which usually possess triangular faces, their idealized geometries can be described accurately as deltahedra or deltahedral fragments. The left-hand column of Figure 1 illustrates the deltahedra containing $n = 6-12$ vertices: the octahedron, pentagonal bipyramid, bisdisphenoid, symmetrically tricapped trigonal prism, bicapped square antiprism, octadecahedron, and icosahedron. These idealized structures are convex deltahedra except for the octahedron, which is not a regular polyhedron. The left-hand column of Figure 1 also represents the class of deltahedral *closo* molecules from which the other idealized structures (deltahedral fragments) can be generated systematically. Any *nido* or *arachno* cluster can be generated from the appropriate deltahedron by ascending a diagonal from left to right. This progression generates the *nido* structure (center column) by removing the most highly connected vertex of the deltahedron, and the *arachno* structure (right column) by removal of the most highly connected atom of the open (nontriangular) face of the *nido* cluster. The structural correlations shown in Figure 1 were formulated in 1971, and subsequently elaborated. The terms

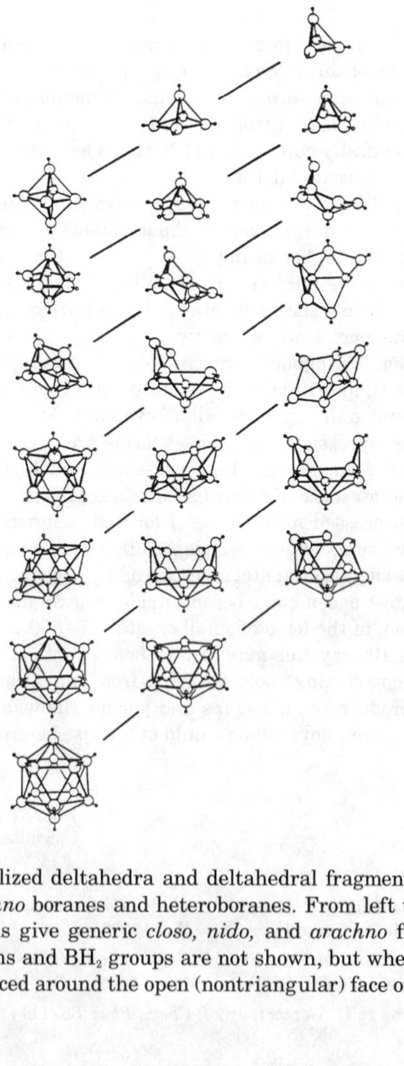

Figure 1. Idealized deltahedra and deltahedral fragments for *closo, nido,* and *arachno* boranes and heteroboranes. From left to right the vertical columns give generic *closo, nido,* and *arachno* frameworks; bridge hydrogens and BH_2 groups are not shown, but when appropriate they are placed around the open (nontriangular) face of the framework (see text).

closo, nido, arachno, and *hypho* are derived from Greek and Latin and imply closed, nestlike, weblike, and netlike structures, respectively. These classifications apply equally well to boranes, heteroboranes, and their metalla analogues, and are intimately connected to a quantity known as the framework, or skeletal, electron count. The partitioning of electrons into framework and exopolyhedral classes allows for predictions of structures in most cases, even though these systematics are not concerned explicitly with the assignments of localized bonds within the polyhedral skeletons of these molecules. That is, the lines depicting the skeletons of the structures illustrated are not electron-pair, or "electron precise," bonds. The lines merely serve to join nearest neighbors and illustrate cluster geometries. However, exopolyhedral lines do represent the usual electron precise bonds.

Placement of Heteroatoms. Many of the deltahedra and deltahedral fragments of Figure 1 have two or more nonequivalent vertices. Nonequivalent vertices are recognized as having a different order; ie, a different number of nearest neighbor vertices within the framework. Heteroatoms generally exhibit a positional preference based on the order of the polyhedral vertex and the electron richness of the heteroatom relative to boron. Electron-rich heteroatom groupings contribute more framework electrons than a :B–H moiety, which has two framework electrons, and generally appear to prefer low order vertices, ie, those having fewer neighbors.

Placement of Extra Hydrogens. The placement of extra hydrogens plays a crucial role in determining the structures adopted by boranes and carboranes. However, the exact position of extra hydrogens sometimes depends on the physical state of the molecule, eg, the tridecahydrodecaborate(1−) anion, $[B_{10}H_{13}]^-$ exhibits different bridge hydrogen placements in the crystal and in solution as can be inferred from experimental evidence, but the solution data are also consistent with a dynamic process of bridge hydrogen tautomerism.

The placement of bridge hydrogens may be the most important variable in the determination of relative isomer stabilities, outranking placement of hetero-atoms.

M–H–B Bridges. Numerous metallaboranes and metallaheteroboranes are known to contain hydrogens bridging between a metal atom and a skeletal boron atom, but complexes containing covalently bound tetrahydroborate(−1), $[BH_4]^-$, constitute the prototypical class. Metal tetrahydroborates have been reviewed.

Exceptions to Structural Systematics. When strong electron-donating or withdrawing groups are present as substituents attached to boron in polyboranes, there is the possibility of structural anomalies. In some cases electron deficiency of boron apparently can be ameliorated by back-donation instead of by the multicenter bonding afforded in a cage framework. Thus it has been suggested that exceptions to the electron-counting paradigms may occur where back-donation from the substituent to a cluster boron is possible. Some metallacarboranes also present anomalies to the electron-counting formalisms.

Bonding

Localized Bonds. Because boron hydrides have more valence orbitals than valence electrons, they have often been called electron-deficient molecules. This electron deficiency is partly responsible for the great interest surrounding borane chemistry and molecular structure.

The elucidation of the structure of diborane(6) led to the description of a new bond type, the three-center bond, in which one electron pair is shared by three atomic centers. The delocalization of a bonding pair over a three-center bond allows for the utilization of all the available orbitals in an electron-deficient system. This key point led to the formulation of a valence–bond description of the bonding in boron hydrides, sometimes termed a topological description. The valence structures of this topological approach give localized bonding descriptions which include delocalized three-center bonds in the basis set of bond types. In addition to the B–H–B three-center bridge bond depicted, a B–B–B three-center bond was introduced to describe bonding in the framework.

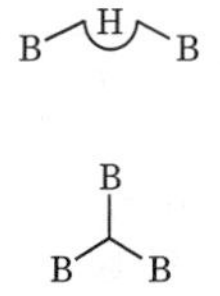

The valence theory includes both types of three-center bonds shown as well as normal two-center, B–B and B–H bonds.

Boranes

Nido and Arachno Boranes. These boranes are generally more reactive and less stable thermally than the corresponding *closo* boranes. The most extensively studied boranes include diborane(6), B_2H_6, tetraborane(10), B_4H_{10}, pentaborane(9), B_5H_9, and decaborane(14), $B_{10}H_{14}$. This subject has been reviewed. A great deal of early work in this area was associated with the government-sponsored high energy fuels programs. Some of this work is summarized. The *nido* and *arachno* boranes smaller than $B_{10}H_{14}$ are quite reactive toward oxygen and water. The properties of selected boranes are given in Table 1.

Molecular Orbital Descriptions. In addition to the localized bond descriptions, molecular orbital (MO) descriptions of bonding in boranes and carboranes have been developed. Early work on boranes helped develop one of the most widely applicable approximate MO methods, the extended Hückel method. Molecular orbital descriptions are particularly useful for *closo* molecules where localized bond descriptions become cumbersome because of the large number of resonance structures that do not accurately reflect molecular symmetry.

Reactions of Boranes with Lewis Bases. Boranes that contain a BH_2 moiety, eg, B_2H_6, B_4H_{10}, B_5H_{11}, hexaborane(12), B_6H_{12}, and nonaborane(15), B_9H_{15}, can generally be cleaved by nucleophiles in two ways termed symmetrical and unsymmetrical bridge cleavage. Using neutral bases, the two modes of cleavage lead to molecular and ionic fragments, respectively.

Certain base adducts of borane, BH_3, such as triethylamine borane, $(C_2H_5)_3N{\cdot}BH_3$, dimethylsulfide borane, $(CH_3)_2S{\cdot}BH_3$, and tetrahydrofuran borane, $C_4H_8O{\cdot}BH_3$, are more easily and safely handled than B_2H_6 and are commercially available. These compounds find wide use as reducing agents and in hydroboration reactions. A wide variety of borane reducing agents and hydroborating agents is available from Aldrich Chemical Company, Milwaukee, Wisconsin.

Proton Abstraction. Although the exopolyhedral hydrogens of *nido* and *arachno* boranes are generally considered hydridic, the bridge hydrogens are acidic as first demonstrated by titration of $B_{10}H_{14}$ and deuterium exchange.

Polyhedral Expansion. The term polyhedral expansion is used to describe a host of reactions in which the size of the polyhedron is increased by the addition of new vertex atoms, whether boron, heteroelements, or metals.

Electrophilic Attack. A variety of boranes, heteroboranes, and metallaboranes undergo electrophilic substitution.

Closo Borane Anions. This group contains a homologous series of very stable polyhedral anions, $[closo - B_nH_n]^{2-}$, $n = 6-12$. Just as the previously known boron hydrides might be considered as analogues of aliphatic hydrocarbons, the *closo* borane anions are analogues of aromatic hydrocarbons. The stability of the *closo* anions is attributable to electron delocalization in a unique three-dimensional aromaticity. Unlike their *nido* and *arachno* counterparts with bridging hydrogens, proton abstraction does not, for practical purposes, occur in *closo* borane chemistry. Instead, acid catalysis is important in their substitution chemistry. The best known members of this series, $[closo - B_{10}H_{10}]^{2-}$ and $[closo - B_{12}H_{12}]^{2-}$, were first reported in 1959 and 1960 and were the subject of detailed studies.

Tetrahydroborates. The tetrahydroboranes constitute the most commercially important group of boron hydride compounds. Tetrahydroborates of most of the metals have been

Table 1. Physical Properties of Boranes

Borane	Molecular formula	Mp, °C	Bp, °C	ΔH°_f, kJ/mol[a]	ΔG°_f, kJ/mol[a]	ΔS°_{298}, J/(K · mol)[a]
diborane(6)	B_2H_6	−164.9	−92.6	35.5	86.6	232.0
tetraborane(10)	B_4H_{10}	−120	18	66.1		
pentaborane(9)	B_5H_9	−46.6	48	73.2	174	275.8
pentaborane(11)	B_5H_{11}	−123	63	103.0		
hexaborane(10)	B_6H_{10}	−62.3	108	94.6		
decaborane(14)	$B_{10}H_{14}$	99.7	213	31.5	216.1	353.0

[a] To convert J to cal, divide by 4.184.

Table 2. Properties of Alkali Metal Tetrahydroborates

	Compound				
Property	$LiBH_4$	$NaBH_4$	KBH_4	$RbBH_4$	$CsBH_4$
mp, °C	268	505	585		
decomp. temp., °C	380	315	584	600	600
density, g/mL	0.68	1.08	1.17	1.71	2.40
refractive index		1.547	1.490	1.487	1.498
lattice energy, kJ/mol[a]	792.0	697.5	657	648	630.1
ΔH°_f, kJ/mol[a]	−184	−183	−243	−246	−264
ΔS°_{298}, J/(mol·K)[a]	−128.7	−126.3	−161	−179	−192

[a] To convert J to cal, divide by 4.184.

characterized and their preparations have been reviewed. The important commercial tetrahydroborates are those of the alkali metals. Some properties are given in Table 2.

The use of tetrahydroborates, as well as the boranes and organoboranes, for organic transformations has proven to be significant because these reduction reactions are highly selective and nearly quantitative. The reducing characteristics of borohydrides may be varied by changing the associated cation and by changing the solvent. Borohydrides are often the reagents of choice for the reduction of aldehydes and ketones to the corresponding alcohols, especially when selective reduction in the presence of other functional groups is required. Many other functional groups, such as acid chlorides, imines, and peroxides, can also be reduced using borohydrides.

Heteroboranes

Heteroboranes contain heteroelements classified as nonmetals. The heteroatoms known to form part of a borane polyhedron include C, N, Si, P, As, S, Se, Sb, and Te either alone or in combination. In principle, most heteroboranes could have a wide range of skeletal sizes. However, with the primary exception of the carbaboranes, extensive chemistry has emerged only for the thiaboranes and azaboranes, which have the greatest availability and demonstrated scope of chemistry.

Carboranes. The term carborane is widely used in American literature as a contraction of the IUPAC-approved nomenclature carbaborane. The first carboranes, isomers of $C_2B_3H_5$, $C_2B_4H_6$, and $C_2B_5H_7$, were prepared in the mid-1950s at Olin Mathiesen. These carboranes were obtained as a mixture in low yield from the reaction of pentaborane(9) with acetylene in a silent electric discharge. The discovery of the icosahedral *closo*-1,2-dicarbadodecaborane(12), 1,2-$C_2B_{10}H_{12}$, came soon after and led to a rapid development of carborane chemistry.

The discovery of the base-promoted degradation of the isomeric *closo*-$C_2B_{10}H_{12}$ cages provided one of the most important carborane anion systems, the isomeric $[nido-C_2B_9H_{12}]^-$ anions,

$$closo-C_2B_{10}H_{12} + RO^- + 2\,ROH \rightarrow [nido-C_2B_9H_{12}]^- + B(OR)_3 + H_2$$

where R = CH_3, C_2H_5, etc. The $[nido-C_2B_8H_{12}]^-$ cages are commonly referred to as dicarbollide ions, derived from the Spanish olla, meaning a bowl. Aside from their extensive use in metallacarborane chemistry, the dicarbollide anions are important intermediates in the synthesis of other carborane compounds.

The arachno carboranes 1,3-$C_2B_7H_{13}$ and 6,9-$C_2B_8H_{14}$ are unusual in that two of the extra hydrogens occur in CH_2 groups. The other two extra hydrogens are present as B–H–B bridges.

As with the simple boranes, the *closo* carboranes are generally more thermally stable than the corresponding *nido* and *arachno* species.

Cage rearrangements in polyhedral carboranes have been studied. Although most carborane cages are stable at room temperature, they frequently undergo rearrangements at elevated temperatures. Many of the carborane isomers obtained by conventional synthetic routes are kinetic products and not the thermodynamically most stable isomers. When subjected to elevated temperatures below the ultimate decomposition temperatures, carboranes often undergo rearrangements to the more stable isomers.

A diversity of polyhedral carborane cage-containing polymers has been prepared. The best known of these are elastomeric polycarboranylsiloxanes which where developed by Olin Corporation and Union Carbide Corporation. These are based on *m*-carborane cages linked by polysiloxane groups with direct C–Si bonds. The properties of these materials can be varied by changing the length and substituents of the polysiloxane linkages as well as their overall molecular weights. Some of these materials have excellent thermal stabilities, chemical resistance, and high temperature elastomeric properties. Polymers of this type, known under the trade name Dexsil, are commercial materials, useful as stationary phases in gas chromatography among other applications. The organic and organometallic chemistry of *closo* carborane derivatives has been reviewed.

Other Heteroboranes. Other well-documented families of heteroboranes include the azaboranes, phosphaboranes, arsenaboranes, stibaboranes, selenaboranes, and telluraboranes.

Metallaboranes

Transition-Element Metallaboranes. The transition-metal hydroborate cluster, $HMn_3(CO)_{10}(BH_3)_2$, containing a B_2H_6 moiety, which is multiply bridging between three manganese carbonyl and manganese carbonyl hydride centers via M–H–B bridges, might be regarded as the first structurally characterized metallaborane cluster. By 1990 a great many *nido* metallaborane clusters has been characterized covering a wide range of sizes and polyhedral fragment geometries.

One of the most extensive classes of metallaboranes are those derived from decaborane, which in most cases produces 11-vertex metallaborane products. The $[B_{10}H_{12}]^{2-}$ anion can also be considered as a bidentate ligand which coordinates metals between boron atoms 2,11 and 3,8, the metal at position 7 such that the metal in effect occupies the position of a bridge hydrogen of the conjugate acid borane. Situations in which a metal vertex may be regarded as equivalent to an H^+, BH^{2+}, or the BH_2^+ moiety have also been discussed.

The first *closo* metallaborane complexes prepared were the nickelaboranes $[closo-(\eta^5-C_5H_5)Ni(B_{11}H_{11})]^-$ and *closo*-1,2-$(\eta^5$-$C_5H_5)_2$-1,2-$Ni_2B_{10}H_{10}$. These metallaboranes display remarkable hydrolytic, oxidative, and thermal stability.

A number of novel products have been isolated from the reaction of $[B_5H_8]^-$ and $CoCl_2$ and $[C_5H_5]^-$ in THF. The predominant product is *nido*-2-(CpCo)-B_4H_8. Also obtained are isomeric clusters containing up to four cobalt atoms, eg, $(\eta^5\text{-}C_5H_5Co)_4B_4H_8$. Characterization of these clusters indicate an unusual $2n$ framework electron count having geometries reminiscent of strictly metallic clusters.

Main Group Element Metallaboranes. A variety of metallaborane clusters, which incorporate main group metals in vertex positions of polyhedral metallaborane clusters, have been reported. Examples are $(BH_4)BeB_5H_{10}$, $MgB_{10}H_{12}\cdot 2O(C_2H_5)_2$, $[(CH_3)HgB_{10}H_{12}]^-$, $[AlB_{10}H_{14}\cdot 2O(C_2H_5)_2]^-$, $[(CH_3)AlB_{11}H_{11}]^{2-}$, $(CH_3)InB_{10}H_{12}$, $[(CH_3)_2TlB_{10}H_{12}]^-$, and $(CH_3)_2MB_{10}H_{12}$ where M = Si, Ge, or Sn.

Exopolyhedral Metallaboranes. Polyboranes may bind exopolyhedral metals in a variety of ways. Most commonly metals are bound via M–H–B interactions. In other cases metals may formally replace bridging hydrogen atoms at edge positions to give B–M–B interactions. Metals may also be attached to polyborane cages by direct M–B ς-bonds.

Metallacarboranes

The isomeric $[nido - C_2B_9H_{11}]^{2-}$ ions, which are commonly known as dicarbollide ions, and many other carborane anions, form stable complexes with most of the metallic elements. Indeed nearly all metals can be combined with polyborane hydride clusters to produce an apparently limitless variety of cluster compounds.

Transition-Metal Metallacarboranes. The first demonstration of the insertion of a transition metal into an open face of a borane cage was with the iron sandwich compound $[commo - Fe(C_2B_9H_{11})_2]^{2-}$. This product is readily air-oxidized to $[commo - (C_2B_9H_{11})_2Fe]^-$, a complex containing a formal Fe^{3+} center. These complexes, as well as those formed from a variety of formally $(d)^3$, $(d)^5$, $(d)^6$, and $(d)^7$ transition metals, have symmetrical sandwich structures of the type shown in Figure 2**a**. By contrast, $(d)^8$ and $(d)^9$ metals form slipped sandwich structures as shown in Figure 2**b**.

Exopolyhedral Metallacarboranes. Many metallacarboranes are known that exhibit exopolyhedral bonding to metals. Most commonly metals are bound via M–H–B interactions in which the B–H group can be regarded as a two-electron donor to the metal center. In other cases, M–B, M–C, or M–M bonding may be involved.

Metallacarboranes in Catalysis. Perhaps the most intensely studied of all metallacarborane complexes is the exopolyhedral metallacarborane *closo*-3,3-$[P(C_6H_5)_3]_2$-3-H-3,1,2-$RhC_2B_9H_{11}$, shown in Figure 3**a** and its cage *C*-substituted derivatives. The three available isomers of *closo*-$[P(C_6H_5)_3]_2(H)Rh\text{-}C_2B_9H_{11}$ are synthesized in high yield by the oxidative addition of $[P(C_6H_5)_3]_2RhCl$ with the appropriate $[nido - C_2B_9H_{12}]^-$ ion, which may also be made chiral by the attachment of a single-alkyl or -aryl group at a carbon position. The resulting hydridorhodacarboranes are quite robust and catalyze a number of reactions, including the isomerization and hydrogenation of olefins, the deuteration of B–H groups, and the hydrosilanolysis of alkenyl acetates. These species function as homogeneous catalyst precursors for the isomerization and hydrogenation of olefins as well as other reactions.

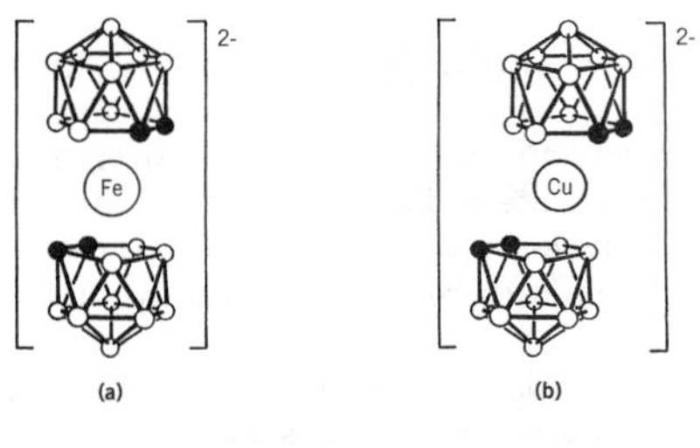

Figure 2. Exemplary structures of (**a**) unslipped and (**b**) slipped metallacarborane dicarbollide sandwich derivatives where ○ represents BH; ●, CH.

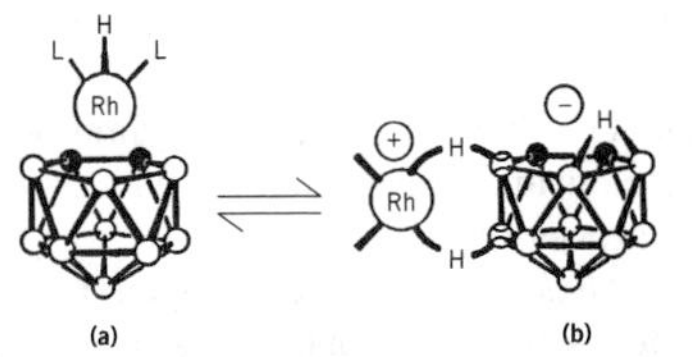

Figure 3. **(a)** The structure of *closo*-3,3-$[(C_6H_5)_3P]_2$-3-H-3,1,2-$RhC_2B_9H_{11}$, and **(b)** one isomer of its tautomer *exo-nido*-$(L_2Rh)_2$-$(\mu\text{-}H)_2$-7,8-$C_2B_9H_{12}$ where L is $(C_6H_5)_3P$. ⊖ represents B; ○, BH; ●, CH.

Main Group Element Carborane Derivatives. Main group element carborane derivatives have been reviewed. Only a few alkaline-earth element metallacarborane derivatives have been characterized.

f-Block Element Metallacarborane Derivatives. The first actinide metallacarborane complex, *commo*-$(C_2B_9H_{11})UCl_2$, was prepared in 1977. The coordination geometry of this complex can be described as distorted tetrahedral with the four positions occupied by two η^5-bound $[7,8 - C_2B_9H_{11}]^{2-}$ cages and two chloride ions. Complexes of this type are often referred to as bent sandwiches because of the configuration of the two-dicarbollide cage about the metal center, which is analogous to the corresponding pentamethyl cyclopentadiene–metal complexes.

Boron Neutron Capture Therapy

One of the most promising applications of polyboron hydride chemistry is boron neutron capture therapy (BNCT) for the treatment of cancers. Boron-10 is unique among the light elements in that it possesses an unusually high neutron capture nuclear cross section (3.8×10^{-25} m^2, 0.02–0.05 eV neutron).

It has been estimated that using available neutron intensities such as 10^2 neutrons/($cm^2 \cdot s$) concentrations of ^{10}B from 10–30 $\mu g/g$ of tumor with a tumor cell to normal cell selectivity of at least five are necessary for BNCT to be practical. Hence the challenge of BNCT lies in the development of practical means for the selective delivery of approximately 10^9 ^{10}B atoms to each tumor cell for effective therapy using short neutron irradiation times. Derivatives of ^{10}B-enriched *closo*-borane anions and carboranes appear to be especially suitable for BNCT because of their high concentration of ^{10}B and favorable hydrolytic stabilities under physiological conditions.

To date, the most extensively studied polyboron hydride compounds in BNCT research have been the icosahedral mercaptoborane derivatives $Na_2[B_{12}H_{11}SH]$ and $Na_4[(B_{12}H_{11}S)_2]$, which have been used in human trials with some, albeit limited, success. New generations of tumor-localizing boronated compounds are being developed.

A related potential medical application of metallacarboranes is based on the highly favorable kinetic stability of many metallacarborane complexes under physiological conditions. This feature makes certain functionalized metallacarboranes containing radiometals ideal choices for use as medical imaging reagents (see IMAGING TECHNOLOGY).

Economic Aspects

Only the simplest of boron hydride compounds, most notably sodium tetrahydroborate, $Na[BH_4]$, diborane, B_2H_6, and some of the borane adducts, eg, amine boranes, are now produced in significant commercial quantities.

Sodium Tetrahydroborate, $Na[BH_4]$. This air-stable white powder, commonly referred to as sodium borohydride, is the most widely commercialized boron hydride material. It is used in a variety of industrial processes including bleaching of paper pulp and clays, preparation and purification of organic chemicals and pharmaceuticals, textile dye reduction, recovery of valuable metals, wastewater treatment, and production of dithionite compounds.

Diborane(6), B_2H_6. This spontaneously flammable gas is consumed primarily by the electronics industry as a dopant in the production of silicon wafers for use in semiconductors. It is also used to produce amine boranes and the higher boron hydrides.

Amine Boranes. Trialkylamine and dialkylamine boranes, such as tri-*t*-butylamine borane and dimethylamine borane, are mainly used in electroless plating processes.

Polyhedral Boron Hydrides. These are used in neutron capture therapy of cancers, and as burn rate modifiers (accelerants) in gun and rocket propellant compositions.

Carboranes. These are used in neutron capture therapy, and as burn rate modifiers in gun and rocket propellants. They are used as high temperature elastomers and other unique materials, high temperature gas–liquid chromatography stationary phases, optical switching devises, and gasoline additives.

Metallacarboranes. These are used in homogeneous catalysis, including hydrogenation, hydrosilylation, isomerization, hydrosilanolysis, phase transfer, burn rate modifiers in gun and rocket propellants, neutron capture therapy, medical imaging, processing of radioactive waste, analytical reagents, and as ceramic precursors.

DAVID M. SCHUBERT
U.S. Borax Inc.

H. C. Brown, *Hydroboration,* Benjamin-Cummings, Reading, Mass., 1980.

R. N. Grimes, *Carboranes,* Academic Press, Inc., New York, 1970.

R. N. Grimes, ed., *Metal Interactions with Boron Clusters,* Plenum Press, New York, 1982.

M. F. Hawthorne, *Angew. Chem. Int. Ed. Engl.* **32,** 950–984 (1993).

BORON HYDRIDES, HETEROBORORANES, AND THEIR METALLA DERIVATIVES (COMMERCIAL ASPECTS)

Tetrahydroborates

Sodium tetrahydroborate, $NaBH_4$, more commonly called sodium borohydride, is the most widely used commercial boron hydride. The largest manufacturer is Morton International Specialty Chemicals Group, which has two plants in the United States and one in Europe.

Sodium borohydride is available as a 12% solution in caustic soda and in solid form either as powder or pellets.

Solid sodium borohydride does not ignite upon contact with moisture and is not shock-sensitive. These characteristics allow it to be handled safely in air. Because it does liberate hydrogen upon contact with water it should be handled with care.

The predominant use for sodium borohydride is in wood pulp (qv) bleaching. The next largest commercial use is as a reducing agent of functional groups in organic synthesis. A significant application in pharmaceutical synthesis is the stereospecific and selective reduction in steroid production.

Sodium borohydride reacts with Lewis acids in nonprotic solvents to yield diborane, B_2H_6, which can then be used to generate other useful organoboranes such as amine boranes, alkyl boranes, and boron hydride clusters.

Borane Complexes

Borane complexes are the most widely used commercial boron compounds, after sodium borohydride. Examples used in organic synthesis are amine borane complexes and borane complexes of tetrahydrofuran and dimethyl sulfide.

Potential Uses of $B_{12}H_{11}SH^{2-}$. Although not commercial as yet, an interesting potential use of borane compounds as therapeutic agents for the treatment of inoperable cancerous tumors is being developed (see CHEMOTHERAPEUTICS, ANTICANCER). Boron neutron capture therapy (bnct) involves the uptake of boron compounds enriched with the boron-10 ^{10}B, isotope, selectively accumulating in cancerous tumors.

JOSEPH BARENDT
BETH DRYDEN
Callery Chemical Company

Sodium Borohydride Digest, Morton International, Specialty Chemicals Group, Danvers, Mass., 1989.

U.S. Pat. 3,017,316 (Jan. 16, 1962), W. H. Rapson (to Hooker Chemical Co.).

H. I. Schlesinger, H. C. Brown, and H. R. Hockstra, *J. Amer. Chem. Soc.* **75,** 199–204 (1953).

H. C. Brown, *Organic Synthesis via Boranes,* John Wiley & Sons, Inc., New York, 1975.

ORGANIC BORON–NITROGEN COMPOUNDS

There are four classes of B–N compounds. (*1*) Amine boranes, $R_3B{-}NR'_3$, have the nitrogen atom which supplies both electrons in the B–N bond; these are isoelectronic with alkanes, $R_3C{-}CR'_3$. (*2*) Aminoboranes, $R_2B{=}NR'_2$, have a covalent bond between B and N. Here, the hybridization of both boron and nitrogen is $(sp)^2$ resulting in a planar $R_2BNR'_2$ unit capable of a π-interaction between nitrogen's free-electron pair and the empty *p* orbital on boron; these are isoelectronic with alkenes, $R_2C{=}CR'_2$. (*3*) Iminoboranes, $RB{\equiv}NR'$, have a two-coordinate boron interacting with the nitrogen via a triple bond; these are isoelectronic with alkynes, $RC{\equiv}CR$. (*4*) Borazines, $(\text{—}RB\text{—}NR'\text{—})_x$, are cyclic compounds containing alternating tricoordinate boron and nitrogen atoms. These compounds are nearly planar and have B–N bond lengths that are substantially shorter than those for single bonds, indicating partial double-bond character. Borazines are isoelectronic with benzene, C_6H_6. There are also B–N ring systems of other sizes. Although these compounds are isoelectronic and nearly isostructural with conventional organic species, properties and reactivity patterns are considerably different. The primary reason for this dissimilarity is the polar nature of the B–N bond.

Reviews covering the literature through 1970 for all areas except iminoboranes and through 1984 are available.

Amine Boranes

Amine–borane adducts have the general formula $R_3N{\cdot}BX_3$ where R = H, alkyl, etc, and X = alkyl, H, halogen, etc. These compounds, characterized by a coordinate covalent bond between boron and nitrogen, form a class of reducing agents having a broad spectrum of reduction potentials.

Synthesis. An efficient, convenient synthesis for the preparation of ammonia borane, the inorganic analogue of ethane, is shown in equation 1 where THF is tetrahydrofuran.

$$(NH_4)_2CO_3 + 2\ NaBH_4 \xrightarrow{\text{anhydrous THF}} 2\ H_3N \cdot BH_3 + 2\ H_2 + Na_2CO_3 \quad (1)$$

Dimethylamine borane, important as a reducing agent in electroless plating (qv), can be synthesized as shown in equation 2.

$$(CH_3)_2NH_2Cl + NaBH_4 \xrightarrow[\text{RT}]{\text{ether}} (CH_3)_2HN{\cdot}BH_3 + H_2 + NaCl \quad (2)$$

Amine–Boronium Cations. The most extensively studied boronium cation is the diammoniate of diborane. In this compound, synthesized in the low temperature reaction between diborane and excess ammonia, the cationic boron is coordinatively saturated in a tetrahedral environment.

Properties and Reactions. Amine boranes are usually colorless, crystalline compounds which exhibit sharp melting points and thermal stability when pure. Primary and secondary amine boranes are generally solids at ambient temperatures. With the exception of trimethylamine borane, the aliphatic *t*-amine boranes are liquids.

Aminoboranes

The aminoboranes are characterized by a normal covalent bond between boron and nitrogen in which an electron from each atom is shared. In this case the hybridization of both boron and nitrogen is $(sp)^2$, resulting in a planar moiety capable of π-interaction by utilizing the nitrogen's free-electron pair and boron's vacant p orbital. There exists a wide variety of aminoborane compounds; among those that have been thoroughly investigated are the monoaminoboranes X_2BNR_2, bisaminoboranes $XB(NR_2)_2$, and trisaminoboranes $B(NR_2)_3$. The substituents X and R may vary widely, but generally R is an alkyl or aryl group, or hydrogen, whereas X can represent a rather wide variety of atoms or groups.

Synthesis. One of the more common routes for the synthesis of aminoboranes involves the aminolysis of the appropriate boron halide.

Properties and Reactions. Monoaminoboranes readily undergo association, the extent of which primarily depends on the steric requirements of the groups attached to boron and nitrogen. The monomers are generally liquids or low melting solids, whereas the dimers and trimers are crystalline solids. The bis- and trisaminoboranes do not show a tendency to dimerize, and are also less sensitive to hydrolysis than the monomeric monoaminoboranes; they are generally high boiling liquids or crystalline solids.

Aminoboranes have been used as ligands in complexes with transition metals.

The reduction of (alkylamino)haloboranes using hydride reagents can provide a convenient route to (alkylamino)boranes.

Iminoboranes

Iminoboranes, SBNR, are isoelectronic with alkynes, XCCR. Whereas the structural and physical properties of these species are rather parallel, this similarity does not hold for reactivity. The polarity of the B–N bond makes iminoboranes much more reactive than the analogous SCCR species. Comprehensive reviews of aminoboranes may be found in the literature.

Synthesis. Iminoboranes, thermodynamically unstable with respect to oligomerization can be isolated under laboratory conditions by making the oligomerization kinetically unfavorable. This is facilitated by bulky substituents, high dilution, and low temperatures. The vacuum gas-phase pyrolysis of (trimethylsilylamino)(alkyl)haloboranes has been utilized as an effective method of generating iminoboranes $RB{\equiv}NR'$. Iminoboranes have been suggested as intermediates in the formation of compounds derived from the pyrolysis of azidoboranes.

Properties and Reactions. The structure of (alkyl)iminoboranes $RB{\equiv}NR'$ is characterized by a linear C–B–N–C geometry and a B–N bond order approaching three. Amino iminoboranes can be described using three resonance structures:

$$\underset{\text{A}}{R_2N{-}B{\equiv}NR'} \leftrightarrow \underset{\text{B}}{R_2N{=}B{=}NR'} \leftrightarrow \underset{\text{C}}{R_2N{-}B{=}NR'}$$

The boron atoms in resonance structures A and B possess a formal negative charge.

The chemistry of these compounds reflects the unsaturated nature of the B–N triple bond. Polar compounds add to iminoboranes, provided the addition proceeds more rapidly than oligomerization of $RB{\equiv}NR'$.

In an analogous fashion to the hydroboration reaction, a variety of boron-containing substrates react with iminoboranes.

A general type of stabilization for iminoboranes is a cyclodimerization, which yields diazadiboretidines $(RBNR')_2$ that are isoelectronic with cyclobutadienes.

Borazines

The largest and most extensively studied family of boron–nitrogen compounds is that of the borazines, characterized by a six-membered ring system containing alternating boron and nitrogen atoms. Borazines are tricoordinated, nearly planar, and have B–N bond lengths considerably shorter than single bonds, indicating partial double-bond character. Because borazine is isoelectronic and isostructural with benzene, it has been called inorganic benzene. The physical properties of borazines tend to confirm π-electron delocalization as in benzenes; however, chemical evidence indicates that the reactions of borazines are dominated by polarization of the B–N bonds.

Synthesis. The parent compound, borazine, is best prepared by a two-step process involving formation of *B*-trichloroborazine followed by reduction with sodium borohydride.

Properties and Reactions. Borazines are liquids or crystalline solids depending on the substitution pattern. Most are sensitive to moisture and must be handled in an inert atmosphere. Borazines are essentially planar except in hexasubstituted cases where there may be some puckering of the borazine ring. Borazines undergo addition reactions rather than electrophilic substitution reactions typical of benzene compounds.

Thermal Stability. Borazine itself shows negligible decomposition at 0–5°C. At ambient temperature 1–2% decomposition has been observed during the first month followed by an increase in rate thereafter, and at higher temperatures appreciable decomposition occurs.

Hydrolysis. Borazine is slowly hydrolyzed by water at ambient or higher temperatures to boric acid.

Addition Reactions. In general, polar molecules such as hydrogen halides add across the B–N bonds, the more electronegative group bonding to boron.

Substitution Reactions. Substitution reactions on borazines are confined mainly to substitution by nucleophilic groups on boron; substituents on nitrogen are inert to most reagents.

Miscellaneous Reactions. Miscellaneous reactions of borazines include photolysis and complex formation.

Fused Rings and Polymers. Borazine analogues of naphthalene and substituted phenalenes are known; the latter compound is formed by reaction of distannylamine and a large excess of tris(methylthio)borane. A general method for the preparation of the polycyclic borazines using fused carbon–heteroatom rings, where X = O, NR and $n = 2, 3$, has been given.

Other B–N Ring Systems. A number of unusual ring systems containing only B–N linkages have been reported.

Manufacture and Uses

Organic boron–nitrogen compounds have not found extensive usage, and therefore very few are manufactured on a large scale. Callery Chemical Company appears to be the largest manufacturer of amine boranes and borazines.

Borazines, particularly polymeric compounds, have been extensively investigated as preceramic materials from which coatings and fibers of boron nitride can be produced upon thermolysis. *B*-Aryl and halogeno–amino borazines are reported to have seen use as fire retardants in cotton and nylon textiles. Other reported uses for borazines are as epoxy resin catalysts, polymerization inhibitors of unsaturated alcohols and esters, and catalysts for polymerization of alkenes.

MARK J. MANNING
T.S. GRIFFIN
U.S. Borax Research Corporation

K. Niedenzu and J. W. Dawson, *Boron–Nitrogen Compounds,* Springer-Verlag, Berlin, 1965; H. Nöth, in R. J. Brotherton and H. Steinberg, eds., *Progress in Boron Chemistry,* Vol. 3, Pergamon Press, Oxford, U.K., 1970, pp. 211–311.

A. Meller, in K. C. Buschbeck and K. Niedenzu, eds., *Gmelin Handbook of Inorganic Chemistry,* 3rd Suppl., Vol. 3, Springer-Verlag, Berlin, 1988, pp. 91–216.

Amine Boranes, Callery Chemical Co., Callery, Pa., 1976.

K. Niedenzu and J. W. Dawson, in E. L. Muetterties, ed., *The Chemistry of Boron and its Compounds,* John Wiley & Sons, Inc., New York, 1967.

BOUNDARY LAYER. See FLUID MECHANICS.

BRAKE FLUIDS. See HYDRAULIC FLUIDS.

BRAKE LININGS AND CLUTCH FACINGS

Brake linings and clutch facings consist of friction materials.

Types of Friction Materials

Prior to the mid-1970s, the most common type of friction materials in use in brakes and clutches for normal duty for original equipment installations and for the aftermarket were termed organics. These materials usually contained about 30–40 wt % of organic components and were asbestos-based.

After the mid-1970s, the downsizing of North American vehicles and the introduction of front wheel drive vehicles brought about the widespread usage of a new class of friction materials called semimetallics, also called semimets and carbon–metallics. Because of the allegedly adverse health effects associated with asbestos (qv) fibers, a second new class of friction materials called nonasbestos organics (NAOs) came about. Such materials are called either asbestos-free or nonasbestos friction materials.

Disk and Drum Materials. Gray cast iron is of reasonably low cost, provides good wear resistance and damping characteristics, and has long been in use as a brake drum or disk material for passenger cars and trucks.

Developments in metal-matrix composites technology has resulted in aluminum matrix materials filled with silicon carbide, SiC, (see CARBIDES, SILICON CARBIDE) particles (15 to 60 vol %) that provide the possibility of weight reduction for brakes. These composite materials are being tested and evaluated.

Friction and Wear

Friction. An analysis of friction mechanisms suggests that a frictional force is likely to consist of several components such as adhesion-tearing, ploughing (or abrasion), elastic and plastic deformation, fracture, shearing of a friction film (glaze), and asperity interlocking, all occurring at the sliding surface. Relative contributions of these mechanisms presumably depend on the normal load and sliding speed as well as the temperature. (Material properties are known to depend on these variables). In the case of automotive friction materials, the coefficient of friction is usually found to decrease with increasing unit pressure and sliding speed at a given temperature, contrary to Amontons' law. This decrease in friction is controlled by the composition and microstructure of friction materials.

Effectiveness, essentially a measure of the stopping efficiency, can be expressed in a number of different ways: as the coefficient of friction, deceleration rate, hydraulic or air line pressure required, torque developed, or distance required to stop a vehicle. Effectiveness levels used by consumers are typically decelerations of 0.15 to 0.30 G achieved at line pressures of 1.2 to 2.5 MPa (12 to 25 bars) in normal braking, increasing up to 0.50 to 0.80 G in panic situations requiring 5.5 to 11.0 MPa (55 to 110 bars).

Wear. For a fixed amount of braking the amount of wear of automotive friction materials tends to remain fairly constant or increase slightly with respect to brake temperature, but once the brake rotor temperature reaches >200 °C, the wear of resin-bonded materials increases exponentially with increasing temperature. This exponential wear is because of thermal degradation of organic components and other chemical changes. At low temperatures the practically constant wear rate is primarily controlled by abrasion, adhesion, and fatigue.

Raw Materials

Binders. Synthetic resins, such as phenolic and cresylic resins (see PHENOLIC RESINS), are the most commonly used friction material binders, and are usually modified with drying oils, elastomer, cardanol, an epoxy, phosphorus- or boron-based compounds, or even combinations of two.

Fibrous Reinforcements. The asbestos usually used in friction materials is chrysotile, $3MgO{\cdot}2SiO_2{\cdot}2H_2O$, the principal mineral of the serpentine group.

Steel, copper, and brass fiber may have a variety of aspect ratios, shape, ie, straight versus curved fibers and cross-sectional geometry, surface roughness, and chemical compositions.

Glass fibers and glassy fibers such as SMF or ceramic fibers are generally more thermally resistant than asbestos. The principal criteria for asbestos substitutes are suitable performance and processing characteristics and that they do not become harmful.

Several types of organic fibers are used: the cellulose-based include cotton (linters), solkafloc, paper (qv), sisal, and other natural fibers; synthetics include acrylics and polyaromatics.

Organic clutch materials contain continuous-strand reinforcements in addition to fibrous reinforcements. These include cotton (primarily for processing), other organic yarns, carbon–graphite yarn, and asbestos yarn, and brass wire or copper wire for high burst strength.

Nonfibrous Reinforcements. Because of the higher costs associated with nonasbestos fibers and the performance requirements needed in replacing asbestos, platy minerals such as mica and talc, and metal powders such as iron and copper, are being used as a portion of the total reinforcement package in NAOs.

Property Modifiers. Property modifiers can, in general, be divided into two classes: nonabrasive and abrasive, and the nonabrasive modifiers can be further classified as high friction or low friction. The most frequently used nonabrasive modifier is a cured resinous friction dust derived from cashew nutshell liquid (see NUTS). Ground rubber is used in particle sizes similar to or slightly coarser than those of the cashew friction dusts for noise, wear, and abrasion control. Carbon black (qv), petroleum coke flour, natural and synthetic graphite, or other carbonaceous materials (see CARBON) are used to control the friction and improve wear, when abrasives are used, or to reduce noise. The abovementioned modifiers are primarily used in organic and semimetallic materials, except for graphite which is used in all friction materials.

Manufacturing

Linings. Most linings are produced from resin wet mix by either an extrusion or a rolling process.

Segments. Segments for heavy-duty use such as for medium-sized trucks are produced by a dry-mix process.

Disk Pads. Organic and semimetallic disks are produced by somewhat similar processes after the mixes are formed. The mix for organics is prepared in an intensive mixer. The mix for semimetallics generally requires a less intensive blender. The mix is then pressed into preforms at room temperature and 28–42 MPa (4–6 kpsi) pressure. These preforms are then hot-pressed at 160–180°C for 5–15 min at 28–55 MPa (4–8 kpsi). Sometimes the preforms can be eliminated, thus going directly to hot-pressing. The pads are cured at 220–300°C for 4–8 h, and then surface-ground to produce the finished disk pads.

Blocks. The mix for organic blocks is prepared as for segments and the mix for semimetallic blocks is prepared as for semimetallic disk pads. Preforms are formed at 10–17 MPa (1.5–2.5 kpsi). To reduce blisters in hot-pressing, the preforms may be heated to 90°C for 15–30 min. The blocks are formed at 130–150°C at 14–21 MPa (2–3 kpsi) for periods of 10–30 min. After slitting to width, the blocks undergo grinding of internal and external radii. Final cure may be in an unconfined form at temperatures as low as 180°C for 15 h or in confined form at temperatures as high as 280°C for 6 h. Grinding, drilling, and chamfering produce the final block.

Clutch Materials. Methods for producing most manual clutch friction materials are concerned with the placement of the reinforcement strand or wire within the matrix using some sort of winding operation. Processing includes molding of mix without strands or wire; molding of mix around strand or wire preforms; or weaving curable preforms.

Woven Bands. Woven bands for heavy-duty operation are produced by an expensive process that begins with an asbestos or nonasbestos fiber cord, which may be reinforced with wire, being passed through a wet mix to pick up resin and modifiers. The saturated cord is then woven into tapes that pass through heated rolls to partially cure the resin. The material can be post-cured at low temperatures (ca 160°C) to remain as a flexible roll lining or postcured at higher temperature (180–230°C) to form rigid segments. Such materials are used in large band brakes used to control large machinery.

Cermets. Cermet materials are manufactured using the powder metallurgy technique (see METALLURGY, POWDER).

Carbon Composites. In this class of materials, carbon or graphite fibers are embedded in a carbon or graphite matrix. The matrix can be formed by two methods: chemical vapor deposition (CVD) and coking.

Evaluation Methods

Chemical, Physical, and Mechanical Tests. Manufactured friction materials are characterized by various chemical, physical, and mechanical tests in addition to friction and wear testing. The chemical tests include thermogravimetric analysis (tga), differential thermal analysis (dta), pyrolysis gas chromatography (pgc), acetone extraction, liquid chromatography (lc), infrared analysis (ir), and x-ray or scanning electron microscope (sem) analysis. Physical and mechanical tests determine properties such as thermal conductivity, specific heat, tensile or flexural strength, and hardness. Much attention has been placed on noise vibration characterization. The use of modal analysis and damping measurements has increased.

Dynamometer and Vehicle Testing. Friction materials are evaluated in the laboratory by a great variety of tests and equipment.

The full brake dynamometer, when properly instrumented and controlled, reflects the actual brake performance in a vehicle with reasonable accuracy. High initial investment is recovered through operation independent of the climatic conditions and by a fully automatic operation for extended periods, minimizing personnel costs.

Numerous vehicle test procedures are used by different organizations. Performance tests are essentially designed to appraise initial effectiveness, burnish and normal effectiveness, fade and recovery, and final effectiveness. Side-to-side and front-to-rear balance can also be determined. Only vehicle tests can determine noise/vibration properties accurately.

Vehicle tests are considered the ultimate in friction material evaluation, but to be accurate these tests must be carefully designed to eliminate variations caused by changing conditions. Controlled-temperature tests and parallel-test controlled vehicles normally perform the function satisfactorily, but at increased cost.

Environmental and Health Considerations

Manufacturing. Asbestos-based friction materials have been virtually phased out for new vehicle installations because OSHA regulations have limited the exposure of workers to airborne asbestos fibers.

Wear Products. Friction material and rotor emissions are generated by normal wear. Because of large-scale usage and the potential health hazard of asbestos, asbestos organic friction materials and wear debris have been extensively studied.

Future Prospects

Organic friction materials continue to serve the drum brake industry, but are being replaced by a trend to 4-wheel disk brakes, which are also preferred for antiskid brake systems. As brakes become smaller, producing higher brake temperatures, the Class A NAOs are expected to become less suitable, requiring Class B NAOs. Trucks and other heavy vehicles are also moving toward more efficient disk brakes. More sintered friction materials are expected to appear in the heavy-vehicle clutch market. At the same time, aircraft continue to move toward light carbon brakes.

Future brakes must satisfy health standards, and most vehicle manufacturers have moved toward removing all asbestos from brakes.

There is much interest and concern for noise/vibration-free brake systems and there is much activity toward friction couples having reduced noise/vibration properties.

M. G. JACKO
S. K. RHEE
Allied-Signal Inc.

M. G. Jacko, C. M. Brunhofer, and F. W. Aldrich, *Proceedings of the National Workshop on Substitutes for Asbestos,* EPA Report 560/3-80-001, U.S. Environmental Protection Agency, Washington, D.C., Nov. 1980, pp. 9–34.

A. E. Anderson, "Brake Systems Performance—Effects of Fiber Types and Concentrations," *Proceedings from Fibers in Friction Materials Symposium,* Asbestos Institute, Atlantic City, N.J., Oct. 1987.

Metals Handbook, 10th ed., Vol. 2, American Society of Metals, 1990, p. 398.

Materially Speaking, Vol. 6, No. 2, Materials Technology Center, Southern Illinois University at Carbondale, Ill., 1989.

A. E. Anderson and co-workers, "Asbestos Emissions from Brake Dynamometer Tests," *SAE Paper* No. 730549, SAE, New York, May 1973.

BROMINE

Bromine, Br_2, is the only nonmetallic element that is a liquid at standard conditions. Bromine, Br, has at no. 35, at wt 79.904, and belongs to Group 17 (VIIA) of the Periodic Table, the halogens. Its electronic configuration is $(1s)^2(2s)^2(2p)^6(3s)^2(3p)^6(3d)^{10}(4s)^2(4p)^5$. The element's known isotopes range in mass number from 74 to 90. Isotopes usable as radioactive tracers are 77, 80, 80m (metastable), and 82. Bromine has two stable isotopes, ^{79}Br and ^{81}Br. The most common valence states are −1 and +5, but +1, +3, and +7 are also observed.

Physical Properties

Bromine is a dense, dark red, mobile liquid that vaporizes readily at room temperature to give a red vapor that is highly corrosive to many materials and human tissues. Bromine liquid and vapor, up to about 600°C, are diatomic (Br_2). Table 1 summarizes the physical properties of bromine.

Bromine is moderately soluble in water, 33.6 g/L at 25°C. Bromine is soluble in nonpolar solvents and in certain polar solvents such as alcohol and sulfuric acid. Bromine can function as a solvent.

Chemical Properties

One of the central features of the chemistry of the halogens is the tendency to acquire an electron to form either a negative ion, X^-, or a single covalent bond, —X, and bromine is no exception.

Reaction with Hydrogen and Metals. Bromine combines directly with hydrogen at elevated temperatures and this is the basis for the commercial production of hydrogen bromide. Heated charcoal and finely divided platinum metals are catalysts for the reaction.

Bromine reacts with essentially all metals, except tantalum and niobium, although elevated temperatures are sometimes required.

Reaction With Other Halides. Bromide ion is oxidized by chlorine to bromine. This is the basic reaction in the production of bromine from seawater, brines, or bitterns.

Reaction with Nonmetals. Bromine oxidizes sulfur and a number of its compounds. Bromine also oxidizes red phosphorus and some phosphorus compounds. Ammonia, hydrazine, nitrites, and azides are oxidized by bromine. Nitrogen is often a product of such reactions. Under certain circumstances bromine reacts with ammonia and amino compounds to form bromamide, NH_2Br, bromimide, $NHBr_2$, or nitrogen bromide, NBr_3. Bromine oxidizes carbon and reacts with carbon monoxide to form carbonyl bromide.

Table 1. Physical Properties of Bromine

Property	Value
mol wt	159.808
freezing point, °C	−7.25
bp, °C	58.8
density, g/mL, at 15°C	3.1396
25°C	3.1055
vapor density, g/L, 0°C, 101.3 kPa[a]	7.139
refractive index at 25°C	1.6475
viscosity, mm^2/s (= cSt), at 20°C	3.14×10^{-1}
40°C	2.64×10^{-1}
surface tension, mN/m (= dyn/cm), 25°C	40.9
critical temperature, °C	311
critical pressure, MPa[b]	10.3
thermal conductivity, W/(m·K)	0.123
specific conductivity, $(\Omega \cdot cm^{-1})$	9.10×10^{-12}
dielectric constant, 25°C, 10^5 Hz	3.33
electrical resistivity, 25°C, Ω·cm	6.5×10^{10}
expansion coefficient from 20–30°C, per °C	0.0011
heat of vaporization, 50°C, J/g[c]	187
heat of fusion, −7.25°C, J/g[c]	66.11
heat capacity, J/mol[c], at 15 K	7.217
60 K	36.33
265.9 K	61.64
288.15 K[d]	78.66

[a] To convert kPa to mm Hg, multiply by 7.50.
[b] To convert MPa to bar, multiply by 10.
[c] To convert J to cal, divide by 4.184.
[d] Liquid bromine.

Reactions in Water. The ionization potential for bromine is 11.8 eV and the electron affinity is 3.78 eV. The heat of dissociation of the Br_2 molecule is 192 kJ (46 kcal). When bromine dissolves in water, it partially disproportionates. The equilibrium constant for this reaction at 25°C is $7.2 \times 10^{-9}\ M^2$. Light catalyzes the decomposition of hypobromous acid to hydrogen bromide and oxygen. In the dark, hypobromous acid decomposes to bromic acid and bromine.

Reactions with Organic Compounds. The addition of bromine to unsaturated carbon compounds occurs readily.

Bromine reacts directly with alkanes but this reaction has little value because mixtures are obtained. However, photochemical bromination of alkyl bromides can be quite selective. The bromination of aromatic hydrocarbons can occur either in a side chain or on the ring, depending on conditions.

In the presence of halogen Lewis acids, such as metal halides or iodine, aromatic hydrocarbons are halogenated on the ring.

Phenols and phenol ethers readily undergo mono-, di-, or tribromination in inert solvents depending on the amount of bromine used. Heterocyclic compounds range from those such as furan, which is readily halogenated and tends to give polyhalogenated products, to pyridine which forms a complex with aluminum chloride that can only be brominated to 50% reaction.

Aliphatic ketones (qv) are readily brominated in the alpha position. Bromination of aldehydes (qv) is more complicated because bromination can take place on the aldehyde carbon as well as the α-carbon. Acids and esters (see ESTERS, ORGANIC) are less easily brominated than aldehydes or ketones. Bromination of α-chloro ethers proceeds readily and often gives 90–95% yields. Bromine can replace sulfonic acid groups on aromatic rings that also contain activating groups.

Organometallic compounds can react with bromine to give bromides, but because organometallic compounds are frequently made from bromides the reaction with iodine to give iodides is of more synthetic significance.

Charge-Transfer Compounds. Similar to iodine and chlorine, bromine can form charge-transfer complexes with organic molecules that can serve as Lewis bases.

Occurrence

Bromine is widely distributed in nature but in relatively small amounts. The only natural minerals that contain bromine are some silver halides, including bromyrite, AgBr, embolite, Ag(Cl, Br), and iodobromite, Ag(Cl,Br,I). Sources of commercial bromine are underground brines in Arkansas, which contain 3–5 g/L bromine, and in China, Russia, and the United Kingdom; bitterns from mined potash in France and Germany; seawater or seawater bitterns in France, India, Italy, Japan, and Spain; and bitterns of potash production from Dead Sea brines, which contain 4–6 g/L bromine, in Israel.

Manufacture

Bromine occurs in the form of bromide in seawater and in natural brine deposits (see CHEMICALS FROM BRINE). Chloride is also present. In all current methods of bromine production, chlorine, which has a higher reduction potential than bromine, is used to oxidize bromide to bromine.

There are four principal steps in bromine production: (*1*) oxidation of bromide to bromine; (*2*) stripping bromine from the aqueous solution; (*3*) separation of bromine from the vapor; and (*4*) purification of the bromine. Most of the differences between the various bromine manufacturing processes are in the stripping step.

Economic Aspects

Facilities for manufacturing bromine are primarily located near sources of natural brines or bitterns containing usable levels of bromine. In 1990, the United States had seven bromine plants owned by four companies. Six of the plants are in southern Arkansas and are operated by two U.S. producers: Great Lakes Chemical Corporation and Ethyl Corporation.

The costs of building and maintaining a bromine plant are high because of the corrosiveness of brine solutions which contain chlorine and bromine and require special materials of construction. The principal operating expenses are for pumping, steam, environmental costs, energy, and chlorine. The plants are very capital intensive.

Requirements and Specifications

Typical specifications for bromine produced in a modern plant generally exceed the ACS requirements for bromine used as a reagent chemical.

Health, Safety, and Handling

Consequences of Exposure. Bromine has a sharp, penetrating odor. The OSHA/ACGIH threshold limit value–time-weighted average for an 8-h workday and 40-h workweek is 0.1 ppm in air. Monitors are available for determining bromine concentrations in air. Symptoms of overexposure include coughing, nose bleed, feeling of oppression, dizziness, headache, and possibly delayed abdominal pain and diarrhea. Pneumonia may be a late complication of severe exposure.

Liquid bromine produces a mild cooling sensation on first contact with the skin. This is followed by a sensation of heat. If bromine is not removed immediately by flooding with water, the skin becomes red and finally brown, resulting in a deep burn that heals slowly. Contact with concentrated vapor can also cause burns and blisters. Bromine is especially hazardous to the tissues of the eyes where severely painful and destructive burns may result from contact with either liquid or concentrated vapor. Ingestion causes severe burns to the gastrointestinal tract.

Protective Equipment. For handling bromine in the laboratory the minimum safety equipment should include chemical goggles, rubber gloves (Buna-*N* or neoprene rubber), laboratory coat, and fume hood.

For handling bromine in a plant, safety equipment should include hard hat, goggles, neoprene full-coverage slicker, Buna-*N* or neoprene rubber gloves, and neoprene boots.

Reactivity. Bromine is nonflammable but may ignite combustibles such as dry grass, on contact. Handling bromine in a wet atmosphere, extreme heat, and temperatures low enough to cause bromine to solidify (−6°C) should be avoided. Bromine should be stored in a cool, dry area away from heat.

Under the Comprehensive Environmental Response, Compensation, and Liability Act (CERCLA)/RCRA regulations in effect at the end of 1986 bromine is regulated as a hazardous waste or material and must be disposed of in an approved hazardous waste facility in compliance with EPA or other applicable local, state, and federal regulations and should be handled in a manner acceptable to good waste management practice. The reportable quantity is 45.4 kg for corrosivity.

Materials of Construction. Glass, lead, tantalum, niobium, nickel, and the fluoropolymers Kynar, Halar, and Teflon are highly resistant to bromine.

Storage and Transportation. Bromine in bulk quantities is shipped domestically in 7570-L and 15,140-L lead-lined pressure tank cars or 6435–6813-L nickel-clad pressure tank trailers.

Uses

An important use of bromine compounds is in the production of flame retardants (qv). Bromine-containing epoxy sealants are used in semiconductor devices (see SEMICONDUCTORS).

Bromine has some use in swimming pools and in bleaching; it is also a disinfectant for cooling water and wastewater. Its main use is as a chemical reactant.

Zinc–bromine storage batteries (qv) are under development as load-leveling devices in electric utilities.

PHILIP F. JACKISCH
Ethyl Corporation

Z. E. Jolles, ed., *Bromine and its Compounds,* Academic Press, New York, 1966.

V. Gutmann, ed., *MTP Int. Rev. Sci.: Inorg. Chem., Ser. One,* 3 (1972).

A. J. Downs and C. J. Adams, in J. C. Bailar, Jr. and co-eds., *Comprehensive Inorganic Chemistry,* Vol. 2, Pergamon Press, Oxford, U.K., 1973, pp. 1107–1594.

V. Gutmann, ed., *Halogen Chemistry,* Academic Press, New York, 1967 (3 vols.).

BROMINE COMPOUNDS

Inorganic Compounds

Bromamines. Bromamide, NH_2Br, is formed from bromine and liquid ammonia:

$$Br_2 + 2\ NH_3 \rightarrow NH_2Br + NH_4^+ + Br^-$$

By evaporating a mixture of bromine and ammonia at low temperatures, nitrogen bromide, NBr_3, is obtained as the ammoniate $NBr_3 \cdot 6NH_3$. This compound decomposes explosively when warmed to room temperature. NH_2Br and bromimide, $NHBr_2$, are formed when bromine reacts with ammonia in ether solution at low temperatures. These products have not been isolated.

In swimming pools disinfected by bromine, bromamide and bromimide can form. These compounds have about half the disinfecting power of HOBr giving bromine an advantage compared to chlorine. Chloramide and chlorimide have 80 to 100 times less disinfecting power than HOCl.

Bromides. *Hydrogen Bromide.* Hydrogen bromide, HBr, (hydrobromic acid), is a colorless gas that fumes strongly in moist air; mp, −86 °C; liquid density, 2.152 g/mL; bp, −67 °C; heat capacity for the solid at −91 °C, 636 J/(kg·K), (152 cal/(kg·K)), liquid, 737 J/(kg·K), (176 cal/(kg·K)), gas at 27°C, 356 J/(kg·K) (85 cal/(kg · K)); heat of fusion at mp, 29.8 kJ/kg (7.12 kcal/kg); heat of vaporization at −66.7 °C, 218 kJ/kg (52 kcal/kg); critical temperature, 89.8°C; and critical pressure, 8510 kPa (84 atm). The gas is highly soluble in water, forming azeotropic mixtures the compositions of which have been determined at various pressures.

The manufacture of HBr gas involves burning a mixture of hydrogen and bromine vapor. Alternatively, platinized asbestos or silica gel can catalyze this reaction.

Hydrobromic acid is one of the strongest mineral acids. It is a more effective solvent for some ore minerals than hydrochloric acid because of its higher boiling point and stronger reducing action.

The liquid and vapors of hydrobromic acid are highly corrosive to tissue. The threshold limit value for HBr gas in an 8-h day is 3 ppm time-weighted average. Inhalation of vapor is so irritating to the nose and throat that a person does not voluntarily remain in an area when vapors are present in hazardous concentrations. Symptoms of overexposure to HBr include coughing, choking, burning in the throat, wheezing, or asphyxia. Ingestion causes severe burns of the mouth and stomach and skin contact can cause severe burns. In the case of liquid or vapor contact with eyes, permanent damage may result. Suitable safety equipment should be used when handling HBr and a safety shower and eye bath should be available.

HBr reacts with metals, producing highly explosive hydrogen gas.

Technical 48% and 62% acids are colorless-to-light yellow liquids available in drums, 15,140-L tank trailers, and 37,850-L tank cars. They are classified under DOT regulations as corrosive materials. Anhydrous hydrogen bromide is available in cylinders, under its vapor pressure of approximately 2.4 MPa (350 psi) at 25°C. It is classified as a nonflammable gas.

A considerable amount of hydrobromic acid is consumed in the manufacture of inorganic bromides, as well as in the synthesis of alkyl bromides from alcohols. The acid can also be used to hydrobrominate olefins (qv).

In the petroleum (qv) industry hydrogen bromide can serve as an alkylation catalyst.

Other Bromides. Alkali and alkaline-earth bromides can be prepared by neutralizing a solution of the corresponding hydroxide or carbonate with hydrobromic acid. Alternatively, bromine and a reducing agent such as ammonia are used in the van der Meulen process:

$$3\ K_2CO_3 + 3\ Br_2 + 2\ NH_3 \rightarrow 6\ KBr + N_2 + 3\ CO_2 + 3\ H_2O$$

Ammonium bromide can be prepared by the direct reaction of bromine with aqueous ammonia:

$$3\ Br_2 + 8\ NH_4OH \rightarrow 6\ NH_4Br + N_2 + 8\ H_2O$$

Anhydrous lithium bromide, LiBr, is a desiccant useful in the industrial drying of air. When it contains sufficient moisture it can be a humectant. Sodium bromide, NaBr, if used in conjunction with an oxidizer, such as chlorine or sodium hypochlorite, is an effective biocide in cooling water systems. Sodium bromide and potassium bromide, KBr, are used to prepare light-sensitive emulsions for photography. Potassium bromide is also used in process engraving. These three bromides have medical applications as sedatives, hypnotics, or anticonvulsants. Rubidium bromide, RbBr, is claimed as a component of x-ray intensifier screens. Cesium bromide, CsBr, finds application in x-ray fluorescent screens, spectrometer prisms, and adsorption-cell windows.

Beryllium bromide, $BeBr_2$, is claimed as an adhesive for poly(vinyl alcohol). Magnesium bromine, $MgBr_2$, is used in organic synthesis. A concentrated solution of calcium bromide, $CaBr_2$, in water has a specific gravity of at least 1.7 and is useful as a completion, workover, and packer fluid in oil-well drilling and maintenance. In medicine magnesium bromide and calcium bromide are sedatives and anticonvulsants. Strontium bromide, $SrBr_2$ has been used as an anticonvulsant. Barium bromide, $BaBr_2$, is a reactant for the manufacture of other bromides and in the preparation of phosphors.

Cerous bromide, $CeBr_3$, and praseodymium bromide, $PrBr_3$, are claimed to be suitable for a molten salt bath used for the reduction of uranium oxide by magnesium. $PrBr_3$ is claimed to be a light filter in a cathode ray tube.

Titanium bromide, $TiBr_4$, is claimed as a catalyst for olefin polymerizations. Chromous bromide, $CrBr_2$, is used in chromizing. Chromic bromide, $CrBr_3$, and tungsten bromide, WBr_6, are catalysts for polymerizing olefins. Manganese bromide, $MnBr_2$, is a catalyst for the formation of aromatic aldehydes from alkylbenzenes and phthalic acids from xylenes. It is also claimed as a catalyst in the ammoxidation conversion of *o*-xylene to phthalimide.

Ferrous bromide, $FeBr_2$, is a polymerization catalyst. Ferric bromide, $FeBr_3$ is a catalyst for organic reactions, particularly in brominations of aromatic compounds. Cobaltous bromide, $CoBr_2$, is used in hydrometers and as a catalyst for organic reactions. Rhodium bromide, $RhBr_3$, is claimed as a catalyst for carbonylating methanol (qv) to acetic acid (see ACETIC ACID AND DERIVATIVES). Nickel bromide, $NiBr_2$, is a catalyst in dimerizing butadiene, in condensing butadiene onto ring systems and benzyl ketones, for oxidizing secondary alcohols to ketones (qv), and for preparing biaryls from aryl iodides. Palladium bromide, $PdBr_2$, is a catalyst for various carbonylation reactions. Platinum bromide, $PtBr_2$, is a general dehydrodimerization catalyst for boron hydrides and carboranes (see BORON COMPOUNDS).

Cuprous bromide, CuBr, is a catalyst for organic reactions. Cupric bromide, $CuBr_2$, is used as an intensifier in photography, as a brominating agent in organic synthesis, as a humidity indicator, as a wood preservative, as a stabilizer for acetylated polyformaldehyde, and in solid- electrolyte batteries. Silver bromide, AgBr, is used in photography, as a topical antiinfective, and as an astringent. Gold tribromide, $AuBr_3$, is claimed as a component of a sensor for halogenated gases. Bromoauric acid, $HAuBr_4$, is a catalyst in the selective oxidation of sulfides to sulfoxides by nitric acid. Sodium tetrabromoaurate, $Na[AuBr_4]$ and potassium tetrabromoaurate, $K[AuBr_4]$, are claimed to be useful in gold-coating solutions.

Zinc bromide, $ZnBr_2$, is consumed to make silver bromide collodion emulsions for photography, in radiation shielding, in gravity separation, as an electrical conductivity fluid, and as an intermediate for the manufacture of various chemicals. Storage batteries with a zinc bromide electrolyte are under development for use as load-leveling devices in electric utilities and as electric car batteries. Mercurous bromide, Hg_2Br_2, reacts with HBr to form hydrogen (qv) quantitatively.

Boron tribromide, BBr_3, is used in the manufacture of diborane and in the production of ultra high purity boron (see BORON, ELEMENTAL; BORON COMPOUNDS). Anhydrous aluminum bromide, $AlBr_3$, is used as an acid catalyst in organic syntheses where it is more reactive and more soluble in organic solvents than $AlCl_3$. Thallium bromide, TlBr, is claimed as a component in radiographic image conversion panels.

Silicon tetrabromide, $SiBr_4$, and tribromosilane, $SiHBr_3$, are used in a process to make high purity silicon. Stannous bromide, $SnBr_2$, is claimed as a catalyst in preparing a lubricant antioxidant. Stannic bromide, $SnBr_4$, is used in the metallurgical separation of minerals.

Ammonium bromide, NH_4Br, is used in photography, process engraving and lithography, fireproofing of wood, corrosion inhibitors, and in medicine as a sedative. Phosphorus tribromide, PBr_3, converts primary alcohols into bromides. Phosphorus pentabromide, PBr_5, is a brominating agent for converting organic acids to acyl bromides. It also can be used to convert phenols and secondary alcohols into bromides. Bismuth bromide, $BiBr_3$, is claimed as a catalyst for dehydrating cyclohexanol to cyclohexene. It is also claimed as part of a solid electrolyte in primary lithium batteries. Selenium tetrabromide, $SeBr_4$, is claimed as a dopant for a photoreceptor for electrophotography (qv) and also as an additive for a rapid bright silver electroplating (qv) bath. Tellurium tetrabromide, $TeBr_4$, is a catalyst for the synthesis of organic acids.

Thionyl bromide, $SOBr_2$, is used to convert alcohols to bromides.

Bromine Halides. Bromine and chlorine react reversibly in the liquid or vapor states to form bromine chloride, BrCl. Bromine chloride is a dark red, fuming, lachrimatory liquid with a boiling point of 5°C. A similar but somewhat more stable compound, iodine bromide, IBr, is formed from bromine and iodine. These compounds are soluble in carbon tetrachloride or acetic acid and are used as halogenating agents for organic substances.

Bromine chloride is used as a brominating agent in the preparation of fire-retardant chemicals, pharmaceuticals, high density brominated liquids, agricultural chemicals, dyes, bleaching agents, and in water treatment, eg, in cooling towers and effluent streams from sewage plants.

Other bromine halides include bromine monofluoride, BrF, bromine trifluoride, BrF_3, and bromine pentafluoride, BrF_5.

Bromine Oxides, Acids, and Salts. *Oxides.* None of the oxides of bromine is stable at ordinary temperatures and none has any practical use.

Acids and Salts. The oxygen acids of bromine are strong oxidants but at ordinary temperatures are stable only in solution. An aqueous solution of hypobromous acid may be prepared by treating bromine water with silver oxide or mercuric oxide.

Hypobromites, the salts of hypobromous acid, do not keep well because they gradually disproportionate to bromide and bromate. Solutions are best prepared as needed from bromine and alkali with cooling. Hypobromites are strong bleaching agents, similar to hypochlorites.

The controlled disproportionation of cold concentrated alkaline hypobromite solutions can yield BrO^-; ammonia or acetone is used to destroy residual BrO^-.

Lithium bromite, $LiBrO_2$, sodium bromite, $NaBrO_2$, potassium bromite, $KBrO_2$, and barium bromite, $Ba(BrO_2)_2$, have been crystallized from such solutions. Sodium bromite is used as a desizing agent in the textile industry.

Bromates are stable in storage. They have various uses based on their oxidizing power. Bromates can be formed by the disproportionation of bromine in basic solution.

Bromates represent a potential fire and explosion hazard if heated, subjected to shock, or acidified. They should not be allowed to contact reactive organic matter, including paper and wood. Industrial quantities are packed in fiber drums with polyethylene liners or in metal drums. For shipment, a yellow oxidizer label is required under DOT regulations.

An important use for sodium bromate, $NaBrO_3$, is as a neutralizer or oxidizer in certain hair-wave preparations. There also is renewed interest in using mixture of sodium bromate and sodium bromide for dissolving gold from its ores. The primary use for potassium bromate, $KBrO_3$, is in flour treatment. In analytical chemistry potassium bromate is used as a primary standard and a brominating agent. Barium bromate, $Ba(BrO_3)_2$, is used in the preparation of rare-earth bromates and as a corrosion inhibitor for low carbon steel.

Perbromic acid, $HBrO_4$, is a strong acid that is completely dissociated in aqueous solutions. It can be prepared by passing sodium perbromate solution through a cation-exchange resin in hydrogen form.

Organic Compounds

Organic compounds of bromine usually resemble their chlorine analogues but have higher densities and lower vapor pressures. The bromo compounds are more reactive toward alkalies and metals; brominated solvents should generally be kept from contact with active metals such as aluminum. On the other hand, they present less fire hazard: one bromine atom per molecule reduces flammability about as much as two chlorine atoms.

Methyl Bromide. Methyl bromide (bromomethane), CH_3Br, is a colorless liquid or gas with practically no odor. Its physical properties are mp, −93.7 °C; bp, 3.56°C; liquid d_4^0, 1.730 g/mL; vapor d_{gas}^{20}, 3.974 g/L; n_D^{20}, 1.4432; vapor pressure at 20°C, 189.3 kPa (1420 mm Hg); and viscosity at −20, 0, and 25°C: 0.475, 0.397, and 0.324 mPa·s(= cP), respectively. The liquid is miscible with most organic solvents and forms a bulky, crystalline hydrate below 4°C. The solubility in water varies with pressure: at normal atmospheric pressure, methyl bromide plus water vapor, the solubility is 1.75 g/100 g solution (20°C).

The bromine atom of methyl bromide is an excellent leaving group in nucleophilic substitution reactions and is displaced by a variety of nucleophiles. Thus, methyl bromide is useful in a variety of methylation reactions, such as the syntheses of ethers, sulfides, esters, and amines. Tertiary amines are methylated by methyl bromide to form quaternary ammonium bromides, some of which are active as microbicides.

Methyl bromide reacts readily with a number of metals to form organometallic reagents useful as catalysts or for the introduction of a methyl group into a variety of organic or organometallic compounds. Reaction with magnesium gives the well-known Grignard reagent (see GRIGNARD REACTION), usually prepared in an ether solvent.

Methyl bromide, when dry (<100 ppm water), is inert toward most materials of construction. Carbon steel is recommended for storage vessels, piping, pumps, valves, and fittings.

Methyl bromide is nonflammable over a wide range of concentrations in air at atmospheric pressure and offers practically no fire hazard.

Commercial manufacture of methyl bromide is generally based on the reaction of hydrogen bromide with methanol.

Because methyl bromide is normally a gas, the liquid evaporates rapidly and can cause frost-type burns when large amounts evaporate from the skin. Vapor trapped next to the skin can cause severe delayed burns. Methyl bromide is corrosive to the eyes. Inhalation of this chemical can cause dizziness, nausea, vomiting, headache, drowsiness, dimming of vision, and death. Respiratory tract inflammation can occur from breathing methyl bromide. These symptoms can be delayed 2–48 h. Repeated and prolonged exposure to lower concentrations (30–100 ppm) causes severe nervous system effects. Methyl bromide has been shown to be mutagenic in laboratory tests. The time-weighted average limit for daily 8-h exposure to the vapor in air is 5 ppm by volume, or 19 mg/m^3.

The primary use for methyl bromide is in the extermination of insect and rodent pests. Except in California, methyl bromide is used in space and structural fumigation. The material is suitable for the fumigation of food commodities such as dried fruits, grain, flour, and nuts and the facilities in which these foods are processed or stored, as well as for tobacco and many kinds of nursery stock.

Methyl bromide finds use as a methylating agent in the syntheses of agricultural and drug chemicals. It is also used in ionization chambers, for degreasing wool, and for extracting oil from nuts, seeds, and flowers.

Other Bromomethanes. Bromochloromethane (methylene chlorobromide), CH_2BrCl, is a clear, colorless liquid with a characteristic sweet odor and very low freezing point, −88.0 °C. Its properties include bp, 68.1°C; $d^{25}{}_4$, 1.9229 g/mL; $n^{25}{}_D$, 1.4808; and heat of vaporization at bp, 232 J/g (55.4 cal/g). The liquid is completely miscible with common organic solvents and soluble in water to the extent of about 0.9 g per 100 g at 25°C. Common methods of preparation involve the partial replacement of chloride in methylene chloride (dichloromethane) by reaction with anhydrous aluminum bromide, treatment with bromine and aluminum, or by reaction with hydrogen bromide in the presence of an aluminum halide catalyst. The principal outlet for bromochloromethane is as a fire-extinguisher fluid. It is also used as an explosion suppression agent and as an intermediate and solvent in the manufacture of pesticides and other products. Its toxicity is also lower than that of many bromine compounds.

Dibromomethane (methylene bromide), CH_2Br_2, is a similar liquid, mp −52.7 °C, bp 96.9°C, $d^{20}{}_4$ 2.4956 g/mL, $n^{20}{}_D$ 1.5419. Water solubility is 1.17 g/100 g at 15°C. It is prepared by the same methods as bromochloromethane, allowing the reaction to proceed to completion. The compound is used as a solvent, as a gauge fluid, and in producing pesticides. Both of these dihalomethanes can be used as dense, readily volatile media for mineral and salt separations.

Tribromomethane (bromoform), $CHBr_3$, is usually sold mixed with up to 3–4% ethanol as a stabilizer. The pure liquid has mp, 7.7°C; bp, 149.5°C; $d^{20}{}_4$, 2.8912 g/mL; and $n^{19}{}_D$ 1.5980. Water solubility is about 0.3 g/100 g at 25°C. Bromoform is prepared from chloroform by the replacement procedures indicated. Uses have been found in syntheses, in pharmacy as a sedative and antitussive, in gauge fluids, and as a dense liquid for separating minerals.

Tetrabromomethane (carbon tetrabromide), CBr_4, is a white to brownish powder, mp 90.1°C; bp, 189.5°C; $d^{20}{}_4$, 3.240 g/mL; $n^{99.5}{}_D$, 1.600. It is prepared by replacement of chlorine in carbon tetrachloride using hydrogen bromide and an aluminum halide catalyst. It is also light-sensitive. A number of patents have covered possible uses in photography and photo-duplicating systems.

Various bromofluoromethanes have been described and proposed for use as fire extinguishing agents. Two that have been recommended highly for this purpose are dibromodifluoromethane, CBr_2F_2, and bromotrifluoromethane, $CBrF_3$.

Ethyl Bromide. Ethyl bromide (bromoethane), CH_3CH_2Br, is a volatile, clear, colorless liquid of mp, −119.3 °C; bp, 38.4°C; $d^{25}{}_4$, 1.4492 g/mL; $n^{25}{}_D$, and 1.421. It is completely miscible with most neutral or acidic organic solvents but may react with bases. The solubility in water is ca 0.9 g/100 g at 25°C. Ordinarily, ethyl bromide is prepared by refluxing ethanol with hydrobromic acid, or with an alkali bromide and sulfuric acid. Ethyl bromide is used mainly as an ethylating agent in syntheses, particularly of pharmaceuticals. Some has been employed as a solvent or refrigerant, and formerly as a local anesthetic applied as a spray. Its toxicity is markedly lower than that of methyl bromide.

Ethylene Dibromide. Ethylene dibromide (ethylene bromide, 1,2-dibromoethane), CH_2BrCH_2Br, is a clear, colorless liquid with a characteristic sweet odor. Its properties include mp, 9.9°C; bp, 131.4°C; $d^{20}{}_4$, 2.1792 g/mL (47); $n^{20}{}_D$, 1.5380; vapor pressure, 1.13 (8.5), 15.98 (119.8), and 38.03 kPa (285.2 mm Hg) at 20, 75, and 100°C, respectively; and viscosity at 20°C, 1.727 mPa·s(= cP). The liquid is completely miscible with carbon tetrachloride, benzene, gasoline, ether, and anhydrous alcohols at 25°C and solubility in water at 20°C is 0.404 g/100 g solution.

Ethylene dibromide is nonflammable and under ordinary conditions is quite stable. Ethylene glycol is produced by high temperature hydrolysis under pressure, and the reaction with zinc in alcohol yields ethylene and zinc bromide.

Ethylene dibromide is a suspected human carcinogen and worker exposure by all routes should be carefully controlled to levels as low as reasonably achievable. Ethylene dibromide causes severe blistering of the skin if contact is prolonged. Eye contact with the liquid will cause pain, irritation, and temporary impairment of vision.

Ethylene dibromide is one of the lowest cost organic compounds of bromine. This compound has found its primary use as an exhaust system scavenger in gasoline containing lead antiknocks and is still used for this purpose in countries where leaded gasolines are sold. Other uses are in the manufacturing of dyes, pharmaceuticals, polymers, and other chemicals. It is a general solvent for resins, waxes, gums, and dyes.

Uses

Bromine compounds are used in dyes and indicators, flame retardants, pesticides, and pharmaceuticals.

PHILIP F. JACKISCH
Ethyl Corporation

Z. E. Jolles, ed., *Bromine and its Compounds,* Academic Press, New York, 1966. V. Gutmann, ed., *MTP Int. Rev. Sci.: Inorg. Chem., Ser. One,* 3 (1972).

A. J. Downs and C. J. Adams, in J. C. Bailar, Jr. and co-eds., *Comprehensive Inorganic Chemistry,* Vol. 2, Pergamon Press, Oxford, U.K., 1973, pp. 1107–1594.

V. Gutmann, ed., *Halogen Chemistry,* Academic Press, New York, 1967 (three vols.).

BRONCHOLYTIC AGENTS. See ANTIASTHMATIC AGENTS; EXPECTORANTS, ANTITUSSIVES, AND RELATED AGENTS; PROSTAGLANDINS.

BRONZE. See COPPER ALLOYS.

BTX PROCESSING

Benzene (B), toluene (T), and the xylenes (X) are the lowest molecular-weight aromatic hydrocarbon homologues. They are each very large-scale chemical feedstocks. Since they are often produced together in the same process, they can be considered as a group, ie, BTX. However, BTX as such is not an article of commerce. It is either an important component of a crude mixture such as reformate or pyrolysis gasoline, or it is separated and purified into its individual components. This article mainly discusses the processes for making those crude mixtures.

Since World War II, the production of BTX has been intimately connected with the production of gasoline. BTX constitutes part of an important gasoline component called reformate, which is discussed below (see GASOLINE AND OTHER MOTOR FUELS). Reformate is highly valued for gasoline because it has a very high octane rating. This results from the high concentration of aromatic compounds, all of which have very high octane values.

Any BTX needed for chemical use is separated from the reformate stream before it is blended into the gasoline pool. Although at a given refinery the total volume of gasoline production (eg, 16,000 m^3/d) usually dwarfs the BTX volume (eg, 800 m^3/d) and may have a higher priority, BTX production is often important enough to support its own reforming facilities and should not be considered simply as a gasoline by-product. This independence from gasoline may be even further emphasized in the future because of restrictions on the allowed level of BTX in gasoline and because new BTX processes may utilize light feeds or natural gas.

However, despite possible dislocations in supply, the demand for chemical uses will still be readily satisfied and the availability and price of BTX for chemical uses probably will not be greatly affected by the change in gasoline composition.

The principal chemical uses of BTX are illustrated in Figure 1.

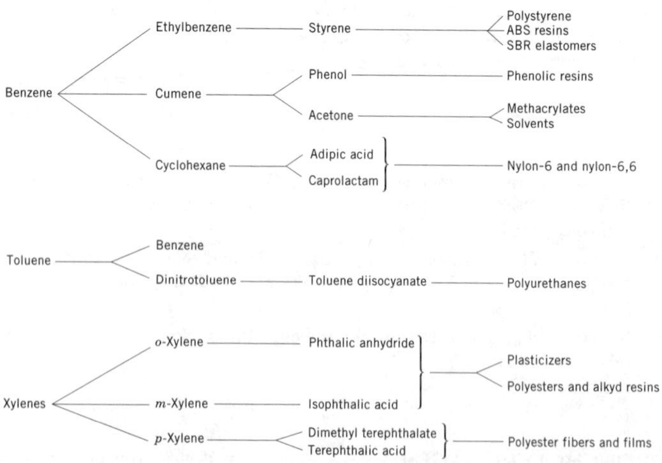

Figure 1. Principal uses of BTX.

Reforming

Reforming, as currently practiced, is a platinum-catalyzed high temperature vapor-phase process which converts a relatively nonaromatic C-6–C-12 hydrocarbon mixture (naphtha) to an aromatic product called reformate. The catalyst often contains less than 1% of platinum, possibly modified with other metals, supported on a high surface area support such as alumina, which provides acidity (see CATALYSIS). The gasoline octane rating of the reformate is directly related to its aromatic content. The aromatic content is higher when the reformer is operated at high severity (high temperature, low space velocity). Some cracking to light products also occurs, and this also increases at high severity. A typical reformate contains BTX in the proportions of 19:49:32, respectively, although these proportions can be varied by tailoring the feed composition. In response to the environmental pressure on benzene in motor gasoline mentioned previously, it is probable that many U.S. refiners will choose to reduce the proportion of benzene in their reformate by raising the cut point on the naphtha feed to their reformers.

Feedstock. Feed for reformers is normally petroleum hydrocarbons that boil roughly in the 70–190°C range.

Reforming Conditions. The main process variables are pressure, 450–3550 kPa (50–500 psig), temperature (470–530°C), space velocity, and the catalyst employed. An excess of hydrogen (2–8 moles per mole of feed) is usually employed. Depending on feed and processing conditions, net hydrogen production is usually in the range of 140–210 m^3/m^3 feed (800–1200 SCF/bbl). The C-1–C-4 products are recovered and normally used as fuels.

A flow diagram for a typical semiregenerative reformer based on the Rheniforming process is shown in Figure 2.

The choice of reforming process depends on the product desired, plant size, and capital availability. If BTX is to be only a coproduct, the refiner might select a semiregenerative process. Severity is usually lower, and the important factor is the yield of gasoline. This yield for some combinations of feeds and catalysts can improve with successive catalyst regeneration cycles. For high BTX yields a swing reactor or continuous regeneration process might be the choice because BTX yields are highest at high severity and low pressure.

Patents cover a new reforming catalyst based on L-zeolite which gives a significantly higher yield of BTX, especially benzene, from light paraffinic feeds. Other new zeolites may also offer advantages over the traditional reforming catalyst supports.

Reforming Chemistry. The main reactions occurring in a reformer are shown in Figure 3; most are reversible, indicating the potential importance of reaction equilibrium.

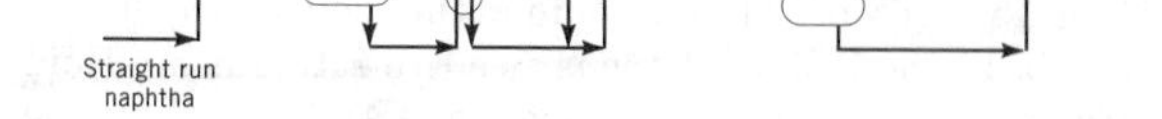

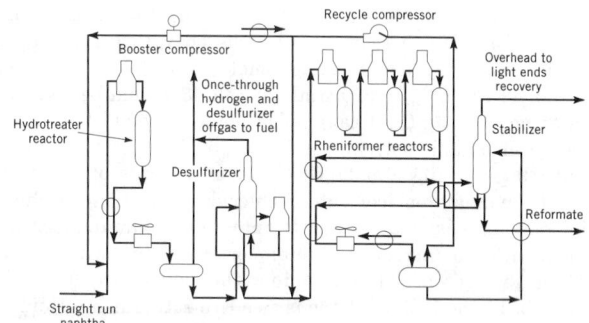

Figure 2. Simplified process flow diagram of a naphtha hydrotreater and rheniformer.

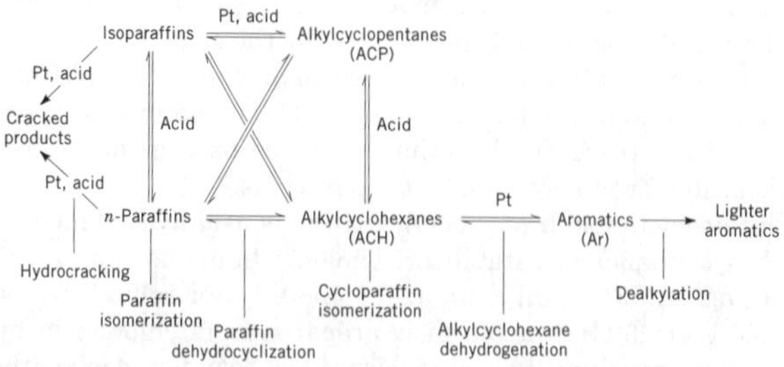

Figure 3. Main reactions of catalytic reforming. Pt and acid refer to predominant active catalytic sites.

Table 1. BTX Yields from Various Pyrolysis Feeds,[a] 10^3 t/yr[b]

Feed	C_2H_6	C_3H_8	n-C_4H_{10}	Mid-crude naphtha	C-6–C-8 raffinate	Gas oil, distilled	
						Atm	Vac
ethylene product rate	450	450	450	450	450	450	450
feed rate	583	1080	1135	1349	1535	1750	2213
ethylene[c]	48.2	34.2	35.8	30.0	26.0	23.0	18.0
benzene	5.0	26.6	34.3	90.0	72.7	105.5	82.5
toluene	0.7	5.8	9.4	45.1	41.5	50.8	64.3
C-8 aromatics			4.0	23.8	18.4	38.0	41.4

[a] Ethane recycled to extinction.
[b] Unless otherwise noted.
[c] Once through yield.

Hydrocarbon Pyrolysis

A large amount of BTX is obtained as a by-product of ethylene manufacture (see ETHYLENE). The amount produced strongly depends on the feed to the ethylene plant. This is illustrated in Table 1 for various feeds to a typical large-scale plant producing 450,000 t/yr of ethylene.

Outside the realm of typical hydrocarbon pyrolysis is the high temperature pyrolysis of methane. In one variant of this process, which has only been commercialized to produce acetylene (with some BTX), methane reacts in an electric arc at about 1500°C with very short contact times. At higher temperatures or with a catalyst and added hydrogen, BTX is produced with fairly high selectivity.

Other BTX Processes

Because of the importance of the petroleum-based processes, only about 1% of the U.S. supply of BTX comes from coal pyrolysis. Outside the United States, coal pyrolysis is more important as a source of BTX. Another interesting route to BTX is the mobil methanol-to-gasoline (MTG) process. Methanol is converted into gasoline containing about 50% aromatics by passing it over a ZSM-5 catalyst.

BTX Recovery

The complexity of separating and purifying the individual BTX components from crude BTX products depends on the amount of nonaromatic impurities present. If the amount is small enough, simple distillation can suffice. If not, distillation alone is not sufficient because the BTX aromatic compounds are close in boiling point to some of the cycloparaffins of the same carbon number or to paraffins of the next higher carbon number. Because of this, extraction or extractive distillation with a polar solvent is widely used to separate the slightly polar aromatic hydrocarbons from the nonpolar nonaromatic hydrocarbons. When olefins are also present, as in pyrolysis gasoline, these interfere with the extraction and must first be hydrogenated or removed by adsorbents.

An option for avoiding the cost of extraction is to increase the severity of the BTX formation step. This reduces the quantity of residual paraffins, and, depending on the BTX formation process, may leave the BTX clean enough to purify by distillation.

Extraction and Extractive Distillation. The choice of an extraction or extractive distillation solvent depends on its boiling point, polarity, thermal stability, selectivity, aromatics capacity, and upon the feed aromatic content (see EXTRACTION). Capacity, defined as the quantity of material that is extracted from the feed by a given quantity of solvent, must be balanced against selectivity, defined as the degree to which the solvent extracts the aromatics in the feed in preference to paraffins and other materials. Most high capacity solvents have low selectivity. The ultimate choice of solvent is determined by economics. The most important extraction processes use either sulfolane or glycols as the polar extraction solvent.

Sulfolane, used in UOP and Shell processes, offers good thermal and hydrolytic stability, high density and boiling point, and a good balance of solvent properties. The UOP Udex Process uses a glycol solvent.

Extractive distillation, using solvents similar to those used in extraction, may be employed to recover aromatics from reformates which have been prefractionated to a narrow boiling range. Extractive distillation is also used to recover a mixed benzene–toluene stream from which high quality benzene can be produced by postfractionation.

Downstream Processing. In addition to extraction, various downstream operations are often carried out on the BTX product to produce products in proportions to fit the market demand.

Benzene and p-xylene are generally in higher demand than toluene, o-xylene, and m-xylene. These are produced from raw BTX by reactive conversion processes (isomerization, disproportionation, dealkylation) and purification processes (adsorption, crystallization).

Environmental Considerations

BTX processing has come under steadily increasing pressure to reduce emissions and workplace exposures (see INDUSTRIAL HYGIENE, TOXICOLOGY). Reductions in the permissible levels of both benzene and total aromatics (BTX) in gasoline have been legislated. Whereas all BTX components are to be controlled, the main focus is on benzene because it is considerably more toxic than the others and is classified as a known carcinogen.

W. A. SWEENEY
P. F. BRYAN
Chevron Research and Technology Company

A. P. Dossett, in E. G. Hancock, ed., *Toluene, the Xylenes, and Their Industrial Derivatives,* Elsevier, Amsterdam, the Netherlands, 1982, Chapt. 5.

D. M. Little, *Catalytic Reforming,* PennWell, Tulsa, Okla., 1985.

R. N. Shreve and J. A. Brink, Jr., *Chemical Process Industries,* 4th ed., McGraw-Hill, New York, 1977, Chapt. 7.

T. Ueno, in T. C. Lo and co-eds., *Handbook of Solvent Extraction,* John Wiley & Sons, Inc., New York, 1983, p. 575.

BUBBLE MEMORY. See MAGNETIC MATERIALS, BULK; MAGNETIC MATERIALS, THIN FILM.

BUILDING MATERIALS

SURVEY

This article discusses traditional building and construction products, ie, not made from synthetic polymers (see BUILDING MATERIALS, PLASTIC), including wood, asphalt, gypsum, glass products, Portland cement, and bricks.

Wood

Wood (qv) is arguably the oldest building material used by humans to construct their dwellings. It is a natural product obtained from trees, used in both structural and decorative applications. The chemical composition of wood is largely cellulose (qv) and lignin (qv). Today there are a variety of composite or reconstituted wood products, such as plywood, particle board, wood fiber boards, and laminated structural beams, where small pieces of wood or wood fiber are combined with adhesives to make larger sheets or boards.

Woods are classified as either hardwoods or softwoods, based on the seed- and leaf-type of the trees from which the wood comes. Softwoods come from trees that are classified as gymnosperms, or naked seeds, whereas the hardwoods are angiosperms, or covered seeds. The common softwoods, such as pine, spruce, fir, cedar, and hemlock, have specific gravities between 0.03 and 0.55, which make them ideal for construction purposes. Because of these properties, softwoods are the common construction wood rather than the generally stronger hardwoods.

Manufacture. The manufacture of lumber or sawn wood starts with cutting the tree and removing the limbs (limbing), followed by sawing the log into slabs parallel to the long direction of the log. Most softwoods are cut either by sawing through and through or around the log (Fig. 1).

The green wood of a fresh cut tree can have a moisture content of 60 to over 150%, depending on species and weather conditions. These moisture levels are too high for most construction applications as the wood can rot and undergo relatively large dimensional and shape changes as it dries out. Most construction wood is dried to below 19% moisture content, often to about 8%, to reduce these problems. As wood dries, it typically shrinks 5–8% perpendicular to, ie, across, the grain and less than 1% with, ie, parallel to, the grain.

Construction lumber is milled, ie, smoothed, after it has a dried to its finished dimensions. These dimensions are less than the nominal size of the lumber, which is based on the size of the green wood cut during the sawing operation.

Finished lumber is visually sorted or graded based on industry organization standards. These standards consider the number of knots, ie, tree-limb locations, straightness, and overall quality of the wood.

Shake Shingle. Whole wood products are used in shake shingles for residential and light commercial application. Cedar is the most common wood used because it splits easily and is rot-resistant, but pine is also used.

Annual Production. The United States is the largest single producer of lumber, with about 25% of all the logs cut in the world. Soft housing and construction markets as well as environmental pressures to limit logging reduce the percentage.

Wood Products or Composite Wood Matierals. The wood products or composite wood materials consist of products where the wood is cut into small or thin pieces and then the cut pieces are glued together to form a larger piece of wood product. The most common example is plywood. This category also includes reconstituted wood products subclassified into wood flake boards, particle boards, or wood fiberboards depending on the size of the wood particle glued together. Products in these categories include low density fiberboard (wood fiberboard), medium-density fiberboard (MDF), high density fiberboard (hardboard), flake board, oriented strand board (OSB), and particle boards.

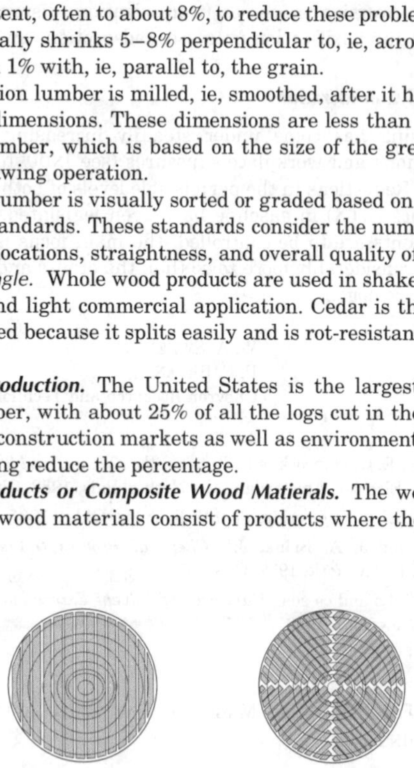

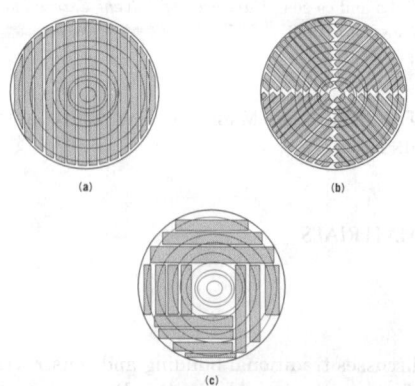

Figure 1. Methods of sawing logs into slabs, where **(a)** is through and through, **(b)** is quartersawn, and **(c)** is sawing around.

Bituminous Products

Bitumen describes a black or dark brown masticlike material that is thermoplastic in nature and softens upon heating. The sources of bitumen are petroleum or coal deposits. The natural product is commonly called gilsonite or pitch, a mineral formed by an old weathered petroleum flow at the surface of the earth that has left behind the larger molecules from the petroleum. A principal source in the past has been Lake Trinidad, a 445,000 m^2 deposit on the island of Trinidad. Bitumen from petroleum or crude oil is called asphalt (qv). It is the material left behind after all the valuable compounds, eg, gasolines, have been distilled out of the crude oil. The amount and quality of asphalt is dependent on the source of the crude oil used in the refining process. Some crude oils have a higher content of asphaltic bitumen left after the distillation process. Bitumen from coal is coal-tar pitch. It remains after the valuable coal oils and tars have been distilled out of the coal tars produced by distractive distillation. Most industrial applications for bitumen products use asphalt or coal-tar pitch because the supply is more uniform and plentiful.

As a construction material, bitumens have principal applications in paving and waterproofing. About 75% of all bitumens are used in paving, and about 20% are used in waterproofing. Of the waterproofing usage, 75% is in roofing applications. Construction applications for natural bitumens are nonexistent and coal-tar application, are less than 20%, with asphalt being the more important material. The use of coal tar is declining because of its listing as a human carcinogen by the U.S. Government.

Asphalt is not an exact chemical composition but rather a mixture of organic compounds whose nature is dependent on many items (see ASPHALT).

Gypsum Products

Gypsum, $CaSO_4 \cdot 2H_2O$, is a naturally occurring mineral found mainly in the western United States and eastern Canada (see CALCIUM COMPOUNDS–CALCIUM SULFATE). The purer deposits require only minimal beneficiation to get a product pure enough for commercial applications. Other deposits require cleaning to remove clay and other impurities.

The principal use of gypsum in construction is as a wall finishing material. The manufacturing process requires that the gypsum be partially dehydrated to the hemihydrate, $CaSO_4 \cdot \frac{1}{2}H_2O$, commonly known as plaster of paris. When mixed with water the hemihydrate dissolves and the gypsum precipitates out as the interlocking crystals that make gypsum a hard monolithic material. The reaction is rapid, exothermic, and autocatalyzed by gypsum itself. For some applications the reaction is slowed by the addition of protein materials to the hemihydrate–water mixture.

In the United States, over 90% of the gypsum products sold are as gypsum board. Gypsum board is used as an interior wall surfacing in both residential and nonresidential construction and is referred to as drywall to differentiate it from the older wet plaster walls. The board is composed of a core of gypsum attached to a facing of heavy paper and is attached to the framing members using either nails or screws made especially for installing drywall. The joints between the board are cosmeticly treated with a reinforcing tape and joint compound composed of mineral fillers and a thermoplastic resin emulsion; some flow control and viscosity modifiers may be added to help get an easy, smooth-applying system. The reinforcing tape, either a

5-cm wide paper or fiber mesh, is embedded into the joint compound, allowed to dry, and then topped with a second or third coat of joint compound to obtain a smooth surface. To help in the taping process most board is produced with a tapered edge on the long edges.

The standard sized sheets are four ft (1.22 m) wide and from 8 to 16 ft long. Sheets are available in thicknesses from $\frac{1}{4}$ up to 1 in. (0.63 to 2.5 cm), but the most common thicknesses are $\frac{1}{2}$ and $\frac{5}{8}$ in. (1.3 and 1.6 cm). There are special products for bathrooms, such as moisture-resistant board (3% of market) and sheathing applications (2% of market), but the vast majority of the product sold is either standard board (60% of market) or the fire-rated Type X board (29% of market). Type X (1.6 cm thick) is the basis of most of the one-hour fire walls in buildings built since the 1960s.

Portland Cement

Portland cement is the most widely used construction material in the world (see CEMENT), especially in Third World nations, because of its availability, ease of use, and versatility.

Portland cement is classified as a hydraulic cement, ie, it sets or cures in the presence of water. Portland cement is not an exact composition but rather a range of compositions, which obtain the desired final properties. The compounds that make up Portland cements are calcium silicates, calcium aluminates, and calcium aluminoferrites (see CALCIUM COMPOUNDS).

Portland cement is not useful by itself to the construction industry. Its value is in the resultant concrete in which it is used as a binder. Concrete is a mixture of smaller particles that coalesce into a solid mass, and the typical particles in a Portland cement concrete are aggregates of rock and sand. In general, the larger the aggregate size the stronger the Portland cement concrete. The crush and shear resistance of the aggregate, and the ratio of Portland cement to sand and aggregate are important to the strength of Portland cement concretes, with higher Portland cement levels yielding stronger concrete.

Applications. The primary market for Portland cement is in building construction, which historically takes approximately 65% of all Portland cement sold. Forty-five percent of the construction market goes for residential construction and 38% goes for commercial construction. These markets use Portland cement for concrete foundations and structural and precast products.

Public works applications take just under 32% of the total Portland cement market, streets and highways represent 68% of this usage, and water and waste account for 23%.

Bricks

Bricks are the oldest manufactured building material in use. Sun-dried bricks were manufactured as early as 6000 BC, and fired bricks were used during the Middle Ages. Today's bricks differ very little except in the efficiency of manufacture; they are still made from clay or shale, a clay-based sedimentary rock that is kiln-fired.

Manufacturing Process. There are three processes used to make bricks. All three start with clay that has been milled and screened to remove coarse materials and impurities. The clay used must have enough plasticity when mixed with water to allow molding and have enough wet and dry tensile strength to maintain the brick shape after forming. Water is added to the prepared clay in an amount appropriate for the brickmaking process that will be used. The clay and water mixture is then kneaded with rotating knives to form a plastic mass, which is molded into a brick shape by one of the three processes. The stiff-mud process is the most common method in use. The soft-mud process is used to make handmade brick. The dry-press process is used to make good quality face brick, ie, brick used on exposed walls.

Glass

The products called glass are in the glassy state and are made mainly from silica (silicon oxide, SiO_2). Sodium oxide, Na_2O, and calcium oxide, CaO, are commonly added, up to 30% or more in combination. Sodium oxide is added to reduce the viscosity or melting temperature of the silica, whereas calcium oxide gives the glass durability against water attack. Glasses can contain a wide range of oxides, depending on the application and properties needed of the glass. Examples of other oxides include the oxides of boron, aluminum, potassium, magnesium, lead, barium, zinc, and lithium.

Sand is the common silica source, but the sand must have only minor impurities in order to make good quality, clear glass. A common source for sodium oxide is sodium carbonate, Na_2CO_3, from soda ash. Upon heating in the glass furnace it forms the oxide. Calcium oxide comes from limestone, which when burnt becomes lime, CaO. Alumina commonly comes from feldspar, which is an alumina silicate rock. Boron oxide is obtained from borax, sodium tetraborate, $Na_2B_4O_7 \cdot 10H_2O$. Other oxides generally are not used in large amounts or at all in glass construction materials.

Uses of glass in construction products fit into three categories: flat glass (window glass); fibrous glass; and specialty glass products. Each is made by different processes and has different applications.

Flat Glass. Flat glass is widely used in applications other than construction, such as automotive glass, etc, but approximately 57% of the flat glass is used in construction applications, mainly as windows.

Fibrous Glass. Fibrous glass is manufactured in two different forms, very fine intermingled fibers called insulation fibrous glass for insulation and fine but coarser fibers called continuous or textile fibers for reinforcement and other textile applications. Both products have construction-related applications.

Specialty Glass Products. Foam glass insulation and glass building blocks are two specialty glass products with construction applications.

THEODORE MICHELSEN
Schuller International

R. B. Hoadley, *Understanding Wood,* The Taunton Press, Inc., Newtown, Conn., 1980.

U.S. Cement Industry Fact Sheet, 9th ed., Portland Cement Association, Skokie, Ill., 1991.

1990 U.S. Industrial Outlook, Dept. of Commerce, International Trade Association, Washington, D.C., 1991, pp. 5-1–7-14.

Current Industrial Reports. Glass Fiber, MA32J(88)-1, Dept. of Commerce, Bureau of the Census, Washington, D.C., 1989.

PLASTIC

Throughout the 1990s, plastics consumption in building materials is forecast to grow at a 4–5% average annual rate. Contributing to this growth will be advances in do-it-yourself and professional remodeling products employing plastics, overall growth in the renovation and home remodeling markets, plus the increased penetration of plastics at the expense of traditional materials in building products because of their superior strength in weight performance, corrosion resistance, environmental stability, lower cost, insulation properties, and ability to fabricate complex designs into a single part, ie, low labor assembly intensity. On the other hand, the overall growth of plastics in building materials will be somewhat limited by concerns over their flammability and smoke toxicity, public perception of their negative environmental impact, resistance by the conservative construction industry in adopting new materials, and competitive actions by producers of traditional building materials, such as wood, concrete, glass, and metal, seeking to defend their market positions.

Polymers and Properties

The physical properties of plastics that are important in building materials are the glass-transition or melt temperature, ease of processing as indicated by the temperatures and pressures needed for molding, heat deflection temperature, uv stability, tensile and impact strength, oxidative degradation, creep set, fatigue, and elongation.

Table 1. Physical Properties of Selected Plastic Building Material

Properties of plastic[a]	LDPE	LLDPE	HDPE	PP	PVC (flexible)	PS	ABS	Polyacrylic (glazing)	Polycarbonate (glazing)	Epoxy (mineral filled)	Acetal homopolymer
melting point, °C	96–115	122–124	130–137	160–175							175–181
glass-transition temp T_g, °C					75–105	74–105	110–125	90–105	140–150		
injection molding temperature, °C	150–232	177–260	176–274	204–288	160–196	177–260	193–260	163–260	270–295		193–243
injection molding pressure, MPa[b]	34–103	34–103	83–103	69–138	7–14	34–138	55–172	34–138	55–140		69–136
tensile strength, MPa[b]	8–31	13–26	22–31	31–41	7–24	35–52	22–55	46–76	65	48–90	67
elongation, %	100–650	100–965	10–1200	500–600	200–400	1–5	2–25	2–10	110	1–3	25–75
Izod impact strength, J/m[c]	no break	no break	21–213	21–64	varies widely	18–24	74–640	16–32	750	16–24	64–123
heat-deflection temperature, °C	40–44		80–91	107–121		66–96	96–118	80–107	138		

Properties of laminates	Melamine	Phenolic, woodbase	Polyester, glass-filled
laminating temperature, °C	145–165	150–160	RT-150
laminating press, MPa[b]	3.5–12	7–14	0–3.5
tensile strength, MPa[b]	69–172	110–220	69–172
Izod impact strength, J/m[c]	16–80	214–427	240–1500
heat-deflection temperature, °C	105	90	

Properties of foams	Polyurethane	Polyisocyanurate	PS board	PS expandable beads
density, kg/m^3	14–42	24–56	24–80	13–15
maximum service temperature, °C	80–170	150	75–80	75–80
thermal conductivity, W/(mK)	0.0125–0.034	0.012–0.02	0.023–0.034	0.03
fire resistance[d]	HF-1		HF-1	HF-1

[a] LDPE = low density polyethylene; LLDPE = linear low density polyethylene; HDPE = high density polyethylene; PP = polypropylene; PVC = poly(vinyl chloride); PS = polystyrene; ABS = poly(acrylonitrile-butadiene-styrene).

[b] To convert MPa to psi, multiply by 145.

[c] To convert J/m to ft·lbf/in., divide by 53.38 (see ASTM D256).

[d] HF-1 = UL Standard 94 for Foam Plastics.

Density, thermal conductivity, and fire resistance are important for foams. The physical properties of the most important plastics building materials are summarized in Table 1.

Applications

Phenolics are consumed at roughly half the volume of PVC, and all other plastics are consumed in low volume quantities, mostly in single application niches, unlike workhorse resins such as PVC, phenolic, urea–melamine, and polyurethane. More expensive engineering resins have a very limited role in the building materials sector except where specific value-added properties for a premium are justified. Except for the potential role of recycled engineering plastics in certain applications, the competitive nature of this market and the emphasis placed on end use economics indicates that commodity plastics will continue to dominate in consumption. The most dynamic growth among important sector resins has been seen with phenolic, acrylic, polyurethane, LLDPE/LDPE, PVC, and polystyrene.

Over 60% of the total plastics volume for building materials is consumed for pipes, fittings, conduit, and wood bonding applications. Other important applications include solar heating, glazing, exterior trim, door and window frames, insulation, panels and siding, and flooring (additional 25%). Ten-year growth has been greatest in profile extrusions, panels and siding, flooring, lighting fixtures, and wood bonding applications.

Toxicity

Indoor Air Quality. With regard to plastics, urea–formaldehyde insulating foams have received the greatest publicity. In the early to mid-1980s, they were studied for their formaldehyde release, ie, outgassing. The U.S. EPA has set an acceptable level of formaldehyde within indoor air at 0.1 ppm. Another potential source of formaldehyde release in buildings is from the binder systems, such as phenol–formaldehyde and urea–formaldehyde, used in pressed wood products such as particle board.

Smoke Toxicity of Burning Plastics. Smoke, not flames, is the primary cause of death in most fires. However, past efforts to determine which building product components, eg, wood, fabric, plastic, etc, generate the most harmful or toxic smoke emissions have proven inconclusive.

Additive Toxicity. Plastic building products almost always incorporate additive such as colorants, plasticizers, uv light stabilizers, and flame retardants. Several families of these various additives are under increasing regulatory scrutiny for their potential risks with regard to smoke toxicity and their effects upon worker health in compounding and processing plants.

Engineering and Recycled Plastics

The plastic building materials industry is facing rapid change and the accelerated use of engineering and recycled plastics is one catalyst for this change. Engineering thermoplastics have better mechanical properties than commodity plastics. Thermoplastics can be combined with fibers, fillers, and traditional building materials like concrete or wood, creating composite materials with special performance features such as durability, fireproofing, and stress resistance. Plastic products account for less than 10% of the materials used to make a typical U.S. home. Despite this limited market penetration, building and construction uses are the second largest market (behind packaging) for plastic resin. Resin demand for building and construction is expected to double by the year 2000, to about 14×10^6 metric tons.

Emerging Developments

A very high price and performance family of polymers called liquid crystal polymers (LCPs) exhibit extremely high mechanical and thermal properties. As their ease of processing and price improve, they may find application in thin-wall, high strength parts such as nails, bolts, and fasteners where metal parts cannot be used for reasons of conductivity, electromagnetic characteristics, or susceptibility to corrosion.

Thermoset polyurethane as a binder material for gravel systems is also under development. Applications could include roofing systems that require a high degree of uv light and abrasion resistance.

Certain state highway authorities are studying the use of fiber-reinforced polymers, typically thermosets such as epoxy or unsaturated polyester, for bridge construction. On an even more futuristic scale, fiber optics that employ polymeric jacketing and, in some cases, optically active polymeric cores, may someday be employed in place of wires for home security systems, climate control, etc.

JAMES A. FINNEGAN
BRIAN P. GERSH
JULIE B. LANG
RONALD LEVY
Arthur D. Little, Inc.

R. Juran, ed., *Modern Plastics Encyclopedia,* Vol. 66, McGraw-Hill Book Co., Inc., New York, 1990, pp. 4–138.

F. W. Billmeyer, *Textbook of Polymer Science,* 2nd ed., John Wiley & Sons, Inc., New York, 1971, Chapt. 8–10.

H. S. Kaufman and J. U. Falcetta, *Introduction to Polymer Science and Technology,* John Wiley & Sons, Inc., New York, 1977, Chapt. 2.

Modern Plastics Encyclopedia, Vol. 56, No. 10A, McGraw-Hill Book Co., Inc., New York, 1979–1980, pp. 4–111, 132, 147, 292–300, 498–521, 528, 626.

BURNER TECHNOLOGY. See COMBUSTION SCIENCE AND TECHNOLOGY.

BUTADIENE

Butadiene, C_4H_6, exists in two isomeric forms: 1,3-butadiene $CH_2{=}CH{-}CH{=}CH_2$, and 1,2-butadiene, $CH_2{=}C{=}CH{-}C_3$. 1,3-Butadiene is a commodity product of the petrochemical industry. Elastomers consume the bulk of 1,3-butadiene, led by the manufacture of styrene–butadiene rubber (SBR). 1,3-Butadiene is manufactured primarily as a coproduct of steam cracking to produce ethylene in the United States, Western Europe, and Japan. However, in certain parts of the world, eg, China, India, Poland, and the former Soviet Union, it is still produced from ethanol.

The other isomer, 1,2-butadiene, a small by-product in 1,3-butadiene production, has no significant current commercial interests. However, there are a number of publications and patents on its recovery and applications, particularly in the specialty polymer area and as a gel inhibitor.

Properties

1,3-Butadiene is a noncorrosive, colorless, flammable gas at room temperature and atmospheric pressure. It has a mildly aromatic odor. It is sparingly soluble in water, slightly soluble in methanol and ethanol, and soluble in organic solvents like diethyl ether, benzene, and carbon tetrachloride. Its important physical properties are summarized in Table 1. 1,2-Butadiene is much less studied. It is a flammable gas at ambient conditions. Some of its properties are summarized in Table 2.

Reactions

The conjugated double bonds of 1,3-butadiene allow a large number of reactions at both the 1,2- and 1,4-positions. These reactions include addition, hydrogenation, oxidation, Diels-Alder, oligomerization, and polymerization reactions. Many of these reactions produce large volumes of important industrial materials.

Table 1. Physical Properties of 1,3-Butadiene

Property	Value
molecular formula	C_4H_6
molecular weight	54.092
boiling point at 101.325 kPa[a], °C	−4.411
freezing point, °C	−108.902
critical temperature, °C	152.0
critical pressure, MPa[b]	4.32
critical volume, cm³/mol	221
critical density, g/mL	0.245
density (liquid), g/mL at	
0°C	0.6452
20°C	0.6211
50°C	0.5818
heat capacity at 25°C, J/(mol·K)[c]	79.538
refractive index, n_D at −25°C	1.4292
solubility in water at 25°C, ppm	735[d]
viscosity (liquid), mPa·s(= cP) at	
−40°C	0.33
40°C	0.20
heat of formation, gas, kJ/mol[c]	110.165
heat of formation, liquid, kJ/mol[c]	88.7
free energy of formation, kJ/mol[c]	150.66
heat of vaporization, J/g[c] at 25°C	389
boiling point	418
flash point, °C	−85
autoignition temperature, °C	417.8

[a] To convert kPa to mm Hg, multiply by 7.5.
[b] To convert MPa to psi, multiply by 145.
[c] To convert J to cal, divide by 4.184.
[d] 245 mol ppm.

Table 2. Physical Properties of 1,2-Butadiene

Property	Value
molecular formula	C_4H_6
molecular weight	54.092
boiling point at 101.325 kPa,[a] °C	10.85
freezing point, °C	−136.19
density (liquid), g/mL at	
0°C	0.676
20°C	0.652
heat of formation at 25°C (gas), kJ/mol[b]	162.21
heat of vaporization at 25°C, kJ/mol[b]	23.426
refractive index at 1.3°C	1.4205

[a] To convert kPa to mm Hg, multiply by 7.5.
[b] To convert kJ to kcal, divide by 4.184.

Handling, Storage, and Shipping

Butadiene reacts with a large number of chemicals, has an inherent tendency to dimerize and polymerize, and is toxic. Therefore, specific handling, storage, and shipping procedures must be followed.

Butadiene is primarily shipped in pressurized containers via railroads or tankers. U.S. shipments of butadiene, which is classified as a flammable compressed gas, are regulated by the Department of Transportation. Most other countries have adopted their own regulations. Other information on the handling of butadiene is also available. As a result of the extensive emphasis on proper and timely responses to chemical spills, a comprehensive handbook from the National Fire Protection Association is also available.

Health, Safety, and Toxicology

Short-term exposure to high concentrations of butadiene may cause irritation to the eyes, nose, and throat. Dermatitis and frostbite may result from exposure to the liquid and the evaporating gas. Long-term physiological reactions to 1,3-butadiene may vary individually.

In several epidemiological studies, elevation in mortality was observed for small subgroups and tumor types. Interpretation of these findings is still incomplete. Based on the assessment of these studies, the American Conference of Governmental Industrial Hygienists (ACGIH) adopted a TLV of 10 ppm for 1,3-butadiene in 1982. Subsequently, in 1989 OSHA proposed a 10-ppm, 15-min exposure level and a 2-ppm TLV. There have been many reviews published on the toxicity of butadiene.

H. N. Sun
J. P. Wristers
Exxon Chemical Company

H. J. Müller and E. Löser, in W. Gerhartz and co-eds., *Ullmann's Encyclopedia of Industrial Chemistry*, Vol. A4, 5th rev. ed., VCH, Weinheim, Germany, 1985, pp. 431–446.

A. K. K. Lee and A. M. Aitani, *Oil Gas J.*, 60 (Sept. 10, 1990), and references therein. *Hazardous Materials Response Handbook*, National Fire Protection Association, 1989.

BUTENE POLYMERS. See Olefin polymers.

BUTYL ALCOHOLS

Butyl alcohols encompass the four structurally isomeric 4-carbon alcohols of empirical formula $C_4H_{10}O$. One of these, 2-butanol, can exist in either the optically active $R(-)$ or $S(+)$ configuration or as a racemic (±) mixture.

Physical and Chemical Properties

The butanols are all colorless, clear liquids at room temperature and atmospheric pressure with the exception of *t*-butyl alcohol which is a low melting solid (mp 25.82°C); it also has a substantially higher water miscibility than the other three alcohols. Physical constants of the four butyl alcohols are given in Table 1.

Physical constants for the optically pure stereoisomers of 2-butanol have been reported as follows:

	d^t_4	n_D^{20}	$[\alpha]^{27}_D$
(*S*)-(+)-2-butanol	0.8025^{27}	1.3954	+13.52
(*R*)-(−)-2-butanol	0.8042^{25}	1.3970	−13.52

Butyl alcohol liquid vapor pressure–temperature responses, which are important parameters in direct solvent applications, are presented in Figure 1.

The butanols undergo the typical reactions of the simple lower chain aliphatic alcohols.

Manufacture

The principal commercial source of 1-butanol is *n*-butyraldehyde, obtained from the Oxo reaction of propylene.

Uses

The largest-volume commercial derivatives of 1-butanol are *n*-butyl acrylate and methacrylate. These are used principally in emulsion polymers for latex paints, in textile applications and in impact modifiers for rigid poly(vinyl chloride).

Isobutyl alcohol has replaced *n*-butyl alcohol in some applications where the branched alcohol appears to have preferred properties and structure.

Health, Safety, and Environmental Considerations

All four butanols are thought to have a generally low order of human toxicity. However, large dosages of the butanols generally serve as central nervous system depressants and mucous membrane irritants.

Table 1. Physical Properties of the Butyl Alcohols (Butanols)

Common Name	*n*-Butyl alcohol	Isobutyl alcohol	*sec*-Butyl alcohol	*t*-Butyl alcohol
systematic name	1-butanol	2-methyl-1-propanol	2-butanol	2-methyl-2-propanol
formula	$CH_3(CH_2)_3OH$	$(CH_3)_2CHCH_2OH$	$CH_3CH(OH)C_2H_5$	$(CH_3)_3COH$
critical temperature, °C	289.90	274.63	262.90	233.06
critical pressure, kPa[a]	4423	4300	4179	3973
normal boiling point, °C	117.66	107.66	99.55	82.42
melting point, °C	−89.3	−108.0	−114.7	25.82
heat of fusion, kJ/mol[b]	9.372	6.322	5.971	6.703
heat of vaporization at normal boiling point, kJ/g[b]	43.29	41.83	40.75	39.07
liquid density, kg/m^3 at 25°C	809.7	801.6	806.9	786.6[c]
refractive index at 25°C	1.3971	1.3938	1.3949	1.3852
flash point, closed cup, °C	28.85	27.85	23.85	11.11
dielectric constant, ϵ	17.5^{25}	17.93^{25}	16.56^{20}	12.47^{30}
solubility in water at 30°C, % by weight	7.85	8.58	19.41	miscible
solubility of water in alcohol at 30°C, % by weight	20.06	16.36	36.19	miscible

[a] To convert kPa to mm Hg, multiply by 7.50.
[b] To convert kJ to kcal, divide by 4.184.
[c] For the subcooled liquid below melting point.

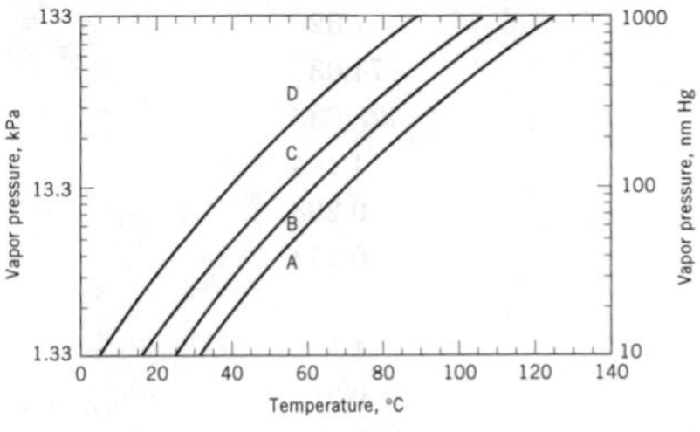

Figure 1. Vapor pressure of butyl alcohols: A, *n*-butyl; B, isobutyl; C, *sec*-butyl; D, *t*-butyl. To convert kPa to mm Hg, multiply by 7.5.

All four butanols are registered in the United States on the Environmental Protection Agency Toxic Substances Control Act (TSCA) Inventory, a prerequisite for the manufacture or importation for commercial sale of any chemical substance or mixture in quantities greater than 454 kg (1000 lbs.). Additionally, the manufacture and distribution of the butanols in the United States are regulated under the Superfund Amendments and Reauthorization Act (SARA), Section 313, which requires that anyone handling at least 4545 kg (10,000 lbs.) a year of a chemical substance report to both the U.S. EPA and the state any release of that substance to the environment.

Storage and Handling

The C-4 alcohols are preferably stored in baked phenolic-lined steel tanks. However, plain steel tanks can also be employed provided a fine-porosity filler is installed to remove any contaminating rust.

Storage under dry nitrogen is also recommended since it limits flammability hazards as well as minimizing water pickup.

ERNST BILLIG
Union Carbide Chemicals
and Plastics Company Inc.

J. B. Cropley, L. M. Burgess, and R. A. Loke, *Chemtech.* **14**, 374–380 (1984).

Chemical Economics Handbook, SRI International, Menlo Park, Calif.

F. E. C. George and D. Clayton, eds., *Patty's Industrial Hygiene and Toxicology,* Vol. 2C, John Wiley & Sons, Inc., New York, 1982, pp. 4571–4586.

BUTYLENES

Butylenes are C_4H_8 mono-olefin isomers: 1-butene, *cis*-2-butene, *trans*-2-butene, and isobutylene (2-methylpropene). These isomers are usually coproduced as a mixture and are commonly referred to as the C-4 fraction. These C-4 fractions are usually obtained as by-products from refinery and petrochemical complexes that crack petroleum fractions and natural gas liquids. Since the C-4 fractions almost always contain butanes, it is also known as the B–B stream. The linear isomers are referred to as butenes.

Physical Properties

For any industrial process involving vapors and liquids, the most important physical property is the vapor pressure. Table 1 presents values for the constants for a vapor-pressure equation and the temperature range over which the equation is valid for each butylene.

Table 2 presents other important physical properties for the butylenes. Thermodynamic and transport properties can also be obtained from other sources.

Chemical Properties

The carbon–carbon double bond is the distinguishing feature of the butylenes and, as such, controls their chemistry. The carbon–carbon bond, acting as a substitute, affects the reactivity of the carbon atoms at the alpha positions through the formation of the allylic resonance structure. This structure can stabilize both positive and negative charges. Thus allylic carbons are more reactive to substitution and addition reactions than alkane carbons. Therefore, reactions of butylenes can be divided into two broad categories: (*1*) those that take place at the double bond itself, destroying the double bond; and (*2*) those that take place at alpha carbons.

The electron-rich carbon–carbon double bond reacts with reagents that are deficient in electrons, eg, with electrophilic reagents in electrophilic addition, free radicals in free-radical addition, and under acidic conditions with another butylene (cation) in dimerization.

Manufacture

The C-4 isomers are almost always produced commercially as by-products in a petroleum refiner–petrochemical process.

There are other commercial processes available for the production of butylenes. However, these are site- or manufacturer-specific, eg, the Oxirane process for the production of propylene oxide; the disproportionation of higher olefins; and the oligomerization of ethylene. Any of these processes can become an important source in the future. More recently, the Coastal Isobutane process began commercialization to produce isobutylene from butanes for meeting the expected demand for methyl-*tert*-butyl ether (MTBE).

New Technology. Several technologies are emerging for the production of isobutylene to meet the expected demand for isobutylene:

Table 1. Vapor-Pressure Equation Constants for the Butanes, Butylenes, and Butadienes[a]

	A	B	C	D	N	Temperature range, K
n-butane	61.5623	−4259.90	−6.20315	3.07575×10^{-7}	2.5	135–423
isobutane	66.7163	−4237.62	−7.08156	4.00506×10^{-7}	2.5	129–408
1-butene	78.8760	−4713.65	−9.05743	1.28654×10^{-5}	2.0	126–416
cis-2-butene	71.9534	−4681.34	−7.87527	1.00237×10^{-5}	2.0	203–358
trans-2-butene	74.3950	−4648.45	−8.33977	1.20897×10^{-5}	2.0	195–358
isobutylene	83.8683	−4822.95	−9.90214	1.51060×10^{-5}	2.0	194–359
1,2-butadiene	49.49.5031	−4021.95	−4.28893	5.13547×10^{-6}	2.0	200–284
1,3-butadiene	73.0016	−4547.77	−8.11105	1.14037×10^{-5}	2.0	164–425

[a] $\ln P = A + B/T + C^{*}\ln T + D^{*}T^{**}N$ where P is in Pa and T is in K.

Table 2. Physical Properties of the Butylenes

Property	Values			
	1-Butene	*cis*-2-Butene	*trans*-2-Butene	Isobutylene
molecular weight	56.11	56.11	56.11	56.11
melting point, K	87.80	134.23	167.62	132.79
boiling point, K	266.89	276.87	274.03	266.25
critical temperature, K	419.60	435.58	428.63	417.91
critical pressure, MPa[a]	4.023	4.205	4.104	4.000
critical volume, L/mol	0.240	0.234	0.238	0.239
critical compressibility factor	0.277	0.272	0.274	0.275
Flammability limits, vol % in air				
lower limit	1.6	1.6	1.8	1.8
upper limit	9.3	9.7	9.7	8.8
autoignition temperature, K	657	598	597	738

[a] To convert MPa to atm, multiply by 9.869.

(*1*) deep catalytic cracking; (*2*) superflex catalytic cracking; (*3*) dehydrogenation of butanes; and (*4*) the Coastal process of thermal dehydrogenation of butanes.

Separation and Purification of C-4 Isomers. 1-Butene and isobutylene cannot be economically separated into pure components by conventional distillation because they are close boiling isomers. 2-Butene can be separated from the other two isomers by simple distillation. There are four types of separation methods available: (*1*) selective removal of isobutylene by polymerization and separation of 1-butene; (*2*) use of addition reactions with alcohol, acids, or water to selectively produce pure isobutylene and 1-butene; (*3*) selective extraction of isobutylene with a liquid solvent, usually an acid; and (*4*) physical separation of isobutylene from 1-butene by absorbents.

There are three important processes for the production of isobutylene; (*1*) the extraction process using an acid to separate isobutylene; (*2*) the dehydration of *tert*-butyl alcohol, formed in the Arco's Oxirane process; and (*3*) the cracking of MTBE.

Handling and Analysis

Storage and Transportation. Handling requirements are similar to liquefied petroleum gas (LPG). Storage conditions are much milder. Butylenes are stored as liquids at temperatures ranging from 0 to 40°C and at pressures from 100 to 400 kPa (1–4 atm). Their transportation is also similar to LPG; they are shipped in tank cars, transported in pipelines, or barged.

Health and Safety

Butylenes are not toxic. The effect of long-term exposure is not known; hence, they should be handled with care. They are volatile and asphyxiants. Care should be taken to avoid spills because they are extremely flammable. Physical handling requires adequate ventilation to prevent high concentrations of butylenes in the air.

Table 3. Butylene Consumption, % of Supply

Use	United States	Western Europe	Japan
Fuel			
alkylate	85.6	22.1	2.0
methyl *tert*-butyl ether	4.5	8.2	0.0
polygas, LPG, blending	3.4	50.3	69.2
Chemicals			
sec-butyl alcohol/MEK	1.5	5.2	7.5
polyethylene copolymer	0.9	0.6	1.6
heptene, octene	0.7	3.4	1.3
butadiene	0.1	0.0	0.0
maleic anhydride	0.0	0.0	1.6
polybutene-polyisobutylenes	1.9	3.1	2.5
butyl rubbers	1.0	3.7	4.8
di- and triisobutylenes	~ 0.0	2.1	2.7
methyl methacrylate	0.0	0.0	4.4
other	0.4	1.3	2.4

Commercial Utilization

Pricing of butylenes determines the end use of butylenes in different geographic areas. The use pattern of butylenes in the United States, Europe, and Japan is shown in Table 3 as a percentage of supply in 1984.

Narasimhan Calamur
Martin E. Carrera
Richard A. Wilsak
Amoco Corporation

DIPPR, Project 801, Data Compilation (July 1990).

R. T. Morrison and R. N. Boyd, *Organic Chemistry,* 4th ed., Allyn and Bacon, Boston, Mass., 1983; D. E. Dorman, M. Jantelot, and J. D. Roberts, *J. Org. Chem.* **36,** 2157 (1971); I. Hirana, O. Kikuchi, and K. Suzuki, *Bull. Chem. Soc. Jpn.* **49,** 3321 (1976).

K. Verscheren, ed., *Handbook of Environmental Data on Organic Chemicals,* Van Nostrand Reinhold Co., New York, 1983, pp. 304, 317.

G. D. Hobson and W. Pohl, *Modern Petroleum Technology,* 5th ed., John Wiley & Sons, Inc., New York, 1984, p. 517.

BUTYL RUBBER. See ELASTOMERS, SYNTHETIC.

BUTYRALDEHYDES

The two isomeric butanals, *n*- and isobutyraldehyde, C_4H_8O, are produced commercially almost exclusively by the Oxo reaction of propylene. They also occur naturally in trace amounts in tea leaves, certain oils, coffee aroma, and tobacco smoke.

Physical Properties

The butanals are highly flammable, colorless liquids of pungent odor. Their physical properties are shown in Table 1.

These aldehydes are miscible with most organic solvents, eg, acetone, ether, ethanol, and toluene, but are only slightly soluble in water.

Table 1. Physical Properties of C-4 Aldehydes

	n-Butyraldehyde	Isobutyraldehyde
formula	$CH_3CH_2CH_2CHO$	$(CH_3)_2CHCHO$
systematic name	butanal	2-methylpropanal
critical temperature, °C	263.95	233.85
critical pressure, kPa[a]	4000	4100
critical specific volume, $m^3/(kg \cdot mol)$[b]	0.258	0.263
melting point, °C	−96.4	−65.0
normal boiling point, °C	74.8	64.1
coefficient of expansion at 20°C	0.00114	
refractive index at 25°C	1.3766	1.3698
liquid density at 20°C, kg/m^3[c]	801.6	789.1
heat of vaporization at normal boiling point, kJ/mol[d]	30.72	31.23
heat of fusion, kJ/mol[d]	11.1	12.0
dielectric constant ϵ at °C	13.426	
solubility parameter at 25°C, $(MJ/m^3)^{0.5}$[e]	18.666	18.446

[a] To convert kPa to mm Hg, multiply by 7.50.
[b] To convert m^3/kg·mol to mL/mol, multiply by 1000.
[c] To convert kg/m^3 to g/mL, divide by 1000.
[d] To convert kJ to kcal, divide by 4.184.
[e] To convert $(MJ/m^3)^{0.5}$ to $(cal/cc)^{0.5}$, multiply by $0.239^{0.5}$.

Reactions

The reactions of *n*- and isobutyraldehyde are characteristic aldehyde reactions of oxidation, reduction, and condensation.

Derivatives and Uses

The majority (92% in 1988) of the butyraldehyde produced in the United States is converted into 1-butanol and 2-ethylhexanol (2-EH). The remaining (8%) *n*-butyraldehyde production of the United States goes into (in decreasing order): poly(vinyl butyral), 2-ethylhexanal, trimethylolpropane, methyl amyl ketone, and butyric acid.

About 69% of the total 1988 U.S. consumption of isobutyraldehyde went into the production of isobutyl alcohol and isobutyraldehyde condensation and esterification products. The other principal isobutyraldehyde derivative markets (as a percentage of total 1988 U.S. isobutyraldehyde consumption) are neopentyl glycol (15%); isobutyl acetate (6%); isobutyric acid (5%); isobutylidene diurea (2.5); and methyl isoamyl ketone (1.7%).

2,2,4-Trimethyl-1,3-pentanediol (TMPD), the hydrogenated aldol condensation product of isobutyraldehyde, is a modifying agent in alkyd resins, high solids coatings, and moisture-set inks.

Economic Aspects

The merchant market for the two aldehydes is relatively insignificant, most of the production being employed captively. The principal U.S. producers of butanals are given in Table 2.

The overall growth for *n*-butyraldehyde depends primarily on *n*-butanol and 2-ethylhexanol. 2-Ethylhexanol is expected to face competition from other alcohols, eg, isodecyl alcohol, as well as from newer production sources.

n-Butanol is the highest volume derivative of *n*-butyraldehyde in the United States with nearly twice the production of 2-EH (56% vs. 36.5%). In sharp contrast, in Western Europe, Japan, and all other countries producing butyraldehydes, 2-EH is dominant.

The most active *n*-butyraldehyde derivatives are expected to be PVB, as more regions require automotive safety glass, and trimethylolpropane.

Low pressure rhodium processes which give higher *n*:iso butyraldehyde ratios (eg, 10:1) have gradually replaced cobalt processes, dramatically affecting the isobutyraldehyde supply.

Health, Safety, and Environmental Factors

Although tests have shown that *n*-butyraldehyde causes some adverse physiological effects, there is no danger to health in normal plant practice. No threshold limit value has been assigned for either butyraldehyde or isobutyraldehyde. Both aldehydes, however, have a pungent, penetrating odor. Their vapors as well as the neat liquids can cause skin, eye, and respiratory organ irritation possibly because of rapid oxidation to the acids on contact with air. Because of the ease of oxidation of the butanals to the corresponding butyric acids, precautions associated with these carboxylic acids must also be noted.

The biological oxygen demand (BOD) in aqueous streams for both butanals is 1.62 wt/wt for five days. The NFPA Hazard classification for both aldehydes are health (blue) 2; flammability (red) 3; and reactivity (yellow) 0.

Both butanals are on the United States Toxic Substances Control Act (TSCA) Inventory, a prerequisite for the manufacture or importation for commercial sale of any chemical substance or mixture in quantities greater than one thousand pounds (455 kg). Additionally, the manufacture and distribution of the butanals in the United States are regulated under the Superfund Amendments and Reauthorization Act (SARA), Section 313, which requires that anyone handling at least ten thousand pounds (4550 kg) a year of a chemical substance report to both the EPA and the state any release of that substance to the environment.

Storage and Handling

Stainless steel, baked phenolic-lined steel, or aluminum are often used for storage and handling of *n*- and isobutyraldehyde. Storage of the aldehydes under nitrogen preserves the integrity of the material. There is some evidence that water stabilizes aldehydes against certain types of exothermic condensation reactions, possibly by precipitating any soluble iron species as hydrous iron oxides.

Table 2. U.S. Producers of Butanals

Plant and location	Capacity, 10^3 t		Catalyst
	n-Butyraldehyde	Isobutyraldehyde	
Aristech Chemical Corp. Pasadena, Texas	114	11.4	Rh
BASF Corp. Freeport, Texas	99	19.5	Rh
Eastman Kodak Co. Longview, Texas	284	166	Co[a]
Hoechst Celanese Bay City, Texas	136	13.6	Rh
Union Carbide Corp. Texas City, Texas	330	33	Rh

[a] Eastman converted to a new, low pressure catalyst in 1989.

ERNST BILLIG
Union Carbide Chemicals and Plastics Company Inc.

Chemical Economics Handbook, SRI International, Menlo Park, Calif. U.S. Pat. 3,527,809 (Sept. 8, 1970), R. L. Pruett and J. A. Smith (to Union Carbide Corp.).

BUTYRIC ACID AND BUTYRIC ANHYDRIDE. See CARBOXYLIC ACID.

BUTYROLACTONE. See ACETYLENE-DERIVED CHEMICALS.

C

CABLE COVERINGS. See INSULATION, ELECTRIC.

CACAO. See CHOCOLATE AND COCOA.

CADMIUM AND CADMIUM ALLOYS

Cadmium, Cd, a Group 12 (IIB) element occurring between zinc and mercury, is a soft, ductile, silver-white metal having a distorted hexagonal close-packed structure (a = 0.29793 nm, c = 0.56181 nm). The crustal abundance of cadmium is somewhere between 0.1 and 0.5 ppm, and several cadmium minerals have been identified, the most common being greenockite, CdS. Cadmium is generally encountered in zinc ores, zinc-bearing lead ores, or complex copper–lead–zinc ores, where, however, it forms an isomorphic impurity in the zinc mineral sphalerite, ZnS, usually in concentrations of 0.1–0.5% cadmium. For this reason, cadmium is almost invariably recovered as a by-product from the processing of zinc, lead, and copper ores.

Properties

Physical properties of cadmium are listed in Table 1. Its electronic structure is $(1s)^2(2s)^2(2p)^6(3s)^2(3p)^6(3d)^{10}(4s)^2(4p)^6(4d)^{10}(5s)^2$, and its oxidation state in almost all of its compounds is +2, although a few compounds have been reported in which cadmium exists in the +1 oxidation state. There are eight natural isotopes:

Mass	Relative abundance, %	Mass	Relative abundance, %
106	1.22	112	24.07
108	0.88	113	12.26
110	12.39	114	28.86
111	12.75	116	7.58

Although it is only slowly oxidized in moist air at ambient temperature, cadmium forms a fume of brown-colored cadmium oxide, CdO, when heated in air. Other elements which react readily with cadmium metal upon heating include the halogens, phosphorus, selenium, sulfur, and tellurium.

Cadmium is rapidly oxidized by hot dilute nitric acid with the simultaneous generation of various oxides of nitrogen.

Table 1. Physical Properties of Cadmium

Property	Value
atomic weight	112.40
melting point, °C	321.1
boiling point, °C	767
specific heat, J/(mol·K),[a] at 20°C	25.9
coefficient of linear expansion at 20°C, μm/(cm·°C)	0.313
electrical resistivity, $\mu\Omega$·cm	
22°C	7.27
600°C	34.8
electrical conductivity, % IACS[b]	25
density, kg/m^3	
26°C	8642
400°C	7930
600°C	7720
thermal conductivity, W/(m·K)	
273 K	98
573 K	89
vapor pressure, kPa[c]	
382°C	0.1013
595°C	10.13
viscosity, mPa·s(= cP)	
340°C	2.37
500°C	1.84
Brinell hardness, kg/mm^2	16–23
tensile strength, MPa[d]	71
elongation, %	50
modulus of elasticity, GPa[e]	49.9
shear modulus,GPa[e]	19.2

[a] To convert J to cal, divide by 4.184.
[b] IACS = International Annealed Copper Standard.
[c] To convert kPa to mm Hg, multiply by 7.5.
[d] To convert MPa to psi, multiply by 145.
[e] To convert GPa to psi, multiply by 145,000.

Production

Cadmium occurs primarily as sulfide minerals in zinc, lead-zinc, and copper–lead–zinc ores. Beneficiation of these minerals, usually by flotation (qv) or heavy-media separation, yields concentrates which are then processed for the recovery of the contained metal values. Cadmium follows the zinc with which it is so closely associated (see ZINC AND ZINC ALLOYS; see also COPPER; LEAD).

Air pollution problems and labor costs have led to the closing of older pyrometallurgical plants, and to increased electrolytic production. On a worldwide basis, 77% of total zinc production in 1985 was by the electrolytic process. In electrolytic zinc plants, the calcined material is dissolved in aqueous sulfuric acid, usually spent electrolyte from the electrolytic cells. Residual solids are generally separated from the leach solution by decantation and the clarified solution is then treated with zinc dust to remove cadmium and other impurities.

Economic Aspects

Cadmium production is dependent on the processing of zinc ores, which often contain 0.2 to 0.4% cadmium. U.S. demand for cadmium normally exceeds the domestic supply and the United States is dependent on imports.

About 30 countries are cadmium producers, led by Russia, Japan, the United States, Canada, Belgium, Germany, and Mexico.

Safety and Handling

Cadmium is classified as a toxic metal. Acute industrial poisoning by cadmium dust or fume can occur during the melting or pouring of cadmium metal; the welding, burning, or heating of cadmium-plated steel; or spraying, brazing, and overheating of cadmium metal. Protection should be provided by a properly designed exhaust ventilation system or, for some intermittent exposures, by a suitable individual filter or air-supplied respirator. Industrial exposure to cadmium fumes and dust has been reported to result in emphysema, hypertension, kidney failure, osteomalacia, and perhaps an increased incidence of cancer.

The 1991 U.S. Occupational Safety and health Administration (OSHA) permissible exposure limit (PEL), 3-h time-weighted average standard (TWA) is 100 μg/m^3, and 200 μg/m^3 of air for cadmium dust, 600 μg/m^3 ceiling concentration.

To help maintain the balance between supply and demand for cadmium, efforts can be made to recycle such cadmium-containing materials as spent nickel–cadmium batteries as well as dusts and other residues from the pigment industry. There are two principal ways to recover cadmium, either by hydrometallurgical or pyrometallurgical processes. Cadmium wastes that cannot be recycled should be discarded in accordance with local regulations governing disposal of hazardous wastes.

Uses

In 1988 the estimated apparent consumption pattern for cadmium was batteries (qv), 32%; coating and plating, 29%, pigments (qv), 15%; plastics and synthetic products, 15%; and alloys and other uses, 9%.

Alloys

Cadmium is an important component in brazing and low melting alloys, used in bearings, solders, and nuclear reactor control rods, and as a hardener for copper (see BEARING MATERIALS). Of interest are two brazing alloys: 20% Ag; 45% Cu; 30% Zn; 5% Cd (mp 615°C; flow point 702°C), and ASTM Ag 2 which is 35% Ag; 26% Cu; 21% Zn; 18% Cd (mp 607°C; flow point 702°C). Other useful brazing compositions are also available (see SOLDERS AND BRAZING ALLOYS).

The commonly used low melting or fusible cadmium alloys are AsarcoLo 158 or Cerrobend (50% Bi; 26.7% Pb; 13.3% Sn; 10% Cd; mp 70°C) for bending pipes and thin sections, glass lens grinding blocks, and fire protection devices; AsarcoLo 158–190 or Cerrosafe (42.5% Bi; 37.7% Pb; 11.3% Sn; 8.5% Cd; mp 70–87°C) for foundry patterns, spotting fixtures, solder, and proofcasting molds; and AsarcoLo 117 or Cerrelow 117 (44.7% Bi; 22.6% Pb; 8.3% Sn; 5.3% Cd; 19.1% In; mp 47°C) for fusible cores, soldering and sealing, holding irregular pieces for machining and plastic lens-grinding blocks.

D. S. CARR
International Lead Zinc Research Organization, Inc.

G. D. Clayton and F. E. Clayton, eds., *Patty's Industrial Hygiene and Toxicology,* Vol. 2A, Toxicology, John Wiley & Sons, Inc., New York, 1981, pp. 1563–1583.

D. M. Chizikov, *Cadmium,* transl. by D. E. Hayler, Pergamon Press, Inc., New York, 1966.

M. C. Sneed and R. C. Brasted, eds., *Comprehensive Inorganic Chemistry,* Vol. IV, D. Van Nostrand Co., Inc., Princeton, N.J., 1955, pp. 65–90.

D. Wilson and R. A. Volpe, eds., "Cadmium 81, Cadmium 83, Cadmium 86 and Cadmium 89," *Proceedings of Third—Sixth International Cadmium Conferences,* Cadmium Council, Inc., Greenwich, Conn., 1981–1989.

CADMIUM COMPOUNDS

Naturally occurring cadmium compounds are limited to the rare minerals, greenockite, CdS, and otavite, an oxycarbonate, but neither is an economically important source of cadmium metal or its compounds. Instead, cadmium compounds are more usually derived from metallic cadmium which is produced as a by-product of lead-zinc smelting or electrolysis (see CADMIUM AND CADMIUM ALLOYS). Typically, this cadmium metal is burnt as a vapor, to produce the brown-black cadmium oxide, CdO, which then acts as a convenient starting material for most of the economically important compounds.

Properties

In general, cadmium compounds exhibit properties similar to the corresponding zinc compounds. Compounds and properties are listed in Table 1.

Uses

The principal areas of cadmium usage in terms of U.S. consumption in 1990 were batteries (qv), 50%; pigments (qv), 20%; plastic stabilizers, 15%; metal finishing, 10%; electronics and optics, <5%; and catalysts (qv), <5%.

Table 1. Physical and Chemical Properties of Selected Cadmium Compounds

Compound	$\Delta H^\circ_{f,298}$, kJ/mol[a]	ΔG°_f298 kJ/mol[a]	S°_{298}, J/mol·K[a]	Density, g/mL	Mp, °C	ΔH_{fus}, kJ/mol[a]	Aqueous solubility, g/100 g H_2O[b]
cadmium antimonide CdSb	−14.4	−13.0	92.9	6.92	452	32.05	
cadmium bromide $CdBr_2$	−316	−296	137.2	5.192	568	20.92	95_{18}
cadmium carbonate $CdCO_3$	−751	−669	92.5	4.26	332 dec		2.8×10^{-6}
cadmium chloride $CdCl_2$	−391	−344	115.3	4.05	568	22.176	128.6_{30}
cadmium fluoride CdF_2	−700	−648	77.4	6.39	1048	22.594	3.45_{25}
cadmium hydroxide $Cd(OH)_2$	−561	−474	96.2	4.79	150 dec		2.6×10^{-4}
cadmium iodide CdI_2, α-form	−203	−201	161.1	5.67	387	33.472	86_{25}
cadmium nitrate $Cd(NO_3)_2$	−456	−255	197.9		350		109_0
cadmium nitrate tetrahydrate $Cd(NO_3)_2 \cdot 4H_2O$	−1649			2.455	59.4	32.636	215_0
cadmium oxide CdO	−258	−228	54.8	8.2	1540 sub	243.509 sub	$9.6 \times 10_{-4}$
cadmium selenide CdSe, α-form	−136	−100	96.2	5.81	1350 dissoc	305.307 dissoc	
cadmium *m*-silicate $CdSiO_3$	−1189	−1105	97.5	4.928	1242		
cadmium sulfate $CdSO_4$	−933	−823	123.0	4.691	1000	20.084	76.6_{20}
cadmium sulfate hydrate $CdSO_4 \cdot H_2O$	−1240	−1069	154.0	3.79	105 trans		
cadmium sulfate hydrate $3CdSO_4 \cdot 8H_2O$	−1729	−1465	229.6	3.09	80 trans		113_0
cadmium sulfide CdS, α-form	−162	−156	64.9	4.82 (4.50)	980 sub in N_2	201.669 subl	$1.3 \times 10^{-4}_{18}$
cadmium telluride CdTe	−92	−92	100.4	6.20	1045		

[a] To convert J to cal, divide by 4.184.
[b] Subscript denotes temperature in °C.

Table 2. Thermodynamic and Stability Constant Data for Selected Aqueous Cadmium Complexes[a]

Complex ion	$\Delta H^\circ_{f,298}$ kJ/mol[b]	$\Delta G^\circ_{f\ 298}$ kJ/mol[b]	Stability constant
$CdCl^+$	−240.5	−224.4	$\log K_1 = 1.32$
$CdCl_3^-$	−561.0	−487.0	$\log K_3 = 0.09$
$Cd(CN)_4^{2-}$	428.0	507.5	$\log K_4 = 3.58$
$Cd(NH_3)_2^{2+}$	−266.1	−159.0	$\log K_2 = 2.24$
$Cd(NH_3)_4^{2+}$	−450.2	−226.4	$\log K_4 = 1.18$
$CdBr^+$	−200.8	−193.9	$\log K_1 = 1.97$
$CdBr_3^-$		−407.5	$\log K_3 = 0.24$
CdI^+	−141.0	−141.4	$\log K_1 = 2.08$
CdI_3^-		−259.4	$\log K_3 = 2.09$
CdI_4^{-2}	−341.8	−315.9	$\log K_4 = 1.59$
$CdSCN^+$		7.5	$\log K_1 = 1.90$
$Cd(SCN)_4^{-2}$			$\log K_4 =$ ca0.1
$Cd(N_3)_4^{-2}$		1,295.0	$\log K_4 = 0.76$

[a] Standard state $M = 1$.
[b] To convert kJ to kcal, divide by 4.184.

Cadmium hydroxide is the anode material of Ag–Cd and Ni–Cd rechargeable storage batteries (see BATTERIES-SECONDARY CELLS). Cadmium sulfide, selenide, and especially telluride find utility in solar cells (see SOLAR ENERGY). Cadmium sulfide, lithopone, and sulfoselenide are used as colorants (orange, yellow, red) for plastics, glass, glazes, rubber, and fireworks (see COLORANTS FOR CERAMICS; COLORANTS FOR PLASTICS; PIGMENTS).

In flexible PVC, cadmium salts of long-chain organic acids, such as stearate and laurate, are used in combination with similar Ba^{2+} salts as heat and light stabilizers (see HEAT STABILIZERS). Cadmium cyanide, acetate, fluoroborate, or sulfate is used as an electrolyte in coating a thin cadmium layer, ie, electroplating (qv), onto other metals thereby imparting enhanced corrosion protection. Cadmium protective overlayers are also deposited by mechanical plating or vapor deposition (see METALLIC COATINGS).

The cadmium chalcogenide semiconductors (qv) have found numerous applications ranging from rectifiers to photoconductive detectors in smoke alarms. Many Cd compounds, eg, sulfide, tungstate, selenide, telluride, and oxide, are used as phosphors in luminescent screens and scintillation counters. Glass colored with cadmium sulfoselenides is used as a color filter in spectroscopy and has recently attracted attention as a third-order, nonlinear optical switching material (see NONLINEAR OPTICAL MATERIALS). Dialkylcadmium compounds are polymerization catalysts for production of poly(vinyl chloride) (PVC), poly(vinyl acetate) (PVA), and poly(methyl methacrylate) (PMMA). Mixed with $TiCl_4$, they catalyze the polymerization of ethylene and propylene.

Demand for cadmium and its compounds in the United States has generally declined after peaking in the 7000–8000 t/yr range during the late 1970s. The declining use of cadmium in plastics and as pigments in the 1980s, largely because of health and environmental concerns and the concomitant introduction of Cd substitutes, has been offset by its increased use in rechargeable storage (Ni–Cd) batteries. Continued and growing concern over the toxicity of cadmium in the environment is expected to lead to alternatives for cadmium compounds in all applications. The outlook for cadmium use, however, seems stable throughout the 1990s.

Inorganic Compounds

Inorganic compounds include cadmium arsenides, antimonides, and phosphides; cadmium borates; cadmium carbonate; cadmium complexes (Table 2); cadmium halides; cadmium hydroxide; cadmium nitrate; cadmium oxide; cadmium phosphates; cadmium selenide and telluride, cadmium silicates, cadmium sulfate, cadmium sulfide, and cadmium tungstate.

Organic Compounds

Many organocadmium compounds are known but few have been of commercial importance. They include dialkyl cadmium compounds, cadmium acetate, and organocadmium soaps.

NORMAN HERRON
E. I. du Pont de Nemours & Co., Inc.

M. Farnsworth, *Cadmium Chemicals,* International Lead Zinc Research Org., New York, 1980.

Cadmium 1990, A Review, Cadmium Association, London, 1990.

Toxicological Profile for Cadmium, U.S. Dept. Commerce, NTIS, Washington, D.C., Mar. 1989.

Nickel Cadmium Battery Update 90, Cadmium Association, London, 1990.

CAFFEINE. See ALKALOIDS.

CALCIUM AND CALCIUM ALLOYS

Calcium, Ca, a member of Group 2 (IIA) of the Periodic Table between magnesium and strontium, is classified, together with barium and strontium, as an alkaline-earth metal and is the lightest of the three. Calcium metal does not occur free in nature; however, in the form of numerous compounds, it is the fifth most abundant element, constituting 3.63% of the earth's crust.

Calcium is mainly used as a reducing agent for many reactive, less common metals; to remove bismuth from lead (qv); as a desulfurizer and deoxidizer for ferrous metals and alloys; and as an alloying agent for aluminum, silicon, and lead. Small amounts are used as a dehydrating agent for organic solvents and as a purifying agent for removal of nitrogen and other impurities from argon and other rare gases (see HELIUM GROUP, GASES).

Physical Properties

Pure calcium is a bright, silvery-white metal, although under normal atmospheric conditions freshly exposed surfaces of calcium quickly become covered with an oxide layer. The metal is extremely soft and ductile having a hardness between that of sodium and aluminum. It can be work-hardened to some degree by mechanical processing. Although its density is low, calcium's usefulness as a structural material is limited by low tensile strength and high chemical reactivity.

Some of the more important physical properties of calcium are given in Table 1. Measurements of the physical properties of calcium

Table 1. Physical Properties of Calcium

Property	Value					
atomic weight				40.08		
electron configuration				$(1s)^2(2s)^2(2p)^6(3s)^2(3p)^6(4s)^2$		
stable isotopes						
atomic weight	40	42	43	44	46	48
natural abundance, %	96.947	0.646	0.135	2.083	0.186	0.18
specific gravity at 20°C, kg/m^3				1.55×10^3		
melting point, °C				839 ± 2		
boiling point, °C				1484		
heat of fusion, ΔH_{fus}, kJ/mol[a]				9.2		
heat of vaporization, ΔH_{vap}, kJ/mol[a]				161.5		
heat of combustion, kJ/mol[a]				634.3		
vapor pressure						
pressure, kPa[b]		0.133	1.33	13.3	53.3	101.3
temperature, °C		800	970	1200	1390	1484
specific heat at 25°C, J/(g·K)[a]				0.653		
coefficient of thermal expansion, 0–400°C, m/(m·K)				22.3×10^{-6}		
electrical resistivity at 0°C, $\mu\Omega$·cm				3.91		
electron work function, eV				2.24		
tensile strength (annealed), MPa[b]				48		
yield strength (annealed), MPa[b]				13.7		
modulus of elasticity, GPa[b]				22.1–26.2		
hardness (as cast)						
HB[c]				16–18		
HR B[d]				36–40		

[a] To convert J to cal, divide by 4.184.
[b] To convert kPa to psi, multiply by 0.145.
[c] Brinell Hardness scale.
[d] Rockwell B Hardness scale.

are usually somewhat uncertain owing to the effects that small levels of impurities can exert.

Chemical Properties

Calcium has a valence electron configuration of $4s^2$ and characteristically forms divalent compounds. It is very reactive and reacts vigorously with water, liberating hydrogen and forming calcium hydroxide, $Ca(OH)_2$. Calcium does not readily oxidize in dry air at room temperature but is quickly oxidized in moist air or in dry oxygen at about 300°C. Calcium reacts with fluorine at room temperature and with the other halogens at 400°C. When heated to 900°C, calcium reacts with nitrogen to form calcium nitride, Ca_3N_2.

Calcium is an excellent reducing agent and is widely used for this purpose.

Commercially produced calcium metal is analyzed for metallic impurities by emission spectroscopy. Carbon content is determined by combustion, whereas nitrogen is measured by Kjeldahl determination.

Manufacture

Although in Western countries the aluminothermic process has completely replaced the electrolytic method, electrolysis is believed to be the method used for calcium production in the People's Republic of China and the Commonwealth of Independent States (CIS). This process likely involves the production of a calcium–copper alloy, which is then redistilled to give calcium metal. For certain applications, especially those involving reduction of other metal compounds, better than 99% purity is required. This can be achieved by redistillation.

Shipment

Because of its extreme chemical reactivity, calcium metal must be carefully packaged for shipment and storage. The metal is packaged in sealed argon-filled containers. Calcium is classed as a flammable solid and is nonmailable. Sealed quantities of calcium should be stored in a dry, well-ventilated area so as to remove any hydrogen formed by reaction with moisture.

Economic Aspects

Calcium metal is produced in the United States by Pfizer Inc. (Canaan, Connecticut), and in Canada by Timminco Metals (Toronto, Ontario). In France it is produced by Pechiney Electrometallurgie. It is also produced in the Commonwealth of Independent States (CIS) and the People's Republic of China.

Health and Safety

Calcium metal and most calcium compounds are nontoxic. Care must be taken, however, to avoid contact with water owing to the exothermic liberation of hydrogen and the resulting explosion hazard.

Calcium Alloys

Calcium alloys can be produced by various techniques. However, direct alloying of the pure metals is normally used in the production of 80% calcium–magnesium, 70% magnesium–calcium, and 75% calcium–aluminum alloys.

Lead alloys containing small amounts of calcium are formed by plunging a basket containing a 77 or 75% calcium–23–25% Al alloy into a molten lead bath or by stirring the Ca-Al alloy into a vortex created by a mixing impellor.

Alloys of calcium with silicon are used in ferrous metallurgy (qv) and are generally produced in an electric furnace from CaO (or CaC_2), SiO_2, and a carbonaceous reducing agent.

STEPHEN G. HIBBINS
Timminco Metals

C. L. Mantell and C. Hardy, *Calcium Metallurgy and Technology*, Reinhold Publishing Corp., New York, 1945.

The Economics of Calcium Metal, 4th ed., Ruskill Information Services Ltd., London, 1990.

CALCIUM CHANNEL ANTAGONISTS.

See CARDIOVASCULAR AGENTS.

CALCIUM COMPOUNDS

SURVEY

The chemical element calcium, Ca, atomic number 20, is an alkaline-earth metal which is fifth in abundance among all elements (ca 4%) and the third most abundant metal found in the earth's crust. It is too reactive to be found naturally in the free state, but its compounds are widespread as the minerals listed in Table 1 indicate.

Inorganic Compounds

Calcium Carbonate. Limestone is the most widely used of all rocks, as such for dimension stone or aggregate in concrete and road building, or as an industrial chemical and precursor of lime and hydrated lime.

Lime and Hydrated Lime. More than 90% of the lime consumed in the United States is used for basic or industrial chemistry. It is produced by thermal decomposition (calcination) of calcium carbonate in various forms including limestone, marble, chalk, oyster shells, and dolomite.

The hydrolysis process, ie, reaction with water, for lime is called slaking and produces hydrated lime, $Ca(OH)_2$. Calcium hydroxide is a strong base but has limited aqueous solubility, 0.219 g $Ca(OH)_2$/100 g H_2O, and is therefore often used as a suspension. It finds widespread industrial application as an alkali because it is cheaper than sodium hydroxide. Lime and hydrated lime are used in mortar, treatment of industrial wastes, and treatment of municipal and industrial water supplies.

Halogen Compounds. *Halides.* Calcium halides are made by reaction of elemental calcium and the halogens directly or more conveniently by the reaction of the corresponding hydrohalic acid and $CaCO_3$, CaO or $Ca(OH)_2$.

Fluorospar, CaF_2, is used as a flux in metallurgical processes such as production of steel in the open-hearth furnace.

Hypochlorites. A common dry form of chlorine used as a bleach or water purifier is made by reaction of gaseous chlorine and high calcium hydrated lime.

Another bleaching agent, calcium hypochlorite, $Ca(OCl)_2$, available chlorine ca 70%, can be made by salting out from a solution of bleaching powder, $CaCl(OCl)\cdot H_2O$, with NaCl. In contrast with bleaching powder, calcium hypochlorite does not decompose on standing (see BLEACHING AGENTS).

Table 1. Calcium-Containing Minerals

Mineral	Molecular formula
marble	$CaCO_3$
limestone	$CaCO_3$
calcite	$CaCO_3$
dolomite	$CaCO_3\cdot MgCO_3$
gypsum	$CaSO_4\cdot 2H_2O$
anhydrite	$CaSO_4$
fluorspar	CaF_2
fluorapatite	$Ca_5F(PO_4)_3$
hydroxylapatite	$Ca_5OH(PO_4)_3$
selenite	$CaSO_4\cdot 2H_2O$
anorthite	$CaAl_2Si_2O_8$

Sulfates and Sulfites. Calcium sulfate occurs in large deposits as $CaSO_4$ and as gypsum, $CaSO_4\cdot 2H_2O$. The dihydrate is a functional additive in Portland cement to control setting time.

Phosphates. The primary constituent of phosphate rock is fluorapatite, $Ca_5FP_3O_{12}$. Industrial phosphates including phosphate fertilizers (qv), phosphoric acid, and calcium phosphates (see PHOSPHORIC ACID AND THE PHOSPHATES) are obtained from the large deposits of fluorapatite found in Florida in the United States, and in Morocco.

Calcium Magnesium Acetate. Calcium magnesium acetate (CMA), suggested formula $CaMg_2(C_2H_3O_2)_6$, is an emerging bulk chemical that has found application as a replacement for salt and calcium chloride as a less corrosive and biodegradable road deicer.

Hydride. Calcium hydride, CaH_2, is an effective reducing agent at high temperatures and has been used to reduce inorganic oxides to their metals. The compound is a convenient drying agent for gases and organic solvents.

Silicates. *Glass.* Ordinary glass (qv), soda-lime glass, is a complex mixture of silicates, chiefly those of sodium and calcium.

Portland Cement. Portland cement is obtained by calcining a mixture of substances to produce an appropriate ratio of the oxides CaO, MgO, Al_2O_3, Fe_2O_3, and SiO_2 (see CEMENT).

Whitewares (Earthenware, China, Porcelain). The chief raw materials of ceramic manufacture are clay, feldspar, and sand. All of the three common types of feldspars are used: soda, potash, $M_2O\cdot Al_2O_3\cdot 6SiO_2$(M = Na, K) and lime, $CaAl_2Si_2O_8$ (see CERAMICS; ENAMELS).

Calcium Silicate Brick. Sand-lime brick is used in masonry in the same way as common clay brick.

Coordination Chemistry of Calcium

Calcium ion shows some tendency to form complexes mainly through coordination with oxygen-containing ligands. A few calcium complexes having nitrogen ligands are known.

Organic Chemistry of Calcium

Calcium Carbide and Its Derivatives. Although hydrocarbon-based acetylene production has become more important, early manufacture of acetylene was based on manufacture of the intermediate, calcium carbide, CaC_2. Calcium carbide is also used to produce calcium cyanamide, $CaCN_2$ (see CYANAMIDES). Calcium cyanamide (lime nitrogen) has been used as a fertilizer.

Calcium cyanamide can be converted to calcium cyanide, used in cyanidation of metallic ores and production of sodium cyanide and ferrocyanides (see CYANIDES). Calcium cyanamide has also been used to make cyanamide, which in turn is the starting material for important industrial organic syntheses.

Salts of Organic Acids. Calcium salts of organic acids may be prepared by reaction of calcium carbonate or hydroxide and the organic acid.

Reagents in Synthesis. Calcium borohydride, $Ca(BH_4)_2$, produced by reaction of $NaBH_4$ and $CaCl_2$, has been used for reductions (see HYDRIDES). Hexaamminecalcium, prepared by passing NH_3 into an ether suspension of calcium, reduces polycyclic aromatic compounds, leaving one isolated aromatic ring. Calcium hydride, CaH_2, and anhydrous calcium sulfate (Drierite), $CaSO_4$, are useful as drying agents.

Organometallic Chemistry. Only a few organocalcium compounds have been reported. Alkyl calcium halides have been prepared by reaction of the halides and calcium in tetrahydrofuran.

Biological Role of Calcium

Biological functions of Ca(II) ion are numerous but may be classified in one of three categories: the formation of solid skeletal material such as bone, teeth, and shell; the stabilizing of protein conformation structure; and the most varied, the ability of Ca(II) to trigger certain physiological activities such as muscle contraction and the release of hormones (qv).

The recommended daily allowances of calcium are for children to 10 years of age, 360–800 mg; teenage children, 1200 mg; adults,

800 mg, increasing to 1200 mg during pregnancy and lactation. Cow's milk supplies ca 1.27 g/L of calcium in available form.

RICHARD L. PETERSEN
MARK B. FREILICH
The University of Memphis

R. S. Boynton, *Chemistry and Technology of Lime and Limestone,* 2nd ed., John Wiley & Sons, Inc., New York, 1980.

G. T. Austin, *Shreve's Chemical Process Industries,* 5th ed., McGraw-Hill Book Co., New York, 1984, pp. 149–212.

R. D. Goodenough and V. A. Stenger, in A. F. Trotman-Dickenson, ed., *Comprehensive Inorganic Chemistry,* Vol. 1, Pergammon Press, Oxford, U.K., 1973, pp. 591–664.

J. J. R. Frausto da Silva and R. J. P. Williams, *The Biological Chemistry of the Elements: The Inorganic Chemistry of Life,* Oxford University Press, Oxford, U.K., 1991, pp. 268–298 and 467–527.

CALCIUM CARBONATE

Calcium carbonate, $CaCO_3$, mol wt 100.09, occurs naturally as the principal constituent of limestone, marble, and chalk. Powdered calcium carbonate is produced by two methods on the industrial scale. It is quarried and ground from naturally occurring deposits and in some cases beneficiated. It is also made by precipitation from dissolved calcium hydroxide and carbon dioxide. The natural ground calcium carbonate and the precipitated material compete industrially based primarily on particle size and the characteristics imparted to a product.

Calcium carbonate is one of the most versatile mineral fillers (qv) and is consumed in a wide range of products, including paper (qv), paint (qv), plastics, rubber, textiles (qv), caulks, sealants (qv), and printing inks (qv). High purity grades of both natural and precipitated calcium carbonate meet the requirements of the *Food Chemicals Codex* and the *United States Pharmacopeia* and are used in dentifrices (qv), cosmetics (qv), foods, and pharmaceuticals (qv).

Properties

Calcium carbonate occurs naturally in three crystal structures: calcite, aragonite, and, although rarely, vaterite. Calcite is thermodynamically stable; aragonite is metastable and irreversibly changes to calcite when heated in dry air to about 400°C.

The commercial grades of calcium carbonate from natural sources are either calcite, aragonite, or sedimentary chalk. In most precipitated grades aragonite is the predominant crystal structure. The essential properties of the two common crystal structures are shown in Table 1.

Economic Aspects

The principal U.S. producers of ground calcium carbonate are Columbia River Carbonates, ECC International, Franklin Limestone Company, Genstar Stone Products, Georgia Marble Company, J.M. Huber Corporation, Calcium Carbonates Division, James River Limestone Company, Inc., OMYA Inc. (Pluess-Staufer), and MTI Inc. The principal U.S. producers of precipitated calcium carbonate are Mississippi Lime Company and MTI Inc.

Health and Safety Factors

Calcium carbonate is listed as a food additive and not considered a toxic material. The exposure to dust is regulated and a Threshold Limit Value–Time-Weighted Average (TLV–TWA) of 10 mg/m^3 is set. Both natural ground and precipitated calcium carbonates can contain low levels of impurities that are regulated.

Table 1. Properties of Calcium Carbonate

Property	Calcite	Aragonite
specific gravity	2.60–2.75	2.92–2.94
hardness, Mohs'	3.0	3.5–4.0
solubility at 18°C, g/100 g H_2O	0.0013	0.0019
melting point, °C	1339[a]	[b]
	dec 900	
index of refraction		
α		1.530
β		1.680
γ		1.685
ω	1.658	
ϵ	1.486	

[a] At 10.38 MPa (102.5 atm).
[b] Decomposes to calcite at temperatures >400°C

F. PATRICK CARR
DAVID K. FREDERICK
OMYA, Inc.

R. J. Reeder, ed., *Carbonates, Mineralogy and Chemistry,* Mineralogical Society of America, Washington, D.C., 1990, p. 191.

C. Klein and C. S. Hurlbut, Jr., *Manual of Mineralogy,* John Wiley & Sons, Inc., New York, 1985, pp. 328, 335.

H. S. Katz and J. V. Milewski, *Handbook of Fillers for Plastics,* Van Nostrand Reinhold Co., New York, 1987, p. 123.

M. O'Driscoll, *Industrial Fillers* **276,** 21 (Sept. 1990).

CALCIUM CHLORIDE

Calcium chloride, $CaCl_2$, is a white, crystalline salt that is very soluble in water. Solutions containing 30–45 wt % $CaCl_2$ are used commercially. Of the alkaline-earth chlorides it is the most soluble in water. It is extremely hygroscopic and liberates large amounts of heat during water absorption and on dissolution. It forms a series of hydrates containing one, two, four, and six moles of water per mole of calcium chloride (Table 1). Another hydrate, $CaCl{\cdot}0.33H_2O$, has been identified, mol wt 116.98; 94.8 wt % $CaCl_2$; heat of solution in water to infinite dilution, −71.37 kJ/mol (−17.06 kcal/mol).

Commercial applications of calcium chloride and its hydrates exploit one or more of its properties with regard to aqueous solubility, hygroscopic nature, the heat gained or lost when one hydrated phase changes to another, and the depressed freezing point of the eutectic solution at a composition of about 30% by weight calcium chloride.

Properties

The properties of calcium chloride and its hydrates are summarized in Table 1.

Calcium Chloride Solutions. Because of high solubility in water, calcium chloride is used to obtain solutions having relatively high densities.

Viscosity is an important property of calcium chloride solutions in terms of engineering design and in application of such solutions to flow-through porous media. Data and equations for estimated viscosities of calcium chloride solutions over the temperature range of 20–50°C are available.

Production and Consumption

Significant quantities of calcium chloride are produced in the United States, Canada, Mexico, Germany, Belgium, Sweden, Finland, Norway, and Japan.

In the United States the principal route for making calcium chloride is by the evaporation of underground brines (see CHEMICALS FROM BRINE). Additional commercial material is available by the action of hydrochloric acid on limestone.

Table 1. Properties of Calcium Chloride Hydrates

Property	$CaCl{\cdot}6H_2O$	$CaCl{\cdot}4H_2O$	$CaCl{\cdot}2H_2O$	$CaCl{\cdot}H_2O$	$CaCl^2$
mol wt	219.09	183.05	147.02	129.00	110.99
composition, wt % $CaCl_2$	50.66	60.63	75.49	86.03	100.00
mp, °C	30.08	45.13	176	187	772
sp gravity, $d^{25}{}_4$	1.71	1.83	1.85	2.24	2.16
heat of fusion or transition, kJ/mol[a]	43.4	30.6	12.9	17.3	28.5
heat of solution in water[b], kJ/mol[a]	15.8	−10.8	−44.05	−52.16	−81.85
heat of formation, at 25°C, kJ/mol[a]	−2608	−2010	−1403	−1109	−795.4
heat capacity, at 25°C, J/(g · °C)[a]	1.66	1.35	1.17	0.84	0.67

[a] To convert J to cal, divide by 4.184.
[b] To infinite dilution.

Uses

Calcium chloride, manufactured for over 100 years, has been used for a variety of purposes. The primary $CaCl_2$ markets have not changed since the 1950s. Significant markets in the United States are for deicing during the winter and roadbed stabilization, and as a dust palliative during the summer. Use as an accelerator in the ready-mix concrete industry is sizable but there is concern about chloride usage because of the possible corrosion of steel in highways and buildings. Calcium chloride is also used in oil and gas well drilling. The size of that market is dependent on the state of the worldwide oil and gas industry.

Food. Food-grade calcium chloride is used in cheese making to aid in rennet coagulation and to replace calcium lost in pasteurization. In the canning industry it is used to firm the skin of fruit such as tomatoes, cucumbers, and jalapenos. It acts as a control in many flocculation, coagulation systems. Food-grade calcium chloride is used in the brewing industry both to control the mineral salt characteristics of the water and as a basic component of certain beers (see BEER).

Toxicity and Environmental

Above certain levels chloride is toxic to plants and animals. Thus, when considering calcium chloride, potentially large concentrations of calcium ion can be tolerated, but at these concentrations the chloride ion becomes toxic.

Calcium chloride solutions, typically employed at 2–5% concentration, are used as antispasmodics, diuretics (see DIURETIC AGENTS), and in the treatment of tetany. Concentrated solutions of calcium chloride cause erythema, exfoliation, ulceration, and scarring of the skin. Injections into the tissue may cause necrosis. If given orally calcium chloride can cause irritation to the gastrointestinal tract unless accompanied by a demulcent. There is no published information on mutagenicity or carcinogenicity caused by calcium ions or calcium chloride. Calcium chloride has been give a toxicity or hazard level of 3. Materials in this classification typically have LD_{50} below 400 mg/kg or an LC_{50} below 100 ppm.

KENNETH I. G. REID
ROGER KUST
Tetra Chemicals

K. K. Meissingset and F. Gronvold, *J. Chem. Thermodynam.* (18), 159–173 (1986).

G. C. Sinke, E. H. Mossner, and J. L. Curnutt, *J. Chem. Thermodynam.* (17), 893–899 (1985).

Chemical Economics Handbook—Chlorine and Alkali Chemicals, SRI International, Menlo Park, Calif., 1990.

N. I. Sax and R. J. Lewis, *Dangerous Properties of Industrial Materials,* 7th ed., Vol. I, Von Nostrand Reinhold, New York, 1989, p. 678.

CALCIUM SULFATE

Calcium sulfate, $CaSO_4$, in mineral form is commonly called and occurs abundantly in many areas of the world. In natural deposits, the main form is the dihydrate. Some anhydrite is also present in most areas, although to a lesser extent. Mineral composition can be found in Table 1. The hemihydrate is normally produced by heat conversion of the dihydrate from which $\frac{3}{2}H_2O$ is removed as vapor. The resulting powder is also known as plaster of Paris. Stucco has the greatest commercial significance of these materials. It is the primary constituent used to fabricate products and in formulated plasters used in job or shop-site applications.

About 23×10^6 t of gypsum are consumed annually. About 80% is processed into the commercially usable hemihydrate. Uses of gypsum are in fabricated or formulated building materials (see BUILDING MATERIALS, SURVEY), Portland cement (qv) set regulation, and agricultural soil conditioning.

Properties

Table 2 lists the physical properties of calcium sulfate.

Sources

The natural, or mineral, form of gypsum is most widely extracted by mining or quarrying and used commercially.

Gypsum is also obtained as a by-product of various chemical processes. The main sources are from processes involving scrubbing gases evolved in burning fuels that contain sulfur (see SULFUR REMOVAL AND RECOVERY), and the chemical synthesis of chemicals, such as sulfuric acid, phosphoric acid, titanium dioxide, citric acid, and organic polymers.

Decomposition Thermodynamics

The thermodynamic properties of gypsum decomposition, which involve two distinct steps, have been the subject of much theoretical and practical study. Two forms of the hemihydrate, α and β, have been identified.

$$CaSO_4{\cdot}2H_2O \xrightarrow{\Delta} CaSO_4{\cdot}\frac{1}{2}\,H_2O + 1\frac{1}{2}\,H_2O$$

$$CaSO_4{\cdot}\frac{1}{2}\,H_2O \xrightarrow{\Delta} CaSO_4 + \frac{1}{2}\,H_2O$$

Manufacture

Natural Gypsum. Gypsum rock from the mine or quarry is crushed and sized to meet the requirements of future processing or removed for direct marketing of the dihydrate as

Table 1. Gypsum Forms and Composition

Common name	Molecular formula	Composition, wt % CaO	SO_3	Combined H_2O
anhydrite	$CaSO_4$	41.2	58.8	
gypsum	$CaSO_4{\cdot}2H_2O$	32.6	46.5	20.9
stucco	$CaSO_4{\cdot}\frac{1}{2}H_2O$	38.6	55.2	6.2

Table 2. Physical Properties of Calcium Sulfate

Property	Dihydrate	Hemihydrate	Anhydrite
mol wt	172.17	145.15	136.14
transition point, °C	128[a]	163[b]	
	163[b]		
mp[c] °C	1450	1450	1450
specific gravity	2.32		2.96
solubility at 25°C, g/100 g H_2O	0.24	0.30	0.20
hardness, Mohs'	1.5–2.0		3.0–3.5

[a] Hemihydrate is formed.
[b] Anhydrous material is formed.
[c] Compound decomposes.

a cement retarder. Once subjected to a secondary crusher, calcining, and drying, the product is fine-ground. Fine-ground dihydrate is commonly called land plaster, regardless of its intended use. The degree of fine grinding is dictated by the ultimate use. The majority of fine-ground dihydrate is used as feed to calcination processes for conversion to hemihydrate.

β-Hemihydrate. The dehydration of gypsum, commonly referred to as calcination in the gypsum industry, is used to prepare hemihydrate, or anhydrite. Kettle calcination continues to be the most commonly used method of producing β-hemihydrate.

α-Hemihydrate. Three processing methods are used for the production of α-hemihydrate. One, developed in the 1930s, involves charging lump gypsum rock 1.3–5 cm in size into a vertical retort, sealing it, and applying steam at a pressure of 117 kPa (17 psi) and a temperature of about 123°C. After calcination under these conditions for 5–7 h the hot moist rock is quickly dried and pulverized.

Another method, first reported in the 1950s, has lower water demand. The dihydrate is heated in a water solution containing a metallic salt such as $CaCl_2$ at pressures not exceeding atmospheric. A third method, developed in 1967, prepares very low water-demand α-hemihydrate by autoclaving powdered gypsum in a slurry. A crystal-modifying substance such as succinic acid or malic acid is added to the slurry in the autoclave to produce large squat crystals.

Anhydrite. In addition to kettle calcination, soluble anhydrite is commercially manufactured in a variety of forms, from fine powders to granules 4.76 mm (4 mesh) in size, by low temperature dehydration of gypsum.

Production and Trade

Crude gypsum is the principal form of calcium sulfate shipped in international trade, although the 1980s saw an increase in the volume of fabricated products moved across international borders.

DONALD J. PETERSEN
NORBERT W. KALETA
LARRY W. KINGSTON
Natural Gypsum

K. K. Kelly, J. C. Southard, and C. T. Anderson, *U.S. Bureau of Mines Technical Papers,* Technical Paper 625, 1941.

Mineral Industry Surveys, U.S. Dept. of Interior, Bureau of Mines, Washington, D.C., Jan. 1990.

CALORIMETRY. See THERMOGRAPHY.

CANCER CHEMOTHERAPY. See CHEMOTHERAPEUTICS, ANTICANCER.

CAPROLACTAM

Caprolactam (2-oxohexamethylenimine, hexahydro-2*H*-azepin-2-one) is one of the most widely used chemical intermediates. However, almost all of the annual production of 3.0×10^6 t is consumed as the monomer for nylon-6 fibers and plastics (see POLYAMIDES, FIBERS; POLYAMIDES, PLASTICS). Cyclohexanone, which is the most common organic precursor of caprolactam, is made from benzene by either phenol hydrogenation or cyclohexane oxidation (see CYCLOHEXANOL AND CYCLOHEXANONE). Reaction with ammonia-derived hydroxylamine forms cyclohexanone oxime, which undergoes molecular rearrangement to the seven-membered ring ϵ-caprolactam.

Physical Properties

Caprolactam, mol wt 113.16, is a white, hygroscopic, crystalline solid at ambient temperature, with a characteristic odor. It is very soluble in water and in most common organic solvents and is sparingly soluble in high molecular weight aliphatic hydrocarbons. Molten caprolactam is a powerful solvent for polar and nonpolar organic chemicals. Selected physical properties are listed in Table 1.

The low melting point of caprolactam and its stability and low viscosity form the basis for commercial transportation practice: caprolactam is handled as a liquid in insulated tank cars or trucks.

Reactions

Caprolactam is an amide and therefore undergoes the reactions of this class of compounds. It can be hydrolyzed, *N*-alkylated, *O*-alkylated, nitrosated, halogenated, and subjected to many other reactions.

Manufacture

All commercial processes for the manufacture of caprolactam are based on either toluene or benzene, each of which occurs in refinery BTX-extract streams (see BTX PROCESSING). Alkylation of benzene with propylene yields cumene (qv), which is a source of phenol and

Table 1. Physical Properties of Caprolactam $CH_2(CH_2)_4CONH$

Properties	Values
melting point, °C	69.3
density (at 77°C), g/cm^3	1.02
bulk density, kg/m^3	600–700
vapor pressure, kPa[a]	
at 270°C	100.6
at 150°C	2.62
at 115°C	0.53
at 25°C (solid)	0.0004
refractive index at 40°C	1.4935
viscosity, mPa·s (= cP)	
at 70°C	12.3
at 80°C	8.5
specific heat, J/(kg·K)[b]	
solid	
at 25°C	1380
at 35°C	1420
liquid	
at 70°C	2117
at 110°C	2412
at 178°C	2608
vapor	
at 100°C	1640
thermal conductivity, W/(m·K)	0.169
heat of fusion, J/g[c]	135.9
heat of vaporization at 80°C, J/g[c]	580
heat of combustion (liquid at 25°C), J/g[c]	−31,900
heat of formation (liquid at 25°C), J/g[c]	−2,840
flash point (closed cup), °C	125
fire point, °C	140

[a] To convert kPa to mm Hg, multiply by 7.5.
[b] To convert J/(kg·K) to cal/(g · °C), divide by 4184.
[c] To convert J to cal, divide by 4.184.

acetone; ca 10% of U.S. phenol is converted to caprolactam. Purified benzene can be hydrogenated over platinum catalyst to cyclohexane; nearly all of the latter is used in the manufacture of nylon-6 and nylon-6,6 chemical intermediates. The five main process routes are the Allied-Signal process, the BASF process, the Dutch State Mines (Stamicarbon) process, the Toray process, and the Snia Viscosa process.

Economic Aspects

Worldwide caprolactam demand increased by about 2% per year during the 1980s, and similar growth is expected into the 1990s. Because of the new capacity coming on-stream in the Far East, production will stay basically flat in the United States and in Europe.

DSM America produces caprolactam only for merchant sales, both domestic and foreign. BASF is a customer, following acquisition of Enka's U.S. fiber and plastics plants, and also a captive producer of caprolactam. Allied-Signal's production is primarily captive for nylon-6 fibers and plastics, but substantial amounts are supplied to the export market.

Health and Safety factors

Caprolactam has a low order of toxicity. It presents no appreciable health hazard if it is handled properly. Prolonged exposure to dust or vapors causes irritation of eyes, mucous membranes, and skin; inhalation may cause irritation of the respiratory tissues. Skin contact, if prolonged, can lead to dermatosis, causing a reddening and tightening of the skin the appearance and sensation of which is similar to sunburn. A thorough wash with water, in which caprolactam is very soluble, or with soap and water, normally is sufficient to remove caprolactam from contaminated parts of the body.

Threshold limit values for caprolactam dust and vapor are 1 mg/m^3 and 4.3 ppm (20 mg/m^3), respectively, although the American Conference of Governmental Industrial Hygienists (ACGIH) has a notice of intended change (1991–1992) of 1 mg/m^3 for dust and 5 ppm (23 mg/m^3) for vapor with short-term exposure limits (STELs) of 3 mg/m^3 and 10 ppm for dust and vapor, respectively (time-weighted averages).

Caprolactam has been extensively tested for its mutagenic potential. Based on the overall weight of evidence, caprolactam would be considered nonmutagenic.

WILLIAM B. FISHER
L. CRESCENTINI
Allied-Signal, Inc.

T. E. Daubert, J. W. Jalowka, and V. Goren, *AIChE Symp. Ser.* **83**(256), 128 (1987).

C. S. Read, *Chemical Economics Handbook,* Marketing Research Report 625-2000, SRI International, Menlo Park, Calif., June 1990.

D. G. Serota, A. M. Hoberman, M. A. Friedman, and S. C. Gad, *J. Appl. Toxicol.* **8**(4), 285 (1988).

CAPRYLIC ACID. See CARBOXYLIC ACIDS.

CARAMEL COLORS. See COLORANTS FOR FOOD, DRUGS, COSMETICS, AND MEDICAL DEVICES.

CARBAMIC ACID

Carbamic acid, NH_2COOH, is the hydrated form of isocyanic acid, H—N═C═O. Carbamic acid is the monoamide of carbonic acid; the diamide is the well-known compound urea, also called carbamide (see UREA).

CARBAZOLE. See AMINES, AROMATIC–DIARYLAMINES.

CARBIDES

SURVEY

Carbon reacts with many elements of the Periodic Table to form a diverse group of compounds known as carbides, some of which are

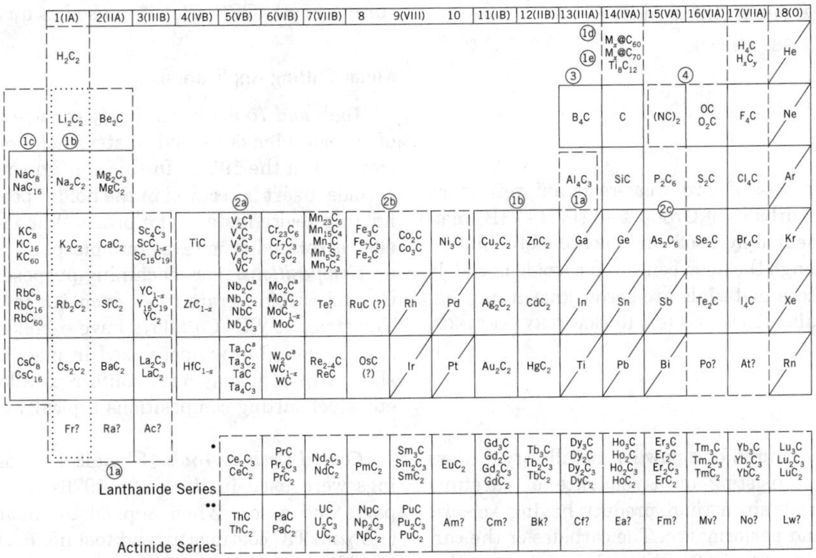

Figure 1. Binary compounds of carbon and their position in the Periodic Table. Numerical designations are ⓐ1a, saltlike carbides; 1b, acetylides; 1c, metal graphite compounds; 1d, metal fullerene compounds; 1e, metallo-carbohedrene clusters; 1f carbonitride 2a, metallic carbides of metals belonging to Groups 4–6 (IVB-VIB); 2b, metallic carbides of the iron metals, including Mn; 2c, metallic carbides of Group 5; 3, diamondlike carbides; and 4, volatile nonmetallic carbides. A line through the box, eg, ⍁, indicates no carbide formation; however, there may be some solubility of carbon in the melt. A question mark, eg, Ac?, indicates the possibility of a carbide. [a]Several modifications exist.

extremely important in technology. For example, calcium carbide, CaC_2, is a source of acetylene; silicon carbide, SiC, and boron carbide, B_4C, are used as abrasives; tungsten carbide, WC, titanium carbide, TiC, and tantalum (niobium) carbide, TaC(NbC) find use as structural materials at extremely high temperatures or in corrosive atmospheres. Cementite, Fe_3C, and the multimetallic complexes $(Co,W)_6C$, $(Cr,Fe,Mo)_{23}C_6$, and $(Cr,Fe)_7C_3$ are the components in tool steels and Stellite-type alloys responsible for their hardness, wear resistance, and excellent cutting performance. There are also emerging applications of these materials as catalysts.

Figure 1 provides a survey of the most important and well-known binary compounds of carbon, according to their position in the periodic system. They are divided into four main groups: the salt-like, metallic, diamond-like, and volatile compounds of carbon. The nature of the bonding is correspondingly of ionic, metallic, semiconductor, or covalent character, but these divisions are not rigid and there are a number of transitional cases. Figure 1 also introduces some additional subdivisions in order to further characterize the compounds.

A number of compounds, although containing carbon, fit only marginally in the category of carbides. This includes the acetylides and the alkali metal–graphite compounds, the coordination and organometallic compounds, ML_y, with ligands, *L,* attached to a metal center, compounds such as M_xC_{60} or M_xC_{70}, formed from diverse elements by association with fullerene structures, and the volatile compounds of carbon. Carbides are generally prepared by the direct reaction of carbon with metals or metal-like materials at elevated temperatures. Their most outstanding properties are extreme hardness and physical strength combined with high temperature stability.

S. TED OYAMA
Clarkson University
RICHARD KIEFFER
Technical University of Vienna

S. T. Oyama, *Catal. Today* **15** (1992).

L. Toth, *Transition Metal Carbides and Nitrides,* Academic Press, New York, 1971.

G. Hägg, *Z. Physik. Chem.* **B6,** 221 (1929); *B12,* 33 (1931).

S. T. Oyama, *J. Solid State Chem.* **96,** 1 (1992).

CEMENTED CARBIDES

Cemented carbides belong to a class of hard, wear-resistant, refractory materials in which the hard carbides of Group 4–6 (IVB–VIB) metals are bound together or cemented by a soft and ductile metal binder, usually cobalt or nickel. Although the term cemented carbide is widely used in the United States, these materials are better known internationally as hard metals (see also REFRACTORIES; REFRACTORY COATINGS; REFRACTORY FIBERS).

Manufacture

Cemented carbides are manufactured by a powder metallurgy process (see METALLURGY, POWDER), consisting of a sequence of carefully controlled steps designed to obtain a final product having specific properties, microstructure, and performance. The carbides or the carbide solid solution powders are prepared and blended with very fine binder metal powder, cobalt or nickel, in ball mills, vibratory mills, or attritors using carbide balls. The powder blends are compacted, presintered, and shaped, and the carbide subjected to sintering and postsintering operations. The sintered product may either be directly put to use or ground, polished, and coated.

Inhalation of extremely fine carbide, cobalt, and nickel powders should be avoided. Efficient exhaust devices, dust filters, and protective masks are essential when handling these powders.

Recycling of Scrap

Recycling or cemented carbide scrap is of growing importance. In one method, the scrap is heated to 1700–1800°C in a vacuum furnace to vaporize some of the cobalt and embrittle the material. After removal from the furnace the material is crushed and screened. Chemical recycling and the zinc reclaim process are also used.

Tool Wear Mechanisms

The performance of a tool material in a given application is dictated by its response to conditions at the tool tip. High temperatures and stresses can cause blunting from the plastic deformation of the tool tip, whereas high stresses alone may lead to catastrophic fracture. In addition to plastic deformation and fracture, the service life of cutting tools is determined by a number of wear processes, including crater wear, flank wear, built-up edge, depth-of-cut-notching, and thermal fatigue.

Evaluation of Properties

In addition to chemical analysis a number of physical and mechanical properties are employed to determine cemented carbide quality. Standard test methods employed by the industry for abrasive wear resistance, apparent grain size, apparent porosity, coercive force, compressive strength, density, fracture toughness, hardness, linear thermal expansion, magnetic permeability, microstructure, Poisson's ratio, transverse rupture strength, and Young's modulus are set forth by ASTM/ANSI and the ISO.

Among the physical properties, cemented carbide density is very sensitive to composition and porosity of the sample and is widely used as a quality control test. Magnetic properties most often measured are magnetic saturation and coercive force.

The properties and performance of cemented carbide tools depend not only on the type and amount of carbide but also on carbide grain size and the amount of binder metal. Information on porosity, grain size, and distribution of carbide particles and the binder phase is obtained by optical and scanning electron microscopy.

Among the mechanical properties, hardness and transverse rupture strength (TRS) are often used as quality control tests.

Metal-Cutting Applications

Tools and Toolholding. Early carbide metal-cutting tools consisted of carbide blanks brazed to steel holders. Indexable inserts were introduced in the 1950s. In this configuration the so-called throwaway carbide insert is secured in the holder pocket by a clamp or some other holding device instead of a braze. When a cutting edge wears, a fresh edge is rotated or indexed into place.

Compositions. For machining purposes, alloys having 3 to 12 wt % Co and carbide grain sizes from 0.5 to >5 μm are commonly used. The straight WC–Co alloys have excellent resistance to simple abrasive wear and are widely used in machining materials that produce short chips, eg, gray iron, nonferrous alloys, high temperature alloys, etc. steel cutting compositions typically contain WC–TiC–(Ta,Nb)C–Co.

Coated Carbide Tools. Chemical vapor deposited (CVD) TiC coatings were used in the early 1970s to combat wear on steel watch parts and cases. When applied to cutting tools, the relatively thin ($\sim$5 μm) TiC coatings extend tool life in steel and cast-iron machining by a factor of two to three by suppressing the crater wear and flank wear. Hard coatings also reduce frictional forces at the chip/tool interface, which in turn reduce the heat generated in the tool resulting in lower tool tip temperatures. CVD coatings have evolved from single layer TiC coatings having narrow application ranges to multilayer hard coatings comprising various combinations of TiC, TiCN, TiN, and Al_2O_3. Multilayered coatings on specially tailored substrates containing cobalt-enriched peripheries have widened the metalcutting application range of the coated tools. In 1992, coated carbides accounted for nearly 65% of all indexable metal-cutting inserts used in the United States.

In the 1980s physical vapor deposition (PVD) emerged as a commercially viable process for applying hard TiN coatings onto cemented carbide tools. A number of factors make PVD process attractive for use with cemented carbide tools: (*1*) lower deposition temperature (<700°C) prevents η-phase formation and produces finer grain sizes in the coating layer; (*2*) PVD coatings are usually crack-free; (*3*) depending on the deposition technique, compressive residual stresses, which are beneficial in resisting crack propagation, may be introduced in the coating; (*4*) PVD coating preserves the transverse rupture strength of the carbide substrate, whereas the CVD process generally reduces the TRS by as much as 30% (20); (*5*) PVD coatings can be applied uniformly over sharp cutting edges.

PVD coated tools are thus successfully employed in operations where sharp edges are most beneficial, including milling, turning, boring, threading, grooving, and cutoff operations. Newer PVD coatings are rapidly becoming commercially available. These include TiCN, TiAlN, CrC, and CrN.

Nonmetal-Cutting Applications

In the early 1990s almost half of the total production of cemented carbides was used for nonmetal-cutting applications such as mining, oil and gas drilling, transportation and construction, metalforming, structural and fluid-handling components, and forestry tools. The majority of compositions used in these applications comprised straight WC–Co grades.

Economic Aspects

There are more than 200 cemented carbide producers in the world. A majority of hard metal production can be attributed to Cerametal Sarl, GTE Valenite Corporation, Kennametal Inc., Krupp Widia GmbH, Mitsubishi Metal Corporation, Plansee Tizit GmbH, Rogers Tool Works Inc., AB Sandvik Coromant, Sumitomo Electric Industries Ltd., Teledyne Firth Stirling, Toshiba Tungaloy Company, Ltd., and Zhuzhou Cemented Carbide Industry Company. Many of the smaller producers have narrow manufacturing capabilities and a limited range of product offerings.

Developments in materials, coatings, and insert geometries have claimed an increasing share of research and development budgets in the cemented carbide industry. An important economic benefit of these effects has been an increase in tool performance. Continuing developments in computer-numerically controlled machining systems have placed a heavy emphasis on tool reliability and consistency, which in turn puts pressure on the industry to invest increasing amounts of capital in developing new materials and processes.

A. T. SANTHANAM
Kennametal Inc.

K. J. A. Brookes, *World Directory and Handbook of Hardmetals and Hard Materials,* 5th ed., International Carbide Data, East Barnet Hertfordshire, U.K., 1992.

G. Schneider, Jr., *Principles of Tungsten Carbide Engineering,* 2nd ed., Society of Carbide and Tool Engineers, ASM International, Materials Park, Ohio, 1989.

E. M. Trent, *Metal Cutting,* 3rd ed., Butterworth-Heinemann Ltd., Oxford, U.K., 1991.

P. Schwarzkopf and R. Kieffer, *Cemented Carbides,* The Macmillan Company, New York, 1960.

INDUSTRIAL HARD CARBIDES

The four most important carbides for the production of hard metals are tungsten carbide, WC, titanium carbide, TiC, tantalum carbide, TaC, and niobium carbide, NbC. The binary and ternary solid solutions of these carbides such as WC–TiC and WC–TiC–TaC (NbC) are also of great importance. Chromium carbide (3:2), Cr_3C_2, molybdenum carbide, MoC, and molybdenum carbide (2:1), Mo_2C, vanadium carbide, VC, hafnium carbide, HfC, and zirconium carbide, ZrC, have minor significance. Carbides and their solid solutions are generally combined with cobalt and used in the form of cemented carbides. The carbides of the actinides, Th, U, Pu, and Np, have gained importance in reactor technology as nuclear fuels (see NUCLEAR REACTORS).

Preparation

In general, the carbides of metals of Groups 4–6 (IVB–VIB) are prepared by reaction of elementary carbon or hydrocarbons and metals and metal compounds at high temperatures. The process may be carried out in the presence of a protective gas, under vacuum, or in the presence of an auxiliary metal (menstruum). Methods include carburization by fusion, carburization by thermal diffusion, carburization by menstruum process, carburization by thermochemical reaction, and reduction.

Physical properties of primary carbides are given in Table 1.

Economic Aspects

Two categories of refining support manufacturers of cemented carbides. The first involves extraction of tungsten in the form of tungstic

Table 1. Physical Properties of Primary Carbides

Property	WC	TiC	TaC	NbC
mol wt	195.87	59.91	192.96	104.92
carbon, wt %	6.13	20.05	6.23	11.45
density, g/cm^3	15.7	4.93	14.48	7.78
mp, °C	2720	2940	3825	3613
microhardness, kg/mm^2	1200–2500	3000	1800	2000
modulus of elasticity, N/mm^{2a}	696,000	451,000	285,000	338,000
transverse rupture strength, N/mm^{2a}	550–600	240–400	350–450	300–400
coefficient of thermal expansion, K − 1	$a = 5.1 \times 10^{-6}$ $c = 7.3 \times 10^{-6}$	7.74×10^{-6}	6.29×10^{-6}	6.65×10^{-6}
thermal conductivity, W/(m·K)	121	21	22	14
heat of formation, $\Delta H_{f,298}$, kJ/mol[b]	−40.2	−183.4	−146.5	−140.7
specific heat, J/(mol·K)[b]	39.8	47.7	36.4	36.8
electrical resistivity, $\mu\Omega \cdot$ cm	19	68	25	35
superconducting temperature, <K	1.28	1.15	9.7	11.1
magnetic susceptibility	+10	+6.7	+9.3	+15.3

[a] To convert N/mm^2 (MPa) to psi, multiply by 145.
[b] To convert J to cal, divide by 4.184.

oxide, WO_3, or ammonium paratungstate, APT, from mineral concentrates. The second converts the WO_3, APT, and other starting materials to primary carbides of tungsten and other metals. Some refining, especially the preparation of primary carbides, is carried out by the cemented carbide manufacturers themselves. Cemented carbide manufacturers also utilize independent refiners specializing in powders of primary metals, carbides, nitrides, carbonitrides, and many lower volume accessory materials used in cemented carbide production. These suppliers include Metallurg Inc., Hermann C. Starck, and Murex Ltd., among others.

A number of cemented carbide producers market some or all primary materials, ranging from monocarbides to binary or ternary solid solutions involving TiC, TaC, NbC, or WC. Among the suppliers offering these materials are Kennametal Inc., GTE Products Corporation, Teledyne Wah Chang, and Tokyo Tungsten Company, Ltd. Kennametal Inc. also manufactures TiN and TiCN powders.

W. M. STOLL
Consulting Metallurgist
A. T. SANTHANAM
Kennametal Inc.

E. K. Storms, *The Refractory Carbides,* Academic Press, New York, 1967.
H. J. Goldschmidt, *Interstitial Alloys,* Butterworth, London, U.K., 1967.
P. Schwarzkopf and R. Kieffer, *Refractory Hard Metals,* The Macmillan Co., New York, 1953.
S. W. H. Yih and C. T. Wang, *Tungsten,* Plenum Press, New York, 1979.

CALCIUM CARBIDE

Chemically pure calcium carbide, Ca_2C, is a colorless solid; however, the pure material can be prepared only by very special techniques. Commercial calcium carbide is composed of calcium carbide, calcium oxide, CaO, and other impurities present in the raw materials. The commercial product's calcium carbide content varies and is sold based on acetylene yield. Industrial-grade calcium carbide contains about 80% as CaC_2, 15% CaO, and 5% other impurities.

In the United States calcium carbide-based acetylene is mainly used in the oxyacetylene welding market, although some continues to be used for production of such chemicals as vinyl ethers and acetylenic alcohols. Calcium carbide is used extensively as a desulfurizing reagent in steel and ductile iron production, allowing steel mills to use high sulfur coke without the penalty of excessive sulfur in the resultant steel (see SULFUR REMOVAL AND RECOVERY). Calcium cyanamide production continues in Europe (see CYANAMIDES).

Properties

Table 1 lists the more important physical properties of calcium carbide.

Table 1. Physical Properties of Calcium Carbide

Property	Value
mol wt	64.10
mp, °C	2300
specific gravity, commercial-grade	
at 15°C	2.34
electrical conductivity, technical-grade, (ohm-cm − 1)	
at 25°C	3,000–10,000
1700°C[a]	0.36–0.47
viscosity at 1900°C, MPa·s(= cP)	
50% CaC_2	6000
specific heat, 0–2000°C, J/mol·K[b]	74.9
heat of formation, $H_{f,298}$, kJ/mol[b]	−59 ± 8
latent heat of fusion, ΔH_{fus}, kJ/mol[b]	32

[a] Material is a liquid.
[b] To convert from J to cal, divide by 4.184.

Reaction With Water. The exothermic reaction of calcium carbide and water-yielding acetylene forms the basis of the most important industrial use of calcium carbide.

$$CaC_2 + 2\ H_2O \rightarrow C_2H_2 + Ca(OH)_2$$

$$\Delta H = -130\text{kJ/mol}(-31.1\text{kcal/mol})$$

Both wet and dry processes are in use for generating acetylene from calcium carbide.

Reaction With Sulfur. An important use of calcium carbide has developed in the iron (qv) and steel (qv) industries, where the carbide has been found to be an effective desulfurizing agent for blast-furnace iron.

Reaction With Nitrogen. Calcium cyanamide is produced from calcium carbide

$$CaC_2 + N_2 \rightarrow CaCN_2 + C$$

$$\Delta H = -295\text{kJ/mol}(-70.5\text{kcal/mol})$$

The reaction is carried out in a refractory oven by passing nitrogen gas through finely pulverized carbide at a temperature of 1000–1200°C.

Manufacture

Calcium carbide is produced commercially by reaction of high purity quicklime and a reducing agent such as coke in an electric furnace at 2000–2200°C.

$$CaO + 3\ C \rightarrow CaC_2 + CO$$

$$\Delta H = 466\ \text{kJ/mol}(111\ \text{kcal/mol})$$

Commercial calcium carbide, containing about 80% CaC_2, is formed in the liquid state. Impurities are mainly CaO and impurities present in raw materials. CO is usually collected for use as a fuel in lime production or drying of the coke used in the process. The liquid calcium carbide is tapped from the furnace into cooling molds.

Environmental Considerations. The principal environmental problem is the prevention of particulate dust emission, which can be handled by cloth filtration equipment.

Specifications and Shipping

Contracts for acetylene-grade carbide are usually based on size and gas yield specification, and include penalties for carbide that fails to meet specified gas yield. The sizes generally available in the trade are based on established U.S. Government specifications.

Calcium carbide is classed as a hazardous chemical under U.S. Department of Transportation regulations. Domestic shipments are mainly in steel tote bins varying in capacity rom 2.5–4.5 t. Containers must be marked "Flammable solid, dangerous when wet" and have the United Nations designation UN 1402.

Calcium carbide for desulfurization is usually sold on the basis of minimum CaC_2 content, minimum levels of various additives, and a specified size distribution. These designations can vary considerably based on the reagent formulation.

Health and Safety

The usual precautions must be observed around the high tension electrical equipment supplying power.

Although acetylene is considered to be a material having a very low toxicity, a threshold limit value (TLV) of 2500 pm has been established by NIOSH. In the presence of a small amount of water carbide may become incandescent and ignition of the evolved air–acetylene mixture may occur. Nonsparking tools should be used when working in the area of acetylene-generating equipment.

Economic Aspects

The calcium carbide industry began a state of decline in the late 1960s when acetylene from carbide was gradually replaced by petrochemical starting materials. Calcium carbide usage for iron desulfurization began in the United States in the mid-1970s and today accounts for 18% of carbide production. Future growth in calcium carbide usage is expected to be modest.

WILLIAM L. CAMERON
Cyanamid Canada, Inc.

D. W. K. Hardie, *Acetylene, Manufacture and Uses*, Oxford University Press, London, U.K., 1965.

F. W. Kampmann and W. Portz, *Chemicals from Coal via the Carbide Route*, Hoechst A.G., Heurth-Knapsack D-5030, Germany, 1991; *Crit. Rep. Appl. Chem.* **14**, 32–44 (1987).

C. J. Macedo, E. A. O. d'Avila, and J. G. Brosnan, *Startup of a Closed Carbide Furnace Using Charcoal as a Reducing Agent*, Vol. 43, *Electric Furnace Proceeding*, Atlanta, Ga., 1985.

G. E. Healy, *Why a Carbide Furnace Erupts*, *Electric Furnace Proceeding*, Pennsylvania State University, University Park, Pa., 1965.

SILICON CARBIDE

Silicon carbide, SiC, is a crystalline material having a color that varies from nearly clear through pale yellow or green to black, depending on the amount of impurities. It occurs naturally only as the mineral moissanite in the meteorite iron of Canon Diablo, Arizona. The commercial product, which is made in an electric furnace, is usually obtained as an aggregate of iridescent crystals. The iridescence is caused by a thin layer of silica produced by superficial oxidation of the carbide. The loose black or green grain of commerce is prepared from the manufactured product by crushing and grading for size.

Traditionally, the metallurgical, abrasive, and refractory industries are the largest users of silicon carbide (see ABRASIVES; METALLURGY; REFRACTORIES). SiC is also used for heating elements in electric furnaces (see FURNACES, ELECTRIC), in electronic devices, and in applications where its resistance to nuclear radiation damage is advantageous. The development of advanced pressureless sintering and complex shape-forming technologies has led to silicon carbide becoming one of the most important structural ceramics. Silicon carbide has found wide acceptance in wear-, erosion-, and corrosion-resistant applications; it has demonstrated excellent performance as a heat-exchanger material; it is also being evaluated for prototype high temperature gas turbine engine component applications.

Properties

The properties of silicon carbide depend on purity, polytype, and method of formation. The measurements made on commercial, polycrystalline products should not be interpreted as being representative of single-crystal silicon carbide. The pressureless-sintered silicon carbides, being essentially single-phase, fine-grained, and polycrystalline, have properties distinct from both single crystals and direct-bonded silicon carbide refractories. Table 1 lists the properties of the fully compacted, high purity material.

Crystal Structure. Silicon carbide may crystallize in the cubic, hexagonal, or rhombohedral structure.

Mechanical Properties. Silicon carbide is a leading candidate material for rotating and static components in many gas turbine engine applications. As is the case for other ceramics, silicon carbide is brittle in nature. It is characterized by low fracture toughness and limited strain-to-failure as compared to metals. The strength of a silicon carbide component is determined by preexisting flaws introduced into the material during processing.

Table 1. Properties of Silicon Carbide

Property	Value
mol wt	40.10
decomposition temperature[a], °C	
α-form	2825 ± 40
β-form	2985
sp gr, g/mL at 20°C	
β-form	3.210
6H polytype[b]	3.211 (3.208)
commercial	3.16
refractive index	
β-form	2.48
α-form[b]	ε ω
4H	2.712 2.659
6H	2.690 2.647
15R	2.687 2.650
heat of formation, $\Delta H^\circ_{f,298}$, kJ/mol[c]	
α-form	−25.73 ± 0.63
β-form	−28.03 ± 2
thermal conductivity at 25°C, W/(m·K)	
commercial, high density	125.6
reaction bonded emissivity[d]	129.7
spectra (3–5 μm)	0.9
coefficient of thermal expansion[d], per °C	
25–200°C	2.97×10^{-6}
700–1500°C	6.08×10^{-6}

[a] The decomposition products are Si, Si_2C, Si_2, SiC, and Si_3.
[b] H = hexagonal, R = rhombohedral.
[c] To convert J to cal, divide by 4.184.
[d] Of the sintered α-form.

Sintered silicon carbide retains its strength at elevated temperatures and shows excellent time-dependent properties such as creep and slow crack growth resistance.

Electrical Properties. The electrical properties of silicon carbide are highly sensitive to purity, density, and even to the electrical and thermal history of the sample.

Resistivity. The temperature coefficient of electrical resistivity of commercial silicon carbide at room temperature is negative. No data are given for refractory materials because resistivity is greatly influenced by the manufacturing method and the amount and type of bond. Manufacturers should be consulted for specific product information.

Semiconducting Properties. Silicon carbide is a semiconductor: it has a conductivity between that of metals and insulators or dielectrics. Because of the thermal stability of its electronic structure, silicon carbide has been studied for uses at high (>500°C) temperature.

Radiation Effects. Alpha silicon carbide exhibits a small degree of anisotropy in radiation-induced expansions along the optical axis and perpendicular to it.

Reactions

Silicon carbide is comparatively stable. The only violent reaction occurs when SiC is heated with a mixture of potassium dichromate and lead chromate. Chemical reactions do, however, take place between silicon carbide and a variety of compounds at relatively high temperatures.

Manufacture and Processing

Powder Preparation. There are several routes to preparing SiC powders having variable purity levels, crystal structure, particle size, shape, and distribution. Methods that have been examined include growth by sublimation from the vapor phase, carbothermic reduction, and crystallization from a melt.

Monolithic Sintered Silicon Carbide. In 1973, it was demonstrated that the simultaneous presence of boron and carbon is required to densify, without pressure, a green compacted body of beta silicon carbide. Pressureless sintering techniques were subsequently established for α-SiC powder made by inexpensive methods such as the Acheson process.

Reinforcements. The high modulus, high intrinsic strength, and temperature stability make SiC, in the form of whiskers, platelets, and fibers, a promising candidate reinforcement material for metal, polymer, and ceramic matrix composites (qv).

In 1988, four firms were producing crude silicon carbide under various trade names at six plants in the United States and Canada: the ExolonESK Company; General Abrasive/Dresser Company; Norton Company; and Superior Graphite Company.

Toxicity

Silicon carbide has been described as a mild inhalation irritant. The threshold limit value for silicon carbide in the atmosphere is 5 mg/m^3. Because of increased interest in SiC whiskers as a reinforcement for composites, the ASTM has established Subcommittee E34.70 on Single-Crystal Ceramic Whiskers to write standards for handling this form of SiC.

R. DIVAKAR
K. Y. CHIA
S. M. KUNZ
S. K. LAU
The Carborundum Company

R. C. Marshall and co-workers, eds., *Silicon Carbide-1973,* University of South Carolina Press, Columbia, S.C., 1974.

H. H. Woodbury and G. W. Ludwig, *Phys. Rev.* **124,** 1083 (1961); W. J. Choyke and co-workers, *Phys. Rev.* **133,** A1163 (1964); R. L. Hartman and P. J. Dean, *Phys. Rev.* **B2,** 951 (1970); L. A. Hemstreet and C. Y. Fong, *Solid State Commun.* **9,** 643 (1971); W. J. Choyke and L. Patrick, *Phys. Rev.* **187,** 1041 (1969); L. A. Hemstreet and C. Y. Fong, *Phys. Rev.* **B6,** 1464 (1972).

U.S. Pat. 4,312,954 (1982), J. A. Coppola, L. N. Hailey, and C. H. McMurtry.

N. I. Sax, *Dangerous Properties of Industrial Materials,* Reinhold Book Corp., New York, 1984, p. 2398.

CARBOHYDRATES

Carbohydrates are found in all plant and animal cells. They are the most abundant of the organic compounds, so abundant that it is estimated that well over half of the organic carbon on earth exists in the form of carbohydrates. Most carbohydrates are produced and found in plants. Carbohydrate molecules make up about three-fourths of the dry weight of plants; most of this is found in cell walls as structural components. Carbohydrates also constitute important energy reserves in plants; one carbohydrate, starch, provides about three-fourths of the calories in the average human diet on a worldwide basis. But the nutritional aspects are only a part of the story of carbohydrates. They have many important industrial uses in such diverse areas as the adhesive, agricultural chemical, fermentation, food, paper and related products, petroleum production, pharmaceutical, and textile industries. Because the basic carbohydrate molecule is functionalized at every carbon atom, and because carbohydrates seldom occur as simple sugars but rather combined with each other or other compounds, the variety of carbohydrates in nature is large, and the number of theoretical possibilities is essentially limitless.

Classification

The basic carbohydrate molecule possesses an aldehyde or ketone group and a hydroxyl group on every carbon atom except the one involved in the carbonyl group. As a result, carbohydrates are defined as aldehyde or ketone derivatives of polyhydroxy alcohols and their reaction products. A look at the formula for glucose ($C_6H_{12}O_6$) shows that it contains hydrogen and oxygen atoms in the ratio in which they are found in water. The name carbohydrate (hydrate of carbon) is derived from the fact that the basic carbohydrate molecule has the formula $C_n(H_2O)_n$. In common practice, all low molecular weight carbohydrates are called sugars.

Numerical prefixes designated the number of carbon atoms are tri-, tetra-, penta-, hexa-, hepta-, etc. In systemic nomenclature, the suffix for the names of aldehyde sugars is -ose and for ketose sugars -ulose. However, common names are routinely used, creating exceptions in both cases.

Monosaccharides are most often joined together in chains. Oligosaccharides are carbohydrate chains that yield 2–20 monosaccharide molecules upon hydrolysis.

Most carbohydrates exist in the form of polysaccharides. Polysaccharides are the principal components of cell walls of land plants (cellulose), seaweeds, and some microorganisms and store energy (starch in plants and glycogen in animals). They are important in the human diet and in many commercial applications.

Chemistry of Saccharides

Most carbohydrates have two kinds of reactive groups: the carbonyl group and primary and secondary hydroxyl groups.

REACTIONS OF THE CARBONYL GROUP

Ring Forms. Aldehydes and ketones react with compounds containing a hydroxyl group (alcohols) to form first hemiacetals and then acetals. Because aldose and ketose molecules have a carbonyl group and hydroxyl groups on the same carbon chain, they can form hemiacetals intramolecularly, as well as by reacting with another molecule. Such an intramolecular reaction forms a ring. The most common rings are the six-membered pyranose ring, a cyclic structure composed of five carbon atoms and one oxygen atom, and the five-membered furanose ring, a cyclic structure composed of four carbon atoms and one oxygen atom.

Glycosides, Oligosaccharides, and Polysaccharides. Few monosaccharides are found free in nature, and these few are usually present in only small amounts. Most monosaccharides occur in combinations, most often with either more of the same sugar or different sugars in the form of polymers (polysaccharides). Less frequently, except in the case of sucrose, they are joined together in oligosaccharide chains. Mono- and oligosaccharides may also be linked to nonsugar organic compounds. These combined forms of sugars are known as glycosides.

For the most part, low molecular weight carbohydrates of commerce are made by depolymerization via enzyme- or acid-catalyzed hydrolysis of polysaccharides. Only sucrose and, to a very much lesser extent, lactose, both disaccharides, are commercial low molecular weight carbohydrates not made in this way.

Oligo- and polysaccharides have reducing and nonreducing ends. A reducing sugar is a carbohydrate that contains an aldehyde or ketone group, either free or in a hemiacetal form. The reducing end of an oligo- or polysaccharide is the one end not involved in a glycosidic linkage and can, therefore, react as an aldehyde or ketone. The sugar units constituting all other ends are attached through glycosidic (acetal) bonds and are, therefore, nonreducing ends. Reducing and nonreducing ends can be demonstrated with the structure of lactose, the reducing disaccharide of milk, β-D-galactopyranosyl-α-D-glucopyranose.

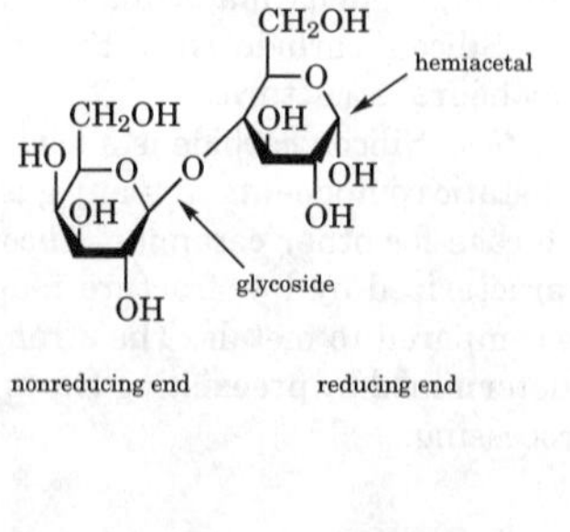

Additional sugar units added to either end of disaccharides form higher oligosaccharides.

By far the most abundant of the naturally occurring oligosaccharides is the disaccharide sucrose, ordinary table sugar from sugar cane or sugar beets (see SUGAR). The two monosaccharide units in sucrose are α-D-glucopyranosyl and β-D-fructofuranosyl units. The two units are joined head-to-head, making sucrose a nonreducing disaccharide, an unusual structure in nature.

Polysaccharides are naturally occurring polymers of monosaccharide (sugar) units. In precise chemical nomenclature, polysaccharides are glycans and are described as being composed of glycosyl units.

In commerce, hydrolysis of glycosidic bonds is far more important than is condensation of sugars with alcohols or other sugar units to form glycosidic bonds. Glycosidic bonds are formed in nature via biosynthetic reactions, and compounds containing them are isolated and used as starting materials for various transformations. Hydrolysis, whether catalyzed by acids or enzymes, follows the same general mechanism.

Synthetic Methods. Although mono- and oligosaccharides are most often made by depolymerization of polysaccharides, oligosaccharides and other compounds with glycosidic bonds can be made synthetically. The classic and still widely used reaction is the Koenigs-Knorr reaction; many modifications of it are known.

Glycoconjugates. Another class of carbohydrates are the glycoconjugates, composed of glycoproteins, proteoglycans, peptidoglycans, and glycolipids.

Oxidation to Sugar Acids and Lactones. When the aldehyde group of an aldose is oxidized, the resulting compound is an aldonic acid (salt form = aldonate)). Some aldonic acids are products of carbohydrate metabolism.

Reduction. Mono- and oligosaccharides can be reduced to polyols (polyhydroxy alcohols) termed alditols (glycitols). Common examples of compounds in this class are D-glucitol (sorbitol) made by reduction of D-glucose and xylitol made from D-xylose.

Cyclitols. Cyclitols are polyhydroxycycloalkanes and -alkenes. They are widely distributed in nature, though never in large quantities. The most abundant of these carbocyclic compounds are the hexahydroxycyclohexanes, commonly called inositols, and their methyl ethers.

REACTIONS OF HYDROXYL GROUPS

Reduction and Oxidation. Hydroxyl groups can be both oxidized to carbonyl groups and removed by reduction. Sugars that have the hydroxyl group missing from one or more of the carbon atoms are called deoxy sugars.

Etherification. Both intramolecular and intermolecular ethers can be formed from carbohydrate hydroxyl groups.

Acetalation. As polyhydroxy compounds, carbohydrates react with aldehydes and ketones to form cyclic acetals.

Replacement of Hydroxyl Groups. Replacement of a hydroxyl group with an amino group at any position produces an aminodeoxysugar.

Isomerization. An aldose, can be converted into the C-2 epimeric aldose and the corresponding ketose, and a ketose can be converted into the two corresponding C-2 epimeric aldoses by base- on enzyme-catalyzed isomerization. It is for this reason that ketoses are reducing sugars.

MODIFICATIONS OF THE CARBON CHAIN

Branched-chain sugars are found in nature, eg, cladinose, ie, 2,6-deoxy-3-*C*-methyl-3-*O*-methyl-L-ribohexose, a component of erythromycin.

Unsaturated sugars are useful synthetic intermediates. The most commonly used are the so-called glycals (1,5- or 1,4-anhydroalditol-1-enes).

Uses of Saccharides

Carbohydrates have widespread utilization, both as low cost, high volume commodities and as low volume specialty chemicals. Significant uses in terms of volume are surveyed here. Not covered are the lower volume uses involving carbohydrates either in the native state or in modified form; these are mainly pharmaceutical applications involving antibiotics, antigens, and synthetic drugs. In the latter case monosaccharides are becoming increasingly important as chiral synthons (chirons) as well as being used more directly to make products such as the nucleoside analogues AraA [9-(β-D-arabinofuranosyl)-9*H*-purin-6-amine], an antineoplastic and antiviral compound known by a number of trade names, and AZT (3′-azido-3′-deoxythymidine,) an antiviral compound also known by a variety of trade names (see ANTIVIRAL AGENTS).

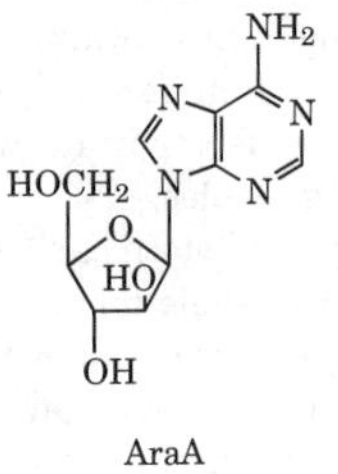

AraA

The considerable uses of carbohydrates as carbon sources for various fermentations or the uses of unrefined carbohydrates, flours for example, are also not described here (see FERMENTATION).

Monosaccharides. Crystalline α-D-glucopyranose is generally sold as dextrose. Glucose is also isomerized to D-fructose to produce high fructose corn syrup (HFCS). Crystalline D-fructose also finds use in the food industry. The annual consumption, in the United States, of dextrose is >600,000 t and of HFCS >8,000,000 t (71% solids basis).

Oligosaccharides. Sucrose is widely used in the food industry to sweeten, control water activity, add body or bulk, provide crispness, give surface glaze or frost, form a glass, provide viscosity, and impart desirable texture. It is used in a wide variety of products from bread to medicinal syrups.

Oligo- and higher saccharides are produced extensively by acid- or enzyme-catalyzed hydrolysis of starch, generally in the form of syrups of mixtures. These products are classified by their dextrose equivalency (DE), which is an indication of their molecular size and is a measure of their reducing power, with the DE value of anhydrous D-glucose defined as 100.

The cyclodextrins or cycloamyloses, a family of cyclic oligosaccharides containing α-D-glucopyranosyl units, most commonly seven (β-cyclodextrin, cycloheptaamylose, cyclomaltoheptaose), with complexing properties, are potentially useful in the food industry to provide stable flavors and fragrances in dry powder form, in the pharmaceutical industry, and in other applications where increased chemical or physical stability, solubility control, or controlled release is desired, eg, with agricultural chemicals (see INCLUSION COMPOUNDS).

Polysaccharides. Since polysaccharides are the most abundant of the carbohydrates, it is not surprising that they comprise the greatest part of industrial utilization. Most of the low molecular-weight carbohydrates of commerce are produced by depolymerization of starch. Polysaccharide materials of commerce can be thought of as falling into three classes: cellulose, a water-insoluble material; starches, which are not water-soluble until cooked; and water-soluble gums.

Cellulose. Cellulose (qv) is the principal cell-wall component of higher plants and the most abundant polysaccharide. Approximately one-half the mass of perennial plants and one-third the mass of annual plants is cellulose. The greatest amount of cellulose used is the purified, but not highly purified, wood pulp that is used in the manufacture of paper (qv), associated products, absorbants, rayons, and nonwovens. A number of derivatives of cellulose are also commercial entities. The water-soluble ones are covered later.

Hemicelluloses and Related Polysaccharides. Hemicelluloses are a large group of polysaccharides that are associated with cellulose in the primary and secondary cell walls of all higher plants, but otherwise have no relationship to cellulose. They are also present in some other plants.

Hemicelluloses (qv) are heteroglycans. They do not comprise a distinct class of chemical structures. Constituent monosaccharides are D-xylose, D-mannose, D-glucose, D-galactose, L-galactose, L-arabinose, D-glucuronic acid, 4-*O*-methyl-D-glucuronic acid, D-galacturonic acid, and to a lesser extent L-rhamnose, L-fucose, and various methyl ethers of neutral sugars, with a limit of perhaps six different glycosyl units per molecule. Both woody and nonwoody tissues contain 20–35% hemicelluloses. Some are neutral polymers, but most are acidic.

Starches. Starch (qv) granules must be cooked before they will release their water-soluble molecules. Starch use permeates the entire economy because it (corn starch in particular) is abundantly available and inexpensive. Another key factor in its widespread use is the fact that it occurs in the form of granules.

All green plants package and store carbohydrate (D-glucose) in the form of starch granules. In granule form, starch is quasi-crystalline (displays spherocrystalline patterns), dense, insoluble in cold water, and only partially hydrated. The sizes and shapes of granules are specific for the plant of origin. Granules can be easily isolated from suspensions by filtration or centrifugation, resuspended, reacted, and recovered.

Normal corn starch is composed of 20–30% of the linear polysaccharide amylose and 70–80% of the branched polysaccharide amylopectin. Through genetic manipulation, corn cultivars with altered starch compositions have been developed. Various modified and derivatized starches are produced by treating a slurry of starch granules with chemicals or enzymes.

Oxidized Starches. Alkaline hypochlorite treatment introduces carboxyl and carbonyl groups, effects some depolymerization, and produces whiter (bleached) products that produce softer, clearer gels. Most of the hypochlorite-oxidized starch and all the ammonium persulfate-oxidized starch is used in the paper industry. The low solution viscosity and good binding and adhesive properties of these products make them especially effective in high solids, pigmented coatings.

Dextrins. Dextrins, like oxidized starches, are so-called converted starches. High solids solutions of some of the more highly converted dextrins produce the tacky, quick-setting adhesives used in paper products.

Acid-Modified Starches. Acid-modified starches are prepared by treating a suspension of starch granules with dilute mineral acid. Acid-modified starches, also called thin-boiling and acid-thinned starches, are used in large quantities as textile warp sizes.

Starch Ethers. A large number of starch ethers have been prepared and patented; only a few are manufactured and used commercially. Commercially available starch ethers are the hydroxyalkyl ethers, hydroxyethylstarch and hydroxypropylstarch, and cationic starches.

Hydroxyethylstarch is widely used with synthetic latexes in the surface sizing of paper and as a coating binder. For these uses, the hydroxyethylstarch is acid-thinned, oxidized, or dextrinized. Hydroxypropylstarch is used in foods to provide viscosity stability and to ensure water-holding during low temperature storage.

Starch Esters. As with the starch ethers, a large number of starch esters have been prepared and patented, but only a few are manufactured and used commercially.

Starch acetates are used in foods to provide paste clarity and viscosity stability at low temperatures. Starch phosphate monoesters are used primarily to make puddings and oil-in-water emulsions.

Cross-Linked Starches. The polymer chains in starch granule can be cross-linked with difunctional reagents that form diethers or diesters. The properties imparted to the starch by such cross-linking are unique and, therefore, these derivatives are considered separately.

Food starches, especially those made from waxy maize, potato, and tapioca starch are usually both cross-linked and phosphorylated, acetylated, or hydroxypropylated to provide appropriate gelatinization, viscosity, and textural properties. Examples of their application are their use in canned foods that are to be retort-sterilized and in the preparation of spoonable salad dressings where products stable to high shear at low pH are required.

Cationic Starches. Commercial cationic starches are starch ethers that contain a tertiary amino or quaternary ammonium group. Cationic starches are used in paper making.

Pregelatinized Starches. Suspensions of starches and starch derivatives can be gelatinized/cooked and dried to yield a variety of products that can be dispersed in cold water to yield pastes comparable to those obtained by cooking granular starch products. These products are made for convenience of use.

Starch Graft Copolymers. A product made by polymerizing acrylonitrile onto gelatinized starch, converting the resulting nitrile groups into a mixture of carbamoyl and alkali metal carboxylate groups by treatment with alkali, and drying will reversibly absorb many hundreds of times its weight of water without dissolving. This product finds use as an additive to absorbent soft goods, such as disposable diapers, incontinent pads, hospitals bed pads, bandages and catamenials, in base root coating of nursery stock, in seed coating, and in hydrogel wound dressing.

Cold-Water Swelling Starches. Special physical treatment produces starch granules that will swell in water without heating. Molecular dispersions can be formed by application of shear to the swollen granules.

General Properties of Starches. Heating a starch in water causes the granules to swell. At sufficient solids concentration the swollen granules occupy most of the space, and a viscous mass called a paste results. Application of shear to these fragile, swollen granules results in formation of a molecular dispersion. The process of granule swelling with concurrent hydration and solubilization of starch molecules is called gelatinization. Gelatinization is accompanied by a loss of birefringence. The temperature at which this occurs is called the gelatinization temperature.

The viscosity obtained by cooking a suspension of starch is determined by the starch type, derivatization or modification, solids concentration, pH, amount of agitation during heating, rate of heating, maximum temperature reached, time held at that temperature, agitation during holding, and the presence of other ingredients.

Water-Soluble Gums. Gums (qv) are polymeric substances which, in an appropriate solvent or swelling agent, form highly viscous dispersions or gels at low dry substance content. Commonly, the term industrial gums refers to water-soluble polysaccharides, glycans in official carbohydrate nomenclature, or polysaccharide derivatives used industrially. They are classified both by structure and by source (Table 1). Particularly in the food industry, the term hydrocolloid is often used interchangeably with gum.

The usefulness of such industrial gums is based on their physical properties, in particular their capacity to thicken or gel aqueous solutions and otherwise to control water. Because all gums modify or control the flow of aqueous solutions, dispersions, and suspensions, the choice of which gum to use for a particular application often depends on its secondary characteristics. These secondary characteristics are responsible for their utilization as adhesives, binders, bodying agents, bulking agents, crystallization inhibitors, clarifying agents, cloud agents, emulsifying agents, emulsification stabilizers, encapsulating agents, film formers, flocculating agents, foam stabilizers, gelling materials, mold release agents, protective colloids, suspending agents, suspension stabilizers, swelling agents, syneresis inhibitors, texturing agents, and whipping agents, in coatings, and for water absorption and binding.

Gums are tasteless, odorless, colorless, and nontoxic. None, except the starches and starch derivatives, are broken down by human digestive enzymes. All are subject to microbiological attack. All can be depolymerized by acid- and enzyme-catalyzed hydrolysis of the glycosidic (acetal) linkages joining the monomeric (saccharide) units.

All native and modified polysaccharides have a range of molecular weights, and the average composition and distribution of molecular weights in a gum sample can vary with the source, the conditions used for isolation or preparation, and any subsequent treatment(s).

Table 1. Classification of Commercial Polysaccharides by Source

Class	Examples
algal (seaweed) extracts	agars, algins, carrageenans, furcellarans, laminarans
higher plants	
insoluble	cellulose
extract	pectins
seeds	corn starches, wheat starch, guar gum, locust bean gum, psyllium seed gum
tubers and roots	potato starch, tapioca starch, konjac mannan
exudates	gum arabics, gum karaya, gum tragacanth
microorganisms (fermentation gums)	curdlan, dextrans, gellan, pullulan, scleroglucan, welan, xanthans
animal	chitins/schitosans
derived	
from cellulose	carboxymethylcelluloses, cellulose acetates, cellulose acetate butyrates, cellulose nitrates, ethylcelluloses, hydroxyalkylcelluloses, hydroxyalkylalkylcelluloses, methylcelluloses
from starch	starch acetates, starch 1-octenylsuccinates, starch phosphates, starch succinates, hydroxyethylstarches, hydroxypropylstarches, cationic starches, oxidized starches, dextrins
from guar gum	carboxymethylguar gum, carboxymethyl(hydroxypropyl)guar gum, hydroxyethylguar gum, hydroxypropylguar gum, cationic guar gum
synthetic	polydextrose

JAMES N. BEMILLER
Purdue University

W. Pigman and D. Horton, eds., *The Carbohydrates: Chemistry and Biochemistry,* Vol. IA, IB, IIA, IIB, 2nd ed., Academic Press, New York, 1970–1972.

G. O. Aspinall, ed., *The Polysaccharides,* Vol. 1, 2, 3, Academic Press, New York/Orlando, 1982–1985.

R. L. Whistler and J. N. BeMiller, eds., *Industrial Gums,* 3rd ed., Academic Press, San Diego, Calif., 1993.

J. W. Rowe, ed., *Natural Products of Woody Plants I. Chemicals Extraneous to the Lignocellulosic Cell Wall,* Springer-Verlag, Berlin, 1989.

R. L. Whistler, J. N. BeMiller, and E. F. Paschall, eds., *Starch: Chemistry and Technology,* 2nd ed., Academic Press, Orlando, Fla., 1984.

Advances in Carbohydrate Chemistry and Biochemistry, Academic Press, Inc., San Diego, Calif.

H. S. El Khadem, *Carbohydrate Chemistry,* Academic Press, Inc., San Diego, Calif., 1988.

P. M. Collins, ed., *Carbohydrates,* Chapman and Hill, London, 1987.

CARBON

CARBON AND ARTIFICIAL GRAPHITE

STRUCTURE AND TERMINOLOGY

Elemental carbon, atomic number six in the periodic table, at wt 12.011, occurs naturally throughout the world in either its crystalline (more ordered) or amorphous (less ordered) form. Carbonaceous materials such as soot or charcoal are examples of the amorphous form, whereas graphite and diamond are crystalline. Carbon atoms bond with other carbon atoms as well as with other elements, principally hydrogen, nitrogen, oxygen, and sulfur, to form carbon compounds, which are the subject of organic chemistry. The manufactured form of carbon and graphite is discussed within this article. In its many varying manufactured forms, carbon and graphite can exhibit a wide range of electrical, thermal, and chemical properties that are controlled by the selection of raw materials and thermal processing during manufacture.

Crystallographic Structure

There are two allotropes of carbon: diamond and graphite. The diamond, or isotropic form, has a crystal structure that is face-centered cubic with interatomic distances of 0.154 nm. Each atom is covalently bonded to four other carbon atoms in the form of a tetrahedron. The nature of the bonding explains the differences in properties of the two allotropic forms. The hardness of diamond is derived from the regular three-dimensional network of σ-bonds; the low electrical conductivity results from fixed-bonding electrons between atoms within the diamond lattice. Graphite, or the anisotropic form, has a structure that is composed of infinite layers of carbon atoms arranged in the form of hexagons lying in planes.

The electronic ground state of carbon is $1s^2$, $2s^2$, $2p^2$, ie, there are four electrons in the outer shell available for chemical bonding. In diamond, the $2s$ and $2p$ electrons mix to form four equivalent covalent σ-bonds. In graphite, three of the four electrons form strong covalent π-bonds with the adjacent in-plane carbon atoms. The fourth electron forms a less strong bond of the van der Waals type between the planes.

Terminology

A wide variety and range of bulk carbon forms are available within the industry. In general, commercial forms are loosely characterized as carbon or graphite, but they are distinctly different. The term *manufactured carbon* (sometimes called formed carbon, amorphous carbon, or baked carbon) refers to a bonded granular carbon body whose matrix has been subjected to a temperature typically between 900 and 2400°C. Manufactured graphite (sometimes called synthetic, artificial graphite, electrographite, or graphitized carbon) refers to a bonded granular carbon body whose matrix has been subjected to a temperature typically in excess of 2400°C and whose matrix is thermally stable below that temperature.

J. C. LONG
UCAR Carbon Company Inc.

T. Ishikawa, T. Nagaoki, and I. C. Lewis, *Recent Carbon Technology,* IEC Press, Cleveland, Ohio, 1983.

H. Marsh, *Introduction to Carbon Science,* Butterworths, London, U.K., 1989.

P. L. Walker, Jr. and P. A. Thrower, eds., *Chemistry and Physics of Carbon; A Series of Advances,* Vol. 1 (and continuing) Marcel Dekker, New York, 1965.

UCAR Carbon Co., Inc., *Industrial Graphite Engineering Handbook,* UCAR Carbon Co., Inc., Danbury, Conn., 1991.

BAKED AND GRAPHITIZED CARBON

Raw Materials

The raw materials used in the production of manufactured carbon and graphite largely control the ultimate properties and practical applications of the final product. This dependence is related to the chemical and physical nature of the carbonization and graphitization processes.

Essentially any organic material can be thermally transformed to carbon. The carbonization process through the elimination of heteroatoms and substituent hydrogen converts the organic precursor into a carbon polymer. This polymer consists of aromatic carbons arranged in large polynuclear aromatic ring systems. With continued heat treatment, this carbon is transformed to a more or less ordered three-dimensional framework approaching the structure of graphite. Differences in the final material depend on the ease and extent of completion of these overall chemical and physical ordering processes.

Filler Materials. Filler materials include petroleum coke, coal-tar pitch coke, natural graphite, carbon blacks and anthracite.

Binders. *Pitches.* Carbon articles are made by mixing a controlled size distribution of coke filler particles with a binder such as coal-tar or petroleum pitch.

Additives. In addition to the primary ingredients, the fillers and binders, minor amounts of other materials are added at various steps in the carbon and graphite manufacturing process. Although the amounts of these additives are usually small, they can play an important role in determining the quality of the final product. Light extrusion oils and lubricants, including petroleum oils, waxes, fatty acids, and esters, are often added to the mix to improve the extrusion rates and structure of the extruded products. Chemical inhibitors are introduced to reduce the detrimental effects of sulfur in high sulfur cokes. Iron oxide is often added to high sulfur coke to prevent puffing, the rapid swelling of the coke caused by volatilization of the sulfur at 1600–2400°C.

Calcining

Nearly all coke utilized in carbon manufacture is calcined. Calcination consists of heating raw coke to remove volatiles and to shrink the coke to produce a strong, dense particle.

I. C. LEWIS
UCAR Carbon Company Inc.

H. Marsh and P. L. Walker, in P. L. Walker and P. Thrower, eds., *Chemistry and Physics of Carbon,* Vol. 15, Marcel Dekker, New York, 1979, p. 230.

C. L. Mantell, *Carbon and Graphite Handbook,* Interscience Publishers, New York, 1968.

PROCESSING OF BAKED AND GRAPHITIZED CARBON

Raw Material Preparation

Crushing and Sizing. Calcined petroleum coke arrives at the graphite manufacturer's plant in particle sizes ranging typically from dust to 50–80 mm diameter. In the first step of artificial graphite production, the run-of-kiln coke is crushed, sized, and milled to prepare it for the subsequent processing steps.

The wide variety of equipment available for the crushing and sizing operations is well described in the literature. Roll crushers are commonly used to reduce the incoming coke to particles that are classified in a screening operation. The crushed coke fraction, smaller than the smallest particle needed, is normally fed to a roll or hammer mill for further size reduction to the very fine (flour) portion of the carbon mix.

Proportioning. The size of the largest particle is generally set by application requirements. Generally, the guiding principle in designing carbon mixes is the selection of the particle sizes, the flour content, and their relative proportions in such a way that the intergranular void space is minimized. If this condition is met, the volume remaining for binder pitch and the volatile matter generated in baking are also minimized.

Mixing. Once the raw materials have been crushed, sized, and stored in charging bins and the desired proportions established, the manufacturing process begins with the mixing operation. The purpose of mixing is to blend the coke filler materials and distribute the pitch binder over the surfaces of the filler grains as it melts or is added as a liquid. The intergranular bond ultimately determines the properties and structural integrity of the graphite. Thus the more uniform the binder distribution is throughout the filler components, the greater the likelihood for a structurally sound product.

Forming. One purpose of the forming operation is to compress the mix into a dense mass so that pitch-coated filler particles and flour are in intimate contact. Two important methods of forming are extrusion and molding.

Baking. In the next stage, the baking operation, the product is fired to 800–1000°C. One function of this step is to convert the thermoplastic pitch binder to solid coke. Another function of baking is to remove most of the shrinkage in the product associated with pyrolysis of the pitch binder at a slow heating rate. This procedure avoids cracking during subsequent graphitization where very fast firing rates are used.

A variety of baking furnaces are in use to provide the flexibility needed to bake a wide range of product sizes and to generate the best possible temperature control. One common baking facility is the pit furnace, so named because it is positioned totally or partially below ground level to facilitate improved insulation. Another common baking facility is the so-called ring furnace. Two equal rows of pit furnaces are arranged in a rectangular ring. A more recent development is the carbottom furnace, which is an above-ground rectangular kiln; the bottom is mounted on wheels and set on tracks so it is movable.

In addition, the development of stainless steel cans, saggers, into which green stock is loaded and then surrounded by packing media, has improved furnace heating effectiveness by reducing the ratio of packing media to green electrode.

Table 1. Effect of One Pitch Impregnation on Graphite Properties

	Unimpregnated[a]		Impregnated[b]	
Property	wg[c]	ag[d]	wg[c]	ag[d]
Young's modulus, GPa[e]	7.4	4.4	11.0	6.3
flexural strength, MPa[f]	10.	7.1	17	13
tensile strength, MPa[f]	5	4.4	8.1	7.3
compressive strength, MPa[f]	21	21	34	33
permeability, μm^2[g]	0.39	0.35	0.19	0.16
coefficient of thermal expansion, 10^{-6}/°C	1.3	2.7	1.5	3.1
specific resistance, $\mu\Omega \cdot m$	8.8	13.0	7.6	11.0

[a] Bulk density = 1.6 g/mL.
[b] Bulk density = 1.7 g/mL.
[c] With the grain, ie, samples cut parallel to the molding plane or extrusion axis.
[d] Across the grain, ie, samples cut perpendicular to the molding or extrusion force.
[e] To convert GPa to psi, multiply by 145,000.
[f] To convert MPa to psi, multiply by 145.
[g] To convert μm^2 to darcys, divide by 0.9869.

Impregnation. In some applications, the baked product is taken directly to the graphitizing facility for heat treatment to 3000°C. However, for many high performance applications of graphite, the properties of stock processed in this way are inadequate. The method used to improve those properties is impregnation with coal-tar or petroleum pitches. Table 1 shows the graphite properties of unimpregnated and impregnated stock 150–300 mm diameter and containing 1.5-mm particles as the largest particle.

Graphitization. Graphitization is an electrical heat treatment of the product to ca 3000°C. The purpose of this step is to cause the carbon atoms in the petroleum coke filler and pitch coke binder to orient into the graphite lattice configuration. This ordering process produces graphite with intermetallic properties that make it useful in many applications.

Puffing. In the temperature range of 1500–2000°C, most petroleum cokes and coal-tar pitch cokes undergo an irreversible volume increase known as puffing. This effect in petroleum cokes has been associated with thermal removal of sulfur and increases with increasing sulfur content.

Many studies of the puffing phenomenon and of means for reducing or eliminating it have been made. As a general rule, puffing increases as particle size increases and is greater across the product grain. Depending on particle size and on the product size, heating rates must be adjusted in the puffing range to avoid splitting the product. Fortunately, the use of puffing inhibitors has eased the problem and has permitted the use of graphitization rates greater than would otherwise be possible.

D. L. TURK
UCAR Carbon Company Inc.

P. LeFrank, G. Schuster, and A. Treugut, *World Steel Metalwork. Exp. Man.* **4,** 110 (1982).

W. L. Root and R. A. Nichols, *Chem. Eng. N.Y.* **80,** 98 (Mar. 19, 1973).

D. McNeil and L. J. Wood, *Industrial Carbon and Graphite,* papers read at the conference held in London, Sept. 24–26, 1957, Society of Chemical Industry, London, U.K., 1958, p. 162.

F. Millhouse and I. W. Gazda, *Iron Steel Int.* **56,** 133 (1983).

PROPERTIES OF MANUFACTURED GRAPHITE

Physical Properties

The graphite crystal, the fundamental building block for manufactured graphite, is one of the most anisotropic bodies known. Properties of graphite crystals illustrating this anisotropy are shown in Table 1. Anisotropy is the direct result of the layered structure with extremely strong carbon–carbon bonds in the basal plane and weak bonds between planes.

Electrical Properties. Manufactured graphite is semimetallic in character with the valence and conduction bands overlapping slightly. Conduction is by means of an approximately equal number of electrons and holes that move along the basal planes. Manufactured graphite is strongly diamagnetic and exhibits a Hall effect, a Seebeck coefficient, and magnetoresistance.

Thermal Conductivity. Compared with other refractories, graphite has an unusually high thermal conductivity near room temperature; above room temperature, the conductivity decreases exponentially to approximately 1500°C and more slowly to 3000°C.

Coefficient of Thermal Expansion (CTE). The volumetric thermal expansion (VTE) of manufactured graphite expressed in equation 1 is anomalously low when compared to that of the graphite single crystal, where wg designates with-grain and cg, cross-grain.

$$VTE = CTE_{wg} + 2\,CTE_{cg} \tag{1}$$

Table 1. Properties of Graphite Crystals at Room Temperature

Property	Value in basal plane	Value across basal plane
resistivity, $\mu\Omega\cdot$m	0.40	ca 60
elastic modulus, GPa[a]	1020	36.5
tensile strength (est), GPa[a]	96	34
thermal conductivity, W/(w·K)	ca 2000	10
thermal expansion, °C^{-1}	-0.5×10^{-6}	27×10^{-6}

[a] To convert GPa to psi, multiply by 145,000.

Table 2. Elastic Constants of Graphite[a]

$c_{11} = 1.06 \pm 0.002$	$s_{11} = 0.98 \pm 0.03$
$c_{12} = 0.18 \pm 0.02$	$s_{12} = -0.16 \pm 0.06$
$c_{13} = 0.015 \pm 0.005$	$s_{13} = -0.33 \pm 0.08$
$c_{33} = 0.0365 \pm 0.0010$	$s_{33} = 27.5 \pm 1.0$
$c_{44} = 0.00018 - 0.00035$	$s_{66} = 2.3 \pm 0.2$

[a] Units c_{ij} (stiffness constant) in TPa, s_{ij} (compliance constant) in (GPa) −1. To convert TPa to psi, multiply by 145,000,000.

At room temperature, the volume coefficient of thermal expansion of a single crystal is approximately 25×10^{-6}/°C, whereas those of many manufactured graphites fall in the range of $4–8 \times 10^{-6}$/°C. There are exceptions and some commercially available, very fine-grain, near-isotropic graphites have a volumetric expansion as high as two-thirds the value for the single crystal. The low value of volume expansion of most manufactured graphite has been related to the microporosity within the coke particles.

Mechanical Properties. The hexagonal symmetry of a graphite crystal causes the elastic properties to be transversely isotropic in the layer plane; only five independent constants are necessary to define the complete set. The self-consistent set of elastic constants given in Table 2 has been measured in air at room temperature for highly ordered pyrolytic graphite.

Chemical Properties. The impurity (ash) content of all manufactured graphite is low, since most of the impurities originally present in raw materials are volatilized and diffuse from the graphite during graphitization.

Graphite reacts with oxygen to form CO_2 and CO, with metals to form carbides, with oxides to form metals and CO, and with many substances to form laminar compounds. Of these reactions, oxidation is the most important to the general use of graphite at high temperatures.

T. R. HUPP
UCAR Carbon Company, Inc.

E. L. Piper, *Soc. Min. Eng. AIME, Preprint Number 73-H-14,* 1973 (properties updated 1990).

W. L. Greenstreet, *U.S. Oak Ridge National Laboratory Report ORNL-4327,* Oak Ridge, Tenn., Dec. 1968.

G. M. Jenkins, in P. L. Walker, Jr. and P. A. Thrower, eds., *Chemistry and Physics of Carbon,* Vol. 11, Marcel Dekker, New York, 1973, p. 189.

J. K. Legg and S. G. Bapat, *Southern Research Institute Technical Report,* AFML-TR-74-161, Birmingham, Ala., 1975.

APPLICATIONS OF BAKED AND GRAPHITIZED CARBON

Aerospace and Nuclear Reactor Applications

Graphite, with its exceptional strength and thermal stability at high temperatures, is a prime candidate material for many aerospace and nuclear applications. Its properties, through process modifications, are tailorable to meet an array of design criteria for survival under extremely harsh environmental operations.

Aerospace and nuclear reactor applications of graphite demand high reliability and reproducibility of properties, physical integrity of product, and product uniformity. The manufacturing processes require significant additional quality assurance steps that result in high cost.

Chemical Applications

Carbon and graphite exhibit excellent resistance to the corrosive actions of acids, alkalies, and organic and inorganic compounds, an attribute that has fostered the use of graphite in process equipment where corrosion is a problem. Other than in the chemical process industries, graphite is used extensively in the steel, food, petroleum, pharmaceutical, and metal finishing industries. The high thermal conductivity and thermal stability of graphite have made it a useful material in heat exchangers and high temperature gas-spray coolers.

Manufactured carbon and graphite exhibit varying degrees of porosity, depending on its method of preparation. Equipment fabricated from these materials must be operated essentially at atmospheric pressure; otherwise, some degree of leakage must be tolerated.

Self-Supporting Structures. Self-supporting structures of carbon and graphite are used in a variety of ways. Water-cooled graphite towers serve as chambers for the burning of phosphorus in air.

The resistance of graphite to thermal shock, its stability at high temperatures, and its resistance to corrosion permit its use as self-supporting vessels to contain reactions at elevated temperatures (800–1700°C), eg, self-supporting reaction vessels for the direct chlorination of metal and alkaline-earth oxides.

Impervious Graphite. For applications where fluids under pressure must be retained, impregnated materials are available. Imperviousness is attained by blocking the pores of the graphite or carbon material with themosetting resins such as phenolics, furans, and epoxies.

Many types of impervious graphite shell and tube, cascade, and immersion heat exchangers are in service throughout the world. Impervious graphite shells and tubes are used in numerous applications for transferring thermal energy, eg, boiling, cooling, or condensing.

Low Permeability Graphite. Graphite manufacturers have developed low permeability graphite materials where permeability is reduced by deposition of carbon and graphite in the pores of the base material. This material is not limited in its operating temperature, except in oxidizing conditions, and it is used to fabricate high temperature interchanger ejectors, fused salt cells, fused salt piping systems, and electric resistance heaters.

Porous Graphite. Several grades of low density porous carbon and graphite are commercially available. Porous carbon and graphite are used in filtration of hydrogen fluoride streams, caustic solutions, and molten sodium cyanide; in diffusion of chlorine into molten aluminum to produce aluminum chloride; and in aeration of waste sulfite liquors from pulp and paper manufacture and sewage streams.

Mechanical Applications

Carbon–graphite possesses lubricity, strength, dimensional stability, thermal stability, and ease of machining, a combination of properties that has led to its use in a wide variety of mechanical applications for supporting rotating or sliding loads in contact. Its principal applications are in bearings, seals, and vanes, which are in sliding contact with a partner material.

Electrode Applications

With the exception of carbon use in the manufacture of aluminum, the largest use of carbon and graphite is as in electric-arc furnaces. In general, the use of graphite electrodes is restricted to open-arc furnaces of the type used in steel production; whereas, carbon electrodes are employed in submerged-arc furnaces used in phosphorus, ferroalloy, and calcium carbide production.

Anode Applications. Graphite has been used as the primary material for electrolysis of brine (aqueous) and fused-salt electrolytes, both as anode and cathode. Technological advances, however, have resulted in a dimensionally stable anode (DSA) consisting of precious metal oxides deposited on a titanium substrate that has replaced graphite as the primary anode (see ALKALI AND CHLORINE PRODUCTS).

Another application for graphite anodes is for cathodic protection. All metal structures placed above ground or underground are subject to corrosion by galvanic action. Current flow, either localized or general, results in oxidation, ie, rusting, of steel. Graphite anodes are used for impressed current protection and a current is induced in the circuit counter to the galvanic current. Since the polarity is reversed, the steel does not corrode.

Metallurgical Applications

Because of their unique combination of physical and chemical properties, manufactured carbons and graphites are widely used in several forms in high temperature processing of metals, ceramics, glass, and fused quartz. A variety of commercial grades is available with properties tailored to best meet the needs of particular applications. Industrial carbons and graphites are available in a broad range of shapes and sizes.

Refractory Applications

Various forms of carbon, semigraphite, and graphite materials have found wide application in the metals industry, particularly in connection with the production of iron, aluminum, and ferroalloys.

J. M. CRISCIONE
(Aerospace and Nuclear Reactor Applications; Chemical Applications; and Mechanical Applications)
R. L. REDDY
(Electrode Applications)
C. F. FULGENZI
(Electrode Applications)
D. J. PAGE
(Electrode Applications, Anode Applications)
F. F. FISCHER
(Metallurgical Applications)
A. J. DZERMEJKO
(Metallurgical Applications, Refractory Applications)
J. B. HEDGE
(Metallurgical Applications)
UCAR Carbon Company Inc.

J. M. Criscione and co-workers, U.S. Air Force Materials Laboratory ML-TDR64-173, Parts I through IV, 1964–1966.

J. M. Criscione, H. F. Volk, and A. W. Smith, *AIAA J.* **4**, 1791 (1966).

U.S. Pat. 4,526,834 (July 2, 1985), R. A. Mercuri and J. M. Criscione (to Union Carbide Corp.).

N. J. Fechter and P. S. Petrunich, *Development of Seal Ring Carbon–Graphite Materials,* NASA Contract Reports CR-72799, Jan. 1971; CR-72986, Aug. 1971; CR-120955, Aug. 1972; and CR-121092, Union Carbide Corp., Parma, Ohio, Jan. 1973.

OTHER FORMS OF CARBON AND GRAPHITE

Flexible Graphite

A useful form of graphite is a flexible sheet or foil. Because of graphite's stability at high temperatures, flexible foil is useful in applications requiring thermal stability in corrosive environments, eg, gaskets and valve packings, and is often used as a replacement for asbestos gaskets.

Carbon and Graphite Foam

Carbon–graphite foam is a unique material that has yet to find a place among the various types of commercial specialty graphites. Its low thermal conductivity, mechanical stability over a wide range of temperatures from room temperature to 3000°C, and light weight make it a prime candidate for thermal protection of new, emerging carbon–carbon aerospace reentry vehicles.

Pyrolytic Graphite

Pyrolytic graphite produced in massive shapes is used for missile components, rocket nozzles, and aircraft brakes for advanced high performance aircraft. Pyrolytic graphite coated on surfaces or infiltrated into porous materials is also used in other applications, such as nuclear fuel particles, prosthetic devices, and high temperature thermal insulators.

Of the many forms of carbon and graphite produced commercially, only pyrolytic graphite is produced from the gas phase via the pyrolysis of hydrocarbons.

Glassy Carbon

Glassy, or vitreous, carbon is a black, shiny, dense, brittle material with a vitreous or glass-like appearance. It is produced by the controlled pyrolysis of thermosetting resins; phenol–formaldehyde and polyurethanes are among the most common precursors. Chemical inertness and low permeability have made glassy carbon a useful material for chemical laboratory crucibles and other vessels.

Carbon and Graphite Paper

Carbon and graphite paper is produced from carbon fibers by conventional papermaking methods. This form of carbon and graphite has outstanding electrical conductivity, corrosion resistance, and moderately high strength. These properties have promoted its use in electrodes for electrostatic precipitators.

J. M. Criscione
UCAR Carbon Company Inc.

R. A. Mercuri, R. A. Howard, and J. J. McGlamery, *Automotive Eng.* **97,** 49–52 (July 1989).

R. A. Mercuri, T. R. Wessendorf, and J. M. Criscione, *Amer. Chem. Soc. Div. Fuel Chem. Prepr.* **12**(4), 103 (1968).

A. W. Moore, *Chem. Phys. Carbon* **11,** 69 (1973).

U.S. Pat. 3,844,877 (Oct. 29, 1974), T. R. Wessendorf and J. M. Criscione (to Union Carbide Corp.).

ACTIVATED CARBON

Activated carbon is a predominantly amorphous solid that has an extraordinarily large internal surface area and pore volume. These unique characteristics are responsible for its adsorptive properties, which are exploited in many different liquid- and gas-phase applications. Through choice of precursor, method of activation, and control of processing conditions, the adsorptive properties of products are tailored for applications as diverse as the purification of potable water and the control of gasoline emissions from motor vehicles.

Physical and Chemical Properties

The structure of activated carbon is best described as a twisted network of defective carbon layer planes, cross-linked by aliphatic bridging groups. X-ray diffraction patterns of activated carbon reveal that it is nongraphitic, remaining amorphous because the randomly cross-linked network inhibits reordering of the structure even when heated to 3000°C. This property of activated carbon contributes to its most unique feature, namely, the highly developed and accessible internal pore structure. The surface area, dimensions, and distribution of the pores depend on the precursor and on the conditions of carbonization and activation.

Functional groups are formed during activation by interaction of free radicals on the carbon surface with atoms such as oxygen and nitrogen, both from within the precursor and from the atmosphere. The functional groups render the surface of activated carbon chemically reactive and influence its adsorptive properties. Activated carbon is generally considered to exhibit a low affinity for water, which is an important property with respect to the adsorption of gases in the presence of moisture. However, the functional groups on the carbon surface can interact with water, rendering the carbon surface more hydrophilic. Surface oxidation, which is an inherent feature of activated carbon production, results in hydroxyl, carbonyl, and carboxylic groups that impart an amphoteric character to the carbon, so that it can be either acidic or basic. The electrokinetic properties of an activated carbon product are, therefore, important with respect to its use as a catalyst support. As well as influencing the adsorption of many molecules, surface oxide groups contribute to the reactivity of activated carbons toward certain solvents in solvent recovery applications.

In addition to surface area, pore size distribution, and surface chemistry, other important properties of commercial activated carbon products include pore volume, particle size distribution, apparent or bulk density, particle density, abrasion resistance, hardness, and ash content.

Manufacture and Processing

Commercial activated carbon products are produced from organic materials that are rich in carbon, particularly coal, lignite, wood, nut shells, peat, pitches, and cokes. The choice of precursor is largely dependent on its availability, cost, and purity, but the manufacturing process and intended application of the product are also important considerations. Manufacturing processes fall into two categories, thermal activation and chemical activation. The effective porosity of activated carbon produced by thermal activation is the result of gasification of the carbon at relatively high temperatures, but the porosity of chemically activated products is generally created by chemical dehydration reactions occurring at significantly lower temperatures.

Forms of Activated Carbon Products. To meet the engineering requirements of specific applications, activated carbons are produced and classified as granular, powdered, or shaped products.

Shipping and Storage. Activated carbon products are shipped in bags, drums, and boxes in weights ranging from about 10 to 35 kg. Bulk quantities of activated carbon products are shipped in metal bins and bulk bags, typically 1–2 m^3 in volume, and in nail cans and tank trucks. Containers can be lined or covered with plastic and should be stored in a protected area both to prevent weather damage and to minimize contact with organic vapors that could reduce the adsorption performance of the product.

Economic Aspects

Over the decade of the 1980s, production capacity in the United States remained essentially unchanged, but minor fluctuations occurred in response to changes in environmental regulations. A similar reaction was noted worldwide. The current demand for activated carbon is estimated at 93% of production capacity. The near-term growth in demand is projected to be approximately 5.5%/yr.

In 1970 the U.S. Congress enacted the Clean Air Act, the Clean Water Act, and the Safe Drinking Water Act. Because of stricter EPA regulations implementing all three acts in 1990, the industry will increase production capacity by 25% during the next several years.

The estimated production capacity of activated carbon in the United States is shown in Table 1 for seven manufacturers. The principal producers are Calgon Carbon (37%), American-Norit (26%),

Table 1. Production Capacity in the United States, Estimated 1990

Company	Location	Capacity, 10^3 t
Acticarb Division, Royal Oak Enterprises	Romeo, Fla.	6.8
American Norit Co.	Marshall, Tex.	38.6
Barneby and Sutcliffe	Columbus, Ohio	3.0
Calgon Carbon Corp.	Catlettsburg, Ky. and Pittsburgh, Pa.	53.5
Ceca Division, Atochem NA	Pryor, Okla.	15.0
Trans-Pacific Carbon	Blue Lake, Calif.	2.3
Westvaco Corp.	Covington, Va.	27.2
Total		*146.4*

Table 2. Gas-Phase Activated Carbon Uses[a]

Application	Consumption, 10^3 t
solvent recovery	4.5
automotive/gasoline recovery	4.1
industrial off-gas control	3.2
catalysis	2.7
pressure swing separation	1.1
air conditioning	0.5
gas mask	0.5
cigarette filters	0.5
nuclear	0.3
Total	*17.4*

[a] In the United States, 1987.

Westvaco (19%), and Atochem (10%). Several other companies purchase activated carbon for resale but do not manufacture products.

Activated carbon is a recyclable material that can be regenerated. Thus the economics, especially the market growth, of activated carbon, particularly granular and shaped products, is affected by regeneration and industry regeneration capacity.

Because powdered activated carbon is generally used in relatively small quantities, the spent carbon has often been disposed of in landfills. However, landfill disposal is becoming more restrictive environmentally and more costly. Thus large consumers of powdered carbon find that regeneration is an attractive alternative.

Health and Safety

Activated carbon generally presents no particular health hazard as defined by the National Institute for Occupational Safety and Health (NIOSH). However, it is a nuisance and mild irritant with respect to inhalation, skin contact, eye exposure, and ingestion. On the other hand, special consideration must be given to the handling of spent carbon that may contain a concentration of toxic compounds.

Liquid-Phase Applications

Activated carbons for use in liquid-phase applications differ from gas-phase carbons primarily in pore size distribution. Liquid-phase carbons have significantly more pore volume in the macropore range, which permits liquids to diffuse more rapidly into the mesopores and micropores. The larger pores also promote greater adsorption of large molecules, either impurities or products, in many liquid-phase applications. Specific-grade choice is based on the isotherm and, in some cases, bench or pilot scale evaluations of candidate carbons. Liquid-phase activated carbon can be applied either as a powder, granular, or shaped form.

Batch-stirred vessels are most often used in treating material with powdered activated carbon. The type of carbon, contact time, and amount of carbon vary with the desired degree of purification. The efficiency of activated carbon may be improved by applying continuous, countercurrent carbon–liquid flow with multiple stages.

Granular and shaped carbons are used generally in continuous systems where the liquid to be treated is passed through a fixed bed.

Liquid-phase applications of activated carbon include potable water treatment, groundwater remediation, industrial and municipal wastewater treatment, sweetener decolorization, chemical processing, mining and the production of food, beverages, cooking oil, and pharmaceuticals.

Gas-Phase Applications

Gas-phase applications of activated carbon include separation, gas storage, and catalysis. Although only 20% of activated carbon production is used for gas-phase applications, these products are generally more expensive than liquid-phase carbons and account for about 40% of the total dollar value of shipments. Most of the activated carbon used in gas-phase applications is granular or shaped. Activated carbon use by application is shown in Table 2.

Separation processes comprise the main gas-phase applications of activated carbon. These usually exploit the differences in the adsorptive behavior of gases and vapors on activated carbon on the basis of molecular weight and size. For example, organic molecules with a molecular weight greater than about 40 are readily removed from air by activated carbon (see ADSORPTION, GAS SEPARATION).

FREDERICK S. BAKER
CHARLES E. MILLER
ALBERT J. REPIK
E. DONALD TOLLES
Westvaco Corporation
Charleston Research Center

P. A. Thrower, ed., *Chemistry and Physics of Carbon,* Marcel Dekker, Inc., New York. *Chemistry and Physics of Carbon,* published in 23 vols. through 1991, is a primary source of excellent review articles on carbon, many relevant to activated carbon.

R. C. Bansal, J.-B. Donnet, and F. Stoeckli, *Active Carbon,* Marcel Dekker, Inc., New York, 1988. A modern treatise on activated carbon based on a comprehensive review of the literature.

P. N. Cheremisinoff and F. Ellerbusch, *Carbon Adsorption Handbook,* Ann Arbor Science Publishers, Inc., Ann Arbor, Mich., 1978. An excellent reference book on activated carbon, ranging from theoretical to applied aspects.

I. H. Suffet and M. J. McGuire, eds., *Activated Carbon Adsorption of Organics from the Aqueous Phase,* Ann Arbor Science Publishers, Inc., Ann Arbor, Mich., 1980. A comprehensive, 2-vol. treatise with many key references.

CARBON BLACK

Carbon black is a generic term for an important family of products used principally for the reinforcement of rubber, as a black pigment, and for its electrically conductive properties. It is a fluffy powder of extreme fineness and high surface area, composed essentially of elemental carbon. Plants for the manufacture of carbon black are strategically located worldwide in order to supply the rubber tire industry, which consumes 70% of production. About 20% is used for other rubber products, and 10% is used for special nonrubber applications, eg, plastics, printing inks, paint, and paper.

A number of processes have been used to produce carbon black, including the oil-furnace, impingement (channel), lampblack, and the thermal decomposition of natural gas and acetylene. Since over 95% of the total output of carbon black is produced by the oil-furnace process, this article emphasizes this process.

Physical Structure of Carbon Black

Molecular and Crystallite Structure. The arrangement of carbon atoms in carbon black has been well-established by x-ray diffraction

Table 1. Chemical Composition of Carbon Blacks, %

Type	Carbon	Hydrogen	Oxygen	Sulfur	Ash	Volatile
rubber-grade furnace	97.3–99.3	0.20–0.40	0.20–1.20	0.20–1.20	0.10–1.00	0.60–1.50
medium thermal	99.4	0.30–0.50	0.00–0.12	0.00–0.25	0.20–0.38	
acetylene	99.8	0.05–0.10	0.10–0.15	0.02–0.05	0.00	<0.40

methods. The diffraction patterns show diffuse rings at the same positions as diffraction rings from pure graphite. The suggested relation to graphite is further emphasized as carbon black is heated to 3000°C. The diffuse reflections sharpen, but the pattern never achieves that of true graphite. Carbon black can have a degenerated graphitic crystallite structure. Whereas graphite has three-dimensional order, carbon black has two-dimensional order. The x-ray data indicate that carbon black consists of well-developed graphite platelets stacked roughly parallel to one another but random in orientation with respect to adjacent layers.

Morphology. In describing carbon black, three terms are used to describe structures of increasing scale and complexity.

Particles (nodules) are the primary structure element. They are roughly spherical elements that are joined in the aggregate structures.

Aggregates are the primary dispersable elements of carbon black in all but thermal blacks. The particles in an aggregate are connected and have grown together.

Agglomerates are undispersed clusters of aggregates held together by van der Waals forces or by binders. The term structure is used to describe both the extent and the complexity with which the particles are interconnected in aggregates. Primary measures of structure focus on the internal space within the aggregate.

Size and shape of the aggregates in composite systems are the principal features that determine the performance of carbon black as a reinforcing agent and as a pigment.

Aggregate Morphology and Structure. The term structure is widely used in the carbon black and rubber industries. It is used to describe the relative void volume characteristics of grades of black of the same surface area. Structure comparisons of grades with different surface areas cannot be made. It is now known that the properties associated with structure are associated principally with the bulkiness of individual aggregates. Aggregates of the same volume, surface area, and number of nodules have high structure in the open bulky and filamentous arrangement and a low structure in a more clustered compact arrangement.

High structure blacks in unvulcanized rubber give higher Mooney viscosities, lower die swell, faster extrusion rates, and better and more rapid dispersion after incorporation. In vulcanized rubber higher modulus is obtained. High structure blacks give lower bulk densities and high vehicle demand in paint systems.

Aggregate Breakdown. Aggregate size analysis by the electron microscope and centrifuge methods are performed on predispersed samples of carbon black. High shear energy, usually ultrasonic, and enough time are employed in these sample preparations to break down microagglomerates to their ultimate aggregates for measurement.

Chemical Composition

Oil-furnace blacks used by the rubber industry contain over 97% elemental carbon. Thermal and acetylene black consist of over 99% carbon. The ultimate analysis of rubber-grade blacks is shown in Table 1.

Carbon Black Formation Mechanisms

Mechanisms of formation must account for the unique morphology and microstructure of carbon black. These features include the presence of nodules, or particles, multiple growth centers within some nodules, the fusion of nodules into large aggregates, and the paracrystalline or concentric layer plane structure of the aggregates. One mechanism of formation involves the decomposition of the aromatic hydrocarbon fuel in a diffusion flame to hydrogen and carbon radicals, and carbon–hydrogen radical fragments. These combine into larger aromatic layer plane units until they are no longer stable and condense out of the vapor phase to form nuclei, or growth centers. Further carbon deposition forms carbon particles that are the precursors of the nodules. The carbon particles collide and coalesce while undergoing further deposition of carbon layer planes and their surface, forming the nodules and aggregates with their characteristic onion microstructure as seen in the micrographs. The various steps in the sequence are not well understood. There are several reviews of carbon formation mechanisms.

Manufacture

The Oil-Furnace Process. The oil-furnace process, based on the partial combustion of liquid aromatic residual hydrocarbons, was first introduced in the United States at the end of World War II. It rapidly displaced the then dominant channel (impingement) and gas-furnace processes because it gave improved yields and better product qualities. It was also independent of the geographical source of raw materials, a limitation on the channel process and other processes dependent on natural gas, making possible the worldwide location of manufacturing closer to the tire customers. Environmentally it favored elimination of particulate air pollution and was more versatile than all other competing processes.

The principal pieces of equipment are the air blower, process air and oil preheaters, reactors, quench tower, bag filter, pelletizer, and rotary dryer. The basic process consists of atomizing the feedstock into the combustion zone of the reactor where the combustion of natural gas and preheated excess air create a high temperature environment of 1200 to 1900°C that almost instantly vaporizes the feedstock and decomposes most of it to carbon black and hydrogen. The remaining feedstock reacts with the excess oxygen in the primary combustion stream to maintain the reaction temperature for carbon formation. In some reactors a number of feedstock streams are atomized radially into the high velocity combustion gases. The reaction products must be quenched rapidly with water sprays to lower the temperature to prevent loss of the carbon black product through reaction with carbon dioxide and water, products of the combustion reactions. The hot, heavy carbon black smoke from the reactors enters the air preheater where thermal energy is transferred to preheat the primary combustion air. From the air preheater the lower temperature combustion products are given a secondary quench for a further lowering of temperature in a tower from which they enter the bag filter that separates the fluffy carbon black product from the tail gases. Since the tail gases are composed mainly of water, nitrogen, carbon monoxide, carbon dioxide, and hydrogen, they have heating value as a fuel to supplement the natural gas used to preheat feedstock and for heating the pellet dryers. Unused tail gas is frequently flared prior to venting to the atmosphere after removal of particulate matter. The fluffy carbon black from the bag filter is mechanically agitated to increase its bulk density and is then conveyed to the wet pelletizers where water is added to transform the product into wet granules. Dry pelletization in rotating drums is practiced for some special applications. The wet pellets are then dried in a rotary dryer after which finished product goes to storage tanks for shipping in bulk or in bags.

Feedstocks. Feedstocks are viscous aromatic hydrocarbons consisting of branched polynuclear aromatics with smaller quantities of paraffins and unsaturates. Preferred feedstocks are high in

Table 2. Applications of Principal Rubber-Grade Carbon Blacks

Designation	General rubber properties	Typical uses
N110, N121	high abrasion resistance	special tire treads, airplane, off-the-road racing
N220, N299, N234	high abrasion resistance, good processing	passenger, off-the-road, special service tire treads
N339, N347, N375, N330	high abrasion resistance, easy processing, good abrasion resistance	standard tire treads, rail pads, solid wheels, mats, tire belt, sidewall, carcass retread compounds
N326	low modulus, good tear strength, good fatigue, good flex cracking resistance	tire belt, carcass, sidewall compounds, bushings, weather strips, hoses
N550	high modulus, high hardness, low die swell, smooth extrusion	tire innerliners, carcass, sidewall, innertubes, hose, extruded goods, v-belts
N650	high modulus, high hardness, low die swell, smooth extrusion	tire innerliners, carcass, belt, sidewall compounds, seals, friction compounds, sheeting
N660	high modulus, high hardness, low die swell, smooth extrusion	carcass, sidewall, bead compounds, innerliners, seals, cable jackets, hose, soling, EPDM compounds
N762	high elongation and resilience, low compression set	mechanical goods, footwear, innertubes, innerliners, mats

aromaticity, free of coke and other gritty materials, and contain low concentrations of asphaltenes, sulfur, and alkali metals.

The principal sources of feedstocks in the United States are the decant oils from petroleum refining operations. These are clarified heavy distillates from the catalytic cracking of gas oils.

Characterization

Carbon blacks differ in particle or nodule size, surface area, aggregate size, and aggregate morphology. Surface activity is also a factor in performance, but this feature has been difficult to define or measure. The ultimate dispersible units are aggregates. Aggregate size distribution and morphology determine such properties as surface area, dibutyl phthalate absorption (DBPA), and testing strength.

Grades and Applications

Table 2 lists the principal rubber grades by their N-number classification, general rubber properties, and typical uses. The behavior of different grades is dominated mainly by surface area and structure (DBPA). High surface area produces high reinforcement as reflected in high tensile and tear strengths, high resistance to abrasive wear, higher hysteresis, and poorer dynamic performance.

The consumption of the various carbon black grades can be divided into tread grades for tire reinforcement and nontread grades for nontread tire use and other rubber applications. Table 3 shows the distribution of production of types for these uses.

Special-Grade Carbon Blacks. Most of the special black grades are manufactured by methods to meet specific product specifications required for their end uses. They sell for a higher average price than the rubber grades. These markets have been growing at an average annual rate twice that of the rubber black grades. Of increasing importance in recent years have been applications in plastics to improve weathering resistance and to impart antistatic and electrically conductive properties. About 42% of special blacks are used in plastics, 35% in printing inks, 7% in paper, and 16% in miscellaneous applications.

Product Safety

The safety aspects of carbon black have been the subject of a number of reviews and articles. The manufacture of carbon results in trace amounts of organic and inorganic impurities. These impurities have been suspected of causing potential health problems. Of particular concern have been the salts of toxic metals and adsorbed polynuclear aromatic hydrocarbons (PNAs). A few of the polyaromatic hydrocarbons are known to be mutagens and/or animal carcinogens. The solvent extract of furnace blacks is in the range of 300–2000 ppm (0.03–0.20%).

Table 3. Carbon Black Production[a] by Grade, 10^3 t

	United States	Western Europe	Japan
N100	35	28	37.1
N200	158	161	118
N300	555	528	300
Total tread grades	*748*	*717*	*418*
percent	55.2	63.8	59.5
N500	120	153	136
N600	326	137	87
N700	129	103	29
N900 (thermal)	23		9
Total nontread grades	*598*	*393*	*261*
percent	44.1	35.0	37.1
other grades			
acetylene	9.1	14	24
Total carbon black	*1355*	*1124*	*703*

[a] 1998.

Most of this extract consists of 10–15 organic compounds, the majority of which are not genotoxic. Statistical studies on the frequency of cancer of long-term employees in a carbon black plant covering a period of 17 years (1939–1956) have been reported. There is no evidence of increased cancer risk from exposure to industrial carbon blacks. OSHA regulations for carbon black dust concentrations call for an average exposure level over a given time period of not more than 3.5 mg/m^3.

Environmental Aspects

The carbon black industry takes extreme efforts to confine product during all stages of manufacture. Highly efficient bag filters are used to collect the product. After collection the fluffy carbon black is densified and pelletized to minimize dusting problems during shipping and use.

The process gases from the filters consist of nitrogen, carbon monoxide, carbon dioxide, hydrogen, water, small amounts of hydrogen sulfide, and other sulfur- and nitrogen-containing gases. In the past the process gases have been flared. Process gas is used as a fuel for in-plant heat needs, and where local conditions warrant, it may be burned to generate steam or power.

ELI M. DANNENBERG
Consultant
LYN PAQUIN
HARRY GWINNELL
Cabot Corporation

ASTM Standards for Carbon Black, American Society of Testing Materials, Philadelphia, Pa., 1987.

J. B. Donnet and A. Voet, *Carbon Black,* Marcel Dekker, Inc., New York, 1976, pp. 1–19.

F. Lyon and K. Burgess, in J. I. Kroschwitz, ed., *Encyclopedia of Polymer Science and Engineering,* John Wiley & Sons, Inc., New York, 1985, pp. 623–640.

DIAMOND, NATURAL

Naturally occurring diamond is a relatively rare polymorphic form of carbon characterized by a three-dimensional arrangement of tetrahedrally coordinated carbon atoms. Both natural and synthesized diamond have the highest hardness of all known materials, the highest thermal conductivity at room temperature, a high refractive index and optical dispersion, a low thermal expansion, and a relatively high inertness to chemical attack. This unique combination of properties permits diamond to be foremost in certain applications: as a highly prized gemstone; industrially as an important abrasive material unsurpassed in certain cutting, drilling, sawing, machining, grinding, and polishing operations for many materials; and in electronic and optical applications as a heat sink and window material, respectively.

Occurrence and Exploration

Upstream exploration has sometimes led to the discovery of the primary source of alluvial stones, namely, kimberlite "pipes." These structures of igneous origin are the principal source of natural diamonds, and there are over 1000 occurrences of them in the world. Only a small number contain a high enough concentration of diamonds to warrant mining. Even in successful mining operations the ratio of diamond to the gangue that has to be removed and crushed is of the order of one part in a million or even less, of which the proportion of gem quality crystals is about 20%. The first key discovery of diamond-containing pipes was in South Africa in 1867, and these structures are well known from the publicity given to the Kimberley, Premier, and other mines there. More recent successful discoveries were a producing pipe in Botswana, Africa in 1967 and the Argyle alluvial deposits and pipes in Australia in 1979. Exploration continues in several countries, including Canada and the United States, for the source of diamonds brought south by the glacier into the United States. Methods include the tracing of certain heavy minerals upstream, drilling, the use of aerial photographs to try to find manifestations of circular pipe structures at the surface, and magnetic anomalies by aerial surveys. There is a pipe in Arkansas where amateurs can dig with a reasonable probability of finding a diamond.

Recovery

Alluvial diamonds are recovered from streambeds by panning or washing techniques as for heavy minerals or, more productively, by removal from pockets where the water has concentrated them. High quality gemstones are recovered from ocean deposits off the coast of Africa by holding back the water with seawalls long enough to remove a considerable overburden of sand and gravel.

The original mining of the pipes begins at the surface in the softer weathered "yellow" ground, but most of the subsequent recovery is by removal from depth of the harder, unweathered "blue" kimberlite. This material is then crushed for separation of the diamonds by a variety of techniques. One of the more interesting sorting methods is the grease table that takes advantage of the fact that most diamonds will stick to grease but not to water that removes the gangue. There are some diamonds that will not stick to grease but can be separated by electrostatic methods. A more modern method for the separation of crystal from rock uses the emission of light by luminescence of diamond when exposed to x-rays. The stones are then themselves separated mostly by hand into two main categories, gem and industrial, with further classification within each of these categories.

Properties

Structure, Density, and Morphology. The lattice constant of the face-centered cubic form of diamond is 0.3567 nm, with a density of 3.52 g/cm^3. Diamond twins readily during growth on the ⟨111⟩ plane, resulting in both interpenetration and flat twins called macles. The growth morphology of natural diamond is often octahedral, but dodecahedral crystals are very common, perhaps because of subsequent solution after growth.

Besides the single crystals usually implied in a discussion of diamond, the latter also occurs in nature in the form of polycrystalline aggregates. These can be classified into two types, based on origin: carbonado and ballas. The former is an aggregate of previously formed grains that have become cemented to each other to varying degrees of bonding. Ballas is an assemblage of grains which grew together simultaneously.

Chemical Composition. Diamond is nominally pure carbon with a ^{12}C:^{13}C ratio of about 99:1. Although other elements are often reported in analyses, most are considered to be present in oxide, silicate, and sulfide phases as inclusions in the diamond. Only boron, nitrogen, and possibly beryllium are considered to be truly substitutional in the lattice. The classification of diamonds into four types (Ia, Ib, IIa, IIb) is based primarily on the presence or absence of boron and nitrogen and their location in the lattice. Hydrogen and oxygen, possibly as OH, may also be important structural contaminants though they may also be present as second-phase gases along with CO, CO_2, H_2O, CH_4, and other species.

Hardness. Diamond is the hardest material known because of its combination of a three-dimensional arrangement of tetrahedrally coordinated C–C bonds with a bond distance of 0.154 nm. The Knoop hardness (*K*) is in the range of 68.7–98.1 kN/m^2 (7,000 to 10,000 kgf/mm^2).

Cleavage. Although hard, diamond is also very brittle and cleaves readily on the ⟨111⟩ plane and also on other planes under certain conditions.

Mechanical Properties. Measurement of the mechanical properties of diamond is complicated, and references should be consulted for the various qualifications. Table 1 compares the theoretical and experimental bulk modulus of diamond to that for cubic BN and for SiC and compares the compressive strength of diamond to that for cemented WC, and the values for the modulus of elasticity *E* to those for cemented WC and cubic BN.

Thermal Conductivity. The value of 2000 W/(m·K) at room temperature for Type IIa natural stones is about five times that of Cu, and recent data on 99.9% isotopically pure ^{12}C Type IIa synthesized crystals are in the range of 3300–3500 W/(m·K).

Thermal Expansion. The averaged value of the coefficient of linear thermal expansion of diamond over the range 20–100°C is 1.34×10^{-6} cm/cm/°C and 3.14×10^{-6} from 20–800°C.

Optical Properties. The high refractive index (2.42 at 589.3 nm) and dispersion (0.044) are the basis for the brilliance and fire of a properly cut gemstone. The optical transmission out to 10.6 μm for Type IIa diamonds makes possible windows for CO_2 lasers and for devices such as were used in the Venus probe.

Electrical and Electronic. Diamond is an electrical insulator ($\sim 10^{16}$ Ω/cm) unless doped with boron when it become a *p*-type semiconductor with a resistivity in the range of 10^{-1} to 100 Ω/cm.

Table 1. Mechanical Properties of Diamond and Other Hard Materials, GPa[a]

	Diamond	Cubic BN	WC[b]	SiC
bulk modulus	440	370		211
compressive strength	8.69		4–6	
modulus of elasticity, *E*	950–1100	890	460–675	

[a] To convert GPa to psi, multiply by 145,000.
[b] Cemented.

ROBERT DEVRIES
Consultant

J. E. Field, ed., *The Properties of Diamond,* Academic Press, London, New York, San Francisco, Calif., 1979, pp. 425–469; 1–469; 595–653; 23–163.

G. Davies, *Diamond,* Adam Hilger Ltd., Bristol, U.K., 1984, pp. 127–245.

Yu. L. Orlov, *The Mineralogy of the Diamond,* John Wiley & Sons, Inc., New York, 1973, pp. 138–148.

A. N. Wilson, *Diamonds from Birth to Eternity,* GIA, Santa Monica, Calif., 1982.

DIAMOND, SYNTHETIC

Reproducible Laboratory Diamond Synthesis

In 1955, a team of research workers at General Electric developed the necessary high pressure equipment and discovered solvent–catalytic processes by which ordinary forms of carbon could be changed into diamond.

Catalyzed Synthesis

In this process, a mixture of carbon (eg, graphite) and catalyst metal is heated high enough to be melted, while the system is at a pressure high enough for diamond to be stable. Graphite is then dissolved by the metal and diamond is produced from it. Effective catalysts are Cr, Mn, Fe, Co, Ni, Ru, Rh, Pd, Os, Ir, Pt, and Ta, and their alloys and compounds.

Apparatus. Many kinds of apparatus have been devised for simultaneously producing the high pressures and temperatures necessary for diamond synthesis. An early successful design is the belt apparatus. In this apparatus, two opposed, conical punches, made of cemented tungsten carbide and carried in strong steel binding rings, are driven into the ends of a short, tapered chamber that is also made of cemented tungsten carbide supported by strong steel rings. A compressible gasket, constructed in a sandwich-fashion of stone, usually pyrophyllite, and steel cones, seals the annular gap between punch and chamber, distributes stress, provides lateral support for the punch, and permits axial movement of the punches to compress the chamber and contents. The reaction zone, usually a cylinder, is buried in pyrophyllite stone in the chamber. The pyrophyllite, a good thermal and electrical insulator, is easily machined and transmits pressure fairly well. The reaction zone is heated electrically with a heavy current.

A belt apparatus is capable of holding pressures of 7 GPa (70 kbar) and temperatures of up to 3300 K for periods of hours. The maximum steady-state temperatures are limited by melting of the refractory near the reaction zone.

Figure 1 shows an arrangement of carbon and catalyst metal. As the sample is heated at high pressure, the metal next to the graphite usually melts and diamond begins to form there. An exceedingly thin film of molten metal (at most a few thousandths of a cm thick) separates the newly formed diamond from the unchanged graphite. This film advances like a wave through the mass of graphite and transforms it to diamond.

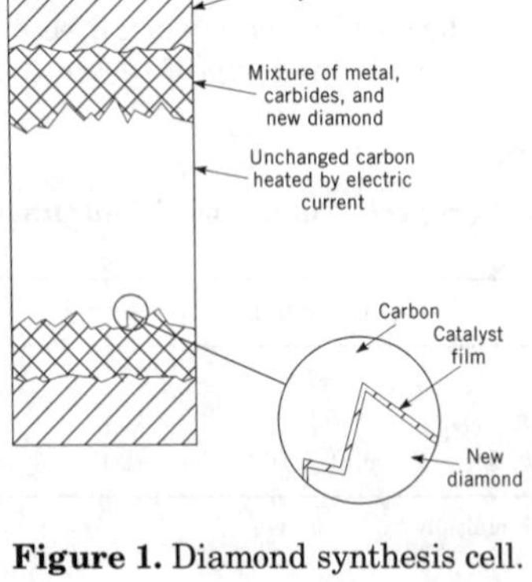

Figure 1. Diamond synthesis cell.

Crystal Morphology. Size, shape, color, and impurities are dependent on the conditions of synthesis. Lower temperatures favor dark-colored, less pure crystals; high temperatures promote paler, purer crystals. Low pressures (5 GPa) and temperatures produce octahedral faces. Nucleation and growth rates increase rapidly as the process pressure is raised above the diamond–graphite equilibrium pressure.

Crystal Growth. If diamond seed crystals are placed in the active diamond growing zone of a typical graphite–catalyst metal apparatus, new diamond usually forms on the seed crystals. Excellent growth can be obtained if pressure and composition are held relatively constant while the change of composition with temperature is employed as a driving force.

Direct Graphite-to-Diamond Process

In this process, diamond forms from graphite without a catalyst. The refractory nature of carbon demands a fairly high temperature (2500–3000 K) for sufficient atomic mobility for the transformation, and the high temperature in turn demands a high pressure (above 12 GPa; 120 kbar) for diamond stability. The combination of high temperature and pressure may be achieved statically or dynamically. During the course of experimentation on this process a new form of diamond with a hexagonal (wurtzitic) structure was discovered.

Shock Synthesis. When graphite is strongly compressed and heated by the shock produced by an explosive charge, some (up to 10%) diamond may form. These crystallite diamonds are small (on the order of 1 μm) and appear as a black powder. The peak pressures and temperatures, which are maintained for a few microseconds, are estimated to be about 30 GPa (300 kbar) and 1000 K. It is believed that the diamonds found in certain meteorites were produced by similar shock compression processes that occurred upon impact.

The annual production of diamond by this process is only a small fraction of total industrial diamond consumption.

Static Pressure Synthesis. Diamond can form directly from graphite at pressures of about 13 GPa (130 kbar) and higher at temperatures of about 3300–4300 K. No catalyst is needed. The transformation is carried out in a static high pressure apparatus in which the sample is heated by the discharge current from a capacitor. Diamond forms in a few milliseconds and is recovered in the form of polycrystalline lumps.

Crystal Structure. Diamonds prepared by the direct conversion of well-crystallized graphite at pressures of about 13 GPa (130 kbar) show certain unusual reflections in the x-ray diffraction patterns. They could be explained by assuming a hexagonal diamond structure (related to wurtzite) with $a = 0.252$ and $c = 0.412$ nm, space group $P6_3/mmc - D_{6h}^4$ with four atoms per unit cell. The calculated density would be 3.51 g/cm^3, the same as for ordinary cubic diamond, and the distances between nearest neighbor carbon atoms would be the same in both hexagonal and cubic diamond, 0.154 nm.

Metastable Vapor-Phase Deposition

Metastable growth of diamond takes place from gases rich in carbon and hydrogen at low pressures where diamond would appear to be thermodynamically unstable.

In a typical use of this method, a mixture of hydrogen and methane is fed into a reaction chamber at a pressure of about 1.33 kPa (10 torr). The substrate upon which diamond forms is at about 950°C and lies about 1 cm away from a tungsten wire at 2200°C. Small diamond crystals, 1 mm or so in size, nucleate and grow profusely on the substrate at a rate around 0.01 mm/h to form a dark, rough polycrystalline layer with exposed octahedral or cubic faces, depending on the substrate temperature. This method has been actively studied since 1993.

The Synthetic Diamond Industry

Soon after the first successful diamond synthesis by the solvent–catalyst process, a pilot plant for producing synthetic diamond was established, the efficiency of the operation was increased, production costs declined, and product performance was improved while the uses of diamond were extended. Today (1993) the price of synthesized diamond is competitive with that of natural diamonds.

The bulk of synthetic industrial diamond production consists of the smaller crystal sizes up to 0.7-mm particle size (25 mesh). This size range has wide utility in industry, and a significant fraction of the world's need for diamond abrasive grit is now met by synthetic production yielding thousands of kilograms per year.

Semiconducting Diamonds. Semiconducting diamonds are prepared by adding small amounts of boron, beryllium, or aluminum to the growing mixture, or by diffusing boron into the crystals at high pressures and temperatures.

Sintered Diamond Masses. Some natural diamonds known as carbonado or ballas occur as tough, polycrystalline masses (see CARBON, DIAMOND, NATURAL). The production of synthetic sintered diamond masses of comparably excellent mechanical properties has only been achieved recently. The essential feature is the presence of direct diamond-to-diamond bonding without dependence on any intermediate bonding material between the diamond grains, since no extraneous bonding material can match the stiffness, thermal conductivity and hardness of diamond.

C_{60} Conversion. Buckminsterfullerene can be crushed to diamond by high pressure applied at room temperature. The process is highly efficient and fast at room temperature, suggesting industrial potential.

Economic Aspects

About 90% of industrial diamond is synthesized at high pressures because its price is relatively low, and it can be tailor-made for efficiency in each application. Total worldwide sales of industrial diamond are currently (1992) about one billion dollars (ca 100 t) per year. They are made in 16 countries; the largest producers are General Electric; De Beers, which also has a large stake in natural diamonds; several Japanese firms; the People's Republic of China; and Russia. The market is very competitive, and manufacturers are reluctant to disclose detailed sales information.

R. H. WENTORF, JR.
General Electric Company

F. P. Bundy, H. M. Strong, and R. H. Wentorf, Jr., in P. L. Walker and P. A. Thrower, eds., *Chemistry and Physics of Carbon,* Marcel Dekker, Inc., New York, 1973, pp. 213–263.

F. P. Bundy and J. S. Kasper, *J. Chem. Phys.* **46,** 3437 (1967).

P. S. DeCarli and J. C. Jamieson, *Science* **133,** 1821 (1961).

R. H. Wentorf, R. C. DeVries, and F. P. Bundy, *Science* **208,** 873–888 (1980).

NATURAL GRAPHITE

Natural graphite, the mineral form of graphitic carbon, occurs worldwide. It differs from the carbon of coal and of diamond in its predominantly lamellar hexagonal crystal structure. The ore usually contains associated silicate minerals that vary in kind and amount with the source. The name natural graphite is used in technical terminology and sometimes elsewhere. It may be simply termed graphite or any of several common names such as plumbago, black lead, silver lead, carburet of iron, and reissblei. The macrophysical form depends on geological genesis, whereas the properties depend on both the macrophysical form and associated mineral suite. The commercial value depends on specific characteristics such as form, percentage and kind of mineral suite, and availability. Graphite occurs in widely distributed places as flakes, lumps, and cryptocrystalline masses referred to commercially as amorphous graphite.

Structure

Parallel layers of condensed planar C_6-rings constitute the graphite crystallite. Each carbon atom joins to three neighboring carbon atoms at 120° angles in the plane of the layer. The C–C distance is 0.1414 nm (this bond is 0.1397 nm in benzene); the width of each C_6-ring is 0.2456 nm. Weak van der Waals forces pin the carbons in adjoining layers, thus accounting in part for the marked anisotropic properties of the graphite crystal.

Physical Properties

Solid articles made of natural graphite always require a binder; ie, they are always composites. The influence of the binder, the processing, and the kind of graphite used, together with graphite's strong anisotropism, influence the properties of the composites. In general, the overall quantities are usually greater than those measured along the *a* axis or along the *c* axis of the graphite under consideration. Table 1 lists some of the physical properties of natural graphite.

Graphite's strength increases as the temperature rises. The thermal conductivity, W/(m · K), along the a axis reaches a maximum of 285 at −100°C and falls rapidly with declining temperature.

Chemical Properties

Graphite burns slowly in air above 450°C, the rate increasing with temperature and exposed area. Above 800°C graphite reacts with water vapor, carbon monoxide, and carbon dioxide. Chlorine has a negligible effect on graphite, and nitrogen none. Many metals and metal oxides form carbides above 1500°C.

Graphite can be regenerated from the crystal compounds because the graphitic structure has not been too greatly altered. The dark crystal compounds are called intercalation compounds, interstitial compound, or lamellar compounds because they are formed by reactants that fit in between the planar carbon networks.

Compounds of graphite with alkali metals or ammonia are electron donors. Compounds with the halogens (except fluorine), metal halides, and sulfuric acid (graphite sulfate) are electron acceptors.

The covalent compounds of graphite differ markedly from the crystal compounds. They are white or lightly colored electrical insulators, have ill-defined formulas, and occur in but one form, unlike the series typical of the crystal compounds. In the covalent compounds, the carbon network is deformed, and the carbon atoms rearrange tetrahedrally as in diamond. Often they are formed with explosive violence.

Table 1. Physical Properties of Natural Graphite

Property	Value
density[a], g/mL, calculated	2.265
compressibility, N/m^2[b], Sri Lanka, average	3.1×10^{-11}
shear modulus, N/m^2[b]	2.3×10^{9}
Young's modulus, N/m^2[b]	1.13×10^{14}
heat of vaporization[c], kJ/mol[c]	711
sublimation point, K	4000–4015
triple point, K	
graphite–liquid–gas, 101.3 kPa[d]	3900 ± 50
graphite–diamond–liquid, 12–13 GPa[d]	4100–4200
surface energy, J/cm^2[c]	ca1.2 $\times 10^{-5}$

[a] The difference between the calculated and experimental values of density is caused by dislocations and imperfections.
[b] To convert N/m^2 to dyn/^{2}, multiply by 10.
[c] To convert J to cal, divide by 4.184.
[d] To convert kPa to atm, divide by 101.3.

Geographic Occurrence

The three physically different forms of natural graphite, which provide essentially different commodities, are found in Sri Lanka (lump), Germany (flake), Norway (flake), Mexico (amorphous and flake), China (amorphous and flake), and the Republic of Korea (amorphous).

Chemical Analysis

There are no generally accepted methods for the complete analysis of natural graphite. Industrial methods usually emphasize either the carbon content or analysis of the ash. Although the carbon percentage is of considerable importance, it is usually true that the mineral suite is more significant for a specific use, eg, fluxing constituents in graphite must be avoided for refractory uses, and abrasive minerals in graphite must be absent for lubricating uses. Associated minerals are seldom reported other than as "ash."

The simplest analytical procedure is to oxidize a sample in air below the fusion point of the ash.

Uses

The many useful properties of graphite give rise to a wide variety of products: unctuous, dry lubricant; marks readily, writing and drafting pencils; combination of lubricity and electrical conductivity, motor and generator brushes; excellent weathering properties and inertness, industrial paint pigment; solubility in molten iron, carbonraiser for steel; poorly wet by most metals and alloys, foundry mold facings; and burns slowly, conducts heat, and retains strength over a large temperature range, refractories such as crucibles, carbon–magnesite brick, continuously casting ware, and stopper heads for steel ladles. Some additional properties of interest include hydrophobicity, forms water-in-oil emulsions, carries a negative charge, has low photoelectric sensitivity, is strongly diamagnetic, and is an infrared absorber.

HAROLD A. TAYLOR, JR.
Bureau of Mines, U.S. Dept. of the Interior

W. N. Reynolds, *Physical Properties of Graphite,* Elsevier Publishing Co., Inc., New York, 1968.

Minerals Yearbook, U.S. Bureau of Mines, Washington, D.C., 1989.

CARBON AND GRAPHITE FIBERS

Carbon fibers are the primary reinforcement used to increase the stiffness and strength of lightweight advanced composites commonly used in aerospace, recreation, and industrial applications. The term carbon fiber generally refers to a variety of filamentary products composed of more than 90% carbon, 5–10 μm in filament diameter, produced via the pyrolysis of polyacrylonitrile (PAN), pitch, or rayon. Carbon fibers are often referred to as graphite fibers; however, only very high modulus carbon fibers with three-dimensional graphite structure can properly be called graphite fibers. Because carbon fibers have extremely high specific strength and modulus when compared to other engineering materials, they are predominantly used in weight-critical applications. In addition to strength and stiffness, carbon fibers possess excellent fatigue strength, vibration damping characteristics, thermal resistance, and dimensional stability. Carbon fibers also have good electrical and thermal conductivity and chemical inertness, except to oxidation.

The physical and mechanical properties of carbon fibers are primarily determined by the axial orientation of the carbon layer planes (Figs. 1 and 2) and the degree of crystalline perfection. Both

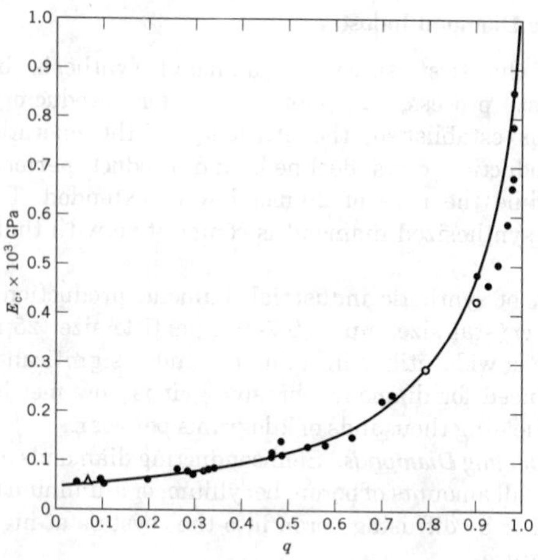

Figure 1. Young's modulus corrected for porosity E_c as a function of preferred orientation q; curve is based on theoretical model where ● = rayon-based fibers; ○ = PAN-based fibers; and △ = pitch-based fibers. To convert GPa to psi, multiply by 145,000.

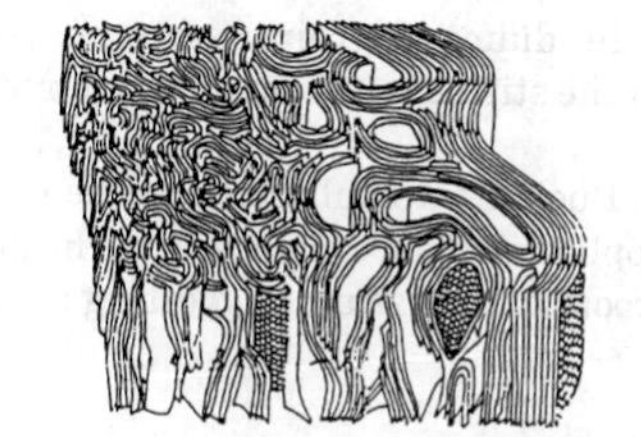

Figure 2. Schematic illustration of PAN-based carbon fiber microstructure based on microscopic observations.

Table 1. Mechanical and Physical Properties of Carbon Fibers[a]

Property	PAN, standard-grade	PAN-IM	PAN-HM	Pitch HM
tensile strength, GPa[b]	3.5–4.8	4.1–7.0	1.7–4.7	2.4–3.0
tensile modulus, GPa[b]	230–240	280–300	350–480	380–520
elongation at break, %	1.5–2.1	1.5–2.4	0.4–1.4	0.4–0.6
density, g/cm^3	1.77–1.81	1.79–1.82	1.70–1.90	2.00–2.14
filament diameter, μm	7	5	4.5–6.5	10
carbon, wt %	92–96	93–96	99+	99+
axial resistivity, $\mu\Omega \cdot$m	14–18	14	6.5–10	7–8.5
axial thermal conductivity, W/(m·K)	8–14	15	70–105	120–185
axial coefficient of thermal expansion at 21°C, ppm/°C	−0.6	−0.75	−1.15	−1.3–1.4
standard filament counts	3K/6K/12K	6K/12K	3K/6K/12K	0.5K/1K/2K/4K

[a] Data from the following technical data sheets: PAN fibers: Amoco Performance Products, Asahi Kasei, BASF Structural Materials, Courtaulds-Grafil, Hercules, Mitsubishi Rayon, Toho Beslon, Toray; Pitch fibers: Amoco Performance Products, Mitsubishi Kasei, Tonen Corp.
[b] To convert GPa to psi, multiply by 145,000.

can be studied through x-ray diffraction. The detailed structural characteristics of carbon fibers have been investigated.

Carbon fibers are generally typed by precursor, such as PAN, pitch, or rayon, and classified by tensile modulus and strength. Tensile modulus classes range from low (<240 GPa) to standard (240 GPa), intermediate (280–300 GPa), high (350–500 GPa), and ultrahigh (500–1000 GPa). Typical mechanical and physical properties of commercially available carbon fibers are presented in Table 1.

Current research to develop low cost carbon fiber includes growing carbon filaments via vapor deposition of carbon from gases such as carbon monoxide, methane, or benzene on metal catalysts.

Producers of PAN-based carbon fiber include Toray, Toho Beslon, Mitsubishi Rayon, and Asahi Kasai Carbon in Japan; Hercules, Amoco Performance Products, Fortafil (Akzo), and Mitsubishi Rayon in the United States; and Akzo, Sigri, and Soficar in Europe. Primary suppliers of high performance pitch-based carbon fibers include Amoco Performance Products, Mitsubishi Kasai, and Tonen Corporation.

Carbon Fiber Manufacturing Processes and Property Relationships

More than 95% of current carbon fiber production for advanced composite applications is based on the thermal conversion of polyacrylonitrile (PAN) or pitch precursors to carbon or graphite fibers. Generally, the conversion of PAN or pitch precursor to carbon fiber involves similar process steps: fiber formation, ie, spinning; stabilization to thermoset the fiber; carbonization–graphitization; surface treatment; and sizing. Schematic process flow diagrams are shown in Figure 3. However, specific process details differ.

Carbon Fiber Property Development

The Young's modulus of carbon fibers is primarily determined by the relative orientation of the hexagonal layer planes to the carbon fiber axis. Increasing precursor molecular orientation, tension applied during processing, or carbonization–graphitization temperatures directly increase carbon fiber modulus.

Density. The density of carbon fibers ranges from approximately 1.5 g/cm^3 to 2.2 g/cm^3. Carbon fiber density is typically less than that

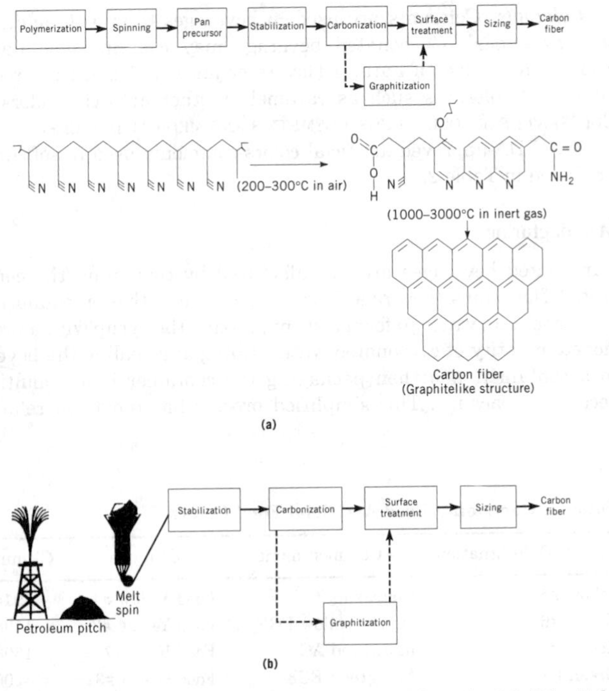

Figure 3. Process flow diagrams for (**a**) PAN-based and (**b**) pitch-based carbon fiber processes.

of single-crystal graphite, 2.26 g/cm^3, because of the imperfect packing of the graphene layer planes and because of internal microporosity present within polymer-derived carbon fibers.

Electrical and Thermal Conductivity. Generally, higher carbonization temperatures increase the concentration of mobile electrons and reduce the energy gap, thereby increasing the electrical and thermal conductivity of PAN and pitch fibers. Carbon fibers exhibiting low and high conductivity are utilized in specialized applications.

Mechanical Properties and Stability at Elevated Temperature. One increasingly important characteristic of carbon fibers is their excellent performance at elevated temperatures. Strength tested in an inert environment remains constant or slightly increases to temperatures exceeding 2500°C.

Although carbon fibers retain their excellent mechanical properties at elevated temperatures in an inert environment or in a vacuum, they are reactive with oxygen at temperatures exceeding 300°C, forming carbon monoxide and carbon dioxide. Oxidative stability is related to the degree of graphitization and impurity level, especially alkali metals that catalyze graphite oxidation reactions. High modulus fibers have improved thermal oxidative stability (TOS) over low modulus fibers because of a lower concentration of reactive edge sites on the fiber surface. Oxidation mechanisms present in graphitized carbons, most of which are also applicable to carbon fibers, have been described.

Quality Control and Specifications

Typically the quality control philosophy used for carbon fibers is based on consistent precursor and tight control of processing parameters such as linespeed, temperature, and gas flow at each process step.

Health and Safety Factors

Safety concerns in handling carbon fibers fall into three categories: dust inhalation, skin irritation, and electrical shorting of equipment. Additionally, the protective finish, or size, which is applied to the fiber may necessitate additional safety precautions.

Applications of Carbon Fibers

With excellent specific stiffness and strength, carbon fibers are ideally suited to applications where weight reduction results in significant performance and cost advantages. Initially military aerospace, space, and recreation were primary markets. More recently carbon fiber composite applications have diversified into automotive, ie, brake linings, drive shafts, springs, wheels, and structural components for racing cars; aircraft engines; recreation, ie, fishing rods and reels, tennis and racquetball racquets, golf club shafts, baseball bats, hockey sticks, oars, paddles, sailing masts, skis, etc; electrical shielding and radar; absorption; medical, ie, x-ray tables and prosthetic devices; and music applications, ie, speaker cones and musical instrument components.

The use of carbon fibers in new applications will continue to grow as fiber costs are reduced further. Development for use in construction, such as cement and carbon reinforcement, and marine applications will result in sustained growth volume through the early twenty-first century.

JOSEPH G. VENNER
BASF Structural Materials, Inc.

W. Johnson, "The Structure of PAN-Based Carbon Fibers and Relationship to Physical Properties," in W. Watt and B. V. Perov, eds., *Handbook of Composites,* Vol. 1, Elsevier Science Publishers, New York, 1985.

M. S. Dresselhaus and co-workers, *Graphite Fibers and Filaments Springer Series in Materials Science,* Vol. 5, Springer-Verlag, New York, 1988.

CARBONATED BEVERAGES

The carbonated beverage has its origin in the study of mineral waters in Europe in the sixteenth century. In the late

Table 1. Nutritive Sweeteners Used in the Preparation of Carbonated Beverages

Types of sweeteners	Sucrose, %	Polysaccharides, %	Dextrose, %	Fructose, %	Solids, %
liquid[a] sucrose	98–100	0	0	0	67
liquid[a] invert sugar	40–60	0	20–30	20–30	76
HFCS-42	0	2–9	49–53	40–44	71
HFCS-55	0	2–5	38–42	54–57	77
HFCS-90	0	1–2	7–8	89–91	77

[a] That is, aqueous solutions.

eighteenth century, artificial mineral waters were investigated for their medicinal properties both in Europe and America. The first commercial artificial mineral water was manufactured in Europe during the 1780s and in America in the early 1800s.

Early bottling of flavored carbonated beverages was limited by spoilage, poor flavor, and color stability. Improvements and innovations in bottling equipment, glass manufacturing, stable flavors and ingredients, crown closures, and transportation resulted in the rapid expansion of the bottled soft drink industry. Soft drinks consist of carbonated water, nutritive or nonnutritive sweeteners, acidulants, preservatives, flavors, juices, and color.

Consumption

The consumption of carbonated beverage has risen steadily since they were first introduced. Annual per capita consumption of soft drinks in the United States was approximately 16 ounces (0.5 L) in 1860 and is estimated to be 47.3 gallons (~180 L) in 1991. Although the consumption of soft drinks has increased, the number of soft drink plants has dropped steadily from 2258 in 1975 to 780 in 1991. Production of soft drinks has increased from 2.6 billion cases in 1975 with a wholesale value of nearly $9 billion to 5.8 billion cases with a wholesale value of $27.5 billion in 1991.

The soft drink industry is dominated by two key players, The Coca-Cola Company and Pepsico. These two companies produce eight of the ten top soft drink brands and comprise over 72% of the soft drink market in the United States.

Colas represent the largest segment of the U.S. soft drink market, followed by lemon-lime brands. Pepper-type, juice-based, root beer, and orange-flavored soft drinks represent two to five percent of the total soft drink market. Diet and caffeine-free categories represent the fastest growing segments of the market.

Ingredients

Water. Water is the largest single ingredient used in carbonated beverages and must be of high purity. Drinking water supplied by local municipalities fails to meet the required purity levels for use in carbonated beverages. It must be treated to remove four types of contaminants that may affect the taste, odor, or appearance of the final beverage. The four contaminants are inorganic material, organic compounds, microbiological contamination, and particulate matter. The water treatment process employed in the soft drink industry varies but may include chemical treatment, reverse osmosis (qv), ultrafiltration (qv), or ion exchange (qv) (see WATER, INDUSTRIAL WATER TREATMENT).

Sweeteners. The sweeteners (qv) used in carbonated beverages may be either nutritive or nonnutritive. The quality of the sweetener is one of the most important parameters affecting the overall quality of the beverage.

Nutritive Sweeteners. These include granulated sucrose, sucrose in solution, invert sugar, dextrose, and high fructose corn syrup (Table 1).

Nonnutritive Sweeteners. Diet or low calorie beverages represent a significant portion 29.8% of the total soft drink market. The diet category is expected to increase at approximately 2% per year. Currently, saccharin and aspartame $C_{14}H_{18}N_2O_5$, are the only nonnutritive sweeteners approved for use in beverages by the U.S. Food and Drug Administration (FDA).

Acidulants. Acidulants give the beverage a tart or sour flavor, adjust pH to facilitate the function of benzoate as a preservative, reduce microbiological susceptibility, and act as a catalyst for the hydrolytic inversion process in sucrose sweetened beverages. The primary carbonated beverage acidulants are phosphoric acid and citric acid. Other acidulants include ascorbic, tartaric, malic, and adipic acid.

Preservatives. The carbonation and acid content in cola and lemon-lime beverages usually act as adequate preservation against microbial growth. Benzoate or sorbate salts are often added to other beverages for protection.

Carbon Dioxide. Carbon dioxide provides soft drinks with a pungent taste, acidic bite, and sparkling fizz. Carbon dioxide also acts as a preservative against yeast, mold, and bacteria. The carbon dioxide used in soft drinks must be food-grade and free of impurities that may affect the taste or odor of the final product.

Flavors. Flavor is the most important attribute of a carbonated beverage. Most carbonated beverages contain complex mixtures of different flavors produced in several commercial forms as alcoholic solutions, emulsions, and concentrates. The majority of flavors used in carbonated beverages are derived from natural sources, eg, caffeine, juice-based flavors, essential oils, oleo resins, alcoholic solutions or extracts, emulsions, concentrates, and concentrated flavor or beverage bases.

Colorants. Colorants are used in beverages to provide additional sensory appeal. Carbonated beverage may contain some natural color from the use of natural flavors or juices but generally require additional colorants such as caramel or other artificial colors (see COLORANTS FOR FOOD, DRUGS, COSMETICS AND MEDICAL DEVICES).

Five FDA-approved artificial colors commonly used in soft drinks are listed in Table 2.

Manufacturing

Carbonated beverages are manufactured by combining the concentrated flavorings (beverage bases) with a nutritive or nonnutritive sweetener and water to form a syrup, mixing the syrup with a proportioned quantity of carbonated water, filling and sealing the beverage in a container, and then packaging the container into a multipack secondary package. This simplified process has remained relatively

Table 2. Commonly Used Artificial Food Colors

FD & C designation	Common name	CI name	CI number
Yellow #5	tartrazine	Food Yellow #4	19140
Yellow #6	sunset yellow FCF	Food Yellow #3	15985
Red #40	allura red AC	Food Red #17	16035
Green #3	fast green FCF	Food Green #3	42053
Blue #1	brilliant blue FCF	Food Blue #2	42090

unchanged, except for modernization and increased efficiency, since the industry began.

Quality Control

Ingredients. Ingredients used in the manufacture of carbonated beverages must meet all *Food Chemical Codex* specifications and be approved for use in soft drinks by the FDA. In addition to the government-stated specifications, manufacturers of carbonated beverages may complete additional analyses based on specific needs or concerns.

Beverages. The quality control for carbonated beverages encompasses all aspects of the product, from actual chemical components to the physical condition of the container. The beverage is evaluated using laboratory tests as well as in-line monitors.

Environmental Issues

The carbonated beverage industry uses environmentally responsible, high quality packaging to protect, transport, display and deliver products to consumers. The industry practices source reduction, uses recycled content in primary packages, and supports recycling efforts. The amount of material in aluminum cans, glass and PET bottles have been significantly reduced since each package was originally introduced. Glass bottles, aluminum and steel cans contain recycled content. In addition, the carbonated beverage industry pioneered the use of recycled PET for food use. Recycling rates for soft drink packages are among the highest of any consumer products packaging. In 1992 in the United States, consumers recycled an impressive 68% of aluminum cans, 30% of glass bottles, 38% of PET (polyethylene terephthalate) soft drink bottles, and 50% of steel cans.

Future Developments

The bulk of all soft drink production is now done by a few large bottlers. The large franchise companies control key portions of their bottling operations and therefore exert management influence with these bottlers. The trend is expected to continue.

R. STEVE MORROW
CHRISTINE M. QUINN
Coca-Cola USA

J. G. Woodroof and G. F. Phillips, *Beverages: Carbonated and Noncarbonated,* rev. ed., AVI Publishing Co., Inc., Westport, Conn., 1981, pp. 109–113, 134–137, 155–159, 208, 217.

Soft Drink Trends, Beverage Industry Annual Manual 1991/92, Beverage Industry, Edgell Communications, Inc., New York, 1991, pp. 20, 24–28, 32.

Beverage World 1990/91 Databank, Keller International Publishing Corp., Great Neck, N.Y., 1991, p. 231.

Carbonated Beverage Quality Control Manual, Vol. II, Beverage Quality Control Department, Coca-Cola USA, Atlanta, Ga., 1989.

CARBON DIOXIDE

Carbon dioxide, CO_2, is a colorless gas with a faintly pungent odor and acid taste first recognized in the sixteenth century as a distinct gas through its presence as a by-product of both charcoal combustion and fermentation. Today carbon dioxide is a by-product of many commercial processes: synthetic ammonia production, hydrogen production, substitute natural gas production, fermentation, limestone calcination, certain chemical syntheses involving carbon monoxide (qv), and reaction of sulfuric acid with dolomite. Generally present as one of a mixture of gases, carbon dioxide is separated, recovered, and prepared for commercial use as a solid (dry ice), liquid, or gas.

Carbon dioxide is also found in the products of combustion of all carbonaceous fuels, in naturally occurring gases, as a product of animal metabolism, and in small quantities, about 0.03 vol %, in the atmosphere. Its many applications include beverage carbonation, chemical manufacture, firefighting, food freezing, foundry-mold preparation, greenhouses, mining operations, oil well secondary recovery, rubber tumbling, therapeutical work, welding, and extraction processes. Although it is present in the atmosphere and the metabolic processes of animals and plants, carbon dioxide cannot be recovered economically from these sources.

Physical Properties

Some values of physical properties of CO_2 appear in Table 1.

Chemical Properties

Carbon dioxide, the final oxidation product of carbon, is not very reactive at ordinary temperatures. However, in water solution it forms carbonic acid, H_2CO_3, which forms salts and esters through the typical reactions of a weak acid.

Although carbon dioxide is very stable at ordinary temperatures, when it is heated above 170°C, the reaction forming CO proceeds to the right to a limited extent in the presence of ultraviolet light and electrical discharges. Carbon dioxide may be reduced by several means. The most common of these is the reaction with hydrogen. Carbon dioxide reacts with ammonia as the first stage of urea manufacture to form ammonium carbamate. The ammonium carbamate then loses a molecule of water to produce urea, $CO(NH_2)_2$. Commercially, this is probably the most important reaction of carbon dioxide and it is used worldwide in the production of urea (qv) for synthetic fertilizers and plastics (see AMINO RESINS AND PLASTICS; CARBAMIC ACID).

Radioactive Carbon. In addition to the common stable carbon isotope of mass 12, traces of a radioactive carbon isotope of mass 14 with a half-life estimated at 5568 yr are present in the atmosphere and in carbon compounds derived from atmospheric carbon dioxide.

Carbon dioxide containing known amounts of ^{14}C has been used as a tracer in studying botanical and biological problems involving carbon and carbon compounds and in organic chemistry to determine the course of various chemical reactions and rearrangements. It has also been used in testing gaseous diffusion theory with mixtures of CO_2 and $^{14}CO_2$ at elevated pressures.

Environmental Chemistry. Carbon dioxide plays a vital role in the earth's environment. It is a constituent in the atmosphere and, as such, is a necessary ingredient in the life cycle of animals and plants.

The balance between animal and plant life cycles as affected by the solubility of carbon dioxide in the earth's water results in the carbon dioxide content in the atmosphere of about 0.03 vol %. However, car-

Table 1. Properties of Carbon Dioxide

Property	Value
sublimation point at 101.3 kPa^a °C	−78.5
triple point at 518 kPa^b °C	−56.5
critical temperature, °C	31.1
critical pressure, kPa^b	7383
critical density, g/L	467
latent heat of vaporization, J/g ^{c}	
at the triple point	353.4
at 0°C	231.3
gas density at 273 K and 101.3 kPa^a, g/L	1.976
liquid density	
at 273 K, g/L	928
at 298 K and 101.3 kPa^a CO_2, vol/vol	0.712
viscosity at 298 K and 101.3 kPa^a, mPa·s(= cP)	0.015
heat of formation at 298 K, kJ/mol^d	393.7

^{a} 101.3 kPa = 1 atm.
^{b} To convert kPa to psia, multiply by 0.145.
^{c} To convert J/g to Btu/lb, multiply by 0.4302.
^{d} To convert kJ/mol to Btu/mol, multiply by 0.9487.

bon dioxide content of the atmosphere seems to be increasing as increased amounts of fossil fuels are burned. There is some evidence that the rate of release of carbon dioxide to the atmosphere may be greater than the earth's ability to assimilate it.

The effects of such an increase, if it occurs, are not known. It could result in a warmer temperature at the earth's surface by allowing the short heat waves from the sun to pass through the atmosphere while blocking larger waves that reflect back from the earth.

Commercial Production

Sources of carbon dioxide for commercial carbon dioxide recovery plants are *(1)* synthetic ammonia and hydrogen plants, in which methane or other hydrocarbons are converted to carbon dioxide and hydrogen ($CH_4 + 2\,H_2O \rightarrow CO_2 + 4\,H_2$); *(2)* flue gases resulting from the combustion of carbonaceous fuels; *(3)* fermentation in which a sugar such as dextrose is converted to ethyl alcohol and carbon dioxide ($C_6H_{12}O_6 \rightarrow 2\,C_2H_5OH + 2\,CO_2$); *(4)* lime-kiln operation in which carbonates are thermally decomposed ($CaCO_3 \rightarrow CaO + CO_2$); *(5)* sodium phosphate manufacture ($3\,Na_2CO_3 + 2\,H_3PO_4 \rightarrow 2\,Na_3PO_4 + 3\,CO_2 + 3\,H_2O$); and *(6)* natural carbon dioxide gas wells.

Methods of Purification. Although carbon dioxide produced and recovered by the methods listed above has a high purity, it may contain traces of hydrogen sulfide and sulfur dioxide, which cause a slight odor or taste. The fermentation gas recovery processes include a purification stage, but carbon dioxide recovered by other methods must be further purified before it is acceptable for beverage dry ice or other uses. The most commonly used purification methods are treatments with potassium permanganate, potassium dichromate, or active carbon.

Methods of Liquefaction and Solidification. Carbon dioxide may be liquefied at any temperature between its triple point (216.6 K) and its critical point (304 K) by compressing it to the corresponding liquefaction pressure and removing the heat of condensation.

Solidification. Liquid carbon dioxide from a cylinder may be converted to "snow" by allowing the liquid to expand to atmospheric pressure. This simple process is used only where very small amounts of solid carbon dioxide are required because less than one-half of the liquid is recovered as solid. Solid carbon dioxide is produced in blocks by hydraulic presses.

Economic Aspects

Throughout the 1970s and 1980s production increased steadily at a rate of 15 to 20% per year. The chief reason for this large predicted increase is the rapid increase in the amount of carbon dioxide used for secondary oil recovery.

Toxicity

Although carbon dioxide is a constituent of exhaled air, high concentrations are hazardous. Ventilation sufficient to prevent accumulation of dangerous percentages of carbon dioxide must be provided where carbon dioxide gas has been released or dry ice has been used for cooling.

Uses

A large portion of the carbon dioxide recovered is used at or near the location where it is generated as an ingredient in a further processing step.

About 51% of the carbon dioxide consumed in the United States is used in the food industry. Approximately 18% of carbon dioxide output is used for beverage carbonation. Both soft drinks and beer production consume the largest quantity of CO_2 for carbonations (see CARBONATED BEVERAGES). About 10% of the carbon dioxide produced is for chemical manufacturing.

RONALD PIERANTOZZI
Air Products and Chemicals, Inc.

L. H. Chen, *Thermodynamic and Transport Properties of Gases, Liquids and Solids,* McGraw-Hill Book Co., Inc., New York, 1959, pp. 358–369.

A. V. Slack and G. R. James, eds., *Ammonia Part II,* Vol. 2, Marcel Dekker, New York, 1974.

J. H. Perry, ed., *Chemical Engineers Handbook,* 3rd ed., McGraw-Hill Book Co., Inc., New York, 1950, p. 702.

CARBON DISULFIDE

Carbon disulfide (carbon bisulfide, dithiocarbonic anhydride), CS_2, is a toxic, dense liquid of high volatility and flammability. It is an important industrial chemical and its properties are well established. Low concentrations of carbon disulfide naturally discharge into the atmosphere from certain soils, and carbon disulfide has been detected in mustard oil, volcanic gases, and crude petroleum. Carbon disulfide is an unintentional by-product of many combustion and high temperature industrial processes where sulfur compounds are present.

Commercial uses grew rapidly from about 1929 to 1970, when the principal applications included manufacturing viscose rayon fibers, cellophane, carbon tetrachloride, flotation aids, rubber vulcanization accelerators, fungicides, and pesticides. Production of carbon disulfide in the United States has declined in recent years. Other chemical fibers and films, as well as environmental and toxicity considerations related to carbon tetrachloride, have had significant impact on the demand for carbon disulfide. Worldwide annual production capacity in 1991 was approximately 1.3 million tons, with actual production estimated at about one million metric tons.

Physical Properties

Pure carbon disulfide is a clear, colorless liquid with a delicate, etherlike odor. Carbon disulfide is slightly miscible with water, but it is a good solvent for many organic compounds. Thermodynamic constants, vapor pressure, spectral transmission, and other properties of carbon disulfide have been determined. Principal properties are listed in Table 1.

Table 1. Properties of Carbon Disulfide

Property	Values
General	
melting point, K	161.11
latent heat of fusion, kJ/kg[a]	57.7
boiling point at 101.3 kPa[b], °C	46.25
flash point at 101.3 kPa[b], °C	−30
ignition temperature in air, °C	
10-s lag time	120
0.5-s lag time	156
critical temperature, °C	273
critical pressure, kPa[b]	7700
critical density, kg/m^3	378
solubility H_2O in CS_2	
at 10°C, ppm	86
at 25°C, ppm	142
dielectric constant	2.641
Thermochemical data at 298 K[a]	
heat capacity, $C°_p$, J/(mol·K)[a]	45.48
entropy, $S°$, J/(mol·K)[a]	237.8
heat of formation, $H°_f$, kJ/mol[a]	117.1
free energy of formation, $G°_f$, kJ/mol[a]	66.9

[a] To convert J to cal, divide by 4.184.
[b] To convert kPa to atm, divide by 101.3.

Carbon disulfide is completely miscible with many hydrocarbons, alcohols, and chlorinated hydrocarbons. Phosphorus and sulfur are very soluble in carbon disulfide.

Chemical Properties

The low flash point temperature of −30°C at atmospheric pressure and wide flammability range of carbon disulfide deserve special attention. The flash point is lowered if the pressure is decreased or the oxygen content enriched. The flammability limits or explosive ranges depend on conditions of temperature, pressure, and geometry of the enclosure. Flammability limits of 1.06–50.0 vol % carbon disulfide in air are reported for upward propagation and 1.91–35.0 vol % for downward propagation in a 75-mm diameter glass tube.

Carbon disulfide chemistry is thoroughly described in several publications which include many references.

Manufacture

The earliest method for manufacturing carbon disulfide involved synthesis from the elements by reaction of sulfur and carbon as hardwood charcoal in externally heated retorts. Safety concerns, short lives of the retorts, and low production capacities led to the development of an electric furnace process, also based on reaction of sulfur and charcoal. The commercial use of hydrocarbons as the source of carbon was developed in the 1950s, and it was still the predominate process worldwide in 1991. That route, using methane and sulfur as the feedstock, provides high capacity in an economical, continuous unit.

Potential Processes. Sulfur vapor reacts with other hydrocarbon gases, such as acetylene or ethylene, to form carbon disulfide. Light gas oil was reported to be successful on a semiworks scale. In the reaction with hydrocarbons or carbon, pyrites can be the sulfur source. With methane and iron pyrite the reaction products are carbon disulfide, hydrogen sulfide, and iron or iron sulfide. Pyrite can be reduced with carbon monoxide to produce carbon disulfide.

Handling, Shipment, and Storage

Transportation of carbon disulfide is controlled by federal regulations. Acceptable shipping containers include drums, tank trucks, special portable tanks, and rail tank cars. Barges have been used in the past. The United States Department of Transportation classifies carbon disulfide as a flammable liquid and a poison. For ship transport, carbon disulfide must be marked as a marine pollutant. All air transport, cargo, or passenger, is forbidden.

Carbon disulfide is normally stored and handled in mild steel equipment.

Contact of carbon disulfide with air should be avoided because the combination of high volatility, wide flammability range, and low ignition temperature results in a readily combustible mixture. Carbon disulfide must be stored in inert-blanketed, closed tanks.

Small carbon disulfide fires can be smothered with carbon dioxide. Large fires can be controlled with certain types of foams or by a fog or spray of water with attention to proper impoundment of the contaminated water runoff.

Economic Aspects

Depending on energy and raw material costs, the minimum economic carbon disulfide plant size is generally in the range of 2,000–5,000 t/yr for an electric furnace process and 15,000–20,000 t/yr for a hydrocarbon-based process. Hydrocarbon–sulfur plants tend to be on the scale of 50,000 to 200,000 t/yr. It is estimated that 53 carbon disulfide plants existed throughout the world in 1991. The U.S. carbon disulfide capacity dropped sharply during 1991 when Akzo Chemicals closed a 159,000 t/yr plant at Delaware City, Delaware. The U.S. carbon disulfide industry still accounts for about 12% of the total worldwide installed capacity.

Production in the United States peaked at 362,000 t/yr in 1969 and had fallen to less than half that amount by 1991. The cost for producing carbon disulfide is sensitive to sulfur and natural gas prices, which have increased greatly since about 1974. The future demand for carbon disulfide is strongly tied to the fate of its principal use, the production of regenerated cellulose products.

Production of carbon disulfide expanded rapidly after World War II to supply the growing needs of the viscose rayon industry, which consumes about 0.31 tons of CS_2 per ton of rayon. Competition from noncellulosic synthetic fibers has caused a drop in rayon production in the United States since the mid-1960s. However, plans have been announce to build a new rayon plant in Louisiana that is expected to achieve improved carbon disulfide utilization and low emissions by recovering and recycling carbon disulfide. This pattern of modern viscose rayon plants replacing aging facilities that cannot be economically upgraded is apt to be repeated in other parts of the world. In a development that could have far-ranging implications, a viscose rayon producer is constructing a solvent spun cellulosic fiber plant using an amine oxide solvent rather than carbon disulfide.

Use of carbon disulfide for manufacture of cellophane has dropped dramatically because of competition from plastic films, and just one cellophane producer, Flexel, Inc., remains in the United States.

Carbon disulfide for manufacture of carbon tetrachloride increased in the 1950s and 1960s to supply the key raw material for chlorofluorocarbon refrigerants and aerosol propellants. Because of ecological and health concerns, carbon tetrachloride consumption began to decline in the mid-1970s. That use for carbon disulfide will suffer under a United Nations proposal to phase out carbon tetrachloride and chlorofluorocarbons to protect the earth's ozone layer. During 1991 the only remaining carbon tetrachloride plant in the United States that employed the carbon disulfide route was permanently shut down. Consumption of carbon disulfide in rubber, agriculture, mining, and specialty industrial applications is anticipated to remain close to 1991 levels for the next several years.

Specifications and Quality Control

Modern plants generally produce carbon disulfide of about 99.99% purity. High product quality is ensured by closely controlled continuous fractional distillation.

Toxicology, Health, and Safety Factors

Care must be exercised in handling carbon disulfide because of both health concerns and the danger of fire or explosions. Occupational exposure potentially may involve as many as 20,000 workers in the United States. Ingestion is rare, but a 10 mL dose can prove fatal. Contact usually occurs by inhalation of vapor. However, vapor and liquid can be absorbed through intact skin and poisoning may occur by the dermal route. Repeated contact of liquid carbon disulfide with the skin can cause inflammation and cracking because carbon disulfide removes protective waxes and oils. Extended skin contact results in blistering and possibly second- and third-degree burns. Precautions should be taken to avoid breathing of vapors or mists that may contain carbon disulfide. Contact with skin or eyes should also be avoided, and adequate safety gear should be worn, including goggles, impervious gloves, and appropriate clothing.

The odor threshold of carbon disulfide is about 1 ppm in air but varies widely depending on individual sensitivity and purity of the carbon disulfide. However, using the sense of smell to detect excessive concentrations of carbon disulfide is unreliable because of the frequent co-presence of hydrogen sulfide that dulls the olfactory sense.

Immediate effects of overexposure to carbon disulfide vapors range from headache, dizziness, nausea, and vomiting to life-threatening convulsions, unconsciousness, and respiratory paralysis. For an exposure time of 30 min, 1150 ppm carbon disulfide in air results in serious symptoms, 3210 ppm is dangerous to life, and 4815 ppm is fatal. Prolonged and repeated exposure to carbon disulfide vapor can affect both the central and peripheral nervous systems. In recent years, previously unrecognized and more subtle toxic effects of repeated lower level exposures became evident. This led OSHA in 1989 to reduce permissible concentration limits to 4 ppm (12 mg/m^3) maximum time-weighted average for 8-h exposure and 12 ppm (36 mg/m^3) maximum for 15-min short-term exposure. Analysis of urine specimens for carbon disulfide metabolites by an iodine-azide test and other methods can indicate overexposure.

Health hazards linked to carbon disulfide are extensively covered in the literature. Also available are epidemiological studies, general reviews containing many references, and a Material Safety Data Sheet.

Uses

United States consumption of carbon disulfide totaled about 108,000 t in 1990 according to SRI International, with the following distribution by end use application: 46,000 t for rayon; 33,000 t for carbon tetrachloride; 12,000 t for rubber; 5,000 t for cellophane; and 12,000 t for agricultural and miscellaneous uses. During 1991 the carbon tetrachloride application disappeared entirely, thereby reducing the annualized carbon disulfide usage to an estimated 75,000 t. Net exports are around 6,000 t, and are expected to increase in the future.

DAVID E. SMITH
ROBERT W. TIMMERMAN
FMC Corporation

L. J. O'Brien and W. J. Alford, *Ind. Eng. Chem.* **43,** 506 (1951).

A. D. Dunn and W. D. Rudorf, *Carbon Disulphide in Organic Chemistry,* Ellis Horwood Ltd., Chichester, UK; Halsted Press, div. of John Wiley & Sons, Inc., New York, 1989.

U.S. Pat. 2,568,121 (Sept. 18, 1951), H. O. Folkins, C. A. Porter, E. Miller, and H. Hennig (to The Pure Oil Co.).

F. A. Patty, *Industrial Hygiene and Toxicology,* Vol. II, 2nd. rev. ed., John Wiley & Sons, Inc., New York, 1962, pp. 901–904.

CARBONIC AND CARBONOCHLORIDIC ESTERS

The reaction of phosgene (carbonic dichloride) with alcohols gives two classes of compounds, carbonic esters and carbonochloridic esters, commonly referred to as carbonates and chloroformates. The carbonic acid esters (carbonates), ROC(O)OR, are the diesters of carbonic acid. The carbonochloridic esters, also referred to as chloroformates or chlorocarbonates, ClC(O)OR, are esters of hypothetical chloroformic acid, ClCOOH.

The reaction proceeds in stages, first producing a carbonochloridic ester (chloroformate), and then a carbonic acid diester (carbonate).

CHLOROFORMATES

Physical Properties

In general, carbonochloridates or chloroformates are clear, colorless liquids with low freezing points and high boiling points. They are soluble in most organic solvents, but insoluble in water, although they do hydrolyze in water. The physical properties of the most widely used chloroformate esters are given in Table 1.

Chemical Properties

Chloroformates are reactive intermediates that combine acid chloride and ester functions. They undergo many reactions similar to those of acid chlorides; however, the rates are usually slower.

Stability. The ester moiety determines thermal stability generally in the following order of decreasing stability: aryl > primary alkyl > secondary alkyl > tertiary alkyl. In terms of mechanistic chemistry, the chloroformates that produce stable carbonium ions on thermal decomposition, eg, benzyl, isopropyl, or tertiary butyl, are unstable.

Reaction with Oxygen Moeities. Chloroformates reach milk water, alkali metal hydroxides, aliphalic alcohols, thiols, heterocyclic alcohols, phenols, and carboxylic acids.

$$\text{RO}\overset{\overset{\text{O}}{\|}}{\text{C}}\text{Cl} + \text{ROH} \longrightarrow \text{RO}\overset{\overset{\text{O}}{\|}}{\text{C}}\text{OR} + \text{HCl}$$

Reactions with Nitrogen Compounds. Chloroformates reach milk ammonia and amines to form carbonates, amino alcohols, and amino phenols to give oxazolidinenes and amino acids to form *N*-protenated amino acids.

$$\text{RO}\overset{\overset{\text{O}}{\|}}{\text{C}}\text{Cl} + 2\,\text{NH}_3 \longrightarrow \text{RO}\overset{\overset{\text{O}}{\|}}{\text{C}}\text{NH}_2 + \text{NH}_4\text{Cl}$$

Acylation. Aryl chloroformates are good acylating agents, reacting with aromatic hydrocarbons under Friedel-Crafts conditions to give the expected aryl esters of the aromatic acid.

Dealkylation. Chloroformates are used to dealkylate tertiary amines.

Manufacture

The reaction of phosgene with alcohols or phenols has been thoroughly discussed. Alkyl chloroformates, especially those of low molecular weight alcohols, are prepared by the reaction of liquid anhydrous alcohols with molar excess of dry, chlorine-free phosgene at low temperature. Corrosion-resistant reactors, lines, pumps, and valves are required. The reactions are most often run in batch reactors, although some of the high volume chloroformates are produced in cascade-type continuous reactors.

Shipping and Storage

Chloroformates are shipped in nonreturnable 208-L (55-gal) polyethylene drums with carbon steel overpacks or high density polyethylene drums.

Chloroformates should be stored in a cool, dry atmosphere, preferably refrigerated, especially for prolonged storage.

Economic Aspects

Most chloroformate production is used captively and production figures are not available. The prices are also not published, but can be obtained by contacting U.S. and other worldwide producers, such as PPG Industries, Inc., BASF, Van Cham, and SNPE.

Toxicity

Chloroformates, especially those of low molecular weight, are lachrimators, vesicants, and produce effects similar to those of hydrogen chloride or carboxylic acid chlorides. They can also irritate the skin and mucous membranes, producing severe burns and possible irreversible tissue damage.

Uses

Chloroformates are versatile, synthetic intermediates, based on the affinity of the chlorine atoms for active hydrogen atoms. Chloroformates should be considered as intermediates for syntheses of pesticides, perfumes, drugs, polymers, dyes, and other chemicals.

CARBONATES

Chloroformates and alcohols or phenols give carbonic diesters. In addition, the higher diesters can be made from the lower ones by alcoholysis or ester interchange by heating the lower diester with a higher alcohol in the presence of acid such as HCl or H_2SO_4, or base such as sodium alcoholate.

More recently, preparation of carbonic esters by nonphosgene routes, such as the reactions of CO or CO_2 with appropriate substances in the presence of catalysts, has been preferred. These

Table 1. Physical Properties of Selected Chloroformates

Chloroformate	Mol wt	Sp gr, d^{20}_4	Refractive index n^{20}_DFlash point, °C[a]	Flash point, °C[a]	Viscosity, mPa·s (= cP), 20°C	Bp, °C at	
						2.67 kPa[b]	101.3 Kpa[b]
methyl	94.5	1.250	1.3864	24.4			71
ethyl	108.53	1.138	1.3950	27.8			94
isopropyl	122.55	1.078	1.3974	27.8	0.65		105
n-propyl	122.55	1.091	1.4045	34.4	0.80	25.3	112.4
allyl	120.5	1.1394	1.4223	27.8	0.71	25	
n-butyl	136.58	1.0585	1.4106	52.2	0.888	44	
sec-butyl	136.58	1.0493	1.4560	35.6	0.897	36	
isobutyl	136.58	1.0477	1.4079	39.5	0.88	39	
2-ethylhexyl	192.7	0.9914	1.4307		1.774	98	
n-decyl	220.7	0.9732	1.4400	118.3	3.00	122	
phenyl	156.57	1.2475	1.5115		1.882	83.5	185
benzyl	170.6	1.2166	1.5175	80.0	2.57	103	152
ethylene bis	186.98	1.4704	1.4512	134.0	4.78	108	
diethylene glycol bis	231.0	1.388	1.4550	160.0	8.76	148	

[a] Tag open cup (TOC).
[b] To convert kPa to mm Hg, multiply by 7.50.

Table 2. Physical Properties of Selected Carbonates

Carbonates	Mol wt	Sp gr, d^{20}_4	Refractive index n^t_D	Flash point, °C	Viscosity[a], mPa·s(= cP)	Bp, °C[b]
dimethyl	90.08	1.073	1.3697	21.7[c] 16.7[d]	0.664 (20)	90.2
diethyl	118.13	0.975	1.3846	46.1[c] 32.8[d]	0.868 (15)	23.8 (1.33) 69.7 (13.33) 126.8
di-*n*-propyl	146.18	0.941	1.4022	64[d]	[e]	165.5–166.6
diisopropyl	146.18					147.0
diallyl	142.15	0.994	1.4280			97 (8.13) 105 (13.33)
di-*n*-butyl	174.14	0.9244	1.4099			166 (97.31)
di-2-ethylhexyl	204.19	0.8974^{20}_{20}	1.4352			207.7
diphenyl	214.08	0.8974^{87}_4				173 (1.33) 302
diethylene glycol bis(allyl)	274.3	1.143	1.4503	177[f]	9 (25)	160 (0.27)
tolyl diglycol	374.4	1.189	1.5229			247–248 (0.27)
ethylene	88.06	1.3218^{39}_4	1.4158			248

[a] At the temperature noted in parentheses (°C).
[b] At 101.3 kPa (= 1 atm) unless otherwise noted in parentheses in kPa.
[c] Tag open cup.
[d] Tag closed cup.
[e] Brookfield no. 1 spindle; rpm (mPa·s 10(5), 20(6.5), 50(8.0), 100(12.0)).
[f] Cleveland open cup.

methods are more economic in many cases and naturally less hazardous than phosgene routes.

Physical Properties

The physical properties of selected carbonates are given in Table 2. The lower alkyl carbonates are neutral, colorless liquids with a mild sweet odor. Aryl carbonates are normally solids.

Chemical Properties

Carbonates undergo nucleophilic substitution reactions analogous to chloroformates, except that in this case, an OR group (rather than chloride) is replaced by a more basic group. Normally these reactions are catalyzed by bases. Carbonates are sometimes preferred over chloroformates because formation of hydrogen chloride as a by-product is avoided, which simplifies handling. However, the reactivity of carbonates toward nucleophiles is considerably less than chloroformates.

Manufacture

Carbonates are manufactured by essentially the same method as chloroformates, except that more alcohol is required in addition to longer reaction times and higher temperatures. The products are neutralized, washed, and distilled. Corrosion-resistant equipment similar to that described for the manufacture of chloroformates is required. Diaryl carbonates are prepared from phosgene and two equivalents of the sodium phenolates or with phenols and various catalysts.

In the 1990s, the trend is to manufacture carbonates without the use of phosgene. This method has the advantage of avoiding the use of highly toxic phosgene as well as a considerably lower cost. A number of different processes for the manufacture of dimethyl carbonate have been patented that use methanol, CO, hydrogen, oxygen, and a variety of catalyst systems. The nonphosgene route has also been utilized for synthesis of aromatic carbonates and ethylene carbonate.

Shipping and Storage

Dimethyl and diethyl carbonates are shipped in nonreturnable 208-L (55-gal) polyethylene drums with carbon steel overpack or high density polethylene drums.

The carbonates should be plainly labeled and stored in cool, dry areas away from sources of ignition. The U.S. Department of Transportation (DOT) Hazardous Materials Regulations control the shipment of carbonates.

Economic Aspects

As in the case of the chloroformates, most of the carbonate production is used captively and production figures are not available.

Health and Safety

Unlike chloroformates, diethyl and dimethyl carbonates are only mildly irritating to the eyes, skin, and mucous membranes. Diethylene glycol bis(allyl carbonate) may be irritating to the skin, but it is not classified as a toxic substance; however, it is extremely irritating to the eyes.

Protective clothing, rubber gloves, safety goggles, and adequate ventilation are recommended for all personnel handling high concentrations of carbonates. In case of fire, foam, carbon dioxide, or dry chemical extinguishing agents should be used. However, it is permissible to use a water spray to cool any drums in the vicinity, thus avoiding any spread of the fire.

Uses

Commercially, the most important carbonate has been the diethyl ester. It is used in many organic syntheses, particularly of pharmaceuticals and pharmaceutical intermediates. It is also used as a solvent for many synthetic and natural resins, and in vacuum tube cathode-fixing lacquers. Dimethyl carbonate is gaining commercial importance because of its lower cost from the nonphosgene route. Dimethyl carbonate is used in the synthesis of pharmaceuticals, agricultural chemicals, dyestuffs, and as a specialty solvent. It has been used in gasoline to improve octane number, as a phase separation inhibitor in liquid hydrocarbon fuel and ethanol mixtures, and as a carbonylating and methylating agent. Dipropyl carbonate is also an organic intermediate, a specialty solvent, and is used in photoengraving as an assist agent for silicon circuiting. Ethylene carbonate is an important solvent for polymers, such as polyacrylonitrile, and may become an important raw material for synthesizing carbonates. The lower alkyl carbonates and diphenyl carbonate are used in the preparation of polycarbonate resins by transesterification. Diethylene glycol bis(allyl carbonate) polymerizes easily because of its two double bonds and is used for colorless, optically clear castings. Polymerization is catalyzed by the use of diisopropyl peroxydicarbonate. Such polymers are used in the preparation of safety glasses, lighweight prescription lenses, glazing cast sheet, and optical cement (see ALLYL MONOMERS AND POLYMERS; POLYCARBONATES).

SURESH B. DAMLE
PPG Industries, Inc.

M. Matzner, R. P. Kurkjy, and R. J. Cotter, *Chem. Rev.* **64**, 645 (1964).

D. N. Kevill, in S. Patai, ed., *The Chemistry of Acyl Halides,* Interscience Publishers, a division of John Wiley & Sons, Inc., New York, 1972.

S. B. Damle, *Hydrolysis of Chloroformates,* unpublished data, 1991.

CARBON MONOXIDE

Carbon monoxide, CO, a colorless, odorless, flammable, toxic gas, is produced by steam reforming or partial oxidation of carbonaceous materials. It is used as a fuel, a metallurgical reducing agent, and a feedstock in the manufacture of a variety of chemicals, notably methanol, acetic acid, phosgene, and oxo alcohols. Increased usage of carbon monoxide from coal in chemicals and fuels manufacture is likely if economic coal gasification technology continues to evolve (see FUELS, SYNTHETIC).

Physical and Thermodynamic Properties

The physical and thermodynamic properties of carbon monoxide are well documented. A summary of particularly useful physical constants is presented in Table 1.

Solid carbon monoxide exists in one of two allotropes, a body-centered cubic or a hexagonal structure.

Extensive listings of thermal conductivity for liquid and gaseous carbon monoxide and relationships of thermal conductivity to pressure appear in the literature.

The solubility of carbon monoxide in a variety of solvents at 928 K is given in Table 2, and a detailed discussion of the solubility of carbon monoxide in water has been provided.

Carbon monoxide burns readily in air or oxygen. Ignition temperatures are 644–658°C and 637–658°C, respectively.

Chemical Properties

According to molecular orbital theory the 10 valence electrons from carbon and oxygen in CO fill the lowest energy molecular orbitals to give the electronic configuration $(\varsigma^b_s)^2(\varsigma^*_s)^2(\pi^b_{xy})^4(\pi^b_z)^2$ (^{b}, bonding; *, antibonding) which predicts one ς- and two π-bonds (24). The π* and ς* orbitals of molecular carbon monoxide remain unfilled but available for bonding with transition-metal atoms.

The bonding between carbon monoxide and transition-metal atoms is particularly important because transition metals, whether deposited on solid supports or present as discrete complexes, are

Table 1. Physical Properties of Carbon Monoxide

Property	Value
mol wt	28.011
melting point, K	68.09
boiling point, K	81.65
ΔH, fusion at 68 K, kJ/mol[a]	0.837
ΔH, vaporization at 81 K, kJ/mol[a]	6.042
density at 273 K, 101.3 kPa[b], g/L	1.2501
sp gr[c], liquid, 79 K	0.814
sp gr[d], gas, 298 K	0.968
critical temperature, K	132.9
critical pressure, MPa[b]	3.496
critical density, g/cm^3	0.3010
triple point	
temperature, K	68.1
pressure, kPa[e]	15.39
$\Delta G°$ formation at 298 K, kJ/mol[a]	−137.16
autoignition temperature, K	882
flammability limits in air[f]	
upper limit, %	74.2
lower limit, %	12.5

[a] To convert J to cal, divide by 4.184.
[b] 101.3 kPa = 1 atm; to convert MPa to atm, multiply by 9.87.
[c] With respect to water at 277 K.
[d] With respect to air at 298 K.
[e] To convert kPa to torr, multiply by 7.5.
[f] Saturated with water vapor at 290 K.

Table 2. Solubility of Carbon Monoxide at 25°C and 101.3 kPa (1 atm) CO Pressure, mol fraction CO×10[4]

Solvent	Solubility	Solvent	Solubility
1-heptane	17.24	chlorobenzene	6.47
cyclohexane	9.91	nitrobenzene	3.72
methylcyclohexane	12.41	methanol	3.76
benzene	6.68	ethanol	4.84
toluene	8.11	2-methyl-1-propanol	6.52
perfluoro-*n*-heptane	38.75	acetone	7.72
perfluorobenzene	21.20	water	0.16
carbon tetrachloride	8.76		

required as catalysts for the reaction between carbon monoxide and most organic molecules.

INDUSTRIALLY SIGNIFICANT REACTIONS OF CARBON MONOXIDE Industrially significant reactions include conversion to hydrogen (water gas shift reaction), oxidation, disproportionation, production of phosgene, production of methanol, production of acetic acid, hydroformylation, production of acrylic acid, Fisher-Tropsch reaction, methanation, and nickel purification.

GENERAL REACTIONS OF CARBON MONOXIDE

With Hydrogen. In a liquid-phase high pressure reaction (60 MPa or 600 atm), a rhodium cluster complex catalyzes the direct formation of ethylene glycol, propylene glycol (see GLYCOLS), and glycerol (qv) from synthesis gas.

With Alcohols, Ethers, and Esters. Carbon monoxide reacts with alcohols, ethers, and esters to give carboxylic acids.

With Unsaturated Compounds. The reaction of unsaturated organic compounds with carbon monoxide and molecules containing an active hydrogen atom leads to a variety of interesting organic products. The hydroformylation reaction is the most important member of this class of reactions.

Oxidative Carbonylation. Carbon monoxide is rapidly oxidized to carbon dioxide; however, under proper conditions, carbon monoxide and oxygen react with organic molecules to form carboxylic acids or esters. With olefins, unsaturated carboxylic acids are produced, whereas alcohols yield esters of carbonic or oxalic acid.

Isocyanate Synthesis. In the presence of a catalyst, nitroaromatic compounds can be converted into isocyanates, using carbon monoxide as a reducing agent.

Dimethylformamide. The industrial solvent dimethylformamide is manufactured by the reaction between carbon monoxide and dimethylamine.

Aromatic Aldehydes. Carbon monoxide reacts with aromatic hydrocarbons or aryl halides to yield aromatic aldehydes (see ALDEHYDES).

Metal Carbonyls. Carbon monoxide forms metal carbonyls or metal carbonyl derivatives with most transition metals (see COORDINATION COMPOUNDS).

Polymerization. Carbon monoxide forms copolymers with ethylene and suitable vinyl compounds.

Manufacture and Purification

Of the four commercial processes for the purification of carbon monoxide, two processes are based on the absorption of carbon monoxide by salt solutions, the third uses either low temperature condensation or fractionation, and the fourth method utilizes the adsorption of carbon monoxide on a solid adsorbent material. All four processes use similar techniques to remove minor impurities. Particulates are removed in cyclones or by scrubbing. Scrubbing also removes any tars or heavy hydrocarbon fractions. Acid gases are removed by absorption in monoethanolamine, hot potassium carbonate, or by other patented removal processes. The purified gas stream is then sent to a carbon monoxide recovery section for final purification and by-product recovery.

Economic Aspects

Carbon monoxide is produced as a component of synthesis gas or as purified gas by many manufacturers, primarily for on-site process applications. Because very few producers engage in merchant sale of purified gases, published production data are not available. By-product hydrogen price, feedstock prices, and delivery charges as well as location, purity, and volume affect carbon monoxide pricing.

Carbon monoxide must bear a red label during shipment because it is flammable. Additional shipping requirements are contained in the Interstate Commerce Commission regulations.

Health and Safety

Occurrence. Carbon monoxide is a product of incomplete combustion and is not likely to result where a flame burns in an abundant air supply, yet may result when a flame touches a cooler surface than the ignition temperature of the gas. Automobile exhaust gas is perhaps the most familiar source of carbon monoxide exposure. The manufacture and use of synthesis gas, calcium carbide manufacture, distillation of coal or wood, combustion operations, heat treatment of metals, fire fighting, mining, and cigarette smoking represent additional sources of carbon monoxide exposure.

Toxicity. Carbon monoxide is the most widespread gaseous hazard to which humans are exposed. The toxicity of carbon monoxide is a result of its reaction with the hemoglobin of blood. The carboxyhemoglobin (COHb) that is formed displaces oxygen and leads to asphyxiation.

The recommended NIOSH limit of 35 ppm is the time-weighted average exposure to carbon monoxide based on a carboxyhemoglobin level of 5%; this amount of COHb is what an employee engaged in sedentary activity would be expected to approach in eight hours of continuous exposure. The standard does not take into account the smoking habits of a worker. First-aid treatment for carbon monoxide poisoning emphasizes elimination of the gas from the body. Elimination of carbon monoxide occurs solely through the lungs, and though rapid at first, the last traces are difficult to remove. The poisoned patients must be removed to fresh air, kept warm, and administered pure oxygen by the best method available.

Prevention of carbon monoxide poisoning is best accomplished by providing good ventilation where contamination is a problem. If good ventilation is not possible, a self-contained breathing apparatus, such as a Scott Air-Pak, must be used.

RONALD PIERANTOZZI
Air Products and Chemicals, Inc.

R. H. Perry and C. H. Chilton, eds., *Chemical Engineers Handbook,* 5th ed., McGraw-Hill Book Co., New York, 1973.

F. A. Patty, in F. A. Patty, ed., *Industrial Hygiene and Toxicology,* Vol. 2, Interscience-Publishers, a division of John Wiley & Sons, Inc., New York, 1963, p. 924.

R. E. Gosselin and co-workers, *Clinical Toxicology of Commercial Products,* Williams & Wilkins Co., Baltimore, Md., 1976, p. 86; M. Sittig, *Hazardous and Toxic Effects of Industrial Chemicals,* Noyes Data Corp., Park Ridge, N.J., 1979, p. 102.

CARBON MONOXIDE–HYDROGEN REACTIONS.

See FUELS, SYNTHETIC.

CARBONYLS

Carbon monoxide (qv), CO, the most important π-acceptor ligand, forms a host of neutral, anionic, and cationic transition-metal complexes. There is at least one known type of carbonyl derivative for every transition metal, as well as evidence supporting the existence of the carbonyls of some lanthanides (qv) and actinides (see ACTINIDES AND TRANSACTINIDES; COORDINATION COMPOUNDS.

Carbonyls are useful in the preparation of high purity metals, as in the Mond process for the extraction of nickel from its ores (see NICKEL AND NICKEL ALLOYS), in catalytic applications, and in the synthesis of organic compounds. Metal carbonyls are employed in the preparation of complexes where the carbon monoxide ligand is replaced by halides, hydrogen, Group 15 (VA) and 16 (VIA) derivatives, arenes, and many chelating ligands (see CHELATING AGENTS).

Structure and Bonding of Metal Carbonyls

Numerous theoretical approaches have been used in predicting and describing the bonding and structure of metal carbonyls. The Sidgwick concept of effective atomic number, or the 18-electron rule, has been particularly useful in predicting formulas, and molecular orbital and ligand field calculations have provided additional insight into the more detailed features of bonding in metal carbonyls.

Bonding. The 18-electron rule requires that each metal interact with a sufficient number of CO molecules and additionally in polynuclear complexes, ie, those containing more than one metal atom, with sufficient neighboring metal atoms to allow the metal to achieve the electronic structure of the subsequent inert gas in the periodic table.

Structure. The CO molecule coordinates in the ways shown diagrammatically in Figure 1.

Mononuclear Carbonyls. The lowest coordination number adopted by an isolable metal carbonyl is four. The only representative of this class is nickel carbonyl, the first metal carbonyl isolated. Representative pentacarbonyls are restricted to the iron, ruthenium, and osmium group.

The neutral complexes of chromium, molybdenum, tungsten, and vanadium are six-coordinate and the CO molecules are arranged about the metal in an octahedral configuration.

Polynuclear Carbonyls. Several structures consist of dinuclear metal carbonyls. The metal atoms in $Mn_2(CO)_{10}$, as also for technetium and rhenium, are held together by a metal–metal bond, and the compound contains 10 terminal CO ligands, five coordinated to each atom.

A great many complexes of nuclearity of three and higher have been discovered since the early 1970s. Many of these contain two or more different metals, other ligands in addition to CO, and heteroatoms such as hydrogen, carbon, sulfur, or phosphorus associated with the cluster. For trinuclear cluster complexes, open (chain) or closed (cyclic) structures are possible. For tetranuclear cluster complexes, three structure types are observed: tetrahedral; open tetrahedral (butterfly); or square planar, for typical total valence electron counts of 60, 62, and 64, respectively.

Heteronuclear Carbonyls and High Nuclearity Carbonyl Clusters

A few of the heteronuclear metal carbonyls that have been reported are presented below. In most cases the structures of the heteronuclear species are similar to their homonuclear analogues.

Heteronuclear carbonyl
$(CO)_5MnCo(CO)_4$
$(CO)_5MnRe(CO)_5$
$[(CO)_5Mn]_2Fe(CO)_4$
$[(CO)_5Mn][(CO)_5Re][(CO)_4Fe]$
$(CO)_5ReCo(CO)_4$
$[(CO)_5Re]_2Fe(CO)_4$
$Ru_2Os(CO)_{12}$
$Fe_2Ru(CO)_{12}$
$Zn[Mn(CO)_5]_2$
$Os(CO)_4[Mn(CO)_5]_2$
$Os(CO)_4[Re(CO)_5]_2$
$Co_2(CO)_7[ZnCo(CO)_4]_2$
$[CoCu(CO)_4]_4$
$[Hg_3Co_3(CO)_9]$-μ-$Hg_3[Hg_3Co_3(CO)_9]$
$Re_2(CO)_8[\mu$-$TlRe(CO)_5]_2$

Syntheses, crystallization, structural identification, and chemical characterization of high nuclearity clusters can be exceedingly difficult. Usually, several different clusters are formed in any given synthetic procedure, and each compound must be extracted and identified. The problem may be compounded by the instability of a particular molecule.

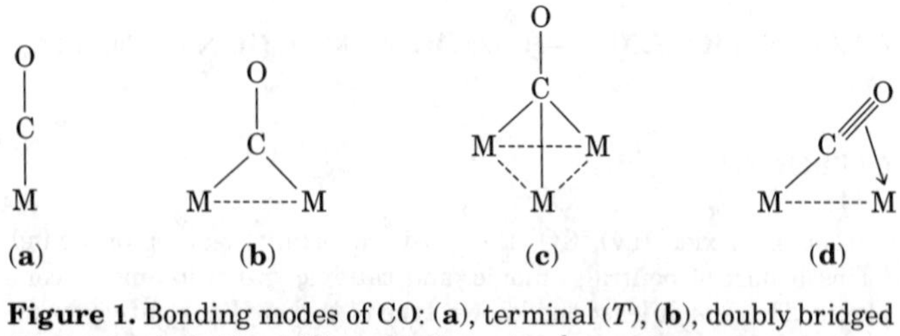

Figure 1. Bonding modes of CO: (**a**), terminal (*T*), (**b**), doubly bridged (*B*); (**c**), triply bridged (*B*); and (**d**), the (μ,ϵ^2–) entity.

Despite the complexity of transition-metal clusters, a basic structural unit is common. A triangular network of metal atoms occurs in most clusters.

In the families of heptanuclear clusters, two geometries are found: the capped octahedron that is typical for 98-valence electrons, and the vertex-sharing open tetrahedral (butterfly) structures typical for 106-valence electrons.

Physical properties of some of the metal carbonyls discussed are summarized in Table 2.

Physical Properties

Most metal carbonyls are volatile solids that sublime easily. The volatility of metal carbonyls coupled with their toxicity is an important safety consideration. The vapor pressure of many metal carbonyls have been tabulated elsewhere.

The thermodynamic properties of simple metal carbonyls have been compiled. Some selected properties are listed in Table 3.

Preparation

Because transition metals even in a finely-divided state do not readily combine with CO, various metal salts have been used to synthesize metal carbonyls. Metal salts almost always contain the metal in a higher oxidation state than the resulting carbonyl complex. Therefore, most metal carbonyls result from the reduction of the metal in the starting material. Such a process has been referred to as reductive carbonylation. Although detailed mechanistic studies are lacking, the process probably proceeds through stepwise reduction of the metal with simultaneous coordination of CO. Processes include syntheses from dry metals and salts, syntheses in solvent systems, condensation of simple metal carbonyls, carbonylation of CO exchange, and low pressure syntheses.

Economic Aspects

Table 4 lists the metal carbonyls that are commercially available and the corresponding U.S. manufacturers or suppliers. Companies producing these compounds on a captive basis are not included.

Toxicology and Safety

Exposure to metal carbonyls can present a serious health threat.

The toxic symptoms from inhalation of nickel carbonyl are believed to be caused by both nickel metal and carbon monoxide. In many acute cases the symptoms are headache, dizziness, nausea, vomiting, fever,

Table 2. Physical Properties of Metal Carbonyls

Name	Molecular formula	Color (solid)	Melting point, °C	Boiling point, °C[a]
vanadium hexacarbonyl	$V(CO)_6$	blue-green	50 dec	$40–50^{2}$ sub
chromium hexacarbonyl	$Cr(CO)_6$	white	149–155	$70–75^{2}$ sub
molybdenum hexacarbonyl	$Mo(CO)_6$	white	150–151 dec	
tungsten hexacarbonyl	$W(CO)_6$	white	169–170	50 sub
decacarbonyldimanganese	$Mn_2(CO)_{10}$	yellow	151–155	$50^{0.001}$ sub
decacarbonylditechnetium	$Tc_2(CO)_{10}$	white	159–160	$40^{0.001}$ sub
decacarbonyldirhenium	$Re_2(CO)_{10}$	white	177	$60^{0.001}$ sub
iron pentacarbonyl	$Fe(CO)_5$	white	−20	103
diiron nonacarbonyl	$Fe_2(CO)_9$	yellow	100 dec	
triiron dodecocarbonyl	$Fe_3(CO)_{12}$	green-black	140 dec	$60^{0.01}$ sub
dodecacarbonyltriruthenium	$Ru_3(CO)_{12}$	orange	150 dec	
dodecacarbonyltriosmium	$Os_3(CO)_{12}$	yellow	224	
dicobaltoctacarbonyl	$Co_2(CO)_8$	orange	50–51	$45^{0.1}$ sub
tri-μ-carbonylnonacarbonyltetracobalt	$Co_4(CO)_{12}$	black	60 dec	
tri-μ-carbonylnonacarbonyltetrarhodium	$Rh_4(CO)_{12}$	red	dec	
hexadecacarbonylhexarhodium	$Rh_6(CO)_{16}$	black	220 dec	
tri-μ-carbonylnonacarbonyltetrairidium	$Ir_4(CO)_{12}$	yellow	210 dec	
nickel carbonyl	$Ni(CO)_4$	white	−25	43

[a] Bp is at 101.3 kPa unless otherwise noted in superscript. To convert kPa to torr, multiply by 7.5.

Table 3. Heats of Combustion and Formation for Metal Carbonyls

Molecular formular	Heat of combustion[a], kJ/mol[b]	Heat of formation, kJ/mol[b]
$Cr(CO)_6$	−1854	−1077
$Mo(CO)_6$	−2123	−982.4
$W(CO)_6$	−2250	−950.6
$Mn_2(CO)_{10}$	−3251	−1677
$Fe(CO)_5$	−1619	−964.0
$Ni(CO)_4$	−1181	−622.2

[a] Determined for the combustion to metal oxides in the presence of oxygen.
[b] To convert J to cal, divide by 4.184.

Table 4. Commercial Availability of Metal Carbonyls

	Company[a]			
Metal carbonyl	Strem Chemicals	Pressure Chemical Co.	Aldrich Chemical Co.	Johnson Matthey (Aesar group)
$V(CO)_6$	A		A	
$Cr(CO)_6$	A	A	A	A
Mo(CO)	A	A	A	A
$W(CO)_6$	A	A	A	A
$Mn_2(CO)_{10}$	A	A	A	A
$Re_2(CO)_{10}$	A	A	A	
$Fe(CO)_5$[b]	A	A	A	
$Fe_2(CO)_9$	A	A	A	A
$Fe_3(CO)_{12}$	A	A	A	A
$Ru_3(CO)_{12}$	A	A	A	A
$Os_3(CO)_{12}$	A	A	A	A
$Co_2(CO)_8$	A	A		A
$Co_4(CO)_{12}$	A			A
$Rh_4(CO)_{12}$	A			
$Rh_6(CO)_{16}$	A	A		A
$Ir_4(CO)_{12}$	A			
$Ni(CO)_4$	A	A		

[a] Where A indicates the carbonyl is available in research-sized quantities of a few grams or kilograms.
[b] GAF Corp. and BASF Corp. are bulk suppliers of $Fe(CO)_5$.

and difficulty in breathing. If exposure is continued, unconsciousness follows with subsequent damage to vital organs and death.

When heated to about 60°C, nickel carbonyl explodes. For both iron and nickel carbonyl, suitable fire extinguishers are water, foam, carbon dioxide, or dry chemical.

FRANK S. WAGNER
Strem Chemicals Inc.
HERBERT D. KAESZ
University of California, Los Angeles

D. M. P. Mingos, in G. Wilkinson, F. G. A. Stone, and E. W. Abel, eds., *Comprehensive Organometallic Chemistry,* Vol. 3, Pergamon Press Ltd., Oxford, U.K., 1982, Chapt. 19, pp. 1–88.

D. M. P. Mingos and A. S. May, D. F. Shriver, H. D. Kaesz, and R. D. Adams, eds., *The Chemistry of Metal Cluster Complexes,* VCH Publishers, New York, 1990, Chapt. 2, pp. 11–120.

I. Wender and P. Pino, eds., *Organic Syntheses Via Metal Carbonyls,* Vol. 2, Wiley-Interscience, New York, 1977.

Registry of Toxic Effects of Chemical Substances (RTECS), U.S. Dept. of Health and Human Services, Washington, D.C., 1990.

CARBOXYLIC ACIDS

SURVEY

Carboxylic acids from the smallest, formic, to the 22-carbon fatty acids, eg, erucic, are economically important; several million metric tons are produced annually. The shorter-chain aliphatic acids are colorless liquids. Each has a characteristic odor ranging from sharp and penetrating (formic and acetic acids), or vinegary (dilute acetic), to the odors of rancid butter (butyric acid), and goat fat (the 6–10-carbon acids). At room temperature, the cis-unsaturated acids through C_{18} are liquids, and the saturated unbranched aliphatic acids from decanoic through the higher acids and trans-unsaturated acids are solids. The latter are higher melting because of their higher degree of linearity and greater degree of crystallinity eg, elaidic acid.

Both odd and even numbered alkanoic acids of molecular formula $C_nH_{2n}O_2$ occur naturally. Until chromatographic techniques provided means to identify minor components in natural mixtures, it was believed that only the even numbered higher acids, most often the C_{18} acids, occurred naturally. Formic acid (qv), acetic acid (qv), propionic, and butyric acids are manufactured in large quantities from

Table 1. Physical Properties of the Straight-Chain Alkanoic Acids, $C_nH_{2n}O_2$

n	Systematic name (common name)[a]	Mol wt	Mp, °C	Bp[b], °C	Density,[c] d^{20}_4	Specific heat, J/g[d]	Viscosity,[e] mPa·s (= cP)
1	methanoic (formic)	46.03	8.4	100.5	1.220		
2	ethanoic (acetic)	60.05	16.6	118.1	1.049		
3	propanoic (propionic)	74.08	−22	141.1	0.992	2.34(l)	1.099
4	butanoic (butyric)	88.11	−7.9	163.5	0.959	2.16(l)	1.538
5	pentanoic (valeric)	102.13	−34.5	187.0	0.942		2.30
6	hexanoic ([caproic])	116.16	−3.4	205.8	0.929	2.23(l)	3.23
7	heptanoic ([enanthic])	130.19	−10.5	223.0	0.922		4.33
8	octanoic ([caprylic])	144.21	16.7	239.7	0.910	2.62(s)	5.74
9	nonanoic (pelargonic)	158.24	12.5	255.6	0.907	2.91(s)	8.08
10	decanoic ([capric])	172.27	31.6	270.0	0.895^{30}		4.30[f]
11	undecanoic ([undecylic])	186.30	29.3	284.0	0.9905^{25}		7.30[f]
12	dodecanoic (lauric)	200.32	44.2	298.9	0.883	1.80(s)	7.30[f]
13	tridecanoic ([tridecylic])	214.35	41.5	312.4	0.8458^{80}		
14	tetradecanoic (myristic)	228.38	53.9	326.2	0.858^{60}	1.60(s)	5.83[g]
15	pentadecanoic ([pentadecylic])	242.40	52.3	339.1	0.8423^{80}		
16	hexadecanoic (palmitic)	256.43	63.1	351.5	0.8534^{60}	1.80(s)	7.80[g]
17	heptadecanoic (margaric)	270.46	61.3	363.8	0.853^{60}		
18	octadecanoic (stearic)	284.48	69.6	376.1	0.847[g]	1.67(s)	9.87[g]
19	nonadecanoic ([nonadecyclic])	298.51	68.6	299_{100}[h]	0.8771^{25}		
20	eicosanoic (arachidic)	312.54	75.3	203_{1}[h]	0.8240^{100}		
22	docosanoic (behenic)	340.59	79.9	306_{20}[h]	0.8221^{100}		
24	tetracosanoic (lignoceric)	368.65	84.2		0.8207^{100}		
26	hexacosanoic (cerotic)	396.70	87.7		0.8198^{100}		
28	octacosanoic (montanic)	424.75					
30	triacontanoic (melissic)	452.81					
33	tritriacontanoic (psyllic)	494.89					
35	pentatriacontanoic (ceroplastic)	522.94					

[a] Brackets signify a trivial name no longer in use.
[b] At 101.3 kPa = 1 atm unless otherwise noted in kPa as a subscript.
[c] At 20°C unless otherwise noted by a superscript number (°C).
[d] To convert J to cal, divide by 4.184.
[e] At 20°C unless otherwise noted.
[f] At 50°C.
[g] At 70°C.
[h] To convert kPa to mm Hg, multiply by 7.5.

Table 2. Physical Properties of the Straight-Chain Alkenoic Acids, $C_nH_{(2n-2)}O_2$

n	Systematic name (common name)	Mol wt	Mp, °C	Bp, °C[a]	Density,[b]	Refractive index,[b] n^{20}_D
3	propenoic (acrylic)	72.06	12.3	141.9	1.0621^{16}	1.4224
4	*trans*-2-butenoic (crotonic)	86.09	72	189	1.018	1.4228$^{79.7}$
4	*cis*-2-butenoic (isocrotonic)	86.09	14	171.9	1.0312^{15}	1.4457
4	3-butenoic (vinylacetic)	86.09	−39	163	1.013^{15}	1.4257^{15}
5	2-pentenoic (β-ethylacrylic)	100.12				
5	3-pentenoic (β-pentenoic)	100.12				
5	4-pentenoic (allylacetic)	100.12	−22.5	188–189	0.9809	1.4281
6	2-hexenoic (isohydroascorbic)	114.14				
6	3-hexenoic (hydrosorbic)	114.14	12	208	0.9640^{23}	1.4935
7	*trans*-2-heptenoic	128.17				
8	2-octenoic	142.20				
9	2-nonenoic	156.23				
10	*trans*-4-decenoic[c]	170.25				
	cis-4-decenoic[c]					
10	9-decenoic (caproleic)	170.25				
11	10-undecenoic (undecylenic)	184.28	24.5	275	0.9075^{25}	1.4464
12	*trans*-3-dodecenoic (linderic)	198.31				
13	tridecenoic	212.33				
14	*cis*-9-tetradecenoic (myristoleic)	226.36				
15	pentadecenoic	240.39				
16	*cis*-9-hexadecenoic (*cis*-9-palmitoleic)	254.41				
16	*trans*-9-hexadecenoic (*trans*-9-palmitoleic)	254.41				
17	9-heptadecenoic	268.44				

Table 2. Physical Properties of the Straight-Chain Alkenoic Acids, $C_nH_{(2n-2)}O_2$ *(continued)*

n	Systematic name (common name)	Mol wt	Mp, °C	Bp, °C[a]	Density,[b] d^{20}_4	Refractive index,[b] n^{20}_D
18	*cis*-6-octadecenoic (petroselinic)	282.47	30		0.8681^{40}	1.4533^{40}
18	*trans*-6-octadecenoic (petroselaidic)	282.47	54			
18	*cis*-9-octadecenoic (oleic)	282.47	13.6	234[d]	0.8905	1.4582^{3}
18	*trans*-9-octadecenoic (elaidic)	282.47	43.7	234[d]	0.8568^{70}	1.4405^{70}
18	*cis*-11-octadecenoic	282.47	14.5			
18	*trans*-11-octadecenoic (vaccenic)	282.47	44		0.8563^{70}	1.4406^{70}
20	*cis*-5-eicosenoic	310.52				
20	*cis*-9-eicosenoic (godoleic)	310.52				
22	*cis*-11-docosenoic (cetoleic)	338.58				
22	*cis*-13-docosenoic (erucic)	338.58	34.7	281[e]	0.85321^{70}	1.44438^{70}
22	*trans*-13-docosenoic (brassidic)	338.58	61.9	265[d]	0.85002^{70}	1.44349^{70}
24	*cis*-15-tetracosenoic (selacholeic)	366.63				
26	*cis*-17-hexacosenoic (ximenic)	394.68				
30	*cis*-21-triacontenoic (lumequeic)	450.79				

[a] At 101.3 kPa = 1 atm unless otherwise noted.
[b] Superscript numbers indicate measurement at a temperature other than 20°C.
[c] The common name for both *cis*- and *trans*-4-decenoic is obtusilic.
[d] At 15 kPa = 113 mm Hg.
[e] At 30 kPa = 225 mm Hg.

Table 3. Some Polyunsaturated Fatty Acids

Total number of carbon atoms	Systematic name (common name)	Mol wt	Mp, °C	Bp, °C at kPa[a]	Refractive index, n^{20}_D
Dienoic acids, $C_nH_{2n-4}O_2$					
5	2,4-pentadienoic (β-vinylacrylic)	98.10			
6	2,4-hexadienoic[b] (sorbic)	112.13	134.5		
10	*trans*-2,4-decadienoic	168.24			
12	*trans*-2,4-dodecadienoic	196.29			
18	*cis*-9,*cis*-12-octadecadienoic (linoleic)	280.45	−5	202 at 1.4	1.4699
18	*trans*-9,*trans*-12-octadecadienoic (linolelaidic)	280.45	28–29		
18	5,6-octadecadienoic (laballenic)	280.45			
22	5,13-docosadienoic	336.56			
Trienoic acids, $C_nH_{2n-6}O_2$					
16	6,10,14-hexadecatrienoic (hiragonic)	250.38			
18	*cis*-9,*cis*-12,*cis*-15-octadecatrienoic (linolenic)	278.44	−10 to −11.3	157 at 0.001	1.4800
18	*cis*-9,*trans*-11,*trans*-13-octadecatrienoic (α-eleostearic)	278.44	48–49	235 at 12	1.5112
18	*trans*-9,*trans*-11,*trans*-13-octadecatrienoic (β-eleostearic)	278.44	71.5		1.5002
18	*cis*-9,*cis*-11,*trans*-13-octadecatrienoic (punicic)	278.44			
18	*trans*-9,*trans*-12,*trans*-15-octadecatrienoic (linolenelaidic)	278.44			
Tetraenoic acids, $C_nH_{2n-8}O_2$					
18	4,8,12,15 octadecatetraenoic (moroctic)	276.42			
18	*cis*-9,*trans*-11,*trans*-13,*cis*-15-octadecatetraenoic (α-parinaric)	276.42			
18	*trans*-9,*trans*-11,*trans*-13,*trans*-15-octadecatetraenoic (β-parinaric)	276.42			
20	5,8,11,14-eicosatetraenoic (arachidonic)	304.47			
Pentaenoic acids, $C_nH_{2n-10}O_2$					
22	4,8,12,15,19-docosapentaenoic (clupanodonic)	330.51			

[a] To convert kPa to mm Hg, multiply by 7.5.
[b] $\Delta; H_{298} = -393.5$ kJ/mol($\frac{94.05}{4.184}$ kcal/mol); flash point (OC) = 127 °C.

petrochemical feedstocks. The higher fatty acids are derived from animal fats, vegetable oils, or fish oils. Some higher saturated fatty acids with significant industrial applications are caprylic, pelargonic, capric, lauric, myristic, palmitic, stearic, and behenic acids (Table 1).

In the alkenoic series of molecular formula $C_nH_{2n-2}O_2$, acrylic, methacrylic, undecylenic, oleic, and erucic acids have important applications (Table 2). Acrylic and methacrylic acids have a petrochemical origin, and undecylenic, oleic, and erucic acids have natural origins (see ACRYLIC ACID AND DERIVATIVES; METHACRYLIC ACID AND DERIVATIVES).

The polyunsaturated aliphatic monocarboxylic acids having industrial significance include sorbic, linoleic, linolenic, eleostearic, and various polyunsaturated fish acids (Table 3). Of these, only sorbic acid (qv) is made synthetically. The other acids, except those from tall oil, occur naturally as glycerides and are used mostly in this form.

Table 4. The Acetylenic Fatty Acids

Total number of carbon atoms	Systematic name (common name)	Mol wt
3	propynoic (propiolic, propargylic)	70.05
4	2-butynoic (tetrolic)	84.07
5	4-pentynoic	98.10
6	5-hexynoic	112.13
7	6-heptynoic	126.16
8	7-octynoic	140.18
9	8-nonynoic	154.21
10	9-decynoic	168.24
11	10-undecynoic (dehydro-10-undecylenic)	182.26
18	6-octadecynoic (tariric)	280.45
18	9-octadecynoic (stearolic)	280.45
18	17-octadecene-9, 11-diynoic (isanic, erythrogenic)	274.40
18	*trans*-11-octadecene-9-ynoic (ximenynic)	278.44
22	13-docosynoic (behenolic)	336.56

Table 5. Some Substituted Acids

Total number of carbon atoms	Systematic name (common name)	Mol wt	Mp, °C	Bp, °C[a]	Density[b] d^{20}_4
4	2-methylpropenoic (methacrylic)	86.09	16	163	1.0153
4	2-methylpropanoic (isobutyric)	88.10	−47	154.4	0.9504
5	2-methyl-*cis*-2-butenoic (angelic)	100.12	45	185	0.9539^{76}
5	2-methyl-*trans*-2-butenoic (tiglic)	100.12	65.5	198.5	0.9641^{76}
5	3-methyl-2-butenoic (β,β-dimethyl acrylic)	100.12			
5	2-methylbutanoic	102.13			
5	3-methylbutanoic (isovaleric)	102.13	−37.6	176	$0.93319^{17.6}$
5	2,2-dimethylpropanoic (pivalic)	102.13	35.5	163	0.905^{50}
8	2-ethylhexanoic	144.21		220	0.9031^{25}
14	3,11-dihydroxytetradecanoic (ipurolic)	260.37			
16	2,15,16-trihydroxyhexadecanoic (ustilic)	304.43			
16	9,10,16-trihydroxyhexadecanoic (aleuritic)	304.43			
16	16-hydroxy-7-hexadecenoic (ambrettolic)	270.41			
18	12-hydroxy-*cis*-9-octadecenoic (ricinoleic)	298.47	5.0,7.7,16.0	226[c]	0.9496^{15}
18	12-hydroxy-*trans*-9-octadecenoic (ricinelaidic)	298.47	52–53	240[c]	
18	4-oxo-9,11,13-octadecatrienoic (licanic)	292.42			
18	9,10-dihydroxyoctadecanoic	316.48	90		
18	12-hydroxyoctadecanoic	300.48	79		
18	12-oxooctadecanoic	298.47	81.5		
18	18-hydroxy-9,11,13-octadecatrienoic (kamlolenic)	294.43	77–78		
18	12,13-epoxy-*cis*-9-octadecenoic (vernolic)	296.45			
18	8-hydroxy-*trans*-11-octadecene-9-ynoic (ximenynolic)	294.43			
18	8-hydroxy-17-octadecene-9,11-diynoic (isanolic)	290.40			
20	14-hydroxy-*cis*-11-eicosenoic (lesquerolic)	326.52			

[a] At 101.3 kPa = 1 atm unless otherwise noted.
[b] Superscript numbers indicate measurement at a temperature other than 20°C.
[c] At 1.3 kPa = 9.75 mm Hg.

The shorter-chain alkynoic or acetylenic acids are common in laboratory organic syntheses, and several long-chain acids occur naturally (Table 4).

Many substituted, ie, branched, fatty acids, particularly methacrylic, 2-ethylhexanoic, and ricinoleic acids, are commercially significant. Several substituted fatty acids exist naturally (Table 5).

Some naturally occurring fatty acids have alicyclic substituents such as the cyclopentenyl-containing chaulmoogra acids, notable for their use in treating leprosy (see ANTIPARASITIC AGENTS, ANTIMYCOTICS), and the cyclopropenyl or sterculic acids (Table 6).

The prostaglandins (qv) constitute another class of fatty acids with alicyclic structures. These are of great biological importance and are formed by *in vivo* oxidation of 20-carbon polyunsaturated fatty acids, particularly arachidonic acid.

Aromatic carboxylic acids are produced annually in amounts of several million metric tons. Several aromatic acids occur naturally, eg, benzoic acid (qv), salicylic acid (qv), cinnamic acid (qv), and gallic acids, but those used in commerce are produced synthetically. These acids are generally crystalline solids with relatively high melting points, attributable to the rigid, planar, aromatic nucleus (see PHTHALIC ACIDS).

Chemical Properties

The alkanoic acids, with the exception of formic acid, undergo typical reactions of the carboxyl group. Formic acid has reducing properties and does not form an acid chloride or an anhydride. The hydrocarbon chain of alkanoic acids undergoes the usual reactions of hydrocarbons except that the carboxyl group exerts considerable influence on the site and ease of reaction. The alkenoic

Table 6. Some Fatty Acids with Alicyclic Substituents

Total number of carbon atoms	Common name	Mol wt
Cyclopentenyl compounds		
6	aleprolic	112.13
10	aleprestic	168.24
12	aleprylic	196.29
14	alepric	224.34
16	hydnocarpic	252.40
18	chaulmoogric	280.45
Cyclopropenyl compounds		
18	malvalic (halphenic)	280.45
19	sterculic	294.48
19	lactobacillic[a]	296.49

[a] Saturated ring.

acids in which the double bond is not conjugated with the carboxyl group show typical reactions of internal olefins. All three types of reactions are industrially important.

Reactions of the carboxyl group include salt and acid chloride formation, esterification, pyrolysis, reduction, and amide, nitrile, and amine formation.

Salt formation occurs when the carboxylic acid reacts with an alkaline substance.

Safety Information

Carboxylic acid dust and vapors are generally described as being destructive to tissues of the mucous membrane, eyes, and skin. The small molecules such as formic, acetic, propionic, butyric, and acrylic acids tend to be the most aggressive. Formic, acetic, propionic, acrylic, and methacrylic acids have time-weighted average exposure limits of 20 ppm or lower.

M. O. BAGBY
United States Department of Agriculture

R. T. O'Connor, in K. S. Markley, ed., *Fatty Acids,* 2nd ed., Pt. 1, Interscience Publishers, Inc., New York, 1960, pp. 285–378.

W. S. Singleton, *ibid.,* pp. 499–607; D. Swern, Pt. 2, pp. 1307–1385.

L. A. Goldblatt, in K. S. Markley, ed., *Fatty Acids,* Pt. 5, Wiley-Interscience, New York, 1968, pp. 3657–3684.

D. Chapman, *The Structure of Lipids,* John Wiley & Sons, Inc., New York, 1965, pp. 221–315.

J. Levy, in E. S. Pattison, ed., *Fatty Acids and Their Industrial Applications,* Marcel Dekker, Inc., New York, 1968, pp. 209–220.

E. N. Frankel and H. J. Dutton, in F. D. Gunstone, ed., *Topics in Lipid Chemistry,* Vol. 1, Logos Press, London, 1970, pp. 161–276.

M. Naudet and E. Ucciani, *ibid.,* Vol. 2, 1971, pp. 99–158.

E. H. Pryde and J. C. Cowan, *ibid.,* pp. 1–98.

MANUFACTURE

Carboxylic acids having 6–24 carbon atoms are commonly known as fatty acids. Shorter-chain acids, such as formic, acetic, and propionic acid, are not classified as fatty acids and are produced synthetically from petroleum sources (see ACETIC ACID AND DERIVATIVES–ACETIC ACID; FORMIC ACID AND DERIVATIVES; OXO PROCESS). Fatty acids are produced primarily from natural fats and oils through a series of unit operations. Clay bleaching and acid washing are sometimes also included with the above operations in the manufacture of fatty acids for the removal of impurities prior to subsequent processing.

The predominant feedstocks for the manufacture of fatty acids are tallow and grease, coconut oil, palm oil, palm kernel oil, soybean oil, rapeseed oil, and cottonseed oil. Another large source of fatty acids comes from the distillation of crude tall oil obtained as a by-product from the Kraft pulping process (see TALL OIL; CARBOXYLIC ACIDS–FATTY ACIDS FROM TALL OIL).

Manufacture of Fatty Acids from Natural Fats and Oils

There are essentially four steps or unit operations in the manufacture of fatty acids from natural fats and oils: *(1)* batch alkaline hydrolysis or continuous high pressure hydrolysis; *(2)* separation of the fatty acids usually by a continuous solvent crystallization process or by the hydrophilization process; *(3)* hydrogenation, which converts unsaturated fatty acids to saturated fatty acids; and *(4)* distillation, which separates components by their boiling points or vapor pressures.

Synthetic Routes to Fatty Acids from Petroleum

Synthetic processes include catalytic oxidation for straight-chain paraffinic hydrocarbons, oxidation of straight-chain 1-olefins, carboxylation of straight-chain 1-olefins, oxidation of straight-chain alcohols, branched-chain carboxylic acids produced by the oxo reactions, and highly branched acids produced from highly branched olefins, carbon monoxide, and an acid catalyst.

R. W. JOHNSON
R. W. DANIELS
Union Camp Corporation

R. H. Potts and V. J. Muckerhoide, in E. S. Pattison, ed., *Fatty Acids and Their Industrial Applications,* Marcel Dekker, Inc., New York, 1968, pp. 21–46.

N. O. V. Sonntag, in R. W. Johnson and E. Fritz, eds., *Fatty Acids in Industry,* Marcel Dekker, Inc., New York, 1989, pp. 40–42.

ECONOMIC ASPECTS

Aliphatic carboxylic acids produced on a reasonably significant commercial scale range from acetic acid (two carbons or C2) through stearic acid (C18). Lesser amounts of commercially available shorter chain-length acids, such as formic (C1), and longer chain-length acids, such as erucic (unsaturated C22) and behenic (saturated C22), are also produced. As a general rule, all of the even chain-length, nonisomeric acids from C6 to C22 are produced from naturally occurring fats and oils. A significant proportion of the lower chain-length (C1–C6) and longer isomeric chain-length (C7–C10) acids are made synthetically.

Nonoleo-Based Carboxylic Acids

Some of the more prominent carboxylic acids that are not fat- or oil-based include acetic, acrylic, and olefin-based propionic, butyric/isobutyric, 2-ethylhexanoic, heptanoic, pelargonic, neopentanoic, and neodecanoic. With the exception of acetic, acrylic, and benzoic, these acids are primarily produced using oxo chemistry (see OXO PROCESS).

Acetic acid (qv) can be produced synthetically (methanol carbonylation, acetaldehyde oxidation, butane/naphtha oxidation) or from natural sources. Oxygen is added to propylene to make acrolein, which is further oxidized to acrylic acid (see ACRYLIC ACID AND DERIVATIVES). An alternative method adds carbon monoxide or water to acetylene. Benzoic acid, (qv) is made by oxidizing toluene in the presence of a cobalt catalyst.

Table 1. U.S. Consumption[a] of Fatty Acids by Market Area in 1987, 10^3 t

household and industrial cleaners	124
coatings and adhesives (including paints and inks)	88
personal care products (excluding toilet soap)	86
industrial lubricants, corrosion inhibitors, and oil field applications	85
plastics (excluding emulsion polymerization)	81
household fabric softeners	50
rubber (excluding emulsion polymerization)	39
emulsion polymerization	29
textiles (including textile fabric softeners)	29
foods	19
asphalt	12
paper	11
mining	8
crayons, candles, and waxes	7
miscellaneous, unassigned, and exported derivatives (including nitriles)	69
Total	*737*

[a] Estimates are based on direct consumption of the acid or ultimate consumption of various derivatives.

Oleo-Based Carboxylic Acids

Worldwide capacity for production of higher aliphatic carboxylic acids (predominantly C8–C18), commonly called fatty acids, approaches 4 million metric tons. U.S. production of C8–C18 linear acids, including tall oil, was reported by the BOC to be approximately 660,000 metric tons, a figure that would appear to be understated. The use of these fatty acids covers many consumer product and industrial applications and historically has correlated well with the GNP in the United States.

Typically, fatty acids make up between 87 and 90% of the fat or oil from which they are made; the remaining 10–13% is glycerol. The most frequently used raw materials are coconut or palm kernel oil (lauric oils) for C8, C10, C12, and C14 acids; tallow, lard, and palm stearine for C16 and C18 acids; and soybean, sunflower, canola, and tall oil for whole cut unsaturated (lower melting point) acids.

With the exception of tall oil and castor oil acids, and acids used for sodium and potassium soaps, U.S. triglyceride-based fatty acids showed a 2–3%/yr growth rate between 1985 and 1990, virtually in line with world projections, with the most significant growth occurring in the stearic and coconut acid segments.

Fatty acids are produced by hydrolysis (splitting), during which the triglyceride (fat or oil) reacts with water, under high temperature and pressure, to yield glycerol and free fatty acids. Split and distilled acids that do not undergo any further processing comprise about 10% of the U.S. market. The majority of whole cut fatty acids are further processed for additional differentiation. Adding hydrogen (hydrogenation) removes double bonds and therefore increases stability and hardness. Partially hardened (intermediate melting point) and fully hardened (high melting point) whole cut acids amount to about 45% of U.S. acid production.

Unhardened whole cut tallow and palm acids contain 40–45% oleic acid, which is typically derived by separation technology. In the 1990s the separation is done using solvents or refrigeration techniques. Oleic and pressed stearics account for about one-third of all U.S. acid production.

Whole cut acids can also undergo fractional distillation to produce fractionated acids with a high percentage content of a specific chain length, eg, 99% lauric, 95% myristic, etc. Fractionated acids fill the remaining 10–12% of U.S. market demand.

Because they are made from renewable natural raw materials, oleo-based fatty acids are completely biodegradable and find widespread usage in a variety of applications and industries.

Table 1 summarizes the consumption of fatty acids by end use industry in 1987, supporting the broad base of usage and explaining the relationship with GNP.

E. T. SAUER
The Procter & Gamble Company

Chemical Economics Handbook, SRI International, Menlo Park, Calif., 1991.

A. J. Kaufman and R. J. Ruebusch, *Oleochemicals: A World Overview,* presented at *World Conference on Oleochemicals—Into the 21st Century,* Kuala Lumpur, Malaysia, Henkel Corp., Emery Group, Oct. 1990; adjusted by author estimates for worldwide tall oil and castor acids.

M2OK Report, U.S. Bureau of the Census, Washington, D.C., 1990.

Glycerine and Oleochemical Division, The Soap and Detergent Association, 1990.

"Natural Fatty Acids Report," in *Chemical Economics Handbook,* SRI International, Menlo Park, Calif., 1988.

FATTY ACIDS FROM TALL OIL

Tall oil fatty acids (TOFA) consist primarily of oleic and linoleic acids and are obtained by the distillation of crude tall oil. Several grades of TOFA are available, depending on rosin, unsaponifiable content, color, and color stability.

Tall oil fatty acids are produced, as of 1993, by six companies using 12 fractionating plants in the United States, one in Canada, 13 in Europe, two in Japan, one in New Zealand, and at least one in Russia. Worldwide crude tall oil fractionating capacity in 1988 was estimated at slightly over 1.4 million metric tons and the fractionating capacity in the United States at just over 900,000 t.

Tall oil fatty acids are used in intermediate chemicals, protective coatings, soaps and disinfectants, and flotation.

ROBERT W. JOHNSON, JR.
ROBERT R. CANTRELL
Union Camp Corporation

D. T. A. Huibers and E. Fritz, "Distillation of Fatty Acids" in R. W. Johnson and E. Fritz, eds., *Fatty Acids in Industry,* Marcel Dekker, Inc., New York, 1989, pp. 85–105.

D. Campbell, *Distilled Tall Oil Products, Summary,* Pulp Chemical Assoc., New York, 1990.

BRANCHED-CHAIN ACIDS

Branched-chain acids contain at least one branching alkyl group attached to the carbon chain, which causes the individual acids to have different physical, and in some cases different chemical, properties than their corresponding straight-chain isomers. Some properties of commercial branched-chain acids are shown in Table 1.

Manufacturing procedures for most branched-chain acids are well known. For example, oxo process acids are manufactured from branched-chain olefins using hydroformylation followed by oxidation (see OXO PROCESS).

Branched-chain acids have a wide variety of industrial uses as paint driers, vinyl stabilizers, and cosmetic products.

The hazards of handling branched-chain acids are similar to those encountered with other aliphatic acids of the same molecular weight. Eye and skin contact as well as inhalation of vapors of the shorter-chain acids should be avoided.

Table 1. Properties and Prices of Branched-Chain Acids

Branched-chain acid (common name)	Molecular weight	Boiling point, °C[a,b]	Melting point, °C	Producers in the United States
2-methylpropanoic (isobutyric)	88	155	−46.1	Hoechst, Eastman
2-methylbutanoic (isopentanoic)	102	180.3	−48	Union Carbide
3-methylbutanoic (isovaleric)	102	175–176.5	ca 30	Hoechst
2, 2-dimethylpropanoic (neopentanoic)	102	163–165	34.4	Exxon
2-ethylhexanoic	144	224–230	<70	Eastman, Union Carbide
mixed isomers (isononanoic)	158	232–246	ca 70	Hoechst
2, 2-dimethyloctanoic (neodecanoic)	172	147–150[b]	<40	Exxon
mixed isomers (isostearic)	284	192–204[c]	ca 7	Emery, Union Camp

[a] At 101.3 kPa.
[b] To convert kPa to mm Hg, multiply by 7.5.
[c] At 20 kPa.[d]
[d] At 5 kPa.[d]

ROBERT W. JOHNSON, JR
ROBERT R. CANTRELL
Union Camp Corporation

Aliphatic Carboxylic Acids, technical bulletin, American Hoechst Corp., Somerville, N.J., 1981.

Neo Acids Properties, Chemistry, and Applications, technical bulletin, Exxon Chemicals Americas, Houston, Tex., 1982.

Oxo Synthesis Products, Farbwerke Hoechst AG, Frankfurt, Germany, 1971.

A. Fisher, *J. Am. Oil Chem. Soc.* **43,** 469 (1966).

TRIALKYLACETIC ACIDS

Trialkylacetic acids are characterized by the following structure:

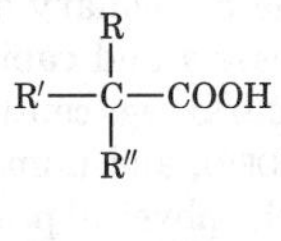

in which R, R′, and R′ are C_xH_{2x+1}, with $x \geq 1$. The lowest member of the series (R = R′ = R′ = CH_3) is the C_5 acid, trimethylacetic acid or 2,2-dimethylpropanoic acid (also, neopentanoic acid, pivalic acid). For higher members in the series, the products are typically mixtures of isomers, resulting from the use of mixed isomer feedstocks and the chemical rearrangements that occur in the manufacturing process.

Trialkylacetic acids have been produced commercially since the early 1960s, in the United States by Exxon and in Europe by Shell, and have been marketed as neo acids (Exxon) or as Versatic Acids (Shell). The principal commercial products are the C_5 acid and the C_{10} acid (neodecanoic acid, or Versatic 10), although smaller quantities of other carbon numbers, such as C_6, C_7, and C_9, are also produced.

The trialkylacetic acids have a number of uses in areas such as polymers, pharmaceuticals, agricultural chemicals, cosmetics, and metal-working fluids. Commercially important derivatives of these acids include acid chlorides, peroxyesters, metal salts, vinyl esters, and glycidyl esters.

Trimethylacetic Acid

Physical Properties. 2,2-Dimethylpropionic acid, $(CH_3)_3CCOOH$, also referred to as neopentanoic acid or pivalic acid, is a solid at room temperature with a pungent odor typical of many lower molecular weight carboxylic acids. Physical properties of a typical commercial sample are given in Table 1.

Chemical Properties. Neopentanoic acid undergoes reactions typical of carboxylic acids, such as acid chloride formation, esterification, and reduction.

Manufacture. Trialkylacetic acids are prepared using variants of the Koch reaction, a two-stage reaction for the preparation of carboxylic acids.

Table 1. Physical Properties of Commercially Available Neopentanoic Acid

Property	Value
mp, °C	34.4
bp, °C	163–165
acid value, mg KOH/g	550
specific gravity at 38/38°C	0.913
viscosity at 60°C, mm^2/s (= cSt)	1.7
flash point, °C (Tag closed cup)	63
vapor pressure, kPa[a] at 60°C	1.33
solubility in water, g/100 mL H_2O at 25°C	2.1

[a] To convert kPa to mm Hg, multiply by 7.5.

Economic Aspects and Shipment. Production worldwide of neopentanoic acid is estimated at 15 thousand metric tons per year. Both Shell and Exxon have announced expansions in capacity. Neopentanoic acid is shipped in heated tank cars, heated tank trucks, and drums.

Health and Safety Factors. Neopentanoic acid possesses low toxicity. The principal hazards associated with neopentanoic acid at ambient temperatures are from eye and skin irritation. At elevated temperatures, where concentrations of the vapor are significant, irritation of the respiratory tract can also occur. Contact with the material should be avoided.

C_{10} Trialkylacetic Acids

Physical Properties. The C_{10} trialkylacetic acids, referred to as neodecanoic acid or as Versatic 10, are liquids at room temperature. Typical physical properties for commercially available material are given in Table 2.

Table 2. Physical Properties of Commercially Available Neodecanoic Acid

Property	Value
mp, °C	>−40
bp, °C	250–257
acid value, mg KOH/g	325
specific gravity at 20/20°C	0.915
viscosity, mm^2/s (= cSt) at 20°C	35.7
flash point, °C (Tag closed cup)	105
solubility in water, g/100 mL H_2O at 25°C	0.017
heat of vaporization, kJ/kg,[a] at the boiling point and 101.3 kPa[b]	249.5

[a] To convert kJ to kcal, divide by 4.184.
[b] To convert kPa to mm Hg, multiply by 7.5.

Chemical Properties. Like neopentanoic acid, neodecanoic acid, $C_{10}H_{20}O_2$, undergoes reactions typical of carboxylic acids including metal salt formation.

Manufacture. The C_{10} trialkylacetic acids are prepared using the same process and catalysts as are used for the preparation of neopentanoic acid.

Economic Aspects and Shipment. The C_{10} trialkylacetic acids are produced in volumes totaling tens of thousands of metric tons per year. The C_{10} acids are shipped in bulk sea vessels, tank cars, tank trucks, and drums.

Health and Safety Factors. The C_{10} trialkylacetic acids have toxicities similar to those for other neo acids: oral LD_{50} in rats is 2.0 g/kg, and dermal LD_{50} in rabbits is 3.16 g/kg.

The primary hazard associated with C_{10} trialkylacetic acids is eye irritation. For skin contact, flush with large quantities of water, using soap if available. To extinguish fires, use foam, dry chemical, or water spray.

Uses. The C_{10} trialkylacetic acids are used in polymers, resins, and coatings, adhesion promoters, metal-working and hydraulic fluids, metal extraction, fuels and lubricants, and electrical and electronic applications.

Glycidyl and Vinyl Esters. Glycidyl neodecanoate, sold commercially as GLYDEXX N-10 (Exxon) or as Cardura E10 (Shell), is prepared by the reaction of neodecanoic acid and epichlorohydrin under alkaline conditions, followed by purification. Physical properties of the commercially available material are given in Table 3. The material is a mobile liquid monomer with a mild odor and is used primarily in coatings.

Total world production of glycidyl neodecanoate is ca 7–10 thousand metric tons per year, with production by Exxon in the United States and by Shell in The Netherlands. The product is shipped in bulk or in drums and must be protected from contact with atmospheric water during storage.

Vinyl neodecanoate is prepared by the reaction of neodecanoic acid and acetylene in the presence of a catalyst such as zinc neodecanoate. Physical properties of the commercially available material, VeoVa 10 from Shell, are given in Table 4. The material is a mobile liquid with a typical mild ester odor used in a number of areas, primarily in coatings, but also in construction, adhesives, cosmetics, and a number of miscellaneous areas.

The only producer of vinyl neodecanoate is Shell in the Netherlands, who markets the material as VeoVa 10. Production is several tens of thousands of metric tons per year. The vinyl ester is shipped in bulk or in lined drums, stabilized with 5 ppm of the monomethyl ether of hydroquinone MEHQ.

Table 3. Physical Properties of Commercially Available Glycidyl Neodecanoates

Property	Cardura E10	GLYDEXX N-10
density at 20°C, g/mL	0.958–0.968	
epoxy equivalent weight, g	244–256	250
flash point, °C	126[a]	128[b]
viscosity, mPa·s(= cP)		
25°C	7.13	
150°C	0.72	
vapor pressure at 37.8°C, kPa[c]	0.9	
boiling range at 101.3 kPa,[c] °C	251–278	
freezing point, °C	<−60	
solubility in water at 20°C	0.01 % mass/mass	

[a] PMCC = Pensky-Martin closed up.
[b] Tagliabue closed cup.
[c] To convert kPa to mm Hg, multiply by 7.5

Table 4. Physical Properties of VeoVa 10

Property	Value
density at 20°C, g/mL	0.875–0.885
vinyl unsaturation, mol/kg	4.85–5.10
kinematic viscosity at 20°C, mm^2/s(= cSt)	
vapor pressure, kPa[a]	2.2
at 30°C	<0.1
at 210°C	101
boiling range, °C at 13.3 kPa[a]	133–136
flash point,[b]	75
freezing point, °C	<−20
solubility in water at 20°C, % mass/mass	<0.1

[a] To convert kPa to mm Hg, multiply by 7.5.
[b] PMCC = Pensky-Martinclosed cup.

M. J. Keenan
M. A. Krevalis
Exxon Chemical Company

H. Bahrmann in J. Falbe, ed., *New Syntheses with Carbon Monoxide,* Springer-Verlag, Berlin, Heidelberg, New York, 1980.

CARDIOVASCULAR AGENTS

The cardiovascular system consists of the heart and a complex network of conduit blood vessels (aorta and vena cavae), arteries, arterioles, capillaries, venules, and veins. The function of the system is to maintain all living cells of the body with an almost constant internal and external environment. To achieve this goal there is a continuous flow of nutrients and oxygen to the cells and a continuous removal of waste products and carbon dioxide from the cells. The heart functions as a pump, distributing its output to the pulmonary and peripheral circulations via this network of blood vessels and capillaries. The heart and the blood vessels, under both neural, eg, central and autonomic nervous systems (see Neuroregulators), and humoral, eg, bloodborne factors control, are also affected by the physical properties of the blood such as volume, viscosity, and the concentration of electrolytes. Under certain conditions the heart and blood vessels may malfunction, causing various disorders or diseases which can lead to irreversible damage and/or death. In the highly developed countries, mortality from cardiovascular disease is extremely high. Cardiovascular agents are used to restore normal function to the cardiovascular system.

ANTIARRHYTHMIC AGENTS

Cardiac arrhythmias are an important cause of morbidity and mortality: approximately 400,000 people per year die from myocardial infarctions (MI) in the United States alone. Individuals with MI exhibit some form of dysrhythmia within 48 h. The goals of antiarrhythmic therapy are to reduce the incidence of sudden death and to alleviate the symptoms of arrhythmias, such as palpitations and syncope. Several excellent reviews of the mechanisms of arrhythmias and the pharmacology of antiarrhythmic agents have been published.

Properties of the Cardiovascular System

The Ionic Basis of Membrane Activity. Almost all living cells maintain specific internal chemical environments that are different from their external environments. In cardiac cells the principal ions involved in maintaining membrane activity are sodium, Na^+; potassium, K^+; chloride, Cl^-; and calcium, Ca^{2+}. The internal (i) and external (o) concentrations of these ions are $Na_o^+ = 140$ mM, $Na_i^+ = 30$ mM; $K_o^+ = 4$ mM, $K_i^+ = 140$ mM; $Cl_o^- = 104$ mM, $Cl_i^- = 30$ mM; $Ca_o^{2+} = 2$ mM, $Ca_i^{2+} = 100$ nM. Ions diffuse through the cell membrane via protein structures called channels, which span the membrane. The resting cell membrane is permeable to K^+ but not to Na^+ and the low concentration of Na_i^+ is maintained by an energy-dependent Na^+/K^+ adenosine triphosphatase (ATPase) pump that exchanges three $Na_2^+\cdot$ for two K_0^+. When the cell depolarizes, there is an increase in the permeability of the cell membrane to Na^+, sodium channels are activated, and Na^+ influx rapidly increases. As the cell membrane potential moves from −90 mV, becoming less negative, there is an increased inward flux of $Ca^{2+}\cdot$ at about −60 to −50 mV, accompanied by the activation of calcium channels. At +10 mV, Na^+ influx is terminated by activation of the Na^+/K^+ ATPase pump and Na^+ efflux begins. The cell begins to repolarize. The first phase of repolarization is thought to result from an outward K^+ conductance. The second phase is a result of the slow Ca^{2+} conductance; the third phase is a result of an outward K^+ current. When the cell is fully repolarized back to its resting membrane potential, phase 4 repolarization, the ionic gradients are reestablished. Calcium is removed from the cell by a Na^+-Ca^{2+} exchange mechanism.

Properties of Excitable Tissues. Various parts of the heart, like most excitable tissues, exhibit automaticity, excitability, conductivity, and refractoriness.

The Cardiac Cycle. The heart performs its function as a pump as a result of a rhythmical spread of a wave of excitation (depolarization) that excites the atrial and ventricular muscle masses to contract sequentially. Maximum pump efficiency occurs when the atrial or ventricular muscle masses contract synchronously. The wave of excitation begins with the generation of electrical impulses within the SA node and spreads through the atria. The SA node is referred to as the pacemaker of the heart and exhibits automaticity, ie, it depolarizes and repolarizes spontaneously. The wave then excites sequentially the AV node; the bundle of His, ie, the penetrating portion of the AV node; the bundle branches, ie, the branching portions of the AV node; the terminal Purkinje fibers; and finally the ventricular myocardium. After the wave of excitation depolarizes these various structures of the heart, repolarization occurs so that each of the structures is ready for the next wave of excitation. Until repolarization occurs the structures are said to be refractory to excitation. During repolarization of the atria and ventricles, the muscles relax, allowing the chambers of the heart to fill with blood that is to be expelled with the next wave of excitation and resultant contraction. This process repeats itself 60–100 times or beats per minute in normal humans in the resting state and represents the basal rate. The electrocardiogram (ECG) is the body surface recording of the wave of electrical excitation of the heart. The P wave is excitation of the atria; the QRS complex and T wave represent depolarization and repolarization of the ventricles, respectively.

Cardiac arrhythmias or dysrhythmias are disturbances of the normal regular rhythm that may be caused by an abnormality in the site of impulse generation, its rate or regularity, or its propagation or conduction.

Class I Antiarrhythmic Agents: The Sodium Channel Blockers

The Class I antiarrhythmic agents inactivate the fast sodium channel, thereby slowing the movement of Na^+ across the cell membrane. This is reflected as a decrease in the rate of development of phase 0 (upstroke) depolarization of the action potential. The Class I agents have potent local anesthetic effects. These compounds have been further subdivided into Classes IA, IB, and IC based on recovery time from blockade of sodium channels. Class IB agents have the shortest recovery times ($t_{1/2} < 1$ s); Class IA compounds have moderate recovery times ($t_{1/2}$ usually <9 s); and Class IC have the longest recovery times ($t_{1/2}$ usually > 9 s).

The Class I agents decrease excitability, slow conduction velocity, inhibit diastolic depolarization (decrease automaticity), and prolong the refractory period of cardiac tissues. These agents have anticholinergic effects that may contribute to the observed electrophysiologic effects. Heart rates may become faster by increasing phase 4 diastolic depolarization in SA and AV nodal cells. This results from inhibition of the action of vagally released acetylcholine, which allows sympathetically released norepinephrine (NE) to act on these structures.

The Class I agents have many similar side effects and toxicities. The anticholinergic side effects include dry mouth, constipation, and urinary hesitancy and retention. Common gastrointestinal (GI) side effects include nausea, vomiting, diarrhea, and anorexia. Cardiovascular adverse effects are hypotension, tachycardia, arrhythmias, and myocardial depression, especially in patients with congestive heart failure. Common central nervous system (CNS) side effects are headache, dizziness, mental confusion, hallucinations, CNS stimulation, paraesthesias, and convulsions.

Class IA Antiarrhythmic Agents. Class IA Antiarrhythmic agents include quinidine, procainamide, and disopyramide.

Class IB Antiarrhythmic Agents. Class IB antiarrhythmic agents include lidocaine, phenytoin, mexilitene, moricizine, and tocainide.

Class IC Antiarrhythmic Agents. Class IC Antiarrhythmic agents include encainide, flecainide, propafenone, and indecainide.

Class I Antiarrhythmic Agents Under Development. These include lorcainide and pirmenol.

Class II Antiarrhythmic Agents: The β-Adrenoceptor Blocking Agents

β-Adrenoceptor blocking agents (β-adrenoceptor blockers) are often called β-blockers. However, the former term more accurately describes the way in which these drugs act. They bind selectively to β-adrenoceptors in a competitive and reversible way to antagonize the responses of neurally released, humoral, or exogenously administered sympathomimetic amines.

β-Adrenoceptors have been subdivided into β_1- and β_2-adrenoceptors. A third subset called nontypical β-adrenoceptors or β_3-adrenoceptors have been described but are still the subject of debate. The focus herein is on the clinically relevant β_1-adrenoceptor-mediated effects on heart and on β_2-adrenoceptor-mediated effects on smooth muscles of blood vessels and bronchioles, the insulin-secreting tissue of the pancreas, and skeletal muscle glycogenolysis for a side effects profile.

Agents that preferentially block β_1-adrenoceptor-mediated effects on the heart, resulting in reduced rate, force, and velocity of cardiac contractility, are classified as cardioselective β-adrenoceptor blockers. Thus these agents would be beneficial in patients having arrhythmias that are present with peripheral vascular disease, asthma, and insulin-dependent diabetes because the β-adrenoceptors in the organs involved are of the β_2-adrenoceptor subtype and blocking the β_2-adrenoceptors would worsen the disease states. However, cardioselectivity may be lost at high doses or concentrations of the β-adrenoceptor blocker. Some β-adrenoceptor blocking agents have local anesthetic properties resulting from membrane stabilizing and have effects on cardiac cells similar to those described for the Class I antiarrhythmic agents, ie, they reduce the rate of rise of the action potential without affecting action potential duration. This effect may be apparent at concentrations in excess of their therapeutic range.

Some β-adrenoceptor blockers have intrinsic sympathomimetic activity (ISA) or partial agonist activity (PAA). They activate β-adrenoceptors before blocking them. Theoretically, patients taking β-adrenoceptor blockers with ISA should not have cold extremities because the drug produces minimal decreases in peripheral blood flow (smaller increases in resistance). In addition, these agents should produce minimal depression of heart rate and cardiac output, either at rest or during exercise.

There have been a number of long-term trials with various β-adrenoceptor blockers in patients surviving acute MI that demonstrated a reduction in mortality, sudden death, and nonfatal reinfarctions. The term cardioprotective has been used to describe this effect for the drugs studied. The mechanism of the cardioprotective effect is not understood and whether this property applies to all β-adrenoceptor blockers is unknown. Timolol, $C_{13}H_{24}N_4O_3S$, metoprolol, atenolol, and propranolol have been shown to possess this cardioprotective property.

In addition to the properties described, β-adrenoceptor blockers may be lipophilic or hydrophilic in nature, depending on the extent to which they partition between water and organic solvents. Lipophilic agents more readily cross the blood brain barrier and are more likely to produce adverse CNS effects, eg, drowsiness, lassitude, impaired concentration, hallucinations, and nightmares. Lipophilic agents are eliminated by the liver and blood levels, duration of action, and toxicological potential may be increased in the presence of liver disease. Hydrophilic agents, on the other hand, are eliminated by the kidneys and blood levels, duration of action, and toxic potential may be increased in the presence of impaired kidney function. These factors have prompted the search for β-adrenoceptor blockers that have a balanced clearance profile, wherein the drugs are excreted equally by the liver and kidneys and elimination is not affected by dysfunction of either organ.

Prolonged exposure of β-adrenoceptor agonists down-regulates β-adrenoceptors, ie, their numbers decrease and they become less responsive. On the other hand, prolonged exposure to β-adrenoceptor antagonists (those without ISA) upregulates β-adrenoceptors, ie, their numbers increase and they become more responsive. Therefore, β-adrenoceptor blocker therapy should be withdrawn gradually from patients on this medication.

The properties of β-adrenoceptor blockers that contribute to antiarrhythmic effects are antagonism of neural/humoral β-adrenergic activity and antagonism of catecholamine-mediated electrophysiological properties, ie, increase in refractory period and decrease in the rate of diastolic depolarization, ie, decrease in automaticity and slow atrioventricular conduction.

β-Adrenoceptor blockers include propranolol, acebutolol, esmolol, carteolol, flestolol, and bopindolol.

Class III Antiarrhythmic Agents

The Class III antiarrhythmic agents markedly prolong action potential duration and effective refractory period of cardiac tissue. The QT interval of the ECG is markedly prolonged. This class includes amiodarone, bretylium, sotalol, and clofilium phosphate.

Class IV Antiarrhythmic Agents: The Calcium Antagonists

The calcium antagonists have also been known as calcium channel blockers, calcium entry blockers, slow channel blockers, or calcium channel antagonists. The term calcium antagonist is the accepted descriptive term for this class of drugs. The World Health Organization (WHO) has subdivided the calcium antagonists into six classes, represented by Roman numerals. The phenylalkylamines, eg, verapamil, constitute Class I; dihydropyridines, eg, nifedipine, class II; benzothiazepines, eg, diltiazem, Class III; flunarizine-like compounds, Class IV; prenylamine-like compounds, Class V; and other agents in Class VI.

The general biochemical mechanism for contraction in cardiac, skeletal, or smooth muscle is dependent on the availability of a critical intracellular calcium concentration. In cardiac cells, calcium enters through slow channels that are 100 times more selective for calcium than sodium. This occurs during phase 2 (plateau phase of repolarization) of the action potential, at which time there is a slow inward current of calcium through these channels. Phase 2 repolarization is depressed or shortened by calcium antagonists. The only calcium antagonist approved for antiarrhythmic indications is verapamil, $C_{27}H_{38}N_2O_4$, although diltiazem, $C_{22}H_{26}N_2O_4S$, and others have similar electrophysiologic properties and may be useful antiarrhythmic agents.

Class V Antiarrhythmic Agents: Selective Bradycardic Agents

Class V agents have antiarrhythmic activity in sinus tachyarrhythmias. The bradycardia produced by these agents results from effects on the SA node to reduce the slope of phase 4 diastolic depolarization and prolonging SA node action potential duration. These agents may restrict anionic currents. They include alinidine.

Autonomic Drugs Used in the Treatment of Arrhythmias

Drugs that mimic or inhibit the actions of neurotransmitters released from parasympathetic or sympathetic nerves innervating the heart may also be used to treat supraventricular bradyarrhythmias, heart block, and supraventricular tachyarrhythmias. They include edrophonium, atropine, isoproterenol, and phenylephrine.

Digital Glycosides in the Treatment of Arrhythmias

The digitalis glycosides play an important role in the treatment of supraventricular tachyarrhythmias, particularly atrial fibrillation. These agents have direct and indirect (vagomimetic or vagal sensitizing effects) actions on various parts of the heart. In general, the therapeutic effect is to decrease AV conduction by increasing refractoriness of the AV node and to slow the heart. Thus the ventricle does not respond to the tachycardic rates of the SA node or ectopic sites in the atrium. Digital glycosides include digoxin and digitoxin.

Unclassified Approved Antiarrhythmic Agents

Adenosine, 6-amino-9-β-ribofuranosyl-9-H-purine, is an endogenous nucleoside found in all cells of the body. Its ubiquitousness suggests that adenosine functions as an autocoid and that its actions are mediated by specific receptors on the plasma membranes of all cells. Adenosine is not active orally, but administered as an iv bolus drug adenosine rapidly eliminates supraventricular tachycardias within 1–2 min after dosing. The drug slows conduction through the AV node.

Adenosine in large doses produces vasodilation, resulting in facial flushing, lightheadedness, dizziness, and hypotension. Shortness of breath and chest pressure is seen in about 10% of the patients.

Antiarrhythmic Agents Under Development. Three potential antiarrhythmic agents (Table 1) are in various stages of development.

Asocainol. Asocainol, a dibenzazonine derivative, has sodium channel (Class I) and calcium channel (Class IV) blocking activity that accounts for the antiarrhythmic activity.

Cifenline. Cifenline succinate, also known as cibenzoline, is an imidazoline derivative that is under development in the United States as an antiarrhythmic agent. It is a sodium channel blocker (Class IA), markedly prolongs action potential duration (Class III), and has calcium channel blocking activity (Class IV).

Achrihellin. Achrihellin, a synthetic cardiac glycoside, is an ester of the corresponding aglycon of hellebrin, a bufodienolide. The compound has electrophysiological properties of the cardiac glycosides and antiadrenergic properties. These properties suggest utility in the treatment of supraventricular tachycardias.

Antiarrhythmic Drug Therapy Problems

Drugs used to treat and prevent cardiac arrhythmias have relatively small therapeutic dose-to-toxic dose ratios. Whereas antiarrhythmic drugs have been shown to be effective in the treatment of a wide variety of arrhythmias, worsening of ventricular arrhythmias (proarrhythmia) is a frequent and serious side effect of antiarrhythmic drugs and may lead to sudden death.

ANTIANGINAL AGENTS

Angina pectoris is the principal symptom of ischemic heart disease, which manifests itself as a dramatic, terrorizing, pressing poststernal pain. It occurs suddenly and can be severe. The location and character of the pain may vary but often radiates from the sternum to the left shoulder and over the flexor surface of the left arm to the tips of the medial fingers. It can be induced by exercise, anxiety, overeating, or stress and is often relieved quickly by rest. Other factors, such as decreased oxygen carrying capacity of the blood, or reduced aortic pressure may be involved. The attack may be transient and damage to the ischemic myocardium may be minimal or it may result in an acute MI and/or death. It is usually accompanied by ST segment changes in the ECG, depending on the condition. The pain results from temporary ischemia of the myocardium when the oxygen supply is insufficient to meet the metabolic demands, either by reason

Table 1. Thrombolytic Agents

Generic name	Trade name
tissue plasminogen activator (alteplase t-PA) streptokinase	Activase
	Streptase, Kabikinase
urokinase	Abbokinase, Breokinase, Win-kinase
anisoylated plasminogen–streptokinase activator (APSAC, anistreplase)	Eminase
prourokinase (scu-PA)	Sandolase

of a decrease in blood supply or an exceedingly large increase in oxygen requirements of the myocardium, or both. For a drug to be efficacious in angina it should improve myocardial oxygen supply (increase blood flow) or reduce myocardial oxygen consumption, or have both actions.

Nitrate Therapy in Angina

Nitrates may produce relief of angina, although they do so differently in the various subsets of angina. For example, in stable exertional angina, the action of nitrates on decreasing myocardial oxygen demand is greater than that on relaxing coronary arteries. In unstable angina, nitrates reduce left ventricular wall tension and dilate eccentric stenotic coronary arteries, whereas in Prinzmetal's angina the primary effect of nitrates is to relax coronary vasospasm.

The precise mechanism of nitrate action is not clearly understood and may be a combination of many factors. The basic pharmacologic action of nitrates is a relaxation of most vascular smooth muscle, eg, vascular, bronchial, gastrointestinal, uretal, and uterine.

Nitrates may release prostacyclin, $C_{20}H_{32}O_5$, from the endothelial lining of the vasculature and thereby relax vascular smooth muscle. Prostacyclin is a potent endogenous vasodilator and inhibitor of platelet aggregation. Nitrates may be effective in relieving anginal pain by redistributing coronary blood flow to the ischemic myocardium following coronary artery dilation, or by improving the perfusion of the subendocardium. Reduction of myocardial oxygen demand has been proposed as part of the mechanism for relieving angina. Nitrate antianginal agents include nitroglycerin, isosorbide, erythrityl, and pentaerythritol.

Calcium Antagonist Therapy in Angina

Calcium antagonists alter the availability of calcium by affecting the entry of this ion into its receptor site(s). Calcium antagonists are used mainly in the treatment of various types of angina, hypertension, and certain arrhythmias but may have potential therapeutic applications in heart failure, acute myocardial infarction, cardioprotection, cardiomyopathy, cerebral vasospasm, and other vasospastic syndromes as well.

Calcium plays a vital role in excitation–contraction coupling, and failure to maintain intracellular calcium homeostasis results in cell death. The availability of the calcium antagonists also provides a powerful tool for basic studies of excitation–contraction coupling, stimulus–secretion coupling, and other specific physiological functions.

Calcium and Vascular Smooth Muscle Contraction. Calcium acts on a number of sites associated with the control of the cytoplasmic calcium concentration. Vascular smooth muscle contraction can be initiated by the opening of the slow calcium channel allowing influx of extracellular calcium through the sarcolemmal membrane into the cytoplasmic compartment.

The calcium antagonists interfere with the entry of calcium through the membrane slow calcium channel and therefore prevent intracellular calcium from reaching the critical concentration necessary to initiate contraction. However, these drugs may be acting through other mechanisms to prevent the elevation of the free calcium ion concentration in the cell.

The calcium antagonists produce beneficial effects in angina by either reducing coronary artery spasm, slowing heart rate, or decreasing contractile force. These effects increase oxygen supply or decrease myocardial oxygen consumption. The clinical effects are increased duration of exercise, reduction in anginal episodes, or may be a reduction in the amount of nitrate therapy consumed to reduce the anginal episode.

The side effects or toxic effects that the calcium antagonists have in common are hypotension, facial flushing, headache, dizziness, weakness, sedation, skin rash, edema, constipation, and abdominal discomfort (nausea, vomiting, and epigastric pressure). Calcium antagonists used in the treatment of angina include verapamil, diltiazem, nifedipine, and nicardipine.

β-Adrenoceptor Blocker Therapy in Angina

The β-adrenoceptor blockers are effective antianginal agents because they reduce myocardial oxygen consumption and improve myocardial perfusion by several mechanisms. A decrease in heart rate increases diastolic filling time, which allows for increased coronary perfusion. By decreasing blood pressure and myocardial contractility, β-adrenoceptor blockers reduce the work load of the heart and ventricular wall tension, thereby decreasing myocardial oxygen demand. The β-adrenoceptor blockers improve exercise tolerance and reduce chest pain in patients having stable exertional angina. Some of the β-adrenoceptor blockers have cardioprotective properties in patients post-MI, reducing the risk of sudden death. Agents used to treat angina include propranolol, metoprolol, nadolol, and atenolol.

AGENTS USED IN THE TREATMENT OF CONGESTIVE HEART FAILURE

Congestive heart failure (CHF) is an ever-present, growing health problem in the United States. Over two million people have been diagnosed as having CHF, and more are expected to be diagnosed as the mean age of the population increases. CHF presents a special problem because of its high annual mortality rate of 20–50%. This high mortality rate is largely the result of the lack of a pharmacologic agent that can improve the prognosis for cure above and beyond symptomatic relief. As of this writing there is a great deal of research centered on improved pharmacotherapeutic agents for the treatment of CHF.

Physiology and Biochemistry of Congestive Heart Failure

When the heart begins to fail as a pump, a number of pathophysiologic processes occur. Perhaps the easiest to perceive is simply that some of the blood within the left ventricle is not completely expelled during muscle contraction or systole. This subsequently requires more work during future systoles as additional blood returns from the pulmonary vein. A sort of congestion begins to develop as the heart cannot develop enough pumping ability to provide adequate oxygenated blood to the arterial circulation or removal of metabolic waste through the venous circulation.

There can be a number of underlying causes of CHF. The most prevalent is the lack of oxygenated blood reaching the heart muscle itself because of coronary artery disease with myocardial infarction. Hypertension and valvular disease can contribute to CHF as well, but to a lesser extent in terms of principal causes for the disease.

Therapy of Congestive Heart Failure

Many of the drugs used to combat congestive heart failure are inotropic agents. Inotrope is a derivation of the Greek *ino* (fiber) and *tropikos* (changing or turning). A positive inotropic agent is therefore one that increases cardiac muscle contractility associated with CHF. They include the cardiac glycosides and nondigitalis inotropic agents, eg, catecholamines, drugs acting on the β-adrenoceptor, phosphodiesterase inhibitors, angiotensin converting enzyme inhibitors, and atrial natriuretic factor.

ATHEROSCLEROSIS AND ANTIATHEROSCLEROSIS DRUGS

Formation of atherosclerotic plaques (atheromas) remains one of the central processes in cardiovascular death, including coronary heart disease. Complications from atheromas in coronary arteries include myocardial infarction and stroke. This latter is a common manifestation of cerebral vascular plaque formation.

Reduction in serum lipids can contribute significantly to prevention of atherosclerosis. Whereas the initial step in reducing serum cholesterol is through reduction of dietary cholesterol intake, a number of drugs are available that can affect serum lipid profile (see FAT REPLACERS). The pathway to cholesterol synthesis is shown in Figure 1.

The primary transporter of cholesterol in the blood is low density lipoprotein (LDL). Once transported intracellularly, cholesterol homeostasis is controlled primarily by suppressing cholesterol synthesis through inhibition of β-hydroxy-β-methyl gluterate–coenzyme A (HMG–CoA) reductase, acyl CoA–acyl transferase (ACAT), and down-regulation of LDL receptors. An important drug in the regulation of cholesterol metabolism is lovastatin, also known as mevinolin, MK-803, and Mevacor, which is an HMG–CoA reductase inhibitor.

Bile Acid Sequestrants. The bile acid binding resins colestipol and cholestyramine are also effective in controlling serum cholesterol levels.

Fibric Acids Derivatives. Fibric acid derivatives have been used since the 1960s for control of blood lipid levels. The four most well-known compounds in this class are fenofibrate, gemfibrozil, clofibrate, and bezafibrate.

Other Cardiovascular Agents Affecting Atherosclerosis. Other cardiovascular agents affecting atherosclerosis include β-adrenoceptor antagonists or blockers, calcium channel blockers, and probucol.

ANTIHYPERTENSIVE AGENTS

Hypertension is one of the two principal risk factors of many cardiovascular diseases, such as coronary heart disease (CHD), stroke, and CHF. Individuals are considered hypertensive if their systolic arterial blood pressure is over 140 mm Hg (18.7 Pa) or their diastolic arterial blood pressure is over 90 mm Hg (12 Pa). Over 60 million people, or one-third of the adult population in the United States are estimated to be hypertensive. About 90% of these patients are classified as primary or essential hypertensive because the etiology of their hypertension is unknown. It is generally agreed that there is a very strong genetic or hereditary component to this disease.

Arterial blood pressure at any given time is determined by the balance of the vasopressor and vasodepressor systems operating in various organs and tissues of the body. Some of the principal vasopressor systems identified are the adrenergic or sympathetic nervous system; the renin–angiotensin–aldosterone system; the thromboxane and the prostaglandin $F_2\alpha$ systems; vasopressor peptides; endothelin, endothelium-derived contracting factors (EDCF); the kidney blood volume regulation system; and the vasopressin system. Some of the primary vasodepressor systems are the cholinergic or parasympathetic nervous system, the kallikrein–kinin system, prostaglandin E_2 and prostacyclin (epoprostenol, PGI_2) atrial natriuretic factor (ANF), endothelium-derived relaxing factors (EDRF), and the renal neutral-lipids system. The central nervous system (CNS) can exert either a vasopressor or vasodepressor influence, depending on the level of homeostasis. Overactivity of the vasopressor systems with normal vasodepressor systems causes hypertension. On the other hand, normal vasopressor systems having underactivity of deficiency of opposing vasodepressor systems theoretically can also lead to hypertension.

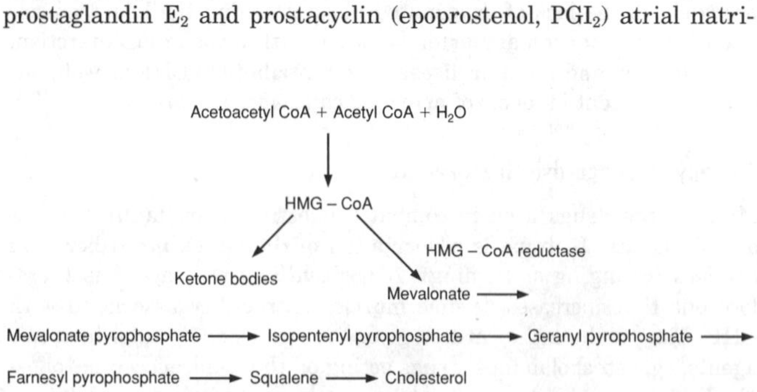

Figure 1. Pathway of cholesterol synthesis.

Treatment of Hypertension

The Joint National Committee on the Detection, Evaluation, and Treatment of High Blood Pressure detailed a recommendation in 1988 (JNC-IV) that emphasized an individualized or patient-oriented (demographic) approach. Factors affecting each patient should be taken into consideration in choosing an antihypertensive drug. The first drug should be either an angiotensin converting enzyme (ACE) inhibitor, a β-adrenoceptor blocker, a calcium channel blocker, or a diuretic. One drug can normalize the blood pressure of about 50–60% of the hypertensive patients and the combination of two complementary drugs normalizes that of over 80% patients. White, young, hyperdynamic, hypertensive patients start with an ACE inhibitor or a β-adrenoceptor blocker; older patients respond well to a calcium channel blocker or a diuretic as do black hypertensive patients; patients having left ventricular hypertrophy respond well to an ACE inhibitor, a calcium channel blocker, or an α_2-adrenoceptor agonist; and patients having concomitant CHF, stroke, diabetes, or hypercholesterolemia can start with an ACE inhibitor. The goal is to give each patient the smallest number of drugs, using the smallest effective amounts having the lowest frequency of dosing and minimum side effects. Thus optimal therapy should reduce all risk factors of coronary heart disease, reverse the hemodynamic abnormalities present by preserving cardiac output and tissue perfusion, and lower total peripheral resistance. The challenge is to choose the best antihypertensive

Agents Affecting the Renin–Angiotensin System

Angiotensin converting enzyme inhibitors

- captopril
- enalapril
- lisinopril
- perindopril, ramipril, spirapril, quinapril, alacepril, delapril, ceranapril, zofenopril, fosinopril, benazepril, libenzapril

Angiotensin II receptor antagonists

- DuP 753, DuP 532
- EXP-9993, EXP-3174
- L-158,809
- diltiazem

Renin inhibitors

- Enalkerin (A-64662)
- CGP 38 560 A
- CP 71362
- KRI-1230, U-71039, ES6864, PD-134672

Agents Affecting the Adernergic Nervous System

Ganglionic blockers

- chlorisondamine
- trimethaphan
- hexamethonium

Neuronal blockers

- guanethidine
- guanadrel, bretylium, debrisoquin, bethanidine

Neuronal norepinephrine depleting agents

- reserpine

α-Adrenoceptor blockers
phentolamine
prazosin
terazosin
doxazosin, bunazosin
Centrally Acting Agents
methyldopa
clonidine
guanabenz
guanfacine
rilmenide
Vasodilators
hydralazine
minoxidil
nitroprusside
cadralazine
Potassium Channel Openers
pinacidil
cromakalim, lemakalim
nicorarndil
diazoxide, minoxidil
β-Adrenoceptor blockers
propanolol
metoprolol
atenolol
bisoprolol
nadolol, acebutolol, pindolol, betaxolol, celiprolol, carvedilol, bucindolol, brefanolol
α/β-Adrenoceptor blockers
labetalol
Calcium Channel Blockers
verapamil
diltiazem
nifedipine
nicardipine
felodipine
amlodipine
nitrendipine
isradipine
lacidipine, benidipine
Diuretics
Low ceiling diuretics
hydrochlorothiazide, other thiazides
chlorthalidone
quinethazone
metolazone
High ceiling diuretics
furosemide
bumetadine
ethacrynic acid
Aldosterone antagonists
spironolactone
Potassium sparing diuretics
triamterene
amiloride

THROMBOLYTIC AGENTS

It has been well documented that the primary cause of acute myocardial infarction (AMI) is coronary arterial thrombosis. Thrombosis formation occurs at sites of ulcerated or fissured atheromatous plaques in the coronary circulation. Reperfusion of the coronaries to the ischemic/infarcted area within 6 h after the onset of AMI can salvage the myocardium and limit infarct size. Furthermore, the better the preservation of the ventricular function, the better the survival rate. Progressive irreversible myocardiac damage occurs 30 min after ischemia starts. Therefore, the faster the coronary arterial thrombus is lyzed, the less irreversible damage done on the myocardium. It is imperative that once an AMI is diagnosed, iv thrombolytic agent should be administered as quickly as possible. Thrombolytic agents available are listed in Table 1.

ECONOMIC ASPECTS

The cardiovascular agents are the fastest growing segment of the world pharmaceutical market. In 1990, the world ethical pharmaceutical market was $77.5 billion. The overall growth rate between 1978–1988 was 18% per annum but was expected to slow to 10% per annum between 1988–1993. Despite the projected slowing of the total market, cardiovascular agents are expected to continue to make substantial progress in all world markets.

PETER CERVONI
DAVID L. CRANDALL
PETER S. CHAN
Lederle Laboratories, American Cyanamid Company

B. R. Lucchesi, in M. Antonacchio, ed., *Cardiovascular Pharmacology,* 3rd ed., Raven Press, N.Y., 1990, p. 369.

J. T. Bigger and B. F. Hoffman, in A. G. Gilman and co-eds., *Goodman and Gilman's, The Pharmacologic Basis of Therapeutics,* 8th ed., Pergamon Press, N.Y., 1990, p. 840.

D. T. Mason and co-workers in *Cardiovascular Drugs: Antiarrhythmic, Antihypertensive and Lipid Lowering Drugs,* Vol. 1, ADIS Press, Sydney, Australia, 1977, p. 75.

CARNALLITE. See POTASSIUM COMPOUNDS.

CARNAUBA WAX. See WAXES.

CARNOTITE. See URANIUM AND URANIUM COMPOUNDS.

CAROTENOIDS. See VITAMINS, VITAMIN A.

CASEHARDENING. See METAL SURFACE TREATMENTS.

CASEIN. See MILK AND MILK PRODUCTS.

CASTOR OIL

Castor oil is derived from the bean of the castor plant *Ricinus communis,* belonging to the family Euphorbiaceae. The castor bean is not a legume as the name implies but is a member of the spurge family. Castor plants occur in practically all tropical and subtropical countries.

They are also grown annually from seed as both ornamental and cultivated plants in temperate zones. There are wide variations in the size, form, and color of the plant, as well as the size and color of the seed.

The seeds of the castor plant are produced in racemes, or clusters of capsules. The capsules each contain three seeds protected by a hull that is removed prior to processing. The seeds are mottled to varying extents, most often with shades of dark brown overlaying shades of light brown. Seeds of commercial varieties range from 250 to 1800/kg. The seeds are toxic, and the ingestion of several seeds can be fatal to humans.

U.S. production was competitive until 1972, when Federal price supports were withdrawn. U.S. production dropped almost to zero by 1974. Interest in the restoration of U.S. castor seed production was initiated in 1989. An aggressive hybrid seed program was developed in Texas, and sufficient hybrid seed was grown in 1990 to plant 40,000 acres of castor in 1991. Mechanization of harvesting equipment for the collection of dwarf (ca 1 m) hybrid plants retaining the seed pods on the plant until harvested, was expected to yield from 800–900 kilograms of seed per acre. Although the U.S. Department of Defense continues to classify castor derivatives as a critical strategic material, in 1990 the United States imported 100% of its castor oil needs, primarily from Brazil and India. Furthermore, fluctuating supplies and prices of imported oil fostered the use of alternative feedstocks.

Properties

Typical of vegetable oils and most fats, castor oil is a triglyceride of various fatty acids (see FATS AND FATTY OILS). Its uniqueness stems from the very high (87–90 wt %) content of ricinoleic acid, $C_{18}H_{34}O_3$, structurally *cis*-12-hydroxyoctadeca-9-enoic acid, $CH_3(CH_2)_5CH(OH)CH_2CH{-}CH(CH_2)_7COOH$, an eighteen-carbon hydroxylated fatty acid having one double bond. Castor oil, sometimes described as a triglyceride of ricinoleic acid, is one of the few commercially available glycerides that contain hydroxyl functionality in such a high percentage of one fatty acid. General properties of castor oil are listed in Table 1.

Castor oil is distinguished from other triglycerides by its high specific gravity, viscosity, and hydroxyl value. Another unique feature is alcohol solubility: one volume of castor oil dissolves in two volumes of 95% ethyl alcohol at room temperature, and the oil is miscible in all proportions with absolute ethyl alcohol. Also the oil is typically soluble

Table 1. Properties of Castor Oil

Property	Value
viscosity,[a] 25°C, mm^2/s(= cSt)	615–790
flash point, Cleveland open cup, °C	285
tag closed cup	230
surface tension at 20°C, mN/m	39.0
pour point, °C	−23
coefficient of expansion, mL/°C	0.00066

[a] Value corresponds to U ± 1/2 in Gardner-Holdt units (see RHEOLOGICAL MEASUREMENTS).

in polar organic solvents and less soluble in aliphatic hydrocarbon and other nonpolar organic solvents.

Oil and Meal Recovery

The commercial process for castor oil recovery from the seed consists of preheating the seed in stack cookers prior to crushing in a hydraulic press or a continuous mechanical screw-type press commonly known as an expeller. The presscake discharged from this mechanical processing contains 10–20 wt % oil and is then processed in solvent extraction units to recover the residual oil. Cold-pressing of castor seed that has not been preheated is still carried out to a limited degree in castor growing countries such as Brazil and India. A light colored, high quality medicinal type oil is recovered. Castor oil recovered from hydraulic or continuous mechanical screw presses requires refining to remove toxic proteins, gums, and foreign matter while improving the color and reducing the free fatty acid content.

Solvent extraction in batch or continuous systems is used to recover most of the residual oil from the presscake.

Chemical Reactions

Castor oil serves as an industrial raw material for the manufacture of a number of complex organic derivatives. Chemical reactions occur at the three basic points of functionality, as shown in Figure 1.

Economic Aspects

Whereas castor oil is available on a worldwide basis from many sources, India, Brazil, and China produced nearly 80% of the world's castor seed for the period 1986–1989.

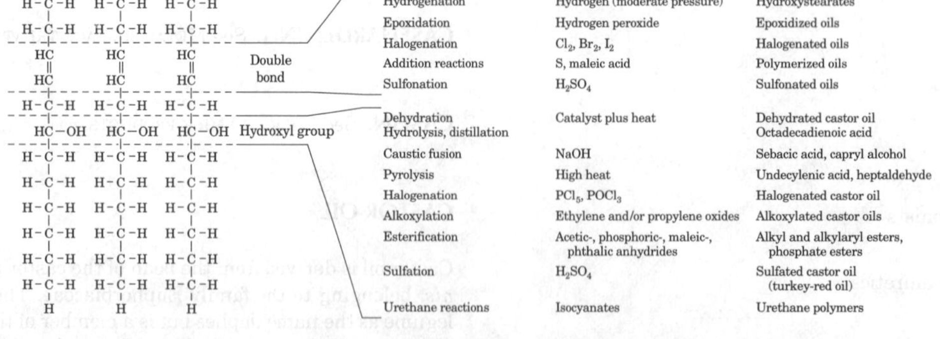

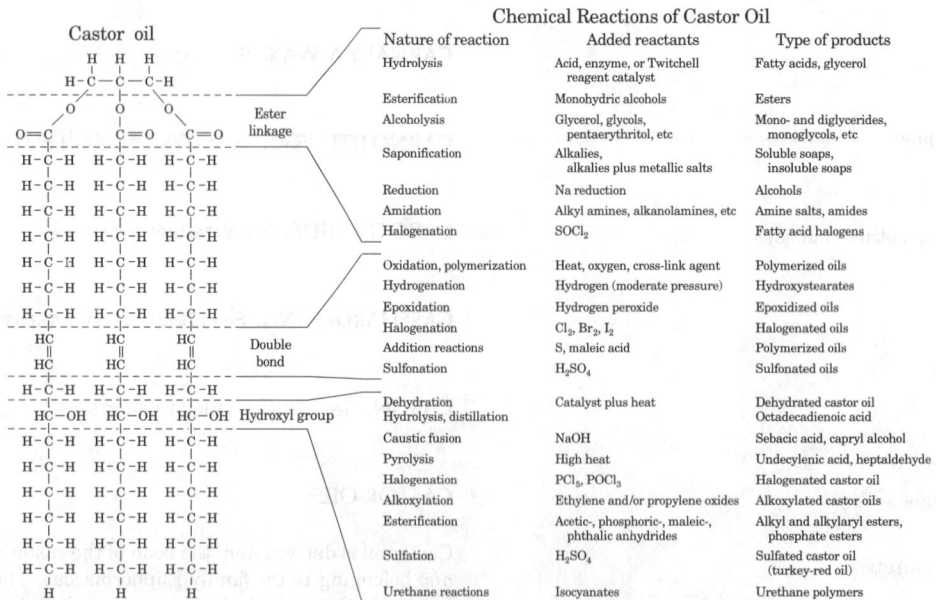

Figure 1. Structure and reactivity of castor oil. Courtesy of CashChem, Inc.

Brazil, China, and India accounted for about 85% of the world exports, and France, the United States, Russia, Germany, and the United Kingdom accounted for about 75% of the world imports from 1986–1989.

Applications

Nylon-11, accounts for the largest single use of castor oil. Castor oil is also used in urethanes, coatings, lubricants, textiles, cosmetics, and surfactants and dispersants.

FRANK C. NAUGHTON
CAMBREX, Inc.

P. W. Sleggs, "Steam Refining of Castor Oil," *Meeting of the International Castor Oil Association,* New York, presented Dec. 6, 1990.

A. M. Altschul, *Processed Plant Protein Foodstuffs,* 1st ed., Academic Press, Inc., New York, 1958, pp. 835–844.

The Processing of Castor Need for Detoxification and Deallergenation, Technical Bulletin No. 1-1989, International Castor Oil Association, 1989.

CATALYSIS

Catalysis is the key to efficient chemical processing. Most industrial reactions and almost all biological reactions are catalytic. The value of the products made in the United States in processes that at some stage involve catalysis approaches a trillion dollars annually, which is more than the gross national products of all but a few nations of the world. Products made with catalysis include food, clothing, drugs, plastics, detergents, and fuels. Catalysis is central to technologies for environmental protection by conversion of emissions.

A catalyst is a substance that increases the rate of approach to equilibrium of a chemical reaction without being substantially consumed itself. A catalyst changes the rate but not the equilibrium of the reaction.

It is well recognized that catalysts function by forming chemical bonds with one or more reactants, thereby opening up pathways to their conversion into products with regeneration of the catalyst. Catalysis is thus cyclic; reactants bond to one form of the catalyst, products are decoupled from another form, and the initial form is regenerated. The simplest imaginable catalytic cycle is therefore depicted as follows:

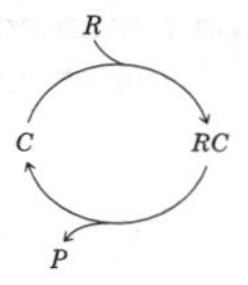

where R is the reactant, P the product, C the catalyst, and RC an intermediate complex. The intermediate complexes in catalysis are often highly reactive and not observable.

Ideally, the catalyst would cycle forever between C and RC without being consumed. But in reality there are competing reactions, and catalysts are converted into species that are no longer catalysts. In practice, catalysts must be regenerated and replaced. Catalyst manufacture is a large industry; catalysts worth some $2–3 billion are sold annually in the United States.

Catalysts may be gases, liquids, or solids. Most catalysts used in technology are either liquids or surfaces of solids. Catalysis occurring in a single gas or liquid phase is referred to as homogeneous catalysis (or molecular catalysis) because of the uniformity of the phase in which it occurs. Catalysis occurring in a multiphase mixture such as a gas–solid mixture is referred to as heterogeneous catalysis; usually this is surface catalysis. Biological catalysts are proteins, ie, poly(amino acids), called enzymes, which may be present in solution or in membranes.

The performance of a catalyst is measured largely by criteria of chemical kinetics, as a catalyst influences the rate and not the equilibrium of a reaction. The catalytic activity is a property of a catalyst that measures how fast a catalytic reaction takes place and may be defined as the rate of the catalytic reaction, a rate constant, or a conversion or temperature required for a particular conversion in a reaction under specified conditions. The selectivity is a measure of the property of a catalyst to direct a reaction to particular products. Because catalysts lose activity and selectivity during operation, they are also evaluated in terms of stability. The stability of a catalyst is a measure of the rate of loss of activity or selectivity. Catalysts that have lost activity are often treated to bring back the activity, ie, regenerated; the regenerability is a measure, often not precisely defined, of how well the activity can be brought back. Technological catalysts are also evaluated in terms of cost.

The solids used as catalysts are typically robust porous materials with high internal surface areas, typically hundreds of square meters per gram. Reaction occurs on the internal catalyst surface. The typical solid catalyst used in industry is a composite material with numerous components and a complex structure.

Catalytic processes are classified roughly according to the nature of the product and the industry of application, and to some degree separate literatures have developed. In chemicals manufacture, catalysis is used to make heavy chemicals, commodity chemicals, and fine chemicals. Catalysis is used extensively in the manufacture of pharmaceuticals. In fuels processing, catalysis is used in almost all the processes of petroleum refining and in coal conversion and related synthesis gas (CO and H_2) conversion. Most of the recent large-scale developments in industrial catalysis have been motivated by the need for environmental protection. Many processes for abatement of emissions are catalytic. Most of the applications of catalysis in biotechnology (qv) are fermentations, often carried out in stirred reactors with gases, liquids, and solids present; the catalysts are enzymes present in living organisms such as yeasts. There are recent applications of whole biological cells and of individual enzymes mounted on supports, ie, carriers, and used in fixed-bed reactors.

Homogeneous Catalysis

Characterization of Solution Processes. There are many important examples of catalysis in the liquid phase, but catalysis in the gas phase is unusual. From an engineering viewpoint, most of the liquid-phase processes have the following characteristics in common.

Pressure and Temperature. The pressure and temperature are relatively low, typically less than about 2 MPa (20 atm) and 150°C.

Corrosiveness. The catalyst solutions are corrosive, and the reactors, separation devices, etc that come in contact with them must be made of expensive corrosion-resistant materials.

Separation Processes. Separation of the catalyst from the products is a significant expense; the process flow diagram and the processing cost are often dominated by the separations. The most common separation devices are distillation columns; extraction is also applied.

Gas–Liquid Contacting. The reactants are often gaseous under ambient conditions. To maximize the rate of the catalytic reaction, it is often necessary to minimize the resistance to gas–liquid mass transfer, and the gases are therefore introduced as swarms of bubbles into a well-stirred liquid or into devices such as packed columns that facilitate gas–liquid mixing and gas absorption.

Exothermicity. The catalytic reactions are often exothermic bond-forming reactions of small molecules that give larger molecules. Consequently, the reactors are designed for efficient heat removal. They may be jacketed or contain coils for heat-transfer media, or the heat of reaction may be used to vaporize the products and aid in the downstream separation by distillation.

There are also a number of generalizations about the chemistry of these processes. Often the reactants are small building blocks, many formed from organic raw materials, namely, petroleum, natural gas, and coal. The reactants include O_2, low molecular-weight olefins, and synthesis gas (CO and H_2). Many reactions are catalyzed by acids and bases, usually in aqueous solution. Many reactions are catalyzed by organometallic compounds, usually in nonaqueous organic solution. The organometallic catalysts used in technology are often highly selective.

Influence of Mass Transport on Reaction Rates. When a relatively slow catalytic reaction takes place in a stirred solution, the reactants are supplied to the catalyst from the immediately neighboring solution so readily that virtually no concentration gradients exist. The intrinsic chemical kinetics determines the rate of the reaction. However, when the intrinsic rate of the reaction is very high and/or the transport of the reactant slow, as in a viscous polymer solution, the concentration gradients become significant, and the transport of reactants to the catalyst cannot keep the catalyst supplied sufficiently for the rate of the reaction to be that corresponding to the intrinsic chemical kinetics.

Acid–Base Catalysis. Inexpensive mineral acids, eg, H_2SO_4, and bases, eg, KOH in aqueous solution, are widely applied as catalysts in industrial organic synthesis. Catalytic reactions include esterifications, hydrations, dehydrations, and condensations. Much of the technology is old and well established, and the chemistry is well understood. Reactions that are catalyzed by acids are also typically catalyzed by bases.

Organometallic Catalysis. Most of the recent innovations in industrial homogeneous catalysis have resulted from discoveries of organometallic catalysts. Thousands of organometallic compounds (complexes), ie, those with metal–carbon bonds, are known, and the rapid development of organotransition metal chemistry in recent years has been motivated largely by the successes and opportunities in catalysis. The chemistry of organotransition metal catalysis is explained by the bonding and reactivity of organic groups (ligands) bonded to the metals. The *d* orbitals of the metals allow bonding of ligands such as H (hydride), CO, and olefins in ways that facilitate their reaction. The important reactions in catalytic cycles are those of ligands bonded in the coordination sphere of the same metal atom. Bonding of ligands such as CO or olefin to a transition metal activates them and facilitates the catalysis. An example of a well-understood catalytic cycle is that of the hydrogenation of olefins shown in Figure 1.

Phase-Transfer Catalysis. When two reactants in a catalytic process have such different solubility properties that they can hardly both be present in a single liquid phase, the reaction is confined to a liquid–liquid interface and is usually slow. However, the rate can be increased by orders of magnitude by application of a phase-transfer catalyst, and these are used on a large scale in industrial processing (see CATALYSIS, PHASE-TRANSFER). Phase-transfer catalysts function by facilitating transport of reactants between the liquid phases. Often most of the reaction takes place close to the interface.

Industrial examples of phase-transfer catalysis are numerous and growing rapidly; they include polymerization, substitution, condensation, and oxidation reactions. The processing advantages, besides the acceleration of the reaction, include mild reaction conditions, relatively simple process flow diagrams, and flexibility in the choice of solvents.

Heterogeneous Catalysis

Characterization of Surface Processes. Most of the largest-scale catalytic processes take place with gaseous reactants in the presence of solid catalysts. From an engineering viewpoint, these processes offer the following advantages, in contrast to those involving liquid catalysts: *(1)* wide ranges of temperature and pressure are economically applied; *(2)* solid catalysts are only rarely corrosive: *(3)* the separation of gaseous or liquid products from solid catalysts is simple and costs little; *(4)* the mixing and mass transport in a fixed- or fluidized-bed reactor are facilitated by the solid catalyst particles through which the reactants and products flow; *(5)* strongly exothermic and strongly endothermic reactions are routinely carried out with solid catalysts.

Properties of Solid Catalysts. Most solid catalysts used on a large scale are porous inorganic materials. A number of these and the reactions they catalyze are summarized in Table 1. Catalysis takes place as one or more of the reactants is chemisorbed (chemically adsorbed) on the surface and reacts there. The activity and selectivity of the catalyst depend strongly on the surface composition and structure.

Important physical properties of catalysts include the particle size and shape, surface area, pore volume, pore size distribution, and strength to resist crushing and abrasion.

Influence of Mass Transport on Catalyst Performance. Reactants must diffuse through the network of pores of a catalyst particle to reach the internal area, and the products must diffuse back. The optimum porosity of a catalyst particle is determined by tradeoffs: making the pores smaller increases the surface area and thereby increases the activity of the catalyst, but this gain is offset by the increased resistance to transport in the smaller pores; increasing the pore volume to create larger pores for faster transport is compensated by a loss of physical strength. If there is a significant resistance to transport of the reactant in the pores, a concentration gradient will exist at steady state, whereby the concentration of the reactant is a maximum at the particle periphery and a minimum at the particle center. The product concentration will be higher at the particle center than at the

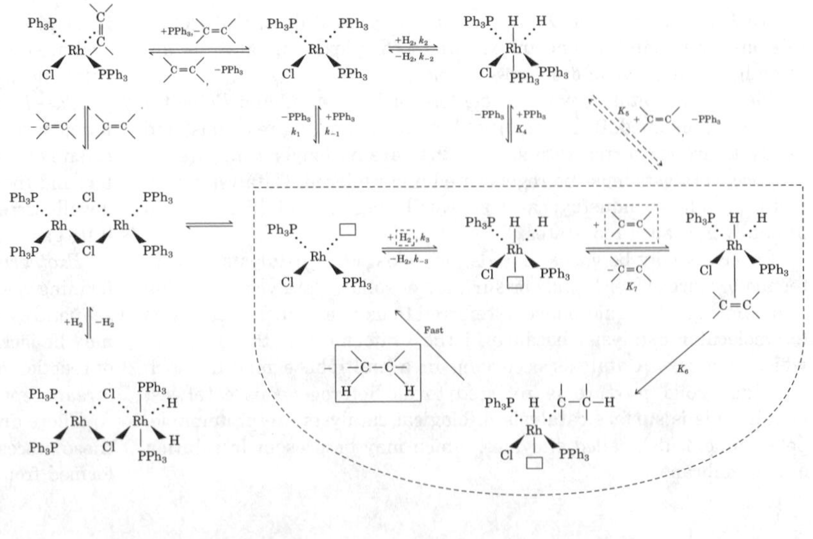

Figure 1. Catalytic cycle (within dashed lines) for the Wilkinson hydrogenation of olefin. Ph represents phenyl (C_6H_5). Values of rate constants and equilibrium constants are as follows: $k = 0.68\ s^{-1}$; $k_{-1} \geq 7 \times 10^4$ L/(mol · s); $K_1 \leq 10^{-5}$ mol/L; $k_{-1}/k_4 \cong 1$; $k_5 = 3 \times 10^{-4}$; $k_2 = 4.8$ L/(mol · s); $k_{-2} = 2.8 \times 10^{-4}\ s^{-1}$; $K_2 = 1.7 \times 10^4$ L/mol; $k_3 \geq 7 \times 10^4$ L/(mol · s); $k_6 = 0.22\ s^{-1}$

Table 1. Some Large-Scale Industrial Processes Catalyzed by Surfaces of Inorganic Solids

Catalyst	Reaction
metals (eg, Ni, Pd, Pt, as powders or on supports) or metal oxides (eg, Cr_2O_3)	C=C bond hydrogenation (eg, olefin + $H_2 \rightarrow$ paraffin)
metals (eg, Cu, Ni, Pt)	C=O bond hydrogenation (eg, acetone + $H_2 \rightarrow$ 2-propanol)
metal (eg, Pd, Pt)	total oxidation of hydrocarbons, oxidation of CO
Fe, Ru (supported and promoted with alkali metals)	$3\ H_2 + N_2 \rightarrow 2\ NH_3$
Ni	$CO + 3\ H_2 \rightarrow CH_4 + H_2O$ (methanation) $CH_4 + H_2O \rightarrow 3\ H_2 + CO$(steam reforming)
Fe or Co (supported and promoted with alkali metals)	$CO + H_2 \rightarrow$ paraffins + olefins + $H_2O + CO_2$ (+ oxygen-containing organic compounds) (Fischer-Tropsch reaction)
Cu (supported on ZnO, with other components, eg, Al_2O_3)	$CO + 2\ H_2 \rightarrow CH_3OH$
Re + Pt (supported on η-Al_2O_3 and promoted with chloride)	paraffin dehydrogenation, isomerization and dehydrocyclization (eg, *n*-heptane $\rightarrow$ toluene + 4 H_2) (naphtha reforming)
solid acids (eg, SiO_2–Al_2O_3, zeolites)	paraffin cracking and isomerization; aromatic alkylation; polymerization of olefins
γ-Al_2O_3	alcohol $\rightarrow$ olefin + H_2O
Pd supported on zeolite	paraffin hydrocracking
metal-oxide-supported complexes of Cr, Ti, or Zr	olefin polymerization (eg, ethylene $\rightarrow$ polyethylene)
metal-oxide-supported complexes of W or Re	olefin metathesis (eg, 2 propylene $\rightarrow$ ethylene + butene)
V_2O_5 or Pt	$2\ SO_2 + O_2 \rightarrow 2\ SO_3$
Ag (on inert support, promoted by alkali metals)	ethylene + $\frac{1}{2}\ O_2 \rightarrow$ ethylene oxide (with $CO_2 + H_2O$)
V_2O_5 (on metal-oxide support)	naphthalene + $\frac{9}{2}O_2 \rightarrow$ phthalic anhydride + 2 CO_2 + 2 H_2O *o* − xylene + 3 $O_2 \rightarrow$ phthalic anhydride + 3 H_2O
bismuth molybdate, uranium antimonate, other mixed metal oxides	propylene + $\frac{1}{2}\ O_2 \rightarrow$ acrolein propylene + $\frac{3}{2}O_2$ + $NH_3 \rightarrow$ acrylonitrile + 3 H_2O
mixed oxides of Fe and Mo	$CH_3OH + O_2 \rightarrow$ formaldehyde (with $CO_2 + H_2O$)
Fe_3O_4 or metal sulfides	$H_2O + CO \rightarrow H_2 + CO_2$ (water gas shift reaction)
Co–Mo/γ − Al_2O_3(sulfided)	olefin hydrogenation
Ni–Mo/γ − Al_2O_3(sulfided)	aromatic hydrogenation
Ni–W/γ − Al_2O_3(sulfided)	hydrodesulfurization
	hydrodenitrogenation

periphery. The concentration gradients provide the driving force for the transport. If these concentration gradients are large, the rate of the catalysts reaction is usually markedly less than it would be otherwise, and the effectiveness factor of the catalyst is small.

Catalyst Components. Industrial catalysts are typically complex in composition and structure. Catalytically active phases, supports, binders, and promoters comprise the components of the catalyst. Typically, the major component is the support, an inexpensive metal oxide such as alumina. The catalytically active components are often expensive, eg, precious metals, and to maximize their utilization, they are dispersed as minute particles on the interior surface of the porous support.

Catalyst Treatment. Catalysts often require activation, redispersion, or regeneration, and their disposal also requires special consideration.

Catalyst Preparation. Catalyst preparation is more an art than a science. Many reported catalyst preparations omit important details and are difficult to reproduce exactly, and this has hindered the development of catalysis as a quantitative science. However, the art is developing into a science and there are now many examples of catalysts synthesized in various laboratories that have nearly the same physical and catalytic properties.

Supports are often prepared first and the catalyst and promoter components added later. Metal oxide supports are usually prepared by precipitation from aqueous solutions. Catalyst and promoter components are usually added as metal salts. Metal oxide supports, eg, silica and alumina, are prepared in the form of hydrogels. Mixed oxides such as silica–alumina are made by cogelation. Careful control of conditions such as pH is important to give uniform products.

Molecular Catalysis on Supports. The term molecular catalysis is commonly applied only to reactions in uniform fluid phases, but it applies nearly as well to some reactions taking place on supports. Straightforward examples are reactions catalyzed by polymers functionalized with groups that closely resemble catalytic groups in solution. Industrial examples include reactions catalyzed by ion-exchange resins, usually sulfonated poly(styrene–divinylbenzene). This polymer is an industrial catalyst for synthesis of methyl *t*-butyl ether (MTBE) from methanol and isobutylene and synthesis of bisphenol A from phenol and acetone, among others. The former application has grown rapidly as MTBE has become a component of high octane number gasoline.

Catalysis by Metals. Metals are among the most important and widely used industrial catalysts. They offer activities for a wide variety of reactions (Table 1). Atoms at the surfaces of bulk metals have reactivities and catalytic properties different from those of metals in metal complexes because they have different ligand surroundings. The surrounding bulk stabilizes surface metal atoms in a coordinatively unsaturated state that allows bonding of reactants. Thus metal surfaces offer an advantage over metal complexes, in which there is only restricted stabilization of coordinative unsaturation. Furthermore, metal surfaces provide catalytically active sites that are stable at high temperatures. For example, supported palladium catalysts have replaced soluble palladium for vinyl acetate synthesis; the advantages of the solid include reduced corrosion and reduced formation of by-products.

Catalysis by Metal Oxides and Zeolites. Metal oxides are common catalyst supports and catalysts. Some metal oxides alone are industrial catalysts; an example is the γ-Al_2O_3 used for ethanol dehydration to give ethylene. But these simple oxides are the exception; mixed metal oxides are more common. The best-understood metal oxide catalysts are zeolites, ie, crystalline aluminosilicates.

Shape-Selective Catalysis. The zeolites are crystalline aluminosilicates that have pores with diameters of only about 1 nm as part of their structures. Consequently, they are molecular sieves; small molecules are able to diffuse through the narrow pores, whereas larger molecules that do not fit are sieved out.

Catalytic processes have been developed to take advantage of the unique transport and molecular sieving properties of zeolites. The zeolite that has found the most applications is the medium-pored HZSM-5. The term shape-selective catalysis is applied to describe the unique effects. There are different kinds of shape selectivity. Mass transport selectivity is a consequence of transport restrictions whereby some species diffuse more rapidly than others in the zeolite pores. In the simplest kind of shape-selective catalysis, small molecules in a mixture enter the pores and are catalytically converted, whereas large molecules pass through the reactor unconverted because they do not fit into the pores where the catalytic sites are located. Similarly, product molecules formed inside a zeolite may be so large that their transport out of the zeolite may be very slow, and they may be converted largely into other products that diffuse more rapidly into the product stream. A different kind of shape selectivity is called restricted transition state selectivity. It is not related to transport restrictions; rather, it is related to the size restriction of the catalyst pore that suppresses the formation of the transition state for a certain reaction, whereas it may not suppress the formation of a smaller transition state for another reaction.

Mass transport selectivity is illustrated by a process for disproportionation of toluene catalyzed by HZSM-5. The desired product is *p*-xylene; the other isomers are less valuable. The ortho and meta isomers are bulkier than the para isomer and diffuse less readily in the zeolite pores. This transport restriction favors their conversion to the desired product in the catalyst pores; the desired para isomer is formed in excess of the equilibrium concentration. Xylene isomerization is another reaction catalyzed by HZSM-5, and the catalyst is preferred because of restricted transition state selectivity. An undesired side reaction, the xylene disproportionation to give toluene and trimethylbenzenes, is suppressed because it is bimolecular and the bulky transition state cannot readily form.

Mixed Metal Oxides and Propylene Ammoxidation. The best catalysts for partial oxidation of hydrocarbons are metal oxides, usually mixed metal oxides. For example, phosphorus–vanadium oxides are used commercially for oxidation of *n*-butane to give maleic anhydride, and oxides of bismuth and molybdenum with other components are used commercially for oxidation of propylene to give acrolein or acrylonitrile.

Catalysis by Supported Metals. Metals used in industrial catalysis are often expensive, and they are predominantly used in a highly dispersed form. Metals dispersed on supports such as amorphous metal oxides or zeolites may be present in clusters of only a few atoms, eg, Pt in zeolite LTL used for naphtha reforming to give aromatics, or in small crystallites, or in larger cystallites, up to about 1 μm in size, eg, Ag supported on α-alumina for ethylene oxidation to ethylene oxide. The catalytic properties for some reactions are sensitive to the structure and size of the metal particles.

Catalysis by Metal Sulfides. Metal sulfides such as MoS_2, WS_2, and many others catalyze numerous reactions that are catalyzed by metals. The metal sulfides are typically several orders of magnitude less active than the metals, but they have the unique advantage of not being poisoned by sulfur compounds. They are thus good catalysts for applications with sulfur-containing feeds, including hydroprocessing of fossil fuels.

Catalyst Development, Testing, and Production

Catalysts are discovered to meet processing needs and opportunities, but the discovery of a catalytic application to take advantage of some newly discovered material almost never occurs. Catalyst development is largely a matter of trial and error testing. The methodology was defined by Haber, Bosch, and Mittasch in the development of the ammonia synthesis process. Catalyst developers benefit from an extensive and diverse literature and often can formulate good starting points in a search for candidate catalysts by learning what has been used successfully for similar reactions. Deeper insights, such as would arise from understanding of the mechanistic details of a catalytic cycle, are usually not attainable; the exceptions to this rule largely pertain to molecular catalysis, usually reactions occurring in solution.

Catalyst testing and evaluation have been revolutionized by computers, automated test reactors, and analytical methods. With modern equipment, researchers can systematically prepare and screen many catalysts in a short time and efficiently determine the initial catalytic activity and selectivity and also the stability and regenerability.

Almost all industrial catalysts are developed by researchers who are motivated to improve existing processes or create new ones. Catalysts are for the most part highly complex specialty chemicals, and catalyst manufacturers tend to be more efficient than others in producing them. Catalyst manufacturing is a competitive industry. Catalyst users often develop close relationships with catalyst manufacturers, and the two may work together to develop and improve proprietary catalysts.

BRUCE C. GATES
University of California, Davis

B. G. Linsen, J. M. H. Fortuin, C. Okkerse, and J. J. Steggerola, eds., *Physical and Chemical Aspects of Adsorbents and Catalysts,* Academic Press, London, 1967.

Advances in Catalysis, Academic Press, New York, continuing series.

J. R. Anderson and M. Boudart, eds., *Catalysis—Science and Technology,* Springer-Verlag, Berlin, Germany, continuing series.

A thorough list of references is given in J. T. Richardson, *Principles of Catalyst Development,* Plenum Press, New York, 1989.

CATALYSIS, PHASE-TRANSFER

Phase-transfer catalysis (PTC) is a technique by which reactions between substances located in different phases are brought about or accelerated. Typically, one or more of the reactants are organic liquids or solids dissolved in a nonpolar organic solvent and the coreactants are salts or alkali metal hydroxides in aqueous solution. The most extensively used catalysts are quaternary ammonium or phosphonium salts, and crown ethers, and cryptates.

Fundamental Process in Displacement Reactions

The mechanistic picture developed by C. M. Starks for liquid–liquid PTC is shown in a graphical form as follows.

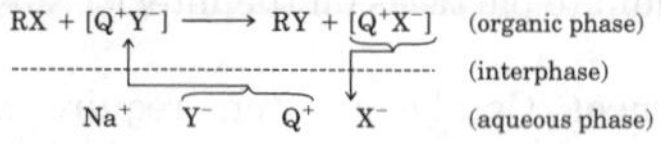

Phase boundary: $[Q^+X^-]_{(org)} + NaY_{(aq)}$ 7 $[Q^+Y^-]_{(org)} + NaX$

Organic phase: $[Q^+Y^-]_{(org)} + RX$ $[Q^+X^-]_{(org)} + RY$

The catalyst cation Q^+ extracts the more lipophilic anion Y^- from the aqueous to the nonpolar organic phase where it is present in the form of a poorly solvated ion pair $[Q^+Y^-]$. This then eacts rapidly with RX, and the newly formed ion pair $[Q^+X^-]$ returns to the aqueous phase for another exchange process: $X^- \rightarrow Y^-$. In practice most catalyst cations used are rather lipophilic and do not extract strongly into the aqueous phase so that the anions are exchanged at the phase boundary.

Types of Phase-Transfer Processes

There are several types of processes, each depends on the nature of the phases, the character of the extracted species, and the direction of extraction.

Phases. Often there are two liquid phases (liquid–liquid PTC) or one solid and one liquid (solid–liquid PTC).

Character of Extracted Species. Most PTC reactions are initiated by the extraction of some anion, be it a nucleophile, a base, or an oxidant, by lipophilic cations. However, relatively hydrophilic cations can be extracted by large anions from an aqueous or solid phase into an organic solvent for electrophilic reactions.

Direction of Extraction. The normal PT process involves the transfer of a reactive agent from a solid or aqueous environment into a nonpolar organic solvent. But the exact opposite can also be executed.

Types of Phase-Transfer Reactions

PTC has an enormous potential for numerous reactions and no single type of mechanism explains everything. Among the more frequently used reactions involving anion extraction, two classes can be distinguished: (*1*) reactions without added bases; and (*2*) reactions in the presence of alkali metal hydroxides, potassium carbonate, or other inorganic bases.

There are two active special fields of phase-transfer applications that transcend classes (*1*) and (*2*): metal–organic reactions both with and without added bases, and polymer chemistry. Certain chemical modifications of side groups, polycondensations, and radical polymerizations can be influenced favorably by PTC.

Factors Influencing the Usefulness of Phase-Transfer Catalysis

In the following typical reaction under phase catalytic conditions,

$$RX + KY_{(water)} \xrightarrow{N(C_4H_9)_4X} RY + KX$$

the relative rate for $X = Cl^-, Br^-, I^-$ is determined by two factors: the actual chemical reaction with I^-, Br^-, or Cl^-; and the magnitude of the competitive extraction of Y^- vs X^-.

Solvent. For For PTC the solvents should be nonmiscible with water and nonhydroxylic; $CHCl_3$ and CH_2Cl_2, chlorobenzene, toluene, acetonitrile, and even petroleum ethers are employed.

Catalyst Cation. The logarithms of extraction constants for symmetrical tetra-*n*-alkylammonium salts ($\log E_{QX}$) rise by ca 0.54 per added C atom. Although absolute numerical values for extraction coefficients are vastly different in various solvents and for various anions, this relation holds as a first approximation for most solvent–water combinations tested and for many anions.

Stability. The stability of the catalysts is another important factor. Reactive nucleophilic anions can attack their own counter ion, leading to Hofmann-type elimination or simple dequaternization.

Selectivity. Steering of reaction directions by the type of catalyst cation, eg, O- vs C-alkylation, substitution vs dihalocarbene addition, as well as enantioselective alkylations by optical active catalysts, have been achieved in some systems. Extensive development is necessary, however, to generate satisfactorily large effects.

Competitive Extraction of Anions. The successful extraction of the necessary anion into the organic phase is crucial for PTC. Often three anions compete for the catalyst cation: the one that is to react, the one formed in the reaction, and the one brought in originally with the catalyst.

Typical Applications

Phase-transfer catalysis is used in displacement reactions, formation of esters, formation of ethers, alkylation of carbanions, nucleophilic aromatic substitution, alkylation of *N*-heterocycles, polymerization of, eg, butyl acrylate, oxidations, reductions, and carbene reactions.

ECKEHARD VOLKER DEHMLOW
Universität Bielefeld, Germany

C. Starks and C. Liotta, *Phase-Transfer Catalysis: Principles and Techniques,* Academic Press, New York, 1978.

A. Brändström, *Preparative Ion Pair Extraction: An Introduction to Theory and Practice,* Apotekarsocieteten, Hässle Läkemedel, Stockholm, Sweden, 1974.

C. Starks, ed., *Phase-Transfer Catalysis: New Chemistry, Catalysts, and Applications,* ACS Symposium Series #326, American Chemical Society, Washington, D.C., 1987.

CATALYSTS

SUPPORTED

Supported catalysts comprise the largest group of materials capable of catalysis, which is broadly defined as the acceleration of chemical reactions (see CATALYSIS). Supported catalysts are extremely useful in virtually all areas of petroleum refining and chemical processing, which accounts for the fact that as a group they were the principal contributor to the \$2.5–3.0 billion worldwide catalyst industry in 1990. Petroleum refining catalysts comprised 37% of the \$1.8 billion U.S. market in 1990, whereas chemical processing made up an additional 34% and emission control catalysts contributed 29%. Of the \$1.0 billion petroleum refining catalyst market, 56% of revenues came from fluid cracking catalysts (FCC), 31.5% from hydrotreating catalysts, 6.5% from hydrocracking catalysts, and 4.5% from reforming catalysts.

An annual review of the worldwide catalyst industry in the *Oil Gas Journal*, identifies current technical and business trends within the catalyst industry and lists virtually all industrial supported (and other) catalysts by manufacturers' designations.

Some novel approaches to using supported catalysts have evolved in the 1990s, among which are inorganic membrane reactors, the automotive catalytic converter, and catalytic distillation technology, used in the manufacture of methyl *t*-butyl ether and certain aromatics employing catalysts such as zeolites contained in fiber glass cloth/stainless steel mesh bales packed in what is essentially a distillation column.

Definition of Supported Catalysts

Supported catalysts are solid or heterogeneous catalysts in which relatively small amounts of catalytically active species, frequently metals, are deposited on the surface of largely inert, porous, shaped support bodies, such as pellets, rings, extrudate, and granules, which are sometimes referred to as carriers. Most materials used in catalyst supports are comprised of single- or mixed-metal oxides, such as alumina, silica, magnesia, titania, zirconia, and aluminosilicates.

Advantages of Supported Catalysts

Separability. One of the greatest advantages of a solid catalyst is that it can be separated easily from the products of reaction.

Cost. The catalytically active components in many supported catalysts are expensive metals. By using a catalyst in which the active component is but a very small fraction of the weight of the total catalyst, lower costs can be achieved.

Catalyst Activity. Of utmost importance in the design of most catalysts is activity, which is a measure of the ability of a catalyst to effect conversion of the reactants to the desired products under specified conditions.

In service, supported catalysts frequently undergo loss of activity over a period of time. In many cases, such catalyst deactivation is accompanied by the loss of accessible surface area of the active phase by sintering, by the accumulation of poisons, or by conversion of active sites to inactive species.

Catalyst Selectivity. Selectivity is the property of a catalyst that determines what fraction of a reactant will be converted to a particular product under specified conditions. A catalyst designer must find ways to obtain optimum selectivity from any particular catalyst. For example, in the oxidation of ethylene to ethylene oxide over metallic silver supported on alumina, ethylene is converted both to ethylene oxide and to carbon dioxide and water. In addition, some of the ethylene oxide formed is lost to complete oxidation to carbon dioxide and water. The selectivity to ethylene oxide in this example is defined as the molar fraction of the ethylene converted to ethylene oxide as opposed to carbon dioxide.

In order to optimize selectivity for any particular system, unwanted by-products must be identified, and reaction conditions and catalyst components that are not favorable to their formation selected. For many reactions, selectivity is found to decrease as the activity increases. Thus sometimes it is necessary to accept a compromise in which some activity or selectivity or both is sacrificed so that the overall product yield or process economics is maximized.

Support Materials

The objective in selecting a support for a catalytic application is to provide a suitable, stable base for the active catalytic component. The support should be chemically inert so that it does not interfere with the role of the catalytic component, and it should possess acceptable physical properties for the intended application. The support should retain its dimensions and chemical integrity under the conditions necessary to operate the catalytic process.

Composition. Among the most commonly used support materials are aluminas, silicas, and aluminosilicates with a wide range of alumina to silica ratios, as well as activated carbon, silicon carbide, selected clays, various ceramics, artificial and natural zeolites, kieselguhr, and pumice. Polymeric materials are also used as supports (see CATALYSIS; ENZYMES; ION-EXCHANGE). Other recognized support materials include calcium carbonate, barium carbonate and sulfate, and magnesium halides.

Size and Shape. It is common practice to form catalyst support materials into convenient sizes and shapes that satisfy the chemical and engineering needs of the overall catalytic process. Some of the more useful shapes are cylinders, rings, and multilobed bodies. Optionally, support materials may be tableted or pelletized to form cylindrical bodies that can be used as is or rolled into spheres.

A catalyst manufactured using a shaped support assumes the same general size and shape of the support. This is an important consideration in the process design because these properties determine packing density and the pressure drop across the reactor.

Surface Area. This property is of paramount importance to catalyst performance because in general catalyst activity increases as the surface area of the catalyst increases. The primary determinant of catalyst surface area is the support surface area, except in the case of certain catalysts where extremely fine dispersions of active material are obtained.

Support surface area can be varied within a wide range of values by a variety of techniques. Alumina, silica, and other metal oxide-containing support materials respond well to calcination conditions, where higher temperatures and longer calcination times can be used to obtain progressively lower surface areas in the base oxide powders. Additional control of surface area in shaped bodies can be obtained by grinding the calcined oxides to specific particle sizes, combining the powder with binders and combustible burnout materials, such as powdered cellulose, wood flour, and polymer powders, prior to shaping and recalcination. Careful control of the particle size of the burnout material powder employed can be used to induce very specific porosities and pore size distributions in the support, which, along with recalcination conditions, can define the surface area of the finished support body. Some catalyst supports rely on a relatively low surface area structural member coated with a layer of a higher surface area support material.

Porosity and Pore Size. The support porosity is the volume of the support occupied by void space and usually is described in units of cm^3/g. This value represents the maximum amount of liquid that may be absorbed into the pore structure, which is an especially important consideration for deposition of metal salts or other active materials on the support surface by liquid impregnation techniques. The concentration of active material to be used in the impregnating solution is determined by the support porosity and the desired level of active material loading on the catalyst.

In most cases, catalyst designers prefer an open-pore structure, that is, pores that have more than one opening, and a pore size as uniform as possible in order to obtain maximum utilization of the available pore volume. This can be achieved by careful choice of raw materials and processing conditions. In some cases, it is desirable to have a bimodal or multimodal pore structure in which there are present both large diameter channels and smaller, intersecting passages.

Attrition Loss. Attrition can occur when support bodies rub against one another and abrade the surface, such as during calcination in a rotary kiln or sizing on moving screens.

Density. This property is of secondary importance to catalyst designers but is of great relevance to manufacturers and users. Density must be taken into account when a catalyst bed is being engineered in order to ensure sufficient strength in the process equipment for safe operation. In addition, the density of a catalyst support affects cost, because a higher density support requires the manufacturer to purchase greater quantities of raw materials.

Cost and Quality. Many factors affect catalyst support cost, including the kind of raw materials used, the purity of the raw materials, the chemical processing steps required, the fabrication method used, the severity of calcination conditions, and the extent of the quality assurance procedure.

Design and Development of Supported Catalysts

Much progress has been made in understanding how to create and use catalysts, but the design and preparation of practical catalysts still relies on the application of known facts and intuition to trial and error methods. Successful catalyst design requires that the very first decisions made be at least partly correct. In a typical design experiment, an element known to be catalytically active toward a certain type of reaction must be correctly combined with a support material considered to be suitable because of its chemical and physical properties. The product catalyst must be evaluated under conditions considered feasible for the reaction. If the experiment is successful, that is, if the catalyst produces conversion of the reactant and adequate selectivity to the desired product, then the experimental catalyst may be a candidate for further development. Much catalyst research in the 1990s is centered on improving or modifying existing catalysts. In addition to developing satisfactory catalysts, the designer may also have to develop a catalyst support or work with a support manufacturer.

Catalyst Preparation Methods. The most common methods used to prepare supported catalysts include extruding, tableting, or pelletizing physically mixed (comulled) solid components, and impregnation of porous, shaped support bodies with the catalyst components dissolved in a solvent, usually water. In addition, some catalysts are produced and used as powders, granules, and the like. The details of most manufacturing methods for industrial catalysts are closely held secrets.

There are other methods of preparation that involve establishing an active phase on a support phase, such as ion exchange, chemical reactions, vapor deposition, and diffusion coating.

Catalysts from Physical Mixtures. Two separate catalysts with different functions may be pulverized to fine powders and mixed to form a catalyst system that accomplishes a reaction sequence which neither of the two individual catalysts alone can achieve. For such catalyst systems, the reaction products of catalyst A become the feedstocks for catalyst B, and vice versa.

Impregnated Catalysts. Preparation of impregnated catalysts usually involves filling the pore structure of a shaped, porous support body with a solution of the catalyst components, removing the solvent, and activating the catalyst.

Dry Impregnation. In the so-called dry impregnation method, the amount of solution added to the solid support is exactly that calculated to fill the pores of the support completely. The solution may be sprayed onto the support in a blending operation and usually is absorbed quickly because of capillary action. The resulting mixture appears dry to the eye, thus the name dry impregnation.

Pore Volume Impregnation. For catalyst supports with extremely small pores or a closed-end pore structure, it may be difficult to fill the support completely with impregnating solution by the dry impregnation method. In this case, immersion of the support in a solution containing the catalytically active components usually is more effective.

Ion-Exchange Catalysts. Catalytically active species can be deposited predictably and reproducibly on support materials capable of ion exchange to produce useful catalysts. The support can be organic, as in the case of ion-exchange resins, or inorganic, as in the case of zeolites, which are by far the largest application for ion-exchange methods.

Testing and Evaluation of Supported Catalysts

In the development phase of catalyst research, testing of the catalyst's chemical and physical properties and evaluation of the catalyst's performance are two essential tasks. In the manufacturing process, many of the same analyses and evaluations are used for quality assurance.

New Developments in Supported Catalysts

Some areas of current interest include methane coupling to C_2+ hydrocarbons, oxidation of methane to useful products, conversion of propane to acrylonitrile, and synthesis gas to oxygenates. A largely unexplored area with vast potential is the supporting of enzymes to effect and control biologically important reactions. New ways to support catalytically active materials are being considered, such as with the use of an immiscible liquid to support a catalyst in solution. The rational application of purely theoretical considerations to catalyst design will be increasingly important.

ROBERT P. NIELSEN
Consultant

G. C. Bond, *Heterogeneous Catalysis: Principles and Applications,* Clarendon Press, Oxford, U.K. 1987.

J. F. LePage, *Applied Heterogeneous Catalysis: Design, Manufacture, Use of Solid Catalysts,* Editions Technip, IFP Publications, Paris, France, 1987.

J. T. Richardson, *Principles of Catalyst Development,* Plenum Publishing Corp., New York, 1989.

REGENERATION

FLUID CATALYTIC CRACKING UNITS

The process of fluid catalytic cracking (FCC) is the central process in a modern, gasoline-oriented refinery. In U.S. refineries, the amount of feed processed by fluid catalytic cracking units (FCCU) is equivalent to 35% of the total crude oil processed in the United States. As of January 1991, installed FCCU capacity in the United States was 8.6×10^5 m^3/d (5.4×10^6 barrel/d.

The popularity of the catalytic cracker stems from its ability to produce large quantities of high octane gasoline and other valuable light products and to use a wide range of refinery process streams as feed materials. The FCCU feeds include high molecular weight vacuum gas oils (VGO), which are traditional feeds; atmospheric and vacuum residues; coker gas oils; gas oils from other thermal and hydrocracking operations; lube oil extracts; and deasphalted oils. In a modern U.S. gasoline refinery, the FCCU typically produces about 35% of the total refinery gasoline pool.

Although the FCCU has long been an important refinery processing tool in the United States, refinery gasoline demands outside the United States have been satisfied largely from the gasoline naturally present in crude oil. This situation is changing, and the need for catalytic cracking is now growing steadily everywhere. Worldwide, excluding North America, Eastern Europe, and mainland China, installed FCCU capacity increased nearly 60% in the 10-year period from 1981 to 1991, and in 1991 stood at $7.3 \times 10^5 m^3/d$ (4.6×10^6 barrel/d).

The FCC process is highly complex but self-contained. A typical unit of mid-1980s design is shown in Figure 1. In the reactor section, hydrocarbon feed is combined with hot fluidizable cracking catalyst at the base of the riser reactor. Catalyst and hydrocarbon vapors pass up the riser in a dilute-phase, plug-type flow and discharge into the reactor vessel where catalyst and hydrocarbon vapors are separated. When in contact with the catalyst, the hydrocarbon feed cracks to form gasoline, middle distillates, liquefied petroleum gas (LPG), and fuel gas. Coke, which is also formed, deposits on the catalyst surface and greatly reduces the catalyst activity. To restore its activity, the catalyst passes into the regenerator section where the coke is burned off. The heat released from coke burning is completely used to provide the heat requirements on the reactor side: to vaporize the feed, to compensate for the endothermic heat of cracking, to bring the reactor to the desired reactor temperature, and to compensate for heat losses to the atmosphere.

Thus the FCCU always operates in complete heat balance at any desired hydrocarbon feed rate and reactor temperature; this heat balance is achieved in units such as the one shown in Figure 1 by varying the catalyst circulation rate. Catalyst flow is controlled by a slide valve located in the catalyst transfer line from the regenerator to the reactor and in the catalyst return line from the reactor to the regenerator.

Prior to entering the regenerator, the spent catalyst from the reactor passes through a steam stripper where the hydrocarbon vapors entrained by the flowing catalyst particles are displaced by steam. The steam also acts to desorb some of the lighter hydrocarbons that are adsorbed during the reaction step. These stripped materials, which represent valuable reaction products, are carried back into the reactor by the upflowing steam from the stripper and go to the product-recovery section with the other reaction products. The removal of these materials from the catalyst stream flowing to the regenerator also reduces the amount of burnable hydrocarbon going to the regenerator, which helps to decrease the regenerator temperature.

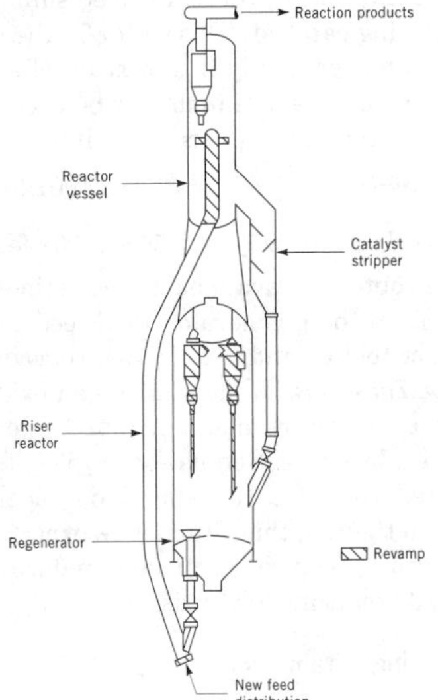

Figure 1. UOP stacked FCC unit revamped to riser cracking.

FCCU Heat Balance

Because of the thermal coupling of reactor and regenerator, any change on the reactor side creates a rapid change on the regenerator side, which, in turn, influences the reactor side, and vice versa. This dynamic interaction rapidly comes to equilibrium, and the catalytic cracker adjusts to a new steady-state, heat-balanced condition. The first law of FCCU catalytic cracking is that the FCCU will always adjust itself to stay in heat balance. If an FCCU is not in heat balance, it is out of control. The operating characteristics of an FCCU regenerator are thus dictated by the constraints of the heat balance.

Environmental Aspects

The FCCU regenerator is one of the principal sources of air pollutants from a refinery. A typical 150,000-barrel-per-day (23,850-m^3/d), modern U.S. refinery with a 50,000-bpd FCCU processing a low sulfur, low metals content VGO feed and operating in complete CO combustion typically emits the following into the atmosphere from the FCCU regenerator flue gas stack: fine catalyst particles, 2–3 t/d; sulfur oxides, 3–4 t/d; nitrogen oxides, 0.5 t/d; and CO, 1.5–2 t/d. Research continues for the development of technologies to reduce these emissions.

Catalyst Emissions. High velocity air passing up through the regenerator bed at velocities typically around 1 m/s carries large quantities of catalyst into the vapor space above the catalyst bed. To contain these catalyst particles within the regenerator vessel, a highly efficient, generally two-stage cyclone system is installed inside the regenerator. The flue gas and the entrained catalyst particles pass through the cyclone system before exiting from the regenerator. With a two-stage cyclone system, 99.995% or more of the catalyst particles are typically returned by the cyclones back into the regenerator. The particles that are not retained and that leave with the exiting flue gas are small, typically 20-μm average particle size. These particles generally represent fines that have been produced as a result of catalyst attrition.

To reduce catalyst losses even further, additional separation equipment external to the regenerator can be installed. Such equipment includes third-stage cyclones, electrostatic precipitators, and more recently the Shell multitube separator, which is licensed by the Shell Oil Co., UOP, and the M. W. Kellogg Co. The Shell separator removes an additional 70–80% of the catalyst fines leaving the first two cyclones. Such a third-stage separator essentially removes from the flue gas stream all particles greater than 10 μm.

SO_x Emissions. The amount of SO_x, which refers to the combination of SO_2 and SO_3, emitted from an FCCU regenerator depends on the type of FCC feed (aromatic, paraffinic, hydrotreated), the amount of (organic) sulfur in the feed, and the conversion level. Depending on these factors, 3–12% (typically 5%) of the feed sulfur deposits with the coke on the circulating catalyst. This sulfur in the coke is oxidized to SO_x in the regenerator, generally in a mixture of about 90% SO_2 and 10% SO_3. The following rule of thumb may be used for estimating the regenerator flue SO_x content in parts per million:

$$\textbf{straight − runfeeds}: \quad SO_x = 800–1100x \text{ feed sulfur in wt}$$

$$\textbf{hydrotreatedfeeds}: \quad SO_x = 2050–2800x \text{ feed sulfur in wt}$$

Three possible routes are available to the refiner for the reduction of SO_x emissions from the regenerator: FCC feedstock hydrodesulfurization; flue gas scrubbing; and use of a SO_x removal catalyst.

Nitrogen Oxide Emissions. Because nitrogen oxide, NO_x, emissions from an FCCU regenerator are much less than SO_x emissions, typically only 100 ppm, much less attention has been given to the development of technology to reduce FCCU NO_x emissions. As environmental constraints continue to tighten, this situation is expected to change.

Feed hydrotreatment can be beneficial in reducing regenerator NO_x emissions as well as reducing SO_x emissions.

Regenerator Operating Parameters

To maximize the performance of an FCCU, most units run at one or more unit constraints. Frequently, one of these constraints is the regenerator temperature, which is set by metallurgical limits for safe operations. Process variables on both the reactor and the regenerator side are thus manipulated to keep the regenerator temperature as close as possible to this regenerator temperature limit.

FCCU Regenerator Configuration and Mechanical Hardware

Since the first fluid-bed catalytic cracking unit was commissioned in 1942, more than 300 additional units have been built. During this time, the process has evolved and has seen considerable improvement in mechanical construction, reliability, and process flow. A modern FCCU typically operates continuously for three to four years between turnarounds, during which time 10^{10} kg of feedstock are processed and 7×10^{10} kg of catalyst circulated. Early FCCU designs were complex compared with the compact configuration of more recent design (Fig. 1).

In the modern unit design, the main vessel elevations and catalyst transfer lines are typically set to achieve optimum pressure differentials because the process favors high regenerator pressure for enhancing power recovery from the flue gas and coke-burning kinetics, and low reactor pressure for enhancing product yields and selectivities.

Influence of Regenerator Design on Catalyst Fluidization

As regenerator design has improved over the years, various fluidization regimes have been used in the regenerator. These designs include the bubbling bed at low gas velocity, the turbulent bed at higher gas velocity, and the fast-fluidized bed at still higher gas velocity.

Solids Mixing Characteristics. Scale-up factors in fluidized systems are notoriously poor. Scale-up is particularly difficult in the area of solid–gas mixing characteristics, because some commercial regenerators exceed 15 m in diameter. The solid–gas mixing characteristics greatly influence thermal gradients, both interparticle and within the bed; residence time distribution; and afterburn in the free board and cyclone area.

The mixing characteristics of bubble- or gas-induced solids in the various fluidized-bed regimes have therefore received considerable attention over the years. Mathematical expressions describing the bubble-induced mixing characteristics of solids in fluidized beds show that the axial mixing in a bubbling-bed regenerator is high (150 kg/m^2·s).

Although this bubble-induced axial turnover rate of solids is extremely high, the radial mixing characteristics are relatively poor on account of the vertical bubble flow, and can lead to distribution problems in the larger-diameter bubbling- or turbulent-bed regenerators.

In the 1970s, UOP developed the high efficiency combustor to overcome these intrinsic problems. The overall benefits of this high efficiency combustor over a conventional bubbling- or turbulent-bed regenerator are enhanced and controlled carbon-burn kinetics (carbon on regenerated catalyst at less than 0.05 wt %); ease of start-up and routine operability; uniform radial carbon and temperature profiles; limited afterburn in the upper regenerator section and uniform cyclone temperatures; and reduced catalyst inventory and air-blower horsepower.

Mechanical Hardware

From a mechanical point of view, the FCCU regenerator can be divided into two main sections: the regenerator vessel and its internals, which include the air distributor and the cyclones, the plenum chamber, and catalyst coolers; and the flue gas handling section, which includes power, heat, and particulate recovery systems. Continuous advances in the mechanical design of both these areas have led to significant improvements in overall FCCU reliability, even though processing conditions, at the same time, have become more difficult as a result of harder and denser catalysts, as well as higher regenerator temperatures.

L. L. Upson
D. A. Lomas
UOP Research Center

L. L. Upson, I. Dalin, and W. R. Wichers, "Heat Balance: The Key to Catalytic Cracking," *Proceedings of Katalistiks, the 3rd Annual FCC Symposium,* Amsterdam, the Netherlands, 1982.

J. A. Sigan and co-workers, "Reducing FCC Emissions with No Capital Costs," *Proceedings of the 1990 AICHE Spring National Meeting,* Miami, Fla., 1990.

D. Kunii and O. Levenspiel, *Fluidization Engineering,* Krieger Publishing, New York, 1977.

D. A. Lomas and co-workers, "Controlled Catalytic Cracking," *Proceedings of the 1990 UOP Technology Conference,* Chicago, Ill., 1990.

NOBLE AND BASE METAL CATALYSTS

Noble Metal Catalysts

Heterogeneous catalysts containing noble metals, such as platinum, are widely used in the petroleum refining industry for a large variety of hydrocarbon conversion processes. The products made by these processes, such as motor fuels and petrochemicals, are of immense economic value. Processes of commercial importance that use platinum catalysts include reforming of low octane naphtha to produce high octane gasoline, reforming to produce aromatics, paraffin isomerization, xylene isomerization, hydrogenation, and dehydrogenation.

Catalysts in this service can deactivate by several different mechanisms, but deactivation is ordinarily and primarily the result of deposition of carbonaceous materials onto the catalyst surface during hydrocarbon charge-stock processing at elevated temperature.

The original performance of the fresh catalyst can be successfully restored by proper regeneration to remove this coke. Regeneration allows continued use of the same catalyst for many years. Thus even expensive and sophisticated catalysts can become economical for commercial use in petroleum refining.

Regeneration Processes. Three different *in situ* techniques are widely used to regenerate naphtha-reforming catalysts: continuous catalyst regeneration (CCR), fixed-bed semiregenerative regeneration, and cyclic, or swing, reactor regeneration.

Base Metal Catalysts

Heterogeneous catalysts containing nonnoble or base metals such as nickel, cobalt, molybdenum, and tungsten are commonly used in the petroleum refining industry for a large variety of hydrocarbon conversion processes. Processes of commercial importance that use such catalysts include hydrotreating, hydrocracking, and residuum cracking, desulfurization, and demetallation.

Hydroprocessing catalysts can deactivate by several different mechanisms, including fouling by carbonaceous deposits, sintering of the metals on the catalytic surface, and deposition of reversible and irreversible poisons on the catalyst. If deactivation is primarily the result of carbonaceous deposits, as in the case of the hydrocracking process, the performance of the original fresh catalyst can be substantially restored by a regenerative oxidation treatment to remove this deposit. The treatment is preferably done *ex situ.*

In many of the other processes that use base metal catalysts, irreversible poisoning of the catalyst occurs as a result of the deposition of metal contaminants from the process feedstock onto the catalyst surface. These catalysts are not considered to be regenerable by ordinary techniques.

D. W. Robinson
UOP

R. L. Peer, R. W. Bennett, and S. T. Bakas, *Proceedings of the 1988 National Refiners Association Annual Meeting,* 1988.

U.S. Pat. 3,751,379 (Aug. 7, 1973), J. C. Hayes (to UOP).

U.S. Pat. 3,935,244 (Jan. 27, 1976), J. C. Hayes (to UOP).

U.S. Pat. 4,832,921 (May 23, 1989), A. R. Greenwood (to UOP).

CAULKING COMPOSITIONS. See Sealants.

CAUSTIC SODA. See Alkali and chlorine products.

CELL CULTURE TECHNOLOGY

Cell culture processes, the *in vitro* growth of animal, insect, or plant cells on a large scale to manufacture biochemicals of commercial importance, have been used for some time for the manufacture of viral vaccines (see Vaccine technology). Significant growth in this technology, primarily because of the advent of recombinant DNA methods (see Genetic engineering) for the production of therapeutic proteins (qv) and hybridoma technology for production of monoclonal antibodies (see Immunoassay), occurred in the late 1980s and early 1990s. The need for cell culture technology stems mainly from the fact that bacteria do not have the capability to perform many of the post-translational modifications that most large proteins require for *in vivo* biological activity. These modifications include intracellular processing steps such as protein folding, disulfide linkages, glycosylation, and carboxylation.

Monoclonals have already found many uses in the diagnostic industry by virtue of extreme specificity of binding to an antigen of interest. They have also found use for immunopurification and as therapeutics.

Recombinant DNA technology has already provided several products of therapeutic interest from mammalian cells. Table 1 gives examples of products from mammalian cells, their use, and the technology used for their production. Although the focus of this article is mainly on mammalian cells, the technologies described herein also apply in principle to insect and plant cells.

Characteristics of Mammalian Cells

Environmental Conditions. Mammalian cells *in vivo* are maintained in a carefully balanced homeostatic environment and thus have evolved to require fairly stringent environmental conditions. These cells differ significantly from bacterial cells in that they lack a rigid cell wall and are hence much more shear-sensitive. Many animal cells are also attachment-dependent, needing a surface to grow on. Many of the cell culture technologies provide the low shear, high surface area environment needed for the mammalian cells. Another approach, is to adapt cells to suspension culture and select cells that are less shear-sensitive, thus permitting the use of fermentation technology for the culture of animal cells. The optimum environmental parameters depend on and are specific to cell type.

Nutritional Requirements. The nutrient requirements for mammalian cells are many, varied, and complex. In addition to typical metabolic requirements such as sugars, amino acids (qv), vitamins) (qv), and minerals, cells also need growth factors and other proteins. Some of the proteins are not consumed, but play a catalytic role in the cell growth process.

Cell growth kinetics of mammalian cells can be described by the typical lag, exponential growth, then stationary and death phases. The exponential phase may be described adequately by a Monod-type kinetic model when the growth rate is much larger than the death rate. At low growth rates, it is necessary to include cell death kinetics to account for the lower viability and to predict the cell viability. Toward the end of the exponential culture, cells are also subject to growth inhibition from metabolic by-products such as lactate and ammonia. Hence, for continuous processes, a comprehensive model should contain terms for cell growth based on the limiting substrate concentration, cell death, and inhibition kinetics.

Table 1. Therapeutic Products from Mammalian Cells

Product	Company	Cell[a]	Process[b]	Therapeutic use
tissue plasminogen activator (tPa)	Genentech	CHO	SC	heart attacks
erythropoietin	Amgen	CHO	ARB	anemia
factor VIII	Cutter/Miles	CHO	SC	hemophilia
OKT-3	Ortho	hybridoma	MA	prevention of transplant rejection
centoxin	Centocor	hybridoma	SC	septic shock

[a] CHO = Chinese hamster ovary.
[b] SC = suspension culture; ARB = automated roller bottles; MA = mouse ascites.

Cell Culture Processes

A wide variety of mammalian cells are used in industrial practice, and, to accommodate the diversity in cell line and product characteristics, a variety of cell culture technologies have evolved. Culture processes include batch suspension culture (Fig. 1), batch microcarrier process, microcarrier perfusion systems (Fig. 2), fluidized-bed systems (Fig. 3), fixed-bed reactors (Fig. 4), and cell recycle systems.

Other Systems. There are several other technologies utilized for specific purposes that do not fall into any of the above categories. For example, erythropoietin is produced in an automated roller bottle plant where cells are grown on the surface of roller bottles in a growth medium. The Technology Partnership (Cambridge, England) offers robotic systems for the production of cell-culture-derived products using roller bottles and T-flasks.

Economic Aspects

As of early 1992, the market for cell-culture-derived products approached $1 billion per year. The market is expected to grow substantially throughout the 1990s.

Process Economics. Relative economics of various cell culture processes depend heavily on the performance of the cell line in a system and on the cost of raw materials, particularly the medium. Often, for high value products, the process that ensures the shortest time to market may be the process of choice because of other economic criteria. This is especially true for pharmaceuticals (qv). Reliability concerns also often outweigh economic considerations in choosing a process for a high value product.

Continuous processes have lower labor costs but higher failure risk. Batch processes can be started back up in a shorter period of time than can a complex continuous process. Batch processes are easier to take through the regulatory process than are continuous processes. Thus batch processes are often chosen for mammalian cell culture systems, even though continuous processes can offer significant cost advantages. Cell culture costs constitute only a small (10–30%) fraction of the overall cost of making a product.

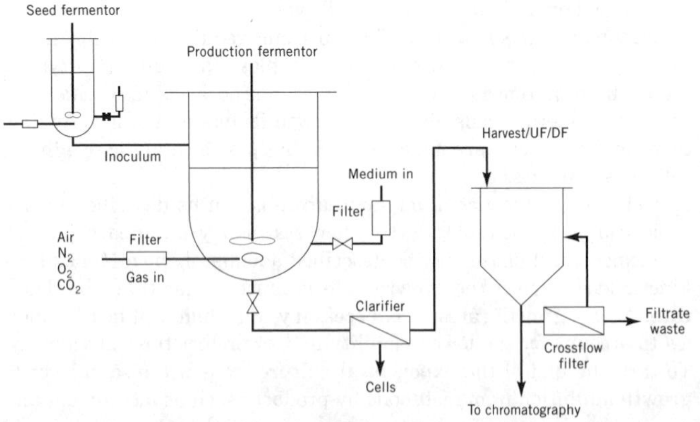

Figure 1. Schematic of a process for batch suspension culture of mammalian cells, where UF is ultrafiltration and DF is diafiltration.

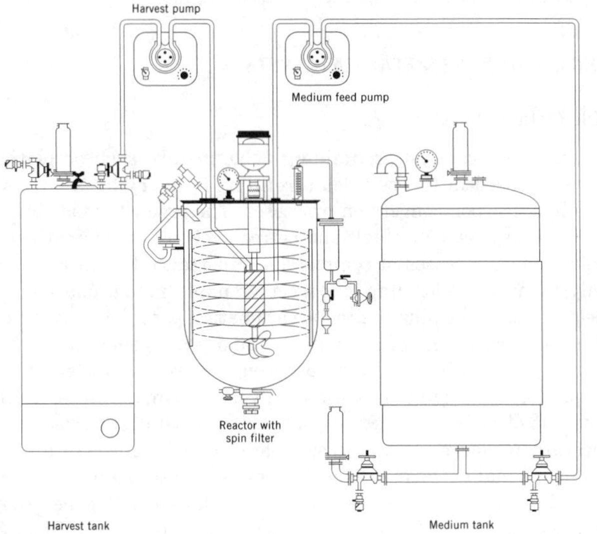

Figure 2. Schematic of a process for microcarrier perfusion culture using a spin filter device.

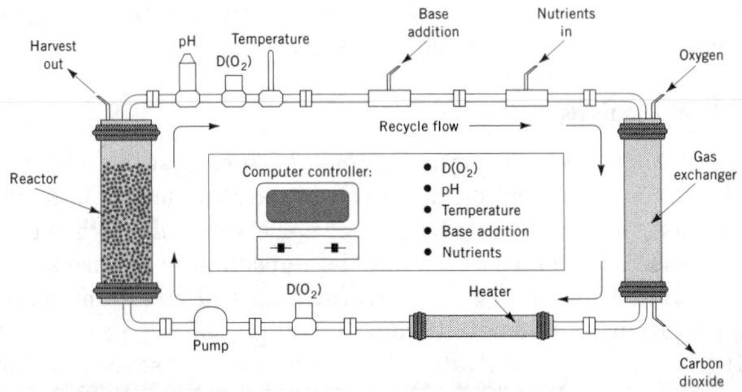

Figure 3. Schematic of a fluidized-bed process for mammalian cell culture, where D(O^2 = dissolved oxygen). Courtesy of Verax Corp.

Regulations and Standards

Most of the products derived from cell culture technologies are for therapeutic or diagnostic applications and manufacture is regulated by the federal government through the Food and Drug Administration (FDA).

The biotechnology (qv) industry has no formal standards for equipment manufacture and quality control as of this writing (1996). The American Society of Mechanical Engineers (ASME) set up a task force in 1989 for developing relevant standards for biotechnology equipment.

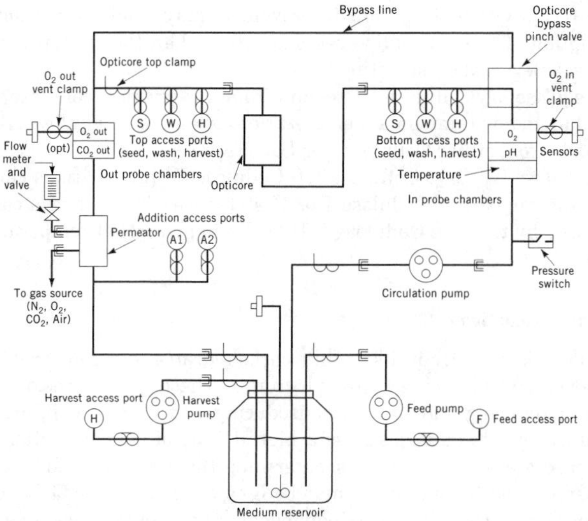

Figure 4. Schematic of a fixed-bed process for culturing mammalian cell using ceramic matrix, where S = seed; W = wash; and H = harvest. Courtesy of Charles River Biotechnical Services.

Safety Considerations

The fact that cell-culture-derived products are often injected into humans as therapeutic agents makes it imperative that there be no component in the final product that can pose a potential health risk to the patient.

Some of the cells used in manufacturing are continuous or "immortal." Many of these have been shown to be tumorigenic in immunosuppressed animals. The cells also contain endogenous materials such as retroviruses and nucleic acids (qv), both of which can induce tumorigenesis, as well as immunogenic foreign proteins. Serum used in media can also introduce adventitious agents such as viruses and mycoplasma into the product. Other process chemicals, including cleaning agents, are low molecular weight compounds that may be hazardous as well. Purification chemicals, such as monoclonals used for affinity purification, can be immunogenic to humans. Some of the potential health risks in mammalian cell culture processes and the methods used for risk reduction include the following:

cells	microfiltration
retroviruses	irradiation, sonication, heat, solvents, etc
nucleic acids	chromatography
cellular proteins	chromatography, ultrafiltration
bacterial contamination	microfiltration
process chemicals	diafiltration with appropriate buffers
serum proteins	affinity/ion-exchange chromatography

Most of these methods are commonly employed in the downstream processing of the desired cell culture technology product. Hence, most of the time it is only necessary to demonstrate that the designed process is reducing the putative risk factors to acceptable levels.

SUBHASH B. KARKARE
Amgen, Inc.

J. P. Mather, ed., *Mammalian Cell Culture: The Use of Serum-Free Hormone-Supplemented Media,* Plenum Press, New York, 1984.

A. S. Lubiniecki, ed., *Large-Scale Mammalian Cell Culture Technology,* Marcel Dekker, New York, 1990.

B. K. Lydersen, ed., *Large-Scale Cell Culture Technology,* Hanser, New York, 1987.

K. L. Nelson, *Biopharm.* **34** (Mar. 1988)

CELLS, ELECTRIC. See BATTERIES.

CELLULOSE

Cellulose is the main component of higher plant cell walls. It is also formed by certain algae, fungi, bacteria, and a group of invertebrate marine animals, the tunicates. It is even produced by humans suffering from the rare disease, scleroderma. About 7.5×10^{10} tons of cellulose grow and disappear each year, establishing it as the most abundant regenerated organic material on earth.

Natural cellulosic materials such as grass are eaten by grazing animals, and various species build nests or dens with wood. Cellulose in wood (qv) or in animal manure, along with lignin (qv), serves directly as fuel; scientists are also striving to develop efficient conversion of cellulose to alcohol and other fuels. After minimal processing, natural cellulosic materials are used as lumber and as textiles based on cotton (qv), jute, ramie, flax (linen), and hemp (see FIBERS, VEGETABLE). After industrial treatment, with and without chemical derivatization, cellulose is made into diverse products, including paper, membranes, explosives, textiles (rayon and cellulose acetate), and dietary fiber.

Native cellulose is fibrous and rather insoluble. Its fibers are relatively strong, having breaking strengths of up to 1 GN/m^2 (145,000 psi) and moduli of elasticity ranging from 70 to 137 GN/m^2 ($10-20 \times 10^6$ psi). The strength and insolubility result because cellulose is an extended, ribbon-like molecule that forms both hydrogen bonds and hydrophobic attractions. Because of the chemical and biochemical stability of cellulose and its combination with lignin, some trees live thousands of years; some bristlecone pines, *Pinus longaeva* (arista), in Colorado are more than 5000 years old.

In the secondary cell wall of plants, cellulose molecules are unbranched chains of up to 17,000 1,4-linked β-D-glucose residues, but shorter chains occur under other circumstances.

Because of the importance of cellulose and the difficulty in unraveling its secrets, several societies (Cellucon, American Chemical Society, and TAPPI) are dedicated to cellulose, lignin, and related molecules, as is at least one journal that is abstracted by *Chemical Abstracts, Cellulose Chemistry and Technology.*

Sources

Cellulose for industrial conversion comes from wood and scores of minor sources such as bagasse, the stalks of sugar cane from which the juice has been pressed. The importance of cellulose recycling is increasing, especially for paper products. On a more biological level, the majority of cellulose is in plant secondary cell walls, where it is usually the principal material. Some cellulose comes from the hairs on seeds, eg, cotton, kapok, and milkweed. The bast fibers such as hemp, ramie (a perennial Asian nettle), linen, and jute are obtained from the stems of plants.

A commercial bacterial cellulose product (Cellulon) has been introduced by Weyerhaeuser. The fiber is produced by an aerobic fermentation of glucose from corn syrup in an agitated fermentor. Because of a small particle diameter (10 μm), it has a surface area 300 times greater than normal wood cellulose, and gives a smooth mouthfeel to formulations in which it is included. Cellulon has an unusual level of water binding and works with other viscosity builders to improve their effectiveness. It is anticipated that it will achieve GRAS status, and is neutral in sensory quality; microcrystalline cellulose has similar attributes.

Only limited success has rewarded attempts to produce cellulose *in vitro.* Enzymes extracted from bacteria do not synthesize cellulose as efficiently as they do *in vivo,* even when the enzymes are activated.

Preparation

Most manufacturers of cellulosic products begin their work with cellulose in the form of pulp (qv). Pulping partially separates cellulose from the lignin (qv) and hemicelluloses (qv), leaving it in a fibrous form that is more susceptible to chemical treatment than the starting material. After pulping, the pulp is purified and otherwise treated to tailor it to the required specifications. Following drying, the pulp is shipped in large rolls.

Microcrystalline Cellulose

Pulverized forms of woodpulp have been widely used as fillers in some foods and pharmaceuticals. However, their utility is limited because the highly fibrous form results in poor mouthfeel. This problem can be overcome by reducing the woodpulp fibers to colloidal microcrystalline cellulose. This is made by hydrolysis with hydrochloric acid to the point of leveling off degree of polymerization. In aqueous suspensoids, these much finer particles have a smooth texture resembling uncolored butter and pseudoplastic properties including stable viscosity over a wide temperature range. It can therefore be used as a low calorie substitute for fat (see FAT REPLACERS). Microcrystalline celluloses are important for their heat stability, ability to thicken with favorable mouthfeel, and flow control. They extend starches, form sugar gels, stabilize foams, and control formation of ice crystals. A few of the foods in which microcrystalline cellulose has been commercially successful are fillings, meringue (cold process), chocolate cake sauce (frozen), cookie fillings, whipped toppings, and imitation ice cream for use as a bakery filling. In the pharmaceutical industry, microcrystalline cellulose is used mostly for tableting.

Over 250,000 metric tons of microcrystalline cellulose have been sold since its commercialization in 1962 and demand continues to increase.

Structure and Its Relation to Chemical and Physical Properties

The chemical and physical properties of cellulose depend in large measure on the spatial arrangements of the molecules. Therefore, cellulose structures have been studied intensively, and the resulting information has been important in helping to understand many other polymers. Despite the extent of work, however, there are still many controversies on the most important details. The source of the cellulose and its history of treatment both affect the structure at several levels. Much of the industrial processing to which cellulose is subjected is intended to alter the structure at various levels in order to obtain desired properties.

Crystal Structure Studies. Cellulose fibers can be studied by diffraction of x-ray, electron, or neutron beams. However, the small quantity of data from fiber diffraction (30–50 spots) is insufficient for complete structural determination. Consequently, structural studies by diffraction are augmented by molecular models. Other techniques, such as solid-state nuclear magnetic resonance and infrared spectroscopy, have furnished qualitative information regarding chain symmetry and hydrogen bonding. Investigations by electron microscopy rely increasingly on techniques such as diffraction contrast.

Unit cells of pure cellulose fall into five different classes, I–IV and x. This organization, with more recent subclasses, is used here, but Cellulose x is not discussed because there has been no recent work on it.

Cellulose I. The majority of celluloses in the native state were previously thought to have the same crystal structure (Cellulose I), varying only in perfection of the crystallites. But in the 1990s, at least two different crystal structures are known for these materials, named Iα and Iβ. These two phases coexist in many natural cellulosic materials. In addition, a few algae and bacteria make only Cellulose II, and some immature or primary wall cellulose has the Cellulose IV form.

Cellulose II. Cellulose II results from the crystallization of dissolved cellulose, with the best results obtained at temperatures near 0°C. Another way to make Cellulose II is by repeated mercerization, the swelling of cellulose in strong alkali, eg, 23% NaOH, followed by rinsing and drying.

Mercerized cellulose fibers have improved luster and do not shrink further. One of the main reasons for mercerizing textiles is to improve their receptivity to dyes. This improvement may result more from the disruption of the crystalline regions rather than the partial conversion to a new crystal structure.

Cellulose III. Cellulose III results from treatment of cellulose with liquid ammonia (ammonia mercerization) or amines. Cellulose III can be made from either Cellulose I or II.

Cellulose IV. Like Cellulose III, Cellulose IV preparations can revert to their parent Cellulose I or II structures. These forms can be prepared by treating Cellulose I, II, or III in glycerol at temperatures ca 260°C.

Chemical Reactivity

Cellulose is chemically like other carbohydrates (qv) that consist of pyranose rings bearing hydroxyl groups. These glucose residues include a reducing end unit, a nonreducing end unit, and intermediate units. Most celluloses have a high degree of polymerization; the intermediate glucose residues determine the chemical and physical properties and the end units may be ignored. The glycosidic bonds in cellulose are strong and this polymer is stable under a wide variety of reaction conditions. It is a generally insoluble, highly crystalline polymer.

Cellulose Solvents

Cellulose is soluble only in unusual and complex solvent systems. Commercially, dissolving pulps, which have lower molecular weights, are used along with strong alkali and derivatization. Cellulose subjected to high temperature and pressure during the steam explosion process can be dissolved in strong base.

Liquid Crystals

Many cellulose derivatives form liquid crystalline phases, both in solution (lyotropic mesophases) and in the melt (thermotropic mesophases). The field has grown rapidly and has been reviewed from different perspectives.

ALFRED D. FRENCH
NOELIE R. BERTONIERE
Southern Regional Research Center
O. A. BATTISTA
(Microcrystalline Cellulose)
Research Services Corporation
JOHN A. CUCULO
(Cellulose Solvents)
North Carolina State University
DEREK G. GRAY
(Liquid Crystals)
Pulp and Paper Research Institute of Canada

T. P. Nevell and S. H. Zeronian, eds., *Cellulose Chemistry and Its Applications,* Ellis Horwood, Chichester, U.K., 1985.

R. A. Young and R. M. Rowell, eds., *Cellulose: Structure, Modification and Hydrolysis,* Wiley-Interscience, New York, 1986.

J. F. Kennedy, G. O. Phillips, and P. A. Williams, eds., *Cellulosics: Chemical, Biochemical and Material Aspects,* Ellis Horwood, Chichester, U.K., 1993.

J. F. Kennedy, G. O. Phillips, and P. A. Williams, eds., *Cellulosics: Pulp, Fibre and Environmental Aspects,* Ellis Horwood, Chichester, U.K., 1993.

CELLULOSE ESTERS

ORGANIC ESTERS

Cellulose esters of almost any organic acid can be prepared, but because of practical limitations esters of acids containing more than four carbon atoms have not achieved commercial significance.

Table 1. Properties of Cellulose Triesters[a]

Cellulose ester	Shrinking point, °C	Mp[c], °C	Water tolerance value[d]	Moisture regain[b], %			Density, g/mL	Tensile strength, MPa[e]
				50% rh	75% rh	95% rh		
cellulose[f]				10.8	15.5	30.5	1.52	
acetate		306	54.4	2.0	3.8	7.8	1.28	71.6
propionate	229	234	26.9	0.5	1.5	2.4	1.23	48.0
butyrate	178	183	16.1	0.2	0.7	1.0	1.17	30.4
valerate	119	112	10.2	0.2	0.3	0.6	1.13	18.6
caproate	84	94	5.88	0.1	0.2	0.4	1.10	13.7
heptylate[g]	82	88	3.39	0.1	0.2	0.4	1.07	10.8
caprate	82	86	1.14	0.1	0.1	0.2	1.05	8.8
caprate[h]	87	88		0.1	0.2	0.5	1.02	6.9
laurate	89	91		0.1	0.1	0.3	1.00	5.9
myristate	87	106		0.1	0.1	0.2	0.99	5.9
palmitate	90	106		0.1	0.1	0.2	0.99	4.9

[a] Courtesy of the American Chemical Society.
[b] At 25% rh moisture regain for cellulose is 5.4%; for the acetate, 0.6%; for the propionate and butyrate, 0.1%; all others are zero.
[c] Char point is 315°C or higher unless otherwise noted.
[d] Milliliters of water required to start precipitation of the ester from 125 mL of an acetone solution of 0.1% concentration.
[e] To convert MPa to psi, multiply by 145.
[f] Starting cellulose, prepared by deacetylation of commercial, medium viscosity cellulose acetate (40.4% acetyl content).
[g] Char point = 290 °C
[h] Char point = 301 °C

Cellulose acetate is the most important organic ester because of its broad application in fibers and plastics; it is prepared in multiton quantities with degrees of substitution (DS) ranging from that of water-soluble monoacetates to those of fully substituted triacetate.

Although cellulose acetate remains the most widely used organic ester of cellulose, its usefulness is restricted by its moisture sensitivity, limited compatibility with other synthetic resins, and relatively high processing temperature. Cellulose esters of higher aliphatic acids, C_3 and C_4, circumvent these shortcomings with varying degrees of success. They can be prepared relatively easily with procedures similar to those used for cellulose acetate. Currently, as of ca 1994, mixed cellulose esters containing acetate and either the propionate or butyrate moieties are produced commercially by Eastman Chemical Company in the United States.

Cellulose esters of aromatic acids, aliphatic acids containing more than four carbon atoms, and aliphatic diacids are difficult and expensive to prepare because of the poor reactivity of the corresponding anhydrides with cellulose; little commercial interest has been shown in these esters. A notable exception, however, is the recent interest in the mixed esters of cellulose succinates, prepared by the sodium acetate-catalyzed reaction of cellulose with succinic anhydride. The additional expense incurred in manufacturing succinate esters is compensated by the improved film properties observed in waterborne coatings.

Mixed cellulose esters containing the dicarboxylate moiety, eg, cellulose acetate phthalate, have commercially useful properties such as alkaline solubility and excellent film-forming characteristics. These esters can be prepared by the reaction of hydrolyzed cellulose acetate with a dicarboxylic anhydride in a pyridine or, preferably, an acetic acid solvent with sodium acetate catalyst. Cellulose acetate phthalate for pharmaceutical and photographic uses is currently, as of ca 1994, produced commercially via the acetic acid–sodium acetate method.

Properties

The properties of cellulose esters are affected by the number of acyl groups per anhydroglucose unit, acyl chain length, and the degree of polymerization (DP) (molecular weight). The properties of some typical cellulose triesters are given in Table 1.

The common commercial products are the primary (triacetate) and the secondary (acetone-soluble, ca 39.5% acetyl, 2.45 DS) acetates; they are odorless, tasteless, and nontoxic. Their properties depend on the combined acetic acid content and molecular weight. Solubility characteristics of cellulose acetates with various acetyl contents are given in Table 2.

In fibers, plastics, and films prepared from cellulose esters, mechanical properties such as tensile strength, impact strength, elongation, and flexural strength are greatly affected by the degree of polymerization and the degree of substitution. Mechanical properties significantly improve as the DP is increased from ca 100 to 250 repeat units.

Stabilization. After hydrolysis, precipitation, and thorough washing of the cellulose esters to remove residual acids, the esters must be stabilized against thermal degradation and color development, which may occur during processing, such as extrusion or injection molding. Thermal instability is caused by the presence of oxidizable substances and small amounts of free and combined sulfuric acid. The sulfuric acid combines with the cellulose almost quantitatively and most of it is removed during the latter stages of hydrolysis. The remaining sulfuric acid can be neutralized with alkali metal salts, such as sodium, calcium, or magnesium acetate, to improve ester stability. The combined sulfate ester may also be removed by treatment in boiling water or at steam temperatures in an autoclave. Treatment with aqueous potassium or calcium iodide reportedly stabilizes the cellulose acetate against thermal degradation.

Cellulose Acetate. Almost all cellulose acetate, with the exception of fibrous triacetate, is prepared by a solution process employing sulfuric acid as the catalyst with acetic anhydride in an acetic acid solvent. The acetylation reaction is heterogeneous and topochemical wherein successive layers of the cellulose fibers react and are solubilized in the medium, thus exposing new surfaces for reaction.

Recent Developments. A considerable amount of cellulose acetate is manufactured by the batch process. In order to reduce produc-

Table 2. Solubility Characteristics of Cellulose Acetates

Acetyl, %	Soluble in	Insoluble in
43.0–44.8	dichloromethane	acetone
37–42	acetone	dichloromethane
24–32	2-methoxymethanol	acetone
15–20	water	2-methoxymethanol
≤13	none of the above	all of the above

tion costs, efforts have been made to develop a continuous process that includes continuous activation, acetylation, hydrolysis, and precipitation.

Demand for cellulose acetate flake in the United States was projected to decline slightly from 1988 to 1993. Cigarette-filter tow for export is the only market projected to grow. Cellulose acetate for textile fibers is expected to decline, as will flake demand for plastics, with the growth of photographic films somewhat offsetting declining markets in other plastics end uses.

Health and Safety Factors

The vapors of the organic solvents used in the preparation of cellulose ester solutions represent a potential fire, explosion, or health hazard.

Cellulose esters are considered nontoxic and may be used in food-contact applications. However, since cellulose esters normally are not used alone, formulators of coatings and films for use in food packaging should ensure that all ingredients in their formulations are cleared by the United States Food and Drug Administration for such use.

Uses

The cellulose esters with the largest commercial consumption are cellulose acetate, including cellulose triacetate, cellulose acetate butyrate, and cellulose acetate propionate. Cellulose acetate is used in textile fibers, plastics, film, sheeting, and lacquers. The cellulose acetate used for photographic film base is almost exclusively triacetate; some triacetate is also used for textile fibers because of its crystalline and heat-setting characteristics.

Cellulose esters, especially acetate propionate and acetate butyrate mixed esters, have found limited use in a wide variety of specialty applications, such as in nonfogging optical sheeting, low profile additives to improve the surface characteristics of sheet-molding (SMC) compounds and bulk-molding (BMC) compounds, and controlled drug release via encapsulation.

STEVEN GEDON
RICHARD FENGL
Eastman Chemical Company

C. J. Malm and co-workers, *Ind. Eng. Chem.* **5,** 81 (1966).

G. D. Hiatt and W. J. Rebel, in N. M. Bikales and L. Segal, eds., *Cellulose and Cellulose Derivatives,* Part V of *High Polymers,* 2nd ed., Vol. 5, Wiley-Interscience, New York, p. 749.

C. J. Malm and L. J. Tanghe, *Ind. Eng. Chem.* **47,** 995 (1955).

C. J. Malm, L. J. Tanghe, and J. T. Schmitt, *Ind. Eng. Chem.* **53,** 363 (1961).

INORGANIC ESTERS

Cellulose esters are cellulose derivatives that result from the esterification of the free hydroxyl groups of cellulose with one or more acids. The esterification can be achieved using mineral acids as well as organic acids or their anhydrides. Cellulose nitrate (CN) is the oldest cellulose derivative and the only inorganic ester of commercial importance. Currently, lacquer finishes constitute the largest market for CN; explosives and propellants are the second-largest market.

Other inorganic esters of cellulose have been frequently described and investigated. They include sulfur-containing and phosphorus-containing cellulose esters.

Cellulose Nitrate

CN is a white, odorless, and tasteless substance. It is manufactured by treating cellulose with nitric acid in the presence of sulfuric acid and water. The amount of water determines the DS attained.

Manufacture. *Raw Material.* Prior to World War II, CN was produced mainly from cotton linters because of their higher degree of

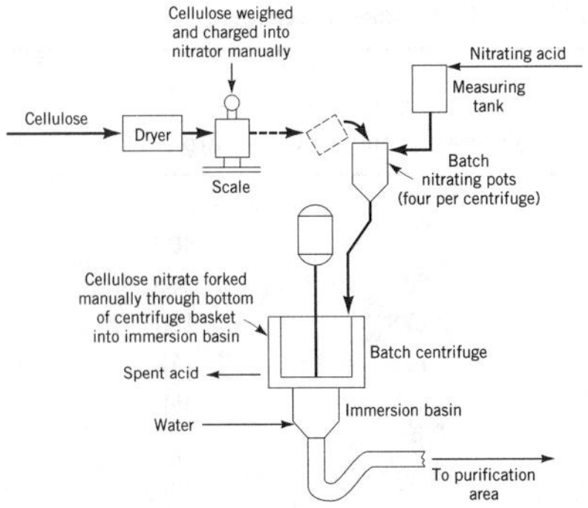

Figure 1. Flow diagram of nitration by batch process.

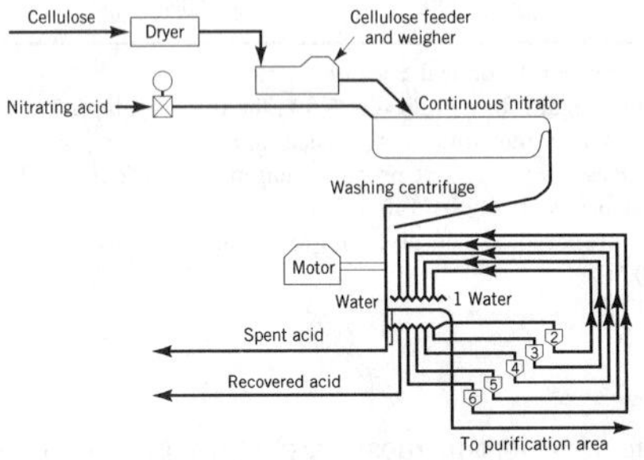

Figure 2. Flow diagram of nitration of cellulose by Hercules continuous process.

purity (alpha cellulose > 98%). The development of highly purified chemical-grade wood pulps has allowed this material to be used in the same manner as are linters.

Manufacturing Methods. Industrial-scale nitration is still frequently carried out using a batch process (Fig. 1)

Continuous nitration processes, which were developed in the 1960s, are more economical and provide a more uniform product (Fig. 2).

RICHARD FENGL
Tennessee Eastman

N. M. Bikales, ed., *Encyclopedia of Polymer Science and Technology,* Vol. 3, Interscience Publishers, a Division of John Wiley & Sons, Inc., New York, 1964.

N. M. Bikales and L. Segal, eds., *High Polymers,* 2nd ed., Vol. 5, Pt. 5, Wiley-Interscience, New York, 1971.

J. F. Kennedy, G. O. Phillips, D. J. Wedlock, and P. A. Williams, eds., *Cellulose and Its Derivatives,* Ellis Horwood Ltd., Chichester, UK, 1985; T. P. Nevell, S. H. Zeronian, eds., *Cellulose Chemistry and Its Applications,* Ellis Horwood, Chichester, UK, 1985.

K. Balser, L. Hoppe, and W. Walsrode, in W. Gerhartz and Y. Yamamoto, eds., *Ullman's Encyclopedia of Industrial Chemistry,* Vol. A5, VCH, Weinheim-New York, 1986, p. 436;

CELLULOSE ETHERS

General Considerations

Alkylation of cellulose yields a class of polymers generally termed cellulose ethers. Most of the commercially important ethers are water-soluble and are key adjuvants in many water-based formulations. The most important property these polymers provide to formulations is rheology control, ie, thickening and modulation of flow behavior. Other useful properties include water-binding (absorbency, retention), colloid and suspension stabilization, film formation, lubrication, and gelation. As a result of having these properties, cellulose ethers have permeated a broad range of industries, including foods, coatings, oil recovery, cosmetics, personal care products, pharmaceuticals, adhesives, printing, ceramics, textiles, building materials, paper, and agriculture.

Cellulose ethers represent a mature industry, with annual sales of over one billion dollars and an annual growth rate averaging 2–3%. The highest-volume cellulose ethers, the industry workhorses, are sodium carboxymethylcellulose, hydroxyethylcellulose, and hydroxypropylmethylcellulose. Cellulose ethers as a class compete with a host of other materials, including natural gums, starches, proteins (qv), synthetic polymers, and even inorganic clays. They provide effective performance at reasonable cost and are derived from a renewable natural resource.

Cellulose ethers are manufactured by reaction of purified cellulose with alkylating reagents under heterogeneous conditions, usually in the presence of a base, typically sodium hydroxide, and an inert diluent. Cottonseed linter fiber and wood fiber are the principal sources of cellulose.

Many cellulose ethers contain mixed substituents (cellulose mixed ethers) in order to enhance or modify the properties of the monosubstituted derivative.

Health and Safety Factors. No adverse toxicological or environmental factors are reported for cellulose ethers in general. Some are even approved as direct food additives, including purified carboxymethylcellulose, methylcellulose, hydroxypropylmethylcellulose, and hydroxypropylcellulose.

The only known hazard associated with cellulose ethers is that they may form flammable dusts when finely divided and suspended in air, a hazard associated with most organic substances.

Commercial Cellulose Ethers

SODIUM CARBOXYMETHYLCELLULOSE

Properties. Sodium carboxymethylcellulose (CMC), also known as cellulose gum, is an anionic, water-soluble cellulose ether, available in a wide range of substitution. The most widely used types are in the 0.7–1.2-DS range. DS is the degree of substitution. It refers to the number of hydroxyl groups substituted per anhydroglucose residue.

Some typical properties of commercial CMCs are given in Table 1.

Manufacture. Common to all manufacturing processes for CMC is the reaction of sodium chloroacetate with alkali cellulose complex.

Economic Aspects. CMC is the most widely used cellulose ether.

In the United States, Aqualon Company, a Hercules Incorporated Company, is the largest producer, followed by Carbose Corporation and MAK Chemical Corporation.

Uses. CMC is an extremely versatile polymer, and it has a variety of applications. A sampling of significant applications is given in Table 2.

Table 1. Typical Properties of Purified CMC

Property	Value
Powder	
appearance	white to off-white
moisture, max %	8.0
charring temp, °C	252
molecular weight, M_w	$9.0 \times 10^4 - 7.0 \times 10^5$
Solution	
viscosity, Brookfield, 30 rpm, mPa·s(= cP)	
at 1% solids (high M_w)	~6000
at 4% solids (low M_w)	~50
sp gr, 2% at 25°C	1.0068
pH, 2%	7.5
surface tension, 1%, mN/m(= dyn/cm)	71

Table 2. Applications for CMC

Industry	Application	Function
foods	frozen desserts	inhibit ice crystal growth
	dessert toppings	thickener
	beverages, syrups	thickener, mouthfeel
	baked goods	water-binder, batter viscosifier
	pet food	water-binder, thickener, extrusion aid
pharmaceuticals	tablets	binder, granulation aid
	bulk laxatives	water-binder
	ointments, lotions	stabilizer, thickener, film-former
cosmetics	toothpaste	thickener, suspension aid
	denture adhesives	adhesion promoter
	gelled products	gellant, film-former
paper products	internal additive	binder, improve dry-strength
	coatings, sizes	water-binder, thickener
adhesives	wallpaper paste	adhesion promoter, water-binder
	corrugating	thickener, water-binder, suspension aid
	tobacco	binder, film-former
lithography	fountain, gumming	hydrophilic protective film
ceramics	glazes, slips	binder (promotes green strength)
	welding rods	binder, thickener, lubricant
detergents	laundry	soil antiredeposition aid
textiles	warp sizing	film-former, adhesion promoter
	printing paste, dye	thickener, water-binder

HYDROXYETHYLCELLULOSE

Properties. Hydroxyethylcellulose (HEC), is a nonionic polymer. Low hydroxyethyl substitutions (MS = 0.05–0.5) yield products that are soluble only in aqueous alkali. Higher substitutions (MS $\geqslant$ 1.5) produce water-soluble HEC. The bulk of commercial HEC falls into the latter category. Water-soluble HEC is widely used because of its broad compatibility with cations and the lack of a solution gel or precipitation point in water up to the boiling point. MS is the molar substitution. It refers to the moles of reagent combined per anhydroglucose residue.

Some typical properties of HEC are given in Table 3.

Table 3. Typical Properties of HEC

Property	Value
Powder	
appearance	white to light tan
ash content (as Na_2SO_4), %	5.5
molecular weight, M_w	$9 \times 10^4 - 1.3 \times 10^6$
Solution	
viscosity, Brookfield, 30 rpm, mPa·s(= cP)	
at 1% solids (high M_w)	5000
at 5% solids (low M_w)	75
sp gr, 2%, g/cm^3	1.0033
pH	7
surface tension, mN/m(= dyn/cm)	
MS 2.5 at 0.1%	66.8
at 0.001%	67.3
refractive index, 2%	1.336

Table 4. Typical Properties of Mixed Ether Derivatives of HEC

Property	CMHEC	Cationic HEC	EHEC	HMHEC
		Powder		
appearance	off-white	light yellow	off-white	off-white
ash content (as Na_2SO_4), %		3	3 (as NaCl)	10 max
		Solution		
pH	6.5–10	7	6–7	6–8.5
flocculation temp in water, °C			~65	
surface tension, mN/m(= dyn/cm)			55	~62

Table 5. Typical Properties of Methylcellulose Ethers

Property	MC	HPMC	HEMC	HBMC
		Powder		
appearance	white	white	white	white
bulk density, g/cm^3		0.25–0.70		
ash content (as Na_2SO_4), %		2.5 max		
		Solution		
viscosity,[a] mPa·s(= cP)	10–15,000	5–70,000	100–70,000	
sp gr, 2% at 20°C		1.0032		
pH, 1%		5.5–9.5		
surface tension, 0.1%, mN/m(= dyn/cm)	47–53	44–56	46–53	49–55
interfacial tension[b], mN/m(= dyn/cm)	19–23	17–30	17–21	20–22

[a] 2% Solution, Brookfield, 20 rpm.
[b] Against paraffin oil.

Manufacture. Purified hydroxyethylcellulose is manufactured in diluent-mediated processes similar to those used to produce carboxymethylcellulose, except that ethylene oxide is used in place of MCA.

Economic Aspects. Aqualon Company and Union Carbide Corporation have manufacturing facilities in the United States and in Western Europe.

Uses. HEC is used as a thickener, protective colloid, binder, stabilizer, and suspending agent in a variety of industrial applications.

Mixed Ether Derivatives of HEC. Several chemical modifications of HEC are commercially available. The secondary substituent is generally of low DS (or MS), and its function is to impart a desirable property lacking in HEC. Derivatives include carboxymethylhydroxyethylcellulose (CMHEC), cationic hydroxyethylcelluloses, and hydrophobic hydroxyethylcelluloses (Table 4).

METHYLCELLULOSE AND ITS MIXED ETHERS

Properties. Methylcellulose (MC) and its alkylene oxide derivatives hydroxypropylmethylcellulose (HPMC), hydroxyethylmethylcellulose (HEMC), and hydroxybutylmethylcellulose (HBMC) are nonionic, surface-active, water-soluble polymers. Each type of derivative is available in a range of methyl and hydroxyalkyl substitutions. The extent and uniformity of the methyl substitution and the specific type of hydroxyalkyl substituent affect the solubility, surface activity, thermal gelation, and other properties of the polymers in solution.

Typical properties of MC, HPMC, HEMC, and HBMC are given in Table 5.

Manufacture. Methylcellulose is manufactured by the reaction of alkali cellulose with methyl chloride.

Economic Aspects. The Dow Chemical Company is the only U.S. manufacturer.

Uses. There are numerous applications for methylcellulose and its derivatives. Some important ones are summarized in Table 6.

ETHYLCELLULOSE AND HYDROXYETHYLETHYLCELLULOSE

Properties. Ethyl cellulose (EC) is a nonionic, organo-soluble, thermoplastic cellulose ether, having an ethyl DS in the range of ~2.2–2.7.

Table 6. Applications for Methylcellulose and its Derivatives

Industry	Application	Function
construction	cements, mortars	thickener, water-binder, workability
foods	mayonnaise, dressing	stabilizer, emulsifier
	desserts	thickener
pharmaceuticals	tablets	binder, granulation aid
	formulations	stabilizer, emulsifier
adhesives	wallpaper paste	adhesive
ceramics	slip casts	binder (promotes green strength)
coatings	latex paints	thickener
	paint removers	thickener
cosmetics	creams, lotions	stabilizer, thickener

Organo-soluble hydroxyethylethylcellulose (HEEC) is highly ethoxylated with small amounts of hydroxyethyl substitution. It is used in coating applications that require solubility in fast-drying aliphatic hydrocarbons. Table 7 gives typical properties for EC and HEEC.

Manufacture. Ethyl chloride undergoes reaction with alkali cellulose in high pressure nickel-clad autoclaves. A large excess of sodium hydroxide and ethyl chloride and high reaction temperatures (up to 140°C) are needed to drive the reaction to the desired high DS values (≥2.0).

Economic Aspects. The Dow Chemical Company and Aqualon Company are the only listed principal producers of EC and HEEC products worldwide.

Uses. A summary of the applications for ethylcellulose is given in Table 8.

HYDROXYPROPYLCELLULOSE

Properties. Hydroxypropylcellulose (HPC) is a thermoplastic, nonionic cellulose ether that is soluble in water and in many organic solvents. HPC combines organic solvent solubility, thermoplasticity, and

Table 7. Typical Properties of EC and HEEC

Property	EC	HEEC
	Powder	
appearance	white	white
bulk density, g/cm^3	0.3–0.35	0.3–0.35
softening point, °C	152–162	
	Film	
specific gravity	1.140	1.120
refractive index	1.470	1.47
tensile strength, MPa[a]	46–72	34–41
elongation, %	7–30	6–10
flexibility[b]	160–2000	500–900

[a] To convert mPa to psi, multiply by 145.
[b] MIT double folds.

Table 8. Applications for EC and HEEC

Industry	Application	Function
coatings	lacquers, varnishes	protective film-former, additive to increase film toughness and durability, shorten drying time
printing	inks	film-former
adhesives	hot melts	additive to increase toughness

Table 9. Typical Properties of HPC

Property	Value
Powder	
appearance	off-white
ash content (as Na_2SO_4), %	0.2–0.5
softening point, °C	100–150
molecular weight, M_w	$8.0 \times 10^4 - 1.15 \times 10^6$
Solution	
viscosity, Brookfield, 30 rpm, mPa·s(= cP)	
at 1% (high M_w)	2500
at 10% (low M_w)	100
surface tension, 0.1%, mN/m(= dyn/cm)	43.6
interfacial tension[a], 0.1%, mN/m	12.5
Film	
tensile strength, MPa[b]	14
elongation, %	50
flexibility[c] (50 μm film)	10,000
refractive index	1.559

[a] Against mineral oil.
[b] To convert MPa to psi, multiply by 145.
[c] MIT double folds.

Table 10. Applications for HPC

Industry	Application	Function
polymerization	PVC suspension polymerization	protective colloid
pharmaceutical	tablets	binder, film-former
coatings	paint remover	thickener
foods	whipped toppings	stabilizer
	processed foods	extrusion aid
ceramics	slip casts	binder (promotes green strength)

surface activity with the aqueous thickening and stabilizing properties characteristic of other water-soluble cellulosic polymers described herein.

Some typical properties of commercial HPC are given in Table 9.

Manufacture. HPC is manufactured by reaction of propylene oxide with alkali cellulose.

Economic Aspects. The Aqualon Company is the only U.S. manufacturer.

Uses. A summary of significant uses for HPC is given in Table 10.

THOMAS G. MAJEWICZ
THOMAS J. PODLAS
Aqualon Company

E. K. Just and T. G. Majewicz, in J. I. Kroschwitz, ed., *Encyclopedia of Polymer Science and Engineering,* 2nd ed., Vol. 3, John Wiley & Sons, Inc., New York, 1985, pp. 224–269.

E. Ott, M. Spurlin, and M. W. Graffin, ed., *Cellulose and Cellulose Derivatives, High Polymers,* Vol. V, Wiley-Interscience, New York, 1954–1955, Parts I–III.

N. M. Bikales and L. Segal, eds., *Cellulose and Cellulose Derivatives, High Polymers,* Vol. V, Wiley-Interscience, New York, 1971, Pts. IV–V.

R. L. Davidson, ed., *Handbook of Water-Soluble Gums and Resins,* McGraw-Hill Book Co., Inc., New York, 1980.

CEMENT

The term "cement" is used to designate many different kinds of substances that are used as binders or adhesives. These can be organic or inorganic materials. The cement produced in the greatest volume and the most widely used in concrete for construction is Portland cement. Portland cement is also found in many specialty cements, such as masonry cement, oil well cement, and expansive cement that are produced for special purposes. Inorganic cements, such as calcium aluminate cement and magnesium phosphate cement not containing Portland cement, are used for special purposes. These cements are distinctly different from polymeric organic cements, such as epoxies. Portland cement is a hydraulic cement. Hydraulic cements set, harden, and gain strength in the presence of water and do not degrade or lose strength in the presence of moisture. Therefore, it is used for construction in underground, marine, and water structures, as well as nonwater applications. They are used where nonhydraulic cements, such as gypsum or lime, cannot be used in water saturated environments. Organic cements, such as epoxys, latexs, and water soluble polymers, do not use water to set and gain strength. The term cement used here is confined to inorganic hydraulic cements, principally Portland and related cements.

Uses

Portland cement is primarily used to make Portland cement concrete. The cement is mixed with water, sand, and coarse aggregate. This combination of materials makes a stone-like mass that is used for construction of buildings, bridges, pavements, sidewalks, dams, housing, and other structures. Portland cement is used in ready-mixed concrete, precast concrete, concrete block and brick, mortar, soil stabilization (soil cement), waste stabilization, and a variety of other applications. Portland cement can be used to make concrete exposed to any natural environment as well as many kinds of severe industrial environments. However, Portland cement products should not be exposed to certain severe chemical conditions such as acid environments or very high temperatures.

Chemistry

Portland cement is a hydraulic cement composed primarily of hydraulic calcium silicates. The primary compounds are as follows: tricalcium silicate, dicalcium silicate, tricalcium aluminate, and tetracalcium aluminoferrite. In the presence of water, the four compounds hydrate to form new ones that are the infrastructure of hardened cement paste in concrete. The calcium silicates, which constitute about 75% of the mass of cement, hydrate to form the compounds calcium hydroxide and calcium silicate hydrate. Hydrated cement contains about 25% calcium hydroxide and 50% hydrated calcium silicate. The strength and other properties of hydrated cement are primarily due to the calcium silicate hydrate. Hydrating tricalcium silicate is primarily responsible for the initial set and early strength of portland cement concrete. Dicalcium silicate hydrates slowly and contributes largely to the strength increase at ages beyond one week. Tricalcium aluminate liberates a large amount of heat during the first few days of hydration and hardening and contributes slightly to early strength development. Gypsum, which is added to the cement in the grinding process, slows down the rapid hydration rate of tricalcium aluminate. Tetracalcium aluminoferrite reduces the clinkering (burning) temperature and assists in the manufacture of cement. It hydrates rapidly but contributes very little to strength development.

As the cement compounds hydrate, Portland cement paste or concrete sets and becomes hard. Hydration continues as long as space for hydration products is available and moisture and temperature conditions are favorable. Most of the hydration and strength development takes place within the first month of concrete's life cycle, but they continue, though much more slowly, for a long time. Strength increases over a 50-year period have been recorded.

Portland cement is generally used at temperatures ordinarily encountered in construction (0–35°C). Temperature extremes should be avoided. The exothermic hydration reaction can play an important part in maintaining adequate temperatures during placement in cold environments; however, the heat generation must be carefully controlled in mass concrete structures in order to prevent excessive temperature rise and cracking due to subsequent cooling.

The initial conditions for the hydration reactions are determined by the concentration of the cement particles in the mixing water (water–cement ratio of 0.2 to 0.7 by mass) and the fineness of the cement (300–500 sq meters per kilogram). Upon mixing with water, the suspension of particles as shown in Figure 1 are surrounded by films of

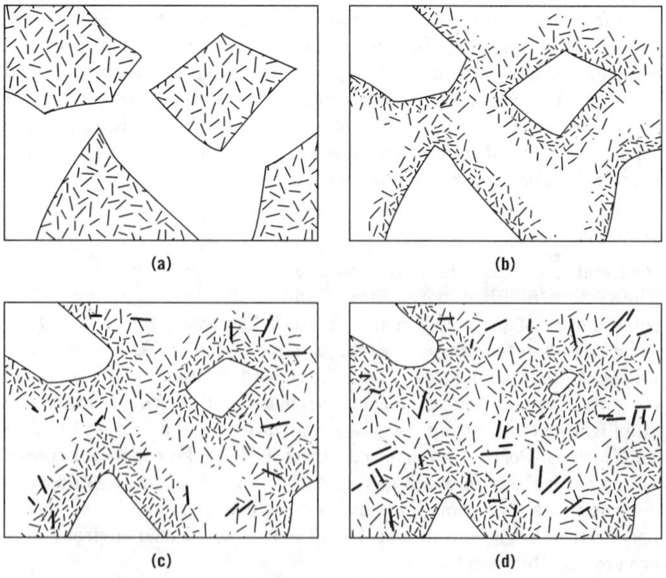

Figure 1. Four stages in the setting and hardening of Portland cement: simplified representation of the sequence of changes. (**a**) Dispersion of unreacted clinker grains in water. (**b**) After a few minutes; hydration products eat into and grow out from the surface of each grain. (**c**) After a few hours; the coatings of different clinker grains have begun to join up, the gel thus becoming continuous (setting). (**d**) After a few days; further development of the gel has occurred (hardening).

water. The anhydride phases initially react by the formation of surface hydration products on each grain, and by dissolution in the liquid phase. The solution quickly becomes saturated with calcium and sulfate ions and the concentration of alkali cations increase rapidly. These reactions consume part of the unhydrated grains, but the reaction products tend to fill that space as well as some of the ordinary water-filled space. The hydration products at this stage are mostly colloidal (less than 0.1 micrometers) but some larger crystals of calcium aluminate hydrates, sulfoaluminate hydrate, and hydrogarnets form. As the reactions proceed, the coatings increase in thickness and eventually form bridges between the original grains. This is the stage of setting. Despite the low solubility and mobility of the silicate ions, growths of the silicate hydrates form in crystals and become incorporated into the calcium compound phases. Upon further hydration, the water filled spaces become increasingly filled with reaction products to produce hardening and strength development.

Manufacture

Portland cement manufacture consists of (*1*) quarrying and crushing of rock, (*2*) grinding the carefully proportioned materials, (*3*) subjecting the raw mix to pyroprocessing at about 1480°C in a rotary kiln, and (*4*) grinding the resulting clinker and a small amount of added gypsum to a fine powder. This fine powder which is smaller than 75 micrometer in size with an average particle size of about 20 micrometer is called Portland cement. Raw materials for making Portland cement must contain appropriate amounts of calcium oxide, silica, alumina and iron components. Raw materials usually consist of limestone, cement rock, clay, iron ore, and other natural raw materials. Figure 2 gives the layout of a dry process plant. Modern installations are equipped with innovations such as preheaters, precalciners and roller mills.

Industrial by-products are becoming more widely used as raw materials for cement. Common examples would be ground granulated blast-furnace slag or coal fly ash. Many industrial by-products are potential raw materials for Portland cement manufacture.

Clinker production requires large quantities of fuel. In the United States, coal and natural gas are the most widely used kiln fuels, but fuels derived from waste materials, such as tires and waste oil, are increasing in importance.

Portland Cement Types

Different types of Portland cement are manufactured to meet various physical and chemical requirements for specific purposes. The American Society for Testing and Materials Specification C150, *Standard Specification for Portland Cement,* provides for the primary types of Portland cement: Type I—Normal; Type II—Moderate sulfate resistant; Type III—High early strength; Type IV—Low heat of hydration, and Type V—High sulfate resistant. Types I, II and III can also be designated as air-entraining cements to provide additional frost resistance for concretes exposed to freezing temperatures.

Type I Portland cement is a general-purpose cement suitable for all uses where special properties of other types are not required. It is commonly used in concrete to make pavements, floors, concrete buildings, bridges, tanks, hydraulic structures, masonry units and precast concrete products.

Type II Portland cement is used for precaution against moderate sulfate attack (when the sulfate content of soil or water in contact with concrete is 0.10–0.20% or 150–1500 ppm, respectively). Type II can also be required to have a moderate heat of hydration to control heat development in massive concrete placements. Cement Types I and II are the most widely available cement types in the United States.

Type III Portland cement provides high strengths at an early period, usually a week or less. It is chemically similar to Type I cement except that the particles have been ground finer to provide a much quicker chemical reaction. It is used when concrete must be put into service as soon as possible or when cold weather requires a shorter curing period.

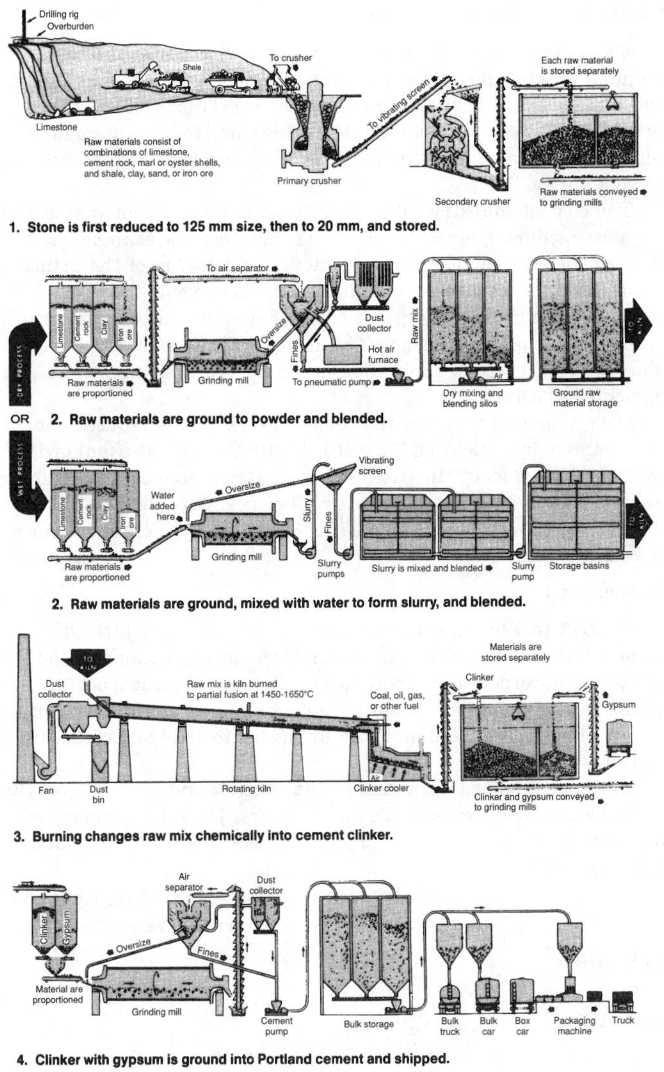

Figure 2. Steps in the manufacture of Portland cement. Courtesy of the Portland Cement Association.

Type IV Portland cement is used where the rate and amount of heat generated from hydration must be minimized. It develops strength slower than other types of portland cement and is intended for massive structures such as large dams where temperature rise from heat generation must be minimized to prevent cracking. Type IV cement is rarely available and is made only for particular applications.

Type V Portland cement is used only in concrete exposed to severe sulfate conditions, principally where soils or ground waters have a high sulfate content (higher than 1500 ppm sulfate in water or 0.20% sulfate in soil).

Blended Hydraulic Cements

Blended hydraulic cements are produced by intimately and uniformly blending two or more types of fine cementitious materials. The primary materials in blended cements are Portland cement, ground granulated blast-furnace slag, fly ash, silica fume, and natural pozzolans. Blended hydraulic cements must conform to the requirements of ASTM C595, *Standard Specification for Blended Hydraulic Cement,* or C1157, *Performance Specification for Blended Hydraulic Cement.* ASTM C595 recognizes five classes of blended cement as follows: Type IS—Portland blast-furnace slag cement; Types IP and P—Portland-pozzolan cements; Type I(PM)—pozzolan-modified-Portland cement; Type S—slag cement; and Type I(SM)—slag-modified-Portland cement. Most of these types can also be designated with air entrainment, moderate or low heat of hydration, and moderate sulfate resistance. Each of these materials has a designated amount or range of slag or pozzolan.

ASTM C1157 recognizes the following types of blended cements: Type GU—cement for general construction; Type HE—high-early strength cement; Type MS—moderate sulfate-resistant cement; Type HS—high sulfate resistant cement; Type MH—moderate heat of hydration cement; and Type LH—low heat of hydration cement. These cements can also be designated with an R suffix designation for use with alkali reactive siliceous aggregates. ASTM C1157 cements do not have restrictions on the particular source of the material ingredients and therefore allow a greater ability to optimize materials to maximize cement and concrete properties.

Blended hydraulic cements are available in most of the United States. Types IP and IS are the most common blended cements. Blended cements often contain by-products from other industries and require less energy to produce than Portland cement. Blended cements provide concrete with engineering and durability properties equal to or greater than that of concrete made with Portland cement. Blended cements can be especially helpful to control alkali–silica reactivity, an expansive condition resulting from a reaction between alkalies in the concrete and certain siliceous aggregates.

Masonry Cements

Masonry cements are hydraulic cements designed for use in mortar for masonry construction. They are composed of one or more of the following: Portland cement, Portland–pozzolan cement, Portland blast-furnace slag cement, slag cement, hydraulic lime, and natural cement. In addition, they usually contain materials such as hydrated lime, limestone, chalk, talc, slag, or clay. Materials are selected for their ability to impart workability, plasticity, and water retention to masonry mortars. Masonry cements meet the requirements of ASTM C91, *Specification for Masonry Cements,* which classifies masonry cements as Type N, Type S, and Type M.

Type N masonry cement is used in ASTM C270, *Specification for Mortar for Unit Masonry,* Type N and Type O mortars. It may also be used with Portland or blended cement to produce Type S and Type M mortars. Type S masonry cement is used in ASTM C270 Type M mortar without the addition of other cements or lime.

Mortar cement, also used to make mortar, is similar to masonry cement except that it has additional bond strength requirements.

Expansive Cements

Expansive cement is hydraulic cement that expands slightly during the early hardening period after setting. It must meet the requirements of ASTM C845, *Standard Specification for Expansive Hydraulic Cement,* in which it is designated as Type E-1. Three varieties of expansive cement are recognized and have been designated as K, M and S. Presently only Type E-1(K) is available in the United States. Type E-1 (K) contains portland cement, anhydrous calcium aluminosulfate, calcium sulfate, and uncombined calcium oxide. The volume increase is due principally to the reaction of the calcium aluminates and calcium sulfates. Expansive cement is used to make shrinkage compensating concrete which can be used to compensate for volume decrease during drying shrinkage, induced tensile stress in reinforcement; and to stabilize the long-term dimensions of post-tensioned concrete structures. One of the main advantages of using expansive cement in concrete is to control and reduce drying shrinkage cracks. Shrinkage compensating concrete is used primarily in floors and bridge decks.

White Portland Cement

White Portland cement is a true Portland cement that differs from gray cement chiefly in color. It is made to conform to ASTM C150 specifications, but the manufacturing process is controlled so that the finished product is white. White cement contains a minimal amount of iron and other dark colored metal oxides. It is primarily used for architectural purposes, such as wall panels, terrazzo, stucco, and decorative concrete. White cement is often combined with pigments to make colored mortar, plaster, or concrete.

Regulated Set Cement

Regulated set cement is a hydraulic cement that can be formulated and controlled to produce concrete with setting times ranging from a few minutes to an hour and with corresponding early-strength development of a 1000 psi within an hour after setting. Its chemistry is similar to that of Portland cement with the addition of calcium fluoroaluminate. It is used to make concretes for rapid repair applications.

Alkali-Activated Alumino–Silicate Cement

Alkali-activated alumino-silicate cement (sometimes called geopolymer cement), is a cement that uses a combination of hydration and alkali activation of alumino silicates to make a cement with high-early-strength properties. These cements can be made from natural or by-product materials and may or may not contain Portland cement. They can be used for rapid-repair applications, such as airport runway or bridge deck repair.

Oil-Well Cements

Oil-well cements, used for sealing oil wells, are usually made from Portland cement and blended hydraulic cement. Generally they must be slow setting and resist high temperatures and pressures. The American Petroleum Institute *Specifications for Materials and Testing for Well Cements,* API Specification 10, includes requirements for eight classes of well cements (Classes A thru H). Each class is applicable for use at a certain range of well depth, temperatures, pressures, and sulfate environments.

Plastic Cements

Plastic cements are inorganic, hydraulic cements made by adding plasticizing agents, up to 12% by total volume, to Type I or Type II Portland cement during the milling operation. Plastic cements are used for making plaster and stucco.

Cements with Functional Additions

Special cements with particular properties can be made by combining Portland or blended cements with functional additions (water reducing, retarding, and accelerating additions). Functional additions, to be interground with clinker, must meet the requirements of ASTM C688, *Standard Specification for Functional Additions for Use in Hydraulic Cements.*

Calcium Aluminate Cement

Calcium aluminate cements are manufactured by heating until molten or by sintering a mixture of limestone and bauxite with small amounts of silica, iron and titanium materials. They do not contain Portland cement. These cements are commonly used in special applications for resistance to certain chemicals or where increased heat resistance is desired. They can also achieve rapid early strength gain. Calcium aluminate cements can be used at low temperatures, down to −10°C. When combined with Portland cement, they can make a mixture with rapid setting and very early strength properties. Care must be taken to properly use this material and avoid conditions where the material undergoes a significant chemical conversion and drastically loses strength. For this reason, it is usually not used in structural applications.

Magnesium Phosphate Cement

Magnesium phosphate cements do not contain Portland cement and do not have a standard. They characteristically have an early strength gain, and therefore, are commonly used for repair and patching of regular concrete.

Environmental and Energy Issues

Portland cement contains the same major and minor elements as are found in the clay, limestone, coal, and other natural materials used as the raw materials or fuel in its manufacture. Trace metals, although present in some amount, occur at levels too small to be meaningful. Extensive studies analyzing the total and leachable elements in Portland cement are available.

The cement industry takes great care to meet all environmental regulations during manufacture. Dust collectors, for example, have a 99.8% efficiency in collecting cement kiln dust. One of the primary waste products of cement manufacturing is cement kiln dust. It is collected from the exhaust gases and either returned to the kiln with other raw materials or disposed of as land fill. Cement kiln dust can also be used in other applications such as synthetic aggregate and agricultural uses.

Over a span of 21 years the cement industry has reduced energy consumption by 28% [from 7.1×10^6 kJ/t(6.7×10^6 Btu/ton) in 1972 to 5.06 kJ/t (4.8×10^6 Btu/ton) in 1993]. Dry process manufacturing has become preferred due to energy efficiency. According to 1993 figures, wet process plants consumed 32% more energy per ton of cement than dry process plants.

Consumption

From 1975 to 1990 cement consumption has changed little. The cement industry has reduced its dependence on bag shipments (54.7% in 1950) and turned to the more labor efficient bulk transport (96% in 1990). In addition, the amount of cement shipped by rail transportation declined from 75% of industry shipments in 1950 to less than 14% in 1990.

In the past 30 years, the ready mix concrete industry became the primary customer for cement manufacturers. The other primary uses are in building materials, concrete products, highway construction, and soil stabilization.

STEVEN H. KOSMATKA
Portland Cement Association

CERAMICS

OVERVIEW

Ceramics may be defined as a class of inorganic, nonmetallic solids that are subjected to high temperature in manufacture or use. The most common ceramics are oxides, carbides (qv), and nitrides (qv), but silicides, borides, phosphides, sulfides, tellurides, and selenides are ceramics, as well as elemental materials such as carbon (qv) and silicon (see SILICON AND SILICON ALLOYS). Ceramic synthesis and processing generally involve high temperatures and the resulting materials are refractory or heat-resistant (see REFRACTORIES). Ceramics are commonly thought to include not only polycrystalline materials, but glasses, which are noncrystalline, and single-crystal materials such as ruby lasers, are classified as ceramics (see GLASS; LASERS). Examples of applications of less common ceramics include quartz optical fibers (see FIBER OPTICS) that are revolutionizing telecommunications, the insulating tiles on the space shuttle (see ABLATIVE MATERIALS), silicon nitride turbocharger rotors used in some passenger automobiles, and the corderite multilayer substrates used as chip carriers in the latest generation of supercomputers (see COMPUTER TECHNOLOGY).

Ceramic components are frequently made by sintering of powders. Alternatives such as melt processing are often uneconomical because many ceramics have very high melting temperatures or decompose before melting. Many ceramic melts are also very reactive with container materials, which imposes additional limitations on available processing methods. A rule of thumb in ceramic processing is that the quality of a ceramic part is no better than the quality of the powder from which it is made. Thus, much effort has been directed to improving the properties of ceramic powders. Improvements include higher purity, finer particle size, less agglomeration, and better control of compositions and distributions of dopants.

One of the many possible ways of classifying ceramics is according to use. One group is the bulk or commodity ceramics that have had relatively little processing beyond the constituent raw materials. These are primarily low value-added materials such as brick, tile, pottery, and abrasive grain (see ABRASIVES). At the other extreme are the engineering or fine ceramics that are characterized as low volume, high value-added, highly processed materials having carefully controlled properties (see ADVANCED CERAMICS). Some of the main types of engineering ceramics include: *(1)* electronic ceramics, which include dielectrics, ferroelectrics (qv), ferromagnetic ceramics, piezoelectrics, and superconductors (see also CERAMICS AS ELECTRICAL MATERIALS); *(2)* structural ceramics, which are strong, fracture-resistant materials such as silicon nitride, Si_3N_4, silicon carbide, SiC, and toughened zirconium dioxide, ZrO_2; *(3)* wear-resistant ceramics, such as the carbides, nitrides, and borides (see BORON COMPOUNDS; TOOL MATERIALS); *(4)* optical ceramics such as Cr doped Al_2O_3 (ruby; silicon dioxide, SiO_2, fiber; lead lanthanum zirconium titanate (PLZT); and yttrium aluminum garnet (YAG); and *(5)* bioceramics, which are low or controlled reactivity materials for in-body use such as aluminum oxide, Al_2O_3, and hydroxyapatite.

RONALD LOEHMAN
Sandia National Laboratories

CERAMIC PROCESSING

In the manufacture of ceramic components, chemical composition and microstructure are specified to optimize the properties (eg, mechanical, electrical, and magnetic) of the finished product for a given application. Optimum properties are achieved by developing and refining processes to produce a target microstructure, and by controlling processes to minimize the concentration and scale of the defects in the finished part. The tolerance of the finished ceramic to defects determines the degree of control that must be exercised during processing. For example, advanced ceramics such as Si_3N_4 turbine blades require greater process control than traditional ceramic electrical porcelain insulators.

Ceramic component fabrication involves simultaneously optimizing multiple processes ranging from raw materials beneficiation to post-sinter machining (Fig. 1). To manufacture reliable ceramic components with reproducible properties, process–microstructure–property relationships must be understood and controlled during processing. These relationships can be determined by characterizing the ceramic raw materials, mixes, and the formed ceramic body intermittently during the various stages of processing and after final thermal consolidation. Established relationships between processing, microstructure, and properties can be applied to control manufacturing processes and optimize properties, and to identify and correct process deficiencies when less than optimal properties are obtained. Examples of some process–microstructure–property relations in advanced ceramics are outlined in Figure 2.

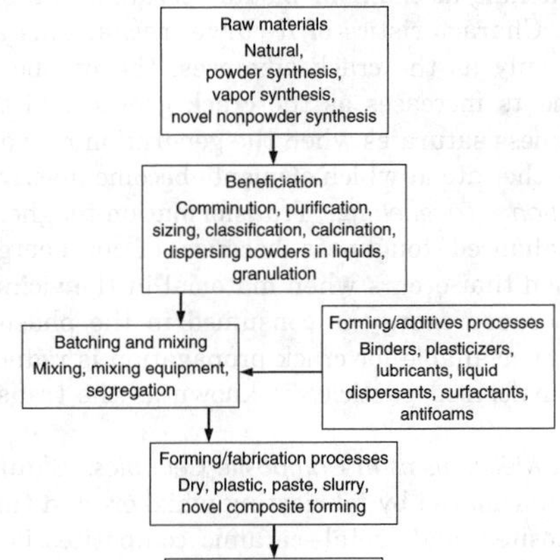

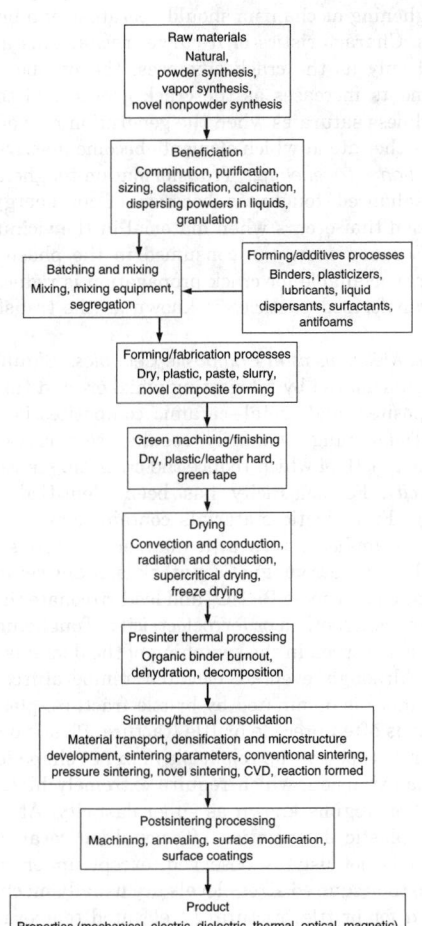

Figure 1. Flow diagram of the steps and processes involved in manufacturing a ceramic.

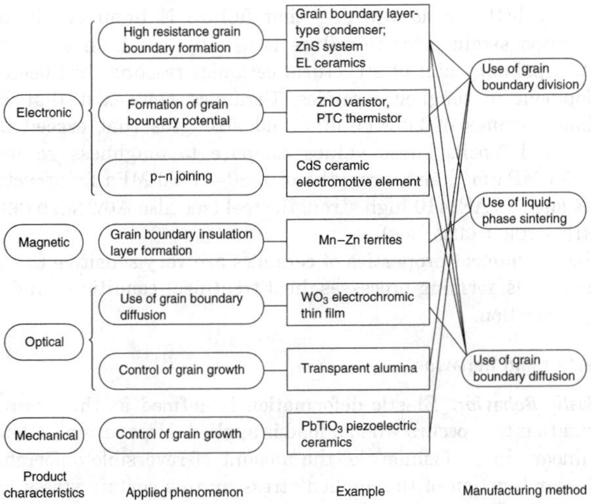

Figure 2. Process—microstructure—property relationships in advanced ceramics.

KEVIN G. EWSUK
Sandia National Laboratories

J. S. Reed, *Introduction to the Principles of Ceramic Processing,* John Wiley & Sons, Inc., New York, 1988.

D. W. Richerson, *Modern Ceramic Engineering,* 2nd ed., Marcel Dekker, Inc., New York, 1992.

Engineering Materials Handbook, Vol 4, *Ceramics and Glasses,* ASM International, Materials Park Ohio, 1991.

N. Ichinose, *Introduction to Fine Ceramics, Applications in Engineering,* John Wiley & Sons, Inc., New York, 1987.

MECHANICAL PROPERTIES AND BEHAVIOR

Structural ceramics are used in applications such as gas turbines, machining, heat exchangers, aerospace components, armor, thermal barrier coatings, and bearings. Advantages of ceramics over metals include dimensional stability; low densities, which translate into weight savings and increased fuel efficiencies; high temperature capabilities; and corrosion resistance.

The overriding concern with regard to the mechanical performance of ceramics is their brittleness and hence sensitivity to flaws. There is usually little or no warning that failure is imminent because deformation strain prior to failure is usually less than 0.1%. As a result, a primary aim of structural ceramics research has been the development of tougher ceramics. Ceramics now exist that have toughness values of 20 $MPa{\cdot}m^{1/2}$ and strengths that exceed 2000 MPa (3×10^5 psi). These values compare to toughness values of 120–153 $MPa{\cdot}m^{1/2}$ and strengths of 1380– 1790 MPa for structural metals such as AF1410 high strength steel (see also ADVANCED CERAMICS–STRUCTURAL CERAMICS).

The mechanical properties of ceramics are very sensitive to starting materials, forming processes, heat treatment conditions, and surface preparation.

Properties and Behavior

Elastic Behavior. Elastic deformation is defined as the reversible deformation that occurs when a load is applied. Most ceramics deform in a linear elastic fashion, ie, the amount of reversible deformation is a linear function of the applied stress up to a certain stress level. If the applied stress is increased any further the ceramic fractures catastrophically. This is in contrast to most metals which initially deform elastically and then begin to deform plastically. Plastic deformation allows stresses to be dissipated rather than building to the point where bonds break irreversibly.

Strength. *Measured Strength versus Theoretical Strength.* The elastic modulus describes how easily atoms in a solid can be moved together or apart for small deformations. If deformation is continued, eventually an atom spacing is reached beyond which there is not enough atomic attraction to hold the atoms together. If entire planes of atoms separate, the ceramic breaks and its theoretical tensile strength has been reached. The theoretical tensile strength of the material, s_{theor}, has been approximated by

$$\sigma_{theor} = \left(\frac{E_\gamma}{a}\right)^{1/2} \cong \frac{E}{10} \qquad (1)$$

where a = equilibrium spacing between planes of atoms, γ = fracture surface energy, and E = Young's modulus.

If all bonds in a material were stressed equally up to the point of failure, the measured strength of a ceramic would be the theoretical strength. Large discrepancies between the theoretical and measured tensile strengths of ceramics result from the presence of microstructural imperfections. These imperfections or flaws can raise the local stress to the point that bonds in the immediate vicinity of the flaw can fail a few at a time, as opposed to every atom in the plane failing simultaneously. Actual fracture strengths depend on the flaw size, c, the fracture toughness, K_{IC}, and a geometrical factor, Y, as follows:

$$\sigma_{fracture} = \frac{K_{IC}}{Y_c^{1/2}} \qquad (2)$$

Statistical Variation in Strength. The wide variety of flaw types and sizes in ceramics produces the large (typically ±25%) variability in strength that has been one of the principal hurdles to the incorporation of ceramics in structural applications. This compares unfavorably with the few percent for variability of the yield stress of a metal. The probability of failure of a ceramic body at a given load depends on the probability of a flaw of a critical size being present in a location where the flaw produces a stress concentration.

Many distribution functions can be applied to strength data of ceramics but the function that has been most widely applied is the Weibull function, which is based on the concept of failure at the weakest link in a body under simple tension. A normal distribution is inappropriate for ceramic strengths because extreme values of the flaw distribution, not the central tendency of the flaw distribution, determine the strength. One implication of Weibull statistics is that large bodies are weaker than small bodies because the number of flaws a body contains is proportional to its volume.

Compressive Strength. Ceramics are much stronger in compression than in tension and are frequently used in applications where they bear compressive loads. Under excessive compressive loads ceramics fail in a brittle manner just as they do in tension; however, the predicted stresses are eight times the predicted tensile stresses. In compression, the failure process appears to begin with microplastic deformation, not the growth of pre-existing flaws. Measured compressive strengths are often far below predicted values because of flaws, twinning, improper loading, and internal stresses resulting from effects such as grain anisotropy.

Fracture Toughness. The fracture criterion is defined by a critical value of the crack tip stress intensity, known as the fracture toughness, K_c. Ceramics often fail in pure tension and in this case K_{IC} is used. In all cases the material fails when the local stress concentration at the crack tip reaches the critical value required for bond breakage. Thus the critical stress intensity, K_{IC}, required for bond rupture is considered to be a material constant. K_{IC} measurements can be made by introducing a crack of a known size in a body having a specific geometry and then loading the body until catastrophic failure occurs.

R-Curve Behavior. Ceramic toughening efforts have focused on the property of some ceramics to exhibit increased apparent fracture toughness as cracks grow. This increase is seen in terms of the far-field value of the stress intensity required to propagate the crack. An important consequence of this effect, which is commonly referred to as R-curve or T-curve behavior, is that the material has increased damage tolerance because there is a crack size regime in which the strength is independent of the crack size. The underlying basis of R-curve behavior is that the crack tip stress is redistributed, either to immediately adjoining material, as in the case of the process zone formed in transformation toughened materials, or to regions far removed from the crack tip, as in the case of fiber-reinforced ceramics. One reason for the interest in crack interface bridging mechanisms is that this toughening mechanism should operate over a broad range of temperatures. Characteristics of R-curve mechanisms are that they are activated only as the crack advances, the number of activated bridging elements increases as the crack grows, and the measured fracture toughness saturates when the generation rate of bridging elements equals the rate at which elements become inactive.

Transformation Toughening. Transformation-toughened materials exhibit enhanced toughness because of an energy-consuming phase transition that occurs when material in the vicinity of a crack is stressed. Because energy is consumed in the phase transformation, the energy available for crack propagation is reduced. The zone containing transformed particles is known as the transformation or process zone.

Toughening Mechanisms in Composite Ceramics. Significant toughening has been achieved by fabricating whisker- and fiber-reinforced ceramic composites, and metal–ceramic composites (see COMPOSITE MATERIALS). Toughening primarily results from crack bridging/or crack deflection, both of which reduce the crack tip stress intensity.

Ferroelasticity. Ferroelasticity has been identified as a source of toughening. Ferroelastic materials contain domains that can be switched by an applied stress in a manner analogous to magnetic domain switching in ferromagnetic materials. Many ceramics that exhibit ferroelasticity, such as $BaTiO_3$ and lead zirconate titanate (PZT), also exhibit ferromagnetism or ferroelectricity. Toughening occurs because energy is absorbed in the switching of the domains.

Plasticity. Although even at elevated temperatures mechanical failure of ceramics is dominated by brittle fracture, plastic deformation mechanisms often precede brittle fracture. Plastic deformation is also important because of the role it plays in net shape forming operations such as extrusion, which require extremely high strain rates in a deformation regime known as superplasticity. At low and high temperatures plastic deformation of crystalline ceramics can occur by slip, but it is not usually observed, except under special conditions, because the required stress levels are usually much higher than those required for brittle fracture. At elevated temperatures plastic deformation can also occur by grain boundary sliding and softening of secondary phases such as glass.

Hardness. Hardness (qv) is a measure of the resistance of a material to deformation, in particular the resistance to plastic deformation during surface penetration. Hardness is related to the bond strength. Because covalent and multivalent ionic bonds are strong and highly directional in nature, slip is very difficult in these cases, and ceramics containing these bonds are generally the hardest materials. The ratio of the distance between planes of atoms, a, and the spacing of the atoms in the plane, b, also plays an important role in how easily planes of atoms slide past each other, and hence in the hardness. Small a/b ratios tend to give harder materials.

Subcritical Crack Growth. Under most environmental conditions, the strength of ceramics, especially glasses, degrades with time. This phenomenon, known as subcritical crack growth (SCG), stress corrosion cracking (SCC), static fatigue, or delayed failure, involves the slow growth of a pre-existing crack at an applied stress intensity lower than that necessary, ie, the critical value, to propagate a crack without environmental influences. Subcritical crack growth is pernicious because a flaw grows slowly at stresses far below the expected failure load until the flaw is large enough to satisfy the Griffith criterion, and then failure occurs catastrophically. The mechanism of subcritical crack growth is the reaction of liquids such as water with highly stressed bonds at the crack tip.

Impact and Erosion. Impact involves the rapid application of a substantial load to a relatively small area. Most of the kinetic energy from the impacting object is transformed into strain energy for crack propagation. Impact can produce immediate failure if there is complete penetration of the impacted body, or if the impact induces a macrostress in the piece, causing it to deflect and then crack catastrophically. Failure can also occur if erosion reduces the cross section and load-bearing capacity of the component, causes a loss of dimensional tolerance, or causes the loss of a protective coating. Detailed information on impact and erosion is available.

Tribological Behavior. Tribological performance of ceramics includes friction, adhesion, wear, and lubricated behavior of two solid materials in contact.

Friction and Adhesion. The coefficient of friction, μ, is the constant of proportionality between the normal force, P, between two materials in contact and the perpendicular force, F, required to move one of the materials relative to the other. Ceramics generally have lower coefficients of friction than metals.

Factors that affect μ include loading geometry, microstructure, crystal orientation, surface chemistry, environment, temperature, and the presence of lubricants.

Wear. Ceramics generally exhibit excellent wear properties. Wear is determined by a ceramic's friction and adhesion behavior, and occurs by two mechanisms: adhesive wear and abrasive wear.

Thermal Stresses and Thermal Shock. Thermal stresses arise when a body is heated or cooled and constrained from expanding or contracting. Thermal stresses can lead to fracture and catastrophic failure when the magnitude of the thermal stress exceeds the strength of the ceramic. Factors that contribute to the generation of large thermal stresses and the failure of ceramics under these stresses are low thermal conductivities, which produce large temperature gradients, and the lack of a stress relief mechanism such as plastic deformation. Approaches used to minimize thermal stresses include matching the expansion of ceramics with the expansions of the materials to which they are joined, minimizing temperature gradients, minimizing cross-sectional thickness changes to ensure that the body heats or cools uniformly, keeping the body at its operating temperature, and heating and cooling slowly.

Cyclic Fatigue. Cyclic fatigue is the weakening and subsequent failure of a material during cyclic loading, often at stress levels significantly less than that required to cause failure under static loading. Cyclic stresses can be produced by repeated heating and cooling, by vibrations, and in applications in which the component is repeatedly loaded and unloaded. In ceramics which show crack interface bridging, cyclic fatigue is largely a result of the loss or destruction of bridges when the crack closes.

Fracture Analysis

Fracture analysis, also known as fractography, plays an important role in understanding the relationships between structure and mechanical properties, and the conditions that lead to failure (see FRACTURE MECHANICS). Systematic examination and interpretation of fracture markings and the crack path can often be used to reconstruct the sequence of events and stresses that led to failure. Fractography also plays an important role in the intelligent use of ceramics because it helps the user differentiate whether failure occurred because the material was weakened by the introduction of processing and handling flaws, or because the applied stress exceeded the design stress.

JILL GLASS
Sandia National Laboratories

B. R. Lawn and T. R. Wilshaw, *Fracture of Brittle Solids,* Cambridge University Press, Cambridge, Mass., 1975.

R. W. Davidge, *Mechanical Behavior of Ceramics,* Cambridge University Press, New York, 1986.

W. D. Kingery, H. K. Bowen, and D. R. Uhlmann, *Introduction to Ceramics,* John Wiley & Sons, Inc., New York, 1976.

D. W. Richerson, *Modern Ceramic Engineering: Properties, Processing and Use in Design,* Marcel Dekker, Inc., New York, 1982.

GLASS STRUCTURE AND PROPERTIES

Inorganic glasses are important ceramic materials and glasses share many of the desirable properties of crystalline ceramics even though they lack long-range order (see GLASS; GLASS-CERAMICS). The structure of glasses is not completely random, however. Glasses do have short-range order in the arrangement of anions around cations.

Glasses, less thermodynamically stable than crystals, must be "captured" in a metastable state, and this is typically accomplished by rapid cooling of a liquid.

Types of Glasses

The simplest inorganic glasses are composed of only one element: B, C, P, As, S, or Se. Glassy selenium in thin-film form has been used extensively in the photocopying industry (see ELECTROPHOTOGRAPHY). Inorganic glasses containing more than one element can be broadly categorized according to the type of anion. There are two main categories: chalcogenide glasses and halide glasses. Chalcogenide glasses contain anions from Group 16 (VI A) of the Periodic Table, O, S, Se, Te; halide glasses contain anions from Group 17 (VII A), F, Cl, Br, I. Although oxide glasses are, strictly speaking, a subset of the chalcogenide glasses, the family of oxide glasses is so large and so important that it is usually considered separately from the other chalcogenides.

Glass Structure

Several criteria for predicting which compounds form glasses have been proposed. The best known guidelines are those originally presented for oxide glasses and known as Zachariasen Rules. They can be summarized as: *(1)* an oxygen or anion must not be linked to more than two cations; *(2)* the oxygen or anion coordination number of the cations must be small; and *(3)* the cation–anion polyhedra must share corners rather than edges or faces.

The radius ratio rule can be used with Zachariasen Rule 1 to predict which compounds tend to form glass. For oxides with the formula MO_2, charge balancing requires that the cations must be four-coordinate if the oxygens are to be two-coordinate. For tetrahedral coordination of a cation by oxygen, the ratio of the cation radius to the oxygen radius must be ≤ 0.414.

In addition to the Zachariasen and radius ratio rules, for oxides the electronegativity of the predominant cation should be between 1.7

and 2.1. If the cation electronegativity is too high, the compound tends to form molecules or discrete polyatomic ions rather than a connected network.

Modifiers in glass are compounds that tend to donate anions to the network whereas the cations occupy "holes" in the disordered structure. These conditions cause the formation of nonbridging anions, or anions that are connected to only one network-forming cation. Modifier compounds usually contain cations with low charge-to-radius ratios (Z/r), such as alkali or alkaline-earth ions.

Optical Properties

Probably the most striking and useful characteristic of common silicate glass is its transparency to visible light. This transparency results from the absence of grain boundaries and delocalized electrons, which tend to scatter and absorb light.

Refractive Index. The refractive index n of a glass is defined as the ratio of the speed of light in vacuum to the speed of light in the glass, v, ie, $n = c/v$. The value of n is always greater than 1, because light travels more slowly in dense media. The refractive index is governed almost exclusively by the nature of the electron-oscillators, with essentially no contribution from atomic vibrations. At visible frequencies, the electrons of an optically transparent glass can follow or move in phase with the applied field, but because of the small vibrational amplitudes of these induced oscillations, the light is reradiated rather than absorbed.

The loosely bound valence electrons make the greatest contribution to n, so large, many-electron atoms, eg, Pb(II) or Bi(III), are added to glass to increase the refractive index.

Dispersion. A glass prism separates white light into its component colors by virtue of the frequency dependence of n. The higher the frequency, the higher the refractive index for a given substance, ie, blue light is slowed to a greater extent than red light and deviated through a larger angle by a prism. It is this dependence of the refractive index on the frequency (color) of light that is known as dispersion. The measure of dispersion is the Abbé number, v_d, which is equal to $(n_d - 1)/(n_F - n_c)$ where n_d, n_F, and n_C are the indexes of refraction at the sodium d-line (587.6 nm), a wavelength in the blue (486.1 nm), and a wavelength in the red (656.3 nm), respectively. The greater the dispersion, the smaller the Abbé number.

Fluorescence and Glass Lasers Some ions absorb light of a certain frequency emitting light of lower frequency. This is known as fluorescence. Examples of ions that fluoresce in glass are Mn(IV), Pb(II), and the lanthanide ions.

The existence of glass lasers (qv) was first reported in 1961. Optically pumped laser action has since been observed for several lanthanide ions in a variety of glass systems. Large, high power neodymium glass lasers have been used for inertial confinement fusion experiments. The best glass laser systems have the following qualities: the absorption spectrum of the lasing ion matches the spectrum of the pump radiation; the absorbed radiation efficiently produces excited-state ions; the excited state has a long lifetime; the probability of radiative decay is high; and the linewidth of the emitted radiation (fluorescence) is narrow. The linewidth of the fluorescence band of the lanthanide ion is affected by the glass matrix. In general, the smaller the field strength of the anions, the less the perturbation of the coordination shell of the fluorescing ion and the narrower the linewidth, ie, fluoride and chloride glasses promote narrower linewidths than those seen in oxide glasses. The stimulated-emission cross sections (efficiencies) of rare-earth sulfide glasses based on Ga_2S_3 are reported to be higher than those of oxide, oxyhalide, and halide glasses.

Electrical Properties

There are many applications in which glass is used as an electrical insulator. One example is glass-to-metal seals. Moreover, other glasses are useful as a result of ionic or electronic conductivity.

Ionic Conductivity. Ionic conductivity in oxide glasses is almost always a result of the movement of monovalent cations, eg, in an alkali silicate glass, the small, univalent alkali ions are much more mobile than the Si^{4+} or O^{2-} ions.

The electrical conductivities of many glasses containing monovalent ions are so high that such glasses are being considered for all-solid-state batteries (qv).

Semiconductivity. Amorphous selenium and other chalcogenide glasses form the basis for the multibillion-dollar electrostatic copying industry. Although amorphous substances possess only short-range and a certain degree of intermediate-range order, they can exhibit extended electronic states similar to those found in crystals. Glassy semiconductors (qv) also possess localized electronic states. Conduction in chalcogenide glasses can occur by the activation of carriers across the mobility gap into the extended states, or by tunneling between the localized states.

Dielectric Constant. The dielectric constant is a measure of the ease with which charged species in a material can be displaced to form dipoles. There are four primary mechanisms of polarization in glasses: electronic, atomic, orientational, and interfacial polarization. Electronic polarization arises from the displacement of electron clouds and is important at optical (ultraviolet) frequencies. At optical frequencies, the dielectric constant of a glass is related to the refractive index: $\kappa = n^2$. Atomic polarization occurs at infrared frequencies and involves the displacement of positive and negative ions.

At lower frequencies, orientational polarization may occur if the glass contains permanent ionic or molecular dipoles, such as H_2O or an Si–OH group, that can rotate or oscillate in the presence of an applied electric field.

Thermal Properties

Glass-Transition Temperature. When a typical liquid is cooled, its volume decreases slowly until the melting point, T_m, is reached, below which the volume decreases abruptly as the liquid is transformed into a crystalline solid.

Unlike the abrupt melting of a crystalline solid, the glass transition is characterized by a continuous change in properties over a small temperature interval. When a solid glass is heated from below its glass-transition temperature, the volume and specific heat increase slowly. As the glass-transition temperature is reached, the rates of change of these quantities become larger, indicating that bonds are being broken and that some parts of the glass have become more mobile; ie, above T_g the behavior of the glass becomes more like that of a liquid, in that there are now opportunities for long-range molecular diffusion.

Working, Softening, Annealing, and Strain Points. Although the glass-transition temperature is important in the scientific study of glass formation, other temperatures are more useful from a technological point of view. For example, the working point refers to the temperature at which the glass can be formed into some desired shape. The working point of a glass corresponds to a viscosity of 10^6 mPa·s(= cP). At the softening point (viscosity of $10^{6.6}$ Pa·s), a glass suspended at two points can begin to sag in a few minutes. The term annealing point is essentially the same as T_g. If annealing is carried out at the strain point, a temperature below T_g at which the viscosity is $10^{13.5}$ Pa·s, the reduction of stresses to acceptable levels takes about four hours.

Thermal Expansion. High silica glasses such as Pyrex have low coefficients of thermal expansion (CTEs) and are used in applications requiring good resistance to thermal shock. In fact, certain ultralow expansion SiO_2–TiO_2 glasses have CTEs of practically zero. Some applications, such as glass-to-metal seals, depend on glasses having high CTEs. Lead-containing glasses are appropriate for such purposes.

Thermal Conductivity. Glasses are good thermal insulators, making them ideal materials for windows in buildings. The thermal conductivity of a glass increases markedly as it is crystallized to form a glass–ceramic. The thermal conductivity of an aerogel is exceptionally low.

Processing

Glass processing techniques include melt processing, sol–gel processing, vapor-phase processing, and thin-film techniques.

Applications

Silicate glass is a familiar, ubiquitous material, used in beverage containers, window panes, and automobile windshields. Increasingly, however, unusual glasses are being employed in high technology applications, such as optical fibers for telecommunications, graded-refractive-index glasses, nonlinear optical glasses, acousto-optic glasses, fast-ion conducting glasses, glass–ceramics, and glass microspheres.

CAROL PHIFER
Sandia National Laboratories

N. J. Kreidl and C. A. Angell, in D. R. Uhlmann and N. J. Kreidl, eds., *Glass: Science and Technology,* Vol. 1, Academic Press, Boston, Mass., 1983, p. 105.

W. H. Zachariasen, *J. Am. Chem. Soc.* **54,** 3841 (1932).

G. H. Beall, in D. C. Boyd and J. F. MacDowell, eds., *Commercial Glasses (Advances in Ceramics)* Vol. 18, American Ceramic Society, Columbus, Ohio, 1986, p. 157.

A. Paul, *Chemistry of Glasses,* 2nd ed., Chapman & Hall, London, 1990, p. 3.

NONLINEAR OPTICAL AND ELECTROOPTIC CERAMICS

Nonlinear Optical Effects

Nonlinear optical (NLO) and electrooptic (EO) processes are events that take place in transparent materials where refractive indexes can vary with electric field intensity. The field elicits a polarization response in which terms dependent on second and higher orders of the field intensity become significant contributions to the overall polarization response. As light propagates through a transparent solid, liquid, or gas, it encounters the electric fields associated with the valence electrons, polarizing them and creating oscillating electric dipoles. These oscillators in turn act as antennas to broadcast or propagate the electric field through the medium. At low optical intensities, the polarizability P thus induced varies linearly with the electric field E:

$$P_i = \overset{(1)}{\chi_{ij}} E_j \tag{1}$$

where $\chi^{(1)}{}_{ij}$ is the linear polarizability. As electric field strength increases, eventually the polarization amplitude is no longer able to follow the electric field in a perfectly linear fashion. The polarizability, P_i must now be expressed as a power series and the higher order terms become more significant with increasing field intensity:

$$P_i = \overset{(1)}{\chi_{ij}} E_j + \overset{(2)}{\chi_{ijk}} E_j E_k + \overset{(3)}{\chi_{ijkl}} E_j E_k E_l + \cdots \tag{2}$$

NLO effects result when the polarization response of the valence electrons becomes significantly anharmonic, usually in intense light beams where the magnitude of E is very large. Ferroelectric ceramics comprise the bulk of technologically useful nonlinear optical materials.

NLO Wavemixing. Second harmonic generation (SHG) can be considered the mixing of two light waves of frequency ω to give a wave of frequency 2ω. It is also known as frequency doubling.

Optical Phase Retardation. A static electric field applied to a ceramic changes its refractive indexes unequally, altering its birefringence, ie, the difference between its minimum and maximum refractive indexes. Such an electrooptic (EO) device acts as a variable phase retarder, and can modulate either the frequency or the amplitude of a transmitted beam.

Photorefractive Effect. In certain crystals, such as barium titanate $BaTiO_3$, and strontium barium niobate (SBN), $Sr_{1-x}Ba_xNb_2O_6$, applying a light pattern that causes alternate portions of the crystal to become illuminated or dark, eg, from the interference of two light beams, causes photoexcited charge carriers to drift from the illuminated to the dark regions, where they are trapped. This results in space charge fields in the crystal, which in turn causes adjacent zones to have different refractive indexes, setting up a diffraction grating within the crystal. Once written, this pattern can remain in the crystal for a considerable length of time until it is erased by strong, uniform illumination. This grating can diffract subsequent beams and is thus capable of storing, processing, and amplifying complex optical information such as images (see HOLOGRAPHY).

Materials and Processing

Materials. NLO materials possess a high degree of optical nonlinearity, physical durability and chemical inertness, and high threshold to optical damage. In addition, it must be possible to process the materials into the form required by the intended application.

Optically nonlinear ceramics all contain highly polarizable bonds. In the case of oxide ceramics, such polarizable bonds are often found between oxygen and early transition metals, eg, $BaTiO_3$, $SrNb_2O_6$, and $KTaO_3$; boron, eg, LiB_3O_5, β-BaB_2O_4; the heavy p-block elements, eg, TeO_2, HIO_3, $Pb_3(PO_4)_2$, and $Bi_{12}SiO_{20}$, and in compounds containing tetrahedral XO_4 groups, eg, $KLiSO_4$, KH_2PO_4, and $Gd_2(MoO_4)_3$. These compounds must crystallize in structures that favor highly directional bonding in order to yield good optical nonlinearities.

Nonoxide NLO ceramics include Si and compound semiconductors (qv) having the silicon structure, eg, GaAs, InP, and InSb, as well as ferrelectrics such as SbSI. These materials tend to be more highly nonlinear than oxide ceramics, although lack of transparency at visible and uv wavelengths prevents them from competing with the oxides for the same applications.

Processing. All-optical wavemixing applications generally require single crystals. The crystals should have relatively low concentrations of point defects, which have been implicated in low thresholds to optical damage. Typical defects in ferroelectrics include Schottky defects and nonstoichiometry, which result in ions residing on the wrong crystallographic positions, eg, siting Nb on Li positions in $LiNbO_3$.

If the ceramic of interest melts congruently, eg, $LiNbO_3$ and $BaTiO_3$, crystals can be grown using the Czochralski method. If the ceramic melts incongruently, eg, $KTiOPO_4$, or is a phase not stable at the composition's melting point, eg, SiO_2, crystals may sometimes be grown hydrothermally, or from a nonaqueous flux.

Techniques and Applications

Frequency Conversion. SHG is most often used to double the frequencies of solid-state and diode lasers operating in the near infrared to yield an output beam in the visible spectrum.

Electrooptic Devices. Electrooptic (EO) materials are optical media in which birefringence can be readily induced or altered by an externally applied electric field. Birefringence, which is the difference between minimum and maximum refractive index (Δn) in an optical medium, occurs in anisotropic crystals, stressed glasses, poled ceramics, or any transparent medium subjected to an external electric field.

As with NLO ceramics, good EO materials tend to be ferroelectrics. Ferroelectric materials are noncentrosymmetric, allowing one to take advantage of the linear response and increased magnitude of the Pockels effect vs the Kerr effect.

Inorganic electrooptic materials can take the form of single crystals or polycrystalline ceramics. For certain applications such as switching and interferometry, single crystals carry certain advantages over polycrystalline compacts. On the other hand, polycrystalline ceramics can be hot-pressed or sintered into a larger variety of shapes and sizes, without regard to crystalographic orientation. In addition, the orientation of the optic axis in such polycrystalline ceramic elements can be controlled by the external electric field.

EO materials are generally used as either bulk optical elements or guided wave devices.

MARK L. F. PHILLIPS
Sandia National Laboratories

J. A. Giordmaine, *Sci. Am.* **38** (1964).

P. Guenter, *Ferroelectrics* **24,** 35–42 (1980).

G. H. Haertling, in L. M. Levinson, ed., *Electronic Ceramics: Properties, Devices and Applications*, Marcel Dekker, New York, 1988, pp. 371–492.

I. P. Kaminow, *An Introduction to Electrooptic Devices,* Academic Press, Inc., New York, 1974.

ELECTRONIC PROPERTIES AND MATERIAL STRUCTURE

The properties of a ceramic material that make it suitable for a given electronic application are intimately related to such physical properties as crystal structure, crystallographic defects, grain boundaries, domain structure, microstructure, and macrostructure. The development of ceramics that possess desirable electronic properties requires an understanding of the relationship between material structural characteristics and electronic properties and how processing conditions may be manipulated to control structural features.

Properties and Applications

Insulators. Ceramic insulators are materials used to support electrical conductors and to prevent the flow of electrical charge between them. Ceramic bodies, such as triaxial porcelains based on compositions in the Al_2O_3–$(K,Na)_2O$–SiO_2 phase diagram, were initially employed as high voltage insulators. Since then, other ceramics have been utilized in higher technology insulation applications, eg, in packaging and very large-scale integration (VLSI) circuits (see INTEGRATED CIRCUITS). VLSI, or multilayer ceramic (MLC), technology involves cofiring conductor and insulator layers to build up multilayer electronic modules, thus increasing device, or circuit, density. Alumina, Al_2O_3, is the material most frequently used for these applications, although because of its very high thermal conductivity, beryllia, BeO, has also been used as a substrate material. Other ceramics, namely, aluminum nitride and glass–ceramics (qv), such as cordierite and cordierite–mullite composites, have also been investigated for use in packaging applications (see PACKAGING, ELECTRONIC MATERIALS). The development of cordierite as a substrate material has progressed to the state where, in conjunction with cosinterable copper electrodes, it is used as the multilayer substrate in high performance computers (see COMPUTER TECHNOLOGY).

The usefulness of ceramics for all of these applications is because of their dielectric, or insulating, characteristics.

Capacitors. Ceramic materials suitable for capacitor (charge storage) use are also dependent on the dielectric properties of the material. Frequently the goal of ceramic capacitors is to achieve maximum capacitance in minimum volume.

Historically, materials based on doped barium titanate were used to achieve dielectric constants as high as 2,000 to 10,000. The high dielectric constants result from ionic polarization and the stress enhancement of k' associated with the fine-grain size of the material.

Because of very high dielectric constants ($k' < 20{,}000$), lead-based relaxor ferroelectrics, $Pb(B_1,B_2)O_3$, where B_1 is typically a low valence cation and B_2 is a high valence cation, have been investigated for multilayer capacitor applications.

Multilayer capacitor devices have become progressively smaller as the dielectric constant of the ceramic layers has increased, and layer thickness has decreased to below 15 μm. There is also growing interest in thin-film dielectric capacitors. For example through the use of processing techniques such as sol–gel solution deposition, thin (~0.25 μm) ceramic layers having dielectric constants ranging from 500 to 2000 in the PZT, $Pb(Zr,Ti)O_3$, and PMN–PT, $Pb(Mn_{1/3}Nb_{2/3})O_3$–$PbTiO_3$, compositional families, respectively, have been prepared.

Piezoelectrics. All ceramics display a slight change in dimension, or strain under the application of an electric field. When the induced strain is proportional to the square of the field intensity, it is known as the electrostrictive effect. Materials that also show the opposite effect, ie, an induced polarization (electric charge) resulting from an applied stress, are referred to as piezoelectric materials, and the effect is referred to as the direct piezoelectric effect.

Certain crystal symmetry requirements are necessary for the existence of piezoelectric behavior in materials. The most widely used class of piezoelectric ceramics possess the ABO_3 perovskite crystal structure, where A is usually a low valence cation and B is a high valence cation. The structure consists of a cubic close-packed array of oxygen and A-site ions, and the B-site ions occupy one-fourth of the octahedral interstices that are not adjacent to the A-site ions. Other useful piezoelectrics possess the tungsten bronze, AB_2O_6, structure.

The piezoelectric electromechanical coupling effects allow for the conversion of mechanical energy into electrical energy or electrical into mechanical energy, thus defining two different classes of transducer applications. Uses range from hydrophones and microphones, to power transducers for ultrasonic cleaning baths. Other applications include filters and resonators. Most (polycrystalline) piezoelectric ceramics for transducer applications are based on compositions in the PZT, $PbZrO_3$–$PbTiO_3$, family.

Pyroelectrics. Pyroelectric ceramics are materials that possess a unique polar axis and are spontaneously polarized in the absence of an electric field. Pyroelectrics are also a subset of piezoelectric materials. Ten of the 20 crystal classes of materials that display the piezoelectric effect also possess a unique polar axis, and thus exhibit pyroelectricity.

Ceramics that display the pyroelectric effect also exhibit a variation in polarization with temperature. The nature of the temperature variation is dependent on the type of crystallographic transformation that the material displays at the Curie point, ie, whether the transition is first or second order.

The most commercially important application that takes advantage of the pyroelectric effect in polycrystalline ceramics is infrared detection, especially for wavelengths in excess of 2.5 μm. Applications range from radiometry and surveillance to thermal imaging, and pyroelectric materials work under ambient conditions, unlike photon detectors, which require cooling.

Ferroelectrics. Ferroelectrics, materials that display a spontaneous polarization in the absence of an applied electric field, also display pyroelectric and piezoelectric behavior. The distinguishing characteristic of ferroelectrics, however, is that the spontaneous polarization must be re-orientable with the application of an electric field of a magnitude lower than the dielectric breakdown strength of the material.

While ferroelectric materials are widely used in capacitor and transducer applications, the number of applications utilizing the ferroelectric response of ceramic materials has been limited.

Ferrites. Magnetic ceramics or ferrites (qv) may be classified according to crystal structure and the type of magnetic properties: *(1)* spinel ferrites; *(2)* hexagonal ferrites, ie, the magnetoplumbites; and *(3)* rare-earth ferrites, which crystallize in the garnet structure. The garnet materials, so named because they crystallize in the structure of garnet, have the chemical formula $3Me_2O_3{\cdot}5Fe_2O_3$, where Me is Y or Gd. These rare-earth ferrites are used in microwave devices (see MICROWAVE TECHNOLOGY) and bubble memory applications.

Hard materials, so named because they are difficult to demagnetize, are used in permanent magnet applications. The general chemical formula for these materials is $MeO{\cdot}6Fe_2O_3$, where Me is a Group 2 (IIA) metal, usually Sr or Ba. The materials crystallize in a hexagonal structure, leading to the name hexagonal ferrites or hexaferrites. They are also isostructural with the naturally occurring mineral, magnetoplumbite.

The most commonly used ferrites, the so-called soft ferrites, are used in soft magnet and low field telecommunication applications, low power transformers, television tube scanning yokes, recording heads, magnetic recording media (see INFORMATION STORAGE MATERI-

ALS–MAGNETIC), antennae, and channel filters. Soft ferrites crystallize in the spinel structure and are characterized by the general formula, $MeFe_2O_4$.

Sensors. One growth area for electronic ceramics is in sensor applications. Sensors (qv) are devices that transform nonelectrical inputs into electrical outputs, thus providing environmental feedback. Smart, or intelligent, sensors also allow for mechanisms such as self-diagnosis, recovery, and adjustment for process monitoring and control (see PROCESS CONTROL).

Ceramic sensor applications are widely varied and likely to grow, in part, because of the increasing emphasis on process control, waste minimization, and environmentally conscious manufacturing. Sensors are used for the measurement of humidity, oxygen gas pressure, carbon monoxide concentration, pressure, temperature, radiation, etc.

The performance characteristics of ceramic sensors are defined by one of more of the following material properties: bulk, grain boundary, interface, or surface. Sensor response arises from the nonelectrical input because the environmental variable affects charge generation and transport in the sensor material.

Varistors. Varistors are devices that exhibit nonlinear current–voltage behavior. At low voltages, current flow is minimal and the device behaves as an ohmic insulator. As the voltage approaches a critical value, the breakdown field (F_{BR}), current flow increases, and the device becomes highly conducting. Because of this characteristic, varistors are typically employed as circuit overvoltage protection devices, ie, devices that protect solid-state circuitry and other components against voltage spikes by diverting the current flow. The most frequently used material is doped zinc oxide, ZnO, which can handle both high current and high energy.

High Temperature Ceramic Superconductors. In 1986, it was reported that the compound $(La, Ba)_2CuO_4$ displayed the transition to the superconducting state at a temperature of ~35K. This extraordinary discovery of superconductivity in a ceramic material led to an explosion of research on other ceramic systems. The most notable include: $YBa_2Cu_3O_{7-\delta}$, T_c ~93 K; $Bi_2CaSr_2Cu_2O_x$, T_c ~80 K; and $Tl_2Ca_2Ba_2Cu_3O_y$, T_c ~125 K. Other bismuth- and thallium-based superconductors having different cation stoichiometries have also been discovered.

Processing Techniques

Traditional Processing Routes. The majority of the electronic ceramics that are used for electronic applications crystallize in the perovskite structure. Many perovskite-type oxides can be prepared by conventional processing routes, ie, by high temperature solid-state reactions between the respective oxide powders.

Chemical Coprecipitation of Powders. Coprecipitation methods have been used extensively for the preparation of barium titanate, lanthanum-doped lead zirconate titanate (PLZT), and oxide superconductor powders.

Solution Deposition of Thin Films. Chemical methods of preparation may also be used for the fabrication of ceramic thin films (qv). Metallo-organic precursors, notably metal alkoxides (see ALKOXIDES, METAL) and metal carboxylates, are most frequently used for film preparation by sol-gel or metallo-organic decomposition (MOD) solution deposition processes.

Alternative Thin-Film Fabrication Approaches. Thin films of electronic ceramic materials have also been prepared by sputtering, electron beam evaporation, laser ablation, chemical beam deposition, and chemical vapor deposition (CVD).

ROBERT SCHWARTZ
Sandia National Laboratories

A. J. Moulson and J. M. Herbert, *Electroceramics,* Chapman & Hall, New York, 1990.

L. E. Cross, *Am. Ceram. Soc. Bull.* **63**(4), 586–590 (1984).

R. R. Tummala, *Am. Ceram. Soc. Bull.* **67**(4), 752–758 (1988).

B. A. Tuttle, *Mater. Res. Soc. Bull.* **12**(7), 40–45 (1987).

CERAMICS AS ELECTRICAL MATERIALS

For a large number of applications involving ceramic materials, electrical conduction behavior is dominant. In certain oxides, borides (see BORON COMPOUNDS), nitrides (qv), and carbides (qv), metallic or fast ionic conduction may occur, making these materials useful in thick-film pastes, in fuel cell applications (see FUEL CELLS), or as electrodes for use over a wide temperature range. Superconductivity is also found in special ceramic oxides, and these materials are undergoing intensive research. Other classes of ceramic materials may behave as semiconductors (qv). These materials are used in many specialized applications including resistance heating elements and in devices such as rectifiers, photocells, varistors, and thermistors. Ceramics (qv) are equally important as electrical insulators.

Electrical Conduction Phenomena

The electrical characteristics of ceramic materials vary greatly, since the atomic processes are different for the various conduction modes. The transport of current may be because of the motion of electrons, electron holes, or ions. Electrical ceramics are commonly used in special situations where refractoriness or chemical resistance are needed, or where other environmental effects are severe (see REFRACTORIES). Thus it is also important to understand the effects of temperature, chemical additives, gas-phase equilibration, and interfacial reactions.

The electrical conductivity is one of the few physical quantities that has been found to vary by many orders of magnitude, ranging from $10^5(\Omega\cdot cm)^{-1}$ for conducting oxides such as rhenium(VI) oxide, ReO_3, or chromium(IV) oxide, CrO_2, to $10^{-14}(\Omega\cdot cm)^{-1}$ for highly insulating materials such as steatite porcelain. Certain compounds exhibit changes in conductivity by several orders of magnitude as a result of doping, temperature, or processing effects.

Ionic Conduction Phenomena

For an ion to move through the lattice, there must be an empty equivalent vacancy or interstitial site available, and it must possess sufficient energy to overcome the potential barrier between the two sites. Ionic conductivity, or the transport of charge by mobile ions, is a diffusion and activated process.

Ions in ceramic crystalline materials constitute potential charge carriers that can contribute to electrical conductivity, but analysis requires a determination of the concentration and mobility of the charge carriers.

In polycrystalline materials, ion transport within the grain boundary must also be considered. For oxides with close-packed oxygens, the O-ion almost always diffuses much faster in the boundary region than in the bulk. In general, second phases at grain boundaries are less close packed and provide a pathway for more rapid diffusion of ionic species. Thus the simplified picture of bulk ionic conduction is made more complex by these additional effects.

Fast Ion Conductors

Several types of compounds showing exceptionally high ionic conductivity have become of technological interest: *(1)* halides and chalcogenides of silver and copper in which the metal atom is disordered over a large number of interstitial sites; *(2)* oxides having the β-alumina, $NaAl_{11}O_{17}$, structure in which conduction channels are formed for the monovalent cation, resulting in high mobilities; and *(3)* oxides of the fluorite, CaF_2, structure in which large concentrations of defects are caused either by a variable-valence cation or by solid solution from a second cation of lower valence; eg, $CaO-ZrO_2$ or $Y_2O_3-ZrO_2$.

Particular interest has been shown in stable materials having high ionic conductivities because of applications as solid electrolytes (Table 1).

Glass Conduction

Electrical conduction in glasses is mainly attributed to the migration of mobile ions such as Li^+, Na^+, K^+, and OH^- under the influence of an

Table 1. Fast Ion Conductors

Compound	Temperature, °C	Conducting ion	$\sigma_{ion}(\Omega \cdot m)^{-1}$
$NaAl_{11}O_{17}$	300	Na	35
$Na_3OZr_2PSi_2O_{12}$	300	Na	20
CeO_2 + 12 mol % CaO	700	O	4
ZrO_2 + 12 mol % CaO	1000	O	0.8
$K_{1.4}Fe_{11}O_{17}$	300	K	2
$Li_2B_4O_7$[a]	150	Li	10^{-4}
	400	Li	0.1
$Li_4B_7O_{12}Cl$[a]	300	Li	0.2
crystal	300	Li	0.8

[a] Glass

applied field. At higher temperatures, >250°C, divalent ions, eg, Ca^{2+} and Mg^{2+}, contribute to conduction, although their mobility is much less. Conduction in glass is an activated process and thus the number of conducting ions increases with both temperature and field. Fast ion conduction may occur in certain glasses such as silver and alkali borates, phosphates, and molybdates. The criterion for such conduction is the existence of a highly ordered structure having channels in which ions can easily move within the sublattice.

Electronic Conduction in Ceramics

The relatively high mobilities of conducting electrons and electron holes contribute appreciably to electrical conductivity. In some cases, metallic levels of conductivity result; in others, the electronic contribution is extremely small. In all cases the electrical conductivity can be interpreted in terms of carrier concentration and carrier mobilities. Including all modes of conduction, the electronic and ionic conductivity is given by the general equation:

$$\sigma_i = \sigma(t_{ionic} + t_{electronic}) \quad (1)$$

where the total electronic conductivity is given as:

$$\sigma = ne\mu_n + pe\mu_p \quad (2)$$

where n and p denote the concentrations of electrons and holes, respectively, and μ_n and μ_p are the corresponding mobilities. The mobility of these electronic defects are generally much higher than those of ionic defects because they are of lower mass and charge density and less confined to particular atomic sites. Scattering of electrons and holes occurs by phonons, point defects, dislocations, and grain boundaries. The conductivity is also determined by the concentration of electrons and holes, and is generally described by energy band structures. Figure 1 schematically illustrates the three and band energy configurations corresponding to intrinsic metal, semiconductor, and insulator conduction. The highest energy band, which is completely filled at

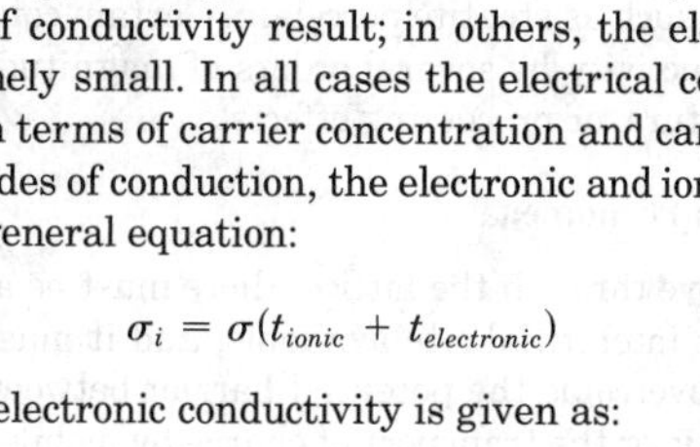

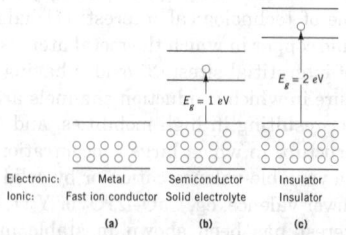

Figure 1. Schematic of band gap energy, E_g, for the three types of electronic and ionic conductors. For electronic conductors the comparison is made of the relative occupancy of valence and conduction bands. For ionic conductors, the bands correspond to the relative occupancy of ionic sublattices. For (**a**), $n = 10^{22} mL^{-1}$ for (**b**), $n = 10^{10} mL^{-1}$; and for (**c**), $n = 1\ mL^{-1}$.

$T = 0$ K, is called the valence band, whereas the next higher band, being empty at this temperature, is the conduction band. These energy levels are separated by an energy or band gap (E_g), which normally is not occupied by electrons.

As shown by Figure 1, at $T > 0$ K, metallic conduction occurs in ceramic materials where there are partially filled valence bands and a corresponding overlap with the unoccupied conduction band states. When a small energy band gap is present with no overlap, the materials become semiconducting. If a large energy gap occurs the materials are insulating, because the energy gap is too great to thermally promote electrons into the conduction band.

Electronic Conducting Ceramics

High Electronic Conductivity. Metals are not the only materials that exhibit metallic conductivity. Some transition-metal oxides also have high levels of conductivity resulting from overlap of the electron orbitals forming wide, unfilled d or f electron energy bands having concentrations of quasi-free electrons of $10^{22}-10^{23} mL^{-1}$. This is equivalent to metallic conduction. The perovskite and rutile oxide structures have been the most studied.

Semiconducting Ceramics. Most oxide semiconductors are either doped to create extrinsic defects or annealed under conditions in which they become nonstoichiometric. Although the resulting defects have been carefully studied in many oxides, the precise nature of the conduction is not well understood. A list os some impurity semiconductors is given in Table 2.

Spinels and Ferrites

The range of published values for the resistivity of ferrite and garnet materials is wide, from about 10^{-4} to 10^9 ohm · cm at room temperature. The low conductivity is typically associated with the simultaneous presence of Fe^{2+} and Fe^{3+} ions on equivalent lattice sites. Ferrite chemistry is complex and it is necessary to consider where the cations within the crystal locate as well as their type and concentration. In the spinel lattice, particular cations display a preference for one of several possible tetrahedrally or octahedrally coordinated lattice sites. The concentration and degree of preference for cations in the lattice modifies the site distribution of any specific cation. Ferrites should have very high electrical resistivities in order to eliminate eddy current and dielectric losses, and allow full penetration of electromagnetic fields throughout the solid. High resistivity is obtained when a cation has only one valence state in the lattice site. For the processing of these materials, high sintering temperatures and reducing atmospheres should be avoided. Such conditions produce mixtures of high and low valence cations. The three main commercial classes of spinels are nickel–zinc, $(NiZn)Fe_2O_4$, manganese–zinc, $(MnZn)Fe_2O_4$, and magnesium–manganese ferrites, eg, $(MnMg)Fe_2O_4$.

Garnets have played an important role in the development of highly sophisticated microwave devices since the development of yttrium–iron garnet, yttrium iron oxide.

Superconductivity

Superconductivity is partly typified by a perfect metallic conductor that has no resistance to current flow below a critical transformation

Table 2. Impurity Semiconductors

			n-Type		
Cds	$BaTiO_3$	Nb_2O_5	Fe_2O_3	WO_3	GeO_2
CdSe	$SrTiO_3$	Ta_2O_5	Tl_2O_3	TiO_2	MnO_2
ZnF_2	$PbCrO_4$	Fe_3O_4	In_2O_3	SnO_2	ZnO
			p-Type		
Se	CuI	Ag_2O	Hg_2O	NiO	PdO
Te	SnS	Cu_2O	MnO	FeO	CoO
			Amphoteric		
Si	Sn	PbSe	Ti_2S	Al_2O_3	Mn_3O_4
Ge	PbS	SiC	PbTe	Co_3O_4	UO_3

temperature (T_c). Besides the disappearance of electrical resistance, there is an expulsion of magnetic flux described as the Meissner effect.

The $YBa_2Cu_3O_{7-x}$ compound is the most intensively investigated high temperature oxide superconductor. It has an oxygen-deficient, distorted orthorhombic 1:1:3 (ABO_3) perovskite structure with *Pmmm* space group. The crystal structure shows the Y and Ba ions located in the center of the unit cell, Cu ions on the corners, and O ions on the edges. Based on neutron diffraction studies, the structure indicates two important features for Cu ions: *(1)* nonplanar CuO_2 planes extend in the crystallographic *ab* planes at $z = 0.36$ and -0.36, and *(2)* fence-like, square planar CuO_3 linear chains extending along the *b* axis at $z = 0$.

Ferroelectrics

A technologically important and often studied ceramic is ferroelectric barium titanate, $BaTiO_3$, which is the base for large volume production of disk and multilayer (MLC) capacitors. Barrier layer and intergranular capacitors (GBBL), as well as nonlinear positive temperature coefficient (PTC) resistors, are also largely based on this material. Ferroelectricity in these materials is a result of dipolar shifts in the cation lattice in going through a temperature transformation range, resulting in crystal symmetry changes and the development of spontaneous polarization charges.

Lead-based materials, PZT, PLZT, and PMN, form a class of ceramics with either important dielectric, relaxor, piezoelectric, or electrooptic properties, and are thus used for applications in actuator and sensor devices. Resistive properties of these materials in film form mirror the conduction processes in the bulk material. Common problems associated with their use are low dielectric breakdown, increased aging, and electrode injection, decreasing the resistivity and degrading the properties.

Varistors

Varistors have a highly nonlinear current–voltage characteristic and are used to protect electronic equipment against voltage surges. Varistors are processed so as to develop a similar type microstructure to that for PTC devices, where conductive grains are surrounded by thin insulating barriers.

To obtain these properties, polycrystalline ZnO is typically doped with antimony(III) oxide, Sb_2O_3, bismuth(III) oxide, Bi_2O_3, and other additives, creating a highly complex oxide ceramic. It is believed that the grains are separated by a thin adsorbed oxide layer that stores a negative charge equal in magnitude to the positive charged depletion layer, which lies wholly within each grain near the grain boundary. A barrier to electron flow at low voltages is created by these depletion regions. Because the varistor action takes place across the ZnO grain boundaries, tailoring devices for specific breakdown voltages is done by fabricating the varistor with the appropriate number of grain boundaries in series between the electrodes.

Insulating Ceramics

Ceramic insulators are materials that do not pass a current when an electric field is applied, owing to the very low concentration of mobile charge carriers, electrons, or ions. These nonconducting materials are insulators. Important ceramic insulators such as SiO_2, Al_2O_3, mullite, BeO, AlN, boron nitride, BN, and Si_3N_4 have resistivities $>10^{14}$ 2Ω · cm. These high values are the result of a large energy gap between a filled valence band and the next available energy level, where the promotion of an electron into a higher state is energetically unfavorable. The conductivity of these ceramics is significantly influenced by both ionic and electronic defects.

The primary function of insulation in electrical circuits is the physical separation of conductors and the regulation or prevention of current flow between them. Ceramic insulators are used in many demanding applications where high electrical resistance is a requirement, together with other important properties such as thermal conductivity, high operating temperatures, high dielectric strength, low dielectric loss, resistance to thermal shock, environmental resistance, thermal expansion, and long-life characteristics. Insulators of this type are known as linear dielectrics.

SiO_2 exhibits the lowest loss properties of any inorganic material. It is commonly used in insulating fibers and in the development of electrical porcelains ($R_2O{\cdot}Al_2O_3{\cdot}SiO_2$). These materials have high dielectric strength with low loss and are therefore suitable for such high voltage applications as transmission line insulators, high voltage circuit breakers, and cutouts.

Al_2O_3 is the most widely used oxide ceramic insulator, particularly as a substrate material for microelectronic devices and for use in severe conditions such as in spark plug insulation.

For electrical insulating sealing applications and heat sinks, Al_2O_3, AlN, SiC, and Si_3N_4 are the most commonly used materials. SiC and Si_3N_4 are also industrially valuable as high temperature heat exchangers because of high thermal conductivity and electrical insulating behavior, high hardness, durability, excellent high temperature, corrosion, and thermal shock resistance.

Processing Effects

Most electrical ceramic properties are ultimately structure-dependent. As indicated, stoichiometry is also very important, especially in ceramic materials having variable valence cations. Thus close control of composition, heat treatment temperature, and ambient atmosphere is needed in processing. Control of microstructure and grain size, particularly in magnetic ceramics, also crucial to the attainment of desired properties. This is equally important in other categories of electrical ceramics, where density changes and the presence of additional phases can grossly affect properties.

R. C. BUCHANAN
R. D. ROSEMAN
University of Cincinnati

W. D. Kingery, H. K. Bowen and D. R. Uhlman, *Introduction to Ceramics*, 2nd ed., John Wiley & Sons, Inc., New York, 1976.

D. W. Richerson, *Modern Ceramic Engineering*, 2nd ed., Marcel Dekker, Inc., New York, 1992.

L. L. Hench and J. K. West, *Principles of Electronic Ceramics*, John Wiley & Sons, Inc., New York, 1990.

R. C. Buchanan, *Ceramic Materials for Electronics*, 2nd ed., Marcel Dekker, Inc., New York, 1991.

CEREALS. See WHEAT AND OTHER CEREAL GRAINS.

CERIUM AND CERIUM COMPOUNDS

Cerium, Ce, at no. 58, is the most abundant member of the series of elements known as lanthanides. Cerium ranks ca 25th in abundance in the earth's crust, and cerium, which occurs at 60 ppm crustal abundance, not lanthanum at 30 ppm, is the most abundant lanthanide.

Cerium, at wt 140.12, electron configuration [Xe] $4f^2 6s^2$, is characterized chemically by having two stable valence states, Ce^{3+}, cerous, and Ce^{4+}, ceric, for which the ionic radii are 114 pm and 97 pm, respectively. The easily accessible tetravalent ion is unique among the Ln series. Indeed, the ceric ion is a powerful oxidizing agent but when associated with the ligand oxygen, it is completely stabilized, and ceric oxide, CeO_2, is the form of cerium most widely used. Cerium still contributes to lighting in the 1990s but is now also to be found in automobiles, televisions, and other technologies.

Resources

Whereas certain rocks of igneous origin formed by melting and recrystallization can include minerals enriched in the lanthanides, cerium is usually present as a trace element rather than as an essential component. Only a few minerals in which cerium is an essential structure-defining component occur in economically significant deposits. Two minerals supply the world's cerium, bastnasite, $LnFCO_3$, and monazite, $(Ln,Th)PO_4$.

Bastnasite has been identified in various locations on several continents. The largest recognized deposit occurs mixed with monazite and iron ores in a complex mineralization at Baiyunebo in Inner Mongolia, China. The mineral is obtained as a by-product of the iron ore mining. The other commercially viable bastnasite source is the Mountain Pass, California deposit where the average Ln oxide content of the ore is ca 9%. This U.S. deposit is the only resource in the world that is mined solely for its content of cerium and other lanthanides.

Several countries supply monazite concentrates for the world market. Extensive deposits along the coast of western Australia are worked for ilmenite and are the primary source of world monazite. Other regions of Australia, along with India and Brazil, also supply the mineral. Because monazite contains thorium, India and Brazil have embargoed its export for many years. In the United States, commerce in the mineral is regulated by the Nuclear Regulatory Commission.

World reserves of contained Ln oxide are estimated as 50,000,000 t with ca 75% of this in China, mostly at Baiyunebo, ca 15% in the United States, and ca 5% in India.

Production

The production of cerium derivatives begins with ore beneficiation and production of a mineral concentrate. Attack on that concentrate to create a suitable mixed lanthanide precursor for later separation processes follows. Then, depending on the relative market demand for different products, there is either direct production of a cerium-rich material, or separation of the mixed lanthanide precursor into individual pure lanthanide compounds including compounds of pure cerium, or both. The starting mineral determines how the suitable mixed lanthanide precursor is formed. In contrast the separation technology, which involves liquid–liquid countercurrent extraction or solvent extraction (SX), for preparing the individual lathanides is essentially independent of the starting mineral (see EXTRACTION, LIQUID–LIQUID). Thus different feedstocks can ultimately be processed by the same separation routines and equipment.

Cerium(IV) Chemistry

The fluorite structure, which has a large crystal lattice energy, is adopted by CeO_2 preferentially stabilizing this oxide of the tetravalent cation rather than Ce_2O_3. Compounds or cerium(IV) other than the oxide, ceric fluoride, CeF_4, and related materials, although less stable, can be prepared.

The tetravalent ceric ion, Ce^{4+}, is the only nontrivalent lanthanide ion, apart from Eu^{2+}, stable in aqueous solution. As a result of the higher cation charge and smaller ionic size, ceric salts are much more hydrolyzed in aqueous solution than those of the trivalent lanthanides. Ceric salt solutions are strongly acidic, basic salts tend to form readily, and there are no stable simple salts of weak acids.

The double salts, ceric ammonium nitrate, $(NH_4)_2[Ce(NO_3)_6]$, and ceric ammonium sulfate, $(NH_4)_2[Ce(SO_4)_3]$, are stable orange compounds prepared by dissolving freshly prepared hydrated oxide in excess of the appropriate acid and adding the correct amount of ammonium salt.

Cerium in the tetravalent state is a strong oxidizing agent and can be reduced by, eg, oxalic acid, halogen acids, ferrous salts, and hydrogen peroxide.

Cerium(III) Chemistry and Compounds

Cerium is strongly electropositive having a low ionization potential for the removal of the three most weakly bound electrons. The trivalent cerous ion, Ce^{3+}, apart from its possible oxidation to Ce(IV), closely resembles the other trivalent lanthanides in behavior.

Ce(III) forms a water-insoluble hydroxide, carbonate, oxalate, phosphate, and fluoride; sparingly soluble sulfate and acetate; and soluble nitrate and chloride (and bromide). In solution the salts are only slightly hydrolyzed. The carbonate is readily prepared and is a convenient precursor for the preparation of other derivatives. The sparingly soluble sulfate and acetate decrease in solubility with an increase in temperature. Calcination of most Ce(III) salts results in CeO_2.

Cerous salts in general are colorless because Ce^{3+} has no absorption bands in the visible. Trivalent cerium, however, is one of the few lanthanide ions in which parity-allowed transitions between $4f$ and $5d$ configurations can take place and as a result Ce(III) compounds absorb in the ultraviolet region just outside the visible.

Metal

In bulk form cerium is a reactive metal. Pure metal is prepared by the calciothermic reduction of CeF_3.

$$2\ CeF_3 + 3\ Ca \rightarrow 2\ Ce + 3\ CaF_2$$

On a fresh surface the metal has a steely lustre but rapidly tarnishes in air as a result of surface formation of oxide and carbonate species. For protection against oxidation the metal is usually stored in a light mineral oil. Cerium reacts steadily with water, readily dissolves in mineral acids, and is also attacked by alkali; it reacts with most nonmetals on heating.

Cerium metal has unique solid-state properties and is the only material known to have a solid–solid critical point. Three allotropes, α, β, γ, are stable at or close to ambient conditions and have complex structural interrelationships.

Mischmetal. Mischmetal contains, in metallic form, the mixed light lanthanides in the same or slightly modified ratio as occurs in the resource minerals. It is produced by the electrolysis of fused mixed lanthanide chloride prepared from either bastnasite or monazite. Although the precise composition of the resulting metal depends on the composition of chloride used, the cerium content of most grades is always close to 50 wt %.

An alternative commercial form of a metallic mixed lanthanide-containing material is rare-earth silicide, produced in a submerged electric-arc furnace by the direct reduction of ore concentrate, bastnasite, iron ore, and quartz.

Biological and Toxicological Behavior

In general the lanthanides, including cerium, have a low toxicity rating, especially when they are present in material having low aqueous solubility.

Table 1. Commercially Available Cerium-Containing Materials and Uses

Type	Material	Use	Cerium content
A	rare-earth chloride, mischmetal	FCC[a] catalysts iron metallurgy	principal component[b]
B	lanthanum concentrate, La-Ln chloride	FCC[a] catalysts	minor component[b]
C	cerium concentrate	glass polishing, glass decolorizing	dominant element[c]
D	oxide, nitrate, metal	autoemission catalysts, etc	>ca 90 wt %
E	oxide, salts	luminescence, catalysts	>ca 99 wt %

[a] FCC = fluid catalytic cracking.
[b] Of mixed-lanthanide composition.
[c] In oxide-type compound.

Economic Aspects

The yearly worldwide production of lanthanides is nearly 70,000 tons measured as contained Ln oxide. For finished products the principal supplying companies are Molycorp (United States), Rhône-Poulenc (France), several Japanese companies, such as Mitsubishi, Santoku, and Shinetsu, and some Chinese organizations. The rise of Chinese lanthanide production during the 1980s has become a significant factor in the global market.

The various cerium-containing derivatives available commercially are summarized in Table 1.

BARRY T. KILBOURN
Molycorp Inc.

J. W. Mellor, *A Comprehensive Treatise on Inorganic and Theoretical Chemistry,* Vol. 5, Longmans, UK, 1924, Chapt. 38; R. C. Vickery, *Chemistry of the Lanthanons,* Academic Press, New York, 1953.

K. A. Gschneidner and L. Eyring, *Handbook on the Physics and Chemistry of Rare Earths,* Vols. 1–13, North-Holland, 1992; *Handbook of Inorganic Chemistry,* Gmelin, System 39, various dates.

J. Scherzer, in R. G. Bautista and M. M. Wong, eds., *Rare Earths, Extraction, Preparation, and Applications,* TMS, 1988, p. 317.

"Rare-Earth Metals," in *Mineral Commodity Summaries 1991,* U.S. Interior Bureau of Mines, Washington, D.C., 1991.

CESIUM AND CESIUM COMPOUNDS

Cesium, Cs, is a member of the Group 1 alkali metals. It resembles potassium and rubidium in the metallic state, and the chemistry of cesium is more like that of these two elements than like that of the lighter alkali metals.

Until the late 1970s cesium continued to be little more than a research element. Much of its limited production was for research into thermionic power conversion, magnetohydrodynamics (qv), and ion propulsion. Although the potential for these applications has not materialized, cesium chemical usage has increased significantly as catalysts in the chemical and petrochemical industries and in biotechnical engineering.

Physical Properties

Pure cesium is a silvery white, soft, ductile metal. Surface alteration by minute traces of oxygen result in the metal taking on a golden hue. Of the stable alkali metals, ie, excluding francium, cesium has the lowest boiling and melting point, the highest vapor pressure, the highest density, and the lowest ionization potential. These properties and the large radius of the monovalent cesium ion have important consequences directly related to applications.

Chemical Properties

The ionization potential of the alkali metals decreases with increasing atomic number; consequently cesium is generally far more reactive than the lower members of the alkali metal group. When cesium is exposed to air, the metal ignites spontaneously and burns vigorously producing a reddish violet flame to form a mixture of cesium oxides. Similarly cesium reacts vigorously with water to form cesium hydroxide, the strongest base known, as well as hydrogen; together with air and water a hydrogen explosion usually occurs as the burning cesium readily ignites the liberated hydrogen gas. Cesium, the most active of the alkali metals toward oxygen and the halogens, is the least reactive toward nitrogen, carbon, and hydrogen.

Cesium salts are, in general, chemically similar to other alkali metal salts. Cesium forms simple alkyl and aryl compounds that are similar to those of the other alkali metals. They are colorless, solid, amorphous, nonvolatile, and insoluble, except by decomposition, in most solvents except diethylzinc.

Occurrence

Cesium is the rarest of the naturally occurring alkali metals, ranking 40th in elemental prevalence. Nevertheless, it is widely distributed in the earth's crust at very low concentrations. By far the most important commercial cesium source is pollucite, ideally $CS_2O{\cdot}Al_2O_3{\cdot}4SiO_2$.

Economic concentrations of pollucite usually occur in highly zoned complex pegmatites, associated with lepidolite, petalite, and spodumene. The Bernic Lake orebody of Tantalum Mining Corp. of Canada Ltd. (Tanco) in southeastern Manitoba, Canada, is the world's largest cesium source containing approximately two-thirds of the known ore. Other significant ore deposits are the Bikita pegmatite in Zimbabwe, and in the Karibib desert of Namibia.

Processing of Pollucite

Pollucite preparation consists simply of mining the ore, crushing it to required size, followed in some instances by hand picking. No other concentration is required. Chemetall Gmbh, Frankfurt, Germany, a Division of Metallgesellschaft AG, is the predominant processor worldwide; the only significant U.S. producer is Cabot Performance Materials Division, at their Revere, Pennsylvania plant.

There are three basic methods of converting pollucite to cesium metal or compounds: direct reduction with metals; decomposition with bases; and acid digestion: The latter is the most common process employed. In each case grinding of the ore to 75 μm precedes conversion.

Production of Cesium Metal

The primary producer of cesium metal in the United States is Cabot Corporation. Cesium is produced by thermochemical methods, thermal decomposition, and electrolytic reduction.

Cesium Alloys

Eutectics melting at about −30, −47, and −40°C are formed in the binary systems, cesium–sodium at about 9% sodium, cesium–potassium at about 25% potassium, and cesium–rubidium at about 14% rubidium. A ternary eutectic with a melting point of about −72°C has the composition 73% cesium, 24% potassium, and 3% sodium. Cesium and lithium are essentially completely immiscible in all proportions.

Cesium Compounds

Cesium compounds are manufactured and distributed by a comparatively large number of companies, considering the size of the total cesium market. Those companies that process pollucite produce their own range of products, some of which are then reprocessed and refined by other, smaller, specialty companies, many of which are located in the United States. Compounds include carbonates (cesium carbonate, Cs_2CO_3 mol wt 325,82; and cesium hydrogen carbonate, $CsHCO_3$, mol wt 193.92), cesium chromate (Cs_2CrO_4, mol wt 381.80), cesium halides (cesium bromide, CsBr, mol wt 212.82; cesium chloride, CsCl, mol wt 168.36; cesium perchlorate, $CsClO_4$, mol wt 232.35; cesium fluoride, CsF, mol wt 151.90; and cesium iodide, CsI, mol wt 259.81), cesium hydroxide (CsOH, mol wt 149.91), cesium nitrate ($CsNO_3$, mol wt 194.91), cesium oxides (cesium monoxide, Cs_2O, mol wt 281.81; the suboxides: cesium heptaoxide, CsO_7, tetracesium oxide, Cs_4O, heptacesium dioxide, Cs_7O_2, and tricesium oxide, Cs_3O; cesium peroxide, Cs_2O_2; and cesium superoxide, CsO_2), cesium permanganate ($CsMnO_4$, mol wt 251.84), and cesium sulfates (cesium sulfate, Cs_2SO_4, mol wt 361.87; and cesium aluminum sulfate (cesium alum) $Cs_2SO_4{\cdot}Al_2(SO_4)_3{\cdot}24H_2O$, mol wt 1136.39).

Economic Aspects

Most current applications require relatively small quantities of cesium, and hence annual world production of cesium metal and compounds is estimated to be on the order of 250 t of cesium chloride equivalents.

Handling, Storage, Transportation, and Occupational Health

Because of the high reactivity of cesium metal, special precautions are required for its storage, transportation, and use. Small quantities are usually contained in evacuated glass ampuls, larger quantities in stainless steel containers which are themselves contained in an outer packing, ensuring that the metal is kept from moisture or air.

The cesium ion is more toxic than the sodium ion but less toxic than the potassium, lithium, or rubidium ion.

The toxicology, occupational health haqards, and transportation regulations of cesium compounds result from the anion rather than the cesium cation. Producers and distributors provide an MSDS as well as detailed shipping requirements for each product.

Uses

The number of commercial uses of both cesium metal and its compounds has grown significantly since the early 1980s.

Electronic Applications. Electronic applications make up a significant sector of the cesium market. The main applications are in vacuum tubes, photoemissive devices, and scintillation counters.

Biotechnology and Medical Applications. Cesium chloride, and to a lesser extent the other halides, cesium trifluoroacetate and cesium sulfate, are used in the purification of nucleic acids (qv), ie, RNA and DNA viruses, and other macromolecules. In medicine, cesium salts have been considered both as an antishock reagent following the administration of arsenical drugs, though a contraindication is the disturbance to heart rhythm, and for the treatment of epilepsy.

Chemical Applications. Cesium metal is used in carbon dioxide purification as an adsorbent of impurities; in ferrous and nonferrous metallurgy (qv) it can be used as a scavenger of gases and other impurities.

One of the major commercial uses of cesium compounds is for catalyst doping. The performance of many metal ion catalysts can be enhanced by doping with cesium compounds, resulting from the low ionization potential of cesium and its ability to stabilize high oxidation states of transition metal oxo anions.

A growing use of cesium compounds is in the field of organic synthesis, for example in esterifications, polymerizations, and in the preparation of organoflourine compounds.

Molten cesium hydroxide (177–343°C) can be used in desulfurizing of heavy oils, recycling the hydroxide by steam hydrolysis.

Energy Related Applications. Much research, with regard to the use of cesium in energy related processes, has resulted in little commercial application.

Cesium is ideally suited for used in magnetohydrodynamic (MHD) power generators. The metal can be used as the plasma seeding agent in closed-cycle (MHD) generators using high temperature nuclear reactors (qv) as the primary heat source.

One alternative method of generating electricity directly from a heat source is by the use of a cesium vapor thermionic convertor, which uses cesium to neutralize the space charge above a hot cathode that is emitting electrons toward a cooler anode.

The use of cesium formate as a high density, low toxicity drilling mud could be commercialized in the near future: if it does so, this could increase total cesium usage by an order of magnitude.

Cesium Isotopes

Naturally occurring cesium and cesium minerals consist only of the stable isotope ^{133}Cs. The radioactive cesium isotopes such as ^{137}Cs are generated in fuel rods in nuclear power plants.

RICHARD O. BURT
Tantalum Mining Corporation

J. D. Downs, *SPE International Symposium on Oilfield Chemistry,* New Orleans, La., March 1993.

P. Černy, *Short Course in Granitic Pegmatites in Science and Industry,* Mineralogical Association, Winnipeg, Canada, 1982, p. 150.

M. Bick, *Cesium Chemicals from the World's Leading Producer,* Chemetall, Frankfurt, Germany.

H. Prinz, *Chemspec. Europe 90 Symposium,* 1990.

CHELATING AGENTS

A chelating agent, or chelant, contains two or more electron donor atoms that can form coordinate bonds to a single metal atom. After the first such coordinate bond, each successive donor atom that binds creates a ring containing the metal atom. This cyclic structure is called a chelation complex or chelate, the name deriving from the Greek word *chela* for the great claw of the lobster.

Chelation is an equilibrium system involving the chelant, the metal, and the chelate. Equilibrium constants of chelation are usually orders of magnitude greater than are those involving the complexation of metal atoms by molecules having only one donor atom.

Chelating agents may be used to control metal ion concentrations. Chelation complexes usually have properties that are markedly different from both the free metal ion and the chelating agent. Consequently, chelating agents provide a means of manipulating metal ions through the reduction of undesirable effects by sequestration or through creating desirable effects such as in metal buffering, corrosion inhibition, solubilization, and cancer therapy.

Chelates and chelation reactions are abundant in nature, ranging from delicately balanced life processes depending on traces of metal ions to extremely stable metal chelates in crude petroleums. Examples of biochemical processes involving chelates include photosynthesis, oxygen transport by blood, certain enzyme reactions, ion transport through membranes, and muscle contraction. Technological applications include scale removal from steam boilers, water softening, ore leaching, textile processing, food preservation, treatment of lead poisoning, chemical analysis, tissue-specific medical procedures, and micronutrient fertilization of agricultural crops.

Structure and Terminology

The structural essentials of a chelate are coordinate bonds between a metal atom or a stable oxo cation, M, which serves as an electron acceptor, and two or more atoms in the molecule of the chelating agent, or ligand, L, which serve as the electron donors. A chelating agent may be bidentate, tridentate, tetradentate, and so on, according to whether it contains two, three, four, or more donor atoms capable of simultaneously complexing with the metal atom. Examples are shown in Figure 1.

Compounds Having Chelating Properties

Compounds with chelating properties can be found in almost any class of structures containing two or more donor atoms spatially situated so that they can coordinate with the same metal atom. The chelate rings formed contain four or more members, but for the same donor atoms, the five- or six-membered chelate rings are usually the most stable and most useful.

Chelating agents may be either organic or inorganic compounds, but the number of inorganic agents is very small. The best known inorganic chelants are polyphosphates. The annual consumption of these compounds exceeds that of all the organic chelating agents combined. Although many hundreds of organic chelating agents are known, only a few members of a few classes of compounds find extensive industrial use. One important class of organic chelating agents is the group of phosphonic acids analogous to the amino- and hydroxycarboxylic acids. These phosphonate chelants possess many of the complexing properties of the inorganic polyphosphates, particularly threshold-scale inhibition, effective at much less than stoichiometric ratios of chelant to metal ion, but unlike the polyphosphates, the phosphonates are stable in water at high temperature and pH.

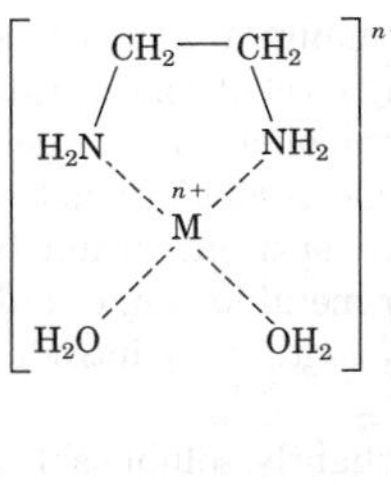

(1)

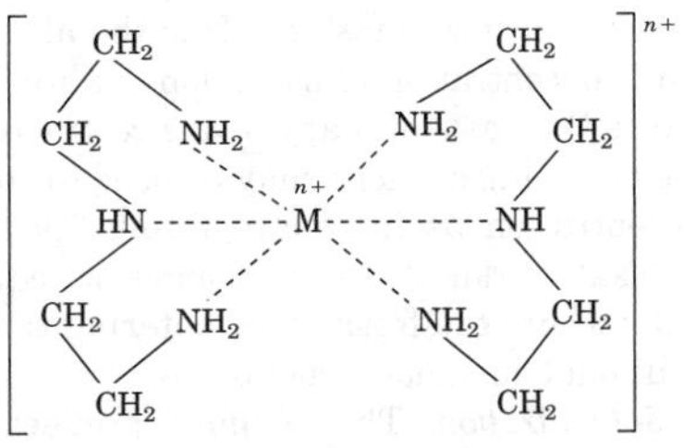

(2)

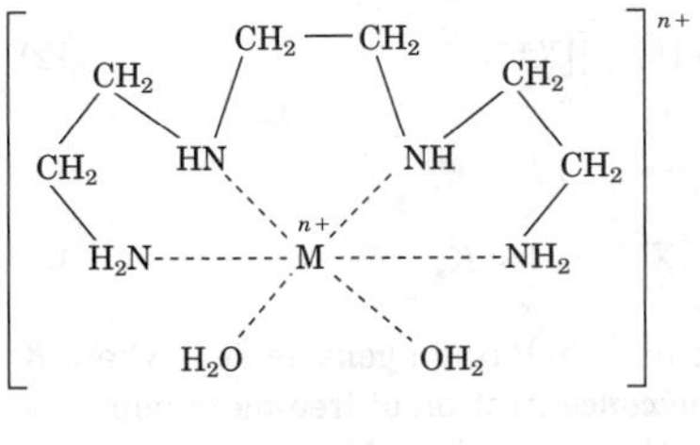

(3)

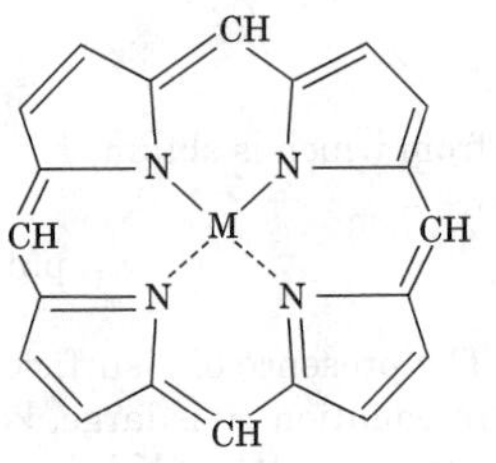

(4)

Figure 1. Types of chelates where **(1)** represents a tetracoordinate metal having the bidentate chelant ethylenediamine and monodentate water; **(2),** a hexacoordinate metal bound to two diethylenetriamines, tridentate chelants; **(3),** a hexacoordinate metal having triethylenetetramine, a tetradentate chelant, and monodentate water; and **(4),** a porphine chelate. The dashed lines indicate coordinate bonds.

Nomenclature and Structural Representation

Chelating Agents. Besides the conventional empirical and structural formulas, chelating compounds and chelates are often represented by type formulas, ie, formulas that show only generalized types of structural features.

For many macrocyclic ligands, simplified names are in common use. For example, crown ether nomenclature consists of four parts: *(1)* the number and type of fused rings on the polyether ring; *(2)* in square brackets the number of atoms in the polyether ring; *(3)* the word crown; and *(4)* the number of oxygen atoms in the macro ring. Ligand structures may be represented by any of the conventional means for depicting structure.

Chelates. Because of length and complexity, systematic names of chelates are little used except for special purposes, such as where unequivocal referencing is essential. Chelates are named in the literature in a variety of ways. The name of the ligand in a chelate is usually given a suffix -o or -ato if it is a negative group but remains unchanged if the ligand is electrically neutral. Prefixes indicate the number of bound ligand molecules. The central atom is given the name of the metal, or a derivative name having the suffix -ate, eg, cuprate and ferrate, if the complex is negatively charged. Oxidation states of the metals are indicated by Roman numerals, eg, iron(III), and ionic charges are shown as part of the name by Ewens-Bassett numbers, eg, (2+) or (1−). Chelates are often named merely as a complex, eg, cadmium complex with acetylacetone.

The Chelation Reaction

Chelate Formation Equilibria. In homogeneous solution the equilibrium constant for the formation of the chelate complex from the solvated metal ion and the ligand in its fully dissociated form is called the formation or stability constant. Whereas the ligand displaces solvent molecules coordinated to the metal, these solvent molecules do not generally enter into the equations. When more than one ligand molecule complexes with a metal atom, the reaction usually proceeds stepwise. For a metal having a coordination number of six and a bidentate chelating agent, the equations representing the equilibria are

$$\mathrm{M} + \mathrm{L} \rightleftarrows K_1 = \frac{[\mathrm{ML}]}{[\mathrm{M}][\mathrm{L}]} \tag{1}$$

$$\mathrm{ML} + \mathrm{L} \rightleftarrows \mathrm{ML_2}K_2 = \frac{[\mathrm{ML_2}]}{[\mathrm{ML}][\mathrm{L}]} \tag{2}$$

$$\mathrm{ML_2L} \rightleftarrows \mathrm{ML_3}K_3 = \frac{[\mathrm{ML_3}]}{[\mathrm{ML_2}][\mathrm{L}]} \tag{3}$$

$$\textit{overall}: \ \mathrm{M} + 3\,\mathrm{L} \rightleftarrows \mathrm{ML_3}K = \frac{[\mathrm{ML_3}]}{[\mathrm{M}][\mathrm{L}]^3} \tag{4}$$

where the square brackets represent concentrations in units of molarity. The overall stability constant is the product of the step stability constants, ie, $K = K_1K_2K_3$, and is often designated by β.

Experimentally determined equilibrium constants are usually calculated from concentrations rather than from the activities of the species involved.

Many experimental approaches have been applied to the determination of stability constants. Techniques include pH titrations, ion exchange, spectrophotometry, measurement of redox potentials, polarimetry, conductometric titrations, solubility determinations, and biological assay.

Displacement Equilibria. Species in solution are generally in formation–dissociation equilibrium, and displacement reactions of any given metal or ligand by another are possible. Thus,

$$\mathrm{ML} + \mathrm{M'} \rightleftarrows \mathrm{M'L} + \mathrm{M} \tag{5}$$

or

$$\mathrm{ML} + \mathrm{L'} \rightleftarrows \mathrm{ML'} + \mathrm{L} \tag{6}$$

If the stability constants for ML and M′L are K and K', respectively, then for the exchange shown in equation 8, the equilibrium constant is K_x.

$$K_x = \frac{[\mathrm{M}][\mathrm{M'L}]}{[\mathrm{ML}][\mathrm{M'}]} = \frac{K'}{K} \tag{7}$$

The extent of displacement depends on the relative stabilities of the complexes and the mass action effect of an excess of M′. For equivalent total amounts of M and M′, K_x must be on the order of 10^4 for 99% complete displacement to occur. Similar considerations apply for the displacement of L from ML by L′. The situation is quite analogous to the familiar competition of two bases for the hydrogen ion.

Metal exchange is the mechanism by which many foods, such as shortenings, shellfish, and dairy products, are stabilized against deleterious effects of trace metals by the addition of $Na_2CaEDTA$ ($\log K = 10.7$). Copper ($\log K = 18.8$) and iron ($\log K = 25.1$) displace calcium and become sequestered so that the remaining concentration of free iron or copper ions is too low for catalytic effects to occur at significant rates. Ligand exchange occurs when ascorbic acid bound to copper ($\log K = 1.57$) is displaced by EDTA, stabilizing this vitamin by disrupting an oxidation mechanism. Dyes and bleaches are similarly protected in the textile industry (see FOOD ADDITIVES; DYES AND DYE INTERMEDIATES; BLEACHING AGENTS).

Rates of Reaction. The rates of formation and dissociation of displacement reactions are important in the practical applications of chelation. Complexation of many metal ions, particularly the divalent ones, is almost instantaneous, but reaction rates of many higher valence ions are slow enough to measure by ordinary kinetic techniques.

Factors Affecting Stability. Many parameters influence the stability of chelates. Several of the stability factors common to all chelate systems are the size and number of rings, substituents on the rings, and the nature of the metal and donor atoms.

pH Effects. Being Lewis bases, the donor atoms of chelating agents react with Lewis acids such as metal and hydrogen ions. In the pH range of aqueous solutions, most of the well-known chelating agents exist as an equilibrium mixture of both protonated and unprotonated forms. Metal ions compete with hydrogen ions for the available donor atoms, and therefore, simultaneous equilibria exist that are treated mathematically by the simultaneous equations for the formation constants of the chelates and the acid dissociation constants of the chelating agents.

Titration Behavior. Protonated chelating agents exhibit titration behavior typical of their respective acidic groups, eg, carboxyl phenolic hydroxyl, ammonium, or sulfhydryl moieties, if they are titrated with bases where the cations have a very weak or no tendency to form chelates. In the presence of a metal ion that coordinates with the donor atom of one of these acidic groups, hydrogen is displaced by the metal, and the acid strength of the group appears to be enhanced. The hydrogen ion concentration is then higher than in the absence of the metal. Strongly chelated metal ions can increase the acidity of an acidic group by several orders of magnitude.

Titration of the hydrogen ion liberated from a strong chelating agent is used to determine the concentration of metal ions in solution. The strength of chelation can also be determined from these data.

Metal Buffering. The equation for the formation constant of the reaction

$$M^{n+} + L^{m} \overset{K}{\rightleftarrows} ML^{n-m} \quad (8)$$

can be rearranged to give

$$\frac{1}{[M^{n+}]} = \frac{K[L^{m-}]}{[ML^{n-m}]} \quad (9)$$

from which, on taking logarithms and defining pM = $-\log[M^{>n+}]$,

$$pM = \log\frac{[L^{m-}]}{[ML^{n-m}]} + \log K \quad (10)$$

The concentration of the metal ion can be controlled by adjusting the ratio of the concentrations of free ligand and metal chelate. If both species are present in appreciable amounts, moderate changes in either concentration have little effect on the ratio. The concentration of the metal ion can thus be buffered in a manner analogous to the buffering of pH by the presence of a weak acid and its anion

$$pH = \log\frac{[A^-]}{[HA]} + pK_a \quad (11)$$

In the equation for pM, log K appears instead of pK because K is a formation constant, the reciprocal of the chelate dissociation constant, which is analogous to the acid dissociation constant K_a.

By choice of chelating agents, and thus log K, pM can be regulated over a wide range. Two or more metals may be selectively buffered at different concentrations by a single chelating agent having different stability constants for the metals. Selective buffering of one metal to a low concentration in the presence of other metals is termed masking. It is the ability to maintain a nearly constant concentration of metal ions at almost any level of concentration that is the basis of many of the commercial uses of chelating agents. The buffering capacity may be used to supply metal ions at a definite concentration as in electroplating (qv) and in nutrient media (see MINERAL NUTRIENTS), or to remove or sequester metal ions in cleaning baths where the fresh stock entering the bath continually introduces additional amounts of metals.

Solubilization. The solubility product of a slightly soluble salt determines the concentration of metal ion that can be present in solution with the anion of that salt. For the salt MX the solubility product is

$$K_{sp} = [M^{n+}][X^{n-}] \quad (12)$$

from which is obtained

$$pM = \log[X^{n-}] - \log K_{sp} \quad (13)$$

The presence of a sufficiently strong chelating agent, ie, one where K in equation 10 is large, keeps the concentration of free metal ion suppressed so that pM is larger than the saturation pM given by the solubility product relation (eq. 13) and no solid phase of MX can form even in the presence of relatively high anion concentrations. The metal is thus sequestered with respect to precipitation by the anion, such as in the prevention of the formation of insoluble soaps in hard water.

Deposits of an insoluble salt can be dissolved as a salt of the metal chelate.

$$MX(s + L^{m-} \overset{K_s}{\rightleftarrows} [ML^{n-m}] + X^{n-}) \quad (14)$$

In the presence of the chelating agent and the insoluble salt, MX, pM of the solution is subject to both the metal buffering and the solubility equilibria.

Electrochemical Potentials. The oxidation potential of a solution-containing a metal in two of its valence states, M^{x+} and M^{x+n}, is given by

$$E = E^0 - \frac{RT}{nF}\ln\frac{[M^{x+n}]}{[M^{x+}]} \quad (15)$$

In the presence of a chelating agent, the concentrations of the two forms of the metal are buffered according to the simultaneous equations

$$[M^{x+n}] = \frac{[M_{ox}L]}{K_{ox}[L]} \text{ and } [M^{x+}] = \frac{[M_{red}L]}{K_{red}[L]} \quad (16)$$

where $M_{ox}L$ and $M_{red}L$ are the chelates of the oxidized and reduced

Table 1. United States Production of Industrial Chelating Agents[a]

Agent	Production, 10^3t/yr[b]	Price range[c] \$/kg	Producers[d]
STPP	277 (241)	0.66–0.88	F, Mon, Oc, Ol, R-P
citric acid	143 (147)	0.88–1.98	HRC, Pfi
aminopolycarboxylic acids	104 (66)		C-G, D, WRG
30–40% solutions		0.66–1.98	
powders/crystals		3.31–7.94	
gluconic acid	7 (11)	0.77–1.32	A, B, Pfa, Pfi, PMP
glucoheptonic acid[e]	9 (8)	0.44–1.54	B, Pfa
organophosphonates	12 (16)	1.54–2.42	Ma, Mon

[a] Estimated for 1990; includes all forms of the compounds.
[b] Thousands of metric tons consumed per year given in parentheses.
[c] Depends on form of compound and size of shipment.
[d] F = FMC Corp.; Mon = Monsanto Co.; Oc = Occidental Petroleum Corp.; Ol = Olin Corp.; R-P = Rhone-Poulenc Inc.; HRC = Haarmann amp; Reimer Corp. (subsidiary of Bayer USA, Inc.); Pfi = Pfizer Inc.; C-G = Ciba-Geigy Corp.; D = The Dow Chemical Company; WRG = W. R. Grace & Co.; A = Akzo America Inc.; B = Belzak Corp.; Pfa = Pfanstiehl Laboratories, Inc.; PMP = PMP Fermentation Products, Inc.; Ma = Mayo Chemical Co.
[e] Glucoheptonic acid, $C_7H_{14}O_8$.

forms of the ions and K_{ox} and K_{red} are the respective formation constants. Substituting these values in the potential equation gives

$$E = E^0 + \frac{RT}{nF} \ln \frac{K_{ox}}{K_{red}} - \frac{RT}{nF} \ln \frac{[M_{ox}L]}{[M_{red}L]} \quad (17)$$

The first two terms of the right-hand side of the equation are sometimes combined and expressed as $E^{0\prime}$, which is called the standard oxidation potential for the chelate system.

Environmental, Health, and Safety Aspects

The primary industrial chelating agents are essentially environmentally benign and nontoxic under the conditions incidental to normal handling and use. With these, eye irritation is mainly a function of the acidity or alkalinity of the form of the product and its solubility. However, use of nitrilotriacetic acid (NTA) is regulated in some jurisdictions. In medical uses the effects of the chelating agents on metal ions balances in the body tissues must be accommodated. Some of the commercial compounds used in smaller amounts as chelants, such as oxalic acid, are toxic, however. The hazards of using any chelant should be determined prior to use.

Solutions of iron chelates can be used to remove hydrogen sulfide and oxides of sulfur and nitrogen in industrial gas scrubbing processes before flue gases are released to the atmosphere.

Economic Aspects

Production and price estimates for the principal industrial chelating agents are given in Table 1.

WILLIAM L. HOWARD
Consultant
DAVID A. WILSON
The Dow Chemical Company

R. M. Smith and A. E. Martell, *Critical Stability Constants,* Vols. 1–6, Plenum Press, New York, 1974–1989.

S. Chaberek and A. E. Martell, *Organic Sequestering Agents,* John Wiley & Sons, Inc., New York, 1959.

A. E. Martell and M. Calvin, *Chemistry of Metal Chelate Compounds,* Prentice-Hall, Inc., Englewood Cliffs, N.J., 1952.

Chemical Economics Handbook Marketing Research Reports, SRI International, Menlo Park, Calif.

M. Salaices with M. Jaeckel and Y. Sakuma, in *Chemical Economics Handbook,* SRI International, Menlo Park, Calif., January 1991.

CHEMICALS IN WAR

Chemicals used in war fit into five categories: flame agents, incendiaries, smokes and obscurants, riot control agents, and toxic agents. Flame and incendiary agents are used to harass and inflict casualties, and to destroy structures and matériel. Smokes and obscurants are employed for screening, signaling, and target marking, in both offensive and defensive applications. Riot control agents are nonlethal tear agents most effective against unprotected personnel. Toxic chemical agents, used to achieve military objectives by producing casualties and discouraging enemy troops from certain areas of the battlefield (terrain denial), may be incapacitating or lethal.

Chemical warfare, a term used since 1917, is of vital interest not only to the world powers, but also to many developing countries. The relative simplicity of production and ease with which existing pesticide or other chemical plants can be converted to a weapons facility, make chemical weapons a continuing threat. Iraq and Libya, each of which possesses chemical weapon production facilities, have been known to use chemicals in war during the latter part of the twentieth century. Chemical weapons are not, as of this writing, banned. Thus whether employed or not, these materials exist as a potential weapon in any conflict. Toxic chemical agents may be defined as chemical substances in the gaseous, liquid, or solid state intended to produce casualty effects ranging from harassment through varying degrees of incapacitation to death. A few such agents are true gases but most are solids or liquids that are converted in use into a gaseous state or disseminated as aerosols (qv). For contamination of terrain the agent can also be disseminated in bulk form with or without additives to modify physical properties.

Toxic chemical agents produce a variety of physiological effects, depending on the nature of the agent. These effects, which range from death to mild incapacitation used for riot control, include blistering, choking, blood poisoning, lacrimation, nerve poisoning, laxation, and various forms of mental and physical disorganization.

Some of the criteria used in the selection of a suitable agent are effectiveness in extremely small concentrations; time to onset of action; effectiveness through various routes of entry into the body, such as the respiratory tract, eyes, and skin; stability in long-term storage; and ease of dissemination in feasible munitions.

Lethal Agents and Incapacitants

Research on chemical agents after World War I led to the elimination of all but a handful of chemicals as being of practical battlefield significance. At the time of World War II the only chemicals considered to be of practical significance included the mustard gases and phosgene.

Discovery of newer agents in Germany led to the availability of a class of compounds at least one order of magnitude more lethal than previously known, where death might occur in a matter of minutes instead of hours.

Mustard and Related Vesicants. Mustard, bis(2-chloroethyl) sulfide (Chemical Agent Symbol HD), $Cl(CH_2)_2S(CH_2)_2Cl$, is a colorless, oily liquid when pure. Most samples have a characteristic garliclike odor. It is primarily a vesicant: blisters are formed by either liquid or vapor contact. Mustard also attacks the eyes and lungs and is a systemic poison, so that protection of the entire body must be provided. It is insidious in its action; there is no pain at the time of exposure, and symptoms usually do not appear until several hours after exposure.

Its primary military application is to restrict the use of terrain or lower the mobility of opposing personnel in a contaminated area. It is volatile enough to be effective as a vapor in warm weather. Relatively modest expenditures of munitions yield severely incapacitating vapor dosages within less than an hour.

The procedure by which mustard is manufactured can be modified to yield either a mixture of mustard and Q (HQ) or a mixture of mustard and T (HT). These mixtures have several advantages over mustard alone, unless the agent is used only for vapor effects. HQ and HT are both more toxic, more vesicant, more persistent, and have lower melting points than mustard alone.

Properties. The physical properties of the mustards are summarized in Table 1.

Uses. The nitrogen mustards are used clinically in the treatment of certain neoplasms. They have been used in treatment of Hodgkin's disease, lymphosarcoma, and leukemia.

Nerve Agents. This term refers to two groups of highly toxic chemical compounds that generally are organic esters of substituted phosphoric acid. The nerve agents inhibit cholinesterase enzymes and thus come within the category of anticholinesterase agents (see ENZYME INHIBITORS). The three most active G-agents are tabun, ethyl phosphorodimethylamidocyanidate (Chemical Agent Symbol GA), $(CH_3)_2NPOCHNOC_2H_5$; sarin, isopropyl methylphosphonofluoridate (GB), $CH_3POFOCH(CH_3)_2$; and soman, pinacolyl methyl-phosphonofluoridate (GD), $CH_3POFOCHOCH_3C(CH_3)_3$.

The G-agent liquids under ordinary atmospheric conditions have sufficiently high volatility to permit dissemination in vapor form. They are generally colorless, odorless or nearly so, and are readily absorbable through not only the lungs and eyes but also the skin and intestinal tract without producing any irritation or other sensation on the part of the exposed individual. These agents are sufficiently potent so that even a brief exposure may be fatal. Death may occur in

Table 1. Properties of Mustard Gases

Property	HD	Q	T	HN1	HN2	HN3
mol wt	159.08	219.08	263.25	170.08	156.07	204.54
bp, °C[a]	$80^{0.67}$		$120.^{0.003}$	$85^{1.3}$	$87^{2.4}$	$144^{2.0}$
mp, °C	14.5	56	10	−34	−60	−4
density at 25°C, g/mL	1.2682		1.24	1.086	1.118	1.2347
volatility at 25°C, mg/m^3	925	0.4	2.8	2.29	3.581	0.120

[a] Pressure in kPa at which boiling point was determined is given as a superscript. To convert kPa to mm Hg, multiply by 7.5.

1–10 min, or be delayed for 1–2 h depending on the concentration of the agent.

Another class of nerve agents, discovered after World War II, is the V-agents. These materials are generally colorless and odorless liquids that do not evaporate rapidly at normal temperatures. In liquid or aerosol form, V-agents affect the body in a manner similar to the G-agents. The advantage is that V-agents produce casualties when absorbed through the skin in concentrations much lower than those required by the G-agents. Aerosolized V-agents are also quite lethal by inhalation.

Properties. Some physical properties of nerve agents are given in Table 2.

Binary Munitions. Binary munitions contain two nonlethal components that are mixed during flight to form a nerve agent. Each component is manufactured separately, remaining in its own container until the munition components are assembled just prior to use. Mixing and subsequent agent formation occurs after firing or launch of the munition. In addition to greatly reduced storage and handling hazards, the binary components can be manufactured in ordinary chemical facilities, which need not be equipped with the stringent safety and environmental controls required for the older nerve agent munitions. The binary technology also overcomes the long-term storage problems associated with the unitary nerve agents.

Agent BZ, 3-quinuclidinyl benzilate, $C_{21}H_{23}NO_3$, is a typical incapacitant. BZ is one of a group of substances, many of them glycolate esters, sometimes known as atropinemimetics. Their action on the central and peripheral nervous systems resembles that of atropine, $C_{17}H_{23}NO_3$. The effects of BZ are those of an anticholinergic psychotomimetic drug. These effects follow about a half hour after exposure to BZ aerosol, reach a peak in four to eight hours, and may then take up to four days to pass. Effects include disorientation with visual and auditory hallucinations. The agent disturbs the higher integrative functions of memory, problem solving, attention, and comprehension. There is a gradual return to normalcy.

Irritants

Riot control agent CS, a modern irritant compound, causes physiological effects that include extreme burning of the eyes accompanied by a copious flow of tears, coughing, difficulty in breathing; chest tightness, involuntary closing of the eyes, stinging sensation of moist skin; runny nose, and dizziness or "swimming" of the head. Heavy concentrations also cause nausea and vomiting.

The effects of agent CS are immediate, even in extremely low concentrations.

Table 2. Properties of Nerve Agents

Property	GA	GB	GD	VX[a]
formula wt	162.13	140.10	182.18	267.38
bp, °C	246	147	167	298
mp, °C	−50	−56	unknown	below −51
density at 25°C, g/mL	1.073	1.0887	1.0222	1.0083
volatility at 25°C, mg/m^3	610	21,900	3,060	10.5

[a] Agent VX, $C_{11}H_{26}NO_2PS$, is a phosphonothioic acid ester.

A water-soluble white crystalline solid, CS is disseminated as a spray, as a cloud of dust or powder, or as an aerosol generated thermally from pyrotechnic compositions. The formulation designated CS1 is CS mixed with an anti-agglomerant; when dusted on the ground, it may remain active for as long as five days. CS2, formulated from CS1 and a silicone water repellent, may persist for as long as 45 days.

The principal uses of CS are in riot control and training; it has limited tactical use in defensive military modes.

Flame

In the modern weapons arsenal flame agents are defined as various hydrocarbons, blends of hydrocarbons, and other readily flammable liquids, usually thickened with additives, that are easily ignited and can be projected to military targets. Although flame agents may be employed against buildings and other flammable targets, their primary role is against personnel in hardened structures or emplacements. In the United States, the principal application of flame agents is now in flame throwers and flame projectors, including flame rockets.

Incendiaries

Incendiary agents are designed for use in the planned destruction of buildings, property, and matériel by fire. Incendiaries burn with an intense, localized heat. They are very difficult to extinguish and are capable of setting fire to materials that normally do not ignite and burn readily. Although there are tactical applications for incendiary agents and munitions, they have played primarily a strategic role in modern warfare.

Incendiary Requirements. The mechanics of starting fires using incendiary agents involve a source of heat to act as a match to initiate combustion in a larger mass; combustible material to serve as kindling; and fuel. The match and the kindling are provided by the incendiary munition; the target is the fuel. All incendiary munitions, except for those containing materials that are spontaneously combustible, must have some sort of initiator such as a fuse or an ignition cup. The second element of the incendiary munition, the kindling, is the important factor, and both the amount and the nature of the combustible material in the munition have been the subject of much research and development.

Metal Incendiaries. Metal incendiaries include those of magnesium in various forms, and powdered or granular aluminum mixed with powdered iron(III) oxide. Magnesium is a soft metal which, when raised to its ignition temperature, burns vigorously in air. It is used in either solid or powdered form as an incendiary filling, and in alloyed form as the casing for small incendiary bombs.

Oil and Metal Incendiary Mixtures. PT1 is a complex mixture composed of magnesium dust, magnesium oxide, and carbon (qv), along with an adequate amount of petroleum (qv) and asphalt (qv) to form the paste.

Smokes

Military smokes are aerosols (qv) of gaseous, liquid, or particulate matter that are tactically employed to defeat enemy surveillance, target acquisition, and weapons guidance devices.

Screening Smokes. Military smoke screens are produced by dispersing either finely divided solids or minute liquid droplets in air. To be useful, a smoke screen must be sufficiently opaque to provide the desired screening power and long-lasting enough to achieve effective military results.

Types of Screening Smokes. The generation of oil smoke is based on the production of minute oil droplets by purely physical means. The most desirable droplet size is 0.5–1.0 μm. The tiny droplets of oil scatter light rays and produce a smoke that appears to be white, and any individual droplet would be transparent under magnification. These droplets are produced as the vaporized oil passes through the nozzle of a generator and is subsequently cooled by the surrounding air.

Another type of smoke mixture, a volatile hygroscopic chloride for thermal generation, has the U.S. Army designation HC, type C. It is composed of ca 6.7 wt % grained aluminum; 46.7 wt % zinc oxide, ZnO; and 46.7 wt % hexachloroethane.

A third screening-smoke type is white phosphorus, P_4, which reacts spontaneously with air and water vapor to produce a dense cloud of phosphorus pentoxide.

Signaling Smokes. Screening smokes also can be adapted for signaling purposes. However, a good signaling smoke must be clearly distinguished from the smoke incident to battle.

Colored signaling smokes are produced by volatilizing and condensing a mixture containing an organic dye. Of the dyes tested by the U.S. Army, the most satisfactory ones are the azo dyes (qv), the azine dyes (qv), the diphenylmethane dyes, and those of anthraquinone (qv) (see DYES AND DYE INTERMEDIATES).

Defense Against Toxic Agents

Defensive measures against toxic agents may be divided into four categories: agent detection and identification, individual and collective protection, decontamination, and medical defense. To these may be added a high degree of training in defensive measures and discipline in using them.

Detection. Many modern toxic chemicals are colorless and odorless; therefore, chemical and physical means must be employed in their detection and identification. Also, because of the extreme toxicity of modern chemical agents, especially the nerve agents, very sensitive and selective detection and identification are necessary.

Individual and Collective Protection. The primary item of individual protection is the protective or gas mask.

The mask alone, however, does not provide protection from substances such as nerve and blister agents that penetrate through the skin. Thus protection for the entire body should be provided. Airtight, impermeable clothing is available for personnel who must enter heavily contaminated areas. Such clothing is cumbersome and enervating because it retards release of body heat and moisture, and personnel efficiency is drastically lowered when it is worn for a long time.

Collective protection enclosures are required for groups of personnel. Such enclosures must be airtight to prevent inward seepage of contamination. They can be independent units or can be formed by adequately treating the interior walls of structures, tents, airplanes, or vehicles. A supply of uncontaminated air, provided by passing ambient air through high efficiency aerosol and carbon filters, must be provided.

Simplified collection protection equipment (SCPE) (U.S. designation M20) provides such protection, using lightweight elements consisting of an inflatable enclosure, a hermetically sealed filter canister, motor blower, protective entrance, and a support kit.

Decontamination. If contaminated equipment or material does not have to be used immediately, natural aeration is an effective decontaminant procedure, as most chemical agents, including the blister and V-agents, are volatile to a certain degree.

If decontamination cannot be left to natural processes, chemical neutralizers or means of physical removal must be employed. In general, the neutralizers are of two types: chlorine-based oxidants or strong bases. Some neutralizers have been especially developed for the decontamination of chemical agents.

Medical Defense. The most important items of U.S. medical defense against organophosphorus nerve agents are atropine and pralidoxime chloride (2-PAM), $C_7H_9ClN_2O_2$. These agents neutralize the effects of the anticholinesterase compounds and are capable of reactivating the inhibited enzymes.

Vesicant agents, such as mustard, require no special treatment once the burns have occurred. Copious washing is quite effective when used early for liquid contamination of the eyes, and soap and water removes the liquid agent from the skin. Burns resulting from mustard agent are treated like any other severe burn. The pulmonary injuries are treated symptomatically; antibiotics are used only IF indicated for the control of infection.

B. L. HARRIS
Consultant
U.S. Army
Chemical Systems Laboratory

W. A. Noyes, Jr., ed., *Science in World War II,* Little, Brown & Co., Boston, Mass., 1948.

The Problem of Chemical and Biological Warfare, Stockholm International Peace Research Institute (SIPRI), Humanities Press, New York, 1973.

Military Chemistry and Chemical Compounds. U.S. Army Field Manual 3-9/U.S. Air Forces Field Manual 355-7, U.S. Government Printing Office, Washington, D.C., Oct. 1975.

R. N. Sterlin, V. I. Yemel'yanov, and V. I. Zimin, "Chemical Weapons and Defense Against Them," *Khim. Oruzhiye i Zashchita ot Nego* (1975).

CHEMICALS FROM BRINE

Almost every country in the world has a source of brine containing usable minerals. Many have underground ore bodies that can be turned into brine by solution mining.

The seas and oceans of the world are the largest sources of brine. A second source of brine is found in terminal lakes. The Dead Sea in Israel and Jordan is an example of a large terminal lake with almost unlimited supplies of magnesium chloride, potassium chloride, and sodium chloride. More than two and a half million tons of potassium chloride are extracted from the Dead Sea each year.

Great Salt Lake, Utah, is the largest terminal lake in the United States. From its brine, salt, elemental magnesium, magnesium chloride, sodium sulfate, and potassium sulfate are produced.

A third source of brine is found underground. Underground brines are primarily the result of ancient terminal lakes that have dried up and left brine entrained in their salt beds.

A fourth source of brine is obtained through solution mining. Potash is mined in Moab, Utah by solution mining. Much of the food-grade sodium chloride in the United States, Europe, and other parts of the world is solution-mined.

The main metals in brines throughout the world are sodium, magnesium, calcium, and potassium. The main nonmetals are chloride, sulfate, and carbonate, with nitrate occurring in a few isolated areas. A significant fraction of sodium nitrate and potassium nitrate comes from these isolated deposits.

All of these metallic and nonmetallic ions join together in a complicated array of salts and minerals called evaporites. Several evaporites usually crystallize simultaneously in a mixture. This often makes separation into pure chemicals difficult.

Recovery Processing

Solar Evaporation. Recovery of salts by solar evaporation is favored in hot dry climates. Solar evaporation is also used in temperate zones where evaporation exceeds rainfall and in areas where seasons of hot and dry weather occur. Other factors affecting solar pond selection are wind, humidity, cloud cover, and land terrain.

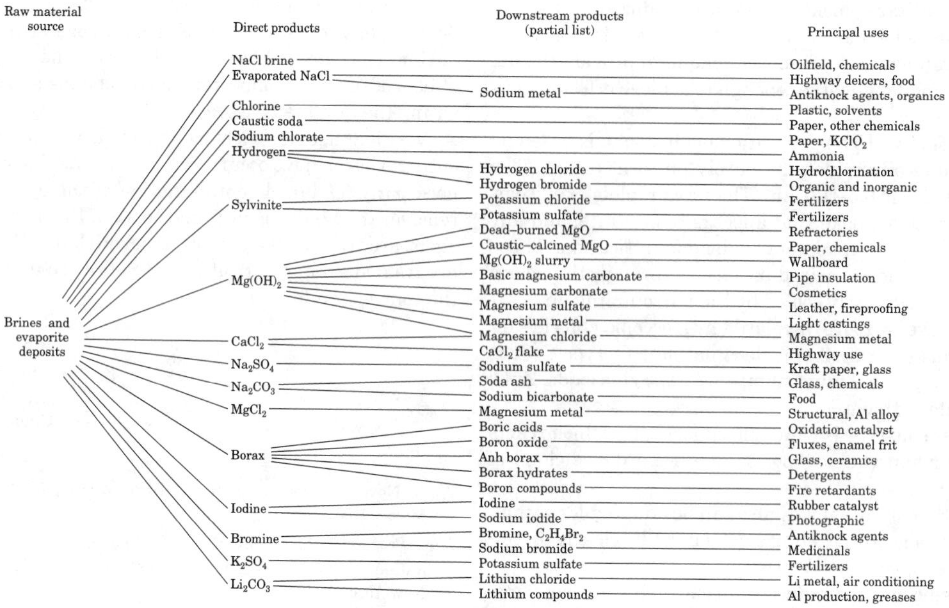

Figure 1. The brine chemical industry and some of its products.

Solar salt operations can be found along the shores of the Great Salt Lake and in the San Francisco Bay area. Salt production from these areas represents 10% of the total salt produced in the United States.

Seawater. Salt extraction from seawater is done by most countries having coastlines and weather conducive to evaporation. Seawater is evaporated in a series of concentration ponds until it is saturated with sodium chloride. At this point over 90% of the water has been removed, and some impurities, $CaSO_4$ and $CaCO_3$, have been crystallized. This brine, now saturated in NaCl, is transferred to crystallizer ponds where salt precipitates on the floor of the pond as more water evaporates.

The Great Salt Lake. All Great Salt Lake mineral extracting facilities have solar ponds as the first stage in processing minerals from brine.

The first salt to saturate and crystallize is halite. This salt is successively followed by epsomite, schoenite, kainite, carnallite, and finally bischofite.

Solution Mining. Solution mining, also known as brining, is the recovery of sodium chloride or any soluble salt in an underground deposit by dissolving it *in situ* and forcing the resultant solution to the surface.

A list of brine evaporates, products, and uses are summarized in Figure 1.

Minerals from Brine

Boron Compounds. *Occurrence.* Brine found in Searles Lake, California is the only brine source where boron is produced commercially.

Boron is found in two underground ores, ulexite and colemanite. Boron is found in many different evaporite deposits.

Recovery Process. Boron is recovered from brine in one of three processes: liquid–liquid extraction, evaporator–crystalizers, or chilling.

Economic Aspects. The principal producers in the United States are U.S. Borax and Chemical Corporation, North American Chemicals Co., and American Borate Corporation.

Bromine. *Occurrence.* Bromine is found in seawater and in underground brine deposits of marine origin. Bromine (qv) is also found in Dead Sea brine and is currently being produced there by the Dead Sea Works.

Recovery Process. Commercial processes depend on the oxidation of bromide to bromine.

Economic Aspects. The United States produced 174,600 tons of bromine in 1989. Over 95% of this was produced in southeastern Arkansas. The remainder was produced from Michigan brines. Bromine is used in gasoline, fire retardants, fumigants, and alkali metals.

Calcium Chloride. *Occurrence.* Brines are the main commercial source of calcium chloride. Some brines of Michigan, Ohio, West Virginia, Utah, and California contain over 4% calcium. Michigan is the leading state in natural calcium chloride production, with California a distant second.

Recovery Process. Because of its high solubility compared to that of other brine constituents, calcium chloride is the final constituent recovered in a multiproduct brine processing operation.

Economic Aspects. Total production of calcium chloride in 1989 was 873,000 tons. Most of this was produced from Michigan brines.

Iodine. *Occurrence.* Iodine is widely distributed in the lithosphere at low concentrations (about 0.3 ppm). It is present in seawater at a concentration of 0.05 ppm. Certain marine plants concentrate iodine to higher levels than occur in the sea brine; these plants have been used for their iodine content. A significant source of iodine is caliche deposits of the Atacama Desert, Chile.

About 40% of the free world's iodine was produced in Japan from natural gas wells, but production from Atacama Desert caliche deposits is relatively inexpensive and on the increase. By 1992, Chile was the primary world producer. In the United States, underground brine is the sole commercial source of iodine.

Recovery Process. Japan and Chile are the leading producers of iodine, producing nearly 7000 metric tons per year.

A large reserve of caliche ore bearing iodine is being processed in the Atacama Desert. Production of iodine there is relatively inexpensive. About 40% of the world supply of iodine is made from these Chilean deposits. The process consists of leaching the caliche with water.

Economic Aspects. Most of the iodine used in the United States comes from Japan and Chile. The United States produces 10% of the world supply but consumes 30%. Production in Chile appears to be relatively low cost, and the product there presently controls prices.

Lithium. *Occurrence.* Numerous brines contain lithium in minor concentrations. Commercially valuable natural brines are located at Silver Peak, Nevada (4000 ppm) and at the Salar de Atacama, Chile.

Lithium brines with commercial potential are found in the Altiplano of Bolivia and Argentina, in salt beds of Chile, and in several salt beds in central and western China.

Recovery Process. Lithium is extracted from brine at Silver Peak Marsh, Nevada, and at the Salar de Atacama, Chile. Lithium brines are concentrated in solar ponds and reacted with soda ash to make lithium carbonate.

Economic Aspects. The lithium concentration at Silver Peak is decreasing. During the 1980s lithium extraction was started at the Salar de Atacama, Chile. This is the largest lithium production plant in the world using brine as its raw material.

Magnesium Compounds. *Occurrence.* Magnesium hydroxide and magnesium chloride are two commercially important magnesium compounds recovered directly from natural brines. From these compounds many other compounds of magnesium are made, such as elemental magnesium and magnesia. Other important compounds containing magnesium are epsomite, schoenite, kainite, and carnallite. The principal magnesium sources are seawater (1300 ppm), Great Salt Lake (1.1% Mg), underground brines near the surface east of Wendover, Utah (1%), subterranean brines in Michigan (0.7–2.5%), and brine from the Yates formation in the Midland Basin of west Texas (3%).

Recovery Process. Magnesium hydroxide can be recovered in relatively pure form either from brine or from an intermediate plant liquor by increasing alkalinity. Better recoveries can be made by replacing calcium hydroxide with dolomite that has been calcined.

Recovery of magnesium chloride is usually economically feasible only as a by-product.

Economic Aspects and Uses. Magnesium hydroxide and magnesium chloride are used as a basic feedstock to make elemental magnesium, MgO refractories, and reactive chemicals.

Potassium Compounds. *Occurrence.* Muriate of potash, KCl, and sulfate of potash, K_2SO_4, are produced from brines in the United States. Three brine potash operations are located in Utah at Moab, Ogden, and Wendover and one in California at Searles Lake. Operations in Searles Lake produce both muriate and sulfate of potash. The Ogden operation produces sulfate of potash. The others produce muriate.

Recovery Process. The Texas Gulf, Cane Creek potash operation of Moab, Utah produces KCl by solution mining. Production of KCl at the Wendover, Utah, operation employs a large 7000 acre complex of solar ponds. Great Salt Lake Minerals Corporation near Ogden, Utah, produces potassium sulfate and several other products from Great Salt Lake brines.

Economic Aspects and Uses. Total world production of potassium products is 29,000,000 tons per year. Its main use is in fertilizers.

Sodium Carbonate. *Occurrence.* The brines of Searles Lake, California, are the sole brine source of sodium carbonate (soda ash) production in the United States. There is a large underground deposit of sodium carbonate brine in the Sua Pan area of Botswana, Africa.

Recovery Process. Presently North American Chemical Co. at Searles Lake is the only producer in the United States of sodium carbonate from brine (1,000,000 t/yr). The process is based on converting all sodium carbonate in the brine to sodium bicarbonate by adding CO_2. The bicarbonate is then converted to carbonate.

Economic Aspects. North American Chemical Co. at Searles Lake is now the only producer of soda ash from naturally occurring brine in the United States. Production from brine represents only about 10% of U.S. production.

Sodium Chloride. *Occurrence.* About half of all the sodium chloride produced in the world is from brine. Approximately one hundred million tons per year are produced from brines of the ocean, terminal lakes, subterranean aquifers, and solution mining. Sodium is found in large quantities in most areas of the world. Its quantity is so large that prices in some locations are only a few dollars per ton.

Recovery Process. There are two main processes. One is to flood solar ponds with brine and evaporate the water leaving sodium chloride crystallized on the pond floor. The other is to artificially evaporate the brine in evaporative crystallizers. Industrial salt is made from solar ponds, whereas food-grade salt, prepared for human consumption, is mostly produced in evaporator-crystallizers.

Economic Aspects. The United States is the largest producer of salt in the world. Salt continues to be one of the most heavily traded chemical ores in the world, representing nearly 65% of all seaborne mineral trade. World consumption is over 200 million tons per year.

Sodium Sulfate. *Occurrence.* In the United States natural sodium sulfate brines are found at Searles Lake, at the shallow castile formation underlying Terry and Gains counties in Texas, and at the Great Salt Lake.

Other natural sodium sulfate brines of commercial importance are found in dry lake beds of southwestern Saskatchewan, Canada; Laguna del Ray in Coahuila, Mexico; the Gulf of Kara-bogaz of the former USSR; and western China.

Recovery Process. The process for making sodium sulfate is different at each facility extracting it from brine. One step common to all facilities is a cooling step to form Glauber's salt, followed by a purification and recrystallization step to form anhydrous sodium sulfate.

In Texas, brine is pumped from underground deposits. At Searles Lake, sodium sulfate is recovered as one of three co-products in a series of complex operations where soda ash and borax are also recovered from the brine. In processing Great Salt Lake brine, Glauber's salt is crystallized in solar ponds by cooling during the winter.

Economic Aspects and Uses. About 50% of all sodium sulfate produced in the United States is from brine. In 1988, 410,000 tons was produced from brine. Most of the production is from North American Chemicals Co., Ozark-Mahoning at Brownfield, and Seagraves, Texas, produce 25% and Great Salt Lake Minerals and Chemicals Corporation produce 5%.

DAVID BUTTS
Great Salt Lake Minerals Corporation

A. C. Bersticker, K. E. Hoekstra, and J. F. Hall, eds., *The Symposia on Salt,* Northern Ohio Geological Society, Cleveland, Ohio. *First Symposium,* 661 pp. in 3 Vols., 1963; J. F. Hall, ed., *Second Symposium,* Vol. I, 443 pp., Vol. II, 1966, 422 pp.; J. L. Rau and L. F. Delling, eds., *Third Symposium,* Vol. I, 474 pp., Vol. II, 1970, 486 pp.; A. H. Coogan, ed., *Fourth Symposium,* Vol. I, 530 pp., Vol. II, 1973, 517 pp.; A. H. Coogan and L. Hauber, eds., *Fifth Symposium,* Vol. I, 485 pp., Vol. II, 1979, 547 pp.; B. C. Schreiber and H. L. Harner, eds., *Sixth Symposium,* Vol. I, 646 pp., Vol. II, 1983, 695 pp., Salt Institute, Alexandria, Va.

W. J. Schlitt, ed., *Salts & Brines '85,* Port City Press, Baltimore, Md., 1985, 209 pp. 501 references.

CHEMOMETRICS

Chemometrics has been defined as the chemical discipline that uses mathematical and statistical methods to design or select optimal measurement procedures and experiments, and to provide maximum chemical information by analyzing chemical data. As a subdiscipline of chemistry, the field emerged in the late 1960s and early 1970s as an outgrowth of pattern recognition and other techniques of artificial intelligence and applied mathematics.

Pattern recognition methods were originally used to address data processing problems in applications such as alphanumeric character recognition, speech analysis, fingerprint identification, and scene analysis. The term pattern recognition refers to a variety of computer tools that can be used to predict a property of an object where the property is not itself measured, but rather is inferred from a set of experimental measurements made on the object that are indirectly related to the property.

Throughout the 1970s, applications of pattern recognition were found in the chemical sciences. Other methods of multivariate mathematics and statistics were borrowed or invented, and a new discipline called chemometrics arose. In 1974, the Chemometrics Society was formed, and the first Chemometrics newsletter came out in 1976.

By about 1980, as advances in computing technology were rapidly expanding the ability of analytical instruments to collect and process data, a new emphasis in chemometrics was emerging. With computers interfaced to spectroscopic instruments, it became possible to collect hundreds of channels of information, ie, to collect intensities at hundreds of wavelengths, creating a data vector of equivalent dimension. That, added to the fact that automated sample handling techniques created a higher sample throughput, resulted in the capability to create large data matrices. By placing aspects of the analytical instrument under computer control and applying chemometrics, it became possible to improve the optimization, resolution, calibration, and other operating parameters in what began to be called intelligent analytical instrumentation (see ANALYTICAL METHODS–TRENDS).

It was recognized that pattern recognition techniques along with other related multivariate methods of mathematics and statistics have much to offer the entire chemical analytical process as shown schematically in Figure 1. This scheme has become the framework for the development of new techniques in the field of chemometrics. Thus the emphasis during the 1970s on applications of pattern recognition to chemical data interpretation was joined by a growing focus during the 1980s on the application of multivariate methods to analytical method development.

Exploratory Data Analysis

Basic Assumptions of Pattern Recognition. Given a k-dimensional data set, that is, a set of samples with k measurements made on each sample, the goal is to learn something about a property or behavior, answer a question, or test a hypothesis about the system. The type of property or question to be investigated influences the selection of data analysis techniques that are appropriate to use. Generally, data analysis problems are one of two types involving either category data or continuous property data. In category-type data analysis, the property of interest about the system relates to the category membership of the data, which is obviously not amenable to the regression approach as used for continuous properties. Pattern recognition is especially designed for category data analysis problems.

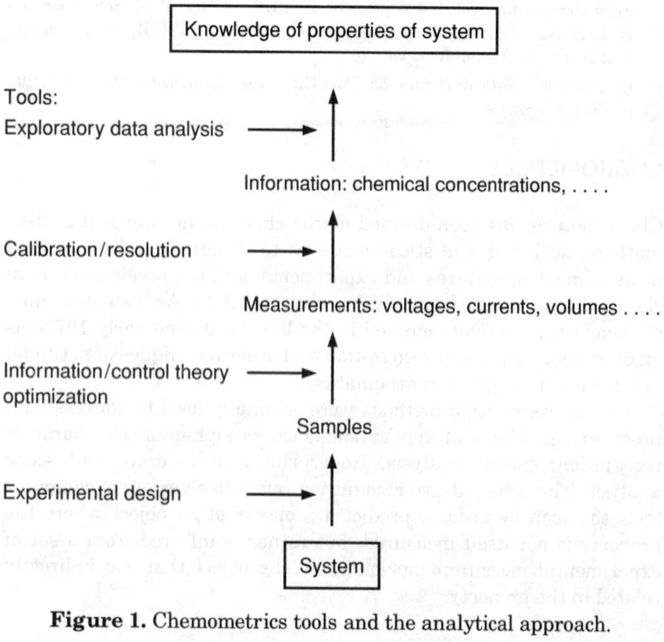

Figure 1. Chemometrics tools and the analytical approach.

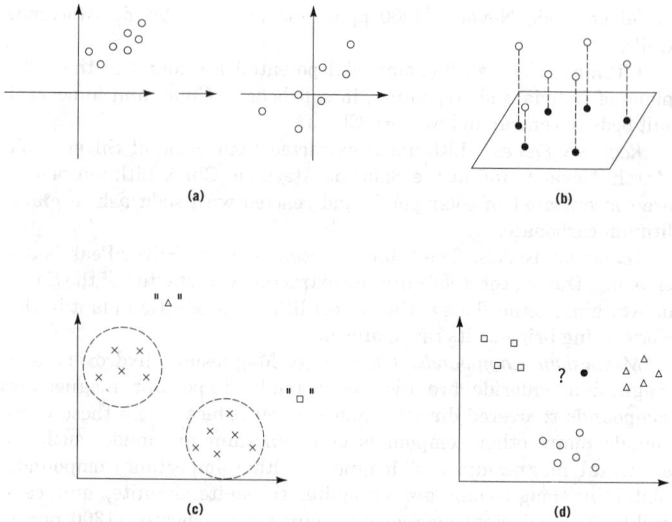

Figure 2. Types of pattern recognition techniques: **(a)** preprocessing, **(b)** display, **(c)** unsupervised learning, and **(d)** supervised learning.

There are four main categories of pattern recognition techniques, as illustrated in Figure 2: preprocessing, display, unsupervised learning, and supervised learning. Preprocessing techniques are utilized to ensure that the form or representation of the data is conformable with the pattern recognition algorithms. They are designed to transform the data into the most informative representation in the context of the study. The data analyst may then choose to display the data in two or three dimensions for human inspection while preserving the maximum amount of information about data structure in the k-dimensional space. Unsupervised learning refers to methods that make no *a priori* assumptions about category-membership of the samples. They are used in uncovering intrinsic clusters or other patterns in the data. Conversely, in supervised learning the computer learns to classify the samples based on *a priori* knowledge about their category designation. The goal is to develop an optimal classification rule in order to test the validity of a hypothesis or classify an unknown sample.

The successful application of pattern recognition methods depends on a number of assumptions. Obviously, there must be multiple samples from a system with multiple measurements consistently made on each sample. For many techniques the system should be overdetermined; the ratio of number of samples to number of measurements should be at least three. These techniques assume that the nearness of points in hyperspace faithfully reflects the similarity of the properties of the samples. The data should be arranged in a data matrix with one row per sample, and the entries of each row should be the measurements made on the sample. The information needed to answer the questions must be implicitly contained in that data matrix, and the data representation must be conformable with the pattern recognition algorithms used.

Calibration Methods

Calibration is an important focus in analytical chemistry. It is the process that relates instrument responses to chemical concentrations. It consists of two basic steps: estimation of the calibration model parameters, and then prediction for new samples of unknown concentration. Calibration refers to the step of the analytical process in Figure 1 where measurements are related to concentrations of chemical species or other chemical information.

In its simplest form, a direct, univariate calibration method proceeds by assuming there is a mathematical model that relates analytical instrument response to concentration. Traditionally in analytical chemistry, the model is assumed to be a linear relation between response r_i and concentration c_i,

$$r_i = kc_i + b$$

where i represents sample number. The slope k and intercept b of the calibration line are commonly estimated using linear regression by least squares using data from a set of standards having known concentrations of the analyte of interest. This is the model for an ideal world, where the analytical sensor is not affected by interfering components in the sample and where the sample matrix itself does not affect the response. In reality, however, the relationship may be nonlinear, and there may be significant interference and matrix effects. Advances in analytical chemometrics have been developed to address statistical considerations, multivariate direct, indirect, internal standard, and standard addition calibration. The statistical aspects include new ways to address nonlinear calibration, to estimate confidence limits and detection limits for the nonlinear case, and to derive expressions for quantities such as limits of detection and determination.

Signal Processing and Resolution

Signal processing pertains to a wide collection of tools used to refine the information contained in a raw analytical signal and to estimate pertinent signal parameters such as peak shape, area, and amplitude. Signal processing applications typically involve either energy-variant or time-variant spectra. Techniques such as smoothing and noise filtering are commonly applied to energy or wavelength spectra, whereas time-variant spectra are often processed using methods such as Fourier transforms, Hadamand transforms, and filtering techniques. Peak detection, second or higher derivative spectroscopy, and methods to enhance the signal-to-noise ratio are common applications in signal processing.

Resolution refers to separating or distinguishing the components of a system and its meaning depends on the nature of the system measurement. In the context of spectroscopy, resolution refers to distinguishing intensities at adjacent wavelengths; in chromatography, resolution is a measure of the separation of chromatographic peaks; in chemical analysis in general, resolution can refer to the estimation of the contributions of components of a mixture to the total analytical response of the sample.

Methods to enhance resolution can be considered in two groups: methods in which assumptions about the system, eg, line shapes and number or identity of constituent components, are made, and those in which there are no *a priori* assumptions except linear behavior, for instance.

Optimization

Optimization refers to the step in the analytical process (see Fig. 1) where some sort of treatment is performed on samples to generate raw data which can be in the form of voltages, currents, or other analytical signals. These data have yet to be calibrated in terms of chemical concentrations.

In the context of chemometrics, optimization refers to the use of estimated parameters to control and optimize the outcome of experiments. Given a model that relates input variables to the output of a system, it is possible to find the set of inputs that optimizes the output.

DEBORAH ILLMAN
University of Washington

P. Geladi and K. Esbensen, *J. Chemometrics* **4**(5–6), 337, 389 (1990).

M. A. Sharaf, D. L. Illman, and B. R. Kowalski, *Chemometrics,* Wiley–Interscience, New York, 1986.

B. R. Kowalski, ed., *Chemometrics: Theory and Application,* ACS Symposium Series 52, American Chemical Society, Washington, D.C., 1977.

S. D. Brown, *Anal. Chem.* **62,** 84R (1990).

CHEMOTHERAPEUTICS, ANTICANCER

Cancer is second only to cardiovascular disease as the principal cause of human mortality. As the median age of populations has risen, total deaths from cancer have increased. Treatment of cancer includes surgery, radiation, and chemotherapy, the last encompassing the use of both cytotoxic agents and relatively nontoxic hormonal agents for the control of tumor growth.

Drugs used in cancer chemotherapy or clinical trials are classified according to primary underlying mechanisms of action. However, many drugs operate through multiple mechanisms. Mechanisms include those of antimetabolites, DNA alkylating and/or cross-linking agents, DNA binding/cleaving agents, DNA topoisomerase interactive compounds, agents that act on tubulin structure, and hormones (qv) (Fig. 1). In addition to those drugs already approved by the FDA, a number of investigational drugs are undergoing clinical evaluation and many others are in the pipeline.

Antimetabolites. Antimetabolites, which represent one of the earliest groups of anticancer agents, are listed in Table 1.

DNA Alkylating/Cross-Linking Agents. This category includes compounds of diverse chemical classes (Table 2).

DNA Binding/Cleaving Agents. DNA binding and/or cleaving agents that have anticancer activity are listed in Table 3. All of the natural products and analogues of natural products in this category (Table 3) are able to bind DNA either as intercalators or as minor groove binders, hence inhibiting DNA-dependant RNA synthesis. Both bleomycin and esperamicin A_1 cleave DNA by forming free radicals in the immediate vicinity of the sugar–phosphate backbone. Activity as antitumor agents is related to the ability to induce irreparable lesions in DNA. Bleomycin generates oxygen free-radical species whereas esperamicin A_1 and a number of related natural products that include neocarzinostatin, dynemicin, and the calicheamicins generate aryl diradical species, which abstract hydrogen atoms directly from the deoxyribose backbone. An analogue of the natural product CC1065 has the unique property of being a DNA alkylating agent which recognizes poly-AT regions of the minor groove of DNA. It remains to be seen if the high potency and unique modes of action ascribed to these novel classes of agents can translate into clinically useful drugs.

Topoisomerase Interactive Drugs. Topoisomerases I and II have emerged as interesting targets for the design of new anticancer agents (Table 4).

Tubulin Active Drugs. Tubulin active drugs are listed in Table 5.

Hormones. Although not strictly cytotoxic, hormones (qv) have been used to control the environment of hormone-dependent tumors such as those of the prostate, breast, and endometrium, ie,

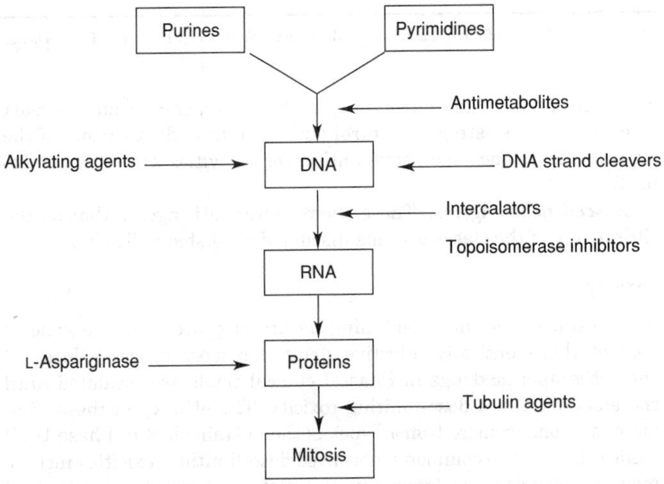

Figure 1. Schematic of nucleic acid and protein synthesis and the steps leading to mitosis showing the common mechanisms of action and various classes of chemotherapeutic agents.

Table 1. Antimetabolites

Drug (trade name)	Molecular formula	Molecular weight	Disease	Toxic effects
5-azacitidine[a] (Mylosar)	$C_8H_{12}N_4O_5$	244.21	acute myelogenous leukemia	nausea, vomiting; hepatic dysfunction; myelosuppression
cytarabine USP[a] (Cytosar)	$C_9H_{13}N_3O_5$	243.22	acute granulocytic leukemia (adults); acute lymphocytic leukemia (children); Hodgkin's disease	bone marrow depression; hepatic toxicity; megaloblastosis; nausea; vomiting; diarrhea
gemcitabine[b]	$C_9H_{11}F_2N_3O_4$	263.20	investigational drug; responses seen in Phase I trials in colon and nonsmall cell lung cancer	myelosuppression observed as dose-limiting toxicity
floxuridine USP[c] (FUDR)	$C_9H_{11}FN_2O_5$	246.21	palliative treatment of gastrointestinal adenocarcinoma with liver metastases	severe hematological toxicity; gastrointestinal hemorrhage; nausea; vomiting; diarrhea; enteritis; stomatitis; erythema
fluorouracil USP[c] (Fluorouracil)	$C_4H_3FN_2O_2$	130.08	palliative treatment of carcinoma of colon, rectum, breast, stomach, and pancreas	bone marrow depression; dermatitis; alopecia; nausea; vomiting; diarrhea; stomatitis; anorexia; GI ulcers; skin pigmentation
mercaptopurine USP[d] (Purinethol)	$C_5H_4N_4S$	152.19	acute leukemia (more effective in children than in adults); chronic granulocytic leukemia	bone marrow depression; hepatic toxicity; anemia; gastrointestinal (GI) ulceration; nausea; vomiting
thioguanine USP[d] (Tabloid)	$C_5H_5N_5S \cdot XH_2O$	167.19	acute leukemia; chronic granulocytic leukemia	bone marrow depression; stomatitis; anorexia; nausea; vomiting
methotrexate USP[e] (Methotrexate)	$C_{20}H_{22}N_8O_5$	454.46	acute lymphocytic leukemia; meningeal leukemia; choriocarcinoma; chorioadenoma destruens; lymphosarcoma; osteogenic sarcoma; cancer of lung, neck, head, cervix; mycosis fungoides; hydatidiform mole high dose MTX followed by leucovorin rescue in nonmetastatic osteosarcoma	bone marrow depression; renal and hepatic toxicity; enteritis; stomatitis; alopecia; abdominal distress; erythematous rash; oral and GI ulceration; diarrhea; nausea; vomiting
leucovorin[e] (calcium USP)	$C_{20}H_{21}CaN_7O_7$	511.51	high dose methotrexate rescue therapy in osteosarcoma	allergic sensitization
DDATHF[b] (Ly237147) (Lometrexol sodium)	$C_{21}H_{23}N_5Na_2O_6$	487.42	investigational drug	
trimetrexate[f]	$C_{19}H_{23}N_5O_3$	369.42	investigational drug; partial remissions in soft tissue sarcomas observed	myelosuppression; mucositis; nausea; vomiting; skin rash
hydroxyurea USP[g] (Hydrea)	$CH_4N_2O_2$	76.05	chronic granulocytic leukemia; melanoma; cancer of ovary, head, neck	vomiting; anorexia; fever; bone marrow depression; nausea; diarrhea

[a] Upjohn. [b] Lilly. [c] Hoffmann-La Roche. [d] Burroughs Wellcome. [e] Lederle. [f] Parke-Davis. [g] Bristol-Myers Squibb.

androgens are used to control the growth of estrogen-dependent breast tumors, whereas estrogens control androgen-dependent tumors of the prostate. Hormones that have anticancer activities are listed in Table 6.

Miscellaneous Agents. Those chemotherapeutic agents that do not fit into any of the classifications discussed are listed in Table 7.

Toxicity

As a result of the life-threatening nature of cancer and the general lack of therapeutically effective drugs for most cancers, doses of chemotherapeutic drugs in Phase I clinical trials are escalated until the emergence of a dose-limiting toxicity. The efficacy of these compounds in one or more tumor types is then established in Phase II/III clinical trials. The commonly observed dose-limiting toxicities include myelosuppression, gastrointestinal upset, and renal, hepatic, and cardiotoxicities.

Toxicity Amelioration. Research efforts to address the problem of toxicity amelioration has progressed in several directions. The three most prominent areas are analogue synthesis, chemoprotection, and drug targeting.

Drug Resistance

The most recognized and studied mechanisms of drug resistance are attributed to multidrug resistance (MDR), gene amplification, DNA repair, topoisomerase II activity, and glutathione and metallothionein levels. Even with advances in understanding the biology and mechanism of drug resistance among different classes of antitumor agents, no real breakthrough appears imminent.

Economics of Cancer Chemotherapy

The increased use of chemotherapy as a modality in the treatment of cancer has caused a corresponding increase in the market for anticancer agents. Only a few anticancer drugs have achieved sales in excess of $100 million per year and many of the drugs discussed

Table 2. DNA Alkylating/Cross-Linking Agents

Drug (trade name)	Molecular formula	Molecular weight	Disease	Toxic effects
carmustine USP[a](BiCNU)	$C_5H_9Cl_2N_3O_2$	214.05	Hodgkin's disease; non-Hodgkin's lymphomas; meningeal leukemia; brain tumor; multiple myeloma	bone marrow depression; hepatic toxicity; nausea; vomiting
lomustine USP[a] (CeeNU)	$C_9H_{16}ClN_3O_2$	233.70	malignant brain tumors; Hodgkin's disease	bone marrow depression; hepatic toxicity
tauromustine[b]	$C_7H_{15}ClN_4O_4S$	286.73	investigational drug responses observed in malignant melanoma	gastrointestinal; thrombocytopenia
streptozocin USP[c](Zanosar)	$C_8H_{15}N_3O_7$	265.22	metastatic islet cell carcinoma of the pancreas	bone marrow depression; renal and hepatic toxicity; nausea; vomiting
busulfan USP[d](Myleran)	$C_6H_{14}O_6S_2$	246.29	chronic granulocytic leukemia; other myeloproliferative disorders	bone marrow depression; hyperuricemia; gynecomastia; amenorrhea; skin hyperpigmentation
cyclophosphamide USP[a] (Cytoxan)	$C_7H_{15}Cl_2N_2O_2P$	279.10	acute and chronic lymphocytic leukemia; lung cancer; rhabdomyosarcoma; neuroblastoma; ovarian and mammary carcinoma; multiple myeloma; lymphosarcoma; Burkitt's lymphoma; Hodgkin's disease; retinoblastoma; mycosis fungoides	bone marrow depression; hepatic toxicity; cystitis; alopecia; nausea; vomiting
ifosphamide USP[a] (Ifex)	$C_7H_{15}Cl_2N_2O_2P$	261.09	germ cell testicular cancer; used in combination with mesna	myelosuppression; urotoxicity; alopecia; nausea; vomiting; CNS toxicities
mesna USP[a] (Mesnex)	$C_2H_5NaO_3S_2$	164.17	prophylactic prevention of hemorrhagic cystitis	
mechlorethamine hydrochloride USP[e](Mustargen)	$C_5H_{11}Cl_2N{\cdot}HCl$	192.52	Hodgkin's disease; non-Hodgkin's lymphomas; lymphosarcoma; cancer of breast, ovary, lung; neoplastic effusion	bone marrow depression; nausea; vomiting; anorexia; diarrhea; local irritation
chlorambucil USP[d]	$C_{14}H_{19}Cl_2NO_2$	304.23	chronic lymphocytic leukemia; cancer of ovary, breast, testis; Hodgkin's disease; non-Hodgkin's lymphomas	bone marrow depression; nausea; vomiting
melphalan USP[d] (Alkeran)	$C_{13}H_{18}Cl_2N_2O_2$	305.20	multiple myeloma; plasmacytic myeloma; cancer of breast and ovary	bone marrow depression; nausea; vomiting; anorexia
thiotepa USP[f](Thiotepa)	$C_6H_{12}N_3PS$	189.21	cancer of breast, ovary, lung, bladder; Hodgkin's disease; non-Hodgkin's lymphomas; neoplastic effusion	bone marrow depression; amenorrhea; anorexia; nausea; vomiting
mitomycin C USP[a] (Mutamycin)	$C_{15}H_{18}N_4O_5$	334.33	chronic myelogenous leukemia; reticulum cell sarcoma; Hodgkin's disease; non-Hodgkin's lymphomas; cancer of stomach, pancreas, lung; epithelial tumors	bone marrow depression; renal toxicity; alopecia; stomatitis; anorexia; nausea; vomiting
BMY-25067[a]	$C_{23}H_{25}N_5O_7S_2$	547.60	investigational drug	
KW2149[g]	$C_{24}H_{34}N_6O_8S_2$	598.7	investigational drug	
cisplatin USP[a] (Platinol)	$Cl_2H_6N_2Pt$	300.06	metastatic testicular tumors; metastatic ovarian tumors; advanced bladder cancer	nephrotoxicity; ototoxicity; myelosuppression; nausea; vomiting; allergic reaction
carboplatin USP[a] (Paraplatin)	$C_6H_{12}N_2O_4Pt$	371.25	recurrent ovarian carcinoma	bone marrow suppression; emesis; allergic reactions
dacarbazine USP[h](DTIC)	$C_6H_{10}N_6O$	182.18	malignant melanoma; Hodgkin's disease; soft tissue sarcomas	bone marrow depression; flulike syndrome; alopecia; nausea; vomiting; anorexia

[a] Bristol-Myers Squibb. [b] Pharmacia. [c] Upjohn. [d] Burroughs Wellcome. [e] Merck Sharp & Dohme. [f] Lederle. [g] Kyowa-Hakko. [h] Dome.

herein sell less than $10 million per year. Those agents that have achieved greater economic importance are newer, frequently used in combination chemotherapy, and of use in the treatment of solid tumors which comprise the bulk of reported cancer incidence. A number of the agents in clinical trials, such as taxol and camptothecin analogues, are expected to have considerable economic impact based on activity in the treatment of the more common human malignancies in, eg, lung, breast, colon, and ovary. In addition to the market for cancer chemotherapeutic drugs, there is also a growing market for biologicals such as interferons.

Table 3. DNA Interactive Agents

Drug (trade name)	Molecular formula	Molecular weight	Disease	Toxic effects
daunorubicin hydrochloride USP[a] (Cerubidine)	$C_{27}H_{29}NO_{10}\cdot HCl$	563.99	acute lymphocytic and granulocytic leukemia; lymphomas	bone marrow depression; cardiac toxicity; alopecia; stomatitis; GI disturbance
doxorubicin USP[b](Adriamycin)	$C_{27}H_{29}NO_{11}$	543.53	soft-tissue and osteogenic sarcomas; Hodgkin's disease; non-Hodgkin's lymphomas; acute leukemia; cancer of thyroid, breast, lung, genitourinary (GU) tract; Wilm's tumor; neuroblastoma	bone marrow depression; cardiac toxicity; alopecia; stomatitis; GI disturbance
idarubicin hydrochloride[b] (Idamycin)	$C_{26}H_{27}NO_9\cdot HCl$	533.96	acute myeloid leukemia in adults	bone marrow suppression; cardiotoxicity; nausea; vomiting; alopecia
mitoxanthrone hydrochloride USP[c](Novantrone)	$C_{22}H_{28}N_4O_6\cdot 2HCl$	517.41	acute nonlymphocytic leukemia, including myelogenous promyelocytic, monocytic, and erythroid acute leukemias	nausea; vomiting; alopecia; mucositis; stomatitus; myelosuppression; cardiotoxicity; allergic reaction; phlebitis
bleomycin sulfate USP[d](Blenoxane)	mixture of bleomycin A_2, B_2 as primary components		squamous cell carcinoma of head, neck, esophagus, skin, GU tract; testicular tumor; Hodgkin's lymphomas	pulmonary fibrosis; skin reactions; alopecia; nausea; vomiting; anorexia; fever; stomatitis
esperamicin A_1[d]	$C_{59}H_{80}N_4O_{22}S_4$	1324.41	investigational drug	
Adozelesin[e](U73,975)	$C_{30}H_{22}N_4O_4$	502.30	investigational drug	
dactinomycin USP[f](Cosmegen)	$C_{62}H_{86}N_{12}O_{16}$	1255.43	Wilm's tumor; Ewing's tumor; choriocarcinoma; testicular carcinoma; rhabdomyosarcoma; neuroblastoma; melanoma; soft-tissue and osteogenic sarcomas	bone-marrow depression; renal and hepatic toxicity; alopecia; mental depression; stomatitis; nausea; vomiting; diarrhea; anorexia; local irritation
plicamycin USP[a] (Mithracin)	$C_{52}H_{76}O_{24}$	1085.16	testicular tumors; hypercalcemia and hypercalciuria associated with advanced malignancies	bone marrow depression; hepatic and renal toxicity; hypocalcemia; hemorrhage; stomatitis; nausea; vomiting; anorexia; diarrhea
procarbazine hydrochloride USP[g](Matulane)	$C_{12}H_{19}N_3O\cdot HCl$	257.76	Hodgkin's disease; non-Hodgkin's lymphomas; lung cancer	bone marrow depression; neurological and dermatological toxicity; nausea; vomiting

[a] Wyeth-Ayerst. [b] Adria. [c] Lederle. [d] Bristol-Myers Squibb. [e] Merck Sharp & Dohme. [f] UpJohn. [g] Hoffmann-La Roche.

Table 4. Topoisomerase Interactive Drugs

Drug (trade name)	Molecular formula	Molecular weight	Disease	Toxic effects
etoposide USP[a] (Vepesid)	$C_{29}H_{32}O_{13}$	588.56	refractory testicular tumors; small cell lung cancer	myelosuppression; mild to moderate nausea and vomiting; transient hypotension; allergic reactions; alopecia
etoposide phosphate[a]	$C_{29}H_{33}O_{16}P$	712.51	investigational drug; prodrug of etoposide	prodrug of etoposide
teniposide[a]	$C_{32}H_{32}O_{13}S$	656.67	refractory acute lymphocytic leukemia in children	myelosuppression; mild to moderate nausea and vomiting; transient hypotension; allergic reactions; alopecia
CPT-11[b]	$C_{33}H_{38}N_4O_6\cdot HCl$	622.78	investigational drug; topoisomerase I inhibitor	
topotecan hydrochloride[c]	$C_{23}H_{23}N_3O_5\cdot HCl$	457.91	investigational drug; topoisomerase I inhibitor	
elsamitrucin tartrate[a]	$C_{33}H_{35}NO_{13}\cdot C_4H_6O_4$	771.37	investigational drug	
amsacrine[d] (Amsidyl)	$C_{21}H_{19}N_3O_3S$	393.46	investigational drug	

[a] Bristol-Myers Squibb. [b] Yakult Honsha. [c] Smith Kline Beecham. [d] Parke-Davis.

Table 5. Tubulin Active Drugs

Drug (trade name)	Molecular formula	Molecular weight	Disease	Toxic effects
vinblastin sulfate USP[a](Velban)	$C_{46}H_{58}N_4O_9{\cdot}H_2SO_4$	909.06	Hodgkin's disease; lymphosarcoma; reticulum cell sarcoma; neuroblastoma; choriocarcinoma; carcinoma of breast, lung, oral cavity, testis, bladder; acute and chronic leukemia; histiocytosis; mycosis fungoides	leukopenia; neurological toxicity (paresthesias, mental depression, loss of deep tendon reflexes, etc); dysfunction of autonomic nervous system (ileus, constipation, urinary retention, etc); alopecia; stomatitis; nausea; vomiting; local irritation
vincristin sulfate USP[a] (Oncovin)	$C_{46}H_{56}N_4O_{10}{\cdot}H_2SO_4$	923.04	acute leukemia in children; lymphocytic leukemia; Hodgkin's disease; non-Hodgkin's lymphomas; Wilm's tumor; neuroblastoma; rhabdomyosarcoma.	neurological toxicity (paresthesias, foot drop, double vision, etc); constipation; ileus; alopecia; leukopenia (occasional);
vindesine sulfate[a] (Eldisine)	$C_{43}H_{55}N_5O_7{\cdot}H_2SO_4$	852.01	investigational drug	
navelbine[b](Vinorelbine)	$C^{45}H^{54}N^{4}O_8$	778.45	investigational drug; nonsmall cell lung cancer	
taxol[c](Paclitaxol)	$C_{47}H_{51}NO_{14}$	853.92	refractory ovarian cancer; refractory breast cancer; melanoma; lung cancer; head and neck cancer	alopecia; neutropenia; hypersensitivity; mucositis; neuropathy
taxotere[d] (Docetaxol)	$C_{43}H_{53}NO_{14}$	807.43	investigational drug	

[a] Lilly. [b] Pierre Fabre. [c] Bristol-Myers Squibb. [d] Rhône-Poulenc.

Table 6. Hormonal Therapy

Drug (trade name)	Molecular formula	Molecular weight	Disease	Toxic effects
tamoxifen citrate USP[a](Nolvadex)	$C_{26}H_{29}NO{\cdot}C_6H_8O_7$	563.65	breast cancer	visual disturbances
diethylstilbestrol diphosphate USP[b](Stilphostrol)	$C_{18}H_{22}O_8P_2$	428.31	prostatic carcinoma	fluid retention; hypercalcemia; common side effects of steroids
chlorotrianisene USP[b] (TACE)	$C_{23}H_{21}ClO_3$	380.87	androgen dependent carcinoma of the prostate	fluid retention; hypercalcemia; common side effects of steroids
estradiol USP[c](Estrace)	$C_{18}H_{24}O_2$	272.39	breast cancer; prostatic carcinoma	fluid retention; hypercalcemia; common side effects of steroids
estramustine phosphate sodium USP[d](Emcyt)	$C_{23}H_{30}Cl_2N{\cdot}Na_2O_6P$	564.35	prostatic carcinoma	side effects because of estradiol; increased dyspnea; nausea; vomiting
medroxyprogesterone acetate[e] (USP Depo-provera)	$C_{24}H_{34}O_4$	386.53	metastatic endometrial carcinoma; renal carcinoma	fluid retention; hypercalcemia; common side effects of steroids
megestrol acetate USP[c] (Megace)	$C_{24}H_{32}O_4$	384.51	carcinoma of the breast or endorometrium	fluid retention; hypercalcemia; common side effects of steroids
testolactone USP[c] (Teslac)	$C_{19}H_{24}O_3$	300.40	breast cancer	fluid retention; hypercalcemia; common side effects of steroids
formestane[f](Lentaron)	$C_{19}H_{26}O_3$	302.41	investigational drug; postmenopausal breast cancer	
goserelin USP[a] (Zoladex)	$C_{59}H_{84}N_{18}O_{14}$	1269.43	prostatic carcinoma	bone pain; common hormonal side effects
leuprolide acetate USP[g](Leupron)	$C_{59}H_{84}N_{16}O_{12}{\cdot}C_2H_4O_2$	1269.47	prostatic carcinoma	common hormonal side effects
octreotide acetate USP[h](Sandostatin)	$C_{49}H_{66}N_{10}O_{10}S_2{\cdot}xC_2H_4O_2$	1019.24	mestastatic carcinoid tumors; vasoactive intestinal peptide-secretory tumors	nausea; diarrhea; loose stools; vomiting; abdominal pain; pain on injection
flutamide[i](Eulexin)	$C_{11}H_{11}F_3N_2O_3$	276.21	metastatic prostatic carcinoma in combination with LHRH agonist	

[a] ICI. [b] Marion-Merrill Dow. [c] Bristol-Myers Squibb. [d] Pharmacia. [e] Upjohn. [f] CIBA-GEIGY. [g] TAP. [h] Sandoz. [i] Schering.

Table 7. Chemotherapeutic Agents

Drug (trade name)	Molecular formula	Molecular weight	Disease	Toxic effects
mitotane USP[a] (Lysodren)	$C_{14}H_{10}Cl_4$	320.05	palliative treatment of inoperable adrenal cortical carcinoma	skin toxicity; vertigo; lethargy; somnolence; anorexia; nausea; vomiting; diarrhea
isotretinoin USP[b] (Accutane)	$C_{20}H_{28}O_2$	300.44	investigational drug	
sulofenur[c]	$C_{16}H_{15}ClN_2O_3S$	350.82	investigational drug; refractory ovarian carcinoma response see in Phase I	anemia; methemoglobinemia
asparaginase[d] (Elspar)			acute lymphocytic leukemia	hepatic, renal, and pancreatic toxicity; neurological effects; hypersensitivity reactions; clotting abnormalities; nausea

[a] Bristol-Myers Squibb. [b] Hoffmann-La Roche. [c] Lilly. [d] Merck Sharp & Dohme.

TERRENCE W. DOYLE
DOLATRAI M. VYAS
Bristol-Meyers Squibb Company

S. K. Carter, M. T. Bakowski, and K. Hellman, *Chemotherapy of Cancer,* John Wiley & Sons, Inc., New York, 1987.

J. D. Fisher and P. A. Aristoff, in *Progress in Drug Research,* Vol. 32, Birkhauser Verlag, Basel, Switzerland, 1988, pp. 411–484.

M. R. Boyd, in *Current Therapy in Oncology,* B. C. Decleer, Inc., Philadephia, Pa., 1993, pp. 11–22.

P. V. Woolley III and K. D. Tew, eds., *Mechanism of Drug Resistance in Neoplastic Cells,* Academic Press, Inc., New York, 1988.

CHEMURGY

Chemurgy, the branch of applied chemistry devoted to industrial utilization of organic raw materials, especially from farm products, was really a social movement during the 1920s and 1930s when there were large surpluses of agricultural materials and severe economic problems in the farm areas. Using farm commodities as chemical or industrial raw materials was seen as a means of solving the economic problems. One significant outgrowth of this movement was the founding of the regional laboratories of the U.S. Department of Agriculture (USDA), which were considered to be chemurgic laboratories. Another success involved the making of strong paper from the fast-growing southern pine, leading to the foundation of the southern pulp (qv) and paper (qv) industry. This industry had been largely centered in the north and northeast and based on slower-growing species.

Some of the early work on the manipulation of proteins (qv) arose out of the chemurgy movement. This technology has found application in synthetic meat production. Other technologies that may be classified as chemurgic include that of the cellulosic fibers, eg, rayon and cellulose acetate (see CELLULOSE ESTERS; FIBERS, CELLULOSE ESTERS); the recovery of turpentine and rosin from paper and pulping processes (see TALL OIL; TERPENOIDS); and the oils and fatty acids business (see CARBOXYLIC ACIDS; FATS AND FATTY OILS). Oils and fatty acids are used in a wide variety of chemical products, including soap (qv) and detergents (see DETERGENCY; SURFACTANTS), cosmetics (qv), and coatings (qv).

Industrial Materials from Renewable Resources

One distinction that can be made in the area of chemurgy is between the use of natural products that are grown solely for industrial purposes on the one hand, and those that are grown primarily for food on the other. In the latter class, industrial materials may be either by-products from food production or substitutes for food uses when the commodity is in surplus.

Industrial Crops as Raw Materials. Trees are by far the largest commodity grown solely for industrial use. About two-thirds of the trees harvested are used for construction or structural uses. Other purely industrial crops include cotton (qv) and flax. Cotton is grown primarily for its fiber, although the protein and oil contained in these seed also contribute to farm income.

There are relatively few other crops that are grown solely for industrial purposes. Among these are tung, a tree nut from which tung oil is produced, and castor, from which is obtained castor oil (qv). Tung oil can be used as a drying oil in coatings. Wild crops being investigated for cultivation include jojoba, crambe, guayule, kenaf, and lesquerella.

Food Crops as Raw Materials. Crops that are grown primarily for food can be used for industrial purposes when the crops are in surplus or are found to be unfit for their intended purpose. Historically, when agricultural surpluses are high, the distinction between industrial and food use is not significant. Additionally, there are large industries based on food commodities, eg, the corn and wheat starch separation processes to give starch (qv) used in paper sizing and textiles (see WHEAT AND OTHER CEREAL GRAINS).

Oilseeds grown primarily for use in salad dressings, margarines, and cooking oils also produce an important by-product in the form of high protein meal. Historically, this meal was fed to animals but increasingly it is refined and fractionated for human consumption.

By-products from meat animals, which may be viewed as a food crop, are also commercially significant. In addition to meat, each of these produces bones, trimmings, fat, hides, and, in the case of sheep, wool (qv).

Wastes or By-Products as Raw Materials. By far the largest volume of natural products for industrial use, aside from the forest products, are wastes or by-products of food processing (qv). The largest use of these wastes is as animal feeds. Because they are used rather than becoming a disposal problem, they are considered to be chemurgic products.

Wastes from the pulp and paper industry are finding increasing applications. These include lignin (qv) and tall oil (qv), as well as sugars in the form of sulfite waste liquor. Although the largest use of lignin is as a fuel in the pulping process, a wide variety of products can be made from it. Lignin is used in dispersants (qv), adhesives (qv), additives to drilling muds, and fillers (qv).

Tall oil has been referred to as the largest and fastest-growing source of extractives such as turpentine and resin.

Sulfite waste liquor from the pulping industry contains a large amount of hydrolyzed sugars. Some sulfite waste liquors are being fermented to give both ethanol (qv) and single-cell protein in the form of yeast.

The trimmings and slash from forest operations increasingly are being used in much the same way as higher quality timber, primarily for pulping of chipping. Agriculture produces large amounts of wastes in the form of animal manures, branches, stems, stalks, and straws. It is possible to recycle animal wastes as animal feed because a large amount of nutritive value is retained in the waste. The enormous volume of animal wastes is also being considered as a potential energy source, probably by anaerobic fermentation to produce methane. Other agricultural wastes, such as straw, have been considered for pulp and papermaking and for digestion to produce fuel. They are also candidates for hydrolysis of the cellulose content to produce glucose. Hydrolysis can be acidic or enzymatic, and products are primarily ethanol but can include other chemicals.

A substantial part of research into uses for wastes from production of food and feeds has been motivated primarily by environmental considerations. For example, kraft black liquor, sulfite waste liquor, and other dilute streams from pulp and papermaking are potential fermentation substrates because of the dissolved sugar, which is also the most serious pollutant. Cereal grain milling produces a variety of fractions that have found historic uses in animal feeds. Research has been conducted on isolation of protein concentrates from these materials for upgrading as human foods. Cellulosic materials, such as farm wastes, can be upgraded for animal feed by simply bringing them into contact with ammonia (qv).

Some of the forest wastes and pulp and paper processing wastes contain dilute concentrations of important acids such as acetic acid (see ACETIC ACID AND DERIVATIVES) and solvents such as methanol (qv), formed from the wood in the high temperature digestion of pulping. It is increasingly attractive to use such processes as liquid–liquid extraction or absorption (qv) on charcoal followed by steam stripping to remove and concentrate the organic chemicals from such wastes.

Processing Methods

The largest class of processes applied to farm commodities are separations, which are usually based on some physical property such as density, particle size, or solubility.

In general, the separation processes used in chemurgy are relatively simple. In contrast, the chemical reactions that are possible with chemurgic materials can be very sophisticated. The chemistry involved is often reductive because so many of the chemurgic materials, such as cellulose and starch, are carbohydrates. Also important are the carboxylic acids and the amine groups of protein. Reactions such as the esterification of sugar (qv) and the epoxidation of soy oil are of commercial significance, as are the acetylation of cellulose to make it soluble for subsequent spinning into fibers, and the solubilization of starch and cellulose as xanthates (qv).

Potential for Renewable Resources

Increasingly, biochemical transformations are used to modify renewable resources into useful materials (see MICROBIAL TRANSFORMATIONS). Fermentation (qv) to ethanol is the oldest of such conversions. Another example is the cell-free enzyme-catalyzed isomerization of glucose to fructose for use as sweeteners (qv). The enzymatic hydrolysis of cellulose is a biochemical competitor for the acid-catalyzed reaction.

Many problems need to be solved before chemurgic materials can be economically used as feedstocks. Among these problems are the recovery, purification, and fractionation of the diverse materials. However, none of these problems are insurmountable. Serious concerns are the supply of the raw material, the relative costs of competitive materials, and competition with other uses for the raw materials.

The economic balance must be considered between recovery, reuse, and modification of a waste material or by-product and its disposal. The future is expected to bring increases in the practice of recycle, recovery, modification, and upgrading of wastes of all sorts, and a reduction in disposal by incineration (qv), biochemical oxidation, or discharge to the environment (see RECYCLING).

J. PETER CLARK
A. Epstein and Sons International, Inc.

U. S. Dept. of Agriculture, *Crops in Peace and War,* U.S. Government Printing Office, Washington, D.C., 1950.

National Research Council, *Renewable Resources for Industrial Materials,* National Academy of Sciences, Washington, D.C., 1976.

I. S. Goldstein, *Science* **189**(4206), 847 (1975).

J. A. Phillips, *Chemtech* **15**(6), 377 (1985).

CHIRAL SEPARATIONS

Chiral separations are concerned with separating molecules that can exist as nonsuperimposable mirror images. Examples of these types of molecules, called *enantiomers* or *optical isomers,* are illustrated in Figure 1. Although chirality is often associated with compounds containing a tetrahedral carbon with four different substituents, other atoms, such as phosphorus or sulfur, may also be chiral. In addition, molecules containing a center of asymmetry, such as hexahelicene, tetrasubstituted adamantanes, and substituted allenes or molecules with hindered rotation, such as some 2,2′ disubstituted binaphthyls, may also be chiral. Compounds exhibiting a center of asymmetry are called *atropisomers.* An extensive review of stereochemistry may be found under PHARMACEUTICALS, CHIRAL.

Although scientists have known since the time of Louis Pasteur that optical isomers can behave differently in a chiral environment (eg, in the presence of polarized light), it has only been since about 1980 that there has been a growing awareness of the implications arising from the fact that many drugs are chiral and that living systems constitute chiral environments. Hence, the optical isomers of chiral drugs may exhibit different bioactivities and/or biotoxicities.

In the case of enantiomerically pure chiral drugs, the possibility of racemization or inversion either *in vivo* or during storage cannot be ruled out. Ibuprofen is an example of a chiral drug which undergoes rapid inversion *in vivo.* In addition, there are several examples of achiral (or *prochiral*) drugs being biotransformed into chiral entities. In some cases, the enantiomeric ratios produced by laboratory animals may differ from that produced in humans. This raises the question of the suitability of laboratory animals as appropriate test models for a certain drug.

For those drugs that are administered as the racemate, each enantiomer needs to be monitored separately yet simultaneously, since metabolism, excretion or clearance may be radically different for the two enantiomers. Further complicating drug profiles for chiral drugs is that often the pharmacodynamics and pharmacokinetics of the racemic drug is not just the sum of the profiles of the individual enantiomers.

Although a great deal of the work currently being done in chiral separations is related to pharmaceuticals, the agricultural and the food and beverage industries are affected as well. For instance, several chiral pesticides are used commercially. It is possible that the enantiomers may differ in their persistence in the environment and their effectiveness against specific pests. In the food and beverage industry, many of the constituents that confer flavor or aroma in foods and beverages are chiral. For instance, the configuration of the 4-alkyl-substituted γ-lactones responsible for much of the flavor in fruits is almost exclusively *R.* Often, the two enantiomers have very different

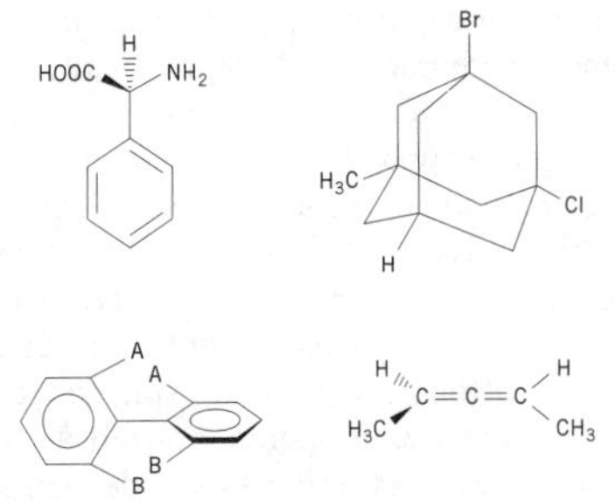

Figure 1. Examples of chiral molecules.

Table 1. Analyte Functional Groups and Chiral Derivatizing Reagents

Analyte functional group	Derivatizing agent	Product	Examples of derivatizing agents
carboxylic acid (acid or base catalyzed)	alcohol	ester	(−)-menthol
	amine	amide	1-phenylethylamine
			1-(1-naphthyl)ethylamine
amine (1°)	aldehyde	isoindole	*o*-phthaldialdehyde–2-mercaptoethanol
amine (1° and 2°)	anhydrides	amide	γ-butyloxycarbonyl-L-leucine anhydride
			O,O-dibenzoyltartaric anhydride
	acyl halides	amide	(*R*)-(−)-methylmandelic acid chloride
			α-methoxy-α-trifluoromethylphenylacetyl chloride
	isocyanates	urea	α-methylbenzyl isocyanate
			1-(1-naphthyl)ethyl isocyanate
	isothiocyanate	thiourea	2,3,4,6-tetra-*O*-acetyl-β-D-glucopyranosyl isothiocyanate
			α-methylbenzyl isothiocyanate
(1°, 2°; can *N*-dealkylate 3°)	chloroformates	carbamate	(−)-menthyl chloroformate
			(+)-1-(9-fluorenyl)ethylchloroformate
alcohols	acyl halides	ester	(−)-menthoxy acid chloride
			(*S*)-*O*-propionylmandelyl chloride
	anhydrides	ester	(*S,S*)-tartaric anhydride
	chloroformate	carbonate	(−)-menthyl chloroformate
	isocyanate	carbamate	α-methylbenzyl isocyanate

aromas or flavors. The presence of any of the "unnatural" enantiomer may confer an "off-flavor" to the substance and may be indicative of racemization under adverse storage conditions, adulteration, or formulation from nonnatural sources.

The growing awareness of the implications of chirality to the pharmaceutical industry has spurred tremendous effort toward stereoselective synthetic strategies and the development of new chiral catalysts. However, the enantiomeric purity of these substances or their chiral precursors needs to be determined. Also, there are many chiral compounds for which no stereospecific synthetic pathways have been devised. Thus, there is a tremendous need not only for analytical scale (<5–10 mg), but bulk-scale chiral separations as well.

Whether analyzing drugs or synthetic precursors for enantiomeric purity, monitoring biological or environmental samples for chiral discrimination or trying to enantioresolve kilogram quantities of a racemic drug, there are a variety of reasons for performing chiral separations. The purpose of the separation dictates, to some extent, the method employed.

Traditionally, chiral separations have been considered among the most difficult of all separations.

A variety of strategies have been devised to obtain them. Although the focus of this article is on chromatographically based chiral separations, other methods include crystallization and stereospecific enzymatic-catalyzed synthesis or degradation. In crystallization methods, racemic chiral ions are typically resolved by the addition of an optically pure counterion, thus, forming diastereomeric complexes.

Enzymatically based methods depend upon the stereospecificity of an enzyme-catalyzed reaction, such as lipase-catalyzed esterification, to degrade enantioselectively the unwanted enantiomer or to produce the desired enantiomer. Because only one enantiomer undergoes the reaction, the subsequent separation is reduced to separating two different species. One disadvantage of enzymatically based methods is that only one enantiomer is obtained and there is usually no analogous method for producing the opposite enantiomer.

An alternative method of creating a chiral environmental is to derivatize a chiral analyte with an optically pure reagent, thus, producing diastereomers. The resultant diastereomers, containing more than one chiral center, have slightly different melting and boiling points and can often be separated using conventional methods. A number of chiral derivatizing agents, as well as the types of compounds for which they are useful, have been developed and are listed in Table 1. Limitations of this approach include lack of suitable functionality in the analyte that can be derivatized with an appropriate enantiomerically pure derivatizing agent, unavailability of a suitable derivatizing agent of sufficiently high or at least known optical purity, difficulty of removing the derivatizing group after the desired separation has been accomplished, enantiodiscrimination during derivatization, potential racemization either during derivatization or removal or the chiral derivatizing group (which is not always possible), and the additional validation required to confirm that the enantiomeric ratio of the final product corresponds to the original enantiomeric ratio.

Use of Chiral Additives

Another method for creating a chiral environment is to add an optically pure chiral selector to a bulk liquid phase. Chiral additives have several advantages over chiral stationary phases and continue to be the predominant mode for chiral separations by tlc and capillary electrophoresis (ce). First of all, the chiral selector added to a bulk liquid phase can be readily changed. The use of chiral additives allows chiral separations to be done using less expensive, conventional stationary phases. A wider variety of chiral selectors are available to be used as chiral additives than are available as chiral stationary phases, thus, providing the analyst with considerable flexibility. Finally, the use of chiral additives may provide valuable insight into the chromatographic conditions and/or likelihood of success with a potential chiral stationary-phase chiral selector. This is particularly important for the development of new chiral stationary phases because of the difficulty and cost involved.

Chiral additives, however, do pose some unique problems. Many chiral agents are expensive or are not commercially available, and therefore, must be synthesized. The presence of the chiral additive in the bulk liquid phase may also interfere with detection or recovery of the analytes. Finally, the presence of enantiomeric impurity in the chiral additive may add analytical complications.

Thin-Layer Chromatography. Thin-layer chromatography (tlc) offers several advantages for chiral separations and in the development of new chiral stationary phases. Besides being inexpensive, tlc can be used to screen mobile-phase conditions rapidly (ie, organic modifier content, pH, etc), chiral selectors, and analytes. Several different analytes may be run simultaneously on the same plate. Usually, no preequilibration of the mobile phase and stationary phase is required. In addition, only small amounts of mobile phase, and therefore, chiral mobile-phase additive, are required. Another significant advantage is that the analyte can always be unambiguously found on the tlc plate.

Two mechanisms for chiral separations using chiral mobile-phase additives, analogous to models developed for ion-pair chromatography, have been proposed to explain the chiral selectivity obtained using chiral mobile-phase additives. In one model, the chiral mobile-phase additive and the analyte enantiomers form "diastereomeric complexes" in solution. As noted previously, diastereomers may have slightly different physical properties such as mobile phase solubilities or slightly different affinities for the stationary phase. Thus, the chiral separation can be achieved with conventional columns.

An alternative model has been proposed in which the chiral mobile-phase additive is thought to modify the conventional, achiral stationary phase *in situ,* thus, dynamically generating a chiral stationary phase. In this case, the enantioseparation is governed by the differences in the association between the enantiomers and the chiral selector in the stationary phase.

Several different types of chiral additives have been used including (1*R*)-(−)-ammonium-10-camphorsulfonic acid, cyclodextrins, proteins, and various amino acid derivatives such as *N*-benzoxycarbonyl-glycyl-L-proline as well as macrocyclic antibiotics.

Chiral separation validation in tlc may be accomplished by recovering the individual analyte spots from the plate and subjecting them to some type of chiroptical spectroscopy such as circular dichroism or optical rotary dispersion. Alternatively, the plates may be analyzed using a scanning densitometer.

Capillary Electrophoresis. Capillary electrophoresis (ce) or capillary zone electrophoresis (cze), a relatively recent addition to the arsenal of analytical techniques, has also been demonstrated as a powerful chiral separation method. Its high resolution capability and lower sample loading relative to hplc makes it ideal for the separation of minute amounts of components in complex biological mixtures.

In a ce experiment, a thin capillary is filled with a run buffer and a voltage is applied across the capillary. The underlying impetus for separations in ce is, in general, derived from the fact that charged species migrate in response to an applied electric field proportionately to their charge and inversely proportionately to their size.

Chiral separations by ce have been performed almost exclusively using chiral additives to the run buffer. The advantages of this approach are identical to the advantages mentioned previously with regard to using chiral mobile-phase additives in tlc. Many of the chiral selectors used successfully as mobile-phase additives in tlc and as immobilized ligands in hplc have been used successfully in ce including proteins, native and functionalized cyclodextrins, various carbohydrates, assorted functionalized amino acids, chiralion pairing agents, and macrocyclic antibiotics.

Although chiral ce is most commonly performed using aqueous buffers, there has been some work using organic solvents such as methanol, formamide, *N*-methylformamide or *N,N*-dimethylformamide with chiral additives such as quinine or cyclodextrins. Nonaqueous ce requires that the background electrolyte be prepared using organic acids (eg, citric acid or acetic acid) and organic bases (eg, tetraalkylammonium halides or tris(hydroxymethyl)aminomethane).

Chiral Stationary Phases

Most chiral chromatographic separations are accomplished using chromatographic stationary phases that incorporate a chiral selector. The chiral separation mechanisms are generally thought to involve the formation of transient diastereomeric complexes between the enantiomers and the stationary phase chiral ligand. Differences in the stabilities of these complexes account for the differences in the retention observed for the two enantiomers. Often, the use of a chiral stationary phase allows for the direct separation of the enantiomers without the need for derivatization. One advantage offered by the use of chiral stationary phases is that the chiral selector need not be enantiomerically pure, only enriched. In addition, for chiral stationary phases having a well understood chiral recognition mechanism, assignment of configuration (eg, *R* or *S*) may be possible even in the absence of optically pure standards. However, chiral stationary phases have some limitations. The specificity required for chiral discrimination limits the broad applicability of most chiral stationary phases; thus there is no "universal" chiral stationary phase. The cost of most chiral columns are typically much higher (~3×) than for conventional columns. In contrast to conventional chromatographic columns, chiral stationary phases are generally not as robust, require more careful handling than conventional columns and usually, once column performance has begun to deteriorate, cannot be returned to their original performance levels. In many cases, chromatographic column choice or mobile phase optimization for chiral stationary phases is not as straightforward as with conventional stationary phases. For many of the chiral stationary phases, adequate chiral recognition models, used to guide selection of the appropriate column for a given separation, have yet to be developed. Column selection, therefore, is often reduced to identifying structurally similar analytes for which chiral resolution methods have been reported in the scientific literature or chromatographic supply catalogues and adapting a reported method for the chiral pair to be resolved.

An additional complication, sometimes arising with the use of chiral stationary phases, may occur when the analytes either exist as *conformers* or can undergo inversion uring the chromatographic analysis.

Thin-Layer Chromatography. Chiral stationary phases in tlc have been primarily limited to phases based on normal or microcrystalline cellulose, triacetylcellulose sorbents or silica-based sorbents that have been chemically modified or physically coated to incorporate chiral selectors such as amino acids or macrocyclic antibiotics into the stationary phase. The cost of many chiral selectors, as well as the accessibility and success of chiral additives, may have inhibited widespread commercialization.

Of the silica-based materials, only the ligand-exchange phases are commercially available (Chiralplate, tlc plates are available through Alltech Associates, Inc.) Supelco, Inc., the Aldrich Chemical Company, and Bodman Industries are all based on ligand exchange. Typically in the case of the ligand-exchange type tlc plates, the ligand-exchange selector is comprised of an amino acid residue to which a long hydrocarbon chain has been attached (eg, (2*S*,4*R*, 2′*RS*)4-hydroxy-1-(2-hydroxydodecyl)proline). The hydrocarbon chain of the functionalized amino acid is either chemically bonded to the substrate or intercalates in between the chains of a reversed phase-stationary phase thus immobilizing the chiral selector. The bidentate amino acid chiral selector is thought to reside close to the surface of the stationary phase and participates as a ligand in the formation of a bi-ligand complex with a divalent metal ion (eg, Cu^{2+}) and the chiral bidentate analyte. Analytes enantioresolvable using ligand exchange are usually restricted to 1,2-diols, α-amino acids, α-amino alcohols, and α-hydroxyacids. Again, differences in the stabilities of the diastereomeric complexes thus formed give rise to the chiral separation.

High Performance Liquid Chromatography. The last decade has seen the commercialization of a large number of different types of chiral stationary phases including the cyclodextrin phases, the chirobiotic phases, the π-π interaction phases, the protein phases as well as the cellulosic and amylosic phases and chiral crown ether phases. Currently, there are over 50 different chiral columns that are commercially available for hplc. Table 2 briefly summarizes the types of columns available as well as typical applications and mobile-phase conditions. Each of these chiral stationary phases are very successful at separating large numbers of enantiomers, which in many cases, are unresolvable using any of the other chiral stationary phases. Unfortunately, despite the large number and variety of chiral stationary phases currently available, there remains a large number of enantiomeric compounds that are unresolvable by any of the existing chiral, stationary phases. In addition, incomplete understanding of the chiral recognition mechanisms of many of these chiral stationary phases limits the realization of the full potential of the existing chiral stationary phases and hampers development of new chiral stationary phases.

Table 2. Classes of Hplc Chiral Stationary Phases

Column chiral selector	Typical mobile phase conditions	Typical analyte features required
pirkle	nonpolar organic; 2-propanol-hexane	π-acid or π-basic moieties for charge transfer complex; hydrogen-bonding or dipole stacking capability near chiral center
protein	phosphate buffers	aromatic near chiral center; organic acids or bases; cationic drugs
cyclodextrin	aqueous buffers; polar organic	good "fit" between chiral cavity or chiral mouth of cyclodextrin and hydrophobic moiety; hydrogen-bonding capability near chiral center
ligand exchange	aqueous buffers	α-hydroxy or α-amino acids near chiral center; can do nonaromatic
chiral crown ether	0.01 *N* perchloric acid	primary amines near chiral center; can do nonaromatic
macrocyclic antibiotics	aqueous buffers, nonpolar and polar organic	amines, amides, acids, esters; aromatic; hydrophobic moiety
cellulosic and amylosic	nonpolar organic	aromatic

Ligand-Exchange Phases

Among the earliest reports of chiral separations by liquid chromatography were based on work done by Davankov using ligand exchange. These types of columns are available from Phenomenex, J. T. Baker, and Regis Technologies, Inc. Although almost any amino acid can form the basis for the chiral selector, proline and hydroxyproline exhibit the most widespread utility. Also, although other metals can be used, copper(II) is usually the metal of choice and is added to the aqueous buffer mobile phase.

The dependence of chiral recognition on the formation of the diastereomeric complex imposes constraints on the proximity of the metal binding sites, usually either an hydroxy or an amine α to a carboxylic acid, in the analyte. Principal advantages of this technique include the ability to: assign configuration in the absence of standards; enantioresolve nonaromatic analytes; use aqueous mobile phases; acquire a stationary phase with the opposite enantioselectivity; and predict the likelihood of successful chiral resolution for a given analyte based on a well-understood chiral recognition mechanism.

Pirkle Phases

Of all of the commercially available chiral stationary phases for liquid chromatography, the chiral recognition mechanism for the "Pirkle" phases are among the best understood. Chiral recognition on Pirkle phases is thought to depend upon complimentary interactions between the analyte and the selector. These interactions may be π-π, steric, hydrogen-bonding, or dipole—dipole interactions and contribute to the overall stability of the diastereomeric association complexes that form between the individual enantiomers and the chiral selector in the stationary phase.

Nonpolar organic mobile phases, such as hexane with ethanol or 2-propanol as typical polar modifiers, are most commonly used with these types of phases. Under these conditions, retention seems to follow normal phase-type behavior (eg, increased mobile phase polarity produces decreased retention). The normal mobile phase components only weakly interact with the stationary phase and are easily displaced by the chiral analytes thereby promoting enantiospecific interactions. Some of the Pirkle-types of phases have also been used, to a lesser extent, in the reversed phase mode.

Reciprocity, an important concept introduced by Pirkle, exploited the notion that analytes that were well resolved using a particular chiral selector would likely be good candidates for chiral selectors to enantioresolve analytes similar to the original chiral selector. This insight spawned a second generation of Pirkle phases based on *N*-(2-naphthyl)-α-amino acids. These phases were very successful at enantio-resolving analytes containing a 3,5-dinitrobenzoyl group, such as 3,5-dinitrophenyl carbamates, and ureas of chiral alcohols and amines. These columns are available through a variety of sources including Phenomenex, Regis Technologies, Inc., J. T. Baker, Inc., and Supelco, Inc.

The structure of the Whelk-O-1 phase, the most recent addition to this type of chiral stationary phase, is illustrated in Figure 2. The presence of both π-acid and π-base features, as well as the inherent rigidity of the chiral selector, confers greater versatility than any of the previous Pirkle-type phases, imposing fewer constraints on both analyte structural features required for successful enantioresolution and mobile phase conditions. Indeed, this chiral stationary phase has demonstrated considerable chiral selectivity for naproxen, warfarin, and its *p*-chloro analogue under nonaqueous reversed-phase conditions and reversed-phase conditions. An additional advantageous feature of this phase is its availability with either the (*R,R*) or (*S,S*) configuration, thus, permitting the enantiomeric elution order to be readily changed. The small size of the chiral selector also promotes fairly high bonded ligand densities in the stationary phase, which coupled with the high enantioselectivities often achieved with these phases, facilitates their use for preparative-scale separations.

Cyclodextrin Phases

Cyclodextrins are macrocyclic compounds comprised of D-glucose bonded through 1,4-α-linkages and produced enzymatically from starch. The greek letter which proceeds the name indicates the number of glucose units incorporated in the CD (eg, $\alpha = 6, \beta = 7, \gamma = 8$, etc). Cyclodextrins are toroidal shaped molecules with a relatively hydrophobic internal cavity (Fig. 3).

Among the most successful of the liquid chromatographic reversed-phase chiral stationary phases have been the cyclodextrin-based phases, introduced by Armstrong and commercially available through Advanced Separation Technologies, Inc. or Alltech Associates. The most commonly used cyclodextrin in hplc is the β-cyclodextrin. In the bonded phases, the cyclodextrins are thought to be tethered to the silica substrate through one or two spacer ligands. The mechanism thought to be responsible for the chiral selectivity observed with these phases is based upon the formation of an inclusion complex between the hydrophobic moiety of the chiral analyte and the hydrophobic interior of the cyclodextrin cavity. Preferential complexation between one optical isomer and the cyclodextrin through stereospecific interactions with the secondary hydroxyls which line the mouth of the cyclodextrin cavity results in the enantiomeric separation. Unlike the Pirkle-type phases, enantiospecific interactions between the analyte

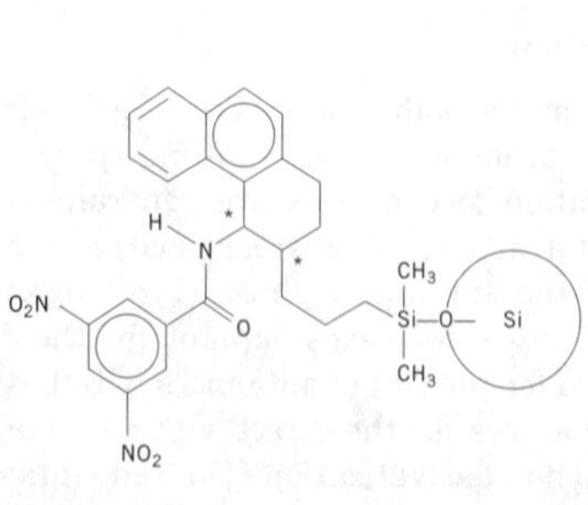

Figure 2. The structure of the chiral selector in the Whelk-O-1 chiral stationary phase.

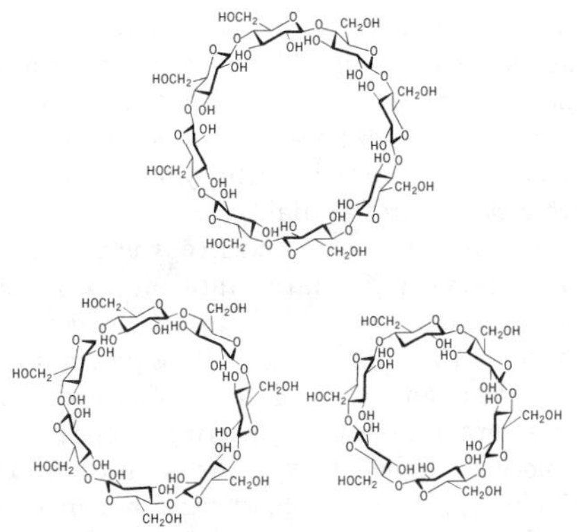

Figure 3. The structure of the three most common cyclodextrins.

and the cyclodextrin are not the result of a single, well-defined association, but more of a statistical averaging of all the potential interactions with each interaction weighted by its energy or strength of interaction.

Vast amounts of empirical data suggest that chiral recognition on cyclodextrin phases in the reversed phase mode require the presence of an aromatic moiety that can fit into the cyclodextrin cavity, that there be hydrogen bonding groups in the molecule, and that the hydrophobic and hydrogen-bonding moieties should be in close proximity to the stereogenic center. Chiral recognition seems to be enhanced if the stereogenic center is positioned between two π-systems or incorporated in a ring.

Most of the chiral separations reported to date using the native cyclodextrin-based phases have been accomplished in the reversed-phase mode using aqueous buffers containing small amounts of organic modifiers. However, polar organic mobile phases have gained in popularity recently because of their ease of removal from the sample and reduced tendency to accelerate column degradation relative to the hydroorganic mobile phases. In these cases, because the more nonpolar component of the mobile phase is thought to occupy the cyclodextrin cavity, the analyte is thought to sit atop the mouth of the cyclodextrin much like a "lid".

Limitations with the chiral selectivity of the native cyclodextrins fostered the development of various functionalized cyclodextrin-based chiral stationary phases, including acetylated, sulfated, 2-hydroxypropyl, 3,5-dimethylphenylcarbamoylated and 1-naphthylethylcaarbamoylated cyclodextrin. Each of the glucose residues contribute three hydroxyl groups to which a substituent may be appended; thus, each cyclodextrin contributes multiple sites for derivatization. The substituents of these functionalized cyclodextrins seem to play a variety of roles in enhancing chiral recognition.

Cellulosic and Amylosic Phases

Cellulose and amylose are comprised of the same glucose subunits as the cyclodextrins. In the case of cellulose, the glucose units are attached through 1,4-β-linkages resulting in a linear polymer. In the case of amylose, the 1,4-α-linkages, as are found in the cyclodextrins, are thought to confer helicity to the polymeric chain.

Cellulosic phases as well as amylosic phases have been used extensively for enantiomeric separations recently. Most of the work in this area has been with various derivatives of the native carbohydrate. The enantioresolving abilities of the derivatized cellulosic and amylosic phases are reported to be very dependent upon the types of substituents on the aromatic moieties that are appended onto the native carbohydrate. Table 3 lists some of the cellulosic and amylosic derivatives that have been used. These columns are available through Chiral Technologies, Inc. and J. T. Baker, Inc.

Protein-Based Phases

Proteins, amino acids bonded through peptide linkages to form macromolecular biopolymers, used as chiral stationary phases for hplc include bovine and human serum albumin, α_1-acid glycoprotein, ovomucoid, avidin, and cellobiohydrolase. The bovine serum albumin column is marketed under the name Resolvosil and can be obtained from Phenomenex. The human serum albumin column can be obtained from Alltech Associates, Advanced Separation Technologies, Inc., and J. T. Baker. The α_1-acid glycoprotein and cellobiohydrolase can be obtained from Advanced Separation Technologies, Inc. or J. T. Baker, Inc.

Table 3. Carbohydrate Derivatives Used as Hplc Chiral Stationary Phases

Cellulosic	Amylosic
triacetate	
tribenzoate	
tribenzylether	
tricinnamate	
triphenylcarbamate	triphenylcarbamate
tris-3,5-dichlorophenylcarbamate	
tris-3,5-dimethylphenylcarbamate	tris-3,5-dimethylphenylcarbamate
tris-1-phenylethylcarbamate	tris-1-phenylethylcarbamate

In most cases, the protein is immobilized onto γ-aminopropyl silica and covalently attached using a cross-linking reagent such as *N,N'*-carbonyldiimidazole. The tertiary structure or three dimensional organization of proteins are thought to be important for their activity and chiral recognition. Therefore, mobile phase conditions that cause protein "denaturation" or loss of tertiary structure must be avoided.

Typically, the mobile phases used with the protein-based chiral stationary phases consist of aqueous phosphate buffers. Often small amounts of organic modifiers, such as methanol, ethanol, propanol, or acetonitrile, are added to reduce hydrophobic interactions with the analyte and to improve enantioselectivity. In some cases, dramatic changes in chiral recognition occur when small amounts of organic modifiers, such as *N,N*-dimethyloctylamine or octanoic acid are added to the mobile phase. As in the case of the cyclodextrin and amylosic and cellulosic phases, the chiral recognition mechanism for these protein-based phases is not well understood. Optimization of chromatographic conditions and selection of analytes that can be successfully resolved on these phases is usually done empirically. In addition, the large molecular weight of these biopolymers dictates that the amount of chiral selector that can be immobilized on the column packing material is very small.

An interesting application of the protein-based phases is various protein binding and displacement experiments which can be done fairly routinely. For instance, differences in the enantioselectivity, toward a particular drug, of a column derived from human serum albumin and a column derived from some other animal serum albumin might be indicative that a particular species might not be a good animal model during drug development, thus, obviating the need for animal testing. Chiral separations on protein-based phases may also provide useful information on drug interactions.

Chirobiotic Phases

The chirobiotic chiral stationary phases are based on macrocyclic antibiotics such as vancomycin and teicoplanin. These chiral selectors, originally used as chiral additives in capillary zone electrophoresis, incorporate aromatic and carbohydrate, as well as peptide and ionizable moieties. The presence of aromatic groups, allowing for π-π interactions, and the macrocyclic rings, offering potential inclusion complexation, give these phases some of the advantages of the protein-based phases (eg, peptide and hydrogen bonding sites) and the carbohydrate-based phases but with greater sample capacity and greater mobile phase flexibility. Indeed, these phases seem to be truly "multimodal" in that they have demonstrated chiral selectivity in the normal, polar organic, and reversed-phase modes. In addition, the use of such well-defined chiral selectors facilitate method development and optimiza-

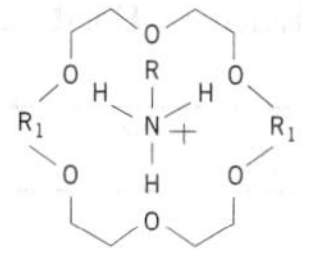

Figure 4. An inclusion complex formed between a protonated primay amine and a chiral crown ether.

tion. These columns are commercially available through Advanced Separation Technologies, Inc. and Alltech Associates.

Chiral Crown Ether Phases

Chiral crown ethers based on 18-crown-6 (Fig. 4) can form inclusion complexes with ammonium ions and protonated primary amines. Immobilization of these chiral crown ethers on a chromatographic support provides a chiral stationary phase which can resolve most primary amino acids, amines and amino alcohols. However, the stereogenic center must be in fairly close proximity to the primary amine for successful chiral separation. Significantly, the chiral crown ether phase is unique in that it is one of the few liquid chromatographic chiral stationary phases that does not require the presence of an aromatic ring to achieve chiral separations.

Mobile phases used with this stationary phase are typically 0.01 *N* perchloric acid with small amounts of methanol or acetonitrile. One significant advantage of these phases is that both configurations of the chiral stationary phase are commercially available and can be obtained from J. T. Baker Inc. and Chiral Technologies, Inc. (Crownpak CR).

Chiral Synthetic Polymer Phases

Chiral synthetic polymer phases can be classified into three types. In one type, a polymer matrix is formed in the presence of an optically pure compound to molecularly *imprint* the polymer matrix. Subsequent to the polymerization, the chiral template is removed, leaving the polymer matrix with chiral cavities. The selectivities achieved with these phases are generally excellent, thus, facilitating semipreparative separations. However, the applicability of these chiral stationary phases are generally limited to the analyte upon which the phase is based and a limited number of analogues. In addition, these types of phases generally exhibit poor efficiency in large part because the polymeric matrix contributes to nonsterespecific binding. Advantages of this approach include the ability to prepare reciprocal phases and the predictability of the enantiomeric elution order.

Another type of synthetic polymer-based chiral stationary phase is formed when chiral catalyst are used to initiate the polymerization. Columns of this type (eg, Chiralpak OT) are available from Chiral Technologies, Inc., or J. T. Baker Inc.

A third type of synthetic polymer-based chiral stationary phase, developed by Blaschke, is produced when a chiral selector is either incorporated within the polymer network or attached as pendant groups onto the polymer matrix. Both are analogous to methods used to produce polymeric chiral stationary phases for gc.

In general, the synthetic polymeric phases seem to have polarities analogous to diol-type phases and a wide range of mobile phase conditions have been used including hexane, various alcohols, acetonitrile, tetrahydrofuran, dichloromethane and their mixtures, as well as aqueous buffers.

Chiral Separation Validation for Hplc

Chiral separations present special problems for validation. Typically, in the absence of spectroscopic confirmation (eg, mass spectral or infrared data), conventional separations are validated by analyzing "pure" samples under identical chromatographic conditions. Often, two or more chromatographic stationary phases, which are known to interact with the analyte through different retention mechanisms, are used. If the pure sample and the unknown have identical retention times under each set of conditions, the identity of the unknown is assumed to be the same as the pure sample. However, often the chiral separation that is obtained with one type of column may not be achievable with any other type of chiral stationary phase. In addition, "pure" enantiomers are generally not available.

Most commonly, uv or uv–vis spectroscopy is used as the basis for detection in hplc. When using a chiral stationary phase, confirmation of a chiral separation may be obtained by either monitoring the column effluent at more than one wavelength or by running the sample more than once. Although not absolute proof of a chiral separation, this approach does provide strong supporting evidence.

As in tlc, another method to validate a chiral separation is to collect the individual peaks and subject them to some type of optical spectroscopy, such as, circular dichroism or optical rotary dispersion. Alternatively, a chiroptical spectroscopy can be used as the basis for detection on-line using commercially available optical rotary dispersion or circular dichroism-based detectors. Another method for validating chiral separations by lc is to couple the chromatographic system to a mass spectrometer.

Chiral Stationary Phases for Gas Chromatography

Gc chiral stationary phases can be broadly classified into three categories: diamide, cyclodextrin, and metal complex.

Diamide Chiral Separations. The first commercially available chiral column was the Chiralsil-val, which was introduced in 1976 for the separation of amino acid type compounds by gas chromatography. It is based on a polysiloxane polymer containing chiral side chains incorporating L-valine-*t*-butylamide. The polysiloxane backbone improved the thermal stability of these chiral stationary phases relative to the original coated columns and extended the operating temperatures up to 220°C. The column is effective for the separation of perfluoroacylated and esterified amino acids, amino alcohols, and some chiral sulfoxides. Another polysiloxane-based chiral stationary phase incorporating L-valine-(*R*)-α-phenylethylamide appended onto hydrolyzed XE-60 was found to be particularly successful at resolving perfluoroacetylated amino alcohol derivatives. Through judicious choice of derivatizing agent, chiral separations were obtained for a wider range of compounds, including amino alcohols, α-hydroxy acids, diols and ketones, than had previously been obtainable using these types of stationary phases.

Metal Complex. Complexation gas chromatography was first introduced by V. Schurig in 1980 and employs transition metals (eg, nickel, cobalt, manganese or rhodium) complexed with chiral terpenoid ketoenolate ligands such as 3-trifluoroacetyl-1*R*-camphorate **(1)**, 1*R*-3-pentafluoro-benzoylcamphorate or 3-heptafluorobutanoyl-(1*R*,2*S*)-pinanone-4-ate. This class of chiral columns is particularly adept at enantioresolving some olefins and oxygen-containing compounds such as ketones, ethers, alcohols, spiroacetals, oxiranes, and esters.

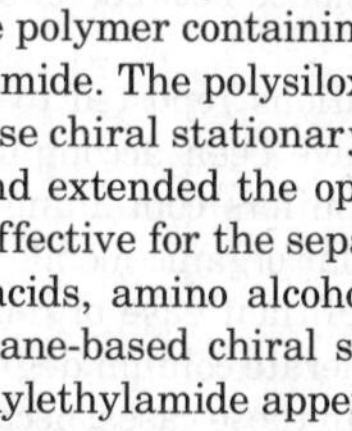

(1)

Cyclodextrins. As indicated previously, the native cyclodextrins, which are thermally stable, have been used extensively in liquid chromatographic chiral separations, but their utility in gc applications was hampered because their highly crystallinity and insolubility in most organic solvents made them difficult to formulate into a gc stationary phase. However, some functionalized cyclodextrins form viscous oils suitable for gc stationary-phase coatings and have been used either neat or diluted in a polysiloxane polymer as chiral stationary phases for gc.

Chiral Separation Validation for Gas Chromatography

The special problems for validation presented by chiral separations can be even more burdensome for gc because most methods of detection (eg, flame ionization detection or electron capture detection) in gc destroy the sample. Even when nondestructive detection (eg, thermal conductivity) is used, individual peak collection is generally more difficult than in lc or tlc. Thus, off-line chiroptical analysis is not usually an option. Fortunately, gc can be readily coupled to a mass spectrometer and is routinely used to validate a chiral separation.

APRYLL M. STALCUP
University of Cincinnati

L. Pasteur, *Comptes Rendus de l'Academie des Sciences* **26,** 535 (1848).

W. J. Wechter, D. G. Loughhead, R. J. Reischer, G. J. Van Giessen, and D. G. Kaiser, *Biochem. Biophys. Res. Comm.* **61,** 833 (1974).

R. A. Kuzel, S. K. Bhasin, H. G. Oldham, L. A. Damani, J. Murphy, P. Camilleri, and A. J. Hutt, *Chirality* **6,** 607 (1994).

S. A. Tobert, *Clin. Pharmacol. Ther.* **29,** 344 (1981).

CHLORAMINES AND BROMAMINES

N-Halamines, inorganic and organic compounds in which halogen is attached to nitrogen, have both research and industrial importance. The *N*-halogen bond is formed by reaction of an amine, imine, amide, or imide with halogen, hypohalous acid, or hypohalite. Numerous *N*-fluoramines have been prepared and are used as selective fluorinating agents. Iodamines are the least stable and least studied. Only the chloro and bromo derivatives are of commercial importance. The chemistry of chloramines and bromamines is diversified not only because nitrogen and halogen act as reaction sites but also because of the different modes by which these functionalities react. In aqueous solution, chloramines and bromamines undergo hydrolysis to varying degrees, forming HOCl and HOBr. Thus these *N*-halamines can be considered as halogen-release agents and many find use in bleaching, disinfecting, and sanitizing applications. Others, such as the halogen derivatives of ammonia, are important because they are industrial process intermediates (monochloramine) or are of significance in water treatment (mono-, di-, and trichloramine).

Properties

Available Halogen. The available halogen in an *N*-halamine is the percentage of *N–X* halogen expressed in terms of equivalent molecular halogen, ie, it is a measure of the oxidizing capacity in terms of elemental halogen. The following equation illustrates the concept of available halogen, in which halamines liberate free halogen upon acidification, one mol of halogen being liberated for each N –Cl or N–Br bond:

$$RR'NX + HX \rightarrow RR'NH + X_2$$

Hydrolysis. Halogen in *N*-halamines is formally positive, ie, the oxidation state is +1. *N*-Halamines hydrolyze, yielding hypohalous acid, which ionizes to hypohalite ion, depending on pH.

Bleaching. A K_h of 10^{-4} is sufficient to provide acceptable performance approaching that of hypochlorite bleaches. Commercial products such as bleaches, dishwasher detergents, and hard-surface cleaners are formulated with alkaline ingredients, such as polyphosphates and silicates, so that chloramines initially hydrolyze to hypochlorite during use.

Disinfection. The disinfection efficiency of *N*-halamines is related to the extent of hydrolysis to hypohalous acid.

In studies with so-called soft *N*-chloramines, the following factors were shown to influence antimicrobial activity significantly: (*1*) the aliphatic chain length; (*2*) the degree of chlorination of the N atom; and (*3*) the nature of a positive charge.

Thermal and Photostability. Some chloramines and bromamines exhibit the lack of stability expected of compounds with bonds between two strongly electronegative elements. Commercial organic *N*-bromamines and *N*-chloramines have good stability, which is a function of temperature, moisture, and impurities.

When chlorine is employed for outdoor swimming pool sanitation, it is relatively rapidly decomposed by sunlight. Isocyanuric acid stabilizes chlorine by formation of photostable chloroisocyanurates. By contrast, bromine is not stabilized by isocyanuric acid.

Inorganic Chloramines and Bromamines

Chloramines are formed by electrophilic attack on ammonia nitrogen via a series of bimolecular reactions, where $k\ (M^{-1}s^{-1})$ and $K\ (M^{-1})$ are the formation and equilibrium constants at 25°C and $\mu = 0.50M$.

Monochloramine. The most important of the ammonia halamines, monochloramine, is prepared by reaction of equimolar solutions of NH_3 and ClO^-. The pure compound decomposes even at low temperatures. Concentrated aqueous NH_2Cl (~70 mol%) is a colorless liquid with a strong pungent odor that decomposes at −50°C to N_2, Cl_2, and NCl_3. However, it is relatively stable when employed in dilute solutions of ether or water. The chemistry of NH_2Cl involves chlorination, amination, addition, condensation, redox, acid–base, and decomposition reactions.

Dichloramine. The least stable chloramine, dichloramine, has not been prepared in pure form. However, it has sufficient stability in dilute organic or aqueous solutions for determination of some physical and chemical properties. It has a pungent odor and can impart an odor or off-taste to water at concentrations above 0.8 ppm. Dichloramine can be produced by reaction of HOCl with a slight excess of NH_3 in the pH range of 4–7 or by disproportionation of NH_2Cl at pH 3.5–4.0.

Trichloramine. Nitrogen trichloride, trichloramine, the only stable pure ammonia halamine, is a shock-sensitive, dense, yellow liquid with a volatility similar to chloroform. Trichloramine has a pungent odor and is a lachrymator. Its solubility in water at room temperature is ~2000 ppm. It is the most irritating of the chloramines and can impart odor or off-taste to water at concentration above only 0.02 ppm. Trichloramine has been used as a bleaching agent for flour and in the manufacture of paper and as a fungicide for the treatment of fruit.

Monobromamine. In organic solvents monobromamine is dark violet. It is formed rapidly and quantitatively by reaction of NH_3 and Br_2 in organic or aqueous media. The solutions are relatively unstable.

Dibromamine. Dibromamine can be prepared in ether by reaction of Br_2 with a slight excess of NH_3. The solution has a strawberry-yellow color and a sharp irritating odor.

Tribromamine. Pure solid nitrogen tribromide is deep red and explodes even at −100°C.

Sulfamates and Imidosulfonates. Sodium, potassium, and bromine analogues of *N*-chlorosulfamic acid, $ClHNSO_3H$, and *N,N*-dichlorosulfamic acid, Cl_2NSO_3H, have been prepared and the kinetics of chlorination of sulfamic acid have been studied. *N*-Chlorosulfamates are useful in dishwashing compositions, textile bleaching, and vat or sulfur dyeing. *N,N*-Dichlorosulfamate can be used as a bleach and disinfectant. *N*-Chloroimidodisulfonates, eg, sodium *N*-chloroimidodisulfonate, $ClN\ (SO_3Na)_2$, can be used for fabric bleaching and stain removal.

Organic Chloramines and Bromamines

Organic chloramines and bromamines can be broadly classified as aliphatic, aromatic, and heterocyclic. Monohalo derivatives of primary amides, carbamates, and sulfonamides are sufficiently acidic to form metal salts. As with ammonia halamines, organo *N*-halamines disproportionate to *N,N*-dihalamines. *N*-Halamines can rearrange under the influence of heat, catalysts, or light, eg, *N*-chloroaniline rearranges to ring-chlorinated anilines. *N*-Halamines are versatile reagents that react with a variety of substrates via radical and polar pathways. They add to olefins and acetylenes providing routes to cyclic compounds and can also cleave certain C–C, C–N, and C–O bonds. They act as aminating, halogenating, dehydrohalogenating, and oxidizing agents and are useful in the preparation of many types of compounds.

Preparation. Substituted *N*-halamines are usually prepared by reaction of RR′NH with halogen, hypohalous acid, or hypohalite, where R is an organic substituent and R′ is either an organic substituent or H.

$$RR'NH + X_2 \rightarrow RR'NX + HX \xrightarrow{OH^-} RR'NX + X^- + H_2O \quad (1)$$

$$RR'NH + HOX \rightarrow RR'NX + H_2O \quad (2)$$

$$RR'NH + XO^- \rightarrow RR'NX + OH^- \xrightarrow{H^+} RR'NX + H_2O \quad (3)$$

$$RR'NH + XO^- \rightarrow RR'NX + OH^- \underset{RR'NH}{\xrightarrow{X_2}} 2\ RR'NX + X^- + H_2O \quad (4)$$

They are generally produced in neutral to slightly acid solution. *N*-Halo-*N*-sodioamidates, *N*-halo-*N*-sodiocarbamidates, and *N*-halo *N*-sodiosulfonamidates are exceptions, being prepared at moderately basic pH. *N*-Chloramines and *N*-bromamines are unstable to excess av Cl_2 or base, which can cleave the C–N bond, forming potentially explosive compounds, eg, NCl_3 and NBr_3. In some cases the acid or base by-product must be neutralized as it is formed, because acid may prevent formation of halamine and base may cause decomposition. The reaction with halogen (eq. 1) is employed in commercial processes such as the production of chloroisocyanurates; the base is added initially in the form of mono-, di-, or tri-sodium cyanurate. Other sources of electropositive halogen have also been used, eg, *N*-halamines, Cl_2O, *t*-C_4H_9OCl, *t*-C_4H_9OBr, and $CH_3C(O)OBr$. Sodium acetate has been used as an acid acceptor. In some cases, eg, in preparation of hexachloromelamine, an HCl acceptor may not be necessary if the reaction is carried out in dilute solution. *N*-Halocarbamidates and *N*-halosulfonamidates are formed directly by reaction of a carbamate or sulfonamide with hypohalite. Neutralization of the Na salts offers a convenient route to *N*-halocarbamates and *N*-halosulfonamides. In the preparation of a bromamine, a combination of Cl_2 or an *N*-chloramine and Br^- or Cl_2 plus Br_2 can be employed in order to eliminate the more expensive Br^- as a by-product: $RR'NH + Cl_2 + Br^- + HO^- \rightarrow RR'NBr + 2\ Cl^- + H_2O$. *N*-Bromo-*N*-chloramines can be prepared from equimolar quantities of the respective halamines: $RNBr_2 + RNCl_2 \rightarrow 2\ RNBrCl$.

Aliphatic Compounds

Amines. Although the chlorination of aqueous alkylamines is analogous to that of ammonia, mono- and dialkylamines generally chlorinate faster because of their higher basicities. Bromination rates are significantly faster than chlorination rates.

Amides. Because amides are less basic, they chlorinate less rapidly than amines. *N*-Halamides are converted to amines in basic solution via intermediate formation of an isocyanate (Hofmann reaction). *N*-Bromoacetamide, mp 102–105°C, is unstable to light and heat. Nevertheless, it is useful in organic synthesis as a brominating and oxidizing agent.

Ureas. Chlorination of aqueous urea yields unstable *N*-chloro compounds. Only two solid derivatives have been isolated: *N*-chlorourea, mp 74–76°C; and *N, N′*dichlorourea, which decomposes at an mp of 83°C with evolution of NCl_3.

Various substituted *N*-bromo- and *N*-chloroureas have also been prepared. These compounds are useful for synthesis of oxazolidinones, and also hydrazine, hydrazo, and azo compounds. *N*-Bromourea is useful for selective oxidation of sugar derivatives.

Cyanamides and Derivatives. *N*-Chloro-*t*-alkylcyanamides, RN (Cl) CN, are oils that have medicinal and other applications. *N,N′*-Dichlorocarbodiimide is useful in the preparation of photothermographic materials. Chlorinated guanylurea (av Cl_2 85%), prepared from dicyanamide, is useful as a bleach for synthetic fabrics. Chloroazodin, dichloroazodicarbonamidine, $NH_2C({=}NCl)N{=}NC(NCl)NH_2$, prepared from guanidine, finds use in vulcanization of rubber. *N*-Chloroamidines, eg, RNHC(CN)=NCL and $CCl_2[C(NHR){=}NCl]_2$, exhibit fungicidal properties. *N*-Chloroamidines are useful for the preparation of biocidal imidazoles and thiadiazolines. *N*-Chloroguanidines, RNHC(=NCl)NHR′, serve as starting materials for the synthesis of imidazoles, oxadiazoles, and thiadiazoles.

Amino Acids. *N*-Halo-α-amino acids decompose to aldehydes and nitriles; the selectivity depends on pH and stoichiometry.

Carbamates. Lower alkyl *N*-halo- and *N,N*-dihalocarbamates are distillable liquids. *N*-Halo-*N*-metallocarbamates are crystalline hygroscopic solids.

Sulfonamidates. *N*-Halo-*N*-alkylsulfonamides, $RSO_2NR'X$, are relatively stable distillable liquids. Under the influence of uv light they form 1:1 adducts with olefins.

Aromatic Compounds

Sulfonamidates. Chloramine-T, *N*-chlor-*N*-sodiomethylbenzenesulfonamidate trihydrate, $CH_3C_6H_4SO_2NClNa{\cdot}3H_2O$, a white to slightly yellow solid, has an mp of 175°C. Chloramine-T is used in analysis and organic synthesis. Bromine analogues are also known, ie, bromamine-T and *N,N*-bromo-*N*-sodio-4-nitrobenzenesulfonamidate. The bromo derivatives are also used in organic synthesis. Chloramine-B, *N*-chlor-*N*-sodiobenzenesulfonamidate, is prepared similarly to chloramine-T. It has been used as a disinfectant in dairies. Polymer-bound *N*-chloro-*N*-sodiobenzenesulfonamidate, prepared via functionalization of poly(styrene-co-divinylbenzene), is useful in water disinfection.

Sulfonamides. Dichloramine-T, *N,N*-dichlor-4-methylbenzenesulfonamide, $CH_3C_6H_4SO_2NCl_2$, mp 80°C, can be prepared by chlorination of the free amine or chloramine-T and has been used as a topical dressing and an antivesicant ointment. It is also used in organic synthesis. Dichloramine-B, *N,N*-dichlorobenzenesulfonamide, $C_6H_5SO_2NCl_2$, has been used as a deodorant and bleach for certain oils.

Heterocyclic Compounds

Glycolurils. Chlorinated glycolurils were developed in the 1950s and 1960s for protection against chemical agents and as bleaches, disinfectants, and foliage protectants. The most important is 2,4,6,8-tetrachloro-2,4,6,8-tetrazobicyclo(3.3.0)octane-3,7-dione. The parent compound, glycoluril, is readily prepared by condensation of urea and glyoxal.

Hydantoins. Chlorinated hydantoins, first introduced in the 1930s, did not find wide use because of its low dissolution rate. It has been used as a chlorinating agent, disinfectant, industrial deodorant, and as a laundry bleach. It is no longer used in home laundering because of changing needs of synthetic fabrics but is used to a small extent in commercial laundries where temperatures of ~ 70°C can be used. 1-Bromo-3-chloro-5,5-dimethylhydantoin (BCDMH) is used in industrial water treatment and in sanitizing spas and hot tubs. 1,3-Dibromo-5,5-dimethylhydantoin is employed in organic synthesis.

Imidazolidinones. Several mono and dichloro isomers have been prepared and tested as disinfectants: 1-chloro-4,4,5,5-tetramethylimidazolidin-2-one; 1,3-dichloro-4,4,5,5-tetramethylimidazolidin-2-one, mp 102–104°C; 1-chloro-2,2,5,5,-tetramethylimidazolidin-4-one, mp 157–158°C; and 1,3-dichloro-2,2,5,5-tetramethylimidazolidin-4-one, mp 69–71°C.

Isocyanurates. Chlorination of cyanuric acid in the presence of base produces dichloroisocyanuric acid (DCCA, HCl_2Cy, where Cy = isocyanurate anion), ie, 1,3-dichloro-*s*-triazine-2,4,6(1*H,* 3*H,* 5*H,*)-trione and trichloroisocyanuric acid (TCCA, Cl_3Cy), ie, 1,3,5-trichloro-*s*-triazine-2,4,6(1*H,* 3*H,* 5*H*)-trione.

Melamines. Hexachloromelamine has a theoretical av Cl_2 of 128%. It is less stable than the trichloro compounds.

Oxazolidinones. 3-Chloro-4,4-dimethyl-2-oxazolidinone has been extensively evaluated as a disinfectant. It is prepared by phosgenation of $(CH_3)_2CH(NH_2)CH_2OH$ followed by chlorination in the presence of caustic. It is a white crystalline solid with a theoretical av Cl_2 of 48.1%, an mp of 71–72.5°C, and a solubility of 1.2% in water.

Succinimides. *N*-Chlorosuccinimide, 1-chloropyrrolidine-2,5-dione, is a white solid with a slight chlorine odor, an mp of 150–151°C, and a solubility in water of 1.4% at 25°C. It is used in organic synthesis for highly selective oxidation of primary and secondary alcohols to carbonyl compounds, providing improved synthesis of prostaglandins, and in conversion of allylic and benzylic alcohols to halides.

Economic Aspects

The estimated 1990 worldwide consumption of monochloramine for hydrazine manufacture was 55,000 t. Consumption data on use of monochloramine in water treatment are not available. The U.S. consumption of chloroisocyanurates and halogenated hydantoins in 1986 was 44,045 and 3,409 t, respectively. Consumption of tetrachloroglycoluril and other specialty *N*-halamines, eg, trichloromelamine, is small.

Safety, Handling, and Toxicity

Chloramines and bromamines react with moisture-releasing, potentially corrosive, toxic, and explosive gases and should be stored under dry conditions at moderate temperatures, segregated from incompatible materials. *N*-Halamines are irritating to the skin, eyes, and mucous membranes. However, they are nonirritating under use conditions in dilute aqueous solution. Monochloramine has been shown to be a weak mutagen and its use in drinking water is under review by the EPA.

JOHN A. WOJTOWICZ
Olin Corporation

C. Morris, in S. D. Faust and J. D. Hunter, eds., *Principles and Applications of Water Chemistry,* John Wiley & Sons, Inc., New York, 1967, pp. 23–53.

J. E. O'Brien, J. C. Morris, and J. N. Butler, in A. J. Rubin ed., *Chemistry of Water Supply, Treatment, and Distribution,* Ann Arbor Science, Ann Arbor, Mich., 1974, p. 338.

J. L. S. Saguinsin and J. C. Morris, in J. D. Johnson, ed., *Disinfection; Water and Wastewater,* Ann Arbor Science, Ann Arbor, Mich., 1975, Chapt. 14.

CHLORAMPHENICOL. See ANTIBIOTICS, CHLORAMPHENICOL AND ANALOGUES.

CHLORINATED BIPHENYLS. See BIPHENYL AND TERPHENYLS; CHLOROCARBONS AND CHLOROHYDROCARBONS, TOXIC AROMATICS.

CHLORINE. See ALKALI AND CHLORINE PRODUCTS.

CHLORINE OXYGEN ACIDS AND SALTS

DICHLORINE MONOXIDE, HYPOCHLOROUS ACID, AND HYPOCHLORITES

Oxidation States

Chlorine has formal positive oxidation numbers (states) in oxychlorine compounds since its appreciable electronegativity (2.83) on the Allred-Rochow scale is exceeded by that of oxygen. References to positive chlorine in this article are strictly in the formal sense unless stated otherwise. All chlorine compounds are strong oxidants because the transfer of electrons to the orbitals of electronegative chlorine is favored in reactions with compounds of less electronegative elements. Chlorine oxides and oxo-acids exhibit the lack of stability expected of compounds having bonds between two strongly electronegative elements. The decomposition reactions of these compounds are, therefore, always energetic and violent in many cases (Table 1). The chemical properties of chlorine oxides and oxo-acids display a trend toward greater thermodynamic and kinetic stability with increasing oxidation state. The reduction potentials of the oxo-acids exhibit a similar trend in that the strongest oxidants have chlorine in its lower states of oxidation. Compounds of chlorine having intermediate oxidation states exhibit a strong tendency to disproportionate.

The chlorine oxides are anhydrides or mixed anhydrides of the chlorine oxo-acids; oxides with an odd number of oxygens are simple anhydrides, whereas those with an even number are mixed anhydrides.

Chlorine in dichlorine monoxide, hypochlorous acid, and ionic hypochlorites is in the +1 formal oxidation state.

The chlorine oxides are anhydrides or mixed anhydrides of the chlorine oxo-acids; oxides with an odd number of oxygens are simple anhydrides, whereas those with an even number are mixed anhydrides.

Chlorine in dichlorine monoxide, hypochlorous acid, and ionic hypochlorites is in the +1 formal oxidation state.

DICHLORINE MONOXIDE

Dichlorine monoxide is the anhydride of hypochlorous acid; the two nonpolar compounds are readily interconvertible in the gas or aqueous phases via the equilibrium: $Cl_2O + H_2O \rightleftharpoons 2\ HOCl$. Like other chlorine oxides, Cl_2O has an endothermic heat of formation and is thus thermodynamically unstable with respect to decomposition into chlorine and oxygen. Dichlorine monoxide typifies the chlorine oxides as a highly reactive and explosive compound with strong oxidizing properties. Nevertheless, it can be handled safely with proper precautions.

Physical Properties

At ordinary temperatures dichlorine monoxide is a brownish yellow gas resembling bromine. The mp of Cl_2O is −120.6°C. It readily dissolves in water to give a solution of hypochlorous acid containing a small equilibrium amount of Cl_2O.

Table 1. Chlorine Oxides[a] and Oxo-Acids

Formula	Oxidation state	Stability
		Oxides
Cl_2O	+1	anhydride of HOCl; yellowish brown gas at 25°C; explodes when heated or sparked
Cl_2O_2	+2	$t_{1/2}$ of gas = 4.8 d at − 78°C
Cl_2O_3	+3	anhydride of $HClO_2$; explodes below 0°C
ClO_2	+4	odd electron molecule; yellow gas; explodes at >6.7 kPa[b]
Cl_2O_4	+1, +7	pale yellow liquid; decomposes to Cl_2, O_2, and Cl_2O_6
Cl_2O_6	+6	red liquid; gas decomposes to Cl_2O_4 + O_2
Cl_2O_7	+7	anhydride of $HClO_4$; colorless liquid; can be distilled under reduced pressure
		Acids
HOCl[c]	+1	very weak acid, pK_a 7.54; cannot be concentrated
$HClO_2$	+3	decomposes rapidly at 25°C; pK_a~2.0; cannot be concentrated
$HClO_3$	+5	decomposes slowly at ~40% and 25°C; cannot be concentrated
$HClO_4$	+7	can be concentrated

[a] Anhydride of chloric acid, Cl_2O_5, is unknown. Oxides with even number of oxygen atoms are mixed anhydrides. Other chlorine oxides such as the radicals ClO, ClO_3, and ClO_4 are known. Chlorine monoxide, ClO, plays a key role in depletion of the ozone layer.
[b] To convert kPa to mm Hg, multiply by 7.5.
[c] The isomeric HClO is an unstable compound formed in the earth's ozone layer.

Dichlorine monoxide reacts with a variety of inorganic substances, eg, its reaction with N_2O_5 is a convenient route to $ClNO_3$. Some reactions of Cl_2O with inorganic compounds are presented below.

$$Cl_2O + 3\ Fe_2 \xrightarrow{CsF,\ -78°C} \underset{>80\%}{ClF_3O} + ClF_3$$

$$2\ Cl_2O + AgF_2 3\ Fe_2 \xrightarrow{65-70°C} ClO_2F + AgF + 3/2\ Cl_2$$

$$5\ Cl_2O + 3\ AsF_5 \xrightarrow{-78\ to\ -50°C} 2\ ClO_2AsOF_3 + 4\ Cl_2$$

$$Cl_2O + N_2O_5 \xrightarrow{0°C} 2\ \underset{90\%}{ClNO_3}$$

$$4\ Cl_2O + 3\ SO_3 \xrightarrow{CFCl_3\ -27°C} (ClO)(ClO_2)[S_3O_{10}] + 3\ Cl_2$$

$$Cl_2O + TiBr_4 \rightarrow TiOBr_2 + Br_2 + Cl_2$$

$$3\ Cl_2O + 2\ BCl_3 \xrightarrow{-78°C} B_2O_3 + 6\ Cl_2$$

$$2\ Cl_2O + P(NCl_2)_3 \rightarrow PO_2Cl + 3\ NCl_3$$

The organic chemistry of Cl_2O has not been extensively explored. Some representative reactions are shown below.

Dichlorine monoxide is a powerful and selective reagent for either side-chain or ring chlorination of deactivated aromatic substrates, providing excellent yields under mild conditions where conventional reagents fail or require harsh conditions.

$$C_3H_7Cl \xrightarrow[CCl_4,\ h\nu]{Cl_2O} \underset{43\%}{CH_3CH_2CHCl_2} + \underset{42\%}{CH_3CHClCH_2Cl} + \underset{15\%}{ClCH_2CH_2CH_2Cl}$$

$$CCl_2{=}CHCl \xrightarrow[CCl_4]{Cl_2O} CCl_3CHCl_2 + CCl_3CHO + (CCl_3CHCl)_2O$$

$$C_6H_5OH \xrightarrow[CCl_4]{Cl_2O} \underset{22\%}{o\text{-}ClC_6H_4OH} + \underset{65\%}{ClC_6H_4OH}$$

$$t\text{-}C_4H_9OOH + Cl_2O \xrightarrow[-30°C]{CCl_3F} t\text{-}C_4H_9OOCl$$

$$+HOCl \xrightarrow[-2\ HOCl]{t\text{-}C_4H_9OOH} t\text{-}C_4H_9OOO\text{-}t\text{-}C_4H_9$$

$$C_2H_5NH_2 + Cl_2O \rightarrow C_2H_5NHCl + HOCl \rightarrow C_2H_5NCl_2 + H_2O$$

$$2\ CH_3C_6H_4NO_2 + 3\ Cl_2O \xrightarrow{CCl_4,\ 75°C} \underset{100\%}{2\ CCl_3C_6H_4NO_2} + 3\ H_2O$$

$$2\ C_6H_5NO_2 + n\ Cl_2O \xrightarrow{H^+,\ 25°C} \underset{95-99\%}{2\ C_6H_{5-n}Cl_nNO_2} + n\ H_2O \quad n = 1-5$$

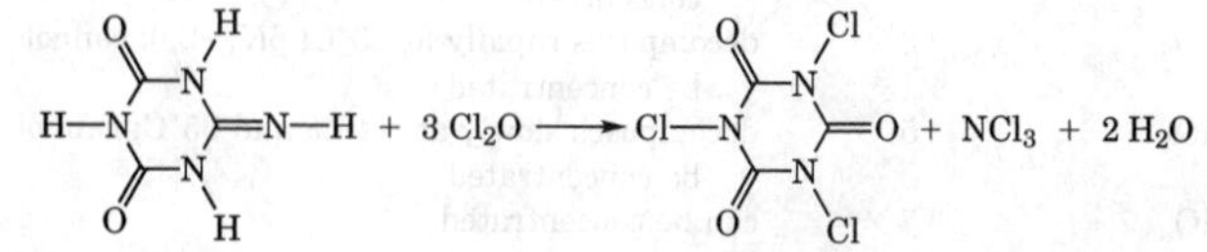

Preparation

On a laboratory scale, gaseous dichlorine monoxide is conveniently generated by reaction of chlorine gas with mercuric oxide in a packed tubular reactor. Routes more amenable to commercial-scale production are based on reaction of chlorine with sodium bicarbonate, carbonate, or sesquicarbonate.

Uses

Dichlorine monoxide is an intermediate in the manufacture of calcium hypochlorite. It has been used in sterilization for space applications (see STERILIZATION TECHNIQUES). Its use in the preparation of chlorinated solvents and chloroisocyanurates has been described. Chlorine monoxide has been shown to be effective in bleaching of pulp (qv) and textiles.

HYPOCHLOROUS ACID

Hypochlorous acid is a highly reactive and relatively unstable compound known both in solution and in the gas phase. In solution it is the most stable and the strongest of the hypohalous acids and is one of the most powerful oxidants among the chlorine oxy-acids. It is an intermediate in the manufacture of hypochlorites. Generated *in situ* via chlorine hydrolysis, it is also an intermediate in the production of chlorohydrins (qv) and chloroisocyanurates. It is the active species that kills bacteria and other microorganisms in municipal water treatment or in swimming pool sanitation when Cl_2, hypochlorites, or chloroisocyanurates are used (see WATER; BLEACHING AGENTS).

Physical Properties

Hypochlorous acid is a weak acid with a dissociation constant $K_a = 2.90 \times 10^{-8}$ at 25°C.

Chemical Properties

Dilute aqueous hypochlorous acid solutions are quite stable if pure, especially if kept cool and in the dark. Aqueous chlorine oxidizes numerous inorganic substrates.

Hypochlorous acid reacts very rapidly and quantitatively with a slight excess of free ammonia, forming monochloramine, NH_2Cl, which reacts at a slower rate with additional HOCl forming dichloramine, $NHCl_2$.

Hypochlorous acid undergoes a variety of reactions with organic substances including addition, oxidation, *C*- and *N*-chlorination, and ester formation.

Addition of HOCl to vinyl chloride yields chloroacetaldehyde; addition to acetylenic compounds produces dichloroketones.

Secondary alcohols are oxidized at room temperature to ketones in high yields by HOCl generated *in situ* from aqueous NaOCl and acetic acid.

Reaction of HOCl, formed from calcium hypochlorite and CO_2, with highly substituted alkenes in CH_2Cl_2 is a convenient route to allylic chlorides.

Stable *N*-chloro compounds are formed by reaction of hypochlorous acid and appropriate N–H compounds

Preparation

Chloride-containing solutions of HOCl are readily prepared by reaction of Cl_2 with H_2O or base and can involve rapid hydrolysis or reaction with OH^-.

Dilute (1–3%), chloride-containing solutions of either HOCl, hypochlorite, or aqueous base can be stripped in a column against a current of Cl_2, steam, and air at 95–100°C and the vapors condensed giving virtually chloride–free HOCl solutions of higher concentration in yields as high as 90%. Virtually chloride-free solutions of HOCl are obtained by reaction of atomized 50% caustic with excess gaseous chlorine at temperatures above the dew point of the system. Reduced pressure distillation of a mixture of chlorine octahydrate and HgO provides a distillate with about 25% HOCl. Preparation of aqueous HOCl substantially free of Cl^- from either aqueous Cl_2 or HOCl–salt solutions has been accomplished by electrodialysis using semipermeable membranes. Organic solutions of HOCl can be prepared in near quantitative yield (98–99%) by extraction of Cl^--containing aqueous solutions of HOCl with polar solvents such as ketones, nitriles, and esters.

Uses

Hypochlorous acid, preformed or generated *in situ* from chlorine and water, is employed in the manufacture of chlorohydrins (qv) from olefins, en route to epoxides, and in the production of chloramines, especially chloroisocyanurates from cyanuric acid (see CYANURIC AND ISOCYANURIC ACIDS).

HYPOCHLORITES

Metal Hypochlorites

Some hypochlorites, either as solutions or solids, are much more stable than hypochlorous acid, and because of their high oxidation potential and ready hydrolysis to the parent acid, find wide use in bleaching and sanitizing applications. One of the novel uses of hypochlorites was for disinfection of Apollo Eleven on its return from the moon.

The only known stable solid neutral hypochlorites are those of lithium, calcium, strontium, and barium. Calcium also forms two stable basic hypochlorites (calcium hydroxide hypochlorites): $(Ca(OCl)_2 \cdot 0.5Ca(OH)_2$ and $Ca(OCl)_2 \cdot 2Ca(OH)_2$.

Calcium hypochlorite is the principal commercial solid hypochlorite; it is produced on a large scale and marketed as a 65–70% product containing sodium chloride and water as the main diluents. A product with a significantly higher available chlorine, av Cl_2, (75–80%) has been introduced by Olin. Calcium hypochlorite is also manufactured to a smaller extent as a hemibasic compound (~60% av Cl_2) and to a lesser extent in the form of bleaching powder (~35% av Cl_2). Lithium hypochlorite is produced on a small scale and is sold as a 35% assay product for specialty applications. Small amounts of NaOCl are employed in the manufacture of crystalline chlorinated trisodium phosphate.

Properties. The solubilities of Li, Na, and Ca hypochlorites in H_2O at 25°C are 40, 45, and 21%, respectively. Thermodynamic properties of the hypochlorite ion are: $\Delta H_f{}^\circ - 107.1$ kJ/mol(-25.6 kcal/mol); $\Delta G_{f^\circ} - 36.8$ kJ/mol(-8.8 kcal/mol); and S°41.8 J/mol·K) (10cal/mol·K))

Chemical Properties. In Solution. Although hypochlorite solutions are much more stable than HOCl, they are subject to decomposition, which is influenced by concentration, ionic strength, pH, temperature, light, and impurities.

Hypochlorites yield HOCl when treated with stoichiometric amounts of acid and are converted to Cl_2 when excess HCl is used. The oxidation of various inorganic anions by hypochlorite has been studied kinetically. Hypochlorite is a strong oxidant capable of oxidizing MnO_4^{2-} to MnO_4^-, OI_3^- to IO_4^-, and Fe^{3+} to FeO_4^{2-}. Its reaction with ammonia to form monochloramine is the basis of the manufacture of hydrazine (qv).

Hypochlorite ion acts as a chlorinating and oxidizing agent toward organic compounds. It readily chlorinates phenols to mono-, di-, and tri-substituted compounds.

Saturated hydrocarbons can be chlorinated in moderate yields under mild conditions in a biphasic system consisting of alkaline hypochlorite solution and CH_2Cl_2 containing Ni(II) bis(salicylidene)ethylenediamine as chlorination catalyst and hexadecyltrimethylammonium bromide as phase-transfer catalyst.

Unsaturated aldehydes, ketones, and nitriles are epoxidized in one step in high yield via nucleophilic attack by hypochlorite ion.

Secondary alcohols are oxidized to ketones in good yields under mild conditions in a triphasic system using an inert solvent, solid $Ca(OCl)_2$, and a hypochlorite resin as catalyst.

Solid State. The stability of neutral calcium hypochlorite is primarily a function of moisture, lime, impurities, and temperature.

Heating calcium hypochlorite in a stream of N_2 during dta results in dehydration at 65–70°C; further heating causes exothermic decomposition at 200–210°C forming $CaCl_2$, O_2, and small amounts of $Ca(ClO_3)_2$.

Anhydrous hypochlorites are oxidized to chlorates by Cl_2O.

Germicidal Properties. The germicidal activity of aqueous chlorine is attributed primarily to HOCl. Although the detailed mechanism by which HOCl kills bacteria and other microorganisms has not been established, sufficient experimental evidence has been obtained to suggest strongly that the mode of action involves penetration of the cell wall followed by reaction with the enzymatic system. The efficiency of destruction is affected by temperature, time of contact, pH, and type and concentration of organisms.

Preparation. Hypochlorite solutions are prepared in near quantitative yield by chlorination of dilute caustic or a lime slurry.

Materials of Construction. Because of the corrosive nature of moist chlorine and hypochlorite solutions, chemically resistant materials are necessary to prevent metallic contamination of the products and to ensure proper functioning of equipment. Chlorination vessels and reactors can be constructed from fiber glass-reinforced polyester (FRP) or carbon steel with a suitable resistant coating or liner made of rubber, Saran, PVC, or poly(vinylidene fluoride) (Kynar) (see VINYL POLYMERS; VINYLIDENE CHLORIDE MONOMER AND POLYMERS).

Hypchlorite Solutions

Sodium Hypochlorite (Liquid Bleach). Commercial strength liquid bleach used by industries, laundries, and in swimming pool sanitation, contains 12–15% av Cl_2 and is sold in 3.8- and 7.6-L polyethylene bottles and 23–57-L carboys, 205-L drums, and tank trucks of about 3-kL capacity and greater. Household bleach contains about 5% av Cl_2 and is sold in 1–5.7-L polyethylene containers. Shipping is limited within a short radius of the plant because of transportation costs. Liquid bleach for use in chemical pulp or textile bleaching is usually prepared on site at concentrations of 30–40 g/L of av Cl_2 (see BLEACHING AGENTS).

Manufacture. Sodium hypochlorite solution is usually prepared by chlorination of NaOH.

Calcium Hypochlorite (Bleach Liquor). Bleach liquor is a solution of $Ca(OCl)_2$ and $CaCl_2$ containing some dissolved lime. It is invariably prepared on site and consumed captively, primarily for chemical pulp bleaching.

Manufacture. The mechanics of the preparation of bleach liquor by chlorination of lime slurry is only slightly different from that of sodium hypochlorite, and the heat liberated per mol of chlorine is approximately the same.

Economic Aspects. U.S. production of sodium hypochlorite has increased by ca 1% per year since 1980. Long-term growth in chemical pulp bleaching is expected to decline because of environmental factors and performance disadvantages compared to alternative bleaching gents. U.S. consumption of calcium hypochlorite bleach liquor amounts to over 2000 t/yr of equivalent Cl_2, ca 53×10^6 L (14 million gallons) of liquor containing 3–3.5% av Cl.

Uses. Sodium hypochlorite is used in chemical pulp and textile bleaching where it is largely produced on site for captive consumption. It is also used as a commercial laundry and household bleach, as a sanitizer for swimming pools, and as a disinfectant for municipal water and sewage. In food processing, sodium hypochlorite is employed as a disinfectant and sanitizer. In the production of oil it is added to water used for flooding to prevent the formation of fungi, which could cause plugging. In oil refineries it is employed as a sweetening agent. Large quantities of sodium hypochlorite are used in the chemical industry, primarily for the manufacture of hydrazine as well as in the synthesis of organic chemicals. Sodium hypochlorite is also used in the manufacture of chlorinated trisodium phosphate (see BLEACHING AGENTS).

Solid Hypochlorites

Sodium Hypochlorite–Trisodium Phosphate Complex. Commercial crystalline trisodium phosphate (TSP) is a complex of the type $(Na_3PO_4 \cdot x\ H_2O)_nNaY$ where n = 4 to 7, x = 11 or 12, and Y is a monovalent anion (see PHOSPHORIC ACIDS AND PHOSPHATES). Chlorinated trisodium phosphate is also a complex of this type with the formula $(Na_3PO_4 \cdot 11H_2O)_4 \cdot NaOCl$. The crystalline, efflorescent product melts at 62°C, which is really a transition to $Na_3PO_4 \cdot 8H_2O$.

Manufacture. Chlorinated TSP is made batchwise by addition of a 15% NaOCl solution containing some NaOH to a hot (75–80°C) concentrated liquor consisting of di- and trisodium phosphates, in a mole ratio of about 1:10, in a suitable reactor, eg, a pan mixer.

The product is shipped in 45-kg polyethylene-lined paper bags, 23-kg Fiberpak drums, and 57- and 137-Leverpak drums.

Economic Aspects. Chlorinated TSP is manufactured by Stauffer (a subsidiary of Rhône-Poulenc, Inc.). The consumption, steadily decreasing since 1980, dropped sharply in 1985 because of reduced use in dishwasher detergents.

Applications. The principal uses are in scouring cleansers and acid metal cleaners for dairy equipment.

Lithium Hypochlorite. High purity, anhydrous lithium hypochlorite, LiOCl, is a white, lightweight, dusty, hygroscopic, and corrosive powder.

Manufacture. Calcined spodumene, a lithium aluminum silicate, $LiAlSi_2O_6$, is treated with sulfuric acid, neutralized to Ph_6, and filtered to remove insolubles. The filtrate containing Li, Na, and K sulfates is treated with caustic (forming LiOH), cooled to 0°C, and filtered to separate crystallized $Na_2SO_4 \cdot 10\ H_2O$.

Economic Aspects. Lithium hypochlorite is produced by Lithium Corporation of America (a subsidiary of FMC). Its total demand is low owing to its relatively high price of about $1.27/lb in truckload quantities.

Applications. Lithium hypochlorite, has limited used in swimming pool and spa sanitation and dry laundry bleaches.

Dibasic Magnesium Hypochlorite. Dibasic magnesium hypochlorite was first synthesized in 1969 by addition of either a NaOCl or $Ca(OCl)_2$ solution to an excess of aqueous $MgCl_2$ or $Mg(NO_3)_2$.

Dibasic magnesium hypochlorite is more thermally stable than neutral or dibasic calcium hypochlorite. In addition, its decomposition, which starts at ~ 325°C, is endothermic rather that exothermic as in the case of the Ca compounds.

Dibasic magnesium hypochlorite can be used as a toilet bowl cleaner, in laundry and textile bleaches, and in scouring cleansers.

Calcium Hypochlorite. High assay calcium hypochlorite is now produced by two additional manufacturers in North America (Table 2). It is similar in composition to anhydrous $Ca(OCl)_2$ except for its higher water content of about 6–12% and a slightly lower available chlorine content. This product has improved resistance to accidental initiation of self-sustained decomposition by a lit match, a lit cigarette, or a small amount of organic contamination. U.S. production in the 1990s consists primarily of partially hydrated $Ca(OCl)_2$, which is sold as a 65% av Cl_2 product mainly for swimming pool use. Calcium hypochlorite is also sold as a 50% av Cl_2 product as a sanitizer used by dairy and food industries and in the home, and as a 32% product for mildew control.

Table 2. Calcium Hypochlorite Producers and Capacities as of 1989

Producer	Plant location	Capacity, 10^3 t/yr
Olin Corp.[a]	Charleston, Tenn.	72.7
PPG Industries, Inc[b]	Natrium, W.Va.	32.7
Saskatoon Chemical	Saskatoon, Sask.	7.0
Total		*103.3*

[a] Olin closed its 27.3 kt/yr Niagara Falls, N.Y. plant in 1982.
[b] PPG opened its Natrium plant in 1984 and closed its 7.7 kt/y Barberton, Ohio plant.

Manufacture. Calcium hypochlorite is made by drying a filter cake of neutral calcium hypochlorite dihydrate containing 30–50% water, depending on the type of filter used. This material is usually prepared from hydrated lime, caustic, and chlorine.

Commercial Processes. Olin's earlier triple salt process, originally commercialized in 1928, was modified in 1983. In the patented process, a slurry of dibasic calcium hypochlorite is mixed with a strong, low salt sodium hypochlorite solution and hypochlorite liquors and chlorinated. The resultant $Ca(OCl)_2 \cdot 2H_2O$ slurry is filtered, the cake going to the dry-end and the filtrate to the dibasic precipitation step where it reacts with lime.

Safety and Handling. Calcium hypochlorite should be stored in a cool, dry, and well-ventilated area. Since calcium hypochlorite can react vigorously and sometimes explosively with certain organic and inorganic materials, they should be kept away from it during shipment, storage, use, or disposal.

Hemibasic Calcium Hypochlorite. This basic hypochlorite is produced by chlorination of lime.

Economic Aspects. The principal producers are Japan and The People's Republic of China; Italy and India are minor producers.

Applications. Because of its high lime content, the use of hemibasic calcium hypochlorite in general sanitation is limited. It is used primarily in Japan and in lesser developed countries as an alternative to bleaching powder.

Bleaching Powder. This material, known since 1798, is made by chlorination of slightly moist hydrated lime, calcium hydroxide, $Ca(OH)_2$. It has the empirical formula $Ca(OCl)_2 \cdot CaCl_2 \cdot Ca(OH)_2 \cdot 2H_2O$.

Manufacture. For bleaching powder, numerous improvements over the original chamber process were developed to deal primarily with heat and moisture removal and mechanical apparatus for batch or continuous manufacture of bleaching powder of improved quality and stability.

Economic Aspects. The production of bleaching powder has been steadily declining. Imports, averaging 1160 t during the period 1980–1987, came from India, Spain, the United Kingdom, Germany, and Canada. It is probably also manufactured in some less developed countries.

Applications. Bleaching powder can be employed for general sanitation and may also be used to disinfect seawater, reservoirs, and drainage ditches where the volume of insolubles is not important. It can be used as a decontaminating agent for areas sprayed with chemical warfare agents such as mustard gas (see CHEMICALS IN WAR). The high insoluble (lime) content is an undesirable feature of bleaching powder owing to the loss of hypochlorite in the sludge that is formed in aqueous slurries.

Organic and Nonmetal Hypochlorites

Alkyl hypochlorites, esters of hypochlorous acid, are nonpolar, volatile liquids with irritating odors and and extremely lachrimatory. The known alkyl hypochlorites (ROCl) are methyl (CH_3) ethyl (C_2H_5), t-butyl (t-C_4H_9), and t-amyl (t-C_5H_{11}). Primary and secondary hypochlorites are very unstable but tertiary hypochlorites exhibit good stability.

Numerous fluorinated and perfluorinated alkyl hypochlorites have been synthesized and characterized, eg, CF_3OCl, C_2F_5OCl, i-C_3F_7OCl, and t-C_4F_9OCl. These nonmetal oxychlorine compounds are much more thermally stable than the corresponding parent compounds and can be prepared by reaction of ClF with the appropriate carbonyl compound or alcohol.

Inorganic nonmetal unipositive oxychlorine compounds include $ClONO_2$, $ClOClO_3$, and the hypochlorites derived from monobasic fluorine-containing oxyacids of the group VIA elements S, Se, and Te; eg, $HOSO_2F$, $HOSO_2CF_3$, $HOSF_5$, $HOSeF_5$, and $HOTeF_5$. In addition, two members of a new class of unipositive chlorine compounds derived from the hydroperoxides CF_3OOH and SF_5OOH have been prepared.

Physical Properties. Data on physical properties of organic hypochlorites is limited. Some physical data on the inorganic nonmetal hypochlorites and the peroxy hypochlorites are available.

Chemical Properties. The primary and secondary alkyl hypochlorites decompose vigorously on warming, and explosively when exposed to light.

In contrast to the alkyl hypochlorites, the fluoroalkyl hypochlorites are extremely susceptible to hydrolysis but are much more thermally stable.

The fluoroalkyl hypochlorites readily react with CO and SO_2 to form the corresponding chloroformates and chlorosulfates in near quantitative yields. They add to olefins giving α-chloroethers. Borate esters are obtained by reaction of perfluoroalkyl hypochlorites with BCl_3. Although the perfluoroacyl hypochlorites are thermally unstable and explosive, CF_3CO_2Cl and $C_3F_7CO_2Cl$ are easily handled and are well characterized.

Uses. *t*-Butyl hypochlorite has been found useful in upgrading vegetable oils and in the preparation of α-substituted acrylic acid esters and esters of isoprene halohydrins. Numerous patents describe its use in cross-linking of polymers (qv), in surface treatment of rubber (qv), and in odor control of polymer latexes. It is used in the preparation of propylene oxide (qv) in high yield with little or no by-products. Fluoroalkyl hypochlorites are useful as insecticides, initiators for polymerizations, and bleaching and chlorinating agents.

JOHN A. WOJTOWICZ
Olin Corporation

F. Aubke and D. D. DesMarteau, in P. Tarrant, ed., *Fluorine Chemistry Reviews,* Vol. 8, Marcel Dekker, Inc., New York, 1977, p. 73.

M. Schmeisser, in S. Y. Tyree, Jr., ed., *Inorganic Synthesis,* Vol. 9, McGraw-Hill Book Co., New York, 1967, p. 127.

G. E. White, *Handbook of Chlorination,* Van Nostrand Reinhold Co., New York, 1972.

CHLOROUS ACID, CHLORITES, AND CHLORINE DIOXIDE

Sodium chlorite, $NaClO_2$, an oxidizing agent, is manufactured and distributed worldwide in commercial quantities as aqueous solution, flake, and powder products. The primary use for sodium chlorite is on-site generation of chlorine dioxide, ClO_2, through oxidation and/or acidification of an aqueous chlorite solution. Sodium chlorate, $NaClO_3$, is the preferred raw material for producing larger quantities of chlorine dioxide.

Chlorine dioxide is finding increasing use as an oxidizing bleaching agent in the pulp (qv) and paper (qv) industry. Chlorine dioxide is also finding increasing use as a disinfectant or biocide in municipal and industrial water (qv) treatment (see DISINFECTANTS AND ANTISEPTICS) as well as an oxidizer in oil field and pollution abatement processes (see PETROLEUM). Chlorous acid, $HClO_2$, is a short-lived chemical intermediate generated in acidified sodium chlorite solutions and is involved in the chemistry of the production and use of chlorine dioxide in oxidation and bleaching processes.

CHLORINE DIOXIDE

Physical Properties

Chlorine dioxide, ClO_2, is a greenish yellow gas having a pungent odor that is distinctive from that of chlorine. Liquid chlorine dioxide has a deep red color and is explosive at temperatures above −40°C. Selected physical and thermodynamic properties of chlorine dioxide are given in Table 1.

Chlorine dioxide exists as a free-radical monomer. It is soluble in water, forming a yellow to yellow-green colored solution that is quite stable if kept cool and in the dark. Various crystalline hydrates of chlorine dioxide have been described, including a hexahydrate, an octahydrate, and an orange-colored decahydrate.

Table 1. Physical and Thermodynamic Properties of Chlorine Dioxide

Property value	Values
mol wt	67.452
critical temperature, K	465
critical pressure, kPa[a]	8621.6
mp, K	213.55
bp at 101.3 kPa[a], K	284.05
density of liquid, g/mL at 0°C	1.640

[a] To convert kPa to mm Hg, multiply by 7.501.

Chemical Properties

Chlorine dioxide gas is a strong oxidizer. The standard $E°$ reversible potential is determined by the specific reaction chemistry. The standard $E°$ potential for gaseous ClO_2 in aqueous solution reactions where a chloride ion is the product is −1.511 V. The potential varies as a function of pH and concentration.

Thermal Decomposition of ClO_2. Chlorine dioxide decomposition in the gas phase is characterized by a long induction period, followed by a rapid autocatalytic phase that may be explosive if the initial concentration is above a partial pressure of 10.1 kPa (76 mm Hg).

In solution, chlorine dioxide decomposes very slowly at ambient temperatures in the dark. The primary decomposition process is the hydrolysis of chlorine dioxide into chlorite and chlorate ions.

Photochemical Reactions. The photochemistry of chlorine dioxide is complex and has been extensively studied. In the gas phase, the primary photochemical reaction is the homolytic fission of the chlorine–oxygen bond to form ClO and O.

Organic Chemistry. The chemistry and mechanisms of chlorine dioxide reactions with organic compounds have been extensively reviewed. A compilation of kinetic data is also available. Chlorine dioxide typically reacts with organics as an oxidant with little or no chlorination. The breaking of organic carbon–carbon bonds is generally not extensive in most of the reactions.

Manufacture

Laboratory Preparation. Pure chlorine dioxide is easily prepared in small quantities by the addition of dilute sulfuric acid into a solution of sodium chlorite.

Small- and Medium-Scale Industrial Production. Sodium chlorite is the preferred raw material for chlorine dioxide production in quantities of less than about 2000 kg/d. This is typical for water treatment and disinfection applications. Other applications not requiring high purity, ie, chlorine-free, chlorine dioxide can use chemical generation methods reacting sodium chlorate with hydrochloric acid or with sulfuric acid in the presence of sodium chloride.

Large-Scale Industrial Production. Large amounts of chlorine dioxide are used in pulp bleaching and smaller quantities are used for the manufacture of sodium chlorite. In these applications, sodium chlorate is the only commercially available raw material. Chlorine dioxide production from sodium chlorate is achieved by the reduction of the chlorate ion in the presence of strong acid.

Chloride Reductant. The use of hydrochloric acid or chloride ion as a reductant produces a chlorine dioxide product containing a substantial quantity of chlorine making it less desirable for bleaching applications. Integrated processes have been developed that separate chlorine from the gas and react it with hydrogen gas from chlorine production to produce hydrochloric acid. The spent chlorine dioxide generator liquor containing NaCl is used as feed for the chlorate plant.

Sulfur Dioxide Reductant. The Mathieson process uses sulfur dioxide, sodium chlorate, and sulfuric acid to produce chlorine dioxide gas with a much lower chlorine content. The sulfur dioxide gas reductant is oxidized to sulfuric acid, reducing the overall acid requirement of the process.

Organic Reductant Processes. A wide range of organic compounds can be employed as reducing agents in the production of chlorine dioxide from sodium chlorate strongly in acidic solutions. Carboxylic acids, alcohols, and carbohydrates have been reported as reducing agents, but methanol is the only organic reductant used in commercial generators.

Electrochemical Reduction and Catalytic Processes. Electrochemical reduction methods for producing pure chlorine dioxide from solutions containing acidified sodium chlorate or pure chloric acid have been developed. The electrolytic reduction process may be catalyzed using graphite, stainless steel, or platinum group metal oxide-coated cathodes.

Other Significant Process Developments. Increasing chlorine dioxide bleaching substitution levels in the pulp mills as well as the need for reduction in sodium sulfate generator salt cane effluent are the driving forces for future process/design changes in commercial generator systems using sodium chlorate. Electrochemical technologies for producing chloric acid as a feedstock or converting sodium sulfate to sulfuric acid and sodium hydroxide are being developed to solve these problems.

Production and Shipment

In all cases, chlorine dioxide is produced at the point of use either from sodium chlorite or sodium chlorate.

Economic Aspects

The costs of manufacturing chlorine dioxide from sodium chlorate vary for different processes, depending on the degree of integration with chlorate manufacture and the need for by-products produced from chlorine dioxide production. The principal costs of chlorine dioxide manufacture are determined by the consumption of sodium chlorate, acid, and reducing agent, together with the fixed costs of operating and maintaining the facility.

The cost of manufacturing chlorine dioxide from sodium chlorite is about two to four times higher than that from sodium chlorate. However, the fixed operating costs of small-scale chlorate-based processes are much higher.

Toxicology

Gaseous chlorine dioxide is explosive at concentrations above 10.1 kPa (76 mm Hg) partial pressure (10 vol %) in air. Explosions may be caused by an electrical discharge including static electricity, hot surfaces, metal oxides, or organics, and may be self-initiated after an induction period.

The threshold limit value–time integrated average, TLV–TWA, of chlorine dioxide is 0.1 ppm, and the threshold limit value–short-term exposure limit, STEL, is 0.3 ppm or 0.9 mg/m^3 of air concentration. Chlorine dioxide is a severe respiratory and eye irritant. Symptoms of exposure by inhalation include eye and throat irritation, headache, nausea, nasal discharge, coughing, wheezing, bronchitis, and delayed onset of pulmonary edema.

Chlorine dioxide solutions are normally stored cold at concentrations of less than 10 g/L in order to keep the concentration of gaseous chlorine dioxide above the aqueous solutions below the explosive limit. Chlorine dioxide solutions are corrosive to the skin and eyes, and solutions must be handled with adequate ventilation to avoid exposure to the gas. Protective equipment required for chlorine dioxide are impervious clothing, neoprene gloves and boots, gas-tight chemical splash goggles and face shields, as well as other appropriate clothing to prevent the possibility of skin contacting aqueous solutions or vapor.

CHLOROUS ACID AND CHLORITES

Physical Properties

Chlorous Acid. The physical properties of $HClO_2$ have been determined using acidified alkali metal chlorite solutions. The short-lived existence of $HClO_2$ is based on spectroscopic evidence.

Sodium Chlorite. Analytical-grade $NaClO_2$ (mol wt = 90.442) is a colorless crystalline solid that decomposes as it melts between 180–200°C. Technical-grade (80 wt % assay) $NaClO_2$ is a white solid that may have a slight greenish tint from trace amounts of ClO_2 or metals such as iron and contains NaCl as a diluent for safety in shipping.

Chemical Properties

Chlorous Acid. The chemical properties of $HClO_2$ have been determined from the acidification of alkali metal chlorites. The K_a acid dissociation constant of chlorous acid at 25°C, reported to be about 1.1×10^{-2}, has been measured using rapid scan spectrophotometric methods and determined to be 1.9×10^{-2}. The standard enthalpy, Gibbs free energy, and standard entropy for $HClO_2$ at 298.15 K are $\delta H^\circ = -51.9$ kJ/mol (-12.4 kcal/mol), $\delta G^\circ = 5.9$ kJ/mol$f$ (1.4 kcal/mol), and $S^\circ = 0.1013$ kJ/(mol·K), (0.024 kcal/(mol·K)), respectively.

Sodium Chlorite. The standard enthalpy, Gibbs free energy of formation, and standard entropy for aqueous chlorite ions are $\delta H^\circ = -66.5$ kJ/mol(-15.9 kcal/mol), $\delta G^\circ = 17.2$ kJ/mol$f$ (4.1 kcal/mol), and $S^\circ = 0.1883$ kJ/(mol·K)) (0.045kcal/(mol · K)), respectively.

Sodium chlorite is used to produce chlorine dioxide by chemical oxidation, electrochemical oxidation, or by acidification with acids. Most of the commercial methods employ chlorine or sodium hypochlorite as chemical oxidants.

Acid–Sodium Chlorite System. The addition of a strong inorganic acid into an aqueous sodium chlorite solution produces chlorous acid, which rapidly disproportionates into chlorine dioxide.

Hypochlorous Acid–Sodium Chlorite System. In this method, chlorine gas is educted into water, forming a hypochlorous acid solution which then reacts with aqueous sodium chlorite to produce chlorine dioxide.

Chlorine Gas–Sodium Chlorite System. In this method, chlorine gas reacts directly with a concentrated sodium chlorite solution under a vacuum and chlorine dioxide gas is removed from the reaction chamber using a water-based eductor.

Sodium Hypochlorite–Acid–Sodium Chlorite System. In this method, hydrochloric or sulfuric acid is added into a sodium hypochlorite, NaOCl, solution before reaction with the sodium chlorite.

$$HOCl \rightleftarrows OCl^- + H^+$$

The acid addition shifts the chlorine solution equilibrium to favor molecular chlorine.

Photochemical Generation of Chlorine Dioxide from Chlorite. Several methods of generating chlorine dioxide by the photochemical oxidation of chlorite have been reported in patents. The chlorite solution is irradiated using a light source and the chlorine dioxide generated is swept from the solution using an inert gas stream.

Electrochemical Generation of Chlorine Dioxide from Chlorite. The electrochemical oxidation of sodium chlorite is an old, but not well-known method of generating chlorine dioxide. Concentrated aqueous sodium chlorite, with or without added conductive salts, is oxidized at the anode of an electrolytic cell having a porous diaphragm or membrane separator between the anode and cathode compartments. The anodic reaction is

$$ClO_2^- \rightarrow ClO_2 + e^-$$

The generated chlorine dioxide must be air stripped from the anode compartment in systems not employing high surface area anodes. In order to achieve high chlorite conversion efficiency.

Mechanisms in Chlorine Dioxide Generation from Chlorite. The reactions between sodium chlorite and chlorine-based oxidizers and acids are complex and involve the formation of a proposed unsymmetrical intermediate, $[Cl_2O_2]$.

Organic Reactions. The chlorite ion, ClO_2^-, is a weak and slow oxidizer in alkaline aqueous solutions. Aldehydes (qv) can be readily oxidized to the corresponding carboxylic acids in neutral or weakly acidic solutions. Mixing solid sodium chlorite with combustive organic materials can result in explosions and fire on shock, exposure to heat, or flames.

Manufacture

The commercial manufacture for sodium chlorite is based almost entirely on the reduction of chlorine dioxide gas in a sodium hydroxide solution containing hydrogen peroxide as the reducing agent. The chlorine dioxide is generated from the chemical or electrochemical reduction of sodium chlorate under acidic conditions.

Production and Shipment

Sodium chlorite solutions with more than 5% available chlorine have a U.S. Department of Transportation (DOT) shipping classification as a class 8 corrosive, identification number UN1908, with a group II packaging requirement. Sodium chlorite dry powder or flake is a class 5.1 oxidizer, identification number UN1496, also with a group II packaging requirement.

Economic Aspects

The specific use applications of sodium chlorite varies from country to country. Important factors are the regulatory and environmental laws in effect for air and water quality standards. Sodium chlorite is generally priced at about four to six times the cost of sodium chlorate.

Toxicology

The toxic effects of sodium chlorite are directly related to its oxidant properties. Details of toxicological studies are summarized in reviews. Sodium chlorite is toxic mainly from ingestion. The acute oral LD_{50} is approximately 180 mg/kg (rat) for the 79 wt % assay product. The dermal LD_{50} exposure is low, greater than 2 g/kg (rabbit). Sodium chlorite produces severe irritation or burns to the skin or eyes.

Dry sodium chlorite decomposes at temperatures above about 175°C, forming sodium chloride, sodium chlorate, and oxygen as by-products. The reaction is extremely exothermic and therefore self-sustaining. Sodium chlorite should be kept away from heat or flames. Contact or mixing of sodium chlorite with acids can produce poisonous and possibly explosive mixtures of chlorine dioxide and chlorine gas. Sodium chlorite is incompatible with all combustibles, reducing agents, and oxidizers.

Protective personal equipment for handling sodium chlorite includes splash-proof goggles, neoprene gloves and boots, waterproof or washable outer clothing, and a dust mask for powder handling.

Uses

More than 80% of all the sodium chlorite produced is used for the generation of chlorine dioxide. Sodium chlorite or the chlorine dioxide generated from it or from sodium chlorate must be registered with the USEPA for each specific application use as a biocide for microbial growth control or disinfection. These regulations are covered under the Federal Insecticide, Fungicide, and Rodenticide Act (FIFRA).

Municipal/Industrial Water Treatment. The principal consumption of sodium chlorite is as the precursor in chlorine dioxide generation for use as an effective disinfectant alternative in municipal and wastewater treatment.

Food Industry Disinfection. ClO_2 solution generated from sodium chlorite is used in the washing of fruits and vegetables as a fungicide and for flume water disinfection.

Industrial Processes. The use of sodium chlorite as an oxidizer in NO_x and SO_x combustion flue gas scrubber systems has been described in the patent art.

Disinfectant Formulations and Sterilization. Hundreds of applications covering disinfectant compositions using sodium chlorite have been described in U.S. and foreign patents.

Metallurgical and Ore Processing. Sodium chlorite is used in solution formulations to oxidize the copper surfaces in multilayer circuit boards for better adhesion as well as in other electronic applications. The use of oxidant gases including chlorine dioxide and chlorine generated in an electrochemical cell has been proposed for the desulfurization of coal (see COAL CONVERSION PROCESSES–CLEANING AND DESULFURIZATION).

Oil Field and Petroleum Processing. Sodium chlorite is finding increasing used as the choice precursor for generating chlorine dioxide for biocidal control in the production of crude oil.

JERRY J. KACZUR
DAVID W. CAWLFIELD
Olin Corporation

G. Gordon, R. G. Kieffer, and D. H. Rosenblatt, *Progr. Inorg. Chem.* **15**, 201–286 (1972).

W. J. Masschelein and R. G. Rice, *Chlorine Dioxide—Chemistry and Environmental Impact of Oxychlorine Compounds*, Ann Arbor Science Publishers, Inc., Ann Arbor, Mich., 1979, pp. 59–87.

Chemical Economics Handbook Product Review, Sodium Chlorate, SRI International, Menlo Park, Calif., Oct. 1991.

AWWA Standard for Sodium Chlorite, B303-88, American Water Works Association, Denver, Colo., Jan. 1, 1992.

R. M. Harrington, D. Gates, and R. R. Romano, "A Review of the Uses, Chemistry and Health Effects of Chlorine Dioxide and the Chlorite Ion," *Chlorine Dioxide Panel of the Chemical Manufacturers Association*, Washington, D.C., Apr. 1989.

CHLORIC ACID AND CHLORATES

Chloric Acid

Chlorates are salts of chloric acid, $HClO_3$.

Physical Properties. Aqueous chloric acid is a clear, colorless solution, stable when cold up to ca 40 wt %. Upon heating, chlorine, Cl_2, and chlorine dioxide, ClO_2, may evolve.

Chloric acid, a strong acid, has $pK_a = -2.7$. It is a strong oxidizing agent, $E_0 = 1.175$ V with ClO_2 as the reduction product. The heat of formation is −99.2 kJ/mol(−23.7 kcal/mol)) and the Gibbs free energy of formation is −3.3 kJ/mol(−0.79 kcal/mol) for both chloric acid and the chlorate ion (eq. 1).

Chemical Properties. Chloric acid is a strong acid and an oxidizing agent. It reacts with metal oxides or hydroxides to form chlorate salts, and it is readily reduced to form chlorine dioxide.

Manufacture. Emerging technologies for the commercial manufacture of chloric acid fall into three categories: (*1*) generation of high purity chloric acid by thermal decomposition of pure solutions of hypochlorous acid, HClO,

$$5\ HOCl \rightarrow HClO_3 + 2\ Cl_2 + 2\ H_2O \qquad (1)$$

(*2*) electrochemical methods employing discrete anion and cation membrane-separated compartment or bipolar membranes, and (*3*) generation of chloric acid by anodic oxidation of hypochlorous acid (eq. 2):

$$4\ HOCl + 2\ H_2O \rightarrow 2\ HClO_3 + 3\ H_2 + Cl_2 \qquad (2)$$

Uses. Chloric acid is the precursor for generation of chlorine dioxide for pulp bleaching and other applications (ee BLEACHING AGENTS). The use of chlorine dioxide for pulp (qv) bleaching applications is growing, and disposal of the by-product solids is a primary environmental concern. Use of chloric acid to generate chlorine dioxide can eliminate this problem.

Chloric acid also has found limited applications as a catalyst for the polymerization of acrylonitrile (qv), C_3H_3N, and in the oxidation of cyclohexanone, $C_6H_{10}O$.

Shipment. Solutions of greater than 10 wt % chloric acid may be shipped using the label, "oxidizing substance, liquid, corrosive, n.o.s.," and using identification number UN3098, packing group II.

Health and Safety Factors. Chloric acid is a strong oxidizing agent and acid must be stored apart from reducing agents and organic materials. Concentrated solutions are corrosive to the skin.

Sodium and Potassium Chlorate

Physical Properties. The physical properties of sodium chlorate and potassium chlorate, $KClO_3$, are summarized in Table 1.

Chemical Properties. On thermal decomposition, both sodium and potassium chlorate salts produce the corresponding perchlorate, salt, and oxygen. Mixtures of potassium chlorate and metal oxide catalysts, especially manganese dioxide, MnO_2, are employed as a laboratory source of oxygen.

Manufacture. Most chlorate is manufactured by the electrolysis of sodium chloride solution in electrochemical cells without diaphragms (eq. 3). Potassium chloride can be electrolyzed for the direct production of potassium chlorate, but because sodium chlorate is so much more soluble, the production of the sodium salt is generally preferred. Potassium chlorate may be obtained from the sodium chlorate by a metathesis reaction with potassium chloride.

The sodium chlorate manufacturing process can be divided into six steps: (*1*) brine treatment; (*2*) electrolysis; (*3*) crystallization and salt recovery; (*4*) chromium removal; (*5*) hydrogen purification and collection; and (*6*) electrical distribution. The overall chemical reaction is

$$NaCl + 3\ H_2O \rightarrow NaClO_3 + 3\ H_2 \qquad (3)$$

Because reaction 3 requires Six Faradays, the production of sodium chlorate is very energy-intrusive, requiring between 5000–6050 k W·h of electricity per metric ton of sodium chlorate produced.

Economic Aspects. Sodium chlorate production has grown at about a 5% rate since the early 1970s and was expected to grow at 8–10% through 1995. The projected rapid growth is related to the increased use of chlorine dioxide in the pulp and paper industry.

Safety and Toxicity. Sodium chlorate is harmful if swallowed, inhaled, or absorbed through the skin. Symptoms of inhalation include burning sensation, coughing, wheezing, and laryngitis. Symptoms of ingestion include nausea, vomiting, abdominal pain, cyanasis, and diarrhea. Acute oral toxicity in laboratory animals for different species are in 1200–1800 mg/kg range. Lethal doses for children are 2 g and for adults are from 15 to 30 g (200–400 mg/kg). Permissible exposure limit for total dust is 15 mg/m^3 and for respirator protection, 5 mg/m^3.

Chlorates are strong oxidizing agents. Dry materials, such as cloth, leather, or paper, contaminated with chlorate may be ignited easily by heat or friction. Extreme care must be taken to ensure that chlorates do not come in contact with heat, organic materials, phosphorus, ammonium compounds, sulfur compounds, oils, greases or waxes, powdered metals, paint, metal salts (especially copper), and solvents. Chlorates should be stored separately from all flammable materials in a cool, dry, fireproof building. Flammable-resistant clothing such as Nomex should be worn when working with chlorates.

Table 1. Physical Properties of Sodium and Potassium Chlorates

Properties	$NaClO_3$	$KClO_3$
molecular weight	106.44	122.55
crystal system	cubic	monoclinic
mp, °C	248–260	356–368
density, g/mL	2.487[a]	2.338[b]
n_D^{20}	1.515	1.440
enthalpy of fusion, ΔH_{fus}, kJ/mol[c]	21.3	
molar heat capacity, J/(mol·K)[c]	100[b,d]	99.8[b]

[a] At 25°C.
[b] At 20°C.
[c] To convert J to cal, divide by 4.184.
[d] From 298 to 533 K.

Uses. The primary (95%) use of sodium chlorate is in the production of chlorine dioxide for bleaching in the pulp (qv) and paper (qv) industry. Sodium chlorate is also used in perchlorates and sodium chlorite manufacture, uranium production, and as an agricultural herbicide.

Chemical wood (qv) pulp bleached with chlorine dioxide has superior brightness over pulps bleached using other reagents.

Potassium chlorate is used mainly in the manufacture of matches (qv) and pharmaceutical preparation. In pyrotechnics, chlorate salts may be mixed with certain organic compounds such as lactose to give a relatively cool flame, so that certain dyes may be incoporated in the mixture to give colored flares.

Other Chlorates

Barium chlorate monohydrate, $Ba(ClO_3)_2{:}H_2O$, has colorless monoclinic crystals; mp or loss of water at 120°C; sp gr, 3.18; $n^{20}{}_D$, 1.562; is prepared by the reaction of barium chloride, $BaCl_2$, and sodium chlorate in solution.

Lithium chlorate, $LiClO_3$, has rhombic needles; mp 124–129°C; decomposes on heating to 270°C. $LiClO_3$ is prepared by adding lithium chloride to sodium chlorate solution. It has limited use in pyrotechnics.

SUDHIR K. MENDIRATTA
BUDD L. DUNCAN
Olin Corporation

A. J. Bard and co-workers, *Standard Potentials in Aqueous Solutions,* Marcel Dekker, Inc., New York, p. 75.

U.S. Pat. 5,108,560 (Apr. 28, 1992), D. W. Cawlfield, R. L. Dotson, S. K. Mendiratta, B. L. Duncan, and K. E. Woodard, Jr., (to Olin Corp.).

B. V. Tilak, E. M. Spore, and J. C. Hanson, paper presented at the *Electrolytic Technology Committee Ad-Hoc Meeting,* U.S. Dept. of Energy and Argonne National Laboratory, Washington, D.C., Mar. 12–13, 1979.

Material Safety Data Sheet Sigma-Aldrich Corp., Milwaukee, Wis., Apr. 1992.

CHLOROCARBONS AND CHLOROHYDROCARBONS

SURVEY

Chlorination of various hydrocarbon feedstocks produces many useful chlorinated solvents, intermediates, and chemical products. The chlorinated derivatives provide a primary method of upgrading the value of industrial chlorine. The principal chlorinated hydrocarbons produced industrially include chloromethane (methyl chloride), dichloromethane (methylene chloride), trichloromethane (chloroform), tetrachloromethane (carbon tetrachloride), chloroethene (vinyl chloride monomer, VCM), 1,1-dichloroethene (vinylidene chloride), 1,1,2-trichloroethene (trichloroethylene), 1,1,2,2-tetrachloroethene (perchloroethylene), mono- and dichlorobenzenes, 1,1,1-trichloroethane (methyl chloroform), 1,1,2-trichloroethane, and 1,2-dichloroethane (ethylene dichloride, EDC).

Ninety-six percent of the EDC produced in the United States is converted to vinyl chloride for the production of poly(vinyl chloride) (PVC) (see VINYL POLYMERS). Chloroform and carbon tetrachloride are used as chemical intermediates in the manufacture of chlorofluorocarbons (CFCs). Methylene chloride, 1,1,1-trichloroethane, trichloroethylene, and tetrachloroethylene have wide and varied use as solvents. Methyl chloride is used almost exclusively for the manufacture of silicone. Vinylidene chloride is chiefly used to produce poly(vinylidene chloride) copolymers used in household food wraps (see VINYLIDENE CHLORIDE MONOMER AND POLYMERS). Chlorobenzenes are important chemical intermediates with end use applications including disinfectants, thermoplastics, and room deodorants.

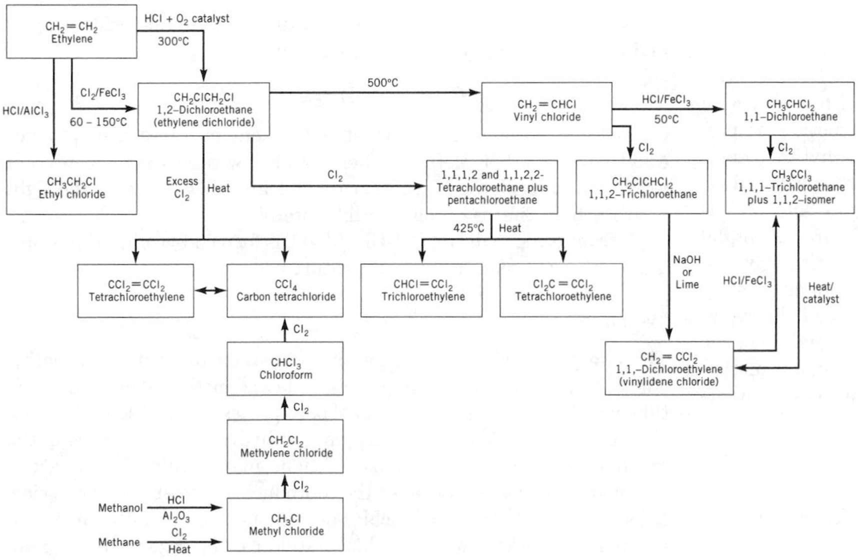

Figure 1. Manufacturing processes for C_1 and C_2 chlorohydrocarbons.

Chlorinated solvents (methylene chloride, methyl chloroform, trichlorethylene and perchloroethylene) have wide-ranging applications including use in adhesives, aerosols, extraction solvents, industrial cleaning solvents, paint and coating solvents, and pharmaceuticals. They are also used in dry cleaning, vapor degreasing, metal cleaning, textile processing, and as reaction media. Methylene chloride is used primarily in paint stripping and aerosol applications. 1,1,1-Trichloroethane has wide use in applications that range from metal cleaning to adhesives, coatings, and aerosols.

Typical manufacturing processes for C_1 and C_2 chlorohydrocarbons are shown in Figure 1.

Environmental Considerations

The use of chlorinated solvents is predicted to decline by the year 2000 as a result of concerns related to smog, groundwater contamination, toxicity, and ozone depletions.

General Properties of Chlorinated Hydrocarbons

Progressive chlorination of a hydrocarbon molecule yields a succession of liquids and/or solids of increasing nonflammability, density, and viscosity, as well as improved solubility for a large number of inorganic and organic materials. Other physical properties such as specific heat, dielectric constant, and water solubility decrease with increasing chlorine content.

All chlorinated hydrocarbons are susceptible to pyrolysis at high temperatures, which liberates hydrogen chloride. Many chlorinated hydrocarbons react readily with aluminum in the so-called bleeding reaction.

Commercial use of many chlorinated derivatives imposes stress on the stability of the solvent. Inhibitors classified as antioxidants (qv), acid acceptors, and metal stabilizers are added to minimize these stresses.

All volatile organic solvents are toxic to some degree. Excessive vapor inhalation of the volatile chlorinated solvents, and the central nervous system depression that results, is the greatest hazard for industrial use of these solvents. Proper protective equipment and operating procedures permit safe use of solvents such as methylene chloride, 1,1,1-trichloroethane, trichloroethylene, and tetrachloroethylene in both cold and hot metal-cleaning operations.

Aliphatic Chlorination Reactions

Substitution Chlorination. The substitution of chlorine for hydrogen atoms in a hydrocarbon is an important commercial chlorination process.

Chlorine free radicals used for the substitution reaction are obtained by either thermal, photochemical, or chemical means.

Chemical initiation generates organic radicals, usually by decomposition of azo or peroxide compounds, to form radicals which then react with chlorine to initiate the radical-chain chlorination reaction.

Addition Chlorination. Chlorination of olefins such as ethylene, by the addition of chlorine, is a commercially important process and can be carried out either as a catalytic vapor- or liquid-phase process. The reaction is influenced by light, the walls of the reactor vessel, and inhibitors such as oxygen, and proceeds by a radical-chain mechanism.

Hydrochlorination. The addition of hydrogen chloride to alkenes in the absence of peroxides takes place by an electrophilic substitution mechanism.

Dehydrochlorination. The primary method for commercially producing vinyl chloride is thermal dehydrochlorination of 1,2-dichloroethane. The gas-phase thermal dehydrochlorination of 1,2-dichloroethane at 350–515°C proceeds by radical chain mechanism. The reaction is accelerated by radical initiators such as chlorine and retarded or inhibited by olefins and alcohols.

Chlorinolysis. Reaction of C_2 and C_3 hydrocarbons with excess chlorine at high temperatures can cleave the C–C bonds of the hydrocarbon to give chlorinated derivatives of shorter-chain length. A well-known commercial process involving this technique is thermal chlorination of propane and chlorinated hydrocarbon feedstocks with chlorine to produce carbon tetrachloride and tetrachloroethylene with hydrogen chloride as by-product.

Oxychlorination. This is an important process for the production of 1,2-dichloroethane which is mainly produced as an intermediate for the production of vinyl chloride. The reaction consists of combining hydrogen chloride, ethylene, and oxygen (air) in the presence of a cupric chloride catalyst to produce 1,2-dichloroethane.

Chlorinated Aromatic Derivatives

Aromatic compounds may be chlorinated with chlorine in the presence of a catalyst such as iron, ferric chloride, or other Lewis acids. The halogenation reaction involves electrophilic displacement of the aromatic hydrogen by halogen.

DANIEL J. REED
Dow Chemical U.S.A.

Z. Sedaghat-Pour, "Ethylene Dichloride," *Chemical Economics Handbook,* 651.5000 S Stanford Research Institute, Menlo Park, Calif., Apr. 1989.

PERP Report #90-4, Chemical Systems, Inc., Sept. 1991, pp. 13–16, 23–29.

METHYL CHLORIDE

Methyl chloride (chloromethane, monochloromethane), CH_3Cl, at ordinary temperatures and pressures is a colorless gas with a very mild odor and sweet taste. Millions of kilograms of methyl chloride are produced naturally every day, primarily in the oceans. Methyl chloride is handled commercially as a liquid. It is miscible with the principal organic solvents and only slightly soluble in water. The dry liquid is stable and noncorrosive; however, in the presence of moisture, the liquid slowly decomposes and becomes corrosive to metals, particularly aluminum, zinc, and magnesium. Gaseous methyl chloride is moderately flammable. Prolonged exposure to high concentrations of the vapor can produce severe toxic effects. Methyl chloride is used mainly in the manufacture of silicones, synthetic rubber, and as a general methylating agent; its refrigerant and extractant applications now have secondary importance.

Physical and Chemical Properties

The physical properties of methyl chloride are listed in Table 1. Methyl chloride is the simplest chlorinated hydrocarbon.

Dry methyl chloride is unreactive with all common metals except the alkali and alkaline-earth metals, magnesium, zinc, and aluminum.

Methyl chloride can be converted into methyl iodide or bromide by refluxing in acetone solution in the presence of sodium iodide or bromide. The reactivity of methyl chloride and other aliphatic chlorides in substitution reactions can often be increased by using a small amount of sodium or potassium iodide as in the formation of methyl aryl ethers.

Methyl chloride reacts with ammonia alcoholic solution or in the vapor phase by the Hofmann reaction to form a mixture of the hydrochloride of methylamine, dimethylamine, trimethylamine, and tetramethylammonium chloride.

Methyl chloride, as a typical aliphatic chloride, may be used in the Friedel-Crafts reaction (qv):

$$CH_3Cl + C_6H_6 \xrightarrow{AlCl_3} C_6H_5CH_3 + HCl$$

Table 1. Physical Properties of Methyl Chloride

Property	Value
mol wt	50.49
mp, °C	−97.7
bp, °C at 101.3 kPa[a]	−23.73
sp gr	
liq, 20/4°C	0.920
gas, 0°C, 101.3 kPa[a] (air = 1)	1.74
density, g/L[b]	2.3045
refractive index	
liq, −23.7 °C	1.3712
gas, 25°C	1.0007
surface tension at 0°C, mN/m (= dyn/cm)	19.5
specific heat, liq, J/g[c], at 20°C	1.599
critical temperature, °C	143.1
critical pressure, kPa[a]	6679.2
critical density, g/cm^3	0.353
critical volume, cm^3/g	2.833
flash point (open cup), °C	−46
autoignition temperature, °C	632
dielectric constant	
liq, −250 °C	12.93
gas, 21°C	1.0109
solubility of methyl chloride in water, 25°C g/100 g H_2O	0.48

[a] To convert kPa to mm Hg, multiply by 7.5.
[b] At sea level, 45 ft latitude, 0°C, 101.3 kPa.[a]
[c] To convert J to cal, divide by 4.184.

When heated to very high temperatures, methyl chloride couples giving ethylene according to the reaction:

$$2\ CH_3Cl \rightarrow CH_2{=}CH_2 + 2HCl$$

Catalytic reactions at somewhat lower temperatures also produce ethylene and other olefins. When coupled with a methane process to methyl chloride, this reaction results in a new route to the light hydrocarbons that is of considerable interest.

A crystalline hydrate, $CH_3Cl \cdot 6H_2O$, is formed by subjecting water and methyl chloride to low temperatures.

Manufacture

The two principal processes for the industrial production of methyl chloride are reaction of hydrogen chloride and methanol, and chlorination of methane. Several variants of both processes are used. Chlorination of methane yields other chlorinated hydrocarbons in substantial amounts; indeed, under certain conditions, methyl chloride may not be the principal product. Because the coproducts, eg, methylene chloride, chloroform, and carbon tetrachloride, are as commercially important as methyl chloride, methane chlorination can be regarded as a multiple product process rather than one with several by-products. The methanol—hydrogen chloride reaction yields methyl chloride as the main product with small amounts of dimethyl ether as the only by-product. It is commercially carried out in both liquid-phase and gas-phase processes.

Handling

All persons who have occasion to use or handle methyl chloride should be thoroughly instructed and adequately supervised in the proper methods of handling the substance to prevent or minimize exposure to the liquid or its vapors and in the proper methods of disposing of this chemical.

Methyl chloride is transported and stored as liquefied gas under pressure in cylinders, tank trucks, and tank cars.

Economic Aspects

Production grew tremendously in the 1960s and again in the late 1980s. The principal U.S. producers and their capacities are shown in Table 2. These capacities do not include the 100+ million kg per year used by The Dow Chemical Company, Occidental, and Vulcan to captively produce other chloromethanes.

The historical growth rate for methyl chloride is 1.7% per year. Methyl chloride is used primarily in the manufacture of silicones.

Toxicity

Methyl chloride is one of the more toxic of the chlorinate hydrocarbons, and there is no adequate warning of the presence of harmful concentrations. The delay in the development of symptoms is characteristic of the toxic effect of methyl chloride. Repeated exposure to low concentrations damages the central nervous system, and, less frequently, the liver, kidneys, bone marrow, and cardiovascular system. Methyl chloride intoxication causes headache, blurred vision, loss of coordination,

Table 2. U.S. Methyl Chloride Producers

Producer	Capacity, 10^3 t/yr
The Dow Chemical Company, Freeport, Tex.	22.7
The Dow Chemical Company, Plaquemine, La.	68.2
Dow Corning, Carrollton, Ky.	90.9
Dow Corning, Midland, Mich.	22.7
General Electric, Waterford, N.Y.	40.9
Occidental, Belle, W. Va.	29.5
Vulcan, Geismen, La.	52.3
Total	*338.6*

and reversible personality change involving moroseness, depression, and anxiety.

The OSHA personal exposure level guideline and the ACGIH TLV are 50 ppm.

Regulation

The manufacturing, storage, and disposal of methyl chloride as well as the other chlorinated methanes are heavily regulated at the national, state, and local level. The potential exposure to methyl chloride by workers may be regulated by OSHA or the state Industrial Hygiene Department. In addition, various state and local regulations may impose other reporting and regulatory standards. Contacting the Environmental or Regulatory Compliance Department before importing, purchasing, selling, using, or disposing of methyl chloride is strongly recommended.

MICHAEL T. HOLBROOK
Dow Chemical U.S.A.

E. T. McBee and co-workers, *Ind. Eng. Chem.* **34**, 296 (1942). An excellent reference paper.

I. I. Kurlyandskaya, S. B. Grinberg, E. B. Svetlanov, and R. M. Flid, *Zh. Fiz. Khim. (Catalysis and Reaction Kinetics)* **47,** 189 (1973); USSR Pat. 558,896 (May 25, 1975), E. B. Svetlanov and co-workers. Svetlanov has produced an excellent series of six papers on gas-phase hydrochlorination.

Chem. Mark. Rep. (Mar. 20, 1989).

METHYLENE CHLORIDE

Methylene chloride (dichloromethane, methylene dichloride), CH_2Cl_2, is a colorless, heavy liquid with a pleasant ethereal odor. Its outstanding solvent properties are the basis of its principal industrial interest. Production reached its zenith in the late 1970s and early 1980s, when the methylene chloride market began to shrink because of the 1985 National Toxicology Program (NTP) study that indicated carcinogenicity in mice.

Physical and Chemical Properties

The physical properties of methylene chloride are listed in Table 1. Methylene chloride is a volatile liquid. Although methylene chloride is only slightly soluble in water, it is completely miscible with other grades of chlorinated solvents, diethyl ether, and ethyl alcohol. It dissolves in most other common organic solvents. Methylene chloride is also an excellent solvent for many resins, waxes, and fats, and hence is well suited to a wide variety of industrial uses.

Dry methylene chloride does not react with the common metals under normal conditions; however, a reaction with aluminum can be initiated, sometimes explosively, by the addition of small amounts of other halogenated solvents or an aromatic solvent.

Methylene chloride is one of the more stable of the chlorinated hydrocarbon solvents. Its initial thermal degradation temperature is 120°C in dry air. This temperature decreases as the moisture content increases. The reaction produces mainly HCl with trace amounts of phosgene.

The most common reaction of methylene chloride is its reaction with chlorine to give chloroform and carbon tetrachloride. This occurs by a free-radical process initiated by heat or light in the gas or liquid phase. Catalytic chlorination to these same products is also known (see CHLOROCARBONS AND CHLOROHYDROCARBONS, CHLOROFORM).

Methylene chloride is easily reduced to methyl chloride and methane by alkali metal ammonium compounds in liquid ammonia.

Bromochloromethane is produced by reaction of an excess of a mixture of methylene chloride and bromine with aluminum at 26 to 30°C.

Manufacture

Methylene chloride is produced industrially in the United States by two methods. The older and currently lesser used method involves a direct reaction of excess methane with chlorine at high temperatures, approximately 400–500°C, or at somewhat lower temperatures either catalytically or photolytically.

Table 1. Properties of Methylene Chloride

Property	Value
mol wt	84.92
bp at 101.3 kPa,[a] °C	39.8
fp, °C	−96.7
sp gr at 20°C	1.320
density at 20°C, kg/m^3	1315.7
diffusivity in air, m^2/s	9×10^{-5}
refractive index at 20°C	1.4244
viscosity at 20°C, mPa·s(= cP)	0.43
surface tension at 20°C, mN/m (= dyn/cm)	0.02812
heat of vaporization at 20°C, kJ/kg [b]	329.23
heat of combustion, MJ/kg[c]	7.1175
critical density, kg/m^3	472
critical temperature, °C	245.0
critical pressure, MPa[d]	6.171
vapor pressure at 0°C, kPa[a]	19.6
solubility in water at 20°C, g/kg	13.2
flash point (ASTM D1310-67)	none
electrical properties at 24°C	
dielectric strength, V/cm[e]	94.488
specific resistivity at 24°C, Ω·cm	1.81×10^8
dielectric constant at 24°C, 100 kHz	10.7

[a] To convert kPa to mm Hg, multiply by 7.5.
[b] To convert kJ/kg to Btu/lb, multiply by 0.4302.
[c] To convert J to cal, divide by 4.184.
[d] To convert MPa to atm, divide by 0.101.
[e] To convert V/cm to V/100 mils, multiply by 0.254.

The predominant method of manufacturing methylene chloride employs as a first step the reaction of hydrogen chloride and methanol to give methyl chloride. Excess methyl chloride is then mixed with chlorine and reacts to give methylene chloride, chloroform, and carbon tetrachloride as coproducts. This reaction is usually carried out in the gas phase thermally but can also be done catalytically or photolytically.

Economic Aspects

Table 2 lists the U.S. producers of methylene chloride and their rated yearly capacities. Since the product mix of a typical chloromethanes process is very flexible, production may be adjusted according to the demand for methylene chloride and chloroform. The demand for methylene chloride has taken a broad downturn as a result of the 1985 NTP carcinogenicity tests.

The historical growth rate (1979–1988), was −2.7% per year. This should have decrease even further to −3 to −5% per year through 1993.

Handling

All persons who have occasion to use or handle methylene chloride should be thoroughly instructed and adequately supervised in the

Table 2. U.S. Methylene Chloride Producers

Producer	Capacity, 10^3 t/yr
Occidental, Belle, W.Va.	40.9
The Dow Chemical Company, Freeport, Tex.	50.0
The Dow Chemical Company, Plaquemine, La.	54.5
Vulcan, Geismar, La.	36.4
Vulcan, Wichita, Kans.	59.1
Total	*264.5*

Table 3. Uses of Methylene Chloride (1989)

paint stripper	28%
aerosols	18%
exports	15%
chemical processing	11%
urethane foam blowing agent	9%
metal degreasing	8%
electronics	7%
other	4%

proper methods of handling the substance to prevent or minimize exposure to the liquid or its vapors and in the proper methods of disposing of this chemical.

Because of its low boiling point, methylene chloride should be stored in a cool place away from direct sunlight. Storage containers may be constructed of mild or plain steel that is galvanized or suitably lined. All bulk storage tanks should be equipped with a vent dryer packed with calcium chloride or other appropriate desiccant to exclude moisture. Methylene chloride is transported in drums, truck transports, rail cars barges, and oceangoing ships.

Toxicity

Methylene chloride is one of the least toxic chlorinated methanes. The ACGIH threshold limit value (TLV) for methylene chloride is 50 ppm by volume for an eight-hour exposure. The OSHA permissible exposure level is 500 ppm time-weighted average, 1000 ppm ceiling with 2000 ppm peak concentration. This is currently under revision.

High levels of methylene chloride vapors have an anesthetic action.

The National Toxicology Program (NTP) reported on tests on mice and rats that there was an increase in the spontaneous incidence of benign tumors in male and female rats and an increase in malignant liver and lung tumors in B6C3F1 mice. NTP concluded that these data demonstrated clear evidence of carcinogenicity in mice and female rats and some evidence in male rats.

Uses

Uses of methyl chloride are listed in Table 3.

Regulation

The manufacturing, storage, and disposal of methylene chloride, as in the case of the other chlorinated methanes, are heavily regulated at the national, state, and local level. The potential exposure to methylene chloride by workers may be regulated by OSHA or the individual state Industrial Hygiene Department. In addition, various state and local regulations may impose other reporting and regulatory standards. Contacting the Environmental or Regulatory Compliance Department before importing, purchasing, selling, using, or disposing or methylene chloride is strongly recommended.

MICHAEL T. HOLBROOK
Dow Chemical U.S.A.

Chem. Mark. Rep. (Feb. 20, 1989).

CHLOROFORM

Chloroform (trichloromethane, methenyl chloride), $CHCl_3$, at normal temperature and pressure is a heavy, clear, volatile liquid with a pleasant, ethereal, nonirritant odor. Although chloroform is nonflammable, its hot vapor in admixture with vaporized alcohol burns with a green tinged flame. Chloroform is miscible with the principal organic solvents and is slightly soluble in water. It is less stable in storage than either methyl or methylene chloride. Chloroform decomposes at ordinary temperatures in sunlight in the absence of air and in the dark in the presence of air. Phosgene is one of the oxidative decomposition products.

Table 1. Physical Properties of Chloroform

Property	Value
mol wt	119.38
refractive index at 20°C	1.4467
flash point, °C	none
mp, °C	−63.2
at 101 MPa[a], °C	−43.4
bp at 101 kPa[b], °C	61.3
sp gr at 0/4°C	1.52637
critical temperature, °C	263.4
critical pressure, MPa[a]	5.45
critical density, kg/m^3	500
critical volume, m^3/kg	0.002
thermal conductivity at 20°C, W/(m·K)	0.130
dielectric constant, 20°C	4.9
heat of combustion, MJ/(kg·mol)[c]	373
solubility of chloroform in water at 0°C, g/kg H_2O	10.62
viscosity, liq, at 0°C, mPa·s(= cP)	0.700

[a] To convert MPa to atm, multiply by 9.87.
[b] To convert kPa to mm Hg, multiply by 7.5.
[c] To convert J to cal, divide by 4.184.

Physical and Chemical Properties

The physical properties of chloroform are listed in Table 1.

Chloroform dissolves alkaloids, cellulose acetate and benzoate, ethylcellulose, essential oils, fats, gutta-percha, halogens, methyl methacrylate, mineral oils, many resins, rubber, tars, vegetable oils, and a wide range of common organic compounds.

Chloroform slowly decomposes on prolonged exposure to sunlight in the presence or absence of air and in the dark in the presence of air. The products of oxidative breakdown include phosgene, hydrogen chloride, chlorine, carbon dioxide, and water.

Chloroform resists thermal decomposition at temperatures up to about 290°C. Pyrolysis of chloroform vapor occurs at temperatures above 450°C, producing tetrachloroethylene, hydrogen chloride, and a number of chlorohydrocarbons in minor amounts.

Chloroform reacts with aniline and other aromatic and aliphatic primary amines in alcoholic alkaline solution to form isonitriles, ie, isocyanides, carbylamines. Many compounds containing either the acetyl (CH_3CO) or $CH_3CH(OH)$ group yield chloroform on reaction with chlorine and alkali or hypochlorite. Many years ago chloroform was almost exclusively produced from acetone or ethyl alcohol by reaction with chlorine and alkali. Although it is no longer practiced commercially, this reaction is possibly an important source of chloroform in the water treating process.

Manufacture

Chloroform can be manufactured from a number of starting materials. Methane, methyl chloride, or methylene chloride can be further chlorinated to chloroform, or carbon tetrachloride can be reduced, ie, hydrodechlorinated, to chloroform. Methane can be oxychlorinated with HCl and oxygen to form a mixture of chlorinated methanes. Methyl chloride chlorination is now the most common commercial method of producing chloroform.

Economic Aspects

The projected growth rate through 1993 was 5 to 6% per year. This high projected growth rate is driven by the expected strong growth of the HCFC-22 and fluoropolymer markets, which account for 90% of the total chloroform market. HCFC-22 is a potential substitute for some applications currently using CFC-11 and CFC-12.

Table 2. U.S. Chloroform Producers

Producer	Capacity, 10^3 t/yr
The Dow Chemical Company, Freeport, Tex.	61.4
The Dow Chemical Company, Plaquemine, La.	68.2
Occidental, Belle, W.Va.	22.7
Vulcan, Geismar, La.	27.2
Vulcan, Wichita, Kans.	50.0
Total	247.7

The U.S. chloroform producers and their capacities are listed in Table 2. With this strong increasing demand and the decline in the methylene chloride market, chloroform producers have had to swing their production away from methylene chloride and toward chloroform. This capability may result in the capacities listed being somewhat understated for chloroform and overstated for methylene chloride, the primary coproduct in chloroform production.

Handling

All persons who have occasion to use or handle chloroform should be thoroughly instructed and adequately supervised in the proper methods of handling the substance to prevent or minimize exposure to the liquid or its vapors and in the proper methods of disposing of this chemical.

Chloroform should be stored in sealed containers in a cool place. Glass containers should be dark green or amber. Bulk storage containers may be constructed of mild or plain steel that is galvanized or suitably lined. All bulk storage tanks should be equipped with a vent dryer packed with calcium chloride or other appropriate desiccant to exclude moisture. Chloroform is transported in drums, truck transports, rail cars, barges, and oceangoing ships.

Health and Safety

The principal hazard in exposure to chloroform is damage to the liver and kidneys resulting from inhalation or ingestion. Inhalation of high concentrations may result in disturbances of equilibrium or loss of consciousness. Chloroform is mildly irritating to skin and mucous membranes upon contact, and to the alimentary tract upon ingestion. It is believed that medically significant quantities are not absorbed through intact skin.

The ACGIH recommended maximum time-weighted average concentration in the workplace atmosphere for eight-hour daily exposure is 10 ppm. OSHA has set the permissible exposure level at 2 ppm. It may be desirable to exclude alcoholics, persons with chronic disorders of the liver, kidneys, and central nervous system, and those with nutritional deficiencies from working with chloroform.

Treatment of chloroform poisoning is symptomatic; no specific antidote is known. Adrenalin should not be given to a person suffering from chloroform poisoning.

Regulation

The manufacturing, storage, and disposal of chloroform and the other chlorinated methanes are heavily regulated at the national, state, and local level. The potential exposure to chloroform by workers may be regulated by OSHA or the state Industrial Hygiene Department. In addition, various state and local regulations may impose other reporting and regulatory standards. Contacting the Environmental or Regulatory Compliance Department before importing, purchasing, selling, using, or disposing of chloroform is strongly recommended.

MICHAEL T. HOLBROOK
Dow Chemical U.S.A.

E. T. McBee and co-workers, *Ind. Eng. Chem.* **34**, 296 (1942).

Chem. Mark. Rep. (Feb. 27, 1989).

CARBON TETRACHLORIDE

Carbon tetrachloride (tetrachloromethane), CCl_4, at ordinary temperature and pressure is a heavy, colorless liquid with a characteristic nonirritant odor; it is nonflammable. Carbon tetrachloride contains 92 wt % chlorine. When in contact with a flame or very hot surface, the vapor decomposes to give toxic products, such as phosgene. It is the most toxic of the chloromethanes and the most unstable upon thermal oxidation. The commercial product frequently contains added stabilizers. Carbon tetrachloride is miscible with many common organic liquids and is a powerful solvent for asphalt, benzyl resin (polymerized benzyl chloride), bitumens, chlorinated rubber, ethylcellulose, fats, gums, rosin, and waxes.

Ca 1950, carbon tetrachloride found a rapidly expanding use as the starting material in the manufacture of fluorinated refrigerants, an application that by 1954 accounted for about half the total demand for carbon tetrachloride and over 95% of the demand today.

Physical and Chemical Properties

The physical properties of carbon tetrachloride ar listed in Table 1. When mixed with excess water and heated to 250°C, carbon tetrachloride decomposes to carbon dioxide and hydrochloric acid; if the quantity of water is limited, phosgene is produced. Dry carbon tetrachloride does not react with most commonly used construction metals, eg, iron and nickel; it reacts very slowly with copper and lead. Like the other chloromethanes, carbon tetrachloride is reactive, sometimes explosively, with aluminum and its alloys. Carbon tetrachloride can be reduced to chloroform using a platinum catalyst or zinc and acid. It is widely employed as an initiator in the dehydrochlorination of chloroethanes at 400–600°C to produce vinyl chloride.

Manufacture

A number of processes have been described for the production of carbon tetrachloride by the chlorinolysis of various hydrocarbon or chlorinated hydrocarbon waste streams, but most literature reports the use of methane as the primary feed. The quantity of carbon tetrachloride produced depends somewhat on the nature of the hydrocarbon starting material but more on the conditions of chlorination.

Table 1. Physical Properties of Carbon Tetrachloride

Property	Value
mol wt	153.82
mp at 101.3 kPa[a], °C	−22.92
bp at 101.3 kPa[a], °C	76.72
refractive index at 15°C	1.46305
sp gr, 0/4°C	1.63195
flash point, °C	none
density of solid at −80°C, g/cm^3	1.809
vapor density (air = 1)	5.32
surface tension at 0°C, mN/m (= dyn/cm)	29.38
specific heat at 20°C, J/kg[b]	866
critical temperature, °C	283.2
critical pressure, MPa[c]	4.6
critical density, kg/m^3	558
average coefficient of volume expansion, 0–40°C	0.00124
dielectric constant, ϵ	
liquid, 20°C	2.205
vapor, 87.6°C	1.00302
heat of combustion, liquid, at constant volume at 18.7°C, kJ/mol[b]	365
solubility of CCl_4 in water at 25°C, g/100 g H_2O	0.08

[a] To convert kPa to mm Hg, multiply by 7.5.
[b] To convert J to cal, divide by 4.184.
[c] To convert MPa to atm, divide by 0.101

Table 2. U.S. Producers of Carbon Tetrachloride

Producer	Location	Capacity, 10^3 t/yr
The Dow Chemical Company	Plaquemine, La.	45.5
Vulcan Materials	Geismar, La.	40.9
Vulcan Materials	Wichita, Kans.	27.3
Total		*271.9*

The principal by-product is perchloroethylene with small amounts of hexachloroethane, hexachlorobutadiene, and hexachlorobenzene. The chlorination of carbon disulfide is a very old method of producing carbon tetrachloride that is still practiced commercially in the United States.

Economic Aspects

In 1978 the use of chlorofluorocarbons in aerosols was essentially banned by the EPA. During the period 1979 to 1988 the compounded growth rate for carbon tetrachloride production was −0.7%/yr. This was projected to decrease to −3%/yr for the years 1989 to 1993.

Production of CFC-11 and -12 is slated to be phased out by the year 2000, though it will probably decrease faster than that as replacement HCFCs become available. The U.S. producers of carbon tetrachloride and their capacities appear in Table 2. All methylene chloride and chloroform producers also make a small amount of carbon tetrachloride as a by-product. Those producers are not listed in Table 2 because the amount of carbon tetrachloride made is small and varies depending on the ratio of methylene chloride to chloroform made.

List prices do not reflect actual market conditions because of overcapacity and the shrinking nature of the market. Although present U.S. carbon tetrachloride capacity of 271,900 t/yr is far in excess of demand, many installations are capable of producing other chlorinated hydrocarbons such as perchloroethylene.

Health and Safety Factors

All persons who have occasion to use or handle carbon tetrachloride should be thoroughly instructed and adequately supervised in the proper methods of handling the substance to prevent or minimize exposure to the liquid or its vapors and in the proper methods of disposing of this chemical.

Carbon tetrachloride is toxic by inhalation of its vapor and oral intake of the liquid. Inhalation of the vapor constitutes the principal hazard. Exposure to excessive levels of vapor is characterized by two types of response: an anesthetic effect similar to that caused by compounds such as diethyl ether and chloroform; and organic injury to the tissues of certain organs, in particular the liver and kidneys.

Organic injury may result from single prolonged exposure to carbon tetrachloride vapor or from repeated short duration exposures. Serious and fatal injuries are usually the result of a single prolonged exposure. Vapor concentrations of only a few hundred parts per million may be sufficient to cause injury. Symptoms of exposure include nausea and vomiting, headache, burning of eyes and/or throat, drowsiness, abdominal pain or discomfort, weakness, and muscle stiffness and soreness. Consequently, a threshold limit value of 5 ppm by volume of carbon tetrachloride in air has been established by ACGIH as a maximum safe concentration for daily eight-hour exposure. The OSHA permissible exposure level is 2 ppm.

Occasional brief contacts of liquid carbon tetrachloride with unbroken skin do not produce irritation, though the skin may feel dry because of removal of natural oils. Prolonged and repeated contacts may cause dermatitis, cracking of the skin, and danger of secondary infection.

In most situations, adequate, usually forced, ventilation is necessary to prevent excessive exposure. Persons who drink alcohol excessively or have liver, kidney, or heart diseases should by excluded from any exposure to carbon tetrachloride. All individuals regularly exposed to carbon tetrachloride should receive periodic examinations by a physician acquainted with the occupational hazard involved.

Handling and Storage

Although in the dry state carbon tetrachloride may be stored indefinitely in contact with some metal surfaces, its decomposition upon contact with water or on heating in air makes it desirable, if not always necessary, to add a small quantity of stabilizer to the commercial product.

Small quantities of carbon tetrachloride can be shipped in 1, 5, and 55 gallon (208 L) metal containers. Larger quantities are shipped by tank truck, tank car, barge, and ship. Special precautions should be taken to prevent contact with aluminum, magnesium metal, and their alloys.

Uses

In 1970, carbon tetrachloride was banned from all use in consumer goods in the United States. Its current principal applications include chlorofluorocarbon production (CFC-11 and -12) and some small use as a reaction medium or chemical intermediate.

Regulation

The manufacturing, storage, and disposal of carbon tetrachloride, as in the cases of the other chlorinated methanes, are heavily regulated at the national, state, and local level. The potential exposure to carbon tetrachloride by workers may be regulated by OSHA or the state Industrial Hygiene Department. In addition, various state and local regulations may impose other reporting and regulatory standards. Contacting the Environment or Regulatory Compliance Department before importing, purchasing, selling, using, or disposing of carbon tetrachloride is highly recommended.

MICHAEL T. HOLBROOK
Dow Chemical U.S.A.

Chem. Mark. Rep. (Feb. 13, 1989).

TOXIC AROMATICS

Chlorinated biphenyls, chlorinated naphthalenes, benzene hexachloride, and chlorinated derivatives of cyclopentadiene are no longer in commercial use because of their toxicity. However, they still impact on the chemical industry because of residual environmental problems. This article discusses the toxicity and environmental impact of these materials.

Polychlorinated Biphenyls

Polychlorinated biphenyls (PCBs) typify halogenated aromatic hydrocarbons (HAHs), industrial compounds or by-products that have been widely identified in the environment and chemical waste dumpsites. PCBs were used in industry as heat-transfer fluids, organic diluents, lubricant inks, plasticizers, fire retardants, paint additives, sealing liquids, immersion oils, adhesives, dedusting agents, waxes, and as dielectric fluids for capacitors and transformers.

Chemistry and Environmental Impact. PCBs are synthesized by the chlorination of biphenyl and the resulting products are designated according to their percent (by weight) chorine content. Over 600 million kg of commercial PCBs were produced in the United States and the estimated worldwide production is approximately double this quantity.

The identification of PCB residues in fish, wildlife, and human tissues has been reported since the 1970s. The results of these analytical studies led to the ultimate ban on further use and production of these compounds.

Commercial PCBs: Toxic and Biochemical Effects. PCBs and related halogenated aromatic hydrocarbons elicit a diverse spectrum of toxic and biochemical responses in laboratory animals dependent on a number of factors including age, sex, species, and strain of the test animal and the dosing regimen (single or multiple). The toxic responses elicited by most PCB preparations are also observed for other classes of HAHs and include a progressive weight loss not simply related to decreased food consumption and accompanied by weakness, debilitation, and ultimately death, ie, a wasting syndrome; lymphoid involution, thymic and splenic atrophy with associated humoral and/or cell-mediated immunosuppression and/or associated bone marrow and hematologic dyscrasia; a skin disorder called chloracne accompanied by acneform eruptions, alopecia, edema, hyperkeratosis, and blepharitis resulting from hypertrophy of the Meibomian glands; hyperplasia of the epithelial lining of the extrahepatic bile duct, the gall bladder, and urinary tract; hepatomegaly and liver damage accompanied by necrosis, hemorrhage, and intrahepatic bile duct hyperplasia; hepatotoxicity also manifested by the development of porphyria and altered metabolism of porphyrins; teratogenesis, developmental and reproductive toxicity observed in several animal species; carcinogenesis as caused by PCBs in laboratory animals and primarily associated with their effects as promoters; and endocrine and reproductive dysfunction, ie, altered plasma levels of steroid and thyroid hormones with menstrual irregularities, reduced conception rate, early abortion, excessive menstrual and postconceptional hemorrhage, and anovulation in females, and testicular atrophy and decreased spermatogenesis in males.

The biochemical responses elicited by PCBs are also numerous and include the induction of CYP1A1 and CYP1A2 gene expression and the associated monooxygenase enzyme activities, ie, aryl hydrocarbon hydroxylase (AHH) and ethoxyresorufin *O*-deethylase (EROD), and several other cytochrome P-450 dependent monooxygenases; the induction of steroid metabolizing enzymes, DT diaphorase, UDP glucuronosyl transferase, epoxide hydrolase, glutathione (*S*)-transferase, and δ-aminolevulinic acid synthetase; increased Ah receptor binding activity; decreased uroporphinogen decarboxylase activity; and decreased vitamin A levels.

Structure–Function Relationships. Since PCBs and related HAHs are found in the environment as complex mixtures of isomers and congeners, any meaningful risk and hazard assessment of these mixtures must consider the qualitative and quantitative structure–function relationships. Several studies have investigated the structure–activity relationships for PCBs that exhibit 2,3,7,8-tetrachlorodibenzo-*p*-dioxin (1) (TCDD)-like activity.

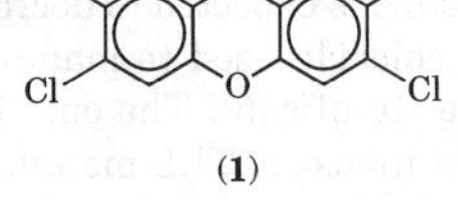

(1)

The data show that 3,3′,4,4′,5-pentaCB is the most toxic coplanar PCB congener and the 2,3,7,8-TCDD/3,3′,4,4′,5-pentaCB potency ratios are 66/1 (body weight loss, rat); 8.1/1 (thymic atrophy, rat); 10/1 (fetal thymic lymphoid development, mouse); 125/1 (AHH induction, rat); 3.3/1 (AHH induction, hepatoma H-4-II E cells, rat); and 100/1 (embryo hepatocytes, chick). Both the 3,3′,4,4′-tetra- and 3,3′,4,4′,5,5′-hexaCB congeners are considerably less toxic than 3,3′,4,4′,5-pentaCB and their relative potencies are highly variable.

Human Health Effects. Any assessment of adverse human health effects from PCBs should consider the route(s) of and duration of exposure; the composition of the commercial PCB products, ie, degree of chlorination; and the levels of potentially toxic PCDF contaminants. As a result of these variables, it would not be surprising to observe significant differences in the effects of PCBs on different groups of occupationally exposed workers.

Chloracne and related skin problems have been observed in several groups of workers and it was suggested that the air concentrations of commercial PCBs >0.2 mg^3 were associated with this effect.

The effects of occupational exposure to PCBs on the concentrations of several serum clinical, chemical, and hematological parameters have been reported. Mildly elevated SGOT and γ-glutamyl transpeptidase (GGTP) suggest some liver damage and induction of hepatic monooxygenase enzymes; these results are similar to those observed in animal studies. A relatively high incidence of pulmonary dysfunction in capacitor-manufacturing workers has been reported, with symptoms including coughing, 13.8%; wheezing, 3.4%; tightness in the chest, 10.1%; and upper respiratory or eye irritation, 48.2%. The pulmonary toxicity of PCBs in laboratory animals has not been widely reported.

It is apparent from most reports that workplace exposure to relatively high levels of PCBs results in limited and moderate toxicity in humans. These toxic symptoms appear to be reversible after exposure to PCBs is terminated and this is accompanied by a decline in serum levels of PCBs.

Environmental exposures to PCBs are significantly lower than those reported in the workplace and are therefore unlikely to cause adverse human health effects in adults. However, it is apparent from the results of several recent studies on children that there was a correlation between *in utero* exposure to PCBs, eg, cord blood levels, and developmental deficits, including reduced birth weight, neonatal behavior anomalies, and cognitive deficits.

Polychlorinated Naphthalenes

Polychlorinated naphthalenes (PCNs) are halogenated aromatic hydrocarbons that are no longer produced. They can be synthesized by the chlorination of naphthalene. The commercial products were graded and sold according to their chlorine content (wt %), and used as waxes and impregnants (for protective coatings), water repellents, and wood preservatives.

Animal and Human Toxicology. The mammalian toxicology of PCNs has not been studied in detail; however, it is believed that these compounds elicit mixture- and structure-dependent biochemical and toxic responses resembling those reported for PCBs and other toxic HAHs.

There have been several reported accidental exposures to commercial PCNs. One of the earliest incidents, the poisoning of cattle, was first reported in 1941 in New York State, and became known as X-Disease because of its unknown etiology. Eventually it was traced to the use of PCNs as high pressure lubricants in feed pelleting machines which resulted in contamination of the feed and ingestion of the PCNs by the animals. The symptoms exhibited by the cattle included a thickening of the skin referred to as hyperkeratosis, excess lacrimation and salivation, anorexia, depression, and a decrease in plasma vitamin A.

Human incidents have been reported in workers involved in the production or uses of PCNs. In humans the inhalation of hot vapors was the most important route of exposure and resulted in symptoms including rashes or chloracne, jaundice, weight loss, yellow atrophy of the liver, and in extreme cases, death.

Lindane and Hexachlorocyclopentadiene

Both lindane (**2**) and hexachlorocyclopentadiene (**3**) are halogenated hydrocarbons; unlike the PCBs and PCNs, they do not contain an aromatic ring.

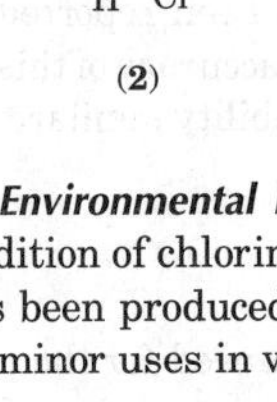

(2)

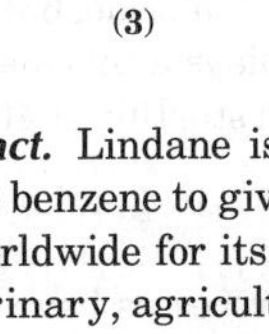

(3)

Chemistry and Environmental Impact. Lindane is produced by the photocatalyzed addition of chlorine to benzene to give a mixture of isomers. Lindane has been produced worldwide for its use as an insecticide and for other minor uses in veterinary, agricultural, and medical products.

The relatively high stability and lipophilicity of lindane and its global use pattern have resulted in significant environmental contamination by this hydrocarbon.

The highly reactive hexachlorocyclopentadiene is rapidly degraded in the environment and is not routinely detected as an environmental pollutant.

Animal and Human Toxicity. The acute toxicity of lindane depends on the age, sex, and animal species, and on the route of administration. Some of the toxic responses caused by lindane in laboratory animals include hepato- and nephrotoxicity, reproductive and embryotoxicity, mutagenicity in some short-term *in vitro* bioassays, and carcinogenicity. The mechanism of the lindane-induced response is not known. Only minimal data are available on the mammalian toxicities of hexachlorocyclopentadiene.

The effects of occupational exposure to lindane have been investigated extensively. These studies indicated that occupational exposure to lindane resulted in increased bodily burdens of this chemical; however, toxic effects associated with these exposures were minimal and no central nervous system disorders were observed.

STEPHEN H. SAFE
Texas A & M University

S. Safe and O. Hutzinger, eds., *Polychlorinated Biphenyls (PCBs): Mammalian and Environmental Toxicology,* Vol. 1, Springer-Verlag Publishing Co., Heidelberg, Germany, 1987.

R. D. Kimbrough and A. A. Jensen, eds., *Halogenated Biphenyls, Terphenyls, Naphthalenes, Dibenzodioxins and Related Products,* 2nd ed., Elsevier/North-Holland, Amsterdam, The Netherlands, 1989.

M. A. Q. Khan and R. H. Stanton, eds., *Toxicology of Halogenated Hydrocarbons,* Pergamon Press, Inc., Elmsford, N.Y., 1981.

K. Baumann, K. Behling, H. L. Brassow, and K. Stapel, *Int. Arch. Occup. Environ. Health* **48,** 165 (1981).

ETHYL CHLORIDE

Ethyl chloride (chloroethane), C_2H_5Cl, is a colorless, mobile liquid of bp 12.4°C, that has a nonirritating ethereal odor and a pleasant taste. It is flammable and burns with a green-edged flame, producing hydrogen chloride fumes, carbon dioxide, and water. Ethyl chloride has primarily been used in the manufacture of tetraethyllead (TEL), an antiknock additive in engine fuel, but it also serves as an ethylating agent, solvent, refrigerant, and local and general anesthetic. It is less toxic than the chloromethanes.

Because of the phasing out of leaded fuels, production of ethyl chloride has decreased steadily since 1979 and imports of ethyl chloride have been essentially zero since 1983. Ethyl chloride demand is expected to continue to diminish.

Physical and Chemical Properties

The physical properties of ethyl chloride are listed in Table 1. At 0°C, 100 g ethyl chloride dissolve 0.07 g water and 100 g water dissolve 0.447 g ethyl chloride. The solubility of water in ethyl chloride increases sharply with temperature to 0.36 g/100 g at 50°C. Ethyl chloride dissolves many organic substances, such as fats, oils, resins, and waxes, and it is also a solvent for sulfur and phosphorus. It is miscible with methyl and ethyl alcohols, diethyl ether, ethyl acetate, methylene chloride, chloroform, carbon tetrachloride, and benzene. Butane, ethyl nitrite, and 2-methylbutane each have been reported to form a binary azeotrope with ethyl chloride, but the accuracy of this data is uncertain. Ethyl chloride displays a thermal stability similar to methyl chloride, with decomposition starting at 400°C.

Manufacture

Three industrial processes have been used for the production of ethyl chloride: hydrochlorination of ethylene, reaction of hydrochloric acid with ethanol, and chlorination of ethane. Hydrochlorination of ethylene is used to manufacture most of the ethyl chloride produced in the United States.

Table 1. Physical Properties of Ethyl Chloride

Property	Value
melting point, °C	−138.3
boiling point at 101 kPa[a], °C	12.4
specific gravity, vapor at 101 kPa[a] (air = 1)	2.23
specific gravity at 0/4°C, liquid	0.92390
refractive index, n_D^{20}	1.3676
specific heat, liquid from −48.8 to 45°[b], J/(kg·K)[c]	$1612 + 2.72\ t + 1.46 \times 10^{-2}\ t^2$
specific heat, vapor at 101 kPa[a], 40°C, J/mol[c]	1.017
critical temperature, °C	186.6
critical pressure, MPa[d]	5.27
thermal conductivity, W/(m·K)	
liquid	0.1467
vapor at bp	0.0095
heat of combustion, kJ/mol[c]	1327
latent heat of evaporation at bp, J/g[c]	383.4
ignition temperature, °C	519
explosive limits in air, vol %	3.16–15
viscosity, mPa·s(= cP)	
liquid, 5°C	0.292
vapor	
12.4°C	0.093×10^{-3}
35°C	0.0165×10^{-3}
vapor pressure, kPa[a]	
0°C	61.86
100°C	1165

[a] To convert kPa to mm Hg, multiply by 7.5.
[b] For example, specific heat at −30°C, 1542.5 J/kg; at 20°C, 1672 J/kg.
[c] To convert J to cal, divide by 4.184.
[d] To convert MPa to atm, divide by 0.101.

Economic Aspects

The economic history of ethyl chloride is entirely dominated by the fact that its principal application is in the manufacture of tetraethyllead (TEL). During the 1980s ethyl chloride production declined steadily. This rate of decline is expected to decrease in the near future as the amount of ethyl chloride used to manufacture products other than TEL becomes more significant. The only important demand for ethyl chloride, other that its use of TEL manufacture, arises from the ethylcellulose industry. As the market for ethyl chloride shrinks, the production of ethyl chloride as a by-product in the manufacture of other chemicals such as vinyl chloride may adversely impact the economics of direct manufacture of ethyl chloride.

Health and Safety Factors

Ethyl chloride is handled and transported in pressure containers under conditions similar to those applied to methyl chloride. In the presence of moisture, ethyl chloride can be moderately corrosive. Carbon steel is used predominantly for storage vessels and prolonged contact with copper should be avoided.

Ethyl chloride is readily absorbed into the body through mucous membranes, lungs, and skin. Recovery of consciousness after exposure to ethyl chloride often entails an unpleasant hangover period. Experiments with animals provide evidence of kidney irritation and promotion of fat accumulation in the kidneys, cardiac muscle, and liver. Concentrations of 15–30 vol % in air are quickly fatal to animals; a concentration of 2% causes some unsteadiness; exposure to 1% concentration has no observable effect. Based on the limited available data, the Environmental Protection Agency has stated that ethyl chloride is one of the least toxic of the chloroethanes. A more recent lifetime inhalation study in rats and mice by the National Toxicology Program showed a high incidence of uterine tumors in female mice exposed to very high (15,000 ppm) concentrations of ethyl chloride.

MATT C. MILLER
Dow Chemical U.S.A.

C. Trabalka and K. Alexandru, *Synth. Ethyl Chloride,* 68354 (1979).

U.S. Pat. 4,849,562 (July 18, 1989), C. Buhs, E. Dreher, and G. McConchie (to The Dow Chemical Company).

Health Effects Assessment for Ethyl Chloride, Report EPA/600/8-88/036, United States Environmental Protection Agency, Environmental Criteria Assessment Office, Cincinnati, Ohio, 1987.

Chem. Business, 32–34 (Nov. 1986).

OTHER CHLOROETHANES

1,1-DICHLOROETHANE

1,1-Dichloroethane, CH_3CHCl_2, ethylidene chloride, ethylidene dichloride, is a colorless liquid with an ethereal odor. It is miscible with most organic solvents and all chlorinated solvents. It is employed as a solvent, but its largest industrial use is as an intermediate in the production of 1,1,1-trichloroethane.

Physical and Chemical Properties

The properties of 1,1-dichloroethane are listed in Table 1.

Manufacture

1,1-Dichloroethane is produced commercially from hydrogen chloride and vinyl chloride at 20–55°C in the presence of an aluminum, ferric, or zinc chloride catalyst. 1,1-Dichloroethane is usually an intermediate in the manufacture of 1,1,1-trichloroethane.

Environmental Concerns

The energy requirements for desorbing 1,1-dichloroethane from activated carbon in a stripping–adsorption process for water purification have been calculated at 112 kJ/kg. Chlorinated hydrocarbons such as 1,1-dichloroethane may easily be removed from water by air or steam stripping.

Toxicity. 1,1-Dichloroethane, like all volatile chlorinated solvents, has an anesthetic effect and depresses the central nervous system at high vapor concentrations. The 1991 American Conference of Governmental Industrial Hygienists (ACGIH) recommends a time-weighted average (TWA) solvent vapor concentration of 200 ppm and a permissible short term exposure level (STEL) of 250 ppm for worker exposure.

Table 1. Properties of Dichloroethanes

Property	1,1-Dichloroethane	1,2-Dichloroethane
melting point, °C	−96.7	−35.3
boiling point; °C	57.3	83.7
density at 20°C, g/L	1.1747	1.2529
n_D^{22}	1.4166	1.4451
viscosity at 20°C, mPa·s(= cP)	0.377	0.84
specific heat at 20°C, J/(g·K)[a]		290
liquid	1.087	5.36
gas	0.824	0.44
critical temperature, °C	261.5	
critical density, g/L		17
critical pressure, MPa[b]	5.06	21
flash point, °C	−12.0	
closed cup		
open cup		
dielectric constant	10.9	10.45
liquid, 20°C		1.0048
vapor, 120°C		5.3
vapor pressure at 10°C, kPa[c]	15.37	

[a] To convert J to cal, multiply by 0.239.
[b] To convert MPa to atm, multiply by 9.87.
[c] To convert kPa to mm Hg, multiply by 7.5.

1,2-DICHLOROETHANE

1,2-Dichloroethane, ethylene chloride, ethylene dichloride, CH_2ClCH_2Cl, is a colorless, volatile liquid with a pleasant odor, stable at ordinary temperatures. It is miscible with other chlorinated solvents and is soluble in common organic solvents, as well as having high solvency for fats, greases, and waxes. It is most commonly used in the production of vinyl chloride monomer.

Physical and Chemical Properties

The physical properties of 1,2-dichloroethane are listed in Table 1.

Pyrolysis. Pyrolysis of 1,2-dichloroethane in the temperature range of 340–515°C gives vinyl chloride, hydrogen chloride, and traces of acetylene and 2-chlorobutadiene.

Hydrolysis. Heating 1,2-dichloroethane with excess water at 60°C in a nitrogen atmosphere produces some hydrogen chloride.

Oxidation. Atmospheric oxidation of 1,2-dichloroethane at room or reflux temperatures generates some hydrogen chloride and results in solvent discoloration.

Corrosion. Corrosion of aluminum, iron, and zinc by boiling 1,2-dichloroethane has been studied.

Nucleophilic Substitution. The kinetics of the bimolecular nucleophilic substitution of the chlorine atoms in 1,2-dichloroethane with NaOH, $NaOC_6H_5$, $(CH_3)_3N$, pyridine, and CH_3COONa in aqueous solutions at 100–120°C has been studied.

Manufacture

1,2-Dichloroethane is produced by the vapor- or liquid-phase chlorination of ethylene or by vapor-phase oxychlorination of ethylene. Most liquid-phase chlorination processes use small amounts of ferric chloride as catalyst.

Economic Aspects and Uses

A significant portion (88%) of U.S. 1,2-dichloroethane production is converted to vinyl chloride monomer. Since it has very few solvent or emissive uses, 1,2-dichloroethane has not faced the regulatory pressures of other chlorinated hydrocarbons. Dramatic growth has been seen for conversion of 1,2-dichloroethane to vinyl chloride. 1,2-Dichloroethane is also a starting material for chlorinated solvents such as 1,1,1-trichloroethane, vinylidene chloride, trichloroethylene, and perchloroethylene. Other uses include its employment as a reactant to prepare ethyleneamines. Most producers of poly(vinyl chloride) resins have back integrated and produce 1,2-dichloroethane for captive use.

Environmental Concerns

Removal of metal chlorides from the bottoms of the liquid-phase ethylene chlorination process has been studied. A detailed summary of production methods, emissions, emission controls, costs, and impacts of the control measures has been made. Residues from this process can also be recovered by evaporation, decomposition at high temperatures, and distillation.

Toxicity. 1,2-Dichloroethane at high vapor concentrations (above 200 ppm) can cause central nervous system depression and gastrointestinal upset characterized by mental confusion, dizziness, nausea, and vomiting. Liver, kidney, and adrenal injuries may occur at the higher vapor levels. The recommended 1991 AGCIH vapor exposure TWA standard for 1,2-dichloroethane was 10 ppm, with a STEL guideline of 40 ppm.

Repeated skin contact should be avoided since the solvent can cause defatting of the skin, severe irritation, and moderate edema. Eye contact may have slight to severe effects.

1,1,1-TRICHLOROETHANE

1,1,1-Trichloroethane, methyl chloroform, CH_3CCl_3, is a colorless, nonflammable liquid with a characteristic ethereal odor. It is miscible with other chlorinated solvents and soluble in common organic solvents.

1,1,1-Trichloroethane is among the least toxic of the chlorinated solvents used in industry today. The commercial metal-cleaning grades contain added inhibitors that make usage acceptable for all common metals including aluminum. It has excellent solvency for various greases, oils, tars, and waxes and a wide range of organic materials (see SOLVENTS, INDUSTRIAL). Emissive uses of 1,1,1-trichloroethane will decline, whereas its use as a chemical intermediate, especially for the production of fluorocarbons, should increase.

Physical and Chemical Properties

The physical properties of 1,1,1-trichloroethane are given in Table 2.

Pyrolysis. The pyrolysis of 1,1,1-trichloroethane at 325–425°C proceeds by a simultaneous unimolecular and radical-chain mechanism to yield 1,1-dichloroethylene and hydrogen chloride.

Hydrolysis. 1,1,1-Trichloroethane heated with water at 75–160°C under pressure and in the presence of sulfuric acid or a metal chloride catalyst decomposes to acetyl chloride, acetic acid, or acetic anhydride.

Oxidation. 1,1,1-Trichloroethane is stable to oxidation when compared to olefinic chlorinated solvents like trichloroethylene and tetrachloroethylene.

Corrosion. The corrosion rates of 1,1,1-trichloroethane with metals in dry and wet environments have been reported.

Inhibition. Organic inhibitors for proprietary grades of 1,1,1-trichloroethane are acid acceptors and metal stabilizers.

Table 2. Properties of 1,1,1- and 1,1,2-Trichloroethane at 20°C kPa

Property	1,1,2-Trichloroethane	1,1,1-Trichlorethane
melting point, °C	−36.5	−33.0
boiling point, °C	113.8	74.0
density at 20°C	1.430[a]	1.3249
n_D^{20}	1.69[b]	1.4377
viscosity at 20°C, mPa·s(= cP)	333	0.858
specific heat at 20°C, J(g·K)[c]	5.141	
liquid	none	1.004
gas		0.782
critical pressure, MPa[d]	2.98[g]	4.48
flash point (closed cup), °C		none
dielectric constant, liquid at 20°C		7.5
vapor pressure[e] at 20°C kPa[f]		13.3

[a] Density at 25°C.
[b] Viscosity at 25°C.
[c] To convert J to cal, divide by 4.184
[d] To convert MPa to atm, multiply by 9.87.
[e] Antoine constants for 1,1,1-trichloroethane: $A = 7.76632$; $B = 1204.66$; $C = 226.671$, where $\log_{10}$ pressure ($kPa = A - B/T + C$); T = temperature in °C.
[f] To convert kPa to mm Hg, multiply by 7.5.
[g] Vapor pressure at 25°C.

Dehydrochlorination. 1,1,1-Trichloroethane over activated alumina or anhydrous aluminum chloride at 0°C gives rapid hydrogen chloride evolution and 1,2-dichloroethylene, which may form polymer.

Economic Aspects

Over 70% of 1,1,1-trichloroethane production is based on the vinyl chloride-1,1-dichloroethane process, 20% on the 1,1-dichloroethylene process, and about 10% on the direct chlorination of ethane.

All nonessential emissive uses of 1,1,1-trichloroethane are expected to be phased out by the year 2000.

Health and Safety Factors

1,1,1-Trichlororethane is among the least toxic of the industrial chlorinated solvents. Vapor inhalation causes depression of the central nervous system (dizziness, light-headedness). The 1991 TWA for 1,1,1-trichloroethane suggested by the ACGIH was 350 ppm, with a recommended STEL of 450 ppm.

Skin exposure to 1,1,1-trichloroethane can cause irritation, pain, blisters, and even burning. Eye exposure may produce irritation, but should not cause serious injury.

Environmental Concerns

1,1,1-Trichloroethane has been targeted for production phaseout by the year 2000 for emissive uses by the . In addition, increasing emphasis is being placed on recovery of spent solvents via carbon adsorption, vacuum distillation, or extraction. Steam or air stripping is a proven method for removing chlorinated hydrocarbons from water solutions. Reducing fugitive emissions to soil and groundwaters and the fate of these solvents in the environment have also been studied.

Uses

Inhibited grades of 1,1,1-trichloroethane are used in hundreds of different industrial cleaning applications.

1,1,2-TRICHLOROETHANE

1,1,2-Trichloroethane, vinyl trichloride, $CH_2ClCHCl_2$, is a colorless, nonflammable liquid with a pleasant odor, miscible with chlorinated solvents, and (as is 1,1,1-trichloroethane) soluble in the other common organic solvents.

Physical and Chemical Properties

The physical properties of 1,1,2-trichloroethane are listed in Table 2.

Dehydrochlorination. 1,1,2-Trichloroethane is easily dehydrochlorinated by a number of catalytic reagents to give 1,1-dichloroethylene and some 1,2-dichloroethylene.

Manufacture

2 1,1,2-Trichloroethane is produced in the United States directly or indirectly from ethylene, eg, by chlorination of 1,2-dichloroethane, a product from ethylene.

1,1,2-Trichloroethane is also a coproduct in the thermal and photochemical chlorination of 1,1-dichloroethane to produce 1,1,1-trichloroethane.

Health, Safety, and Environmental Factors

1,1,2-Trichloroethane is much more toxic than 1,1,1-trichloroethane in acute exposure studies. The 1991 ACGIH recommended TWA value for 1,1,2-trichloroethane is 10 ppm.

1,1,2-Trichloroethane may be removed from water by several methods such as evaporation or air or steam stripping. Elastomeric thin-film pervaporation membranes may also be used.

Uses

The principal use of 1,1,2-trichloroethane is as a feedstock intermediate in the production of 1,1-dichloroethylene. 1,1,2-Trichloroethane is also used where its high solvency for chlorinated rubbers, etc, is needed, as a solvent for pharmaceutical preparation, and in the manufacture of electronic components.

1,1,1,2-TETRACHLOROETHANE

1,1,1,2-Tetrachloroethane, CCl_3CH_2Cl, is used primarily as a feedstock for the production of solvents such as trichloroethylene and tetrachloroethylene.

Physical and Chemical Properties

The physical properties of 1,1,1,2-tetrachloroethane are listed in Table 3.

Pyrolysis Thermal decomposition of 1,1,1,2-tetrachloroethane produces tetrachloroethylene (by disproportionation), hydrogen chloride, and trichloroethylene via dehydrochlorination.

Oxidation. Oxidation of 1,1,1,2-tetrachloroethane in the presence of ionizing radiation gives dichloroacetyl chloride, $Cl_2CHCOCl$.

Manufacture

1,1,1,2-Tetrachloroethane is often an incidental by-product in the manufacture of chlorinated ethanes. It can be prepared by heating the 1,1,2,2-isomer with anhydrous aluminum chloride or chlorination of 1,1-dichloroethylene at 40°C. Hydrochlorination of trichloroethylene using a $FeCl_3$ catalyst may also be used.

Toxicity

Rats exposed to 1000 ppm of vapors for 4–7 h/d for eight days showed ataxia, decreased body weight and growth rate, and minimal central fatty metamorphosis of the liver.

1,1,2,2-TETRACHLOROETHANE

1,1,2,2-Tetrachloroethane, acetylene tetrachloride, $CHCl_2CHCl_2$, is a heavy, nonflammable liquid with a sweetish odor. It is miscible with the chlorinated solvents and shows high solvency for a number of natural organic materials. It is also a solvent for sulfur and a number of inorganic compounds, eg, sodium sulfite.

Physical and Chemical Properties

The physical properties of 1,1,2,2-tetrachloroethane are listed in Table 3.

Pyrolysis. 1,1,2,2-Tetrachloroethane, like the 1,1,1,2-isomer, is thermally degraded with or without a catalytic agent to give trichloroethylene, tetrachloroethylene, and hydrogen chloride.

Dehydrochlorination and Chlorination. The simultaneous chlorination and dehydrochlorination of 1,1,2,2-tetrachloroethane proceeds via formation of labile intermediate, Cl_3CCHCl_2.

Manufacture

1,1,2,2-Tetrachloroethane is produced by direct chlorination or oxychlorination utilizing ethylene as a feedstock.

Toxicity and Environmental Concerns

1,1,2,2-Tetrachloroethane has a TWA of 1 ppm as recommended by the ACGIH (1991). Skin adsorption may also pose an exposure hazard. 1,1,2,2-Tetrachloroethane is one of the most toxic chlorinated hydrocarbons. The liver is most affected.

Silica gel has been used to remove acid and water from 1,1,2,2-tetrachloroethane. Steam and air stripping readily remove chlorinated hydrocarbons such as 1,1,2,2-tetrachloroethane from water solutions.

Uses

The only significant use of 1,1,2,2-tetrachloroethane is as a feedstock in the manufacture of trichloroethylene, tetrachloroethylene, and 1,2-dichloroethylene.

PENTACHLOROETHANE

Pentachloroethane, $CHCl_2CCl_3$, is a colorless, heavy, nonflammable liquid with a chloroform-like odor; it is miscible with common organic solvents.

Physical and Chemical Properties

Physical properties of pentachloroethane are listed in Table 4.

Manufacture

Pentachloroethane can be made by chlorinating 1,1,2,2-tetrachloroethane under ultraviolet light, or trichloroethylene at 70°C in the presence of ferric chloride, sulfur, or ultraviolet light. Oxychlorination of ethylene gives pentachloroethane as well as lower chlorinated hydrocarbons. Chlorination of trichloroethylene can also give pentachloroethane in good yield.

Toxicity

The toxicity of pentachloroethane is similar to that of the tetrachloroethanes. The strong narcotic effect of pentachloroethane is even greater than that of chloroform. Significant pathological changes

Table 3. Properties of 1,1,1,2- and 1,1,2,2-Tetrachloroethane

Property	1,1,2,2-Tetrachloroethane	1,1,1,2-Tetrachloroethane
melting point, °C	−42.5	−68.7
boiling point, °C	146.3	130.5
density at 20°C,	1.593	1.5465
n_D^{20} g/L	1.4942	1.4822
flash point, °C	1.77	none
viscosity at 20 °C, mPa·s(= cP)	388	1.509
critical temperature, °C	3.99	
critical pressure, MPa[a]	0.176	

[a] To convert MPa to atm, multiply by 9.87.

Table 4. Properties of Hexachloroethane and Pentachloroethane

Property	Hexachloroethane	Pentachloroethane
melting point, °C	186.0	−29
boiling point, °C	2.094	161.95
density at 20°C, g/L	0.728	1.678
n_D^{20}	3.073	1.5035
viscosity at 20°C, mPa·s(= cP)	0.028	2.45
specific heat, liquid at 25°C, J/(g·C)[a]	194	182.4
heat of combustion, kJ/g[a]	3.073	4.25
vapor pressure at 20°C, kPa[b]	0.28	0.44

[a] To convert J to cal, divide by 4.184.
[b] To convert kPa to mm Hg, multiply by 7.5.

in the liver, lungs, and kidneys of cats were observed at vapor concentrations of 121 ppm given 8–9 h daily for 23 days.

Uses

Pentachloroethane is still used as an intermediate in some tetrachloroethylene processes.

HEXACHLOROETHANE

Hexachloroethane, perchloroethane, CCl_3CCl_3, is a white crystalline solid with a camphor-like odor. Hexachloroethane is nonflammable and has a number of minor industrial uses which are limited because of its toxic nature. Crystalline hexachloroethane is a minor product in many industrial chlorination processes of saturated and unsaturated C_2 hydrocarbons.

Physical and Chemical Properties

Physical properties of hexachloroethane are listed in Table 4.

Trichloromethyl Free Radical. Degradation of carbon tetrachloride by photochemical, x-ray, or ultrasonic energy produces the trichloromethyl free radical which on dimerization gives hexachloroethane.

Manufacture

Hexachloroethane is formed in minor amounts in many industrial chlorination processes designed to produce lower chlorinated hydrocarbons, usually via a sequential chlorination step.

Toxicity

Hexachloroethane is considered to be one of the more toxic chlorinated hydrocarbons. The 1991 ACGIH recommended time-weighted average (TWA) for hexachloroethane was 1 ppm or 10 mg/m^3 of air. Skin adsorption is a route of possible exposure hazard. The primary effect of hexachloroethane is depression of the central nervous system.

Uses

Hexachloroethane, like carbon tetrachloride and 1,1,1-trichloroethane, can be used to formulate extreme pressure lubricants. It has been used as a chain-transfer agent in the radiochemical emulsion preparation of propylene tetrafluoroethylene copolymer and as a chlorinating agent in the production of methyl chloride from methane.

Other uses of hexachloroethane are as moth repellent, plasticizer for cellulose esters, anthelmintic in veterinary medicine, rubber accelerator, and as a component in fungicidal and insecticidal formulations.

GAYLE SNEDECOR
The Dow Chemical Company

R. W. Gallant, *Hydrocarbon Process.* **45**(7), 111 (1966).

L. E. Horsley, *Azeotropic Data, Advances in Chemistry Series,* No. 6, American Chemical Society, Washington, D.C., 1952; *Azeotropic Data-II,* No. 35, 1962.

J. A. Key, C. W. Stuewe, and R. L. Standifer, *Technical report,* U.S. Environmental Protection Agency, Office of Air Quality Planning and Standards, Washington, D.C., EPA-450/3-80-028c, 363 pp. 1980.

D. D. Irish, in F. A. Patty, ed., *Industrial Hygiene and Toxicology,* 3rd rev. ed., John Wiley & Sons, Inc., New York, 1963, pp. 3491–3497.

DICHLOROETHYLENE

1,1-Dichloroethylene is more commonly known as vinylidene chloride (see VINYLIDENE CHLORIDE MONOMER AND POLYMERS).

1,2-Dichloroethylene (1,2-dichloroethene) is also known as acetylene dichloride, dioform, α,β-dichloroethylene, and *sym*-dichloroethylene. It exists as a mixture of two geometric isomers: *trans*-1,2-dichloroethylene (**1**) and *cis*-1,2-dichloroethylene (**2**).

Table 1. Physical Properties of the Isomeric Forms of 1,2-Dichloroethylene

Property	Trans	Cis
mol wt	96.95	96.95
mp, °C	−49.44	−81.47
bp, °C	47.7	60.2
density at 15°C, g/mL	1.2631	1.2917
	1.44903	1.45189
viscosity at 0°C, mPa·s(= cP)	0.498	0.577
surface tension at 20°C, mN/m (= dyn/cm)	25	28
vapor pressure at 0°C, kPa[a]	15.1	8.7
flash point, °C	4	6

[a] To convert kPa to mm Hg, multiply by 7.5.

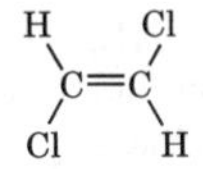

(**1**)

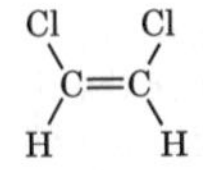

(**2**)

The isomeric mixture is a colorless, mobile liquid with a sweet, slightly irritating odor resembling that of chloroform. It decomposes slowly on exposure to light, air, and moisture. The mixture is soluble in most hydrocarbons and only slightly soluble in water.

Physical and Chemical Properties

1,2-Dichloroethylene consists of a mixture of the cis and trans isomers, as manufactured. The physical properties of both isomeric forms are listed in Table 1.

Manufacturing and Processing

1,2-Dichloroethylene can be produced by direct chlorination of acetylene at 40°C. It is often produced as a by-product in the chlorination of chlorinated compounds and recycled as an intermediate for the synthesis of more useful chlorinated ethylenes. 1,2-Dichloroethylene can be formed by continuous oxychlorination of ethylene by use of a cupric chloride–potassium chloride catalyst, as the first step in the manufacture of vinyl chloride.

Storage and Handling

1,2-Dichloroethylene is usually shipped in 208-L (55 gal) and 112-L (30 gal) steel drums. Because of the corrosive products of decomposition, inhibitors are required for storage.

Health and Safety

1,2-Dichloroethylene is toxic by inhalation and ingestion and can be absorbed by the skin. It has a TLV of 200 ppm. Thorough ventilation is essential whenever the solvent is used for both worker exposure and flammability concerns. Symptoms of exposure include narcosis, dizziness, and drowsiness.

Uses

1,2-Dichloroethylene can be used as low temperature extraction solvent for organic materials such as dyes, perfumes, lacquers, and thermoplastics. It is also used as a chemical intermediate in the synthesis of other chlorinated solvents and compounds.

JAMES A. MERTENS
Dow Chemical USA

C. Marsden, *Solvents Guide,* 2nd ed., Interscience Publishers, New York, 1963.

L. Scheflan, *The Handbook of Solvents,* D. Van Nostrand Co., New York, 1953.

G. Hawley, *The Condensed Chemical Dictionary,* Van Nostrand Reinhold Co., Inc., New York, 1977.

TRICHLOROETHYLENE

Trichloroethylene, trichloroethene, $CHCl{=}CCl_2$, commonly called "tri," is a colorless, sweet smelling (chloroform-like odor), volatile liquid and a powerful solvent for a large number of natural and synthetic substances. It is nonflammable under conditions of recommended use. In the absence of stabilizers, it is slowly decomposed (autoxidized) by air. The oxidation products are acidic and corrosive. Stabilizers are added to all commercial grades. Trichloroethylene is moderately toxic and has narcotic properties.

By the mid-1950s, perchloroethylene had replaced trichloroethylene in dry-cleaning, and metal cleaning became the principal use for trichloroethylene.

The restrictions on production of 1,1,1-trichloroethane from the 1990 Amendments to the Montreal Protocol on substances that deplete the stratospheric ozone and the U.S. Clean Air Act 1990 Amendments will lead to a phase-out of 1,1,1-trichloroethane by the year 2005, which in turn will likely result in a slight resurgence of trichloroethylene in vapor-degreasing applications. The total production, however, will probably stay relatively low because regulations will require equipment designed to assure minimum emissions.

Physical and Chemical Properties

The physical properties of trichloroethylene are listed in Table 1. Trichloroethylene is immiscible with water but miscible with many organic liquids; it is a versatile solvent. It does not have a flash or fire point. However, it does exhibit a flammable range when high concentrations of vapor are mixed with air and exposed to high energy ignition sources. The most important reactions of trichloroethylene are atmospheric oxidation and degradation by aluminum chloride.

Shipping and Storage

Shipment of trichloroethylene is usually by truck or rail car and also in 208-L (55-gal) steel drums Seamless black iron pipes are suitable for transfer lines, gasketing should be of Teflon, Viton, or other solvent impermeable material.

Manufacture

Most trichloroethylene is made from ethylene or 1,2-dichloroethane.

Containers should bear warning labels against breathing vapors, ingesting the liquid, splashing solvent in eyes or on skin and clothing, and using it near an open flame, or where vapors will come in contact with hot metal surfaces (>176°C). Precautions in handling any waste products in conformance with federal, state, and local regulations should be included.

For worker exposure to trichloroethylene vapor, OSHA set a maximum eight-hour time-weighted average (TWA) concentration of 100 ppm. This severely restricted certain applications, and many organizations converted to other chlorinated solvents. As a result, U.S. production of trichloroethylene declined about 70% from a peak in 1970. In 1989, OSHA lowered the permissible exposure limit (PEL) from 100 ppm eight-hour TWA to 50 ppm eight-hour TWA. This added further pressure for some users to consider changing to alternative solvents. No new production capacity is planned in the United States for the foreseeable future.

Shortages, together with rapidly escalating fuel and feedstock prices, have led to a dramatic increase in the price of trichloroethylene, which more than doubled between 1972 and 1976 and doubled again between 1975 and 1985. The price stayed flat during the late 1980s. During the 1990s, the price will likely depend on energy demands and the availability of trichloroethylene.

Table 1. Properties of Trichloroethylene

Property	Value
molecular weight	131.39
melting point, °C	−86.5
boiling point, °C	87.3
vapor density at bp, kg/m^3	4.61
viscosity, mPa·s(= cP) liquid at 20°C	0.57
critical temperature, °C	300.2
critical pressure, MPa[a]	4.986
dielectric constant, liquid at 16°C	3.42
heat of combustion, kJ/g[b]	−6.56

[a] To convert MPa to atm, divide by 0.101.
[b] To convert J to cal, divide by 4.184.

Health and Safety Factors (Toxicity)

Trichloroethylene is acutely toxic, primarily because of its anesthetic effect on the central nervous system. Exposure to high vapor concentrations is likely to cause headache, vertigo, tremors, nausea and vomiting, fatigue, intoxication, unconsciousness, and even death. Because it is widely used, its physiological effects have been extensively studied.

Ingestion of large amounts of trichloroethylene may cause liver damage, kidney malfunction, cardiac arrhythmia, and coma; vomiting should not be induced, but medical attention should be obtained immediately.

Protective gloves and aprons should be used to prevent skin contact, which may cause dermatitis. Eyes should be washed immediately after contact or splashing with trichloroethylene.

During the 1980s a significant amount of work was done on developing methods for treatment of contaminated groundwater and also on setting standards for trichloroethylene under the Safe Drinking Water Act. The EPA has set a maximum contaminant level goal (MCLG) at 0 based on the animal carcinogenic effects. The maximum contaminant level (MCL) is currently set at 5 μg/L.

Uses

Approximately 85% of the trichlorethylene produced in the United States is consumed in the vapor degreasing of fabricated metal parts; the remaining 15% is divided equally between exports and miscellaneous applications.

JAMES A. MERTENS
Dow Chemical U.S.A.

E. Linak with H. J. Lutz and E. Nakamura, "C_2 Chlorinated Solvents," in *Chemical Economics Handbook,* Stanford Research Institute, Menlo Park, Calif., Dec. 1990, pp. 632.30000a–632.3001Z.

L. M. Elkin, *Process Economics Program,* Chlorinated Solvents, Report No. 48, Stanford Research Institute, Menlo Park, Calif., Feb. 1969.

D. M. Avaido and co-workers, *Methyl Chloroform and Trichloroethylene in the Environment,* CRC Press, Cleveland, Ohio, 1976.

L. P. Brown, D. G. Farrar, and C. G. DeRooij, *Health Risk Assessment of Environmental Exposure to Trichloroethylene, Regulatory, Toxicol. Pharmacol.* **11,** 24–41 (1990).

TETRACHLOROETHYLENE

Tetrachloroethylene, perchloroethylene, $CCl_2{=}CCl_2$, is commonly referred to as "perc" and sold under a variety of trade names. It is the most stable of the chlorinated ethylenes and ethanes, having no flash point and requiring only minor amounts of stabilizers. These two properties combined with its excellent solvent properties account for its dominant use in the dry-cleaning industry as well as its application in metal cleaning and vapor degreasing.

Demand for tetrachloroethylene peaked in the 1980s. The decline in demand can be attributed to use of tighter equipment and solvent recovery in the dry-cleaning and metal cleaning industries and the phaseout of CFC 113 (trichlorotrifluoroethane) under the Montreal Protocol.

Physical and Chemical Properties

The physical properties of tetrachloroethylene are listed in Table 1. It dissolves a number of inorganic materials including sulfur, iodine, mercuric chloride, and appreciable amounts of aluminum chloride. Tetrachloroethylene dissolves numerous organic acids, including benzoic, phenylacetic, phenylpropionic, and salicylic acid, as well as a variety of other organic substances such as fats, oils, rubber, tars, and resins. It does not dissolve sugar, proteins, glycerol, or casein. It is miscible with chlorinated organic solvents and most other common solvents. Tetrachloroethylene forms approximately sixty binary azeotropic mixtures.

Manufacture

The following processes are commonly used today: chlorination of ethylene dichloride, chlorination of C1-C3 hydrocarbons of partially chlorinated derivatives, and oxychlorination of C2 chlorinated hydrocarbons.

Economic Aspects

Several United States manufacturers have shut down facilities since around 1980. Current manufacturers and their capacities include Dow Chemical Co. (Plaquemine, La.), 40.8; Occidental Chemical Corp. (Deer Park, Tex.), 81.6; PPG Industries, Inc. (Lake Charles, La.), 90.7; Vulcan Materials Co. in Geismar, La., 68.0, and Wichita, Kans., 22.7.

Toxicity

Overexposure to tetrachloroethylene by inhalation affects the central nervous system and the liver. At concentrations in excess of 1000 ppm the anesthetic and respiratory depression effects can cause unconsciousness and death. Alcohol consumed before or after exposure may increase adverse effects.

The OSHA permissible exposure limit (PEL) for tetrachloroethylene is 25 ppm (8-h TWA). The American Conference of Governmental Industrial Hygienists (ACGIH) threshold limit value (TLV) is 50 ppm. In addition they recommend a 15-minute, short-term exposure limit (STEL) of 200 ppm.

Repeated exposure of skin to liquid tetrachloroethylene may defat the skin, causing dermatitis. Tetrachloroethylene can cause significant discomfort if splashed in the eyes. Although no serious injury results, it can cause transient, reversible corneal injury. If contact with skin or eyes occurs, follow standard first-aid practices.

Tetrachloroethylene is classified in Group 2B, a "possible human carcinogen" by the International Agency for Research on Cancer (IARC). The National Toxicology Program (NTP) lists tetrachloroethylene as "reasonably anticipated to cause cancer in humans."

Table 1. Properties of Tetrachloroethylene

Property	Value
molecular weight	165.83
melting point, °C	−22.7
boiling point at 101 kPa[a], °C	121.2
viscosity, mPa·s(= cp), liquid, °C	
15	0.932
critical temperature, °C	347.1
critical pressure, MPa[b]	9.74
dielectric constant at 1 kHz, 20°C	2.20

[a] To convert kPa to mm Hg, multiply by 7.5.
[b] To convert MPa to atm, divide by 0.101.

Environmental Regulations

Tetrachloroethylene is subject to inventory and release reporting under Title III of the Superfund Amendments and Reauthorization Act of 1986 (SARA). Tetrachloroethylene waste is considered hazardous waste under the Resource Conservation and Recovery Act of 1984 (RCRA). The EPA revised the reportable quantity (RQ) for tetrachloroethylene to 100 lb. in 1989. Under the Clean Air Act Amendment of 1990, tetrachloroethylene is considered a hazardous air pollutant. Under this act, the EPA will develop standards to control tetrachloroethylene emissions in dry-cleaning and metal cleaning applications. Under the Safe Drinking Water Act, EPA has established a maximum contaminant level (MCL) of 0.005 mg/L and a goal of 0 mg/L for tetrachloroethylene. Packed tower aeration and granular activated carbon are considered the best available technologies for removal of tetrachloroethylene from drinking water.

J. C. HICKMAN
The Dow Chemical Company

1990–1991 Threshold Limit Values for Chemical Substances and Physical Agents, American Conference of Governmental Industrial Hygienists, 1990.

Specialty Chlorinated Solvents Product Stewardship Manual, 1991 ed, The Dow Chemical Company, Midland, Mich., form 100-6170-90HYC.

L. W. Rampy, J. F. Quast, B. K. J. Leong, and P. J. Gehring, *Proc. 1st Int. Cong. Toxicol.,* Academic Press, New York, 1978.

Final Report EPA/600/8-83/0005F, U.S. EPA, Washington, D.C., 1985.

ALLYL CHLORIDE

Efficient and economical synthesis of allyl chloride or 3-chloropropene was made possible by the discovery in the late 1930s of a direct high temperature (300–500°C) chlorination reaction by the Shell Development Company and commercialized in 1945. This synthesis allows good yields and use of common inexpensive raw materials such as propylene and chlorine. A second method for synthesis of allyl chloride is thermal dehydrochlorination, ie, cracking, of 1,2-dichloropropane.

Physical Properties

Allyl chloride is a colorless liquid with a disagreeable, pungent odor. Although miscible in typical compounds such as alcohol, chloroform, ether, acetone, benzene, carbon tetrachloride, heptane, toluene, and acetone, allyl chloride is only slightly soluble in water. Other physical properties are given in Table 1.

Chemical Properties

Allyl chloride exhibits reactivity as an olefin and as an organic halide. Its activity as a chloride is enhanced by the presence of the double bond, but its activity as an olefin is somewhat less than that of propylene. Allyl chloride participates in most types of reactions characteristic of either functional group; reactions can be directed by control of conditions, selection of reagents, and provision of suitable catalysts. Allyl chloride does not polymerize well by free-radical techniques (see ALLYL MONOMERS AND POLYMERS).

Manufacture and Processing

Substitutive chlorination of propylene is the commercial route to allyl chloride. For this reaction $\delta H^{\circ}_{298} = -113$ kJ/mol(0 − 27 kcal/mol).

$$CH_2{=}CH{-}CH_3 + Cl_2 \rightarrow CH_2{=}CH{-}CH_2Cl + HCl$$

Storage and Shipment

Storage. Purified and dry allyl chloride can be safely stored in carbon steel vessels.

Shipment. The use of vapor balancing or closed-loop systems is recommended when transferring to or from transportation containers in order to minimize vent flow and use of fresh pad gas.

Economic Aspects

Producers. In the years since 1945, production capacities and the number of producing companies have substantially increased; however the high temperature chlorination reaction has remained the exclusive technique for commercial production of allyl chloride. Production facilities thought to be in existence in 1993 are listed here in order of estimated production capacities: The Dow Chemical Co. (Freeport, Tex. and Stade, Germany), Shell Chemical Co. (Pernis, Holland and Norco, La.), Solvay & Cie (Tavaux, France and Rheinberg, Germany), Kashima Chemical Co. (Kashima, Japan), Qilu Petrochemical Complex (Qilu, China), Daiso (Mizushima, Japan), Han Yang Chemical Corp. (Yochon, South Korea), Organika-Zachem (Bydgoszcz, Poland), Sumitomo Chemical Co. (Niihama, Japan), MTT Co. (Tokuyama, Japan), and Spolek Pro Chemickou (Usti nad Labem, Czech Republic).

Approximately 90% of allyl chloride production is used captively to synthesize epichlorohydrin. The remainder is sold on the merchant market.

Health and Safety Factors

Health Hazards. Allyl chloride is a toxic, highly flammable compound that is severely irritating to the skin and mucous membranes. Exposure to large amounts (oral, dermal, or inhalation) can cause injury and even death. The Material Safety Data Sheet should be consulted for further safety and health information prior to handling allyl chloride.

Exposure Limits. The American Conference of Governmental Industrial Hygienists (ACGIH) has recommended a threshold limit value (TLV) of 1 ppm allyl chloride in air based on a time-weighted average (TWA) of an eight-hour work day, with a short-term exposure limit (STEL) of 2 ppm. OSHA has established its permissible exposure limit (PEL) at this same level.

Personal Protective Equipment. Personal protective and emergency safety equipment should not be relied on as the primary protection from allyl chloride. Prevention of exposure should be considered the preferred precautionary measure.

Emergency Response to Fires, Spills, and Leaks. Vapors of allyl chloride are heavier than air and may travel a considerable distance to sources of ignition. Combustion products of allyl chloride may be more toxic than the allyl chloride itself. Emergency response personnel should wear full protective clothing and self-contained breathing apparatus. Alcohol foam, carbon dioxide, and dry chemicals are effective extinguishing agents for allyl chloride fires. Water may be used to keep fire-exposed containers cool, and water spray may be used to flush burning spills away from exposure to ignition sources. Allyl chloride floats on water, making water alone potentially inadequate for firefighting.

Spills should be confined and prevented from entering water sources.

Table 1. Physical Properties of Allyl Chloride

Property	Value
molecular weight	76.53
freezing point, °C	−134.5
boiling point at 101.3 kPa[a], °C	45.1
specific gravity at 20/4°C	0.938
flash point (tag closed cup), °C	−29
flammable limits (by volume in air), %	3.3–11.1
critical temperature, °C	240.7
critical pressure, kPa[a]	4710

[a] To convert kPa to atm, divide by 101.3.

Uses and Derivatives

Allyl chloride is typically used to make intermediates for downstream derivatives such as resins and polymers. Allyl chloride is very important in the production of epichlorohydrin, which is used as a basic building block for epoxy resins (qv). Synthetic glycerol is also a very important derivative of allyl chloride with epichlorohydrin, an intermediate in this process (see CHLOROHYDRINS; GLYCEROL). Allyl chloride is a starting material for allyl ethers of phenols, bisphenol A, novolak phenolic resins (qv), and the like.

CHRIS KNEUPPER
LESTER SAATHOFF
The Dow Chemical Company

A. W. Fairbairn, H. A. Cheney, and A. J. Cherniavsky, *Chem. Eng. Progr.* **43**(6), 280–290 (June 1947).

Allyl Chloride, Technical Bulletin 296-676-86, The Dow Chemical Company, Midland, Mich., 1986.

Allyl Chloride, Material Safety Data Sheet, The Dow Chemical Company, Midland, Mich., June 14, 1990.

Allyl Chloride, Material Safety Data Sheet, Shell Oil Co., Houston, Tex., Jan. 10, 1991.

CHLOROPRENE

Chloroprene (2-chloro-1,3-butadiene is prepared by caustic dehydrochlorination of 3,4-dichlorobutene-1. It is used almost exclusively for polymerization to form specialty synthetic elastomers.

Physical Properties

Selected physical properties of chloroprene are listed in Table 1. When pure, the monomer is a colorless, mobile liquid with slight odor, but the presence of small traces of dimer usually give a much stronger, distinctive odor similar to terpenes and inhibited monomer may be colored from the stabilizers used.

Chemical Properties

Chloroprene is significantly less reactive than butadiene toward attack by nucleophilic reagents, slightly less reactive toward electrophilic reagents, eg, chlorine or maleic anhydride, but more reactive toward attack by polymerization initiators or other free-radical reagents. Spontaneous polymerization and dimerization at room temperature seriously limit its use for purposes other than polymerization. Chloroprene is normally polymerized with free-radical catalysts in aqueous emulsion, limiting the conversion of monomer to avoid formation of cross-linked insoluble polymer, at a typical temperature of 40°C.

Table 1. Physical Properties of Chloroprene

Property	Value
mol wt	88.54
melting point, °C	−130 2
boiling point at 101 kPa[a]°C	59.4
critical temperature, °C	261.7
viscosity at 25°C, mPa·s(= cP)	0.394
flash point (ASTM, open cup), °C	−20
latent heat of vaporization, kJ/g[b]	
0°C	0.3328
60°C	0.3027
dielectric constant at 27°C	4.9

[a] To convert kPa to mm Hg, multiply by 7.5.
[b] To convert J to cal, divide by 4.184.

Manufacture

Chloroprene is prepared commercially by chlorinating butadiene to a mixture of 3,4-dichlorobutene-1 and 1,4-dichlorobutene-2, separation of the desired, lower boiling 3,4-isomer, isomerizing the higher boiling 1,4-isomer, and dehydrochlorinating the 3,4-isomer with aqueous sodium hydroxide. The older process involving dimerization of acetylene and addition of hydrogen chloride is seldom used today.

Storage, Handling, and Shipment

Uninhibited chloroprene suitable for polymerization must be stored at low temperature (<10°C) under inert atmosphere to maintain both safety and quality. It is a flammable, polymerizable liquid with significant toxicity and should be handled with care even in the laboratory. When transportation of the monomer is required, it is inhibited and loaded cold into sealed, insulated vessels with exclusion of air and careful monitoring of loading and arrival temperatures and duration of transit. It is then purified by distillation at the site where it is to be used.

Economic Aspects

Chloroprene production, polychloroprene production, and polymer sales are approximately equal, recently about 0.3 million metric tons per year, excluding Russia, China, and former Soviet-dominated states. U.S. producers are DuPont and Mobay.

Safety and Health Factors

Uncontrolled polymerization and fire are the most important acute hazards in handling chloroprene. Flammable limits in air are 1.9–10%. It is detectable by odor at about 1 ppm in air, or lower if appreciable dimer impurities are present. It is physiologically active and exposure should be minimized. The maximum exposure limit cited (DuPont Company, OSHA) is 10 ppm in air.

CLARE A. STEWART, JR.
Consultant

W H. Carothers, I. Williams, A. M. Collins, and E. J. Kirby, *J. Am. Chem. Soc.* **53,** 4203 (1931).

CHLORINATED PARAFFINS

The principal feedstocks used today are the normal paraffin fractions C10–C13, C12–C14, C14–C17, and C18–C20, together with paraffin wax fractions of C24–C30; precise compositions may vary depending on petroleum oil source. Chlorination extent generally varies from 30 to 70% by weight. The choice of paraffinic feedstock and chlorine content is dependent on the application.

Chemical and Physical Properties

The physical and chemical properties of chlorinated paraffins are determined by the carbon chain length of the paraffin and the chlorine content. This is most readily seen with respect to viscosity and volatility; increasing carbon chain length and increasing chlorine content lead to an increase in viscosity but a reduction in volatility.

Chlorinated paraffins vary in their physical form from free-flowing mobile liquids to highly viscous glassy materials. Physical properties of some commercially available chlorinated paraffins are listed in Table 1.

A key property associated with chlorinated paraffins, particularly the high chlorine grades, is nonflammability, which has led to their use as fire-retardant additives and plasticizers in a wide range of polymeric materials. The fire-retardant properties are considerably enhanced by the inclusion of antimony trioxide.

Chlorinated paraffins are relatively inert and exhibit excellent resistance to chemical attack and are hydrolytically stable. They are soluble in chlorinated solvents, aromatic hydrocarbons, esters, ketones, and ethers but only moderately soluble in aliphatic hydrocarbons and virtually insoluble in water and lower alcohols.

Table 1. Physical Properties of Selected Commercial Chlorinated Paraffins

Paraffin carbon chain length	Viscosity,[a] mPa·s(= cP)	Density, g/mL	Volatility,[b] %w/w	Refractive index
C10–C13	80	1.19	16.0	1.493
	3500	1.36	4.4	1.516
	30,000	1.44	2.5	1.525
C14–C17	70	1.10	4.2	1.488
	1600	1.25	1.4	1.508
C18–C20	1700	1.21	0.8	1.506
Wax	2500	1.16	0.4	1.506

[a] At 25°C unless otherwise noted.
[b] Measured in a standard test for four hours at 180°C.

Manufacture

Chlorinated paraffins are manufactured by passing pure chlorine gas into a liquid paraffin at a temperature between 80 and 100°C, depending on the chain length of the paraffin feedstock.

Storage and Transportation

Liquid chlorinated paraffins are shipped in drums usually lacquer-lined mild steel or polyethylene and in road or rail barrels. Where appropriate larger quantities can be shipped by sea either in deck tanks of conventional cargo ships or in chemical parcel tankers for larger consignments.

Economic Aspects

The global market for chlorinated paraffins excluding the former Soviet Union and the People's Republic of China is around 300,000 t.

The largest single market is the United States at approximately 40,000 t. Europe as a whole is approximately two and a half times greater than the United States mainly because of the extensive use of chlorinated paraffins as secondary plasticizers in plasticized PVC, which is virtually absent in the United States.

In the United States approximately 50% of the 40,000 t of chlorinated paraffins consumed domestically are used in metal-working lubricants. Approximately 20% are consumed as plastic additives, mainly fire retardants, and similarly 12% in rubber. The remainder is used as plasticizers in paint (9%) and in caulks, adhesives, and sealants at 6%.

Health and Safety Factors

A substantial body of information on the toxicological and environmental effects of chlorinated paraffins has been compiled over the past 20 years, and research is still continuing in both areas.

Toxicity. The acute toxicity of chlorinated paraffins has been tested in a range of animals and was found to be very low.

Environmental. In general, chlorinated paraffins biodegrade; the rate is determined by chlorine content and carbon chain length. Microorganisms previously acclimatized to specific chlorinated paraffins show a greater ability to degrade the compounds than nonacclimatized organisms. Mammals and fish have been shown to metabolize chlorinated paraffins.

In the United States further information and advice is readily available from the Chlorinated Paraffin Manufacturers Association (CPIA) based in Washington, D.C.

KELVIN L. HOUGHTON
ICI Chemicals and Polymers Ltd.

H. J. Caesar "Chlorinated Paraffins as Secondary Plasticizers in PVC," *Chem. Ind.* (Aug. 1978).

H. J. Caesar and P. J. Davis "Flame Retardant Vinyl Compounds," *33rd Annual Technical Conference,* Atlanta, Ga., May 6, 1975.

K. L. Houghton and M. E. Moss "Chlorinated Paraffins as Plasticizers in Polymer Sealant Systems," *ASC Supplier Short Course,* Nashville, Tenn., May 14–17, 1990.

R. D. N. Birtley and co-workers, *Toxicol. Appl. Pharmacol.* **54,** 514 (1980).

CHLORINATED BENZENES

The chlorination of benzene can theoretically produce 12 different chlorobenzenes. With the exception of 1,3-dichlorobenzene, 1,3,5-trichlorobenzene, and 1,2,3,5-tetrachlorobenzene, all of the compounds are produced readily by chlorinating benzene in the presence of a Friedel-Crafts catalyst (see FRIEDEL-CRAFTS REACTIONS). The usual catalyst is ferric chloride either as such or generated *in situ* by exposing a large surface of iron to the liquid being chlorinated. With the exception of hexachlorobenzene, each compound can be further chlorinated; therefore, the finished product is always a mixture of chlorobenzenes. Refined products are obtained by distillation and crystallization.

With the discontinuation of some herbicides, eg, 2,4,5-trichlorophenol, based on the higher chlorinated benzenes, and DDT, based on monochlorobenzene, both for ecological reasons, the production of chlorinated benzenes has been reduced to just three with large-volume applications of (mono)chlorobenzene, *o*-dichlorobenzene, and *p*-dichlorobenzene. Monochlorobenzene remains a large-volume product, considerably larger than the other chlorobenzenes, in spite of the reduction demanded by the discontinuation of DDT.

Physical and Chemical Properties

The important physical properties of chlorobenzenes appear in Table 1. Only limited information is available for some chlorobenzenes. Nitration of chlorobenzenes, mostly monochlorobenzene in the United States, with nitric acid has wide industrial applications.

Table 1. Physical Properties of Chlorobenzenes

Property	Chloro-benzene	1,2-Dichloro-benzene	1,3-Dichloro-benzene	1,4-Dichloro-benzene	1,2,4-Trichloro-benzene
mol wt	112.56	147.005	147.005	147.005	181.45
mp, °C	−45.34	−16.97	−24.76	53.04	17.15
bp at 101.3 kPa[a] °C	131.7	180.4	173.0	174.1	213.8
critical temperature, °C	359.2	417.2	415.3	407.5	453.3
critical pressure, kPa[a]	4519	4031	4864	4109	3718
critical density, kg/L	0.3655	0.411	0.458	0.411	0.447
viscosity, mPa·s(= cP)	0.756	1.3018	1.0254		
flash point[b], °C	28	71	67	99	

[a] To convert kPa to mm Hg, multiply by 7.5.
[b] ASTM method D56-70, closed cup.

Table 2. Toxicity of Chlorinated Benzenes

Chlorobenzene	Fish[a] toxicity, mg/L[b]	LD_{50}, g/kg	TLV (inhal), ppm[c]	Saturated concentration ppm by vol at 20°C
monochloro, rat	<3[d,e]	2.9	75	11,900
rabbit	16	2.8		
1,2-dichloro, guinea pig	3	0.8 – 2.0	50	1,125
1,4-dichloro, rat	0.7[e,f]	3.8	75	1,570
guinea pig	5	2.8		
1,2,4-trichloro, rat	2	1	[g]	260

[a] Fathead unless otherwise noted. [b] No observed adverse effect at this concentration in H_2O; 72 h static test unless otherwise noted. [c] Volume per volume of air. [d] Rainbow trout [e] 96 h dynamic test. [f] Bluegill. [g] No TLV suggested.

Storage, Shipment, and Handling

Chlorobenzenes are stored in manufacturing plants in liquid form in steel containers.

Health and Safety Aspects

In general, all of the chlorobenzenes are less toxic than benzene. Liquid chlorobenzenes produce mild to moderate irritation upon skin contact.

Contact with eye tissue at normal temperatures causes pain, mild to moderate irritation, and possibly some transient corneal injury. Prompt washing with large quantities of water is helpful in minimizing the adverse effects of eye exposure.

The threshold limit value (TLV), the vapor concentration in ppm by volume to which humans may be exposed for an eight-hour working day for many years without adverse effects, is also reported in Table 2.

Economic Aspects

Total production of chlorobenzenes in the three principal producing regions of the world amounted to approximately 400 thousand metric tons in 1988: the United States, 46%, Western Europe, 34%, and Japan, the remainder. Monochlorobenzene accounted for over 50% of the total production of chlorinated benzenes. The largest use of monochlorobenzene worldwide is for the production of nitrochlorobenzene: 41% for the United States' demand, 70% for the Western European demand, and 89% for the Japanese demand in 1988. Nitrochlorobenzenes are used to make dye and pigment intermediates, rubber processing chemicals, pesticides, pharmaceuticals, and other organic intermediates.

With the exception of use in the manufacture of polymers, markets for chlorobenzenes are mature, and demand is expected to show little if any growth in the next few years.

The chlorobenzene operations in the United States were developed primarily for the manufacture of phenol, aniline, and DDT. However, with the process changes in the production of phenol and aniline, the phase-out of DDT production, and changes in the herbicide and solvent markets, the U.S. production of chlorinated benzenes has shrunk by more than 50% since the total production peaked in 1969. Commercial chlorination of benzene today is carried out as a three-product process (monochlorobenzene and *o*- and *p*-dichlorobenzenes.

The principal use of *o*-dichlorobenzene is to manufacture 3,4-dichloroaniline, which is a raw material for several herbicides and for the production of 3,4,4′-trichlorocarbanilide (TCC), a bacteriostat used in deodorant soaps.

The largest single market and a growing outlet for *p*-dichlorobenzene in the United States is the production of poly(phenylene sulfide) (PPS) resin. The second largest consumption in the United States of *p*-dichlorobenzene (16%) is the room deodorant market which is static and likely to remain unchanged. Moth control (11%) is also expected to remain static.

JAMES G. BRYANT
Standard Chlorine of Delaware, Inc.

W. K. Johnson with A. Leder and Y. Sakuma, "CEH Product Review", *Chlorobenzenes Chemical Economics Handbook,* SRI International, Menlo Park, Calif., Oct. 1989.

J. A. Barter and R. S. Nair, *Review of the Scientific Evidence on the Human Carcinogenic Potential of Para-Dichlorobenzene,* Chlorobenzene Producers Association, Washington, D.C., 1990.

J. B. Cohen and P. Hartley, *J. Chem. Soc.* **87**, 1360 (1905).

M. Campbell and H. Hatton, *Herbert H. Dow: Pioneer in Creative Chemistry,* Appleton-Century-Crosts, Inc., New York, 1951.

RING-CHLORINATED TOLUENES

The ring-chlorinated derivatives of toluene form a group of stable, industrially important compounds. Many chlorotoluene isomers can be prepared by direct chlorination. Other chlorotoluenes are prepared by indirect routes involving the replacement of amino, hydroxyl, chlorosulfonyl, and nitro groups by chlorine and the use of substituents, such as nitro, amino, and sulfonic acid, to orient substitution followed by their removal from the ring.

Mono- and dichlorotoluenes are used chiefly as chemical intermediates in the manufacture of pesticides, dyestuffs, pharmaceuticals, and peroxides, and as solvents. Worldwide annual production of *o*- and *p*-chlorotoluene is estimated at several tens of thousands of metric tons. Yearly production of polychlorotoluenes is in the range of 100–1000 tons.

MONOCHLOROTOLUENES

Physical Properties

o-Chlorotoluene (1-chloro-2-methylbenzene, OCT) is a mobile, colorless liquid with a penetrating odor similar to chlorobenzene. It is miscible in all proportions with many organic liquids such as aliphatic and aromatic hydrocarbons, chlorinated solvents, lower alcohols, ketones, glacial acetic acid, and di-*n*–butylamine; it is insoluble in water, ethylene and diethylene glycols, and triethanolamine.

p-Chlorotoluene (1-chloro-4-methylbenzene, PCT) and *m*-chlorotoluene (1-chloro-3-methylbenzene, MCT) are mobile, colorless liquids with solvent properties similar to those of the ortho isomer.

Physical properties of the monochlorotoluene isomers, mol wt 126.59, appear in Table 1.

Chemical Properties

The monochlorotoluenes are stable to the action of steam, alkalies, amines, and hydrochloric and phosphoric acids at moderate temperatures and pressures. Three classes of reactions, those involving the aromatic ring, the methyl group, and the chlorine substituent, are known for monochlorotoluenes.

Reactions of the Aromatic Ring.

Ring chlorination of *o*-chlorotoluene yields a mixture of all four possible dichlorotoluenes, the 2,3-, 2,4-, 2,5-, and 2,6-isomers.

Table 1. Physical Properties of the Monochlorotoluenes, C_7H_7Cl

	Isomer		
Property	Ortho	Meta	Para
mp, °C	−35.6	−47.8	7.5
bp, °C	159.2	161.7	162.4
flash point, °C	47	47	49
density, kg/m^3, at 20°C	1082.5	1072.2	1069.7

Reactions of the Methyl Group. Monochlorotoluenes are widely used to synthesize compounds derived from reactions of the methyl group.

Halogen Reactions. Hydrolysis of chlorotoluenes to cresols has been effected by aqueous sodium hydroxide.

Preparation

Monochlorotoluenes have been prepared by chlorinating toluene with a wide variety of chlorinating agents, catalysts, and reaction conditions. The ratio of ortho and para isomers formed can vary over a wide range. Particular attention has been given to studies aimed at increasing the para-isomer content owing to its greater commercial significance. The meta-isomer must be prepared by indirect means since only a small amount, <1%, is formed by direct chlorination.

Handling and Shipment

Monochlorotoluenes are shipped in bulk in steel tank cars and tank trucks. Under DOT regulations, for transport of over 415 L (110 gal) of monochlorotoluenes, freight classification is combustible liquid NOS, and for truck transport, chemical NOI.

Health and Safety Factors

Inhalation of high concentrations of monochlorotoluenes will cause symptoms of central nervous system depression. *o*- and *p*-Chlorotoluene are both considered moderately toxic by ingestion.

HIGHER CHLOROTOLUENES

Dichlorotoluenes

There are six possible dichlorotoluene isomers, $C_7H_6Cl_2$, (mol wt 161.03) all of which are known. Only the 2,4-, 2,5-, and the 3,4-isomers are available from direct chlorination of monochlorotoluenes. Physical properties of the dichloro- and other higher chlorotoluenes are given in Table 2.

Trichlorotoluenes

The chlorination of toluene and *o*-and *p*-chlorotoluenes produces a mixture of trichlorotoluenes, $C_7H_5Cl_3$, (mol wt 195.48): the 2,3,6-isomer (1,2,4-trichloro-3-methylbenzene) and 2,4,5-trichlorotoluene (1,2,4-trichloro-5-methylbenzene) containing small amounts of 2,3,4-trichlorotoluene (1,2,3-trichloro-4-methylbenzene) and 2,4,6-trichlorotoluene (1,3,5-trichloro-2-methylbenzene).

Tetra- and Pentachlorotoluenes

2,3,4,6-Tetrachlorotoluene, $C_7H_4Cl_4$ (mol wt 229.93) (1,2,3,5-tetrachloro-4-methylbenzene), is prepared from the Sandmeyer reaction on 3-amino-2,4,6-trichlorotoluene. Pentachlorotoluene (pentachloromethylbenzene), $C_7H_3Cl_5$ (mol wt 246.37), is formed in 90% yield by the ferric chloride-catalyzed chlorination of toluene in carbon tetrachloride or hexachlorobutadiene solution.

Table 2. Physical Properties of the Higher Chlorotoluenes

Toluene	Mp, °C	Bp, °C	n_D^t	Density at 20°C, kg/m^3
2,3-dichloro	5	208.3	1.5511^{20}	
2,5-dichloro	5	201.8	1.5449^{20}	1253.5
3,4-dichloro	−15.3	208.9	1.5471^{20}	1256.4
2,3,4-trichloro	43–44	244		
2,3,6-trichloro	45–46	118[a]		
2,4,6-trichloro	38			
2,3,4,5-tetrachloro	98.1			
2,3,5,6-tetrachloro	93–94			

[a] At 2.4 kPa (18 mm Hg).

HENRY C. LIN
RAMESH KRISHNAMURTI
Occidental Chemical Corporation

U.S. Pat. 4,851,596 (July 25, 1989), M. Franz-Josef, F. Helmut, R. Kai, and W. Karlfried (to Bayer A-G).

D. R. Thielen, P. S. Foreman, A. Davis, and R. Wyeth, *Environ. Sci. Technol.* **21,** 145 (1987).

I. P. Ulanova, P. N. Dyachkov, and A. I. Khalepo, *Pharmacochem. Libr.* **8** (QSAR Toxicol. Xenobiochem.), 83 (1985).

U.S. Pat. 4,006,195 (Feb. 1, 1977), S. Gelfand (to Hooker Chemicals & Plastics Corp.).

BENZYL CHLORIDE, BENZAL CHLORIDE, AND BENZOTRICHLORIDE

Nearly all of the benzyl chloride, benzal chloride, and benzotrichloride manufactured is converted to other chemical intermediates or products by reactions involving the chlorine substituents of the side chain. Each of the compounds has a single primary use that consumes a large portion of the compound produced. Benzyl chloride is utilized in the manufacture of benzyl butyl phthalate, a vinyl resin plasticizer; benzal chloride is hydrolyzed to benzaldehyde; benzotrichloride is converted to benzoyl chloride. Benzyl chloride is also hydrolyzed to benzyl alcohol, which is used in the photographic industry, in perfumes (as esters), and in peptide synthesis by conversion to benzyl chloroformate (see BENZYL ALCOHOL AND β-PHENETHYL-ALCOHOL; CARBONIC AND CARBONOCHLORIDIC ESTERS).

Several related compounds, primarily ring-chlorinated derivatives, are also commercially significant. *p*-Chlorobenzotrichloride is converted to *p*-chlorobenzotrifluoride, an important intermediate in the manufacture of dinitroaniline herbicides.

Physical Properties

Benzyl chloride [(chloromethyl)benzene, α-chlorotoluene], $C_6H_5CH_2Cl$, is a colorless liquid with a very pungent odor. Its vapors are irritating to the eyes and mucous membranes, and it is classified as a powerful lacrimator. The physical properties of pure benzyl chloride are given in Table 1. Benzyl chloride is insoluble in cold water, but decomposes slowly in hot water to benzyl alcohol. It is miscible in all proportions at room temperature with most organic solvents.

Benzal chloride [(dichloromethyl)benzene, α,α-dichlorotoluene, benzylidene chloride], $C_6H_5CHCl_2$, is a colorless liquid with a pungent, aromatic odor. Benzal chloride is insoluble in water at room temperature but is miscible with most organic solvents.

Benzotrichloride [(trichloromethyl)benzene, α,α,α-trichlorotoluene, phenylchloroform], $C_6H_5CCl_3$, is a colorless, oily liquid with a pungent odor. It is soluble in most organic solvents, but it reacts with water and alcohol.

Table 1. Physical Properties of Benzyl Chloride, Benzal Chloride, and Benzotrichloride

Property	Benzyl chloride	Benzal chloride	Benzotrichloride
mol wt	126.58	161.03	195.48
freezing point, °C	−39.2	−16.4	−4.75
boiling point, °C	179.4	205.2	220.6
density, kg/m^3	1113.5^4_4	1256^{14}_{14}	1374^{20}_4
refractive index, n^t_D	1.5392^{20}	1.5502^{20}	1.55789^{20}
heat of combustion, kJ/mol^a	3708^b	3852^b	3684^c

^{a} To convert J to cal, divide by 4.184.
^{b} At constant volume.
^{c} At constant pressure.

Manufacture

Benzyl chloride is manufactured by the thermal or photochemical chlorination of toluene at 65–100°C. Reaction is limited to 50% conversion of toluene to minimize benzol chloride formation.

Benzyl chloride is manufactured in 70% yield by chlorination with 2.0–2.2 moles of chlorine per mole of toluene. The product is purified by distillation.

Further chlorination at a temperature of 100–140°C with ultraviolet light yields benzotrichloride. Yields of 95% are obtained.

Competing reactions in these reactions lead to undesired by-products. Addition of chlorine to the aromatic ring occurs at temperatures below 40°C and high concentration of chlorine. Electrophilic substitution of chlorine on the ring is promoted by traces of iron.

Handling and Shipment

As is the case during manufacture, contact with those metallic impurities that catalyze Friedel-Crafts condensation reactions must be avoided. The self-condensation reaction is exothermic and the reaction can accelerate, producing a rapid buildup of hydrogen chloride pressure in closed systems.

Benzyl chloride is available in both anhydrous and stabilized forms. Both forms can be shipped in glass carboys, nickel and lined-steel drums, and nickel tank trucks and tank cars.

Benzyl chloride is classified by DOT as chemicals NO1BN, poisonous, corrosive and a hazardous substance (100 lbs, 45.45 kg). Benzal chloride is classified as poisonous and a hazardous substance (5000 lbs, 2270 kg). Benzotrichloride is classified under DOT regulation as a corrosive liquid NOS and a hazardous substance (10 lbs, 4.5 kg). The Freight Classification Chemical NOI applies. It is shipped in lacquer-lined steel drums and nickel-lined tank trailers. Benzal chloride is handled in a similar fashion.

Economic Aspects

Plant capacities for the production of benzyl chloride in the western world totaled 144,200 t/yr in 1989. Monsanto is the largest producer.

In the United States, in addition to Monsanto, Akzo, which took over part of Stauffer, is the only other producer of benzyl chloride (9000 t/yr). Total western world production in 1988 was approximately 92,700 t, with U.S. production at 26,500 t or 54% of capacity.

Benzyl chloride and butyl alcohol react with phthalic anhydride in one step to yield benzyl butyl phthalate a plasticizer made by Monsanto and known by its trade name Santicizer 160.

Benzotrichloride is a chemical intermediate used to produce two significant products. Partial hydrolysis or reaction with benzoic acid yields benzoyl chloride, whereas chlorination and subsequent reaction with hydrogen fluoride yields *p*-chlorobenzotrifluoride.

Health and Safety Factors

Benzyl chloride is a severely irritating liquid and causes damage to the eyes, skin, and respiratory tract, including pulmonary edema. Other possible effects of overexposure to benzyl chloride are CNS depression and liver and heart damage.

IARC states there is limited evidence that exposure to benzyl chloride is carcinogenic in experimental animals; epidemiological data was inadequate to evaluate carcinogenicity to humans.

Other toxicological effects that may be associated with exposure to benzyl chloride based on animal studies are skin sensitization and developmental embryo or fetal toxicity. A 1980 OSHA regulation has established a national occupational exposure limit for benzyl chloride of 5 mg/m^3 (1 ppm). Vapors of both benzal chloride and benzotrichloride are strongly irritating and lacrimatory.

For all three compounds, biological data relevant to the evaluation of carcinogenic risk to humans are summarized in the World Health Organization International Agency for Research on Cancer monograph.

Table 2. Physical Constants of the Main Ring-Chlorinated Derivatives of Benzyl Chloride, Benzal Chloride, and Benzotrichloride

Benzene derivative (methyl)	Mp, °C	Bp, °C	n_D^{20}	Density, kg/m^3
1-chloro-2-(chloro-)	−17	217	1.5330	1270
1-chloro-3-(chloro-)		215–216[a]		1269.5
1-chloro-4-(chloro-)	31	222	1.5554	
1-chloro-2-(dichloro-)		228.5	1.5670[b]	1399
1-chloro-3-(dichloro-)		235–237		
1-chloro-4-(dichloro-)		236[c]		
2,4-dichloro-1-(chloro-)	−2.6	248	1.5761	1407
1,3-dichloro-2-(chloro-)	39–40	117–119[d]		
1,2-dichloro-4-(chloro-)	37–37.5	241		1412
1-chloro-2-(trichloro-)	29.4	264.3	1.5836	1519
1-chloro-3-(trichloro-)		255	1.4461	1495
1-chloro-4-(trichloro-)		245	1.4463	1495
1,3-dichloro-2-(dichloro-)		250		
1,2-dichloro-4-(dichloro-)		257		1518
2,4-dichloro-1-(dichloro-)	47–48	155–159[e]		
1,2-dichloro-4-(trichloro-)	25.8	283.1	1.5886	1591

[a] At 100.4 kPa (753 mm Hg). [b] At 16°C [c] At 100.7 kPa (755 mm Hg). [d] At 1.87 kPa (14 mm Hg). [e] At 2.67 kPa (20 mm Hg).

Derivatives

Ring-Substituted Derivatives. The ring-chlorinated derivatives of benzyl chloride, benzal chloride, and benzotrichloride are produced by the direct side-chain chlorination of the corresponding chlorinated toluenes or by one of several indirect routes if the required chlorotoluene is not readily available. Physical constants of the main ring-chlorinated derivatives of benzyl chloride, benzal chloride, and benzotrichloride are given in Table 2.

Side-Chain Chlorinated Xylene Derivatives. Only a few of the nine side-chain chlorinated derivatives of each of the xylenes are available from direct chlorination. All three of the monochlorinated compounds, α-chloro-*o*-xylene (1-(chloromethyl)-2-methylbenzene), α-chloro-*m*-xylene (1-(chloromethyl)-3-methylbenzene), α-chloro-*p*-xylene (1-(chloromethyl)-4-methylbenzene) are obtained in high yield from partial chlorination of the xylenes. 1,3-Bis(chloromethyl)benzene can be isolated in moderate yield from chlorination mixtures.

The fully side-chain-chlorinated products, 1,3-bis(trichloromethyl)-benzene and 1,4-bis(trichloromethyl)benzene, are manufactured by exhaustive chlorination of meta- and para-xylenes.

1-(Dichloromethyl)-2-(trichloromethyl)benzene, the end product of exhaustive side-chain chlorination of *o*-xylene, is an intermediate in the manufacture of phthalaldehydic acid.

HENRY C. LIN
JOSEPH F. BIERON
Occidental Chemical Corporation

J. S. Ratcliffe, *Br. Chem. Eng.* **11,** 1535 (1966).

Faith, Keyes, and Clark's Industrial Chemicals 4th ed., John Wiley & Sons, Inc., New York, 1975, pp. 145–148.

H. G. Haring and H. W. Knol, *Chem. Process. Eng.* **45,** 540, 619, 690 (1964).

IARC Monogr. Eval. Carcinog. Risk Chem. Man. **29,** 59 (1982).

TOXIC AROMATICS

Chlorinated biphenyls, chlorinated naphthalenes, benzene hexachloride, and chlorinated derivatives of cyclopentadiene are no longer in commercial use because of their toxicity. However, they still impact on the chemical industry because of residual environmental problems. This article discusses the toxicity and environmental impact of these materials.

Polychlorinated Biphenyls

Polychlorinated biphenyls (PCBs) typify halogenated aromatic hydrocarbons (HAHs), industrial compounds or by-products that have been widely identified in the environment and chemical waste dumpsites. PCBs were used in industry as heat-transfer fluids, organic diluents, lubricant inks, plasticizers, fire retardants, paint additives, sealing liquids, immersion oils, adhesives, dedusting agents, waxes, and as dielectric fluids for capacitors and transformers.

Chemistry and Environmental Impact. PCBs are synthesized by the chlorination of biphenyl and the resulting products are designated according to their percent (by weight) chorine content. Over 600 million kg of commercial PCBs were produced in the United States and the estimated worldwide production is approximately double this quantity.

The identification of PCB residues in fish, wildlife, and human tissues has been reported since the 1970s. The results of these analytical studies led to the ultimate ban on further use and production of these compounds.

Commercial PCBs: Toxic and Biochemical Effects. PCBs and related halogenated aromatic hydrocarbons elicit a diverse spectrum of toxic and biochemical responses in laboratory animals dependent on a number of factors including age, sex, species, and strain of the test animal and the dosing regimen (single or multiple). The toxic responses elicited by most PCB preparations are also observed for other classes of HAHs and include a progressive weight loss not simply related to decreased food consumption and accompanied by weakness, debilitation, and ultimately death, ie, a wasting syndrome; lymphoid involution, thymic and splenic atrophy with associated humoral and/or cell-mediated immunosuppression and/or associated bone marrow and hematologic dyscrasia; a skin disorder called chloracne accompanied by acneform eruptions, alopecia, edema, hyperkeratosis, and blepharitis resulting from hypertrophy of the Meibomian glands; hyperplasia of the epithelial lining of the extrahepatic bile duct, the gall bladder, and urinary tract; hepatomegaly and liver damage accompanied by necrosis, hemorrhage, and intrahepatic bile duct hyperplasia; hepatotoxicity also manifested by the development of porphyria and altered metabolism of porphyrins; teratogenesis, developmental and reproductive toxicity observed in several animal species; carcinogenesis as caused by PCBs in laboratory animals and primarily associated with their effects as promoters; and endocrine and reproductive dysfunction, ie, altered plasma levels of steroid and thyroid hormones with menstrual irregularities, reduced conception rate, early abortion, excessive menstrual and postconceptional hemorrhage, and anovulation in females, and testicular atrophy and decreased spermatogenesis in males.

The biochemical responses elicited by PCBs are also numerous and include the induction of CYP1A1 and CYP1A2 gene expression and the associated monooxygenase enzyme activities, ie, aryl hydrocarbon hydroxylase (AHH) and ethoxyresorufin *O*-deethylase (EROD), and several other cytochrome P-450 dependent monooxygenases; the induction of steroid metabolizing enzymes, DT diaphorase, UDP glucuronosyl transferase, epoxide hydrolase, glutathione (*S*)-transferase, and δ-aminolevulinic acid synthetase; increased Ah receptor binding activity; decreased uroporphinogen decarboxylase activity; and decreased vitamin A levels.

Structure–Function Relationships. Since PCBs and related HAHs are found in the environment as complex mixtures of isomers and congeners, any meaningful risk and hazard assessment of these mixtures must consider the qualitative and quantitative structure–function relationships. Several studies have investigated the structure–activity relationships for PCBs that exhibit 2,3,7,8-tetrachlorodibenzo-*p*-dioxin (**1**) (TCDD)-like activity.

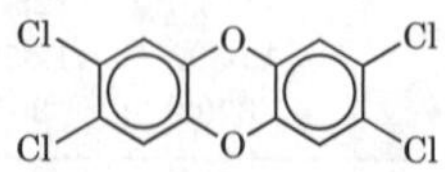

(**1**)

The data show that 3,3′,4,4′,5-pentaCB is the most toxic coplanar PCB congener and the 2,3,7,8-TCDD/3,3′,4,4′,5-pentaCB potency ratios are 66/1 (body weight loss, rat); 8.1/1 (thymic atrophy, rat); 10/1 (fetal thymic lymphoid development, mouse); 125/1 (AHH induction, rat); 3.3/1 (AHH induction, hepatoma H-4-II E cells, rat); and 100/1 (embryo hepatocytes, chick). Both the 3,3′,4,4′-tetra- and 3,3′,4,4′,5,5′-hexaCB congeners are considerably less toxic than 3,3′,4,4′,5-pentaCB and their relative potencies are highly variable.

Human Health Effects. Any assessment of adverse human health effects from PCBs should consider the route(s) of and duration of exposure; the composition of the commercial PCB products, ie, degree of chlorination; and the levels of potentially toxic PCDF contaminants. As a result of these variables, it would not be surprising to observe significant differences in the effects of PCBs on different groups of occupationally exposed workers.

Chloracne and related skin problems have been observed in several groups of workers and it was suggested that the air concentrations of commercial PCBs >0.2 mg^3 were associated with this effect.

The effects of occupational exposure to PCBs on the concentrations of several serum clinical, chemical, and hematological parameters have been reported. Mildly elevated SGOT and γ-glutamyl transpeptidase (GGTP) suggest some liver damage and induction of hepatic monooxygenase enzymes; these results are similar to those observed in animal studies. A relatively high incidence of pulmonary dysfunction in capacitor-manufacturing workers has been reported, with symptoms including coughing, 13.8%; wheezing, 3.4%; tightness in the chest, 10.1%; and upper respiratory or eye irritation, 48.2%. The pulmonary toxicity of PCBs in laboratory animals has not been widely reported.

It is apparent from most reports that workplace exposure to relatively high levels of PCBs results in limited and moderate toxicity in humans. These toxic symptoms appear to be reversible after exposure to PCBs is terminated and this is accompanied by a decline in serum levels of PCBs.

Environmental exposures to PCBs are significantly lower than those reported in the workplace and are therefore unlikely to cause adverse human health effects in adults. However, it is apparent from the results of several recent studies on children that there was a correlation between *in utero* exposure to PCBs, eg, cord blood levels, and developmental deficits, including reduced birth weight, neonatal behavior anomalies, and cognitive deficits.

Polychlorinated Naphthalenes

Polychlorinated naphthalenes (PCNs) are halogenated aromatic hydrocarbons that are no longer produced. They can be synthesized by the chlorination of naphthalene. The commercial products were graded and sold according to their chlorine content (wt %), and used as waxes and impregnants (for protective coatings), water repellents, and wood preservatives.

Animal and Human Toxicology. The mammalian toxicology of PCNs has not been studied in detail; however, it is believed that these compounds elicit mixture- and structure-dependent biochemical and toxic responses resembling those reported for PCBs and other toxic HAHs.

There have been several reported accidental exposures to commercial PCNs. One of the earliest incidents, the poisoning of cattle, was first reported in 1941 in New York State, and became known as X-Disease because of its unknown etiology. Eventually it was traced to the use of PCNs as high pressure lubricants in feed pelleting machines which resulted in contamination of the feed and ingestion of the PCNs by the animals. The symptoms exhibited by the cattle included a thickening of the skin referred to as hyperkeratosis, excess lacrimation and salivation, anorexia, depression, and a decrease in plasma vitamin A.

Human incidents have been reported in workers involved in the production or uses of PCNs. In humans the inhalation of hot vapors was the most important route of exposure and resulted in symptoms including rashes or chloracne, jaundice, weight loss, yellow atrophy of the liver, and in extreme cases, death.

Lindane and Hexachlorocyclopentadiene

Both lindane (**2**) and hexachlorocyclopentadiene (**3**) are halogenated hydrocarbons; unlike the PCBs and PCNs, they do not contain an aromatic ring.

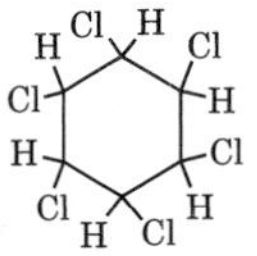

(**2**)

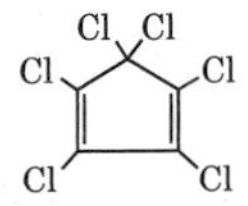

(**3**)

Chemistry and Environmental Impact. Lindane is produced by the photocatalyzed addition of chlorine to benzene to give a mixture of isomers. Lindane has been produced worldwide for its use as an insecticide and for other minor uses in veterinary, agricultural, and medical products.

The relatively high stability and lipophilicity of lindane and its global use pattern have resulted in significant environmental contamination by this hydrocarbon.

The highly reactive hexachlorocyclopentadiene is rapidly degraded in the environment and is not routinely detected as an environmental pollutant.

Animal and Human Toxicity. The acute toxicity of lindane depends on the age, sex, and animal species, and on the route of administration. Some of the toxic responses caused by lindane in laboratory animals include hepato- and nephrotoxicity, reproductive and embryotoxicity, mutagenicity in some short-term *in vitro* bioassays, and carcinogenicity. The mechanism of the lindane-induced response is not known. Only minimal data are available on the mammalian toxicities of hexachlorocyclopentadiene.

The effects of occupational exposure to lindane have been investigated extensively. These studies indicated that occupational exposure to lindane resulted in increased bodily burdens of this chemical; however, toxic effects associated with these exposures were minimal and no central nervous system disorders were observed.

STEPHEN H. SAFE
Texas A & M University

S. Safe and O. Hutzinger, eds., *Polychlorinated Biphenyls (PCBs): Mammalian and Environmental Toxicology,* Vol. 1, Springer-Verlag Publishing Co., Heidelberg, Germany, 1987.

R. D. Kimbrough and A. A. Jensen, eds., *Halogenated Biphenyls, Terphenyls, Naphthalenes, Dibenzodioxins and Related Products,* 2nd ed., Elsevier/North-Holland, Amsterdam, The Netherlands, 1989.

M. A. Q. Khan and R. H. Stanton, eds., *Toxicology of Halogenated Hydrocarbons,* Pergamon Press, Inc., Elmsford, N.Y., 1981.

K. Baumann, K. Behling, H. L. Brassow, and K. Stapel, *Int. Arch. Occup. Environ. Health* **48,** 165 (1981).

CHLOROHYDRINS

A chlorohydrin has been defined as a compound containing both chloro and hydroxyl radicals, and chlorohydrins have been described as compounds having the chloro and the hydroxyl groups on adjacent carbon atoms. Common usage of the term applies to aliphatic compounds and does not include aromatic compounds. Chlorohydrins are most easily prepared by the reaction of an alkene with chlorine and water, though other methods of preparation are possible. The principal use of chlorohydrins has been as intermediates in the production of various oxirane compounds through dehydrochlorination.

Properties

Ethylene chlorohydrin, $HOCH_2CH_2Cl$, is the simplest chlorohydrin. It is a liquid at 15°C and 101.3 kPa (1 atm) (Table 1), and is miscible with water and ethanol and slightly soluble in ethyl ether.

Table 1. Physical Properties of Ethylene Chlorohydrin

Property	Value
molecular weight	80.51
boiling point at 101.3 kPa[a], °C	128.7
melting point, °C	−67.5
viscosity at 20°C, mPa·s(= cP)	3.43
refractive index, n_D^{20}	1.4418–1.442
flash point, °C	55

[a] To convert kPa to mm Hg, multiply by 7.5.

Table 2. Physical Properties of Propylene Chlorohydrins, C_3H_7ClO

Property	2-Chloro-1-propanol	1-Chloro-2-propanol
mol wt	94.54	94.54
boiling point, °C	133–134	126–127
flash point, °C	44[a]	52

[a] ASTM D3278.

Table 2 gives physical property data for the propylene chlorohydrins, 2-chloro-1-propanol and 1-chloro-2-propanol.

3-Chloro-1,2-propanediol, $HOCH_2CHOHCH_2Cl$, has a mol wt of 110.48 and a specific gravity at 20°C of 1.3218. It is a liquid with $n_D^{20} = 1.4831$, boils at 213°C and 101.3 kPa (1 atm) with decomposition, but it can be distilled at 114–120°C at 1.87 kPa (14 mm Hg). This compound, commonly known as glycerol monochlorohydrin, is miscible in water, ethanol, ethyl ether, and acetone and is soluble in hot benzene.

3-Chloro-1,2-propanediol has a flash point of 135°C. Its heat of formation at 298 K is −525.8 kJ/mol(−125.7 kcal/mol) and the heat of combustion at constant volume is 15.2 kJ/g (3.63 kcal/g).

Physical property data for dichloropropanols or dichlorohydrins, $C_3H_6Cl_2O$, appear in Table 3. 1,2-Dichloro-3-propanol it is miscible in ethanol, ether, acetone, and benzene and is slightly soluble in H_2O.

1,3-Dichloro-2-propanol, $ClCH_2CHOHCH_2Cl$, has a vapor pressure at 28°C of 0.13 kPa (0.98 mm Hg).

Chemistry

Synthesis of Chlorohydrins. Chlorohydrins are produced by hypochlorination of olefins, from the hydrochlorination of epoxides, via enzyme technology, and from chromyl chloride reaction with olefins.

Reactions of Chlorohydrins. Reactions include dehydrochlorination to epoxides, formation of mustard gas from ethylene chlorohydrin, hydrolysis to glycols, formation of cyclic carbonates, esterification, etherification, oxidation, and quaternization.

Manufacture and Processing

For many years ethylene chlorohydrin was manufactured on a large industrial scale as a precursor to ethylene oxide, but this process has been almost completely displaced by the direct oxidation of ethylene to ethylene oxide over silver catalysts. However, because other commercially important epoxides such as propylene oxide and epichlorohydrin cannot be made by direct oxidation of the parent olefin, chlorohydrin intermediates are still important in the manufacture of these products.

Propylene Chlorohydrin. Propylene chlorohydrin is made by chlorohydrination of propylene in chlorine and water, with nonaqueous hypochlorous acid, and with *tert*-alkyl hypohalites.

Manufacture of Glycerol Monochlorohydrins. Glycerol monochlorohydrins are manufactured from allyl alcohol and from glycerol.

Manufacture of Glycerol Dichlorohydrins. Glycerol dichlorohydrins are manufactured from allyl chloride and from allyl alcohol.

Table 3. Physical Properties of Dichloropropanols

Property	1,2-Dichloro-3-propanol	1,3-Dichloro-2-propanol
mol wt	128.99	128.99
boiling point, °C	183–185	174.3
mp, °C		−4

Economic Aspects

The most important chemical reaction of chlorohydrins is dehydrochlorination to produce epoxides. In the case of propylene oxide. The Dow Chemical Company is the only manufacturer in the United States that still uses the chlorohydrin technology. Recently, Dow Europe SA announced a decision to expand its propylene oxide capacity by 160,000 metric tons per year at the Stade, Germany, site. This represents about a 40% increase over the current capacity.

Epichlorohydrin (chloromethyloxirane), which has a production capacity in the United States of 291,000 t/yr, is manufactured by the chlorohydrination of allyl chloride and subsequent dehydrochlorination of the glycerol dichlorohydrin isomers. Dow and Shell Chemical are the two producers of epichlorohydrin in the United States.

The merchant market for chlorohydrins is small, primarily for specialty applications.

Health and Safety Factors

In general, chlorohydrins are relatively toxic irritants. They are harmful if swallowed, inhaled, or absorbed through the skin. They cause irritation to the eyes, skin, mucous membrane, and upper respiratory tract.

For handling chlorohydrins, chemical safety goggles, chemical-resistant gloves, OSHA/MSHA approved respirators, and other protective clothing are required. In case of contact, one should immediately flush eyes or skin with copious amounts of water for at least 15 minutes and remove contaminated clothing and shoes. If chlorohydrins are inhaled, the person should be moved to fresh air.

Chlorohydrins are combustible and should be stored away from heat and open flame in a cool, dry place. These materials are generally incompatible with strong oxidizing agents and strong bases. Under fire conditions toxic fumes of hydrogen chloride, phosgene, and carbon monoxide may be generated.

W. Frank Richey
The Dow Chemical Company

P. Sherwood, *Petroleum Refiner* **28,** 120 (1949).

A. J. Gait, in E. G. Hancock, ed., *Propylene and Its Industrial Derivatives,* John Wiley & Sons, Inc., New York, 1973, pp. 274–279.

A. C. Fyvie, *Chem. Ind.,* 384–388 (Mar. 7, 1964).

H. H. Szmant, *Organic Building Blocks of the Chemical Industry,* John Wiley & Sons, Inc., New York, 1989, p. 281.

CHLOROPHENOLS

The chlorophenols make up an important class of industrial chemical compounds. They are used as either intermediates in the synthesis of agrochemicals, dyestuffs, and pharmaceuticals or directly in formulations.

Physical Properties

The main characteristics and physical properties of the chlorophenols are brought together in Table 1. With the exception of *o*-chlorophenol, they are all solids at room temperature.

Table 1. Physical Properties of the Chlorophenols

Compound	Mp, °C	Bp, °C	pK_a water[a]
2-chlorophenol[b]	8.7	175–176	8.5–8.52
3-chlorophenol	32.8	215–217	8.97–9
4-chlorophenol[c]	40–41	219	9.37–9.44
2,3-dichlorophenol	58	206	7.4–7.71
2,4-dichlorophenol	42.8	210	7.9–7.9
2,5-dichlorophenol	58	212–213	7.5–7.51
2,6-dichlorophenol	67	219–220	6.8–6.80
3,4-dichlorophenol	65	253	8.6–8.62
3,5-dichlorophenol	68	233	8.2–8.25
2,3,4-trichlorophenol	83.5		6.97–6.97
2,3,5-trichlorophenol	62	255	6.43
2,3,6-trichlorophenol	101	272	5.8–5.80
2,4,5-trichlorophenol	68	245–246	6.72–7.3
2,4,6-trichlorophenol	68	244.5	5.99–6.2
3,4,5-trichlorophenol	101	275	7.55–7.8
2,3,4,5-tetrachlorophenol	115–117		5.64–5.64
2,3,4,6-tetrachlorophenol	69–70		5.22–5.22
2,3,5,6-tetrachlorophenol	115		5.02–5.03
pentachlorophenol	190	309–310	4.74–4.8

[a] At 25°C.
[b] pK_a in pyridine = 12.1.
[c] pK_a in DMSO = 16.1.

Preparation

Chlorination of Phenols. Industrially, the phenols are chlorinated without solvent. Chlorine reacts rapidly with phenol and with the chlorophenols, which makes it difficult to determine the relative reaction rates because of the superchlorination that sometimes results from an unsatisfactory chlorine dispersion.

Hydrodechlorination. The polychlorophenols can be broken back down into lighter chlorophenols by catalytic hydrogenation with Pd, CO in liquid or in gaseous phase.

Sandmeyer Reaction. This general reaction allows the phenol function to be introduced.

Polyhalogenobenzene Hydrolysis. The chlorobenzenes can be transformed into chlorophenols by hydrolysis in a liquid-phase basic medium.

Sulfonation–Desulfonation of Chlorobenzenes. Sulfonation of chlorobenzenes can also be used to produce chlorophenols. Sulfonation is carried out at 60–80°C using oleum at 15–20%. The subsequent desulfonation usually calls on aqueous alkali solutions at 15–20% at temperatures between 170 and 230°C.

Health and Safety

Effects in Humans. In chlorophenol production, irritation symptoms of the nose, eyes, respiratory tract, and skin resulting in chloroacne have been observed. The results of epidemiology studies on the long-term effects of chlorophenols are quite contradictory and have not allowed the experts to reach any firm conclusions.

Economic Aspects

Overall, the chlorophenol market is in decline. Rhône-Poulenc, with a capacity of around 20,000 t/yr, is the world's leading producer of light chlorophenols. Excluding the unknown factors for which no statistics are available (China, Russia), the market for pentachlorophenol can be estimated at ~25,000 t/yr. The principal producers of pentachlorophenol are Vulcan (USA), Idacon (USA), Rhône-Poulenc, and Chapman.

JEAN-ROGER DESMURS
Rhône-Poulenc Recherches
SERGE RATTON
Rhône-Poulenc Specialités Chimiques

J. Drahonovsky and Z. Vacek, *Collect. Czech. Chem. Commun.* **36,** 3431 (1971).
S. Li, M. Paleologou, and W. C. Purdy *J. Chromatogr. Sci.* **29,** 66 (1991).
M. Nigretto and M. Josefowicz, *Electrochim. Acta* **18,** 148 (1973).

CHLOROPHYLL. See DYES, NATURAL.

CHLOROPRENE. See CHLOROCARBONS AND CHLOROHYDROCARBONS.

CHLOROSULFURIC ACID

Chlorosulfuric acid, $ClSO_3H$, is a clear to straw-colored liquid with a pungent odor. It is a highly reactive compound that reacts violently with water to produce heat and dense white fumes of hydrochloric acid and sulfuric acid. Chlorosulfuric acid reacts with most organic materials, in some cases with charring. It is used principally in organic synthesis as a sulfating, sulfonating, or chlorosulfonating agent. The main application for chlorosulfuric acid is as an intermediate in the production of synthetic detergents, drugs, and dyestuffs. This acid is preferred in many applications because it yields the desired isomers. It has also been used as a smoke-forming agent in warfare.

Physical Properties

The physical property values given in Table 1 are considered to be the most reliable available.

Chemical Properties

Chlorosulfuric acid is a strong acid containing a relatively weak sulfur–chlorine bond. Many salts and esters of chlorosulfuric acid are known, most of them are relatively unstable or hydrolyze readily in moist air.

Strong dehydrating agents such as phosphorous pentoxide or sulfur trioxide convert chlorosulfuric acid to its anhydride, pyrosulfuryl chloride.

In organic reactions, chlorosulfuric acid is a powerful sulfating and sulfonating agent, a fairly strong dehydrating agent, and a specialized chlorinating agent.

Manufacture

Modern plants manufacture chlorosulfuric acid by direct union of equimolar quantities of sulfur trioxide and dry hydrogen chloride gas.

Table 1. Physical Properties of Chlorosulfuric Acid

Property	Value
mol wt	116.531
mp, °C	−81 to −80
bp, °C	151–152
vapor pressure,[a] in Pa at 30°C	1.2×10^2
specific gravity at 15.6°C	1.752
viscosity, mPa·s(= cP)	
at −31.6 °C	10.0
at 15.6°C	3.0
dielectric constant at 15°C	60 ± 10

[a] To convert Pa to mm Hg, multiply by 0.0075.

Economic Aspects

There are 20 manufacturers of chlorosulfuric acid in Europe, Asia, and Australia, plus manufacturers in Brazil and Mexico. The two United States manufacturers are E. I. du Pont de Nemours & Co., Inc. having a capacity in excess of 30,000 t/yr, and Gabriel Chemical Co. having a capacity of 13,600 t/yr. The United States and Canadian consumption is about 27,000 t/yr.

Detergent and other surfactants manufacturing is the leading consumer of chlorosulfuric acid at approximately 40%; pharmaceuticals is next at 20%.

Health and Safety Factors

Safety. Chlorosulfuric acid is a strong acid and the principal hazard is severe chemical burns when the acid comes into contact with body tissue. The vapor is also hazardous and extremely irritating to the skin, eyes, nose, and respiratory tract. Exposure limits for chlorosulfuric acid have not been established by OSHA or ACGIH. However, chlorosulfuric acid fumes react readily with moisture in the air to form hydrochloric and sulfuric acid mists, which do have established limits. The OSHA 8-h TWA limits and ACGIH TLV–TWA limits are sulfuric acid = 1 mg/m^3; hydrochloric acid = 5 ppm or 7 mg/m^3 (ceiling limit).

Fire Hazard. Although chlorosulfuric acid itself is not flammable, it may cause ignition by contact with combustible materials because of the heat of reactions.

Storage and Handling. All lubricants and packing materials in contact with chlorosulfuric acid must be chemically resistant to the acid.

Spills and Waste Disposal. Depending on the magnitude of the spill, control can be achieved by absorption into absorbents such as diatomaceous earth, sand, or limestone.

Fuming is reduced by dilution with large volumes of water or by use of foam to blanket the spill.

C. E. McDONALD
E. I. du Pont de Nemours & Co., Inc.

J. W. Mellor, *A Comprehensive Treatise on Inorganic and Theoretical Chemistry,* Vol. 10, Longmans, Green & Co., London, 1920, pp. 684–692.

CHOCOLATE AND COCOA

In the United States, chocolate and cocoa are standardized by the U.S. Food and Drug Administration under the Federal Food, Drug, and Cosmetic Act. The current definitions and standards resulted from prolonged discussions between the U.S. chocolate industry and the Food and Drug Administration (FDA). The definitions and standards originally published in the *Federal Register* of December 6, 1944, have been revised only slightly.

The FDA announced in the *Federal Register* of January 25, 1989 a proposal to amend the U.S. chocolate and cocoa standards of identity. The proposed amendments respond principally to a citizen petition submitted by the Chocolate Manufacturers Association (CMA) and, to the extent practicable, will achieve consistency with the Codex standards. The proposed amendments would allow for the use of nutritive carbohydrate sweeteners, neutralizing agents, and emulsifiers; reduce slightly the minimum milkfat content and eliminate the nonfat milk solids-to-milkfat ratios in certain cocoa products including milk chocolate; update the language and format of the standards; and provide for optional ingredient labeling requirements. FDA has also received a proposal to establish a new standard of identity for white chocolate. Comments regarding the proposal amendments are under review by FDA, and a final ruling is expected to be issued in the near future.

White Chocolate. There is at present no standard of identity in the United States for white chocolate.

White chocolate has been defined by the European Economic Community (EEC) Directive 75/155/EEC as free of coloring matter and consisting of cocoa butter (not less than 20%); sucrose (not more than 55%); milk or solids obtained by partially or totally dehydrated whole milk, skimmed milk, or cream (not less than 14%); and butter or butter fat (not less than 3.5%).

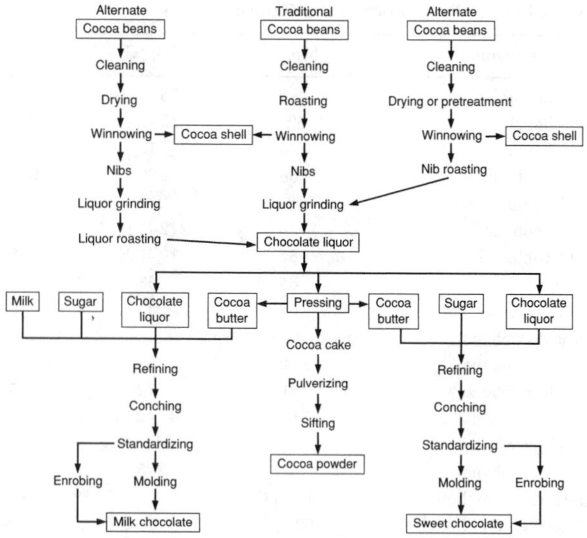

Figure 1. Flow diagram of chocolate and cocoa production.

Cocoa Beans

The cocoa bean is the basic raw ingredient in the manufacture of all cocoa products. The beans are converted to chocolate liquor, the primary ingredient from which all chocolate and cocoa products are made. Figure 1 depicts the conversion of cocoa beans to chocolate liquor, and in turn to the chief chocolate and cocoa products manufactured in the United States, ie, cocoa powder, cocoa butter, and sweet and milk chocolate.

Significant amounts of cocoa beans are produced in about 30 different localities. These areas are confined to latitudes 20° north or south of the equator. Although cocoa trees thrive in this very hot climate, young trees require the shade of larger trees such as banana, coconut, and palm for protection.

Fermentation (Curing). Prior to shipment from producing countries, most cocoa beans undergo a process known as curing, fermenting, or sweating. These terms are used rather loosely to describe a procedure in which seeds are removed from the pods, fermented, and dried. Unfermented beans, particularly from Haiti and the Dominican Republic, are used in the United States.

Commercial Grades. Most cocoa beans imported into the United States are one of about a dozen commercial varieties that can be generally classified as Criollo or Forastero. Criollo beans have a light color, a mild, nutty flavor, and an odor somewhat like sour wine. Forastero beans have a strong, somewhat bitter flavor and various degrees of astringency. The Forastero varieties are more abundant and provide the basis for most chocolate and cocoa formulations. The main varieties of cocoa beans imported into the United States, usually named for the country or port of origin, are Ivory Coast, Accra (Ghana), Lagos, Nigeria, Fernando Po, and Sierra Leone (from Africa); Bahia (Brazil), Arriba (Ecuador), and Venezuelan (from South America); Malaysia, New Guinea, Indonesia, and Samoa (in the Pacific); and Sanchez (Dominican Republic), Grenada, and Trinidad (in the West Indies).

Blending. Most chocolate and cocoa products consist of blends of beans chosen for flavor and color characteristics.

Production. Worldwide cocoa bean production ranged from 2.3–2.5-million t and cocoa bean production was stagnant between 1988 and 1993. Indonesia was the only country significantly expanding production. Production in Brazil and Malaysia actually dropped.

Consumption. Worldwide cocoa bean consumption increased by 14% between 1988 and 1993 from approximately 2.1 million t in the 1988–1989 crop year to almost 2.4 million t today. North America and Western Europe increased grind by approximately 26% over this time period, whereas in Russia and Eastern Europe grind dropped by 46%.

Marketing. Most of the cocoa beans and products imported into the United States are done so by New York and London trade houses. The New York Sugar, Coffee, and Cocoa Exchange provides a mechanism by which both chocolate manufacturers and trade houses can hedge their cocoa bean transactions.

Chocolate Liquor

Chocolate liquor is the solid or semisolid food prepared by finely grinding the kernel or nib of the cocoa bean. It is also commonly called chocolate, unsweetened chocolate, baking chocolate, or cooking chocolate. In Europe chocolate liquor is often called chocolate mass or cocoa mass.

Cocoa Powder

Cocoa powder (cocoa) is prepared by pulverizing the remaining material after part of the fat (cocoa butter) is removed from chocolate liquor. The U.S. chocolate standards define three types of cocoas based on their fat content. These are breakfast, or high fat cocoa, containing not less than 22% fat; cocoa, or medium fat cocoa, containing less than 22% fat but more than 10%; and low fat cocoa, containing less than 10% fat.

Cocoa powder production today is an important part of the cocoa and chocolate industry because of increased consumption of chocolate-flavored products. Cocoa powder is the basic flavoring ingredient in most chocolate-flavored cookies, biscuits, syrups, cakes, and ice cream. It is also used extensively in the production of confectionery coatings for candy bars.

Cocoa Butter

Cocoa butter is the common name given to the fat obtained by subjecting chocolate liquor to hydraulic pressure. It is the main carrier and suspending medium for cocoa particles in chocolate liquor and for sugar and other ingredients in sweet and milk chocolate.

The FDA has not legally defined cocoa butter, and no standard exists for this product under the U.S. Chocolate Standards. For the purpose of enforcement, the FDA defines cocoa butter as the edible fat obtained from cocoa beans either before or after roasting. Cocoa butter as defined in the *U.S. Pharmacopeia* is the fat obtained from the roasted seed of *Theobroma cacao Linne.*

Composition and Properties. Cocoa butter is a unique fat with specific melting characteristics. It is a solid at room temperature (20°C), starts to soften around 30°C, and melts completely just below body temperature. Its distinct melting characteristic makes cocoa butter the preferred fat for chocolate products.

Cocoa butter is composed mainly of glycerides of stearic, palmitic, and oleic fatty acids (see FATS AND FATTY OILS). The triglyceride structure of cocoa butter has been determined as tri-saturated, 3%; mono-unsaturated (oleo-distearin), 22%; oleo-palmitostearin, 57%; oleo-dipalmitin, 4%; di-unsaturated (stearo-diolein), 6%; palmito-diolein, 7%; and tri-unsaturated, tri-olein, 1%.

Although there are actually six crystalline forms of cocoa butter, four basic forms are generally recognized as alpha, beta, beta prime, and gamma.

Substitutes and Equivalents. In the past 25 years, many fats have been developed to replace part or all of the added cocoa butter in chocolate-flavored products. These fats fall into two basic categories commonly known as cocoa butter substitutes and cocoa butter equivalents.

Cocoa butter substitutes and equivalents differ greatly with respect to their method of manufacture, source of fats, and functionality; they are produced by several physical and chemical processes. Cocoa butter substitutes are produced from lauric acid fats such as coconut, palm, and palm kernel oils by fractionation and hydrogenation; from domestic fats such as soy, corn, and cotton seed oils by selective hydrogenation; or from palm kernel stearines by fractionation. Cocoa butter equivalents can be produced from palm kernel oil and other specialty fats such as shea and illipe by fractional crystallization; from glycerol and selected fatty acids by direct chemical synthesis; or from edible beef tallow by acetone crystallization.

In the early 1990s, the most frequently used cocoa butter equivalent in the United States was derived from palm kernel oil but a synthesized product was expected to be available in the near future.

Sweet and Milk Chocolate

Most chocolate consumed in the United States is consumed in the form of milk chocolate and sweet chocolate. Sweet chocolate is chocolate liquor to which sugar and cocoa butter have been added. Milk chocolate contains these same ingredients and milk or milk solids.

U.S. definitions and standards for chocolate are quite specific. Sweet chocolate must contain at least 15% chocolate liquor by weight and must be sweetened with sucrose or mixtures of sucrose, dextrose, and corn syrup solids in specific ratios. Semisweet chocolate and bittersweet chocolate, though often referred to as sweet chocolate, must contain a minimum of 35% chocolate liquor. The three products, sweet chocolate, semisweet chocolate, and bittersweet chocolate, are often simply called chocolate or dark chocolate to distinguish them from milk chocolate.

Sweet chocolate can contain milk or milk solids (up to 12% max), nuts, coffee, honey, malt, salt, vanillin, and other spices and flavors as well as a number of specified emulsifiers. Many different kinds of chocolate can be produced by careful selection of bean blends, controlled roasting temperatures, and varying amounts of ingredients and flavors.

The most popular chocolate in the United States is milk chocolate. The U.S. Chocolate Standards state that milk chocolate shall contain no less than 3.66 wt % of milk fat and not less than 12 wt % of milk solids. In addition, the ratio of nonfat milk solids to milk fat must not exceed 2.43:1 and the chocolate liquor content must not be less than 10% by weight.

Production. The main difference in the production of sweet and milk chocolate is that in the production of milk chocolate, water must be removed from the milk. Many milk chocolate producers in the United States use spray-dried milk powder. Others condense fresh whole milk with sugar, and either dry it, producing milk crumb, or blend it with chocolate liquor and then dry it, producing milk chocolate crumb. These crumbs are mixed with additional chocolate liquor, sugar, and cocoa butter later in the process. Milk chocolates made from crumb typically have a more caramelized milk flavor than those made from spray-dried milk powder.

Theobromine and Caffeine

Chocolate and cocoa products, like coffee, tea, and cola beverages, contain alkaloids. The predominant alkaloid in cocoa and chocolate products is theobromine, though caffeine is also present in smaller amounts. Concentrations of both alkaloids vary depending on the origin of the beans. Published values for the theobromine and caffeine content of chocolate vary widely because of natural differences in cocoa beans and differences in analytical methodology.

Nutritional Properties of Chocolate Products

Chocolate and cocoa products supply proteins, fats, carbohydrates, vitamins, and minerals. The Chocolate Manufacturers' Association of the United States (McLean, Virginia) completed a nutritional analysis from 1973 to 1976 of a wide variety of chocolate and cocoa products representative of those generally consumed in the United States.

Economic Aspects

Chocolate consumption (wholesale Dollar value) on a global basis was approximately \$23 billion in 1992. In the United States, Hershey, Mars, and Nestlé control about 70% of the market.

The leading chocolate companies continue to pursue a global confectionery business strategy with an increase in the early 1990s of confectionery business activity in the Eastern Bloc countries, Russia, China, and South America. Generally, as per capita income increases, chocolate consumption increases and sugar consumption decreases. Consumer demographics, the declining child population, and the increase in consumer awareness of health issues play important roles in the economics of chocolate consumption. Chocolate confectionery business trends during the early 1990s include product down-sizing leading to snack size finger foods, increased emphasis on specialty chocolates with concentration on dessert chocolates, and chocolate brand equity spread into beverages, baked goods, frozen novelties, and even sugar confections.

B. L. ZOUMAS
J. F. SMULLEN
Hershey Foods Corporation

W. T. Clarke, *The Literature of Cacao,* ACS, Washington, D.C., 1954.

L. R. Cook, *Chocolate Production and Use,* Magazines for Industry, Inc., New York, 1972.

B. W. Minifie, *Chocolate, Cocoa, and Confectionery: Science and Technology,* AVI, Westport, Conn., 1970.

E. M. Chatt, in Z. J. Kertesz, ed., *Economic Crops,* Vol. 3, Interscience Publishers, Inc., New York, 1953, p. 185.

CHOLINE

Choline base, $[(CH_3)_3NCH_2CH_2OH]^+OH^-$, trimethyl(2-hydroxyethyl)-ammonium hydroxide, derives its name from bile (Greek *cholē*) which it was first obtained. This so-called free-choline is a colorless, hygroscopic liquid with an odor of trimethylamine. The quarternary ammonium compound (1) choline or a precursor is needed in the diet as a constituent of certain phospholipids universally present in protoplasm.

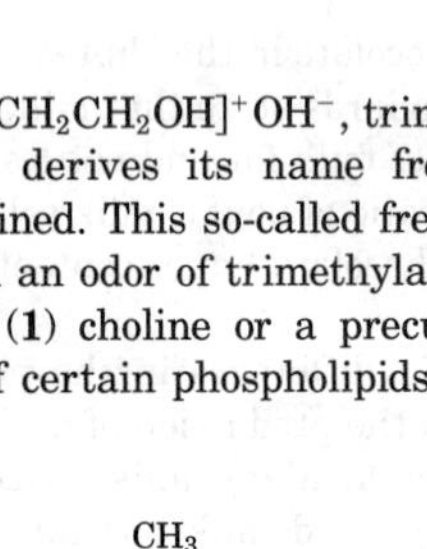

(1)

This makes choline an important nutritional substance. It is also of great physiological interest because one of its esters, acetylcholine, appears to be responsible for the mediation of parasympathetic nerve impulses and has been postulated to be essential to the transmission of all nerve impulses. Choline is used clinically in liver disorders and as a constituent in animal feeds.

Physical and Chemical Properties

Choline is a strong base ($pK_B = 5.06$ for 0.0065–0.0403 M solutions). It crystallizes with difficulty and is usually known as a colorless deliquescent syrupy liquid, which absorbs carbon dioxide from the atmosphere. Choline is very soluble in water and in absolute alcohol but insoluble in ether.

Biological Functions

In nutrition, the most important function of choline appears to be the formation of lecithin (phosphatidylcholine) (2) and other choline-containing phospholipids.

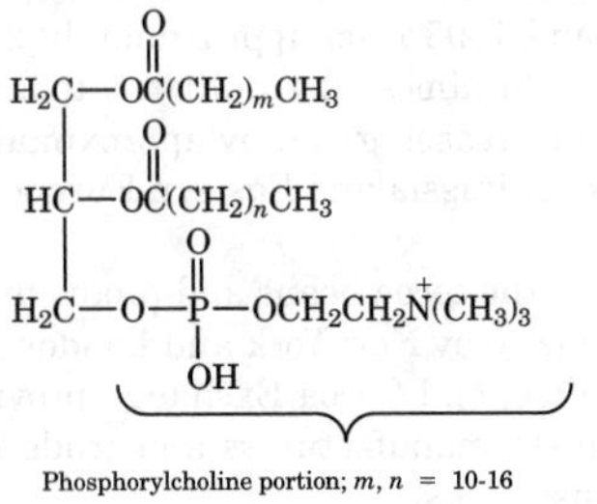

(2)

Occurrence

Choline occurs widely in nature and, prepared synthetically, it is available as an article of commerce. Soybean lecithin and egg-yolk lecithin have been used as natural sources of choline for supplementing the diet. Other important natural-food sources include liver and certain legumes.

Preparation

Choline is not usually encountered as the free base but as a salt, most commonly, the chloride, $[(CH_3)_3N(CH_2CH_2OH)]^+Cl^-$.

Choline is made from the reaction of trimethylamine with ethylene oxide or ethylene chlorohydrin.

Choline Salts

Choline salts include choline chloride, choline dihydrogen citrate, tricholine citrate concentrate, and choline bitartrate.

Economic Aspects

The world market for choline chloride used in animal feeds is estimated at 113,000 t on a 100% basis. The market for good grade choline chloride is a small market by comparison and is utilized mainly in the supplementation of infant formulas. Other choline salts are utilized solely in the human vitamin supplementation markets and are also small compared to animal feed usage.

There are nine primary producers of choline chloride within the world: DuCoa, (U.S.), Bioproducts, Inc. (U.S), Chinook, (Canada), I.C.I. (U.K.), U.C.B. (Belgium), Akzo (the Netherlands), BASF (Germany), and Mitsubishi (Japan).

Derivatives

Important derivatives of choline are acetylcholine, acetyl-β-methylcholine, and carbamylcholine. Many other choline derivatives have been synthesized and studied, but have not been found satisfactory for clinical use.

CURTIS E. GIDDING
DuCoa

S. Budavari, ed., *The Merck Index,* 11th ed., Merck & Co., Inc., Rahway, N.J., 1989, p. 342.

The United States Pharmacopeia, 22nd rev. ed. (USP XXII), The United States Pharmacopeial Convention, Inc., Rockville, Md., 1990, p. 1736.

Choline: Functions and Requirements, DuCoa, Highland, Ill., 1992.

CHROMATOGRAPHY

Chromatography is a technique for separating and quantifying the constituents of a mixture. Separation techniques are essential for the characterization of the mixtures that result from most chemical processes. Chromatographic analysis is used in many areas of science and engineering: in environmental studies, in the analysis of art objects, in industrial quality control, in analysis of biological materials, and in forensics (see BIOPOLYMERS–ANALYTICAL TECHNIQUES; FINE ART EXAMINATION AND CONSERVATION; FORENSIC CHEMISTRY). Most chemical laboratories employ one or more chromatographs for routine analysis.

Chromatographic separations rely on fundamental differences in the affinity of the components of a mixture for the phases of a chromatographic system. Thus chromatographic parameters contain information on the fundamental quantities describing these interactions and these parameters may be used to deduce stability constants, vapor pressures, and other thermodynamic data appropriate to the processes occurring in the chromatograph.

Principles

The principle of chromatographic separation is quite straightforward. A mixture is allowed to come into contact with two phases, one referred to as the stationary phase and the other as the mobile phase. The stationary phase is contained in a column or sheet through which the mobile phase moves in a controlled manner relative to the stationary phase, carrying with it any material that may prefer to mix with it.

Because of differences in affinity of the mixture's constituents for the mobile phase, as compared to the stationary phase, these constituents tend to be swept along with the mobile phase at different rates. This selective interaction is known as partitioning. In adsorption chromatography the constituents in the dissolved sample compete with the mobile phase for the active sites on the stationary phase. To remove constituents adsorbed on the stationary phase, the mobile-phase chromatographic strength is increased by modifying the mobile phase to have a greater affinity for the stationary phase than the adsorbed sample constituents. To determine the effects of adsorption or partitioning on retention of substances on the column, a detector measures either the time required to travel a given distance or the distance traveled in a fixed time. The detector may be as simple as the human nose or the human eye or as complex as a microsensor. A plot of detector response versus time of travel for a fixed distance is called a chromatogram. In preparative chromatography a device may be attached to the end of a column to collect the separated components of a mixture.

The nature of the stationary and mobile phases in a particular chromatographic experiment determines the efficacy of component separation in a particular mixture. A wide variety of combinations of stationary and mobile phases is used as is shown in Figure 1. The stationary phase may be a solid or a liquid supported on a solid. The mobile phase may be a gas, a liquid, or a material such as a supercritical fluid. A particular chromatographic technique is specified by naming the mobile phase, followed by the stationary. Thus, gas–liquid chromatography (glc) is a system that uses a gaseous mobile phase in contact with a film of liquid stationary phase.

Development of the Chromatogram. The term development describes the process of performing a chromatographic separation. There are several ways in which separation may be made to occur,

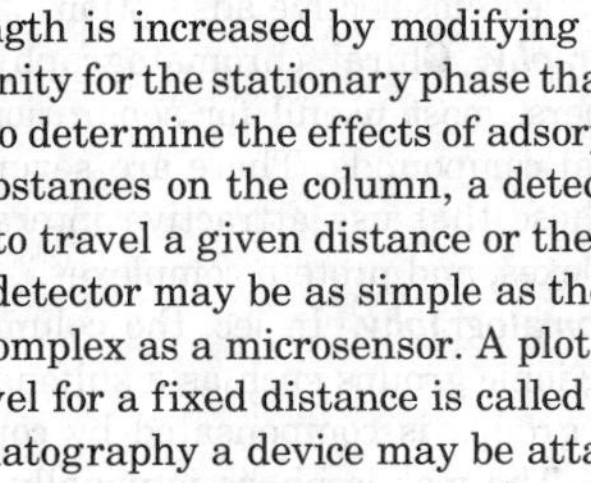

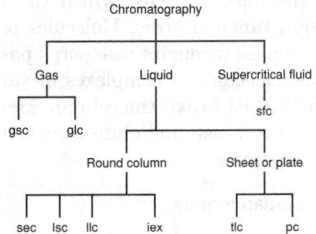

Figure 1. Classification of chromatographic systems where gsc is gas–solid chromatography; glc, gas–liquid chromatography; sec, size-exclusion chromatography; lsc, liquid–solid chromatography; llc, liquid–liquid chromatography; iec, ion-exchange chromatography; tlc, thin-layer chromatography; pc, paper chromatography; and sfc, supercritical-fluid chromatography.

eg, frontal, displacement, and elution chromatography. Frontal chromatography uses a large quantity of sample and is usually unsuited to analytical procedures. In displacement and elution chromatography, much smaller amounts of material are used.

Of the three principal types of chromatography, elution is by far the most common. Frontal chromatography is a technique in which the sample is introduced onto a column continuously. In essence the sample that is collected at the end of the column is the mobile phase, free of materials that adsorb/absorb on the stationary phase. Once the bed, ie, the stationary phase, is saturated and can no longer remove the impurities, the material coming off the column contains these materials. Using an appropriate detector, the condition at which this transition occurs can be determined, thereby determining the capacity of the column. This technique is called frontal chromatography or frontal analysis. Figure 2**a** shows an integral chromatogram from a frontal analysis.

In displacement chromatography a small sample is displaced by the much more strongly held mobile phase, so that the sample is gradually pushed through the column as the mobile phase advances. As this happens, the components are dispersed into bands that can then either be excised to obtain the pure material or displaced from the column. Such techniques are useful for the generation of quantities of pure material, particularly in application to problems in environmental analyses. Figure 2**b** shows an integral chromatogram obtained by displacement analysis.

Both frontal and displacement chromatographies suffer a significant disadvantage in that once a column has been used, part of the sample remains on the column. The column must be regenerated before reuse. In elution chromatography all of the sample material is usually removed from the column during the chromatographic process, allowing reuse of the column without regeneration. Most analytical applications of chromatography employ elution methods where a small sample is put onto the column, at the column head as a plug or a band. The sample is applied, sometimes by injection, while a stream of eluent, the mobile phase, is moving through the column. Because of the difference in affinities of the sample's components for the stationary phase, constituents travel through the column at different rates and elute at different times. Figure 2**c** shows a typical differential elution chromatogram.

Gas Chromatography

The most frequently used chromatographic technique is gas chromatography (gc).

Gas chromatography, depending on the stationary phase, can be either gas–liquid chromatography (glc) or gas–solid chromatography (gsc). The former is the most commonly used. Separation in a gas–liquid chromatograph arises from differential partitioning of the sam-

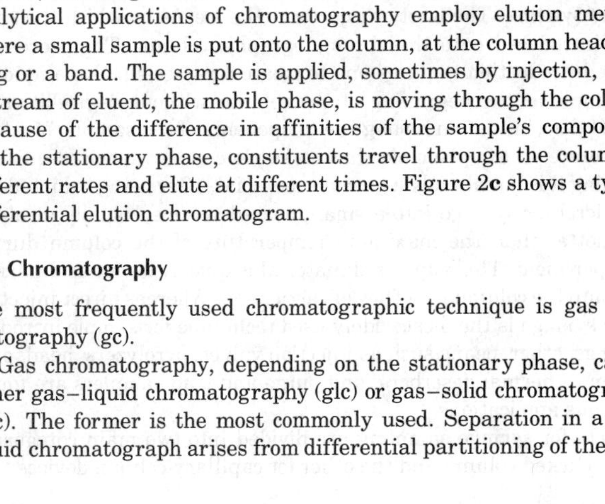

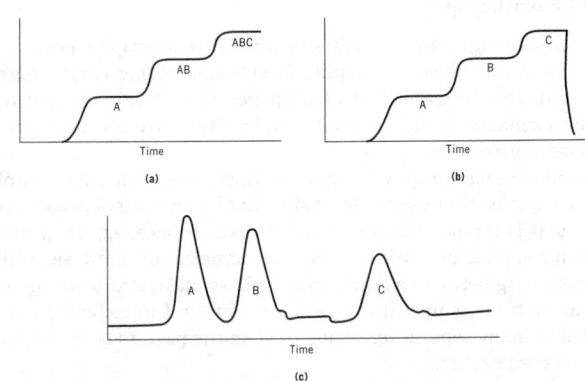

Figure 2. Chromatograms of a mixture containing three components A, B, and C, where A is less sorbed than B, and B is less sorbed than C; (**a**) frontal analysis; (**b**) displacement analysis; and (**c**) differential elution chromatogram.

ple's components between the stationary liquid phase adsorbed on a porous solid, and the gas phase. Separation in a gas–solid chromatograph is the result of preferential adsorption on the solid or exclusion of materials by size.

A second way of classifying gas chromatographic separations is by use. If the desired result is an analysis of the sample, the technique is analytical gas chromatography. If the desired result is the production of purer materials through fractionation in the chromatographic column, the technique is known as preparative gas chromatography. If the analytical chromatography is performed as part of the control of a manufacturing process, the technique is known as process gas chromatography.

Topic. Most theoretical models of gas chromatographic processes are based on analogy to processes such as distillation (qv) or countercurrent extraction experiments. The separation process is viewed as a type of successive partitioning of the components of a mixture between the stationary and mobile phases similar to the partitioning that occurs in distillation columns. In those experiments an important parameter is the number of theoretical plates of which the column may be considered to be composed; the greater the number of theoretical plates, the greater the efficiency of the column for achieving separations of similar components. In gas chromatography, the equivalent measure of efficacy is the height equivalent to theoretical plates (HETP), which measures the ultimate ability of the column to separate like components. This quantity depends on many instrumental parameters such as wall or particle diameter, type of carrier gas, flow rate, liquid-phase thickness, etc. The theoretical expression relating these various parameters is called the van Deemter equation, which relates the change in efficiency to the flow: $HETP = A + B/\mu + C\mu$, where A is the eddy diffusion or multipath effect term, B is the molecular diffusion term, μ is the linear gas velocity or flow rate, and C is the resistance to mass transfer term.

Inlet Systems. The inlet or injector is the means by which the sample is introduced onto the chromatographic column. The process of sample introduction requires one to create a representative aliquot of the sample at the beginning of the column without degradation or without discrimination among the components of the sample. Most inlets operate on the principle that a sample can be vaporized quickly, assuming it is not already a gaseous material, after being squirted out of a microliter syringe into a small, heated volume, usually at about 50°C hotter than the maximum temperature of the column during the experiment. This vaporized material is quickly swept as a narrow band onto the column by a flow of carrier gas. Whereas direct injection from a syringe is the most widely used technique for sample introduction in gc, other means such as injection valves, pyrolyzers, headspace samplers, thermal desorbers, and purge-and-trap samplers are found in various applications.

Inlets for syringe sampling are divided into two main categories: one for packed-column and the other for capillary-column devices.

Liquid Chromatography

Liquid chromatography (lc) refers to any chromatographic process in which the mobile phase is a liquid. Traditional column chromatography, thin-layer chromatography (tlc), paper chromatography (pc), and high performance liquid chromatography (hplc) are all members of this class of processes.

Liquid chromatography is complementary to gas chromatography because samples that cannot be easily handled in the gas phase, such as nonvolatile compounds or thermally unstable ones, eg, many natural products, pharmaceuticals, and biomacromolecules, are separable by partitioning between a liquid mobile phase and a stationary phase, often at ambient temperature. Developments in the technology of lc have led to many separations, done by gc in the past, to be carried out by liquid chromatography.

An advantage of liquid chromatography is that the composition of the mobile phase, and perhaps of the stationary phase, can be varied during the experiment to provide greater efficacy of the separation.

In classical column chromatography the usual system consists of a polar adsorbent, or stationary phase, and a nonpolar solvent, mobile phase, such as a hydrocarbon. In practice, the situation is often reversed, in which case the technique is known as reversed-phase lc.

Paper chromatography originated in the 1940s and tlc in the 1950s. In these techniques a chamber is usually used to isolate the column, which is a piece of filter paper in pc and a glass plate coated with an adsorbent such as silica gel in tlc, from the laboratory environment. The chromatogram is developed by allowing a mobile phase to creep through the column, carrying with it materials soluble to various extents. After this process has proceeded sufficiently, the column is removed from the solvent tank and the mobile phase evaporated. The separated components are visualized elsewhere, for example under an ultraviolet lamp in which various fluorescent bands indicate how fluorescent materials are separated by the movement of the solvent. Paper chromatography is sometimes described as a type of liquid–liquid chromatography, because the paper inherently contains bound water that acts as a stationary liquid phase. In tlc the usual mechanism for separation is partitioning resulting from adsorption on the stationary phase.

Paper and thin-layer chromatography may be further classified as either one- or two-dimensional, by direction, eg, ascending, descending, ascending/descending, and by capacity as either analytical or preparative.

Affinity Chromatography. This technique involves the use of a bioselective stationary phase placed in contact with the material to be purified, the ligate. Because of its rather selective interaction, sometimes called a lock-and-key mechanism, this method is more selective than other lc systems based on differential solubility. Affinity chromatography is sometimes called bioselective adsorption.

Chiral Chromatography. Chiral chromatography is used for the analysis of enantiomers, most useful for separations of pharmaceuticals and biochemical compounds. There are several types of chiral stationary phases: those that use attractive interactions, metal ligands, inclusion complexes, and protein complexes.

Ion-Exchange Chromatography. In iec, the column contains a stationary phase having ionic groups such as a sulfonate or carboxylate. The charge of these groups is compensated by counterions such as sodium or potassium. The mobile phase is usually an ionic solution, eg, sodium chloride, having pH and salt concentrations that act as the separation variables, having ions similar to the counterions.

Ion chromatography (ic), a novel form of ion-exchange chromatography, is a technique in which a weak ion-exchange column is used for separation. Detection is usually done conductimetrically.

Ion-pair chromatography (ipc), a variant of iec, is also sometimes called paired-ion chromatography (pic), soap chromatography, extraction chromatography, or chromatography with a liquid ion exchanger. In this technique the mobile phase consists of a solution of an aqueous buffer and an organic cosolvent containing an ion of charge opposite to the charge on the sample ion. The sample ion and the solvated ion form an ionic pair that is soluble in the stationary phase. Thus retention is determined by the ability to form the ion pair as well as the solubility of the complex in the stationary phase.

In size-exclusion chromatography (sec) or gel-permeation chromatography (gpc), the material with which the column is packed has pores in a certain range of sizes. Molecules or solvent-molecule complexes too large to pass through these pores pass rapidly through the column, whereas molecules or complexes of suffiently small size are retained and are the last to exit the column. Molecules of intermediate size are partially retained and elute from the chromatographic column at intermediate times.

Supercritical-Fluid Chromatography

Supercritical-fluid chromatography (sfc), is the link between gc and lc, because its mobile phase, a supercritical fluid, has physicochemical properties intermediate between a gas and a liquid. The physicochemical properties of the mobile phase are strong factors determining the selectivity, sensitivity toward a component, and efficiency of separation in the chromatographic process. Carbon dioxide is the mobile phase most often used in sfc.

Sfc can be performed with either capillary or lc-like packed columns. Carbon dioxide is compatible with chromatographic hardware, is readily available, and is noncorrosive. The most important detector for sfc is the flame-ionization detector because the mobile phase does not give a significant background signal. Recent applications of sfc include separations in fields as diverse as natural products, drugs, foods, pesticides, herbicides, surfactants, and polymers. These are a direct result of the advantages that sfc has over other forms of chromatography because of low operating temperature, selective detection, and sensitivity to molecular weight.

MARY A. KAISER
E. I. du Pont de Nemours & Co., Inc.
CECIL DYBOWSKI
University of Delaware

L. R. Snyder and J. J. Kirkland, *Introduction to Modern Liquid Chromatography,* Wiley-Interscience, John Wiley & Sons, Inc., New York, 1989.

B. Fried and J. Sherma, *Thin-Layer Chromatography, Techniques and Applications,* Marcel Dekker, New York, 1986.

H. F. Walton and R. D. Rocklin, *Ion Exchange in Analytical Chemistry,* CRC Press, Boca Raton, Fla., 1990.

R. L. Grob, *The Modern Practice of Gas Chromatography,* 3rd ed., John Wiley & Sons, Inc., New York, 1994.

CHROMIUM AND CHROMIUM ALLOYS

Chromium, Cr, also loosely called chrome, is the twenty-first element in relative abundance with respect to the earth's crust, ranking with V, Zn, Ni, Cu, and W, yet is the seventh most abundant element overall because Cr is concentrated in the earth's core and mantle. It has atomic number 24 and belongs to Group 6 (VIB) of the Periodic Table and is positioned between vanadium and manganese. Other Group 6 members are molybdenum and tungsten. On a tonnage basis, chromium ranks fourth among the metals and thirteenth of all mineral commodities in commercial production.

Occurrence and Mining

The only commercial ore, chromite, which is also called chromite ore, chrome ore, and chrome, has the ideal composition $FeO\cdot Cr_2O_3$, ie, 68 wt % Cr_2O_3, 32 wt % FeO, or ca 46 wt % chromium. Actually the Cr:Fe ratio varies considerably and the ores are better represented as $(Fe,Mg)O\cdot(Cr,Fe,Al)_2O_3$.

Chromite deposits occur in olivine- and pyroxene-type rocks and derivatives. Geologically these appear in stratiform deposits several feet thick covering a very wide area and are usually mined by underground methods in such countries as South Africa, Zimbabwe, India, and Finland.

Decreasing world supplies of high grade lumpy ore and increasing availability of high grade fines and concentrates have led to an increased use of three agglomeration methods: *(1)* briquetting with a binder; *(2)* production of an oxide pellet by kiln firing; and *(3)* production of a prereduced pellet by furnace treatment.

Properties

Chemical Properties. The valence states of chromium are +2, +3, and +6, the latter two being the most common. The +2 and +3 states are basic, whereas the +6 is acidic, forming ions of the type $(CrO_4)^{2-}$ (chromates) and $(Cr_2O_7)^{2-}$ (dichromates). The blue-white metal is refractory and very hard.

Chromium is highly acid-resistant and is only attacked by hydrochloric, hydrofluoric, and sulfuric acids. It is also resistant to other common corroding agents including acetone, alcohols, ammonia, carbon dioxide, carbon disulfide, foodstuffs, petroleum products, phenols, sodium hydroxide, and sulfur dioxide.

Table 1. Physical Properties of Chromium

Property			Value	
at no.			24	
at wt			51.996	
isotopes				
mass	50	52	53	54
relative abundance, %	4.31	83.76	9.55	2.38
crystal structure			bcc	
lattice parameter, a_0, nm			0.2888–0.2884	
density at 20°C, g/mL			7.19	
mp, °C			1875	
bp, °C			2680	
elastic modulus, GPa[a]			250	

[a] To convert GPa to psi, multiply by 145,000.

Physical and Mechanical Properties. The physical properties of chromium are listed in Table 1.

Perhaps more so than any other common metal, the mechanical properties of chromium depend on purity, history, grain size, strain rate, and surface condition. Chromium metal of ordinary purity is extremely brittle; and even the purest (iodide) chromium, while quite ductile at room temperature in wrought form, is readily embrittled by recrystallization or by nitrogen exposure at elevated temperature (Table 2).

Production

Very little chromite is processed all the way to bulk metal; most can be used in an intermediate form, eg, as alloying agent, chromium source for coating processes, or to form metallurgically or chemically important chromium compounds.

Ferrochromium. Ferrochromium, also called ferrochrome and typically classified as low carbon, high carbon, or charge-grade (charge chrome), is usually made by reduction of chromite with coke in a three-phase electric submerged arc furnace. This process leads inevitably to a high carbon ferrochromium, the use of which was historically restricted to high carbon steels. However, it can now also be used in argon–oxygen decarburization (AOD) and similar alloy and stainless steels processing.

Chromium Metal by Pyrometallurgical Reduction. The principal pyrometallurgical process for commercial chromium metal is the reduction of Cr_2O_3 by aluminum: $Cr_2O_3 + Al \rightarrow 2\ Cr + Al_2O_3$.

Electrowinning of Chromium. In the chrome alum process typified by the 2000 tons per year Elkem Metals Company plant at Marietta, Ohio, high carbon ferrochromium is leached with a hot solution of reduced anolyte plus chrome alum mother liquor and makeup sulfuric acid to form ammonium chrome alum, from which chromium is recovered by electrolysis.

Alternatively, chromium metal may be produced electrolytically or pyrometallurgically from chromic acid, CrO_3, obtained from sodium dichromate by any of several processes. Small amounts of an ionic catalyst, specifically sulfate, chloride, or fluoride, are essential to the

Table 2. Mechanical Properties of Room Temperature Swaged Iodide Chromium[a]

Condition	Yield strength, 0.2% offset, MPa[b]	Ultimate strength, MPa[b]	Elongation, %[c]	Reduction of area, %
wrought	362	413	44	78
recrystallized		282	0	0

[a] Pure chromium made by the iodide process (99.996% Cr).
[b] To convert MPa to psi, multiply by 145.
[c] Percent elongation in a 6 mm gauge length.

electrolytic production of chromium. Fluoride and complex fluoride catalyzed baths have become especially important in recent years.

Purification. The metal obtained from both electrolytic processes contains considerable oxygen, which is believed to cause brittleness at room temperature. For most purposes the metal as plated is satisfactory. However, if ductile metal is desired, the oxygen can be removed by hydrogen reduction, the iodide process, calcium refining, or melting in a vacuum in the presence of a small amount of carbon.

Consolidation and Fabrication. Chromium metal may be consolidated by powder metallurgy techniques or by arc melting in an inert atmosphere.

The initial cast structure of arc-melted ingots must be carefully broken down by hot working in order to permit subsequent warm working. Forging, swaging, and extrusion are all possible.

Electroplating, Chromizing, and Other Chromium-Surfacing Processes

Electroplating. Chromium is electroplated onto various substrates in order to realize a more decorative and corrosion- or wear-resistant surface. About 80% of the chromium employed in metal treatment is used for chromium plating; over 50% is for decorative chromium plating (see METAL SURFACE TREATMENTS). Hard chromium plating differs from decorative plating mostly in terms of thickness. Hard chromium plate may be 10 to several 100 μm thick, whereas the chromium layers in a decorative plate may be as thin as 0.25 μm, which corresponds to about two grams Cr per square meter of surface.

Chromizing. The other principal method of obtaining a chromium-rich surface on steel is by chromizing. The material to be treated is embedded in a mixture of ferrochromium powder, a chromium halide, alumina, and sometimes NH_4Cl. The chromium is diffused in by a furnace treatment at about 1100°C to produce an effective stainless steel surface where the mean composition is about 18 wt % Cr and the thickness is controlled by the time of treatment.

Other Surface Processes. Whereas sputtering, ion implantation, chemical vapor deposition (CVD), metal spraying, cladding, and weld overlayment have been used for chromium-based protective coatings, only the last two have commercial significance for protecting steels.

Economic Aspects

In the latter part of the twentieth century, the United States has become completely dependent on imports from South Africa and Turkey (chromite); South Africa, Zimbabwe, Turkey, and Yugoslavia (ferrochromium); and the Philippines (chromite for refractory brick).

A 3–4% annual growth in chromium consumption leading to a total U.S. primary chromium demand of 1,000,000 metric tons in the year 2000 has been projected. Chromium is absolutely essential to the production of stainless steel, which accounts for the largest use of this metal. Moreover, a chromium-free stainless steel is unlikely. Thus chromium importance can be judged in view of the importance of stainless steel to the U.S. industrial economy.

The cost of chromium ore is determined by operating, ie, mining and beneficiation, and transportation costs, whereas the price of the ore is affected by chromium and carbon contents and particle size. Lumpy or coarse grades usually command a premium price.

CHROMIUM ALLOYS

Metallurgically Important Chromium Compounds

There are a number of metallic compounds of chromium that are used either as the compound itself or as metallurgical constituents in Cr-bearing alloys. Trichromium dicarbide, Cr_3C_2, is important as a wear-resistant gauge material; chromium boride, (1:1), CrB, for oil well drilling; chromium dioxide, CrO_2, for magnetic tape; and $Cr_xMn_{2x}Sb$ as a magnetic material having unique characteristics. The intermetallic compounds trichromium aluminum, Cr_3Al, trichromium silicon, Cr_3Si, and dichromium titanium, Cr_2Ti, are encountered in developmental oxidation-resistant coatings. The carbides chromium carbide, (23:6), $Cr_{23}C_6$, and chromium carbide (7:3), Cr_7C_3; chromium iron (1:1), CrFe (ς phase), and chromium iron molybdenum (12:36:10), $Cr_{12}Fe_{36}Mo_{10}$ (χ phase), are found as constituents in many alloy steels; Cr_2Al_{13} and CoCr are found in aluminum and cobalt-based alloys, respectively. The chromium-rich interstitial compounds, Cr_2H, chromium nitride (2:1), Cr_2N, and carbide, $Cr_{23}C_6$, play an important role in the effect of trace impurities on the properties of unalloyed chromium. The intermetallics and the interstitial compounds of chromium are stabilized by electronic and/or spatial factors and are not to be regarded as simple ionic or covalent compounds.

Stainless Steel

The stainless quality is conferred on steels that contain enough chromium to form a protective surface film. About 12 wt % chromium is required for protection in mild atmospheres or in steam. At 18–20 wt % chromium, sufficient protection is achieved for satisfactory performance in a wide variety of more destructive environments, including those occurring in the chemical, petrochemical, and the power-generating industries. Stainless grades having 25 wt % chromium or more and containing other alloying elements such as molybdenum provide even higher corrosion resistance. In certain stainless steels, the chromium depresses the martensite transformation below room temperature. By thus stabilizing austenite the chromium permits achievement of desired mechanical properties without loss of corrosion resistance (see STEEL).

Stainless steels are classified in terms of their microstructures as austenitic, martensitic, ferritic, duplex (austenite + ferrite), and precipitation-hardening (PH). The microstructure type is determined by base composition and heat treatment and, in turn, dictates the properties, especially strength, toughness, and corrosion resistance.

Other Alloy Steels

In low alloy steels chromium contributes more to hardenability, tempering resistance, and toughness than to solid-solution hardening or oxidation resistance. The marked effect of small additions of chromium on hardenability is shown in Figure 1. Whereas other alloying elements may show a similar or greater effect, chromium is among the cheapest. In the high chromium tool steel compositions, chromium carbides contribute high abrasion resistance and improve the high hot hardness.

Wrought alloy steels, alloyed cast irons and steels, and tool steels account for 22–25% of the annual U.S. consumption of chromium. The

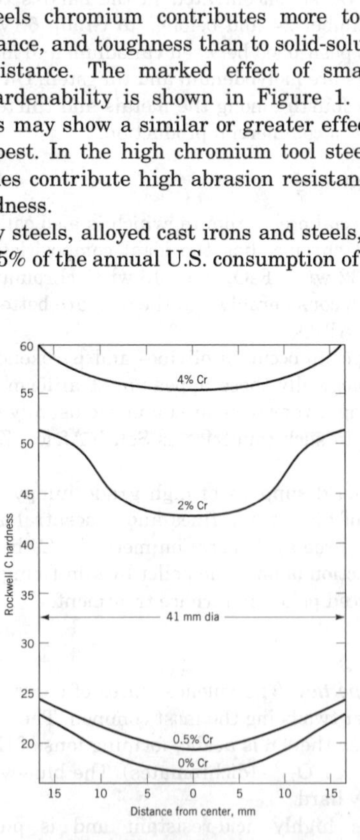

Figure 1. Effect of chromium on hardenability of steel as indicated by hardness distribution across 41 mm rounds of oil-quenched 0.35% C steel (from H. W. Paxton and E. D. Bain, *Alloying Elements in Steel*, 2nd ed., ASM, 1966).

wrought alloy steels are the largest category. In terms of chromium consumption, the most important classes are, in order, the CrMo, NiCrMo, and Cr steels. These may be further subdivided into carburizing and through-hardening grades. Such steels are extensively used in machinery, construction, and other structural work, and in machine parts such as bearings, gears, rolls, springs, and shafting.

A very different chromium-bearing alloy family is the Cr–Co–Fe magnetic alloys discovered in 1971. These alloys, containing about 30 wt % Cr, have magnetic properties comparable to the brittle Alnicos yet are cold formable. The magnetic properties are superior to the cobalt magnet steels, CuNiFe's, and Vicalloys. The most extensive application is as a cold-formed, cup-shaped magnet for telephone receivers.

Nonferrous Alloys

Nonferrous alloys account for only about 2 wt % of the total chromium used in the United States. Nonetheless, some of these applications are unique and constitute a vital role for chromium. For example, in high temperature materials, chromium in amounts of 15–30 wt % confers corrosion and oxidation resistance on the nickel-base and cobalt-base superalloys used in jet engines; the familiar electrical resistance heating elements are made of Ni–Cr alloy; and a variety of Fe–Ni and Ni-based alloys used in a diverse array of applications, especially for nuclear reactors, depend on chromium for oxidation and corrosion resistance.

Recovery and Reuse. As of this writing (ca 1993) only about 20% of the chromium consumed in the United States is recycled, and this amount comes largely from stainless steel scrap. Improved recovery of the substantial chromium losses incurred in the past from refractory and foundry applications of chromite grain is being investigated.

Toxicity and Environmental Aspects

Chromium is generally recognized as being essential to human health; however, hexavalent chromium compounds are also known to be toxic, significantly more so than trivalent ones. Thus, chromium releases into the environment are regulated by the Environmental Protection Agency, and workplace exposure is regulated by OSHA. The exposure limit set by OSHA for chromium metal is 1 mg/m^3 in an 8-h time-weighted average, and for chromic acid and chromates it is 0.1 mg/m^3, ceiling exposure limit.

JACK H. WESTBROOK
Brookline Technologies

M. J. Udy, *Chromium*, Vol. 1; *Chemistry of Chromium and its Compounds;* Vol. 2; *Metallurgy of Chromium and its Alloys,* Reinhold Publishing Corp., New York, 1956.

Roskill Information Services, *The Economics of Chromium,* 4th ed., Roskill Information Services, London, 1982.

E. R. Parker and co-workers, *Contingency Plans for Chromium,* NMAB Report 335, 1977.

D. Pechner and I. M. Bernstein, *Handbook of Stainless Steels,* McGraw-Hill Book Co., Inc., New York, 1977.

CHROMIUM COMPOUNDS

Russia and the Republic of South Africa account for more than half the world's chromite ore production. Almost all of the world's known reserves of chromium are located in the southeastern region of the continent of Africa. South Africa has 84% and Zimbabwe 11% of these reserves. The United States is completely dependent on imports for all of its chromium. The chromite's constitution varies with the source of the ore, and this variance can be important to processing. Typical ores are from 20 to 26 wt % Cr, from 10 to 25 wt % Fe, from 5 to 15 wt % Mg, from 2 to 10 wt % Al, and between 0.5 and 5 wt % Si. Other elements that may be present are Mn, Ca, Ti, Ni, and V. All of these elements are normally reported as oxides; iron is present as both Fe(II) and Fe(III).

Properties

Chromium compounds number in the thousands and display a wide variety of colors and forms. Examples of these compounds and the corresponding physical properties are given in Table 1.

Chromium is able to use all of its 3*d* and 4*s* electrons to form chemical bonds. It can also display formal oxidation states ranging from Cr(−II) to Cr(VI). The most common and thus most important oxidation states are Cr(II), Cr(III), and Cr(VI). Although most commercial applications have centered around Cr(VI) compounds, environmental concerns and regulations in the early 1990s suggest that Cr(III) may become increasingly important, especially where the use of Cr(VI) demands reduction and incorporation as Cr(III) in the product.

Low Oxidation State Chromium Compounds. Cr(0) compounds are π-bonded complexes that require electron-rich donor species such as CO and C_6H_6 to stabilize the low oxidation state. A direct synthesis of $Cr(CO)_6$, from the metal and CO, is not possible. Normally, the preparation requires an anhydrous Cr(III) salt, a reducing agent, an arene compound, carbon monoxide that may or may not be under high pressure, and an inert atmosphere (see CARBONYLS).

Chromium(II) Compounds. The Cr(II) salts of nonoxidizing mineral acids are prepared by the dissolution of pure electrolytic chromium metal in a deoxygenated solution of the acid. It is also possible to prepare the simple hydrated salts by reduction of oxygen-free, aqueous Cr(III) solutions using Zn or Zn amalgam, or electrolytically. These methods yield a solution of the blue $Cr(H_2O)_6^{2+}$ cation. The isolated salts are hydrates that are isomorphous with Fe^{2+} and Mg^{2+} compounds.

Chromium(III) Compounds. Chromium(III) is the most stable and most important oxidation state of the element. The $E°$ values (show that both the oxidation of Cr(II) to Cr(III) and the reduction of Cr(VI)

Table 1. Physical Properties of Chromium Compounds

Compound	Density[a], g/cm^3	Mp, °C	Bp, °C
chromium(0) hexacarbonyl	1.77_{18}	148.5	210[b]
dibenzene chromium(0)	1.519	284–285	sub 150[c]
bis(biphenyl) chromium (I) iodide	1.617_{16}	178	dec
chromium(II) acetate dihydrate	1.79		
chromium(II) chloride	2.88	815	1300
ammonium chromium(II) sulfate hexahydrate			
chromium(III) chloride	2.76_{15}	877	sub 947
chromium(III) acetylacetonate	1.34	216	340
potassium chromium(III) sulfate dodecahydrate	1.826_{25}	89[d]	400[e]
chromium(III) chloride hexahydrate	1.835_{25}	95	
chromium(III) oxide	5.22_{25}	2330	3000
chromium(IV) oxide	4.98[f]	dec	dec 300
chromium(IV) fluoride	2.89	ca 277	ca 400
chromium(VI) oxide	2.7_{25}	197	dec
chromium(VI) dioxide dichloride	1.9145_{25}	−96.5	115.8
ammonium dichromate(VI)	2.155_{25}	dec 180	
potassium dichromate(VI)	2.676_{25}	398	dec 500
sodium dichromate(VI) dihydrate	2.348_{25}	356; 84.6[e]	dec 400
potassium chromate(VI)	2.732_{18}	975	
sodium chromate(VI)	2.723_{25}	792	
potassium chlorochromate(VI)	2.497_{39}	dec	
silver chromate(VI)	5.625_{25}		
barium chromate(VI)	4.498_{25}	dec	
strontium chromate(VI)	3.895_{15}	dec	

[a] Measurement taken at temperature in °C noted in subscript. [b] Explodes. [c] In vacuum. [d] Incongruent. [e] Loses all water at temperature indicated. [f] Calculated value.

to Cr(III) are favored in acidic aqueous solutions. The preparation of trivalent chromium compounds from either state presents few difficulties and does not require special conditions. In basic solutions, the oxidation of Cr(II) to Cr(III) is still favored. However, the oxidation of Cr(III) to Cr(VI) by oxidants such as peroxides and hypohalites occurs with ease. The preparation of Cr(III) from Cr(VI) in basic solutions requires the use of powerful reducing agents such as hydrazine, hydrosulfite, and borohydrides, but Fe(II), thiosulfate, and sugars can be employed in acid solution. Cr(III) compounds having identical counterions but very different chemical and physical properties can be produced by controlling the conditions of synthesis.

Chromium(IV) and Chromium(V) Compounds. The formal oxidation states Cr(IV) and Cr(V) show some similarities. Both states are apparently intermediates in the reduction of Cr(VI) to Cr(III). Neither state exhibits a compound that has been isolated from aqueous media, and Cr(V) has only a transient existence in water. The majority of the stable compounds of both oxidation states contain either a halide, an oxide, or a mixture of these two. As of this writing, knowledge of the chemistry is limited.

Chromium(VI) Compounds. Virtually all Cr(VI) compounds contain a Cr–O unit. The chromium(VI) fluoride, CrF_6, is the only binary Cr^{6+} halide known and the sole exception. The primary Cr–O bonded species is chromium(VI) oxide, CrO_3, which is better known as chromic acid.

Manufacture

The primary industrial compounds of chromium made directly from chromite ore are sodium chromate, sodium dichromate, and chromic acid. Secondary chromium compounds produced in quantity include potassium dichromate, potassium chromate, and ammonium dichromate. The secondary trivalent compounds manufactured in quantity are chrome acetate, chrome nitrate, basic chrome chloride, basic chrome sulfate, and chrome oxide.

Sodium Chromate, Dichromate, and Chromic Acid. The basic chemistry used to process chromite ore has not changed since the early nineteenth century. However, modern technologies have added many refinements to the manufacturing techniques, and plants have been adapted to meet health, safety, and environmental regulations. A generalized block flow diagram for the modern chromite ore processing plant is given in Figure 1.

Other Chromates and Dichromates. Potassium and ammonium dichromates are generally made from sodium dichromate by a crystallization process involving equivalent amounts of potassium chloride or ammonium sulfate. In each case the solubility relationships are favorable so that the desired dichromate can be separated on cooling, whereas the sodium chloride or sulfate crystallizes out on boiling.

Potassium chromate is prepared by the reaction of potassium dichromate and potassium hydroxide. Sulfates are the most difficult impurity to remove, because potassium sulfate and potassium chromate are isomorphic.

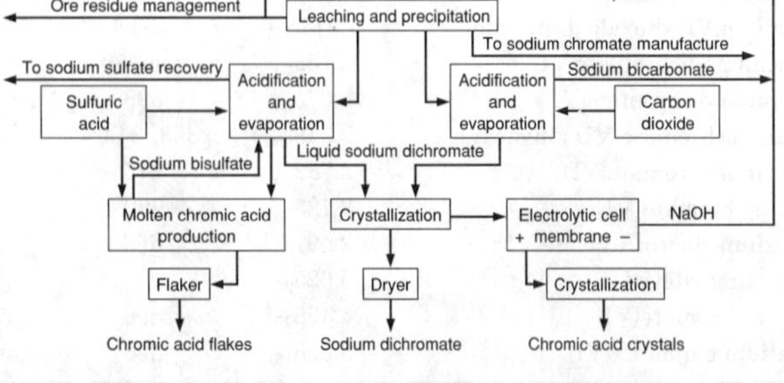

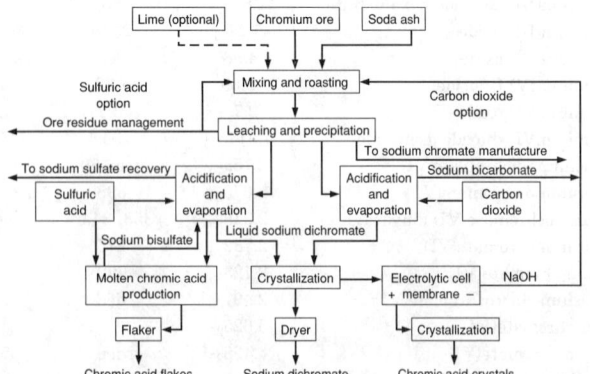

Figure 1. Flow diagram for the production of sodium chromate, sodium dichromate, and chromic acid flake and crystals.

Water-Soluble Trivalent Chromium Compounds. Most water-soluble Cr(III) compounds are produced from the reduction of sodium dichromate or chromic acid solutions. This route is less expensive than dissolving pure chromium metal, it uses high quality raw materials that are readily available, and there is more processing flexibility. Finished products from this manufacturing method are marketed as crystals, powders, and liquid concentrates.

Economic Aspects

In 1993, the total weight of chromite ore imported into the United States was 288,000 metric tons. The total consumption of chromite ore for the chemical, metallurgical (the principal use), and refractory industries was 319,000 metric tons.

Health and Safety Factors

Acute and Chronic Toxicity. Only Cr(III) and Cr(VI) compounds are produced in large quantities and are accessible to most of the population. Therefore, the toxicology of chromium compounds has been historically limited to these two states, and virtually all of the available information is about compounds of Cr(III) or Cr(VI). However, there is some indication that Cr(V) may play a role in chromium toxicity.

The primary routes of entry for animal exposure to chromium compounds are inhalation, ingestion, and, for hexavalent compounds, skin penetration. This last route is more important in industrial exposures. Most hexavalent chromium compounds are readily absorbed, are more soluble than trivalent chromium in the pH range 5 to 7, and react with cell membranes. Although hexavalent compounds are more toxic than those of Cr(III), an overexposure to compounds of either oxidation state may lead to inflammation and irritation of the eyes, skin, and the mucous membranes associated with the respiratory and gastrointestinal tracts. Skin ulcers and perforations of nasal septa have been observed in some industrial workers after prolonged exposure to certain hexavalent chromium compounds, ie, to chromic acid mist or sodium and potassium dichromate.

Acute systemic poisoning by chromium compounds is rare, and only hexavalent compounds have been implicated. It has been suggested that the principal routes of exposure allow for a detoxification by reduction of Cr(VI) to Cr(III) by the body's sulfur-containing proteins, eg, glutathione, or the oxidizable compounds contained in the gastric juices and saliva. The target organ for acute systemic toxicity is the kidney. Usually, poisoning by Cr(VI) results in acute tubular necrosis of the kidney, the reported cause of death.

Carcinogenicity, Mutagenicity, and Genotoxicity. There is evidence that hexavalent chromium may be a carcinogen, but there is some doubt about which Cr(VI) compounds are implicated. The National Institute for Occupational Safety and Health (NIOSH) has classified the chromate and dichromate salts of lithium, potassium, ammonium, rubidium, cesium, and hydrogen, plus chromic acid, as noncarcinogenic Cr(VI) compounds. NIOSH considers all of the other Cr(VI) compounds carcinogenic. Manufacturing processing practices have reduced worker exposure to hexavalent chromium. No carcinogenic potency has been demonstrated for Cr(III) compounds.

Nutrition. Chromium, in the trivalent oxidation state, is recognized as an essential trace element for human nutrition, and the recommended daily intake is 50 to 200 micrograms.

Chromium Exposure Levels and U.S. Government Regulations. The Occupational Safety and Health Administration (OHSA) has established workplace permissible exposure limits (PEL) for chromium metal and three forms of chromium compounds. OSHA's PEL for chromic acid and chromates is 0.1 mg/m^3 CrO_3 as both a ceiling, ie, no exposure above this concentration is allowed, and an 8-h time-weighted average (TWA). Chromium metal and insoluble chromium salts have an 8-h TWA PEL of 1.0 mg/m^3 Cr, and the same standard is 0.5 mg/m^3 Cr for soluble Cr(III) and Cr(II) compounds.

The NIOSH recommended exposure limit for carcinogenic hexavalent chromium is 1 $\mu g/m^3$ Cr(VI) as a 10-h TWA, and for noncarcinogenic Cr(VI) the 10-h TWA is 25 $\mu g/m^3$ Cr(VI), including a 15-min maximum exposure of 50 $\mu g/m^3$ Cr(VI). According to NIOSH, the noncarcinogenic Cr(VI) compounds are chromic acid and the chromates and dichromates of sodium, potassium, lithium, rubidium, cesium, and ammonia. NIOSH considers any hexavalent chromium compound that does not appear on the preceding list carcinogenic.

There are no federal standards proclaimed for ambient air concentrations of chromium, but the EPA has published a notice of intent to list Cr(VI) or total Cr as a toxic air contaminant. Although natural sources may be responsible for part of the chromium value, all of the Cr(VI) and some of the Cr(III) measured is probably from anthropogenic sources. The most likely origins of Cr(VI) values are industrial emissions, and the most likely forms are aqueous aerosol fog, mists, or droplets and aerosol powders.

The EPA has set the National Interim Primary Drinking Water Standard at 50 $\mu g/L$ total chromium and the current Maximum Contamination Level (MCL) is 120 $\mu g/L$. This agency has also issued a Cr(VI) ambient water quality standard of 50 $\mu g/L$ and has proposed a Maximum Contamination Level Goal (MCLG) of 0.1 $\mu g/L$. Industrial discharges of total Cr(VI) are regulated by National Pollutant Discharge Elimination System (NPDES) permits, specific for the area that receives the waste or discharge.

Chromium containing solids from manufacturing and wastewater treatment sludges are classified as hazardous wastes and must be handled as such.

The EPA has established exposure levels for both Cr(III) and Cr(VI) for the general population. For exposures of short duration that constitute an insignificant fraction of the lifespan the acceptable intake subchronic (AIS) by ingestion is 979 mg/d for trivalent chromium and 1.75 mg/d for hexavalent chromium. There was insufficient data to calculate an AIS by inhalation for Cr(III), and the EPA believes this type of standard is inappropriate for hexavalent chromium. For lifetime exposures, an acceptable intake chronic (AIC) of 103 mg/d Cr(III) and 0.35 mg/d Cr(VI) is established for ingestion. The inhalation AIC is estimated to be 0.357 mg/d Cr(III). The EPA has calculated an inhalation cancer potency for Cr(VI) of 41 [mg/(kg·d)]−1 risk for a lifetime exposure to 1 $\mu g/m^3$ hexavalent chromium.

Waste Management

Despite modern engineering designs, production and consumption waste by-products containing chromium are generated. These wastes have been traditionally managed by burial of dewatered sludges that are mainly the result of the reduction of Cr(VI) to Cr(III) and the latter's precipitation as a hydrated oxide from the treatment of wastewaters. Scrap iron, ferrous sulfate, sodium bisulfite, sulfur dioxide, sodium hydrosulfite, and sulfide wastes have all been employed as reducing agents for waste streams containing hexavalent chromium. Following hexavalent chromium's reduction, lime or other alkali is added to raise the pH, causing hydrated chromium(III) oxide to precipitate. The slurry is allowed to segregate and the clear, purified supernatant is decanted and either recycled or discharged. Often this type of procedure allows several metals, such as Cu and Ni, to be removed with the chromium, because the hydrous oxide is a good collector.

Where appropriate, the direct precipitation of hexavalent chromium with barium, and recovery of the Cr(VI) value can be employed. Another recycling option is ion exchange a technique that works for chromates and Cr^{3+}. Finally, recovery of the chromium as the metal or alloy is possible by a process similar to the manufacture of ferrochromium alloy and other metals.

Uses

Chromium compounds are essential to many industries. The percentage distribution of consumption for chromium compounds is wood preservation, 38; metal finishing, 15; leather tanning, 10; pigments, 8; chemical manufacturing, 8; oil drilling muds, 4; textiles, 3; magnetic tapes, 2; and other uses such as for catalysts, photography, etc, 11.

BILLIE J. PAGE
GARY W. LOAR
McGean-Rohco, Inc.

M. J. Udy, ed., *Chromium,* Vol. 1, Reinhold Publishing Co., New York, 1956.

J. C. Bailer, Jr., H. J. Emeléus, R. Nyholm, and A. F. Trotman-Dickenson, eds., *Comprehensive Inorganic Chemistry,* Vol. 3, Pergamon Press, Oxford, U.K., 1973.

J. O. Nriagu and E. Nieboer, eds., *Chromium in the Natural and Human Environments,* John Wiley & Sons, Inc., New York, 1988.

CHROMOGENIC MATERIALS

ELECTROCHROMIC

Electrochromism is a reversible and visible change in transmittance and/or reflectance that is associated with an electrochemically induced oxidation–reduction reaction. This optical change is effected by a small electric current at low d-c potential. The potential is usually on the order of 1 V, and the electrochromic material sometimes exhibits good open-circuit memory. Unlike the well-known electrolytic coloration in alkali halide crystals, the electrochromic optical density change is often appreciable at ordinary temperatures.

Coloration occurs both cathodically and anodically, as well as in both organic and inorganic materials. Compounds of all types may be classified within one or the other of two general groups based on the nature of charge balancing. In one group, an electrolyte separates a cathode–anode pair, one or both of which may be chromogenically active. Typically, the chromogenic material is a thin film on the cathode or anode. As charge neutrality must be preserved, and the electrochromic cathode or anode is a solid, insertion/extraction of ions, often H^+ or alkali, accompanies reduction–oxidation within the electrode surface layer. Insertion/extraction in the cathode or anode is the distinguishing feature of this group. The second group is best described by referring to the viologens, a family of halides of quaternary bases derived from the 4,4′-bipyridinium structure. Viologens are recognized as the first important organic electrochromic materials. Some of these color deeply within solution by simple reduction; others are distinguished by their deep coloration when electrodeposited from solution onto a cathode. These colorations typify the noninsertion group, though incidental insertion may accompany electrodeposition.

Members of the ion-insertion/extraction group, as inorganic or organic thin films, especially the former, have attracted the widest interest most recently. Tungsten trioxide was the earliest exploited inorganic compound, even before the mechanism of its electrochromic response was understood. It is still the best known of the important ion-insertion/extraction group.

Ongoing research with a wide variety of materials from both groups has been reviewed frequently. Much of the earliest published work followed from research on displays, but opportunities for switchable mirrors, windows, and sunglasses, have been highlighted as well. With one noteworthy exception, however, there has been no remarkable commercial success. In part, this is because the technology involves many complex scientific and engineering principles. Also, the competing liquid crystal technology has evolved successfully in some display applications. The one commercial exception is an electrochromic automotive rearview mirror which has been gaining popularity since 1988. The mirror contains an encapsulated solution of viologen, which undergoes optical switching without electrodeposition.

CHARLES B. GREENBERG
PPG Industries, Inc.

B. W. Faughnan and R. S. Crandall, in J. I. Pankove, ed., *Display Devices, Topics in Applied Physics,* Vol. 40, Springer-Verlag, Berlin, Germany, 1980, p. 181.

C. M. Lampert, *Sol. Energy Mater.* **11,** 1 (1984).

N. Baba, M. Yamana, and H. Yamamoto, eds., *Electrochromic Display,* Sangyo Tosho Co. Ltd., Tokyo, 1991.

PHOTOCHROMIC

For the purposes of this article the definition of photochromism is restricted to a reversible change in the color or darkening of material caused by absorption of ultraviolet or visible light. The change in color, or darkening of the material, implies a change in the absorption spectrum in the visible range of light (400–700 nm). The change must be reversible, although there is no implication about the kinetics or speed of the reversion to the original state.

Schematically, the photochromic reaction can be stated by the simple equation 1:

$$A \underset{\delta \text{ or } h\nu'}{\overset{h\nu}{\rightleftarrows}} B \qquad (1)$$

Substance A has an absorption spectrum in one or more regions of the ultraviolet or visible spectral range. Irradiation of A at a wavelength corresponding to one of the absorption bands results in formation of substance B, which has a visible absorption spectrum different from A. Most commonly, substance A is uncolored or only slightly colored, whereas substance B is colored or appears darker than A.

The reverse reaction, B returning to A, can be driven either by thermal or photochemical energy, or both. When the reversion is photochemically driven, the process is called optical bleaching. Optical bleaching is a general characteristic and is a factor in almost all photochromic systems, even those normally thought of as being thermally reversible.

Another important concept in the discussion of photochromic systems is fatigue. Fatigue is defined as a loss in photochromic activity as a result of the presence of side reactions that deplete the concentration of A and/or B, or lead to the formation of products that inhibit the photochemical formation of B. The inhibition can result from quenching of the excited state of A or screening of active light. Fatigue, therefore, is caused by the absence of total reversibility within the photochromic reaction (eq. 2).

$$A \underset{\delta \text{ or } h\nu'}{\overset{h\nu}{\rightleftarrows}} B \rightarrow C \qquad (2)$$

Photochromic systems can be separated into two broad categories, organic and inorganic. The two types are vastly different in their observable characteristics and mechanisms, but there are several examples of both which fit the definition of photochromism given. The purpose of the discussion is to define, with the help of the examples, the principal characteristics of each photochromic system.

Inorganic Photochromic Systems

Silver Halide-Containing Glasses. The most important examples of inorganic systems are those containing silver halide crystallites dispersed throughout a glass matrix. In general, these systems are characterized by broad absorption of visible light by the colored species and excellent resistance to fatigue.

The principle behind the generation of a photochromic glass with silver halide is the controlled formation of silver halide particles or crystallites suspended throughout the glass matrix. The formation of crystallites of the correct size and concentration is the key to a useful photochromic system. The general procedure involves the initial melting of a glass-forming mixture which is then cooled to a solid glass shape. Rapid cooling to room temperature results in a nonphotochromic glass. Holding the solid at a temperature in the range of 500–600°C for several minutes to hours causes the nucleation and growth of silver halide crystallites, the active photochromic species.

Organic Photochromic Systems

The organic photochromic systems that have been studied are numerous and it is helpful to classify them into a few categories by way of the general mechanism of the photochromic reaction in each category. They include photochromism based on geometric isomerism, photochromism based on cycloaddition reactions, photochromism based on tautomerism, photochromism based on dissociation processes, photochromism based on triplet formation, photochromism based on redox reactions, and photochromism based on electrocyclic reactions.

Applications

Although the proposed applications for photochromic systems are numerous, few have received broad use. By far, the most successful commercial application is the use of photochromic silver halide-containing glasses in prescription eyewear.

Besides the use of photochromic systems in light filters, their color development has also received considerable attention. For example, the introduction of photochromic components into product labels, tickets, credit cards, etc adds a mechanism for verification of authenticity. The active components are invisible until activated with an ultraviolet light source, after which they are easily detected.

The color development of photochromic compounds can also be utilized as a diagnostic tool. An example is their use in a nondestructive inspection technique for honeycomb aerospace structures.

Photochromic compounds that can be thermally faded have also been used in engineering studies to visualize flows in dynamic fluid systems.

JOHN C. CRANO
PPG Industries, Inc.

W. H. Armistead and S. D. Stookey, *Science* **144,** 150 (1964).

H. Rau, in H. Durr and H. Bouas-Laurent, eds., *Photochromics, Molecules and Systems,* Elsevier, Amsterdam, the Netherlands, 1990, p. 165.

PIEZOCHROMIC

In its most general sense piezochromism is the change in color of a solid under compression. There are three aspects of the phenomenon. The first is, in a sense, trivial, but it is very general. The color of a solid results from the absorption of light in selected regions of the visible spectrum by excitation of an electron from the ground electronic state to a higher level. If the two electronic energy levels are perturbed differently by pressure, compression results in a color change. This is the basic definition of pressure tuning spectroscopy. Examples include, among others, increased splitting of the *d* orbitals of transition-metal ions in complexes with pressure, the shift to higher energy of a color center (a vacancy containing an electron) in an alkali halide or glass environment, and a change in the relative energy of bonding and antibonding orbitals as pressure increases.

Phase Transitions. A second aspect involves a discontinuous change of color when a crystalline solid undergoes a first-order phase transition from one crystal structure to another. The most obvious example is the change of the absorption edge.

Changes in Molecular Geometry

The phenomenon of most interest is a change in color of a solid as a result of a change in the molecular geometry of the molecules that make up the solid. The color change takes place because the change in geometry alters the relative energy of different electronic orbitals, and therefore the electronic absorption spectrum. Frequently it rearranges the order of these orbitals or provides new combinations of atomic orbitals because of symmetry changes. The rearrangements may be discontinuous at a given pressure, may occur over a modest range of pressures, or may occur gradually over the whole range of available pressure as for chemical equilibria in solution.

The systematic study of piezochromism is a relatively new field. It is clear that, even within the restricted definition used here, many more systems will be found which exhibit piezochromic behavior. It is quite possible to find a variety of potential applications of this phenomenon. Many of them center around the estimation of the pressure or stress in some kind of restricted or localized geometry, eg, under a localized impact or shock in a crystal or polymer film, in such a film under tension or compression, or at the interface between bearings. More generally it conveys some basic information about inter- and intramolecular interactions that is useful in understanding processes at atmospheric pressure as well as under compression.

H. G. DRICKAMER
University of Illinois

H. G. Drickamer, in R. Pucci and J. Picatto, eds., *Molecular Systems under High Pressure,* North Holland Press, New York, 1991.

H. G. Drickamer and K. L. Bray, *Intern. Rev. of Phys. Chem.* **8,** 41 (1989).

H. G. Drickamer and K. L. Bray, *Accts. Chem. Res.* **23,** 55 (1991).

J. F. Rabolt and co-workers, *Polym. Prep.* **31,** 262 (1991).

THERMOCHROMIC

Thermochromism is the reversible change in the spectral properties of a substance that accompanies heating and cooling. Strictly speaking, the meaning of the word specifies a visible color change; however, thermochromism has come to also include some cases for which the spectral transition is either better observed outside of the visible region or not observed in the visible at all. Primarily, thermochromism occurs in solid or liquid phase, but it also describes a thermally dependent equilibrium between brown nitrogen dioxide, NO_2, and colorless dinitrogen tetroxide, N_2O_4, a rare example in the gas phase.

There are many materials, especially organic and metal-organic materials, which exhibit true thermochromism, with a variety of sometimes debatable structural transition mechanisms; it is difficult to summarize the whole with any continuity. For this reason, an effort is made to delineate the scope of the field by listing several thermochromic transitions (Table 1).

Table 1. Thermochromic Compounds and Transitions

Thermochromic material	Thermochromic transition[a]
Co^{2+} solutions and glasses	equilibrium shift, two coordinations
$[(C_2H_5)_2NH_2]_2CuCl_4$	square planar to tetrahedral[b]
$[(CH_3)_2NH_2]_3CuCl_5$	variation in bandwidth
$Cu_4I_4(Py)_4$	fluorescence variations
$Al_{2-x}Cr_xO_3$ (ruby)	lattice expansion/contraction
VO_2	monoclonic/tetragonal[c]
Cu_2HgI_4	order/disorder[c]
di-β-naphthospiropyran	close/open spiro ring
poly(xylylviologen dibromide)	hydration/dehydration[d]
ETCD polydiacetylene[e]	side group rearrangement[f]

[a] When applicable, expressed as a change upon heating; various colors have been reported, often qualitatively. [b] Transition temperatures, T_t, for sharp transition = 50°C. [c] T_t = 68°C. [d] T_t = 100°C. [e] Urethane-substituted polymer of $(=\underset{R}{C}-C=C-\underset{R}{C}=)_n$ where R = $(CH_{24}OCONHCH_2CH_3)$. [f] T_t =~115°C.

CHARLES B. GREENBERG
PPG Industries, Inc.

K. Sone and Y. Fukuda, *Inorganic Thermochromism,* Vol. 10, Springer-Verlag, New York, 1987.

K. Nassau, *The Physics and Chemistry of Color,* John Wiley & Sons, Inc., New York, 1983.

J. H. Day, *Chem. Rev.* **63,** 65 (1963); *ibid.,* **68,** 649 (1968).

A. Jayaraman, P. D. Dernier, and L. D. Longinotti, *Phys. Rev. B* **11,** 2783 (1975); *High Temp.-High Press.* **7,** 1 (1975).

CINNAMIC ACID, CINNAMALDEHYDE, AND CINNAMYL ALCOHOL

The earliest references to cinnamic acid, cinnamaldehyde, and cinnamyl alcohol are associated with their isolation and identification as odor-producing constituents in a variety of botanical extracts. It is now generally accepted that the aromatic amino acid L-phenylalanine, a primary end product of the Shikimic Acid Pathway, is the precursor for the biosynthesis of these phenylpropanoids in higher plants.

The widespread use of cinnamic derivatives has led to the pursuit of reliable methods for their direct synthesis. Commercial processes have focused on condensation reactions between benzaldehyde and a number of active methylene compounds for assembly of the requisite carbon skeleton. The presence of a disubstituted carbon–carbon double bond in the sidechain of these chemicals also gives rise to the existence of two distinct stereoisomers, the cis or (*Z*)- and trans or (*E*)-isomers:

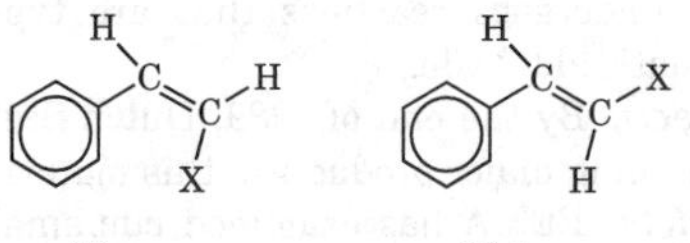

(*Z*)-isomer (*E*)-isomer

where X = COOH, CHO, or $CHOH_2$.

A considerable range of products, including flavors, fragrances, agrochemicals, pharmaceuticals, and polymers, has been developed using these chemicals as either synthetic intermediates or ingredients.

Cinnamic Acid

3-Phenyl-2-propenoic acid, commonly referred to as cinnamic acid, is a white crystalline solid having a low intensity sweet, honey-like aroma.

Physical and Chemical Properties. Cinnamic acid is generally encountered as the thermodynamically favored (*E*)-isomer. (*E*)-Cinnamic acid is an off-white solid having the properties outlined in Table 1.

For (*Z*)-cinnamic acid, three distinct polymorphic forms have been characterized. The most stable form, referred to as allocinnamic acid, has a melting point of 68°C, and the two metastable forms, isocinnamic acids, have melting points of 58°C and 42°C, respectively. (*E*)-Cinnamic acid can be converted to the (*Z*)-isomer photochemically through irradiation of a solution with ultraviolet light. Cinnamic acid undergoes reactions that are typical of an aromatic carboxylic acid.

Economic Aspects. There are no published production figures for cinnamic acid. Most of the manufactured acid is consumed internally to generate a series of cinnamate esters for flavor and fragrance applications. With this in mind, it was possible to estimate a 1990 usage in the range of 175 metric tons.

Health and Safety. The Flavor and Extract Manufacturers' Association (FEMA) expert panel has given cinnamic acid GRAS (generally recognized as safe) status and FEMA No. 2288 has been assigned to this material. As a consequence, the FDA has approved it for food use. The acid is likewise devoid of any significant dermal irritation or sensitization and has been approved for fragrance use.

Table 1. Properties of (*E*)-Cinnamic Acid

Property	Value
molecular formula	$C_9H_8O_2$
mol wt	148.2
melting point, °C	133
boiling point, °C at 101.3 kPa[a]	300

[a] To convert kPa to mm Hg, multiply by 7.5.

Table 2. Properties of (*E*)-Cinnamaldehyde

Property	Value
molecular formula	C_9H_8O
mol wt	132.2
boiling point, °C at 101 kPa[a]	252

Cinnamaldehyde

3-Phenyl-2-propenal, also referred to as cinnamaldehyde, is a pale yellow liquid with a warm, sweet, spicy odor and pungent taste reminiscent of cinnamon. It is found naturally in the essential oils of Chinese cinnamon (*Cinnamomum cassia,* Blume) (75–90%) and Ceylon cinnamon (*Cinnamomum zeylanicum,* Nees) (60–75%) as the primary component in the steam distilled oils. It also occurs in many other essential oils at lower levels.

Physical and Chemical Properties. The (*E*)- and (*Z*)-isomers of cinnamaldehyde are both known. (*E*)-Cinnamaldehyde is generally produced commercially and its properties are given in Table 2. Cinnamaldehyde undergoes reactions that are typical of an α,β-unsaturated aromatic aldehyde.

Economic Aspects. By the end of 1989, Dutch State Mines (DSM) was the only remaining major producer of this material.

Health and Safety. FEMA has examined cinnamaldehyde and established its GRAS status (No. 2286). The material has been used in some fragrance compositions, but RIFM has noted its potential for sensitization and limited the use in perfumes for skin contact at 1% in the formula.

Cinnamyl Alcohol

3-Phenyl-2-propen-1-ol, commonly referred to as cinnamyl alcohol, is a colorless crystalline solid with a sweet balsamic odor that is reminiscent of hyacinth. Its occurrence in nature is widespread as, for example, in Hyacinth absolute (*Hyacinthus orientalis*), the leaf and bark oils of cinnamon (*Cinnamomum cassia, Cinnamomum zeylanicum,* etc), and Guava fruit (*Psidium guajava L.*). In many cases it is also encountered as the ester or in a bound form as the glucoside.

Physical and Chemical Properties. Although both the (*E*)- and (*Z*)-isomers of cinnamyl alcohol are known in nature, (*E*)-cinnamyl alcohol is the only isomer with commercial importance. Its properties are summarized in Table 3.

When heated in the presence of a carboxylic acid, cinnamyl alcohol is converted to the corresponding ester. Oxidation to cinnamaldehyde is readily accomplished under Oppenauer conditions with furfural as a hydrogen acceptor in the presence of aluminum isopropoxide.

Table 3. Properties of (*E*)-Cinnamyl Alcohol

Property	Value
molecular formula	$C_9H_{10}O$
mol wt	134.2
boiling point, °C at 1.33 kPa	117–118
melting point, °C	33

Manufacture. A limited amount of natural cinnamyl alcohol is produced by the alkaline hydrolysis of the cinnamyl cinnamate present in Styrax oil.

The commercial production of cinnamyl alcohol is accomplished exclusively by the reduction of cinnamaldehyde.

$$C_6H_5CH{=}CHCHO \rightarrow C_6H_5CH{=}CHCH_2OH$$

Economic Aspects. As of this writing, DSM is the only significant supplier for this material.

Health and Safety. Cinnamyl alcohol has been evaluated by FEMA and given GRAS status (FEMA No. 2294). Two of its esters, cinnamyl cinnamate (FEMA No. 2298) and cinnamyl acetate (FEMA No. 2293), are also used extensively in flavor and fragrance compositions. Cinnamyl alcohol has also been tested by RIFM and found to be safe for use. There have been reported cases of irritation and several manufacturers market a desensitized alcohol for use in fragrance applications.

ROBERT G. EILERMAN
Givaudan-Roure Corporation

P. Schreier, *Chromatographic Studies of Biogenesis of Plant Volatiles,* A. Hüthig Verlag, Heidelberg, 1984, pp. 53, 84–88.

M. Luckner, *Secondary Metabolism in Microorganisms, Plants and Animals,* Springer-Verlag, Heidelberg, Berlin, 1984.

Beilstein's Handbuch der Organische Chemie, Vol. 6, 4th ed., Springer-Verlag, Berlin.

J. A. Dean, ed., *Lange's Handbook of Chemistry,* 13th ed., McGraw-Hill Book Co., New York, 1985, p. 7–240.

T. E. Furia and N. Bellanca, *Fenaroli's Handbook of Flavor Ingredients,* 2nd ed., Vol. 2, CRC Press, Cleveland, Ohio, 1975, p. 92.

R. O. B. Wijesekera, *CRC Crit. Rev. Food Sci. Nutr.* **10**(9), 1–30 (1978).

D. L. Opdyke, *Food Cosmet. Toxicol.* **17,** 253 (1979).

D. L. Opdyke, *Food Cosmet. Toxicol.* **12,** 855 (1974).

CITRIC ACID

Citric acid (2-hydroxy-1,2,3-propanetricarboxylic acid), is a natural component and common metabolite of plants and animals. It is the most versatile and widely used organic acid in foods, beverages, and pharmaceuticals.

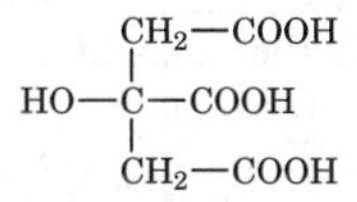

Because of its functionality and environmental acceptability, citric acid and its salts (primarily sodium and potassium) are used in many industrial applications for chelation, buffering, pH adjustment, and derivatization. These uses include laundry detergents, shampoos, cosmetics, enhanced oil recovery, and chemical cleaning.

Citric acid specifications are defined in a number of compendia including *Food Chemicals Codex* (FCC), *United States Pharmacopeia* (USP), *British Pharmacopeia* (BP), *European Pharmacopeia* (EP), and *Japanese Pharmacopeia* (JP).

Physical Properties

Citric acid, anhydrous, crystallizes from hot aqueous solutions as colorless translucent crystals or white crystalline powder. Its crystal form is monoclinic holohedra. Citric acid is deliquescent in moist air. Some physical properties are given in Table 1.

Aqueous solutions of citric acid make excellent buffer systems when partially neutralized because citric acid is a weak acid and has three carboxyl groups, hence three pK_as.

Citric acid monohydrate has a molecular weight of 210.14 and crystallizes from cold aqueous solutions. When gently heated, the crystals lose their water of hydration at 70–75°C and melt in the range

Table 1. Physical Properties of Citric Acid, Anhydrous

Property	Value
molecular formula	$C_6H_8O_7$
mol wt	192.13
gram equivalent weight	64.04
melting point, °C	153

of 135–152°C. Citric acid monohydrate is available in limited commercial quantities since most applications now call for the anhydrous form.

Chemical Properties

Citric acid undergoes most of the reactions typical of organic hydroxy polycarboxylates including decomposition, esterification, oxidation, reduction, salt formation, chelate formation, and corrosion.

Occurrence

Citric acid occurs widely in the plant and animal kingdoms. It is found most abundantly in the fruits of the citrus species, but is also present as the free acid or as a salt in the fruit, seeds, or juices of a wide variety of flowers and plants. The citrate ion occurs in all animal tissues and fluids. The total circulating citric acid in the serum of humans is approximately 1 mg/kg body weight. Normal daily excretion in human urine is 0.2–1.0 g.

Physiological Role of Citric Acid. Citric acid accurs in the terminal oxidative metabolic system of virtually all organisms. This oxidative metabolic system (Fig. 1), variously called the Krebs cycle (for its discoverer, H. A. Krebs), the tricarboxylic acid cycle, or the citric acid cycle, is a metabolic cycle involving the conversion of carbohydrates, fats, or proteins to carbon dioxide and water. This cycle releases energy necessary for an organism's growth, movement, luminescence,

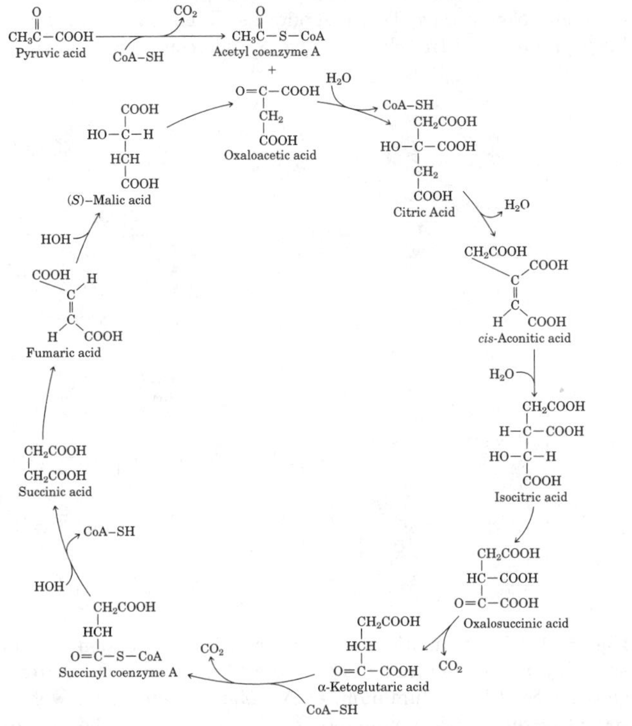

Figure 1. Krebs (citric acid) cycle. Coenzyme A is represented CoA–SH. The cycle begins with the combination of acetyl coenzyme A and oxaloacetic acid to form citric acid.

chemosynthesis, and reproduction. The cycle also provides the carbon-containing materials from which cells synthesize amino acids and fats. Many yeasts, molds, and bacteria conduct the citric acid cycle, and can be selected for their ability to maximize citric acid production in the process. This is the basis for the efficient commercial fermentation processes used today to produce citric acid.

Shipment and Storage

Crystalline citric acid, anhydrous, can be stored in dry form without difficulty, although conditions of high humidity and elevated temperatures should be avoided to prevent caking. Storage should be in tight containers to prevent exposure to moist air. Several granulations are commercially available with the larger particle sizes having less tendency toward caking.

Liquid citric acid is commercially available in a variety of concentrations with 50% w/w being most common. Grades are available that vary in appearance, purity, and color. Packaging is usually in drums, tank trucks, or rail cars. Liquid citric acid should be kept above 0°C to prevent crystallization.

Recommended materials of construction for pipes, tanks, and pumps handling citric acid solutions are 316 stainless steel, fiber glass-reinforced-polyester, polyethylene, polypropylene, and poly(vinyl chloride). At elevated temperatures, 304 stainless steel is not recommended.

Although not as corrosive as the acid, the sodium and potassium salts of citric acid should be handled in the same type of equipment as the acid to avoid corrosion problems.

Economic Aspects

Citric acid is manufactured in over 20 countries with 1990 worldwide production estimated at approximately 550,000 t. Most of this production is used for foods and beverages; however, industrial applications, eg, detergents and metal cleaning, of citric acid are becoming more important on a worldwide basis.

It was estimated that 1990 U.S. citric acid and citrate salt consumption was 152,000 t. Citric acid represents approximately 90% of this volume. This citric acid/citrate use and its historical distribution in various markets is described in Table 2.

Health, Safety, and Environmental Considerations

Citric acid, as well as its common sodium and potassium salt forms, are Generally Recognized As Safe (GRAS) by the U.S. Food and Drug Administration as Multiple Purpose Food Substances. Citric acid is also approved by the Joint FAO/WHO Expert Committee on Food Additives for use in foods without limitation.

Tests have shown that citric acid is not corrosive to skin but is a skin and ocular irritant. For these reasons it is recommended that individuals use appropriate personal protection to cover the hands, skin, eyes, nose, and mouth when in direct contact with citric acid solutions or powders.

Citric acid is biodegraded readily by many organisms under aerobic and anaerobic wastewater treatment conditions and in the natural environment.

Table 2. U.S. Citric Acid/Citrate Distribution by End Use, %

Product	1990	1988	1986
beverages	45	43	44
foods	21	20	22
pharmaceuticals and cosmetics	8	8	8
household detergents and cleaning products	19	23	20
misc. nonfood	7	6	6
Total annual consumption	*152,000*	*148,000*	*128,000*

Derivatives

Salts. The trisodium citrate salt is made by dissolving citric acid in water at a concentration of 50% w/w or higher. A solution of sodium hydroxide is carefully added to pH 8.0–8.5. The reaction is exothermic and cooling may be necessary. The tripotassium salt of citric acid is made in a similar manner using potassium hydroxide. The product crystallizes as the monohydrate. Ammonium salts of citric acid are made by adding either aqueous or anhydrous ammonia to citric acid dissolved in water.

Esters. The significant esters of citric acid are trimethyl citrate, triethyl citrate, tributyl citrate, and acetylated triethyl- and tributyl citrate. Many other esters are available but have not been used on a commercial scale. Citric acid esters are made under azeotropic conditions with a solvent, a catalyst, and the appropriate alcohol.

GARY BLAIR
PHILIP STAAL
Haarmann & Reimer Corporation

G. T. Blair and M. F. Zienty, *Citric Acid: Properties and Reactions,* Miles Laboratories, Inc., 1979.

Scientific Literature Review on GRAS Food Ingredients-Citrates, PB-223 850, National Technical Information Service, Springfield, Va., Apr. 1973; *Scientific Literature Review on GRAS Food Ingredients-Citric Acid,* PB-241 967, National Technical Information Service, Springfield, Va., Oct. 1974.

Code of Federal Regulations, Title 21, §182.1033, 182.6033, U.S. Government Printing Office, Washington, D.C., 1990.

Ecological Effects of Non-Phosphate Detergent Builders-Final Report on Organic Builders Other than NTA, International Joint Commission, Windsor, Ontario, Canada, July 21, 1980.

CLAYS

SURVEY

The terms *clay* or *clays* commonly refer to either rocks that are consolidated or unconsolidated sediments, or a group of minerals having unique properties. Traditionally, clays (rocks) are distinctive in at least two properties that render them technologically useful: plasticity and composition. Clays are predominantly composed of hydrous phyllosilicates, referred to as clay minerals. These are hydrous silicates of Al, Mg, K, and Fe, and other less abundant elements. Clay minerals are extremely fine crystals or particles, often colloidal in size and usually plate-like in shape. The nonclay mineral portion of clays (rocks) may consist of other minerals, portions of rocks, and organic compounds.

The very fine particles yield very large specific surface areas that are physically sorptive and chemically surface reactive. Many clay mineral crystals carry an excess negative electric charge owing to internal substitution by lower valent cations, and thereby increase internal reactivity in chemical combination and ion exchange. Clays, which may have served as substrates selectively absorbing and catalyzing amino acids in the origin of life, apparently catalyze petroleum formation in rocks (see PETROLEUM).

A clay deposit usually contains nonclay-like minerals as impurities and these impurities may actually be essential in determining the unique and specially desired properties of the clay. Both crystalline and amorphous minerals and compounds may be present in a clay deposit.

A broad definition of clays includes the following properties: *(1)* Crystalline hydrated silicates of aluminum, iron, and magnesium comprise the majority of clay minerals; however, amorphous hydrated aluminum compounds are also included. Distinctions among clay minerals are made by chemical and structural parameters. The chemical variations range from kaolinite, which is relatively uniform in chemical composition, to smectite minerals, which vary widely in chemical composition, base exchange properties, and expanding crystal lattice. Clay minerals are excellent examples of mixed layering, both random and regular, in layer-structure silicates. These mixed-layer clays are among the most ubiquitous of the various clay minerals. The structural differences among clay minerals are related to the arrangement of tetrahedral (T) and octahedral (O) layers, and the manner in which electrostatic charge imbalances, created by chemical substitution, are neutralized. Figure 1 shows several examples. *(2)* The possible content of hydrated alumina and iron. *(3)* The extreme fineness of individual clay particles, which may be of colloidal size in at least one dimension. *(4)* The property of thixotropy in various degrees of complexity. *(5)* The possible content of quartz, SiO_2, sand and silt, feldspars, mica, chlorite, opal, volcanic dust, fossil fragments, high density so-called heavy minerals, sulfates, sulfides, carbonate minerals, zeolites, and many other rock and mineral particles ranging upward in size from colloids to pebbles.

Geology and Occurrence

Clays may originate through several processes: *(1)* hydrolysis and hydration of a silicate, ie, alkali silicate + water → hydrated aluminosilicate clay + alkali hydroxide; *(2)* solution of a limestone or other soluble rock containing relatively insoluble clay impurities that are left behind; *(3)* slaking and weathering of shales (clay-rich sedimentary rocks); *(4)* replacement of a preexisting host rock by invading guest clay where the constituents are carried in part or wholly by solution; *(5)* deposition of clay in cavities or veins from solution; *(6)* bacterial and other organic activity, including the extraction of metal cations as nutrients by plants; *(7)* action of acid clays, humus, and inorganic acids on primary silicates; *(8)* alteration of parent material or diagenetic processes following sedimentation in marine and freshwater environments; and *(9)* resilication of high alumina minerals.

Clays or shales that may be utilized in the manufacture of bricks, tiles, and other heavy clay products exist in every state in the United States (see BUILDING MATERIALS–SURVEY). Glacial clays, as unassorted glacial till or secondarily deposited melt water are abundant in the United States north of the Missouri and Ohio Rivers. Fire clays are those that resist fusion at a relatively high temperature, usually around 1600°C. Missouri, Pennsylvania, Ohio, Kentucky, Georgia, Colorado, New Jersey, Texas, Arkansas, Illinois, and Maryland are large producers of fire clays. Adobe, a calcareous, sandy to silty clay

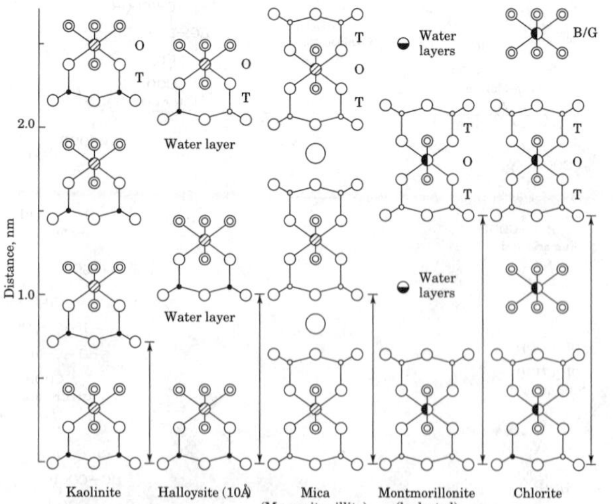

Figure 1. Diagrammatic representation of the succession of layers in some layer lattice silicates (12) where ○ is oxygen; ◎, hydroxyl; •, silicon; , Si–Al; ⊘, aluminum; ◐, Al–Mg; ○, potassium; ◒, Na–Ca. Sample layers are designated as O, octahedral; T, tetrahedral; and B/G, brucite- or gibbsitelike. The distance depicted by arrows between repeating layers in nm are 0.72, kaolinite; 1.01, halloysite (10 Å); 1.00, mica; ca 1.5, montmorillonite; and 1.41, chlorite.

used extensively for making sundried brick, is available in the more arid southwestern states. Slip clay for glazing pottery is produced near Albany, New York.

Bentonite, widely distributed geographically and geologically, also varies widely in properties.

Each continent has clays of almost every type; however, certain deposits are outstanding.

The commercial value of a clay deposit depends on market trends, competitive materials, transportation facilities, new machinery and processes, and labor and fuel costs. Naturally exposed outcrops, geological area and structure maps, aerial photographs, hand and power auger drills, core drills, earth resistivity, and shallow seismic methods are used in exploration for clays. Clays are mined primarily by open-pit operation, including hydraulic extraction; however, underground mining is also practiced.

Mineralogy

The development of apparatus and techniques, such as x-ray diffraction, contributed greatly to research on clay minerals. Crystalline clay minerals are identified and classified primarily on the basis of crystal structure and the amount and locations of charge (deficit or excess) with respect to the basic lattice. Amorphous (to x-ray) clay minerals are poorly organized analogues of crystalline counterparts.

The structural variations among the clay minerals can be understood by considering various physical combinations of tetrahedral and octahedral sheets and the electrostatic effect chemical substitution has on the structural units. The tetrahedral sheets are composed primarily of Si^{4+} and oxygen, but minor amounts of Al^{3+} or Fe^{3+} may substitute for Si^{4+}. The substitution of M^{3+} for Si^{4+} leaves the tetrahedral sheet negatively charged. The cations of the octahedral sheet are composed primarily of Al^{3+}, Fe^{3+}, Mg^{2+}, and Fe^{2+}, but all other transition elements, except Sc, may be included. The anions of the octahedral sheet are O^{2-}, OH^-, and F^-. The smallest unit of the octahedral sheet contains three octahedral having an ideal net charge of negative six, ie, three O^{2-}. If the negative charge is balanced by two trivalent cations, the layer is referred to as a dioctahedral layer; if balanced by three bivalent cations, the layer is referred to as a trioctahedral layer. Substitution of bivalent cations for trivalent cations, univalent cations (Li^+) for bivalent cations or unfilled octahedral sites, leaves the ocahedral layer a net negative charge. The tetrahedral apical oxygen is shared with the ocahedral layer to join the two types of layers.

The least complicated clay minerals are the 1:1 clay minerals composed of one tetrahedral (T) layer and one octahedral (O) layer.

Clay minerals that are composed of two tetrahedral layers and one octahedral layer are referred to as 2:1 clay minerals or TOT minerals.

The multitude of variation in clay minerals is caused by substitution in the octahedral and tetrahedral layers, resulting in charge deficits. The manner in which the charge deficit is balanced leads to many of the useful and unique properties of clay minerals.

Crystalline and Paracrystalline Groups

Kaolins. The kaolin minerals include kaolinite, dickite, and nacrite which all have composition $Al_2O_3 \cdot 2\ SiO_2 \cdot 2\ H_2O$; halloysite (7 Å), $Al_2O_3 \cdot 2\ SiO_2 \cdot 2\ H_2O$; and halloysite (10 Å), $Al_2O_3 \cdot 2\ SiO_2 \cdot 4\ H_2O$. The structural formulas for kaolinite and halloysite (10 Å), which are shown in Figure 1, are $Al_4Si_4O_{10}(OH)_8$ and $Al_4Si_4O_{10}(OH)_8 \cdot 4\ H_2O$, respectively. The so-called fire clay mineral is a *b*-axis disordered kaolinite; halloysite (7 Å) and halloysite (10 Å) are disordered along both the *a*- and *b*-axes. Indeed, most variations in the kaolin group originate as structural polymorphs, related to variations in layer stacks.

Halloysite, a mineral in the kaolin family, has a chemical composition similar to, but physical properties that differ greatly from, kaolinite. Halloysite differs from kaolinite in tetrahedral Al content, layer stacking sequence, and configuration of the six-fold rings. Four basic morphologies of halloysite are recognized: tubular (long and short), spheroidal, platy, and prismatic.

Kaolin most commonly originates by the alteration of feldspar or other aluminum silicates via an intermediate solution phase, usually by surface weathering or by rising warm (hydrothermal) waters.

Large deposits of relatively pure kaolinite have developed from parent, feldspar-rich pegmatites, whereas others are secondarily deposited in sedimentary beds after transportation.

The textures of kaolin (rock) include varieties similar to examples observed in igneous and metamorphic as well as sedimentary rocks. Kaolin grains and crystals may be straight or curved, sheaves, flakes, face-to-face or edge-to-edge floccules, interlocking crystals, tubes, scrolls, fibers, or spheres.

Serpentines. Substituting 3 Mg^{2+} for the 2 Al^{3+} in the kaolin structure results in the serpentine minerals, $Mg_3Si_2O_5(OH)_4$. In serpentines all three possible octahedral cation sites are filled. Most serpentine minerals are tubular to fibrous in structure presumably because of misfit between Mg octahedral and tetrahedral layers.

Talc and Pyrophyllite. Talc and pyrophyllite are 2:1 layer clay minerals having no substitution in either the tetrahedral or octahedral layer. These are electrostatically neutral particles ($x = 0$) and may be considered ideal 2:1 layer hydrous phyllosilicates. Talc and pyrophyllite are found in metamorphic rocks that are rich in Mg and A1, respectively.

Smectites (Montmorillonites). Smectites are the 2:1 clay minerals that carry a lattice charge and characteristically expand when solvated with water and alcohols, notably ethylene glycol and glycerol.

Smectites are structurally similar to pyrophyllite or talc, but differ by substitutions mainly in the octahedral layers.

The minerals of the smectite group have been formed by surface weathering, low temperature hydrothermal processes, alteration of volcanic dust in stratified beds, action of circulating water of uncertain source among fractures and in veins, and laboratory synthesis.

Illite. Illite is a general term for the clay mineral constituents of argillaceous sediments that strongly resemble mica minerals. Other names that have been used for illite include bravaisite, degraded mica, hydromica, hydromuscovite, hydrous illite, hydrous mica, K-mica, micaceous clay, and sericite. Illite and the mica minerals have a 2:1 sheet structure similar to the smectite minerals except that the maximum charge deficit in mica is typically in the tetrahedral layers and contains potassium held tenaciously in the interlayer space, which contributes to a 1.0-nm basal spacing.

The formula of illite can be expressed as $2K_2O \cdot 3MO \cdot 8R_2O_3 \cdot 24SiO_2 \cdot 12H_2O$, and the crystal structure by the formula $K_y[Al_{4-x}(Fe,Mg)_x](Si_{(8-y)+x}Al_y)O_{20}(OH)_4$ where y refers to the K^+ ions that satisfy the excess charges resulting when about 15% of the Si^{4+} positions are replaced by Al^{3+}.

Illite was defined as the most abundant clay mineral in Paleozoic shale and is widespread in many other sedimentary rocks; it is common in soils, slates, certain alteration products of igneous rocks, and recent sediments. Its origin has been attributed to alteration of silicate minerals by weathering and hydrothermal solutions, reconstitution, wetting and drying of soil clays, and diagenesis involving other three-layer minerals and potassium during geologic time and pressure under deep burial. Illitization of smectite via illite–smectite mixed-layer intermediates is a very common and an important reaction in the formation of shales during burial diagenesis.

Glauconite. Glauconite is a green, dioctahedral, micaceous clay rich in ferric iron and potassium. The generally accepted formula for glauconite is $(Na,K)_{0.78}(Fe^{3+}_{1.01}Al_{0.45}Mg_{0.39}Fe^{2+})_{2.05}(Si_{3.65}Al_{0.35})O_{10}(OH)_2$. Glauconite has many characteristics common to illite, but much glauconite contains random mixed expanding layers, and can be referred to as interstratified glauconite–smectite minerals. In addition, glauconite found in Late Cenozoic rocks tends to have less crystallographic order than older glauconite; therefore, the modifiers ordered (well crystalline) and disordered (poorly crystalline) are commonly used.

Glauconite occurs abundantly in sand-size or bigger pellets, or in pellets within fossils, notably foraminifera, giving it an organic connotation.

Celadonite is an iron-rich dioctahedral micaceous mineral that is similar to glauconite. Celadonite has a composition of: $(Na, K)_{0.83}(Fe^{3+}_{0.72}Al_{0.49}Mg_{0.63}Fe_{2+0.20})_{2.05}(Si_{3.81}Al_{0.19})O_{10}(OH)_2$ (39) and, like glauconite, has well crystalline, poorly crystalline, and interstratified varieties.

Chlorite and Vermiculite. Chlorite is a 1.4-nm (14Å) clay mineral that cannot be expanded or collapsed by traditional laboratory procedures. Structurally, the unit layer of chloride is composed of a 2:1 layer combined with a 0.4-nm Mg or Al interlayer or hydroxide sheet.

Palygorskite and Sepiolite. Palygorskite (attapulgite) and sepiolite are clay minerals in which the 2:1 layers are linked together in chain-like or a combination of chain-sheet structures.

Palygorskite and sepiolite are different from other clay minerals in the manner in which the 2:1 layers are joined. Rather than being joined in a continuous manner, the tetrahedral sheets are joined to an adjacent inverted tetrahedral layer, making the octahedral layers noncontinuous and leaving an open channel in the mineral structure.

Palygorskite has an ideal formula that approximates $MgAl_3Si_8O_{20}(OH)_3(OH_2)_4 \cdot x[R^{2+}(H_2O)_4]$; the ideal formula for sepiolite is $Mg_8Si_{12}O_{30}(OH)_4(OH_2)_4 \cdot x[R_{2+}(H_2O)_8]$. The chemical composition of a specific sample may vary widely because there is substitution of Na, Fe, Mn, Al, and Ni in the octahedral sheets of sepiolite, and substitution of Na, Fe, and Mn in paylgorskite.

These clays have distinctive uses and properties not shown by platy clay minerals.

Mixed-Layer Minerals. In addition to polymorphism resulting from the disordering and proxying of one element for another, clay minerals exhibit ordered and random intercalation sandwiches with one another.

Mixed-layer clays, particularly illite–smectite, are very common minerals and illustrate the transitional nature of the 2:1 layered silicates. The transition from smectite to illite occurs when smectite, in the presence of potassium from another mineral such as potassium feldspar, or from thermal fluids, is heated and/or buried. With increasing temperature smectite plus potassium is converted to illite.

The physical structure of mixed-layer minerals is open to question. In the traditional view, the MacEwan crystallite is a combination of 1.0-nm (10-Å) nonexpandable units (illite) that forms as an epitaxial growth on 1.7-nm expandable units (smectite) that yield a coherent diffraction pattern. This view is challenged by the fundamental particle hypothesis which is based on the existence of fundamental particles of different thickness.

Amorphous and Miscellaneous Groups

Allophane and Imogolite. Allophane is an amorphous clay that is essentially an amorphous solid solution of silica, alumina, and water. Allophane has been found most abundantly in soils and altered volcanic ash. It usually occurs in spherical form but has also been observed in fibers.

Imogolite is an uncommon paracrystalline clay mineral assigned the formula $1.1SiO_2 \cdot Al_2O_3 \cdot 2.3-2.8H_2O$. The morphology of imogolite has been reported as thread-shaped and as hollow spheres. Imogolite is generally viewed as an intermediate between allophane and kaolinite. In modern environments both allophane and imogolite are associated with volcanic material in areas of high rainfall.

High Alumina Clay Minerals. Several hydrated alumina minerals should be grouped with the clay minerals because the two types may occur so intimately associated as to be almost inseparable. Diaspore (α-AlO(OH)) and boehmite (γ-AlO(OH)), both $Al_2O_3 \cdot H_2O$ (Al_2O_3, 85%; H_2O, 15%) are the chief constituents of diaspore clay, which may contain over 75% Al_2O_3 on the raw basis. Gibbsite, $Al_2O_3 \cdot 3\ H_2O$ (Al_2O_3, 65.4%; H_2O, 34.6%), and cliachite, the so-called amorphous alumina hydrate (much cliachite is probably cryptocrystalline), as well as the monohydrates, occur in bauxite, bauxitic kaolin, and bauxitic clays.

The hydrated alumina minerals usually occur in oolitic structures (small spherical to ellipsoidal bodies the size of BB shot, about 2 mm in diameter) and also in larger and smaller structures. High alumina minerals are found where intense weathering and leaching has dissolved the silica. It is generally believed that a very humid, subtropical climate is required for this (lateritic) stage of weathering.

T. DOMBROWSKI
Engelhard Corporation

C. E. Weaver and L. D. Pollard, *The Chemistry of Clay Minerals,* Elsevier, New York, 1973.

R. E. Grim. *Clay Mineralogy,* McGraw-Hill Book Co., Inc., New York, 1968.

D. D. Carr, Sr. ed., *Clays: Industrial Minerals and Rocks,* 6th ed., AIME, pp. 229–277.

R. F. Giese, *Hydrous Phyllosilicates (exclusive of micas), Miner. Soc. Am. Rev. Mineral.* **19,** 29–62 (1988).

USES

Clays are composed of extremely fine particles of clay minerals which are layer-type aluminum silicates containing structural hydroxyl groups. Clay particles generally give well-defined x-ray diffraction patterns from which the mineral composition can readily be determined. Clay particles are so finely divided that their properties are often controlled by the surface properties of the minerals rather than by their chemical composition. Particle size distribution, particle shape, the nature and amount of soluble and mineral impurities, the presence of organic matter, exchangeable ions, and degree of crystal perfection are all known to affect the properties of clays profoundly.

Ceramic Products

A large proportion of the annual production of ball clay, fire clay, and common clay and shale are used for ceramics. Ceramic products are made from fine-grained oxides, silicates and other naturally occurring or synthetic minerals through the application of high temperature. The suitability of clays as ceramic raw materials is determined by their specific properties as outlined in the following section.

Properties. Plasticity is defined as the property of a material that permits it to be deformed under stress without rupturing and to retain the shape produced after the stress is removed. The amount of water to reach a plastic state ("water of plasticity") is typically used with ceramic clays as one measure of their suitability for ceramic applications. In general, a relatively low value for water of plasticity is desired, hence kaolinite, illite, and chlorite clays have better plasticity characteristics than attapulgite or montmorillonite (smectite).

Green strength is the transverse breaking strength measured while the plasticizing water is still present. The values obtained relate to the ease of formation of molded shapes of the wet clay and to the durability of the shape formed.

The reduction in length or volume that takes place on drying is termed drying shrinkage. In the production of ceramic ware, minimal drying shrinkage is desired. Illite, chlorite, and kaolinite give less drying shrinkage than montmorillonite.

Dry strength is measured as the transverse breaking strength of test piece after drying, usually at 150°C. High dry strength is usually desirable since it gives greater durability to the unfired piece.

Heating clay materials to a sufficiently high temperature results in the fusion of the material and the desirable ceramic properties are obtained. Varying the firing time and temperature can control to some extent the properties of the finished ceramic.

Raw Materials. Depending on the specific ceramic desired, raw materials vary considerably. Almost any clay composition is satisfactory for the manufacture of brick unless the clay contains a large percentage of coarse material that cannot be eliminated or ground to an adequate fineness.

Roofing and structural tiles are usually made from the same material as brick. Drain tiles are often made from clays having about 75% of fine-grained nonclay mineral material in addition to components that provide a high green and dry strength and a low fusion point.

Clays composed of mixtures of clay minerals containing 25–50% fine-grained unsorted quartz are well suited to the manufacture of stoneware.

Porcelain and dinnerware are made up of about equal amounts of kaolin, ball clay, flint (ground quartz), feldspar, or some other white-firing fluxing material such as talc and nepheline.

The slurry used in porcelain enameling is commonly composed of ball clay, frits (finely ground glass with a low fusion temperature) and coloring pigments.

Refractory products are prepared from a wide variety of naturally occurring materials such as chromite, magnesite, and clays containing a large proportion of kaolinite. Increasingly, higher purity synthetic materials are being used to obtain special properties.

Paper

The paper industry is the largest consumer of processed clays, nearly all of which is kaolin. Kaolin has two main uses in paper: as a filler where kaolin is mixed with the pulp fibers; and as a coating where it is combined with water, binders, and various additives and coated onto the surface of the paper sheet. As a filler, kaolin improves the appearance and opacity of the sheet, while as a coating it markedly improves the print quality, especially in the case of multicolor printing.

Types of Kaolin. Kaolins for paper are generally classified into three groups according to how they are processed: air-floated, water-washed, and calcined. Kaolins that are air-floated are processed by selecting appropriate crudes, drying, crushing, pulverizing, and removal of oversize particles by air separation. Water-washed kaolins, as the name infers, are processed as aqueous slurries. Wet processing consists of mining selected crudes, dispersing in water, degritting to remove oversize particles, centrifugation into different particle size fractions, chemical bleaching, filtration, redispersion, and drying. Additional steps to improve product whiteness such as magnetic separation, froth flotation and selective flocculation are often carried out. In some cases, the kaolin is calcined by heating to 1000°C to further improve its whiteness and opacity.

Sources. The largest sources of kaolin for the paper industry are Cornwall in the United Kingdom and Georgia in the United States. Smaller, but important, sources of production are located in Australia and Brazil.

Properties. Kaolin is useful in paper because it is white, of fine particle size, and gives low viscosity in aqueous slurries at high solids. The properties of kaolin vary considerably according to the source of crude clay, the method of processing, and its particle size. Water-washed kaolins are generally higher in brightness and more uniform than air-floated kaolins. Calcined kaolins are of still higher brightness.

Since paper kaolins are nearly white, a blue reflectance measurement, or "brightness" is used to specify color. Air-floated kaolins range from 81 to 85 in brightness, water-washed kaolins cover the brightness range of 82 to 90; calcined kaolins are of 93 brightness or slightly higher. Higher brightness products are more costly because of the more elaborate processes used in their production.

The ability of kaolin to give a low viscosity slurry at up to 70% solids is important in paper coating. Since the kaolin is coated on the sheet at high speed, the highest possible solids are desired to minimize the cost of evaporating water after the coating operation.

All coating kaolins must be free of oversize particles (greater than 44 micrometers) and typically have a fairly broad particle size distribution. Finer particle clays are more suitable for coating paper than are coarser particle kaolins. Finer particle clays give improved opacity, gloss, and print quality as compared to coarser particle materials. When clays are used as fillers, particle size is less important. Some clays are selectively ground or "delaminated" to give relatively large diameter but thin platelets that are especially valuable in obtaining good paper quality, even at low coating weights.

Molding Sands

Molding sands, composed essentially of sand and clay, are used extensively in the metallurgical industry for the shaping of metal by the casting process. The molding sand may be a natural sand containing clay or a synthetically prepared mixture of clean quartz sand and clay. The bentonites, composed essentially of montmorillonite, are the most-used clay for this application.

Plastics and Rubber

Clay is used in plastics and rubber as a reinforcing agent or as a filler to extend the polymer. Kaolin is by far the most used clay filler in polymers and various grades are produced for this industry. The kaolins may be air-floated, water-washed, or calcined. In many cases, the kaolins have been surface treated to improve compatibility or reinforcing power. Attapulgite is being used as a replacement for fibrous asbestos in some applications.

Drilling Fluids

Clays are an important ingredient of most water based drilling fluids since they give a higher density to the fluid and sufficient viscosity that cuttings from the drill bit may readily be brought to the surface. Montmorillonite (in the form of bentonite), attapulgite, and sepiolite are the main clays used in this application. In addition to clays, weighting agents, organic thickeners, and other materials are part of the drilling fluid composition.

Paint

Clays are widely used in both oil-based and water-based paints since they perform several important functions: they extend the much higher cost titanium dioxide opacifying pigment, control viscosity so as to prevent settling during storage, provide thixotropy so that the paint does not sag after application, improve gloss retention, promote film integrity, and aid in tint retention. Both regular kaolin and calcined kaolin are used in paint although the latter is preferred because it contributes significantly to opacity. Bentonite and attapulgite are also used but as a viscosity control agent rather than as an extender pigment. Organic treated bentonites are used as suspending and antisag agents in oil-based paints.

Miscellaneous

Clays are used in a vast number of products and, in a few cases, as a chemical raw material such as for the production of zeolites or aluminum.

Adsorbents. Acid activated and raw clays have been used to decolorize and remove off-tastes from edible oils, as kitty litter, and in sweeping compounds. Clays used as adsorbents include fuller's earth (largely montmorillonite), attapulgite, and kaolin.

Adhesives. Clays, especially kaolin and attapulgite, are used in various adhesive formulations. In addition to serving as an extender, clays can also increase viscosity to reduce dripping and sagging, improve smoothness of the surface, and slow the penetration of the adhesive into the substrate. Both air-floated and water-washed kaolins are used.

Catalysts. Although crude clays have been used in petroleum refining in the past, more recently processed clays have replaced crude clays as a raw material for catalyst production. Various proprietary processes are used and numerous patents have been issued.

Cement. Kaolin, and in some cases, other clays are used to provide controlled amounts of silica and alumina in the product.

Chemical Raw Material. In addition to a raw material for catalyst and zeolite production, kaolin has also been studied as a raw material for aluminum production. Kaolin, with a much lower alumina content than bauxite, has not yet been shown to be economically competitive. Kaolin, usually air floated, is an essential ingredient in continuous filament fiber glass.

Inks. Refined kaolin is a common ingredient in printing inks where it serves as an extender, controls rheology, and improves adhesion. Kaolin for this application must be of extreme fineness and free from larger particles.

Pelletizing. Bentonite is used as a pelletizing agent in several industries to agglomerate fine particle products into larger sizes. The fertilizer and iron ore industries consume substantial quantities for this application.

Pesticides. Highly concentrated pesticides are diluted with fine particle inert clays to allow their application as dusts or granules. Attapulgite, montmorillonite, and kaolin are used for this application.

Other Clay Uses. Other applications for clays include use as a suspending agent, eg, montmorillonite and attapulgite in liquid fertilizers and dishwasher detergents; in pharmaceuticals, eg, kaolinite and attapulgite for diarrhea control; in cosmetics; montmorillonite and attapulgite; and in water impedance where bentonite linings are used for reservoirs and waste disposal areas.

PAUL SENNETT
Engelhard Corporation

H. H. Murray, ed., *Applied Clay Science,* Vol. 5, No. 5 and 6, Mar. 1991.

R. E. Grim, *Applied Clay Mineralogy,* McGraw-Hill, New York, 1962.

Mineral Facts and Problems, Clays, U.S. Dept. of the Interior, Bureau of Mines, Washington, D.C., 1985.

CLINICAL CHEMISTRY. See AUTOMATED INSTRUMENTATION, CLINICAL CHEMISTRY.

CLUTCH FACINGS. See BRAKE LININGS AND CLUTCH FACINGS.

COAGULANTS AND ANTICOAGULANTS. See BLOOD, COAGULANTS AND ANTICOAGULANTS.

COAL

Coal is usually a dark black color, although geologically younger deposits of brown coal have a brownish red color (see LIGNITE AND BROWN COAL). The color, luster, texture, and fracture vary with rank, type, and grade. Coal is the result of combined biological, chemical, and physical degradation of accumulated plant matter over geological ages. The relative amounts of remaining plant parts leads to different types of coal, which are sometimes termed banded, splint, nonbanded (cannel and boghead); or hard or soft; or lignite, subbituminous, bituminous, or anthracite. In Europe the banded and splint types are generally referred to as ulmic or humic coals. Still other terms refer to the origins of the plant parts through maceral names such as vitrinite, liptinite, and inertinite. The degree of conversion of plant matter or coalification is referred to as rank. Brown coal and lignite, subbituminous coal, bituminous coal, and anthracite make up the rank series with increasing carbon content. The impurities in these coals cause differences in grade.

Coal consists primarily of carbon, hydrogen, and oxygen, and contains lesser amounts of nitrogen and sulfur and varying amounts of moisture and mineral matter. The mode of formation of coal, the variation in plant composition, the microstructure, and the variety of mineral matter indicate that there is a mixture of materials in coal. The nature of the organic species present depends on the degree of biochemical change of the original plant material, on the historic pressures and temperatures after the initial biochemical degradation, and on the finely divided mineral matter deposited either at the same time as the plant material or later. The principal types of organic compounds have resulted from the formation and condensation of polynuclear and heterocyclic ring compounds containing carbon, hydrogen, nitrogen, oxygen, and sulfur. The fraction of carbon in aromatic ring structures increases with rank.

Nearly all coal is used in combustion and coking (see COAL CONVERSION PROCESSES). At least 80% is burned directly in boilers for generation of electricity or steam for industrial purposes. Small amounts are used for transportation, space heating, firing of ceramic products, etc. The rest is essentially pyrolyzed to produce coke, coal gas, ammonia, coal tar, and light oil products from which many chemicals are produced (see FEEDSTOCKS–COAL CHEMICALS). Combustible gases and chemical intermediates are also produced by the gasification of coal, and different carbon products are produced by various heat treatments. A small amount of coal is used in miscellaneous applications such as fillers, pigments, foundry material, and water filtration.

World reserves of bituminous coal and anthracite are ca 5.6×10^{12} t of coal equivalent, ie, 29.3 GJ/t (12.6×10^3 Btu/lb), and subbituminous and lignite are 2.9×10^{12} t of coal equivalent. For economic and environmental reasons coal consumption has been cyclic.

Coal Petrography

The study of the origin, composition, and technological application of these materials is called coal petrology, whereas coal petrography involves the systematic quantification of the amounts and characteristics by microscopic study. The petrology of coal may involve either a macroscopic or a microscopic scale.

On the macroscopic scale, two coal classifications have been used: humic or banded coals and sapropelic or nonbanded coals.

Macerals. Coal parts derived from different plant parts are referred to as macerals. The maceral names end in "-inite" as do the mineral forms of rocks. For example, for the maceral group there is vitrinite; the macerals include telinite (with submaceral telocollinite) and collinite (with submacerals gelocollinite, desmocollinite, and corpocollinite).

Vitrinite Reflectance. The amount of light reflected from a polished plane surface of a coal particle under specified illumination conditions increases with the aromaticity of the sample and the rank or the coal or maceral. Precise measurements of reflectance, usually expressed as a percentage, are used as an indication of coal rank.

Application of Coal Petrology and Petrography Petrographic analysis is frequently carried out for economic evaluation or to obtain geologic information. Samples are usually lumps or more coarsely ground material that have been mounted in resins and polished. Maceral analysis involves the examination of a large number (usually >500) of particles during a traverse of a polished surface to identify the macerals at specified intervals. A volume percentage of each of the macerals present in a sample is calculated.

Seam correlations, measurements of rank and geologic history, interpretation of petroleum formation with coal deposits, prediction of coke properties, and detection of coal oxidation can be determined from petrographic analysis.

Classification Systems

Prior to the nineteenth century, coal was classified according to appearance, eg, bright coal, black coal, or brown coal. A number of classification systems have since been developed. These may be divided into two types, which are complementary: scientific and commercial. Both are used in research, whereas the commercial classification is essential industrially. In the scientific category, the Seyler chart has considerable value.

Systems include the Seyler classification (Fig. 1), the ASTM classification, the National Coal Board classification for British coals, and International classification.

Composition and Structure

The constitution of a coal involves both the elemental composition and the functional groups that are derived therefrom. The structure of the coal solid depends to a significant extent on the arrangement of the functional groups within the material.

Composition. The functional groups within coal contain the elements C, H, O, N, or S. The significant oxygen-containing groups

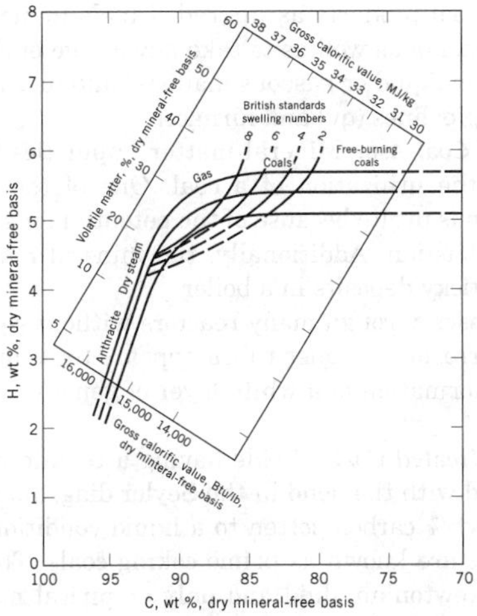

Figure 1. Simplified form of Seyler's coal classification chart. Note that ASTM uses the free-swelling index.

found in coals are carbonyl, hydroxyl, carboxylic acid, and methoxy. The nitrogen-containing groups include aromatic nitriles, pyridines, carbazoles, quinolines, and pyrroles. Sulfur is primarily found in thiols, dialkyl and aryl-alkyl thioethers, thiophene groups, and disulfides. Elemental sulfur is observed in oxidized coal. The relative and absolute amounts of the various groups vary with coal rank and maceral type.

Coal Structure. Conclusions regarding the chemical structure of the macromolecules within coal are generally based on experimental measurements and an understanding of structural organic chemistry.

Several requirements must be met in developing a structure. Not only must elementary analysis and other physical measurements be consistent, but limitations of structural organic chemistry and stereochemistry must also be satisfied. Mathematical expressions have been developed to test the consistency of any given set of parameters used to describe the molecular structure of coal and analyses of this type have been reported.

The macromolecules that make up the coal structure are held together by a variety of forces and bonds. The coal network model is one approach to describing the three-dimensional structure of the solid. Aromatic clusters are linked by a variety of connecting bonds, through oxygen, methylene or longer aliphatic groups, or disulfide bridges, and the proportions of the different functional groups change as the rank of the coal is progressively increased.

Coal Constitution. Chemical composition studies indicate that brown coals have a relatively high oxygen content. About two-thirds of the oxygen is bonded carboxyl, acetylatable hydroxyl, and methoxy groups. Additionally, unlike in bituminous coals, some alcoholic hydroxyl groups are believed to exist.

The anthracites, which approach graphite in composition (see CARBON–CARBON AND ARTIFICIAL GRAPHITE), are classified higher in rank, have less oxygen and hydrogen, and are less reactive than bituminous coals.

Mineral Matter in Coal. The mineral matter in coal results from several separate processes. Some comes from the material inherent in all living matter; some from the detrital minerals deposited during the time of peat formation; and a third type from secondary minerals that crystallized from water which has percolated through the coal seams.

Properties

Pieces of coal are mixtures of materials somewhat randomly distributed in differing amounts. The mineral matter can be readily distinguished from the organic, which is itself a mixture. Coal properties reflect the individual constituents and the relative proportions. By analogy to geologic formations, the macerals are the constituents that correspond to minerals that make up individual rocks. For coals, macerals, which tend to be consistent in their properties, represent particular classes of plant parts that have been transformed into coal. Most detailed chemical and physical studies of coal have been made on macerals or samples rich in a particular maceral, because maceral separation is time consuming.

The predominant maceral group in U.S. coals is vitrinite. The other important maceral groups include inertinite consisting of micrinite, a dull black amorphous material, fusinite, a dull fibrous material similar to charcoal, and the liptinite group including sporinite, which is relatively fusible and volatile.

In the United States the commercial classification of coals is based on the fixed carbon (or volatile matter) content and the moist heating value. Table 1 indicates the usual range of composition of commercial coals of increasing rank.

Physical Methods of Examination. Physical methods used to examine coals can be divided into two classes which, in the one case, yield information of a structural nature such as the size of the aromatic nuclei, ie, methods such as x-ray diffraction, molar refraction, and calorific value as a function of composition; and in the other case indicate the fraction of carbon present in aromatic form, ie, methods such as ir and nuclear magnetic resonance spectroscopies, and density as a function of composition.

Physical Properties. Most of the physical properties discussed herein depend on the direction of measurement as compared to the bedding plane of the coal. Additionally, these properties vary according to the history of the piece of coal. Properties also vary between pieces because of coal's brittle nature and the crack and pore structure.

The specific electrical conductivity of dry coals is very low and the specific resistance $10^{10}-10^{14}$ ohm·cm, although it increases with rank. Coal has semiconducting properties.

The dielectric constant is also affected by structural changes on strong heating. Also, the value is very rank-dependent, exhibiting a minimum at about 88 wt % C and rising rapidly for carbon contents over 90 wt %.

Density values of coals differ considerably, even after correcting for the mineral matter, depending on the method of determination. The true density of coal matter is most accurately obtained from measuring the displacement of helium after the absorbed gases have been

Table 1. Composition of Humic Coals

Type of coal	Composition, wt %[a] C	H	O	N	Moisture as found	Volatile matter	Calorific value, kJ/g[b]
peat	45–60	3.5–6.8	20–45	0.75–3.0	70–90	45–75	17–22
brown coals and lignites	60–75	4.5–5.5	17–35	0.75–2.1	30–50	45–60	28–30
bituminous coals	75–92	4.0–5.6	3.0–20	0.75–2.0	1.0–20	11–50	29–37
anthracites	92–95	2.9–4.0	2.0–3.0	0.5–2.0	1.5–3.5	3.5–10	36–37

[a] Dry, mineral-matter-free basis except for moisture value.
[b] To convert kJ/g to Btu/lb, multiply by 430.2.

removed from the coal sample. Density values increase with carbon content or rank for vitrinites.

Thermal conductivity and thermal diffusivity are also dependent on pore and crack structure. Thermal conductivities for coals of different ranks at room temperature are in the range of 0.23–0.35 W/(m·K).

The specific heat of coal can be determined by direct measurement or from the ratio of the thermal conductivity and thermal diffusivity. The latter method gives values decreasing from 1.25 J/(g·K) (0.3 cal/(g·K)) at 20°C to 0.4 J/(g·K) (0.1 cal/(g·K)) at 800°C. The specific heat is affected by the oxidation of the coal.

Ultrafine Structure. Coal contains an extensive network of ultrafine capillaries that pass in all directions through any particle. The smallest and most extensive passages are caused by the voids from imperfect packing of the large organic molecules. Vapors pass through these passages during adsorption, chemical reaction, or thermal decomposition. The rates of these processes depend on the diameters of the capillaries and any restrictions in them. Most of the inherent moisture in the coal is contained in these capillaries. The porous structure of the coal and products derived from it have a significant effect on the absorptive properties of these materials.

Mechanical Properties. Mechanical properties are important for a number of steps in coal preparation, from mining through handling, crushing, and grinding. The properties include elasticity and strength as measured by standard laboratory tests and empirical tests for grindability and friability, and indirect measurements based on particle-size distributions.

Properties Involving Utilization. Coal rank is the most important single property for application of coal. Rank sets limits on many properties such as volatile matter, calorific value, and swelling and coking characteristics. Other properties of significance include grindability, ash content and composition, and sulfur content.

Chemistry

Coal reactions, which on heating are important to the production of coke and synthetic fuels, are complicated by its structure.

Mature (>75 wt % C) coals are built of assemblages of polynuclear ring systems connected by a variety of functional groups and hydrogen-bonded cross-links. The ring systems themselves contain many functional groups. These polynuclear coal molecules differ one from another to some extent in the coal matter. For bituminous coal, a tarlike material occupies some of the interstices between the molecules. Generally coal materials are nonvolatile except for some moisture, light hydrocarbons, and contained carbon dioxide. The volatile matter produced on carbonization reflects decomposition of parts of the molecule and the release of moisture. Rate of heating affects the volatile matter content such that faster rates give higher volatile matter yields.

Coal composition is denoted by rank. Rank increases with the carbon content and decreases with increasing oxygen content. Partial oxidation as carried out in gasification produces carbon monoxide, hydrogen gas, carbon dioxide, and water vapor. The first two are used to make a wide range of synthetic fuels and chemicals. Surface oxidation short of combustion, or using nitric acid or potassium permanganate solutions, produces regenerated humic acids similar to those extracted from peat or soil. Further oxidation produces aromatic acids and oxalic acid, but at least half of the carbon forms carbon dioxide.

Treatment with hydrogen at 400°C and 12.4 MPa (1800 psi) increases the coking power of some coal and produces a change that resembles an increase in rank. Treatment of coal with chlorine or bromine results in addition and substitution reactions. Hydrolysis using aqueous alkali has been found to remove ash material including pyrite.

The pyritic sulfur in coal can undergo reaction with sulfate solution to release elemental sulfur. Processes to reduce the sulfur content of coal have been sought.

Many of the products made by hydrogenation, oxidation, hydrolysis, or fluorination are of industrial importance. Concern about stable, low cost petroleum and natural gas supplies is increasing the interest in some of the coal products as upgraded fuels to meet air pollution control requirements as well as to take advantage of the greater ease of handling of the liquid or gaseous material and to utilize existing facilities such as pipelines (qv) and furnaces.

Reactions of Coal Ash. Mineral matter impurities have an important effect on the utilization of a coal. One of the constituents of greatest concern is pyrite because of the potential for sulfur oxide generation on combustion. Additionally, the mineral matter has a tendency to form sticky deposits in a boiler.

Coal ash passes through many reactors without significant chemical change. Corrosion of boiler tubes appears to be initiated in some cases with the formation of a white layer of general composition $(Na, K)_3Al(SO_4)_3$.

Plasticity of Heated Coals. Coals having a certain range of composition associated with the bend in the Seyler diagram (see Fig. 1) and having 88–90 wt % carbon soften to a liquid condition when heated. These materials are known as prime coking coals. The coal does not behave like a Newtonian fluid and only empirical measurements of plasticity can be made.

Pyrolysis of Coal. Most coals decompose below temperatures of about 400°C, characteristic of the onset of plasticity. Moisture is released near 100°C, and traces of oil and gases appear between 100–400°C, depending on the coal rank. As the temperature is raised in an inert atmosphere at a rate of 1–2°C/min, the evolution of decomposition products reaches a maximum rate near 450°C, and most of the tar is produced in the range of 400–500°C. Gas evolution begins in the same range but most evolves above 500°C. If the coal temperature in a single reactor exceeds 900°C, the tars can be cracked, the yields are reduced, and the products are more aromatic. Heating beyond 900°C results in minor additional weight losses but the solid matter changes its structure. The tests for volatile matter indicate loss in weight at a specified temperature in the range of 875–1050°C from a covered crucible. This weight loss represents the loss of volatile decomposition products rather than volatile components.

A predictive macromolecular network decomposition model for coal conversion based on results of analytical measurements has been developed called the functional group, depolymerization, vaporization, cross-linking (FG-DVC) model. Data are obtained on weight loss on heating (thermogravimetry) and analysis of the evolved species by Fourier transform infrared spectrometry. Separate experimental data on solvent swelling, solvent extraction, and Gieseler plastometry are also used in the model. Volatile matter yields decrease with increasing coal rank. An overall picture of the pyrolysis process is generally accepted but the detailed mechanism is controversial.

The mechanism of coal pyrolysis has been discussed in the literature. The early stages involve formation of a fluid through depolymerization and decomposition of coal organic matter containing hydrogen. Around 400–550°C aromatic and nonaromatic groups may condense after releasing hydroxyl groups. The highest yields of methane and hydrogen come from coals having 89–92 wt % C. Light hydrocarbons other than methane are released most readily below 500°C; methane is released at 500°C. The highest rate for hydrogen occurs above 700°C.

Resources

World Reserves. Amounts of coal of some specified minimum deposit thickness and some specified maximum overburden thickness existing in the ground are termed resources. There is no economic consideration for resources, but reserves represent the portion of the resources that may be recovered economically using conventional mining equipment.

Comprehensive reviews of energy sources are published by the World Energy Conference, formerly the World Power Conference, at six-year intervals. The 1986 survey includes reserves and also gives total resources. In 1986 the total proven reserves of recoverable solid fuels were given as 6×10^{11} metric tons. One metric ton is defined as 29.2×10^3 MJ (27.7×10^6 Btu) to provide for the variation of calorific value in different coals. The total estimated additional reserves recoverable and total estimated additional amount in place are 2.2×10^{12} and 7.7×10^{12} metric tons, respectively. These figures are about double the 1913 estimates, primarily because significantly increased reserves have been indicated for Russia.

The part of the resource that is economically recoverable varies by country. The estimates made in the survey show that the proven recoverable reserves would last about 1200 years at the 1988 annual rate of production and that the estimated additional amount in place represents almost 1700 years' supply at 1988 annual consumption.

Coal is widely distributed in the United States as indicated in Figure 2.

Coal Production. World coal and lignite production rose to about 4.7×10^9t in 1988. Coal production in the United States has increased with fluctuations to ca 1000×10^6t in 1992. The demand for energy is continually increasing and the highest energy consumption in the world occurs in the United States. Estimated coal consumption reduces the known recoverable reserves at about 1%/yr. Whereas the use of bituminous coal is expected to continue to increase in terms of tonnage, the percentage of coal used in the United States has stabilized.

Sample Sources

Basic coal research requires a variety of coal samples of different ranks that workers may access using a minimum of effort. Coal sample banks fill this need. Moreover, over the past decades it has become evident that the quality of samples degrades from atmospheric oxidation and the degradation has limited the ability of researchers to compare results. The U.S. Department of Energy Office of Basic Energy Sciences has sponsored the Argonne Premium Coal Sample Program to permit the acquisition of ton-sized samples of each of eight different coals representing a range of coal ranks, chemical composition, geography, and maceral content.

Mining and Preparation

Mining. Coal is obtained by either surface mining of outcrops or seams near the surface or by underground mining depending on geological conditions, which may vary from thick, flat seams to thin, inclined seams that are folded and need special mining methods. Coal mining has changed from a labor intensive activity to one that has become highly mechanized.

Strip or open-pit mining involves removal of overburden from shallow seams, breaking of coal by blasting or mechanical means, and loading the coals. The two methods of underground mining commonly used are room-and-pillar and longwall.

Preparation. Coal preparation is of significant importance to the coal industry and to consumers. Preparation normally involves some size reduction of the mined coal and the systematic removal of some ash-forming material and very fine coal.

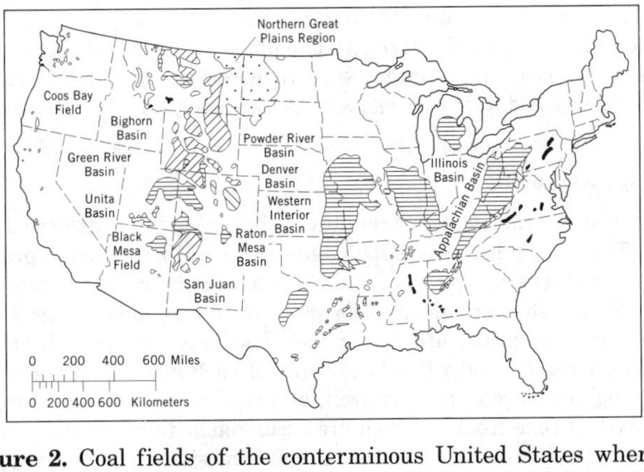

Figure 2. Coal fields of the conterminous United States where ■ represents anthracite and semianthracite; ▨, low volatile bituminous coal; ▤, medium and high volatile bituminous coal; ▧, subbituminous coal; and ▩, lignite.

In some areas, run-of-mine coal is separated into three products: a low gravity, premium-priced coal for metallurgical or other special use, a middlings product for possible boiler firing, and a high ash refuse. The complete preparation of coal usually requires several precesses.

Jig washing is the most widely used of all cleaning methods. Froth flotation is the most important method for cleaning fine coal because very small particles cannot be separated by settling methods.

Draining on screens removes substantial amounts of water from larger coal, but other dewatering methods are required for smaller sizes having larger surface areas. Vibrating screens and centrifuges are used for dewatering.

For utilities, two types of storage are used. A small amount of coal in storage meets daily needs and is continually turned over. This coal is loaded into storage bins or bunkers. However, long-term reserves are carefully piled and left undisturbed except as necessary to sustain production.

Transportation

The usual means of transporting coal are railroad, barge, truck, conveyer belt from mine to plant, and slurry pipelines.

Economic Aspects

Of the 1989 total, 65,128,000 t were metallurgical coal and 34,910,000 t were steam coal. Exports of coke from the United States in 1989 were 1,169,120 t, whereas anthracite exports were 745,749 t. Lignite exports were 163,628 t. In 1989 Canada produced 77,727,000 t, imported 16,160,000 t and exported 36,094,000 t. Japan is the principal recipient of Canadian exports.

Health and Safety Factors

Coal mining has been a relatively dangerous occupation. In the seven years after the passage of the Federal Coal Mine Health and Safety Act of 1969, the average fatality rate decreased to 0.58, and by 1989 the rate was 0.25.

The principal causes of fatalities are falling rock from mine roofs and faces, haulage, surface accidents, machinery, and explosions. For disabling injuries the primary causes are slips and falls, handling of materials, use of hand tools, lifting and pulling, falls of roof rock, and haulage and machinery.

Regulations. The U.S. Bureau of Mines, Mining Enforcement and Safety Administration (MESA) studies hazards and advises on accident prevention. MESA also administers laws dealing with safety in mines. Individual states may also have departments of mines to administer state standards.

The Federal Coal Mine Health and Safety Act set standards for mine ventilation, roof support, coal dust concentration levels, mine inspections, and equipment. As a part of this comprehensive act, miners must receive medical examinations at employer expense, and payments are made from the U.S. government to miners who cannot work because of black lung disease.

Uses

Coal As Fuel. Coal is used as a fuel for electric power generation, industrial heating and steam generation, domestic heating, railroads, and coal processing. About 87% of the world's coal production is burned to produce heat and derived forms of energy. The balance is practically all processed thermally to make coke, fuel gas, and liquid by-products.

Coal Processing to Synthetic Fuels and Other Products. The primary approaches to coal processing or coal conversion are thermal decomposition, including pyrolysis or carbonization, gasification, and liquefaction by hydrogenation. The hydrogenation of coal is not currently practiced commercially.

In the United States the Clean Coal Technology program was created to develop and demonstrate the technology needed to use coal in a more environmentally acceptable manner. Activities range from basic research and establishing integrated operation of new processes in pilot plants through demonstration with commercial-scale equipment.

Bioprocessing and Biotreatment of Coal. The use of biotechnology to process coal to make gaseous and liquid fuels is an emerging field. Bacteria and enzymes have been studied to establish the technical feasibility of conversion.

Biological processes are also being studied to investigate the ability to remove sulfur species in order to remove potential contributors to acid rain. These species include benzothiophene-type materials, which are the most difficult to remove chemically, as well as pyritic material.

KARL S. VORRES
Argonne National Laboratory

M. C. Stopes, *Proc. R. Soc. London Ser. B.* **90** 470 (1919); *Fuel* **14,** 4 (1935).

International Coal 1990 Edition, National Coal Association, Washington, D.C., 1990.

"Gaseous Fuels, Coal and Coke," *Annual Book of ASTM Standards,* Vol. 5.05, American Society for Testing and Materials, Philadelphia, Pa., published annually; *British Standards 1016,* parts 1–16, British Standards Institute, London, published annually.

Surveys of Energy Resources 1986, World Energy Conference, Central Office, London, 1986.

COAL CHEMICALS. See FEEDSTOCKS–COAL CHEMICALS.

COAL CONVERSION PROCESSES

CARBONIZATION

Coal carbonization is the process for producing metallurgical coke for use in iron-making blast furnaces and other metal smelting processes. Carbonization of coal (qv) entails heating coal to temperatures as high as 1100°C in the absence of oxygen in order to distill out tars and light oils. A gaseous by-product referred to as coke oven gas (COG), along with ammonia, water, and sulfur compounds, is also thermally removed from the coal. The coke that remains after this distillation largely consists of carbon, in various crystallographic forms, but also contains the thermally modified remains of various minerals that were in the original coal. These mineral remains, commonly referred to as coke ash, do not combust and are left as a residue after the coke is burned. Coke also contains a portion of the sulfur from the coal. Coke is principally used as a fuel, reductant, and support for other raw materials in ironmaking blast furnaces (see FURNACES, FUEL-FIRED; IRON). A much smaller tonnage of coke is similarly used in cupola furnaces in the foundry industry. The carbonization by-products are usually refined, within the coke plant, into commodity chemicals such as elemental sulfur (qv), ammonium sulfate, benzene, toluene, xylene, and naphthalene (see also AMMONIUM COMPOUNDS; BTX PROCESSING). Subsequent processing of these chemicals produces a host of other chemicals and materials. The COG is a valuable heating fuel used mainly within steel plants for such purposes as firing blast furnace stoves, soaking furnaces for semifinished steel, annealing furnaces, and lime kilns as well as heating the coke ovens themselves.

Supply and Demand

The vast majority of coke is produced from slot-type by-product coke ovens. Total coke production worldwide was about 378×10^6 t in 1990. This tonnage has remained relatively stable for the last two decades. In 1990, the former USSR (CIS) was the largest coke producer, producing 80×10^6 t, followed closely by the People's Republic of China, producing 73×10^6 t. Japan produced 53×10^6 t and the United States produced about 27×10^6 t. The United States dropped from being the No. 2 producer in 1970 to being the No. 4 producer in 1990.

Worldwide demand for blast furnace coke has decreased over the past decade. Increased technical capabilities, although not universally implemented, have allowed for about a 10% decrease in coke rate, ie, coke consumed per pig iron produced, because of better specification of coke quality and improvements in blast furnace instrumentation, understanding, and operation methods. As more blast furnaces implement injection of coal into blast furnaces, additional reduction in coke rate is expected. In some countries that have aggressively adopted coal injection techniques, coke rates have been lowered by 25%.

Production of coke is expected to continue to decrease unless new cokemaking facilities are constructed, because the effective production capacities of coke plants decrease as the coke plants age. This situation is particularly acute in North America where the majority of coke plants are over 25 years in age and the economic life spans of conventional coke batteries average about 20 years.

Coals for Cokemaking

Known world coal reserves in 1990 were estimated to be about 1000–1600 billion metric tons. The geographic distribution of these reserves is widespread, but about two-thirds of this coal resides in the United States, People's Republic of China, and the Commonwealth of Independent States. Some 637–1075 billion metric tons is classified as anthracite and bituminous coals, of which 10% is estimated to be suitable for cokemaking. Thus this 60–108 billion metric tons of coking coal, if recovered in a fully usable form, represents enough coal to supply coke plants at 1990 consumption rates for about 100–200 years. North America is estimated to possess about 130 billion metric tons of bituminous coal, of which perhaps one-tenth would be classified as coking coal, most of which resides in the United States.

The United States possesses a wealth of good quality coking coals in the Appalachian states as well as in locations in some southern and western states (see COAL). Coal blends normally consist of higher rank (more metamorphosed) coals in minor proportion relative to certain lower rank coals. The higher rank coals are referred to as medium volatile and low volatile. Similarly, the lower rank coking coals are referred to as high volatile. The reference to volatility reflects the relative amounts of by-products derived from the coals. High volatile coals generate more gas and tar during coking than do the medium and low volatile coals. Coals having either very low or very high volatile contents are not extensively used in cokemaking, for technical reasons.

Coal arrives at the coke plant by ship, rail, conveyor, or truck. Each type of coal is unloaded into a separate stockpile in the coal field. Reclaiming of coal from the stockpile can be accomplished using mobile equipment or bridge-mounted hoists. Coal of each type is moved to its coal bunker. In some plants, it is possible to crush each coal independently prior to its reaching the coal bunker.

Coals are not usually stored at coke plants for lengthy time periods. Besides the costs to maintain such inventories, coal undergoes low temperature oxidation that can adversely impact its coking behavior.

Coking Mechanism

There are several necessary conditions for coal to be transformed into coke. These include a heat supply, enclosure, or blanketing to prevent oxygen contact with the coal, and close contact between the coal particles during the carbonization process. In conventional vertical coke ovens these conditions are readily met. The heat is supplied from gas-burning flues located within the walls of each oven. Loading, called charging, of the coal into the oven, and retrieving, called pushing, of the product coke from the oven are accomplished via openings in the oven that can be sealed to prevent the incursion of air. Finally, the coal particles are charged into the oven by being dropped from a height so that the particles are packed together between the opposite walls of the oven, causing contact between particles. In this configuration, even for cubic particles and excluding contact with gas pockets, each particle, on average, comes into contact with six other particles.

Upon heating, coal molecules undergo many reactions. The primary reactions involve pyrolysis and formation of radicals having lower molecular weight than in the original coal. Some of the radicals, enriched with hydrogen, form liquid and gaseous products. In other reactions some radicals form more stable substances of higher molecular weight and less hydrogen content. Surface tension on the liquid components promotes additional contact between these components to further facilitate fusing the coal particles as they are heated.

Coke Ovens and Battery Operation

Individual coke ovens are constructed of interlocking silica bricks that are produced in numerous shapes for special purposes. It is not uncommon for modern coke oven batteries to contain 2000 different shapes and sizes of brick. Typical coke ovens are 12–14 m in length, 4–6 m in internal height, and have less than 0.5 m internal width. On each side of the oven are heating flues, also constructed of silica brick. Batteries of adjacent ovens, where ovens share heating flues, contain as many as about 85 ovens. At each end of each oven, refractory-lined steel doors are removed and reseated for each oven charge and push. Beneath each oven is a refractory substructure of the heating system regenerator checkers and sole flues for heating the oven floors. Combustion air, and sometimes combustion gases, are preheated in the regenerator. Reversing equipment periodically, perhaps twice per hour, reverses gas and air flow to the oven flues in order to maintain uniform temperature distributions in all of the flues. A concrete basement in which various battery equipment and portions of the heating system are contained is also standard. Coke batteries are generally heated with part of the coke oven gas generated in the cokemaking process; however, they can be heated using blast furnace gas and natural gas also. Once heated, the battery generally remains hot for its entire life because cooling of the silica brick causes a mineralogical change in the silica that lowers the strength of the brick. This effect is not reversed upon reheating; thus unexpected, sustained loss of heat can be catastrophic to the ovens.

Above the ovens is a roof system capable of supporting the moving Larry car from which coal is discharged into each oven through 3–5 charging holes in the roof of each oven. The Larry car itself is filled, for each oven charge, from a large blended coal silo that is constructed above the travel of the Larry car, usually at one end of the coke battery. Modern Larry car technology includes telescopic charging chutes for minimizing dust emissions during charging.

Coke ovens in a battery are charged with coal and pushed according to planned schedules. These schedules are an attempt to ensure that wall pressures generated in adjacent ovens are balanced to prevent wall movement, and that heat utilization and movement of charging and pushing equipment are optimized.

Coke Properties and Use

Coke, used in ironmaking blast furnaces, provides three primary functions. First, coke is the fuel that is burned in the blast furnace. The heat generated by combustion of coke provides the energy needed for the various reactions in the process including the melting of iron raw materials and the elevation of liquid iron to the temperatures required for downstream processing into steel. Second, the gases produced from combustion and gasification of coke are the reducing agents to remove oxygen from the iron raw materials in the blast furnace. Third, coke is the only blast furnace material that remains solid at the high temperatures that exist in the lower portions of the furnaces. This means that the coke can support the rest of the materials and provide passageways for the ascending gases and descending iron droplets to meet and interact. The ability to maintain permeability in the blast furnace is perhaps the most important function of coke.

The combustion and reduction functions of coke are enhanced by maximization of the carbon in the coke. This means that the other chemical constituents of coke derived from the coal sulfur and coal minerals should be as low as possible, as the presence of these impurities dilutes the amount of carbon in coke. Moreover, these impurities must be melted and prevented from entering the molten product iron, and they must then be removed from the blast furnace. This is accomplished by maintaining a molten slag layer that floats on the molten iron and periodically “tapping” this slag from the furnace. The slag chemistry and properties are continually adjusted to drive the coke impurities, as well as impurities from other raw materials, into the slag rather than into the pig iron. The materials added to the blast furnace in order to effect this slag volume, chemistry, and properties must also be melted.

Other Cokemaking Technology

Owing to the importance of coke to the steel industry, means for improving the quality of coke, lowering its cost of production, and developing cleaner cokemaking processes are always under investigation.

One modification to the vertical coke oven process, coal preheating, gained attention and acceptance in the late 1960s and early 1970s. Over the years, however, most of these facilities shut down as it was realized that costs of operating and maintaining the preheaters and additional battery maintenance negated the benefits of shortened coking cycles.

Another modification to the vertical coke oven process, dry quenching of the coke as first developed in the CIS, started becoming of interest in the 1960s and has continued to grow steadily. Extensions to the size of vertical coke ovens have been developed and this idea continues to be evaluated.

Formcoke Processes. A completely different approach to making coke is embodied in the various types of formcoke processes which produce coke briquettes in a series of reactors and vessels.

Nonrecovery Cokemaking. Another cokemaking technology that is being practiced in various forms in several countries is nonrecovery cokemaking. This technology centers around the use of horizontal ovens somewhat similar to those that were used historically, along with beehive ovens.

Other types of cokemaking technology include both batch and continuous processes, and processes that use electrical induction as the heat-transfer mechanism.

DENNIS KAEGI
VALERY ADDES
HARDARSHAN VALIA
MICHAEL GRANT
Inland Steel Company

IISI, *Future Supplies of Coking Coal,* Brussels, 1992.

IISI, *Western World Cokemaking Capacity,* Brussels, 1989.

R. Loison, R. Foch, and A. Boyer, *Coke Quality and Production,* Butterworth's, London, 1989.

M. A. Elliott, *Chemistry of Coal Utilization,* 2nd Suppl. Vol., John Wiley & Sons, Inc., New York, 1981.

CLEANING AND DESULFURIZATION

Coal (qv) is a primary source of energy for the United States and is expected to continue to be so into the twenty-first century (see FUEL RESOURCES). However, combustion of raw coal directly in the furnace of an electric power generating plant yields flue gases containing sulfur oxides (SO_x), nitrogen oxides (NO_x), and compounds of toxic metals. These materials, which are principally derived from impurities in the coal feed, may be reduced below permissible emission levels by various processes: the coal can be cleaned before it is fed to the furnace; the contaminants can be captured in a solid sorbent during, or immediately following, combustion as the hot product gases pass through the boiler; or the cool flue gases can be cleaned after leaving the heat exchange region (see AIR POLLUTION CONTROL METHODS; EXHAUST CONTROL,

INDUSTRIAL; FUELS, SYNTHETIC–GASEOUS FUELS; SULFUR REMOVAL AND RECOVERY).

Coal Cleaning

In 1990 coal production in the United States reached 0.9 billion metric tons and worldwide production was estimated to be over four billion metric tons. In 1982 it was estimated that at least 50% of the world coal production was cleaned in some manner before use. As higher quality coal reserves are depleted and more stringent environmental regulations on pollutants, particularly sulfur oxides, are enacted, this percentage is expected to increase.

Impurities. The three categories of potential pollutants in coal are sulfur, nitrogen, and ash. Sulfur and ash are associated with both the mineral and organic portions of coal, whereas nitrogen is mainly associated with the organic matter.

Concern over the release of hazardous trace elements from the burning of coal has been highlighted by the 1990 Clean Air Act Amendments. Most toxic elements are associated with ash-forming minerals in coal. As shown in Table 1, levels of many of these toxic metals can be significantly reduced by physical coal cleaning.

Conventional Coal Preparation Plants. Coal cleaning (preparation) is based principally on size and density differences, with the exception of flotation (qv). In this manner physical impurities, ie, ash and pyrite, may be removed from coal. Four general categories of coal preparation plants can be defined based on levels or degrees of cleaning: level 1 involves crushing and screening only; level 2, coarse coal cleaning only; level 3, coarse coal and partial fine coal cleaning; and level 4, total cleaning, ie, all size fractions are cleaned. At each successive level, the process design becomes increasingly more sophisticated. Conventional coal cleaning processes can remove about 50% of pyritic sulfur and 30% of total sulfur.

Advanced Coal Cleaning Technologies. As the easy-to-remove relatively clean coals are gradually mined out and as fuel specifications become more stringent in order to meet environmental regulations, the need for advanced fine coal cleaning processes has grown. For any fine coal cleaning process, two characteristics tend to dominate. As coal particles are crushed into finer size, the specific surface area increases and the mass of each particle becomes smaller. This leads to the development of surface force-controlled processes or advanced density-based processes that are quite different from the specific gravity-controlled processes found in coarse coal cleaning. In general, advanced processes are capable of producing a deep-cleaned coal product having low ash and low sulfur content. However, as of this writing, most of these processes are in the small-scale demonstration stage and have not been tested on a commercial scale. They include column flotation, agglomeration-based fine coal cleaning, heavy-medium and heavy-liquid cycloning processes, and dry coal cleaning.

Chemical and Biological Coal Cleaning. Whereas physical coal cleaning is capable of removing most of the ash (mineral matters) and inorganic (pyritic) sulfur, it cannot be used to remove organic sulfur. For most bituminous coals in the United States, 40 to 60% of the sulfur content is bound in organic matter. Organic sulfur can be classified into four types: thiols, sulfides, disulfides, and heterocyclic thiophenes, such as dibenzothiophene, $C_{12}H_8S$. The sulfur from these compounds can only be removed using chemical or biological methods. Some of these processes have progressed to the miniplant stage. However, most are still at laboratory scale. They include the oxidative desulfurization process, coal cleaning by reactive leaching, microwave desulfurization, chlorinalysis, self-scrubbing coal, and microbial coal cleaning.

Combustion of Synthetic Fuels. Sulfur may also be removed from coal before combustion by converting the coal to a synthetic gaseous fuel (syngas). Gasifying the coal yields H_2S, which is then removed from the syngas.

Table 1. Effect of Coal Cleaning on Trace Elements[a]

Coal	Cd	Cr	Cu	F	Hg	Mn	Ni	Pb
	Trace element content, ppm							
feed	3.15	55	25	156	0.20	53	26	18
product[b]	0.05	28	10	71	0.09	7.9	11	3.0
reduction, %	98	49	60	54	55	85	58	83

[a] Upper Freeport Coal, W.Va., 0.075 mm top size, ie, all particles ≤0.075 mm.
[b] Float at 1.40 specific gravity.

Coal Desulfurization in the Furnace and Ductwork

Sulfur dioxide in the hot gases from a coal-fired combustor may be transferred to a solid reaction product that usually contains the calcium ion, at any point in the furnace. In a fluidized-bed combustor the SO_2-sorbent is intimately mixed with the coal throughout the region where combustion occurs. In entrained-bed systems the sorbent is introduced in finely divided form into the hot exhaust gases as they pass through the heat exchange system inside the boiler or through the ductwork leading to the particulate removal equipment following the boiler.

Sulfur Removal form Flue Gases

As of 1992, more than 150 coal-fired boilers in the United States operated with flue gas desulfurization (FGD) systems. The total electrical generating capacity of these plants has risen to 72 gigawatts. FGD processes are classified into *(1)* wet-throwaway, *(2)* dry-throwaway, *(3)* wet-regenerative, and *(4)* dry-regenerative processes.

Clean Coal Choices

When the Clean Air Act of 1990 was signed into law, electric utilities were required to establish plans and initiate projects to comply with that Act's Title IV. Each utility had to evaluate how the various commercial and emerging clean coal systems fit into the utility's technical and business environment, resulting in strategies to utilize fuel switching and wet throwaway FGD processes almost exclusively.

SHIAO-HUNG CHIANG
JAMES T. COBB, JR.
University of Pittsburgh

F. T. Princiotta, *Chem. Eng. Prog.* **74**(2), 58–64 (Feb. 1978).

J. W. Leonard and B. C. Hardinge, eds., *Coal Preparation,* 5th ed., Society for Mining, Metallurgy and Exploration, Inc., Littleton, Colo., Chapts. 7 and 14.

D. J. Smith, *Power Eng.* **95**(12), 18–23 (Dec. 1991).

R. McInnes and R. Van Royen, *Chem. Eng.* **97**(9), 124–127 (Sept. 1990).

GASIFICATION

Interest in coal gasification was renewed in 1973 when international oil and gas prices increased sharply. Extensive development efforts over the past 20 years by companies such as Shell, Texaco, Destec, Krupp Koppers, British Gas, and Lurgi have led to high temperature, high pressure slagging processes that offer high efficiencies, improved economics, and excellent environmental performance. Simultaneously, the development of high firing-temperature gas turbines has created a new and potentially very large market for coal-derived syngas as a fuel for combined cycle power generation.

Coal Gasification Combined Cycle Power Generation

Electricity Demand. The U.S. Department of Energy (DOE) forecasts that, even when comprehensive energy conservation programs are taken into account, electricity consumption in the United States by the year 2000 is expected to be more than 20% above 1990 levels. The DOE expects the demand for electricity to grow at almost twice the rate of total energy demand; electricity demand in 2010 is predicted to be almost 50% greater than the demand in 1990.

Electric utilities are therefore expected to build new power plants or to extend the lives of existing, older ones. Ready availability, secure supply, and low price are expected to make coal the fuel of choice for most of the new baseload generating capacity.

Coal Gasification Combined Cycle. Coal gasification combined cycle (CGCC) integrates two commercially proven technologies: the manufacture of a clean-burning fuel gas from coal and the highly efficient use of that gas to produce electricity in a combined cycle power generation system. The combined cycle system has two basic components: *(1)* high efficiency gas turbines, which burn the clean fuel gas to produce electricity, and *(2)* exhaust heat, which is recovered to power traditional high efficiency steam turbines to generate additional electricity. The overall system is shown in Figure 1.

Demonstration Projects. The principal developers of advanced coal gasification technologies constructed and successfully operated large-scale demonstration plants during the 1980s.

As of this writing (ca 1994), Destec Energy, Inc. has been operating a CGCC power plant using its coal gasification technology at a Dow Chemical Company plant in Plaquemine, Louisiana, since April 1987. This plant is designed for 1435 t/d of subbituminous coal on a dry basis or 1835 t/d of low quality lignites on a dry basis (see LIGNITE AND BROWN COAL). The syngas is fed to two Westinghouse WD 501 gas turbines that, in conjunction with the steam cycle, produce a net power output of 161 MW.

Shell's demonstration unit, called SCGP-1, was placed in service in 1987 at Shell's Deer Park manufacturing complex near Houston, Texas. SCGP-1 has a capacity ranging from 227 t/d on bituminous coal to 364 t/d on high moisture, high ash lignite. During four years of operation, SCGP-1 logged over 14,000 h while providing engineering and environmental data on 18 significantly different feedstocks, including Texas lignite, a subbituminous Powder River Basin coal, a wide variety of bituminous coals, and petroleum coke.

Commercialization of Coal Gasification Technologies. The successful operation of the demonstration plants is expected to lead to widespread commercialization of coal gasification technologies in the 1990s (Table 1).

Table 1. Coal Gasification Combined Cycle Projects

Location	Capacity, MW	Design[a]	Estimated start-up
	Europe		
Buggenum (Demkolec)	250	Shell	1993
Berrenrath (Rheinbraun)	300	HTW	1995
Spain	330	Prenflo	1996
Italy	350	undecided	1996
	North America		
Tennessee (TVA)	265[b]	Shell	1998
Wabash River	265	Destec	1995
Cool Water, Calif.	100[c]	Texaco	1992
Lakeland, Fla.	260	Texaco	1996
Delaware City, Del.	125	Texaco	1995
Springfield, Ill.	60	ABB CE	1996
Reno, Nev.	80	KRW	1997
Coeburn, Va. (Tamco)	107	U-Gas	1996
Canada	250	undecided	1996

[a] Terms are defined in text.
[b] Urea is also to be produced.
[c] Methanol is also a product.

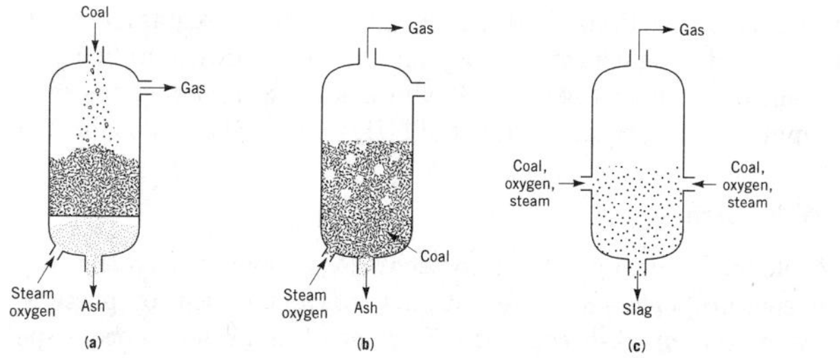

Figure 2. Types of gasifiers: (**a**) moving-bed (dry-ash), (**b**) fluidized-bed, and (**c**) entrained-flow.

Types of Gasifiers

There are essentially three types of coal gasifiers: moving-bed or countercurrent reactors; fluidized-bed or back-mixed reactors; and entrained-flow or plug-flow reactors. The three types are shown schematically in Figure 2.

Commercial Processes

The Shell Coal Gasification Process is based on a dry feed, entrained-bed, high pressure, high temperature slagging design. The Texaco Coal Gasification Process incorporates a single-stage slagging, pressurized, entrained-bed downflow gasifier. Dow Coal Gasification Process (Destec) is a two-stage, slurry feed, entrained-flow, slagging gasifier. High Temperature Winkler Process (HTW) developed by Rheinbraun and especially targeted for the gasification of brown and hard coals, peat, and biomasses, is a fluidized-bed gasifier. The British Gas/Lurgi Slagging Gasifier technology, developed by the British Gas Corporation (BGC) and Lurgi, started with the dry ash Lurgi gasifier and incorporated enhancements such as operation at a higher temperature that melts the coal ash to slag. A significant efficiency advantage is gained by reducing the steam requirement to only about 15% of that required by the dry-ash Lurgi gasifier.

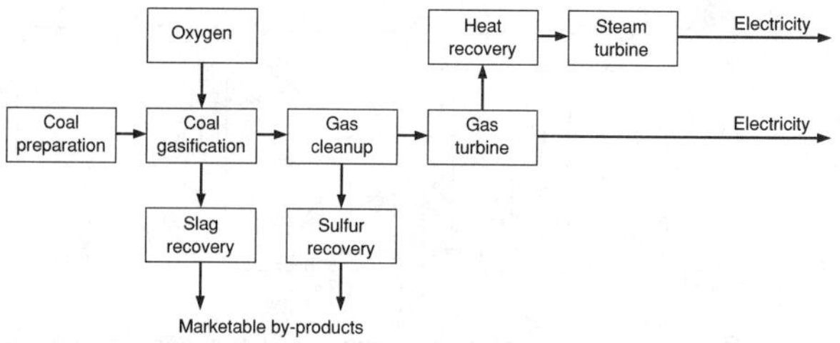

Figure 1. Schematic of coal gasification combined cycle power generation.

Coal Gasification Chemistry

Gasification involves incomplete combustion in an oxygen-deficient or reducing atmosphere. Gasification uses 20–30% of the oxygen theoretically required for complete combustion to carbon dioxide and water. Carbon monoxide and hydrogen are the principal products, and only a fraction of the carbon in the coal is oxidized completely to carbon dioxide. The heat released by partial combustion provides the bulk of the energy necessary to drive the gasification reactions.

In fixed-bed gasifiers the coal is first dried by evaporation of the surface and inherent moisture, then devolatilized, which breaks the weaker chemical bonds, forming tars, oils, phenols, and hydrocarbon gases. These devolatilization products exit the gasifier with the syngas, because of low temperatures and lack of oxygen in fixed-bed gasifiers. Fluidized-bed gasifiers provide better mixing and uniform temperatures that allow oxygen to react with the devolatilization products. These products also undergo thermal cracking, primarily on hot char surfaces, reacting with steam and H_2. In dry fluidized-bed gasifiers, temperatures have to be maintained below the ash melting point, which leads to incomplete carbon conversion for unreactive coals. Agglomerating ash gasifiers operate at higher temperatures, near the ash softening point, which provides improved carbon conversion. Entrained-bed slagging gasifiers provide uniform high temperatures, resulting in complete conversion of all coals to hydrogen, carbon monoxide, and carbon dioxide, and producing no tars, oils, or phenols. Thus the principal gasification reactions are

$$C + O_2 \rightarrow CO_2 \ (exothermic)$$

$$C + 1/2\ O_2 \rightarrow CO \ (exothermic)$$

$$CO + H_2O \rightarrow CO_2 + H_2 \ (exothermic, shift\ reaction)$$

$$C + H_2O \rightarrow COOH_2 \ (endothermic)$$

$$C + CO_2 \rightarrow 2\ CO \ (endothermic)$$

$$CO + 3\ H_2 \rightarrow CH_4 + H_2O \ (exothermic;\ methanation)$$

$$C + 2\ H_2 \rightarrow CH_4 \ (exothermic;\ methanation)$$

Trace elements such as sulfur and nitrogen are also involved in the gasification reactions. Sulfur in coal is converted primarily to H_2S under the reducing conditions of gasification. Approximately 5 to 15% of the sulfur is converted to COS, whereas the coal nitrogen is converted primarily to N_2; trace amounts of NH_3 and HCN are also formed.

Coal Properties

Developers of coal gasification technology have studied the impact of key coal properties on different parts of the gasification process.

Ash Content. Ash content affects gasifier performance, especially for most slagging gasifiers, because molten ash in the form of slag provides an insulating coverage on the wall of the gasifier, which reduces the heat transferred during the gasification reaction. Ash content also influences the requirements of the slag tap and the slag handling system.

Ash Melting Point/Slag Viscosity. For gasification technologies utilizing a slagging gasifier, slag flow behavior is an important parameter. For coals having high slag viscosities, slag behavior can be modified by the addition of a flux such as limestone (calcium carbonate).

Fouling Characteristics. Fouling of heat transfer surfaces can result from constituents such as chlorine, sodium, potassium, and calcium.

Reactivity. Reactivity is used to describe the relative degree of ease with which a coal undergoes gasification reactions. The primary property affecting the ease of conversion is the oxygen content of the coal, which in turn reflects its age or rank.

Corrosion. The primary coal properties affecting corrosion are sulfur and chlorine levels.

Gasifier Performance

Operating Parameters. The primary gasifier operating parameters are coal composition, coal throughput, oxygen/coal ratio, and steam/oxygen ratio. The amount of oxygen and steam fed to the gasifier depends on the coal composition. In general, low rank coals are very reactive and require less oxygen and little to no steam, whereas high rank coals are relatively unreactive, requiring more oxygen and a moderate amount of steam. Steam provides an alternative source of oxygen for the gasification reaction and helps to moderate the gasification temperature. Gasifier performance is evaluated in terms of syngas production and composition, carbon conversion, and cold gas efficiency.

Heat Balance. Mass and heat balances are calculated around the gasification block, which includes the gasifier, quench, syngas cooler, and solids removal systems.

Environmental Performance of Coal Gasification Technology

One advantage of modern coal gasification combined cycle systems is excellent environmental performance. Not only are regulatory standards met, but emissions and effluents are well below accepted levels.

Acid Rain Emissions. Coal gasification combined cycle (CGCC) represents a superior technology for controlling SO_2 and NO_x emissions. Emissions are much lower than those from traditional coal combustion technologies. During gasification, the sulfur in the coal is converted to reduced sulfur compounds, primarily H_2S and a small amount of carbonyl sulfide, COS. Because the sulfur is gasified to H_2S and COS in a high pressure concentrated stream, rather than fully combusted to SO_2 in a dilute-phase flue gas stream, the sulfur content of the coal gas can be reduced to an extremely low level using well-established acid gas treating technology. The sulfur is recovered from the gasification plant as salable, elemental sulfur. A small quantity of sulfur can also be captured in the slag as sulfates. Figure 3 compares emissions from a coal gasification plant with a modern pulverized coal (PC) power plant. New technologies are being developed for removing sulfur and other contaminants at high temperature. One hot-gas cleanup process uses metal oxide sorbents to remove H_2S + COS from raw gas at high (> 500 °C) temperature and system pressure.

During coal gasification the nitrogen content of coal is converted to molecular nitrogen, N_2, ammonia, NH_3, and a small amount of hydrogen cyanide, HCN. In moving-bed gasifiers, some of the nitrogen also goes into tars and oils. The NH_3 and HCN can also be removed from the coal gas using conventional (cold) gas treating processes. Other techniques are being investigated in hot-gas cleanup technologies. After removal of HCN and NH_3, combustion of the coal gas in the gas turbine produces no fuel-based NO_x. Only a small amount of thermal NO_x is formed, and this can be controlled to low levels through turbine combustor design and, if necessary, steam or nitrogen addition. Based on tests using SCGP-type coal gas fired in a full-scale GE-frame 7F combustor, a NO_x concentration of no more than 10 ppm in the gas turbine flue gas is attainable. See Figure 3 for a comparison of NO_x emissions from a PC plant equipped with low NO_x burners.

Criteria Air Pollutants. Moving-bed gasifiers produce tars, oils, phenols, and heavy hydrocarbons, the concentrations of which are controlled by quenching and water scrubbing. Fluidized-bed gasifiers produce significantly lower amounts of these compounds because of higher operating temperatures. Entrained-flow gasifiers operate at even higher temperatures, typically in excess of 1650°C. SCGP-1 experience has confirmed that carbon conversions of greater than 99.5% are easily attainable for any coal and that essentially no organic compounds heavier than methane are produced. Emissions of volatile organic compounds (VOC) from a CGCC plant are expected to be approximately 300 times lower than those from a similarly sized coal-fired steam plant equipped with low NO_x burners and an FGD unit.

Hazardous Air Pollutants. The total emissions of hazardous air pollutants from a CGCC plant having wet cleanup are expected to be at least an order of magnitude lower than those achievable from a modern coal-fired steam plant. Metals removal in hot-gas cleanup systems is still under development.

Water Consumption and Effluent Characterization. Another advantage of CGCC power generation is derived from lower water require-

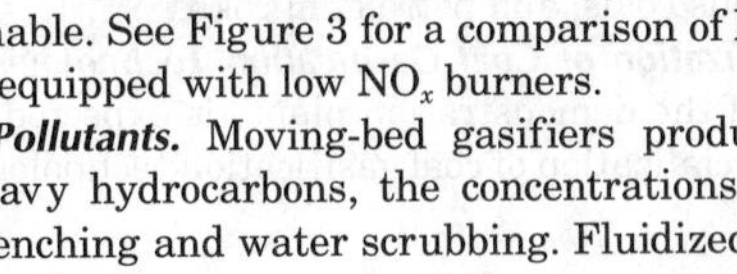

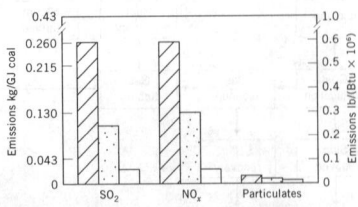

Figure 3. Environmental emissions, where ▨ represents new source performance standards (NSPS) requirements; ▩ represents a pulverized coal (PC) plant; and □ represents SCGP-1.

ments. Because more than half of the power generated in a CGCC plant comes from the gas turbine, the water requirement is only 70–80% of that required for a coal-fired power plant, where all of the power is generated from steam turbines.

Whereas moving-bed gasifiers require complex water-treatment systems to address tars, phenols, and metals, this complexity is mostly alleviated for fluidized-bed gasifiers and is eliminated for entrained-flow gasifiers.

Solid By-Products. Coal gasification power generation systems do not produce any scrubber sludge, a significant advantage over both direct coal combustion processes that use limestone-stack gas scrubbers and fluidized-bed combustion processes that use solid absorbents for sulfur capture. In coal gasification, the sulfur in the coal is recovered as bright yellow elemental sulfur, for which there are several commercial applications, the largest being in the phosphate fertilizer industry. The ash in the coal is converted to slag, fly slag, or fly ash. Environmental characterization of SCGP-1 slag and fly slag was performed for several coals using the extraction procedure (EP) toxicity tests and the toxicity characteristic leaching procedure test (TCLP), confirming that toxic trace metal concentrations in the leachate were well below Resource Conservation and Recovery Act (RCRA) requirements.

As part of a solids utilization program at SCGP-1, gasifier slag has been used as a principal component in concrete mixtures (Slagcrete) to make roads, pads, and storage bins. Other applications of gasifier slag and fly slag that are expected to be promising are in asphalt aggregate, Portland cement kiln feed, and lightweight aggregate (see CEMENT). Compressive strength and dynamic creep tests have shown that both slag and fly slag have excellent construction properties.

CO_2 Emissions and Global Warming. The high coal-to-busbar efficiency of a CGCC system provides a significant advantage in responding to CO_2 emissions and thus to global warming concerns. High efficiency translates to lower coal consumption and lower CO_2 production per unit of electricity generated.

Syngas Chemistry

Whereas near-term application of coal gasification is expected to be in the production of electricity through combined cycle power generation systems, longer-term applications show considerable potential for producing chemicals from coal using syngas chemistry. Products could include ammonia, methanol, synthetic natural gas, and conventional transportation fuels.

UDAY MAHAGAOKAR
A. B. KREWINGHAUS
Shell Development Company

D. R. Simbeck, R. L. Dickenson, and E. D. Oliver, *EPRI Rep. AP-3109,* June 1983.

"Cool Water Coal Gasification Program," *Final Rep. EPRI GS-6806,* Dec. 1990.

R. H. Fisackerly and D. G. Sundstrom, "The Dow Syngas Project—Project Overview and Status Report," presented at the *Sixth Electric Power Research Institute Gasification Contractors Conference,* Palo Alto, Calif., Oct. 1986.

U. Mahagaokar and co-workers, "Shell's SCGP-1 Test Program—Final Overall Results," *Tenth Annual EPRI Conference on Gasification Power Plants,* San Francisco, Calif., Oct. 1991.

LIQUEFACTION

Coal (qv) can be converted to liquid and gaseous fuels by two different processing routes normally termed "direct" and "indirect" and "indirect" (see also FUELS, SYNTHETIC–GASEOUS FUELS; FUELS, SYNTHETIC–LIQUID FUELS). Direct liquefaction processes include those that normally proceed to liquids in a single processing sequence, using solid coal as the primary reactant. Some direct liquefaction schemes also involve chemical pretreatment steps. Indirect liquefaction processes normally involve gasification of coal as the first conversion step, followed by catalytic recombination of the resulting synthesis gas mixture ($CO + H_2$) to form hydrocarbons and oxygenates.

Direct Liquefaction

Coal liquefaction involves raising the atomic hydrogen-to-carbon ratio from approximately 0.8/1.0 for a typical bituminous coal, to 2/1 for liquid transportation fuels or 4/1 for methane. In this process, molecular weight reduction and removal of mineral matter and heteroatoms such as sulfur, oxygen, and nitrogen may need to be effected.

Hydrogenation or hydroliquefaction and pyrolysis are the two means used for direct liquefaction. In direct hydrogenation the primary reactions are a combination of homogeneous thermal cracking, ie, free-radical generation, and heterogeneous hydrogenation involving hydroaromatics in the slurry vehicle or the coal itself as hydrogen transfer agents. Rapid and efficient capping of the primary free radicals generated by thermolysis is thought to be necessary in order to prevent regressive reactions leading to formation of char. Other theories of coal liquefaction suggest that hydrogen can engender reactions involving scission of strong bonds in the coal macromolecule, and hence can act as an active bond cleaving agent rather than simply a passive radical quencher.

Process schemes that apply pyrolysis chemistry normally involve thermolysis in an inert or reducing atmosphere and produce two principal products from coal: a tar and char. The relative proportion of char to tar can be quite high; hence the rationale for liquefaction by pyrolysis is often not production of coal-derived distillate materials. Hydropyrolysis processing schemes involving thermolysis in the presence of hydrogen or pyrolysis under conditions of rapid heating can, however, generate yields of distillate products significantly in excess of the volatile matter content of the starting coal.

Hydrogenation. Examples of hydrogenation processes include the Solvent Refined Coal process (SRC), the Exxon donor solvent (EDS) process, and the H-Coal process. Both the SRC and EDS processing schemes are primarily thermal, while the H-Coal process uses an added heterogeneous hydrogenation catalyst.

Pyrolysis and Hydropyrolysis. The second category of direct liquefaction aimed at producing distillate materials from coal involves the processes of pyrolysis and hydropyrolysis. Pyrolysis, sometimes called destructive distillation, essentially involves heating the coal in an inert atmosphere, followed by recovery of coal-derived tars and distillates in the off-gas stream. Depending on the coal type and processing conditions, yields of condensables can equal or even exceed the volatile matter content of the parent coal. Pyrolysis processes are, however, usually burdened with poor liquid yield, relative to direct hydrogenation, and the coal-derived liquids are high in heteroatoms and inorganic particulate matter. When high yields of coal-derived distillates are desired, pyrolysis is usually carried out in a hydrogen atmosphere, eg, hydropyrolysis, or at extremely rapid heating rates, eg, flash pyrolysis.

Indirect Liquefaction

The second category of coal liquefaction involves those processes which first generate synthesis gas, a mixture of CO and H_2, by steam gasification of coal, $C(s) + H_2O \rightarrow CO + H_2$, followed by production of solid, liquid, and gaseous hydrocarbons and oxygenates via catalytic reduction of CO in subsequent stages of the process. Whereas coal is usually the preferred feedstock, other carbon-containing materials such as coke, biomass, or natural gas can also be used. This concept is the basis for the well-known Fischer-Tropsch process.

Processes which operated at relatively low pressure, in the range of 100–200 kPa (1–2 atm) dominated commercial applications of the Fischer-Tropsch process in Germany prior to 1939. Large-scale commercial developments for synthesis of hydrocarbons from coal have been carried out in South Africa (SASOL).

Much of the research and process development on indirect liquefaction of coal in the 1990s is aimed at matching the synthesis conditions

Table 1. Synthesis Gas Compositions from Gasifiers Operating on Western U.S. Coals

Material	Dry gasifier product, vol %	
	Shell	Lurgi
H_2	30	50
CO	66	25
CO_2	3	10
CH_4 + higher hydrocarbons		15
inerts	1	
H_2/CO ratio, vol basis	0.45	2
net efficiency to synthesis gas, %[a]	78–80	61

[a] Net efficiency includes thermal losses in reforming methane to synthesis gas.

with modern, efficient coal gasifiers such as those developed by Texaco, Dow, and Shell (see COAL CONVERSION PROCESSES–GASIFICATION). A comparison of the gas product mix from a Shell gasifier with that from the older standard Lurgi system is shown in Table 1. Whereas the newer gasifiers are considerably more efficient than the older design, there is the drawback that the gas produced is much lower in hydrogen content. This problem may be solved by shift conversion of the raw synthesis gas, a process that is expensive, or it may be obviated by design of synthesis reactors and catalysts that can utilize the low H_2/CO gas directly. Because of the exothermic nature of the synthesis reactions, design efforts have focused on the development of slurry-phase reactors to replace conventional fixed-bed and fluid-bed systems.

ROBERT M. BALDWIN
Colorado School of Mines

E. E. Donath, "Hydrogenation of Coal and Tar," in H. H. Lowry, ed., *Chemistry of Coal Utilization, Supplementary Volume,* John Wiley & Sons, Inc., New York, 1963.

S. B. Alpert and R. H. Wolk, in M. A. Elliott, ed., *Chemistry of Coal Utilization, Second Supplementary Volume,* John Wiley & Sons, Inc., New York, 1981.

H. D. Schindler; "Coal Liquefaction: A Research Needs Assessment," Vol. 2, Technical Background, *Final Report on DOE Contract No. DE-AC01-87ER30110,* 1989.

C. Y. Wen and S. Dutta, in C. Y. Wen and E. S. Lee, eds., *Coal Conversion Technology,* Addison Wesley, Inc., Reading, Mass., 1979.

COATED FABRICS

A coated fabric is an engineered product derived by a combination of textile and polymer coating technology.

Textile Component

Industrial-grade fabrics, as opposed to apparel-grade fabrics, typically constitute the substrate classification from which the coated fabrics producer selects a construction to fulfill specific end use requirements. A comprehensive listing of industrial fabric suppliers has been compiled.

Fibers. For many years cotton (qv) and wool (qv) were the primary textile components, contributing the properties of strength, elongation control, and aesthetics to the finished product. As cotton again became an important fiber for the apparel industry during the 1980s, the supply and pricing situation changed significantly and polyester–cotton blends and 100% polyesters became the fabrics of choice. Polyester is now the most widely used industrial coating and laminating fiber. When moisture absorbance and glueability are the most important properties, polyester–cotton blends and in some cases 100% cotton textiles still see significant use, ie, wallcovering and case goods covering.

Textile Construction. There are many choices in textile construction. The original and still the most commonly used is the woven fabric. Woven fabrics have four basic constructions: the plain weave, the drill weave, the satin weave, and the twill weave. For shoe uppers and other applications where strength is important, woven cotton fabrics are used.

Knitted fabrics are used where moderate strength and considerable elongation are required.

Many types of nonwoven fabrics (qv) are utilized as substrates for coated fabrics, including products made by the wet web method, saturated nonwovens, spunbonded nonwovens, and needled nonwovens. Today's nonwovens are engineered fabrics designed for specific end uses. The most common nonwovens used for coated fabric substrates are the lightly needled, low density nonwovens which are typically prepared with either polypropylene or polyester fibers.

Post-Finishes of the Textile Component. The construction that results from either weaving or knitting is called greige good. In many cases, other steps are required before the fabric can be coated. This often includes scouring to remove surface impurities and finishes added to the yarns to improve weaving, and heat setting to correct the width, stabilize the textile, and minimize shrinkage during coating.

Polymeric Coating Component

Rubber and Synthetic Elastomers. For many years nondecorative coated fabrics consisted of natural rubber on cotton cloth. Natural rubber is possibly the best all-purpose rubber but some characteristics, such as poor resistance to oxygen and ozone attack, reversion and poor weathering, and low oil and heat resistance, limit its use to special application areas (see ELASTOMERS, SYNTHETIC; RUBBER, NATURAL).

Polychloroprene (Neoprene), introduced in 1933, rapidly gained prominence as a general purpose synthetic elastomer having oil, weather, and flame resistance. The introduction of new elastomers in solid or latex form was accelerated by World War II. In addition to natural rubber and polychloroprene, other rubbery polymers in use include: styrene–butadiene (SBR), polyisoprene, polyisobutylene (Vistanex), isobutylene–isoprene copolymer (Butyl), polysulfides (Thiokol), polyacrylonitrile (Paracril), silicones, chlorosulfonated polyethylene (Hypalon), poly(vinyl butyral), acrylic polymers, ethylene–propylene–diene monomer (EPDM), fluorocarbons (Viton), polybutadiene, polyolefins, and many more. Copolymerization makes the number of variations available staggering. The number of commercially available elastomers is large, with many producers of those polymers offering several variations that provide a wide range of properties.

Poly(vinyl chloride). By far the most important polymer used in coated fabrics is poly(vinyl chloride). This relatively inexpensive polymer resists aging processes readily, resists burning, and is durable. It can be compounded readily to improve processing, aging, burning properties, softness, etc. In addition, it can be decorated to fit nearly any required use including leather prints, textile looks, or detailed patterns. PVC-coated fabrics are used for furniture, marine and automotive upholstery, window shades, automotive trim, wallcoverings, book covers, convertible topping, shoe uppers and liners, and many other uses. Two of the largest uses of PVC-coated fabrics are in vinyl wallcoverings and upholstery.

Processing

Coated fabrics can be prepared by lamination, direct calendering, direct coating, or transfer coating.

Post-Treatment. Coated fabrics can be decorated and protected by applying inks and coatings to the surface.

Economic Aspects

In the total market for coated fabrics, the first consideration is the required performance of the finished product, followed by cost considerations. Because of their higher cost, the rubber and specialty elastomers are only used for producing coated fabrics that require oil or chemical resistance, low air permeability, or other unique properties. Likewise, urethane-coated fabrics are often used for the production of style- and design-oriented coated fabrics for the apparel, shoe, and handbag market because it is easy to produce small-run sizes using transfer coating methods. In addition, the finished products have light weight and good abrasion and scuff resistance and are significantly less expensive than the natural leather hides that they replace. Because of performance and cost considerations, PVC-coated fabrics are the workhorse coated fabrics. They are easily processed by either calender coating or transfer coating, are easily decorated by a combination of printing and embossing, and depending on the choice of compound formulation and textile backing can be designed for many different uses.

Health and Safety Factors

Some materials used in coating operations have been identified by the federal government as being hazardous to workers' health. Because of this, manufacturers of coated fabrics are subject to OSHA standards relating to acceptable exposure to these chemicals. In most cases, depending on the individual chemical, this required engineering changes to the process, protective equipment, personnel monitoring, and extensive record-keeping. One change that many manufacturers have made or are presently working on is the development of formulations for coatings, inks, and finishes that do not contain heavy metals. This is particularly true in the vinyl industry where there is a movement away from stabilizer systems containing cadmium, mildewcides that contain arsenic, and pigment systems containing lead chromates and lead molybdates.

Most exposure problems are related to solvents and dusts, so particular attention should be given to raw material mixing areas and solvent exposure during coating and post-printing and finishing operations.

BRUCE BARDEN
GenCorp Polymer Products

CFFA Standard Test Methods, Chemical Films & Fabrics Association, Inc. (CFFA), Cleveland, Ohio, 1983.

Modern Plastics Encyclopedia 1993, McGraw-Hill Inc., New York, 1993.

1993 Rubber Red Book, Communication Channels, Inc., Atlanta, Ga., 1993.

1993 Plastics Directory, The Cahners Publishing Co., Newton, Mass., 1990.

ASTM Annual Book of Standards, American Society of Testing and Materials, Philadelphia, Pa., 1990.

COATING PROCESSES

SURVEY

Coatings technology covers a wide variety of products and processes. Typical are paints (see PAINT) and the diverse surface coatings (qv) used to protect houses, bridges, appliances, and automobiles. These coatings fulfill functional needs such as these provided by waterproofing, flameproofing, and corrosion protection, as well as having decorative aesthetic qualities (see CORROSION AND CORROSION CONTROL). Coating processes are also important to the production of such coated products as photographic films for medical, industrial, graphic arts, and consumer use (see COLOR PHOTOGRAPHY; PHOTOGRAPHY); magnetic media for audio and visual use and for data storage (see INFORMATION STORAGE MATERIALS; MAGNETIC MATERIALS); adhesives (qv); printing plates; and paper (qv).

Coating is defined herein as replacing air at a substrate interface with a new material. The replacing is the coating process and includes application techniques, the importance of solution parameters, the selection of coating method, and the mechanisms involved. Because most coating solutions are applied from some solvent that must be removed for the coating to be functional, the drying step is also an integral part of the coating process.

Coating Methods Selection

The application of a liquid to a traveling web or substrate is accomplished via a large number of diverse coating methods. Steps in the coating process involve: preparing the solution to be coated; metering the coating solution to the desired coating weight; applying the coating to the support with an applicator; followed by removing the solvent. The coating method selected depends on several factors which include: the nature of the support to be coated; the coating composition rheology; coating solvent; wet-coating weight or coverage desired; coating uniformity; the desired coating speed; the number of layers; and whether the coating is to be continuous or intermittent.

The primary substrates or support include many types of paper and paperboard, polymer films such as polyethylene terephthalate, metal foils, woven and nonwoven fabrics, fibers, and metal coils. Although the coating process is better suited to continuous webs than to short individual sheets, it does work very well for intermittent coating, such as in the printing process. In general, there is an ideal coater arrangement for any given product. However, most available coating machines are required to produce a great many different products and coating thicknesses, and the machine chosen is therefore usually a compromise one, ie, is used for several applications. Table 1 describes the capabilities of some of the principal coating processes.

Rheology. Because the coating process imparts shear and extensional stresses to the solution, the rheological properties of coating liquids are important factors in the selection and successful running of a coating operation.

Coating solutions often exhibit a mixture of viscous and elastic behavior with the response of a particular system, depending on the structure of the material and the extent of deformation. Because the coating process involves the formation and maintenance of interfaces, interfacial or surface rheology must also be considered.

Coating Processes

Many different coating processes are used in the coating industry. Continuous web coating methods include both single layers and multilayers. Dip, rod, knife, blade, air knife, gravure, and forward and reverse roll, methods are all single-layer; slide extrusion, slot, and curtain coating methods may be either single- or multilayer. For discrete surface coating, spray, dip, spin, vacuum, or curtain coating methods can be used.

The control of the coverage or coating weight of the coating and its resulting uniformity is an important characteristic of the coating

Table 1. Summary of Coating Methods[a]

Process	Viscosity, Pa·s[b]	Wet thickness, μm[c]	Coating accuracy, %	Speed max, m/min	Effect of web roughness
		Single layer			
rod, wire wound	0.02–1	5–50	10	250	large
reverse roll	0.1–50	12–1200	5	300	slight
forward roll	0.02–1	10–200	8	150	
air knife	0.005–0.5	2–40	5	500	large
knife over roll	0.1–50	25–750	10	150	large
blade	0.5–40	1–30		1500	large
gravure	0.001–5	1–50	2	700	
slot	0.005–20	15–250	2	400	slight
extrusion	50–5000	15–750	5	700	
		Multilayer			
slide	0.005–0.5	15–250	2	300	slight
curtain, precision	0.005–0.5	2–500	2	300	slight

[a] Values given are meant to be guideline values.
[b] Pa·s = 1000 cP.
[c] 1μm = 1 cm^3/m^2 of wet coating, and 1μm = 1 g/m^2 for a density of 1 g/cm^3.

method. There are two basic classes: premetered, which deliver a set flow rate of solution per unit width to the coated web, and post-metered, in which the coverage is a function of the liquid properties, the system geometry, and the web and roll speeds.

The precision or uniformity of the coating is very important for some products such as photographic or magnetic coatings. Some processes are better suited for precise control of coverage. When properly designed, slot, slide, curtain, gravure, and reverse-roll coaters are able to maintain coverage uniformity to within 2%. In many of the other coating processes the coverage control may be only 10%. Table 1 lists attainable control.

Limits of Coatability. In any coating process there is a maximum coating speed above which coating does not occur. Above this coating speed air is entrained resulting in many bubbles in the coating, or in ribs and finally rivulets, or wet and dry patches. Lower viscosity liquids can be coated faster and thinner. Polymer solutions can be coated at higher speeds than Newtonian liquids.

Discrete Surface-Coating Methods

Coatings may be applied by spraying on irregularly shaped and compound curved or sharp-edged surfaces. Many coating materials of suitable dielectric constant may be electrically polarized so that the powders are attracted to a grounded or oppositely charged surface. Techniques include dip and spin coating and vacuum deposition techniques.

Coating Process Mechanisms

One of the principal advances in the coating process area in the 1980s was the development of techniques to understand and define basic coatings mechanisms. This has led to improved quality and a wider range of utility for most coating techniques. Advances have been in the computer modeling of the coating process and the development of visualization techniques to enable users to actually see the flows in the coating process. The flow patterns predicted by the computer models can be verified by the visualization techniques.

Drying

The drying (qv) process after the application of the coating is as important as the coating process itself. Coatings applied in the fluid state are not complete until dried. The coated film or web is transported through the dryer where the properties of the coating can either be enhanced or made to deteriorate by the drying process. Drying of coatings involves the removal of the inert inactive solvent (vehicle) used to suspend, dissolve, or disperse the active ingredients of the coating, ie, the polymer, binder, pigments, dyes, slip agents, hardener, coating aids, etc. After drying, only the desired coating on the substrate should be retained. Coating solvents can range from the easy-to-handle water to flammable and toxic organic materials. The drying process is a thermal one in which the solvent evaporates. Drying must occur without adversely affecting the coating formulation, maintaining the desired physical uniformity of the coating.

Whereas drying can be a physical process involving only solvent removal, chemical reactions can also be used to help solidify the coating. Solidification, or cross-linking, can be accelerated by catalysts or be accomplished by an electron beam or uv radiation. When this happens, the process is called curing. Practically, most coatings undergo both drying and some form of curing in the dryer, and the distinction is in the degree of curing relative to drying.

Air Impingement Dryers. Air impingement dryers, the most widely used dryers, for coated webs, basically consist of a heat source and heat exchangers, fans to move the air, nozzles or air delivery devices positioned close to the web, and solvent-removal devices in the return airstream.

In these single-sided dryers, the air impinges only one of the coated sides, heating and drying from the coated side only. The two-sided or floater dryer is the most modern dryer design. In this configuration the roll transport system in the dryer is replaced with air nozzles on the rear of the web so that the air transports and supports the web as well as heating and drying the web from both sides.

Contact Dryers. Coatings on webs as well as sheets of newly formed paper can be dried by contacting the web around the surface of a heated drum.

Convection Drying Modeling. Models of the drying process have been developed to estimate whether a particular coating can dry under the conditions of an available dryer. These models can be run on desktop personal computers. To model convection drying, both the heat transfer to the coated web and the mass transfer from the coating must be considered.

Commercial Availability

All of the many types of coaters and dryers discussed herein are commercially available from many different vendors. These vendors usually have pilot facilities so that new coating and drying techniques can be easily tested. Contract coating companies, specializing only in coating, also exist.

EDWARD D. COHEN
E. I. du Pont de Nemours & Co., Inc.
EDGAR B. GUTOFF
Consultant

E. D. Cohen and E. B. Gutoff, eds., *Modern Coating and Drying Technology,* VCH Publishers, New York, 1992, pp. 64, 67, 83, 89, 92, 105, 120.

D. Satas, ed., *Web Processing and Converting Technology and Equipment,* Van Nostrand Reinhold, New York, 1984, pp. 32, 44.

D. Satas, ed., *Coatings Technology Handbook,* Marcel Dekker, Inc., New York, 1991.

H. L. Weiss, *Coating and Laminating Machines,* Converting Technology Co., Milwaukee, Wis., 1977.

POWDER TECHNOLOGY

In the fluidized-bed coating process, the coating powder is placed in a container having a porous plate as its base. Air is passed through the plate, causing the powder to expand in volume and fluidize. In this state, the powder possesses some of the characteristics of a fluid. The part to be coated, which is usually metallic, is heated in an oven to a temperature above the melting point of the powder and dipped into the fluidized bed where the particles melt on the surface of the hot metal to form a coating. Using this process, it is possible to apply coatings ranging in thickness from ca 250 to 2500 μm. It is difficult to obtain coatings thinner than 250 μm and, therefore, fluidized-bed applied coatings are generally referred to as thick-film coatings, differentiating them from most conventional thin-film coatings applied from solution or as a powder at thicknesses of 20–75 μm (see THIN FILMS).

In the electrostatic spray process, the coating powder is dispersed in an air stream and passed through a corona discharge field where the particles acquire an electrostatic charge. The charged particles are attracted to and deposited on the grounded object to be coated. The object, usually metallic and at room temperature, is then placed in an oven, where the powder melts and forms a coating. Using this process, it is possible to apply thin-film coatings comparable in thickness to conventional solution coatings, ie, 20–75 μm. A hybrid process based on a combination of high voltage electrostatic charging and fluidized-bed application techniques (electrostatic fluidized bed) has evolved, as well as triboelectric spray application methods.

Compared to other coating methods, powder technology offers a number of significant advantages. These coatings are essentially 100% nonvolatile, ie, no solvents or other pollutants are given off during application or curing. They are ready to use, ie, no thinning or dilution is required. Additionally, they are easily applied by unskilled operators and automatic systems because they do not run, drip, or sag as do liquid coatings. The reject rate is low and the finish tougher and more abrasion-resistant than that of most conventional paints (see PAINT). Thicker films provide electrical insulation, corrosion protection, and other functional properties. Powder coatings cover sharp edges for better corrosion protection.

Coating powders are frequently separated into decorative and functional grades. Decorative grades are generally finer in particle size, and color and appearance are important.

Coating powders are based on both thermoplastic and thermosetting resins. For use as a powder coating, a resin should possess low melt viscosity, which affords a smooth continuous film; good adhesion to the substrate; good physical properties when properly cured, eg, high toughness and impact resistance; light color, which permits pigmentation in white and pastel shades; good heat and chemical resistance; and good weathering resistance.

The volume of thermosetting powders sold exceeds that of thermoplastics by a wide margin. Thermoplastic resins are almost synonymous with fluidized-bed applied thick-film functional coatings whereas thermosetting powders are used almost exclusively in electrostatic spray processes and applied as thin-film decorative coatings.

Thermoplastic Coating Powders

As a coating powder, a thermoplastic resin must melt and flow at the application temperature without significant degradation. The principal polymer types are based on plasticized poly(vinyl chloride) (PVC), polyamides (qv), and other specialty thermoplastics.

Thermosetting Coating Powders

Thermosetting coating powders, with minor exceptions, are based on resins that cure by addition reactions. Thermosetting resins are more versatile than thermoplastic resins in that many types are available both in varying molecular-weight ranges and having different functional groups; numerous cross-linking agents are available, and thus the properties of the applied film can be modified; the resins possess a low melt viscosity during application, allowing application of thin films, and addition of pigments and fillers required to achieve opacity in the thin films can be incorporated without adversely affecting flow; gloss, textures, and special effects can be produced by modifying the curing mechanism or through the use of additives; and manufacturing costs are lower because compounding is carried out at lower temperatures and the resins are friable and can be ground to a fine powder without using cryogenic techniques.

Formulation. In addition to the resin and curing agents or hardeners, a variety of other ingredients are normally present in coating powder formulations. Catalysts and accelerators are used to modify the reaction rate and curing characteristics. Flow control additives are employed to prevent cratering and promote leveling of the molten polymer film and wetting of the substrate. The most widely used types are low molecular-weight polymers of butyl acrylate and copolymers of ethyl acrylate and 2-ethylhexyl acrylate (see ACRYLIC ESTER POLYMERS).

Matting or flattening agents are employed to control gloss, which is dependent on microscopic surface smoothness. Colorants are used in most powder coating formulations. Carbon black, titanium dioxide, iron oxides, and other inorganic pigments (qv) are widely used.

Specialty Coatings. Clear coatings are formulated using curing agents and flow control additives, which have a high degree of compatibility with the resin. Conventional uv stabilizers can be added to improve exterior durability, and metallic finishes can also be prepared.

Thermosetting coating powders include epoxy coating powders, epoxy–polyester hybrids, urethane polyesters, unsaturated polyester resins, and Acrylic resins.

Manufacture

Coating powders are either melt-mixed or dry-blended, as shown in Figure 1.

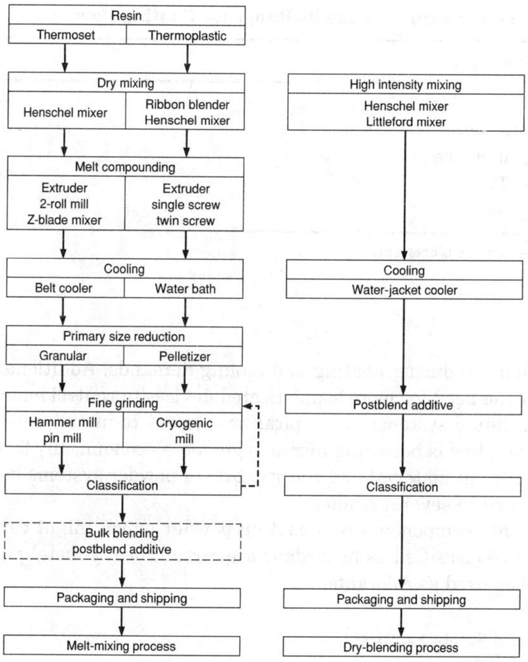

Figure 1. Flow diagram for powder coating manufacture.

Application Methods

Application methods include fluidized-bed coating, electrostatic fluidized-bed coating, electrostatic spray coating, and hot flocking.

Metal Cleaning and Preparation. As in any finishing operation, the surface of the object to be coated must be clean, dry, and free from rust, mill scale, grease, oil, drawing compounds, rust inhibitors, or any soil that might prevent good wetting of the surface by the coating powder. Steel should be sandblasted or centrifugally blast-cleaned to give a near white finish (see METAL SURFACE TREATMENTS). Phosphate and chromate conversion coatings improve performance in harsh environments.

Economic Aspects

A significant factor in the growth of coating powders has been the increasingly stringent environmental regulations governing volatile organic compounds (VOCs) and disposal of hazardous waste. A breakdown of worldwide coating powder consumption and estimated growth rates shows that Europe consumes 52.7% with a growth rate of 9.5%; followed by North America (22.4%) and Far East/Australia (21.6), with growth rates of 12.0% and 15+%, respectively.

The polyester–epoxy hybrids account for 50% of the European market, yet have only a ≤25% market share in the rest of the world. Similarly, the polyester urethanes, which account for only a small market share in Europe, find significant usage in the United States and Japan. Acrylics hold a significant share only in the Japanese market.

Thermoset decorative coatings are by far the largest powder coatings market segment, amounting to 67.641 t in North America in 1992. However, in the United States in 1992, the market for functional thermoset powders (all epoxy) was estimated to be about 13,612 t, primarily going into the pipe and rebar markets.

Environmental and Energy Considerations

In additional to the environmental advantages, the low volatile emissions of powder coatings during the baking operation has economic and energy saving advantages. Further, in the coating operation almost all powder is recovered and reused, resulting in higher material utilization, and waste minimization. The air used in the coating booths during application is filtered and returned to the workplace

Table 1. Lower Explosive Limits Range for Coating Powders

Resin type	g/m^3
epoxy	45–78
epoxy–polyester	33–35[a]
polyester–urethane	65–71
polyester–TGIC	40–70
polypropylene	32

[a] Unfilled—organic binder only.

atmosphere, reducing heating and cooling demands. Additionally, because of the need for more sophisticated devices to control emission of VOCs in liquid systems, the capital investment to install a new powder coating line is becoming increasingly more economically favorable. The savings in material and energy costs of powder systems has been documented in several studies.

The only components in a coating powder which might cause the waste to be classified as hazardous are certain heavy-metal pigments sometimes used as colorants.

Health and Safety Factors

The most significant hazard in the manufacture and application of coating powders is the potential of a dust explosion. The severity of a dust explosion is related primarily to the material involved, its particle size, and concentration in air at time of ignition. The lower explosive limit (LEL), the lowest concentration of a material dispersed in air that explodes when ignited, is essentially the same as the minimum explosive concentration (MEC). The LEL values for a number of coating powders are given in Table 1.

In powder coating installations, the design of the spray booth and duct work, if any, should be such that the powder concentration in the duct is always kept below the LEL, employing a wide margin of safety. General safety considerations are detailed.

The health hazards and risk associated with the use of powder coatings must also be considered (see HAZARD ANALYSIS AND RISK ASSESSMENT). Practical methods to reduce employee exposure to powder, such as the use of long-sleeved shirts to prevent skin contact, should be observed. Furthermore, exposure can be minimized by good maintenance procedures to monitor and confirm that the spray booths and dust collection systems are operating as designed. Ovens should also be properly vented so that any volatiles released during curing, such as caprolactam in the case of urethane polyesters, do not enter the workplace atmosphere.

In general, the raw materials used in the manufacture of powder coatings are relatively low in degree of hazard. None of the epoxy, polyester, or acrylic resins normally used in the manufacture of thermoset powder coatings are defined as hazardous materials by the OSHA Hazard Communication Standard. Most pigments and fillers used in powder coatings generally have no hazards other than those associated with particulate nature.

Coating powders nonetheless should be treated as nuisance dusts and the concentration in air kept below 10 mg/m^{3}TLV–TWA, primarily through environmental controls.

DOUGLAS S. RICHART
Morton International, Inc.

D. S. Richart, "Coating Methods, Powder Coating," in J. I. Kroschwitz, ed., *Encyclopedia of Polymer Science and Engineering,* 2nd ed., Vol. 3, John Wiley & Sons, Inc., New York, 1985, pp. 575–601.

D. A. Bate, *The Science of Powder Coatings—Chemistry Formulation and Application,* Vol. 1, Scholicem Int., Port Washington, N.Y., 1990.

T. Misev, *Powder Coatings in Chemistry and Technology,* John Wiley & Sons, Inc., New York, 1991.

D. L. Ulrich, ed., *User's Guide to Powder Coatings,* 3rd ed., AFP/Society Manufacturing Engineering Dearborn, Mich., 1993.

SPRAY COATING

A coating may be applied to articles, ie, workpieces, by spraying. This application method is especially attractive when the articles have been previously assembled and have irregularly shaped and curved surfaces. The material applied is frequently a paint (qv), ie, a combination of resin, solvent, diluent, additives, and pigment. The material can also be a hot thermoplastic, an oil, or a polymer dissolved in a solvent. Many types of spray equipment are available. Methods can be used in combinations, and most of the techniques can be used for simple one-applicator manual systems or in highly complex computer-controlled automatic systems having hundreds of applicators. In an automatic installation, the applicators can be mounted on fixed stands, reciprocating or rotating machines, or even robots.

Atomization

Airless Atomization. In airless or pressure-atomizing systems, the coating is atomized by forcing the coating (or the liquid) through a small-diameter nozzle under high pressure. The fluid pressure is typically between 5 and 35 MPa (700–5000 psi); fluid flow rates are between 150–1500 cm^3/min. In most commercial applications, a pump designed for the type of material sprayed is used to develop the high pressure.

A variation of airless atomization is called air-assisted airless. A small amount of compressed air at 35–170 kPa (5–25 psi) is introduced adjacent to the airless nozzle and impinges upon the thin sheet of fluid as it exits from the nozzle. This air aggravates the turbulence in the fluid and results in improved atomization at lower fluid pressures. Often, material that cannot be properly atomized using straight airless atomization can be using the air-assisted airless method. In some cases, the introduction of the air allows some control of the fan size.

Air Atomization. In an air atomizer, an external source of compressed air, usually supplied at pressures of 70–700 kPa (10–100 psi), is used to atomize the liquid. Air atomization is perhaps the most versatile of all the atomization methods. It is used with liquids of low to medium viscosity, and flow rates of 50–1000 cm^3/min are common. Medium-to-fine particle sizes are produced, and the resulting surface finish is very good. It is sometimes difficult to penetrate small recessed areas, however, because the atomization air forms a barrier in the recess that the coating particles must then penetrate. When higher air pressures are used, air atomization produces considerable misting and overspray, which can be a disadvantage under some conditions. Air-atomizing devices can be of internal- or external-mix design.

Electrostatic Atomization. The atomization of the coating material by electrostatic forces occurs when an electrical charge is placed on a filament or thin sheet of coating, and the mutual repulsion of the charges tears the coating material apart. For this process to produce acceptable atomization, the physical properties of the coating material must be within a relatively narrow range. The material is charged by an external source of high voltage, either prior to or as it is forced to flow over a knife edge, through a thin slot or orifice, or it is discharged from the edge of a slowly rotating disk or bell (cup). The thin sheet or small-diameter stringers or cusps of coating material are torn apart by the mutual repulsion of the charges on the material.

Rotary Atomization. In rotary atomization, a bell (cup) or disk rotates at a speed of 10,000–60,000 rpm. In contrast to electrostatic atomization, mechanical forces dominate. The coating material is introduced near the center of the rotating device, and centrifugal force distributes it to the edge, where the material has an angular velocity close to that of the rotating member. As the coating material leaves the surface its main velocity component is tangential, and it is spun off in the form of a thin sheet or small cusps. The material is then atomized by turbulent or aerodynamic disintegration, depending on exact conditions. Rotary atomization produces the most uniform atomization of any of the aforementioned techniques, and produces the smallest maximum particle size. It is almost always used with electrostatics to aid deposition, and at lower rotational speeds the electrostatics assist the atomization.

Recent developments in rotary atomization include the use of semiconductive composites for the rotary cup, permitting the construction of a unit that does not produce an ignition spark when brought close to a grounded workpiece yet has the transfer efficiencies associated with a rotary atomizer.

Electrostatic Spraying

Use of electrostatic spraying or electrostatic deposition increases the efficiency of material transfer to the workpiece (see COATING PROCESSES, POWDER TECHNOLOGY). The cost of solvents and coating materials and the emphasis on reducing emissions to the atmosphere have both increased dramatically since the late 1970s. These factors have effected an emphasis on increased transfer efficiency, ie, the fraction of the material removed from the coating bucket that is placed on the workpiece. The transfer efficiency is affected by the painting technique, workpiece geometry, the coating material, how the workpiece is presented to the atomizer, the ambient air movement, and other variables.

Electrostatic forces can be very effective in increasing the transfer efficiency. An electrical charge, usually negative, is placed on the coating material before atomization or as the coating particles are being formed. This is accomplished either by direct charging, where the coating material comes in contact with a conductor at high voltage, or by an indirect method, where the air in the vicinity of the coating particles is ionized and these ions then attach themselves to the coating particles.

An electrostatic force is exerted on each coating particle equal to the product of the charge it carries and the field gradient. The trajectory of the particle is determined by all the forces exerted on the particle. These forces include momentum, drag, gravity, and electrostatics. The field lines influencing the coating particles are very similar in arrangement to the alignment of iron particles when placed between two magnets. Using this method, coating particles that would normally pass alongside the workpiece are attracted to it, and it is possible to coat part or all of the back side of the workpiece.

All of the atomization techniques that produce a spray can be used with electrostatic spraying.

Economic Aspects

Spray equipment is marketed in a variety of ways in the United States. Several large manufacturers have broad product lines and sell equipment both directly to the user and through distributors. These companies can also provide custom-engineered automatic systems for specific applications. The largest U.S. manufacturers include Binks, Graco, ITW (DeVilbiss and Ransburg), and Nordson.

K. J. COELING
Nordson Corporation

F. A. Robinson, Jr., G. Pickering, and J. Scharfenberger, *Paint Con. '84,* Hitchcock Publishing Co., Carol Stream, Ill., 1984.

J. M. Lipscomb, *Surface Coating '83,* Chemical Coaters Association, *Finishing '83 Conference Proceedings,* Association for Finishing Processes of SME, pp. 10-1, 10-10.

J. Schrantz and J. M. Bailey, *Indust. Finish.* (June 1989).

COATINGS

Coatings are ubiquitous in an industrialized society. Coatings are used for decorative, protective, and functional treatments of many kinds of surfaces (see PAINT). Coatings such as interior flat wall paints are used for decoration but incidentally serve to reflect light diffusely to provide more uniform illumination. Some coatings, such as those on undersea pipelines, are only for protective purposes (see also COATINGS, MARINE). Others, such as exterior automobile coatings, fulfill both decorative and protective functions. Still others provide friction control on boat decks or car seats. Some coatings control the fouling of ship bottoms, others protect food and beverages in cans (see FOOD PACKAGING). Silicon chips, printed circuit panels, coatings on waveguide fibers for signal transmission, and magnetic coatings on videotapes and computer disks are among many so-called high-tech applications for coatings (see also ELECTRONIC COATINGS; INFORMATION STORAGE MATERIALS; MAGNETIC MATERIALS–THIN FILMS AND PARTICLES).

Each year tens of thousands of coating types are manufactured. In general, these are composed of one or more resins or polymers, a mixture of solvents (except in powder coatings), commonly one or more pigments (qv), and frequently one or more additives. Coatings can be classified into thermoplastic and thermosetting coatings. Thermoplastic coatings contain at least one polymer having sufficiently high molecular weight to provide the required mechanical strength properties without further polymerization. Thermosetting coatings contain lower molecular-weight polymers that are further polymerized after application in order to achieve desired properties. The term thermosetting is used not only when the coating is cured or baked in an oven after application, but also when it is dried or cured at ambient temperature (see also COATING PROCESSES).

Film Formation

Most coatings are manufactured and applied as liquids and are converted to solid films after application to the substrate. In the case of powder coatings, the solid powder is converted after application to a liquid, which in turn forms a solid film (see COATING PROCESSES–POWDER TECHNOLOGY). The polymer systems used in coatings are amorphous materials and therefore the term solid does not have an absolute meaning, especially in thermoplastic systems such as lacquers, most plastisols, and most latex-based coatings. A useful definition of a solid film is that it does not flow significantly under the pressures to which it is subjected during testing or use. Thus a film can be defined as a solid under a set of conditions by stating the minimum viscosity at which flow is not observable in a specified time interval. For example, it is reported that a film is dry-to-touch if the viscosity is greater than about 10^6 mPa·s(= cP). However, if the definition of dry is that the film resists blocking, that is, sticking together, when two coated surfaces are put against each other for two seconds under a mass per unit area of 1.4 kg/cm^3 (20 psi), the viscosity of the film has to be greater than 10^{10} mPa·s(= cP).

For practical coatings, it is not sufficient just to form a film; the film must also have a minimum level of strength, depending on product use. Film strength depends on many variables, but one critical factor is molecular weight. This required molecular weight varies according to the chemical composition of the polymer and the mechanical properties required for a particular application.

Solvent Evaporation from Solutions of Thermoplastic Polymers. A solution of a copolymer of vinyl chloride (chloroethene), C_2H_3Cl, vinyl acetate (acetic acid ethenyl ester), $C_4H_6O_2$, and a hydroxy-functional vinyl monomer having a number average molecular weight ($\overline{M}_n$) of 23,000 and a T_g of 79°C, gives coatings having good mechanical properties without cross-linking. A simple coating having only the resin and 2-butanone (methyl ethyl ketone, MEK), C_4H_8O, as the sole solvent would give a polymer concentration of about 19 wt % solids or approximately 12 vol % in order to have a viscosity of about 100 mPa·s(= cP) for spray application (see COATING PROCESSES–SPRAY COATING). Because of the relatively high vapor pressure under application conditions, MEK evaporates rapidly and a substantial fraction of the solvent evaporates in the time interval between the coating leaving the orifice of the spray gun and its arrival on the surface being coated. As the solvent evaporates, the viscosity increases and the coating reaches the dry-to-touch stage very rapidly after application and does not block under the conditions discussed. However, if the film is formed at 25°C, the dry film contains several percent retained solvent.

In the first stages of solvent evaporation from such a film the rate of evaporation depends on the vapor pressure at the temperatures encountered during the evaporation, the ratio of surface area to volume of the film, and the rate of air flow over the surface and is essentially independent of the presence of polymer. However, as the solvent evaporates the T_g increases, free volume decreases, and the rate of loss of solvent from the film at some point becomes dependent not on how fast the solvent evaporates, but on how rapidly the solvent molecules can diffuse to the surface of the film. In this diffusion-control stage, solvent molecules must jump from free-volume hole to free-volume hole to reach the surface where evaporation can occur. As solvent loss continues, T_g increases, and free volume decreases. When the T_g of the remaining polymer solution approaches the temperature at which the film is being formed, the rate of solvent loss becomes very slow.

The rate of solvent diffusion through the film depends not only on the temperature and the T_g of the film, but also on the solvent structure and solvent–polymer interactions.

Film thickness is an important factor in solvent loss and film formation. In the first stage of solvent evaporation, the rate of solvent loss depends on the first power of film thickness. However, in the second state when the solvent loss is diffusion rate-controlled, it depends on the square of the film thickness.

Thermoplastic polymer-based coatings have low solids contents because the relatively high molecular weight requires large amounts of solvent to reduce the viscosity to that required for application. Air pollution (qv) regulations limiting the emission of volatile organic compounds (VOC) and the increasing cost of solvents has led increasingly to replacement of such coatings with types that require less solvent for application (see COATING PROCESSES–SURVEY).

Film Formation from Solutions of Thermosetting Polymers. Substantially less solvent is required in formulating a coating from a low molecular weight resin that can be further polymerized to a higher molecular weight after application to the substrate and evaporation of the solvent. Theoretically, difunctional reactants could be used. However, in contrast to polymers for fibers and some plastics, this is not feasible for coatings where the close control of stoichiometric ratio and purity required to achieve a desired molecular weight reproducibly with difunctional reactants is impractical. Therefore, the average functionality must be over two in order to ensure that the molecular weight of the final cured film is high enough for good properties. Not only should the average functionality be over two, but it is usually preferable for the number of monofunctional molecules to be at a minimum because these terminate polymerization. If any of the resin molecules have no functional groups, they cannot react and remain in the film as a plasticizer (see PLASTICIZERS). The reactions are commonly called cross-linking reactions. A cross-linked film not only has very high molecular weight; it is also insoluble in solvents. For many but not all applications, this solvent resistance is an advantage of thermosetting coatings over thermoplastic coatings.

The mechanical properties of the cross-linked film depend strongly on many factors; two of the most important are the lengths of the segments between cross-links and the T_g of the cross-linked resin. Segment length depends on the average equivalent weight and the average functionality of the components and the fraction of cross-linking sites actually reacted.

Film Formation by Coalescence of Polymer Particles. Latex paints have low solvent emissions as well as many other advantages. A latex is a dispersion of high molecular weight polymer particles in water (see LATEX TECHNOLOGY). The dispersion is stabilized by charge repulsion and entropic repulsion. (The terms steric repulsion or osmotic repulsion are sometimes used instead of entropic repulsion.) Because the latex polymer is not in solution, the rate of water loss by evaporation is almost independent of concentration until near the end of the evaporation process. When a dry film is prepared from a latex, the forces that stabilize the dispersion of latex particles must be overcome and the particles must coalesce into a continuous film. As the water evaporates, the particles come closer and closer together. As they approach each other, they can be thought of as forming the walls of capillary tubes in which surface tension leads to a force striving to collapse the tube. The smaller the diameter of the tube, the greater the force. When the particles get close enough together so that the force pushing them together exceeds the repulsive forces holding them apart, coalescence is possible. A surface tension driving force also promotes coalescence because of the decrease in surface area when the particles coalesce to form a film. Both factors have been shown to be important in film formation from latexes. Coalescence, however, also requires that the polymer molecules in the particles be free to intermingle with those from adjoining particles. This movement can occur only if there are sufficient number and size of free-volume holes in the polymer particles into which the polymer molecules from other particles can move. In other words, the T_g of the latex particles must be lower than the temperature at which film formation is being attempted.

The rate of coalescence is controlled by the free-volume availability, which in turn depends mainly on $(T - T_g)$. The viscosity of the coalesced film is also dependent on free volume.

Powder coatings form films by coalescence. Because the powder must not fuse or sinter during storage, the free volume at storage temperature must be sufficiently low to avoid coalescence at this stage.

Resins

Alkyds. Although no longer the principal class of resins used in coatings in the United States, alkyds are still very important and a wide range of types of alkyds are manufactured (see ALKYD RESINS). Whereas some nonoxidizing alkyds are used as plasticizers in lacquers and cross-linked with melamine–formaldehyde resins in baking enamels, the majority are oxidizing alkyds for use in coatings for air-dry and force-dry applications. The principal advantages of alkyds are low cost and relatively foolproof application characteristics, resulting primarily from low surface tensions. The principal shortcomings of these resins are embrittlement and discoloration upon aging and relatively poor hydrolytic stability.

Polyesters. The term polyester is used in the coatings field almost entirely for low molecular-weight hydroxy, or sometimes carboxylic acid, terminated oil-free polyesters. Polyesters have been one important class of replacements for alkyd resins in melamine–formaldehyde cross-linked baking enamels. Hydroxy-terminated polyesters are also used with polyfunctional isocyanates in making air-dry and force-dry coatings, as well as with blocked isocyanates in coatings for higher baking temperature and in powder coatings. Carboxylic acid-terminated polyesters are used predominantly with epoxy cross-linkers in powder coatings (see EPOXY RESINS). When adhesion directly to metal is required, polyesters are generally preferred over acrylic resins.

Amino Resins. Melamine–formaldehyde (MF) resins are the most widely used cross-linking agents for baking enamels (see AMINO RESINS AND PLASTICS). Unlike the MF resins used in plastics, textile, and paper (qv) treatment, the MF resins used in coatings are ether derivatives. The resins are made by reacting melamine (1,3,5-triazine-2,4,6-triamine), $C_3H_6N_6$, and formaldehyde, CH_2O, followed by etherification of the methylol groups using an alcohol. Two classes of MF resins are widely used. Class I resins are made using excess formaldehyde and a high fraction of the methylol groups are etherified with alcohol.

Class II MF resins are made using a lower ratio of formaldehyde to melamine and a significant fraction of the nitrogens have one alkoxymethyl group and a hydrogen.

Acrylic Resins. Acrylic resins are the largest volume class of coatings resins. The term acrylic is used in a general sense to mean resins where a significant fraction of the comonomers are acrylic or methacrylic esters (see ACRYLIC ESTER POLYMERS; METHACRYLIC POLYMERS). Other monomers, such as styrene (qv) (ethenylbenzene), C_8H_8, vinyl acetate, and others, may be included in the copolymer. High molecular weight thermoplastic acrylics, which were widely used in acrylic lacquers for new automobiles, are now restricted to repair and aftermarket lacquers because of the need to control VOC emissions from automobile assembly plants. As VOC controls become stricter, use of acrylic lacquers in repair shops can be expected to be eliminated. Use of thermosetting acrylic resins continues to expand, with increasing emphasis on high solids and water-reducible types. In architectural coatings, acrylic latexes and styrene or vinyl acetate acrylic copolymer latex-based paints have replaced drying oil and alkyd-based paints almost entirely except in gloss enamels. The consumption of acrylic latexes in industrial coatings applications is small but growing. The main advantages of acrylic coatings involve the high degree of resistance to thermal and photoxidation and to hydrolysis, giving coatings that have superior color retention, resistance to embrittlement, and exterior durability.

Hydroxy-functional thermosetting acrylics are widely used in baking enamels for automobile and appliance top coats, exterior can coatings, and coil coating.

Epoxy Resins. Epoxy resins are used to cross-link other resins with amine, hydroxyl, and carboxylic acid (or anhydride) groups. The epoxy group, properly called an oxirane, is a cyclic three-membered ether group. By far the most widely used epoxy resins in coatings are bisphenol A (BPA) (4,4′-(1-methylethylidene)bisphenol), $C_{15}H_{16}O_2$, epoxy resins.

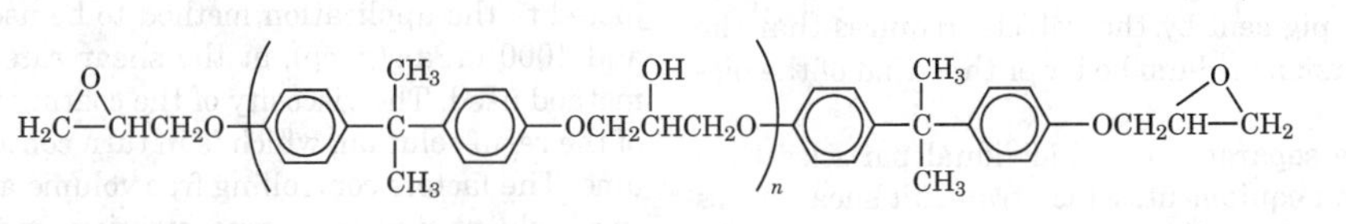

bisphenol A (BPA) epoxy resin

A principal use for epoxy resins is as a component in two-package primers for steel. One package contains the epoxy resin and the other a polyfunctional amine. Epoxy/amine coatings are particularly effective as corrosion-protective primers for steel, because the amine groups resulting from the cross-linking reaction promote adhesion to the steel in the presence of water and because the cross-linked resins are completely resistant to hydrolysis.

Epoxy resins also are used to cross-link phenolic resins. Such coatings are widely used in interior can linings. The hydrolytic stability and adhesion of the coatings are critical in terms of selection for this application. Adhesion is still further improved by the incorporation of a small amount of epoxy phosphate in the coatings.

Epoxy resins are widely used in powder coatings. Probably the largest volume usage is of BPA epoxy resins cross-linked with dicyandiamide (cyanoguanidine), $C_2H_4N_4$. Because BPA epoxy resins are easily photoxidized, they are not useful in coatings requiring exterior exposure. Triglycidylisocyanurate (TGIC) (1,3,5-tris(oxiranylmethyl)-1,3,5-triazine-2,4,6(*1H,3H,5H*)-trione), $C_{12}H_{15}N_3O_6$, is widely used in powder coatings that require exterior durability.

Another use for epoxy resins is as a raw material to make epoxy esters. Epoxy esters are made by reaction of BPA epoxy resins and drying oil fatty acids.

Urethane Systems. Isocyanates react with a wide variety of functional groups to give cross-links. The most widely used coreactants are hydroxy-functional polyester and acrylic resins. The isocyanate group reacts with the hydroxyl group to generate a urethane cross-link. The reaction proceeds relatively rapidly at ambient or modestly elevated temperatures. In addition to the low curing temperatures, significant advantages of urethane coatings are generally excellent abrasion and impact resistance, combined with solvent resistance.

Other Coatings Resins. A wide variety of other resin types are used in coatings. Phenolic resins, ie, resins based on reaction of phenols and formaldehyde, have been used in coatings for many years. Silicone resins provide coatings having outstanding heat resistance and exterior durability. Polyfunctional 2-hydroxyalkylamides can serve as cross-linkers for carboxylic acid-terminated polyester or acrylic resins, and a range of acetoacetylated resins has been introduced.

Volatile Components

In most coatings, solvents are used to adjust the viscosity to the level required for the application process and to provide for proper flow after application. Most methods of application require coating viscosities of 50–1000 mPa·s(= cP). Many factors must be considered in the selection of the solvent or, more commonly, solvent mixtures. Except for water-borne systems, solvents that dissolve the resins in the coating formulation are usually chosen. Solubility parameters have been recommended as a tool for selecting solvents that can dissolve the resins. However, the concept of solubility parameters for the prediction of polymer solubility is a gross oversimplification and the old principle that like dissolves like is the most useful selection criterion. For resins that are soluble in esters (see ESTERS, ORGANIC) and ketones, costs can generally be decreased by using mixtures that contain hydrocarbons and alcohols to replace part of the esters and ketones.

The problem of solvent selection is most difficult for high molecular-weight polymers such as thermoplastic acrylics and nitrocellulose in lacquers. As molecular weight decreases, the range of solvents in which resins are soluble broadens. Even though solubility parameters are inadequate for predicting all solubilities, they can be useful in performing computer calculations to determine possible solvent mixtures as replacements for a solvent mixture that is known to be satisfactory for a formulation.

An important characteristic of solvents is rate of evaporation. Rates of solvent loss are controlled by the vapor pressure of the solvent(s) and temperature, partial pressure of the solvent over the surface, and thus the air-flow rate over the surface, and the ratio of surface area to volume. Tables of relative evaporation rates, in which *n*-butyl acetate is the standard, are widely used in selecting solvents.

Final adjustment of solvent selection must be done under actual field conditions. Toxic hazards, environmental considerations, flammability, odor, surface tension, and viscosity also affect solvent selection.

Pigments

Pigments (qv) in coatings provide opacity and color. Pigment content governs the gloss of the final films and can have important effects on mechanical properties. Some pigments inhibit corrosion. Pigmentation affects the viscosity and hence the application properties of coatings. An important variable determining the properties of pigments is particle size and particle-size distribution. Pigment manufacturing processes are designed to afford the particle size and particle-size distribution that provide the best compromise of properties for that pigment. In the process of drying the pigment, the particles generally aggregate. The coatings manufacturer must disperse these dry pigment aggregates in such a way as to achieve a stable dispersion where most, if not all, of the pigment is present as individual particles.

Pigments can be divided into four broad classes: white, color, inert, and functional pigments. The ideal white pigment, when dispersed in the coating, would absorb no visible light and would efficiently scatter light entering the film.

Color pigments selectively absorb some wavelengths of light more strongly than others. A wide variety of color pigments are used in coatings.

Inert pigments, also called extenders and fillers do not exhibit significant absorption or scattering of light when incorporated into coatings. In most cases, inert pigments are used to occupy volume in the coating composition. A wide variety of clays, silica, and carbonates are used as inert pigments. The most widely used functional pigments are the corrosion-control pigments. These pigments inhibit the corrosion of steel.

Pigment Dispersion. The dispersion of pigments involves wetting, separation, and stabilization. Wetting, ie, displacement of air and water from the surface of the pigment by the vehicle, requires that the surface tension of the dispersion medium be lower than that of the pigment surface.

Pigment aggregates are separated into individual particles by a variety of types of dispersion equipment, which transmit shear stress of sufficient magnitude to break up the aggregates.

Pigment dispersions are stabilized by charge repulsion and entropic, ie, steric or osmotic, repulsion. Although both types of stabilization force may be present in most cases, for pigment dispersions in solvent-borne coatings entropic repulsion is usually the most important mechanism for stabilization.

Dispersion of pigments for latex paints is the principal application of aqueous dispersions in coatings. Commonly, three surfactants are used in preparing the dispersion of the white and inert pigments: potassium tripolyphosphate, an anionic surfactant, and a nonionic surfactant. The final colored latex paint formulations, which are very complex, commonly contain seven pigments, some having high surface energies (inorganic pigments) and some having low surface energies (organic pigments). The latex polymer is present as a dispersion that must be stabilized against coalescence and flocculation, and latexes themselves commonly contain two or more surfactants and a water-soluble polymer.

Pigment Volume Relationships. Pigmentation can have profound effects on the properties of coating films, depending on the level of pigmentation of these films. Variations in these effects are best interpreted in terms of volume relationships rather than weight relationships. Pigment volume concentration (PVC) is defined as the volume of pigment in a dry film divided by the total volume of the dry film, commonly expressed as a percentage. The term pigment volume content is sometimes used for pigment volume concentration.

As the volume of pigment in a series of formulations is increased, properties change, and at some PVC there is a fairly drastic change in a series of properties that is a function of PVC. This PVC is defined as the critical pigment volume concentration (CPVC) for that system. The CPVC is the maximum PVC that can be present in a dry film of that system, having sufficient solvent-free resin to adsorb on all the pigment surfaces and fill all the interstices between the pigment particles. In other words, when the PVC is above the CPVC, there are voids in the film. Gloss decreases as PVC increases. Hiding generally increases as PVC increases, but above CPVC there is a rapid increase in the rate of increase of hiding because of the presence of voids. Tinting strength of white paints behaves like hiding, increasing rapidly above CPVC. Tensile strength of films increases with increasing PVC, passing through a maximum at CPVC. Stain resistance is poorer and the ease of removing stains becomes more difficult for coatings above CPVC, compared to those with PVC below CPVC. Blistering of films on wood is less likely to occur above CPVC. The intercoat adhesion to a primer is improved when the primer has a PVC greater than CPVC. The CPVC effects result from the voids in dry films having a PVC greater than CPVC.

Color Matching. Most pigmented coatings must be color matched to a certain standard. In many cases, a significant part of the time involved in formulating a new coating is in establishing the color match, and a significant part of the manufacturing cost is the color matching operation. Poor color matching is a common source of customer complaints, and problems and costs can be minimized if established specifications for the initial color match and for judging the acceptability of production batches are made.

Flow

Rheological properties, that is, flow and deformation, of coatings have significant impacts on application and performance properties. The application and film formation of liquid coatings require control of the flow properties at all stages. The mechanical properties of the applied coating films are controlled by the viscoelastic responses of the films to stress and strain.

Viscosity of Resin Solutions. The viscosity of coatings must be adjusted to the application method to be used. It is usually between 50 and 1000 mPa·s(= cp), at the shear rate involved in the application method used. The viscosity of the coating is controlled by the viscosity of the resin solution, which is in turn controlled mainly by the free volume. The factors controlling free volume are temperature, resin structure, solvent structure, concentration, and solvent–resin interactions.

Viscosity of Systems with Dispersed Phases. A large proportion of coatings are pigmented and, therefore, have dispersed phases. In latex paints, both the pigments and the principal polymer are in dispersed phases. The viscosity of a coating having dispersed phases is a function of the volume concentration of the dispersed phase and can be expressed mathematically by the Mooney equation, a convenient form of which is $\ln \eta = \ln \eta_e + K_E V_i/(1 - V_i/\phi)$, where η_e is the viscosity of the external phase, K_E is a shape constant (2.5 for spheres), V_i is the volume fraction of internal phase, and ϕ is the packing factor, ie, the volume fraction of internal phase when the V_i is at the maximum close-packed state possible for the system. The Mooney equation assumes rigid particles having no particle–particle interaction. It fits pigment dispersions and latexes that exhibit Newtonian flow.

Flow During Application. Many methods of film application to a substrate give an uneven surface. For example, application by brush gives brush marks that are not related to the bristles of the brush, but rather to the fact that the film of wet paint is split between the substrate and the brush. Similarly, when paint is applied by a roller, the film is split such that part of the paint remains on the roller and part transfers to the substrate. Film splitting gives an uneven surface. It is generally desirable for these nonuniformities to level out. Leveling occurs most rapidly if the viscosity of the coating on the substrate is low. In order to keep the viscosity low for a longer period, it is common to formulate brush-applied coatings using relatively slow-evaporating solvents. However, low viscosity coatings applied to a vertical wall sag, that is, they run down the wall unevenly. The viscosity requirements to promote leveling and to minimize sagging are in conflict with each other. Sometimes solvents can be chosen that evaporate slowly enough to permit reasonable leveling, but evaporate rapidly enough so that the viscosity increases before serious sagging occurs.

Film Properties

The film properties required for some applications can only be determined by the performance of the applied coating in practice. Because requirements and exposure conditions vary widely, devising laboratory tests to predict film performance is difficult and frequently not possible. Data banks of actual field performance as functions of coating compositions, application variables, and environmental factors can be very useful.

Adhesion

In most cases, it is desirable to have a coating that is difficult to remove from the substrate to which it has been applied. An important factor controlling this property is the adhesion between the substrate and the coating (see ADHESIVES). In formulating a coating, it is important to remember that difficulty in removing a coating also can be affected by how difficult it is to penetrate through the coating and how much force is required to push the coating out of the way as the coating is being removed from the substrate as well as the actual force holding the coating onto the substrate. Furthermore, the difficulty of removing the coating can be strongly affected by the roughness of the substrate. If the substrate has undercut areas that are filled with cured coating, a mechanical component makes removal of the coating more difficult. This is analogous to holding two dovetailed pieces of wood together.

Surface roughness affects the interfacial area between the coating and the substrate. The force required to remove a coating is related to the geometric surface area, whereas the forces holding the coating onto the substrate are related to the actual interfacial contact area. Thus the difficulty of removing a coating can be increased by increasing the surface roughness. However, greater surface roughness is only of advantage if the coating penetrates completely into all irregularities, pores, and crevices of the surface. Failure to penetrate completely can lead to less coating-to-substrate interface contact than the corresponding geometric area and leave voids between the coating and the substrate, which can cause problems.

Adhesion to Metals. For interaction between coating and substrate to occur, it is necessary for the coating to wet the substrate. Somewhat oversimplifying it may be stated that the surface tension of the coating must be lower than the surface tension of the substrate. In the case of metal substrates, clean metal surfaces have very high surface tensions and any coating wets a clean metal substrate.

Penetration of the vehicle of the coating as completely as possible into all surface pores and crevices is critical to achieving good adhesion. The critical viscosity is that of the continuous phase because many of the crevices in the surface of metal are small compared to the size of pigment particles. Because penetration takes time, the initial viscosity of the external phase should be low and the viscosity should be kept as low as possible for as long as possible. Slow evaporating solvents are best for coatings that are to be applied directly on metal.

Adhesion is strongly affected by the interaction between coating and substrate. On a clean steel substrate, hydrogen bond or weak acid–weak base interactions between the surface layer of hydrated iron oxide that is present on any clean steel surface and polar groups on the resin of the coating provide such interaction.

Fracture mechanics affect adhesion. Fractures can result from imperfections in a coating film which act to concentrate stresses. In some cases, stress concentration results in the propagation of a crack through the film, leading to cohesive failure with less total stress application. Propagating cracks can proceed to the coating/substrate interface; then the coating may peel off the interface, which may require much less force than a normal force pull would require.

Adhesion of coatings is also affected by the development of stresses as a result of shrinkage during drying of the film.

The formation of covalent bonds between resin molecules in a coating and the surface of the substrate can enhance adhesion.

Adhesion to Plastics and Coatings. In contrast to the application of coatings on clean steel, wetting can be a serious problem for adhesion of coatings to plastics. Some plastic substrates have such low surface tension that it may be difficult to formulate coatings having a sufficiently low surface tension to wet the substrate. Polyolefin plastics, in particular, are difficult to wet. Frequently the surface of the polyolefin plastic must be oxidized to increase surface tension and provide groups to interact with polar groups on the coating resin. Adhesion to plastics can be enhanced if resin molecules from the coating can penetrate into the surface layers of the plastic. Intercoat adhesion to other coatings is a specialized case of adhesion to plastics.

Testing for Adhesion. Because of the wide range of exposures to stresses in actual use, there is no really satisfactory laboratory test for the adhesion of a coating film to the substrate during use. Probably the most useful guide for an experienced coatings formulator is the use of a penknife to see how difficult it is to remove the coating and to observe its mode of failure. Many tests for adhesion have been devised.

Exterior Durability

In many cases, an important performance requirement for coatings is exterior durability. There are many potential modes of failure when coatings are exposed outdoors. Although many failure mechanisms are involved, the two most common modes are hydrolysis and photochemical oxidation by free-radical chain reactions. In general, resins that have backbone linkages that cannot hydrolyze provide better exterior durability than systems having, for example, ester groups in the backbone.

Susceptibility to free-radical-induced photoxidation varies with structure of the resins and pigments and, in some cases, with the interactions between pigment and resin. In general terms, resistance of resins to photochemical failure is related to the ease of abstraction of hydrogens from the resin molecules by free radicals. The greatest resistance is shown by fluorinated resins and silicone resins, especially methyl-substituted silicone resins. The greatest sensitivity to degradation is shown by resins having methylene groups between two double bonds; methylene groups adjacent to amine nitrogens, ether oxygens, or double bonds; and methine groups.

Pigment selection can also be critical in formulating for high exterior durability. Some pigments are more susceptible to color change on exposure than others. Some pigments act as photosensitizers to accelerate degradation of resins in the presence of uv and water.

The exterior durability of relatively stable coatings can be enhanced by use of additives. Ultraviolet absorbers reduce the absorption of uv by the resins and hence decrease the rate of photodegradation. Further improvements can be gained by also adding free-radical trap antioxidants such as hindered phenols and especially hindered amine light stabilizers (HALS).

Corrosion Control

An important function of many coatings is to protect metals, especially steel, against corrosion. Corrosion protection is required in two different situations: in one case, the steel is protected against corrosion with intact coating films; in the other case, the objective is to protect the steel against corrosion even when the film has been ruptured.

Protection by Intact Films. In the case of intact films, the key factors responsible for corrosion protection are adhesion of the coating to the steel in the presence of water, oxygen and water permeability, and the resistance of the resins in the film to saponification. If the resins in the coating are adsorbed to cover the steel surface completely and if the adsorbed groups cannot be displaced by water, oxygen and water permeating through the film cannot contact the steel and corrosion does not occur. In effect, only a monolayer is required to protect against corrosion, if the monolayer stays in place over the period of exposure.

Protection by Nonintact Films. It is also possible to achieve corrosion control by coatings after a film has been ruptured. Because the coatings used to achieve this control generally give poorer protection when their films are not ruptured, such systems should be used only when film rupture must be anticipated or when complete coverage of the steel interface cannot be achieved. There are two techniques used on a large scale: primers containing corrosion-inhibiting pigments and zinc-rich primers.

Mechanical Properties

Hundreds of tests of the mechanical properties of coatings have been developed. Many are suitable only for quality control work, but a few have been sufficiently correlated with actual field performance and hence have value for predicting performance. The ASTM provides detailed procedures for many paint tests. A monograph on mechanical properties of coatings has been published that gives an excellent presentation of a limited range of tests. Basic mechanical properties such as tensile strength-to-break, elongation-to-break, work-to-break, loss and storage modulus, and loss tangent (tan delta) can be more readily related to structure and correlated with actual performance than most paint tests. Dynamic mechanical thermal analysis has become a powerful tool for the assessment of cure and the study of the effect of extent of sure on film properties and performance.

Coating films are viscoelastic; therefore, mechanical properties depend on temperature and the rate of application of stress. Behavior is shifted toward the elastic mode with increased tensile strength and decreased elongation-to-break and with a more nearly constant modulus as a function of stress as temperature is decreased or as the rate of application of stress is increased.

In designing coatings that must withstand large deformations without cracking, such as in container and coil coating applications, the T_g, as determined at a rate of application of stress comparable to that which is experienced in the forming operation, should be lower than the temperature at which forming takes place.

Abrasion resistance of floor coatings has been related to work-to-break of coatings. There is often no correlation to the most widely used specification test for abrasion, the Tabor Abraser test. Polyurethane coatings, in general, provide superior abrasion resistance. This may result from the presence of what might be thought of as two types of cross-links, the covalent cross-links and cross-links resulting from intermolecular hydrogen bonds between urethane groups on different molecules.

In other cases, abrasion resistance is related to the coefficient of friction between the surfaces in contact with each other. Abrasion resistance is also related to the tendency for stresses to be concentrated or dissipated within the film. If there are imperfections that concentrate the stress, the film can tear more easily and abrasion resistance is poorer. On the other hand, if there are particles within the film that serve to dissipate stress, the film is less likely to tear and abrasion resistance is improved. Abrasion is one of several types of wear to be considered in a broad approach to fracture energetics and surface energetics of polymer wear.

Architectural Coatings

Roughly half the volume of all coatings sold in the United States are classified as architectural coatings. These coatings are designed to be applied to residences and offices, and for other light-duty building purposes. In contrast to most product coatings, they are designed to be applied in the field, in some cases by contractors but in large measure by do-it-yourself consumers. A wide range of products is involved.

Flat Wall Paint. The largest volume of architectural coatings is flat wall paint. In the United States, virtually all flat wall paint is latex-based. Latex paints have the advantages of low odor, fast drying, easy clean-up when wet, durability of color and film properties, and lower VOC emissions as compared to oil- or alkyd-based paints. They are manufactured primarily as white base paints, which are tinted by the retailer to the color selected by the customer from large collections of color chips. Two or three base whites are made: one for pastel shades, one for deep shades, and sometimes one for medium color shades. Bases for pastels require a relatively high content of TiO_2 to provide hiding.

Exterior House Paints. Latex paints dominate the exterior house paint market in the United States because of superior performance in almost all cases. As compared to oil or alkyd paints, the exterior durability of latex paints, that is, resistance to chalking, checking, and cracking, is superior particularly when using latexes having high methyl methacrylate (2-methyl-2-propenoic acid methyl ester), $C_4H_6O_2$, content. Methyl methacrylate has excellent resistance to photodegradation and hydrolysis. Another advantage of latex paints used on wood surfaces is the fact that, because of high moisture vapor permeability, they are much less likely to blister than oil or alkyd paints.

An application where latex paints show outstanding performance is over masonry such as stucco or cinder block construction. This performance results from saponification resistance in the presence of the alkali from the cement.

There are limitations to the applicability of exterior latex house paints providing a small continuing market for oil or alkyd exterior house paints. Because film formation from latex paints occurs by coalescence, there is a temperature limit, below which the paint should not be applied. If painting must be done when the temperature is below 5–7°C, oil or alkyd paint is preferable.

Another limitation is that latex paints do not give good adhesion when applied over a chalky surface such as weathered oil or alkyd paint.

Gloss Enamels. In contrast to exterior and flat wall paint, about half of the gloss paint or enamels sold are based on alkyd resins. Professional painters particularly favor the continued use of alkyd gloss paints. The need for reduction of VOC emission levels, especially in California, has led to efforts to increase the solids content of alkyd paints or overcome the disadvantages of latex gloss paints.

The principal limitation of gloss latex paints is not gloss; rather, it is the greater difficulty of getting adequate hiding from one coat. Probably the largest factor affecting hiding by gloss latex paints is poor leveling.

The rheological properties of gloss latex paints greatly influence leveling, and therefore, hiding (see RHEOLOGICAL MEASUREMENTS). Latex paints have exhibited a much higher degree of shear thinning than alkyd gloss paints, leading to paints having viscosity that is too low at high shear rate, and a subsequent applied film thickness that is too thin.

Latex paints formulated with associative thickeners have increased high shear viscosity, allowing the application of thicker wet films. Furthermore, the thickeners afford reduced low shear rate viscosity and a slower rate of recovery of the low shear rate viscosity, which improves leveling at the same time.

Another shortcoming of latex, which is particularly evident in gloss paints, is the time required to develop full film properties. Latex paints dry to touch and even to handling much more rapidly than do alkyd paints, but the latex requires a much longer time to reach the full dry properties.

Another important potential problem for use of gloss latex paints is adhesion to an old gloss paint surface when water is applied to the new dry paint film.

Product Coatings

About 32% of the total volume of coatings are applied in factories to a very large variety of products ranging from automobiles to toys. These are called product or, commonly, industrial coatings, or industrial finishes. These coatings are all proprietary. Additionally, because government regulations of air quality, effluent, waste disposal, and toxic hazards are continually changing, any specific information that is available as of this writing may already be outdated.

A large fraction of the product coatings are applied to metal. The essential first step in metal coating is the preparation of the metal. Oil and related contaminants must be removed by detergent or solvent washing. For best adhesion and corrosion resistance, the surface of the metal should be treated. In the case of steel, phosphate conversion coating treatments are used: for aluminum, chromate conversion treatments are available. The surface should be carefully rinsed before applying paint. Surface contaminants can result in crawling or blistering of the coating.

Primers for Metal. If reasonably high performance is required in the end product and unless cost is of paramount importance, a minimum of two coats, usually a primer and a top coat, should be applied to metal. For highest performance, primer vehicles should provide good wet adhesion, be saponification-resistant, and have low viscosity to permit penetration of the vehicle into microsurface irregularities in the substrate. Color, color retention, exterior durability, and other such properties are generally not important in primers. Resin systems such as those including bisphenol A epoxy resins that provide superior wet adhesion can thus be used in spite of their poor exterior durability.

Primers for automobiles and other products and a significant part of primers for household appliances are applied by electrodeposition. Almost all electrodeposition primers are now cationic. The compositions are proprietary, but the vehicles in some primers are epoxy/amine resins neutralized with volatile organic acids such as lactic acid;

an alcohol-blocked isocyanate is used as a cross-linking agent. The pigments are dispersed in the resin system and the coating is reduced with water.

Top Coats. The selection of a top coat depends on cost, method of application, and product use and performance requirements, among other factors. As a result of increasingly stringent air quality standards and increased solvent costs, approaches to reduction of solvent emissions are being sought.

There is no agreement on a definition of high solids coatings. For any particular use, this term means higher solids than previously used in that application. In automotive metallic coatings, high solids usually is taken to be about 45 vol %. In the case of clear or high gloss polyester coatings, volume solids of over 70% can be achieved with reasonable film properties.

Whereas the main driving force behind the development of higher and higher solids coatings has been the reduction of VOC emissions, solvent cost is also a factor. A further advantage is that the same dry film thickness can be applied in less time. High solids coatings are made using lower molecular weight resins having fewer average functional groups per molecule compared to conventional coatings. As a result, more complete reaction of the functional groups is necessary to achieve good film properties.

Sagging of spray-applied high solids coatings is more difficult to control than for conventional coatings. Little solvent is lost between the spray gun and the surface being coated during the application of high solids coatings. Therefore, there is little increase in viscosity before the coating droplets arrive on the object being sprayed, thus increasing the probability of sagging. In many cases, the only available method of controlling sagging is the incorporation of an additive that imparts thixotropic flow properties. In many applications, fine particle size silicon dioxide provides the desired flow properties.

Microgel additives are useful to control sagging and also minimize the problem in using high solids automotive clear top coats over wet base coats.

In industrial coatings two classes of water-borne systems are used: water-reducible and latex systems. As of this writing (ca 1993), water-reducible systems are used on a much larger scale, but the consumption of latex product coatings is increasing.

Because the molecular weight and average functionality of the resins used in water-reducible coatings are comparable to those of conventional solution thermosetting coatings, film properties obtained after curing are fully equivalent.

Until the 1980s, latex-based coatings were infrequently used in product coating applications. High gloss coatings cannot be made using latexes, and transparent coatings are difficult if not impossible to make. Furthermore, the rheological properties of the coatings limit utility. The availability of associative thickeners that minimize flocculation of latex particles permits formulation of industrial coatings that are less thixotropic and hence level better after application. Low VOC emissions and excellent film properties resulting from the high molecular weight of the polymer are expected to lead to increases in the use of latexes in product coatings.

An important segment of the product coatings market is sold for application to coiled metal, both steel and aluminum. In this process, the metal is first coated and then fabricated into the final product rather than fabricating first into product. The method offers a considerable economic advantage because coatings are applied by direct or reverse roll-coating at high (up to 400 m/min) speeds to wide (up to 3 m) coils in a continuous process. The metal is cleaned, conversion-coated, and coated with primer (if desired) and top coat all in an in-line operation. In many cases, coatings are applied to both sides of the metal during the same run through the coil coating line. The labor cost of coating application is much less than for application to a previously fabricated product. There is essentially 100% effective utilization of coating, and the loss by overspray involved in coating fabricated products is avoided.

A wide range of resin compositions are used in coil coatings, depending on the product performance requirements and cost limitations. The lowest cost coatings are generally alkyd coatings. They are appropriate for indoor applications where color requirements are not stringent. Alkyd–melamine coatings are sometimes used outdoors when long-term durability is not needed. Greater durability and better color retention on overbaking are obtained using polyester–melamine systems.

For greater exterior durability, silicone-modified polyesters or silicone-modified acrylic resins are used.

In the can industry, large-volume three-piece cans, used for packing many fruits and vegetables, are made from coil-coated stock. In this case, generally, oil-modified phenolic resins are used for coating tinplated steel coils on the side that becomes the interior of the can. Coil stock for making ends for two-piece beverage cans is coated with epoxy–melamine, epoxy–phenolic, or a cationic uv-cure epoxy coating.

Radiation has several significant advantages over heat as the source of energy to carry out cross-linking reactions. Coatings can be designed that cure rapidly, in a second or less, at room temperature yet have relatively long storage lives. The energy requirements for curing are much lower than for thermal baking systems. Rather than volatile solvents, reactive monomers can be used to provide for the low viscosities needed for application. Thus VOC emissions can be very low. In some cases, the energy source is high energy electron, but more commonly ultraviolet radiation curing systems are used.

There are many advantages to uv curing, but there are also limitations. Only flat surfaces that can be passed under the focused uv source, or cylindrical surfaces that can be rotated under the source, can be practically cured by this method. Radical initiated uv-cure systems are poor candidates for coatings requiring good exterior durability, because the residual photoinitiator is also a photoinitiator for photodegradation reactions. Solvent-free pigmented uv-cure coatings are limited not only by the depth of penetration of the uv radiation but also by the effect of the pigmentation on flow properties.

Cationic systems have fewer limitations. They are not air-inhibited, the photoinitiators do not initiate photodegradation reactions, there is less shrinkage during curing, and the stability of the acids generated permits migration if initiator through a pigmented film, allowing cure of somewhat thicker pigmented films. However, commercial adoption has been slow, possibly because of higher costs and slower cure rates at ambient temperatures.

Uses that have developed for uv curing reflect the special advantages of the system rather than replacement to reduce VOC emissions or energy consumption. Clear coatings on heat-sensitive flat plastic substrates, where rapid cure is needed, is an area where uv curing has found application. Other examples of application to heat-sensitive substrates are coatings for thin wood veneers used for door skins and coatings for paper.

Another coating technique that substantially reduces VOC emissions is powder coating. A wide variety of coating types and application methods are used, and powder coatings have been the most rapidly growing part of the coatings industry. The two most widely used methods of application involve passing heated objects to be coated into a fluidized bed of powder particles suspended in air and electrostatically spraying grounded articles at ambient temperature with powder, then baking. Triboelectric charging of the powder particles has improved spray application of powders (see COATING PROCESSES–POWDER TECHNOLOGY).

Examples of applications for powder coatings include pipe lining, coatings for rebars, underbody automotive parts, wheels, garden tractors, home appliances, playground equipment, metal furniture, fire extinguishers, and many others. As VOC restrictions become more stringent, powder coatings can be expected to increase in usage.

Wood Product Coatings. Furniture is one important class of wood (qv) products that is industrially coated. In most cases, the appearance standards are set by the fine furniture industry where the flat areas are composed of plywood having high quality top veneer; the legs, rails, and so forth are solid wood; and there are frequently carved wood decorative additions (see also WOOD-BASED COMPOSITES AND LAMINATES). The finishing process for fine furniture is long and requires significant artistic skill. If the final overall color of the furniture is lighter than the color of any part of the wood, the first step is bleaching, using a solution of hydrogen peroxide in methanol, CH_4O. The bleached wood, or unbleached for darker color finishes, is given a wash coat of size to stiffen the fibrils so they can be cleanly removed by sanding. The wood is then coated with stain, that is, a solution of substantive acid dyes in methanol, to give a desired overall color tone to the piece of furniture. A wash coat, generally a low solids vinyl chloride copolymer lacquer, is applied over the stain, partially to seal the stain in place but also to prepare the surface for the next operation, filling. The filler, a dilute dispersion of pigments, usually in linseed oil with mineral spirits solvent, is sprayed over the whole piece of furniture. It is then wiped off, leaving filler only in the pores of the wood. The colors of fillers are commonly dark brown; the filler serves to emphasize the grain pattern in the hardwood veneer. Next, shading stains are selectively sprayed on the wood to give different colors to various sections.

It is also common to distress the surface for resemblance to antique furniture. When this "art work" is completed, a sanding sealer is applied to immobilize the lower layers of the finish and to provide a surface that can be sanded smooth. Finally, a top coat is applied and polished smooth.

The primary binder used in the sanding sealer and the top coat is nitrocellulose. Top coats generally include fine particle-size silicon dioxide to reduce the gloss. In making a low gloss top coat, it is essential to retain the transparency.

Considerable effort has been invested in developing water-borne coatings for wood furniture. The two most promising approaches seem to be water-reducible acrylic resins with urea–formaldehyde cross-linkers and hydroxy-functional vinyl acetate copolymer latexes, again with urea–formaldehyde cross-linking.

Uv-cure coatings are widely used in European furniture manufacture but have found more limited applications in the United States.

The other principal component of industrial wood finishing is the panel industry. The highest cost segment of this industry, fine hardwood veneer paneling for executive office walls, is comparable to the fine furniture industry, and similar nitrocellulose lacquer finishing systems are used. However, the bulk of the industry requires less expensive finishing operations. Some plywood is stained followed by roll coating with a low gloss nitrocellulose lacquer top coat. It is common to print grain patterns from woods such as walnut onto inexpensive, relatively featureless veneers such as luan before applying a lacquer top coat. Lacquer top-coats are being replaced by alkyd–urea coatings to reduce VOC.

Special Purpose Coatings

Special purpose coatings include those coatings that do not fit under the definition of architectural or product coatings. About 16% of the volume (6.1×10^5 m^3) of United States coatings falls into this category. They include heavy-duty maintenance and marine coatings and automotive refinish paints.

Economic Aspects

The value of United States coatings consumption in 1991 reached a record $13.6 billion, continuing the slow but steady growth in dollar value that has occurred over many years.

A primary factor affecting both price and volume has been changes resulting largely from the effect of regulations reducing permissible VOC emissions.

Coating volume is also affected by other technologies which can substantially reduce the need for coatings. For example, in many cases coatings are not required on molded plastics products which have replaced coated metal products. High pressure laminates are increasingly used as furniture tops.

The effect of environmental regulation and the increasing recognition of immediate or potential toxicity hazards of coating components has led to a technical revolution in the coatings field in the 1970s and 1980s. Further drastic changes are expected in the 1990s.

A substantial fraction of the industry is concentrated in chemical companies and a few large independent companies. However, there are a large number of small and medium-sized companies that generally serve specialized segments of the business or restricted geographical areas. It is estimated that 45% of the companies have fewer than 20 employees.

ZENO W. WICKS, JR.
Consultant

Monograph series published by the Federation of Societies for Coatings Technology, Blue Bell, Pa. Titles issued thus far: *Introduction to Coatings Technology, Introduction to Polymers and Resins, Film Formation, Organic Pigments, Inorganic Primer Pigments, Solvents, Mechanical Properties of Coatings, Corrosion Protection by Coatings, Application of Paints and Coatings, Coating Film Defects, Rheology, Aerospace and Aircraft Coatings, Marine Coatings, Coil Coatings, Automotive Coatings, Radiation Cure Coatings, Cationic Radiation Curing, Sealants and Caulks,* and *Powder Coatings.*

Z. W. Wicks, Jr., F. N. Jones, and S. P. Pappas, *Organic Coatings: Science and Technology,* John Wiley & Sons, Inc., New York, Vol. I, 1992; Vol. II, 1994.

COATINGS, MARINE

The selection and application of marine coatings has become a highly specialized discipline in which governmental regulations are a dominant influence. Many paints and painting procedures that were used in the past are no longer permitted or are extremely costly to use.

Corrosion Control Plan. A corrosion control plan for each ship or structure, which is designed to control deterioration in the most economical and practical manner and to include all appropriate mechanisms for corrosion control, must be developed before construction begins. For steel in the marine environment, the chief methods available are protective coatings and cathodic protection.

Protective Coatings. Each coating in the protective coating system is designed to perform a specific function and to be compatible with the total system. The effectiveness of a coating is directly related to its ability to maintain adhesion to the substrate, its integrity, and its thickness. Areas that cannot be easily or safely repaired, especially those which require drydocking for repair, need to be given the best available coating.

Fouling organisms attach themselves to the underwater portions of ships and have a severe impact on operating costs. Because fouling is controlled best by used of antifouling paints, it is important that these paints be compatible with the system used for corrosion control and become a part of the total corrosion control strategy.

Environmental Concerns

Local environmental regulations have significantly affected the production, transportation, use, and disposal of coatings.

Volatile Organic Compounds. As coatings dry, solvents are released into the atmosphere, where they undergo chemical reactions in sunlight and produce photochemical smog and other air pollutants (see AIR POLLUTION). As a general rule, the volatile organic compound (VOC) content of marine coatings is restricted to 340 g/L. In the locations where ozone (qv) levels do not conform to the levels established by the Environmental Protection Agency, regulations require an inventory of all coatings and thinners from the time they are purchased until they are used.

Heavy-Metal Pigments. Lead and chromate pigments, used for many years as corrosion inhibitors in metal primers and topcoats for marine coating systems, have been linked to adverse health and environmental effects. Because of these concerns, lead- and chromate-containing pigments are not used in marine coatings.

Organotins. In the mid-1970s compounds based on derivatives of triphenyl- or tributyltin (see TIN COMPOUNDS) known generically as organotins, were found to be much more effective than cuprous oxide paints in controlling fouling, and numerous products were introduced. These fell into two classes. Free-association coatings contains a tributyltin salt, eg, acetate, chloride, fluoride, or oxide, physically mixed into the coating. Copolymer coatings contain organotin which is covalently bound to the resin of the coating and is not released until a tin–oxygen bond hydrolyzes in seawater. This controlled hydrolysis produces a low and steady leach rate of organotin and creates hydrophilic sites on the binder resin. This layer of resin subsequently washes away and exposes a new layer of bound organotin.

However, there is now considerable evidence that sufficiently high concentrations of organotins kill many species of marine life and affect the growth and reproduction of others. Thus many nations restrict organotin coatings to vessels greater than 25 meters in length. In the United States, laws prohibit the retail sale of copolymer paints containing greater than 7.5% (dry weight) of tin, and of free-association paints containing greater than 2.5% (dry weight) of tin, but do not restrict the size of the ship to which the paints may be applied.

Abrasive Blast Cleaning. Removal of paint by abrasive blasting may lead to adverse health effects for workers who breathe dust formed during the operation. Regulations restrict blasting operations to such procedures as blasting within enclosures, using approved mineral abrasives, using a spray of water to reduce dust, and blasting with alternative materials such as ice, plastic beads, or solid carbon dioxide.

Debris from the removal of paint may contain lead, chromium, or other heavy metals. Collection of such debris is required to prevent release of these metals into the environment and to avoid exposure and contamination of workers.

Reactive Coatings. Coatings that cure by chemical reaction of two component parts are the most widely used in marine applications, and protection of workers from the reactive ingredients is required.

Surface Preparation for Marine Painting

Surface preparation, always important in obtaining optimal coatings performance, is critical for marine coatings (see METAL SURFACE TREATMENTS). Surface preparation usually comprises about half of the total coating costs, and if inadequate may be responsible for early coating failure. Proper surface preparation includes cleaning to remove contaminants (eg, abrasive blasting) and roughening the surface to facilitate adhesion.

Types of Coatings

Coatings ingredients fall into four principal classes. Resins form a continuous solid film after curing, bind all ingredients within the film, and provide adhesion to the substrate. The properties of a coating are determined principally by the resins it contains. Pigments are metals or nearly insoluble salts that impart opacity, color, and chemical activity to the coating. Solvents are used primarily to facilitate manufacture and application, but are lost by evaporation after application and are not a permanent part of the coating. Additives used in small (1–50 ppt) amounts give the coating such desirable additional properties as ease of manufacture, stability in shipment and storage, ease of application, or increased performance of the dried film.

These ingredients may be formulated to give coatings that protect against corrosion in different ways. Barrier coatings physically separate oxygen, water, ions, and other corrosive agents from the steel surface. Inhibitive coatings prevent corrosion by absorbing or neutralizing corrosive agents, or by slowly releasing protective ions. Sacrificial coatings contain a metal (usually zinc) that is oxidized more rapidly than steel, thereby providing protection for the substrate by electrochemical action. Conversion coatings chemically oxidize the surface of the substrate to a depth of 7–10 μm, producing a passive layer which resists corrosion better than the metal itself.

Modern marine coatings fall into eight generic categories: epoxies, urethanes, alkyds, inorganic silicate coatings, vinyls, chlorinated rubber, coal tar, and powder coatings.

Application Methods

The application of marine coatings is a critical factor in achieving maximum performance. Protective clothing and breathing equipment should be worn during application. Because of large surface areas, ships are usually spray-painted (see COATING PROCESSES–SPRAY COATINGS). Three techniques are widely used: air, airless, and electrostatic spraying. Manual painting occurs mostly during touch-up or repair, and is best suited for piping, railings, and other hard-to-spray places.

Transfer Efficiency. Many components of ships and marine structures are now coated in the shop under controlled conditions to reduce the amount of solvents released into the atmosphere, improve the quality of work, and reduce cost. Regulations designed to limit the release of volatile organic compounds into the air confine methods of shop application to those having transfer efficiencies of 65%. Transfer efficiency is defined as the percentage of the mass or volume of solid coating that is actually deposited on the item being coated.

Selection of Coatings

Underwater Hull. Hull coatings consist of two layers of an anticorrosive coating topped with one layer of an antifouling coating. The coating system must resist marine fouling, severe corrosion, the cavitation action of high speed propellers, and the high current densities near the anodes of the cathodic protection system. Epoxy and coal-tar epoxy systems are commonly used as anticorrosive coatings.

Anticorrosive systems require an antifouling topcoat. Arsenic, cadmium, and mercury are proscribed and organotins are severely restricted. Cuprous oxide has always been and remains the most widely used toxicant.

The system of hull coatings, including antifouling paint, must be compatible with the cathodic protection system. Thus the coating system must have good dielectric properties to minimize cathodic protection current requirements and must be resistant to the alkalinity produced by the electric current. The cathodic protection system should prevent corrosion undercutting of coatings that become damaged, and the current density should be able to be increased easily to meet the increased electrical current needed as the coating deteriorates.

Boottop and Freeboard Areas. The boottop suffers mechanical damage from tugs, piers, and ice, and experiences intermittent wet and dry periods with nearly constant exposure to sunlight. Thus coatings for this area require resistance to sunlight and mechanical damage, good adhesion, and flexibility. Frequently the hull coating system is used in the boottop area, and one or two extra topcoats are applied for added strength. In the freeboard areas, commercial ships use organic zinc-rich primers extensively and usually topcoat them with a two- or three-coat epoxy system.

Weather Decks. Coatings for decks must be resistant to abrasion by pedestrians and small vehicles, and must be slip-resistant. Inorganic zinc primers overcoated with epoxy coatings for additional corrosion protection perform well on steel decks, or a multiple-coat all-epoxy system can be used. Nonskid coatings are used on aircraft carrier landing and hangar decks and in passageways of all ships to maintain traction during wet and slippery conditions. The coatings contain epoxy resins and a coarse grit and are applied using a roller over epoxy primers to produce a textured finish. Nonskid coatings are 6–10 mm thick when dry.

Superstructure. Deck hardware and machinery, masts, and booms are coated with an inorganic zinc primer, an intermediate coat of epoxy, and a finish coat of aliphatic polyurethane or silicone-alkyd enamel. Powder coatings are used effectively on antennas and other equipment on the superstructure. This equipment, as well as exhaust stacks, steam riser valves and piping, and other hot surfaces, can also be coated using thermal-sprayed aluminum.

Tanks. Coatings for liquid cargo tanks are selected according to the materials that the tanks are to contain. Tank coatings protect the cargo from contamination and must be compatible with the material carried. Epoxy systems are most frequently selected because they perform well with both aqueous and organic products.

Machinery Spaces, Bilges, and Holds. Machinery spaces and bilges are so inaccessible that surface preparation is a significant problem and damage to machinery that cannot be removed must be avoided. Chemical cleaning by aqueous citric acid solutions, followed by degreasing using a nonflammable solvent, is widely used. Surfaces are best protected using a two- or three-coat epoxy–polyamide system having a total thickness of 250–300 μm. Alkyd enamels perform well in dry machinery spaces. Holds for carrying cargo may be painted with either of these systems, but the epoxy system is preferred for chemical and abrasion resistance.

Living Areas. Living areas are generally painted with nonflaming coatings, or with intumescent coatings which foam when heated and produce a thick char that lessens damage to the substrate. Epoxy systems are generally used in damp areas such as galleys, washrooms, and showers where moisture causes enamels to deteriorate.

ROBERT F. BRADY, JR.
U. S. Naval Research Laboratory
RICHARD W. DRISKO
U. S. Naval Civil Engineering Laboratory

H. R. Bleile and S. D. Rogers, *Marine Coatings,* Federation of Societies for Coatings Technology, Blue Bell, Pa., 1989.

R. F. Brady, Jr., "Marine Applications," in J. I. Kroschwitz, ed., *Encyclopedia of Polymer Science and Engineering,* Vol. 9, John Wiley & Sons, Inc., New York, 1988, pp. 295–300.

Journal of Coatings Technology, published monthly by the Federation of Societies for Coatings Technology, Blue Bell, Pa.

Journal of Protective Coatings and Linings, published monthly by the Steel Structures Painting Council, Pittsburgh, Pa.

COBALT AND COBALT ALLOYS

COBALT

Cobalt, a transition series metallic element having atomic number 27, is similar to silver in appearance.

Cobalt and cobalt compounds (qv) have expanded from use as colorants in glasses and ground coat frits for pottery (see COLORANTS FOR CERAMICS) to drying agents in paints and lacquers (see DESICCANTS), animal and human nutrients (see MINERAL NUTRIENTS), electroplating (qv) materials, high temperature alloys (qv), hardfacing alloys, high speed tools (see TOOL MATERIALS), magnetic alloys (see MAGNETIC MATERIALS, BULK; MAGNETIC MATERIALS, THIN-FILM), alloys used for prosthetics (see PROSTHETIC AND BIOMEDICAL DEVICES), and uses in radiology (see RADIOACTIVE TRACERS). Cobalt is also used as a catalyst for hydrocarbon refining from crude oil for the synthesis of heating fuels (see FUELS, SYNTHETIC; PETROLEUM).

Occurrence

Cobalt is the thirtieth most abundant element on earth and comprises approximately 0.0025% of the earth's crust. It occurs in mineral form as arsenides, sulfides, and oxides; trace amounts are also found in other minerals of nickel and iron as substitute ions. Cobalt minerals are commonly associated with ores of nickel, iron, silver, bismuth, copper, manganese, antimony, and zinc. The world's largest cobalt reserves are in Zaire, Zambia, Morocco, Canada, and Australia. Together the ores of these countries contain well over one-half of the world cobalt supply. The reserves of Canada and Australia comprise approximately one-fourth of the world supply. Smaller but commercially practical ore bodies also existin Russia, Finland, Uganda, and the Philippines. Leteritic ores are becoming increasingly important as a source of nickel, and cobalt is a by-product.

Table 1. Properties of Cobalt

Property	Value		
at wt	58.93		
mp, °C	1493		
bp, °C	3100		
coefficient of thermal expansion, °C^{-1}			
cph at RT	12.5		
fcc at 417°C	14.2		
thermal conductivity at RT, W/(m·K)	69.16		
hardness[a], diamond pyramid, of % Co	99.9	99.98[b]	
at 20°C	225	253	
at 900°C	22	17	
strength of 99.9% cobalt, MPa[c]	as cast	annealed	sintered
tensile	237	588	679
compressive	841	808	

[a] Vickers.
[b] Zone refined.
[c] To convert MPa to psi, multiply by 145.

Properties

The electronic structure of cobalt is [Ar] $3d^7 4s^2$. At room temperature the crystalline structure is hexagonal close-packed with lattice parameters of a = 0.25.1 nm and c = 0.4066 nm. Cobalt is magnetic up to 1123°C and at room temperature the magnetic moment is parallel to the c-direction. Physical properties are listed in Table 1.

Metallic cobalt dissolves readily in dilute H_2SO_4, HCl, or HNO_3 to form cobaltous salts. Cobalt cannot be classified as an oxidation-resistant metal. Scaling and oxidation rates of unalloyed cobalt in air are 25 times those of nickel.

Economic Aspects

During the period of 1978–1981 the price and availability of cobalt were erratic. In May 1978, the production of cobalt in Zaire came to a temporary standstill and the world price of the metal immediately increased at least fourfold. Many users sought alternative materials. The supply and price have stabilized somewhat since 1978.

COBALT ALLOYS

Mechanical properties depend on the alloying elements. Addition of carbon to the cobalt base metal is the most effective choice. The carbon forms various carbide phases with the cobalt and the other alloying elements (see CARBIDES). The presence of carbide particles is controlled in part by alloying elements such as chromium, nickel, titanium, manganese, tungsten, and molybdenum that are added during melting. The distribution of the carbide particles is controlled by heat treatment of the solidified alloy.

Cobalt alloys are strengthened by solid-solution hardening and by the solid-state precipitation of various carbides and other intermetallic compounds. Minor phase compounds, when precipitated at grain boundaries, tend to prevent slippage at those boundaries, thereby increasing creep strength at high temperatures. Aging and service under stress at elevated temperatures induce some of the carbides to precipitate at slip planes and at stacking faults, thereby providing barriers to slip. If carbides are allowed to precipitate to the point of becoming continuous along the grain boundaries, they often initiate fracture (see FRACTURE MECHANICS).

Cobalt-Base Alloys

As a group, the cobalt-base alloys may generally be described as wear-resistant, corrosion-resistant, and heat-resistant, ie, strong even at high temperatures. Many of the alloy properties arise from the crystallographic nature of cobalt, in particular its response to stress;

the solid-solution-strengthening effects of chromium, tungsten, and molybdenum; the formation of metal carbides; and the corrosion resistance imparted by chromium. Generally, the softer and tougher compositions are used for high temperature applications such as gas-turbine vanes and buckets. The harder grades are used for resistance to wear.

Cobalt-Base Wear-Resistant Alloys

The main differences in the Stellite alloy grades of the 1990s vs those of the 1930s are carbon and tungsten contents, and hence the amount and type of carbide formation in the microstructure during solidification. Carbon content influences hardness, ductility, and resistance to abrasive wear. Tungsten also plays an important role in these properties.

Types of Wear. There are several distinct types of wear that can be divided into three main categories: abrasive wear, sliding wear, and erosive wear. The type of wear encountered in a particular application is an important factor influencing the selection of a wear-resistant material.

Alloy Compositions and Product Forms. Stellite alloys 1, 6, and 12, derivatives of the original cobalt–chromium–tungsten alloys, are characterized by their carbon and tungsten contents. Stellite alloy 1 is the hardest, most abrasion-resistant, and least ductile.

Stellite alloy 21 differs from the first three alloys in that molybdenum rather than tungsten is used to strengthen the solid solution. Stellite alloy 21 also contains considerably less carbon. Each of the first four alloys is generally used in the form of castings and weld overlays. Haynes alloy 6B differs in that it is a wrought product available in plate, sheet, and bar form.

The Tribaloy alloy T-800 is from an alloy family developed by DuPont in the early 1970s, in the search for resistance to abrasion and corrosion. Excessive amounts of molybdenum and silicon were alloyed to induce the formation during solidification of hard and corrosion-resistant intermetallic compounds, known as Laves phase.

The physical and mechanical properties of the six commonly used cobalt wear alloys are presented in Table 2.

Cobalt-Base High Temperature Alloys

For many years, the predominant user of high temperature alloys was the gas-turbine industry. The use of high temperature alloys in the 1990s is more diversified, as more efficiency is sought from the burning of fossil fuels and waste, and as new chemical processing techniques are developed.

Cobalt-base alloys are not as widely used as nickel and nickel–iron alloys in high temperature applications. Nevertheless,

Table 2. Mechanical and Physical Properties of Cobalt-Base Wear-Resistant Alloys

Property	Alloy[a]					
	1	6	12	21	6B	T-800
hardness, Rockwell	55	40	48	32	37[b]	58
yield strength, MPa[b]	541		649	494	619[b]	
ultimate tensile strength, MPa[c]	618	896	834	694	998[b]	
elongation, %	<1	1	<1	9	11	
specific gravity	8.69	8.46	8.56	8.34	8.39	8.64
melting range, °C						
solidus	1255	1285	1280	1186	1265	1288
liquidus	1290	1395	1315	1383	1354	1352

[a] Refers to stellites 1, 6, 12, and 21, and Haynes Alloy 6B and Tribaloy T-800.
[b] 3.2 mm (1/8 in.) thick sheet.
[c] To convert MPa to psi, multiply by 145.

cobalt-base high temperature alloys play an important role because of excellent resistance to sulfidation and strength at temperatures exceeding those at which the γ'- and γ''-precipitates in the nickel and nickel–iron alloys dissolve. Cobalt is also used as an alloying element in many nickel-base high temperature alloys.

Alloy Compositions and Product Forms. Stellite 21, an early type of cobalt-base high temperature alloy, is used primarily for wear resistance. The use of tungsten rather than molybdenum, moderate nickel contents, lower carbon contents, and rare-earth additions typify cobalt-base high temperature alloys of the 1990s.

Cobalt-Base Corrosion-Resistant Alloys

To satisfy the industrial need for alloys which exhibit resistance to aqueous corrosion yet share the attributes of cobalt as an alloy base, ie, resistance to various forms of wear and high strength over a wide range of temperatures, several low carbon, wrought cobalt–nickel–chromium–molybdenum alloys have been produced. In addition, the cobalt–chromium–molybdenum alloy Vitallium is widely used for prosthetic devices and implants, owing to excellent compatibility with body fluids and tissues.

Economic Aspects

With cobalt historically having a cost approximately twice that of nickel, cobalt-base alloys for both high temperature and corrosion service tend to be much more expensive than competitive alloys. In some cases of severe service their performance increase is, however, commensurate with the cost increase and they are a cost-effective choice.

F. Galen Hodge
Haynes International, Inc.

C. T. Sims, N. S. Stoloff, and W. G. Hagel, eds., *Superalloys II,* John Wiley & Sons, Inc., New York, 1987, p. 135.

N. J. Grant and J. R. Lane, *Trans. ASM* **41**, 95 (1949).

R. D. Gray, *A History of the Haynes Stellite Company,* Cabot Corp., Kokomo, Ind., 1974.

ASM Metals Handbook, Vol. 1, 10th ed., ASM Internationad, Materials Park, Ohio, 1990.

COBALT COMPOUNDS

Cobalt forms numerous compounds and complexes of industrial importance.

Cobalt exists in the +2 and +3 valence states for the majority of its compounds and complexes. A multitude of complexes of the cobalt(III) ion exist, but few stable simple salts are known. Cobalt(II) forms numerous simple compounds and complexes, most of which are octahedral or tetrahedral in nature; cobalt(II) forms more tetrahedral complexes than other transition-metal ions. Because of the small stability difference between octahedral and tetrahedral complexes of cobalt(II), both can be found in equilibrium for a number of complexes. Typically, octahedral cobalt(II) salts and complexes are pink to brownish red; most of the tetrahedral Co(II) species are blue.

Preparation and Properties

Cobalt(II) Salts. Cobalt(II) acetate tetrahydrate, $Co(C_2H_3O_2)_2 \cdot 4H_2O$, occurs as pink, deliquescent, monoclinic crystals. It can be prepared by reaction of cobalt carbonate or hydroxide and solutions of acetic acid, by reflux of acetic acid solutions in the presence of cobalt(II) oxide, or by oxygenation of hot acetic acid solutions over cobalt metal. It is used as a bleaching agent and drier in inks and varnishes, and in pigments, catalysis, agriculture, and the anodizing industries.

The mauve-colored cobalt(II) carbonate of commerce is a basic material of indeterminate stoichiometry, $(CoCO_3)_x \cdot (Co(OH)_2)_y \cdot zH_2O$, that contains 45–47% cobalt. It is prepared by adding a hot solution of cobalt salts to a hot sodium carbonate or sodium bicarbonate solution.

Cobalt(II) acetylacetonate, cobalt(II) ethylhexanoate, cobalt(II) oleate, cobalt(II) linoleate, cobalt(II) formate, and cobalt(II) resinate can be produced by metathesis reaction of cobalt salt solutions and the sodium salt of the organic acid, by oxidation of cobalt metal in the presence of the acid, and by neutralization of the acid using cobalt carbonate or cobalt hydroxide.

Cobalt(II) chloride hexahydrate, $CoCl_2 \cdot 6H_2O$, is a deep red monoclinic crystalline material that deliquesces. It is prepared by reaction of hydrochloric acid with the metal, simple oxide, mixed valence oxides, carbonate, or hydroxide.

Cobalt(II) hydroxide, $Co(OH)_2$, is a pink, rhombic crystalline material containing about 61% cobalt. The material is prepared by mixing a cobalt salt solution and a sodium hydroxide solution. The hydroxide is a common starting material for the preparation of cobalt compounds. It is also used in paints and lithographic printing inks and as a catalyst.

Cobalt(II) nitrate hexahydrate, $Co(NO_3)_2 \cdot 6H_2O$, is a dark reddish to reddish brown, monoclinic crystalline material containing about 20% cobalt. Cobalt nitrate can be prepared by dissolution of the simple oxide or carbonate in nitric acid, but more often it is produced by direct oxidation of the metal with nitric acid. The nitrate is used in electronics as an additive in nickel–cadmium batteries, in ceramics, and in the production of vitamin B_{12}.

Cobalt(II) oxalate, CoC_2O_4, is a pink to white crystalline material that absorbs moisture to form the dihydrate. It precipitates as the tetrahydrate on reaction of cobalt salt solutions and oxalic acid or alkaline oxalates. It is used in the production of cobalt powders for metallurgy and catalysis, and is a stabilizer for hydrogen cyanide.

Cobalt(II) phosphate octahydrate, $Co_3(PO_4)_2 \cdot 8H_2O$, is a red to purple amorphous powder. The product is obtained by reaction of an alkaline phosphate and solutions of cobalt salts. The phosphate is used in glazes, enamels, pigments, and plastic resins, and in certain steel phosphating operations.

Cobalt(II) sulfamate, $Co(NH_2SO_3)_2$, is generally produced and sold as a solution containing about 10% cobalt. The product is formed by reaction of sulfamic acid and cobalt(II) carbonate or cobalt(II) hydroxide, or by the aeration of sulfamic acid slurries over cobalt metal. Cobalt(II) sulfamate is used in the electroplating (qv) industry and in the manufacture of precision molds for record and compact discs (see INFORMATION STORAGE MATERIALS).

Cobalt(II) sulfate heptahydrate, $CoSO_4 \cdot 7H_2O$, is a reddish pink monoclonic crystalline material that effloresces in dry air to form the hexahydrate. Cobalt(II) sulfate can be prepared by solution of cobalt(II) carbonate, cobalt(II) hydroxide, or cobalt(II) oxide in sulfuric acid. Cobalt sulfate heptahydrate and cobalt(II) sulfate monohydrate are the most economical sources of cobalt ion and are used in feed supplements as well as in the electroplating industry, in storage batteries, in porcelain pigments and glazes, and as a drier for inks.

Cobalt Oxides. Cobalt(II) oxide, CoO, is an olive-green, cubic crystalline material. The simple oxide is most often produced by oxidation of the metal at temperatures above 900°C. It is used in glass decorating and coloring and is a precursor for the production of cobalt chemicals.

Cobalt(II) dicobalt(III) tetroxide, Co_3O_4, is a black cubic crystalline material containing about 72% cobalt. It is prepared by oxidation of cobalt metal at temperatures below 900°C or by pyrolysis in air of cobalt slats, usually the nitrate or chloride. It is used in enamels, semiconductors, and grinding wheels. Both oxides adsorb molecular oxygen at room temperatures.

Cobalt Carbonyls. Dicobalt octacarbonyl, $Co_2(CO)_8$, is an orange-red solid that decomposes in air. It is prepared by heating cobalt metal to 300°C under 20–30,000 kPa (3–4000 psi) of carbon monoxide, by reduction of cobalt(II) carbonate with hydrogen under pressure at high temperatures, or by heating a mixture of cobalt(II) acetate and cyclohexane to 160°C in the presence of carbon monoxide and hydrogen at 30,000 kPa (4000 psi) pressure.

Economic Aspects

Prices of cobalt compounds are directly related to the cost of cobalt metal, which fluctuates widely.

Health and Safety

Cobalt is one of twenty-seven known elements essential to humans. It is an integral part of the cyanocobalamin molecule, ie, vitamin B_{12}, the only documented biochemically active cobalt component in humans.

Cobalt compounds can be classified as relatively nontoxic. There have been few health problems associated with workplace exposure to cobalt. The primary workplace problems from cobalt exposure are fibrosis, also known as hard metal disease, asthma, and dermatitis. The exposure level established by NIOSH for the workplace is 0.1 mg/m^3. ACGIH has recommended a TLV of 0.05 mg/m^3 for cobalt.

Uses

Cobalt in Catalysis. Over 40% of the cobalt in nonmetallic applications is used in catalysis. About 80% of those catalysts are employed in three areas: (*1*) hydrotreating/desulfurization in combination with molybdenum for the oil and gas industry (see SULFUR REMOVAL AND RECOVERY); (*2*) homogeneous catalysts used in the production of terphthalic acid or dimethylterphthalate (see PHTHALIC ACIDS AND OTHER BENZENEPOLYCARBOXYLIC ACIDS); and (*3*) the high pressure oxo process for the production of aldehydes and alcohols (see ALCOHOLS, HIGHER ALIPHATIC; ALCOHOLS, POLYHYDRIC).

Cobalt in Driers for Paints, Inks, and Varnishes. The cobalt soaps, eg, the oleate, naphthenate, resinate, linoleate, ethylhexanoate, synthetic tertiary neodecanoate, and tall oils, are used to accelerate the natural drying process of unsaturated oils such as linseed oil and soybean oil.

Cobalt as a Colorant in Ceramics, Glasses, and Paints. Cobalt(II) ion displays a variety of colors in solid form or solution ranging from pinks and reds to blues or greens. It has been used for hundreds of years to impart color to glasses and ceramics (qv) or as a pigment in paints and inks (see COLORANTS FOR CERAMICS).

H. WAYNE RICHARDSON
CP Chemicals, Inc.

J. D. Donaldson, S. J. Clark, and S. M. Grimes, *Cobalt in Medicine, Agriculture and the Environment,* monograph series, Cobalt Development Institute, London, 1986.

J. D. Donaldson, S. J. Clark, and S. M. Grimes, *Cobalt in Chemicals, The Monograph Series,* Cobalt Development Institute, London, 1986.

B. Delmon, *Proceedings of the International Conference on Cobalt Metallurgy and Uses,* ATB Metallurgy, Brussels, Belgium, 1981.

COFFEE

Commercial coffees are grown in tropical and subtropical climates at altitudes up to ca 1800 meters; the best grades are grown at high elevations. Most individual coffees from different producing areas possess characteristic flavors. Commercial roasters obtain preferred flavors by blending or mixing the varieties before or after roasting. Colombian and washed Central American coffees are generally characterized as mild, winey-acid, and aromatic; Brazilian coffees as heavy body, moderately acid, and aromatic; and African robusta coffees as heavy body, neutral, slightly acid, and slightly aromatic. Premium coffee blends contain higher percentages of Colombian and Central American coffees.

Green Coffee Processing

The coffee plant is a relatively small tree or shrub belonging to the family Rubiaceae. It is often controlled to a height of 3 to 5 meters. *Coffea arabica* (milds) accounts for 69% of world production; *Coffea canephora* (robustas), 30%; and *Coffea liberica* and others, 1%.

Green coffee processing is effected by either the dry or wet method. The dry method produces so-called natural coffees. The wet method usually produces the more uniform and higher quality washed coffees.

The dry method is used in most of Brazil and in other countries where water is scarce in the harvesting season. The cherries from strip picking are spread on open drying ground and turned frequently to permit thorough drying by the sun and wind. Sun-drying usually takes two to three weeks depending on weather conditions. Some producing areas use hot air, indirect steam, and other machine-drying devices. When the coffee cherries are thoroughly dry, they are transferred to hulling machines which remove the skin, pulp, parchment shell, and silver skin in a single operation.

In the wet method, as practiced in Colombia, freshly picked ripe coffee cherries are fed into a tank for initial washing. Stones and other foreign material are removed. The cherries are then transferred to depulping machines which remove the outer skin and most of the pulp. However, some pulp mucilage clings to the parchment shells that encase the coffee beans. Fermentation tanks, usually containing water, remove the last portions of the pulp. Fermentation may last from twelve hours to several days. Because prolonged fermentation may cause development of undesirable flavors and odors in the beans, some operators use enzymes to accelerate the process. The beans are subsequently dried either in the sun or in mechanical dryers.

Coffee prepared by either the wet or dry method is machine-graded by sieves, oscillating tables, and airveyors into large, medium, and small beans. Damaged beans and foreign matter are removed by hand-picking, machine separators, electronic sorters, or a combination of these techniques. Commercial coffee is graded according to the number of imperfections present, such as black beans, damaged beans, stones, pieces of hull, or other foreign matter. Processors also grade coffee by color, roasting characteristics, and cup quality of the beverage. After all processing, coffee will maintain acceptable quality for approximately one year.

Coffee Chemistry

Chemical Composition of Green Coffee. The chemical composition of green coffee can affect roasted flavor quality. The chemistry can vary according to species, variety, growing environment, post-harvest handling including wet or dry processing, and storage time, temperature, and humidity. The composition data in Table 1 are given as a range compiled from literature.

Effects of Roasting on Components. Green coffee has no desirable taste or aroma; these are developed upon roasting. Many complex physical and chemical changes occur during roasting, including the obvious change in color from green to brown. In the first stage of roasting, loss of free water (typically 11% of the bean) occurs. In the second stage, chemical dehydration, fragmentation, recombination, and polymerization reactions occur. Many of these are associated with the Maillard reaction and lead to the formation of lower molecular-weight compounds such as carbon dioxide, free water, and those associated with flavor and aroma; and higher molecular-weight colored materials, both soluble and insoluble in water.

Table 2 indicates some of the chemical changes that occur in green coffee as a result of roasting.

Chemistry of Brewed Coffee. The chemistry of brewed coffee is dependent on the extraction of water-soluble and hydrophobic aromatic components from the coffee cells and lipid phase, respectively. Factors that affect extraction and flavor quality of brewed coffee are degree of roast; blend composition; grinding technique; particle size and density; water quality; water-to-coffee ratio; and brewing technique or device, such as drip filter, percolator, or espresso, which defines the water temperature, steam pressure, brewing time, water recycle, etc. Extraction yields upon home brewing range from about 9 to 28% and typically about 23% dry basis roasted and ground (R&G). The trend in the United States with some notable exceptions has been toward weaker brew strengths, with typical recipes of about 5 g of R&G coffee per 6 oz cup (brew solids concentration about 0.7%), compared to about 10 g a generation ago (brew solids about 1.2%).

Table 1. Typical Analyses of Green Coffee, %

Constituent[b]	Type[a]	
	Robusta	Arabica
moisture	11(10–13)	12.5(10–13)
lipids	10(7–11)	15(14–17)
ash	4.2(3.9–4.5)	4.0(3.5–4.5)
caffeine	2.0(1.5–2.6)	1.3(1.1–1.4)
chlorogenic acid	9(7–10)	7(5–8)
carboxylic acids	2(1–3)	2.5(1.5–3.5)
trigonelline	0.7(0.3–0.9)	1.1(0.9–1.2)
protein	11(9–13)	11(9–13)
free amino acids	0.8	0.5
sucrose	4(3–6)	8(5–9)
reducing sugars	0.5(0.4–0.6)	0.1(0.1–0.2)
others[c]	8.8(5–10)	5.5(3–8)
polymeric carbohydrate		
mannan	22	22
arabinogalactan	17(16–18)	15(14–16)
cellulose	8(7–9)	7(7–8)

[a] Typical value and (range).
[b] Dry basis (db) (ex moisture; as is basis).
[c] Others by difference.

Table 2. Approximate Analyses of Roasted, Brewed, and Instant Coffee, %[a]

Constituent[b]	Roasted	Brewed	Instant
moisture	4 (1–5)		3.5 (2–5)
oil	17 (16–20)	0.8 (0.2–1.0)	(0.1–0.6)
ash	4.5(4–5)	14	9(7–11)
caffeine	1.2 (1.0–1.6)	4.8	3.5 (2–5)
chlorogenic acid	1.5 (1–3.5)	14.8	(3–9)
carboxylic acids	3 (2–4)	3.0	5.5 (4–8)
trigonelline	0.5–1.0	1.6	(0.5–2)
protein	8–10	6	(1–6)
reducing sugar	0.2	0.4	(1–5)
sucrose	0.2 (0–0.5)	0.8	(0.6–0)
aroma compounds	~0.1	~0.1–2	~0.05
browning products and others	(22–11)	29.4	(20–35)
polymeric carbohydrate	38	24	(30–50)
mannan	20		
arabinogalactan	12		
cellulose	6		

[a] Typical value and (range).
[b] Dry basis (ex moisture; as is basis).

Instant Coffee. The chemistry of instant or soluble coffee is dependent on the R&G blend and processing conditions. This is indicated in Table 2 by the wide range of constituents. In addition to the atmospherically extractable solids found in brewed coffee, commercial percolation generates water-soluble carbohydrate by hydrolysis, which contributes to the yield.

Roasted and Ground Coffee Processing and Packaging

The main processing steps in the manufacture of roast and ground coffee products are blending, roasting, grinding, and packaging.

Roasting Technology. Roasting is usually by hot combustion gases in rotating cylinders or fluidized-bed systems; infrared- and microwave-roasted coffees have recently (ca 1993) come to market in Japan.

Grinding. Grinding of the roasted coffee beans is tailored to the intended method of beverage preparation. Average particle-size distributions range from very fine (500-μm or less) to very coarse (1100-μm).

Instant Coffee Processing and Packaging

Instant coffee is the dried water-extract of ground, roasted coffee.

Green beans for instant coffee are blended, roasted, and ground similarly to those for roasted and ground products. A concentrated coffee extract is normally produced by pumping hot water through the coffee in a series of cylindrical percolator columns. The extracts are further concentrated prior to a spray- or freeze-drying step, and the final powder is packaged in glass or other suitable material. Some soluble coffees, both spray- and freeze-dried, are manufactured in producing countries for export. Spray-dried instant coffees are most often sold in an agglomerated form.

Decaffeinated Coffee Processing

Decaffeinated coffee products represented 18% of the coffee consumed in 1991 in the United States.

Until the 1980s, synthetic organic solvents commonly were used in the United States to extract the caffeine, either by direct contact or by an indirect secondary water-based system. In each case, steaming or stripping was used to remove residual solvent from the beans, and the beans were dried to their original moisture content (10–12%) prior to roasting.

In the 1980s, manufacturers' commercialized processes which utilized either naturally occurring solvents or solvents derived from natural substances to position their products as naturally decaffeinated. The three most common systems use carbon dioxide under supercritical conditions, oil extracted from roasted coffee, or ethyl acetate, an edible ester naturally present in coffee.

In all the above-mentioned processes of coffee decaffeination, changes occur that affect the roast flavor development.

To make an instant decaffeinated coffee product, the decaffeinated roast and ground coffee is extracted in a manner similar to nondecaffeinated coffee. Alternatively, the caffeine from the extract of untreated roasted coffee is removed by using the solvents described previously.

Economic Importance of Coffee

Coffee has been a significant factor in international trade since the early 1800s. It is among the leading agricultural products in international trade, along with wheat, corn, and soybeans.

Historically, any factors that affect the balance of supply and demand, eg, political, climatic, etc, have contributed to the high volatility of green coffee prices. The International Coffee Organization (ICO), consisting of seventy-two producer and consumer nations, developed the International Coffee Agreement (ICA) in 1962 to achieve stable prices through export quotas adjusted by indicator price change. The breakdown of the ICA in 1989 led to the historically low prices of the early 1990s. In 1993 the major producing countries unilaterally instituted a voluntary holdback program in an attempt to bolster price levels.

Coffee Regulations and Standards

Various standards and regulations for green coffee are set either by legislation in the producing and consuming countries, eg, the United States, France, or the European Economic Community, or by voluntary standards set by various trade organizations and associations in the consuming countries, eg, National Coffee Association or London Terminal Market. These standards and regulations define what is being purchased by contract and protect consumers against fraud and potential health risks.

Decaffeination Regulation. In the United States, decaffeination usually signifies that 97% of the caffeine has been removed. Permissible solvents for decaffeination processes are defined by national legislation, eg, FDA or EEC directive.

Caffeine and Health. Caffeine is listed in the U.S. Code of Federal Regulations as a multipurpose, generally recognized as safe (GRAS) substance and in Food Chemicals Codex, published by the U.S. National Academy of Sciences/National Research Council, as a flavoring agent and stimulant. A consensus has developed among knowledgeable scientists that consumption of caffeine in moderate amounts is not associated with any serious health consequences.

Coffee Substitutes

Coffee substitutes, which include roasted chicory, chick peas, cereal, fruit, and vegetable products, have been used in all coffee consuming countries.

New Technology

Coffee Biotechnology. In South and Central America, breeding programs are in progress to improve coffee disease resistance, productivity, and cup quality. Micropropagation methods based on multiplying cells in a bioreactor have been reported and will allow for the propagation of identically cloned plants, thus eliminating nonuniformity in coffee plant material. Coffee plant regeneration studies from protoplasts, ie, coffee cells without cell walls, have been reported and may serve as a key part of the methodology involved in transforming coffee plants via recombinant DNA techniques. Coffee bioconversions through enzymatic hydrolysis have been used to modify green coffee and improve the finished product. Potential consumer benefits from biotechnology are lower cost and improved quality.

Liquid Coffee Products. Liquid coffee products, generally presweetened and ready to drink with milk added, represent a significant part of Japanese and Korean coffee markets. Much of this product is sold in vending machines, hot and cold, for immediate consumption. Recently the U.S. market has seen entry of several similarly processed flavored products.

GERALD WASSERMAN
HOWARD D. STAHL
WARREN REHMAN
PETER WHITMAN
Kraft General Foods Corporation

R. J. Clarke and R. Macrae, eds., *Coffee,* Vol. 1, *Chemistry,* 1985; Vol. 2, *Technology,* 1987; Vol. 3, *Physiology,* 1988; Vol. 4, *Agronomy,* 1988; Vol. 5, *Related Beverages,* 1987; Vol. 6, *Commercial and Technico-Legal Aspects,* 1988; Elsevier Applied Science Publishers, Ltd., Barking, U.K. An excellent reference series.

M. Sivetz and N. Desrosier, *Coffee Technology,* AVI Publishing Co., Inc., Westport, Conn., 1979. Somewhat dated but good reference for the basic coffee processing technology.

ASIC (Association Scientifique Internationale Du Café) Colloquium Series—Proceedings of the Biannual Worldwide Coffee Technology Symposium.

COLLOIDS

A colloid is a material that exists in a finely dispersed state. It is usually a solid particle, but it may be a liquid droplet or a gas bubble. Typically, colloids have high surface-area-to-volume ratios, characteristic of matter in the submicrometer-size range. Matter of this size, from approximately 100 nm to 5 nm, just above atomic dimensions, exhibits physicochemical properties that differ from those of both the constituent atoms or molecules and the macroscopic material. The differences in composition, structure, and interactions between the surface atoms or molecules and those on the interior of the colloidal particle lead to the unique character of finely divided material, specifics of which can be quite diverse (see FLOCCULATING AGENTS).

Today, colloids are encountered in a wide variety of industries, and many important technological problems relate to the behavior of dispersions, the mixtures obtained when colloids are distributed, ie, dispersed, in a medium, such as a liquid or a gaseous fluid (see DISPERSANTS). Commercial examples include metals, magnetic powders, catalysts, ceramics, minerals, oil recovery, technical glasses, paints and pigments, polymers, pulp and paper, prepared foods, pharmaceuticals, fibers, detergents, and purified water (see LATEX TECHNOLOGY; PLASTICIZERS; PRINTING PROCESSES). Natural and biological systems may also depend, to some extent, on the behavior of colloids. Phenomenological examples are found in soil science and plant nutrition, meteorology, hematology, membrane science, and antigen–antibody reactions in medical technology (see MEMBRANE TECHNOLOGY).

General Properties

Dimensions. Most colloids have all three dimensions within the size range ~100 nm to 5 nm. If only two dimensions (fibrillar geometry) or one dimension (laminar geometry) exist in this range, unique properties of the high surface area portion of the material may still be observed and even dominate the overall character of a system. The non-Newtonian rheological behavior of fibrillar and laminar clay suspensions, the reactivity of catalysts, and the critical magnetic properties of multifilamentary superconductors are examples of the numerous systems that are ultimately controlled by such colloidal materials.

A dispersion factor, defined as the ratio of the number of surface atoms to the total number of atoms in the particle, is commonly used to describe highly dispersed systems that do not exhibit a particularly high surface-area-to-volume ratio.

In view of the facts that three-dimensional colloids are common and that Brownian motion and gravity nearly always operate on them and the dispersing medium, a comparison of the effects of particle size on the distance over which a particle translationally diffuses and that over which it settles elucidates the colloidal size range.

Nomenclature. Colloidal systems necessarily consist of at least two phases, the colloid and the continuous medium or environment in which it resides. The properties of such systems greatly depend on the composition and structure of each phase. Therefore, it is useful to classify colloids according to their states of subdivision and agglomeration, and with respect to the dispersing medium. The possible classifications of colloidal systems are given in Table 1.

Behavior. Diffusion, Brownian motion, electrophoresis, osmosis, rheology, mechanics, and optical and electrical properties are among the general physical properties and phenomena that are primarily important in colloidal systems. Of course, chemical reactivity and adsorption often play important, if not dominant, roles. Any physical and chemical feature may ultimately govern a specific industrial process and determine the final product characteristics.

Physical and Chemical Properties

Formation. Colloid formation involves either nucleation and growth phenomena or subdivision processes. The former case requires a phase change; the latter one pertains to the comminution or atomization of coarse particles (solids) or droplets (liquids). Three nucleation and growth processes by which colloids have been synthesized in the laboratory or in nature are condensation of vapor to yield liquid or solid directly, condensation of vapor to form colloidal liquid droplets (aerosols, emulsions) that may subsequently solidify, and precipitation from liquid or solid solutions. Chemical reactions often induce these phase changes; however, they are not essential for producing either homogeneously or heterogeneously nucleated colloids. The various mechanisms of colloid formation and the resultant powder properties of industrially produced single- and multicomponent colloidal solids are listed in Table 2. Processes based on subdivision contrast these chemical processes and are widely used.

Specific advancements in the chemical synthesis of colloidal materials are noteworthy. Many types of generating devices have been used to produce colloidal liquid aerosols and emulsions; among them are atomizers and nebulizers of various designs. A unique feature of producing liquid or solid colloids via aerosol processes (Table 2) is that material with a relatively narrow size distribution can be routinely prepared. Another significant approach makes colloidal powders prepared as liquid suspensions increasingly available. These powders have uniform chemical and phase composition, particle size, and shape.

Characterization. The proper characterization of colloids depends on the purposes for which the information is sought because the total description would be an enormous task. The following physical traits are among those to be considered: size, shape, and morphology of the primary particles; surface area; number and size distribution of pores; degree of crystallinity and polycrystallinity; defect concentration; nature of internal and surface stresses; and state of agglomeration. Chemical and phase composition are needed for complete characterization, including data on the purity of the bulk phase and the nature and quality of adsorbed surface films or impurities.

Hazards of Colloidal Systems

Two situations exist for which the chemical reactivity of colloidal materials may be detrimental or even physically harmful. The first one is a result of their high specific surface area, which typically ensures that the colloidal chemical reactivity differs considerably from that of the identical macroscopic material with much lower specific surface area. This aspect is particularly important if the colloidal surface is easily and rapidly oxidized. The conditions make dust explosions and spontaneous combustion potential dangers whenever certain materials exposed to oxidizing environments exist as finely divided dry matter (see POWDERS, HANDLING).

A second hazard is that many colloidal substances, including fibrillar matter, are often inhaled to the extent that physiological problems may arise if they are retained by bodily tissues such as the lungs. This hazard may be quite acute if the particulates are ca 1 μm. This problem does not occur with exceedingly small particulates because these may be exhaled; particulates that are significantly larger than 1 μm also do not normally present such problems because they may settle and therefore are not often inhaled into the lungs. Short-term allergic

Table 1. Classifications of Two-Phase Colloidal Systems

Dispersing medium	Dispersed (colloidal) matter	Names[a]
gas	liquid	aerosol, fog, mist
	solid	aerosol, smoke
liquid	gas	foam
	liquid	emulsion
	solid	sol[b] gel, dispersion, suspension
solid	gas	solid foam
	liquid	gel, solid emulsion
	solid	solid sol, alloy

[a] The appropriate name depends on the specific properties of the system.
[b] Aqueous sols are commonly called hydrosols.

Table 2. Industrially Produced Colloidal Materials and Related Processes

Mechanism	Examples
vapor → liquid → solid ↓ → → → ↑	oxides, carbides via high intensity arc; metallic powders via vacuum or catalytic reactions
vapor + vapor → solid	chemical vapor deposition, radio frequency-induced plasma, laser-induced precipitation
liquid → solid	ferrites, titanates, aluminates, zirconates, molybdates via precipitation
solid → solid	oxides, carbides via thermal decomposition

conditions derived from colloidal substances are common, eg, those induced by pollen and household dust, asthma, and hay fever. The long term effects of such matter may be fatal, as is the case with silicosis, asbestosis, and black lung disease.

Specific potential hazards have with certainty been associated with a diverse spectrum of colloidal materials, eg, chemicals, coals, minerals, metals, pharmaceuticals, plastics, and wood pulp. Limits for human exposure for many particulate, hazardous materials are published. If the material is a colloidal solid or liquid, the exposure limits, typically expressed in mg/m^3, may be misleading, owing to the fact that problems arise from the high specific surface area rather than the usual concentration-related chemical reactions of hazardous materials. One salient feature is that large fractions of readily hydrolyzable, toxic metals may exist as adsorbed species on suspended (colloidal) solids in fresh and marine water systems and can also be anticipated in industrial wastewater. Similarly, hazardous elements such as lead, zinc, and vanadium that are released into the atmosphere as vapors may subsequently condense or are removed as solid particulates by rain, thereby becoming highly mobile. Liquid droplets may also constitute a hazard, in much the same manner as solid colloids. For instance, smog often contains sulfuric acid aerosols.

The cost to protect the general population from the hazardous colloids that act as pollutants is, of course, highly variable, owing to transient economic issues and governmental restrictions. Although the latter constraints may set a rigorous schedule for technologically isolating and containing waterborne and airborne pollutants, they are largely unenforced. Yet, economic and governmental forces have helped to foster quite large water- and air-treatment industries. The increasing cost of research on these particulate pollutants is possibly as great as 60% of the increasing annual cost of research and acutely affects the progress made in controlling colloidal pollutants.

New Developments

Important problems in colloid science remain to be addressed if the potential of colloids is to be fully exploited. The extension of existing understanding to more concentrated suspensions, the testing of predictions using model colloidal powders, and the examination of relaxation phenomena in ordered colloids are among those receiving increased attention.

ALAN BLEIER
Oak Ridge National Laboratory

H. R. Kruyt, *Colloid Science, Volume I, Irreversible Systems*, Elsevier, Amsterdam, the Netherlands, 1952.

R. J. Hunter, *Foundations of Colloid Science*, Vol. 1, Clarendon Press, Oxford, U.K., 1987.

E. Matijević, *Chem. Technol.* **3,** 656 (1973); *Ibid.* **21,** 176 (1991); *Ann. Rev. Mater. Sci.* **15,** 483 (1985).

C. R. Veale, *Fine Powders*, John Wiley & Sons, Inc., New York, 1972; J. K. Beddow, *Particulate Science and Technology*, Chemical Publishing, New York, 1980.

COLOR

Color Fundamentals

An immediate complexity is illustrated in the two early and apparently incompatible theories of color vision. Trichromatic theory, first proposed in 1801 by Thomas Young and later refined by Hermann von Helmholtz, postulated three types of color receptors in the eye. This explained many phenomena, such as various forms of color blindness, and was confirmed when three types of blue-, green-, and red-sensitive cones were reported to be present in the retina in 1964. Yet Ewald Hering's 1878 opponent theory, which used three pairs of opposites, light-dark, red-green, and blue-yellow, also offered much insight, including the explanation of contrast and afterimage effects and the absence of some color combinations such as reddish greens and bluish yellows. In the modern zone theories it is now recognized that the data from three trichromatic detectors in the eye are processed on their way to the brain into opponent signals, thus removing the apparent incompatibility.

In color technology and measurement, both types of approaches are used. Color printing, for example, generally employs three primary colors (usually plus black), and the ever useful CIE system was founded on experiments in which colors were matched by mixtures of three primary colors, often blue, green, and red. Yet transmitted television signals are based on the opponent system, with one intensity and two color-balance signals, as are the modern representations of color, such as the CIELAB and related color spaces based on red-green and yellow-blue opponent axes.

Light and Color. Visible light is that part of the electromagnetic spectrum, shown in Figure 1, with wavelengths between the red limit at about 700 nm and the violet limit of about 400 nm. Depending on the observer, light intensity, etc, typical values for the spectral colors are red, 650 nm; orange, 600; yellow, 580; green, 550 and 500; and blue, 450.

The appearance of color depends significantly on the exact circumstances. Normally one thinks of viewing a colored object under some type of illumination, the object mode (Table 1). In viewing a light source, there is the illuminant mode. Finally, in viewing through a hole in a screen there is the aperture mode. Color perception differs significantly in these modes.

Interactions of Matter with Light. In the most generalized interaction of light with matter the many phenomena of Figure 2 are possible.

Color Vision

The Eye. Light passes through the cornea, the transparent outer layer of the eye, through the lens and the aqueous and vitreous humors, and is focused onto the retina. The iris, forming the pupil, acts as a variable aperture to control the amount of light that enters the eye, varying from about f/2.5 to f/13 with a 30:1 light intensity ratio. The two humors serve merely as neutral transmission media and to keep the eyeball distended. The retina is a layer about 0.1 mm thick that contains the light-sensitive rods and cones. Only the rods function in low levels of illumination of about less than 1 lux, providing

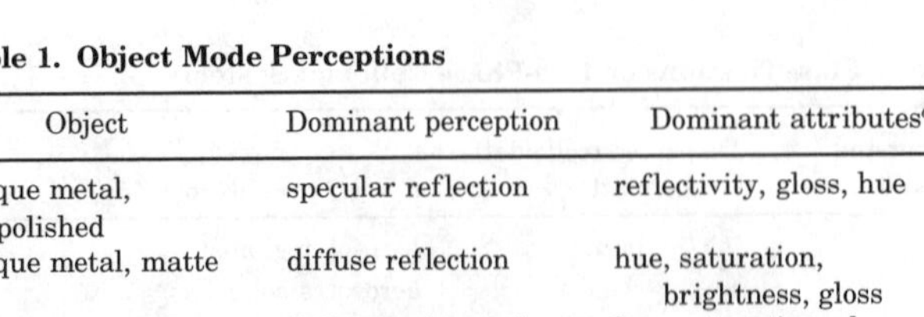

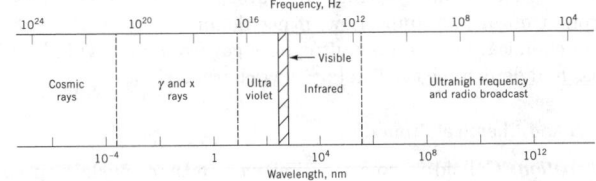

Figure 1. The electromagnetic spectrum.

Table 1. Object Mode Perceptions

Object	Dominant perception	Dominant attributes[a]
opaque metal, polished	specular reflection	reflectivity, gloss, hue
opaque metal, matte	diffuse reflection	hue, saturation, brightness, gloss
opaque nonmetal, glossy	diffuse and specular reflections	hue, saturation, gloss, brightness
opaque nonmetal, matte	diffuse reflection	hue, saturation, brightness
translucent nonmetal	diffuse transmission	translucency, hue, saturation
transparent nonmetal	transmission	hue, saturation, clarity

[a] In approximate sequence of importance.

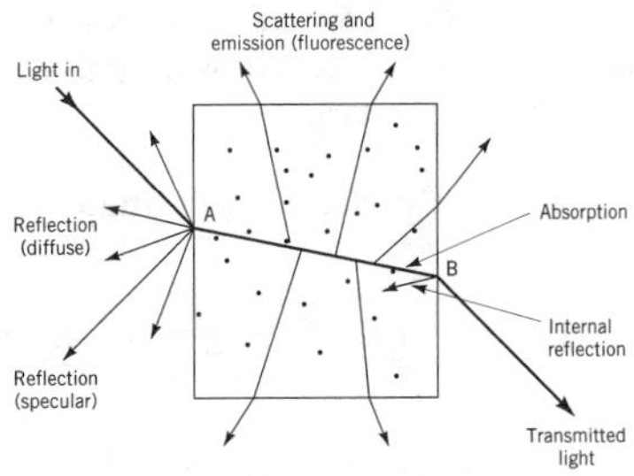

Figure 2. The adventures of a beam of light passing through a block of partly transparent substance.

an achromatic, noncolor image. Each set of cones is sensitive to a wide range of wavelengths, with extensive overlap. Appropriate designations are short-, medium-, and long-wavelength sensitive cones.

The trichromatic theory, subsequently confirmed by the existence of the three sets of cones, must be combined with the opponent theory, which is involved in the retinal pathway. A third approach, the appearance theory or the retinex theory, must be added to explain color constancy and other effects. As one example of this last, consider an area perceived as red in a multicolored object such as a Mondrian painting when illuminated with white light. If the illumination is changed so that energy reflected by this same area is greater at shorter wavelengths than the energy reflected at longer wavelengths, this area is still perceived as red within the overall visual context. If all other colors are now covered so that only our "red" area is visible, corresponding to the aperture mode, then this area is perceived as green. Clearly, all three approaches must be melded to give a full description of color perception, a process that is not yet complete.

Color Vision Defects. Anomalous color vision is present, eg, if one of the three sets of cones is inoperative (dichromacy) or defective (anomalous trichromacy). There are other color vision defects.

Color Order Systems

Many one-, two-, and three-dimensional systems have been developed over the years to order colors in a systematic way and provide specimen colors for visual comparison. Coordination has now been achieved with computer programs between essentially all of these systems and the CIE systems described below, and conversions can easily be made between them.

The Munsell System. The best known and most widely used is the Munsell system, developed by the artist A. H. Munsell in 1905 and modified over the years. This is a three-dimensional space. With interpolations some 100,000 colors can be distinguished within the Munsell system by visual comparison under carefully standardized viewing conditions.

Other Color Order Systems. The Natural Color System, (NCS) developed in Sweden is an outgrowth of the Hesselgren Color Atlas, and uses the opponent color approach. There are also the Colorcurve, the Coloroid, and the German DIN systems, among others. The Optical Society of America has published the OSA Uniform Color Scale System with 558 equally spaced color chips. For many purposes it is convenient to use a much simpler set of 267 color regions as provided by the ISCC-NBS Centroid System, a joint project of the Inter-Society Color Council and the USA National Bureau of Standards, now NIST.

Basic Colorimetry

The International Commission on Illumination (abbreviated CIE from the French expression) over the years has recommended a series of methods and standards in the field of color.

When considering light of a certain spectral energy distribution falling on an object with a given spectral reflectance and perceived by an eye with its own spectral response, to obtain the perceived color stimulus it is necessary to multiply these factors together. Standards are clearly required for both the observer and the illuminant.

The CIE Standard Observer. The CIE standard observer is a set of curves giving the spectral color matching functions $\bar{x}$, $\bar{y}$, and $\bar{z}$, of an imaginary observer representing an average population. The 1931 CIE standard observer was determined for 2° foveal vision, while the later 1964 CIE supplementary standard observer applies to a 10° vision; a subscript 10 is usually used for the latter. The standard observers were defined in such a way that of the three primary responses, the value of $\bar{y}$ corresponds to the spectral photopic luminous efficiency, ie, to the perceived overall lightness of an object. The tristimulus values X, Y, and Z are derived from these color matching functions.

Chromaticity Diagrams. The CIE 1931 chromaticity diagram uses the chromaticity coordinates:

$$x = X/(X+Y+Z);\ y = Y/(X+Y+Z);\ z = Z/(X+Y+Z)$$

It is not actually necessary to specify z, since $x + y + z = 1$. Here Newton's pure spectral colors in fully saturated form follow a horseshoe-shaped curve with wavelengths from 400 nm violet to 700 nm red.

One of several defects of the chromaticity diagram is that the minimum distinguishable colors are not equally spaced, that is, that equal changes in x, y, and Y, the three parameters required to specify fully any color, do not correspond to equally perceived color differences.

Standard Illuminants. Clearly, standardized light sources are desirable for color matching, particularly in view of the phenomenon of illuminant metamerism described herein. Over the years CIE has defined several standard illuminants, some of which can be closely approximated by practical sources. In 1931 there was Source A, defined as a tungsten filament incandescent lamp at a color temperature of 2854 K. Sources B and C used filtering of A to simulate noon sunlight and north sky daylight, respectively. Subsequently a series of D illuminants was established to better represent natural daylight. Of these the most important is Illuminant D_{65}.

Gloss and Opacity. Attributes such as gloss, transparency, translucency, opacity, haze, and luster may apply to some materials (see Table 1), and these are relevant in that they may influence the judgment of color differences. When present, these attributes can be measured using specialized approaches.

Light Mixing. Light or additive mixing applies to light beams. White results when any suitable set of three-color beams of the appropriate intensity are mixed.

Colorant Mixing. A colorant, whether a dye dissolved in a medium or pigment particles dispersed in it, produces color by absorbing or scattering part of the transmitted light. If only absorption is present, the Beer-Lambert law applies:

$$A(\lambda) = \log 1/T(\lambda) = \sum^{i} a_i(\lambda) b c_i$$

where A is the absorbance, T is the transmittance, $a_i(\lambda)$ is the absorptivity or specific absorbance of absorber i at wavelength λ, b is the length of the absorbing path, and c_i the concentration of the absorber. When colorants are mixed, they function by each independently absorbing light and the subtractive mixing rules merely specify this additivity, a mixture of the three primary colorants producing black.

When both absorption and scattering are present, the Beer-Lambert Law must be replaced by the Kubelka-Munk equation employing the absorption and scattering coefficients K and S, respectively. This gives the reflectivity R_∞

$$(1 - R_\infty)^2/2R_\infty = K/S = \sum^{i} c_i K_i(\lambda) \Big/ \sum^{i} c_i S_i(\lambda)$$

where c_i is the concentration and $K_i(\lambda)$ and $S_i(\lambda)$ are the specific absorbance and scattering parameters, respectively, of absorber and scatterer i at wavelength λ.

The color of an opaque paint depends both on the size of the pigment particles and on refractive index considerations.

The ready availability of computers has led to the detailed analysis of the colorant formulation problems faced every day by the textile, coatings, ceramics, polymer, and related industries. The resulting computer match prediction has produced improved color matching and reductions in the amounts of colorants required to achieve a specific result, with accompanying reductions of cost.

Metamerism. There are several types of metamerism, the phenomenon in which two objects perceived as having a perfect color match under one set of conditions are found to differ in color under other conditions. Most common is illuminant metamerism that occurs when a change in illuminant is the cause. This originates from the situation that a given visually identified hue can be caused by many different stimuli. Another, related cause would be the presence of an ultraviolet component in one of the sources (usually actual daylight), causing a fluorescent object to emit light in addition to that reflected, compared to a visible-light-only source such as an incandescent lamp. The standard illuminant D_{65} specifies significant intensity at wavelengths less than 400 nm in the ultraviolet.

Observer metamerism derives from the significant differences in spectral response found even among persons with normal color vision. Finally, there is geometric metamerism observed with a change in apparent color and with a change in viewing angle, as with some metallic paints.

Advanced Colorimetry

Two colorimetry systems have been ultimately agreed upon, designated the CIELUV and CIELAB color spaces, and have come into widespread use.

The 1976 CIELUV Color Space. Properly designated CIE $L^*u^*v^*$, this uses a white object or light source designated by the subscript n as the reference standard and employs the transformations:

$$L^* = (116Y/Y_n)^{1/3} - 16; Y/Y_n > 0.008856$$

$$L^* = 903.3Y/Y_n; Y/Y_n \geq 0.008856$$

$$u^* = 13L^*(u' - u'_n)$$

$$v^* = 13L^*(v' - v'_n)$$

The CIELUV space preserves a property of the CIE 1931 chromaticity space which is important in the field of color reproduction, eg, in the television, photographic, and the graphic arts industries. This is the characteristic that the chromaticities of additive mixtures of color stimuli lie on the straight line connecting the chromaticities of the component stimuli; this is true of the 1976 metric chromaticity diagram but not of the CIELAB space that follows.

The 1976 CIELAB Color Space. Defined at the same time as the CIELUV space, the CIELAB space, properly designated CIE $L^*a^*b^*$, is a nonlinear transformation of the 1931 CIE X, Y, Z space. It also uses the metric lightness coordinate L^*, together with:

$$a^* = 50[(X/X_n)^{1/3} - (Y/Y_n)^{1/3}]$$

$$b^* = 200[(Y/Y_n)^{1/3} - (Z/Z_n)^{1/3}]$$

These equations apply for X/X_n, Y/Y_n, and Z/Z_n all >0.008856. For $X/X_n \leq 0.008856$, the term $[(X/X_n)^{1/3}]$ is replaced by $[7.787(X/X_n) + 16/116]$ and similarly for Y and for Z in these two equations.

This transformation results in a three-dimensional space that follows the opponent color system with $+a^*$ as red, $-a^*$ as green, $+b^*$ as yellow, and $-b^*$ as blue. CIELAB is closely related to the older Adams-Nickerson, modified Adams-Nickerson, and other spaces of the L, a, b type, which it replaced.

The 1976 CIE Metric Color Spaces. Both the CIELUV and CIELAB spaces can have their Cartesian coordinates converted to cylindrical coordinates, called metric or hue-angle coordinates, with L^* unchanged.

Hunter L, a, b and Other Color Spaces. This was the earliest practical opponent-based system which is still widely used. In this system, for illuminant C and the 2° standard observer:

$$L = 10Y^{1/2}\text{(lightness coordinate)}$$

$$a = 17.5(1.02X - Y)Y^{1/2}\text{(red-green coordinate)}$$

$$b = 7.0(Y - 0.847Z)/Y^{1/2}$$

There are other equations for other illuminants, other observers, and for special conditions.

Color Difference Assessment. Color difference scales include those of Judd-Hunter, Macadam, Adams-Nickerson, ANLAB, and ANLAB40. All of these have limitations in some way or another.

Color Measuring Instruments

There has been a tremendous change in the last two decades as computers have taken over the tedious calculations involved in color measurement. Indeed, microprocessors either are built into or are connected to all modern instruments, so that the operator may merely need to specify, for example, x, y, Y or L^*, a^*, b^* or L^*, C^*, h, either for the 2° or the 10° observer, and for a specific standard illuminant, to obtain the desired color coordinates or color differences, all of which can be stored for later reference or computation. The use of high intensity filtered Xenon flash lamps and array detectors combined with computers has resulted in almost instantaneous measurement in many instances.

Spectrophotometers and Spectrocolorimeters. Spectrophotometers are the most sophisticated color measuring instruments and provide the most detailed and accurate information. They may provide continuous spectral data of reflectance and transmittance against wavelength or use up to 20-nm wavelength steps, with high precision in reflectance (down to 0.01%) and tristimulus values (down to 0.01). There may be dual beams with one for the sample and a second as reference for the most stable and precise operation. Many spectrophotometers use an integrating sphere, although some use other geometries or permit alternative ones.

Slightly less sophisticated are spectrocolorimeters that determine spectral response curves for further computation but from which the spectral curve itself is not available.

Colorimeters. Also known as tristimulus colorimeters, these are instruments that do not measure spectral data but typically use four broad-band filters to approximate the $\bar{y}$, $\bar{z}$, and the two peaks of the $\bar{x}$ color-matching functions of the standard observer curves.

The Fifteen Causes of Color

No fewer than 15 distinct chemical and physical mechanisms explain the various causes of color, ordered into five groups as in Table 2. In the first group, covered by quantum theory, there are incandescence, simple electronic excitations, and vibrational and rotational excitations. Most chemical compounds contain only paired electrons that require very high energies to become unpaired and form excited energy levels. This requires ultraviolet; hence, there is no visible absorption and no color. Absorption color can, however, be derived from the easier excitation of unpaired electrons in transition-metal compounds and impurities, covered by ligand field theory in the second group. Absorptions from paired electrons can be shifted into the visible by increasing the size of the region over which the electrons are localized, as in organic compounds covered by molecular orbital theory in the third group; this also explains various forms of charge transfer. In the fourth group there is color in metals and in semiconductors such as yellow cadmium sulfide, both pure and doped, covered by band theory; this also covers color centers. In the final group there are four color-causing mechanisms explained by geometrical and physical optics.

Table 2. Fifteen Causes of Color

Cause	Examples
	Vibrations and simple excitations
incandescence	flames, lamps, carbon arc, limelight
gas excitations	vapor lamps, flame tests, lightning, auroras, some lasers
vibrations and rotations	water, ice, iodine, bromine, chlorine, blue gas flame
	Transitions involving ligand field effects
transition-metal compounds	turquoise, chrome green, rhodonite, azurite, copper patina
transition-metal impurities	ruby, emerald, aquamarine, red iron ore, some fluorescence and lasers
	Transitions between molecular orbitals
organic compounds	most dyes, most biological colorations, some fluorescence and lasers
charge transfer	blue sapphire, magnetite, lapis lazuli, ultramarine, chrome yellow, Prussian blue
	Transitions involving energy bands
metals	copper, silver, gold, iron, brass, pyrite, ruby glass, polychromatic glass, photochromic glass
pure semiconductors	silicon, galena, cinnabar, vermillion, cadmium yellow and orange, diamond
doped semiconductors	blue and yellow diamond, light-emitting diodes, some lasers and phosphors
color centers	amethyst, smoky quartz, desert amethyst glass, some fluorescence and lasers
	Geometrical and physical optics
dispersive refraction	prism spectrum, rainbow, halos, sun dogs, green flash, fire in gemstones
scattering	blue sky, moon, eyes, skin, butterflies, bird feathers, red sunset, Raman scattering
interference	oil slick on water, soap bubbles, coating on camera lenses, some biological colors
diffraction	diffraction gratings, opal, aureole, glory, some biological colors, most liquid crystals

KURT NASSAU
Consultant

R. S. Hunter and R. W. Harold, *The Measurement of Appearance*, 2nd ed., John Wiley & Sons, Inc., New York, 1987; Table A7.

R. W. G. Hunt, *Measuring Color,* 2nd ed., John Wiley & Sons, Inc., New York, 1992.

F. W. Billmeyer, Jr. and M. Saltzman, *Principles of Color Technology,* 2nd ed., John Wiley & Sons, Inc., New York, 1981.

K. Nassau, *The Physics and Color,* John Wiley & Sons, Inc., New York, 1983.

COLORANTS FOR CERAMICS

Any product that depends on aesthetics for consideration for purchase and use will be improved by the use of color. Hence, many ceramic products, such as tile, sanitary ware, porcelain enameled appliances, tableware, and some structural clay products and glasses, contain colorants.

For both economic and technical reasons, the most effective way to impart color to a ceramic product is to apply a ceramic coating that contains the colorant. The most common coatings, glazes, and porcelain enamels are vitreous in nature. Hence, most applications for ceramic colorants involve the coloring of a vitreous material.

There are a number of ways to obtain color in a ceramic material. First, certain transition-metal ions can be melted into a glass or dispersed in a ceramic body when it is made. Although it is suitable for bulk ceramics, this method is rarely used in coatings because adequate tinting strength and purity of color cannot be obtained in this way.

A second method used to obtain color is to induce the precipitation of a colored crystal in a transparent matrix. Certain materials dissolve to some extent in a vitreous material at high temperatures, but when the temperature is reduced, the solubility is also reduced and precipitation occurs. This method is used to disperse nonoxide precipitates of gold, copper, or cadmium sulfoselenide in bulk glass. In coatings it is used for opacification, ie, the production of an opaque white color. Normally, some or all of the opacifier added to the coating slip dissolves during the firing process and recrystallizes upon cooling. For oxide colors other than white, however, this method lacks the necessary control for reproducible results and is seldom used.

The third method used to obtain color in a vitreous matrix is to disperse in that matrix an insoluble crystal or crystals that are colored. The color of the crystal is then imparted to the transparent matrix. This method is the one most commonly used to introduce color to vitreous coatings.

Ceramic Pigments

The principal method of coloration of ceramic coatings is dispersal of a ceramic pigment in a vitreous matrix. To be suitable as a ceramic pigment, a material must possess a number of properties which fall in two categories: strength of pigmentation and stability. Another desirable property for a ceramic color is a high refractive index.

Manufacturing. Although a number of different pigment systems exist, most are prepared by similar manufacturing methods. The first step in pigment manufacture is close control over the selection of raw materials. The raw materials are weighed and then thoroughly blended. The reaction forming the pigment crystal occurs in a high temperature calcining operation. Following calcination, the product may require milling to reduce particle size to that necessary for use. It has been found that there are several advantages to modification of pigments by addition to the calcined product of a small quantity of a dispersing agent.

Pigment Systems. Most of the crystals used for ceramic pigments are complex oxides, owing to the great stability of oxides in molten silicate glasses. The one significant exception to the use of oxides is the family of cadmium sulfoselenide red pigments. This family is used because the colors obtained cannot be obtained in oxide systems; thus it is necessary to sustain the difficulties of a nonoxide system.

There are no oxides that can be used to give a true red pigment which is stable to the firing of ceramic coatings. Hence, orange, red, and dark red colors are obtained by the use of cadmium sulfoselenide pigments. The cadmium sulfoselenides are a group of pigments based on solid solutions of cadmium selenide, cadmium sulfide, or zinc sulfide.

Although red is not available in oxide systems, pink and purple shades are obtained in several ways. One such system is the chrome alumina pinks. Chrome alumina pinks are combinations of ZnO, Al_2O_3, and Cr_2O_3. Depending on the concentration of zinc, the crystal structure may be either spinel or corundum.

The most stable pink pigment is the iron-doped zircon system. This pigment is made from a mixture of ZrO_2, SiO_2, and iron oxide, and produces shades from pink to coral.

The chrome–tin system is the only family known to produce purple and maroon shades, as well as pinks. The system can be defined as pigments that are produced by the calcination of mixtures of small amounts of chromium oxide with substantial amounts of tin oxide. In addition, most formulations contain substantial amounts of silica and calcium oxide.

The most important brown pigments used in ceramic coatings are the zinc iron chromite spinels. This pigment system produces a wide palette of tan and brown shades.

There are several systems for preparing a yellow ceramic pigment. Moreover, there are valid technical and economic reasons for the use of a particular yellow pigment in a given application. The pigments of greatest tinting strength, the lead antimonate yellows, cadmium sulfide, and the chrome titania maples, do not have adequate resistance to molten ceramic coatings. Thus other systems must be used if the firing temperature is above 1000°C.

Zirconia vanadia yellows are prepared by calcining ZrO_2 with small amounts of V_2O_5. They are economical pigments for use with a broad range of coatings firing at temperatures above 1000°C. They are brighter and stronger in glazes low in PbO and B_2O_3.

Tin vanadium yellows are prepared by introducing small amounts of vanadium oxide into the cassiterite structure of SnO_2. Tin vanadium yellows develop a strong color in all ceramic coating compositions.

The praseodymium zircon pigments are formed by calcination of about 5% of praseodymium oxide with a stoichiometric zircon mixture of ZrO_2 and SiO_2 to yield a bright yellow pigment.

Just as there are several alternative yellow pigments, there are several alternative green pigments. Formerly, the chromium ion was the basis for green pigments. Higher quality results are obtained if chromium oxide is used as a constituent in a calcined pigment. One such system is the cobalt zinc alumina chromite used to produce blue-green pigments. Victoria green is prepared by calcining silica and a dichromate with calcium carbonate to form the garnet $3CaO–Cr_2O_3–3SiO_2$. This pigment gives a transparent bright green color. Because of all the difficulties with the use of chromium-containing pigments and because there is a definite limitation on the brilliance of green pigments made of chromium, many green ceramic glazes are now made of a mixture of yellow and blue zircon pigments. The use of copper compounds to make a green pigment is of little interest to most industrial manufacturers, but the colors obtained from them are of great interest to artists because of the many subtle shades that can be obtained.

The traditional way to obtain blue in a ceramic coating is with cobalt, which has been used as a solution color since antiquity. Today, cobalt may react with Al_2O_3 to produce the spinel $CoAl_2O_4$ or with silica to produce the olivine Co_2SiO_4. The silicate involves higher concentrations of cobalt, with only modestly stronger color. In glazes, the cobalt pigments have largely been replaced by pigments based on vanadium-doped zircon. These pigments are turquoise in shade and are less intense than the cobalt pigments. Therefore, they are not applicable when the greatest tinting strength is required or when a purple shade is called for. Where they are applicable they give vastly improved color stability.

Black ceramic pigments are formed by calcination of several oxides to form the spinel structure. The prototype black is a cobalt iron chromite.

It is easiest to obtain a uniform gray color when a calcined pigment is used which is based on zirconia or zircon as a carrier for various ingredients of blacks such as Co, Ni, Fe, and Cr oxides. This pigment is called cobalt nickel gray periclase.

Use of Pigments in Coatings

There are several additional factors that must be considered in selecting pigments for a specific coating application. These factors include processing stability requirements, pigment uniformity and reproducibility, particle size distribution, dispersibility, and compatibility of all materials to be used.

Health and Safety Factors

Properly handled, ceramic colorants should not cause unacceptable problems affecting health and safety. Preventive measures to avoid inhalation of fine particulate matter should invariably be used. Care should be taken to avoid ingestion of pigments by thorough washing before eating or smoking. Particular care should be taken in handling cadmium sulfoselenide pigments and lead antimonate pigments, which are highly toxic if ingested or inhaled.

When these pigments are used with lead-containing glazes, care should be exercised to use lead-safe glaze materials (see LEAD COMPOUNDS, INDUSTRIAL TOXICOLOGY).

Economic Aspects

Owing to the limited market and the variety and complexity of the products, ceramic pigments are manufactured by specialist firms, not by the users. The principal producers are Ceramic Color and Chemical Corp., Drakenfeld Colors, CGRDEC, Englehard Corp., Ferro Corp., General Color and Chemical Corp., Mason Color and Chemical Corp., and Miles, Inc. Estimated annual production is about 2500–3000 metric tons. This figure does not include some of the same and similar products manufactured for use in nonceramic applications. The costs of ceramic pigments range from $9/kg to $50/kg or higher, depending on the elemental composition and the required processing. The most expensive pigments are those containing gold and the cadmium sulfoselenides.

RICHARD A. EPPLER
Eppler Associates

A. Burgyan and R. A. Eppler, *Am. Ceram. Soc. Bull.* **62**(9), 1001–1003 (1983).

R. A. Eppler, *Am. Ceram. Soc. Bull.* **56**(2), 213–215, 218, 224 (1977).

R. A. Eppler, *Am. Ceram. Soc. Bull.* **66**(11), 1600–1604 (1987).

COLORANTS FOR FOODS, DRUGS, COSMETICS, AND MEDICAL DEVICES

Colorants currently in use and their status follow. Colorants permitted in foods that are (*1*) subject to certification: FD&C Blue No. 1, FD&C Blue No. 2, FD&C Green No. 3, FD&C Red No. 3, FD&C Red No. 40, FD&C Yellow No. 5, FD&C Yellow No. 6, Citrus Red No. 2, and Orange B. (*2*) exempt from certification: annatto extract, β-apo-8′-carotenal, canthaxanthin, caramel, β-carotene, carrot oil, cochineal extract (carmine), corn endosperm oil, dehydrated beets (beet powder), dried algae meal, ferrous gluconate, fruit juice, grape color extract, grape skin extract, paprika, paprika oleoresin, riboflavin, saffron, synthetic iron oxide, tagetes meal and extract, titanium dioxide, toasted partially defatted cooked cottonseed flour, turmeric, termeric oleoresin, ultramarine blue, and vegetable juice. Colorants permitted in drugs (including colorants permitted in foods) that are (*1*) subject to certification: FD&C Red No. 4, D&C Blue No. 4, D&C Blue No. 9, D&C Green No. 5, D&C Green No. 6, D&C Green No. 8, D&C Orange No. 4, D&C Orange No. 5, D&C Orange No. 10, D&C Orange No. 11, D&C Red No. 6, D&C Red No. 7, D&C Red No. 17, D&C Red No. 21, D&C Red No. 22, D&C Red No. 27, D&C Red No. 28, D&C Red No. 30, D&C Red No. 31, D&C Red No. 33, D&C Red No. 34, D&C Red No. 36, D&C Red No. 39, D&C Violet No. 2, D&C Yellow No. 7, D&C Yellow No. 8, D&C Yellow No. 10, D&C Yellow No. 11, and Ext. D&C Yellow No. 7; (*2*) exempt from certification: alumina, aluminum powder, annatto extract, bismuth oxychloride, bronze powder, calcium carbonate, canthaxanthin, caramel, β-carotene, chromium–cobalt–aluminum oxide, chromium hydroxide green, chromium oxide green, cochineal extract (carmine), copper powder, dihydroxyacetone, ferric ammonium citrate, ferric ammonium ferrocyanide, ferric ferrocyanide, guanine, logwood extract, mica, potassium sodium copper chlorophyllin, pyrogallol, pyrophyllite, synthetic iron oxide, talc, titanium dioxide, and zinc oxide, β-carotene, chromium hydroxide green, chromium oxide greens, copper powder, dihydroxyacetone, disodium EDTA-copper, ferric ammonium ferrocyanide, ferric ferrocyanide, guaiazulene, guanine, henna, lead acetate, manganese violet, mica, potassium sodium copper chlorophyllin, pyrophyllite, silver, synthetic iron oxides, titanium dioxide, ultramarine blue, ultramarines (green, pink, red, violet), and zinc oxide. These lists are accurate as of January 1993 but are subject to change. Such changes as well as any changes in the regulation of color additives are routinely published in the *Federal Register*. The FDA, Division of Colors and Cosmetics also provides additional regulatory information.

Coloring Food

Colorless Foods. The principal use of color additives in food is in products containing little or no color of their own. These include many liquid and powdered beverages, gelatin desserts, candies, ice creams, sherbets, icings, jams, jellies, and snack foods.

Process and Storage Difficulties. Often the process used to prepare a food results in the formation of a color in the product, the depth of which depends largely on the time, temperature, pH, air exposure, and other parameters experienced during processing. It is deemed necessary to supplement the color of the product to ensure its uniformity from batch to batch. Items that fall into this category include certain beers, blended whiskies, brown sugars, table syrups, toasted cereals, and baked goods.

During storage, natural pigments frequently deteriorate with time because of exposure to light, heat, air, and moisture or because of interaction of the components of the food with each other or with the packaging material. The color of maraschino cherries, for example, fares so poorly with storage that they are routinely bleached, then artificially colored.

Regional and Seasonal Problems. The problems of the dairy and citrus fruit industries are typical of those encountered with products produced in different areas of the country or at different times of the year.

Most varieties of Florida oranges tend to be green, suggesting immaturity, even though they contain the proper ratio of solids to acid for fully nutritious, mature fruit. The necessity of coloring these oranges to make them comparable in appearance and thus as commercially acceptable as naturally orange-colored fruit was recognized years ago and began on a commercial scale about 1934.

In milk approximately 90% of the yellow color is a result of the presence of β-carotene, a fat-soluble carotenoid extracted from feed by cows. Summer milk is more yellow than winter milk because cows grazing on lush green pastures in the spring and summer months consume much higher levels of carotenoids than do cows barn-fed on hay and grain in the fall and winter. Various breeds of cows and even individual animals differ in the efficiency with which they extract β-carotene from feed and in the degree to which they convert it into colorless vitamin A. The differences in the color of milk are more obvious in products made from milk fat, since here the yellow color is concentrated. Thus, unless standardized through the addition of colorant, products like butter and cheese show a wide variation in shade and in many cases appear unsatisfactory to the consumer.

Other products having natural color that varies enough to make standardization of their color desirable include the shells of certain kinds of nuts, the skins of red and sweet potatoes, and ripe olives.

Miscellaneous Uses. Inks used by inspectors to stamp the grade or quality on meat must, by law, be made from food-grade colors. Dyes used in packaging materials that come in direct contact with a food must also be food-grade or, if not, it must be established that no part of the colorant used migrates into the food product. Pet foods, too, if colored, must contain only those colorants recognized by the FDA as suitable for the purpose.

Coloring Drugs

Compared with the food and cosmetic industries, pharmaceuticals are a minor though important consumer of colorants. Originally, dyes were used in drugs to make them more appealing to the consumer by adding color to otherwise colorless products, by masking unsatisfactory natural colors, and by standardizing the appearance of drugs the color of which varied from batch to batch as a consequence of the manufacturing process, a difference in the color of the raw materials used, or both. Some drugs, of course, contain added color for cosmetic purposes, as in the case of the skin-tone dyes added to certain creams and ointments used to treat disorders such as acne.

Although colors are still added to drugs for these purposes, the principal use of colorants in pharmaceuticals currently is to provide the manufacturer with a simple means of identifying products so that they are not inadvertently mixed during production and shipment.

Coloring Cosmetics

Compared with foods and drugs, cosmetics usually contain much higher amounts of colorants. Although foods and drugs seldom contain more than a few to several hundred parts per million (ppm) of colorant, cosmetics often contain several percent.

Coloring Medical Devices

Color additives are routinely added to medical devices such as surgical sutures, surgical cements, and contact lenses. Sutures are usually colored to make them more visible during surgery and, depending on the application, during removal of the suture after the sutured area has healed. Surgical cements, too, are colored to make them more visible during use. Colorants are used in contact lenses for several reasons.

Regulations Governing Use

Listed and Provisionally Listed Colorants. Colorants can be divided into two groups: those listed for use and those provisionally listed. Listed additives are colors that have been sufficiently evaluated to convince FDA of their safety for the applications intended. These colorants are also known popularly as permanently listed colorants; however, they in fact can be delisted for sufficient cause. Provisionally listed colorants, on the other hand, are dyes and pigments that are not considered unsafe but that have not undergone all the tests required by the Color Additives Amendments of 1960 to establish their eligibility for permanent listing. Currently, these colors can still be used in those applications in which they were used prior to enactment of the 1960 amendments, unless newer temporary regulations restrict their use further.

Certification of Colorants. A further distinction between color additives is made relative to whether there is requirement for FDA certification. In general, only synthetic organic colorants are now subject to certification, whereas natural organic and inorganic colorants, such as turmeric and titanium dioxide, are not.

Use Restrictions. There are numerous restrictions on the use of color additives. They cannot, for example, be employed to deceive the public by adding weight or bulk to a product or by hiding quality. In addition, special permission is needed to use colorants or products containing them in the area of the eyes, in injections, in surgical sutures, and in foods for which standards of identity have been promulgated under Section 401 of the Federal Food, Drug, and Cosmetic Act.

Other restrictions pertaining to the areas of use and the quantities of colorants allowed in products are specified in regulations for particular additives.

Certified Colors

Presently (ca 1997), all certified colors are factory-prepared materials belonging to one of several different chemical classes. Although a few, such as D&C Blue No. 6 (indigo), are known to exist in nature, certified colors owe their commercial importance to their synthetic production.

Compared to noncertified color additives, certified colors are a cheap, brighter, more uniform, and better characterized group of dyestuffs with higher tinctorial strengths and a wider range of hues. They are available singly (primary colors) and in admixture with other certified colors (secondary mixes).

Chemical Classifications. Azo colors comprise the largest group of certified colorants. They are characterized by the presence of one or more azo bonds ($-N{=}N-$) and are synthesized by the coupling of a diazotized primary aromatic amine to a coupling component, usually a naphthol. Certifiable azo colors can be subdivided into four groups: insoluble unsulfonated pigments, soluble unsulfonated dyes, insoluble sulfonated pigments, and soluble sulfonated dyes.

Lakes. Lakes are a special kind of color additive prepared by precipitating a soluble dye onto an approved insoluble base or substratum. In the case of D&C and Ext. D&C lakes, this substratum may be alumina, blanc fixe, gloss white, clay, titanium dioxide, zinc oxide, talc, rosin, aluminum benzoate, calcium carbonate, or any combination of two or more of these materials. Currently (ca 1997), alumina is the only substratum approved for manufacturing FD&C lakes.

FD&C lakes were first approved for use in 1959. Today, they are the most widely used type of lake. To make a lake, an alumina substrate is first prepared by adding sodium carbonate or sodium hydroxide to a solution of aluminum sulfate. Next, a solution of certified colorant is added to the resulting slurry, then aluminum chloride is added to convert the colorant to an aluminum salt, which then adsorbs onto the surface of the alumina. The slurry is then filtered, and the cake is washed, dried, and ground to an appropriate fineness, typically 0.1–40 μm.

Production and Use. The primary FD&C colors account for 80% or more of the total weight of colorant certified during any one year.

Based on maximum color concentrations and the total annual production of food in each food category, the total certified food color that might be ingested per person per year is estimated to be 19.5 g. Based on recent annual colorant production figures and current total population, this figure is closer to 11 g/yr.

Colorants Exempt from Certification

The Commissioner of Food and Drug has the authority to exempt particular color additives from the batch-certification procedure when it is believed that, because of their nature, certification is not needed to protect the public health.

Exempt colorants are made up of a wide variety of organic and inorganic compounds representing the animal, vegetable, and mineral kingdoms. In general, exempt colorants have less coloring power than certified colorants and thus have to be used at higher concentration.

Exempt colorants are inherently neither more nor less safe than certified colorants. However, they are viewed as having been obtained from nature (natural) and thus imagined as less of a health hazard than certified colorants. In fact, like all color additives, they are fabricated products.

DANIEL MARMION
Consultant

S. H. Hochheiser, *Synthetic Foods Colors in the United States: A History Under Regulation,* University Microfilms International, 83-04269, Ann Arbor, Mich., 1986. An excellent history of the development of legislation to control colorants used in foods, drugs, and cosmetics.

U. S. Code of Federal Regulations, Title 21, Pts. 70–82.

D. M. Marmion, *Handbook of U.S. Colorants, Foods, Drugs, Cosmetics and Medical Devises,* John Wiley & Sons, Inc., New York, 1991.

B. C. Hesse, *Coal-Tar Colors Used in Food Products,* Bureau of Chemistry, Bulletin No. 147, Feb. 10, 1912. Results of the Hesse study made at the turn of the century.

COLORANTS FOR PLASTICS

The initial uses of colorants in plastics were as extenders and additives. Carbon black and titanium dioxide were and are still used as fillers because of their low cost. Almost from plastics' inception the limitation of black and white did not offer sufficient color choices for end users looking to differentiate their products. The increase in aesthetic requirements along with different performance requirements and resin compatibilities led to a great expansion in the number of different chemical classes of colorants and forms in which these colorants are available in today's market.

Traditionally, colorants are divided into three classes: inorganic pigments, organic pigments, and dyes. Dyes are soluble under conditions of use but must be completely dissolved in order to show little or no haze in the final product. Dyes generally make very good transparent colors and colors that are very bright with high chroma attributes. In most cases, dyes often do not weather as well as pigments, especially in conjunction with an opacifying agent such as titanium dioxide. Pigments are insoluble and consist of finite particles that must be dispersed by physical means. Because pigments are not soluble in the resin, they often influence the physical properties of the mixtures.

The thermoplastic or thermoset nature of the resin in the colorant–resin matrix is also important. For thermoplastics, the polymerization reaction is completed, the materials are processed at or close to their melting points, and scrap may be reground and remolded, eg, polyethylene, propylene, poly(vinyl chloride), acetal resins, acrylics, ABS, nylons, cellulosics, and polystyrene. In the case of thermoset resins, the chemical reaction is only partially complete when the colorants are added and is concluded when the resin is molded. The result is a nonmeltable cross-linked resin that cannot be reworked, eg, epoxy resins, urea–formaldehyde, melamine–formaldehyde, phenolics, and thermoset polyesters.

There is the possibility of a chemical reaction between a plastic and a colorant at processing temperatures. Thermal stability of both the polymer and colorant plays an important role. Furthermore, the performance additives that may have been added to the resin such as antioxidants, stabilizers, flame retardants, ultraviolet light absorbers, and fillers must be considered. The suitability of a colorant in a particular resin must be evaluated and tested in the final application after all processing steps to ensure optimum performance.

Available Colorant Forms

Colorants can be added to plastics by several methods. The incorporation of the colorant often is a balance between a particular end property requirement and inventory control. The typical forms in which colorants are added are raw colorant, dispersed colorant, dry concentrate, and liquid concentrate. Furthermore, resins can be purchased that have the colorants dispersed in them, thus the resin is precolored.

Product Properties

Weathering. Colorants need to be evaluated in the resin in which they are incorporated for a particular application before the colorants can be judged acceptable. Weathering performance can vary widely from resin to resin and among color types; thus, correlations and predictions are only valid within the constrained set of testing conditions that have been developed.

Optical Properties. Haze is the most common optical property problem that depends on colorants. In transparent or translucent applications haze development becomes an important criterion in colorant evaluation.

Mechanical and Chemical Properties. Colorants, especially pigments, can affect the tensile, compressive, elongation, stress, and impact properties of a polymer. The colorants can act as an interstitial medium and cause microcracks to form in the polymer colorant matrix. This then leads to degradation of the physical properties of the system.

Thermal and Electrical. Colorants can also affect the thermal and electrical properties of the resin and thus cause differences in melt flow of the polymer or shrinkage of the final plastic part.

Inorganic Pigments

White. Titanium dioxide is by far the most widely used white pigment for plastics. It is available in several particle sizes and coatings for use in a varied group of applications.

Black. Carbon black is another outstanding pigment for plastics. It has the unique property of blocking uv, visible, and ir radiation. A common use is to improve weatherability of polyethylene sheeting for outdoor use.

Iron Oxides. In addition to the black iron oxide, there are several natural and synthetic yellow, brown, and red oxides. As a class, they provide inexpensive but dull, lightfast, chemically resistant, and nontoxic colors. The natural products are known as ocher, sienna, umber, hematite, and limonite.

Chromium Oxide Greens. Chromium oxide, Cr_2O_3, is dull but is the most weatherable green pigment.

Iron Blue. There are three common varieties of iron blue: Milori, Chinese, and Prussian (they are sometimes called toning blues). The three types differ chiefly in color, ease of dispersion, and reactivity characteristics.

Violet. There are several inorganic violet pigments. As a class, the inorganic violets are weak and difficult to disperse.

Ultramarine Pigments. Ultramarine is a reddish blue with a clean, brilliant shade having good durability except to acids.

Blue, Green, Yellow, and Brown Metal Combinations. There are a number of inorganic oxide pigments that are indispensable for coloring transparent or translucent high temperature plastics. Their properties have been summarized. The ingredients are mixtures of metals that are calcined at 800–1300°C. The resulting oxides are ground, washed, dried, and pulverized to fine powders. The products are insoluble in solvents, and all resins have excellent resistance to heat, light, and chemical attack as well as low oil absorption, good dispersibility, and little or no tendency to migrate. Compared with organic pigments and dyes, the oxide pigments are at a serious disadvantage in terms of brightness and, usually, cost.

The inorganic blues include cobalt aluminate, which has the nominal formula $CoO{\cdot}Al_2O_3$. A wide variety of greens ranging from blue to yellow in shade are based on cobalt in combination with chromium, aluminum, titanium, nickel, magnesium, antimony, or zinc. These are brighter than the chromium oxides. Inorganic yellow oxide combinations may contain lead, antimony, tin, nickel, or chromium. They are classed as yellows rather than brown, but they are dull compared with the cadmium yellows. Brown combinations usually contain iron with chromium, zinc, titanium, or aluminum.

Lead Chromates and Molybdates. The lead chromates appear in several shades of yellow. These pigments have the advantages of opacity, brightness, and low cost. They are sensitive to acids, alkalies, and hydrogen sulfide. Pigments containing lead are banned from any use, and hexavalent chromium is a suspected carcinogen. Use is confined to vinyls and low temperature polyolefin and polystyrene resins, except for chrome orange, which is too sensitive to acids for use in vinyls.

Cadmium Pigments. The cadmiums constitute a continuous series from greenish yellow through orange and red to maroon. They may be used in most plastics and are stable at 325°C and up to 550°C for short periods. Resistance to chemicals is good. Lightfastness is satisfactory indoors, improving from yellow to red. The full tones are strong and bright, but light tints, especially reds, become dull.

Titanate Pigments. These pigments are relatively weak but have extreme heat resistance and outdoor weatherability, eg, the yellow is used where a light cadmium could not be considered. They are compatible with most resins.

Pearlescent Pigments. The preferred process of producing pearlescent pigments is to coat mica with layers of titanium dioxide.

Metallic Pigments. For paint use, aluminum flakes are coated with stearic acid to produce a leafing grade that tends to line up with the platelets parallel to the surface. The metal protects plastics by reflecting uv light.

Bronze powders are available for used in plastics in various compositions and particle sizes. Copper is seldom used alone; all bronzes contain zinc.

Organic Pigments

The composition of inorganic pigments is well known. However, structures of organic pigments are often not disclosed (see PIGMENTS–ORGANIC). Most suppliers identify their pigments by CI designations and common names.

Monoazo Pigments. In combination with other groups, the azo linkage, —N=N—, imparts color to many dyes and pigments (see AZO DYES). The simplest of these, ie, the Hansa yellows, toluidine reds, and naphthol reds, do not have the lightfastness and heat stability required for plastics. Permanent Yellow FGL and Permanent Red 2B are stable enough for vinyls, polyethylene, polypropylene, and cellulosics. Permanent Red 2B is available as the calcium, barium, or manganese salt.

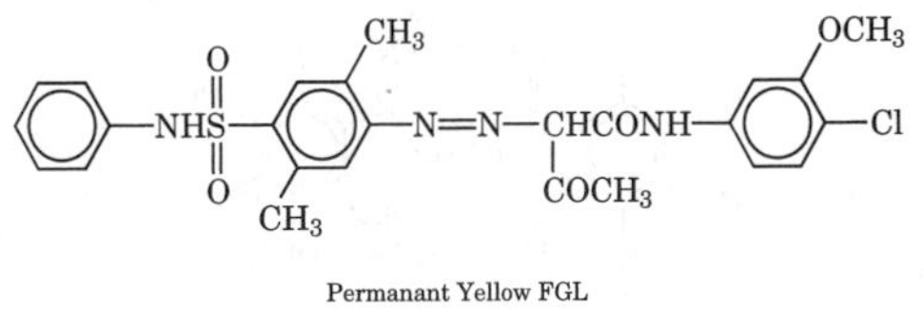

Permanant Yellow FGL

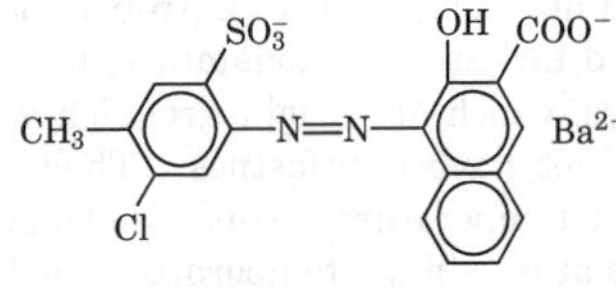

Permanant Red 2B

Disazo Pigments. The diarylide yellow and oranges also known as benzidines are derivatives of benzidine coupled to two moles of substituted acetoacetanilide. Benzidine Yellows AAMX, AAOT, AAOA, and HR (PY 13, 14, 17, and 83) are examples. Yellows AAMX and AAOT are used in flexible vinyls. AAOA also colors polyethylene and polypropylene. These three differ only slightly in shade. Benzidine Yellow HR is redder.

Disazo Condensation Pigments. In the manufacture of diarylide pigments, both amino groups are diazotized and coupled at the same time. CIBA-GEIGY Corp. has developed a series of disazo colors in which two azo dyes are combined with benzidine into one much larger molecule. These disazo pigments are trade-marked Cromophtal. The resulting red, orange, or yellow pigments are recommended for vinyls, acrylics, ABS, polyethylene, and similar thermoplastics as well as for epoxy and phenolic thermosets.

Quinacridone Pigments. The quinacridones have the following formula:

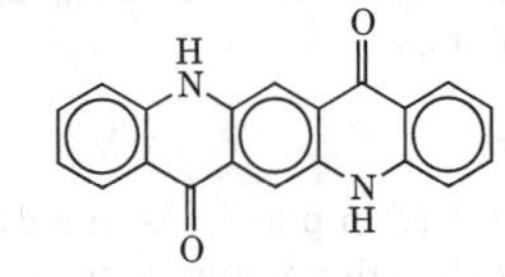

Pigment Violet 19

Practical methods for synthesis and elucidation of the optimum physical forms were developed at DuPont. The violets fill the void in the color gamut when the inorganics are inadequate. The quinacridones may be used except polymers such as nylon-6,6, polystyrene, and ABS.

Vat Pigments. A few of the many anthraquinone vat dyes have been exploited as pigments, first in automotive finishes and more recently in plastics. As a class, the vat pigments have good light and heat fastness and virtually no tendencies to migrate or bleed. However, they are expensive and are not stable under reducing conditions.

Perylene Pigments. The perylenes are a class of red and maroon pigments. These pigments are recommended for most plastic systems because of their excellent stability to chemicals, bleeding, and light.

Thioindigo Pigments. The thioindigos are red and violet pigments developed for textiles.

Phthalocyanine Pigments. Copper Phthalocyanine Blue and its related greens present the best combination of properties available in any pigment class. The bright, clean hues and outstanding fastness make them the colorants of choice in many media.

Tetrachloroisoindolinones. The tetrachloroisoindolines are yellow, orange, and red pigments marketed by CIBA-GEIGY under the trade name Irgazin. These pigments have excellent resistance to bleeding and light and are stable up to 290°C. They are recommended for ABS, polypropylene, and thermosets.

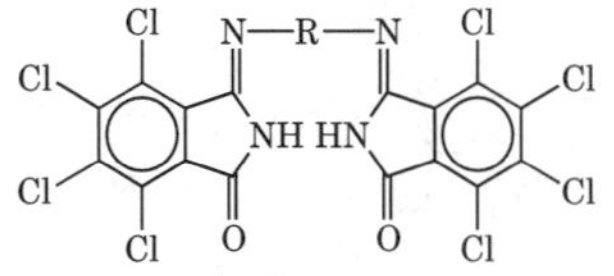

tetrachloroisoindolinones

Fluorescent Pigments. Fluorescent pigments or dyes depend on their ability to absorb light at one wavelength and to reemit it in a narrow intense band at a longer wavelength (see LUMINESCENT MATERIALS). The dyes used include the rhodamines, which emit pink, and aminonaphthalimides, which are bright greenish yellow. Traditional fluorescent dyes do not have lightfastness. Their use in plastics is confined to the lower temperature resins, vinyls, polyethylene, polystyrene, and acrylics at maximum temperatures of 200°C.

Dyes

Dyes should be checked for migration, sublimation, and heat stability before use. These precautions are particularly important for plasticized resins.

Azo Dyes. In general, the azo colors are useful for coloring polystyrene, phenolics, and rigin poly(vinyl chloride).

Anthraquinone Dyes. These dyes have much superior weatherability and heat stability compared with the azos, but at higher cost.

Xanthene Dyes. This class is best represented by Rhodamine B. It has high fluorescent brilliance but poor light and heat stability; it may be used in phenolics.

Azine Dyes. Azine dyes include induline and nigrosines. They produce jet blacks unobtainable with carbon black. The nigrosines are used in ABS, polypropylene, and phenolics.

Economic Aspects

The cost of the individual colorants plus the method of addition (concentrate, dispersion, or raw colorant) may be a significant portion of a colored part's costs. These costs often can be rapidly changing because of raw material availabilities.

Health and Safety Factors

The toxicity of colorants used in plastics is in a dynamic environment today. With the rapidly changing regulations, it would not be appropriate to identify the acceptabilities of particular colorants. The regulations regarding heavy-metal compounds represent the largest area of concern. As organic dye and pigment replacements are developed to replace heavy metals, there may be additional concerns regarding the decomposition of these chemicals with incineration (see DYES, ENVIRONMENTAL CHEMISTRY).

Traditionally, one of the biggest problems with colorants is their dustiness. Several producers offer low dusting or encapsulated products to improve this situation. Dyes may also sublime under operating conditions and show on mold surfaces.

GARY BEEBE
Rohm and Haas Company

Colour Index, and its Additions and Amendments, 3rd ed., Society of Dyers and Colourists, London, and American Association of Textile Chemists and Colorists, Durham, N.C. It now consists of seven volumes. The *Colour Index* was originally written by and for the textile industry, but pigments are receiving increasing attention.

P. A. Lewis, ed., *Pigment Handbook,* 3 vols., John Wiley & Sons, Inc., New York. Vol. 1: *Properties & Economics,* 2nd ed., 1988. Vol. 2: *Applications & Markets,* 2nd ed., 1988. Vol. 3: *Characterization & Physical Relationships,* 1973. John Wiley & Sons, Inc.

T. C. Patten, ed., *Pigment Handbook,* 2nd ed., 3 vols., Wiley-Interscience, New York, 1979. Written by a number of specialists in the pigment field; the three volumes describe pigment chemistry and applications.

K. Venkataramam, ed., *The Chemistry of Synthetic Dyes,* 8 vols., Academic Press, Inc., New York, 1951–1978. The most comprehensive treatise on dyes; emphasis is on chemistry, not applications.

COLOR PHOTOGRAPHY

Color photography is a technology by which the visual appearance of a three-dimensional subject may be reproduced on a two-dimensional surface having a pleasing balance of brightness, hue, and color saturation. The physical record of the image is expected to have reasonable permanence. It may be viewed directly as a color print, by projection as a color transparency (slide), or by back-illumination as a display transparency. There are two essentials in the practice of color photography: the camera and the film. The task of the camera is to present an undistorted image to the plane of the photographic film with an intensity level and exposure time appropriate to the sensitivity of the film being used. The technology of the light-sensitive film is discussed herein. (see PHOTOGRAPHY; COLOR PHOTOGRAPHY, INSTANT).

The Light-Recording Element

The primary element for light capture in color photography is the silver halide crystal. By common usage, the term emulsion has come to denote in the photographic literature what is actually a dispersion of silver halide crystals (grains) in gelatin. Four discrete steps can be identified in the formation of a colored image: *(1)* light absorption by the crystal; *(2)* the solid-state processes leading to the formation of a latent image; *(3)* the reduction of all or part of a crystal bearing latent image to metallic silver by a mild reducing agent; and *(4)* use of the by-products of the silver reduction to create a colored image in register by dye formation, dye destruction, or dye-transfer processes.

Sensitivity, or photographic speed, is one of the most important attributes of the light-sensitive element. Practical color photography using a handheld camera is possible from conditions of bright sunlight to night street lighting. These conditions span a factor of about 10^5 in illuminance, and must be accommodated for by the combination of camera shutter speed, lens aperture, and choice of film speed. Most commercial emulsions contain a population of silver halide crystals varying widely in size and shape. Although the structure of the AgBr and AgCl lattice is simple face-centered cubic, an enormous variety of crystal shapes can be obtained, depending on the number and orientation of twin planes present in the crystal, the silver ion concentration during growth, and the presence of growth modifiers. To add to this complexity, the crystals in commercial emulsions usually contain mixed-halide phases. Films suitable for a handheld camera generally contain silver bromoiodide, in which iodide ions are incorporated into the AgBr lattice during crystal growth.

The relationship between crystal size and photographic speed can be understood using simple geometric arguments. For an individual crystal, sensitivity may be defined as the reciprocal of the minimum light absorption required to generate a developable latent image. For a silver halide crystal without any sensitizing dye, blue light absorption is proportional to volume. If it is assumed that the crystal is a sphere and that the latent image can be formed with equal efficiency at all grain sizes, the relationship shown in Figure 1 is obtained. However, the adsorption of dyes is necessary to confer sensitivity to the green and red regions of the spectrum; this is frequently called "minus-blue" speed. To a first approximation, minus-blue speed depends on the surface area available for dye adsorption. Again, assuming sphericity, the line shown in Figure 1 is the expected change of minus-blue speed with crystal size. Even for the highest speed films, the crystals do not usually exceed 3 μm in linear dimension.

Over the years emphasis has been placed on obtaining greater uniformity in silver halide crystal size and habit in the grain population, in the belief that the chemical sensitization process can then yield a higher average imaging efficiency.

Sensitizing dyes are essential to the practice of color photography because they permit the separate color-forming layers to react to the

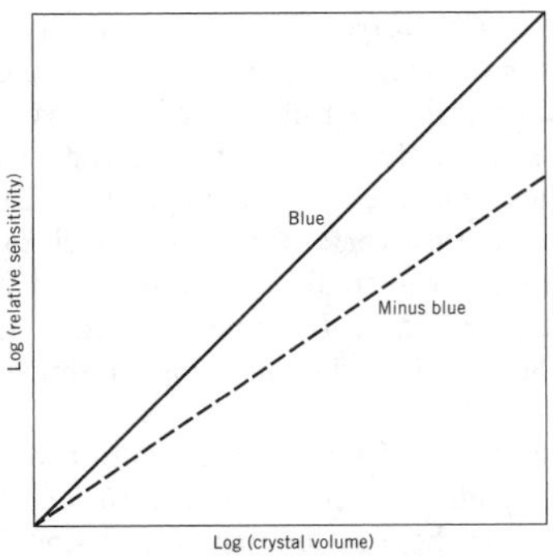

Figure 1. Calculated relationship between log (relative sensitivity) and log (crystal volume) for (—) intrinsic response (blue) and (----) dyed response (minus blue), assuming crystals are spherical.

corresponding wavelengths of light from the scene. The most widely used of these materials are the cyanine dyes, which are heterocyclic moieties linked by a conjugated chain of atoms. More than 20,000 dyes of this class have been synthesized. In the example shown in Figure 2, when $n = 0$, the dye is yellow and provides sensitization to blue light. Dyes which sensitize in the blue are particularly important for AgCl emulsions, which lack intrinsic blue sensitivity. When $n = 1$, the dye, a carbocyanine, is magenta and absorbs green light. For $n = 2$, a dicarbocyanine, the dye appears cyan and sensitizes silver halide to red light. Extending the conjugation further produces dyes which absorb in the infrared, producing films useful for aerial photography and thermal analysis.

There are several requirements for a good sensitizing dye. A good dye is adsorbed strongly to silver halide. A good sensitizing dye absorbs light of the desired wavelength range with high efficiency and efficiently transfers that energy to form the silver halide latent image. A good sensitizing dye also does not interfere with other system properties.

Color Processes

Additive Mixing. The first commercially successful color photographic systems were based on the additive primaries, red, green, and blue.

In the 1990s, the additive process is used in color television, in which light emitted from a tiny regular mosaic of red, green, and blue phosphors blends to give the colored image. Another modern additive color system is Polaroid's Polachrome 35-mm transparency film, which consists of a positive silver image overlying an additive screen having 394 triplets of red, green, and blue lines per centimeter of film. However, because additive photographic systems are inherently wasteful

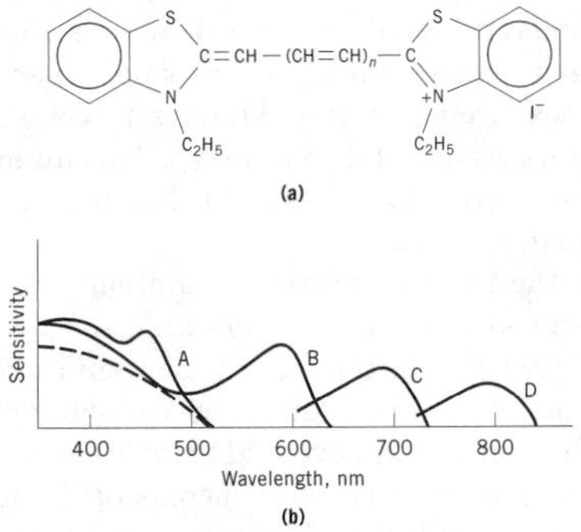

Figure 2. (**a**) Cyanine sensitizing dye structure. (**b**) Sensitivity curves (---) intrinsic silver halide; and (—) for the dye in (**a**) where A corresponds to $n =$ O; B to $n = 1$; C to $n = 2$ and D to $n = 3$.

of light (each additive filter absorbs two-thirds of the light energy), most modern systems rely on the subtractive primaries.

Subtractive Mixing. There are mechanical difficulties in separating a photographic image into three images to record red, green, and blue information, only to combine them later. Perfect registration of the color information can be preserved if the three-color records are stacked on top of each other on the same support. This film structure is known as an integral tri-pack. Photographic systems have been designed in which the three subtractive primary dyes, cyan, magenta, and yellow, can be formed in register, destroyed in register, or transferred in register to create the full-color image.

In the most widely used films, the uppermost image-forming layer contains silver halide sensitized to blue light along with a yellow-dye former or coupler. Below that is a layer of green-sensitized silver halide with a magenta-forming coupler and on the bottom a red-sensitized record containing a cyan-forming coupler. Between the blue and green records is a filter layer to remove blue light that would otherwise form latent images on the underlying records which retain their intrinsic blue-light sensitivity even where they are spectrally sensitized to green or red light.

There are two ways in which the developed tri-pack can be processed. One leads directly to a positive image; the other leads to a negative image that can be subsequently printed on a negative-working paper.

Chromogenic Chemistry

Developers. The detection and amplification of the latent image on the silver halide crystal occurs through the intervention of a mild reducing agent called a developer. The developer converts the silver halide to silver metal in an auto catalytic process with an amplification factor of about 10. In chromogenic development, the oxidized developer has the further function of reacting with the dye-forming coupler to produce the color image.

For modern color photographic systems, the developing agent of choice is an *N,N*-disubstituted *p*-phenylenediamine (PPD).

Practical developers must possess good image discrimination; that is, rapid reaction with exposed silver halide, but slow reaction with unexposed grains. Developing agents must also be soluble in the aqueous alkaline processing solutions. Some of the color developers in commercial use are CD-2, CD-3, and CD-4.

In addition to the developer, sulfite, and pH buffer, commercial developer solutions often contain antifoggants or restrainers that reduce the rate of development of unexposed silver halide relative to exposed grains. A common restrainer is the halide ion of the same type as in the film's silver halide emulsion.

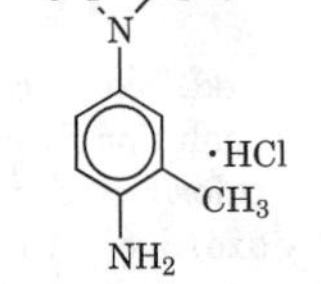

CD-2

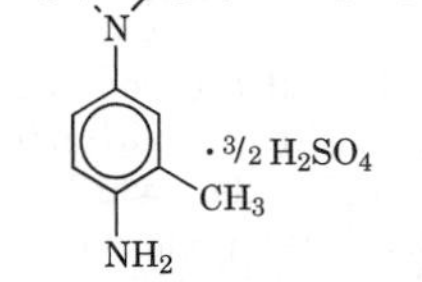

CD-3

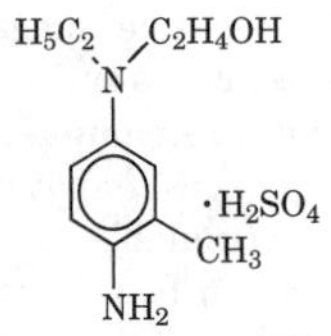

CD-4

Coupler Types. A photographic coupler is a weakly acidic organic compound that reacts with an oxidized *p*-phenylenediamine to produce a dye, usually one of the subtractive primaries, cyan, magenta, or yellow (Fig. 3). In addition to the dye-forming portion of the molecule, most couplers also bear an organic ballast, a long aliphatic chain or

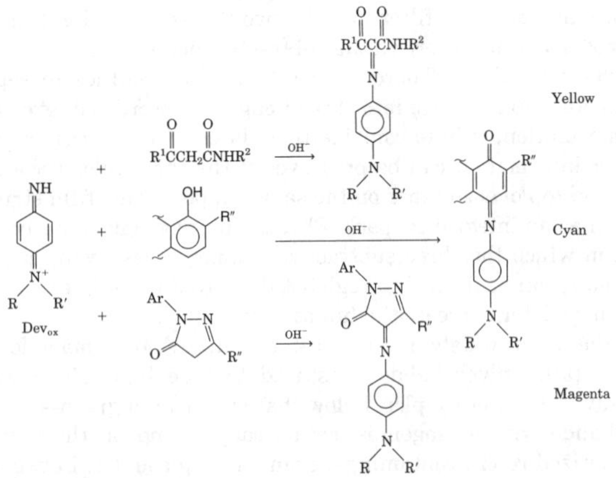

Figure 3. In each image-forming layer, developer oxidized by the exposed silver halide (Dev_{ox}) reacts with the appropriate coupler to form the corresponding subtractive primary dye, yellow, cyan, or magenta. Ar represents an aryl group and the various R's are undefined organic segments.

combination of aliphatic and aromatic groups. This allows the coupler to be suspended in droplets of a high boiling organic liquid, called the coupler solvent, which serves to anchor the coupler in its appropriate film layer.

Cyan Couplers. Substituted phenols and α-naphthols are the primary classes of cyan dye-forming couplers.

Yellow Couplers. The most important classes of yellow dye-forming couplers are derived from β-ketocarboxamides, specifically the benzoylacetanilides and the pivaloylacetanilides.

Magenta Couplers. For many years the most widely used magenta couplers have been derived from the 1-aryl-2-pyrazolin-5-ones. Substituents in the aryl ring or at the 3-position have been used to alter dye hue and stability as well as to control coupler reactivity. More recently, pyrazolo-(3,2,-c)-5-triazoles and related isomers have appeared in color films and papers.

Colored Masking Couplers. The dyes produced in chromogenic development have unwanted absorptions. For example, the cyan dye is expected to control or modulate red light alone and thus should absorb only between 600 and 700 nm, but it shows lesser absorptions in the blue (400–500 nm) and green (500–600 nm) regions as well. Thus exposure of the red-sensitive layer of the film produces not only the desired density to red light in the negative, but also undesirable densities to blue and green light, resulting in desaturation or "muddying" of the color.

For materials that are not directly viewed, like a color negative film, masking couplers provide an ingenious solution. Unlike a normal cyan dye-forming coupler, which is itself colorless, a cyan masking coupler bears a colored, preformed (usually azo) dye in the coupling-off position. The hue of this dye is chosen to match the unwanted blue-green absorptions of the cyan dye that is generated. When coupling occurs, the preformed dye is released and washes out of the film or is destroyed. The result is a negative image formed by the cyan dye with its unwanted absorptions and an entirely complementary positive image left by the preformed dye remaining on the residual coupler.

DIR Couplers. Masking couplers cannot be used for directly viewed materials because of the objectionable color of the mask itself. But similar advantages and more can be achieved by using development-inhibitor-releasing (DIR) couplers. These materials are usually image couplers that carry a silver development inhibitor (In) linked directly or indirectly to the coupling site. When released as a function of dye formation, the development inhibitor or its carrier migrates to the silver halide grain to either slow or stop further development. In addition to correction of unwanted dye absorptions, DIR couplers can be used to improve sharpness and reduce granularity of films.

Post-Development Chemistry. The silver and silver halide remaining in the film after color development must be removed, both to improve the appearance of the color image and to prevent the appearance from changing as silver halide is slowly photoreduced to silver metal. This is generally accomplished in two steps. The first, called bleaching, is an oxidation that converts silver metal back to silver salts. The second, called fixing, is the solubilization of the silver salts by complexation with a silver ligand. In some processes, particularly those used for color paper, the two steps can be combined in a single step called a bleach-fix or blix.

The most common color film bleaches contain a ferric complex as the oxidant and a halide ion, often bromide, to complex the silver ion being formed and drive the reaction to its conclusion. Thiosulfate, usually as its sodium or ammonium salt, is almost universally employed as the fixing agent for color films.

Film Quality

Speed. Standards for photographic speed are now coordinated worldwide by the International Standards Organization (ISO) in Geneva, Switzerland. ISO speed is determined under specified conditions of exposure, processing, and measurement. Standards are published for color negative and color reversal films. In amateur color photography, the most popular film speeds are in the ISO 100–400 range.

Color Reproduction. Color has three basic perceptual attributes: brightness, hue, and saturation. Saturation relates to how much of the hue is exhibited. The primary influences on the color quality of the final image are the red, green, and blue spectral sensitivities of the film or paper, and the spectral absorption characteristics of the image dyes.

Dye Stability. The dyes used in photographic systems can degrade over time, both by thermal reactions and, if the image is displayed for extended periods of time, by photochemical processes. The relative importance of these two mechanistic classes, known as dark fade and light fade, respectively, depends on how the product is to be used.

Modern photographic products have stabilities vastly improved over those of the past. For example, the magenta light stability of color negative print papers has improved by about two orders of magnitude between 1942 and the present.

Image Structure. Because the primary photographic sensors are a population of silver halide crystals having a random spatial distribution, the final image is also particulate. In chromogenic photography using incorporated couplers, the image is formed by the coupler-containing hydrophobic droplets dispersed with the silver halide. The droplet size, typically 0.2 μm in diameter, is usually less than the crystal size. Dye is formed in a cloud of droplets around each developing crystal as oxidized developer is released. Individual droplets cannot be resolved under usual viewing conditions, but the dye clouds can be seen under magnification and convey a visual sensation of nonuniformity or graininess. The objective correlate of graininess is granularity, which is the spatial variation in density observed when numerous readings are taken on a uniformly exposed patch using a densitometer having a very small aperture. The distribution of such measurements approximates to Gaussian and can be characterized by its standard deviation, SD. This quantity, called root mean square (RMS) granularity, is published by film manufacturers as a figure of merit.

Sharpness is the ability to discriminate edge detail in the final image. The sharpness of a film is often assessed by the modulation transfer function (MTF), which measures how sinusoidal test patterns of different frequencies are reproduced by the photographic material. A perfect reproduction would have a MTF of 100% at all frequencies. In practice, a decrease of MTF with increasing frequency occurs as a result of optical degradation. Films having couplers that release development inhibitors can display MTF values above 100% at low frequencies, because of edge effects that increase the output signal amplitude beyond that of the reference signal.

Environmental Aspects

Photographic processing, ie, photofinishing, is a geographically dispersed chemical industry. Processing machines are rarely emptied and refilled with fresh solutions, but do require replenishment because of chemical use, evaporation, and overflow. New films have been designed that make more efficient use of coated silver and thus require less in terms of the chemicals for processing. The solution overflow that does occur is usually chemically regenerated and returned to the tank. During the past decade, some components in the effluent have been reduced to near zero, whereas an overall reduction in effluent concentration of as much as 80% has been achieved.

Economic Aspects

The number of photographs taken annually in the United States now exceeds 16 billion. In 1989, the breakdown was 14.9 billion color prints, 0.8 billion color slides, and 0.4 billion black-and-white prints. The 1980s saw an increasing dominance of 35-mm color negative at the expense of color slide, because of its greater exposure latitude, the convenience of viewing reflection prints, and the ease of obtaining duplicate prints. The growth in popularity of picture-taking has been fueled by the increasing sophistication of 35-mm point-and-shoot cameras. These offer features such as auto-exposure, auto-focus, motor drive, and built-in flash, all within a very compact camera body. The 35-mm single-use camera in which the film box itself is equipped with a lens and shutter has also been very successful.

The photofinishing industry has been growing at more than 5% annually. Since the mid-1970s a shift has occurred to favor local minilabs over large centralized processing laboratories. Although minilabs are generally more expensive, consumers appreciate the convenience, rapid-access, and personal service. Within this environment, hardware is now available to enable the customer to personally zoom and frame selected areas of the negative for enlargement.

The professional segment of color photography includes portrait and wedding photography, advertising photography, and photojournalism. Products tailored to these markets are offered by the principal manufacturers. The motion picture industry enjoyed continued growth in the 1980s. In spite of video cassette recorders, there continues to be strong demand for the theatrical experience of first-run movies. This experience is being enhanced by the improved picture quality of 70-mm origination and projection. Color negative film is the preferred medium for prerecorded television shows; telecine transfer converts the images to electronic form for transmission.

Although camcorders have displaced home movie films, electronic still photography as a viable consumer product appears still to be some distance in the future. Besides poorer picture quality, the equipment cost and the relative lack of convenient hard copy has so far limited this technology to the enthusiast. With the opening of previously closed large foreign markets such as Eastern Europe and China, where there is an unsatisfied demand for high quality personal imaging, the prospects for continued growth in conventional color photography, with its relatively low initial investment, appear excellent.

JON KAPECKI
JAMES RODGERS
Eastman Kodak Company

R. W. G. Hunt, *The Reproduction of Colour,* 5th ed., Fountain Press, UK, 1995.

T. H. James, ed., *Theory of the Photographic Process,* 4th ed., Macmillan Publishing Co., New York, 1977.

J. Sturge, V. Walworth, and A. Shepp, eds., *Imaging Processes and Materials—Neblette's Eighth Edition,* Van Nostrand Reinhold, New York, 1989.

P. Kowaliski, *Applied Photographic Theory,* John Wiley & Sons, Inc., New York, 1972.

COLOR PHOTOGRAPHY, INSTANT

The term "instant color photography" originally referred to one-step processes that provide finished color photographs within a minute after exposure of the film. The processing of each film unit is initiated in the camera immediately after the film has been exposed. Processing proceeds rapidly under ambient conditions. Such processes are outwardly dry, and the reagent is usually provided as a part of the film unit. The reagent is applied by mechanical action of the camera, film holder, or other processing device (see also COLOR PHOTOGRAPHY; PHOTOGRAPHY).

The technology of instant photography has been extended to include the application of similar chemistry to films wherein processing is delayed for a time after exposure. Whereas the multistep darkroom processes used for noninstant color films require precise time and temperature control, the instant processes require little or no timing and operate over a wide range of temperatures.

Principles of Instant Photography

Film and Process Design. Handheld camera use requires a film of sufficient photographic speed to permit short exposures at small apertures. Silver halide emulsions have high sensitivity and, upon development, enormous amplification. The original system involved a one-step process in which a viscous reagent was spread between two sheets, one bearing an exposed silver halide emulsion and the other an image-receiving layer. Both sheets were drawn through a pair of pressure rollers, and a sealed pod attached to one of the sheets ruptured to release the viscous reagent, which was spread to form a thin layer between the two sheets, temporarily bonding them together. The action of the reagent produced concomitantly a negative image in the emulsion layer and a positive image in the image-receiving layer. After about a minute, the two sheets were stripped apart to reveal the positive image.

Reagents for Instant Photography. An essential component of each instant film is the reagent system. Essentially dry processing is realized by using a highly viscous fluid reagent and restricting the amount to just that needed to complete the image-forming reaction for a single picture.

The high viscosity of the reagent is provided by water-soluble polymeric thickeners. Suitable polymers include hydroxyethyl cellulose, the alkali metal salts of carboxymethyl cellulose and carboxymethyl hydroxyethyl cellulose.

In addition to high molecular weight polymer, reducing agents, and alkali, the viscous reagent may contain reactive components that participate in image formation, deposition, and stabilization. Some reactants may also be incorporated in coatings (qv) on either of the two sheets.

The reagent-containing pod must be carefully designed for both containment and discharge of its contents (Fig. 1).

One-Step Cameras and Processors. The earliest one-step cameras used roll film and completed processing inside a dark chamber within the camera. The first instant color film, Polacolor, was provided in roll film format to fit these cameras. Flat-pack film cameras (Fig. 2), introduced in 1963, permitted the film to be drawn between processing rollers and out of the camera before processing was completed. Film holders for instant 10 × 13 cm(4 × 5 in.) film packets contain retractable rollers that permit the film to be loaded without rupturing the pod. For 20 × 25 cm(8 × 10 in.) films, the processing rollers are part of a tabletop processor. The exposed film, contained in a protective black envelope, and a positive sheet with pod attached are inserted into separate slots of a tray that leads into the processor. The film passes through the rollers into a covered compartment within which processing is completed.

Fully automatic processing was introduced in 1972 with the SX-70 camera, which ejected each integral picture unit automatically, passing it between motorized processing rollers and out of the camera immediately after exposure. Kodak instant cameras, introduced

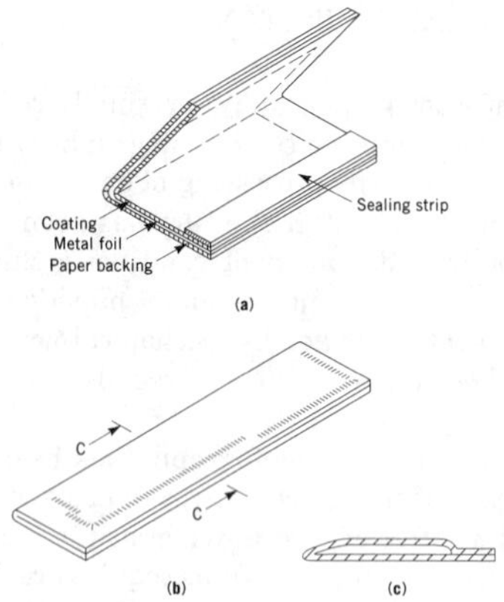

Figure 1. Patent illustration showing a typical pod structure. (**a**) Unfilled pod; (**b**) filled, sealed pod; (**c**) cross-section of filled pod along line C–C in (**b**).

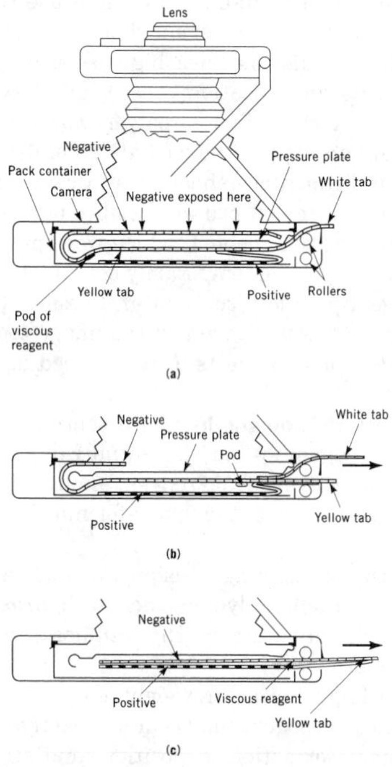

Figure 2. Schematic section of Model 100 camera (1963), illustrating processing of pack film outside the camera. The outer surfaces of both negative and positive sheets are opaque. (**a**) Position of each element during exposure; (**b**) following exposure, pulling the white paper tab leads the negative into position for processing; (**c**) pulling the yellow paper tab draws the entire film unit between pressure rollers, rupturing the pod and spreading the reagent, and leads the film unit out of the camera.

in 1976 and now discontinued, included both motorized and hand-cranked models. Fuji instant cameras for integral films are motorized.

Film Configuration. Using peel-apart materials, two separate sheets are laminated together as reagent is spread at the start of processing. The sheets are then peeled apart to terminate processing and permit viewing of the image. An integral print film comprises two sheets permanently secured as a single unit. The image-forming layers are located on the inner surfaces of the two sheets, at least one of which is transparent. The image is usually viewed through the transparent sheet against a reflective white pigment layer within the film unit.

Integral films are designed to provide exposure of the emulsion layers and viewing of the print either through the same surface or through opposite surfaces of the film unit. Polaroid integral color film units are exposed and viewed through the same surface, and correct image orientation is obtained using a single mirror reversal within the camera, an arrangement chosen for compactness in optical design. Fuji integral films, like the earlier Kodak integral films, are exposed and viewed through opposite surfaces, so that image orientation is correct either without optical reversal in the camera or with two mirror reversals.

Formation of Instant Images. A series of complementary positive and negative images, one or more of which can serve as a starting point for a transfer process that leads to a useful positive image, includes exposed grains, unexposed grains; developed silver, undeveloped silver halide; oxidized developer, unoxidized developer; neutralized alkali, alkali not neutralized; hardened gelatin, unhardened gelatin.

In instant color processes that utilize dye-developers, that is, molecules that are both image dyes and photographic developers, an image in terms of unoxidized dye-developer transfers to form a positive dye image in the image-receiving layer. In dye-release systems both negative-working and positive-working emulsions have been used. Dye release is effected by one or more of the initial images in terms of silver or developer.

Methods of Color Reproduction

The reproduction of color requires the selective recording and presentation of principal regions of the visible spectrum, which extends roughly from 400 to 700 nm. For most processes three records are used, corresponding to blue (400–500 nm), green (500–600 nm), and red (600–700 nm) (see COLOR PHOTOGRAPHY).

In subtractive color photography, the three color records are formed in separate silver halide emulsions sensitive to blue, green, and red light, and coated in a multilayer structure. Processing produces positive images in terms of complementary dyes: the blue record is transformed to an image in yellow, or minus blue, dye; the green record to an image in magenta, or minus green, dye; and the red record to an image in cyan, or minus red, dye. When white light passes through the superposed set of yellow, magenta, and cyan images, the dyes absorb blue, green, and red components of the white light so that the light that is not absorbed represents the original colors.

In additive color photography, the three color records are separated laterally by an array, or screen, of blue, green, and red filter elements superposed over a panchromatic silver halide emulsion. That is, there is an emulsion sensitive to blue, green, and red light. The exposed emulsion is processed to form a positive transparency comprising black-and-white records in silver of the respective blue, green, and red components of the original scene. When these silver images are projected in registration with the superposed color screen, the images add to reproduce the full color image. Additive color photography is used only for transparencies because the minimum density is too high for reflection prints.

Both subtractive and additive color reproduction are utilized in instant color films. Subtractive systems include all of the instant print and large format transparency materials except Polachrome 35-mm slide films, which are additive.

Dye Developer Processes. The first instant color film, Polacolor, introduced the dye developer, a bifunctional molecule comprising both a preformed image dye and a silver halide developer. The multilayer Polacolor negative comprises a set of three negative-working emulsions, blue-, green-, and red-sensitive, respectively, each overlying a layer containing a dye developer complementary in color to the emulsion's spectral sensitivity. During processing, development of exposed grains in each of the emulsion layers results in oxidation and immobilization of a corresponding portion of the contiguous dye developer. Dye developer that is not immobilized migrates through the layers of the negative to the image-receiving layer to form the positive image.

In addition to having suitable diffusion properties, dye developers must be stable and inert in the negative before processing. After completion of the process, the dye developer deposited in the image-receiving layer must have suitable spectral absorption characteristics and stability to light. The requirements of a developer moiety for incorporation into a dye developer are well fulfilled by hydroquinones.

Although as of this writing none of the commercialized films incorporate color-shifted dye developers, such compounds have been investigated extensively. These materials offer the option of incorporating the dye developer and the silver halide in the same layer without losing speed through the unwanted absorption of light by the dye developer.

Other types of dyes that have been studied as chromophores in dye developers include rhodamine dyes, azamethine dyes, indophenol dyes, and naphthazarin dyes. Cyanine dyes, although not generally stable enough for use as image dyes, have also been incorporated in dye developers.

Dye-Release Processes. Dyes or dye precursors that do not participate directly in the development process and are initially immobile or of low mobility in alkali may be released by agents generated imagewise during development. One of the advantages of such a process is that the released species may themselves be unreactive with respect to other components of the negative. Dyes may thus diffuse through layers of the negative to reach the image-receiving layer without undergoing unwanted reactions.

Dye release may relate either directly or inversely to the image-related reduction of silver halide. Release of the dye or dye precursor may be accomplished or initiated by the oxidized developing agent or the unoxidized developing agent or by alkali or silver salts.

Stability of Dye Images. Unlike the dye images produced by bath processes, the instant image is formed in a single processing step and is not washed or treated afterward. Therefore the instant process must incorporate both image-formed reactions and reactions that provide a safe environment for the finished image.

The subtractive instant color films provide for image stabilization by the inclusion of polymeric acid layers that operate in conjunction with timing layers to produce a carefully timed reduction in pH following image formation. Reducing the pH terminates development reactions and stops the migration of dyes and other alkali-mobile species, thus stabilizing both the dye image and its environment.

Another concern is the stability of the image dyes to prolonged exposure to light. Many dyes that would otherwise be suitable as image dyes undergo severe degradation upon such exposure and are particularly sensitive to uv irradiation. Important considerations include the dye structures and the structure of the mordant in the image-receiving layer, as well as the chemical environment following the termination of processing.

Subtractive Instant Color Films

Table 1 provides a summary of instant color camera-speed films introduced prior to January 1993. Instant color reprographic films in this period were Ektaflex PCT negative (Kodak), for printing from color negative film; Ektaflex PCT reversal (Kodak), Copycolor CCN (Agfa-Gevaert), Agfachrome-Speed (Agfa-Gevaert), and Colorcopy DC (Fuji), for printing from color prints or transparency films; and Fujix Pictrography 1000 (Fuji), for printing from electronic records. Of these reprographic films, only the Copycolor CCN, Colorcopy DC, and Pictrography films are commercially available as of this writing.

Polacolor. The first instant color film, Polacolor, was introduced by Polaroid Corporation in 1963. Polacolor was replaced in 1975 by Polacolor 2, a film with improved light stability, which utilized metallized dye developers. An extended range version, Polacolor ER, introduced in 1980, utilizes the cyan and yellow metallized dye developers together with a magenta dye developer that incorporates a xanthene dye having reduced blue absorption.

The Polacolor process produces subtractive multicolor prints comprising positive images in terms of yellow, magenta, and cyan dye developers. The dye developers form the positive image concomitantly with the formation of negative silver images and negative dye images within the layers of the negative. Image formation is based on the immobilization of dye developers by oxidation in areas where exposed silver halide grains are developed and on the diffusion of dye developers from areas that are unexposed.

Table 1. Instant Color Films

Classification, film configuration[a]		ISO speed
Subtractive color films		
Polaroid		
peel-apart print films	Polacolor 64T	64
	Polacolor[b] (types 38, 48, 58, 88, 108, 636)	75
	Polacolor 2 (types 58, 88, 108, 668, 808)	80
	Polacolor ER (types 59, 559, 669, 809)	80
	Polacolor 100	100
	Polacolor Pro 100	100
peel-apart transparency	Colorgraph (types 691, 891)	80
integral print films	SX-70	150
	Time-Zero, type 778	150
	600b[b], 600 Plus, type 779	640
	Autofilm 339	640
	Spectra, type 990	640
	Vision 95[c]	600
Kodak		
integral print films	PR-10[b]	150
	Instant Color[b]	150
	Kodamatic[b]	320
integral/peel print	Trimprint[b]	320
integral/peel transparency	Instagraphic[b]	64
Fuji		
integral print films	FI-10	150
	FI-800, FI-800G, FI800GT	800
peel-apart print film	FP-100	100
Additive color films		
Polaroid: additive color screen; silver transfer process		
Super-8 film	Polavision[b]	40
35-mm films	Polachrome CS	40
	Polachrome HC	40

[a] Polaroid film type, where shown, denotes format as well as photographic characteristics, eg, 50 series are 10 × 13 cm sheet films; image area 8.9 × 11.4 cm. 80 series are 8.3 × 8.1 cm pack films; image area 7 × 7.3 cm. 100, 660, and 690 series are 8.3 × 10.8 cm pack films; image area 7.3 × 9.5 cm. 300 series are 11.4 × 10.8 cm integral pack films; image area 10 × 7.5 cm. 500 series are 10 × 13 cm pack films; image area 8.9 × 11.7 cm. 800 series are 20 × 25 cm sheet films; image area 19 × 24 cm. Polacolor 100 and Polacolor Pro 100 are pack and sheet films of the same formats as Polacolor ER. SX-70, Time-Zero, 600, 600 Plus, and 700 Series are 8.9 × 10.8 cm; integral pack films; image area 8 × 8 cm. Spectra and Type 990 are 10.8 × 11.4 cm integral pack films; image are 7.2 × 9.1 cm. Vision 95 is 6.4 × 11 cm integral pack film; image area 2.29 × 5.46 cm. Fuji Films include both 8.3 × 10.8 cm packs, image area 7.3 × 9.5 cm, and 10 × 12.7 cm packs, image area 8.5 × 10.8 cm.

[b] Discontinued films.

[c] Vision 95 designates film introduced in Europe in 1992; it may have a different designation when introduced in the United States.

Polaroid Integral Films. In 1972 the SX-70 automatic camera and integral film system were introduced. The SX-70 film provided images that required no timing and no peeling apart. Each film unit was ejected through processing rollers immediately after exposure. The entire development process took place within the film unit under ambient conditions.

The integral film format required different processing chemistry and film components from those used previously. The processed film

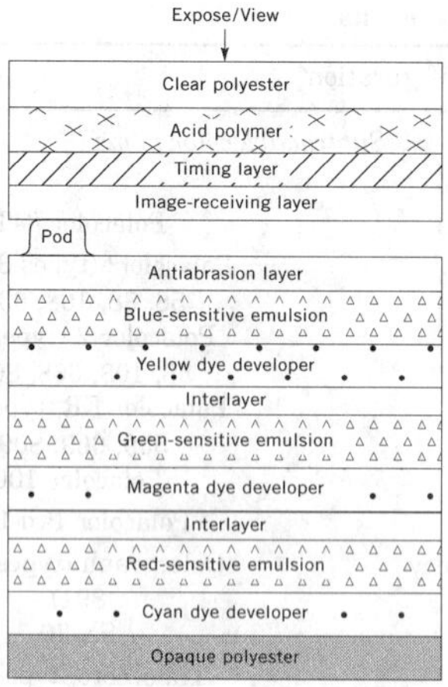

Figure 3. Schematic cross section of SX-70 integral film unit. The film is exposed through the clear upper sheet. When the reagent from the pod is spread, it forms a white pigmented layer. The image formed in the image-receiving layer is viewed against the reflective pigment layer. Courtesy of Van Nostrand Reinhold.

unit needed to contain all of the reaction products along with the final color image. Polaroid followed the original SX-70 film with Time-Zero SX-70 film (1979), 600 film (1981), Spectra film (1986), and 600 Plus film (1988).

The SX-70 picture unit is an integral multilayer structure having no air spaces after processing. The sequence of dye developers, silver halide emulsions, and other layers is shown schematically in Figure 3.

Kodak Instant Films. Kodak entered the instant photography market in 1976 with the introduction of PR-10 integral color print film, rated at ISO 150, and a series of instant cameras designed for this film. These films and cameras were incompatible with the Polaroid instant photographic systems. Later Kodak integral films included Kodak Instant Color Film (ISO 150), Kodamatic and Kodamatic Trimprint films (ISO 320), Instagraphic print film (ISO 320), and Instagraphic color slide film (ISO 64). The films were balanced for daylight exposure. Kodak discontinued the production of instant films and cameras in 1986.

Fugi Instant Films. The first Fuji instant system, Fotorama (1981), provided an integral color film, FI-10, rated at ISO 150, and cameras compatible with the Kodak instant system. The film is similar in structure and processing to Kodak PR-10 film but has thinner emulsion layers, as well as a different set of dye-releaser compounds. The thinner negative may account for the observation that the FI-10 image appeared to reach completion more quickly than the PR-10 image, although the faster image completion may also relate to the specific dyes and dye-releaser compounds used.

A high speed integral color film, FI-800 (ISO-800), and a new series of cameras suited to this higher speed, were introduced in 1984. The FI-800 film structure includes spacer layers that separate each of the light-sensitive emulsions from the dye-releaser layer that it controls.

In 1984 Fuji introduced FP-100, a peel-apart instant color film rated at ISO 100. The FP-100 system uses a dye-release process similar to that used in the Fuji integral films.

Reprographic Films. Both Kodak and Agfa have marketed reprographic films based on dye release and transfer, using liquid activators in film processors.

Photothermographic Color Films. Several photothermographic materials fall within the speed class of reprographic films.

Redox dye-release chemistry similar to that used in the Fuji instant films was utilized with negative-working emulsions in the photothermographic Fujix Pictrography 1000 film (1987). This film is designed to transform digital data into three-color prints. Film layer sensitivity is matched to the output of a light-emitting diode (LED) printer, rather than the more conventional blue, green, and red spectral regions (see LIGHT GENERATION–LIGHT-EMITTING DIODES).

A positive-working photothermographic material for reprography, designed to produce either color prints or color transparencies, has been described. The thermal development technology is similar to that used for Fujix Pictrography, but with 15-s processing at 80°C. The Colorcopy DC film incorporates a set of negative-working emulsions sensitive to blue, green, and red light, respectively, along with dye releasers of the ROSET type, an electron donor, and an electron-transfer agent. The interlayers of the donor film contain a development-inhibitor-releasing (DIR) compound, described as Hyper-DIR, which is said to function very rapidly, providing enhanced color saturation. The dye releasers undergo reduction and release dye in unexposed areas, thus providing positive transfer images.

The Colorcopy DC film is used in a Fuji Colorcopier DC3000, which is designed to photograph reflection copy, transparencies, or solid objects and to produce either color prints or transparencies for overhead projection.

A dry photothermographic system based on dye transfer from a multilayer negative donor sheet to a hydrophobic receiving sheet, using only materials contained within the two sheets, has been described (Konica). The negative sheet components include silver halides, organic silver salts, couplers immobilized on a polymer, a color developer precursor, and solvents. Following exposure the two sheets are laminated together and heated to activate the processing. The dyes formed by color coupling are hydrophobic, low molecular-weight compounds that are displaced from the polymer and diffuse readily into the receiving sheet. The process is negative-working, the dye released in exposed areas in proportion to the amount of oxidized color developer generated.

The development of thermally processed dry silver color materials is based on the use of photosensitive silver halide–silver behenate layers. The multilayer film comprises red-, green-, and blue-sensitized layers, each including the precursor of an appropriate dye, and barrier layers to prevent cross talk between the image-forming layers. The process is negative-working; dye images are formed *in situ* in exposed regions. Both transparencies and reflection prints may be produced. Processing is similar to that of 3M black-and-white dry silver products, for example, 10 s at 132°C.

Instant Additive Color Films

Polaroid introduced Polavision, a Super-8-mm instant motion picture system, in 1977. Polachrome CS 35-mm slide film followed in 1982, and a high contrast version, Polachrome HCP, appeared in 1987. Each of the films comprises a very fine additive color screen and an integral silver image transfer film. The Polavision system, which included a movie camera and a player that processed the exposed film and projected the movie, is no longer on the market. The Polavision film was provided in a sealed cassette, and the film was exposed, processed, viewed, and rewound for further viewing without leaving the cassette.

Polachrome is provided in standard-size 35-mm cassettes. Its processing is carried out in a Polaroid Autoprocessor using a processing pack that contains a reagent pod and a strip sheet.

Polachrome Films. The Polachrome film structure and the formation of a Polachrome color image employ a color screen separated from the image-receiving layer by an alkali-impermeable barrier layer.

Economic Aspects

Although the cost per print for instant color prints is greater than for conventional film, there is significant value in the immediate availability of instant prints and the fact that it is not necessary to process a whole roll when only a few images are desired.

The technical and industrial markets for instant photography continue to grow in diversity. In 1991 such applications accounted for approximately one-third of the worldwide sales, 60% of which were outside the United States.

Health and Safety Factors; Toxicology

No toxicological hazards have been associated with the normal use of instant color films. However, direct contact with the highly alkaline processing fluids can cause alkali burns.

As solid-waste management becomes a more critical concern, there is increasing attention to materials that are incinerated or added to landfills. The negatives discarded from peel-apart color films include silver, although at much lower concentrations than in conventional color films, and there are also small amounts of chromium and copper; the pod includes a layer of aluminum foil. The zinc–carbon Leclanché batteries provided in Polaroid's integral film packs at one time contained a small concentration of mercury, but its use has been eliminated. As of this writing, some used film components and batteries are being recycled.

Uses

Instant film formats and corresponding apparatuses have been developed to fit a variety of specialized needs. Many laboratory instruments and diagnostic machines include built-in instant-film camera backs (see ANALYTICAL METHODS; AUTOMATED INSTRUMENTATION; IMAGING TECHNOLOGY). Digital film recorders produce color prints, slides, or overhead transparencies from computer output and from cathode-ray tube displays. There are also direct screen cameras for photographing still cathode ray tube (CRT) displays. Polaroid Freeze-Frame video image recorders provide color prints and transparencies from a variety of video sources, including VCRs, laser discs, and video cameras.

Instant photographs are widely used for identification purposes, for example, for drivers' licenses, student identification cards, and credit cards that can be issued immediately.

In studio photography, instant color slides exposed simultaneously with conventional color films are used to provide proofs that can be projected immediately for viewing by the customer. Professional photographers also use instant films as proof material to check composition and lighting. Large format Polacolor films are often used directly for exhibition prints.

Instant color film has a unique application in the generation of images by transfer from a partially developed Polacolor negative to a material other than the usual receiving sheet, eg, plain paper, fabric, or vellum. The transferred image may then be modified for artistic purposes by reworking with watercolors or other dyes. The final images, generally known as transfer images, may be displayed directly or reproduced for commercial use.

In the instant reprographic field, Copycolor materials are used extensively in Europe but are not distributed in the United States. Principal markets are in seismological charts and maps for the oil industry, mapmaking, and reproduction of large graphs, charts, and engineering drawings. The films are also used for small color stats and for position proofs in layout work.

VIVIAN K. WALWORTH
STANLEY H. MERVIS
Polaroid Corporation

E. H. Land, H. G. Rogers, and V. K. Walworth, in J. M. Sturge, ed., *Neblette's Handbook of Photography and Reprography,* 7th ed., Van Nostrand Reinhold, New York, 1977, pp. 258–330.

V. K. Walworth and S. H. Mervis, in J. Sturge, V. Walworth, and A. Shepp, eds., *Imaging Processes and Materials: Neblette's Eighth Edition,* Van Nostrand Reinhold, New York, 1989, pp. 181–225.

S. Fujita, K. Koyama, and S. Ono, *Nippon Kagaku Kaishi* **1991,** 1 (1991).

E. H. Land, *Photogr. Sci. Eng.* **21,** 225 (1977).

S. Fujita, K. Koyama, and S. Ono, *Rev. Heteroat. Chem.* **7,** 229–267 (1992).

COMBUSTION SCIENCE AND TECHNOLOGY

Fuel combustion is a complex process, the understanding of which involves knowledge of chemistry (structural features of the fuel), thermodynamics (feasibility and energetics of the reactions), mass transfer (diffusion of fuel and oxidant molecules), reaction kinetics (rate of reaction), and fluid dynamics of the process. Therefore, the design of combustion systems involves utilizing information and data generated in a range of disciplines. Often, the design of practical combustion systems is based on experience rather than on fundamental mechanistic understanding. However, for certain fuels such as methane, the combustion mechanism is better understood than for other more complex fuels such as coal. To accommodate the variety of approaches used to solve practical problems, this article is divided into two subsections: combustion science and combustion technology.

Combustion Science

Higher Heating Value. The higher heating value of a fuel is the heat of combustion at constant pressure and temperature (usually ambient) determined by a calorimetric measurement in which the water formed by combustion is completely condensed.

FAR or AFR. The composition of a mixture of fuel and air or oxidant is often specified according to the Fuel to Air Ratio (FAR), and can be expressed on a mass, molar, or volume basis.

Flammability Limits. Any given mixture of fuel and oxidant is flammable (explosive) within two limits referred to as the upper (rich) and lower (lean) limits of flammability (Fig. 1).

Flash Point. The flash point is the lowest temperature at which vapor, given off from a liquid, is in sufficient quantity to enable ignition to take place.

Ignition. To understand the phenomenon of ignition it is necessary to consider the following concepts: ignition source, gas temperature, flame volume, and presence of quench wall surfaces. In general, there are two main methods of igniting a flammable mixture. In the self-ignition method, the mixture is heated slowly so that the vapor released as the temperature is raised ignites spontaneously at a particular temperature. In the forced ignition method, a small quantity of combustible mixture is heated by an external source and the heat released during the combustion of this portion results in propagation of a flame. The external ignition source can be an electric spark, pilot flame, shock wave, etc. For ignition to take place the following conditions should be satisfied: (*1*) the amount of energy supplied by the ignition source should be large enough to overcome the activation energy barrier; (*2*) the energy released in the gas volume should exceed the minimum critical energy for ignition; and (*3*) the duration of the spark or other ignition source should be long enough to initiate flame propagation, but not too long to affect the rate of propagation. Ignition models fall into two categories: the thermal model explains the ignition as resulting from supplying the mixture with the amount of heat sufficient to initiate reaction. In the chemical diffusion model the main role of the ignition source is attributed to the formation of a large number of free radicals in the preheat zone, where their diffusion to the surrounding region initiates the combustion process. The thermal model

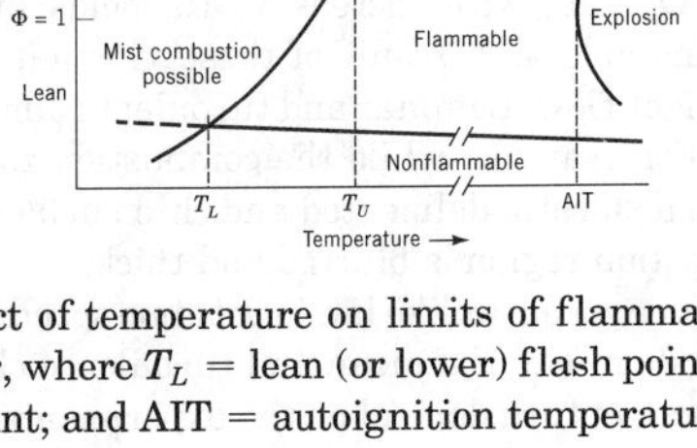

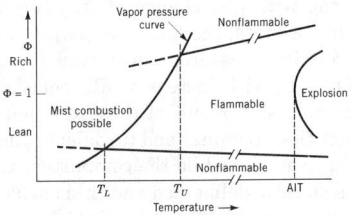

Figure 1. Effect of temperature on limits of flammability of a pure liquid fuel in air, where T_L = lean (or lower) flash point; T_U = rich (or upper) flash point; and AIT = autoignition temperature.

is applied more widely in the literature and shows better agreement with experimental data.

Cool Flames. Under particular conditions of pressure and temperature, incomplete combustion can result in the formation of intermediate products such as CO. As a result of this incomplete combustion, flames can be less exothermic than normal and are referred to as cool flames. An increase in the pressure or temperature of the mixture outside the cool flame can produce normal spontaneous ignition.

Flame Temperature. The adiabatic flame temperature, or theoretical flame temperature, is the maximum temperature attained by the products when the reaction goes to completion and the heat liberated during the reaction is used to raise the temperature of the products. Flame temperatures, as a function of the equivalence ratio, are usually calculated from thermodynamic data when a fuel is burned adiabatically with air.

Flame Types and Their Characteristics. There are two main types of flames: diffusion and premixed. In diffusion flames, the fuel and oxidant are separately introduced and the rate of the overall process is determined by the mixing rate. Examples of diffusion flames include the flames associated with candles, matches, gaseous fuel jets, oil sprays, and large fires, whether accidental or otherwise. In premixed flames, fuel and oxidant are mixed thoroughly prior to combustion. A fundamental understanding of both flame types and their structure involves the determination of the dimensions of the various zones in the flame and the temperature, velocity, and species concentrations throughout the system.

The structure of a one-dimensional premixed flame is well understood. By coupling the rate of heat release from chemical reaction and the rate of heat transfer by conduction with the flow of the unburned mixture, an observer moving with the wave would see a steady laminar flow of unburned gas at a uniform velocity, S_u, into the stationary wave or flame. Hence S_u is defined as the burning velocity of the mixture based on the conditions of the unburned gas. The thickness of the preheating zone and the equivalent reaction zone is found to be inversely proportional to the burning velocity. By considering the heat release from the chemical reaction, it is possible to calculate the thickness of the effective reaction zone.

There are a number of sources of instability in premixed combustion systems. Laminar flame instabilities are dominated by diffusional effects that can only be of importance in flows with a low turbulence intensity, where molecular transport is of the same order of magnitude as turbulent transport. Flame instabilities do not appear to be capable of generating turbulence. They result in the growth of certain disturbances, leading to orderly three-dimensional structures which, though complex, are steady.

Combustion processes and flow phenomena are closely connected and the fluid mechanics of a burning mixture play an important role in forming the structure of the flame. Laminar combusting flows can occur only at low Reynolds numbers, defined as

$$Re = \frac{\rho u d}{\mu} \tag{1}$$

where, ρ = density, kg/m^3; u = velocity, m/s; d = diameter, m; and μ = kinematic viscosity, kg/ms. When $Re > Re_{cr}$ the laminar structure of a flow becomes unstable and when the Reynolds number exceeds the critical value by an order of magnitude, the structure of the flow changes. Along with this change of structure from an orderly state to a more chaotic state, the following parameters begin to fluctuate randomly: velocity, pressure, temperature, density, and species concentrations. Overall, with increasing Reynolds number, laminar flow becomes unstable as a result of these fluctuations and breaks down into turbulent flow. Laminar and turbulent flames differ greatly in appearance. For example, while the combustion zone of a laminar Bunsen flame is a smooth, delineated and thin surface, the analogous turbulent combustion region is blurred and thick.

Turbulent flame speed, unlike laminar flame speed, is dependent on the flow field and on both the mean and turbulence characteristics of the flow, which can in turn depend on the experimental configuration.

In high speed dusted premixed flows, where flames are stabilized in the recirculation zones, the turbulent flame speed grows without apparent limit, in approximate proportion to the speed of the unburned gas flow.

In the reaction zone, an increase in the intensity of the turbulence is related to the turbulent flame speed.

The balanced equation for turbulent kinetic energy in a reacting turbulent flow contains the terms that represent production as a result of mean flow shear, which can be influenced by combustion, and the terms that represent mean flow dilations, which can remove turbulent energy as a result of combustion. Some of the discrepancies between turbulent flame propagation speeds might be explained in terms of the balance between these competing effects.

To analyze premixed turbulent flames theoretically, two processes should be considered: (*1*) the effects of combustion on the turbulence, and (*2*) the effects of turbulence on the average chemical reaction rates.

A unified statistical model for premixed turbulent combustion and its subsequent application to predict the speed of propagation and the structure of plane turbulent combustion waves is available.

Many different configurations of diffusion flames exist in practice. Laminar jets of fuel and oxidant are the simplest and most well-understood diffusion flames.

The discussion of laminar diffusion flame theory addresses both the gaseous diffusion flames and the single-drop evaporation and combustion, as there are some similarities between gaseous and liquid diffusion flame theories. A frequently used model of diffusion flames has been developed, and despite some of the restrictive assumptions of the model, it gives a good description of diffusion flame behavior.

The Displacement Distance theory suggests that since the structure of the flame is only quantitatively correct, the flame height can be obtained through the use of the displacement length or "displacement distance" (eq. 2), where h = flame height, m; V = volumetric flow rate, m^3/s; and D = diffusion coefficient.

$$h = \frac{V}{2\pi D} \tag{2}$$

As the velocity of the fuel jet increases in the laminar-to-turbulent transition region, an instability develops at the top of the flame and spreads down to its base. This is caused by the shear forces at the boundaries of the fuel jet. The flame length in the transition region is usually calculated by means of empirical formulas of the form (eq. 3): where l = length of the flame, m; r = radius of the fuel jet, m; v = fuel flow velocity, m/s; and C_1 and C_2 are empirical constants.

$$l = \frac{r}{C_1 - \frac{C_2}{v}} \tag{3}$$

Laminar diffusion flames become turbulent with increasing Reynolds number. Some of the parameters that are affected by turbulence include flame speed, minimum ignition energy, flame stabilization, and rates of pollutant formation. Changes in flame structure are believed to be controlled entirely by fluid mechanics and physical transport processes.

The various studies attempting to increase our understanding of turbulent flows comprise five classes: moment methods disregarding probability density functions, approximation of probability density functions using moments, calculation of evolution of probability density functions, perturbation methods beginning with known structures, and methods identifying coherent structures.

Fundamentals of Heterogeneous Combustion. The discussion of combustion fundamentals so far has focused on homogeneous systems. Heterogeneous combustion is the terminology often used to refer to the combustion of liquids and solids. From a technological viewpoint, combustion of liquid hydrocarbons, mainly in sprays, and coal combustion are of greatest interest.

Most theories of droplet combustion assume a spherical, symmetrical droplet surrounded by a spherical flame, for which the radii of the droplet and the flame are denoted by r_d and r_p, respectively. The flame is supported by the fuel diffusing from the droplet surface and the oxidant from the outside. The heat produced in the combustion zone ensures evaporation of the droplet and consequently the fuel supply. Other assumptions that further restrict the model include (*1*) the rate of chemical reaction is much higher than the rate of diffusion and hence the reaction is completed in a flame front of infinitesimal thickness; (*2*) the droplet is made up of pure liquid fuel; (*3*) the composition of the ambient atmosphere far away from the droplet is constant and does not depend on the combustion process; (*4*) combustion occurs under steady-state conditions; (*5*) the surface temperature of the droplet is close or equal to the boiling point of the liquid; and (*6*) the effects of radiation, thermodiffusion, and radial pressure changes are negligible.

In order to obtain an expression for the burning rate of the droplet, the following parameters are needed: physical constants such as the specific heat, and the thermal conductivity of the droplet, the radius of the flame, and the temperature of the flame. To determine these quantities, heat conduction, diffusion, and the kinetics of the chemical processes associated with droplet combustion need to be analyzed. This is achieved mathematically by solving the equations of mass continuity, mass continuity for components, and the energy equation. The solving of these equations can be facilitated if the following simplifying assumptions are made: the flame surrounding the droplet is a diffusion flame and, by definition, is formed where the fuel and oxidant meet in stoichiometric proportions; the temperature of this flame is very close to the adiabatic flame temperature; and the heat required for evaporation of the droplet and the heat loss to the surroundings through the burned gas are small and can therefore be neglected. These equations are usually solved in spherical coordinates for a one-dimensional case. However, since the flame is relatively thick, and the droplet is relatively small, the one-dimensional model of the process may not be a particularly accurate representation. Nevertheless, the values obtained for burning rates provide useful information.

The amount of data available on droplet combustion is extensive. However, the results can be easily summarized, because the burning rate constants for the majority of fuels of practical interest fall within the narrow range of 7 to 11×10^{-3} cm^2/s. An increase in oxygen concentration results in an increase in the burning rate constant. If the burning takes place in pure oxygen, the values for burning rate are increased by a factor of about 2.0, compared to when the burning takes place in air.

The convective gas flow around a burning particle affects its burning rate. It has been postulated that in the absence of convection, the burning rate is independent of pressure. Forced convection, on the other hand, is believed to increase the burning rate.

During the final stages of the combustion of a droplet, coke remains, and although it represents a relatively small percentage of the mass of the original oil droplet, the time taken for the heterogeneous reaction between the oxygen-depleted combustion air and the coke particle is generally the slowest of all the combustion steps.

The reaction between a porous solid, such as a coke sphere, and a gas, such as oxygen, occurs in the following stages: (*1*) the main reactant species diffuse thoroughly through the boundary layer toward the solid surface and the products of reaction diffuse away from the surface; (*2*) diffusion and simultaneous chemical reaction take place within the pores of the solid proceeding from the external surface toward the interior, and gaseous products diffuse in the opposite direction; and (*3*) at the participating surfaces, the reacting gas chemisorbs, some intermediate species are formed, then the final products of the reaction desorb from the surface. Thus the observed reaction rate is a function of the individual resistances–boundary layer diffusion, pore diffusion, and chemical kinetics, and the rate controlling process is the slowest of them or a combination of these processes. Even though a number of gas reactions may take place at the surface of a burning carbonaceous solid, the reaction forming CO is most often assumed, $C + 1/2\ O_2 \rightarrow CO$. Coke combustion is treated mathematically like char combustion.

In a practical combustion chamber, the droplets tend to burn in the form of sprays, hence it is important to understand the fundamentals of sprays. In the most simple case, a fuel spray suspended in air will support a stable propagating laminar flame in a manner similar to a homogeneous gaseous mixture. In this case, however, two different flame fronts are observed. If the spray is made up of very small droplets they vaporize before the flame reaches them and a continuous flame front is formed. If the droplets are larger, the flame reaches them before the evaporation is complete, and if the amount of fuel vapor is insufficient for the formation of a continuous flame front, the droplets burn in the form of isolated spherical regions. Flames of this type are referred to as heterogeneous laminar flames. Experimental determination of burning rates and flammability limits for heterogeneous laminar flames is difficult because of the motion of droplets caused by gravity and their evaporation before the arrival of the flame front.

In modern liquid-fuel combustion equipment the fuel is usually injected into a high velocity turbulent gas flow. Consequently, the complex turbulent flow and spray structure make the analysis of heterogeneous flows difficult and a detailed analysis requires the use of numerical methods.

The combustion of a coal particle occurs in two stages: (*1*) devolatilization during the initial stages of heating with accompanying physical and chemical changes, and (*2*) the subsequent combustion of the residual char. The burning rate or reactivity of the residual char in the second stage is strongly dependent on the process conditions of the first stage.

The ignition mechanism is rather complex and is not well understood in terms of actually defining the ignition temperature and reaction mechanisms. The ignition temperature is known, however, not to be a unique property of the coal and depends on a balance between heat generated and heat dissipated to the surroundings around the coal particle. Measuring the ignition characteristics is complicated by the fact that they are strongly dependent on the physical arrangement of the particles, eg, single particle, clouds of coal dust, or coal piles. Reported ignition temperatures range from 303 to 373 K in the case of spontaneous ignition of coal piles at ambient temperature to 1073–1173 K in the case of single coal particle ignition. Characteristics such as coal type, particle size and distribution of mineral matter, and experimental conditions such as gas temperature, heating rate, oxygen, and coal dust concentration are some of the important factors that influence values obtained for ignition temperatures.

A variety of techniques has been used to determine ignition temperatures: fixed beds, the crossing point method, the critical air blast method, photographic techniques, entrained flow reactors, electric spark ignition, luminous glow observations, plug flow reactors, shock tubes, and thermogravimetric analysis. The techniques mostly used are constant temperature methods, in which coal particles are introduced into a preheated furnace maintained at a fixed temperature.

The structure of residual char particles after devolatilization depends on the nature of the coal and the pyrolysis conditions such as heating rate, peak temperature, soak time at the peak temperature, gaseous environment, and the pressure of the system.

The rate limiting step in the combustion of char is either the chemical kinetics (adsorption of oxygen, reaction, and desorption of products) or diffusion of oxygen (bulk and pore diffusion). Variations in the reaction rate with temperature for gas–carbon reactions have been grouped into three main regions or zones depending on the rate limiting resistance.

Combustion Technology

Technology addresses the more applied, practical aspects of combustion, with an emphasis on the combustion of gaseous, liquid, and solid fuels for the purpose of power production. In an ideal fuel burning system (*1*) there should be no excess oxygen or products of incomplete combustion, (*2*) the combustion reaction should be initiated by the input of auxiliary ignition energy at a low rate, (*3*) the reaction rate between oxygen and fuel should be fast enough to allow rapid rates of heat release and it should also be compatible with acceptable nitrogen and sulfur oxide formation rates, (*4*) the solid impurities introduced with the fuel should be handled and disposed of effectively,

(5) the temperature and the weight of the products of combustion should be distributed uniformly in relation to the parallel circuits of the heat absorbing surfaces, (6) a wide and stable firing range should be available, (7) fast response to changes in firing rate should be easily accommodated, and (8) equipment availability should be high and maintenance costs low.

Combustion of Gaseous Fuels. In any gas burner some mechanism or device (flame holder or pilot) must be provided to stabilize the flame against the flow of the unburned mixture. This device should fix the position of the flame at the burner port. Although gas burners vary greatly in form and complexity, the distribution mechanisms in most cases are fundamentally the same. By keeping the linear velocity of a small fraction of the mixture flow equal to or less than the burning velocity, a steady flame is formed. From this pilot flame, the main flame spreads to consume the main gas flow at a much higher velocity. The area of the steady flame is related to the volumetric flow rate of the mixture by equation 4:

$$V_{mix} = A_f S_u \quad (4)$$

where $\dot{V}_{\text{mix}}$ = volumetric flow rate, m^3/s; A_f = area of the steady flame, m^2; and S_u = burning velocity, m/s.

The volumetric flow rate of the mixture is, in turn, proportional to the rate of heat input (eq. 5):

$$\dot{V}_{\text{mix}} \cdot (HHV) = \dot{Q} \quad (5)$$

where $\dot{V}_{\text{mix}}$ = volumetric flow rate, m^3/s; HHV = higher heating value of the fuel, J/kg; and Q = rate of heat input, J/s.

Most of the commercial gas–air premixed burners are basically laminar-flow Bunsen burners and operate at atmospheric pressure.

Atmospheric pressure industrial burners are made for a heat release capacity of up to 50 kJ/s (12 kcal/s), and despite the varied designs, their principle of stabilization is basically the same as that of the Bunsen burner.

Gas burners that operate at high pressures are usually designed for high mixture velocities and heating intensities and therefore stabilization against blowoff must be enhanced. This can be achieved by a number of methods such as surrounding the main port with a number of pilot ports or using a porous diaphragm screen.

It is often desired to substitute directly a more readily available fuel for the gas for which a premixed burner or torch and its associated feed system were designed. Satisfactory behavior with respect to flashback, blowoff, and heating capability, or the local enthalpy flux to the work, generally requires reproduction as nearly as possible of the maximum temperature and velocity of the burned gas, and of the shape or height of the flame cone. Often this must be done precisely, and with no changes in orifices or adjustments in the feed system.

Turbulence in the flow of a premixture flattens the velocity profile and increases the effective burning velocity of the mixture; eg, at a pipe-Reynolds number of 40,000 the turbulent burning velocity is several times the laminar burning velocity and it can be perhaps fifty times larger at very high Reynolds numbers.

Combustion of Liquid Fuels. There are several important liquid fuels, ranging from volatile fuels for internal combustion engines to heavy hydrocarbon fractions, sold commercially as fuel oils. The technology for the combustion of liquid fuels for spark-ignition and compression-ignition internal combustion engines is not described here. The emphasis here is primarily on the combustion of fuel oils for domestic and industrial applications.

In general, the combustion of a liquid fuel takes place in a series of stages: atomization, vaporization, mixing of the vapor with air, and ignition and maintenance of combustion (flame stabilization). Recent advances have shown the atomization step to be one of the most important stages of liquid fuels combustion. The main purpose of atomization is to increase the surface area to volume ratio of the liquid. This is achieved by producing a fine spray. The finer the atomization spray the greater the subsequent benefits are in terms of mixing, evaporation, and ignition. The function of an atomizer is twofold: atomizing the liquid and matching the momentum of the issuing jet with the aerodynamic flows in the furnace.

Atomizers for large boiler burners are usually of the swirl pressure jet or internally mixed twin-fluid types, producing hollow conical sprays. Less common are the externally mixed twin-fluid types.

Combustion of fuel oil takes place through a series of steps, namely vaporization, devolatilization, ignition, and dissociation, which finally lead to attaining the flame temperature.

The study of the combustion of sprays of liquid fuels can be divided into two primary areas for research purposes: single-droplet combustion mechanisms and the interaction between different droplets in the spray during combustion with regard to droplet size and distribution in space. The wide variety of atomization methods used and the interaction of various physical parameters has made it difficult to give general expressions for the prediction of droplet size and distribution in sprays. The main fuel parameters affecting the quality of a spray are surface tension, viscosity, and density, with fuel viscosity being by far the most influential parameter.

Combustion of Solid Fuels. Solid fuels are burned in a variety of systems, some of which are similar to those fired by liquid fuels. In this article the most commonly burned solid fuel, coal, is discussed. The main coal combustion technologies are fixed-bed, eg, stokers, for the largest particles; pulverized-coal for the smallest particles; and fluidized-bed for medium size particles (see COAL).

Fixed-Bed Technology. Fixed-bed firing of coal by means of stokers consists of a solid bed of large (2–3-cm) coal particles on grates with combustion air passing through the grates and ash removal from the end of the grate.

Pulverized-Coal Firing. This is the most common technology used for coal combustion in utility applications because of the flexibility to use a range of coal types in a range of furnace sizes. Nevertheless, the selection of crushing, combustion, and gas-cleanup equipment remains coal-dependent.

Prior to being fed to a pulverized fuel burner, coal is ground to a size generally specified such that at least 70% passes a 200 mesh screen (75 μm) and less than 2% is retained on a 50 mesh screen (300 μm). The top size is determined by the classifying component of the crushing mill, oversize material being retained for further grinding.

Suspensions of pulverized coal or coal dust in air can be explosive; hence, it is essential to have adequate guidelines and procedures to ensure safe and stable operation during pulverized-coal (PC) firing.

As for oil and gas, the burner is the principal device required to successfully fire pulverized coal. The two primary types of pulverized-coal burners are circular concentric and vertical jet-nozzle array burners.

The self-igniting characteristics of pulverized coal vary from one coal to another, but for most coals it is possible to maintain ignition without auxiliary fuel when firing above the capacity of the boiler. The igniters may have to be activated in the following cases: (1) when firing pulverized coal with volatile matter less than about 25%, (2) when firing excessively wet coal, and (3) when feeding coal sporadically into the pulverizer.

Compared to natural gas and oil, complete combustion of coal requires higher levels of excess air, about 15% as measured at the furnace outlet at high loads, and this also serves to avoid slagging and fouling of the heat absorption equipment.

As pulverized-coal combustion potentially has a significant impact on the environment, the 1980s saw the employment of techniques such as coal washing and beneficiation to reduce the emissions of fly-ash, SO_x, and water-soluble metallic oxides.

The environmental impact associated with pulverized-coal firing has given rise to efforts to develop other combustion technologies such as fluidized beds or the use of coal-water slurry fuels (CWSF), which can be burned as substitutes for certain liquid fuels. CWSFs were

Table 1. Comparison of Design Parameters for Fossil Fuel Boilers

Parameter	Gas	Oil	Coal		
			Grate	Fluid bed	Pulverized coal
heat rate, mW (t)	0.03–3000	0.03–3000	0.3–30	up to 30	30–3000
volumetric combustion intensity, kW/m^3	250–450	250–450	250–750[a]	up to 2000[a] (based on bed volume)	150–250
area combustion intensity, kW/m^2	280–500	280–500	2000	3000	7500
fuel firing density					
kg/m^3h			30–100	≈250	15–30
kg/m^2h	6–11	6–11	40–250	up to 500	up to 1000
practical combustion temperature, °C	1000–1600	1100–1700	1200–1300	850–950	1600–1700
combustion time, s	10×10^{-1}	$20°25 \times 10^{-1}$	up to 5000	100–500	≈1–2
particle heating rate, °C/s			>1	10^3–10^4	10^4–10^5

[a] Based on the total combustion volume which includes space between the bed and the convective tubes.

developed as alternatives to more expensive and increasingly scarce conventional hydrocarbon fuels. The main challenge in the utilization of CWSFs is obtaining stable mixtures that can be successfully atomized and burned.

Fluidized-Bed Technology. In fluidized-bed combustion of coal, air is fed into the bed at a sufficiently high velocity to levitate the particles. This velocity is referred to as the minimum fluidizing velocity, u_{mf}. At this velocity, the volume occupied by the bed increases abruptly and the bed exhibits some of the characteristics of a fluid. The two predominant designs of fluidized beds are bubbling and recirculating, with most theories of fluidization being based on the simpler bubbling bed concept.

Fluidized combustion of coal entails the burning of coal particles in a hot fluidized bed of noncombustible particles, usually a mixture of ash and limestone. Once the coal is fed into the bed it is rapidly dispersed throughout the bed as it burns. The bed temperature is controlled by means of heat exchanger tubes. Elutriation is responsible for the removal of the smallest solid particles and the larger solid particles are removed through bed drain pipes. To increase combustion efficiency the particles elutriated from the bed are collected in a cyclone and are either reinjected into the main bed or burned in a separate bed operated at lower fluidizing velocity and higher temperature.

Fluidized beds are ideal for the combustion of high sulfur coals since the sulfur dioxide produced by combustion reacts with the introduced calcined limestone to produce calcium sulfate. The chemistry involved can be simplified and reduced to two steps, calcination and sulfation.

Calcination $CaCO_3 \rightarrow CO_2 + CaO$

Sulfation $SO_2 + CaO + 1/2\ O_2 \rightarrow CaSO_4$

The main steps associated with coal combustion (heating, devolatilization, volatiles combustion, and char burnout), occur sequentially to some extent; however, there is always some overlap between the stages.

Design Considerations in Fossil Fuel Combustion Systems. One of the most important considerations in the design of a combustion chamber for a boiler is the fuel that is to be burned in the chamber (see FURNACES, FUEL-FIRED). Although all fuels burn and release heat during combustion, the rate at which a fuel burns and releases heat, and the impurities associated with the fuel have to be considered.

Furnaces for Oil and Natural Gas Firing. Natural gas furnaces are relatively small in size because of the ease of mixing the fuel and the air; hence, there is a relatively rapid combustion of gas. Oil also burns rapidly with a luminous flame. To prevent excessive metal wall temperatures resulting from high radiation rates, oil-fired furnaces are designed slightly larger in size than gas-fired units in order to reduce the heat absorption rates.

Furnaces for Pulverized Coal Firing. The main differences between boilers fired with coal and those fired with oil or natural gas result from the presence of mineral matter in coals. The volume of the coal-fired furnace is higher because of the longer residence time required for the complete combustion of coal particles, the requirement of a controlled combustion rate to reduce NO_x formation, the provision for a larger heat-transfer surface area resulting from decreased heat-transfer rates because of ash deposits on the surfaces, and increased spacing of heat-transfer tubes to reduce flue gas velocities and thereby erosion of heat-transfer surfaces. Even when firing coal, depending on the reactivity of the coal (rank), the size of the combustion chamber required can vary. Table 1 provides some design parameters for fossil fuel burners.

Environmental Considerations. Atmospheric pollutants released by combustion of fossil fuels fall into two main categories: those emitted directly into the atmosphere as a result of combustion and the secondary pollutants that arise from the chemical and photochemical reactions of the primary pollutants (see AIR POLLUTION).

The main combustion pollutants are nitrogen oxides, sulfur oxides, carbon monoxide, unburned hydrocarbons, and soot. Combustion pollutants can be reduced by three main methods depending on the location of their application: before, after, or during the combustion. Techniques employed before and after combustion deal with the fuel or the burned gases. A third alternative is to modify the combustion process in order to minimize the emissions.

Diffusion Flame Chemistry. Since most combustion systems employ mixing-controlled diffusion flames, which are characterized by very high pollutant emissions, it is imperative to look into the chemistry occurring in diffusion flames. In a typical diffusion flame the mixture composition in the reaction zone is close to the stoichiometric proportion and the temperature is at a maximum resulting from the large volume of this zone, thus NO_x production is favored. If, however, the surrounding gas cools the combustion products rapidly, further reactions of CO and NO are eliminated. This fixes the concentrations of these pollutants at unfavorable levels. Furthermore, the fuel diffuses into the combustion zone through the burned gases and thus is heated in the absence of oxygen. This creates ideal conditions for the formation of soot and the reduction of the CO_2 produced in the combustion zone to CO. Additionally, diffusion flames have low combustion intensity and efficiency and hence release large amounts of unburned hydrocarbon emissions. In general, despite the fact that the structure of the diffusion flame is more complex and difficult to analyze, the same basic description of soot formation and oxidation should apply to diffusion flames as for premixed flames.

Emissions Control. More advanced techniques for emissions control include electrical or plasma jet augmentation of flames based on radical production. Because in two-phase, heterogeneous combustion the flames are always diffusion flames on the microscale, ie, the individual droplets or particles burn as diffusion flames, and because at the characteristic times for evaporation, decomposition and burning of individual particles can be comparable to the characteristic times for mixing and pollutant formation, prevaporization, or gasification of the fuel can reduce pollutant emissions. For this reason catalytic systems for liquid-fuel decomposition and coal gasification are being considered seriously as alternatives to conventional combustion technology.

REZA SHARIFI
SARMA V. PISUPATI
ALAN W. SCARONI
Pennsylvania State University

N. Chigier, "Energy," *Combustion and Environment,* McGraw-Hill, Inc., New York, 1981.

J. M. Beer and N. A. Chigier, *Combustion Aerodynamics,* Applied Science, London; John Wiley & Sons, Inc., New York, 1972.

R. H. Essenhigh, in M. A. Elliot, ed., *Fundamentals of Coal Combustion,* in *Chemistry of Coal Utilization,* 2nd Suppl. Vol., John Wiley & Sons, Inc., New York, 1981.

S. C. Stultz and J. B. Kitto, eds., *Steam, Its Generation and Use,* 40th ed., Babcock and Wilcox Co., Barberton, Ohio, 1992.

COMPOSITE MATERIALS

SURVEY

The term *composite material* is used to describe macroscopic combinations of two or more materials. Macroscopic combinations are specified to exclude alloys that consist of materials combined on a microscopic scale.

The fundamental goal in the production and application of composite materials is to achieve a performance from the composite that is not available from the separate constituents or from other materials. The concept of improved performance is broad and includes increased strength of reinforcement of one material by the addition of another material.

In the early 1960s there was a significant increase in interest in the science of composite materials when very strong and stiff but brittle ceramic, boron, and carbon fibers became available. These new fibers were at least as strong as the glass fibers of the same diameter but approximately five to ten times stiffer. The term *advanced composites* has been coined to cover those reinforced plastics containing continuous strong, stiff fibers such as carbon (graphite), boron, aramid, or glass. Advanced composites have been developed primarily for the aerospace industry, in which the demand for strong and stiff lightweight structures overcame the prohibitive costs of early composite material systems. Currently, advanced composite materials, mainly carbon-reinforced epoxy resins, are used widely in military aircraft and increasingly in civil aircraft. The largest volume usage of these materials has, however, been in the recreational area in applications such as tennis rackets and golf club shafts.

Composite Classification and Terminology

Composites are generally classified as fibrous, laminated, and particulate.

Fibrous Composites. These composites consist of fibers in a matrix. The fibers may be short or discontinuous and randomly arranged; continuous filaments arranged parallel to each other; in the form of woven rovings (collections of bundles of continuous filaments); or braided.

Laminates. Two or more layers of material bonded together form a laminated composite.

Plastic laminated sheets produced in 1913 led to the formation of the Formica Products Company and the commercial introduction, in 1931, of decorative laminates consisting of a urea–formaldehyde surface on an unrefined (kraft) paper core impregnated with phenolic resin and compressed and heated between polished steel platens. Since 1937, the surface layer of most decorative laminates has been fabricated with melamine–formaldehyde, which can be prepared with mineral fillers, thus offering improved heat and moisture resistance and allowing a wide range of decorative effects.

Plywood is a laminate consisting of thin sheets of wood arranged with the grain in alternate sheets at right angles.

Fiber composite laminates often consist of unidirectional (parallel) continuous fibers in a polymer matrix with the individual layers, plies, or laminae stacked with selected fiber angles so as to produce specific laminate stiffness and strength values. Laminates are also made by stacking layers of mats or woven fabric, which have many possible weave configurations.

Particulate Composites. These composites encompass a wide range of materials. As the word particulate suggests, the reinforcing phase is often spherical or at least has dimensions of similar order in all directions. Examples are concrete, filled polymers, solid rocket propellants, and metal and ceramic particles in metal matrices.

Reinforcements

The choice of reinforcement for a particular engineering application is likely to depend on a large number of parameters, including strength, stiffness, environmental stability, long-term characteristics, and cost. At present, carbon, glass, and aramid fibers account for over 95% of the industrial market, with increasing use in areas such as the aerospace, automotive, construction, biomedical, and sport sectors. No single fiber type can be said to be truly superior to another; each has its own merits as well as shortcomings. The stress–strain responses obtained by bending typical carbon, glass, and aramid fibers are summarized in Figure 1. Carbon fibers generally offer the highest strengths and stiffnesses but can be brittle, failing at relatively low applied strains. Glass fibers offer intermediate strengths and higher failure strains but exhibit lower moduli. Aramid fibers such as Kevlar, a proprietary material developed by DuPont, have lower strengths but are capable of absorbing considerable energy without fracture. Rather than using strength and stiffness as the performance parameters for material selection, specific properties obtained by normalizing the property with respect to the density, ρ, of the fiber are often quoted. For tensile or compressive members, the specific strength, σ/ρ, yields the largest load-carrying capacity for a given mass and the specific stiffness, E/ρ, produces the smallest deflection for a given mass. A convenient way of comparing fibers is to plot specific strength against specific modulus.

Glass Fibers. Glass fibers represent the most frequently used reinforcement in modern polymer composites. This popularity results from their relatively low cost and high tensile strength. In bulk form, glass is brittle, having a relatively low strength. However, when extruded and drawn into fine fibers, the strength of glass increases enormously, by as much as two orders of magnitude. Glass-fiber-reinforced composites are currently used in a wide variety of applications, including boat construction and the automotive and aerospace industries. Many types of glass fiber are available, each having specific properties and characteristics. The most commonly used fibers, E-glass, contain approximately 14% Al_2O_3, 18% CaO, 5% MgO, 8% B_2O_3, and 1% $Na_2O + K_2O$ as well as silica (see GLASS).

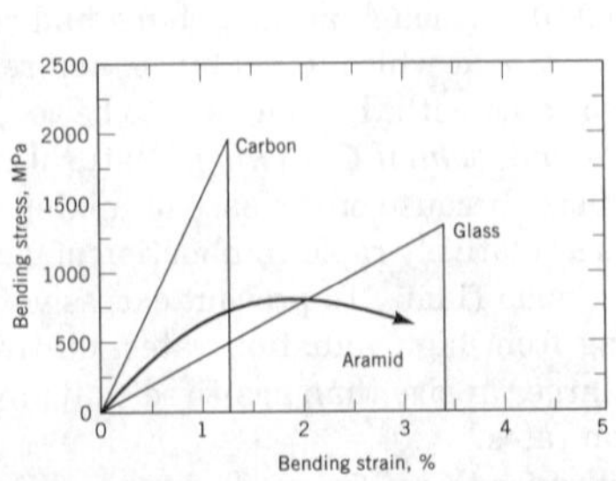

Figure 1. Bending stress versus bending strain for typical carbon, glass, and aramid fibers. To convert MPa to psi, multiply by 145.

Glass fibers are produced by dry mixing the individual components and then heating them to form a melt. The temperature of the melt depends on the glass composition but is typically about 1250°C. The molten glass then passes into the fiber-drawing furnace and subsequently flows through a large number of tiny orifices, forming fine filaments. It is common practice, therefore, to apply a size, a surface coating, to protect the fibers during handling and prevent damage during any subsequent processing stages, such as weaving.

Carbon Fibers. The current technology for manufacturing carbon fibers is based on the thermal decomposition of organic precursor materials such as polyacrylonitrile (PAN) and pitch. Generally, the stages in the production of carbon fibers from the various organic precursors can be identified as spinning, stabilization, carbonization, and graphitization.

Surface treatment of fibers is an important stage in the manufacturing process. The primary aim of such treatment is to improve the adhesion between the fiber and the matrix and to improve handleability. The types of treatment are oxidative gas or dry oxidation, wet oxidation, and aqueous electrolytic or anodic oxidation.

Carbon-fiber-reinforced plastic composites are currently used in primarily and secondary aircraft structures, helicopter rotor blades, sporting goods such as fishing rods and tennis rackets, and certain biomedical applications.

Aramid Fibers. Aramid fibers are formed by mixing a polymer, poly(*p*-phenylene terephthalamide), with a strong acidic solution and extruding the mixture through spinnerets at a temperature between 50 and 100°C. The fibers are cooled and then washed thoroughly before being dried on bobbins. The properties of the aramid fibers can be modified by using solvent additives or by using a post-spinning heat treatment. Aramid fibers with relatively low moduli, such as Kevlar 29, find use in energy-absorbing applications such as bullet-resistant and other protective clothing, helmets, and ropes. The higher modulus counterpart, Kevlar 49, is used in high performance engineering applications such as the manufacture of load-bearing components for the aerospace industry where it is used in filament-wound rocket motor cases.

Other Fibers. Currently boron fibers are used in metal matrices such as aluminum and magnesium. Boron-reinforced plastics, however, still find use as repair patches for damaged aircraft structures as well as in certain sporting goods, such as fishing rods and golf club shafts.

Polyethylene fibers are attracting considerable interest owing to their high specific strength and stiffness as well as their excellent energy-absorbing capability. Two methods are used to manufacture polyethylene fibers (see FIBERS–OLEFIN). The first involves the extrusion and drawing of a medium molecular weight polyethylene. The second method involves dissolving the polymer in a solvent and spinning at a temperature of around 140°C.

Silicon carbide fibers exhibit high temperature stability and therefore find use as reinforcements in certain metal matrix composites. Silicon carbide fibers can be made in a number of ways, eg, by vapor deposition on carbon fibers.

Matrix Materials

The mechanical properties of composites based on the fibers discussed depend not only on the characteristics of the fibers but also on those of the matrix itself as well as on the fiber–matrix interface.

Polymer Matrices. The matrix in a polymer composite serves both to maintain the position and orientation of the fibers and to protect them from potentially degrading environments. Polymer matrices may be thermosets or thermoplastics. Thermosetting polymers are rigid, cross-linked materials that degrade rather than melt at high temperatures; thermoplastics are linear or branched molecules that soften upon heating. A comparison of the mechanical properties and relative cost of various thermosetting and thermoplastic materials is given in Table 1. Thermoset-based composites are somewhat less expensive than thermoplastic-based composites, but have lower heat distortion temperatures and poorer toughness when tested in an interlaminar mode. The majority of present-day composite components are still based largely on thermosetting matrices, such as unsaturated polyesters, epoxies, and phenolic resins.

Table 1. Mechanical Properties and Relative Costs of Thermosetting and Thermoplastic Composites

Matrix material	Young's modulus, GPa[a]	Tensile strength, MPa[b]	Heat distortion temperature, °C	Relative cost[c]
		Thermosets		
polyester	3.6	60	95	1
vinyl ester	3.4	83	110	1.8
epoxy	3.0	85	110	2.3
phenolic	3.0	50	120	0.8
		Thermoplastics		
PES	2.8	84[d]	203	6
PEI	3.0	105[d]	200	7
PEEK	3.7	92[d]	140	25

[a] To convert GPa to psi, multiply by 145,000.
[b] To convert MPa to psi, multiply by 145.
[c] Relative to the 1988 cost of isophthalic polyester.
[d] Yield stress.

Other Matrix Materials. Advanced materials, eg, structural components, in aerospace vehicles also employ ceramics and metals as composite matrices (see COMPOSITE MATERIALS, CERAMIC-MATRIX; METAL-MATRIX COMPOSITES).

Fabrication Methods

A large number of methods are presently available for manufacturing long fiber composites. The cost of finished components depends not only on the price of the raw materials but also on labor costs and energy requirements. Many composite components are manufactured by hand, involving relatively long manufacturing cycles. Large engineering components, such as boat hulls, are frequently manufactured using hand lay-up or spray-up techniques. The whole cycle may last several months, involving a large work force. Considerable effort is being made to find cheaper, more efficient ways to manufacture composite parts and structures. Some of the more commonly employed techniques used to manufacture such parts are fabrication methods such as hand and spray lay-up, filament winding, autoclave molding, compression molding, pultrusion, and resin transfer molding.

Theories of Reinforcement

The advantages of composites include improved stiffness and strength of the composite material system compared with the base line or unreinforced material. In composites, enhanced toughness can be achieved by increasing the energy required to initiate and propagate a crack through the brittle matrix. Increased work of fracture has been obtained in composites through debonding at the fiber–matrix interface; by frictional interaction between the fiber and the matrix as fibers bridging a matrix crack are pulled out of the matrix as the crack extends and opens; by deformation of the fibers; and by fiber fracture. In the case of brittle matrix composites, small quantities of reinforcing material are needed to achieve the desired performance.

Fiber volume fraction is a quantitative measure of degree of reinforcement of the matrix material in a fiber-reinforced composite.

Strength predictions of composites are in general quite complex and somewhat limited. This is particularly true of compressive and shear strengths, which are needed, together with the tensile strengths, in composite failure prediction.

The tensile strength of a unidirectional lamina loaded in the fiber direction can be estimated from the properties of the fiber and matrix for a special set of circumstances. If all of the fibers have the same tensile strength σ_f and the composite is linear elastic until failure of the fibers, then the strength of the composite is given by

$\sigma_c = \sigma_f v_f + \epsilon_{mf} E_m v_m$, where ϵ_{mf} is the strain in the matrix when the fibers fail. For carbon fibers in epoxy resin the tensile strength of the composite is predicted to be approximately proportional to the fiber volume fraction. The assumption of a constant failure stress of the fibers is unrealistic. This strength prediction relates to a fiber-dominated model. However, at low values of fiber volume fraction the fiber-dominated model is invalid. For very low values of fiber volume fraction the composite tensile strength is given approximately by the matrix-dominated model: $\sigma_c = \sigma_m(1 - v_f)$. The stress–strain relationship is used in conjunction with the rules for determining the stress and strain components with respect to some angle θ relative to the fiber direction to obtain the stress–strain relationship for a lamina loaded under plane strain conditions where the fibers are at an angle θ to the loading axis. Classical laminated plate theory is used to determine the stiffness of laminated composites. Details of the Kirchoff-Love hypothesis on which the theory is based can be found in standard texts.

The strength of laminates is usually predicted from a combination of laminated plate theory and a failure criterion for the individual lamina. A general treatment of composite failure criteria is beyond the scope of the present discussion. Broadly, however, composite failure criteria are of two types: noninteractive, such as maximum stress or maximum strain, in which the lamina is taken to fail when a critical value of stress or strain is reached parallel or transverse to the fibers in tension, compression, or shear; or interactive, such as the Tsai-Hill or Tsai-Wu type, in which failure is taken to be when some combination of stresses occurs.

Economic Considerations

In the form of fiber-reinforced unidirectional and multidirectional composites, very high values of strength and stiffness can be achieved with fiber volume fractions of about 60%. Excellent fatigue properties can also be obtained. Transverse impact damage tolerance, which was an early limiting factor in carbon fiber composites, has been improved greatly through the development of high strain to failure fibers and tough matrix and interleaf materials. However, these improvements in properties have been associated with significant increases in the material costs. There have been many applications in which cost has been a secondary factor to performance, such as in the military aerospace fields. However, cost has become the critical issue in the continued development and application of advanced composite materials. Composite components can be fabricated through various routes, depending on the quality of the end product, and economies in the finished cost can be achieved by reducing the number of parts and attachments and assembly operations. In comparing costs, it is important to include the life-cycle costs. In the oil and gas industry, significant reductions in the life-cycle costs can be achieved by replacing steel with fiber-reinforced plastics. Another very large market with enormous potential for the application of medium-technology composites is the automotive industry.

J. MORTON
Virginia Polytechnic Institute and State University
W. J. CANTWELL
Ecole Polytechnique Fédérale de Lausanne

R. M. Jones, *Mechanics of Composite Materials,* Scripta Book Co., Washington, D.C., 1975.

R. A. Baker and co-workers in S. M. Lee, ed., *International Encyclopedia of Composites,* Vol. 2, VCH Publishing, Inc., 1990, pp. 182–222.

G. Lubin, ed., *Handbook of Composites,* Van Nostrand Reinhold Co., Princeton, N.J., 1982.

W. J. Cantwell, P. Davies, P.-Y. Jar, P.-E. Bourban, and H. H. Kausch, in H. Hornfeld, ed., *Plastics—Metals—Ceramics,* SAMPE (European), 1990, pp. 411–427.

POLYMER-MATRIX THERMOSETS

Polyester Resins

Unsaturated polyester resins predominate among fiber-reinforced composite matrices for several reasons. A wide variety of polyesters is available and the composites fabricator must choose the best for a particular application. The choice involves evaluation of fabrication techniques, temperatures at which the resin is to be handled, cure time and temperature desired, and required cured properties (see POLYESTERS–UNSATURATED).

Manufacture. Polyester resins are manufactured by the reaction of dibasic acids with glycols.

Application. Polyesters are cured by free radicals, most commonly produced by the use of peroxides.

Because the elevated-temperature curing resin systems are thermally activated, they provide a very long time at lower temperatures for fabrication (pot life). Long pot lives make applications such as sheet molding compounds possible; it is the attribute responsible for most of the elevated temperature cure applications.

Some characteristics restrict the application of polyesters. Polyesters have a limited shelf life, polymerizing slowly over a period of months at room temperature. The shelf life can be extended by cold storage or by the addition of polymerization inhibitors by the manufacturer.

Ease of cure, easy removal of parts from mold surfaces, and wide availability have made polyesters the first choice for many fiber-reinforced composite molders. Sheet molding compound, filament winding, hand lay-up, spray up, and pultrusion are all well adapted to the use of polyesters. Table 1 lists the desirable properties for a number of fiber-reinforced composite fabrication methods.

To optimize the resin system for a given process and part, consideration should be given to fillers that can greatly affect the cost and performance of the composite. Fillers are often much cheaper than the resin they displace, and they can improve the heat resistance, stiffness, and hardness of the composite. Certain fillers, such as fumed silica, impart thixotropy to the resin, increasing its resistance to drainage.

Health, Safety, and Environment. Manufacturers of fiber-reinforced polyester composites need to be concerned with proper handling of hazardous wastes, emissions of volatile organic compounds, and a host of recent laws and regulations. Of primary concern is worker exposure to, and plant emissions of, styrene. OSHA permissible exposure limits on air contaminants for 1990 placed an 8-h time-weighted average of 50 ppm for styrene. The listing of styrene as a probable carcinogen has led to increased regulation.

The organic peroxides used to cure polyester resins need to be stored separately from the polyester resins and promotors.

Table 1. Resin Properties Required for Various Fabrication Methods

Process	Viscosity, mPa·s(= cP)	Cure temperature, °C	Thixotropy	Filler, %	Glass, %
SMC/BMC[a]	200–2500	150	no	25–50	25–50
hand lay-up	400–800	RT	yes		20–40
filament winding	600–2000	RT–150	yes and no		40–70
pultrusion	400–2000	100–150	no	0–20	60–80
prepreg	50,000+	100–150	no		60–80
spray up	200–1000	RT	yes	0–20	20–40

[a] SMC = sheet molding compound; BMC = bulk molding compound.

Phenolic Resins

Most processors of fiber-reinforced composites choose a phenol formaldehyde (phenolic) resin because these resins are inherently fire-retardant, are highly heat-resistant, and are very low in cost. Phenolic resins (qv) are often not chosen, however, because the resole types have limited shelf stability, both resole and novolac types release volatiles during their condensation cure, formaldehyde emissions are possible during both handling and cure, and the polymers formed are brittle compared with other thermosetting resins.

Manufacture. Phenolic resins are diverse in structure and functionality. A phenol and less than a stoichiometric amount of formaldehyde are condensed with an acid catalyst resulting in a thermoplastic resin termed a *novolac.* These resins are oligomers that are terminated by phenol groups. They require the addition of a curing agent to effect curing. Novolacs are usually solid resins. Condensing phenol and a stoichiometric amount of formaldehyde with an alkaline catalyst and stopping the reaction while the resin is still a thermoplastic results in a resole. Resoles are terminated by methyl groups. They can be cured by heating or by the addition of a catalyst and are normally produced as liquids. The bulk of the phenolic resins used in fiber-reinforced composites are mixtures of novolacs and curing agents.

Applications. Curing agents for novolacs are aldehydes or methylene donors.

Resoles can be cured by the addition of base or by heat alone. Resoles are often used in unreinforced applications in electronics and high moisture areas.

Heat-cured phenolic prepregs have almost completely replaced all other thermosetting resins in aircraft and other transit interiors, where flammability and toxicity of combustion products is a prime consideration.

Health and Safety. Free phenols may be present in phenolic novolacs and resoles. Phenol is poisonous and caustic, irritating the skin and mucous membranes. Formaldehyde and ammonia are often emitted during the cure of novolacs and must be properly vented. Formaldehyde is listed as a human carcinogen; worker exposure and emissions are controlled by OSHA and the EPA.

Epoxy Resins

Currently, epoxy resins (qv) constitute over 90% of the matrix resin material used in advanced composites. The total usage of advanced composites is expected to grow to around 45,500 t by the year 2000, with the total resin usage around 18,000 t in 2000. Epoxy resins are expected to still constitute about 80% of the total matrix-resin-systems market in 2000. The largest share of the remaining market will be divided between bismaleimides and polyimide systems (12 to 15%) and what are classified as other polymers, including thermoplastics and thermoset resins other than epoxies, bismaleimides, cyanate esters, and polyimide systems (see COMPOSITES, POLYMER-MATRIX–THERMOPLASTICS).

The earliest and still the most widely used matrix resins in high performance composites are the bisphenol A-based epoxy resin systems. Probably the most widely recognized property of cured epoxy resin systems is their excellent adhesion to a very broad range of substrates and reinforcements. A contributing factor is the low shrinkage exhibited by epoxy resin systems during cure, which results in lower stress levels in the composite than is found in other polymer systems with higher shrinkage. Another factor contributing to the excellent strength of articles produced from epoxy resins is that no by-products are formed during cure. Thus, there are no volatiles liberated that can lead to voids nor are nonvolatiles generated that can act as plasticizers.

Epoxy resins can be generically characterized as a group of commercially available oligomeric materials that contain one or more epoxy (oxirane) groups per molecule. The value of epoxy resins is that they can be processed into a variety of useful products, such as protective coatings, adhesives, and structural components of almost any size and shape, by reaction of the epoxy groups with an appropriate curing agent. The products obtained from epoxy resins that contain more than one epoxy group per molecule are thermosetting polymers.

Liquid resins such as glycidyl esters and bisphenol A epoxy resins are used mainly in ambient temperature cure coatings, electrical castings, flooring, electrical laminates, and fiber-reinforced composites. These applications require low viscosity materials for good flow and are cured through the epoxide ring. The higher n value resins, particularly those above 3000 molecular weight, are normally used in solution and find their greatest application in heat-cured finishes.

For more demanding uses at higher temperatures, for example, in aircraft and aerospace and certain electrical and electronic applications, multifunctional epoxy resin systems based on epoxy novolac resins and the tetraglycidyl amine of methylenedianiline are used. The tetraglycidyl amine of methylenedianiline is currently the epoxy resin most often used in advance composites.

Curing Agents. The two principal classes of curing agents used in epoxy matrix resins for advanced composites are aromatic diamines and anhydrides.

Manufacture. Liquid epoxy resins are generally manufactured from bisphenol A and epichlorohydrin. Typically, the bisphenol A reacts with epichlorohydrin to give a bischlorohydrin, which is dehydrohalogenated with caustic to give the desired epoxy resin. Glycidylamines such as tetraglycidyl methylenedianiline (TGMDA) are manufactured by a similar process.

High Performance Epoxy Resins

Tetraglycidyl methylenedianiline cured with diaminodiphenyl sulfone has been the principal resin system used in high performance composite applications. Although the specific details of most formulated matrix resins are proprietary, formulations of some of the more popular early systems for carbon fiber prepreg based on TGMDA–DDS contain bisphenol epoxy novolac resin and glycidyl ester are believed to be formulated into the systems to modify the curing characteristics and tack and drape, respectively. Boron trifluoride complex is added as an accelerator. These matrix resin systems offer a combination of ease of melt processing into prepreg as well as tack and drape of the prepreg and out time.

Current TGMDA systems, however, do not meet all the requirements for advanced composites for future aircraft. Specifically, TGMDA systems lack the necessary hot–wet performance and they are too brittle.

New products designed to overcome these shortcomings are beginning to appear. The Dow Chemical Company has introduced a glycidyl ether based on a hydrocarbon epoxy novolac under the trade name TACTIX 556. The properties of neat resin castings are given in Table 2. This multifunctional epoxy resin has a glass-transition temperature of over 223°C when cured with DDS and outstanding moisture resistance and thermal oxidative resistance.

Shell Chemical Company has introduced a new tetraglycidyl amine and a new stiff backbone diglycidyl ether under the respective trade names EPON HPT Resin 1071 and EPON HPT Resin 1079, which have superior hot–wet performance compared with TGMDA (Table 3). These two materials are chemically *N,N,N′,N′*-tetraglycidyl-α, α'-bis (4-aminophenyl)-*p*-diisopropylbenzene and diglycidyl-9,9-bis(4-hydroxyphenyl)fluorene.

Although changes in the structure of highly functional epoxy resins have resulted in improved hot–wet performance, brittleness is an inherent property of highly cross-linked systems. Improvements in ductility, fracture resistance, and impact strength, however, can be achieved without substantially degrading the thermal and mechanical properties of the epoxy matrix. For example, carboxyl terminated butadiene–acrylonitrile copolymer, which is initially soluble in conventional bisphenol A epoxy resins, forms a second phase of dispersed particles on cure, resulting in an epoxy resin system with improved toughness. The degree to which any epoxy resin system can be toughened depends on many factors, including the type of modifier, heterophase size distribution, interfacial adhesion between the heterophase and the matrix, combination of resin and curing agent, cure

Table 2. Mechanical Properties of High Performance Epoxy Resins[a] in Unreinforced Resin Castings

Properties	TGMDA	EPON HPT[b] 1071[c]	EPON HPT[b] 1079[c]	TACTIX 556[d,e]
T_g, °C		249	279	
flexural properties RT/dry				
strength, MPa[f]	138	117	124	136
modulus, GPa[g]	3.9	3.9	3.3	3.1
elongation, %	5.0	3.7	4.7	
flexural properties hot/wet[h]				
strength, MPa[f]	76	90	90	
modulus, GPa[g]	2.5	3.4	3.0	3.0[h]
elongation, %	4.7	3.6	4.1	
moisture gain, wt%[i]	5.7	3.6	2.8	2.4[j]

[a] Resin cured with 100% diaminodiphenylsufone.
[b] Trademark of Shell Chemical Co.
[c] Cured 2 h at 150°C and 4 h at 200°C.
[d] Trademark of The Dow Chemical Company.
[e] Cured 3 h at 177°C and 2 h 232°C.
[f] To convert MPa to psi, multiply by 145.
[g] To convert GPa to psi, multiply by 145,000.
[h] Tested in water at 93°C after 2 weeks' immersion at 93°C.
[i] Specimens conditioned 200 h in boiling water.
[j] Conditioned 2 weeks in 93°C water.

conditions, and molecular weight between cross-links in the matrix. A thorough and critical review of all the variables affecting the ability to toughen epoxy resin systems is beyond the scope of this discussion.

The use of elastomeric modifiers for toughening thermoset resins generally results in lowering the glass-transition temperature, modulus, and strength of the modified system. More recently, ductile engineering thermoplastics and functional thermoplastic oligomers have been used as modifiers for epoxy matrix resins and other thermosets.

Applications. Epoxy resins constitute over 90% of the matrix resin material used in advanced composites. In addition, epoxy resins are used in all the various fabrication processes that convert resins and reinforcements into composite articles. Liquid resins in combination, mainly, with amines and anhydride are used for filament winding, resin transfer molding, and pultrusion. Parts for aircraft, rocket cases, pipes, rods, tennis rackets, ski poles, golf club shafts, and fishing poles are made by one of these processes with an epoxy resin system.

Bismaleimides

Bismaleimides (BMI) are a relatively young class of thermosetting polymers that are gaining acceptance by the industry because they combine a number of unique features including excellent physical property retention at elevated temperatures and in wet environments, almost constant electrical properties over a wide range of temperatures, and nonflammability properties. Their excellent processibility and balance of thermal and mechanical properties have made them popular in advanced composites and electronics.

Bismaleimides are best defined as low molecular weight, at least difunctional monomers or prepolymers, or mixtures thereof, that carry maleimide terminations.

The principal concern with BMA resins has been their inherent brittleness owing to their high cross-link density. Ciba-Geigy, however, demonstrated that BMI-*o,o'*-diallylbisphenol A copolymers are tougher than TGMDA-DDS epoxy resins. Experience has shown that a useful BMI-resin system comprises both a bismaleimide part (the BMI resin) and a comonomer part (reactive diluent).

Building Blocks and Systems. A standard synthesis of *N,N'*-arylene bismaleimide involves the chemical dehydration of *N,N'*-arylene bismaleamic acid with acetic anhydride and sodium acetate as a catalyst at temperatures below 80°C. The yield of pure recrystallized BMI is usually 65–75%. Various by-products, such as isoimides and acetanilides, are responsible for the relatively low yield of pure BMI. Almost every aromatic amine (diamine) can be converted to the corresponding maleimide (bismaleimide). The most widely used building block is 4,4'-bismaleimidodiphenylmethane, because the precursor diamine is readily available and cheap. For reasons of processibility BMIs with low melting points are the preferred building blocks.

The most important property of a bismaleimide is its ability to undergo a temperature-induced polymerization. The maleimide double bond is highly activated owing to the adjacent carbonyl groups of the imide ring; therefore, heating the BMI above its melting point effects polymerization.

Besides low molecular weight building blocks, long-chain maleimide-terminated oligomers have been synthesized for molding, adhesive, and composite applications. The key step is the preparation of an amine-terminated intermediate needed to introduce the maleimide group.

Other backbone structures that have generated a great interest are the polyether ketones. An attempt was made to synthesize amino-terminated arylene ether ketones, which were subsequently converted into the corresponding maleimide-terminated oligomers. The aim of this approach was to obtain tough, solvent-resistant, high temperature thermosets.

The common synthetic route to bismaleimides or maleimide functionalized oligomers is the condensation of diamines or amino-terminated oligomers with maleic anhydride. Another possibility is the use of an AB-type monomer of the following general formula to build the polymaleimide, where X represents a functional group that can be employed in condensation reactions.

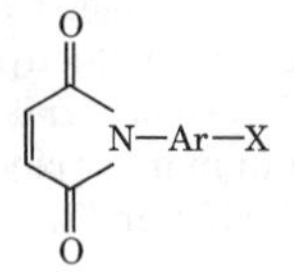

The maleimide is prebuilt into the molecule in a separate step.

Bismaleimide Resin Concepts. The bismaleimide building block is not the final resin product. Although building blocks usually make up 50–75% by weight of the resin, other ingredients, such as comonomers, reactive diluents, processing additives, elastomers, and catalysts, are combined with BMI so as to obtain a product suitable for the application considered. The application areas for bismaleimide resins are reinforced composites for printed circuit boards (with glass fabric), structural laminates (with glass, carbon, and aramid fibers), and moldings (with short fibers and particulate filters).

In order to fulfill the processing requirements, bismaleimide building blocks have to be formulated into products that enable their use as highly concentrated solutions, powders, or hot melts.

Michael Additions. The reaction of a bismaleimide with a functional nucleophile (diamine, bisthiol, etc) via the Michael addition reaction converts a BMI building block into a polymer.

Bismaleimide Resins via ENE Reaction. The copolymerization of a BMI with *o,o'*-diallylbisphenol A (DABA) is a resin concept that has been widely accepted by the industry because BMI–DABA blends are tacky solids at room temperature and therefore provide all the desired properties in prepregs, such as drape and tack, similar to epoxies. Crystalline BMI can easily be blended with DABA, which is a high viscosity fluid at room temperature. Upon heating BMI–DABA blends copolymerize via complex ENE and Diels-Alder reactions.

Diels-Alder Copolymers. The Diels-Alder reaction can also be employed to obtain thermosetting polyimides. If bismaleimide (the bisdienophile) and the bisdiene react nonstoichiometrically, with bismaleimide in excess, a prepolymer carrying maleimide terminations is formed as an intermediate, which can then be cross-linked to yield a temperature-resistant network.

Table 3. Commercially Available BMI Resins

Company	Resin name	Description
CIBA-GEIGY Corp.	Matrimide 5292 A,B	two-component resin system comprising 4,4′-bismaleimidodiphenylmethane (M5292A) and diallylbisphenol A (M5292B)
	RD85-101	BMI building block based on diaminodiphenylindane; designed for hot-melt prepregging
	Araldite XU5292	bismaleimide resin solution for PCB applications
DuPont Co.	MVA-2	1,3-bismaleimidabenzene building block
Mitsubishi Gas Chemicals Co.	BT resins	blends of bismaleimide(B) and triazine(T) resins; used primarily in printed circuit boards
Mitsui Toatsu Chemicals, Inc.	Bismaleimide-S	4,4′-bismaleimido-diphenylmethane. (The company also offers other BMI building blocks.)
Rhône Poulenc Chimie	Kerimide 601	soluble BMI powder for multilayer board application
	FE70000	series of research and development products
	Rhodimid M3	4,4′-bismaleimido-diphenylmethane
Shell Chemical Co.[a]	COMPIMIDE MDAB	high purity 4,4′-bismaleimidodiphenylmethane
	COMPIMIDES 353, 353A, 796	unformulated BMI resins
	COMPIMIDE 15 MRK	for injection and compression molding
	COMPIMIDE 65 FWR	for filament winding
	COMPIMIDE 121	liquid bis(allylphenyl) compound
	COMPIMIDE 123	[bis-(*o*-propenyl)phenoxy] benzophenone
	COMPIMIDE 1206-F55	55–60 wt % solution of bismaleimide for prepregging

[a] Also, Technochemie GmbH-Verfahrenstechnik, Dossenheim, Germany, is a member of the Royal Dutch/Shell Group of Companies.

Styrene can react with bismaleimide via a complex Diels-Alder–ENE route in a 1:2 stoichiometric ratio. Other vinylbenzene compounds, such as propenylphenoxydiphenyl sulfone and bis(*o*-propenylphenoxy)benzophenone, react in a similar way with bismaleimide.

Propenylphenoxy compounds have attracted much research. BMI-propenylphenoxy copolymer properties can be tailored through modification of the backbone chemistry of the propenylphenoxy comonomer. Bis[3-(2-propenylphenoxy)phthalimides] have been synthesized from bis(3-nitrophthalimides) and *o*-propenylphenol sodium involving a nucleophilic nitro displacement reaction. They copolymerize with bismaleimide via Diels-Alder and provide temperature-resistant networks.

Under appropriate thermal conditions, the strained four-membered ring of benzocyclobutene undergoes electrocyclic ring opening to generate, *in situ*, *o*-chinodimethane, which, in the presence of bismaleimide, reacts via a Diels-Alder reaction. Certain bisbenzocyclobutene–BMI (COMPIMIDE 353) blends form compatible mixtures in a wide range of rations which, after cure, show remarkable thermal oxidative stability and high glass-transition temperatures.

Modified Bismaleimides. Bismaleimide resins may be further modified and blended with other thermoset resins or reactive diluents to achieve either specific end-use properties or processibility. Thermoset resins that can be used for modification are unsaturated polyesters, vinyl esters, cyanate esters, and epoxies.

Toughened Bismaleimide Resins. Bismaleimide homopolymers are brittle thermosets and laminates made from them display low impact damage tolerance. The techniques used to toughen brittle bismaleimides include: the use of monomers or monomer–comonomer systems that provide inherently tough networks through reduced cross-link density; the use of elastomeric materials as modifiers to achieve second-phase toughening similar to rubber-modified polystyrene; and the use of engineering thermoplastics instead of elastomers to maintain the high elastic modulus of the thermoset in the toughened system.

Commercial BMI Resins. Table 3 indicates the BMI resins available at the end of 1994. No reference is made to prepreg systems available from various sources. A complete list of available products and information on any particular material should be obtained from the suppliers.

RONALD S. BAUER
STEVEN L. STEWART
Shell Development Company
HORST D. STENZENBERGER
Technochemie GmbH-Verfahrenstechnik

Modern Plastics **68**(1), 104, 108 (1991).

Bismaleimides for Advanced Printed Circuit Boards, N. N. Rhône-Poulenc.

B. Ellis, *Chemistry and Technology of Epoxy Resins,* Blackie Academic and Professional, 1993.

C. A. May, *Epoxy Resins,* 2nd ed., Marcel Dekker, Inc., New York 1988.

THERMOPLASTICS

Thermoplastic resins have received considerable attention in the past few years as the matrix material in organic resin-based composites. Although their use in advanced composites is not widespread, thermoplastic composites are used extensively in commercial applications ranging from automobiles to durable goods.

Thermoplastic polymers are usually linear molecules with no chemical linkage between the molecules. The molecules are held together by weak secondary forces, such as van der Waals or hydrogen bonding, and as such are deformed by the application of heat or pressure. Thermoplastic resins can be amorphous, that is, structureless, or semicrystalline, in which some of the molecules form an ordered array. A material is usually considered semicrystalline if as little as 5% of the polymer is in the crystalline form. Semicrystalline resins exhibit a higher modulus, but amorphous materials are tougher; amorphous materials are usually more solvent sensitive but can be processed at lower temperatures. One of the most important advantages of thermoplastic resins is their toughness, that is, high impact strength and fracture resistance, which, unfortunately, is not linearly translated into properties of the composite. Other advantages of thermoplastic polymers include long shelf life at room temperature; postformability, that is, thermal reforming; ease of repair by thermal welding or solvents; and ease of handling, that is, they are not tacky.

One of the principal advantages of true thermoplastic polymers is their ability to consolidate or flow at elevated temperatures; however, this quality also limits their upper-use temperature.

Thermoplastic composites can be classified according to use, cost, performance, or processing methods. In the following discussion of the chemistry of the resin systems utilized in composites, three classes are considered: (*1*) Resins used in conventional composites (usually commodity materials) and typically containing 10–40 wt % reinforcing agent. (*2*) Resins used in advanced or high performance composites, which contain more than 50 wt % reinforcing agent and are the typical aerospace materials. (*3*) Pseudothermoplastic resin systems, which are formed as conventional thermoplastic materials and then cured or postcured in a manner similar to that used for thermosetting resins to enhance high temperature properties.

Commodity Resins

Most of the resin systems used in commodity composites are slight modifications of the standard commercial molding grade material. Usually certain selected properties, such as purity or molecular weight range or distribution, are enhanced or carefully selected. In addition, special additives, such as flow controllers, thermal stabilizers, or antioxidants, are often added by the resin manufacturer prior to shipment. Many of the conventional or commodity-type resins used in thermoplastic composites are listed in Table 1.

Polyamides. Nylon was the first commercial thermoplastic polymer produced on a large scale. Nylons contain repeating amide functionality (see POLYAMIDES). The aliphatic polyamide nylon is one of the most important thermoplastic materials. It is used extensively as a molding compound and in fiber glass (both short and long fibers) reinforced composites.

Many modifiers and additives have been described for use with nylon composites, but generally a small amount, 0.05–1 wt %, of a lubricity aid, such as sodium or zinc stearate is added to enhance both resin flow during processing and removal from the mold after consolidation.

Polyolefins. The most common polyolefin used to prepare composites is polypropylene, a commodity polymer that has been in commercial production for almost 40 years. A Ziegler catalyst consisting of titanium tetrachloride and an aluminum alkyl was used to achieve controlled polymerization to produce isotactic polypropylene directly from propylene. Polypropylene is available with many different reinforcing agents or fillers, such as talc, mica, or calcium carbonate; chopped or continuous strand fiber glass is the most common reinforcing agent used for composites. Many additives have been developed to enhance the thermal stability of polypropylene to minimize degradation during processing. One of the most important requirements of the polypropylene used in the manufacture of composites is that it be relatively pure and free of residual catalyst. Recent developments to form copolymers of polypropylene and polyethylene have great promise for relatively inexpensive, tough, thermoplastic composite applications (see OLEFIN POLYMERS).

Acetals. Acetal resins (qv) are polymers of formaldehyde and are usually called polyoxymethylene.

Precise amounts of chain terminators are added during polymerization to produce final polymers having various molecular weights. Stabilizers and antioxidants are added to both the homopolymer and the copolymer to create the basic polymer grades. Although several fillers, such as talc or calcium carbonate, can be added to polyoxymethylene, the most common reinforcing agent for engineering applications is fiber glass.

Polycarbonates. Polyarylates are aromatic polyesters commonly prepared from aromatic dicarboxylic acids and diphenols. One of the most important polyarylates is polycarbonate, a polyester of carbonic acid. Polycarbonate composite is extensively used in the automotive industry because the resin is a tough, corrosion-resistant material. Polycarbonates can be prepared from aliphatic or aromatic materials by two routes: reaction of a dihydroxy compound with phosgene accompanied by liberation of HCl or transesterification of a dihydroxy compound with dialkyl or diaryl carbonates.

Polyesters. Polyesters are widely used as the matrix for conventional composites. Two resins of particular importance because of the large amounts used are poly(ethylene terephthalate) (PET) and poly(butylene terephthalate) (PBT). The most practical methods for the preparation of high molecular-weight polyesters involves ester-exchange reactions.

Table 1. Conventional Thermoplastic Resins

Chemical class	Repeat unit	Polymer	Morphology
polyamides	—C(=O)—NH—	nylon-6,6	semicrystalline
		nylon-6	semicrystalline
polyolefins	—CH₂—CH(X)—	polypropylene	semicrystalline
		polystyrene and its copolymers	amorphous
polycarbonates	—O—C(=O)—O—	polycarbonate	amorphous
polyesters	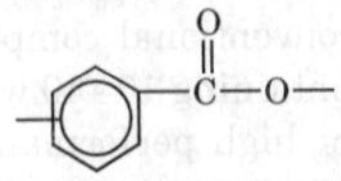	poly(ethylene terephthalate)	semicrystalline
		poly(butylene terephthalate)	semicrystalline
acetals	$—CH_2O—$	polyoxymethylene	semicrystalline

Advanced Composites

The term *advanced composite* normally implies a high volume fraction reinforcing agent, of the order of 60 wt %. Much research into the development of high temperature advanced thermoplastic composites has been conducted since the late 1970s. The primary goal of this work has been the development of high temperature tough materials. This development presents a twofold problem: (*1*) the synthesis of resins that are stable and do not degrade at the high temperatures of interest, and (*2*) incorporation of the resin system into a reinforcing agent so that a structural composite can be fabricated. The latter is the more difficult of the two problems.

Resins for advanced composites can be classified according to their chemistry; typical resins are polyaryl-ether ketones, polysulfides, polysulfones, and a very broad class of polyimides containing one or more additional functional groups (Table 2) (see also ENGINEERING PLASTICS).

It is important to distinguish between true thermoplastic and pseudothermoplastics. Pseudothermoplastics are systems in which the chemistry continues during processing or an extra postcure step is added after the resin is consolidated. Many of the so-called advanced thermoplastic resins are actually pseudothermoplastics. Although pseudothermoplastics are not completely reprocessible as are the true thermoplastics, usually they can be thermoformed a second or

Table 2. Resins for Advanced Composites

Chemical class	Resin	Morphology	T_g, °C	T_m, °C
polyetherketones	polyether ether ketone	semicrystalline	143	345
	polyether ketone ketone	semicrystalline	170	370
polysulfones	polysulfone	amorphous	200	
	polyethersulfone	amorphous	230	
polysulfides	polyphenylene sulfide[a]	semicrystalline	85	285
polyimides	polyimide[a]	semicrystalline	250–300	
	polyetherimide	amorphous	220	
	polyamide imide[a]		275	
	polybenzimidazoles[a]		230–290	
	fluorinated polyimides	amorphous	340	

[a] These materials are also referred to as pseudothermoplastics.

possibly a third time if necessary to correct defects in a structural component.

Polyarylether Ketones. The aromatic polyester ketones are true thermoplastics. Although several are commercially available, two resins in particular, poly ether ether ketone (PEEK) from ICI and poly ether ketone ketone (PEKK) from DuPont, have received the most attention. Tough, semicrystalline PEEK is prepared by the condensation of bis(4-fluorophenyl) ketone with the potassium salt of bis(4-hydroxyphenyl) ketone in a diaryl sulfone solvent such as diphenyl sulfone. A commercial prepreg of PEEK and carbon fibers (manufactured and sold by ICI as APC-2) can be made by the hot melt process.

PEEK can also be spun into fibers, which are commingled with a reinforcing fiber to form a yarn.

Poly(phenylene ether). The only commercially available thermoplastic poly(phenylene oxide) PPO is the polyether poly(2,6-dimethylphenol-1,4,-phenylene ether). PPO is prepared by the oxidative coupling of 2,6-dimethylphenol with a copper amine catalyst.

Polysulfones. The most common polysulfone is actually a sulfone ether, polyethersulfone (PES). It is prepared from the polycondensation of the disodium salt of bisphenol A and 4,4-dichlorodiphenyl sulfone in a polar aprotic solvent such as dimethyl sulfoxide.

Polyimides. Polyimides represent a very broad class of materials ranging from thermosets to both amorphous and semicrystalline thermoplastics. Polyimides are used extensively as high temperature adhesives, as insulation for electrical wire, and for printed circuit boards. They have great potential for high temperature applications because of their thermooxidative stability, solvent resistance, and general ease of preparation. However, their use in resin matrix composites has been rather limited in spite of their great potential.

Aromatic polyimides are generally produced by the reaction of aromatic dianhydrides with aromatic diamines.

Polyether Imides. Polyether imides (PEIs) are amorphous, high performance thermoplastic polymers that have been in use since 1982. The first commercial polyether imides were the Ultem series developed by the General Electric Company. The first, Ultem 1000, is prepared from phthalic anhydride, bisphenol A, and *meta*-phenylenediamine. PEI resins reinforced with up to 40 wt % fiber glass are available.

Another series of PEI resins has been introduced by DuPont, the Avimid series. The Avimid K-III resin was formulated to be a high temperature thermoplastic polyimide that could be processed with existing autoclave equipment designed for thermosets.

Polyamide Imides. Polyamide imides (PAIs) are formed from the condensation of trimellitic anhydride and aromatic diamines. One commercial PAI is Torlon, supplied by Amoco.

Polybenzimidazoles. The polybenzimidazoles (PBIs) are generally produced by the high temperature, melt polycondensation reaction of aromatic bis-*ortho*-diamines and aromatic dicarboxylates (acids, esters, or amides). The particular polymer used in many commercial and developmental applications is poly(2,2′-*m*-phenylene)-5,5′-dibenzimidazole).

Fluorinated Polyimides. Often the substitution of fluorine atoms for hydrogen atoms in a polymer chain markedly increases the thermal stability of the base polymer; this is true for polyimides. A typical fluorinated polyimide is prepared from the reaction of 2,2-bis(3,4-dicarboxyphenyl) hexafluoropropane dianhydride and 2,2-bis-(4-amino phenyl)hexafluoropropane. This material is manufactured by DuPont and sold as resin NR-150B2.

Semicrystalline Polyimides. Semicrystalline polyimides containing carbonyl and ether groups have been synthesized the NASA Langley Research Center.

Poly(phenylene sulfide). Poly(phenylene sulfide) (PPS) is a semicrystalline thermoplastic with a T_g of 85°C and a T_m of 285°C. The nominal degree of crystallinity is 50–65 vol %. It is produced commercially by the direct reaction of *p*-dichlorobenzene with sodium sulfide in a polar organic solvent.

Novel Methods

One of the principal problems facing the use and further development of thermoplastic matrices for composites is the need to wet the reinforcing agent with the matrix to obtain a good strong bond. Amorphous thermoplastics must be heated above their T_g and crystalline materials must be heated above their crystalline melting points to flow. The upper temperature limit to which the polymer can be heated is controlled by rate of thermal degradation; thus, a relatively narrow processing window is imposed by the need to flow without degrading. In the usable temperature region, most thermoplastics are quite viscous (several hundred Pa · s) and high pressures are required for the resin to flow. Even then, in many cases the resin wets the fiber poorly. Recently, new methods have been described to circumvent this problem and still prepare a true thermoplastic composite. In one case relatively low molecular-weight, low viscosity cyclic compounds are injected directly onto preforms and the polymerization conducted rapidly *in situ*. This requires a special low viscosity monomer coupled with a unique catalytic system that rapidly polymerizes the monomer but does not lead to excess thermal degradation at elevated temperatures.

Another method that has great potential for the preparation of advanced prepregs and has been explored extensively requires fine powders. The reinforcing fibers are coated with fine particles of the resin and, when heated, the resin flows over the fiber.

CLARENCE J. WOLF
Washington University

"Thermoplastic Resins," in *Engineered Materials Handbook: Engineering Plastics,* Vol. 2, ASM International, Metals Park, Ohio, 1988, pp. 98–221.

E. J. Degrup, in R. B. Seymour and E. S. Kirshenbaum, eds., *High Performance Polymers: Their Origin and Development,* Elsevier, New York, 1986, p. 81.

D. Wilson, H. D. Stenzenberger, and P. M. Hergenrother, *Polyimides—Chemistry and Applications,* Blackie and Sons Ltd., Glasgow, Scotland, 1990.

P. M. Hergenrother, "Polyimides Condensation" in S. M. Lee, ed., *International Encyclopedia of Composites,* Vol. 4, VCH Publishers, New York, 1990, p. 180.

CERAMIC-MATRIX

Monolithic ceramics are brittle and are thus very sensitive to intrinsic flaws and damage produced by use. Failure of these materials occurs in a catastrophic manner and at low strain-to-failure ratios. However, the problem can be alleviated by reinforcing monolithic ceramics with a second phase which is itself capable of operating at high temperatures. Such systems are designated as ceramic matrix composites (CMC).

The reinforcing phases in ceramic matrix composites are usually also ceramic and have many possible morphologies: particulate, platelet, whisker, short-fiber, or continuous-fiber. Reinforcing entities are typically added to ceramic matrices to produce tough composites. In comparison, high strength reinforcements are added to polymer-based composites to increase strength and stiffness. To enhance toughness high strength reinforcements with high elastic modulus and weak interfaces with the matrix are required; to produce high strength and stiffness, strong interfaces along with high stress transfer are needed to allow efficient load transfer or shedding from the matrix to the reinforcement.

Ceramic Composites Systems

With the appropriate choice of composite properties such as, reinforcement and matrix materials, reinforcing geometry and composite interface, an otherwise brittle mode of failure of a ceramic becomes more "ductile" and noncatastrophic in nature. Thus, the choice of the component materials is an important aspect of designing ceramic matrix composites. Two questions need to be addressed when making these choices. First, if a matrix crack encounters a potential bridging entity, will it deflect along the reinforcement/matrix interface or will fracture of the reinforcement occur? Second, if interface debonding occurs, will the interfacial sliding shear resistance, τ, be low enough to allow the bridge to slip in the matrix or will fracture of the bridging-reinforcement occur?

A partial answer to the first question has been provided by a theoretical treatment that examines the conditions under which a matrix crack will deflect along the interface between the matrix and the reinforcement. The calculations indicate that, for any elastic mismatch, interface failure will occur when the fracture resistance of the bridge is at least four times greater than that of the interface.

About the second question, concerning the relative strengths of the bridge and the interfacial sliding resistance, little is known a priori. The general recommendation is to have a high bridge strength and a low interfacial sliding shear resistance.

Various combinations of ceramic–matrix composites have been manufactured at the research level. Their properties are given in Table 1 for oxide-based matrices and in Table 2 for nonoxide matrices. Some commercial products are identified for information only. Such identification does not imply recommendation or endorsement by NIST, nor does it imply that the products are the best available for the purpose.

Table 1. Oxide-Based Ceramic-Matrix Composites

Reinforcement					
Type[a]	Amount, vol %	Strength,[b] MPa	Toughness,[c] Mpa$\sqrt{\text{in.}}$	Modulus, GPa[d]	Density, g/cm^3 or % td[e]
Al_2O_3 matrix					
B_4C_p	50		4.5	380	3.28
SiC_w	20		2.5	400	
SiC_w/SiO_{2i}[f]	20		6.0	420	
SiC_w/Si_3N_{4w}	20	203	3.4		95% td
SiO_{2f}		6.3	28		
TiC_p	30		4.0	400	4.26
BN_p		24.6		490	91.1% td
Al_p	20		8.4		
$ZrO_2(t)_p$[g]		2000	5–8	333	4.54
Aluminosilicate glass matrix					
SiC_w			0.8	80	
Al_2O_{3f}		311	3.3		
Cordierite glass matrix					
SiC_f		128	1.6		2.44
Pyrex glass matrix					
Al_2O_{3f}		305	3.7		
SiC_p	30	171	1.79		
SiC_w	30	180	3.04		
SiC_p/SiC_w[h]		159	2.73		
Soda-lime silicate matrix					
SiC_w	20		0.7	72	
LASIII glass ceramic matrix					
SiC_w	35	327	5.1		
$3Al_2O_3 \cdot 2SiO_2$ matrix					
SiC_w	10	274	2.7	197	2.84
SiC_p		262	2.35	240	
$ZrO_2(t)_p$[g]		250	4.0	150	98.9% td
ZrO_2 matrix					
$ZrO_2(t)_p$[g]		400–600	10	200	6.08

[a] Subscripts denote reinforcement morphology; p = particulate, l = platelet, w = whisker, f = fiber, i = interlayer between reinforcement and matrix.
[b] Strength as measured in a four-point flexure test (modulus of rupture); to convert MPa to psi, multiply by 145.
[c] Fracture toughness; to convert MPa $\sqrt{\text{m}}$ to psi $\sqrt{\text{in.}}$, multiply by 910.048.
[d] To convert GPa to psi, multiply by 145,000.
[e] % td = percentage of theoretical density.
[f] 20% SiC_w.
[g] Tetragonal.
[h] 10% each.

Composite Reinforcements

The structure of reinforcements can be either equiaxed or acicular. The nature of their placement within a composite, the composite architecture, is critical to the resultant composite properties. Possible architectures are summarized in Figure 1.

Equiaxed particles which are well dispersed in the ceramic matrix, tend to produce isotropic composite behavior. The particles, either ceramic or metallic, may be single crystal or polycrystalline in nature.

Acicular reinforcements such as whiskers and platelets tend to produce rather more anisotropic composite properties. Whiskers and platelets are usually single crystals with aspect ratios up to 100 and with tensile strengths near their theoretical value. Composite processing can be tailored to produce either an aligned microstructure with the principal axis of all reinforcements lying in the same direction; a textured microstructure in which the principal axis is randomly arranged within a single plane; or an isotropic microstructure in which the reinforcements are randomly arranged in three dimensions. Aligned reinforcements produce a composite with highly unidirectional properties in the alignment direction, but with properties that are isotropic in the transverse direction. Such microstructures are produced when whisker-reinforced composites are fabricated by extrusion or when platelet-reinforced composites are fabricated by tape-casting or hot-pressing techniques. Textured microstructures produce composites that have isotropic properties in the reinforcement plane. This tends to be the most common type of microstructure produced when a whisker-reinforced composite is fabricated by hot pressing or tape-casting techniques. Fabrication techniques required to produce a completely random microstructure with resulting isotropic properties are extremely difficult. Hence, most ceramic composites reinforced with acicular particles tend to have some form of texture and thus, some anisotropy of properties.

Fiber reinforcements can be amorphous, single crystal or polycrystalline in structure. They can be either short fibers producing similar composite architectures to those of whiskers or they can be continuous. Continuous fiber reinforced composites tend to have orthotropic properties. For unidirectional composites properties transverse to the fiber direction are significantly different from those parallel to the fiber direction.

The role of reinforcements in a ceramic-matrix composite system is to transfer stress from the matrix to the reinforcement, thereby shielding the crack tip from the applied load and providing an additional dissipative energy sink to resist crack propagation. This function is usually achieved via strong reinforcements with a weak interface between the reinforcement and the matrix. This combination allows ligament debonding and energy dissipation via frictional sliding of the reinforcement in the matrix. The matrix and reinforcement are usually chosen to allow weak interface debond stress. However, in practice, it is difficult to achieve this state because most ceramic systems react chemically. In fiber and whisker reinforced ceramic composite systems an interlayer coating of pyrolitic carbon is usually incorporated at the interface to facilitate easy debonding.

An alternative to the weak debond coatings is to create a mechanically weak debond interface. One that shows much promise is a porous coating of the matrix itself on the reinforcement. The coating is well-bonded to the reinforcement and the matrix but is mechanically much weaker than either because of its degree of porosity. A debond crack will thus run preferentially through the coating.

Table 2. Nonoxide-Based Ceramic-Matrix Composites

Reinforcement Type[a]	Amount, vol %	Strength,[b] MPa	Toughness,[c] $MPa\sqrt{m}$	Modulus, GPa[d]	Density, g/cm^3
AlN matrix					
BN_p		65.5		480	
SiC matrix					
SiC_w			19.9	240	
TiB_{2p}	16		4.5	430	3.30
TiC_p	25		6.0	450	3.36
Si_3N_4 matrix					
SiC_w	10	620	7.8		
SiC_w	20		4.0	350	
SiC_w	10	436	5.7		
Si_3N_{4p}		680	7.6–8.6	160	
TiC_p		578	7.2	328	
TiC_p	30		4.5	350	3.7
TiN matrix					
Al_2O_{3p}/AlN[e]		229	10.2		
WC matrix					
Co	20		16.9	442	

[a] Subscripts denote reinforcement morphology; p = particulate, l = platelet, w = whisker, f = fiber, i = interlayer between reinforcement and matrix.
[b] Strength as measured in a four-point flexure test; to convert MPa to psi, multiply by 145.
[c] Fracture toughness; to convert MPa $\sqrt{m}$ to psi $\sqrt{in.}$, multiply by 910.048.
[d] To convert GPa to psi, multiply by 145,000.
[e] 30% each.

An alternative to the weak debond interface approach may lie in a ductile interface that is well-bonded to both the reinforcement and the matrix. Debonding of the interface then entails ductile yielding and shearing of the interface. Such a process potentially dissipates more energy than debonding and frictional interfacial sliding alone.

A related and important issue in choosing a reinforcing material is the chemical compatibility of the reinforcement with the matrix. The reinforcement must also have high strength that is retained to elevated temperatures. If the environment has access to the reinforcement, either at the surface or through matrix cracking, then the reinforcement must be sufficiently chemically inert in the service-environment. Table 3 presents the properties of a few of the currently available platelets, whiskers, and fibers for use as reinforcements in ceramic composites.

Ceramic Matrices

Ceramic matrices are usually chosen on their merits as high temperature materials; reinforcements are added to improve their toughness, reliability, and damage tolerance. The matrix imparts protection to the reinforcements from chemical reaction with the high temperature environment. The principal concerns in choosing a matrix material are its high temperature properties, such as strength, oxidation resistance, and microstructural stability, and chemical compatibility with the reinforcement.

Another consideration is the difference in thermal expansion between the matrix and the reinforcement. Composites are usually manufactured at high temperatures. On cooling any mismatch in the thermal expansion between the reinforcement and the matrix results in residual mismatch stresses in the composite. These stresses can be either beneficial or detrimental: if they are tensile, they can aid debonding of the interface; if they are compressive, they can retard debonding, which can then lead to bridge failure.

Compressive interfacial stresses increase the interfacial shear resistance. Although usually detrimental to toughening, these stresses can enhance toughening if bridge pullout is the operative toughening process.

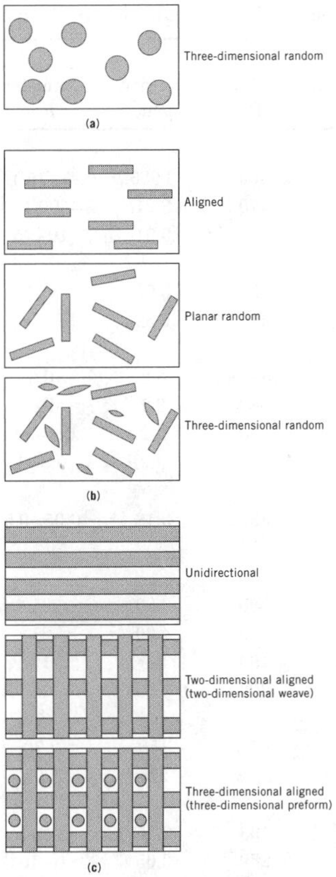

Figure 1. Reinforcement architectures for ceramic–matrix composites and corresponding composite properties. (**a**) Spherical particles; (**b**) platelets, whiskers, short fibers; and (**c**) continuous fibers.

Toughening Processes

The toughness induced in ceramic matrices reinforced with the various types of reinforcements, that is, particles, platelets, whiskers, or fibers, derives from two phenomena: crack deflection and crack-tip shielding. These phenomena usually operate in synergism in composite systems to give the resultant toughness and noncatastrophic mode of failure.

Crack-Resistance Behavior. The goal of composite reinforcement is to produce tough, flaw-insensitive materials that fail in a "ductile" manner. Such materials are more damage tolerant than the monolithic ceramics because they can withstand larger cracks without fracture and the fracture strength may be independent of crack size within a certain flaw size range. This important property of flaw tolerance and stable crack growth results from a fracture resistance behavior known as $\mathcal{R}$-curve or T-curve behavior, in which the fracture resistance rises with crack extension. Fracture resistance can be formulated in terms of either stress intensity factor T or strain energy release rate $\mathcal{R}$ (or J). If stress intensity factor is used, then the ordinate is the square root of crack length and the plot is termed a T-curve. If, however, strain energy-release rate is used, $\mathcal{R}$, (or J_c) is plotted directly as a function of crack length and the curve is termed an $\mathcal{R}$-curve.

Crack Deflection Contribution to Toughening. Crack deflection is a phenomenon that leads both to toughening and to the formation of bridges that shield the crack tip from the applied stress. Little is known of the bridge formation process, but its effect, that is, crack-tip shielding, is considered in the following section.

The condition for propagation of a mode I edge crack, that is, a crack that is subjected to pure opening (tensile) stresses applied perpendicular to the crack plane, is given by:

$$K_a = Y\sigma_a\sqrt{c} = K_{IC} \tag{1}$$

Table 3. Reinforcements for Ceramic-Matrix Composites

	Tensile strength, GPa[a]	Modulus, GPa[a]	Density, g/cm^3	Diameter, μm	Maximum use temperature, °C
Platelets					
Al_2O_3[b]		400	3.986	5–15/1	2040
SiC[c]	3	470	3.21	5–500/1–15	1600
SiC	0.5	470	3.21	10–15	1600
Whiskers					
Al_2O_3[d]	20	450	3.96	4–7	2040
B_4C	14	490	2.52		2450
SiC					
Silar SC[e]	7	340–690	3.2	0.6	1760
VLS[f]	8.3	580	3.2	4–7	1400
Tokamax[g]		600	3.2	0.1–1.0	1400
SiC[h]		600	3.2	0.5–1	1400
Si_3N_4					
SNWB[i]	14	385	3.18	0.05–0.5	1900
Fibers					
Al_2O_3					
FP[j]	1.38	380	3.90	21	1316
PRD166[j]	2.07	380	4.20	21	1400
Sumitomo[k]	1.45	190	3.9	17	1249
Safimax[l]	2.0	300	3.30	3	1250
mullite					
Nextel312[m]	3.12	1.55	150	2.70	1204
Nextel480[m]	4.80	2.28	224	3.05	1200
SiC					
Nicalon[n]	2.62	193	2.55	10	1204
SCS[o]	2.80	280	3.05	6–10	1299
Sigma[p]	3.45	410	3.40	100	1259
SiTiCO					
Tyrrano[q]	2.76	193	2.5	10	1300
Si_3N_4					
TNSN[r]	3.3	296	2.5	10	1204
SiO_2					
Astroquartz	3.45	69	2.2	9	993
Graphite					
T300R[s]	2.76	2.76	1.8	10	1648
T40R[s]	3.45	276	1.8	10	1648

[a] To convert GPa to psi, multiply by 145,000.
[b] Atochem, Centre de Recherche, France.
[c] C-Axis, Jonquiere, Quebec.
[d] Catapal XW, Vista Chemical Co., United States.
[e] Advanced Composite Materials Corp. (ACMC), Greer, S.C.
[f] Los Alamos National Lab, Los Alamos, N. Mex.
[g] Tokai Carbon Co., Japan.
[h] J.M. Huber, Corp., Nacagdoches, Tex.
[i] UBE Industries, Japan.
[j] E.I. du Pont de Nemours & Co. Inc., Wilmington, Del.
[k] Sumitomo Chemical America, New York
[l] ICI Advanced Materials, Wilmington, Del.
[m] 3M Co., St. Paul, Minn.
[n] Nippon Carbon Co., Tokyo.
[o] AVCO Specialty Materials/Textron Inc., Lowell, Mass.
[p] Berghoff, Tubingen, Germany.
[q] Dow-Corning/Celanese, Midland, Mich.
[r] Toa Nevyo Kogyo K. K, Tokyo.
[s] Amoco Performance Products, Ridgefield, Conn.

where Y is a dimensionless geometry term, $ς_a$ is the applied stress, and c is the crack length. Once a crack is deflected from its original plane, further crack extension requires a higher driving force to accommodate the mode II (shear) or mode III (tearing shear) contribution to the stress intensity factor on the new crack plane.

Crack-Tip Shielding. Crack-tip shielding has two origins: process-zone shielding and crack-wake bridging. Process-zone shielding derives from mechanisms occurring in a zone around the crack tip which extend to the crack wake as the crack advances, indirectly applying closure forces to the crack flanks. Crack-wake bridging derives from intact bridging elements in the wake of the crack, directly applying closure forces to the crack flanks.

Mechanical Performance

Particle Reinforcement. Particle reinforcement is an excellent method for toughening brittle ceramic matrices. The toughness imparted to such composites is due to multiple toughening mechanisms including crack deflection, crack pinning, microcracking, residual stress, frictional bridging, particle pullout, and transformation toughening. The mechanisms important to any specific system depend on the physical properties of the particles: size; morphology; thermal expansion mismatch with the matrix; and strength, toughness, and ductility.

Brittle Particles. Reinforcement via small brittle particles exploits the toughening mechanisms of crack deflection, microcracking, crack pinning, and crack bowing. The toughening contribution from the mechanisms of crack bridging and frictional pullout may be significant if the reinforcing particles are of the order of the matrix grain size or larger. All of these mechanisms arise from, or are strongly enhanced by, thermal expansion mismatch stresses in the composite.

Ductile Particles. Ductile particle reinforced ceramic composites show promise as composite material for high strength–high toughness applications.

Ductile particles can act as bridging sites in the crack wake. Instead of fracturing in a brittle manner, they undergo plastic yielding as the crack opens up. A second toughening mechanism that operates simultaneously with crack bridging is the ductile yielding of particles in the crack-tip stress field within a process zone.

Transforming Particles. A special type of particulate-composite are those based on the tetragonal form of zirconia. Tetragonal zirconia has the ability to undergo a stress-induced martensitic phase-transformation from its tetragonal crystal form to a monoclinic form with an accompanying dilatation of 4% unconstrained. The toughening owing to the phase-transformation-induced dilatation in terms of a stress intensity factor approach has been calculated to be that shown in equation 2,

$$K_a = K_o + K_p = T = K_o + 0.3Ee^T V_f w^{1/2} \quad (2)$$

where E is the modulus of the matrix, e^T is the transformation strain per particle, V_f is the volume fraction of transforming particles, and w is the width of the process zone.

Whisker Reinforcement. Toughening for whisker-reinforced composites has been shown to arise from two separate mechanisms: frictional bridging of intact whiskers, and pullout of fractured whiskers, both of which are crack-wake phenomena.

Whisker reinforcement is a viable method of toughening composites. However, health considerations associated with the aspiration of fine, high aspect-ratio whiskers raise serious concern about their widespread use.

Platelet Reinforcement. Ceramic composites reinforced with crystalline platelets show similar values of toughness as whisker-reinforced ceramic matrices. Platelets have the additional advantages of being at least one tenth the cost of whiskers, easier to process, and have higher thermal stability and none of the health hazards associated with the aspiration of whiskers. Toughness comes from a combination of crack deflection, frictional bridging and platelet pullout.

Fiber Reinforcement. The whiskers bridging mechanics apply also to short random fiber bridging mechanisms.

Composites reinforced by continuous fibers can fail in one of several possible modes depending on the interface properties and the fiber strength. When the distribution of fiber strengths is broad (as characterized by a low Weibull modulus) in the regime of low fiber strength/high shear resistance, fiber fracture in the crack wake occurs away from the crack mid-plane and the fibers pull out. The majority of toughening is a result of the frictional pullout mechanism, although there may be some contribution from frictional bridging before fiber fracture occurs.

A transition to a different mode of composite failure, which is still within this low fiber strength/high shear resistance regime, occurs when the fiber strengths have a much tighter strength distribution as characterized by a high Weibull modulus. Initially the fiber strength is sufficient to allow the formation of a bridging zone in the crack wake before fiber fracture occurs at the point of highest stress in the fiber, that is, in the crack mid-plane. There is little to no fiber pullout contribution to toughening, and the contribution from fractional bridging predominates.

In the region of high fiber strength and low interfacial shear resistance, matrix cracks can propagate around the fibers, leaving them intact in the wake of the crack. The matrix can be completely cracked through, with the fibers supporting all the load before fiber failure begins. In such a material the toughness is primarily a result of the bridging contribution, rather than fiber pullout.

Chemical and Thermal Stability

Ceramic-matrix composites are a class of materials designed for structural applications at elevated temperature. Exposure at these temperatures will be for many thousands of hours. Therefore, the composite microstructure must be stable to both temperature and environment. Relatively few studies have been conducted on the high temperature mechanical properties and thermal and chemical stability of ceramic composite materials.

Reinforcement Integrity. Strength degradation with increasing temperature occurs to a much greater extent with ceramic reinforcements, particularly those of continuous fibers, than it does with monolithic materials. Reinforcements have high surface areas to volume so that they are more susceptible to strength degradation resulting from surface reactions with the atmosphere. These reactions can also decrease the toughness of the composite if crack-wake bridging and pullout are the predominant toughening mechanisms.

Composite Response. A majority of ceramic-matrix composites show strong trends in the manner in which the mechanical properties are affected by temperature. If the interface degrades, allowing strong bonding to occur between the reinforcement and the composite matrix, the toughness is considerably reduced. If the interface remains weak enough to allow debonding and pullout, composite strength and elastic modulus are reduced.

Studies on creep resistance of particulate reinforced composites seem to indicate that such composites are less creep-resistant than are monolithic matrices. In contrast to the particulate-reinforced composites, all other reinforcement morphologies appear to provide enhanced creep resistance.

E. P. BUTLER
E. R. FULLER, JR.
National Institute of Standards and Technology

R. Warren, *Ceramic-Matrix Composites,* Blackie, Glasgow and London, 1992.

B. R. Lawn, *Fracture of Brittle Solids,* The Cambridge Press, Cambridge, 1992.

I. N. Sneddon and M. Lowengrub, *Crack Problems in the Classical Theory of Elasticity,* John Wiley & Sons, Inc., New York, 1969.

E. P. Butler, H. Cai, and E. R. Fuller, Jr., in *Engineering Ceramics Division, 16th Annual Conference on Composites & Advanced Ceramics,* ACerS, Cocoa Beach, Fla., 1992, pp. 475–482.

COMPUTER-AIDED DESIGN AND MANUFACTURING (CAD/CAM)

With the aid of computers, design processes can be expedited and implemented with greater accuracy. Computers can simulate the shape of an object, make changes, and display a three-dimensional perspective of the object on a terminal monitor. Numerous computer-aided design (CAD) software packages are commercially available. In automating chemical processes and computations, companies must decide whether to use commercial software packages or to develop in-house CAD software.

Except for the physical processes, such as cutting, forging, and mixing, used to manufacture items, the commonly discussed topics in computer-aided manufacturing (CAM), such as material planning, process and procedure designs, numerical control, and scheduling, might also be considered CAD because design is involved. Computer-integrated manufacturing systems (CIMS) programs have played an important role in the development of CAM activities. Flexible manufacturing systems (FMS) refers to a more recent concept of adapting basic manufacturing modules to meet a particular manufacturing need.

Solid modeling now plays an important role in CAD/CAM. It allows a design item to be simulated on a display monitor, making possible a trial assembly of various design parts before they are actually fabricated.

Computer Application in Chemical Technology

The *Software Directory,* published by the American Institute of Chemical Engineers as a supplement to the journal *Chemical Engineering Progress,* shows the scope of computer utilization in chemical technology. Selection and evaluation of available software packages have been discussed.

Steady-state chemical process simulation has been reviewed. Uses of computers manufactured by Apple, Control Data, Gould, Telex, Lear-Siegler, MINC, ICS, IMLAC, Vector General, IBM, Megtek, and Pet for applications in process design, kinetics, heat transfer, thermodynamics, and mass transfer have been reported, based on two surveys conducted by CACHE (Computer Aids for Chemical Engineering) Corporation.

Supercomputers, such as the CRAY X-MP, CRAY Y-MP, and CRAY-2, are partially available and used for flow-sheet and optimization studies.

Networking provides easy access to software worldwide. NSFNet (National Science Foundation network), Ethernet (Xerox Corporation), ProNet (Proteon, Inc,), and DECnet (Digital Equipment Corporation) are among the most notable networks.

Many commercially available software packages, such as AutoCAD and VersaCAD, can be used to draw chemical processes in block-diagram representation. To manipulate block diagrams, numerous automated procedures are also available. Various pointer devices (mouse, light pen, joystick, and others) are used for interactive selection of particular blocks or branches for simplification and derivation of the system's transfer function. The fundamental steps that need to be taken in such manipulations, using a program called IMPROVISA, have been explained. The number of more elaborate software packages specially designed for chemical processing is rapidly increasing.

Computer Graphics in Instruction

The use of graphic displays as an essential element of computer-based instructional systems has been exploited in a number of ways. Molecular modeling and visualization techniques have supplemented the traditional set of stick models in courses on organic and inorganic chemistry, and animation of molecular motion and of the progress or mechanism of chemical reactions has been a useful classroom tool.

Computer Graphics and Chemical Structural Analysis

CAD/CAM techniques have provided the framework for using the computer as a tool in the drawing and analysis of chemical structures and, more recently, in the use of chemical structures to design reaction pathways and new products. The essential elements in these applications of CAD/CAM are that the possible structures are relatively deterministic and that allowable changes in structure through reaction are governed by thermodynamic, stoichiometric, and steric constraints.

Molecular Modeling for Reaction Path Synthesis and Molecular Design

Applications of CAD/CAM to drawing and modeling chemical structures are essentially passive; that is, although the user decides which compound to draw, strict rules govern what that compound will look like. A reasonable goal for the application of molecular modeling capabilities is the ability to design chemical syntheses based on the structure of a desired product, a set of potential reaction pathways, perhaps defined on the basis of a connection table and a Gibbs energy minimization routine.

An area that has used chemical structures for predictive purposes quite successfully is the estimation of thermophysical properties of compounds. There has been an extensive compilation of estimation methods, and prediction of physical properties has been automated using these techniques.

Pattern Recognition and Interactive Graphics Applications

The graphics capabilities of the CAD/CAM environment offer a number of opportunities for data manipulation, pattern recognition, and image creation. The direct application of computer graphics to the automation of graphic solution techniques, such as a McCabe-Thiele binary distillation method, or to the preparation of data plots are obvious examples. Graphic simulation has been applied to the optimization of chemical process systems as a technique for energy analysis.

The interactive features of the modern computer allow the use of graphic methods to be further expanded. An early effort to couple data fitting with interactive computing was the VIPER program. In the VIPER system the user changes parameters within a curve-fitting routine with a light pen, after viewing the results from the previous fit.

The graphic operating framework has been effective as a basis for image processing applications.

The use of graphic simulators for flow visualization has grown from relatively simple models, such as the generation of natural convection streak lines, to sophisticated commercial programs for the analysis of complex flow regimes.

CAD/CAM Applications in Process Flow-Sheet Development

The use of the computer in the design of chemical processes requires a framework for depiction and computation completely different from that of traditional CAD/CAM applications. For this reason, most practitioners use computer-aided process design to designate those approaches that are used to model the performance of individual unit operations, to compute heat and material balances, and to perform thermodynamic and transport analyses. Typical process simulators have, at their core, techniques for the management of massive arrays of data, computational engines to solve sparse matrices, and unit-operation-specific computational subroutines.

The introduction of menu- and icon-driven interactive programs has permitted improvements in computer-user interfacing. Two programs, PFG for process flow sheets and PIG for piping and instrumentation diagrams, use icon and menu input.

A future goal for the integration of graphics and process design simulators is to be able to use an interactive graphics program to prepare the input to the process simulator. This capability would allow true on-line process modification, flow-sheet optimization, and process optimization, and is likely to be one of the key developments in this field in the 1990s.

Graphics Applications in Process Optimization and Control

The capabilities for computer-aided process design and for utilization of CAD/CAM techniques in association with design have progressed further than similar applications in the design of process control systems.

The goal of integrating automated process optimization with automated design of process control and instrumentation diagrams will likely be more difficult to attain than the integration of graphic input for process simulation programs. The computational engines for optimization are under development at this time, as are effective dynamic simulators.

There are, however, two areas in which graphic methods have had a significant effect on both process design and process control. The synthesis of heat-exchange networks is an exercise in examination of multiple combinations of hot and cold streams transferring heat across heat-exchanger nodes. The network synthesis methods are intrinsically graphic in approach. An interactive software package, RESHEX has been developed to implement this approach. The HEXTRAN program by Simulation Sciences, Inc., is another example of the automation of heat-exchanger network design.

The second area, the implementation of a modern process monitoring and control system, is the most dramatic current application of CAD/CAM technology to the chemical process industry. The state of the art is the use of computer graphics to display the process flow diagram for sections of the process, current operating conditions, and controller-set points.

The process monitors and controllers typically also have the capability for data logging, analysis, and display. This capability has made on-line control of pilot plants as well as commercial-scale processes desirable. A number of commercially available process control programs that run on microprocessors have been reviewed. Virtually all of them incorporate graphic display as an integral part of the interactive capability of the program.

Essential Elements for Developing In-house, Special-Purpose CAD Software

Any program developed in-house must be easy to use, or user-friendly. If the program has various options for input, analysis, computation, and output, then it must provide the user with a fast way to select them. To meet this need, the system is likely to be menu-driven. The peripheral interactive devices such as mice, joysticks, light pens, graphic tablets, and templates are helpful and often used to expedite the selection process.

Modular Approach and Three-Dimensional Solid Modeling

Commonly used entities such as the pipe fittings are increasingly being created and stored as graphic files for quick retrieval. In the construction of three-dimensional piping systems, they can be retrieved, rescaled, and positioned by use of a pointing device for connecting to the other pipe elements.

Other Capabilities

The creation and analysis of process flow sheets has become much easier because of the availability of automated systems to draw and revise them. The goal of the use of the flow sheet as the input for process simulation and for process control is likely to be achieved reasonably soon. The use of interactive graphic displays for process monitoring and control is pervasive today.

Computer graphics will continue to provide an effective tool for drawing chemical structures. Great improvement in software packages can also be expected.

The use of color graphics is also an effective means for displaying chemical structures. This method is far better than typesetting the three-dimensional architecture of complex multimolecule assembly.

CAD/CAM capabilities have had a subtle, but significant impact on chemical technology. Molecular modeling (qv) capabilities have led to the ability to describe chemical behavior and are showing promise in the design of new molecules. Pattern recognition techniques are being used to automate the analysis of chromatograms and spectral information.

Y. C. Pao
L. D. Clements
University of Nebraska, Lincoln

Y. C. Pao, *Elements of Computer-Aided Design and Manufacturing,* John Wiley & Sons, Inc., New York, 1984.

M. F. Hordeski, *CAD/CAM Techniques,* Reston Publishing, Reston, Va., 1986.

M. P. Groover and E. W. Zimmers, Jr., *CAD/CAM Computer-Aided Design and Manufacturing,* Prentice-Hall, Englewood Cliffs, N.J., 1984.

Y. C. Pao and M. Foltz, *Engineering Drafting and Solid Modeling with Silver Screen,* CRC Press, Inc., Boca Raton, Fla., 1993.

COMPUTER-AIDED ENGINEERING (CAE)

Whenever a computer is used for engineering computations, several layers of programming are involved. Figure 1 represents those layers and their relationships with each other.

Process Simulation

At the simplest level, process simulation involves making material and energy balances for a process flow sheet. Generally, it involves much more detailed modeling of any type of process for which there is a continuous flow of materials and energy from one processing unit to the next. Simulators have been used to model processes in chemical and petrochemical industries, petroleum refining, oil and gas processing, synthetic fuels, power generation, metals and minerals, pulp and paper, food, pharmaceuticals, and biotechnology.

Flow-sheet models are used at all stages in the life cycle of a process plant during process development, for process design and retrofits, and for plant operations. Input to the model consists of information normally contained in the process flow sheet. Output from the model is a complete representation of the performance of the plant, including the composition, flow, and properties of all intermediate and product streams and the performance of the process units.

During process development, a model can be developed as soon as a conceptual flow sheet has been formulated. This model can be updated as more information about the process is obtained. Even at an early stage in the project, the model can be used to assess the preliminary economics of the process and the effect of technological changes on these economics. The model can aid in interpreting pilot-plant data and allows the study of many process alternatives.

During process design, once the decision has been made to build a new plant or to modernize an existing plant, simulation models can be used to study trade-offs, to investigate off-design operation, and to analyze the flexibility of the plant to handle a range of feedstocks.

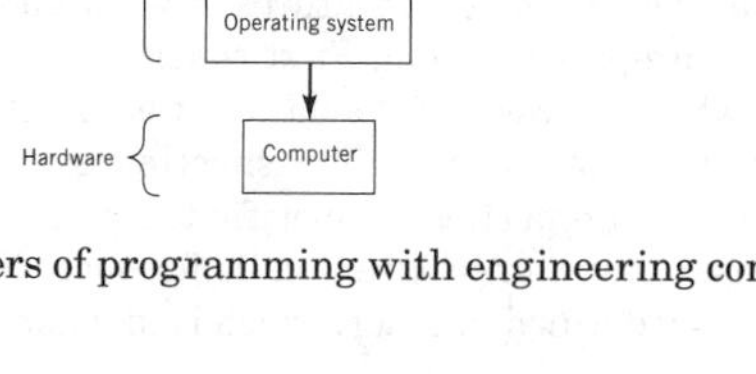

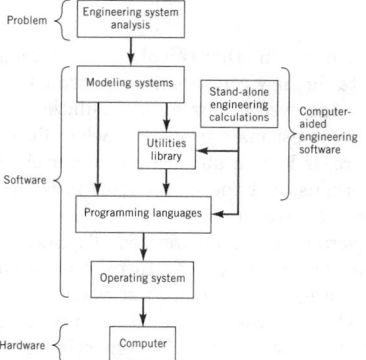

Figure 1. Layers of programming with engineering computation software.

For an existing plant, a model can serve as a powerful tool for plant engineers to improve plant operations, to enhance yield and throughput, and to reduce energy use.

Using simulation offers several advantages. Simulation makes it possible to investigate and experiment with the complex internal interactions of the system being studied whether a single process, an overall plant, or a complete company. It allows investigation of the sensitivity of a system to small changes in internal parameters or environmental conditions and thus helps determine the accuracy to which these factors must be known. Simulation also allows the testing of the model system in regimes of operation that would be too costly, too dangerous, too time-consuming, or beyond the operating ranges of any one particular physical system, ie, systems sizes, construction materials, etc. Because they obviously cause no disturbances in the real system being studied, investigations at the limits of operation of the real system can be made with impunity, without danger to the system or its human operators.

The steps involved in developing a flow-sheet model are as follows. (*1*) Define the process flow sheet to be modeled and the purpose of the model. (*2*) Select the units of measurement for input data and output reports. (*3*) Specify the chemical components present in the streams of the flow sheet. (*4*) Specify the methods and models to be used for calculating physical properties. (*5*) Break the process flow sheet into unit operations and choose an appropriate model for each unit. (*6*) Specify the performance of each unit operation to represent the design and operating conditions of the process. (*7*) Impose design specifications. (*8*) Set up sensitivity analyses or case studies.

Equations-Oriented Simulators. In contrast to the sequential-modular simulators that handle the calculations of each unit operation as an input–output module, the equations-oriented simulators treat all the material and energy balance equations that arise in all the unit operations of the process flow sheet as one set of simultaneous equations. In some cases, the physical properties estimation equations also are included as additional equations in this set of simultaneous equations.

Historically, sequential-modular simulators were developed first. They were also developed primarily in industry. They continue to be widely used. In terms of unit operations, each module can be made as simple or complex as needed. New modules can be added as needed. Equation-oriented simulators, on the other hand, are able to handle arbitrary specifications and limitations for the entire process flow sheet more flexibly and conveniently than sequential-modular simulators, and process optimization can also be carried out with less computer effort.

Convergence Methods. In the simplest terms, convergence means finding a solution to a trial-and-error calculation. The simplest procedure is direct substitution of the recalculated value in the trial-and-error calculation. There are several mathematical and algorithmic techniques (studied in the field of numerical analysis) of speeding up this convergence process by providing a better guess for each succeeding iteration than direct substitution. Newton's method is most commonly the first one attempted, because it is simple and, in many situations, effective. It consists of linearizing the equations based on derivatives at the current point of iteration and then solving the linearized simultaneous equations. The solution provides the guess for the next iteration. For highly nonlinear situations and when the functions have many inflections, Newton's method could run into problems of cycling, and Wegstein's and other methods are used instead.

Steady-State Simulation Programs. The earliest process simulators were developed by petroleum (Exxon and Chevron) and chemical (Union Carbide, Monsanto, and Imperial Chemical) companies and by engineering construction contractors (Kellogg). Many of these simulators contained proprietary information about the properties of some of their materials and were customized to handle proprietary models of some of their processes. A number of independent software companies now develop and support high quality simulation software, and the trend has been for companies that previously developed their

Table 1. Comparison of Simulator[a] Capabilities

	Simulator							
Characteristics	ASPEN PLUS	Chemcad II	Design-II	Hysim	Pro-II	Quasilin	Span	SPEEDUP
company	Aspen Technology	Coade Chem Stations	Chem-Share Corp.	Hyprotech Inc.	Simulation Sciences Inc.	Lynxvale Ltd.	Kesler Eng. Inc.	Aspen Technology UK Ltd.
hardware[a]	1–5	1	1–4	1	1–5	2–4	1,3	2–5
Physical property capabilities[b]								
methods and data available	1,4,5,10–13	1,4,5,7,10–13	1,3–5,7,10–13	1,4,5,7,10–12	1,10–13	7,9	10–12	6,8
handle electrolytes?	Y	Y	Y	N	Y	N	N	Y
handle petrol. fractions?	Y	Y	Y	Y	Y		Y	Y
add own estim. methods?	Y	Y		Y	N		N	Y
data regression?	Y	Y	Y	Y	Y	N	Y	Y
Unit process modeling								
reactor models available[c]	A,B,C	A,B,C	A,B	A,B	A,B,C	A	A	A,B
incorporate user modules?	Y	Y	Y	Y	Y	Y	N	Y
handle batch reactor (R) or batch distillation (D)?	Y	Y	Y	N	D only	Y	N	Y
handle streams with solids?	Y	Y	N	Y	Y	N	N	Y
Input/output features								
accept graphical input as flow diagram?	Y	Y	Y	Y	Y		N	Y
produce PFD[d] as output?	Y	Y	Y	N	Y	N	N	Y
plot curves of simulation results and sensitivity studies	Y	Y	Y	Y	Y	Y	N	Y
System modeling features								
optimizer available?	Y	N	Y	Y	Y	Y	Y	Y
dynamic stimulator available?	N	N	Y	Y	Y	Y	N	Y
algorithm used[e]	SMD	SMD	SM	HS	SM	EO	SM	EO
Other features								
has project design database?	Y	N	Y	Y	Y	Y	Y	Y

[a] Computer platforms: where 1 represents IBM PCs and compatibles, 386 or better; 2, Unix-based work stations (or equivalent) including HP/Apollo, SUN, and IBM RS6000; 3, DEC VMS; 4, IBM mainframe, MVS or VM, or compatible; and 5, others including Cray, Data Gen., etc.
[b] Physical property methods and data include 1, activity coefficient correlation; 2, ChemAdat; 3, ChemTran; 4, DIPPR; 5, Group Contribution; 6, PPDS; 7, Prausnitz; 8, AspenTech's PROPERTIES PLUS; 9, Thermopack; 10, UNIFAC; 11, UNIQUAC, Van Leer, Wilson, and NRTL; 12, Soave-Redlich-Kwong, Peng-Robinson Eq.'s of S; and 13, Chao-Seader, Grayson Streed plus one or more of Braun K-10, B-W-R, and other Eq.'s of S.
[c] Reactor types modeled: A, stoichiometric conversion; B, equilibrium/free-energy minimization, continuous stirred tank, and plug flow; C, reactive distillation. Some vendors have special models for special reactions; also, private company simulators usually have reactors of specific interest to their company.
[d] PFD = process flow diagram.
[e] Algorithmic structure of simulator: EO, equation oriented; HS, hybrid system; SM, sequential modular; and SMD, sequential modular, suitable for design.

own simulators to rely on these independently produced simulation packages. Table 1 lists these companies and their software.

Physical Properties

A process-simulation program almost always contains a physical property service, because the quality of process design ultimately depends on the way in which the laws of physics and chemistry are applied to the problem. Accordingly, the quality of this service is an important consideration to the user of a flow-sheeting system.

The physical property service must perform a number of tasks, but the most useful of these are the following: (*1*) To supply estimates repetitively for a number of different physical properties while the simulation is in progress. (*2*) To provide the user with the values of properties of interest during the calculation and at simulation completion, for subsequent use in other calculations. (*3*) To allow the user to input special data for new components and transform them into the form required by the system during a simulation. (*4*) To supply the user with a means to estimate properties when little, except perhaps chemical structure, is known about a particular chemical compound; again the system must put these estimates into the form needed by the simulation during execution.

General Properties of Computerized Physical Property System. Figure 2 summarizes the general features of what one may expect to find in a physical property system. First consider the central column of the diagram, which represents the basic set of facilities.

The simulation models of the flow-sheeting system must make frequent requests for properties at specific temperatures, pressures, and compositions. Computer-program calls for such data are usually made in a rigorously defined manner, which is independent of both the

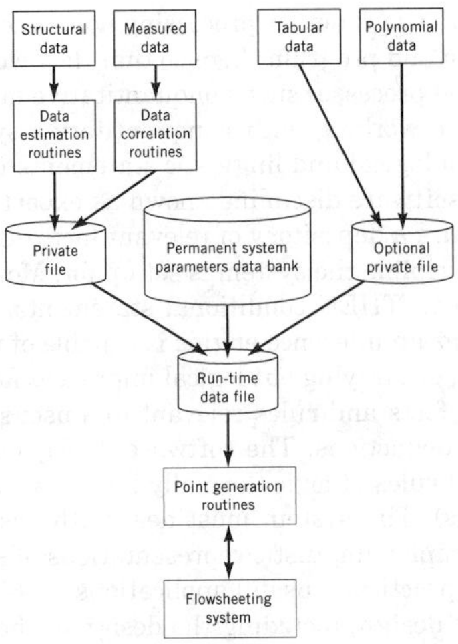

Figure 2. General features of a physical property system.

point data generation models and the particular components. These point generation routines provide the property values, using selected methods that base their calculations on a set of parameters for each component.

At the time of a computer run of a simulation, it is desirable to store only the parameters for those particular components involved in the simulation, and so this set of basic data is normally copied over from the permanent system parameters data bank into run-time data locations. In typical flow-sheeting problems this involves the collection of particular parameters for 5 to 20 components from a data bank with extensive sets of data for up to several thousand components.

Use of such a data system is easy, provided that the permanent system data bank has entries for the components of interest. Thus it is often important to the user that the data bank is extensive and not restricted to a small class of compounds. However, the number of chemical species is enormous and expanding at a rapid rate. Accordingly, for wide application it is necessary that the physical property system embodies facilities for the use of user-supplied data.

This requirement of a physical property system is generally accommodated by enabling the user to create the equivalent of the permanent system data bank by explicitly entering data in the same format. Such private data banks can be used independently or in conjunction with the system data bank. Data supplied in this way normally require that the user has access to expertise in physical property data and possibly in computer use as well. In these circumstances the user may also wish to provide data in tabular or polynomial form for use by an appropriate set of interpolative point generation routines. This facility is shown at the top right of Figure 2.

In addition to these facilities for supply of data in an explicit form for direct use by the system, there also are options designed for the calculation of the parameters used by the system's point generation routines. The first of these applies to the correlation of raw data and is most commonly applied to the estimation of binary interaction parameters.

The second category differs from those discussed above in that it relates, in the main, to those situations for which no data or only characterizing data exist. In such cases, this small set of characterizing data or, in its absence, structure data are used to estimate a set of parameters of the type required by point generation routines.

All the foregoing facilities form part of the spectrum of options that, in addition to the permanent system data bank, enable the engineer to get the most out of a flow-sheeting system. Physical properties that are often required for process simulation are equations of state, density, compressibility, thermal expansion coefficients, heat capacity, enthalpy, heat of mixing, entropy, Gibbs free energy, heats of formation, entropy of formation, heat of reaction, free energy of reaction, K-values (vapor–liquid equilibrium), activity coefficients, fugacity coefficients, chemical potential, fugacity, viscosity, thermal conductivity, diffusion coefficients, and surface tension.

Process Equipment Design and Rating

There is wide array of computer programs that have been developed for the sizing and detailed mechanical design of processing equipment such as distillation and other separation columns, heat exchangers, compressors, etc.

Some process simulators have sizing calculations built into the unit operation models. For example, having determined the flows and internal refluxes in a distillation column, the user can then calculate the diameter of a bubble cap or a packed column. Generally, the sizing calculations do not interact with the material and energy balances and are invoked by the engineer through calling special routines after the material and energy balances have been completed.

For many pieces of equipment, such as heat exchangers and distillation columns, stand-alone programs are available that calculate material and energy balances around that piece of equipment, size the equipment, and calculate or rate its performance.

Process simulators stop generally at the process specifications for the equipment. For the detailed mechanical design of the equipment, such as heat exchangers and distillation columns, stand-alone programs are often used. They make process calculations, size the equipment, calculate thermal and mechanical stresses, design mechanical support of the parts of the equipment, design inlet and outlet nozzles, etc.

Heat exchange packages include HETRAN (Simulation Sciences, Inc.), ADVENT (Aspen Technology, Inc.), HEXNAN (Kesler Engineering, Inc.), HETRAN teams (B-Jac International), Chemcalc 6 (Gulf Publishing Co.), HTC-STX (Heat Transfer Consultants, Inc.), PCI-HEXN (PCI Consultants, Inc.), Energy analyst (Thermal Analysis Systems Co.), SUPER-TARGET (Linnhoff March, Inc.), Chemcalc 5 (Gulf Publishing Co.)

Some computer programs can both design and rate equipment, whereas some can only design or rate the performance of the equipment. For the design of equipment, the process conditions are given as input to the program, which then produces all the size parameters of the equipment as the output. In rating, on the other hand, the physical dimensions or the size of existing equipment are input into the program, which then predicts the performance of the equipment and the outlet conditions of the process streams from the equipment. Rating programs are useful for studying alternative uses of existing equipment for purposes other than what it might have originally been designed for, and for monitoring the internal condition of equipment in service, such as the fouling of a heat exchanger or proper operation of trays or packing in separation columns. They can also be used to identify material and energy losses arising from internal breakdowns of equipment such as tube ruptures.

Optimization

Finding the best solution when a large number of variables are involved is a fundamental engineering activity. Exhaustive enumeration is the most straightforward method of finding an optimum when the number of possibilities is limited and manageable with the computer power available to an engineer. Many stand-alone design programs and most process simulators have a way of making and recording case studies, varying certain parameters so that the optimum can be found by the user conveniently.

Many process simulators come with optimizers that vary any arbitrary set of stream variables and operating conditions and optimize an objective function.

There are several mathematical methods for producing new values of the variables in this iterative optimization process. Mathematical methods that provide continual improvement of the objective function in the iterative process can be classified into the following categories: mountain-climbing procedures, iterative uses of linear programming, methods using penalty functions, and successive quadratic programming.

Dynamic Models and Process Control

Mathematically speaking, a process simulation model consists of a set of variables (stream flows, stream conditions and compositions, conditions of process equipment, etc) that can be equalities and inequalities. Simulation of steady-state processes assume that the values of all the variables are independent of time; a mathematical model results in a set of algebraic equations. If, on the other hand, many of the variables were to be time dependent (in the case of simulation of batch processes, shutdowns and startups of plants, dynamic response to disturbances in a plant, etc), then the mathematical model would consist of a set of differential equations or a mixed set of differential and algebraic equations.

There are special numerical analysis techniques for solving such differential equations. Issues related to the stability and convergence of a set of differential equations must be addressed. The differential equation models of unsteady state process dynamics and a number of computer programs model such unsteady-state operations. They are of paramount importance in the design and analysis of process control systems (see PROCESS CONTROL).

Data Reconciliation

The purpose of data reconciliation is to produce one set of consistent data that can be used for materials accounting, for monitoring the operations of a plant, and for determining the parameters of a model of a process unit.

Process Synthesis

Process synthesis is the step in design when the chemical engineer selects component parts and the interconnection between them to create the flow sheet. This formal approach to design includes developing a representation of the synthesis problem, using a means to evaluate alternatives, and following a strategy to search the almost infinitely large space of possible alternatives. Effective solutions depend heavily on the nature of the synthesis problem being addressed.

Whereas process simulation includes quantitative analysis of a design given the structure of the design, process synthesis involves determining the structure that will meet the requirements of the design as well as finding the best structure for the requirements.

Any formal approach to process synthesis of an industrial problem soon grows into a computational problem of developing an astronomical number of alternatives and then finding a way of selecting among them. The two main approaches to these problems are heuristic and mixed-integer programming. The heuristic approach involves making up different case studies and evaluating them. Mixed-integer programming is a mathematical tool wherein all the defined alternate structures are modeled as a mathematical programming problem and special computer codes produce an optimal solution with the structure and the values of all the variables that produce the best value of the objective function.

Typically, process synthesis involves the synthesis of the following: heat-exchanger networks, separation systems, chemical reaction paths, complete flow sheets, and control systems.

Computer-aided process synthesis systems do not mean completely automated design system. Process synthesis should be carried out by interactive systems, in which the engineer's role is to carry out synthesis and the machine's role is to analyze the performance of synthesized systems. Computer applications in the future will probably deal with the knowledge-based system in applied artificial intelligence.

Artificial Intelligence and Expert Systems

Computers that started out as machines to carry out programmed calculations were soon used as information-processing machines. The next progression was to use them for processing nonnumerical, nonquantitative information and program them so that they would simulate the way a human mind processes such nonquantitative information.

An approach to working with nonquantitative symbolic information in the form of logical and linguistic statements using a computer developed into a software discipline known as expert systems. Expert systems (qv) require a depository of relevant facts (known by experts) for the field or discipline the system is set up for. Most of the facts are in the form of IF ... THEN conditional statements, known as rules. Software known as an inference engine is capable of using the library of facts and rules, of carrying out logical inferences forward and backward, or relating facts and rules relevant to a user's input inquiries, and of reporting deductions. The software deploys and stays within the confines of all rules of logic (formally known as propositional and predicate calculus). The system must deal with vast volumes of information and complex linguistic representations of such information and inquiries for practically useful applications.

The process of design, including the design of chemical processes, is a complex nonquantitative process requiring a vast amount of information, deep knowledge of the field, and expert experience. Artificial intelligence will most likely be applied in developing process designs in the future. This is a promise and a possibility that is a target of current research.

Fault Tree Analysis

Fault tree analysis (FTA) is a widely used computer-aided tool for plant and process safety analysis. One of the primary strengths of the method is the systematic, logical development of the many contributing factors that might result in an accident. This type of analysis requires that the analyst have a complete understanding of the system and plant operations and the various equipment failure modes.

FTA breaks down an accident into its contributing equipment failures and human errors. The method therefore is a reverse-thinking technique, ie, the analyst begins with an accident or undesirable event that is to be avoided and identifies the immediate cause of that event. Each of the immediate causes is examined in turn until the analyst has identified the basic causes of each event. The fault tree is a diagram that displays the logical interrelationships between these basic causes and the accident.

The result of the FTA is a list of combinations of equipment and human failures that are sufficient to result in the accident. These combinations of failures are known as minimal cut sets. Each minimal cut set is the smallest set of equipment and human failures that are sufficient to cause the accident if all the failures in that minimal set exist simultaneously. Thus a minimal cut set is logically equivalent to the undesired accident stated in terms of equipment failures and human errors.

Spreadsheet Programs

Spreadsheet programs are used by engineers to perform a variety of computer design and analysis functions. The applications that are implemented on them include process design and optimization, equipment design, batch distillation, process control, process economics, and many other functional areas of interest.

A spreadsheet program is intuitive in its operation and can immediately show the effect of any one change throughout the whole spreadsheet. It is thus a subtle form of data validation; an error may be spotted immediately. Furthermore, when a material balance is set up, the total effect caused by a change in one variable may be seen at once.

Basically, a spreadsheet program is one that operates on many rows and columns (8196 rows and 256 columns is currently the minimum for any respectable program). A caption or a value may be placed in any row—column location (which is usually referred to as a cell); alternatively, a cell may contain a formula for how its value is to be computed as a function of values stored in other cells. Rows or columns (or portions thereof) may be summed or be involved in other mathematical calculations. Logic tests for conditional results can be put into a cell, regression analyses can be performed on a series of cells, averages can be computed, and many other mathematical functions (rounding, exponentiation, calculation of net present value, etc) are available. A series of cells may also be plotted graphically in many ways (line, bar, pie charts, etc) so that presentation of results is simplified for the user while giving the output a professional appearance. Among the available spreadsheet programs, there is *Lotus 1-2-3* (Lotus Development Corporation), *Symphony* (Lotus Development Corporation), *Full Impact* (Ashton Tate), *Supercalc* (Computer Associates), *Excel* (Microsoft Corporation), and *Quattro Pro* (Borland International, Inc.).

Another implicit advantage to use of spreadsheets, besides their ubiquity and ease of use, is the fact that so many engineers and scientists already are familiar with them. Their use then obviates the need to learn another programming language.

A useful feature of spreadsheet programs is the ability to copy and replicate portions of the spreadsheet that have already been entered. Another important feature of spreadsheet programs is the ability to generate graphical output.

Design and analysis programs involving the spreadsheet approach are available for a wide array of engineering problems. These programs are called templates, and they load the prepackaged spreadsheet into the spreadsheet program. The template contains the column and row headings, built-in constants, and built-in processing rules for all the relevant cells. The user may use the template "as is" or edit it to adapt it to the special situation at hand. If the template exactly fits the problem the user wishes to solve, only those cells that represent input variables are entered into the spreadsheet.

M. T. TAYYABKHAN
Tayyabkhan Consultants, Inc.
H. BRITT
Aspen Technology, Inc.

J. F. Boston, H. I. Britt and M. T. Tayyabkhan, *Chem. Eng. Prog.* **89**(11), 38–49 (Nov. 1993).

R. G. E. Franks, *Modeling and Simulation in Chemical Engineering,* John Wiley & Sons, Inc., New York, 1972.

V. E. Jenson and G. V. Jeffreys, *Mathematical Methods in Chemical Engineering,* Academic Press, Inc., Orlando, Fla., 1990.

C. L. Dym and R. E. Levitt, *Knowledge-Based Systems in Engineering,* McGraw-Hill Book Co., Inc., New York, 1991.

COMPUTER TECHNOLOGY

Personal Computers

Microprocessor based personal computers are now common throughout the chemical community. At the time of writing, the preferred desktop computer for most chemists is either an Intel-based system running Microsoft DOS, Windows or Windows 95, or an Apple MacIntosh system. Currently, Intel/Microsoft systems enjoy an overall market share in excess of 80%, with Apple gaining less than a 10% market share on the desktop. MacIntosh's however, have been considerably more popular within the chemical community and remain the preferred platform in many chemistry focused organizations.

At the time of writing, most chemical information is delivered to the desktop via either terminal emulation to legacy, host-based systems, or, increasingly, via one of several recently emerged, Chemically aware, client-server systems. Computer networking, via both Local Area Networks (LAN) and the Internet is currently facilitating rapid and profound changes in the delivery of chemical information, workgroup computing and access to high performance computing resources. In many organizations, analytical instruments are interfaced to computers, so that results can be viewed, manipulated, archived and recalled from the chemist's desktop. Robot control of instruments and repetitive chemical operations is now widespread.

The current generation of Intel processors (Pentium), and the new PowerPC processors used by Apple, have considerably narrowed the performance gap between "personal computers" and "workstations". Given extremely favorable price/performance ratios, and the emergence of more fully featured operating systems such as Windows NT and Linux, there is an increasing trend towards performing many kinds of numerically intensive computations on "commodity" hardware rather than on more expensive workstations. Many excellent graphics and statistics packages, together with numerous structure drawing and manipulation tools are now available for personal computers, further enhancing their utility to the chemical community.

Supercomputers

Although there may be no universally accepted definition of a supercomputer, there are some characteristics that all supercomputers have. A supercomputer is expensive. Its cost has stayed relatively constant over the years, typically between $1 and $30 million. Performance is the primary goal of the designers of a supercomputer; cost is a secondary consideration at best. Supercomputers utilize the fastest electronic components available, connected in ways designed to minimize transmission delays. With a large number of electronic components driven at very fast rates, the typical supercomputer generates enormous amounts of heat. Most require extensive power-supply and liquid-cooling support systems.

Supercomputers are found in many government research laboratories, intelligence agencies, universities, and a small number of industrial companies.

Vector Computers. Most computers considered supercomputers are vector-architecture computers. The concept of vector architecture has been a source of much confusion.

Most computers use pipelining to maximize performance. Pipelining is analogous to an automobile assembly line, with the finished product being the result of many sequential but otherwise independent operations on the object being produced.

There are many reasons that it might be difficult to keep the pipelines full. The most obvious is a data dependency, where a previously initiated computation must pass through all stages of the pipeline before the next operation can be commenced.

The best way to ensure that the pipelines are full is to gather together a complete series of inputs before starting the pipeline. These are the vector registers of a supercomputer. Vector supercomputers typically contain at least eight vector registers, each holding 64 or more floating-point numbers.

A significant amount of machine overhead is involved in setting up a vector operation, with maximum benefit accruing if there are a full 64 elements to compute. For short loops, typically eight elements or fewer, vector operations are often slower than doing the computations one at a time in nonvector mode.

Other Performance Considerations. Even if a program allows main memory to supply operands at peak rate, it may not be fast enough to keep the CPU operating at its peak rate.

Because of the relative slowness of main memory (compared with the CPU), most computers have a much smaller, but much faster cache memory subsystem that augments main memory. The size of the cache memory and the extent to which a program can utilize the cache can be critical determinants of performance.

The hierarchy of disk, main memory, and cache, each one faster than the one before, is a general one. On the top of this pyramid are registers. Supercomputers typically contain a small number of ultrafast registers within the CPU. The registers hold scalar variables, such as loop counters, or accumulator variables. At all stages of the hierarchy, performance tuning involves maximizing the use of the faster components. The compiler usually decides which variables should be kept in registers.

Performance. The most commonly cited performance measure in the scientific computing world is the LINPACK benchmark. The benchmark involves diagonalizing a 100×100 double precision matrix. The 100×100 problem is moderately vectorizable in that good speedups over scalar execution can be achieved, but it generally does not permit vector computers to perform near peak performance. Because LINPACK is such a well-known benchmark, a great deal of effort is devoted to optimizing its performance on many computers. Successful efforts tuning LINPACK may or may not correlate with increased performance for other programs.

A common acronym is MFLOPS, millions of floating-point operations per second. Because most scientific computations are limited by the speed at which floating point operations can be performed, this is a common measure of peak computing speed. Supercomputers of 1991 offered peak speeds of 1000 MFLOPS (1 GFLOP) and higher.

Performance achieved on single-processor vector computers is governed by Amdahl's law. Once started, vector operations can be performed much faster than single arithmetic operations. Thus, only when substantial parts of the program can be run in full vector mode can significant overall speed improvements be achieved, and regardless of how efficiently the part of the program that can be run in vector mode is run, the nonvector part will still limit the execution time. Most real world, as opposed to the synthetic benchmark, programs contain significant nonvectorizable portions.

Peak Performance. Every computer has a theoretical peak speed, the speed that would be achieved if all the pipelined functional units of the machine could be kept fully supplied with operands. It is a speed that is guaranteed never to be exceeded. These peak processing speeds make for interesting discussion and speculation; however, few programs allow vector computers to run at anything approaching peak speed. Recent LINPACK reports contain a section that describes the performance of a program that does allow vector computers to perform near peak speeds.

Chemistry. Many computational chemistry programs have been adapted for vector processing, often with good gains in speed. The work on adapting the programs has been done both by the original developers and by staffs at the National Science Foundation (NSF) centers. Many of the optimizations are related to the use of optimized matrix manipulation routines or to recasting data structures into forms that are amenable to vectorization.

Ab initio quantum chemistry programs can easily be made to occupy any computer fully, and so are good candidates for a supercomputer.

Supercomputer Costs and Benefits. Given the great expenses associated with supercomputers, not only the initial purchase and facilities cost but also the large support staffs typically required, they are most often shared by large numbers of people, with the undesirable effect of providing to each user only a small fraction of the computer's performance. Sometimes in these situations no user gets supercomputer-class throughput. On widely shared supercomputers, large jobs are often run overnight only in order to maximize the availability of resources for daytime users. There is a great deal of debate about the ultimate value of widely shared supercomputers. Well-funded, grand-challenge, or mission-critical applications can often justify their own dedicated supercomputers; however, such projects seem to be more the exception than the rule.

Minicomputers

The minicomputer became the standard for departmental scientific computing during the 1980s, when computers still cost much more than the labor of the people using them. Painstaking efforts were undertaken to optimize computational chemistry programs for the minicomputer architectures. With computations taking days or more to run, even small percentage gains could translate to time savings of hours.

The Digital VAX rose to prominence as a departmental minicomputer and became a virtual standard in the world of chemistry. The VAX offered a user-friendly flexible environment, together with what was then considered good computational throughput. Much computational chemistry methodology was developed on the VAX.

Chemistry. The widespread availability of minicomputers and the advent of robust programs led to greatly enhanced use of *ab initio* and semiempirical quantum chemistry techniques.

As the minicomputer rose to prominence so did the more widespread use of molecular mechanics as a computational technique. Whereas the quantum chemical programs deal with molecules as nuclei and electrons, the molecular mechanics paradigm treats each atom as a classical ball of a certain mass. The bonds connecting the balls are treated as classical, generally harmonic springs, and bond angles are described by similar classical terms. Through space (London), rather than through bond, interactions are typically described by Lennard-Jones potential functions.

The strength of molecular mechanics is that by treating molecules as classical objects, fully described by Newton's equations of motion, quite large systems can be modeled. Computations involving enzymes with thousands of atoms are done routinely. As computational capabilities have advanced, so have the size and complexity of the systems modeled with molecular mechanics. Unfortunately, the execution time can scale as the square of the number of particles, but the method does hold interesting possibilities for parallel processing, which will be discussed later.

The preeminent offerings in this crowded market include MacroModel (Columbia University, New York), Insight (Biosym Technologies, California), Sibyl (Tripos, Missouri), ChemX (Chemical Design, UK), BioGraf (Molecular Simulations, California), Charmm/Quanta (Polygen Corp., Massachusetts), PC Model (Serena Software, Indiana), ChemLab (ChemLab Inc., Illinois), and a large number of personal computer-based packages.

Work Stations

The mid-1980s was a turning point for minicomputers as microprocessor-based UNIX work stations began to appear. These systems offered large, high resolution, multiwindow monitors, together with computational speed that rivaled or exceeded that available on the minicomputers of the time. There began a transition that later became a stampede.

Vendors such as SUN and MIPS introduced lines of computers based on RISC (reduced instruction set computer) chips. These computers offered significant performance advantages over the CISC (complex instruction set computer) minicomputers, at least for CPU-bound work. Although there are still active debates about what RISC and what CISC are, the essence of RISC is simplicity.

The philosophy of RISC is that the CPU performs a very small number of very simple operations. Whereas a CISC-based computer might have an instruction that fetches a number from memory and updates a counter, a RISC system implements such an operation with multiple, but simple, instructions. By keeping the CPU simple, it can be more readily scaled up to ever greater speeds. The idea is that, although it might execute many more instructions than a CISC machine, it can perform its simple instructions so much faster that it gets more work done in a given time period.

An additional advantage of the RISC microprocessor computers is that their implementors laid plans with a view of semiconductor and compiler technology of the 1990s. Most of the earlier CISC systems were defined with a view of what the technology of the 1980s would bring. RISC-microprocessor-based computers hold an increasing performance advantage over CISC-based systems.

The widespread confusion about millions of instructions per second (MIPS) and MFLOPS, together with the existence of programs that exhibit widely differing behavior on different computers, led to the formation of the SPEC group, Systems Performance Evaluation Cooperative. This group maintains the SPEC benchmark, a suite of programs having differing characteristics. Some are limited by integer performance, others by floating-point performance; some benefit from vectorization, others do not. By reporting a performance metric that is an average over several programs, it is anticipated that the SPEC rating will be more generally predictive than are benchmarks based on a single program. Of course, for those who use primarily floating-point intensive programs, the integer intensive benchmarks in the SPEC benchmark may not be of great interest; however, the purpose of the SPEC suite is to provide a general purpose rating. Note, however, that compilation is

Table 1. SPECmark Ratings[a] for Popular RISC Computers

Computer	SPEC int	SPEC fp	Rating[b]
Hewlett Packard model 730	51	100	76
IBM RS/6000 model 550	34	119	72
Hewlett Packard model 720	39	78	59
MIPS RC6280			42
IBM RS/6000 model 320	16	53	32
		Silicon Graphics 4D/ 35	31
SUN SparcStation II	20	21	21
DEC Decstation 5000/200	18	20	19
DEC Decstation 3100			10
SUN SparcStation I			8

[a] Where available, the SPEC integer (int) rating and the SPEC floating-point (fp) rating are reported separately.
[b] Ratings are normalized to a VAX 11/780. As compilers change, SPEC mark ratings may change.

an integer intensive operation. The SPECmark rating of several well-known RISC work stations is included in Table 1.

RISC work stations have been doubling in performance every two years since the mid-1980s. This growth is much more aggressive than that observed in other technologies (supercomputers, mainframes, minicomputers, etc). This trend must slow down eventually, as limits imposed by constraints such as the speed of light are approached.

As CPU performance increases, the gap between CPU and disk and memory speeds will continue to widen. As limits of technology are approached, other techniques will be needed to gain performance advantages; more functional units, multiple processors, and so on. These approaches are discussed in the sections on minisupercomputers and parallel processing.

Technical RISC work stations have revolutionized computational science in five years.

Visualization

Although humans are not efficient when reading numbers sequentially, they are extremely efficient at visual interpretation because visual information is processed in a parallel fashion. Thus, if a mass of data can be presented in a meaningful visual form, the analysis of computational science can be greatly facilitated. Visual presentation not only speeds up interpretation, but also makes readily discernible insights into the results that may be virtually unobtainable from a pile of paper. Visualization aids in the initial interpretation of the data and also in its communication to others.

Two successful and widespread applications of visualization techniques in the field of chemistry are the visualization of molecular orbitals and the visualization of molecules in molecular mechanics studies.

The challenges for visualization are at least twofold. Faster graphics hardware will be required to display and manipulate more complex data displays. More importantly, the human effort required to develop visualization systems must be reduced. It is the realm of the expert programmer to implement a usable visualization system. General purpose tools that allow the nonexpert to import data in different formats into robust visualization systems are just beginning to appear.

Minisupercomputers

The so-called minisupercomputers that emerged and then declined during the 1980s were for many years some of the most interesting computers in terms of architecture. Many of the more successful aspects of their designs are expected to be incorporated into general computer design practice. The minisupercomputers were typically minicomputer-sized systems that offered performance levels of one-quarter to one-third of that available on the supercomputers of the time. As their prices were close to minicomputer prices, they were an attractive alternative for many who needed supercomputer performance but did not have supercomputer budgets. A dedicated minisupercomputer might provide more throughput than a crowded, shared supercomputer.

For all the excitement and enthusiasm of the computer architects, these computers did not meet with great success in the marketplace, and few companies remain as viable entities. One of the primary reasons for their demise seems to have been the simultaneous rise of the RISC work stations, which killed off numerous other architectural initiatives, whence the term *killer micros*.

Parallel Processing

The vast majority of the speed increases previously described in the uniprocessor world are, in fact, parallel in nature: multiple functional units, multiple pathways to and from memory, pipelined operations, and so on. These have always been within the context of a single, tightly integrated CPU unit that executed a single sequential stream of instructions. With the speeds of such uniprocessors approaching insurmountable physical limitations, the next great leap in computational throughput can only come from parallel processing, that is, having more than one processor cooperatively working on the same problem at the same time. However, there usually comes a point at which adding an extra processor to a problem does not increase throughput.

Mutual Exclusion (MUTEX). The idea of multiple entities all working on the same piece of work raises the issue of coordination and communication among the individual processes.

The notion of an atomic operation is important for synchronization. An atomic operation is one that is indivisible. Once initiated, it will continue to completion. There are usually a large number of synchronization primitives in a parallel computer, most commonly test and set primitives, or semaphores implemented in hardware. A test and set operation tests the current value of a variable and optionally sets a new value, all in one indivisible operation. A semaphore forces the serialization of multiple processes around a critical section, a part of a program that must be executed by only one process at a time, such as changing the balance of a bank account, for example. When multiple processes request the same semaphore, they are automatically serialized. Of course, in systems with multiple semaphores, deadlock conditions can easily arise. These situations are either avoided by careful design or must be detected and resolved as they occur. This is generally a nontrivial problem that could consume significant resources.

Parallel Languages. The computer science community generally abhors FORTRAN and has been predicting its demise for some time. A relatively new realization is that the scientific community will not be abandoning FORTRAN any time soon and that parallel computing must be fully available within a FORTRAN context. The scientific community has been slow in changing to languages that inherently express parallelism. Nevertheless, there are vigorous ongoing efforts in developing parallel programming environments and languages.

Performance Boundaries. Some fundamental laws work against parallel computers. For example, the same program will always run slower on a two-processor parallel computer than on a uniprocessor having double the processor speed, because the parallel computer must spend processing time synchronizing the work of its two processors, a task that the uniprocessor does not need to perform. There is also no guarantee that a program can be broken down into two computationally equal parts.

Speedup. The term *good performance* merits discussion. The ideal parallel computer has as many as an infinite number of processors, as much as an infinite amount of zero-latency shared memory, and all interprocessor communications require zero time. If an infinitely parallelizable problem takes T seconds to execute on one of the processors, then it will take T/N seconds to execute on N processors. More commonly this is referred to as an N-fold speedup. Real computers and problems are different, and full N-fold speedup with N processors is impossible. One measure of the efficiency of a parallel computer is how close it comes to achieving N-fold speedup with a given program. This is a strong function of both the program and how well the parallelism inherent to the problem can be mapped onto the parallelism of the computer. In many cases problems may be too large to run on a single processor, making speedup measurements difficult.

The other kind of performance scaling is, for a constant number of processors, how does the running time change as the size of the problem increases?

Cache Coherence. Another effect that leads to reduced throughput from computers with multiple independent processors is the issue of cache coherence. The performance of many of these machines is critically dependent on the cache utilization of the program. For efficiency, individual processors may store frequently used variables in cache only and not write the value to memory immediately (a so-called write-back cache system). A problem arises when another processor needs to access the current value of that variable. The most current value for that variable may be in another processor's cache memory, rather than in main memory. Multiprocessors with cache memories require explicit mechanisms for ensuring cache coherency among the processors.

Types of Computers. Computers can be classified by Flynn's taxonomy. The three important classes are SISD: single instruction, single data; SIMD: single instruction, multiple data; and MIMD: multiple instructions, multiple data.

Unfortunately, Flynn's classification, although commonly used, is quite restrictive when discussing parallel-architecture computers. There have been several attempts to formulate more detailed classification schemes for the great variety of parallel computers now available. None of these efforts have been entirely successful, and none appear to be in general use.

Coarse-Grained Parallelism. The coarsest grained parallelism is the situation in which separate networked computers cooperate to work on a single problem.

MIMD Multicomputers. Probably the most widely available parallel computers are the shared-memory multiprocessor MIMD machines. Examples include the multiprocessor vector supercomputers, IBM mainframes, VAX minicomputers, Convex and Alliant minisupercomputers, and Silicon Graphics server machines. Most of these computers have several CPUs sharing a common memory and usually a common bus.

Hypercube and Massively Parallel Systems. The NCUBE computer is one of a large class of hypercube topology computers. A hypercube of dimension n contains 2^n-nodes. Hypercube topology offers significant advantages in parallel-computer design. Hypercubes represent a reasonable trade-off between the number of connectors at each node, with the maximum number of interprocessor "hops" a message will need in order to pass from any one node to another being $\log_2(n)$. As this value grows slowly with n, such systems are scalable. Many other kinds of processor topology can be mapped onto a hypercube, with less-connected schemes, a mesh for example, achieved by simply not using existing connections. In order to double the number of processors in a hypercube array, it is only necessary to add one extra connection to each processor. This is a favorable implication for large-scale implementations.

The NCUBE and similar computers often exhibit very good performance on grid-based problems. There is a natural mapping from the spatial nature of the grid to the interconnectivity of the hypercube.

Molecular mechanics simulations can be readily mapped onto such kinds of machine architecture by using the spatial locality of the atoms to determine their allocation to processors. Short-range van der Waals forces can usually be accurately modeled with a cut-off distance of less than one nm, so interprocessor communication requirements can also be localized.

Whereas the NCUBE and Intel iPSC MIMD computers utilize relatively complex chips at each node, the Connection Machine combines as many as 65,536 simple processors, each of which has 8K or more of local memory (a system maximum of 8 gigabytes in a fully configured system). When operating at full capacity, a Connection Machine can perform 3500 million 32-bit integer operations per second. If equipped with floating-point processors, its peak speed is 21 GFLOPS (double precision). Like the NCUBE, the Connection Machine can be logically subdivided into independent subcubes, thereby facilitating sharing by multiple users and experiments with differing numbers of processors.

The massively parallel approach adopted in the Connection Machine has been termed data parallel. Whereas a uniprocessor must sequentially step through large amounts of data, a data parallel machine moves processors to the data. Aggregate memory to processor bandwidth in the Connection Machine is more than 700 megabytes per second.

The Connection Machine computers have been quite successful, finding application in a wide variety of fields despite relatively high costs. Many image processing algorithms exhibit good data locality and perform well on a Connection Machine as do grid-based fluid dynamics calculations.

IAN WATSON
Consultant

K. Dowd, *High Performance Computing,* O'Reilly & Associates, 1993.

CONTACT LENSES

Contact lenses are small, hemispherical-shaped optical devices placed in contact with the transparent tissue at the front of the eye called the cornea. The capillary attraction of the liquid tear layer between the contact lens and the cornea and the partial coverage of the lens by the eyelids prevent contact lenses from being dislodged during eye movement and blinking. Contact lenses are used to correct vision deficiencies such as myopia (nearsightedness), hyperopia (farsightedness), presbyopia (loss of near focusing power with age), and astigmatism. An increasingly common use of contact lenses is to change the color of a normal eye or to improve the appearance of an eye disfigured by injury or disease. Contact lenses can also be used as therapeutic devices in the medical treatment of certain eye diseases or injuries.

The number of contact lens wearers has grown to an estimated 24 million in the United States and 50 million worldwide. Concurrently, there has been a proliferation of contact lens manufacturers and products. The 1980s saw the widespread introduction of lens products made of more oxygen-permeable materials, ie, rigid gas-permeable (RGP) materials that made PMMA lenses virtually obsolete and high water content hydrogels that competed with HEMA-based lenses.

Research continues for materials having improved oxygen permeability, deposit resistance, and comfort. In addition, there is the search for breakthrough lens designs, such as bifocal contact lens products. The challenge for contact lens manufacturers is to capture more than a 15% share of the vision correction population, now dominated by traditional eyeglasses. Furthermore, competition includes developing new surgical techniques, eg, excimer laser refractive surgery that modifies the shape of the cornea to correct vision, potentially without the use of eyeglasses or contact lenses. Detailed information on historical and clinical aspects of contact lenses and overviews of refractive surgery techniques are available.

Clinical Aspects. No particular contact lens type or product is considered universally superior. In some regions of the world hard lenses dominate the market, eg, some European countries and Japan; in other regions, eg, North America and Scandinavia, soft lenses dominate. Contact lens practitioners select their preferred type of lens using criteria other than just lens material properties. However, among soft lenses, HEMA-based lenses are prescribed most often, and among hard lenses, silicone–acrylate RGP lenses are most common.

To remain safe and efficacious on the eye, contact lenses must maintain clear and wetted surfaces, provide an adequate supply of atmospheric oxygen to and adequate expulsion of carbon dioxide from the cornea, allow adequate flow of the eye's tear fluid, and avoid excessive abrasion of the ocular surface or eyelids, all under a variety of environmental conditions. The clinical performance of a contact lens is controlled by the nature of the lens material; the lens design; the method and quality of manufacture; the lens parameters or specifications prescribed by the practitioner; and the cleaning, disinfection, and wearing procedures used by the patient.

Corneal edema occurs with almost every type of contact lens as a result of corneal hypoxia during lens wear. It is most noticeable when lenses are worn during sleep. Although corneal edema is reversible, excessive and prolonged edema can lead to changes in corneal curvature, growth of blood vessels into the central cornea across the line of sight, or vision reduction through corneal haziness. Excessive edema may predispose the cornea to abrasion, inflammation, or infection, leading to a corneal ulcer. Consequently, a primary goal of contact lens manufacturers has been to develop materials with higher oxygen permeability to reduce corneal hypoxia.

All contact lenses tend to accumulate debris, deposits, and discolorations on and sometimes below the surfaces. The deposits are usually proteinaceous or lipid but may also be inorganic compounds and microorganisms, eg, fungi and bacteria. The source of these deposits is usually the eye itself, the tear film that bathes the surface tissues, the eyelid glands, the mucoid substance secreted by certain conjunctival cells, or the immunoglobulins secreted through the vascular system. Although regular cleaning and disinfection can control these accumulations and reduce the likelihood of clinical problems such as reduced vision, inflammation or red eye, abrasion, and infection, the deposits are considered undesirable.

Clinical experience has shown that certain types of lens materials are more prone to deposit problems. In general, lenses with negatively charged moieties at the surface accumulate greater amounts of lysozyme, the principal tear film protein. The introduction and use of disposable lenses make these deposits and their clinical problems less significant.

Classification of Contact Lenses. All contact lenses can be divided according to the wearing modality. Daily wear lenses are worn during the day only, being placed on the eye in the morning and removed, cleaned, and disinfected before sleep at night. Extended wear or flexible wear lenses can be worn continuously for several days and nights, including during sleep. They have a higher oxygen transmissibility than daily wear lenses, achieved either by making the lens thinner or by making the lens from a more oxygen-permeable material.

It is convenient to classify lenses according to whether they are rigid (hard) or flexible (soft). Some newer materials give rise to lenses that can be termed semirigid, semisoft, and the combination soft–rigid. Hard lenses, including RGP lenses, are invariably nonhydrogels, although their surfaces are wettable on the eye. Soft lenses are usually hydrogels. One notable exception is the silicone elastomer lens that is quite flexible but is not a hydrogel.

The flexibility of a contact lens material dictates, to a large extent, the design of the lens. Hard lenses are generally 8–10 mm in diameter, cover only some of the corneal surface, and fit mostly in between the upper and lower eyelids. Soft lenses are generally 13–15 mm in diameter, cover the entire corneal surface, extend onto the white of the eye, and fit well underneath the upper and lower eyelids.

Properties of Contact Lenses

Not every polymer can be manufactured successfully into a contact lens. Several important properties for both ocular physiology and patient handling are required of a material for a contact lens application. In addition, the type of lens application, ie, rigid, flexible, or soft, will dictate the range and importance of the key properties. Key properties include oxygen permeability, water content and refractive index, mechanical properties (strength and modulus), light transmittance, wettability, and deposition.

Rigid (Hard) Lenses

Hard lenses can be defined as plastic lenses that contain no water, have moduli in excess of 5 MPa (500 g/mm^2), and have Tg well above the temperature of the ocular environment. Poly(methyl methacrylate) (PMMA) has excellent optical and mechanical properties and scratch resistance and was the first and only plastic used as a hard lens material before higher oxygen-permeable materials were developed.

During the 1980s, a series of hard lenses with higher oxygen permeability, the RGPs, were developed and introduced to the market. As a result, the popularity of PMMA lenses decreased even further.

Rigid Gas-Permeable Lens Products. From the late 1970s to the early 1990s, a long series of RGP lens products were introduced, offering higher oxygen permeability, better wettability, and more deposit resistance. Table 1 gives a representative list of RGP products introduced to the marketplace.

Hydrogel Contact Lenses

Hydrogels are water-containing polymers, hydrophilic in nature, yet insoluble. In water, these polymers swell to an equilibrium volume and maintain their shape. The hydrophilicity of hydrogel is a result of the presence of functional groups such as $—NH_2$, —OH, —COOH, $—CONH_2$, —CONH—, $—SO_3H$, etc. The insolubility and stability of hydrogels are caused by the presence of a three-dimensional network. The scope, preparation, and characterization of hydrogels has been reviewed.

Hydrogels used for contact lens applications can be classified based on the chemical structure of hydrophilic monomers and on the water content of the hydrogel lens.

The U.S. Food and Drug Administration, which has the functional authority to regulate contact lenses, classifies soft contact lens according to the water content and ionic character of the soft lens, ie, group I, low water (<50%) nonionic; group II, high water (>50%) nonionic; group III, low water (<50%) ionic; and group IV, high water (>50%) ionic. This classification is particularly useful for clinicians and the FDA in determining care system efficacy.

Typical Hydrogel Components. Regardless of water content, a hydrogel lens formulation consists of hydrophilic monomers, cross-linkers, and initiators. Hydrophilic monomers provide the water needed in a hydrogel, and the cross-linkers help to hold the hydrogel's shape. The initiator starts the polymerization and curing and is removed from the hydrogel after the polymerization. In addition to the above three components, a hydrogel formulation also may contain a hydrophobic monomer as a physical property modifier and a solvent to facilitate the processing.

The water content of a hydrogel depends on the hydrophilicity of the monomer.

Mechanical properties of a hydrogel lens also are affected by the use of a hydrophobic monomer, such as a low alkyl methacrylate.

Table 1. Partial List of Rigid Gas-Permeable Lens Materials

USAN name[a]	Material composition	*Dk*, barrer[b]	Trade name	Manufacturer
airfocon	*t*-butylstryene	25	Airlens	Wesley-Jessen
fluorofocon	fluoropolymer	95	Advent	3M Corp.
itafluorofocon	fluorosilicone	55	Equalens	Polymer Technology
itafocon	silicone acrylate	14	Boston II	Polymer Technology
		32	Optacryl K	Optacryl
		54	Optacryl EXT	Optacryl
pasifocon	silicone acrylate	49	Paraperm EW	Paragon
siflufocon	silicone acrylate	55	Quantum I	Bausch & Lomb

[a] U.S. Adopted Name Council designation for polymeric materials.
[b] 1 barrer = 10^{-11} cm^3O_2 (at STP)·cm/s·cm^2·mm Hg.

Hydrogel lenses are obtained by photo or thermal polymerization of monomers in bulk or in solution. A variety of thermal and photo catalysts have been used in hydrogel formulations.

Because of the many choices of hydrophilic monomers, cross-linkers, and hydrophobic monomers, a large number of formulations have been developed and manufactured into hydrogel lenses. The water content of these hydrogel lenses ranges from about 38%, for HEMA-based lenses, to 80%, for poly(vinyl alcohol) and partially hydrolyzed acrylonitrile lenses.

Oxygen permeability depends exclusively on water content. To increase oxygen transport to the cornea through a hydrogel lens, the lenses should either be made thinner or be fabricated from hydrophilic monomer with higher hydrophilicity. Some hydrogel lenses, such as etafilcon, are manufactured from formulations containing an acid monomer, such as methacrylic acid, to gain higher water content. Unfortunately, the presence of methacrylate anion invites large amounts of lysozyme uptake, which reduces the water content of the lens and may affect clinical performance.

Silicone Hydrogels. Based on the success in RGP applications, polysilox-anylalkyl acrylates–methacrylates, such as TRIS, have been employed in the preparations of silicone hydrogels. These materials were used in combination with hydrophilic monomers such as HEMA and *N,N*-dimethylacrylamide to form hydrogel lenses with high oxygen permeability. They also have been used with methacrylate-based prepolymers, such as those with polyurethane linkages, to give hydrogel lenses with high oxygen permeability as well as excellent mechanical properties.

Fluorohydrogels. In addition to applications in hard lenses, fluorine-based monomers, such as fluoroalkyl acrylates–methacrylates and fluorostyrene and fluorosulfonamide monomers give hydrogel lenses after copolymerizing with hydrophilic monomers such as substituted acrylamide and NVP. In general, these hydrogels claim to have good oxygen permeability and to be useful as extended wear contact lenses.

Miscellaneous Hydrogel Classes. Because of unique structure and flexibility in manipulating the mechanical properties, polyurethanes have been evaluated for contact lens applications.

Flexible Nonhydrogel Lenses

Silicone–Fluorosilicone Lenses. Efforts to modify the silicone lens surface for improved wettability have achieved limited success. These efforts include grafting hydrophilic monomers, such as HEMA, GM, and NVP, to the lens surface and plasma treatments of finished lenses.

Other polysiloxane-containing polymerizable systems, such as prepolymers described earlier, also claim to have application as flexible nonhydrogel lenses. Prepolymers containing block copolymers of polysiloxane and poly(alkylene oxide) give highly oxygen-permeable flexible lenses with good wettability. Monomers with fluoroalkyl groups, or polysiloxane prepolymers with fluoroalkyl modified siloxane side chains, are used in addition to a wetting monomer to minimize lipidlike deposits.

Nonsilicone Flexible Lenses. To avoid wettability and lipid deposit problems associated with silicone, nonsilicone-containing elastomeric materials have been investigated as lens materials. The compositions studied have shown good oxygen permeability.

Tinted Lenses

Contact lenses are tinted for cosmetic reasons, ie, to modify or change eye color, and for visibility, ie, to locate the lens more easily when it is out of the eye. Both rigid and hydrogel lenses can be visibility tinted. However, only hydrogel lenses are cosmetically tinted, because the rigid lens is rarely fit to cover the entire colored portion of the eye. Cosmetic tints were introduced in the early 1980s as an added feature to soft contact lenses. A wearer's natural eye color was combined with the tint in the lens to either enhance the existing color, ie, blue on blue, or to change the eye color, ie, a yellow-tinted lens over blue eyes results in green eyes. Lenses are tinted after lens manufacture with a masking technique to provide a clear annulus around the lens periphery. Because a standard soft contact lens extends past the limbal junction and the iris, the clear annulus was required to prevent the tinted portion from being visible against the white of the eye.

In the late 1980s, a new type of cosmetic-tinted lens appeared on the market. Called an opaque tint, the lens blocks the natural color of the iris and provides a totally new color to the eye.

Visibility tints are similar to cosmetic tints except that they are significantly lighter in intensity. Visible light transmission losses are typically 1–2%. With these tints, the lens is visible in the lens package and against a variety of surfaces, and the lens is easier to handle. The tint can be added as cosmetic tints, with a masking technique to produce a clear annulus, or the whole lens can be tinted.

Lens Colorants. Colorants used with rigid lenses are generally pigments cast into the solid matrix during polymerization. The vast majority of tints used for soft contact lenses use dyes originally developed for the textile industry, and classified according to the American Association of Textile Chemists and Colorists (AATCC). Vat, pigment, and reactive dyes are routinely used (see DYES, REACTIVE).

Surface-Modified Lenses

Surface modification of a contact lens can be grouped into physical and chemical types of treatment. Physical treatments include plasma treatments with water vapor (silicone lens) and oxygen and plasma polymerization for which the material surface is exposed to the plasma in the presence of a reactive monomer. Surfaces are also altered with exposure to uv radiation or bombardment with oxides of nitrogen. Ion implantation (qv) of RGP plastics can greatly increase the surface hardness and hence the scratch resistance without seriously affecting the transmission of light.

Chemical treatments provide a wider variety of applications, depending on the chemical functionality at the original surface and the type of functionality desired as the replacement. Most treatments replace hydrophobic and nonwetting surfaces with hydrophilic and wetting surfaces. Chemical modifications by acylation, esterification, hydroxylation, and treatment with organic anhydrides and alkali have been reported.

Manufacture of Contact Lenses

Today, most contact lens materials, including xerogels, ie hydrogels in their dry state, can by produced in a disklike form (button) that allows lathe cutting for accurate manufacture of contact lenses.

The first hydrogel poly(HEMA) lenses were manufactured using the technique of spin casting. Today, cast molding is an increasingly used manufacturing process for both rigid gas-permeable and hydrogel contact lenses.

YU-CHIN LAI
ALAN C. WILSON
STEVE G. ZANTOS
Bausch & Lomb

R. B. Mandell, *Contact Lens Practice*, 4th ed., Charles C Thomas, Springfield, Ill., 1988.

E. S. Bennett and B. A. Weissman, eds., *Clinical Contact Lens Practice*, J. B. Lippincott. Philadelphia, Pa., 1991.

V. Kudela, in J. I. Kroschwitz, ed., *Polymers: Biomaterials and Medical Applications*, Encyclopedia Reprint Series, John Wiley & Sons, Inc., New York, 1989.

CONTAINERS. See PACKAGING, CONTAINERS and INDUSTRIAL MATERIALS.

CONTRACEPTIVES

Today, millions of people throughout the world use a variety of methods to regulate human fertility, including chemical methods, eg, oral contraceptives, vaginal contraceptives, injectable/implantable contraceptives, and contragestational agents; intrauterine devices; intravaginal barrier methods; male and female condoms; surgical sterilization; induced abortion; and natural family planning. Each of these has its advantages and disadvantages, and it is generally accepted that none of these methods represents an ideal method of fertility regulation for everyone.

Oral Contraceptives

Almost 150 million women have used oral contraceptives sometime during their reproductive lives. The commercial market for oral contraceptives is large and expanding.

Second and Third Generation Oral Contraceptives. Most oral contraceptives are combinations of an estrogenic agent and a progestational agent (progestogen). Estrogens are found in the ovary, and are important in preventing pregnancy; they work in conjunction with the progestogen to suppress ovulation.

The estrogenic and progestational components provide their primary contraceptive effect by blocking ovulation, ie, preventing the selection of a dominant follicle in the ovary by a negative feedback action on the hypothalamus and pituitary. This inhibits pituitary secretion of Follicle-Stimulating Hormone (FSH) and Luteinizing Hormone (LH), with the resultant inhibition of ovulation. The estrogen component also provides stability to the endometrium so that unwanted breakthrough bleeding can be avoided. This combination also provides several ancillary contraceptive mechanisms by interfering with fertilization and implantation processes should ovulation occur.

Early oral contraception formulations contained up to 100–150 μg of estrogen. By the early 1990s the majority of oral contraceptives contained only 30 or 35 μg of estrogen (Table 1). These low dose products are commonly referred to as second-generation oral contraceptives, ie, new products which have almost totally replaced the original high dose oral contraceptives. Numerous review articles and reports on cohort studies describing the use of oral contraceptives and the incidence of side effects are available.

As companies continued to conduct studies to find the lowest doses of estrogen and progestogen effective as contraceptives, women began to experience increased levels of intermenstrual bleeding. This observation led to the development of multiphasic oral contraceptives, a new approach to low-dose contraception (Table 2). Multiphasic oral contraceptives vary the dose of active ingredients or the ratio of progestogen to estrogen throughout the cycle, instead of remaining constant as in conventional combination oral contraceptives, to utilize the lowest effective dose of active ingredients yet still control intermenstrual bleeding. Some, but not all, of the low dose multiphasic oral regimens studied significantly reduce intermenstrual bleeding and spotting. The success of this approach was seen in the marketplace when the first triphasic oral contraceptive was introduced in the United States (Ortho-Novum 7/7/7) in 1984. It is now the most commonly prescribed oral contraceptive in that country. The products in Tables 1 and 2 are representative of products being sold throughout the world.

Studies published in the 1970s and 1980s demonstrate that androgenicity associated with the progestational component of some of the original progestogens is associated with changes in lipid metabolism. These changes may impact cardiovascular morbidity. Hence, oral contraceptives that do not disturb the various blood lipid fractions and do not lower HDL may be preferable to the ones that shift the lipid profile in an undesirable direction. Three companies, ie, Ortho Pharmaceutical Corporation, Organon, and Schering AG, have attempted to dissociate the androgenicity and progestational activities of steroidal progestogens using medicinal chemistry approaches. Norgestimate, desogestrel, and gestodene emerged from this research. These more selective progestogens, in combination with ethinyl estradiol, composes the third generation oral contraceptives. During the 1990s and beyond, the usage of oral contraceptives containing these progestins will continue to grow.

Table 1. Low Dose Monophasic Estrogen–Progestogen[a] Combination Oral Contraceptives Marketed in the United States

Trade name	Estrogen, μg	Progestogen, mg	Launch date	Manufacturer
Bevicon	35	0.50	10/75	Syntex
Demulen 1/35	35	1.00[b]	1/82	Searle
Demulen 1/50	50	1.00[b]	12/70	Searle
Desogen	30	0.15	1/93	Organon
Genora 0.5/35	35	0.50	10/89	Rugby Labs
Genora 1/35	35	1.00	12/86	Rugby Labs
Genora 1/50	50[c]	1.00	10/86	Rugby Labs
Levlen	30	0.15[d]	1/86	Berlex
Loestrin 1.5/30	30	1.50[e]	10/73	Parke-Davis
Loestrin 1/20	20	1.00[e]	10/73	Parke-Davis
Lo-ovral	30	0.30[d]	3/75	Wyeth
M.E.E.	35	1.00	3/88	Lexis Pharm
Modicon	35	0.50	12/74	Ortho
Nelova 0.5/35E	35	0.50	11/87	Warner-Chilcott
Nelova 1/35E	35	1.00	11/87	Warner-Chilcott
Nelova 1/50M	50[c]	1.00	11/88	Warner-Chilcott
Norcept-E 1/35	35	1.00	8/89	Gynopharma
Nordette	30	0.15[d]	5/82	Wyeth
Norinyl +35	35	1.00	12/83	Syntex
Norinyl +50	50[c]	1.00	12/83	Syntex
OrthoCept	30	0.15	1/93	Ortho
Ortho Novum 1/35	35	1.00	1/80	Ortho
Ortho Novum 1/50	50[c]	1.00	4/67	Ortho
Ortho-Cyclen	35	0.25[f]	10/92	Ortho
Ovcon 35	35	0.40	4/76	Mead Johnson
Ovcon 50	50	1.00	9/78	Mead Johnson
Ovral	50	0.50[d]	9/76	Wyeth

[a] Ethinyl estradiol-norethindrone unless otherwise noted.
[b] Ethynodiol diacetate.
[c] Mestranol.
[d] Norgestrel.
[e] Norethindrone acetate.
[f] Norgestimate.

Chemical Classification. Chemically, the various progestogens can be classified in many ways. Estranes are 19-nortestosterone derivatives; gonanes are 19-nortestosterone derivatives with a C-13 ethyl group; and progenanes are 17-alpha-OH progesterone derivatives similar in structure to progesterone itself.

Progestogen-Only Oral Contraceptives. Progestogen-only oral contraceptives, ie, minipills, are available but are not used as extensively as combination oral contraceptives. These preparations contain 19-norsteroids. The contraceptive effectiveness of these products is not as high as that of combination oral contraceptives; intermenstrual or breakthrough bleeding and spotting occur more frequently with these products. Progestogen-only oral contraceptives are prescribed for breast-feeding women because progestins do not influence milk production, and for women for whom estrogens are contraindicated.

Areas of Continued Research. Research continues in many academic and pharmaceutical laboratories throughout the world with the objective of improving oral contraceptives and better understanding their pharmacological and clinical actions.

Long-Acting Contraceptives Long-acting contraceptives avoid compliance issues, and are useful in countries with fewer health professionals. Their popularity is based on simplicity of administration and a relatively high degree of effectiveness.

Table 2. Low Dose Multiphasic Estrogen–Progestogen Combination Oral Contraceptives Marketed in the United States

	Ethinyl estradiol		Progestogen[a]			
Trade name	Regimen, days	Dosage, μg	Regimen, days	Dosage, mg	Launch date	Manufacturer
			Levonorgestrel			
Triphasil	1–6	30	1–6	0.050	12/84	Wyeth
	7–11	40	7–11	0.075		
	12–21	30	12–21	0.125		
Tri-levlen	1–6	30	1–6	0.050	1/86	Berlex
	7–11	40	7–11	0.075		
	12–21	30	12–21	0.125		
			Norethindrone			
Jenest	1–21	35	1–7	0.50		Organo
			8–21	1.00		
Ortho Novum 10/11	1–21	35	1–10	0.50	3/82	Ortho
			11–21	1.00		
Ortho Novum 7/7/7	1–21	35	1–7	0.50	4/84	Ortho
			8–16	0.75		
			17–21	1.00		
Tri-norinyl	1–21	35	1–7	0.50	4/84	Syntex
			8–16	1.00		
			17–21	0.50		
			Norgestimate			
Ortho Tri-cyclen	1–21	35	1–7	0.180	10/92	Ortho
			8–16	0.215		
			17–21	0.250		

[a] Products are grouped by progestogen.

Injectable Contraceptives. Long-acting activity of number of different approaches. Steroid activity can be extended over a period of time by chemical modifications, such as esterification, which result in products that require enzymatic hydrolysis over time for conversion to active products. Long-acting activity is also achieved by the presentation of drugs, with limited aqueous solubility, to the injection site in a microcrystalline aqueous suspension.

Two well-known injectable long-acting products are medroxyprogesterone acetate (Depo Provera) and norethindrone enanthate (NET EN). Both of these products are progestational in nature. Depo Provera (Upjohn Company) is a once-every-three-months injectable administered as a microcrystalline aqueous suspension. NET EN (Schering AG) is a two-month injectable administered as solution in caster oil.

The acceptability of long-acting injectable contraceptives is tempered by the effects of these drugs on endometrial function and the presence of nuisance side effects such as intermenstrual spotting and bleeding.

The importance of predictable withdrawal bleeding has led to the development of combined progestin–estrogen injectable formulations. These types of formulations are administered on a monthly basis. Regularized bleeding occurs approximately one week after each injection and is the result of decreasing levels of estrogen. This is termed an estrogen withdrawal bleeding. Once-a-month injectables are popular in some Latin and South American countries; their introduction into other developing countries is being attempted by WHO.

Implanted Contraceptives. Controlled release of contraceptive progestins also can be accomplished by incorporating the drug into an implantable cylinder or rod. The best known implant is Norplant, developed by the Population Council. Norplant is composed of six matchlike silastic cylinders with the progestogen levonorgestrel incorporated on the inside of each cylinder. After implantation under the skin, the product can provide effective contraception for a period of five years. Because silastic is not biodegradable, the implant can be removed from the patient at any time.

Biodegradable Implants. The Capronor device has walls composed of sigma caprolactone and releases levonorgestrel. It is a single implant with a projected life span of 12–19 months. Capronor I had the drug dissolved in ethyl oleate. Since the use of excipients such as ethyl oleate is undesirable, investigators have attempted to regulate the rate of drug release by reducing the wall thickness. Studies in animals indicate that the thinner-walled Capronor II may have an effective life of at least 12 months.

Fused Pellets. Another form of an implant is a fused pellet. The pellets may be composed of either the drug alone, or the drug fused with cholesterol, and are formed as small cylinders by melting the drug and then solidifying it under pressure. Clinical studies with norethindrone pellets have been in progress for a number of years.

Vaginal Rings. Contraceptive vaginal rings consist of silastic shells or core rings of various sizes and membrane thicknesses. They have been developed for delivery of progestins alone or progestins combined with estrogens. Rings releasing both a progestin and an estrogen resemble combined oral contraceptives. After the ring is removed, a predictable withdrawal bleeding takes place. Vaginal rings have been under development for nearly fifteen years, by the World Health Organization and Population Council.

Contragestational Drugs

Pharmacological substances that either inhibit implantation or interrupt pregnancy after implantation have been investigated during the 1980s and 1990s. A number of different terms have been used to describe these compounds, including antiimplantive agents, postcoital contraceptives (eg, Danazol $C_{22}H_{22}NO_2$, RU-486, and Centchroman 3,4-*trans*-7-methoxy-2,2-dimethyl-3-phenyl-4-[4(2-pyrrolidino ethoxy) phenyl]-chromane), morning-after pills, once-a-week pills, interceptives, abortifacients (eg, Prostaglandins, RU-486, and progesterone synthesis inhibitors), and contragestational agents. This medical approach to fertility regulation presents several principal advantages, including potentially fewer long-term side effects as a result of short-term periodic administration, and greater convenience.

However, in the decision to develop and market contragestational drugs, social, political, legal, ethical, and religious factors have been of critical importance.

Vaginal Contraceptives

The active agents employed in vaginal contraceptive products may be classified as weak acids, organometallic compounds, and surfactants. Because of the inferior spermicidal potency of the weak acids, and the growing concerns over the potential toxicity of mercury-containing compounds, surface active agents constitute the most important class of spermicidal compounds in vaginal contraceptive products. A review of the history of vaginal contraceptives that provides description of selected products and reviews the literature on clinical efficacy and various aspects of use and distribution is available. Surface active spermicidal agents include nonoxynol-9 (7-nonylphenoxypolyethoxyethanol, Triton N), octoxynol (4-diisobutylphenoxypolyethoxy ethanol, Triton X-100), and menfegol (4-menthanylphenylpolyoxyethylene ether).

In the last several decades, physical properties of vaginal contraceptive formulations have been improved to deliver spermicide more effectively and enhance consumer compliance. The formulation that delivers the spermicide can affect the efficacy of vaginal contraceptives. Formulations currently available include jellies, creams, suppositories, aerosol foams, and foaming tablets. Each consists of a relative inert base material that serves as a carrier for the chemically active spermicide and blocks to some extent the passage of sperm.

Intrauterine Devices

Intrauterine devices are medical products that prevent conception when placed in the uterus. In spite of their ancient origins, modern intrauterine devices (IUDs) have been widely used only in the last

30 years. The two generic subclasses of IUDs are nonmedicated (inert) devices and medicated IUDs, ie, progestin-releasing and copper IUDs.

Complications associated with IUDs include uterine perforation and pelvic inflammatory disease. Uterine bleeding and cramping are the most common causes for discontinuation of this method.

Inert Devices. Inert IUDs act by creating an environment hostile to sperm or fertilized ova and by blocking implantation. The exact mechanism of action of IUDs is not totally clear, but convincing evidence is mounting to support the idea that IUDs act primarily as contraceptives and not abortifacients.

There are eight types of inert IUDs used around the world; two are unmedicated and six are copper. Outside of China, the Lippes Loop, made of polyethylene, is the most widely used unmedicated IUD. The other main type of unmedicated IUD, used mostly in China, is a flexible steel ring, ie, the Chinese IUD.

Medicated Devices. Medicated IUDs consist of an inert base reservoir for a uterus-affecting or spermicidal agent. Medicated IUDs as of this writing are either metal-bearing or progestogen-bearing devices.

Sterilization

Sterilization is the most used contraceptive method in the world, predominantly because of usage in developing countries, including China, and is the second leading contraceptive method in the U.S. for contraceptors ages 15–44. Although sterilization procedures in the male and female can be reversed under certain circumstances, the procedure is irreversible in most cases. Worldwide, an estimated 42 million couples rely on vasectomy; nearly 140 million rely on female sterilization.

Physical Barrier Methods

Various physical barrier devices are available for contraceptive use by men and women. Modern barrier methods such as diaphragms, condoms, and cervical caps were made possible by the discovery of the vulcanization of rubber.

Natural Methods

Natural methods, ie, natural family planning, are methods based on awareness of the fertile and infertile segments of the menstrual cycle. This awareness can be utilized to avoid pregnancy or to become pregnant.

Recent findings may lead to better identification of the fertile period. Changes in the quality and quantity of cervical mucus occurring 5–6 days prior to the mid-cycle surge of luteinizing hormone (LH), which initiates ovulation, can be used to predict the fertile period. However, it is sometimes difficult to recognize the changes in mucus. Various devices have been marketed to automate the daily temperature monitoring or to assist in cervical mucus collection. However, such devices produce only marginal improvement because of the inherently variable nature of temperature and mucus texture as fertility markers. Colorimetric enzyme immunoassays have been developed for the measurement of LH in urine. Since LH is rapidly excreted, the increase in urine LH levels can be used as a marker to predict impending ovulation. However, because of the life span of sperm, one must be cautious in predicting the usefulness of this method. Research on methods of consistently and accurately predicting ovulation is ongoing.

Breast-Feeding

In many societies, it is believed that women who are breast-feeding are incapable of becoming pregnant. Suckling leads to a release of prolactin and endorphins that interfere with the hormones necessary for ovulation. Although breast-feeding does affect ovulation, the duration of lactational amenorrhea and infertility is variable, and lactation appears to be unreliable as the sole method of fertility regulation.

New Approaches

Contraceptive Vaccines. Major research efforts involve immunological approaches to fertility control. The development of contraceptive vaccines is directed towards the immunoneutralization of reproductive process or the interference of fertilization by inducing antibodies against oocytes and spermatozoa. Attempts have been made to develop vaccines against leutinizing hormone releasing hormone (LHRH) (also known as gonadotropin releasing hormone, GnRH), LH, follicle stimulating hormone (FSH), human chorionic gonadotropin (hCG), placenta antigen, the zona pellucida of the ovum, and different sperm antigens.

Luteinizing Hormone-Releasing Hormone (LHRH). Scientists also continue to study LHRH analogues for contraception. Treatment with LHRH analogues blocks ovulation in women and spermatogenesis in men. However, it also results in a loss of estrogen and testosterone and causes other related side effects.

Inhibin and Activin. Although the spectrum of functions of inhibin and activin are not completely understood at present, this peptide family has already demonstrated, by the nature of its differential subunit association, a powerful mechanism for the generation of dimers with opposing biologic actions. These characteristics of the inhibin peptide family warrant further study and evaluation as alternative approaches to fertility control.

Progesterone Antagonists as Contraceptives. Another area of antifertility research involves progesterone antagonists or inhibitors. Inhibitors of progesterone synthesis, such as epostane, and inhibitors of progesterone-receptor binding, such as RU-486, have been investigated for termination of pregnancy. Studies in the nonhuman primate indicate that progesterone antagonist may have antifertility potential other than as an abortifacient.

Male Fertility Control. The ideal male contraceptive would produce azoospermia without compromising libido and sexual potency. While not totally fulfilling the criteria for a perfect male contraceptive, GnRH antagonists hold a greater potential than GnRH agonists. Unlike the agonists, GnRH antagonists inhibit gonadotropin secretion, decrease androgen levels, and induce azoospermia in male primates. Similar effects on hormone secretion have been reported in men.

D. W. Hahn
J. L. McGuire
R. W. Johnson Pharmaceutical Research Institute
Gabriel Bialy
National Institutes of Health

W. H. W. Inman, M. P. Vessey, B. Westerholm, and A. England, *Br. Med. J.* **2,** 203 (1970).

H. W. Ory, A. Rosenfield, and L. C. Landman, *Fam. Plann Perspect.* **12,** 278 (1980).

S. Ramcharan, R. A. Pellegrin, R. M. Ray, and co-workers, *The Walnut Creek Contraceptive Drug Study: A Prospective Study of the Side Effects of Oral Contraceptives,* Vol. 3, U.S. Dept. of Health and Human Services, Government Printing Office, 1981.

"Oral Contraceptives", in *Popul. Rep.* [*A*] **6** (May–June 1982).

R. J. Aitken and M. J. K. Harper, *Contraception* **16,** 227 (1977).

S. J. Segal, *Frontiers in Reproduction and Fertility Control,* MIT Press, Cambridge, 1977, p. 170.

Out Look **8**(3), (1990).

CONTROLLED RELEASE TECHNOLOGY

AGRICULTURAL

Controlled release technology in agriculture encompasses the controlled delivery of plant nutrients, ie, fertilizers, as well as control chemicals, eg, herbicides (qv), insecticides, fungicides (qv), etc, to a target in a manner which maximizes its use efficiency, minimizes potential negative effects associated with overdosage, and/or extends the time in which sufficient dosages are delivered. Controlled release fertilizer (CRF) technologies are emphasized here.

The benefits of controlled release fertilizers come at a cost premium, eg, the cost of controlled release nitrogen (CRN) can be anywhere from 2.5 to 10 times the cost of nitrogen from a conventional fertilizer such as urea. Because of the large price differential with conventional fertilizers, usage of CRFs has been limited. In 1990, controlled released nitrogen sources accounted for only 1% of total U.S. nitrogen fertilizer consumption. Usage of CRFs has been limited primarily to specialty markets where their advantages justify the increased cost, ie, high to medium value agricultural crops, consumer lawn and garden, lawn and landscape care companies, golf courses, nurseries and greenhouses, and professional turf applications.

Numerous technologies have been developed with varying degrees of success to maximize the various benefits of controlled release. Heightened attention to environmental concerns has prompted increased activities worldwide in the development and commercialization of controlled release fertilizers.

Synthetically produced CRFs are categorized either as nitrogen reaction products or coated fertilizers. Nitrogen reaction products are produced by the chemical reaction of water-soluble nitrogen compounds, such as urea or ammonia, to create nitrogen fertilizers possessing more complex molecular structures. These complex molecules have limited water solubility and are slowly broken down in the soil environment to nitrogen forms which the plant can assimilate. The limited solubility and the rate of decomposition controls the nitrogen availability to the plant.

Coated fertilizers achieve controlled release by coating a soluble fertilizer core (substrate) with a water-insoluble barrier which limits the access of water to the fertilizer and thus limits its dissolution rate.

Nitrogen Reaction Products

Urea–Formaldehyde Reaction Products. Urea–formaldehyde (UF) reaction products represent one of the older controlled release nitrogen technologies. The reaction of urea with formaldehyde results in a distribution of polymers (oligomers) of varying polymer chain length, with their solubility in water varying inversely with molecular weight.

Granular compositions of UF reaction products are divided into three classes, each based on their degree of water solubility as affected by their polymer distributions, ie, ureaform, the least water-soluble class; methylene ureas; and methylene diurea/dimethylene triurea (MDU/DMTU) compositions, the shortest-chain MU oligomers and the most water-soluble.

Urea–formaldehyde reaction products also are available commercially as liquids that can be categorized into two classes, ie, water suspensions and water solutions.

Physical properties of the methylene urea polymers which have been isolated are compared to urea in Table 1.

Urea–formaldehyde fertilizers are prepared from the reaction of formaldehyde with excess urea (U/F mole ratio greater than 1) (Table 2).

Urea–Other Aldehyde Reaction Products. Urea can also react with other aldehydes to form slow release nitrogen fertilizers. However, cost constraints associated with higher aldehydes have either precluded or limited broad commercial development of these products. Two exceptions are isobutylidene diurea (IBDU), registered trademark of Vigoro Industries, and crotonylidene diurea (CDU), registered trademark of Chisso-Asahi Fertilizer Company.

Table 1. Physical Properties of Methylene Urea Polymers

Product	Melting point, °C	Water solubility, g/100 g	Temperature, °C
urea	132.7	100	17
MDU	205–207	2.5	25
		7.0	50
		0.1	25
DMTU	231–232	4.4	100

Table 2. U.S. Manufacturers of UF Reaction Products

UF reaction product	Manufacturer	Trade name
	Granular products	
ureaform	NOR-AM Chemical Co.	Nitroform
	Omnicology, Inc.	Organiform
	The Scotts Company	Granuform
methylene ureas	The Scotts Company	ProTurf
		HiTech
		Scotts
		ProGrow
	NOR-AM Chemical Co.	Nutralene
	Lebanon Chemical Corp.	Country Club Greens Keeper
MDU/DMTU compositions	The Scotts Company	Scotts
		Triaform
	Liquid products	
methylol urea solutions	CoRoN Corp.	CoRoN, Folocron
	Georgia Pacific Corp.	GP 4340
		GP 4341
	Hickson Korlen, Inc.	Formalene Plus
MDU solutions	C.P. Chemical Co., Inc.	Nitro 26-CRN
urea–triazone solutions	Hickson Kerley, Inc.	N-Sure
		Trisert
UF suspensions	Marral Chemical Co	UWIN
	Georgia Pacific Corp.	RESI-GROW

IBDN, the reaction product of urea/and isobutyraldehyde, is produced by Mitsubishi Kasei Corporation in Japan, BASF in Germany, and IB Chemical in the United States. It is marketed in the United States by Vigoro Industries for use in the turfgrass, nursery, and specialty (medium to high value) agriculture market.

CDU, the reaction product of urea and acetaldehyde, is manufactured by Chisso Corp. (Japan) and marketed as CDU by Chisso Asahi Fertilizer Co., Ltd. It is also produced by BASF (Germany) under Japanese license. It is sold under the trade name Crotodur in Europe.

CDU has very limited use in the United States. It is used primarily in Japan and Europe where it is produced. It serves the turf and specialty agriculture markets and is typically formulated into granulated N–P–K fertilizers.

Other Reaction Products. Nitrogen reaction fertilizers are commercially available that do not involve reactions between urea and aldehydes. These are oxamide and melamine.

Coated Fertilizers

Coated fertilizers represent the fastest growing segment in controlled release fertilizer technology. During the 1980s, coated fertilizer products grew at a rate of about 10% per year, UF product usage grew at about 2% per year, and all manufactured controlled release fertilizers, coated and nitrogen reaction products, grew at a rate of about 4% per year. The growth of coated fertilizers is the result of more favorable economics, increased flexibility in nutrient release patterns as compared to nitrogen reaction products, and the flexibility in controlling the release of other nutrients in addition to nitrogen.

The three categories of coated fertilizers are those using sulfur as the coating material, those that employ a polymeric material, and hybrid products that utilize a multilayer coating of sulfur and polymer.

Sulfur-Coated Fertilizers. Sulfur-coated ureas (SCUs) typically have bulk densities of 720–800 kg/m^3 (45–50 lb/ft^3). The color ranges from brown to tan to bright yellow depending on the source of urea, whether or not a sealant is used, and the type sealant employed. Soft wax sealants are typically used as a secondary coating over the sulfur coating to fill in imperfections in the sulfur coating and to provide handling integrity to the brittle sulfur coat.

The total nitrogen content of SCUs vary with the amount of coating applied; SCUs available in the early 1990s range from 30 to 40% nitrogen.

The process for sulfur-coated urea, as developed by The Tennessee Valley Authority (TVA), involves the precision application of three coating materials to a soluble fertilizer core, usually granular urea. The coatings include molten sulfur, a hydrocarbon/oil (soft wax) sealant, and a flow conditioner. The process involves preheating urea prior to coating with molten sulfur. Preheating the urea allows better adhesion of the sulfur and more time for the molten sulfur to wrap the urea prior to solidifying. The quality of the urea substrate as well as the sulfur are essential in order to produce a quality product.

Polymer-Coated Fertilizers. Polymer-coated fertilizers (PFC) represent the most technically advanced state of the art in terms of controlling product longevity and providing nutrient efficiency. Because most polymer-coated products release by diffusion through a semipermeable membrane, the rate of release can be altered by composition of the coating and the coating thickness. Polymer coatings can be categorized as either thermoset resins or thermoplastic resins (see COMPOSITE MATERIALS–POLYMER-MATRIX). Because of the relative high coatings cost of most polymer-coated products compared to SCU, their use has been primarily restricted to high value applications. Polymer-coated fertilizer technologies vary greatly between suppliers.

Polymer/Sulfur-Coated Fertilizers. Polymer/sulfur coated fertilizers (PSCF) are hybrid products that utilize a primary coating of sulfur and a secondary polymer coat. These fertilizers were developed to deliver control release performance approaching polymer-coated fertilizers, but at a much reduced cost. Sulfur is employed as the primary coating because of its low cost. Low levels of a polymer surcoat are used to control nutrient release rate.

HARVEY M. GOERTZ
The Scotts Company

New Developments in Fertilizer Technology, 14th Demonstration, TVA: National Fertilizer Development Center, Muscle Shoals, Ala., Oct. 5–6, 1983, pp. 30–33.

Farm Chemicals Handbook, Meister Publishing, Willoughby, Ohio, 1990.

Proceedings of the Symposium of Fertilizer, Present and Future, Japanese Society of Soil Science and Plant Nutrition, Sept. 25–26, 1989.

S. P. Landels, M. M. Smart, J. Bakker, and J. Shimosato, "Controlled Release Fertilizers and Nitrification Inhibitors," *Chemical Economics Handbook,* SRI International, Menlo Park, Calif., 1994.

PHARMACEUTICAL

Controlled-release drug delivery not only provides a predictable, patterned action but controls the rate of drug release for a predetermined period.

Controlled-release dosage forms enhance the safety, efficacy, and reliability of drug therapy. They regulate the drug release rate to control drug action and reduce the frequency of drug administration to encourage patients to comply with dosing instructions.

A principal focus of research and development in controlled release technology is the delivery of proteins and peptides, owing in part to biotechnological advances that allow the large-scale production of therapeutically important natural proteins, such as insulin and analogues. Because these substances tend to be broken down and inactivated in the gastrointestinal tract before they can be absorbed in therapeutic amounts, researchers are looking into enhancing oral delivery, eg, using permeation enhancers to increase absorption, as well as using less conventional delivery sites and routes of administration including the rectum, skin, oral cavity, nose, eyes, lungs, uterus, and vagina.

Delivery Sites, Portals, and Systems

Controlled-release pharmaceuticals provide local or systemic treatment, depending on their use and route of administration.

Gastrointestinal Tract. The gastrointestinal (GI) tract is the most familiar and widely used portal for drug delivery, largely because of the ease with which medications can be administered.

Stomach, Small Intestine, and Colon. These organs provide accessible portals for local delivery of therapeutic agents, including laxatives, antidiarrheal medications, gastroprotective agents such as sucralfate, and anthelmintics. They are also important portals for systemic therapy.

Dosage forms have been designed to optimize the localized actions or systemic absorption of drugs by targeting specific GI tract areas for drug release. Many dosage forms have pH-sensitive enteric coatings that prevent digestion in the stomach, reducing local irritation and ensuring that the drug core does not dissolve until it reaches the small intestine or colon. Permeation enhancers can be used to improve absorption through the gastric mucosa.

The design and fabrication of numerous commercially available controlled-release oral dosage forms have been widely reviewed.

Rectum. Local treatment of conditions and diseases of the rectum, including hemorrhoids and local inflammation, usually involves drug-containing suppositories or enemas. As a portal for systemic drug delivery, the rectum has the advantage of bypassing the liver and the upper GI tract. Drugs delivered rectally include sedatives, antihistamines, and headache medications.

Skin. Dosage forms that use the skin as a reliable portal for systemic drug delivery are available. Despite skin's limited permeability, the transdermal route is appealing because it avoids first-pass metabolism and minimizes the dose necessary to achieve therapeutic serum-drug levels. To be suitable for systemic therapy from a passive transdermal delivery system, a drug must be nonirritating and potent and have a short half-life, low molecular weight, and low melting point.

Passive transdermal delivery systems on the market tend to be either matrix- or membrane-controlled. In matrix devices, the structural and molecular characteristics of the drug–polymer matrix determine drug release.

Membrane-controlled diffusional systems enhance transdermal delivery by increasing control over serum-drug concentrations. They are particularly suitable for delivering compounds with a narrow therapeutic index. Because it is predominantly the membrane rather than the skin that governs drug release, the system reduces the problem of differing skin absorption rates.

Various strategies are available to overcome the barrier properties of skin. Permeation enhancers can facilitate transdermal transport of drugs with low percutaneous flux. Numerous potential permeation enhancers have been reported in the literature, including various alcohols, propylene glycol, and sodium lauryl sulfate. Ultrasound (phonophoresis), electrical charge (electrotransport), and bioconvertible drug precursors, or pro-drugs, also are being investigated as ways to enhance drug transport through the skin.

Oral Cavity. The oral mucous membranes are subject to many diseases and chronic lesions, for which a variety of local treatments are available. Rinses and swabbed liquids are effective but must be applied frequently. Slow-dissolving lozenges, eg, Mycostatin, which delivers the antifungal antibiotic nystatin, and chewing gum that delivers the antifungal miconazole can sustain drug release in the oral cavity, reducing the need for frequent reapplication. Incorporation of drugs in inert polymerized acrylics prolongs drug delivery for several days. Oral mucosal membranes provide a port for systemic therapy as well.

Nose. Intranasal delivery of drugs, such as antihistamines and decongestants, to alleviate local symptoms is common. There is now increasing interest in using the nose as a portal for systemic drug delivery, particularly for the delivery of proteins and peptides. Several classes of nasal permeation enhancers have been investigated, including surfactants, such as polyoxyethylene-9-lauryl ether, bile salts and bile salt derivatives, such as sodium taurodihydro-fusidate, and mucoadhesive systems. Controlled delivery from bioadhesive microspheres has been proposed to avoid the rapid clearance of therapeutic agents by nasal cilia. Such microspheres would have to be larger than 10 to 15 μm in order to remain in the nasal cavity and not travel to the lungs.

Eye. The eye is an easily accessible but problematic portal for local therapy, that is, for eye diseases or trauma, because constant tear flow and lacrimal–nasal drainage result in extensive drug loss. The most familiar local treatments (eye drops, suspensions, and ointments) require frequent application to achieve steady-state drug levels. Numerous approaches enhance local treatment by improving corneal permeation with ionophores, ion pairs, liposomes, or pro-drugs, or by reducing loss from drainage with viscosity-enhancing agents, suspensions, emulsions, ointments, or erodible or nonerodible matrices.

Additional research activity includes development of an ocular polymeric drug delivery system for local therapy. The system utilizes the prepolymer α, ω-bis-(4-methacryloxybutyl)-poly(dimethylsiloxane) (M_2D_x) copolymerized with acrylate-based monomers for local therapy.

No ocular products for systemic therapy are commercially available, but research is under way on ocular systems for the systemic delivery of therapeutic agents such as insulin.

Lungs. The lungs are easily accessible for both local and systemic treatment. Local treatment of various lung and bronchial diseases typically involves the inhalation of drugs such as sodium cromoglycate, corticosteroids, pentamidine, and beta adrenoceptor agonists such as salbutamol and terbutaline. Solutions of these drugs are converted into aerosols and delivered with nebulizers or metered dose inhalers. Strategies to lengthen the duration of drug release into the respiratory tract include modifying a drug's molecular design, eg, salbutamol is modified to form salmeterol; changing the form of a drug, eg, from a salt suspension to a parent free base; or incorporating drugs into liposomes.

The lungs' large, permeable surface makes systemic delivery possible. For example, an inhaler delivers 360 μg per dose of aerosolized ergotamine tartrate for migraine, and inhalant systems deliver anesthetic gases. Research is under way on the systemic delivery of proteins and peptides through the lungs.

Uterus. As a portal for drug delivery, the uterus has had fairly limited use. The only local treatment is through intrauterine devices (IUDs) of contraceptive metals or steroid hormones, which provide long-term contraception. No uterine dosage forms are available for systemic drug delivery.

Vagina. The vagina is an accessible but little used route of drug administration. All commercially available vaginal delivery systems are for local therapy. Vaginal douches, suppositories, creams, foams, and ointments effectively deliver drugs for vaginal infections, contraception (spermicides), and vaginal dryness (estrogens).

The permeability of the vaginal mucosa makes the vagina a route for systemic delivery as well. Vaginal rings that deliver contraceptive steroids for up to 21 days are being tested clinically, as are rings delivering estradiol for postmenopausal estrogen replacement.

Intramuscular, Intravenous, and Subcutaneous Portals. These routes require injections, infusion systems, or implants. Injections for local treatment are limited to a few common drugs, including procaine or cortisone to ease pain and stiffness in joints and bursas. For systemic therapy, a wide variety of drugs are available in intramuscular or subcutaneous injections for intermittent dosing. Sustained systemic treatment can, of course, be achieved by intra-arterial or intravenous infusions with catheters attached to containers or pumps. Several miniaturized infusion pumps that attach to the patient, allowing prolonged ambulatory care, are now available, including the Baxter Infusor, the Mill Hill Infuser, and the Auto-Syringe. Sustained systemic treatment is also available through implants such as the Norplant system, a 5-yr contraceptive recently approved for marketing in the United States. For agents such as insulin, a vapor-pressure-powered implantable system about the size and shape of a hockey puck provides long-term continuous infusion.

Energy Sources for Controlled-Release

Biodegradation. The design of a therapeutic system varies with its energy source as well as its route of administration. Some overlap exists; biodegradable devices often employ biodegradation and diffusion processes for drug release.

Three mechanisms for the biodegradation of polymers have been discussed. Mechanism I involves water-soluble polymers that are cross-linked by covalent bonds to make them insoluble. Hydrolytic cleavage of either the cross-links (type IA) or the backbone (type IB) yields water-soluble polymers or polymer fragments whose size depends on the density of the bonds being hydrolyzed. Mechanism II involves water-insoluble polymers that become soluble when pendent groups are hydrolyzed, ionized, or protonated. In Mechanism III, cleavage of the hydrolyzable bonds in the polymer backbone produces low molecular weight, water-soluble fragments. Actual biodegradation may be a combination of these mechanisms.

Biodegradable polymers have several advantages as drug carriers. They allow controlled release over a designated period of time, can target a specific body site, cause little or no tissue reaction, and cause less discomfort and inconvenience than multiple injections. Implants of biodegradable polymers need not be removed. Finally they enable the use of drugs with short *in vivo* half-lives and may improve the bioavailability of drugs having low aqueous solubility.

In order to become useful drug delivery devices, biodegradable polymers must be formable into desired shapes of appropriate size, have adequate dimensional stability and appropriate strength-loss characteristics, be completely biodegradable, and be sterilizable. The polymers most often studied for biodegradable drug delivery applications are carboxylic acid derivatives such as polyamides; poly(α-hydroxy acids) such as poly(lactic acid) and poly(glycolic acid); cross-linked polyesters; poly(orthoesters); polyanhydrides; and poly(alkyl 2-cyanoacrylates).

Homopolymers and copolymers of lactic acid and glycolic acid, originally developed for use in absorbable sutures (qv) (for example, Dexon and Vicryl), are particularly appealing for controlled drug delivery because of their demonstrated safety. Many types of pharmaceutical agents, including narcotic antagonists, contraceptive steroids, antiinflammatory steroids, anticancer agents, antimalarials, antibiotics, peptides, and local anesthetics, have been incorporated into these polymers. Systems based on the polymers have been prepared in several forms, including cylinders and rods, particles and powders, microcapsules and microspheres, beads, films, fibers, and needles.

Drug release from biodegradable devices occurs by degradation of the polymer, by diffusion, or by a combination of these two processes.

Diffusion. Diffusional drug delivery systems utilize the physicochemical energy resulting from concentration differentials. Drug molecules diffuse through a polymer matrix or through a polymer membrane film from a region of high concentration to one of low concentration.

Elasticity. A portable infusion controlled-release device, the Baxter Infusor, utilizes energy stored in a distended, elastomeric rubber tube to deliver drug solutions as constant rates of flow through fixed resistances. The pharmacist inflates the drug reservoir with a loaded syringe inserted through a septum at the filling port, and the liquid drug is metered through a glass capillary. A description of the physicochemical properties of the elastomeric component and the design of the elastomeric reservoir is available.

Electrokinetics. Electrokinetics, the motion produced by an electrical field, is expected to play an increasingly important role in enhancing transdermal drug delivery. In an electrical field, transdermal transport of ionized compounds occurs primarily through the sweat glands and hair follicles, the so-called shunt pathways through the stratum corneum, significantly increasing transport rates and allowing delivery of large molecules such as proteins and peptides that do not cross the skin in passive systems. Electrically assisted transdermal drug delivery, ie, electrotransport or iontophoresis, involves the three key transport processes of passive diffusion, electromigration, and electroosmosis.

A simple electrotransport therapeutic system is composed of an electrical power source, an anode, and a cathode. The anode and cathode are typically composed of multiple layers.

Electrotransport technology offers a number of benefits for therapeutic applications, including systemic or local administration of a wide variety of therapeutic agents with the potential administration of peptides and proteins; long-term noninvasive administration, improving convenience and compliance; controlled release, providing a desired delivery profile over an extended period with rapid onset of efficacious plasma drug levels and in some cases reduced side effects; and a transport rate relatively independent of skin type or site. Additional benefits include easy inception and discontinuation of treatment, patterned and feedback-controlled delivery, and avoidance of first-pass hepatic metabolism.

Future electrotransport therapeutic systems will draw on advances in microelectronics and transdermal system technology to provide transdermal therapy for compounds with low passive permeation rates, patterned or pulsed drug delivery, on-demand drug administration such as patient-controlled analgesia, or closed-loop drug delivery using a biosensor. Electrotransport systems may be completely integrated and will likely be the size of a conventional transdermal system, that is, smaller than 50 cm^2, and easy to wear for extended periods.

Osmosis. Osmosis, a natural process in which molecules of a solvent move through a semipermeable membrane from a region of low to high solute concentrations, is the energy source for several commercially available therapeutic systems. Osmotic systems for human therapy have a solid core, usually shaped like a standard tablet, that contains the drug and often the osmotic agents. This core is coated with a semipermeable, rate-controlling membrane containing one or more laser-drilled orifices. In an aqueous environment such as the gastrointestinal tract, the osmotic activity of the core components establishes an osmotic activity gradient across the membrane, drawing water into the system at a rate controlled by the membrane's composition, thickness, and area. Drug delivery begins when the water enters the system to dissolve or suspend the drug; the drug solution or suspension flows out of the delivery orifice or orifices at a rate equal to the rate of water inflow through the membrane.

Numerous marketed oral dosage forms utilize osmosis to control drug delivery.

A drug-dedicated osmotic implant for human and veterinary use has been developed to deliver hormones, peptides, and proteins that are digested or rendered inactive after oral administration.

pH Sensitivity. Dosage forms that employ pH sensitivity to trigger the release of active agents have been designed with both physical and chemical mechanisms. Systems with physical mechanisms are discussed here; a description of chemical mechanisms, such as pH-sensitive hydrogels, is available.

Delivery systems that respond to changes in pH have been known to the pharmaceutical industry for more than a century. The pH-sensitive enteric coating is probably the oldest controlled-release technology.

Enteric coatings are generally formulated from anionic polymers with pendent carboxyl groups and typically have a pK_a of 4 to 6.

Other dosage forms can be designed to use pH sensitivity. Examples include microcapsules or liposomes that release drugs by changes in pH and may be used to treat tumor cells, which reportedly have a pH substantially lower than that of healthy tissues; and a dosage form incorporating pH-sensitive electrodes in a feedback system with mechanical pumps that continuously monitors gastric pH values with a nasogastric electrode and intravenously infuses appropriate levels of antiulcer medication in response to changes in intragastric pH.

Vapor Pressure. The Shiley Infusaid implantable infusion pump utilizes energy stored in a two-phase fluorinated hydrocarbon fluid. The pump consists of a refillable chamber that holds the drug and a chamber that holds the fluid. The equilibrium vapor pressure of the fluid, a constant 60 kPa (450 mm Hg), compresses the bellows, pumping the drug through a bacterial filter, a capillary flow restrictor, and an infusion cannula to the target body site.

Emerging Applications of Controlled-Release Technology

Patterned Delivery. Although steady serum–drug concentrations are the preferred mode of therapy for many drugs, patterned delivery of some compounds, including the peptides calcitonin, growth hormone, and corticotropin-releasing hormone, mimics the body's own secretion patterns or responses to certain compounds or diseases. Osmotic pumps can be programmed to deliver drugs in a variety of patterns. Patterned delivery is also possible with electrically assisted transdermal systems that apply an intermittent, and sometimes varying, electrical charge to enhance drug permeation at programmed intervals.

Targeted Delivery. By delivering drugs to the specific cells or organs where action is required, targeted dosage forms may enhance therapeutic outcome and avoid many toxic effects. One approach to targeted delivery is to couple a pharmacological agent to a macromolecular vector specifically taken up by target cells. Vectors include antibodies, nanoparticles, polysaccharides, synthetic polypeptides, liposomes, synthetic and natural cells, polymer-based microcapsules, albumin, and glycoprotein conjugates. Recently, companies have begun developing systems to circumvent the blood–brain barrier and target the brain.

Triggered and Closed-Loop Delivery. Triggered delivery systems release a drug in response to a signal from the body, for example, the presence of a substance to be controlled, such as morphine, or from the patient.

The ultimate controlled-release device, a closed-loop system, delivers a metered amount of a drug each time it encounters a signal from the body, eg, the plasma concentration of a metabolite or a substance such as glucose, until the system is depleted. Closed-loop insulin delivery has been the focus of much research.

DAVID EDGREN
HAROLD LEEPER
KIRSTIN NICHOLS
JEREMY WRIGHT
ALZA Corporation

J. Kost and R. Langer, *Adv. Drug Deliv. Revs.* **6,** 19–50 (1991).

L. F. Prescott and W. S. Nimmo, *Novel Drug Delivery and its Therapeutic Application,* John Wiley & Sons, Ltd., Chichester, U.K., 1989.

J. R. Robinson and V.H.L. Lee, eds., *Controlled Drug Delivery: Fundamentals and Applications,* 2nd ed., Marcel Dekker, Inc., New York, 1987.

P. Tyle, ed., *Drug Delivery Devices: Fundamentals and Applications,* Marcel Dekker, Inc., New York, 1988.

CONVEYING

Conveying is a term used for the transport of bulk solids. Bulk solids conveyors are in large part made up of components that are dimensionally standardized, but fabricated in varying classes of construction to meet operating conditions ranging from light-duty intermittent operation to heavy-duty continuous operation. Conveyors are volumetric machines that transport material fed to them at a controlled rate. The selection of the type of conveyor for a specific application is dependent first on the required capacity; second, on the conveying path, ie, horizontal, vertical, or a combination of both; and third, on the handling characteristics of the material.

Characterization of Bulk Material

The basis of all bulk conveyor engineering is the precise definition and accurate classification of materials according to individual characteristics under a specific combination of handling conditions. Since the late 1960s there has been an extraordinary growth in research into the fundamental properties and behavior of particulate solids. However, as of this writing, it is not possible to predict the handling behavior of a bulk solids material relevant to conditions in a specific conveyor, merely on the basis of the discrete particle properties.

Industry practice is to describe flow behavior by descriptive terms derived by combining data from empirical bench-top tests and observations of actual conveying systems in operation. This information is then used first, as a guide in the selection of the particular type of conveyor, and second, for determining the design features required for the particular application. Collections of most of this information can be found in the guides published by manufacturers' engineering associations in various countries. These guides contain extensive listings of measured and observed handling characteristics for specific materials. Typical are the publications of the Conveyor Equipment Manufacturers Association (CEMA) in the United States, used primarily for mechanical conveying.

Information on the behavior of bulk materials in conveying equipment is being developed through a number of sources. Extensive research on the characterization of bulk solids and the flow behavior of bulk solids in conveyors and feeders has been published.

Conveyor Types

Belt Conveyors. A belt conveyor is made up of an endless fabric or elastomer covered belt that traverse between two or more pulleys, and is supported at intermediate points by idler rolls. These conveyors can handle a wide range of materials, from fine powders to large, lumpy stone and coal. Material can be transported at rates of over 5000 t/h and the conveyors operated at belt speeds ranging from 20 to 300 m/min over very long distances. Versatility, reliability, and range of capacities have made belt conveyors the most commonly used bulk handling conveyors in industry.

A typical belt conveyor arrangement is shown in Figure 1.

Belt-conveying design technology is changing rapidly. Newer methods using computers for dynamic analysis can take the viscoelastic properties of the conveyor belt and the distribution of stresses during startup and shut down into consideration. Analysis is being applied to long conveyor belts and belts with horizontal curves. A description of these design techniques is available.

Material Characteristics Influencing Design. Materials ranging from fine powders to large lumpy materials, nonabrasive to very abrasive, free-flowing to cohesive, and nonfriable to friable can be handled on properly designed belt conveyors. Very sticky materials can be a problem, however, if these cannot be continuously cleaned from the belt surface.

Material characteristics typically used as criteria for determining the required belt width, the carrying capacity of a particular belt, and the maximum inclination at which the belt can be operated are (*1*) The angle of repose and the angle of surcharge. The angle of surcharge is a basic design parameter for determining the transport capacity of a belt conveyor. (*2*) The size and proportion of lumps. The larger the lumps and the greater the number, the wider the belt must be to prevent the lumps from spilling over the edge of a horizontal conveyor, and the more likely they are to roll back, or fall over the edge, on an inclined conveyor. (*3*) Fluidizing or air retention properties. Materials that are easily aerated and have long air retention times impose limitations on allowable belt inclinations and belt speeds.

Carrying Idlers. The most commonly used troughed carrying idlers have the outer rolls inclined at either 20, 35, or 45° from the horizontal (see Fig. 1**b**). Carrying capacity of the belt increases as the idler angle increases. The newer belts, having carcasses of various synthetic fiber blends, are more flexible and can be designed to trough at the higher angles. As a result, the 35 and 45° idlers are gaining in popularity. Troughed idlers can be mounted on pivoted supports and spaced at intervals along the conveyor for training or guiding the belt, ie, to keep it centered on the carrying idlers during transient upsets.

Idler life is determined by a combination of factors such as bearings, seals, shell thickness, load density, and the operating environment.

Conveyor belts are manufactured in widths up to 2500 mm. The belt represents a substantial part of the initial cost of a belt conveying system and it is the component most susceptible to damage. The typical belt consists of two principal elements: the covers (top and bottom), and the carcass. The primary purpose of the covers is to protect the carcass against environmental effects, wear, and cutting. Covers are natural or synthetic rubber, thermosetting elastomers, and thermoplastic materials. The carcass provides the tensile strength required to start and move the loaded belt, the transverse and longitudinal flexibility needed to allow the belt both to support the load and conform to the shape of the idlers when running empty and to properly wrap around pulleys, and the strength to resist impact forces.

Carcass and cover ratings for belts manufactured in the U.S. have been established by the Rubber Manufacturers Association (RMA). Similar, but not the same belt rating systems have been established in most other countries.

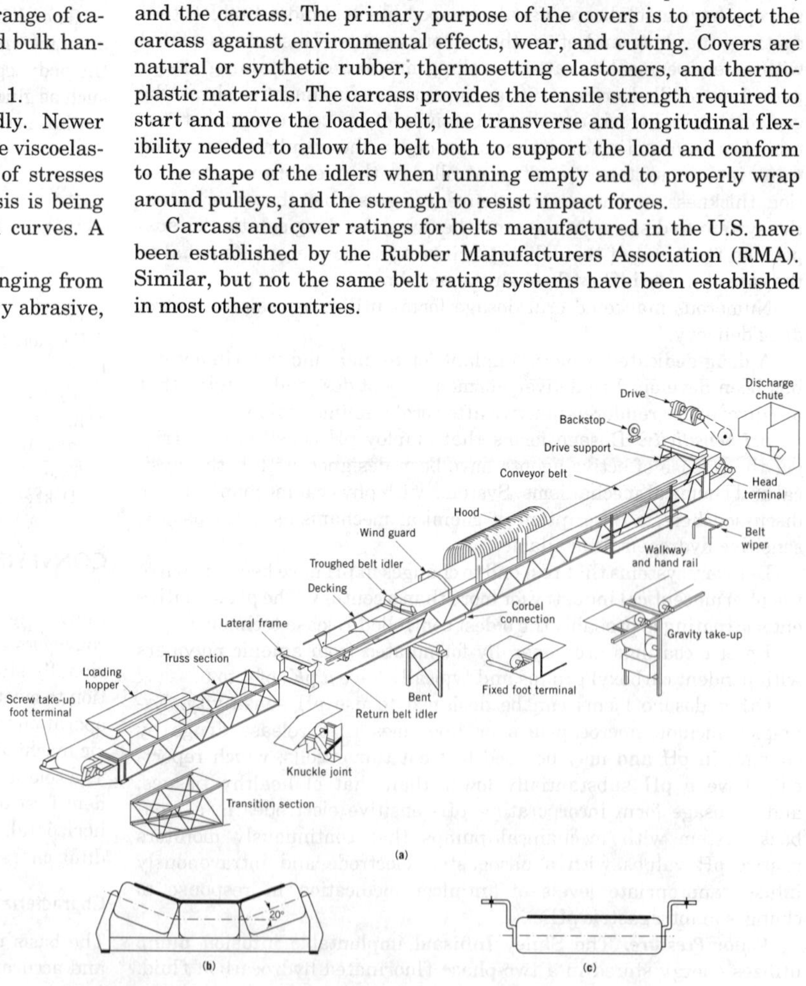

Figure 1. (**a**) Belt conveyor and support assembly, (**b**) 20° troughing idler, and (**c**) return idler.

Carcass Construction. Carcasses are made of one or more plies of a woven-fabric bonded together with an elastomeric compound. Woven materials that are used include cotton, rayon, nylon, polyester, aramids, and glass, in the pure form or in blends. The fabrics are constructed with warp yarns that run lengthwise along the belt, and filling (weft) yarns that run crosswise. There are a variety of fabric weaves available for specific applications.

A more recent development in belt technology has been the solid-woven carcass belts impregnated and covered with poly(vinyl chloride) (PVC) or urethane plastisol.

Take-Ups. A take-up is required on a belt conveyor to ensure the proper belt tension at the drive pulley and along the conveyor, as well as to ensure the proper troughing contour between idlers. A take-up is also needed to compensate for changes in belt length caused by elastic stretch during start-up, and any elongation characteristics of the belt that occur over a period of time.

Backstops. A backstop is a device that permits rotation of the pulley in the forward direction but automatically prevents rotation in the opposite direction.

Belt Cleaning. There are available a variety of single and multiple-blade belt scrapers having spring or counterweighted supports, motor driven or belt powered blade, or brush cleaners. Each has had varying success in thoroughly cleaning the belt surface. A variety of self-cleaning idlers constructed using rubber-disk, spiral, and beater rolls for use on the belt return run have been moderately successful in dislodging material from the belt surface.

Power. The power required to drive a belt conveyor is derived from the tensile forces required to propel or restrain the belt at the design speed. These include the tensile forces produced by the frictional resistance of the drive, conveyor components, and material; the acceleration of the material; and the gravitational forces required to lift or lower the material.

Newer Designs. A number of belt conveyor designs that depart from troughed belt technology have come onto the market since the mid-1980s. These are expected to have a significant impact on belt conveying technology. They include flexible sidewall belt conveyors, serpentix three-dimensional continuous path conveyors, air cushion conveyors, sandwich belt conveyors, enclosed tubular-type belt conveyors, folding belt designs, and enclosed pocket belt conveyors.

Screw Conveyors. A screw conveyor consists of a helical flight fastened around a pipe or solid shaft, mounted within a tubular-shaped or U-shaped trough. As the screw rotates, material heaps up in front of the advancing flight and is pushed through the trough. Particles in the heap, adjacent to the flight surface, are carried part way up the flight surface and then cascade down on the forward-moving side of the heap. Screw conveyors are of simple, relatively low cost construction, and are comprised of highly standardized component parts. These conveyors can handle a wide variety of solid particles ranging from lumps to powders within a completely enclosed housing at temperatures up to 275°C or higher if they are liquid cooled. The flights can be configured for particle mixing during transport, and flights and housing configured for particle cooling. Lumpy, sticky, or fiberous materials may cause problems in a screw conveyor. Conveying distances are limited by the torque capacity of available drive shafts. Power requirements are relatively high and conveying efficiency is considerably reduced when screws are inclined or mounted vertically.

Construction. In a typical screw assembly, the flights are fabricated, then welded to a pipe that has bushings press-fitted or welded into each end to provide reinforcing for the conveyor couplings. There are two types of flights: helicoid and sectional.

Screw conveyors are assembled by connecting screw sections in sequence within a trough enclosure, using screwed, flanged, or bolted connections.

Considerations for Sizing. The amount, size, and character of lumps in the material to be handled is the first consideration when selecting the proper diameter of a screw for a particular application. If the lumps are expected to be friable, and easily broken in passing through the conveyor, or if the material contains no lumps, there is no limitation on the diameter of the conveyor, and selection can be made on conveying capacity alone. If the lumps are hard and not liable to be broken up during transit through the conveyor, then the screw diameter must have enough clearance between the pipe shaft and trough to accommodate these lumps.

In general, screw conveyors can handle all free-flowing materials.

The flight pitch effects screw conveying efficiency. A pitch-to-diameter ratio of one has been found to provide the best combination of efficiency and cost effectiveness for horizontal screw conveyors.

The angle at which the screw is inclined to the horizontal affects conveying efficiency. As a screw is inclined, the slope of the flight surface becomes closer to horizontal and becomes less effective in moving material forward.

Vertical Screw Conveyors. Many free-flowing materials can be conveyed vertically with a screw. Screw elevators, designed for this purpose, have tubular housings, run at speeds ranging from 200 to 400 rpm, and have volumetric efficiencies of about 25% of an equivalent horizontal screw. The vertical screws are not self-cleaning; ie, if material feed to the inlet at the bottom stops even though the vertical screw is rotating, some material remains at the bottom until more material is fed into it.

Screw Conveyor Capacity. The volumetric capacity of a horizontal screw conveyor is calculated on the assumption that all material contained within one screw pitch moves one pitch distance in one screw revolution. Volumetric conveying capacity is calculated as $(A_s - A_p)P \cdot K \cdot N$, where A_s is the cross-sectional area of the screw flight, A_p is the cross-sectional area of the pipe shaft, P is the flight pitch, N, the rotational speed, and K is the percentage of the cross-sectional annular space between flight and pipe shaft occupied by the material.

Power to Operate a Screw Conveyor. The power required to operate a screw conveyor is dependent, to a large extent, on the handling characteristics of the material to be transported. Formulas for calculating power use empirically derived factors to account for the conveying characteristics of specific materials, the configuration of the screw, and the bearing friction. These formulas have been developed by CEMA and can be found in the literature and in engineering handbooks. It is assumed that the total power is equal to the sum of the power required to overcome friction and the power required to transport the material.

Bucket Elevators. In a bucket elevator, a series of buckets attached to an endless belt or chain are filled with material and lifted vertically to a head pulley or sprocket, where the material is dumped. The buckets are then returned back down to a tail pulley or sprocket at the bottom. Bucket elevators are not self-feeding. They must be fed at a controlled rate to avoid overfilling the buckets and damaging the machinery.

There are four broad classifications of bucket elevators: centrifugal, continuous, positive, and internal discharge. Centrifugal and continuous discharge elevators are by far the most commonly used. Two specialized versions used for high capacity handling are the super-capacity elevator and the cement mill elevator. The positive discharge and the internal discharge elevators are used for special applications.

Buckets. Elevator capacity is a function of bucket volume as well as speed and spacing on the chain or belt. There are a variety of bucket geometries, and selection is based on the material to be handled.

Casing Enclosure. Steel casings are normally furnished in intermittently welded, dust-tight construction, which is the lowest cost construction. Weather-tight, watertight, or gastight can be specified.

Chain. Chain, which can be used in all types of elevators, must be used for handling hot (>120°C) materials. Several types of chains are commonly used. The three types of chains most commonly used in industrial elevators are *(1)* steel side bar, bushed chain, or steel knuckle chain made entirely of alloy steel. *(2)* Welded steel chain which is dimensionally similar to the knuckle chain. *(3)* Combination chain is made up of cast malleable iron links alternating with steel side bars.

Belting. In a belt elevator, the buckets are fastened to the belt with special flat-headed bolts. Belts used for bucket elevator service have the same type of construction as those used on belt conveyors but the selection criteria differs in one important respect: the belt must have sufficient body strength to prevent the bolt heads from pulling through the belt carcass.

Take-Up and Hold-Back. A take-up adjustment is needed to maintain tension between the head and foot shafts. A manually adjusted screw take-up that moves the tail shaft or head shaft of a self-adjusting weighted take-up that maintains a constant gravity force on the tail shaft may be used.

The weight of material in the buckets on the loaded side of an elevator chain causes the elevator to momentarily run backwards if, during operation, the power is interrupted or there is a failure in the driving system. Because this could be a hazard to operating personnel, as well as damage to the elevator, a backstop, similar to that described for a belt conveyor, should be used.

Vibrating Conveyors. A vibrating conveyor consists of a trough supported by tuned springs and/or hinged links having a drive system. The drive system is arranged to oscillate the trough, causing solid particles to be moved along the trough. Thus these conveyors are sometimes called oscillating conveyors. There are two types of oscillating conveyors: reciprocating and vibrating.

Material must be fed onto a vibrating conveyor at a controlled rate. These conveyors are not designed to operate under a head load from solids in a storage silo or hopper. In addition to horizontal conveying, these conveyors can be used to perform other functions such as elevating, heating, drying, cooling, fluidizing, agglomeration, screening, and dewatering (qv).

Vibratory conveyors are capable of handling a wide range of materials. They provide gentle handling of food products such as friable flakes and pellets, pharmaceuticals, powde2red and granular chemicals and minerals, and discrete metal parts. They are uniquely suited for handling abrasive, hot, and dusty materials and can be designed to withstand heavy impact loads from materials such as rocks, iron and steel castings, and metal and wood scrap. They operate at frequencies ranging from 5 to 15 Hz; strokes range from 50 to 5 mm; and lengths are up to 50 m.

Drive Systems. Positive, fixed displacement mechanical drives are the most commonly used.

Compressed air or hydraulically driven reciprocating piston or rotary exciters are sometimes used in short conveyors. They are particularly useful where explosion hazards limit the use of electrical drives.

Design Variations. Vibrating conveyor troughs can be custom-designed for particular needs. Troughs can be rectangular, with or without covers, and they can be tubular.

Isolation Mounting. The static and dynamic forces that are imposed on a supporting foundation or supporting structure by a vibrating conveyor are an important consideration in conveyor selection. Static forces are vertical forces resulting from the weight of the conveyor assembly and material in the trough. The dynamic forces act in the line of action of the conveyor and can be resolved into vertical and horizontal components. The structures supporting these conveyors must be rigidly designed to withstand cycling forces. The conveyors can be designed to minimize the forces transmitted to the supporting structure. This is particularly desirable when the conveyors are to be mounted on elevated supports, or on the upper floors of a building. Methods to reduce transmission of forces to supporting structures include: hanging the conveyor on flexible cables, mounting the conveyor on a tuned spring supported weighted base, or using a counterbalance mounted on a tuned spring system identical to that on the conveyor trough and driving it 180° out of phase with the trough. The dynamic forces transmitted into supporting structures have been reduced by up to 95% by these methods.

En-Masse Conveyors. An en-masse conveyor consists of an endless chain or cable pulling a series of spaced skeleton or solid plug flights through the enclosed casing or housing. Material is introduced through an opening in the casing, where it is captured by the flights and drawn through the casing until it reaches an opening in the housing, where is discharges by gravity. Several types exist, including chain and tubular en-masse conveyors.

En-masse conveyors offer unique advantages over other types of conveyors: they are compact in cross section and totally enclose the bulk material; they can be made vapor and gas tight; they can handle many materials with little particle attrition; they can have an L-shaped or Z-shaped path, thereby eliminating transfer points that are required by conventional straight line conveyors; they can combine feeding, conveying, and elevating in one machine; and they can have multiple inlet and discharge openings.

Material Characteristics. Performance of an en-masse conveyor is very dependent on the characteristics of the bulk materials handled. This conveyor is best suited for nonabrasive, free-flowing, granular or powder materials. Sticky or smearing materials can build up in the clearance between flight and casing causing a mechanical overload.

Air Activated Gravity Conveyor. The air activated gravity conveyor is a very simple, inexpensive, maintenance-free, and low power-using device for conveying fine, easily fluidized, and aerateable powders. Originally developed as the Air Slide conveyor, it consists of a downward-sloped rectangular trough bisected by a porous membrane that defines a lower and upper channel in the trough. When air is supplied to the lower channel, it permeates through the membrane, and aerates powder resting in the upper channel, causing the powder to flow like a liquid down the surface of the inclined membrane. Powder flow occurs even if the membrane is only partially covered with material, because the pressure drop through the membrane is such that the air flow is uniformly distributed across the membrane surface independent of the depth of material above.

Pneumatic Conveyors. In a pneumatic conveyor, bulk solids particles are transported through a closed duct in a gas stream. A wide range of particle sizes, from powders to large chunks and from fibers to chopped sheet waste, can be handled through these systems. The only significant restrictions are that the conveying pipe should be able to accommodate the largest particles, and the material should not be sticky. Pneumatic conveyors have a number of important advantages over mechanical conveyors that have led to widespread use: they provide great flexibility and compactness in system arrangement; they can be arranged with a multiplicity of solids pick-up and discharge points; they can provide a complete enclosure for protection from contamination of the operating areas, or contamination of the materials being handled; and they can be designed to provide heating, cooling, drying, or mixing during transport. Recirculating inert gases can be used for safely conveying explosive, toxic, or other sensitive materials.

Classifications. Pneumatic conveyors are commonly classified as being either a dilute phase, or a dense phase conveyor.

F. M. THOMSON
Consultant

Belt Conveyors for Bulk Materials, 4th ed., Manassas, Va.

Screw Conveyors, CEMA Book No. 300, Conveyor Equipment Manufacturers Association, Manassas, Va., 1988.

H. Colijn, *Mechanical Conveying for Bulk Solids,* Elsevier, Amsterdam, the Netherlands, 1985, pp. 236–328.

R. D. Marcus, G. E. Klinzing, and F. Rizk, *Pneumatic Conveying of Solids,* Chapman and Hall, New York, 1991.

COORDINATION COMPOUNDS

A coordination compound typically consists of a metal atom or ion (Lewis acid) surrounded by a number of electron-pair donors (Lewis bases) called ligands. Coordination compounds, also known as metal complexes, are pervasive throughout chemistry, biochemistry, and chemical technology. The metallic elements, which constitute 80% of the Periodic Table, exhibit predominantly coordination chemistry. Whereas the ligands may have charges equal in magnitude and opposite in sign to the charge of the metal ion, in which case a neutral coordination compound or inner complex results, often the metal plus ligands result in a charged entity or complex ion. In these situations, the coordination compound may include neutral ligands or counterions, either simple or complex. Generally coordination compounds have properties which are unique relative to both the ligands and the metal ion itself.

Coordination compounds are used as catalysts in nature, ie, metal enzymes, and in industry in situations in which the metal ion or ligand would not work alone. The ligands often modify the properties of the metals or metal ions. For example, the metal deactivators in gasoline modify the chemistry of dissolved copper to the extent that the copper does not promote gum formation (see GASOLINE AND OTHER MOTOR FUELS). Some ligands react with different metal ions such that selective analysis of metallic elements is possible through coordination followed by solvent extraction, spectroscopy, gravimetry, electrochemistry, etc. Furthermore, complexes of copper and zinc in water allow brass to be electroplated (see COPPER ALLOYS; ELECTROPLATING). Conversely, metals are sometimes used to modify the properties of ligands. For example, azo dyes (qv) are metallated to give more permanence and/or alter color tones; the coordination of zinc to bactericides modifies the properties of the bactericides. The modification includes template and neighboring group effects, promotion of nucleophilic substitution, enhanced ligand acidity, and strain modification. Common geometries for coordination numbers from two through nine are shown in Figure 1.

Other Types of Compounds

A coordination compound that has a chelating ligand in a ring around the metal ion is termed a macrocycle. Heme (**1**), the iron porphryn derivative in human blood that carries oxygen, is an example. Other types of macrocyles such as dibenzo-[18]-crown-6 (**2**) known as crown ether and cryptate (**3**) complexes, as well as metal cluster coordination compounds, have been known for many years.

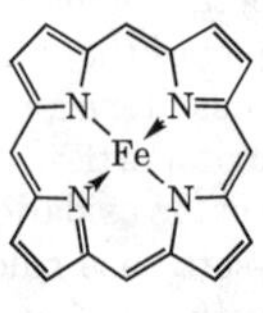

(**1**)

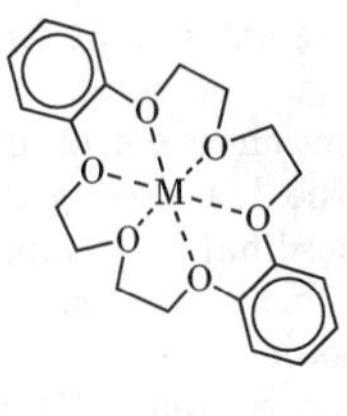

(**2**)

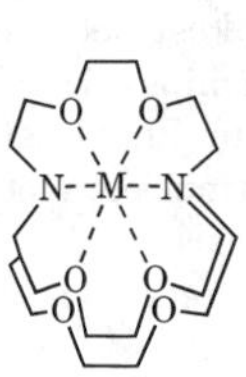

(**3**)

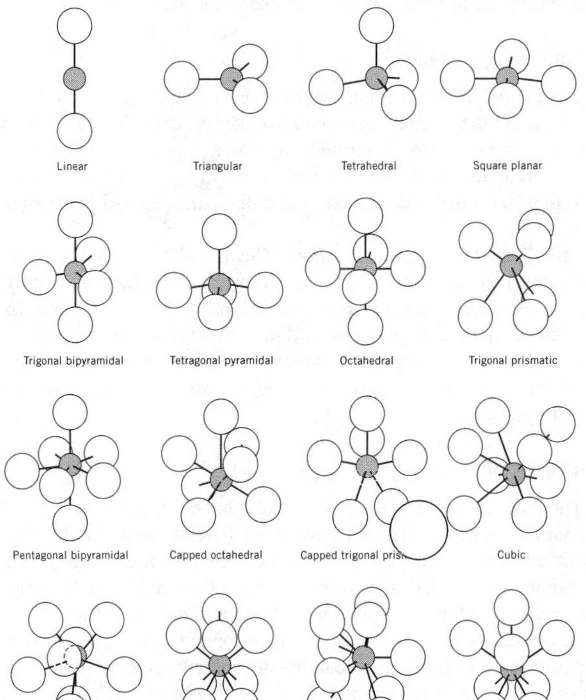

Figure 1. Common geometries for metal coordination numbers from two through nine, where ● represents the metal and ○, a ligand donor atom. The principal axis orientation is vertical.

Newer are the metal fullerenes. A large number of metal encapsulated fullerenes, ie, three-dimensional C_x polytopal cage derivatives, have been observed.

Coordination Theories and Bonding

The quest for a comprehensible theory of coordination chemistry has given rise to the use of valence-bond, crystal-field, ligand-field, and molecular-orbital theories. Ligand field theory incorporates covalency with the electrostatic crystal field. The symmetry-induced separation of energy levels that is perceived from the chemical and physical properties of coordination compounds is inherent in all four theories, but symmetry effects are more apparent in the latter three. Molecular orbital treatments of complexes range from the very simple semiempirical, such as Hückel, angular overlap, etc, through so-called *ab initio* methods.

Properties

Stability. The thermodynamic stability of coordination compounds in solution has been extensively studied. The equilibrium constants may be reported as stability or formation constants: $M + nL \rightleftarrows ML_n$. This compound, ML_n, has a cumulative stability constant β_n related to the activities a of the species by $\beta_n = a_{MLn}/(a_M a_L^n)$; and the stepwise constant $K_n = a_{MLn}/(a_{MLn-1} a_L)$. Alternatively, instability or dissociation constants are sometimes used to describe compounds, and caution is necessary when comparing values from different sources.

Steric Selectivity. In addition to the normal regularities that can be rationalized by electronic considerations, steric factors are important in coordination chemistry.

Coordination stereochemistry (including various forms of isomerization) is an area of significant research interest. This aspect of coordination is important for stereospecific catalytical applications.

Reactions

Substitution. Coordination species are often categorized in terms of the rate at which they undergo substitution reactions. Complexes that react with other ligands to give equilibrium conditions almost as fast as the reagents can be mixed by conventional techniques are termed labile. Included are most of the complexes of the alkali metals, the alkaline earths, the aluminum family, the lanthanides, the actinides, and some of the transition-metal complexes. On the other hand, numerous transition-metal complexes that kinetically resist substitution reactions are termed inert. These terms refer to substitution reactivity and not to thermodynamic properties.

Oxidation–Reduction. Redox or oxidation–reduction reactions are often governed by the hard–soft base rule. For example, a metal in a low oxidation state (relatively soft) can be oxidized more easily if surrounded by hard ligands or a hard solvent. Metals tend toward hard-acid behavior on oxidation. Redox rates are often limited by substitution rates of the reactant so that direct electron transfer can occur.

Photochemistry. Substitution rates of many complexes are enhanced by irradiation of the low energy d–d transitions, such as $t_{2g} \rightarrow e_g$ in octahedral coordination compounds. Quantum yields, Φ, defined as the ratios of moles of product formed (or reactant depleted) to the moles of photons absorbed, vary from very good, eg, chromium(III) ca 0.5, to poor, eg, cobalt(III) < 0.01, for ligand substitution. The substituted ligand is normally the stronger ligand of the two on the axis with the weakest net pair of ligands as determined by spectrochemical relationships, ie, $CN^- > NO_2 > NH_3 > H_2O$, $F^- > Cl^-$. Exceptions do occur. Photochemical ligand dissociation is useful in the synthesis of multinuclear metal complexes such as diiron nonacarbonyl from iron pentacarbonyl.

The use of photochemical redox for practical energy-transfer is being actively pursued.

Applications

Coordination compounds are used in bactericides and fungicides, catalysis, coordination polymers, dyes and pigments, photography, electroplating, petroleum additives, and therapeutic chelates.

RONALD D. ARCHER
University of Massachusetts, Amherst

F. Basolo and R. C. Johnson, *Coordination Chemistry,* 2nd ed., Sci. Revs., Northwood, UK, 1986. Easy-to-read paperback on coordination chemistry suitable for anyone who has studied basic chemistry.

F. A. Cotton and G. Wilkinson, *Advanced Inorganic Chemistry,* 5th ed., Interscience Publishers, a division of John Wiley & Sons, Inc., New York, 1988.

R. G. Wilkins, *Kinetics and Mechanism of Reactions of Transition Metal Complexes,* 2nd ed., VCH, Weinheim, Germany, 1991. A critical and selected compilation of kinetics and mechanism data.

G. Wilkinson, R. D. Gillard, and J. A. McCleverty, eds., *Comprehensive Coordination Chemistry,* Pergamon Press, Oxford, UK, 1987, 7 Vols., Vol. 1, theory and background; Vol. 2, ligands; and Vol. 6, applications.

COPOLYMERS

Synthetic polymers have become extremely important as materials over the past 50 years and have replaced other materials because they possess high strength-to-weight ratios, easy processability, and other desirable features. The emphasis in research has shifted from developing new synthetic macromolecules toward preparation of cost-effective multicomponent systems (ie, copolymers, polymer blends, and composites). These multicomponent systems can be "tuned" to achieve the desired properties (within limits, of course) much easier than through the total synthesis of new macromolecules.

Homopolymers and Copolymer Structures

Homopolymers are high molecular-weight molecules prepared by linking a large number of smaller molecules called monomers (eq. 1).

$$\underset{\text{monomer}}{n\text{A}} \longrightarrow \underset{\text{polymer}}{+\!\text{A}\!+_n} \qquad (1)$$

Macromolecules in which two or more different monomers (comonomers) are incorporated in the same polymer chain are copolymers (eq. **2**).

$$\underset{\text{comonomers}}{n\text{A} + n\text{B}} \longrightarrow \underset{\text{copolymer}}{+\!\text{A–B}\!+_n} \qquad (2)$$

Copolymers can be further described by specifying the number and distribution of monomer units within the copolymer molecule. Thus a polymer with a statistical placement of monomer units, eg, ABBABABAAB is called a random copolymer. An alternating copolymer consists of an alternating arrangement of the comonomers, ABABABABA. On the other hand, a block copolymer has a long segment of one monomer followed by a long segment of another monomer, AAAABBBB. A graft copolymer is a type of block copolymer in which a polymer chain (main chain or backbone) has chains (branched chains) of another polymer attached at intervals along the backbone. A network copolymer is a cross-linked or three-dimensional copolymer.

```
   AAAAA
    B
    B
    B
    B


AAAAABAAAA
    C
    C
    C
    C
AAAAABAAAA
```

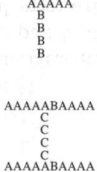

A polymer blend is a physical or mechanical blend (alloy) of two or more homopolymers or copolymers.

Copolymers extend the number and range of available materials, enabling the polymer scientist to achieve combinations of material properties (eg, tensile strength, solubility, solvent resistance, low temperature flexibility, etc) unattainable from the simple constituent homopolymers. As a result, a large number of copolymers have become commercially important. Table 1 lists some of them.

Copolymerization Reactions

The mutual polymerization of two or more monomers is called copolymerization. This topic has been comprehensively reviewed. Monomers frequently show a different reactivity toward copolymerization than toward homopolymerization. In fact, some monomers that can be homopolymerized only with great difficulty can be readily copolymerized.

Chain-Growth Copolymerization Theory. The theory of chain-growth (eg, radical, anionic, etc) copolymerization has received more attention than that of step-growth or other copolymerizations. In the case of chain-growth copolymerization, growing polymer chains must choose between more than one monomer. Such a choice or relative reactivity has been quantitatively treated by the reactivity ratio and the Q–e schemes.

Synthetic Methods

Free-radical copolymerizations have been performed in bulk (comonomers without solvent), solution (comonomers with solvent), suspension (comonomer droplets suspended in water), and emulsion (comonomer emulsified in water). On the other hand, most ionic and coordination copolymerizations have been carried out either in bulk or solution, because water acts as a poison for many ionic and coordination catalysts. Similarly, few condensation copolymerizations involve emulsion or suspension processes. The following reactions exemplify the various copolymerization mechanisms: free-radical copolymerization, anionic copolymerization, cationic copolymerization, coordination copolymerization, step-growth copolymerization, group-transfer polymerization, ring-opening polymerization, ring-opening metathesis polymerization, telechelic polymers and macromonomers, and postpolymerization reactions.

Table 1. Some Commercially Important Copolymers

Comonomers	Nomenclature	Generic name, trade name, and/or abbreviation
butadiene–styrene	poly(butadiene-*co*-styrene)	synthetic rubber; GRS, SBR
ethylene–propylene	poly(ethylene-*co*-propylene)	EPM, EPR rubber
ethylene–propylene–diene	poly(ethylene-*co*-propylene-*co*-5-ethylidene-2-norbornene)	EPDM rubber
butadiene–acrylonitrile	poly(butadiene-*co*-acrylonitrile)	NBR rubber
styrene–butadiene–styrene (triblock)	polystyrene-*block*-polybutadiene-*block*-polystyrene	SBS thermoplastic rubber; Kraton
isobutylene–isoprene	poly(isobutylene-*co*-isoprene)	butyl rubber, GR-1
vinyl chloride–vinyl acetate	poly(vinyl chloride-*co*-vinyl acetate)	Vinylite flooring, Tygon tubing, coatings
vinyl chloride–acrylonitrile	poly(vinyl chloride-*co*-acrylonitrile)	Dynel fibers
vinyl chloride–vinylidene chloride	poly(vinyl chloride-*co*-vinylidene chloride)	Saran packaging, fibers
acrylonitrile–vinyl acetate	poly(acrylonitrile-*co*-vinyl acetate)	Orlon, Acrilon acrylic fibers
acid–glycol–diisocyanate (step growth multiblock)	poly(acid-*co*-glycol-*co*-diisocyanate-*co*-diamine)	Spandex fibers
acrylonitrile–butadiene–styrene[a]	polybutadiene-*graft*-poly(styrene-*co*-acrylonitrile) + poly(styrene-*co*-acrylonitrile)	ABS plastics
butadiene–styrene[a]	polybutadiene-*graft*-polystyrene + polybutadiene + polystyrene	IPS impact styrene plastics
styrene–acrylonitrile	poly(styrene-*co*-acrylonitrile)	SAN plastics
styrene–maleic anhydride	poly(styrene-*alt*-maleic anhydride)	SMA resins

[a] Graft copolymer and polymer blend.

Effect of Monomer Unit Arrangement on Physical Properties

Random Arrangement. The primary incentive for preparing copolymers is to attain certain properties in the products. The effect of random copolymerization on polymer properties is easily shown by differences in polymer crystallinity, melting point (T_m), glass-transition temperature (T_g), and solubility between a copolymer and the corresponding homopolymers. Because random comonomer enchainment tends to reduce symmetry and modify intermolecular forces, it is not surprising that random copolymers have different melting behavior than the corresponding constituent crystalline homopolymers.

Block (Linear) Arrangement. By far the most interesting block copolymers are those in which there is little or no mixing of the block phases. Such heterogeneous block copolymers tend to show the properties of the components rather than the averaging of homopolymer properties. Heterogeneous copolymers also tend to be soluble in a wide variety of solvents. In fact, they may act like surfactants if the respective blocks have widely different solubilities.

Not only do block copolymers have properties different from those of homopolymers, random copolymers, and polymer blends, but the properties of block copolymers themselves differ depending on the length and chemical composition of the blocks. This results in interesting morphologies and structures.

Block (Star) Arrangement. The known star polymers, like their linear counterparts, exhibit microphase separation. In general, they exhibit higher viscosities in the melt than their analogous linear materials. Their rheological behavior is reminiscent of network materials rather than linear block copolymers.

Graft Arrangement. Graft copolymers typically exhibit some microphase separation, especially in cases in which the backbone polymer has different solubility compared with the chains grafted onto it. The physical properties lie between those of a heterogeneous diblock or triblock copolymer and a random copolymer.

Random Copolymers

Many random copolymers have found commercial use as elastomers and plastics. For example, SBR, poly(butadiene-*co*-styrene), has become the largest volume synthetic rubber. It can be prepared in emulsion by use of free-radical initiators, such as $K_2S_2O_8$ or Fe^{2+}/ROOH, or in solution by use of alkyl lithium initiators. Emulsion SBR copolymers are produced under trade names by such companies as American Synthetic Rubber (ASPC), Armtek, B.F. Goodrich (Ameripool), and Goodyear (Plioflex); solution SBR is manufactured by Firestone (Stereon).

Although most of the SBR produced in the United States has been used in tires, use in nonautomotive applications has been on the rise. For example, SBR use in wire and cable materials, mechanical goods, footwear, and foam products has increased.

Poly(butadiene-*co*-acrylonitrile), NBR, is another commercially significant random copolymer. This rubber is manufactured by free-radical emulsion polymerization.

Another important class of random copolymers are the ethylene–propylene elastomers. The saturated hydrocarbon backbone provides good ozone resistance and better weathering and aging characteristics than diene rubbers. Even though these materials are synthesized by Ziegler-Natta catalysts, they do not exhibit much stereospecificity. There are basically two types of these copolymers, ethylene–propylene (EPM) and ethylene–propylene–diene (EPDM) materials. They have become important in automotive uses (radiator and heater hoses, weather stripping, and door and window seals), tires, and tubes.

SAN resins are random, noncrystalline copolymers of styrene and acrylonitrile. These materials are manufactured by emulsion, suspension, or continuous mass polymerization. Polymer properties are adjusted by varying molecular weight and styrene–acrylonitrile ratio. Significant producers are Dow (Tyril) and Monsanto (Lustran SANZI). SAN copolymers are typically tougher and stronger than polystyrene. SAN is used primarily in housewares (tumblers, salad bowls, serving trays, etc) and appliances (refrigerator drawers, vacuum cleaner parts, washing machine detergent dispensers).

Random copolymers of vinyl chloride and other monomers are important commercially. Most of these materials are produced by suspension or emulsion polymerization using free-radical initiators. Important producers for vinyl chloride–vinylidene chloride copolymers include Borden, Inc. and Dow.

Poly(vinyl chloride-*co*-vinyl acetate) has found application in flooring, phonograph records, protective coatings, fibers, and some films and sheeting.

Poly(ethylene-*co*-vinyl alcohol) is made by saponification of ethylene–vinyl acetate copolymers. They are used commercially as barrier resins for packaging. Important producers include DuPont and EVALCA.

Fluoroelastomers are copolymers based on two or more of the following monomers: vinylidene fluoride (VF_2), hexafluoropropylene (HFP), chlorotrifluoroethylene (CTE), tetrafluoroethylene (TFE), and perfluoromethyl vinyl ether (FVE). They are made by a high pressure, free-radical initiated emulsion polymerization process. The excellent properties of these fluoroelastomers come with a high price tag. These materials are used in automotive applications (seals, gaskets, fuel hose lines, engine parts, etc), where they can withstand under-hood temperatures.

Alternating Copolymers

Poly(styrene-*alt*-maleic anhydride) is a classic and commercial example of an alternating copolymer. This material is manufactured by free-radical bulk, solution, or emulsion copolymerization. Important producers are ARCO (SMA) and Monsanto (Lytron). These resins and their derivatives are seldom used alone but are used as dispersants (to increase the pigment concentration) and floor polishes (to act as emulsifiers and protective colloids).

Block Copolymers

Block copolymers have become commercially valuable commodities because of their unique structure–property relationships. They are best described in terms of their applications such as thermoplastic elastomers (TPE), elastomeric fibers, toughened thermoplastic resins, compatibilizers, surfactants, and adhesives (see ELASTOMERS, SYNTHETIC–THERMOPLASTIC ELASTOMERS).

A thermoplastic elastomer is a material that combines the processability of a thermoplastic with the performance of a thermoset rubber. A thermoplastic elastomer results when block copolymers have an ABA, $(AB)_nX$, or +AB+_n but not an AB diblock arrangement of A (thermoplastic) and B (rubbery) blocks. Important block copolymers include Kraton (Shell), Vector (Dexco), Estane (B. F. Goodrich), Texin (Mobay), and Hytrel (Du Pont).

The physical properties of block copolymer TPE depend on the type and arrangement of the blocks. Table 2 compares the property advantages of various block copolymer thermoplastic elastomers.

The properties and prices of the various block copolymer TPE greatly affect their markets. For example, the low cost butadiene–styrene block copolymers have found utility in footwear (sneakers, tennis shoes), injection-molded or extruded goods (automotive sight shields, fender extensions, toys, housewares), and adhesives (solvent cement and hot-melt types). The principal commercial supplier of styrenic block copolymers is Shell (Kraton).

In contrast, the copolyester–ether block copolymer TPE are relatively expensive, with high performance characteristics. They are produced by a melt-transesterification polymerization process and can be processed by conventional techniques such as injection, blow, transfer, or rotational molding. Important commercial products are produced by DuPont (Hytrel), Eastman Kodak (Ecdel), General Electric (Lomod), and Hoechst Celanese (Riteflex). Their most important uses are in wire cable materials (eg, the coiled stretch telephone cords), injection-molded articles (eg, small mechanical parts), and high pressure hoses.

Table 2. Property Advantages of Various Block Copolymer TPE

Property	Styrene–diene	Hydrogenated styrene–diene	Ester–ether	Urethane–ester
tensile			+	+
recovery	+	+		
upper use temperature			+	+
lower use temperature	+	+		
aging stability		+		
acid–base resistance	+	+		
oil resistance			+	+
electrical	+	+		
abrasion resistance				+
melt processability			+	+
cost	+			

[a] A designation of + indicates a performance strong point.

Graft Copolymers

Two commercially significant graft copolymers are acrylonitrile–butadiene–styrene (ABS) resins and impact polystyrene (IPS) plastics.

ABS is the sixth largest volume thermoplastic resin and the principal engineering (structural or load-bearing) plastic. ABS is a terpolymer manufactured by copolymerizing acrylonitrile and styrene in the presence of polybutadiene rubber. Important produces of ABS plastics include General Electric, Monsanto (Lustran), and Dow (Abtec).

The properties of ABS can be modified by varying the relative proportions of the basic components, the degree of grafting, and the molecular weight. The principal markets for ABS are automotive (25% of total consumption), business machines (24%), and pipes and fittings (13%).

Impact polystyrene (IPS) is one of a class of materials that contains rubber grafted with polystyrene. This composition is usually produced by polymerizing styrene (by mass or solution free-radical polymerization) in the presence of a small amount (ca 5%) of dissolved elastomer. Some of the important producers of impact-resistant polystyrenes are BASF (Polystyrol), Dow (Styron), and Monsanto (Lustrex).

Impact polystyrenes have found use in the manufacture of refrigerator door liners and in packaging.

Characterization of Copolymers

The characterization of copolymers must distinguish copolymers from polymer blends and the various types of copolymers from each other. In addition, the exact molecular structure, architecture, purity, supermolecular structure, and sequence distribution must be determined.

Assessing whether a material is a copolymer or a mixture of homopolymers can sometimes be accomplished by extracting the prospective copolymers with solvents selective for the component homopolymers. However, this method is effective only when the copolymer segments differ significantly in solubility behavior. The situation is further complicated by the fact that block copolymers can themselves act as compatibilizing agents for homopolymers and can confound extraction experiments. Alternatively, solution fractionation has been used to test copolymers versus homopolymers or polymer blends.

An indication of whether block or copolymer architecture is AB, ABA, or +AB+_n can often be seen in its rheological behavior.

The molecular structure of the copolymers is also important. Molecular-weight measurements (osmometry, gpc) and functional group analysis are useful.

Economic Aspects

The economic importance of copolymers can be clearly illustrated by a comparison of U.S. production of various homopolymer and copolymer elastomers and resins. SBR, a random copolymer, constitutes the bulk of the entire U.S. production. Copolymers of ethylene and propylene, and nitrile rubber (a random copolymer of butadiene and acrylonitrile) are manufactured in smaller quantities. Nevertheless, the latter copolymers approach the volume of elastomeric butadiene homopolymers.

The relative U.S. production of styrene homopolymer and copolymer resins is also noteworthy. The impact polystyrene (graft and polymer blend) copolymers are produced in nearly the same quantities as styrene homopolymers. The ABS resins are synthesized in lesser, yet significant, quantities.

Future Trends

Copolymer technology is progressing along two "fronts." First, new applications for copolymers are being found to increase the volume of materials that are already commercially available. One example of this is the rapid growth of styrenic block copolymers sold as asphalt (qv) and polymer modifiers over the past 10 years. Another is the increased interest in graft and block copolymers as compatibilizers for polymer blends and alloys. Of particular interest are compatibilizers for recycled polymer scrap.

The second front originates in the polymer synthesis community. Efforts are mainly directed toward production of monodisperse block copolymers by living polymerizations. These structures typically result in microphase separated systems; if one block is a high T_g material and the other is elastomeric in nature, then the overall system will be thermoplastic.

Furthermore, increased governmental scrutiny of chemical substances will make it more difficult to bring a new product to market. The choice of comonomers and copolymers may be based partly on EPA, FDA, OSHA, and TSCA rulings. In addition to these regulations, the impetus toward the recycling of polymers is expected to impact copolymer production.

If new copolymers cannot be competitively introduced, new ways must be developed for improving old ones. This might involve copolymer blending or improved processing techniques.

There will also be a growing need for macromolecules for special end use requirements. One method for obtaining these specialty materials is through copolymerization. Biomedical materials are of special interest.

CHRISTINE A. COSTELLO
DONALD N. SCHULZ
Exxon Research and Engineering Company

G. Odian, *Principles of Polymerization,* 3rd ed., John Wiley & Sons, Inc., New York, 1991, pp. 453–523 and references therein.

W. Saltman, in M. Morton, ed., *Rubber Technology,* 2nd ed., Van Nostrand Reinhold Co., Inc., New York, 1973, Chapt. 7.

L. E. Forman, in J. P. Kennedy and E. G. Tornquist, eds., *Polymer Chemistry of Synthetic Elastomers II, High Polymers,* Vol. 23, Wiley-Interscience, New York, 1969.

C. A. Harper, in C. A. Harper, ed., *Handbook of Plastics and Elastomers,* McGraw-Hill Book Co., New York, 1975, Chapt. 1.

COPPER

Copper, Cu, critically important to the development of civilization, is the only common metal found naturally in the metallic state.

Copper, the first element of Group 11 (IB) of the Periodic Table, is immediately above silver and gold. It is classed with silver and gold as a noble metal and can be found in nature in the elemental form. Copper occurs as two natural isotopes, ^{63}Cu and ^{65}Cu.

Large-scale mining of low grade ores is a development of the twentieth century. The exploitation of large ore bodies in Chile and Peru has made South America the world's largest producer of copper. The United States is the second largest, followed by Zaire and Zambia, and the CIS. Other important deposits are found in southern Oceania (Papua New Guinea, the Philippines, and Indonesia), Canada, Mexico, and Poland.

Occurrence

Cosmically, copper is relatively abundant: 100–400 ppm are found in the metal phase of meteoric iron (see EXTRATERRESTRIAL MATERIALS). The high affinity of copper for sulfur is the principal factor in determining the manner of occurrence in the earth's crust. Copper–iron sulfides are the last minerals to crystallize, and these fill the interstices between other minerals in igneous rocks, which contain an average of about 60–70 ppm copper. Other copper compounds occurring in nature are oxides and silicates. The strong affinity of copper for sulfur is the prime factor in separating copper from iron in the pyrometallurgical reduction of copper from sulfide ore.

Copper ore minerals may be classified as primary, secondary, oxidized, and native copper. Most copper deposits are (*1*) porphyry deposits and vein replacement deposits, (*2*) strata-bound deposits in sedimentary rocks, (*3*) massive sulfide deposits in volcanic rocks, (*4*) magmatic segregates associated with nickel in mafic intrusives, or (*5*) native copper typified by the lava-associated deposits of the Keweenaw Peninsula, Michigan.

Exploration. Because it takes years to bring a mine into production, significant new copper deposits are sought and known reserves are expanded more or less continually. These exploration expenditures are highly sensitive to metal market conditions, and, as of this writing, gold is at a better price level than copper. Worldwide exploration continues, however, even after discovery of a deposit and the start of mining.

Ocean Nodules. A less conventional copper resource consists of deep-sea ferromanganese nodules. Although a number of companies are studying methods for recovering values from this source, copper resources from nodules must be considered tentative. World resources are estimated at 0.7 billion metric tons.

Properties

The properties of copper, silver, and gold are compared in Table 1. Like silver and gold, copper has an atomic structure that results in outstanding electrical and thermal conductivities and a high

Table 1. Some Properties of Pure Copper, Silver, and Gold[a]

Property	Copper	Silver	Gold
atomic weight	63.54	107.87	196.97
atomic volume, cm^3/mol	7.11	10.27	10.22
mass numbers stable isotopes (relative abundance, %)	63 (69.1)	107 (51.35)	197 (100)
	65 (30.9)	109 (48.65)	
oxidation states	1, 2, 3	1, 2, 3	1, 2, 3
standard electrode potential, 25°C, V	Cu/Cu^+ = 0.520	Ag/Ag^+ = 0.799	Au/Au^+ = 1.692
	Cu/Cu^{2+} = 0.337		
density, kg/m^3	8.96×10^3	10.49×10^3	19.32×10^3
metallic radius, nm	0.1276	0.1442	0.1439
ionic radius, M^+, nm	0.096	0.126	0.137
covalent radius, nm	0.138	0.153	0.150
electronegativity	2.43	2.30	2.88
ionization energy, kJ/mol[b]			
1st	745	732	891
2nd	1950	2070	
heat of atomization, kJ/mol[b]	339	286	354
thermal conductivity, W/(m·K)	394	427	289
electrical resistivity at 20°C, $\mu\Omega$/cm	1.6730	1.59	2.35
temperature coeff. of electrical resistivity[c]	0.0068	0.0041	0.004
melting point, °C	1083	960.8	1063
heat of fusion, kJ/kg[b]	212	102	67.4
boiling point, °C	2595	2212	2970
heat of vaporization, kJ/kg[b]	7369	2400	1860
specific heat at 20°C, J/(kg·°C)[b]	384	233	131 (18°C)
linear coeff. of expansion $\times 10^6$ per °C at 20°C	16.5	10.68	14.2
tensile strength, annealed metal, MPa[d]	230	280	170

[a] The crystal structure of each of the pure metals is fcc.
[b] To convert J to cal, divide by 4.184.
[c] From 0–100°C.
[d] To convert MPa to psi, multiply by 145.

degree of malleability (see GOLD AND GOLD COMPOUNDS; SILVER AND SILVER ALLOYS). Although the ground state electronic configuration, $1s^22s^2p^63s^23p^63d^{10}4s^1$, implies a stable closed shell of 18 electrons, the shell is not inert. Rather, the underlying *d* orbitals appear to participate in metallic bonding by promotion of at least one *d* electron into a higher energy orbital of the outermost principal quantum level. There this electron is available for participation in electrical and thermal conduction. Although commercial copper is of excellent purity, differences in data obtained for pure copper and for electrolytic copper are significant.

The unique nature of the electronic configuration of copper, which contributes to its high electrical and heat conductivity, also provides chemical properties intermediate between transition and 18-shell elements. Copper can give up the 4*s* electron to form the copper(I) ion or release an additional electron from the 3*d* orbitals to form the copper(II) ion.

The higher ionization energy and smaller ionic radius of copper contribute to its forming oxides much less polar, less stable, and less basic than those of the alkali metals. Because of the relative instability of its oxides, copper joins silver in occurring in nature in the metallic state.

Elemental copper is resistant to aerated alkaline solutions except in the presence of ammonia. Copper does not displace hydrogen from acid but dissolves readily in oxidizing acids such as nitric acid (qv) or in acid solutions that contain an oxidizing agent, such as sulfuric acid solution containing ferric sulfate. Because of corrosion resistance to salt solutions, copper is used in marine applications (see COATINGS, MARINE; CORROSION AND CORROSION CONTROL). Resistance to oxidation by water vapor at high temperatures has made copper a material of choice in cooling systems.

Recovery and Processing

Most copper is processed using a combination of mining, concentrating, smelting, and refining, or by leaching waste and solvent extraction–electrowinning. Open-pit mining is more common than underground mining, and the overburden, or waste, contains some copper. Frequently the waste is leached to extract the copper, which may be recovered by contacting the solution with an organic solvent extraction and then recovering the copper by electroplating (qv) it on copper sheets (electrowinning).

Secondary Recovery. Metal returning from the store of metal in use is referred to as old scrap, in contrast with scrap generated within the copper fabrication process, which is called new scrap (see RECYCLING). In 1990 the amount of the U.S. copper supply derived from old scrap was 24% of the total copper consumed. About 40% of old scrap is used for producing refined copper; most of the remainder is used in the production of brass and bronze ingots (see COPPER ALLOYS). About 75% of new scrap is consumed by brass mills, with most of the remainder used in the production of refined copper. Some estimates suggest that as much as 60% of the copper produced is ultimately recycled for reuse. Old scrap combined with new scrap from fabricating plants accounts for about 40% of the metallic input to domestic copper furnaces.

Effluents from Production. Sulfur dioxide is produced when copper and iron sulfide minerals are oxidized, an essential step in copper production. Stringent limitations have been placed on sulfur emissions by many industrialized countries. In the United States, the EPA sets rules for plant emissions, whereas OSHA is concerned with sulfur dioxide concentrations in working areas. The source standards of the EPA were established by the Clean Air Act of 1977 and amended in 1990. The U.S. copper industry has made significant investments to meet environmental regulations, resulting in a dramatic reduction in SO_2 emissions. In 1991 some 90% of these emissions were under control. U.S. copper smelters are now a minor source of SO_2 emissions and are no longer considered a significant national contributor to pollution.

Copper mining operations have always been faced with a large solid waste disposal problem caused by large tonnages of waste rock or overburden with open-pit mines, tailings from the concentration step, and slag and dust piles associated with smelters. In the United States, the Solid Waste Disposal Act, amended in 1976 by the Resources Conservation and Recovery Act (RCRA), requires solid and hazardous wastes to be managed in accordance with regulations adopted by the EPA. Storage and disposal are organized to comply with government regulations.

U.S. government regulations on water discharge require the installation of water treatment plants for smelter effluents carrying significant levels of heavy metals as well as for discharges from the concentrator tailings ponds. Copper concentrators, smelters, and refineries in the United States practice maximum water recycling. Such action has required new approaches to internal water treatment and modification of flotation reagent systems to compensate for buildup in circulating loads of organic and inorganic constituents. Total recycling of water from dump-leaching operations is common practice. Actions are needed to keep leach solutions from entering groundwater systems (see GROUNDWATER MONITORING). Uncontrolled runoff from mining operations requires impoundment with appropriate treatment before release into surface systems.

Copper Sulfate. Copper sulfate, $CuSO_4$, is produced from copper scrap, blister copper, copper precipitates, electrolytic refinery solutions, and spent electroplating solutions (see COPPER COMPOUNDS). Data from domestic producers for 1993 indicated that 59% of shipments went for agricultural uses, 31% for industrial uses, including wood preservatives, and 10% for water treatment. In agriculture, copper sulfate is principally used as a fungicide for treatment of citrus and vegetable crops (see FUNGICIDES, AGRICULTURAL).

Energy. Mining uses about 20% of the total energy requirement of copper production; milling, around 40%; smelting, from 10 to 20%; and converting and refining, the remaining 20 to 30%. Actual requirements vary widely, depending on the mine characteristics and type of smelter. Pollution control accounts for a large percentage of the energy demand for smelting.

The energy consumption per unit of copper production is expected to increase as copper grades decline and as facilities are added for environmental control. A significant reduction in energy consumption could be achieved by installing new facilities; however, energy cost is only a minor component of the total cost of copper production.

Economic Aspects

The United States is largely self-sufficient with respect to copper, meeting any shortfall by imports. Australia and the CIS consume most of their production on the domestic market. Japan and Western Europe import substantial quantities of copper in the form of concentrates, blister, and refined copper. Copper industries in Chile, Peru, Zaire, and Zambia are largely nationalized. The United States is the world's largest copper producer. Most of the copper came from 25 mines.

A detailed statistical picture of the supply and consumption of copper and copper alloys in the United States is available annually. The statistics trace the flow of copper from mining and scrap collection through smelting, refining, and ingot making to the wire mills, brass mills, and foundries and then on to the final markets. There are strong indications that the future demand for copper will exceed production capacity.

In 1988, a comprehensive report on the technology and competitiveness of the U.S. copper industry was issued. This report concludes that the revitalized U.S. copper industry could compete in all but the worst foreseeable markets and that the industry's turnaround came entirely from its own efforts, with little governmental assistance. The U.S. copper industry is a world leader in smelter and refinery production, applying modern technology and measures to improve productivity.

Purity. Electrolytic copper is one of the purest of the materials of commerce. The average copper content of ETP copper, for instance, is over 99.95%, and even the highest level of impurities other than oxygen are found only to the extent of 15–30 ppm. Up to 0.05% oxygen is present in the form of copper(I) oxide. Even at these low impurity levels, properties of interest to fabricators are affected in varying degree.

Electrical Conductivity. All dissolved impurities lower the conductivity of copper.

Fabricability. Impurities in electrolytic copper are of such low levels as to have little effect on hot- or cold-working operations.

Annealability. The first studies of impurity effects were made early on in conjunction with research on the production of spectrographically pure copper and involved iron and nickel in the 0.7–500 ppm range and cobalt in the 20–500 ppm range. It was concluded that in oxygen-bearing copper, that is, oxygen in the ETP range 0.02–0.05%, these impurities did not change the recrystallization temperature. In ETP copper, phosphorus had no effect on the softening temperature up to 200 ppm because of compound formation with oxygen. Arsenic (up to 5 ppm) and silver (up to 35 ppm) had no appreciable effect, and neither tin nor cadmium showed any effect because they react with oxygen. Antimony, sulfur, tellurium, and selenium were found highly effective in increasing the softening temperature of copper, subject, however, to heat treatment.

The effects of trace quantities of silver, iron, sulfur, lead, selenium, and phosphorus on the conductivity and recrystallization behavior of copper were also investigated. Measurements of the annealing behavior of copper rod and wire made from wire bars containing various levels and combinations of tellurium, selenium, antimony, bismuth, lead, and silver showed tellurium to have the greatest unit effect. Selenium, antimony, and bismuth also raised the softening temperature. Silver decreased the recrystallization temperature. The effect of lead depends on the thermal history of the copper, decreasing the softening temperature under certain annealing conditions.

Effect of Thermal History. Many of the impurities present in commercial copper are in concentrations above the solid solubility at low (eg, 300°C) temperatures. Other impurities oxidize in oxygen-bearing copper to form stable oxides at lower temperatures. Hence, because the recrystallization kinetics are influenced primarily by solute atoms in the crystal lattice, the recrystallization temperature is extremely dependent on the thermal treatment prior to cold deformation.

Quality Control. The spectrometer is the most suitable instrument for determining most low level residual impurities.

Health and Safety

Copper is required for all forms of aerobic and most forms of anaerobic life. In humans, the biological function of copper is related to the enzymatic action of specific essential copper proteins. Lack of these copper enzymes is considered a primary factor in cerebral degeneration, depigmentation, and arterial changes. Because of the abundance of copper in most human diets, chemically significant copper deficiency is extremely rare.

Accidental ingestion of large amounts of copper salts from food or beverages contaminated by copper can cause gastrointestinal disturbances, and inhalation of copper fumes can cause metal fume fever. However, no chronic copper poisoning has been reported. The human metabolic system is very efficient in promoting a discriminating copper absorption. Copper is bound to albumin in blood plasma, and large amounts can be stored in and eliminated through the liver. Therefore, although systemic effects such as hemolysis, liver damage, and renal damage have been reported after ingestion of large amounts of copper salts, recovery has usually been rapid upon treatment.

About 50% of copper in food is absorbed, usually under equilibrium conditions, and stored in the liver and muscles. Excretion is mainly via the bile, and only a few percent of the absorbed amount is found in urine. The excretion of copper from the human body is influenced by molybdenum. A low molybdenum concentration in the diet causes a low excretion of copper, and a high intake results in a considerable increase in copper excretion. This copper–molybdenum relationship appears to correlate with copper deficiency symptoms in cattle. It has been suggested that, at the pH of the intestine, copper and molybdenum ions react to form biologically unavailable copper molybdate.

Humans tolerate fairly large oral doses of copper without harmful effects, and it is used in various therapies. Copper sulfate is a powerful emetic and has been used clinically in the treatment of intoxications.

Copper has been employed as a bactericide, molluscicide, and fungicide for a long time and is of importance in the control of schistosomiasis (see also ANTIPARASITIC AGENTS–ANTHELMINTICS). Addition of copper to lake water acts as an efficient deterrent to transmittal of the disease by eliminating snails that act as hosts for the responsible parasite. Copper is commonly utilized at ca 0.1 mg/L as an algicide. In fresh water, acute toxicosis in fish is unusual if the copper concentration is below 0.025 mg/L.

Uses

The properties of wrought and cast copper and its alloys, which result in countless industrial uses, are high electrical and thermal conductivity, corrosion resistance, formability, ease of joining, machinability, low temperature strength, moderate high temperature strength, long service, recyclability, and aesthetics.

In statistical analyses of the copper, brass, and bronze industry, there are five primary markets: *(1)* building construction (41%), including building wiring, plumbing, and heating; air conditioning and commercial refrigeration (see REFRIGERATION); builders' hardware, and architectural materials (see BUILDING MATERIALS–SURVEY); *(2)* electrical and electronic products (23%), for example, power utilities, telecommunications, business electronics, and lighting and wiring devices (see also EELECTRICAL CONNECTORS; ELECTRONIC MATERIALS); *(3)* industrial machinery and equipment (14%), such as in-plant equipment, industrial valves and fittings, nonelectrical instruments, off-highway vehicles, and heat exchangers (see HEAT EXCHANGE TECHNOLOGY); *(4)* transportation equipment (12%), including automobiles, trucks and buses, railroads, marine vehicles, aircraft, and aerospace vehicles; and *(5)* consumer and general products (10%), such as appliances, cord sets, ordnance, consumer electronics, fasteners, coinage, utensils and cutlery, and miscellaneous items.

DAVID B. GEORGE
Kennecott Corporation

A. K. Biswas and W. G. Davenport, *Extractive Metallurgy of Copper,* Pergamon Press, New York, 1980.

Eng. Min. J., Copper USA Supplement, **191** (Jan. 1990).

U.S. Bureau of Mines, annual *Minerals Yearbook,* Washington, D.C.

COPPER ALLOYS

WROUGHT COPPER AND WROUGHT COPPER ALLOYS

Typically, copper is alloyed with other elements to provide a broad range of mechanical, physical, and chemical properties that account for widespread use. The principal characteristics of copper alloys are moderate-to-high electrical and thermal conductivities combined with good corrosion resistance, good strength, unique decorative appearance, and moderate cost. Most copper alloys are readily hot and cold formed, joined (soldered, brazed, and welded), and plated.

Chief consumers of copper and copper alloys are the building construction industry for electrical wire, tubing, builder's hardware, plumbing, and sheathing (see BUILDING MATERIALS–SURVEY); electrical and electronic products for motors, connectors, printed circuit copper foil, and leadframes (see ELECTRICAL CONNECTORS; ELECTRONIC MATERIALS; INTEGRATED CIRCUITS); and the transportation sector for radiators and wiring harnesses. Other industries include ordinance, power utilities, and coinage (see EXPLOSIVES AND PROPELLANTS; POWER GENERATION).

Alloy Designations

Elements typically added to copper are zinc, tin, nickel, iron, aluminum, silicon, chromium, and beryllium.

Table 1. UNS Designation for Copper and Copper Alloys

Alloy group	UNS designation	Principal alloy elements
coppers[a]	C10100–C15999	Ag, As, Mg, P, Zr
high coppers[b]	C16000–C19999	Cd, Be, Cr, Fe, Ni, P, Mg, Co
brasses	C20500–C28580	Zn
leaded brasses	C31200–C38590	Zn–Pb
tin brasses	C40400–C40980	Sn, Zn
phosphor bronzes	C50100–C52400	Sn–P
leaded bronzes	C53200–C54800	Sn–P–Pb
phosphorus–silver	C55180–C55284	P, Ag–P
aluminum bronze	C60600–C64400	Al, Fe, Ni, Co, Si
silicon bronze	C64700–C66100	Si, Sn
modified brass	C66400–C69950	Zn, Al, Si, Mn
cupronickels	C70100–C72950	Ni, Fe, Sn
nickel silvers	C73150–C77600	Ni–Zn
leaded nickel silvers	C78200–C79900	Ni–Zn–Pb

[a] Contains a minimum of 99.3 wt % copper.
[b] Contains a minimum of 96 wt % copper.

Copper and its alloys are classified in the United States by composition according to the Unified Numbering System (UNS) for metals and alloys. The designations of wrought copper alloys are given in Table 1. Designations that start with numeral 8 or 9 are reserved for cast alloys.

Most wrought alloys are provided in conditions that have been strengthened by various amounts of cold work or heat treatment. Cold worked tempers are the result of cold rolling or drawing by prescribed amounts of plastic deformation from the annealed condition. Alloys that respond to strengthening by heat treatment are referred to as precipitation or age hardenable. Cold worked conditions can also be thermally treated at relatively low temperatures to affect a slight decrease in strength (stress relief annealed) to benefit other properties, such as corrosion resistance and formability.

Temper. The system for designating material condition, whether the product form is strip, rod, or wire, is defined in ASTM Recommended Practice B601. The ASTM system uses an alpha-numeric code for each of the standard temper designations. This system replaces the historical terminology of half hard, hard, spring, etc.

Product Forms and Processing

The output from brass mills in the United States is split nearly equally between copper and the alloys of copper. Copper and dilute copper alloy wrought products are melted and processed from electrically refined copper so as to maintain low impurity content. Copper alloys are commonly made from either refined copper plus elemental additions or from recycled alloy scrap. Copper alloys can be readily manufactured from remelted scrap while maintaining low levels of nonalloy impurities. A greater proportion of the copper alloys used as engineering materials are recycled than are other commercial materials.

Wrought alloy product forms are varied and include plate, sheet, strip and foil, round and special cross-section bars, rod, and wire.

Alloying for Strengthening

Copper alloys can also be grouped according to how the principal elemental additions affect properties.

Solid Solution Alloys. Copper dissolves other elements to varying degrees to produce a single-phase alloy that is strengthened relative to unalloyed copper. The contribution to strengthening from an element depends on the amount in solution and by its particular physical characteristics such as atom size and valency. Tin, silicon, and aluminum show the highest strengthening efficiency of the common solute additives, whereas nickel and zinc are the least efficient.

Dispersed Phase Alloys. The presence of finely dispersed second-phase particles in copper alloys contributes to strength, through refined grain size and increased response to hardening from cold working.

Precipitation (Age) Hardening Alloys. Only a few copper alloys are capable of responding to precipitation or age hardening. Those that do have the constitutional characteristics of being single-phase (solid solution) at elevated temperatures and are able to develop into two or more phases at lower temperatures that are capable of resisting plastic deformation. The copper alloy systems of commercial importance are based on individual additions of Be, Cr, or Ni + X where X = Al, Sn, Si, and Zr.

Special Addition Alloys. The most notable of the special additives to copper alloys are those added to enhance machinability. Lead, tellurium, selenium, and sulfur are within this group of additives. Because of increasing concern over lead toxicity, interest has centered on use of bismuth which is nearly as effective as lead for improving machinability. The alloys that contain such additives are limited because of the difficulty they cause to hot and cold working.

Properties

Strength. There is an increase in strength and an accompanying decrease in conductivity that derives from alloying of copper. This trend is clearly apparent among solid solution strengthened alloys, namely those that contain zinc, tin, nickel, and zinc plus nickel as their principal alloying constituents. Notable exceptions to the association of low conductivity with high strength are some alloys that are precipitation strengthened, namely C1751, C182, and C7025. For the latter group, both high strength and moderate conductivity are possible.

Alloy selection is not made from only consideration of strength and conductivity. For example, the cupronickels have about the same strength as do copper–zinc brasses, and also have much lower conductivity. However, the corrosion resistance of the cupronickels far exceeds that of brass and is worth the higher cost if needed in the application. Similar trade-offs exist between these properties and formability, softening resistance, and other properties.

Formability. Copper and most of the wrought alloys are readily formed by bending, drawing, upset forging, stamping, and coining. The maximum formability condition is the fully soft or annealed condition. When additional strength or hardness is desired in the final part, the forming step is done starting with a cold worked temper or the part is not annealed between forming steps. Cold forming operations always work-harden alloys and in many cases sufficient strengthening is produced.

Softening Resistance. The ability of being readily annealed or softened in a controlled manner to restore ductility is beneficial in mill processing, but resistance to softening of the wrought product is often preferred during fabrication and subsequent service. Joining operations such as welding, brazing, and soldering are prime examples where softening resistance is essential. Most additions of alloying elements and impurities increase softening resistance. The amount by which softening resistance is increased is specific to the element being added and also depends on whether the element remains in solid solution or forms a second-phase particle.

Stress Relaxation. The rate at which elastic stress is reduced depends on the alloy, its temper, the temperature of exposure, and the duration. Resistance to stress relaxation of copper is improved by alloying with solid solution elements, as well as by dispersion and precipitation strengthening. Changing temper to higher strength for a given alloy results in some loss in stability. Relief annealing where yield strength is decreased slightly while causing little or no change in tensile strength is used to improve relaxation resistance.

The highest stress relaxation resistance, at both high strength and moderate conductivity, is available from precipitation hardened alloys.

Fatigue. Imposed cyclic stressing of metals may result in localized cracking that leads to fracture. The pattern of stressing can be in reversed bending, reversed torsion, and tension–compression, or half cycles of these such as bending in only one direction. The number of cycles of stressing that can be endured without fracture depends on the

Table 2. Fatigue Strengths of Copper Alloys[a]

Alloy	Average 0.2% yield strength, MPa[b]	Fatigue strength 10^8 cycles, MPa[b]
C172[c]	760	275
C194	480	150
C260	615	185
C510	690	235
C762	725	205

[a] All are H08, spring temper unless indicated.
[b] To convert MPa to psi, multiply by 145.
[c] TMO2 ($\frac{1}{2}$ HM).

magnitude of the peak applied stress, the pattern of stressing, and the alloy's mechanical properties.

Fatigue strengths for several copper alloys are listed in Table 2. Generally, fatigue strength increases with tensile strength of the material. The rule-of-thumb is that the fatigue strength of a copper alloy is around one-third of its tensile strength.

Corrosion. Copper and selected copper alloys perform admirably in many hostile environments. Copper alloys with the appropriate corrosion resistance characteristics are recommended for atmospheric exposure (architectural and builder's hardware), for use in fresh water supply (plumbing lines and fittings), in marine applications (desalination equipment and biofouling avoidance), for industrial and chemical plant equipment (heat exchangers and condensers), and for electrical/electronic applications (connectors and semiconductor package leadframes) (see PACKAGING).

C260 shows the most susceptibility; C706, C194, and C110 show the least.

Hydrogen Embrittlement. Copper alloys that contain cuprous oxide in their microstructures, as in C110, are potentially susceptible to embrittlement when heated in hydrogen-containing gases. Accordingly, susceptible alloys are annealed during processing in nitrogen or very low hydrogen-containing gas, at low temperatures, and for short times, to avoid embrittlement.

Solderability. Most copper alloys have good solderability, meaning that they are wet by molten tin, tin–lead, and modifications of these to produce a continuous coating that has few to no pinhole sized nonwet areas. This characteristic of copper and its alloys accounts for the significant use of tin and solders to provide corrosion resistance and in joining (see SOLDERS AND BRAZING ALLOYS).

Brazing. Brazing is, by definition, elevated temperature soldering, that is, soldering above the arbitrarily defined temperature of 425°C. Copper and its alloys are readily brazed and often are brazed in order to take advantage of the stronger and more stable brazed joint compared to the soldered joint. In addition, it is easy to match the properties of the filler metal to the copper alloy to be brazed because many of the brazing alloys are themselves copper-base alloys. Filler alloys for brazing of copper alloys are usually copper base. These include copper–zinc alloys (RBCuZn), copper–phosphorus alloys (BCuP-1), and copper–silver (Zn, Cd, or Li) alloys (BAg).

Weldability. There are three primary considerations for successful welding of copper. The first and foremost is the high thermal conductivity of copper and its resulting ability to conduct heat away from the weld zone. The second consideration is the chemical and in some cases toxic properties of the typical elements alloyed with copper. Adequate ventilation must be provided for the lead vapor, the zinc fumes, or the toxic compounds of Be, As, or Sb that can be emitted from the copper alloys containing these elements. The third concern is the ready solubility of oxygen in copper at elevated temperatures and the subsequent precipitation of cuprous oxide particles at grain boundaries during solidification and cooling, which if uncontrolled cause reduced strength and ductility in the weld zone.

Arc welding has long been used to join copper alloys. The gas tungsten-arc welding (GTAW) and the inert gas metal arc welding (GMAW) methods are the preferred arc welding methods for copper alloys.

Machinability. Copper and its alloys can be machined with differing degrees of ease. Special additives such as lead, tellurium, selenium, and sulfur, are added, to enhance machinability, although other properties, such as formability and tensile ductility, normally suffer. These particular alloying elements form second-phase particles that promote chip fracture and the development of lubricative films at the tool-to-chip interface. Smaller, easier to handle chips and lower cutting forces having longer tool life result from use of these additives. Notable uses for special alloys having high machinability are rod, for high production rate screw machine items such as fasteners and plumbing components, and strip for keys.

Alloy Specific Properties

Copper. The physical properties of pure copper are given in Table 3. The mechanical properties of pure copper are essentially the same as those for C101 and C110. The coppers represent a series of alloys ranging from the commercially pure copper, C101, to the dispersion hardened alloy C157. The difference within this series is the specification of small additions of phosphorus, arsenic, cadmium, tellurium, sulfur, zirconium, as well as oxygen. To be classified as one of the coppers, the alloy must contain at least 99.3% copper.

Trace elements added to copper exert a significant influence on electrical conductivity. Effects on conductivity vary because of inherent differences in effective atomic size and valency. The decrease in conductivity produced by those elements appearing commonly in copper, at a fixed atomic concentration, rank as follows: Zn (least detrimental), Ag, Mg, Al, Ni, Si, Sn, P, Fe (most).

The mechanical properties of coppers having UNS designations between C10100 and C13000 are listed in Table 4. The coppers include high purity copper (C101, C102), electrolytic tough pitch (C110), phosphorus deoxidized (C120, C122), and silver-bearing copper (C115, C129, etc). The mechanical properties of these alloys are essentially the same. Other coppers, C142 through C157, offer higher strength at usually lower conductivity.

Excellent resistance to saltwater corrosion and biofouling are notable attributes of copper and its dilute alloys. High resistance to atmospheric corrosion and stress corrosion cracking, combined with high conductivity, favor use in electrical/electronic applications.

High Copper Alloys. The high copper alloys contain a minimum of 96 wt % copper; most contain about 98.5 wt %. These alloys are found within the UNS classification of C162 to C197. As a group, they offer higher strength and better softening resistance than the specialty coppers. Like the coppers, the high copper alloys have excellent resistance to general and stress corrosion cracking. However, the high copper alloys have lower electrical conductivities than the coppers, they are more expensive, and special processing may be required to incorporate dispersoids or precipitates for optimal effectiveness.

Table 3. Physical Properties of Copper (C10100)[a]

Parameter	Value
mp, °C	1083
density, g/mL	8.94
electrical conductivity, % IACS min	101
electrical resistivity, $\mu\Omega\cdot$cm	1.71
thermal conductivity, W/(m·K)	391
coefficient of thermal expansion from 20–300°C, μm/(m·K)	17.7
specific heat, J/(kg·K)[b]	385
elastic modulus, GPa[c]	
tension	115
shear	44

[a] All properties at 20°C unless otherwise noted.
[b] To convert J to cal, divide by 4.184.
[c] To convert GPa to psi, multiply by 145,000.

Table 4. Tensile Properties of Copper (C10100–C13000)

Temper	Tensile strength, MPa[a]	Yield strength,[b] MPa[a]	Elongation, %	Hardness, HRF[c]
OS 025[d]	235	76	45	45
H01	260	205	25	70
H02	290	250	14	84
H04	345	310	6	90
H08	380	345	4	94
H10	395	365	4	95

[a] To convert MPa to psi, multiply by 145.
[b] 0.5% offset.
[c] Hardness is on the Rockwell-F scale.
[d] Annealed to average grain size of 0.025 mm.

Copper–Zinc Brasses. Copper–zinc alloys have been the most widely used copper alloy during the 1990s.

Brass alloys fall within the designation C205 to C280 and cover the entire solid solution range of up to 35 wt % zinc in the Cu–Zn alloy system. Zinc, traditionally less expensive than copper, does not too greatly impair conductivity and ductility as it solution hardens copper.

Brass alloys are highly formable, either hot or cold, and provide moderate strength and conductivity. Moreover, the alloys have a pleasing yellow "brass" color at zinc levels above 20 wt %. The material is amenable to polishing, buffing, plating, and soldering. By far, the best known and most used composition is the 30 wt % zinc alloy, Cartridge Brass. Door knobs and bullet cartridges are the best known applications and illustrate the material's excellent formability and general utility.

Brasses are susceptible to dezincification in aqueous solutions when they contain >15 wt % zinc. Stress corrosion cracking susceptibility is also significant above 15 wt % zinc. Over the years, other elements have been added to the Cu–Zn base alloys to improve corrosion resistance. For example, a small addition of arsenic or phosphorus helps prevent dezincification to make brasses more useful in tubing applications.

Tin Brasses. The tin brass series of alloys consists of various copper– (2.5 to 35 wt %) zinc alloys to which up to about 4 wt % tin has been added. These are solid solution alloys that have their own classification as the C40000 series of alloys. Tin provides better general corrosion resistance and strength without greatly reducing electrical conductivity. As with all the brasses, these alloys are strengthened by cold work and are available in a wide range of tempers. These alloys offer the combined strength, formability, and corrosion resistance required by their principal applications, namely, fuse clips, weather stripping, electrical connectors, heat exchanger tubing, and ferrules.

Tin Bronzes. Whereas bronze is still used for statuary, these alloys are found in many modern applications, such as electrical connectors, bearings, bellows, and diaphragms. The wrought tin bronzes are also called phosphor bronzes because 0.03 to 0.35 wt % phosphorus is commonly added for deoxidation and improved melt fluidity.

The several wrought alloys of commercial importance span the range of 1.0 to 10 wt % tin and are mostly used in work-hardened tempers. These alloys are single-phase and offer excellent cold working and forming characteristics. Strength, corrosion resistance, and stress relaxation resistance increase with tin content. Unfortunately, conductivity decreases, cost increases (tin has been historically more costly than copper), and the capability of being hot processed is impaired as tin level is increased.

Aluminum Bronzes. Aluminum bronze alloys comprise a series of alloys (C606 to C644) based on the copper–aluminum (2–15 wt %) binary system, to which iron, nickel, and/or manganese are added to increase strength.

Aluminum bronze alloys are used for their combined good strength, wear, and corrosion-resistance properties where high electrical and thermal conductivity are not required. Corrosion resistance results from the formation of an adherent aluminum oxide that protects the surface from further oxidation. Mechanical damage to the surface is readily healed by the redevelopment of this oxide. The aluminum bronzes are resistant to nonoxidizing mineral acids such as sulfuric or hydrochloric acids, but are not resistant to oxidizing acids such as nitric acid. However, these alloys must be properly heat treated to be resistant to dealloying and general corrosion.

The aluminum oxide surface layer that provides wear and corrosion resistance is not without drawbacks. This adherent film is difficult to remove during industrial cleaning of the alloys. Furthermore, excessive tool wear in stamping and shearing equipment is caused by the presence of this film. Soldering and brazing are also made difficult by this oxide film.

Two single-phase, binary alloys are used commercially: C606, containing 5 wt % Al, and C610, 8 wt % Al. Both alloys have a golden color and are used in rod or wire applications, such as for bolts, pump parts, and shafts.

Silicon Bronzes. Silicon bronzes have long been available for use in electrical connectors, heat exchanger tubes, and marine and pole line hardware because of their high solution hardened strength and resistance to general and stress corrosion. As a group, these alloys also have excellent hot and cold formability. Unlike the aluminum bronzes, the silicon bronzes have moderately good soldering and brazing characteristics. Their compositions are limited to below 4.0 wt % silicon because above this level, an extremely brittle phase (kappa) is developed that prevents cold processing. Electrical conductivities of silicon bronzes are low.

Modified Copper–Zinc Alloys. The series of copper–zinc base alloys identified as C664 to C698 have been modified by additions of manganese (manganese brasses and the manganese bronzes), aluminum, silicon, nickel, and cobalt. Each of the modifying additions provides some property improvement to the already workable, formable, and inexpensive Cu–Zn brass base alloy. Aluminum and silicon additions improve strength and corrosion resistance. Nickel and cobalt form aluminide precipitates for dispersion strengthening and grain size control. The high zinc-containing alloys are formulated to facilitate hot processing by transforming to a highly formable (beta) phase at elevated temperature. C674 and C694 are commonly used in rod and wire forms.

Copper–Nickels. The copper–nickel alloy system is essentially single-phase across its entire range. Alloys made from this system are easily fabricated by casting, forming, and welding. They are noted for excellent tarnishing and corrosion resistance. Commercial copper alloys extend from 5 to 40 wt % nickel.

Iron is added in small (usually 0.5–1.0 wt %) amounts to increase strength. More importantly, iron additions also enhance corrosion resistance, especially when precautions are taken to retain the iron in solution. A small (up to 1.0 wt %) amount of manganese is usually added both to react with sulfur and to deoxidize the melt. These copper alloys are most commonly applied where corrosion resistance is paramount, as in condenser tube or heat exchangers.

Nickel–Silvers. Nickel–silver alloys, once called German silver, are a series of solid solution Cu–Ni–Zn alloys, the compositions of which encompass the ranges of 3 to 30 wt % Zn and 4 to 25 wt % Ni. This family of alloys falls within the UNS designation numbers C731 to C770. Leaded nickel–silvers that contain from 1.0 to 3.5 wt % lead for improved machining characteristics are designated as C782 to C799.

Nickel–silver alloys are not readily hot worked but have excellent cold fabricating characteristics. Because of the high nickel and zinc contents, these alloys exhibit good resistance to corrosion, good strength, and usually adequate formability, but have low electrical conductivity. Wire and strip are the dominant forms used for hardware, rivets, nameplates, hollowware, and optical parts.

Precipitation Hardening Alloys. Copper alloys that can be precipitation hardened to high strength are limited in number. In addition to the metallurgical requirement that the solubility of the added element(s) decrease with temperature, the precipitated phase that forms during aging must be distributed finely and have characteristics that act to resist plastic deformation.

Commercial precipitation hardening copper alloys are based on beryllium, chromium, and nickel, this last in combination with silicon or tin. The principal attributes of these alloys are high strength in association with adequate formability. Electrical conductivity varies according to alloy and ranges from around 20 to 80% IACS.

JOHN F. BREEDIS
RONALD N. CARON
Olin Corporation

Properties and Selection: Nonferrous Alloys and Pure Metals, Vol. 2, *Metals Handbook,* 10th ed., ASM International, Materials Park, Ohio, 1990.

Source Book on Copper and Copper Alloys, ASM International, Materials Park, Ohio, 1979.

Application Data Sheet—Standard Designations for Wrought and Cast Copper and Copper Alloys, Copper Development Association, Greenwich, Conn., 1992.

Annual Book of ASTM Standards, Vol. 02.01, *Copper and Copper Alloys,* American Society for Testing and Materials, Philadelphia, Pa., 1992.

CAST COPPER ALLOYS

Copper (qv) alloy castings are used for their generally superior corrosion resistance (see COROSION AND CORROSION CONTROL), high electrical and thermal conductivities, and good bearing and wear qualities. Some of the alloys are heat-treatable and couple high strength with good electrical and thermal conductivity. Irregular and complex external and internal shapes can be produced by various casting methods. The production of the same configurations by other methods may be mechanically impractical or too costly.

Properties and Characteristics

Cast copper alloys can be classified into two main groups: single-phase alloys, characterized by moderate strength, high ductility (except for leaded varieties), moderate hardness, and good impact strength; and polyphase alloys, having high strength, moderate ductility, and moderate impact strength.

Typical mechanical properties and electrical conductivity for various cast alloys are shown in Table 1.

Stresses and Stress Relieving. Nonuniform cooling leads to unbalanced residual stress patterns in the casting. Exposure of stressed castings to environments containing ammonia, ammoniacal compounds, or mercury can cause cracking of the alloys. Castings stressed to a level of ca 80 MPa (12,000 psi) and greater may crack in mercury environments, and stress levels of only a few MPa may produce cracks when exposed to ammonia. The coppers and copper–nickel–iron alloys usually have very high resistance to stress-corrosion cracking and do not require stress relieving.

Stress relieving to safe stress levels can be accomplished by a thermal treatment, sometimes called stress-relief annealing even though annealing may not be involved. The following alloys are all effectively stress relieved by heating at 260°C for 24 min/cm of section: high copper alloys, red brass, semired brass, yellow brass, manganese–bronze, silicon–bronze, tin–bronze, and nickel–bronze. Aluminum–bronze alloys are treated at 316°C for 24 min/cm of section.

Machinability. The cast copper alloys can be placed in three groups relative to machinability and are rated in the same general manner as wrought copper alloys. Group one contains the leaded alloys and is considered to be free-machining. Lead causes chip breakage during machining operations, permitting higher cutting speeds, decreased tool wear, and improved surface finish.

Alloys in the second group are polyphase alloys having a second phase generally harder than the matrix, which can cause brittleness and some chip breakage. This group comprises leaded tin bronze, silicon bronze, high tin bronze, aluminum bronze, and manganese bronze. Group three, the most difficult to machine, consists of high strength manganese and aluminum–bronze high in iron or nickel content.

Electrical and Thermal Conductivities. Wrought alloys have higher conductivities than the comparable casting alloys; wrought alloys usually have lower concentrations and fewer alloying elements.

Bearing and Wear Properties. Copper alloys have been used as bearing materials (qv) because of the combination of moderate to high strength, very good corrosion resistance wear resistance, and self-lubricating characteristics. Bearing alloys containing copper can be placed into three groups: phosphor–bronze alloys; copper–tin–lead alloys; and aluminum–bronze and silicon–bronze alloys.

Joining Characteristics. Cast copper alloys may be joined by welding (qv), brazing, and soldering techniques with varying degrees of ease and success.

Mechanical Properties. Most alloys containing tin, lead, or zinc have moderate tensile and yield strengths and high elongation. Higher tensile or yield strengths are available through the use of aluminum and manganese bronzes, silicon brasses and bronzes, and some nickel alloys. Some of these alloys, such as beryllium–copper, chromium–copper, and some aluminum bronze, are heat-treatable to attain maximum tensile strengths. Mechanical and physical properties for copper-base casting alloys are summarized in Table 1.

Production Methods

Precautions. Care should be taken not to impair the quality of the metal as a result of the melting operation. The best practice is to select scrap of known good quality. The use of miscellaneous scrap may not be economical, despite the lower cost of such material; however, it is not common practice to make much use of virgin metals. Production methods include sand casting, permanent-mold castings, die casting, plaster-mold casting, centrifugal casting, investment casting, and continuous casting.

Comparison or Casting Methods. Table 2. compares several factors of various casting procedures. The factors are rated A through E; letter A is the most advantageous.

Alloy Chemistry

Effect of Various Alloying Elements. The mechanical properties of cast copper alloys are a function of alloying elements and their concentrations. The specific effects of a number of these alloying elements are given in the following sections.

Zinc is added to copper as a predominating alloying element in concentrations of 5–40%, forming the alloy series known as the brasses. Zinc increases the tensile strength at a significant rate up to a concentration of ca 20%, whereas the tensile strength increases only slightly more for additions of zinc of 20–40%. The Rockwell F scale hardness is substantially an increasing straight-line function with zinc additions of 5–35%.

Tin added to copper in concentrations of 5–20% forms the tin–bronze alloy series. Leaded tin bronze is also produced. Tin imparts strength and hardness to the copper-base alloys, making them tough and wear-resistant. Tin also enhances the corrosion resistance of copper-base alloys in nonoxidizing media. Tin lowers both the tensile strength and the ductility of the alloy.

Lead is added to copper in amounts up to 40%. Because it imparts a certain degree of brittleness to the structure, it enhances machining operations by causing the alloy to break into chips as cutting tools are thrust into the matrix.

Silicon added to copper forms alloys of high strength and toughness along with improved corrosion resistance, particularly in acidic media. Silicon in small amounts can improve fluidity. Silicon is a very harmful impurity in leaded tin bronze alloys, however, because it contributes to lead sweat and unsoundness.

Aluminum added as the predominating alloying element to copper forms a series of high strength alloys called aluminum bronzes.

Table 1. Properties of Selected Cast Copper Alloy[a]

Common name	UNS designation	0.5% Yield strength, MPa[b]	Compressive yield strength, MPa[b]	Tensile strength, MPa[b]	Elongation in 5 cm, %	Brinell hardness, 500-kg load	Electrical conductivity, % IACS	Thermal conductivity at 20°C, W/(m·K)[c]
copper	C80100	62		172	40	44	100	391
chromium–copper								
cast	C 81500	83		214	35	63	45	
precipitation-hardened		296		379	18	110	82	315
beryllium–copper	C 81700	469	551[d]	620	8	217[e]	48	188
ASTM B22								
high strength yellow brass	C 86300	124	413[f]	820	83	225	8.0	35.4
gun metal	C 90500	138–158	276[g]	310	25	75	11.0	74
tin–bronze 84:16	C 91100	172		214	2	135[e]	8.5	
high leaded tin–bronze	C 93700	124	90[f]	214	20	60	10.0	47
ASTM B61								
steam–bronze	C 92200	137	262[d]	2276	20	65	14	69
ASTM B62								
leaded red brass	C 83600	103	96[f]	241	32	62	15	73
ASTM B66								
phosphor–bronze	C 94400	110	303[d]	221	18	55	10	52
high leaded tin–bronze	C 93800	110	83[f]	208	16	55	11.5	52
medium bronze	C 94500	83	248[d]	172	12	50	10	52
ASTM B148								
aluminum–bronze 9A	C 95200	172–208	186–214[f]	482–600	22–38	110–140	11	50
aluminum–silicon–bronze	C 95600	234		517	18	140[e]8.5	38	
manganese–aluminum–bronze	C 95700	310	1034[d]	655	26	180[e]	3.0	12
nickel–aluminum–bronze	C 95800	262	689[d]	655	25	159[e]	7.0	35
ASTM B176								
die-casting yellow brass	C 85800	207[h]		379	15		20	
die-cast silicon–brass	C 87800	345[h]		586	25		6.7	28
silicon–yellow brass	C 87900	241[h]		482	25		15.0	
ASTM B584								
leaded red brass	C 83600	103	96	255	32	60	15.0	71.9
leaded semired brass	C 84400	90–117		200–269	18–30	50–60	18.0	73
leaded yellow brass	C 85200	83–96	55–96[f]	241–276	25–40	40–50	15–22	83.9
commercial no. 1								
yellow brass	C 85400	76–103	62[f]	208–262	20–35	40–60	18–25	88
yellow brass	C 85700	96–138		276–310	15–40	50–75	20—26	83.9
high strength yellow brass	C 86200	331	345[f]	655	20	180[e]	7.5	35.4
leaded high strength yellow brass	C 86400	172[h]	158[f]	448	20	105[e]	19.0	88
high strength yellow brass	C 86500	193[h]	165[f]	489	30	130[e]	22	86
leaded high strength yellow brass	C 86700	289		586	20	155[f]	16.7	
silicon–bronze	C 87200	172	124[f]	379	30	85	6.0	28.4
silicon–brass	C 87400	165		379	30	70	6.7	27.7
tin–bronze	C 90300	145	90[f]	310	30	70	12	74.7
gun metal	C 90500	152	276[g]	310	25	75	11	74.7
leaded tin–bronze	C 92300	138	241[d]	276	25	70	12	74.7
nickel–tin bronze								
cast	C 94700	138–158		310–345	25–35	85	12	53.9
heat-treated	C 94700	345–413	0	517–551	5–10	180[e]		
leaded nickel–tin bronze								
cast	C 94800	138–158		310–345	20–35	80	12	38.6
heat-treated		207		413	8	120[e]		
	C 94900	103		262	15			
copper–nickel 90:10	C 96200	172		310	20		11	45
copper–nickel 70:30	C 96400	221		413	20	140[e]	5	29
12% nickel–silver	C 97300	117		241	20	55	5.7	28.5
20% nickel–silver	C 97600	117		262	20	70	5	22

[a] Mechanical property data were developed from separately cast test bars; values shown are based on technical literature. [b] To convert MPa to psi, multiply by 145. [c] To concert W/(m·K) to Btu − ft^2/(ft·h·°F, multiply by 0.578. [d] 0.1 cm set/cm. [e] 3000 kg load. [f] 0.001 cm set/cm. [g] 0.01 cm set/cm. [h] Offset of 0.2%.

Table 2. Comparison of Casting Methods[a]

Factor	Casting method						
	Sand	Die	Investment	Permanent mold	Plaster	Centrifugal	Continuous
tolerance	C	A	A	B	A	D	C
surface finish	D	B	A	C	A	D	C
thickness of section	C	D	D	C	D	A	A
pattern cost	A	E	B	C	B	A	A
ease of getting into production	A	D	B	C	B	C	C
production rate	B	A	C	C	D	B	C
cost per piece	B	A	C	C	D	A	C
flexibility as to alloy	A	D	A	D	C	A	A
limitation of size	A	D	C	B	D	A	A

[a] Comparisons are all relative based on letters A through E; A is most advantageous.

Aluminum present in leaded tin bronze alloys promotes unsoundness.

Iron added to copper alloys adds strength to the silicon, aluminum, and manganese bronzes.

Phosphorus is used principally as a deoxidizer in copper and high copper alloys.

Boron is a commercial deoxidizer.

Manganese is added as an alloying element in high strength brasses, where it forms compounds with other elements such as iron and aluminum. Manganese may also be used as a deoxidizer, although that is not a common usage.

Nickel added to copper markedly whitens the resulting alloy. Iron, up to a nominal 1.4%, added along with nickel significantly enhances the resistance toward cavitation or impingement corrosion. Added to bronzes, nickel refines the cast grain structure and adds toughness. Nickel improves strength and corrosion resistance.

Beryllium added to copper forms a series of age- or precipitation-hardenable alloys. These heat-treatable alloys are the strongest of all known copper-bases alloys.

Chromium also forms heat-treatable copper alloys.

Arsenic, antimony, and phosphorus can be added in small quantities (0.05%) to the all alpha-phase brass alloys containing less than 80% copper to inhibit the dezincification type of corrosion in yellow brass alloys.

Economic Aspects

Casting is used for irregular external and internal shapes that are impractical, impossible, or too costly to produce using other methods. The choice of an alloy for any casting usually depends on four factors: metal cost, castability, properties, and final cost. A cost analysis determines the most economical method of producing a casting, although frequently the choice can be based on experience.

Health and Safety Factors

During melting and pouring, certain metals, such as zinc, volatilize, enter the atmosphere, and immediately oxidize to solid particulates, forming a smoke. Some of these fumes or smokes can be hazardous to health, depending on the chemical composition of the particulate, its concentration in the off-gas, and the duration of exposure. More and more melt and casting shops are significantly controlling the atmospheric conditions in working areas by the use of adequate ventilating hoods, ducts, and exhaust fans in strategic locations.

Local and national agencies have set limits on the quantities of particulates permissible in casting-shop atmospheres and in the exhaust from collection systems. Agencies in the United States are the EPA, NIOSH, and OSHA.

Daniel L. Twarog
American Foundrymen's Society

P. R. Beeley, *Foundry Technology,* Halsted Press, New York, 1972.

G. J. Cook, *Engineered Castings,* McGraw-Hill Book Co., New York, 1961.

Metals Handbook, 9th ed., vol. 2, American Society for Metals, Metals Park, Ohio, 1979.

J. L. Morris, *Metal Castings,* Prentice-Hall Inc., Englewood Cliffs, N.J., 1957.

COPPER COMPOUNDS

Copper compounds, which represent only a small percentage of all copper production, play key roles in both industry and the biosphere. Copper, mol wt = 63.546, $[Ar]3d^{10}4s^1$, is a member of the first transition series and much of its chemistry is associated with the copper(II) ion, $[Ar]3d^9$. Copper forms compounds of commercial interest in the +1 and +2 oxidation states.

Manufacture of Commercially Important Compounds

Copper(II) Carbonate Hydroxide. Basic copper carbonate, also named copper(II) carbonate hydroxide, occurs in nature as the green monoclinic mineral malachite. The approximate stoichiometry is $CuCO_3{\cdot}Cu(OH)_2$. There are two grades available commercially, the light and the dense. The light grade is produced by adding a copper salt solution to a concentrated solution of sodium carbonate, usually at 45–65°C. The blue, voluminous azurite, $C_2H_2Cu_3O_8$, forms initially and converts to the green malachite within two hours. The dense product can be produced by boiling an ammoniacal solution to copper(II) carbonate or by addition of a copper salt solution to sodium bicarbonate at 45–65°C. A dense product can also be produced by simultaneous addition of copper(II) salt solutions and soda ash solutions at controlled pH.

Basic copper carbonate is essentially insoluble in water, but dissolves in aqueous ammonia or alkali metal cyanide solutions. It dissolves readily in mineral acids and warm acetic acid to form the corresponding salt solution.

Copper Chloride. Copper(I) chloride, CuCl, is a colorless or gray cubic crystal and occurs in nature as the mineral nantokite. The commercial product is white to gray to brown to green and of variable purity. Copper(I) chloride is usually produced at 450–900°C by direct combination of copper metal and chlorine gas to yield a molten product. The molten product is variously cast, prilled, flaked, or ground, depending on final use.

Copper(I) chloride is insoluble to slightly soluble in water. Solubility values between 0.001 and 0.1 g/L have been reported. CuCl is insoluble in dilute sulfuric and nitric acids, but forms solutions of complex compounds with hydrochloric acid, ammonia, and alkali halide. Copper(I) chloride is fairly stable in air at relative humidities of less than 50%, but quickly decomposes in the presence of air and moisture.

Cupric chloride or copper(II) chloride, $CuCl_2$, is usually prepared by dehydration of the dihydrate at 120°C. The anhydrous product is a deliquescent, monoclinic yellow crystal that forms the blue-green orthohombic, bipyramidal dihydrate in moist air. Both products are available commercially.

Copper(II) oxychloride, $Cu_2Cl(OH)_3$, is found in nature as the green hexagonal paratacamite or rhombic atacamite. It is usually precipitated by air oxidation of a concentrated sodium chloride solution of copper(I) chloride.

Copper Hydroxide. Copper(II) hydroxide, $Cu(OH)_2$, produced by reaction of a copper salt solution and sodium hydroxide, is a blue, gelatinous, voluminous precipitate of limited stability. Usually ammonia or phosphates are incorporated into the hydroxide to produce a color-stable product.

Copper hydroxide is almost insoluble in water (3 μg/L) but readily dissolves in mineral acids and ammonia forming salt solutions or copper ammine complexes.

Copper Nitrates. The most common commercial forms for copper nitrate are the trihydrate and solutions containing about 14% copper. Copper nitrate can be prepared by dissolution of the carbonate, hydroxide, or oxides in nitric acid.

The trihydrate is very soluble in water and ethanol.

Copper Oxides. Copper(I) oxide is a cubic or octahedral naturally occurring mineral known as cuprite. It is red or reddish brown in color. Commercially prepared copper(I) oxides vary in color from yellow to orange to red to purple as particle size increases. Usually copper(I) oxide is prepared by pyrometallurgical methods.

Copper(I) oxide is stable in dry air, but reacts with oxygen to form copper(II) oxide in moist air. Cu_2O is insoluble in water, but dissolves in ammonia or hydrochloric acid.

Copper(II) oxide, CuO, is found in nature as the black triclinic tenorite or the cubic or tetrahedral paramelaconite. The black product of commerce is most often prepared by evaporation of $Cu(NH_3)_4CO_3$ solutions or by precipitation of copper(II) oxide from hot ammonia solutions by addition of sodium hydroxide.

Copper(II) oxide is insoluble in water, but readily dissolves in mineral acid or in hot formic or acetic acids. CuO slowly dissolves in ammonia solution, but alkaline ammonium carbonate solubilizes it quickly.

Copper(II) Sulfates. Copper(II) sulfate pentahydrate $CuSO_4{\cdot}5H_2O$, occurs in nature as the blue triclinic crystalline mineral chalcanthite. It is the most common commercial compound of copper.

Copper(II) sulfate can be prepared by dissolution of oxides, carbonates, or hydroxides in sulfuric acid solutions.

Copper(II) sulfate monohydrate, $CuSO_4{\cdot}H_2O$, which is almost white in color, is hygroscopic and packaging must contain moisture barriers. This product is produced by dehydration of the pentahydrate at 120–150°C.

Anhydrous copper(II) sulfate is a gray to white rhombic crystal and occurs in nature as the mineral hydrocyanite. $CuSO_4$ is hygroscopic. It is produced by careful dehydration of the pentahydrate at 250°C.

The basic copper(II) sulfate that is available commercially is known as the tribasic copper sulfate, $CuSO_4{\cdot}3Cu(OH)_2$, which occurs as the green monoclinic mineral brochantite. This material is essentially insoluble in water, but dissolves readily in cold dilute mineral acids, warm acetic acid, and ammonia solutions. Tribasic copper sulfate is usually prepared by reaction of sodium carbonate and copper sulfate.

Economic Aspects

It has been estimated that 100,000 t of copper is used in the production of copper compounds annually, representing less than 1% of the total production of primary metal. As much as 70,000 t/yr (as copper) is used in agriculture for fungicides, animal feeds, and crop nutrients. Algicides make up the second largest group of copper compounds, followed by wood (qv) treatment and antifouling pigments (qv), mining flotation (qv), electroplating (qv), colorants, electronics, and catalysts.

Health and Safety

Copper is one of the twenty-seven elements known to be essential to humans. The daily recommended requirement for humans is 2.5–5.0 mg. Copper is probably second only to iron as an oxidation catalyst and oxygen carrier in humans. Copper aids in photosynthesis and other oxidative processes in plants.

Copper is toxic in exceedingly low concentrations to most fungi, algae, and certain bacteria and can be lethal to higher life forms in relatively high doses. The acute oral toxicity in humans, LD_{Lo}, is 100 mg/kg, but recovery from ingestion of 600 mg/kg has occurred. The symptoms of copper poisoning are nausea, vomiting, cramps, gastric disturbances, apathy, anemia, convulsions, coma, and death.

Inhalation of dusts can cause metal fume fever, and ulceration or perforation of the nasal septum. The workplace standard (TLV) for copper dusts or mist is 1 mg/m^3 and 0.2 mg/m^3 for copper fume.

H. WAYNE RICHARDSON
CP Chemicals, Inc.

W. Hatfield and R. Whyman, in R. Carlin, ed., *Transition Metal Chemistry,* Vol. 5, Marcel Dekker, New York, 1969, pp. 47–179.

H. W. Richardson, ed., *Copper Compounds Application Handbook,* Marcel Dekker, New York, 1992.

COPYRIGHTS AND TRADEMARKS

COPYRIGHTS

Copyright has been grouped with other forms of legal protection under the general term *intellectual property*. It is a means of protecting that particular form of creativity which has been variously referred to as originality of authorship, expression of ideas, or writings of an author. It is distinct from forms of intellectual property that do not protect original expression of authorship, eg, patents, which protect novel inventions or discoveries; trademarks, which protect terms identifying the source of origin of goods and services; and trade secrets which protect confidential, proprietary information.

Copyrightability

Under U.S. law, a work is either protected, ie, copyrighted, or unprotected and free for all to use, ie, in the public domain. Once a work enters the public domain, it cannot thereafter be recovered and protected again.

The Copyright Act specifies that copyright extends to "original works of authorship fixed in any tangible medium of expression, now known or later developed, from which they can be perceived, reproduced, or otherwise communicated, either directly or with the aid of a machine or device". (17 U.S.C. § 102(a) (1988). Many of the requirements for copyrightability may be gleaned from this provision.

The copyrightable work must be an original work of authorship. Thus originality, and not novelty as for patent rights, constitutes the touchstone of protection. The work must also be the product of an author. Finally, the work must be fixed in a tangible medium of expression. To a very limited extent, there are some works that are not so fixed, such as purely improvised and unrecorded pieces of music or choreography; extemporaneous speeches; or live, unrecorded, and ephemeral broadcasts. Unfixed works are protected by state common law copyright, and not the federal statute.

The Subject Matter of Copyright. The Copyright Act specifies, by way of example, the types of works that are covered, ie, literary works; musical works, including lyrics; dramatic works, including accompanying music; pantomimes and choreographic works; pictorial, graphic, and sculptural works; motion pictures and other audiovisual works; sound recordings; and architectural works. This list is nonexhaustive.

The Idea/Expression Dichotomy. Copyright protects the expression of ideas, but not ideas themselves. Thus copyright will not protect any procedure, process, system, method of operation, concept, principle, or discovery, regardless of the form in which it is described, explained, illustrated, or embodied.

Utilitarian Works. These works may be copyrightable, but only to the extent that they contain copyrightable subject matter. That copyrightable subject matter must be physically or conceptually separable from the purely utilitarian object.

Compilations. Copyright extends not only to works that can exist on their own but also to compilations of such works or even of public domain material. The Copyright Law imposes a three-step test for such copyrightable compilations. They must first constitute the collection and assembling of preexisting data or materials. Second, those materials must be selected, coordinated, or arranged in a particular fashion. Third, that selection, coordination, or arrangement must itself possess sufficient originality and creativity to constitute an original work of authorship.

Copyright Ownership

Copyright is a property right. Although it differs from most other forms of property in that it is intangible, it nevertheless has the essential elements of property and is governed by the principles of property ownership.

The intangible nature of copyright requires a distinction between the intangible property of the copyright, called a work, and the material object in which the copyrighted work is embodied, ie, a copy or phonorecord, which includes such diverse media as paper and ink, computer disks, and audiotapes. Ownership of the copyrighted work does not constitute ownership of the material object in which it is embodied, and vice versa. Copyright ownership vests initially in the author or authors of the work.

Joint Authorship. When more than one author has created a work, the work is said to be a joint work. Under the law, such a joint work is one prepared by two or more authors with the intention that their contributions be merged into inseparable or interdependent parts of a unitary whole.

Joint Ownership. Joint ownership of copyright occurs when there is joint authorship, but it may also occur in other ways, for example, by transfer of a copyright to two or more individuals, such as when an author bequeaths a copyright to two children.

Works Made for Hire. In many instances one person has created a work at the behest of another. In such circumstances, the creator does not own the copyright. As an easy example, consider a company that manufactures an appliance and has one of its employees write an instruction manual for the appliance. Logically, the company, and not the employee, should own the copyright.

Such situations are governed by the work-made-for-hire doctrine of the Copyright Law. Under the Copyright Law, copyright ownership vests initially in the author of the work. In cases of works made for hire, the law specifies that the employer or other person for whom the work is prepared is deemed to be the author. Thus the appliance company would be deemed to be the author, and hence the initial copyright owner, of the copyrighted instruction manual.

Transfers and Licenses of Copyright. Like other forms of property, copyright may be freely transferred. However, there are certain special rules governing the transfer of copyrights, and certain aspects of the law concerning transfer of property are of special importance to copyright.

Copyright Formalities

Changes to the Copyright Law, starting with the 1976 Copyright Act and continuing with the Berne Convention Implementation Act of 1988, have radically changed U.S. copyright law regarding copyright formalities; many formalities previously of paramount importance have been eased or entirely eliminated.

Copyright Notice. In the past, the law contained an absolute requirement that each copy of a published work bear a proper copyright notice. An amendment to the law abolished the notice requirement for all works published on or after March 1, 1989. For such works, no copyright notice is required. Notice, however, is still required on all copies of works first published before that date. In addition, notice is still widely used even when it is not required so as to inform the world of the copyright status of the work.

Copyright Deposit. The law requires that copies of every published work be submitted to the United States Copyright Office, which is a branch of the Library of Congress.

Copyright Registration. The term copyrighting a work is usually misused. Although it refers to registering a work with the Copyright Office, copyright registration is not required for copyright protection. Federal copyright protection exists from the moment a work is created, ie, fixed in a tangible medium of expression, even if it is never registered. Registration is entirely permissive, but does have certain procedural and substantive benefits in litigation.

Copyright Duration. Two different regimes of copyright duration apply in the United States. One is for works first created, published, or registered for copyright on or after January 1, 1978 (new law works); the other is for works published or registered before that date (old law works). In all cases, copyright terms run through December 31 of their anniversary year.

New Law Works. The basic copyright term for new law works is the life of the author plus 50 years after the author's death. In the case of joint authors, the life in question is that of the longest surviving author.

Old Law Works. Protection for works registered or published prior to 1978 endures for a dual-term system. There is an initial term of 28 years from the earlier of publication or registration, followed by a renewal term of an additional 47 years, for a total of 75 years of protection. Under a 1992 amendment to the Copyright Act, renewal is automatic, but there are procedural advantages to obtaining a renewal registration in the last year of the initial term.

Copyright Rights

The Copyright Act grants copyright owners five exclusive rights. These rights include not only the right to do specified actions, but also to authorize them: the right to reproduce in copies, the right to prepare derivative works, the right of public distribution, the right of public performance, and the right of public display.

Author's Rights. The Copyright Law was amended on June 1, 1991, to grant very limited additional rights to authors of certain types of works, even if they had parted with copyright ownership. These author's rights, sometimes referred to as moral rights, are applicable only to works of visual art that exist in single copies or multiples of up to 200. Even within this limited category of works, there are many exceptions, for example, author's rights do not apply to works made for hire. The author's rights are those of attribution, that is, right to have the author's name attached to or deleted from the work, and integrity, that is, the right to prevent mutilation or distortion of the work that would prejudice the author's honor or reputation.

Limitations and Exemptions

Fair Use. The Copyright Law contains several limitations on copyright rights and exemptions for certain uses. The best-known exemption is the fair use doctrine. Certain uses of copyrighted works that would otherwise be infringements are excused from liability because they are fair. The law gives examples of uses for purposes such as criticism, comment, news reporting, teaching, scholarship, and research.

First Sale Doctrine. Although the copyright owner has the exclusive right to distribute copies to the public, the bona fide possessor of a particular copy, in most circumstances, may further dispose of that copy without the copyright owner's consent. Thus, for example, the purchaser of a book may freely resell that copy. Hence, used book stores do not violate copyright law. The Copyright Act has been amended to prohibit the rental of sound recordings or computer software, even though that rental would have been permitted by the first sale doctrine.

Infringement

Anyone who violates the exclusive rights of a copyright owner is liable for infringement in a lawsuit brought in federal court. There is a three-year statute of limitations on copyright infringement actions.

The Test for Infringement. It is rare that actual evidence of copying exists. Proof of copying is usually circumstantial and is shown by a two-part test. First, the alleged infringer must be shown to have had access to the copyrighted work. Second, the two works must appear to their hypothetical intended audience to be substantially similar.

Remedies. A copyright owner successfully proving infringement has three remedies available: recovery of monetary damages; injunctive relief; and, in the court's discretion, recovery of costs, including attorneys' fees.

International Copyright

Because copyright easily transcends national boundaries, several copyright conventions have been developed to protect copyrights internationally. The best known and most widely effective conventions are the Berne Convention for the Protection of Literacy and Artistic Works and the Universal Copyright Convention; the United States is a signatory to both. In varying degrees, the treaties specify minimum standards that each member country's copyright law must meet.

I. FRED KOENIGSBERG
White & Case

P. Goldstein, *Copyright,* Little Brown & Co., Boston, 1989. A new treatise by a highly regarded copyright scholar.

M. B. Nimmer and D. Nimmer, *Nimmer on Copyright,* Matthew Bender, New York, 1991, 4-vol treatise. The most often cited and detailed copyright treatise, updated semiannually.

The United States Copyright Office furnishes a wide variety of circulars and publications on all aspects of copyright, as well as registration forms. All may be obtained from the United States Copyright Office, Library of Congress, Washington, D.C., 20559.

TRADEMARKS

A trademark, as defined in the federal statutes, includes any word, name, symbol, device, or combination thereof, adopted and used by a manufacturer or merchant to identify his or her goods and distinguish them from those manufactured or sold by others. Related to trademarks are service marks, certification marks, and collective marks.

Ownership of a trademark confers the exclusive right to use it or authorize its use in connection with the goods of its owner or goods made by others to the owner's quality standards. Conversely, it also confers the right to prevent others from using the mark, or another mark so similar as to be confused for the original, on similar or related goods. Trademark protection is a type of restraint afforded by the courts against unfair competition.

Although distinct from trademarks, trade names also are entitled to protection against unfair competition. Trade names and trademarks should not be confused. A trademark identifies the particular goods of a business; a trade name identifies the business itself. Trade names include individual names and surnames, firm names, and other names used by manufacturers, industrialists, merchants, agriculturists, and others to identify their businesses, vocations, or occupations. Depending on the manner of its use, a trade name may also function as a trademark. For instance, the name General Electric, when used to refer to the corporation, is a trade name; but when applied to merchandise, eg, General Electric refrigerators, it is a trademark.

Trademarks, Patents, Copyrights. A trademark is acquired by use on or in connection with the goods of its owner. It is accorded legal protection at common law independent of statutory provisions or registration. Registration, available under the laws of the states as well as under federal statute, enhances or facilitates protection of the trademark right. Acquisition of the trademark right by use in connection with the goods is a prerequisite to obtaining registration. Under the 1988 Trademark Revision Act, applications for trademark registration may be filed on the basis of a bona fide intent to use the mark, but proof of actual use is still required before a registration is issued.

Patents afford the owner the right to exclude others from making, using, or selling an invention, and are entirely dependent on statutory registration. They are acquired by disclosing an invention in an application duly filed and prosecuted in accord with the patent laws (see PATENTS AND TRADE SECRETS).

Copyrights afford in most cases an exclusive right to control distribution, reproduction, adaptation, public performance, and public display of literary or artistic works. They arise automatically upon creation of an eligible work, but the exercise of such rights is governed exclusively by federal statute.

Authority for protection of patents and copyrights is set out in the Constitution and is the exclusive province of the federal government. Federal legislation to protect trademarks is based on the authority of Congress, under the commerce clause of the Constitution, to regulate interstate and foreign commerce of the United States; protection afforded by individual states is based on their power to regulate intrastate commerce.

Selection of Trademarks

Trademarks distinguish the merchandise of the owner of the mark from the merchandise of others. Any designation commonly used by the public in describing the merchandise, its functions, or its properties cannot be preempted by one manufacturer to the exclusion of the public. Thus a mark for which exclusive protection is sought cannot be merely descriptive of the merchandise. Preferably, a mark should be fanciful or arbitrary. A trademark must also be adequately distinguishable from other trademarks for similar merchandise to avoid confusing or deceiving customers about the origin of the goods. To determine whether a prospective mark is distinct enough from other marks, a search in the fields of commerce in which the mark's use is contemplated is customary.

Trademark Registration

Trademarks' function may be characterized as an identification of origin, a guarantee of quality, and an advertising device.

Registration Rights. The right to trademark protection arises through exclusive use of the mark in connection with the goods of the owner. This right is protected by common law independent of registration. Trademark registration is a statutory creation; it affords a means of publicizing a claim to a trademark right and facilitating its protection. The federal statute, the Lanham Act, provides for two separate systems of registration: a Principal Register for marks fulfilling all requirements for full registration, and a Supplemental Register for those lacking certain of the requirements for registration on the Principal Register. Registration on the Principal Register constitutes constructive notice of the claim of ownership of the mark. Federal registration on either register extends protection to all parts of the United States and its territories and possessions. A common law trademark right, on the other hand, is effective only in the region where the mark has become known through use. Federal registration on either register also confers jurisdiction of trademark actions on the federal courts and affords the trademark owner additional rights against infringers. Substantial advantages can thus be derived from federal trademark registration.

Registration Procedure. The procedure for registering a trademark under the Lanham Act involves filing an application with the Patent and Trademark Office requesting registration on either the Principal Register or the Supplemental Register, and declaring either an actual, present use of the mark in interstate or foreign commerce or a bona fide intention to use the mark in commerce.

After examination, any objections, including rejection in view of prior registrations, are communicated to the applicant in an office action, requiring a response within a limited period. The response may involve amendment of the application, an argument to overcome the grounds for rejection, or a showing of further facts. Final rejections of applications may be appealed to the Trademark Trial and Appeal

Board, if necessary or desired, as described in the following case of adversary proceedings.

When the application is found to be allowable, it is published in the *Official Gazette* of the Patent and Trademark Office, and within 30 days, or within such extension thereof as may be granted to a potential opposer, any person may file an opposition alleging the belief that he or she will be damaged by the registration, and the grounds of that belief. The application is then transferred to the jurisdiction of the Trademark Trial and Appeal Board, which fixes terms for taking testimony by the opposer and the applicant. A final hearing is held before the Trademark Trial and Appeal Board, which renders a decision based on the testimony presented by the parties. This decision may be appealed to the Court of Appeals for the federal Circuit, or the applicant may have a remedy by civil action in the federal district courts. If the parties are able to negotiate a settlement while the matter is pending, the opposition can be withdrawn.

If no opposition is filed, or if a decision favorable to the applicant is made in an opposition, the application is allowed, and the registration is issued.

Pursuant to the amendments of the 1988 Act, an intent-to-use application may be filed in advance of actual use and, subject to approval, remain pending for a specified period of time until a statement of use is filed.

All marks registered under the Lanham Act are subject to cancellation by petition of a third party at any time within the five years following registration. The same time period applies after publication of a notice in the *Official Gazette* converting a registration under a prior trademark statute to one entitled to protection under the Lanham Act. Such conversions are allowed except in those instances in which the mark has become the common descriptive name of the article or substance, has been abandoned, or is disqualified by one of the five criteria listed earlier precluding registration.

For maintenance of a registration, the Lanham Act requires that an affidavit or declaration under penalty or perjury be filed within the sixth year of registration showing that the mark is still in use in commerce subject to regulation by Congress or that its nonuse is the result of circumstances excusing nonuse and not to an intention to abandon the mark. Marks registered only under earlier acts are not incontestable and do not require the filing of an affidavit of use within six years after registration or renewal. All marks, however, require renewal at ten-year intervals after registration. Such renewal is effected by filing a complete application within six months before expiration or within three months after expiration, identifying the goods or services for which the mark is still in use or showing, that nonuse is the result of circumstances excusing such nonuse and not indicating an intent to abandon the mark.

Effect of Registration. Under the federal statues, the registrant of a mark can bring civil action in the federal courts against the unauthorized use of a reproduction, counterfeit, copy, or colorable imitation of the mark in connection with any goods or services for which such use is likely to cause confusion or mistake or to deceive. A certificate of registration on the Principal Register of the Lanham Act, or republished from the act of 1881 or 1905, is *prima facie* evidence of the validity of the registered mark and the registration, the distinctiveness of the mark, the registrant's ownership of the mark, and exclusive right of the registrant to use the mark in commerce.

If the mark has become incontestable because of proper filing of a statement of five years of continuous use, the registration is conclusive evidence of the validity and ownership of the mark and registration, of the distinctiveness of the mark, and of the exclusive right of the registrant to use the mark on or in connection with goods or services specified in the declaration conferring incontestability. Exceptions are made in certain well-defined circumstances, for example, fraudulent registration, abandonment, and generic use of the mark.

The registrant of a trademark can enforce his or her rights by civil action in federal court against one who makes an unauthorized use of the mark that is likely to cause confusion or mistake or to deceive. In cases of contributory infringement, where the unauthorized use is unknowingly made by one engaged solely in the business of printing the infringing mark or of reproducing for others an advertisement displaying the mark in a periodical or an electronic communication, the remedy is limited to an injunction against future printing or advertising of the infringing matter. Thus, although printers and publishers who innocently reproduce an infringing work may be liable for producing the infringing advertisements, labels, and so on, they do not suffer any monetary liability. The mark's owner can obtain only injunctive relief, that is, a prohibition against further printing or publication. Profits or damages are recoverable only if the infringement was committed with knowledge or an intent to cause confusion or mistake or to deceive.

A registrant prevailing in a federal court action against an infringer is granted an injunction that is enforceable in any District Court of the United States. In addition, the plaintiff may recover the defendant's profits, damages sustained by the plaintiff, and costs. The court may, in its discretion, assess up to triple damages. The court may also order the destruction of all labels, signs, prints, packaging, wrappers, receptacles, and advertising in the possession of the defendant bearing the infringing mark and all plates, molds, matrices, or the like for producing them.

Registration of a mark on the Principal (but not the Supplemental) Register, as well as registration under the Acts of 1881 or 1905, constitutes constructive notice of the owner's claim to the mark. Damages and profits are available only from the time of actual notice to an infringer, unless the owner displays with the mark as used a notice of registration in the form specified in the statute, that is, the phrase *Registered in United States Patent and Trademark Office,* the abbreviation *Reg. U.S. Pat. & Tm. Off.,* or the letter *R* enclosed in a circle. In those cases not precluded by the statute of limitations, damages and profits may be recovered for infringing acts that occurred within three years prior to institution of a civil action for infringement.

Registration of a mark on the Principal Register prohibits importation of merchandise that copies the registered mark. Under the regulations of the Department of the Treasury, the registration may be recorded with the Customs authorities and merchandise bearing the infringing mark is subject to seizure and exclusion. The recording procedure includes supplying the Bureau of Customs with a specified number of copies of the registration. These are distributed to the various ports of entry and border stations for reference and notice to the Customs inspectors.

Transfer of Trademark Rights

A trademark and its registration may be assigned, but only together with the good will of the business in connection with which the mark is used or that part of the good will of the business connected with the use of and symbolized by the mark. An assignment of an intent-to-use application for registration is further limited to instances where the business in which the mark is to be used exists and is ongoing and is transferred to the assignee of the application along with the application and rights therein.

Proper Use of Trademarks

Proper use of a trademark by the owner or a licensee under a valid license is essential to preservation of the exclusive trademark right. The Lanham Act provides, for example, that nonuse for a period of two years constitutes *prima facie* abandonment of the trademark right. A trademark may also be lost through other circumstances. For instance, if the mark comes to be understood by the public, or the class of persons constituting the usual customers for merchandise sold under the mark, as a generic name for the product for which it is used, then the mark enters the public domain, and the exclusive right of the owner is lost. The danger of losing the exclusive right increases as a mark becomes more popular. Examples of marks that have been lost or severely endangered in this manner are Aspirin for acetylsalicylates, Escalator for moving stairways, and Cellophane for regenerated cellulose film.

State Registration

Registration of a trademark under state law is appropriate when the mark is to be used primarily in intrastate commerce. Procedure for registration generally parallels, but is simpler than, federal procedure.

Foreign Trademark Registration

The trademark laws of foreign countries differ in many fundamental respects from the laws of the United States. Countries deriving their system of laws from the United Kingdom generally recognize use of a mark as the basis for the trademark right. Registration is permitted in some of these countries based on an allegation of intent to use the mark, and continuance of the registration is dependent on demonstrating within a specified period of time that the intended use has occurred. Countries that derive their legal systems from continental Europe generally permit registration without use, although nonuse may be a ground for cancellation after a period of years.

Under the International Convention for Protection of Industrial Property, which has some 100 member countries, including the United States, if an application for registration of a trademark has been filed in one member country, a subsequent application for registration of the same trademark in another member country will be treated as if it were filed on the same date as the first application for purposes of priority, as long as the later application is filed within six months of the first and convention priority is claimed. A U.S. application may also be based on foreign registration or a foreign application for registration cojoined with a bona fide intention to use the mark in commerce in the United States.

Twenty-nine countries of the convention are also parties to a treaty for trademark registration known as the Madrid Arrangement. Under this treaty, a mark registered in the member home country of a trademark owner may be registered in the other member countries by filing with the International Trademark Bureau in Berne an application showing the home country registration.

LILE H. DEINARD
NICHOLAS J. STATHIS
MARY E. RASENBERGER
White & Case

International Bureau of the World Intellectual Property Organization, *Major Provisions of Trademark Legislation in Selected Countries,* Geneva, 1977.

S. Kane, *Trademark Law—A Practitioner's Guide,* Practicing Law Institute, New York, 1987.

Thomas Register of American Manufacturers, Thomas Publishing Co., New York, 1992.

Patent and Trademark Office, U.S. Department of Commerce, *Official Gazette,* Washington, D.C., published weekly. The trademark section, available separately, lists marks accepted, subject to opposition, with details of the application and all marks actually registered.

Patent and Trademark Office, U.S. Department of Commerce, *Trademark Rules of Practice of the Patent and Trademark Office with Forms and Statutes,* Washington, D.C., 1968. This publication can be purchased from the Superintendent of Documents, U.S. Government Printing Office, Washington, D.C.

CORROSION AND CORROSION CONTROL

Corrosion is the natural degradation of materials in the environment through electrochemical or chemical reaction. Traditionally, the definition of corrosion refers to the degradation of metals and has not included the degradation of nonmetals such as wood (rotting) or plastics (swelling or crazing), but increasingly, natural degradation of any engineering material is being regarded as corrosion. The vast majority of the technologically significant corrosion involves the deterioration of metallic materials, and only the corrosion of metallic materials is discussed here.

Electrochemical Nature of Corrosion

Ores are mined and are then refined in an energy intensive process to produce pure metals, which in turn are combined to make alloys (see METALLURGY; MINERALS RECOVERY AND PROCESSING). Corrosion occurs because of the tendency of these refined materials to return to a more thermodynamically stable state. The key reaction in corrosion is the oxidation or anodic dissolution of the metal to produce metal ions and electrons $M \xrightarrow{k_1} M^{n+} + ne^-$. The ions, M^{n+}, formed by this reaction at a rate, k_1, may be carried into a bulk solution in contact with the metal, or may form insoluble salts or oxides. In order for this anodic reaction to proceed, a second reaction which uses the electrons produced, ie, a reduction reaction, must take place. This second reaction, the cathodic reaction, occurs at the same rate, ie, $I_c = I_a$, where I_c and I_a are the cathodic and anodic currents, respectively. The cathodic reaction, in most cases, is hydrogen evolution or oxygen reduction.

The three elements necessary for corrosion are an aggressive environment, an anodic and a cathodic reaction, and an electron conducting path between the anode and the cathode. Other factors such as a mechanical stress also play a role. The thermodynamic and kinetic aspects of corrosion determine, respectively, if corrosion can occur, and the rate at which it does occur.

Manifestations of Corrosion

The most common form of corrosion is uniform corrosion, in which the entire metal surface degrades at a near uniform rate. Often the surface is covered by the corrosion products. The rusting of iron (qv) in a humid atmosphere or the tarnishing of copper (qv) or silver alloys in sulfur-containing environments are examples (see also SILVER AND SILVER ALLOYS). High temperature, or dry, oxidation, is also usually uniform in character. Uniform corrosion, the most visible form of corrosion, is the least insidious because the weight lost by metal dissolution can be monitored and predicted.

An especially insidious type of corrosion is localized corrosion which occurs at distinct sites on the surface of a metal while the remainder of the metal is either not attacked or attacked much more slowly. Localized corrosion is usually seen on metals that are passivated, ie, protected from corrosion by oxide films, and occurs as a result of the breakdown of the oxide film. Generally the oxide film breakdown requires the presence of an aggressive anion, the most common of which is chloride. Localized corrosion can cause considerable damage to a metal structure without the metal exhibiting any appreciable loss in weight. Localized corrosion occurs on a number of technologically important materials such as stainless steels, nickel-base alloys, aluminum, titanium, and copper (see ALUMINUM AND ALUMINUM ALLOYS; NICKEL AND NICKEL ALLOYS; STEEL; and TITANIUM AND TITANIUM ALLOYS).

Two types of localized corrosion are pitting and crevice corrosion. Pitting corrosion occurs on exposed metal surfaces, whereas crevice corrosion occurs within occluded areas on the surfaces of metals such as the areas under rivets or gaskets, or beneath silt or dirt deposits. Crevice corrosion is usually associated with stagnant conditions within the crevices. A common example of pitting corrosion is evident on household storm window frames made from aluminum alloys.

Another type of corrosion is dealloying which has also been called parting or selective leaching. Dealloying is the preferential removal of one of the alloying elements from an alloy resulting in the enrichment of the other alloying element(s). Common examples are the loss of zinc from brasses (dezincification) and the loss of iron from cast irons (graphitization) (see COPPER ALLOYS).

Corrosion may also appear in the form of intergranular attack, ie, preferential attack of the grain boundary region. Often, intergranular attack is the result of stress corrosion, but it can also occur because the grain boundary and the grain have different corrosion tendencies, ie, different potentials. In the latter case, the grain boundary acts as the anode and the grain as the cathode. Intergranular corrosion can lead to a loss in strength or ductility of the metal.

Corrosion also occurs as a result of the conjoint action of physical processes and chemical or electrochemical reactions. The specific manifestation of corrosion is determined by the physical processes involved. Environmentally induced cracking (EIC) is the failure of a metal in a corrosive environment and under a mechanical stress. Examples are the failure of brasses in ammonia environments and stainless steels in chloride or caustic environments.

When a stress is cyclic rather than constant, the failure is termed corrosion fatigue. Fretting corrosion results from the relative motion of two bodies in contact, one or both being a metal. The motion is small such as a vibration. Erosion corrosion results from the action of a high velocity fluid impinging on a metal surface. Metals and alloys can also experience cracking in liquid metal environments. This form of corrosion is referred to as liquid metal cracking (LMC).

Galvanic corrosion occurs as a result of the electrical contact of different metals in an aggressive environment. The driving force is the electrode potential difference between the two metals. One metal acts principally as a cathode and the other metal as the anode.

Microbiologically influenced corrosion, which results from the interaction of microorganisms and a metal, is receiving increased emphasis. The action of microorganisms is at least one of the reasons why natural seawater is more corrosive than either artificial seawater or sodium chloride solutions. Microorganisms attach to the surfaces of metals and can, for example, act as diffusion barriers; produce metabolites that enhance or initiate corrosion; act as sinks or sources for species involved in cathodic reactions, such as oxygen and hydrogen; increase the pH at the surface as a result of photosynthesis; or decrease the pH by production of acid metabolites. A more detailed discussion of the various forms of corrosion may be found in the literature.

Electrochemical Equilibrium Diagrams

The thermodynamic data pertinent to the corrosion of metals in aqueous media have been systematically assembled in a form that has become known as Pourbaix diagrams. The data include the potential and pH dependence of metal, metal oxide, and metal hydroxide reactions and, in some cases, complex ions. The potential and pH dependence of the hydrogen and oxygen reactions are also supplied because these are the common corrosion cathodic reactions.

Kinetics of Electrochemical Reactions

Even in uniform corrosion, a corroding metal surface has numerous local anodes and cathodes. The sites of these local reactions may be fixed by microstructural features or may change as corrosion proceeds. The oxidation reaction at anodic sites on the metal surface can be represented as in equation 1. A corresponding reduction reaction must be occurring at cathodic sites. The potentials of these two reactions would be moved toward each other, away from the respective equilibrium potentials, and the metal surface would assume an overall uniform potential.

Galvanic Corrosion and Cathodic Protection

Galvanic corrosion can be used to a corrosion advantage. If, for example, a metal such as zinc or magnesium, having a low corrosion potential in most environments, is coupled with a steel component the zinc or magnesium pulls the potential of the steel down causing the steel to corrode less. When a sacrificial metal or alloy, called a sacrificial anode, is attached to a structure having a higher corrosion potential to intentionally pull the potential of the higher potential metal down and thus decrease the corrosion rate, it is called cathodic protection. This method of corrosion mitigation is common for underground pipelines (qv), residential hot water heaters, and hulls and tanks (qv) of ships. The lowering of potential can also be achieved by the application of external electrical current, ie, impressed current cathodic protection.

The lower potential metal in a galvanic couple does not always have its corrosion rate accelerated. For metals that form a passive film, coupling with another metal of higher potential can cause the film-forming metal's potential to shift from a value at which it corrodes to one at which it passivates and therefore corrodes less. When this is done intentionally the procedure is referred to as anodic protection, ie, achieving protection by intentionally shifting the potential in the positive direction. Anodic protection can also be brought about by adding oxidizers to the electrolyte.

Environmental Effects

The environment plays several roles in corrosion. It acts to complete the electrical circuit, ie, supplies the ionic conduction path; provide reactants for the cathodic process; remove soluble reaction products from the metal surface; and/or destabilize or break down protective reaction products such as oxide films that are formed on the metal. Some important environmental factors include: the oxygen concentration; the pH of the electrolyte; the temperature; and the concentration of anions.

Reduction of oxygen is one of the predominant cathodic reactions contributing to corrosion.

Very often the environment is reflected in the composition of corrosion products, eg, the composition of the green patina formed on copper roofs over a period of years. The determination of the chemical composition of this green patina was one of the first systematic corrosion studies ever made (see COPPER). The composition varied considerably depending on the location of the structure as shown in Table 1.

Metallurgical Factors

The primary determining factor in corrosion behavior of metals and alloys is usually chemical composition. The amount of chromium in a stainless steel is a good example. There are, however, other metallurgical factors, such as crystallography, grain size and shape, grain heterogeneity, second phases, impurity inclusions, and residual stress, that influence corrosion. The technologically important structural materials are polycrystalline aggregates. Each individual crystal is referred to as a grain. Grain orientation can affect corrosion resistance as evidenced by metallographic etching rates and pitting behavior. Grain shape may likewise vary greatly depending on the alloy and processing history.

Stainless Steels. The most common and serious metallurgical factor affecting the corrosion resistance of stainless steels is termed sensitization. This condition is caused by the precipitation of chromium-rich carbides (qv) at the grain boundaries, giving rise to chromium depleted grain boundary areas. These areas are anodic to the grain interior and tend to dissolve thereby causing intergranular corrosion.

There are several measures available to mitigate sensitization. Low carbon grades such as AISI 304L and 316L are available which have much less tendency toward sensitization. Additionally, stabilized grades such as AISI 321 and 347 that contain titanium and niobium, respectively, to tie up the carbon are available.

Copper Alloys. Brasses are susceptible to dealloying in the form of dezincification, ie, the preferential loss of zinc from the alloy. Brasses having zinc concentrations of 15% or greater are prone to dezincification and dezincification is generally more severe in brasses that have two metallurgical phases.

Table 1. Composition of Green Patina on Copper from Different Locations

Location of structure	Age of structure, yr	Composition of green patina, %			
		$CuCO_3$	$CuCl_2$	$Cu(OH)_2$	$CuSO_4$
urban	30	14.6		9.6	49.8
rural	300	1.4		58.5	25.6
marine	13	12.8	26.7	52.5	2.5
urban–marine	38		4.6	61.5	29.7

Conditions that favor dezincification include stagnant solutions, especially acidic ones, high temperatures, and porous scale formation. Additions of small amounts of arsenic, antimony, or phosphorus can increase the resistance to dezincification.

Aluminum Alloys. Copper, silicon, magnesium, zinc, and manganese are some common alloying additions to aluminum. Most of the alloying elements are added to aluminum to produce alloys having improved mechanical properties. However, the strengthening phases which result from the alloying can disrupt the passive oxide layer on aluminum and lead to localized corrosion in the form of pitting attack. Also, these strengthening phases can result in intergranular corrosion.

Nickel Alloys. Nickel-based alloys have great technological importance in the development of high strength, high temperature alloys (qv), in part because of the occurrence of unique intermetallic phases. However, the precipitation of certain intermetallic phases can decrease the resistance to corrosion because this can deplete the matrix of alloying elements.

Environmentally Induced Cracking

Environmentally induced cracking (EIC) is a brittle fracture process caused by the conjoint action of a mechanical stress and a corrosive environment. There are several different types of EIC: stress corrosion cracking (SCC), corrosion fatigue cracking (CFC), and hydrogen-induced cracking (HIC).

The cracking in EIC can proceed either intergranularly, ie, between the grains, or transgranularly, ie, across the grains.

Copper Alloys. Copper alloys under an applied or residual stress are susceptible to stress-corrosion cracking in environments containing ammonia (qv) or ammonium compounds (qv). Trace quantities of nitrogen oxides may also cause SCC. The nitrates settle as dust on the brass parts. The susceptibility to SCC can be minimized by: (*1*) proper alloy selection; (*2*) thermal stress relief; (*3*) avoiding contact with ammonia and ammonia compounds; and (*4*) using an inhibitor.

Aluminum Alloys. Both the 2000 and the 7000 series aluminum alloys have experienced significant SCC in service. The alloys are most vulnerable to SCC if stressed parallel to the short transverse grain direction, ie, parallel to the thinnest dimension of the grain; most resistant if stressed only parallel to the longest grain dimension; and show intermediate susceptibility if stressed parallel to the long transverse direction. Hence prudent practice is to avoid designs in which high sustained stresses are imposed across the short transverse direction.

In addition to the possibility of selecting alloys having minimum SCC susceptibility while retaining other properties as needed, there are other steps possible to reduce the SCC probability: (*1*) avoid designs that permit water to accumulate; (*2*) avoid conditions in which salts, especially chlorides, can concentrate; and (*3*) where available and otherwise acceptable, use a clad alloy (see METAL SURFACE TREATMENTS).

High Strength Steels. Steels that owe their strength to heat treatment, whether martensitic, precipitation hardened, or maraging, and whether stainless or not, are susceptible to SCC in aqueous environments, including water vapor. The primary factor in determining the degree of SCC susceptibility of a given steel is its strength.

Stainless Steels. Austenitic stainless steels undergo SCC when stressed in hot aqueous environments containing chloride ion. The oxygen level of the environment can be important, probably through its effect in establishing the electrode potential of the steel. The standard methods for avoiding chloride SCC in austenitic stainless steels include: avoidance of fabrication stresses; minimizing chloride ion level; and minimizing oxygen concentration in the environment.

Corrosion-Resistant Materials

Alloys having varying degrees of corrosion resistance have been developed in response to various environmental needs. At the lower end of the alloying scale are the low alloy steels. These are iron-base alloys containing from 0.5–3.0 wt % Ni, Cr, Mo, or Cu and controlled amounts of P, N, and S. The exact composition varies with the manufacturer. The corrosion resistance of the alloy is based on the protective nature of the surface film, which in turn is based on the physical and chemical properties of the oxide film. As a rule, this alloying reduces the rate of corrosion by 50% over the first few years of atmosphere exposure. Low alloy steels have been used outdoors with protection.

Several copper alloys are exceptionally resistant to certain atmospheres. The copper–nickel alloys, 90% Cu–10% Ni and 70% Cu–30% Ni, have very good resistance to corrosion in seawater, if the iron content is properly controlled, ie, from 0.5 to 1% Fe.

For most environments quantitative studies have been reported describing the corrosion rate of various materials including a number of corrosion-resistant alloys.

Inhibitors

Corrosion inhibitors are substances which slow down or prevent corrosion when added to an environment in which a metal usually corrodes. Corrosion inhibitors are usually added to a system in small amounts either continuously or intermittently. The effectiveness of corrosion inhibitors is partly dependent on the metals or alloys to be protected as well as the severity of the environment.

Inhibitors act and are classified in a variety of ways. Types of inhibitors include (*1*) anodic, (*2*) cathodic, (*3*) organic, (*4*) precipitation, and (*5*) vapor-phase inhibitors.

Coatings for Corrosion Prevention

Coatings (qv) are applied to metal substrates to prevent corrosion. Generally the coating protects the metal by imposing a physical barrier between the metal substrate and the environment. However, the coating can also act to provide corrosion protected by cathodic protection or by serving as holding reservoirs for inhibitors. Coatings may be divided into organic, inorganic, and metallic coatings (see COATINGS MARINE; METAL SURFACE TREATMENTS).

PATRICK J. MORAN
United States Naval Academy
PAUL M. NATISHAN
Naval Research Laboratory

M. G. Fontana, *Corrosion Engineering,* 3rd ed., McGraw-Hill Book Co., Inc., New York, 1986.

D. A. Jones, *Principles and Prevention of Corrosion,* Macmillan Publishing Co., New York, 1992.

L. S. Van Delinder, ed., *Corrosion Basics: An Introduction,* National Association of Corrosion Engineers, Houston, Tex., 1984.

Metals Handbook, 9th ed., Vol. 13 (*Corrosion*), ASM International, Metals Park, Ohio, 1987.

COSMETICS

Cosmetics are products created by the cosmetic industry and marketed directly to consumers. The cosmetic industry is dominated by manufacturers of finished products but also includes manufacturers who sell products to distributors as well as suppliers of raw and packaging materials. Cosmetics represent a large group of consumer products designed to improve the health, cleanliness, and physical appearance of the human exterior and to protect a body part against damage from the environment. Cosmetics are promoted to the public and are available without prescription.

Cosmetics, regardless of form, can be grouped by product use into the following seven categories: (*1*) skin care and maintenance, including products that soften (emollients and lubricants), hydrate (moisturizers), tone (astringents), protect (sunscreens), etc, and repair (antichapping, antiwrinkling, antiacne agents); (*2*) cleansing, including soap, bath preparations, shampoos, and dentifrices (qv); (*3*) odor improvement by use of fragrance, deodorants, and antiperspirants; (*4*) hair removal, aided by shaving preparations, and depilatories; (5) hair care and maintenance, including waving, straightening, antidandruff, styling and setting, conditioning, and coloring products (see HAIR PREPARATIONS); (6) care and maintenance of mucous membranes by use of mouthwashes, intimate care products, and lip antichapping products; and (7) decorative cosmetics, used to beautify eyes, lips, skin, and nails.

Economic Aspects

Numerous cosmetic trade organizations exist. Foremost among them are the Cosmetic, Toiletry, and Fragrance Association (CTFA), formerly the Toilet Goods Association; the European Cosmetics Industries Federation (COLIPA); and the Japanese Cosmetic Industry Association (JCIA). These organizations provide member companies with regulatory and technical information and supply documentation on the industry's practices to governments and consumers.

Regulation of the Cosmetic Industry

In the United States, the 1938 revision of the Federal Food and Drug Act regulates cosmetic products and identifies these materials as:

> *(1) articles intended to be rubbed, poured, sprinkled, or sprayed on, introduced into, or otherwise applied to the human body or any part thereof for cleansing, beautifying, promoting attractiveness, or altering the appearance, and (2) articles intended for use as a component of any such articles, except that such term shall not include soap.*

This definition establishes the legal difference between a drug and a cosmetic. It is clearly the purpose of, or the claims for, the product, not necessarily its performance, that legally classifies it as a drug or a cosmetic in the United States. For example, a skin-care product intended to beautify by removing wrinkles may be viewed as a cosmetic because it alters the appearance and a drug because it affects a body structure.

The FDA is responsible for enforcing the 1939 act as well as the Fair Packaging and Labeling Act. In light of the difficulty of differentiating between cosmetics and drugs, the FDA has in recent years implemented its regulatory power by concluding that certain topically applied products should be identified as OTC drugs. As a group, these OTC drugs were originally considered cosmetics and remain among the products distributed by cosmetic companies. They include acne, antidandruff, antiperspirant, astringent, oral-care, skin-protectant, and sunscreen products.

The use or presence of poisonous or deleterious substances in cosmetics and drugs is prohibited. The presence of such materials makes the product "adulterated" or "misbranded" and in violation of good manufacturing practices (GMP), which are applicable to drugs and, with minor changes, to cosmetics.

In contrast to prescription drugs, OTC drugs and cosmetics are not subject to preclearance in the United States. However, the rules covering OTC drugs preclude introduction of untested drugs or new combinations. A "new chemical entity" that appears suitable for OTC drug use requires work-up via the new drug application (NDA) process. In contrast, the use of ingredients in cosmetics is essentially unrestricted and may include less well known substances.

Color Additives. The FDA has created a unique classification and strict limitations on color additives (see also COLORANTS FOR FOODS, DRUGS, COSMETICS, AND MEDICAL DEVICES). Certified color additives are synthetic organic dyes that are described in an approved color additive petition. Each manufactured lot of a certified dye must be analyzed and certified by the FDA prior to usage.

Hair colorants, the fourth class of color additives, may be used only to color scalp hair and may not be used in the area of the eye. Use of these colorants is exempt, that is, coal-tar hair dyes may be sold with cautionary labeling, directions for preliminary (patch) testing, and restrictions against use in or near the eye.

Under the Fair Packaging and Labeling Act, the FDA has instituted regulations for identifying components of cosmetics on product labels. To avoid confusion, the CTFA has established standardized names for about 6000 cosmetic ingredients.

European Regulations. Regulations for cosmetics differ from country to country but, in general, are similar to or patterned after U.S. regulation.

Japanese Regulation. Japanese regulations of cosmetics are similar to those already discussed. The safety and quality of cosmetic products are regulated under the Pharmaceutical Affairs Law with detailed requirements for approval and licensing of manufacturing and import, for labeling and advertising standards, and for reporting safety data to the Ministry of Health and Welfare.

Product Requirements

Safety. Cosmetic products must meet acceptable standards of safety during use, must be produced under sanitary conditions, and must exhibit stability during storage, shipment, and use.

For many years the safety of cosmetic ingredients has been established using a variety of animal safety tests. More recently, animal welfare organizations have urged that this type of safety testing be abandoned. Despite widespread use of cosmetics without professional supervision, the incidence of injury from cosmetic products is rare.

In vitro safety testing technology is becoming more common. Validation of these methods is based on comparisons with early animal safety data. Whether these *in vitro* tests can ensure the safety of all products that reach the consumer cannot be predicted. As of this writing, the principle that *in vitro* testing may be substituted for *in vivo* testing for complete safety substantiation has not been accepted by regulatory agencies. In the United States, the CTFA created the Cosmetic Ingredient Review (CIR) for the purpose of evaluating existing *in vitro* and *in vivo* data and reviewing the safety of the ingredients used in cosmetics.

In addition to the CIR process the cosmetic industry has instituted a second, important self-regulatory procedure: the voluntary reporting of adverse reactions, which is intended to provide data on the type and incidence of adverse reactions noted by consumers or by their medical advisors.

Safety testing of a finished cosmetic product should be sufficient to ensure that the product does not cause irritation when used in accordance with directions, neither elicits sensitization nor includes a sensitizer, and does not cause photoallergic responses.

Production Facilities. The manufacture of acceptable cosmetic products requires not only safe ingredients but also facilities that maintain high standards of quality and cleanliness. Most countries have established regulations intended to assure that no substandard product or batch is distributed to consumers. Good Manufacturing Practices (GMP) represent workable standards that cover every aspect of drug manufacture, from building construction to distribution of finished products. GMPs in the United States that have been established for drug manufacture are commonly used in cosmetic production.

Contamination. Manufacturers of cosmetics must be careful to guard against chemical and microbial contamination. Compendial specifications and publications by the CTFA and other professional societies form the basis of most intracompany raw material specifications. Moreover, all packaging components must meet not only physical and design specifications but also such chemical requirements as extractables and absence of dust and similar contaminants.

Stability. An additional mandatory requirement for cosmetic products is chemical and physical stability. Interactions between ingredients that lead to new chemical entities or decomposition products are unacceptable. In the absence of an expiration date, a cosmetic product or an OTC drug should be stable for 60 months at ambient temperature.

Performance. Consumer acceptance is a criterion on which cosmetic marketers cannot compromise. Performance is tested by *in vitro* techniques during formulation, but the ultimate test of a product's performance requires in-use experience with consumers and critical assessment by trained observers.

Ingredients

Manufacturers of cosmetics employ a surprisingly large number of raw materials. Some of these ingredients are active constituents that have purported beneficial effects on the skin, hair, or nails, for example, acting as moisturizers or conditioners. These substances are generally used in limited quantities. Other ingredients are used to formulate or create the vehicle. These are bulk chemicals used in comparatively large amounts. The resulting combination of various substances affects the nature (viscosity, oiliness, etc) of the finished cosmetic. As a rule numerous combinations and permutations are tested to optimize textural characteristics and to match these to consumers' preferences. Finally, cosmetics may include substances added primarily to appeal to consumers. These ingredients need not contribute appreciably to product performance.

About 6000 different cosmetic ingredients have been identified. These can be divided into smaller groups according to chemical similarity or functionality. Table 1 represents a breakdown by functionality on the skin or in the product.

Antioxidants. The operant mechanisms of antioxidants are interference with radical propagation reactions, reaction with oxygen, or reduction of active oxygen species. Antioxidants are intended to protect the product but not the skin against oxidative damage resulting from ultraviolet radiation or singlet oxygen formation.

Preservatives. Several microorganisms can survive and propagate on unpreserved cosmetic products. Preservatives are routinely added to all preparations that can support microbial growth.

Lipids. Natural and synthetic lipids are used in almost all cosmetic products. Lipids serve as emollients or occlusive agents, lubricants, binders for creating compressed powders, adhesives to hold makeup in place, and hardeners in such products as lipsticks. In addition, lipids are used as gloss-imparting agents in hair-care products. The primary requirements for lipids in cosmetics are absence of excessive greasiness and ease of spreading on skin.

Solvents. Solvents can be added to cosmetics to help dissolve components used in cosmetic preparations. Water is the most common solvent and is the continuous phase in most suspensions and water/oil (w/o) emulsions. Organic solvents are required in the preparation of colognes, hair fixatives, and nail lacquers. Selected solvents are used to remove soil, sebum, and makeup from skin.

Surfactants. Substances commonly classified as surfactants (qv) or surface active agents are required in a wide variety of cosmetics. Prolonged contact with anionic surfactants can cause some swelling of the skin. Although this is a temporary phenomenon, skin in this swollen condition allows permeation of externally applied substances. Nonionic surfactants as a group are generally believed to be mild even under exaggerated conditions. The more hydrophobic nonionics, those that are water dispersible (not water-soluble), can enhance transdermal passage. Amphoteric surfactants as a group exhibit a favorable safety profile. Finally, cationic surfactants are commonly rated as more irritating than the anionics, but the evidence for generalized conclusions is insufficient.

Colorants. Color (qv) is used in cosmetic products for several reasons: the addition of color to a product makes it more attractive and enhances consumer acceptance; tinting helps hide discoloration resulting from use of a particular ingredient or from age; and finally, decorative cosmetics owe their existence to color.

The importance of coal-tar colorants cannot be overemphasized. The cosmetic industry, in cooperation with the FDA, has spent a great deal of time and money in efforts to establish the safety of these dyes (see COLORANTS FOR FOOD, DRUGS, COSMETICS, AND MEDICAL DEVICES). Contamination, especially by heavy metals, and other impurities arising from the synthesis of permitted dyes are strictly controlled. Despite this effort, the number of usable organic dyes and of pigments derived from them has been drastically curtailed by regulatory action.

In addition to the U.S. certified coal-tar colorants, some noncertified naturally occurring plant and animal colorants, such as alkanet, annatto, carotene, $C_{40}H_{56}$, chlorophyll, cochineal, saffron, and henna, can be used in cosmetics. In the United States, however, natural food colors, such as beet extract or powder, turmeric, and saffron, are not allowed as cosmetic colorants.

The terms FD&C, D&C, and External D&C (Ext. D&C), which are part of the name of colorants, reflect the FDA's colorant certification. FD&C dyes may be used for foods, drugs, and cosmetics; D&C dyes are

Table 1. Cosmetic Functions and Representative Ingredients

Function	Ingredient[a]	Molecular formula
Biologically active agents		
antiacne	salicylic acid	$C_7H_6O_3$
anticaries	monosodium fluorophosphate	Na_2HPO_3F
antidandruff	zinc pyrithione	$C_{10}H_8N_2O_2S_2Zn$
antimicrobial	benzalkonium chloride	
antiperspirant	aluminum chlorohydrate	$Al_2ClH_5O_5$
biocides	triclosan	$C_{12}H_7Cl_3O_2$
sunscreen	octyl methoxycinnamate	$C_{18}H_{26}O_3$
skin protectant	dimethicone	$(C_2H_6OSi)_nC_4H_{12}Si$ $(C_2H_6OSi)_n$
external analgesic	methyl salicylate	$C_8H_8O_3$
Nonbiologically active agents		
abrasive		
skin	oatmeal	
teeth	dicalcium phosphate	$Ca_2(HPO_4)_2$
antifoam	simethicone	
antioxidant	ascorbic acid	$C_6H_8O_6$
antistatic agent	dimethylditallow alkylammonium chlorides	
binder	hydroxypropylcellulose	
chelator	hydroxyethyl ethylenediamine triacetic acid (HEDTA)	$C_{10}H_{18}N_2O_7$
colorant		
pigment	ultramarine	$Na_7Al_6Si_6O_{24}S_2$
dye	FD&C Red No. 4	$C_{18}H_{16}N_2O_7S_2 \cdot 2$ Na
emulsion stabilizer	xanthan gum	
film former	PVP	$(C_6H_9NO)_x$
hair colorant	*p*-phenylenediamine	$C_6H_8N_2$
hair conditioner	sodium lauroamphoacetate	$Na_2C_{18}H_{35}N_2O_3 \cdot HO$
humectant	glycerol	$C_3H_8O_3$
deodorant		
mouth	zinc chloride	$ZnCl_2$
external	cetylpyridinium chloride	$C_{21}H_{38}ClN$
preservative	propylparaben	$C_{10}H_{12}O_3$
emollient	octyl stearate	$C_{26}H_{52}O_2$
skin-conditioning agent		
general	pyrrolidinone carboxylic acid (PCA)	$C_5H_7NO_3$
occlusive	petrolatum	C_nH_{2n+2}
film forming	hyaluronic acid	
solvent	ethanol	C_2H_6O
cleansing agent	sodium lauryl sulfate	$C_{12}H_{25}NaO_4S$
emulsifying agent	polysorbate 65	
foam booster	cocamide DEA	
suspending agent	sodium lignosulfonate	
hydrotrope	sodium toluenesulfonate	
viscosity-controlling agent		
decrease	propylene glycol	$C_3H_8O_2$
increase	hydroxypropylmethyl cellulose	

[a] CTFA adopted names are used; this notation is used for cosmetic labeling.

allowed in drugs and cosmetics; and Ext. D&C dyes are permitted only in topical products.

In addition to various white pigments, other inorganic colorants are used in a number of cosmetic products. These usually exhibit excellent lightfastness and are completely insoluble in solvents and water.

For many years nacreous pigments were limited to guanine (from fish scales) and bismuth oxychloride. Mica, gold, copper, and silver, in flake form, can also provide some interesting glossy effects in products and on the face. An entirely new set of colored, iridescent, inorganic pigments, which may be described as mixtures of mica and titanium dioxide (sometimes with iron oxides), has been created by coating mica flakes with titanium dioxide.

Skin Preparation Products

Products for use on the skin are designed to improve skin quality, to maintain (or restore) skin's youthful appearance, and to aid in alleviating the symptoms of minor diseases of the skin. Many of these products are subject to different regulations in different countries. Skin products are generally formulated for a specific consumer purpose.

Skin-Care Products. Preparations are generally classified by site of application and purpose. The smoothing or emollient properties of creams and lotions are critical for making emulsions the preferred vehicles for facial skin moisturizers, skin protectants, and rejuvenating products. On the body, emollients provide smoothness and tend to reduce the sensation of tightness commonly associated with dryness and loss of lipids from the skin. Although a wide variety of plant and animal extracts have been claimed to impart skin benefits, valid scientific evidence for efficacy has been provided only rarely.

Emulsion components enter the stratum corneum and other epidermal layers at different rates. Most of the water evaporates, and a residue of emulsifiers, lipids, and other nonvolatile constituents remains on the skin. Some of these materials and other product ingredients may permeate the skin; others remain on the surface. If the blend of nonvolatiles materially reduces the evaporative loss of water from the skin, known as the transepidermal water loss (TEWL), the film is identified as occlusive.

The ability to moisturize the stratum corneum has also been claimed for the presence of certain hydrophilic polymers, for example, guar hydroxypropyl trimonium chloride, on the skin. By far the most popular way to moisturize skin is with humectants. It is claimed that humectants attract water from the environment and thereby provide moisture to the skin.

Antiacne Preparations. Antiacne products are designed to alleviate the unsightly appearance and underlying cause of juvenile acne.

As of 1991 in the United States, OTC antiacne preparations may contain only a few active drugs, for example, sulfur, resorcinol acetate, resorcinol, salicylic acid, and some combinations. OTC antiacne constituents may be included in a variety of conventional cosmetic preparations, which then become OTC drugs.

Sunscreens. The use of uv light absorbing substances is accepted worldwide as a means of protecting skin and body against damage and trauma from uv radiation. These colorless organic substances are raised to higher energy levels upon absorption of uv light, but little is known about mechanisms for the disposal of this energy. These substances can be classified by the wavelengths at which absorbance is maximal. Absorption throughout the incident uv range (285 to about 400 nm) affords the best protection against erythema.

It is also possible to deflect uv radiation by physically blocking the radiation using an opaque makeup product. A low particle-size titanium dioxide can reflect uv light without the undesirable whitening effect on the skin that often results from products containing, for example, zinc oxide or regular grades of titanium dioxide.

A list of uv absorbing substances found useful in protective sunscreen products is provided in Table 2. Some information on the levels permitted in products in both the United States and the EEC is included. Descriptions and specifications of sunscreens have been published.

Table 2. Cosmetic Uv Absorbers

Ingredient	Quantity approved, % U.S.	EEC
uvA Absorbers		
benzophenone-8	3	
menthyl anthranilate	3.5–5	
benzophenone-4	5–10	5
benzophenone-3	2–6	10
uvB Absorbers		
p-aminobenzoic acid (PABA)	5–15	5
pentyl dimethyl PABA	1–5	5
cinoxate	1–3	5
DEA *p*-methoxycinnamate	8–10	8
digalloyl trioleate	2–5	4
ethyl dihydroxypropyl PABA	1–5	5
octocrylene	7–10	
octyl methoxycinnamate	2–7.5	10
octyl salicylate	3–5	5
glyceryl PABA	2–3	5
homosalate	4–15	10
lawsone (0.25%) plus dihydroxyacetone (33%)		
octyl dimethyl PABA	1.4–8	8
2-phenylbenzimidazole-5-sulfonic acid	1–4	8
TEA salicylate	5–12	2
sulfomethyl benzylidene bornanone		10
urocanic acid (and esters)		2
Physical barriers		
red petrolatum	30–100	
titanium dioxide	2–25	

In principle, emulsified sunscreen products are similar to emollient skin-care products in which some of the emollient lipids are replaced by uv absorbers.

Facial Makeup. This classification applies to all products intended to impart a satinlike tinted finish to facial skin and includes liquid makeups, tinted loose or compressed powders, rouges, and blushers.

In modern liquid makeups and rouges, the required pigments are extended and ground in a blend of suitable cosmetic lipids. This magma is then emulsified, commonly as o/w, in a water base. Soaps, monostearates, conditioning lipids, and viscosity-increasing clays are primary components. Nonionic emulsifiers can replace part or all of the soap.

Tinted dry powders form the second type of facial makeup. Commonly, the blended solids are compressed into compacts.

Skin Coloring and Bleaching Preparations. Products designed to simulate a tan, to lighten skin color in general, or to decolorize small hyperpigmented areas such as age spots either impart to or remove color from the skin. Skin stains are intended to create the appearance of tanned skin without exposure to the sun. The most widely used ingredient is dihydroxyacetone (2–5% at pH 4 to 6) which reacts with protein amino groups in the stratum corneum to produce yellowish brown Maillard products. Lawsone and juglone are known to stain skin directly. Stimulation of melanin formation is another approach to artificial tanning. Commercialization, which is limited, depends primarily on topical application of products containing tyrosine or a tyrosine precursor.

The number of cosmetically acceptable bleaching ingredients is very small, and products for this purpose are considered drugs in the United States. The most popular ingredient is hydroquinone at 1–5%; the addition of uv light absorbers and antioxidants reportedly helps to reduce color recurrence.

Astringents

Astringents are designed to dry the skin, denature skin proteins, and tighten or reduce the size of pore openings on the skin surface. These products can have antimicrobial effects and are frequently buffered to lower the pH of skin. They are perfumed, hydro-alcoholic solutions of weak acids, such as tannic acid or potassium alum, and various plant extracts, such as birch leaf extract. In the United States, some astringents, depending on product claims, are considered OTC drugs. Only three ingredients, aluminum acetate, aluminum sulfate, and hamamelis, are considered safe and effective.

Antiperspirants and Deodorants. There are many forms of antiperspirants and deodorants: liquids, powders, creams, and sticks. Deodorants do not interfere with the delivery of eccrine or apocrine secretions to the skin surface but control odor by reodorization or antibacterial action. Deodorant products, regardless of form, are antimicrobial fragrance products. An important antimicrobial or cosmetic biocide used in many products is triclosan. Other active agents include zinc phenolsulfonate, *p*-chloro-*m*-xylenol, and cetrimonium bromide. There have been claims that ion-exchange polymers and complexing agents provide protection against unpleasant body odors. In addition, delayed-release, that is, liposomal or encapsulated, substances of diverse activity have been employed.

The mechanism of antiperspirant action has not been fully established but probably is associated with blockage of ducts leading to the surface by protein denaturation by aluminum salts. The FDA has mandated that an antiperspirant product must reduce perspiration by at least 20% and has provided some guidelines for testing finished products.

Cleansing Preparations

Cleansing preparations are products, based on surfactants or abrasives, that are designed to remove unwanted oil and debris from skin, hair, and the oral cavity. Soaps (qv) are the best-known cleansers but are not considered cosmetic products unless they are formulated with agents that prevent skin damage or contain antimicrobial agents.

Hair Cleansers. Except for a few specialty preparations, hair cleansers, or shampoos, are based on aqueous surfactants. The most popular surfactants in shampoos are alkyl sulfates and alkylether sulfates, commonly used at about 10–15% active. These ingredients by themselves do not provide the dense, copious foam desired by consumers, and additives are required, especially for use on oily hair or scalps. The foam boosters usually found in finished shampoos are fatty acid alkanolamides, fatty alcohols, and amine oxides.

Excessive degreasing by shampoos can be overcome by treatment with an after-shampoo (cream) rinse or a hair dressing. A wide variety of hair-conditioning additives have been recommended and tested. Only a few have gained wide acceptance, for example: dialkyl (C_{12}–C_{18}) dimethylammonium chloride, hydrolyzed collagen polymeric quaternary derivatives, potassium cocoyl hydrolyzed collagen, sodium cocoamphoacetate, sodium lauroyl glutamate, and stearamidopropyl betaine.

Dandruff, a benign scaling skin disease of the scalp, is commonly viewed as a hair problem. The etiology and therapy of dandruff are similar to those of seborrheic dermatitis. Antidandruff shampoos are formulated using antimicrobial or desquamating agents to reduce the lipophilic yeasts (qv) widely believed to be the cause of scalp flaking. In the United States, shampoos for which antidandruff claims are made are OTC drugs. The choice of active agents is limited to coal tar, zinc pyrithione, salicylic acid, selenium sulfide, and sulfur, which can be added to shampoos or other scalp preparations.

Oral Cleansing Products. Toothpastes and mouthwashes are considered cosmetic oral cleansers as long as claims about them are restricted to cleaning or deodorization. Because deodorization may depend on reduction of microbiota in the mouth, several antimicrobial agents, either quaternaries, such as benzethonium chloride, or phenolics, such as triclosan, are permitted. Products that include anticaries or antigingivitis agents or claim to provide such treatment are considered drugs.

Mouthwashes are hydro-alcoholic preparations in which flavorants, essential oils (see OILS, ESSENTIAL), and other agents are combined to provide long-term breath deodorization. Palatability can be improved by including a polyhydric alcohol such as glycerin or sorbitol (see ALCOHOLS, POLYHYDRIC).

Dentifrices (qv), or toothpastes, depend on abrasives to clean and polish teeth. The principal ingredients in toothpastes or powders are 20–50% polishing agents, such as calcium carbonate, di- or tricalcium phosphate, insoluble sodium metaphosphate, silica, and alumina; 0.5–1.0% detergents, for example, soap or anionic surface-active agents; 0.3–10% binders (gums); 20–60% humectants, such as glycerol, propylene glycol, and sorbitol; sweeteners (saccharin, sorbitol); preservatives, such as benzoic acid or *p*-hydroxybenzoates; flavors, for example, essential oils; and water. The most widely used surfactant is sodium lauryl sulfate, which is available with high purity. It produces the desired foam during brushing, acts as a cleansing agent, and has some bactericidal activity.

Shaving Products

Cosmetic shaving products are preparations for use before, during, or after shaving.

Preshaves. Preshave products are used primarily for dry (electric) shaving. Solid preshaves are usually compressed-powder sticks based on lubricating solids, such as talc or zinc or glyceryl stearate. Liquid preshaves are intended to remove perspiration residues and tighten and lubricate the skin. The alcohol content is relatively high (50–80%) to accelerate drying. The remaining ingredients may be polymeric lubricants, such as 1–2% polyvinylpyrrolidinone (PVP), emollients, such as 1–5% diisopropyl adipate, and up to about 5% propylene glycol.

Shaving Creams. Shaving creams and soaps are available as solids, that is, bars; creams, generally in tubes; or aerosols.

The principal ingredients of shaving creams and aerosols are liquid soaps, usually a blend of potassium, amine, and sodium salts of fatty acids, formulated to create a foam with the desired consistency and rinsing qualities. The soap blend may include synthetic surfactants, skin-conditioning agents, and other components.

The objectives of shaving creams include protecting the face from cuts by cushioning the razor. Beard-softening qualities are attributable almost exclusively to hair hydration, which also depends on pH.

After-Shaves. After-shave preparations serve the same function as and are formulated similarly to skin astringents.

Nail-Care Products

Over the years the cosmetic industry has created a wide variety of products for nail care. Some of these, such as cuticle removers and nail hardeners, are functional; others, such as nail lacquers, lacquer removers, and nail elongators, are decorative.

Functional Nail-Care Products. Cuticle removers are solutions of dilute alkalies that facilitate removal, or at least softening, of the cuticle.

Nail hardeners have been based on various protein cross-linking agents. Only formaldehyde is widely used commercially. Contact with skin and inhalation must be avoided to preclude sensitization and other adverse reactions.

Decorative Nail-Care Products. Nail lacquers, or nail polishes, consist of resin, plasticizer, pigments, and solvents. The most commonly used resin is nitrocellulose. Ethyl acetate, butyl acetate, and toluene are typical solvents. Toluenesulfonamide–formaldehyde resin and similar polymers are the resins of choice as secondary film formers for optimal nail adhesion.

Camphor, dibutyl phthalate, and other lipidic solvents are common plasticizers. Nail lacquers require the presence of a suspending agent because pigments have a tendency to settle. Most tinted lacquers contain a suitable flocculating agent, such as stearalkonium hectorite.

The blend of pigments used to create a particular shade must conform to regulations covering pigments and dyes in cosmetics.

Nail lacquer removers are simply acetone or blends of solvents similar to those used in nail lacquers. It is commonly accepted that solvents have a drying effect on nails, and nail lacquer removers are often fortified with various lipids such as castor oil or cetyl palmitate.

Nail elongators are products intended to lengthen nails. These have become extremely popular. Nail elongation is achieved by adhering a piece of nonwoven nylon fabric (referred to as nail wrap) to the nail with a colorless lacquer. This process may be repeated until the desired nail thickness has been reached. After shaping, the artificial nail is further decorated.

Hair Products

Hair Conditioners. Hair conditioners are designed to repair chemical and environmental damage, replace natural lipids removed by shampooing, and facilitate managing and styling hair. The classical hair-conditioning products were based on lipids, which were deposited on hair either directly, with oils or pomades, or from emulsions.

An entirely different, and in the 1990s more popular, type of hair conditioning is achieved by treating hair with substantive quaternary compounds or polymers. Quaternaries are sorbed by hair, retained despite rinsing with water, and removed only by shampooing. The most widely used quaternary is stearalkonium chloride, which has been used at 3–5% concentration in cream rinses for many years.

A third type of hair-conditioning product relies on the use of proteins, amino acids (qv), botanicals, and amphoterics. Many ingredients have been identified as hair conditioners. Some of them are claimed to be substantive to the hair, whereas others are claimed to penetrate into the hair and repair previously incurred damage. Some of these hair-conditioning substances have been incorporated into newer delivery systems, such as mousses.

Hair Fixatives. These products are designed to assist in hair styling and in maintaining the style for a period of time. In contrast with hair dressings, hair fixatives do not leave an oily residue on the hair but tend to coat the hair with film-forming residues after drying. As in the case of hair dressings, style-holding qualities depend primarily on fiber–fiber adhesion and to a minor extent on fiber coating. The products may be conveniently divided into two groups: those that are applied to damp or wet hair, hair-setting products, and those that are applied to hair after styling, hair sprays.

Wave-setting products can be applied to wet hair and should not interfere with or delay drying. Wave sets can be formulated with water-soluble polymeric substances or with polymers that show solubility only in hydro-alcoholic media. Some of the preferred hair-fixative polymers are combined with lubricants or emollients and other excipients. The viscosity of these products can vary from that of a water-thin fluid to a rather firm gel.

Hair-spray products containing little or no water are preferred, because the presence of significant levels of water tends to soften a preexisting style. Environmental regulations today preclude the use of propellant solvents. Thus higher levels of alcohols are now used and the propellants of choice are low concentrations of hydrocarbons.

Hair Colorants. Hair colorants are commonly divided into temporary, semipermanent, and permanent types. Decolorizing (bleaching) represents a fourth type of hair coloring.

Hair bleaching removes the pigment melanin from the hair shaft by oxidative destruction. Alkaline hydrogen peroxide is the agent of choice. Thickening is required in order to retain the blended oxidizing mixture on the hair. Surfactants are required to assure that every hair fiber is thoroughly wet by the blended moisture. Bleaching by hydrogen peroxide is enhanced by the presence of a peroxydisulfate, such as potassium persulfate. Bleaching damages the hair by converting some cystine to cysteic acid. In addition, the high pH induces swelling and cuticular damage. These adverse effects are counteracted by conditioning after treatment or by including some protectants in the hair-bleach product.

Temporary hair colorants are removed from the hair by a single shampoo. Temporary hair colorants usually employ certified dyes that have little affinity for hair. They are incorporated into aqueous solutions, shampoos, or hair-setting products.

Semipermanent hair colorants employ dyes that are absorbed directly by the hair. These dyes add color to the preexisting (natural) hair color and are useful primarily for blending in gray fibers. These dyes may fade significantly owing to exposure to sunlight and also are gradually removed by shampooing. Typically, temporary hair colorings are distributed as pourable lotions. Formulations may include alkanolamides, polymeric substances, fatty alcohols, thickeners, and conditioners commonly employed in all hair cosmetics. The CTFA lists temporary hair dyes with other substances as "Color Additives—Hair Colorants".

Permanent hair colorants, frequently identified as oxidation dyes, show much greater resistance to fading and shampoo loss than do semipermanent hair colorants. As a rule, these dyes remain on the hair; it is common practice to dye only that portion of the hair shaft that has emerged from the scalp since the last application. In permanent dyeing, a lotion containing developers and couplers is blended with hydrogen peroxide and then applied to the hair. The objective is uniform penetration of the various components into the hair, oxidation of the developer to a reactive intermediate, and formation of a colored dye stuff with the coupler. The dyes are synthesized within the fiber and migrate outward only slowly because of their size. In addition, the reactions occur in alkaline media, and some of the peroxide bleaches the hair. Thus it is possible to generate colored hair lighter than the original shade.

Hair-Waving and Straightening Products. The development of hair-waving and hair-straightening products requires a careful balance between product performance and hair damage. The hair-waving process essentially depends on converting some cystine cross-links in keratin to cysteine residues, which are reoxidized after the configuration of the hair has been changed. Sometimes hair straightening can also be achieved by a similar, relatively innocuous chemical change in the hair. As a rule, however, much more chemical destruction is required to achieve rapid and permanent straightening than to achieve permanent waving.

Permanent waving depends on the metathesis of a mercaptan and the cystine in hair while the hair is held in a curly pattern on a suitable device (curler). The most commonly used mercaptan is thioglycolic acid, although some other nonvolatile mercaptans can be employed. The active species is the mercaptide anion. Thus, adjustment to a pH between about 8.8 and 9.5 using amines or especially ammonia is required. A typical hair-waving product may consist of a 0.5–0.75 N solution of thioglycolic acid adjusted to a pH of about 9.1 with ammonia. The product generally includes a nonionic surfactant, to ensure thorough wetting of the wound hair tress, and a fragrance. Opaque lotion products can be created by adding the actives to mineral oil or other lipid-containing emulsions. The thioglycolate lotion is allowed to remain on the hair for about 10–30 min; then the hair is rinsed with water. Next, an oxidizing solution consisting of a dilute (1.5–3%) acidified hydrogen peroxide solution or of a potassium or sodium bromate is applied to the hair. This so-called neutralizing solution oxidizes the cysteine residues to cystine (without bleaching) in a new configuration within about 5–10 min. The neutralizer may contain a variety of hair conditioners and is removed from the hair by thorough rinsing with water after unwinding.

Hair straightening is more difficult than hair waving. Kinky hair has a tight crimp that cannot be straightened by winding over a rod or curler. Two processes for straightening exist. One, based on thioglycolates, effects the same chemical change as that occurring during permanent waving. The other, more aggressive process is based on (1–8%) sodium hydroxide (or guanidine). In order to hold the hair straight, hair-straightening products are viscous. The chemical reactions with sodium hydroxide involve formation of cysteine and dehydroalanine residues in the hair with some loss of sulfur. The cysteine and dehydroalanine can subsequently react to form the thioether, lanthionine, which helps repair the mechanical strength of the fiber to some extent. Similar chemical reactions occur when steam is allowed to interact with hair, such as during hot pressing, which was an earlier technique for straightening hair.

Conditioners, lipids, acid rinses, and related cosmetics have been developed to minimize hair damage from these rather destructive processes.

Hair Removers. Hair removers are designed to remove hair from the skin surface without cutting in order to avoid undesirable stubble. Cosmetic products have been developed for chemical destruction of hair, that is, depilation, and for facilitating mechanical hair removal, that is, epilation.

Depilatories epitomize the chemical destruction of hair and allow hair removal by scraping with a blunt instrument or by rubbing with terry cloth. Chemical depilatories are based on 5–6% calcium thioglycolate in a cream base (to avoid runoff) at a pH of about 12. The pH is maintained with calcium or strontium hydroxide. This type of treatment does not destroy the dermal papilla, and the hair grows back.

Epilation is required for permanent hair removal. The most effective epilation process is electrolysis or a similar procedure. Epilation can also be achieved by pulling the fibers out of the skin. For this purpose, wax mixtures (rosin and beeswax) are blended with lipids, for example, oleyl oleate, which melt at a suitable temperature (about 50–55°C).

Decorative Cosmetics

Decorative cosmetics are products intended to enhance appearance by adding color or by hiding or deemphasizing physical defects. In Western cultures, most decorative cosmetics are for use on the face. Products in this category are various types of powders, facial makeups, and lip- and eye-coloring products.

MARTIN M. RIEGER
M & A Rieger Associates

CTFA International Cosmetic Ingredient Dictionary, 5th ed., CTFA, Washington, D.C., 1993.

CTFA International Cosmetic Ingredient Handbook, 2nd ed., CTFA, Washington, D.C., 1992.

E. Jungermann, ed., *Cosmetic Science and Technology Series,* Marcel Dekker, New York, continuing series.

COTTON

Cotton is the most important vegetable fiber used in spinning (see FIBERS, VEGETABLE). Its origin, breeding, morphology, and chemistry have been described in many publications. It is a member of the Malvaceae, or mallow family, is a plant of the genus *Gossypium,* and is widely grown in warm climates the world over.

The average cotton plant is a herbaceous shrub with a normal height of 1.2 to 1.8 m, although some tree varieties reach a maximum height of 4.5 to 6.0 m. The most important species included in the genus *Gossypium* are ***hirsutum, barbadense, aboreum,*** and ***herbaceum.***

The most favorable growing conditions for cotton include a warm climate (21 to 30°C, mean temperature) for at least 150 days. Fairly moist and loamy soil produces the highest yields. Under normal climatic conditions, cotton seeds germinate in 7 to 10 days. Flower buds (known as squares) appear in 35 to 45 days followed by open flowers 21 to 25 days later. One day after the flower opens the cotton boll begins to grow rapidly, if the flower has been fertilized. The mature boll opens 45 to 90 days after flowering, depending on variety and environmental conditions. Within the boll are three to five divisions called locks, each of which normally has seven to nine seeds that are covered with both lint and fuzz fibers. The fuzz fibers form a short, shrubby undergrowth beneath the lint hairs on the seed. At least 10,000 fibers are attached to each seed and there are close to 500,000 fibers in each boll. The usual range of planting time in the United States extends from the beginning of March to the end of May; harvest time is in the late summer or early fall.

The cotton fiber is a single cell that originates in the epidermis of the seed coat at about the time the flower opens; it first emerges on the broad, or chalazal, end of the seed and progresses by degrees to the sharp, or micropylar, end. As the boll matures, the fiber grows until it attains its maximum length, which averages about 2500 times its width.

The seed hairs of cultivated cottons are divided into two groups (fuzz and lint) that differ in length, width, pigmentation, and strength of adherence to the seed. The growth of fuzz fibers is much the same as that of lint, but they are usually about 0.33 cm long compared with the 2.5 cm average length of lint fibers and are twice as thick, or about 32 μm. Their color ranges from greenish brown to gray. After lint fibers have been ginned off the seed, the fuzz fibers (which are also called linters) remain. Removal of linters requires a machine similar to the regular cotton gin. Linters are an important source of cellulose for chemical purposes and are used in upholstery and batting.

Cultivation and Production

At present, the chief cotton-growing countries of the world are the People's Republic of China (23%), the United States (17%), the former USSR (15%), India (11%), Pakistan (8%), Brazil (4%), Turkey (3%), and Egypt (2%) (5). In the Western Hemisphere, cotton is cultivated between about 37° N and 32° S latitude; in the Eastern Hemisphere, the limits extend to 47° N and 30° S.

In the United States management and harvesting practices have increased the yield so only about half as much land is needed today to produce the same amount of cotton as in 1930. Cultivation of cotton differs markedly from one country to another, depending on methods of mechanization. Approximately 70% of the U.S. cotton is rain grown, but western states (California and Arizona) only grow irrigated cotton. The use of supplemental irrigation is increasing in Texas and the Mid-South states.

Through cotton-breeding research, the United States has developed varieties that today remain the leading fine-quality cottons. Two such varieties, Deltapine 50 and Paymaster HS26, were planted on 30% of the U.S. cotton acreage.

Fertilizers. On average in 1990 U.S. cotton was fertilized with 31 kg of nitrogen, 10 kg of P_2O_5, and 6.8 kg of K_2O. The type and concentration of fertilizer required for high yield depends on many factors such as soil type, previous fertilization rate, cropping system, and irrigation. Therefore, an efficient fertilization program must be based on results of soil and tissue tests and the yield desired from the crop.

Pests and Insecticides. The most destructive pests of the cotton plant are the boll weevil and the bollworm/budworm complex. The domestic cotton crop lost to the weevil is worth $200 million a year. In addition, about $75 million a year is spent for pesticides to control this destructive pest. Unfortunately, some insecticides used to control the weevil kill many beneficial insects. Among the undesired casualties are insects that help to control the bollworm and the tobacco budworm, pests that cause another $200 million loss in cotton.

In the mid-1970s, a test program to eradicate the boll weevil, carried out on an 8100-ha (81-km^2) region of Mississippi, Alabama, and Louisiana, led to plans for beltwide eradication of this devastating pest. The pest could be eliminated from the United States by the year 2015.

Another destructive insect is the pink bollworm, which overwinters as diapausing (hibernating) larvae in the soil.

A technique to combat this pest has been developed by agricultural research scientists in cooperation with the Arizona Agricultural Experiment Station in Phoenix. The number of overwintering pink bollworms is limited by reduction of their food supply late in the year. Chemical growth regulators are used to prevent late boll formation; this has no effect on the existing bolls and causes no loss in harvesting, as most of the late bolls fail to mature anyway. Along with the chemical treatments, other eradication practices include early stalk shredding, early and deep tillage, and winter irrigation that drowns diapausing larvae. Other insects injurious to the cotton plant include aphids, leafhoppers, lygus bugs, mites, whiteflies, fleahoppers, thrips, cutworms, and leaf miners.

Harvesting. Except for the cotton gin, the introduction of the mechanical harvester has probably had a greater effect on cotton production than any other single event. Commercial mechanical harvesters were introduced into the United States after World War II. In the early 1990s more than 99% of the U.S. cotton crop was mechanically harvested, although cotton was still hand-harvested in some other countries.

Once the plant is ready, the cotton is mechanically harvested by using either a spindle picker or cotton stripper. The spindle picker selectively harvests seed cotton from open bolls. Two types of cotton strippers are currently in use in the United States: the finger-type stripper and the roll-type stripper.

Ginning. Gin equipment is designed to remove foreign matter, moisture, and cottonseed from raw seed cotton. Typical types of gin equipment are cylinder cleaners, stick machines, and lint cleaners for cleaning; shelf-type driers for removing moisture; and gin stands for separating the fiber from the cottonseed. The gin stand is actually the only item of equipment required to gin cotton, the other equipment is for trash removal and drying.

Greige Processing. The cotton from the gin is received by the textile mill in the form of highly compressed bales, weighing approximately 227 kg. The lint cotton still contains various forms of trash, including stem, leaf, and seed coat fragments that must be removed in the manufacturing process. Before it can be spun into yarn, the cotton must be opened, blended, cleaned, parallelized, and reduced to the proper size.

The final process in the yarn manufacturing operation is spinning. Ring spinning is the mainstay of the textile industry and accounts for over 50% of all cotton yarn produced in the United States.

The newer spinning methods produce yarn directly from drawing sliver. Rotor, or open-end spinning is a relatively new method of yarn formation that can produce coarser yarns at three to five times the rate of ring spinning.

Some other new methods of producing spun cotton yarns at exceptionally high production rates have been developed. Air-jet spinning is one of the newest yarn formation techniques and can spin yarns at speeds of up to 200 yards (183 m) per minute.

Most woven and knitted cotton fabrics are produced from single yarns as supplied by the spinning process. However, for the manufacture of industrial fabrics such as canvas, it is necessary to combine, or ply twist, several strands of single yarns together to obtain increased strength and resilience. Sewing thread and cordage are also produced from multiple plies of single yarns twisted together.

Physical Properties

Fiber length is universally accepted as the most important fiber property, because it greatly affects processing efficiency and yarn quality. The recognized reference machine method for fiber length information is the Suter-Webb Comb Sorter. Variations of length are unique to specific varieties of cotton and range from less than 2.5 cm for short-staple Upland varieties to 2.6 to 2.8 cm for medium-staple Uplands, to >2.85 cm for long-staple varieties (Pima, Egyptian, and Sea Island).

Next to length, fiber strength is the most important physical property that relates to fiber and yarn quality. The recognized reference method for fiber strength is based on measurements made on bundles of parallel fibers. Fiber strength is expressed as breaking stress or force to break per linear density of the bundle. These units are newtons (or gram force (gf)) per linear density (tex), where 1.0 tex = 1 g/1000 m. Variations in fiber strength are also unique to specific varieties of cotton and range from 0.176–0.216 N/tex (18 to 22 gf/tex) for short-staple Upland varieties to 0.235–0.275 N/tex (24 to 28 gf/tex) for some medium-staple Uplands to 0.314–0.373 N/tex (32 to 38 gf/tex) for long-staple varieties (Pima, Egyptian, and Sea Island).

Another important characteristic property of cotton fibers is their fineness, or linear density, or weight per unit length. The normal units for fineness are millitex. Fineness is directly related to the amount of cellulose in the fiber, which is a function of the fiber wall area, excluding the hollow center (lumen), and the fiber length. Developments in computerized microscopic image analysis allow for rapid and accurate measurements of fiber wall area and perimeter.

In addition to fiber length, strength, and fineness, two other properties that have significant bearing on fiber and yarn properties are color and trash measurements, which have been accomplished by instrumentation such as the Colorimeter and the Shirley Non-Lint Analyzer.

High Volume Instrument Systems. In the mid-1960s, a cooperative effort between the USDA and instrument manufacturers began that was aimed at developing instruments for high volume use such as would be necessary for cotton classification. This led to the development of high volume instrument (HVI) systems. Modern HVI systems make use of the latest advances in electronic instrumentation and space-age technology to rapidly and inexpensively measure the more important fiber properties, including length, length uniformity, strength, fineness, color, and trash.

Morphology

The cotton fiber is a single biological cell, 15 to 24 μm in width and 12 to 60 mm long; it has a central canal, or lumen, down its length except at the tip. It is tapered for a short length at the tip, and along its entire length the dried fiber is twisted frequently; the direction of twist reverses occasionally. These twists are referred to as convolutions, and it is believed that they are important in spinning because they contribute to the natural interlocking of fibers in a yarn.

Cotton is essentially 95% cellulose. The noncellulosic materials, consisting mostly of waxes, pectinaceous substances, and nitrogenous matter, are located to a large extent in the primary wall, with small amounts in the lumen.

Of the noncellulosic substances in cotton, protein normally occurs in the largest amounts. This protein is apparently the protoplasmic residue left behind on the gradual drying up of the living cell. It occurs almost entirely in the lumen.

Chemical modification of the cotton fiber must be achieved within the physical framework of this rather complicated architecture. Uniformity of reaction and distribution of reaction products are inevitably influenced by rates of diffusion, swelling and shrinking of the whole fiber, and by distension or contraction of the fiber's individual structural element during finishing processes.

Structure and Reactivity

Chemical Structure. The raw cotton fiber is predominantly cellulose, and its structures and chemical properties are essentially those of the cellulose polymer.

Supramolecular Structure. The chains of cellulose molecules associate with each other by forming intermolecular hydrogen bonds and hydrophobic bonds. They coalesce to form microfibrils also called crystallites. In cotton, the microfibrils can organize into macrofibrils 60 to 300 nm wide. The macrofibrils are organized into fibers. Cotton fibers have a complex, reversing, helical arrangement of macrofibrils. There are several different forms or polymorphs (cellulose I to IV and X with recent subclasses Iα and Iβ), depending on the source and treatment. There are both different unit cells and different packing arrangements in the unit cell. Native cotton is cellulose I.

Pore Structure and Affinity for Water. The cotton fiber is a porous, hydrophilic material that accounts for the comfort of cotton clothing. Moisture is retained tenaciously in cotton. The moisture absorbed from the atmosphere and held under ambient conditions is expressed either as moisture content (amount of moisture as a percentage of original sample mass) or more commonly as moisture regain (amount of moisture as a percentage of oven-dry sample). Under ordinary atmospheric conditions, moisture regain is 7 to 11%.

Pores accessible to water molecules are not necessarily accessible to chemical agents. Many uses of cotton, eg, easy care fabric, depend on chemical modification to impart the desired properties. Knowledge of its accessibility under water-swollen conditions to dyes and other chemical agents of various sizes is required for better control of the various chemical treatments applied to cotton textiles.

Availability of Hydroxyl Groups. The regular occurrences of intermolecular and intramolecular hydrogen bonds in the crystalline regions of cotton cellulose render the involved hydroxyl groups unavailable to chemical agents under mild reaction conditions. Chemical agents that have access to the interior pores of the cotton fiber thus find potential reactive sites unavailable for reaction. The order of decreasing availability of hydroxyl groups in cotton is $2 - OH > 6 - OH \gg 3 - OH$.

Reactions for Practical Objectives

Chemical modification has assisted in building cotton's position in the marketplace despite the advent of synthetic fibers.

Mercerization. One of the earliest known modifications of cotton that had commercial potential was mercerization. Traditionally, the process employed a cold concentrated sodium hydroxide treatment of yarn or woven fabric followed by washing and a mild acetic acid neutralization. Maintaining the fabric under tension during the entire procedure was integral to the expected properties. The resultant mercerized cotton of commerce has improved luster and dyeability and, to a lesser extent, improved strength. If the cotton is allowed to shrink freely during contact with mercerizing caustic, slack mercerization takes place; this technique produces a product with greatly increased stretch (stretch cotton) that has found application in both medical and apparel fields.

Etherification. The accessible, available hydroxyl groups on the 2, 3, and 6 positions of the anhydroglucose residue are quite reactive and provide sites for much of the current modification of cotton cellulose to impart special or value-added properties. The two most common classes into which modifications fall include etherification and esterification of the cotton cellulose hydroxyls as well as addition reactions with certain unsaturated compounds to produce cellulose ethers (see CELLULOSE ETHERS). One large class of cellulose-reactive dyestuffs in commercial use attaches to the cellulose through an alkali-catalyzed etherification by nucleophilic attack of the chlorotriazine moiety of the dyestuff.

Cross-Linking. By far the most important commercial modifications of cotton cellulose remain those that occur through etherification. For example, commercial modification of cotton to impart durable-press, smooth drying, or shrinkage resistance properties involves cross-linking adjacent cellulose chains through amidomethyl ether linkages.

Increasing Resiliency. Base-catalyzed reactions of cotton cellulose with either monoepoxides or diepoxides to form cellulose ethers also result in fabrics with increased resiliency.

Other Cellulose Ethers. Other cotton cellulose ethers include carboxymethyl, carboxyethyl, hydroxyethyl, carbamoylethyl, cyanoethyl, sulfoethyl, and aminoethyl (aminized cotton) products. Most, with the exception of cyanoethylated and aminized cotton, are of interest in applications requiring solubility or swellability in water or alkali.

Flameproofing. The flameproofing of cotton is discussed elsewhere (see FLAME RETARDANTS FOR TEXTILES). Although certain cellulose esters such as the ammonium salt of phosphorylated cotton and cellulose phosphate are flame resistant, the attachment of most currently used durable polymeric flame retardants for cotton is through an ether linkage to the cellulose at a relatively low degree of substitution (DS). Nondurable flame retardants based on liquid or vapor-phase applications of boric acid or methyl borate are used in treatment of cotton batting for upholstery, bedding, and automotive cushions.

Water Repellency. The development of water-repellent cellulose ethers has been reviewed (see WATERPROOFING). A typical example of a commercial etherification for waterproofing cotton is with stearamidomethylpyridinium chloride:

$$C_{17}H_{36}\overset{O}{\overset{\|}{C}}-NCH_2-\overset{+}{N}\bigcirc + cell-OH \longrightarrow$$

$$C_{17}H_{36}\overset{O}{\overset{\|}{C}}-NCH_2O-cell + HCl + N\bigcirc$$

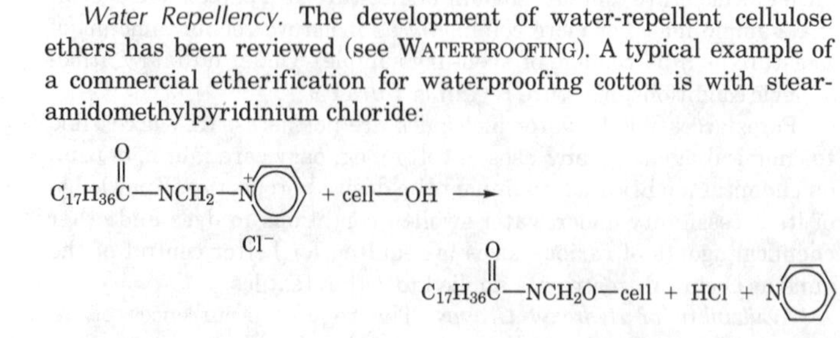

Cyanoethylation. One of the earliest examples of etherification of cellulose by an unsaturated compound through vinyl addition is the cyanoethylation of cotton. Cyanoethylation can impart a wide variety of properties to the cotton fabric such as rot resistance, heat and acid resistance, and receptivity to acid and acetate dyes.

Irradiation. The effects of high energy radiation on cotton properties have also been investigated, eg, the effects of gamma radiation on cotton have been described. Depolymerization of the cellulose occurs with increasing energy absorption; carbonyl formation, carboxyl formation, and chain cleavage occur in the ratio of 20:1:1. With these chemical changes, there is a corresponding increase in solubility in water and alkali and a decrease in fiber strength. The induction of cellulose free radicals by near uv irradiation forms the basis for photofinishing with vinyl monomers to produce graft polymers on the cotton.

Another useful reaction of cotton cellulose occurs with an ionized atmosphere. This is essentially a surface reaction. Glow discharge treatment of cotton yarn in air increases water absorbency and strength, and surface-dependent properties of cotton fabric are drastically changed by exposure to low temperature–low pressure plasma generated by radio-frequency radiation.

Insolubilization. With the exception of fiber-reactive dyestuffs, all of the cotton dyes, ie, substantive, vat and sulfur, are insolubilized within the fiber, the latter two after an oxidizing step.

Insolubilization. Insolubilization is also used to impart antimicrobial properties to cotton. Insolubilization of poly(ethylene glycol)s cross-linked with methylolamides renders fabrics temperature adaptable; such fabrics absorb and release heat in repeatable cycles.

Esterification. There are both inorganic and organic esters of cellulose. Of the three most common inorganic esters, cellulose nitrate, phosphate, and sulfate, only the cellulose sulfate is soluble in water.

Acetylation of cotton to an acetyl content slightly greater than 21% produces a material with greatly increased resistance to fungal and microbiological degradation, in addition to a tolerance of high temperatures not exhibited by native cotton.

In the 1960s, esterification of cotton cellulose by polycarboxylic acids to produce smooth-drying fabrics was investigated. In the late 1980s, better catalysis was discovered for the ester cross-linking of cellulose; inorganic salts of phosphorus-containing acids were found to give ester cross-links that are durable to multiple home launderings. Because of the improved catalysis, certain tricarboxylic and tetracarboxylic acids have shown promise for commercialization.

Economic Aspects

Marketing. There are several routes by which cotton fiber in the United States changes ownership from the grower to its final destination at the spinning mill. The grower may sell cotton directly to a spinning mill under a grower contract or the grower may sell cotton to a gin, broker, commission firm, or shipper. Some growers, after ginning their cotton, may sell through a cooperative organization or may place the cotton in a depository as collateral under the Commodity Credit Corporation Loan Program, to be either withdrawn on repayment of the loan plus interest and storage or forfeited for sale by the government. These intermediate buyers then sell the cotton to foreign or domestic mills that have not purchased the cotton directly from the grower.

Health and Safety Factors

Byssinosis. Byssinosis is a controversial form of occupational lung disease that can affect a small number of textile workers from repeated inhalation of the dust generated during the processing of cotton and some other vegetable fibers (eg, flax and soft hemp). Views vary radically between those who believe byssinosis to be a fully reversible disease of the airways not associated with any disability to those who believe it may cause progressive and disabling airway narrowing. Appropriate engineering controls in cotton textile processing areas for the most part can eliminate incidence or workers' reaction to cotton dust. The U.S. Occupational Safety and Health Administration (OSHA) issued revised standards for occupational exposure to cotton dust in 1985.

Formaldehyde. Formaldehyde is a component of resins used to impart durable-press and other properties to cotton fabrics. It is classified as a "probable human carcinogen" because it has been shown to be an animal carcinogen, and there is limited evidence to indicate that it is a carcinogen in humans. Exposure to formaldehyde from cotton textiles is controlled by the chemical technology on low emitting formaldehyde resin technology and nonformaldehyde finishes and by increased ventilation. OSHA has issued standards for control of occupational exposure to formaldehyde.

NOELIE R. BERTONIERE
Southern Regional Research Center, ARS, USDA
DEVRON P. THIBODEAUX
GEORGE F. RUPPENICKER, JR.
WILTON R. GOYNES, JR.
BETHLEHEM K. ANDREWS
DICKY D. HARDEE
J. RAY WILLIFORD
WILLIAM S. ANTHONY
CHARLES K. BRAGG
Agricultural Research Service, USDA
JESS BARR
KATER HAKE
PHIL WAKELYN
National Cotton Council

U. S. Cotton Handbook, Cotton Council International and National Cotton Council of America, Washington D.C., 1976.

H. B. Brown and J. O. Ware, eds., *Cotton,* 3rd ed., McGraw-Hill Book Co., Inc., New York, 1958.

T. P. Nevel and S. H. Zeronian, eds., *Cellulose Chemistry and Its Applications,* Ellis Horwood, Chichester, UK, 1985.

J. T. Marsh, ed., *An Introduction to Textile Finishing,* 2nd ed., Chapman and Hall, London, 1966.

COUMARIN

Coumarin, 2*H*-1-benzopyran-2-one; 1,2-benzopyrone, $C_9H_6O_2$ (**1**), is one of the most important aroma chemicals having unique characteristics not only because of its haylike bittersweet odor, but also because of its quality as a perfume fixative.

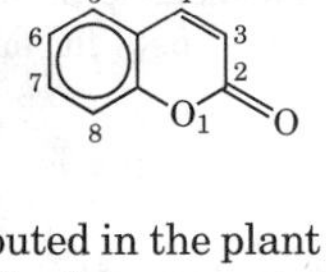

Coumarin is widely distributed in the plant kingdom but most of it has been produced synthetically for many years for commercial uses. In addition to its use in the perfumery, cosmetic, and related industries, coumarin has several other industrial applications. Coumarin is the parent substance of a large group of derivatives, many of which occur naturally and some of which are of economic significance.

Occurrence

Coumarin's better known occurrences are in touka beau (*Dipteryx odoratiae*), sweet clover (*Melilotus officinalis* and *alba*), woodruff (*Asperula odorata*), cassia (*Cinnamorum cassia*), melilot (*Melilotus officinalus*), lavender (*Lavandum officinalus*), and balsam of Peru (*Myroxylon pereirae*).

Physical Properties

Coumarin is usually sold in the form of colorless shiny leaflets or rhombic crystals. Physical constants appear in Table 1.

Table 1. Physical Constants of Coumarin

Constant	Value
mol wt	146.14
mp, °C	71
bp at 100 kPa[a], °C	301.1
density at 100°C, g/cm^3	1.178

[a] To convert kPa to mm Hg, multiply by 7.5.

Chemical Properties

The chemical properties of coumarin are those of the lactone of an α,β-unsaturated aromatic acid.

The lactone is easily hydrolyzed by alkalies to the corresponding salts of coumarinic acid or *o*-hydroxy-*cis*-cinnamic acid. Hydrogenation of coumarin with a Rancy catalyst gives 3,4-dihydrocoumarin or octahydrocoumarin. Coumarin is reduced to *o*-hydroxycinnamyl alcohol by reaction with lithium aluminum hydride. Coumarin combines readily with sodium bisulfite solutions to form soluble sodium 3- and 4-hydrosulfonates. Coumarin reacts with bromine under moderate conditions to give 3,4-dibromocoumarin. Coumarin is not readily oxidized by chromic acid but, by action of the Fenton's reagent, it is converted into 7-hydroxycoumarin (umbelliferone). Fuming sulfuric acid reacts with coumarin to give coumarin-6-sulfonic acid at moderate temperature and coumarin-3,6-disulfonic acid at higher temperature. Fuming nitric acid forms 6-nitrocoumarin. Methylating agents such as methyl sulfate and methyl iodide react with coumarin in the presence of sodium hydride to give methyl 2-methoxycinnamate. A coumarin dimer is formed by prolonged exposure of coumarin to sunlight or uv radiation.

Methods of Preparation

Until the late 1890s, coumarin was obtained commercially from only natural sources by extraction from tonka beans and deer tongue. Then synthetic methods of preparation and industrial manufacturing processes were discovered and developed starting principally from *o*-cresol, phenol, and salicylaldehyde. Various methods can be used to obtain coumarin from each of these starting materials. The most current industrial process uses the Perkin reaction, in which salicylaldehyde is reacted with acetic anhydride.

Purification and Shipment

In order to be suitable for perfumary uses, synthetic coumarin must be treated to a high degree of purity. Normal purification processes include fractional distillation under high vacuum and crystallization from suitable solvents such as methanol or ethanol.

Health and Safety

The acute and subchronic toxicity of coumarin have been well described. A subchronic oral toxicity study of coumarin administered was conducted for the National Toxicology Program (NTP) in 1981. The results of this study, conducted for 13 weeks using gavage doses from 19 to 300 mg/kg/d, indicated significant decreases in body weight gain and dose-related changes in clinical pathology parameters indicative of hepatic injury at the two highest dose levels, 150 and 300 mg/kg/d. Increased liver weight and a histopathologic diagnosis of toxic hepatitis were reported at these two dose levels.

In contrast, the chronic toxicity and potential carcinogenicity of coumarin have been surrounded by scientific debate. In 1987, the International Agency for Research on Cancer (IARC) classified coumarin as a Group 3 chemical, indicating limited evidence of carcinogenicity in animals and insufficient evidence in humans. Since 1954, coumarin has been classified by the FDA as a toxic substance and its use in foods was banned.

Uses

Because of its unique sweet note and stability, coumarin has long been recognized as an important raw material in the fragrance industry. It is widely used in hand soaps, detergents, lotions, and perfumes at concentrations usually extending from 0.01 to 0.8%.

In other fields, coumarin has a significant use in the electroplating industry, mostly in the automotive area, to provide high polished quality to chrome plated steel (see ELECTROPLATING) but this use is presently declining. Coumarin and some of its derivatives have been tested in pharmacology for treatment of schizophrenia, or of microcirculation disorders and angiopathic ulcers, and also for treatment of high protein edemas in animals.

Derivatives

A large number of coumarin derivatives have been identified in plants and many of them have been synthesized and studied for their physiological activity. Only a few are mentioned here because of their economic significance.

3,4-Dihydrocoumarin is prepared by catalytic hydrogenation of coumarin. It is also used in the perfumery industry for its hay-like odor. It has GRAS status and can be used as a food flavor ingredient with a sweet caramel-like taste.

3-Methylcoumarin and 6-methylcoumarin have some use in the perfume industry. The 6-methyl derivative is permitted in flavor compositions.

4-Hydroxycoumarin can be synthesized by cyclization of acetyl methyl salicylate. It is a coumarin metabolite occurring in spoiled hay. Derivatives of 4-hydroxycoumarin such as dicoumarol, warfarin, cyclocoumarol, ethylbis–coumaracetate, and bis-4-hydroxycoumarin are synthetic blood anticoagulants.

7-Hydroxycoumarin, known as umbelliferone, occurs naturally in gum resins of umbelliferae and is an important coumarin metabolite. It is readily manufactured from resorcinol and maleic or fumaric acid. Umbelliferone and β-methylumbelliferone are used as fluorescent brighteners.

Dicoumarol was isolated from spoiled sweet clover hay. It is prepared synthetically by reaction of 4-hydroxycoumarin with formaldehyde. It is used in anticoagulant therapy often associated with heparin.

PAUL M. BOISDE
WALTER C. MEULY
Rhône-Poulenc, Inc.

J. R. Johnson, *Organic Reactions,* Vol. 1, John Wiley & Sons, Inc., New York, 1942, p. 210.

IARC Monograph, *Evaluation of the Carcinogenic Risks to Man, Supplement* **7,** 56 (1987).

COUMARONE-INDENE RESINS. See HYDROCARBON RESINS.

COUPLING AGENTS. See SILICON COMPOUNDS, SILYLATING AGENTS.

CRACKING. See CATALYSIS; PETROLEUM.

CREAM. See MILK AND MILK PRODUCTS.

CRESIDINE. See AMINES, AROMATIC AMINES—ANILINE AND ITS DERIVATIVES.

CRESOLS. See ALKYLPHENOLS.

CROWN ETHERS. See ANTIBIOTICS, POLYETHERS; CATALYSIS, PHASE-TRANSFER; CHELATING AGENTS.

CRYOGENICS

In present-day usage, the word cryogenics refers to "all phenomena, processes, techniques or apparatus occurring or using temperatures below 120 K". (B. A. Hands, Cryogenic Engineering).

Cryogenic technology has contributed greatly to scientific research and is widely used in many industrial applications. The ability to condense a gas such that it can be stored and shipped as a cryogenic liquid rather than a pressurized gas has found several applications. Natural gas is stored and transported as LNG to many countries in the world. Liquid hydrogen is used as fuel for space vehicles. The use of slush hydrogen, a solid–liquid mixture of hydrogen, is being considered for transonic and hypersonic aircraft. Oxygen, nitrogen, and argon are shipped as liquids by truck and rail car to the point of end use.

Another significant use of cryogenic technology has been to produce low cost, high purity gases through fractional condensation and distillation. Cryogenic air separation is used for the production of pure oxygen, nitrogen, and argon. These gases are used in primary metals manufacturing (eg, steel), chemical manufacturing, electronic industries, partial oxidation and coal gasification processes, enhanced oil recovery and many other applications. Cryogenic methods are also used for the purification of hydrogen, helium and carbon monoxide. Hydrogen and carbon monoxide gases are used in chemical manufacturing and some metal industries. Helium is used in welding, medicine, gas chromatography, and diving.

Cryogenic processes that provide low temperatures to refrigerate other materials or to alter their properties have been used in many applications. Liquid nitrogen is used for freezing food such as hamburgers and shrimp. Rubber tires and scrap metal from old cars are reclaimed using cryogenic cooling techniques to make them brittle for easier fracturing and component separation. Biological materials such as bone marrow, blood, animal semen, tissue cultures, tumor cells and skin are preserved by cryogenic freezing and storage.

Magnetic resonance imaging (MRI) uses cryogenics to cool high conductivity magnets for nonintrusive body diagnostics. Low temperature infrared detectors are utilized in astronomical telescopes. Cryogenic technology is being used to increase the speed of computers. Cryogenic refrigerators have been applied industrially for cryopumping to yield high pumping speeds and ultrahigh vacuum. With the recent advent of high temperature superconductivity, it is anticipated that applications of superconductivity at near liquid nitrogen temperature have great potential for electric power transmission, magnetic transportation systems, and magnets for energy generation in fusion processes.

Refrigeration Methods

Refrigeration for cryogenic applications is produced by absorbing or extracting heat at low temperatures and rejecting it to the atmosphere at higher temperatures. Three general methods for producing cryogenic refrigeration in large-scale commercial applications are the liquid vaporization cycle, the Joule-Thomson (J-T) expansion cycle, and the engine expansion cycle. The first two are similar in that they both utilize irreversible isenthalpic expansion of a fluid, usually through a valve. Expansion in an engine approaches reversible isentropic expansion with the performance of work.

Equipment

Machinery. Compressor selection for a cryogenic process plant depends on the fluid, the volumetric flow rate, the pressures involved, the compressor efficiency, and the cost of energy and capital. Centrifugal compressors are lower in installed cost than reciprocating machines, and are preferred if the volumetric flows and pressures of the process allow them to be applied. For large volumetric flows, axial compressors are used. At very high pressures with small volumetric flows, reciprocating machines are required. Sometimes a combination of more than one kind of compression stage is used to yield the most cost-effective and efficient system.

Expanders provide refrigeration by extracting work from a fluid, thereby reducing its temperature. Gas expanders can be either reciprocating or centrifugal, and can be loaded (braked) by electric generators, gas blowers or compressors, oil-film cups, or oil pump brakes. The extracted work is usefully recovered in an electric generator or compressor. Reverse-running liquid pumps have been used as liquid expanders. Isentropic efficiencies for gas expanders can be as high as 85–90% for machines with high discharge volumetric flow.

Heat Exchangers. The two most prominent types of heat exchangers used in cryogenic service are the wound coil tube-in-shell exchanger and the plate and fin (core) exchanger.

Distillation Columns. In a cryogenic air separation plant, distillation (qv) accounts for a major fraction of the total energy consumption. The low relative volatility characteristic of many cryogenic separations requires the use of many stages. Columns operating at low pressures are consequently designed for a low pressure drop per theoretical stage of separation.

In some cryogenic hydrocarbon separation plants such as a nitrogen rejection unit, both flow rate and feed composition may change over the life of the plant. This requires the use of valve or bubble cap trays with high turndown capability for distillation.

Insulation. Cryogenic insulation should economically reduce heat leak into the system so that its impact on the overall refrigeration requirement is minimized. Insulation can be categorized as unevacuated bulk type (eg, purged rock-wool or perlite), rigid foam (eg, foam–glass or urethane), vacuum-jacketed (VJ), evacuated powder (eg, perlite), and multilayer insulation (MLI) (eg, evacuated aluminimized Mylar) (see INSULATION, THERMAL).

Safety

The possibility of an uncontrolled release of a cryogenic fluid such as liquid oxygen, liquid methane or liquid hydrogen from storage and during handling must be carefully considered during the design of a cryogenic facility. The level of risk is often reduced by providing dikes for secondary containment of liquid spills. Procedures to protect personnel from cryogenic burns and asphyxia and to protect the nearby equipment from embrittlement failure must be carefully considered and followed. When trace impurities in the feed streams can lead to the combination of an oxidant with a flammable cryogen (eg, solid oxygen in liquid hydrogen) or a combustible with an oxidant (eg, acetylene in liquid oxygen), special precautions must be taken to eliminate them. Many materials react with pure oxygen, so care must be taken in the selection of materials in contact with oxygen and in the cleaning of oxygen systems prior to use. Potential ignition sources must be minimized, particularly in oxygen compression and in systems for handling oxygen at elevated pressures.

RAKESH AGRAWAL
HOWARD C. ROWLES
GLENN E. KINARD
Air Products and Chemicals, Inc.

B. A. Hands, ed., *Cryogenic Engineering*, Academic Press, New York, 1986.

K. D. Timmerhaus and T. M. Flynn, *Cryogenic Process Engineering*, Plenum Press, New York, 1989.

R. E. Latimer, *Chem. Eng. Prog.* **63**(2), 35 (1967).

R. B. Scott, *Cryogenic Engineering*, Met-Chem Research Inc., Boulder, Colo., 1988.

CRYSTALLIZATION

Crystallization is one of the oldest unit operations in the portfolio of industrial and/or laboratory separations. Almost all separation techniques involve formation of a second phase from a feed, and processing conditions must be selected that allow relatively easy segregation of the two or more resulting phases. This is a requirement for crystallization also, and there are a variety of other properties of the solid product that must be considered in the design and operation of a crystallizer. Interactions among process, function, product, and phenomena important in crystallization are illustrated in Figure 1.

Several possible functions that can be achieved by crystallization: separation, purification, concentration, solidification, and analysis.

Products. In all of the instances in which crystallization is used to carry out a specific function, product requirements are a central component in determining the ultimate success of the process. These requirements grow out of how the product is to be used and the processing steps between crystallization and recovery of the final product. Key determinants of product quality are the size distribution (including mean and spread), the morphology (including habit or shape and form), and purity. Of these, only the last is important with other separation processes.

Process. In each of the systems noted above there is a need to form crystals, to cause the crystals to grow, and to separate the crystals from residual liquid. There are various ways to accomplish these objectives, leading to a multitude of processes designed to meet requirements of product yield, purity, and, uniquely, crystal size distribution.

Phenomena. The critical phenomena in crystallization are, as shown in Figure 1, nucleation and growth kinetics, interfacial phenomena, breakage, and agglomeration.

Solid–Liquid Equilibria and Mass and Energy Balances

Solubility. Solid–liquid equilibrium, or the solubility of a chemical compound in a solvent, refers to the amount of solute that can be dissolved at constant temperature, pressure, and system composition; in other words, the maximum concentration of the solute in the solvent at static conditions. In a system consisting of a solute and a solvent, specifying system temperature and pressure fixes all other intensive variables. In particular, the composition of each of the two phases is fixed, and solubility diagrams of the type shown for a hypothetical mixture of R and S in Figure 1 can be constructed. Such a system is said to form an eutectic, ie, there is a condition at which both R and S crystallize into a solid phase at a fixed ratio that is identical to their ratio in solution. Consequently, there is no change in the composition of residual liquor as a result of crystallization.

Several features of the hypothetical system in Figure 1 can be used to illustrate proper selection of crystallizer operating conditions and limitations placed on the operation by system properties. Suppose a

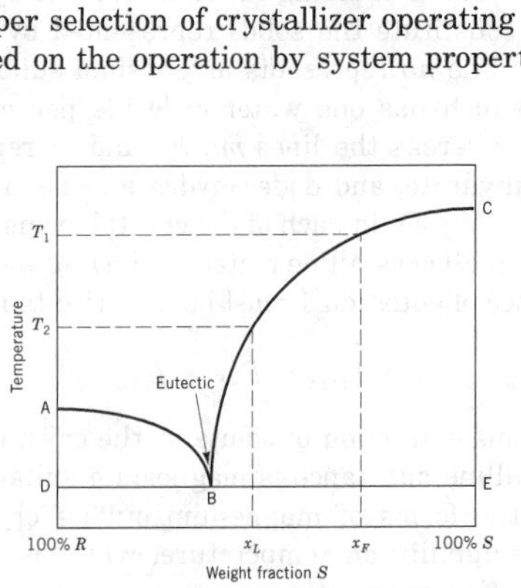

Figure 1. Solubility diagram for a hypothetical system. The curves AB and BC represent solution compositions that are in equilibrium with solids whose compositions are given by the lines AD and CE. If AD and CE are vertical along the respective axes, the crystals are pure *R* and *S*, respectively. Crystallization from any solution whose composition is to the left of the vertical line through point B produces crystals of pure *R*, whereas solutions to the right of the line produce crystals of pure *S*. A solution whose composition falls on the line through B produces a solid mixture that has a composition identical to the liquid solution.

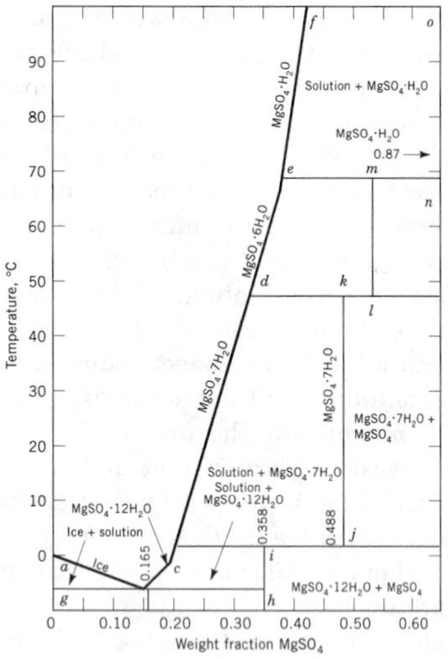

Figure 2. Solubility diagram for magnesium sulfate in water.

saturated solution at temperature T_1 is fed to a crystallizer operating at temperature T_2. Because the feed is saturated, the weight fraction of S in the feed is given as shown in Figure 2. The maximum crystal production rate $P_{\max}$ from such a process depends on the value of T_2 and is given by

$$P_{\max} = Fx_F - Lx_L \tag{1}$$

where F is the feed rate to the crystallizer, and L is the solution flow rate leaving the crystallizer. No other stream is fed to or removed from the crystallizer. Note that the lower limit on T_2 is given by the eutectic point B.

Figure 2 presents the equilibrium behavior of magnesium sulfate in water, and it is illustrative of systems that form hydrated salts. Equilibrium solution concentrations are plotted as curves *ab, bc, cd, de,* and *ef;* the solid phases that are in equilibrium with these solutions have compositions given by the lines *ag, hi, jk, lm,* and *no,* respectively. Ice is the solid phase whose composition is given by *ag,* and crystals containing differing ratios of water of hydration to magnesium sulfate constitute the solids represented by the other lines. Specifically, the line *no* represents magnesium sulfate monohydrate ($MgSO_4{\cdot}H_2O$), which has one water molecule per molecule of magnesium sulfate, whereas the lines *ml, kj,* and *ih* represent the hexahydrate, heptahydrate, and dodecahydrate forms, respectively. The weight fraction of $MgSO_4$ in each of the crystal forms is shown in Figure 2, and as with all crystalline materials having water of hydration, the solute balance of equation 1 must be modified to read

$$x_c P_{\max} = Fx_F - Lx_L \tag{2}$$

where x_c is the mass fraction of solute in the crystal, eg, x_c is 0.488 when the crystalline substance is magnesium sulfate heptahydrate. Differences in the forms of magnesium sulfate crystals affect the dependence of solubility on temperature, which is reflected by the slopes of the solution composition curves.

Supersaturation. The thermodynamic driving force for both crystal nucleation and growth is the key variable in setting the mechanisms and rates by which these processes occur. Supersaturation is defined rigorously as the deviation from thermodynamic equilibrium, which is the difference between the chemical potential of the solute at the existing conditions of the system μ and the chemical potential of the solute equilibrated at the system conditions μ^*. Less abstract definitions involving measurable system properties such as temperature, concentration, or mass or mole fraction also have been used to express supersaturation.

Mass and Energy Balances. The formulation of mass and energy balances follows procedures outlined in many basic texts. The use of solubilities to calculate crystal production rates from a cooling crystallizer was demonstrated by the discussion of equations 1 and 2. Subsequent to determining the yield, the rate at which heat must be removed from such a crystallizer can be calculated from an energy balance:

$$F\hat{H}_F = P\hat{H}_C + L\hat{H}_L + Q \tag{3}$$

where F, P, and L are feed rate, crystal production rate, and mother liquor flow rate, respectively; $\hat{H}$ is the specific enthalpy of the stream corresponding to the subscript; and Q is the required rate of heat transfer. As F, P, and L are known or can be calculated from a simple mass balance, determination of Q requires methods of estimating specific enthalpies.

If appropriate enthalpy data are unavailable, estimates can be obtained by first defining reference states for both solute and solvent. Often the most convenient reference states are crystalline solute and pure solvent at an arbitrarily chosen reference temperature. The reference temperature selected usually corresponds to that at which the heat of crystallization $\Delta\hat{H}_C$ of the solute is known. The heat of crystallization is approximately equal to the negative of the heat of solution.

The mass balance on a crystallizer is related to the growth kinetics that occur within the unit. This may be illustrated by considering systems in which crystal growth kinetics are sufficiently fast to use essentially all of the supersaturation provided by the crystallizer. Under such conditions (referred to in the crystallization literature as class II or fast-growth behavior), the solute concentration in the mother liquor can be assigned a value corresponding to saturation. Alternatively, should supersaturation in the mother liquor be so great as to affect the solute balance, the operation is said to follow class I or slow-growth behavior. An expression coupling the rate of growth to a solute balance must be used to describe such a system.

Crystallization Kinetics

Along with operating variables of the crystallizer, nucleation and growth determine such crystal characteristics as size distribution, purity, and shape or habit.

Nucleation. Crystal nucleation is the formation of an ordered solid phase from a liquid or amorphous phase. Nucleation sets the character of the crystallization process, and it is, therefore, the most critical component in relating crystallizer design and operation to crystal size distributions.

Mechanisms. Classical nucleation theory is based on homogeneous and heterogeneous mechanisms, both of which call for the formation of crystals through a process of sequentially combining the constituent units that form a crystal. Heterogeneous and homogeneous mechanisms are referred to as primary nucleation because existing crystals play no role in the nucleation.

Both homogeneous and heterogeneous mechanisms require relatively high supersaturation, and they exhibit a high order dependence on supersaturation. These factors often lead to production of excessive fines in systems where primary nucleation mechanisms are important. The classical theoretical treatment of primary nucleation results in the expression:

$$B^0 = A\exp\left(-\frac{16}{\pi\sigma^3\nu^2}3k^3T^3[\ln(s+1)]^2\right) \tag{4}$$

where k is the Boltzmann constant, ς is surface energy per unit area, ν is molar volume, and A is a constant. This equation can be simplified by recognizing that $\ln(s + 1)$ approaches s as s approaches 0. So for small supersaturations,

$$B^0 = A\exp\left(-\frac{16\pi\sigma^3\nu^2}{3k^3T^3s^2}\right) \tag{5}$$

The most important variables affecting nucleation rate are shown by equations 4 and 5 to be interfacial energy, temperature, and supersaturation.

Secondary nucleation is crystal formation through a mechanism involving the solute crystals; crystals of the solute must be present for secondary nucleation to occur. Thorough reviews have been given.

Several features of secondary nucleation make it more important than primary nucleation in industrial crystallizers. First, continuous crystallizers and seeded batch crystallizers have crystals in the magma that can participate in secondary nucleation mechanisms. Second, the requirements for the mechanisms of secondary nucleation to be operative are fulfilled easily in most industrial crystallizers. Finally, low supersaturation can support secondary nucleation but not primary nucleation, and most crystallizers are operated in a low supersaturation regime that improves yield and enhances product purity and crystal morphology.

Secondary nucleation can occur as the result of several mechanisms that have been identified in selected systems and include initial breeding, contact nucleation, and shear breeding.

Process Variables Affecting Contact Nucleation. Pioneering studies elucidated many factors affecting contact nucleation. The number of crystals produced by a controlled impact of an object with a seed crystal depends on energy of impact, supersaturation at impact, supersaturation at which crystals mature, hardness of the impacting object, area of impact, angle of impact, and system temperature.

Crystal Growth. At least two resistances determine growth kinetics, those associated with integration or incorporation of the crystalline unit (for example, solute molecules) into the crystal surface (lattice) and molecular or bulk transport of the unit from the surrounding solution to the crystal face. The primary concern here is with surface incorporation.

Numerous models have been proposed to describe surface reaction kinetics, including those that assume crystals grow by layers and others that consider growth to occur by the movement of a continuous step. Each model results in a specific relationship between growth rate and supersaturation, but none can be used for a priori predictions of growth kinetics. Insights regarding the roles of certain process variables can be obtained, however, and with additional research predictive capabilities may be achieved.

Models used to describe the growth of crystals by layers call for a two-step process: (*1*) formation of a two-dimensional nucleus on the surface and (*2*) spreading of the solute from the two-dimensional nucleus across the surface. The relative rates at which these two steps occur give rise to the mononuclear two-dimensional nucleation theory and the polynuclear two-dimensional nucleation theory. In the mononuclear two-dimensional nucleation theory, the surface nucleation step occurs at a finite rate, whereas the spreading across the surface is assumed to occur at an infinite rate. The reverse is true for the polynuclear two-dimensional nucleation theory.

The screw dislocation theory, often referred to as the BCF theory (after its formulators), shows that the dependence of growth rate on supersaturation can vary from a parabolic relationship at low supersaturation to a linear relationship at high supersaturation.

All the models described above indicate the importance of system temperature on growth rate. Dependencies of growth kinetics on temperature are often expressed in terms of an Arrhenius expression:

$$k_G = k_G^0 \exp\left(\frac{\Gamma E_G}{RT}\right) \qquad (6)$$

where k_G is a growth rate coefficient, $k^0{}_G$ is a constant, and ΔE_G is an activation energy.

The presence of impurities usually decreases the growth rates of crystalline materials, and problems associated with the production of crystals smaller than desired are commonly attributed to contamination of feed solutions. Accordingly, monitoring the composition of recycle streams so as to detect possible accumulation of impurities is important. Furthermore, crystallization kinetics used in scaleup should be obtained from experiments on solutions as similar as possible to those expected in the full-scale process.

The effects of a solvent on growth rates have been attributed to two sets of factors: one has to do with the effects of solvent on mass transfer of the solute through adjustments in viscosity, density, and diffusivity; the second is concerned with the structure of the interface between crystal and solvent.

Multicrystal magma studies usually involve examination of the rate of change of a characteristic crystal dimension or the rate of increase in the mass of crystals. The characteristic dimension depends on the method used in the determination of size; eg, the second-largest dimension is measured by sieve analyses, whereas both electronic-zone-sensing and laser-light-scattering instruments provide estimates of an equivalent spherical diameter.

Anomalous growth means that growth rates of crystals in a magma are not identical or that the growth rate of an individual crystal or mass of crystals is not constant. Two theories have been used to explain growth rate anomalies: size-dependent growth and growth rate dispersion. Both alter the form of the population density function obtained from perfectly mixed continuous crystallizers; unfortunately, such behavior cannot be used to distinguish between size-dependent growth and growth rate dispersion, as both have the same qualitative effects on population density.

Crystal Characteristics

The morphology (including crystal shape or habit), size distribution, and purity of crystalline materials can determine the success in fulfilling the function of a crystallization operation.

Morphology. A crystal is highly organized, and constituent units, which can be atoms, molecules, or ions, are positioned in a three-dimensional periodic pattern called a space lattice. A characteristic crystal shape results from the regular internal structure of the solid with crystal surfaces forming parallel to planes formed by the constituent units. The surfaces (faces) of a crystal may exhibit varying degrees of development, with a concomitant variation in macroscopic appearance.

If atoms, molecules, or ions of a unit cell are treated as points, the lattice structure of the entire crystal can be shown to be a multiplication in three dimensions of the unit cell. Only 14 possible lattices (called Bravais lattices) can be drawn in three dimensions. These can be classified into seven groups based on their elements of symmetry. Moreover, examination of the elements of symmetry (about a point, a line, or a plane) for a crystal shows that there are 32 different combinations (classes) that can be grouped into seven systems. The correspondence of these seven systems to the seven lattice groups is shown in Table 1.

The general shape of a crystal is referred to as its habit. The appearance of the crystalline product and its processing characteristics (such as washing and filtration) are affected by crystal habit. Relative growth rates of the faces of a crystal determine its shape; faster-growing faces become smaller than slower-growing faces and, in the

Table 1. The 14 Bravais Lattices

Type of symmetry	Lattice	Crystal system
cubic	cube	regular
	body-centered cube	
	face-centered cube	
tetragonal	square prism	tetragonal
	body-centered square prism	
orthorhombic	rectangular prism	orthorhombic
	body-centered rectangular prism	
	rhombic prism	
	body-centered rhombic prism	
monoclinic	monoclinic parallelepiped	monoclinic
	clinorhombic	
triclinic	triclinic parallelepiped	triclinic
rhomboidal	rhombohedron	triclinic
hexagonal	hexagonal prism	hexagonal

extreme, may disappear from the crystal altogether. Growth rates depend on the presence of impurities, rates of cooling, temperature, solvent, mixing, and supersaturation. Furthermore, the importance of each of these factors may vary from one crystal face to another.

A number of studies have shown that various additives can be included in a process stream to alter crystal habit. Prediction of such behavior is difficult and extensive laboratory or bench-scale experiments may be required to evaluate the effectiveness of habit modifiers. More recently, some measure of success has been achieved with altering the habit of organic crystals based on the molecular structure and characteristics of the crystallizing species.

Polymorphism is a condition in which a specific chemical substance may crystallize into different forms. Transitions from one polymorphic form to another may be accompanied by changes in specific volume, which may lead to destruction of the crystal and containers in which the substance is stored.

Agglomeration. Many of the analyses of industrial crystallizers require that the particle recovered from the crystallizer consist of a single crystal. It is only with this type of system that single growth rates are likely to be exhibited and, moreover, many of the properties of the crystal are affected deleteriously by agglomeration. Purity, for example, typically is diminished when agglomeration occurs. Countering the negative aspects of agglomeration is recognition that in many systems the single crystals produced by normal crystal growth would be too small to be separable using conventional solid–liquid separation equipment. In such instances, there would be no product without agglomeration.

Purity. Although crystallization has been employed extensively as a separation process, purification techniques using crystallization have always been important. Mechanisms by which impurities can be incorporated into crystalline products include adsorption of impurities on crystal surfaces; solvent entrapment in cracks, crevices, and agglomerates; and lattice substitution and inclusion of pockets of liquid. It has been noted that the key to producing high purity crystals was to maintain the supersaturation at a low level so that large crystals were obtained. Others have found that reducing the size of ammonium perchlorate crystals resulted in a substantial decrease in moisture due to inclusion.

Crystal Size Distributions. Particulate matter produced by crystallization has a distribution of sizes that varies in a definite way over a specific size range. A crystal size distribution (CSD) is most commonly expressed as a population (number) distribution relating the number of crystals at each size to the size or as a mass (weight) distribution expressing how mass is distributed over the size range. The two distributions are related and affect many aspects of crystal processing and properties, including appearance, solid–liquid separation, purity, reactions, dissolution, and other properties involving surface area.

Population Balances and Crystal Size Distributions

Population balances and crystallization kinetics may be used to relate process variables to the crystal size distribution produced by the crystallizer. Such balances are coupled to the more familiar balances on mass and energy. It is assumed that the population distribution is a continuous function and that crystal size, surface area, and volume can be described by a characteristic dimension L. Area and volume shape factors are assumed to be constant, which is to say that the morphology of the crystal does not change with size.

Determination of Crystallization Kinetics. From a series of runs at different operating conditions, a correlation of nucleation and growth kinetics with appropriate process variables can be obtained; the resulting correlation can then be used to guide either crystallizer scaleup or the development of an operating strategy for an existing crystallizer. The variables affecting nucleation and growth kinetics include temperature, supersaturation, magma density, and external stimuli, such as agitation or circulation rate of the magma. Empirical power-law functions are used most frequently in correlating nucleation and growth rates, but the choice of the independent variables can be justified from a mechanistic perspective.

Mass Balance Constraints. The following mass balance on solute can be constructed for a continuous crystallizer:

$$Q_i c_i = Q_0 c_0 + Q_0 M_T \tag{7}$$

where Q_i and Q_0 are input and output volumetric flow rates, respectively, C_i, C_0 and M_T are inlet and outlet solute concentrations and mass of crystals per unit volume, and c_0 is determined by system kinetics and constrained by a solid-liquid equilibrium (solubility) relationship, which gives the equilibrium concentration c^* at the system conditions. The system (solute–solvent and crystallizer) is characterized by the magnitude of the supersaturation $(c_0 - c^*)$ remaining in the solution exiting the crystallizer. If the mass balance is closed by substituting c^* for c_0 in equation 7, the system is said to be a fast-growth or class II system. If the mass balance is not closed, significant supersaturation remains in the solution and the system is said to be a slow-growth or class I system.

CSD Characteristics for MSMPR Crystallizers. The population density function, n, is used to formulate population balances and express key CSD characteristics. It is defined as

$$n = \lim_{\Delta N \to 0} \frac{\Delta N}{\Delta L}$$

where ΔN is the number of crystals per unit volume in the size range L to $L_t\,\Delta L$. The perfectly mixed crystallizer described in the preceding discussion is highly constrained and the form of crystal size distributions produced by such systems is fixed. Such distributions have the following characteristics.

(*1*) Moments of the distribution can be calculated for MSMPR crystallizers by the simple expression

$$m_j = j!n^0(G\tau)^{j+1} \tag{8}$$

Properties of the distribution such as total number of crystals per unit volume, total length of crystals per unit volume, total area of crystals per unit volume, and mass of crystals per unit volume can be evaluated from the moments.

(*2*) The spread of the mass density function about the dominant size gives a coefficient of variation (cv) of 50% for an MSMPR crystallizer. Such a cv may be too large for certain commercial products, which means either the crystallizer must be altered or the product must be screened to separate the desired fraction.

(*3*) The magma density M_T (mass of crystals per unit volume of slurry or liquor) may be obtained from the third moment of the population density function and is given by

$$M_T = 6\rho/k_v n^0(G\tau)^4 \tag{9}$$

(*4*) A pair of kinetic parameters, one for nucleation rate and another for growth rate, describe the crystal size distribution for a given set of crystallizer operating conditions. Variation in one of the kinetic parameters without changing the other is not possible. Accordingly, the relationship between these parameters determines the ability to alter the characteristic properties (such as dominant size) of the distribution obtained from an MSMPR crystallizer.

Preferential Removal of Crystals. Crystal size distributions produced in a perfectly mixed continuous crystallizer are highly constrained; the form of the CSD in such systems is determined entirely by the residence time distribution of a perfectly mixed crystallizer. Greater flexibility can be obtained through introduction of selective removal devices that alter the residence time distribution of materials flowing from the crystallizer. The functions of classified removal are best described in terms of idealized models of clear-liquor advance, classified-fines removal, and classified-product removal.

Clear-liquor advance is simply the removal of mother liquor from the crystallizer without simultaneous removal of crystals. The primary objective of *fines removal* is preferential withdrawal from the crystallizer of crystals whose size is below some specified value. Such crystals may be redissolved and the resulting solution returned to the crystallizer. *Classified-product removal* is carried out to remove preferentially those crystals whose size is larger than some specified value.

Batch Crystallization. Crystal size distributions obtained from batch crystallizers are affected by the mode used to generate supersaturation and the rate at which supersaturation is generated.

Crystallizers and Crystallization Operations

Crystallization equipment can vary in sophistication from a simple stirred tank to a complicated multiphase column, and the operation can range from allowing a vat of liquor to cool through exchanging heat with the surroundings to the complex control required of batch cyclic operations. In principle, the objectives of these systems are all the same: to produce a pure product at a high yield with an acceptable crystal size distribution. However, the characteristics of the crystallizing system and desired properties of the product often dictate that a specific crystallizer be used in a particular operating mode.

Crystallization from Solution. Crystallization techniques are related to the methods used to induce a driving force for solids formation and to the medium from which crystals are obtained. Approaches include cooling crystallizers, evaporative crystallizers, evaporative-cooling crystallizers, salting-out crystallization, reactive crystallization, and the use of supercritical fluid solvents.

Crystallizers. The basic requirements of a system involving crystallization from solution are as follows: (*1*) a means of generating supersaturation in a fashion commensurate with the requirements of producing satisfactory crystal size distribution, (*2*) a vessel to provide sufficient residence time for crystals to grow to a desired size, and (*3*) mixing to provide a uniform environment for crystal growth. There are numerous manufacturers of crystallization equipment; in addition, many chemical companies design their own crystallizers based on expertise developed within their organizations. Crystallizers include the forced-circulation crystallizer, the draft-tube-baffle (DTB) crystallizer, and the Oslo crystallizer.

Melt Crystallization. The use of a solvent can be avoided in some systems. In such cases, the system operates with heat as a separating agent, as do several processes involving crystallization from solution, but formation of crystalline material is from a melt of the crystallizing species rather than a solution.

For the following reasons, melt crystallization holds great promise in situations in which it can be substituted for crystallization from solution: (*1*) Without the need to recover and maintain the purity of a solvent, processing costs are reduced substantially. (*2*) Because there is no contaminated solvent to handle, melt crystallization may be more environmentally benign. (*3*) Energy costs found in evaporative crystallization obviously would be reduced if it is possible to produce a desired solid without the need to evaporate solvent. (*4*) Melt crystallization may be a reasonable alternative to other separation and purification processes, because the heat of vaporization of most volatile organic materials is between two and five times their heat of fusion. An analysis of the energy requirements in melt processes concludes that such processes can compete with other thermal separation techniques only if the plant is well designed and the process precisely controlled.

RONALD W. ROUSSEAU
Georgia Institute of Technology

A. S. Meyerson, *Handbook of Industrial Crystallization,* Butterworth-Heinemann Ltd., Boston, Mass., 1993.

J. W. Mullin, *Crystallization,* 3rd ed., Butterworth-Heinemann Ltd., Boston, Mass., 1993.

A. D. Randolph and M. A. Larson, *Theory of Particulate Processes,* 2nd ed., Academic Press, New York, 1988.

R. W. Rousseau and C. G. Moyers, in R. W. Rousseau, ed., *Handbook of Separation Process Technology,* John Wiley, New York, 1987, Chapt. 11.

CUMENE

The cumene molecule can be visualized as a straight-chain propyl group having a benzene ring attached at the middle carbon.

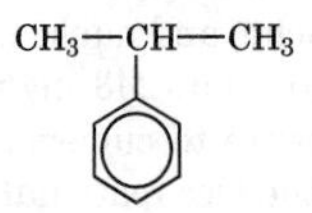

Thus cumene (1-methylethylbenzene, 2-phenylpropane, isopropylbenzene), C_9H_{12}, is a substituted aromatic compound in the benzene (qv), toluene (qv), and ethylbenzene series (see BTX PROCESSING; XYLENES AND ETHYLBENZENE). Cumene is a clear liquid at ambient conditions. It is the principal chemical used in the worldwide production of phenol and its coproduct acetone.

Properties

Some physical, chemical, and thermodynamic properties of cumene are listed in Table 1. Useful health and safety data have been included.

Manufacture

Cumene as a pure chemical intermediate is produced in modified Friedel-Crafts reaction processes that use acidic catalysts to alkylate benzene with propylene (see ALKYLATION; FRIEDEL-CRAFTS REACTIONS). The majority of cumene is manufactured with a solid phosphoric acid catalyst. Much of the remainder is made with aluminum chloride catalyst.

Economic Aspects

Cumene production follows the demand for phenol and its derivatives. Based on a trend from 1985 on, the demand is projected to increase at about 3% per year through 1995.

Health and Safety Factors

Cumene is a significant fire hazard when exposed to flame or sparks and is in the class of liquids that can be ignited under almost all

Table 1. Some Properties of Cumene

Property	Value
freezing point, °C	−96.03
boiling point, °C	152.39
density at 0°C, g/cm^3	0.8786
refractive index, n_D	1.4915
thermal conductivity at 25°C, W/(m·K)	0.124
viscosity at 0°C, mPa·s(= cp)	1.076
surface tension at 20°C, mN/m(= dyn/cm)	28.2
vapor pressure,[a] kPa[b]	
35°C	1
180°C	185
flash point, °C	44
relative molar mass	120.2
critical temperature, °C	351.4
critical pressure, kPa[b]	3220
critical density, g/cm^3	0.280
heat of vaporization at bp, J/g[c]	312
heat of formation (liquid) at 25°C, J/mol[c]	−44,150
heat capacity (liquid) at 25°C, J/(mol·K)[c]	197
odor threshold, ppmv	1.2
threshold limit value, ppmv	50

[a] Calculated from the equation: $\ln P = A - B/(t + C)$; where t = temp, °C; P = vapor pressure, kPa[b]; $A = 13.99$; $B = 3400$; and $C = 207.78$.
[b] To convert kPa to mm Hg, multiply by 7.5, to atm, divide by 101.3.
[c] To convert from J to cal, divide by 4.184.

normal temperature conditions. Cumene fires should be extinguished with foam, carbon dioxide, or dry chemical. Because cumene vapors are heavier than air and may travel long distances, they could encounter an ignition source.

Cumene is a primary skin and eye irritant. Exposure may result in significant narcosis, headache, and nausea. The recommended threshold limit value (TLV) is 50 ppm (243 mg/m^3), which is an 8-h time-weighted average for exposure to cumene. The permissible exposure limit for cumene given by the Occupational Safety and Health Administration is also 50 ppm (245 mg/m^3), again with a skin notation.

Studies of the effects of cumene on fresh and saltwater fish indicate the lowest reported toxic concentration (LC_{50}) for fishes was 20 to 30 mg/L. The solubility of cumene is about 50 mg/L. Among invertebrates, the lowest reported concentration that was toxic to test organisms was 0.012 mg/L after 18 hours. The only available data on the effect of cumene on aquatic plants indicate that the photosynthesis of several species was inhibited at concentrations from 9 to 21 mg/L.

R. C. SCHULZ
P. J. VAN OPDORP
D. J. WARD
UOP

Z. Sedaghat-Pour, *Chemical Economics Handbook*, Stanford Research Institute International, Menlo Park, Calif., Mar. 1989.

Cumene, Hazardous Substances Databank No. 172, TOXNET (Toxicology Data Network), National Library of Medicine, Bethesda, Md., 1989.

C. L. Yaws, *Chem. Eng. (N.Y.)* **82**(20), 73 (1975).

R. C. Schulz and co-workers, "Cumene Technology Improvements," paper presented by UOP at the *AIChE Summer Meeting*, Denver, Colo., 1988.

CYANAMIDES

It has been suggested that under primordial conditions, cyanamide could have acted as the original peptide-forming and phosphorylating reagent at the beginning of life on earth. Structural formulas of cyanamide, CH_2N_2, and its dimer, $C_2H_4N_4$, and trimer, $C_3H_6N_6$, are given as follows:

$H_2NC{\equiv}N$

(1)

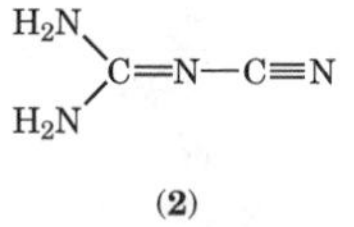

(2)

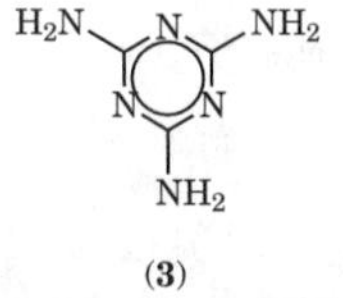

(3)

In North America, calcium cyanamide is no longer used as fertilizer, but it has limited use in special agricultural applications for defoliants, fungicides, herbicides, and as a weed killer. The primary industrial use is as a chemical intermediate for the manufacture of calcium cyanide, hydrogen cyanamide solution, and dicyandiamide. Calcium cyanamide is also used to add nitrogen to steel.

Cyanamide

Properties. Cyanamide, also called carbamodiimide or carbamic acid nitrile, crystallizes from a variety of solvents as somewhat unstable, colorless, orthorhombic, deliquescent crystals. The properties of cyanamide are listed in Table 1.

Table 1. Properties of Cyanamide

Property	Value
molecular weight	42.04
mp, °C	46
bp at 101 kPa[a], °C	dec
density[b], g/mL	1.282
specific heat at 0–39°C, J/(g K)[c]	2.288
heat of formation at 25°C, kJ/mol[c]	58.77
heat of solution[d] at 15°C, kJ/mol[c]	−15.05
heat of combustion at 25°C, kJ/mol[c]	−737.9

[a] To convert kPa to mm Hg, or Pa to μm Hg, multiply by 7.5.
[b] Calculated.
[c] To convert J to cal, divide by 4.184.
[d] In 1000 parts H_2O.

Cyanamide is a weak acid with a very high solubility in water. It is highly soluble in polar organic solvents, such as the lower alcohols, esters, and ketones, and less soluble in nonpolar solvents.

Reactions. Reactions of cyanamide are either additions to the nitrile group or substitutions at the amino group. Both are involved in the dimerization to dicyandiamide.

Manufacture. The basic process for the manufacture of cyanamide comprises four stages. The first three steps produce calcium cyanamide: *(1)* lime is made from high grade limestone (see LIME AND LIMESTONE); *(2)* calcium carbide is manufactured from lime and coal or coke (see CARBIDES); *(3)* calcium cyanamide is produced by passing gaseous nitrogen through a bed of calcium carbide with 1% calcium fluorspar, which is heated to 1000–1100°C in order to start the reaction. The heat source is then removed and the reaction continues because of its strong exothermic character: $CaC_2 + N_2 \rightarrow CaCN_2 + C$.

In the final fourth step cyanamide is manufactured from calcium cyanamide by continuous carbonation in an aqueous medium.

Other methods include reaction of lime, (CaO) with hydrogen cyanide and reaction of limestone, ($CaCO_3$), with ammonia.

Economic Aspects. A peak in calcium cyanamide production was probably reached in 1962 when the world production for fertilizer use was of the order of 1,000,000 metric tons of calcium cyanamide per year, and for industrial use approximately 300,000 t (excluding the then USSR). In 1990, the total production of cyanamide products was about half that of 1962. The largest producers are in Japan, Germany, and Canada.

Handling and Storage. Cyanamide should be stored under refrigeration and the pH tested periodically. Stabilized cyanamide can be kept at ambient temperature for a few weeks.

Health and Safety Factors (Toxicology). Manufacture of cyanamide and calcium cyanamide does not present any serious health hazard. Ingestion of alcoholic beverages by workmen within several hours of leaving work sometimes results in a vasomotor reaction known as cyanamide flush.

Commercial grades of calcium cyanamide contain lime and are moderate skin irritants where contact is repeated or prolonged.

Contact or ingestion of cyanamide must be avoided, and precautions taken to prevent inhalation of dust or spray mist. The compound is considered to be moderately toxic both by ingestion in single doses and by single-skin applications.

Dicyandiamide

Properties. Dicyandiamide **(2)** (cyanoguanidine) is the dimer of cyanamide and crystallizes in colorless monoclinic prisms. It is amphoteric, and generally soluble in polar solvents and insoluble in nonpolar solvents. Its properties are listed in Table 2.

Reactions. The reactions of dicyandiamide resemble those of cyanamide. However, cyclizations take place easily and the nitrile group is less reactive.

Table 2. Properties of Dicyandiamide

Property	Value
mol wt	84.08
mp, °C	208
bp, °C	dec
heat of formation at 25°C, kJ/mol[a]	24.9
heat of combustion at 25°C, kJ/mol[a]	−1382
heat of solution at 15°C, kJ/mol[a]	−24.1

[a] To convert J to cal, divide by 4.184.

Manufacture. Dicyandiamide is manufactured by dimerization of cyanamide in aqueous solution.

Uses. Dicyandiamide is used as a raw material for the manufacture of several chemicals, such as guanamines, biguanide and guanidine salts, and various resins. Since 1975, it has also been used in the manufacture of potassium or sodium dicyanamide which are used as insecticides and in chemotherapy. Melamine has extensive applications in the resin and plastic industry; guanamines are used as copolymers (qv) in many resin compositions. Guanidine phosphate is employed as a fire retardant in applications where water solubility is not a drawback.

Melamine

Properties. The outstanding characteristic of melamine, usually a white crystalline matrial, is its insolubility in most organic solvents. This property is also evident in melamine resins after they are cured. On the other hand, melamine is appreciably soluble in water, its solubility increasing with increased temperature.

The chemistry of melamine has been reviewed. Melamine, although moderately basic, is better considered as the triamide of cyanuric acid than as an aromatic amine (see CYANURIC AND ISOCYANURIC ACIDS). Its reactivity is poor in nearly all reactions considered typical for amines. In part, this may be a result of its low solubility (see AMINO RESINS AND PLASTICS).

Manufacture. Dicyandiamide is converted into melamine by heating.

Toxicity. Extensive toxicity investigations performed with melamine in experimental animals suggest that the compound may have a low order of biological activity.

Human subjects were given patch tests with melamine. No evidence of either primary irritation or sensitization was found. Such results suggest that melamine crystal may be handled in ordinary industrial use without special hygienic precautions.

Uses. Most of the melamine produced is used in the form of melamine-formaldehyde resins (see AMINO RESINS AND PLASTICS). Other applications include the use of melamine pyrophosphate in fire retardant textile finishes, chlorinated melamine as a bactericide, and melamine as a tarnish inhibitor in detergent compositions, in papermaking, and manufacture of adhesives.

BALDEV K. PATEL
Cyanamid Canada Inc.

CYANIDES

HYDROGEN CYANIDE

Hydrogen cyanide (hydrocyanic acid, prussic acid, formonitrile), HCN, is a colorless, poisonous, low viscosity liquid having an odor characteristic of bitter almonds. The compound has been known and used as a poison for decades. Today, hydrogen cyanide is used in the manufacture of many important chemicals.

It is theorized that hydrogen cyanide played a key role in the origin of plant and animal life on earth via formation of amino acids. Hydrogen cyanide is present in the normal human being's blood. People who smoke or who consume vegetables having relatively high cyanide content have slightly higher blood concentrations.

Hydrogen cyanide is a basic chemical building block for such chemical products as adiponitrile to produce nylon, methyl methacrylate to produce clear acrylic plastics (see ACRYLIC ESTER POLYMERS), sodium cyanide for recovery of gold (see GOLD AND GOLD COMPOUNDS), triazines for agricultural herbicides, methionine for animal food supplement (see FEEDS AND FEED ADDITIVES), chelating agents for water treatment, and many more.

Properties

The physical properties of hydrogen cyanide are listed in Table 1.

Chemical Properties. Hydrogen cyanide is a weak acid. Its structure is that of a linear, triply bonded molecule, HC≡N.

Hydrogen cyanide, as the nitrile of formic acid, CH_2O_2, undergoes many of the typical nitrile reactions.

Hydrogen cyanide can be oxidized by air at 300–650°C over silver or gold catalyst to give yields of up to 64% cyanic acid, HOCN, and 26% cyanogen, $(CN)_2$. Reaction with chlorine in the liquid phase gives cyanogen chloride, CClN, which is the basic route to triazines of which melamine, $C_3H_6N_6$, is an important derivative (see CYANAMIDES; UREA). Bromine reacts similarly, but the reaction with iodine is incomplete.

Hydrogen cyanide adds to an olefinic double bond most readily when an adjacent activating group is present in the molecule, eg, carbonyl or cyano groups.

Table 1. Physical Properties of Hydrogen Cyanide

Property	Value
molecular weight	27.03
melting point, °C	−13.24
boiling point, °C	25.70
density, g/mL	
0°C	0.7150
20°C	0.6884
specific gravity of aqueous solutions[a]	
10.04% HCN	0.9838
60.23% HCN	0.829
vapor pressure, kPa[b]	
−29.5°C	6.697
27.2°C	107.6
vapor specific gravity, at 31°C[c]	0.947
surface tension at 20°C, mN/m(= dyn/cm)	19.68
liquid viscosity at 20.2°C, mPa·s(= cP)	0.2014
specific heat, J/mol[d]	
−33.1°C, liquid	58.36
27°C, gas	36.03
heat of fusion at −14°C, kJ/mol[d]	$7.1; ts10^3$
heat of formation, ΔH_f, kJ/mol[d]	
gas at 25°C	−130.5
liquid at 25°C	−105.4
heat of combustion, net, kJ/mol[d]	642
heat of vaporization, kJ/mol[d]	25.2
heat of polymerization, kJ/mol[d]	42.7
flash point, closed cup, °C	−17.8
explosive limits in air at 100 kPa[b] and 20°C, vol %	6–41
autoignition temperature, °C	538

[a] Measured at 18°C, compared to water at 18°C.
[b] To convert kPa to mm Hg, multiply by 7.5.
[c] Air = 1.
[d] To convert J to cal, divide by 4.184.

Hydrogen cyanide adds across the carbonyl group of aldehydes and ketones and opens the oxirane ring of epoxides, both under mildly basic conditions. Several of these cyanohydrins are commercially important.

Hydrogen cyanide reacts with formaldehyde and aniline to form *N*-phenylglycinonitrile, and with formaldehyde alone to form glycolonitrile. Hydrogen cyanide reacts with NaOH, KOH, and $Ca(OH)_2$ to form the corresponding cyanides. Amines can be derived from olefins and hydrogen cyanide via the Ritter reaction.

Cyanohydrins (qv) are formed by the reaction of glucose and similar compounds with hydrogen cyanide. The corresponding aminonitrile from methyl isobutyl ketone can be formed with ammonia and hydrogen cyanide.

Dimethylformamide can be produced from the reaction of hydrogen cyanide and methanol. Adenine can be prepared from hydrogen cyanide in liquid ammonia. Thioformamide can be produced from hydrogen cyanide and hydrogen sulfide.

Under certain conditions hydrogen cyanide can polymerize to black solid compounds, eg, hydrogen cyanide homopolymer and hydrogen cyanide tetramer, $C_4H_4N_4$.

Although hydrogen cyanide is a weak acid and is normally not corrosive, it has a corrosive effect under two special conditions: *(1)* water solutions of hydrogen cyanide cause transcrystalline stress cracking of carbon steels under stress even at room temperature and in dilute solution and *(2)* water solutions of hydrogen cyanide containing sulfuric acid as a stabilizer severely corrode steel (qv) above 40°C and stainless steels above 80°C.

Manufacture and Processing

Hydrogen cyanide has been manufactured from sodium cyanide and mineral acid, and from formamide by catalytic dehydration. As of this writing, primarily because of high raw material costs, only one manufacturer uses the formamide route and one plans to use the sodium cyanide route for small quantities.

Two synthesis processes account for most of the hydrogen cyanide produced. The dominant commercial process for direct production of hydrogen cyanide is based on classic technology involving the reaction of ammonia, methane (natural gas), and air over a platinum catalyst; it is called the Andrussow process. The second process involves the reaction of ammonia and methane and is called the Blausäure-Methan-Ammoniak (BMA) process; it was developed by Degussa in Germany. Hydrogen cyanide is also obtained as a by-product in the manufacture of acrylonitrile (qv) by the ammoxidation of propylene (Sohio process).

The Shawinigan process uses a unique reactor system. The heart of the process is the fluohmic furnace, a fluidized bed of carbon heated to 1350–1650°C by passing an electric current between carbon electrodes immersed in the bed. Feed gas is ammonia and a hydrocarbon, preferably propane. High yield and high concentration of hydrogen cyanide in the off-gas are achieved. This process is presently practiced in Spain, Australia, and South Africa.

Health and Safety Factors

The cyanides are true noncumulative protoplasmic poisons, ie, they can be detoxified readily. Cyanide combines with those enzymes at the blood tissue interfaces that regulate oxygen transfer to the cellular tissues. Unless the cyanide is removed, death results through insufficient oxygen in the cells. The warning signs of cyanide poisoning include dizziness, numbness, headache, rapid pulse, nausea, reddened skin, and bloodshot eyes. More prolonged exposure can cause vomiting and labored breathing followed by unconsciousness; cessation of breathing; rapid, weak heart beat; and death. Severe exposure by inhalation can cause immediate unconsciousness. Hydrogen cyanide can enter the body by inhalation, oral ingestion, or skin absorption.

First Aid and Medical Treatment. Action should be fast and efficient. With the protection of a gas mask remove or drag the victim to fresh air. Remove contaminated clothing and rinse contaminated body areas. Keep victim warm. If the victim is conscious and speaking, no treatment is necessary. If the victim is unconscious but breathing, break an ampul of amyl nitrite in a cloth and hold it under the victim's nose for 15 s. Repeat five or six times. Use a fresh ampul every 3 min. Continue until the victim regains consciousness. If the patient is not breathing, apply artificial respiration; this can best be done using an oxygen resuscitator. The amyl nitrite antidote should also be administered during resuscitation. Mouth-to-mouth resuscitation is the next-best method followed by the Holger-Mielsen arm-lift method. Notify a physician immediately.

Disposal. Small quantities of concentrated hydrogen cyanide can be burned in a hood in an open vessel. Large-scale burning in outdoor pans can be performed, but special safety precautions must be employed. A cyanide solution can be decontaminated by making the solution strongly basic (pH 12) with caustic and pouring it into ferrous sulfate solution. The resulting ferrocyanide is relatively nontoxic. Cyanide solution can be converted to less toxic cyanate by treatment with chlorine, sodium or calcium hypochlorite, or ozone at pH 9 to 11. A solution of 10% hypochlorite maximum should be used. The final solution should be checked for absence of free cyanide. The hypochlorite or Cl_2 + NaOH method is by far the most widely used commercially.

Environmental. The toxicity of cyanide in the aquatic environment or natural waters is a result of free cyanide, ie, as HCN and CN^-. Much work has been done to establish stream and effluent limits for cyanide to avoid harmful effects on aquatic life, as fish are extremely sensitive to very low concentrations.

Another important environmental issue is the fate of cyanide. Hydrogen cyanide, if spilled, evaporates quite readily, but is not accumulating in the atmosphere. That which does not evaporate is soon decomposed or rendered nonhazardous by complexing with iron in the soil, biological oxidation, or polymerization.

General Safety Aspects. Laboratory work with hydrogen cyanide should be carried out only in a well-ventilated fume hood. Special safety equipment such as air masks, face masks, plastic aprons, and rubber gloves should be used. A chemical-proof suit should be available for emergency. Where hydrogen cyanide is handled inside a building, suitable ventilation must be provided. The people involved should be thoroughly trained in first aid. The most important rule when working with hydrogen cyanide is never to work alone. A second person must be in view at all times about 9 to 10 m away, must be equipped to make a rescue, and must be trained in first aid for hydrogen cyanide exposure.

Besides toxicity, hydrogen cyanide presents other hazards. Hydrogen cyanide undergoes an exothermic polymerization at conditions of pH 5 to 11. This polymerization can become explosively violent, especially if confined. The reaction is between hydrogen cyanide and cyanide ions, so the presence of water and heat contribute to the onset of this polymerization. Stored hydrogen cyanide should contain less than 1 wt % water, should be kept cool, and should be inhibited with sulfuric, phosphoric, or acetic acid. Manufacturers recommend a maximum of 90-day storage even for inhibited hydrogen cyanide.

Explosively violent hydrolysis can occur if an excess of a strong acid (H_2SO_4, HNO_3, or HCl) is added to hydrogen cyanide. Because of its low boiling point, hydrogen cyanide can be a fire and explosion hazard.

SODIUM CYANIDE

Sodium cyanide, NaCN, is a white cubic crystalline solid commonly called white cyanide.

Sodium cyanide is made by the neutralization or wet process in which liquid hydrogen cyanide and sodium hydroxide solution react and water is evaporated. The resulting crystals are briquetted or made into granular form. The principal applications of sodium cyanide are gold and silver extraction, electroplating, synthesis of iron blues, and synthesis of a large number of chemicals.

Properties

Physical Properties. The physical properties of sodium cyanide are listed in Table 2.

Table 2. Physical Properties of Sodium Cyanide

Property	Value
molecular weight	49.015
melting point, °C	562
boiling point, °C	1530
density of 30% NaCN solutions, at 25°C, g/mL	1.150
heat capacity[a], 25–72°C, J/(g·K)[b]	1.40
heat of fusion, J/g[b]	179
heat of formation, ΔH°_f, NaCN(c), J/mol[b]	-89.9×10^3
heat of solution,[c], ΔH_{soln}, J/mol[c]	−1548

[a] The heat capacity of sodium cyanide has been measured between 100 and 345 K.
[b] To convert J to cal, divide by 4.184.
[c] In 200 mol H_2O.

Sodium cyanide is soluble in liquid ammonia. At 15°C, 100 g anhydrous methanol dissolves 6.44 g anhydrous sodium cyanide; at 67.4°C, it dissolves 4.10 g. Sodium cyanide is slightly soluble in formamide, ethanol, methanol, SO_2, furfural, and dimethylformamide.

Sodium chloride and sodium cyanide are isomorphous and form an uninterrupted series of mixed crystals.

Chemical Properties. When heated in a dry CO_2 atmosphere, sodium cyanide fuses without much decomposition. A brown-black color appears when water vapor and CO_2 are present at temperatures of 100°C below the fusion point. This color is presumably from hydrogen cyanide polymer.

In the presence of a trace of iron or nickel oxide, rapid oxidation occurs when cyanide is heated in air, first to cyanate and then to carbonate. Case hardening of steels using a sodium cyanide molten bath depends on these reactions, wherein the active carbon and nitrogen are absorbed into the steel surface; hence the names carburizing and nitriding.

When sodium cyanide and sodium hydroxide are heated in the absence of water and oxygen above 500°C, sodium carbonate, sodium cyanamide, sodium oxide, and hydrogen are produced. In the presence of small amounts of water at 500°C decomposition occurs with the formation of ammonia and sodium formate, and the latter is converted into sodium carbonate and hydrogen by the caustic soda. In the presence of excess oxygen, sodium carbonate, nitrogen, and water are produced.

Molten sodium cyanide reacts with strong oxidizing agents such as nitrates and chlorates with explosive violence. In aqueous solution, sodium cyanide is oxidized to sodium cyanate by oxidizing agents such as potassium permanganate or hypochlorous acid. The reaction with chlorine in alkaline solution is the basis for the treatment of industrial cyanide waste liquors.

Sodium cyanide, when fused with sulfur or a polysulfide, is converted into sodium thiocyanate; this compound is also formed when a solution of sodium cyanide is boiled with sulfur or a polysulfide. A solution of sodium cyanide shaken with freshly precipitated ferrous hydroxide is converted to a ferrocyanide.

Aqueous solutions of sodium cyanide are slightly hydrolyzed at room temperature. At temperatures above 50°C, irreversible hydrolysis to formate and ammonia becomes important.

Hydrogen cyanide is a weak acid and can readily be displaced from a solution of sodium cyanide by weak mineral acids or by reaction with carbon dioxide, eg, from the atmosphere; however, the latter takes places at a slow rate.

In the presence of oxygen, aqueous sodium cyanide dissolves most metals in the finely divided state, with the exception of lead and platinum. This is the basis of the MacArthur process for the extraction of gold and silver from their ores.

Economic Aspects

Sodium cyanide is sold as granular or powder, pillow-shaped briquettes of 15-g and 30-g sizes, tablets of 30 g, and 30% aqueous solution. Sodium cyanide is packed in mild steel or fiber drums and in 1.4-t Flo-bins. Dry sodium cyanide is also shipped in wet-flo tank cars and trucks of up to 32 t net. At destination, water is circulated through the wet-flo car or trailer to dissolve the dry sodium cyanide at delivery. This type of shipment reduces freight costs and reduces environmental risks compared with 30% aqueous solution shipment. Safety regulations are imposed by the various shipping lines and by the countries in which cyanide is transported.

Health and Safety Factors

Handling, storage, and the use of alkali metal cyanides must be carried out by trained people. Most serious injuries and fatalities have been caused by inadvertently mixing these cyanides with acids, thereby releasing hydrogen cyanide. The present threshold limit value for 8-h exposure to cyanide dust is 5 mg/m^3 calculated as CN. Cyanide salts also must be protected from large concentrations of carbon dioxide to avoid hydrogen cyanide liberation. Carbon dioxide fire extinguishers should not be used. Cyanide salts as solids or solutions must be stored in tightly closed containers that must be protected from corrosion or damage.

Rubber gloves should be worn when handling dry salts. In addition, the following protective items should be used with solution or dusty salts: protective sleeves, aprons, shoes, boots or overshoes made of rubber, chemical safety goggles, full-face shield, and filter-type respirator (where dust is present). Cyanide spills should be flushed to a contained area where treatment to destroy the cyanide can be carried out. In the event that cyanide salts or solutions contact the eyes, they should be flushed for 15 min with a copious, gentle flow of water followed by immediate medical attention. Eating, smoking, and chewing should be forbidden in areas where cyanide salts are handled. Employees should be required to wash carefully after working with cyanide salts and before eating, smoking, or chewing.

POTASSIUM CYANIDE

Potassium cyanide, KCN, a white crystalline, deliquescent solid, was initially used as a flux, and later for electroplating, which is the single greatest use in the 1990s. With the decline in the use of alkali cyanides for plating the demand for potassium cyanide continues to decline. The total U.S. production in 1993 was estimated at about 300 t.

Commercial potassium cyanide made by the neutralization or wet process contains 99% KCN; the principal impurities are potassium carbonate, formate, and hydroxide. To prepare 99.5 + %KCN, high quality hydrogen cyanide and KOH must be used.

Properties

Physical Properties. The physical properties of potassium cyanide are given in Table 3. Unlike sodium cyanide, potassium cyanide does not form a dihydrate.

The solubility of potassium cyanide in nonaqueous solvents is as follows: in anhydrous liquid ammonia, 4.55 g/100 g NH_3 at −33.9°C; 4.91 g/100 g methanol at 19.5°C; 0.57 g/100 g ethanol at 19.5°C; 146 g/L solution in formamide at 25°C; 41 g/100 g hydroxylamine at 17.5°C; 24.24 g/100 g glycerol of specific gravity 1.2561 at 15.5°C; 0.73 g/L solution in phosphorus oxychloride at 20°C; 0.017 g/100 g liquid sulfur dioxide at 0°C; and 0.22 g/100 g dimethylformamide at 25°C.

At room temperature, potassium cyanide has fcc crystal structure.

Chemical Properties. Potassium cyanide is readily oxidized to potassium cyanate by heating in the presence of oxygen or easily reduced oxides, such as those of lead or tin or manganese dioxide, and in aqueous solution by reaction with hypochlorites or hydrogen peroxide.

Dry potassium cyanide in sealed containers is stable for many years. An aqueous solution of potassium cyanide is slowly converted to ammonia and potassium formate.

Many reactions can be carried out between potassium cyanide and organic compounds with the alkalinity of the KCN acting as a catalyst; these reactions are analogous to reactions of sodium cyanide. The reactions of potassium cyanide with sulfur and sulfur compounds are also

Table 3. Physical Properties of Potassium Cyanide

Property	Value
molecular weight	65.11
melting point, °C	634
density, g/mL	
cubic	1.55
orthorhombic at −60°C	1.62
specific heat, 25–72°C, J/g[a]	1.01
heat of fusion, J/mol[a]	14.7×10^3
heat of formation, $\Delta H°_f$, J/mol[a]	-113×10^3
heat of solution, ΔH_{soln}, J/mol[a]	−12550
solubility at 25°C, g/100 g H_2O	71.6

[a] To convert J to cal, divide by 4.184.

analogous to those of sodium cyanide. Potassium cyanide is reduced to potassium metal and carbon by heating it out of contact with air in the presence of powdered magnesium. Beryllium, calcium, boron, and aluminum act in a similar manner. Malonic acid is made from monochloroacetic acid by reaction with potassium cyanide, followed by hydrolysis. The acid and the intermediate cyanoacetic acid are used for the synthesis of polymethine dyes, synthetic caffeine, and for the manufacture of diethyl malonate, which is used in the synthesis of barbiturates. Most metals dissolve in aqueous potassium cyanide solutions in the presence of oxygen to form complex cyanides.

OTHER CYANIDES

Lithium, Rubidium, and Cesium Cyanides

Lithium cyanide, rubidium cyanide, and cesium cyanide are white or colorless salts, isomorphous with potassium cyanide. In physical and chemical properties these cyanides closely resemble sodium and potassium cyanide. As of this writing these cyanides have no industrial uses.

All of these alkali metal cyanides may be prepared by passing hydrogen cyanide into an aqueous solution of the hydroxide or by precipitating a solution of barium cyanide with lithium, rubidium, or cesium sulfate. A product with fewer contaminants may be obtained by the reaction of the base in absolute alcohol or dry ether with anhydrous hydrogen cyanide. In another method of preparation, a suspension of rubidium, cesium, or lithium metals in anhydrous benzene is treated with anhydrous hydrogen cyanide, and the benzene subsequently removed by evaporation under reduced pressure.

These cyanides are all soluble in water.

Lithium cyanide melts at 160°C. In the fused state the specific gravity at 18°C is 1.075. It is highly hygroscopic. Rubidium cyanide is not hygroscopic and is insoluble in alcohol or ether. Cesium cyanide is highly hygroscopic.

Ammonium Cyanide

Ammonium cyanide, NH_4CN, a colorless crystalline solid, is relatively unstable, and decomposes into ammonia and hydrogen cyanide at 36°C. Ammonium cyanide reacts with ketones (qv) to yield aminonitriles. Reaction of ammonium cyanide with glyoxal produces glycine. Because of its unstable nature, ammonium cyanide is not shipped or sold commercially.

Ammonium cyanide may be prepared in solution by passing hydrogen cyanide into aqueous ammonia at low temperatures. It may also be prepared from barium cyanide and ammonium sulfate, or calcium cyanide with ammonium carbonate.

Calcium Cyanide

Crude calcium cyanide, about 48 to 50 eq % sodium cyanide, is the only commercially important alkaline-earth metal cyanide, and output tonnage has been greatly reduced. This product, commonly called black cyanide, is marketed in flake form as a powder or as cast blocks under the trademarks Aero and Cyanogas of the American Cyanamid Company.

Physical and Chemical Properties. Because of decomposition, the melting point of calcium cyanide can only be estimated by extrapolation to be 640°C.

Calcium cyanide diammoniate, $Ca(CN)_2 \cdot 2NH_3$, is formed in liquid ammonia by reaction of calcium hydroxide or nitrate with ammonium cyanide. Deammoniation under heat and high vacuum yields calcium cyanide, a white powder, which is readily hydrolyzed to hydrogen cyanide.

Aqueous solutions of calcium cyanide prepared even at low temperature turn yellow or brown owing to the formation of HCN polymer. Calcium cyanide hydrolyzes readily.

Ferrocyanides are produced by reaction of ferrous salts. With sulfur in aqueous medium, calcium cyanide forms calcium thiocyanate.

Manufacture. Calcium cyanide is made commercially from lime, CaO, coke, and nitrogen. The reactions are carried out in an electric furnace.

Safety Precautions. Precautions similar to those used for sodium cyanide should be used for black cyanide.

Uses. The extraction or cyanidation of precious metal ores was the first, and is still the largest, use for black cyanide.

LAWRENCE D. PESCE
E. I. du Pont de Nemours & Co., Inc.

R. J. Cicerone and R. Zellner, *J. Geophys. Res.* **88**(C15), 10689 (1983).

D. Hasenberg, *HCN Synthesis on Polycrystalline Platinum and Rhodium,* dissertation, University of Minnesota, 1984.

J. L. Huiatt and co-workers, eds., *Proceedings of a Workshop—Cyanide From Mineral Processing,* University of Utah, Salt Lake City, 1982.

Sodium Cyanide Material Safety Data Sheet, E.I. du Pont de Nemours & Co., Inc., Wilmington, Del., 1990.

CYANINE DYES

This large class of dyes shows absorptions that cover the ultraviolet to the infrared region and, as a group, cover a wider span of the spectrum than any other dye class. The cyanine dyes are also among the oldest known class of synthetic dyes; the first dye was discovered in 1856. The great usefulness of cyanines was discovered in photography. Important reasons for the cyanines' prominence as photographic spectral sensitizers include *(1)* high light absorption per molecule, coupled in many cases with a single absorption band in the visible or infrared spectral region, which gives very color-selective absorption of light; *(2)* a tendency to form dye aggregates that have even narrower, more color-selective absorptions than the monomeric dyes themselves; and *(3),* high chemical and photochemical reactivity for dyes adsorbed to silver halides, which leads to efficient participation in photographic sensitization processes.

The cyanine class of dyes is also useful in biological, medical, laser, and electro-optic applications. Dyes marketed as Povan and Dithiazanine are useful anthelmintics, and Indocyanine Green is an infrared-absorbing tracer for blood-dilution medical diagnoses. "Stains-All" is a well-studied biological stain, and Merocyanine 540's photochemotherapeutic activity is known in some detail. Many commercially available red and infrared laser dyes are cyanines.

The cyanine dye literature has been reviewed extensively. References prior to 1980 are best for the synthetic heterocyclic chemistry of cyanine and related dyes. Later reviews (1977 to date) combine syntheses with compilations of physical and photophysical data.

Well-developed synthetic methods allow cost-effective manufacture of cyanines for commercial applications as well as a high degree of dye-structure design for new and innovative studies, such as solar cells, electrophotography, Langmuir-Blodgett nonlinear optical layers, and photoreceptors for processes activated by infrared solid-state lasers. Thus, tailoring the many characteristics of a dye is a well-practiced art in the cyanine class. Combinations of heterocycles, substituents, and chromophore lengths can yield a series of dye structures having parallel shifts in oxidation-reduction potentials almost independent of absorption wavelength. Steric features of substitutents either enhance or decrease aggregation. Solubility in either aqueous or hydrocarbon solvents can be provided by other substituents, almost independently from those that change redox potentials or steric properties. Controlling the number of conformations of the polymethine chain, achieved by several synthetic routes, is important to enhanced infrared absorption strength.

Color and Constitution

The color and constitution of cyanine dyes may be understood through detailed consideration of their component parts, ie, chromophoric systems, terminal groups, and solvent sensitivity of the dyes. Resonance theories have been developed to accommodate significant trends very successfully. For an experienced dye chemist, these are useful in the design of dyes with a specified color, band shape, or solvent sensitivity. More recently, quantitative values for reversible oxidation–reduction potentials have allowed more complete correlation of these dye properties with organic substituent constants.

Chromophoric Systems. The primary types of chromophores for cyanine dyes are the amidinium-ion system (A), the carboxyl-ion system (B), and the dipolar amidic system (C). For each system two extreme resonance structures are shown in (A), (B), and (C) where the formal charges are located at the ends of the chromophore. Intermediate resonance structures, with the charges closer to the center of the chromophore or with additional dipoles, are less important in the resonance picture of dyes. However, structural changes that favor intermediate forms have significant effects on the color of symmetrical dyes containing (A). For the amidic dyes (C), structural features stabilizing both neutral and dipolar extreme resonance forms in an equivalent manner give dyes absorbing at longer wavelengths.

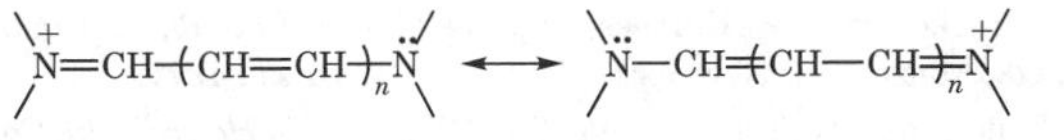

(A)

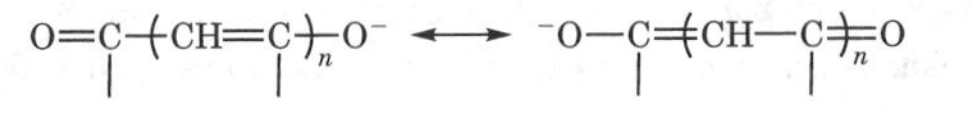

(B)

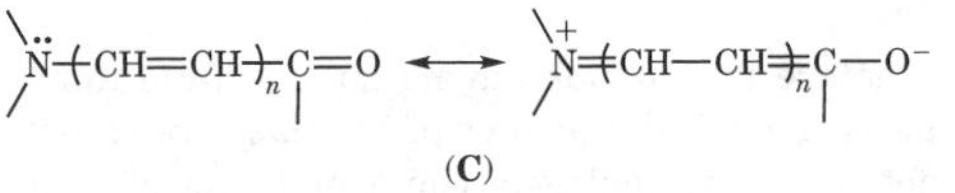

(C)

The important characteristics that influence the absorption wavelengths for these dyes are the length of the conjugated chain and the nature of the terminal group.

Terminal Groups. The cyanines (A), oxonols (B), and merocyanines (C) were originally considered as polymethine dyes with two heterocyclic terminal groups. Hemicyanines were defined as dyes with one heterocyclic and one noncyclic terminal group. Currently, dyes without heterocyclic terminal groups are designated as either cyanines or polymethine dyes. In fact, almost any atom or group of atoms can function as a terminal group for dyes if the nitrogens and oxygens in the primary chromophores (A), (B), and (C) are replaced by electronically equivalent atoms. Dyes from novel terminal groups are quite numerous. However, the fundamental concepts and perhaps the largest class of useful cyanines are derived from dyes with heterocyclic terminal groups. The heterocycles are of two principal types: basic or electron-donating, and acidic or electron-accepting.

Solvent Sensitivity. There is a large influence of solvents on absorption spectra for dyes, particularly for the merocyanines, (C). Merocyanines with widely different terminal groups illustrate the general pattern of solvent effects on the absorption maxima and peak intensities: Weakly polar dyes are red-shifted (longer wavelength) and show increased extinction coefficients (ϵ_{max}) as solvent polarity increases; and highly polar dyes are blue-shifted and show decreased ϵ_{max} values for increasingly polar solvents. The large effects of the environment (solvent) around a dye on the absorption spectra led to the synthesis of hundreds of dyes to clearly document relations between structure and spectra.

Both merocyanine and cyanine dye classes can exhibit solvent sensitivity. For dyes with long chromophoric chains of =CH— (methine) groups, the conformation and thus the absorption of charged dyes may change as a function of solvent.

Synthesis of Cyanines and Related Dyes

Nonoxidative syntheses are the most versatile and employ varied combinations of nucleophilic and electrophilic reagents. Reviews dating from between 1964 and 1977 tabulate these reagents and their uses in dye syntheses.

Nonoxidative Syntheses. Dye syntheses are characterized as condensations, "two intermediates characterized as condensations, "two intermediates reacting under suitable conditions with elimination of some simple molecule" (Brooker in *Theory of the Photographic Process*). Nucleophilic methylene bases are derived from the deprotonation of quaternary salts like 3-ethyl-2-methylbenzothiazolium iodide. The methylene bases react with a variety of electrophilic reagents to form dyes in which the nucleophilic reagent is the terminal heterocyclic group for the chromophore. The quaternary salts are also converted with ortho esters or anilides to generally useful electrophilic reagents, eg, the intermediate 3-ethyl-2-[2-(acetylphenylamino) ethenyl]benzothiazolium iodide.

Ring-Closure Reactions. Some interesting dyes are prepared by ring-closure reactions at or near the dye-forming step of a synthetic sequence. The structural identity of thiacyanine was originally established by the reaction of diethyl malonate and *o*-aminothiophenol.

Highly fluorescent, rigidized dyes form readily. Quaternary salts (with reactive *N*-ethylformyl groups) are formed readily from acrolein and nitrogen heterocycles. Direct formation of dyes from the uncyclized quaternary salt lead to the thiacarbocyanine. The reactive *N*-ethylformyl groups in this dye can be ring-closed to give a completely rigid and highly fluorescent thiacarbocyanine.

Photophysical Properties

The Electronic Structure of Sensitizers. Large conjugated molecules such as the cyanines and related dyes are well described by general quantum theories as well as by resonance concepts. Both prove useful in the design and understanding of dyes. More recently, the alternating pi-electronic charge characteristic of a polymethine chain was used to define a "polymethine state," as distinct from olefinic and aromatic conjugated systems. Considering the complete electronic structure of dye molecules, the filled orbitals of the ground states contain sigma (ς), pi (π), and lone-pair (n) electrons. Antibonding orbitals are typically pi* (π^*) and sigma* (ς*). The long-wavelength transitions for cyanine dyes are generally high-extinction transitions involving the π-electrons and show long-axis polarizations in cyanine dyes, as determined by dichroic absorption of polarized light (stretched polymer films with dye) and by fluorescence polarization. Shorter-wavelength transitions are polarized both parallel and perpendicular to the long axis.

Most cyanines show prominent, hypsochromic vibrational shoulders at shorter wavelength associated with the long-wavelength electronic transition. Excited-state properties of the cyanine and related dyes are complex. Most cyanine dyes exhibit small Stokes shifts for fluorescence maxima.

Aggregation. The most notable spectral property of the cyanine dye class is the ability to self-associate or aggregate. For the more symmetrical cationic dyes, these aggregates can exhibit exceptionally narrow,

Table 1. Acute Toxicity Data for Cyanine Dyes[a]

Dye name	Oral LD_{50}		Dermal LD_{50}		Intraperitoneal LD_{50}	
	Mouse, mg/kg	Rat, mg/kg	Mouse, mg/kg	Guinea pig, mg/kg	Mouse, mg/kg	Rat, mg/kg
indocyanine green					60[b,c]	
cryptocyanine	25			>50	1[b]	
DODC iodide	25			>1000	5–10	
DTTC iodide	400–800		>100	>1000	1	
HIDC iodide		178 (M) >100 (F)		>1000		
HPTS	2200	800–1600		1000		
IR-125	>100					700
Q-switch 1[d]				>1000	1600–3200	

[a] Interchemical comparisons of acute toxicity (LD_{50}) are valid only for the same species and route of administration. Except where noted in the table, all data are previously unpublished, Health and Environment Laboratories, Eastman Kodak Company, Rochester, N.Y.

[b] Material Safety Data Sheet, Sigma-Aldrich Corporation, Milwaukee, Wis.

[c] This LD_{50} was derived from an intravenous injection and not an intraperitoneal injection.

[d] For neodymium laser.

high extinction absorptions at longer wavelength than the monomeric dye absorption. Such aggregates, termed J-aggregates for Jelley, who first observed them, provide exceptionally color-selective and efficient sensitization for color imaging systems. Hypsochromically shifted aggregates (dimers, trimers, H-aggregates) also occur, but their absorptions are typically as broad as the monomer.

The π-systems of dyes can overlap strongly if the long dimensions of the chromophores are parallel and the methine chain/heterocycles are stacked face-to-face (as in a deck of cards) and closely spaced.

In a more general description of the relation between these structures and their tendency to aggregate, compact dyes were defined as structures that were planar but with tightly packed substituents from a steric hindrance viewpoint; these dyes aggregated readily. Less hindered loose dyes (perhaps having several conformations) or more hindered crowded dyes (probably non-planar) do not aggregate as well. Of course, other groups such as aromatic ring or nitrogen substituents also influence aggregation, by changing electronic and solubility properties.

Merocyanine and hemicyanine dyes are often components of thin-film multilayers such as Langmuir-Blodgett films, where tight packing of dye molecules within a layer reinforces the self-association tendencies of the dyes. Chromophores, which orient in the films due to the method of preparation, can exhibit very strong second-order nonlinear optical properties. Symmetrical cyanine dyes in polymer films show large third-order nonlinear optical properties associated with their large conjugated π-electron systems.

Reactivity of Cyanine and Merocyanine Dyes

Photoreactions. Photooxidation and photoreduction of dyes have been less studied for the cyanines and merocyanines for noncyanine dye classes. Perhaps this is due to the low singlet-triplet intersystem crossing yields and the high propensity of cyanines to undergo rapid $S_1 \rightarrow S_0$ radiationless deactivation. Low quantum yield photofading reactions in solutions of cyanine dyes are observed under both oxidative conditions (oxygen, reducible metal ions) and reductive conditions (ascorbic acid, gelatin).

Chemical Reactions of Dyes. Decolorization is important for cyanines used in imaging materials. Understanding decolorization provides clues to dye reactions that may cause degradation of imaging materials during preparation and storage. For many dyes, protonation of the methine chain occurs readily and reversibly. Highly basic carbocyanine dyes like those from benzimidazole protonate so readily that this provides a practical decolorization method.

Innovative Uses of Dyes

Recent reviews cite both patents and scientific articles that indicate a widening spectrum of uses for the cyanine dye classes (and other dyes) in innovative technological areas. A long-chain indolenine-type cyanine dye has been described as the infrared sensitizer in optical disk memories. Several hundred compounds have been applied as voltage-sensitive, optical probes of neuronal activity in the brain. Other dyes are designed to improve detection of ribonucleic acid (RNA) when a preparation is stained with a dye and to improve bioassays by reaction of isothiocyanate-substituted, near-infrared dyes with amino moieties or proteins.

Optical properties of cyanines can be useful for both chiral substituents/environments and also third-order nonlinear optical properties in polymer films.

Suppliers. Suppliers of cyanine dyes include manufacturers of other specialty organic and photographic chemicals: Aldrich Chemical Company (Milwaukee, Wisconsin), Eastman Chemical Company (Kingsport, Tennessee), Japanese Institute for Photosensitizing Dyes (Okayama, Japan), Molecular Probes (Eugene, Oregon), NK Dyes (Japan), Pfaltz and Bauer (Stamford, Connecticut), and Riedel de-Haen (Karlsruhe, Germany). More importantly, these firms provide sources of generally useful reagents which, in two or three synthetic steps, lead to many of the commonly used cyanine dyes.

Health and Safety Information

A wide variety of structures exist in the cyanine, merocyanine, and oxonol classes of dyes. Properties that may affect toxicity vary widely also. These include solubility, propensity to be oxidized or reduced, aggregation tendency, and diffusion through membranes. Specific acute toxicity data are listed in Table 1, and the LD_{50} data vary widely with the test used.

DAVID M. STURMER
Eastman Kodak Company

L. G. S. Brooker, in C. E. K. Mees, ed., *The Theory of the Photographic Process*, 1st ed., Macmillan, New York, 1942, p. 987; 3rd ed., 1977.

F. M. Hamer, in A. Weissberger, ed., *The Chemistry of Heterocyclic Compounds*, Vol. 18, Wiley-Interscience, New York, 1964.

D. M. Sturmer, in E. C. Taylor and A. Weissberger, eds., *The Chemistry of Heterocyclic Compounds*, Vol. 30, Wiley-Interscience, New York, 1977.

M. Matsuoka, ed., *Infrared Absorbing Dyes*, Plenum Press, New York, 1990.

CYANOCARBONS

Cyanocarbons are compounds having such a large number of cyano groups that the chemical reactions of the class are essentially new in kind and not shared by analogous compounds free of such groups.

Tetracyanoethylene. Tetracyanoethylene ethenetetracarbonitrile (TCNE) has a high positive heat of formation (Table 1), but is stable to shock. It has good thermal stability and can thus be sublimed unchanged through a tube at 600°C. Tetracyanoethylene resists oxidation, but once ignited in oxygen it burns with an intensely hot flame that may be above 4000 K and is, at any rate, hotter than the oxygen–acetylene flame.

Tetracyanoethylene is colorless but forms intensely colored complexes with olefins or aromatic hydrocarbons. TCNE is conveniently prepared in the laboratory from malononitrile by debromination of dibromomalononitrile with copper powder. The debromination can also be done by pyrolysis at ca 500°C.

Tetracyanoethylene undergoes two principal types of reactions, addition to the double bond and replacement of a cyano group.

Hexacyanobutadiene. Hexacyanobutadiene, 1,3-butadiene-1,1,2,-3,4,4-hexacarbonitrile, is prepared in good yield by a two-step process from the disodium salt of tetracyanoethane. It is like TCNE is forming colored π-complexes and an anion radical.

Table 1. Physical Properties of Tetracyanoethylene

Property	Value
mol wt	128.1
mp, °C	200–202
bp, °C	223
density, g/mL	1.318
heat of combustion, kJ/mol[a]	−3022
heat of formation, kJ/mol[a]	628

[a] To convert kJ/mol to kcal/mol, divide by 4.184.

Tetracyanoquinodimethane. Tetracyanoquinodimethane, 2,2′-(2,5-cyclohexadiene-1,4-diylidene)bispropanedinitrile (TCNQ), is prepared by condensation of 1,4-cyclohexanedione with malononitrile to give 1,4-bis(dicyanomethylene)cyclohexane, which is oxidized with bromine. It resembles tetracyanoethylene in that it adds reagents such as hydrogen, sulfurous acid, and tetrahydrofuran to the ends of the conjugated system of carbon atoms; suffers displacement of one or two cyano groups by nucleophilic reagents such as amines or sodiomalononitrile; forms π-complexes with aromatic compounds; and takes an electron from iodide ion, copper, or tertiary amines to form an anion radical.

Tetracyanomethane. Tetracyanomethane, methanetetracarbonitrile, is prepared by heating silver tricyanomethanide in liquid cyanogen chloride. It is a very strong cyanating agent.

Hexacyanoethane. Hexacyanoethane, ethanehexacarbonitrile, is quite unstable and readily decomposes to TCNE and cyanogen (NC–CN). It is prepared as follows:

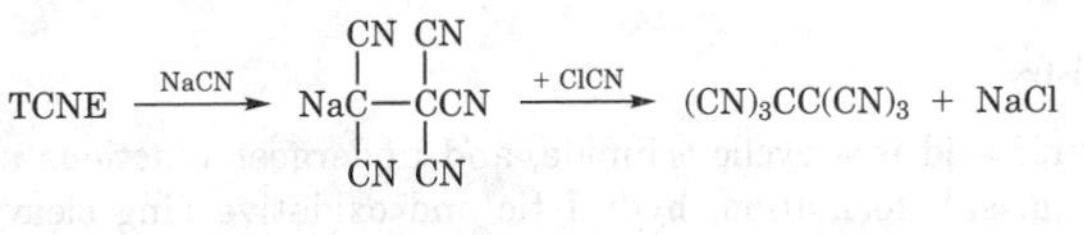

Dicyanoacetylene. Dicyanoacetylene, 2-butynedinitrile, is obtained from dimethyl acetylenedicarboxylate by ammonolysis to the diamide, which is dehydrated with phosphorus pentoxide. It burns in oxygen to give a flame with a temperature of 5260 K, the hottest flame temperature known. Alcohols and amines add readily to its acetylenic bond.

Hexacyanobenzene. Hexacyanobenzene, benzenehexacarbonitrile, is prepared from 2,4,6-trifluorobenzene-1,3,5-tricarbonitrile by substitution with calcium cyanide. It forms colored π-complexes with aromatic hydrocarbons.

Tetracyanobenzoquinone. Tetracyanobenzoquinone, 3,6-dioxo-1,4-cyclohexadiene-1,2,4,5-tetracarbonitrile, is a remarkable strong oxidizing agent for a quinone. It is a stronger π-acid than TCNE because it forms more deeply colored π-complexes with aromatic hydrocarbons.

Cyanocarbon Acids. These acids are organic molecules that contain a plurality of cyano groups and are readily ionized to hydrogen ions and resonance-stabilized anions. Many of these acids rival mineral acids in strength and are usually isolable only as salts with metal or ammonium ions.

Oxacyanocarbons. Tetracyanoethylene oxide, oxiranetetracarbonitrile, is the most notable member of the class of oxacyanocarbons. It is made by treating TCNE with hydrogen peroxide in acetonitrile. In reactions unprecedented for olefin oxides, it adds to olefins to form 2,2,5,5-tetracyanotetrahydrofurans via cleavage of the ring C–C bond.

With pyridine, reaction takes place at the nitrogen rather than at a double bond, and an ylid is formed. Sulfides react similarly to give sulfilidenes and carbonyl cyanide. This is the most convenient synthesis of carbonyl cyanide.

Thiacyanocarbons. The thiacyanocarbons, tetracyano-1,4-dithiin and tetracyanothiophens, are derived from "Bahr's Salt", the disodium salt of 2,3-dithiol-maleonitrile, prepared from carbon disulfide and sodium cyanide.

Azacyanocarbons. Hydrogen cyanide tetramer (*Z*-),2,3-diamino-2-butenedinitrile, an azacyanocarbon, is produced by Nippon Soda in pilot-plant quantities for development as a chemical intermediate.

Health and Safety Factors (Toxicology)

Unless specifically tested, all cyanocarbons should be considered as toxic as sodium cyanide or hydrogen cyanide. They should be used only in a fume hood, and rubber gloves should be worn.

OWEN W. WEBSTER
E. I. du Pont de Nemours & Co., Inc.

E. Ciganek, W. J. Linn, and O. W. Webster, in Z. Rapport, ed., *The Chemistry of the Cyano Group,* Interscience Publishers, London, 1970, pp. 423–638. An extensive review of cyanocarbon chemistry.

K. Wallenfels and co-workers, *Agnew. Chem. Int. Ed. Engl.* **15,** 261 (1976).

W. R. Hertler and co-workers, *Mol. Cryst. Liq. Cryst.* **171,** 205 (1989). Cyanocarbon history.

CYANOHYDRINS

A cyanohydrin is an organic compound that contains both a cyanide and a hydroxy group on an aliphatic section of the molecule. Cyanohydrins are usually α-hydroxy nitriles which are the products of base-catalyzed addition of hydrogen cyanide to the carbonyl group of aldehydes and ketones. The IUPAC name for cyanohydrins is based on the α-hydroxy nitrile name. Common names of cyanohydrins are derived from the aldehyde or ketone from which they are formed.

The outstanding chemical property of cyanohydrins is the ready conversion to α-hydroxy acids and derivatives, especially α-amino and α,β-unsaturated acids. Because cyanohydrins are primarily used as chemical intermediates, data on production and prices are not usually published. The industrial significance of cyanohydrins is waning as more direct and efficient routes to the desired products are developed.

Properties

Cyanohydrins are usually colorless to straw yellow liquids with an objectionable odor akin to that of hydrogen cyanide. Table 1 list physical properties of some common cyanohydrins.

Cyanohydrins can react either at the nitrile group or at the hydroxyl group.

Preparation

Cyanohydrins can be formed by the acid- or base-catalyzed reaction of hydrogen cyanide with an aldehyde or ketone, the displacement of bisulfite ion by cyanide ion on the bisulfite addition compounds of aldehydes and ketones, or the exchange of cyanide ion between a ketone cyanohydrin and an aldehyde to give the usually more stable aldehyde cyanohydrin.

Shipping, Storage, and Handling

Cyanohydrins should be stabilized with acid to pH 3–4 to prevent decomposition to hydrogen cyanide and carbonyl compound. When cyanohydrins are shipped, steel drums, carboys, tank cars, and barges are used. In general, cyanohydrins are combustible liquids and many decompose upon heating. They should be stored in a cool, dry place, preferably outside and separated from other storage. Containers should be protected against physical damage.

Health and Safety

Cyanohydrins are highly toxic by inhalation or ingestion, and moderately toxic through skin absorption. Special protective clothing should be worn and any exposure should be avoided. The area should be adequately ventilated. Immediate medical attention is essential in case of cyanohydrin poisoning.

Table 1. Physical Properties of Some Cyanohydrins

Name	Mol wt	Mp, °C	Bp,[a] °C	Specific gravity	n_D^{20}	Flash point, °C
formaldehyde cyanohydrin	57.06	<−72	119 at 3.2 kPa[b]	1.104	1.4117	
acetaldehyde cyanohydrin	71.03	−40	182–184, dec	0.988	1.4050	77
acetone cyanohydrin	85.10	−19	85 at 3.1 kPa[b]	0.927	1.3992	74
cyclohexanone cyanohydrin	121.17	29	109–113 at 1.2 kPa[b]	1.032	1.4576	60
benzaldehyde cyanohydrin	133.15	−10	170, dec	1.117	1.5315	
ethylene cyanohydrin	71.08	−46.2	228	1.059	1.4256	129
propylene cyanohydrin	85.11		207		1.4280	

[a] At 101.1 kPa[b] unless otherwise noted.
[b] To convert kPa to mm Hg, multiply by 7.5.

Specific Compounds

Formaldehyde Cyanohydrin. This cyanohydrin, also known as glycolonitrile, is a colorless liquid with a cyanide odor. It is soluble in water, alcohol, and diethyl ether. Equimolar amounts of 37% formaldehyde and aqueous hydrogen cyanide mixed with a sodium hydroxide catalyst at 2°C for one hour give formaldehyde cyanohydrin in 79.5% yield.

Acetaldehyde Cyanohydrin. This cyanohydrin, commonly known as lactonitrile, is soluble in water and alcohol, but insoluble in diethyl ether and carbon disulfide. Lactonitrile is used chiefly to manufacture lactic acid and its derivatives, primarily ethyl lactate. Lactonitrile is manufactured from equimolar amounts of acetaldehyde and hydrogen cyanide containing 1.5% of 20% NaOH at −10–20°C. The product is stabilized with sulfuric acid.

Acetone Cyanohydrin. This cyanohydrin, also known as α-hydroxyisobutyronitrile and 2-methyllactonitrile, is very soluble in water, diethyl ether, and alcohol, but only slightly soluble in carbon disulfide or petroleum ether. Acetone cyanohydrin is the most important commercial cyanohydrin as it offers the principal commercial route to methacrylic acid and its derivatives, mainly methyl methacrylate (see METHACRYLIC ACID AND DERIVATIVES). The principal U.S. manufacturers are Rohm and Haas Company, DuPont, CyRo Industries, and BP Chemicals.

Acetone cyanohydrin is manufactured by the direct reaction of hydrogen cyanide with acetone, catalyzed by base, generally in a continuous process.

Benzaldehyde Cyanohydrin. This cyanohydrin, also known as mandelonitrile, is a yellow, oily liquid, insoluble in water, but soluble in alcohol and diethyl ether. Mandelonitrile is a component of the glycoside amygdalin, a precursor of laetrile found in the leaves and seeds on most *Prunus* species (plum, peach, apricot, etc). It is commercially prepared from benzaldehyde and hydrogen cyanide.

Ethylene Cyanohydrin. This cyanohydrin, also known as hydracrylonitrile or glycocyanohydrin, is a straw-colored liquid miscible with water, acetone, methyl ethyl ketone, and ethanol, and is insoluble in benzene, carbon disulfide, and carbon tetrachloride. Ethylene cyanohydrin differs from the other cyanohydrins discussed here in that it is a β-cyanohydrin. It is formed by the reaction of ethylene oxide with hydrogen cyanide.

MICHAEL S. CHOLOD
Rohm and Haas Company

V. Migrdichian, *The Chemistry of Organic Cyanogen Compounds*, Reinhold Publishing Co., New York, 1947, Chapt. 9.

CYANURIC AND ISOCYANURIC ACIDS

The chemistry of cyanuric acid is diversified because of multiple reaction sites. *N*-Chlorination of cyanuric acid produces chloroisocyanurates that are widely used as disinfectants, sanitizers, and bleaches. The triallyl- and tris(2-hydroxyethyl) derivatives are employed as cross-linking and curing agents, and tris(2,3-epoxypropyl) isocyanurate is used in weather-resistant powder coatings. Melamine cyanurate finds significant use as a fire retardant in plastics. Considerable interest has developed in the use of cyanuric acid for reduction of nitrogen oxides in exhaust gases from combustion of oil, gas, and coal.

Properties

Cyanuric acid (CA) is an odorless, white, crystalline solid that does not melt up to 330°C; at higher temperatures it sublimes and dissociates to isocyanic acid (HNCO). The vapor pressure of CA at 167–200°C is given by $\log P(kPa) = -5552/T + 11.54$, where T is in K. The equilibrium constants for dissociation of CA in the gas phase are 0.02 and 0.76 MPa^2 (1.9 and 74.3 atm^2) at 365°C and 434°C, respectively. Below 350°C, isocyanic acid tends to polymerize, whereas above 350°C it can form by depolymerization of CA.

Cyanuric acid is a titrable weak acid ($pK_{a1} = 6.88$, $pK_{a2} = 11.40$, $pK_{a3} = 13.5$). Cyanuric acid is only slightly soluble (≤0.1%) at room temperature in common organic solvents such as acetone, benzene, diethyl ether, ethanol, and hexane. Solubility is significant in basic nitrogen compounds (eg, dimethylformamide 7.2%) or unusual solvents such as DMSO (17.4%). Solubility in water is only 0.2% at 25°C but increases to 2.6% at 90°C and 10.0% at 150°C. In aqueous alkali (eg, NH_4OH, NaOH, KOH) solubility increases due to salt formation.

Chemistry

Cyanuric acid is a cyclic triimide, and undergoes reactions at N, O, or C, eg, salt formation, hydrolytic and oxidative ring cleavage, *C*-halogenation, *N*-halogenation, and alkylation. Reaction at nitrogen produces isocyanurates $R_3(NCO)_3$, whereas reaction at oxygen forms cyanurates $(RO)_3(NC)_3$. Mixed derivatives are possible, as in the case of the sodium salt of dichloroisocyanuric acid.

Organic Derivatives. Although numerous mono-, di-, and trisubstituted organic derivatives of cyanuric and isocyanuric acids appear in the literature, many are not accessible via cyanuric acid. Cyanuric chloride 2,4,6-trichloro-*s*-triazine, is generally employed as the intermediate to most cyanurates. Tri-substituted isocyanurates can also be produced by trimerization of either aliphatic or aromatic isocyanates with appropriate catalysts (see ISOCYANATES, ORGANIC). Virtually all of the organo derivatives of CA are produced by reactions characteristic of a cyclic imide, wherein isocyanurate nitrogen (frequently as the anion) nucleophilically attacks a positively polarized carbon of the second reactant.

Preparation

A convenient laboratory synthesis of high purity CA is hydrolysis of cyanuric chloride. On a commercial scale, CA is produced by pyrolysis of urea.

Manufacture

The majority of the cyanuric acid produced commercially is made via pyrolysis of urea (mp 135°C), primarily employing either directly or indirectly fired stainless steel rotary kilns.

Beside continuous horizontal kilns, numerous other methods for dry pyrolysis of urea have been described, eg, use of stirred batch or continuous reactors, ribbon mixers, ball mills, etc, heated metal surfaces such as moving belts, screws, rotating drums, etc, molten tin or its alloys, dielectric heating and fluidized beds (with preformed urea cyanurate). All of these modifications yield impure CA.

Economic Aspects

Monsanto and Olin, in the United States, are the world's largest bulk suppliers of cyanuric acid and chloroisocyanurates, ie, trichloroisocyanuric acid (TCCA), sodium dichloroisocyanurate (SDCC) and its dihydrate (SDCC-H), potassium dichloroisocyanurate (PDCC), and the double salt (TCCA · 4PDCC). They have a combined CA capacity of ~38,000 t/yr, 90% of which is converted into chlorinated derivatives. The main foreign producers of CA are Shikoku Kasei Kogyo Company, Nissan Chemical Industries Ltd., and Nippon Soda in Japan and Atochem in France. Swimming pool sales represent 76% of the total U.S. consumption.

Health and Safety

Acute toxicities (LD_{50} g/kg, rats) for CA and chloroisocyanurates are CA >5.0, SDCC 1.67, PDCC 1.22, and TCCA 0.75. Toxicological studies of CA and its chlorinated derivatives show that the compounds are safe for use in swimming pool and spa/hot tub disinfection, sanitizing, and bleaching applications when handled and used as directed. Most uses of chloroisocyanurates are regulated by the EPA under FIFRA.

Cyanuric acid is stable and relatively inert. Chloroisocyanurates are also stable when dry, uncontaminated, and kept away from fire or a source of high heat. They are, however, active chlorine compounds and thus are strong oxidizing agents; care must be exercised to prevent hazardous contamination.

Uses

Most of the CA produced commercially is chlorinated to produce SDCC, SDCC-H, PDCC, TCCA, and the double salt TCCA·4PDCC. These have become standard ingredients in formulations for scouring powders, household bleaches, institutional and industrial cleansers, automatic dishwasher compounds, and general sanitizers, and most importantly, in swimming pool and spa/hot tub disinfection.

J. A. WOJTOWICZ
Olin Corporation

J. Elguero and co-workers, in A. R. Katritzky and A. J. Boulton, eds., *Tautomerism of Heterocycles,* Suppl. 1, Academic Press, Inc., New York, 1976, pp. 138–139.

Pool/Spa Operators Handbook, National Swimming Pool Foundation, San Antonio, Tex., 1988.

C. E. Schildknecht, in J. I. Kroschwitz, ed., *Encyclopedia of Polymer Science and Engineering,* Vol. 4, John Wiley & Sons, Inc., New York, 1986, pp. 802–811; *Triallyl Isocyanurate,* product bulletin, Allied Chemical Corp., Morristown, N.J., 1973.

CYCLOHEXANE. See HYDROCARBONS.

CYCLOHEXANOL AND CYCLOHEXANONE

Cyclohexanol is a colorless, viscous liquid with a camphoraceous odor. It is used chiefly as a chemical intermediate, a stabilizer, and a homogenizer for various soap detergent emulsions, and as a solvent for lacquers and varnishes. Cyclohexanol was first prepared by the treatment of 4-iodocyclohexanol with zinc dust in glacial acetic acid, and later by the catalytic hydrogenation of phenol at elevated temperatures and pressures.

Cyclohexanone is a colorless, mobile liquid with an odor suggestive of peppermint and acetone. Cyclohexanone is used chiefly as a chemical intermediate and as a solvent for resins, lacquers, dyes, and insecticides. Cyclohexanone was first prepared by the dry distillation of calcium pimelate, $-OOC(CH_2)_5COO^-Ca^{2+}$, and later by Bouveault by the catalytic dehydrogenation of cyclohexanol.

Table 1. Properties of Cyclohexanol and Cyclohexanone

Property	Cyclohexanol	Cyclohexanone
	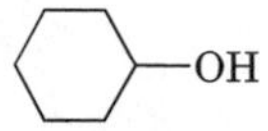	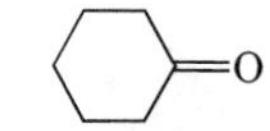
mp, °C	25.15	−47
bp, °C	161.1	156.7
d_4^{20}, g/mL	0.9493	0.9478
n_D^{25}	1.4648	
sp heat, 15–18°C, J/g[a]	1.75	1.81
viscosity, 25°C, mPa·s(= cP)	4.6	2.2
flash point, open cup, °C	67.2	54

[a] To convert J to cal, divide by 4.184.

Physical Properties

Important physical properties of cyclohexanol and cyclohexanone are shown in Table 1. Cyclohexanol is miscible in all proportions with most organic solvents, including those customarily used in lacquers. It dissolves many oils, waxes, gums, and resins.

Cyclohexanone is miscible with methanol, ethanol, acetone, benzene, *n*-hexane, nitrobenzene, diethyl ether, naphtha, xylene, ethylene glycol, isoamyl acetate, diethylamine, and most organic solvents. This ketone dissolves cellulose nitrate, acetate, and ethers, vinyl resins, raw rubber, waxes, fats, shellac, basic dyes, oils, latex, bitumen, kaure, elemi, and many other organic compounds.

Reactions

Cyclohexanol shows most of the typical reactions of secondary alcohols. Cyclohexanone shows most of the typical reactions of aliphatic ketones.

Economic Aspects

Estimated annual cyclohexanone production capacities were 665 × 10^3 t in 1992; the production is greater than 90% captive for caprolactam production. The annual cyclohexanol production is only 10 thousand metric tons. These production figures do not include KA-oil (cyclohexanol–cyclohexanone) production for adipic acid. Worldwide annual capacity for cyclohexanone is approximately 3.0 million metric tons, also primarily for caprolactam production. Projected new capacity for caprolactam could add 0.5 million metric tons worldwide in this decade.

Health and Safety Factors

Cyclohexanol is slightly toxic by the oral route of exposure and is slightly irritating to the skin. It can cause severe eye irritation and transient corneal injury.

The ACGIH threshold limit value (TLV), time-weighted average for an 8-h workday, 40-h workweek, was set at 50 ppm (~ 200 mg/m^3) with a notation for skin absorption.

Cyclohexanone has only slight toxicity by the oral, dermal, and inhalation routes of exposure. Liquid or vapor exposures may result in transient corneal injury. Primary irritation and defatting of the skin can result from substantial or prolonged contact with cyclohexanone. Exposure to high vapor concentrations can cause central nervous system (CNS) depression, an effect which can also occur after ingestion or repeated dermal exposure.

The time-weighted average OSHA permissible exposure limit (PEL), as well as the ACGIH threshold limit value (TLV), for cyclohexanone is 25 ppm (100 mg/m^3) with a notation for skin absorption.

The precautions usually observed when handling volatile solvents should be observed as a matter of course with cyclohexanone and cyclohexanol. These include adequate and proper ventilation, avoidance of prolonged breathing of vapor or contact of the liquid with the skin, avoidance of internal consumption, and protection of the eyes against splashing liquids.

WILLIAM B. FISHER
JAN F. VANPEPPEN
Allied Signal Inc.

U.S. Pat. 4,092,360 (May 30, 1978), J. F. VanPeppen and W. B. Fisher (to Allied Chemical Corp.).

Chemical Economics Handbook, SRI International, Menlo Park, Calif., Sept. 1990; Allied Signal Co., internal data, 1992.

1991–1992 Threshold Limit Values for Chemical Substances and Physical Agents and Biological Exposure Indices, American Conference of Government Industrial Hygienists, Cincinnati, Ohio, 1991.

Registry of Toxic Effects of Chemical Substances (RTECS), NIOSH Database, Cincinnati, Ohio, 1992.

CYCLOPENTADIENE AND DICYCLOPENTADIENE

Cyclopentadiene (CPD), C_5H_6, (1) and its more stable and available dimer form, dicyclopentadiene (DCPD), $C_{10}H_{12}$, (2) are two well-established and versatile chemical building blocks.

(1)

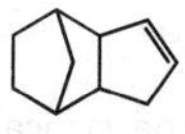

(2)

The monomer, CPD, obtained via cracking of the dimer, DCPD, and the dimer both have extensive uses. Cyclopentadiene is probably the most widely studied conjugated, cyclic diolefin system.

Physical Properties

Dicyclopentadiene exists in two stereoisomeric forms, the endo and exo isomers. Commercial DCPD, 3a,4,7,7a-tetrahydro-4,7-methano-1*H*-indene, is predominantly the endo isomer (exo:endo 6:953 by capillary gas chromatography). The dimer is the form in which CPD is sold commercially.

The physical properties of CPD and DCPD are given in Table 1.

Table 1. Physical Properties of Cyclopentadiene and Dicyclopentadiene

Physical properties	Cyclopentadiene	Dicyclopentadiene[a]
mol wt	66.1	132.2
bp, 101.3 kPa[b], °C	41.5	170[c]
mp, °C	−85	33.6
physical form	colorless liquid	colorless crystals
odor	sweet terpenic	camphoraceous
heat of combustion, kJ/mol[d]	2929	5767
heat of vaporization, kJ/mol[d]	29.3	38.5
heat of cracking, kJ/mol[d]		102.9
heat of fusion, kJ/mol[d]		2.1
specific heat, kJ/(kg·K)[d]		1.7
spontaneous ignition temp, °C		510
in oxygen		680
in air	640	
dielectric constant at 40°C	2.43	

[a] Endo isomer.
[b] To convert kPa to mm Hg, multiply by 7.5.
[c] Depolymerizes at boiling point to form two molecules of cyclopentadiene.
[d] To convert kJ to kcal, divide by 4.184.

Chemical Reactions

Cyclopentadiene contains conjugated double bonds and an active methylene group and can thus undergo a Diels-Alder diene addition reaction with almost any unsaturated compound. The number of its derivatives is extensive.

Source and Production

No commercial process for the sole, deliberate, synthetic production of cyclopentadiene exists. It is obtained as a by-product from thermal operations. In the carbonization of coal, the tar, light oil, and coke-oven gas contain cyclopentadiene and dicyclopentadiene in very low concentrations. Steam-cracking (thermal cracking in the presence of steam) of hydrocarbons, ethane, propane, and particularly gas oil and naphtha, represents the principal source of CPD and also methylcyclopentadienes (MCPD).

Cyclopentadiene is recovered from the other hydrocarbons as follows: The total cracked product is initially separated in a primary fractionation tower. Compounds boiling lower than 230°C are sent to a splitter tower, whose cut point (final boiling point of the overhead stream) is determined by the overall economics and dispositions of the steam-cracked products. Usually this is set at the boiling point of benzene. Splitter tower bottoms are processed into benzene, toluene, xylenes, and a wide boiling stream whose reactive components are comprised of styrene, alkylstyrenes, indene, and alkylindenes. This latter stream is called resin oils. Cyclodienes (CPD and MCPD) are sometimes present as dimers and codimers in these resin oils, which are used in hydrocarbon resin syntheses.

The splitter tower overhead, consisting of C_5-hydrocarbons and lighter components, is processed through a series of distillation towers. In the course of this process, the distillate is heated in the reboilers of the towers to a temperature of about 100°C to convert monomeric CPD to DCPD. The DCPD, which boils higher than the unreacted hydrocarbons of the distillate, is recovered as distillation bottoms.

Handling and Storage

Cyclopentadiene monomer spontaneously dimerizes at room temperature. Because of this spontaneous dimerization and the fact that DCPD is readily reconverted to the monomer, the dimer forms a safe and convenient way of handling CPD commercially. Generally, DCPD is moved in barges, tank cars, and trucks, all of which may be equipped with heating coils to maintain a liquid, pumpable product. However, DCPD should not be left in heated, idle lines and pumps for long times, ie, several weeks, as it will slowly polymerize and plug them. The consumer of CPD monomer must thermally crack DCPD to obtain it.

Health and Safety Factors

Toxicological studies conducted on DCPD indicate that it is a moderately toxic material and, to some extent, an irritant and a narcotic.

The 1978 TLV for DCPD was established at 5 ppm by the ACGIH. This is for a daily 8-h exposure on a time-weighted average.

Uses of Cyclopentadiene and Dicyclopentadiene

Because of the unusual reactivity of the DCPD molecule, there are a number of wide and varying end use areas. The primary uses in the United States are: DCPD-based unsaturated polyester resins (36%); hydrocarbon type resins, based on DCPD alone or with other reactive olefins (39%); EPDM elastomers via a third monomer; ethylidenenorbornene or DCPD (16%); and miscellaneous uses (9%), including polychlorinated pesticides, polyhalogenated flame retardants, and polydicyclopentadiene for reaction injection molding.

Specialty Derivatives and Uses

Fuels. Because of their high density, both DCPD and CPD produce high heat when burned. This property would make them excellent high energy fuels except for their reactive characteristics and, therefore, poor storage stability. The tetrahydrogenated products of these reactive diolefins have been claimed to be high energy fuels for racing cars, missiles, etc (see FUELS SYNTHETIC). A current use for the tetrahydrogenated DCPD compounds is as fuel (JP-10) for cruise missiles.

Dicyclopentadienedicarboxylic Acid. The dicyclopentadienedicarboxylic acid can be used in making alkyd resins (qv) that have excellent air-drying and baked-film properties. It can also be converted into many other products characteristic of dibasic acids.

MICHAEL J. KEENAN
Exxon Chemical Company

J. H. Wells and P. J. Wilson, *Chem. Rev.* **34,** 1 (1944).

A. S. Onishchenko, *Diene Synthesis,* Old Bourne Press, London, 1964.

D

DAIRY SUBSTITUTES

Dairy products (see MILK AND MILK PRODUCTS) have been staple items of the diet for many centuries, and have long been the target for imitation. The development of nutritional guidelines emphasizing the need to reduce total dietary fat, dietary cholesterol, $C_{27}H_{46}O$, and saturated fatty acids (see FATS AND FATTY OILS; FAT REPLACERS), has increased the interest in imitation dairy foods. However, with the exception of butter and cream the market penetration of dairy substitutes has been limited.

Ingredients

The characteristics and stability of imitation dairy products depend largely on the characteristics of the main ingredients, ie, fat, protein, and carbohydrates, together with the minor functional ingredients that stabilize the fat and protein systems. The ingredients generally include fats or oils and their stabilizing emulsifiers, proteins and their stabilizing gums (qv), and salts and carbohydrates. Many substitute dairy products have been reformulated in the past few years to meet consumers' interest in lower fat foods.

Fats and Oils. In industrial use, fats are considered solid and oils liquid at room temperature. The latter are liquids because of a higher content of unsaturated fatty acids. Hydrogenation converts oils to fats. Fats and oils consist mainly of the triglyceride-rich lipid fraction, which contains small amounts of other lipids, eg, sterols and phospholipids. For imitation dairy products, fats are generally selected to have low and narrow melting point ranges, generally 32–36°C. The fats and oils are characterized in a number of ways, including iodine number, polyunsaturated fatty acid content, saturated fatty acid content, Wiley melting point, and by solid–fat content at different temperatures.

The physical characteristics of a fat or oil for imitation dairy products are not necessarily dictated by the fat being replaced, but by the composition, processing methods, and conditions of use of the substitute product.

Fat Replacers. The reduction of fat in substitute dairy products results in an increase in water and a stress on the food system both in respect to body and texture, and to flavor. There is no universal fat replacer, but microparticulated proteins having particle sizes $<10\ \mu m$ and/or starch derivatives, and gums have been used as fat replacers.

Emulsifiers. In food systems, two general types of emulsions exist: oil-in-water and water-in-oil. An emulsifier is a product that has a water-soluble and a fat-soluble portion in the same molecule and stabilizes fat-containing food products. Emulsifiers may be natural products, eg, phospholipids and proteins, or derived from natural products, eg, esters of long-chain fatty acids and a polyhydric alcohol. Types of emulsifiers used in dairy substitutes include mono- and diglycerides, lactic acid and fatty acid esters of glycerol, polyglycerol esters of fatty acids, sorbitan esters of fatty acids, polyoxyethylene esters of fatty acids, and propylene esters of fatty acids.

Proteins. Proteins are especially significant in imitation dairy foods with respect to nutritional and physical properties of the product. The relative significance of the nutritional quality of the protein depends upon product type and the extent to which the product contributes to the total protein intake of a given population.

The role of protein with respect to physical properties varies with the type of product. Proteins contribute to a number of functions in an imitation food. These include emulsification, gelation, melting, water binding, and whipping.

A wide number of protein sources are available for use in dairy substitutes. These include animal proteins, ie, skim milk in liquid, condensed, or dry form (filled products); casein, caseinates, and coprecipitates; whey proteins; oil-seed proteins, fish proteins; and blood proteins. Oil-seed protein sources include soybean protein concentrates and isolates, groundnut protein, cottonseed protein, and sunflower seed, rapeseed, coconut, and sesame seed proteins (see SOYBEANS AND OTHER OIL SEEDS). Other sources are leaf and single-cell proteins (see FOODS, NONCONVENTIONAL). Of these protein sources, milk and soybean proteins are most widely used. Protein usage is based on economics, flavor, functionality, and availability.

Whey has been used in some substitute dairy products but not as a source of protein. Functionality of whey protein concentrates varies with whey type and concentration. Concentrates are used in a limited number of products: ice cream and other frozen desserts, fermented products, coffee whiteners, and whipped toppings.

Gum Stabilizers. Gums are used in substitute dairy products for one of the following reasons: to provide viscosity control and improve mouth feel; to improve whipping (aerating) properties of whipped products; to provide a protective colloid to stabilize proteins to heat processing; to modify the surface chemistry of fat surfaces to minimize creaming; to provide acid stability to protein systems; to increase freeze–thaw stability; to provide a hard-water resistant coffee whitener; and to provide desired melting characteristics to imitation cheese. Gums can be classified as neutral and acidic, straight- and branched-chain, gelling and nongelling. The principal gums and the functions in imitation dairy products are listed in Table 1.

Stabilizing Salts. Citrates and phosphates are used in substitute dairy products for one or more of the following purposes: to alter buffering capacity of the system; to improve the stability of the protein to calcium ions; to improve the heat stability of the protein; to minimize the age gelation of ultra-high temperature-(UHT) processed substitute products; to serve as emulsifying salts in imitation cheese manufacture; and to modify the water-binding capacity of proteins and improve solubility. Common phosphate salts used in these foods are monosodium phosphate (MSP), disodium phosphate (DSP), trisodium phosphate (TSP), disodium pyrophosphate, tetrasodium pyrophosphate (TSPP), sodium tripolyphosphate (STP), $Na_5O_{10}P_3$, sodium hexametaphosphate (SHMP), sodium trimetaphosphate, $Na_3O_9P_3$, and sodium tetrametaphosphate. Although there are similarities in some functional uses, phosphates are more broadly used than citrates in substitute dairy foods.

Composition and Processing

The composition of dairy substitutes is highly variable and generally represents the least-cost formulation consistent with consumer acceptance of the product. These imitations invariably have lower fat and protein levels than the dairy products that they are made to resemble. A comparison of the composition of certain dairy products and their substitutes is presented in Table 2.

Processes used for the manufacture of dairy substitutes are essentially the same as for dairy product counterparts. The processes common to all fluid products include ingredient blending, pumping,

Table 1. Gums and Associated Functions in Dairy Substitutes

Gum	Imitation product use	Function
alginates	ice cream	aeration, reduce whip time
	cheese	texture modifier, prevent oil separation
carrageenan	milk puddings	gelation
	ice cream	prevent wheying off
	infant formula	stabilize proteins to heat
	evaporated milk	stabilize proteins to heat
locust bean gum	ice cream	smooth melt down, freeze–thaw resistance
	cheese spread	texture modifier
guar gum	ice cream	water binding, body control
carboxymethyl cellulose	ice cream	prevent ice crystal growth
	whipped topping	aeration, protective colloid
xanthan gum	sour cream	prevent wheying off

Table 2. Comparison of Dairy Products and Substitutes

	Gross composition, g/100 g[a]				Minerals, mg/100 g[a]				Vitamins		
	Carbohydrates	Fat	Protein	Ash	Calcium	Phosphorus	Sodium	Potassium	A, IU	Riboflavin, μg	Total solids, g/100 g[a]
milk (fluid), whole	38.9	27.8	27.8	5.6	936	738	396	1142	1112	1350	12.6
milk substitutes	52.8	31.7	7.4	4.0	25	248					12.1
liquid	61.8	27.7	6.8	4.5	136	673	636	3270			11.0
light cream	14.6	72.7	10.5	2.2	353	280	182	331	3020	509	27.5
coffee whiteners	48.5	39.9	4.9	2.8	12	718	293	788	110	0	98.9
dry	49.1	36.5	5.0	2.7	16	625	258	768	110	0	98.9
	46.1	37.2	5.0	2.7	46	561	290	606	440	108	98.9
	48.7	35.8	4.9	3.0	12	62	146	1040	200	219	98.5
liquid	50.2	47.9	3.0	1.5	23	30	543	6	2170	0	26.7
	42.0	48.0	5.0	2.0	70	155	245	300	250	0	20.0
	49.1	40.5	8.6	2.7	72	212	496	121	0	0	22.2
	37.9	52.6	2.6	1.1	52	57					21.1
heavy cream	7.8	85.3	5.6	1.2	190	149	98	134	3510	268	41.0
whipped toppings	40.8	43.2	4.6	1.0	17	32	100	48	800	0	99.9
dry	40.6	45.4	5.7	0.6	12	46	97	6	1370	0	98.9
liquid	25.7	58.5	0.0	0.5	15	3	198	13	1040	0	39.3
	31.6	55.3	2.0	0.2	14	20	20	2	2220	0	49.1
aerosols	22.0	58.0	10.7	2.3	310	276	255	310	2540	145	35.5
dairy-based	20.4	63.3	8.5	2.3	274	267	413	209	2650	189	41.2
nondairy	29.2	58.7	0.0	0.5	10	3	243	11	1130	0	39.1
	23.9	67.7	7.6	0.5	23	71	173	6	1370	0	39.3

[a] Dry matter.

pasteurization or sterilization, homogenization, cooling, and packaging. Products can be manufacture by batch or by continuous processes.

Margarine. Margarine or oleomargine, originally marketed as an imitation butter, has a recognized identity of its own. The proportion of fat blend and other ingredients varies with the type of margarine and with the country of manufacture. Within a given country, variations in fat blends, aqueous phases, and other ingredients are fixed by legislation.

Skim milk was initially used as the aqueous phase in margarine. Where the law allows, margarines may contain caseinates, whey proteins, or soy proteins as the proteins component in the aqueous phase. The addition to margarine of 0.01–0.1 wt % sodium caseinate in place of milk has been proposed to eliminate sticking during frying. Substituting soy proteins for milk would have the same effect.

Cheese. The evolution of imitation cheese has come from the substitution of milk components in the development of filled and nondairy cheese and development of a synthetic cheese based on the Chinese food sufu, a form of tofu, which is based on soybean curd.

Coffee Creams/Whiteners. Coffee-cream substitutes have been given the generic name of coffee whiteners and are available in liquid, frozen, and dry forms. The emulsifiers and proteins are key stabilizing factors with respect to the stability of the product in coffee. In addition, the method of processing has a significant effect on the stability and whitening power of the product.

Ice Cream and Frozen Desserts. Imitation frozen desserts include ice cream, sherbets, and specialty products, eg, milkshake bases. Filled ice cream, called mellorine in the United States, is the dominant product in the world market and is sold both as a soft-serve and as a hardened product. The composition and processing is as for ice cream, except for the substitution of vegetable fat for milkfat.

Milk. Imitation milks fall into three broad categories: filled products based on skim milk, buttermilk, whey, or combinations of these; synthetic milks based on soybean products; and toned milk based on the combination of soy or ground nut (peanut) protein with animal milk. Milk is the one area where nutrition is of primary concern, especially in the diets of the young. Substitute milks are being made for human and animal markets. In the latter area, the emphasis is for products to serve as milk replacers for calves.

Guidelines for fluid milk substitutes include recommendations regarding raw materials, methods of preparation, composition, sanitation, control of toxic substances, heat destruction of trypsin inhibitors, and recommendations for the distribution of filled or toned milk and products containing non-cow-milk components. Most of the cow-milk substitutes are based on soy bean or groundnut (peanut).

Soy-Based Beverages. Soy milk is a traditional oriental product and has been the basis for development of dairy substitutes for western markets. Soybean-based beverages may be classified as: *(1)* traditional soy milks, unfermented, starting with uncomminuted full-fat beans, full-fat or defatted soy flours, or soy protein concentrates; *(2)* traditional fermented yogurt-like products; or *(3)* simulated milks based on soy protein concentrates and isolates, ie, of the dairy type, fluid single strength, concentrated, or sterile, nonfat milk replacers such as dry blends of soy protein isolates with dry whey, infant-feeding beverages simulating human milk, and fermented yogurt-like products.

Infant Formulas. Milk-based infant formula are sometimes referred to as humanized milk. These products are formulated to resemble the composition of human milk and contain a lower amount of carbohydrate than cow's milk. They are sold in three forms: liquid ready-to-use, liquid concentrates, and dry powders. A typical processing scheme involves adding skim milk to a mixing vat that is adjusted to 60°C; adding oil-emulsifier, minerals, and vitamins; pasteurizing; homogenizing at 21 MPa (3000 psi); standardizing composition; packaging; sterilizing; cooling; and storing.

Most vegetable-protein-based infant formulas are soybean-based. Soybean-based formulas have served as useful replacements for milk in the nutritional management of infants who are allergic to cow's milk. Allergies to soybean milk also exist, so the substitution of soy milk for cow's milk does not always solve allergy problems. This concern is lessened by hydrolyzing the protein to reduce potential allergenicity.

Whipping Creams and Whipped Toppings. Whipped topping is the term generally used for imitation whipping cream. Products are available as liquids to be whipped, frozen prewhipped products, and as liquids in aerosol cans that form whips upon release to the atmosphere. Dry powdered products that are reconstituted and whipped are also available. Toppings were first made from nonfat dry milk or sodium caseinate as the protein source and vegetable fat as the source of fat.

Most of these products are heavily dependent upon these constituents to impart the desired characteristics and to provide uniformity in the whipped topping. Generally, the toppings have a lower fat content than traditional whipping cream.

Miscellaneous. Miscellaneous products having market significance include sour cream, chip dips, milkshake bases, puddings, yogurt, and fat-reduced forms of all substitute dairy foods. These products have been formulated from both caseinates and soybean proteins.

Economic Aspects

Of the filled and imitation products, margarine has made the greatest penetration into the sale of dairy foods.

Other dairy substitutes have penetrated the U.S. market to the extent of 1% for fluid whole milk, 58% for creams, <1% for low-fat milk, 6–7% for cheeses, 10% for evaporated and condensed milks, and 2% for ice cream. About 60% of the substitute and imitation cheese sold in the United States is being used as the cheese material for pizza.

Based on brand name products, the total number of branded substitute and imitation dairy products worldwide is estimated to exceed 1000. Almost all multinational food companies market one or more dairy substitutes.

Cost is a significant factor in the consumer's acceptance of substitute dairy foods. In all cases the prices for the substitutes are lower than the prices of the respective diary product. The smallest price margin is in the area where the substitute products are advertised as fat-reduced or cholesterol-free.

Regulatory Aspects

The legal status of dairy substitutes varies widely among countries. In the United States, the FDA has held that filled products should be nutritionally equivalent to the products they resemble. In the case of imitation milks, the FDA proposed regulations for nutritional equivalency. A section of the Food, Drug, and Cosmetic Act defines misbranded foods, and the FDA has set up standards of identity for foods under this part of the law which includes standards for imitation milks, cheese, and creams.

The definition of margarine varies from country to country. The U.S. Code provides regulations on margarine and on imitation ice cream, called mellorine.

W. JAMES HARPER
Ohio State University

F. Winklemann, *Imitation Milk and Milk Products,* Food and Agricultural Organization of the United Nations, New York, 1974, pp. 1–117.

M. K. Switzer, *Margarine and Other Food Fats,* Leonard Hill, Ltd., London, 1956, pp. 240–273.

21 FCR 120, Chapt. 1, Part 45.

21 FCR 120, Chapt. 8, Part 20.

DATABASES

Chemically related database searches can be used to establish concepts and patentable ideas. For instance, searches have identified researchers using particular monomers in a potentially patentable latex formulation; found precedents for a polymeric emulsifier; summarized publications of people being considered as consultants, expert witnesses, employees or speakers to an industrial group; and provided market description information for a new pigment manufacturing firm to identify target markets.

Among the reasons for undertaking a search are those based on an interest in the competitive and strategic factors related to market needs. Information from databases might help determine the markets for an emerging technology; how amenable the technology is to production scale-up; the patents held by competitors; what regulatory issues apply; what toxicity data are available and what is known about safe-handling or shipping; who manufactures the chemicals required, and what the physical and chemical properties are; what the manufacturing costs and pricing implications are; where company headquarters, subsidiaries, and executive information are located; and what information can be retrieved from molecular structure searches.

Growth in Databases

Size of the database industry can be measured in terms of the number of database records, databases, database entries in CRDB, database producers, database vendors, or on-line searches. The slowest growth is at the vendor level, the fastest growth is in database files.

Many database producers provide on-line access to their databases or distribute their databases on compact disk read-only-memory (CD-ROM) and so are also considered vendors or producer/vendors. Whereas numerical growth in vendors is indicated, the success of the database industry is largely a result of the transition of the information industry from paper-based to computer-based services. Thus industry growth can also be measured in terms of the increase in use of computer-readable databases as exemplified by the number of searches.

The database industry is dynamic, changing and growing every year. There is no sign of leveling-off in the 1990s. New media for distribution and access to databases have increased the potential for attracting users. In particular, the development and distribution of CD-ROM databases has greatly increased the use of computer-readable databases in universities and colleges where cost has often been a barrier to access. At the same time, there has been a decrease in the use of online searching of those databases that are also distributed on CD-ROM. Decreases have been noticed particularly on ERIC, PsycINFO, NTIS, and MEDLINE, which are all databases well used in academia.

Classification of Databases

Form of Data. Databases can be classified in many ways. One method is by form of data representation, ie, data may be in the form of words, numbers, images, or sounds. The corresponding databases may then be considered to be word-oriented, number-oriented, image-oriented (video), or sound-oriented (audio). Data representation affects file structures and software for search and data retrieval. Thus the structures and search techniques vary considerably among these four basic classes.

Many databases can be classified in multiple ways because of multiple type data, eg, text and numeric data, text and image data, image and audio data, etc. Two additional classes of databases are electronic services and software. Both of these data types could also be classed by form of representation because of use of words and numbers. However, the way in which these databases are used is different and they have special characteristics. Thus they are presented as additional classes. Whereas electronic information services such as bulletin boards, electronic mail, and electronic conferencing contain data that are transitory and nonarchival, these must be included among databases because several of the principal vendors sell these services in the same way as database search services are sold.

Subject Categories. The determinant for user selection of a database is usually subject matter. That is, when chemical information is desired, a chemical database is selected. The form or media of the database is of secondary importance. The type of search may dictate the need for a full-text or statistical database.

Business databases remain number one among all database categories, followed by science/technology/engineering and then by health/life science.

Medium of Distribution. Another way of looking at databases is in terms of the recording medium used for distribution or access, ie, on-line, batch, CD-ROM, diskette, and magnetic tape.

Producers. The producers of databases are sometimes called database publishers. Some producers publish hard-copy counterparts to databases and so are publishers in the traditional sense; others publish data only in electronic form. Database producers are responsible both for the determination of content and for database production. Most producers offer their databases for lease or license to private organizations or database vendors. Vendors offer database search services to the marketplace on a fee basis. An increasing number of producer/vendors such as Mead Data Central, U.S. National Library of Medicine, and DRI/McGraw-Hill (formerly Data Resources), offer search services (batch or online) from their own databases as well as from the databases of other products.

Producers of databases can be classified in terms of a country's infrastructure. Identifiable groups are the government; commerce/industry; not-for-profit (NFP), which includes academe; and mixed status.

Vendors. Database vendors provide value added processing of databases and offer search services, on-line and/or batch, or distribute CD-ROM products to database users. Vendors also provide the technology for accessing databases.

Value is added to databases in several ways: through database preparation (loading); by means of the special capabilities of the vendor's search software; and through related services such as online document ordering and selective dissemination of information (SDI)/current awareness.

Connecting to Databases

Online Connections. One highly reliable telecommunications network is DIALNET, offered by DIALOG. DIALNET offers dial-up access from over 50 principal U.S. cities at 300, 1200, 2400, and higher baud rates to 9600. Other networks for users in the United States and Canada include SprintNet or TYMNET networks. INWATS and access is also available.

Using a Personal Computer. A personal computer can be used to communicate with on-line services when the following three items are available: (*1*) some form of communications software which must be synchronous and compatible with the particular computer; (*2*) a modem which must also be compatible with the computer; and (*3*) a dedicated telephone line which allows no transfers or call-waiting. Call-waiting options may be disabled by computer command in dial-out code in some areas.

Using a Terminal. Dial-up terminals may also be used to access a vendor's databases; however, the terminal, personal computer, word processor, or microcomputer must all be compatible with specific functionalities. In addition, one also needs an EIA-compatible (within the United States) or CCITT-compatible (non-U.S.) modem for dial-up.

MARTHA E. WILLIAMS
University of Illinois
JAMES L. GRANT
DIALOG information Services, Inc.

DIALOG Focus on Biotechnology, DALOG Information Services, Inc., Palo Alto, Calif., Sept. 1989.

CASSI Cumulative (Chemical Abstracts Service Source Index), Chemical Abstracts Service, American Chemical Society, Columbus, Ohio, 1991.

M. E. Williams, "The State of Databases Today: 1993," in K. Y Marcaccio, K. L. Norlan, and G. E. Tureecki, eds., *Gale Directory of Databases,* Gale Research, Inc., Detroit, Mich., 1993.

K. E. Marcaccio, K. Hillstrom, C. Tomassini, G. E. Turecki, and M. E. Williams, eds., *Computer Readable Databases: A Directory and Data Sourcebook,* 8th ed., Gale Research, Inc., Detroit, Mich., 1992.

M. E. Williams, *Information Market Indicators: Information Center/Library Market,* issue 39, Information Market Indicators, Monticello, Ill., 1992.

DDT. See INSECT CONTROL TECHNOLOGY.

DECARBOXYLASES. See ENZYME APPLICATIONS, INDUSTRIAL; MICROBIAL TRANSFORMATIONS.

DEFLOCCULATING AGENTS. See CLAYS; DISPERANTS; FLOCCULATING AGENTS.

DEFOAMERS

The control or elimination of the foam that occurs in many industrial processes is often a key factor in their efficient operation. The additives that are used in low concentration to achieve this effect are known variously as defoamers, antifoaming agents, foam inhibitors, and foam controllers. Defoaming implies breaking a pre-existing foam whereas antifoaming or foam inhibition indicates prevention of the formation of that foam. Such distinctions call for different product features. A defoamer is expected to exhibit rapid knockdown of a foam, whereas longevity of action might be the key requirement in many antifoam applications. Despite these varying performance features, many applications require both preventive and control functions, and in practice the same types of materials are used both for antifoaming and defoaming.

Many industries now rely on the efficient and economical use of defoamers both as a process aid in product manufacture and to increase the quality of the finished product in its subsequent application. The most obvious use of defoamers as process aids is to increase holding capacity of vessels and improve efficiency of distillation or evaporation equipment. They are also used to improve filtration, dewatering, washing, and drainage of suspensions, mixtures, or slurries. Among the finished products that are improved in quality or efficacy by the proper inclusion of defoamers are lubricants, particularly cooling lubricants in metal working; diesel fuel, and hydraulic and heat-transfer fluids; paints and other coatings; adhesives; inks; detergents; and antiflatulence tablets.

Most modern defoamers are complex, formulated specialty chemicals. They are usually proprietary products and the patent literature is the best guide to their probable composition. In addition to control of foam and associated features such as rate of foam knockdown and the persistence of the effects, other frequently needed application requirements of these specialty materials include adequate shelf life, absence of adverse effects on and by the products being treated, ease of handling, lack of toxicity to manufacturing personnel and users, environmental acceptability, and cost-effectiveness. Defoamers range from relatively inexpensive mineral oils to costly fluorinated polymers, but it is not the cost per kilogram of defoamer that matters, but rather the cost per unit produced using this processing aid. Another factor that strongly influences the choice of a specific defoamer is its ancillary surface properties, such as wetting, dispersion, and leveling.

Active Ingredients

Liquid-Phase Components. Because lowering of surface tension is a key defoamer property and this effect is a function of the nonpolar portion of the liquid-phase component, it is preferable to classify organic liquids by the hydrophobic, nonpolar portion. This approach identifies four liquid-phase component classes: hydrocarbons, polyethers, silicones, and fluorocarbons.

Solid-Phase Components. Dispersed solids are vital ingredients in commercial antifoam formulations. Much of the current theory on antifoaming mechanism ascribes the active defoaming action to this dispersed solid phase with the liquid phase primarily a carrier fluid, active only in the sense that it must be surface-active in order to carry the solid particles into the foam films and cause destabilization. Using the same classification adopted for liquids gives three solid-phase component classes: hydrocarbons, silicones, and fluorocarbons.

In most cases, these active defoaming components are insoluble in the defoamer formulation as well as in the foaming media, but there are cases which function by the inverted cloud-point mechanism. These products are soluble at low temperature and precipitate when the temperature is raised. When precipitated, these defoamer-surfactants function as defoamers; when dissolved, they may act as foam stabilizers.

Ancillary Agents

Surface-Active Materials. Since defoaming is a surface effect, all active components are necessarily surface active materials in that sense, but in this ancillary category the surfactants are incorporated in the formulation for other effects such as emulsification or to enhance dispersion. Emulsifiers are essential in the common oil-in-water emulsion systems but they are also required where mixtures of active liquid components are used. These additives increase the speed of foam decay by promoting rapid dispersion of the defoamer throughout the foaming media. Examples of emulsifying agents used in defoamer compositions are fatty acid esters and metallic soaps of fatty acids; fatty alcohols and sulfonates, sulfates and sulfosuccinates; sorbitan esters; and ethoxylated products such as ethoxylated octyl or nonylphenols.

Carriers. The function of the carrier is to provide an easily handleable, readily dispersible system for delivering the active defoamer components to the foaming system and also to tie the complex defoamer formulation together, ie, coupling agents, compatibilizers, or solubilizers. Sometimes the carrier is used simply as an extender to lower the cost of the final product. Many of the low viscosity organic solvents that are used also exhibit some antifoaming properties in cases where they are both insoluble and of lower surface tension than the medium to which they are applied. Any of the usual paraffinic, naphthenic, aromatic, chlorinated or oxygenated organic solvents can be used, but aliphatic hydrocarbons are the most common. Water is often used as a carrier fluid. In these cases, the defoamer product is typically an oil-in-water emulsion. With growing concern over unrecovered solvents, this has become a preferred type of defoamer formulation. Such products usually require preservatives to prevent bacterial spoilage in the drum or other shipping and storage container.

Sometimes the defoamer is required in a solid form; for example, to be suitable for incorporation into a low-sudsing detergent powder or agricultural chemical composition. Water soluble inorganic sorbent carriers such as sodium sulfate, sodium carbonate or sodium tripolyphosphate are used as well as organic polymers such as methylcellulose.

Defoaming Theory

Foams are thermodynamically unstable. To understand how defoamers operate, the various mechanisms that enable foams to persist must first be examined. There are four main explanations for foam stability: *(1)* surface elasticity; *(2)* viscous drainager retardation effects; *(3)* reduced gas diffusion between bubbles; and *(4)* other thin-film stabilization effects from the interaction of the opposite surfaces of the films.

All these mechanisms except high bulk viscosity require a stabilizer in the surface layers of foam films. Accordingly, most theories of antifoaming are based on the replacement or modification of these surface-active stabilizers. This requires defoamers to be yet more surface active; most antifoam oils have surface tensions in the 20 to 30 mN/m range whereas most organic surfactant solutions and other aqueous foaming media have surface tensions between 30 and 50 mN/m(= dyn/cm).

Based on this low surface tension feature and the commonly observed insolubility of defoamers, two related antifoam mechanisms have been introduced: *(1)* The agent dispersed in the form of fine drops enters the liquid film between bubbles and spreads as a duplex film. The tensions created by this spreading lead to the rupture of the original liquid film. *(2)* A droplet of the agent enters the liquid film between bubbles, but rather than spreading produces a mixed monolayer on the surface. This monolayer, if of less coherence than the original film-stabilizing monolayer, causes destabilization of the film.

These theories adequately account for agents consisting only of insoluble liquids such as silicone fluids used in the defoaming of crude oil. However, practical experience also shows that dispersed hydrophobic solids can greatly enhance defoamer effectiveness in certain cases, particularly in aqueous foam systems. Materials such as hydrophobic silica or high melting-point hydrocarbon amides such as ethylenediamine distearamide are notably effective. A strong correlation between defoamer action and contact angle for silicone-treated silica in a hydrocarbon oil has been demonstrated. Most recent publications on the mechanism of defoamers have concentrated on the role of the dispersed hydrophobic solids. For example, dewetting of the hydrophobic silica by the bubble film is believed to cause foam collapse by the direct mechanical shock of the event. Such dewetting helps thin the film and promote instability, and is particularly effective when sizes are such that the particle occupies both surfaces of the film.

Applications

Defoamers are used in adhesives and sealants, chemical processing, cleaning compounds, the construction industry, fermentation processes, fertilizers, food and beverage preparation, leather processing, metal working, oil recovery and petrochemical operations, coatings, polymer production, pulp and paper manufacture, textile processing, and wastewater treatment.

Health and Safety Factors

Defoamers are usually added at low bulk concentrations ranging from a few to a thousand parts per million of the foaming medium. Often the health risk posed by such additives is negligible compared to that of the material being defoamed. Sometimes a specific defoamer type/foaming medium combination presents a particular problem, so the supplier should always be involved in defoamer selection. Health and safety concerns arise primarily in applications in the food and drug industries. U.S. government regulations governing the use of additives such as defoamers in food and drugs are listed in the *Code of Federal Regulations*.

MICHAEL J. OWEN
Dow Corning Corporation

R. Hofer and co-workers, in B. Elvers, J. F. Rounsaville, and G. Schulz, eds., *Ullmann's Encyclopedia of Industrial Chemistry,* Vol. A11, 5th ed., VCH Publishers, New York, 1988, pp. 465–490.

M. J. Owen, in J. I. Kroschwitz, ed., *Encyclopedia of Polymer Science and Engineering,* Vol. 2, 2nd ed., John Wiley & Sons, Inc., New York, 1985, pp. 59–72.

P. R. Garrett, *Defoaming: Theory and Industrial Applications,* Surfactant Science Series No. 45, Marcel Dekker Inc., New York, 1993.

DENTAL MATERIALS

Dental therapy includes the replacement of hard and soft oral tissues lost through disease using inert materials that may be metallic, ceramic, or organic, or composites employing combinations of these three classes. The operative restorations and prostheses are made of amalgam, precious and nonprecious alloys, special cements, synthetic polymers, porcelain, and glass–ceramics (see CEMENT; GLASS; ENAMELS, PORCELAIN OR VITREOUS). All must withstand the rigors of the oral environment (see also PROSTHETIC AND BIOMEDICAL DEVICES). The accessory materials needed in the fabrication procedures include synthetic polymers, synthetic and natural gums (qv) and waxes (qv), hydrocolloids, gypsums, and refractories (qv).

Specifications. The ADA maintains a list of certified dental materials and devices based upon the certification by the maker that the item complies with ADA specification and that the testing for specification compliance of the item is procured in Association laboratories. The ADA also maintains a list of classified dental materials and devices which prove to be acceptable or provisionally acceptable to the Association based upon data submitted by the applicant and data available in the literature.

The Food and Drug Administration (FDA) has responsibility for enforcement of the Federal Food, Drug, and Cosmetic Act of 1938 and its various amendments, eg, May, 1976, in which dental materials, instruments, and equipment are included. Premarketing clearance requirements apply for establishing the safety and effectiveness of new products. There is a close liaison between the FDA and the ADA standards and certification programs.

Biocompatibility. Many of the dental materials are used in contact with body tissues, eg, tooth, bone, soft tissue; and body fluids, eg, saliva and blood; usually for extended periods of time, ie, years. As knowledge of the toxic, allergic, and carcinogenic responses to substances foreign to the body has increased, the need for evaluation of biological responses to materials used in specific locations and for specific functions has become clear. In order to address the need for evaluation, ADA/ANSI specification #41, Biological Evaluation of Dental Materials, has been developed. This specification is revised and updated regularly. For materials for which an ANSI/ADA specification does not yet exist, the demonstration of acceptable tolerance by specification #41 helps provide to the patient the highest level of protection against injury by the new material.

Dental Ceramics

Ceramic materials are well suited for biomedical applications requiring tissue compatibility, compressive strength, durability, radiopacity, and inertness towards sterilization chemicals and conditions (see CERAMICS). In addition, glass-matrix ceramics, containing appropriate crystalline fillers and colorants, can richly mimic many of the aesthetic and optical properties of natural teeth (see COLORANTS FOR CERAMICS). Limitations to the use of ceramics include design restrictions based upon the often low and variable tensile strengths of ceramic products; susceptibility to stress corrosion, eg, crack extension in the presence of water or other low molecular weight species; and shrinkage during processing.

The principal use of ceramics in dentistry is for the aesthetic restoration of missing teeth or tooth structure, ie, crowns and bridges, primarily as amorphous (glassy) ceramic coatings fused to an underlying metal framework. Additional aesthetic use is made of ceramics not supported by metal substructures. These uses include single-tooth crowns, ceramic fillings, veneers, denture teeth, inserts, and orthodontic brackets. Some semi-aesthetic, generally more highly crystalline, ceramics are utilized as core materials in place of cast metal substructures, to be veneered with more aesthetic ceramics.

Aesthetic dental ceramics are essentially glass-matrix materials with varying volume fractions of crystalline fillers. Crystalline fillers are used in the glass matrix both for dispersion strengthening, usually at volume fractions of 40–70%, and for altering optical properties, usually at low volume fractions. Dental ceramics are generally manufactured from two distinct classes of materials, ie, beneficiated feldspathic minerals and glass–ceramics.

Dental Cements

Dental cements are composites, ie, they have a continuous or matrix phase linked to a discontinuous or reinforcing phase by an interphase. Based on their chemistry, dental cements can be divided into acid–base cements, acrylic or resin cements, and resin-modified acid–base cements, ie, hybrid cement composites. Acid–base cements are either aqueous-based cements or nonaqueous-based cements, although small amounts of water or protoic agents are essential to the acid–base setting mechanism of the latter. Acrylic cements are resin-based composites modified for cement applications. The monomer(s) used in these cements can be simply methyl methacrylate or various types of multifunctional monomers, eg, BIS-GMA, triethylene glycol dimethacrylate, etc. The setting mechanism is by free-radical addition polymerization, either chemical, photochemical, or a combination of the two. The hybrid cement-composites include a vinyl resin system as well as the usual acid–base components of the cement. They have a dual, ie, ionic and free radical, setting mechanism.

Although used in relatively small quantities, dental cements are essential in a number of dental applications as temporary, intermediate and, in some cases, more permanent restoratives; cavity liners and bases; luting agents to bond preformed restorations and orthodontic devices; pulp capping agents and endodontic sealers; components in periodontal packs; and impression pastes. Because this wide range of dental uses makes it virtually impossible for one type of cement to have all the necessary properties demanded, tailored dental cements exist to meet certain specific objectives. Generally, cements are sought that excel in strength, oral environmental resistance, durability, adhesiveness, and biocompatibility.

Research is focused on the improvement of the polyelectrolyte cements, eg, glass–ionomer, and the development of durable, nonshrinking resin or polymer-based adhesive materials with improved biocompatibility.

Dental Plasters

Gypsum is widely distributed naturally as calcium sulfate dihydrate, $CaSO_4 \cdot 2H_2O$. Plaster is the rehydrated calcined gypsum. The American Dental Association classifies five types of dental plaster according to the physical properties: type I, impression plaster; type II, model plaster; type III, dental stone; type IV, high strength dental stone; and type V, high strength, high expansion dental stone. These different types are the result of various calcining methods.

Although plaster has been a very successful and serviceable material, it is seriously lacking in hardness, edge strength, chip resistance, abrasion resistance, and strength to fulfill many needs of dentistry. Some of these requirements have been partially filled by the development of the type III and type IV plasters. Table 1 lists the compression strength of dental plasters.

To form plaster, gypsum is ground and subjected to temperatures of 110–120°C in open kettles to drive off part of the water of crystallization. The crystals thus formed are large and porous and are the type I and type II plasters. These crystals require a 2:1 powder–gauging water ratio for proper consistency. Type I, impression plaster, and type II, model plaster, differ in additives that control working and setting times. When gypsum is formed these crystals form the type III plaster commonly called dental stone and require 28–35 mL of water for 100 g of powder. Type IV, high-strength dental stone is formed by calcining gypsum in a 30% solution of calcium chloride. The crystals resulting from this process are slightly larger and denser than the type II crystals and require even less gauging water (20–22 mL/100 g of powder). Type V dental stone has gypsum that is formed by a process

Table 1. Compressive Strength of Dental Plasters, Investments

Materials	Compressive strength, MPa[a,b]	ISO standard no.
Plasters		
impression	4.0[c]–8.0[d]	6873
model	9.0[c]	6873
dental stone	20[c]	6873
high-strength dental stone	35[c]	6873
Investments		
inlay and crown[e]	2.3 (1.7)	7490
partial denture[e]	5.8 (3.3)	7490
phosphate-bonded[f]	3.0	7490
ethyl silicate-bonded[f]	1.5	proposed
Dies		
gypsum refractory	13.0	proposed
phosphate refractory	13.0	proposed

[a] To convert MPa to psi, multiply by 145.
[b] Value given is at room temperature. Value in parenthesis is at casting temperature.
[c] Value given is minimum value.
[d] Value given is maximum value.
[e] Low temperature casting investments.
[f] High temperature casting investments.

similar to that for the type IV stone, with the use of calcium chloride. Additives, such as fillers and surface tension reducing chemicals, result in higher setting expansion and higher strength. Hence, less gauging water is required, ie, only 18–20 mL/100 g is needed.

Dental Investments

Dental investments are comprised of refractory materials capable of withstanding elevated temperatures. They are used as casting investments and model investments.

Casting Investments. Casting investments are used to form molds into which molten metal may be cast. The cavity for receiving the metal is formed by the lost wax process. The composition of investments used for alloys cast from low (≤1100°C) temperatures are different from those used for alloys cast from higher (≥1300°C) temperatures.

A casting investment must provide sufficient expansion to compensate for shrinkage (up to 0.4%) of the wax (or plastic) pattern during its fabrication, and shrinkage (1.2–2.0%) of the cast alloys resulting from solidification and cooling. The higher the solidification temperature of an alloy the greater is its casting shrinkage.

Investments for Low Temperature ($T \leq 1100°C$) Casting Alloys. Low temperature casting alloys are usually comprised of gold, silver, and copper and are used for inlays, crowns, and removable partial dentures. At the temperature tested, the compressive strengths specified for inlay or partial denture investments are adequate to prevent fracture during handling and in casting. The strength may vary at casting temperature. The strengths of gypsum-bonded investments at high temperatures vary with the additive used to reduce shrinkage during heating. The type of binder also influences the effect of the temperature on strength. Table 1 presents the mechanical property limits defined by ISO standards for casting investment for low-temperature dental casting alloys.

These investments consist of a binder of calcined gypsum, ie, calcium sulfate hemihydrate, $CaSO_4 \cdot 1/2H_2O$; to which is added silica and modifying agents such as finely ground gypsum, soluble sulfates, and chlorides, ie, accelerators of setting; and low solubility salts, ie, retarders of setting (111). The refractory base consists of the silica in some crystalline form; quartz and crystobalite are the main constituents.

Investments for High Temperature Casting Alloys. These investments are used for alloys that require high (≥1300°C) casting temperatures. The alloys vary greatly in nobility. Some are rich in gold and/or palladium, and others are predominately cobalt, nickel, and chromium. The noble-based alloys for fixed partial dentures require casting temperatures in the neighborhood of 1300°C, and phosphate bonded investments are used. The cobalt, nickel, and chromium alloys require casting temperatures of ca 1400–1500°C, and investments that use ethyl silicate or stabilized silica solutions are used.

Model Investments. Model investments are materials used for noncasting operations in the fabrication of dental protheses. They differ from casting investments in various ways depending on the prosthetic device being constructed. For low temperature operations, such as soldering, gypsum is used; phosphate-bonded materials are employed for higher solder temperatures or for the fabrication of porcelain veneers.

Investment Casting Rings and Liners. Casting rings and ring liners are needed prior to investing a wax pattern. Casting rings are cylindrical tubes that mold and retain the casting investment. Ring liners are paper or fabric sheets adapted to the inside of the rings to allow the investment to expand within the ring.

Three types of castings rings commonly used are metal rings, split rings, and disposable rings.

Liners are either cellulose, which readily absorbs water, or ceramic, which does not absorb water or investment liquid.

Dental Waxes

Pattern Waxes. The pattern waxes, ie, inlay casting, base-plate, and sheet and shape waxes, are used to construct the prototype or pattern from which a finished dental restoration is produced.

Inlay Casting Waxes. The three types of inlay casting waxes, ie, types A, B, and C, are used to produce wax patterns for the lost wax casting process in the production of cast gold inlays, crowns, and bridges. Some inlay wax is also used to produce patterns for acrylic restorations.

Type A waxes are hard waxes used when an extra hard wax is preferred by the dentist in making patterns in the mouth. Type B waxes are medium waxes used by most dentists to make patterns in the mouth. Type C waxes are soft waxes used for making patterns outside the mouth.

The exact formulations for inlay casting waxes are considered trade secrets, and little has been published on the subject.

Inlay waxes are generally dark blue, green, black, or purple; however, inlay waxes for acrylic work are uncolored and are a natural ivory. The waxes are generally produced in stick form, about 75 mm long and 6.5 mm across; the cross section may be round or hexagonal.

Base-Plate Waxes. Base-plate waxes are used to substitute for, or be used in conjunction with, a base plate to form a pattern for the production of complete or partial dentures and certain orthodontic appliances to be molded of acrylic resin, modified vinyl resins, or other denture-base polymers.

Base-plate waxes are formulated for specific uses or working conditions into types I, II, and III. Consequently, the flow requirements differ. Type I waxes are soft waxes for building contours and veneers, type II waxes are medium waxes used for pattern production in the mouth in temperate weather, and type III waxes are hard waxes used for production in the mouth in hot weather.

Base-plate wax compositions are generally regarded as trade secrets.

Sheet and Shape Waxes. Sheet and shape waxes are used to produce patterns from which complete or partial dentures are cast of gold or base metal alloys. They are used to fabricate the restoration prototype directly upon a refractory investment cast.

The flow of sheet and shape wax is much higher than that of inlay wax and base-plate wax, reflecting increased pliability, ductility, and plasticity.

The compositions of sheet and shape waxes are also trade secrets. However, they are blends of various proportions of paraffin, microcrystalline waxes, carnauba wax, ceresin, beeswax, gum dammar, mastic gum, and possibly other resins. Sheet waxes are marketed in square sheets approximately 80 by 90 mm. Various thicknesses are available from 32 gauge (0.5 mm) to 14 gauge (1.63 mm).

Shape wax is similar in composition and properties to the nontacky sheet waxes. It is processed into preformed shapes or definite shape and gauge to facilitate the waxing-up of partial-denture patterns. The shapes and sizes conform to those most needed for lingual bars, palatal bars, clasps, saddle construction, retainers, etc.

Impression Waxes. Impression waxes include those waxes used to obtain a negative cast of the mouth structure (impression waxes), waxes used to establish tooth articulation (bite-registration waxes), and waxes used to detect tooth interference and high spots or improper fit of denture bases (disclosing waxes). They must be plastic and moldable at mouth temperatures, and chill to a still nonplastic mass upon cooling within a few degrees below mouth temperature.

The materials that have been identified in the compositions of impression waxes are paraffin, ceresin, vegetable waxes, rosin, mastic gum, and spermaceti. Bite-registration waxes are generally compounded from high flow, low melting paraffins, microcrystalline waxes, and resins. Disclosing waxes include soft paraffins, petrolatums, coconut oil, zinc oxide, titanium dioxide, and suitable dyes.

Processing Waxes. The extensive amount of handwork and craftsmanship necessary in the fabrication of most dental restorations and appliances has created a need for several types of wax compositions. These are known as boxing, sticky, utility, or study waxes.

Boxing wax, as the name implies, was developed to box impressions, ie, to construct a retaining ring of wax around an impression to confine the plaster or stone when the cast is poured. These waxes may have melting points of 65.5–71.1°C, and a flow of 60–80% at 37.5°C.

Boxing waxes are generally formulated from various microcrystalline waxes to give the desired tackiness and pliability. Various resinous additions may also be made to increase strength and tackiness.

Sticky waxes are used as thermoplastic cements. Sticky wax is useful in almost any operation where it is desired to position and hold several small pieces in a temporary relationship. They are generally composed of resins and wax.

Utility waxes are generally useful for sealing, filling, contouring, building up areas, positioning, and many other functional purposes. Their compositions, considered trade secrets, are probably microcrystalline waxes or blends of beeswax, petrolatum, and rosin.

Study waxes, ie, waxes for carving, are useful in the study and modeling of tooth forms and the teaching of anatomical detail. Carving wax compositions include paraffin, ceresin, ozokerite, carnauba wax, montan waxes, and Acrawax C. Fillers and pigments may be added.

Wax Manufacture. Wax compositions for dental usage are usually compounded in simple melting and blending equipment.

Wax compositions are processed into rods, sheets, cakes, and special forms by a variety of processes. Casting, roll forming, and extrusion, by mechanical or hydraulic-piston-type extruders or by continuous screw extrusion, are operations in practice. Stamping, roll cutting, or molding are also employed.

Dental Alloys and Metals

The chemical and physical properties of some metals and alloys, such as hardness, strength, stiffness, toughness, corrosion resistance, and biocompatibility have provided materials capable of withstanding the most severe demands of restorative dentistry, namely the harsh corrosive environment of the mouth and high stresses on the small cross-sectional areas of teeth.

Amalgams. Dental amalgam is a novel alloy. Within a few seconds of the start of trituration, the alloy and mercury must amalgamate to a smooth plastic mass. Within 3–5 min it must set to a carvable mass and remain so for 15 minutes. Within 2 h it must develop sufficient strength, hardness, and toughness to resist mild biting and chewing forces. It should not tarnish or corrode in the mouth, nor react to produce toxic or soluble salts, and must maintain its color.

The mercury content in dental amalgam is generally restricted to about 50% or less. Excessive mercury in the final restoration results in a low strength amalgam with poor creep resistance, excessive setting expansion, and poor corrosion resistance. Properly proportioned, mixed, and processed amalgams will have compressive strengths in the range of 210–345 MPa (30,000–50,000 psi) in 24 hours.

Composition. The composition of powdered alloys used in preparing dental amalgams usually includes 66.7–75.5% silver; 25.3–27.0% tin; 0.0–6.0% copper; and 0–1.9% zinc. These are commonly referred to as conventional alloys, and the amalgams made from them are conventional amalgams. This composition range was in use for almost a century until two Canadian researchers designed an alloy consisting of a mixture of two powders; one having the customary flakes of the aforementioned composition range and the other, spherical shapes of a silver–copper eutectic, ie, 72% Ag, 28% Cu. The ratio of the customary powder to the spherical one was approximately 2:1 by weight, giving an approximate composition of 70% silver, 18% tin, 11% copper, and 1% zinc. Amalgams made with this powder mix have high compressive strengths and low creep, have better clinical performance than conventional amalgams, and are sometimes called dispersed amalgams.

Other powders, made from alloys of uniform composition, appeared on the market with approximately the same overall composition but with a high compressive strength of 462 MPa (67,000 psi), high tensile strength of 55 MPa (8000 psi), and low creep (0.2%) when compared to the amalgams made from the mixture of powdered alloys. Other alloys having copper content higher than 18% were subsequently developed. The range of composition of some of these alloys is 42.2–70.3% silver, 17.7–30.0% tin, 12.0–27.8% copper, and 0–0.3% zinc. In addition, one of the powdered alloys contained 3.4% indium.

The copper-rich amalgams have performed well in clinical trials in which they were compared with alloys having lower copper content. An improved marginal stability was observed, which may be associated with a longer clinical lifetime. These amalgams have also been called non-γ_2 amalgams, where γ_2 refers to the compound $Sn_{7-8}Hg$ comprised of Sn_8Hg and Sn_7Hg.

Manufacturing. In the general procedure manufacturing processes used to produce alloy powders for dental amalgams, metals of the required purity are melted together in their proper proportions under nonoxidizing conditions. Either atomization is used to produce spherical particles or ingots are cast and cooled at a rate that produces some Widmanstätten structure, a geometrical pattern caused by the formation of a new phase along certain crystallographic planes of the parent solid solution.

Gallium-Based Alloys. A gallium-based alloy has been introduced commercially in Japan as a substitute for dental amalgam, but similar alloy types have previously been associated with abnormal cellular reactions and are not much used elsewhere; nickel–gallium alloys have produced carcinomas in rats. The corrosion resistance of the gallium alloys is also marginal.

Gold and Gold Alloys. Gold foil, crystal powder, a gold-calcium alloy, and combinations thereof in the noncohesive states are used in dentistry as direct-filling materials; an adsorbed layer of ammonia renders the gold noncohesive for packaging, and heating returns it to a cohesive state for use. Gold alloys are used for cast restorations and prosthetic devices. Wrought gold alloys are used for wire clasps and fabricated orthodontic appliances (see GOLD AND GOLD COMPOUNDS).

Gold and gold alloys serve the needs of dentistry better than any other metals or alloy systems. Gold alloys have a broad range of working characteristics and physical properties, coupled with excellent resistance to tarnish and corrosion in the mouth.

Gold Casting and Wrought Alloys. Gold alloys useful in dentistry may contain gold, silver, platinum, palladium, iridium, indium, copper, nickel, tin, iron, and zinc. Other metals occasionally are found in minor amounts. The effect of each of the constituents is empirical, but some observations have been made.

Gold Alloys, Cast Types. Four types of gold alloys have been recognized for cast dental restorations. They provide desired material for specific uses. The appropriate specifications for these alloys is ANSI/ADA specification no. 5.

Type I, soft alloys (20–22-carat golds), are used for inlays of simpler non-stress-bearing types.

Type II, medium-hard alloys, are harder, stronger, and have lower elongation than type I alloys. They are used for moderate stress application, eg, three-quarter crowns, abutments, pontics, full crowns, and saddles.

Type III, hard alloys, are the hardest, strongest, and least ductile of the inlay casting alloys. Their use is indicated for restorations required to resist large forces such as three-quarter crowns, abutments, pontics, supports for appliances, and precision-fitting inlays.

Type IV, extra hard (partial denture) alloys, are indicated where high strength, hardness, and stiffness are required. These partial-denture alloys are used for cast removable partial dentures, precision-cast fixed bridges, certain three-quarter crowns, saddles, bars, arches, and clasps.

Gold Alloys, Wrought Type. Two types of wrought gold alloys were formerly recognized by the ADA specification no. 7 for the fabrication of orthodontic and prosthetic dental appliances. Alloys of this type are seldom used in the United States; they have been replaced by stainless steels and nickel–titanium alloys.

Platinum and Platinum Alloys. Platinum has excellent resistance to strong acids and, at elevated temperatures, to oxidation. Under reducing conditions at high temperatures it must be protected from low fusing elements or their oxides. Easily reduced metals at high temperatures may form low fusing alloys with platinum.

Platinum, a bluish-white metal, is soft, tough, ductile, and malleable. It has a melting point of 1773°C. The coefficient of thermal expansion is $9 \times 10^{-6}/°C$.

Platinum has many uses in dentistry. Pure platinum foil serves as the matrix in the construction of fused-porcelain restorations. Platinum foil may be laminated with gold foil for cold-welded foil restorations. Platinum wire has found use as retention posts and pins in crown and bridge restorations. Heating elements and thermocouples in high-fusing porcelain furnaces are usually made of platinum or its alloys (see PLATINUM-GROUP METALS).

Platinum, as an alloying element, is used in many dental casting golds to improve hardness and elastic qualities. Platinum in combination with palladium and iridium has limited use for dental pins and wires.

Palladium and Palladium Alloys. Palladium is not used in the pure state in dentistry. However, it is a useful component of many gold casting alloys.

Alloys based on Ag–Pd have been used for a number of years and are available from most gold alloy manufacturers. The palladium content is 22–50 wt %; silver content is from 35 to 66 wt %.

Base-Metal Alloys. Base-metal casting alloys are inferior to gold-based casting alloys in some dental applications but are superior in others. Base-metal alloys and low noble alloys have made inroads into dentistry at the expense of the noble metal alloys. Base-metal alloys containing sufficient chromium to make them passive have essentially displaced gold-based alloys in the casting of skeletons for partial dentures. Some inlays, crowns, or bridges are made of nonprecious alloys. Wrought stainless steels of the 18Cr–8Ni type are used infrequently as denture bases but have supplanted gold alloys in orthodontic appliances for preventing and correcting malocclusion and associated dental and facial disharmonies. Chromium-containing alloys and titanium-based alloys are the only base metal alloys that can be used with dental techniques and have almost no tarnish or corrosion in the mouth.

Cobalt–Chromium Alloys. Co–Cr and Ni–Cr alloys are used predominately for the casting of removable partial dentures; fixed partial dentures (bridges), crowns, and inlays are also cast. Because of high hardness, corrosion resistance, and wear resistance cobalt-chromium alloys are used for bite adjustments and as serrated inserts in plastic teeth used in full dentures. These alloys are well tolerated by the body and also are used for dental implants and orthopedic implant alloys.

Nickel–Chromium Alloys. Because gold is so expensive, there was an intensive development in the 1970s and 1980s of alloys based mostly on nickel–chromium with as many as eight modifying elements. Nickel–chromium alloys cannot be cast as easily or as accurately as gold–based alloys, nor can they be fabricated as easily by soldering. Ni–Cr alloys primarily contain nickel with ca 12% or more of chromium. They are used mainly for the casting of fixed partial dentures (bridges), crowns, or inlays, either with or without aesthetic porcelain or plastic veneers. However, some of them are used for removable partial dentures.

Titanium-Based Casting and Wrought Alloys. Titanium-based alloys offer an attractive alternative to gold alloys and to the base-metal alloys that contain nickel or chromium.

Many of the technical problems of fabrication that formerly inhibited the use of titanium alloys in dental castings have been effectively solved, and titanium castings may now be obtained for virtually any type of dental appliance at prices that are increasingly competitive.

The property most frequently cited in connection with the use of Ti dental or medical appliances is titanium's unique biocompatibility. This helps practitioners avoid occasional allergic reactions that occur with nickel or chromium alloys, and removes concerns about the toxic or carcinogenic potential of appliances that contain nickel, chromium, or beryllium. Wrought alloys of titanium are used for orthodontic wires because of their unique elastic properties, and are used to fabricate most dental implants.

Properties. The casting shrinkage of titanium alloys is comparable, but not identical, to that of chromium-containing alloys, and greater than that of gold alloys. The ability of commercially pure titanium to be cast in relatively thin sections, together with its low density, permits light-weight appliances in bulky restorations.

The Vickers hardness of Ti alloys is strongly dependent on the presence of elements such as oxygen, nitrogen, and carbon that dissolve in hot or molten titanium through contact with residual air molecules or with the surface of refractory molds. Under optimum conditions, castings of commercially pure titanium show yield strengths of 400–500 MPa and ultimate strengths of 490–650 MPa with elongations of 20–30%. Yield strengths and tensile strengths are increased by the presence of dissolved elements.

The stiffness of pure titanium can be increased slightly by alloying; alloys such as Ti–6Al–4V may be specified for partial dentures requiring additional rigidity. Titanium appliances do not tarnish or corrode in the mouth, have no metallic taste, and are easy to clean because plaque and calculus do not adhere to them. The relatively low thermal conductivity of titanium (relatively close to that of tooth enamel) gives these appliances a seemingly natural feel in the mouth and minimizes thermal sensitivity.

Composition. Acceptable composition limits for titanium dental castings have not yet been established, but current practice favors the use of commercially pure or unalloyed titanium.

Oxygen, nitrogen, and carbon promote brittle behavior in all titanium alloys, and it is important to control the cumulative effect of these three elements during casting operations.

Stainless Steels (Iron-Based Alloys). Today, stainless steels find their primary use in wrought form for temporary applications such as orthodontic wires, brackets, and temporary crowns. The temporary crowns are obtained in preformed sizes/shapes and then are trimmed by the dentist with shears to fit over prepared teeth that are awaiting the fabrication of permanent cast crowns.

Orthodontic wires and brackets owe their strength to work hardening during their formation, contrary to the heat treatment process used for wrought gold alloy wires. The 18-8 stainless steels are the most commonly used alloys; their strength, ductility, and elastic modulus are generally higher than the wrought gold alloys.

Aluminum. Aluminum and aluminum alloys play a limited role in dental applications. Aluminum is used to make lightweight impression trays, articulator facebows, and radiograph alignment holders. Soft aluminum is used as preformed temporary crowns where shaping and bending can be easily done at chairside.

Electrodeposited Metals. Electrodeposited, electroplated, or electroformed metals are used to produce an accurate metal-clad die, or cast, from a compound polysulfide rubber, or silicone rubber impression (see ELECTROPLATING). Copper and silver are used to give accurate dies and casts with an improved working surface having strength, abrasion resistance, and chip resistance.

Solders and Fluxes. Dental solders, like all dental alloys, must be biologically tolerated in the oral environment. They are specifically designed or employed for the purpose of fusing two pieces of dental alloy through the use of an intermediate low temperature filler metal.

The soldering process involves the use of solders and fluxes. Solders are materials that fuse and join two dental alloys together. Fluxes coat the surfaces of alloys to be joined in order to produce clean metal surfaces at the temperatures of joining. This enables the molten solder to fuse with those surfaces.

Solders. Modern dental solders are made from mostly corrosion-resistant, nontoxic metals. Minimal quantities of tin and other elements are often added, some of which could produce toxic effects in the unalloyed state. Each solder is used for specific applications; typical solders used in dentistry are low karat, general purpose, high karat, nonprecious alloy, silver, and procelain–gold.

Fluxes. Fluxes, composed mostly of salts or oxides of metals, serve to protect underlying metal from the air. This prevents the formation of surface oxides that impede fusion and the formation of a strong solder joint. Fluxes may also act to selectively leach elements from the surface of the underlying metal. The result is a surface free of obstacles to fusion, and of a composition readily wetted by the solder.

Fluxes also contain agents that facilitate application, such as petroleum jelly or alcohol. These burn or evaporate at elevated temperatures before the soldering temperature is attained, leaving behind a uniform coating of flux. Fluxes become molten before the joining process is reached. The molten flux flows or spreads to form a continuous coating over the surfaces to which they have been applied.

Antifluxes are materials used to coat certain surfaces to prevent solder from flowing to those regions. Antifluxes are not wetted by the solder. They should be removed at the time when the flux is removed.

Nonconventional Solder Systems. Nonconventional solder systems are developed for use with newer alloys, especially base metal alloys. They are few in number and will probably remain the exception rather than the rule.

Polymeric Dental Materials

Acrylic Resins. Because of the unique combination of properties, eg, aesthetics and ease of fabrication, acrylic resins based on methyl methacrylate and its polymer and/or copolymers have received the most attention since their introduction in 1937. However, deficiencies include excessive polymerization shrinkage and poor abrasion resistance. Polymers used in dental application should have minimal dimensional changes during and subsequent to polymerization; excellent chemical, physical, and color stability; processability; and biocompatibility and the ability to blend with contiguous tissues.

Denture Bases. Dentures require accurate fit, reasonable chewing efficiency, and lifelike appearance. The chewing efficiency of artificial dentures is one-sixth that of natural dentition. Acrylic resins are generally used as powder/liquid formulations for denture base, bone cement, and related applications. Polymerization is achieved thermally using initiators; photochemically using photoactive chemicals and either uv or visible light irradiation; and at ambient temperatures using initiator/activator systems.

Properties of Denture-Base Materials. Physical properties of acrylic denture-base materials are given in Table 2. Mechanical properties of denture bases can vary considerably, and depend on composition, mode of polymerization, and degree or interaction with the oral environment.

Special-Purpose Resins, Repair Resins. Fractured acrylic dentures can be repaired with materials similar in composition to cold-cured denture resins. These materials generally cure more rapidly because of the relatively simple manipulations involved. The process is quick and there is little dimensional change, but the strength of the repaired denture may be only half that of the original appliance.

Denture Reliners. A denture can be readapted to the changing contours of soft tissue by relining it with rigid or resilient materials.

Resilient Liners. Resilient liners reduce the impact of the hard denture bases on soft oral tissues. They are designed to absorb some of the energy produced by masticatory forces that would otherwise be transmitted through the denture to the soft basal tissue. The liners should adhere to but not impair the denture base. Other critical properties include total recovery from deformation, retention of mechanical properties, good wettability, minimal absorption of fluids, nonsupport of bacterial or fungal growth, and ease of cleaning. At present, no material fulfills all of these requirements.

Resilient liner materials are generally supplied as powders, liquids, or ready-to-use sheets. The commercial materials currently available are plasticized acrylics or silicones (see SILICON COMPOUNDS–SILICONE).

Tissue Conditioners. Tissue conditioners are gels designed to alleviate the discomfort from soft-tissue injury, eg, extractions. Under a load, they exhibit viscous flow, forming a soft cushion between the hard denture and the oral tissues. The polymer in tissue conditioners is often the same as that used for resilient liners. The liquid is a plasticizer containing an alcohol of low volatility.

Crown and Bridge Resins. These materials, based on methyl methacrylate, higher molecular weight methacrylates, and the epimine resin system, are used as interim tooth coverage during fabrication of permanent prostheses. They are used to maintain the correct biting relationship, stop teeth drifting, and protect the prepared tooth against pulpal irritation and fracture. The epimine resin system requires cationic polymerization rather than free radical polymerization common to acrylic materials.

Resins are also used for permanent tooth-colored veneers on fixed prostheses, ie, crown and bridges. Compositions for this application include acrylics, vinyl–acrylics, and dimethacrylates, as well as silica- or quartz-micro-filled composites.

Plastic Teeth. Plastic teeth are manufactured by an injection- or transfer-molding process.

In addition to acrylics, vinyl resins, polycarbonate, and polysulfone have been suggested for teeth molding. Newer compositions contain very finely dispersed spheres such as pyrogenic silica as reinforcing fillers, ie, microfillers; a urethane dimethacrylate resin; nonfilled highly crosslinked copolymers with interpenetrating networks; and a layered tooth with an exterior containing a BIS–GMA-type resin and silica filler. Pigments impart a natural appearance. The fluorescence of human teeth under uv light is simulated by the incorporation of additives, and a more natural appearance is obtained by painting on stains and striations.

Acrylic teeth have a compressive strength of ca 76 MPa (11020 psi) and a Knoop hardness of 18–20 KHN. These values are lower than those of alloys used for dentures and those of human enamel or dentin. Acrylic teeth have a more natural appearance and are more fracture-resistant than porcelain teeth because of their higher impact strength. The low modulus of elasticity reduces the clicking often experienced by porcelain denture wearers. The deficiencies of plastic teeth are their relatively high wear rate and ultimate loss of occlusal relationship.

Mouth Protectors. The widespread use of protective mouth guards in contact sports has greatly reduced the incidence of orofacial injuries. Guards are produced from ready-made stock or are custom-fabricated. Natural rubber, poly(vinyl chloride), poly(vinyl acetate-*co*-ethylene), polyurethane, and silicone elastomers are the materials of choice.

Maxillofacial Prosthetic Materials. Extraoral or external maxillofacial prosthetics (EMFP) is the science of using polymeric biomaterials for the restoration of missing and/or defective facial tissues. The synthesis of materials that can be easily fabricated into lifelike facial devices has long challenged researchers. Ideally, maxillofacial polymeric materials should mimic as closely as possible the mechanical, physical, chemical, and aesthetic properties of the natural tissues they replace.

Maxillofacial polymers include the chlorinated polyethylenes, polyetherurethanes, polysiloxanes (see ELASTOMERS SYNTHETIC), and conventional acrylic polymers. These are all deficient in a number of critical performance and processing characteristics. It is generally agreed that there is a need for improved maxillofacial polymers that can be conveniently fabricated into a variety of prostheses.

Other Uses. Epoxy die materials are used for the fabrication of cast prostheses in some commercial dental laboratories. Cold-curing plastics are used in making contoured impression trays. These resins contain substantial quantities of fillers that increase the rigidity of the materials after setting. Occlusal night guards are made from poly(methyl methacrylate) or another thermoplastic to protect teeth

Table 2. Physical Properties of Denture-Base Resins

Property	Poly(methyl methacrylate)	Vinyl–acrylics
tensile strength, MPa[a]	48–62	51
compressive strength, MPa[a]	75	61–75
elongation, %	1–2	7–10
elastic modulus, MPa[a]	3.8×10^3	2.3×10^3
impact strength, N·m	1050	3150
transverse strength, MPa[a]	41–55	41–55
flexural strength, MPa[a]	83–117	69–110
Knoop hardness	16–22	14–20
thermal coefficient expansion, °C^{-1}	81×10^{-6}	71×10^{-6}
heat-distortion temp, °C	160–195	130–170
polymerization shrinkage, %	6	6
24-h water sorption	0.3–0.4	0.07–0.4

[a] To convert MPa to psi, multiply by 145.

against excessive grinding. Orthodontics makes extensive use of plastics in retainers, splints, temporary space maintainers, and bite plates.

Elastomer Impression Materials. Dentistry requires impression materials that are easily handled and accurately register or reproduce the dimensions, surface details, and interrelationship of hard and soft oral tissues. Flexible, elastomeric materials are especially needed to register intraoral tooth structures that have undercuts. The flexibility of these elastomers allows their facile removal from undercut areas while their elasticity restores them to their original shape and size.

Impression materials based on natural polymers, eg, reversible hydrocolloids (agar), irreversible hydrocolloids (alginates), combinations of agar/alginate, and, to a lesser extent, zinc oxide–eugenol cements, are still employed. However, the growing use of elastic impression materials by the dental profession has spurred a continuous search for better elastomers. The primary emphasis has been in the development of nonaqueous elastomeric impression materials.

Polysulfide Impression Materials. Significant improvements in strength, toughness, and especially dimensional stability of the set polysulfide elastomers over the aqueous elastic impression materials made these materials popular.

The materials are available in three basic grades, ie, light-bodied or syringe, regular, and heavy-bodied types. The products are supplied as a two-part paste system, usually packaged in collapsible tubes.

The polysulfide impression materials can be formulated to have a wide range of physical and chemical characteristics by modifying the base (polysulfide portion), and/or the initiator system. Further changes may be obtained by varying the proportion of the base to the catalyst in the final mix. Characteristics varied by these mechanisms include viscosity control from thin fluid mixes to heavy thixotropic mixes, setting-time control, and control of the set-rubber hardness from a Shore A Durometer scale of 20 to 60. Variations in strength, toughness, and elasticity can also be achieved.

A number of modifying agents are desirable. Sulfur, in fractional percentages, is a powerful activator. Stearic acid and oleic acid may be used as retarding agents. Additions of acid must be carefully selected because strong acid attacks the methylenedioxy linkages and degrades the polymer. Buffering agents for pH control and perfumes for odor-masking also are used frequently.

The second catalyst paste of the two-paste product is a curing agent. A wide variety of materials convert the liquid polysulfide polymers to elastomeric products. Alkalies, sulfur, metallic oxides, metallic peroxides, organic peroxides, and many metal–organic salts, ie, paint driers, are all potential curing agents.

Only two types of curing systems have found application in dentistry. Lead peroxide is the curing agent most frequently used for the polysulfide polymers that serve as dental impression materials.

The catalyst paste is usually dark brown owing to the high concentration (20–78%) of lead peroxide. A nonlead peroxide curing system offers curing pastes in yellow and red that produce lighter shades of set-rubber colors.

Condensation Silicones. Odor, color, and stickiness of the polysulfide rubbers have deterred universal acceptance. The development of room-temperature-vulcanizing (RTV) silicone rubbers, or siloxanes, has made another acceptable elastomeric system available to dentistry.

The silicone bases have a mild, pleasant odor, and are generally white, although some manufacturers add a pink pigment. The materials are nontacky and can be wiped away from instruments, hands, etc, at any stage of the mix or set. The silicones offer a selection of setting rates and mix viscosities. Regardless of base viscosity, the set rubbers are about the same hardness, ie, a Shore A Durometer value of about 55.

Both the silicone base and the catalyst compositions are moisture-sensitive and subject to deterioration when exposed to the atmosphere.

Condensation silicone materials are based on hydroxyl-terminated polydimethylsiloxane and are made in the form of two paste or paste–liquid catalyst systems.

Addition Silicones. Perhaps the most important development in the area of elastic impression materials has been the addition siloxane (or silicone) system. Several reviews have been published on the materials.

Addition siloxanes are the kind most frequently used in the dental clinic and are supplied as a two-paste system. One paste contains a low molecular weight silicone with terminal vinyl groups and reinforcing filler, and the other consists of a hydrogen-terminated siloxane oligomer, filler, and chloroplatinic acid catalyst; mixing gives a crosslinked elastomer. Hydroxyl-containing silicones evolve hydrogen in the setting reaction. Some formulations contain palladium to absorb this gas. Addition silicones have the best elastic properties and lowest dimensional change on setting of all elastomeric impression materials.

Polyether Impression Materials. A polyether-base polymer elastomeric impression material was introduced in 1964. This material is related to the epimine resin and cured by the reaction between aziridine rings, which are at the ends of branched polyether molecules. The main chain is probably an ethylene oxide–tetrahydrofuran copolymer. Setting is by cross-linking brought about by an aromatic sulfonate ester catalyst that produces cross-linking by cationic polymerization via the imine end groups. These polyethers are supplied as two pastes.

A new type of impression material that has a polyether backbone with urethane–acrylic end groups and that cures free-radically by visible light irradiation, has been developed.

Restoratives. Polymer resins were introduced as tooth restorative materials in the early 1940s. These materials can be classified as unfilled tooth-restorative resins, composite or filled restorative resins, and pit and fissure sealants.

Composite or Filled Tooth Restorative Resins. Improvements in the properties of resin-based restoratives, brought about by the addition of silane-treated inorganic fillers to unfilled resins, has made these the primary anterior restorative material used today.

They are used by direct placement in tooth cavities or by custom fabricating using composites on a gypsum model of the cavity and then cementing the restoration into the tooth cavity.

The addition–reaction product of bisphenol A and glycidyl methacrylate is a compromise between epoxy and methacrylate resins. This BIS–GMA resin polymerizes through a free-radical induced covalent bonding of methacrylate rather than the epoxide reaction of epoxy resins. Mineral fillers coated with a silane coupling agent, which bond the powdered inorganic fillers chemically to the resin matrix, are incorporated into BIS–GMA monomer diluted with other methacrylate monomers to make it less viscous. A second monomer commonly used to make composites is urethane dimethacrylate.

Composite resins can be cured using a variety of methods. Intraoral curing can be done by chemical means, where amine–peroxide initiators are blended in the material to start the free-radical reaction. Visible light in the blue (470–490 nm) spectrum is used to intraorally cure systems containing amine–quinone initiators. Ultraviolet systems were used in some early materials but are no longer available. Laboratory curing of indirect restorations can be done by the above methods as well as the additional application of heat and pressure.

The composites are dispensed as paste systems in one or two parts.

Pit and Fissure Sealants. The BIS–GMA or urethane dimethacrylate portion of the composite restorative material has been further diluted with methyl methacrylate monomer or other low viscosity monomers and used to seal developmental pits and fissures on natural tooth enamel.

Adhesives. The fundamentals of adhesion have an important role in modern restorative dentistry (see ADHESIVES). The retention of restorative materials to tooth structure, in addition to holding the restoration to the teeth, seals the interface between the tooth and restorative material. Maintenance of this seal reduces pulpal irritation and the potential for recurrent decay. The application of high performance adhesives reduces the need for mechanically retentive cavity designs, thereby minimizing removal of healthy tooth structure. Adhesion also plays an important role in holding prosthetic materials to one another. Polymer adhesive materials can be classified as composite or filled resin cements, porcelain or ceramic coupling agents, metal coupling agents, or enamel and dentin adhesives.

Abrasives

Dental abrasives range in fineness from those that do not damage tooth structure to those that cut tooth enamel. Abrasive particles should be irregular and jagged so that they always present a sharp edge, and should be harder than the material abraded. Another property of an abrasive is its impact strenth, ie, if the particle shatters on impact it is ineffective; if it never fractures, the edge becomes dull. Other desirable characteristics include the ability to resist wear and solvation.

Dental abrasives can be classified either according to their use or according to the degree of their ability to abrade. They include aluminum oxide silicas and silicates.

Therapeutic Dental Materials

Fluorides. Most worldwide reductions in dental decay can be ascribed to fluoride incorporation into drinking water, dentifrices, and mouth rinses. Numerous mechanisms have been described by which fluoride exerts a beneficial effect. Fluoride either reacts with tooth enamel to reduce its susceptibility to dissolution in bacterial acids or interferes with the production of acid by bacterial within dental plaque. The multiple modes of action with fluoride may account for its remarkable effectiveness at concentrations far below those necessary with most therapeutic materials. Fluoride release from restorative dental materials follow the same basic pattern. Fluoride is released in an initial short burst after placement of the material, and decreases rapidly to a low level of constant release. The constant low level release has been postulated to provide tooth protection by incorporation into tooth mineral.

Composite Resins. Many composite restorative resins have incorporated fluoride into the filler particles.

Glass-Ionomers. Glass-ionomers show fluoride release at levels that are usually higher than those found in composite materials.

Other Fluoride Containing Materials. Many forms of fluoride are used strictly as preventive or therapeutic materials. Varnishes containing sodium fluoride are available to place in a cavity prior to filling. These varnishes can also be applied as protective agents on tooth crown and root surfaces. Cavity rinses containing stannous fluoride and/or acidulated phosphate fluoride are applied topically to a cavity just before filling. The largest group of therapeutic fluorides are the topical formulations used for treatment of both primary and permanent teeth. These materials are topically applied to tooth surfaces as gels or rinses. Prescription fluoride supplements are supplied as rinses, drops, and tablets containing sodium fluoride or acidulated phosphate fluoride in varying doses. These are used as both topical treatments and dietary supplements. Fluoride is also available at varying concentrations in many over-the-counter mouth rinses, toothpastes, and chewable vitamins. The most common and widespread form of fluoride is that contained in drinking water supplies.

Chlorhexidine Gluconate. Chlorhexidine gluconate (1,1′-hexamethylene bis[5-(*p*-chlorophenyl) biguanide] di-D-gluconate) is used as an antimicrobial against both aerobic and anaerobic bacteria in the oral cavity. It is used as a therapeutic supplement in the treatment of gingivitis, periodontal disease, and dental caries. A mouth rinse form is available as a 0.12 wt% aqueous solution.

Calcium Phosphate Materials

Hydroxyapatite. The mineral of teeth and bone comprises impure forms of hydroxyapatite (HA), $Ca_{10}(PO_4)_6(OH)_2$. Because of its excellent biocompatibility, synthetic HA, in ceramic forms, has been used clinically for filling bony defects since the mid-1970s. Dense and porous types of ceramic HA are available. Dense HA ceramics are made by compressing calcium phosphate solids from a very fine powder form, known as the green state, into a pellet which is then subjected to a heat treatment that causes the powder particles to fuse together by means of solid-state diffusion. Porous HA ceramics are derived from certain species of coral in the genus *Protes* which have regularly patterned 200 μm diameter pores.

Other Ceramic Calcium Phosphate Materials. Other ceramic calcium phosphate materials for repairing bony defect include β-tricalcium phosphate (β-TCP), β-$Ca_3(PO_4)$, and biphasic calcium phosphate (BCP) ceramics which consist of both β-TCP and HA.

Calcium Phosphate Cements. Self-setting calcium phosphate cements have been a subject of considerable interest in recent years. Materials that are totally biocompatible and also harden like a cement at the site of application are highly desirable in a wide range of biomedical applications. Data in the literature show that cementation can occur in mixtures containing a variety of calcium phosphate compounds. The products formed in these systems included a dibasic calcium phosphate known as dicalcium phosphate dihydrate (DCPD). $CaHPO_4{\cdot}2H_2O$, octacalcium phosphate (OCP), $Ca_8H_2(PO_4)_6{\cdot}10H_2O$, and HA.

Dental Implants

The use of dental implants is increasing due to improved success rates. Success for dental implants is based on immobility, no radiolucency between bone and implant, no bone loss, and no other adverse symptoms such as violations of the mandibular nerve canal. A number of present-day implants have means for mechanical fixation such as grooved and porous surfaces. Modern dental implants can function for 10 years or more.

Regulation. Dental implants are regulated by the Food and Drug Administration.

Requirements. Requirements for dental implant materials are the same as those for orthopedic uses. The first requirement is that the material used in the implant must be biocompatible and not cause any adverse reaction in the body. The material must be able to withstand the environment of the body, and not degrade and be unable to perform the intended function.

Materials. Titanium is the most widely used metallic implant today, and titanium–6aluminum–4vanadium alloy is widely used. The cobalt–chromium–molybdenum was used successfully for many years and still is used for casting individual implants and for some of the blade implants. Ceramic materials including hydroxyapatite, high density alumina, and single-crystal sapphire (alpha alumina) also are being studied and/or used as dental implants.

Hydroxyapaite, the mineral constituent of bone, is applied to the surfaces of many dental implants for the purpose of increasing initial bone growth. Some investigators believe that an added benefit is that the hydroxyapatite shields the bone from the metal. However, titanium and its alloy, Ti-6Al-4V, are biocompatible and have anchored successfully as dental implants without the hydroxyapatite coating.

Surface preparation of the dental implant prior to implantation will have an effect on corrosion behavior, initial metal ion release, and interface tissue response. The titanium and titanium alloy dental implants in present use have many forms to assist bone ingrowth attachment including cylinders with holes, screw threaded surfaces, porous surfaces, and other types of roughened surfaces. Methods used to produce porous surfaces include arc plasma spraying of metal onto the implant and the sintering of spheres to the implant surface.

Types of Dental Implants. Indications for a specific type of implant are based primarily on the amount of bone available to support the implant. Also to be considered is the implant proven most successful. Three types of implants are discussed here.

Endosseous. The implants are anchored in the bone and osseointegrated, and are the most successful implants to date. Endosseous implant types are root form, ie, cylinder, screw, cone; bladeform, ie, plate; and ramus frame.

Subperiosteal. The subperiosteal implants are placed on the residual bony ridge and are not osseointegrated. This implant is most commonly used in the mandible but sometimes is used in the maxilla.

Transosteal and Staple Implants. This implant goes through he bone and is for the purpose of attaching dentures. The mandibular bone staple plate has replace the transosteal pin.

Certain commercial materials and equipment may be identified in the article for adequate definition of subject matter. In no instance does such identification imply recommendation or endorsement by the National Institute of Standards and Technology, or that the material or equipment is necessarily the best available for the purpose.

JOHN A. TESK
JOSEPH M. ANTONUCCI
FRED C. EICHMILLER
J. ROBERT KELLY
NELSON W. RUPP
RICHARD W. WATERSTRAT
ANNA C. FRAKER
LAWRENCE C. CHOW
LAURIE A. GEORGE
JEFFREY W. STANSBURY
EDWARD E. PARRY
National Institute of Standards and Technology (NIST)

The Standards Coordinator, The American Dental Association, Chicago.

W. J. O'Brian ed., *Dental Materials, Properties and Selection,* Quintessence Publishing Co., Chicago, 1989.

R. W. Phillips, *Science of Dental Materials,* W. B. Saunders Co., Philadelphia, 1991, pp. 217–235.

National Institutes of Health, *Consensus Development Conference Statement on Dental Implants,* **7** (1988).

DENTIFRICES

The three forms of marketed dentifrices are pastes (toothpastes), liquid concentrates, and powders. Pastes account for an overwhelming portion of the dentifrice market. Liquid concentrates and powders account for only a minor portion because they have disadvantageous dispensing characteristics and they do not provide a means to accurately and consistently dispense a dose of active agent.

Dentifrices and dental rinses have many common purposes and ingredients, thus dental rinses are discussed briefly herein.

Toothpastes

Toothpastes are packaged in flexible tubes, other flexible containers, and mechanically operated pump dispensers. They are usually extruded as cylindrical ribbons of a cohesive, smooth paste, approximately 2.54 cm in length and weighing approximately 1.5 g. New or modified dispensing devices are continually introduced to increase consumer interest.

General Toothpaste Formulation. The key functions of dentifrices, to remove dental stains and to freshen the oral cavity, are accomplished by abrasive cleaning and the masking or elimination of unpleasant oral odors. Materials designed to deliver antitartar, antiplaque, or anticaries benefits must be compatible with the ability of the dentifrice to fulfill those two functions.

Toothpaste contains an abrasive (qv), flavor, a humectant system, a surfactant, a binding and thickening agent, color, and one or more therapeutic or cosmetic agents.

Abrasive. An abrasive is usually chemically inert, neither interacting with other dentifrice ingredients nor dissolving in the paste or the mouth. Substances used as dentifrice abrasives include amorphous hydrated silica, dicalcium phosphate dihydrate, anhydrous dicalcium phosphate, insoluble sodium metaphosphate, calcium pyrophosphate, α-alumina trihydrate, and calcium carbonate.

Flavor. Generally recognized as safe (GRAS) flavors or flavors from approved lists are used. The most popular flavors are peppermint, spearmint, cinnamon, and mixtures of these. Synthetic sweeteners are usually added, although regulatory concerns limit their selection; for example, cyclamate is not used in the United States.

Humectant System. The humectant system comprises one or more liquids in which the other toothpaste ingredients are dispersed to provide a stable paste. The liquids are usually glycerol and 70% aqueous sorbitol. Propylene glycol and polyethylene glycols are used occasionally.

Surfactant. The surfactant most commonly used is the anionic detergent sodium lauryl sulfate. Other surfactants that have been used include sodium dodecylbenzene sulfonate, sodium *N*-lauroyl sarcosinate or Gardol, and sodium cocomonoglyceride sulfonate.

Binding and Thickening Agent. Gums and resins are employed to obtain the desired thickening and binding.

Color. Colorants used in dentifrices are regulated by the Food and Drug Administration. They are FD&C (food, drug, and cosmetic) or D&C (drug and cosmetic).

Special Active Agents. Fluoride added to a compatible dentifrice base at a level of 1000 ppm has been clinically proven to reduce the incidence of dental caries by about 25% on average, even in areas where the water supply is fluoridated. Elevation to 1500 ppm increases the protection. Sources of fluoride approved for use in dentifrices are sodium fluoride (0.22%), sodium monofluorophosphate (0.76%), and stannous fluoride (0.41%).

Several agents delivered via toothpaste inhibit the accumulation of dental calculus. Pyrophosphate salts, with or without a methoxyethylene–maleic acid copolymer, and zinc salts have given positive results in clinical trials.

Dentifrices are also vehicles for agents that alleviate dentinal hypersensitivity. Among the materials that have given positive results in clinical tests are potassium nitrate (5%) and strontium chloride (10%).

Claims for safety and efficacy of therapeutic dentifrices are regulated by the Food and Drug Administration. The Council on Dental Therapeutics of the American Dental Association reviews products and grants a Seal of Acceptance to those products deemed worthy.

Specific Toothpaste Formulations. Two types of toothpaste formulation predominate. Type 1 is a low abrasive–high solvent toothpaste; type 2 is a high abrasive–low solvent toothpaste. The most important differences are the ratio of humectant to abrasive and the nature of the abrasive.

Gels. Amorphous hydrated silicas of a purity and structure typical of those used in type 1 dentifrices and the liquid portion (humectant system) of type 1 dentifrices both have approximately the same refractive index, ie, about 1.47. As a result, the type 1 dentifrices are inherently transparent or translucent. In the marketplace it has become popular to refer to such dentifrices as gels.

Dental Rinses

Dental rinses, as discussed here, do not include gargles and other liquids that are indicated for inflammation and diseases of the throat and are unrelated to dental plaque.

The primary function of a dental rinse is to clean and refresh the mouth. With few exceptions dental rinses marketed in the early 1990s also fulfill important secondary functions. Some contain agents that inhibit or destroy oral microbial populations that aid in generating dental plaque or oral malodor. Other rinses deliver fluoride to the teeth to prevent the development of caries. Dental rinses can also deliver active agents that cannot be provided by toothpaste because of chemical incompatibility between the agent and the toothpaste ingredients.

An oral dental rinse generally consists of water, alcohol, a humectant, an emulsifier, flavor, color, and an active agent. Water is the primary vehicle. The alcohol provides bite and is also a formulation aid. The humectant improves the feel in the mouth and also prevents locking of the cap to the container between uses; glycerin or noncrystallizing sorbitol may be satisfactory. The emulsifier is a nonionic type, for example, a polyoxyethylene–polyoxypropylene block copolymer or a polyoxyethylene sorbitan fatty acid ester. Flavors are generally a type of mint or cinnamon. Colors are FD&C or D&C.

Active agents vary according to use. For controlling bad breath, zinc salts, sodium lauryl sulfate, and flavors are used. To destroy oral microorganisms, chlorhexidine, cetylpyridinium chloride, and benzalkonium chloride are valuable. Essential oils, such as thymol, eucalyptol, menthol, and methyl salicylate reduce plaque-related gingivitis (see OILS, ESSENTIAL). Sodium fluoride aids in caries control.

Chlorhexidine is the most potent oral antimicrobial agent available. It has side effects and is sold only with a prescription.

A dental rinse containing sodium lauryl sulfate, marketed to be used before toothbrushing, removes some plaque directly and makes residual plaque easier to remove by brushing.

Claims for oral dental rinses are regulated by the Food and Drug Administration whether they are marketed as drugs or cosmetics. The Council on Dental Therapeutics of the American Dental Association reviews oral rinses, and may authorize use of the Seal of Acceptance for a product.

Economic Aspects

The strongly research-oriented dentifrice market is dominated by a few primary companies. The industry funds efforts to develop products with superior cosmetic and therapeutic performance. The four principal manufacturers of toothpaste worldwide are Procter & Gamble, Unilever, Colgate-Palmolive, and Beecham. Numerous smaller companies and private-label houses also exist.

Tooth Whitening Agents

Tooth whitening preparations are available by dentist's prescription and over the counter. Many products are marketed as systems comprising toothpastes and treatment pastes or gels. The whitening agent in most is 10% carbamide peroxide, which reacts with water to release hydrogen peroxide. Some preparations contain calcium peroxide as the bleaching agent. The products have undisputed efficacy. Some dental authorities have questioned their safety, hypothesizing that long-term chronic exposure to hydrogen peroxide can result in ill effects.

MORTON PADER
Consumer Products Development Resources, Inc.

M. Pader, *Oral Hygiene Products and Practice,* Marcel Dekker, New York, 1988.

American Dental Association "Guide to Dental Health", *J. Am. Dental Assoc.* **116,** special issue (1988).

M. Pader, *Chemistry and Technology of the Cosmetics and Toiletries Industry,* Blackie Academic & Professional, London, 1992, Chap. 7.

M. Pader, *Rheological Properties of Cosmetics and Toiletries,* Marcel Dekker, New York, 1993, Chap. 7.

DESICCANTS

Some materials have sufficient capacity for water or efficiency for drying as well as appropriate physical and chemical properties that they are classed as drying agents, or desiccants. These substances are widely used for removing water from gases, liquids, and solids. Desiccants may be liquids or solids. They may be used repetitively by regenerating the desiccant after use to return it to its active state, or they may be used only once. If the desiccant is used only once, it may last the life of the article being dried or may be discarded when spent. Drying agents are used either in a static (batchwise) or dynamic (continuous or semicontinuous) mode. Their use may be further classified as open system, if fluid flows through the system, or closed system if it does not. Examples of the industrial uses of desiccants, designated as dynamic or static and open- or closed-system applications, are given in Table 1. The list is not all inclusive and ignores various laboratory uses.

Desiccants have varied fundamental characteristics in terms of water capacity and the rate of water sorption. The degree of water removal achieved, or efficiency, is usually given in terms of the water content remaining in the substance that has been dried. The effectiveness of any drying agent can be measured in terms of its water capacity.

Table 1. Applications of Desiccants

Industry	Application	Classification
compressed air	prevent freeze-up and corrosion in air-actuated components	dynamic, open
air separation	prevent ice formation in heat exchangers before cryogenic distillation	dynamic, open
natural gas	prevent corrosion and hydrate formation in pipelines, remove water before cryogenic hydrocarbon recovery, dry liquefied petroleum gas (LPG) to prevent freeze-ups during vaporization	dynamic, open
petrochemical	remove moisture before low temperature fractionation, remove moisture during the rejuvenation or burnoff of spent catalysts, prevent side reactions during catalytic refining	dynamic, open
chemical	remove water that is a diluent or contaminant of some finished product, remove water prior to or during polymerization reactions, prevent caking and corrosion	static or dynamic, closed or open
storage and shipping	prevent food deterioration and corrosion of equipment by relative humidity control	static or dynamic, closed or open
moisture vapor control	lower the dew point in sealed spaces where condensation could occur	static, closed
vapor compression refrigeration	remove moisture from circulating refrigerants	dynamic, closed
space cooling	dry air to permit cooling by evaporation of water	dynamic, open
dehumidification	dry ambient air for air-conditioning or for storage, manufacture, or drying of moisture-sensitive parts or materials	dynamic, open
corrosion control	reduce the dew point in automobile exhaust systems during cool down to reduce internal cold condensate corrosion	static, semiclosed
absorption refrigeration	cyclic absorption and stripping of water with liquid desiccants to produce chilled water for air conditioning[a] and process cooling	dynamic, open

[a] See AIR CONDITIONING

Mechanism

The drying mechanisms of desiccants may be classified as follows: Class 1: chemical reaction, which forms either a new compound or a hydrate; Class 2: physical absorption with constant relative humidity or vapor pressure (solid + water + saturated solution); Class 3: physical absorption with variable relative humidity or vapor pressure (solid or liquid + water + diluted solution); and Class 4: physical adsorption.

These mechanisms are characterized by the relative magnitudes of the heats of reaction, solution, or adsorption (see ADSORPTION, SEPARATION). All useful drying mechanisms are exothermic.

Compatibility

Desiccants must be chemically compatible with the material being dried. Ideally, the desiccant and the material should not react because such a reaction may produce harmful or undesirable by-products.

Static Drying

Many liquids are dried batchwise rather than continuously. The drying agent is added to the liquid and sufficient time is allowed to dry the product. The liquid is then separated from the drying agent by filtration, decantation, or distillation. Drying agents employing Class 1 or 2 mechanisms are generally used for these applications.

Desiccants Used in Static Drying. The most commonly used desiccants are barium oxide, calcium chloride, calcium oxide, calcium sulfate, lithium chloride, perchlorates, and sodium and potassium hydroxides. Activated alumina, silica gel, and molecular sieves, which are listed under dynamic, solid drying agents, are also widely used in static or batch-drying situations.

Capacity and Efficiency. The higher capacity desiccants go through the various hydrate levels to yield fairly broad ranges of constant relative humidities as moisture is picked up. However, these compounds do not produce very low relative humidities or dew points. The best performance, or lowest dew point, occurs with excess drying agent.

From a study of the efficiency of some 25 desiccants for drying several families of laboratory solvents and reagents it was concluded that molecular sieves are the desiccants of choice in most cases.

Closed-System Drying. Equilibrium capacity is the principal consideration in the design of closed nonregenerative, relatively static drying systems. The total amount of moisture to be removed must first be calculated from the volume of the system and the initial water concentration. Depending on the final moisture content desired, a drying agent can be selected based on its compatibility with the material to be dried and its ability to produce the final dew point.

Closed systems are usually nonregenerative: the desiccant charge is designed for the life of the system or is replaceable.

Because the system likely is nonisothermal, the analysis of a closed-desiccant system requires knowledge of the temperature of the desiccant as well as the dew point (ice point) or water concentration (partial pressure) specification. Indeed, the whole system may undergo periodic temperature transients that may complicate the analysis.

Another aspect to consider in the design of closed-drying systems is the dry-down time. The drydown time is the period required for the system to dry down from its initial water concentration (or partial pressure) to a concentration that approaches equilibrium with the desiccant. During this time, the system is not fully protected from the negative effects of the moisture that the desiccant is designed to remove. In such a system, the instantaneous drying rate is proportional to the water content at any time.

If required, the drydown can be hastened by increasing desiccant mass, particle surface area, or mass-transfer coefficient. The mass-transfer coefficient can be altered to some extent by the design of the desiccant container.

Dynamic Desiccants

Continuous drying is employed when drying a volume of gas or liquid in a batchwise fashion is not practical. When a solid, dynamic desiccant is used, the fluid stream is passed over a fixed bed of the drying agent, which must have the physical properties to allow the fluid to pass readily through. When liquid desiccants are used, the drying is usually achieved by countercurrent contact of the gas (flowing up) against the liquid (flowing down). The desiccants are usually regenerable.

Glycols and sulfuric acid are the principal examples of liquid desiccants.

The solid desiccants used in dynamic applications fall into a class called adsorbents. Because they are used in large packed beds through which the gas or liquid to be treated is passed, the adsorbents are formed into solid shapes that allow them to withstand the static (fluid plus solid head) and dynamic (pressure drop) forces imposed on them. The most common shapes are granules, extruded pellets, and beads. Solid desiccants include activated alumina, silica gel, and molecular sieves.

Design of Dynamic Adsorption Drying Systems

Adsorbent drying systems are typically operated in a regenerative mode with an adsorption half-cycle to remove water from the process stream and a desorption half-cycle to remove water from the adsorbent and to prepare it for another adsorption half-cycle. Usually, two beds are employed to allow for continuous processing. In most cases, some residual water remains on the adsorbent after the desorption half-cycle because complete removal is not economically practical. The difference between the amount of water removed during the adsorption and desorption half-cycle is termed the differential loading, which is the working capacity available for dehydration.

The two most common types of drying systems operate on either a pressure-swing cycle or a thermal-swing cycle to take advantage of the difference in water loading on the desiccant with changes in pressure and temperature.

Adsorption Plots. Isotherm plots are the most common method of presenting adsorption data. An isotherm is a curve of constant temperature: the adsorbed water content of the adsorbent is plotted against the water partial pressure in equilibrium with the adsorbent. An isostere plot shows curves of constant adsorbed water content: the vapor pressure in equilibrium with the adsorbent is plotted against temperature.

Mass Transfer and Useful Capacity. The term useful capacity, or breakthrough capacity, differs from the equilibrium capacity. The useful capacity is a measure of the total moisture taken up by a packed bed of adsorbent at the point where moisture begins to appear in the effluent. Thus the drying process cycle must be stopped before the adsorbent is fully saturated. The portion of the bed that is not saturated to an equilibrium level is called the mass-transfer zone.

The parameters affecting the size and shape of a mass-transfer zone are adsorbent type, adsorption isotherm shape, flow rate, packed-bed depth, adsorbent particle size, physical properties of the carrier fluid, temperature, pressure, and the concentration of water in the carrier fluid.

Economic Aspects

The largest U.S. manufacturer of molecular sieves for adsorbent and desiccant use is UOP, which has a production capacity of 18–20 million kg/year. W.R. Grace and Zeochem have about 7 and 2 million kg/year capacity, respectively. W.R. Grace is the largest producer of silica gel desiccants. Activated alumina for use as adsorbent and desiccant is produced by LaRoche Chemicals (formerly Kaiser) and by Aluminum Company of America. About one-third of the U.S. supply of activated alumina adsorbent and desiccant is imported by Rhône-Poulenc.

The largest users of molecular sieve desiccants are the natural gas processing, insulating glass, and refrigeration industries. Silica gel dominates the packaging industry, where the material is used to protect electronic equipment and pharmaceuticals, for example, from moisture. Silica gel is also used in dehumidification of buildings. The principal uses of activated alumina are in refrigeration, where its primary function is to adsorb organic acids rather than water, air drying, and alkylation feed stream drying in oil refineries.

Future Applications

Energy Storage. Reactivating a desiccant stores the reactivation energy in the dehydrated desiccant. This energy-storage feature is useful if the energy source is intermittent or seasonal, such as solar energy, or interruptable. Research suggests that this energy-storage feature is especially useful if the desiccant is used to dry air for agricultural applications.

Desiccant Cooling. Considerable work is now being done in the field of desiccant-based cooling systems. In these systems, a desiccant is used to produce an extremely dry air stream. The dry air is then cooled in a heat exchanger and humidified with a water spray. The evaporation of the water absorbs heat from the air and produces a cold, almost saturated air stream for air conditioning. The desiccant is thermally regenerated with exhaust air, which is heated with solar energy, natural gas, or waste heat from power generation. Both liquid and solid desiccant systems have been studied.

ALAN P. COHEN
UOP

O. A. Hougen and F. W. Dodge, *The Drying of Gases,* J. W. Edwards, Ann Arbor, Mich., 1947.

C. Misra, *Industrial Alumina Chemicals,* ACS Monograph 184, American Chemical Society, Washington, D.C., 1986, pp. 107–120.

R. K. Iler, *The Chemistry of Silica,* John Wiley & Sons, Inc., New York, 1979, pp. 462–621.

D. W. Breck, *Zeolite Molecular Sieves: Structure, Chemistry, and Use,* Wiley-Interscience, New York, 1974, 715 pp.

DESIGN OF EXPERIMENTS

The main reason for designing an experiment statistically is to obtain unambiguous results at a minimum cost. The need to learn about interactions among variables and to measure experimental error are added reasons.

Many chemists and engineers think of experimental design mainly in terms of standard plans for assigning treatments to experimental units, such as the Latin square, factorial, fractional factorial, and central composite (or Box) designs. These designs are described in books, such as those summarized in the General References, and catalogued in various reports and articles. Important as such formal plans are, the final selection of test points represents only the proverbial tip of the iceberg, the culmination of a careful planning process.

Statistically planned experiments are characterized by *(1)* the proper consideration of extraneous variables; *(2)* the fact that primary variables are changed together, rather than one at a time, in order to obtain information about the magnitude and nature of interactions and to gain improved precision in estimating the effects of these variables; and *(3)* built-in procedures for measuring the various sources of random variation and for obtaining a valid measure of experimental error against which one can assess the impact of the primary variables and their interactions.

A well-planned experiment is often tailor-made to meet specific objectives and to satisfy practical constraints. The final plan may or may not involve a standard textbook design.

Purpose and Scope of the Experiment

Designing an experiment is like designing a product. Every product serves a purpose; so should every experiment. This purpose must be clearly defined at the outset.

In addition to defining the purpose of a program, one must decide on its scope. An experiment is generally a vehicle for drawing inferences about the real world. Since it is highly risky to draw inferences about situations beyond the scope of the experiment, care must be exercised to make this scope sufficiently broad.

Experimental Variables

An important part of planning an experimental program is the identification of the variables that affect the response and deciding what to do about them. Controllable or independent variables in a statistical experiment can be dealt with in four different ways. The assignment of a particular variable to a category often involves a trade-off among information, cost, and time.

Primary Variables. The most obvious variables are those whose effects on performances are to be evaluated directly. Such variables may be quantitative or qualitative. When there are two or more variables, they might interact with one another.

An important purpose of a designed experiment is to obtain information about interactions among the primary variables. This is accomplished by varying factors simultaneously rather than one at a time.

Background Variables and Blocking. In addition to the primary controllable variables there are those variables, though not of primary interest, that cannot, and perhaps should not, be held constant. It is of crucial importance that such background variables are not varied in complete conjunction with the primary variables. If the background variables do not interact with the primary variables, they may be introduced into the experiment in the form of experimental blocks. An experimental block represents a relatively homogeneous set of conditions within which different conditions of the primary variables are compared.

A main reason for running an experiment in blocks is to ensure that the effect of a background variable does not contaminate evaluation of the effects of the primary variables. However, blocking removes the effect of the blocked variables from the experimental error as well, thus allowing more precise estimation of the experimental error and, as a result, more precise estimates of the effects of the primary variables.

Uncontrolled Variables and Randomization. A number of further variables, such as ambient conditions (temperature, pressure, etc), can be identified but not controlled, or are only hazily identified or not identified at all but affect the results of the experiment. To ensure that such uncontrolled variables do not bias the results, randomization is introduced in various ways into the experiment to the extent that this is practical.

Variables Held Constant. Finally, some variables should be held constant in the experiment. Holding a variable constant limits the size and complexity of the experiment but, as previously noted, can also limit the scope of the resulting inferences.

Experimental Environment and Constraints

The operational conditions under which the experiment is to be conducted and the manner in which each of the factors if varied must be clearly spelled out.

Practical considerations enter into the experimental plan in various other ways. They include prior knowledge, response variables, types of repeat information, preliminary estimates of repeatability, and consistent data-recording procedures.

Stagewise Experimentation

Whether or not to conduct a particular experiment in stages depends on the program objectives and the specific experimental situation; a stagewise approach is recommended when units are made in groups or one at a time and a rapid feedback of results is possible. Running the experiment in stages is also attractive in searching for an optimum response, because it might permit a move closer to the optimum from stage to stage. On the other hand, a single-stage experiment may be desirable if there are large start-up costs at each stage or if there is a long waiting time between fabricating the units and measuring their performance.

Formal Experimental Plans

After the preceding considerations have been taken into account, a test plan is developed to best meet the goals of the program. This might involve one of the standard plans, developed by statisticians. Such plans are described in various texts. Sometimes, combinations of plans are encountered, such as a factorial experiment conducted in blocks or a central composite design using a fractional factorial base. They include: blocking designs, complete factorial and fractional factorial designs, response surface designs, mixture designs, and taguchi designs.

Computer-Aided Design of Experiments

In recent years, numerous computer software packages have been developed to help generate experimental designs. They are especially useful in performing the mechanical tasks associated with the design and analysis of experiments. They can also be helpful in generating "statistically optimum" test plans for special situation.

GERALD J. HAHN
General Electric Company

G. E. P. Box, W. G. Hunter, and J. S. Hunter, *Statistics for Experimenters: An Introduction to Design, Data Analysis, and Model Building,* John Wiley & Sons, Inc., New York, 1978.

D. C. Montgomery, *Design and Analysis of Experiments,* 3rd ed., John Wiley & Sons, Inc., New York, 1991.

R. D. Moen, T. W. Nolan, and L. P. Provost, *Improving Quality through Planned Experimentation,* McGraw-Hill, New York, 1991.

DETERGENCY

The term detergency is limited to systems in which a liquid bath is present and is the main cleaning component of the system. The cleaning is enhanced primarily by the presence in the bath of a special solute, the surfactant, which alters interfacial effects at the various phase boundaries within the system.

In the cleaning or washing process in a typical detersive system the soiled substrate is immersed in or brought into contact with a large excess of the bath liquor. Enough bath is used to provide a thick layer over the whole surface of the substrate. During this stage, air is displaced from soil and substrate surfaces, ie, they are wetted by the bath. The system is subjected to mechanical agitation, either rubbing or shaking, which provides the necessary shearing action to separate the soil from substrate and disperse it in the bath. Agitation also promotes mass-transfer in the system, just as in a heterogeneous chemical reaction. The bath carrying the removed soil is drained, wiped, squeezed, or otherwise removed from the substrate. The substrate is rinsed free of the remaining soiled bath. This rinsing step determines the final cleanliness of the substrate. The cleaned substrate is dried or otherwise finished.

A meaningful discussion of detergency requires a definition of clean. In the physiochemical sense, a surface is clean if it contains no molecular species other than those in the interior of the two adjoining phases. It is difficult to achieve such a state even under the most exacting laboratory conditions. Practically, a surface is clean if it has been brought to a desired state with regard to foreign matter present upon it, as judged by agreed upon criteria. Most standards for cleanness involve a visual or optical judgment for the presence of foreign matter.

Components of Detersive Systems

Substrates. Solid objects to be cleaned vary widely in chemical composition and surface configuration. With few exceptions, however, they can be divided into fabrics and fibrous materials, and hard surfaces.

Soils. Soils vary greatly. They may be a single solid or liquid phase but usually are two or more phases, intimately and randomly mixed and irregularly disposed over the substrate. In a large number of important detersive systems, the nature of soil and the quantity present are well known.

As a result of many painstaking investigations, the soils on apparel encountered in laundering have been shown to be complex mixtures containing both oily and finely divided solid material. The oily material consists largely of fatty acids and polar fatty material but a considerable proportion of neutral nonpolar oil is also present. The solid components vary widely with the locale in which samples are taken, and resemble local street dust in composition.

Particle size is one of the most important factors in determining the ease with which solid soil can be removed from a substrate. Particles of >5 μm dia are generally easily removed. Particles of $<10-$ nm dia cannot be removed by ordinary detersive processes once they are attached to a typical textile fabric. Such particles are responsible for the gradual irreversible graying of white goods with continued wear and laundering. Particles in this size range tend to form clumps and clusters before they reach the fiber surface. These clusters behave like individual large particles. Particles or clusters in the range of 100 nm dia resist removal by simple agitation in liquids that are not surface active, but these particles are removable by normal detersive processes. This is the size range of greatest interest.

Soil may include material that is soluble in the bath, such as encrusted sugar residues and molecularly dispersed material such as fruit juice stains. Removal of these soils is an important aspect of cleaning but is not generally considered in discussions of detergency.

Baths. The baths discussed herein are aqueous solutions. Some nonaqueous systems, however, are true detersive systems. Modern dry cleaning baths, for example, contain solutes that are surface active in the conventional hydrocarbon or chlorinated hydrocarbon medium and aid soil removal. The physical chemistry of such systems differs considerably from that of aqueous systems. Among bath components the solute that is effective in cleaning, usually a mixture of several components, is called the detergent. The term detergent is also used frequently in the restricted sense of a surfactant of high detersive power. In many hard-surface systems, however, nonsurfactants such as alkaline silicates and phosphates exert a true detersive effect. They are, in fact, the principal detergents in these systems even in the complete absence of any surfactants. In the cleaning of fabric systems, the most important detersive component in the bath is the surfactant. Nonsurfactant components that augment the cleaning effect of surfactants are called builders. Many materials that act as builders in fabric systems, eg, phosphates and silicates, are the primary detergents in hard-surface systems, although their primary contribution to the cleaning process may differ in the two cases.

Formulation

Detergents are formulated to clean a defined set of soiled substrates under an expected range of washing conditions. Some detergents, the familiar bar or toilet soap, for example, consist essentially of only one component. There are few systems, however, in which a suitably formulated detergent consisting of several components does not outperform the best single-component system. Although detergents for hand dishwashing rarely contain builders, those currently used in the U.S. contain at least three surfactants, and may contain up to six. Ingredients of laundering detergent formulations for fabrics may be divided into the following groups: surfactants, including soap and various others; the inorganic salts, acids, and bases, including builders, and other compounds that do not contribute to detergency but provide other functions, such as regulating density and assuring crispness of powdered formulations; organic additives that enhance detergency, foaming power, emulsifying power, or soil-suspending effect of the composition; and special purpose additives, such as bleaching agents, fluorescent whitening agents, antimicrobial agents, blueing agents, or starch, which provide desirable performance functions but have no direct effect on soil removal (see also INDUSTRIAL ANTIMICROBIAL AGENTS).

Fabric detergent formulations for special applications, such as the various specific operations within the textile mill, are frequently much simpler. They tend to contain little if any builder or special-purpose additive. The indispensable ingredient in fabric detergency is the organic surfactant. Formulations for hardsurface detergency such as those used in automatic machine dishwashing, are simpler than fabric-washing compositions. An organic surfactant is frequently not needed and inorganic salts are the detersive ingredients.

Surfactants. The most important components of detersive systems are, of course, the surfactants described elsewhere in the *Encyclopedia.*

Builders. Builders are substances that augment the detersive effects of surfactants. Most important is the ability to remove hardness ions from the wash liquor (ie, soften the water) and thus to prevent them from interacting with the surfactant. They include phosphates, sodium carbonate, silicates, zeolites, clays, nitrilotriacetic acid, alkalies, and neutral soluble salts.

Organic Additives. Certain nonsurfactant organic additives improve cleaning performance and exhibit other desirable properties.

Antiredeposition agents contribute to the appearance of washed fabrics. Sodium carboxymethylcellulose, NaCMC is the most widely used, and on cotton fabrics, the most effective. With the advent of synthetic fabrics, other cellulose derivatives, eg, methylcellulose, hydroxybutylcellulose, hydroxypropyl- and mixed methyl and hydroxybutycellulose ethers have been shown to be more effective than NaCMC (see CELLULOSE ETHERS).

Fluorescent whitening agents (qv) absorb ultraviolet radiation and subsequently emit some of the radiation energy in the blue part of visible spectrum. As a result, they confer enhanced whiteness to the appearance of washed articles. For synthetic fabrics such as polyester, it has proved to be more effective to prebrighten the fabric by incorporating the fluorescent whitening agent in the spin-melt during manufacture rather than depend on adsorption from the detergent bath. As a result, the usage of fluorescent whitening agents in formulated laundry products has decreased in recent years.

Blueing agents, which are dyes, provide another approach to maintaining fabric whiteness by a mechanism in which a yellow cast of washed fabrics is covered by the blue dye. Since this approach reduces reflectance, it is less desirable than the use of fluorescent whitening agents that increase reflectance.

Proteolytic enzymes in particular are widely used in premium products. They degrade proteinaceous stains and aid the cleaning performance of other formulation ingredients. Amylases and lipases have been used in a few U.S. detergents, the former to remove starches and the latter, fatty esters and triglycerides.

One U.S. detergent manufacturer has introduced detergents with the bleach activator sodium nonanoyloxybenzene sulfonate. This activator forms the surface active species pernonanoic acid that does provide a bleach benefit under U.S. conditions. In automatic-dishwashing formulations, bleaching agents are needed to remove food stains from dishware and break down proteinaceous soil. Chlorine is the most cost-effective agent available for this purpose and is present in all U.S. products as chlorinated isocyanurate.

Foam regulators such as amine oxides, alkanolamides, and betaines are present in products where high foam value is functionally or esthetically desirable, mainly hand-dishwashing liquids and shampoos. In automatic dishwashing products, on the other hand, copious foam volumes interfere with the efficiency of the mechanical rotors during operation. In this type of product, a foam depressant is often present.

Organic sequestering agents serve the same purpose as the sequestering phosphates, ie, to remove interfering metal ions from the detergent bath. They would appear to also provide some benefits through ionic strength and soil suspending effects. They are used where the less expensive phosphates are, for one reason or another, not applicable.

Factors Influencing Detergency

Detergency is mainly affected by the concentration and structure of surfactant, hardness and builders present, and the nature of the soil and substrate. Other important factors include wash temperature; length of time of washing process; mechanical action; relative amounts of soil, substrate, and bath, generally expressed as the bath ratio, ie, the ratio of the bath weight to substrate weight; and rinse conditions.

Mechanisms

Even the simplest detersive system is surprisingly complex and heterogeneous. It can nevertheless be conceptually resolved into simpler systems that are amenable to theoretical treatment and understanding. These simpler systems are represented by models for substrate-solid soil and substrate-liquid soil. In practice, many soil systems include solid–liquid mixtures. However, removal of these systems can generally be analyzed in terms of the two simpler model systems. Although these two systems differ markedly in behavior and structure, and require separate treatment, there are certain overriding principles that apply to both.

The first principle is that soil systems can be regarded and treated as classical systems of colloid and surface chemistry. A second principle applying to these model systems is derived from their colloidal nature. With the usual thermodynamic parameters fixed, the systems come to a steady state in which they are either agglomerated or dispersed. No dynamic equilibrium exists between dispersed and agglomerated states. In the solid-soil systems, the particles (provided they are monodisperse, ie, all of the same size and shape) either adhere to the substrate or separate from it. In the liquid-soil systems, the soil assumes a definite contact angle with the substrate, which may be anywhere from 0° (complete coverage of the substrate) to 180° (complete detachment). The governing thermodynamic parameters include pressure, temperature, concentration of dissolved components, and electrical conditions.

In applying this concept, the factor of particle size must be continuously borne in mind. A heterodisperse system can reach a steady state wherein the smaller particles are agglomerated and the larger particles are dispersed, giving the apparent effect of an equilibrium. In ideal monodisperse systems under steady conditions, however, no such effects are noted.

Purely mechanical disturbances (which are not usually considered thermodynamic variables) may influence the state of aggregation of a colloidal system; for example, floc size in carbon and iron oxide suspensions varies with the degree of agitation being imposed on the system.

A final consideration in resolving practical detersive systems into their simpler components relates to soil removal versus redeposition. Superficially, it would appear that the redeposition phenomenon contradicts the all-or-nothing concept that the system must exist in either the agglomerated or the dispersed state. Keeping in mind both the composite nature and the kinetics of a practical system, it is readily shown that no such contradiction exists. The soil particle that redeposits is in a different state from what it was during its initial removal from the substrate. Thus, the initial group of agglomerated systems, which composes the soil–substrate–bath complex before soil removal, is quite different from the agglomerated system composed of substrate and redeposited soil.

Solid-Soil Detergency

Adsorption. Many studies have been made of the adsorption of soaps and synthetic surfactants on fibers in an attempt to relate detergency behavior to adsorption effects. Relatively fewer studies have been made of the adsorption of surfactants by soils. Plots of the adsorption of sodium soaps by a series of carbon blacks and charcoals show that the fatty acid and the alkali are adsorbed independently, within limits, although the presence of excess alkali reduces the sorption of total fatty acids. No straightforward relationship was noted between detergency and adsorption.

In a study of the adsorption of soap and several synthetic surfactants on a variety of textile fibers, it was found that cotton and nylon adsorbed less surfactant than wool under comparable conditions. Among the various surfactants, the cationic types were adsorbed to the greatest extent, whereas nonionic types were adsorbed least. The adsorption of nonionic surfactants decreased with increasing length of the polyoxyethylene chain. When soaps were adsorbed, the fatty acid and the alkali behaved more or less independently just as they did when adsorbed on carbon.

Adsorption of bath components is a necessary and possibly the most important and fundamental detergency effect. Adsorption (qv) is the mechanism whereby the interfacial free energy values between the bath and the solid components (solid soil and substrate) of the system are lowered, thereby increasing the tendency of the bath to separate the solid components from one another. Furthermore, the solid components acquire electrical charges that tend to keep them separated, or acquire a layer of strongly solvated radicals that have the same effect.

Mass Transfer near the Substrate Surface. Mechanical action has a great effect on soil removal, probably by influencing mass transfer, ie, the diffusion of soluble material away from the immersed fibers. Mechanical action tends to maintain a high concentration gradient near the fiber, and the resulting increased diffusion causes stronger diffusion currents to flow. These diffusion currents are presumably responsible for carrying away the soil particles that have already been detached or loosened from the fiber surface by physicochemical action.

Colloidal Stabilization. Surfactant adsorption reduces soil–substrate interactions and facilitates soil removal. For a better understanding of these interactions, a consideration of colloidal forces is required.

The model solid-soil detersive system is advantageously treated as a sol-agglomerate colloid system or, in more general terms, a lyophobic colloid. In the typical lyophobic colloid, consisting of a single disperse phase in an aqueous suspending medium, only one type of liquid-solid interface and one type of solid–solid interface is present. The simplest detersive system, however, has an added degree of complexity in the presence of two types of liquid–solid interface: soil–bath and substrate–bath. Also present are two effective types of solid–solid interface: soil–substrate and soil–soil. The soil–soil interface relates to flocculation or dispersion of soil particles remote from the substrate, and is not of primary concern in the present discussion. The soil–soil interface is, of course, important in practical detergency since soil aggregates can be regarded as large single particles.

There are two general theories of the stability of lyophobic colloids, or, more precisely, two general mechanisms controlling the dispersion and flocculation of these colloids. Both theories regard adsorption of dissolved species as a key process in stabilization. However, one theory is based on a consideration of ionic forces near the interface, whereas the other is based on steric forces. The two theories complement each other and are in no sense contradictory. In some systems, one mechanism may be predominant, and in others both mechanisms may operate simultaneously. The fundamental kinetic considerations common to both theories are based on Smoluchowski's classical theory of the coagulation of colloids.

Oily-Soil Detergency

Roll-up. The principal means by which oily soil is removed is probably roll-up. The applicable theory is simply the theory of wetting.

Solubilization. The role of micellar solubilization (as the term is used in the physical chemistry of surfactants) in oily-soil removal has been debated for many years. The amount of oily soil that could be present in a normal wash load could not all be removed and held in micellar solution by anionic surfactants. On the other hand, nonionic surfactants could do so, because of their greater solubilizing ability. High solubilizing power is definitely linked with good detergency. Thus, a very direct relationship between the solubilizing power of a surfactant for the test dyestuff Orange OT and its ability to remove polar solid from steel surfaces was established.

Phase Changes at the Soil–Bath Interface. Closely related to solubilization is a phenomenon that involves polar organic soils and surfactant solutions. If a complete phase diagram is plotted for a ternary system containing sodium dodecyl sulfate (or glycerol oleate), and water, several important and unusual features are noted. A large area represents a liquid phase consisting of a microemulsion, where the dispersed particles are so small that the system is isotropic, like the familiar soluble oils. Also, over another large area, a liquid crystalline phase is formed, containing all three components. This liquid crystalline phase flows like a liquid, at least in one direction. Flow perpendicular to the oriented planes is accomplished by folding the planes cylindrically, but the physical flow is still of the purely viscous type, with no yield point evident. These two phases, particularly the liquid–crystal phase, play an important part in detergency. Furthermore, liquid–crystal formation lowers interfacial tension. Although this phenomenon was demonstrated in tertiary oil recovery, the principles could also apply to oily-soil detergency.

When the polar organic component is a solid at ordinary temperatures, the addition of detergent and water markedly lowers the melting point; more specifically, as the temperature is raised, a point is reached where surfactant and water penetrate the solid. Thus, the ternary liquid–crystal phase might form spontaneously at room temperature by mixing the components, or, more precisely, an aqueous detergent solution can literally melt and liquefy a relatively large proportion of solid polar fatty matter (see LIQUID CRYSTALLINE MATERIALS).

In a detersive system containing a dilute surfactant solution and a substrate bearing a solid polar soil, the first effect is adsorption of surfactant at the soil–bath interface. This adsorption is equivalent to the formation of a thin layer of relatively concentrated surfactant solution at the interface, which is continuously renewable and can penetrate the soil phase. Osmotic flow of water and the extrusion of myelin forms follows the penetration, with ultimate formation of an equilibrium phase. This equilibrium phase may be microemulsion rather than liquid crystalline, but in any event it is fluid and flushable from the substrate surface. This phase change effect explains the detersive behavior of sucrose fatty esters in admixture with alkylarenesulfonates.

Measurement of Detergency

The measurement of detergency can be approached from two different points of view. The theoretical approach is concerned with the relative quantity of soil bound to the substrate before and after washing. In this case, measurement is a necessary analytical procedure in the study of the detergency mechanism. The second approach emphasizes the development of reproducible laboratory methods that predict the results of practical cleaning operations.

The measurement of detergency in the laboratory requires the following components: a means of measuring or estimating the amount of soil on the substrate or the degree of cleanness both before and after washing; satisfactory substrates and soiling composition; a means of applying soil to substrate in a realistic manner; and a realistic and reproducible cleaning device. These fundamental requirements apply regardless of the particular type of substrate that is being cleaned.

Fabric Detergency

Laundering. Reflectance is the most commonly used measurement for the whiteness of fabrics, although the transmittance of light by fabric specimens can also be used. The most commonly used instrument for reflectance measurement is the Gardner colorimeter, although the Zeiss Elrepho is also used. For general detergency, the grayness of the fabric is measured. Color effects can also be measured, and fabric yellowing is especially important. It is masked by fluorescent whitening agents (FWA). Special filters are available to eliminate this effect, and whitening caused by soil removal can be distinguished from that of FWA deposition.

Textile Mill Operations. Detergency is important in textile finishing because small quantities of foreign matter on the goods can interfere seriously with dyeing and other finishing treatments. Furthermore, the goods are expected to be uniformly and thoroughly clean when sold. Many detergency tests in this area are of the semipractical type, ie, test swatches are analyzed for soil content. This analysis generally consists of gravimetric determination of the soil content either directly from the fabric weight, or by extraction (see TEXTILES).

Hard Surface Detergency

Despite the variety of hard-surface objects that are purposefully cleaned at regular intervals, detergency has been studied quantitatively in relatively few cases only. The small-scale user normally judges washing results as satisfactory or unsatisfactory. If satisfactory results are obtained with the amount of detergent and the degree of mechanical action employed, the user is not interested in minor qualitative differences. In those areas where specifications are important and where differences among detergents or mechanical washing equipment are readily perceivable, quantitative methods for measuring detergency have been developed.

In specific cases of metal cleaning where small amounts of residual soil must be detected and are difficult to measure by conventional means, radiotracer methods have been employed. Interest in these techniques has been stimulated by the development of methods for decontaminating hard surfaces subjected to atomic fallout.

Quantitative measurements have been obtained for ceramics and glass, metals, and organic surfaces such as painted and plastic tile.

Glassware and Dishwashing. Dishes are washed either by hand or in an automatic dishwashing machine. Hand-dishwashing detergents are generally high foaming compositions containing organic surfactants as the main ingredient. The consumer judges efficiency not only by the cleanness of plates but also by foam persisting throughout the operation. Evaluation of hand-dishwashing products by manufacturers simulates this procedure. The number of plates that can be washed clean, judged visually without or with a color or fluorescence indicator, and the number of plates necessary to kill the foam in the dishpan is taken as the measure of detersive efficiency.

Metal Cleaning. The purpose of cleaning steel is to remove dirt and leave the article in a state where it can be delivered for use without further finishing (see METAL SURFACE TREATMENTS). The surface must therefore be covered with a tenacious corrosion-resistant coating as it emerges from the cleaning bath. Many emulsion cleaners remove lubricants and other unwanted dirt while depositing an anticorrosive coating on the metal. The primary test for efficacy in this situation is a corrosive test of the cleaned article.

Organic Surfaces. Tests for detergency on organic surfaces such as painted walls and plastic tile generally include a rubbing or sponging step corresponding to the manner in which such surfaces are cleaned in practice.

Detergent Manufacture

Liquid Products. The manufacture of liquid detergent products is generally a straightforward process requiring batch equipment with provisions for metered addition of individual ingredients, agitation, and if needed, heating and cooling. Capital cost can vary depending on the degree of automation.

Spray-Dried Products. The manufacture of powdered product is more complicated. High-pressure spray-drying of an aqueous slurry has replaced the earlier process in which a solidified cake of the product had to be broken up mechanically. Spray-drying equipment requires a relatively high capital outlay.

The first step in preparing spray-dried products involves producing a slurry of liquid and soil ingredients. Two processes are available, a batch process or a continuous process.

Acids such as fatty acids and alkylbenzenesulfonic acids are neutralized with NaOH during slurry preparation to form soap and sodium alkylbenzenesulfonate, respectively.

After preparation, the slurry is transported to an aging vessel. During residence time of 20–30 min, the neutralization process, hydration of sodium tripolyphosphate and structural changes in the slurry are completed to provide a homogeneous composition. By means of a high pressure pump, the slurry is conveyed to the spray-drying tower under a pressure of ca 10 MPa (100 atm).

The slurry, at 80–100°C, is forced through nozzles of 2.5–3.5 mm dia arranged on a nozzle ring. In the tower, the slurry encounters hot air that has entered the tower at 250–350°C. Upon exit, the powder is conditioned during passage via a belt conveyer and an airlift to the packaging machinery.

Because of stringent air pollution rules the exit gases are wet-scrubbed in brine because high NaCl concentrations reduce foam formation. NaOH scrubs out SO_2 from sulfur-containing fuel. The scrubbing solution, saturated with detergent fines, is recycled to the water tank for slurry preparation.

Although spray-drying accommodates relatively high content of surfactants, certain types, such as the alkanolamides and some nonionic surfactants are best added to the product after spray-drying.

Dry-Blended Product. In addition to lower capital outlay, dry-blending requires considerably less processing energy. Final product density, which is usually near unity, depends on the density of the starting materials and the nature of equipment used to blend these materials. Modern mixing and blending equipment, if properly controlled, can give product density and particle sizes comparable to spray-dried products.

Agglomerated Products. The process of agglomeration is intermediate between spray-drying and dry-blending. In agglomeration, a liquid is sprayed onto a continuously agitated powder. Equipment designs include stationary mixers, rotating mixers with spray nozzles, and rotating blenders with a liquid dispersion bar, either twin shell or continuous zigzag. Automatic dishwashing detergents are manufactured mostly by agglomeration.

Health and Safety Factors

As a class, surfactants and detergent products are among the most widely used chemical compositions. Almost everyone is exposed to these products on a daily basis in situations that range from ingestion of food-grade emulsifiers to intimate contact of skin and eyes with personal-care and laundry products. Safety is therefore a matter of great importance (132,133). Ranges of surfactant LD_{50} values are shown in Table 1.

Under conditions of normal use, detergent products are not hazardous to users. Nonetheless, surfactants possess some toxicity, and they are mild irritants.

The manufacture of surfactants and of detergent products is regulated by OSHA. Dust concentration in detergent plants as well as factory noise levels are the primary areas of relevance, since the individual components in these products are essentially nonhazardous. Of more immediate concern to the detergent industry is the Federal Hazardous Substances Act (FHSA) of 1960 and the Consumer Products Safety Act (CPSA) of 1972. The FHSA defines specific labeling requirements, such as "Danger" for extremely flammable, corrosive, or highly toxic substances, and "Warning" or "Caution" for less hazardous materials.

Environmental Considerations

The introduction of surfactant products into the environment, after use by consumers or as part of waste disposed during manufacture, is regulated by the Clean Water Act, and Clean Air Act, and the Resource Conservation and Recovery Act. In this respect, surfactants are subject to the same regulations as chemicals in general. There are, however, two areas of specific relevance to surfactants and detergent products, ie, biodegradability and eutrophication.

Extensive investigations led to the conclusion that a branched hydrophobe impedes the rate and extent of degradation of surfactants by microorganisms. The most immediately apparent remedy, therefore, was to replace the propylene tetramer in ABS with a straight hydrocarbon chain giving straight-chain alkylbenzenesulfonate, so-called linear alkanesulfonate (LAS). At the same time, commercialization of the Ziegler process for the oligomerization of ethylene provided another route to straight-chain hydrophobes that could easily be converted to detergent alcohols and straight-chain nonionic

Table 1. Rat Oral LD_{50} Values of Surfactant

Type of compounds	Oral LD_{50}, mg/kg
alkylbenzenesulfonates	700–2,480
alcohol ethoxylates	1,600 to greater than 25,000
sulfated alcohol ethoxylates	7,000 to greater than 50,000
alcohol sulfates	5,000–15,000

surfactants. By 1965, the U.S. detergent industry had completed a voluntary switch from hard to soft surfactants at a cost that has been estimated at ca 150×10^6. In addition to ABS, other surfactants based on propylene oligomers, such as alkylphenol derivatives, have largely disappeared from U.S. consumer laundry products.

Even though the biodegradability problem has been solved for all practical purposes, the subject continues to receive considerable attention. The biodegradation of LAS has been studied intensively, and several mechanistic pathways have been identified such as β- and ω-oxidation as well as reductive and oxidative desulfonation. Investigation of the biodegradation of LAS, alcohol ethoxylates, and alkylphenol ethoxylates in the laboratory and under sewage plant operating conditions showed that LAS and straight-chain alcohol ethoxylates and their sulfates degrade to CO_2 and H_2O.

Eutrophication. This term, which denotes excessive nutrition or overfertilization, has been applied to the contribution excessive amounts of phosphorus may make to the growth of algae under certain conditions. Phosphorus in water supply originates from run-off of agricultural fertilizers, human excrement, and sodium tripolyphosphate present in detergent formulations. It has been estimated that 25–30% of the phosphorus in waste water comes from laundry detergents, and that detergents contribute about 3% of the phosphorus annually entering U.S. surface waters. In the United States, and later in Western Europe, detergent phosphates were singled out in the early 1960s as the cause of eutrophication, and their removal from consumer laundry formulations was proposed as a feasible approach to improvement of environmental water quality. Many states and a number of local jurisdictions have banned detergent products containing phosphate.

The efforts of the detergent industry toward solution of its part of the eutrophication problem are, at this point, less complete than its response to the biodegradability problem. Soda ash, Na_2CO_3, sodium silicate, and, to a lesser extent, sodium citrate formed the basis of the early formulations marketed in the areas where phosphates were banned. Technically, these substances are considerably less effective than sodium tripolyphosphate. As a precipitant builder, soda ash can lead to undesirable deposits of calcium carbonate on textiles and on washing machines.

JESSE L. LYNN, JR.
Lever Brothers Company

A. M. Schwartz, in E. Matijevic, ed., *Surface and Colloid Science,* Vol. 5, John Wiley & Sons, Inc., New York, 1972.

A. W. Adamson, *Physical Chemistry of Surfaces,* 5th ed., John Wiley & Sons, Inc. New York, 1990.

M. J. Rosen, *Surfactants and Interfacial Phenomena,* 2nd ed., Wiley-Interscience, New York, 1989.

McCutcheon's Emulsifiers & Detergents, North Am. & International Ed., Glen Rock, N.J., 1991.

DETONATING AGENTS. See EXPLOSIVES AND PROPELLANTS.

DEUTERIUM AND TRITIUM

The element hydrogen has three known isotopes. These hydrogen species are identical in atomic number, ie, they have identical extranuclear electronic configurations ($1s^1$), but they differ in nuclear mass. Over 99.98% of the hydrogen in nature has a nucleus consisting of a single proton and, therefore, has mass 1 (symbol ^1H). Two heavier isotopes of hydrogen, present in small amounts in nature, of mass 2 and 3 are also known; these have nuclei consisting of one proton and one neutron (deuterium) or two neutrons (tritium). The three isotopes of hydrogen resemble each other closely in chemical and physical properties, but because the ratios of their masses are the largest for any set of isotopes in the Periodic Table, differences in chemical and physical properties exist that are larger than those encountered in any other set of isotopes. Whereas ordinary hydrogen and deuterium are stable isotopes, tritium is unstable and its nucleus undergoes radioactive decay (see RADIOACTIVE TRACERS; RADIOISOTOPES). The two heavy isotopes of hydrogen are always present to a small extent in any compound or substance containing the light isotope of hydrogen. Because both isotopically pure deuterium and tritium can be manufactured on a large industrial scale, a large variety of isotopically pure deuterium and tritium compounds is readily available. Heavy water, D_2O, is the most important compound of deuterium and the only form of deuterium produced and used on a large scale.

DEUTERIUM

Physical Properties

As in the case of hydrogen and tritium, deuterium exhibits nuclear spin isomerism. However, the spin of the deuteron is 1 instead of $\frac{1}{2}$ as in the case of hydrogen and tritium. As a consequence, and in contrast to hydrogen, the ortho form of deuterium is more stable than the para form at low temperatures, and at normal temperatures the ratio of ortho- to para-deuterium is 2:1 in contrast to the 3:1 ratio for hydrogen. The physical and thermodynamic properties of elemental hydrogen and deuterium and of their respective oxides illustrate the effect of isotopic mass differences.

Properties of Light and Heavy Hydrogen. The equilibrium state for these substances is the low temperature ortho–para composition existing at 20.39 K, the normal boiling point of normal hydrogen. The normal state is the high (above 200 K) temperature ortho–para composition, which remains essentially constant.

Thermodynamic data on H_2, the mixed hydrogen–deuterium molecule, HD, and D_2, including values for entropy, enthalpy, free energy, and specific heat, have been tabulated. Some physical properties of liquid H_2 and D_2 at 20.4 K are presented in Table 1.

Properties of Light and Heavy Water. Selected physical properties of light and heavy water are listed in Table 2. At room temperature both light and heavy water appear to be highly structured, ie, extensively hydrogen bonded. Heavy water is the more structured.

Deuterium as Neutron Moderator. Deuterium has very desirable properties as a moderator for neutrons. As illustrated in Table 3, heavy water has a much greater moderating ratio than any of the other materials commonly used as moderators.

Kinetic Isotope Effects. The principal difference in chemical behavior between H and D derives from the generally greater stability of chemical bonds formed by D. The most important factor contributing to the difference in bond energy and the kinetic isotope effect is the lower (5.021–5.275 kJ/mol (1.2–1.5 kcal/mol)) zero-point vibrational energy for D bonds.

Table 1. Properties of Liquid H_2 and D_2 at 20.4 K

Property	H_2[a]	D_2
Equilibrium states		
density, g/L	70	169
viscosity, mPa·s(= cP)	1.4×10^{-2}	4.0×10^{-2}
surface tension, mN/m (= dyn/cm)	2.1_7	3.7_2
thermal conductivity, W/(cm·K)	11.6	12.6_4
dielectric constant	1.22_6	1.27_5
Normal states		
molar volume at 20 K, mL	28.3	23.5
heat of vaporization, J/mol[c]	904	1226
heat of fusion, J/mol[c]	117	197

[a] The equilibrium state of hydrogen has 0.21% o-H_2O; the normal state has 75% o-H_2O.
[b] The equilibrium state of deuterium has 97.8% o-D_2; the normal state has 67% o-D_2.
[c] To convert J to cal, divide by 4.184.

Table 2. Physical Properties of Light and Heavy Water

Property	H_2O	D_2O
molecular weight	18.015	20.028
melting point, T_m, °C	0.00	3.81
normal boiling point, T_b, °C	100.00	101.42
critical constants		
temperature, °C	374.1	371.1
pressure, MPa[a]	22.12	21.88
volume, cm^3/mol	55.3	55.0
density at 25°C, g/cm^3	0.99701	1.1044
vapor pressure, liquid at 25°C, kPa[b]	3.166	2.734
coefficients of thermal expansion, $°C^{-1}$		
solid at T_m	1.39×10^{-4}	1.39×10^{-4}
liquid at T_m	-5.9×10^{-5}	-3.2×10^{-5}
compressibility at 20°C, Pa^{-1}[d]	4.45	4.59
length of the hydrogen bond, nm	0.2765	0.2760
dielectric constant at 25°C	78.304	77.937
refractive index, $n^{20}{}_D$	1.3330	1.3283
viscosity at 25°C, mPa·s(= cP)	0.8903	1.107
surface tension at 25°C, mN/m(= dyn/cm)	71.97	71.93
ion product constant at 25°C	1.01×10^{-14}	1.11×10^{-15}
heat of ion product at 25°C, kJ/mol[c]	56.27	60.33

[a] To convert MPa to atm, multiply by 10.1.
[b] To convert kPa to mm Hg, multiply by 7.5.
[c] To convert Joules to calories, divide by 4.184.

Table 3. Properties of Neutron Moderators

Moderator	Slowing-down power $\xi \times \Sigma_s$, cm^{-1}	Macroscopic absorption cross section Σ_a, cm^{-1}	Moderating ratio, $\xi \times \Sigma_a/\Sigma_a$
water	1.28	2.2×10^{-2}	58
heavy water	0.18	8.5×10^{-6}	21,000
helium	10^{-5}	2.2×10^{-7}	45
beryllium	0.16	1.2×10^{-3}	130
graphite	0.065	3.3×10^{-4}	200

Biological Effects of Deuterium. Replacement of more than one third of the hydrogen by deuterium in the body fluids of mammals or two thirds of the hydrogen in higher green plants has catastrophic consequences for the organisms. At lower deuterium levels in higher plants, growth is markedly slowed. In mice and rats low levels of deuterium result in sterility, and at higher concentrations neuromuscular disturbances, fine muscle tremors, and a tendency to convulsions can be noted. Impairment of kidney function, anemia, disturbed carbohydrate metabolism, central nervous system disturbances, and altered adrenal function have been found in deuterated mice. Hemoglobin and red blood cell count, serum glucose, and cholesterol all decrease in deuterated dogs.

Extensive replacement of H by D in living organisms is, however, not invariably fatal to the organisms. Numerous green and blue-green algae have been grown in which >99.5% of the hydrogen has been replace by D. Numerous varieties of bacteria, molds, fungi, and even a protozoan have been successfully grown in fully deuterated form. These organisms of unnatural isotopic composition and the deuterated compounds that can be extracted from them have found uses in many areas of biological research.

Kinetic isotope effects are an important factor in the biology of deuterium. Isotopic fractionation of hydrogen and deuterium in plants occurs in photosynthesis. The lighter isotope ^1H is preferentially incorporated from water into carbohydrates and lipids formed by photosynthesis. Hydrogen isotopic fractionation has thus become a valuable tool in the elucidation of plant biosynthetic pathways.

Deuterium isotope effects provide information about the mechanisms of enzyme action. Fully deuterated griseofulvin and benzylpenicillin have antifungal and antibiotic potencies at least as great as their ordinary hydrogen analogues, and are probably metabolized *in vivo* more slowly because of the enhanced stability of C—D relative to C—H bonds. Synthetic deuterated drugs have also been considered as therapeutic agents. Considering the well-documented toxic effects of deuterium in mammals, long-term administration of D_2O to humans is not likely to become a routine procedure. However, D_2O confers significant protection to yeast against hydrostatic pressure damage, and pretreatment with D_2O has been claimed to protect cultured cells against x-ray damage.

Isotope Effects on Superconductivity. Substitution of hydrogen by deuterium affects the superconducting transition temperature of palladium hydride, PdH_2, palladium silver hydride, $Pd_{1-y}Ag_yH_zD_x$, and vanadium–zirconium–hydride, $V_2ZrH_x(D_x,T_x)$.

Production of Heavy Water

Because of the low natural abundance of deuterium, very large amounts of starting material, which is water, must be processed to produce relatively small amounts of highly enriched deuterium. No water or other hydrogen compound has been found either in nature or as a by-product of an industrial operation that is significantly enriched in deuterium. The cost of subsequent enrichment to 99% is negligible compared to the costs incurred in the initial enrichment from natural abundance to 1%. For small-scale preparations, a highly efficient but very expensive process such as electrolysis can be used. For large-scale use, however, a high enrichment factor per stage is of only secondary importance to the overall costs of operation both in power and in capital investment. The isotope separation methods that have attracted the greatest interest include chemical exchange between water and hydrogen sulfide, hydrogen and water, and hydrogen and ammonia; distillation of water or hydrogen; and electrolysis of water in combination with other procedures.

Economic Aspects. The principal market for deuterium has been as a moderator for nuclear fission reactors fueled by unenriched uranium. The decline in nuclear reactor construction has sharply reduced the demand for heavy water.

JOSEPH J. KATZ
Argonne National Laboratory

D. Kritchevsky, ed., *Ann. N.Y. Acad. Sci.* **84,** 573 (1960).

P. A. Rock, ed., *Isotopes and Chemical Principles,* American Chemical Society Symposium Series No. 11, 1975.

C. J. Collins and N. S. Bowman, eds., *Isotope Effects in Chemical Reactions, American Chemical Society Monograph 167,* Van Nostrand Reinhold Co., New York, 1971.

TRITIUM

Tritium, the name given to the hydrogen isotope of mass 3, has symbol ^3H or more commonly T. Its isotopic mass is 3.0160497. Molecular tritium, T_2, is analogous to the other hydrogen isotopes. The tritium nucleus is energetically unstable and decays radioactively by the emission of a low-energy β particle. The half-life is relatively short (~12 yr), and therefore tritium occurs in nature only in equilibrium with amounts produced by cosmic rays or man-made nuclear devices.

Physical Properties

Tritium is the subject of various reviews, and a book provides a comprehensive survey of the preparation, properties, and uses of tritium compounds. Selected physical properties for molecular tritium, T_2 are given in Table 1. All components appear miscible in both liquid and

Table 1. Physical Properties of Molecular Tritium

Property	Value
melting point, at 21.6 kPa[a], K	20.62
boiling point, K	25.04
critical temperature, K	40.44
critical pressure, MPa[b]	1.850
critical volume, cm^3 mol	57.1[c]
heat of sublimation, J/mol[d]	1640
heat of vaporization, J/mol[d]	1390
entropy of vaporization, J/(mol·K)[d]	54.0
molar density of liquid, mol/L	
20.62 K[c]	45.35
29 K	39.66

[a] Value represents the triple point (162 mm Hg).
[b] To convert MPa to psi, multiply by 145.
[c] Value calculated.
[d] To convert J to cal, divide by 4.184.

solid phases from 17 to 22 K. The T–T bond energy has been estimated at 4.5881 *e*V. The entropy of T_2 at 298.15 K is 164.8562 kJ/mol (39.4016 kcal/mol), the specific heat is 29.1997 J/(mol·°C) (6.9789 cal/(mol·°C)), and the Gibbs free energy is 135.9083 kJ/mol (32.4829 kcal/mol).

Ortho–Para Tritium. As in the case of molecular hydrogen, molecular tritium exhibits nuclear spin isomerism. The spin of the tritium nucleus is $\frac{1}{2}$, the same as that for the hydrogen nucleus, and therefore H_2 and T_2 obey the same nuclear isomeric statistics. Below 5 K, molecular tritium is 100% para at equilibrium. At high (100°C) temperatures the equilibrium concentration is 25% para and 75% ortho. The kinetic parameters of conversion for T_2 at low temperatures are faster than rates at corresponding temperatures for H_2. In the solid phase the conversion of molecular tritium to a state of ortho–para equilibrium is 210 times as fast as that for molecular hydrogen.

Properties of T_2O. Some important physical properties of T_2O are listed in Table 2. Tritium oxide can be prepared by catalytic oxidation of T_2 or by reduction of copper oxide using tritium gas. T_2O, even of low (2–19% T) isotopic abundance, undergoes radiation decomposition to form HT and O_2.

Nuclear Properties

Radioactivity. Tritium decays by β-emission, $3\ T \xrightarrow{3} He + \beta^-$.

Nuclear Fusion Reactions. Tritium reacts with deuterium or protons (at sufficiently high temperatures) to undergo nuclear fusion. Nuclear fusion using tritium can be initiated and sustained at the lowest temperature (at least in principle) of any nuclear fusion reaction known. Tritium thus becomes the key element both for controlled thermonuclear energy sources and in the uncontrolled release of thermonuclear energy in the hydrogen bomb.

Nuclear Magnetic Resonance. All three hydrogen isotopes have nuclear spins, $I \neq 0$, and consequently can all be used in nmr spectroscopy. Tritium is an even more favorable nucleus for nmr than is ^1H, which is by far the most widely used nucleus in nmr spectroscopy.

Table 2. Physical Properties of T_2O

Property	Value
mol wt	22.032
temperature of maximum density, °C	13.4
boiling point, °C	101.51
density at 25°C, g/mL	1.2138
ionization constant at 25°C	ca 6×10^{-16}

Chemical Properties

Most of the chemical properties of tritium are common to those of the other hydrogen isotopes. However, notable deviations in chemical behavior result from isotope effects and from enhanced reaction kinetics induced by the β-emission in tritium systems. Isotope exchange between tritium and other hydrogen isotopes is an interesting manifestation of the special chemical properties of tritium.

Production

Nuclear Reactions. The primary reaction for the production of tritium is $6\ Li + ^1n \xrightarrow{3} T + ^4He + 4.78$ MeV.

Production in Target Elements. Tritium was produced on a large scale by neutron irradiation of ^6Li. The principal U.S. site of production was the Savannah River plant near Aiken, South Carolina where tritium was produced in large heavy-water moderated, uranium-fueled reactors.

Production in Heavy Water Moderator. A small quantity of tritium is produced through neutron capture by deuterium in the heavy water used as moderator in the reactors.

Production in Fission of Heavy Elements. Tritium is produced as a minor product of nuclear fission.

Production-Scale Processing. The tritium produced by neutron irradiation of ^6Li must be recovered and purified after target elements are discharged from nuclear reactors. In the recovery process the gaseous constituents of the target are evolved and the hydrogen isotopes separated from other components of the gas mixtures.

Isotopic Concentration. A number of techniques have been reported for concentrating tritium from naturally occurring sources, eg, low (20–25 K) temperature distillation, concentration by gas chromatography, separation via a cryogenic thermal diffusion column, by diffusion through a palladium–silver–nickel membrane, and by chromatography on coated molecular sieves, and laser separation of tritium.

Health Aspects

Hazards. Because tritium decays with emission of low energy radiation ($E_{av} = 5.7$ keV), it does not constitute an external radiation hazard. However, tritium presents a serious hazard through ingestion and subsequent exposure of vital body tissue to internal radiation.

Monitoring and Control. Detailed descriptions of methods used for handling and monitoring tritium at Savannah River and the European Tritium Handling Program have been published.

A widely used instrument for air monitoring is a type of ionization chamber called a Kanné chamber. Uptake of tritium by personnel is most effectively monitored by urinalyses normally made by liquid scintillation counting on a routine or special basis. Environmental monitoring includes surveillance for tritium content of samples of air, rainwater, river water, and milk.

The radiological hazard of tritium to operating personnel and the general population is controlled by limiting the rates of exposure and release of material. Personnel are protected in working with tritium primarily by containment of all active material. Several new technologies were in development at the Savannah River plant that would have considerably enhance safety in handling large amounts of tritium. Metal hydride technology has been developed to store, purify, pump, and compress hydrogen isotopes. Conversion to or extraction from metal triteride would offer flexibility and size advantages compared to conventional processing methods that use gas tanks and mechanical compressors, and should considerably reduce the risk of tritium gas leaks (see HYDRIDES).

Personnel who must work in areas in which tritium contamination exceeds permitted levels are safeguarded by protective clothing, such as ventilated plastic suits.

Uses

Nuclear fusion is an approach to the ever increasing global demands for energy. All nuclear fusion reactions require very high temperatures for initiation. The thermal threshold is lowest for light ions, and the nuclear reaction $D(T,n)^4He$, involving the fusion of deuterium and tritium nuclei is considered to be the most practical approach to the realization of nuclear fusion energy. Whereas the technology for large-scale production of deuterium exists, the production and handling of tritium is one of the key problems in the achievement of practical nuclear fusion.

The development of a tritium fuel cycle for fusion reactors is likely to be the focus of tritium chemical research into the twenty-first century. Tritium is widely used as a tracer in molecular biology (see RADIOACTIVE TRACERS).

JOSEPH J. KATZ
Argonne National Laboratory

E. A. Evans, *Tritium and Its Compounds,* 2nd ed., John Wiley & Sons, Inc., New York, 1974.

K. Wilzbach, *J. Am. Chem. Soc.* **79,** 1013 (1957).

G. R. Choppin and J. Rydberg, *Nuclear Chemistry,* Pergamon Press, Oxford, U.K., 1980.

M. Garcia-Leon and G. Madurga, eds., *Low-Level Meas. Man-Made Radionuclides Environ., Proc. Inc. Summer Sch., 2nd,* World Science, Singapore, 1991.

DEWATERING

Dewatering is the last process applied to separate water from a solid, unless thermal drying (qv) is used. Dewatering is usually a mechanical process that presses residual water from solids or displaces the water with a gas, and the energy required is negligible compared with the heat required for drying. Thus there is a significant incentive for adding a dewatering step to a process.

Dewatering processes are normally concerned with removing water bound in capillaries, and function by changing the size distribution of capillary radii, reducing adhesion of water to the solids, displacing water from the capillaries, and reducing the energy required to cause flow through and out of the capillaries. These effects are achieved by three methods. (*1*) The particulate matrix may be compacted by applying stress, that is, forces can be induced by frictional drag of the liquid as it flows through the pores. Body forces can arise from gravity or centrifugal motion, and boundary stresses can be applied with rolls, membranes, pistons or screw presses, or acoustic energy (see MEMBRANE TECHNOLOGY). (*2*) The water may be displaced, usually with a gas, by application of vacuum or pressure. Centrifugal force in a pusher centrifuge also results in the water being displaced by a gas. (*3*) An electrical field may be applied to a slurry of charged particles.

Although all the techniques are effective, in industrial applications there is rarely time to achieve an equilibrium reduced saturation state (see FILTRATION), so variables that affect only the kinetics of dewatering and not the equilibrium and residual moisture are also very important. The most important kinetic variables in displacing the liquid from the solid are increases in pressure differentials and viscosity reduction.

Cake Dewatering

The most important function of filtration is the formation of a filter or centrifuge cake from a slurry (see also SEPARATION–CENTRIFUGAL). The most important function of dewatering is removal of the liquid from the resulting cake. Theories of cake dewatering are covered elsewhere.

Ways to reduce the final moisture content of a centrifuge cake include the use of steam, surfactants, or flocculants (see FLOCCULATING AGENTS), as well as pretreatments by pelletizing, oil agglomeration, thermal treatment, and freeze–thaw processes. The main dewatering variable in the centrifuge itself is centrifugal force sufficient to expel the liquid from the pores in the cake through the filter medium.

Cake dewatering is related to cake formation; in a filter cake, the final moisture content is dependent on many variables that also control cake formation. Pretreatment processes, for example, affect both cake formation and cake dewatering. Cake dewatering is achieved by compacting the solids; displacing the residual liquid in the cake with another phase, usually a gas; or applying an electrical current to remove the liquid. Each process relies on different properties of the cake.

Compaction and expression are two widely used methods for dewatering filter cakes. Compaction is the reduction of water content using hydraulic flow to consolidate particles closer together in the cake. High pressure filters like plate and frame filters use this method. Expression is the mechanical squeezing of the solids against each other to reduce space and eliminate water. Equipment for expression dewatering has improved greatly over the past decade and expression is widely used. Typical equipment are variable-chamber plate and frame filters, belt-filter presses, and screw presses. Screw presses are the traditional equipment for dewatering fibrous solids.

Improving Cake Dewatering. Equations relating the rate of liquid flow through a filter cake can be simplified to

$$\frac{V}{A} = \frac{K\Delta P}{\mu l}$$

where V in units of m^3/s is the flow rate through the cake, A in m^2, is the area of the cake, K, m^2, is the cake permeability, ΔP, Pa, is the pressure drop across the cake, l in m, is the thickness of the cake, and μ is the cake viscosity (Darcy's law). Viscosity, a kinetic variable, does not appear in equations describing reduced saturation levels in a cake. Because most practical filtration is limited by the time available for the steps of cake forming and dewatering, an increase in the flow rate of filtrate during dewatering translates directly to lower cake moisture levels. If the liquid is water, the viscosity can drop by a factor of three as the temperature rises from 15°C to 80°C, and consequently the flow rate of water through the cake is tripled.

Although the use of steam to improve dewatering is consistently beneficial, the effects of surfactants on residual moisture are highly inconsistent. Additions of anionic, nonionic, or sometimes cationic surfactants of a few hundredths weight percent of the slurry, 0.02–0.5 kg/t of solids, are as effective as viscosity reduction in removing water from a number of filter cakes, including froth-floated coal, metal sulfide concentrates, and fine iron ores. A few studies have used both steam and a surfactant on coal and iron ore and found that the effects are additive, giving twice the moisture reduction of either treatment alone.

Virtually all solids become charged in water, either by reaction with the water to form surface hydroxyl groups that can ionize or by adsorption of ions from the water. The charges on the particles affect how much water is bound to the particle. Reducing the charge on the particle increases the amount of dewatering possible in conventional dewatering equipment.

Flocculants are capable of improving filtration rates up to 100 times, especially of fine clay, sludges, or tailings. One of the main uses of flocculants in filtration is to make extremely slow filtering slurries filterable at reasonable rates. In addition, flocculants are critical in making belt-filter presses and dewatering centrifuges effective.

Sludge conditioning is the chemical, physical, or heat treatment of wastewater sludges to improve dewatering. Coagulants such as ferric and ferrous chloride, alum, and lime have been added which chemically react in the sludge and improve dewatering. These are often used with polymer flocculants. Diatomaceous earth, fly ash, and ash derived from incinerating dewatered sludge or bark have been added as body feeds to improve sludge dewatering.

Biological processing of sludge reduces the amount of sludge needing dewatering and changes the dewatering properties. The changes are not always helpful. A comparison of high biological oxidation of wastewater sludges, to minimize the amount of sludge that needed disposal, showed higher overall disposal costs compared with the costs of less oxidation and twice as much "wasting" (removal and disposal) of sludge. In an unusual application of biological conditioning, enzymes have been used to aid in the dewatering of phosphatic clay ponds.

Use of Mechanical Vibration. Vibration of the sludge or cake can further release entrapped moisture. Pretreatment of organic sludges and highly hydrated inorganic sludges using ultrasonic energy can significantly reduce the amount of polyelectrolyte polymer needed.

Less Common Commercial Dewatering Processes

When solids dewatering is known to be a problem early in the process-design stage of a plant or is serious enough to warrant consideration of a range of dewatering alternatives, two approaches are available, and both can be used together. First, processes that begin the dewatering process while in the original suspension may be used. A second approach is to extract water or apply unusual desaturating forces to water present in sludges and cakes. Processes include agglomeration of suspended solids (such as pelletizing flocculation or oil agglomeration), extraction processes (oil-phase extraction of solids from water or solvent extraction of water from the solids), thermal processes (thermal drying, thermal treatment of organic sludges, freeze-thaw dewatering, freeze crystallization, clathrate freezing), capillary suction processes to withdraw water from sludges, electromagnetic processes (electrical enhancement of dewatering, electroosmosis with vibration), and magnetic enhancement of dewatering.

Economic Aspects

About 500 U.S. companies manufacture liquid–solid separation equipment. Some of the manufacturers in this worldwide business include Ametek, Anderson International Corp., Arus-Andritz Inc., Ashbrook-Simon-Hartley, Bird Machine Co., Inc., Bepex Corp., Black Clawson, Centrifugal and Mechanical Industries, Denver Process Equipment Co., Dorr-Oliver Inc., Ebara Infilco, Envirex Inc., Eimco Process Equipment Co., Hitachi Plant Engineering and Construction Co., Ltd., Infilco-Degremont Inc., Ingersoll Rand Co., JWI Inc., KHD Humboldt Wedag, Krauss-Maffei Corp., Komline-Sanderson Engineering Corp., Kurita Machinery Manufacturing Co., Ltd., Larox Oy, Mitsubishi Kakoki Kaisha, Ltd., Rosenmund Inc., Sharples Division, Alfa-Laval Separation Inc., Sparkler Filters Inc., U.S. Filter, Wemco, and Zimpro-Passavant Environmental Systems, Inc.

BOOKER MOREY
SRI International

O. E. Albertson and co-workers, *Dewatering Municipal Wastewater, Sludge Design Manual,* EPA/625/1-87/014, Sept. 1987.

K. Fouhy and S. Moore, *Chem. Eng.,* 33 (July 1994).

D. B. Purchas, *Solid–Liquid Separation Technology,* Uplands Press, 1981.

D. B. Purchas and R. J. Wakeman, *Solid–Liquid Separation Suck-Up,* 2nd ed., Uplands Press, 1986.

DIALYSIS

Dialysis is a membrane separation process in which one or more dissolved species flow across a selective barrier in response to a difference in concentration. It is the earliest molecularly separative membrane process to be identified and described. The mode of transport is diffusion, and separation occurs because small molecules diffuse more rapidly than larger ones, and also because the degree to which membranes restrict solute transport usually increases with permeant size. The basic principles are illustrated in Figure 1. Solute c is present at concentrations c' and c'' on opposite sides of a membrane. In the absence of differences in pressure, temperature, or electrical potential, Fick's phenomenological first-order description of diffusion, published in 1855, states that solute will move from region of greater to lesser concentration and at a rate proportional to the difference. In equation 1, ϕ = unit solute flux in g/cm²·s; D = diffusion coefficient, cm²/s; c = concentration in g/cm³; x = distance in cm; and the minus sign accounts for the convention that flux is considered positive in the direction of decreasing concentration.

$$\phi = -D\frac{\delta c}{\delta x} \tag{1}$$

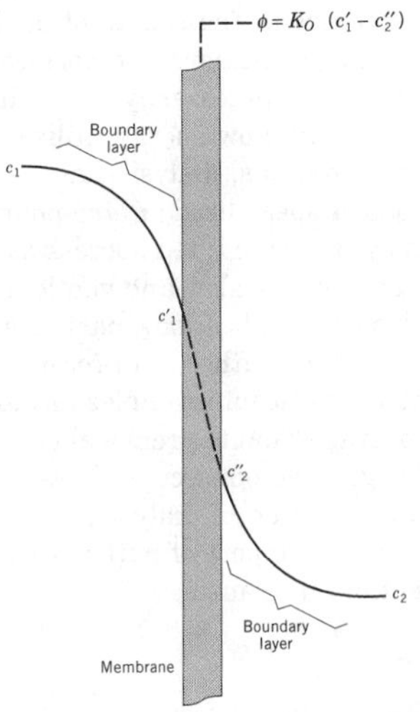

Figure 1. General dialysis is a process by which dissolved solutes move through a membrane in response to a difference in concentration and in the absence of differences in pressure, temperature, and electrical potential. The rate of mass transport or solute flux, ϕ, is directly proportional to the difference in concentration at the membrane surfaces (eq. 1). Boundary layer effects, the difference between local and wall concentrations, are important in most practical applications.

Diffusion coefficients decrease roughly in proportion to the square root of molecular weight, are widely tabulated for aqueous solutions, or may be estimated from the Stokes Einstein equation. Ignoring boundary layer effects for the moment, and by assuming that diffusion within the membrane is analogous to that in free solution, equation 1 can be integrated across a homogeneous membrane of thickness d to yield the following equation, where S represents the dimensionless solute partition coefficient, ie, the ratio of solute concentration in external solution to that at the membrane surface, and D_M represents solute diffusion within the membrane and is assumed independent of solute concentration.

$$\phi = \frac{SD_M\Delta c}{d} \tag{2}$$

The product of SD_M is often termed permeability; if two or more solutes are dialysing at the same time, the degree of separation or enrichment is proportional to the ratio of their permeabilities. The closer the permeability of a membrane is to that of an equivalent thickness of free solution, the more rapid is the resultant dialytic transport. Considerable effort has been devoted to understanding how the physical and chemical properties of a membrane determine its permeability. The simplest approaches are geometric and consider the membrane to comprise a series of parallel pores that provide a topographic obstacle to hard noninteracting permeant molecules; far more complex analyses are also available. As a general rule, permeability for a particular species increases with porosity (solute content) of the membrane and with the diameter of its pores. Equation 2 also states that the mass flow rate of solute is inversely proportional to membrane thickness, but the degree of separation (selectivity) is independent of thickness. For this reason, membranes are always made as thin as possible consistent with the requirements of mechanical strength and reliability.

Solutions adjacent to the membranes are rarely well mixed, and the resistance to transport resides not just in the membrane but also in the fluid regions, termed boundary layers, on both the dialysate and feed side. Boundary layer effects typically account for from 25 to 75% of overall resistance. They are minimized by rapid convective flow tangential to the surface of the dialysing membrane.

Dialysis is a highly constrained process. Molecular diffusion is slow in the context of industrial dimensions. The driving force is set by the system itself, decreases in the course of purification, and is not amenable to extrinsic augmentation. The permeant species is not recovered in pure form, and is necessarily more dilute in the dialysate than in the starting stream. Low energy utilization is offset by high capital costs. For these reasons, dialysis has been largely limited to laboratory separations or specialized *in vivo* pharmacological investigations, and has enjoyed very limited success as a broad-based commercial unit operation. But the slow and gentle nature of dialysis has a special appeal for biologic applications, particularly when partial purification of the feed stream, rather than recovery of a product, is intended. Commercially significant examples include the adjustment of alcohol content of beverages and the removal of salts from solutions of proteins or other biologic macromolecules. However, the most successful and widespread application of dialysis, or for that matter of any membrane process, is the support of patients with kidney failure by repeated intermittent blood cleansing.

MICHAEL J. LYSAGHT
CytoTherapeutics, Inc.
ULRICH BAURMEISTER
Akzo Faser AG

W. Pusch, *Desalination* **59,** 105–198 (1986).

C. K. Colton and E. G. Lowrie, in B. M. Brenner and F. C. Rector, eds., *The Kidney,* 2nd ed., Saunders Publishing Co., Philadelphia, 1981.

M. J. Lysaght and P. C. Farrell, *J. Membr. Sci.* **44,** 5–33 (1989).

H. Strathmann and H. Goehl, *Contrib. Nephrol.* **78,** 119–141 (1990).

DIAMINES AND HIGHER AMINES, ALIPHATIC

The aliphatic diamine and polyamine family encompasses a wide range of multifunctional, multireactive compounds. This family includes ethylenediamine (EDA) and its homologues, the polyethylene polyamines (commonly referred to as ethyleneamines), the diaminopropanes and several specific alkanediamines, and analogous polyamines. The molecular structures of these compounds may be linear, branched or cyclic, or combinations of these.

The ethyleneamines have found the broadest commercial application and are the primary focus of this article. The lower molecular weight ethylenediamines, ie, EDA, diethylenetriamine (DETA), piperazine (PIP), and *N*-(2-aminoethyl)-piperazine (AEP), are available commercially as industrially pure products. The tetramine (TETA), pentamine (TEPA), hexamine (PEHA), and higher polyamine products are commercially available as boiling point fractions consisting of natural mixtures of linear, branched, and cyclic compounds. Their compositions are largely determined by the chemical processes used in their production. The individual components in these higher ethyleneamines are generally not available in industrial quantities.

The predominant commercial diaminopropanes are 1,2-propylenediamine (1,2-PDA), 1,3-diaminopropane (1,3-PDA), iminobispropylamine (IBPA), and dimethylaminopropylamine (DMAPA). Other commercially important products include other higher alkylenediamines, such as hexamethylenediamine (HMDA); certain cyclic amines, such as triethylenediamine (TEDA); and various alkyl- and hydroxyalkyl-derivatives.

Physical Properties

Physical properties of some commercially available polyamines appear in Table 1. Generally, they are slightly to moderately viscous, water-soluble liquids with mild to strong ammoniacal odors. Although completely soluble in water initially, hydrates may form with time, particularly with the heavy ethyleneamines (TETA, TEPA, PEHA, and higher polyamines), to the point that gels may form or the total solution may solidify under ambient conditions. The amines are also completely miscible with alcohols, acetone, benzene, toluene and ethyl ether, but only slightly soluble in heptane. Piperazine, the lowest mol wt cyclic diamine, freezes above room temperature. As such, it is available commercially as either the anhydrous solid or an aqueous solution.

Chemical Properties

The aliphatic alkyleneamines are strong bases exhibiting behavior typical of simple aliphatic amines. Additionally, dependent on the location of the primary or secondary amino groups in the alkyleneamines, ring formation with various reactants can occur. This same feature allows for metal ion complexation or chelation. The alkyleneamines are somewhat weaker bases than aliphatic amines and much stronger bases than ammonia as the pK_b values indicate.

Alkylene Oxides and Aziridines. Alkyleneamines react readily with epoxides, such as ethylene oxide or propylene oxide, to form mixtures of hydroxyalkyl derivatives. Aziridines react in an analogous fashion to epoxides.

Aliphatic Alcohols and Alkylene Glycols. Simple aliphatic alcohols, such as methanol, can be used to alkylate alkyleneamines.

Organic Halides. Alkyl halides and aryl halides, activated by electron withdrawing groups (such as NO_2) in the ortho or para positions, react with alkyleneamines to form mono- or disubstituted derivatives.

Aldehydes. Alkyleneamines react exothermically with aliphatic aldehydes. The products depend on stoichiometry, reaction conditions, and structure of the alkyleneamine.

Organic Acids and Their Derivatives (Anhydrides, Nitriles, Ureas). Alkyleneamines react with acids, esters, acid anhydrides or acyl halides to form amidoamines and polyamides. Various diamides of EDA are prepared from the appropriate methyl ester or acid at moderate temperatures.

Sulfur Compounds. EDA reacts readily with two moles of CS_2 in aqueous sodium hydroxide to form the bis sodium dithiocarbamate.

Environmentally Available Reactants. Under normal conditions ethyleneamines are considered to be thermally stable molecules. However, they are sufficiently reactive that upon exposure to adventitious water, carbon dioxide, nitrogen oxides, and oxygen, trace levels or byproducts can form and increased color usually results.

Manufacture

Ethyleneamine Processes. Present industrial processes are based on ethylene and ammonia. The sixty-year-old ethylene dichloride (EDC) process is still the most widely practiced industrial route for producing ethyleneamines.

Alternative processes for the manufacture of ethyleneamines have been actively sought since the late 1960s. The catalytic reductive amination of monoethanolamine (MEA), which was the first such process to appear, produces the lighter ethyleneamines (EDA, DETA, PIP, and AEP) and coproduct aminoethylethanolamine (AEEA), and hydroxyethylpiperazine (HEP).

The condensation of MEA with EDA over heterogeneous catalysts to form primarily DETA represents the newest commercial technology for making ethyleneamines.

A process for the production of ethylenimine, a suspected carcinogen, by the vapor phase dehydration of monoethanolamine has been developed.

Diaminopropane Processes. 1,2-Propylenediamine can be produced by the reductive amination of propylene oxide, 1,2-propylene glycol, or monoisopropanolamine. 1,3-Propanediol can be used to make 1,3-diaminopropane. Various propaneamines are produced by reducing the appropriate acrylonitrile–amine adducts. Polypropaneamines can be obtained by the oligomerization of 1,3-diaminopropane.

Economic Aspects

Worldwide growth for most ethyleneamines is expected to parallel GDP. Some regional and certain applications demands will show somewhat higher growth rates.

Table 1. Properties of Commercial Diamines and Higher Amines

Commercial name	Molecular weight	Freezing point, °C	Bp[a], °C	ΔH^{a}_{vap} kJ/mol	Refractive index, n^{20}_{D}	Viscosity at 20°C, mPa·s(= cP)
ethylenediamine	60.1	10.8	117.0	40.7	1.4565	1.8
diethylenetriamine	103.2	−39	206.9	54.0	1.4859	7.2
triethylenetetramine	146.2[b]	−35	277.4	56.4	1.4986	26
tetraethylenepentamine	189.3[b]	>−40	315		1.5067	76
pentaethylenehexamine[c]	232.4[b]	−30	180–280[d]			100–300
aminoethylpiperazine	129.2	−17	221	41.2	1.5003	15
piperazine	86.1	109.6	144.1			
1,2-propylenediamine	74.1	−27	120–123	38.2	1.4455	1.6
1,3-diaminopropane	74.1	−12	137–140	46.4[e]	1.4555	2.0
iminobispropylamine	131.2	−16	110–120	76.2[e]	1.4791	9.6
N-(2-aminoethyl)-1,3-propylenediamine	117.2		80			
N,N'-bis-(3-aminopropyl)-ethylenediamine	174.3		170			
dimethylaminopropylamine	102.2	−56	134.9	35.6	1.4350	1.1
menthanediamine	170.3	−45	107–126[f]		1.479	17.5[g]
triethylenediamine	112.2	158	174	61.9[h]		
hexamethylenediamine[i]	116.2	41	204.0	51.0		

[a] At 101.3 kPa = 1 atm unless otherwise noted. [b] Linear component. Commercial product consists of a mixture of linear, branched, and cyclic structures with the same number of nitrogen atoms. [c] Commercial higher polyamine products contain up to about 40% PEHA. [d] At 0.67 kPa, 10–60% distills in this range. [e] At 93.3°C. [f] At 1.3 kPa. [g] At 25°C. [h] Heat of sublimation, below 78°C. [i] For manufacture of HMDA in preparation of nylon-6,6 see POLYAMIDES.

Of the worldwide ethyleneamines capacity, over 50% is EDC-based; the balance is monoethanolamine-derived.

Storage and Handling

By virtue of their unique combination of reactivity and basicity, the polyamines react with, or catalyze the reaction of, many chemicals, sometimes rapidly and usually exothermically. Some reactions may produce derivatives that are explosives (eg, ethylenedinitramine). The amines can catalyze a runaway reaction with other compounds (eg, maleic anhydride, ethylene oxide, acrolein, and acrylates), sometimes resulting in an explosion.

As commercially pure materials, the ethyleneamines exhibit good temperature stability, but at elevated temperatures noticeable product breakdown may result in the formation of ammonia and lower and higher mol wt species.

Like many other combustible liquids, self-heating of ethyleneamines may occur by slow oxidation in absorbent or high surface-area media. In some cases, this may lead to spontaneous combustion; either smoldering or a flame may be observed. These media should be washed with water to remove the ethyleneamines, or thoroughly wet prior to disposal in accordance with local and Federal regulations.

Since ethyleneamines react with many other chemicals, dedicated processing equipment is usually desirable. Amines slowly absorb water, carbon dioxide, nitrogen oxides, and oxygen from the atmosphere, which may result in the formation of low concentrations of by-products and generally increase color. Storage under an inert atmosphere minimizes this sort of degradation.

Galvanized steel, copper, and copper-bearing alloys are unacceptable for all ethyleneamine service. A 300 series stainless steel or aluminum are recommended for storage of lighter amines (particularly EDA and DETA) to maintain product quality. Carbon steel generally can be used for storage of the heavier ethyleneamines without noticeable impact on product quality if the storage temperature is modest (<60°C), nitrogen blankets are maintained to exclude air, and the material is anhydrous. A 300 series stainless steel is often specified for heating coils, transfer lines, and small agitated tanks, because carbon steel can suffer enhanced corrosion as a result of the erosion of the passive film by the product velocity. Similar logic suggests cast 316 stainless steel for pumps and valves in ethyleneamine service.

Baked phenolic-lined carbon steel is acceptable for storage of many pure ethyleneamines, except EDA. Gaskets utilized in ethyleneamine service generally are made of Grafoil flexible graphite or polytetrafluoroethylene (TFE).

Most common thermal insulating materials are acceptable for ethyleneamine service. However, porous insulation may introduce the hazard of spontaneous combustion if saturated with ethyleneamines from a leak or external spill.

Certain ethyleneamines require storage above ambient temperature to keep them above their freezing points (EDA and PIP) or to lower the viscosity (the heavy amines). As a result, the vapors "breathing" from the storage tank can contain significant concentrations of the product. Water scrubbers may be used to capture these vapors.

Solid ethyleneamine carbamates, formed by the reaction of the amines with carbon dioxide, can foul tank vents and pressure relief devices. Vent fouling can be minimized by using a nitrogen blanket that prevents atmospheric CO_2 from being drawn in, or by steam-tracing the vents (>160°C) to decompose the carbamates.

Although the ethyleneamines are water soluble, solid amine hydrates may form at certain concentrations that may plug processing equipment, vent lines, and safety devices. Hydrate formation usually can be avoided by insulating and heat tracing equipment to maintain a temperature of at least 50°C. Water cleanup of ethyleneamine equipment can result in hydrate formation even in areas where routine processing is nonaqueous. Use of warm water can reduce the extent of the problem.

Health and Safety Factors

Ethyleneamine vapors are painful and irritating to the eyes, nose, throat, and respiratory system. Extremely high vapor concentration may cause lung damage. Prolonged or repeated inhalation may lead to kidney, liver, and respiratory system injury. Contact with the liquids will severely damage the eyes and may cause serious burns to the skin. When swallowed, the concentrated liquid materials may produce considerable local injury. Both vapors and liquid can cause sensitization in some individuals, resulting in contact dermatitis and/or the development of an asthmatic respiratory response. This may occur in certain susceptible individuals following exposure to extremely low concentrations of ethyleneamines, even below the irritation threshold.

The ACGIH has adopted TLVs of 10 ppm (25 mg/m^3) and 1 ppm (4 mg/m^3) for EDA and DETA, respectively. Strict precautions should be observed to prevent direct contact with ethyleneamines, including eye, skin, and respiratory protection. If contact is made, medical treatment should be obtained immediately, in addition to flushing and washing with copious amounts of water. Vomiting is not to be induced following ingestion.

Applications

Polyalkylene polyamines find use in a wide variety of applications by virtue of their unique combination of reactivity, basicity, and surface activity. With a few significant exceptions, they are used predominantly as intermediates in the production of functional products. End-use profiles for the various ethyleneamines include fungicide, oil and fuel additives, polyamides/epoxy curing, paper resins, chelating agents, fabric softeners/surfactants, petroleum production, bleach activator, and anthelmintics/pharmaceuticals.

RICHARD G. CARTER
ARTHUR R. DOUMAUX, JR.
STEVEN W. KAISER
PAMLA R. UMBERGER
Union Carbide Chemicals and Plastics Company Inc.

DIATOMITE

Diatomite is a naturally occurring, porous, high surface area form of hydrous silica that is used as a filter aid and as a mineral filler. Diatomite products may be classified according to manufacturing method into three categories: natural diatomite, calcined diatomite, and flux-calcined diatomite. Products from all three categories find widespread use in industrial filtration applications as a filter aid for achieving higher clarity, longer filter cycles, and removing high solids concentrations (Table 1). Products from all three categories are also used as functional fillers where diatomite properties add to the performance of paints, plastics, rubber, catalysts, agricultural chemicals, pharmaceuticals (qv), toothpastes, polishes, and other chemicals.

Table 1. Property Ranges of Diatomite Products[a]

Property	Filter aids: Natural	Filter aids: Calcined	Filter aids: Flux-calcined	Fillers, all types
permeability range, μm^2 [b]	0.06	0.5–2.0	1.0–29.6	
density, kg/m^3				
wet cake	240–350	270–350	290–380	
bulk	112	120–128	144–336	104–160
particle size distribution, μm				
10% <	1.5–3.6	2.5–4.4	7–11	2–4
50% <	7.0–13.4	10.0–16.1	25–37	6–20
90% <	25–44.5	30.0–58.9	65–97	14–30
approx. pressure differential[c], kPa[d]	36.5	2.33–4.56	1.11–0.058	
specific gravity	2.00	2.25	2.33	2.0–2.3
porosity, by vol, %	65–85	65–85	65–85	65–85
median pore size range, μm	1.5	3.5–5.0	7–22	[e]
surface area, m^2/g	10–20	4–6	1–4	0.7–30
pH	6.0–8.0	6.0–8.0	8.0–10.0	6.0–10.0
refractive index	1.42	1.44	1.48	1.40–1.49
oil absorption, %				100–210

[a] Values typical or estimated, not specifications. [b] To convert from μm^2 to d'Arcys, multiply by 1.013. [c] Measurement at 0.034 cm/s and 0.1 g/m^2 precoat. [d] To convert from kPa to psi, multiply by 0.145. [e] Hegman gauge readings, useful for paint manufacture, run from 0–55.

Diatomite, also known as diatomaceous earth, or kieselguhr, consists mainly of accumulated shells or frustules of intricately structured amorphous hydrous silica secreted by diatoms, which are microscopic, one-celled golden brown algae of the class Bacillariophyceae. Diatoms exist in many different environments and are abundant in regions of oceanic upwelling.

Origins of Deposits

Most commercial marine diatomite deposits exploit accumulations resulting from large blooms of diatoms that occurred in the oceans during the Miocene geological epoch. Marine deposits must have been formed on the bottom of protected basins or other bodies of quiet water, undisturbed by strong currents, in an environment similar to the existing Santa Barbara Channel or Gulf of California.

The main deposits of freshwater diatomite were laid down in large lakes. Many of these deposits in the western United States formed during glacial times, when the local climate was wetter. Several tens of square kilometers in Nevada west of Tonopah are covered with diatomite as are other large areas in the Great Basin.

Location of Deposits. Deposits of diatomite are known to exist on every continent and in nearly every country. Most of the deposits are not large enough or sufficiently pure to have commercial value. Production figures show the location of the deposits that meet commercial standards in both respects.

In the United States the most extensive commercial deposits are located in California, Nevada, Oregon, and Washington. The U.S. Bureau of Mines also reports the commercial operation of diatomite deposits in Arizona.

Physical and Chemical Properties

Chemically, diatomite consists primarily of silicon dioxide, $SiO_2 \cdot nH_2O$, and is essentially inert. It is attacked by strong alkalies and by hydrofluoric acid but is virtually unaffected by other acids. The silicon dioxide has a unique structure, resulting from the intricate form of the diatom skeletons. The chemically combined water content varies from 2 to 10%. Impurities that are often found mixed with the diatomite are other aquatic fossils such as sponge residues, Radiolaria, silicoflagellata, sand, clay, volcanic ash, mineral aerosols, calcium carbonate, magnesium carbonate, soluble salts, and organic matter.

The color of pure diatomite is white, or near white, but impurities such as carbonaceous matter, clay, iron oxide, volcanic ash, etc may darken it.

Individual diatom frustules are porous. The diatoms are highly variable in shape and size, having particles that range in effective diameter from 0.75 to 1000 μm, but most are 50 to 100 μm in diameter. Diatom shapes can range from simple cylinders and disks to complex, highly variable, but always punctate, forms.

The bulk density of powdered diatomite varies from 112 to 320 kg/m^3. The true specific gravity of diatomite is 2.1 to 2.2, the same as for opaline silica, or opal. The thermal conductivity of bulk quantities of diatomite is low but increases with higher percentages of impurities and a higher density. The fusion point depends on the purity but averages about 1430°C for pure material, which is slightly less than for pure silica. The addition of chemical agents, such as soda ash, reduces the fusion point.

Diatomite has only weak adsorption (qv) powers but shows excellent absorption (qv) because of its structure and high surface area. Acids, liquid fertilizers (qv), alcohol, water, oils, and other fluids are absorbed by diatomite.

Mining and Processing

Diatomite deposits are usually discovered by observation of outcrop, and the value of the deposits is determined by geological prospecting and exploration.

Mining. Most diatomite is mined by open-pit methods.

Economic Aspects Owing to the low bulk density of diatomite, freight and trucking rates (on a weight basis) are high. Domestic finished products are packed and shipped in laminated kraft-paper bags, usually containing 22.5 kg, or the product is shipped in bulk or semibulk bags. Bagged products are shipped by truck or rail boxcar.

Domestic Producers. A principal company mining diatomite and processing it into finished products is Celite Corp. (Lompoc, California), which has wholly owned mines and processing facilities in Lompoc, California; Quincy, Washington; Jalisco, Mexico; Murat, France; Alicante, Spain; and a joint venture mine in Iceland.

Production. After the United States, Romania, the former USSR, and France are the largest producers of diatomite. Combined with the United States, these countries account for more than 75% of the world's production.

KENNETH R. ENGH
Sandkuhl Clay Works

F. L. Kadey, in S. J. Lefond, ed., *Industrial Minerals and Rocks,* 5th ed., AIME, New York, 1983.

ASTM D604-81; D719-86, American Society for Testing and Materials, Philadelphia, Pa., 1989.

C. W. Cain, Jr., in J. J. McKetta, ed., *Encyclopedia of Chemical Processing and Design,* Vol. 21, Marcel Dekker, Inc., New York, 1984.

J. Kiefer, *Brauwelt Int.* **6,** 300 (1991).

DICARBOXYLIC ACIDS

The diacids are characterized by two carboxylic acid groups attached to a linear or branched hydrocarbon chain. Aliphatic, linear dicarboxylic acids of the general formula $HOOC(CH_2)_nCOOH$ and branched dicarboxylic acids are the subject of this article. The bifunctionality of the diacids makes them versatile materials, ideally suited for a variety of condensation polymerization reactions. Several diacids are commercially important chemicals that are produced in multimillion-kg quantities and find application in a myriad of uses.

Nomenclature

Unsubstituted aliphatic dicarboxylic acids, $HOOC(CH_2)_nCOOH$, are most often referred to by their trivial names for $n = 2$ to 10 (Table 1). Higher homologues are named using the IUPAC system by adding the suffix dioic to the parent hydrocarbon.

Physical Properties

Detailed summaries of physical properties are given. The diacids are colorless, crystalline solids that melt somewhat higher than monoacids of the same molecular weight. For diacids of even carbon number, melting points decrease sharply for numbers 2–10 and remain relatively constant for numbers 12–20 (see Table 1). There is a marked alternation in melting point and other physical properties with changes in carbon number from even to odd within the series. Odd members exhibit lower melting points, and higher solubility. Theoretical treatments have been developed to correlate these physical properties. The alternating effects are the result of the inability of odd carbon number compounds to assume an in-plane orientation of both carboxyl groups with respect to the hydrocarbon chain. Other properties showing these alternations are decarboxylation temperature and index of refraction. These effects have practical consequences in the selection of material for a given preparation because acid melting point, decarboxylation temperature, and solubility are often key considerations. The effects persist in derivatives based on the diacids, particularly polyamides, polyurethanes, and polyesters.

The temperature at which decarboxylation occurs is of particular interest in manufacturing processes based on polymerization in the molten state where reaction temperatures may be near the point at which decomposition of the diacid occurs. The diacids become more heat stable at carbon number four, with even-numbered acids always more stable. Thermal decomposition is strongly influenced by trace constituents, surface effects, and other environmental factors.

Table 1. Physical Properties of C_2–C_{21} Aliphatic Dicarboxylic Acids

IUPAC name	Mp, °C	Bp, °C[a]	Water solubility,[b] g/100 mL	Density, g/mL
ethanedioic	187 (dec)		9.5	
propanedioic	134–136 (dec)		152	1.619
butanedioic	187.6–187.9		8.35	1.572
pentanedioic	98–99	200[c]	130	1.424
hexanedioic	153.0–153.1	265	3.08[d]	1.345
heptanedioic	105.7–105.8	272	5.0	1.287
octanedioic	143.0–143.4	279	0.16	1.270
nonanedioic	107–108	286.5	0.214	1.235
decanedioic	134.0–134.4	294.5	0.10	1.231
undecanedioic	110.5–112		0.003	
dodecanedioic	128.7–129.0	254[e]	0.004	1.16
tridecanedioic	114			
tetradecanedioic	126.5			
pentadecanedioic	114.7			
hexadecanedioic	125			
heptadecanedioic	117–118			
octadecanedioic	124.6–124.8			
nonadecanedioic	118–119.5			
eicosanedioic	124–125			
heneicosanedioic	118–120			

[a] At 13.3 kPa = 100 mm Hg unless otherwise noted.
[b] At 20–25°C unless otherwise noted.
[c] At 2.7 kPa = 20 mm Hg.
[d] At 34.1°C.
[e] At 2.0 kPa = 15 mm Hg.

Lower members of the series are water soluble; solubility falls off sharply above adipic acid. Alternating effects are again expressed with acids of odd carbon numbers being the most soluble (see Table 1). Dibasic acids are ionized in aqueous solution to varying degree depending upon the proximity of the carboxyl groups within the individual structures. The carboxyl group, being electron-withdrawing, causes the neighboring carboxyl hydrogen to be more readily dissociated.

Chemical Properties

The dibasic acids undergo the reactions typical of monocarboxylic acids.

Manufacture, Preparation, and Processes

Glutaric Acid. Until 1990–1991 glutaric acid was available commercially from DuPont as a by-product in the production of adipic acid. It is no longer available, but DuPont produces dimethyl glutarate and mixtures of dimethyl succinate and dimethyl glutarate, as well as mixtures of dimethyl glutarate and dimethyl adipate.

Several procedures for making glutaric acid have been described in *Organic Syntheses* starting with trimethylene cyanide, methylene bis (malonic acid), γ-butyrolactone, and dihydropyran.

Pimelic Acid. This acid is manufactured by Tateyama Chemical Company in Japan in quantities of about 1000–2000 kg/yr, and by Heinrich Mock Nachf in Germany. The method or process they are using has not been disclosed. Pimelic acid is available in small quantities with purities of 98% from laboratory chemical supply companies. The preparation of pimelic acid has been described in *Organic Syntheses;* cyclohexanone condenses with diethyl oxalate, followed by decarboxylation to ethyl 2-keto-hexahydrobenzoate, and then cleavage of the β-keto ester with strong alkali.

Suberic Acid. This acid is not produced commercially at this time. However, small quantities of high purity (98%) can be obtained from chemical supply houses. If a demand developed for suberic acid, the most economical method for its preparation would probably be based on one analogous to that developed for adipic and dodecanedioic acids; air oxidation of cyclooctane to a mixture of cyclooctanone and cyclooctanol. This mixture is then further oxidized with nitric acid to give suberic acid.

Azelaic Acid. This acid is produces by the Emery Group of Henkel Corporation in Cincinnati, Ohio, in multimillion kg quantities. The process that is currently used is based on the ozonolysis of oleic acid (from grease or tallow) followed by the decomposition of the ozonide with oxygen.

Sebacic Acid. This acid is produced commercially by Union Camp in Dover, Ohio, by Hokoku Oil Company in Japan, and by a state enterprise in the People's Republic of China. The process used in each case is based on the caustic oxidation of castor oil or ricinoleic acid in either a batch or continuous process.

Dodecanedioic Acid. Dodecanedioic acid (DDDA) is produced commercially by Du Pont in Victoria, Texas, and by Chemische Werke Hüls in Germany. The starting material is butadiene which is converted to cyclododecatriene using a nickel catalyst. Hydrogenation of the triene gives cyclododecane, which is air oxidized to give cyclododecanone and cyclododecanol. Oxidation of this mixture with nitric acid gives dodecanedioic acid.

Brassylic Acid. This acid is commercially available from Nippon Mining Company (Tokyo, Japan). It is made by a fermentation process.

C-19 Dicarboxylic Acids. The C-19 dicarboxylic acids are generally mixtures of isomers formed by the reaction of carbon monoxide on oleic acid. Because the reaction produces a mixture of isomers, no single chemical name can be used to describe them. There are currently no commercial producers of C-19 dicarboxylic acids.

C-20 Dicarboxylic Acids. These acids have been prepared from cyclohexanone via conversion to cyclohexanone peroxide followed by decomposition by ferrous ions in the presence of butadiene. Okamura Oil Mill (Japan) produces a series of commercial acids based on a modification of this reaction.

C-21 Dicarboxylic Acids. C-21 dicarboxylic acids are a mixture of predominately 5-(6)-carboxy-4-hexyl-2-cyclohexene-1-octanoic acid, 5-isomer and 6-isomer. C-21 dicarboxylic acids are produced by Westvaco Corporation in Charleston, South Carolina in multimillion-kg quantities. The process involves reaction of tall oil fatty acids (TOFA) (containing about 50% oleic acid and 50% linoleic acid) with acrylic acid and iodine at 220–250°C for about 2 h.

Dicarboxylic Acids via Microorganisms

During the 1980s a number of patents were issued describing the preparation of dicarboxylic acids or esters using microorganisms. The α, ω-n-alkanedioic acids that have been prepared generally have 5–25 carbons. One of the first methods described the preparation of dimethyl 1,16-hexadecanedioate using a nutrient solution, n-hexadecyl bromide, and certain strains of *Torulopsis.* Other methods have been described to give dibasic acid of 8–22 carbons using n-alkanes or n-alcohols and various organisms.

Derivatives and Uses

Diacids, owing to their ready incorporation into polymers, are components in a wide variety of materials. The diacids are important industrial intermediates for the manufacture of diesters, polyesters, and polyamides. These derivatives find application as plasticizing agents, lubricants, heat transfer fluids, dielectric fluids, fibers, copolymers, inks and coatings resins, surfactants, fungicides, insecticides, hot-melt coatings, and adhesives. Of the higher diacids, azelaic, sebacic, and dodecanoic find the greatest application. Derivatives of glutaric and C-21 diacids also enjoy significant commercial applications.

Economic Aspects

In addition to azelaic, sebacic, dodecanedioic, eicosanedioic (C_{20} diacids), and C_{21} diacids, undecanedioic, brassylic, tetradecanedioic, hexadecanedioic, docosanedioic, and tetracosanedioic acids are available, expensive, and in limited quantity from research chemical supply houses.

Health and Safety Factors

In general, the higher diacids are essentially nontoxic. There are no indications that the dicarboxylic acids detailed here are carcinogenic or teratogenic in animals or humans. It is generally recognized that these diacids are ocular irritants and that the inhalation of the dust of these diacids is irritating to the mucous membranes and the respiratory tract. The water solubility of glutaric acid fosters its toxicity. Glutaric acid is a known nephrotoxin.

Environmental Effects. In general, the higher diacids do not pose substantial environmental risk; however, releases of significant quantities into surface or ground waters may be reportable under the Clean Water Act. The low biotoxicity of the higher diacids results, in part, from their limited water solubility. Glutaric acid is significantly more water soluble than the other diacids described herein, and the aquatic biotoxicity of glutaric acid and dimethyl glutarate is established.

ROBERT W. JOHNSON
CHARLES M. POLLOCK
Union Camp Corporation
ROBERT R. CANTRELL
Union University

E. H. Pryde and J. C. Cowan in J. K. Stille and T. W. Campbell, eds., *Condensation Monomers,* Wiley-Interscience, New York, 1972, pp. 1–153.

R. W. Johnson in R. W. Johnson and E. Fritz, eds., *Fatty Acids in Industry,* Marcel Dekker, Inc., New York, 1989, Chapt. 13.

DIETARY FIBER

Cell-wall dietary fiber is a complex system composed of variable amounts of cellulose, other polysaccharides such as hemicellulose and pectin, and lignin. The precise composition and proportion of polysaccharide types is related to the plant source, stage of maturity, and growing conditions. The composition and physical properties of dietary fiber also may be affected by both postharvest physiological changes and food processing. Cereal grains, legumes, vegetables, and fruits are primary sources of dietary fiber (see WHEAT AND OTHER CEREAL GRAINS). A smaller proportion of total dietary fiber comes from polysaccharides (gums and mucilages) added for their functionality in processed foods.

The variation in water solubility among polysaccharides results in varied physiological roles. Plant cell-wall polysaccharides and lignin provide insoluble dietary fiber (IDF); nondigestible storage polysaccharides, some pectic polysaccharides, and most of the functional additives contribute soluble dietary fiber (SDF).

A resurgence of interest in dietary fiber has been stimulated by epidemiological evidence of differences in colonic disease patterns between cultures with diets containing large quantities of fiber, and Western cultures having more highly refined diets. Many African countries, for example, are relatively free of diverticular disease, ulcerative colitis, hemorrhoids, polyps, and cancer of the colon. Whereas most interest has focused on the beneficial role of dietary fiber, there is also concern that high fiber diets may cause disturbances in the absorption of nutrients such as minerals (see MINERAL NUTRIENTS) and vitamins (qv).

Terminology. Dietary fiber is the accepted terminology in the United States for nutritional labeling. Total dietary fiber (TDF) and its subfractions, insoluble dietary fiber (IDF) and soluble dietary fiber (SDF), are defined analytically by official methods.

Sources, Composition, and Structure of Dietary Fiber

Natural sources of fiber in the diet include fruits, vegetables, legumes, and cereal grain products. The insoluble fiber content of some processed foods and breads is supplemented by incorporation of purified cellulose, cereal brans, or other plant fiber preparations. Cellulose and its chemically modified derivatives; seaweed polysaccharides, alginates and carrageenans; seed mucilaginous polysaccharides, guar and locust bean galactomannans; highly complex plant exudate polysaccharides, gum arabic, tragacanth, and others; microbially synthesized polysaccharides, xanthan and gellan gum; pectins; and other plant polysaccharides are added to foods for a variety of purposes. Because these materials are also nondigestible, they contribute to the total effect of dietary fiber and are encompassed by its definition.

Dietary fiber is a mixture of simple and complex polysaccharides and lignin. In intact plant tissue these components are organized into a complex matrix, which is not completely understood. The physical and chemical interactions that sustain this matrix affect its physicochemical properties and probably its physiological effects. Several of the polysaccharides classified as soluble fiber are soluble only after they have been extracted under fairly rigorous conditions.

Any starch escaping digestion in the upper gastrointestinal tract also contributes to dietary fiber effects. Some food starches, and the amylose fraction in particular, are readily converted into a nondigestible or slowly digestible physical form under certain food processing conditions. These resistant starches are readily fermented by colonic bacteria. Small amounts of waxes, cutin, and minerals in fruits and vegetables contribute to total dietary fiber values but may be physiologically inert.

Physiological Properties

The beneficial effects of dietary fiber, including both soluble and insoluble fiber, are generally recognized. Current recommendations are for daily intakes of 20–35 g in a balanced diet of cereal products, fruits, vegetables, and legumes. However, the specific preventive role of dietary fiber in certain diseases has been difficult to establish, in part because dietary risk factors such as high saturated fat and high protein levels are reduced as fiber levels increase.

Dietary fiber is important in the functioning of the entire gastrointestinal (GI) tract and affects the structure and morphology of the intestine. The process of chewing insoluble fiber-rich foods increases salivation and the flow of gastric secretions. Fiber components that increase viscosity or gel, for example, guar gum and pectin, delay gastric emptying. Fibers exert buffering action and may alter gastric pH by their effect on gastrointestinal hormones. A reduced glycemic response, probably resulting from delayed absorption of glucose, is associated with various fiber fractions, particularly viscosity-enhancing fibers.

Dietary fiber has a pronounced effect on the characteristics of the fecal mass and on the rate of passage of digest through the GI tract. High fiber diets also play a role in the excretion of bile acids and cholesterol.

Physicochemical Properties

Several physicochemical properties of dietary fiber contribute to its physiological role. Water-holding capacity, ion-exchange capacity, solution viscosity, density, and molecular interactions are characteristics determined by the chemical structure of the component polysaccharides, their crystallinity, and surface area.

Sources of Dietary Fiber for Processed Foods

An increasing number of fiber sources are available for food processing, representing a diversity of sources and technological advances. Commercial food-grade purified cellulose products and cereal brans are used in food products to enhance the fiber content or for other functional purposes. Many purified or partially purified nondigestible polysaccharides are used in food systems for their physicochemical properties, for example, for viscosity or as suspending agents. These polysaccharides contribute to the dietary fiber content even though they are not used for that purpose. Fiber sources include commercial cellulose products (powdered cellulose, microcrystalline cellulose, cellulose derivatives, and bacterial cellulose), bran, sugar-beet pulp, soybean cotyledons, and pea fiber. Sources of functional polysaccharides contributing to SDF include seaweed or algal, microbial, plant exudate, legume seeds, and tubers.

B. A. LEWIS
Cornell University

J. W. Anderson, D. A. Deakins, T. L. Floore, B. M. Smith, and S. E. White, *Crit. Rev. Food Sci. Nutr.* **29,** 95 (1990).

M. L. Dreher, *Handbook of Dietary Fiber, An Applied Approach,* Marcel Dekker, New York, 1987.

I. Furda and C. J. Brine, eds., *New Developments in Dietary Fiber,* Plenum Press, New York, 1990.

D. Kritchevsky, C. Bonfield, and J. W. Anderson, eds., *Dietary Fiber, Chemistry, Physiology and Health Effects,* Plenum Press, New York, 1990.

DIFFUSION SEPARATION METHODS

In chemical engineering, the term diffusional unit operations normally refers to the separation processes in which mass is transferred from one phase to another, often across a fluid interface, and in which diffusion is considered to be the rate-controlling mechanism. A number of special processes have been developed for difficult separations, such as the separation of the stable isotopes of uranium and those of other elements. Two of these processes, gaseous diffusion and gas centrifugation, are used by several nations on a multibillion dollar scale to separate partially the uranium isotopes and to produce a much more valuable fuel for nuclear power reactors. Because separation in these special processes depends upon the different rates of diffusion of the components, the processes are often referred to collectively as diffusion separation methods.

The most important industrial application of the diffusion separation methods has been for the enrichment of uranium-235, ^{235}U.

The United States was the first country to employ the gaseous diffusion process for the enrichment of ^{235}U, the fissionable natural uranium isotope. During the 1940s and 1950s, this enrichment application led to the investment of several billion dollars in process facilities. The original plants were built in 1943–1945 in Oak Ridge, Tennessee, as part of the Manhattan Project of World War II.

As of the early 1990s, diffusion separation methods are being employed or developed internationally in *(1)* Argentina, which has a gaseous diffusion project; *(2)* Brazil, where work is ongoing on gas centrifuges; *(3)* China, which has gaseous diffusion and gas centrifuges under development; *(4)* France, including Eurodif, owned by France, Italy, Spain and Belgium, which has a gaseous diffusion plant at Tricastin, plus a topping plant at Pierrelatte; *(5)* Germany, which has a large-scale centrifuge plant (Urenco); *(6)* India, which has gas centrifuges; *(7)* Japan, which has gas centrifuges; *(8)* the Netherlands, where there is a large-scale Urenco gas centrifuge plant; *(9)* Pakistan, which has gas centrifuges; *(10)* South Africa, which has a version of an advanced vortex tube process and has been working on the gas centrifuge process; *(11)* the former USSR, which has large-scale gaseous diffusion and gas centrifuge plants; and *(12)* the UK, which has gaseous diffusion and Urenco gas centrifuge plants.

In the United States, a group of domestic investors, including Duke Power, and Urenco, have applied for permission to construct a new gas centrifuge plant in Louisiana.

General Process and Design Selection

For difficult separations, such as isotope separations that involve the separation of molecules having very similar physical and chemical properties, the enrichment that can be obtained in a single equilibrium stage or transfer unit of the process is quite small. Hence, an extremely large number of these elementary separating units must be connected to form a separation cascade in order to achieve most desired separations. Consequently, very large separation systems requiring large amounts of energy are needed, and the total energy requirement is one of the most important cost considerations.

The energy or power required by any separation process is related more or less directly to its thermodynamic classification. There are, broadly speaking, three general types of continuous separation processes: reversible, partially reversible, and irreversible. Among the diffusion separation methods discussed herein, the centrifuge process (pressure diffusion) constitutes a theoretically reversible separation process with a relatively low energy requirement. Chemical exchange exemplifies a partially reversible process. All of the other diffusion methods discussed herein are essentially irreversible processes having high energy requirements.

Cascade Design

Less conventional diffusional separation operations are characterized by the relatively small separations that can be obtained by the elementary separation mechanism. That is, the changes in fluid composition attained in gaseous diffusion across the barrier, in thermal diffusion between the hot and cold walls, in mass diffusion between the inlet and the condensing surface for the sweep vapor, and in the centrifuge between the axis of the rotor and its periphery, are all quite small. Thus, a large number of separating units must be employed. Cascade is the term given to the aggregation of separating units that have been interconnected so as to be able to produce the desired material. The optimum arrangement of the separating units in a separation cascade generally minimizes the unit cost of product, and its design is a problem common to all separation processes. In a stagewise separation process such as gaseous diffusion, each unit of equipment consists of one separation stage.

Separation Stage. A fundamental quantity, α, exists in all stochastic separation processes, and is an index of the steady-state separation that can be attained in an element of the process equipment. The numerical value of α is developed for each process under consideration in the subsequent sections. The separation stage, which in a continuous separation process is called the transfer unit or equivalent theoretical plate, may be considered as a device separating a feed stream, or streams, into two product streams, often called heads and tails, or product and waste, such that the concentrations of the components in the two effluent streams are related by the quantity, α.

Separative Capacity. The separation stage is characterized not only by the separation factor α but also by its capacity or throughput of which the upflow L is a measure, and in the case of the continuous process, also by the length S. It is therefore desirable to define and determine a quantity indicative of the amount of useful separative work that can be done per unit time by a single stage. Such a quantity is called the separative capacity of the stage. It is postulated that the separation stage does useful work on the streams it processes, hence increasing their net value. The value of a stream must be a function of its concentration; let this value function be designated by $v(x)$. Then the separative capacity of the stage by definition is set equal to the increase in value it creates. The separative capacity of a unit is a very useful concept and permits comparisons to be made between different separation processes.

The separative capacity of a single stage operating with a cut of one-half is seen to be $\delta U = 1/4L(\alpha - 1)^2$. Thus, the separative capacity of a stage is directly proportional to the stage upflow as well as to the square of the separation effected.

The expression for the value function is $v(x) = (2x - 1) \times \ln[x/(1 - x)]$.

Cascade Gradient Equations. An arrangement of separation stages to form a simple cascade is shown in Figure 1. A simple cascade is one that divides a single cascade feed stream into a product stream and a waste stream. Additional side streams, however, could easily be handled. To be consistent with the conventions given for the single stage, the desired component is assumed to be enriched in the

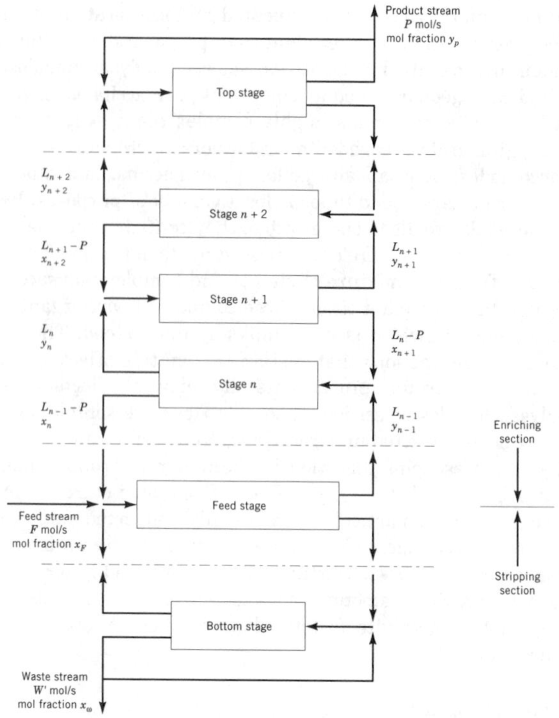

Figure 1. Separation stages arranged to form a simple cascade.

product stream at the top of the cascade. The cascade feed is introduced at some intermediate stage between the top and bottom of the cascade. The portion of the cascade that lies above the feed point is termed the enriching section; that which lies below the feed point is termed the stripping section. The gradient equations for the cascade are obtained from a combination of the material balance equations, frequently called the operating-line equations, and the α relationship, usually called the equilibrium-line equation. From a material balance around the top of the cascade down to, but not including, stage n of the enriching section, is obtained the operating-line equation, $L_n y_n = (L_n - P)x_{n+1} + Py_P$, which can be combined with the equilibrium-line, $y - x(\alpha - 1)x(1 - x)$ to give the following:

$$x_{n+1} - x_n = \frac{L_n}{L_n - P}[(\alpha - 1)x_n(1 - x_n) - (P/L_n)(y_P - x_n)]$$

For the case under consideration, where the value of $\alpha - 1$ is quite small, it follows that everywhere in the cascade, except possibly at the extreme ends, the stage upflow is many times greater than the product withdrawal rate. Thus, $L/(L - P)$ can be set equal to unity. Furthermore when the value of $\alpha - 1$ is small, the stage enrichment $x_{n+1} - x_n$ can be approximated by the differential ratio dx/dn without appreciable error. The gradient equation for the enriching section of a simple cascade therefore takes the form

$$dx/dn = (\alpha - 1)x(1 - x) - (P/L)(y_P - x) \quad (1)$$

Similarly, one obtains a gradient equation for the stripping section that has the form

$$dx/dn = (\alpha - 1)x(1 - x) - (W/L)(x - x_w) \quad (2)$$

Equations 1 and 2 are the basic equations for cascade design. Although these equations were derived from a consideration of a cascade composed of discrete separation stages, equations of the same form are also obtained for cascade designs based on continuous or differential separation processes. For use in the case of continuous separation processes, however, the term dx/dn, which is the enrichment per stage, is usually replaced by the equivalent terms $S\,dx/dz$, where S is the stage length and dx/dz the enrichment per unit length of process equipment.

These equations may then be used to calculate the output from a given cascade configuration, that is, from a cascade for which the variation of $\alpha - 1$ and L is known as a function of the stage number.

The Ideal Cascade. A cascade of particular interest to design engineers is the ideal cascade: a continuously tapered cascade (ie, L is a continuously varying function of x or n) that has the property of minimizing the sum of the stage upflows of all the stages required to achieve a given separation task. Because, in general, the total volume of the equipment required and the total power requirement of the cascade are directly proportional to the sum of the stage upflows, a consideration of the ideal plant requirements often permits a good economic estimate of the unit cost of product to be made without having to resort to the much more painstaking labor of designing a real (as opposed to ideal) cascade to accomplish the separation job.

Equations for Large Stage Separation Factors. The preceding results have been obtained with the use of an approximate equilibrium line equation and by replacing the finite difference, $x_{n+1} - x_n$, by the differential, dx/dn, both of which are valid only when the quantity $(\alpha - 1)$ is very small compared with unity. However, there has been renewed interest, partly because of the development of the gas centrifuge process to commercial status, in the design of cascades composed of stages with large stage separation factors. When the stage separation factor is large, the number of stages required in an ideal cascade in which all stages have the same separation factor is given by

$$N_{\text{ideal}} = \frac{2}{\ln \alpha} \ln\left(\frac{y_P}{1 - y_P} \Big/ \frac{x_W}{1 - x_W}\right) - 1 \tag{3}$$

Real Cascades. Although the ideal cascade minimizes the volume of equipment and the energy requirements, the cost of the cascade is generally not minimized because production economies are realized in the manufacture of the process equipment when a large number of identical units are produced. Thus, a minimum-cost cascade consists of a number of square cascade sections rather than uniformly tapered nonidentical stages. A first approximation to the optimum practical cascade, once the size (length and width) of the individual separating units available is known, is obtained by fitting the ideal plant shape with square sections in some intuitively appealing manner, as illustrated in Figure 2.

In the final analysis the problem of determining the optimum practical cascade is rather complex. Equipment performance and costs need to be related to the selected independent process variables. The whole process is provided with services, and auxiliary systems and feed and withdrawal facilities. It is usually enclosed by a building and surrounded by land of the proper type. Sizes and cost of these important items must be related to the process variables.

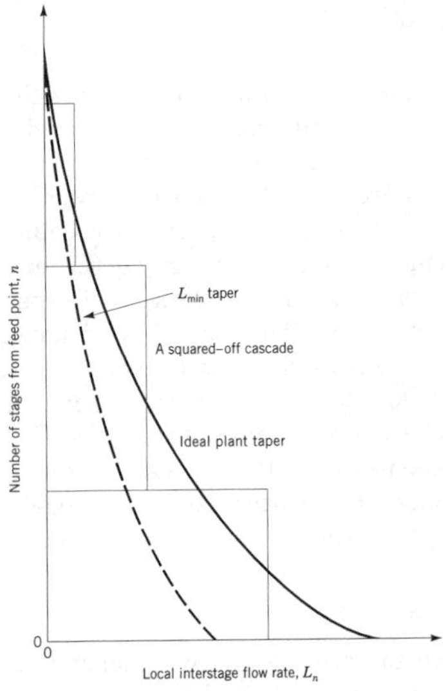

Figure 2. Design of a real cascade obtained by squaring off an ideal enriching section.

Time-Dependent Cascade Behavior. The period of time during which a cascade must be operated from start-up until the desired product material can be withdrawn is called the equilibrium time of the cascade. The equilibrium time of cascades utilizing processes having small values of $\alpha - 1$ is a very important quantity. Often a cascade may prove to be quite impractical because of an excessively long equilibrium time. An estimate of the equilibrium time of a cascade can be obtained from the ratio of the enriched inventory of desired component at steady state, H, to the average net upward transport of desired component over the entire transient period from start-up to steady state, τ.

The Gaseous Diffusion Process

The gaseous diffusion separation process depends on the separation effect arising from the phenomenon of molecular effusion (that is, the flow of gas through small orifices). When a mixture of two gases is confined in a vessel and is in thermal equilibrium with its surroundings, the molecules of the lighter gas strike the walls of the vessel more frequently, relative to its concentration, than the molecules of the heavier gas. This is caused by the greater average thermal velocity of the lighter molecules. If the walls of the container are porous with holes large enough to permit the escape of individual molecules, but sufficiently small so that bulk flow of the gas as a whole is prevented (that is, with pore diameters approaching mean-free-path dimensions of the gas), then the lighter molecules escape more readily than the heavier ones, and the escaping gas is enriched with respect to the lighter component of the mixture. The equation for the separation factor, α, for this process reflects the relative ease of light versus heavy molecules in escaping through the pores. Indeed, α^*, the ideal separation factor, is the ratio of the two molecular velocities. Because the kinetic energies, $1/2\ mv^2$, of the two species are the same, α^*, the ratio of the two velocities is equal also to the square root of the inverse ratio of the two molecular weights.

The large plants built for the separation of uranium isotopes following World War II are outstanding applications of the gaseous diffusing process. In the United States this work culminated in the construction of the gaseous diffusion cascade called the K-25 plant at Oak Ridge, Tennessee.

Process Description. The basic unit of a gaseous diffusion cascade is the gaseous diffusion stage. The main components are the converter holding the barrier in tubular form, motors, and compressors moving the gas between stages, a heat exchanger removing the heat of compression introduced by the stage compressors, the interstage piping, and special instruments and controls to maintain the desired pressures and temperatures.

The feed stream to a stage consists of the depleted stream from the stage above and the enriched stream from the stage below. This mixture is first compressed and then cooled so that it enters the diffusing chamber at some predetermined optimum temperature and pressure. Within the diffusion chamber the gas flows along a porous membrane or diffusion barrier. Approximately one-half of the gas passes through he barrier into a region of lower pressure. This gas is enriched in the component of lower molecular weight. The enriched fraction, upon leaving the diffusion chamber, is directed to the stage above where it is recompressed to the barrier high side pressure. The gas that does not pass through the barrier is depleted with respect to the light component. This depleted fraction, upon leaving the chamber, passes through a control valve, and is directed to the stage below where it too is recompressed to the barrier high side pressure. However, because it is necessary in this case to compensate only for the frictional losses and the control valve pressure drop, the compression ratio may not need to be as high as that for the enriched fraction.

Stage Design. The important parameters of a separation cascade employing gaseous diffusion stages are the stage separation factor and the size of a stage required to handle the desired stage flows. Both of these parameters depend on the characteristics of the barrier.

Plant Operation and Costs. Information on the design of new gaseous diffusing plants is available.

The cost of enriched material from a gaseous diffusion plant depends both on the cost of separative work and of feed material.

Pressure Diffusion Processes

Several devices have been developed for the purpose of producing such pressure gradients. The best known is the gas centrifuge (see also SEPARATION, CENTRIFUGAL). High speed centrifuges can develop gravitational fields equal to many thousand times that of the earth. Thus relatively large pressure gradients can exist between the axis and periphery of a centrifuge, giving rise to appreciable separation effects. By moving streams of gas at the periphery and at the axis countercurrently, the centrifuge can be made equivalent to a multistage separating column.

A second type of apparatus based on the pressure diffusion effect is the separation nozzle. Pressure gradients in a curved expanding jet produce an isotopic separation similar to that in a centrifuge. The separation effect obtained with a single jet is relatively small, and separation nozzle stages, similar to gaseous diffusion stages, must be used in a cascade to realize most of the desired separations.

A third device that utilizes pressure diffusion is the vortex chamber. Here, as in the centrifuge, angular acceleration effects in a rapidly rotating gas provide the pressure gradient. The vortex chamber may be considered as a centrifuge with a stationary outer wall. The mechanical difficulties of high-speed rotating machinery are avoided at the expense of friction effects between the gas and the stationary wall.

A vortex tube process has been developed in South Africa and is being used there for the enrichment of uranium. It appears that cascades of this type are characterized by an extremely high power consumption.

The Gas Centrifuge

The gas centrifuge process is today the most attractive alternative to gaseous diffusion as the means of providing new uranium enrichment capacity. The gas centrifuge is essentially a hollow, vertical cylinder which is spun about its axis at a high angular velocity. Gaseous uranium hexafluoride is fed to the centrifuge and is accelerated to approximately the angular velocity of the rotor. Centrifugal forces cause steep pressure gradients to be built up in the gas and ensuing pressure diffusion results in a radial concentration with the light isotope being slightly enriched in the vicinity of of the axis, the heavy isotope being slightly enriched in the vicinity of the rotor wall. Countercurrent gas centrifuges in which an axial convective circulation of the process gas is induced within the rotor in order to produce large end-to-end separation effects are the type of most interest. The centrifuge spins inside of an evacuated casing. The introduction of the feed gas and the withdrawal of enriched and depleted streams are accomplished by means of stationary tubes at the axis of the rotor with stationary scoops at the ends of the rotor which extend into the rotating gas. The countercurrent circulation may be induced by the mechanical action of the unshielded scoop or by temperature differences between the ends of the rotor. The bowl is supported at the top by a magnetic bearing, at the bottom by a needle bearing, and is driven by an electric motor.

The separative efficiency of a gas centrifuge used for isotope separation is best defined in terms of separative work. Thus, the separative efficiency E is defined by

$$E = \frac{\delta U(\text{experimental})}{\delta U(\text{max})} \tag{4}$$

where δU (experimental) is the actual separative work produced per unit time by the centrifuge under consideration and δU (max) is the maximum theoretical separative capacity of the machine. The maximum separative capacity of a gas centrifuge is given by

$$\pi U(\text{max}) = \frac{\pi Z c D_{AB}}{2}\left(\frac{\Delta M V^2}{2RT}\right)^2 \tag{5}$$

where δU is the separative capacity in mols per unit time, Z is the length of the rotor, ΔM is the difference in the mol wts of the components being separated, and V is the peripheral velocity of the centrifuge ($V = or_2$). The expression for the maximum separative capacity of a centrifuge indicates a desirability for (*1*) low-temperature operation because the theoretical maximum separative capacity of a centrifuge varies inversely as the temperature; (*2*) long centrifuge bowls because the theoretical maximum separative capacity varies directly as Z and that δU (max) is independent of the radius of the bowl; and (*3*) high peripheral velocity because the theoretical maximum separative capacity varies as the fourth power of the peripheral speed. At the higher speeds the predicted separative capacity increases with increasing peripheral speed much more slowly than the fourth-power relationship. Nevertheless, over the entire range of speeds investigated there is still an appreciable gain in separative capacity to be realized from an increase in speed.

The separative efficiency E of a countercurrent gas centrifuge may be considered to be the product of four factors, all but one of which can be evaluated on the basis of theoretical considerations. In this formulation the separative efficiency is defined by $E \equiv e_c e_I e_F e_E$, where e_c designates the circulation efficiency, e_I designates the ideality efficiency, e_F designates the flow pattern efficiency, and e_E designates the experimental efficiency and includes all phenomena such as turbulence and end effects not taken into account by the preceding terms.

E. VON HALLE
Martin Marietta Energy Systems
J. SHACTER
Consultant

M. Benedict, T. Pigford, and H. Levi, *Nuclear Chemical Engineering,* 2nd ed., McGraw-Hill Book Co., New York, 1981, Chapts. 12 and 14.

H. R. C. Pratt, *Countercurrent Separation Processes,* American Elsevier Publishing Company, Inc., New York, 1967.

S. Villani, *Isotope Separation,* American Nuclear Society Monograph, ANS Publications, 1976.

S. Villani, ed., *Topics in Applied Physics,* Vol. 35, Springer-Verlag, New York, 1979.

DIMENSIONAL ANALYSIS

Dimensional analysis is a technique that treats the general forms of equations governing natural phenomena. It provides procedures of judicious grouping of variables associated with a physical phenomenon to form dimensionless products of these variables; therefore, without destroying the generality of the relationship, the equation describing the physical phenomenon may be more easily determined experimentally. It guides the experimenter in the selection of experiments capable of yielding significant information and in the avoidance of redundant experiments, and makes possible the use of scale models for experiments (see DESIGN OF EXPERIMENTS). The method is particularly valuable when the problems involve a large number of variables. On such occasions, dimensional analysis may reveal that, whatever the form of the inaccessible final solution, certain features of it are obligatory. The technique has been utilized effectively in engineering modeling.

Units and Dimensions

The concepts used to describe natural phenomena are based on the precise measurement of quantities. The quantitative measure of anything is a number that is found by comparing one magnitude with another of the same type. It is necessary to specify the magnitude of the quantity used in making the comparison if the number is to be meaningful. The statement that "the length of a car is 6 meters" implies that a length has been chosen, namely, one meter, and that the ratio of the length of the car to the chosen length is 6. The chosen magnitudes, such as the meter, are called *units* of measurement.

Classical physics is built on the foundation of the laws of motion. It was felt at the time that the entire subject could be based on the laws of classical mechanics, and further work would undoubtedly make electromagnetism another branch of mechanics. Under these circumstances, it was natural to regard length l, mass m, and time t as the fundamental, primary, or reference dimensions. However, such designations lead to dimensional ambiguity in that two distinct concepts may possess the same dimensions.

Over the years the number of reference dimensions in physics has evolved from the original three, to four, to five, and then gradually downwards to an absolutely necessary one, and then upwards again through an understanding that, though only one is absolutely necessary, a considerable convenience can stem from using three, to four, or five reference dimensions depending on the problem at hand. There is nothing sacrosanct about the number of reference dimensions, and dimensional analysis is merely a tool that may be manipulated at will. This principle of free choice of the reference dimensions has been widely accepted, although one still finds references to true dimensions. Thus, an important step in dimensional analysis is the selection of reference dimensions in such a way that the others, called the secondary or derived dimensions, can be expressed in terms of them. The relation between reference and derived dimensions is generally established either through the fundamental law or equation governing the phenomenon or through definitions.

Dimensional Matrix and Dimensionless Products

An appropriate set of independent reference dimensions may be chosen so that the dimensions of each of the variables involved in a physical phenomenon can be expressed in terms of these reference dimensions. In order to utilize the algebraic approach to dimensional analysis, it is convenient to display the dimensions of the variables by a matrix. The matrix is referred to as the dimensional matrix of the variables and is denoted by the symbol $\boldsymbol{D}$. Each column of $\boldsymbol{D}$ represents a variable under consideration, and each row of $\boldsymbol{D}$ represents a reference dimension. The ith row and jth column element of $\boldsymbol{D}$ denotes the exponent of the reference dimension corresponding to the ith row of $\boldsymbol{D}$ in the dimensional formula of the variable corresponding to the jth column. As an illustration, consider Newton's law of motion, which relates force F, mass M, and acceleration A by (eq. 1):

$$F = \text{constant} \times MA \tag{1}$$

If length l, mass m, and time t are chosen as the reference dimensions, the dimensional formulas for the variables F, M, and A are as follows:

Variables	Dimensional formulas
F	$m^1 l^1 t^{-2}$
M	$m^1 l^0 t^0$
A	$m^0 l^1 t^{-2}$

The dimensional matrix associated with Newton's law of motion is obtained as (eq. 2)

$$\boldsymbol{D} = \begin{matrix} m \\ l \\ t \end{matrix} \overset{\begin{matrix} F & M & A \end{matrix}}{\begin{bmatrix} 1 & 1 & 0 \\ 1 & 0 & 1 \\ -2 & 0 & -2 \end{bmatrix}} \tag{2}$$

The validity of the method of dimensional analysis is based on the premise that any equation that correctly describes a physical phenomenon must be dimensionally homogeneous. An equation is said to be dimensionally homogeneous if each term has the same exponents of dimensions. Such an equation is of course independent of the systems of units employed provided the units are compatible with the dimensional system of the equation. It is convenient to represent the exponents of dimensions of a variable by a column vector called dimensional vector represented by the column corresponding to the variable in the dimensional matrix. In equation 2, the dimensional vector of force $\boldsymbol{F}$ is $[1,1,-2]'$ where the prime denotes the matrix transpose.

Suppose that there are n variables $Q_1\ Q_2, \ldots, Q_n$ that are involved in a physical phenomenon whose dimensional vectors are $\boldsymbol{D}_1, \boldsymbol{D}_2, \ldots, \boldsymbol{D}_n$, respectively. This phenomenon can generally be expressed by (eq. 3):

$$f(Q_1, Q_2, \cdots, Q_n) = 0 \tag{3}$$

Where such a function is established or assumed, it will still exist even after the variables are intermultiplied in any manner whatsoever. This means that each variable in the equation can be combined with other variables of the equation to form dimensionless products whose dimensional vectors are the zero vector. Equation 3 can then be transformed into the nondimensional form as (eq. 4):

$$f(\pi_1, \pi_2, \cdots, \pi_n) = 0 \tag{4}$$

where the dimensionless products π_i ($i = 1, 2, \ldots, n$) can generally be expressed as the power products of the form (eq. 5):

$$\pi_i = Q_1^{x1i} Q_2^{x2i} \cdots Q_n^{xni} \tag{5}$$

Let $R_1, R_2, \ldots, R_m$ be a set of chosen reference dimensions. Then the dimensional formulas for the variables Q_i are given by (eq. 6):

$$R_1^{d1i} R_2^{d2i} \cdots R_m^{dmi} \tag{6}$$

where the exponents of dimensions are represented by the dimensional vectors as (eq. 7):

$$\mathbf{D}_i' = [d_{1i}, d_{2i}, \cdots, d_{mi}],\ i = 1, 2, \cdots, n \tag{7}$$

Using equation 6 the dimensional formulas for π_i of equation 5 can be written to give (eq. 8):

$$\left[R_1^{d11} R_2^{d21} \cdots R_m^{dm1}\right]^{x1i} \left[R_1^{d12} R_2^{d22} \cdots R_m^{dm2}\right]^{x2i} \cdots \left[R_1^{d1n} R_2^{d2n} \cdots R_m^{dmn}\right]^{xni} \tag{8}$$

Because π_i are dimensionless products having dimensional vectors equal to the zero vector, the exponents of the R_j ($j = 1, 2, \ldots, m$) must add up to zero, giving (eq. 9):

$$\begin{aligned} d_{11}x_{1i} + d_{12}x_{2i} + \cdots + d_{1n}x_{ni} &= 0 \\ d_{21}x_{1i} + d_{22}x_{2i} + dx_{2i} + \cdots + d_{2n}x_{ni} &= 0 \\ \cdots\cdots\cdots \\ \cdots\cdots\cdots\cdots\cdots\cdots \\ d_{m1}x_{1i} + d_{m2}x_{2i} + \cdots + d_{mn}x_{ni} &= 0 \end{aligned} \tag{9}$$

In terms of the dimensional vectors of equation 8, equation 10 can be written as (eqs. 10–12):

$$[\mathbf{D}_1, \mathbf{D}_2, \cdots, \mathbf{D}_n]\mathbf{X}_i = \mathbf{0}, i = 1, 2, \cdots, n \tag{10}$$

where

$$X_i' = [x_{1i}, x_{2i}, \cdots, x_{ni}] \tag{11}$$

or more compactly

$$\mathbf{DX} = \mathbf{0} \tag{12}$$

where $\mathbf{X} = \mathbf{X}_i$, $i = 1, 2, \ldots, n$.. Thus, the product of a set of variables is dimensionless if, and only if, the exponents of these variables are a solution of the homogeneous linear algebraic equation. A vector $\boldsymbol{X}$ is said to be a B-vector of $\boldsymbol{D}$ if it is a solution of equation 12. The corresponding dimensionless product associated with the variables of a B-vector is called a B-number.

WAI-KAI CHEN
University of Illinois at Chicago

J. F. Douglas, *An Introduction to Dimensional Analysis for Engineers,* Sir Isaac Pitman & Sons, London, 1969.

W. W. Happ, *J. Appl. Phys.* **38,** 3918 (1967).

W. K. Chen, *J. Franklin Inst.* **292,** 403 (1971).

E. R. Van Driest, *J. Appl. Mech.* **13,** A-34 (1946).

DIMER ACIDS

The dimer acids, 9- and 10-carboxystearic acids, and C-21 dicarboxylic acids are products resulting from three different reactions of C-18 unsaturated fatty acids. These reactions are, respectively, self-condensation, reaction with carbon monoxide followed by oxidation of the resulting 9- or 10-formylstearic acid (or, alternatively, by hydrocarboxylation of the unsaturated fatty acid), and Diels-Alder reaction with acrylic acid. The starting materials for these reactions have been almost exclusively tall oil fatty acids or, to a lesser degree, oleic acid, although other unsaturated fatty acid feedstocks can be used.

Physical Properties

The physical properties of polymerized fatty acids are influenced by the basestock, by the dimerization conditions and catalysis, and by the degree to which monomer, dimer, and higher oligomers are separated following the dimerization.

Dimer acids are relatively high mol wt (ca 560) and yet are liquid at 25°C. This liquidity is a consequence of the many isomers present, most with branching or cyclic structures.

Most of the products listed in Tables 1 and 2 are based on manufacture from tall oil fatty acids. Dimer acids based on other feedstocks (eg, oleic acid) may have different properties.

Chemical Properties

Structure and Mechanism of Formation. Thermal dimerization of unsaturated fatty acids has been explained both by a Diels-Alder mechanism and by a free-radical route involving hydrogen transfer. The Diels-Alder reaction appears to apply to starting materials high in linoleic acid content satisfactorily, but oleic acid oligomerization seems better rationalized by a free-radical reaction.

Chemical Reactions. The reactions of dimer acids were reviewed fully in 1975. The most important is polymerization; the greatest quantities of dimer acids are incorporated into the non-nylon polyamides. Other reactions of dimer acids that are applied commercially include polyesterification, hydrogenation, esterification, and conversion of the carboxy groups to various nitrogen-containing functional groups.

Table 1. Properties of Dimer and Distilled Dimer Acid[a]

Physical property	Distilled dimer acid: Hydrogenated[b]	Distilled dimer acid: Unhydrogenated[c]	Dimer acid
composition %			
dimer acids	97	95	82–83
trimer acids (and higher)	3	4	14–16
monobasic acids	trace	1	1–5
acid number	191–197	190–196	189–197
viscosity at 25°C, mm^2/s (= cSt)	~5200	7000–8000	7500–9000
specific gravity, 25/25°C	0.94	0.9	~0.95

[a] Hystrene series of dimer acids, Humko Chemical Division of Witco Corporation.
[b] Empol series of dimer acids, Henkel Corp., Emery Group (oleic-based, thus of lower viscosity).
[c] Hystrene series of dimer acids, Humko Chemical Div. of Witco Corp.

Table 2. Properties of Trimer Acids[a]

Physical property	Value
composition, %	
dimer	40
trimer	60
monobasic acids	trace
acid number	170–190
viscosity at 25°C, mm^2/s (= cSt)	~30,000

[a] Hystrene 5460, Humko Chemical Div. of Witco Corp.

Manufacture

Clay-catalyzed oligomerization and thermal oligomerization are the two commercial processes used to manufacture dimer acids.

Process Modification. Dimer acid process modifications have fallen into three categories; those claiming higher dimer:trimer ratios, those utilizing varying types of clays, and those purporting to result in improved yields. Another aspect of process improvement is color improvement.

Other Polymerization Methods. Experimental alternatives include the use of peroxides, hydrogen fluoride, a sulfonic acid ion-exchange resin, and corona discharge.

Energy Requirements. The production of dimer acids is quite energy-intensive. A standard operation sequence normally results in the expenditure of about 18.6 MJ (17,600 Btu) (equivalent to 0.67 kg coal or 0.33 kg natural gas of fuel oil) to produce each kg of crude dimer and to separate it into monomer, dimer, and trimer.

Storage and Handling

Since dimer acids, monomer acids, and trimer acids are unsaturated, they are susceptible to oxidative and thermal attack, and under certain conditions they are slightly corrosive to metals. Special precautions are necessary, therefore, to prevent product color development and equipment deterioration.

Economic and Market Aspects

According to one estimate, the current capacity for manufacturing dimer acids in the U.S. is around 55,000 t per year. Current demand is estimated at about 33,600 t per year, and is expected to grow at about 2–3% per year to 35,000 t in 1993.

The current market situation for dimer acids includes relatively high raw material costs, high energy costs, slow growth, and relatively low prices. It is generally recognized as a mature market, with hopes for future growth hinging on factors such as increased polyamide use and a resurgence of oil drilling, where dimers are used for corrosion inhibition.

Health and Safety Aspects

The acute oral toxicity and the primary skin and acute eye irritative potentials of dimer acids, distilled dimer acids, trimer acids, and monomer acids have been evaluated based on the techniques specified in the *Code of Federal Regulations* (CFR). Based on these evaluations, monomer acids, distilled dimer acids, dimer acids, and trimer acids are classified as nontoxic by ingestion, are not primary skin irritants or corrosive materials, and are not eye irritants as these terms are defined in the federal regulations.

Uses

Nonreactive Polyamide Resins. Dimer-based polyamide resin markets are divided into those for reactive polyamides and those for nonreactive polyamides. The largest-volume commercial application of dimer acids is in nonreactive polyamide resins. Dimer acids impart flexibility, corrosion resistance, chemical resistance, moisture resistance, and adhesion to nonreactive polyamides. Hot-melt adhesives, the largest commercial application of nonreactive polyamide resins, are thermoplastics that have fairly sharp melting ranges. Flexographic printing inks utilize nonreactive polyamides from dimer acids as resin binders. The most important coating application for the nonreactive polyamide resins is in producing thixotropy.

Reactive Polyamide Resins. Reactive polyamide resins are used extensively to react with epoxy or phenolic resins. The amount used in epoxy applications far exceeds the use with phenolic resins.

THOMAS E. BREUER
Humko Chemical Division of Witco Corporation

E. C. Leonard, ed., *The Dimer Acids,* Humko Sheffield Chemical, Memphis, Tenn., 1975.

R. W. Johnson in E. H. Pryde, ed., *Fatty Acids,* American Oil Chemists Society, Champaign, Ill., 1979.

E. C. Leonard in E. H. Pryde, ed., *Fatty Acids,* American Oil Chemists Society, Campaign, Ill., 1979.

R. W. Johnson in R. W. Johnson and E. Fritz eds., *Fatty Acids in Industry,* Marcel Dekker, Inc., New York, 1989.

DISINFECTANTS AND ANTISEPTICS

The U.S. Environmental Protection Agency (EPA) and the Centers for Disease Control (CDC) reported in the early 1990s that diseases caused by viruses and parasites are on the increase. Bacterial diseases that had been under control have reappeared with large numbers of cases. Cholera has been epidemic in South American countries and tuberculosis has revived in the United States and Russia and has become a global problem again. Furthermore, new virulent strains and antibiotic-resistant forms puts greater stress on prevention of infection by employing disinfectants and antiseptics. Disinfectants find additional use in preventing spoilage of products such as food (see FOOD ADDITIVES), pharmaceuticals (qv), cosmetics (qv), paints (see PAINT), wood (qv), cloth (see TEXTILES), and even in helping to keep office buildings from becoming uninhabitable.

Definitions

Disinfectant. A disinfectant is a chemical or physical agent that frees from infection, that kills bacteria, fungi, viruses, and protozoa, but may not kill or inactivate bacterial spores, and is used only on inanimate objects, not on or in living tissue. A bactericide, fungicide, virucide, etc, is a disinfectant intended to kill the organisms indicated in the term. A germicide claims to kill pathogenic microorganisms, or germs.

Antiseptic. An antiseptic is a chemical substance that prevents or inhibits the action or growth of microorganisms but may not necessarily kill them, and is used topically on living tissue. The distinction between a disinfectant and an antiseptic is that the former is expected to kill all vegetative cells and is used only on inanimate objects, whereas the latter may not kill all cells and is used on the body.

Use of Disinfectants and Antiseptics

The disinfectant market is about $1 billion annually at the retail level in the United States. Disinfectants are used in janitorial supplies for hospitals and the home to treat toilet bowls, floors and walls in sick rooms, operating rooms, and wherever infective microorganisms are a problem. Instruments such as scalpels, scissors, catheters, and endoscopes used to invade tissues are treated with disinfectants, as are dental instruments. Laws require that hospital waste must be disinfected so that bacteria and viruses, such as the hepatitis virus and the AIDS virus, do not infect hospital workers and people in the community.

Disinfecting chemicals also experience wide application in the food industry in the growth and production of plant and animal foods. Finally, disinfectants are employed in a range of industrial applications such as prevention of paint mildewing, microbial contamination of pharmaceuticals and cosmetics, bacterial clogging of oil wells, biocorrosion of oil storage tanks, and decay of timber.

Antiseptics are used in the home for simple cuts and wounds, and in hospitals for treating patients' skin and surgeons' hands prior to operative procedures. They also are used for preparation of the skin prior to insertion of items such as intravascular lines, chest tubes, temporary pacemakers, and catheters of all kinds.

Dyes. No group of compounds is so intimately tied to the history and fabric of microbiology as are the dyes. Dyes have been used to selectively stain microorganisms for microscopic examination and identification, and also as antiseptics to inhibit growth. However, dyes are generally no longer used, in part because of stains to clothing, bedding, and noninfected body parts.

Halogens

Chlorine, Hypochlorites, and Chlorine Dioxide. Chlorine and its compounds are not only among the oldest disinfectants, but are used in greatest amount because these materials are cheap and effective (see ALKALI AND CHLORINE PRODUCTS–CHLORINE AND SODIUM HYDROXIDE). They find use in treatment of drinking water, wastewater, swimming pools, and general sanitation in commercial plants and the home although there is a move to restrict the use of chlorine owing to the production of carcinogens with some organic compounds. Chlorine has the greatest antimicrobial activity of the halogens. Fluorine (qv) is too reactive to be used as a disinfectant, but fluoride and silico-fluoride salts have found use as mold-control agents (see ANTIPARASITIC AGENTS–ANTIMYCOTICS) and wood preservatives (see also BLEACHING AGENTS).

N-Chloramines. *N*-Chloramines comprise the derivatives of amines in which one or two valences of trivalent nitrogen are taken up by chlorine. These compounds, like hypochlorites, are used in the cleaning and sanitizing of equipment and utensils in the food and dairy industries, for water and sewage treatment, and for bleaching and sanitizing in commercial laundries. Like chloramine, they are slower in effecting germicidal action than chlorine and hypochlorous acid. In swimming pools this is desirable because the germicidal effect lasts longer where chlorine is lost from sunlight and aeration. As dry powders, chloramines are stable in storage.

Iodine. Iodine has been important for many years, primarily as an antiseptic. Tincture of iodine and aqueous iodine are not as popular as they used to be because they stain skin and clothes a brown color and also because of their toxicity. These problems have been considerably reduced, but not completely resolved, in the production of iodophors.

Iodophors are the product of the chemical reaction of iodine and surface-active agents or polymers to produce complexes that retain the germicidal activity but not the undesirable properties of iodine. These are water-soluble, nonstaining, less irritating to the skin and other tissues, nonirritating to the eyes and mucous membranes, and do not cause a burning sensation when applied to raw skin. For these reasons, iodophores have virtually replaced tincture of iodine. An organic carrier of the iodine may be poly(vinylpyrrolidine) (Povidone–iodine), a nonionic surfactant, or a cationic detergent. Iodophors are important as broad-spectrum antiseptics for the skin, although they do not have the persistent action of some other antiseptics. They are also used as disinfectants for clinical thermometers that have been used by tuberuclous patients, for surface disinfection of tables, etc, and for clean equipment in hospitals, food plants, and dairies, much as chlorine disinfectants are used.

Alcohols

Alcohols, particularly ethanol and 2-propanol, are important disinfectants and antiseptics. In the aliphatic series of straight-chain alcohols, the antimicrobial activity increases with increasing molecular weight up to a maximum, depending on the organism tested. The order of bactericidal activity for alcohols is primary > secondary > tertiary. Alcohols are bactericidal, fungicidal, and virucidal but not sporicidal.

Ethanol and 2-propanol find application as skin antiseptics for personnel handwashing, surgical scrub, and preoperative skin preparations because these alcohols evaporate leaving no residue and are rapid in action.

Ethanol and 2-propanol have also found use in disinfecting clinical thermometers, and as preservatives to prevent microbial deterioration of cosmetics and medicinals. They are sometimes combined with other disinfectants, namely formaldehyde, phenolics, chlorhexidine, hypochlorite, and phenols.

Phenolic Compounds

As a disinfectant or antiseptic, phenol (carbolic acid) is mostly of historical interest. However, its extensive use continues in both investigative and analytical microbiology, eg, as in the AOAC phenol coefficient and use-dilution methods.

Phenol Derivatives. Derivatives of phenol have been investigated more thoroughly than any other group of disinfectants, and have found use as antiseptics, as the active ingredient in germicidal soaps and lotions, as hard-surface disinfectants, as preservatives for toiletries, and as antimicrobial products for institutions and the household. In industry they have varied uses, such as wood preservatives, mildewcides for leather, and to control algae, slime-forming bacteria, fungi, and sulfate-reducing bacteria in oil fields.

Studies of the effect of the structure of phenol derivatives on antimicrobial activity have generated a number of general observations. *(1)* In a homologous series of monoalkyl phenols, the potency against four common organisms, ie, *Salmonella typhosa, Staphylococcus aureus, Mycobacterium tuberculosis,* and *Candida albicans,* increases with the increase in molecular weight until the *n*-amyl derivative is reached. *(2)* Halogenation increases the antimicrobial activity of phenols. *(3)* Greater activity is found where the alkyl group is ortho to the phenolic moiety, and the halogen is para, rather than the reverse. *(4)* Nitro group substitution in phenols also increases antimicrobial activity, believed to result from uncoupling of oxidative phosphorylation.

Because of lower toxicity and high antimicrobial activity, the phenols having the greatest use in disinfections are *o*-phenylphenol (Dowicide 1), $C_{12}H_{10}O$; *o*-benzyl-*p*-chlorophenol (Santophen 1), $C_{13}H_{11}ClO$; and *p-tert*-amylphenol, $C_{11}H_{16}O$. They possess similar general characteristics; ie, broad-spectrum antimicrobial activity toward gram-negative and gram positive bacteria, fungi, *Mycobacterium tuberculosis,* and protozoa; virucidal activity against lipophilic and intermediate but not hydrophilic viruses; tolerance for organic loading and hard water; residual activity; and biodegradability.

Phenols are considered to be low-to-intermediate level disinfectants, appropriate for general disinfection of noncritical and semicritical areas. They are not sporicidal and should not be used when sterilization is required.

Bisphenols. Bisphenols are a group of phenols with outstanding antimicrobial properties. The outstanding member is hexachlorophene has found extensive use because it does not lose activity in the presence of soap and greatly reduces the number of skin flora when used in antimicrobial soaps. It has also been incorporated in soapless detergent bases, eg, pHisohex, and in deodorant soaps. It is especially effective against *Staphylococcus aureus,* active at 1–1.5 ppm. Unfortunately, the action of hexachlorophene is reversed upon contact with blood. This is deemed significant for any prophylactic application to broken skin against infection in injury, or for preoperative use. Skin that has been degermed using hexachlorophene is therefore not protected against subsequent contamination from transient pathogens or against lesions of bacterial origin. Although reports of the extensive use of hexachlorophene do not show toxic symptoms in humans, neurotoxicity has been demonstrated in rats using large doses, in monkeys, and in premature infants. The FDA has banned over-the-counter sale of soaps, cosmetics, and drugs containing more than 0.1% hexachlorophene. All products of higher percentage have been put on a prescription basis. All products must be labeled to prevent use by pregnant women and children. The United Kingdom has initiated similar restrictions.

Another important bisphenol is dichlorophene (dichlorophane, methylene-bis(4-chlorophenol)). Whereas it is not as active against bacteria as hexachlorophane, it has found miscellaneous applications, eg, as a rot preservative for textiles; as a treatment for athlete's foot and for tapeworm in humans and domestic animals; as a slimicide in paper manufacture; and as an antibacterial agent in water-cooling systems.

Fentichlor (bithionol, thiobis-(4,6-dichlorophenol)) is a bisphenol having a sulfur bridge rather than a methylene bridge. Like hexachlorophene, it is more inhibitory to gram-positive than gram-negative bacteria, but it is highly active toward fungi and yeasts. Its chief application is in the treatment of dermatophytic conditions; however it has shown photosensitivity in humans in some cosmetic and soap applications. A comprehensive review of bisphenols is available.

Triclosan (Irgasan, trichlorohydroxydiphenyl ether) is a bisphenol with an oxygen bridge. It has been recommended for soaps and washing products, and many antibacterial soaps contain triclosan. It inhibits staphylococci at 0.03–0.1 ppm, and some *E. coli* in the same range, but Pseudomonas aeruginosa requires 100–1000 ppm. Molds are inhibited at 1–30 ppm.

Coal-Tar Disinfectants. Coal-tar disinfectants formerly constituted the most important category of disinfectants for general use. These are obtained from coal tar (see COAL; COAL CONVERSION PROCESSES; TAR AND PITCH) which, when fractionated, yields a group of chemicals including phenols, organic bases, and neutral hydrocarbon oils.

In addition to coal tar, petroleum has been a source of the same chemicals, and many of the individual phenols have been produced in the pure state by synthetic processes.

Acids

Hydroxybenzoic Acids and Esters. The phenolic hydroxyl group of benzoic acid derivatives is largely responsible for their antimicrobial activity. Benzoic acid, itself effective as a preservative in foods and cosmetics, is relatively weak as a disinfectant as compared to phenols. *ortho*-Hydroxybenzoic acid (salicylic acid) has its phenolic group masked by hydrogen bonding to the oxygens of the carboxyl group. It has keratinolytic activity, and is used with benzoic acid in a successful athlete's foot treatment, ie, Whitfield's ointment. The methyl, ethyl, propyl, and butyl esters of vanillic acid (4-hydroxy-3-methoxy benzoic acid) have antifungal properties at 0.1–0.2% concentration, and have been used in the preservation of foods and food packaging materials.

The most successful compounds in this group are the esters of *p*-hydroxybenzoic acid, known as parabens, which have been in continuous use since the 1920s. Their antimicrobial activity increases from the methyl to the benzyl ester, but water solubility limits use mainly to the esters. These compounds are active against gram-positive and gram-negative bacteria, yeasts, and fungi. The low order of toxicity, lack of irritation, and absorption and excretion characteristics in both humans and animals make these compounds well suited as preservatives for pharmaceuticals, cosmetics, and food. Their application in antimycotic therapy has been considered by several investigators and tests on their use in a cosmetic lotion show that in combination with another antimicrobial preservative, imidazolindyl urea, they are ideal in providing a broad spectrum preservative system.

Inorganic Acids. Strong inorganic acids have little antimicrobial activity in themselves but inhibit microorganism growth by lowering the pH. Disinfectant toilet bowl cleaners that contain 9.5% HCl or more are antimicrobial. Carbonic acid in soft drinks provides some antibacterial preservation. Sulfurous acid is an effective preservative used to preserve wines (see WINE), fruit juices (qv), and dried fruits.

Organic Acids. Among the organic acids, acetic acid, as vinegar (qv), is an effective food preservative and has been used medically since ancient times for treating open wounds. Propionic acid is a preservative for cheeses and food wrappers and prevents the growth of microbial rope and mold in bread. Lactic acid preserves pickles, sauerkraut, and silage. Benzoic acid has been used for many years to control mold growth in ketchup and other foods. Sorbic acid (2,4-hexadienoic acid) inhibits yeasts (qv) and molds at 25–500 ppm, and finds application as a preservative in foods, cosmetics (qv), and medicines. Dehydroacetic acid has similar activity for bacteria and fungi, and has had similar applications. Undecenoic acid (undecylenic acid) alone, or with its zinc salt, is used in a popular treatment for athlete's foot fungus and other superficial skin dermatophytoses. Glutaric acid and glutaric acid analogues have been reported to be virucidal against rhinoviruses.

Aldehyde Antimicrobials

Two aldehydes (qv) have made their mark in the field of disinfection, namely formaldehyde and glutaraldehyde. Gaseous formaldehyde is effective for the sterilization of heat-sensitive medical instruments and hospital supplies, such as bedding and blankets.

Formalin is an aqueous solution of 34–38% formaldehyde. As an alkylating agent, formaldehyde reacts with proteins and nucleic acids through the amino acid groups and the sulfhydryl, phenolic, or indole residues. Microbiologically, it is active against bacteria, fungi, bacterial spores, and many viruses. Formaldehyde has had varied uses in embalming fluids, as a preservative for laboratory specimens, for inactivating poliovirus, and for the disinfection of isolators, ion-exchange columns, and soils. Used at 8% concentration, formaldehyde is a much more active microbiocide in 70% isopropanol than in water. The alcohol solution is said to be rapidly bactericidal, tuberculocidal, and sporicidal. A 6–8% formaldehyde solution is rated as a sterilant, and at 1–8% as a low-to-high level disinfectant. Controversy in the early 1990s regarding formaldehyde as a potential occupational carcinogen has limited its use.

A number of disinfectants apparently owe their activity to formaldehyde, although there is argument on whether some of them function by other mechanisms. In this category, the drug with the longest history is hexamethylenetetramine (hexamine, urotropin), which is a condensation product of formaldehyde and ammonia that breaks down by acid hydrolysis to produce formaldehyde. Hexamine was first used for urinary tract antisepsis. Other antimicrobials that are adducts of formaldehyde and amines have been made; others are based on methylolate derivations of nitroalkanes. The applications of these compounds are widespread, including inactivation of bacterial endotoxin; preservation of cosmetics, metal working fluids, and latex paint; and use in spin finishes, textile impregnation, and secondary oil recovery.

The other aldehyde of importance in disinfection is glutaraldehyde. It has two aldehyde groups in its five-carbon molecule, and is a more powerful germicide than other dialdehydes and formaldehyde. The 2% alkaline solution of glutaraldehyde is rapid in action, reportedly killing most bacteria in less than 1 min, the tubercle bacillus and viruses in less than 10 min, and bacterial spores in 3 h or less; however, EPA registration requires 45 min for the tubercle bacillus and 10 h for spores. At 2% concentration glutaraldehyde is classified as a high level germicide, capable of producing sterility.

The primary application for glutaraldehyde is in the disinfection or sterilization of heat-sensitive medical and surgical instruments like endoscopes. Glutaraldehyde also has been recommended for cold sterilization of hemostats, cystoscopes, food containers, anesthetic equipment, dental equipment, urological equipment, and gynecologic laproscopy equipment.

Caution should be taken when using glutaraldehyde. Gloves and aprons should be worn and adequate ventilation provided. It has been reported to produce contact dermatitis, eye irritation, nausea, headache, rashes, and asthmatic reaction.

Peroxygen Compounds

Hydrogen Peroxide. Hydrogen peroxide, H_2O_2, a long-time favored antiseptic in the home and in medical circles, fell out of favor once it was discovered that peroxide is quickly decomposed by the catalase in tissues. There was also a problem with the stability of peroxide preparations, but this has been overcome and stable solutions having a long shelf-life are available. Peroxide has found important applications as a disinfectant and sporicide rather than as an antiseptic. Applications include purifying drinking water; treating contaminated water supplies in hospitals; treating raw milk; sterilizing spacecraft; and disinfecting contact lenses (qv) and acrylic resin sections of surgical implants. These applications use peroxide in concentrations of 10–25% rather than the 3% solution used as an antiseptic. Another important application of peroxide is in the sterilization of the contact surfaces of food packaging (qv) for nonrefrigerated milk and fruit juices (qv).

Peracetic Acid. Peracetic acid (peroxyacetic acid), $C_2H_4O_3$, the peroxide of acetic acid, is a disinfectant having the desirable properties of hydrogen peroxide, ie, broad-spectrum activity against microorganisms, lack of harmful decomposition products, and infinite water solubility. Peracetic acid also has greater lipid solubility and is free from deactivation by catalase and peroxidase enzymes. However, it is corrosive, and degradation products may have to be rinsed from the surface of disinfected materials. Peracetic acid is a more powerful antimicrobial agent than hydrogen peroxide and most other disinfectants. It has advantages for disinfection and sterilization not found in any other agent.

The strong antimicrobial activity of peracetic acid makes it valuable in maintaining sterile conditions in the production of germ-free animals. It has been accepted worldwide in the food processing and beverage industries as an ideal for clean-in-place systems; it does not require rinsing where the breakdown product, acetic acid, is not objectionable in high dilution. Peracetic acid is more toxic than hydrogen peroxide and is a weak carcinogen but can be used with safety when diluted. Like all peroxides, it is a powerful oxidizer and should be handled with proper safety precautions. It is more corrosive to metals and plastics than is hydrogen peroxide.

Surface-Active Agents

Quaternary Ammonium Compounds. The quats may be regarded as long-chain alkyl relatives of ammonium salts, NH_4X, where the hydrogens are replaced by organic groups, R_4NX. They have been termed invert soaps, the counterpart of soaps. Like fatty acid soaps, these are salts containing a long-chain hydrocarbon group, and are surface-active substances. In soap the anion contributes the hydrophobic portion; in quats the cation is hydrophobic.

The unique property of quats is the ability to produce bacteriostasis in very high dilution. The quats have a narrower antibacterial spectrum than the phenols, afd are much more active against gram-positive bacteria than gram-negative bacteria. Acid-fast bacteria and bacterial spores are resistant to quats. Quats, moderately active against most fungi and lipophilic viruses but not against hydrophilic viruses, are very active against algae, which makes them useful in those industrial water processes where algae are a problem.

Further developments have brought forth polymeric quats having antimicrobial properties. Different kinds of polyquats have been described with molecular weight from 2,000 to 60,000. Polymeric quats have two characteristics that make them uniquely different from the monomeric quats. One is the absence of foaming, even at high concentrations. The other is their remarkably low toxicity in skin and eye irritant tests and oral ingestion tests. Yet, even at this low toxicity, these polymer quats have significant antibacterial and antifungal activity. Polyquaternium 1 is considered a candidate preservative for ophthalmic preparations, and is registered with FDA for contact lens solutions; WSCP, LD_{50} of 2.77 g/kg, is registered with EPA as a swimming pool algaecide; and the biguanides, another group of polymers, are known under such trademarks as Vantocil and Cosmocil CQ. A further development to extend the use of quats in medicine is the so-called soft drugs or labile antimicrobial quats; ie, those that would be broken down *in vivo* to nontoxic fragments.

Quats are used as surface disinfectants for floors, walls, and equipment in hospitals and nursing homes, breweries, food plants, and the home. They may be used as sanitizers in rinsewater for glasses and dishes in restaurants. In combination with compatible nonionic wetting agents, quats are employed in products giving one-step cleaning and disinfection for environmental surfaces.

An application that takes advantage of the charge attraction to negatively charged fabrics is in treatment of diapers to suppress bacterial diaper rash and to prevent the liberation of ammonia produced by bacterial breakdown of urea in urine. For laundry where hot water is not used, a final rinse with 200 ppm of quats based on the dry weight of the fabric provides residual bacteriostatic activity to the fabric. An antibacterial and antimildew treatment of carpeting, underwear, socks, mattress ticking, etc, was introduced as Slygard and Biogard; it resists washing, and results from an organic silicon quat.

Considerable work has been done to try to explain why quats are antimicrobial. The following sequence of steps is believed to occur in the attack by the quat on the microbial cell: *(1)* adsorption of the compound on the bacterial cell surface; *(2)* diffusion through the cell wall; *(3)* binding to the cytoplasmic membrane; *(4)* disruption of the cytoplasmic membrane; *(5)* release of cations and other cytoplasmic cell constituents; *(6)* precipitation of cell contents and death of the cell.

Acid–Anionic Sanitizers. This group, like the cationic disinfectants, are surface active agents but, like common soaps, these compounds have the hydrophobic principle in the anion rather than the cation. Whereas the cationic agents perform best in an alkaline environment, the anionics work best under acid conditions, usually between pH 1 and 3. The acidic groups donating the hydrophilic groups are carboxylic acid, sulfonic acid, sulfuric acid ester, phosphoric acid ester, or phosphonic acid. The hydrophobic portion is contributed by alkyl chains, which may be substituted with aromatic rings such as benzene or naphthalene. The acid–anionic sanitizers combine low levels of an anionic surface-active agent and an organic or inorganic acid, a solubilizer, and in some cases, a small amount of nonionic surfactant. Whereas these preparations do not possess the high bacteriostatic activity of quaternary ammonium germicides, they have the alternative advantage of being rapidly functional in acid solution.

The acid–anionics are particularly valuable in the dairy industry, where they have had the greatest acceptance. The acid helps remove calcium deposit from milk (milkstone), thus enabling the anionic to do a more efficient job of destroying the bacteria. The detergency of acid–anionic surfactants is useful for cleaning equipment in dairies and food plants, and they are not corrosive to the stainless steel equipment used in these plants. Because they contain common detergents and food-grade acid, the sanitizers have very low toxicity, and are used to disinfect cows' udders prior to milking as a precaution against mastitis infection. Foaming may be a problem, but new low-foam acid–anionic surfactants have been introduced. Another use proposed for these sanitizers is in hard surface disinfection of walls, floors, and equipment in hospitals, nursing homes, schools, hotels, restaurants, etc, where the surfaces are not adversely affected by the acid solution.

Amphoteric Surfactant Disinfectants. Amphoteric surfactants (ampholytes) differ from cationic and anionic surfactants because they ionize in water to give zwitterions, ie, ions with both a positive and a negative charge in the same molecule, and, depending on pH, may act more like anions or cations. They are typically amino acids (qv) with a long-chain alkyl group, and are more basic than acidic in nature because they have two or three amine groups but only one carboxylic acid group. As in the case of the other surfactants, the long-chain hydrocarbon group donates the hydrophobic part of the molecule and the surface activity. The ampholytes, which must be used in higher concentrations than many other disinfectants when employed at room temperature, are greatly improved as the temperature is increased.

Whereas tests indicated that ampholytes were effective in skin cleansing for preoperative use, for wound cleansing, and as an antiseptic in the oral cavity, as well as other medical applications, the food and beverage industries have proved to be the principal employers of these compounds. Ampholytes are used as sanitizers and disinfectants, not as food preservatives. Low toxicity, absence of skin irritation, and noncorrosiveness, along with antimicrobial activity, has given ampholytes acceptance in dairies, meat plants, and the brewing and soft drink industries. These disinfectants have been manufactured and distributed in Europe and Japan, but not in the United States.

Chelating Agents. 8-Hydroxyquinoline (8-quinolinol, oxine) might be thought to function as a phenol, but of the 7 isomeric hydroxyquinolines only oxine exhibits significant antimicrobial activity, and is the only one to have the capacity to chelate metals. If the hydroxyl group is blocked so that the compound is unable to chelate, as in the methyl ether, the antimicrobial activity is destroyed.

Certain halogen derivatives of 8-hydroxyquinoline have a record of therapeutic efficacy in the treatment of cutaneous fungus infections and also of amebic dysentery. Among these are 5-chloro-7-iodo-8-quinolinol (iodochlorhydroxyquin, Vioform), 5,7-diiodo-8-hydroxyquinoline (diiodohydroxyquin), and sodium 7-iodo-8-hydroxyquinoline-5-sulfonate (chiniofon).

Copper 8-quinolinolate (copper oxinate), the copper compound of 8-hydroxyquinoline, is employed as an industrial preservative for a variety of products, which include wood, textiles, and interior paints for food plants. More recent concern that hydroxyquinolines are carcinogenic has resulted in reduced interest in this group.

Another compound, the antimicrobial action of which is associated with chelation, is 2-pyridinethiol-*N*-oxide (Omadine). In the form of its zinc chelate it is found in shampoos to control seborrheic dermatitis. Other applications of this useful chemical include preservation of adhesives, plastics, latex paints, polyurethane foam, and metal working fluids.

Many compounds capable of chelation have been tested for antimicrobial properties. Those showing positive results include salicylaldoxime, 1-nitroso-2-naphthol, mercaptobenzothiazol, dimethylglyoxime, salicylaldehyde, cupferron, phenanthroline, isoniazid, thiosemicarbazones, the sulfur analogue of oxine, and numerous antibiotics including tetracyclines. Whether these compounds function exclusively, partially, or at all by virtue of their ability to chelate is open to debate.

Among the compounds capable of chelation that have proved especially useful are the dithiocarbamates, which for years have controlled fungus diseases on tomatoes, potatoes, and other plants. They also have been used as industrial preservatives and sodium-*N*-methyldithiocarbamate is an effective soil fumigant (see FUNGICIDES, AGRICULTURAL).

Ethylenediaminetetraacetic acid (EDTA), an important chelating agent, is not considered a bactericide in its own right, as it has generally no effect on gram positive bacteria. However, it can act alone or with other agents to cause lysis of some gram-negative bacteria, eg, *Pseudomonas aeruginosa*. It potentiates the activity of chemically unrelated antibacterial compounds against gram-negative bacteria. It does so by chelating cations in the outer membrane of the bacteria, thus disrupting the membrane, and causing the release of lipopolysaccharide.

Biguanidine Compounds. Chlorhexidine(1,6-di(4-chlorophenyldiguanidino)hexane, Hibitane), a leading skin antiseptic and a cationic bisguanide, is a strong base that reacts with acids to form salts, most of which are water-insoluble. However, the salt with gluconic acid is soluble and is dispensed as a 20% solution that is colorless and odorless. Chlorhexidine is moderately surface active, stable in the range of pH 5 to 8, and is generally compatible with other cationic germicides. Some disinfectant products incorporate both chlorhexidine and a quaternary, eg, Presept Liquid, which has 1.25% chlorhexidine gluconate and 0.1% quaternary ammonium compounds in 70% alcohol. Chlorhexidine is incompatible with organic anions such as soaps, anionic detergents, and many dyes.

Unlike the quats, chlorhexidine is highly active against both gram-positive and gram-negative bacteria; yeasts and fungi are also sensitive to this agent, which has high activity against the lipid viruses, including many respiratory viruses, herpes, cytomegalovirus, and HIV, but not the hydrophilic viruses, such as poliovirus and the enteric viruses. Acid-fast bacteria are inhibited but not killed by aqueous chlorhexidine solutions, but are killed by alcoholic solutions. Prolonged use of the drug has not led to the development of resistant bacteria.

Chlorhexidine achieved its outstanding position as a skin disinfectant because of its apparent low toxicity to most body parts, absence of skin irritation, and ability to reduce hospital nosocomial skin infections. Its low toxicity was demonstrated when used on the intact skin of newborns, but in the U.S. restrictions are made for its use around ears.

Chlorhexidine has found other medical applications, eg, in urology in preventing urinary tract infections, in obstetrics and gynecology, in controlling infection in burns and wounds, and in the prevention of oral disease. Hypersensitivity to chlorhexidine has been reported in Japan but 0.05% concentration is considered to be safe.

Another important biguanide disinfectant is polyhexamethylene biguanide (PHMB, Vantocil IB). This compound is a polymer containing a spread of polydispersed oligomers of molecular weight between 500 and 6000, for which the tetramer is the predominant species. Each oligomer can have an amine or a cyanoguanidine group at either end position.

PHMB is one of the few biologically active synthetic polymers. It is active against most gram-positive and gram-negative bacteria, yeasts, many fungi, and aquatic algae, but is not active against Mycobacteria and bacterial spores. It is water-soluble, and is used as a solid surface disinfectant at concentrations of 1000–2000 ppm. It is registered with EPA for the control of pathogenic bacteria in bottle washwater, cannery cooling water, synthetic adhesives, leather processing liquors, and metalworking fluids, and has also been proposed for the preservation of food and cosmetics.

Anilides. Salicylanilide, the product of salicylic acid and aniline, was developed in 1930 under the name Shirlan for preventing mildew on stored and shipped woolen goods. It has been used as a mildewcide in plastics, paints, lacquers, leather, and paper. Later, halogenated salicylanilides were introduced as antimicrobial soap additives for degerming skin.

Nitrogen Heterocycles

Imidazole and imidazoline derivatives have provided some useful antimicrobial compounds. Metronidazole (2-methyl-5-nitroimidazole-1-ethanol) inhibits the growth of pathogenic protozoa, *Trichomonas vaginalis,* and *Eistamoeta histolyticum* in urogenital infections, and is effective against infections resulting from anaerobic bacteria and facultative anaerobes. Other derivatives, ie, clotrimazole, miconazole, econazole, and ketoconazole, demonstrate a broad antimycotic spectrum of activity. Glyodin (2-heptadecyl-2-imidazoline acetate) protects apple and cherry trees from leaf spot disease. It functions as a cationic surfactant. Dantoin (DMDMH-55, Glydant (1,3-di(hydroxymethyl)-5,5-dimethylhydantoin)) has a wide spectrum of activity against bacteria and fungi at 250–500 ppm. It is water-soluble and active over a wide pH range. It can be used as a preservative for resins, adhesives, and emulsions. Germall 115, *N,N″*-methylene bis(5′-(1-hydroxylmethyl)-2,5-dioxo-4-imidazolidinylurea), is an antimicrobial preservative tested for the cosmetic industry. It is more active against bacteria than fungi.

Some triazines and other nitrogen heterocycles are produced by reacting amines and formaldehyde and may exert antimicrobial activity by virtue of the slow release of formaldehyde. These compounds find use as preservatives for latex paints, adhesives, and cosmetic products.

Mercaptobenzothiazole is an effective agent against antibacterial and antifungal properties. It is used in combination with dimethyldithiocarbamate (Vancide 51) as a general industrial preservative. A related compound, Busan 1030, 2-(thiocyanomethythio)benzothiazole, is a preservative for caulking compounds, vinyl acetate wallcovering adhesives, particle board, and various paints and protective coatings. It is also sold as a protectant for lumber against sapstain and mold.

Mylone (tetrahydro-3,5,dimethyl-2*H*-1,3,5-thiadiazine-2-thione), a preservative with various trade names, ie, Metasol D3T, Busan 1058, and Biocide-N-521, has a range of suggested applications in leather, paint, glue, casein, starch, pigment slurries, and paper manufacture. Antimicrobial thiazole compounds include Proxel CRL, 1,2-benzoisothiazolin-3-one; Kathon 886 (Kathon CT, Kathon CG), a mixture of 5-chloro-2-methyl-3(2*H*)-isothiazalone and 2-methyl-3(2*H*)-isothiazalone, Skane M-8, 2-*n*-octyl-4-isthiazolin-3-one, and Metasol TK-100, (2-(4-thiazolyl)benzimidazole). These chemicals are used as preservatives for paint, leather, fabrics, hydraulic fluids, cooling tower water, etc.

Sulfur Compounds

Sulfur has long been known for its properties as a pesticide and a curative agent. Sulfur is selective in its antimicrobial activity, and there is debate whether it has any toxicity of its own or if it acts through chemical conversion to some other form. However, colloidal sulfur, a mixture of 70% sulfur and 30% polythionic acids, has a phenol coefficient against common gram-positive and gram-negative bacteria of 1.5–5, and against plant pathogens of 4–276. Colloidal sulfur has been reported fungicidal to *Tinea corporis* and *Trichophyton interdigitale* at concentrations of 1%, but not to *Monilia tropicalis,* even at 5%.

Sulfur dioxide, sulfites, and metabisulfites have had extensive use as antimicrobial preservatives in the food industry. In pharmaceuticals they have had a dual role, acting as preservatives and antioxidants. The sulfa drugs, or sulfonamides, the first effective chemotherapeutic agents to be employed systemically for the prevention and cure of bacterial infections, have a wide range of activity against gram-positive and gram-negative organisms in a concentration of about 50–100 ppm. Unlike the more highly active antibiotics that have largely replaced them, however, sulfas are inhibited by blood, pus, and tissue breakdown products. Sulfas are employed to some degree in combination with antibiotics or other drugs.

The toxicity of sulfur to microorganisms is taken advantage of in several industrial biocides. Methylene bisthiocyanate (Metasol TK, Biosperse 284) has powerful action against slime-forming bacteria, spore formers, and fungi, and is used in paper mills and water cooling systems, where high activity and low cost are primary considerations.

Metal Compounds

Metal compounds, particularly compounds of the heavy metals, have a history of importance as antimicrobial agents. Because of regulations regarding economic poisons in the environment they are no longer widely used in this application. The metals whose compounds have been of primary interest as antimicrobials are mercury, silver, and copper.

Mercury. Mercury is remarkable for its great bacteriostatic activity. Organic mercurials with popular names like mercurochrome, metaphen, and merthiolate (thimerosal), appeared in the decade after World War I.

Sulfides, thiols, and proteinacious organic matter, particularly plasma and whole blood, seriously depress and may even abolish the germicidal action of mercury compounds (qv). As of this writing approved uses for mercurials are limited to contact lens cleaning fluids, spoilage prevention of stored water-based paints, mildew control in finished paints, and a retardant for sapstain in lumber. It is uncertain, however, whether even these applications will be permitted to continue. A comprehensive review of mercurial antimicrobials is available.

Silver. The outstanding bactericidal properties of highly diluted silver metal, as apart from silver ions in salts like silver nitrate, was termed oligodynamic action. Silver nitrate is astringent and a protein precipitant, which is not medically desirable. Other forms of silver have been used to avoid this problem, including colloidal silver, silver-protein preparations, and finely divided silver metal called Katadyn silver.

Penicillin and other antibiotics have since replaced the caustic silver nitrate for anti-infection treatment of the eyes of newborn infants, but silver, which is effective against both *Pseudomonas aeruginosa* and other gram-negative bacteria associated with burned tissue, is still employed in that therapy. Silver also may be expected to continue to find use in water purification.

Copper. Copper, like mercury and silver, has a long history of use as an antimicrobial agent. Its use has been mainly as a fungicide and algicide. Copper fungicides for plants have been largely replaced by organic compounds like the dithiocarbamates. The cupric ion is algicidal at 0.5–2.9 ppm, and is used to prevent algae growth in swimming pools. Copper naphthenate and copper-8-quinolinolate, $C_{18}H_{12}CuN_2O_2$, are still used to treat tents, ropes, tarpaulins, and gun covers to prevent rotting. The latter has also proved to be an effective mildewcide with many applications.

Other Metals. Other metals having antimicrobial properties include zinc, chromium, arsenic, boron, and tin. Zinc, as zinc oxide, has been used in paint to enhance mildew resistance, and zinc chloride is a wood preservative. Chromated zinc chloride and copper chromate are wood preservatives that combine zinc and copper with chromium. Arsenic, in the form of lead arsenate, was formerly used as a fungicide–insecticide for protecting plants. Sodium arsenate is a wood preservative; an organic arsenical, oxybisphenoxyarsine (Durotex), is effective in combating fungi in adhesives, wall paper, paint, plastics, and emulsions (see ARSENIC COMPOUNDS).

Boron, as barium metaborate, is marketed as a mildew preventative for paints. Borax is used as a wood preservative, and an organic boron, 2,2′-(1-methyltrimethylenedioxy)-bis(4,4,6-trimethyl)-1,3,2-dioxaborinane (Biobor JF) is a biocide for jet fuel. Whereas tin metal is used to coat steel cans used as food containers, organic tin in the form of tributyl tin compounds have proven to be powerful antimicrobials, and have found use in antifouling coatings for ship bottoms, paints, and wood preservatives.

Gaseous Sterilants

Ethylene Oxide. Ethylene oxide C_2H_4O, is the gas most used in hospitals to sterilize items that cannot be sterilized at high temperatures. Items sterilized include most of the plastics, medical and biological preparations, surgical implants, medical instrumentation, hospital bedding, and oxygen tents. Ethylene oxide is also used to sterilize dry powders and food spices. It reacts with nucleic acids, which is thought to be the primary cause of this biocidal activity. Ethylene oxide is able to inactivate all microorganisms including spores and viruses. Ethylene oxide was also shown to be virucidal.

Ozone. Ozone, O_3, is an allotropic form of oxygen having outstanding properties as an oxidant. It is also a powerful germicide, being lethal to all forms of microorganisms. The odor of ozone can be detected with as little as 0.02–0.04 ppm in air, and 20–30 ppm causes irritation to the eyes, nose, and throat. Prolonged exposure to a concentration of 1000 ppm can cause death. The toxicity, instability, and corrosiveness have limited the practical applications of ozone. In addition, it is necessary to generate it electrically as it is used. Because ozone is very reactive, organic matter greatly reduces its germicidal effectiveness. Its main use over the years has been for the disinfection of drinking water and process water for manufacturing pharmaceuticals, cosmetics, and household products. However, it has been suggested that because of its high oxidative properties, ozone could be used in hospitals for resterilizing instruments composed of materials such as noble metals, titanium, stainless steel, silicone rubber, ceramics, poly(vinyl chloride), and polyurethane.

Chlorine Dioxide. Like ozone, chlorine dioxide is a powerful oxidant. It is usually generated as used. It has been used for disinfecting drinking water and bleaching paper pulp. Unlike chlorine and other chlorine oxidants, it does not produce carcinogenic chloromethanes.

SEYMOUR S. BLOCK
University of Florida, Gainesville

S. S. Block, ed., *Disinfection, Sterilization, and Preservation,* 4th ed., Lea & Febiger, Philadelphia, Pa., 1991.

A. D. Russell, W. B. Hugo, and G. H. J. Ayliffe, eds., *Principles and Practice of Disinfection, Preservation, and Sterilization,* 2nd ed., Blackwell, Boston, Mass., 1992.

J. F. Gardner and M. M. Peel, *Introduction to Sterilization, Disinfection, and Infection Control,* 2nd ed., Churchill Livingstone, New York, 1991.

DISPERSANTS

Dispersants are materials that help maintain fine solid particles in a state of suspension, and inhibit their agglomeration or settling in a fluid medium. With the help of mechanical agitation, dispersants can also break up agglomerates of particles to form particle suspensions. Another use of dispersants is to inhibit the growth of crystallites in a supersaturated solution. This characteristic is also known as precipitation inhibition, threshold inhibition, or antinucleation. Overall, dispersants are useful in preventing settling, deposition, precipitation, agglomeration, flocculation, coagulation, adherence, or caking of solid particles in a fluid medium.

Physical Chemistry of Dispersants

A convenient way to understand particle dispersion is to consider the process in four successive parts: the nature of particles and surfaces, adsorption onto particles, interface properties, and forces of attraction and repulsion.

Particles and Surfaces. Dispersants are primarily used to increase stability (prevent settling) of solid particles in liquid media, whereas surfactants are used more frequently to stabilize liquid (including polymer latex) surfaces within liquids. When the surface of a liquid is increased (stressed), molecules of the liquid flow to the surface to lower its energy, "healing" it. In contrast, solids exhibit no significant flow to the surface. Any stresses applied therefore remain in the form of a higher energy surface. Thus the history of a particle is important to its surface properties. Treatments that alter particle surface properties include freshly cleaving a surface along lowest energy crystal faces, adsorption of molecules and ions, heating or cooling, friction, corrosion, and grinding or polishing. The process by which a particle is formed also affects its surface properties. Examples of this include screw and spiral dislocations, missing layers, and other defects due to contamination or stress during formation.

Adsorption onto Particles. The Gibbs Adsorption law relates how adsorption (qv) onto surfaces affects interfacial tension, $d\gamma = -RT\Gamma d\ln c$, where γ = interfacial or surface tension, in N/m (1N/m = 1000 dyn/cm); R = gas constant; T = absolute temperature; Γ = interfacial or surface concentration, in mol/unit area (ie, adsorption); and c = dimensionless concentration ($d\ln c = dc/c$, thus units cancel).

If adsorption occurs ($\Gamma > 0$), then increasing the concentration of dispersant in the bulk water reduces interfacial tension.

Most adsorption processes are exothermic (ΔH is negative). Adsorption processes involving nonspecific interactions are referred to as physical adsorption, a relatively weak, reversible interaction. Processes with stronger interactions (electron transfer) are termed chemisorption. Chemisorption is often irreversible and has higher heat of adsorption than physical adsorption. Most dispersants function by chemisorption, in contrast to surfactants, which tend to physically adsorb.

Interface Properties. A polymeric dispersant may have segments extended into the solution, or the segments may be coiled, depending on whether the solvent is good (polymer–solvent interactions energetically favored) or poor (polymer–polymer and solvent–solvent contacts favored). Between these two solvent–polymer interactions is a θ (theta) solvent, in which neither condition is favored. If the polymer–solvent interaction is better than θ conditions, the extending chains or segments will repel chains adsorbed on other particles, as well as making the distance between two particles greater and enhancing steric repulsion. If the interaction is worse than θ conditions, the particles may flocculate due to mutual attraction of the polymer layers. On the other hand, if the polymer–solvent interaction is too strong, the polymeric dispersant may be adsorbed only weakly or not at all. This can lead to depletion flocculation, which is due to desorbed chains that are squeezed out from between two approaching particles. The desorbed chains can cause solvent to flow from between the particles by osmotic forces, leaving a bare area so that attraction between the particles is increased.

Attractive and Repulsive Forces. The force that causes small particles to stick together after colliding is van der Waals attraction. There are three van der Waals forces: *(1)* Keesom-van der Waals, due to dipole–dipole interactions that have higher probability of attractive orientations than nonattractive; *(2)* Debye-van der Waals, due to dipole-induced dipole interactions (ie, uneven charge distribution is induced in a nonpolar material); and *(3)* London dispersion forces, which occur between two nonpolar substances.

As the distance between two approaching particles decreases, their electrical double layers begin to overlap. As a first approximation, the potential energy of the two overlapping double layers is additive, which is a repulsive term since the process increases total energy. Electrostatic repulsion can also be considered as an osmotic force, due to the compression of ions between particles and the tendency of water to flow in to counteract the increased ion concentration.

The overall stability of a particle dispersion depends on the sum of the attractive and repulsive forces as a function of the distance separating the particles. DLVO theory, named for Derjaguin and Landau and Verwey and Overbeek, encompasses van der Waals attraction and electrostatic repulsion between particles, but does not consider steric stabilization. The net energy, ΔG_T, between two particles at a given distance is the sum of the repulsive and attractive forces: ΔG_T = (electrostatic repulsive forces) − (van der Waals attractive forces). The electrostatic repulsive forces are a function of particle kinetic energy (kT), ionic strength, zeta potential, and separation distance. The van der Waals attractive forces are a function of the Hamaker constant and separation distance.

Although some progress has been made in calculating steric repulsive forces, the theory concerning them is not as completely developed as DLVO theory. The adsorbed polymer layers of two particles (in a good solvent) begin to interpenetrate as the particles approach each other. The interaction between these polymer layers can have an osmotic effect due to an increase in the local concentration of the adsorbed polymer layers, and can have an entropic or volume restriction effect due to crowding of the interacting chains. In both cases, entropy decreases, which is unfavorable. Moreover, the osmotic effect can create unfavorable enthalpic changes due to desolvation of closely packed chains. To regain lost entropy, the particles must separate to allow the chains more freedom of movement, while the solvent moves in to resolvate the polymer layer. As with electrostatic repulsion, an energy barrier is created. A common approximation used is that the strength of the energy barrier rises steeply at slightly less than the adsorbed layer thickness. Some of the practical differences between sterically and electrostatically stabilized dispersions may be summarized as follows:

Steric stabilization	Electrostatic stabilization
insensitive to electrolyte	coagulation occurs with increased electrolyte
effective in aqueous and nonaqueous media	more effective in aqueous media
effective at high and low concentrations	more effective at low concentrations
reversible flocculation common	coagulation often irreversible
good freeze–thaw stability	freezing often induces irreversible coagulation

Dispersant Materials

Dispersant materials include condensed phosphates, organic polymeric dispersants, poly(meth)acrylates, polymaleates, condensed phosphates, polysulfonates, sulfonated polycondensates, and tannins, lignins, glucosides, and alginates.

Uses

Dispersants are used in recirculating cooling water, boiler water, geothermal fluids, seawater distillation, reverse osmosis, sugar processing, oilfields, drilling muds, cement, paints and pigments, mineral processing, caulks, sealants, roof coatings, pesticides, animal feeds, detergents, and cleaners.

Environmental Considerations

Biodegradability of Dispersants. Most reviews on biodegradable polymers suggest that, with the exception of poly(vinyl alcohol) and poly(ethylene glycol)s, most synthetic organic dispersants are recalcitrant in the environment. More recently developed dispersants displaying biodegradability are polymers containing ester linkages and ether linkages. There is currently a great deal of research activity to develop dispersants that are both effective and biodegradable. Consequently, developments in this area should be rapid and changing into the year 2000.

WILLIAM M. HANN
Rohm and Haas Company

DISTILLATION

Distillation is a method of separation that is based on the difference in composition between a liquid mixture and the vapor formed from it. This composition difference arises from the dissimilar effective vapor pressures, or volatilities, of the components of the liquid mixture. Distillation as normally practiced involves condensation of the vaporized material, usually in multiple vaporization/condensation operations, and thus differs from evaporation (qv), which is usually applied to separation of a liquid from a solid but which can be applied to simple liquid concentration operations.

Distillation is the most widely used industrial method of separating liquid mixtures and is at the heart of the separation processes in many chemical and petroleum plants (see SEPARATION SYSTEMS SYNTHESIS). The most elementary form of the method is simple distillation in which the liquid is brought to boiling and the vapor formed is separated and condensed to form a product. If the process is continuous with respect to feed and product flows, it is called flash distillation. If the feed mixture is available as an isolated batch of material the process is a form of batch distillation and the compositions of the collected vapor and residual liquid are thus time dependent. The term fractional distillation, which may be contracted to fractionation, was originally applied to the collection of separate fractions of condensed vapor, each fraction being segregated. In modern practice the term is applied to distillation processes in general, where an effort is made to separate an original mixture into several components by means of distillation. When the vapors are enriched by contact with counterflowing liquid reflux, the process is often called rectification. When fractional distillation is accomplished with a continuous feed of material and continuous removal of product fractions, the process is called continuous distillation. When steam (qv) is added to the vapors to reduce the partial pressures of the components to be separated, the term steam distillation is used.

Most distillations conducted commercially operate continuously, with a more volatile fraction recovered as distillate and a less volatile fraction recovered as bottoms or residue. If a portion of the distillate is condensed and returned to the process to enrich the vapors, the liquid is called reflux. The apparatus in which the enrichment occurs is usually a vertical, cylindrical vessel called a still or distillation column.

This apparatus normally contains internal devices for effecting vapor–liquid contact; the devices may be categorized as plates or packings.

Vapor–Liquid Equilibria

The equilibrium distributions of mixture component compositions in the vapor and liquid phases must be different if separation is to be made by distillation. It is important, therefore, that these distributions be known. The compositions at thermodynamic equilibrium are termed vapor–liquid equilibria (VLE) and may be correlated or predicted with the aid of thermodynamic relationships.

Thermodynamic Relationships. A closed container with vapor and liquid phases at thermodynamic equilibrium may be depicted as in Figure 1, where at least two mixture components are present in each phase. The components distribute themselves between the phases according to their relative volatilities. A distribution ratio for mixture component i may be defined using mole fractions:

$$K_i = y_i^*/x_i \quad (1)$$

where the asterisk is used to denote an equilibrium condition. This K term, known as the vapor–liquid equilibrium ratio, or often the K value, is widely used, especially in the petroleum (qv) and petrochemical industries. For any two mixture components i and j, their relative volatility, often called the alpha value, is defined as

$$\alpha_{ij} = \frac{K_i}{K_j} = \frac{y_i x_j}{x_i y_j} = \frac{y_i(1 - x_i)}{x_i(1 - y_i)} \quad (2)$$

The relative volatility, α, is a direct measure of the ease of separation by distillation. If $\alpha = 1$, then component separation is impossible, because the liquid-and-vapor-phase compositions are identical. Separation by distillation becomes easier as the value of the relative volatility becomes increasingly greater than unity.

When both phases form ideal thermodynamic solutions, ie, no heat of mixing, no volume change on mixing, etc, Raoult's law applies:

$$p_i^V = x_i P_i^0 \quad (3)$$

where P_i^0 is the vapor pressure of i at the equilibrium temperature. Combining this expression with Dalton's law of partial pressures, K values and relative volatilities may be obtained:

$$K_i = P_i^0/P \quad (4)$$

$$\alpha_{ij} = P_i^0/P_j^0 \quad (5)$$

The development and thermodynamic significance of activity coefficients is discussed in most chemical engineering thermodynamics texts. The liquid-phase coefficients are strong functions of liquid composition and temperature and, to a lesser degree, of pressure. A system with positive deviation, ie, the two components having activity coefficients greater than one such that the logarithm of the coefficient is positive, is shown in Figure 2**a**; a system with negative deviation, the coefficients less than unity and logarithms negative, is shown in Figure 2**b**. In a few cases one component of a binary mixture has a positive deviation and the other a negative deviation. Most commonly, however, both coefficients have positive deviations.

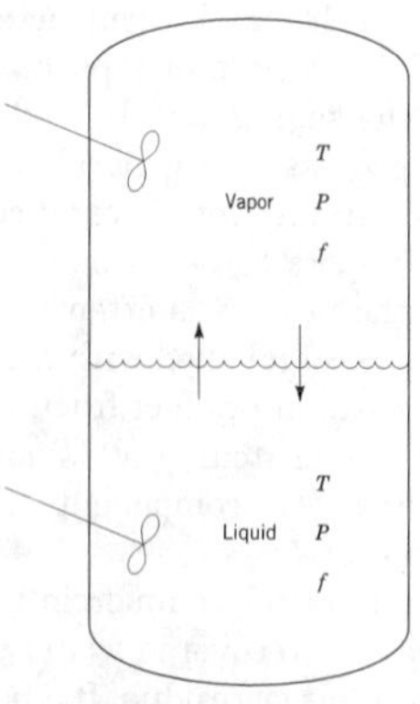

Figure 1. Equilibrium between vapor and liquid. The conditions for equilibrium are $T^V = T^L$ and $P^V = P^L$. For a given T and P, phase fugacities are equal, ie, $f^V = f^L$ and $f_i^V = f_i^L$.

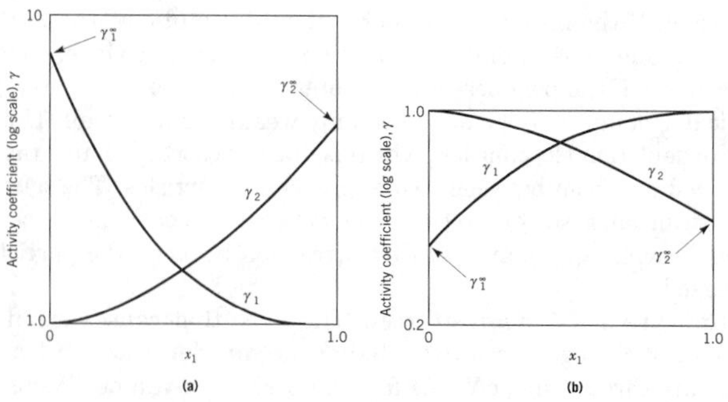

Figure 2. Binary activity coefficients for two component systems having (**a**) positive and (**b**) negative deviations from Raoult's law. Conditions are either constant pressure or constant temperature and terminal coefficients, γ^∞_i, are noted.

Terminal activity coefficients, γ^∞_i, are noted in Figure 2. These are often called infinite dilution coefficients and for some systems are given in Table 1.

If the molecular species in the liquid tend to form complexes, the system will have negative deviations and activity coefficients less than unity, eg, the system chloroform-ethyl acetate. In azeotropic and extractive distillation (see DISTILLATION, AZEOTROPIC AND EXTRACTIVE) and in liquid–liquid extraction, nonideal liquid behavior is used to enhance component separation (see EXTRACTION, LIQUID–LIQUID).

Azeotropic Systems. An azeotropic mixture is one that vaporizes without any change in composition. In homogeneous azeotropic systems, positive activity coefficients tend to produce minimum boiling azeotropes, and negative coefficients tend to produce maximum boiling azeotropes.

Heterogeneous azeotropes are formed when the positive activity coefficients are sufficiently large to produce two liquid phases which exist at the boiling point, and a constant boiling mixture which is formed at some composition, generally within the liquid immiscibility composition range.

Distillation Processes

For ease of presentation and understanding, the initial discussion of distillation processes involves binary systems. Examining the binary boiling point (Fig. 3**a**) and phase (Fig. 3**b**) diagrams, the enrichment

Table 1. Terminal Activity Coefficients at Atmospheric Pressure[a]

Component 1	Component 2	γ^∞_1	γ^∞_2
chloroform	ethyl acetate	0.3	0.3
chloroform	benzene	0.9	0.7
n-hexane	*n*-heptane	1.0	1.0
ethyl acetate	ethanol	2.5	2.5
ethanol	toluene	6.0	6.0
benzene	methanol	9.0	9.0
ethanol	isooctane	11.0	8.0
methyl acetate	water	20.0	7.0
ethyl acetate	water	100.0	15.0
water	water	>100.0	>100.0

[a] Values are approximate.

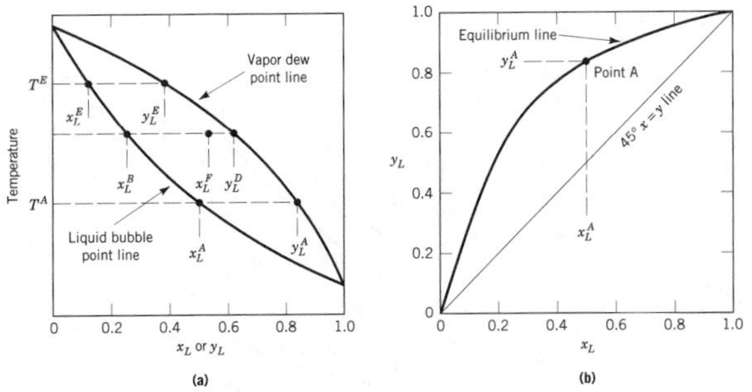

Figure 3. Isobaric VLE diagrams: (**a**) dew and bubble point; (**b**) vapor-liquid ($y - x$) equilibrium.

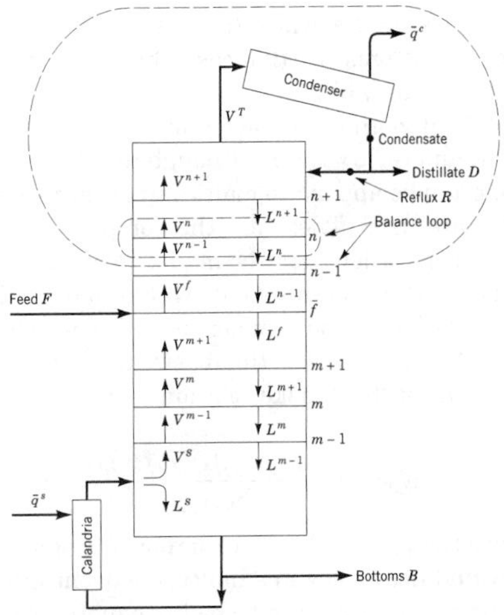

Figure 4. Distillation column with stacked multiple equilibrium stages. Terms are defined in text.

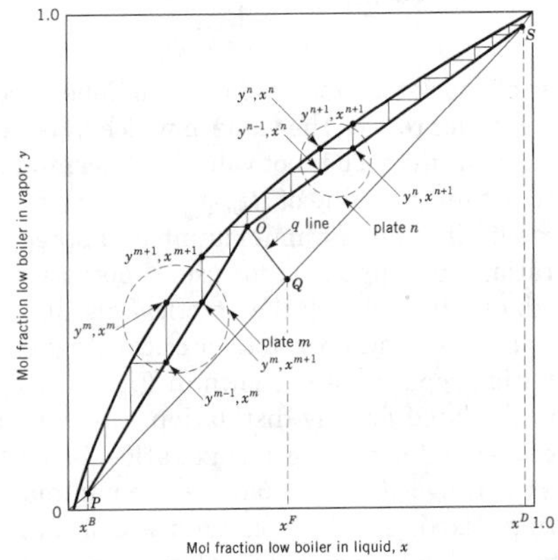

Figure 5. McCabe-Thiele diagram. Terms are defined in text.

from liquid composition x_L to vapor composition y_L represents a theoretical step, or equilibrium stage.

Simple Distillations. Simple distillations utilize a single equilibrium stage to obtain separation. Simple distillation, also called differential distillation, may be either batch or continuous, and may be represented on boiling point or phase diagrams. In Figure 3**a**, if the batch distillation begins with a liquid of composition $x_L{}^A$ the initial distillate vapor composition is $y_L{}^A$. As the distillate is removed, the remaining liquid becomes less rich in the low boiler, *L*, and the boiling liquid composition moves to the left along the bubble point line. If the distillation is continued until the liquid has a composition of $x_L{}^E$, the last vapor distillate has a composition of $y_L{}^E$. Simple batch distillation is not widely used in industry, except for the processing of high valued chemicals in small production quantities, or for distillations requiring regular sanitization.

Simple continuous distillation, also called flash distillation, has a continuous feed to a single equilibrium stage; the liquid and vapor leaving the stage are considered to be in phase equilibrium. On the boiling point diagram (Fig. 3**a**), the feed is represented by $x_L{}^F$, the bottoms liquid by $x_L{}^B$, and the equilibrium vapor distillate by $y_L{}^D$. The overall mass balance is $F = D + B$ and the component L balance is $x_L^F F = y_L^D D + x_L^B B$. Flash distillations are widely used where a crude separation is adequate.

Multiple Equilibrium Staging. The component separation in simple distillation is limited to the composition difference between liquid and vapor in phase equilibrium. To overcome this limitation, multiple equilibrium staging is used to increase the component separation. Figure 4 schematically represents a continuous distillation that employs multiple equilibrium stages stacked one upon another. The feed, *F*, enters the column at equilibrium stage *f*. The heat $\overline{q}^s$ required for vaporization is added at the base of the column in a reboiler or calandria. The vapors V^T from the top of the column flow to a condenser from which heat $\overline{q}^c$ is removed. The liquid condensate from the condenser is divided into two streams: the first, a distillate *D*, which is the overhead product (sometimes called heads or make), is withdrawn from the system, and the second, a reflux *R*, which is returned to the top of the column. A bottoms stream *B* is withdrawn from the reboiler. The overall separation is represented by feed *F* separating into a distillate *D* and a bottoms *B*.

The graphical McCabe-Thiele design method facilitates a visualization of distillation principles while providing a solution to the material balance and equilibrium relationships. Here, the subscripts *L* and *H* are not used and *x* and *y* refer to the lower boiler, ie, more volatile component, in the binary system. A McCabe-Thiele diagram is given in Figure 5 where *P*, *Q*, and *S* are the x^B, x^F, and x^D compositions on the $y = x$, 45° construction line, respectively. Line *OP* is the stripping operating line and line *OS* is the rectifying operating line.

The McCabe-Thiele method employs the simplifying assumption that the molal overflows in the stripping and the rectification sections are constant. This method is based on the simplifying assumption that the molal overflow is constant in both the rectifying and stripping sections. For many problems this assumption is not valid and more precise calculations are necessary. For the more general case, detailed enthalpy balances are made around individual stages or groups of stages. Standard distillation texts discuss the internal enthalpy calculations by algebraic balances or by graphical procedures, eg, the stage-to-stage mass and enthalpy balances with equilibrium calculations and also by means of the graphical Ponchon-Savarit procedure. Hand algebraic and graphical methods requiring internal enthalpy calculations have been largely superseded by simulations performed on modern computing devices, including personal computers (see COMPUTER TECHNOLOGY).

There are infinite combinations of reflux ratios and numbers of theoretical stages for any given distillation separation. The larger the reflux ratio, the fewer the theoretical stages required. For any distillation system with its given feed and its required distillate and bottoms compositions, there are two constraints within which the variables of reflux ratio and number of theoretical stages must lie: the minimum number of theoretical stages and the minimum reflux ratio. The minimum reflux ratio occurs when the reflux ratio is reduced so that the upper and lower operating lines and the *q* line are coincident at a single point on the equilibrium curve. The slope of the *q* line is $q/(q - 1)$, where

$$q = \frac{\text{heat needed to vaporize one mole of feed}}{\text{molal latent heat of feed}} \tag{6}$$

When this condition exists, an infinite number of theoretical stages would be required to make the separation. The minimum number of theoretical stages occurs when the system is at total reflux: no feed, distillate, or bottoms. In this case the two operating lines are coincident with the diagonal line.

Simple analytical methods are available for determining minimum stages and minimum reflux ratio. Although developed for binary mixtures, they can often be applied to multicomponent mixtures if the two key components are used. These are the components between which the specification separation must be made; frequently the heavy key is the component with a maximum allowable composition in the distillate and the light key is the component with a maximum allowable specification in the bottoms. On this basis, minimum stages may be calculated by means of the Fenske relationship:

$$N_{\min} = \frac{\ln[(y_i/y_j)D(x_j/x_i)_B]}{\ln \alpha_{ij,\text{avg}}} \tag{7}$$

where i and j are the light and heavy components of a binary mixture, or the light key and heavy key in a multicomponent mixture.

For minimum reflux ratio, the following equations may be used:

$$\sum_i \frac{\alpha_i x_{if}}{\alpha_i - \phi} = 1 - q \tag{8}$$

$$\sum_i \frac{\alpha_i (x_{id})}{\alpha_i - \phi} = R_{\min} + 1 \tag{9}$$

where the value of q is determined as in the McCabe-Thiele procedure. Equation 8 is solved for root ϕ, the value of which must lie between 1.0 and the light key volatility. The root value so determined is then used in equation 9 to obtain the value of $R_{\min}$.

Both of these limits, the minimum number of stages and the minimum reflux ratio, are impractical for useful operation, but they are valuable guidelines within which the practical distillation must lie. A representative plot of the number of theoretical stages vs reflux ratio for some distillation separation is shown in Figure 6. Both minimum limits may be calculated for any distillation, thereby bracketing the practical design. Actual operating reflux ratios for most commercial columns are in the range of 1.1 to 1.5 times the minimum reflux ratio.

The operating, fixed, and total costs of a distillation system are functions of the relation of operating reflux ratio to minimum reflux ratio. Figure 7 shows a typical plot of costs; as the operating to minimum reflux ratio increases, the operating cost (principally energy cost for the boil-up) increases almost linearly. Similarly, the fixed costs at first decrease from the infinite number of stages, pass through a minimum, and then increase again as the diameter of column increases with increased reflux ratio. These costs for typical distillations have been calculated; the ratio of the economic optimum reflux to the minimum reflux is often 1.2 or less.

The calculations that determine the reflux and stage requirements are more difficult to make for multicomponent systems than for binary systems. When the concentration of a component in the distillate and

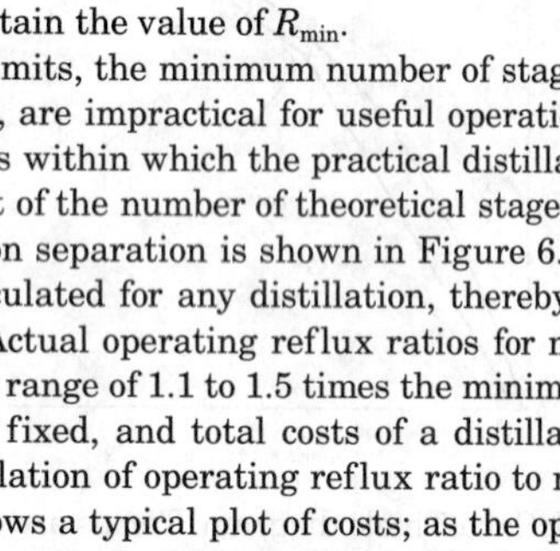

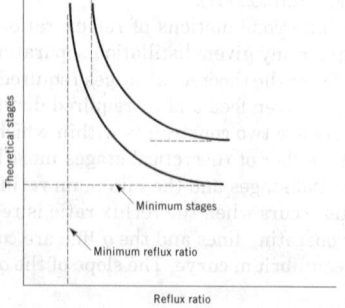

Figure 6. Representative plot of theoretical stages vs reflux ratio for a given separation. Each curve is the locus of points for a given separation. Note the limiting conditions of minimum reflux and minimum stages.

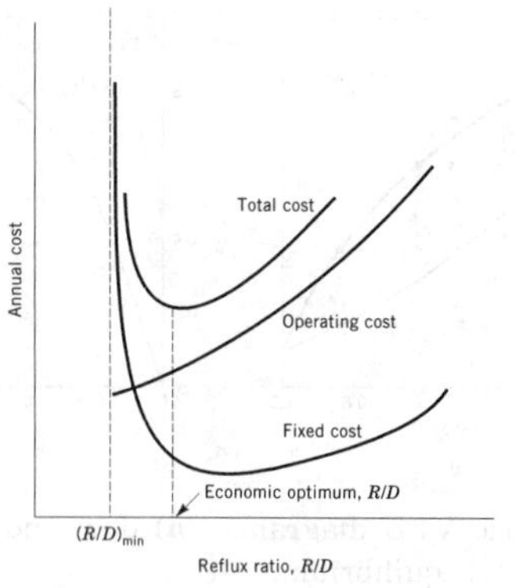

Figure 7. Fixed, operating, and total costs of a typical distillation, as a function of reflux ratio.

in the bottoms is specified for the overall solution of a binary distillation, the component balance around the column also is completely specified. In the multicomponent case, only a single high boiling key component can be specified in the distillate and a single low boiling key component in the bottoms; the split of other components can be determined only by detailed calculations. These require a series of trial and error computations to obtain the solution at any given reflux ratio and number of stages. As the number of components and number of stages become large, the mathematical problem becomes formidable. Two approaches may be followed: use of approximate, ie, shortcut, methods, or use of a suitable computer.

Multiple Products. If each component of a multicomponent distillation is to be essentially pure when recovered, the number of columns required for the distillation system $N^* - 1$, where N^* is the number of components. Thus, in a five-component system, recovery of all five components as essentially pure products requires four separate columns. However, those four columns can be arranged in 14 different ways.

The number of columns in a multicomponent train can be reduced from the $N^* - 1$ relationship if side-stream draw-offs are used for some of the component cuts. The feasibility of multicomponent separation by such draw-offs depends on side-stream purity requirements, feed compositions, and equilibrium relationships.

Distillation Columns

Distillation columns are vertical, cylindrical vessels containing devices that provide intimate contacting of the rising vapor with the descending liquid. This contacting provides the opportunity for the two streams to achieve some approach to thermodynamic equilibrium. Depending on the type of internal devices used, the contacting may occur in discrete steps, called plates or trays, or in a continuous differential manner on the surface of a packing material. The fundamental requirement of the column is to provide efficient and economic contacting at a required mass-transfer rate. Individual column requirements vary from high vacuum to high pressure, from low to high liquid rates, from clean to dirty systems, and so on. As a result, a large variety of internal devices has been developed to fill these needs.

Packed vs Plate Columns. Relative to plate towers, packed towers are more useful for multipurpose distillations, usually in small (under 0.5 m) towers or for the following specific applications: severe corrosion environment where some corrosion-resistant materials, such as plastics, ceramics, and certain metallics, can easily be fabricated into packing but may be difficult to fabricate into plates; vacuum operation where a low pressure drop per theoretical plate is a critical requirement; high (eg, above 49,000 kg(h·m^2) (~10,000 lb/(h·ft^2))) liquid rates; foaming systems; or debottlenecking plate towers having plate spacings that are relatively close, under 0.3 m.

Plate columns have the advantage of lower fabrication cost, less dependence on good liquid and gas distribution, and protection against vapor bypassing the liquid in critical zones, eg, regions of extremely low impurities. Further, methods for the design on plate columns are somewhat more reliable than those for many of the packings, especially those packings of a proprietary nature.

There are notable cases where plate columns have been converted to packed columns to gain advantage of the low pressure drop exacted from the vapor stream. More recently the packings have been largely of the structured type. Illustrative of this is the trend toward the use of structured packing in ethylbenzene–styrene fractionators, some of which have diameters of 10 m or higher.

Molecular Distillation

Molecular distillation occurs where the vapor path is unobstructed and the condenser is separated from the evaporator by a distance less than the mean-free path of the evaporating molecules. This specialized branch of distillation is carried out at extremely low pressures ranging from 13–130 mPa (0.1–1.0 μm Hg) (see VACUUM TECHNOLOGY). Molecular distillation is confined to applications where it is necessary to minimize component degradation by distilling at the lowest possible temperatures. Commercial usage includes the distillation of vitamins (qv) and fatty acid dimers (see DIMER ACIDS).

Distillation as a Separation Method

Distillation is the most important industrial method of separation and purification of liquid components. Liquid separation methods in less common use include liquid–liquid extraction (see EXTRACTION, LIQUID–LIQUID), membrane diffusion (see DIALYSIS; MEMBRANE TECHNOLOGY), ion exchange, and adsorption. However, distillation does not require a mass separating agent such as a solvent, adsorbent, or membrane, and distillation utilizes energy in a convenient heating medium (often steam). Also, a wealth of experience with design and operations makes distillation column performance prediction more reliable than equivalent predictions for other methods. At times distillation also competes indirectly with methods involving solid–liquid separations such as crystallization. An extensive discussion of the selection of alternative separation methods is available (see SEPARATION SYSTEMS SYNTHESIS).

The suitability and economics of a distillation separation depend on such factors as favorable vapor–liquid equilibria, feed composition, number of components to be separated, product purity requirements, the absolute pressure of the distillation, heat sensitivity, corrosivity, and continuous vs batch requirements. Distillation is somewhat energy-inefficient because in the usual case heat added at the base of the column is largely rejected overhead to an ambient sink. However, the source of energy for distillations is often low pressure steam which characteristically is in long supply and thus relatively inexpensive. Also, schemes have been devised for lowering the energy requirements of distillation and are described in many publications.

Column Control

Distillation columns are controlled by hand or automatically. The parameters that must be controlled are *(1)* the overall mass balance, *(2)* the overall enthalpy balance, and *(3)* the column operating pressure. Modern control systems are designed to control both the static and dynamic column and system variables.

JAMES R. FAIR
The University of Texas at Austin

K. C. D. Hickman, in R. H. Perry and C. H. Chilton, eds., *Chemical Engineers' Handbook,* 5th ed., McGraw-Hill Book Co., Inc., New York, 1973, section 13.

J. R. Fair, in Y. A. Liu, H. A. McGee, and W. R. Epperly, eds., *Recent Developments in Chemical Process and Plant Design,* John Wiley & Sons, Inc., New York, 1987, Chapt. 3.

E. J. Henley and J. D. Seader, *Equilibrium-Stage Separation Operations in Chemical Engineering,* John Wiley & Sons, Inc., New York, 1981.

M. Van Winkle, *Distillation,* McGraw-Hill Book Co., Inc., New York, 1967.

DISTILLATION, AZEOTROPIC AND EXTRACTIVE

This article describes special distillation techniques for economically separating low relative volatility and azeotropic mixtures. Whereas there is extensive literature on design methods for azeotropic and extractive distillation, much less has been published on operability and control. It is, however, widely recognized that azeotropic distillation columns are difficult to operate and control because these columns exhibit complex dynamic behavior and parametric sensitivity. In contrast, extractive distillations do not exhibit such complex behavior and even highly optimized columns are no more difficult to control than ordinary distillation columns producing high purity products.

Most methods for distilling azeotropic and low relative volatility mixtures rely on the addition of specially chosen chemicals to facilitate the separation. These separating agents can be divided into distinct classes which define the principal distillation techniques used to separate mixtures containing azeotropes. The five methods for separating azeotropic mixtures are *(1)* extractive distillation and homogeneous azeotropic distillation where the liquid separating agent is completely miscible. For extractive distillation, separating agents are variously known as solvents, extractive agents, entrainers, or extractants. *(2)* Heterogeneous azeotropic distillation or, more commonly, azeotropic distillation where the liquid separating agent, called the entrainer, forms one or more azeotropes with the other components in the mixture and causes two liquid phases to exist over a broad range of compositions. This immiscibility is the key to making the distillation sequence work. *(3)* Distillation in the presence of ionic salts. The salt dissociates in the liquid mixture and alters the relative volatilities sufficiently that the separation becomes possible. *(4)* Pressure-swing distillation where a series of columns operating at different pressures are used to separate binary azeotropes which change appreciably in composition over a moderate pressure range or where a separating agent which forms a pressure-sensitive azeotrope is added to separate a pressure-insensitive azeotrope. *(5)* Reactive distillation where the separating agent reacts preferentially and reversibly with one of the azeotropic constituents. The reaction product is then distilled from the nonreacting components and the reaction is reversed to recover the initial component.

Of these five methods all but pressure-swing distillation can also be used to separate low volatility mixtures. It is also possible to combine distillation and other separation techniques such as liquid–liquid extraction (qv), adsorption (qv), melt crystallization (qv), or pervaporation to complete the separation of azeotropic mixtures.

Residue Curve Maps

The least complicated of all distillation processes is the simple distillation, or open evaporation, of a mixture from an open vessel. The liquid is boiled in such a way that the vapor is removed from contact with the liquid as soon it is formed. The composition of the liquid changes continuously with time because the vapor is always richer in the more volatile components than the liquid from which it came. The path of liquid compositions starting from some initial point is called a simple distillation residue curve or simply, a residue curve. The collection of all such curves for a given mixture is called a residue curve map. These maps contain exactly the same information as the corresponding phase diagram for the mixture, but represent the information in a way that it is much more useful for understanding and designing distillation systems.

Mixtures that do not contain azeotropes have residue curve maps that all look the same. For ternary mixtures of this sort, the map looks like the one shown in Figure 1. The residue curves all start at the lowest boiling pure component, they then move towards the intermediate boiling component, and finally end at the highest boiling component. These curves indicate that when such a mixture is boiled, the liquid initially gets richer in each of the heavier components but eventually gets richer in only the heaviest component until only the pure heavy component remains. One of the most important properties

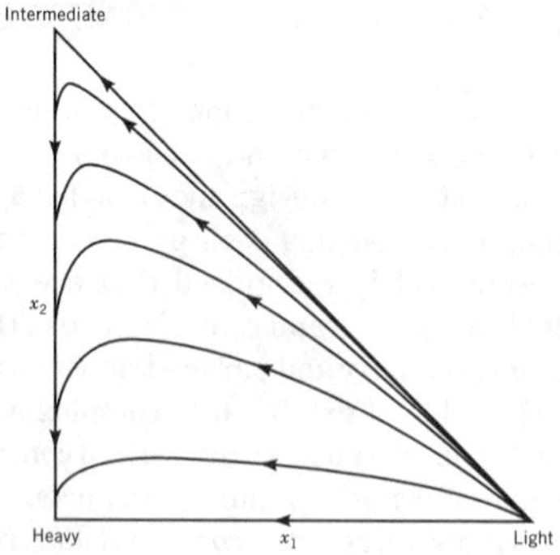

Figure 1. Residue curve map for a ternary nonazeotropic mixture.

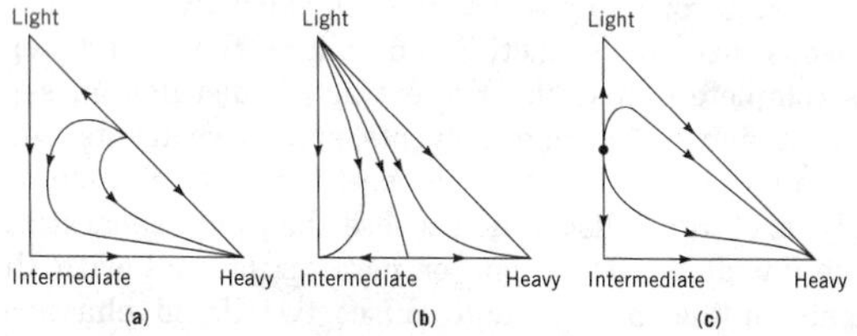

Figure 2. Residue curves for **(a)** where the separating agent is intermediate and does not introduce a new azeotrope; **(b)** for the methanol (light)–2-propanol (intermediate)–water (heavy) system; and **(c)** for ethanol (light)–water (intermediate)–ethylene glycol (heavy) system.

of residue curves is that they must move in such a way that the boiling temperature of the mixture increases along every curve. The arrows on the edges point in the direction of increasing boiling temperature.

The presence of even one binary azeotrope destroys the structure shown in Figure 1. If the mixture contains a single minimum boiling binary azeotrope, there are three possible residue curve maps depending on whether the azeotrope is between the lightest and heaviest components (Fig. 2**a**), between the intermediate and heaviest components (Fig. 2**b**), or between the intermediate and lightest components (Fig. 2**c**). Notice that the residue curves in Figures 2**a** and **c** all end up at the heaviest boiling pure component, but in Figure 2**b**, the curves end at either the intermediate vertex or the heavy vertex depending on the initial composition. The special curve that divides these two distillation regions is called a distillation boundary. Mixtures with more than one azeotrope exhibit more complex residue curve maps with either single or multiple distillation boundaries.

Heterogeneous Azeotropic Distillation

Heterogeneous azeotropic distillation, or simply azeotropic distillation, is widely used for separating nonideal mixtures. The technique uses minimum boiling azeotropes and liquid–liquid immiscibilities in combination to overcome the effect of other azeotropes or tangent pinches in the mixture that would otherwise prevent the desired separation. The azeotropes and liquid heterogeneities that are used to make the desired separation feasible may either be induced by the addition of a separating agent, usually called the entrainer, or they may be intrinsically present, in which case the mixture is sometimes called self-entrained. The most common case is the former; it includes such classic separations as ethanol dehydration using either benzene, heptane, ethyl ether, etc, as the entrainer, and acetic acid recovery from water using either ethyl acetate, 1-propyl acetate, or 1-butyl acetate as the entrainer. In ethanol dehydration the entrainer breaks the homogeneous minimum boiling azeotrope between ethanol and water; in the acetic acid recovery process the entrainer is used to overcome the tangent pinch between acetic acid and water.

MICHAEL F. DOHERTY
University of Massachusetts, Amherst
JEFFREY P. KNAPP
E. I. du Pont de Nemours & Co., Inc.

M. F. Doherty and G. A. Caldarola, *Ind. Eng. Chem. Fund.* **24,** 474–485 (1985).

C. Black and D. E. Ditsler, in R. F. Gould, ed., *Azeotropic and Extractive Distillation,* Advances in Chemistry Series No. 115, American Chemical Society, Washington, D.C. 1972, pp. 1–15.

C. S. Robinson and E. R. Gilliland, *Elements of Fractional Distillation,* 4th ed., McGraw-Hill Publishing Co., New York, 1950.

P. J. Ryan and M. F. Doherty, *AIChE J.* **35,** 1592–1601 (1989).

DIURETIC AGENTS

Diuretics are agents that increase urine output or flow. The term is generally used to describe all drugs that act on the kidney to increase the production of urine. More specifically, the terms saliuretic or natriuretic are used to describe those agents that exert diuretic effects by primarily increasing the excretion of sodium chloride. Aquaretics are agents that increase urine output by producing a water diuresis but do not promote the urinary excretion of electrolytes. Both the use of diuretics in the treatment of hypertension, and the effects of diuretics on the kidney to promote the excretion of urine to normalize derangements in body fluid distribution leading to edematous states are discussed herein.

Disturbances in body fluid distribution may occur at three principal sites: *(1)* within the interstitial space, as occurs in peritonitis or cirrhosis with ascites; *(2)* between the interstitial space and the vascular tree, as in the nephrotic syndrome; or *(3)* within the vascular tree, as in congestive heart failure. Thus, the underlying disease may be of cardiac, hepatic, or renal origin. These derangements provide what amounts to a low blood volume signal to the kidneys that activates renal mechanisms to retain salt and water. If the retained salt and water do not terminate the low volume signal, the kidneys continuously retain salt and water, resulting in edema. The clinical outcome of excessive accumulation of salt and water depends on the particular sector of the extracellular space to which the retained fluid is relegated. Clinical signs of edema appear when the volume of the extracellular space is exceeded by several liters. Diuretics are used in the treatment of edematous states because, in most cases, they produce satisfactory mobilization and subsequent prevention of fluid accumulation in the interstitial space, the abdominal cavity, the lungs, and/or thoracic cavity. However, diuretic therapy is symptomatic in nature, and unless the underlying pathology is corrected, the kidneys continue to retain salt and water and the retained fluid and electrolytes are redistributed to the various compartments described. The principal indications of diuretics in the treatment of edema are in congestive heart failure; renal disease; hepatic cirrhosis with ascites; obesity, where salt and water retention are prominent (see ANTIOBESITY DRUGS); premenstrual tension; edema of pregnancy, including toxemia; and steroid administration. Edema may also be associated with other clinical conditions such as inflammation or hypersensitivity reactions.

Pharmacology and Mechanism of Action

Low Ceiling Diuretics. The designation of low ceiling diuretics denotes that the total excretion of the filtered sodium ion load is less than 10% compared to about 30% for the high ceiling diuretics. There are many chemical classes in this category, ie, thiazides, quinazoline sulfonamides, chlorthalidone, indapamide, etc, but their site of action in the kidney is similar, and they are grouped as thiazide-type diuretics for general discussion.

The most popular diuretics in this class are hydrochlorothiazide and chlorthalidone; there are more potent low ceiling diuretics available. The long duration of action of chlorthalidone, 24 to 72 h, makes once a day dosing possible, and achieves good patient compliance.

Cyclothiazide, polythiazide, and trichlormethiazide are about 15 to 30 times more potent than hydrochlorothiazide, and about 500 to 1000 times more potent than chlorothiazide, the first member of the thiazide family marketed.

The low ceiling diuretics increase urinary sodium excretion by acting directly on the Na^+-Cl^- transport mechanism in the convoluted distal tubules of the kidney.

Indapamide has been shown to possess diuretic and independent vasodilatory effects. It lowers the elevated blood pressure and reduces total peripheral resistance without an increase in heart rate. Indapamide antagonizes the vasoconstricting effects of the catecholamines and angiotensin II, a property not shared by other thiazide-type diuretics. Tripamide is also reported to have direct vasodilatory effects.

Weight loss is a good indicator of fluid loss and excretion. The first wave of fluid mobilized is from the periphery. The excretion of chloride and water is considered passive, and the excretion of potassium and magnesium is increased. In long-term use, the excretion of calcium is decreased.

In long-term treatment, the thiazides may produce hypokalemia, hyperglycemia, hyperuricemia, and a 5% increase in plasma cholesterol; indapamide has been shown not to increase plasma cholesterol or lipids at therapeutic doses. Thiazides can cause hyponatremia in patients with large water intake while on the drug; hyponatremia may be associated with nausea, vomiting, and headaches.

Paradoxically, the thiazides are efficacious, especially if combined with a prostaglandin synthetase inhibitor such as indomethacin or aspirin, in the treatment of nephrogenic diabetes insidipus, in which the patient's renal tubules fail to reabsorb water despite the excessive production of ADH. Thiazides can decrease the urine volume up to 50% in these patients.

High Ceiling (Loop) Diuretics. The principal action of the loop diuretics is inhibition of sodium and chloride reabsorption in the thick ascending limb of the loop of Henle. They can produce an excretion of 20 to 30% of the filtered sodium ion load. Most loop diuretics have a rapid onset of action, steep diuretic dose-response curves, and usually are short acting.

The most commonly used loop diuretics are furosemide, bumetanide, and ethacrynic acid. Newer agents available in some countries include torasemide and piretanide. The potency ratio for furosemide:ethacrynic acid:torasemide:piretanide:bumetanide is 1:1:2:4:40.

After long-term use of a loop diuretic, the extracellular fluid volume contracts and water reabsorption in the proximal and distal tubules increases, thus overriding and diminishing the diuretic's effects. A second high ceiling diuretic may sometimes induce diuresis again. This may be due to additional mechanisms. The loop diuretics increase urinary excretion of potassium and magnesium, as do the thiazides, but loop diuretics also increase urinary excretion of calcium; the excretion of chloride ion is greater than that of sodium ion, suggesting the inhibition of active chloride transport by the high ceiling diuretics in the loop of Henle. Alkalosis may develop in patients treated with the loop diuretics, and plasma renin activity (PRA) is markedly elevated. The high ceiling diuretics inhibit the ability of the kidney to concentrate urine even in the presence of high concentrations of vasopressin.

Ototoxicity, as evidenced by transient or permanent hearing loss, is a serious side effect of ethacrynic acid, and occurs less frequently with furosemide. Bumetanide is claimed to have only 20% of the ototoxic potential of furosemide. It has been reported that patients treated with torasemide at high doses for four weeks did not suffer hearing loss.

Potassium-Sparing Diuretics. Potassium-sparing diuretics act on the aldosterone-sensitive portion of cortical collecting tubules, and partially in the distal convoluted tubules of the nephron. The commonly used potassium-sparing diuretics are triamterene, amiloride, and spironolactone. Spironolactone is a competitive aldosterone receptor antagonist, whereas triamterene and amiloride are not.

Amiloride is far more soluble than triamterene, and is the most widely studied potassium-sparing diuretic. Its natriuretic effect is minimal because only 2 to 3% of the filtered sodium ion load reaches the collecting tubules of the nephron. Etozolin is a newer, long-lasting agent that has a gradual onset of action.

Spironolactone antagonizes the effects of aldosterone by binding at the aldosterone receptor in the cytosol of the late distal tubules and renal collecting ducts. Side effects of spironolactone are gynecomastia, decreased libido, and impotency.

Potassium-sparing by diuretic agents, particularly spironolactone, enhances the effectiveness of other diuretics because the secondary hyperaldosteronism is blocked. This class of diuretics decreases magnesium excretion. The most important and dangerous adverse effect of all potassium-sparing diuretics is hyperkalemia, which can be potentially fatal; the incidence is about 0.5%. Therefore, blood potassium concentrations should be monitored carefully.

Natriuretic Peptide Diuretics. Atrial natriuretic peptide (ANP), an endogenous diuretic, natriuretic, and vasodilator, is a peptide hormone primarily synthesized and stored by atrial cardiocytes, and secreted by the atria in response to mechanical stretch of the atria. ANP is also known as anaritide, $C_{112}H_{175}N_{39}O_{35}S_3$; atrial natriuretic factor (ANF); auriculin; cardionatrin; and atriopeptide. Its primary action is in the kidney and the vascular system.

It has been suggested that both the increased glomerular filtration rate (GFR) caused by ANP and the direct epithelial action in the collecting ducts by ANP are necessary to explain the diuretic effects of ANP. It appears that ANP may increase GFR by relaxing the glomerular mesangial cells resulting in increased surface area for filtration. Due to the augmented GFR, ANP increases the delivery of sodium and water to the renal tubules beyond the distal convoluted tubule. In the collecting ducts, ANP reduces sodium and free-water reabsorption by antagonizing the action of vasopressin. Therefore, the increased loads of sodium and water passing through the collecting ducts without the increased compensatory reabsorption result in profound diuresis and natriuresis.

Atrial Natriuretic Peptide Potentiator Diuretics. Neural endopeptidase inhibitors or atrial peptidase inhibitors are compounds that inhibit the enzyme that degrades ANP, resulting in higher plasma concentrations, and longer duration of action, of ANP. The diuretic effects of this class of compounds resemble those resulting from administration of ANP. Compounds such as thiorphan, candoxatril, SCH-34826, and SCH-39370 have been studied in hypertension and congestive heart failure in humans with only limited success.

Osmotic Diuretics. An effectiveosmotic diuretic is a nonionic compound, freely filterable at the glomerulus, not reabsorbed by the tubules of the nephron, and biologically inert except for its osmotic properties. One of the best examples of an osmotic diuretic is mannitol, used to prevent acute renal failure in many major surgeries and traumatic injuries. An osmotic diuretic increases urine flow, rather than the excretion of sodium, by maintaining a high osmotic gradient resulting from the presence of large amounts of nonreabsorbable solutes in the luminal side of the proximal tubules. Under such conditions, the reabsorption of water is impaired along the descending loop of Henle as well as the collecting ducts, and urine flow increases. At high concentrations of an osmotic diuretic, the urinary excretion of sodium is also increased. This is due to the reduced reabsorption of sodium in the proximal tubules.

Carbonic Anhydrase Inhibitor Diuretics. Carbonic anhydrase accelerates the hydration of carbon dioxide to carbonic acid in aqueous solution, up to 7500-fold as compared to the nonenzymatic reaction. The hydrogen ions liberated from carbonic acid in the epithelial cells of the proximal tubules of the nephron are exchanged for sodium ions in the renal tubular lumen. When the generation of hydrogen ions is inhibited by a carbonic anhydrase inhibitor, the exchange of hydrogen for sodium ions is greatly diminished and the diuretic effect ensues. The site of action of carbonic anhydrase inhibitors is in the proximal tubules. In addition to sodium, the excretion of bicarbonate and potassium is also increased. Owing to the increased urinary bicarbonate excretion, the urine becomes alkaline and the blood becomes acidotic. The diuretic effect ceases once metabolic acidosis occurs.

Acetazolamide, the best example of this class of diuretics, is rarely used as a diuretic since the introduction of the thiazides. Its main use is for the treatment of glaucoma and some minor uses, eg, for the alkalinization of the urine to accelerate the renal excretion of some weak acidic drugs, and for the prevention of acute high altitude mountain sickness.

Methylxanthine Diuretics. The methylxanthines are of very limited efficacy when used as diuretics. The excretion of sodium and chloride ions are increased, but the potassium excretion is normal. Even though the methylxanthines have been demonstrated to have minor direct effects in the renal tubules, it is believed that they exert their diuretic effects through increased renal blood flow and GFR.

Organomercurial Diuretics. Before the advent of the thiazide diuretics, mercurial and organomercurial diuretics were the mainstay therapy for the treatment of edema. They have become obsolete and are of historical value only.

Water Diuretics (Aquaretics). A water diuretic, ie, aquaretic, decreases urinary osmolality by influencing the kidney to excrete water selectively without a concomitant proportionally increased excretion of sodium ions. A water diuretic should be efficacious for the treatment of hyponatremia, ie, low plasma sodium concentration, and the syndrome of inappropriate antidiuretic hormone secretion (SIADH). In many diseases and conditions, when water is retained to a greater extent as related to sodium ions, hyponatremia results. This is seen in many edema cases arising from congestive heart failure (CHF), hepatic cirrhosis, renal failure, and nephrotic syndrome. In the treatment of these conditions, the conventional diuretics will lose their effectiveness once hyponatremia occurs. Diseases of the brain and the lung, certain surgeries, and some tumors also will cause hyponatremia; in these conditions, with any given plasma osmolality, the plasma antidiuretic hormone (ADH) concentrations are inappropriately high. When this occurs, the patient is inferred to have SIADH.

There is no specific water diuretic marketed as of this writing. Demeclocycline has been used clinically with only limited success.

Economic Aspects

The sales of oral diuretics are declining, and are forecast to continue their decline in constant dollars during the 1990s. Several possible explanations can be offered for these trends. The patents of market leaders are expiring, leading to the introduction of generic brands at ca 40% below the cost of the branded market leaders; physicians are switching to newer treatments for hypertension, eg, calcium channel blockers and angiotension-converting enzyme inhibitors; and concerns are growing about the possible adverse effects of diuretics, eg, hypokalemia, the progression of atherosclerosis, and the increase in mortality, serum cholesterol, glucose tolerance, and diabetes.

PETER CERVONI
PETER S. CHAN
American Cyanamid Company

I. M. Weiner, in A. G. Gilman and co-workers, eds., *The Pharmacological Basis of Therapeutics,* 8th ed., Pergamon Press, Inc., Elmsford, N.Y. 1990.

N. D. Larkin and D. D. Fanestil, in J. B. West, ed., *Physiological Basis of Medical Practice,* 12th ed., Williams and Wilkins Co., Baltimore, Md., 1991.

J. B. Hook and R. Z. Gussin in M. Antonaccio, ed., *Cardiovascular Pharmacology,* 2nd ed., Raven Press, New York, 1984.

E. J. Cragoe, Jr., *Diuretics, Chemistry, Pharmacology, and Medicine,* John Wiley & Sons, Inc., New York, 1983.

DRIERS AND METALLIC SOAPS

Metal soaps as a class of compounds have been defined as the reaction products of alkaline, alkaline-earth, or transition metals with monobasic carboxylic acids containing 6–30 carbons. Commercially important metal soaps include those of aluminum, barium, cadmium, calcium, cobalt, copper, iron, lead, lithium, magnesium, manganese, potassium, nickel, zinc, and zirconium. Their solubility or solvation in a variety of organic solvents accounts for their many and varied uses. Significant application areas for metal soaps include lubricants and heat stabilizers in plastics as well as driers in paint,paint varnishes, and printing inks. Other uses are as processing aids in rubber, fuel and lubricant additives, catalysts, gel thickeners, emulsifiers, water repellents, and fungicides.

A significant advance in metal soap technology occurred in the 1920s with the preparation of the metal naphthenates. Naphthenic acids are not of precise composition, but rather are mixtures of acids isolated from petroleum. Because the mixture varies, so does acid number, or the combining equivalent of the acid, so that the metal content of the drier would not always be the same from lot to lot. The preparation of solvent solutions of these metal naphthenates gave materials that were easy to handle and allowed the metal content to be standardized. Naphthenates soon became the standard for the industry.

Octoates were the next drier development. Because these driers are produced from synthetic 2-ethylhexanoic acid, the chemical composition can be controlled and uniformity assured. Also, other synthetic acids, eg, isononanoic and neodecanoic, became available and are used for metal soap production. Compared to naphthenic acid, these synthetic acids have high acid values, are more uniform, lighter in color, and do not have its characteristic odor. It is also possible to produce metal soaps with much higher metal content by using synthetic acids.

More recently, so-called overbased driers with even higher metal contents have become available. These driers are made with combinations of monocarboxylic acids and carbon dioxide or polyfunctional acids.

Composition and Properties

Metal soaps are composed of a metal and acid portion supplied as solutions in solvent or oil. The general formula for a metal soap is $(RCOO)_x M$. In the case of neutral soaps, x equals the valence of the metal M. Acid soaps contain free acid (positive acid number) whereas neutral (normal) soaps contain no free acid (zero acid number); that is, the ratio of acid equivalents to metal equivalents is greater than one in the acid soap and equal to one in the neutral soap. Basic soap is characterized by a higher metal-to-acid equivalent ratio than the normal metal soap. Particular properties are obtained by adjusting the basicity.

Properties are furthermore determined by the nature of the organic acid, the type of metal and its concentration, the presence of solvent and additives, and the method of manufacture. Higher melting points are characteristics of soaps made of high molecular-weight, straight-chain, saturated fatty acids. Branched-chain unsaturated fatty acids form soaps with lower melting points.

The anion used to prepare the metal soap determines to a large extent whether it will meet fundamental requirements, which can be summed up as follows: solubility and stability in various kinds of vehicles (this excludes the use of short-chain acids); good storage stability; low viscosity, making handling the material easier; optimal catalytic effect; and best cost/performance ratio.

Manufacture

Metallic soaps are manufactured by one of three processes: a fusion process, a double decomposition or precipitate process, or a direct metal reaction (DMR). The choices of process and solvent depend on the metal, the desired form of the product, the desired purity, raw material availability, and cost.

Health and Safety Factors

The hazards encountered in the manufacture, processing, handling, and use of metal soaps are largely associated with the inherent toxicity of the metals and solvents. In general, the acid portion of the metal soap is low in toxicity. Material Safety Data Sheets (MSDS) are available from the commercial suppliers of these metal soaps specifying the inherent hazards. The Hazardous Material Identification System (HMIS) rating for liquid metal soaps may be summarized as follows, where the hazard rating index rates 0–4 as minimal to extreme.

Factor		Rating
health	2	moderate toxicity, may be harmful if inhaled or absorbed
flammability	2	combustible, requires moderate heating to ignite flash point 38 to 93°C
reactivity	0	normally stable, does not react with water

Metal soaps may cause skin irritation or sensitization. They are harmful if swallowed or ingested, which could result in gastrointestinal irritation and vomiting. Inhalation of concentrated vapors can lead to headaches and incoordination.

Solid metal soaps, when finely divided, may present an explosion hazard and are capable of spontaneous combustion. Inhalation of the dust can cause eye and/or respiratory irritation, so they require adequate ventilation.

Uses

Metal Stearates. More than half the metal stearates produced in the United States are applied as lubricants and heat stabilizers in plastics, particularly in the processing of poly(vinyl chloride) (PVC) resins.

Metal 2-Ethylhexanoate (Octoates). The principal applications of metal 2-ethylhexanoates are as paint driers. Types of driers are listed below.

Oxidative	Polymerizing	Auxiliary
Co(II)	Pb(II)	Ca(II)
Mn(II)	Zr(IV)	Zn(II)
Ce(III)		modifiers
Fe(II)	through-dry	
surface-dry		

Metal Naphthenates. Naphthenates of cobalt, manganese, calcium, copper, iron, zinc, and zirconium are used as driers in printing inks. Their use in coatings is declining as a result of the use of higher metal content synthetic driers and the overall trend to latex paint in architectural coatings.

MARVIN LANDAU
Hüls America Inc.

W. S. Stewart, in W. H. Madison, ed., *Paint Driers and Additives, Federation Series of Coatings Technology,* Federation of Societies for Paint Technology, Philadelphia, Pa., 1969, Unit 11.

Stearate Product Specifications, Tenneco Chemicals, Inc., Piscataway, N.J.; *Witco Metallic Stearates, Bulletin 55-4R-5-63,* Witco Chemical Corp., New York, May 1963.

ASTM D1544, American Society for Testing and Materials, Philadelphia, Pa., 1992.

DRUG DELIVERY SYSTEMS

The range of plasma levels for drugs that provides efficacy and avoids side effects in most of the patient population is known as the therapeutic window or range. Design of a dosage form that inputs drug into the physiological system at a specified rate profile for the longest convenient dosing interval is the goal and definition of controlled release drug delivery.

The two most common temporal input profiles for drug delivery are zero order (constant release), and half order, ie, release that decreases with the square root of time. These two profiles correspond to diffusion through a membrane and desorption from a matrix, respectively. In practice, membrane systems have a period of constant release, ie, steady-state permeation, preceded by a period of either an increasing (time lag) or decreasing (burst) flux. This initial period may affect the time of appearance of a drug in plasma on the first dose, but may become insignificant upon multiple dosing.

Design of a controlled release dosage form requires sufficient knowledge of both the desired therapy to specify a target plasma level and the pharmacokinetics.

Permeation and Desorption Studies

To design a drug delivery system, release from the device and its component materials should be investigated, and absorption across the relevant biological membrane for the selected route of access into the body must be measured. The two most common experimental paradigms for evaluating drug absorption or materials for drug delivery devices are membrane permeation and desorption (or dissolution) (see MEMBRANE TECHNOLOGY). A gradient in chemical potential or activity is the driving force for diffusion. In both experiments, the temperature is controlled, the receiver solution is well-stirred, and the chemical activity in the receiver solution is maintained near zero concentration by either very large volumes, frequent replacement, or flow through diffusion cells.

Physiological Routes for Drug Delivery

Design of a drug delivery device is dictated by the properties of the physiological barrier, the effective plasma levels, and the total dosage.

Oral. The oral route for drug delivery includes the gastrointestinal (GI) tract and the oral cavity including the buccal mucosa. The buccal mucosa is considered separately because of differences in the approach to drug delivery via this route.

The primary function of the GI tract is the digestion and absorption of food. Thus, drugs entering the GI tract are exposed to a wide range of pH values, from 1–2 in the stomach to 5.0–6.5 in the small intestine, as well as high levels of various enzymes involved in the digestion of proteins, fats, and carbohydrates.

The transit of a dosage form through the GI tract can have a profound influence on its performance. Total GI transit time is between 24 and 48 h on average.

Absorption of drugs across the wall of the GI tract is primarily the result of passive diffusion. Absorption is believed to take place by partitioning of the drug from the aqueous GI environment into the lipoidal membrane, diffusion through the membrane, and partitioning into the blood and body fluids.

Drugs, such as opiates, may undergo metabolism both in the intestinal wall and in the liver (first-pass metabolism). The metabolism may be extensive and considerably reduce the amount of drug reaching the systemic circulation. Alternatively, the metabolite may be metabolically active and contribute significantly to the action of the parent drug. Some compounds undergo enterohepatic circulation in which they are secreted into the GI tract in the bile and are subsequently reabsorbed. Enterohepatic circulation prolongs the half-life of a drug.

For those compounds absorbed from a small part of the intestine, the amount of drug absorbed can be increased by extending the residence time of the dosage form in the GI tract. The two basic approaches used are gastric flotation or retention devices, and bioadhesive delivery systems.

Drug absorption from the colon has become the subject of much attention. The development of dosage forms that release drug for 16 to 24 h depends on the drug being absorbed from the colon, because the bulk of the delivery period may be spent there. Only those compounds that exhibit good colonic absorption are suitable for extended delivery dosage forms, eg, metoprolol. Protein and peptide drugs are more readily available since the advent of recombinant DNA technology, and the oral delivery of these compounds has become the holy grail of drug delivery. However, proteins (qv) present several challenges because of size, hydrophilicity, and susceptibility to hydrolysis and degradation by proteases. Compared to the upper GI tract, proteolytic activity is lower in the colon and the residence time is longer, which has led to interest in the development of dosage forms targeted to the colon. However, it appears unlikely that significant absorption of proteins occurs from the colon in the absence of either protease inhibitors, absorption enhancers, or both. Targeting drugs to the colon has followed two basic approaches, ie, delayed release and exploitation of microbial enzymes found in the colonic flora. Delayed release generally relies upon enteric coating to ensure safe passage through the stomach, and a delay of 4 to 6 h before drug release.

Rectal. The rectal route for drug delivery is an extremely unpopular one in the United States, but may present advantages in certain situations. Enemas containing either steroids or 5-acetylsalicylic acid for the treatment of proctitis, ie, inflammation of the rectum, offer good therapy in inflammatory bowel disease. The rectal route may be used when gastric stasis or vomiting is present, making the oral route of drug delivery untenable, eg, ergotamine for the treatment of migraine. The vascular drainage of the rectum may partially avoid firstpass metabolism which offers definite advantages for those drugs undergoing extensive metabolism.

Transdermal. The skin offers a formidable barrier to the entry of foreign compounds, including drugs, into the body, both in terms of a physical barrier and an immunological one. The principal barrier to drug diffusion lies in the outer few layers of the epidermis, the stratum corneum, which is 10–20 μm thick in humans and consists of sheets of keratinized epithelial cells joined by tight junctions. The remainder of the epidermis, which is about 100 μm thick in humans, consists of living cells that are metabolically active. A drug applied to the skin must therefore diffuse through the epidermis to reach the blood capillaries in the dermis for distribution to the systemic circulation. Blood supply to the skin can vary tremendously from 200 to 4000 $mL/(m^2 \cdot min)$ as a result of its role in the control of body temperature. Drug delivery by the transdermal route avoids presystemic metabolism in the gastrointestinal tract or first-pass metabolism in the liver. The permeability of skin is low, which limits the usefulness of this route to highly permeable, potent compounds. Permeability varies somewhat with regions of the body. The greatest permeability is in the scrotum.

The use of absorption enhancers for transdermal delivery may be necessary as a result of the low permeability of a drug through skin. As of this writing ethanol is the only enhancer in use in a commercially available system, and the flux of estradiol and nitroglycerin is linearly correlated with the flux of ethanol. Other absorption enhancers such as 1-dodecylazacycloheptan-2-one (Laurocapram), terpenes, oleic acid, pyrrolidones, n-alkanols, and alkyl esters, are under evaluation.

Dermal irritation and sensitization are issues specific to the transdermal route of drug delivery and can result in the cessation of therapy. Transdermal drug delivery is associated with a relatively long time lag before the onset of efficacy, and removal of the system is followed by a correspondingly extended fall in plasma concentration, which probably results from formation of a drug depot in the skin that dissipates slowly. The time lag is approximately 3 to 5 h for many drugs that have low binding in the skin, but may be considerably longer. In contrast, plasma drug levels may be obtained between 2 and 5 min by the oral, buccal, or nasal routes.

Despite the limitations imposed by the physiology of the skin, several marketed controlled release transdermal drug delivery systems are available in the United States; for example, scopolamine for the treatment of motion sickness, nitroglycerin for angina, estradiol for the relief of postmenopausal symptoms and osteoporosis, clonidine for the treatment of hypertension, fentanyl as an analgesic, and nicotine as an aid to smoking cessation. These systems are designed to deliver drug for periods of one to seven days.

Buccal. Buccal mucosa has a high blood flow of 20–30 mL/min for each 100 g of tissue and good lymphatic drainage. Vascular drainage is directly into the systemic circulation, and thus first-pass metabolism is avoided. The buccal mucosa is readily accessible to the patient for self-administration of drugs, as well as rapid removal of the dosage form should it be necessary.

The buccal route may prove useful for peptide or protein delivery because of the absence of protease activity in the saliva. However, the epithelium is relatively tight, based on its electrophysiological properties. Absorption of proteins and peptides is generally low and somewhat erratic. The judicious use of absorption enhancers may be necessary and can be accomplished in a very controlled manner in this area.

Commercially available buccal or sublingual dosage forms include nitroglycerin for angina, buprenorphine for pain relief, ergotamine for the treatment of migraine, methyltestosterone for hypogonadism, captopril for hypertensive emergencies, and nifedipine for hypertensive emergencies and acute angina. Nicotine gum is available as a smoking cessation aid. Absorption is predominantly from the oral cavity, with a minor contribution from intestinal absorption of swallowed drug. These dosage forms are essentially tablets that dissolve rapidly over a few minutes. An alternative approach is the use of a bioadhesive, polymeric system that would provide sustained drug delivery over an extended period of time. The use of a backing material that is impermeable to the drug and saliva directs the drug toward the mucosa and prevents drug loss because of swallowing. The feasibility of this approach has been demonstrated in clinical trials.

Nasal. The nose has good vascular drainage and an estimated blood supply of 40 mL/min for each 100g of tissue. The nasal cavity is obviously accessible, absorption is very rapid, and first-pass metabolism in the liver is avoided. A potential disadvantage is the rapid mucociliary clearance rate for removal of trapped particles from the nose. The estimated turnover time is 15 min. Both the common cold and conditions such as allergic rhinitis can affect clearance as well as the extent of absorption. The nasal route is used primarily for topical delivery of drugs, generally in aerosol form, for the treatment of allergic rhinitis and cold/flu symptoms. This route may have utility for rapid delivery of proteins or peptides, ie, compounds which may require pulsatile rather than sustained delivery.

The permeability of the nasal mucosa is similar to that of the ileum, and it is therefore a leaky epithelium. The structural requirements for drug absorption from the nasal cavity have been analyzed. Examination of data for 24 compounds has shown that the nasal route is suitable for the efficient, rapid delivery of many drugs having mol wts <1000. Mean bioavailability is 70%, without the use of adjuvants. This limit may be extendable to compounds of at least 6000 mol wt using adjuvants. Another approach is to prolong the residence time using agents such as methylcellulose, carboxymethylcellulose, hydroxypropyl cellulose, and polyacrylic acid, or bioadhesive microspheres that also protect proteins from degradation.

The effects of drugs and adjuvants must be assessed, both in short-term administration and during chromic treatment. Local effects include changes in mucociliary clearance, cell damage, and irritation. Chronic erosion of the mucous membrane may lead to inflammation, hyperplasia, metaplasia, and deterioration of normal nasal function.

Pulmonary. Drug delivery to the lung is currently primarily for local therapy, but pulmonary delivery may offer opportunities for systemic delivery of compounds, including vaccines.

Ocular. Drug delivery to the eye presents several challenges based on anatomy and physiology. Drugs have to cross two lipid layers and an aqueous layer to enter the eye, and compounds such as acetazolamide that are readily absorbed elsewhere cannot effectively cross the corneal barrier. The epithelium is rate-limiting for most drugs; the aqueous region, ie, the stroma, is rate-limiting for very lipophilic drugs.

The eye is highly innervated, and patient comfort is of paramount importance in order to achieve good compliance. The eye is designed to keep the surface free of foreign bodies by blinking, tear production, and rapid drainage into the nasolacrimal duct. Two approaches used to increase the residence time of drugs in the eye, and consequently the amount of drug absorbed, are increasing the viscosity of the solution and the use of an implant, such as Ocusert, or hydrogel contact lenses (qv) loaded with drug. Polymers that undergo a phase change from a liquid to a gel in response to temperature, pH, or ionic strength also show promise in this field.

Vaginal. The vaginal mucosa consists of stratified squamous epithelium, thrown into numerous transverse folds or rugae. The area is well supplied with both blood and lymphatic drainage. Although vaginal drug delivery is currently used primarily for the treatment of local infections and contraceptive purposes, this route may offer opportunities for systemic delivery for the treatment of diseases, such as osteoporosis, in which the patient population is predominantly female.

Classes of Controlled Drug Delivery Systems

Controlled release drug delivery systems include those that are diffusion-controlled, chemically controlled, swelling-controlled, and externally controlled. Osmotically controlled systems are a subset of diffusion-controlled systems and are often classified separately.

Diffusion-Controlled. When two materials are in contact and contain components at different activities, the components diffuse in such a way as to equalize the activities. This driving force provides a method of controlling drug delivery systems. The flux of the drug mass is defined by Fick's First law, which states that the drug flux is equivalent to the concentration gradient in the system multiplied by the diffusivity of the drug in the medium. This expression incorporates the driving force of concentration difference and the drug's diffusivity, ie, mobility in the medium. A reservoir system and a monolithic system are the two basic types of drug delivery systems designed from this principle. The Norplant System is a long-term system for birth control based on a reservoir system. The Minitran System (3M Health Care) is a monolithic transdermal system that delivers nitroglycerin at a continuous rate for a day.

Chemically Controlled. These systems depend on the hydrolysis or enzymatic cleavage of a chemical bond that allows delivery of the drug. There are two main types of systems, ie, pendent chain systems and bioerodible systems. Zoladex and Lupon Depot are two commercially available bioerodible systems used in the treatment of prostate cancer.

Swelling-Controlled. The mechanism of swelling-controlled release of drugs is derived from the glassy/rubbery nature of polymers. A penetrant may lower the glass transition temperature of the polymer below the ambient conditions, thus changing it from the glassy state to its rubbery state. Some polymers when dry are in a glassy state and as water migrates into the polymer, the polymer relaxes and swells. The swollen polymer is in a rubbery state and allows the drug to diffuse toward areas of lower activity. The polymer must be chosen very carefully to achieve zero order release as this occurs only when the drug diffuses more rapidly than the swelling front, and the swelling front is moving with constant velocity

Osmotic Control. Several oral osmotic systems (OROS) have been developed by the Alza Corporation to allow controlled delivery of highly water-soluble drugs. The elementary osmotic pump consists of an osmotic core containing drug surrounded by a semi-permeable membrane having a laser-drilled delivery orifice. The system looks like a conventional tablet, yet the outer layer allows only the diffusion of water into the core of the unit. The rate of water diffusion into the system is controlled by the membrane's permeability to water and by the osmotic activity of the core. Because the membrane does not expand as water is absorbed, the drug solution must leave the interior of the tablet through the small orifice at the same rate that water enters by osmosis. The osmotic driving force is constant until all of the drug is dissolved; thus, the osmotic system maintains a constant delivery rate of drug until the time of complete dissolution of the drug. Procardia XL is an example of this technology used in the management of hypertension.

External Control. The use of external control to govern the release of drugs from delivery systems has largely been experimental. A number of mechanisms have been explored, and include external sources such as electrical currents, magnetism, ultrasound, temperature changes, and irradiation. Each of these systems relies upon an external trigger to activate drug release, eg, iontophoretic devices that depend on release of drug by an electric field or polymer-drug conjugates that are light-activated.

System Characterization and Manufacturing

A drug delivery system is a vehicle that provides a stable environment to store an active ingredient prior to usage, and controls the release of the drug during usage. Typically, it is desirable for the system to be stable for a period of at least two years from the date of manufacturing.

Material Characterization. Drug delivery systems are made from polymeric materials. Chemical and physical characterization needs to be performed at each stage of development and manufacturing to prove the safety and efficacy of the final product. Chemical analyses are used to identify and quantify the compositions of the polymer, and the minor constituents.

Material Processing. There are many types of drug delivery systems and the manufacturing of each system consists of fabricating a device as well as loading the drug and other excipients into this device. The drug can be introduced into various components of the system by physical and chemical means, and at various stages of system fabrication.

The first processing step is the handling of raw materials. These materials need to be transported from storage units to the next processing station. Next, these materials are converted into a deformable state by melting, plasticizing, or forming a solution.

When a material is in a deformable state, it can be shaped into desirable configurations such as films, disks, rods, or microspheres. The primary shaping processes are extrusion and molding. Drug loading can be accomplished by dispersion or absorption.

System Characterization. After all the components are assembled to produce a drug delivery system, extensive testing is performed to ensure that the performance and stability of the drug delivery system meet specifications. The extent of chemical and physical interactions among the components of a drug delivery system are characterized. The performance of the drug delivery system needs to be characterized. The rate of drug release and the total amount of drug loaded into a drug delivery system can be determined in a dissolution apparatus or in a diffusion cell. In addition to transport properties, the adhesive properties are characterized by tensile measurements.

ELIZABETH QUADROS
ANN R. COMFORT
STEVEN M. DINH
BRET BERNER
Ciba-Geigy Corporation

J. Crank, *The Mathematics of Diffusion,* 2nd ed., Clarendon Press, Oxford, 1975.

A. Kydonieus, *Treatise on Controlled Drug Delivery,* Marcel Dekker, Inc., New York, 1991.

K. Heilmann, *Therapeutic Systems: Rate-Controlled Drug Delivery, Concept and Development,* 2nd rev. ed., Thieme-Stratton, Inc., New York, 1984.

R. W. Duncan and L. W. Seymour, *Controlled-Release Technologies,* Elsevier Advanced Technology, Oxford, N.Y., 1989.

DRYING

Drying is an operation in which volatile liquids are separated by vaporization from solids, slurries, and solutions to yield solid products. In dehydration, vegetable and animal materials are dried to less than their natural moisture contents, or water of crystallization is removed from hydrates. In freeze drying (lyophilization), wet material is cooled to freeze the liquid; vaporization occurs by sublimation. Gas drying is the separation of condensable vapors from noncondensable gases by cooling, adsorption, or absorption (see also ADSORPTION, GAS SEPARATION). Evaporation differs from drying in that feed and product are both pumpable fluids.

Reasons for drying include user convenience, shipping cost reduction, product stabilization, removal of noxious or toxic volatiles, and waste recycling and disposal. Environmental factors, such as emission control and energy efficiency, increasingly influence equipment choices. Drying operations involving toxic, noxious, or flammable vapors employ gas-tight equipment combined with recirculating inert gas systems having integral dust collectors, vapor condensers, and gas reheaters (see EXHAUST CONTROL, INDUSTRIAL).

Drying is an applied science; ie, drying theory is based on the laws of physics, physical chemistry, and the principles underlying the transfer processes of chemical and mechanical engineering: heat (see HEAT EXCHANGE TECHNOLOGY–HEAT TRANSFER), mass and momentum transfer, vaporization, sublimation, crystallization, fluid mechanics, mixing (see MIXING AND BLENDING), and material handling. Drying is one of several unit operations involving simultaneous heat and mass transfer. However, drying is complicated by the presence of solids that interfere with heat, liquid, and vapor flow and retard the transfer processes, at least during the final drying stages or when a solids phase is continuous.

Because all drying operations involve processing of solids, equipment material handling capability is of primary importance. In fact, most industrial dryers are derived from material handling equipment designed to accommodate specific forms of solids. If possible, liquid separation from solids as liquid, by dewatering in a mechanical separation operation, should precede drying.

Several methods are employed to classify commercial dryers by process application. Mode of heat transfer is the conventional choice. The principal heat-transfer mechanisms in drying are (*1*) convection from a hot gas that contacts the material, used in direct-heat or convection dryers; (*2*) conduction from a hot surface that contacts the material, used in indirect-heat or contact dryers; (*3*) radiation from a hot gas or hot surface that contacts or is within sight of the material, used in radiant-heat dryers; and (*4*) dielectric and microwave heating in high frequency electric fields that generate heat inside the wet material by molecular friction, used in dielectric, or radio frequency, and microwave dryers. In the last group, high internal vapor pressures develop and the temperature inside the material may be higher than at the surface.

In order of priority, the factors that govern the selection of industrial dryers are (*1*) personnel and environmental safety; (*2*) product moisture and quality attainment; (*3*) material handling capability; (*4*) versatility for accommodating process upsets; (*5*) heat- and mass-transfer efficiency; and (*6*) capital, labor, and energy costs.

Costs are determined by energy, labor, capacity, and equipment materials of construction. Continuous dryers are less expensive than batch dryers and drying costs rise significantly if plant size is less than 500 t/yr. Vacuum batch dryers are four times as expensive as atmospheric-pressure batch dryers and freeze dryers are five times as costly as vacuum batch dryers. Once-through air dryers are half as costly as recirculating inert-gas dryers. Per unit of liquid vaporization, freeze and microwave dryers are the most expensive. The cost difference between direct- and indirect-heat dryers is minimal because of the former's large dust recovery requirement. Drying costs for particulate solids at rates of $1 \times 10^3 - 50 \times 10^3$ t/yr are about the same for rotary, fluid-bed, and pneumatic conveyor dryers, although few applications are equally suitable for all three.

Psychrometry

Before drying can begin, a wet material must be heated to such a temperature that the vapor pressure of the contained liquid exceeds the partial pressure of vapor already present in the surrounding atmosphere. The effect of a dryer's atmospheric vapor content and temperature on performance can be studied by construction of a psychrometric chart for the particular gas and vapor. Figure 1 is a standard chart for water vapor in air.

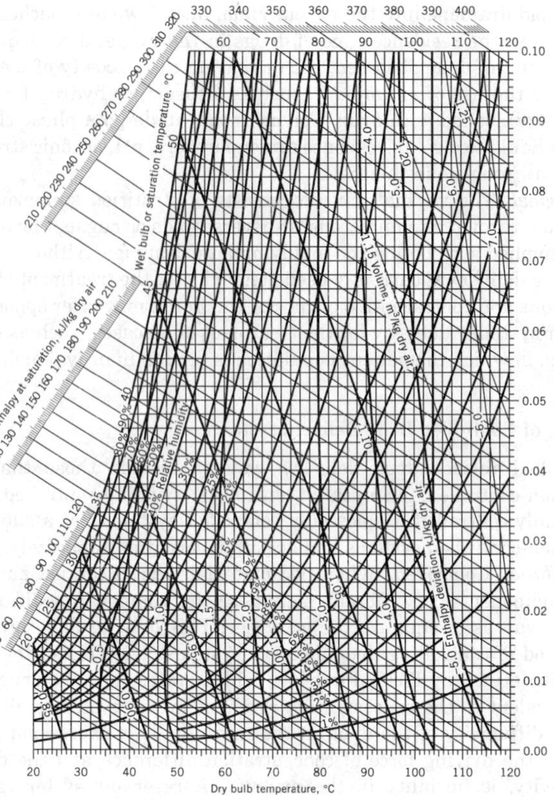

Figure 1. Carrier psychrometric chart for air and water vapor at 101.325 kPa (1 atmosphere) total pressure. Courtesy of Carrier Corp.

The wet bulb or saturation temperature curve indicates the maximum weight of vapor that can be carried by a unit weight of dry gas. For any temperature on the abscissa, saturation humidity is found by reading up to the saturation temperature curve, then across to the ordinate, kg/kg dry air.

Drying Mechanisms

Drying Periods. The goal of most drying operations is not only to separate a volatile liquid, but also to produce a dry solid of a desirable size, shape, porosity, density, texture, color, or flavor. An understanding of liquid and vapor mass-transfer mechanisms is essential for quality control. Mass-transfer mechanisms are best understood by measuring drying behavior under controlled conditions in a prototypic, pilot-plant dryer. No two materials behave alike and a change in material handling method or any operating variable, such as temperature or gas humidity, also affects mass transfer.

Figure 2**a** shows drying time profiles for one material dried under three conditions. Corresponding rate profiles are in Figure 2**b.** Three products having uniquely different characteristics were produced by three different kinds of agitation. Other controllable drying conditions were constant. These profiles show that during drying several distinct periods may occur, which depend on how the material is handled. These are (*1*) an induction period during which wet material is heated to drying temperature; (*2*) a constant rate drying period indicated by the horizontal portions of the profiles in Figure 2**b;** (*3*) a period of decreasing rate shown by the sloping portions of two rate profiles during which the drying rate appears proportional to moisture content; and (*4*) a period of decreasing rate shown by the curved portions of two rate profiles during which the drying rate is evidently a more complex function of moisture content than simple proportionality.

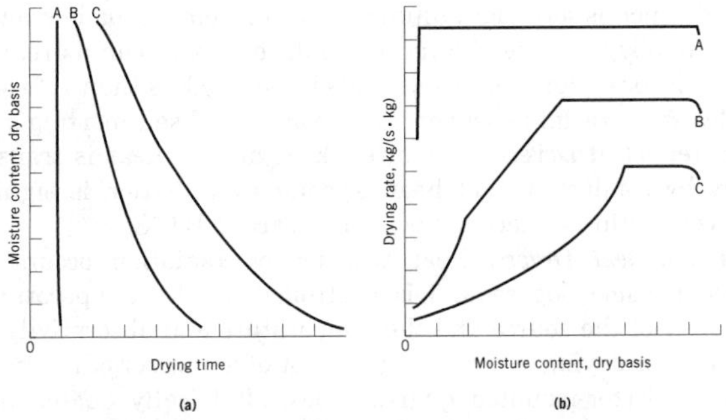

Figure 2. (**a**) Profiles of moisture content vs drying time; and (**b**) drying rate vs moisture content for a slightly soluble, water-wet organic powder centrifuge cake at 4.0 kPa (0.58 psi) absolute pressure using 120°C indirect heat. Profile A was produced in a continuous, high speed agitator dryer provided with scrapers to maintain a clean heating surface. Drying time was 45 s and at an almost entirely constant rate because of high solids surface exposure to the heating surface; the product particle size was 100% less than 150 μm. Profile B was produced in a paddle-agitated batch dryer also having scrapers. Drying time was 70 min, including periods of constant rate, capillary, and diffusion drying. Because of the much slower agitator, the product was a porous, 100–500 μm powder having some dust. Profile C was produced in a double-cone batch dryer using some dry recycle. Drying time was 120 min, and was almost entirely liquid and vapor diffusion-controlled because turning of the double-cone pelletized the wet material early in the cycle; the product was composed of rather dense (200–800 μm) spheres, having negligible dust.

The moisture content at the end of constant rate drying is the critical moisture content. Drying periods following are falling rate periods. The curved portion of profile B in Figure 2**b** is a second falling rate period; moisture content at the second break is the second critical moisture content. Profile C shows that drying may occur almost entirely in a falling rate period; a slight change in specified product moisture content can have a significant effect on drying time.

Constant Rate Drying. During constant rate drying, vaporization occurs from a liquid surface of constant composition and vapor pressure. Material structure has no influence except moisture movement from within the material must be fast enough to maintain the wet surface. The vaporization rate is controlled by the heat-transfer rate to the surface. The mass-transfer rate adjusts to the heat-transfer rate and the wet surface reaches a steady-state temperature. The drying rate remains constant, therefore, as long as external conditions are constant. If heat is supplied solely by convection, the steady-state temperature is the gas wet bulb temperature. When conduction and radiation contribute, eg, the material contacts and/or receives radiation from a warm surface, a liquid surface temperature between the wet bulb temperature and the liquid's boiling point is obtained. In indirect-heat and radiant-heat dryers, where conduction and radiation predominate, surface liquid may boil regardless of ambient humidity and temperature. During constant rate drying, material temperature is controlled more easily in a direct-heat dryer than in an indirect-heat dryer because in the former the material temperature does not exceed the gas wet bulb temperature as long as all surfaces are wet. For convection, all principles relating to simultaneous heat and mass transfer between gases and liquids apply.

Contact Drying. Contact drying occurs when wet material contacts a warm surface in an indirect-heat dryer. A sphere resting on a flat heated surface is a simple model. The heat-transfer mechanisms across the gap between the surface and the sphere are conduction and radiation.

Critical Moisture Content. Critical moisture content is the average material moisture content at the end of constant rate drying. Critical moisture content cannot be determined except by a prototypic drying test.

Particle size distribution determines surface-to-mass ratios and the distance internal moisture must travel to reach the surface. Large pieces thus have higher critical moisture contents than fine particles of the same material dried under the same conditions. Pneumatic-conveyor flash dryers work because very fine particles are produced during initial dispersion and these have low critical moisture contents.

Case hardening refers to a circumstance in which a mass of nonporous, soluble, or colloidal material is dried at such a high rate during initial constant rate drying that the surface overheats and shrinks. Because liquid diffusivity decreases with moisture content, the barrier formed by the overdried surface prevents moisture flow from the interior of the mass to the surface. Case hardening of nonporous materials can be minimized by initially maintaining a high relative humidity environment and consequently a high surface equilibrium moisture content until internal moisture has time to escape.

Equilibrium Moisture Content. Equilibrium moisture content is the steady-state equilibrium reached by the gain or loss of moisture when material is exposed to an environment of specific temperature and humidity for a sufficient time. The equilibrium state is independent of drying method or rate. It is a material property. Only hygroscopic materials have equilibrium moisture contents. Clean beach sand is nonhygroscopic and has an equilibrium moisture content of 0. The same rules apply to organic vapors. Hygroscopic material retains a constant fraction of moisture under specific ambient humidity and temperature conditions. At constant temperature, if ambient humidity increases or decreases, an increase or decrease in moisture content follows. This is called equilibrium moisture because it is held in vapor pressure equilibrium with the partial pressure of vapor in the atmosphere. The reason it is retained even when the atmosphere is quite dry is that the retention mechanism reduces effective liquid vapor pressure. It is bound moisture because it is bound to material in solution or by adsorption and bound moisture behaves as if the atmosphere were saturated even when the atmosphere is not saturated relative to the unbound liquid's normal vapor pressure. Chemically combined liquid may behave like bound moisture depending on the nature of the chemical bond. Because equilibrium is influenced by partial vapor pressure in the atmosphere and the effective vapor pressure of the bound liquid, temperature and humidity are both important. For many materials in the 15–50°C temperature range, equilibrium moisture content can be plotted vs relative humidity as an essentially straight line.

Falling Rate Drying. Heat transfer is limited by material conductivity, but the drying rate usually is controlled by internal liquid and vapor mass transfer. The principal mass-transfer mechanisms are (*1*) liquid diffusion in continuous, homogeneous materials; (*2*) vapor diffusion in porous and granular materials; (*3*) capillarity in porous and fine granular materials; (*4*) gravity flow in granular materials; (*5*) flow caused by shrinkage-induced pressure gradients; and (*6*) pressure flow of liquid and vapor when porous material is heated on one side, but vapor must escape from the other.

Usually, only one mass-transfer mechanism predominates at any given time during drying, although several may occur together. In most materials, the mechanisms of internal liquid and vapor flow during falling rate drying are complex. Simultaneous heat transfer is a factor and falling rate drying rarely can be described with mathematical precision. Computer models for some materials are published, but most employ data from actual drying tests. In the absence of tests, the falling rate drying periods usually are studied on the assumption that internal mass transfer is controlled either by diffusion or capillarity depending on whether the material is porous or nonporous, soluble or not.

Drying Profiles. An application of diffusion principles to falling-rate drying is exemplified by Figure 3. Single drops of whole milk were dried by suspension in a warm-air stream. Because of the rapid formation of surface films, drying was mostly by vapor diffusion.

Dryers

Industrial dryers may be broadly classified by heat-transfer method as either direct or indirect heat. Dryers evolved from material han-

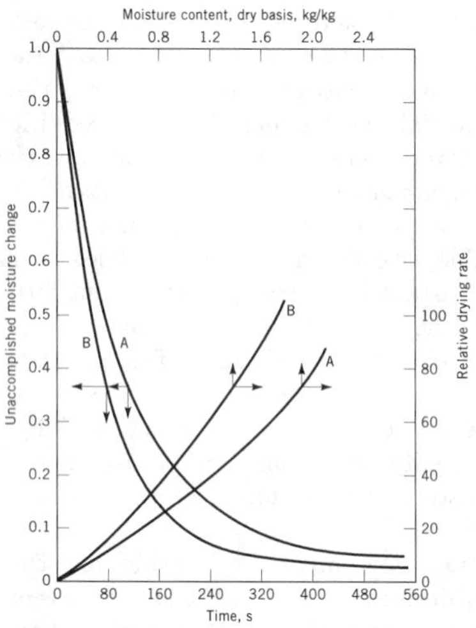

Figure 3. Drying profiles for single drops of whole milk at 94°C and 0.6 m/s relative air flow. The initial diameter of drop $A = 1900$ μm, initial moisture content = 2.6 kg/kg, dry basis; drop $B = 1470$ μm initial diameter and 2.4 kg/kg moisture.

dling equipment, and thus most types of industrial dryers are specially suited for certain forms of material. Dryers are also classified as batch or continuous.

The material suitability of industrial dryers is summarized below.

Dryer	Material form
spray dryer	pumpable, heat-sensitive pastes, slurries, and solutions; all pumpables at high capacities
indirect-heat drum dryer	pumpable, heat-insensitive pastes, slurries, and solutions
pneumatic conveyor dryer	materials instantly dispersible into discrete particles in the drying gas
fluid-bed dryer	fluidizable particulate materials
spouted-bed dryer	particulate materials too coarse or uniform in size to fluidize adequately
hopper dryer	preheated coarse or uniform materials and beds pervious to gas throughflow
direct-heat rotary dryer	particulate materials too coarse, sticky, or unpredictable to be fluidized or spouted
indirect-heat rotary dryer	fine, dusty materials
double-cone (vacuum) dryer	particulate materials that do not stick together, ball, or pelletize during drying
agitator (vacuum) dryer	particulate materials that may stick together, ball, or pelletize until almost dry
through-circulation or band dryer	materials that can be formed into static beds pervious to gas throughflow
continuous conveyors	continuous webs, paper, fabric, film, and fiber tow
batch, cabinet, and tray dryers	small lots, batch identification, and single-product plants less than 500 t/yr

Direct-Heat Dryers. In direct-heat dryers, steam-heated, extended-surface coils are used for gas heating up to about 200°C. Electric and hot oil or vapor heaters are added for higher temperatures. Diluted combustion products are used for all temperatures. An increasingly popular technique for producing inert gas is to recycle the dryer exit gas and vapor as secondary dilution gas for incoming combustion products. Thereby the oxygen level in the dryer gas stream is reduced to safe levels for organic materials. This is usually less than 10% oxygen, but always material-dependent. These are called self-inerting heaters.

Indirect-Heat Dryers. In indirect-heat dryers, heat is transferred mostly by conduction, but heat transfer by radiation is significant when conducting surface temperatures exceed 150°C.

Radiant-Heat Dryers. Heat transfer by radiation occurs in all dryers to some degree and is controlled by the temperature and emissivity of the source and the temperature and absorptivity of the receiver. For drying, sources may consist of a number of incandescent lamps, reflector-mounted quartz tubes, electrically heated ceramic surfaces, and ceramic-enclosed gas burners. Usual source temperatures are 800–2500 K. Radiant energy does not penetrate most material surfaces. Heat penetration below the surface is dependent on material conductivity. Radiant heaters are most suitable for the drying of thin films, eg, paint films.

Dielectric and Microwave Dryers. Dielectric, also called radiofrequency, dryers operate in the frequency range of 1–100 MHz. Microwave dryers in the United States operate at 915 MHz and 2450 MHz.

Microwave applicators are single, like a microwave oven, or multimode cavities in which material is placed or through which it is conveyed, or rectangular waveguides which in effect surround material as it is conveyed. Rapid reversal of electrode polarity generates heat in the material. In a mechanism called dipole rotation, dipoles, which normally are in random orientation, become ordered in the electrical field. As the field dies, they return to random orientation; as the field reverses, they again become ordered but in the opposite direction. Electrical energy is converted to potential energy, to random kinetic energy, and to heat. In ionic conduction, ions are accelerated by the electrical field. They collide with nonionized molecules in random billiard ball fashion. Electrical energy is converted to kinetic energy and to heat. These are two primary mechanisms of energy conversion.

Industrial applications of dielectric and microwave energy for drying are many; however, response to high frequency electromagnetic radiation depends on a material's dielectric constant and dissipation factor, the product of which is its loss factor. A material having a loss factor greater than 0.05 is a potential drying candidate.

The cost of microwave equipment per kilowatt output is about twice that of the dielectric. For irregular shapes, microwaves are preferable because to avoid hot spots during heating, dielectric electrodes are needed that conform to the material shape. Industrial dielectric dryers are employed for lumber drying, plywood bonding and drying, furniture parts drying, textile skeins and package drying, paper moisture leveling, tire cord drying, and many food products. Dielectric heating frequently is combined with radiant heat and hot air for print and coating drying. Microwave dryers are employed for drying cloth, lumber, and foods. Microwaves are used as an energy source in vacuum and freeze dryers.

Dielectric and microwave heating are generally more costly than alternative methods. Thus many applications involve material preheating and second-stage drying where energy demand is low and cycle times can be reduced significantly. Dielectric and microwave heating are chosen mostly when other methods will not work or are impractical.

PAUL Y. MCCORMICK
Drying Unincorporated

D. W. Green, ed., *Perry's Chemical Engineers' Handbook,* 6th ed., McGraw-Hill, Inc., New York, 1984, pp. 20–14.

C. W. Hall and A. S. Mujumdar, eds., *Drying Technology,* Vols. 1–12 Marcel Dekker, Inc., New York, 1983–1994.

A. S. Mujumdar, ed., *Handbook of Industrial Drying,* Marcel Dekker, Inc., New York, 1987.

R. B. Keey, *Drying of Loose and Particulate Materials,* Hemisphere Publishing Corp., New York, 1991.

DRYING OILS

Drying oils oxidize upon exposure to air from a liquid film to a solid, dry film. Consumption of drying oils in the United States peaked in the late 1940s or early 1950s; in the early 1990s much smaller, but still significant, amounts are used. The use of synthetic drying oil-based resins has exceeded the use of natural drying oils; however, their consumption is declining owing to discoloration and embrittlement caused by continued oxidation after film formation.

Natural Oils

Occurrence and Isolation. Most drying oils are derived from plant seeds. The largest volume drying oil, linseed oil, is obtained from flaxseed, *Linum usitatissimum.* In the United States flax is grown in North Dakota, Minnesota, and South Dakota. Flax for oil is also raised in Canada, Argentina, India, and parts of the former USSR. Soybean oil, the second most important oil, is obtained from soybeans, the seed of *Glycine max* (L) Merrill. It is produced in the United States, Brazil, Argentina, and China. Without modification, it is semidrying oil, not a drying oil. Perilla, safflower, sunflower, and walnut oils have limited uses as drying oils. Tung oil, also called wood oil or chinawood oil, is obtained from the seed kernels of the tung tree, *Aleurites fordii.* The principal source of the oil is China. Limited quantities of another conjugated oil, oiticica oil from the oiticica tree, *Licania rigida,* also are used. The only animal oils used as drying oils on a significant scale are marine fish oils, primarily from anchovy, menhaden, pilchard, and sardines. Fish oil is isolated by steam treatment of the fish.

Castor oil, derived from the beans of *Ricinus communis,* is converted to a drying oil by heating with catalysts to yield dehydrated castor oil.

Trees, especially conifers, contain tall oils. Tall oil is not isolated directly; tall oil fatty acids are isolated from the soaps generated as by-products of the sulfate pulping process for making paper.

A wide variety of other oils have been investigated and, in many cases, used commercially over the years. More complete listings are available.

Composition and Analysis. Naturally occurring drying oils are triglycerides. The reactivity of the oils results from the presence of esters of fatty acids with two or more nonconjugated double bonds separated by single methylene groups, $—CH{=}CHCH_2CH{=}CH$ hose with two or more conjugated double bonds. Typical compositions of some of the more important oils are listed in Table 1.

Autoxidation. Oils are classified as drying oils, which form solid films on exposure to air; semidrying oils, which form tacky, sticky films; and nondrying oils, which do not undergo marked increase in viscosity on exposure to air. Drying oils are further classified as nonconjugated and conjugated oils, depending on whether the double bonds in the predominant fatty acids are separated by one methylene group or are conjugated.

A general statement, useful in considering synthetic drying oils as well as natural oils, is that if the average number of methylene groups between two double bonds per molecule is greater than 2.2, the oil is a drying oil; if less than 2.2, the oil is a drying oil. There is no sharp dividing line between semidrying and nondrying oils.

The reactivity of nonconjugated drying oils is related to the average number of methylene groups between double bonds per molecule. Such methylene groups are allylic to two double bonds, and show much greater reactivity than methylene groups allylic to only one double bond.

The reactions taking place during drying are complex, with many side reactions. Films form from a drying oil, such as linseed oil, in the following steps: an induction period during which naturally present antioxidants, mainly tocopherols, are consumed; a period of rapid oxygen uptake with a weight gain of about 10% (ftir shows an increase in hydroperoxides and appearance of conjugated dienes during this stage); a complex sequence of autocatalytic reactions in which hydroperoxides are consumed and the cross-linked film is formed; and cleavage reactions to form low mol wt by-products.

The rates at which nonconjugated drying oils dry are slow. Metal salts (driers) are known to catalyze the drying rate. The most widely used are the oil-soluble cobalt, manganese, lead, zirconium, and calcium salts of 2-ethylhexanoic acid or naphthenic acids (see DRIERS AND METALLIC SOAPS).

Combinations of metal salts are almost always used. Although mixtures of lead with cobalt and/or manganese are particularly effective, toxicity regulations ban the use of lead driers in consumer paints sold in interstate commerce in the United States. Combinations of cobalt and/or manganese with zirconium, and frequently also with calcium, are commonly used. The amounts of driers needed are very system specific. Their use should be kept to the minimum possible level since they not only catalyze drying but also catalyze the post-drying embrittlement and discoloration reactions.

Oils containing conjugated double bonds, such as tung oil, dry more rapidly than nonconjugated drying oils. Free-radical polymerization of the conjugated diene systems can lead to chain-growth polymerization rather than just combination of free radicals to form cross-links. In general, the water and alkali resistance of films formed using conjugated oils are superior, presumably because more of the cross-links are stable carbon-to-carbon bonds. However, because α-eleostearic acid in tung oil has three double bonds, discoloration on baking or aging is severe.

Both nonconjugated and conjugated drying oils can be polymerized by heating under an inert atmosphere to form so-called bodied oils. Bodied oils have higher viscosities and are often used in oil paints to improve application and performance characteristics.

Table 1. Typical Fatty Acid Composition of Drying Oils From Seeds,[a] %

Oil	Saturated[b]	Oleic	Linoleic	Linolenic
linseed	10	22	16	52
perilla	7	14	16	63
safflower	10	13	77	
soybean	16	24	51	9
sunflower[c]	14	14	72	
sunflower[c]	9	72	19	
tung[d]	6	4	8	
walnut	8	16	72	

[a] Proportions shown are approximate; actual compositions can vary greatly. [b] Palmitic and stearic acids. [c] Examples of the especially large variations in composition of available sunflower oils. [d] Also 82% α-eleostearic acid.

Synthetic and Modified Drying Oils

Varnishes. The drying rate of drying oils can be increased by dissolving a solid resin in the oil and diluting with a hydrocarbon solvent. Such a solution is called a varnish. The solid resin serves to increase the glass-transition temperature, T_g, of the solvent-free film so that film hardness is achieved more rapidly.

In varnish manufacture, the drying oil, ie, linseed oil, tung oil, or mixtures of the two, and the resin are cooked together to high temperatures to yield a homogeneous solution for the proper viscosity. The varnish is then thinned with hydrocarbon solvents to application viscosity. Varnishes were widely used in the nineteenth and early twentieth centuries. They have been replace almost completely by a wide variety of other products, especially alkyds, epoxy esters, and urethane oils.

Synthetic Conjugated Oils. Tung oil dries rapidly, but is expensive, and its films discolor rapidly owing to the presence of three double bonds. These defects led to efforts to synthesize conjugated oils, especially those containing esters of fatty acids with two conjugated double bonds.

Esters of Higher Functionality Polyols. When oil-derived fatty acids react with polyols having more than three hydroxyl groups per molecule, the average number of cross-linking sites per molecule increases proportionally to the functionality of the polyol. Because the number of reactive sites in soybean oil with the composition listed in Table 1 is 2.07, soybean oil is a semidrying oil. However, the pentaerythritol (2,2-bis(hydroxymethyl)-1,3-propanediol) ester of soybean fatty acids with 2.76 reactive sites per molecule is a drying oil.

Oxidizing alkyds can be considered as still-higher functionality drying oils. Their drying speed is faster owing to both the higher functionality and the higher T_g of the rigid aromatic rings in the phthalate esters.

Oils Modified with Maleic Anhydride (Maleated Oils). Oils, with either conjugated or nonconjugated double bonds, react with maleic anhydride (2,5-furandione) to form adducts. The products of these reactions with maleic anhydride, termed maleated oils, react with polyols to give moderate mol wt derivatives that dry faster than the unmodified oils. Such products have not found significant commercial use, but similar reactions with alkyds and epoxy esters are used on a large scale to make water-dilutable derivatives.

Vinyl-Modified Oils. Both conjugated and nonconjugated drying oils react in the presence of free-radical initiators with such vinyl monomers as styrene, vinyltoluene, acrylic esters, and cyclopentadiene. High degrees of chain transfer cause the formation of wide varieties of products, including low mol wt homopolymers of the vinyl monomer, short-chain graft copolymers, and dimerized drying oil molecules. The reaction products with drying oils, except cyclopentadiene, are not commercially important, but the same principle is widely used in making modified alkyds.

Uses

Although some drying oils continue to be used as drying oils, the largest use of drying and semidrying oils in the early 1990s is as raw materials, either directly or as the fatty acids obtained by saponification, in the manufacture of oxidizing alkyds, epoxy esters, urethane oils, and synthetic drying oils. Tall oil fatty acids also are used to manufacture dimer acids (qv) (see CARBOXYLIC ACIDS–FATTY ACIDS FROM TALL OIL).

Since drying oils are considered a renewable resource, they may again become important in paints and printing inks, depending on the cost of drying oils compared with petroleum-derived raw materials. However, in many cases, the properties that can be obtained with synthetic binders, especially retention of flexibility, gloss, and color, are superior to those that can be obtained with a drying oil or drying oil-derived binder.

Paints. Although most drying oils have been replaced as paint vehicles by latexes and other synthetic resins, oils are still being used to a degree in paint and allied products. In exterior house paints, linseed oil or oxidizing alkyds are used when paint must be applied at temperatures as low as 4 to 5°C, ie, temperatures at which latexes do not coalesce satisfactorily. They also are used in primers over chalky surfaces where latex paints do not provide adequate adhesion.

Most stains used for finishing shingles and other natural wood exterior products are made with pigmented linseed oil, diluted with hydrocarbon solvents for penetration.

Red lead-in-oil primers have been used for many years for corrosion protection of steel where it is not practical to remove oily rust particles from the surface of the steel; concern about lead content has resulted in use reduction.

Printing Inks. The use of drying oils in printing inks has decreased. Some inks based on drying oils are still used for sheet-fed letterpress printing. The principal remaining use is in lithographic printing, particularly sheet-fed lithographic printing.

ZENO W. WICKS, JR.
Consultant

M. W. Formo, in D. Swern, ed. *Bailey's Industrial Oil and Fat Products,* John Wiley & Sons, Inc., New York, Vol. I, 1979, pp. 177–232 and 687–816; Vol. II, 1982, pp. 343–406.

A. E. Rheineck and R. O. Austin, in R. R. Myers and J. S. Long, eds., *Treatise on Coatings,* Vol. I, No. 2, Marcel Dekker, New York, 1968, pp. 181–248.

Z. W. Wicks, Jr., F. N. Jones, and S. P. Pappas, in *Organic Coatings. Science and Technology,* Wiley-Interscience, New York, Vol. I, 1993, pp. 133–143.

DYE CARRIERS

Dye carriers are needed for complete dye penetration of polyester fibers. Carriers cause the glass-transition temperature, T_g, of the polyester polymer to become lower and allow the penetration of water-insoluble dyes into the fiber.

Dye carriers, occasionally called dyeing accelerants, are used on cellulose triacetate fibers, but have found their greatest use in the dyeing of polyester.

Carrier Properties

Most carriers are aromatic compounds, and have similar solubility parameters to the poly(ethylene terephthalate) fibers and to some disperse dyes.

Table 1 lists the four main groups of compounds most commonly used as dye carriers. In order for these compounds to act effectively as carriers, they must be homogeneously dispersed in the dyebath. Because the carrier-active compounds have little or no solubility in water, emulsifiers are needed to disperse these compounds in the dyebath.

Carrier Formulation

The formulation of a carrier depends on four considerations: *(1)* the carrier-active chemical compound; *(2)* the emulsifier; *(3)* special additives; and *(4)* environmental concerns. Additional parameters to be considered in the formulation of a carrier product with satisfactory and repeatable performance arise from the equipment in which the dyeing operation is to be carried out. The choice of equipment is usually dictated by the form in which the fiber substrate is to be processed, eg, loose fiber, staple, continuous or texturized filament, woven or knot fabric, yarn on packages or in skeins (see TEXTILES).

Carrier Selection

A carrier is selected by the dyer according to various criteria. The type of equipment and conditions under which it is to be used have

Table 1. Compounds Most Commonly Used as Dye Carriers

Compounds	Mol wt	Bp,°C
phenolics		
o-phenylphenol	170.2	280–284
p-phenylphenol	170.2	305–308
methyl cresotinate	166.0	240
chlorinated aromatics		
o-dichlorobenzene	147.0	172–178
1,3,5-trichlorobenzene	181.45	214–219
aromatic hydrocarbons and ethers		
biphenyl	154.2	255.9
methylbiphenyl	168.24	255.3
diphenyl oxide	170.0	259.0
1-methylnaphthalene	142.2	244.6
2-methylnaphthalene	142.2	241
aromatic esters		
methyl benzoate	136.14	198–200
butyl benzoate	178.22	250
benzyl benzoate	212.24	323–324
phthalates		
dimethyl phthalate	194.18	298
diethyl phthalate	212.18	298
diallyl phthalate	246.25	290
dimethyl terephthalate	194.18	284

already been mentioned. Other considerations include color yield, dye migration, and product and emulsion stability.

Health and Safety Factors

Most carrier-active compounds are based on aromatic chemicals with characteristic odor. An exception is the phthalate esters, which are often preferred when ambient odor is objectionable or residual odor on the fabric cannot be tolerated. The toxicity of carrier-active compounds and of their ultimate compositions varies with the chemical or chemicals involved. The environment surrounding the dyeing equipment where carriers are used should always be well-ventilated, and operators should wear protective clothing (eg, rubber gloves, aprons, and safety glasses or face shields, and possibly an appropriate respirator). Specific handling information can be obtained from the supplier or manufacturer.

OSHA and EPA have established exposure limits that must be carefully considered in relation to the waste disposal method available and the environment in which dye carriers are to be used.

The increasingly stringent government regulations and the introduction of carrierless–dyeable polyester have not substantially affected the use of carriers. The factor with the greatest impact on their use has been provided by the spectacular technological advances in dyeing equipment that have taken place since the early 1980s. High temperature dyeing has greatly reduced the time element (dyeing cycle) and the need for carriers.

ERNESTO DE GUZMAN
BOYCE SUTTON, JR.
Sybron Chemicals Inc.

K. Tandy, Jr., "Characteristics of Polyester Homopolymer Fiber Which Affect Dyeing Properties," paper presented at the *14th AATCC New England Regional Technical Conference,* New Hamsphire, May 19–21, 1977.

S. Salvin and co-workers, *Am. Dyest. Rep* (22) (Nov. 2, 1959).

C. M. Hansen, *J. Paint Technol* **39**, 104 (1967).

M. C. Keen and R. J. Thomas, "Absorption Properties of Latyl Disperse Dyes on Application to Dacron Polyester Fibers," *Dyes and Chemicals Technical Bulletin,* E. I. du Pont de Nemours & Co., Inc., Organic Chemicals Dept., Wilmington, Del., 1992.

DYES AND DYE INTERMEDIATES

Classification Systems for Dyes

Dyes may be classified according to chemical structure or by their usage or application method. The former approach is adopted by practicing dye chemists who use terms such as azo dyes, anthraquinone dyes, and phthalocyanine dyes. The latter approach is used predominantly by the dye user, the dye technologist, who speaks of reactive dyes for cotton and disperse dyes for polyester. Very often, both terminologies are used, for example, an azo disperse dye for polyester and a phthalocyanine reactive dye for cotton.

Classification of Dyes by Use or Application Method

The classification of dyes according to their usage is summarized in Table 1, which is arranged according to the *Colour Index* (CI) application classification. It shows the principal substrates, the methods of application, and the representative chemical types for each application class.

Although not shown in Table 1, dyes are also being used in high technology applications, such as in the medical, electronics, and especially the reprographics industries.

Nomenclature of Dyes

Dyes are named either by their commercial trade name or by their *Colour Index* (CI) name. In the *Colour Index* these are cross-referenced.

Table 1. Usage Classification of Dyes[a]

Class	Principal substrates
acid	nylon, wool, silk, paper, inks, and leather
azoic components and compositions	cotton, rayon, cellulose acetate, and polyester
basic	paper, polyacrylonitrile-modified nylon, polyester, and inks
direct	cotton, rayon, paper, leather, and nylon
disperse	polyester, polyamide, acetate, acrylic, and plastics
fluorescent brighteners[b]	soaps and detergents, all fibers, oils, paints, and plastics[c]
food, drug, and cosmetic[d]	foods, drugs, and cosmetics
mordant[e]	wool, leather, and anodized aluminum
natural[f]	food
oxidation bases	hair, fur, and cotton
pigments[g]	paints, inks, plastics, and textiles
reactive[h]	cotton, wool, silk, and nylon
solvent	plastics, gasoline, varnish, lacquer, stains, inks, fats, oils, and waxes
sulfur	cotton and rayon
vat	cotton, rayon, and wool

[a] *Encyclopedia* articles on specific chemical types of dyes are AZINE DYES; AZO DYES; CYANINE DYES; DYES, ANTHRAQUINONE; PHTHALOCYANINE COMPOUNDS; POLYMETHINE DYES; STILBENE DYES; SULFUR DYES; TRIPHENYLMETHANE AND RELATED DYES; XANTHENE DYES.
[b] See FLUORESCENT WHITENING AGENTS.
[c] See COLORANTS FOR PLASTICS.
[d] See COLORANTS FOR FOOD, DRUGS, COSMETICS, AND MEDICAL DEVICES.
[e] See DYES, APPLICATION AND EVALUATION.
[f] See DYES, NATURAL.
[g] See PAINT; PIGMENTS; INKS.
[h] See DYES, REACTIVE.

The commercial names of dyes are usually made up of three parts. The first is a trademark used by the particular manufacturer to designate both the manufacturer and the class of dye, the second is the color, and the third is a series of letters and numbers used as a code by the manufacturer to define more precisely the hue, and also to indicate important properties the dye possesses.

Classification of Dyes by Chemical Structure

The two overriding trends in dyestuffs research for many years have been improved cost-effectiveness and increased technical excellence. Improved cost-effectiveness usually means replacing tinctorially weak dyes such as anthraquinone, the second largest class after the azo dyes, with tinctorially stronger dyes such as heterocyclic azos, triphendioxazines, and benzodifuranones. This theme will be pursued throughout this section discussing dyes by chemical structure.

Azo Dyes. These dyes are by far the most important class, accounting for over 50% of all commercial dyes, and having been studied more than any other class (see AZO DYES). Azo dyes contain at least one azo group (—N═N—) but can contain two (disazo), three (trisazo), or, more rarely, four or more (polyazo) azo groups. The azo group is attached to two radicals of which at least one, but, more usually, both are aromatic.

In monoazo dyes, the most important type, the A radical often contains electron-accepting groups, and the E radical contains electron-donating groups, particularly hydroxy and amino groups.

Almost without exception, azo dyes are made by diazotization of a primary aromatic amine followed by coupling of the resultant diazonium salt with an electron-rich nucleophile.

In theory, azo dyes can undergo tautomerism: azo/hydrazone for hydroxyazo dyes; azo/imino for aminoazo dyes, and azonium/ammonium for protonated azo dyes. A more detailed account of azo dye tautomerism can be found elsewhere.

The three metals of importance in azo dyes are copper, chromium, and cobalt. The most important copper dyes are the 1:1 copper(II): azo dye complexes of formula; they have a planar structure.

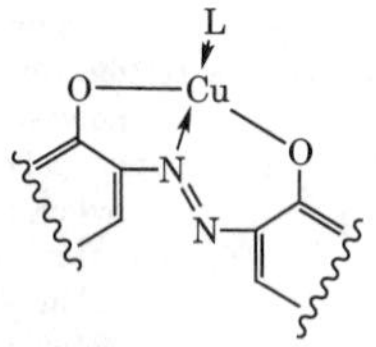

In contrast, chromium(III) and cobalt(III) form 2:1 dye:metal complexes that have nonplanar structures. Geometrical isomerism exists.

Premetallized dyes are now used widely in various outlets to improve the properties of the dye, particularly lightfastness. However, this is at the expense of brightness, because metallized azo dyes are duller than nonmetallized dyes.

Carbocyclic azo dyes are the backbone of most commercial dye ranges. Based totally on benzene and naphthalene derivatives, they provide yellow, red, blue, and green colors for all the major substrates such as polyester, cellulose, nylon, polyacrylonitrile, and leather.

The carbocyclic azo dye class provides dyes having high cost-effectiveness combined with good all-around fastness properties. However, they lack brightness, and consequently, they cannot compete with anthraquinone dyes for brightness. This shortcoming of carbocyclic azo dyes is overcome by heterocyclic azo dyes.

One long-term aim of dyestuffs research has been to combine the brightness and high fastness properties of anthraquinone dyes with the strength and economy of azo dyes. This aim is now being realized with heterocyclic azo dyes, which fall into two main groups: those derived from heterocyclic coupling components, and those derived from heterocyclic diazo components.

All the heterocyclic coupling components that provide commercially important azo dyes contain only nitrogen as the hetero atom. They are indoles, pyrazolones, and especially pyridones; they provide yellow to orange dyes for various substrates.

In contrast to the heterocyclic coupling components, virtually all the heterocyclic diazo components that provide commercially important azo dyes contain sulfur, either alone or in combination with nitrogen (the one notable exception is the triazole system). These S or S/N heterocyclic azo dyes provide bright, strong shades that range from red through blue to green, and therefore complement the yellow-orange colors of the nitrogen heterocyclic azo dyes in providing a complete coverage of the entire shade gamut.

Anthraquinone Dyes. Anthraquinone dyes are based on 9,10-anthraquinone, which is essentially colorless. To produce commercially useful dyes, powerful electron-donor groups such as amino or hydroxy are introduced into one or more of the four alpha positions (1,4,5, and 8). The most common substitution patterns are 1,4-, 1,2,4-, and 1,4,5,8-. To maximize the properties, primary and secondary amino groups (not tertiary) and hydroxy groups are employed.

Anthraquinone dyes are prepared by the stepwise introduction of substituents on to the performed anthraquinone skeleton or ring closure of appropriately substituted precursors.

The principal advantages of anthraquinone dyes are brightness and good fastness properties, but they are both expensive and tinctorially weak. However, they are still used extensively, particularly in the red and blue shade areas, because other dyes cannot provide the combination of properties offered by anthraquinone dyes, albeit at a price.

Benzodifuranone Dyes. BDFs are unusual in that they span the whole color spectrum from yellow through red to blue, depending on the electron-donating power of the R group on the phenyl ring of the aryl acetic acid, ie, Ar = $=C_6H_4R$ (R = H, yellow–orange; R = alkoxy, red; R = amino, blue). The first commercial BDF, Dispersol Red C-BN, a red disperse dye for polyester, is already making a tremendous impact. Its brightness even surpasses that of the anthraquinone reds, while its high tinctorial strength (ca 3–4 times that of anthraquinones) makes it cost-effective.

Polycyclic Aromatic Carbonyl Dyes. Structurally, these dyes contain one or more carbonyl groups linked by a quinonoid system. They tend to be relatively large molecules built up from smaller units, typically anthraquinones. Since they are applied to the substrate (usually cellulose) by a vatting process, the polycyclic aromatic carbonyl dyes are often called the anthraquinonoid vat dyes.

Although the colors of the polycyclic aromatic carbonyl dyes cover the entire shade gamut, only the blue dyes and the tertiary shade dyes, namely, browns, greens, and blacks, are important commercially. As a class, the polycyclic aromatic carbonyl dyes exhibit the highest order of lightfastness and wetfastness.

Indigoid Dyes. Like the anthraquinone, benzodifuranone, and polycyclic aromatic carbonyl dyes, the indigoid dyes also contain carbonyl groups. They are also vat dyes.

Indigoid dyes represent one of the oldest known classes of dyes. Although many indigoid dyes have been synthesized, only indigo itself is of any importance today. Indigo is the blue used almost exclusively for dyeing denim jeans and jackets and is held in high esteem because it fades in tone to give progressively paler blue shades.

Polymethine and Related Dyes. Cyanine dyes are the best known polymethine dyes. Nowadays, their commercial use is limited to sensitizing dyes for silver halide photography. However, derivatives of cyanine dyes provide important dyes for polyacrylonitrile. They include azacarbocyanines, hemicyanines, and diazahemicyanines.

Styryl Dyes. The styryl dyes are uncharged molecules containing a styryl group C_6H_5—CH=C usually in conjugation with an *N,N*-dialkylaminoaryl group. Styryl dyes were once a fairly important group of yellow dyes for a variety of substrates. They are synthesized by condensation of an active methylene compound, especially malononitrile with a carbonyl group, especially an aldehyde. As such, styryl dyes have small molecular structures and are ideal for dyeing densely packed hydrophobic substrates such as polyester.

Yellow styryl dyes have now been largely superseded by superior dyes such as azopyridones, but there has been a resurgence of interest in red and blue styryl dyes. The addition of a third cyano group to produce a tricyanovinyl group causes a large bathochromic shift: the resulting dyes are bright red rather than the greenish yellow color of the dicyanovinyl dyes. These tricyanovinyl dyes have been patented by Mitsubishi for the transfer printing of polyester substrates. Two synthetic routes to the dyes are shown: one is by the replacement of a cyano group in tetracyanoethylene, and the second is by oxidative cyanation of a dicyanovinyl dye with cyanide. The use of such toxic reagents could hinder the commercialization of the tricyanovinyl dyes (see CYANOCARBONS).

Di- and Triaryl Carbonium and Related Dyes. As a class, these dyes are bright and strong, but are generally deficient in lightfastness. Consequently, they are used in outlets where brightness and cost-effectiveness, rather than permanence, are paramount, for example, the coloration of paper. Many dyes of this class, especially derivatives of pyronines (xanthenes), are among the most fluorescent dyes known.

Resurgence of interest in triphendioxazine dyes arose through successful modification of the intrinsically strong and bright triphendioxazine chromogen to produce blue reactive dyes for cotton. These blue reactive dyes combine the advantages of azo dyes and anthraquinone dyes. Thus they are bright, strong dyes with good fastness properties.

Phthalocyanines. Apart from the recent discoveries of benzodifuranone dyes and diketopyrrolopyrrole pigments, phthalocyanine is the only novel chromogen of commercial importance discovered since the nineteenth century.

Phthalocyanines are analogues of the natural pigments chlorophyll and heme. However, unlike these natural pigments, which have extremely poor stability, phthalocyanines are probably the most stable of all the colorants in use today. Substituents can extend the absorption to longer wavelengths, into the near infrared, but not to shorter wavelengths, and so their hues are restricted to blue and green.

Of all the metal complexes evaluated, copper phthalocyanines give the best combination of color and properties and consequently the majority of phthalocyanine dyes are based on copper phthalocyanine.

Besides being extremely stable, phthalocyanines are bright and tinctorially strong ($\epsilon_{max} \sim 100,000$); this renders them cost-effective. Consequently, phthalocyanines are used extensively in printing inks and paints.

Quinophthalones. Like the hydroxy azo dyes, quinophthalone dyes can, in theory, exhibit tautomerism. The dyes are synthesized by the condensation of quinaldine derivatives with phthalic anhydride. Quinophthalones provide important yellow dyes for the coloration of plastics and for the coloration of polyester.

Sulfur Dyes. These dyes are synthesized by heating aromatic amines, phenols, or nitro compounds with sulfur or, more usually, alkali polysulfides. Sulfur dyes are used for dyeing cellulosic fibers. They are insoluble in water and are reduced to the water-soluble leuco form for application to the substrate by using sodium sulfide solution. The sulfur dye proper is then formed within the fiber pores by atmospheric oxidation. Sulfur dyes constitute an important class of dye for producing cost-effective tertiary shades, especially black, on cellulosic fibers.

Nitro and Nitroso Dyes. These dyes are now of only minor commercial importance, but are of interest for their small molecular structures. The most important nitro dyes are the nitrodiphenylamines. Their small molecules are ideal for penetrating dense fibers such as polyester, and are therefore used as disperse dyes for polyester. All the important dyes are yellow. Although the dyes are not terribly strong ($\epsilon_{max} \sim 20,000$), they are cost-effective because of their easy synthesis from inexpensive intermediates.

Nitroso dyes are metal-complex derivatives of *o*-nitrosophenols or naphthols. Tautomerism is possible in the metal-free precursor between the nitrosohydroxy tautomer and the quinoneoxime tautomer. The only nitroso dyes important commercially are the iron complexes of sulfonated 1-nitroso-2-naphthol. These inexpensive colorants are used mainly for coloring paper.

Miscellaneous Dyes. Other classes of dyes that still have some importance are the stilbene dyes and the formazan dyes. Stilbene dyes are in most cases mixtures of dyes of indeterminate constitution that are formed from the condensation of sulfonated nitroaromatic compounds in aqueous caustic alkali either alone or with other aromatic compounds, typically arylamines. The sulfonated nitrostilbene is the most important nitroaromatic, and the aminoazobenzenes are the most important arylamines.

Formazan dyes bear a formal resemblance to azo dyes, since they contain an azo group. The most important formazan dyes are the metal complexes, particularly copper complexes, of tetradentate formazans. They are used as reactive dyes for cotton.

Dye Intermediates

The precursors of dyes are called dye intermediates. They are obtained from simple raw materials, such as benzene and naphthalene, by a variety of chemical reactions. Usually, the raw materials are cyclic aromatic compounds, but acyclic precursors are used to synthesize heterocyclic intermediates. The intermediates are derived from two principal sources, coal tar and petroleum (qv).

Intermediates Classification. Intermediates may be conveniently divided into primary intermediates (primaries) and dye intermediates. Large amounts of inorganic materials are consumed in both intermediates and dyes manufacture.

Inorganic materials include acids (sulfuric, nitric, hydrochloric, and phosphoric), bases (caustic soda, caustic potash, soda ash, sodium carbonate, ammonia, and lime), salts (sodium chloride, sodium nitrite, and sodium sulfide) and other substances such as chlorine, bromine, phosphorus chlorides, and sulfur chlorides. The important point is that there is a significant usage of at least one inorganic material in all processes, and the overall tonnage used by, and therefore the cost to, the dye industry is high.

Primary intermediates are characterized by one or more of the following descriptions, which associate them with raw materials rather than with intermediates.

1. Manufactured in a dedicated plant.
2. At least 1000 t/yr capacity from a single plant.
3. Manufacturing process and/or operation is continuous or semi-continuous.
4. A primary intermediate has established usage in basic industries such as rubber, polymers, or agrochemicals in addition to dyes.

All the significant primaries, about 30 different products, are derived from benzene, toluene, or naphthalene. The primaries are listed here with a reference to the *Encyclopedia* article that covers them in detail.

The following amines are covered under the title AMINES, AROMATIC: aniline, *p*-nitroanilineaniline, *o*-toluidine, *p*-toluidine, dimethylaniline, *m*-phenylenediamine, and *p*-phenylenediamine. The article NITROBENZENE AND NITROTOLUENES covers the primaries: nitrobenzene, *p*-chloronitrobenzene, *o*-chloronitrotoluene, and *p*-nitrotoluene.

Some primaries have articles devoted to them and their derivatives, ie, BENZOIC ACID, PHENOL, SALICYCLIC ACID, and phthalic anhydride as one derivative of PHTHALIC ACIDS. The primary β-naphthol is discussed in NAPHTHALENE DERIVATIVES.

Dye intermediates are defined as those precursors to colorants that are manufactured within the dyes industry, and they are nearly always colorless. Colored precursors are conveniently termed color bases. As distinct from primaries they are only rarely manufactured in single-product units because of the comparatively low tonnages required. Fluorescent brightening agents (FBAs) are neither intermediates nor true colorants.

There are at least 3000 different intermediates in current manufacture (over half that number are specifically mentioned in the *Colour Index*), and in addition there is a comparatively small number of products manufactured by individual companies for their own specialties.

Intermediates vary in complexity, usually related to the number of chemical and operational stages in their manufacture, and therefore cost. Prices may be classed as cheap (less than \$1500/t, as with primaries), average (\$1500 to \$5000/t) or expensive (more than \$5000/t).

The Chemistry of Dye Intermediates

The chemistry of dye intermediates may be conveniently divided into the chemistry of carbocycles, such as benzene and naphthalene, and the chemistry of heterocycles, such as pyridones and thiophenes.

Chemistry of Aromatic Carbocycles. Benzene and naphthalene are by far the most important aromatic carbocycles used in the dyes industry. The hundreds of benzene and naphthalene intermediates used can be prepared from these parent compounds by the sequential introduction of a variety of substituents eg, NO_2, NR^1R^2, Cl, SO_3H, etc. Introduction of these groups are known as unit processes. The substituents are introduced into the aromatic ring by either electrophilic or nucleophilic substitution. In general, aromatic rings, because of their inherently high electron density, are much more susceptible to electrophilic attack than to nucleophilic attack. Nucleophilic attack only occurs under forcing conditions unless the aromatic ring already contains a powerful electron-withdrawing group, eg, NO_2. In this case, nucleophilic attack is greatly facilitated because of the reduced electron density at the ring carbon atoms.

Unit Processes. The unit processes encountered in intermediate and dye chemistry are summarized in Table 2.

Chemistry of Aromatic Heterocycles. In contrast to the benzenoid intermediates, it is unusual to find a heterocyclic intermediate that is synthesized via the parent heterocycle. They are synthesized from acyclic precursors.

The most important heterocycles are those with five- or six-membered rings; these rings may be fused to other rings, especially a benzene ring. Nitrogen, sulfur, and to a lesser extent oxygen, are the most frequently encountered heteroatoms. They are often considered

Table 2. Unit Processes in Dyes Manufacture

Process	Primaries[a]	Intermediates (common usage)	Colorants (common usage)
nitration	6	✓	
reduction	8	✓	
sulfonation	4	✓[b]	✓
oxidation	5	✓	
fusion/hydroxylation	3	✓	
amination	3	✓[c]	
alkylation	2	✓	✓
halogenation	2	✓	✓
hydrolysis	2	✓	
condensation	1	✓	✓
alkoxylation	1	✓	
esterification	1	✓	
carboxylation	1	✓	
acylation	1	✓	✓
phosgenation	1	✓	✓
diazotization	1	✓	✓
coupling (azo)	1	✓	✓

[a] Number of occurrences within 30 identified product manufactures.
[b] Includes chlorosulfonation.
[c] Includes the Bucherer reaction.

in two groups: those containing only nitrogen, such as pyrazolones, indoles, pyridones, and triazoles which, except for triazoles, are used as coupling components in azo dyes, and those containing sulfur (and also optionally nitrogen), such as thiazoles, thiophenes, and isothiazoles, that are used as diazo components in azo dyes. Triazines are treated separately since they are used as the reactive system in many reactive dyes.

Equipment and Manufacture

The basic types of dye (and intermediate) manufacture are shown in Figure 1. There are usually several reaction steps or unit processes.

The reactions for the production of intermediates and dyes are carried out in bomb-shaped reaction vessels made from cast iron, stainless steel, or steel lined with rubber, glass (enamel), brick, or carbon blocks. Wooden vats are also still used in some countries, eg, India. These vessels have capacities of 2–40 m^3 (ca 500–10,000 gal) and are equipped with mechanical agitators, thermometers or temperature recorders, condensers, pH-probes, etc, depending on the nature of the operation. Jackets or coils are used for heating and cooling by circulating through them high boiling fluids (eg, hot oil, or Dowtherm), steam, or hot water to raise the temperature, and air, cold water, or chilled brine to lower it. Unjacketed vessels are often used for aqueous reactions, where heating is affected by direct introduction of steam, and cooling by addition of ice or by heat exchangers. The reaction vessels normally span two or more floors in a plant to facilitate ease of operation (see REACTOR TECHNOLOGY).

Products are transferred from one piece of equipment to another by gravity flow, pumping, or by blowing with air or inert gas. Solid products are separated from liquids in centrifuges, on filter boxes, on continuous belt filters, and perhaps most frequently, in various designs of plate-and-frame or recessed plate filter presses. The presses are dressed with cloths of cotton, Dynel, polypropylene, etc. Some provide separate channels for efficient washing, others have membranes

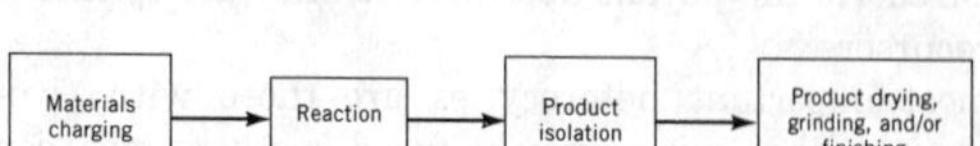

Figure 1. Operation sequence in dye and intermediate manufacture.

for increasing the solids content of the presscake by pneumatic or hydraulic squeezing.

The plates and frames are made of wood, cast iron, or now usually hard rubber, polyethylene, and polyester.

When possible, the intermediates are taken for the subsequent manufacture of other intermediates or dyes without drying because of savings in energy costs and handling losses. There are, however, many cases where products, usually in the form of pastes discharged from a filter, must be dried. Where drying is required, air or vacuum ovens (in which the product is spread on trays), rotary dryers, spray dryers, or less frequently, drum dryers (flakers) are used. Spray dryers have become increasingly important.

The final stage in dye manufacture is grinding or milling. Dry grinding is usually carried out in impact mills (Atritor, KEK, or ST); considerable amounts of dust are generated, and well-established methods are available to control this problem. Dry grinding is an inevitable consequence of oven drying, but more modern methods of drying, especially continuous drying, allow the production of materials that do not require a final comminution stage. Wet milling has become increasingly important for pigments and disperse dyes.

In the past the successful operation of batch processes depended mainly on the skill and accumulated experience of the operator. This operating experience was difficult to codify in a form that enabled full use to be made of it in developing new designs. The gradual evolution of better instrumentation, followed by the installation of sequence control systems, has enabled much more process data to be recorded, permitting maintenance of process variations within the minimum possible limits.

Full computerization of multiproduct batch plants is much more difficult than with single-product continuous units because the control parameters vary fundamentally with respect to time. The first computerized azo and intermediates plants were brought on stream by ICI Organics Division (now Zeneca Specialties) in the early 1970s, and have now been followed by many others. The additional cost (ca 10%) of computerization has been estimated to give a saving of 30 to 45% in labor costs. However, highly trained process operators and instrument engineers are required.

Economic Aspects

With the state of the world economy as it stands in 1993, the growth rate for the industry is likely to be as low as 2–3%, but even this low figure represents something like an additional 20,000 t/yr.

Health and Safety Aspects

Toxicology and Registration. The toxic nature of some dyes and intermediates has long been recognized. Acute, or short-term, effects are generally well known. They are controlled by keeping the concentration of the chemicals in the workplace atmosphere below prescribed limits and avoiding physical contact with the material. Chronic effects, on the other hand, frequently do not become apparent until after many years of exposure.

The positive links between benzidine derivatives and 2-naphthylamine with bladder cancer prompted the introduction of stringent government regulations to minimize such occurrences in the future. Currently, the three principal regulatory agencies worldwide are European Core Inventory (ECOIN) and European Inventory of Existing Commercial Substances (EINECS) in Europe, Toxic Substances Control Act (TOSCA) in the United States, and Ministry of Technology and Industry (MITI) in Japan. Each of these has its own set of data and testing protocols for registration of a new chemical substance.

Environmental Concerns. Dyes, because they are intensely colored, present special problems in effluent discharge; even a very small amount is noticeable. However, the effect is more aesthetically displeasing rather than hazardous. Of greater concern is the discharge of toxic heavy metals such as mercury and chromium.

Effluents from both dye works and dyehouses are treated both before leaving the plant, eg, neutralization of acidic and alkaline liquors and heavy metal removal, and in municipal sewage works. Various treatments are used.

Biological treatment is the most common and most widespread technique used in effluent treatment, having been employed for over 140 years. There are two types of treatment, aerobic and anaerobic.

Removal of color by adsorption using activated carbon is also employed. Activated carbon is very good at removing low levels of soluble chemicals, including dyes. Its main drawback is its limited capacity. Consequently, activated carbon is best for removing color from dilute effluent (see CARBON–ACTIVATED CARBON).

Chemical treatment of the effluent with a flocculating agent is the most robust and generally most efficient way to remove color.

Chemical oxidation is a more recent method of effluent treatment, especially chemical effluent. This procedure uses strong oxidizing agents like ozone, hydrogen peroxide, chlorine, and potassium permanganate in order to force degradation of even some of the more resilient organic molecules. As of this writing (ca 1993), these treatments remain very expensive and are of limited size, thought they may have some promise in the future.

Additional strategies being implemented to minimize dye and related chemical effluent include designing more environmentally friendly chemicals, more efficient (higher yielding) manufacturing processes, and more effective dyes, eg, reactive dyes having higher fixation.

PETER GREGORY
Zeneca Specialties

K. Venkataraman, *The Chemistry of Synthetic Dyes*, Vols. I–VIII, Academic Press, Inc., New York, 1952–1974.

Colour Index, 3rd ed., The Society of Dyers and Colorists, Bradford, U.K., 1971.

H. Zollinger, *Color Chemistry: Synthesis, Properties and Applications of Organic Dyes and Pigments*, 2nd ed., VCH, 1991.

P. F. Gordon and P. Gregory, *Organic Chemistry in Color*, Springer-Verlag, Berlin, 1983.

DYES, ANTHRAQUINONE

The synthesis of an anthraquinone dye generally involves a large number of steps. Highly toxic metals such as mercury or chromium(VI) are sometimes required. Some processes need to employ a large amount of organic solvent, and others involve a great quantity of waste acids. With the increasing demand for environmental protection, the regulation of pollutant effluents has become more stringent year after year, which has caused a sharp increase in the costs for wastewater treatment. This situation has led to intensive improvement of conventional methods and the development of new synthetic routes as well.

Efforts have also been made to overcome complicated processes. Methods to reduce the number of steps or to use new starting materials have been studied extensively.

Because of their small extinction coefficients anthraquinone dyes have less tinctorial strength than azo dyes; that is the intrinsic disadvantage of anthraquinones. This fact and the complexity of preparation have made their production costs higher than those of azo dyes (qv). However, the anthraquinone dyes have excellent properties that are not attainable by azo dyes, such as brilliancy of color, fastness, and excellent dyeing properties (leveling and dye bath stability). Thus the anthraquinone dyes have been widely used in the areas where these properties are required. Cotton or polyester–cotton blend fibers for military wear and working wear that require extreme fastness are dyed mainly with anthraquinone vat dyes. Most polyester fabrics for automobile seats are dyed with anthraquinone disperse dyes, since the requirement for lightfastness is extremely high and, simultaneously, bright shades are needed.

World dye manufacturers have already begun to develop new types of dyes that can replace the anthraquinones technically and economically. Some successful examples can be found in azo disperse red and blue dyes. In the reactive dye area intensive studies have continued to develop triphenodioxazine compounds to replace anthraquinone blues.

Color and Structure

The uv–vis spectrum of anthraquinone shows an absorption maximum at 323 nm ($\epsilon = 4500$) due to a π-π^* transition and very weak absorption in the visible range, 405 nm ($\epsilon = 60$) due to a n-π^* transition. Thus anthraquinone is almost colorless. Introduction of electron-donating substituents causes a bathochromic shift. This is due to the charge-transfer band from the lone pair of amino or hydroxyl groups to the oxygen atom of the carbonyl group. By increasing the electron-donating ability of substituents, the bathochromic shifts are enhanced (Table 1). In the case of the same substituent, the bathochromic shift is larger when the substituent is in the 1-position rather than in the 2-position. The introduction of an electron-withdrawing group has little effect on the absorption maximum of the spectrum.

The absorption maximum of a disubstituted anthraquinone greatly depends on the substituents and their positions. The 1,4-disubstituted compound shows a remarkable bathochromic shift. Larger bathochromic shifts are observed with increasing electron-withdrawing ability of β-substituents.

1,4,5,8-Tetrasubstituted anthraquinones give a slightly reddish blue tint to greenish blue color, depending on the substituents and their positions.

In addition to the color and the tinctorial strength, which are very important factors for the molecular design of anthraquinone dyes, affinity for fibers, various kinds of fastness (light, wet, sublimation, nitrogen oxides (NO_x) gas, washing, etc), and application properties (sensitivity for dyeing temperature, pH, etc) must be considered thoroughly as well.

Method of Synthesis

Anthraquinone dyes are derived from several key compounds called dye intermediates, and the methods for preparing these key intermediates can be divided into two types: *(1)* introduction of substituent(s) onto the anthraquinone nucleus, and *(2)* synthesis of an anthraquinone nucleus having the desired substituents, starting from benzene or naphthalene derivatives (nucleus synthesis). The principal reactions are nitration and sulfonation, which are very important in preparing α-substituted anthraquinones by electrophilic substitution. Nucleus synthesis is important for the production of β-substituted anthraquinones such as 2-methylanthraquinone and 2-chloroanthraquinone. Friedel-Crafts acylation using aluminum chloride is applied for this purpose. Synthesis of quinizarin (1,4-dihydroxyanthraquinone) is also important.

Key Intermediates

1-Aminoanthraquinone and Related Compounds. 1-Aminoanthraquinone is the most important intermediate for manufacturing acid,

Table 1. Spectral Data for Some Monosubstituted Anthraquinones[a] in Methanol

	1-position		2-position	
Substituent	λ_{max}, nm	ϵ	λ_{max}, nm	ϵ
Electron-donating groups				
OCH_3	378	5200	363	3950
OH	402	5500	368	3900
$NHCOCH_3$	400	5600	367	4200
NH_2	475	6300	440	4500
$NHCH_3$	503	7100	462	5700
$N(CH_3)_2$	503	4900	472	5900
Electron-withdrawing groups				
NO_2	325	4300	323	5200
Cl	333	5000	325	3900

[a] Unsubstituted anthraquinone λ_{max} = 323 nm; ϵ = 4500.

reactive, disperse, and vat dyes. It has been manufactured from anthraquinone-1-sulfonic acid by ammonolysis of the sulfo group with aqueous ammonia in the presence of an oxidizing agent such as nitrobenzene-3-sulfonic acid. In this process the starting material can only be obtained by mercury-catalyzed sulfonation of anthraquinone with oleum. For improved ecology, the alternative route based on 1-nitroanthraquinone was established. 1-Nitroanthraquinone is prepared from anthraquinone by nitration in sulfuric acid or organic solvent. 1-Aminoanthraquinone is prepared from 1-nitroanthraquinone by reduction with sodium hydrogen sulfide or by catalytic hydrogenation. Highly purified product is manufactured by continuous vacuum distillation.

1-Amino-4-bromoanthraquinone-2-sulfonic acid (bromamine acid) is the most important intermediate for manufacturing reactive and acid dyes. Bromamine acid is manufactured from 1-aminoanthraquinone-2-sulfonic acid by bromination in aqueous medium, or in concentrated sulfuric acid.

1-Amino-2-bromo-4-hydroxyanthraquinone is one of the most important intermediates for manufacturing red disperse dyes. It is prepared by dibrominating 1-aminoanthraquinone in concentrated sulfuric acid and subsequent hydrolysis in the presence of boric acid.

1-Amino-2-chloro-4-hydroxyanthraquinone is another important intermediate in red disperse dye manufacture. 1-Amino-2-chloro-4-hydroxyanthraquinone is prepared via a route from chlorobenzene and phthalic anhydride as the raw materials.

1,4-Dihydroxyanthraquinone. This anthraquinone, also known as quinizarin, is of great importance in manufacturing disperse, acid, and vat dyes. It is manufactured by condensation of phthalic anhydride with 4-chlorophenol in oleum in the presence of boric acid.

1,4-Diaminoanthraquinone and Related Compounds. Leuco-1,4-diaminoanthraquinone (leucamine) is an important precursor for 1,4-diaminoanthraquinone and is prepared by heating 1,4-dihydroxyantraquinone with sodium dithionite in aqueous ammonia under pressure. 1,4-Diaminoanthraquinone is an important intermediate for vat dyes and disperse dyes, and is prepared by oxidizing leuco-1,4-diaminoanthraquinone with nitrobenzene in the presence of piperidine.

1,4-Diamino-2,3-dichloroanthraquinone (CI Disperse Violet 28) is an important compound as an intermediate for CI Disperse Blue 60 and CI Disperse Violet 26, and is prepared by chlorination of leuco-1,4-diaminoanthraquinone with chlorine gas or sulfuryl chloride in an inert organic solvent such as nitrobenzene.

1,4-Diamino-2,3-dicyanoanthraquinone is the key intermediate for manufacturing CI Disperse Blue 60. 1,4-Diamino-2,3-dicyanoanthraquinone is manufactured by reaction of 1,4-diaminoanthraquinone-2,3-disulfonic acid with alkali metal cyanide.

1,4,-Diaminoanthraquinone-2,3-dicarboxyimide is the intermediate for CI Disperse Blue 60 and is prepared by hydrolysis of 1,4-diamino-2,3-dicyanoanthraquinone in concentrated sulfuric acid.

Anthraquinone-1-Sulfonic Acid and Its Derivatives. Anthraquinone-1-sulfonic acid has become less competitive than 1-nitroanthraquinone as the intermediate for 1-aminoanthraquinone. However, it still has a great importance as an intermediate for manufacturing vat dyes via 1-chloroanthraquinone.

Anthraquinone-1-sulfonic acid is prepared from anthraquinone by sulfonation with 20% oleum in the presence of mercury catalyst, a Hg(II) salt such as $HgSO_4$ or HgO, at 120°C.

1-Chloroanthraquinone is an intermediate for manufacturing vat dyes such as CI Vat Brown 1. 1-Chloroanthraquinone is prepared by chlorination of anthraquinone-1-sulfonic acid with sodium chlorate in hydrochloric acid at elevated temperature.

1-Methylaminoanthraquinone is an important intermediate for manufacturing solvent dyes and acid dyes, and is prepared from anthraquinone-1-sulfonic acid by replacing the SO_3H group with methylamine.

Anthraquinone-α,α'-Disulfonic Acids and Related Compounds. Anthraquinone-α,α'-disulfonic acids and their derivatives are important intermediates for manufacturing disperse blue dyes (via 1,5- or 1,8-dihydroxyanthraquinone, or 1,5-dichloroanthraquinone) and vat dyes (via 1,5-dichloroanthraquinone).

Anthraquinone-1,5-disulfonic acid and anthraquinone-1,8-disulfonic acid are produced from anthraquinone by disulfonation in oleum.

1,5-Dichloroanthraquinone is an important intermediate for vat dyes and disperse blue dyes. 1,5-Dichloroanthraquinone is prepared by the reaction of anthraquinone-1,5-disulfonic acid with $NaClO_3$ in hot hydrochloric acid solution.

1,5-Dihydroxyanthraquinone (anthrarufin) is an important intermediate for manufacturing disperse blue dyes and is prepared from anthraquinone-1,5-disulfonic acid by heating with an aqueous suspension of calcium oxide and magnesium chloride under pressure at 200–250°C.

α,α'-Dinitroanthraquinones and Related Compounds. 1,5- and 1,8-Dinitroanthraquinone are the key intermediates for manufacturing disperse blue dyes via dinitrodihydroxyanthraquinone and vat dyes via diaminoanthraquinones. 1,5-Dinitroanthraquinone and 1,8-dinitroanthraquinone are prepared by nitration of anthraquinone with nitric acid in sulfuric acid. α,β'-Dinitroanthraquinones are also formed in the reaction.

1,5-Diaminoanthraquinone is prepared from 1,5-dinitroanthraquinone by ammonolysis, by catalytic hydrogenation, or by reduction with sodium sulfide. It is also prepared from anthraquinone-1,5-disulfonic acid by ammonolysis. 1,5-Diaminoanthraquinone is an important intermediate for manufacturing vat dyes.

1,5-Dihydroxy-4, 8-dinitroanthraquinone is an important dye precursor for CI Disperse Blue 56, and is prepared from 1,5-diphenoxyanthraquinone by hexanitration in sulfuric acid and subsequent hydrolysis with aqueous alkali. 1,5-Dinitro-4,8-dihydroxyanthraquinone is also prepared from dimethoxyanthraquinone. High purity of 1,5-dimethoxyanthraquinone is required for manufacturing disperse blue dyes.

2-Methylanthraquinone and Related Compounds. 2-Methylanthraquinone and its derivatives are important as intermediates for manufacturing various kinds of vat dyes and brilliant blue (turquoise blue) disperse dyes. 2-Methylanthraquinone is prepared from phthalic anhydride and toluene via a benzoylbenzoic acid. 1-Nitroanthraquinone-2-carboxylic acid is of great importance as an intermediate for manufacture of vat dyes as well as disperse dyes. It is conventionally prepared from 2-methyl-1-nitroanthraquinone, by oxidation in sulfuric acid with sodium dichromate. 1-Aminoanthraquinone-2-carboxylic acid is also an important intermediate for vat dyes and disperse dyes and is prepared from 1-nitroanthraquinone-2-carboxylic acid by reaction with ammonia.

2-Chloroanthraquinone and Its Derivatives. 2-Chloroanthraquinone and its derivatives are the most important intermediates for vat dyes and high performance organic pigments. 2-Chloroanthraquinone is prepared by Friedel-Crafts reaction of chlorobenzene and phthalic anhydride in the presence of aluminum chloride, followed by ring closure in concentrated sulfuric acid. 2-Amino-3-hydroxyanthraquinone is prepared by heating 5-benzoylbenzoxazolone-2′-carboxylic acid in sulfuric acid. This compound is an intermediate for CI Vat Red 10.

Benzanthrone and Related Compounds. Benzanthrone is prepared by the reaction of anthraquinone with glycerol, sulfuric acid, and a reducing agent such as iron. Benzanthrone is an important intermediate for manufacturing vat dyes.

N-Methylanthrapyridone and Its Derivatives. 6-Bromo-3-methylanthrapyridone is an important intermediate for manufacturing dyes soluble in organic solvents. These solvent dyes are prepared by replacing the bromine atom with various kinds of aromatic amines. 6-Bromo-3-methylanthrapyridone is prepared from 1-methylamino-4-bromoanthraquinone by acetylation with acetic anhydride followed by ring closure in alkali. The starting material of this route is anthraquinone-1-sulfonic acid.

Reactive Dyes

Most of the anthraquinone reactive dyes are derived from bromamine acid. These dyes give a bright blue shade and excellent lightfastness.

A great number of reactive groups have been proposed; typical examples include sulfatoethylsulfone, dichlorotriazine, monochlorotriazine, monofluorotriazine, and other heterocyclic groups (see DYES, REACTIVE).

Disperse Dyes

Disperse dyes are water-insoluble, aqueous dispersed materials that are used for dyeing hydrophobic synthetic fibers, including polyester, acetate, and polyamide.

By introducing amino, hydroxy, or methyl groups onto the anthraquinone moiety as the principal auxochromes, dyes that have yellow through greenish blue shades are obtained. Among these dyes many that have brilliant red, violet, blue, and greenish blue shades have great industrial importance in view of their affinity for polyester or cellulose acetate fibers and lightfastness and sublimation resistance. On the contrary, yellow or orange dyes are not satisfactory because of the rather simple molecular structure. Therefore these shades are obtained from other chromophores.

On the basis of the kind and the position of their substituents and their color range, the anthraquinoid disperse dyes may be classified as follows:

Color range	Chemical description
red	1-amino-4-hydroxyanthraquinones
blue, greenish blue	1,1,4,5,8-substituted anthraquinones
greenish blue	1,4-diaminoanthraquinone-2,3-dicarboxyimides
violet, blue	1,4-diaminoanthraquinone derivatives
violet, blue	*N*-substituted 1-amino-4-hydroxyanthraquinones

Acid Dyes

Acid dyes are used for dyeing wool, synthetic polyamides, and silk in aqueous media. Anthraquinone acid dyes give brilliant reds, violets, blues, and greens and exhibit excellent lightfastness. Because of their relatively high cost, they are used to dye high grade textiles in pale and moderate shades. Various kinds of anthraquinone acid dyes have been developed so far mainly by IG-Farbenindustrie in Germany applying chemical reactions that were studied in developing vat dyes. However, the number of commercial products has declined because of poor properties or unavailable raw materials. Anthraquinone acid dyes may be classified into two groups: bromamine acid derivatives and quinizarin derivatives.

Vat Dyes

Anthraquinone vat dyes have been used to dye cotton and other cellulose fibers for many decades. Despite their high cost, relatively muted colors, and difficulty in application, anthraquinone vat dyes still form one of the most important dye classes of synthetic dyes because of their all-around superior fastness.

Anthraquinone vat dyes are water-insoluble dyes. They are converted to leuco compounds (anthrahydroquinones) by reducing agents such as sodium hydrosulfite in alkaline conditions. These water-soluble leuco compounds have an affinity to cellulose fibers and penetrate them. After reoxidation by means of air or other oxidizing agents, the dye becomes water-insoluble again and fixes firmly on the fiber.

The anthraquinone vat dyes can be classified into several groups on the basis of their chemical structures: *(1)* benzanthrone dyes, *(2)* indanthrones, *(3)* anthrimides, *(4)* anthrimidocarbazoles, *(5)* acylaminoanthraquinones, *(6)* anthraquinoneazoles, *(7)* anthraquinone acridones, *(8)* anthrapyrimidines, and *(9)* highly condensed ring systems. Recently, research and development efforts have focused on improved manufacturing of traditional vat dyes.

Mordant Dyes

Mordant dyes have hydroxy groups in their molecular structure that are capable of forming complexes with metals. Although a variety of metals such as iron, copper, aluminum, and cobalt have been used, chromium is most preferable as a mordant. Alizarin or CI Mordant Red 11 (CI 58000), the principal component of the natural dye obtained from madder root, is the most typical mordant dye. Many mordant dyes have given way to the vat or the azoic dyes, which are applied by much simpler dyeing procedures.

Acid–mordant dyes have characteristics similar to those of acid dyes which have a relatively low molecular weight, anionic substituents, and an affinity to polyamide fibers and mordant dyes. In general, brilliant shades cannot be obtained by acid–mordant dyes because they are used as their chromium mordant by treatment with dichromate in the course of the dyeing procedure. However, because of their excellent fastness for light and wet treatment, they are predominantly used to dye wool in heavy shades (navy blue, brown, and black). In terms of chemical constitution, most of the acid–mordant dyes are azo dyes; some are triphenylmethane dyes; and very few anthraquinone dyes are used in this area. CI Mordant Black 13 is one of the few examples of currently produced anthraquinone acid–mordant dyes.

Functional Dyes

The investigation of new dyes has always been focused on the development of fast, brilliant, inexpensive, and easy applicable dyes. Because a great emphasis has been placed especially on fastness, the dyes with poorer fastness have been ignored in the past. However, in recent years new needs for dyes that change color in response to low energy stimuli including light, electricity, or heat have arisen in the electronics industry. This new application includes information recording, information display, and energy conversion. The term *functional dye* has been applied to dyes that are used in advanced technologies based on optoelectronics since 1981 when the book entitled *The Chemistry of Functional Dyes* was published in Japan.

In order to develop the dyes for these fields, characteristics of known dyes have been re-examined, and some anthraquinone dyes have been found usable. One example of use is in thermal-transfer recording where the sublimation properties of disperse dyes are applied. Anthraquinone compounds have also been found to be useful dichroic dyes for guest-host liquid crystal displays when the substituents are properly selected to have high order parameters. These dichroic dyes can be used for polarizer films of LCD systems as well. Anthraquinone derivatives that absorb in the near-infrared region have also been discovered, which may be applicable in semiconductor laser recording.

Economic Aspects

Production Capacity and Demand. The production capacity for each dye or dye intermediate has rarely been announced officially by the individual manufacturers. Principal manufacturers of anthraquinone dyes and their intermediates are as follows: Bayer, BASF, and Hoechst (Germany); Hoechst Mitsubishi Kasei, Sumitomo Chemical, Mitsui Badische Dyes, and Nippon Kayaku (Japan); Ciba-Geigy and Sandoz (Switzerland); Zeneca and Holliday (U.K.); Crompton & Knowles (U.S.); and IDI (India).

Recent increases in environmental costs have become a serious problem, and future prospects for the anthraquinone dye industry are not optimistic. Some traditional manufacturers have stopped the production of a certain dye class or dye intermediates that were especially burdened by environmental costs, eg, vat dyes and their intermediates derived from anthraquinone-1-sulfonic acid and 1,5-disulfonic acid. However, several manufacturers have succeeded in process improvement and continue production, even expanding their capacity. In the forthcoming century the worldwide framework of production will change drastically.

Consumption

Anthraquinone dyes are the most important dye class after azo dyes. The consumption of each dye class or set of classes is approximately parallel to the consumption of fibers to which they are applied.

Among these dye classes, anthraquinone dyes are in an important position in reactive dyes and vat dyes for cellulose fibers, disperse dyes for polyester, and acid dyes for polyamide. Applications for high performance organic pigments for plastics and paints are also important areas.

Health and Safety Information

In general, anthraquinone dyes and their intermediates have not been reported as strongly toxic substances, but for many compounds safety data have not been evaluated. 1-Nitroanthraquinone, 1-chloroanthraquinone, and benzanthrone are reported to cause mild skin irritation in a test with rabbits, 500 mg/24 h. Some eye irritation data have been reported.

There are some tumorigenic data for anthraquinone dyes and intermediates which have been evaluated thoroughly. Data for 2-aminoanthraquinone and 2-methyl-1-nitroanthraquinone are available. 2-Aminoanthraquinone has been assessed by the United Nations International Agency for Research on Cancer (IARC) from studies in animals, and is judged to fall into the *Animal: Limited Evidence* group. 2-Aminoanthraquinone has been evaluated by EPA (Genetic Toxicology program) and a positive carcinogenic effect for rat and mouse is designated. 2-Methyl-1-nitroanthraquinone has been assessed by IARC and judged as belonging in the *Animal: Sufficient Evidence* group. 2-Methyl-1-nitroanthraquinone has been evaluated by the National Cancer Institute (NCI) and clear evidence of carcinogenicity for rat and mouse is demonstrated.

Most anthraquinone dyes and their intermediates are handled in a powder form. Their dust poses the threat of contact to eyes and skin or contamination of surroundings. Attention must be paid to avoid these hazards. Special attention should be paid to avoid contact with compounds that are recognized to have probable carcinogenicity.

In the case of handling in relatively small quantities, ie, for laboratory use, normal personal equipment, ie, dust masks, safety glasses, and gloves, and hoods with local exhaust ventilation should be used. In plant operations, special technical handling measures should be taken because the possibility of contact is extremely high, especially when charging the raw materials or isolating or packaging the intermediates or final products (see INDUSTRIAL HYGIENE; PLANT SAFETY).

MAKOTO HATTORI
Sumitomo Chemical Company

K. Venkataraman, *The Chemistry of Synthetic Dyes,* Vol. 2, Academic Press, Inc., New York, 1952.

F. B. Stilmar and co-workers in H. A. Lubs, ed., *The Chemistry of Synthetic Dyes and Pigments,* Reinhold Publishing Corp., New York, 1955, pp. 335–550.

H. R. Schweizer, *Künstriche Organische Farbstoffe und ihre Zwischenproducte,* Springer-Verlag, Berlin, 1964, pp. 301–320.

G. Hallas in J. Shore, ed., *Colorants and Auxiliaries,* Society of Dyers and Colourists, Bradford, U.K., 1990, pp. 230–267.

DYES, APPLICATION AND EVALUATION

The global consumption of textiles is estimated at around 30 million t, and this is expected to grow at 3% per year. The coloration of this amount needs ca 700,000 t of dye, with a value of $4400 million. The principal reasons for coloring textiles are for aesthetic appearance and decoration or for utilitarian purposes, and unless there is an unpredicted change in human behavior the majority of textiles will continue to be dyed to produce colored apparel, home furnishings, carpets, etc. Among the aesthetic uses are fashion garments and household articles such as drapes, towels, and carpets. In the utilitarian group are uniforms (military and civil), and work wear.

In order for a colored substance to be regarded as a dyestuff, a number of requirements must be satisfied. A dyestuff must be substantive for a textile and exhaust from an aqueous solution into the fiber; have a high exhaustion; exhaust at a rate allowing economic processing; give a uniform level dyeing; and have satisfactory fastness for the particular end use the textile is intended for. The process of dyeing is therefore a combination of chemistry, application technology, economics, and customer needs.

Classification of Dyestuffs According to Application

The *Colour Index* categorizes all coloring matters according to application characteristics. The following are the main types currently of interest as dyes.

Acid Dyes. These are anionic dyes, usually containing sulfonic acid groups, that are substantive to wool, other protein fibers, and polyamides when dyed from an acidic dyebath. The lower the pH the more rapid the dyeing, and exhaustion efficiency is enhanced by increased acidity.

Mordant Dyes. This group includes many natural as well as synthetic dyes. They have no or low substantivity for textile fibers and are therefore applied to cellulosic or protein fibers that have been treated (mordanted) with metallic oxides to give points of attraction for the dye. The dye forms a complex with the metal and depending on the metal and fiber can simply form a large macromolecule incapable of desorbing, or a dye molecule bound to the fiber resulting from chelation with the metal. An important subgroup is chrome dyes where wool is treated with Cr^{3+} with which it reacts; dye is then applied which in turn complexes with the chromium.

Metals such as chromium and cobalt can be introduced into dye molecules to give larger molecules. They can be regarded as being a special form of mordant dye. The complexes can be formed by chelating one or two molecules of dye with metal. They are applied in a similar manner to acid dyes.

Direct Dyes. These are defined as anionic dyes, again containing sulfonic acid groups, with substantivity for cellulosic fibers. They are usually azo dyes (qv) and can be mono-, dis-, or polyazo, and are in general planar structures. They are applied to cellulosic fibers from neutral dyebaths, ie, they have direct substantivity without the need of other agents. Salt is used to enhance dyebath exhaustion. Some direct dyes can be applied to wool and polyamides under acidic conditions, but these are the exception.

Fiber-Reactive Dyes. These dyes can enter into chemical reaction with the fiber and form a covalent bond to become an integral part of the fiber polymer. They therefore have exceptional wetfastness. Their main use is on cellulosic fibers where they are applied neutral and then chemical reaction is initiated by the addition of alkali. Reaction with the cellulose can be by either nucleophilic substitution, using, for example, dyes containing activated halogen substituents, or by addition to the double bond in, for example, vinyl sulfone, $—SO_2CH{=}CH_2$, groups.

Basic Dyes. These are usually the salts of organic bases where the colored portion of the molecule is the cation. They are therefore sometimes referred to as cationic dyes. They are applied from mild acid, to induce solubility, and applied to fibers containing anionic groups. Their main outlet is for dyeing fibers based on polyacrylonitrile (see FIBERS–ACRYLIC).

Vat Dyes. The basic mechanism of vat dye application is the conversion of an insoluble complex polycylic molecule based on the quinone structure into a soluble leuco form by treatment with alkaline-reducing agents. This leuco form is then absorbed onto cellulose. Once the dye has been exhausted into the cellulose it is reconverted *in situ* to the insoluble pigment form which is trapped within the fiber. These dyes have high wet- and lightfastness. A subgroup of vat dyes is the solubilized vat dyes which are temporarily solubilized to allow easy application without reducing agents followed by regeneration of the insoluble dye after dyeing. These dyes are no longer of commercial importance.

Sulfur Dyes. These are complex molecules containing sulfur obtained from the reaction between selected organic intermediates such as 4-aminophenol, or *p*-phenylenediamine and molten sulfur or polysulfide. The actual structures of sulfur dyes are largely unknown although it is considered that they possess sulfur-containing heterocyclic rings. They are applied like vat dyes with the leuco form being generated by using a reducing agent such as sodium sulfide.

Disperse Dyes. These are substantially water-insoluble dyes applied from aqueous dyebath in a finely dispersed form. They are the most important class of dye for dyeing hydrophobic synthetic fibers such as polyester and acetates.

Ingrain Dyes/Azoic Dyes. These are dyes that are formed in the fiber by applying precursors. An example of this class are the azoic dyes. With these dyes a coupling component is applied to the fabric followed by a diazonium compound to form the insoluble dyes. Alternatively, a stabilized mix of the two can be applied and the insoluble azoic dye created on the fiber in a separate treatment, eg, acid steaming. This is a traditional method for obtaining bright heavy shades cheaply but has lost some popularity as a result of poor rubbing fastness and the decreasing availability of the amines and diazonium salts. Other dyes in this group are phthalocyanine compounds which still have commercial importance, particularly in textile printing.

Other Dyes. Other dye classes listed in the *Colour Index* include dyes for leather, solvents, paper, and food. Leather dyes are those acid, direct, mordant, and basic dyes that show substantivity for leather, good diffusion into it, and acceptable fastness. They are essentially applied in an analogous manner to acid or basic dyes. Paper is colored by both inorganic pigments and natural and synthetic organic colorants. The main dyes used are basic, acid, and direct dyes. Solvent dyes can be regarded as similar to disperse dyes. They are small, unsulfonated molecules, plus a few basic dyes that show high solubility in solvents. Finally food dyes, nontoxic colored substances that can be added to food, are included (see COLORANTS FOR FOOD, DRUGS, COSMETICS, AND MEDICAL DEVICES).

Fluorescent Whitening Agents. These are fluorescent substances that transform invisible ultraviolet light into visible blue light. Fluorescent whitening agents change the appearance of substrates in two ways: by emitting light and therefore increasing the luminosity (brightness); and by changing the shade from yellowish white to bluish white. They are unfavorably influenced by factors such as low uv content light and high uv light absorption of the substrate or other chemicals present on the substrate. Like dyes, fluorescent whitening agents are available in classes analogous to acid, basic, direct, and disperse dyes for application to all substrates, including paper.

Physical and Organic Chemistry of Dyes and the Dyeing Process

The practical characteristic of a dyestuff is that when a textile is immersed in a solution containing a dye, the dye preferentially adsorbs onto and diffuses into the textile. The thermodynamic equations defining this process have been reviewed in detail. The driving force for this adsorption process is the difference in chemical potential between the dye in the solution phase and the dye in the fiber phase. In practice it is only necessary to consider changes in chemical potential and to understand that the driving force is the reduction in free energy associated with the dye molecule moving from one phase to the other, as the molecule always moves to the state of lowest chemical potential.

Influence of the Fiber. In order for a dye to move from the aqueous dyebath to the fiber phase the combination of dye and fiber must be at a lower energy level than dye and water. This in turn implies that there is a more efficient, lower energy sharing of electrons or intramolecular energy forces, and there are a number of mechanisms that allow this to happen.

Fibers exist as natural, or synthetic, hydrophilic, hydrophobic, nonionic, and ionic. Natural fibers have complex chemical structures with a multitude of possible points of attraction for a dyestuff and are difficult to characterize because of the structure being strongly influenced by regional, climatic variations and the species of plant or animal. Dyeing of natural fibers is therefore much more complex than dyeing synthetic fibers where structures can be characterized and the availability of points of attraction can be deliberately engineered into the fiber's molecular chain. The various types of fiber are summarized in Table 1. The fiber type dictates the type of dye needed.

Modes of Attraction

The force of attraction between a dye and fiber results from the usual electronic interactions. They include ionic forces (coulombic attraction), ion-dipole forces, hydrogen bonds, charge-transfer forces, van der Waals forces, hydrophobic interaction, and covalent bonds.

Dyestuff Organic Chemistry

Dyestuffs impart color to textiles because of their ability to absorb electromagnetic radiation in the wavelengths visible to the human eye (400–650 nm). When white light strikes a dyestuff molecule certain wavelengths are absorbed, depending on the molecular construction, and others are reflected. The wavelengths of the reflected light give the specific color of the dyestuff.

Dyestuff organic chemistry is concerned with designing molecules that can selectively absorb visible electromagnetic radiation and have affinity for the specified fiber, and balancing these requirements to achieve optimum performance. To be colored the dyestuff molecule must contain unsaturated chromophore groups, such as azo, nitro, nitroso, carbonyl, etc. In addition, the molecule can contain auxochromes, groups that supplement the chromophore. Typical auxochromes are amino, substituted amino, hydroxyl, sulfonic, and carboxyl groups.

There is little correlation between classifications according to chemical type and application properties. Application classifications are of most practical usefulness to the dyer.

The Dyeing Process

The physical chemistry associated with dyeing has been described elsewhere both in detail and in summary. The purpose of this treatment is to outline those basic concepts that have a direct impact on dyeing in order to appreciate the fundamental processes taking place.

Table 1. Fiber–Dye Property Requirements

Fiber name[a]	Type/general classification	Chemical constitution	Ionic nature in dyebath
cotton, linen, and other vegetable fibers	natural, hydrophilic	cellulose	anionic
viscose rayon	synthetic,[b] hydrophilic	regenerated cellulose	anionic
wool, silk, hair	natural, hydrophilic	complex proteins	cationic
nylon	synthetic, somewhat hydrophobic	polyamide	usually cationic
acrylics	synthetic, hydrophobic	modified polyacrylonitriles	anionic
acetate	synthetic,[b] hydrophobic	acetylated cellulose	nonionic
triacetate	synthetic,[b] hydrophobic	acetylated cellulose	nonionic
polyester	synthetic, hydrophobic	polyester	usually nonionic
polypropylene	synthetic, hydrophobic	polyolefin	nonionic

[a] See FIBERS–SURVEY.
[b] Some references distinguish between synthetic fibers made from synthetic polymers and those made by modification of cellulose (man-made fibers).

Zeta Potential. When a textile is immersed in water a negative charge is developed on its surface. This is called the zeta potential. This happens even with ionic fibers in neutral dyebaths. Negatively charged dyes therefore are coulombically repelled.

Internal and External Phases. When dyeing hydrated fibers, for example, hydrophilic fibers in aqueous dyebaths, two distinct solvent phases exist, the external and the internal. The external solvent phase consists of the mobile molecules that are in the external dyebath so far away from the fiber that they are not influenced by it. The internal phase comprises the water that is within the fiber infrastructure in a bound or static state and is an integral part of the internal structure in terms of defining the physical chemistry and thermodynamics of the system. Thus dye molecules have different chemical potentials when in the internal solvent phase than when in the external phase. Further, the effects of hydrogen ions (H^+) or hydroxyl ions (OH^-) have a different impact. In the external phase acids or bases are completely dissociated and give an external or dyebath pH. In the internal phase these ions can interact with the fiber polymer chain and cause ionization of functional groups. This results in the pH of the internal phase being different from the external phase and the theoretical concept of internal pH.

Isotherms. When a fiber is immersed in a dyebath, dye moves from the external phase into the fiber. Initially the rate is quick but with time this slows and eventually an equilibrium is reached between the concentration of dye in the fiber and the concentration of dye in the dyebath. For a given initial dyebath concentration of a dye under given dyebath conditions, eg, temperature, pH, and conductivity, there is an equilibrium concentration of dye in fiber, D_f, and dye in the dyebath external solution, D_s. Three models describe this relationship: simple partition isotherm, Freundlich isotherm, and Langmuir isotherm.

With simple partition the situation is comparable to the partition of a solute between two solvents. The bonding forces involved between uncharged dye and uncharged fiber, and uncharged dye and uncharged solvent are considered to be the same. The dye is sometimes referred to as in solid solution in the fiber. The simple partition isotherm is found in practice with disperse dyes on cellulose acetate and polyester. It represents the dyeing situation with the minimum restrictions for the dye to enter the fiber; the only restriction is when the fiber solution becomes saturated.

The Freundlich isotherm, where the dye in fiber D_f is directly proportional to $(D_s)^x$ and a plot of log D_f against log D_s gives a straight line, is generally found with cellulosic and other ionic hydrophobic fibers.

In a Langmuir isotherm, the reciprocal of dye in fiber $1/DF_f$ is directly proportional to the reciprocal of dye in the dyebath $1/DF_s$. A plot of $1/D_f$ against $1/D_s$ therefore gives a straight line. Langmuir isotherms are typically found with ionic synthetic fibers and ionic dyes, eg, dyeing polyacrylonitrile with modified basic dyes, and on hydrophilic fibers in situations when the number of sites becomes very low. This may arise when the internal pH is such that only a small number of sites ionize.

Aggregation, Activity, and Solution. Theoretical treatments use the term activity which assumes that the dyestuff is present in a monomolecular dissociated state in solution. Dyes are not generally in this state except at very dilute concentrations; some molecular interaction is more likely. Thus practical situations are not fully described or characterized by theoretical treatments assuming monomolecularity, but the errors involved are usually of no practical consequence. However, when gross aggregation takes place there is significant interference as the dyestuff available for absorption is removed and previous theoretical considerations become invalid. Where aggregation takes place less dye is absorbed than predicted by theory. In general, aggregation is therefore to be avoided, except in exceptional situations where it is introduced to deliberately slow down the rate of dyeing. Dyebath conditions are usually adopted to minimize the potential for aggregation.

Rate of Diffusion. Diffusion is the process by which molecules are transported from one part of a system to another as a result of random molecular motion. This eventually leads to an equalization of chemical potential and concentration throughout the system, and in the case of dyeing an equilibrium between dye in the fiber and dye in the dyebath. In dyeing there are three stages to diffusion: diffusion of dye through the bulk solution of the dyebath to the fiber surface, diffusion through this surface, and diffusion of dye from the surface into the body of the fiber to allow for more dye to diffuse through the surface layer. These processes have been summarized elsewhere.

Level Dyeing. The concept of obtaining a level dyeing in reasonable time is fundamental in practical processes. Dyes that have a low affinity are likely to diffuse to the fiber surface slowly (a low rate of strike), and once in the fiber quickly diffuse through the fiber to give a uniform distribution of dye molecules. Because of the low affinity of these dyes there are no strong forces of attraction restricting movement. These dyes exhibit good migration and are likely to give level dyeings.

Level dyeings can be obtained by applying dyes of low affinity that dye quickly if unevenly, and allow them to migrate to give uniform level dyeings by extending the time of dyeing. This simple approach has the disadvantage that such low affinity dyes produce dyeings of relatively low wetfastness. Good wetfastness requires high affinity dyes. Because these exhibit low migration it is necessary to ensure that a level dyeing is obtained from the start and maintained throughout the dyeing process.

In order to ensure level dyeings from high affinity dyes it is necessary to make them behave as if they were of low affinity. This is done by reducing the difference between their chemical potential in solution and fiber. The techniques used include: higher temperature for more diffusion; change in pH to control the number of sites available; addition of an electrolyte to either compete with the dye for sites, or neutralize the sites preventing ionic attraction; and addition of auxiliary agents that either compete with the dye for the fiber by lowering the chemical potential in the dyebath phase thus making it a more attractive environment for the dye, or by removing dye from the equilibrium by forming a temporary complex or aggregate in the dyebath. In essence, in order to obtain level dyeings with higher affinity dyes the objective is to slow down the rate of dyeing.

Compatibility. To produce a desired shade more than one dye is usually needed. Often combinations of three dyes are used, eg, yellow, red, and blue, in order to obtain the maximum number of shades available from the minimum number of dyestuffs. In order to give uniform coloration it is necessary to apply such mixtures of dyes from the same dyebath. As the dyeing proceeds the textile takes up more of the dyes. If the hue of the fabric is the same throughout all stages of dyeings, simply becoming stronger with time, and the hue of the final textile is reproducibly uniform both on its surface and within its interior, eg, for a wound package of yarn, then the dyes are said to be compatible under the dyeing conditions used. Compatibility is mainly a function of exhaustion rate, but can also be influenced by migration.

Dyeing of Cellulosic Fibers

Preparation for Dyeing. Cotton fibers are coated with natural waxes and pectins. These can be removed by aqueous alkalies at 80°C or above or by solvent treatment to improve the absorbancy and dyeability of the fibers. Cotton may be made suitable for dyeing in a variety of forms, such as raw stock, yarn, or piece goods. Raw stock is normally dyed without thorough dewaxing, since the natural waxes aid in subsequent spinning operations. Careful preparation of cotton piece goods is essential to achieve suitable dye penetration, fastness, and general appearance. Fabric construction dictates whether the fabrics will be processed in rope or open-width forms. Heavy piece goods, and those which are subject to rubs and crease marks, are handled in open width.

Before dyeing in light or bright shades, the goods should be bleached with hydrogen peroxide and caustic soda to bleach the motes. This operation also helps in the removal of trace impurities that remain after boil-off.

Mercerizing is accomplished by passing the cotton fabric through 15–30% caustic soda under tension. Improved luster and increased dye affinity result. With knitted fabrics it is necessary to remove the knitting oils by either alkali treatment or solvents. Viscose rayon, because of its low wet strength, must be processed under minimum tension at all stages of preparation. Skeins contain few impurities and require only light scouring.

The Dyeing Process. When cotton fiber is immersed in water it develops a negative charge. In order for dyes to show good buildup on cotton the dyes must be soluble, planar, aromatic structures. Solubility is obtained by incorporating negatively charged sulfonic acid groups, ie, the anions. Therefore the dyes show long-range, coulombic (ionic) repulsion, but very strong short-range van der Waals forces of attraction. Thus there is a potential barrier that the dyestuff molecule has to overcome. Natural and introduced thermal agitation tend to equalize the distribution of ions.

The use of salt or similar electrolyte is critical in the dyeing of cellulose. When sodium chloride is added to the dyebath, sodium ions (Na^+) diffuse to the negative charges on the cellulose and neutralize them. With the introduction of the sodium ions (Na^+) the coulombic repulsion force between fiber and negatively charged dye is removed and only the strong attraction forces exist. It is then possible for negatively charged dye to diffuse unhindered to the fiber surface.

Direct Dyes

The simplest way of coloring cellulosic fibers is with direct dyes. The dyeing mechanism follows exactly the outline just described where the addition of salt is used to allow dyestuff to be absorbed on the fiber. This is done carefully to ensure that level dyeing is achieved, especially during the early stages of dyeing.

Leveling Power. Direct dyes are classified according to their leveling characteristics. Class A direct dyes migrate well and have high leveling power, ie, they have low affinity and high diffusion. Class B direct dyes have poor leveling power and exhaustion must be brought about by controlled salt addition. Class C direct dyes are dyes of poor leveling power which exhaust well in the absence of salt and the only way of controlling the rate of exhaustion is by temperature control.

In all these application methods the same procedure is adopted. At the beginning the rate of dyeing must be as slow as needed to give levelness; once this has been achieved the rate of dyeing is systematically increased to give complete exhaustion.

Wetfastness. Class A direct dyes offer the most trouble-free process for dyeing cellulose. However, they do not always provide sufficient wetfastness.

The Class B and C dyes show better resistance to desorption, ie, they show higher wetfastness, but they do not overcome it fully, and even the Class C direct dyes show inadequate wetfastness and poor staining of adjacents in fastness tests as a result of the reversible nature of the dyeing process. Attempts to overcome this problem have concentrated on chemical treatment of the direct dyes after they were applied. These treatments essentially make the direct dye molecules already on the fiber much bigger and thereby increase the nonionic forces of attraction or reduce solubility in order to reduce desorption and give good wetfastness. Methods used include applying dyes containing free amino groups, diazotizing them and coupling with a base; aftertreatment with formaldehyde; applying dyes containing hydroxyl groups ortho to the azo group and then aftertreating with metallic salts, eg, Cu; and applying 1–4% of a cationic surface-active agent (over 15–30 min at 25–60°C) to form a sparingly soluble complex with the dye. Today only the latter two processes are used.

Fiber-Reactive Dyes

Because of the limitations of direct dyes and the ability to use simple acid dye chromophores to give bright washfast dyeings, fiber-reactive dyes have become a well-established, popular way of dyeing cellulose. The growth rate of reactive dye consumption of 3.9% per annum is four times the growth rate of other dyes for cellulosic fibers.

A reactive dye for cellulose contains a chemical group that reacts with ionized hydroxyl ions in the cellulose to form a covalent bond. When alkali is added to a dyebath containing cellulose and a reactive dye, ionization of cellulose and the reaction between dye and fiber is initiated. As this destroys the equilibrium more dye is then absorbed by the fiber in order to re-establish the equilibrium between active dye in the dyebath and fiber phases. At the same time the addition of extra cations, eg, Na^+ from using Na_2CO_3 as alkali, has the same effect as adding extra salt to a direct dye. Thus the addition of alkali produces a secondary exhaustion.

At the end of the dyeing process there is fixed dye on the fiber, and hydrolyzed dye in both the dyebath and fiber. All active dye disappears. In order to take advantage of the fastness offered by covalently bonding the dye to the cellulose it is necessary to remove all the hydrolyzed dye from the fiber. Unfortunately, the hydrolyzed dye does exhibit some affinity for the cellulose and removing it is a desorption process rather than a rapid physical removal. Thus the removal is a relatively difficult procedure that is a critical part of the total dyeing process. Once a dye is fixed it cannot migrate and therefore dyeings achieved using fiber-reactive dyes must be level before they are fixed.

The Ideal Fiber-Reactive Dye Profile. Figure 1 shows the general profile for the application of a reactive dye. In addition to showing the rate profile of fixation between dye and fiber, three other practical parameters (A–C) are noted.

The overall objective is to make the fixation (C) as high as possible for economic reasons. The closer the fixation (C) is to the total dye on the fiber (B) then the smaller the concentration of [dye–OH] and the less that needs to be removed in the washing-off process after dyeing.

Application Methods. There are many detailed application methods used for applying reactive dyes, and all have been described in detail. Examples of the main methods include cold exhaust dyeing fiber-reactive dyes, warm, hot exhaust dyeing dyes, migration exhaust technique for less than 0.5% depth of shade, all-in method, continuous dyeing, and cold pad-batch dyeing.

Chemical Types. A wide range of reactive groups have been investigated, with 20–30 used commercially and over 200 patented. These have been described in detail elsewhere. Because these reactive groups differ chemically the activation of the reactive systems is different as are the rates of reaction with cellulose, from one reactive system to another. This rate of reaction with cellulose, or reactivity, dictates the temperature and pH needed for dyeing.

The most important reactive groups are those based on halotriazine or halopyrimidine systems, where an activated halogen substituent undergoes a nucleophilic substitution reaction with ionized cellulose, or dyes based on sulfatoethylsulfonyl groups.

Bifunctional fiber-reactive dyes have been developed. The concept behind bifunctional dyes is that if two distinct reactive groups are used, the probability of obtaining dye covalently bonded to the cellulose at the end of dyeing instead of being hydrolyzed is increased because each molecule must react twice with OH to be fully hydrolyzed. The claimed benefits of bifunctional reactive dyes are a generally

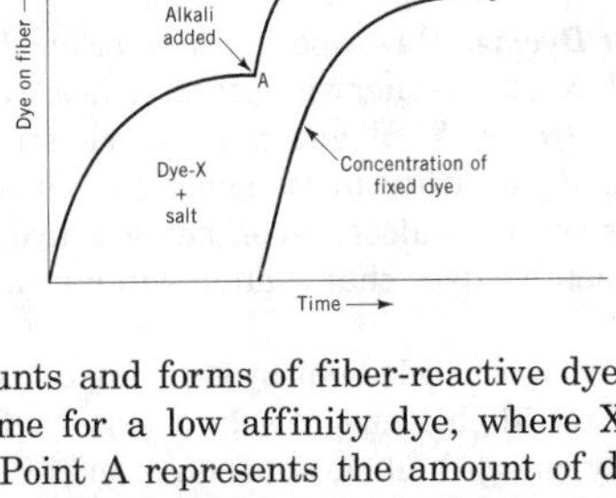

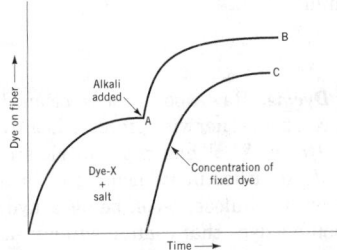

Figure 1. Amounts and forms of fiber-reactive dye on the fiber as a function of time for a low affinity dye, where X represents the reactive group. Point A represents the amount of dye exhausted in neutral conditions; B is the total amount of dye exhausted at the end of the dyeing process, ie, [dye-OH] + [dye-X] + [dye-O-cell]; and C is the amount of dye fixed [dye-O-cell].

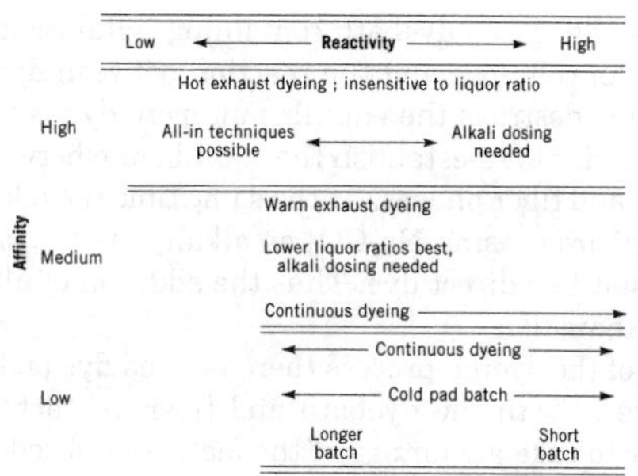

Figure 2. Summary of dyeing techniques related to dye reactivity and affinity characteristics.

higher level of fixation, and in the case of mixed reactive groups, suitability for application over a range of temperatures and methods.

Correlation of Application, Affinity, and Reactivity. Figure 2 correlates fiber-reactive dye application suitability to reactivity and affinity.

Vat Dyes on Cellulose

Most are based on the quinone structure and are solubilized by reduction with alkaline reducing agents such as sodium dithionite. Conversion back to the insoluble pigment is achieved by oxidation. The dyes are applied by either exhaust or continuous dyeing techniques. In both cases the process is comprised of five stages: preparation of the dispersion, reduction, dye exhaustion, oxidation, and soaping.

Uses of Vat Dyes. The main characteristic of vat dyes is their excellent fastness to light, water, and other agents, eg, chlorine. Vat dyes are therefore widely used in outlets demanding high lightfastness such as outerwear, furnishings, drapes, etc; high wetfastness and fastness to repeated washing such as workwear; high chlorine fastness such as institutional articles; or where general excellent fastness is required as in the case of sewing threads where it is impossible to know the use of final garment they will be used to construct. The majority of vat dyes used worldwide are applied by continuous dyeing; polyester–cotton blends are the most important substrate.

Sulfur Dyes. These are a special case of vat dyes and behave in an analogous manner except that the reducing agent used is sodium sulfide. In order to obtain rapid oxidation chemical oxidizing agents are used. The main outlet for these dyes is in the economic production of navy and black shades on woven fabrics by continuous dyeing, often applying the pre-reduced form of the sulfur dye.

Indigo. Indigo is similar to vat dyes in its application, however, it is not based on a quinone structure but on indigotin. In the presence of alkaline-reducing agents the C=O group is reduced to CH—OH and the dye rendered soluble. As with vat dyes the reaction is reversible, oxidation being achieved with atmospheric oxygen. The principal use for indigo is in denim.

Dyeing of Wool

Preparation for Dyeing. Raw wool must be cleaned before it can be efficiently carded, combed, otherwise processed, or dyed.

Mechanism of Dyeing. Wool has a polymeric structure based on amino acids. It is dyed either in its neutral or in its net positively charged form. As with cellulose, wool, being a hydrophilic fiber, is dyed with water-soluble dyes that contain sulfonic acid groups to impart solubility.

The dyeing of wool is carried out by applying a negatively charged dye to a neutral (or slightly negatively charged substrate) or to a strongly positively charged fiber, depending on pH. Strong ionic attraction exists that results in high affinity and rapid uptake of dye, so rapid that it is essential to control this rate of uptake if level dyeing is to be achieved. The wool dyeing processes are therefore designed around methods of obtaining level dyeings under practical application conditions.

Acid Dyes

Classes. There are three classes of acid dyes: acid leveling, acid milling, and super milling. Acid leveling dyes are molecular dispersions at low temperatures (true solutions) and are simple molecules. Acid milling dyes are colloidal dispersions at low temperatures and true solutions at high temperatures. Super milling dyes are colloidal dispersions at both low and high temperatures and are complex molecules containing low alkyl hydrophobic chains to enhance fastness.

Controlling Dyeing Behavior. As with all other dyes the dyeing process concentrates on obtaining level dyeings within an economic time period, and once again slower dyeing means better control and level dyeing is enhanced. When dyeing wool with acid dyes four factors control the dyeing behavior: pH of the dyebath, presence and concentration of electrolyte, temperature of the dyebath, and choice of dyestuff class.

Practical Processes. With acid leveling dyes no real problems exist because the dyes show good migration, electrolyte is added from the beginning, and rather like Class A direct dyes on cotton, level dyeing is achieved by prolonging the times at the boil.

The other extreme is found with super milling dyes when at the start ammonium acetate, sulfate, or an organic ester is present without any electrolyte. Dyeing is carried out more slowly taking some 60 min to reach the boil, and often the dye is applied with a cationic leveling agent.

Acid milling dyes are intermediate in behavior, being applied with acetic or formic acid in the presence of sodium sulfate. A disadvantage of acid dyes is that their wetfastness depends on the formation and maintenance of a salt linkage between the charged wool and dye. This requires an acidic internal pH to be maintained in the wool.

Mordant Dyes/Metal Complex Dyes

Certain acid dyes can have their fastness properties improved by combining the dye with a metal atom (chelation). The most common metal is chromium, although cobalt is sometimes used, and this can be introduced in a number of ways. The basic mechanism is donation of electron pairs by groups in the dye (ligands) to a metal ion.

Methods of Introducing Metal. Methods of introducing metal include chroming and dyeing together, afterchroming, and metal introduced into the dyestuff molecule in manufacture.

Dyeing Wool with Fiber-Reactive Dyes

Fiber-reactive dyes are by no means as popular for dyeing wool as they are for cotton because the fastness of fiber-reactive dyes on wool is not that much better than other dyes. They are difficult to apply and the severe washing treatments needed to remove unfixed dye can damage the wool itself. They are used on specially treated wools that have been made suitable for washing in automatic washing machines by treating with a polymer. The dye reacts with cations in the polymer. These dyes are not very commercially important, as the need for bright high fast shades on wool is not as high as on cotton. The dyeing mechanism has been described in detail elsewhere.

Silk

Because it is also a protein, silk can be dyed as wool, but in practice the dyes used are generally acid dyes in view of the fiber not being treated to any severe washing in its life. The main difference between wool and silk is in the preparation of the fiber for dyeing.

Silk in its raw state is coated with sericin. It is necessary to remove this gum in order to develop the silk luster and dyeability. Synthetic detergent systems, such as higher alcohol sulfates, and soda ash and boric acid have replaced soap to a large extent for degumming.

Dyeing of Synthetic Polyamides

Dyeing Mechanism. Nylon is similar in its general chemical structure to the natural fiber wool, and therefore all the previously described processes for wool are applicable to dyeing nylon with acid, metallized, and other dyes. There are, however, significant differences. Nylon is synthetic, it has defined chemical structure depending on the manufacturing process, and it is hydrophobic.

Chemically there are important differences. There are no side chains and unlike wool the number of amino and carboxylic groups differs; there is an excess of carboxylic groups. The numbers of amino groups can be changed by chemical modification, eg, in deep dyeing nylon, but for the most part nylon fibers can be considered to have a limited number of sites, which can differ from one chemical type to another.

Physically there are differences. Like all polymeric fibers nylon contains crystalline and noncrystalline areas. Only amino groups in the noncrystalline regions are accessible.

Finally, because the fiber is synthetic, polymer formation followed by drawing into a yarn presents the likelihood of chemical and physical variations in the yarn. It is usual to stabilize the fibers by a heat-setting process before dyeing, and again further physical variation can be introduced at this stage. The manufacturing history of the polymer therefore plays a role in determining the dyeing performance. In order to obtain level dyeings it is necessary to consider not only the basic chemical reactions taking place but also the relative sensitivities of the dyes to physical and chemical variation in the fiber.

Acid Dyes. The majority of acid dyes is applied to nylon rather than to wool. There are three groups of dyes: Group 1 includes dyes with little affinity at neutral or acidic pH but which exhaust under strongly acidic conditions; Group 2, the largest group of dyes which exhaust onto nylon in the pH range 3.0–5.0; and Group 3, dyes with a high affinity for nylon under neutral or weakly acidic pH. Only dyes within one group should be used together, and dyestuff manufacturers assist in this by having different nomenclature for dyes in each group.

Tanning Agents. It is possible to improve the wetfastness of acid dyes by aftertreatment. The original method was to apply tannic acid, tartar emetic, and formic acid.

Metal Complex Dyes. The 1:2 metal–dye complexes are of commercial interest because of their excellent lightfastness in pale shades. These macromolecules are difficult to apply level and are sensitive to both chemical and physical variations. In their application they are treated as the Group 3 acid dyes.

Other Soluble Hydrophilic Dyes. Some direct dyes have profiles on nylon very similar to Group 3 dyes and therefore, to supplement the range of shades available, they are sometimes applied with Group 3 dyes.

Disperse Dyes. The insoluble, hydrophobic disperse dyes readily dye nylon, and because their mode of attraction is completely nonionic they are completely insensitive to chemical variations and pH. Small molecular-sized disperse dyes (ca mol wt 400) show very high rates of diffusion and excellent migration properties and they are insensitive to physical variations in the nylon. As the molecular size of disperse dyes increases they show increasing sensitivity to physical variation.

Although when using disperse dyes on nylon they are readily absorbed at temperatures up to the boil, they are also readily desorbed when the dyed fabric is immersed in wash liquors.

The main use for disperse dyes is where excellent coverage of fibers likely to have physical and chemical variations is needed, and where wetfastness is not critical. The small molecular-weight dyes are therefore widely used for pale shades on continuous filament yarns used in hosiery. There is also some use made in exhaust dyeing of carpets made from continuous bulk filament nylon to give good coverage.

Dyeing is relatively simple. The disperse dye is added to a dyebath containing a nonionic dispersing agent, sodium hexametaphosphate, and sometimes acetic acid is added to give pH 5.5 to prevent decomposition of some disperse dyes. Dyeing is carried out by bringing the dyebath to the boil, and continuing until exhaustion is completed.

Dyeing of Acrylic Fibers

In order to make fibers of commercial interest acrylonitrile is copolymerized with other monomers such as methacrylic acid, methyl methacrylate, vinyl compounds, etc, to improve mechanical, structural, and dyeing properties. Fibers based on at least 85% of acrylonitrile monomer are termed acrylic fibers; those containing between 35–85% acrylonitrile monomer, modacrylic fibers. The two types are in general dyed the same, although the type and number of dye sites generated by the fiber manufacturing process have an influence (see FIBERS, ACRYLIC).

Basic dyes are the most popular class applied to acrylic fibers. Like nylon, acrylic can be dyed with disperse dyes, but with the same reservations of fastness. Disperse dyes are therefore only used for pale shades where excellent levelness is needed or difficult to obtain by any other method owing to variations in the fiber.

Preparation for Dyeing. Fabrics are scoured with a synthetic detergent at 45–65°C and are rinsed before further processing to remove tints, size, wax, grease, spinning oils, or other impurities that were applied or picked up during the manufacturing operation. Bleaching, when required, is usually accomplished by means of a sodium chlorite bleach, a selected optical brightener, or a suitable combination of the two. Acrylic-blend fabrics may require other bleaching agents if chlorine-sensitive fibers are present. Most acrylic fibers require a presetting in open-width in boiling water to avoid dimensional stability problems during subsequent wet-processing steps.

Dyeing Mechanism. As the importance of acrylic fibers grew, basic dyes were developed having localized charge in one specific part of the molecule, allowing stronger salt links to be formed than with the delocalized type. These newer dyes are often referred to as modified basic dyes. Essentially, their structure is that of a disperse dye that has been protonated. These dyes therefore have high rates of diffusion into the fiber, and their mode of attraction is almost entirely ionic.

The effect of pH depends on the fiber type. The SO_3^- groups on fibers are so strong that they are deprotonated even in neutral dyebaths. The dyebath pH therefore has no influence on the availability of these sites in the fiber and therefore pH cannot be used to control the uptake and level dyeing behavior of the dye. For carboxylic acid groups the pK_a is about 5.5. At lower pH values there are considerably fewer sites available for the dye and at higher pH values considerably more COO^-.

As with wool and nylon when applying dyes to the fiber where dye and site are oppositely charged the need is to control the rate of exhaustion to promote level dyeing. With acrylic this need is made all the more important by two additional factors: first, the modified basic dyes show poor migration because they form very strong salt bonds especially between dyes with delocalized charges and fibers with strongly negative SO_3^- sites; secondly acrylic fibers do not readily dye below their glass-transition temperature, T_g, which is usually around 80°C.

Compatibility Values. The need to apply dyes in admixture to give more shades necessitates a way of measuring the compatibility of dyes. Depending on charge, the degree of localization, and molecular shape and size, dyes have different affinities and behavior and hence different dyeing rates. A qualitative testing procedure has been defined where dyes are applied with a range of known dyes and the unknown dye is ascribed the same compatibility value as that already given to the known dye with which it dyes compatibly under all practical exhaust-dyeing conditions except in the presence of anionic dyes or auxiliaries. There are five values and in combinations the dye with the lowest value exhausts most rapidly. For best results dyes should be mixed with dyes having the same compatibility value, or at least no more than one value different.

Level Dyeing Techniques. It is exceptionally difficult to obtain level dyeings on acrylic, and temperature and pH control depend on fiber type and are not always adequate. Sodium sulfate in limited amounts can be used to some effect.

The more popular method to control leveling is to use cationic products that act as colorless dyes competing with the colored cationic dyes for the fiber sites. If amounts of colored modified basic dye and colorless modified basic dye equal to the saturation value of the fiber are uniformly dissolved in the dyebath, then level dyeing behavior is promoted.

Dyeing of Polyester

Polyester fibers are based on poly(ethylene terephthalate) (PET); some modified versions are formed by copolymerization, eg, basic dyeable polyester. The modified forms dye in analogous manner to other fibers of similar charge.

Preparation for Dyeing. A hot alkaline scour with a synthetic surfactant and with 1% soda ash or caustic soda is used to remove size, lubricants, and oils. Sodium hypochlorite is sometimes included in the alkaline scouring bath when bleaching is required. After bleaching, the polyester fabric is given a bisulfite rinse and, when required, a further scouring in a formulated oxalic acid bath to remove rust stains and mill dirt which is resistant to alkaline scouring.

Dyeing Mechanism. Unmodified polyester fibers are very hydrophobic and absorb only minimal amounts of water, and are therefore only dyeable with hydrophobic disperse dyes. The mechanism of dyeing is by simple partition, the so-called solid solution mechanism. Disperse dyes are only sparingly soluble and therefore high temperatures are needed to increase the amount of soluble dye in the system. Disperse dyes on polyester generally have good fastness as a result of the fiber being below its glass-transition temperature in wash treatments and the slow rate of desorption. The degree of crystallinity, the drawing, and heat-setting temperatures of polyester all play a role in determining the rate and amount of dye uptake.

Disperse Dyes. There is a general correlation between heat fastness, the propensity to desorb under conditions of dry heat onto a white piece of polyester, and the dyeing properties of disperse dyes. Low energy dyes are not usually used in thermofixation, as their low heat fastness at the thermofixation temperatures used (200–210°C) results in the subliming of them from the hot fabric.

Medium energy dyes are based on larger sized molecules than the low energy dyes. They have slower rates of dyeing, better heat fastness, and generally higher wetfastness. They are not suitable for carrier dyeing. Their main application methods are exhaust dyeing at temperatures of 125–135°C, and for continuous dyeing by thermofixation at around 30–60 s at 190–210°C. Because of their medium molecular size these dyes dye rapidly (15–30 min) at 125°C.

High energy dyes are based on large molecules with polar groups. These dyes have excellent heat fastness resulting from extremely low rates of sublimation. Their main use is in dyeing fabrics that are to be given a subsequent high temperature heat treatment. Dyeing with these dyes requires either longer times or temperatures than with medium energy dyes to achieve full exhaustion, eg, 45–60 min at 125–135°C in exhaust dyeing.

Dyeing Processes. Polyester yarns and fabrics are usually dyed by exhaust techniques; continuous dyeing is largely used only for blends with cellulose. The basic dyeing process is relatively simple. The dyebath is set with disperse dye and dispersing agent (a nonionic or anionic surface-active agent) at pH 5.5 obtained with, for example, acetic acid. The temperature is slowly raised up to the dyeing temperature (125–135°C) and kept there to complete exhaustion and promote migration followed by cooling to below the boil for removal of the dyed material.

During the cooling process after dyeing, the solubility of the disperse dye remaining in the dyebath decreases rapidly and it can precipitate on the surface of the polyester fibers. If it is not removed, the resulting dyeing will exhibit both poor fastness to rubbing and poor wetfastness. This precipitated dye is removed by a combined chemical decomposition and stripping. The dye is destroyed by reduction using hot (70°C) caustic soda and sodium hydrosulfite, optionally in the presence of a detergent.

This process, based on strong reducing agents, can be avoided by the use of disperse dyes that are removed by aqueous alkali alone. Two types of dye are used: dyes containing diesters of carboxylic acid and dyes destroyed by mild alkali.

Thermal Migration. In any subsequent heat treatment of the polyester such as heat-setting to stabilize the fiber or fabric, or in the application of a finishing agent such as a softener or antistat, the polyester is again taken above its glass-transition temperature, and dyestuff molecules again have mobility within the fiber.

Some general observations are that low energy dyes thermally migrate more than medium or high energy dyes, presumably because of their tendency to sublime out of the fiber at high temperature; the behavior of medium and high energy dyes has no correlation to their heat fastness; the more polar the dye the greater the likelihood of thermal migration; the higher the temperature the greater the risk; and the presence of hydrophobic finishing agents increases the likelihood of thermal migration.

Dyeing of Cellulose Esters

Acetate fibers are dyed usually with disperse dyes specially synthesized for these fibers. They tend to have lower molecular size (low and medium energy dyes) and contain polar groups presumably to enhance the forces of attraction by hydrogen bonding with the numerous potential sites in the cellulose acetate polymer (see FIBERS–CELLULOSE ESTERS). Other dyes can be applied to acetates such as acid dyes with selected solvents, and azoic or ingrain dyes can be applied especially for black colorants. However their use is very limited.

Cellulose Diacetate. When preparing cellulose diacetate for dyeing, strong alkalies must be avoided in the scouring of acetate because the surface of the cellulose acetate would be saponified by such treatment.

Very small quantities of acetate staple are dyed; however, large quantities of acetate filament are found in satin, taffeta, and tricot fabrics. These are usually dyed open-width on a jig owing to their inclination to crease or crack easily.

Cellulose Triacetate. Cellulose acetate having 92% or more of the hydroxyl groups acetylated is referred to as triacetate. This fiber is characteristically more resistant to alkali than the usual acetate and may be scoured, generally, in open-width, with aqueous solutions of a synthetic surfactant and soda ash. Triacetate is a hydrophobic fiber, as compared to secondary acetate, and consequently does not dye rapidly. It is necessary to increase the rate of diffusion of the disperse dye into the fiber by increasing the dyeing temperature to 110–130°C or using a dye accelerant or carrier, or both.

Dyeing of Fiber Blends

Fiber blends combine the advantageous properties of two or more fibers into one fabric. They are available as blends of natural fibers, synthetic fibers, or natural fibers blended with synthetic. The differences in dyeability between the many fibers on the market open a wide field of multicolored yarns and fabrics to the stylist.

Fiber blends can be dyed into union shades (tone-on-tone) or multicolor effects can be obtained by coloring the individual components in different shades or by maintaining one fiber in an undyed state (reserving). A complete reserving of a fiber is not possible in all cases.

When dyeing fiber blends it must be decided whether the fibers can be dyed simultaneously from the same dyebath, or separately and in what order from different dyebaths.

With respect to fiber components that are dyed with completely different dye classes, the ability to use single-bath techniques (exhaust and continuous) depends on the interaction between the dyes and the compatibility of their dyeing procedures.

Cellulosic Fiber Blends

Cellulosic–Polyester Fibers. One of the most important fiber blends on the market is the mix of 35/65 or 50/50 cotton–polyester. High tenacity viscose fibers are sometimes used instead of cotton. Although the knitgoods are dyed in exhaust dyeing procedures, most of the woven fabrics are dyed according to one of the continuous dyeing processes. The choice of dyes and hence dyeing method is determined by the fastness properties required.

Cellulosic–Acrylic Fibers. Commonly this blend is used in knitgoods, woven fabrics for slacks, drapery, and upholstery fabrics. Since anionic direct dyes are used for the cellulosic fiber and cationic dyes for the acrylics, a one-bath dyeing process is only suitable for light to medium shades. Auxiliaries are needed to prevent precipitation of any dye complexes.

In two-bath processes either the cotton or the acrylic can be dyed first. Heavy shades are best dyed by first dyeing the acrylic and then dyeing the cotton under alkaline conditions. In order to prevent desorption of the cationic dye the dyeing temperature for the cotton dyeing must be below the glass-transition temperature for the acrylic of 80°C.

Cotton–acrylic fiber blends are also used for high quality upholstery pile fabrics. Besides the one-bath exhaust dyeing procedure involving a very high ratio of liquor to bath, a continuous pad-steam process is used to dye these fabrics.

Cellulosic Fiber–Nylon Blends. These blends are used in fabrics for apparel, corduroy, and swimwear. If wetfastness requirements are relatively low, the nylon portion can be dyed with disperse dyes and the cellulosic fiber with direct dyes and a one-bath procedure can be employed. For better wetfastness, the nylon portion is dyed with level dyeing acid colors together with the direct dyes in one bath at 95°C using a reserving agent to prevent the direct dyes from dyeing the nylon. An aftertreatment with a cationic fixative improves the wetfastness properties. For swimwear, the cotton portion is dyed with fiber-reactive dyes. After rinsing hot and cold and soaping at the boil, the nylon portion is dyed with a phosphate buffer system. Selected acid and/or acid milling colors are applied. An aftertreatment with a phenolsulfonic acid condensation product results in best wetfastness properties.

Wool Blends

Wool–Cellulosic Fibers. One of the oldest fiber blends in the textile market is the combination of wool and cotton or wool and viscose. In a one-bath process, selected direct and acid dyes are applied at pH 4.5–5.0 at 98–100°C. A phenolsulfonic acid condensation product is added as a reserving agent, to prevent the direct dyes from dyeing the wool under acid conditions. If optimum wetfastness properties are required, fiber-reactive dyes can be applied to both fibers by use of a two-bath process.

Wool–Nylon. Nylon has been blended with wool in order to give additional strength to the yarn or fabric. It is used mainly in the woollen industry for coats and jackets and, to a lesser extent, for socks and carpet yarns. Both fibers are dyed with the same products, however the fibers have different affinity to them. Generally level dyeing acid dyes are applied.

Wool–nylon upholstery fabrics and carpet yarns require higher light- and wetfastness properties. Neutral premetallized dyes are used in these cases. However, they have a much higher affinity to the nylon than the wool. Therefore, stronger retarding agents have to be employed, eg, phenolsulfonic acid condensation products.

Wool–Acrylic Fibers. This blend is being used for industrial and hand knitting yarns. Special precautions are necessary because the two fibers are colored with dyes of opposite ionic type. Usually, level dyeing acid dyes are used for the wool portion in combination with the cationic dyes for acrylic fiber.

Wool–Polyester Fibers. The wool–polyester blend is the most common fiber combination in the worsted industry. Disperse dyes for polyester and acid or neutral premetallized dyes for wool are employed in a one-bath process.

Blends of Synthetic Fibers

Polyester Fiber Blends. Disperse dyeable and cationic dyeable polyester fibers are frequently combined in apparel fabrics for styling purposes. Whereas the disperse dyes dye both fibers, but in different depths, selected cationic dyes reserve the disperse dyeable fiber completely, resulting in color/white effects.

Polyester Fiber–Nylon Blends. This fiber blend is used in apparel fabrics as well as in carpets. Disperse dyes dye both fibers, however they possess only marginal fastness properties on nylon. Therefore it is important to select those disperse dyes that dye nylon least under the given circumstances. The nylon is dyed with acid dyes, selected according to the fastness requirements.

Polyester Fiber–Acrylic Fiber Blends. This fiber blend is dyed in a similar fashion to that of the blends of the different polyester fibers. The selection of cationic dyes is substantially larger for the acrylic blend.

Nylon Blends. Differential dyeing nylon types and cationic dyeable nylon blends are used primarily in the carpet industry. The selection of cationic dyes for nylon is rather limited; most products have very poor fastness to light. These blends are dyed in a one-bath procedure at 95–100°C. Selected acid dyes are used for differential dyeing. Disperse dyes will dye all different types in the same depth.

Elastomeric Fibers. Elastomeric fibers are polyurethanes combined with other nonelastic fibers to produce fabrics with controlled elasticity (see FIBERS–ELASTOMERIC). Processing chemicals must be carefully selected to protect all fibers present in the blend.

Dyeing is carried out by the method best suited to the fiber used as the outer sheath, eg, acid or premetallized dyes for nylon-based, reactive or direct dyes for cotton-based.

Other Application Procedures

Pigment Dyeing. Many dyers do not look upon this form of coloration as dyeing; nevertheless millions of meters of fabric are dyed by this system each year. A finely dispersed (0.5–5.0 μ diameter) organic pigment is applied by padding together with organic binders and, depending on the binder system, a catalyst. After drying, the fabric is cured at 170–175°C when polymerization and optionally cross-linking of the binder takes place. The typical binder systems used are acrylic and butadiene resins. The pigments used cover the range of azoics, carbon black, phthalocyanines, triphenylmethanes, and dioxazine derivatives. They give excellent lightfastness and good washfastness in pale to medium shades. Fabrics being used range from lightweight poplins and sheetings to corduroy of cellulosic or fabrics of polyester–cellulosic blends. In addition to polyester–cellulosic and cellulosic fabrics, pigments may also be applied to 100% synthetic fibers of special construction for unique uses.

Solvent Dyeing. Solvent dyeing generally refers to dyeing in nonaqueous media. In the early 1970s, solvent dyeing was expected to become the dyeing process of the future and was discussed and researched extensively. This interest did not materialize into practical acceptance and the technique has not achieved importance.

Dyeing Machinery

In the application of dyes three techniques are used: the dye liquor is moved as the material is held stationary, the textile material is moved without mechanical movement of the liquor, or both move.

Transportation of Dye Liquor through the Textiles. Regardless of the form of the textile, raw stock, sliver, yarn, or cloth, the principle is generally the same. A large stainless steel kier, capable of withstanding sufficient pressure to reach a maximum operating temperature of 145°C, has one or more perforated spindles through which the dye liquor is pumped. Around this spindle the textile is packed tightly in the form of a cake in a perforated basket, as yarn, or in a package as a sliver or tow in a can, or as cloth around a beam barrel. The dye liquor is pumped through the textile, then flows to the bottom of the machine and into the return side of the pump.

Transportation of the Textile Material with No Mechanical Movement of the Liquor. The chain warp dyeing procedure is widely used in the dyeing of indigo on warp yarns which are in the form of ropes or

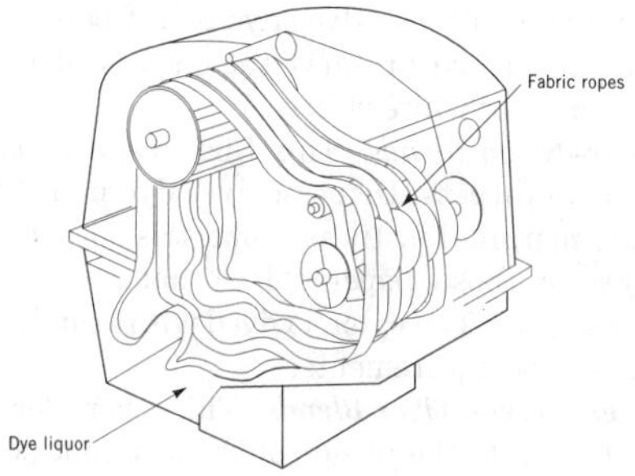

Figure 3. Winch or beck dyeing machine.

chains. In this procedure, several warp ropes are pulled through a series of tubs containing the dye liquor and gradually are dyed to the desired shade. Piece goods are dyed by means of a batch process on a dye jig. Winch or beck dyeing is one of the oldest forms for dyeing mechanized piece goods (fabric) (Fig. 3). In continuous dyeing, the equipment at hand may be simple padder or a complete dye range. Most dye padders consist of a medium density rubber roller across the width of which pressure can be applied. This roller presses against a stainless steel or a hard rubber roller. These rollers may be mounted either vertically or horizontally. The cloth is passed through a stainless steel pad box, down under a rod or a roller which is below the dye liquor level, between the squeeze rollers.

Dye ranges can be of different configuration depending on the composition of the fabrics being processed and dye system used. All contain similar units; differences are mainly in the method of heating.

The following sequence is typical for polyester–cotton blended fabric. The infrared units reduce moisture content 20–30% and greatly minimize uneven migration of the dye on the wet goods when they go onto the drying cylinders. The dried cloth progresses into the thermosol unit where the dyestuff for the synthetic portion of the fabric is fixed. Goods then continue through the chemical pad for immersion into an alkaline (or reducing) solution, depending on whether fiber-reactive or vat dyes are applied to the cellulosic fibers. They then pass through a steamer and finally through 8–10 wash boxes, which contain various chemicals depending on the class of dyestuff applied.

Machines Based on Movement of Both Dye Liquor and Material. One example of a machine in which both yarn and dye liquor are moved is the Klauder-Weldon skein dye machine; not only do the skeins turn, but the liquor is pumped in small streams over the yarn as the threads pass over the spindles. This process assures maximum uniformity and levelness.

Another extremely popular machine of this type is the jet dyeing machine which conserves energy by reducing the cloth-to-liquor ratio to 1:10 or lower as compared to 1:20 for the winch. In this machine, the fabric which is in a rope form is transported by movement of the dye liquor through a Venturi jet. This method provides intimate contact between the dye liquor and each meter of material. The machine operates at 40–135°C.

Overflow machines may be thought of as hybrids of jets and winches. They usually feature a winch reel which, unlike the cloth-guiding roller in a jet, is normally driven and provides motive power to the fabric. There is also some driving force on the cloth from the circulation of the liquor through the overflow tube down which they both pass. Both pressurized and nonpressurized versions exist and are available from a large number of different makers. The principal advantage that overflow machines possess over jets is a gentler action on the fabric.

Control of Dyeing Equipment. Over the years, the dyer and machinery manufacturer have applied any mechanical or electrical equipment that would enable them, day after day, to produce repeatable dyeings of top quality. First, thermometers were installed in dye lines; these soon evolved into thermocouples with remote recording. Other improvements were soon developed, such as automatic four-way valves with variable-interval controls, flow controls, pressure recorders, hydraulic and air pressure sets on rollers, pH controls, etc.

Textile Printing

The term textile printing is used to describe the production of colored designs or patterns on textile substrates through a combination of various mechanical and chemical means. In printing on textiles, a localized dyeing process takes place, whereby in general the chemical and physical parameters of dyeing apply.

The process of textile print coloration can be divided into three steps. First, the colorant is applied as pigment dispersion, dye dispersion, or dye solution from a vehicle called print paste or printing ink, containing in addition to the colorant such solutions or dispersions of chemicals as may be required by the colorant or textile substrate to improve and assist in dye solubility, dispersion stability, pH, lubricity, hygroscopicity, rate of dye fixation to the substrate, and colorant-fiber bonding. The second step is the fixation process. During the afterscouring, the third step, the prints are rinsed and scoured in a detergent solution.

Colorants for Textile Printing

Pigments. Pigment-printed textiles represent the highest percentage of all printed textiles, accounting for between 40 and 50% of all cellulose and over 90% of polyester–cotton blend prints.

Disperse Dyes. Disperse dyes are used in powder or pasted form, or ready-to-prepare aqueous dispersions for incorporation into a thickener solution.

Acid Dyes. These dyes have their greatest importance in printing of polyamide.

Premetallized Dyes. This dye group is applied to the same textile fibers and with the same procedures as those with acid dyes. The premetallized dyes offer better fastness properties, but lack brilliancy of shade.

Direct Dyes. A few selected direct dyes are used to complement the acid dyes in printing of polyamide.

Fiber-Reactive Dyes. This dye class represents, next to pigments, the main dye group for cellulosic fibers, ie, cotton and rayon.

Basic (Cationic) Dyes. The use of basic dyes is confined mainly to acrylic textile fibers, acetate, and as complementary dyes for acid-modified polyester fibers that accept this class of dyes.

Vat Dyes. Applied to cellulosic fibers, vat dyes yield prints with excellent fastness properties. They are used to print furnishings, drapes, and camouflage where their infrared reflectance resembles natural terrain and foliage.

Azoic Dyes. These are used to produce cost-effective heavy yellow, orange, red, maroon, navy blue, brown, and black shades and are printed alongside other dye classes to extend the coloristic possibilities for the designer.

Phthalocyanine Dyes. These dyes are synthesized as the metal complex on the textile fiber from, eg, phthalonitrile and metal salts.

Dye Combinations. In certain cases it is desirable to print fiber blends with combinations of the appropriate dye classes, rather than with pigments. Only polyester–cellulose blends are of commercial importance and the following dye systems have been developed for them. The dyes of the different classes are contained in the same print paste and, therefore, are applied simultaneously in one print operation. They include disperse–reactive combinations, disperse–pigment combinations, and disperse–vat combinations.

Styles of Printing

Styles of printing include direct printing, discharge printing, resist printing, and wax printing.

Printing Machinery

Textile materials can be printed at different steps of the textile manufacturing process. Woven fabrics comprise the largest percentage of printed goods. In recent years, knitted textile fabrics have considerably increased in importance. However, printing can also be done on yarns in skein form, or on warps being passed from a warp beam to another beam, or as yarn strands. Space printing is a process where a yarn, temporarily knitted into a loose fabric, is printed and then deknitted. Carpets can be printed in woven or tufted constructions. Vigoureux printing is the printing of woolen slubbing. Regardless of the state of the textile material, any printing process makes use of one of the following methods: screen printing or roller printing.

Paper Coloring

Colorants for Paper. Among the colorants that have been and are being used for the dyeing of paper are natural inorganic pigments (ochre, sienna, etc); synthetic inorganic pigments (chromium oxides, iron oxides, carbon blacks, etc); natural organic colorants (indigo, alizarin, etc); and synthetic organic colorants. The last is the largest and most important group.

Basic Dyestuffs. Basic dyestuffs are usually used for dyeing of unbleached pulp in mechanical pulp such as wrapping paper, kraft paper, box board, news, and other inexpensive packaging papers. Their strong and brilliant shades also make them suitable for calendar staining and surface coloring where lightfastness is not critical.

Acid Dyestuffs. Because of poor affinity and good solubility, acid dyestuffs have poor bleedfastness and form colored backwater and are therefore suitable for paper that does not require wetfastness, such as construction grades. Acid dyestuffs are most suitable for calendar staining or surface coloring because of their solubility and brightness of shade.

Direct Dyestuffs. Direct dyestuffs' bonding ability to nonligneous pulps and excellent fastness properties to light and bleeding make them useful for all fine papers. The shades of direct dyestuffs are not as bright as those of acid or basic dyestuffs and in blended furnishes (bleached–ligneous pulps) mottling or graniting may occur.

Pigments. Synthetic organic pigments are replacing the use of some inorganic pigments for ecological reasons. Pigments do not react chemically with the fiber, but are fixed physically and are dependent on filtration, absorption, occlusion, and flocculation. paper dyeings with pigments have outstanding fastness properties, but poor affinity, low tinctorial strength, and two-sidedness problems limit their application to paper.

Fluorescent Whitening Agents. Fluorescent whitening agents (qv) change the appearance of paper in two ways: by emitting light and therefore increasing the luminosity (brightness); and by changing the shade from yellowish white to bluish white.

Dyeing Processes

Paper may be colored by dyeing the fibers in a water suspension by batch or continuous methods. The classic process is by batch dyeing in the beater, pulper, or stock chest. Continuous dyeing of the fibers in a water suspension is adaptive to modern paper machine processes with high production speeds in modern mills. Solutions of dyestuffs can be metered into the high density or low density pulp suspensions in continuous operation.

Nonimpact Printing. Interest is growing in the use of nonimpact styles because of the quickness of color changeover and the ability to interface these machines to computer-aided design systems. Two basic types exist: drop on demand and constant drop techniques.

Dyeing of Leather

Not only may the compound used to convert hide substance into leather vary chemically over a wide range, but the quantities used, the method of application, and the physical condition of the hide prior to tanning or dyeing may vary, with each factor in turn affecting the dyeing properties of the resultant leather. Also, leather retains many of the properties originally associated with the parent substance, and these affect profoundly and, in many ways, limit the dyeing properties of the final product. Chief among these properties are sensitivity to extremes of pH, thermolability, and the tendency to combine with acidic or basic compounds.

Leather Dyes. The main classes of dyes employed in the coloring of leather are the acid, acid/direct, direct, and basic types. On chrome leather, the direct dyes usually have greater affinity and produce fuller or heavier shades than do acid or chrome dyes. Acid/direct dyes as well as the metallized-type dyestuffs may be classified for the purpose of leather dyeing as the main types in use. Basic dyes color chrome leather weakly and unevenly, unless the leather is first mordanted or retanned with suitable materials, such as vegetable tannin, syntans, or previously applied acid and/or direct dyes. They may be used alone on vegetable tanned leather to produce full shades or, as is done more frequently, following a preliminary coloring with acid or acid/direct dyes. In the latter case, basic dyes are used to impart fullness of shade with minimum coloring matter and cost.

Leather Dyeing Methods and Equipment. The methods used in dyeing leather are quite simple and they obtain their names from the equipment employed, such as drum, wheel, paddle, brush, tray, or spray dyeing. Most leather is dyed in drums.

Fastness Tests for Textiles

The principal active bodies in the field of colorfastness testing have been the American Association of Textile Chemists and Colorists (AATCC) and the Europaische-Convention für Echtheitprüfung/ Groupement d'Etudes des Commissions Européenes pour la Soliditié (ECC). The ISO subcommittee concerned with colorfastness tests is ISO TC 38/SCI. This meets every two or three years to coordinate developments in standard testing methods and to seek international agreement on proposed new tests and modifications to existing tests. The purpose of ISO is solely to produce useful standard test methods. The setting of specifications and levels of acceptance on the basis of such test methods is a matter to be resolved between buyer and seller.

Testing of Dyes. At the 1989 meeting of ISO/TC38/SC1 in Williamsburg, Va., a new work group (WG11), Characterization of Dyestuffs, was established. The following tests are significant, together with alternative techniques currently being considered for introduction as ISO standards: evaluation of dyestuff migration, thermal fixation properties of disperse dyes on polyester–cotton, transfer of disperse dye on polyester, transfer of basic dyes on acrylics, transfer of acid and premetallized dyes in nylon, dispersion stability of disperse dyes at high temperature, foaming propensity of disperse dyes, and evaluation of the dusting properties of powder (or other solid dyes).

Other New Methods. Because the values obtained are dependent on the conditions of measurement, standard test procedures are under review by ISO for: determination of cold-water solubility of water-soluble dyes; determination of the solubility and solution stability of water-soluble dyes; and determination of the electrolyte stability of reactive dyes.

Safe Handling of Dyes

The Ecological and Toxicological Association of Dyes and Organic Pigments Manufacturers (ETAD), an international body of all primary manufacturers based in Europe but also with standing committees in the United States, Brazil, and Japan, issues clear guidelines for the safe handling of dyes. In December 1991 the United States Operating Committee of ETAD joined with the United States Environmental Protection Agency in publishing a pollution prevention guidance manual for the dye manufacturing industry (see also DYES, ENVIRONMENTAL CHEMISTRY).

Colour Index Generic Names

The *Colour Index* assigns CI generic names to commercial dyes. This CI name is defined as "a classification name and serial number which when allocated to a commercial product allows that product to be uniquely identified within any *Colour Index* Application Class." This enables the particular commercial products to be classified along with other products whose essential colorant has the same chemical constitution.

BRIAN GLOVER
Zeneca Colours

Colour Index, 3rd ed. (4th edition in preparation), Society of Dyers and Colourists (SDC), U.K., in collaboration with American Association of Textile Chemists and Colourists (AATCC), Research Triangle Park N.C., 1971.

J. Shore, ed., *Colourants and Auxiliaries, Organic Chemistry and Application Processes,* Vol. 1, *Colourants,* Society of Dyers and Colourists, U.K., 1990.

Textiles—Tests for Colour Fastness, ISO 105, AATCC, Research Triangle Park, N.C., 1990.

J. R. Aspland, *Textile Chemist Colourist, A Series on Dyeing,* AATCC, Research Triangle Park, N.C., Oct. 1991–Nov. 1993, Chapts. 1–15.

DYES, ENVIRONMENTAL CHEMISTRY

Effluent Treatment Methods

Methods of effluent treatment for dyes may be classified broadly into three main categories: physical, chemical, and biological.

Physical	Chemical	Biological
adsorption	neuralization	stabilization ponds
sedimentation	reduction	aerated lagoons
flotation	oxidation	trickling filters
flocculation	electrolysis	activated sludge
coagulation	ion exchange	anaerobic digestion
foam fractionation	wet-air oxidation	bioaugmentation
polymer flocculation		
reverse osmosis/ ultrafiltration		
ionization radiation		
incineration		

There are four stages: preliminary, primary, secondary, and tertiary treatment processes, which differ mainly by the number of operations performed on the waste steams.

Preliminary treatment processes of dye waste include equalization, neutralization, and possibly disinfection. Primary stages are mainly physical and include screening, sedimentation, flotation, and flocculation. The objective is to remove debris, undissolved chemicals, and particulate matter. Secondary stages are used to reduce the organic load, which essentially is a combination of physical/chemical separation and biological oxidation. Tertiary stages are important because they serve as a polishing of effluent treatment. These methods are adsorption, ion exchange, chemical oxidation, hyperfiltration (reverse osmosis), electrochemical, etc.

Fate of Dyes

Recent estimates indicate 12% of the synthetic textile dyes used yearly are lost to waste streams during dyestuff manufacturing and textile processing operations. Approximately 20% of these losses enter the environment through effluents from wastewater treatment plants.

With few exceptions, the normal use of organic colorants poses few problems in terms of acute ecological effects. On the other hand, certain dyestuffs exhibit toxic effects toward microbial populations and can be toxic and/or carcinogenic to animals. Also, the possible contamination of drinking water supplies is of concern because certain classes of dyes are known to be enzymatically degraded in the human digestive system, producing carcinogenic substances.

Until recently, few papers appeared on the fate of dyes in the environment. But because of the importance of this subject, work is being done primarily by the U.S. Environmental Protection Agency (U.S. EPA) and the Ecological and Toxicological Association of the Dyestuff Manufacturing Industry (ETAD).

One of the reasons for lack of literature was probably because environmental analysis depends heavily on gas chromatography/mass spectrometry, which is not suitable for most dyes because of their lack of volatility. However, significant progress is being made in analyzing nonvolatile dyes by newer mass spectral methods such as fast atom bombardment (FAB), desorption chemical ionization, thermospray ionization, etc.

Dyestuffs in general, and azo dyes in particular, are likely to undergo substantial primary biodegradation in an anaerobic environment through reductive cleavage of azo bonds into aromatic amines. Lipophilic aromatic primary amines are aerobically degradable, but depending on their precise structure, some sulfonated aromatic amines may not be degradable.

The fate of one dye that has been thoroughly studied is the azo dye Disperse Blue 79 which may be designated 6-bromo-2, 4-dinitroaniline → 3-(*N*, *N*-diacetoxyethylamino)-4-ethoxyacetanilide (see AZO DYES).

Disperse Blue 79 is the largest volume dye on the market today. It has been estimated that during the manufacture of Disperse Blue 79 there should be released the following amounts of dye: 4.5–14 t per year at a total of nine sites with an estimated 3–20 kg per day.

Of particular importance is the degradation, cleavage, or reduction of the dye into aromatic amines, one of which is 6-bromo-2,4-dinitroaniline. This amine is toxic, mutagenic, and was selected for carcinogenic study. The precursor for preparation of 6-bromo-2,4-dinitroanaline, is 2,4-dinitroaniline, which has been extensively evaluated and found to be highly toxic, mutagenic, and was selected for carcinogenic study.

Small amounts of the following aromatic compounds were found to be present in the effluent after manufacture of Disperse Blue 79: 6-bromo-2,4-dinitroaniline, 6-bromo-2,4-dinitrophenol, 3-diacetoxyethylamino-4-ethoxy acetanilide, 3-dihydroxyethylamino-4-ethoxyacetanilide, 3-[(*N*-hydroxyethyl, *N*-acetoxyethyl)amino]-4-ethoxyacetanilide, and 3-acetoxyethylamino-4-ethoxy acetanilide. However, after a neutralization and heat stabilization step is added after coupling in manufacture, no diazotizable amine or phenol and only traces of the coupling component and its impurities were found.

The fate study of Disperse Blue 79 in anaerobic sediment–water systems shows the following degradation products:

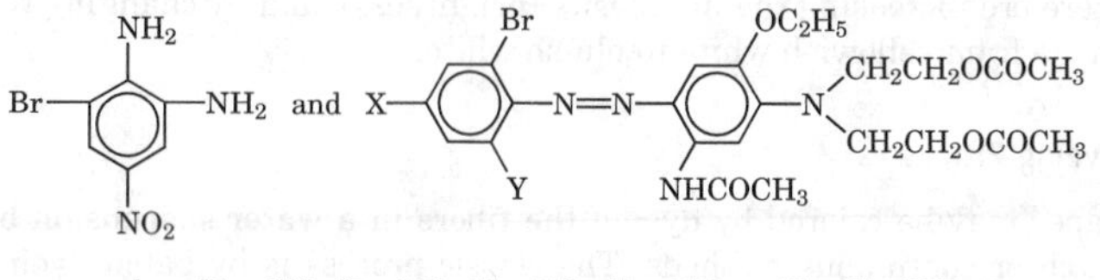

where $X = NH_2$, $Y = NO_2$; $X = NO_2$, $Y = NH_2$; or $X = Y = NH_2$

These products suggest that this dye may undergo reduction in bottom sediments in the environment, resulting in the subsequent release of potentially hazardous aromatic amines into water.

A large study was done to determine the fate of Disperse Blue 79 in a conventionally operated activated sludge process and in an anaerobic sludge digestion system. The results showed no degradation in the activated sludge system, but did show degradation in the anaerobic digester of which no positive identification of compounds were made.

Pollution Prevention

Cooperation between industry, government, academia, and private environmental groups to implement these pollution prevention acts has begun in earnest.

The dye and dyeing industries have begun to give pollution prevention and its other forms of lessening or eliminating waste generation, such as waste minimization and source reduction, a high priority. The USEPA and the Ecological and Toxicological Association of the Dyestuffs Manufacturing Industry (ETAD) have jointly set up a program for pollution prevention in the dyestuff industry. The original goals were to develop a pollution prevention guidance manual and conduct a baseline survey of industry prevention practices for dye manufacture. Recently the manual was published and the data for the baseline survey collected by USEPA and ETAD.

There was a number of papers and patents on recycling dye and textile industry wastewater for reuse of dye, textile auxiliaries, and water. Recycling is considered a part of pollution prevention.

Heavy Metals

The heavy metals, copper, chromium, mercury, nickel, and zinc, which are used as catalysts and complexing agents for the synthesis of dyes and dye intermediates, are considered priority pollutants.

A number of papers have appeared on the removal of heavy metals in the effluents of dyestuff and textile mill plants. The methods used were coagulation, polymeric adsorption, ultrafiltration, carbon adsorption, electrochemical, and incineration and landfill. Of interest is the removal of these heavy metals, especially copper by chelation using trimercaptotriazine and reactive dyed jute or sawdust.

Toxicity

The past experience of the dyestuff industry in its use of dye intermediates such as β-naphthylamine and benzidine, known human bladder carcinogens, have led to studies as to whether or not handlers of dyes are exposed cancer, dermatitis, and other disorders.

The National Institute of Occupational Safety and Health (NIOSH) and the Occupational Safety and Health Administration (OSHA) reported in 1978 that the three primary benzidine-based azo dyes, namely Direct Black 38, Direct Blue 6, and Direct Brown 95, were carcinogenic in animals as a result of being converted to benzidine. These dyes are characterized by having a biphenyl diazo linkage:

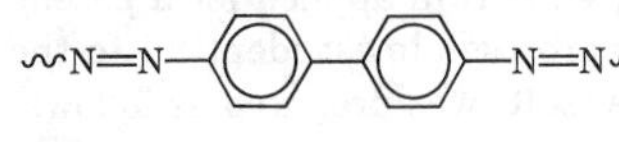

This has led to concern about possible carcinogenicity from these dyestuffs, and therefore benzidine-based dyes and pigments are no longer produced by the large dyestuff manufacturers.

Two large studies were done for the selection of azo, nitro, and anthraquinone dyes for carcinogen bioassay. Based on previous information or testing, a total of 30 dyes were selected based on chemical structure, potential exposure, and suspicion of carcinogenicity.

Because of the large number of dyestuffs and the fact that most of these colorants have not been tested for carcinogenicity, structure–activity theory may help predict possible candidates for study.

In order to minimize the possible toxicity and damage to humans and the environment arising from the production and applications of colorants, an international association, the Ecological and Toxicological Association of the Dyestuff Manufacturing Industry (ETAD), was established in 1974. ETAD coordinates the ecological and toxicological efforts of synthetic organic dyes and pigment manufacturers. To date, ETAD consists of 32 members in Western Europe, North America, Japan, and India. The purpose of ETAD's toxicological work is to identify and assess risks caused by colorants and their intermediates with respect to their potential acute toxicity and their chronic effects on human health. This is accomplished by recommended methods to member firms following appraisal and development by appropriate ETAD committees, lectures, and publications. One of the projects of ETAD was a survey of acute oral toxicity, as measured by LD_{50}, the 50% lethal dose, which showed that of 4461 colorants tested, only 44 had a $LD_{50} < 250$ mg/kg, but 3669 exhibited practically no toxicity ($LD_{50} > 5$ g/kg). The evaluation of these colorants by chemical and coloristic classification showed that the most toxic ones are found among diazo and cationic dyes. Pigments and vat dyes, on the other hand, have a low acute toxicity, presumably because of their low solubility in water and in lipophilic systems.

Table 1. Important Environmental Legislation

Law	Worker	Consumer	Public and environment
United States			
Toxic Substances Control Act (TSCA)	X	X	X
Occupational Safety and Health Act (OSHA)	X		
Food, Drug, and Cosmetic Law		X	
Consumer Product Safety Act		X	
Labeling of Hazardous Materials		X	X
Federal Water Pollution Control Act			X
Clean Air Act			X
Resource Conservation and Recovery Act (RCRA)			X
Superfund Amendments and Reauthorization Act (SARA) Plus Title III	X		X
Superfund (CERCLA)			X
Hazardous Materials Transportation (DOT)			X
CONEG (Heavy Metals)			X
USDA			X
California Proposition 65			X
EEC			
Control of Certain Industrial Activities	X	X	X
Classification, Packaging, and Labeling of Dangerous Substances		X	X
UK			
Health and Safety at Work Act	X		
Carcinogenic Substances Regulations	X		
Toy (Safety) Regulations		X	
Pencil and Graphic Instruments (Safety) Regulations		X	
Poison Act		X	
Clean Air Act			X
Switzerland			
Environment Protection Law		X	X
Poison Law		X	
France			
Control of Chemical Products		X	X
Consumer Protection and Information		X	
Pollution Control Law			X
Germany			
Environmental Chemicals Law	X	X	X
Japan			
Chemical Control Law			X

Legislation

There has been a tremendous increase in regulatory activities worldwide aimed at achieving safer manufacture, use, and disposal of chemicals, including colorants. Table 1 is a summary of important United States, European, and Japanese Environmental Legislation affecting workers, consumers, and the public and environment.

The two most important pieces of chemical control legislation enacted affecting the dye and pigment industries are the United States' toxic Substance Control Act (TSCA) and EEC's Classification, Packaging, and Labeling of Dangerous Substances and its amendments.

Besides the federal laws, all 50 U.S. states have also passed environmental laws. As previously mentioned, both the United States and the state of New Jersey have passed Pollution Prevention Acts.

There is a difference between the United states, European, and Japanese environmental laws in introducing a new chemical compound. The EEC and Japan require animal and other toxicological and ecological testing, whereas the United States encourages but does not require these tests. This could result in rejection of a U.S. chemical exported product. Subsequently, greater harmonization is needed in many federal versus state versus overseas regulatory issues impacting the dye industries.

ABRAHAM REIFE
CIBA-GEIGY Corporation

E. A. Clarke and R. Anliker, *Environmental Chemistry, Anthropogenic Compounds,* Vol. 3, Part A, Springer-Verlag, New York, 1980, p. 181.

C. T. Helmes and co-workers, *J. Env. Sci. Health* **A19**(2), 97 (1984).

R. Anliker and E. A. Clarke, *J. Soc. Dyer. Color.* **98,** 42 (1982).

J. Houk, M. J. Doa, M. Dezube, and J. M. Rovinski, "Evaluation of Dyes Submitted Under the Toxic Substance Control Act New Chemicals Programme," *Colour Chemistry,* Elsevier Applied Science, London and New York, 1991.

DYES, NATURAL

Natural dyes were replaced by synthetic dyes, although lately there has been a revival of the use of natural dyes for coloring foods, and some textile manufacturers are using natural dyes for dyeing their products. This article discusses those natural dyes formerly manufactured.

Anthraquinones

The anthraquinone structure occurs in both the plant and animal kingdom. Those natural dyes having this structure surpass all other natural dyes in fastness properties (see DYES, ANTHRAQUINONE).

Alizarin. There is only one significant plant anthraquinone dye, alizarin (CI Natural Red 6, 8, 9, 10, 11, and 12; CI 75330).

Alizarin is a mordant dye forming various colored coordination complexes with different metallic salts. Based on analytical results, a structural formula has been proposed for the alizarin complex.

Anthracene was oxidized to anthraquinone, dibrominated, and the dibromo derivative subjected to a caustic fusion. Alizarin was obtained in an impure form and in low yield. This represented the first synthesis of a natural dye.

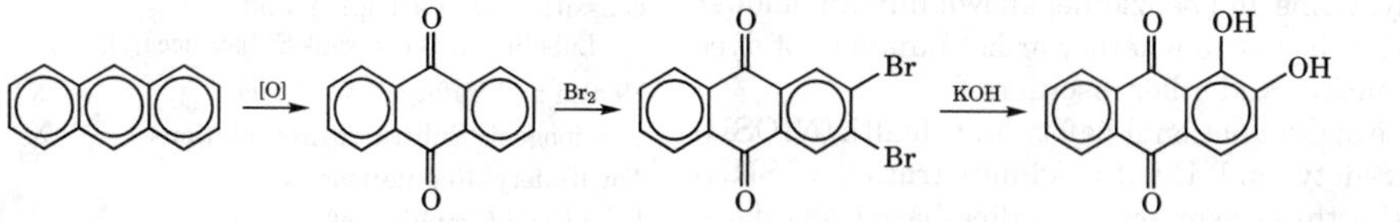

(1)

Later, at BASF, a process was developed for the manufacture of alizarin by the caustic fusion of anthraquinone-2-sulfonic acid (so-called silver salt) which was patented in England on the 25th of June, 1869. One day later, W. Perkin applied for a patent for the manufacture of alizarin by a process almost identical to the German process except that the "silver salt" was prepared as follows:

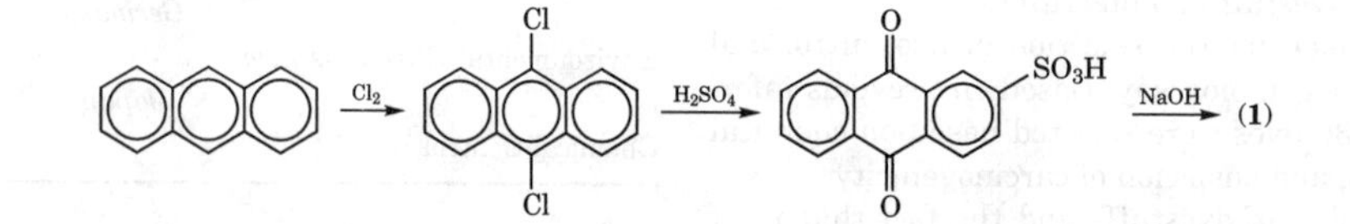

Later, improvements were made in the process: use of oleum for sulfonating anthraquinone and the addition of an oxidizing agent to the caustic melt.

For years this was the process used to manufacture alizarin, although it was claimed that a more economical process would result if 2-chloroanthraquinone were to be used instead of silver salt.

Just as synthetic alizarin forced natural alizarin out of the market, synthetic alizarin has been replaced by azoic dyes because they are easier to apply.

Animal Anthraquinone Dyes. Kermisic Acid. Many accounts claim that kermisic acid (CI Natural Red 3; CI 75460) is the oldest dyestuff ever recorded. The dye produces a brilliant scarlet color with an alum mordant. Although expensive, it was cheaper than its rival Tyrian Purple. It was in great demand until the sixteenth century, when it was displaced by carminic acid.

The structure of kermisic acid is 1,3,4,5-tetrahydroxy-7-carboxy-8-methylanthraquinone. Carminic acid (CI Natural Red 4; CI 75470), is a red dye occurring as a glycoside in the body of the cochineal insect *Dactylopius coccus* of the order Homoptera, family Coccidae. Until the advent of synthetic dyes, the principal use for carminic acid was for dyeing tin-mordanted wool or silk. Its aluminum lake, carmine, finds use in the coloring of foods. The structural formula of carminic acid is (**2**).

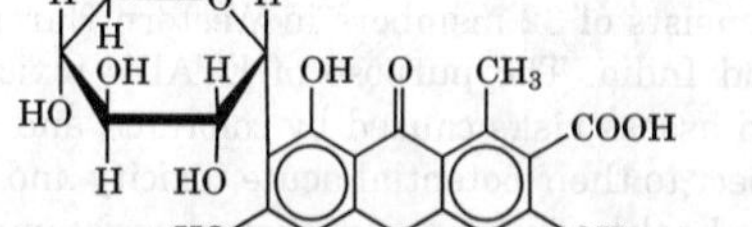

(**2**)

Laccaic acid has been designated (CI Natural Red 25; CI 75450). Lac dye ranks as the most ancient of animal dyes. It is found in lac, the resinous secretion of a very small insect, *Coccus laccae,* found growing in India and Southeast Asia. Lac dye is actually a mixture of acids derived from 2-phenylanthraquinone

Naphthoquinone Dyes

Although naphthoquinones represent the largest group of naturally occurring quinones, only a small number of these achieved importance as dyestuffs.

Lawsone (CI Natural Orange 6; CI 75420), also known as henna and isojuglone, occurs in the shrub henna (*Lawsone alba*). Lawsone has been identified as 2-hydroxy-1,4-naphthoquinone. It has been synthesized by the Thiele acetylation of 1,4-napththoquinone followed by hydrolysis and oxidation.

Lapacol (CI Natural Yellow 16; CI 75490) (lapachic acid, taiguie acid, tecomin) is a yellow pigment occurring in the wood of trees of the genus Tecoma, native to the West Indies and tropical South America.

Juglone (CI Natural Brown 7; CI 75500) was isolated from the husks of walnuts in 1856). Juglone occurs in walnuts as a glycoside of its reduced form, 1,4,5-trihydroxynaphthalene. Its structure is (**3**).

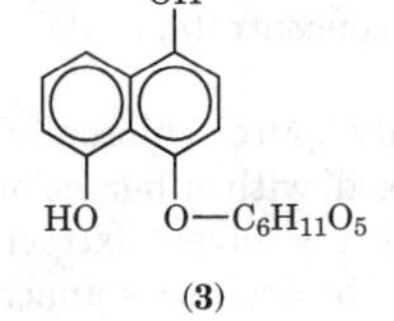

(**3**)

Juglone is most readily synthesized by Bernthsen's method. Although it no longer has any commercial value as dye, it is a fungicide and as such finds use in the treatment of skin diseases. Its toxic properties have been made use of in catching fish. Juglone has been used to detect very small amount of nickel salts since it gives a deep violet color with such salts.

Alkannin, shikonin, and shikalkin are grouped together because the first two are enantiomers and the last one is their racemate. Alkannin (CI Natural Red 20; CI 75530) (*Anchusa tinctoria* or *alkanna tinctoria*) is a member of the Boraginaceae family. It is found in the roots of alkanet, a perennial shrub native to Southern Europe.

Alkannin occurs in the roots of the plant as the alkali-sensitive ester of angelic acid. It may be extracted from the roots by using boiling light petroleum ether. Treatment of this extract with dilute sodium hydroxide gives a blue solution from which the dye is precipitated by the addition of acid. The crude product is purified by vacuum sublimation. Its structure is a hydroxylated naphthoquinone with a long, unsaturated side chain; it has the (*S*)-configuration.

Shikonin (CI 75535) occurs as an acetyl derivative in the Japanese shikone, *Lithospermum erythrorhizon,* another member of the Boraginaceae family. It is the (*R*)-optical isomer of alkannin. Tissue cultures of *L. erythrorhizon* are used in Japan to manufacture shikonin mainly for cosmetic use. Both alkannin and shikonin are mordant dyes producing violet to gray colors on fabrics. Shikalkin the racemate, has been synthesized.

Flavones. These compounds are the most widely distributed natural coloring matter formerly used as dyestuffs. Flavone-type dyes occur in all the higher plants: in the leaves, roots, bark, fruits, pollen, and flower petals. The most widespread flavone dye are quercetin and kaempferol. In general, the dyes occur as glycosides, the most common sugar being glucose.

The basic unit of the flavone-type dyes is 2-phenylbenzopyrone (**4**) which unsubstituted is flavone (**4**); isoflavone is (**5**); and flavonol (**6**) is

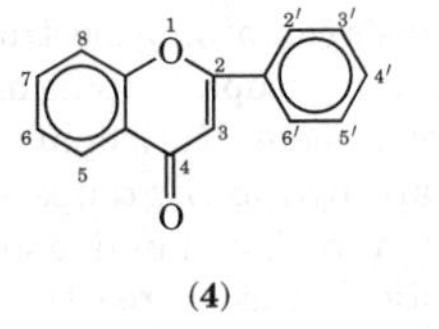

(**4**)

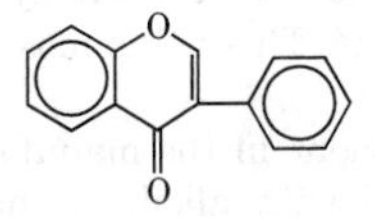

(**5**)

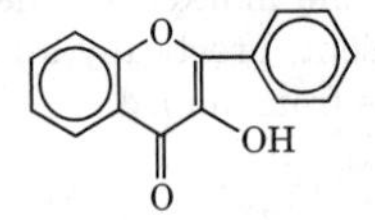

(**6**)

Flavone dyes having these structures are hydroxylated and methoxylated derivatives. Those dyes containing not more than three hydroxyls are generally termed flavones, whereas those containing up to and including six are flavonols.

The flavone, isoflavone, and flavonol-type dyes owe their importance to the presence of an *o*-hydroxy carbonyl structure within the molecule. Positions 4 and 5 can chelate with different metallic salts to give colored insoluble complexes. In other words, these dyes require a mordant in order to fix them onto the fiber.

Anthocyanins. Like the flavones, the anthocyanins are found throughout nature. This class of polyphenolic compounds is responsible for the pink, red, violet, and blue colors found in plants. Like many other natural phenolic substances, anthocyanins occur in plants as glycosides; the sugar-free anthocyanins are known as anthocyanidins.

All anthocyanidins have the 2-phenylbenzopyrylium or flavylium cation structure, a resonance hybrid of oxonium forms and carbenium forms. There are three fundamental groups of anthocyanidins to which all the other anthocyanidins could be referred. In the following structure R = R′ = H designates pelargonidin R = OH, R′ = H is cyanidin; and R = R′ = OH is delphinidin.

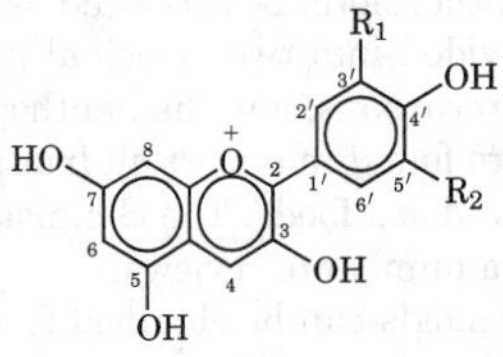

The anthocyanins are pH sensitive. Their color, in part, is determined by the pH of the sap.

A convenient method for synthesizing anthocyanidins involves the condensation of an *o*-hydroxybenzaldehyde with an acetophenone.

Indigoid Dyes

Tyrian Purple. The ancient kingdom of Tyre owed its fame and fortune to the purple dye produced from the lowly mollusks found on its shores. The dye became known as Tyrian Purple (CI 75800) reflecting its place of origin. These mollusks belong to the Muricidae family and the genera *Murex* and *Purpura* which include *M. brandaris* and *M. trunculus,* the principal sources of the dye. The dye is 6,6′-bromoindigotin.

Indigotin. The blue dye of the ancient world was derived from indigo and woad (CI Natural Blue; CI 75780). Indigo belongs to the legume family. The two most important species are *Indigo tinctoria* and *I. suffruticosa,* found in India and the Americas, respectively. The leaves of the indigo plant do not contain the dye as such, but in the form of its precursor, a glycoside known as indican.

Woad, *Isatis tinctoris,* belongs to a genus that comprises some thirty species.

In 1890, it was observed that treatment of ω-bromoacetanilide with alkali produced oxindole. Based on this observation, K. Heumann treated *N*-phenylglycine with alkali and obtained indoxyl (keto form), which on aerial oxidation converted to indigotin. Later, a variation of the original Heumann process was made: aniline, formaldehyde, and hydrogen cyanide react to form phenylglycinonitrile, which is hydrolyzed to phenylglycine. This is the most widely used process for manufacturing indigotin.

The greatest improvement in the manufacture of indigotin came when sodamide was used with alkali in the conversion of phenylglycine to indoxyl.

Although there is still demand for indigotin for dyeing blue jeans, it has lost a good part of the market to other blue dyes with better dyeing properties. At present, practically all the indigotin consumed in the United States comes from abroad.

Natural Food Colors

In the 1970s, decertification of the important food colors FD&C Reds 2 and 4 caused much concern among manufacturers of food dyes. With the possibility that other synthetic dyes would be banned present, attention was turned to the use of natural dyes as food colorants. Many such dyes had been in use for hundreds of years until they were replaced by synthetic dyes.

The yellow dye curcumin (CI Natural Yellow 3; CI 75300), also known as tumeric, occurs in the roots of the plant *Curcuma tinctoria,* found growing wild in Asia. The dye is an oil-soluble bright yellow material, and is the only natural yellow dye that requires no mordant. It finds use as a colorant for baked goods such as cakes. Carmine is the trade name for the aluminum lake of the red anthraquinone dye carminic acid obtained from the cochineal bug. The dye is obtained from the powdery form of cochineal by extraction with hot water, the extracts treated with aluminum salts, and the dye precipitated from the solution by the addition of ethanol. This water-soluble bright red dye is used for coloring shrimp, pork sausages, pharmaceuticals, and cosmetics. It is the only animal-derived dye approved as a colorant for foods and other products.

Carotenoids. The carotenoids are a group of widely distributed, highly colored, fat-insoluble, naturally occurring organic compounds. They owe their color to the four repeating isopyrene units found in the molecule and may, therefore, be classified as tetraterpenoids. The carotenoids may be divided into two principal groups, the carotenes, which are strictly hydrocarbons, and the xanthophylls, which contain oxygen. Carotenoids are found in almost all fruits and vegetables, egg yolk, dairy products, and sea foods. The chemistry of the carotenoids has been described in a number of reviews.

Although the carotenoids can be obtained from natural sources, it is far more economical to manufacture them for commercial use. Three have been manufactured for many years: β-carotene, canthaxanthin and β-apo-8′-carotenal.

In general, the low solubility of the carotenoids creates a problem when they are applied as food dyes. To overcome this, a microcrystalline powder form has been prepared. This is then mixed into an edible fat. In this form it finds use for coloring margarine, butter, cake mixtures, and other fat-containing foods. β-Carotene is available also as an emulsion, in a water-dispersed form, and as a liquid suspension. Canthaxanthin is commercially available as a 10% water-dispersable beadlet or spray-dried powder. Because of its exceptionally good tomato color-enhancing properties, it finds use in tomato-based products such as pizza and spaghetti sauce. It is useful in water-based foods such as peach ice cream and pink grapefruit beverages. β-Apo-8′-carotenal has high tinctorial strength; because of this, it is marketed in several different strengths, usually as a dispersion in vegetable oils. Its main use is for coloring process cheese and French dressing.

Carotenoids have two general characteristics of importance to the food industry: they are not pH sensitive in the normal 2–7 range found in foods, and they are not affected by vitamin C, making them especially important for beverages.

Bixin (CI Natural Orange 4; CI 75120) is found in the seed of the plant *Bixa orellana,* native to India. Later it was found growing in South America, where the Indians used the red dye from the seeds as a body paint. An extract of the seeds appears on the market as annatto. This extract is used in coloring butter, margarine, and cheese such as Leicester cheese. In Mexican and South American cuisine, it finds special use as a flavor and coloring matter. Annato is available as an aqueous solution, as an oleaginous dispersion, and a spray-dried powder.

Crocetin (CI Natural Yellow 6; CI 75100) occurs in saffron as crocin, the digentiobiose ester of crocetin. Saffron is found in the pistils of the plant *Crocus sativus.*

Betalaines. In 1968, the term betalaines was used to describe collectively two groups of plant pigments: the red betacyanins and the yellow betaxanthins. The red and yellow dyes found in beets, *Beta vulgaris,* fall into this category.

The color of betalaines is barely affected by the pH range normally found in foods. However, the dyes are heat-sensitive, which places some limitations on their use as food dyes.

Beet juice contains about 80% of fermentable carbohydrates and nitrogenous compounds. To remove these compounds, a yeast fermentation utilizing *Candida utillis* has been suggested. By so doing, a more concentrated form of the dye becomes available. The red dye from beets is sold as beet juice concentrate, as dehydrated beet root, and as a dried powder.

Chlorophyll. Chemically pure chlorophyll is difficult to prepare, since it occurs mixed with other colored substances such as carotenoids. Commercially it is solvent extracted from the dried leaves of various plants such as broccoli or spinach. Chlorophyll is water-insoluble. It has none of the characteristics of a dye in that it has no affinity for the usual fibers such as cotton or wool. Chlorophyll is properly classified as a pigment (CI Natural Green 3; CI 75810). As such, it finds use for coloring soaps, waxes, inks, fats, or oils. Chlorophyll is an ester composed of an acidic part, chlorophyllin, esterified by an aliphatic alcohol known as phytol. Hydrolysis of chlorophyll using sodium hydroxide produces the moderately water-soluble sodium salts of chlorophyllin, phytol, and methanol. The magnesium in chlorophyllin may be replaced by copper. The sodium copper chlorophyllin salt is heat-stable, and is ideal for coloring foods where heat is involved, such as in canning.

Health, Safety, and Environmental Factors of Natural Dyes

Natural dyes comprise those colors derived from plant or animal matter without chemical processing. Modern tests have verified the safety of natural dyes as food colorants; many of these dyes are on the FDA's list of approved food dyes.

Natural dyes processed for the market do not undergo any chemical operations. Those operations involved are purely physical, such as grinding, spray or vacuum drying, and water or solvent extractions. None of these operations create any great environmental problems.

A. J. COFRANCESCO
Consultant

A. G. Perkin and A. E. Everest, *The Natural Organic Colouring Matters,* Longmans, Green and Co., New York, 1981.

W. F. Leggett, *Ancient and Medieval Dyes,* Chemical Publishing Co., New York, 1964.

R. H. Thomson, *Naturally Occurring Quinones,* 3rd ed., Academic Press, Inc., New York, 1987.

DYES, REACTIVE

Reactive dyes are those dyes containing electrophilic functional groups capable of reacting with a nucleophile to form a covalent bond either through addition or displacement. Nucleophiles within fibers that typically react with dyes are hydroxyl groups in cellulose, amino, hydroxyl, and thiol groups in wool, and amino groups in polyamide. The outstanding characteristic feature of reactive dyes is their high wetfastness properties attributed to covalent bonding, an advantage over those dyes fixed through adsorption or mechanical entrapment. Unlike large bulky direct dyes, since reactive dyes are chemically bonded to the substrate, smaller molecules giving brighter colors are possible. The principal use of reactive dyes is far and away greatest for cellulose (cotton and rayon), followed by wool, with polyamide (nylon) a very distant third. Applications to silk and leather represent only very minor uses (see DYES, APPLICATION AND EVALUATION).

Reactive Dye Structure

Reactive dyes consist basically of three components: a dye, a bridging group (B), and the reactive group (R), dye—B—R. The reactive group may be considered in two parts as a carrying group and the reactive component.

Most commonly used chromophores parallel those of other dye classes. Azo dyes (qv) represent the largest number with anthraquinone and phthalocyanine making up most of the difference. Metallized azo and formazan dyes are important and have gained in importance as a chromophore for blue dyes during recent years (see DYES AND DYE INTERMEDIATES).

Yellow dyes are generally monoazo, and most are pyrazolone or pyridone couplings. Orange dyes are generally monoazo derived from couplings to pyrazolones or of slightly substituted phenyl and naphthyl groups.

Many red dyes are based on H-acid (**1**), eg, Reactive Reds 2, 24, and 218. Others are substituted phenyl and naphthyl or metallized systems. Violet dyes are also metallized monoazo dyes.

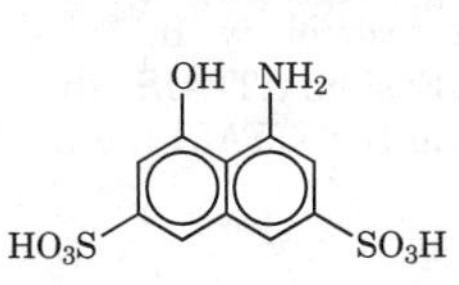

(**1**)

Blue dyes are derived from anthraquinone, phthalocyanine, or metallized formazan (see DYES, ANTHRAQUINONE) (Figs. 1 and 2). There are also oxazine and thiazine dyes reported (see AZINE DYES) (Fig. 2).

Brown and black dyes are generally disazo with exceptions for metallized or polycyclic structures. Two disazo dyes are Reactive Brown 11 (**9**) and Reactive Black 5 (CI 20505) (**10**).

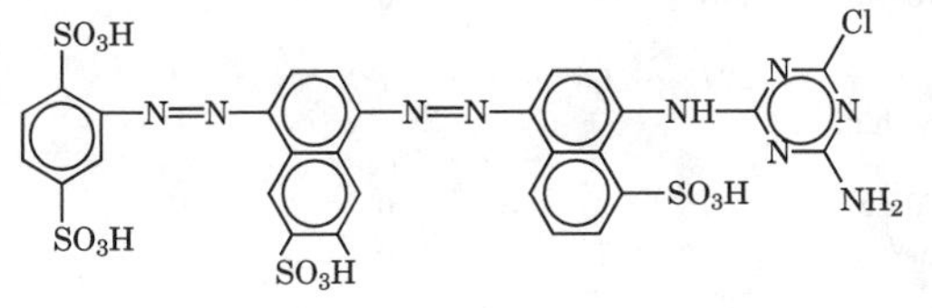

(**9**)

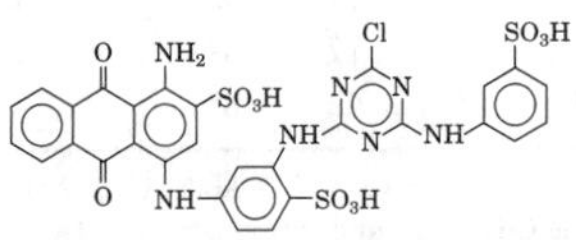

(**2**)

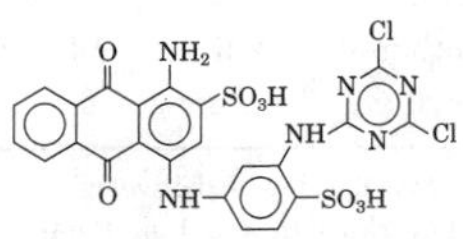

(**3**)

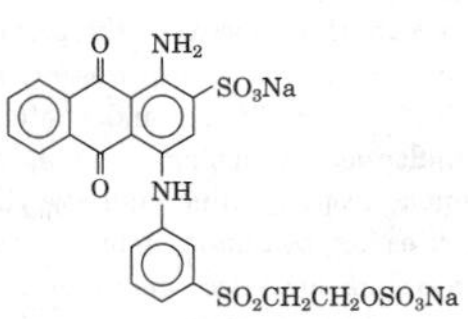

(**4**)

Figure 1. Blue reactive dyes from anthraquinone. Reactive Blue 5 (CI 61210) (**2**), Reactive Blue 4 (CI 61205) (**3**), eg, Procion Blue MX-R, Reactive Blue 19 (CI 61200) (**4**), eg, Remazol Brilliant Blue R.

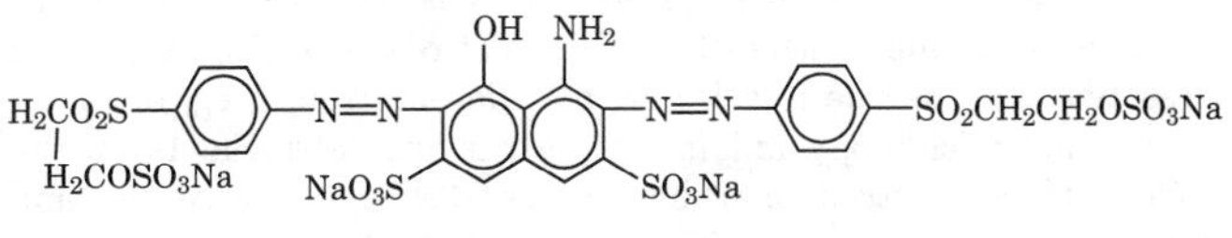

(**10**)

Green dyes are obtained by bridging an anthraquinone blue chromogen with a yellow chromogen, as in the following reactive green (**11**) or from phthalocyanine.

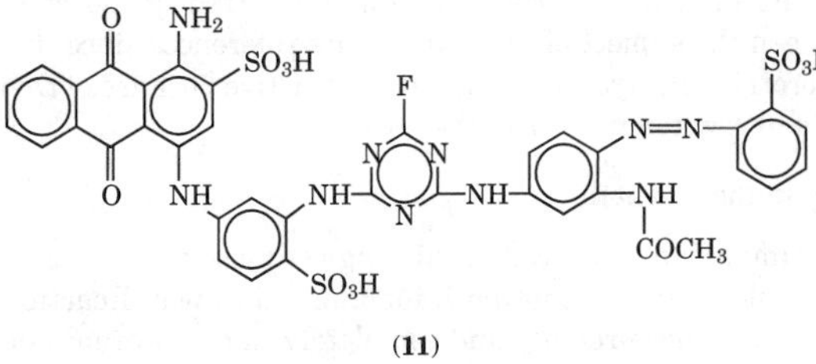

(**11**)

Properties

Reactive Groups. Although fastness of reactive dyes to washing is good, once fixed on the substrate, hydrolysis during application leads to low dye fixation rate, ie, dye that hydrolyzes in the application process is not fixed on the fiber. Reactive dyes are highly sulfonated, and as such are very water-soluble. Fairly large amounts of salt are required to force the dye into the substrate during application. The hydrolyzed dye is even more soluble than the dye, and presents concerns for the dyer because of effluent color and low biodegradability as well as economic losses. A very large part of the development effort spent on reactive dyes during recent years has been toward improving fixation on the substrate. Approaches taken have been to find different reactive groups as well as to include more than one reactive group in the dye structure and select reactive groups less sensitive to hydrolysis or more reactive to the substrate under application conditions.

During the 1980s, there was a revived interest in bireactive dyes. Every principal dye manufacturer has introduced dyes with more than

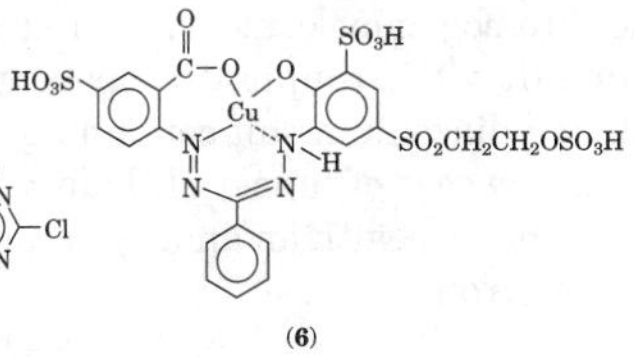

(**6**)

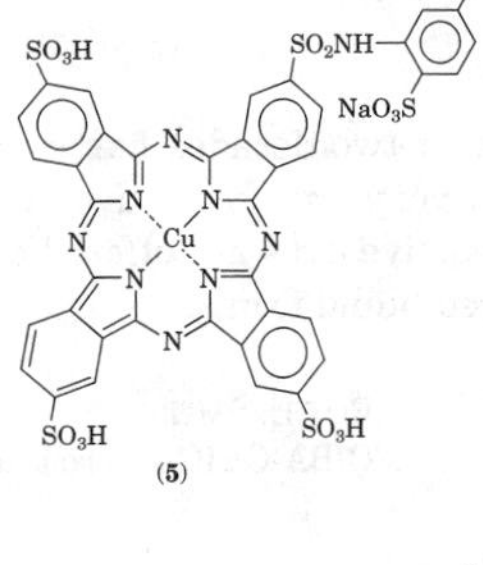

(**5**)

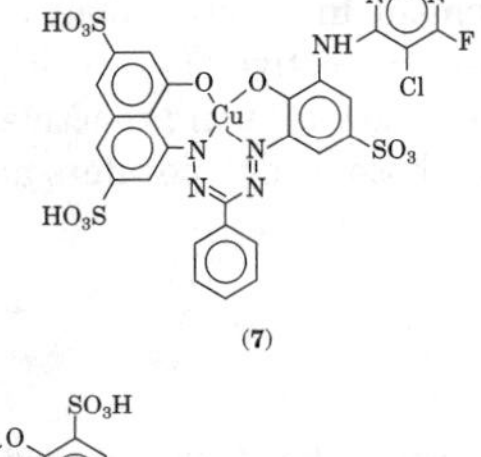

(**7**)

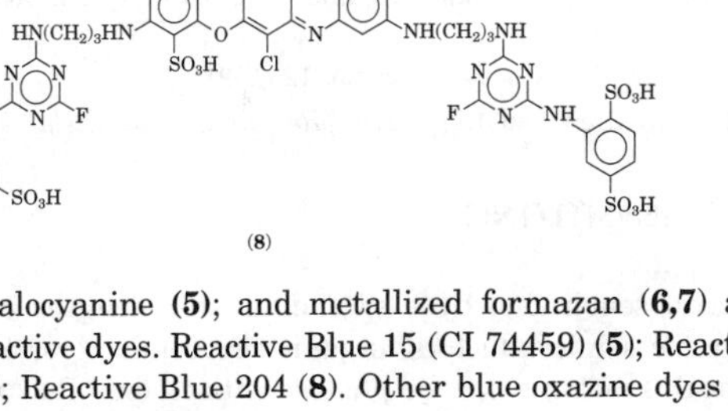

(**8**)

Figure 2. Phthalocyanine (**5**); and metallized formazan (**6,7**) and azine (**8**) blue reactive dyes. Reactive Blue 15 (CI 74459) (**5**); Reactive Blues (**6**) and (**7**); Reactive Blue 204 (**8**). Other blue oxazine dyes are other fluorotriazines and a dichlorotriazine.

one reactive group. The ones offering highest fixation have two or more reactive groups with different rates of reaction. Different rates of reaction may be due to selection of different groups, eg, dichlorotriazine and trichloropyrimidine, or by varying substituents on the triazine ring; electron-donating groups decrease reactivity, and electron-withdrawing groups increase reactivity.

Reactive groups have minimal auxochrome effect on color intensity, and color yield per molecular weight decreases with increasing numbers of reactive groups. Increased dye fixation and reduced environmental impact of hydrolyzed dye more than compensate for color reduction of additional reactive groups.

The fluorotriazine reactive group was reported in the mid-1970s, and has been the subject of many literature references since that time. Monofluorotriazine dyes are especially attractive because of their high color yield at low temperatures (below 40°C).

Reactivity in the Dyebath

The most important discovery in dyeing cellulose with reactive dyes was the application of Schotten-Baumaun principles. Reaction of alcohols proceeds more readily and completely in the presence of dilute alkali, and the cellulose anion (cell$-O^-$) is considerably more nucleophilic than is the hydroxide ion. Thus the fixation reaction (eq. 1) competes favorably with hydrolysis of the dye (eq. 2).

$$\text{dye—B—R} + \text{cell} - O^- \rightarrow \text{dye—B—O—cell} \quad (1)$$

$$\text{dye—B—R} + OH^- \rightarrow \text{dye—B—OH} \quad (2)$$

The most important reactive groups today are monochlorotriazine (high energy), vinyl sulfone (medium energy), and monofluorotriazine (low to medium energy), although dichlorotriazine, 2,3-dichloroquinoxaline and 2,4-difluoro-5-chloropyrimidine have a significant presence. α-Bromoacrylamide and vinyl sulfone from *N*-methyltaurine are the most important reactive dyes for wool.

Bifunctional Dyes. There are many examples of dyes with two or more reactive groups, including many mixed reactive systems. Dye fixation is increased significantly with increasing number of reactive groups. Some multiple reactive dyes are claimed to have as high as 95% fixation.

Methods of Synthesis

Reactive dyes are synthesized by (*1*) condensation of an amine function in the chromogen molecule with a reactive group; (*2*) by coupling a diazonium salt with a coupling component that has a reactive group, or by coupling a diazonium salt containing a reactive group with a coupler; or (*3*), in the case of copper phthalocyanine (CPC), by condensing CPC sulfonyl chloride with an amino-containing bridging group attached to a reactive group.

Economic Aspects

Development in reactive dyes over the past two decades has clearly been the most active of any class of dyes. Every significant dye manufacturer is now offering reactive dyes. Reactive dyes are offered commercially both as dry powders and buffered liquid forms.

ROY E. SMITH
CIBA-GEIGY Corporation

M. J. Bradbury, P. S. Collishaw, and D. A. S. Phillips, *J. Soc. Dyers Colourists,* 108, (1992).

H. Zollinger, *Text. Chem. Color.* **23,** 12 (1991).

A. H. M. Renfrew and J. A. Taylor, *Rev. Prog. Color. Relat. Top.,* **20** (1990).

DYES, SENSITIZING

Spectral sensitizing dyes extend the wavelengths of light to which inorganic semiconductors, organic semiconductors, and chemical (biological) reactions can be photosensitized (see PHOTOCHEMICAL TECHNOLOGY). Spectral sensitizers are needed for the blue, green, and red portions of the visible spectrum (see COLOR PHOTOGRAPHY). For infrared photographic, electrophotographic, and biological applications, sensitizing dyes are also needed to match output wavelengths of solid-state lasers (optical data storage, laser printing, range finding, and data transmission), to provide color-selective infrared photography (an effective environmental survey method), or to match transmission wavelengths of body tissue.

Spectral sensitizing dyes are considered "functional" dyes to distinguish them from conventional colorants. The absorption of radiation by a functional dye causes some additional function(s) to occur, and in many cases, the sensitizing dyes can exhibit more than one type of sensitization reaction. Many texts and reviews provide extensive details about the diversity of spectrally sensitized processes.

The detection of spectral sensitizing action often depends on amplification methods such as photographic or electrophotographic development or, alternatively, on chemical or biochemical detection of reaction products. Separation of the photosensitization reaction from the detection step or the chemical reaction allows selection of the most effective spectral sensitizers. Prime considerations for spectral sensitizing dyes include the range of wavelengths needed for sensitization and the absolute efficiency of the spectrally sensitized process. Because both sensitization wavelength and efficiency are important, optimum sensitizers vary considerably in their structures and properties.

Structural Classes of Spectral Sensitizers

A useful classification of sensitizing dyes is the one adopted to describe patents in image technology. In Table 1, the Image Technology Patent Information System (ITPAIS), dye classes and representative patent citations from the ITPAIS file are listed as a function of significant dye class.

Table 1. Dyes and Semiconductors, Patent Citations[a]

ITPAIS classification	Silver halide	ZnO	TiO_2	ZnS	Se	CdS, CdSe	Misc. organic semiconductor
dyes, cyanine[b]	420	33	27	10	45	16[c]	11
dyes, merocyanine[b]	118	2	1	0	27	2	0
all other polymethine dyes[b]	277	5	4	1	34	3	4
dyes, acridine	0	2	0	0	1	0	0
dyes, azine[d]	4	4	3	0	7	5	0
dyes, azo	107	18	28	3	28	10	4
dyes, arylmethane	3	3	1	2	7	2	0
dyes, quinone type[e]	48	10	16	1	21	2	1
dyes, porphine[f]	12	8	17	1	24	2	1
dyes, xanthene[g]	6	14	11	3	17	3[h]	3
dyes, pyrylium[i]	1	0	1	0	42	3[h]	4

[a] ITPAIS was developed between 1975–1985 by Eastman Kodak Co., Agfa-Gevaert (Antwerp/Leverkusen), and Fuji Photo Film Co., Ltd., and encompasses selected patents and literature references related principally to the chemical aspects of image technology. Search terms used for this table were the same as in the previous edition, and the Derwent patent database was used for the search data presented here. [b] Dyes, polymethine: used for dyes having at least one electron donor and one electron acceptor group linked by methine groups or aza analogues; allopolar cyanine, dye bases, complex cyanine, hemicyanine, merocyanine, oxonol, streptocyanine, and styryl. Supersensitization has been reported for these types—18 cites for cyanines, 3 for merocyanine, and 6 for all other polymethine types. [c] Also 3 citations for misc. inorganic semiconductors. [d] Dyes, azine: used for azines, thiazines, and oxazines (see AZINE DYES). [e] Dyes, quinone type: used for anthraquinones, indamines, indoanilines, indophenols, and miscellaneous quinones (see DYES, ANTHRAQUINONE). [f] Dyes, porphine derivative: used for chlorophyll, phthalocyanines, and hemin. [g] Dyes, xanthene: used for eosin, fluorescein-type phthaleins, rhodamines, and rose bengal. [h] Also 1 cite for misc. inorganic semiconductors. [i] Dyes pyrylium: also used for thiapyrylium and benzothiapyrylium.

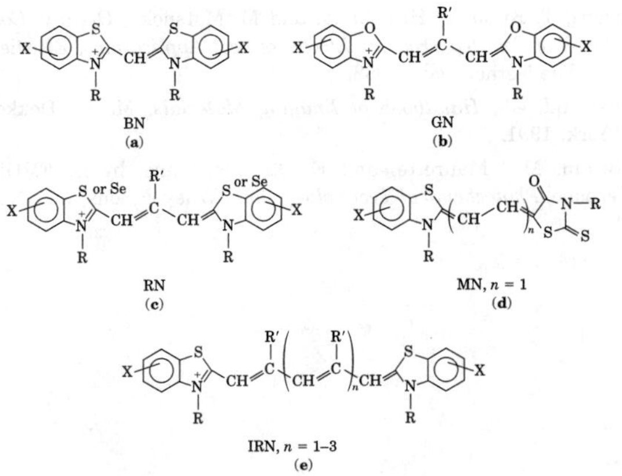

Figure 1. Spectral sensitizing dyes for silver halides (**a**) Blue sensitizers (400–500 nm) are designated BN; (**b**) green sensitizers (500–600 nm) are designated GN (the ring oxygen may be replaced by N(R)); (**c**) red sensitizers (600–700 nm) are designated RN; (**d**) MN designates a merocyanine dye; and (**e**), infrared sensitizers (>700 nm) are designated IRN.

Spectral Sensitization of Silver Halides

The large number of patents for spectral sensitizers of silver halides indicates their extensive use in photographic plates, films, and papers. Color films that give good color reproduction under a variety of illuminants (daylight, electronic flash, tungsten) and exposure times (short for electronic flash, long for available light) have required more extensive tailoring of spectral sensitizers than imaging systems based on other semiconductors. Early reviews cover spectral sensitizing dyes specifically for silver halides, including synthetic methods, general sensitization mechanisms, electrochemical potentials and dye efficiency, dye absorption and aggregation, and supersensitization of dyes.

Photographic products are often subjected to wide variations in temperature, humidity, and time before processing. Consequently, high priorities are also given to spectral sensitizers that do not degrade either the film or image quality, as well as provide efficient spectral sensitization at the desired wavelengths. Although there are many available dyes and pigments, commercial silver halide films, papers, and plates are efficiently sensitized by just a few types of chromophores in the cyanine and merocyanine dye classes (Fig. 1).

The suitability of these chromophores rests in large measure in being able to simultaneously optimize three properties (electrochemical potentials, J-aggregation, and solubility) by choosing various substituent and heteroatom combinations.

Spectral Sensitization of Inorganic and Organic Solids

Relatively few spectral sensitizing dyes continue to serve the large-volume color (imaging and printing) photographic markets, where silver halide is the dominant semiconductor. Technology to serve both color printing and office copy markets will employ not only silver halides but other inorganic and organic semiconductors as well. Inorganic semiconductors include selenium, germanium, CdS, HgO, HgI_2, ZnO, PbO, Cu_2O, thallium halides, and TiO_2. As noted by patent citations and the published literature, both zinc oxide and titanium dioxide have been extensively investigated. Spectrally sensitized photoconduction has also been observed for organic semiconductors, ie, anthracene, poly(*N*-vinylcarbazole), polyacetylene, copper phenylacetylide, phthalocyanines, and other solid dyes.

Continued commercial development of nonsilver office copying into color printing and optical data storage to higher densities provides many opportunities for spectral sensitizer improvement among competing technologies. Spectral sensitizers for these other materials are already quite varied (Table 2). Recent handbooks review both sensitizing dye technology and imaging systems that utilize dyes.

Table 2. Spectral Sensitizers for Inorganic and Organic Semiconductors

Semiconductor	Sensitizer
CdS (Cu doped)	thiacarbocyanine
	benzothiazolylrhodanines
	rhodamine B
ZnO	thiacarbocyanine
	erythrosin
	eosin
	phthalocyanine
	rhodamine B
	rose bengal
	methylene blue
	fluorescein
TiO_2	benzothiazolylstyryl dye
	thiacarbocyanine
poly(*N*-vinylcarbazole)	rose bengal
	2,4,7-trinitrofluorenone
	thiapyrylium dyes
solid dye particles	phthalocyanines
photoelectrophoresis	thioindigo
	flavanthrone
	quinacridone
dye electrodes	benzothiazolylrhodanines
	phthalocyanines

Spectral Sensitization in Photochemical Technology

Functional dyes of many types are important photochemical sensitizers for chemical reactions involving oxidation, polymerization, (polymer) degradation, isomerization, and photodynamic therapy. Often, dye structures from several classes of materials can fulfill a similar technological need, particularly for laboratory or small-scale reactions where production efficiency may be of secondary importance. Commercial photochemical technology, however, is more selective and requires photochemical efficiency, ease of product separation, and lack of unwanted side reactions to an extent similar to that required by imaging processes. In addition, reusability of the spectral sensitizer is also preferred in commercial photochemical reactions.

Uses and Suppliers

Sensitizing dyes are used primarily for specialty purposes: photographic sensitizers, electrophotographic sensitizers, laser dyes, infrared (optical disk, etc) imaging, and certain medicinal applications. Because of this, their manufacture is limited to significantly smaller quantities than for fabric dyes or other widely used coloring agents. However, the photographic, laser, and medicinal uses place high demands on the degree of purity required, and the reproducibility of synthetic methods and purification steps is very important. Suppliers of cyanine dyes include manufacturers of other specialty organic and photographic chemicals: Aldrich Chemical Co. (Milwaukee, Wisconsin), Eastman Chemical Company (Kingsport, Tennessee), Japanese Institute for Photosensitizing Dyes (Okayama, Japan), Molecular Probes (Eugene, Oregon), NK Dyes (Japan), Pfaltz and Bauer (Stamford, Connecticut), Riedel deHaen (Karlsruhe, Germany), and H. W. Sands. More importantly, these firms provide sources of generally useful reagents that, in two or three synthetic steps, lead to many of the commonly used sensitizers.

Sensitizing Dye Toxicology

Nearly every significant class of dyes and pigments has some members that function as sensitizers. Toxicological data are often included in surveys of dyes, reviews of toxic substance identification programs, and in material safety data sheets provided by manufacturers of dyes. More specific data about toxicological properties of sensitizing dyes are contained in the *Encyclopedia* under the specific dye classes.

DAVID M. STURMER
Eastman Kodak Company

D. M. Sturmer, in E. C. Taylor and A. Weissberger, eds., *Special Topics in Heterocyclic Chemistry, The Chemistry of Heterocyclic Compounds,* Vol. 30, Wiley-Interscience, New York, 1977, p. 441.

M. Okawara, T. Kitao, T. Hirashima, and M. Matsuoka, *Organic Colorants, Handbook of Selected Dyes for Electro-optical Applications,* Elsevier, Amsterdam, The Netherlands, 1988.

A. S. Diamond, ed., *Handbook of Imaging Materials,* Marcel Dekker, Inc., New York, 1991.

A. M. Braun, M-T Maurette, and E. Oliveros, trans. by D. F. Ollis and N. Serpone, *Photochemical Technology,* John Wiley & Sons, Inc., New York, 1991.

E

ECONOMIC EVALUATION

Economic evaluation is an assessment of the probable benefit or reward of a proposed course of action, relative to other choices. This type of evaluation is a primary planning activity from the pilot-plant to the corporate level, involving research, engineering, manufacturing, finance, sales, marketing, and other business groups (see also MARKET AND MARKETING RESEARCH; RESEARCH TECHNOLOGY MANAGEMENT). Such evaluation is an evolutionary process that is repeated endlessly from the research stage through the manufacturing cycle. Economic studies are employed to assess the expected profitability as the result of various choices such as plant size, location, pricing, product grades, and engineering features, or to make a comparison with other ventures, including the competition. Statistical methodologies can be employed to make use of probabilistic data. The four essential parts of any economic evaluation are problem definition, cost estimation, revenue estimation, and profitability analysis.

Cost Estimation

The three general types of cost estimates needed are equipment cost, capital investment cost, and product cost. Equipment cost estimates are needed as part of the capital investment estimate, which indicates the amount of money that is needed to start the venture. Both of these estimates are reflected in the product cost estimate, which is important to both management and marketing groups.

Equipment Costs. Equipment costs include the purchased cost of process and materials handling equipment, storage facilities, waste treatment equipment, structures, and site service facilities. Installation costs such as insulation, piping, painting and finishing, foundations, process structures, instrumentation, and electrical service connections are estimated or factored separately. Actual quoted prices from suppliers are the best data, but these are not usually available when estimates are made. The quick, inexpensive cost estimates are based largely on personal cost files, internal company cost data, or published cost correlations.

Capital Investment Cost. The capital investment involved in a proposed project is important because it represents the money that must be raised to get the project started, is used in profitability forecasts, and is reflected in the estimated manufacturing cost of a product. The capital investment is classified herein as fixed capital, working capital, and land cost. Sample capital investment estimate forms provide for separate materials (M) and labor (L) categories, or just combined M&L figures. Estimates include those for fixed capital, order-of-magnitude estimates, predesign estimates, overall factor estimates, category factor estimates, module factor estimates, estimates for unproven technology, and estimates of working capital.

Product Cost. An estimate of total product cost is an important part of economic evaluation and management planning. Total product cost can be viewed as the sum of the manufacturing cost and the general expense.

Typical operating cost data in dollars per weight or volume basis of product are frequently available. These data, which usually do not include capital costs, can be scaled directly over a moderate range of annual capacities to give rough estimates of annual operating costs as a function of annual capacity. Capital costs can be annualized using a capital charge factor, which converts the capital cost to equivalent annual costs.

Unit cost data should be carefully assessed to ensure that process type, size, and raw materials are similar to the proposed venture. Operating cost data sometimes are reported for separate categories such as operating labor, maintenance labor, supervision, and utilities.

A more detailed product cost estimation method is to relate manufacturing cost items to a few calculated items, such as raw materials, labor, and utilities by means of simple factors. Internal accounting groups often develop factors to use with this method. This factor method is very popular.

Revenue Estimation

Revenues are money inflows from venture activities. Product sales, based on the unit product price ($/unit) and the yearly sales volume (units/yr), are the only type of revenue considered herein. The price and sales volume are related to each other and typically contribute much of the uncertainty in economic forecasts of chemical process systems. Whereas the estimation of price and sales volume is a marketing or sales function, technical factors are frequently part of the estimation process.

Product Price. If a need is not met by any other available product, then price can be set as high as the need can support. However, a very high price tends to limit market growth and encourage the introduction of competitive or substitute products. The preferred strategy is to establish a moderately high initial price, but to plan on future price reductions to help expand the market and meet any competitive pressures.

If a need is met by a variety of products, then the price should reflect the price of these competitive products and any unique or advantageous features of the product being priced.

If the product is essentially identical to that produced by other manufacturers, then the price is determined principally by the commodity market price. However, contract features such as guaranteed delivery schedules can influence price.

Sales Volume. The quantity of annual sales is often called the sales volume, although the units may be mass quantities instead of volume units. The estimation of the annual sales volume over the expected life of a production facility is extremely difficult. It requires an estimate of the total market, production capabilities and costs of the competition, market share expected, selling strategy to be employed, and future economic conditions.

The expected annual sales volume is important not only for estimating sales revenue, but also for the selection of plant capacity or process type.

Profitability Analysis

Profitability analysis involves the generation of criteria that relate to the financial return to be expected from a proposed investment choice, as well as a comparative assessment with other choices. It provides quantitative measures that aid in the subjective decision-making process; this is an art that depends on both experience and luck.

Essence of Profitability. If an investor purchases a computer in the morning and sells it in the afternoon for a larger sum, then a return on the investment is realized. This is profit; a reward for the effort (investment) made. A somewhat more difficult situation is the case of three choices having different purchase costs and expected selling prices.

Before a decision is made, all three items, ie, investment, return, and rate of return, would be examined, as would the current cash position, perceived risk, other venture opportunities, and a variety of subjective criteria. For this elementary situation, economists would also employ an incremental approach analogous to the above, based on the tenet that each increment of investment should itself make an adequate return. Rarely is there a unique correct decision. Only future events determine the wisdom of the selection; even then, the results that another decision would have produced are rarely known. This is the essence of profitability analysis.

In multiyear process ventures, the money flows are more complicated and must be discounted to a common point in time before they can be combined. However, the three basic parameters, investment, return, and rate of return, should be retained in some logical and consistent manner.

Multiyear Process Ventures. The economics of a proposed process venture, from initial development through final shutdown, should be examined. A venture timetable and schedule of money flows is part of the problem definition. All money flows are typically tabulated as end-of-year flows for simplification. Investments and other outflows are negative; sales revenues and other inflows are positive.

Profitability Analysis. Modern profitability analysis is based on the annual cash flows of the venture, where the annual cash flow is defined as:

$$\text{net annual cash flow} = \text{after-tax income} + \text{depreciation tax-basis} + \text{end-of-life items} - \text{yearly investment}$$

These cash flows are discounted to a common time point to account for the time value of money. The principal parameters for the profitability analysis are the Net Present Value (discounted return), Net Return Rate (discounted annual return rate), and Discounted Total Capital (discounted investment). Other discounted criteria that still find use include the Internal Return Rate and Overall Return Rate. Simplified criteria that do not involve discount considerations are the Return on Investment and Payout Time.

A simplified annual cost view is sometimes employed when comparing alternatives. The approach is to convert the capital cost to an equivalent annual expense using a fixed charge factor or an amortization factor. Then the total annual costs or profits of the choices can be compared. This type of analysis is commonly employed for components where one choice might have a higher capital cost, but a lower operating cost. The fixed charge factor is a rate that represents the annual cost of capital and an annual component of capital recovery; typical values range from 12 to 20%.

In energy production ventures, the unit energy costs (cents per kilowatt-hour) are estimated as a measure of venture feasibility. An averaging approach widely used for such projects includes converting the cost items occurring in the various years to an equivalent present worth using a discount factor, and then converting the present worth to an equivalent uniform annual amount using the capital recovery factor

$$\left[\frac{i(1+i)^n}{(1+i)^n - 1}\right]$$

where i is the annual discount rate in decimal form and N is the number of years involved in the translation, for the project lifetime. The result is a levelized annual amount.

When mutually exclusive ventures having different levels of investment are compared, an attractive concept is that each increment of investment must itself yield a satisfactory return. This concept has led to a variety of incremental approaches for profitability analysis. Because the risk can vary with investment level and cloud the meaning of satisfactory, any incremental approach to multiyear investment analysis should be viewed with caution.

Break-Even Charts. A break-even chart is a visual tool for analyzing operating profitability at various levels of production. In this type of diagram, annual expenses are separated into fixed, variable, and semivariable categories.

A typical break-even chart is used with production models to predict optimum production levels, break-even points, and shutdown conditions under various scenarios. These models tend to involve a reasonable amount of approximation.

Inflationary Effects. Inflation can have a significant effect on the profitability of a venture. However, the U.S. federal tax laws do not allow for indexing the inflationary effects on depreciation schedules, salvage values, replacement costs, or taxable income. Inflation rates can vary unpredictably with time and can differ for certain revenues or expenditures.

Prevailing interest rates probably tend to reflect an estimate of future inflation and contain a component that can be attributed loosely to inflationary expectations. However, the classical treatment is to assume that an inflation-free interest rate, r_e, and average inflation rate, r_i, over the project lifetime can be identified. A discount factor $(1 + r)^{-n}$ can be modified so that $(1 + r)^{-n} = [(1 + r_e)^{-n}(1 + r_i)^{-m}]$, where m is the number of years to the reference or constant dollar year. It is often assumed that $m = n$. Various treatments of inflationary effects have been reported.

Statistical Criteria. In order to treat probability, statistical measures are employed to characterize the probability distributions. Because most distributions in profitability analysis are not accurately known, the common assumption is that normal distributions are adequate. The distribution of a quantity then can be characterized by two parameters, the expected value and the variance. These usually have to be estimated from meager data.

The two principal approaches of interest for situations involving risk and uncertainty are decision trees and Monte Carlo simulation. These find wide application in general business analysis but are less widely used in engineering economic analysis.

THOMAS J. WARD
Clarkson University

T. Au and T. P. Au, *Engineering Economics for Capital Investment Analysis,* 2nd ed., Prentice-Hall, Englewood Cliffs, N.J., 1992.

O. Axtell and J. Robertson, *Economic Evaluation in the Chemical Process Industries,* John Wiley & Sons, Inc., New York, 1986.

T. J. Ward, *Chem. Eng.* **101**, 1, 102 (1994).

J. A. White, M. H. Agee, and K. Case, *Principles of Engineering Economic Analysis,* 3rd ed., John Wiley & Sons, Inc., New York, 1989.

EGGS

Eggs are defined herein as eggs from chickens, and refer to both shell eggs and/or egg products. Egg products in liquid, frozen, or dried form contain egg as the principal ingredient. An egg product can be anything from a frozen or dried product made from 100% egg white to a scrambled egg mix that has 51% whole egg. Eggs are primarily used as food. Shell eggs used in the home, restaurants, and institutions are fried, hard-cooked, poached, etc, or are used as ingredients in other foods. Egg products generally are utilized in the food industry.

Eggs contribute important proteins (qv), fats, vitamins (qv), and minerals to the diet. Comprehensive reviews of the chemistry and biology and marketing of eggs are available.

Properties

Physical Properties. The egg is composed of three basic parts: shell, whites (albumen), and yolk. Each of these components has its own membranes to keep the component intact and separate from the other components. The vitelline membrane surrounds the yolk, which in turn is surrounded by the chalaziferous layer of albumen, keeping the yolk in place. Egg white (albumen) consists of an outer thin layer next to the shell, an outer thick layer near the shell, an inner thin layer, and finally, an inner thick layer next to the yolk. Table 1 shows the various physical properties for components of eggs.

Functional Properties. Eggs function in different ways to give food products certain desirable characteristics. Functions include coagulating and thickening, whipping or beating, and emulsifying.

Chemical Properties. Egg white contains mostly proteins having the physical and chemical characteristics given in Table 2. Some proteins in egg white have biological activities that protect from microbiological growth, eg, lysozyme lyses certain bacteria, conalbumin ties up iron, and avidin binds biotin. Most of these activities are destroyed when the egg white is cooked. pH of liquid egg white is normally about 9.0. However, egg white from freshly laid eggs has a pH of about 7.6. pH increases quite rapidly as carbon dioxide escapes during storage. The high 9.0 pH of natural egg white retards the growth of many bacteria.

The yolk is separated from the white by the vitelline membrane, and is made up of layers that can be seen upon careful examination.

Table 1. Physical Properties of Liquid Egg Products

Property	Whites	Yolks	Whole
solids, %	12.1	44.0	24.5
specific gravity	1.035	1.035	1.035
specific heat	0.940	0.780	0.880
freezing point, °C	−0.4	−0.4	−0.4
specific heat below freezing	0.500	0.500	0.500
latent heat of freezing, kJ/kg[a]	531.4	338.9	451.9
viscosity, mPa(= *cP*)			
5°C	12	260	20
50°C	5		
60°C		45	7

[a] To convert kJ to kcal, divide by 4.184.

Table 2. Proteins in Egg White

Protein	Amount of albumen, %	pH[a]	Mol wt
ovalbumin[b]	54	4.5	45,000
ovotransferrin[c]	12	6.1	76,000
ovomucoid[d]	11	4.1	28,000
ovomucin[e]	3.5	4.5–5.0	$5.5-8.3 \times 10^6$
lysozyme[f]	3.4	10.7	14,300
G_2 globulin[g]	4.0	5.5	$3.0-4.5 \times 10^4$
G_3 globulin	4.0	4.8	
ovoinhibitor	1.5	5.1	49,000
ficin inhibitor	0.05	5.1	12,700
ovoglycoprotein	1.0	3.9	24,400
ovoflavoprotein	0.8	4.0	32,000
ovomacroglobulin	0.5	4.5	$7.6-9.0 \times 10^5$
avidin	0.05	10.0	68,300

[a] Isoelectric point. [b] Denaturation = 84.0°C. [c] Also known as conalbumin; denaturation = 61.0°C. [d] Denaturation = 70.0°C. [e] Denaturation (sialoprotein) is viscous. [f] Denaturation = 75.0°C. [g] Denaturation = 92.5C.

Egg yolk is a complex mixture of water, lipids, and proteins. Lipid components include glycerides, 66.2%; phospholipids, 29.6%; and cholesterol, 4.2%. The phospholipids consist of 73% lecithin, 15% cephalin, and 12% other phospholipids. Of the fatty acids, 33% are saturated and 67% unsaturated, including 42% oleic acid and 7% linoleic acid. Fatty acids can be changed by modifying fatty acids in the laying feed. (see CARBOXYLIC ACIDS).

Yolk can be separated into two fractions, granules and plasma by high speed centrifugation (see SEPARATION–CENTRIFUGAL). Granules contain a high percentage of high density lipoproteins (HDL) and lesser amounts of low density lipoproteins (LDL) and water-soluble proteins (phosvitins). The plasma contains water-soluble proteins (livetins) and finely dispersed LDL; most of the glycerides reside in this LDL fraction. The glycerides apparently form the inner core of the LDL, which is surrounded by a phospholipid shell, with protein wrapped around the shell. Half of the water in egg yolk is bound to the proteins and lipoproteins; half is free. The pH of yolk is normally about 6.6, and in freshly laid eggs it is 6.0.

Eggs, considered to be one of the most nutritious foods, have the highest quality protein of any food, and are important as a source of minerals and certain vitamins. Lipids in eggs are easily digested, and the amount of unsaturated fatty acids is greater than in most animal products.

Cholesterol has received the most attention of the components in eggs. Considerable controversy surrounds the role of dietary cholesterol in eggs and the part it plays in the development of arterioschlerosis. The concentration of blood cholesterol is not affected strongly by dietary cholesterol, but rather is dependent on the degree of saturation of dietary triglycerides. The USDA has found that the average large egg contains 22% less cholesterol than previously believed. Cholesterol is an important part of the animal tissue and cells, eg, it is necessary in the production of Vitamin D, certain hormones, and bile salts. It is carried to the tissues by blood. The body maintains a certain cholesterol level, and synthesizes any additional amount that is not supplied by the diet. There is evidence that high levels of egg in the diet do not present a greater risk of heart disease in a normal individual. However, as a precautionary measure, diets low in cholesterol may be advised for persons having higher than normal blood cholesterol levels, and for persons who may be prone to heart disease.

The egg shell is 94% calcium carbonate, $CaCO_3$, 1% calcium phosphate, and a small amount of magnesium carbonate. A water-insoluble keratin-type protein is found within the shell and in the outer cuticle coating. The pores of the shell allow carbon dioxide and water to escape during storage. The shell is separated from the egg contents by two protein membranes. The air cell formed by separation of these membranes increases in size because of water loss. The air cell originally forms because of the contraction of the liquid within the egg shell when the temperature changes from the body temperature of the hen at 41.6°C to a storage temperature of the egg at 7.2°C.

Shell Eggs

Production. Most production is carried out on farms having 30,000 hens or more per flock or house. Almost everything in the house is automated, eg, feeding, watering, ventilation, and gathering and sorting or eggs. Eggs are put on filler-flats and placed on racks to be transferred to the processing area. The racks are brought into a refrigerator at about 10°C until they are picked up for transfer to the shell egg or egg products plant. The eggs are usually tempered at 10°C before going to the egg products plant, where they are broken and separated into whites, yolks, and mix, ie, standardized whole egg solids.

Grading. Eggs are graded and sorted according to size and to quality factors, which include both shell and interior quality. In modern processing, eggs are flash-candled on a continuous conveyor within a short time after being laid. Because of their freshness, most eggs have uniformly high interior quality where the proportion of thick to thin egg white is relatively high. The USDA has an egg grading program which is run on a voluntary basis in cooperation with each state. Most states also have their own egg grading laws, usually patterned after USDA regulations. The United States size or weight classifications for eggs are listed below.

Size or weight class	Minimum net weight per dozen, kg (oz)
extra large	0.77 (27)
large	0.68 (24)
medium	0.60 (21)
small	0.51 (18)

Eggs are downgraded according to specific conditions of the shell; Table 3: the shell is unbroken and has adhering dirt or foreign materials, prominent stains, or moderate stains covering more than one-fourth of the shell surface; the individual egg has a broken shell or a crack in the shell, but the shell membrane is intact and its contents do not leak; the shell and membrane are broken so that the contents are leaking (USDA regulations prohibit use of this type of egg for human consumption).

Processing. Methods for handling shell eggs are highly automated. This includes collecting, sorting, washing, sanitizing, drying, candling, and packing.

Egg Products

Manufacturing. The first step in making egg products is breaking and separating the whites from the yolks or breaking out the eggs as whole egg.

Table 3. United States Standards for Quality of Individual Shell Eggs

Factor	AA	A	B
shell	clean, unbroken, practically normal	clean, unbroken, practically normal	clean to slightly stained,[a] unbroken, abnormal
air cell	3.2 mm or less in depth, unlimited movement, and free or bubbly	4.8 mm or less in depth, unlimited movement, and free or bubbly	over 4.8 mm in depth, unlimited movement, and free or bubbly
white	clear, firm	clear, reasonably firm	weak and watery, small blood and meat spots present[b]
yolk	outline slightly defined, practically free from defects	outline fairly well defined, practically free from defects	outline plainly visible, enlarged and flattened, clearly visible germ development but no blood, other serious defects

[a] Moderately stained areas permitted, ie, $\frac{1}{32}$ of surface if localized, or $\frac{1}{16}$ if scattered.
[b] If they are small, ie, aggregating not more than 3.2 mm in diameter.

Three components, ie, whites, yolks, and mix, flow away from the breaking machines to small inspection vats. After inspection, the liquids are pumped through filters or centrifuges, and then through cooling plates to a storage vat, where they are held for further processing. The solids of yolk and whole egg (or mix) are usually standardized at this point by the addition of whites or yolks.

The egg products are finally processed and spray-dried. Sometimes liquid egg whites are concentrated before spray-drying by ultrafiltration or reverse osmosis procedures.

Pasteurization. All egg products must be pasteurized to render them *Salmonella*-negative. Conventional plate-type pasteurizers having the usual attachments, including holding tubes, flow diversion valve, regeneration cycle, etc, are used (see STERILIZATION TECHNIQUES). Minimum pasteurization requirements are based on the bacterial kill obtained when heating whole egg to 60°C for a holding time of 3.5 min.

Liquid Egg Products. Liquid egg products include egg white; egg yolk; whole egg; extended shelf life refrigerated liquid egg products, ie, whites, yolks, whole; and concentrated sugared whole egg. These products, generally consumed by large users such as large bakeries who have the necessary handling equipment, are usually transported by refrigerated tank truck holding approximately 20 t.

Liquid egg products must be of excellent microbiological quality with very low total bacteria counts. Pasteurization conditions are more severe than conventional methods for pasteurizing egg products, and aseptic packaging is usually necessary for the success of these products.

Frozen Egg Products. Frozen egg products include egg white, plain whole egg, whole egg with yolk added (ie, fortified), plain egg yolk, fortified whole egg with corn syrup, sugared egg yolk, salted egg yolk, salted whole egg, and scrambled eggs and omelets. Egg products are frozen in a blast freezer at −40°C for up to 72 h, and then held for storage at −24°C (see REFRIGERATION). They are used by large and small bakeries and for other uses.

Dried Egg Products. Dried egg products are listed as follows:

Egg white	Whole egg	Egg yolk
spray-dried egg white solids (whipping and nonwhipping)	standard whole egg solids	standard egg yolk solids
flake albumen (pan-dried)	stabilized (glucose-free) whole egg solids	stabilized (glucose-free) egg yolk solids
instant egg white with sugar	blends of whole egg with sugar	free-flowing egg yolk solids
	blends of whole egg with corn syrup	blends of egg yolk with sugar
	free-flowing whole egg solids	blends of egg yolk with corn syrup

Most dried egg products are made by spray-drying, which produces a powder form. For almost all egg whites, and some whole egg and yolk products, the natural glucose is removed before spray-drying. This gives dried egg white products excellent stability under almost any storage conditions, and dried whole egg and yolk products good stability under room temperature as well as refrigerated storage conditions. The reducing group of the glucose reacts with the amino groups of the protein leading to browning, poor solubility, off-flavor, and off-odor developments.

Health and Safety Factors

The interior of shell eggs is mostly sterile at the time of lay. A few eggs may be contaminated inside the shell because of infection of the birds at the time the egg is being formed. However, contamination of the outside of the shell occurs after lay from fecal matter, nesting material, floor litter, dust, etc. Although shell eggs have several physical and chemical barriers that protect the contents from bacterial contamination, eg, shell membrane and antibacterial factors in egg whites, eggs have been implicated in food poisoning outbreaks. *Salmonella enteriditis* is the microorganism that has caused most of the problems in shell eggs. However, because one contaminated egg can contaminate the entire batch of eggs mixed with it, it is recommended that all eggs be cooked thoroughly before serving.

Egg products are relatively free of *Salmonella* because pasteurization and testing for *Salmonella* is required.

Certain individuals are allergic to eggs and egg products.

Economic Aspects

Following are figures, in metric tons, for the amount of liquid, frozen, and dried egg products produced in 1990 and 1991. Dried egg products are given as the liquid equivalent.

Product, t	1990	1991
liquid eggs	213,000	239,000
frozen eggs	184,000	182,000
dried eggs	201,000	227,000

An important aspect of economic consideration is the prevention of egg and egg product loss to the drain or the atmosphere.

DWIGHT H. BERGQUIST
Henningsen Foods, Inc.

R. W. Burley and D. V. Vadehra, *The Avian Egg Chemistry and Biology,* John Wiley & Sons, Inc., New York, 1989.

W. J. Stadelman and O. J. Cotterill, eds., *Egg Science and Technology,* 3rd ed., Haworth Press, Binghamton, N.Y., 1986.

Regulations Governing the Grading of Shell Egg and United States Standards, Grades, and Weight Classes for Shell Eggs, USDA 7 CFR, Part 56, U.S. Government Printing Office, Washington, D.C., May 1, 1991.

Regulations Governing the Inspection of Eggs and Egg Products, USDA 7 CFR, Part 59, U.S. Government Printing Office, Washington, D.C., May 1, 1991.

ELASTOMERS, SYNTHETIC

SURVEY

The purpose of this article is to provide a brief overview of the materials designated synthetic elastomers and the elastomeric or rubbery state. Subsequent entries describe the individual classes of elastomers in detail. Table 1 provides a fundamental description of the principal classes of synthetic elastomers.

Definition of Elastomers

The term elastomer is the modern word to describe a material that exhibits rubbery properties, ie, that can recover most of its original dimensions after extension or compression. Ever since the pioneering work of Staudinger in the early 1900s, it has been accepted that such rubbery behavior results from the fact that the material is composed of cross-linked, long-chain, flexible polymer molecules. When such a material is extended or stretched, the individual long-chain molecules are partially uncoiled, but will retract or coil up again when the force is removed because of the kinetic energy of the segments of the polymer chain. The flexibility of such polymer-chain molecules is actually the result of the ability of the atoms comprising the chain to rotate around single bonds. Theories of rubberlike elasticity are well-developed.

The properties of elastomeric materials are also greatly influenced by the presence of strong interchain, ie, intermolecular, forces which can result in the formation of crystalline domains. Irregular polymer chains will not crystallize, have weak interchain interaction and have the properties of an amorphous material.

A most interesting class of materials is comprised of elastomers with chain regularity that undergo a temporary crystallization when stretched to a high extension, thus virtually becoming fibers, but that retract to their original dimension when the force is removed. Such strain-crystallizing rubbers can thus demonstrate unusually high tensile strength, but revert to the amorphous state when the force is relaxed because of relatively weak interchain forces.

Effect of Temperature on Polymer Properties

There are two principal forces that govern the ability of a polymer to crystallize: the interchain attractive forces, which are a function of the

Table 1. Elastomers[a] and Their Characteristics

Name	Chemical name	Repeat unit structure	Vulcanizing agent	Stretching crystallization	Gum tensile strength
natural rubber	*cis*-1,4-polyisoprene (>99%)	[b]	sulfur	good	good
styrene–butadiene rubber	poly(butadiene-*co*-styrene)	$-(CH_2-CH=CH-CH_2)_m-(CH_2-CH(C_6H_5))_n-$	sulfur	poor	poor
butadiene rubber	polybutadiene (>97% *cis*,-1,4)	$-(CH_2-CH=CH-CH_2)-$	sulfur	poor to fair	poor to fair
isoprene rubber	*cis*-1,4-polyisoprene (>97%)	$-(CH_2-C(CH_3)=CH-CH_2)-$	sulfur	good	good
EP(D)M	poly(ethylene-*co*-propylene-*co*-diene)	$-(CH_2-CH_2)_m-(CH(CH_3)-CH_2)_n-(\text{diene with }CH=CH_2)_o-$	sulfur	poor	poor
butyl rubber	poly(isobutyene-*co*-isoprene)	$-(CH_2-C(CH_3)_2)_{50}-(CH_2-C(CH_3)=CH-CH_2)-$	sulfur	good	good
nitrile rubber	poly(butadiene-*co*-acrylonitrile)	$-(CH_2-CH=CH-CH_2)_m-(CH_2-CH(CN))_n-$	sulfur	poor	poor
chloroprene rubber	polychloroprene (mainly trans)	$-(CH_2-C(Cl)=CH-CH_2)-$	MgO or ZnO	good	good
silicones	polydialkylsiloxane (mainly poly-dimethyl siloxane)	$-(SiR_2-O)-$	peroxides	poor	poor
fluorocarbon elastomers	poly(vinylidene fluoride-*co*-hexafluoropropene)	$-(CH_2-CF_2)_x-(CF_2-CF(CF_3))_y-$	diamines	poor	poor
polysulfide rubber	poly(alkylene sulfide)	$-(CH_2-CH_2-S_{2-4})-$	metal oxides	fair	poor
polyurethanes	polyurethanes	$HO-(R-OCONHR'NHCOO)_x-R-OH$	diisocyanate	fair	good

[a] Not inclusive; see also ACRYLIC ELASTOMERS, PHOSPHAZENES, CHLOROSULFONATED POLYETHYLENE, ETHYLENE-ACRYLIC ELASTOMERS, and POLYETHERS under the title ELASTOMERS, SYNTHETIC.

[b] See ISOPRENE.

chain structure, and the countervailing kinetic energy of the chain segments, which is a function of the temperature. The fact that polymers consist of long-chain molecules also introduces a third parameter, ie, the imposition of a mechanical force, eg, stretching, which can also enhance interchain orientation and favor crystallization.

In addition to the phenomena of crystallization and melting, which both represent a change of state in the material, there is a third transition which plays a strong role in the behavior of polymers, although it is by no means absent in the behavior of simple liquids. This is the glass-transition temperature, T_g.

Polymers fall into two classes, those that are capable of crystallization and those that are not. A noncrystalline (amorphous) polymer is considered a liquid (although a highly viscous one), which becomes a glass at reduced temperatures. Thus, atactic polystyrene is always amorphous, but it is in a "glassy" state at room temperature, since its T_g is about 105°C. Above its T_g it becomes rubbery although its chemical stability in air at that temperature is so poor as to render it useless. On the other hand, polyethylene and isotactic polypropylene are crystalline, to a greater or lesser extent, at room temperature, and hence do not exhibit rubbery behavior even though above T_g.

The class of rubbers that show the ability to crystallize when stretched represent a special class of rubbers.

Compounding and Vulcanization of Elastomers

In order to "cure" or "vulcanize" an elastomer, ie, cross-link the macromolecular chains (Fig. 1), chemical reactants are mixed or compounded with the rubber, depending on its nature. The mixing process depends on the type of elastomer: a high viscosity type, eg, natural rubber, requires powerful mixers (such as the Banbury type or rubber mills), while the more liquid polymers can be handled by ordinary rotary mixers, etc (see RUBBER COMPOUNDING).

Typical tire rubbers, eg, natural, SBR, or polybutadiene, being unsaturated hydrocarbons, are subjected to sulfur vulcanization, and this process requires certain ingredients in the rubber compound, besides the sulfur, eg, accelerator, zinc oxide, and stearic acid. Accelerators are catalysts that accelerate the cross-linking reaction so that reaction time drops from many hours to perhaps 20–30 min at about 130°C. There are a large number of such accelerators, mainly organic compounds, but the most popular are of the thiol or disulfide type. Zinc oxide is required to activate the accelerator by forming zinc salts. Stearic acid, or another fatty acid, helps to solubilize the zinc compounds.

In addition to the ingredients that play a role in the actual vulcanization process, there are other components that make up a typical rubber compound (see RUBBER CHEMICALS). Softeners and extenders, generally inexpensive petroleum oils, help in the mastication and mixing of the compound. Antioxidants (qv) are necessary because the unsaturated rubbers can degrade rapidly unless protected from atmospheric oxygen. They are generally organic compounds of the amine or phenol type (see also ANTIOZONANTS). Reinforcing fillers, eg, carbon black or silica, can help enormously in strengthening the rubber against rupture or abrasion. Nonreinforcing fillers, eg, clay or chalk, are used only as extenders and stiffeners to reduce cost.

Figure 1. Vulcanization of rubber macromolecules: (**a**), before cross-linking; (**b**), after cross-linking.

Styrene–Butadiene Rubber (SBR). This is the most important synthetic rubber and represents more than half of all synthetic rubber production (Table 2) (see STYRENE–BUTADIENE RUBBER). It is a copolymer of 1,3-butadiene, $CH_2{=}CH{-}CH{=}CH_2$, and styrene, $C_6H_5CH{=}CH_2$, and is a descendant of the original Buna S first produced in Germany during the 1930s. The polymerization is carried out in an emulsion system where a mixture of the two monomers is mixed with a soap solution containing free-radical initiators. The final product is an emulsion of the copolymer, ie, a fluid latex (see LATEX TECHNOLOGY).

Polybutadiene. The homopolymer polybutadiene (PB) is next in importance to SBR, as shown by its world production (Table 2). It is prepared by polymerization of butadiene in solution, using organometallic initiators, either of the Ziegler-Natta type or lithium compounds. It is, therefore, a result of the discovery of stereospecific polymerization during the 1950s and 1960s (see ELASTOMERS, SYNTHETIC–POLYBUTADIENE).

Polyisoprene (Synthetic). Polyisoprene has four possible chain unit geometric isomers: *cis*- and *trans*-1,4-polyisoprene, 1,2-vinyl, and 3,4-vinyl.

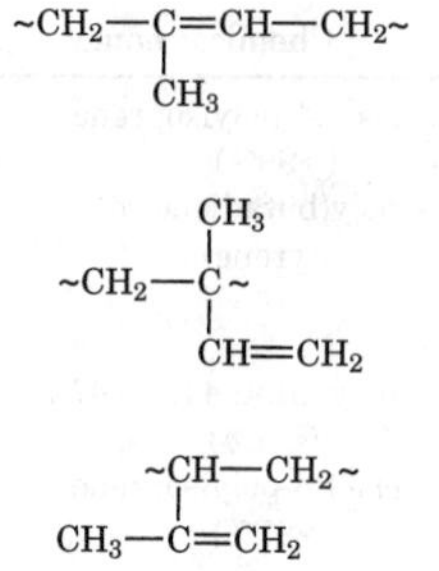

As in the case of polybutadiene, the Ziegler-Natta type of initiators are used to produce a polymer of high *cis*-1,4 structure (~98–99%) (natural rubber is 100% *cis*-1,4).

Age-Resistant Elastomers

Ethylene–Propylene (Diene) Rubber. The age-resistant elastomers are based on polymer chains having low unsaturation not in the backbone. This is sufficient for sulfur vulcanization but low enough to reduce oxidative degradation and ozone attack does not cause chain cleavage and subsequent cracking. Because of its irregular chain structure, EPDM is amorphous at <60 wt % ethylene and shows no crystallization on stretching. Hence, as is the case with all amorphous

Table 2. World Synthetic Rubber Consumption[a,b], 10^3t

Rubber type	1987	1988	1989[c]	1993[c]
SBR solid	2324	2382	2405	2616
SBR latex	241	245	250	265
carboxylated latex	984	1049	1088	1189
polybutadiene	1044	1098	1117	1215
ethylene–propylene	498	540	557	642
polychloroprene	249	251	251	258
nitrile, solid and latex	221	237	240	262
other synthetics[d]	958	1022	1040	1163
Total	*6519*	*6824*	*6948*	*7610*

[a] From *Chem. Eng.*, (Apr. 17, 1989). Courtesy of the Inter. Institute of Synthetic Rubber Producers. [b] Excludes Communist countries. [c] Estimated. [d] Includes butyl and polyisoprene.

elastomer, it exhibits poor strength and requires carbon black reinforcement. The presence of unsaturation in the side chain makes it possible to use sulfur vulcanization, but a higher proportion of accelerators must be used because of the low unsaturation. Its excellent aging and low temperature properties make it ideal for use as a sheet rubber for roofing applications, where it has shown a rapid growth.

Butyl Rubber. Butyl rubber is a copolymer of isobutylene and isoprene, with just enough of the latter to provide cross-linking sites for sulfur vulcanization. The polymerization system is of the cationic type, using coinitiators such as $AlCl_3$ and water at very low temperatures (−100°C) and leading to an almost instantaneous polymerization.

Solvent-Resistant Elastomers

Nitrile Rubber (NBR). This is the most hydrocarbon oil swelling resistant of the synthetic elastomers. This elastomer is prepared by emulsion polymerization, similar to that used for SBR, but generally carried out to high conversion. As for SBR, the chain irregularity leads to a noncrystallizing rubber, so that this polymer requires carbon black reinforcement for strength.

Acrylic Elastomers. These materials are based principally on an acrylate chain structure, as follows:

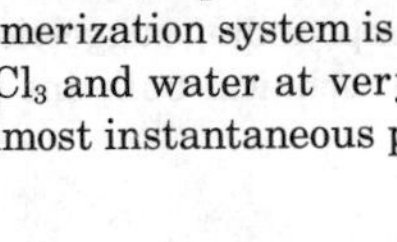

where R is generally an alkyl group. Because of the absence of unsaturation, vulcanization is carried out by amine compounds instead of sulfur, but this absence of unsaturation also confers good aging properties. This rubber also shows good resistance to hydrocarbon solvents because of the polar acrylic ester groups present. Practically all commercial acrylic elastomers are produced by free-radical polymerization. Of the four processes available, ie, bulk, solution, suspension, and emulsion, only aqueous suspension and emulsion polymerization are used to produce the acrylic elastomers present in the market (see ELASTOMERS, SYNTHETIC–ACRYLIC ELASTOMERS; ETHYLENE–ACRYLIC ELASTOMERS).

Chloroprene Rubber. Polychloroprene can be represented by the formula:

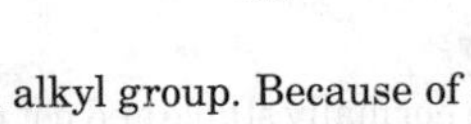

The polymer is prepared by emulsion polymerization of chloroprene. Polychloroprene has a variety of uses, both in latex and dry rubber form. In addition to its excellent solvent resistance, it is also much resistant to oxidation or ozone attack than natural rubber. It is also more resistant to chemicals and has the additional property of flame resistance from the chlorine atoms.

Temperature-Resistant Elastomers

Silicone Rubber. These polymers are based on chains of Si—O rather than carbon atoms, and owe their temperature properties to their unique structure. The most common types of silicone rubbers are specifically and almost exclusively the polysiloxanes, eg, polydimethylsiloxane. The Si—O—Si bonds can rotate much more freely than the C—C bond, or even the C—O bond, so the silicone chain is much more flexible and less affected by temperature (see SILICON COMPOUNDS, SILICONES). The polymer chain is formed by a ring-opening reaction caused by the action of alkalies on the monomer, a cyclic siloxane.

Fluorocarbon Elastomers. These elastomers are the most resistant elastomers to heat, chemicals, and solvents known, but they are also the most expensive, ie, between $22 and $35 per kg. The most common types are copolymers of vinylidene fluoride and hexafluoropropene. Emulsion polymerization is used, but the latex is too unstable for use and all the latex is coagulated to dry rubber.

Liquid Rubber Technology

An entirely new concept was introduced into rubber technology with the idea of "castable" elastomers, ie, the use of liquid, low molecular-weight polymers that could be linked together (chain-extended) and cross-linked into rubbery networks. This was an appealing idea because it avoided the use of heavy machinery to masticate and mix a high viscosity rubber prior to molding and vulcanization. In this development three types of polymers have played a dominant role, ie, polyurethanes, polysulfides, and thermoplastic elastomers.

MAURICE MORTON
The University of Akron

J. P. Queslel and J. E. Mark, in J. I. Kroschwitz, ed., *Encyclopedia of Polymer Science and Engineering*, 2nd ed., Vol. 5, John Wiley & Sons, Inc., New York, 1986, pp. 365–408.

ACRYLIC ELASTOMERS

Acrylic elastomers have the ASTM designation ACM for polymers of ethyl acrylate and other acrylates. Conventionally, the M indicates a polymer having a saturated chain of the polymethylene type. The repeat structure of this polymer family shows both the presence of a saturated backbone, which is responsible for the high heat and oxidation resistance, and the ester side groups, which contribute to the marked polarity of this elastomer chain. These two main properties are not present in general-purpose rubbers. As a result, ACMs are designated as specialty elastomers.

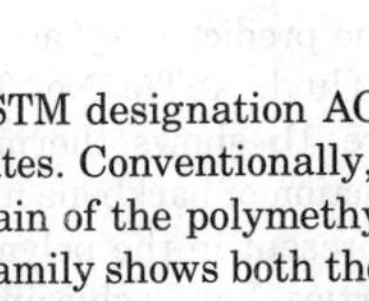

Manufacture

Practically all commercial acrylic elastomers are produced by free-radical polymerization. Of the four methods available, ie, bulk, solution, suspension, and emulsion polymerization, only aqueous suspension and emulsion polymerization are used to produce the ACMs present in the market.

Monomers

Two kinds of monomers are present in acrylic elastomers: backbone monomers and cure-site monomers. Backbone monomers are acrylic esters that constitute the majority of the polymer chain (up to 99%), and determine the physical and chemical properties of the polymer and the performance of the vulcanizates. Curesite monomers simultaneously present a double bond available for polymerization with acrylates and a moiety reactive with specific compounds in order to facilitate the vulcanization process.

Backbone Monomers. Ethyl acrylate (EA), butyl acrylate (BA), and 2-methoxy ethyl acrylate (MEA) constitute the building blocks of current acrylic elastomers.

Cure-Site Monomers. A large variety of cure-site monomers have been proposed, but only a few have achieved commercial significance. Two of the most important classes are labile chlorine-containing monomers and epoxy/carboxyl-containing monomers.

Structure–Property Relationships

The modern approach to the development of new elastomers is to satisfy specific application requirements. Acrylic elastomers are very powerful in this respect, because they can be tailor-made to meet certain performance requirements. Even though the structure-property studies are proprietary knowledge of each acrylic elastomer manufacturer, some significant information can be found in the literature.

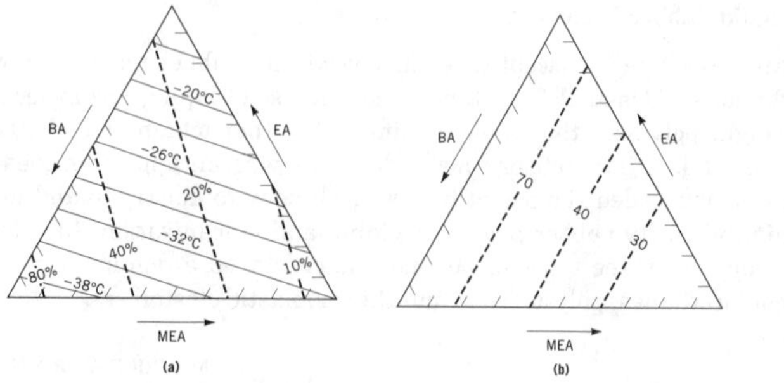

Figure 1. Elastomer properties as a function of monomer composition, butyl acrylate (BA), ethyl acrylate (EA), and methoxyethyl acrylate (MEA). (**a**), (–) glass-transition temperature; (––––) swelling in ASTM No. 3 oil; (**b**), (––––) residual elongation at break, %, after heat aging.

Figure **1a** shows the predicted T_g, according to GCT, and the volume swell in reference fluid, ASTM No. 3 oil, related to each monomer composition. Figure **1b** shows thermal aging resistance of acrylic elastomers as a function of backbone monomer composition.

The cure sites present in the polymer also significantly influence the expected properties. Labile chlorine cure sites generally give good elongation at break retention after heat aging, and show low sensitivity to acidic compounds in the mix recipe. Epoxy/carboxyl cure sites give good compression set and good hydrolysis resistance. Dual cure site-based polymers vulcanized by quaternary ammonium salts may behave in different ways compared to conventional cure sites.

Applications

Because at their high oil and temperature resistance, acrylic elastomers are extensively used in highly demanding applications such as automotive underhood components. Some significant examples are uses as lip and shaft seals, O-rings, oil pan and cover valve gaskets, and hoses.

Compounding

A rubber compound typically consists of many additives and ingredients, although in the case of acrylic rubber compound, the recipes are usually simpler. General recipes include the following:

Ingredient	phr
polymer	100
stearic acid	1–3
processing aid	1–3
antioxidant	1–2
filler (black and/or white)	20–100
plasticizer	0–5
cure system	0.5–8

The compound recipe is a function both of cure site(s) present in the polymer and of application requirements.

Vulcanization Systems

Because of the different vulcanization chemistry involved in each commercial ACM, a vulcanization system specific to the cure site present has to be adopted. Many cure systems for labile chlorine containing ACM have been proposed. Among these the alkali metal carboxylate–sulfur cure system, or soap–sulfur as it is called in the United States, became the mainstay of acrylic elastomer technology in the early 1960s, and continues to be widely used.

New efficient vulcanization systems have been introduced in the market based on quaternary ammonium salts to vulcanize epoxy/carboxyl cure sites. They have been found effective in chlorine containing ACM dual cure site with carboxyl monomer. This accelerator system together with a retarder (or scorch inhibitor) based on stearic acid and/or guanidine can eliminate post-curing. More recently, in the United States a proprietary vulcanization package based on zinc diethyldithiocarbamate has been reported that offers a superior balance between cure rate at molding temperatures and safety at process temperatures. It is proposed as a curative system that does not require post-vulcanization.

More detailed compounding suggestions and additives information can be found in the technical bulletins published by ACM and additives suppliers.

Economic Aspects

Acrylic rubbers, as is the case for most specialty elastomers, are characterized by higher price and smaller consumption compared to general-purpose rubbers. As a first approximation, the ACM consumption in 1991 was estimated to be 7000 t distributed among the United States, Western Europe, and Japan/Far East, where automotive production is significantly present.

The price of ACM is approximately four to five times greater than that of general-purpose rubbers like SBR or NR, and about three times the price of standard engineering rubbers such as NBR and EPDM. However, they cost about four to six times less than other specialty elastomers like fluorocarbon elastomers FKM or HNBR.

Health and Safety

Acrylic elastomers are normally stable and not reactive with water. The material must be preheated before ignition can occur, and fire conditions offer no hazard beyond that of ordinary combustible material.

Overexposure to acrylic rubbers is not likely to cause significant acute toxic effects. ACM, however, may contain residual monomers, mainly acrylate monomers, vapors of which are known to cause eye and/or skin irritation.

According to ACGIH (1991–1992 edition) the Threshold Limit Values (TLV) for a normal 8-h workday and 40-h workweek exposure (TWA) are 5 ppm (20 mg/m^3) for ethyl acrylate, and 10 ppm (55 mg/m^3) for butyl acrylate. Furthermore these monomers are codified A2 (Suspect Human Carcinogen). Therefore, according to good practice of industrial hygiene, these types of rubber should be used in a well-ventilated area. The use of eye protection and gloves is recommended when handling acrylic elastomers.

A. L. SPELTA
EniChem Elastomeri

A. L. Spelta, G. Cantalupo, and L. Gargani, paper presented at *The German Rubber Conference,* Nuremberg, July 4–7, 1988.

D. W. van Krevelen, *Properties of Polymers,* 2nd ed., Elsevier Science Publishing Co., New York, 1976.

P. H. Starmer, *Prog. Rubber Plastics Technol.* **3**(1) (1987).

E. Chang and E. Mazzone, "A New Non-Postcure Curative Package for Polyacrylate Elastomers", paper presented at *ACS Rubber Division,* Detroit, Mich., Oct. 17–20, 1989.

R. D. De Marco, *Rubber Chem. Technol.* **52**, 173 (1979).

BUTYL RUBBER

Butyl rubber and other isobutylene polymers of technological importance include various homopolymers and isobutylene copolymers containing unsaturation achieved by copolymerization with isoprene. Bromination or chlorination of the unsaturated site is practiced commercially, and other modifications are being investigated.

The first use for butyl rubber was in inner tubes, the air-retention characteristics of which contributed significantly to the safety and convenience of tires. Good weathering, ozone resistance, and oxidative stability have led to applications in mechanical goods and elastomeric sheeting.

Halogenated butyl rubber greatly extended the usefulness of butyl rubber by providing much higher vulcanization rates and improving the compatibility with highly unsaturated elastomers. Moreover, the halogenated elastomers can undergo vulcanization by different mechanisms than those for SBR or NR. These properties permitted the production of tubeless tires with chlorinated or brominated butyl inner liners. Tire durability was extended by the retention of air pressure and low intercarcass pressure.

Polyisobutylene is produced in a range of mol wts, and has found a host of uses. The low mol wt liquid polybutenes have applications as adhesives, sealants, coatings, lubricants, and plasticizers, and for the impregnation of electrical cables. Moderate mol wt polyisobutylene was one of the first viscosity-index modifiers for lubricants. High mol wt polyisobutylene is used to make uncured rubbery compounds, and as an impact additive for thermoplastics.

Synthesis

Monomers for manufacture of butyl rubber are 2-methylpropene (isobutylene) and 2-methyl-1,3-butadiene (isoprene). Polybutenes are copolymers of isobutylene and *n*-butenes from mixed-C_4 olefin-containing streams.

Isobutylene Polymerization Mechanism. The mechanism of cationic polymerization of isobutylene and copolymerization of isobutylene with isoprene with Lewis acids is highly complex. Friedel-Crafts Lewis acid and Brønsted acid coinitiators at low temperature give an extremely high polymerization rate in hydrocarbon or halogenated hydrocarbon diluants.

Halobutyls. Chloro- and bromobutyls are commercially the most important butyl rubber derivatives. The halogenation reaction is carried out in hydrocarbon solution using elemental chlorine or bromine (equimolar ratio with enchained isoprene).

Conjugated-Diene Butyl. CDB can be obtained by the controlled dehydrohalogenation of halogenated butyl rubber.

Isobutylene–Isoprene–Divinylbenzene Terpolymer. A partially cross-linked isobutylene–isoprene–divinylbenzene terpolymer containing some unreacted substituted vinylbenzene appendages is commercially available. It is employed primarily in the manufacture of sealant tapes and caulking compounds.

Liquid Butyl Rubber. This material is commercially produced by degradation of a conventional high mol wt butyl rubber, most likely via extrusion at high shear rate and temperature. The principal areas of use are in sealant, caulks, potting compounds, and coating, where advantage can be taken of its relatively low viscosity in formulating high solids compounds that can be poured, sprayed, troweled, or spread.

New Materials

Grades of polyisobutylene, butyl rubber, halogenated butyl rubber, and partially cross-linked isobutylene–isoprene–divinylbenzene terpolymer have been developed to meet specific processing and property needs. Recently, two new polyisobutylene-based elastomers have been developed.

Star-Branched (SB) Butyl. Star-branched butyl has a bimodal mol wt distribution with a high mol wt branched mode and a low mol wt linear component. The polymer is prepared by a conventional cationic copolymerization of isobutylene and isoprene in methyl chloride below −80°C with a Friedel-Crafts coinitiator, eg, $AlCl_3$, in the presence of a polymeric branching agent. Star-branched butyl rubbers offer a unique balance of viscoelastic properties resulting in significant processability improvements. Dispersion in mixing and mixing rates are improved. Compound extrusion rates are higher, die swell is lower, shrinkage is reduced, and surface quality is improved. The balance between green strength and stress relaxation at ambient temperature is improved, making shaping operations such as tire building easier. Several grades of Exxon SB Butyl Polymers including copolymer, chlorinated, and brominated copolymers are now commercially available.

Brominated Poly(isobutylene-co-p-methylstyrene). *para*-Methylstyrene (PMS) can be readily copolymerized with isobutylene via classical cationic copolymerization using strong Lewis acid, eg, $AlCl_3$ or alkyl aluminum in methyl chloride at low temperature. These new high mol wt copolymers encompass an enormous range of properties from polyisobutylene-like elastomers to poly(*p*-methylstyrene)-like tough hard plastic materials with T_gS above 100°C, depending on monomer ratio. The brominated copolymer with reactive benzyl bromide functionality can be crosslinked with a variety of crosslinking systems and easily converted by nucleophilic substitution reactions to other functional groups and graft copolymers.

Manufacturing

The bulk of the world production of butyl rubber is made by a precipitation (slurry) polymerization process in which isobutylene and a minor amount of isoprene are copolymerized using aluminum chloride in methyl chloride at −100 to −90 °C. General description of the process and patent information are found in the literature. Halogenated butyl rubbers are produced commercially by dissolving butyl rubber in hydrocarbon solvent and contacting the solution with elemental halogen.

Polymer Properties

Polyisobutylene. Isobutylene polymerizes in a regular head-to-tail sequence to produce a polymer having no asymmetric carbon atoms. The glass-transition temperature is about −70°C.

Polybutenes. Copolymerization of mixed isobutylene and 1-butene containing streams with a Lewis acid catalyst system yields low mol wt (several hundred to a few thousand) copolymers that are clear, colorless, viscous liquids. The chain-ends are unsaturated, and they are often chemically modified through this functionality.

Butyl Rubber. In butyl rubber, isoprene is enchained by 1,4-addition in the trans configuration.

Halogenated Butyl Rubber. Halogenation at the isoprene site in butyl rubber proceeds by a halonium ion mechanism, leading to a double-bond shift and formation of an exomethylene alkyl halide.

Polyisobutylene has the chemical properties of a saturated hydrocarbon. The unsaturated end groups undergo reactions typical of a hindered olefin and are used, particularly in the case of low mol wt materials, as a route to modification; eg, the introduction of amine groups to produce dispersants for lubricating oils. The in-chain unsaturation in butyl rubber is attacked by atmospheric ozone, and unless protected can lead to cracking of strained vulcanizates. Oxidative degradation, which leads to chain cleavage, is slow, and the polymers are protected by antioxidants.

Cross-linking reactions for the polyisobutylene-type polymers depend on adding a reactive site, usually an allylic hydrogen or halogen. These reactive sites allow vulcanization with sulfur and accelerators or metal oxides.

Polyisobutylene is readily soluble in nonpolar liquids. The solution properties of polyisobutylene, butyl rubber, and halogenated butyl rubber are very similar. Cyclohexane is an excellent solvent, benzene a moderate solvent, and dioxane a nonsolvent for polyisobutylene polymers.

The most important physical properties of the elastomeric polyisobutylenes are those exhibited by vulcanized compounds, especially high loss modulus and low permeability air.

Elastomeric Vulcanizates. As with almost all rubbers, the final properties are determined by compounding and subsequent vulcanization or cross-linking. Various fillers, processing aids, plasticizers, tackifiers, cure systems, and antidegradants are used.

Fillers and Plasticizers. Of the compounding ingredients,fillers most significantly influence stress–strain and dynamic properties. Polymer chains interact with the surface of reinforcing fillers altering chain dynamics. Carbon black (qv) because of its high surface area (small particle size) enhances tensile properties and abrasion resistance. Butyl rubber has a low affinity for carbon black when compared to highly unsaturated rubbers such as SBR. This can be an important consideration in elastomeric blends since the carbon black can accumulate in one phase. The effect of carbon black on butyl and halobutyl rubbers is similar to that on other elastomers. Mineral fillers are used for light-colored compounds. Petroleum-based oils are commonly used as plasticizers.

Vulcanization

Vulcanization or curing is accomplished via chemical cross-linking reactions involving allylic hydrogen or halogen sites along the polymer backbone to produce a polymer network. Sulfur, resin, and metal oxide or amine cures are used.

Health and Safety Factors

Polyisobutylene and isobutylene–isoprene copolymers are considered to have no chronic hazard associated with exposure under normal industrial use.

Economic Aspects

Butyl and halogenated butyl rubbers are manufacture by Exxon Chemical Co., Bayer AG (formed Polysar Ltd.), and in the former USSR. Polybutenes are manufactured by 10 companies throughout the world: Amoco Chemical Co. (Texas City, Texas; Whiting, Indiana; Antwerp, Belgium); Chevron Chemical Co. (Richmond, California); Cosden Petroleum Co. (Big Spring, Texas); Exxon Chemical Co. (Baytown, Texas; Bayway, New Jersey; Koln, Germany); Lubrizol (Deer Park, Texas; Rouen, France); Petrofina (Montreal, Canada); Polybutenos Argentinos (Ensenada, Argentina); BASF (Ludwigshafen, Germany); British Petroleum Co. (Grangemouth, U.K.; Naphtachemie, France), and Nipon Sekiyu Kagaku, Ltd. (Kawasaki, Japan).

EDWARD KRESGE
H-C. WANG
Exxon Chemical Company

E. N. Kresge, R. H. Schatz, and H-C Wang, in J. I. Kroschwitz, ed., *Encyclopedia of Polymer Science and Engineering*, Vol. 8, 2nd ed., John Wiley & Sons, Inc., New York, 1987.

F. P. Baldwin and I. J. Gardner, *Chemistry and Properties of Crosslinked Polymers*, Academic Press, Inc., New York, 1977.

J. V. Fusco and P. Hous in R. F. Ohm, ed., *Vanderbilt Rubber Handbook*, R. T. Vanderbilt Co., Norwalk, Conn., 1990.

H-C. Wang and K. W. Powers, *Elastomerics*, Jan. 1992.

J. P. Kennedy and M. Marechal, *Carbocationic Polymerization*, John Wiley & Sons, Inc., New York, 1982.

CHLOROSULFONATED POLYETHYLENE

Chlorosulfonated polyethylene, CSM, represents a family of curable polymers, ranging from soft and elastomeric to hard and plastic, containing pendent chlorine and sulfonyl chloride groups. Chlorosulfonated base resins, other than polyethylene, are closely related and are therefore considered a part of this family. Addition of chlorine and sulfonyl chloride groups onto these base resins enhances solubility in common solvents and, when properly compounded and cured, gives resistance to light discoloration, oil, flame, and oxidizing chemicals. Resistance to thermal degradation and ozone attack result from the absence of unsaturation in the polymer backbone. The additional functionality of comonomers and grafts contribute to adhesion and polymer mechanical reinforcement. This combination of value-added properties promotes special end use applications in coating, adhesives, roofing membranes, electrical wiring insulation, automotive and industrial hose, tubing and belts, and in molded goods.

Physical Properties

The rubbery character of the chlorosulfonated polyethylene is derived from the flexibility of the polyethylene chain in the absence of crystallinity. Introduction of chlorine onto the polymer chain provides sufficient molecular irregularity to prevent crystallinity in the relaxed state. The sulfonyl chloride groups provide cross-linking sites for nonperoxide curing procedures.

CSM products may be divided into three groups depending on the type of precursor resin: low density (LDPE), high density (HDPE), and linear low density (LLDPE). LDPE is made by a high pressure free-radical process, while HDPE and LLDPE are made via low pressure, metal coordination catalyst processes (see OLEFIN POLYMERS).

The uncured physical properties of polymers within each group depend on the molecular weight, molecular-weight distribution, and the extent and distribution of chlorination and chlorosulfonation. The molecular weight, molecular weight distribution (MWD), and chain branching are generally set by the choice of parent polyethylene resin, ie, neither chain scission nor cross-linking take place during chlorosulfonation.

These conclusions are supported by expected physical properties of dried film of chlorosulfonated polyethylene from the different types of polyethylene (Table 1).

These values are given for polymers of narrow molecular-weight distribution, with number-average molecular weights (M_n) of about 20,000 prior to chlorination.

The uncured property most often used for CSM in dry applications is Mooney viscosity, a low shear bulk viscosity (ca 1.6 s^{-1}) determined at 100°C. Mooney viscosity is a rubber industry standard used to predict raw rubber and compound processibility, ie, mixing, extrusion, molding, etc.

Chemical Properties

The known chemistry of the functional groups in CSM, ie, chlorine and sulfonyl chloride groups, make reactions predictable from their functions in low molecular-weight substances. Acid, ester, and amide derivatives of the sulfonyl chloride group have been prepared and their infrared spectra have been studied.

The sulfonyl chloride group is the cure site for CSM and determines the rate and state of cure along with the compound recipe.

Vulcanization

Most end use applications of CSM, as with other thermoset polymers, involve mixing with various fillers, plasticizers, processing aids, curatives, etc, then shaping and cross-linking in its final form. Acid acceptors are required in CSM compounds because acidic by-products of curing reactions interfere with the cure and cause equipment corrosion. Acceptable acid acceptors include magnesia, litharge, organically bound lead oxide, calcium hydroxide, synthetic hydrotalcite, and epoxy resins.

For some applications, ie, nuclear power cable, high filler loadings are undesirable because of their contribution to radiation leakage. Unlike most other elastomers, large quantities of reinforcing fillers are not required to achieve good mechanical properties for high density

Table 1. Properties of Uncured CSM

Property	CSM HDPE		CSM LLDPE		CSM LDPE	
Cl, wt %	21.5	27.7	22	30	20.8	29
sulfur, wt %	1.7	0.51	1.0	0.98	1.3	1.5
tensile strength, MPa[a]	13.4	10.9	9.3	12.9	5.4	0.48
elongation at break, %	420	880	1259	1935	2100	25
ΔH^b J/g[c]	38	14	12	1.5	11	2

[a] To convert MPa to psi, multiply by 145.
[b] Heat of fusion.
[c] To convert J to cal, divide by 4.184.

CSM vulcanizates. Carbon black fillers give the best reinforcement of physical properties and the best resistance to chemical degradation. Small amounts of carbon black can also give significant improvement in weatherability. Mineral fillers are used to take advantage of CSM's nondiscoloring characteristics. Clays, silicas, and calcium carbonate augment flame and heat resistance. White compounds include titanium dioxide for light stability. Plasticizers are added to reduce viscosity during processing and increase low temperature flexibility of the final product. Petroleum oils are widely used because of low cost. Ester plasticizers provide the best combination of low temperature flex, heat resistance, and mechanical property retention.

Three different covalent cure systems are commonly used: sulfur-based or sulfur donor, peroxide, and maleimide. These systems rely on a cross-linking agent and one or more accelerators to develop high cross-link density.

Manufacture

Unlike most elastomeric polymers, which are made by direct polymerization of monomers or comonomers, chlorosulfonated polyethylene, as the name implies, is made by chemical modification of a preformed thermoplastic polymer. The chlorination and chlorosulfonation reactions are usually carried out simultaneously but may be carried out in stages.

A simplified form of the reaction may be represented by

$$\mathrm{-\!\!\left(CH_2\right)_x\!\!-} + (y+z)\mathrm{Cl_2} + (z)\mathrm{SO_2} \xrightarrow{\text{initiator}} \mathrm{-\!\!\left(CH_2\right)_{x-y-z}\!\left(\underset{\displaystyle Cl}{CH}\right)_y\!\left(\underset{\displaystyle SO_2Cl}{CH}\right)_z\!\!-} + (y+z)\,\mathrm{HCl}$$

or

$$\mathrm{-\!\!\left(CH_2\right)_x\!\!-} + (y+z)\mathrm{SO_2Cl_2} \xrightarrow[\text{catalyst}]{\text{initiator +}} \mathrm{-\!\!\left(CH_2\right)_{x}\!\left(\underset{\displaystyle Cl}{CH}\right)_y\!\left(\underset{\displaystyle SO_2Cl}{CH}\right)_z\!\!-} + (y+z)\,\mathrm{HCl} + y\,\mathrm{SO_2}$$

Economic Aspects

Production of chlorosulfonated polyethylene products on a worldwide basis is estimated to be approximately 50,000 t/yr. The Du Pont Co. is the primary manufacturer with one plant in the United States having a capacity of about 33,000 t and one plant in Northern Ireland with about 13,000 t capacity. The remaining world capacity is provided by Toyo Soda Manufacturing Ltd. in Japan. The Du Pont Co. manufactures all CSM types under the trade names of Hypalon and Acsium Synthetic Rubber at each of its plant sites. Toyo Soda makes closely related products under the trade name CSM-CP, or Ts. Since the precursor material is primarily ethylene, materials costs are related to petroleum prices. Costs of environment control procedures surrounding the use of carbon tetrachloride solvent have escalated extensively because it was placed on the Montreal Protocol of potential ozone depleting agents in 1989. Because carbon tetrachloride is to be completely phased out by the year 1996, significant investment in new equipment designed to handle replacement solvents is anticipated.

Health and Safety Factors

Hypalon may contain small amounts of carbon tetrachloride residue and a much lesser amount of chloroform. These chemicals are toxic and carcinogenic with TWA exposure limits of 5 ppm. Both are regulated as air contaminants in the United States under the Occupational Safety and Health Act (OSHA). Significant amounts of sulfur dioxide and hydrogen chloride may also be evolved during mixing or processing.

Hypalon raw polymer compounds or cured product may be disposed of in an approved landfill.

ROYCE ENNIS
DuPont-Beaumont Works

P. J. Canterino, in N. M. Bikales, ed., *Encyclopedia of Polymer Science and Technology,* Vol. 6, John Wiley & Sons, Inc., New York, 1967, p. 442.

Vanderbilt Rubber Handbook, R. T. Vanderbilt Co., Norwalk, Conn., 1987, pp. 185–189.

ETHYLENE–ACRYLIC ELASTOMERS

Ethylene–acrylic elastomers are best known for their excellent heat and oil resistance, but they also possess a good balance of compression set resistance, flex resistance, physical strength, low temperature flexibility, and weathering resistance. Special compounded attributes include uniquely temperature-stable vibrational damping properties and the ability to produce flame-resistant compounds with combustion products having an exceptionally low order of smoke density, toxicity, and corrosiveness. Because of this balance of properties, ethylene–acrylic elastomers have found ready acceptance in many high performance applications, especially in the automotive market.

Polymer Properties

Polymer Composition. Ethylene–acrylic elastomer terpolymers are manufactured by the addition copolymerization of ethylene and methyl acrylate, in the presence of a small amount of an alkenoic acid to provide sites for cross-linking with diamines.

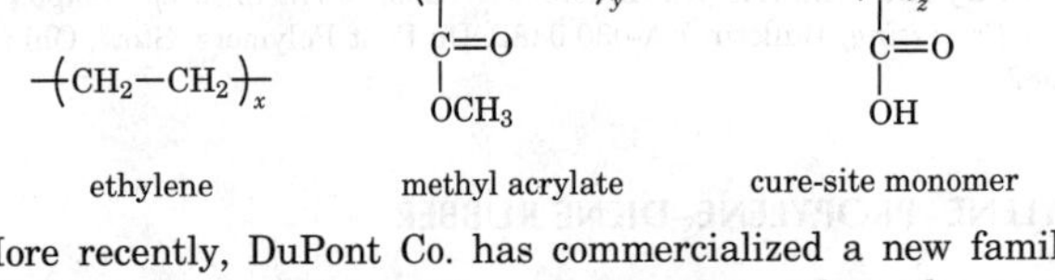

ethylene methyl acrylate cure-site monomer

More recently, DuPont Co. has commercialized a new family of copolymers of just ethylene and methyl acrylate, where the cure-site monomer has been removed from the polymer backbone.

The process yields a random, completely soluble polymer that shows no evidence of crystallinity of the polyethylene type down to −60°C. The polymer backbone is fully saturated, making it highly

Table 1. VAMAC Ethylene-Acrylic Elastomer Polymers

Commercial designation	Monomers[a]	Methyl acrylate level	Type of cure system
VAMAC G	E/MA/CS	average	amine
VAMAC LS	E/MA/CS	high	amine
VAMAC D	E/MA	average	peroxide
VAMAC DLS	E/MA	high	peroxide

[a] *E* is ethylene; MA, methyl acrylate; and CS, proprietary cure-site monomer.

resistant to ozone attack even in the absence of antiozonant additives. The fluid resistance and low temperature properties of ethylene–acrylic elastomers are largely a function of the methyl acrylate to ethylene ratio.

Commercial Forms. Four different base polymers of VAMAC ethylene–acrylic elastomer are commercially available (Table 1).

Processing

Mixing. Ethylene–acrylic elastomers are processed in the same manner as other elastomers. An internal mixer is used for large-scale production and a rubber mill for smaller scales.

Extrusion and Calendering. Most compounds of ethylene–acrylic elastomers have low nerve and yield smooth extrusions or calendered sheets.

Molding. Parts can be produced from ethylene–acrylic elastomers using compression, transfer, and injection molding techniques.

Post-Curing. Whenever production techniques or economics permit, it is recommended that compounds based on terpolymer grades be post-cured. Relatively short press cures can be continued with an oven cure in order to develop full physical properties and maximum resistance to compression set.

Adhesion. Commercially available 1- or 2-coat adhesive systems produce rubber failure in bonds between ethylene–acrylic elastomer and metal. Adhesion to nylon, polyester, or aramid fiber cord or fabric is greatest when the cord or fabric have been treated with carboxylated nitrile rubber latex.

Economic Aspects

The growth rate for ethylene–acrylic elastomers has been greater than 10% since the late 1980s. Over 50% of ethylene–acrylic elastomers are sold in Europe.

THERESA M. DOBEL
DuPont Chemical Company

R. G. Peck, *Ethylene/Acrylic Elastomer—Meeting The Challenges of a Demanding Market,* Bulletin EA-020.0185, DuPont Polymers, Stow, Ohio, Jan. 24, 1985.

T. M. Dobel, *New Development in Ethylene/Acrylic Elastomers,* Paper No. 28, American Chemical Society Rubber Division, Detroit, Mich., Oct. 1991.

C. Williams, *VAMAC Ethylene/Acrylic Elastomer, A Survey of Properties, Compounding and Processing,* Bulletin H-34753, DuPont Polymers, Stow, Ohio, Jan. 1992.

J. W. Crary, *Ethylene/Acrylic Elastomer—Basic Principles of Compounding and Processing,* Bulletin EA-030.0482, Du Pont Polymers, Stow, Ohio, Apr. 1982.

ETHYLENE–PROPYLENE–DIENE RUBBER

Copolymers of ethylene and propylene (EPM) and terpolymers of ethylene, propylene, and a diene (EPDM) as manufactured today are rubbers based on the early work of G. Natta and co-workers. A generic formula for EPM and EPDM may be given as follows, where $m = \sim 1500$ (~60 mol%), $n = \sim 975$ (~39 mol%), $o = \sim 25$ for EPDM (~ 1 mol%), and zero for EPM in an average amorphous molecule, and the comonomers are statistically distributed along the molecular chain.

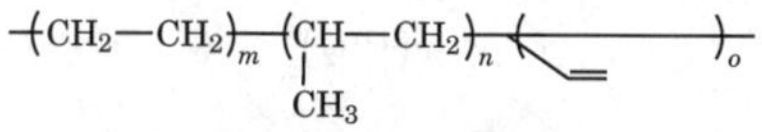

EPM can be vulcanized radically by means of peroxides. A small amount of built-in third diene monomer in EPDM permits conventional vulcanization with sulfur at the pendent sites of unsaturation.

Among the variety of synthetic rubbers, EPM and EPDM are the fastest growing elastomers, particularly by virtue of their excellent ozone resistance in comparison with natural rubber (*cis*-1,4-polyisoprene) and its synthetic counterparts IR (isoprene rubber), SBR (styrene-butadiene rubber), and BR (butadiene rubber). Secondly, EPDM rubber can be extended with fillers and plasticizers to an extremely high level in comparison with the other elastomers mentioned, and still give good processibility and properties in end articles. This gives it a price advantage.

Even though EPM and EPDM rubbers have been commercially available for more than 30 years, the technology concerning these products, both their production and their applications, is still very much under development.

Polymer Properties

The properties of EPM copolymers are dependent on a number of structural parameters of the copolymer chains: the relative content of comonomer units in the copolymer chain, the way the comonomers are distributed in the chain, the variation in the comonomer composition of different chains, average mol wt, and mol wt distribution. The structure of EPM shows it to be a saturated synthetic rubber. Because EPM does not contain any carbon–carbon unsaturation, it is inherently resistant to degradation by heat, light, oxygen, and, in particular, ozone. The properties of typical EPDM rubbers are shown in Table 1.

Table 1. Properties of Raw Ethylene—Propylene–Diene Co- and Terpolymers

Property	Value
specific gravity	0.86–0.87
appearance	glassy white
ethylene/propylene ratio by wt	
amorphous types	50/50
crystalline or sequential types	75/25
onset of crystallinity, °C	
amorphous types	below −50
crystalline types	below ca 30
glass-transition temperature,[a] °C	−45 to −60
heat capacity, kJ/(kg·K)[b]	2.18
thermal conductivity, W/(m·K)	0.335
thermal diffusivity, m/s	1.9×10^{-5}
thermal coefficient of linear expansion per °C	1.8×10^{-4}
Mooney viscosity, ML (1 + 4) 125°C[c]	10–90

[a] All types dependent on third monomer content.
[b] To kJ to kcal, divide by 4.184.
[c] Oil extended grades, when viscosity >100% of the raw polymer.

Manufacture

The two principal raw materials for EPM and EPDM, ethylene and propylene, both gases, are available in abundance at high purity.

EPM and EPDM rubbers are produced in continuous processes. Most widely used are solution processes, in which the polymer produced is in the dissolved state in a hydrocarbon solvent, eg, hexane.

Noteworthy developments in the field of EPDM-manufacture include the development of highly active catalyst species, achieved by supporting titanium halide on a magnesium chloride carrier. Another development is replacement of the steam-stripping operation by a progressive series of degassing operations, vessels, and extruders, for cost reasons and to do away with water in the whole operation. Special reactor designs with multiple feeding locations to achieve special molecular structures for specific purposes have been developed. Gas-phase polymerization of EPDM is possible as an extension of the well-known gas-phase processes for polyethylene and polypropylene.

Compounding

EPM/EPDM grades have to be compounded with reinforcing fillers if high levels of mechanical properties are required. Carbon blacks are usually used as fillers. The semireinforcing types, such as FEF (Fast Extrusion Furnace) and SRF (Semi-Reinforcing Furnace) give the best performance. The most widely used plasticizers are paraffinic oils.

Although EPM can only be cross-linked with peroxides, peroxide or sulfur plus accelerators or even other vulcanization systems like resins can be used for EPDM. The choice of chemicals used in an EPDM vulcanizate depends on many factors, such as mixing equipment, mechanical properties, cost, safety, and compatibility.

Modern legislation puts much emphasis on the prevention of the formation of carcinogenic secondary nitrosamines as by-products of sulfur vulcanization. This requires the choice of specific accelerators.

Processing

Only compounds of low Mooney EPM or EPDM grades can be mixed on open mills. EPM/EPDM compounds are therefore almost exclusively mixed in internal mixers.

In general, EPM/EPDM compounds can be extruded easily on all commercial rubber extruders. Furthermore, EPDM compounds can be calendered both as unsupported sheeting and onto a cloth substrate.

EPM/EPDM compounds are cured on all of the common rubber-factory equipment: press cure, transfer molding, steam cure, hot-air cure, and injection molding are all practical.

Properties of EPM and EPDM Vulcanizates

Mechanical properties depend considerably on the structural characteristics of the EPM/EPDM and the type and amount of fillers in the compound. A wide range of hardnesses can be obtained with EPM/EPDM vulcanizates. The elastic properties are by far superior to those of many other synthetic rubber vulcanizates, particularly of butyl rubber, but they do not reach the level obtained with NR or SBR vulcanizates. The resistance to compression set is surprisingly good, in particular for EPDM with a high ENB content.

The resistance to heat and aging of optimized EPM/EPDM vulcanizates is better than that of SBR and NR. EPM/EPDM vulcanizates have an excellent resistance to chemicals, such as dilute acids, alkalies, alcohol, etc. This is in contrast to the resistance to aliphatic, aromatic, or chlorinated hydrocarbons. EPM/EPDM vulcanizates sell considerably in these nonpolar media.

The electrical-insulating and dielectric properties of the pure EPM/EPDM are excellent, but in compounds they are also strongly dependent on the proper choice of fillers. The electrical properties of vulcanizates are also good at high temperatures and after heat-aging.

Health and Safety Factors

EP(D)M is not classified as hazardous. It is not considered carcinogenic according to OSHA Hazard Communications Standard and IARC Monographs.

In handling EP(D)M, normal industrial hygienic procedures should be followed. The use of EP(D)M is permitted for food contact under the conditions given in the respective FDA-paragraphs: §177.1520 for Olefin polymers, and §177.2600 for rubber articles intended for repeated use.

Uses

Contrary to the general-purpose elastomers such as NR, SBR, and BR, EPM and EPDM still show a steady growth over the years. Part of this growth still comes from replacement of these commodity rubbers by virtue of their better ozone and thermal resistance.

The main uses of EPM of EPDM are in automotive applications as profiles, (radiator) hoses, and seals; in building and construction as profiles, roofing foil, and seals; in cable and wire as cable insulation and jacketing; and in appliances as a wide variety of mostly molded articles.

Another important application for EPDM is in blends with general-purpose rubbers. Considerable amounts of EPM and EPDM are also used in blends with thermoplastics.

Substantial amounts of EPM are also used as additives to lubrication oils because of its excellent heat and shear stability under the operating conditions of automobile engines.

JACOBUS W. M. NOORDERMEER
DSM Elastomers Europe

S. Cesca, *Macromol. Rev.* **10,** 1–230 (1975).

F. P. Baldwin and G. VerStrate, *Rubber Chem. Technol.* **45,** 709–881 (1972).

FLUOROCARBON ELASTOMERS

Fluorocarbon elastomers are synthetic, noncrystalline polymers that exhibit elastomeric properties when cross-linked. They are designed for demanding service applications in hostile environments characterized by broad temperature ranges and/or contact with chemicals, oils, or fuels. Table 1 lists the typical commercial fluorocarbon elastomers.

Properties

Table 2 summarizes general characteristics of vulcanizates prepared from commercially available fluorocarbon elastomer gumstocks.

Table 1. Commercial Fluorocarbon Elastomers

Copolymer	Trademark	Supplier
poly(vinylidene fluoride-*co*-hexafluoropropylene)	Dai-el	Daikin
	Fluorel	3M
	Technoflon	Ausimont
	Viton	DuPont
plus cure-site monomer[a]	Fluorel	3M
poly(vinylidene fluoride-*co*-hexafluoropropylene-*co*-tetrafluoroethylene) with and without cure-site monomer[a]	Dai-el	Daikin
	Fluorel	3M
	Tecnoflon	Ausimont
	Viton	DuPont
poly(vinylidene fluoride-*co*-tetrafluoroethylene-*co*-perfluorovinyl ether) plus cure-site monomer?[a]	Viton	DuPont
	Tecnoflon	Ausimont
poly(tetrafluoroethylene-*co*-perfluoro(methyl vinyl ether) plus cure-site monomer[b]	Kalrez	DuPont
poly(tetrafluoroethylene-*co*-propylene)[a]	Aflas	Asahi Glass
	Aflas	3M
poly(vinylidene fluoride-*co*-chlorotrifluoroethylene)[a]	Kel-F	3M
poly(vinylidene fluoride-*co*-tetrafluoroethylene-*co*-propylene)[b] plus cure site	Fluorel II	3M
	Aflas	Asahi Glass
	Aflas	3M

[a] Peroxide curable.

[b] Proprietary cure system.

Table 2. Fluorocarbon Elastomers Physical Property Ranges

Property	Value
Physical properties	
tensile strength, MPa[a]	7.00–20.00
100% modulus, MPa	2.00–16.00
elongation at break, %	100–500
hardness range, Shore A	50–95
compression-set[b]	
70 h at 25°C	9–16
70 h at 200°C	10–30
1000 h at 200°C	50–70
specific gravity (gumstock)	1.54–1.88
low temperature flexibility, °C[c]	0 to −30
brittle point (ASTM D746), °C	0 to −50
thermal degradation temperature, °C	400 to 550
General characteristics	
gas permeability	very low
flammability	self-extinguishing or nonburning (when properly formulated)
radiation resistance	good to fair
abrasion resistance	good and satisfactory for most uses
weatherability and ozone resistance	outstanding (unaffected after 200 h exposure to 150 ppm ozone)

[a] To convert MPa to psi, multiply by 145.
[b] ASTM method B, 3.5 mm O-ring.
[c] Highly dependent on grade of material used.

Manufacture and Processing

Manufacture of Fluorocarbon Elastomers. These elastomers are typically prepared by high pressure, free-radical, aqueous emulsion polymerization techniques. The initiators (qv) can be organic or inorganic peroxy compounds such as ammonium persulfate. The emulsifying agent is usually a fluorinated acid soap, and the temperature and pressure of polymerization ranges from 30 to 125°C and 0.35 to 10.4 MPa (50–1500 psi).

Cross-Linking Chemistry. Like other thermosetting elastomers, fluorocarbon elastomers must be cured in order to get useful properties. Three distinct cross-linking systems have been developed to achieve this goal: diamine, bisphenol–onium, and peroxide curing agents. Over the years, the bisphenol–onium cure system, which is the most practical in terms of processing latitude and cured properties, has become the most widely used.

Compounding. Owing to the number of ingredients required in a conventional rubber recipe, fluorocarbon elastomer compounding seems simple compared to typical hydrocarbon elastomer recipes. However, the apparent simplicity of such formulations makes a selection of appropriate ingredients especially important in order to obtain the excellent properties inherent in available gumstocks.

With a clear idea of use requirements and rubber response to specific additives, a formulation may be selected. Uses generally fall into one of three classes: O-rings, molded goods, and extruded forms.

Formulation Parameters. Gum viscosity is of primary importance to the determination of processibility, as this factor affects vulcanizate properties, especially compression set.

Compound stability and safety must also be considered when determining processibility, as they are strongly affected by compounding ingredients and cure systems. Typical fluorocarbon elastomer O-ring compounds and properties are shown in Table 3.

Mixing. Fluorocarbon elastomer formulations can be compounded by any standard rubber mixing technique. Open mill mixing can be used since most commercial gums mix well. Internal mixing is widely used with fluorocarbon elastomers.

Preforming. Extrusion preforming is easily accomplished if relatively cool barrel temperatures are used with either a screw or piston type extruder (Barwell).

Table 3. Fluorocarbon Elastomer O-Ring Compounds

Typical formulation	I[a]	II[b]	III[c]
Compound ingredients, phr			
MT black (N-990)	30	30	25
calcium hydroxide	6	6	
magnesium oxide	3	3	
hexafluoroisopropylidenediphenol	2.1	2.1	
triphenylbenzylphosphonium chloride	0.45	0.45	
triallyl isocyanurate			5
α,α-bis(t-butylperoxy)diisopropylbenzene			1
sodium stearate			1
Physical properties[d]			
tensile strength, MPa[e]	15.0	15.0	16.0
elongation at break, %	200	200	300
hardness, Shore A	75	75	71
compression-set (3.5 mm O-rings), % for 70 h at 200°C, (ASTM D395)	15	10	35
specific gravity	1.8	1.8	1.6

[a] 100 phr FKM2230 (ML1 + 10 at 121°C = 40) where FKM2230 is poly-(vinylidene fluoride-*co*-hexafluoropropylene); available from 3M.
[b] 100 phr FKM2178 (ML1 + 10 at 121°C = 100) where FKM2178 is poly-(vinylidene fluoride-*co*-hexafluoropropylene); available from 3M.
[c] 100 phr FKM100S (ML1 + 10 at 121°C = 90) where FKM100S is poly-(tetrafluoroethylene-*co*-propylene); available from 3M.
[d] Press cure: 5 min at 177°C; post-cure: 24 h at 230°C.
[e] To convert MPa to psi, multiply by 145.

Calendering operations are done routinely, and warm rolls (40–90°C) are recommended for optimum sheet smoothness.

Molding. Compression molding, transfer molding, and injection molding are used.

Extrusion. Extrusion techniques are used in the preparation of tubing, hose, O-ring cord, preforms, and shaped gaskets.

Post-Curing. Post-curing at elevated temperatures develops maximum physical properties (tensile strength and compression-set resistance) in fluorocarbon elastomers.

Economic Factors

Annual worldwide fluorocarbon elastomer usage totaled about 7300 metric tons in 1993. Approximately 40% of this usage was in the United States, 30% in Europe, and 20% in Japan.

Health and Safety Factors

In general, under normal handling conditions, the fluorocarbon elastomers have been found to be low in toxicity and irritation potential. Specific toxicological, health, and safe handling procedures are provided by the manufacturer of each fluorocarbon elastomer product upon request.

Uses

About 60% of the United States usage is in ground transportation. Typical components include engine oil seals, fuel system components such as hoses and O-rings, and a variety of drive train seals. Growth in this area is expected to continue with the general strength of the U.S. automotive industry coupled with increased demands from higher underhood temperatures, alcohol containing fuels, and more aggressive lubricants. Other major U.S. segments include petroleum/petrochemical, industrial pollution control, and industrial hydraulic and pneumatic applications.

WERNER M. GROOTAERT
GEORGE H. MILLET
ALLAN T. WORM
3M Company

W. W. Schmiegel, *Makromekulare Chemie* **76/77**, 39 (1979).
A. L. Logothetis, *Prog. Polym. Sci.* **14**, 251–296 (1989).

NITRILE RUBBER

Nitrile rubber is a synthetic polymer made from 1,3-butadiene and acrylonitrile using emulsion polymerization techniques. Nitrile rubber is classified as a specialty rubber and is well known for its resistance to various oils, fuels, and chemicals. After mixing with other ingredients (fillers, plasticizers, antidegradants, curatives) and curing, NBR compounds commonly see use in various seals, gaskets, hose, and roll applications.

Chemical Properties

Nitrile rubbers are high molecular-weight amorphous copolymers of 1,3-butadiene, $CH_2{=}CH{=}CH_2$, and acrylonitrile, $CH_2{=}CH{-}CN$. Average molecular weights of commercial products have been reported to be between 250,000 and 600,000, with a wide distribution around the average in any single product. Structure of the polymer can vary widely from largely linear to highly branched to cross-linked, depending on the conditions of polymerization.

Chemically modified nitrile rubbers are also produced commercially with the objective of changing their chemistry in such a way that specific properties are enhanced. The oldest of the chemically modified NBRs are the carboxylated varieties, made by copolymerizing methacrylic or acrylic acid with the butadiene and acrylonitrile. The resultant products, which typically contain 2–6% by weight of the acid monomer, can be vulcanized with polyvalent metals in addition to the normal sulfur or peroxide cure systems. The most outstanding result of this modification is a large improvement in abrasion resistance. Another chemical modification currently in use is the copolymerization of a monomer which causes an antioxidant structure to be attached to the polymer. The result is improved resistance to oxidation, particularly after the rubber has been exposed to a hydrocarbon fluid that would normally extract a conventional antioxidant from the polymer, leaving it unprotected. The most recent and potentially the most important chemical modification is hydrogenation of the polymer so that little of the unsaturation remains. This results in a product with much improved resistance to oxidation and weathering, but with little or no sacrifice in other useful properties.

Other chemical modifications such as attachment of isocyanate or hydroxyl functionality have been reported, but have not become commercially significant.

Physical Properties

Nitrile rubbers are produced over a wide range of monomer ratios and molecular weights, so their physical constants and basic polymer properties also cover a range of values. Some of the more widely used properties are listed in Table 1.

Table 1. Physical Properties of Nitrile Rubber

Property	Acrylonitrile content, %	Value
specific gravity	15	0.94
	20	0.95
	35	0.99
	45	1.02
	50	1.03
T_g,°C	15	−49
	22	−40
	30	−30
	40	−19
	50	−9
thermal conductivity, kJ/(m·h·°C)[a]	28	0.90
	33	0.90
	38	0.92
thermal expansion coefficient $\times 10^6$/°C	28	175
	33	170
	38	150
specific heat, J(g·°C)[a]	40	$0.00283T + 1.126$

[a] To convert J to cal, divide by 4.184.

Manufacturing

Virtually all nitrile elastomers are manufactured by emulsion polymerization. The primary steps in the process include polymerization into a latex form, coagulation of the latex into a wet crumb, and then drying into a final product. In practice, latex is usually collected in large blend tanks, where it is sampled and tested before "finishing" to a dry product.

Economic Aspects

Nitrile rubber is generally considered a mature product and growth of production and sales have been relatively low.

There are five nitrile rubber producers in North America. Details of production facilities for nitrile rubber as provided by the International Institute of Synthetic Rubber Producers (IISRP) are shown in Table 2.

Table 2. North American Production Facilities for Nitrile Rubber

Producer	Plant location	Trade name	Capacity, 10^3 t
	North America		
Zeon Chemicals	Louisville, Ky.	Nipol	35
Goodyear Tire & Rubber	Houston, Tex.	Chemigum	28
Uniroyal Chemical Co.	Painesville, Ohio	Paracril	20
Polysar/Bayer	Sarnia, Ontario	Krynac	20
Copolymer Rubber/DSM	Baton Rouge, La.	Nysyn	10
Total North America			*113*

Grades of Nitrile Rubber

There are many grades of nitrile rubber available on the market today; for example, Zeon Chemicals offers approximately 75 different grades of NBR. The principal variables that are commonly changed include the following: acrylonitrile content, Mooney viscosity, emulsifier type, antioxidant type, third monomers, preplasticizer, blends with PVC, masterbatches with carbon black, and physical form.

Health and Safety Factors

Nitrile rubber presents no unusual hazards when processed and handled in areas with good housekeeping and ventilation. Like other elastomers, nitrile rubber emits fumes and vapors when heated to high temperatures during processing and curing. Acrylonitrile and 1,3-butadiene monomers are known or suspected carcinogens, thus producers of NBR must follow strict Occupational Safety and Health Administration (OSHA) standards that regulate permissible exposure of personnel to these materials during the nitrile rubber manufacturing process.

Uses

Common applications for nitrile rubber take advantage of the chemical resistance of the rubber and are as follows: seals, O-rings, gaskets, oil field parts, diaphragms, printing supplies, gloves, pump stators, belts, wire and cable insulation, hose tubes/covers, rolls, weather stripping, footwear/shoe products, milking inflations, and miscellaneous molded rubber goods.

Nitrile rubber is also used in many applications that do not involve the mixing of a traditional rubber compound as described above. Some of these areas include the following: modification of plastics, adhesives/cements/sealants/coatings, and friction materials.

Hydrogenated Nitrile Rubber

Hydrogenated nitrile rubber is a high performance polymer with significantly better high temperature properties, compared to nitrile rubber, owing to the largely saturated backbone of the rubber. It is produced by a solution process that leads to a selective hydrogenation of the butadiene unsaturation in the polymer.

As in nitrile rubber there are different grades of hydrogenated nitrile rubber that vary in acrylonitrile content and Mooney viscosity. In addition, various grades with different extents of hydrogenation are available. Finally, grades of hydrogenated nitrile rubber are available containing zinc oxide and methacrylic acid. Extremely high tensile strength, up to 58 MPa (8400 psi), and excellent abrasion resistance have been reported for hydrogenated nitrile rubber compounds containing zinc oxide and methacrylic acid.

Applications for hydrogenated nitrile rubber are similar to those of nitrile rubber in that they take advantage of the excellent chemical and oil resistance of the polymer. However, the increased high temperature performance of the hydrogenated nitrile rubber makes it far superior to standard grades of nitrile rubber.

DONALD MACKEY
AUGUST H. JORGENSEN
Zeon Chemicals USA, Inc.

W. Hoffmann, *Rubber Chem. Technol.* **37**(2), part 2 (Apr.–June 1964).

R. C. Klingender, in M. Morton, ed., *Rubber Technology,* 3rd ed., Van Nostrand Reinhold, New York, 1987.

J. R. Purdon, in *The Vanderbilt Rubber Handbook,* 13th ed., R. T. Vanderbilt Co., Inc., New York, 1990.

D. A. Seil and F. R. Wolf, in M. Morton, ed., *Rubber Technology,* 3rd ed., Van Nostrand Reinhold, New York, 1987, Chapt. 11.

PHOSPHAZENES

Polyphosphazenes have a backbone of alternating nitrogen and phosphorus atoms with two substituents on each phosphorus atom. The backbone is isoelectronic with that of silicones; these polymer backbones share the characteristics of thermal stability and high flexibility. The normal synthesis route provides a large range of possible derivatives. Coatings, fluids, elastomers, and thermoplastics can be produced by varying the polymer molecular weight and the substituents on the phosphorus. These materials have been suggested for use in a wide variety of areas including solid electrolytes for advanced batteries, and biomedical devices, including implants and drug carriers. Other reviews are available covering various aspects of polyphosphazene chemistry and applications (see INORGANIC HIGH POLYMERS).

Two elastomers have been commercialized with unique property profiles. One has fluoroalkoxy substituents that provide resistance to many fluids, especially to hydrocarbons. This material also has a broad use temperature range and useful dynamic properties. Aryloxy substituents provide flame retardant materials without halogens.

Synthesis

Phosphazene polymers are normally made in a two-step process. First, hexachlorocyclotriphosphazene, trimer, is polymerized in bulk to poly(dichlorophosphazene), chloropolymer. The chloropolymer is then dissolved and reprecipitated to remove unreacted trimer. After redissolving, nucleophilic substitution on with alkyl or aryloxides provides the desired product.

Toxicology

Hexachlorocyclotriphosphazene (cyclic trimer) is a respiratory irritant. Nausea has also been noted on exposure. Linear chloropolymer is also believed to be toxic. Upon organic substitution, the high molecular weight linear polymers have been shown to be inert.

Fluoroalkoxyphosphazene Elastomers

FZ Characterization. FZ elastomer is a translucent pale brown gum with a glass-transition temperature, T_g, of −68 to −72°C. The gum can be cross-linked using peroxides such as dicumyl peroxide and α,α'-bis(*t*-butylperoxy)diisopropylbenzene. It is provided in compounded form for specific applications, or in masterbatch form to allow custom compounding. Surface treated silica and semireinforcing carbon black are preferred fillers (qv). Clays, silicates, and nonreinforcing blacks can be used as extending fillers. Compounding is performed using an internal mixer, and completed using a roll mill. Compounds can be processed using most conventional rubber processing equipment. The range of typical compound property values is given in Table 1. FZ elastomers have excellent resistance to hydrocarbons and inorganic acids, as is expected for a fluorinated elastomer. They are strongly affected by polar solvents, but are more resistant to amines than most other fluorinated elastomers.

Table 1. Properties of FZ Compounds

Property	Value
density, g/mL	1.75–1.85
durometer hardness, Shore A	35–90
tensile strength, MPa[a]	6.9–13.8
100% modulus, MPa[a]	2.8–14.8
ultimate elongation, %	75–250
compression set, 70 h at 150°C, %	15–55
tear resistance, die B, kN/m[b]	to 26
retraction temperature, TR-10, °C	−56
brittle point, °C	−68
weatherability and ozone resistance	excellent
flame resistance, LOX[c]	39–46

[a] To convert MPa to psi, multiply by 145.
[b] To convert kN/m to ppi, divide by 0.175.
[c] Limiting oxygen index.

Economic Factors. Annual usage was under 10 metric tons, and was largely consumed in the United states. Ethyl Corp. has exited this business. It had been the sole supplier.

Applications. Initial applications have been largely in military and aerospace areas.

Aryloxyphosphazene Elastomers

Properties and Applications. Aryloxyphosphazene elastomers using phenoxy and *p*-ethylphenoxy substituents have found interest in a number of applications involving fire safety. This elastomer has a limiting oxygen index of 28 and contains essentially no halogens. It may be cured using either peroxide or sulfur. Peroxide cures do not require the allylic cure monomer. Gum physical properties are as follows:

T_g °C	−18
Mooney viscosity, $ML1 + 4$, 100°C	40–70
specific gravity	1.25

The gum is soluble in tetrahydrofuran, toluene, and cyclohexane.

Economic Considerations. Gum production by Ethyl Corp. was under 20 metric tons per year. Products were sold primarily for use in U.S. and N.A.T.O. naval applications. Ethyl Crop. exited this business. Atochem developed their technology and constructed a pilot plant to produce aryloxyphosphazenes, but has also discontinued production of these polymers.

JEFFREY T. BOOKS
Albemarle Corporation

D. F. Lohr and H. R. Penton in A. K. Bhowmick and H. L. Stephens, eds., *Handbook of Elastomers,* Marcel Dekker, Inc., New York, 1988, p. 535.

R. E. Singler, G. L. Hagnauer, and R. W. Sicka in J. E. Mark and J. Lal, eds., *Elastomers and Rubber Elasticity,* ACS Symposium Series, Vol. 193, Washington, D.C., 1985, p. 229 (good general review).

H. R. Allcock, *Phosphorus–Nitrogen Compounds,* Academic Press, Inc., New York, 1972.

M. Zeldin, KJ Wynne, and H.R. Allcock, eds., *Inorganic and Organometallic Polymers, Macromolecules Containing Silicon, Phosphorus and Other Inorganic Elements,* ACS Symposium Series, Vol. 360, Washington, D.C., 1988, Chapts. 19–25, pp 250–313.

POLYBUTADIENE

1,3-Butadiene can be polymerized to produce various resinous and elastomeric polymers. The basic microstructural units in polybutadiene include *cis*-1,4, *trans*-1,4, and 1,2 units. A variety of polymers with different properties can be produced by changing the ratio of these units and their tacticity.

The preparation and characterization of 1,3-butadiene monomer is discussed extensively elsewhere.

Microstructures of Polybutadiene

The conjugated structure of 1,3-butadiene gives it the ability to accept nucleophiles at both ends and distribute charge at both carbon 2 and 4. The initial addition of nucleophiles leads to transition states of π-allyl complexes in both anionic and transition-metal polymerizations.

High Vinyl Polybutadiene. These 1,2 addition products are categorized into three main groups: syndiotactic, isotactic, and atactic 1,2-polybutadiene. The 1,2 vinyl products follow a stereospecific addition in which the chiral carbon carrying the pendent vinyl group leads to the formation of ldl, lll, or ddd configurations. The ldl structure is called syndiotactic polybutadiene, and the lll or ddd structures are called isotactic polybutadiene. A mixed structure (dldlldldl) is called atactic polybutadiene. Each of these structures gives polymers with unique physical, mechanical, and rheological properties.

Syndiotactic Polybutadiene. Syndiotactic polybutadiene melts at high (150–220°C) temperatures, depending on the degree of crystallinity in the sample, and it can be molded into thin films that are flexible and have high elongation. The unique feature of this plastic-like material is that it can be blended with natural rubber. *cis*-1,4-Polybutadiene and the resulting blends exhibit a compatible formulation that combines the properties of plastic and rubber.

There are several methods described in the literature using various cobalt catalysts to prepared syndiotactic polybutadiene.

By using different catalysts and polymerization techniques, syndiotactic polybutadiene can be prepared with various melting points. An extensive review of high melting syndiotactic polybutadiene has been published. Two types of syndiotactic polybutadiene are most relevant to industrial applications; a high melting syndio-polybutadiene (T_m = 190–216 °C) and a low melting syndio-polybutadiene (T_m = 70–90 °C). The low melting type is commercially available from the Japan Synthetic Rubber Co. (JSR). The physical properties of syndiotactic polybutadiene are controlled by its melting point, degree of crystallinity, and molecular weight. The physical properties of low melting point (60–105°C) syndiotactic polybutadienes commercially available from JSR are shown in Table 1.

Table 1. Physical Properties of Low Melting Syndiotactic 1,2-Polybutadiene

Properties	Measured value JSR RB805	JSR RB810	JSR RB820	JSR RB830
density, g/cm^3	0.898	0.901	0.906	0.909
microstructure, % of 1,2 bonds	90	90	92	93
refractive index	1.508	1.510	1.512	1.517
thermal properties, °C				
vicat softening point		39	52	68
brittle point	−42	−40	−37	−35
melting point	59	71	95	105
tensile properties, 300% modulus, MPa	2.4	3.9	5.9	7.8
strength at break, MPa	4.9	6.4	10.3	12.7
elongation at break, %	800	750	700	670
hardness, Shore D	19	32	40	47
izod impact strength, notched, (N·cm)/cm	21.6–22.6	—does not break—		
mold shrinkage, %	2.3–3.3	0.7–0.9	0.3–0.5	0.3–0.6

Amorphous High 1,2-Polybutadiene. In recent years, high vinyl polybutadiene has become increasingly important to the tire industry. Amorphous, high vinyl polybutadiene is useful in tread formulations developed for energy conservation owing to it low hysteresis and good wetgrip characteristics. An amorphous, high 1,2-polybutadiene has been commercialized by the Nippon Zeon Co. under the trade names BR1240 and BR 1245.

The preparation of amorphous high (99%) 1,2-polybutadiene was first reported in 1981. The use of a heterocyclic chelating diamine such as dipiperidine ethane in the polymerization gave an amorphous elastomeric polymer of 99.9% 1,2 units and a glass-transition temperature of +5 °C. Preparation and characterization of low, medium, and high vinyl polybutadienes used polar modifiers. The vinyl content of polybutadiene is controlled by the ratio of the polar modifier to the active lithium catalyst as well as the polymerization temperature. Raising the modifier/lithium ratio increases the vinyl content and T_g of the polybutadiene (Fig. 1). At a 2:1 modifier:lithium ratio, the vinyl content decreases with increasing polymerization temperature up to about 60°C (Fig. 2).

Microstructural variation can give polymers with a wide range of glass-transition temperature. Using this method, one can develop a single polymer in which the desired combination of physical properties can be obtained.

Based on this variety of properties, amorphous polybutadiene has found a niche in the rubber industry. Moreover, it appears that the anionically prepared polymer is the only polymer that can be functionalized by polar groups. The functionalization is done by using aromatic substituted aldehydes and ketones or esters. Functionalization has been reported to dramatically improve polymer-filler interaction and reduce tread hysteresis.

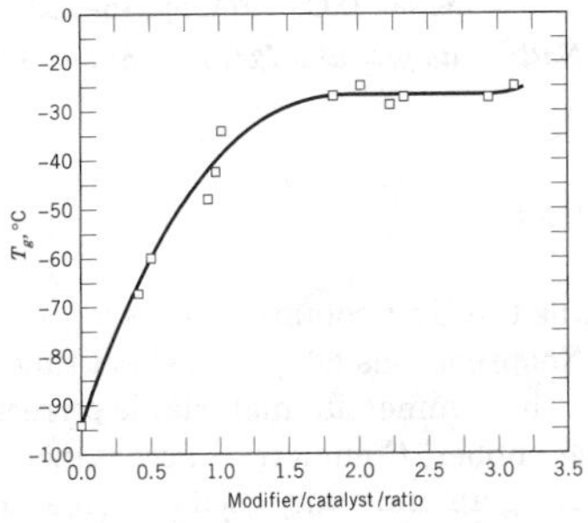

Figure 1. Effect of modifier ratio on polymer T_g. Courtesy of Goodyear Tire & Rubber Co.

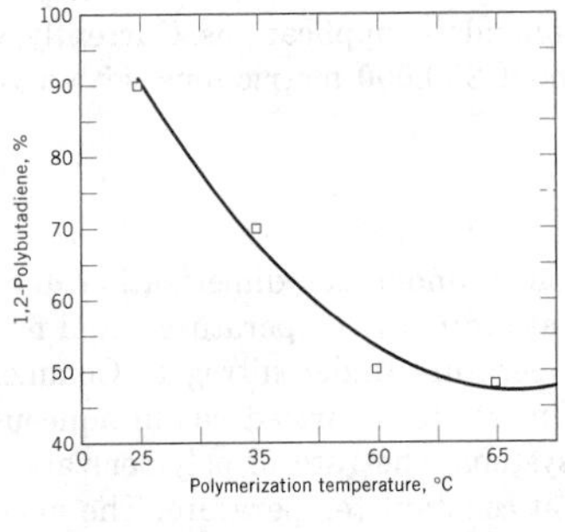

Figure 2. Effect of polymerization temperature on vinyl content at a 2:1 modifier:lithium ratio. Courtesy of Goodyear Tire & Rubber Co.

***cis*-1,4-Polybutadiene.** There are numerous references in the literature on the preparation of *cis*-1,4-polybutadiene. These authors have used transition metals in the presence of a reducing agent such as organoaluminum compounds or its chloride or hydride derivatives.

***trans*-1,4-Polybutadiene.** *trans*-1,4-Polybutadiene can be prepared using transition-metal catalysts or a nontransition-metal catalyst system based on Group I and II metals. The transition metals used include titanium, vanadium, chromium, rhodium, iridium, cobalt, and nickel.

The *trans*-1,4-polybutadiene made by transition-metal catalysis is a resin-like material that has two melting temperatures, 50 and 150°C. This solid resinous material has not found much application because it is difficult to stabilize.

A *trans*-1,4-polybutadiene that is useful as a tire rubber and can be stabilized and processed using conventional equipment has been made using an alkoxide of group II, reduced with organolithium or organomagnesium compounds in the presence of lithium alkoxide salts and aluminum alkyls.

Mixed Microstructure Polybutadiene. An amorphous polybutadiene of 10% vinyl, 35% *cis*-1,4, and 55% *trans*-1,4 structure can be made with a living anionic catalyst. This polymer is currently sold by Firestone under the name of Diene, and by Asahi under the name of Taktene. It has poor green strength, but shows excellent physical properties when compounded in a tire tread. This tread rubber, made in a living polymerization using a lithium catalyst, shows excellent tread wear as well as low hysteresis. Moreover, this type of polymer can be functionalized since the end of the polymer chain carries an active carbon metal linkage. This carbon-bound metal has been functionalized with ketones, aldehydes, acid chlorides, and metal halides of silicon and tin.

ADEL F. HALASA
J. M. MASSIE
Goodyear Tire & Rubber Company

A. F. Halasa and co-workers, in F. G. A. Stone and R. West, eds., *Advances in Organometallic Chemistry,* Vol. 18, Academic Press, Inc., New York, 1980.

Y. Takeuchi, A. Sekimoto, and M. Abe, *Am. Chem. Soc. Symp. Ser. 4,* American Chemical Society, Washington, D.C., 1974, pp. 15–26.

J. Boor, Jr., *Zeigler-Natta Catalysts and Polymerizations,* Academic Press, Inc., New York, 1979.

POLYCHLOROPRENE

Polychloroprene was the first commercially successful synthetic elastomer. The name Neoprene was adopted a short time later, and is now in general use, but the commercial material is correctly designated as CR, or chloroprene rubber. Commercial acceptance of CR was based on the material having an unusually good combination of environmental resistance and toughness, particularly in dynamic applications involving heat buildup and resistance to flex cracking. The unusual chemistry of chloroprene monomer and the versatility of emulsion polymerization have led to a broad spectrum of products, often optimized for particular end use applications. Currently, worldwide annual consumption is about 350,000 metric tons with a value approaching $1.4 billion.

Polymerization

Chloroprene monomer undergoes dimerization and autopolymerization when stored at ordinary temperatures, and for that reason it is stored at low temperature under nitrogen. Commercial chloroprene polymerization is most often carried out in aqueous emulsion using an anionic soap system. The rate of polymerization is controlled by catalyst addition at constant temperature. The major soap system is a rosin acid which both provides colloidal stability and plasticizes the rubber.

There are only a limited number of commercial applications of bulk or solution polymerization of chloroprene. These involve either graft polymerization of adhesives or production of liquid polymers.

Polymer Structure and Properties

Branching and Gel Formation. Long-chain branching and gel formation are caused by chain-transfer and double-bond addition reactions with polymer during polymerization. Monomeric conversion is limited to about 70% to control the rheology and processability of the polymer.

Molecular Weight Control. Polymer bulk viscosity, eg, Mooney viscosity, is widely used to predict processibility. It is usually controlled by addition of a chain-transfer agent to the polymerization system. Normally a mercaptan is used, but other materials such as xanthogen disulfides have also been used to provide reactive end groups.

Sulfur Copolymers. Molecular weight may also be controlled by copolymerization of the monomer with sulfur. The copolymer contains polysulfide units along the polymer chain that are subsequently cleaved chemically to control molecular weight.

Copolymerization. Over the years, almost every vinyl and diene monomer has been tested with chloroprene in free-radical polymerization. Few monomers are sufficiently reactive to compete with chloroprene. As a result, copolymers usually contain only a limited amount of random comonomer with much unreacted comonomer. If the polymerization is driven to completion, the second monomer may form a combination of homopolymer and graft polymer. Sometimes, however, a minor amount of comonomer provides beneficial functional groups. Methacrylic acid, for example, promotes adhesion and increases cohesive strength. An exception to the reactivity problem is 2,3-dichlorobutadiene, which is even more reactive than chloroprene and is often used to improve the crystallization resistance of the polymer. A large number of graft polymers of polychloroprene have been described, but the only ones of commercial significance are those made with acrylates and methacrylates. These are particularly useful for adhesion to plasticized vinyl. The graft polymers may be made either in solution or emulsion polymerization.

Interpenetrating networks have been made by co-curing polychloroprene with copolymers of 1-chloro-1,3-butadiene. The butadiene copolymer with 1-chloro-1,3-butadiene and an octyl acrylate copolymer improved the low temperature brittleness of polychloroprene. The acrylate also improved oil resistance and heat resistance.

Finally, block copolymers have been made in a two-step process.

Microstructure. Whereas the predominate structure of polychloroprene is the head to tail *trans*-1,4-chloroprene unit (**1**) other structural units (**2,3,4**) are also present. The effects of these various structural units on the chemical and physical properties of the polymer have been determined. The high concentration of structure (**1**) is responsible for crystallization of polychloroprene and for the ability of the material to crystallize under stress. Structure (**3**) is quite important in providing a cure site for vulcanization, but on the other hand reduces the thermal stability of the polymer. Structures (**3**), (**4**), and especially (**2**) limit crystallization of the polymer.

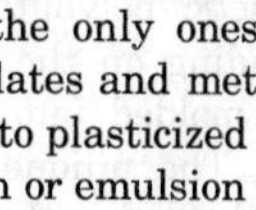

trans-1,4-unit
(**1**)

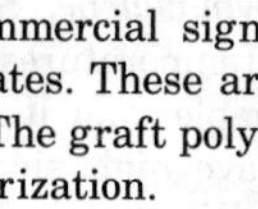

cis-1,4-unit
(**2**)

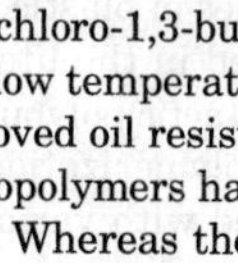

1,2-unit
(**3**)

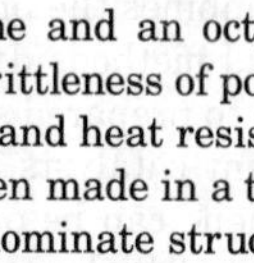

3,4-unit
(**4**)

A further structural irregularity was found to be inverted alignment of the structural units to give tail to tail and head to head arrangements in addition to the predominant head to tail arrangement. The total amount of structures (**2**), (**3**), and (**4**) varies with polymerization temperature from 5% at −40°C to about 30% at 100°C. The amounts produced of the various structures at −150 to +90°C have been determined by ^{13}C nmr (Table 1).

When the polymerization is run at less than about 20°C, the *trans*-1,4 content is high, and the resulting polymer is hard and tough. As such, it has valuable properties as an adhesive. Polymer made at higher temperature, with more chain irregularities, tends to be much slower-crystallizing, and is more suitable for mechanical goods applications.

Manufacture of Chloroprene Rubber

Chloroprene rubber is usually manufactured by either batch or continuous emulsion polymerization and isolated either by freeze coagulation or drum drying of a polymer film.

Properties of CR

General Vulcanizate Properties. CR has a good balance of toughness and environmental resistance. The high strength of CR vulcanizates that do not contain reinforcing filler is a result of its tendency to crystallize under stress. Few elastomers other than natural rubber do this.

Crystallization. The rate of crystallization and the melting point of polychloroprene both increase as the polymerization temperature decreases.

Heat Aging and Degradation. The resistance of polychloroprene to air oxidation, ozone attack, and weathering is enhanced, compared to other diene elastomers, by the presence of an electronegative chlorine atom on the repeat unit double bond. However, antidegradants are required for long-term exposure.

Compounding and Processing Chloroprene Rubber

Compound processibility is a key factor in the optimization of new polychloroprene types. As a result, commercial compounds can be mixed, shaped, and cured by virtually all the methods used in the rubber industry. A typical polychloroprene compound includes a variety of additives designed to improve compound rheology, cure rate, and vulcanizate properties.

Curing Systems. Polychloroprene can be cured with many combinations of metallic oxides, organic accelerators, and retarders. The G family of polymers, containing residual thiuram disulfide, can be cured with metallic oxides alone, although certain properties, for example compression set, can be enhanced by addition of an organic accelerator. The W, T, and xanthate modified families require addition of an organic accelerator, often in combination with a cure retarder, for practical cures.

Antidegradants. Degradation of CR at elevated temperatures is clearly related to air exposure. The addition of even a small amount of a free-radical inhibitor has a significant effect on the resistance of vulcanizates to oxidative attack.

Table 1. Microstructure of Polychloroprene by ^{13}C nmr

	1,4 Addition					
Polymerization temperature, °C	trans	inverted	cis	1,2 Addition	Isomerized 1,2	3,4 Addition
+90	85.4	10.3	7.8	2.3	4.1	0.6
+40	90.8	9.2	5.2	1.7	1.4	0.8
+20	92.7	8.0	3.3	1.5	0.9	0.9
0	95.9	5.5	1.8	1.2	0.5	1.0
−20	97.1	4.3	0.8	0.9	0.5	0.6
−40	97.4	4.2	0.7	0.8	0.5	0.6
−150	~100	2.0	<0.2	<0.2	>0.2	<0.2

Reinforcing Fillers. Polychloroprene and natural rubber are both self-reinforcing rubbers, and do not require reinforcing fillers to achieve high tensile strength. However, reinforcing fillers are normally used to improve certain properties, such as hardness and modulus, to improve processibility and to reduce cost. Carbon black is the most widely used filler. The principal non-black mineral fillers are hard clay, silica, and calcium silicate. Hard clay provides a good degree of reinforcement and tear strength, and is used to extend more expensive black compounds. Calcined clays are used for best electrical properties. Hydrated alumina is used for improved flame resistance, and platy talcs for good extrusion and electrical properties.

Plasticizers. These are used to improve compound processibility, modify vulcanizate properties, and reduce cost. For many applications, where cost and processibility are the objective, naphthenic and aromatic oils are preferred. If flammability is an issue, liquid chloroprene polymers (eg, DuPont FB or Denki LCR-H-050) can be used.

Polyester plasticizers are much more effective than hydrocarbon oils in reducing the brittle point and the glass-transition temperature of vulcanizates. There is a trade-off, however. The plasticizers are most effective in that lowering brittle point also promotes crystallization. The trade-off can be adjusted by using blends of dioctyl sebacate with aromatic oil or be going to an unsaturated vegetable oil such as rapeseed oil.

Mixing. CR may be mixed with the various types of equipment used for general-purpose elastomers. The key factor is to minimize mix time and temperature to avoid scorch. This is often best done in an internal mixer.

Calendering and Extrusion. Friction compounds are used to build up composite structures of fabric and rubber. The surface of the calendered fabric must have good green tack, and for that reason compounds to be calendered are best made from a slow crystallizing peptizable polymer such as DuPont GRT or Baypren 610.

Extrusion compounds use FEF carbon black to obtain good die definition, low nerve, and a good level of physical properties.

Molding. Molding is used widely for fabricating CR into belting, hose, sponge, and a variety of industrial products. All of the standard molding techniques have been used successfully with commercially available equipment. Molding methods include compression, transfer and injection, blow molding, vacuum molding, and tubing mandrel wrap.

Commercial Polymer Types

The *Synthetic Rubber Manual* (Institute of Synthetic Rubber Producers, Houston, Texas, 1989) lists 93 chloroprene rubber dry polymer types manufactured by six producers, not including those types produced in China or those for adhesive applications. For the most part, the polymers fall into the categories in Table 2.

Freeze-Resistant Polymers. Chloroprene homopolymers made at conventional polymerization temperatures of 40–50°C are not sufficiently freeze resistant for some applications. In particular, automotive parts such as belts, boots, and air springs are used in dynamic applications and need substantial freeze resistance during cold weather. Certain polychloroprene manufacturers make use of 2,3-dichloro-1,3-butadiene as a comonomer to retard crystallization, while others polymerize at higher temperature.

Applications

Dry Polymer Types. Polychloroprene has a well-balanced combination of properties including processibility, strength, flex and tear resistance, weatherability, flame resistance, and adhesion, together with sufficient heat and ozone resistance for most applications. Typical applications involve power transmission and timing belts, automotive boots, airsprings, and truck engine mounts. Other applications involve the proven long-term dependability of the material in adverse environments.

Solvent Adhesive Applications. A large number of CR types are involved, both standard and specially designed, for adhesive uses. The

Table 2. Distinguishing Features of CR Grades

Standard types	Advantages	Limitations
mercaptan modified W types	heat resistance compression set shelf stability nonpeptizable	cure rate flex resistance
xanthate modified types	tensile properties cure rate tan delta compression set rheology	heat resistance flex resistance
sulfur copolymers G types	peptizable building tack flex resistance tan delta cure chemistry	heat resistance compression set shelf stability
precross-linked grades T types	low nerve fast extrusion low die swell collapse resistance	tensile properties
freeze-resistant copolymers		
2,3-dichlorobutadiene styrene	slow crystallization	slightly higher T_g
polymers made at higher temperatures	slow crystallization	heat resistance

special polymer types are designed to have high uncured strength either through polymer crystallization or by ionomer formation.

Latex Adhesive Applications. Polychloroprene latex adhesives have a long history of use in foil laminating adhesives, facing adhesives, and construction mastics. Increasingly stringent restrictions on the emission of photoreactive solvents has heightened interest in latex compounds for broader applications, particularly contact bond adhesives.

Polychloroprene Latexes Polychloroprene latexes can often be used where either dry or solvent-based processes would be impractical. The water base systems have high solids, minimal solvent emissions, are unaffected by polymer rheology, and have extremely small particle size. The main problem areas are the slow evaporation rate of water and the need to ensure that drying conditions favor film formation.

Latex Types. Latexes are differentiated both by the nature of the colloidal system and by the type of polymer present. Nearly all of the colloidal systems are similar to those used in the manufacture of dry types. That is, they are anionic and contain either a sodium or potassium salt of a rosin acid or derivative. In addition, they may contain a strong acid soap to provide additional stability.

Latex Compounding. Latex compounding must take into account the stability of the latex both before and after compounding. Where consideration of solids concentration permits, the additives are best predispersed in a compatible aqueous surfactant before addition to the latex. The volume of additives, especially if clay fillers are involved, may easily be enough to starve the system for soaps and flocculate the compound. Dry powders or molten resins may often be added directly to the nonionic latex to avoid dilution of the compound.

All latex compounds should contain at least 5 phr of zinc oxide. This is needed to absorb evolved hydrochloric acid either in the compound or finished part.

Accelerators may be added to improve the physical properties of the polymer when needed. Mineral fillers, for example clay, can be added to reduce cost. A light process oil, free of polycyclic aromatics, can be used to improve the flexibility of films. An ester plasticizer can be used to improve low temperature properties.

Processing and Applications. Latexes are used in a variety of special ways to provide useful products. Applications include binders, coatings, dipped goods, elasticizers, and foam.

Economic Aspects

CR is currently manufactured in five countries in Western Europe, North America, and Japan with a combined capacity of 385,000 metric tons. DuPont is the largest supplier, with 49% of the total capacity and plants in the United States, U.K., and Japan.

Although CR continues to have about the same production volume, its share of total synthetic elastomer production has dropped from around 5% in 1975 to 3% in 1989. In part, this is related to the introduction of new materials that may not match CR's balance of properties but work quite well in specific areas and are less expensive.

On the other hand, CR is a versatile material that is able to take advantage of new opportunities. At this writing (ca 1993) these are in dynamic applications such as airbags, airsprings, and heat-resistant engine mounts.

Health, Safety, and Environmental Factors

Fire and uncontrolle runaway polymerization are a concern in the handling of chloroprene monomer. The refined monomer is ordinarily stored refrigerated under nitrogen and inhibited to avoid the latter.

After polymerization, excess monomer is stripped, condensed, and recycled. The residual monomer content of the stripped emulsion does not normally represent an acute hazard in subsequent operation. Worker exposure to monomer is monitored, and sources of exposure identified and corrected.

Polychloroprene latexes contain 0.5 to 0.02% residual chloroprene depending on the specific latex type. The amount of free alkali in the water phase of latexes varies from 0.1 to 0.08% depending on type and age of the material. Eye protection and appropriate skin protection have been recommended for use in situations where splashes or spills are possible. Manufacturers should be contacted for precise handling requirements.

Although polychloroprene itself has not been shown to have potential health problems, it should be understood that many rubber chemicals that may be used with CR can be dangerous if not handled properly. This is particularly true of ethylenethiourea curatives and, perhaps, secondary amine precursors often contained in sulfur-modified polychloroprene types. Material safety data sheets should be consulted for specific information on products to be handled.

The principal atmospheric emissions from a chloroprene polymerization facility result from gaseous venting during monomer transfer, from incomplete condensation of monomer after distillation and stripping operations, and from the polymer dryer vents. Back-up condensers are used to capture monomer venting from monomer-transfer operations. Chilled brine cooling is used to maximize condenser efficiency. Recipes have been modified to minimize dryer emissions. Equipment and emissions control procedures are ordinarily arrived at by agreement with appropriate governmental agencies.

The effluent from the isolation wash belt is the principal wastewater stream from the polymerization process. It contains highly diluted acetic acid and a surfactant that may not be biodegradable. The wastewater streams are sent to sewage treatment plants where BOD is reduced to acceptable levels. Alternative biodegradable surfactants have been reported in the literature.

Polymer that does not meet specifications is sold at reduced price. What cannot be sold is buried in nonhazardous landfill operations.

W. K. WITSIEPE
Consultant

L. J. Op, P. J. Hwan, and B. H. On, *Hwahak Kwa Hwahak Kongop* **20,** 119 (1977).

Y. Miyata and S. Matsunaga, *Polymer* **28,** 2233 (1987); Y. Miyata and M. Sawada, *Polymer* **29,** 1495 and 1683 (1988).

M. M. Coleman, D. L. Tabb, and E. G. Brame, *Rubber Chem. Technol.* **51,** 49 (1978); M. M. Coleman and E. G. Brame, *Rubber Chem. Technol.* **51,** 668 (1978).

C. A. Hargreaves II, in J. P. Kennedy and E. G. Tornqvist, eds., *Polymer Chemistry of Synthetic Elastomers,* Wiley-Interscience, New York, 1986.

C. A. Stewart, Jr., T. Takeshita, and M. L. Coleman in J. I. Kroschwitz, ed., *Encyclopedia of Polymer Science and Engineering,* Vol. 3, 3rd ed., John Wiley & Sons, Inc., New York, 1985.

POLYETHERS

Structure

Epichlorohydrin Elastomers without AGE. ECH homopolymer, poly-epichlorohydrin (**1**), and ECH–EO copolymer, poly(epichloro-hydrin-*co*-ethylene oxide) (**2**), are linear and amorphous. Because it is unsymmetrical, ECH monomer can polymerize in the head-to-head, tail-to-tail, or head-to-tail fashion. The commercial polymer is 97–99% head-to-tail, and has been shown to be stereorandom and atactic. Only low degrees of crystallinity are present in commercial ECH homopolymers; the amorphous product is preferred.

$$+CH_2CHO\overline{)_n}$$
$$\quad\quad | $$
$$\quad\quad CH_2Cl$$

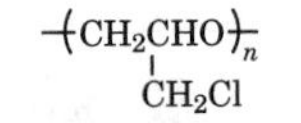

(**1**)

$$+CH_2CHO\overline{)_m}(CH_2CH_2O\overline{)_n}$$
$$\quad\quad | $$
$$\quad\quad CH_2Cl$$

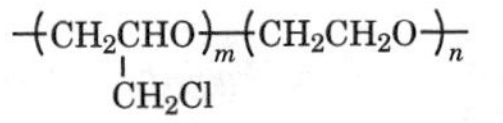

(**2**)

AGE-Containing Elastomers. ECH–AGE, poly(epichlorohydrin-*co*-allyl glycidyl ether) (**3**), ECH-EO-AGE, poly(epichlorohydrin-*co*-ethylene oxide-*co*-allyl glycidyl ether) (**4**), ECH–PO–AGE, and PO–AGE are also amorphous polymers.

$$+CH_2CHO\overline{)_m}(CH_2CHO\overline{)_n}$$
$$\quad | \qquad\qquad | $$
$$\quad CH_2Cl \qquad CH_2OCH_2CH{=}CH_2$$

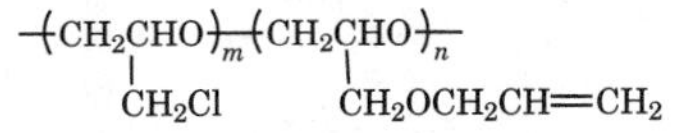

(**3**)

$$+CH_2CHO\overline{)_m}(CH_2CH_2O\overline{)_n}(CH_2CHO\overline{)_o}$$
$$\quad | \qquad\qquad\qquad\qquad | $$
$$\quad CH_2Cl \qquad\qquad CH_2OCH_2CH{=}CH_2$$

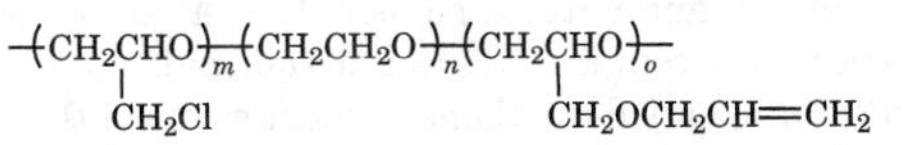

(**4**)

Crystallinity is low; the pendent allyl group contributes to the amorphous state of these polymers.

Properties

Properties of the uncompounded elastomers are listed in Table 1 and properties of the compounded polymers in Table 2.

Epichlorohydrin Elastomers without AGE. The ECH homopolymer has very low gas permeability (two to three times greater than that of butyl rubber), outstanding ozone resistance, good building tack, and

Table 1. Commercial Polyether Elastomers

Elastomer	ECH, %	Chlorine, %	Ethylene oxide, %	Sp gr	ML^a	T_g, °C
ECH	100	38	0	1.36	40–80	−22
ECH–EO	68	26	32	1.27	40–130	−40
ECH–AGE	92	35	0	1.24	60	−25
ECH–EO–AGE	48–70	24–29	20–50	1.27	50–100	−38
ECH–PO–AGE	40	15	0	1.12	60–80	−48
PO–AGE	0	0	0	1.01	^{b}	−62

^{a} Mooney viscosity at 100°C, ASTM 1646.
^{b} Oscillating disk rheometer (ODR) viscosity is 21–26.

Table 2. Vulcanizate Properties of Polyether Elastomers

Properties	CO	ECO	GCO	GECO	GPCO	GPO
Mooney viscosity at 100°C						
polymer	78	80	62	80	78	72
compound	73	93	60	86	75	83
originals cured at 170°C						
100% modulus, MPa^a	7.6	5.4	7.2	5.5	5.6	3.4
tensile strength, MPa^a	16.5	15.0	15.5	15.1	11.4	10.2
elongation, %	225	275	205	280	210	365
hardness, Shore A	74	74	74	73	77	68
air oven aged 70 h at 125°C						
100% modulus, MPa^a	7.6	5.9	7.5	5.7	0	6.2
tensile strength, MPa^a	15.3	14.4	13.6	14.7	11.8	10.6
elongation, %	195	240	165	250	80	180
hardness, Shore A	75	72	74	73	90	76
compression set, 70 h at 100°C	13	19	18	19	39	39
brittleness, temp^b °C	−20	−40	−24	−40	−46	−61
tear strength, kN/m^c	32	33	26	35	46	51
ozone resistance, 100 ppm at 49°C, h	>168	>168	>168	>168	>168	>168

^{a} To convert MPa to psi, multiply by 145.
^{b} ASTM 2137.
^{c} To convert kN/m to lb/in., multiply by 5.71.

low heat buildup. The polymer is flame retardant as a result of its high chlorine content. It has poor resilience at room temperature, but this improves upon heating. The ECH–EO copolymer is less flame retardant due to its lower chlorine content. It has some impermeability to gases similar to that of the medium high acrylonitrile rubbers. It has low temperature flexibility to −40°C and exhibits good heat resistance. In contrast with ECH homopolymer, ECH–EO has poor tack.

AGE-Containing Elastomers. ECH–AGE copolymer shows excellent ozone resistance and good resistance to softening upon heat aging. Like ECH and ECH–EO, vulcanizates of ECH–AGE and ECH–EO–AGE are resistant to ASTM oils, aliphatic solvents, and aromatic containing fuels.

Two propylene oxide elastomers have been commercialized, PO–AGE and ECH–PO–AGE. These polymers show excellent low temperature flexibility and low gas permeability. After compounding, PO–AGE copolymer is highly resilient, and shows excellent flex life and flexibility at extremely low temperatures (ca −65°C). It is slightly better than natural rubber in these characteristics. Resistance to oil, fuels, and solvents is moderate to poor. Wear resistance is also poor. Unlike natural rubber, PO–AGE is ozone resistant and resistant to aging at high temperatures. The properties of compounded ECH–PO–AGE lie somewhere between those of ECH–EO copolymer and PO–AGE copolymer. As the ECH content of the terpolymer increases, fuel resistance increases while low temperature flexibility decreases. Heat resistance is similar to ECH–EO; fuel resistance is similar to polychloroprene. The uncured rubber is soluble in aromatic solvents and ketones.

Molecular Weight Determination and Solution Behavior. Molecular weight determinations using dilute solution viscosity measurements have been reported for ECH. Intrinsic viscosity is related to molecular weight by the Mark-Houwink equation: $[\eta] = KM^a$, where K and a are measured experimentally.

Molecular weight determinations of ECH–EO, ECH–AGE, ECH–EO–AGE, ECH–PO–AGE, and PO–AGE have not been reported. Some solution studies have been done on poly(propylene oxide), and these may approximate solution behavior of the PO–AGE copolymer.

Polymer Preparation

Epichlorohydrin Elastomers without AGE. Polymerization on a commercial scale is done as either a solution or slurry process at 40–

130°C in an aromatic, aliphatic, or ether solvent. Typical solvents are toluene, benzene, heptane, and diethyl ether. Trialkylaluminum–water and trialkylaluminum–water–acetylacetone catalysts are employed. A cationic, coordination mechanism is proposed for chain propagation. The product is isolated by steam coagulation. Polymerization is done as a continuous process in which the solvent, catalyst, and monomer are fed to a back-mixed reactor.

AGE-Containing Elastomers. The manufacturing process for ECH–AGE, ECH–EO–AGE, ECH–PO–AGE, and PO–AGE is similar to that described for the ECH and ECH–EO elastomers.

Processing and Fabrication

All of the polyether elastomers, like other vulcanizable elastomers, can be compounded with processing aids, fillers, plasticizers, stabilizers, and vulcanizing agents to make useful rubber products.

Economic Aspects

Polyether elastomers are moderately costly when compared with other synthetic rubbers because of their specialty, small-volume nature.

Health and Safety Factors; Toxicity

Monomers. The monomers used for commercial production of these elastomers are suspected carcinogens. These monomers, in addition to being potential human carcinogens, also produce acute toxic effects upon exposure. They are harmful when ingested or inhaled, or when they come in contact with skin. AGE, ECH, and PO are absorbed through the skin and may cause systemic toxic effects.

The American Conference of Governmental Industrial Hygenists (ACGIH) has set threshold limit values (TLVs) for airborne concentrations in the workplace (ppm, time-weighted average, 8-h day) as follows: AGE, 5; ECH, 2; EO, 1; and PO, 20. However, the ACGIH has reexamined data on ECH, and has proposed lowering the TLV to 0.1 ppm; it has categorized this monomer as Group A2, a suspected human carcinogen.

Polymers. A subacute vapor inhalation toxicity study in which animals were exposed to emission products from compounded Parel 58 suggests that no significant health effects would be expected in workers periodically exposed to these vapors.

Compounding Ingredients. Ethylene thiourea (ETU), the most commonly used curing agent for epichlorohydrin elastomers, has been determined to be carcinogenic, teratogenic, and goitrogenic in animal studies.

The common acid acceptors, red lead oxide and barium carbonate, are both toxic when inhaled or ingested. They are, and should be, used in industry as dispersions in EPDM or ECO.

Uses

Epichlorohydrin Elastomers without AGE. Vulcanizates of ECH homopolymer and ECH–EO copolymer are important in automotive applications such as fuel, air, and vacuum hoses, vibration mounts, and adhesives. Used as an additive, ECH–EO copolymer can also impart antistatic properties to plastics. Other industrial applications of these polymers include drive and conveyor belts; hoses, tubing, and diaphragms; pump parts including inner coatings, seals, and gaskets; printing rolls and blankets; fabric coatings for protective clothing; pond liners, and membranes in roofing material. In oil well drilling equipment, uses include drill-pipe protectors, packers, pipe scrubbers, and submersible power-cable jacketing.

AGE-Containing Elastomers. ECH–EO–AGE vulcanizates find applications in constant velocity boots, mounting isolators, and hose and wire covers. Automotive applications of ECH–PO—AGE include dust and fuel hose covers and rubber boots for suspension and transmission systems. This polymer is also used in some automotive antivibration applications as well as in covering for cable.

Epichlorohydrin Elastomer Derivatives

The principal route of chemical modification of epichlorohydrin elastomers is nucleophilic substitution on the pendent chloromethyl group. Reported nucleophilic substitution products include acetate, glycolate, hydrazine, α-pyrrolidonate, azide, phosphinyl, thiosulfate, α-mercaptoacetic acid, thiosalicylic acid, thioethers, dithiocarbamate, isothiouronium, imides, α,β-unsaturated carboxylic acids, cinnamate, carbazole, and methoxide. Applications of these modified polyethers include water thickeners, breaking emulsions, flocculating agents, drainage aids in paper manufacture, selectively permeable membranes, photosensitive material, flame retardants, and shrinkproofing wool.

Modifications of epichlorohydrin elastomers by radical-induced graft polymerization have been reported. Incorporated monomers include styrene and acrylonitrile, styrene, maleic anhydride, vinyl acetate, methyl methacrylate, and vinylidene chloride, acrylic acid, and vinyl chloride. When the vinyl chloride-modified epichlorohydrin polymers were used as additives to PVC, impact strength was improved.

KATHRYN OWENS
VERNON L. KYLLINGSTAD
Zeon Chemicals USA, Incorporated

K. E. Steller in E. J. Vandenberg, ed., *Polyethers,* American Chemical Society, Washington, D.C., 1975.

D. A. Berta and E. J. Vandenberg, in A. K. Bhowmick and H. L. Stephens, eds., *Handbook of Elastomers: New Developments and Technology,* Marcel Dekker, Inc., New York, 1988.

R. W. Body and V. L. Kyllingstad, in J. I. Kroschwitz, ed., *Encyclopedia of Polymer Science and Engineering,* Vol. 6, John Wiley & Sons, Inc., New York, 1986.

POLYISOPRENE

Physical Properties

In Table 1 some of the properties of raw synthetic *cis*-1,4-polyisoprene (Goodyear's Natsyn) and natural rubber (Hevea) are presented along with references that contain additional thermal, optical, electrical, and mechanical property data. Some properties of synthetic *trans*-1,4-polyisoprene (Kuraray TP-301) are also given. Molecular weights and mol wt distribution are determined by gel-permeation chromatography (gpc).

Chemical Properties

Polymer Structure. Isoprene can undergo 1,4-, 1,2-, or 3,4-addition polymerization depending on the catalyst type and conditions, resulting in several structures:

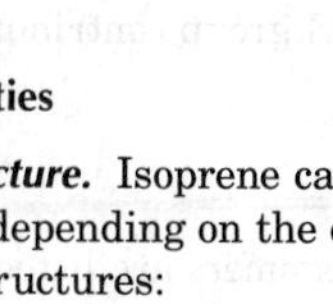

1,2-polyisoprene

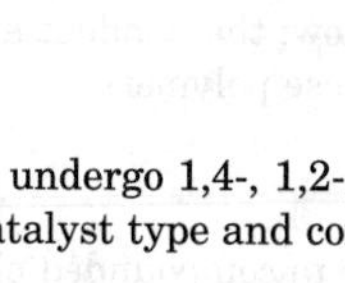

3,4-polyisoprene

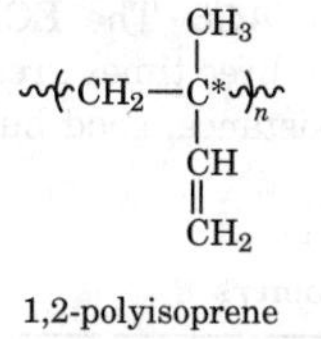

cis-1,4-polyisoprene

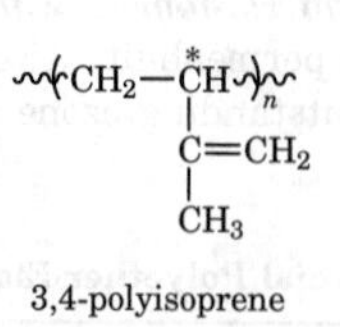

trans-1,4-polyisoprene

Polymerization

Ziegler-Natta. In 1954, the B. F. Goodrich Company used the newly discovered Ziegler catalyst for polymerizing ethylene to polyethylene

Table 1. Properties of Polyisoprenes

	cis-1,4-Polyisoprene		*trans*-1,4-Polyisoprene	
Property	Natural rubber (Hevea)	Natsyn 2200[a]	Kuraray TP-301[b] Gel content = 0%;	Natural balata
density, g/mL	0.913	0.91	0.96	
refractive index, n_D	1.5191			
Mooney viscosity, ML 1 + 4, 100°C		70–90	30	25–33
ash content, % max	0.5–1.5	0.60	0.15	0.5
T_g, °C	−72	−72		
T_m °C			67	67
$M_w \times 10^{-5}$		7.55–9.55		
$M_n \times 10^{-5}$		2.4–3.5		
microstructure, nmr				
cis-1,4, mol %	~100%	98.3[c]		
trans-1,4, mol %		1.4	99	~100
3,4, mol %		0.3	0.2	0.2

[a] The Goodyear Tire & Rubber Co., Ziegler Ti–Al catalyst.
[b] volatile matter = 0.3%.
[c] Cariflex, from Shell Nederland Chemie, and an alkyllithium catalyst, has 90.9% *cis*-1,4, 5.2% *trans*-1,4, and 3.9% 3,4.

as a catalyst for making *cis*-1,4-polyisoprene. The catalyst consisted of titanium tetrachloride and a trialkylaluminum. Aliphatic or aromatic hydrocarbons were used as the polymerization solvents with the exclusion of oxygen and moisture.

Alfin Catalysts. Alfin catalysis give polyisoprenes of high *trans*-1,4 microstructure.

Alkali Metal Catalysts. In hydrocarbon solvent or bulk, the polymerization of isoprene with alkali metals occurs heterogeneously, whereas in highly polar solvents the polymerization is homogeneous. Of the alkali metals, only lithium in bulk or hydrocarbon solvent gives over 90% *cis*-1,4 microstructure.

Anionic Polymerization. Polyisoprenes of predictable molecular weight and uniform chain length can be prepared by anionic polymerization, which is characterized by living growing chains and the absence of chain-termination.

Cationic Polymerization. A small amount of isoprene is cationically copolymerized with isobutylene in the commercial process for making butyl rubber, wherein the isoprene provides the unsaturation required for sulfur vulcanization.

Free-Radical Polymerization. The best method for polymerizing isoprene by a free-radical process is emulsion polymerization. Since emulsion polyisoprene has low *cis*-1,4 microstructure, it has poorer physical properties than the high *cis*-1,4-polyisoprene and has not been commercialized.

Manufacture and Processing

Dry prepurified isoprene monomer and dry aliphatic hydrocarbon solvent along with a Ziegler catalyst consisting of triisobutylaluminum and titanium tetrachloride are utilized in one commercial solution process for *cis*-1,4-polyisoprene. Isoprene, the Ziegler catalyst, and recycled aliphatic hydrocarbon solvent in an inert atmosphere (nitrogen) are continuously fed into a continuous polymerization reactor system, and the isoprene is polymerized to the desired conversion.

Compounding. Another important part of the use of polyisoprene rubber is its compounding into formulations that can be utilized to make finished goods and articles. The raw polyisoprene rubber is mixed in a Banbury (internal) mixer or on a mill with ingredients such as fillers (carbon black, clays, or silicates), plasticizers, antioxidants, accelerators, and sulfur. The filler, eg, carbon black, gives durability and reduces cost of the compound. The sulfur functions as a cross-linking agent for the highly unsaturated polyisoprene. The accelerators, eg, sulfenamides, thiazoles, guanidines, thiuram sulfides, and thiocarbamates, help reduce the vulcanization time. The antioxidants (qv), eg, arylamines, hindered phenols, etc, retard the deterioration of the finished article due to aging. Antiozonants (qv), eg, *p*-phenylenediamine, and waxes are also used in tire formulations to retard cracking under stress. The resulting rubber compound is calendered, molded, extruded, or fabricated into the desired shape and vulcanized or cured at around 150°C.

Economic Aspects

In 1995, synthetic *cis*-1,4-polyisoprene was manufactured commercially in the United States only by The Goodyear Tire and Rubber Co. at Beaumont, Texas. This plant has a capacity of 61,000 metric tons/yr and utilizes a Ziegler catalyst. Since 1985, annual production of polyisoprene has been flat at about 48,000 to 49,000 metric tons. The only Western European producer of synthetic *cis*-1,4-polyisoprene is Shell Nederland Chemie (member of the Royal Dutch/Shell Group), with a plant at Pernis (near Rotterdam), Netherlands. The plant capacity has been reduced to about 45,000 metric tons per year.

trans-1,4-Polyisoprene is not produced commercially in the United States, although the 1990 consumption was estimated to be around 2000 metric tons (synthetic and natural). In Japan, Kuraray Co., Ltd. produces about 400 metric tons/yr of *trans*-1,4-polyisoprene.

Health and Safety Factors

Polyisoprene rubber is relatively nonhazardous, but must be kept away from sparks, open flames, or excessive heat because it will burn. The current Material Safety Data Sheet (MSDS) should always be checked for known hazards before using polyisoprene or any other chemical materials.

Uses

cis-1,4-Polyisoprene is used in tires and tire products, belts, gaskets, hoses, foam rubber, molded and mechanical goods, bottle nipples, gloves, caulking, sealants, footwear and sporting goods, rubber bands, erasers, and rubber sheeting.

Because *trans*-1,4-polyisoprene is a crystalline thermoplastic, it resists abrasion, scuffing, and cutting, and is used mainly in high quality golf ball covers and orthopedic devices and splints, and to some extent in transmission belts, cable covering, and adhesives.

3,4-Polyisoprene is finding some applications in specialty tire rubber compounds.

MICHAEL SENYEK
The Goodyear Tire & Rubber Company

E. Ceausescu, *Stereospecific Polymerization of Isoprene*, Pergamon Press Inc., New York, 1983.

M. Morton, *Anionic Polymerization: Principles and Practice*, Academic Press, New York, 1983.

E. Schoenberg, H. A. Marsh, S. J. Walters, and W. M. Saltman, *Rubber Chem. Technol.* **52**, 533 (1979).

J. F. Auchter, *Polyisoprene Elastomers, Chemical Economics Handbook*, CEH Product Review, SRI International, Menlo Park, Calif., Sept. 1993.

THERMOPLASTIC ELASTOMERS

Thermoplastic elastomers have many of the physical properties of rubbers, ie, softness, flexibility, and resilience; but in contrast to conventional rubbers, they are processes as thermoplastics. Rubbers must be cross-linked to give useful properties. In the terminology of the plastics industry, vulcanization is a thermosetting process. Like other thermosetting processes, it is slow and irreversible and takes place upon heating. With thermoplastic elastomers, on the other hand, the transition from a processible melt to a solid, rubberlike object is rapid and reversible and takes place upon cooling. Thermoplastic elastomers can be processed using conventional plastics techniques, such as injection molding and extrusion; scrap is usually recycled.

Because of increased production and the lower cost of raw material, thermoplastic elastomeric materials are a significant and growing part of the total polymers market. World consumption in 1995 is estimated to approach 1,000,000 metric tons. However, because the melt to solid transition is reversible, some properties of thermoplastic elastomers, eg, compression set, solvent resistance, and resistance to deformation at high temperatures, are often not as good as those of the conventional vulcanized rubbers. Applications of thermoplastic elastomers are, therefore, in areas where these properties are less important, eg, footwear, wire insulation, adhesives, polymer blending, and not in areas such as automobile tires.

The classification given in Table 1 is based on the process, ie, thermosetting or thermoplastic, by which polymers in general are formed into useful articles and on the mechanical properties, ie, rigid, flexible, or rubbery, of the final product. All commercial polymers used for molding, extrusion, etc, fit into one of these six classifications; the thermoplastic elastomers are the newest.

Structure

Thermoplastic elastomers are often multiphase compositions in which the phases are intimately dispersed. In many cases, the phases are chemically bonded by block or graft copolymerization. In others, a fine dispersion is apparently sufficient. In these multiphase systems, at least one phase consists of a material that is hard at room temperature but becomes fluid upon heating. Another phase consists of a softer material that is rubberlike at RT. A simple structure is an A–B–A block copolymer, where A is a hard phase and B an elastomer, eg, poly(styrene-*b*-elastomer-*b*-styrene).

Four block copolymers are of major commercial importance: poly(styrene-*b*-elastomer-*b*-styrene), thermoplastic polyurethanes, thermoplastic polyesters and thermoplastic polyamides. All but the first have the multiblock $(A–B)_n$ structure. The morphology of the poly(styrene-*b*-elastomer-styrene) polymer is shown diagramatically in Figure 1. That of polyurethane, polyester, and polyamide block copolymers is shown diagramatically in Figure 2. It has some similarities to that of poly(styrene-*b*-elastomer-*b*-styrene) equivalents and also some important differences: (*1*) the hard domains are more interconnected; (*2*) they are crystalline; and (*3*) these long $(A–B)_n$ molecules may run through several hard and soft phases.

Not all thermoplastic elastomers are block copolymers. Those that are not are usually combinations of a hard thermoplastic with a softer, more rubberlike polymer. Usually, the components are mechanically mixed together, although it is sometimes possible to produce the rubber component *in situ* during polymerization. Typically, the two components form interdispersed multiphase systems as shown diagrammatically in Figure 3. Blends with ethylene–propylene copolymer (EPR) are now more important commercially, and propylene copolymers often replace polypropylene homopolymer as the hard phase. However, some combinations of a hard thermoplastic with a rubberlike polymer are claimed to be single-phase systems. In some cases, the elastomer phase is cross-linked while the mixture is being highly sheared. This process if often referred to as "dynamic vulcanization" and gives a finely dispersed and cross-linked elastomer phase, as shown diagramatically in Figure 4.

Table 1. Classification of Polymers According to Properties and Processing

Property	Thermoset	Thermoplastic
rigid	epoxy, melamine–formaldehyde, sheet molding compounds	polypropylene, high density polyethylene, polystyrene
flexible	highly vulcanized rubber	low density polyethylene, ethylene–vinyl, acetate copolymer, plasticized PVC
rubbery	vulcanized rubber	thermoplastic rubbers

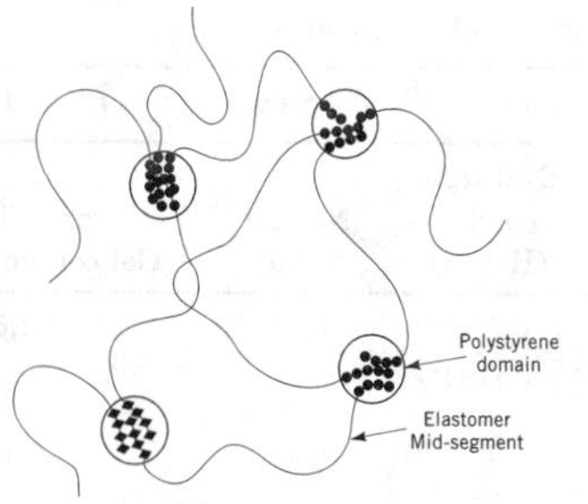

Figure 1. Phase arrangement in styrenic block copolymers.

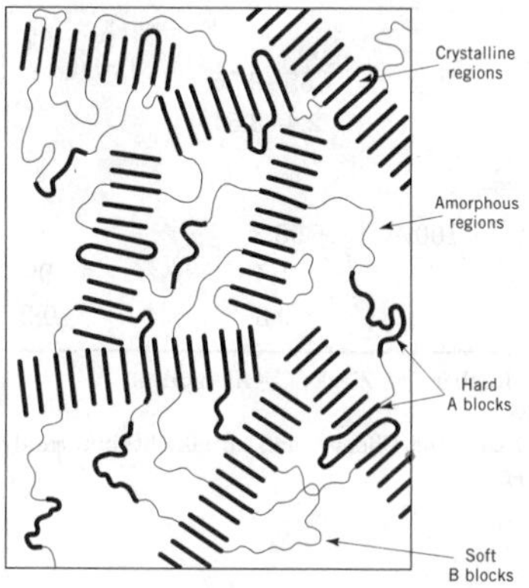

Figure 2. Phase arrangement in crystalline block copolymers.

Other thermoplastic elastomer combinations, in which the elastomer phase may or may not be cross-linked, include blends of polypropylene with nitrile, butyl, and natural rubbers, blends of PVC with nitrile rubber, and blends of halogenated polyolefins with ethylene interpolymers. Collectively, thermoplastic elastomers of this type are referred to herein as hard polymer/elastomer combinations.

Thermoplastic elastomers based on blends of a silicone rubber (cross-linked during processing) with block copolymer thermoplastic elastomers have also been described.

Property–Structure Relationships

Effects of variations in the structure of styrenic A–B–A block copolymers and similar materials were described in early work and have been reviewed.

Molecular Weight. Compared with homopolymers of similar molecular weight, styrenic block copolymers have very high melt viscosities which increase with increasing molecular weight.

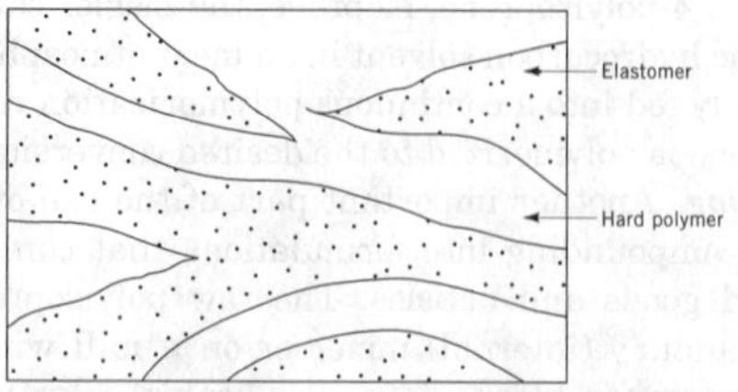

Figure 3. Phase arrangement in hard polymer/elastomer blends.

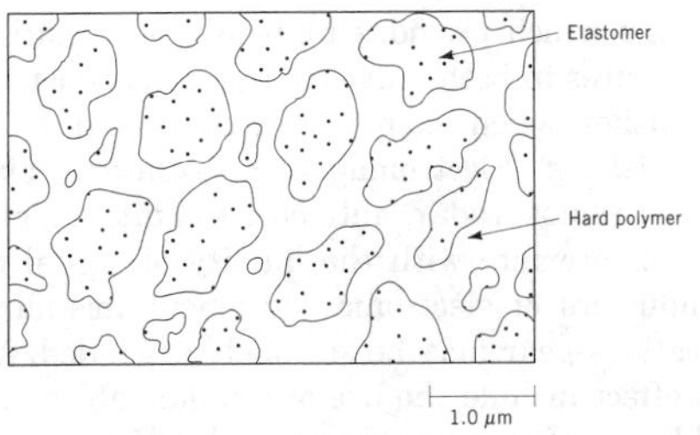

Figure 4. Phase arrangement in hard polymer/elastomer combinations in which the elastomer phase has been dynamically vulcanized.

Proportion of Hard Segments. As expected, the modulus of styrenic block copolymers increases with the proportion of the hard polystyrene segments. As the styrene content is increased, the products change from very weak, soft, rubberlike materials to strong elastomers, then to leathery materials, and finally to hard glassy thermoplastics. The latter have been commercialized as clear, high impact polystyrenes under the trade name K-Resin (Phillips Petroleum Company). Other types of thermoplastic elastomers show similar behavior; that is, as the ratio of the hard to soft phase is increased, the product in turn becomes harder.

Elastomer Segment. The choice of elastomer segment has a profound effect on the properties of styrenic block copolymers. Four are commercially important: polybutadiene, polyisoprene, poly(ethylene-*co*-butylene), and poly(ethylene-*co*-propylene). The corresponding styrenic triblock copolymers are referred to as S–B–S, S–I–S, S–EB–S, and S–EP–S, respectively. In thermoplastic polyurethanes, polyesters, and polyamides, the elastomer segments may be either polyesters or polyethers. The former type are tougher and have better oil resistance; the latter are more flexible at low temperature and have better hydrolytic stability.

Hard Segment. The choice of the hard segment determines the upper service temperature and also influences the solvent resistance. In styrenic block copolymers, those based on poly(α-methylstyrene) have higher upper service temperature and tensile strength than analogues based on polystyrene; both are soluble in common solvents. Replacing the polystyrene end segments in S–EB–S by polyethylene (giving E–EB–E block copolymer) improves solvent resistance; the phases are not separated in the melt.

In thermoplastic polyurethanes, polyesters, and polyamides, the crystalline end segments, together with the polar center segments, impart good oil resistance and high upper service temperatures. The hard component in most hard polymer/elastomer combinations is crystalline and imparts resistance to solvents and oils, as well as providing the products with relatively high upper service temperatures.

Synthesis

Block copolymers are synthesized by a variety of methods; most important are sequential polymerization and step growth. Thermoplastic elastomers that are hard polymer/elastomer combinations are often not truly synthesized. Instead, the two polymers that form the hard and soft phases are intimately mixed on high shear equipment.

Commercial Production

The poly(styrene-*b*-elastomer-*b*-styrene) materials are made by anionic polymerization.

Commercially, anionic polymerization is limited to three monomers: styrene, butadiene, and isoprene; therefore only two useful A–B–A block copolymers, S–B–S and S–I–S, can be produced directly. In both cases, the elastomer segments contain double bonds which are reactive and limit the stability of the product. To improve stability, the polybutadiene mid-segment can be polymerized as a random mixture of two structural forms, the 1,4 and 1,2 isomers, by addition of an inert polar material to the polymerization solvent; ethers and amines have been suggested for this purpose. Upon hydrogenation, these isomers give a copolymer of ethylene and butylene.

Thermoplastic polyurethane elastomers are produced from prepolymers by polycondensation. Commercial thermoplastic polyesters are synthesized in a similar way by the reaction of a relatively high molecular-weight polyether glycol with butanediol and dimethyl terephthalate. The polyether chain becomes the soft segment in the final product, whereas the terephthalic acid–butanediol copolymer forms the hard crystalline domains.

There are several types of thermoplastic polyamides. Their synthesis is similar to that of the polyurethane and polyester equivalents. In some cases, only the soft segments are prepolymers; whereas in others, prepolymers are used to give both segmental types. Both polyesters and polyethers are used for the soft segments, and this choice affects the final properties of the product. Various polyamides, including those based on aromatic groups, may be used for the hard segments. These all have different melting points and degrees of crystallinity and so give products with a wide range of properties.

The production of the hard polymer/elastomer combinations is more simple. The two components are mixed together under conditions of intensive shear. In some cases, grafting may occur. In a variation of this technique, the elastomer can be cross-linked while the mixing is taking place, a process described as dynamic vulcanization.

A large number of hard polymer/elastomer combinations made by the last technique have been investigated. In some cases, the components are technologically compatibilized by use of a grafting reaction, but usually a fine dispersion of the two phases is formed that is sufficient to give the product the properties of a thermoplastic elastomer.

Economic Aspects

Global consumption of thermoplastic rubbers of all types is estimated at about 600,000 t/yr. Of this, 42% was estimated to be consumed in the United States, 39% in Western Europe, and 19% in Japan. At present, the worldwide market is estimated to be divided as follows: styrenic block copolymers, 48%; hard polymer/elastomer combinations, 26%; thermoplastic polyurethanes, 12%; thermoplastic polyesters, 4%; and others, 9%. The three largest end uses were transportation, 23%; footwear, 18%; and adhesives, coatings, etc, 16%.

Health and Safety

Most thermoplastic elastomers are stable materials and decompose only slowly under normal processing conditions. If decomposition does occur, the products are usually not particularly hazardous and should not present a problem if good ventilation is provided. Extra caution should be exercised when processing polyurethanes, especially those containing polycaprolactone segments. In these cases the decomposition products may include isocyanates and caprolactam, both of which are potential carcinogens.

Of course, all materials that are processed in the molten state can cause burns if the hot material comes in contact with the skin. Care must be taken to avoid this. In all cases, it is recommended that the manufacturer's Material Safety Data Sheet be consulted before working with any of these materials.

Reprocessing

Easy reprocessing is one of the great advantages that thermoplastic elastomers have over conventional vulcanized rubbers. The scrap can be reground and is usually blended with virgin material before being reworked. Thermoplastic elastomers can also be used to "sweeten" regrind; that is, they can be blended with reground scrap from conventional thermoplastics to restore impact strength and reduce brittleness. Many applications, eg, coextrusion, generate mixed scrap, which usually has very poor properties. Thermoplastic elastomers can often convert this into useful material.

GEOFFREY HOLDEN
Consultant

N. R. Legge, G. Holden, and H. E. Schroeder, eds., *Thermoplastic Elastomers—A Comprehensive Review*, 2nd ed., Carl Hanser Verlag and Oxford University Press, Munich, New York.

B. M. Walker and C. P. Rader, eds., *Handbook of Thermoplastic Elastomers*, 2nd ed., Van Nostrand Reinhold, New York, 1988.

ELECTRICAL CONNECTORS

Electrical connectors are mechanical devices that connect wires, cables, printed circuit boards, and electronic components to each other and to related equipment. Connector designs include miniature units for microelectronic applications; specialized cable; rack and panel designs for incorporating combinations of a-c, d-c, and radio-frequency conducting contacts; and high current connectors for industrial application and for transmission and distribution of electrical power in overhead and underground networks. Further categorization of connectors can be made according to: application, whether connectors permanently join conductors and components or permit separation and rejoining; the means used to effect connection, whether by fusion (welding, soldering) or by pressure, the values of which can be small or great enough to severely deform metal; the distribution type, whether of power or of low (signal) levels of current; and the conductor size. The term electrical contact describes the junction between two or more current-carrying members that provide electrical continuity at their interfaces. Connector contacts ordinarily remain stationary in active circuits, eg, they are not mated or separated.

Many connectors for single conductors have an insulating sleeve, and almost all connectors that join two or more conductors have a plastic body, or dielectric, which separates the contact elements (see INSULATION, ELECTRIC). Metal or plastic shells with mechanical aids, such as screws, levers, and other coupling devices to facilitate joining and separation of the contacts, also may surround a connector. The shell may have mounting features for securing the connector to a chassis, supports for the wires and cables, and polarizing keys for prevention of improper mating.

Connector Configurations

Electronic Connectors. The complexity and size of many electronic systems necessitate construction from relatively small building blocks which are then assembled with connectors. An electronic connector is a separable electrical connector used in telecommunications, apparatus, computers, and in signal transmission and current transmission $\leq$5 A. Separable connectors are favored over permanent or hard-wired connections because the former facilitate the manufacture of electronic systems; also, connectors permit assemblies to be easily demounted and reconnected when inspection, replacement, or addition of new parts is called for.

Electronic connectors may connect internally or externally. Internal connections may be between a component and a printed circuit board or wire; a printed circuit board and a wire or another printed circuit board which is in a chassis; and between chassis in the same cabinet. External connectors join separate pieces or equipment.

Surface mount refers to a method of securing connectors to the conductors of a printed circuit board by soldering appropriately shaped contacts to the board surface. Higher contact densities can be achieved and the need to drill holes in the board is avoided.

The contacts of an electronic connector have spring elements which press the mating surfaces together with a predetermined force, usually in the range of 0.25–10 N (0.056–2.24 lbf) for plated contacts; this range depends on the connector design and the materials from which the contact is made. Mating of the connector is usually by sliding the surfaces together. Less frequently, the contacts are butted after they have been positioned in zero insertion force (ZIF) connectors.

Another design which provides unusually low mating forces employs bundles of wires in both halves of the connector that intermesh, like two hair brushes, when the parts are connected.

Connector Shielding. Electromagnetic radiation, either human-made, ie, from tv, radios, radar, automotive ignitions, etc; or natural, ie, lightning, can interfere with the quality of signal transmission. Signals in conductors of electronic equipment including computers and communications gear may be seriously degraded. Methods used to control this effect include the use of coaxial cables and connectors in which shielding and grounding are used, twisted pair conductors, and filter connectors.

Joining to Electronic Connectors. The most widely used techniques for the termination of wires to separate contacts are the soldering (see SOLDERS AND BRAZING ALLOYS), welding (qv), crimping, solderless wrapping, and slotted-beam methods. Except for crimping and welding, it is usually possible to replace wires to a contact a limited number of times if repair or wiring changes are necessary.

Splicing Connectors. Splicing connectors are used to permanently join wire to wire. Some are simple sleeve barrels that are crimped to bare wire; others are preinsulated where the crimp is made by compressing the sleeve and its positioned insulation onto wires which may or may not have insulation.

Terminals. Terminals are connectors having individual wires that are designed to be screwed down at separable ends, and to which conductors are permanently joined at the back end, usually by crimping.

Utility and Industrial Connectors. Connectors used in power distribution systems are nearly always of the permanent type and are usually made for single conductors. Sleeve barrels are used to splice cable by crimping. Insulating covers may be applied to the connector after the joint is made between the connector and the insulated cable. Clamp-type connectors are used with one or more clamping bolts and may have additional resilience when used with dished washers made of spring steel (Belleville washers).

Methods of Application. Attachment of separable contacts to conductors may by achieved using automated machinery or specialized hand tools.

Contact Principles

The design of electrical contacts and the selection of contact materials are based on a body of interrelated physical, metallurgical, and chemical principles.

Nature of Mechanical Contact. The surfaces of solids are irregular on a microscopic scale. When two metallic bodies are placed in contact at a light load, the bodies touch at only a few small spots, or asperities. As the load is increased, more and more asperities come into contact and the surfaces move together. The true area of contact depends, therefore, on normal load and on the hardness of the metal. The real area of contact is only a fraction of the apparent area in most cases, except at very high loads.

Nature of Electrical Contact. If metallic surfaces are covered by a nonconducting layer, such as an oxide or a sulfide tarnish film, the area of metallic contact is zero provided that the film is unbroken. Significant current does not flow between such surfaces except when the film is less than 2–3 nm in thickness. At or less than these thicknesses, electron tunneling (tunnel conduction), which is voltage independent, occurs by a wave-mechanical effect analogous to the transmission of light through metal foil of thickness comparable to the wave length. If the nonconductive layer on a surface is discontinuous or is punctured, the mechanical load is borne by both film and metal. Current then flows through the metallic spots. The lines of electric flow converge at these spots. Constriction resistance, the increase of resistance beyond that of a continuous solid, ie, not having an interface, originates at this convergence.

In the simplest case, for a single circular contact spot between identical metals having a uniform film, contact resistance R has the relationship: $R = R_c + R_f$ and $R = \frac{\rho}{2a} + \frac{\sigma}{\pi a^2}$, where R_c = constriction resistance, R_f = film resistance, a = radius of the a spot, ρ = bulk resistivity of the contact metal, and σ = film resistance (ohms per unit area). The radius of the a spot can be calculated from the hardness of the metal and the applied load.

Because a length of metal associated with the connector contact is ordinarily in the path between the contact end to which a wire is terminated and the contact interface, its resistance (bulk resistance) must be added to contact resistance when considering the connector as a circuit element. This overall resistance is sometimes erroneously called contact resistance.

Resistance Heating of Contacts. The contact material, contact area, and heat dissipating ability, as well as the heat dissipating ability of the structure to which the material is attached, limit the amount of current that a contact can transport. Excessive current heats and softens the metal contact. This softening results in an increase in the surface area of the contact and a corresponding reduction in contact resistance.

Voltage Breakdown of Films. If a significant voltage can be passed by a circuit across a film-covered contact, the film, depending on thickness and composition, may break down electrically. This action, called the coherer effect or fritting, results in the formation of minute metallic conductive paths through the film. Puncturing is of the order of 0.1 V/nm of film. The potential drop across the film after puncturing is the melting point voltage.

Overview. Metallic contact between surfaces of separable connectors usually is obtained either by using noble metals, which are essentially film-free, or by designing the contact so that any films that are present are broken before or as the surfaces are brought together.

The contact resistance of any electrical connector in a circuit must be stable and generally low for proper functioning of that circuit. Low voltage circuits which are common in modern electronic systems, have open-circuit voltages of not more than a few volts. Low contact resistance should be achieved using noble metal contacts or base metals concurrently with methods that mechanically perforate any insulating films that are present.

Materials and Processing

Contact Substrates. The substrate must be able to be terminated readily as well as be a good electrical conductor. In electronic connectors the substrate may serve as a spring element. The most widely used spring materials for connectors are the copper alloys: 98.1 wt % Cu; 1.9 wt % Be (Unified Numbering System (UNS) designation C17200); 94.8 wt % Cu, 5.0 wt Sn, 0.19 wt % P (phosphor bronze, C51000); and 88.2 wt % Cu, 9.5 wt % Ni, 2.3 wt % Sn (C72500) (see COPPER ALLOYS). Sometimes springs made of metals that have poor conductivity, such as stainless steel, are used as inserts in connector barrels of brass or similar metals which are inexpensive and can be terminated easily.

Contact Finishes. Coatings of other metals commonly are used to obtain corrosion resistance, to provide conductivity, or to facilitate termination to conductors by soldering, wire wrapping, or by other means. Application of finishes is achieved by electroplating (qv), cladding, and by hot-dipping when low melting metals such as tin are used. The principal noble contact finishes are gold, palladium, rhodium, and alloys having a high gold or palladium content. Palladium-based finishes usually have a thin (0.05–0.1 μm) top coating of gold. The non-noble finishes are tin, silver, and nickel. Alloys of these metals, such as 50 wt % Sn, 50 wt % Pb, also are widely used.

Conductive Elastomers. Conductive elastomers (qv), rubbers that are made conductive by molding metal or carbon powders in them, have characteristics of both a contact material and a spring. Silicone rubbers, neoprene, polyurethane, and other elastomers have been used; however, silicones are the most popular because these have a low compressive set and operate over a wide temperature range, from ca − 65 to 200°C (see also ANTISTATIC AGENTS; ELECTRICALLY CONDUCTIVE POLYMERS).

However, conductive elastomers have only ca $\leq 10^{-3}$ of the conductivity of solid metals. Also, the contact resistance of elastomers changes with time when they are compressed. Therefore, elastomers are not used where significant currents must be carried or when low or stable resistance is required. Typical applications, which require a high density of contacts and easy disassembly for servicing, include connection between liquid crystal display panels and between printed circuit boards in watches.

Contact Lubricants. Debilitating wear of separable connector contacts, which may occur if the metallic coating is thin or if forces normal to the contact surfaces are high, can be minimized by coating the contacts with thin films (qv) or organic lubricants (see LUBRICATION AND LUBRICANTS). Viscous mineral oils, poly(phenyl ethers), per(fluoroalkyl) polyethers, soft microcrystalline waxes (qv), and petrolatum have been used on electronic connector contacts.

Joint Aids. Good metallic contact with aluminum connectors or aluminum conductors, including those with severely deformed surfaces, is difficult if aluminum oxide is present. It has been found that finely divided zinc incorporated in grease can be coated on the interior surfaces of an aluminum connector to provide lower initial contact resistance and better long-term resistance stability. The grease is also able to retard ingress of air and water which can seriously degrade reliability of aluminum connectors, especially in exposed out-of-doors service in power distribution networks. Other soft metal coatings are effective, such as indium, which has been proposed as a plating on copper alloy connectors intended for permanent joints to aluminum communications cables having slotted-beam connectors.

Insulators. Molded plastics serve as insulators for multicontact connectors and glass is used in hermetic connectors intended for bulkhead mounting. Environmentally sealed circular connectors combine both elastomeric technology with plastics by bonding the seals and grommets to the plastic insert. A wide variety of plastics are employed in electronic connector bodies depending on the size, strength requirements, complexity of the design, and service environment. Plastic materials are often reinforced using about 30 to 40% glass fibers. Thermoplastics are favored by a wide margin over thermosets for electronic connectors because of manufacturing economies and generally superior performance.

Reliability and Testing

Mechanisms of Failure. The causes of connector contact failure can be of a thermal, chemical, or mechanical nature, in addition to misapplication and physical abuse.

Thermal. Heat cycling forms the basis of much connector testing, eg, 500 cycles of heating by current flow to 100°C above ambient followed by cooling to room temperature for aluminum conductor overhead power distribution connectors. A connector is acceptable if this stress does not cause significant change in overall resistance.

Chemical. Chemical degradation can be avoided by using closed structures, which may have protective covers or which may be fully hermetic, as well as by using unreactive metals. Barrier coatings are effective in some cases, and polyethylene–polybutene grease is used in some splicing connectors for telecommunications cable.

Many techniques have been developed for the accelerated testing of connector contacts. These include elevated relative humidity exposure, cycling temperature–humidity procedures, and aging in chambers containing gaseous pollutants.

Mechanical. Underplatings, contact lubricants, and hard materials reduce mechanical wear.

Fiber Optic Connectors

Optical connectors are used to terminate and interconnect fiber optic cables.

Economic Aspects

Most electronic connectors are designed and made by one of the hundreds of companies devoted entirely to these products. The larger organizations both manufacture and sell internationally. In addition, this industry supports materials suppliers who provide piece parts such as metal stampings and molded connector bodies or offer a metal finishing service. The 10 largest connector companies have world sales ranging from about 300 to 2500 million U.S. dollars. These include AMP, MOLEX, AMPHENOL, 3M, ITT Cannon, DuPont, Framatome, Hirose, Japan Aviation Electronics (JAE), and Japan Solderless Terminal (JST). Some large electrical connector users, such as those in the automotive fields, have captive connector manufacturing facilities and also offer products to the general trade.

The market is expected to increase slowly. Japan and Pacific Rim suppliers are growing in terms of their world share at the expense of North America and European suppliers.

MORTON ANTLER
Contact Consultants, Inc.

R. Holm, *Electric Contacts Handbook*, 4th ed., Springer-Verlag, New York, 1967; comprehensive treatment of electric contact theory.

Connectors and Interconnection Handbook, 5 vols., International Institute of Connector and Interconnection Technology, Inc., Westfield, N.J.; emphasis on electronic connector design.

Cumulative Index of the Proceedings of the IEEE Holm Conferences on Electrical Contacts and the International Conferences on Electrical Contacts, IEEE Service Center, Piscataway, N.J.

ELECTRICALLY CONDUCTIVE POLYMERS

Generally speaking, electrically conductive polymers are composed of conjugated polymer chains with π-electrons delocalized along the backbone. In the neutral, or undoped, form the polymers are either insulating or semiconducting. The polymers are converted to the electrically conductive, or doped, forms via oxidation or reduction reactions which form delocalized charge carriers. Charge balance is accomplished by the incorporation of an oppositely charged counterion into the polymer matrix. The conductivity is electronic in nature and no concurrent ion motion occurs in the solid state. The redox doping processes are reversible and can be accomplished electrochemically. During electrochemical switching, ions do move into and out of the polymers as charge-balancing species for the charge carriers on the polymer backbone. Present and potential applications of conducting polymers utilize both their electronic (electrically conducting and optical) properties and their ionic properties (eg, as battery or sensor electrodes).

Synthesis of Electrically Conductive Polymers

A number of synthetic routes have been developed for the preparation of conjugated polymers. The diversity has been driven by the desire to examine many different types of conjugated polymers and attempts to improve material properties. Material property enhancement has centered on the synthesis of polymers that are processible in various forms. Five primary classes of conjugated polymers have been shown to exhibit high levels of electrical conductivity in the doped state. These include polyacetylenes, polyarylenes and polyheterocycles, poly(arylene vinylenes), and polyanilines. In addition, a number of multicomponent materials, usually polymer blends and composites, have been prepared in which at least one of the components is a conducting polymer.

Polyacetylenes. Despite improvement in properties of polyacetylenes prepared from acetylene, the materials remained intractable. To avoid this problem, soluble precursor polymer methods for the production of polyacetylene have been developed. The most highly studied system utilizing this method is the Durham technique.

Polyphenylenes. Poly(*p*-phenylene) (PPP), synthesized using direct polymerization methods, yields oligomers of a completely intractible material. Although it is generally difficult to use intractable polymers in practical applications, sintering techniques are available which may make this polymer technologically useful.

Polyheterocycles. Heterocyclic monomers such as pyrrole and thiophene form fully conjugated polymers with the potential for doped conductivity when polymerization occurs at the 2,5 positions. The heterocycle monomers can be polymerized by an oxidative coupling mechanism, which can be initiated by either chemical or electrochemical means. Similar methods have been used to synthesize poly(*p*-phenylenes). The electrochemical polymerization of pyrrole is generally believed to follow a radical step-growth mechanism.

Poly(arylene vinylenes). The use of the soluble precursor route has been successful in the case of poly(arylene vinylenes) containing either benzenoid or heteroaromatic species as the aryl groups. The simplest member of this family is poly(*p*-phenylene vinylene) (PPV).

Polyanilines. Initial preparations of polyaniline (PANI) led to insoluble materials that were difficult to characterize. Poly(*p*-phenylene amineimine) (PPAI) was synthesized directly to demonstrate that PANI is purely para-linked. Comparison of the properties of PPAI and PANI showed PPAI to be an excellent model both structurally and electronically.

PANI is more commonly prepared by polymerization of aniline using $(NH_4)_2S_2O_8$ in HCl. As prepared, it has a structure known as emeraldine hydrochloride. In this form, PANI is highly conductive but completely insoluble. When emeraldine hydrochloride is deprotonated with NH_4OH, the highly soluble emeraldine base is produced. As was the case for polythiophenes, substitution along the polyaniline backbone has been utilized as a means of improving processibility. Many derivatives are possible for PANI because of the possibility of substitution of the monomers at either the main-chain nitrogen atoms, or on the aromatic ring. Substituents studied have included alkyl, aryl, sulfonyl, and amino groups.

Conducting Polymer Blends, Composites, and Colloids. Incorporation of conducting polymers into multicomponent systems allows the preparation of materials that are electroactive and also possess specific properties contributed by the other components. Dispersion of a conducting polymer into an insulating matrix can be accomplished as either a miscible or phase-separated blend, a heterogeneous composite, or a colloidally dispersed latex. When the conductor is present in sufficiently high composition, electron transport is possible.

There are several approaches to the preparation of multicomponent materials, and the method utilized depends largely on the nature of the conductor used. In the case of polyacetylene blends, *in situ* polymerization of acetylene into a polymeric matrix has been a successful technique.

Because of the aqueous solubility of polyelectrolyte precursor polymers, another method of polymer blend formation is possible. The precursor polymer is co-dissolved with a water-soluble matrix polymer, and films of the blend are cast. With heating, the fully conjugated conducting polymer is generated to form the composite film. This technique has been used for poly(arylene vinylenes) with a variety of water-soluble matrix polymers, including polyacrylamide, poly(ethylene oxide), poly(vinylpyrrolidinone), methylcellulose, and hydroxypropylcellulose. These blends generally exhibit phase-separated morphologies.

The true thermoplastic nature of poly(3-alkylthiophenes), ie, solubility and fusibility, allows the use of compounding methods commonly used in the plastics industry for the preparation of composites of these polymers. The polymers can be co-dissolved with a matrix polymer, then processed from organic solution. Again the resulting blends are phase separated, but if the composition of conducting polymer is high enough, the conducting component forms the matrix with the insulating polymer dispersed within it, and high conductivity is possible.

Electrochemical polymerization of heterocycles is useful in the preparation of conducting composite materials. Conducting polymer composites have also been formed by co-electrodeposition of matrix polymer during electrochemical polymerization. The preparation of molecular composites by electropolymerization of heterocycles in solution with polyelectrolytes is an extremely versatile technique, and many polyelectrolyte systems have been studied.

Redox Doping

In order to induce high electrical conductivity in organic conjugated polymers, charge carriers must be introduced. These charge carriers are created by removing electrons from, or adding electrons to, the delocalized π-electron network of the polymer, creating a conducting unit that is now a polymeric ion rather than a neutral species. The charges introduced are compensated by ions from the reaction medium. This process is called doping by analogy to the changes that occur in inorganic semiconductors upon addition of small quantities of electronic defects. However, it proceeds through a different mechanism and is more precisely termed a redox reaction. The doping level, or ratio of charge carriers per polymer repeat unit, is generally between 0.2 and 0.4 in most polyarylenes and can be determined by measuring the content of charge-balancing counterions. The ability to control the electrical properties of conducting polymers over wide ranges, by adjusting the redox doping level, has created interest in these materials for a number of emerging applications.

Charge Carriers in Conducting Polymers. The mechanism for the conductivity increase resulting from doping in inorganic semiconductors involves the formation of unfilled electronic bands. Electrons are removed from the top of the valence band during oxidation, called *p*-type doping, or added to the bottom of the conduction band during reduction, termed *n*-type doping. Extension of this argument to the case of conjugated organic polymers was found to be inaccurate as the conductivity in many conducting polymers was found to be associated with spinless charge carriers. *In situ* epr/electrochemistry techniques have shown that the conducting entity in polyacetylene, polypyrrole, polythiophene, and poly(*p*-phenylene) can be spinless, although evidence exists for mixed valence charge carriers as well.

The conductivity increase following doping in conjugated polymers is explained in terms of local lattice distortions and localized electronic states. In this case the valence band remains full and the conduction band remains empty so that there is no appearance of metallic character. When the polymer chain is redox doped, a lattice distortion results and the equilibrium geometry for the doped state is different than the ground state geometry.

Doping Processes. Redox doping can be accomplished through both chemical and electrochemical means. Vapor-phase doping of neutral polyacetylene has been carried out using AsF_5 and I_2 as oxidants.

Solution doping of many conjugated polymers in suspension or, in the case of processible derivatives, in solution, has been accomplished using a variety of oxidizing and reducing agents. Typical chemical dopants include $FeCl_3$, $NOPF_6$, and sodium naphthalide. Conducting polymers that are synthesized using oxidative coupling techniques are obtained in the oxidized form.

These conjugated polymers can be chemically and electrochemically reduced and reoxidized in a reversible manner. In all cases the charges on the polymer backbone must be compensated by ions from the reaction medium which are then incorporated into the polymer lattice. The rate of the doping process is dependent on the mobility of these charge compensating ions into and out of the polymer matrix.

Electrogenerated conducting polymer films incorporate ions from the electrolyte medium for charge compensation. Electrochemical cycling in an electrolyte solution results in sequential doping and undoping of the polymer film.

Optical Properties. The energy difference between the valence band and the conduction band in conjugated polymers is referred to as the band gap and can be determined using optical spectroscopy. In the neutral (or insulating) form, conducting polymers exhibit a single electronic absorption in either the visible or ultraviolet region attributed to the electronic transition between the HOMO and LUMO levels.

In the doped form, transitions between the band edges and newly formed intragap electronic states are observed in the optical spectra of conducting polymers. When polarons are present as charge carriers, an additional transition is apparent which corresponds to the electronic transitions between the two gap states. Since the intragap electronic states are taken from the band edges, the bandgap increases with increasing doping levels. Also, since a bipolaron creates a larger lattice distortion than a polaron, the gap states are further away from the band edges in the bipolaron model.

The changes in the optical absorption spectra of conducting polymers can be monitored using optoelectrochemical techniques.

Electrical Properties

The electrical conductivity, ς, of a material is equal to the inverse of its specific resistivity, ρ, and is a measure of a material's ability to transport electrical charge.

In simplistic terms, the conductivity of a material is controlled by both the density, n, and mobility, μ, of the charge carriers, having a charge of e: $\sigma = ne\mu$. One of the benefits of conductive polymers is that the number of charge carriers can be quite high because of the high density of redox sites along the conjugated backbones. The mobility of the charges are controlled by both intrachain and interchain interactions. Defect-free conjugated polymers that are highly oriented in the direction of electrical transport, can exhibit high mobilities along the chains. At the same time, strong π-overlap interchain interactions allow charge carriers to "hop" from chain to chain and ultimately high electrical conductivities are possible.

Table 1 shows the present state-of-the-art for the electrical conductivity of doped conjugated polymers.

Stability of Conducting Polymers

Although polyacetylene has served as an excellent prototype for understanding the chemistry and physics of electrical conductivity in organic polymers, its instability in both the neutral and doped forms precludes any useful application. In contrast to polyacetylene, both polyaniline and polypyrrole are significantly more stable as electrical conductors. When addressing polymer stability it is necessary to know the environmental conditions to which it will be exposed; these conditions can vary quite widely.

Applications of Conducting Polymers

The novel combination of electrical, electrochemical, and chemical properties of conductive polymers may lead them to be used in a number of applications. At this time (1993), the systems do not have appropriate electronic properties and stabilities to use conditions to be considered as replacement materials for metals or semiconductors. Of course, this does not preclude further advances from being made in the near future. Because of the present advanced commercial development of Versicon by AlliedSignal, Inc. relative to the other conducting polymers, it is being examined for many applications. Some of those suggested by the manufacturer include lightly colored coatings for antistatic films, and blends for electrostatic dissipation, and as corrosion-preventive paints. Conductive hosings, gaskets, cable shields, and radar-absorbing coatings are being tested for emi shielding applications. The conductive properties are being exploited in adhesives, inks, electrodes, and resistive heaters. In electronics applications PANI has been used as a discharge layer for electron-beam

Table 1. Electrical Conductivity Ranges for Conductive Polymers

Polymer	S/cm
polyacetylene	10^3–10^5
poly(*p*-phenylene)	10^3
polyazulene	10^0
poly(*p*-phenylene sulfide)	10^1–10^2
poly(*p*-phenylene vinylene)	10^3–10^4
poly(thienylene vinylene)	10^3
polythiophene	10^2
poly(3-alkyl thiophene)	10^2–10^3
polyisothianaphthene	10^1
polypyrrole	10^2–10^3
polyfuran	10^2
polyaniline	10^2–10^3

lithography, as a removable sem discharge layer, and as a conducting electrode for electrolytic metallization of copper on through-holes.

Many of the applications of conductive polymers utilize their unique properties and advantages over other material systems, for example low density and controllable electrical properties. Examples include charge-storage batteries, conductive textiles, chemical and biochemical sensors, electromagnetic shielding and antistatic coatings, gas separation membranes, and electrooptics and electrochromism.

JOHN R. REYNOLDS
ANDREW D. CHILD
University of Florida
MELINDA B. GIESELMAN
3M Specialty Adhesives and Chemicals

A. J. Epstein, in T. A. Skotheim, ed., *Handbook of Conducting Polymers*, Vol. 2, Marcel Dekker, New York, 1986, p. 1041.

A. O. Patil, A. J. Heeger, and F. Wudl, *Chem. Rev.* **88**, 183 (1988).

J. R. Reynolds, *Chemtech* **18**, 440 (1988).

J. R. Reynolds and M. Pomerantz, in T. A. Skotheim, ed., *Electroresponsive Molecular and Polymeric Materials*, Marcel Dekker, New York, 1991.

ELECTROANALYTICAL TECHNIQUES

Electroanalysis employs electrochemical cells. The most common electrochemical cells are batteries which consist of two electrodes: an anode and a cathode. The electrochemical principles governing batteries and electroanalytically useful cells are the same. The typical electroanalytical cell has at least two electrodes, a working electrode and a reference electrode. The working electrode may serve as either anode or cathode depending on applied voltage, or as a source or sink for ion exchange as in the case of ion-selective electrodes. The reference electrode invariably serves as a source or sink for ions at its interface with the solution, and as a source or sink for electrons at its interface with an external circuit. The solution or electrolyte in batteries has a very high ionic strength in order to carry enormous amounts of current. In electroanalytical cells the solution is the sample and must typically contain supporting electrolyte, in the form of electrochemically inert salts, in order to support much lower current densities. Supporting electrolytes may also be used to define ionic strength.

Cells useful for electroanalysis typically consist of two or more electrodes dipping into the sample, ie, the solution to be analyzed. Sample solutions can range from water to blood but can also include virtually all organic solvents. Analyte concentrations are usually in the picomolar to millimolar range. Numerous choices exist for working electrodes, ranging from electron exchangers to in exchangers. These may be metals of many kinds, eg, Pt, Au, Ag, and Hg; semiconductors, eg, Si, Ge, and TiO_2; or plastics, ion exchangers and sequesterers in poly(vinyl chloride) (PVC). Choices for reference electrodes are more limited. The problems involving reference electrodes are often more profound, and availability of the right reference electrode may ultimately dictate the feasibility of an assay.

Reference electrodes are inherently unstable. These electrodes drift, leak, become foul or plugged, and frequently need to be replaced.

Electrochemical cells may be used in either active or passive modes, depending on whether or not a signal, typically a current or voltage, must be actively applied to the cell in order to evoke an analytically useful response. Electroanalytical techniques have also been divided into two broad categories, static and dynamic, depending on whether or not current flows in the external circuit. In the static case, the system is assumed to be at equilibrium. The term dynamic indicates that the system has been disturbed and is not at equilibrium when the measurement is made. These definitions are often inappropriate because active measurements can be made that hardly disturb the system and passive measurements can be made on systems that are far from equilibrium. The terms static and dynamic also imply some sort of artificial time constraints on the measurement. Active and passive are terms that nonelectrochemists seem to understand more readily than static and dynamic.

Active Techniques

Active techniques are classified by the method of collection and display of data. If a voltage is applied to the cell and the resultant current measured and displayed as a function of time, the technique is called chronoamperometry. If a current is applied to the cell and the resultant voltage is measured and displayed as a function of time, the technique is called chronopotentiometry. Similarly, if current is measured and displayed as a function of applied potential, the technique is called voltammetry. Subcategorizing depends on such things as the geometry and size of the working electrode, its physical treatment, usually its rotation, and the functionality of the applied waveform. Therefore there are rotating disk voltammetry, cyclic voltammetry, pulse voltammetry, etc. Even post-collection treatment of the data may result in a renaming of the technique. Integration of the current, either during analogue data acquisition or digitally from computer stored currents, results in chronocoulometry.

Passive Techniques

Perhaps the most precise, reliable, accurate, convenient, selective, inexpensive, and commercially successful electroanalytical techniques are the passive techniques, which include only potentiometry and use of ion-selective electrodes, either directly or in potentiometric titrations. Whereas these techniques receive only cursory or no treatment in electrochemistry textbooks, the subject is regularly reviewed and treated. There is a journal, *Ion-Selective Electrode Reviews,* devoted solely to the use of ion-selective electrodes.

According to the definition, a passive technique is one for which no applied signal is required to measure a response that is analytically useful. Only the potential (the equilibrium potential) corresponding to zero current is measured. Because no current flows, the auxiliary electrode is no longer needed. The two-electrode system, where the working electrode may or not be an ion-selective electrode, suffices.

The equilibrium potential may be forced to change by the addition of a titrant, as in potentiometric titrations employing nonspecific working electrodes. But the equilibrium potential can also be analytically useful in its own right if the response of the electrode is highly selective to the analyte. Chemically modified electrodes are being researched that may provide this high selectivity, eg, for use in active techniques, such as the glucose sensor. However, ion-selective electrodes, of which the pH glass electrode is probably the most common example, dominate applications involving high selectivity even though their response mechanisms have nothing to do with oxidation and reduction. Potentiometry employing ion-selective electrodes without the addition of titrants is termed direct potentiometry.

Static and Dynamic Measurements

The definition herein of static electroanalytical measurements implies a time independent response, regardless of whether or not that response is generated by an applied signal. If an applied signal is needed, the method is active; if not, it is passive. These terms should not be confused with other definitions where static more or less equates to passive and dynamic to active.

Economic Aspects

The reference electrode contributes heavily to the economics of electroanalytical chemistry. Companies that sell and service electroanalytical instrumentation are few in number and small in size, or they are parts of much larger companies. One supplier of electroanalytical instrumentation is Princeton Applied Research Corporation (PARC) of Princeton, New Jersey. PARC is a subsidiary of EG&G Instruments, Inc. Among the many suppliers of ion-selective electrodes are Orion (Boston, Massachusetts), Corning (Corning, New York), and Ingold

(Wilmington, Massachusetts). Brinkmann Instruments, Inc. (Westbury, New York) is a useful supplier of titration equipment.

Electroanalytical chemistry does not yet generate the kind of revenue that battery technology and electrowinning (see METALLURGY-EXTRACTIVE METALLURGY) do. Electroanalysis, even at the most sensitive limits, can be performed using simple instruments that cost less than $100. However, assays generally require highly skilled technicians and are thus correspondingly expensive even if the equipment used is not.

In the area of consumer products, amperometric glucose sensors hold high potential. Industrially, process monitors for the manufacture of consumer chemicals are under development. However, replacement of defective reference electrodes, which in a laboratory environment may be trivial, may be prohibitively difficult *in vivo* or in an industrial process environment.

JAMES R. SANDIFER
Eastman Kodak Company

P. T. Kissinger and W. R. Heineman, eds., *Laboratory Techniques in Electroanalytical Chemistry,* Marcel Dekker, New York, 1984.

A. J. Bard and L. R. Faulkner, *Electrochemical Methods, Fundamentals and Applications,* John Wiley and Sons, Inc., New York, 1980.

W. E. Morf, *The Principles of Ion-Selective Electrodes and of Membrane Transport,* Elsevier, New York, 1981.

W. Matuszewski, S. A. Rosario, and M. E. Meyerhoff, *Anal. Chem.* **63,** 1906 (1991).

ELECTROCHEMICAL PROCESSING

INTRODUCTION

Electrochemical systems convert chemical and electrical energy through charge–transfer reactions. These reactions occur at the interface between two phases. Consequently, an electrochemical cell contains multiple phases, and surface phenomena are important. Electrochemical processes are sometimes divided into two categories: electrolytic, where energy is supplied to the system, eg, the electrolysis of water and the production of aluminum; and galvanic, where electrical energy is obtained from the system, eg, batteries and fuel cells.

The industrial economy depends heavily on electrochemical processes. Electrochemical systems have inherent advantages such as ambient temperature operation, easily controlled reaction rates, and minimal environmental impact. Electrosynthesis is used in a number of commercial processes. Batteries and fuel cells, used for the interconversion and storage of energy, are not limited by the Carnot efficiency of thermal devices. Corrosion, another electrochemical process, is estimated to cost hundreds of millions of dollars annually in the United States alone (see CORROSION AND CORROSION CONTROL). Electrochemical systems can be described using the fundamental principles of thermodynamics, kinetics, and transport phenomena.

Thermodynamics of Electrochemical Cells

Consider the cell

α	β	γ		ε	β′	α′
Pt(s)	Fe(s) H_2	NaOH in H_2O	membrane	NaCl in H_2O	TiO_2(s) Cl_2	Pt(s)

for which the electrode reactions are

$$2\ Cl^- \rightarrow Cl_2 + 2e^- \tag{1}$$

$$2\ H_2O + 2e^- \rightarrow H_2 + 2\ OH^- \tag{2}$$

The electrode where oxidation occurs is the anode; the electrode where reduction occurs is called the cathode. Electrons released at the anode travel through an external circuit and react at the cathode. The vertical lines denote phase separation; the squiggly lines separate a junction region. Although adjacent phases are in equilibrium, not all species are present in every phase. The membrane provides an ionic path for sodium ions that are transported from the anode to the cathode, but also separates the chlorine and hydrogen gases. Within the junction region, ie, the membrane, transport processes occur (see also ALKALI AND CHLORINE PRODUCTS; MEMBRANE TECHNOLOGY).

Determining the cell potential requires knowledge of the thermodynamic and transport properties of the system. The analysis of the thermodynamics of electrochemical systems is analogous to that of neutral systems. For ionic species, however, the electrochemical potential replaces the chemical potential.

Kinetics and Interfacial Phenomena

The rate of an electrochemical process can be limited by kinetics and mass transfer. Before considering electrode kinetics, however, an examination of the nature of the interface between the electrode and the electrolyte, where electron-transfer reactions occur, is in order.

Because some substances may preferentially adsorb onto the surface of the electrode, the composition near the interface differs from that in the bulk solution. If the cell current is zero, there is no potential drop from ohmic resistance in the electrolyte or the electrodes. The question from where this potential arises can be answered by considering the interface.

At the interface between phases, there is a region called the electrical double layer where potential variation occurs. Figure 1 shows this region. Although the electrode and electrolyte are overall electrically neutral, the metal electrode may have a net charge near the surface. The solution layer closest to the electrode contains specifically adsorbed ions and is called the inner Helmholtz plane (IHP). Ions that are hydrated, generally the cations, can approach the metal surface only to a finite distance, and comprise the outer Helmholtz plane (OHP). The nonspecifically adsorbed ions are distributed by thermal agitation; this region is called the diffuse double layer and lies just outside the OHP.

Even in the absence of Faradaic current, ie, in the case of an ideally polarizable electrode, changing the potential of the electrode causes a transient current to flow, charging the double layer. The metal may have an excess charge near its surface to balance the charge of the specifically adsorbed ions. These two planes of charge separated by a small distance are analogous to a capacitor. Thus the electrode is analogous to a double-layer capacitance in parallel with a kinetic resistance.

In electrode kinetics a relationship is sought between the current density and the composition of the electrolyte, surface overpotential, and the electrode material. This microscopic description of the double layer indicates how structure and chemistry affect the rate of charge-transfer reactions. Generally in electrode kinetics the double layer is regarded as part of the interface, and a macroscopic relationship is sought. For the general reaction

$$O + ne^- \rightleftarrows R \tag{3}$$

the cathodic and anodic reaction rates can be written as

$$r_a = k_a c_R \exp\left[(1-\beta)\frac{nF}{RT}V\right] \tag{4}$$

and

$$r_c = k_c c_O \exp\left[-\beta\frac{nF}{RT}V\right] \tag{5}$$

β is a symmetry factor equal to the fraction of the potential that promotes the cathodic reaction. The reaction rate and current are related through Faraday's law:

$$\frac{i}{nF} = r_a - r_c \tag{6}$$

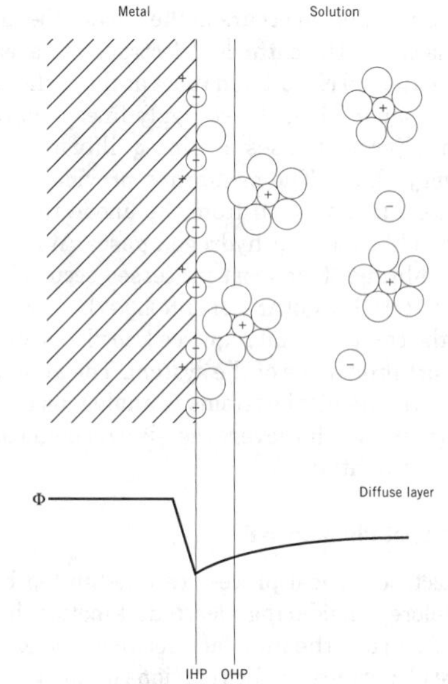

Figure 1. The structure of the electrical double layer where ○ represents the solvent; ⊖, specifically adsorbed anions; ⊖, anions; and ⊕, cations. The inner Helmholtz plane (IHP) is the center of specifically adsorbed ions. The outer Helmholtz plane (OHP) is the closest point of approach for solvated cations or molecules. Φ, the corresponding electric potential across the double layer, is also shown.

These two reactions may be generalized to the Butler-Volmer equation:

$$i = i_o\left\{\exp\left(\frac{\alpha_a F \eta_s}{RT}\right) - \exp\left(-\frac{\alpha_c F \eta_s}{RT}\right)\right\} \quad (7)$$

The exchange current density, i_o, depends on temperature, the composition of the electrolyte adjacent to the electrode, and the electrode material. The exchange current density is a measure of the kinetic resistance. High values of i_o correspond to fast or reversible kinetics. The three parameters, α_a, α_c, i_o, are determined experimentally. The surface overpotential, η_s, is the difference in potential of the metal and the potential of an electrode of the same kind in the electrolyte measured adjacent to the electrode, ie, just outside the double layer, but passing no current. The surface overpotential appears in the exponential terms for both the anodic and cathodic reactions, and can be considered the driving force for the electrochemical reaction.

Transport Processes

In addition to electrode kinetics, the rate of an electrochemical reaction can be limited by the rate of mass transfer of reactants to and from the electrode surface. In dilute solutions, four principal equations are used. The flux of species i is

$$\mathbf{N}_i = -z_i u_i F c_i \nabla\Phi - D_i \nabla c_i + c_i v \quad (8)$$

These three terms represent contributions to the flux from migration, diffusion, and convection, respectively. The bulk fluid velocity is determined from the equations of motion. Equation 8, with the convection term neglected, is frequently referred to as the Nernst-Planck equation. In systems containing charged species, ions experience a force from the electric field. This effect is called migration. The charge number of the ion is z_i, F is Faraday's constant, u_i is the ionic mobility, and Φ is the electric potential. The ionic mobility and the diffusion coefficient are related: $D_i = RTu_i$ (eq. 9). This relation, discovered by Nernst and Einstein, applies in the limit of infinite dilution.

A material balance on an element of the fluid gives

$$\frac{c_i}{t} = -\nabla \cdot \mathbf{N}_i + R_i \quad (10)$$

where R_i is the homogeneous reaction rate. Except near the diffuse double layer, to a good approximation the solution is electrically neutral

$$\sum_i z_i c_i = 0 \quad (11)$$

The current density is given by

$$\mathbf{i} = F \sum_i z_i \mathbf{N}_i \quad (12)$$

Equations 8, 10–12 using the appropriate boundary conditions, can be solved to give current and potential distributions, and concentration profiles. Electrode kinetics would enter as part of the boundary conditions. The solution of these equations is not easy and often involves detailed numerical work.

THOMAS F. FULLER
JOHN NEWMAN
University of California, Berkeley

J. Newman, *Electrochemical Systems*, Prentice-Hall, Inc., Englewood Cliffs, N.J., 1991.

A. J. Bard and L. R. Faulkner, *Electrochemical Methods*, John Wiley & Sons, Inc., New York, 1980.

G. Prentice, *Electrochemical Engineering Principles*, Prentice-Hall, Inc., Englewood Cliffs, N.J., 1991.

K. J. Vetter, *Electrochemical Kinetics*, Academic Press, Inc., New York, 1967.

INORGANIC

The electrochemical production of inorganic chemicals and metals in the United States consumes about 5% of all the electricity generated annually, and about 16% of the electric power consumed by industry. This includes the production of such commodity chemicals as sodium hydroxide, NaOH, and chlorine, Cl_2.

Hardware for Electrochemical Processing

Power supplies and electrolytic cells are the distinguishing features of electrochemical processes. Nearly all electric power is generated and transported as high voltage multiple-phase alternating current. Industrial electrochemical processes require direct current; thus transformers are needed to decrease voltage, and rectifiers are required to convert the alternating current to direct current. Rectifiers may be rated for voltages up to 400 V or more and for any amperage up to hundreds of thousands of amperes. Rectifier efficiencies generally increase as voltages increase. Very high amperages are achieved by connecting rectifier units in parallel. State-of-the-art rectifiers use thyristers which are semiconductor devices that conduct only when a triggering potential is applied (see SEMICONDUCTORS). Thyristers can be made to conduct for half of an alternating current cycle and not conduct the other half-cycle, rectifying alternating current to direct current.

Electrochemical processes require feedstock preparation for the electrolytic cells. Additionally, the electrolysis product usually requires further processing. This often involves additional equipment, as is demonstrated by the flow diagram shown in Figure 1 for a membrane chlor-alkali cell process (see ALKALI AND CHLORINE PRODUCTS). Only the electrolytic cells and components are discussed herein.

Design possibilities for electrolytic cells are numerous, and the design chosen for a particular electrochemical process depends on factors such as the need to separate anode and cathode reactants or products, the concentrations of feedstocks, desired subsequent chemical reactions of electrolysis products, transport of electroactive species to electrode surfaces, and electrode materials and shapes. Cells may be arranged in series and/or parallel circuits. Some cell design possibilities for electrolytic cells are

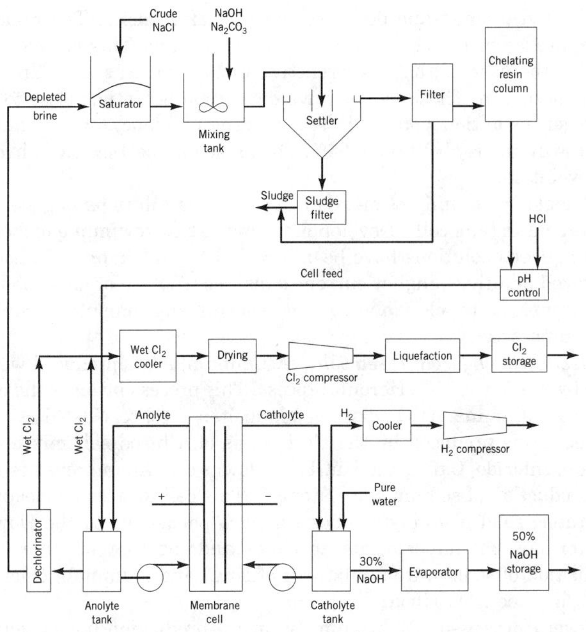

Figure 1. Flow diagram for chlor-alkali production by a membrane cell process.

Cell type	Process
one-compartment cells	
open-top inert tank	
monopolar electrodes	MnO_2
bipolar electrodes	chlorate
open-top cathodic tank	chlorate and perchlorate
enclosed horizontal liquid metal cathode	chlorine–caustic and aluminum
two-compartment cells	
diaphragm cells	
monopolar electrodes	chlorine–caustic and Mn metal
bipolar electrodes	chlorine–caustic
membrane cells	
monopolar electrodes	chlorine–caustic
bipolar electrodes	chlorine–caustic
pressurized cells	
filter press	
bipolar electrodes	hydrogen–oxygen
rotating cylindrical cathode	continuous production of metal
particle-bed electrodes	removal of low concentration of metals from waste streams

Electrode materials and shapes may have a profound effect on cell designs. Anode materials encountered in electrochemical processes are

Anode material	Process
carbon	fluorine aluminum
graphite	chlorine–caustic chlorate
DSA	chlorine–caustic chlorate
Pt	perchlorates persulfates
lead dioxide, PbO_2, on graphite or titanium	chlorate perchlorate
lead alloy (Pb, 1% Ag)	zinc electrowinning manganese electrowinning

Industrial Process Conditions

Electrolysis of Chloride Solutions. Chloride may be oxidized electrochemically to chlorine or hypochlorite, chlorate, and perchlorate.

The electrolysis of NaCl brine for the production of chlorine and caustic soda is one of the oldest and certainly one of the most important industrial electrochemical processes. The overall reaction is

$$2\ NaCl + 2\ H_2O \xrightarrow[\text{energy}]{\text{electrical}} 2\ NaOH + Cl_2 + H_2 \qquad (1)$$

There are three main technologies available for carrying out this process: diaphragm cells, mercury cells, and membrane cells. Membrane cells are the most recent development, and are generally chosen for new production capacity.

Chlorates. Sodium chlorate is produced by the electrolysis of sodium chloride at pH 6.5–7.5 in a one-compartment cell. DSA anodes and steel cathodes are generally used in chlorate cells. The electrolysis products, hypochlorous acid and hypochlorite ions, react chemically to produce chlorate.

Sodium dichromate, $Na_2Cr_2O_7$, is added to chlorate cell electrolytes. The chromate minimizes the reduction of hypochlorite and chlorate at the cathode. Chromate also minimizes the corrosion of cathodes when or where cathodic current densities are inadequate to cathodically protect the steel cathodes from the very corrosive electrolyte.

Nearly 95% of the sodium chlorate produced in North America is used to produce chlorine dioxide, ClO_2, for pulp (qv) bleaching. Minor amounts are used to produce other chemicals such as $KCLO_3$, $NaClO_2$, $NaClO_4$, etc, to recover uranium, U, and for agricultural uses as a defoliant or herbicide.

Perchlorates. Concentrated solutions of sodium chlorate are electrolyzed to produce sodium perchlorate, $NaClO_4$. Most of the $NaClO_4$ produced is converted to ammonium perchlorate, NH_4ClO_4.

$$NaClO_4 + HCl + NH_3 \rightarrow NH_4ClO_4 + NaCl \qquad (2)$$

The products of equation 2 are separated by controlled crystallizations to produce high purity crystalline anhydrous ammonium perchlorate and sodium chloride. The main use for ammonium perchlorate is as an oxidizer in the propellant of rockets and missiles (see EXPLOSIVES AND PROPELLANTS).

Manganese Dioxide. High performance alkaline batteries and some lithium batteries require pure electrolytic manganese dioxide (EMD), MnO_2. The production of EMD involves the following unit operations: first, high quality manganese ores are reduced. Reduced ore is then leached using acidic electrolyte from the cells and makeup sulfuric acid, H_2SO_4. Crude manganese sulfate, $MnSO_4$, solution is treated to remove impurities. Purified manganese sulfate solution is fed into an electrolytic cell where MnO_2 is deposited on the anode. Deposited MnO_2 is periodically harvested, milled, neutralized, and packaged for market.

Water Electrolysis. The electrolysis or water for hydrogen and oxygen production is economically attractive only in those areas where electric power is available at very low cost. Hydrogen is usually the primary product (see HYDROGEN ENERGY); oxygen (qv) is a co-product. Hydrogen in large quantities is produced by steam (qv) reforming methane, ie, natural gas (see GAS, NATURAL). Hydrogen produced in chlor-alkali and chlorate cells is available in some areas, and in some cases is compressed and sold in cylinders and as liquid hydrogen. Oxygen is commonly produced by the separation of air (see CRYOGENICS).

Cells for the electrolysis of water are available from several sources. Cells are usually of a filter-free design incorporating bipolar electrodes, porous diaphragms or ion-exchange membranes, alkaline electrolyte, KOH, and catalyzed electrodes.

Recent developments in water electrolysis cells include the use of ion-exchange membranes, optimization of pressure and temperature, and suggestions for better electrolysis cell designs. Interest in water electrolysis is associated with load leveling and space exploration. Electric power available during periods of low usage, ie, off-peak periods, can be used to electrolyze water. The hydrogen and oxygen produced can be stored and used in fuel cells to produce electric power when peak power demands occur.

Heavy water, D_2O, was produced by a combination of electrolysis and catalytic exchange reactions. Some nuclear reactors (qv) require heavy water as a moderator of neutrons.

Cold fusion has been reported to result from electrolyzing heavy water using palladium, Pd, cathodes. Experimental verification of the significant excess heat output and various nuclear products are still under active investigation.

Fluorine. Fluorine is the most reactive product of all electrochemical processes. The principal use of fluorine continues to be the production of UF_6 from UF_4.

$$UF_4 + F_2 \rightarrow UF_6$$

The electrolyte used in fluorine cells is KF–HF in a ratio that minimizes melting point, HF vapor pressure, and corrosion of materials. Fluorine has been compressed, liquified, and shipped. However, most fluorine is produced and used on-site. Fluorine production in the United States is based on electrolytic cells developed in the 1940s. Modern type "E" cells are rated for 6 kA.

Permanganate. Potassium permanganate, $KMnO_4$, is produced in commercial quantities. It may be prepared from manganese dioxide or directly from manganese metal, ferromanganese, or other manganese alloys. The Carus Chemical Company produces potassium permanganate from manganese dioxide in the United States.

Hydrogen Peroxide. Peroxydisulfuric acid, $H_2S_2O_8$, is one of the strongest oxidizing agents known. It and other peroxydisulfates are produced electrochemically. The production of peroxydisulfates was once important for the manufacture of hydrogen peroxide (qv), H_2O_2. The development of the autoxidation of alkyl anthraquinones led to a rapid increase in the production of H_2O_2 but a sharp decline in the importance of the electrolytic process. No H_2O_2 is produced by the electrolytic peroxydisulfate process. The last plant using this process closed in 1983.

Rapid growth in demand for H_2O_2 in the late 1980s created widespread interest in new processes for H_2O_2 production. New process development efforts were generally focused either on direct combination of hydrogen and oxygen to produce H_2O_2 or the electrolytic reduction of oxygen to produce H_2O_2. HD Tech Inc., a joint venture of Dow Chemical Canada Inc. and Huron Technologies Inc., has developed and commercialized an electrolytic reduction of oxygen process.

Electrowinning of Metals

The metals that are produced by electrolysis are included in Table 1. Fused salt processes are used when the reactivity of the metal does not allow electrowinning from aqueous solutions. Manganese is the most reactive metal that is produced by electrolysis of an aqueous solution.

Electrowinning from Aqueous Solutions. The aqueous processes for electrowinning of metals from ores have the following common unit operations or steps: (*1*) the metal in the ore is converted to an acid-soluble form and this may be an oxidizing roast or a reduction; (*2*) ores from step 1 are leached, usually in sulfuric acid; (*3*) metal solutions from step 2 are purified and in some cases concentrated; (*4*) purified metal solutions are electrolyzed in cells where the metal is deposited on the cathode; and (*5*) acid is produced at the anode and recycled to the leaching step 2. Some acid values are lost, usually in the purification step, 3. Makeup acid is added in the leaching step, *2*. In most cases the metal solution from leaching step *2* contains impurities, other metals. Many of these metals have the characteristic of low hydrogen overvoltage. Codeposition of the impurity metals causes contamination of the desired product and decreases current efficiencies. The removal of impurities before electrolysis is very important. This is especially true in the case of the more reactive metals such as zinc, Zn, and manganese, Mn. These metals have deposition potentials close to the hydrogen evolution potential. The current efficiency of manganese electrowinning is about 60 to 68%. The principal inefficiency is hydrogen evolution.

The electrowinning of metals from aqueous solutions is generally carried out in tank cells. Developments in the electrowinning of metals form aqueous solutions have been directed toward improved anodes, improved additives, higher current densities, the use of ion-exchange membranes, better electrolyte quality control, and computer modeling of the processes.

Electrowinning from Fused Salts. Aluminum, Al, is produced worldwide by the Bayer-Hall-Heroult process. This process involves the electrolysis of alumina, Al_2O_3, dissolved in molten cryolite, Na_3AlF_6.

Sodium is produced by the electrolysis of a fused salt mixture of calcium chloride, $CaCl_2$, and NaCl in a Downs cell. An improved cell for the production of sodium was patented. This cell utilizes an electrode separator, solid electrolyte tubes that are permeable to the flow of sodium ions but impermeable to fluids and the flow of other ions. Production of sodium from mixtures of NaCl and aluminum chloride, $AlCl_3$, has been described.

Several processes for lithium, Li, metal production have been developed. The Downs cell with LiCl–KCl electrolyte produces lithium in much the same manner as sodium is produced. Lithium metal or lithium–aluminum alloy can be produced from a mixture of fused chloride salts. Granular Li metal has been produced electrochemically from lithium salts in organic solvents.

There are three electrolytic processes for magnesium, Mg, production: the Dow process, a process developed by I.G. Farbenindustrie in Germany, and an Alcan process. All processes involve the electrolysis of magnesium chloride, $MgCl_2$.

Research and development efforts have been directed toward improved cell designs, theoretical electrochemical studies of magnesium cells, and improved cathode conditions. A stacked-type bipolar electrode cell has been operated on a lab scale. Electrochemical studies of the mechanism of magnesium ion reduction have determined that it is a two-electron reversible process that is mass-transfer controlled.

Beryllium Be, metal is produced by electrolysis of $KCl–NaCl–BeCl_2$ melts.

Electrochemical Waste Treatment

In many instances the metal processing industry produces aqueous effluents containing dissolved metals. Many of these metals are toxic and controlled by the U.S. Environmental Protection Agency (EPA) regulations. Most water treatment plants do not remove toxic metals, or they concentrate the toxic metals in sludges that are classified as hazardous waste. Concentrations of the metals in wastewater are generally very low. The flow rates may be relatively high. Electrodes with large areas operating at low current densities are desirable for this application. Various porous electrodes, particulate-bed electrodes, fluidized-bed electrodes, and roll cells have been developed for metals recovery from dilute wastewater streams. Electrochemical processing has advantages over other chemical processes because the electrochemical process usually requires no addition of materials. Electrochemical processes for wastewater treatment often provide recovery of metal resources that partially offset processing costs. Silver has been recovered electrolytically from spent photographic liquors for many years. Silver recovery was motivated by economics. More recently, heavy-metal recovery has been motivated by environmental regulations.

In addition to metal recovery, which generally involves cathodic reduction, some waste may be treated by anodic oxidation processes. Many organic contaminants in wastewater can be oxidized by electrochemical treatments. High over-voltage anodes have been developed to improve the efficiency of these oxidations. Electrochemical oxidation of

Table 1. Production of Metals by Electrolysis of Aqueous Solutions

Metal	Anode	Diaphragm	Cathode	Cell feed[a], g/L	Electrolyte[a], g/L	Temperature, °C	Cell voltage, V	Cathode current density, A/m^2	Energy requirement, kW·h/kg	Current efficiency, %
Cd	Pb–Ag	no	Al	90–200 Cd, 20–40 Zn	10–20 Cd, 20–40 Zn, 60–140 H^+	30–35	2.5–2.7	80	1.5	90
Cr	Pb–Ag	yes	316 stainless steel		[b]	53	4.2	700	18	45
Cu	Pb–Sb–Ag	no	Cu	20–70 Cu, 20–70 H^+		30–35	2.0–2.2	130	2.2	80–90
Mn	Pb–Ag	yes	stainless steel or Ti	30–40 Mn, 125–150 NH_4^+, 0.1 SO_2 + glue	[c]	35	5.1	400–600	8–9	60–68
Ni	Pb	yes	Ni 99.9%		[d]	52	3.4	180		91–96
Zn	Pb–Ag	no	Al	100–200 Zn	100–200 H^+, 20–40 Zn	35	3.2–3.6	350–1000	3.3	90

[a] As sulfates unless otherwise noted.
[b] Anolyte in g/L: 13 Cr(VI), 2 Cr(III), 24 NH_3, 280 H_2SO_4. Catholyte in g/L: 11.5 Cr(III), 12.5 Cr(II), 84 NH_3; pH = 2.1–2.4.
[c] Anolyte in g/L: 10–20 Mn, 25–40 H_2SO_4, 125–150 NH_4^+. Catholyte in g/L: pH 6–7.2.
[d] Catholyte in g/L: 70 Ni(II) + H_3BO_3, Na_2SO_4; pH = 3.0–3.5. Anolyte in g/L: 40 H_2SO_4.

halogenated hydrocarbons has been achieved using barium peroxide, BaO_2, in aqueous NaCl solutions containing cationic surfactants.

Electrolytically generated hypochlorite may by used for the oxidative destruction of cyanides (qv) or the sterilization of domestic wastes. Several on-site systems for swimming pool sterilization and municipal waste treatment works have been developed. On-site production and immediate use of chlorine is considered safer than the transportation of chlorine.

Other electrochemical processes such as electrodialysis, electroflotation, and electrodecantation are also used in waste treatment.

Safety and Environmental Considerations

The electrochemical process industries are confronted with a wide range of hazards. These include electrical hazards, various explosion hazards, and the hazards associated with exposure to reactive chemicals.

Economic Aspects

Most large electrochemical processing facilities are located where raw materials, including electric power, are readily available at reasonable costs. Other factors influencing the location of electrochemical plants are proximity to markets, established transportation facilities, availability of water, and a source of labor. Large electrochemical plants are capital intensive, requiring large capital investment cost per employee.

The total annual value of products produced by electrochemical processing is roughly $18 billion.

MORRIS P. GROTHEER
Kerr-McGee Corporation

J. B. Talbot and S. D. Fritts, *J. Electrochem. Soc.* **139**, 2981–3018 (1992).

J. O'M. Bockris and co-workers, eds., *Comprehensive Treatise of Electrochemistry*, Vol. 2, Plenum Press, New York, 1981, pp. 3–9.

N. M. Prout and J. S. Moorhouse, eds., *Modern Chlor-Alkali Technology*, Vol. 4, Society of Chemical Industry, Elsevier Applied Science Publishers, Ltd., London, 1990, pp. 40–42, 97, and 178.

A. T. Kuhn, *Industrial Electrochemical Processes*, Elsevier Publishing Co., Amsterdam, The Netherlands, 1971, Chapt. 3.

"Report of the Electrolytic Industries," published annually in Aug./Sept. by The Electrochemical Society, Inc.

ORGANIC

An electroorganic reaction is a heterogeneous process having similarities to heterogeneous catalysis. In both cases selectivity is influenced by materials used, ie, the electrode, mass transport, mixing rates, and surface and bulk regimes.

The electrochemical cell simultaneously provides for oxidation and reduction processes, which occur at the anode and cathode, respectively. An organic molecule on the surface of an electrode can undergo electron removal (anode) or electron addition (cathode), resulting for uncharged organic reactants in the formation of a radical cation or radical anion, respectively. These reactive species, at essentially ambient temperatures, are open to a wide range of reaction paths depending on the species, medium, and adsorption at the electrode. Reviews of the wide range of electroorganic reactions investigated are available.

Oxidation and reduction reactions can be carried out using reformer hydrogen and oxygen from the air. To decide when electroorganic synthesis is likely to be a viable option for a desired product, some opportunity factors are use of cheaper feedstock; elimination of process step(s) or a difficult reaction; avoidance of waste disposal, toxic materials, and/or ability to recycle reagent; and ability to obtain products from anode and cathode.

For a profitable electrochemical process some general factors for success might be listed as high product yield and selectivity; current efficiency >50%, electrolysis energy <8 kW·h/kg product; electrode, and membrane in divided cells, lifetime >1000 hours; simple recycle of electrolyte having >10% concentration of product; simple isolation of end product; and the product should be a key material and/or the company should be comfortable with the electroorganic method.

Cell Design

Materials. *Cell Structure.* Because electroorganic syntheses in the main are carried out at close to room temperature, a wide range of plastics are available for cell construction. A full range of readily available polymers, eg, polyethylene, poly(vinyldene fluoride) (PVDF), polypropylene, polytetrafluoroethylene (PTFE), and other thermoplastic materials, is used for cell frames or vessels, and elastomers (qv), eg, neoprene, Viton, ethylene–propylene–diene monomer (EPDM), and PTFE encased, are used to seal cell parts.

Electrodes. At least three factors need to be considered in electrode selection as the technical development of an electroorganic reaction moves from the laboratory cell to the commercial system. First is the selection of the lowest cost form of the conductive material that both produces the desired electrode reactions and possesses structural integrity. Second is the preservation of the active life of the electrodes. The final factor is the conductivity of the electrode material within the context of cell design. An in-depth discussion of electrode materials for electroorganic synthesis as well as a detailed discussion of the influence of electrode materials on reaction path (electrocatalysis) are available. A general account of electrodes for industrial processes is also available.

The widely used cathode materials include Hg, Pb, Al, Zn, Ni, Fe, Cu, Sn, Cd, C, and alloys.

Generally, steel or nickel is used in making anodes for oxygen evolution in alkaline electrolytes. The so-called dimensionally stable anode (DSA) iridium oxide also makes a good oxygen evolve at low pH. Graphite or nickel is frequently employed as anodes for anodic halogenation of organics. Platinum makes a good, if expensive, anode material. Lead dioxide is also used but requires careful fabrication. Mixed-metal oxides and spinels have found some application.

In general, the best view of the electrode is as a catalyst in the process. Thus it should be both costed and assessed accordingly, eg, fractions of an electrode per kg of product. With time, the nature of the active electrode surface is likely to change either chemically or physically. Electrode surface contamination can be circumvented by *(1)* periodic mechanical and/or chemical cleaning; *(2)* current reversal; *(3)* stringent raw material specifications with regard to problem impurities; *(4)* purging and processing a recycle electrolyte stream; or *(5)* replacing electrodes.

Electrical conductivity is important and should be arranged to be as high as possible.

Diaphragms. In many electroorganic systems, the use of a diaphragm to separate anolyte and catholyte is necessary to minimize the side reactions caused by the incompatibility of a cell's two half-cell reactions. In the development of a commercially feasible process, the selection of a diaphragm for large-scale cells often presents a serious obstacle. The preferred electrolysis diaphragm would have low cost, low electrical resistance, limited interdiffusion of anolyte and catholyte constituents other than carrying ions, long operating life, good dimensional stability, and be insusceptible to plugging and fouling. The probability of finding all of these qualities in a single material is, unfortunately, small. Diaphragm materials are available in two types: permeable and permselective.

Electrolyte. The ideal electrolyte, ie, the fluid part of the cell, for organic synthesis would give high solubility to the organic, possess good conductivity, have low cost, contain easy recovery and purification, and be noncorrosive. Quaternary ammonium salts provide many of the above criteria in aqueous systems. A concise compilation of solvents and salts used in electroorganic chemistry is available.

Transport Phenomena. Electrochemical reactions are heterogeneous and are governed by various transport phenomena, which are important features in the design of a commercial electroorganic cell system. As for other heterogeneous reactions, the electrochemical reaction is impacted by heat and mass transport. The electrochemical reaction, however, is unique in that it also has charge transport characteristics. More comprehensive works on transport phenomena are available.

Mass Transport. Probably the most investigated physical phenomenon in an electrode process is mass transfer in the form of a limiting current. A limiting current density is that which is controlled by reactant supply to the electrode surface and not the applied electrode potential.

Methods proposed for improving mass-transfer rates in large-scale cells are *(1)* rotation of cylindrical or disk electrode including wiping of surface; *(2)* use of turbine or propeller agitators; *(3)* fluidized bed of electrode particles; *(4)* fluidized bed of nonconducting particles; *(5)* vibration of electrode; *(6)* gas sparging; and *(7)* external pumping of electrolyte in open channels or channels having turbulence promoters.

Charge Transport. Side reactions can occur if the current distribution (electrode potential) along an electrode is not uniform. The side reactions can take the form of unwanted by-product formation or localized corrosion of the electrode. The problem of current distribution is addressed by the analysis of charge transport in cell design. The path of current flow in a cell is dependent on cell geometry, activation overpotential, concentration overpotential, and conductivity of the electrolyte and electrodes. Three types of current distribution can be described when these factors are analyzed, a nontrivial exercise even for simple geometries. The three factors are primary current distribution (electrical), secondary current distribution (polarization), and tertiary current distribution (limiting current).

Heat Transfer. Heat removal for the commercial electrolysis system is generally carried out by internal or external evaporative cooling; circulation of electrolytes through external heat exchangers; or internal cooling with coils, jackets, or tubes that may act as electrodes.

Reaction Engineering. Electrochemical reaction engineering considers the performance of the overall cell design in carrying out a reaction. The joining of electrode kinetics with the physical environment of the reaction provides a description of the reaction system. Both the electrode configuration and the reactant flow patterns are taken into account. More in-depth treatments of this topic are available.

Electrochemical Cells as Reactors. The electrochemical reactor readily parallels the chemical reactor of which there are two types used as ideal models. The first is the plug flow reactor (PFR) where reactants and products move through the reactor in a plug-like manner. No mixing occurs in the PFR, which presumes that concentration changes occur only in the direction of electrolyte flow. The second model is referred to as a back-mix or stirred tank reactor (STR). The STR assumes perfect mixing and uniform composition in all zones of the reactor. An E is inserted for the electrochemical version of these abbreviations to produce PFER and STER. Idealized models are not attained in practice: mixing occurs in the PFER, whereas imperfect mixing occurs in the STER. Thus reactors are often described as having either PFER or STER features.

Scale-Up of Electrochemical Reactors. The intermediate scale of the pilot plant is frequently used in the scale-up of an electrochemical reactor or process to full scale. Dimensional analysis (qv) has been used in chemical engineering scale-up to simplify and generalize a multivariant system, and may be applied to electrochemical systems, but has shown limitations. It is best used in conjunction with mathematical models. Scale-up often involves seeking a few critical parameters. For electrochemical cells, these parameters are generally current distribution and cell resistance. The characteristics of electrolytic process scale-up have been described.

Cell Geometries. Uniform electrode potential, short interelectrode gaps, and good mixing and mass transport benefit many electrochemical reactions. It is difficult to depart from the narrow-spaced rectangular plates with turbulent flow electrolyte to achieve this. Reviews of electrolytic cell design are available in the literature. Cell designs include tank cells, two-dimensional electrode flow cells, and three-dimensional electrode flow cells.

Commercially Available Cells. A significant obstacle to electrode process development was the design and construction of the electrochemical cell. This has changed because a number of companies fabricate a range of cells for general use in electroorganic synthesis. These are parallel plate cells that may be divided or undivided. All use electrolyte recirculation for convection, and may include turbulence promoters to enhance mass transport. They follow the plate and frame design with external or internal electrolyte manifolding. A variety of electrode materials is offered, eg, pure metals, PbO_2, DSA, graphite, and alloys, and the latest polymers are used for materials of construction, eg, polypropylene, poly(vinylidenedifluoride), Teflon, and a range of elastomers. The peripherals of pumps, piping, etc, are also made available. Brochures on these cells are available from the respective companies, which also offer some consulting and support services. Cell fabricators include: ElectroCell AB (Sweden), the Electrocatalytic (New Jersey), ICI (UK), and the Aquanautics (California).

Product Recovery. Comparison of the electrochemical cell to a chemical reactor shows the electrochemical cell to have two general features that impact product recovery. Cell product is usually liquid, can be aqueous, and is likely to contain electrolyte. In addition, there is a second product from the counter electrode, even if this is only a gas. Electrolyte conservation and purity are usual requirements. Because product separation from the starting material may be difficult, use of reaction to completion is desirable; cells would be run batch or plug flow. The water balance over the whole flow sheet needs to be considered, especially for divided cells where membranes transport a number of moles of water per Faraday. At the inception of a proposed electroorganic process, the product recovery and refining should be included in the evaluation to determine true viability. Thus early cell work needs to be carried out with the preferred electrolyte/solvent and conversion. The economic aspects of product recovery strategies have been discussed. Some process flow sheets are also available.

Economic Aspects. Several publications probe the various areas of electroorganic process cost. Cells, overall process costs, economic optimization, and a comparison between the chemical and electrochemical methods are all discussed.

Industrial Electroorganic Processes

Large-Scale Processes. *Adiponitrile.* The most significant commercial electroorganic synthesis process is Monsanto's electrohydrodimerization (EHD) of acrylonitrile to adiponitrile. The importance of adiponitrile is as a precursor to hexamethylenediamine which is used in the manufacture of nylon-6,6. The cost of manufacturing nylon-6,6 is critically dependent on the cost of the intermediates used, and this has maintained the pressure to produce improvements in the EHD process.

Main cathode reaction

$$2\ CH{=}CHCN + 2\ H_2O + 2\ e \rightarrow NC(CH_2)_4CN + 2\ OH^-$$

Anodic water decomposition

$$2\ H_2O \rightarrow 4\ H^+ + O_2 + 4\ e^-$$

Two major cathode by-products

$$2\ CH_2{=}CHCN + 2\ H_2O + 2\ e^- \rightarrow CH_3CH_2CN + 2\ OH$$

$$3\ CH_2{=}CHCN + 2\ H_2O + 2e^- \longrightarrow NCCH(CH_2)_3CN\ (\text{with } (CH_2)_2CN \text{ on CH}) + 20H^- \quad \text{trimer}$$

The original process as pioneered by Monsanto in the 1960s used a membrane divided cell. Considerable energy savings were achieved when the undivided cell technology was introduced in the early 1970s. Also, the concentrated quaternary ammonium salt electrolyte was replaced by a dilute quaternary ammonium salt in 10 weight percent phosphate alkali metal salt solution. Yields of adiponitrile close to 90% can be achieved with this process. Asahi Chemical operate their version of the EHD process.

Tetraalkyllead. The Nalco Process. The second greatest success story for electroorganic processing in terms of total tons of product is that of tetraalkyllead. However, because of the curtailment of the used of lead-based antiknock additives in gasoline, Nalco closed its commercial plant in the early 1980s.

Smaller-Scale Processes. *Fluorination.* Perfluorinated organic compounds are important industrial surfactants (qv) and textile treating agents.

The discovery that a number of organic compounds could be fluorinated by electrolysis in a solution of anhydrous hydrogen fluoride, HF, was made around 1940 (see FLUORINE COMPOUNDS, INORGANIC–HYDROGEN). In the Simons process, fluorination of the dissolved organics takes place at a nickel anode, generally without generation of free fluorine; hydrogen is evolved at the iron cathode of diaphragmless cell. Alkali fluorides may be added for improved electrolyte conductivity. Electrochemical fluorinations of a wide range of compounds including hydrocarbons, alcohols, ketones, carboxylic acids, and amines have been studied. A comprehensive review of the field has been published.

Sebacic Acid. Sebacic acid, $C_{10}H_{18}O_4$, is an important intermediate in the manufacture of polyamide resins (see POLYAMIDES). It has an estimated demand worldwide of approximately 20,000 t/yr. The alkaline hydrolysis of castor oil (qv), which historically has shown some wide fluctuations in price, is the conventional method of preparation. Because of these price fluctuations, there have been years of considerable interest in an electrochemical route to sebacic acid based on adipic acid (qv) as the starting material. The electrochemical step involves the Kolbé-type of Brown-Walker reaction where anodic coupling of the monomethyl ester of adipic acid forms dimethyl sebacate.

BASF, Asahi Chemical Industry, and a Russian group have carried out pilot-plant studies of the electrochemical route to sebacic acid. Yields of dimethyl sebacate reported for the BASF and Russian processes are in the range of 80–85%. Asahi claims product yields as high as 92% at current efficiencies in the range of 85–90%.

Maltol. Otsuka Chemical Company in Japan has operated several electroorganic processes on a small commercial scale. It has used plate and frame and annular cells at currents in the range of 4500–6000 A. The process for the synthesis of maltol, a food additive and flavor enhancer, starts from furfural. The electrochemical step is the oxidation of α-methylfurfural to give a cyclic acetal. The remaining reaction sequence is acid-catalyzed ring expansion, epoxidation with hydrogen peroxide, and then acid-catalyzed rearrangement to yield maltol.

Fenoprofen. Fenoprofen is one of a series of antiinflammatory drugs of the arylpropionic acid family (see ANALGESICS, ANTIPYRETICS, AND ANTIINFLAMMATORY AGENTS). As is common with many complex drugs, its synthesis involves many steps of sometimes difficult chemistry. France's Société Nationale des Poudres et Explosifs (SNPE) uses an electrochemical step to replace a seven-step process by one of four steps. This is made possible by the availability of 3-phenoxybenzaldehyde, which is used as the starting point for the synthesis. The electrochemical step is an electrocarboxylation of the organic halide using carbon dioxide at a stainless steel cathode having some lead coverage. An aprotic solvent, such as dimethylformamide made conductive with a small amount of terabutylammonium halide is used, and the anode process uses the dissolution of magnesium.

Propylene Oxide. An electrochemical route to 1,2-propylene oxide has received considerable attention, but as yet has not been commercialized. The electrochemical route involves passing gaseous propylene through the anode compartment of a divided cell in which both anolyte and catholyte are a dilute solution of sodium chloride. Hypochlorous acid produced in the anode compartment combines with the propylene to form propylene chlorohydrin, which diffuses through the porous diaphragm and is decomposed in the alkaline environment generated by the cathodic reduction of water.

There have been a number of cell designs tested for this reaction. Undivided cells using sodium bromide electrolyte have been tried. These have had electrode shapes for in-cell propylene absorption into the electrolyte. The chief advantages of the electrochemical route to propylene oxide are elimination of the need for chlorine and lime, as well as avoidance of calcium chloride disposal (see CALCIUM COMPOUNDS–CALCIUM CHLORIDE; LIME AND LIMESTONE). An indirect electrochemical approach meeting these same objectives employs the chlorine produced at the anode of a membrane cell for preparing the propylene chlorohydrin external to the electrolysis system. The caustic made at the cathode is used to convert the chlorohydrin to propylene oxide, reforming a NaCl solution which is recycled. Attractive economics are claimed for this combined chlor-alkali electrolysis and propylene oxide manufacture.

CHRIS J. H. KING
Monsanto Chemical Group

M. M. Baizer and H. Lund, eds., *Organic Electrochemistry*, Marcel Dekker, Inc., New York, 1991.

N. L. Weinberg and B. V. Tilak, eds., *Technique of Electroorganic Synthesis*, Part III, A. Weissberger, series ed., *Techniques of Chemistry*, Vol. 5, John Wiley & Sons, Inc., New York, 1982.

J. D. Genders and D. Pletcher, eds., *Electrosynthesis, From Laboratory, to Pilot, to Production*, Electrosynthesis Co., E. Amherst, N.Y., 1990.

F. Walsh, *A First Course in Electrochemical Engineering*, The Electrosynthesis Co., East Amherst, N.Y., 1993.

ELECTROLESS PLATING

Electroless plating is defined as the controlled autocatalytic deposition of a continuous film at a catalytic interface by the reaction in solution of a metal salt and a chemical reducing agent. Electroless deposition can produce films of metals, alloys, compounds, and composites on both conductive and nonconductive surfaces.

The growth of electroless plating is directly traceable to *(1)* the discovery that some alloys produced by electroless deposition, notably nickel phosphorus, have unique properties; *(2)* the growth of the electronics industry, especially the development of printed circuits; and *(3)* the large-scale introduction of plastics into everyday life.

Theory of Electroless Plating

The theory and practice of electroless plating parallels that of electrolytic plating.

The actual metal reduction and film development occurs at the interface of the solution and the item being plated in both the electrolytic and electroless processes. The main difference is that the electrons in electroless plating are supplied by a chemical reducing agent present in solution. This means that electroless solutions are not thermodynamically stable, because the reducing agent and the metal salt are always present and ready to react.

Electroless solutions contain a metal salt, a reducing agent, a pH adjuster or buffer, a complexing agent, and one or more additives to control stability, film properties, deposition rates, etc.

Of the large number of potential reducing agents, the principal commercial materials are formaldehyde (qv) for copper and silver, hypophosphite for nickel and palladium, and organoboron compounds for gold, nickel, palladium, and copper. The latter two reducing agents produce phosphorus- or boron-containing alloys. The detailed theory of electroless plating has been discussed thoroughly in a number of works.

Most reducing agents are too slow, giving insufficient plating rates, or too fast, resulting in bulk decomposition. Each combination of metals and reducing agent requires a specific pH range and bath formulation. The metal salt and reducing agent must be replenished at periodic intervals. Buffers, complexers, and other additives are added to compensate for drag-out losses, and to counter bath aging effects such as total salts accumulation. Stabilizers are typically used at a few to a few tens ppm, and may affect deposition rate, deposit color, or metal ductility, or control spontaneous decomposition reactions.

The ideal electroless solution deposits metal only on an immersed article, never as a film on the sides of the tank or as a fine powder. Room temperature electroless nickel baths closely approach this ideal; electroless copper plating is beginning to approach this stability when carefully controlled. Any metal that can be electroplated can theoretically also be deposited by electroless plating. Only a few metals, ie, nickel, copper, gold, palladium, and silver, are used on any significant commercial scale.

Electrolytic plating rates are controlled by the current density at the metal–solution interface. The current distribution on a complex part is never uniform, and this can lead to large differences in plating rate and deposit thickness over the part surface. Uniform plating of blind holes, re-entrant cavities, and long projections is especially difficult.

A primary advantage of electroless solutions is the ability to produce conductive metallic films on properly prepared nonconductors, along with the ability to uniformly coat any platable object.

Electroless plating rates are affected by the rate of reduction of the dissolved reducing agent and the dissolved metal ion which diffuse to the catalytic surface of the object being plated. When an initial continuous metal film is deposited, the whole surface is at one potential determined by the mixed potential of the system. The current density is the same everywhere on the surface as long as flow and diffusion are unrestricted so the metal deposited is uniform in thickness over the whole surface. However, maximum plating rates are usually lower for electroless plating than those possible for electrolytic plating. Extremely thin films of electroless coatings are not uniform, because the initial deposition is confined to discrete nucleation sites that grow and coalesce into a film.

Catalysts for dielectric surfaces are more complex than the simple salts used on metals. One-step catalysts consist of a stabilized, prereacted solution of the palladium and stannous chlorides. A separate acceleration or activation solution removes loose palladium and excess tin before the catalyzed part is placed in the electroless bath, prolonging bath life and stability.

Equipment

An electroless plating line consists of a series of lead-lined (for plastics etching) or plastic-lined tanks equipped with filters and heaters, separated by rinse tanks.

The preferred methods for heating are a double-jacket heated tank, nonmetallic heat exchangers, quartz heaters, or Teflon-coated low watt density stainless steel heaters.

Some hot nickel and flash electroless copper solutions are plated to the point of exhaustion and then discarded. Most baths are formulated to give bath lives of ≥6 turnovers of the bath constituents; some reach steady-state buildup of the by-products and can be used indefinitely. All regenerable solutions should be filtered to remove particulates that can cause deposit roughness and bath instability.

Racks are of stainless steel, titanium, or plastic-coated steel. Tank and rack design are less important than tank loadings and rack coatings, especially when plating nonconductors. The parts may be closely racked without harm if reracking is done after electroless plating; otherwise, racking should be the same as for electrolytic plating. However, more racks may be placed in a given electroless tank than in a similar electrolytic tank. The critical points are to use air sparging to maximize metal transfer rates; filter to remove catalytic nuclei before they grow; and not to overload the plating bath.

Safety and Waste Disposal

Electroless plating operations must comply with EPA and local waste treatment guidelines. The most troublesome chemicals are electroless nickels and coppers, and the chromic acid etchant. Most solutions are considered hazardous, as they are usually highly acidic or basic, and usually contain metals (see WASTE TREATMENT hazardous waste). All reducing agents should be stored separately from oxidizing substances such as chromic acid or nitric acid. Proper ventilation and fume scrubbing are extremely important, as is the use of proper safety equipment. Formaldehyde reducing agent is a suspect carcinogen. All personnel must have appropriate training in case of spills or accidents.

Waste minimization (see WASTES, INDUSTRIAL) techniques must always be implemented as part of waste disposal and treatment. Countercurrent rinsing reduces water flow and facilitates waste treatment.

The electroless baths, especially copper and nickel, may be difficult to treat by conventional precipitation because of the presence of strong chelating agents (qv). Ion-exchange and accelerated plateout systems are among the best techniques, although ozonolysis and other oxidative treatments also destroy chelating agents. The metal is tightly chemically bound by organic complexers or chelating agents. Specific techniques for breaking the complex and precipitating the metal include treatment using lime, ferrous sulfate, sodium borohydride, and organic sulfides. Suppliers of the proprietary baths can make specific recommendations, depending largely on the type of chelate used.

Plating on Metals

Electroless Nickel. Properties of electroless nickel deposits vary greatly depending on the reducing agent: hydrazine gives a practically pure nickel; the organoboron reducing agents give very hard nickel–boron alloys; and the most widely used sodium hypophosphite deposits a range (from 1–15 wt % P) of nickel–phosphorus alloys having unique properties. Acidic (pH 4.0–5.5) baths are preferred, but alkaline (pH 8–10) baths are also used. Operating temperatures are 70–95°C. Typical formulations use sodium hypophosphite as the reducing agent, nickel sulfate or nickel chloride, and ammonium or organic salts as buffers. Mild complexing agents such as citrate or glycolate, and trace quantities of proprietary stabilizers including heavy metals or sulfur compounds, are used. A large number of commercial baths are available.

The engineering properties of electroless nickel have been summarized. The Ni–P alloy has good corrosion resistance, lubricity, and especially high hardness. This alloy can be heat-treated to a hardness equivalent to electrolytic hard chromium, and the lubricity is also comparable. The wear characteristics are extremely good, especially with composites of electroless nickel and silicon carbide or fluorochloropolymers. Thus the main applications for electroless nickel are in replacement of hard chromium.

Other Metal Processes. The only other types of electroless processes commonly used on metals are electroless gold and palladium. These are restricted to specialized uses because of extremely high cost. The properties of these coatings are not significantly different from electroplated coatings. Typical applications include solder mask-coated printed circuit boards, insertion tabs, and surface mount devices. New electroless silver processes are available for specialized uses.

Plating on Nonconductors

Plating on nonconductors comprises two technologically very different categories: plating of plastics and printed circuit production.

Plating of Plastics. Plastics can be coated with metals for decorative and/or functional purposes.

Electroless films have two functions: they provide an electrically conductive layer that allows further coating by electroplating; and they provide a secure bond between the plastic and the electroplated layer. Plated plastics, however, have several disadvantages. Plating normally lowers impact strength. The coefficient of thermal expansion is much higher for plastics than for metals, so stress buildup and adhesion loss can occur on severe thermal cycling. Blistering can occur during corrosion. The relatively low heat distortion temperature of most plated plastics can also limit applications. Molding cycles are relatively long.

One of the most important advantages of plated plastics is that weight savings can be as much as 60% as compared to an equivalent all-metal part. The molded plastic parts need no buffing or other finishing step before plating. Plated plastics have improved tensile strength, elasticity, flexural strength, a reduced total coefficient of thermal expansion, and improved abrasion and weathering resistance.

Applications and Economic Aspects. There are various reasons for coating plastics with metals. In the automotive industry, the light weight of plastics is combined with the consumer appeal of metal. The largest single application for electroless plating is for automotive parts such as grilles and headlamp bezels. This market segment has dropped in size because cars are smaller, and painted parts have replaced painted finishes. The electroless coating serves only as the base for electrolytic final metal coatings. Other large market segments include hardware, appliances, marine applications, plumbing fixtures, knobs, closures, and novelties; RFI/EMI-shielding applications are a newer market opportunity. Plated ABS has about 80% of the market. Most of the remainder is ABS–polycarbonate alloys (especially for computer housings) and modified poly(phenylene oxide).

Chemistry. Successful electroless plating depends on the optimized interaction of five separate complex chemical solutions to clean, roughen, and catalyze the surface before plating. These steps are critical for formation of an adherent continuous electroless coating, and for optimum durability after electrolytic plating.

Etching is necessary to give good metal-to-plastic adhesion. Solutions of strongly oxidizing chromic acid–sulfuric acid–water, or chromic acid–water are operated near the point of mutual saturation. The etchant both physically roughens the surface and chemically modifies it to give a very hydrophilic surface.

Neutralizing removes the large amount of hexavalent chromium from the surface of the part. Hexavalent chromium shortens the life of the catalyst, and trace amounts completely inhibit electroless nickel deposition.

Catalysis is done by an acidic solution for the stabilized reaction product of stannous chloride and palladium chloride.

Acceleration modifies the surface layer of palladium nuclei, and stannous and stannic hydrous oxides and oxychlorides. Any acid or alkaline solution in which excess tin is appreciably soluble and catalytic palladium nuclei become exposed may be used. The activation or acceleration step is needed to remove the excess tin from the catalyzed surface, which otherwise would inhibit electroless plating.

Electroless metal is provided by either copper or nickel. A number of reports have indicated that copper may be better for corrosion resistance. A nickel–copper–phosphorus alloy of low copper content is also available.

Printed Circuit Boards. Electroless copper metallization is an indispensable part of the modern electronics revolution. Printed circuit board production uses electroless copper to coat nonconductive plastic surfaces exposed when drilling through-holes for internal electrical paths, and to define the circuit patterns on the board surfaces. Printed circuit production is much more complicated than POP, involving up to 80 or more process and inspection steps. Automated electroless plating lines are often used, but separated from the electrolytic processing cycle. Thus the boards are processed and plated in the semibulk mode, where large numbers of boards are spaced closely together on special racks. The electroless plated panels are reracked for electrolytic plating after the intermediate processing steps. Printed circuit production also includes electrolytic plating of copper, tin, tin–lead, nickel, and gold.

Numerous variations exist in the electroless plating solutions, processes, and techniques employed both in laboratory and commercial form, to create a great variety of products. All produce a layer of highly conductive copper in specified areas. Modern electroless copper films have a ductility and conductivity identical to that of electrolytic copper. The three basic classes of copper baths are

Copper bath	Thickness desired, μm	Processing type
flash	0.5–1	subtractive
fast	1–2.5	semiadditive
heavy build	25–40	additive

Economic Aspects. Printed circuit board production is a multibillion dollar business in the United States alone. An estimated 800 shops are involved in production, ranging from small companies specializing in one aspect of production, to huge organizations containing both research and production facilities capable of producing any type of board.

The cost of the chemicals used in electroless copper plating is very low, rarely exceeding $2.78/m^2, except for fully additive processes. The principal costs of printed circuit board production arise mainly from handling steps and other operations.

In 1990 the majority of U.S. PCB production resulted from subtractive or print-and-etch processing; additive processes were less than 6% of the total; multilayer boards accounted for 55.8%.

Chemistry. Many of the chemicals and process steps used for PCB processing are similar to those used in the POP industry; however, electroless copper is used exclusively.

Application Methods. The three PCB processing techniques may be summarized as:

Subtractive process	Semiadditive process	Additive process
copper-clad board	thin copper-clad board	unclad board
form holes	form holes	adhesive coat
catalyze	catalyze	form holes
accelerate	accelerate	etch surface
electroless copper	electroless copper	catalyze
positive resist coat	negative resist coat	negative resist coat
electrolytic copper	electrolytic copper	accelerate
dissimilar metal electroplate	dissimilar metal electroplate	electroless copper
strip resist	strip resist	
etch copper	etch copper	

Additive and semiadditive processing can give material savings and higher circuit densities, whereas subtractive processing is technologically easier.

Additional steps used for multilayer production include application of black or red oxide to foil; application of photoresist or screened etch resist; imaging and development of photoresist; copper etching; removal of resist; assembling of layers, checking registration and lamination; drilling; and removal of drill smear on copper inner layers. Rinses and additional cleaning, neutralizing, light etching, and developing steps are required.

Electroless Copper. Electroless copper, introduced in the mid-1950s, is available commercially in great variety. Formaldehyde is usually the reducing agent, copper sulfate (occasionally copper nitrate or copper chloride) is the metal salt, and sodium hydroxide is used to control pH. The complexers may be tartrate, ethylenediaminetetraacetic acid (EDTA), tetrakis(2-hydroxypropyl)ethylenediamine, nitrilotriacetic acid (NTA), or some other strong chelate. Numerous proprietary stabilizers, eg, sulfur compounds, nitrogen heterocycles, and cyanides (qv) are used. These formulated baths differ in deposition rate, ease of waste treatment, stability, bath life, copper color and ductility, operating temperature, and component concentration. Most have been developed for specific processes; all deposit nearly pure copper metal.

Printed Circuit Etchants. Two types of etchants are used in printed circuit board processing for entirely different purposes. Etchants used for etchback or desmear attack only organic materials, whereas copper etchants dissolve only copper. Etchback refers to massive removal of the organic board material perpendicular to the drilled hole, to expose annular copper rings from the inner layers of multilayer boards to ensure good conductivity after plating. Desmear is a less aggressive type to roughen drilled hole surfaces and remove organic residues from the drill-exposed copper inner layers. Older chromic acid and sulfuric acid etchants have been replaced by regenerable permanganate systems and by plasma etching. These etchants directly contribute to electroless plating through microroughening the surface and removing loosely bonded debris, thereby improving adhesion and performance.

Emerging Printed Circuit Technologies. Much research has been devoted to developing safer, cheaper, and more reliable alternatives to formaldehyde reduced electroless coppers, and many processes are in the advanced stages of development. These include hypophosphite-reduced electroless coppers; substitution of conductive graphite colloids for electroless coppers; direct electroplating over catalyzed or catalyzed and sulfide treated substrates; and non-noble metal catalysts. Three-dimensional printed circuit boards, usually on engineering quality plastics, are becoming more common (see ENGINEERING PLASTICS). High performance boards of novel types continue to be developed with controlled impedance layers; ultrathin and narrow copper conductors; ceramic substrates; high temperature plastic laminates; and Invar or Kovar inner layers in place of copper or dielectric.

Other Processes

Silver, the first metal used for metallizing nonconductors, is rarely employed because of high cost, low stability, and inability to plate a thick deposit autocatalytically. Formulations typically consist of ammoniacal silver nitrate solutions and a reducing agent of sugar, formaldehyde, hydrazine, or dimethylamineborane. These baths are generally unstable, and are used until exhausted.

Most electroless silver applications are for silvering glass or metallizing record masters. Mirror production is the principal usage for electroless silver. Silver is also used in the jewelry and button industry.

The other important commercial glass-plating application is for production of architectural reflective glasses. Translucent metal films are used for decoration and for reduction of environmental heat gain. Electroless plating is used by one producer for this type of product.

Ceramic plating is similar to glass plating in that a two-step catalyst is used. The surface is often mechanically roughened or chemically etched to improve the metal adhesion.

An electroless nickel matrix can be used to securely bond diamonds to cutting tools, and electroless nickel–diamond composites are also used (see TOOL MATERIALS).

Electroless gold is commercially used on printed circuit boards, flex circuits, ceramics, three-dimensional circuits, metals, and highly specialized substrates. The newest type is extremely stable and cost-effective; older baths were highly unstable. Electroless palladium is also used in specialty applications in the electronics industry where the need for pore-free or discontinuous coatings justifies the high price. Both coatings are normally applied over electroless nickel, Kovar, or other materials, but cannot be used directly on copper.

GERALD A. KRULIK
Applied Electroless Concepts, Inc.

W. Riedel, *Electroless Nickel Plating*, Finishing Publications, Ltd., Hertfordshire, U.K., 1991.

G. O. Mallory and J. B. Hajdu, eds., *Electroless Plating: Fundamentals and Applications*, American Electroplaters and Surface Finishers Society, Orlando, Fla., 1990.

Metal Finishing Guidebook & Directory, Metals and Plastics Publications, Inc., Hackensack, N.J., 1993.

The Engineering Properties of Electroless Nickel Deposits, International Nickel Co., New York.

ELECTRONIC MATERIALS

Electronic materials are those exquisitely pure crystal structures which form the basis for the information technology of the 1990s. The chemistry and chemical engineering required for making electronic materials, ie, these very specific inorganic chemical bonding structures, have come about by techniques employing chemical vapor deposition to effect material growth and through an increased control over surface and interfacial chemistry.

Semiconductor Energy Levels

Semiconductor materials are rather unique and exceptional substances. The entire semiconductor crystal is one giant covalent molecule. In semiconductors, the electron wave-functions are delocalized, in principle, over an entire macroscopic crystal. Because of the size of these wave functions, no single atom can have much effect on the electron energies, ie, the electronic excitations in semiconductors are delocalized.

Good semiconductors are drawn from the central columns, Groups 13, 14, and 15 (III, IV, and V), of the Periodic Table, where the atoms tend to be nonpolar. For this reason, and because of the giant size of the wave functions, the electron–atom interaction is very weak. The electrons move as if in free space, colliding with the atomic lattice rather infrequently.

In a semiconductor the available energy levels are the valence and conduction bands, which are generally filled and empty, bonding and antibonding, respectively, separated by a forbidden gap. The electrons in the conduction and valence bands act as two separate subsystems. Not only do the electrons ignore the crystallographic lattice of atoms, the electrons in one band tend to ignore those in the other subsystem. This property is unique to electronic materials.

Electrons excited into the conduction band tend to stay in the conduction band, returning only slowly to the valence band. The corresponding missing electrons in the valence band are called holes. Holes tend to remain in the valence band. The conduction band electrons can establish an equilibrium at a defined chemical potential, and electrons in the valence band can have an equilibrium at a second, different chemical potential. Chemical potential can be regarded as a sort of available voltage from that subsystem. Instead of having one single chemical potential, ie, a Fermi level, of all the electrons in the material, the possibility exists for two separate quasi-Fermi levels in the same crystal.

The possibility of two separate electronic equilibria, ie, the establishment of two quasi-Fermi levels or two different chemical potentials, requires a very slow decay of electrons from the conduction band back into the valence band. The very weak electron–atom coupling, resulting from the large delocalized wave functions in nonpolar materials, permits the slow decay. Electron-hole recombination requires getting rid of the electronic band gap energy, which is usually around one volt, and dumping it off as heat of atomic motion. In relation to characteristic energies of atomic motion, one volt is a huge amount of energy to dissipate in a single step. The weak electron–atom coupling makes nonradiative decay an extremely unlikely event and the low probability for semiconductor materials to dissipate electronic energy as heat is probably their most unique property. By contrast, in organic molecules, decay by nonradiative recombination is sufficiently likely to occur that it is given the name internal conversion.

In fact, nonradiative recombination does occur in semiconductors, but primarily as a result of chemical defects that introduce new energy levels into the forbidden gap. These defect levels act as stepping stones, permitting conduction electrons to cascade down to the valence band in two smaller steps rather than one improbable leap.

Thus a principal goal of semiconductor materials science has been to create chemically perfect semiconductor structures. Any defects that disturb the perfect valence bonding structure, allowing energy levels in the forbidden gap, must be eliminated as far as possible. Even the utmost extrinsic chemical purity is insufficient, however, because intrinsic defects such as broken bonds, self-interstitials, and vacancies are also proscribed. In particular, unsaturated chemical bonds on the material surface, or in the bulk, contribute nonbonding orbitals having unwanted energy levels in the forbidden gap. The rigid, tetrahedrally coordinated semiconductor crystal structures of silicon, Si; germanium, Ge; and gallium arsenide, GaAs, have a tendency to reject both extrinsic and intrinsic defects, contributing to their technological success.

Semiconductor Surfaces

Semiconductor surfaces are the most likely location for intrinsic defects such as dangling or weak bonds to occur. The bulk chemical defect densities that can be tolerated in solid-state electronics range from 1 in 10^6 to 1 in 10^{11}, depending on the specific application. Corresponding surface defect densities that can be tolerated range form 1 in 10^4 to 1 in 10^7, ie, nearly all of the semiconductor surface atoms must be cleanly saturated with strong covalent bonds, because defects introduce energy levels into the forbidden gap. These requirements for semiconductor surfaces and interfaces give rise to a chemical figure of merit, ie, equivalent to a surface chemical reaction having a 99.99% to 99.99999% yield.

The Role of Silicon

The Si–SiO$_2$ Interface. Beginning in the mid-1950s, thermal oxidation of silicon at high temperatures in oxygen was begun, coating the silicon with a thin layer of silicon dioxide, SiO_2, glass. The thermal oxidation recipe was gradually perfected and by the late 1960s the figure of merit for the Si–SiO_2 interface had been improved to 1 defective chemical bond in 10^6. The Si–SiO_2 interface is an amorphous/crystalline heterojunction. The interfacial bonds can be 99.9999% saturated. Thus, in short order the microprocessor (1969), the memory chip, and the pocket calculator, were developed. In 1992, microchip production was a $\sim\$6 \times 10^{10}$ annual industry worldwide, supporting a $\sim 5 \times 10^{11}$ systems, software, and communications industry, employing millions of highly educated people.

Purification of Silicon. Chemical purity plays an equally important role in the bulk of materials as on the surface. To approach the goal of absolute structural perfection and chemical purity, semiconductor Si is purified by the distillation of trichlorosilane, $SiHCl_3$, followed by chemical vapor deposition (CVD) of bulk polycrystalline silicon (at 1100°C), $SiHCl_3 + H_2 \rightarrow Si(s) + 3HCl(g)$. Purified polycrystalline CVD silicon from this reaction is then melted and a single-crystal boule weighing as much or more than 50 kg, and having a diameter up to 20 cm, is pulled from the melt by Czochralski growth. Metallurgical-grade silicon is not sufficiently pure for applications in electronics (see SILICON AND SILICON ALLOYS).

III–V Semiconductors

For optoelectronics the binary compound semiconductors drawn from Groups 13 and 15 (III and V) of the Periodic Table are essential. These often have direct rather than indirect band gaps, which means that, unlike Si and Ge, the lowest lying absorption levels interact strongly with light. The basic devices of optical communications, light-emitting diodes (LEDs) (see LIGHT GENERATION–LIGHT-EMITTING DIODES) and semiconductor lasers, are made of these III–V semiconductors. Aluminum arsenide, AlAs, GaAs, and the alloys of these compounds have historically been the most important III–V material system. The reason once again derives from the need to control interfacial chemical bonding structures.

The double heterostructure, invented in the early 1960s, is essentially a crystalline sandwich: the bread is made of AlAs and the filling of GaAs. Because the band gaps of AlAs and GaAs are 2.2 eV and 1.4 eV, respectively, the GaAs wave functions are sandwiched in by the 2.2 eV potential barriers. Although the electrons and holes are prevented from seeing any external surface, they do see the AlAs–GaAs interface. However, the cubic unit cell dimensions of GaAs and AlAs are 0.56533 nm and 0.56605 nm, respectively. Thus the mismatch at the interface is less than 0.1%, meaning that the crystal structures can match up nearly perfectly, leaving only a few unsaturated chemical bonds at the interface. Generally, the interfacial bonds in the $Ga_{1-x}Al_xAs$ system are 99.999% saturated. Although this is not as good as the best Si–SiO_2 interfaces, it is excellent nonetheless. The growth of successive atomic layers of semiconductor material, in perfect registry with atoms in the underlying crystal, is called epitaxy. The perfect atomic registry between layers of differing composition is called heteroepitaxy.

Physics and chemistry researchers approach III–V synthesis and epitaxial growth differently. The physics approach, known as molecular beam epitaxy (MBE), is essentially the evaporation of the elements. The chemistry approach, organometallic chemical vapor deposition (OMCVD) is exemplified by the typical chemical reaction: (at 580°C) $Ga(CH_3)_3 + AsH_3 \rightarrow GaAs(s) + 3CH_4(g)$. Thin-film epitaxy by OMCVD is generally more flexible, faster, lower in cost, and more suited for industrial production than MBE. An OMCVD system usually consists of two principal components, a gas manifold for blending the gas composition, and a graphite substrate holder which is usually inductively heated.

ELI YABLONOVITCH
University of California at Los Angeles

A. S. Grove, *Physics and Technology of Semiconductor Devices*, John Wiley & Sons, Inc., New York, 1967.

F. L. Carter, ed., *Molecular Electronic Devices*, Marcel Dekker, Inc., New York, 1982.

M. A. Herman and H. Sitter, *Molecular Beam Epitaxy: Fundamentals and Current Status*, Springer-Verlag, Berlin, 1989.

G. B. Stringfellow, *OMVPE: Theory and Practice*, Academic Press, Boston, Mass., 1989.

ELECTRONICS, COATINGS

Coating technology can be defined as replacing air at the surface of a substrate to give a film structure having varying layers of different properties. This technology covers a wide range of products and processes (see COATING PROCESSES; COATINGS; PAINT). Electronics coatings refers to the wide variety of coated structures used in electronic devices.

These coatings have some unique aspects of formulation, production, and use. The ability to produce high resolution coatings, at high volume and low cost, has been the driving force for the revolutionary advances in a wide variety of electronic components, integrated circuits (qv), memory chips, magnetic storage devices, cathode ray tubes, optical storage disks, etc (see also ELECTRONIC MATERIALS; INFORMATION STORAGE MATERIALS). Improved functionality of these components has led to advances in a wide range of consumer, industrial, and military products. Typical devices are television sets, video cassette recorders, cameras, fax machines, cellular phones, personal and laptop computers, fly-by-wire control systems for airplanes, and the guidance system for such military ordnance as smart bombs and cruise missiles.

Significant advances have also been made in the more traditional area of wire and cable used to transmit the electricity needed for these devices (see ELECTRICAL CONNECTORS). The optical cables (see FIBER OPTICS; NONLINEAR OPTICAL MATERIALS) that are replacing the traditional copper (qv) cables for telephone usage require very sophisticated coatings to protect cable from light loss and deterioration. The ribbon cable and multiple coaxial cable assemblies used for cable television, stereo systems, and burglar alarms all depend on the cable coatings reducing interference with each other so that signal quality is maintained.

Economic Aspects

Electronic coatings are of significant economic importance, as are the finished products in which they go. The worldwide total value of the resulting products is $500 billion. The annual electronic coatings market value is estimated to be $5 billion. These coatings are manufactured in several countries. Some of the principal manufacturers of electronic coatings are

Company	Products
Minnesota Mining and Manufacturing	conformal coatings
W. R. Grace & Co.	dielectric and conductive adhesives, encapsulants, heat dissipating materials, manufacturing aid coatings
Union Carbide Corp.	Parylene conformal coatings, photoresists, developers, etchants, solder masks, potting compounds
Conap unit of American Cyanamid	potting compounds, conformal coatings
DuPont Co.	conformal coatings, photoresists, manufacturing aids, electronic functional coating compounds
OGG Microelectronic Materials, Olin/CIBA-GEIGY venture	photosensitive polyimides, high performance semiconductors
Dexter	encapsulants
Dow Corning	potting and encapsulating compounds
Asahi	packing materials
Nippon Kayaku	packing materials
General Electric	packaging materials
Unilver	adhesives, dielectric coatings, protective coatings, dielectric interlayers, electronic functional coatings

Functions of Electronics Coatings

Electronics coatings can generally be classified as (*1*) functional, ie, coatings that provide a variety of electrical, magnetic, optical, chemical, mechanical, and/or thermal properties enabling a device to function as prescribed; (*2*) manufacturing aids, ie, coatings that serve as an integral part of the manufacturing process to make a specific component or device; or (*3*) protective/decorative, ie, coatings that protect the component or device from environmental damage during its normal use cycle.

Coatings Properties

Material property specifications must be written by design and material engineers to control engineering requirements and to control incoming raw material quality. Material property requirements depend on various in-use functional needs in terms of electrical, mechanical, thermal, chemical, optical, and magnetic properties.

Electronic coatings can be classified as either organic, inorganic, or metallic. Organic coatings are typically polymeric in nature. A wide variety of polymer types are used. Although polymeric properties vary considerably less over the full range than those of the inorganics and metallics, both chemistry and molecular structure play a role because these dictate properties and coating behavior. A list of polymer uses in electronic coatings follows. More detail can be found in the literature.

Polymer	Uses
alkyd	polyesters for protective and decorative coatings; paints
acrylic	general purpose for molding, casting, and coating formulations; dip and sprayable PC board coatings
epoxy	protective and electrical-grade adhesives, electronic encapsulation, composites, and PC boards; solder masks
fluorocarbons	good chemical and electrical properties, protective coatings, release coatings, barrier coatings
phenolic	work-horse thermoset for adhesives; parts
polyimide	high temperature applications, glass-reinforced PC boards, flexible film and cable, interlayer dielectrics
polyurethane	tough, abrasion resistant for casting, potting, and encapsulation of connectors and modules; conformal coatings for PC assemblies
polyvinyls	chlorides, fluorides, vinylidene chlorides and fluorides, vinyl aldehydes, and polystyrene; used for moisture barriers, primary-wire insulation, corrosion protection, dielectric impregnants, and baking enamels
polyxylylene	conformal coatings for PC assemblies; insulation

silicone	available as solvent solutions, room-temperature vulcanizable (RTV) rubbers, and solventless resins; used as insulation for high temperature and high voltage parts

Components

Printed Circuit Functional Coatings. Coatings are used in all aspects of printed circuits, from the base substrate materials, through fabrication, to final protection/encapsulation of the finished product.

Screenable Resists. Screenable resists or inks (qv) are applied to the metal-clad substrate through a silk (qv), nylon, or stainless steel screen on which a circuit pattern has been defined. Organic soluble resists have almost entirely been replaced by aqueous alkali strippable resists.

Screenable inks have a resin or polymer base and are of three types: organic solvent soluble, aqueous alkali soluble, and permanent. Primarily because of pollution requirements and higher solvent costs, the aqueous types have come into greater use. The permanent types are used as solder masks or for marking the boards. Uv-curable inks are also in use.

Photoresists. For high resolution circuit lines $\leq 125\ \mu m$ (5 mils) photosensitive resists are used. These can either be applied as liquids and dried or laminated as dried films supported on a polyester support. When exposed to light, typically ultraviolet radiation, these coatings change chemically, and their solubility to certain solvents or developers changes. There are two types of photoresist: negative-acting and positive-acting. Negative resists typically consist of a mixture of acrylate monomers, a polymeric binder, and a photoinitiator. Upon uv exposure through a mask, the resist polymerizes and becomes insoluble to the developer. Unexposed areas remain soluble and are washed away. Positive resists function in the opposite way with exposed areas becoming soluble in the developing solvent. As for screenable inks, photoresists are available commercially both as solvent or aqueous developable, as well as permanent.

Protective Coatings. Solder masks are resists designed to define or mask areas of a finished printed circuit board where components are to be attached by selective deposition of molten solder (see SOLDERS AND BRAZING ALLOYS). In contrast to etching or plating resists, the cured solder mask film remains permanently.

Many electronic component assemblies are inexpensive and designed to be nonrepairable and throwaway. For these, if a protective coating is used at all, it is an inexpensive varnish or polyester. For more expensive modules, specialty coatings are used to ensure long-term reliability. These are typically acrylic, epoxy, polyurethane, silicone, or polyxylylene in composition and can be applied by numerous methods, eg, dip-coating, spraying, fluidized-bed coating, casting, or vacuum deposition.

Hybrid Microelectronic Coatings. Hybrid circuits are fabricated using two complementary technologies: thick- and thin-film printed circuits.

Thick-film technology is an additive process and involves screen printing of functional inks, also called pastes or simply compositions. The paste formulation consists of a high (20,000−250,000 mPa·s(= cP)) viscosity thixotropic dispersion of fine particles of an electrically functional phase and a glass and/or ceramic frit binder. Also included for screen printing are a polymeric binder, a solvent or vehicle, and modifiers such as wetting and flow-control aids.

Polymer thick films also perform conductor, resistor, and dielectric functions, but here the polymeric resins remain an integral part after curing.

Thin-film technology offers the highest circuit/feature resolution, down to micrometer level geometries or even below, and film thicknesses that range from 0.01 to several micrometers (see THIN FILMS). Although used in hybrid manufacturing, these processes are more typically found in the fabrication or manufacture of microelectronic, ie, integrated circuit devices. Coating is primarily done by either of two vacuum techniques: evaporation or sputtering.

Integrated Circuit Coatings. The integrated circuit (IC) is the basic building block in microelectronics. It is a semiconductor device having both passive and active components built both into and onto a silicon wafer.

Polyimides, both photodefinable and nonphotodefinable, are coming into increased use. Applications include planarizing interlayer dielectrics on integrated circuits and for interconnects, passivation layers, thermal and mechanical stress buffers in packaging, alpha particle barriers on memory devices, and ion implantation (qv) and dry etching masks.

A number of polymeric coatings, primarily polyimides, epoxies, and silicones, are used as adhesives and in other aspects of packaging, as well as for final encapsulation and protection of the integrated circuit.

Coating Application Methods

There are a wide variety of coating applications processes in use. The majority of these techniques are similar to those used in other coatings industries, and the same basic operating principles apply to these uses as to coating a photographic film or a coil of metal for a refrigerator.

In addition, cleanliness is far more important at all stages of the electronics coating process than generally necessary for nonelectronics coatings. There is also a wider range of coating compositions used, and because most of the processes are fluid coating application processes, many diverse solvents are in use. This contrasts with the paper and photographic industries where water is almost always the exclusive solvent used. The wide use of high volatility solvents in a coating application is of concern, however, because of the increasing environmental needs.

EDWARD COHEN
HOWARD E. SIMMONS III
DuPont Central Science and Engineering

J. E. Sturge, V. Walworth, and A. Shepp, eds., *Imaging Process and Materials*, 8th ed., Van Nostrand Rheinhold, New York, 1989, p. 252.

J. J. Locari and L. A. Hughes, *Handbook of Polymer Coatings for Electronics*, 2nd ed., Noyes Publications, Park Ridge, N.J., 1990, p. 224.

C. F. Coombs, Jr., *Printed Circuit Handbook*, 2nd ed., McGraw-Hill Book Co., New York, 1979, pp. 6-9–6-31.

D. S. Soane and Z. Martynenko, *Polymers in Microelectronics: Fundamentals and Applications*, Elsevier, Amsterdam, the Netherlands, 1989.

ELECTROPHOTOGRAPHY

The combination of electrostatic charging and development using the phenomenon of photoconductivity, first discovered in selenium crystals in the 1870s, led to the invention of electrophotography. Photoconductivity, the enhancement of the electrical conductivity of a material by illumination with light, allows the image of an entire document to be copied by simply projecting its optical image onto a page-size photoconductive layer, known as the photoreceptor, uniformly charged with ions. Reflected light from the document then produces selective photodischarge in the photoreceptor proportional to the incident light intensity. The resultant image, consisting of the remaining surface charge, replicates the information content of the document. This latent electrostatic image can then be developed by utilizing its electrostatic attraction for charged powder. The highly photosensitive amorphous form of selenium was discovered in the mid-1940s. Other improvements involving ion charging processes, electrostatic transfer, and dry-ink or toner materials followed. Commercialization became feasible and the Greek words for dry, xeros, and writing or drawing, graphein, were combined to give the name xerography and Xerox as a trade name.

The heart of a xerographic machine is the photoreceptor which consists of a thin film, 20 to 50 μm thick, coated onto a grounded aluminum drum or metallized belt (see THIN FILMS). The film material must have a high ($>10^{12}$ Ω·cm) resistivity in the dark and yet be photosensitive on exposure to visible light. The photoreceptor drum or belt is mechanically rotated with respect to the other necessary subsystems. The first of these is a corona charging device, such as a corotron, which produces either positive or negatively charged air ions depending on the polarity of a high voltage applied to a wire. These ions are accelerated by the applied voltage and deposited on the surface of the photoreceptor, which thus charges as a parallel plate capacitor. The resultant voltage across the film is proportional to the number of ions deposited per unit surface area and to the inverse of the material's dielectric constant. Typically, voltages of a few hundred volts are used to provide the necessary development fields. This surface potential, divided by the film thickness, determines the applied electric field within the photoreceptor. The charged area of the photoreceptor then rotates, with minimal loss in the initial voltage because of its high resistivity in the dark, into the exposure position.

In the exposure position an optical image, produced by scanning the document to be copied with a slit light source, is focused by lenses onto the photoreceptor surface. Where light reflected from white areas of the document strikes the photoreceptor, photogeneration of free electrons and holes occurs within the absorption depth of the light, which is typically much less than the film thickness. The dark characters on the original document reflect less light and proportionally fewer photocarriers are created. Assuming the initial voltage is positive, photogenerated holes move across the photoreceptor film and electrons move the shorter distance back to the top surface under the influence of the electric field within the photoreceptor. This causes the initial voltage to selectively discharge to a degree depending on the amount of light locally absorbed by the photoreceptor. In laser printers or digital copiers, the optical image is directly written on the photoreceptor surface by a scanning laser or LED array where the light output is controlled in an on–off fashion using digital electronic signals.

Leaving the exposure zone, the photoreceptor now carries an invisible distribution of surface charge or voltage which replicates the document image. This latent electrostatic image is then developed by cascading charged dry ink or toner, consisting of polymer particles containing carbon black or colored pigment, over the photoreceptor surface. The toner charge which is produced by controlled triboelectric charging is opposite in sign, negative in this example, to the original charge deposited on the photoreceptor. The negatively charged toner particles are attracted by mutual coulomb electrostatic force to the remaining positive surface charges of the latent electrostatic image, and become attached to the photoreceptor surface.

After leaving the development zone, the original document charge image is now a visible toner image. This is next transferred from the photoreceptor to plain paper by using a transfer corotron. This charges the paper with ions of the same polarity as that used to charge the photoreceptor, and toner now moves from the photoreceptor onto the paper. Because only a fraction of the toner is transferred in this step, the remainder must be removed from the photoreceptor surface, for example, by mechanical brushing. Finally, any residual latent-image surface charge must be discharged. This is done by uniformly exposing the entire photoreceptor to light from an intense erase lamp. After the transfer step, the paper moves to a fixing or fusing station where the toner image is softened using heat. In this way the toner adheres to the paper, the image is fixed, and the copy is ready. At this point the process cycle is completed and can automatically restart. In modern high speed xerographic products, these steps occur automatically to produce over 100 copies or prints per minute.

Photoreceptor Materials

The photoreceptor employs photoconductive materials and basically serves the same function as silver halide film in conventional photography (qv). The photoreceptor is designed to transform an optically input image into an electrostatic latent image on the photoconductor surface. The photoconductor remains charged if unilluminated, or even when exposed to unabsorbed wavelengths of light, discharging in the regions where light is absorbed.

The most desirable characteristics of photoreceptors for use in electrophotography are the following:*(1)* The ability to accept and hold the electrostatic charge in the darkness.*(2)* The photoreceptor must lose most of its surface charge upon exposure to the white light reflected from the original or upon exposure to an attenuated laser beam.*(3)* The photogenerated charge carriers should exit the device in a time shorter than the time between exposure and the development of the latent image.*(4)* The photoreceptor must be stable in performance.*(5)* The photoreceptor should be mechanically robust.*(6)* Lastly, the photoreceptors must be inexpensive and easy to fabricate into defect-free, large-area thin films with uniform thickness of all layers.

Photoconductors. Only the most recently developed organic photoconductors (OPC) approach optimum behavior.

Amorphous (vitreous) selenium, vacuum-deposited on an aluminum substrate such as a drum or a plate, was the first photoconductor commercially used in xerography. It is highly photosensitive, but only to blue light. Its light absorption falls off rather rapidly above 550 nm. Because of the lack of photoresponse in the red or near infrared regions, selenium photoreceptors cannot be used in laser printers having He–Ne lasers (632.8 nm), or solid-state lasers (680–830 nm).

The useful spectral range of selenium can, however, be extended. For example, the addition of substantial amounts of tellurium shifts the absorption tail to the red or eventually to the near infrared. Se–Te alloys in layered configurations have been used in a number of copiers.

Another disadvantage of selenium is that it crystallizes readily when heated above about 55°C. Crystallization can be retarded by the addition of small amounts of arsenic (several wt % based on selenium). The addition of larger amounts of As also extends the spectral response into the red region. Selenium alloy photoreceptors having about 40 wt % As (nearly stoichiometric As_2Se_3) are highly sensitive throughout the visible region of light and have been used in high speed laser printers in conjunction with the He–Ne lasers.

Zinc oxide was also quite successful in xerographic photoreceptors, particularly in Electrofax. Other photoconductive pigments (qv) which found commercial application are cadmium sulfide, CdS, or the alloy with Se, CdS_xSe_{1-x}.

Amorphous Silicon. Significant effort has been expended in the development of an amorphous silicon photoreceptor. The main advantage of α-Si:H photoreceptors is not the reduction of thickness by wear and abrasion, but a catastrophic failure attributable to microscopic scratches caused by foreign objects, dust, or paper, or to corona damage. It is necessary to overcoat α-Si:H using a thin hard shell of nonstoichiometric silicon carbide, SiC, or silicon nitride, Si_3N_4. Also, because the α-Si:H photoreceptors are charged positively, spontaneous dark injection of electrons from the conductive substrate must be prevented because such dark-injected electrons would neutralize the surface charge before the latent image could be developed. Electron injection is blocked by a thin layer of α-Si:H heavily doped with boron. As shown in Figure 1 about 1000 ppm boron is inserted between the electrode and the bulk of α-Si:H. Boron doping reduces the electron range.

α-Si:H has sufficient sensitivity to He–Ne lasers (632.8 nm) but lacks the sensitivity to solid-state lasers (780–830 nm) (see LASERS). The spectral response of α-Si:H can be extended to the near ir range by adding germanium. This is accomplished by codeposition of silane

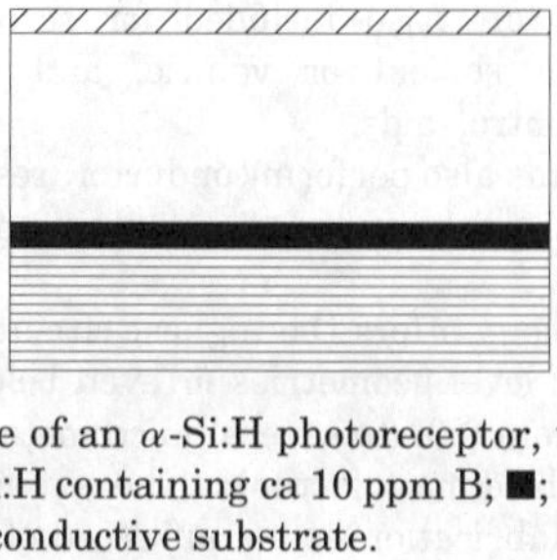

Figure 1. Structure of an α-Si:H photoreceptor, where ▨ represents SiC_x or SiN_x; □, α-Si:H containing ca 10 ppm B; ■; α-Si:H with ca 1000 ppm B; and ▤, the conductive substrate.

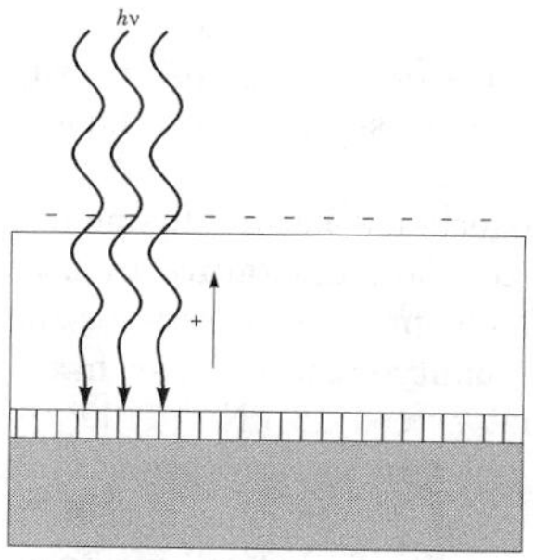

Figure 2. Schematic of an organic layered photoreceptor, where the − signs represent the corona-deposited charge, which is typically negative; □, the CTL; ▥, the CGL; and ■, the conductive base. Terms are defined in text.

in the presence of germane, GeH_4. Doping with Ge, however, causes an undesirable increase of dark conductivity. This effect is partially mitigated by forming a separate thin layer of α-Si:Ge:H in conjunction with Ge-free α-Si:H, as in the case of alloying selenium, Se, and tellurium, Te.

Organic Photoconductor-Based Photoreceptors (OPCs). Most modern xerographic copiers, duplicators, and printers use OPCs.

Most of the currently used OPCs are, however, two-layer photoconductors schematically illustrated in Figure 2. In these devices, the charge-photogeneration function is separated from the charge-transport function. This separation enables independent optimization of each layer. The charge-generation layer (CGL) is optimized for the spectral response and the carrier-supply efficiency (sensitivity), and the charge-transport layer (CTL) is optimized for wear resistance, high charge carrier mobility, flexibility, corona resistance, etc. The CGL is usually thin, between 0.3 and ~3 μm, ie, only thick enough to absorb most of the incident light. The CTL is typically 10–25 μm thick. The charge-generation layers are typically sandwiched between the conductive substrate, generally in drum or belt form, and the polymeric charge-transport layer which protects the CGL from the corrosive and abrasive xerographic environment.

Most of the known charge-transport layers are *p*-type or hole transporting. Thus this type of layered photoconductor must be charged negatively.

The CGLs contain highly absorbing photoconductive pigments, either dispersed in binder resins or vapor-deposited (sublimed) directly onto the conductive substrate. Photogenerating pigments and dyes that are most frequently mentioned in the literature are the phthalocyanines, perylenes, squaraine and azo-type pigments, pyrylium and thiopyrylium dyes and pigments, and inorganic pigments such as trigonal selenium in conjunction with an organic CTL. For imaging processes that employ visible, broad-spectrum light sources, the most popular are the bisazo dyes and pigments, primarily because these are relatively easy to synthesize, sufficiently sensitive, and easy to fabricate into photoreceptors. The two-layer photoreceptors may be fabricated by successive solvent coatings of CGL and CTL, although in some cases the CGL pigment is deposited directly by evaporation.

Charge-Generation Layers. For xerographic laser printers, photoconductors must respond to the wavelengths of coherent light emitted either by He–Ne lasers at 632.8 nm, or by solid-state lasers which emit in the near infrared (ir) region, between 650 and 830 nm. Four groups of compounds that show the near infrared response are derivatives of squaraine, phthalocyanines, bisazo and trisazo pigments, and thiapyrylium or pyrylium pigments. Among these groups of materials, phthalocyanines are most popular because of high sensitivity. Particularly sensitive appears to be the Phase 3 titanyl phthalocyanine which requires only ~0.25 $\mu J/cm^2$ of light energy at 5.3×10^5 V/cm to discharge the photoreceptor to one-half of the original potential of 800 V. This is roughly the same as the sensitivity of α-Se in white light. The issues plaguing ir-sensitive photoreceptors are not the lack of sensitivity, but rather the thermodynamic stability of the most sensitive form, excessive dark conductivity, and the stability of the xerographic performance in general.

The design of photogenerating materials is still largely empirical but some tentative design criteria appear to be emerging. The CGL must typically be in an aggregated or crystalline form to have high absorptivity and high photogeneration efficiency; have recognizable donor and acceptor groups separated by a conjugated system of double bonds or an aromatic structure to support polarization of the excited state; have reversible redox for stable operation; and have a broad absorption spectrum for light-lens imaging, or, if near ir-sensitivity is desired, must have extended π-conjugation.

Charge-Transport Layers. The charge-transport layer materials are largely responsible for the overall mechanical characteristics of the photoreceptors. These can be classified into four principal classes: (*1*) polymers having pendent transport-active groups, such as PVK, where the main chain can be vinylic, acrylic, or similar, and the active pendent group is typically a complex aromatic amine or a large polynuclear aromatic group such as anthracene; (*2*) molecularly doped polymers, which are essentially solid solutions of transport-active molecules (20–50 wt %) in tough polymers such as polycarbonate or polyester; (*3*) condensation polymers having transport-active groups acting as building blocks of the polymer chain and (*4*) polymers having a transport-active chain or at least segments of the chain, such as π-conjugated or σ-conjugated polymers represented by polyphenylenevinylene or polysilylenes.

Even though numerous patents describing all four classes have been issued, almost all existing OPCs now in use in copiers and xerographic laser printers employ the molecularly doped polymeric charge-transport layers.

The trend in the xerographic copying and printing industry is to expand the use of organic photoconductors, which are typically inexpensive and have a useful life. However, the α-Si:H photoreceptors, even though more expensive to manufacture, may ultimately have an advantage in greater durability.

Developer Materials

The Development Process. The physical basis of development, which may be simple in principle, is complex in practice. A uniform surface charge on a photoreceptor of infinite extent produces associated electric fields, the lines of force of which are confined within the photoreceptor, because these lines terminate on the induced charges of opposite sign in the conductive substrate. This is not true if the photoreceptor has finite dimensions or if a nonuniform distribution of surface charge exists, because discontinuities exist in the surface charge density. At these discontinuities, force lines extend out of the photoreceptor into space. These so-called fringe-fields, associated with nonuniform distribution of surface charge produced by the imagewise exposure of a photoreceptor, though quite strong, are confined to a narrow region at such a boundary. External to the photoreceptor, these fringe-fields determine the development field E_D and the electrical force, $F_D = Q \times E_D$, on a toner particle having total charge Q. Because such fields at the center of solid areas are very small, fringe-field development is such that only narrow lines or edges of large-area black images are adequately developed. This problem was overcome by using a biased-development electrode positioned close to the photoreceptor surface so that the lines of force are pulled out of the photoreceptor, even if the receptor is uniformly charged.

The development fields, which fall off rapidly over distances of micrometers moving away from the photoreceptor surface, exert an electrostatic force on a triboelectrically charged particle of these dimensions. The average toner particle is only 5–10 μm in diameter, and the challenge is to control and direct large numbers of such particles. The coulombic forces between charged toner and the charges on the photoreceptor surface are used, but to come under the influence of these electrostatic fields the toner must be brought within micrometers of the photoreceptor surface.

A variety of methods can be employed to achieve the required proximity, most involving an intermediary component called a carrier. These different development processes include powder-cloud, cascade, and magnetic-brush development. Liquid development, in which an insulating hydrocarbon liquid is used as the carrier, allows the use of much finer toner than can be typically controlled as a dry powder and therefore inherently offers the possibility of much higher resolution development. Powder-cloud or aerosol development uses air as the carrier. There is also the inability to manage the majority of the toner cloud, because it is only the toner within micrometers of the photoreceptor surface that experiences the development fields. For certain specialized applications, however, such development can be the process of choice. One example is in xeroradiography where small changes in transmitted x-ray intensity need to be detected.

Other processes use solid carriers for toner delivery, so that the developer consists of both toner and carrier, and is known as two-component development. The process called cascade development is characterized by a dependence on gravitational forces for its operation. In the predominantly used development process known as magnetic-brush development, the carrier is made of a magnetic material such as iron, so that magnetic forces are the dominant delivery force. A single carrier particle or bead, at about 100 micrometers in diameter, is much larger in size than typical toner and delivers the toner to the surface of the photoreceptor. It also controls and determines the sign and magnitude of the toner charge, for which purpose it is usually coated with a thin polymer film.

Toner. Almost all black toners consist of a thermoplastic polymer or polymer blend with carbon black as the pigment in a weight ratio of about 85% polymer, 15% carbon black. Styrene–acrylic, styrene–butadiene and polyester polymers and copolymers are largely used for negatively charged toners for use with positively charged photoreceptors such as amorphous selenium and its alloys; polyamides (qv), polyethylene, and ethylene–vinyl acetate copolymers are used as positively charged toners for polymeric-based photoreceptors, which are typically used with negative charge.

Carrier Materials. Carrier composition includes the core material and various carrier coatings that may be employed. A variety of carrier-core materials have been used such as spherical or nonspherical steel shot, sponge or porous iron, and ferrites. Processes used to produce carrier materials include omizing or electrolytic and chemical reduction methods to produce steel or iron carriers, or spraying techniques for ferrites. The carrier beads are typically coated with a thin polymer film to provide a uniform carrier surface, which also serves as a low surface energy layer to reduce impact adhesion of toner to the carrier, thus extending developer and photoreceptor life. These carrier coatings are also used to control the charging characteristics such as polarity and charge-to-mass ratio.

Imaging Media. Various physical properties of paper (qv) are critically important not only in the transfer process, but also in its stripping from the photoreceptor surface, high speed transport through the machine, and in the fusing step. Paper, an interpenetrating network of cellulose fibers, possesses a significant amount of free volume occupied by air, whereas xerographic-grade paper also contains fillers to enhance its mechanical strength and smoothness, and to minimize the shedding of fibers. All of these factors are important in minimizing paper jams.

Fusing. As the toner image is transferred from the photoreceptor to the paper, the image must be permanently bonded to the paper, rather than attached only by electrostatic and dispersion forces. Otherwise the toner images are easily smudged. In the fusing step, toner particles are made to coalesce with each other, and interpenetrate the paper fibers. The thermoplastic toner may be softened or melted using heat, pressure, or a combination of both. This must happen using minimum energy both for economic reasons and to avoid damaging the paper substrate. The energy needed to soften a thermoplastic is related to rheological properties, specifically the glass-transition temperature, T_g (see RHEOLOGICAL MEASUREMENTS). T_g, the temperature at which the polymer chains of the thermoplastic acquire sufficient thermal molecular motion that it begins to soften, is also characterized by a rapid drop in viscosity through the transition. Toners are generally designed to have glass-transition temperatures lying between 50°C and 65°C.

Modern copier products use hot-roll fusing in which a combination of pressure, shear force, and temperature are used.

An alternative fixing process is cold-pressure fixing. This uses steel rollers of high quality surface finish having nip pressures between the rollers that exceed 18 kN/m (100 ppi). The temperature rise associated does not lead to significant softening of the toner so that the fixing is mostly a consequence of cold flow. This means that softer toners than those used in hot-roll fusing must be employed, eg, polyethylene-type polymers.

The Cleaning Step. The cleaning process employs direct physical contact with the toner and photoreceptor. One approach involves the use of brushes; the second uses a blade.

MILAN STOLKA
JOSEPH MORT
Xerox Corporation

J. Mort, *The Anatomy of Xerography*, McFarland & Co., Publishers, Jefferson, N. C., 1989.

R. M. Schaffert, *Electrophotography*, Focal Press, New York, 1975.

E. M. Williams, *The Physics and Technology of Xerographic Processes*, Wiley-Interscience, New York, 1984.

ELECTROPLATING

Electroplating is the process of applying a metallic coating to an article by passing an electric current through an electrolyte in contact with the article. The ASTM adds some quality restriction by defining electroplating as electrodeposition of an adherent metallic coating on an electrode such that a surface having properties or dimensions different from those of the basis metal is formed.

Progress in electroplating is linked to improvements in materials of construction, power supplies and other plating equipment, purer industrial chemicals and anodes, and improved analytical test and control methods. The quality of electroplating is dependent on the basis metal surface. Cleaner, less porous castings and better casing alloys, and improved steel (qv) and steel finishes have helped significantly.

Materials

Surfaces. Essentially any electrically conductive surface can be electroplated, although special techniques may be required to make the surface electrically conductive. Many techniques are used to metallize nonconductive surfaces. These are well-covered in the literature and can range from coating with metallic-loaded paints or reduced-silver spray, to autocatalytic processes on tin–palladium activated surfaces or vapor-deposited metals. Preparation steps must be optimized and closely controlled for each substrate being electroplated.

Although metals and alloy substrates account for much of the volume in electroplating, there is a large and growing amount of plastic surfaces being plated, both for decorative trim and for electronic shielding applications. On a smaller scale, other materials that are plated include wood (qv), plaster, fibers (qv) and cloth materials, and plant and animal tissue, such as leaves, leather (qv), paper (qv), and seashells.

Electroplated Metals and Alloys. The metals electroplated on a commercial scale from specially formulated aqueous solutions include cadmium, chromium, cobalt, copper, gold, indium, iron, lead, nickel, platinum-group metals, silver, tin, and zinc. Although it is possible to electroplate some metals, such as aluminum, from nonaqueous solutions as well as some from molten salt baths, these processes appear to have achieved little commercial significance.

In addition to the metals listed above, many alloys are commercially electroplated: brass, bronze, many gold alloys, lead–tin, nickel–iron, nickel–cobalt, nickel–phosphorus, tin–nickel, tin–zinc, zinc–nickel, zinc–cobalt, and zinc–iron. Electroplated alloys in lesser use include lead–indium, nickel–manganese, nickel–tungsten, palladium alloys, silver alloys, and zinc–manganese. Whereas tertiary and many other alloys can feasibly be electroplated, these have not found commercial applications.

Composites. Another type of electrodeposit in commercial use is the composite form, in which insoluble materials are codeposited along with the electrodeposited metal or alloy to produce particular desirable properties. Polytetrafluoroethylene (PTFE) particles are codeposited with nickel to improve lubricity (see LUBRICATION AND LUBRICANTS). Silicon carbide and other hard particles including diamond are co-deposited with nickel to improve wear properties or to make cutting and grinding tools (see CARBIDES; TOOL MATERIALS).

Economic Aspects

Electroplating is done both in job shops, where a customer's work is plated, and in captive (in-house) shops. There were reported to be about 7500 plating plants in the United States in 1992. This is a decrease from the ca 12,000 reported by the same source in 1980. The reduction, particularly in the number of smaller job shops, is related to the problems in meeting the waste regulations imposed on plating shop effluents.

Uses

Electroplated materials are generally employed for a specific property or function. There is, of course, some overlap; for example, decorative use certainly requires some degree of corrosion resistance. The various usages and the principal plating metals employed are as listed. There are also smaller amounts of other metals and alloys used for specific applications.

Property/function	Principal plating metals
decorative	chromium, copper, nickel, brass, bronze, gold, silver, platinum-group, zinc
corrosion resistance	nickel, chromium, electroless nickel, zinc, cadmium, copper and copper alloys
wear, lubricity, hardness	chromium, electroless nickel, bronze, nickel, cadmium, metal composites
bearings	copper and bronze, silver and silver alloys, lead–tin
joining, soldering, brazing, electrical contact resistance, conductivity nickel, electroless nickel, electroless copper, copper, cadmium, gold, silver, lead–tin, tin, cobalt	
barrier coatings, antidiffusion, heat-treat, stop-off	nickel, cobalt, iron, copper, bronze, tin–nickel
electromagnetic shielding	copper, electroless copper, nickel or electroless nickel, zinc
paint/lacquer base, rubber bonding	zinc, tin, chromium, brass
manufacturing; electroforming	copper, nickel
manufacturing; electronic circuitry	electroless copper, copper, electroless nickel, nickel
dimensional buildup, salvage of worn parts	chromium, nickel, electroless nickel, iron

Functional Plating. *Plating Method.* Materials such as strip steel are plated on machines where coils of steel are unrolled on a continuous, high production basis, fed through a sequence of preparation steps, and into the plating tank(s) on a series of rolls and rollers. Short plating times, high current densities, and relatively thin deposits are the rule.

Wire. In electroplating, wire is uncoiled from spools or reels, passed through the processing and plating steps as individual strands, and then recoiled. Several spools can be plated simultaneously through multistrand machines. Wire is plated commercially with several metals. Among these are copper and copper alloys, zinc, iron and iron alloys, nickel and nickel alloys, gold, and silver.

Stampings. Stampings, moldings, and castings are usually mounted onto specially designed plating racks to allow rigid and proper positioning of the part.

Barrel Plating. When the parts to be plated are small enough, bulk plating methods may be used. Although some work may be electroplated in dipping baskets, plating barrels are more effective. Plating barrels come in many shapes, sizes, and styles. All are capable of turning while going through the process, are shaped to tumble small parts so these are continually mixed, and are equipped with holes to allow for passage of current and exchange of solutions. Plating barrels are made of inert plastic materials with means for electrical contact to the parts and the d-c power source.

Brush Plating. When parts are large and only smaller areas of the part require plating brush plating is employed. Other terms used for this method include selective, contact, swab, and out-of-tank plating. Specially designed plating tools, which are essentially shaped-anode materials covered with some absorbent material saturated with a plating solution, are used.

Automation. Handling of racks or barrels through the process tanks in a plating line can be done manually, or it can be fully or partially automated. A variety of automated styles are used: a return-type conveyor system transports the racks or barrels from a loading station through the sequence of tanks without backtracking; the conveyor returns through the second half of the process cycle off to the side of the first half so that in no case does the work drip solution drainage into previous process tanks in the cycle. A straight-line cycle could provide the same sequence, but the load and unload stations would be widely separated and require additional load–unload equipment or additional personnel if manually loaded and unloaded. Another style tries to place process tanks such that those most sensitive to drippage problems and cross-contamination are placed in parts of the in-line layout where they would be least troublesome. The transport system is then programmed to take the work in a non-sequential order, skipping over stations and backtracking in order to complete the plating cycle.

Fundamentals of Electroplating

The essential components of an electroplating process are an electrode to be plated (the cathode); a second electrode to complete the circuit (the anode); an electrolyte containing the metal ions to be deposited; and a d-c power source. The electrodes are immersed in the electrolyte such that the anode is connected to the positive leg of the power supply and the cathode to the negative. As the current is increased from zero, a minimum point is reached where metal plating begins to take place on the cathode. The physics of this process has been the topic of many studies, and several theories havebeen proposed. A discussion of these theories can be found elsewhere.

Plate Thickness. In plating processes, plate thickness can be predicted knowing the cathode efficiency of a particular plating solution, the current density, and time of plating.

Plate thickness is an important factor in electroplating, in terms of both performance and economics. Corrosion resistance, porosity, wear, appearance, and several other properties are proportional to plate thickness. Minimum plate thicknesses are, or should be, specified as should the location, or check-point, where the thickness is to be measured. In some applications, such as threaded fasteners, maximum thicknesses should be specified. Root diameters of finer machine threads can be adversely affected by as little as 10 μm of plating.

Cathode Efficiency. Faraday's law relates the passage of current to the amount of a particular metal being deposited; ie, 96,485 coulombs, equal to one Faraday, deposits one gram-equivalent weight of a metal at 100% efficiency. The cathode efficiency, an important factor in commercial electroplating, is the ratio of the actual amount of metal deposited to that theoretically calculated multiplied by 100%.

Throwing Power. It has long been observed that plate thickness distribution does not always follow the primary current distribution. Throwing power, a term used to describe the relative plate thickness distribution, was initially defined as the improvement in percentage of the metal distribution ratio above the primary current distribution ratio. A test method for the measurement and mathematical expression of throwing power, called the Haring Cell test, is available. Plating solutions vary widely in ability to exhibit good throwing power. As current is increased in a given plating process, a point is reached where the metal ion being deposited is not replaced in the solution film nearest the surface fast enough and a concentration polarization occurs, shifting some of the current to the unpolarized, lower current density areas. The effect is to reduce the plating rate in these higher current areas.

A plating solution that exhibits a decrease in cathode efficiency with increasing current density exhibits better throwing power, ie, more even plate thickness distribution, than other solutions. Generally, the solutions having positive throwing power are those that contain some complexing agent for the metal ion being deposited.

Examples of plating solutions having good throwing power include cyanide plating baths such as copper, zinc, cadmium, silver, and gold, and noncyanide alkaline zinc baths. Examples of poorer throwing power baths are acid baths such as copper, nickel, zinc, and hexavalent chromium.

Covering Power. Covering power refers to the lowest current density area where plating appears on a part. There are qualitative and semiquantitative tests for covering power, but no standard method is described by ASTM. One method simulates plating a tubular part by rolling a metal foil into a tubular shape, plating the rolled foil, unrolling, and measuring the distance that the plating covered on the inside of the foil.

Current Density. For a given electroplating process, as the current density is increased on the work, the point where the deposit becomes rough, coarse-grained, and takes on what is termed a burned and generally unacceptable appearance is eventually reached. The range from the minimum covering power to just below this burn producing current is the usable current density range. In electroplating, this range is not the same for all metal plating baths, and is influenced by metal ion concentration, chemical compositions, temperature, agitation, and anode-to-cathode spacing, as well as by the shape of the work and how it is positioned. In production plants, it is usually preferred to plate at the highest possible current density to shorten production time. This is not always done, however, especially when using lower throwing power plating baths, because increasing current decreases throwing power even further. For parts that require more even deposit thicknesses, lower current densities and longer plating times are used.

The Plating Tank. Materials of construction for the plating tank and the required auxiliary equipment are chosen either to be totally inert to the plating solution, or lined with inert material to protect the tank. In all cases, it is preferable to have a nonconductive inner surface of the tank to provide electrical as well as chemical resistance. A more recent requirement in many states is a containment surrounding the plating tank area to prevent accidental chemical spills from entering the environment (see TANKS AND PRESSURE VESSELS). For alkaline plating solutions, mild steel materials are used; in acid plating solutions, other materials are used, depending on the chemical composition of the plating bath. Titanium and various stainless steel alloys, polytetrafluoroethylene (PTFE), Karbate, Hastalloys, zirconium alloys, and others are among the choices.

Sizing. There is no limit to plating tank size. Commercial electroplating lines have varied from liter-size tanks in benchtop wire plating lines to those holding 100,000 L and more.

Busbars. Fitting the tank for d-c power is usually accomplished using round copper busbars, both for supporting the anodes and the work or cathodes.

Filters. Continuously filtered plating solutions are becoming much more common, except for hexavalent chromium plating solutions. Baths are filtered to remove any fine particulate matter that could codeposit with the electroplate and cause roughness.

Heating and Cooling. In some plating tanks heating or cooling is required preferably using a means of automatic control. Heating coils, plate coils, and other forms of immersion heaters are placed along a tank wall, preferably in an area that has good solution agitation. Heating can also be accomplished using external heat exchangers where the solution is pumped from the plating tank.

Cooling is required in some plating solutions. Cooling can be accomplished using cooling water coils, if demand is not high, or refrigeration units, commonly called chillers.

Temperature control is important to all electroplating baths where good quality is required, because many properties are affected with only a few degrees of temperature change.

Anodes. There are two types of anodes: soluble and insoluble. Most electroplating baths use one or the other specifically; however, a few baths use either or both. Soluble anodes are designed to dissolve efficiently with current flow and preferably, not to dissolve when the system is idle. A plating solution having the anode efficiency close to the cathode efficiency provides a balanced process that has fewer control problems and is less costly. If the anode efficiency is much greater than the cathode efficiency and there are only small solution losses, the dissolved metal concentration rises until at some time the bath has to be diluted back or the excess metal has to be reduced by some other means. If the anode efficiency is less than the cathode efficiency, the dissolved metal decreases, pH decreases, and eventually metal salt additions and other solution corrections are required. Based on the cost of metal, it is usually considerably more economical to plate from the anode rather than add metal salt.

Faraday's law applies to the anode as well as to the cathode; ie, the total reaction at the anode is proportional to the current, and much like the cathode, the anode efficiency varies with the current density. As the current on the anode is increased, the anode efficiency decreases, slightly at first, until it reaches a point at which the anode metal cannot dissolve fast enough through the anode film. The first stage of dissolution for the soluble anode is the oxidation of the metal followed by dissolution of the oxide. When the oxide dissolution rate is less than the oxidation rate, polarization of the anode takes place. The oxide film builds up in sufficient thickness to form an insulating coating, and the current decreases rapidly. The thick anode films can dislodge at the reduced current and remain as particulate matter until they slowly redissolve. As particulate matter, the dislodged anode films form a significant source of roughness because they can codeposit with the metal on the work before redissolving. While the anode is polarized, oxygen and other gasses can be given off from the anode. Solutions containing chlorides or bromides can emit chlorine or bromine, which are hazardous. Anode current densities should be maintained well below the point at which polarization occurs. This can be accomplished by increasing the anode area and lowering the current.

Agitation of the solution around the anodes and some chemical variations can increase the anode current density range before polarization. Some platers use anode bags to prevent particulate anode and anode film material from reaching the work. However, the solution flow through the bag material may be inadequate and add to the polarization problem. The weave opening of the bag material is important to maximize solution flow without passing excessive particulates. Anode bags must be replaced once plugged.

Insoluble anodes are used exclusively in some plating baths. Chromium plating solutions utilize lead–tin, lead–antimony, or lead anodes. Gold and other precious metal plating processes use stainless steel anodes, keeping inventory costs down.

Whenever insoluble anodes are used, the pH of the plating solution decreases along with the metal ion concentration. In some plating baths, a portion of the anodes is replaced with insoluble anodes in order to prevent metal ion buildup or to reduce metal ion concentration.

The use of insoluble anodes can also result in side effects. In alkaline cyanide solutions, the generation and buildup of carbonates is accelerated remarkably, along with a significant reduction in alkalinity. In acid solutions the pH decreases as well, requiring frequent adjustments. In sulfamate nickel plating solutions, insoluble anodes, and even slightly passive soluble anodes, partially oxidize the sulfamate ion to form sulfur-bearing compounds which change the character and performance of the deposit.

d-c Power Supplies. The Daniell cells and batteries of the mid-nineteenth century, replaced by motor-generator sets in the first half of the twentieth century, have given way to solid-state silicon rectifiers. A variety of modifications in power supplies has become available. Stepless controls, constant current, and constant voltage are popular. Automatic current manipulations are also getting more use. Current interruption cycles where the current is periodically discontinued have been used in alkaline baths. The newer forms of current manipulation are called pulse cycles because the cycle is typically on the order of milliseconds. A number of variations are possible.

The d-c output from a rectifier can have a portion of modulation from the a-c input, called ripple. Modern rectifiers usually have 5–10% or less ripple at full load. High ripple can have some bad effects, especially in chromium plating where ripple can cause dullness and poor coverage. In all cases, the d-c power supply needs a means of control, stepless preferred, and appropriate ammeter and voltmeter. Ampere-hour recorders are convenient for making additions and keeping records.

Preparation for Plating

Preparation cycles vary depending on the particular substrate being electroplated. High carbon steels, low carbon steels, free-machining steels (lead or bismuth), HSLA steels, stainless steels, cast-iron, and high nickel alloys all require special considerations in preplate treatments, as do the many nonferrous alloys and various plastic substrates. Often only slight variations in substrate composition significantly influence the preparation process. Heat treating variations also contribute to the complications in preparation. Determining an optimum preparation process for a given material often becomes a matter of trial and error. Some guidelines can be found in the literature and from suppliers of proprietary preparation products. Examples of ASTM guides are (B253) *Preparation of Aluminum Alloys for Electroplating*; (B322) *Cleaning Metals Prior to Electroplating; Practices for preparation of and electroplating on*: (B630) *Chromium (Electrodeposits) on Chromium*; (B281) *Copper and Copper-Base Alloys*; and (B727) *Plastic Materials*.

Poor preparation of the substrate can result in loss of adhesion, pitting, roughness, lower corrosion resistance, smears, and stains. Because electroplating takes place at the exact molecular surface of a work, it is important that the substrate surface be absolutely clean and receptive to the plating. In the effort to get the substrate into this condition, several separate steps may be required, and it is in these cleaning steps that most of the problems associated with plating arise.

Cleaning Methods. Some plating baths can clean surfaces and thus tolerate minimally cleaned surfaces; others need surfaces cleaned to near perfection. Chromium plating from chromic acid solutions is a good example of the former. Another plating bath that possesses good cleaning properties is the high cyanide zinc plating solution.

The majority of plating baths require excellent cleaning. Substrate surfaces have to be free of even the slightest of soil-containing films or oxides as the work enters the plating solution. No simple, universal cleaning cycle exists for electroplating. Several methods and cleaning solutions may be used in a single-plating process. Cleaning methods include mechanical cleaning, solvent cleaning and vapor degreasing, diphase cleaning, spray cleaning, ultrasonic cleaning, soak cleaning, acid-soak cleaners, electrocleaning, and alkaline derusting.

Soak Cleaning. A common visual test to determine cleanliness of the work is to observe how the surface rinses; if the surfaces drain evenly when rinsed using clean water, holding a sheet of water, it is said to be water-break-free. If rinse water beads up in droplets on the part surface or does not hold a sheet of water as it drains, it is said to water-break and is in need of additional cleaning. However, if this test is run on surfaces having surfactant-solubilized oil films, the water-break test is not definitive. Thus, the part has to be thoroughly rinsed, acid dipped in clean acid, and thoroughly rinsed again before using the water-break test. Neither does the water-break test reliably detect inorganic films. Thus, the water-break test indicates when a part is dirty, but not necessarily when it is clean.

Cleaner formulations can be classified with respect to the ability to emulsify oils, keep them in solution; or to reject oils, split them out of solution.

Cleaning Cycles. Cleaning cycles usually start with a soak cleaner followed by electrocleaning. If work is heavily soiled, however, a pre-cleaning step may be required. Buffing compound residues fall into this class. Solvents, diphase mixes, power washers, and other presoaks are used. If instead work is received in a fairly clean state, the high surfactant soak cleaner is omitted from the cleaning cycle. The electrocleaner formulations have less tendency toward leaving substantive films and is preferred. In some plating lines, there are more problems removing the soil-remover than in removing the soil.

Combination soak-and-electrocleaner products are touted for use when normal soak cleaner is too much and a normal electrocleaner is too little. These combination cleaners are often electrocleaners with additional surfactants. The results are often hard-to-rinse electrocleaner or a low capacity soak. The term heavy-duty, often used to describe cleaners, can refer to high caustic content or to good soil-removing, high soil-load capacity.

Rinsing. Rinsing is extremely important to the overall operation of a plating line. Work traveling from a processing solution carries a film on its surface from that solution. This amount of drag-out varies with the shape and surface area of the part, along with the nature of the solution. Drag-out can reach considerable quantities, eg, from 20–80 mL/m^2.

To reduce costs of the more expensive plating solutions and decrease the amount of hazardous or regulated material in a waste stream, recovery and reuse of the drag-out is a common practice. This is done by closing off the water flow to the first rinse tank following the process tank, and periodically returning the accumulated solution to the process tank.

Recycling drag-out losses can have the unwanted effect of concentrating impurities in the plating tank. Thus, it is good practice to purify the drag-out before returning it to the process tank. Filtration (qv) through activated carbon is helpful on most nonchromium solutions; cation exchange has been useful with chromic acid drag-out solutions, before it is evaporated.

Good rinsing requires good water, not too cold, vigorous agitation, and time.

Dual- or triple-rinse tanks follow most process solutions. Additionally, many lines utilize counterflow or cascade rinsing with the multiple tanks plumbed so freshwater flows from one tank to the next counter to the direction of the work. This method of conserving water is recommended procedure for all plating lines.

Spray rinsing can be efficient because of the impingement; misting at the exit of a processing tank can reduce drag-out. Barrel plating lines need much more rinsing time and special consideration because of the relatively high drag-out rates.

Acid Dips and Acid Pickles. Acids are used to remove inorganic soils. A distinction is made between acid dips and acid pickles. Where there is no rust or scales, a dilute acid dip is used for activation and as a rinse aid. Caustic residues from cleaners are notoriously difficult to remove with water rinsing alone. By contrast, strong acid pickles are used to remove rust and scale.

The most commonly used acids in preplate processes are hydrochloric acid, HCl, more common for rust and scale removal, and sulfuric acid, H_2SO_4 (see HYDROGEN CHLORIDE; SULFURIC ACID AND SULFUR TRIOXIDE).

Electropickling. At times, electrolytic pickling is used; steel can be treated anodically in strong sulfuric acid, 70%.

Properties, Specifications, and Test Methods

Standard test methods are required to measure the properties of electroplated materials. Documents on plating specifications for many phases of the plating process are published by such organizations as the Federal government, the military, ASTM, ISO, SAE, etc. An excellent cross-index of these is available in the literature.

Environmental Aspects

No topic is as important to the plating industry as waste management, which has become the key to economic survival (see WASTES, INDUSTRIAL). Waste management includes recycling (qv), waste reduction, and waste treatment and disposal. Metal finishers (electroplaters) are regulated under a Federal EPA program, National Pollutant Discharger Elimination System (NPDES), often administered by the State.

The Resource Conservation and Recovery Act (RCRA), is the most comprehensive pollution control law for metal finishers. One requirement assigns cradle to grave responsibility for waste to the generator, ie, the plater. Unless the waste is made into another form or recycled as another product, the original generator is still responsible for future treatment or liability costs. Other requirements of RCRA include a generator's license, proper waste containers, labels, and manifest shipping documents, licensed haulers, record maintenance, and an annual waste minimization report. This act was expanded in 1984 and now includes everyone that generates more waste than 100 kg/mo. The Clean Water Act was sufficiently strengthened by the Water Quality Act of 1987.

For the foreseeable future, the plating industry has to expect tighter restrictions and more regulation of recovery and recycling operations (see also REGULATORY AGENCIES). Solid wastes, the sludge from waste treatment processes, have to pass stringent leaching tests to be allowed in landfills, and costs for the disposal of solid wastes are increasing dramatically. More recent legislation, the Clean Air Act of 1990, is concerned with air pollutants. This act restricts chromium in air exhausts from chromium-plating and chromic acid anodizing plants to 0.01–0.03 mg/m^3. The Pollution Prevention Act of 1990 stresses reduction of chemicals that could enter the waste stream before treatment.

Plating Bath Formulations

Formulations of plating baths can be flexible in some systems and very sensitive to variation in others. Many of the more recent changes have resulted from waste treatment and safety requirements. Besides the ability to deposit a coating having acceptable appearance and physical properties, the desired properties of a plating bath would include: high metal solubility, good electrical conductivity, good current efficiencies for anode and cathode, noncorrosivity to substrates, nonfuming, stable, low hazard, low anode dissolution during down-time, good throwing power, good covering power, wide current density plating range, ease of waste treatment, and economical to use. Few formulas have all these attributes. Only a few plating solutions are commercially used without special additives, but chemical costs often constitute a relatively low percentage of the total cost of electroplating. Additives are used to brighten, reduce pitting, or otherwise modify the character of the deposit or performance of the solution. Preferred formulations are normally specified by the suppliers of the proprietary additives.

Plating Bath Purification. Purification, often needed once a plating bath is made, is used periodically to maintain the plating solutions. Alkaline zinc plating solutions are sensitive to a few mg/L of heavy-metal contamination, which can be precipitated using sodium sulfide and filtered out. Nickel plating solutions may contain excess iron and unknown organic contaminants. Iron is removed by peroxide oxidation, precipitation at a pH of about 5, then filtered out. The more complex, less water-soluble organic contaminants along with some trace metals are removed with activated carbon treatments in separate treatment tanks.

Another common purification treatment used both on new and used plating solution is dummying. Heavy-metal impurities are removed by electrolyzing, usually at low current densities, using large disposable steel cathodes. Good agitation and lower pH speed the process.

Testing and Control. Analysis and testing are required whenever a new plating solution is made up, and thereafter at periodic intervals. The analyses are relatively simple and require little equipment. Trace metal contaminants can be analyzed using spot tests, colorimetrically, and with atomic absorption spectrophotometry (see TRACE AND RESIDUE ANALYSIS). Additives, chemical balance, impurity effects, and many other variables are tested with small plating cells, such as the Hull cell.

Table 1. Cadmium-Plating Solutions[a]

Parameter	Cyanide	Acid sulfate	Fluoborate
Cd metal, g/L	20–30	15–30	75–150
NaCN, g/L	90–150		
Na_2CO_3, g/L	30–60		
H_2SO_4, g/L		45–90	
NH_4BF_4, g/L[b]			60–120
NaOH, g/L	10–20		
pH	12.5–13.5	very acid	2.5–3.0
anodes	Cd and steel	Cd and graphite	Cd
temperature, °C	15–40	18–23 max	10–40
current density, A/m^2	50–900	300–800[c]	100–600

[a] Addition agents for all three plating solutions are as required.
[b] Some baths contain 20–30 g/L $NaBF_4$ and 20–30 g/L H_3BO_3 instead of NH_4BF_4.
[c] Current densities from 100–200 A/m^2 are used for barrel plating.

Individual Plating Baths

Cadmium. Compositions of cadmium plating baths are shown in Table 1. Whereas cadmium provides better corrosion resistance to steel and other substrates than zinc when exposed to marine environments, zinc is better in industrial, sulfur-bearing atmospheres. Resistance of cadmium is improved with chromate conversion coatings; bright, yellow iridescent, olive drab, and black finishes are used. Cadmium is readily solderable, provides lubricity on threaded fasteners, and has been used as a high temperature protective coating in the aircraft industry when diffused into nickel plating. Olive-drab chromated cadmium plate is used on nickel-plated aluminum electrical connectors to extend salt spray resistance. For diffusion processes, for low hydrogen embrittlement, and for plating directly on cast iron, cadmium–cyanide plating solutions are used without brighteners; in other applications, brighteners produce attractive lustrous deposits.

The future of cadmium plating is in question because of cadmium toxicity. Cyanide-free cadmium plating systems have experienced some growth. Acid cadmium, based on cadmium sulfate compositions, is replacing some cyanide baths in the United States. The fluoborate cadmium is reported in use in the U.K., especially in barrel plating. Cadmium plating is covered by ASTM, U.S. government, and ISO specifications.

Chromium. Applications of chromium plating can be separated into two areas: hard chromium, also called functional, industrial, or engineering chromium, and decorative chromium. The plating bath compositions may be the same for both. In most cases, the differentiating factor is plate thickness. Decorative chromium is usually less than about 1 μm; hard chromium can be from about 1 μm to 500 μm or more.

Table 2. Chromium-Plating Solutions[a]

Parameter	Conventional baths		Cocatalyzed baths	
	Range	Typical	Range	Typical
CrO_3, g/L	150–400	240–260	150–400	150–180
SO_4^{2-}, g/L	1.25–5	2.4–2.6	0.6–1.3	0.9–1.0
SiF_6^{2-}, g/L			0.3–0.8	0.5–0.6
$CrO_3 : SO_4^{2-}$ ratio	80–120:1	90–110:1	150–200:1	170–180:1
anodes	Pb–7% Sn or Pb–6% Sb		Pb-7% Sn	
current density, A/m^2	400–4000	1000–3000	400–4500	1000–3500

[a] Most decorative chromium baths are run at 38–43°C; hard chromium baths at 40–60°C.

Formulations and conditions for operating hard chromium plating solutions are shown in Table 2.

Chromic acid-based plating solutions differ from the other common metal-plating baths in that chromium solutions have poor current efficiencies, poor covering power, and poor throwing power, as well as a need to use much higher current densities to plate. Modifications developed in attempts to improve these weaknesses incorporate a second catalyst, and are called cocatalyzed baths. The second catalyst is often fluoride, or more commonly, the fluosilicate ion.

Impurities. Chromic acid plating solutions are affected by metallic impurities that usually are accompanied by chromium reduction increasing electrical resistance. Iron is a common impurity that increases rapidly when the chromium plating tank is used as the reverse current chromium etch tank. A separate chromic acid etch tank is used to extend the life of the plating solution. Trivalent chromium can be treated by electrolytic oxidation treatments at temperatures of about 80°C and high (5000–5500 A/m^2) cathode and low (2000 A/m^2) anode current density. Treatment using the electrolytic porous pot is also used to purify chromium solutions.

Other purification treatments include: cation-exchange treatments on diluted solutions followed by evaporation back to working bath strength; and electrodialysis directly in the working solution.

Cobalt. Cobalt plating is little used because most applications can be satisfied using nickel at considerably less cost. Cobalt has been used to coat tungsten carbide components that needed to be brazed and exposed to high temperatures in use; and as a barrier coating to inhibit migration of copper into gold. Cobalt is sputtered onto computer memory disks for its magnetic properties (see INFORMATION STORAGE MATERIALS). Cobalt is used in nickel electroforming solutions to increase strength and hardness. Cobalt plating solutions are so similar to nickel plating solutions in composition that in most cases the equivalent cobalt salts can be substituted for nickel salts in published nickel bath formulations.

Copper. Copper plating used as a final finish in some applications, is primarily employed as an undercoat for other deposits. Of the several types of copper baths, the two most popular in the United States are the acid sulfate and the cyanide baths (Tables 3 and 4).

Cyanide copper baths are a target for replacement, and alkaline, noncyanide baths are being advertised as replacements for cyanide copper. Care is needed in choosing a noncyanide copper system to avoid trading one waste treatment problem for another.

The acid sulfate bath is by far the most widely used copper plating bath, both for plating and for electroforming and electrowinning.

Brass, the first alloy to be electroplated, is a popular decorative finish. Plated as an undercoat for nickel–chrome, brass has given good corrosion protection compared to copper, especially on aluminum and zinc substrates. There has been considerable use of brass plating on steel wire and other shapes because of improved rubber bonding characteristics. Several copper–zinc alloys are commercially plated; the more common are yellow brass (ca 30% zinc), red brasses (10–15% zinc), and white brass (ca 72% zinc).

Bronze electrodeposits have been used for decorative applications as a final finish, with and without subsequent coloring or antiquing treatments. A clear lacquer helps preserve the deposit. On a limited basis bronze has also been used as an undercoat for nickel–chromium finishers on steel, zinc, and aluminum. In functional applications, plated bronze is used as a stop-off material in the selective nitriding of steel parts. It also finds good use as a nongalling bearing material.

Table 3. Cyanide Copper-Plating[a]

Parameter	Solution		
	Strike	Rochelle	High speed
Cu metal, g/L	15–22	22–36	56–71
CuCN, g/L	21–31	31–51	79–100
KCN			
total[b]	31–70	55–89	130–165
free[c]	10–15	12–18	20–25
KOH, g/L	3–18[d]	15–25	15–25
Rochelle salts, g/L	10–20[e]	30–45	35–55
Na_2CO_3, g/L[f]	10–15[e]	15–30[e]	15–30[e]
temperature, °C	42–49	54–60	60–70
current density, A/m^2			
cathode[g]	160–320	up to 270	up to 380
anode[h]	50–100	110–160	110–160

[a] Wetting agents and other additives are added as required.
[b] Total KCN = 1.4(CuCN) + free KCN; total NaCN = 1.1(CuCN) + free NaCN.
[c] Free KCN = surplus KCN over that required to complex the CuCN; the titratable KCN. NaCN can be substituted on an equimolar basis with the CN content. Current density range is less and throwing power is less but NaCN costs less.
[d] KOH in strikes for zinc die casting is kept below about 3 g/L; for Cu strike on zincated Al, sodium bicarbonate is added to lower pH to about 9.8–10.2.
[e] Optional additive.
[f] CO_3^{2-} forms in quantity and increases with use. Sometimes CO_3^{2-} is added to new baths to break them in.
[g] Current densities cited are average based on racked work, not Hull cell derived.
[h] Higher currents polarize anodes; current density is critical.

Table 4. Acid Copper-Plating Baths[a]

Parameter	Sulfate			Fluoborate	
	Typical	Average range	High throw	Low	High
Cu metal, g/L	57	38–64	15–23	60	120
$CuSO_4 \cdot 5H_2O$, g/L	225	150–250	60–90		
H_2SO_4, g/L	60	30–75	170–220		
Cl^-, g/L	0.05	0.20–0.12	0.05–0.10		
$Cu(BF_4)_2$, g/L				225	450
HBF_4, g/L				15	30
H_3BO_3, g/L				15	30
pH	<0	<0	<0	1.2–1.7	0.2–0.6
temp., °C	25	20–45	20–30	18	65
cathode current density, A/m^2	300–700	300–700	300–700	800–1400	1400–3800
anodes[b,c]		0.02– 0.08% PCu		OFHC Cu	OFHC Cu

[a] For all baths the anode to cathode ratio is 2:1; the baths are mechanically, hydraulically, or air agitated, and there is continuous filtration.
[b] Dynel or polypropylene anode bags are used for both types of anodes.
[c] OFHC = oxide-free high conductivity.

Gold. Electroplated gold is used both for decorative and industrial or engineering purposes. The important properties of industrial gold are corrosion resistance, low electrical contact resistance, and good solderability.

Gold is usually plated over electroplated (or electroless) nickel-plated substrates. A small amount of gold is used in electroforming with substantial thicknesses. Aerospace applications for both thin and thick deposits are newer uses for gold.

There are literally hundreds of gold-plating bath formulations; most of which are proprietary. Gold flash, gold strikes, soft golds, hard golds, bright golds, and golds of all purities and colors account for the wide selection. Formulations can be classified as alkaline, neutral, or acidic.

Indium. Indium diffuses readily into copper, silver, and lead, increasing hardness and imparting good antifriction properties. Indium plating is used for bearing materials and some is used in electronics for contact pads for flip chip bonding. Although indium can be plated from cyanide, fluoborate, and sulfate baths, the standard bath recommended is based on sulfamate.

Iron. Iron plating is not widely used. However, copper alloy soldering iron tips are plated with thick (175–200 μm) iron plating to prevent solder from dissolving the copper. Iron has also been used to plate stereotype printing plants, and in plating of aluminum pistons and cylinders in automotive engine blocks. Iron has been plated from chloride, sulfate, mixed chloride–sulfate, fluoborate, and sulfamate solutions; chloride is the most common.

Lead and Lead–Tin Alloys. Lead is plated from fluoborate solutions. Lead is used in battery parts for its good resistance to sulfuric acid. Low tin alloys of lead are used in bearings. Lead–tin alloys of 3–15% tin are called terne; for hot-dip coatings, terneplate can be 20% tin. Terneplate has been used to protect steel in gasoline tanks. Solder deposits are also plated. Lead–tin baths are similar to lead baths. Stannous fluoborate is added to supply the tin. Lead is under strict regulation in the environment and waste streams. The future of lead plating is uncertain.

Nickel. Nickel plating continues to be very important. Many plating baths have been formulated, but most of the nickel plating is done in either Watts baths or sulfamate baths. Typical bath compositions and conditions are shown in Table 5.

Table 5. Nickel-Plating Baths

Parameter	Watts	Watts high chloride	Sulfamate
Ni metal,[a] g/L	82	77	75
$NiSO_4 \cdot 6H_2O$,[b] g/L	300	135[c]	
$NiCl_2 \cdot 6H_2O$,[b] g/L	60	190	[d]
$Ni(SO_2NH_2)_2 \cdot 4H_2O$,[b] g/L			410
H_3BO_3,[e] g/L	35–45	35–45	35–45
pH	2–4	2–4	3.5–4.5[f]
temperature, °C[g]	50–60	50–60	40–60
cathode current density,[h] A/m^2	up to 430	up to 500	up to 550
anodes	Ni, electrolytic or S-bearing	electrolytic Ni	S-bearing Ni[i]

[a] As Ni is increased, the H_3BO_3 should be reduced.
[b] Nickel salts are commercially available as liquid concentrates.
[c] Proportions of Cl^- and SO_4^{2-} can be varied; total metal is usually maintained at 70–90 g/L.
[d] Cl^- is kept low in sulfamate, concentrations run from zero to about 3 g/L as Cl^-; $NiCl_2 \cdot 6H_2O$, $MgCl_2 \cdot 6H_2O$, and KCl have been used.
[e] H_3BO_3 is kept at near-saturation; solubility varies directly with temperature.
[f] Higher temperatures and lower pH should be avoided.
[g] Temperature should be kept constant to maintain H_3BO_3 level. Low (45°C) temperature Watts baths can be run if Ni is reduced to about 45 g/L total and H_3BO_3 is increased to about 50 g/L.
[h] Higher currents are attained using higher agitaion, higher Ni, and higher temperatures.
[i] If using other anodes, sulfamate ion can be oxidized.

Both Watts and sulfamate baths are used for engineering application. The principal difference in the deposits is in the much lower internal stress obtained, without additives, from the sulfamate solution.

Bright Nickel. Nickel as electroplated from pure nickel salts is a dull grey, satin-like deposit that has to be buffed to obtain a bright finish. Brighteners are used to obtain bright deposits directly from the bath. The additives currently used fall into two classes, which have variously been labeled primary and secondary, first class and second class, and carrier and brightener. The last is more commonly used in plating plants.

The additives that go into the nickel bath are almost always proprietary; carrier portions may include the sodium salts of benzene disulfonic acid, naphthalene 1,5-disulfonic acid, saccharin, and allyl sulfonic acid. Generally, these compounds are characterized by a $=C—SO_2$-group. Brightener–leveler portions include a large number of possible materials. A few are based on small amounts of zinc, cadmium, lead, selenium, or tellurium, etc, and many are based on organic compounds containing unsaturated groups such as C═O, C═C, C≡C, C═N, C≡N. Nickel brighteners and the theories of brightening are well-covered in patents and other literature.

Dual Nickel. Semibright nickel is a term given to sulfur-free systems. The more well-known of these were based on coumarin (qv). At first, these soft, hazy deposits were buffed bright and a layer of bright nickel deposited for appearance. It was later observed that better corrosion protection was obtained using this dual nickel system; not all of which could be attributed to the buffing. This improvement is ascribed to the sulfur-free nickel being much more noble than the more anodic top layer. Improvements in sulfur-free additives have resulted in much brighter deposits and the term semibright could be replaced by sulfur-free.

Triple Nickel and More. As an extension to the dual nickel, a thin, higher sulfur-containing nickel strike is deposited between the sulfur-free and the bright nickel. A fourth nickel deposit has shown improved protection by the effects is has on subsequent chromium deposits. Highly stressed, these nickel strikes have been used to aid in producing microcracked chromium.

Special Nickel-Plating Solutions. Black nickel for nonreflective, decorative, or solar absorptive uses is plated from nickel solutions containing ammonium, zinc, and thiocyanate salts. Nickel composites containing hard particles for wear, or abrasive particles for cutting and grinding tools, are plated from conventional Watts or sulfamate solutions to which the particles are added.

Nickel Alloys. Nickel alloys containing 10–35% iron are used in decorative applications as a cost-saving substitute for all-nickel deposits. One Watts-based variation contains about 45 g/L nickel, 3 g/L iron, and boric acid and additives such as citrates and gluconates. Hydroxy carboxylic acids act as stabilizers for the ferrous iron.

Alloys of nickel–cobalt, are used for engineering properties, especially in electroforming. Deposits having 1–50% cobalt have been obtained from various baths. Nickel–cobalt plating solutions are similar to those used for engineering nickel with cobalt salts substituted from some of the nickel to add strength and hardness to nickel deposits.

Interest in electrodeposited nickel–phosphorus came with realization of the benefits of the electroless nickel-plated alloy. The properties of the alloys appear to be the same from either process and are related to the phosphorus content. Low (2%) phosphorus deposits, can be obtained simply using phosphorus acid additions to a Watts bath and operating the bath at 0–2 pH with a temperature of 75–95°C and current of 500–4000 A/m^2.

Alloys can be deposited from acid Watts baths that contain sodium tungstate producing deposits with up to about 4% tungsten. Alkaline baths having ammonium salts in the nickel solution, along with Rochelle salts, citrates, or glycolates can deposit 10–20% tungsten. No significant commercial uses of nickel-tungsten alloys have been publicized, although the properties may be of interest in special applications.

Platinum-Group Metals. Rhodium is the most commonly plated platinum-group metal. In addition to its decorative uses, rhodium has useful properties for engineering applications. It has good corrosion resistance, stable electrical contact resistance, wear resistance, heat resistance, and good reflectivity.

Platinum plating has found application in the production of platinized titanium, niobium, or tantalum anodes, which are used as insoluble anodes in many other plating solutions.

Palladium is used in telephone equipment and in electronics applications as a substitute for gold in specific areas. Palladium is plated from ammoniacal and acid baths; available along with chelated variations as proprietary processes.

Ruthenium, the least expensive of the platinum group, is the second best electrical conductor, has the hardest deposit, and has a high melting point.

Silver Plating. Cyanide baths are the standard for silver plating. Only minor modifications and improvements in brighteners have occurred since silver electroplating was first patented in 1840. Typical plating baths are shown in Table 6. Silver strikes are necessary to obtain adhesion on most metals. Although a large proportion of silver plating is used for decorative purposes, such as tableware, jewelry, art work, musical instruments and the like, a growing amount is found in industrial applications. These engineering uses include bearings, reflective surfaces, contact areas on busbar and other electrical and electronic contacts, solderable surfaces, in thermocompression bonding, selective plating on semiconductor components to replace gold on many devices, and other applications where the high thermal and good electrical conductivity are useful.

Tin Plating. Tin is plated from alkaline stannate and acid sulfate or fluoborate baths. Deposits are matte when plated without brighteners. Brighteners for use in sulfate baths have been vastly improved, and very bright, decorative deposits can be obtained directly from the bath.

Tin Alloys. Tin–lead alloys are plated from fluoborate baths and, more recently, from methanesulfonate solutions. Solder 60% tin plate is used on contacts and as a solderable etch resist on printed circuit boards. Higher (96–98% tin) tin alloys are used in semiconductor and other electronic applications.

The 65% tin alloy exhibits good resistance to chemical attack, staining, and atmospheric corrosion, especially when plated copper or bronze undercoats are used. This alloy has a low coefficient of friction. Deposits are solderable, hard (650–710 HV_{50}), act as etch resists, and find use in printed circuit boards, watch parts, and as a substitute for chromium in some applications. The most common plating bath uses fluoride to complex the tin.

Although alloys of all concentrations are possible, 80% tin–20% zinc gives the best combination of properties. This alloys has a low coefficient of friction, low electrical contact resistance, is solderable,

Table 6. Silver-Plating Baths[a]

Parameter	Decorative	Industrial	Strike[b]
AgCN, g/L[c]	30–35	45–50	1.5–5.0
KCN, g/L[d]	50–78	65–72	75–90
free	35–50	45–50	
K_2CO_3, g/L	15[e]	40–80	
KNO_3, g/L		40–60	
KOH, g/L		1–14	
temperature, °C	20–28	42–45	22–30
current density, A/m^2	50–150	500–1000	130–300

[a] Anodes are normally pure silver, 999+, often bagged.
[b] For strike on nonferrous metals and a second strike on steel, first strike uses 1.5 g/L AgCN, 75–90 g/L KCN, and 10–15 g/L copper cyanide.
[c] Insoluble in water; dissolves in aqueous solution of KCN.
[d] Some KCN is used to form the silver complex; the excess is called free CN^-. NaCN has been used, but has some deficiencies.
[e] Value given is minimum.

slightly anodic to steel, and does not form voluminous corrosion products. In addition, the tin–zinc alloy has good paint adhesion qualities, good ductility, and its easily spotwelded. A typical bath is based on stannate and cyanide.

Zinc. Zinc plating continues to be a popular, cost-effective way to protect steel from atmospheric corrosion. Very bright decorative, chromium-like coatings can be obtained. Post-plate treatments in chromate conversion coatings are used to extend corrosion protection and aid in paint adhesion.

Plating solutions for continuous strip, called electrogalvanizing, sheet, and wire are usually simple zinc sulfate solutions, although chloride and mixed variations have found some use.

Chloride Baths. Chloride zincs are brighter, can plate directly on cast iron, and operate at high efficiency, but have poor throwing power, require acid-proof equipment, and are more corrosive to surrounding steel materials. A typical chloride zinc may contain about 35 g/L zinc, 135 g/L Cl^-, 30 g/L boric acid, and proprietary brighteners and wetting agents.

Zincate Baths. Zincates are simple, economical, noncorrosive to steel and have superior throwing power. Newer systems are showing better tolerance to operating variables, with less tendency toward flaking or delayed blistering, and good brightness. Zincate baths, however, require good cleaning, frequent analysis, close chemical control, and have low cathode efficiency. Zinc plating is discussed in detail in the literature.

Cyanide Baths. Cyanide zincs have excellent covering power and throwing power, good cleaning ability, deposit relatively pure zinc, are capable of thick deposits, and require little control. Zinc deposits from cyanide baths are purer than from the other baths, but still contain traces of the metal contaminants, brightener components, sulfur, hydrogen, and other gases. Pure, sulfur-free zinc is resistant to hydrochloric acid.

Zinc Alloys. There has been considerable worldwide activity in the area of plating zinc alloys. This interest results from efforts to improve the corrosion resistance of automobiles and automotive components without using cadmium. Three zinc alloys dominate the interest: zinc–nickel, zinc–iron, and zinc–cobalt. Europe produces predominantly zinc–nickel. In Japan, Zinc–nickel and zinc–iron are more popular. The annual nickel consumption for zinc–nickel alloy plating has been estimated at 2700–3175 metric tons. In the United States, consumption is estimated at about 225 metric tons for this purpose. Alloys are generally 6–12% nickel. Usage is expected to increase.

Steel has the best salt spray resistance when the nickel is 12–13% of the alloy.

Alloys of Zn–Co usually contain 0.3–0.8% cobalt. Higher cobalt alloys, from 4–8%, have shown better salt spray resistance, but the commonly plated alloy is 0.3–0.8%.

The Zn–Fe alloy is plated from an alkaline bath. Deposits are 0.3–0.8% iron and can be given attractive, resistant, black, silver-free chromate coatings. Corrosion protection requires the heavier, darker chromates. Zinc–iron baths are the most economical of the zinc alloys.

Post-Treatments. Although many post-treatments have been used over plated metals, chromate conversion coatings remain the most popular. Chromates are used to improve corrosion resistance, provide good paint and adhesive base properties, or to produce brighter or colored finishes.

Electroforming

Electroforming is the production or reproduction of articles by electrodeposition upon a mandrel or mold that is subsequently separated from the deposit. The separated electrodeposit becomes the manufactured article. Of all the metals, copper and nickel are most widely used in electroforming. Mandrels are of two types: permanent or expendable. Permanent mandrels are treated in a variety of ways to passivate the surface so that the deposit has very little or no adhesion to the mandrel, and separation is easily accomplished without damaging the mandrel. Expendable mandrels are used where the shape of the electroform would prohibit removal of the mandrel without damage. Low melting alloys, metals that can be chemically dissolved without attack on the electroform, plastics that can be dissolved in solvents, are typical examples.

JACK HORNER
Consultant

Metal Finishing Guidebook and Directory, '92, Metal Finishing, Hackensack, N.J.

"Metallic and Inorganic Coatings," *ASTM Annual Book of Standards*, Vol. 02.05, 1992. Issued annually; note that volume numbers are subject to change.

H. Geduld, *Zinc Plating*, ASTM Int'l. and Finishing Publications Ltd., Materials Park, Ohio, 1988.

A. Brenner, *Electrodeposition of Alloys*, Vol. 1, Academic Press, Inc., New York, 1963,

ELECTROSEPARATIONS

ELECTRODIALYSIS

Electrodialysis (ED) is a process for moving ions across a membrane from one solution to another under the influence of a direct electric current (see DIALYSIS and MEMBRANE TECHNOLOGY). In 1940, a multicompartment ED was invented by K. Meyer and W. Strauss in Zürich. Such process and apparatus is what is now meant by ED. In such an apparatus, many membranes *A* selective to anions alternate with membranes *C* selective to cations between a single pair of electrodes (Fig. 1). When a d-c potential is applied, cations M^+ move toward the negatively charged cathode and are able to permeate the cation-selective membranes, but not the anion-selective membranes, if the latter are perfectly selective. Similarly, anions X^- move toward the positively charged anode and are able to permeate the anion-selective membranes, but not the cation-selective ones. As a result, the odd-numbered compartments in the figure become depleted in the electrolyte, the even-numbered compartments enriched. It is shown in Figure 1 that the choice of which electrode is positive and which negative is arbitrary. If the left hand electrode is negative and the right hand one positive, then the odd numbered compartments become enriched in electrolyte, the even-numbered ones are depleted. This is the basis of reversing type ED (EDR) invented by W. McRae in 1956, in which the polarities of the electrodes are reversed regularly at intervals of 15 minutes to 1 week depending on the details of the application.

The depletion compartments of the figure (and in some applications, the enrichment compartments as well) may be filled with ion-exchange beads, fibers, or fabric (see ION EXCHANGE), leaving interstices through which fluid may flow. This type of filled cell ED was invented by W. Walters, D. Weiser, and L. Marek in 1955. They filled both depletion and enrichment compartments with a mixture of anion exchange (AX) and cation exchange (CX) beads. This process is called electrodeionization (EDI).

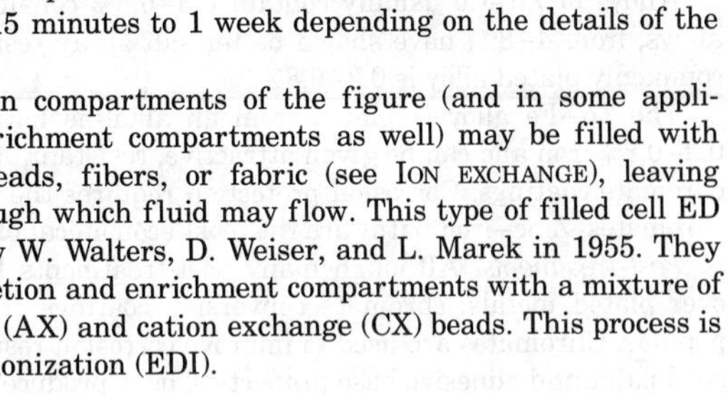

Figure 1. Principle of multicompartment electrodialysis. See text.

Ion-Exchange Membranes.

In 1948, ion-selective membranes having high selectivity, low electrical resistance, good mechanical strength, and good chemical stability were invented by W. McRae. They were essentially insoluble, synthetic, polymeric, organic ion-exchange resins in sheet form. Typical chemical structures for modern membranes of this type are shown schematically in Figure 2. The cation-selective membranes (Fig. 2**a**) consist of cation-exchange (CX) resin composed of polystyrene having negatively charged sulfonate groups chemically bonded to most of its phenyl groups. The charges of the sulfonate groups are electrically balanced by positively charged cations (counterions). Sulfonated polystyrene swells greatly in water. The amount of swelling is typically controlled by including cross-linking agents in the polymer, eg, divinyl benzene; by incorporating electrically neutral polymers; or by having extensive regions (blocks) in the polymer which lead to substantial microcrystallinity. The positively charged counter-ions, eg, Na^+, Ca^{2+}, or Mg^{2+}, are appreciably dissociated from the chemically bound negatively charged groups once the membrane is exposed to water. Thus the counter-ions are mobile and may be exchanged for other cations from an ambient solution, maintaining the electrical neutrality of the membrane. This high (typically >1 meq/cm^3 of membrane) concentration of counter-ions in ion-exchange resins is responsible for the low electrical resistance of the membrane. The high concentration of bound negatively charged groups tends to exclude mobile, negatively charged ions (co-ions) from an ambient solution and is responsible for the high ion selectivity of the membranes.

The anion-selective (AX) membranes (Fig. 2**b**) may also consist of cross-linked polystyrene, but have positively charged quaternary ammonium groups chemically bonded to most of the phenyl groups in the polystyrene instead of the negatively charged sulfonates. In this case, the counter-ions are negatively charged, eg, Cl^-, HCO_3^-, NO_3^-, or SO_4^{2-}.

The membranes described above called homogeneous ion-exchange membranes, represent about half the world production of ion-exchange membranes and are made in the United States and in Japan. The other half (made principally in the CIS and in the People's Republic of China) are made by bonding about 75 parts of AX or CX powder with 25 parts of, eg, polyethylene and are called heterogeneous. Whether homogeneous or heterogeneous, the ion-exchange membranes tend to transfer divalent ions (such as Ca^{2+}, Mg^{2+}, SO_4^{2-}) relatively more rapidly than univalent ions. Skinned homogeneous membranes are made in Japan and on the contrary tend to transfer univalent ions much more rapidly than divalent. CX membranes based on perfluorinated polymers and having carboxylate CX groups on one surface and sulfonate groups on the other are used in membrane chloralkali cells (see ALKALI AND CHLORINE PRODUCTS). Such membranes are made principally in the United States and Japan.

A large quantity of homogeneous ion-exchange membranes are made annually for separators in alkaline batteries (see BATTERIES PRIMARY CELLS; BATTERIES, SECONDARY CELLS).

Commercial membranes are usually reinforced with woven, synthetic fabrics to improve the mechanical properties. Several hundred thousand square meters of ion-exchange membranes are now produced annually. The mechanical and electrochemical properties are varied

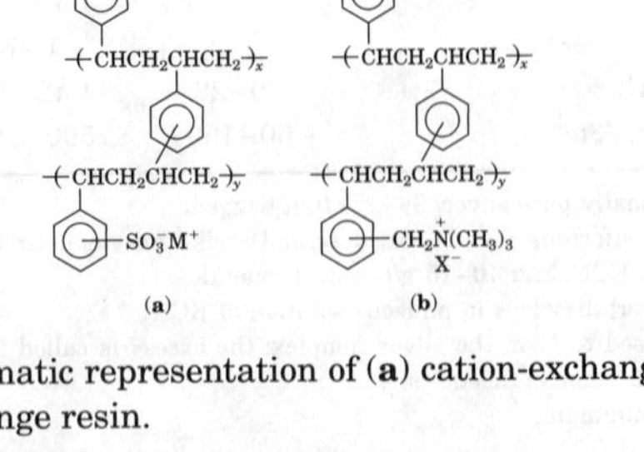

Figure 2. Schematic representation of (**a**) cation-exchange resin, and (**b**), anion-exchange resin.

by the manufacturers to suit the proposed applications. The electrochemical properties of most importance for ED are *(1)* the electrical resistance per unit area of membrane; *(2)* the ion transport number, related to current efficiency; *(3)* the electrical water transport, related to process efficiency; and *(4)* the back-diffusion, also related to process efficiency.

Apparatus

The apparatus for ED is fundamentally an array of alternating AX and CX membranes terminated by electrodes. The membranes are separated from each other by gaskets which form the fluid compartments between the membranes. The enrichment and depletion compartments also alternate through the array. Holes in the gaskets and membranes register with each other to provide two pairs of internal hydraulic manifolds to carry fluid into and out of the compartments. One pair communicates with the depletion compartments and the other with the enrichment compartments. Much effort has been expended on the design of the entrance and exit channels from the manifolds to the compartments to prevent unwanted cross-leak of fluid intended for one class of compartment into the other class. This effort is made increasingly difficult by the continuing trend to thinner membranes and gaskets, the latter determining membrane spacing and the thickness of the fluid compartments; such trends are intended to reduce energy consumption. A contiguous group of two membranes and the associated two fluid compartments is called a cell pair. A group of cell pairs and the associated end electrodes is called a stack or pack. Generally 100–600 cell pairs are arranged in a single stack, the choice being made on the basis of ED capacity desired, the uniformity of flow distribution achieved among the several compartments of the same class in a stack, and the maximum total direct current potential desired. One or more stacks may be arranged in a press, designed to compress the membranes and gaskets against the force of fluid flowing through the compartments thereby preventing fluid leaks to the outside and internal cross-leaks between compartments. For small presses such compression is usually provided by tie-rods; for larger presses, hydraulic rams are frequently used.

Commercial membranes have typical thicknesses of ca 0.15–0.5 mm; the compartments between the membranes have typical thicknesses of ca 0.3–2 mm. The thickness of a cell pair is therefore in the 1.3–5.0 mm range, commonly about 3.0 mm. One hundred cell pairs have a combined thickness of about 300 mm. The effective area of a cell pair for current conduction is generally on the order of 0.2–2 m^2.

Polarization. In concentrated electrolytes the electric current applied to a stack is limited by economic conditions, the higher the current I the greater the power consumption W in accordance with the equation $W = I^2R_s$ where R_s is the electrical resistance of the stack. In relatively dilute electrolytes, the electric current which can be applied is limited by the ability of ions to diffuse to the membranes across a convection free, laminar flow, diffusion layer contiguous with the depletion faces of the membranes. The need for such diffusion is due to the fact that almost all the current passing through a membrane is carried by its counter-ions, whereas, in the ambient solution roughly only half the current is carried by such counter-ions, the remainder by the co-ions. Therefore, there is a deficit of counter-ions at the membrane-solution interface. When the concentration c of electrolyte at the interface between an AX membrane and the solution in the depletion compartment is negligible compared to the concentration c_b in the bulk solution (ie, outside the laminar flow layer), then the rate at which ions can diffuse to the interface is a maximum and the AX membrane and the apparatus are said to be concentration polarized or simply polarized. The applied current density corresponding to such maximum rate of diffusion is the limiting current density i_{lim}. As the applied current density approaches i_{lim}, the electrical resistance of the ED stack increases substantially even though the bulk solution in the depletion compartments may still contain an appreciable concentration of electrolyte. This is a sign of polarization; i_{lim} is not a wall. At applied current densities above i_{lim} much of the increase in current through the AX membrane is carried by hydroxide ions (OH^-) resulting from dissociation of water at the interface. As a result, the enrichment compartments become somewhat alkaline and the depletion compartments somewhat acidic. This is also a sign of polarization.

The rate at which ions can diffuse to the interface, and therefore, i_{lim}, is inversely proportional to δ (the thickness of the convection free, laminar flow, diffusion layer). The latter is typically reduced by including turbulence promoting structures (such as screens) in the depletion compartments, and generally also in the enrichment compartments. In such case, i_{lim} in A/cm^2 is, in order of magnitude, equal numerically to the concentration c_b in gram-equivalents/L. i_{lim}, measured in terms of the AX membrane area, can also be increased (by a factor of 10 or more) by including ion-exchange beads, fibers, or fabrics in the depletion compartments as in EDI.

Applications

ED is used principally in the production of high purity water, in demineralization of brackish water (about 1/10th the concentration of seawater), in concentration of seawater (to 18–20%), and in deashing cheese whey. For the production of high purity water it is typical to use EDI. When EDI is operated at currents well above i_{lim} for the filled cells, then substantial water dissociation occurs and EDI behaves in many ways as a continuously electrically regenerated mixed-bed ion exchanger producing ultrapure water equivalent to that produced by chemically regenerated mixed beds. EDI is rapidly replacing chemically regenerated ion-exchange for the production of high purity water.

EDR is typically used for demineralization of brackish water, which often contains poorly soluble minerals such as calcium bicarbonate and calcium sulfate, as well as colloids such as humic and fulvic acids and iron hydroxides. The periodic reversal of the direction of the electric current avoids scaling and fouling of the membranes by such substances.

Univalent ion selective membranes are used in nonreversing ED for seawater concentration. Both EDR and nonreversing ED are used for whey deashing.

Bipolar ED. Bipolar ion-exchange membranes have one surface consisting of CX resin and the opposite surface of AX resin. The interface between the CX and AX resins may be regarded as a zero gap ED compartment. When a direct current is passed through such a membrane in a direction to pull anions out of such interface and through the AX resin, the interface rapidly becomes depleted of all ions other than those resulting from the dissociation of water. Dilute alkali can therefore be produced at the outer surface of the AX region and dilute acid at the outer surface of the CX layer.

The bipolar membranes are used in a more or less conventional ED stack together with conventional unipolar membranes. Such a stack has many acid–alkali producing membranes between a single pair of end electrodes. The advantages of the process compared to direct electrolysis seem to be that because only end electrodes are required, the cost of the electrodes used in direct electrolysis is avoided, and the energy consumption at such electrodes is also avoided.

The disadvantages appear to be that the bipolar membranes are comparatively expensive, and the economic life is limited to about one year. Such short lifetime appears to result from the very high ($\sim 10^6$ V/cm) potential gradients at the interface between the AX and CX regions. Additionally, practical current densities are limited to about 1000 A/m^2 available area.

Economic Aspects

About 5000 ED plants of all types have been installed worldwide for the demineralization of brackish and potable water. These range in capacity from a few to more than 10,000 m^3/d. The total installed capacity is greater than 1.5×10^6 m^3/d and uses more than 3×10^6 m^2 of ion-exchange membranes. Ionics, Incorporated (Watertown, Massachusetts) the leading supplier outside the CIS and the People's Republic China has sold more than 2000 ED plants all using homogeneous membranes. Most of their plants are of the EDR type. Their total installed capacity is greater than 600,000 m^3/d and uses more than 1.2 million m^2

ion-exchange membrane. Most of the remaining 3000 ED plants were built by CIS and the People's Republic of China entities using heterogeneous membranes and are used almost exclusively in the countries in which they were made.

About 10 ED plants for concentrating seawater have been built, exclusively using nonreversing ED stacks and univalent ion-selective membranes made by the Japanese companies Asahi Chemical Industry Company, Asahi Glass Company, and Tokuyama Corporation. The combined capacity is about 1,700,000 metric tons of salt per year using about 850,000 m^2 of ion-exchange membranes. Obviously each plant is very large. Seven of the plants are in Japan as a result of the decision by the Japanese government that all domestic salt had to be made in Japan. Salt produced by ED is not economic compared to imported solar salt. A recent decision by the government removing the above monopoly probably means such ED plants will be phased out.

More than 150,000 metric tons per year of 90% demineralized, dry, whey solids are made by ED or by ED followed by chemically regenerated ion-exchange.

WAYNE A. MCRAE
Consultant

J. R. Wilson, ed., *Demineralization by Electrodialysis*, Butterworths, London, 1960.

L. H. Shaffer and M. S. Mintz, in K. S. Spiegler, ed., *Principles of Desalination*, Academic Press, Inc., New York, 1966, pp. 199–289.

R. Rautenbach and R. Albrecht, *Membrane Processes*, John Wiley & Sons, Inc., New York, 1989.

H. Strathmann, *Membrane Separation Systems—A Research Needs Assessment*, U.S. Dept. of Energy, Washington, D.C., p. 8–1.

ELECTROPHORESIS

Electrophoresis is a separation technique most often applied to the analysis of biological or other polymeric samples. It has frequent application to analysis of proteins (qv) and DNA fragment mixtures (see NUCLEIC ACIDS). The high resolution of electrophoresis has made it a key tool in the advancement of biotechnology (qv). Variations of this methodology are being used for DNA sequencing (see GENETIC ENGINEERING), isolating active biological factors associated with diseases such as cystic fibrosis, sickle-cell anemia, myelomas, and leukemia, and establishing immunological reactions between samples on the basis of individual compounds (see also CHEMOTHERAPEUTICS, ANTICANCER; IMMUNOASSAY). Electrophoresis is an extremely effective analytical tool because it does not affect a molecule's structure, and it is highly sensitive to small differences in molecular charge and mass.

The term electrophoresis refers to the movement of a solid particle through a stationary fluid under the influence of an electric field. The study of electrophoresis has included the movement of large molecules, colloids (qv), fibers (qv), clay particles (see CLAYS), latex spheres (see LATEX TECHNOLOGY), basically anything that can be said to be distinct from the fluid in which the substance is suspended. This diversity in particle size makes electrophoresis theory very general.

The fundamental principle behind electrophoresis is the existence of charge separation between the surface of a particle and the fluid immediately surrounding it. An applied electric field acts on the resulting charge density, causing the particle to move, the fluid around the particle to move, or both. An applied electric field also generates heat, through resistive heating, and gases, through electrolysis reactions. Each is important in understanding and designing working electrophoresis equipment.

There are three distinct modes of electrophoresis: zone electrophoresis, isoelectric focusing, and isotachophoresis. These three methods may be used alone or in combination to separate molecules on both an analytical (μL of a mixture separated) and preparative (mL of a mixture separated) scale. Separations in these three modes are based on different physical properties of the molecules in the mixture, making at least three different analyses possible on the same mixture.

Distinction is also made among electrophoretic techniques in terms of the type of matrix employed for analysis. Matrices include polymer gels such as agarose and polyacrylamide, paper, capillaries, and flowing buffers. Each matrix is used for different types of mixtures, and each has unique advantages.

There are a variety of techniques for detecting separated sample compounds using chemical stains, photographic media, and immunochemistry. Each detection technique also gives different information about the identity, quantity, and physical properties of the molecules in the mixture. Detection is often the focus of electrophoresis, and usually yields basic information about the mixture being studied.

Electrolysis Reactions. The electrodes in electrophoresis equipment are typically constructed from platinum wire, and sodium chloride generally carries the bulk of the current in any electrophoretic medium. This results in the reactions at the cathode of $2\,H_2O + 2e^- \rightarrow 2\,OH^- + H_2$, and $H_2O + OH^- \rightleftarrows A^- + H_2O$; at the anode: $H_2O \rightarrow 2\,H^+ + 20.5\,O_2 + 2e^-$, $H^+ + A^- \rightleftarrows HA$. That is, water is electrolyzed. The hydrogen gas produced at the cathode can be hazardous, especially because it is in the vicinity of an electrode that is also producing heat. For this reason, electrode chambers are usually open to the atmosphere so that gases can vent.

The other reactions at the electrodes produce acid (anode) and base (cathode) so that there is a possibility of a pH gradient throughout the electrophoresis medium unless the system is well buffered (see HYDROGEN-ION ACTIVITY). Buffering must take the current load into account because the electrolysis reactions proceed at the rate of the current. Electrophoresis systems sometimes mix and recirculate the buffers from the individual electrode reservoirs to equalize the pH.

Modes of Electrophoretic Separations

Electrophoresis Equipment. Most electrophoresis equipment shares a basic design, a diagram of which is shown in Figure 1.

Zone Electrophoresis. In zone electrophoresis multicomponent samples are applied to an electrophoretic medium, most commonly a gel, an electric field is applied, and after a predetermined length of time or after a certain level of power, current, or voltage has been applied, the electrophoretic medium is inspected for resolution of the sample components.

Disc Electrophoresis. Resolution in zone electrophoresis depends critically on getting sample components to migrate in a focused band, thus some techniques are employed to concentrate the sample as it migrates through the gel. The most common technique is referred to as discontinuous pH or disc electrophoresis. Disc electrophoresis employs a two-gel system, where the properties of the two gels are different.

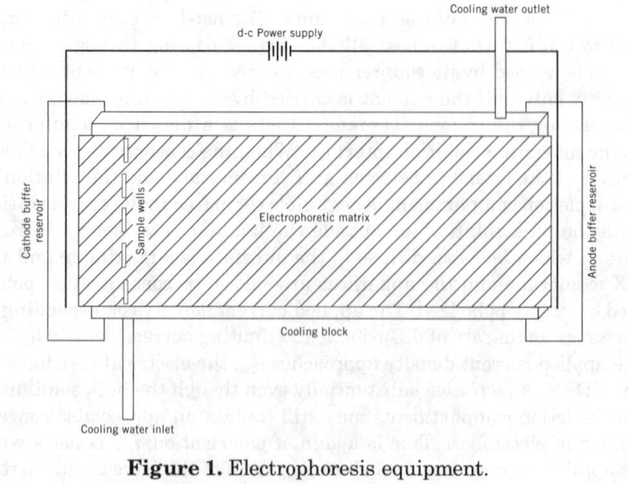

Figure 1. Electrophoresis equipment.

Reduced SDS Electrophoresis. The combination of sodium dodecyl sulfate (SDS), $CH_3(CH_2)_{10}CH_2OSO_3Na$, also known as lauryl sulfate, treatment of samples and polyacrylamide gel electrophoresis was first described in the late 1960s. SDS is an ionic surfactant which solubilizes and denatures proteins (see SURFACTANTS). The surfactant coats a protein through hydrophobic interactions with the polypeptide backbone, effectively separating most proteins into their polypeptide subunits. The majority of proteins to which SDS binds then unfold into linear molecules having a similar surface potential. Nonreduced proteins bind approximately 0.9–1.0 grams of SDS per gram of protein.

SDS–polyacrylamide gel electrophoresis (SDS–page) allows separation of molecules strictly on the basis of size, ie, molecular weight. When SDS-treated samples migrate into a gel and are electrophoresed, the principal difference is size or length.

The way proteins behave in disc electrophoresis systems depends primarily on differences in the pH in the two gels. The pH of the buffers in both the second gel (the separating gel) and the electrode reservoir are similar, whereas the buffers of the first gel (the stacking gel) and the sample itself are of a lower pH. This difference in pH allows for different sample–gel interactions. The stacking gel is only about 5 cm, including distance for the sample wells, and is stacked on top of the separating gel which is about 20 cm in length. As with most electrophoretic methods, a current is applied and the molecules in the sample wells begin to migrate anodally. Here, the migration is electrochemically different because the buffer in the upper reservoir chamber has a higher pH than that of the samples and stacking gel. This difference in pH allows the molecules in the samples to migrate rapidly through the stacking gel. When the sample compounds enter the separating gel, movement is slowed because of the pH change. This focuses the molecules into narrow bands and allows more bands to be resolved from one another.

Another difference between other types of electrophoresis and disc electrophoresis is that the molecules in a sample do not start to significantly separate until entering the separating gel. A discontinuous gel system may be used with almost any type of zone electrophoresis application.

Native Zone Electrophoresis. In some cases, good resolution between sample species can be obtained with little or no sample pretreatment. In these cases, the gels are said to be native gels. In this method, the charge on the individual sample component is primarily responsible for its differential migration.

Pulsed Field Gel Electrophoresis. In pulsed field electrophoresis, the direction of the field is intermittently changed, either forward and backward or from side to side. A small molecule notices no real difference in its electrophoresis because it can completely reorient in each field direction. However, the redirectioning of the electric field causes larger molecules to travel in a zigzag pattern, putting kinks in to the length of the molecule. The longer the molecule, the more kinks in its length, and the slower it travels down the length of the gel. The larger molecule finds itself traveling backward along some sections of its length with respect to the direction of the electric field. This has allowed resolution of megabase size strands of DNA, and has made the Human Genome Project feasible.

Isoelectric Focusing. Isoelectric focusing is a technique used for protein separation, by driving proteins to a pH where they have no mobility. Resolution depends on the slope of a pH gradient that can be achieved in a gel.

Ampholytes or Zwitterions. An ampholyte is a molecule that can be either positively or negatively charged, depending on the pH. These molecules are also called zwitterions. All amino acids (qv) and proteins are ampholytes, or amphoteric.

A special class of ampholytes has been synthesized for the purpose of isoelectric focusing. These ampholytes have an amino end and a carboxyl end that are separated by varying numbers of methylene groups. The further apart the amino and carboxyl groups, the less one affects the ionization of the other; thus a different isoelectric point (pI) is established for each molecule. These ampholytes, which may be added to an electrophoretic medium, migrate in one direction or another, under the influence of an applied electric field, until they reach a zone in which the pH is the same as that ampholyte's isoelectric point. The ampholyte molecules buffer themselves and establish the local pH as they migrate through the gel. As the ampholytes reach an isoelectric pH, they establish a stationary spatial pH gradient in the electrophoretic medium.

Isotachophoresis. Isotachophoresis takes advantage of the fact that electroneutrality must be maintained in an electrophoretic system in order to support an electric field. If a current passes through a medium, that current must be constant from one electrode to the other, regardless of the local ion concentration or mobility; ie, dilute ions must move faster to keep up with a zone of more concentrated ions. Electric fields compensate for this because the electric field strength does not have to be constant along the length of the medium. The electric field strength is lowest where the ions are most concentrated and most mobile. Isotachophoresis takes advantage of this phenomenon by lining up the ions of interest, fastest (most mobile) to slowest. This is a highly specialized technique that requires detailed knowledge of the properties of the sample to be separated, and is generally not applicable to analytical separations.

Electrophoretic Materials and Matrices

Agarose Electrophoresis. Agarose is produced from the processing of red seaweed. To prepare a gel for electrophoresis a combination of agarose and buffer is heated until the agarose solid is dissolved and boiling. The solution is cooled sufficiently and then poured into a warmed gel casting apparatus which forms the shape of the gel as it cools. After the cooled solution is poured into the casting apparatus, it is allowed to gel and can then be used in an agarose electrophoresis method.

The use of agarose as an electrophoretic method is widespread. The advantages of agarose electrophoresis are that it requires no additives of cross-linkers for polymerization, it is not hazardous, low concentration gels are relatively sturdy, it is inexpensive, and it can be combined with many other analytical methods.

Polyacrylamide Electrophoresis. Polyacrylamide gels are synthesized through the combination of acrylamide (qv), $CH_2{=}CHCONH_2$, monomer and a cross-linking comonomer. Cross-linking comonomer of choice is *N,N′*-methylenebisacrylamide.

The most commonly used combination of chemicals to produce a polyacrylamide gel is acrylamide, bisacrylamide, buffer, ammonium persulfate, and tetramethylenediamine (TEMED).

Polyacrylamide gel electrophoresis is one of the most commonly used electrophoretic methods. Analytical uses of this technique center around protein characterization, for example, purity, size, or molecular weight, and composition of a protein. Polyacrylamide gels can be used in both reduced and nonreduced systems as well as in combination with discontinuous and isoelectric focusing (ief) systems.

Paper Electrophoresis. Besides being easy to obtain, paper is a good medium because it does not contain many of the charges that interfere with the separation of different compounds. Two types of paper employed in this type of electrophoresis are Whatman 3 MM (0.3 mm) and Whatman No. 1 (0.17 mm).

In paper electrophoresis, the sample is placed directly onto chromatographic or filter paper and then exposed to a buffer solution at each end and an electric field is applied. As in most electrophoretic techniques, charged dyes are combined with samples and standards to see the progress of the electrophoresis.

Capillary Electrophoresis. The glass capillaries used are typically 20 to 200 μm in diameter, may be filled with buffer or gel, and are frequently coated on the inside. Capillaries are used because of the high surface-to-volume ratio which allows high voltages without heating effects. The only limitations associated with capillaries are limits of detection and clearance of sample components.

Capillary electrophoresis is a commercially available technique, and has been integrated with most automated lab equipment such as autosamplers, computer peak analysis (the charts generated are called electropherograms), temperature control, and recirculating buffers.

The use of capillaries as a support medium for electrophoresis is advantageous because it avoids the effects of heating that occur in a gel, can be very rapid, and produces a chart recording rather than a stained gel for archiving.

Free-Flow Electrophoresis. Free-flow electrophoresis is the most common technique for scaling up electrophoresis for commercial application. In this technique, sample compounds are injected into a curtain of buffer which flows between two flat plates, with electrodes parallel to the flow at each end. The electric field is then applied perpendicularly to the flow direction, so that as compounds flow down between the electrodes they separate horizontally and exit the flow field at different locations.

Detection Techniques

Most sample components analyzed with electrophoretic techniques are invisible to the naked eye. Thus methods have been developed to visualize and quantify separated compounds. These techniques most commonly involve chemically fixing and then staining the compounds in the gel. Other detection techniques can sometimes yield more information, such as detection using antibodies to specific compounds, which gives positive identification of a sample component either by immunoelectrophoretic or blotting techniques, or enhanced detection by combining two different electrophoresis methods in two-dimensional electrophoretic techniques.

SCOTT RUDGE
KATHLEEN MARKEY
Synergen, Inc.

R. C. Allen, C. A. Saravis, and H. R. Maurer, *Gel Electrophoresis and Isoelectric Focusing of Proteins*, Walter de Gruyter, New York, 1984.

A. Chrambach, *The Practice of Quantitative Gel Electrophoresis*, VCH Publishers, New York, 1985.

B. A. Baldo and E. R. Tovey, *Protein Blotting: Methodology, Research, and Diagnostic Applications*, Karger, Switzerland, 1989.

J. F. Robyt and B. J. White, *Biochemical Techniques: Theory and Practice*, Waveland Press, Inc., Prospect Heights, Ill., 1987.

EMBEDDING

Advances in electronic technology have had great technological and economic impact on the electronic industry throughout the world. The rapid growth of the number of components per chip, the rapid decrease of device dimension, and the steady increase in integrated circuit (IC) chip size have imposed stringent requirements, not only on IC physical design and fabrication, but also in electronic packaging and embedding. Electronic embedding is one of the most common processes used to encapsulate and protect these electronic components.

Materials for Electronic Embedding

Silicones. Polydimethylsiloxanes, polydiphenylsiloxanes, and polymethyl-phenylsiloxanes are generally called silicones (see SILICON COMPOUNDS, SILICONES). With a repeating unit of alternating silicon–oxygen, the siloxane chemical backbone structure, silicone possesses excellent thermal stability and flexibility that are superior to most other materials. Polydimethylsiloxane provides a very low glass-transition temperature T_g material but is suitable for use at temperatures up to 200°C. The basis of commercial production of silicones is that chlorosilanes are readily hydrolyzed to give disilanols which are unstable and condense to form siloxane oligomers and polymers. Depending on the reaction conditions, a mixture of linear polymers and cyclic oligomers is produced. The cyclic components can be ring-opened by either acid or base to become linear polymers and it is these linear polymers that are of commercial importance. The linear polymers are typically liquids of low viscosity and, as such, are not suited for use as encapsulants. These must be cross-linked (or vulcanized) in order to increase the molecular weight sufficiently to achieve useful properties. For electronic applications, only the high purity room temperature vulcanized (RTV) condensation cure silicone, which uses an alkoxide-cure system with noncorrosive alcohol by-products, and platinum-catalyzed addition heat-cure (hydrosilation) silicone systems are suitable for device encapsulation.

Epoxies. The unique chemical and physical properties such as excellent chemical and corrosion resistances, electrical and physical properties, excellent adhesion, thermal insulation, low shrinkage, and reasonable material cost have made epoxy resins (qv) very attractive in electronic applications. The commercial preparation of epoxies is based on bisphenol A, which upon reaction with epichlorohydrin produces diglycidyl ethers. The reactant ratio (bisphenol A:epichlorohydrin) determines the final viscosity of the epoxies. In addition to the bisphenol A resins, the novolak resins with multifunctional groups which led to higher cross-link density and better thermal and chemical resistance have gained increasing acceptance in electronic applications.

Polyurethanes. Recent work has focused on the use of intermediates which are low molecular weight polyethers with reactive functional groups such as hydroxyl or isocyanate groups able to further cross-link, chain extend, or branch with other chain extenders to become higher molecular weight polyurethanes. Diamine and diol are chain extended with the prepolymer, either polyester or polyether, to form polyurethanes with urea or urethane linkages, respectively. The morphology of polyurethane is well characterized. Hard and soft segments from diisocyanates and polyols, respectively, are the key to the excellent physical properties of this material (Fig. 1).

High performance polyurethane elastomers are used in conformal coating, potting, and in reactive injection molding (or reaction impingement molding) of IC devices. Furthermore, rigid polyurethane foams, most often in free-foam densities of 128–288 kg/m^3 (8–18 lbs/ft^3 (pcf)), are useful for embedding complex electronic systems.

Polyesters. Polyester is used in embedding resins for electronic components because of its low cost compared to silicones and epoxides. Polyesters (qv) are condensation products of dicarboxylic acids and dihydroxy alcohols; the reaction provides a wide range of viscosities for polyesters.

$$\text{HOOC—R—COOH} + \text{HO—R'—OH} \xrightarrow{-\text{H}_2\text{O}} \left(\text{O—}\overset{\text{O}}{\overset{\|}{\text{C}}}\text{—R—}\overset{\text{O}}{\overset{\|}{\text{C}}}\text{—O—R'}\right)_n$$

There are electronic-grade polyesters with relatively low viscosities that are suitable for embedding of coils, as in transformers. Furthermore, free-radical-cured polyesters are popularly used for display castings.

Polysulfides. Polysulfides are organic polymers that contain sulfur in disulfide linkages (S–S), mercaptans (S–H), or thiol groups (S–R), (see POLYMERS CONTAINING SULFUR). Low molecular weight polysulfides (3000–4000) may oxidatively react with free-radical reactants such as lead, tin, cobalt octoate, *p*-quinone, dicumene hydroperoxide, lead and manganese dioxides to yield rubbery flexible polysulfides with good water, gas, and moisture sealer properties. Most polysulfides have excellent adhesion to coated substrates and resistance to oxidation, solvents, and ozone. However, the electrical properties of this material

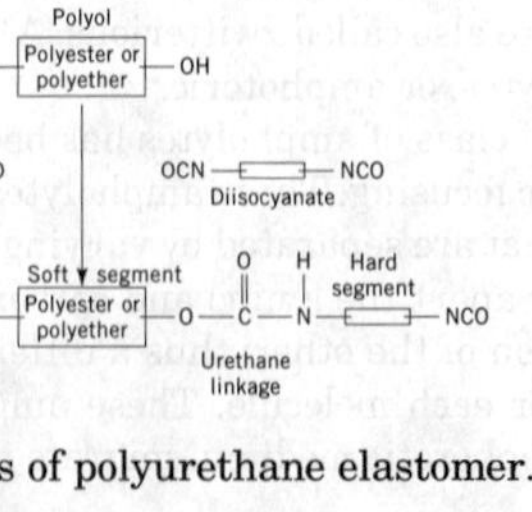

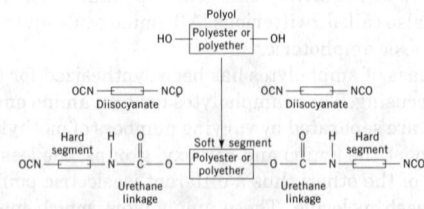

Figure 1. Synthesis of polyurethane elastomer.

are marginal as an insulator, and the dielectric constant of the material is relatively high ($\epsilon_1 \leq 7$). It is a low cost embedding material for transformers and connector sealing, but not too common for microelectronic embedding.

Advanced Thermoplastics Materials. Thermoplastics and linear plastics of finite molecular weight that can be fabricated into very complex structures by hot melt or injection molding are different from the thermoset materials that require cross-linking to build up infinite molecular weight to form network (cross-link) structures. Advances in thermoplastic engineering materials include amorphous thermoplastics, crystalline thermoplastics, liquid crystal thermoplastics, and fluorinated thermoplastics (see ENGINEERING PLASTICS).

Embedding Materials Properties

The ability of a given material to perform as an electronic embedding encapsulant depends largely on its properties. Ultrapure chemical properties with a low level of mobile ions such as sodium, potassium, and chloride are essential. Furthermore, the material's electrical, mechanical, and rheological properties are critical.

Material Processes

Material processes consist of cavity-filling and saturation coating. The cavity-filling process involves molding, potting, and coating.

Material Curing

In order to optimize each embedding material property, complete cure of the material is essential. Various analytical methods are used to determine the complete cure of each material. Differential scanning calorimetry, Fourier transform-infrared (ftir), and microdielectrometry provide quantitative curing processing of each material.

C. P. WONG
PAUL D'AMBRA
AT&T Bell Labs

C. P. Wong, in C. P. Wong, ed., *Polymers for Electric and Photonic Applications*, Academic Press, San Diego, Calif., 1993, Chapt. 4, pp. 167–214.

C. A. Harper, ed., *Electronic Packaging and Interconnect Handbooks*, McGraw-Hill Book Co., New York, 1991.

R. R. Tummala and E. J. Rymaszewski, eds., *Microelectronic Packaging Handbook*, Van Nostrand Reinhold Co., Inc., New York, 1989.

C. P. Wong, in J. H. Lai, ed., *Polymers in Electronics*, CRC Press, Boca Raton, Fla., 1989, Chapt. 3, pp. 63–92.

L. T. Manzione, *Plastic Packaging of Microelectronic Devices*, Van Nostrand Reinhold, New York, 1990.

EMULSIONS

Simple Emulsions

A (macro)emulsion is formed when two immiscible liquids, usually water and a hydrophobic organic solvent, an oil, are mechanically agitated so that one liquid forms droplets in the other one. A microemulsion, on the other hand, forms spontaneously because of the self-association of added amphiphilic molecules. During the emulsification agitation both liquids form droplets, and with no stabilization, two emulsion layers are formed, one with oil droplets in water (o/w) and one of water in oil (w/o). However, if not stabilized the droplets separate into two phases when the agitation ceases. If an emulsifier (a stabilizing compound) is added to the two immiscible liquids, one of them becomes continuous and the other one remains in droplet form.

During emulsification new surfaces are created between the two phases. Such a process requires energy; the surface free energy, numerically identical to the easily measure surface tension, reflects this amount.

There are two fairly common misconceptions about the conclusions that may be drawn from the value of the surface free energy. First, the energy input to enlarge the interface during emulsification is not a significant part of the total energy needed for this process. The viscous resistance during the agitation absorbs most of the energy, giving a small temperature rise to the system. Secondly, the emulsion is stabilized by the addition of a compound adsorbed to the interface, termed an emulsifier. Emulsifiers are molecules with nonpolar and polar parts that reside at the interface. Their presence causes a reduction of the surface tension and a stabilization of the emulsion, but these two features are not directly related. The interfacial tension reflects the amount of emulsifier at the interface.

Any conclusion that a low interfacial tension per se is an indication of enhanced emulsion stability is not reliable. In fact, very low interfacial tensions lead to instability. The stability of an emulsion is influenced by the charge at the interface and by the packing of the emulsifier molecules, but the interfacial tension at the levels found in the common emulsion has no influence on stability.

Emulsification

Much commercial equipment is available for emulsification (Fig. 1) and has been well described. The following discusses the relations between energy input and emulsion droplet size.

The emulsification process in principle consists of the break-up of large droplets into smaller ones due to shear forces. The simplest form of shear is experienced in lamellar flow, and the droplet break-up may be visualized according to Figure 2. The phenomenon is governed by two forces, ie, the Laplace pressure, which preserves the droplet, and the stress from the velocity gradient, which causes the deformation. The ratio between the two is called the Weber number, We, where η_c is the viscosity of the continuous phase, G the velocity gradient, r the droplet radius, and γ the interfacial tension: $We = \eta_c \cdot G \cdot r/\gamma$.

As an approximate rule, break-up of droplets occurs for a Weber number in excess of one, a rule of thumb that is actually valid for the range of viscosity ratios of the dispersed phase to the continuous phase of less than approximately five. Higher viscosities of the disperse phase lead to serious difficulties with emulsification because the shear energy is then dispersed in rotation of the droplets.

As may be expected, turbulent flow is more efficient for droplet formation in low viscosity liquids.

Finally, some general rules for the amount of surfactant appear to be valid. For anionic surfactants the average size of droplets is reduced for an increase of surfactant concentration up to the critical micellization concentration, whereas for nonionic surfactants a reduction occurs also for concentrations in excess of this value. The latter case may reflect the solubility of the nonionic surfactant in both phases, causing a reduction of interfacial tension at higher concentrations, or may reflect the stabilizing action of the micelles per se.

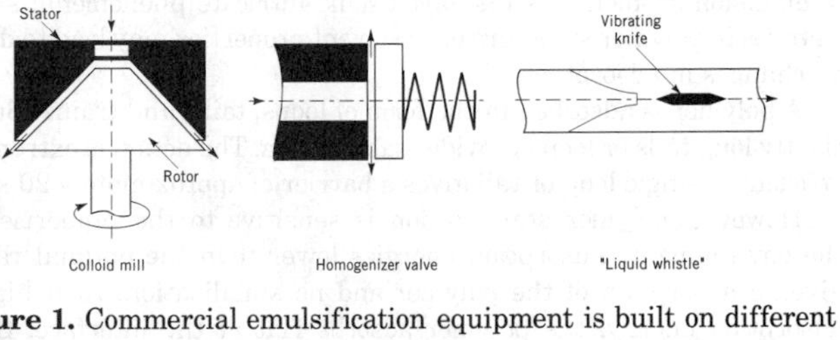

Figure 1. Commercial emulsification equipment is built on different principles for mixing one liquid into another.

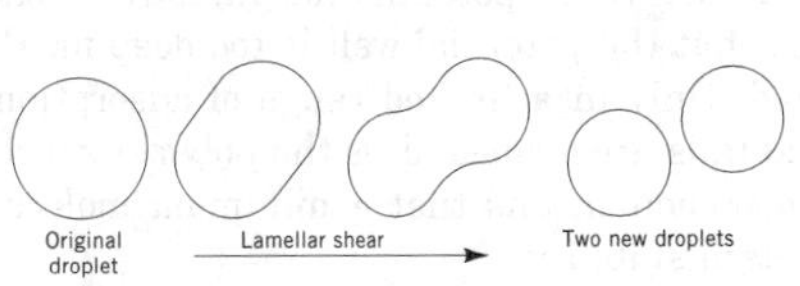

Figure 2. A droplet is broken into two droplets in lamellar shear.

Stability of Two-Phase Emulsions

Before determining the degree of stability of an emulsion and the reason for this stability, the mechanisms of this destabilization should be considered. When an emulsion starts to separate, an oil layer appears on top, and an aqueous layer appears on the bottom. This separation is the final state of the destabilization of the emulsion; the initial two processes are called flocculation and coalescence. In flocculation, two droplets become attached to each other but are still separated by a thin film of the liquid. When more droplets are added, an aggregate is formed, in which the individual droplets cluster but retain the thin liquid films between them. The emulsifier molecules remain at the surface of the individual droplets during this process.

In the coalescence step, the thin liquid film between the droplets is destabilized, and a large droplet is formed. Hence, the coalescing emulsion is characterized by a wide size distribution of the droplets, but no clusters are present. Finally, the droplets achieve such a size that they are recognized by the naked eye as a separate phase. A fully separated emulsion consists of an oil layer and an aqueous layer.

The sequence, flocculation → coalescence → separation, is complicated by the fact that creaming or sedimentation occurs and that this process is determined by the droplet size.

The definition of stability and the appropriate measure of the rate of destabilization of an emulsion depend on the application. For an application such as the use of a fluorocarbon as a blood substitute the destabilization stage of importance is obviously the flocculation; aggregates of droplets would clog blood vessels. At the other extreme are beverages such as soda or reconstituted fruit juices in which aggregation of flavor droplets is not noticed by the consumer and is not an essential disadvantage. The consumer instead would notice any separated oil forming a "greasy"ring in the bottle neck. In this case creaming must be prevented, a distinctly different problem from stabilization against flocculation.

Stabilization Mechanisms. The stabilization of an emulsion involves slowing the destabilization, primarily the flocculation process. This may be achieved in two principal manners: by reducing the mobility of droplets through enhanced viscosity or by inserting an energy barrier between them.

Two kinds of barriers are important for two-phase emulsions: the electric double layer and steric repulsion from adsorbed polymers. The repulsion from the electric double layer is famous because it played a decisive role in the theory for colloidal stability that is called DLVO, after its originators Derjaguin, Landau, Vervey, and Overbeek. The theory provided substantial progress in the understanding of colloidal stability, and its treatment dominated the colloid science literature for several decades.

Polymer Stabilization. Polymers have so far been used comparatively less than the common surfactants to stabilize emulsions in spite of the fact that excellent stabilization by them can be achieved. Application probably has been limited because the adsorption of polymers to emulsion droplets has displayed some intricate phenomena; small changes in polymer structure or in solvent properties may lead to drastic changes in adsorption.

A polymer is adsorbed in the form of loops, tails, and trains. Sufficiently long tails or loops provide stabilization. The action is extremely efficient; a single loop or tail gives a barrier of approximately 20 kT.

However, polymer stabilization is sensitive to the properties of the environment. Adsorption energies lower than the optimal range give no adsorption of the polymer and no stabilization. At a higher adsorption energy, the polymer adsorbs flat at the interface. Such an adsorption is also without stabilization effect because at short distances the van der Waals potential has already reached such large negative values that the potential well is too deep for the droplets to be deaggregated. Only in a limited range of adsorption energies, in which loops and tails are formed, does the polymer serve. In addition, the same phenomenon means that a minimum molecular weight is necessary to obtain stability.

These problems may be overcome by using block copolymers. The polymer blocks are chosen to be selectively soluble in the aqueous and oil phase, respectively.

Stability of Three-Phase Emulsions

In the simplest emulsions just described, the final separation is into two liquid phases upon destabilization. The majority of emulsions are of this kind, but in some cases the emulsion is divided into more than two phases. One obvious reason for such a behavior is the presence of a material that does not dissolve in the oil or the water. One such case is the presence of solid particles, which is common in emulsions for food, pharmaceuticals, and cosmetics. Another less trivial reason is that the surfactant associates with the water and/or the oil to form a colloidal structure that spontaneously separates from the two liquid phases. This colloidal structure may be an isotropic liquid or may be a semisolid phase, a liquid crystal, with long-range order.

In the case of emulsions with three liquids the presence of the third phase results in a reduction of the energy input for the emulsification process, whereas the emulsion with a liquid crystal as the third phase shows interesting stabilization mechanisms. Finally, the emulsion with added particles illustrates the importance of liquid–solid wetting for stability.

STIG E. FRIBERG
STEVEN JONES
Clarkson University

P. Becher, *Emulsions: Theory and Practice*, Reinhold, New York, 1965.

P. Sherman, *Emulsion Science*, Academic Press, Inc., New York, 1969.

International Union of Pure and Applied Chemistry, *Manual on Colloid and Surface Science*, Butterworths, London, 1972.

D. E. Napper, *Polymeric Stabilization of Colloidal Dispersion*, Academic Press, Inc., New York, 1983.

ENAMELS, PORCELAIN OR VITREOUS

Porcelain enamel is a glassy coating applied to various metal substrates through a high (usually >425 °C) temperature fusion process (see also COATINGS). The resultant chemically bonded glass–metal composite exhibits properties which reflect the chemical, physical, electrical, and aesthetic properties of the glass while combining the strength, ease of fabrication, and durability of the metal (see also COMPOSITE MATERIALS). Porcelain enamels are formulated to develop specific properties on metals or alloys.

The porcelain enamel is composed of various inorganic metal oxides fused between 1100 and 1400°C to form an alkali borosilicate glass (qv). The glass is rapidly quenched to produce either flakes by rapid chilling using water-cooled rollers, or glass granules by water quenching a narrow stream of molten glass. The resultant product is known as frit. An enameled article is produced by applying either wet or dry finely ground frit particles to the metal substrate. The enamel is fired to form one or more glassy layers to achieve a desired surface finish and property group for the article.

Modern appliances were developed utilizing the chemical and physical properties of the glass-on-metal coatings on cast iron (qv), sheet steel (qv), and, more recently, aluminum. Enamels have been used on a wide array of kitchen utensils, cooking devices such as ovens and range tops, sanitary ware such as bathtubs and lavatories, water heater tanks, specially built industrial chemical vessels, cast-iron piping, storage silos for agricultural and industrial products, as well as architectural interior and exterior surfaces.

Porcelain enamels are used in modern mass transit facilities because of inherent fire resistance as well as ease of maintenance and durability in the face of highly corrosive atmospheres resulting from vehicular traffic. Newer applications of porcelain enamels include electronic substrates (see ELECTRONIC MATERIALS), pyrolytic self-cleaning oven interiors, microwave oven interiors, outdoor cookers, and fireplace liners as well as wood-stove exteriors, writing boards for erasable markers, institutional surfaces such as bathroom stalls and hospital operating room walls that need to be easily disinfected, and subway car interiors and elevator walls that are durable and easy to clean. Porcelain enamel surfaces are frequently used in food contact applications. The glassy nature of porcelain enamels has a significant effect on encapsulating and minimizing any solubility of constituents. Enamels can also be formulated to have specific acid, alkali, and abrasion-resistant properties.

Metals for Enameling. Sheet steel is usually bought in precut sheets or coils for subsequent stamping and pressing into shapes. The steels are chemically formulated to make them suitable for the fabrication and porcelain enameling operations. The ASTM has classified enameling-grade steels into Type I and Type II (A or B) as well as specifying various qualities such as commercial, drawing, and drawing quality special killed. Type I has an extremely low carbon level, commonly produced by decarburizing in an open-coil process. This material is suitable for direct cover coat enameling practice. The less expensive Type II has moderately low carbon and is suitable for ground coat enameling operations.

The basic steel types are undergoing gradual modifications to adapt the steels to the continuous casting process. This has led to changes in the minor constituents of steel such as boron, nitrogen, titanium, and other alloying elements.

Cast Iron. Cast irons for enameling contain between 2.8 and 3.7 wt % carbon; the more usual content is between 3.25 and 3.6 wt %. The carbon is usually found in two forms: graphitic carbon and combined carbon. Some additional elements in the cast iron are silicon, manganese, sulfur, and phosphorus. The cast iron known as gray cast iron is the most widely used for enameling purposes. Before enameling, a casting must be cleaned, usually by abrasively blasting the surface with sand or steel shot.

Aluminum. The most commonly used aluminum alloys for enameling are 3003 P.E. grade and 6061 P.E. grade. It is important that the magnesium and copper contents of the alloys be kept to a minimum; otherwise, spalling of the enamel can occur.

Other Metals. Metals such as the austenitic series, Types 301–347, and the ferritic series, Types 409–446, of stainless steels may be enameled, as well as a number of other alloys. The metal preparation usually consists of degreasing and grit blasting. Copper, gold, and silver are also enameled. These metals are usually prepared for application by degreasing.

Metal Preparation. Enameling cannot be successful unless the metal is thoroughly cleaned and kept clean until the final coat is fired. Simply touching the surface with a hand can cause defects. Cast iron, thick steel parts, and aluminum castings may be sandblasted. The thickness precludes the danger of deformation resulting from metal loss. Sand, silicon carbide, and steel grit are satisfactory abrasives (qv). Products made from thin sheet material are most satisfactorily and most economically cleaned by chemical methods that require alkali and soap solutions to remove grease and dirt, and acid solutions to remove oxidized metal.

Firing of Enamels. Firing can be carried out in intermittent box-type furnaces or continuous furnaces. The dryer and the furnace form one continuous unit or function as separate units in the continuous firing process. Most industrial furnaces are fiber-lined (low thermal mass), which lowers cost and downtime between firing schedules.

Energy Requirements. The energy needed to heat 0.45 kg of enamel ware and tooling to the firing temperature ranges from 400–1000 kJ/kg (300–450 Btu/lb) and depends on such factors as furnace loading, tool weight, and, in gas-fired furnaces, flue-gas losses. Advances in furnace design, firing systems, and low thermal mass insulating materials have reduced the total energy needed over the years. Thus overall energy requirements for enameling have dropped as much as 50% since the 1970s. Technologies such as electrostatic frit powder application, various two-coat–one-fire processes (eg, dry-over-dry, dry-over-wet, and wet-over-wet), as well as reductions in the metal pretreatment requirements have also contributed.

Composition

Porcelain enamels are basically alkali borosilicate glasses. These enamels are complex, however, because of the large number and types of oxides which are needed to develop proper adherence and functional properties. Network-forming ingredients and modifiers are used as in normal glass (qv) making practice. The principal network formers are SiO_2, B_2O_3, and P_2O_5. Modifiers include the alkali metal oxides (Na_2O, K_2O, and Li_2O) and alkaline-earth metal oxides (CaO, BaO, and SrO). Other common oxides include Al_2O_3, MgO, ZrO_2, ZnO, TiO_2, Sb_2O_3, and the halide F^-. Transition-metal oxides such as Fe_2O_3, CoO, NiO, CuO, and MnO_2 are used for adherence to the sheet-steel substrate and color development in ground coats. Continuous-cleaning (catalytic) oven enamels have high percentages of the transition-metal oxides to achieve cleaning effectiveness. Less frequently used oxides of cerium, cadmium, lead, and tin are used in sheet-steel, cast-iron, and aluminum enamels.

Enamels used on cast iron and aluminum have traditionally been composed of SiO_2, B_2O_3, P_2O_5, and PbO. The lead oxide produces good surface quality, fusibility, and acid resistance when properly formulated with other oxides. More recently some nonlead-bearing compositions have been developed for both cast-iron and aluminum metals. Glasses containing lead oxide are not recommended for food contact surfaces.

Porcelain enamels meet a variety of performance characteristics required for different applications. The common characteristics of all enamels include good adherence to the substrate and good thermal expansion fit to the metal. Specific properties depend on usage; for example, acid and alkali resistance, hot water resistance, abrasion resistance, thermal shock resistance, high gloss, high reflectance, specific color, heat resistance, and cleanability.

Titanium Dioxide. The recrystallization of titanium dioxide in a cover-coat glass is very important to the development of thin, highly opaque finish coats. Titania, TiO_2, is the primary opacifying agent for white finish coats. Two polymorphic forms of titania, anatase and rutile, may be present in the enamel (see TITANIUM COMPOUNDS–INORGANIC). Anatase is preferred because anatase crystals are present in the size range (0.1–0.2 μm) for maximum reflectance, and therefore generate the most desirable bluish white color.

Properties

Thermal Fit and Residual Stresses. Thermal expansion measurements are typically carried out on dilatometric equipment consisting of a fused quartz pusher rod inside a tube of the same material.

A porcelain enamel glass becomes less viscous as the temperature is increased during firing. Above the softening point, the enamel is relatively fluid and conforms to the metal surface. As the porcelain enamel is cooled from the firing temperature of 750–800°C to the softening point, the fluid glass does not retain stress. However, as cooling proceeds below the softening point, the expansion (or contraction) or the coating exceeds that of the steel, and tensile stresses begin to develop in the coating. On further cooling, the stress increases until the temperature at which the expansion of the glass equals that of the metal. With still further cooling, the coefficient of expansion of the glass is less than that of the metal, coating tensile stresses decrease, and compressive stresses develop and are retained at room temperature.

Thermal expansion comparisons of the coating and metal have often been used to determine residual stresses in the coatings.

Composite Modulus of Elasticity. The modulus of elasticity of the enamel glass–steel composite system has been shown to lie between the modulus of the glass and that of the metal. The composite modulus can be calculated by $E_c = (E_m - E_e)Q^3 + E_e$, where E_c, E_m, and E_e are the modulus of elasticity of the composite, the metal, and the enamel, respectively; and Q = thickness of the metal divided by total thickness of the composite.

Residual Compressive Stress. Residual compressive stress in commercial ground coat enamels varies with enamel thickness.

Thinner coatings have higher compressive stresses, other factors being equal. Higher residual compressive stress in the coating also can be obtained by using enamel glass having a lower thermal expansion coefficient, or a metal having a higher expansion coefficient or a higher modulus of elasticity.

Maximum Strain. Strain in enamels that leads to failure is on the order of 0.002–0.003 cm/cm. Thinner enamels having higher residual compressive stresses are more flexible and can be strained to a greater degree.

Some other physical properties of enamel glass are density, from 2.5–3.5 g/mL; Mohs' hardness, 5–6; tensile strength, 34–103 MPa (4,900–15,000 psi); compressive strength, 1380–2760 MPa ($2-4 \times 10^5$ psi); modulus of elasticity, 55–83 GPa ($8-12 \times 10^6$ psi); and dielectric constant, 5–10.

Appearance and Color. Porcelain enamel allows the designer great variety with regard to color, texture, and aesthetic appeal. The enamels exhibit exceptional color stability whether used in domestic interiors or architectural exteriors. Colors should be grouped according to the type and the usage of enamel. Categories are *(1)* ground coats, which can generally be used by themselves as a finish coat or a base coat; *(2)* cover coats that are self-opacified; and *(3)* clear and semiopaque enamels, used for developing a wide range of colors. Transition-metal oxides used principally to develop a bond between the glass and the substrate metal also produce specific colors. Those range in shades of grays from blue to yellow and red to green. Combinations of each are also possible. Cobalt oxide is the principal colorant for blue, nickel oxide, and iron oxide, Fe_2O_3, for green, manganese dioxide for red, and nickel oxide with other oxides for yellow.

Cover coats such as self-opacified titanium enamels derive their color (qv) from titanium dioxide crystals nucleated in the glass during firing of the coating.

Textures in enamels are developed by the use of semicrystalline glasses or by the addition of refractory materials such as quartz, alumina, zirconia and zircon, feldspar, various clays, and titania. More refractory glasses are also added to impart a lower gloss and a texture.

Colored enamels are produced by tinting the titania enamels during the smelting operation through the addition of colorant oxides. Control of enamel color is most popularly done using computer-controlled color measuring systems consisting of spectrophotometric detectors (380–720 nm) and hardware to measure, integrate, evaluate, report, and store the color data. The most common systems in use are the Commission Internationale de l'Eclairage (CIE) $L^*a^*b^*$ and Rd,a,b color scales and various illuminates.

Decorating. The decoration of enamels is primarily done by silk screening. Other methods include indirect printing such as use of decals (decalcomania), and indirect lithographic, thermoplastic, and total transfer printing. Photographic printing is also done as a specialty.

Microstructure and Thickness. The microstructure of enamels is of importance to understanding and thus being able to control the macroproperties of the enamel. The microstructure is also related to thickness of the enamel and its firing history. Porcelain enamels typically have a bubble structure which is a result of gas evolution during firing.

The thickness of the enamel layer varies with the type of use and metal. However, typical thicknesses are from 75 to 150 μm (3 to 6 mils) for sheet steel, 175 to 359 μm (7 to 14 mils) for hot-rolled steel, 100 to 125 μm (4 to 5 mils) for each coat of wet process cast iron (760 to 788°C fire), 15 to 25 μm (0.5 to 1 mil) for dry process cast-iron base coats and 750 μm (average) to nearly 2250 μm (30 mils to 90 mils) for the dusted cover coats (898 to 955°C fires), and 25 to 50 μm (1 to 2 mils) for aluminum alloys. Stainless steel and copper may have enamel coatings from 40 μm to 175 μm (1.5 to 7 mils) thick.

The study of cross sections is extremely valuable as a way of tracking defects and determining sources of contaminant once the enameled article has been produced.

Enamel Testing

Standards. The development of standards for porcelain enamel coatings is shared by several national and international organizations. The American Society for Testing and Materials (ASTM), the Porcelain Enamel Institute (PEI), and the American National Standards Institute (ANSI), as well as the Association of Home Appliance Manufacturers (AHAM) are active in developing, collecting, and disseminating information to interested organizations. Cooperation with the International Standards Organization (ISO) is also fostering the development of internationally unified standards for vitreous enamel coatings. Enamel is tested for abrasion resistance, adherence, impact resistance, thermal shock resistance, resistance to chemical attack, and enamel defects.

Economic Aspects

The porcelain enameling market has been gradually declining since 1974 in the United States in terms of tonnage. Self-smelters represent about half of the total production and specialty producers the remainder. Substitution of less expensive, and often less durable, organic coatings and plastics has been mainly responsible for the decline in the usage of porcelain enamel.

William D. Faust
Ferro Corporation

A. I. Andrews, *Porcelain Enamels*, 2nd ed., Garrard Press, Champaign, Ill., 1961.

J. F. Wright, C. G. Bergeron, and J. C. Oliver, "Porcelain Enamel," in S. J. Schneider, Jr., vol. chrmn., *Engineering Materials Handbook*, Vol. 4, ASM International, Materials Park, Ohio, 1991, pp. 937–942.

J. Watril, *Vitreous Enamels*, Borax Holdings Ltd., London, 1984.

A. H. Dietzel, *Emaillierung: Wissenschaftliche Grundlagen und Grundzuge der Technologie*, Springer-Verlag, Berlin, 1981.

ENERGY MANAGEMENT

In the chemical industry, which is inherently energy intensive, energy costs, including feedstock, average approximately 8% of the value added. For large-volume chemicals these costs represent a much higher fraction. For example, for nitrogen-based fertilizer the energy costs are approximately 70% of the value added (see Fertilizers).

Energy management includes energy conservation, but also encompasses utility system reliability; the intermesh of process design with utility systems; purchasing, including plant location for minimum energy cost; environmental impacts of energy use; tracking energy performance; and the optimization of energy against capital in equipment selection (see also Economic evaluation; Power generation; Process energy conservation).

Energy and the Chemical Industry

The chemical industry used 21% of the energy consumed by the U.S. industrial sector, and the other three related process industries, paper (qv), petroleum (qv), and primary metals, combined for an additional 50% of the industrial consumption.

Feedstocks. A separate breakdown between fuels and feedstocks (qv) for the chemical industry shows that the quantity of hydrocarbons (qv) used directly for feedstock is about as great as that used for fuel. Much of this feedstock is oxidized accompanied by the release of heat, and in many processes, by-product energy from feedstock oxidation dominates purchased fuel and electricity.

Fuels. Two-thirds of the fuel used by the United States chemical industry in 1988 was natural gas, which is clean and easy to combust (see Gas, natural). Although relatively inexpensive at the wellhead, natural gas is costly to transport. Hence the chemical industry is concentrated in regions where natural gas is produced, keeping the average price paid by the U.S. chemical industry for natural gas in 1988 to only 80% of the average U.S. industrial price. Similarly the movement of chemical commodity production to the Middle East is driven by the desire to obtain low cost natural gas.

Waste Fuel Utilization. It is always preferable to minimize or upgrade by-products for sale as chemicals, however when this is not feasible the fuel value can still be recovered. Increased combustion of by-product gases and liquids was one of the principal components in the improvement in energy efficiency that has occurred in the industry.

Electricity. Electricity, including the losses associated with production, represents 24% of the total energy used by the chemical industry. On a cost basis, electricity represents a higher share at 29% of the energy bill including feedstocks. Increases in electrical costs have provided the driving force for increased cogeneration, ie, the recovery of power as a by-product of other process plant operations.

Energy Efficiency Improvements. Efficiency improvement is driven by two distinct forces: technological progress which is the long-term trend, and cost optimization which is the short-term response to price swings. The baseline, long-term trend has been in the range of 2 to 3% per year improvement. The forecast for the 1990s is 1 to 2% per year, reflecting a more mature industry.

Whereas energy conservation is an important component of cost reduction for the chemical industry, conservation is rarely the only driving force for technological change. Much of the increased energy efficiency comes as a by-product of changes made for other reasons such as higher quality, increased product yield, lower pollution, increased safety, and lower capital.

Energy and the Environment. The impact of energy usage on gaseous emissions has emerged as a primary environmental issue, and regulatory action has required emission reductions in NO_x and SO_2 (see AIR POLLUTION; AIR POLLUTION CONTROL METHODS; EXHAUST CONTROL, INDUSTRIAL). Control of NO_x is achieved by limiting the temperature of combustion and limiting oxygen. In some cases this involves running the first part of the combustion in a reducing atmosphere. Control of SO_2 either requires changing fuel or flue gas scrubbing. Because the preferred fuel of the chemical industry is already predominantly low sulfur natural gas, the primary impact on the chemical industry of SO_2 regulation is expected to be to raise the price of electricity derived from coal (qv).

Issues related to gases such as CO_2, which contribute to the global greenhouse effect, are also rising in importance (see ATMOSPHERIC MODELING). Energy conservation directly reduces CO_2 emissions. The elimination of fugitive hydrocarbon emissions as a result of improved maintenance procedures is also a tangible step that the industry is taking.

Energy Technology

Energy management requires the merging of such technologies as thermodynamics, process synthesis, heat transfer, combustion chemistry, and mechanical engineering (see also COMBUSTION SCIENCE AND TECHNOLOGY; HEAT-EXCHANGE TECHNOLOGY; and THERMODYNAMICS).

Thermodynamics. The first law of thermodynamics, which states that energy can neither be created nor destroyed, dictates that the total energy entering an industrial plant equals the total of all of the energy that exits. Feedstock, fuel, and electricity count equally, and a plant should always be able to close its energy balance to within 10%. If the energy balance does not close, there probably is a big opportunity for saving.

The second law of thermodynamics focuses on the quality, or value, of energy. The measure of quality is the fraction of a given quantity of energy that can be converted to work. What is valued in purchased energy is the ability to do work.

Unlike the conservation guaranteed by the first law, the second law states that every operation involves some loss of work potential, or energy. The second law is a very powerful tool for process analysis, because this law tells what is theoretically possible, and pinpoints the quantitative loss in work potential at different points in a process. Typically, the biggest loss that occurs in chemical processes is in the combustion step. The second law can also suggest appropriate corrective action. For example, in combustion, preheating the air or firing at high pressure in a gas turbine, as is done for an ethylene (qv) cracking furnace, improves energy efficiency by reducing the lost work of combustion.

Converting Heat to Work. There has been a historic bias in the chemical industry to think of energy use in terms of fuel and steam (qv) systems. A more fundamental approach is to minimize the input of work potential embedded in the fuel and feedstock, as well as work purchased directly as electricity.

Economics of Energy Levels and Power Recovery. In a steam system, steam at high pressure can be let down through a turbine for power. The shaft work developed by the turbine is sometimes referred to as by-product power, and the process is referred to as cogeneration.

The by-product power takes only 40% as much energy to produce as on-purpose firing for power only. As a result, by-product power is much cheaper than power purchased as electricity.

There is, however, only a limited quantity of by-product power available, and for large process operations the demand for power is usually far greater than the simple steam cycle can produce. Many steam system design decisions fall back to the question of how to raise the ratio of by-product power to process heat. One simple approach is to limit the turbines that are used to extract power to large sizes, where high efficiency can be obtained. Another way to raise the power/heat ratio is by raising the pressure of the steam system.

The combined cycle first fires fuel into a gas turbine and greatly increases the power extracted per unit of steam produced. The big advantage of the gas turbine in cogeneration is that it permits a much higher ratio of power to heat. This ratio, which is routinely >0.8, gets bigger as the unit size of the turbine increases. The ratio of power to heat is also larger for aero-derivative systems, which are basically jet engines exhausting into power recovery turbines.

Gas turbine cogeneration is inherently relatively low in capital cost. The absence of heat exchange surface in the gas turbine part of the cycle provides the basic capital advantage, and standardized equipment, prepackaged as skid mounted components, adds to the capital advantage. When these factors are coupled with low priced natural gas, a situation results in which petrochemical plants have become exporters of cogenerated power to utilities. The gas turbine also has advantages that are firmly rooted in thermodynamics. It utilizes energy directly at a high temperature level, without large driving forces for pressure drop and temperature difference.

Most gas turbine applications in the chemical industry are tied to the steam cycle, but the turbines can be integrated anywhere in the process where there is a large requirement for fired fuel. The combined cycle is also applicable to dedicated power production.

Steam is by far the biggest opportunity for power recovery from pressure letdown, but others such as tailgas expanders in nitric acid plants and on catalytic crackers, also exist.

Heat Recovery, Energy Balances, and Heat-Exchange Networks. The goal of heat recovery is to be sure that energy does the maximum useful work as it cascades to ambient. An energy balance is a summary of all of the energy sources and all of the energy sinks for a unit operation, a process unit, or an entire manufacturing plant. Table 1 gives an energy balance for a simple propane-fired dryer. The energy balance is almost as important to understanding how a process works as the material balance. The energy balance is the basic tool for analyzing an operation for energy conservation opportunities. When incorporated into a computer program, the energy balance becomes the base for a model of the process. Operational changes, system configuration, and equipment alterations can be evaluated via the model.

The heat exchanger network analysis, sometimes called pinch technology, is a special kind of model that has been developed into a sophisticated way of attacking heat recovery problems. This type of analysis defines the optimum interchange between all the heat sinks and heat sources.

In most chemical process plants, the steam system is the integrating energy system. Recovering waste heat by generating steam makes the heat usable in any part of the plant served by the steam system.

Table 1. Product Dryer Analysis Heat Balance

Material	Mass, kg/h	Energy, MJ/h[a]
Inputs		
fuel, C_3H_8	130	6,553
air		
combustion	6,817	106
secondary	14,846	232
in-leakage	4,289	67
water with product	1,731	354
dry product solids	4,478	249
Totals in	*32,291*	*7,561*
Outputs		
water vapor		
from product	1,445	3,808
from combustion of H_2	212	560
air and combustion products	25,870	1,933
dry product solids	4,478	458
water with product out	286	146
heat losses		656
Totals out	*32,291*	*7,561*

[a] To convert J/h to Btu/h, multiply by 0.95×10^{-3}.

Heat exchange is commonly used to cool the product of a thermal process by preheating the feed to that process, thus providing a natural stabilizing, feed forward type of process integration. Product-to-feed interchange is common on reactors as well as distillation (qv) trains.

Flue gas to air exchange, a type of product-to-feed heat exchange is extremely important because of the large loss associated with the combustion of unpreheated air. This exchange process has generated fairly unique types of hardware such as the Ljungstrom or rotary wheel regenerator; the brick checkerwork regenerators used in metallurgical furnaces, hot oil, or hot water belts (also called "liquid runarounds"); and heat pipes. Liquid runaround systems make it practical to use finned surface on both gas exchange surfaces. These are particularly useful for retrofit because of the ability to move the heat to physically separated units.

The use of heat pumps adds a compressor to boost the temperature level of rejected heat. It can be effective in small plants having few opportunities for heat interchange.

Heat Recovery Equipment. Factors that limit heat recovery applications are corrosion, fouling, safety, and cost of heat-exchange surface. Most heat interchange utilizes shell and tube-type units because of the rugged construction, ease of mechanical cleaning, and ease of fabrication in a variety of materials. However, there are a rich assortment of other heat exchangers. Examples found in chemical plants in special applications include the following: plate heat exchangers, brazed-fin aluminum cores, and spiral plate construction.

Utility Systems

Steam. The steam system serves as the integrating energy system in most chemical process plants. Steam holds this unique position because it is an excellent heat-transfer medium over a wide range of temperatures. Water gives high heat-transfer coefficients whether in liquid phase, boiling, or in condensation. In addition, water is safe, nonpolluting, and if proper water treatment is maintained, noncorrosive to carbon steel.

The steam balance is usually the most important plant-wide energy balance. It shows each service requirement, including the use of steam as a working fluid to develop power.

Because of increased emphasis on maximizing cogenerated power, newer plants are trying to utilize back-pressure turbines only in applications where efficiencies above 70% can be attained. This typically means limiting the applications to the large (>1000 kW) drives, and using small machines only where they are necessary for the safe shutdown of the unit. Multistage turbines are used even on the smaller loads. Most large plants also have some condensing turbines to handle process and seasonal swings and provide some flexibility to the steam balance.

In a process plant, steam traps are used to drain and return condensate. Given proper application and continuous maintenance, these can operate with minimal steam leakage.

Electrical. The plant electrical system is sometimes more important than the steam system. The electrical system consists of the utility company's entry substation, any in-plant generating equipment, primary distribution feeders, secondary substations and transformers, final distribution cables, and various items of switch-gear, protective relays, motor starters, motors, lighting control panels, and capacitors to adjust power factor.

Other Energy Systems. Chemical plants usually require cooling water, compressed air, and fuel distribution systems. Sometimes also included are refrigeration, pressurized hot water, or specialized heat-transfer fluids such as Therminol liquid or condensing vapor. Each of these systems serves the process and reliability is the most important characteristic. Thus a project in any of them that achieves a 10% reduction in energy cost at the expense of a 1% loss of reliability loses money for the operation.

Capital and Equipment Areas

Virtually all chemical processing is energy driven, but in separations such as distillation, drying (qv), and evaporation (qv), this is particularly clear. All three of these processes are simple thermal operations that involve separation through vaporization, and only a minor change in the chemical energy of the products. The capital related costs of those operations are typically three to five times as large on a lifetime basis as the energy costs. In almost all cases there is a balance between capital and energy costs, and typically one is traded against the other to achieve the lowest overall cost. Much can be said about this energy/capital trade based on first principles and engineering correlations.

Insulation. A surprisingly important capital element of energy management is insulation. On large projects the capital cost of insulation is in the same range as that for heat exchangers or distillation towers and trays. At the optimum insulation thickness, the lifetime value of the insulation approximates the lifetime value of the heat loss; ie, insulation is as costly as the heat loss that it prevents.

Insulation provides other functions in addition to energy conservation. A key role of insulation is safety. It protects personnel from burns and minimizes hot surfaces that could ignite inflammables. It also protects equipment, piping, and contents in event of fire.

Compressors. Compression equipment accounts for a large fraction of power use as well as a large fraction of installed capital. Usually the energy bill for a compressor is large enough to warrant a very visible monitor of the driver, such as a control room electric meter.

Pumps. Energy use for pumps (qv) can best be controlled by design for the proper flow and discharge pressure.

Vacuum Systems. The basic question in vacuum systems is what can be done to cut design inert loading. One factor driving toward greater use of vacuum pumps is the large reduction they achieve in effluent to be treated.

Boilers and Process Furnaces. Boilers and process fired heaters are the entry point for the energy released from burning fuel. Fuel combustion is irreversible, and fired heaters are typically the principal loss point for work potential. Air preheat cuts energy losses by cutting fuel firing and increasing the flame burst temperature.

A more obvious energy loss is the heat to the stack flue gases. The sensible heat losses can be minimized by reduced total air flow, ie, low excess air operation. Flue gas losses are also minimized by lowering the discharge temperature via increased heat recovery in economizers, air preheaters, etc.

Distillation. The optimum reflux rate for a distillation column depends on the value of energy, but is generally between 1.05 times and

1.25 times the reflux rate, which could be used with infinite trays. At this level, excess reflux is a secondary contributor to column inefficiency. However, when designing to this tolerance, correct vapor–liquid equilibrium data and adequate controls are essential.

Energy savings for improved control are surprisingly high. A 2 to 20% savings from a series of control projects has been reported. Improved control achieves this by permitting a reduction in the margin of safety that the operators use to handle changes in feed conditions. The key element is the addition of feed-forward capability, which automatically handles changes in feed flow and composition.

The real work used in a distillation varies with the temperature difference between the heating medium and the cooling medium.

There are a number of ways to provide the heating or cooling medium at temperatures closer to the optimum level. One is by use of double-effect distillation, which uses the overhead vapor from one column as the heat source for another column such that the second column's reboiler becomes the first column's condenser. This basically cuts the temperature differential in half, and shows up as an energy saving because external heat is supplied to only one of the units.

Another element that sets the temperature differential across the distillation is the pressure drop in the column and its auxiliaries. One of the more recent changes is the introduction of special structured packings which give extremely low (10% of an equivalent column with trays) pressure drop. This energy benefit can show up in an overhead temperature high enough to permit generation of by-product steam. It can also show up in a variety of other ways including lower bottoms temperature, yielding less fouling and product degradation to by-products, as in the styrene–ethylbenzene separation.

Drying. A typical dryer mass and energy balance shows that the heat loss is 10% of the fuel input. Improving insulation is one of the simplest ways to reduce energy input. Another simple way to reduce energy input is improving the dewatering (qv) of the feed.

Some of the other energy conservation approaches applicable to dryers are interchange between the stack vapor and the incoming dryer air; recovering sensible heat from the product; use of waste heat from another operation for air preheat; and using less, but hotter drying air. This last is limited to nonheat-sensitive materials.

A single-effect evaporator produces slightly less than a kilogram of water vapor per kilogram of steam. By using the vapor produced by the first-effect as the heat source for a second-effect evaporator, steam use can be essentially halved. The performance can be improved almost in proportion to the number of effects employed.

Much as reverse osmosis (qv) can compete with evaporation in desalination applications, osmosis should also be considered as an alternative for process evaporation.

Energy Accounting and Improvement

Energy Accounting. Long-term costs at plant gate meters are the costs that should be considered in project evaluations. The real benefit of a technical action can be quite different from the savings allocated by the accounting system.

Measurements and Audits. The enabling element of continuous improvement is measurement. Metering of the cost elements at each unit in a chemical plant provides effective accountability. Measurements should be linked via computer software to production as well as to weather to result in maximum feedback.

DAN STEINMEYER
Monsanto Company

W. F. Kenney, *Energy Conservation in the Process Industries*, Academic Press, Inc., New York, 1984.

R. Smith and B. Linnhoff, *Chem. Eng. Res. Des.* **66**, 195 (1988).

D. E. Steinmeyer, *CHEMTECH* **188** (Mar. 1982).

ENGINEERING, CHEMICAL DATA CORRELATION

It is a familiar scenario in the chemical sciences: a straightforward set of calculations becomes bogged down for lack of one or more pieces of data, particularly physical property data. Whereas the use of experimentally obtained data is usually preferred over values obtained from data prediction methods, such data are often unavailable or are of dubious quality. In such cases, correlations to obtain the required values must be employed. Because it is not sufficient to provide just a solution, but the best solution given the constraints in time, funding, and information, a large portion of scientific and engineering investigation is devoted to the assessment of the degree to which mathematical constructs faithfully mimic real systems. It is perhaps equally important to be able to determine the level of confidence to which a solution is ascribed.

Theoretically based correlations (or semitheoretical extensions of them), rooted in thermodynamics or other fundamentals are ordinarily preferred. However, rigorous theoretical understanding of real systems is far from complete, and purely empirical correlations typically have strict limits on applicability. Many correlations result from curve-fitting the desired parameter to an appropriate independent variable. Some fitting exercises are rooted in theory, eg, Antoine's equation for vapor pressure; others can be described as being semitheoretical. These distinctions usually do not refer to adherence to the observations of natural systems, but rather to the agreement in form to mathematical models of idealized systems. The advent of readily available computers has revolutionized the development and use of correlation techniques (see CHEMOMETRICS; COMPUTER TECHNOLOGY; DIMENSIONAL ANALYSIS).

For engineering and physical property data correlations, it is important to *(1)* determine the accuracy of the experimental data to which the correlation is to be applied; *(2)* understand the limitations of the correlation technique employed: *(3)* minimize mathematical complexity (additional fitting parameters may improve data regressions, but may also generate local minima/maxima which do not exist beyond the mathematical model); *(4)* perform sufficient sensitivity analyses to be able to assess the confidence level of final results; and *(5)* to present solutions within the context of confidence levels and assumptions.

Thermodynamic Correlations for Property Estimation

In the broadest sense, thermodynamics is concerned with mathematical relationships that describe equilibrium conditions as well as transformations of energy from one form to another. Many chemical properties and parameters of engineering significance have origins in the mathematical expressions of the first and second laws and accompanying definitions. Particularly important are those fundamental equations which connect thermodynamic state functions to real-world, measurable properties such as pressure, volume, temperature, and heat capacity (see also THERMODYNAMIC).

The phase rule specifies the number of intensive properties of a system that must be set to establish all other intensive properties at fixed values, without providing information about how to calculate values for these properties. The field of applied engineering thermodynamics has grown out of the need to assign numerical values to thermodynamic properties within the constraints of the phase rule and fundamental laws. In the engineering disciplines there is a particular demand for physical properties, both for pure fluids and mixtures, and for phase equilibrium data.

Data compilations, the first recourse for an engineering calculation requiring physical property or parameter data, are often incomplete or do not contain data within the appropriate range of temperature or pressure. For this reason, correlation and estimation methods play an important role in applied thermodynamics.

The common theme in the evolution of methods for property and parameter prediction is the development of equations, either theoretical or empirical, containing quantities that can be calculated from theoretical considerations or experimental data. Mathematical expressions for correlating thermodynamic data may take several forms, eg, limiting laws, equations of state, fundamental property relation, semiempirical relationships, mathematical consistency, chemical potential, and generalized correlations.

Theoretical and Empirical Correlating Forms. *Fundamental Property Relation.* For homogeneous, single-phase systems the fundamental property relation, is a combination of the first and second laws of thermodynamics that may be written as $d(nU) = Td(nS) - Pd(nV) + \epsilon(\mu_i dn_i)$ (eq. 1) where U = internal energy, T = absolute temperature, S = entropy, P = pressure, μ_i = chemical potential, V = system volume, n_i = moles of i, and n = total moles in the system.

The definitions of enthalpy, H, Helmholtz free energy, A, and Gibbs free energy, G, also give equivalent forms of the fundamental relation which apply to changes between equilibrium states in any homogeneous fluid system.

Clausius-Clapeyron Equation. Derived from equation 1, the Clapeyron equation is a fundamental relationship between the latent heat accompanying a phase change and pressure–volume–temperature (PVT) data for the system:

$$\Delta H = T\Delta V\frac{dP^{\text{sat}}}{dT} \tag{2}$$

The Clapeyron equation is most often used to represent the relationship between the temperature dependence of a pure liquid's vapor pressure curve and its latent heat of vaporization.

Equations of State. An equation of state can be an exceptional tool for property prediction and phase equilibrium modeling. The term equation of state refers to the equilibrium relation among pressure, volume, temperature, and composition of a substance. This substance can be a pure chemical or a uniform mixture of chemicals in gaseous or liquid form.

Theoretical equation forms may be derived from either kinetic theory or statistical mechanics. However, empirical and semitheoretical equations of state have had the greatest success in representing data with high precision over a wide range of conditions.

The Virial Expansion. Many equations of state have been proposed for gases, but the viral equation is the only one having a firm basis in theory. The pressure-explicit form of the virial expansion is

$$Z \equiv \frac{PV}{RT} = 1 + B'P + C'P^2 + D'P^3 + \cdots \tag{3}$$

where the ratio PV/RT is called the compressibility factor and is given the symbol Z. An equivalent expression for Z, explicit in volume, is

$$Z = 1 + \frac{B}{V} + \frac{C}{V^2} + \frac{D}{V^3} + \cdots \tag{4}$$

Equations 3 and 4 are known as virial expansions, and the coefficients $B', C', D', \ldots B, C, C, \ldots$ are called virial coefficients. For a given substance, these coefficients are functions of temperature only.

Liquid-Phase Models. The Tait equation for the pressure–volume behavior of liquids correlates data accurately, and is expressed mathematically as

$$V = V_0 - D\ln\left(\frac{P + E}{P_0 + E}\right) \tag{5}$$

where the parameters D and E are empirical constants for a given temperature, and V_0 and P_0 are molar volume and pressure values at some reference state of the liquid.

Another empirical model for liquid pressure–volume behavior is the generalized equation for the molar volumes of saturated liquids given by the Rackett equation: $V^{\text{sat}} = V_c Z_c^{(1-Tr)^{0.2857}}$ (eq. 6) which employs critical volume, V_c, the compressibility factor at the critical point, Z_c, and reduced temperature, T_r. The Rackett equation produces results accurate to about 2%.

Mixing Rules. Kay's method is a mixing rule for determining the pseudocritical temperature, T_{cm}, and pressure P_{cm}, of a mixture of gases. In this method, pseudocritical values for mixtures of gases are calculated on the assumption that each component in the mixture contributes to the pseudocritical value in the same proportion as the number of moles of that component. The resulting values are referred to as linearly weighted mole-average pseudocritical properties of pressure, temperature, and volume.

Correlations for Enthalpy of Vaporization. Enthalpy or heat of vaporization, which is an important engineering parameter for liquids, can be predicted by a variety of methods which focus on either prediction of the heat of vaporization at the normal boiling point, or estimation of the heat of vaporization at any temperature from a known value at a reference temperature.

Limiting Laws. *Ideal Gas Behavior.* Expressed mathematically, the ideal gas law is $pV = nRT$ (eq. 7), where p is the pressure exerted by the gas, V is the volume of the gas, n is the number of moles of gas, R is the universal gas constant, and T is the absolute temperature of the gas. Real gases closely obey this pressure–volume–temperature relationship at low pressures and high temperatures.

Ideal Liquid Solutions. Two limiting laws of solution thermodynamics that are widely employed are Henry's law and Raoult's law, which represent vapor–liquid partitioning behavior in the concentration extremes. These laws are used frequently in equilibrium problems and apply to a variety of real systems.

In general, equality of component fugacities, ie, chemical potentials, in the vapor and liquid phases yields the following relation for vapor–liquid equilibrium:

$$y_i\phi_i P_T = x_i\gamma_i P_i^{\text{sat}}\phi_i^{\text{sat}}\pi_i \tag{8}$$

in which the Poynting correction factor, π_i, is defined as:

$$\pi_i = \exp\left(\frac{V_i^l(P_T - P_i^{\text{sat}})}{RT}\right) \tag{9}$$

and where y_i and x_i are the equilibrium mole fractions of i in the vapor and liquid phases, respectively, V^l_i is the liquid molar volume or pure i, ϕ_i is the vapor-phase fugacity coefficient, and P_T is the total pressure.

Partial Molar Properties. Perhaps the most significant of the partial molar properties, because of its application to equilibrium thermodynamics, is the chemical potential, μ_i. This fundamental property, and related properties such as fugacity and activity, are essential to mathematical solutions of phase equilibrium problems. The natural logarithm of the liquid-phase activity coefficient, $\ln\gamma_i$, is also defined as a partial molar quantity. For liquid mixtures, the activity coefficient, γ_i, describes nonideal liquid-phase behavior.

Heat Capacities. The heat capacities of real gases are functions of temperature and pressure, and this functionality must be known to calculate other thermodynamic properties such as internal energy and enthalpy.

Mathematical Consistency Requirements. Theoretical equations provide a method by which a data set's internal consistency can be tested or missing data can be derived from known values of related properties. The ability of data to fit a proved model may also provide insight into whether that data behaves correctly and follows expected trends.

From the definition of a partial molar quantity and some thermodynamic substitutions involving exact differentials, it is possible to derive the simple, yet powerful, Duhem data testing relation. Stated in words, the Duhem equation is a mole-fraction-weighted summation of the partial derivatives of a set of partial molar quantities, with respect to the composition of one of the components.

The well-known Gibbs-Duhem equation is a special mathematical redundance test which is expressed in terms of the chemical potential. The general Duhem test procedure can be applied to any set of partial molar quantities. It is also possible to perform an overall consistency test over a composition range with the integrated form of the Duhem equation.

In some cases, reported data do not satisfy a consistency check, but these may be the only available data. In that case, it may be possible to smooth the data in order to obtain a set of partial molar quantities that is thermodynamically consistent. The procedure is simply to reconstruct the total molar property by a weighted mole fraction average of the n measured partial molar values and then recalculate normalized partial molar quantities. The new set should always be consistent.

For interrelated properties, a method of assessing data quality is to see if data for those properties are consistent with theoretical equations.

Many additional consistency tests can be derived from phase equilibrium constraints. From thermodynamics, the activity coefficient is known to be the fundamental basis of many properties and parameters of engineering interest. Therefore, data for such quantities as Henry's constant, octanol–water partition coefficient, aqueous solubility, and solubility of water in chemicals are related to solution activity coefficients and other properties through fundamental equilibrium relationships. Accurate, consistent data should be expected to satisfy these and other thermodynamic requirements. Furthermore, equilibrium models may permit a missing property value to be calculated from those values that are known.

Phase Equilibria and Chemical Potential. *Models of Phase Nonideality.* The correlation and prediction of phase equilibria at low (near-vacuum) to moderate (several hundred kPa) pressures is typically based on a form of equation 8. The key is the degree to which the various parameters for phase nonideality are known or can be estimated.

An exact equation that is widely used for the calculation of fugacity coefficients and fugacities from experimental pressure–volume–temperature (PVT) data is

$$\ln \hat{\phi}_i = \int_0^P (\overline{Z}_i - 1)\frac{dP}{P} \qquad \text{(constant } T, x) \tag{10}$$

where ϕ_i = fugacity coefficient of i in a vapor mixture, and

$$\overline{Z}_i = P\overline{V}_i/RT \tag{11}$$

which leads to another form of equation 10:

$$\ln \hat{\phi}_i = \frac{-1}{RT}\int_0^P \left(\frac{RT}{P} - \overline{V}_i\right)dP \qquad \text{(constant } T, x) \tag{12}$$

Here a suitable equation of state is required to provide a mathematical expression for the mixture molar volume, $\overline{V}_i$. For some equations of state, it is better to use a form of equation 12 in which the integral is volume explicit. Note also that for an ideal gas $Z_i = \overline{Z}_i = 1$, and $\phi_i = 1$.

Activity coefficients in liquid mixtures are directly related to the molar excess Gibbs energy of mixing, ΔG^E, which is defined as the difference in the molar Gibbs energy of mixing between the real and ideal mixtures. It is typically an assumed function. Various functional forms of ΔG^E give rise to many of the different activity coefficient models found in the literature. Typically, the liquid-phase activity coefficient is a function of temperature and composition; explicit pressure dependence is rarely included.

There are many simple two-parameter equations for liquid mixture constituents, including the Wilson, Margules, van Laar, nonrandom two-liquid (NRTL), and universal quasichemical (UNIQUAC), equations.

Other algorithms are much more complex and require computer software, such as group contribution models for estimating liquid-phase activity coefficients. These models include analytical solution of groups (ASOG), modified separation of cohesive energy density (MOSCED), and UNIQUAC functional-group activity coefficient (UNIFAC). Each model is applicable for some, but not all, chemical systems and conditions, and the choice of the functional form for the Gibbs excess energy leads to specialization and limitation.

The solvophobic model of liquid-phase nonideality takes into account solute–solvent interactions on the molecular level. In this view, all dissolved molecules expose microsurface area to the surrounding solvent and are acted on by the socalled solvophobic forces. These forces, which involve both enthalpy and entropy effects, are described generally by a branch of solution thermodynamics known as solvophobic theory. The solvophobic model has been used to deduce a functional form for a Henry's constant correlation based on molecular connectivity index and polarizability.

Regular Solution Theory. The key assumption in regular-solution theory is that the excess entropy, *SE*, is zero when mixing occurs at constant volume. This idea of a regular solution leads to the equations:

$$G^E = \sum_k [x_k V_k(\delta_k - \bar{\delta})^2] \tag{13}$$

and

$$\ln \gamma_i = \frac{V_i}{RT}(\delta_i - \bar{\delta})^2 \tag{14}$$

where

$$\bar{\delta} = \frac{\Sigma_k(x_k V_k \delta_k)}{\Sigma_k(x_k V_k)} \tag{15}$$

In these equations, *Vk* is the molar volume and δ_k is the solubility parameter of the pure liquid *k*. The solubility parameter is the square root of an energy density, defined as

$$\delta_i = \left(\frac{\Delta U_i^{\text{vap}}}{V_i}\right)^{1/2} \tag{16}$$

where $\Delta U^{\text{vap}}{}_i$ is the internal energy change of vaporization of pure liquid *i*. Values are determined from heats of vaporization.

Thermodynamics of Vapor–Liquid Equilibrium. Assuming ideal vapor and choosing the Lewis-Randall standard state, a chemical distributed between a vapor and liquid in equilibrium often behaves according to: $y_i P_T = p_i = x_i \gamma_i P_i^{\text{sat}}$ (eq. 17), where *yi* is the mole fraction of *i* in the vapor phase; *xi* is the mole fraction of *i* in the liquid phase; P_T is the total system pressure; p_i is the partial pressure of *i* in the vapor mixture; γ_i is the liquid-phase activity coefficient of *i* (Lewis-Randall basis); and $P^{\text{sat}}{}_i$ is the saturation pressure of pure liquid *i* at the system temperature.

Henry's Law Constant. Henry's law for dilute concentrations of contaminants in water is often appropriate for modeling vapor–liquid equilibrium (VLE) behavior. At very low concentrations, a chemical's Henry's constant is equal to the product of its activity coefficient and vapor pressure. Activity coefficient models can provide estimated values of infinite dilution activity coefficients for calculating Henry's constants as a function of temperature.

Thermodynamics of Liquid–Liquid Equilibrium. Phase splitting of a liquid mixture into two liquid phases (I and II) occurs when a single liquid phase is thermodynamically unstable. The equilibrium condition of equal fugacities (and chemical potentials) for each component in the two phases allows the fugacities f'_i and f''_i in phases I and II to be equated and expressed as: $x'_i\gamma'_i = x''_i\gamma''_i$ (eq. 18). The same reference (standard) state, $f^0{}_i$, is chosen for the two phases, so that it cancels on both sides of equation 18. The products $x'_i\gamma'_i$ and $x''_i\gamma''_i$ are referred to as activities. Because equation 18 holds for each component of a liquid–liquid system, it is possible to predict liquid–liquid phase splitting when the activity coefficients of the individual components in a multicomponent system are known. These values can come from vapor–liquid equilibrium experiments or from prediction methods developed for phase-equilibrium problems.

Aqueous Solubility. Solubility of a chemical in water can be calculated rigorously from equilibrium thermodynamic equations. Because activity coefficient data are often not available from the literature or direct experiments, models such as UNIFAC can be used for structure–activity estimations. Phase-equilibrium relationships can then be applied to predict miscibility. Simplified calculations are possible for low miscibility; however, when there is a high degree of miscibility, the phase-equilibrium relationships must be solved rigorously.

Octanol–Water Partition Coefficient. In environmental calculations, the octanol–water partition coefficient, K_{ow}, is related to a chemical's lipophilicity and correlates well with many properties.

Based on liquid–liquid equilibrium principles, a general model of octanol–water partitioning is possible if accurate activity coefficients can be determined.

Generalized and Reduced Equations

Reduced conditions are corrected, or normalized, conditions of temperature *T*, pressure *p*, and specific volume $\hat{V}$; and are expressed mathematically as $T_r = T/T_c$ (eq. 19), $p_r = p/p_c$ (eq. 20), $\hat{V}_r = \hat{V}/\hat{V}_c$ (eq. 21) where the subscript c denotes the critical state, and r the reduced state.

Principle of Corresponding States. *Generalized Correlations.* A simple and reliable method for the prediction of vapor–liquid behavior has been sought for many years to avoid experimentally measuring the thermodynamic and physical properties of every substance involved in a process. Whereas the complexity of fluids makes universal behavior prediction an elusive task, methods based on the theory of corresponding states have proven extremely useful and accurate while still retaining computational simplicity. Methods derived from corresponding states theory are commonly used in process and equipment design.

A generalized correlation is a functional relationship between dimensionless or reduced variables. Most generalized correlations may be expressed by the following function: $z = \psi(T_r, P_r)$ (eq. 22), which is the keystone to the theory of corresponding states. This form readily lends itself to the use of a generalized chart, where the compressibility factor (or any other property) is plotted as a function of the reduced pressure for each reduced isotherm. A great advantage of the generalized chart is its ease of use and fair accuracy (within 10%) obtained from limited information, ie, usually the critical pressure and critical temperature. Drawbacks to the generalized chart include the inability to predict properties with high precision (< 1% error) and the inconvenience of the form to computer application.

Reduced Properties. One of the first attempts at achieving an accurate analytical model to describe fluid behavior was the van der Waals equation, in which corrections to the ideal gas law take the form of constants a and b to account for molecular interactions and the finite volume of gas molecules, respectively.

$$\left(P + \frac{a}{\hat{V}^2}\right)(\hat{V} - b) = RT \tag{23}$$

This equation bridged the gap between gases and liquids by showing each could be described by a single cubic equation. Although the van der Waals equation's accuracy fails at the critical point and in the liquid region, it has had a great effect on the understanding of fluids. It is from this simple equation that the theory of corresponding states was proposed.

All fluids, when compared at the same reduced temperature and reduced pressure have approximately the same compressibility factor and deviate from ideal gas behavior to the same extent, giving

$$\left(P_r + \frac{3}{\hat{V}_r^2}\right)(3\hat{V}_r - 1) = 8T_r \tag{24}$$

which is theoretically equivalent for all substances.

The theory of corresponding states only works for certain classes of fluids, thus fluids have been placed into generalized groups based on behavior. These groupings include the following: simple fluids, normal fluids, slightly polar fluids, hydrogen-bonded, associating, polar fluids, and quantum fluids.

Reduced Equations of State. A generalized cubic equation is the Redlich-Kwong equation:

$$Z = \frac{1}{1-h} - \frac{4.9340}{T_r^{1.5}}\left(\frac{h}{1+h}\right) \tag{25}$$

where $h = 0.08664P_r/ZT_r$. This equation is useful for gases above the critical point. Only reduced pressure, P_r, and reduced temperature, T_r, are needed. In the form represented by equation 25, iteration quickly gives accurate values for the compressibility factor, Z. However, this two-parameter equation only gives accurate values for simple and nonpolar fluids. Unless the Redlich-Kwong equation (eq. 25) is explicitly solved for pressure in nonreduced variables, it does not give accurate liquid volumes.

An example of a more complex equation that exhibits better accuracy is the Benedict-Webb-Rubin equation:

$$P = \frac{RT}{\hat{V}} + \frac{B_0RT - A_0 - C_0/T^2}{\hat{V}^2} + \frac{bRT - a}{\hat{V}^3} + \frac{a\alpha}{\hat{V}^6} + \frac{c}{\hat{V}^3T^2}\left(1 + \frac{\gamma}{\hat{V}^2}\right)e^{-\left(\frac{\gamma}{\hat{V}^2}\right)} \tag{26}$$

where A_0, B_0, C_0, a,b,c, α, and γ are all empirical constants for specific substances. The complexity and lack of constant values for an equation of this sort is a weakness of some generalized equation correlations.

Three Parameter Models. Most fluids deviate from the predicted corresponding states values. Thus the acentric factor, ω, was introduced to account for asymmetry in molecular structure. The acentric factor is defined as the deviation of reduced vapor pressure from 0.1, measured at a reduced temperature of 0.7. In equation form this becomes: $\omega = -\log P_r - 1.000$ (eq. 27), where P_r is evaluated at $T_r = 0.7$. This definition was chosen because of the abundance and accuracy of vapor pressure data near the normal boiling point. Acentric factors for various pure compounds are given in many references.

The acentric factor, ω, was the third parameter used in an equation based on the second virial coefficient. This equation was further modified and is suitable for reduced temperatures above 0.5.

$$\left(\frac{BP_c}{RT_c}\right) = \left(\frac{B^{(0)}P_c}{RT_c}\right) + \omega\left(\frac{B^{(1)}P_c}{RT_c}\right) \tag{28}$$

where

$$\left(\frac{B^{(0)}P_c}{RT_c}\right) = 0.1445 - \frac{0.330}{T_r} - \frac{0.1385}{T_r^2} - \frac{0.0121}{T_r^3} - \left(\frac{0.000607}{T_r^8}\right)$$

$$\left(\frac{B^{(1)}P_c}{RT_c}\right) = 0.073 + \frac{0.46}{T_r} - \frac{0.50}{T_r^2} - \frac{0.097}{T_r^3} - \frac{0.0073}{T_r^8}$$

The second virial coefficient is related to the compressibility factor:

$$Z = 1 + B^{(0)}\frac{P_r}{T_r} + \omega B^{(1)}\frac{P_r}{T_r} \tag{29}$$

Four Parameter Models. Two- and three-parameter theories are only accurate for simple, normal, and some slightly polar fluids. In order to accurately predict polar fluid behavior a fourth parameter is needed. The Stiel polarity factor, χ, is one such fourth parameter and follows from the equation

$$\log P_r = (\log P_r)^{(0)} + \omega(\log P_r)^{(1)} + \chi(\log P_r)^{(2)} \tag{30}$$

where $(\log P_r)^{(0)}$ and $(\log P_r)^{(1)}$ are the same as those obtained for a normal fluid by equation 28.

Application to Other Properties. Extension of Generalized Charts. In 1975, the usefulness of generalized charts was extended upon the publication of extensive tables of residual enthalpy, entropy, and heat capacity. This tabular data has also been converted into graphical form. The corresponding equations incorporate the acentric factor. Another property which can be represented by generalized charts is fugacity, ϕ.

Generalized Surface Tension Correlations. Use of the principle of corresponding states has provided a practical and accurate method for the estimation of surface tensions. The functional relationship for the surface tension of a pure substance is $\sigma_r = \sigma_r(T_r; \alpha_k)$ (eq. 31) where

$$\sigma_r = \frac{\sigma}{P_c^{2/3}(xT_c)^{1/3}} \tag{32}$$

x is a constant that depends on the units of P_c and T_c and

$$\sigma = 4.6 \times 10^{-4}[P_c^{2/3}T_c^{1/3}\alpha_k(1 - T_r)^{11/9}] \tag{33}$$

$$\alpha_k = 0.1207\left[1 + \frac{T_b}{T_c}\frac{(\ln P_c - 11.526)}{\left(1 - \frac{T_b}{T_c}\right)}\right] - 0.281$$

For 84 simple inorganic substances and a variety of organic compounds, equation 33 produced an average error of 3.0%. The above correlations should not be applied to the quantum substances, associating substances, or highly polar substances.

In 1971 the Stiel polar factor was incorporated in an equation that accounts for polar fluids

$$\sigma = P_c^{2/3}T_c^{1/3}(\sigma_r\,\vdots_{T_r=0.6})\left(\frac{1 - T_r}{0.4}\right)^m \tag{34}$$

where

$$\sigma_r\,\vdots_{T_r=0.6}\ 0.1574 + 0.359\omega - 1.769\chi - 13.69\chi^2 - 0.510\omega^2 + 1.298$$

and

$$m = 1.210 + 0.5385\omega - 14.61\chi - 32.07\chi^2 - 1.65\omega^2 + 22.03\omega\chi$$

This equation was empirically derived from 16 polar fluids and has an average error of 2.9%. A technique for estimating surface tension using nonretarded Hamaker constants (89) has also been presented.

Generalized Correlations for Viscosity. Gas viscosity has also been predicted by corresponding states theory using

$$\eta\xi = [0.807T_r^{0.618} - 0.357e^{(-0.449T_r)} + 0.340e^{(-4.058T_r)} + 0.018]F_P^0F_Q^0 \tag{35}$$

where ξ is the reduced, inverse gas viscosity and is defined as

$$\xi = 0.176\left(\frac{T_c}{M^3P_c^4}\right)^{1/6} \tag{36}$$

where M is the molecular weight of the gas. For nonpolar, nonquantum gases the values of the low pressure polarity and quantum factors, $F^0{}_P$ and $F^0{}_Q$, are equal to 1.

Miscellaneous Generalized Correlations. Generalized charts and corresponding states equations have been published for many other properties in addition to those presented. Most produce accurate results over a wide range of conditions. Some of these properties include *(1)* transport properties; *(2)* second virial coefficients; *(3)* third virial coefficients; *(4)* liquid mixture activity coefficients; *(5)* Henry's constant; and *(6)* diffusivity.

Reference Substances. Use of a reference substance has its origins in the work of Clausius-Clapeyron and equation 73,

$$\ln P^{sat} = -\frac{\Delta H^{vap}}{RT} + I \tag{37}$$

where I is an integration constant. If the natural log of the vapor pressure, P^{sat}, is plotted vs the inverse of absolute temperature, T, then the slope of the line becomes the heat of vaporization, ΔH^{vap}, divided by the universal gas constant. A plot called a Cox chart can be constructed by plotting the ln P^{sat} of an unknown compound vs the ln P^{sat} of a well-documented substance at equal temperatures. By recording only the temperature at which the reference vapor pressure was measured on the x axis, the Cox chart becomes a valuable tool for estimating vapor pressures at temperatures for which no experimental values are available.

By use of Othmer plots of reference substances, large tables of thermodynamic data can be expressed as simple correlations which are extremely accurate and easy to use. The real power of these correlations is the ability to interpolate and extrapolate the correlations beyond the experimental values with considerable accuracy.

Values for many properties can be determined using reference substances, including density, surface tension, viscosity, partition coefficient, solubility, diffusion coefficient, vapor pressure, latent heat, critical properties, entropies of vaporization, heats of solution, colligative properties, and activity coefficients.

Compared to the theory of corresponding states, the reference substance method gives highly accurate results for compounds having sparse experimental data. The corresponding states method gives moderate accuracy for numerous compounds even without actual data.

Empirical Representations of Data

Fitting Simple Functions. In many engineering applications there may be no apparent theoretical basis for the relationship of two variables or the relationship may be too complex to apply. Thus the search for a correlating equation form may at first be along empirical lines. A simple plot of the data in ordinary Cartesian coordinates gives an immediate indication of the essential form of the data.

Each of the graphs shown in Figure 1 suggests an equation given in Table 1 by which a data set may be correlated. For most of these equations, the method of reduction to a linear form may be apparent, possibly through a transformation of variables. Typically, it is desirable to reduce the information to a linear form because the constants of the resulting linear equation can be found either directly from the slope and intercept of the straight line graph, or by means of analytical techniques.

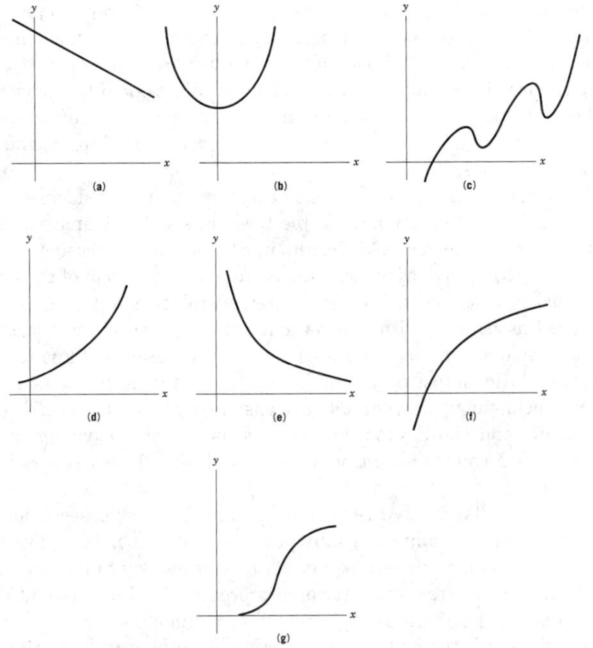

Figure 1. Some common curves: (**a**) linear; (**b**) parabolic; (**c**) polynomial; (**d**) exponential; (**e**) hyperbolic; (**f**) logarithmic; and (**g**) logistic. See Table 2 for corresponding equations.

Table 1. Curve Equations and Linear Reduction Methods

Curve	Equation	Method of reduction
linear	$y = a + bx$	plot y vs x; $a = y$ − intercept; b = slope
parabolic	$y = a + cx^n$	first obtain $a = y$ − intercept of y vs x plot; then plot log($y - a$) vs log x; c = antilog of new y − intercept at $x = 1$; n = slope
polynomial quadratic[a]	$y = a + bx + cx^2$	first obtain $a = y$ − intercept onyvsxplot; then plot $y - y_n/x - x_n$ vs x where y_n, x_n are the coordinates of any point on a smooth curve through the experimental points; c = slope; $b + cx_n = y$ − intercept
exponential	$y = ab^x$	plot log y vs x; a = antilog of y − intercept; b = antilog of slope
	$y = ae^{bx}$	plot ln y vs x; a = antilog of y − intercept; b = slope
hyperbolic	$y = a + b/x$	plot y vs $1/x$; $a = y$ − intercept; b = slope
logarithmic	$y = a \log x$	plot y vs log x; a = slope
logistic	$1/y = a + be^x$	plot $1/y$ vs e^x; $a = y$ − intercept and b = slope or plot1/yvsx; $a = y$ − intercept; then plot $\log(\frac{1}{y} - a)$ vs log x; b = antilog of y − intercept

[a] For the general polynomial quadratic equation, $y = z + bc + cx^2 + dx^3 + \cdots$; graphical procedures are almost impossible.

Once the form of the correlation is selected, the values of the constants in the equation must be determined so that the differences between calculated and observed values are within the range of assumed experimental error for the original data. However, when there is some scatter in a plot of the data, the best line that can be drawn representing the data must be determined. If it is assumed that all experimental errors (ϵ) are in the y values and the x values are known exactly, the least-squares technique may be applied. In this method the constants of the best line are those that minimize the sum of the squares of the residuals, ie, the difference, α, between the observed values, y, and the calculated values, Y.

The least-squares procedure can be applied to the transformed variables of any of the equations in Table 1, where a simple transformation of one or both of the variables results in a linearized expression.

Curve fitting to data is most successful when the form of the equation used is based on a known theoretical relationship between the variables associated with the data points, eg, use of the Clausius-Clapeyron equation for vapor pressure. In the absence of known theoretical relationships, polynomials are one of the most useful forms to describe a curve. Polynomials are easy to evaluate; the coefficients are linear; and the degree, ie, the highest power appearing in the equation, is a convenient measure of smoothness. Lower orders yield smoother fits.

Some formulas are not readily linearized. In these cases a nonlinear regression technique, usually computational in nature, must be applied. For such nonlinear equations it is necessary to use an iterative or trial-and-error computational procedure to obtain roots to the set of resultant equations. Most of these techniques are well developed and include methods such as successive substitution, variations of Newton's rule, and continuation methods.

Nomographs. The word *nomograph*, from the Greek, means the image of the law, so literally a nomograph is any diagram that represents a mathematical function. The function is evaluated by drawing lines through axes. These lines intersect a result curve. There are three types of nomographs, the Cartesian graph, slide rule, and alignment chart.

Computer Techniques. Basic to most engineering work stations are databases (qv) for efficient data retrieval and management, and spreadsheets in which numerous routine calculations can be quickly correlated, regressed, or otherwise manipulated.

Molecular-Based Methods

In the most general sense, a molecular-based property prediction method can be thought of as any method in which the physical features of a molecule (or molecules for multicomponent properties) are correlated to a property. Some methods, such as molecular thermodynamics, generalize characteristics of whole molecules to correlate them to bulk properties consistent with classical thermodynamics. Other methods correlate molecular fragments, bonds, atomic compositions, molecular topological features, and solution interactions.

Molecular Thermodynamics, the Kinetic Theory of Gases, and Statistical Thermodynamics. Molecular thermodynamics attempts to explain observed physical properties via individual molecular properties. It uses classical thermodynamics, as well as concepts from statistical thermodynamics and chemical physics. Statistical thermodynamics, or equilibrium statistical mechanics, provides the mathematical glue between the kinetic theory and macrosystems, ie, systems containing large numbers of molecules. It takes advantage of the fact that molecules are very numerous and that average properties can be estimated through probability and statistical analysis.

The kinetic theory attempts to describe the individual molecules' energies and interactions; statistical thermodynamics attempts to fundamentally develop the equation of state from considerations of groupings of molecules. These approaches are complementary in many ways. A well-referenced text covering molecular thermodynamics is also available.

Basic Problem in Statistical Thermodynamics and Quantum States. All problems in statistical thermodynamics have been postulated to be essentially forms of the same problem, that is, given a constant energy level, E, for a closed set of identical systems, eg, molecules, determine the distribution of these systems over all possible states in which the set can exist. This set is often referred to as an assembly or ensemble. Implicit in this description is that whereas there is interaction between the systems, the interaction is so weak that it can be disregarded. Thus, from a mathematical point of view, each system is thought of as containing its own distinct energy and E is obtained by summing all distinct system energies.

Distinct molecular energies are found at discrete, or quantized, levels. One method to determine the energy associated with a quantum state of a system is by describing the spins, or angular-momentums of the components and subcomponents. The wave functions which are used in these descriptions, eg, Schrödinger's wave equation, may be symmetric or antisymmetric depending on the spins.

Ensembles and Postulates. Once estimations of time-averaged energies for specific quantum states of identical systems have been made, a large number of identical systems called an ensemble is defined. An ensemble is typically defined according to a set of boundary conditions encountered in a real system. The ensemble is a mathematical construct that facilitates summation of all possible states. The typical method is to assign a state to each system in an ensemble. If the number of systems is great enough, summing an ensemble of systems with fixed states arrives at the same total as averaging all possible states of a single system and multiplying by the number of systems. The microcanonical and canonical ensembles are the two most frequently encountered in chemical thermodynamics. A microcanonical ensemble represents an isolated system having independent parameters, E; volume, V; and N molecules. A canonical ensemble represents a closed isothermal system with independent parameters N, V, and temperature T. An ensemble encountering fewer mathematical difficulties is the grand canonical ensemble in which restrictions on E and N are relaxed. T, V, and the chemical potentials μ_1, μ_2, ... of components 1,2, ... are the usual independent parameters.

Mechanical–thermodynamic property values for each system are assigned based on quantum theory and the following postulates; *(1)* the most probable distribution of systems among all possible quantum states is analogous to the equilibrium property value from classical thermodynamics, and *(2)* the most probable distribution of systems among all possible quantum states is that distribution for which the number of quantum states of the systems and their surroundings is a maximum. Mechanical–thermodynamic properties are those for which a system in a quantum state would have a well-defined value, eg, pressure, volume, energy, and mass. Nonmechanical–thermodynamic properties, eg, temperature and entropy, do not have well-defined values for a system, and therefore can only be evaluated through a comparison with classical thermodynamic relationships.

Applications. Statistical thermodynamics holds great promise as a means to characterize molecules, bonds, reactions, and energy fields in ways which are consistent with both observed and calculated classical thermodynamic properties. However, it is exceedingly difficult, if not impossible, to solve analytically for the trajectories of just three mutually interacting bodies from quantum or classical mechanics, and the problem at hand involves on the order of 10^{23} bodies. Few physical property prediction methods are available that are based solely on statistical thermodynamics. However, its methodology is fundamental to group contribution and most other molecular-based physical property prediction methods. Some insight into the applications of statistical thermodynamics to nonequilibrium problems is available. An alternative approach in which field theory is applied to replace volume and pressure with strain and stress tensors, respectively, involves continuum mechanics. Areas in which molecular thermodynamics goes beyond classical thermodynamics is of interest.

Group Contribution Methods. It has been shown that many macroscopic physical properties, ie, those derived from experimental measurements of bulk solutions or substances, can be related to specific constituents of individual molecules. These constituents, or functional groups, are usually composed of commonly found combinations of atoms. One procedure for correlating functional groups to a property is as follows. *(1)* A set of substances is selected for which the property has been determined experimentally. *(2)* Groups are determined for each substance based on a previously determined set of rules. *(3)* Property contributions for each group are determined based on a previously determined summation equation. This is done with a variety of computer optimization algorithms. *(4)* Property values are calculated based on the group contributions and compared to the experimental results. As a further test, property values are sometimes estimated for a selection of substances not included in the optimization.

Group contribution methods routinely predict properties of substances, both real and imagined, for which the only information known is structural.

Several group contribution methods for predicting liquid solubilities have been developed. These methods as well as other similar methods are often called quantitative structure-activity relationships (QSARs). This field is experiencing rapid development.

Group contribution methods to estimate physical properties vary in reliability and usage. Almost all are restricted to organic compounds. Before a method is used in studies requiring more than rough estimation, the literature should be reviewed. Concise descriptions of many of these methods including comparisons to nongroup contribution methods are available.

Linear Free Energy–Linear Solvation Energy Relationships. Linear free energy (LFER) and linear solvation energy (LSER) relationships are used to develop correlations between selected properties of similar compounds. These are fundamentally a collection of techniques whereby properties can be predicted from other properties for which linear dependency has been observed.

Kamlet-Taft Linear Solvation Energy Relationships. Most recent works on LSERs are based on a powerful predictive model, known as the Kamlet-Taft model, which has provided a framework for numerous studies into specific molecular thermodynamic properties of solvent–solute systems. This model is based on an equation having three conceptually explicit terms. $Prop = cavity\ term + dipolar\ term + hydrogen - bonding\ terms$ (eq. 38), where *Prop* is the physical property of interest, the *cavity term* describes the energy associated with creation and stabilization of a hole of correct size and shape in a continuous solvent phase large enough to accept a solute molecule, and the remaining terms relate electrostatic and hydrogen-bonding interactions between the molecules to stabilize the solute molecule in the hole.

A sampling of applications of Kamlet-Taft LSERs include the following.*(1)* The Solvatochromic Parameters for Activity Coefficient Estimation (SPACE) method for infinite dilution activity coefficients where improved predictions over UNIFAC for a database of 1879 critically evaluated experimental data points has been claimed. *(2)* Observation of inverse linear relationship between log 1-octanol-water partition coefficient and liquid (or subcooled liquid) solubility. *(3)* Correlation of fish bioconcentration factors with solvatochromic parameters. *(4)* The distribution of organic solutes between water and immiscible organic solvents. *(5)* Carbon adsorption of organic compounds. A similar, alternative approach is one based on solvophobic theory. *(7)* Correlation of octanol–water partition coefficients of organic nonelectrolytes (including strong HBD solutes).

Molecular Connectivity Indexes and Graph Theory. Computational procedures are simplified when the number of parameters used to describe the salient features of a problem is reduced. Because many properties of molecules correlated well with structures, parameters have been developed which grossly quantify molecular structural characteristics. These parameters, or connectivity indexes, are usually based on the numbers and orientations of atoms and bonds in the molecule.

As computing capability has improved, the need for automated methods of determining connectivity indexes, as well as group compositions and other structural parameters, for existing databases of chemical species has increased in importance. New naming techniques, such as SMILES, have been proposed which can be easily translated to these indexes and parameters by computer algorithms. Discussions of the more recent work in this area are available. SMILES has been used to input Contaminant structures into an expert system for aquatic toxicity prediction by generating LSER parameter values.

Perhaps a more flexible approach to incorporating structural information into a series of rule-based parameters is found in the incorporation of graph theory into chemical representation studies. The enhanced flexibility of graph theory stems form the fact that it retains molecular structural information in a matrix format which is essentially a two- (or more) dimensional representation of a molecule's spatial orientation. Through the use of large computer capacity and sophisticated regression techniques, parameters of optimum correlation to the property of interest may be obtained. An excellent introduction to the application of graph theory to structure–property relationships is available.

MICHAEL E. MULLINS
TONY N. ROGERS
Michigan Technological University
PETER P. RADECKI
Center for Clean Industrial and
Treatment Technologies (CenCITT)

ENGINEERING PLASTICS

Engineering plastics account for 7% of all plastics production and about 9% of the thermoplastics market. Engineering plastics are those "plastics which lend themselves to use for engineering design, such as gears and structural members" (*Dictionary of Scientific and Technical Terms*, 1978, p. 542).

Plastics are either crystalline or amorphous. Acetals, most nylons, some polyesters, poly(aromatic sulfides), and polyetherketones are crystalline resins. Nonoriented crystalline polymers do not transmit light, whereas amorphous polymers have some degree of transparency. Besides light transmission, there are other properties distinguishing amorphous from crystalline polymers (Table 1).

Polymers with differing morphologies respond differently to fillers (qv) and reinforcements. In crystalline resins, heat distortion temperature (HDT) increases as the aspect ratio and amount of filler and reinforcement are increased. In fact, glass reinforcement can result in the HDT approaching the melting point. Amorphous polymers are much less affected. Addition of fillers, however, interrupts amorphous polymer molecules' physical interactions, and certain properties, such as impact strength, are reduced.

Engineering thermoplastics are priced between the very expensive resins and the high volume, low priced commodities, eg, poly(phenylene oxide)–polystyrene alloys at $2/kg to polyetheretherketones at $60/kg. Many specialty and commodity resins are addition polymers. Their prices are primarily governed by raw material costs and the complexity of the manufacturing process. Engineering polymers tend to be composed of aromatic monomers, except for acetals and nylons, and their monomers are usually linked by condensation (ester, carbonate, amide, imide), substitution (sulfide) or oxidative coupling (ether), and in a few cases, addition (polyolefins). High monomer costs and complex polymerization processes force higher prices. Larger volumes of production take advantage of economy of scale. Key producers include multinational companies, with plants

Table 1. Morphology Effects and Polymer Properties

Property	Crystalline	Amorphous
organic solvent resistance	high	low
light transmission	none to low (random)	high
lubricity	high	low
dimensional stability	high[a]	moderate
mold shrinkage	high	low

[a] With a high degree of crystallinity and no water absorption.

in many countries. Recent political events such as the opening of Eastern Europe, the North American Free Trade Zone, the advancing European Economic Union, and sharpening competition are increasing the multinational efforts to seek local pricing advantages by use of cheaper local labor, elimination of transport charges, and local tax incentives.

Processing

Processing methods include injection molding, blow molding, extrusion, and reaction-injection molding and reactive casting.

Properties

For physical, thermal, electrical, and mechanical properties, ASTM test methods are employed. Flammability ratings are often based on Underwriters Laboratories' (UL) standards. UL flammability ratings given in this article are not intended to reflect the hazards presented by the resins under use conditions.

Physical Properties. Physical properties include specific gravity, water absorption, mold shrinkage, transmittance, haze, and refractive index. Specific gravity affects performance and has commercial implications. The price of the material divided by the specific gravity gives the yield in cost per unit volume. Comparison of yields gives an evaluation of raw material costs.

Thermal Properties. Thermal properties include heat-deflection temperature (HDT), specific heat, continuous use temperature, thermal conductivity, coefficient of thermal expansion, and flammability ratings.

Electrical and Mechanical Properties. Electrical properties include dielectric strength, dielectric constant, dissipation factor, and volume resistivity; these properties can change with temperature and absorbed water.

Chemical Resistance. Chemical resistance is less rigidly defined than the previously discussed properties. Measurement methods include immersion in selected vapors or liquid and response to various surface treatments, including painting. Elevated temperature, temperature cycling, wet–dry cycling, application of stress and post-treatments are often employed. Parts tested are typically ASTM flexural, tensile, or impact specimens, or a commercial part. Resistance to chemical treatment is rated by comparing mechanical and optical properties before and after stress by organic chemicals; aqueous acids, bases, salts, and buffers; light of various wavelengths; or combinations of the foregoing agents.

Rheological Properties. Rheological properties of engineering thermoplastics are determined in both solid and molten states. Melt-flow behavior is important for the design of plastic parts and molds and affects formation of knit lines, locations of internal stresses, mold fill, melt strength, and the conditions necessary for extrusion and molding. Useful rheological information is gained from viscosity as a function of shear rate.

Mechanical Properties. Along with thermal properties, mechanical properties are key to the utility of engineering plastics. In the majority of applications the materials are subject to mechanical stress, often in conjunction with thermal stress. Stiffness and impact strength are among the most desirable properties of a plastic. The first allows for easy and economical design, coming closest to the classical design tradition based on metal properties; the second guarantees that the inevitable sudden stresses which everyday articles experience in everyday use will not destroy a given article or prematurely shorten its useful lifespan. Frequently, fillers such as glass and other fibers are compounded into plastics to increase their stiffness. Especially in the case of brittle resins this also results in an increase in the tensile and impact strength. To improve impact, special rubber additives of small particle size are employed. Typically these impact modifiers are produced by emulsion polymerization and they are almost always copolymers whose outermost layer is composed of a polymer designed to be compatible with the future host resin. Another approach frequently used in impact modification is the chemical reaction of an elastomer, eg, ethylene–propylene–diene, suitably modified to react with the amine functional groups of the polyamide molecule. Impact modifiers are believed to act by reducing formation of large cracks through craze formation, thereby absorbing energy.

Mainstream Engineering Resins

Acetal Resins. Acetal resins (qv) are poly(methylene oxide) or polyformaldehyde homopolymers and formaldehyde copolymerized with aliphatic oxides such as ethylene oxide. About six firms, not including their subsidiaries, make acetal resins; Hoechst-Celanese, DuPont, and their subsidiaries are the principal producers.

Polyacetals are translucent white and crystalline. They are solvent-resistant and self-lubricating. Unlike nylons, they have low moisture absorption and maintain dimensional stability under varying humidity. Melt-flow characteristics are excellent, allowing for short mold cycles. Good fatigue allows applications such as ski-binding parts. On the other hand, impact resistance is low. Mold design must allow for shrinkage above 25% for unfilled resins. Specific gravities are high. Effective flame retardants have not been developed. For these reasons, applications are limited to small parts or those for which mold warpage, flammability, and toughness are not important.

Acetal resins are processed mainly by injection molding. Most resin products are virgin or pigmented neat resins; uv-light stabilized, glass-reinforced, impact modified, metal-plateable, and conductive grades are available.

Worldwide acetal resin capacity is ca 228,000 t/yr. Applications are tied to the consumer, appliance, construction, and automotive markets and will probably grow only as fast as the world economy. Growth rates of 2–10% are projected for various areas.

Nylon Resins. Nylon engineering thermoplastic resins have the following polyamide structures:

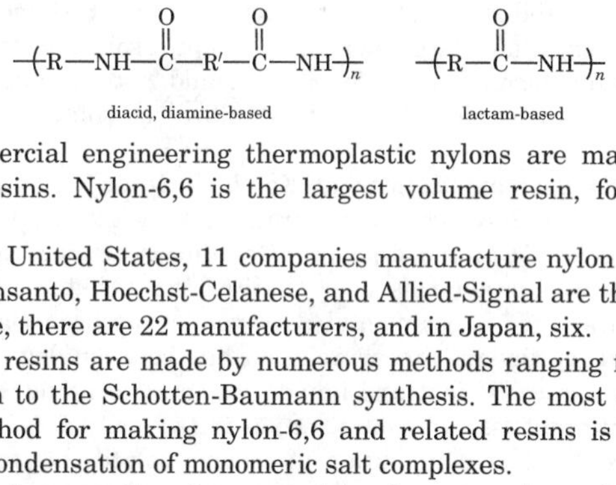

diacid, diamine-based lactam-based

Commercial engineering thermoplastic nylons are mainly crystalline resins. Nylon-6,6 is the largest volume resin, followed by nylon-6.

In the United States, 11 companies manufacture nylon resin; Du Pont, Monsanto, Hoechst-Celanese, and Allied-Signal are the leaders. In Europe, there are 22 manufacturers, and in Japan, six.

Nylon resins are made by numerous methods ranging from ester amidation to the Schotten-Baumann synthesis. The most commonly used method for making nylon-6,6 and related resins is the heat-induced condensation of monomeric salt complexes.

Physical properties of commercial nylons are shown in Table 2. Crystalline nylons are white and chemically and hydrolytically stable. With good melt flow and rapid crystallization rates, nylons are easy to process. On the other hand, they exhibit high mold shrinkage, and dimensions and properties vary with the amount of absorbed water, which can be as high as 10% by weight.

Amorphous nylons are transparent. Heat-deflection temperatures are lower than those of filled crystalline nylon resins, and melt flow is stiffer; hence, they are more difficult to process. Mold shrinkage is lower and they absorb less water. Warpage is reduced and dimensional stability less of a problem than with crystalline products. Chemical and hydrolytic stability are excellent. Amorphous nylons can be made by using monomer combinations that result in highly asymmetric structures which crystallize with difficulty or by adding crystallization inhibitors to crystalline resins such as nylon-6.

Crystalline nylons are processed by injection molding and extrusion. The extrusion products are mostly films. Coated or laminated with moisture-barrier resin, eg, PVDC or PE polymers, they are used for meat wrappings and shaped food containers. Injection-molded parts are used for small mechanical, electrical, and building construction applications. In the automotive area, nylon resins are used for

Table 2. Properties of Nylon Resin

Property	ASTM method	Nylon-6,6 Zytel 101[a]	Nylon-6 Zytel 211[a]	Trogamid T[b]
HDT at 1.82 N/mm^2, °C[c]	D648	194	129	124
melting point, °C	D2117	270	228	
notched Izod at 20°C, 3.2-mm thickness, J/m[d]	D256	53	80	65
flex modulus at 20°C, N/mm^2[c]	D790	540–2800	1000	2700
specific gravity	D792	1.14	1.13	1.12
flammability rating, UL 94		V-2	HB	V-0

[a] As molded; crystalline.
[b] Amorphous yellow resin; light transmission 85–90%.
[c] To convert N/mm^2 to psi, multiply by 145.
[d] To convert J/m to ftlbf/in., divide by 53.38.

interior, exterior, and under-the-hood applications. The construction industry uses injection molded parts for fasteners, hardware, and power tools.

Over 565,000 t/yr of nonfiber crystalline nylons is sold worldwide. Because markets are controlled by the economy, a modest growth of 5–8%/yr is expected. Although currently only ca 900 t/yr of amorphous nylons is sold worldwide, a growth rate of 10% is expected because of increased research activity. Currently, the amorphous nylon resins compete with PEI and polyesters in many applications.

Polyester Resins. The general formula of engineering thermoplastic polyester resin may be given as follows:

$$\left[\overset{\displaystyle O}{\overset{\|}{C}} - R - \overset{\displaystyle O}{\overset{\|}{C}} - O - R' - O \right]_n$$

Most polyesters are based on phthalates. They are referred to as aromatic-aliphatic or aromatic according to the copolymerized diol.

Aromatic–Aliphatic Polyester Resins. Unlike most other classes of engineering plastics, which are made by only a few manufacturers, aromatic–aliphatic polyester resins are produced and compounded by several dozen firms. The aliphatic polyester resin marketplace is characterized by wide product differentiation and competition.

Low molecular weight PET and PBT resins are made by melt processes. For higher molecular weight resins, both melt processes or solid-state polymerization are used.

The engineering applications of PET resins include blow-molded bottles, films, molding, and extrusion. Resins made for the latter two uses and related purposes are called molding resins in this article. The PBT resins are mainly used for molding and related applications.

Both PET and PBT resins are used for moldable and extrudable engineering resins.

PET and PBT molding resins are available as neat resins or compounded with other resins (including each other), pigments, filler, flame retardants, and reinforcements. Filled and reinforced resins are most commonly used for large parts and those with flat surfaces.

PET and PBT resins often compete for the same applications. The PET resins are considered in cases where higher heat-deflection temperatures and greater rigidity are desired. The PBT resins have the advantages of faster molding cycles and reduced drying time. Typical applications include automotive parts. Blends of PBT with PC resins are used for auto bumpers. Higher ductile, specially formulated polyester resins will likely compete for other large exterior parts, including doors, fenders, and quarter panels. Other markets are electrical and electronic parts (which use flame-retarded UL 94 V-0 grade), home appliances, furniture, plumbing valves, pump housings, brackets, medical components, and parts for sports equipment.

Processing is similar to other engineering plastic resins. Drying is necessary before extrusion or molding. Special drying precautions are required for PET products to prevent degradation and splay.

Both PET and PBT resin serve some of the same markets as other plastics. Nylon, unsaturated polyester fiber glass, phenolic, PC, and polyarylate resins sometimes compete for the same molded part.

Combined PET and PBT polyester resins are sold at a rate of 800,000 t/yr worldwide and are expected to have an average annual volume growth rate of 7–12%; in Europe, bottle-grade PET is growing at a rate of 23%. Growth rates for molding resins could increase if usage for large exterior automotive parts increases.

Aromatic Polyester Resins. A number of attempts to commercialize aromatic polyesters have been made. However, for reasons such as moldability, clarity, and color stability, significant market share has never been achieved.

Commercial aromatic polyester resins or polyarylates are a combination of bisphenol A with isophthalic acid or terephthalic acid. The resins are made commercially by solution polymerization or melt transesterification.

Hydrolytic stability and resistance to organic solvents are fair.

Polyarylates have been employed for electrical and electronic components, firefighter helmets, and applications requiring higher heat-deflection temperatures than PC resins.

Although several companies have produced polyarylates, Unitika in Japan and Hoechst-Celanese are the only ones active in the marketplace. Worldwide sales are quite small, probably less than 1000 t/yr.

Polycarbonate (PC) Resins. Polycarbonates (qv) based on bisphenol A are sold in large quantities. Small quantities of PC based on tetramethylbisphenol A are used as blending resins, and polyester carbonate copolymers are used for applications requiring heat-deflection temperatures above those of standard PC resins.

Bisphenol A Polycarbonate Resins. These resins are manufactured by interfacial polymerization.

Although some 12 firms produce PC resin worldwide, Bayer AG along with its subsidiary Miles, Inc., General Electric Company, and The Dow Chemical Company are the principal producers.

PC resins are glassy, amorphous polymers with little color and excellent optical properties. Properties are given in Table 3.

PC resins materials are sold mainly as injection-moldable pellets available in transparent, flame-retarded, mineral-filled, reinforced, and impact-modified grades. PC resins are also used for foam molding. Sheet products are available in forms ranging from uv-stabilized construction grades to multilayer laminates designed for bullet-resistant glazing and light weight twin-wall extrusions. Both blown and extruded film are produced.

PC resins find applications in many areas. Sheet products are used for glazing and thermoformed products such as signs, cases, and furniture. PC resin film is suitable for silk-screen printing and finds uses in illuminated consoles and soft-touch controls. PC resins are utilized for packaging as components of multilayer extruded sheet and film and in coextruded blow-molded containers. Injection molding consumes 80% of production. Applications include electrical and electronic

Table 3. Properties of Polycarbonate (PC) Resins

Property	ASTM method	Lexan 101	Merlon 8320[a]	Lexan 4701[b]
HDT at 1.82 N/mm^2[c], °C	D648	132	142	163
notched Izod at 20°C, 3.2-mm thickness, J/m[d]	D256	600	100	500
flex modulus at 20°C, N/mm^2[c]	D790	2300	3380	2300
specific gravity	D792	1.21	1.27	1.2
flammability rating, UL 94		V-2	V-1	HB
light transmission T, %		85	opaque	85

[a] 20% glass-reinforced.
[b] Polyester carbonate.
[c] To convert N/mm^2 to psi, multiply by 145.
[d] To convert J/m to ftlbf/in., divide by 53.38.

parts, appliance parts, power-tool housing, parts for the transportation industries ranging from automobile dashboards to headlamps. Sports equipment requiring high impact and other uses might be categorized under metal or glass replacement. PC blow-molding resins are used for containers, including large water bottles.

Current (ca 1993) worldwide PC resin sales are ca 290,000 t/yr; annual growth rates are expected to be near 10%.

Polyester Carbonate Copolymers. Polyester carbonate resins have molecular structures composed of iso- and terephthalate units in conjunction with the standard bisphenol A PC moieties.

Polyester carbonate resins are made by the interfacial process described for standard PC resins. At present, Bayer, GE, and Miles produce polyester carbonate resins; sales volume is low, probably ca 100 t/yr. Polyester carbonates are used primarily in applications requiring 5–25°C higher heat-deflection temperature and better hydrolytic performance than are provided by standard PC resins. Properties are given in Table 3.

Poly(phenylene ether) Alloys. Poly(phenylene ether) resins, composed of phenolic monomers, have a very high T_g. The commercial resins are based on 2,6-dimethylphenol. The resins is produced by oxidative polymerization in toluene solution over an amine catalyst.

With the glass-transition temperature above 260°C, neat poly(phenylene ether) resins are difficult to process. Poly(phenylene ether) resins are soluble in all proportions with polystyrene resin and other styrenic polymers, which allow blends or alloys to be made.

Typical poly(phenylene ether) resin–styrene resins are cream colored and are characterized by excellent melt flow. They are easily foamed either by an internal heat-activated blowing agent or by inline injection of a volatile material. Flame retardance is conferred by halogen-free compounds such as aryl phosphates. Electrical properties are outstanding and moisture absorbance is low. With excellent dimensional stabilities, the alloys can be made into large parts. Although hydrolytic stability is good, organic solvent resistance is poor. This has limited applications, such as large automotive parts. Final finishing must take chemical resistance into consideration. Another problem area is color stability to uv radiation.

Pellets for injection-molding, extrusion, and blow-molding applications include mineral-filled, glass-filled, pigmented, and flame-retarded grades. Foamed products are used in cabinets for business machines, computers, and copiers. Automotive seat backs are made by blow molding and replacements for metal parts by injection molding.

Worldwide sales of poly(phenylene ether)–styrene resin alloys are 100,000–160,000 t/yr; annual growth rates are ca 9%. Other resin, particularly acrylonitrile–butadiene–styrene (ABS) polymers and blends of these resins with PC resins, compete for similar applications.

High Performance Engineering Resins

Polysulfone Resins. Commercially important polysulfones are aromatic, ie, in the generalized formula for the repeating unit R and R′ both contain aromatic rings (see POLYMERS CONTAINING SULFUR–POLYSULFONE RESINS). They are possess ether linkages as well, so that use of the designations polysulfone, polyarylsulfone (PAS), and polyethersulfone (PES) is somewhat arbitrary.

The resins are made by batch processes employing Friedel-Crafts reactions or nucleophilic aromatic substitution.

Udel is a slightly yellow but transparent engineering thermoplastic. It has low flammability and smoke emission and good electrical properties. It has excellent resistance to water, steam, and alkaline solutions. Specific uses for Udel include microwave cookware, beverage dispensers, coffee brewers, cookware, hair dryers, corn poppers, and steam table trays. Its steam resistance makes it particularly fit for a dishwasher environment. Properties of polysulfone resins are given in Table 4.

Blends of the polysulfone resin have been made with ABS, poly(ethylene terephthalate), polytetrafluoroethylene (PTFE), and polycarbonate. These are sold by Amoco under the Mindel trademark. Additional materials are compounded with mineral filler, glass, or carbon fiber to improve properties and lower price.

Table 4. Properties of Polysulfone Resins

Property	PES Victrex 200P	PAS Radel A-400	PSO Udel
specific gravity	1.37	1.37	1.24
tensile strength, MPa[a]	82.8	82.8	70.3
elongation at break, %	40–80	40	50–80
flexural modulus, GPa[a]	2.55	2.69	2.69
notched Izod, J/m[b]	85.4	85.4	69.4
heat deflection temperature, °C at 1.82 MPa[a]	203	204	174
flammability rating, UL 94	V-0	V-0	V-2

[a] To convert MPa to psi, multiply by 145.
[b] To convert J/m to ftlbf/in., divide by 53.38.

Amoco Performance Products (Atlanta, Ga.), is the sole supplier of polysulfone, Udel, polyarylsulfone, Radel A, and polyphenylsulfone, Radel R. ICI Advanced Materials (Exton, Pennsylvania), is the sole domestic supplier of the polyethersulfone, Victrex PES, but announced in 1991 it was withdrawing from the business.

In 1989 BASF announced its intention to market in the United States a polysulfone, Ultrason S, and a polyethersulfone, Ultrason E. Both materials would likely be made in Germany.

Medical uses for Udel resin include surgical trays, nebulizers, flow controllers for blood, and respiration regulators. Transportation applications center around automotive fuse housings, electrical connectors, and switches. Electrical and electronic end uses include coil bobbins, housings, connectors, bushings, capacitor film, and business machine parts. Finally, water, heater dip tubes, milking machine parts, pollution control equipment, and some filtration membranes are made.

Victrex (PES) and Radel (PAS) serve the same end uses and are, like Udel, transparent, slightly yellow polymers exhibiting heat deflection temperatures about 28°C higher than Udel PSO. Their main advantage in electrical and electronic end use is high heat resistance. They also have excellent water and chemical resistance; like all polysulfones they are steam sterilizable. Electrical and electronic end uses represent the largest market segment. Applications include connectors, switches, bobbins, and sleeves.

In automotive and aerospace end uses, the applications are also often electrical. Polysulfones do not have as good solvent resistance as poly(phenylene sulfide). They perform well in hydrocarbons like gasoline and oil, or in antifreeze, but are attacked by the alcohol-blend fuel mixtures. This may limit their under-the-hood applications.

Polyetherimide Resins. Polyetherimide resins (PEI) were commercialized during the 1980s (see POLYIMIDES). They are produced by an unusual nucleophilic substitution process or by a conventional condensation of diamines and dianhydrides.

Ultem PEI resins are amber and amorphous, with heat-distortion temperatures similar to polyethersulfone resins. In spite of the high use temperature, they are processible by injection molding, structural foam molding, or extrusion techniques at moderate pressure between 340 and 425°C. They are inherently flame retardant and generate little smoke; dimensional stabilities are excellent. Large flat parts such as circuit boards or hard disks for computers can be injection-molded to maintain critical dimensions.

PEI polymers exhibit minimal creep under load. PEI copolymers with silicone rubber allow for flame-retardant, high temperature applications such as plenum wire coatings (Siltem). Reinforcement of PEI resins improves their temperature performance. Table 5 compares unreinforced and 20% glass-reinforced Ultem resins.

The Ultem PEI resins compete with PAI, polyarylethersulfone, nylon, and polyester resins in certain markets. General Electric Company is the sole U.S. manufacturer of PEI resins. High cost coupled with stiff competition from metals and ceramics have limited growth.

Poly(phenylene sulfide) Resins. Poly(phenylene sulfide) (PPS) resin is manufactured from *p*-dichlorobenzene and sodium sulfide in a dipolar aprotic solvent.

Table 5. Properties of Polyetherimide (PEI) Resin

Property	ASTM method	Ultem 6200[a]	Ultem 1000[b]
HDT at 1.82 N/mm^{2c}, °C	D648	223	200
notched Izod at 20°C, 3.2-mm thickness, J/m^d	D256	80	50
flex modulus at 20°C, N/mm^{2c}	D790	6550	3240
specific gravity	D792	1.43	1.7
flammability rating, UL 94		V-0	V-0

[a] 20% glass-reinforced.
[b] Neat.
[c] To convert N/mm^2 to psi, multiply by 145.
[d] To convert J/m to ftlbf/in., divide by 53.38.

The resin is recovered by precipitation as white powder, which is then purified by extraction with 2-propanol, dried, and then generally compounded with fiber glass and mineral filler.

The polymer (PPS) possessing high crystallinity exhibits greater resistance to creep than the amorphous material. On the other hand, the latter has higher elongation. Most poly(phenylene sulfide) resin is sold as a compound of glass or mineral filler. The main advantage of PPS lies in its superior heat deflection and continuous service temperatures. PPS is an excellent electrical insulator and is inherently flame resistant. Its chemical resistance is outstanding even at elevated temperatures, except in oxidizing agents, strong acids, and chlorinated hydrocarbons (see POLYMERS CONTAINING SULFUR–POLY(PHENYLENE SULFIDE)).

PPS resins are chiefly used for injection molding. Applications include chemical-resistant parts for pumps and processing equipment such as impellers. The high heat resistance of glass-reinforced grades allows application in large exterior light reflectors and automotive engine components.

PPS resins must compete with PEI and phenolics. There are two domestic manufacturers of poly(phenylene sulfide): Phillips and Fortron Industries. Worldwide there is currently large overcapacity. Prices for PPS resins and compounds range from $8.80/kg for unreinforced resin to $3.30/kg for 65% filled resins.

Electrical and electronics are the largest end uses for PPS.

Liquid Crystalline Polymers. Liquid crystalline molding resins are, by and large, special aromatic resins possessing rigid linear molecular structures. These rigid rods form ordered crystal-like areas in the liquid state that coexist with amorphous areas. This molecular order can be improved by such processes as orienting a film or stretching a fiber as the polymer rods are then dragged against each other and are forced into a parallel array and achieve the close packing needed to "lock." Order is preserved by cooling, thus locking the structure in the solid state. LCP's rod-like molecular structure provides the strength and stiffness normally associated with fiber-reinforced thermoplastics. At temperatures of 200°C and above, LCPs maintain stiffness comparable with that of some engineering plastics at room temperature. These polymers are intended as metal replacement in electronic, automotive, and industrial markets.

Four companies (trade name in parentheses), Amoco (Xydar), Hoechst-Celanese (Vectra), DuPont, and Granmont (Granlar), make thermotropic LCPs for various types of extrusion and molding processes.

Chemically all the currently commercial materials are linear aromatic esters. Markets have developed more slowly than initially anticipated in electronic, automotive, and industrial areas. The thermotropic LCP market was estimated at 1300 t in 1990 with a growth rate at 14%/yr. Prices, once $22 + /kg have fallen to $17–20/kg.

Polyetheretherketone Resin (PEEK). The resin was commercialized as Victrex PEEK by Imperial Chemical Industries, Ltd. (ICI) in the late 1970s and by Amoco Chemicals Corporation in the middle 1980s under the trade name Kadel. It is produced by both companies in the United States. Kadel is believed made by the displacement reaction of 4,4′-difluorodiphenyl ketone by the potassium salt of hydroquinone. Other companies offering similar materials are DuPont, BASF, and Hoechst-Celanese. PEEK is a difficult resin to manufacture because of batch procedures and insolubility.

The PEEK resin is gray, crystalline, and has excellent chemical resistance; T_g is ca 185°C, and it melts at 288°C. The unfilled resin has an HDT of 165°C, which can be increased to near its melting point by incorporating glass filler. The resin is thermally stable, and maintains ductility for over one week after being heated to 320°C; it can be kept for years at 200°C. Hydrolytic stability is excellent. The resin is flame retardant, has low smoke emission, and can be processed at 340–400°C. Crystallinity is a function of mold temperature and can reach 30–35% at mold temperatures of 160°C. Recycled material can be safely processed.

The PEEK resin is marketed as neat or filled pellets for injection molding, as powder for coatings, or as preimpregnated fiber sheet and tapes. Applications include parts that are exposed to high temperature, radiation, or aggressive chemical environments. Aerospace and military uses are prominent. At present, polyamideimide (PAI) resin and poly(arylene sulfides) are the main competitors for applications requiring service temperatures of 280°C. At lower temperatures, polyethersulfones, amorphous nylons, and polyetherimides (PEI) can be considered.

Production of PEEK resin is less than 500 t/yr worldwide. Growth is expected to exceed 10%/yr. Because of chemical and manufacturing constraints, the price, ca $55/kg for neat resin, is likely to remain high.

Polyamideimide Resin (PAI). Amoco introduced the first commercial polyamideimide (PAI) resin, Torlon, in the early 1970s. It is produced by the solution condensation of trimellitic trichloride with methylenedianiline.

Torlon resin is an amber-to-gray amorphous polymer with an HDT of 274°C, flexural modulus of 4600 MN/m^2 (667,000 psi), and good impact resistance. Although it contains amide linkages, it is dimensionally much more stable under various humidity conditions than the standard crystalline nylons because of its elevated glass-transition temperature. Also, like PPS, Torlon appears to undergo "curing" with additional heat treatment. Like aromatic polyamides and imides, Torlon is flame retardant and has low smoke emission. Creep under load is low. Electrical, chemical, hydrolytic, and radiation stability are excellent. Because of these properties, PAI replaces metal in gears, connectors, and other mechanical parts. It is also used in electrical applications where high temperature and chemical resistance are required.

Resins competing with PAI include PEI, polyethersulfones, and thermoset resins.

Other Polyimide Resins. The polyimide category, which includes both the polyetherimide and the polyamide resins discussed earlier, can be broken down into three subdivisions.

Nonthermoprocessible Condensation Polyimides. These are obtained from condensation of aromatic dianhydrides with aromatic diamines.

These resins have extensive applications in the high performance insulation markets (traction motor, wire cable) and flexible printed circuits. DuPont supplies Kapton film and also supplies Verpel stock shapes made from the same resin by powder sintering methods.

Thermoplastic Condensation Polyimides. These include General Electric's Ultem Resin and Amoco's Torlon polyamideimide, although the latter is no longer offered as injection moldable pellets, but rather as compression moldable powder and in solution.

Additive Polyimides. Rhône-Poulenc's Kinel molding compound and Kerimid impregnating resin, Mitsubishi's BT resins, and Toshiba's Imidalov Resin are based on bismaleimide technology. Although these resins have been on the market for a number of years, volumes and applications remain low in spite of favorable properties including high temperature resistance. Excellent wear and friction properties, good electrical properties, chemical inertness, and inherent nonflammability. Processibility and price may be the limiting factors.

Specialty Polyolefins

As recently as 1986 almost all addition polymers were excluded from the ranks of engineering plastics. However, progress since then has been made in the development of addition polymeric resins such as polymethylpentene and polycyclopentadiene and its copolymers (see CYCLOPENTADIENE AND DICYCLOPENTADIENE).

Blends and Alloys

The differentiation between polymer blends (qv) and polymer alloys is blurred. The terms (often used interchangeably), refer to multiresin compositions. Such systems have become increasingly important. For thermodynamic reasons (low entropy of mixing), polymers are usually not miscible with each other; mutual solubility in all proportions is the exception rather than the rule. Mutually compatible mixtures are often referred to as alloys to distinguish them from others where the component resins exist as separate entities in well-defined if random geometrical arrangements, with good adhesion between them. These latter systems are more frequently referred to as blends.

Alloys and blends are of great commercial significance. The archetype of "alloys" is the poly(phenylene oxide)–polystyrene resin discussed earlier. Companies produce alloys and blends in order to extend their existing resin capacity and product lines without high capital investment and to achieve property profiles required for specific applications.

Currently, over 110,000 t/yr of engineering resin blends are consumed worldwide, primarily in the transportation, business-machine, hardware, electrical, and appliance industries. Annual growth is projected to be ca 17%/yr. New blends based on PC, terephthalate, and nylon resins are experiencing the greatest expansion. These projections could be surpassed if large-volume metal applications such as automotive panels are replaced by engineering resin blends which are currently being field-tested.

DONALD CLAGETT
Provisional Technical Secretariat Organisation for the Prohibition of Chemical Weapons
RAYMOND NARR
General Electric Company

O. H. Fenner in C. A. Harper, ed., *Handbook of Plastics and Elastomers*, McGraw-Hill, Inc., New York, 1975, pp. 4-1–4-34.

H. Faig in J. Agranoff, ed., *Modern Plastics Encyclopedia 1982–1983*, Vol. 60, No. 10A, McGraw-Hill, Inc., New York, 1983, pp. 248–262, 268.

Polymeric Materials—Long Term Property Evaluations, UL 746B, 2nd ed., Underwriters Laboratories, Inc., Northbrook, Ill., 1979, pp. 6–26.

W. F. Christopher and D. W. Fox, *Polycarbonates*, Reinhold Publishing Corp., New York, 1962, pp. 161–173.

ENVIRONMENTAL IMPACT

The significant impact of environmental regulations in the 1990s on industry in general, and the chemical industry in particular, is indisputable. There is a growing tendency to merge environmental issues and economics in decision making within the chemical industry (see CATALYSTS–REGENERATION; ECONOMIC EVALUATION; FUELS FROM WASTE; HERBICIDES; RECYCLING; SOLVENTS, INDUSTRIAL; WASTES, INDUSTRIAL). Moreover, because environmental issues are pervasive in the chemical industry, these concerns have brought about change in chemical technology. In many fields, environmental concerns and regulations are driving technological development (see COATING PROCESSES; DYES, ENVIRONMENTAL CHEMISTRY; EXHAUST CONTROL, AUTOMOTIVE; EXHAUST CONTROL, INDUSTRIAL).

ENZYME APPLICATIONS

INDUSTRIAL

Enzymes, like other proteins (qv), are composed of up to 20 different amino acids (qv). They accelerate hundreds of reactions taking place simultaneously in the cell and its immediate surroundings, and are essential for the development and maintenance of life.

Industrial applications of enzymology form an important branch of biotechnology. Enzymatic processes enable natural raw materials to be upgraded and turned into finished products. They offer alternative ways of making products previously made only by conventional chemical processes.

Genetic Engineering of Enzymes

Genetic engineering makes it possible to take DNA out of a donor cell, modify the DNA is a test tube, and introduce the modified recombinant DNA into a new host cell in such a way that the recombinant DNA becomes a stable part of the genetic material of the host. Thus it is possible to transfer the gene encoding a particular enzyme from any exotic organism into a well-known production organism like a *Bacillus* or an *Aspergillus*; ie, a production organism may be obtained that both produces the desired enzyme of the exotic donor and has all the desirable properties of a safe industrial microorganism.

Host Microorganisms. As of this writing (ca 1993), only microorganisms are used as recombinant production organisms for industrial enzymes. The choice of host microorganism for production of industrial enzymes is often critical for the commercial success of the product. Potential hosts should give sufficient yields, be able to secrete large amounts of protein, be suitable for industrial fermentations, produce a large cell mass per volume quickly and on cheap media, be considered safe based on historical experience or evaluation by regulatory authorities, and should not produce harmful substances or any other undesirable products. Few microorganisms fulfill all the above criteria and are used for production of industrial enzymes. Important hosts include *Escherichia coli, Bacilli, Aspergilli*, and yeast.

Engineering of New Enzymes. It is possible to modify and improve natural enzymes with protein engineering. Because the properties of any enzyme are determined by its three-dimensional structure, which in turn is determined by the linear combination of amino acids, it is possible, by a precise and site-specific modification of a gene, to change any amino acid into any other, or design a synthetic gene of any composition, and hence design a protein of any three-dimensional structure as long as basic physical and chemical rules are obeyed. Protein engineers can only predict the structural consequences of minor changes in enzymes (ca 1993). Protein-engineered enzymes on the market include a detergent protease that has been stabilized against chemical oxidation by replacement of two amino acids of the native enzyme.

It is likely that any new enzymes isolated by screeners will be quickly and routinely cloned by genetic engineers, and be sequenced and expressed as almost pure proteins. Protein chemists can then evaluate the properties of the new enzyme and determine its three-dimensional structure. This vast amount of information allows the protein engineers and their computers to design the enzymes of the future.

Abzymes. To design an improved enzyme, protein engineers must know the structure of a protein resembling the desired enzyme. Since the number of different known types of enzymes is less than 1000, only a limited number of chemical reactions can be catalyzed by known naturally occurring enzymes. Catalytic antibodies, or abzymes, may overcome this limitation. Abzymes are antibodies having the ability to complex with the transition state of a chemical reaction, thereby lowering the free energy of the intermediate. Since the reaction rate is determined by the difference in free energy between the reactants and the transition state, the lowering of the energy of the transition state speeds up the process until chemical equilibrium, which is independent of catalysis, is reached. Since the immune system of mammals is able to produce approximately 100 million antibodies, it is likely that an antibody that recognizes any transition-state molecule, or a stable analogue of this, can be raised. The concept has been shown to work with at least seven different types of reactions, including a specific ester hydrolysis which was stimulated one millionfold by an abzyme. A cost-effective industrial abzyme must still be developed.

Catalytic Activity

Enzymatic Catalysis. Enzymes are biological catalysts. They increase the rate of a chemical reaction without undergoing permanent change and without affecting the reaction equilibrium.

The characteristics of enzymes are their catalytic efficiency and their specificity. Enzymes increase the reaction velocities by factors of at least one million compared to the uncatalyzed reaction. Enzymes are highly specific, and consequently a vast number exist. An enzyme usually catalyzes only one reaction involving only certain substrates.

Another characteristic of enzymes is their frequent need for cofactor. A cofactor is a nonprotein compound that combines with the otherwise inactive enzyme to give the active enzyme.

Enzymes accelerate reactions by stabilizing the transition states, the highest energy species on the reaction pathway, and thereby decreasing the activation barrier.

Enzyme Kinetics. A simple enzyme catalyzed reaction can be described: $E + S \rightleftarrows ES \rightarrow E + P$. The enzyme, E, and the substrate, S, initially combine to form an enzyme–substrate complex, ES, which is held together by physical forces. In the second step the chemical process occurs, whereby the enzyme and the product, P, are liberated. This step is controlled by a first-order rate constant k_{cat}, called the catalytic constant or the turnover number. When deriving kinetic expressions, it is generally assumed that the concentration of enzyme is negligible compared to the concentration of substrate. Furthermore, it is assumed that what is being measured is the initial velocity v for the formation of products, ie, the rate of formation for the first few percent of the product. Under these conditions the products have not accumulated, the substrates have not been depleted, and the reaction velocity is generally linear with time. It is found experimentally that v is directly proportional to the concentration of enzyme $[E_0]$, but varies with the substrate concentration $[S]$. At low $[S]$, v is directly proportional to $[S]$. At higher $[S]$, however, this relation begins to break down, and at sufficiently high $[S]$ v tends toward a limiting value v_{max}. The Michaelis-Menten equation expresses this relation quantitatively.

Kinetic parameters are often derived from a set of experimental data by rearranging the data into a linear form, eg, plotting $1/v$ as a function of $1/[S]$ (a Lineweaver Burk, or double-reciprocal, plot) allows the determination of k_{cat} and K_m from the intercepts of the axes.

Enzyme Inhibition. Enzyme inhibitors (qv) are reagents that bind to the enzyme and cause a decrease in the reaction rate. Irreversible inhibitors bind to the enzyme by an irreversible reaction, and consequently cannot dissociate from the enzyme or be removed by dilution or dialysis.

Reversible inhibition is characterized by an equilibrium between enzyme and inhibitor. Many reversible inhibitors are substrate analogues, and bear a close relationship to the normal substrate. When the inhibitor and the substrate compete for the same site on the enzyme, the inhibition is called competitive inhibition. The effect of adding a competitive inhibitor is illustrated in Figure 1.

Effect of Temperature and pH. The temperature dependence of enzymes often follows the rule that a 10°C increase in temperature doubles the activity. However, this is only true as long as the enzyme is not deactivated by the thermal denaturation characteristic for enzymes and other proteins. Most enzymes have temperature optima between 40 and 60°C. However, thermostable enzymes exist with optima near 100°C.

The pH dependency of enzyme-catalyzed reactions also exhibits an optimum. The pH optima for enzyme-catalyzed reactions cover a wide range of pH values. The nature of the pH profile often gives clues to the elucidation of the reaction mechanism of the enzyme-catalyzed reaction. The temperature at which an experiment is performed may affect the pH profile and vice versa.

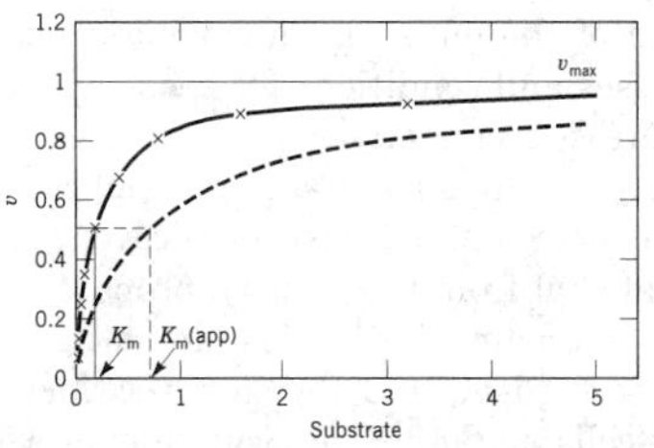

Figure 1. The effect on kinetic parameters of adding a competitive inhibitor. Reaction velocity v as a function of [S] is shown. (– × –) Uninhibited reaction; (––––) inhibited reaction. As indicated on the figure, the parameter K_m is increased by adding the competitive inhibitor; both curves eventually reach the same v_{max}.

Enzyme Assays. An enzyme assay determines the amount of enzyme present in sample. However, enzymes are usually not measured on a stoichiometric basis. Enzyme activity is usually determined from a rate assay and expressed in activity units. As mentioned above, a change in temperature, pH, and/or substrate concentration affects the reaction velocity. These parameters must therefore be carefully controlled in order to achieve reproducible results.

Spectrophotometry, a simple and reliable technique, is often used in rate assays. Potentiometry is another useful method for determining enzyme activity in cases where the reaction liberates or consumes protons.

Enzyme Nomenclature. Enzymes are classified on the basis of the reactions they catalyze. Despite its apparent complexities, the system is precise and very descriptive, accommodating existing enzymes and serving as a systematic basis for the naming of new enzymes. All enzymes are placed in one of the six principal classes.

Number	Class	Type of reaction catalyzed
1	oxidoreductases	transfer of electrons
2	transferases	group-transfer reactions
3	hydrolases	transfer of functional groups to water
4	lyases	addition of groups to double bonds or the reverse
5	isomerases	transfer of groups within molecules to yield isomeric forms
	ligases	formation of C–C, C–S, C–O, and C–N bonds by condensation reactions coupled to ATP cleavage

Each class is divided into groups or subclasses according to the type of reaction catalyzed. These groups are further subdivided according to the nature of the substrate involved. Each enzymes is then assigned a four-digit classification number (EC number) based on this division, and a systematic name to identify the reaction catalyzed. A trivial name may be used if the systematic name is too long or cumbersome.

Production of Industrial Enzymes

With the exception of trypsin, chymosin, papain, and a few others, industrial enzymes are now produced by microorganisms grown in aqueous suspension in large vessels, ie, by fermentation (qv). A small (5%) fraction is obtained by surface culture, ie, solid-state fermentation, of microorganisms.

Enzymes are usually sensitive to harsh physical and chemical conditions, and care must be taken during recovery and purification to avoid inactivation of the enzyme. This demands careful selection of production processes and conditions for each individual enzyme. Different methods are subsequently applied to assure the stability and activity of the enzymes during storage and application.

Recovery. The principal purpose of recovery is to remove nonproteinaceous material from the enzyme preparation. Enzyme yields vary, sometimes exceeding 75%. Most industrial enzymes are secreted by a microorganism, and the first recovery step is often the removal of whole cells and other particulate matter by centrifugation or filtration. In the case of cell-bound enzymes, the harvested cells can be used as is or disrupted by physical (eg, bead mills, high pressure homogenizer) and/or chemical (eg, solvent, detergent, lysozyme, or other lytic enzyme) techniques. Enzymes can be extracted from disrupted microbial cells, and ground animal (trypsin) or plant (papain) material by dilute salt solutions or aqueous two-phase systems.

Ultrafiltration (uf) is increasingly used to remove water, salts, and other low molecular-weight impurities; water may be added to wash out impurities, ie, diafiltration.

Purification. Enzyme purity, expressed in terms of the percent active enzyme protein of total protein, is primarily achieved by the strain selection and fermentation method. In some cases, however, removal of nonactive protein by purification is necessary. The key purification method is selective precipitation of the product or impurities by addition of salt, eg, sodium sulfate, or solvent, eg, ethanol or acetone; by heat denaturation; or by isoelectric precipitation, ie, pH adjustments.

Product Requirements. When an enzyme is recovered from fermentation broth, it is usually present in an aqueous solution or processed to a dried state. Both types of preparation have to be formulated to comply with requirements appropriate to their final application.

Requirements related to the storage of enzymes from the time of manufacture to the time of application include enzyme stability, ie, the catalytic activity of the enzyme must remain after prolonged storage at relevant temperatures; microbial stability, ie, industrially recovered enzymes in aqueous solution are potentially excellent growth media for microorganisms, so it is usually necessary to prevent microbial growth; and physical stability, ie, the enzyme must remain in solution to avoid inhomogeneous dosage. Some applications demand special requirements. These can be of a technical nature such as having no precipitate, off-odor, and off-color when mixed into a liquid detergent; and/or an absence of side activities, eg, transferase activity must be absent in saccharifying enzymes like amyloglucosidase, and protease must be absent in cell wall-degrading enzymes for the upgrading of vegetable proteins. Other requirements in certain applications derive from approval considerations, eg, only food-grade ingredients, absence of certain microorganisms, and kosher restrictions on enzymes for food applications.

Any formulation is a compromise among the previously mentioned requirements. For example, the fermentation broth may contain enzyme-stabilizing substances, but the application of the enzyme or precipitation problems in the formulation may demand a high degree of purification that eliminates the stabilizers.

The formulation of any enzyme is normally considered a way to store and transport the enzyme until its application. One common exception is that of immobilized enzymes where formulation is an active part of their application.

Immobilization. Cost reduction is the primary argument for using immobilized enzymes, especially when comparing this method with soluble enzyme or nonenzymatic methods; nevertheless, satisfactory technical solutions can be found among the latter two alternatives. When immobilized, expensive enzymes can be reused, and this may compensate for the cost of immobilization.

A significant advantage of immobilized enzymes is the total absence of catalytic activity in the product. Moreover, the degree of substrate-to-product conversion can be controlled during processing, eg, by adjusting the flow rate through a paced-bed column reactor of immobilized enzyme.

Choice of Method. Because enzymes can be intracellularly associated with cell membranes, whole microbial cells, viable or nonviable, can be used to exploit the activity of one or more types of enzyme and cofactor regeneration, eg, alcohol production from sugar with yeast cells. Viable cells may be further stabilized by entrapment in aqueous gel beads or attached to the surface of spherical particles. Otherwise cells are usually homogenized and cross-linked with glutaraldehyde to form an insoluble yet penetrable matrix. This is the method upon which the principal industrial applications of immobilized enzymes are based.

Extracellular microbial enzymes can be immobilized in the form of proteins purified to varying degrees. Other immobilization methods are based on chemical and physical binding to solid supports, eg, polysaccharides, polymers, glass, and other chemically and physically stable materials, which are usually modified with functional groups such as amine, carboxy, epoxy, phenyl, or alkane to enable covalent coupling to amino acid side chains on the enzyme surface.

Membrane reactors, where the enzyme is adsorbed or kept in solution on one side of an ultrafiltration membrane, provides a form of immobilized enzyme and the possibility of product separation.

Microemulsions or reverse micelles are composed of enzyme-containing, surfactant-stabilized aqueous microdroplets in a continuous organic phase. Such systems may be considered as a kind of immobilization in enzymatic synthesis reactions.

The choice of a suitable immobilization method for a given enzyme and application is based on a number of considerations including previous experience, new experiments, enzyme cost and productivity, process demands, chemical and physical stability of the support, approval and safety issues regarding support, and chemicals used. Enzyme characteristics that greatly influence the approach include intra- or extracellular location; size; surface properties, eg, charge/pI, lysine content, polarity, and carbohydrate; and active site, eg, amino acids or cofactors. The size, charge, and polarity of the substrate should also be considered.

Industrial-Scale Applications. The introduction of immobilized penicillin G acylase and then immobilized glucose isomerase (IGI) for the production of high fructose corn syrup (HFCS) demonstrated the enormous industrial potential of immobilized enzymes.

Commercial IGIs of the 1990s are based on various enzyme sources, largely *Streptomyces* spp., and include immobilization techniques incorporating the adsorption of purified glucose isomerase.

The second most important group of immobilized enzymes is still the penicillin G and V acylases. These are used in the pharmaceutical industry to make the intermediate 6-aminopenicillanic acid, which in turn is used to manufacture semisynthetic penicillins, in particular ampicillin and amoxicillin.

During the 1980s, molecular biology techniques were used successfully to reduce enzyme production costs; this is part of the reason the number of immobilized enzyme processes used in industry are relatively few compared with soluble enzyme processes, and certainly less than had been expected in the 1970s. It is expected that the pharmaceutical, organic chemical, and oils and fats industries will increase their use of this technology. Proteases are used in peptide synthesis and nitrilase in the synthesis of acrylamide. Of particular interest is the use of immobilized lipases for optical resolution, ester synthesis, and the production of specific lipids by interesterification. These have already demonstrated the feasibility of using immobilized enzyme technology in nonaqueous systems of low water activity.

Granulation of Enzymes. Although the trend is to market industrial enzymes as liquid products, solid enzyme is needed. Examples of this are enzymes for solid detergents, animal feed, and flour improvement.

Several different methods have been used for the granulation of enzymes. In general, the development of granulation methods focuses on parameters such as the cost of the process and solubility of the granulate. The focus for detergent enzymes is directed onto a single parameter, ie, a dust-free product.

To reduce the enzyme dust level in detergent factories to an absolute minimum, the majority of detergent enzyme granulates are coated with a layer of inert material.

For granulation of enzymes for purposes other than incorporating into detergents, the same methods can be used as for the detergent enzymes, although some of the additives needed for the production of a rigid granulate may not be accepted, ie, in certain enzymes for the food industry. Fortunately for such applications the handling of the enzyme is more gentle, and the requirement for physical stability less. For these enzymes, a fluid-bed granulation performed as an agglomeration of powder with a liquid binder, or as a coating of the enzyme onto inert carrier particles of selected size, often gives the desired quality of product. These enzyme preparations may also be coated.

Industrial Applications

The largest industrial use of enzymes is in detergents. Enzymes are also used in starch conversion, in textile finishing, in the tannery, in pulp and paper production, in the animal feed industry, in the modification of lipids, in food applications (dairy, baking, brewing, and protein modification, eg, bone cleaning, and modification of wheat gluten), and extraction processes of material.

Economic Aspects

Worldwide consumption of industrial enzymes amounted to approximately $720 million in 1990; about one-third was accounted for by the U.S. market. The detergent and starch conversion industries are by far the most important, and account for 60% of total enzyme sales. Five principal industries account for around 80% of enzyme sales, whereas the remaining sales are spread over many different industries.

Three proteases account for almost all sales to the dairy industry, ie, chymosin extracted from calves' stomachs, chymosin produced by fermentation, and substitutes also produced by fermentation. Four different types of enzyme are used in the detergent industry, ie, proteases, amylases, cellulases, and lipases. Cellulases and lipases have only recently been introduced.

Environmental and Safety Aspects

The industrial use of microbial enzymes produced by modern biotechnology is an important contribution to the development of green technology. Enzymes have a positive impact on the environment because they replace conventional chemical-based technologies and conventional energy-intensive manufacturing processes, originate from natural biological systems, are totally biodegradable, and leave no harmful residues.

The safety and environmental impact of the production of industrial enzymes can be evaluated on three different levels, ie, the potential risk if the microorganisms, their products, or both are released into the environment; the possible health hazards to staff working with the microorganisms, their products, or both; and safety when products are used by the consumer.

Enzymes are totally biodegradable, and their release into the environment does not cause problems. The release of the production organism itself is controlled by two categories of safety measures which are complementary. The first is physical containment in a fermenter system and recovery plant with a high standard of hygiene. The second is biological containment. Being specially bred, either by traditional techniques or by modern genetic engineering techniques, to produce one specific substance, the production organisms are adapted to grow optimally only under the defined conditions during fermentation. The growth of strains of production organisms in nature is handicapped in comparison with microorganisms already existing in the environment. Their chances of survival in the environment are extremely limited.

Like other proteins, enzymes are potential allergens. In addition, proteases may act as skin and eye irritants. However, during the production and handling of industrial enzymes, the occupational health risks entailed by these properties can be avoided by protective measures, and by the form in which the enzyme preparations are supplied. In order to reduce dust generation, enzymes are supplied as liquids, encapsulates, or immobilized preparations.

To guarantee that enzymes can be used safely by the consumer, microbial enzymes are obtained from nonpathogenic and nontoxinogenic microorganisms grown on raw materials that do not contain compounds hazardous to health.

Genetically engineered microorganisms can be used under the same conditions of containment, and the same security rules apply as for equivalent, naturally occurring microorganisms. Provided an enzyme is produced by a harmless host, the contained use of recombinant microorganisms does not warrant any special provisions concerning production conditions, worker protection, environmental assessment, field monitoring, or product approval.

Regulatory Aspects. National authorities have preferred to use or adapt existing legislation and regulations. For the adaptation of the food additive regulations to fit the processing aids applications of enzymes, guidance has been available in the recommendations of the Joint FAO/WHO Expert Committee on Food Additives (JECFA), and the Food Chemicals Codex (FCC). The enzyme manufacturers associations, the Association of Microbial Food Enzyme Producers (AMFEP) in Europe and the Enzyme Technical Association (ETA) in the United States, work nationally as well as internationally for a harmonization of regulation. The Codex Committee on Food Additives and Contaminants (CCFAC) plays an important role in this work. In the European Community, common guidelines are being developed for the evaluation of enzymes for food and feed uses. This is a result of the increased attention on new biotechnology rather than on enzymes *per se*. The contained use of genetically modified microorganisms in the manufacture of enzymes has been one of the first cases of a product of recombinant DNA technology reaching the industrial marketplace. A working group on food safety and biotechnology is also considering the use of food enzymes made from genetically modified organisms.

PEDER HOLK NIELSEN
HENRIK MALMOS
TURE DAMHUS
BOERGE DIDERICHSEN
HENRIK KIM NIELSEN
MERETE SIMONSEN
HANS ERIK SCHIFF
ANDERS OESTERGAARD
HANS SEJR OLSEN
PETER EIGTVED
TAGE KJAER NIELSEN
Novo Nordisk A/S

J. E. Bailey and D. F. Ollis, *Biochemical Engineering Fundamentals*, 2nd ed., McGraw-Hill Book Co., Inc., New York, 1986, Chaps. 3–4.

A. Fersht, *Enzyme Structure and Mechanism*, 2nd ed., W. H. Freeman and Co., New York, 1988.

T. Palmer, *Understanding Enzymes*, 1st ed., Ellis Horwood, New York, 1991.

D. V. Roberts, *Enzyme Kinetics*, 1st ed., Cambridge University Press, London, 1977.

L. Stryer, *Biochemistry*, 3rd ed., W. H. Freeman and Co., New York, 1988.

D. Freifelder, *Physical Biochemistry*, 2nd ed., W. H. Freeman and Co., New York, 1982.

R. J. Leatherbarrow, *TIBS* **15**, 455–458 (1990).

THERAPEUTIC

Physical and Chemical Properties

Enzymes are protein catalysts of remarkable efficiency and specificity. Lipid, carbohydrate, nucleotide, or metal-containing prosthetic groups may be attached to these enzymes and serve as essential components of their catalyses by enhancing specificity and/or stability. Each enzyme has a specific temperature and pH range where it functions to its optimal capacity; the optima for these proteins usually lie between 37–47°C, and pH optima range from acidic, ie, 1.0 in the case of gastric pepsin, to alkaline, ie, 10.5 in the case of alkaline phosphatase.

Table 1. Therapeutic Enzymes

Enzyme	Catalysis	Use
neuraminidase	hydrolysis of terminal acylneuraminyl residues	antineoplastic
ribonuclease	RNA → oligoribonucleotides	antineoplastic
L-α-arabinofuranosidase	L-α-arabinofuranoside → alcohol + L-arabinose	antineoplastic
brinase	fibrinogen → fibrin	fibrinolytic
α-glucosidase	D-1, 4α-glucoside → α-D-glucose	metachromatic leukodystrophy
β-glucosidase	D-1, 4β-glucoside → β-D-glucose	type A glycogenosis
arylsulfatase	phenolsulfate → phenol + sulfate	metachromatic leukodystrophy
α-galactosidase	α-D-galactoside → α-D-galactose	Fabry's disease
β-galactosidase	β-D-galactoside → β-D-galactose	Fabry's disease
bromelain	protein → amino acids and peptides	antiinflammatory
collagenase	collagen → amino acids and peptides	dermal ulcers
papain	protein → amino acids and peptides	reduction of edema after dental surgery
L-asparaginase	L-asparagine → L-aspartic acid + NH_3	antineoplastic
streptokinase	plasminogen → plasmin	thrombolytic
arvin	fibrinogen → fibrin	fibrinolytic
urokinase or tissue plasminogen activator	plasminogen → plasmin	thrombolytic
coagulation factor VIII	prothrombin → thrombin	hemophilia
glucocerebrosidase	glycolipid glucocerebroside → α-D-glucose + ceramide	Gaucher's disease
lipase	carboxylic ester → alcohol + carboxylic acid	pancreatic deficiency
L-glutaminase	L-glutamine → L-glutamic acid + NH_3	antineoplastic
L-arginase	L-arginine → L-ornithine + urea	antineoplastic
L-tyrosinase	L-tyrosine + O_2 → dihydrophenylalanine + H_2O	antineoplastic
L-serine dehydratase	L-serine → pyruvate + NH_3	antineoplastic
L-threonine deaminase	L-threonine → 2-ketobutyric acid + NH_3	antineoplastic
L-tryptophanase	L-tryptophan → indole + pyruvate + NH_3	antineoplastic
deoxyribonuclease	DNA → oligodeoxyribonucleotides	chronic bronchitis
trypsin	protein → peptides	athletic injuries
chymotrypsin	protein → peptides	athletic injuries
superoxide dismutase	$O_2^{[[mindo]]} + O_2^{[[mindo]]} + 2 H^+ \rightarrow O_2 + H_2O_2$	antiinflammatory

However, enzymes from extremely thermotolerant bacteria have become available; these can function at or near the boiling point of water, and therapeutic use of these ultrastable proteins can be anticipated.

Hydrolases represent a significant classes of therapeutic enzymes (Table 1). Another group of enzymes with pharmacological uses has built-in cofactors, eg, in the form of pyridoxal phosphate, flavin nucleotides, or zinc.

Commercial enzymes are available in oral form, sometimes formulated with appropriate stabilizers and excipients. However, such preparations are seldom suitable for parenteral use. Therefore, dry preparations devoid of high salts, excipients, or reducing agents have been adopted for the final formulation of therapeutic enzymes involves lyophilization in the presence of mannitol and a physiological buffer. Since many enzymes can be denatured by heat even in the dry state, refrigeration or freezing during transit or storage is customary.

The therapeutic utility of an enzyme preparation is largely dependent on its stability as finally formulated. A number of chemical modifications have been employed and include binding to inert surfaces and encapsulation to increase the resistance of these intrinsically labile macromolecules to decomposition. Unfortunately, some of these modifications can interfere with the optimal kinetic performance of the enzyme. The development and deployment of recombinant DNA technologies have made significant contributions toward meeting the goal of mass production of human enzymes for human use.

Uses

Therapeutic enzymes have a broad variety of specific uses, ie, as oncolytics, thrombolytics, or replacements for inherited deficiencies. Additionally, there is a growing group of miscellaneous enzymes of diverse function.

Immobilized, Derivatized, and Entrapped Enzymes

Immobilized Enzymes. With the development of techniques for binding enzymes to insoluble supports, immobilized enzymes have been used not only in clinical analysis but also for therapeutic purposes.

Enzyme Conjugates. One approach used to prolong residence time of a given enzyme in the circulation is to conjugate that enzyme with albumin, a natural plasma protein.

Erythrocyte Entrapment of Enzymes. Erythrocytes have been used as carriers for therapeutic enzymes in the treatment of inborn errors. Exogenous enzymes encapsulated in erythrocytes may be useful both for delivery of a given enzyme to the site of its intended function and for the degradation of pathologically elevated, diffusible substances in the plasma.

Chemically or enzymatically modified preparations of enzymes present another potential immunologic problem since modifications designed to stabilized enzyme activity may produce new antigenic determinants.

As of this writing (ca 1993) there are eight methods of erythrocyte entrapment. Six methods depend on loading via hypotonic exchange, one method depends on chlorpromazine-induced endocytosis, and one depends on voltage-step induced transitory permeation.

Health and Safety Factors

Repetitive doses of foreign proteins may produce severe immunologic reactions, ranging from mild allergy to anaphylactic shock and death. Parenteral administration of metabolically active enzymes can be acutely toxic on account of their biochemical effects. However, large metabolically effective doses of these products may be less toxic than small doses, due to immune paralysis, wherein large doses of antigen repress the expression of the complementary antibody. It is essential to circumvent such allergic responses in the enzyme therapy of genetic diseases, because replacement enzyme therapy may be required over a lifetime. By contrast, treatment of cancer with enzymes may be somewhat less problematic because of the attenuation for the immunologic reactivity of patients in many cases. A significant effort is being directed toward the use of human sources of enzymes, or of human-type enzymes produced in cultures of prokaryotic or eukaryotic organisms, with a goal of reducing or eliminating the problems of antigenicity.

For enzymes intended for parenteral use, the manufacturer must assure that the enzyme preparation is essentially pure and free of endotoxins. All preparations of enzymes intended for parenteral use are tested for safety in lower animals under the conditions anticipated in clinical trials.

HIREMAGULAR N. JAYARAM
Indiana University School of Medicine
GURPREET S. AHLUWALIA
Gillette Research Institute
DAVID A. COONEY
National Cancer Institute

D. M. Goldberg, *Clin. Chim. Acta* **206,** 45–76 (1992).

D. B. Kohn and co-workers, *Human Gene Ther.* **2,** 101–105 (1991).

M. J. Cowan, *Clin. Biochem.* **24,** 375–381 (1991).

J. S. Holcenberg and J. Roberts, eds., *Enzymes as Drugs,* Wiley-Interscience, New York, 1981.

ENZYME INHIBITORS

The development of potent enzyme inhibitors has led to an increased understanding of enzyme mechanisms and has provided effective therapeutic agents for the treatment of diseases. Enzymes are natural biocatalysts that promote specific reactions essential for the viability of living organisms. They have unique recognition sites that allow them to select their substrates out of the vast pool of biologically important compounds in living cells. The substrate binds to the active site, that portion of the enzyme responsible for promoting the chemistry involved in converting the substrate to product. Inhibitors of enzymes prevent this chemical reaction from occurring by altering the active site and thereby rendering the enzyme at least temporarily inactive.

Applications for Enzyme Inhibitors

Enzyme inhibitors are often used to further the understanding of enzyme mechanisms. Inhibitors serve as probes for kinetic and chemical processes during catalysis. Inhibitors are also used for *in vivo* studies to localize and quantify enzymes in organs or to mimic certain genetic diseases that involve the absence of an enzyme in a given biosynthetic pathway. In pharmacological research, enzyme inhibitors are used to inactivate specific enzymes or groups of enzymes, leading to the treatment of many diseases.

An important area of drug design is the development of drugs against microorganisms and parasites in humans. A powerful strategy consists of choosing a target enzyme that is essential for the existence of the invader, but not the host. Therefore, these drugs usually exhibit low toxicity toward the host. An alternative approach is the choice of a target enzyme that exists as different species, called isozymes, in the invader and the host.

Another class of therapeutic agents is used for the treatment of certain genetic diseases or other enzymatic disorders caused by the dysfunction or absence of one particular enzyme. This often leads to an unwanted accumulation or imbalance of metabolites in the organism.

A third class of enzyme inhibitors consists of antitumor agents. Their design is quite challenging because tumor cells do not appear to contain enzymes that are very different from those in normal cells. Most antitumor drugs, called antiproliferative agents, take advantage of the fact that tumor cells generally grow and divide much faster than normal cells.

Classification of Enzyme Inhibitors

Enzyme inhibitors may be classified as follows:

Noncovalent inhibitors	Covalent inhibitors
rapid reversible inhibitors	mechanism-based inhibitors
tight, slow, slow-tight binding inhibitors	affinity labels
transition-state analogues	pseudoirreversible inhibitors
multisubstrate analogues	

All of these enzyme inhibitors are active site-directed.

Computer-Aided Inhibitor Design

Computer-aided inhibitor design is a relatively new and powerful approach for the development of novel, potentially potent, nonsubstrate-analogue enzyme inhibitors. Computer-aided methods and biological screening can each lead to new classes of novel inhibitors. However, computer-aided design methods can focus the search for inhibitors, thereby circumventing much of the time-consuming synthetic and natural product purification procedures for those compounds they find unlikely to function as inhibitors.

Design of Inhibitors from an Enzyme Structure. As high quality, three-dimensional structures of enzymes have become increasingly available, they have been used for the structure-based design of inhibitors. The enzyme structure provides a map showing the shape of available spaces, the locations of potential hydrogen-bonding groups, and the locations of potentially changed groups.

Computers are integral to structure-based design processes. Graphics workstations and a variety of computer algorithms (procedures) are used to help visualization of three-dimensional structures. Molecular graphics programs can provide tools for displaying, building, and energy-minimizing three-dimensional structures, in addition to docking structures, bringing them together in three-dimensional space like an inhibitor into the active-site of an enzyme. Visual display requires molecular surface contours, color codes, depth perception cues, and tools for three-dimensional manipulation. Currently available program packages that contain molecular graphics capabilities include BIOGRAPH (from BioDesign), CHEM-X (from Chemical Design, Ltd.), FRODO, INSIGHT (from Biosym Technologies, Inc.) MACROMODEL, MIDAS, MOGLI (from Evans and Sutherland Computer Corp.) QUANTA (from Polygen Corp.), and SYBYL (from Tripos Association) (see MOLECULAR MODELING).

In addition to assisting with structure visualization, computer algorithms are used to automate the search for inhibitors by performing tasks such as screening databases and analyzing enzyme active sites. DOCK, a computer program package, brings molecules together in space to score the fit of a single potential inhibitor in an enzyme active site, or to screen small molecule databases for compounds that score well. Another program, GRID, finds regions in the active site that should have favorable interactions with a specific chemical group. A third program, HINT, identifies and maps potential hydrophobic and polar interactions.

Design of Inhibitors from a Series of Structurally Related Compounds. Traditionally, medicinal chemists have used trial-and-error methods to study the relationship between structure and function for the development of inhibitors in a fashion more rational than biological screening. After measuring the inhibitory effect of an original compound, a small chemical modification is made to create a structural analogue, which is tested for a change in inhibition. It is assumed that the structural change is small enough so that the mechanism of action is not drastically altered. By testing several structural analogues, the spatial limits of the enzyme's binding site are explored without knowing the structure of the enzyme. Modern, quantitative approaches use the speed and computational power of computer algorithms to determine specific structure–function relationships for the design of more potent analogues of known inhibitors. Methods include quantitative structure–activity relationship (QSAR) methods, comparative molecular field analysis methods (CoMFA), and perturbation free-energy calculations.

Design of Inhibitors from a Pharmacophore. Pharmacophore-based inhibitor design is a method for developing inhibitors that combines the structural elements known to be important for binding with new structural templates. From a series of compounds that bind to the enzyme, the pharmacophore, the structural unit containing the chemical elements in the orientation required for binding, must first be identified. Once the pharmacophore has been defined, it can be used in many ways to develop inhibitors. For example, structural databases can be searched for existing molecules that contain the proposed pharmacophore. Alternatively, potential inhibitors can be built from structural pieces that connect the elements of the pharmacophore in the proper fashion.

ANGELIKA MUSCATE
CYNTHIA L. LEVINSON
GEORGE L. KENYON
University of California, San Francisco

D. V. Santi and G. L. Kenyon in M. E. Wolff, ed., *Burger's Medicinal Chemistry*, 4th ed., Wiley-Interscience, New York, 1980, pp. 349–391.

A. Fersht, *Enzyme Structure and Mechanism*, 2nd ed., W. H. Freeman and Co., New York, 1985.

I. D. Kuntz, *Science* **257**, 1078–1082 (1992).

Y. C. Martin, *Methods Enzymol.* **203**, 587–613 (1991).

ENZYMES, IMMOBILIZED.

See ENZYME APPLICATIONS; ENZYMES IN ORGANIC SYNTHESIS.

ENZYMES IN ORGANIC SYNTHESIS

The application of enzymes in organic synthesis is a modern and rapidly growing area in synthetic organic chemistry. The rapid expansion of this area in recent years has been brought about by a number of factors, most importantly that a large number of enzymes have become commercially available. Of about 2500 enzymes identified thus far about 300 are available in a partly purified form. Moreover, because of advances in molecular biology, fermentation, and purification techniques, the cost of some enzymes has been reduced to less than $500/kg.

The synthetic utility of enzymes for the preparation of organic molecules is tremendous. Enzymes catalyze virtually all types of chemical reactions with the exception of Diels-Alder condensation. They possess remarkably high catalytic power (up to 10^{12} rate acceleration compared to the nonenzymatic reactions) and unsurpassed stereo- and regioselectivity. From a technological standpoint they also offer a number of advantages: the mild temperatures, neutral pH, and atmospheric pressure under which most enzymes operate result in processes that are environmentally acceptable, low energy consuming, and usually do not require high capital investments.

Biotransformations are carried out by either whole cells (microbial, plant, or animal) or by isolated enzymes. Both methods have advantages and disadvantages. In general, multistep transformations, such as hydroxylations of steroids, or the synthesis of amino acids, riboflavin, vitamins, and alkaloids that require the presence of several enzymes and cofactors are carried out by whole cells. Simple one- or two-step transformations, on the other hand, are usually carried out by isolated enzymes. Compared to fermentations, enzymatic reactions have a number of advantages including simple instrumentation; reduced side reactions, easy control, and product isolation.

The principal emphasis of this article is on reactions carried out by isolated enzymes that have the broadest synthetic utility, ie, hydrolases, oxidoreductases, and lyases. Biotransformations catalyzed by living cells are considered to a lesser extent. For more detailed information on biotransformations the reader may consult several books and numerous reviews.

Hydrolases

Quantitative Analysis of Selectivity. One of the principal synthetic values of enzymes stems from their unique enantioselectivity, ie, ability to discriminate between enantiomers of a racemic pair. Detailed quantitative analysis of kinetic resolutions of enantiomers relating the extent of conversion of racemic substrate (*c*), enantiomeric excess (*ee*), and the enantiomeric ratio (*E*) has bene described elsewhere.

Enzyme-Catalyzed Asymmetric Synthesis. The extent of kinetic resolution of racemates is determined by differences in the reaction rates for the two enantiomers. At the end of the reaction the faster reacting enantiomer is transformed, leaving the slower reacting enantiomer unchanged. It is apparent that the maximum product yield of any kinetic resolution cannot exceed 50%.

The situation is different if the substrate is a prochiral or meso compound. Since these molecules have a center or plane of symmetry the binding of pro-S or pro-R forms is equivalent. The chirality appears only as a result of the transformation. Hence, at least theoretically, the compound can be converted to one enantiomer quantitatively.

Hydrolytic enzymes such as esterases and lipases have proven particularly useful for asymmetric synthesis because of their abilities to discriminate between enantiotopic ester and hydroxyl groups. A large number of esterases and lipases are commercially available in large quantities; many are inexpensive and accept a broad range of substrates.

Dicarboxylic Acid Monoesters. Enzymatic synthesis of monoesters of dicarboxylic acids by hydrolysis of the corresponding diesters is a widely used and thoroughly studied reaction. It is catalyzed by a number of esterases, lipases, and proteases and is usually carried out in an aqueous buffer, pH 6–8 at room temperature. Organic cosolvents may be added to increase solubility of the substrates. The pH is maintained at a constant level by the addition of aqueous hydroxide. After one equivalent of base is consumed the monoesters are isolated by conventional means. Recent developments in the area of enzymatic catalysis in nonaqueous media have significantly broadened the repertoire of hydrolytic enzymes.

Monoacyl Diols. Enzymatic synthesis of chiral monoacyl diols can be carried out either by direct enzymatic acylation of prochiral diols or by hydrolysis of chemically synthesized dicarboxylates.

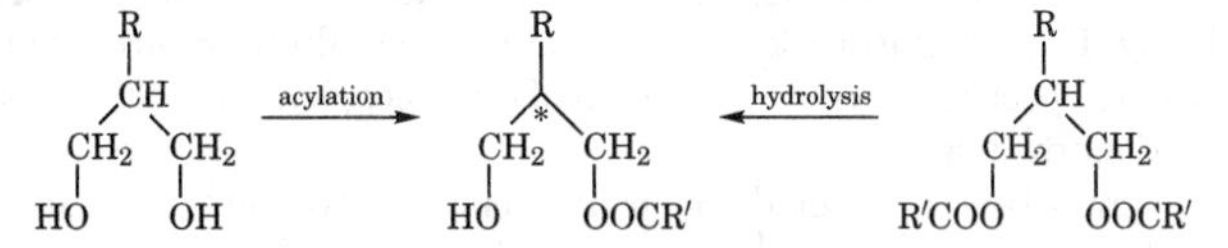

Generally, these two methods complement each other. With some rare exceptions an enzyme that produces an S ester in the hydrolysis reaction produces an R isomer in acylation reaction and vice versa.

Kinetic Resolutions. From a practical standpoint the principal difference between formation of a chiral molecule by kinetic resolution of a racemate and formation by asymmetric synthesis is that in the former case the maximum theoretical yield of the chiral product is 50% based on a racemic starting material. In the latter case a maximum yield of 100% is possible. If the reactivity of two enantiomers is substantially different the reaction virtually stops at 50% conversion, and enantiomerically pure substrate and product may be obtained in close to 50% yield. Conveniently, the enantiomeric purity of the substrate and the product depends strongly on the degree of conversion so that even in those instances where reactivity of enantiomers is not substantially different, a high purity material may be obtained by sacrificing the overall yield.

The variety of enzyme-catalyzed kinetic resolutions of enantiomers reported in recent years is enormous. Similar to asymmetric synthesis, enantioselective resolutions are carried out in either hydrolytic or esterification–transesterification modes. Both modes have advantages and disadvantages. Hydrolytic resolutions that are carried out in a predominantly aqueous medium are usually faster and, as a consequence, require smaller quantities of enzymes. On the other hand, esterifications in organic solvents are experimentally simpler procedures, allowing easy product isolation and reuse of the enzyme without immobilization.

Optically Active Acids and Esters. Enantioselective hydrolysis of esters of simple alcohols is a common method for the production of pure enantiomers of esters or the corresponding acids. Lipases, esterases, and proteases accept a wide variety of esters and convert them to the corresponding acids, often in a highly enantioselective manner. Lipase-catalyzed kinetic resolutions are often practical for the preparation of optically active pharmaceuticals.

Optically Active Alcohols and Esters. In addition to the hydrolysis of esters formed by simple alcohols described above, lipases and esterases also catalyze the hydrolysis of a wide range of esters based on more complex and synthetically useful cyclic and acyclic alcohols. Although the hydrolysis of acetates often gives the desirable resolution, to achieve maximum selectivity and reaction efficiency, comparison of various esters is recommended. Both saturated and unsaturated derivatives are easily accepted by lipases and esterases.

Hydrolysis of Enol Esters. Enzyme-mediated enantioface-differentiating hydrolysis of enol esters is an original method for generating optically active α-substituted ketones.

Chiral Lactones and Polyesters. Similar to intermolecular reactions described previously, lipases also catalyze intramolecular acylations of hydroxy acids; the reaction results in the formation of lactones.

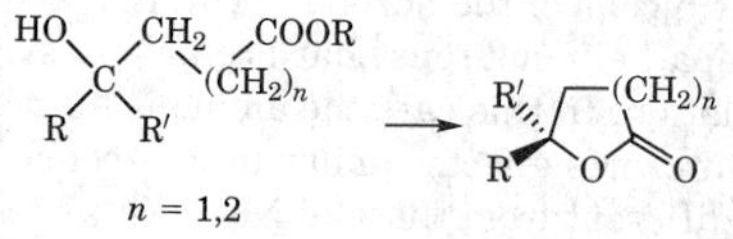

Regioselective Acylation of Hydroxy Compounds. Aliphatic diols can be selectively acylated at the primary and secondary positions by a number of lipases in nonaqueous solvents.

Resolution of Racemic Amines and Amino Acids. Acylases (EC 3.5.1.14) are the most commonly used enzymes for the resolution of amino acids. Porcine kidney acylase (PKA) and the fungal *Aspergillus* acylase (AA) are commercially available, inexpensive, and stable. Amino alcohols can be resolved by a number of pathways, including hydrolysis, esterification, and transesterification.

Unprotected racemic amines can be resolved by enantioselective acylations with activated esters. This approach is based on the discovery that enantioselectivity of some enzymes strongly depends on the nature of the reaction medium.

Two interesting approaches for resolution of racemic amino acids have been reported. Both are based on enantioselective ring opening with *in situ* racemization of the nonreactive isomer (Fig. 1).

Hydrolysis of Nitriles. The chemical hydrolysis of nitriles to acids takes place only under strong acidic or basic conditions and may be accompanied by formation of unwanted and sometimes toxic by-products. Enzymatic hydrolysis of nitriles by nitrile hydratases, nitrilases, and amidases is often advantageous since amides or acids can be produced under very mild conditions and in a stereo- or regioselective manner.

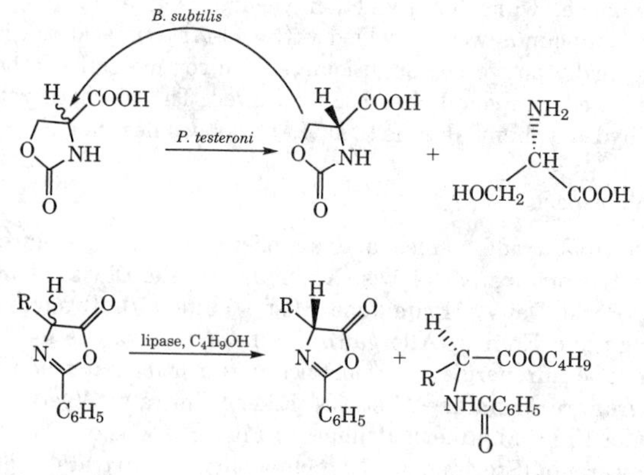

Figure 1. Enzymatic resolution of amino acids by ring-opening reaction.

There are two distinct classes of enzymes that hydrolyze nitriles. Nitrilases (EC 3.5.5.1) hydrolyze nitriles directly to corresponding acids and ammonia without forming the amide. In fact, amides are not substrates for these enzymes. Nitriles also may be first hydrated by nitrile hydratases to yield amides which are then converted to carboxylic acid with amidases. This is a two-enzyme process, in which enantioselectivity is generally exhibited by the amidase, rather than the hydratase.

Peptide Synthesis. Despite significant progress in the chemical synthesis of peptides and proteins by both liquid- and solid-phase methodologies, a number of shortcomings still exist. The principal limitation of chemical methods stems from the formation of by-products at each individual condensation step that accumulate during the course of the repeated reactions. Moreover, even the small amount of racemization that often occurs in the presence of highly activated coupling reagents reduces the purity of final product and complicates purification dramatically. In this respect the use of enzymes for amino acid or peptide coupling has a number of advantages. Mild reaction conditions and the excellent stereo-and regioselectivity of enzymes require only minimal protection, precludes racemization, and guarantees the structural fidelity of the product.

There are two basic strategies for enzyme-catalyzed peptide synthesis: equilibrium- and kinetically controlled synthesis. The former is the direct reversal of proteolysis and involves the condensation of an amino component with unactivated carboxyl component. The latter proceeds by the aminolysis of an activated peptide ester.

Lyases

Aldol Additions. These reactions catalyzed by lyases are perhaps the most synthetically useful enzymatic reactions for carbon–carbon bond formation.

There are two distinct groups of aldolases. Type I aldolases, found in higher plants and animals, require no metal cofactor and catalyze aldol addition via Schiff base formation between the lysine ϵ-amino group of the enzyme and a carbonyl group of the substrate. Class II aldolases are found primarily in microorganisms and utilize a divalent zinc to activate the electrophilic component of the reaction. A great variety of aldol additions have been carried out, allowing the synthesis of numerous nitrogen-containing sugars, deoxysugars, and fluorosugars, based, among others, on D-threose, L-threose, D-ribose, L-arabinose, D-xylose, D-glucose-6-P, and D-mannose-6-P, where P = phosphate.

Cyanohydrin Synthesis. Another synthetically useful enzyme that catalyzes carbon–carbon bond formation is oxynitrilase (EC 4.1.2.10). This enzyme catalyzes the addition of cyanides to various aldehydes that may come either in the form of hydrogen cyanide or acetone cyanohydrin (Fig. 2). The reaction constitutes a convenient route for the preparation of α-hydroxy acids and β-amino alcohols. Acetone cyanohydrin can also be used as the cyanide carrier, and is considered to be superior since it does not involve hazardous gaseous HCN and also virtually eliminates the spontaneous nonenzymatic reaction.

Oxidoreductases

Biocatalytic redox reactions offer great synthetic utility to organic chemists. The majority of oxidase-catalyzed preparative bioconversions are still performed using a whole-cell technique, despite the fact that the presence of more than one oxidoreductase in cells often leads to product degradation and lower selectivity. Fortunately, several efficient cofactor regeneration systems have been developed, making some cell-free enzymatic bioconversions economically feasible.

The two oxidoreductase systems most frequently used for preparation of chiral synthons include baker's yeast and horse liver alcohol dehydrogenase (HLAD). The use of baker's yeast has been recently reviewed in great detail and therefore will not be covered here. The emphasis here is on dehydrogenase-catalyzed oxidation and reduction of alcohols, ketones, and keto acid, oxidations at unsaturated carbon, and Bayer-Villiger oxidations.

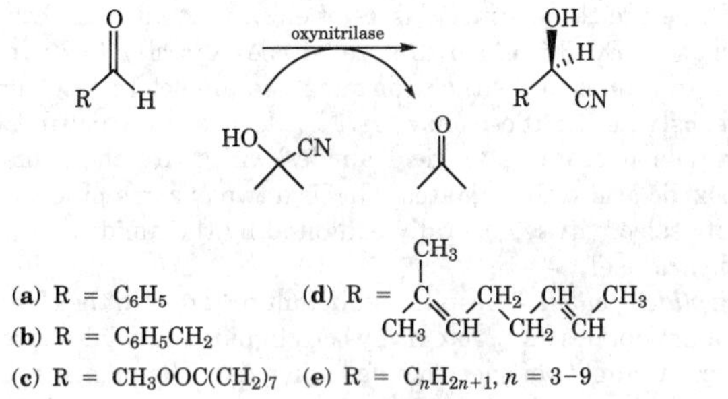

Figure 2. Cyanohydrin formation.

Chiral Alcohols and Lactones. HLAD has been widely used for stereoselective oxidations of a variety of prochiral diols to lactones on a preparative scale. In most cases pro-(*S*) hydroxyl is oxidized irrespective of the substituents. The method is applicable to, among others, *cis*-1,2-bis(hydroxymethyl) derivatives of cyclopropane, cyclobutane, cyclohexane, and cyclohexene. Resulting γ-lactones are isolated in 68–90% yields and of 100% *ee*.

Although alcohol dehydrogenases (ADH) also catalyze the oxidation of aldehydes to the corresponding acids, the rate of this reaction is significantly lower. The systems that combine ADH and aldehyde dehydrogenases (EC 1.2.1.5) (AldDH) are much more efficient.

Alcohol dehydrogenase-catalyzed reduction of ketones is a convenient method for the production of chiral alcohols. HLAD, the most thoroughly studied enzyme, has a broad substrate specificity and accommodates a variety of substrates.

Hydroxy and Amino Acids. Reduction of 2-oxoacids with NADH, catalyzed by lactate dehydrogenase, is an established method for the preparation of homochiral α-hydroxy acids of both S and R configurations. The enzyme is found in all higher organisms and can be easily isolated from a variety of mammalian and bacterial sources.

Baeyer-Villiger Oxidations. The biological equivalent of the Baeyer-Villiger reaction is a useful transformation in providing a mild technique for converting ketones into esters or lactones.

Lipoxygenase-Catalyzed Oxidations. Lipoxygenase-1 catalyzes the incorporation of dioxygen into polyunsaturated fatty acids possessing a 1(*Z*),4(*Z*)-pentadienyl moiety to yield (*E*),(*Z*)-conjugated hydroperoxides. A highly active preparation of the enzyme from soybean is commercially available in purified form.

Summary

The use of enzymatic catalysis inorganic synthesis has grown tremendously in recent years and continues to grow. Greater availability of enzymes, development of new methodologies for their utilization, use of nonconventional environments, and the design and synthesis of new biocatalysts with altered selectivity and increased stability were essential for the successful development of this field. As more is learned about selectivity of enzymes toward unnatural substrates, the choice of an enzyme for a particular transformation will become easier to predict. It will simplify a search for an appropriate catalyst and help to establish biocatalytic procedures as a useful supplement to classical organic synthesis.

ALEKSEY ZAKS
Schering-Plough Research Institute

C.-H. Wong and G. M. Whitesides, *Enzymes in Synthetic Organic Chemistry*, Tetrahedron Organic Chemistry Series, Vol. 12, Elsevier Science Inc., Tarrytown, N.Y., 1994.

H. G. Davis, R. H. Green, D. R. Kelly, and S. M. Roberts, eds., *Biotransformations in Preparative Organic Chemistry*, Academic Press Ltd. London, 1989.

E. Santaniello, P. Ferraboschi, P. Grisenti, and A. Manzocchi, *Chem. Rev.* **92**, 1071 (1992).

H. L. Holland, *Organic Synthesis with Oxidative Enzymes*, VCH Publishers, Inc., New York, 1992.

EPINEPHRINE AND NOREPINEPHRINE

(−)-Epinephrine (E), (−)-norepinephrine (NE), and dopamine (DA) belong to a class of compounds known as catecholamines, and are neurotransmitters and/or hormones in peripheral tissues and the central nervous system. NE is the principal postganglionic sympathetic neurotransmitter in the periphery and in adrenergic neurons in the central nervous system. It is released from nerve terminals, and acts to transmit a nerve impulse across the synapse (synaptic cleft) between the nerve terminal and postsynaptic site of action. Many neurotransmitters, including E and NE, can be both inhibitory and excitatory depending on which tissue or organ the neurotransmitter is acting on. For example, NE may be excitatory, as in the heart and vascular, inhibitory, as in the intestine, or both excitatory and inhibitory, as in the central nervous system depending on location.

The chromaffin cells of the adrenal medulla may be considered to be modified sympathetic neurons that are able to synthesize E from NE by *N*-methylation. In this case the amine is liberated into the circulation, where it exerts effects similar to those of NE; in addition, E exhibits effects different from those of NE, such as relaxation of lung muscle (hence its use in asthma). Small amounts of E are also found in the central nervous system, particularly in the brain stem where it may be involved in blood pressure regulation. DA, the precursor of NE, has biological activity in peripheral tissues such as the kidney, and serves as a neurotransmitter in several important pathways in the brain.

Physical Properties

Some of the physical properties of E and NE are summarized in Table 1.

Stability. E is sensitive to air, light, heat, and alkalies. Metals, notably copper, iron, and zinc, destroy its activity. In solution with sulfite or bisulfite, it slowly forms an inactive sulfonate. The red color that forms when neutral or alkaline solutions are exposed to air is caused by adrenochrome.

NE is unstable in light and air, especially at neutral and alkaline pH. Oxidation to noradrenochrome occurs in the presence of oxygen and such divalent metal ions as copper, manganese, and nickel.

Chemical Synthesis

Friedel-Crafts acylation of catechol with chloroacetyl chloride yields chloroacetocatechol. Displacement of the chlorine by methylamine yields the methylamine derivative, adrenalone, which on catalytic reduction yields (±)-epinephrine. Substitution of ammonia for methylamine in the sequence yields the amino derivative noradrenalone, which on reduction yields (±)-norepinephrine. The racemic compounds were resolved with (+)-tartaric acid to give the physiologically active (−)-enantiomers. The commercial synthesis of E and related compounds has been reviewed. The synthetic route for L-3,4-dihydroxyphenylalanine (L-DOPA) has been described.

Economic Aspects

The principal trade names and suppliers of (−)E, an adrenergic vasoconstrictor, are as follows: Adrenalin (Parke-Davis); Adrenalin in Oil (Parke-Davis); Primatene Mist (Whitehall); Bronkaid Mist (Sterling); and Epifrin (Allergan). The bitartrate salt is used as an ophthalamic adrenergic vasoconstrictor; the principal trade names (suppliers) are Asmatane Mist (3M Riker); Epitrate (Wyeth-Ayerst); Medihaler-Epi (3M Riker); Primatene Mist Suspension (Whitehall); and Suprarenin (Sterling). (−)Norephinephrine bitartrate, used similarly to (−)E, is supplied by Sterling under the trade name Levophed. Dopamine hydrochloride, a cardiovascular agent, is supplied under

Table 1. Physical Properties of Epinephrine and Norepinephrine[a]

Property	D-(−)-Epinephrine	D-(−)-Norepinephrine
IUPAC name	(−)-3,4-dihydroxy-α-[(methylamino)methyl]benzyl alcohol	(−)-α-(aminomethyl)-3,4-dihydroxy benzyl alcohol
formula	$C_9H_{13}NO_3$	$C_8H_{11}NO_3$
formula wt	183.2	169.2
appearance	colorless microcrystals[b]	colorless microcrystals
mp	ca 209–210°C (dec)	ca 212°C (dec)
solubility	in water, ca 0.1 g/100 mL; insoluble in alcohol and most other organic solvents	slightly soluble in water
absorption spectrum	0.1 *M* HCl λ_{max} 221 nm, ϵ ca 6100; λ_{max} 280 nm, ϵ ca 2700	in 0.1 *M* HCl, λ_{max} 221 nm, ϵ ca 5860; λ_{max} 279 nm, ϵ ca 2560
fluorescence spectrum	in 0.05 *M* sodium acetate buffer, pH 4, λ_{ex} 283 nm, λ_{em} 337 nm	in 0.05 *M* sodium acetate buffer, pH 4, λ_{ex} 283 nm, λ337 nm; iodine oxidation in alkaline ascorbate, λ_{ex} 412 nm, λ_{ex} 505 nm
source	resolution of (±)-epinephrine with (+)-tartaric acid	
likely impurities	(+)-epinephrine	(+)-norepinephrine

[a] Data given in: *Specifications and Criteria for Biochemical Compounds, Supplement: Biogenic Amines and Related Compounds*. Courtesy of the National Academy of Sciences.
[b] Acceptable preparations may be slightly off-white.

the name Dopestat from Parke-Davis and Intropin from DuPont Pharmaceuticals. Levodopa, an antiparkinsonian agent, is supplied under the following trade names (manufacturers) as Bendopa (ICN), Dopan (Norwich-Eaton), Larodopa (Hoffman-LaRoche), and Levopa (SmithKline-Beecham), and is a component of Madopa (Hoffman-LaRoche) and Sinemet (Merck).

Regulation of Synthesis, Storage, Release, and Metabolism

Catecholamine biosynthesis begins with the uptake of the amino acid tyrosine into the sympathetic neuronal cytoplasm, and conversion to DOPA by tyrosine hydroxylase. This enzyme is highly localized to the adrenal medulla, sympathetic nerves, and central adrenergic and dopaminergic nerves. Tyrosine hydroxylase activity is subject to feedback inhibition by its products DOPA, NE, and DA, and is the rate-limiting step in catecholamine synthesis; the enzyme can be blocked by the competitive inhibitor α-methyl-*p*-tyrosine.

DOPA in the bloodstream can be taken up into neural tissue and into tissue devoid of tyrosine hydroxylase, thus bypassing the rate-limiting enzymatic synthetic step. Uptake of DOPA by the brain is the basis of the therapeutic effect of DOPA in the treatment of Parkinson's disease (a disease characterized by depletion of DA in basal ganglia due to loss of nerve cells in the substantia nigra). In addition, the neurotoxin 1-methyl-4-phenyl-1,2,3,6-tetrahydropyridine (MPTP) depletes DA in the striatum and produces a clinical syndrome like Parkinson's disease that responds to DOPA.

DA in the cytoplasm of neurons is taken up into storage vesicles by an energy requiring process that can be inhibited by the alkaloid reserpine. The depletion by reserpine is long lasting, and the storage granules are irreversibly damaged. Cytoplasmic NE can also be taken up stereoselectively into the vesicles. The vesicles contain dopamine-β-oxidase that catalyzes the conversion of DA to NE. In the vesicle NE can be stored, can leak back into the cytoplasm, or can be released into the synapse. Finally, at some sites, notably in adrenal medulla and brain stem, NE is *N*-methylated to form E by phenylethanolamine *N*-methyltransferase.

The storage vesicles play a dual role: they maintain a ready supply of catecholamines at the terminal (also in the adrenal medulla) available for release, and they mediate the process of release. Efficient regulatory mechanisms operate to modulate the rate of synthesis and release of catecholamines, depending on the need.

Unlike the cholinergic system where the action of acetylcholine is terminated by cholinesterase, the action of catecholamine neurotransmitters released into the synaptic cleft is mainly brought to an end not by enzymatic degradation but by reuptake back into the nerve terminals (uptake 1) that released the catecholamines. Each of the three types of catecholamine neurons has its own uptake mechanism which is an energy-requiring, stereoselective process that can be blocked by certain classes of drugs such as tricyclic antidepressants (desipramine), cocaine, and inhibitors of the Na–K ATPase such as ouabain. Nonneuronal uptake of catecholamines (uptake 2) also occurs, and is an energy-requiring, nonstereoselective process that can be blocked by adrenocorticosteroids such as corticosterone.

Two important pathways for catecholamine metabolism are *O*-methylation by catechol-*o*-methyl transferase (COMT), which is cytoplasmically localized, and oxidative deamination by the mitochondrial localized enzyme monoamine oxidase (MAO).

Catecholamine Receptors

When catecholamines are released from adrenergic neuronal terminals or adrenal medulla, they are recognized by and interact with specific receptor proteins (adrenergic) on plasma membranes of effector cells or tissues that mediate the action of the agonist, resulting in a physiological response. The diverse actions of E and NE led to the 1948 proposal of the concept of distinct subtypes of adrenergic receptors, classified as alpha- and beta-adrenergic receptors.

Second Messengers

Adrenergic receptors comprise integral membrane proteins responsible for binding extracellular E and NE and initiating membrane signals through interaction with one of a series of guanine nucleotide binding regulatory proteins (*G*-proteins) to activate second messenger systems. In this respect there are two significant mechanisms involved in the interaction of E and NE with their receptors, the first involving changes in calcium conductance and the phosphoinositol (PI) response (alpha-) and the second related to inhibition (alpha-2, DAD-2) or activation (beta, DAD-1) of adenylyl cyclase.

Molecular Biology

Extensive pharmacological data together with the emergence of molecular biological approaches in the study of receptor structure and function, have yielded new information concerning the classification of alpha- and beta-adrenergic and dopaminergic receptors. A feature of these receptors is that they are members of a large superfamily of genes encoding numerous receptors that couple to guanine nucleotide regulatory proteins during signal transduction and show common structural features, eg, seven putative hydrophobic membranes. This new information has resulted in the identification of a number of distinct subtypes. The following subdivision of receptor subtypes is based on data generated on endogenous receptors and cloned receptors. Three distinct alpha-1 (designated 1A, 1B, and 1C), three distinct alpha-2 (designated 2A, 2B, and 2C), three beta (designated beta-1, beta-2, and beta-3), and five or six DA (designated D-1, ie, D-1A, D-1B, and D-1C, also known as D-5, and D-2, ie, D-2A [2L], D-2B [2L], D-3, and D-4) receptors have been characterized. The D-2 receptor exists as two isoforms (2A and 2B), which are generated through alternative splicing of *m* RNA and differ by the presence or absence of a 29-amino acid insert in the third cytoplasmic loop.

Therapeutic Uses

As described, E, NE, DA, and related compounds exhibit a wide diversity of physiological functions. Patients in clinical shock are usually treated with an iv-administered pressor amine, including isoproterenol (beta agonist selective) NE, E, DA, and dobutamine. E is also the primary treatment for anaphylactic shock. E is commonly included in local anesthetic solutions to promote hemostasis, and by vasoconstriction to reduce absorption resulting in prolongation of anesthesia. Several related sympathomimetic vasoconstrictor amines (eg, phenylephrine hydrochloride) are used for nasal congestion. Because of their relaxation of bronchial smooth muscle, E and selected beta-2 agonists are used to antagonize the bronchospasm observed in asthma. NE is used for treating hypotension during anesthesia when tissue perfusion is good.

Alpha-adrenergic blocking agents such as prazosin (alpha-1 selective), which causes vasodilation in both arteries and veins without usually causing reflex tachycardia, are used to treat mild to moderate hypertension. Nonselective beta-adrenergic antagonists such as propranolol are used in the treatment of hypertension (usually with a diuretic), as prophylaxis in angina pectoris, and for prophylaxis of supraventricular and ventricular arrhythmias and other selected disorders. Selective beta-1 adrenergic antagonists such as metoprolol are used mainly for the treatment of hypertension. In addition, clonidine (an alpha-2 agonist) and methyldopa (metabolized to alpha-methylnorepinephrine in brain) act centrally on vasomotor centers of the brain to reduce sympathetic outflow to the peripheral vessels and thus are used, but to a lesser extent, in the treatment of hypertension.

In Parkinson's disease, treatment with the amine precursor DOPA (with the decarboxylase inhibitor carbidopa), has been shown to ameliorate the symptoms and signs of the condition and prolong life.

There are several other disorders of the central nervous system in which catecholamines have been shown to be involved and drugs that affect the actions of catecholamines have a therapeutic action. Dopamine receptor antagonists that encompass several chemical classes such as phenothiazines (eg, chlorpromazine, butyrophenones (eg, haloperidol, and thioxanthene derivatives (eg, chlorprothixene, are prescribed for the management of both acute and chronic psychoses and in nonpsychotic individuals who are delusional or excited (eg, mania). In the treatment of depression, most antidepressants are believed to improve mood by increasing catecholamine and/or serotonin concentrations.

Besides behavior and blood pressure, catecholamine neurons also have important roles in other brain functions. Regulation of neuroendocrine function is a well-known action of catecholamines; for example, DA agonists reduce serum prolactin concentration, especially in conditions of hypersecretion. Ingestive behavior can be modulated by brain catecholamines, and some appetite-suppressing drugs are believed to act via catecholaminergic influences. Catecholamines also participate in regulation of body temperature.

Toxicity

Untoward effects of both E and NE (usually to a lesser degree) are anxiety, headache, cerebral hemorrhage (from vasopressor effects), cardiac arrhythmias, especially in presence of digitalis and certain anesthetic agents, and pulmonary edema as a result of pulmonary hypertension. The minimum subcutaneous lethal dose of E is about 4 mg, but recoveries have occurred after accidental overdosage with 16 mg subcutaneously and 30 mg intravenously, followed by immediate supportive treatment.

THOMAS A. PUGSLEY
Warner-Lambert/Parke-Davis

U. S. von Euler in H. Blaschko and E. Muscholl, eds., *Catecholamines, Handbuch der Experimentellen Pharmakologie*, Vol. 33, Springer-Verlag, Berlin, 1972, p. 186.

J. R. Cooper, F. E. Bloom, and R. H. Roth, *The Biochemical Basis of Neuropharmacology*, Oxford University Press, New York, 1991, p. 224.

H. Winkler in U. Trendelenberg and N. Weiner, eds., *Handbook of Experimental Pharmacology*, Vol 90/I, Springer-Verlag, Berlin, 1988, p. 43.

D. Bylund and co-workers, *Pharmacological Rev.* **46**, 121 (1994).

EPOXY RESINS

Epoxy resins are characterized by the presence of a three-membered ring known as the epoxy, epoxide, oxirane, or ethoxyline group. Commercial epoxy resins contain aliphatic, cycloaliphatic, or aromatic backbones. The capability of the epoxy ring to react with a variety of substrates imparts versatility to the resins. Treatment with curing agents gives insoluble and intractable thermoset polymers. In order to facilitate processing and modify cured resin properties, other constituents may be included in the compositions: fillers, solvents, diluents, plasticizers, accelerators, curatives, and tougheners.

Resin Properties

Epichlorohydrin and Bisphenol A-Derived Resins. The most widely used epoxy resins are diglycidyl ethers of bisphenol A (**1**) derived from bisphenol A and epichlorohydrin.

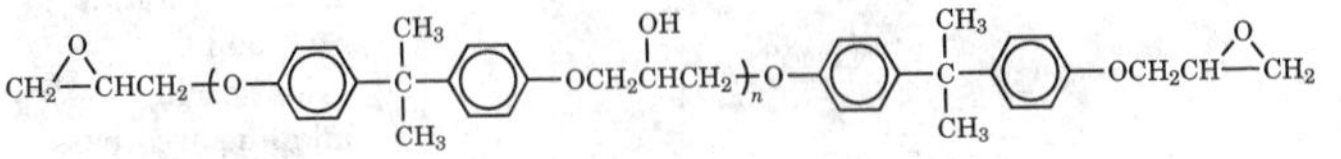

(**1**)

The outstanding performance characteristics of the resins are conveyed by the bisphenol A moiety (toughness, rigidity, and elevated temperature performance), the ether linkages (chemical resistance), and the hydroxyl and epoxy groups (adhesive properties and formulation latitude, or reactivity with a wide variety of chemical curing agents) (see also PHENOLIC RESINS).

The bisphenol A-derived epoxy resins are most frequently cured with anhydrides, aliphatic amines, or polyamides.

Diluents are commonly used to reduce the viscosity of epoxy systems to aid handling, improve ease of application, and to facilitate higher filler loading to reduce formulation cost. This, however, is achieved at the expense of other properties. To achieve a balance of properties, careful selection of diluent is needed.

Specialty Epoxy Resins. In addition to bisphenol, other polyols such as aliphatic glycols and novolaks are used to produce specialty resins. Epoxy resins may also include compounds based on aliphatic, cycloaliphatic, aromatic, and heterocyclic backbones. Glycidylation of active hydrogen-containing structures with epichlorohydrin and epoxidation of olefins with peracetic acid remain the important commercial procedures for introducing the oxirane group into various precursors of epoxy resins.

Epoxy Cresol–Novolak Resins (ECN). The cresol–novolak epoxy resins (**2**) are multifunctional, solid polymers characterized by low ionic and hydrolyzable chlorine impurities, high chemical resistance, and good thermal performance. ECN resins are widely used as base components in high performance electronic and structural molding compounds, high temperature adhesives, castings and laminating systems, tooling applications, and powder coatings.

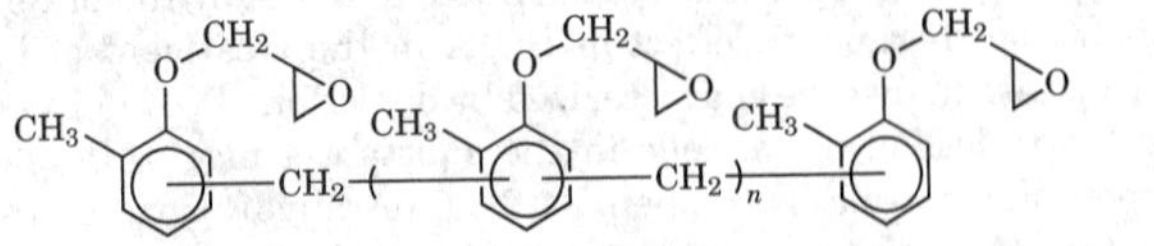

(**2**)

The epoxy cresol–novolak resins (**2**) are prepared by glycidylation of *o*-cresol–formaldehyde condensates in the same manner as the phenol–novolak resins.

Bisphenol F Resin. Bisphenol F epoxy resin is of the same general structure as the epoxy phenol novolaks. Bisphenol F is 2,2′-methylene bisphenol.

Owing to relatively low viscosity, these resins offer advantages for 100% solids (solvent-free) systems. Higher filler levels are possible because of the low viscosity. Faster bubble release is also achieved. Higher epoxy content and functionality of bisphenol F epoxy resins can provide improved chemical resistance compared to conventional epoxies.

Bisphenol F epoxy resins are used in high-solids-high-build systems such as tank and pipe linings, industrial floors, road and bridge deck toppings, structural adhesives, grouts, coatings, and electrical varnishes. Bisphenol F epoxy resins are manufactured in Europe and Japan.

Epoxy Phenol–Novolak Resins. Epoxy phenol–novolak resins are represented by the general idealized structure (**3**) whereby multifunctional products are formed containing a phenolic hydroxyl group per phenyl ring in random para–para′, ortho–para′, and ortho–ortho′ combinations.

Subsequent epoxidation with epichlorohydrin yields the highly functional epoxy novolak. The product can range from a high viscosity liquid of $n = 0.2$ to a solid of n value greater than 3.

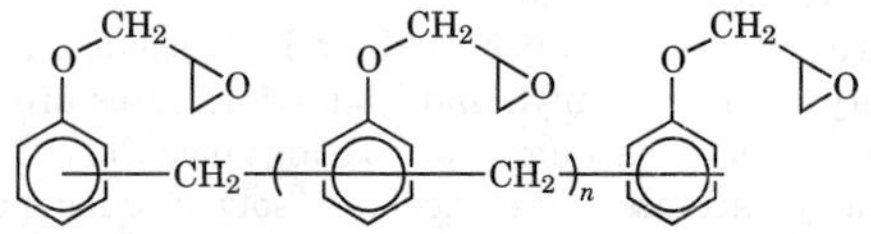

(**3**)

The thermal stability of epoxy phenol–novolak resins is useful in adhesives, structural and electrical laminates, coatings, castings, and encapsulations for elevated temperature service. Filament-wound pipe and storage tanks, liners for pumps and other chemical process equipment, and corrosion-resistant coatings are typical applications using the chemically resistant properties of epoxy novolak resins.

Curing agents that give the optimum in elevated temperature properties for epoxy novolaks are those with good high temperature performance such as aromatic amines, catalytic curing agents, phenolics, and some anhydrides.

Polynuclear Phenol–Glycidyl Ether-Derived Resins. This is one of the first commercially available polyfunctional products. Its polyfunctionality permits upgrading of thermal stability, chemical resistance, and electrical and mechanical properties of bisphenol A–epoxy systems. It is used in molding compounds and adhesives.

Cycloaliphatic Epoxy Resins. This family of aliphatic, low viscosity epoxy resins consists of two principal varieties, cycloolefins epoxidized with peracetic acid and diglycidyl esters of cyclic dicarboxylic acids.

The nonaromatic nature of these materials provides for improved uv resistance and arc-track resistance compared to conventional epoxies. The best properties are generally achieved with anhydride and phenolic curing agents.

Recommended applications include transformers, insulators, bushings, wire and cable coatings, generators, motors and switchgear, additives for adhesives, vinyl stabilization, and as viscosity depressants.

Aromatic and Heterocyclic Glycidyl Amine Resins. Among the specialty epoxy resins containing an aromatic amine backbone, the following are commercially significant.

Tetraglycidylmethylenedianiline-Derived Resins. Resins from aromatic glycidyl amines can be formulated into hot-melt or solution-binder systems with various reinforcements, eg, glass, graphite, boron, or aramid. They are utilized for graphite-reinforced composites in aerospace and leisure products, structural adhesives, laminates, tooling and casting applications, and structures such as wings and fuselages.

Triglycidyl p-Aminophenol-Derived Resins. Resins derived from triglycidyl *p*-aminophenol, originally developed by Union Carbide Corp., are currently marketed by CIBA-GEIGY. Synthesis is conducted by reaction of epichlorohydrin with the phenolic and amino groups followed by dehydrohalogenation. The product is a viscous liquid (1.5–5 Pa·s (15–50 P) at 25°C) which is considerably more reactive toward amines than standard bisphenol A-derived resins.

Used to increase heat resistance and cure speed of bisphenol A epoxy resins, it has utility in such diverse applications as adhesives, tooling compounds, and laminating systems.

Triazine-Based Resin. Triglycidyl isocyanurate is a solid resin that provides superior thermal, electrical, and mechanical properties and is recommended for laminates, insulating varnishes, coatings, and adhesives. Widely used as a curing agent for special polyester-based weatherable powder coatings, it is also used in electronic applications owing to its retention of optical transparency after aging at temperatures up to 150°C and minimal smoke evolution on thermal decomposition (see EMBEDDING).

The triazine ring-containing product 1,3,5-triglycidyl isocyanurate is synthesized by glycidylation of cyanuric acid with epichlorohydrin.

Resin Synthesis and Manufacture

Epichlorohydrin and Bisphenol A-Derived Resins. Liquid epoxy resins may be synthesized by a two-step reaction of an excess of epichlorohydrin to bisphenol A in the presence of an alkaline catalyst. The reaction consists initially in the formation of the dichlorohydrin of bisphenol A and further reaction by dehydrohalogenation of the intermediate product with a stoichiometric quantity of alkali.

In recent years, production of liquid resins of higher purity, ie, higher monomer content and fewer side-reactions, has been accomplished. This is in response to more stringent product quality requirements.

Aliphatic Glycidyl Ethers. Aliphatic epoxy resins have been synthesized by glycidylation of difunctional or polyfunctional polyols such as a 1,4-butanediol, 2,2-dimethyl-1,3-propanediol (neopentyl glycol), polypropylene glycols, glycerol, trimethylolpropane, and pentaerythritol.

The epoxidation is generally conducted in two steps: *(1)* the polyol is added to epichlorohydrin in the presence of a Lewis acid catalyst (stannic chloride, boron trifluoride) to produce the chlorohydrin intermediate, and *(2)* the intermediate is dehydrohalogenated with sodium hydroxide to yield the aliphatic glycidyl ether. Solid epoxy resins are prepared by the Taffy or Advancement processes.

Taffy Process. Bisphenol A reacts directly with epichlorohydrin in the presence of a stoichiometric amount of caustic. The molecular weight of the product is governed by the ratio of epichlorohydrin–bisphenol A. In practice, the taffy process is generally employed for only medium molecular-weight resins ($n = 1-4$).

Advancement Process. In the advancement process, sometimes referred to as the fusion method, liquid epoxy resin (crude diglycidyl ether of bisphenol A) is chain-extended with bisphenol A in the presence of a catalyst to yield higher polymerized products. The advancement process is more widely used in commercial practice.

In recent years, proprietary catalysts for advancement have been incorporated in precatalyzed liquid resins. Thus only the addition of bisphenol A is needed to produce solid epoxy resins. Use of the catalysts is claimed to provide resins free from branching which can occur in conventional fusion processes. Additionally, use of the catalysts results in rapid chain-extension reactions because of the high amount of heat generated in the processing.

The preparation of flame-retardant epoxy resins is accompanied by inclusion of tetrabromobisphenol A in the advancement process (see FLAME RETARDANTS). Products containing ca 20 wt % Br are extensively employed in the printed circuit board industry.

Liquid resins containing bromine (ca 49 wt %) can also be prepared directly from tetrabromobisphenol A and epichlorohydrin and are used for critical applications where a high degree of flame retardancy is required.

Curing Reactions

A variety of reagents has been described for converting the liquid and solid epoxy resins to the cured state, which is necessary for the development of the inherent properties of the resins. Liquid epoxy resins contain mainly epoxy groups and solid resins are composed of both epoxy and hydroxyl curing sites. The curing agents or hardeners are categorized as either catalytic or coreactive and the functional groups of the resins are terminal epoxy together with a pendent hydroxyl per repeat unit of the polymer chain.

Economic Aspects

Epoxy resin sales increased rapidly in the 1970s and continued the increase into the 1980s as new applications were developed.

High performance multifunctional epoxy resins have achieved a higher compounded growth rate than the significantly larger volume DGEBPA conventional resins. Although epoxy resins cost more than competitive products, their longer service life and high performance capabilities provide a better cost-performance ratio.

The fastest growth over the past decade has been in laminates and composites. Tremendous growth in the electronics market has markedly increased the demand for epoxy laminating resins for the manufacture of printed wiring boards and epoxy-molding compounds for semiconductor encapsulation. Use of epoxy composites in the transportation industry, including the military and commercial aircraft and automotive fields, where a high strength-to-weight ratio is required, is growing at a steady rate that is expected to increase through the end of the century (see COMPOSITE MATERIALS–POLYMER-MATRIX).

As with other petrochemical-based products, prices of epoxies have risen rapidly during the 1980s and 1990s as oil prices have escalated.

Trademarks of the principal epoxy resin producers are as follows: Shell Chemical Co. (EPON, EPONOL, Epikote); Dow Chemical Co. (D.E.R., D.E.N., D.E.H., Tactix, Quatrex); CIBA-GEIGY Corp. (Araldite); Dainippon Ink and Chemicals (Epotuf, Kelpoxy); and Union Carbide Co. (Unox).

Health and Safety Factors

There have been many investigations of the toxicity of various classes of epoxy-containing materials (glycidyloxy compounds). The use and interpretation of the vast amount of data available has been obscured by two factors: *(1)* proper identification of the epoxy systems in question and *(2)* lack of meaningful classification of the epoxy materials. In general, the toxicity of many of the glycidyloxy derivatives is low but the diversity of compounds found within this group does not permit broad generalizations for the class.

Applications

Epoxy resins are used in protective coatings, eg, waterborne coatings, high solids coatings, and powder coatings, structural composites, electrical laminates, electrical, electronics, and structural applications (molding components and casting and encapsulation), and adhesives.

JOHN GANNON
Consultant

H. Lee and K. Neville, *Handbook of Epoxy Resins,* McGraw-Hill, Inc., New York 1967, reprinted 1982.

W. G. Potter, *Epoxide Resins,* Springer-Verlag, New York, 1970.

C. A. May, ed., *Epoxy Resins Chemistry and Technology,* 2nd ed., Marcel Dekker, Inc., New York, 1988.

L. V. McAdams and J. A. Gannon, in J. I. Kroschwitz, ed., *Encyclopedia of Polymer Science and Engineering,* 2nd ed., Vol. 6, Wiley-Interscience, New York, 1986, pp. 322–382.

ESTERIFICATION

This article describes methods for the production of carboxylic esters:

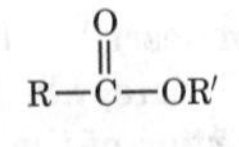

For the properties of these compounds, see ESTERS, ORGANIC. For esters of inorganic acids, see the articles on nitric acid, phosphoric acids, sulfuric acid, etc.

Esters are most commonly prepared by the reaction of a carboxylic acid and an alcohol with the elimination of water. Esters are also formed by a number of other reactions utilizing acid anhydrides, acid chlorides, amides, nitriles, unsaturated hydrocarbons, ethers, aldehydes, ketones, alcohols, and esters (via ester interchange).

On the basis of bulk production, poly(ethylene terephthalate) manufacture is the most important ester producing process. This polymer is produced by either the direct esterification of terephthalic acid and ethylene glycol, or by the transesterification of dimethyl terephthalate with ethylene glycol. Dimethyl terephthalate is produced by the direct esterification of terephthalic acid and methanol.

Other large-volume esters are vinyl acetate (VAM), methyl methacrylate (MMA), and dioctyl phthalate (DOP). VAM is produced for the most part by the vapor-phase oxidative acetoxylation of ethylene. MMA and DOP are produced by direct esterification techniques involving methacrylic acid and phthalic anhydride, respectively.

The acetates of most alcohols are also commercially available and have diverse uses. Because of their high solvent power, ethyl, isopropyl, butyl, isobutyl, amyl, and isoamyl acetates are used in cellulose nitrate and other lacquer-type coatings (see CELLULOSE ESTERS). Butyl and hexyl acetates are excellent solvents for polyurethane coating systems (see COATINGS; URETHANE POLYMERS). Ethyl, isobutyl, amyl, and isoamyl acetates are frequently used as components in flavoring (see FLAVORS AND SPICES), and isopropyl, benzyl, octyl, geranyl, linalyl, and methyl acetates are important additives in perfumes (qv).

Effect of Structure. The rate at which different alcohols and acids are esterified as well as the extent of the equilibrium reaction are dependent on the structure of the molecule and types of functional substituents of the alcohols and acids.

In making acetate esters, the primary alcohols are esterified most rapidly and completely, ie, methanol gives the highest yield and the most rapid reaction. Under the same conditions, the secondary alcohols react much more slowly and afford lower conversions to ester products; however, wide variations are observed among the different members of this series. The tertiary alcohols react slowly and the conversions are generally low (1–10% conversion at equilibrium).

The introduction of a nitrile group on an aliphatic acid has a pronounced inhibiting effect on the rate of esterification.

Substitutions that displace electrons toward the carboxyl group of aromatic acids diminish the rate of the reaction.

Kinetic Considerations. Extensive kinetic and mechanistic studies have been made on the esterification of carboxylic acids since Berthelot and Saint-Gilles first studied the esterification of acetic acid. A number of mechanisms for acid- and base-catalyzed esterification have been proposed. One possible mechanism for the bimolecular acid-catalyzed ester hydrolysis and esterification is shown below.

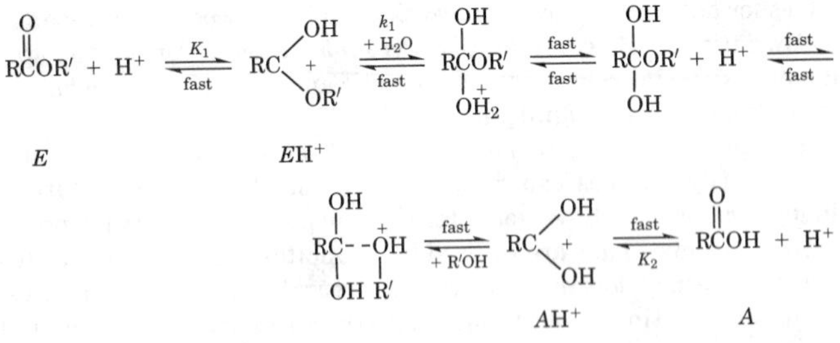

This mechanism leads to the rate equation for hydrolysis and to an analogous expression for the esterification:

$$-\frac{d[E]}{dt} = \frac{k_1K_1[E][H_2O][H^+]}{1+\alpha} - \frac{k_2K_2[A][R'OH][H^+]}{1+1/\alpha} \quad (2)$$

In this expression, α depends on those rate coefficients in the above mechanism whose values are assumed to be high. Other mechanisms for the acid hydrolysis and esterification differ mainly with respect to the number of participating water molecules and possible intermediates.

Applications of kinetic principles to industrial reactions are often useful. Initial kinetic studies of the esterification reaction are usually conducted on a small scale in a well stirred batch reactor. In many cases, results from batch studies can be used in the evaluation of the esterification reaction in a continuous operating configuration.

Equilibrium Constants. The reaction between an organic acid and an alcohol to produce an ester and water is expressed as:

$$\mathrm{RC(=O)OH} + \mathrm{R'OH} \rightleftharpoons \mathrm{RC(=O)OR'} + H_2O$$

This was first demonstrated in 1862 by Berthelot and Saint-Gilles, who found that when equivalent quantities of ethyl alcohol and acetic acid were allowed to react, the esterification stopped when two-thirds of the acid had reacted. Similarly, when equal molar proportions of ethyl acetate and water were heated together, hydrolysis of the ester stopped when about one-third of the ester was hydrolyzed. By varying the molar ratios of alcohol to acid, yields of ester >66% were obtained by displacement of the equilibrium. The results of these tests were in accordance with the mass action law shown. $K = [ester][water]/[acid][alcohol]$. However, in many cases the equilibrium constant is affected by the proportion of reactants. The temperature as well as the presence of salts may also affect the value of the equilibrium constant.

Completion of Esterification. Because the esterification of an alcohol and an organic acid involves a reversible equilibrium, these reactions usually do not go to completion. Conversions approaching 100% can often be achieved by removing one of the products formed, either the ester or the water, provided the esterification reaction is equilibrium limited and not rate limited. A variety of distillation methods can be applied to afford ester and water product removal from the esterification reaction (see DISTILLATION). Other methods such as reactive extraction and reverse osmosis can be used to remove the esterification products to maximize the reaction conversion. In general, esterifications are divided into three broad classes, depending on the volatility of the esters:(*1*) Esters of high volatility, such as methyl formate, methyl acetate, and ethyl formate, have lower boiling points than those of the corresponding alcohols, and therefore can be readily removed from the reaction mixture by distillation.(*2*) Esters of medium volatility are capable of removing the water formed by distillation. Examples are propyl, butyl, and amyl formates, ethyl, propyl, butyl, and amyl acetates, and the methyl and ethyl esters of propionic, butyric, and valeric acids.(*3*) Esters of low volatility are accessible via several types of esterification.

Use of Azeotropes to Remove Water. With the aliphatic alcohols and esters of medium volatility, a variety of azeotropes is encountered on distillation. Removal of these azeotropes from the esterification reaction mixture drives the equilibrium in favor of the ester product.

Use of Desiccants and Chemical Means to Remove Water. Another means to remove the water of esterification is calcium carbide supported in a thimble of a continuous extractor through which the condensed vapor from the esterification mixture is percolated (see CARBIDES). A column of activated bauxite (Florite) mounted over the reaction vessel has been used to remove the water of reaction from the vapor by adsorption.

Catalysts. The choice of the proper catalyst for an esterification reaction is dependent on several factors. The most common catalysts used are strong mineral acids such as sulfuric and hydrochloric acids. Lewis acids such as boron trifluoride, tin and zinc salts, aluminum halides, and organo–titanates have been used. Cation-exchange resins and zeolites are often employed also.

Acid-Regenerated Cation Exchangers. The use of acid-regenerated cation resin exchangers (see ION EXCHANGE) as catalysts for effecting esterification offers distinct advantages over conventional methods. Several types of cation-exchange resins can be used as solid catalysts for esterification. In general, the strongly acidic sulfonated resins comprised of copolymers of styrene, ethylvinylbenzene, and divinylbenzene are used most widely. With the continued improvement of ion-exchange resins, such as the macroporous sulfonated resins, esterification has become one of the most fertile areas for use of these solid catalysts.

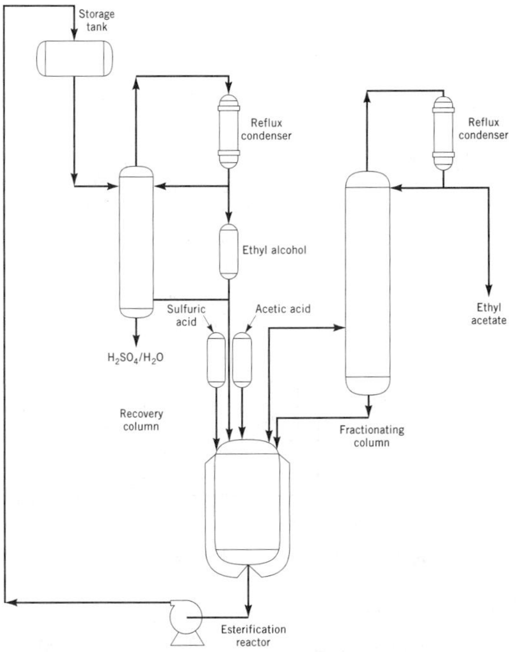

Figure 1. Batch ethyl acetate process.

Despite the higher cost compared with ordinary catalysts such as sulfuric or hydrochloric acid, the cation exchangers present several features that make their use economical. The ability to use these agents in a fixed-bed reactor operation makes them attractive for a continuous process. Cation-exchange catalysts can be used also in continuous stirred tank reaction (CSTR) operation.

Batch Esterification

Batch esterification is used to produce ethyl acetate (Fig. 1) and *n*-butyl acetate.

Continuous Esterification

The law of mass action, the laws of kinetics, and the laws of distillation all operate simultaneously in a process of this type. Esterification can occur only when the concentrations of the acid and alcohol are in excess of equilibrium values; otherwise, hydrolysis must occur. The equations governing the rate of the reaction and the variation of the rate constant (as a function of such variables as temperature, catalyst strength, and proportion of reactants) describe the kinetics of the liquid-phase reaction. The usual distillation laws must be modified, since most esterifications are somewhat exothermic and reaction is occurring on each plate. Since these kinetic considerations are superimposed on distillation operations, each plate must be treated separately by successive calculations after the extent of conversion has been determined (see DISTILLATION).

Continuous esterification of acetic acid in an excess of *n*-butyl alcohol with sulfuric acid catalyst using a four-plate single bubblecap column with reboiler has been studied. The rate constant and the theoretical extent of reaction were calculated for each plate, based on plate composition and on the total incoming material to the plate. Good agreement with the analytical data was obtained.

A continuous distillation process has been studied for the production of high boiling esters from intermediate boiling polyhydric alcohols and low boiling monocarboxylic aliphatic or aromatic acids. The water of reaction and some of the organic acid were continuously removed from the base of the column.

Continuous esterification is used to produce methyl acetate (Fig. 2).

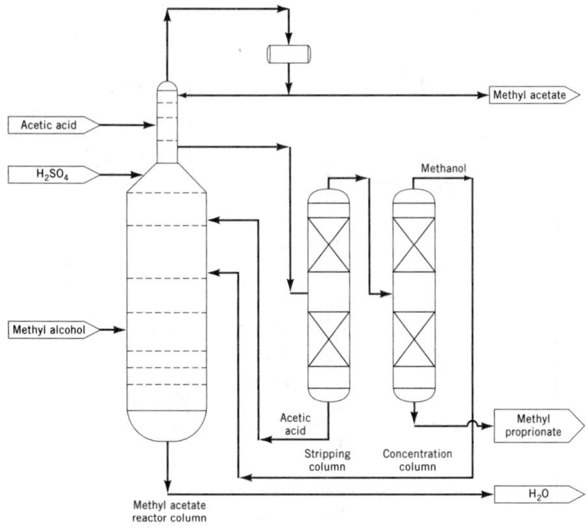

Figure 2. Continuous methyl acetate process.

Vapor-Phase Esterification

Catalytic esterification of alcohols and acids in the vapor phase has received attention because the conversions obtained are generally higher than in the corresponding liquid-phase reactions.

Physicochemical Considerations. The determination of the equilibrium constant K_G for the reaction $C_2H_5OH + CH_3COOH \rightarrow C_2H_5OOCH_3 + H_2O$ has been the subject of a number of investigations over the temperature range of 40–300°C. The values of the equilibrium constant range from 6–559 with 71–95% ester as the equilibrium concentration from an equimolar mixture of ethyl alcohol and acetic acid, depending on the technique used. A study of the reaction mechanism indicates that adsorption of acetic acid is the rate-controlling step; the molecularly adsorbed acetic acid then reacts with alcohol in the vapor phase.

Ethyl Acetate. Catalysts proposed for the vapor-phase production of ethyl acetate include silica gel, zirconium dioxide, activated charcoal, and potassium hydrogen sulfate. More recently, phosphoric-acid-treated coal and calcium phosphate catalysts have been described.

Esterification of Other Compounds

Acid Anhydrides. Acid anhydrides react with alcohols to form esters (in high yields in many cases) with a carboxylic acid formed as by-product:

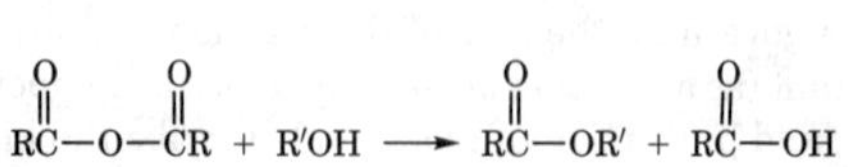

However, this method is applied only when esterification cannot be effected by the usual acid–alcohol reaction because of the higher cost of the anhydrides. The production of cellulose acetate (see FIBERS–CELLULOSE ESTERS), phenyl acetate (used in acetaminophen production), and aspirin (acetylsalicylic acid) (see SALICYLIC ACID AND RELATED COMPOUNDS) are examples of the large-scale use of acetic anhydride.

Formic anhydride is not stable. However, formate esters of alcohols and phenolics can be prepared using formic–acetic anhydride. Dibasic acid anhydrides such as phthalic anhydride and maleic anhydride readily react with alcohols to form the monoalkyl ester. Ketene, like acid anhydrides, reacts with alcohols to form (acetate) esters.

Acid Chlorides. Acid chlorides react with alcohols to form esters.

Amides. Alcoholysis of amides provides another method for synthesizing esters.

Other methods of converting amides to esters have been described. Alkyl halides can be treated with amides to give esters. Also, esters can be synthesized from *N*-alkyl-*N*-nitrosoamides, which are derived from the corresponding amides.

Nitriles. Alcoholysis of nitriles offers a convenient way to produce esters without isolating the acid. Catalysts such as hydrogen chloride, hydrogen bromide, and sulfuric acid have been employed. One of the most important applications of this process is that of methyl methacrylate manufacture.

Unsaturated Hydrocarbons. Olefins from ethylene through octene have been converted into esters via acid-catalyzed nucleophilic addition.

Most of the vinyl acetate produced in the United States is made by the vapor-phase ethylene process.

Ethers. In the presence of anhydrous agents such as ferric chloride, hydrogen bromide, and acid chlorides, ethers react to form esters. Esters can also be prepared from ethers by an oxidative process.

Unsaturated esters can be prepared from the corresponding acetylenic ethers with yields in most cases of >50%. β-Hydroxyethyl esters can be prepared from carboxylic acids and ethylene oxide.

Aldehydes and Ketones. Esters are obtained readily by condensation of aldehydes in the presence of alcoholate catalysts such as aluminum ethylate, $Al(OC_2H_5)_3$, by the Tishchenko reaction.

Alcohols. The direct synthesis of esters by dehydrogenation or oxidative hydrogenation of alcohols offers a simple method for the preparation of certain types of esters, such as ethyl acetate.

Technical Preparation of Esters

Esterification is generally carried out by refluxing the reaction mixture until the carboxylic acid has reacted with the alcohol and the water has been split off. The water of the ester is removed from the equilibrium by distillation. The choice of the esterification process to obtain a maximum yield is dependent on many factors, ie, no single process has universal applicability.

Methyl Esters. Methyl esters are obtained in good yield using methylene dichloride or ethylene dichloride as solvent.

Medium Boiling Esters. Esterification of ethyl and propyl alcohols, ethylene glycol, and glycerol with various acids, eg, chloro- or bromoacetic, or pyruvic, by the use of a third component such as benzene, toluene, hexane, cyclohexane, or carbon tetrachloride to remove the water produced is quite common.

High Boiling Esters. The following procedure can be used for making diethyl phthalate and other high boiling esters. Phthalic anhydride (1 equiv) and 2.5 equivalents of ethanol are refluxed for 2 h in the presence of 1% of concentrated H_2SO_4.

Difficulty Esterifiable Acids. The sterically hindered acids, such as 2,6-disubstituted benzoic acids, cannot usually be esterified by conventional means. Several esters of sterically hindered acids such as 2,4,6-triisopropylbenzoic acid have been prepared by dissolving 2 g of the acid in 14–20 mL of 100% H_2SO_4. After standing a few minutes at room temperature, when presumably the acylium cation is formed, the solution is poured into an excess of cold absolute methanol. Most of the alcohol is removed under reduced pressure, about 50 mL of water is added, and the distillation is continued under reduced pressure to remove the remainder of the methanol. The organic matter is extracted with ether and treated with sodium carbonate solution. The ester is then distilled. Yields of esters made in this manner are 57–81%.

Ester Interchange

Ester interchange (transesterification) is a reaction between an ester and another compound, characterized by an exchange of alkoxy groups or of acyl groups, and resulting in the formation of a different ester. The process of transesterification is accelerated in the presence of a small amount of an acid or a base.

Three types of transesterification are known: *(1)* exchange of alcohol groups, commonly known as alcoholysis, *(2)* exchange of acid groups, acidolysis, and *(3)* ester–ester interchange. These reactions are reversible and ordinarily do not involve large energy changes.

Applications. Transesterifications via alcoholysis play a significant role in industry as well as in laboratory and in analytical chemistry. The reaction can be used to reduce the boiling point of esters by exchanging a long-chain alcohol group with a short one, eg, methanol, in the analysis of fats, oils, and waxes.

An industrial example of acidolysis is the reaction of poly(vinyl acetate) with butyric acid to form poly(vinyl butyrate). Often a butyric acid–methanol mixture is used and methyl acetate is obtained as a coproduct.

MOHAMMAD ASLAM
G. PAULL TORRENCE
EDWARD G. ZEY
Hoechst Celanese Corporation

E. E. Reid in P. Grotggins, *Unit Processes in Organic Synthesis*, 5th ed., McGraw-Hill Book Co., Inc., New York, 1958.

S. Patai, *The Chemistry of Carboxylic Acids and Esters*, Wiley-Interscience, New York, 1969.

K. S. Markley in K. S. Markley, ed., *Fatty Acids*, part 2, Wiley-Interscience, New York, 1961, p. 757.

R. C. Larock, *Comprehensive Organic Transformations*, VCH Publishers, Inc., New York, 1989.

ESTERS, ORGANIC

Esters are compounds that, on hydrolysis, yield alcohols or phenols and acids according to the equation: $RA + H_2O \rightleftarrows ROH + HA$, where R is a hydrocarbon fragment and A is the anion portion of an organic acid.

Nomenclature

The names of esters consist of two words that reflect their formation from an alcohol and a carboxylic acid. According to the IUPAC rule, the alkyl or aryl group of the alcohol is cited first followed by the carboxylate group of the acid with the ending -ate replacing the -ic of the acid. For example, $CH_3CH_2COOCH_3$, the methyl ester of propanoic acid, is called methyl propanoate (or methyl propionate, if the trivial name, propionic acid, is used for the carboxylic acid).

Physical Properties

The physical properties of organic esters vary according to the molecular weight of each component. Lower molecular weight esters are colorless, mobile, and highly volatile liquids that usually have pleasant odors. As the molecular weight increases, volatility decreases and the consistency becomes waxy, then solid, and eventually even brittle, often with formation of lustrous crystals. The melting point of an ester is generally lower than that of the corresponding carboxylic acid. However, the boiling point depends on the chain length of the alcohol component and eventually exceeds that of the acid. Lower molecular weight esters are relatively stable when dry and can be distilled without decomposition. Organic esters are generally insoluble in water, but soluble in various organic liquids. Lower esters are themselves good solvents for many organic compounds. The physical properties of commercially important aliphatic and aromatic organic esters are listed in Table 1.

Chemical Properties

The reactions of esters have been reviewed. Because of the large number of possible acid and alcohol moieties, the chemical properties of esters may differ considerably. Only typical reactions applicable to the majority of esters are described in the following sections.

Hydrolysis. Esters are cleaved (hydrolyzed) into an acid and an alcohol through the action of water. This hydrolysis is catalyzed

Table 1. Physical Properties of Some Common Esters

Ester	Mol wt	n_D^{20}	Bp, °C[a]	Freezing point, °C
methyl formate	60.05	1.344	32	−99.8
ethyl formate	74.08	1.3598	54.3	−80
butyl formate	102.13	1.3889	106	−91.9
methyl acetate	74.08	1.3594	57	−98.1
ethyl acetate	88.1	1.3723	77.1	−83.6
vinyl acetate	86.1	1.3959	72.2	−93.2
propyl acetate	102.13	1.3844	101.6	−92.5
isopropyl acetate	102.13	1.3773	90	−73.4
butyl acetate	116.16	1.3951	126	−73.5
isobutyl acetate	116.16	1.3902	117.2	−98.6
sec-butyl acetate	116.16	1.3877	112	
t-butyl acetate	116.16	1.3855	97	
pentyl acetate	130.18	1.4023	149.3	−70.8
isoamyl acetate	130.18	1.4000	142	−78
sec-hexyl acetate	144.22	1.4014	157	0
2-ethylhexyl acetate	172.26	1.4204	199.3	−93
ethylene glycol diacetate	146.14	1.415	191	−31
2-methoxyethyl acetate	118.13	1.4019	145	−65.1
2-ethoxyethyl acetate	132.16	1.4058	156.4	−61.7
2-butoxyethyl acetate	160.12	1.42	187.8	−32
2-(2-ethoxyethoxy)ethyl acetate	176.21	1.423	217.4	−25
2-(2-butoxyethoxy)ethyl acetate	204.27	1.4265	247	−32.2
benzyl acetate	150.18	1.5232	215.5	−51.5
glyceryl triacetate	218.23	1.4296	258	−78
ethyl 3-ethoxypropionate	146.19		165–172	−50
glyceryl tripropionate	260.3	1.4318	176	−58
methyl acrylate	86.09	1.4040	80.5	<−75
ethyl acrylate	100.11	1.4068	99.8	<−72
butyl acrylate	128.17	1.4185	69	−64.6
2-ethylhexyl acrylate	184.28		130[i]	−90
methyl methacrylate	100.12	1.4119	100	−48
methyl butyrate	102.13	1.3878	102.3	−84.8
ethyl butyrate	116.16	1.4000	121.6	−100.8
butyl butyrate	144.22	1.4075	166.6	−91.5
methyl isobutyrate	102.13	1.3840	92.6	−84.7
ethyl isobutyrate	116.16	1.3870	110	−88
isobutyl isobutyrate	144.22	1.3999	148.7	−80.7
methyl stearate	298.5	1.457	215	40
ethyl stearate	312.52	1.429	213–215	33.7
butyl stearate	340.58		343	27.5
dodecyl stearate	440.8	1.433		28
hexadecyl stearate	496.91	1.441		57
dimethyl maleate	144.13	1.4409	204	
dimethyl oxalate	111.09	1.4096	185	−41
dimethyl adipate	174.2	1.4283	115	10.3
diethyl adipate	202.25	1.4372	245	−19.8
di(2-ethylhexyl) adipate	370.58	1.4472	214	−60
methyl benzoate	136.15	1.517	199.5	−12.5
ethyl benzoate	150.18	1.505	212.9	−34.2
methyl salicylate	152.15	1.536	223.3	−8.6
ethyl salicylate	166.18	1.522	231.5	1.3
dimethyl phthalate	194.19	1.515	282	−2
diethyl phthalate	222.24	1.499	295	−33
dibutyl phthalate	278.35	1.4911	340	−35
di(2-ethylhexyl) phthalate	390.56	1.486	231	−50
dimethyl isophthalate	194.19	1.5168	124	67
dimethyl terephthalate	194.19		288	140
methyl anthranilate	151.17	1.584	132	24
benzyl cinnamate	238.29		244	39
dimethyl carbonate	90.08	1.3682	90	3
diethyl carbonate	118.13	1.3854	127	−43

[a] At 101.3 kPa = 760 mm Hg unless otherwise stated.

by acids or bases. The mechanistic aspects of ester hydrolysis have received considerable attention and have been reviewed.

Enzymatic Hydrolysis. Enzymatic hydrolysis has received enormous attention. The enzymes generally employed are lipases from microorganisms, plants, or mammalian liver. The great advantage of the enzymatic process is its high chemo- and stereoselectivity.

Transesterification. When esters are heated with alcohols, acids, or other esters in the presence of a catalyst, the alcohol or acid groups are exchanged. This process is called transesterification. It is accelerated by the presence of a small amount of acid or alkali. Three types of transesterification are known: (*1*) exchange of alcohol groups (alcoholysis), (*2*) exchange of acid groups (acidolysis), and (*3*) ester–ester interchange (see ESTERIFICATION).

Ammonolysis and Aminolysis. Esters and ammonia react to form amides and alcohols.

If primary or secondary amines are used, *N*-substituted amides are formed. This reaction is called aminolysis.

Reduction. Esters can be reduced to alcohols by catalytic hydrogenation using molecular hydrogen or by chemical reduction.

Reaction of Enolate Anions. In the presence of certain bases, eg, sodium alkoxide, an ester having a hydrogen on the α-carbon atom undergoes a wide variety of characteristic enolate reactions. Mechanistically, the base removes a proton from the α-carbon, giving an enolate that then can react with an electrophile.

Grignard and Related Reactions. Esters react with alkyl magnesium halides in a two-stage process to give alcohols. The reaction involves nucleophilic substitution of R^3 or OR^2 and addition of R^3MgX to the carbonyl group. With 1,4-dimagnesium compounds, esters are converted to cyclopentanols. Lactones react with Grignard reagents and give diols as products. Many other organometallic compounds also react with carbonyl.

α-Halo esters react with aldehydes or ketones in the presence of zinc to form β-hydroxy esters. This is known as the Reformatsky reaction.

Preparation of Acyloins. When aliphatic esters are allowed to react with metallic sodium, potassium, or sodium–potassium alloy in inert solvents, acyloins (α-hydroxyketones) are formed.

Pyrolysis. The pyrolysis of simple esters of the formula $RCOOCR^1R^2CHR^3{}_2$ to form the free acid and an alkene is a general reaction that is used for producing olefins.

Carbonylation Reaction. The carbonylation of methyl acetate is an important industrial reaction for producing acetic anhydride.

Substitution, Alkylation, and Rearrangement. The reaction of alkaline phenoxides with alkyl *S*-2-(chloro)- or *S*-2- (mesyloxy)propionate gives optically active *R*-2-aryloxyalkanoic acid esters in good chemical and optical yields (>97%*ee*). The reaction is utilized in the synthesis of several phenoxy herbicides.

Optically active 2-arylalkanoic acid esters have been prepared by Friedel-Crafts alkylation of arenes with optically active esters, such as methyl *S*-2-(chlorosulfonoxy)- or *S*-2-(mesloxy)propionate, in the presence of aluminum chloride.

The Fries rearrangement of phenol esters gives a mixture of 2- and 4-acylphenols. Similarly, enol esters undergo rearrangement to give the corresponding 1,3-diketones.

Occurrence and Preparation

Currently, most of the simple esters used commercially are of synthetic origin, although esters occur naturally in large quantities in fats, oils, and waxes. Microorganisms produce a complex array of compounds containing the ester linkage, ranging from simple esters to macrocyclic lactones, such as erythromycin, which are important because of their antibacterial properties.

Recovery of naturally occurring esters is accomplished by steam distillation, extraction, pressing, or by a combination of these processes. Synthetic esters are generally prepared by reaction of an alcohol with an organic acid in the presence of a catalyst such as sulfuric acid, *p*-toluenesulfonic acid, or methanesulfonic acid.

Stability and Storage

All organic esters are unstable in the presence of acid or base and nucleophiles such as water or alcohols. However, if stored anhydrous, they are stable. Storage vessels can be constructed of steel, aluminum, or other metallic materials, but plastic storage tanks are unsuitable because the highly lipophilic esters can sometimes permeate into the container boundary and soften or even dissolve it.

The properties of flash point, autoignition temperature, and flammable limit should be considered when an ester is to be handled in any fashion.

Health and Safety Factors

Toxicity. The degree of toxicity of organic esters covers a wide range. These toxicities are usually described in terms of threshold limiting values (TLV), or permissible exposure limits (PEL). Both the PEL and the TLV describe the average concentration over an 8-h period to which a worker may be exposed without adverse effects. The lethal dosages for 50% of the exposed animals, LD_{50}s, are also used as an indicator of the relative toxicity. The LD_{50}s of organic esters for small mammals range between 0.4 and 16 g/kg. The TLVs of organic esters range between 5 and 400 ppm.

When ingested or absorbed, organic esters are likely to be hydrolyzed to the corresponding alcohols and carboxylic acids. Therefore the toxicities of the hydrolysis products should also be considered. Some organic esters are highly volatile and can act as asphixiant or narcotic. Also, skin absorption and inhalation are among the hazards associated with esters that are volatile or have good solvent action. Because of the high solubility of fats and oils in organic esters, prolonged or repeated exposure to skin can cause drying and irritation.

Acetates generally do not cause any physiological effects unless high exposure occurs since they are usually converted into or occur naturally as metabolites. However, large enough exposure to acetate esters can cause narcotic effects.

Propionates and higher aliphatic esters generally become less toxic as the size of the alkyl carboxylate increases.

The acrylate esters are more physiologically hazardous than their saturated homologues. They are usually lachrymators and irritants, and their toxicities decrease with increasing molecular weight.

Among adipates, oxalates, malonates, and succinates, the adipates are the least toxic.

Benzoate esters, like most organic esters, are not very toxic. They are not absorbed through the skin as rapidly as alkyl esters but are more potent physiologically. They are also moderate skin irritants.

The phthalate esters are one of the most widely used classes of organic esters, and fortunately they exhibit low toxicity.

More information on the toxicities of a range of organic esters is available in the literature.

Exposure Limits. The Occupational Safety and Health Act (OSHA) of 1990 lists a multitude of acetates, phthalates, formates, and acrylates along with the corresponding permissible exposure limits and threshold limit values. If there is potential for exposure to an organic ester for which PEL or TLV data has been identified, then an exposure limit lower than that listed

Regulation and Waste

Waste from production of organic esters is usually not a problem since the method of synthesis often involves a carboxylic acid condensation with an alcohol and the only by-product is water. Any organic remnants lost to the process water can usually be biologically degraded. The biochemical oxygen demand (BOD) or chemical oxygen demand (COD) should be measured if biological treatment is used on the process waste from ester production. Organic ester vapor emitted in processing usually can be burned.

Extensive federal environmental regulations exist that govern organic esters as well as many other substances. These regulations must always be consulted for complete information before using large amounts of organic esters. State and local regulations must also be met, which in some cases are more stringent than federal regulations.

Uses

Organic esters are used as solvents, plasticizers, in resins, plastics, and coatings, as lubricants, in perfumes, flavors, cosmetics, and soap, as surface-active agents, as medicinals, and as herbicides and pesticides.

KWOLIANG D. TAU
VARADARAJ ELANGO
JOSEPH A. MCDONOUGH
Hoechst Celanese Corporation

S. Patai, ed., *The Chemistry of Acid Derivatives,* Suppl. B, Parts 1 and 2, John Wiley & Sons, Inc., New York, 1979.

R. W. Johnson and E. Fritz eds., *Fatty Acids in Industry,* Marcel Dekker, Inc., New York, 1989.

B. M. Trost and I. Fleming, eds., *Comprehensive Organic Synthesis: Selectivity, Strategy and Efficiency in Modern Organic Chemistry,* Vol. 1–9, Pergamon Press, Inc., Elmsford, N.Y., 1991.

J. March, *Advanced Organic Chemistry,* 4th ed., John Wiley & Sons, Inc., New York, 1992.

ESTROGENS. See HORMONES.

ETHANOIC ACID. See ACETIC ACID AND DERIVATIVES, ACETIC ACID.

ETHANOL

Ethanol or ethyl alcohol, CH_3CH_2OH, is one of the most versatile oxygen-containing organic chemicals because of its unique combination of properties as a solvent, a germicide, a beverage, an antifreeze, a fuel, a depressant, and especially as a chemical intermediate for other organic chemicals. The use of fermentation-derived ethanol as an automotive fuel additive to enhance octane and reduce emissions has seen explosive growth since about 1980 so that it now accounts for over 80% of the 6.0×10^9 L ethanol market in the United States.

Physical Properties

A summary of physical properties of ethyl alcohol is presented in Table 1. Detailed information on the vapor pressure, density, and viscosity of ethanol can be obtained from many references. Ethanol forms binary and ternary azeotropes with a wide variety of other organic chemicals.

Table 1. Physical Properties of Ethanol

Property	Value
freezing point, °C	−114.1
normal boiling point, °C	78.32
density, d_4^{20}, g/mL	0.7893
refractive index, n_D^{20}	1.36143
$\Delta n_D/\Delta t$, 20–30°C, per °C	0.000404
surface tension, at 25°C, mN/m (= dyn/cm)	23.1
viscosity, at 20°C, mPa·s(= cP)	1.17
solubility in water, at 20°C	miscible
heat of combustion, at 25°C, J/g[a]	29676.69
flammable limits in air, vol %	
lower	4.3
upper	19.0
autoignition temperature, °C	423.0
specific heat, at 20°C, J/(g·°C)	2.42
thermal conductivity, at 20°C, W/(m·K)	0.170

[a] To convert J to cal, divide by 4.184.

Chemical Properties

The chemistry of ethyl alcohol is largely that of the hydroxyl group, namely, reactions of dehydration, dehydrogenation, oxidation, and esterification.

Manufacture

Industrial ethyl alcohol can be produced synthetically from ethylene, as a by-product of certain industrial operations, or by the fermentation of sugar, starch, or cellulose. The synthetic route supplies most of the industrial market in the United States.

The synthetic process employing direct hydration has since the early 1970s completely supplanted the old sulfuric acid process in the United States. This process, the catalytic vapor-phase hydration of ethylene, is now practiced by only three U.S. companies: Union Carbide Corporation (UCC), Quantum Chemical Corporation, and Eastman Chemical Company UCC imports crude industrial ethanol, CIE, from SADAF (the joint venture of SABIC and Pecten [Shell]) in Saudi Arabia, and refines it to industrial grade.

Other Methods of Preparation. In addition to the direct hydration process, the sulfuric acid process, and fermentation routes to manufacture ethanol, several other processes have been suggested. These include the hydration of ethylene by dilute acids, the hydrolysis of ethyl esters other than sulfates, the hydrogenation of acetaldehyde, and the use of synthesis gas. None of these methods has been successfully implemented on a commercial scale, but the route from synthesis gas has received a great deal of attention since the 1974 oil embargo.

Recovery and Purification

Various distillation and equipment modifications are used to ensure a pure water azeotrope of ethanol (95% by volume ethanol).

Purification schemes have generally emphasized the following techniques: (*1*) extractive distillation using water reflux to distill a large share of the impurities and concentrate the crude alcohol–water mixture; (*2*) efficient fractionation to produce approximately 190 proof alcohol; (*3*) hydrogenation to convert aldehyde impurities to alcohols, together with the use of chemicals such as inorganic bases and sodium sulfite; and (*4*) ion-exchange resins or azeotropic distillation to dehydrate 190-proof to 200-proof or absolute alcohol.

Manufacturing Costs

The cost of producing industrial ethanol depends on the location of the manufacturing plant; the design, type, and degree of modernization of equipment; the kind of raw material used; the price paid for the raw material; the relative labor costs represented; the scale of production; and the total investment. On a raw material basis, each kilogram of ethanol requires 0.6 kg ethylene, 2 kg sugar, 3.3 kg corn, or 4 kg molasses. If all other cost factors were equivalent, the economics of fermentation ethanol derived from corn would compare favorably with synthetic ethanol if corn prices were 5.5 times less than ethylene on a weight basis. With ethylene at $0.44/kg, as it was in 1991, corn would have to be about $0.08/kg (about $2/bushel). Corn by-products that are produced by some fermentation plants offset the price of corn to a varying degree depending on their market value. It is expected that ethylene will continue to be the primary raw material for industrial ethanol, because during the early to mid-1990s most of the fermentation ethanol will go to fuel due to the impact of the Clean Air Act, and ethylene prices are not expected to significantly increase because of additional ethylene capacity coming onstream (see ETHYLENE).

Shipment

Commercial ethyl alcohol is shipped in railroad tank cars, tank trucks, 208-L (55-gal) and 19-L (5-gal) drums, and in smaller glass or metal containers having capacities of 0.473 L (one pint), 0.946 L (one quart), 3.785 L (one U.S. gal), or 4.545 L (one Imperial gal). All containers, of course, must comply with the specifications of the U.S. Department of Transportation. Both 190 proof and 200 proof ethyl alcohol are considered red label (flammable) materials by the DOT, as both have flash points below 37.8°C by the Tag closed-cup method.

Economic Aspects

Worldwide ethanol demand in 1994 was about 22×10^9 L, of which the industrial demand was about 5.3×10^9 L. The majority of the worldwide demand was for fuel use, with Brazil consuming 14×10^9 L and the United States consuming 5×10^9 L. The worldwide synthetic ethanol capacity amounted to 2×10^9 L, with the United States having a capacity of 0.8×10^9 L, 39% of the world total. Total ethanol capacity in the United States was 6.4×10^9 L, with an industrial demand of 1.0×10^9 L. Fermentation alcohol and imports supplied the remainder of the industrial market not supplied by synthetic alcohol.

Health and Safety Considerations

Ethyl alcohol is a flammable liquid. Vapor concentrations between 3.3 and 19.0% by volume in air are explosive. Liquid ethyl alcohol can react vigorously with oxidizing materials. Ethyl alcohol has found wide application in industry, and experience shows that it is not a serious industrial poison. If proper ventilation of the work environment is maintained, there is little likelihood that inhalation of the vapor will be hazardous.

The threshold limit value for ethyl alcohol vapor in air has been set at 1000 ppm for an 8-h time-weighted exposure by the ACGIH (1989 listing). Exposure to concentrations of 5,000–10,000 ppm result in irritation of the eyes and mucous membranes of the upper respiratory tract and, if continued for an hour or more, may result in stupor or drowsiness.

Ethyl alcohol is oxidized completely to carbon dioxide and water in the body, and thus it is not a cumulative poison. Alcohol poisoning and alcohol intoxication are almost invariably the result of using alcohol as a beverage, rather than inhalation as a vapor.

Uses

Industrial ethanol is one of the largest-volume organic chemicals used in industrial and consumer products. The main uses for ethanol are as an intermediate in the production of other chemicals and as a solvent. As a solvent, ethanol is second only to water. Ethanol is a key raw material in the manufacture of drugs, plastics, lacquers, polishes, plasticizers, perfumes, and cosmetics.

Denatured Ethanol. For hundreds of years alcoholic beverages have been taxed all over the world to generate government revenue. When ethanol emerged as a key industrial raw material, the alcohol tax was recognized as a burden to many essential manufacturing industries. To lift this burden, the Tax-Free Industrial and Denatured Alcohol Act of 1906 was passed in the United States. The U.S. Treasury, Bureau of Alcohol, Tobacco, and Firearms (BATF), now oversees the production, procurement, and use of ethanol in the United States.

The concern of the government is to prevent tax-free industrial ethanol from finding its way into beverages. To achieve this end, the regulations call for a combination of financial and administrative controls (bonds, permits, and scrupulous record keeping) and chemical controls (denaturants that make the ethanol unpalatable). Regulations establish four distinct classifications of industrial ethanol. The classifications with the most stringent financial and administrative controls call for little or no chemical denaturants. The classifications that call for the most effective chemical denaturants require the least financial and administrative controls. For a list of denaturants currently authorized, consult the appropriate Code of Federal Regulations.

Chemicals Derived From Ethanol

Ethanol's use as a chemical intermediate suffered considerably from its replacement in the production of acetaldehyde, butyraldehyde, acetic acid, and ethylhexanol. The switch from the ethanol route to those products has depressed demand for ethanol by more than 300×10^6 L (80×10^6 gal) since 1970. This decrease reflects newer technologies for the manufacture of acetaldehyde and acetic acid, which is the largest use for acetaldehyde, by direct routes using ethylene, butane, and methanol. Oxo processes (qv) such as Union Carbide's Low Pressure Oxo process for the production of butanol and ethylhexanol have totally replaced the processes based on acetaldehyde.

Ethylene. Where ethylene is in short supply and fermentation ethanol is made economically feasible, such as in India and Brazil, ethylene is manufactured by the vapor-phase dehydration of ethanol.

Glycol Ethers. The addition of one mole of ethylene oxide to ethanol gives ethylene glycol monoethyl ether. Addition of two moles of oxide gives the monoethyl ether of diethylene glycol. The oxide–alcohol route is the only commercially important route to glycol ethers now in use. Anhydrous alcohols must be used; otherwise the water present forms contaminating glycols.

Vinegar. Dilute solutions of alcohol as fermented worts are oxidized by air at 30–40°C in the presence of various organisms such as *Mycoderma aceti, B. aceti,* and *B. xylinus,* to produce dilute acetic acid as vinegar (qv). Vinegar based on synthetic ethanol has a fully acceptable aroma and taste and has gained a healthy share of the market for vinegar used in such products as pickles, ketchup, and mustard. However, vinegar based on fermentation ethanol is gaining back some of this market because of the "all natural ingredients" trend in advertising.

Ethylamines. Mono-, di-, and triethylamines, produced by catalytic reaction of ethanol with ammonia, are a significant outlet for ethanol.

Ethyl Acrylate. The esterification of acrylic acid is a primary use for ethanol. Acrylic acid can also react with either ethylene or ethyl esters of sulfuric acid. These processes have supplanted the condensation reaction of ethanol, carbon monoxide, and acetylene as the principal method of generating ethyl acrylate.

Ethyl Ether. Most ethyl ether is obtained as a by-product of ethanol synthesis via the direct hydration of ethylene.

Ethyl Vinyl Ether. The addition of ethanol to acetylene via a vapor-phase reaction gives ethyl vinyl ether.

Ethyl tert-Butyl Ether. Ethanol can react with isobutylene to form ETBE much the same way as methanol is now processed into MTBE, methyl *tert*-butyl ether (see ETHERS).

Ethyl Acetate. The esterification of ethanol by acetic acid was studied in detail over a century ago, and considerable literature exists on determinations of the equilibrium constant for the reaction. The usual catalyst for the production of ethyl acetate is sulfuric acid, but other catalysts have been used, including cation-exchange resins, α-fluoronitrites, titanium chelates, and quinones and their partly reduced products. Ethyl acetate is made industrially by both batch and continuous processes.

Ethyl Chloride. Previously a significant use for industrial ethanol was the synthesis of ethyl chloride for use as an intermediate in producing tetraethyllead, an antiknock gasoline additive. Ethanol is converted to ethyl chloride by reaction with hydrochloric acid in the presence of aluminum or zinc chlorides. However, since about 1960, routes based on the direct addition of hydrochloric acid to ethylene or ethane have become more competitive.

JOHN E. LOGSDON
Union Carbide Corporation

L. F. Hatch, *Ethyl Alcohol*, Enjay Chemical Co., a Division of Humble Oil and Refining Co., New York, 1962.

S. A. Miller, ed., *Ethylene and Its Industrial Derivatives*, Ernst Benn Ltd., London, 1969.

American Institute of Chemical Engineers, *Design Institute for Physical Property Data*, (DIPPR File), University Park, Pa., 1989.

A. Chauvel and G. Lefebvre, *Petrochemical Processes, Technical and Economic Characteristics*, Gulf Publishing, Houston, Tex., 1989.

ETHERS

Ethers are compounds of the general formula Ar—O—Ar′, Ar—O—R, and R—O—R′ where Ar is an aryl group and R is an alkyl group.

Physical Properties

In general, ethers are neutral, pleasant-smelling compounds that have little or no solubility in water, but are easily soluble in organic liquids. Their boiling points approximate those of hydrocarbons having comparable molecular weights and geometries. Detailed physical properties for the ethers most commonly used as solvents are listed in Table 1.

Chemical Properties

Most ethers, particularly dialkyl ethers, are comparatively unreactive compounds because the carbon–oxygen bond is not readily cleaved. For this reason, ethers are frequently employed as inert solvents in organic synthesis. However, within the ether family, the cyclic ethers are more reactive, the most reactive being the olefin oxides or epoxides. Epoxides are generally used as intermediates for producing other small molecules or polymers (see EPOXY RESINS; POLYETHERS). Ethers do react with exceptionally powerful basic reagents, particularly certain alkali metal alkyls, to give cleavage products. Ethers react with less powerful bases to give the same cleavage products, but only under the forcing conditions of high temperature and pressure. The ether linkage can also be cleaved by strong acids, generally at high temperatures.

Most ethers are potentially hazardous chemicals because, in the presence of atmospheric oxygen, a radical-chain process can occur, resulting in the formation of peroxides that are unstable, explosion-prone compounds. One of the exceptions to the peroxide forming tendency of ethers is methyl *tert*-alkyl ethers such as methyl *tert*-butyl ether (MTBE) and *tert*-amyl methyl ether (TAME). Both have shown little tendency if any to form peroxides.

Ethers are weakly basic and are converted to unstable oxonium salts by strong acids such as sulfuric acid, perchloric acid, and hydrobromic acid; relatively stable complexes are formed between ethers and Lewis acids such as boron trifluoride, aluminum chloride, and Grignard reagents (qv). Like other aromatic compounds, aromatic ethers can undergo substitution in the aromatic ring with electrophilic reagents. They also undergo Friedel-Crafts (qv) alkylation and acylation.

Fuel Properties

In addition to MTBE, two other ethers commonly used as fuel additives are *tert*-amyl methyl ether (TAME) and ethyl *tert*-butyl ether (ETBE). There are a number of properties that are important in gasoline blending (see GASOLINE AND OTHER MOTOR FUELS) (Table 2).

Table 1. Typical Physical and Chemical Properties of Commonly Used Ethers

Property	Ethyl	MTBE	THF	Isopropyl	*n*-Butyl
chemical formula	$C_4H_{10}O$	$C_5H_{12}O$	C_4H_8O	$C_6H_{14}O$	$C_8H_{18}O$
molecular weight	74.12	88.15	72.10	102.17	139.22
boiling point at 101.3 kPa,[a] °C	34.5	55.0	66.0	68.4	142.0
vapor pressure at 20°C, kPa[a]	56	27	17	16	0.67
evaporation rate[b]	11.8	8.5	5.7	8.0	0.66
viscosity at 20°C, mPa·s(= cP)	0.23	0.35	0.48	0.38	0.65
flash point	−40	−30	−17	−13	25
autoignition temp, °C	160	426	321	440	185
flammability limits in air, vol %					
lower	1.9	1.6	1.8	1.0	0.9
higher	48.0	8.4	11.8	21.0	8.5

[a] To convert kPa to mm Hg, multiply by 7.5.
[b] *n*-Butyl acetate = 1.

Table 2. Typical Gasoline-Related Properties for Ethers Used in Gasoline

Property	MTBE	ETBE	TAME
research octane number (RON)	118	119	112
motor octane number (MON)	102	104	99
(RON + MON)/2	110	111	105.5
boiling point			
°F	131	161	187
°C	55	72	86
Reid vapor pressure, kPa[a]	55	28	7
oxygen, wt %	18.2	15.7	15.7

[a] To convert kPa to psi, multiply by 0.145.

Preparation

The most versatile method of preparing ethers is the Williamson ether synthesis, particularly in the preparation of unsymmetrical alkyl ethers. Alkyl tertiary alkyl ethers can be prepared by the addition of an alcohol or phenol to a tertiary olefin under acid catalysis. Commercially, sulfonic acid ion-exchange resins are used in fixed-bed reactors to make tertiary alkyl ethers.

Health and Safety Factors

Although ethers are not particularly hazardous, their use involves risks of fire, toxic effects, and several unexpected reactions.

Flammability. Because almost all ethers burn in air, an assessment of their potential hazards depends on flash points and ignition temperatures.

Toxicity. The effect of ethers owing to ingestion, skin contact, or inhalation may range from drowsiness and lack of coordination to serious injury or death.

Prolonged or repeated contacts of ethers with skin cause tissue defatting and dehydration leading to dermatitis. Some compounds penetrate the skin in harmful amounts. 1,4-Dioxane and 1-allyl-3,4-methylenedioxybenzene (safrole) have been listed as Category I carcinogens by OSHA.

Inhalation is the most common means by which ethers enter the body. The effects of various ethers may include narcosis, irritation of the nose, throat, and mucous membranes, and chronic or acute poisoning. In general, ethers are central nervous system depressants, eg, ethyl ether and vinyl ether are used as general anesthetics.

Peroxide Formation. Except for the methyl *tert*-alkyl ethers, most ethers tend to absorb and react with oxygen from the air to form unstable peroxides that may detonate with extreme violence when concentrated by evaporation or distillation, when combined with other compounds that give a detonable mixture, or when disturbed by heat, shock, or friction.

Applications

Alkyl ethers are used for organic reactions and extractions, as plasticizers, as vehicles for other products, as anesthetics (qv) and octane (and oxygen) enhancers in gasoline, and in paint and varnish removers; as high boiling solvents for gums, resins, and waxes; in lubricating oils; as an extraction solvent in the fragrance industry; and as an inert reaction medium in the pharmaceutical industry. The vapors of certain ethers are toxic to insects and are useful as agricultural insecticides and industrial fumigants.

Aryl ethers have distinctive, pleasant odors and flavors which make them valuable to the perfume (qv) and flavor industries. Because of their heat stability, they are useful as heat-transfer fluids. Other aryl ethers are useful as food preservatives and antioxidants.

Commercially Important Ethers

Ethyl Ether. Diethyl ether is one of the more important members of the ether family. It is a colorless, very volatile, highly flammable liquid with a sweet, pungent odor and burning taste. As a commercial product it is available in several grades; it is used in chemical manufacture, as a solvent, extractant, or reaction medium, and as a general anesthetic.

Much of the diethyl ether manufactured is obtained as a by-product when ethanol (qv) is produced by the vapor-phase hydration of ethylene (qv) over a supported phosphoric acid catalyst.

The handling of ethyl ether is hazardous because of its highly flammable properties. The area in which ethyl ether is handled should be considered a Class I hazardous location as defined by the National Electrical Code. Special containers have been developed for anesthetic ether to prevent deterioration before use.

Though 1990 U.S. production capacity was estimated at 25.5 × 10^6 kg, production was estimated as only 12 × 10^6 kg in 1991. Much of the decrease has been the result of a decline in arsenal demand (smokeless gun powder). List prices for ether have been steadily increasing, and reached $1.12/kg by 1989, refined, tanks (fob).

The toxicity of ethyl ether is low and its greatest hazards in industry are fire and explosion.

Methyl Tertiary Butyl Ether. MTBE production grew during the 1980s as lead-based octane enhancers were being phased out of gasoline.

In November 1990, the U.S. government revised the Clean Air Act (CAA). For the 39 metropolitan areas that exceeded the federal standard for carbon monoxide, the gasoline would have to contain at least 2.7 wt % oxygen for at least four winter months beginning November 1992. This represented nearly 30% of the U.S. gasoline market.

Starting in January 1995, the nine metropolitan areas with the most severe ozone problem must sell a "reformulated" gasoline all year long. These nine cities represent approximately 20% of the U.S. gasoline market. In addition to many other very restrictive specifications, this gasoline must also contain at least 2.0 wt % oxygen. Other metropolitan areas exceeding the federal ozone standard are also allowed to choose the program when sufficient capacity exists to make the reformulated gasoline.

MTBE is easily made by the selective reaction of isobutylene and methanol over an acidic ion-exchange resin catalyst, in the liquid phase and at temperatures below 100°C. To be economically competitive, MTBE's use as an octane enhancer in gasoline has been dependent on low cost isobutylene.

MTBE production capacity has grown steadily, usually at an annual rate of 10 to 20% per year. By 2000, MTBE may be the second largest organic chemical produced in the United States, second only to ethylene.

Economic Aspects

Fuel-grade ethers are generally produced in crude oil refineries for internal use in the manufacture of gasoline or in petrochemical plants for resale to the refiners. The value of these ethers fluctuates with the price of gasoline and octane in the industry. The two largest MTBE markets are in the U.S. Gulf Coast region and in Northern Europe. MTBE is typically sold to refiners in large barges in quantities of ten to one hundred thousand U.S. barrels (1–12 10^3 t). It is also sometimes shipped in batches on gasoline product pipelines.

Because there does not appear to be sufficient MTBE capacity to satisfy the oxygen requirements for the 1990 CAA revision, there has been a general upward movement in MTBE market prices that should provide the incentive for additional capacity to be built.

Uses. MTBE and related ethers are used to add octane to gasoline. MTBE also adds oxygen to the gasoline, which allows for more efficient combustion, and therefore less carbon monoxide and unburned hydrocarbon in the exhaust emissions. A refined grade of MTBE is used in the solvents and pharmaceutical industries. Its higher autoignition temperature and narrower flammability range also make it relatively safer to use compared to other ethers.

One other unique use of MTBE is a medical procedure for the removal of gallstones.

Tetrahydrofuran. Except for the special case of the epoxides, THF represents the largest use of a heterocyclic ether. Unlike the dialkyl ethers, THF is totally miscible in water at ambient conditions. Its cyclic structure also allows it to be more reactive than the dialkyl ethers. More than half of the THF produced is used as an intermediate in making other chemicals of elastomers.

Almost all the THF in the United States is currently produced by the acid-catalyzed dehydration of 1,4-butanediol. Only one plant in the United States still makes THF by the hydrogenation of furfural.

In 1990, approximately 105 × 10^6 kg of THF was produced from a U.S. annual capacity of about 123 × 10^6 kg. Consumption was expected to grow by 5–6% per year through 1995. THF is generally shipped in tank cars, tank trucks, or drums.

Others Ethers. *n*-Butyl ether is prepared by dehydration of *n*-butyl alcohol by sulfuric acid or by catalytic dehydration over ferric chloride, copper sulfate, silica, or alumina at high temperatures. It is an important solvent for Grignard reagents and other reactions that require an anhydrous, inert medium. *n*-Butyl ether is also an excellent extracting agent for use with aqueous systems owing to its very low water-solubility.

Isopropyl ether is manufactured by the dehydration of isopropyl alcohol with sulfuric acid. It is obtained in large quantities as a by-product in the manufacture of isopropyl alcohol from propylene by the sulfuric acid process. Isopropyl ether is of moderate importance as an industrial solvent. It is also being promoted as another possible ether to be used in gasoline.

Like MTBE, TAME is produced by the simple reaction of methanol and isoamylenes (2-methyl-1-butene and 2-methyl-2-butene). It is used for gasoline blending, and also as a feedstock to make high purity isoamylene.

Similar to methanol in the MTBE reaction, ethanol can react with isobutylene to produce ETBE.

Vinyl ether is manufactured by the pyrolytic dehydrochlorination of 1,1′-dichloroethyl ether. Vinyl ether is used as a general inhalation anesthetic for procedures of short duration.

This ether is prepared by methylating 1,2-dihydroxybenzene. It is useful as an antioxidant for fats, oils, and vitamins, and as a polymerization inhibitor.

2- and 3-*tert*-Butyl-4-methoxyphenol (butylated hydroxyanisole (BHA)) is prepared from 4-methoxyphenol and *tert*-butyl alcohol over silica or alumina at 150°C or from hydroquinone and *tert*-butyl alcohol or isobutene, using an acid catalyst and then methylating. It is widely used in all types of foods.

LAWRENCE KARAS
W. J. PIEL
ARCO Chemical Company

E. W. Flick, *Industrial Solvents Handbook*, 3rd ed., Noyes Data Corp., Park Ridge, N.J., 1985.

A. G. Davies, *Organic Peroxides*, Butterworths, London, 1961, p. 79.

The Pharmacopeia of the United States of America, 19th ed., USP XIX, Mack Publishing Co., Easton, Pa., 1975; USP XX-NFXF, 1980.

W. J. Piel, *Fuel Reformulation* **2**(6) (1992).

ETHYLENE

Ethylene (ethene), $H_2C{=}CH_2$, is the largest volume building block for many petrochemicals. This olefin is used to produce many end products such as plastics, resins, fibers, etc. Ethylene is produced mainly from petroleum-based feedstocks by thermal cracking, although alternative methods are also gaining importance.

Table 1. Physical Properties of Ethylene

Property	Value
mol wt	28.0536
normal freezing point temperature, °C	−169.15
normal boiling point temperature, °C	−103.71
density of liquid	
mol/L	20.27
d_4^{-104}	0.566
specific heat of liquid, J/(mol·K)[a]	67.4
viscosity of the liquid, mPa·s(= cP)	0.161
limits of flammability at atmospheric pressure and 25°C	
lower limit in air, mol %	2.7
upper limit in air, mol %	36.0
autoignition temperature in air at atmospheric pressure, °C	490.0
solubility in water at 0°C and 101 kPa[b], mL/mL H_2O	0.226

[a] To convert kPa to mm Hg, multiply by 7.5
[b] To convert J to cal, divide by 4.184.

Physical Properties

Ethylene is the lightest olefin. It is a colorless, flammable gas with a slightly sweet odor. Physical and thermodynamic properties are given in many publications and are briefly summarized in Table 1.

Chemical Properties

Structure. Ethylene is a planar molecule with a carbon–carbon bond distance of 0.134 nm, which is shorter than the C—C bond length of 0.153 nm found in ethane.

Reactivity. Ethylene is a very reactive intermediate, and hence is involved in many chemical reactions. The ethylene double bond reacts readily to form saturated hydrocarbons, their derivatives, or polymers. Ethylene reacts with electrophilic reagents like strong acids (H^+), halogens, and oxidizing agents, but not with nucleophilic reagents such as Grignard reagents and bases. Some of the reactions have commercial significance and others have only academic interest.

The principal reactions with commercial significance include polymerization, oxidation, and addition including halogenation, alkylation, oligomerization, hydration, and hydroformylation.

Polymerization. Very high purity ethylene (>99.9% plus) is polymerized under specific conditions of temperature and pressure in the presence of an initiator or catalyst.

Oxidation. Ethylene oxide (qv) is produced by oxidizing ethylene.

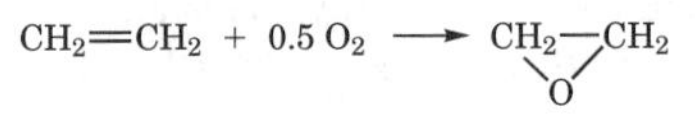

Addition. Addition reactions of ethylene have considerable importance and lead to the production of ethylene dichloride, ethylene dibromide, and ethyl chloride by halogenation–hydrohalogenation; ethylbenzene, ethyltoluene, and aluminum alkyls by alkylation; α-olefins by oligomerization; ethanol by hydration; and propionaldehyde by hydroformylation.

Biological Properties

Ethylene is slightly more potent as an anesthetic than nitrous oxide, and the smell of ethylene causes choking.

The controlled ripening of various fruits and vegetables by ethylene is of considerable importance.

Manufacture by Thermal Cracking

Although ethylene is produced by various methods as follows, only a few are commercially proven: thermal cracking of hydrocarbons, catalytic pyrolysis, membrane dehydrogenation of ethane, oxydehydrogenation of ethane, oxidative coupling of methane, methanol to ethylene, dehydration of ethanol, ethylene from coal, disproportionation of propylene, and ethylene as a by-product.

Thermal cracking of hydrocarbons is the principal route for the industrial production of ethylene. The chemistry and engineering of thermal cracking has been reviewed. In thermal cracking, valuable by-products including propylene, butadiene, and benzene are also produced. Commercially less valuable methane and fuel oil are also produced in significant proportions. An important parameter in the design of commercial thermal cracking furnaces is the selectivity to produce the desired products.

Mechanism. The thermal cracking of hydrocarbons proceeds via a free-radical mechanism.

Conversion. The terms severity or conversion are used to measure the extent of cracking. Conversion can easily be measured for a single component (C) feed and is defined as follows, where the quantities are measured in weight units. conversion $= \frac{C_{\text{in}} - C_{\text{out}}}{C_{\text{in}}}$ When a mixture is cracked, one or more components in the feed may also be formed as products.

Instead of conversion, some producers prefer to use other identifications of severity, including coil outlet temperature, propylene to methane ratio, propylene to ethylene ratio, or cracking severity index. Of course, all these definitions are somewhat dependent on feed properties, and most also depend on the operating conditions.

Industrial Furnaces. Thermal cracking of hydrocarbons is accomplished in tubular reactors commonly known as cracking furnaces, crackers, cracking heaters, etc.

Environmental. Stringent environmental laws require that nitrogen oxides (NO_x) and sulfur oxides emission from furnaces be drastically reduced. In many parts of the world, regulations require that NO_x be reduced to 70 vol ppm or lower on a wet basis. Conventional burners usually produce 100 to 120 vol ppm of NO_x. Many vendors (McGill, John Zink, and North American) are supplying low NO_x burners.

Product Distribution. In addition to ethylene, many by-products are also formed. The product distribution is strongly influenced by residence time, hydrocarbon partial pressure, steam-to-oil ratio, and coil outlet pressure.

Kinetic Models Used for Designs. Numerous free-radical reactions occur during cracking; therefore, many simplified models have been used.

Many researchers have correlated the overall decomposition as an nth order reaction, with most paraffins following the first order and most olefins following a higher order. In general, isoparaffin rate constants are lower than normal paraffin rate constants. To predict the product distribution, yields are often correlated as a function of conversion or other severity parameters.

Instead of radical reactions, models based on molecular reactions have been proposed for the cracking of simple alkanes and liquid feeds like naphtha and gas oils. However, the validity of these models is limited, and cannot be extrapolated outside the range with confidence.

With the introduction of Gear's algorithm for integration of stiff differential equations, the complete set of continuity equations describing the evolution of radical and molecular species can be solved even with a personal computer. Many models incorporating radical reactions have been published.

Run Length. Coke is produced as a side product that deposits on the radiant tube walls. This limits the heat transfer to the tubes, and increases the pressure drop across the coil. The coke deposition not only limits the heat transfer, but also reduces the olefin selectivity. Periodically, the heater has to be shut down and cleaned. Typical run lengths are 40 to 100 days between decokings. Prediction of run length of a commercial furnace is still an art, and various mechanisms are postulated in the literature.

Recovery and Purification. The pyrolysis gas leaves the transferline exchanger at 300 to 400°C for gaseous and light naphtha feeds and at 550 to 650°C for heavy liquid feeds. In order to minimize any further cracking for liquid feeds, the temperature must be quickly reduced. This is achieved by spraying quench oil directly into the effluent. For naphtha-based plants, quenching is performed before primary fractionation. In gas oil plants, quenching is done immediately after the transferline exchanger, resulting in two-phase flow in the transferline.

For all feeds the effluent is separated into desired products by compression in conjunction with condensation and fractionation at gradually lower temperatures.

Energy Efficiency Improvement. Modern plants are more energy efficient than those designed in the 1980s. Reduction in energy consumption was made possible by improvements in cracking coil technology and improvements in recovery section design. Some improvements may not only reduce energy, but can also increase the capacity of an existing plant. They include quench oil viscosity control, feed saturation, predemethanization, demethanizer overhead expander and multifeed fractionation, tower internals and equipment modification, and dephlegmaters.

Other Routes to Ethylene Production

In addition to conventional thermal cracking in tubular furnaces, other thermal methods and catalytic methods to produce ethylene have been developed. None of these are as yet commercialized. Other routes include the advanced cracking reactor, the adiabatic cracking reactor, fluidized-bed cracking, catalytic pyrolysis, the membrane reactor, oxidative coupling of methane, methanol to ethylene, ethanol to ethylene, ethylene from coal, dehydrogenation, oxydehydrogenation, propylene disproportionation, and ethylene as a by-product.

Advanced Computer Control Systems and Training Simulators

An ethylene plant contains more than 300 equipment items. Traditionally, operators were trained at the site alongside experienced coworkers. With the advent of modern computers, the plant operation can be simulated on a real-time basis, and the results displayed on monitors. Computers are used in a modern plant to control the entire operation, eg, they are used to control the heaters and the recovery section.

Shipment and Storage

In the United States, the Gulf Coast produces and consumes the majority of the U.S. ethylene production. The plants are located along the Texas southeast coast extending into Louisiana. The plants are served by a system of pipelines connecting the production and consuming plants.

Safety and Environmental Factors

Although ethylene is a colorless gas with a mild odor that is not irritating to the eyes or respiratory system, it is a hydrocarbon and therefore a flammable gas. All vessels must be designed for handling the liquids and gases during operation at the temperatures and pressures that exist, and safety and depressuring valves must be provided to relieve excessive pressure. Releasing of hydrocarbons into the air in large amounts must be avoided because of health and fire hazards. If hydrocarbons must be released into the air, it is done under a blanket of steam. To protect the plant and personnel in case of fire, a complete fire fighting system is provided with tanks grouped to minimize fire and provided with foam makers and deluge systems. Reviews at various stages of a project assure safety is given constant attention in the plant design (see HAZARD ANALYSIS AND RISK ASSESSMENT). An ethylene plant produces liquid, gaseous, and solid wastes that must be disposed of in an environmentally safe manner.

Uses

Almost all ethylene produced is consumed as feedstock for manufacturing other petrochemicals. Only a very small amount has been used in the agricultural industry for ripening fruits.

Although some ethylene is shipped across the oceans in large quantities, the preference is to ship first-generation products such as polyethylene, ethylbenzene, etc.

Economic Aspects

In 1992, world capacity for the production of ethylene was approximately 69×10^6 t. The U.S. production capacity accounted for almost 29% of the world capacity, or approximately 20×10^6 t, followed by Western Europe with almost 26%, or 18×10^6 t.

Although the United States and Western Europe account for more than half of the world's production capacity, the most rapid growth in capacity has been occurring in the developing areas of the world. Growth in capacity in the developing countries is almost twice that of North America and Western Europe. These differing growth rates will lead to a shift in exports and imports among the developed and developing countries.

The economics for the production of ethylene depend to a large extent on the prices for feedstocks and coproducts. In the United States the feedstocks of choice have been the lighter feeds of ethane and propane, as opposed to Western Europe and the Far East, which favor naphtha.

A significant factor affecting capital investment is plant capacity. By 1990 it was theoretically possible to build a single-train (no duplication of compressors or other equipment except for cracking heaters) ethylene plant with a capacity of approximately 900,000 t/yr. The limitation above this capacity becomes the suction volume to the charge gas compressor. There are reports, however, of one plant operating at 950,000 t/yr with as yet no limitation.

K. M. SUNDARAM
M. M. SHREEHAN
E. F. OLSZEWSKI
ABB Lummus Crest, Inc.

K. M. de Reuck and co-workers, eds., *Ethylene (Ethene)-International Thermodynamic Tables of the Fluid State-10,* Blackwell Scientific Publishers, Oxford, UK, 1988.

TRC Thermodynamic Tables, Thermodynamic Research Center, The Texas A & M University System, College Station, Tex., 1986.

L. Kniel, O. Winter, and K. Stork, *Ethylene-Keystone to the Petrochemical Industry,* Marcel Dekker, Inc., New York, 1980.

G. F. Froment, *Chem. Eng. Sci.* **36,** 1271 (1981).

1992–1993 Ethylene Annual, Dewitt & Co. Inc., Houston, Tex.

ETHYLENE OXIDE

Today about 9.6×10^6 t of ethylene oxide are produced each year worldwide. The primary use for ethylene oxide is in the manufacture of derivatives such as ethylene glycol, surfactants, and ethanolamines.

Physical Properties

Ethylene oxide is a colorless gas that condenses at low temperatures into a mobile liquid. It is miscible in all proportions with water, alcohol, ether, and most organic solvents. Its vapors are flammable and explosive. The physical properties of ethylene oxide are summarized in Table 1.

Clathrate Formation. Ethylene oxide forms a stable clathrate with water molecules of water in the unit cell. The maximum observed melting point is 11.1°C.

Chemical Properties

Ethylene oxide is a highly reactive compound, and so is used industrially as an intermediate for many chemical products. The three-membered ring is opened in most of its reactions. These reactions are very exothermic because of the tremendous ring strain in ethylene oxide, which has been calculated.

Polymerization. The reaction of ethylene oxide with a nucleophile introduces the hydroxyethyl group:

Table 1. Some Physical Constants of Ethylene Oxide, C_2H_4O

Property	Value
molecular weight	44.05
bp, °C at 101.3 kPa[a]	10.4
dielectric constant at 0°C	14.50
explosive limits in air, %	
upper	100
lower	3
flash point, Tag open cup, °C	<−18
freezing point, °C	−111.7
heat of combustion at 25°C, kJ/mol[b]	−1218

[a] To convert kPa to mm Hg, multiply by 7.5.
[b] To convert kJ to kcal, divide by 4.184.

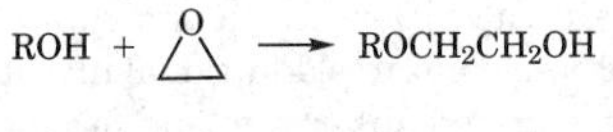

The product of this reaction can also react with ethylene oxide; if this process is repeated many times, a polymer is formed.

Low molecular weight polymers of ethylene oxide, poly(ethylene glycol), are formed by allowing ethylene oxide to react with water or alcohols under the proper conditions. The average molecular weight can be varied from 200 to 14,000.

Crown Ethers. Ethylene oxide forms cyclic oligomers (crown ethers) in the presence of fluorinated Lewis acids such as boron trifluoride, phosphorus pentafluoride, or antimony pentafluoride. Hydrogen fluoride is the preferred catalyst.

Other Chemical Reactions. Reaction with water is slow at ambient temperatures and neutral conditions, but is much faster with either acid or base catalyst. By-products, namely diethylene and triethylene glycol, are also formed in this reactions.

Reactions with alcohol parallel those of ethylene oxide with water. The primary products are monoethers of ethylene glycol; secondary products are monoethers of poly(ethylene glycol). Most are appreciably water-soluble.

The carboxyl group of an organic acid reacts with ethylene oxide to give the corresponding ethylene glycol monoester. Ethylene glycol diesters may be obtained directly by the reaction of ethylene oxide with the acid anhydride.

Ethylene oxide reacts with ammonia to form a mixture of mono-, di-, and triethanolamines.

Complex nitrogen compounds are formed from the reaction of alkylamines with ethylene oxide. Primary and secondary aromatic amines react with ethylene oxide to give the corresponding arylaminoethanols.

Ethylene oxide reacts with hydrogen sulfide to yield 2-mercaptoethanol and thiodiglycol (bis-2-hydroxyethyl sulfide). The reaction of ethylene oxide with long-chain alkyl mercaptans yields polyoxyethylene mercaptans, some of which are nonionic surfactants.

Ethylene oxide reacts with Grignard reagents, RMgX, to yield the corresponding two carbon homologue, RCH_2CH_2OH.

Ethylene oxide reacts with acetyl chloride at slightly elevated temperatures in the presence of hydrogen chloride to give the acetate of ethylene chlorohydrin.

Compounds containing active $—CH_2—$ or $—CH—$ groups, such as malonic and monosubstituted malonic esters, ethyl cyanoacetate, and β-keto esters, react with ethylene oxide under basic conditions.

The 2-hydroxylethyl aryl ethers are prepared from the reaction of ethylene oxide with phenols at elevated temperatures and pressures. 2-Phenoxyethyl alcohol is a perfume fixative. The water-soluble alkyl-phenol ethers of the higher poly(ethylene glycol)s are important surface-active agents.

Ethylene oxide reacts readily with hydrogen cyanide in the presence of alkaline catalysts, such as diethylamine, to give ethylene cyanohydrin.

Autodecomposition of ethylene oxide vapor occurs at ~500 °C at 101.3 kPs (1 atm) to give methane, carbon monoxide, hydrogen, and ethane. Isomerization of ethylene oxide to acetaldehyde occurs at elevated temperatures in the presence of catalysts such as iron oxide.

Manufacture

At the present time (ca 1993), all the ethylene oxide production in the world is achieved by the direct oxidation process. The direct oxidation technology, as the name implies, utilizes the catalytic oxidation of ethylene with oxygen over a silver-based catalyst to yield ethylene oxide. The process can be divided into two categories depending on the source of the oxidizing agent: the air-based process and the oxygen-based process.

Several companies have developed technologies for direct oxidation plants. Union Carbide Corp. and Dow Chemical use their own technologies. Shell Development, Scientific Design, and more recently, Nippon Shokubai, license ethylene oxide technology, and over 70% of present world capacity is based on their processes. All the ethylene oxide plants that have been built during the last 15 years were oxygen-based processes, and a number of existing ethylene oxide plants were converted from the air to the oxygen-based process during the same period. There are 13 producers of ethylene oxide in the United States: BASF, Dow, Eastman, Hoechst-Celanese, Olin, Oxy Petrochemicals, PD Glycols, Quantum, Shell, Sun Refining, Huntsman, and Union Carbide, and Fermosa.

Process Technology Considerations. Innumerable complex and interacting factors ultimately determine the success or failure of a given ethylene oxide process. Those aspects of process technology that are common to both the air- and oxygen-based systems are reviewed below, along with some of the primary differences.

Of all the factors that influence the utility of the direct oxidation process for ethylene oxide, the catalyst used is of the greatest importance. There are four basic components in commercial ethylene oxide catalysts; the active catalyst metal; the bulk support; catalyst promoters that increase selectivity and/or activity and improve catalyst life; and inhibitors or anticatalysts that suppress the formation of carbon dioxide and water without appreciably reducing the rate of formation for ethylene oxide.

Silver-containing catalysts are used exclusively in all commercial ethylene oxide units although the catalyst composition may vary considerably.

The chemical and physical properties of the support strongly dictate the performance of the finished catalyst. Although nonsupported silver catalysts have been advocated in some patents, it is unlikely that they are used commercially since pure silver tends to sinter at reaction temperatures with a resultant activity loss. For commercial operation, the preferred supports are alundum (a-alumina) and silicon carbide. Other supports are glass wool, quartz, carborundum, and ion-exchange zeolites (see MOLECULAR SIEVES).

Silver alone on a support does not give rise to a good catalyst. However, addition of minor amounts of promotor enhance the activity and the selectivity of the catalyst, and improve its long-term stability. The most commonly used promoters are alkaline-earth metals, such as calcium or barium, and alkali metals such as cesium, rubidium, or potassium.

Many organic compounds, especially the halides, are very effective for suppressing the undesirable oxidation of ethylene to carbon dioxide and water, although not significantly altering the main reaction to ethylene oxide. These compounds, referred to as catalyst inhibitors, can be used either in the vapor phase during the process operation or incorporated into the catalyst manufacturing step.

Temperature is used to control two related aspects of the reaction: heat removal from the reactor bed and catalyst operating temperature.

Space velocity has a strong effect on the process economics. It establishes the reactor size and pressure drop, affecting compression costs. The optimum space velocity is a function of energy costs, reaction rate, and selectivity.

Operating pressure has a marginal effect on the economics of the ethylene oxide process. High pressure increases production due to higher gas density, increases heat transfer, increases ethylene oxide and carbon dioxide recovery in the absorbers, and lowers compression costs. Also, since the total number of moles decreases in the formation of ethylene oxide from ethylene and oxygen, high pressure is consistent with high conversion. However, high pressures reduce the flammable limit of the process gas as well as increase equipment costs. Typical commercial pressures are 1–2 MPa (10–20 atm).

Air process technology uses nitrogen as the diluent gas.

The choice of diluent for the oxygen process is based on the thermal properties of the gas. The small process purge makes it economically possible for the process to operate under a wide variety of ballast gases. Several gases have been proposed in the patent literature, including methane and ethane.

The oxygen process has four main raw materials: ethylene, oxygen, organic chloride inhibitor, and cycle diluent. The purity requirements are established to protect the catalyst from damage due to poisons or thermal runaway, and to prevent the accumulation of undesirable components in the recycle gas. The latter can lead to increased cycle purging, and consequently higher ethylene losses.

Typical ethylene specifications call for a minimum of 99.85 mol % ethylene. The primary impurities are usually ethane and methane. Impurities that strongly affect catalyst performance and reactor stability include acetylene, propylene, hydrogen, and sulfur.

Oxygen specifications can vary depending on the economics of the process. The dominant impurity in oxygen is argon. As the argon concentration in the reactor feed increases, the flammable limit of the gas decreases.

Organic chloride and cycle diluent specifications are less critical since the flows are significantly less. The organic chloride specifications must prevent gross contamination as well as the potential of solids that would lead to plugging. The cycle diluent must also be free of gross contamination as well as significant catalyst poisons such as sulfur.

The air process has similar purity requirements to the oxygen process. The ethane content of ethylene is no longer a concern, due to the high cycle purge flow rate. Air purification schemes have been used to remove potential catalyst poisons or other unwanted impurities in the feed.

An economic recovery scheme for a gas stream that contains less than 3 mol % ethylene oxide (EO) must be designed. It is necessary to achieve nearly complete removal since any ethylene oxide recycled to the reactor would be combusted or poison the carbon dioxide removal solution. Commercial designs use a water absorber followed by vacuum or low pressure stripping of EO to minimize oxide hydrolysis. Several patents have proposed improvements to the basic recovery scheme. Other references describe how to improve the scrubbing efficiency of water or propose alternative solvents.

The main impurities in ethylene oxide are water, carbon dioxide, and both acetaldehyde and formaldehyde. Water and carbon dioxide are removed by distillation in columns containing only rectifying or stripping sections. Aldehydes are separated from ethylene oxide in large distillation columns.

Process Safety Considerations. Unit optimization studies combined with dynamic simulations of the process may identify operating conditions that are unsafe regarding fire safety, equipment damage potential, and operating sensitivity.

The safe operating ranges of the unit are dependent on all of the process parameters: temperature, pressure, residence time, gas composition, unit dynamic responses, instrumentation system, and the presence of ignition sources. The ethylene oxide reaction cycle operates close to the flammable limit. Owing to the reactivity and explosibility of ethylene oxide, another sensitive area is the final refining of ethylene oxide. The selection of the safe operating conditions and design of effective process safety systems is a complex task that requires extensive laboratory testing to determine the effect of the various process parameters on explosibility as well as proven commercial experience.

Environmental Considerations. A detailed study of the environmental considerations in the manufacture of ethylene oxide by the direct oxidation of processes is available. The primary air emissions from the formation of ethylene oxide by direct oxidation are ethylene, ethylene oxide, carbon dioxide, and ethane. Traces of NO_x and SO_x from pollution control and process machinery operations have also been reported.

Liquid emissions from ethylene oxide units originate in the recovery section. The water of reaction from complete combustion of ethylene must be purged from the oxide absorber water cycle. This stream contains glycol, organic salts, aldehydes, and ethylene oxide. The location of the purge stream is selected to minimize ethylene oxide and glycol emissions. This stream is readily biodegradable. Several technologies have been proposed to reduce the amount of organics in the waste, either by distillation or the use of membranes to recover the contained glycol.

Air vs Oxygen Process Differences and Economics. The relative economics of the air vs oxygen process have been reported. Two process characteristics dictate the difference in the capital costs for the two processes. The air process requires additional investment for the purge reactors and their associated absorbers, and for energy recovery from the vent gas. However, this is offset by the need for an oxygen production facility and a carbon dioxide removal system for an oxygen-based unit. In a comparison of necessary investments for medium to large capacity units (>20,000t/yr), oxygen-based plants have a lower capital cost even if the air-separation facility is included. However, for small- to medium-scale plants, the air process investment is smaller than that required for the oxygen process and the air-separation unit, unless the oxygen is purchased from a large air-separation unit serving many customers.

Other Processes. Other processes for producing ethylene oxide include the chlorohydrin process, the arsenic-catalyzed liquid-phase process, the thallium-catalyzed epoxidation process, the Lummus hypochlorite process, the liquid-phase epoxidation with hydroperoxides, the electrochemical process, the unsteady-state direct oxidation process, the fluid-bed direct oxidation process, and biological routes.

Economic Aspects

United States production of ethylene oxide in 1990 was 2.86×10^6 metric tons. Total world capacity in 1992 was ca 9.6×10^6 metric tons.

More than 98% of the total production of ethylene oxide is converted to derivatives.

Shipment and Storage

Small shipments of ethylene oxide are made in either compressed gas cylinders up to ~0.1 m^3 (30 gal) or in 1A1 steel drums (61 gal). Very large shipments >40 m^3 (10,000–25,000 gal) are made in insulated, type 105J100W or other DOT approved tank cars.

Storage. Carbon steels and stainless steel should be used for all equipment in ethylene oxide service.

Ethylene oxide storage tanks are pressurized with inert gas to keep the vapor space in a nonexplosive region and prevent the potential for decomposition of the ethylene oxide vapor.

Worldwide capacity for ethylene oxide (ca 1993) was projected to rise by ca 2×10^6 t/yr to a total of 11.6×10^6 t/yr by 1995. Additions have been announced by Union Carbide in the United States and Canada, Formosa Plastics and Shell in the United States, BASF in Belgium (rebuild of an existing unit), and SABIC in Saudi Arabia. Expansions have also been announced in South Korea and India, and may occur in other countries as well.

Health and Safety Factors

Toxicology. An excellent review of the toxicity and health assessment of ethylene oxide has been compiled. Ethylene oxide (EO) can be relatively toxic as both a liquid and gas. Inhalation of ethylene oxide in high concentrations may be fatal.

Inhalation exposure to high concentrations of ethylene oxide has been reported to result in respiratory system irritation and edema.

There is some evidence that occupational exposure to high levels of ethylene oxide can result in cataracts. Neurological effects have also been reported in association with recurrent human and animal inhalation exposures to ethylene oxide. Again, depending on the degree of exposure, headache, nausea, vomiting, diarrhea, dizziness, loss of coordination, convulsion, or coma may occur. The onset of illness is rapid in severe exposures, but may be delayed after moderate exposure. In the reports of human peripheral neurotoxic effects or central nervous system toxicity, most cases have shown a marked improvement on removal from further exposure.

Ethylene oxide has been shown to produce mutagenic and cytogenic effects in a variety of test systems.

When the data as a whole are reviewed for studies on humans exposed to ethylene oxide, no conclusion can be made that there is an increase in mortality associated with those exposed to ethylene oxide.

Developmental toxicity inhalation tests have been conducted using rats or rabbits. No teratogenic effects were observed. The only developmental effects noted were decreased fetal weights and delayed ossification.

Dermal exposure information has been collected from case reports of industrial accidents. Concentrated ethylene oxide evaporates rapidly from the skin and produces a freezing effect, often compared to frost bite, leaving burns ranging from first- to third-degree severity.

OSHA considers that, at excessive levels, ethylene oxide may present reproductive, mutagenic, genotoxic, neurologic, and sensitization hazards. In addition, ethylene oxide is considered by OSHA, IARC, and NTP as a potential human carcinogen.

Explosibility and Fire Control. As in the case of many other reactive chemicals, the fire and explosion hazards of ethylene oxide are system-dependent. Each system should be evaluated for its particular hazards including start-up, shut down, and failure modes. Storage of more than a threshold quantity of 5000 lb (~2300 kg) of the material makes ethylene oxide subject to the provisions of OSHA 29 CFR 1910 for "Highly Hazardous Chemicals."

Liquid Hazards. Pure liquid ethylene oxide will deflagrate given sufficient initiating energy either at or below the surface, and a propagating flame may be produced. Liquid mists of ethylene oxide will decompose explosively in the same manner as the vapor.

Liquid ethylene oxide under adiabatic conditions requires a temperature of about 200°C before a self-heating rate of 0.02°C/min is observed. However, in the presence of contaminants such as acids and bases, or reactants possessing a labile hydrogen atom, the self-heating temperature can be much lower.

Ethylene oxide is an electrically conductive liquid that does not accumulate static electricity in grounded equipment. Static electricity can, however, accumulate in liquid mist produced by splashing and spraying. Although the vapor alone has a large minimum ignition energy, mixtures with oxidants such as air can be very sensitive to ignition.

Vapor Hazards. Ethylene oxide vapor can decompose explosively and propagate a decomposition flame when its pressure is greater than about 300 mm Hg (depending on temperature). To prevent decomposition, dilution using an inert gas (N_2 or CO_2) or an extinguishant such as a halocarbon have been used. The amount of dilution required depends on temperature, pressure, and the anticipated ignition source.

Hazards of Mixtures with Air. Pools of liquid ethylene oxide will continue to burn until they are diluted with at least 22 parts of water by volume. This ratio must be increased to about 100 parts water if the vapor is confined, such as in a sewer.

Mixtures of ethylene oxide with air are far easier to ignite and burn much faster than the pure vapor.

When ethylene oxide vapor or liquid leaks into porous refractory insulation such as mineral wool or calcium silicate, it reacts with water contained in the insulation forming low molecular weight poly(ethylene glycol)s. Whereas ethylene oxide is volatile, the glycols can accumulate and self-heat in the presence of air. The event may lead to a fire in the insulation that might overheat small diameter lines causing internal decomposition. To prevent this, cellular glass insulation may be used since it is nonporous and cannot accumulate glycols.

Catalysts such as iron oxides cause isomerization of the ethylene oxide to acetaldehyde with the evolution of heat. The acetaldehyde has a much lower autoignition temperature in air than does ethylene oxide, and the two effects may lead to hot-spot ignition.

Uses

Ethylene oxide is an excellent fumigant and sterilizing agent. Ethylene oxide has been studied for use as a rocket fuel and as a component in munitions. It has been reported to be used as a fuel in FAE (fuel air explosive) bombs.

Derivatives

Derivatives include ethylene glycol, di-, tri-, and tetraethylene glycols, nonionic surface-active agents, ethanolamines, glycol ethers, poly(ethylene glycol)s, and poly(ethylene oxide).

J. P. Dever
K. F. George
W. C. Hoffman
H. Soo
Union Carbide Corporation

Ethylene Oxide, Brochure F-ICD23, Union Carbide Corp., Danbury, Conn., 1993.

I. Kiguchi, T. Kumazawa, and T. Nakai, *Hydrocarbon Process.* **55**(3), 69 (1976).

B. J. Ozero and J. V. Procelli, *Hydrocarbon Process.* **63**(3), 55 (1984).

G. H. Twigg, *Trans. Faraday Soc.* **42**, 284, and 657 (1946); *Proc. Roy. Soc. (London) Ser. A* **188**, 92 (1946).

L. G. Britton, *Plant/Operations Prog.* (Apr., 1992).

EVAPORATION

In the chemical engineering sense, evaporation is the removal of volatile solvent from a solution or a relatively dilute slurry by vaporizing the solvent. In nearly all industrial applications the solvent is water, and in most cases the nonvolatile residue is the valuable constituent. Evaporation differs from distillation (qv) in that when the volatile stream consists of more than one component no attempt is made to separate those components. Thus, production of distilled water from impure feedwater utilizes an evaporator rather than a distillation unit even though small-capacity units are usually called stills. Although the product of an evaporator system may be a solid, the heat required for vaporization of the solvent must be transferred to a solution or a slurry of the solid in its saturated solution in order for the device to be classified as an evaporator rather than a dryer. It is not unusual for an evaporator to be used to produce a solid as its only product.

The highly varied purposes for which evaporators are used industrially include: (*1*) reducing the volume to economize on packaging, shipping, and storage costs, eg, of salt, sugar (qv), caustic soda (see Alkali and chlorine products), orange juice (see Fruit juices), and milk (see Milk and milk products); (*2*) obtaining a product in its most useful form, eg, salt from brine or sugar from cane juice; (*3*) eliminating minor impurities, eg, salt, sugar; (*4*) removing major contaminants from a product, eg, diaphragm cell caustic soda solutions contain more salt than caustic when produced but practically all the salt can be precipitated by concentrating to a 50% NaOH solution; (*5*) concentrating a process stream for recovery of resources, eg, pulp (qv) mill spent cooking liquor, if concentrated sufficiently in an evaporator, can be burned in a boiler to produce steam, yielding also an ash that can be used to reconstitute fresh cooking liquor; (*6*) concentrating wastes for easier disposal, such as nuclear reactor wastes (see

NUCLEAR REACTORS), dyestuff plant effluents, and cooling tower blowdown streams; (*7*) transforming a waste into a valuable product, such as spent distillery slop after alcohol recovery, which can be concentrated to produce an animal feed (see FEEDS AND FEED ADDITIVES); and (*8*) recovering distilled water (qv) from impure streams such as sea water and brackish waters.

Steam-Heated Evaporators

The three principal requirements of steam-heated evaporators are (*1*) transfer to the liquid of the large amounts of heat needed to vaporize the solvent, (*2*) efficient separation of the evolved vapor from the residual liquid, and (*3*) the accomplishment of these aims with the least expenditure of energy justifiable by the capital cost involved.

Natural Circulation Evaporators. Natural circulation evaporators (Fig. 1) were the first developed commercially and still represent probably the largest number of units in operation. These evaporators utilize the density difference between the liquid and the generated vapor to circulate the liquid past the heating surface and thereby give good heat-transfer performance.

Forced Circulation Evaporators. The forced circulation evaporator, suitable for the largest variety of applications, is usually the most expensive type. It usually consists of a shell-and-tube heat exchanger, a vapor–liquid separator (variously called vapor head, vapor body, separator, flash chamber, or body), and a pump to circulate the liquor from the body through the heater and back to the body. The system is usually arranged so that there is no boiling in the heater. The heat input is therefore absorbed as sensible heat, and vapor liberation does not occur until the liquor enters the flash chamber.

Several configurations of forced-circulation evaporators are shown in Figure 2. The most common arrangement, shown in Figure 2**a**, is one having an external vertical single-pass heater and a tangential inlet to the body.

Film-Type Evaporators. Film-type evaporators are illustrated in Figure 3. Figure 3**a** shows the rising film or long tube vertical (LTV) evaporator that is widely used in the United States.

Energy Conservation

Most of the complexity and cost of an evaporator installation is a result of attempts to reduce energy consumption, which is usually by far the most important element of operating cost. Evaporators are not normally rated directly in efficiency of energy usage because the separation of solvent from the solution requires very little theoretical energy in an ideal system.

Steam-heated evaporators are generally rated in terms of steam economy, ie, weight of water evaporated per weight of steam used, also called gained output ratio or performance ratio, and frequently standardized as pounds of water evaporated per 1000 Btu extracted from the steam (kg evaporated per 2324 kJ).

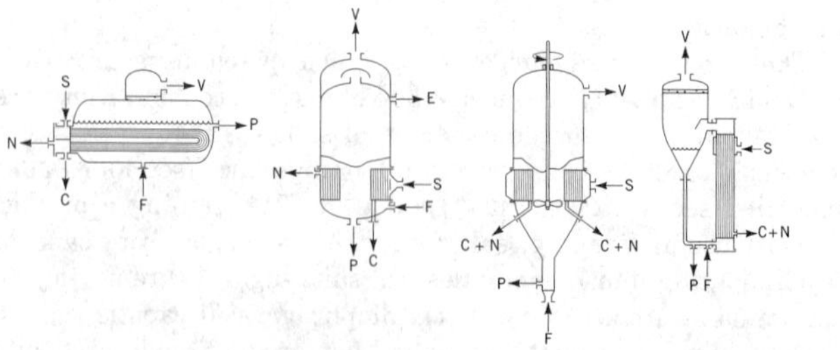

Figure 1. Natural circulation evaporators where C = condensate, E = entrainment return, F = feed, N = noncondensibles' vent, P = product or concentrate, S = steam, V = vapor, and ▥ = knitmesh separator: (**a**) horizontal-tube, (**b**) short-tube vertical, (**c**) propeller calandria, and (**d**) long-tube recirculating.

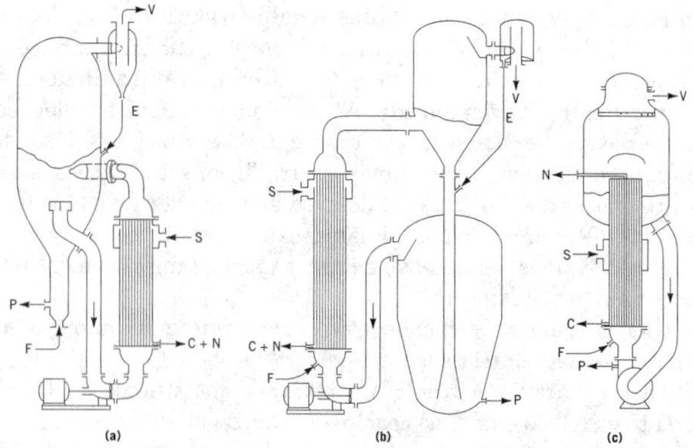

Figure 2. Forced circulation evaporators: (**a**) submerged-tube, shown as circulating magma crystallizer; (**b**) submerged-tube, shown as suspension type crystallizer; and (**c**) boiling type. Terms are defined in Figure 1.

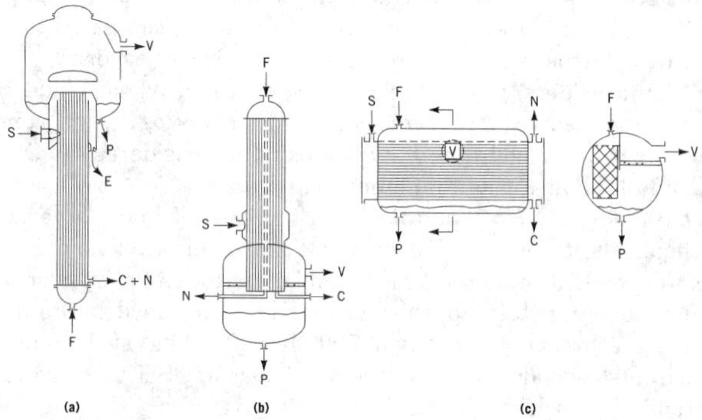

Figure 3. Film-type evaporators: (**a**) long-tube vertical, (**b**) falling film, and (**c**) horizontal tube. Terms are defined in Figure 1. ⊠ represents end view of tube bundle in (**c**).

The single-effect evaporator is the simplest arrangement. It uses steam from an outside source and exhausts its vapor to the atmosphere or to an air- or water-cooled condenser. Such an evaporator requires about a 1-to-1 ratio of steam to water evaporated, but somewhat more if the feed is colder than the product and heat cannot be recovered by preheating the feed with concentrate and condensate. The high steam consumption limits the use of single effects to small capacities or to materials requiring an expensive type of evaporator, such as the wiped-film type, or having a very high boiling point, or to cases where the vapor is contaminated by materials that would cause excessive fouling or corrosion of heating surfaces when condensed, eg, rayon spin-bath liquor. Such evaporators may be operated on a continuous, batch, or semibatch basis with very little difference in heat requirements. In both batch and semibatch operation, final concentration is not reached until the end of the cycle, and these methods are therefore used primarily when the heat-transfer properties become markedly poorer as final concentration is approached. Semibatch operation is the more common. Feed is added continuously during most of the cycle in order to maintain a liquid inventory large enough to permit the evaporator to operate properly.

The single-effect evaporator produces almost as much vapor as the amount of steam used, the only difference being that the vapor is at a lower temperature and hence lower pressure. Compressing the vapor for reuse as the heat source was put into operation in the nineteenth century. This thermocompression or vapor recompression operation can be accomplished by either mechanical or steam-jet compressors. Mechanical compressors are by far the more efficient and may be driven by electric motors, gas or diesel engines, or steam or gas turbines.

The ideal mechanical power requirement of a thermocompression evaporator is given by the Carnot equation: $W = Q\Delta T/T$, where W is the work done, Q is the heat received at absolute suction temperature T, and ΔT is the difference in saturation temperature at compressor discharge and suction pressures. To minimize the work required ΔT is kept low so the evaporator must have a large heating surface area. The optimum balance between power consumption and evaporator cost is usually at net a ΔT in the range of 3–10 K. The need to operate at low net ΔT is a disadvantage of some types of evaporator, such as natural circulation and LTV units, where coefficients fall off at low ΔT; and the submerged, forced circulation types, which lose ΔT as a result of temperature rise in the heating element and short-circuiting in the vapor head. The most advantageous evaporator type for vapor compression operation, when suitable for the liquor handled, is the falling film, which has very little ΔT loss.

The largest number of vapor-compression evaporators have been built for producing potable water from seawater or brackish water. These are usually relatively small-capacity units for military use, offshore drill rigs, marine vessels, and the like, and use engine-driven compressors and short-tube vertical evaporators. Another use has been for citrus juice concentration, which must be done at low temperatures to avoid degradation. Since the 1973 energy crisis, and as a result of higher efficiency, thermocompression evaporators are used in fields where they were not previously considered, such as for paper-mill waste cooking liquors and for disposal of cooling-tower blowdown wastes.

If steam instead of power is the source of energy, multiple-effect evaporation is the principal means of energy conservation. In this operation, the vapor from one effect is used to heat another effect boiling at a lower temperature and the vapor from this effect is used to heat yet another effect boiling at still lower temperature. In such evaporators it is desirable to use an initial steam temperature as high as possible in the first effect and a heat sink temperature as low as possible to condense the vapor from the last effect, in order to develop the highest practical total temperature difference for heat transfer.

Another means of gaining multiple-effect steam economy is by multistage flash operation, developed originally as the Alberger salt process. In this process, cool liquor is preheated in stages by vapor condensing at successively higher temperatures in a recovery section and finally by prime steam in a brine heater. The heated liquor is then flashed down in stages to generate the vapor used for preheating. The condensate from this vapor is also flashed down in the same manner so that the total flow being flashed is the same as the flow being heated. As a result, the temperature rise and flashing range in all heaters and flashers is about the same. Flashing through a 75 K range evaporates only about 12% of the feed; thus it is necessary to recirculate the liquor through the heaters and flashers when an appreciable degree of concentration is required. This then requires a rejection section in the flash train, containing one or more stages in which liquor is flashed to reject heat to cooling water before being recirculated to the recovery section. The principal advantages of the multistage flash evaporator are that only one circulating pump is needed and that a number of stages, comprising both flash space and heating elements, can be combined in one vessel.

It is possible, of course, to combine the different types of evaporator as well as the various methods of energy conservation into a single system.

Temperature Difference and Heat Transfer

The capacity of a steam-heated evaporator is governed by the amount of heat it can transfer, which is determined in turn by the amount of heating surface area and temperature difference available, and the heat-transfer coefficients achieved (see HEAT EXCHANGE TECHNOLOGY–HEAT TRANSFER). There are common conventions in the industry for these terms but they are not universally used; therefore, care must be exercised in interpreting and applying available data. Heating surface area is almost always taken as the total tubing area on the side in contact with the liquor being evaporated, which is different from normal heat exchanger practice, where it is always the outside area of the tubes. The definition of temperature difference is open to the widest interpretation and is the most frequent cause of misapplication of data. Considering only one effect of an evaporator, it is the difference between saturated steam temperature on one side of the heating surface and the liquor temperature on the other side that is effective for heat transfer. However, for the effect as a whole, with its associated liquor circuits, vapor piping, and entrainment separators, it is the difference between saturated steam temperature at the inlet to the heating element and the condensing temperature of the evolved vapor at the exit of the vapor discharge piping that determines how the evaporator effect reacts with its surroundings. Certain losses in temperature difference are inherent, including boiling point rise, vapor circuit losses (those from friction and acceleration and deceleration heads of steam and vapor flowing into and through the heating element, entrainment separator, and vapor piping) and liquor circuit losses (the difference between the temperature at which heat is absorbed at the heating surface and the temperature at which it is released in the vapor head). Boiling point rise, or boiling point elevation (BPE), is usually the most important of these losses and is the difference between the boiling point of the solution in the evaporator and the boiling point of pure solvent at the same pressure. Because the vapor cannot condense and give up the bulk of its heat content until it has cooled to the boiling point of the pure solvent, BPR represents a loss of available ΔT and is deducted from the overall ΔT before computing heat-transfer coefficients.

Any pressure drop losses of steam in the heater or of vapor in the entrainment separator and vapor piping also reduce condensing temperatures, and therefore represent a loss in ΔT available for heat transfer. These losses become more serious at the lower temperatures, because of the increasing difficulty of handling lower density vapor and the decreasing slope of the vapor pressure curve.

Losses of ΔT in the liquor circuit may also be substantial but are not always taken into consideration. Only in the forced circulation evaporator are at least part of these losses usually calculated.

When designing a new evaporator or analyzing the operation of an existing evaporator, a temperature distribution table first should be prepared. Such a table compares the magnitude of the various losses in ΔT and hence indicates the principal obstructions to capacity.

When used in design, a temperature distribution table serves as the basis for calculating a heat and material balance around the evaporator to determine heat loads in the individual effects, vapor flows, and liquor concentrations, from which vapor-circuit losses or line sizes and BPR estimates can be reconfirmed.

Actual heat-transfer coefficients encountered in evaporators cover a wide range, depending on the physical properties of the solution, its fouling characteristics, the type of evaporator employed, the boiling temperature, and the temperature difference. Only in the submerged-tube forced circulation and the falling-film evaporator can heat-transfer coefficients easily be calculated from theory. For other types, performance estimates are usually based on earlier experience with the same or a similar liquor. In general, coefficients range from a low of about 175 $J/(m^2 \cdot s \cdot K)$ (100 $Btu/(h \cdot ft^2 \cdot {}^\circ F)$) to a high of about 3500 (2000).

Vapor–Liquid Separation

The heating surface usually determines the evaporator cost and the vapor head the space requirements. The vapor–liquid separator must have enough horizontal plan area to allow the bulk of the initial entrainment to settle back against the rising flow of vapor and enough height to smooth out variations in vapor velocity and to prevent splashing directly into the vapor outlet. Separators are usually sized on the basis of the Souders-Brown expression: $U = K((\rho_l - \rho_g)/\rho_g)^{1/2}$, where U = vapor velocity, ρ_l = density of liquid, ρ_g = density of gas. For most types of evaporator, the decontamination factor *DF* (DF = kg of vapor per kg of entrained liquid) decreases as *K* increases.

In general, it is not economically attractive to provide the full degree of entrainment separation desired in the vapor head alone. Instead, the vapor head is sized for a decontamination factor on the order of 200 and reliance is place on supplementary separators such as shown in Figures 1–3 for removal of residual entrainment.

Heat Removal and Noncondensible Gases

A single- or multiple-effect evaporator does not consume heat; it merely degrades the heat input, and means must be provided for removing the waste heat. Heat is usually rejected to a river, well water, or a cooling tower. The most common means of heat rejection is by a barometric condenser in which the vapor from the last effect is condensed by direct, countercurrent contact with water cascading over weirs or trays. Noncondensible gases in the vapor accumulate at the top of the condenser and are cooled close to the temperature of the incoming water, thereby reducing the amount of water vapor associated with the gases. These gases are removed by either a steam-jet ejector (usually of several stages) or a mechanical vacuum pump.

Noncondensible gases are more prevalent in an evaporator system than in most other steam-heated equipment and must be properly handled to avoid serious impairment of heat transfer or reduction in steam economy.

Other Evaporative Methods

Solar evaporation, the first evaporative method developed, is still in widespread use, primarily for production of salt from seawater (ie, at coastal locations where groundwater contamination is not a problem).

There are other evaporator types that are not steam-heated. When the BPR is extremely high (as for manufacture of anhydrous NaOH), the evaporator may be heated by molten salt or other high temperature heat-transfer fluids instead of steam. The LTV-type evaporator is normally used in this service. Wiped-film evaporators, because of their high unit cost of heating surface, also sometimes use these high temperature, low pressure heating media to achieve high ΔTs without need for very heavy heat-transfer wall thicknesses. Another method, submerged combustion, involves direct contact with gases from a burner immersed in the liquid. These evaporators are impractical for all except very low capacities or the most refractory scale-forming or corrosive liquids.

Reverse osmosis (qv) is being used for some of the work previously only feasible by evaporation, such as concentrating dilute solutions that do not deposit solids on being concentrated. The limit is approximately that of recovering fresh water from seawater, where osmotic pressure to be overcome reaches about 3600 kPa (520 psi).

Increasing energy costs have pushed optimum steam economy of evaporators higher and higher. This in turn has aided introduction of falling-film evaporators into more services because the heat-transfer performance does not degrade at the lower temperature differences then available.

Some use is being made of lower grade heat sources, such as moist air from driers, and the construction of auxiliaries, such as condensers, integral with the evaporator body. A further step is elimination of the conventional condenser–cooling tower–vacuum pump circuit by recirculating last-effect liquor over the equivalent of a cooling tower built as an integral part of the evaporator body.

FERRIS C. STANDIFORD
Consultant

D. W. Green, ed., *Chemical Engineers Handbook*, 6th ed., McGraw-Hill Book Co., New York, 1984, Section 10, pp. 34–38; Section 11, pp. 11–40.

Energy Conservation Tools for the Process Engineer: Upgrading Existing Evaporators to Reduce Energy Consumption, Report COO/2870-2, National Technical Information Service, U.S. Department of Commerce, Springfield, Va., 1977.

Equipment Testing Procedures—Evaporators, AIChE, New York, 1978.

N. Lior, ed., *Measurement and Control in Water Desalination*, Academic Press, Inc., New York, 1986.

EXHAUST CONTROL, AUTOMOTIVE

Automobiles and trucks consume large amounts of gasoline, producing commensurately large amounts of gaseous exhaust consisting primarily of carbon dioxide (qv), water, unburned hydrocarbons (qv), carbon monoxide (qv), and oxides of nitrogen, NO_x (see GASOLINE AND OTHER MOTOR FUELS). The latter three atmospheric pollutants have been regulated since the 1970s by the U.S. government and more stringently by the State of California (see AIR POLLUTION). Automobile companies have developed fuel metering and exhaust systems using the catalytic converter to meet emission regulations. Carbon dioxide emissions are indirectly controlled by corporate average fuel economy (CAFE) standards for passenger cars and small trucks.

By 1994, when the Tier I exhaust emission standards mandated by the took effect in the United States, the degree of cleanup for automobiles and small trucks was required to be 97.4% for hydrocarbon emissions, 96.0% for carbon monoxide, and 90% for oxides of nitrogen as compared to pre-control of exhaust emissions. For areas in the United States that do not reach minimum ambient air standards by the end of the 1990s, the U.S. Congress has set conditional Tier II standards for vehicles to take effect as early as the year 2003. Also, California, because of unusually severe atmospheric pollution conditions, has established the most stringent automobile exhaust regulations in the world to control exhaust emissions to the absolute minimum levels. These California standards, called Low Emission Vehicle (LEV) Standards, started to take effect in 1996. Other states are expected to adopt the California standards.

Key elements for achieving the degree of exhaust emission control during the early 1990s were the monolithic catalytic converter, the three-way catalyst, and the closed loop system based on the oxygen (qv) sensor. Over 200 million vehicles had been equipped with the catalytic converter by ca 1993, enabling it to be rated among the top 10 engineering breakthroughs of the twentieth century. Emission control is achieved without negatively affecting fuel economy or performance. The shift to alternative transportation fuels such as methanol, ethanol, natural gas, liquefied petroleum gas (LPG), and reformulated gasoline in accordance with the Clean Air Act and the National Energy Policy Act of 1992 were expected to produce further modifications to the catalytic converter.

Diesel engine emission control technology is not discussed herein. As of early 1993 only one passenger car manufacturer was marketing diesel fuel cars. Emission control technology for diesel engines, used in some light-duty trucks and in medium- and heavy-duty trucks, is evolving at a rapid pace. This technology includes engine modifications such as high pressure fuel injection, variable valve timing, and intercooled turbochargers. Catalytic aftertreatment that is quite different from that discussed herein has also been developed.

Exhaust Gas Composition

The exhaust composition from gasoline/air combustion is dependent on many factors. Total combustion in the engine is not possible even when excess oxygen is present. Formation of the air/fuel mixture as well as the design of the combustion chamber influence the combustion process, as do engine power and ignition system timing. However, for emission control, the main factor affecting the composition of the exhaust gas is the air/fuel mixture or ratio.

Unburned hydrocarbons in the exhaust originate primarily from crevices in the combustion chamber, such as gaps between the piston and cylinder wall, where the combustion flame cannot burn. The composition of unburned hydrocarbons is dictated primarily by the composition of the fuel. Carbon monoxide results from areas of insufficient oxygen. Oxides of nitrogen are produced in the high temperature zones during combustion by the reaction of nitrogen molecules and oxygen atoms thermally produced from oxygen and oxygen-containing species, according to the Zeldovich mechanism.

Hydrocarbons and carbon monoxide emissions can be minimized by lean air/fuel mixtures, but lean air/fuel mixtures maximize NO_x emissions. Very lean mixtures (>20 air/fuel) result in reduced CO and NO_x, but in increased HC emissions owing to unstable combustion. The turning point is known as the lean limit. Improvements in lean-burn engines extend the lean limit.

Emission Control System

A typical 1993 model year automobile emission control system contained a multipoint fuel injection fuel metering system which metered the fuel in response to a measured amount of air. Incorporated into the exhaust stream prior to the catalytic converter is an oxygen sensor which indicates whether the exhaust air/fuel mixture is rich or lean of the stoichiometric point (defined as neither air nor fuel in excess). The catalytic converter is located in the exhaust line leading from the exhaust manifold, upstream of the acoustic muffler. The signal generated by the oxygen sensor is sent to the computer controller as is a signal from an air flow measurement device. The computer controller regulates the fuel metered by the fuel injection system in response to the air measurement signal. Air flow varies in response to the throttle position and load (inlet vacuum). The oxygen sensor quickly detects any change in oxygen concentration in the exhaust and the controller adjusts for this change. Thus the air/fuel ratio is constantly being adjusted back and forth slightly rich and slightly lean of the stoichiometric mixture. The three-way conversion (TWC) catalyst therefore receives exhaust gas that reflects this constant change back and forth in air/fuel mixture and is designed to operate under those conditions to convert NO_x by reduction and HC and CO by oxidation of at least 80 to 90%.

Catalytic Converter. The converter consists of a catalytic unit contained in a metal canister which surrounds the fragile ceramic catalytic unit with a steel shell. In between the steel shell and the exterior of the catalytic unit is a compliant layer that grips the catalytic unit with sufficient force to prevent movement of the catalytic unit within the canister, and which compensates for the differences in thermal expansion between the catalyst and the metal shell.

The catalytic unit is designed to provide enough surface area so that all exhaust gases contact the catalyst surfaces as they pass from the engine to the tailpipe. In order to function quickly after the engine is started, the catalytic unit must rapidly heat up to operating temperature. It therefore must possess good heat-exchange properties to extract the necessary heat from the exhaust gas. Once the minimum catalytic operation temperature is reached, the catalytic unit is designed to maximize transfer of the pollutants from the exhaust gas to the surface of the catalytic unit. Heat transfer and mass transfer are driven by temperature difference and concentration difference, respectively. At operating temperatures above 300 or 350°C, the catalytic reactions are so fast that only the exterior surfaces of the catalyst are utilized for the catalytic function.

Automobile exhaust catalysts have been developed that maximize the catalyst surface area available to the flowing exhaust gas without incurring excessive pressure drop. Two types have been extensively studied: the monolithic honeycomb type and the pellet type.

Catalytic Unit. The catalytic unit consists of an activated coating layer spread uniformly on a monolithic substrate. The catalyst predominantly used in the United States and Canada is known as the three-way conversion (TWC) catalyst, because it destroys all three types of regulated pollutants: HC, CO, and NO_x. When the air/fuel mixture is near the stoichiometric point, an optimum value of conversion for all three components is achieved.

Activated Coating. The activated coating layer of a TWC is applied to the high geometric surface area monolithic honeycomb body or substrate. Precious metals are dispersed within this catalytic layer, which contains particles as small as 10 μm. The small (10-μm) coating particles are typically aluminum oxide, Al_2O_3. Alumina is used because it is relatively inert and provides the high surface area needed to efficiently disperse the expensive active catalytic components.

The activated coating layer must possess two additional properties. It must adhere tenaciously to the monolithic honeycomb surface under conditions of rapid thermal changes, high flow, and moisture condensation, evaporation, or freezing. It must have an open porous structure to permit easy gas passage into the coating layer and back into the main exhaust stream. It must maintain this porous structure even after exposure to temperatures exceeding 900°C.

Precious Metal Catalysts. Precious metals are deposited throughout the TWC-activated coating layer. Rhodium plays an important role in the reduction of NO_x, and is combined with platinum and/or palladium for the oxidation of HC and CO. Only a small amount of these expensive materials is used (see PLATINUM-GROUP METALS).

Catalytic Support Body Monolithic Honeycomb Unit. The terms substrate and brick are also used to describe the high geometric surface area material upon which the active coating material is placed. Monolithic honeycomb catalytic support material comes in both ceramic and metallic form. Both are used in automobile catalysts and each possesses unique properties. A common property is a high geometric surface area which is inert and does not react with the catalytic layer.

Ceramic. The ceramic substrate is made from a mixture of silicon dioxide, talc, and kaolin to make the compound cordierite. Cordierite possesses a very low coefficient of thermal expansion and is thermal-shock resistant.

Metallic. Metallic substrates are also used as a support for the activated coating layer. A class of metal alloys containing Fe, Cr, and Al, when stabilized with Y or Ce, have excellent oxidation resistance at the extreme temperatures found in automobile exhaust. Melting temperatures are about in the same range as that of the ceramic cordierite.

Mass Transfer. Exhaust gas catalytic treatment depends on the efficient contact of the exhaust gas and the catalyst. The process of catalyst heating and initiation of the catalytic function is shown in Figure 1, where there are three or four distinct regions. Depending on the location of the catalytic unit in the exhaust system and the thermal mass present prior to the catalytic unit, it can take from 30 to 120 seconds for the catalyst unit to reach the catalyst ignition temperature of approximately 250–300°C (Region I). At temperatures above this ignition point, the activity of the catalyst increases rapidly with temperature (Region II). Some catalyst reaches a point where the sharp rise with temperature abruptly takes on a mild positive slope (Region III). Then, a point is reached at which catalytic performance improves only slightly to an increase in temperature (Region IV).

Catalyst Function. Automobile exhaust catalysts are perfect examples of materials that accelerate a chemical reaction but are not consumed. Reactions are completed on the catalyst surface and the products leave. Thus the catalyst performs its function over and over again. The catalyst also permits reactions to occur at considerably lower temperatures.

Concerning the reduction of NO_x, automobile three-way catalysts exhibit a property called selectivity. Catalyst selectivity occurs when

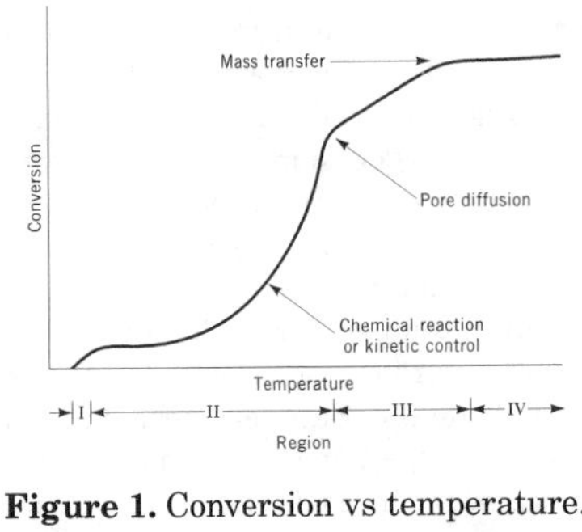

Figure 1. Conversion vs temperature.

several reactions are thermodynamically possible but one reaction proceeds at a faster rate than another. In the case of a TWC catalyst, CO, HC, and H_2 are all potential reductants of NO. On the other hand, O_2 is present, which oxidizes the CO, HC, and H_2. If these oxidation reactions are too rapid, no reductant is available to convert NO. Using modern TWC catalysts, however, NO reduction is fast enough that it is substantially completed before the reductants are consumed by O_2.

Two classes of metals have been examined for potential use as catalytic materials for automobile exhaust control. These consist of some of the transitional base metal series, for instance, cobalt, copper, chromium, nickel and manganese.

The precious metals possess much higher specific catalytic activity than do the base metals. In addition, base metal catalysts sinter upon exposure to the exhaust gas temperatures found in engine exhaust, thereby losing the catalytic performance needed for low temperature operation. Also, the base metals deactivate because of reactions with sulfur compounds at the low temperature end of auto exhaust. As a result, a base metal automobile exhaust catalyst would need to be considerably larger than a precious metal one and, even if a large bed were used, it would not heat up quickly enough to achieve the catalytic performance demanded of the emission control systems.

Catalyst function in an exhaust gas stream can be understood by examining the catalyst performance. As the temperature increases from ambient to about 200°C, there is no apparent action on the part of the catalyst to consume any of the reactants. The carbon monoxide present strongly chemisorbs on the surface of the catalyst, and prevents oxygen access. As the inlet gas temperature approaches 200°C, the CO bonds to the metal surface are relaxed. Oxygen molecules are now able to chemisorb, and catalytic ignition occurs.

Beyond the catalytic ignition point there is a rapid increase in catalytic performance with small increases in temperature. A measure of catalyst performance has been the temperature at which 50% conversion of reactant is achieved.

Catalyst Durability. Automobile catalysts last for the life of the vehicle and still function well at the time the vehicle is scrapped. However, there is potential for decline in total catalytic performance from exposure to very high temperatures, accumulation of catalyst poisons, or loss of the active layer.

Oxygen Sensor and the Closed Loop Fuel Metering System

The first commercial application of the TWC catalyst and closed loop fuel metering system using an oxygen sensor came with the introduction of the 1977 Volvo for the California market. This catalyst was developed by Engelhard. Other car companies introduced the system the following year, and in the 1990s almost 100% of U.S. and Canada passenger cars and light-duty vehicles utilize it. The fully developed system is described in the literature (see SENSORS).

The function of the oxygen sensor and the closed loop fuel metering system is to maintain the air and fuel mixture at the stoichiometric condition as it passes into the engine for combustion; ie, there should be no excess air or excess fuel. The main purpose is to permit the TWC catalyst to operate effectively to control HC, CO, and NO_x emissions. The oxygen sensor is located in the exhaust system ahead of the catalyst so that it is exposed to the exhaust of all cylinders (see Fig. 1). The sensor analyzes the combustion event after it happens. Therefore, the system is sometimes called a closed loop feedback system. There is an inherent time delay in such a system and thus the system is constantly correcting the air/fuel mixture cycles around the stoichiometric control point rather than maintaining a desired air/fuel mixture.

Oxygen Sensor. The oxygen sensor is also known as the lambda sonde or lambda sensor (Fig. 2), from the greek letter used to denote the air/fuel ratio, and as an exhaust gas oxygen sensor (EGO). The sensor consists of a ceramic porous solid electrolyte that is ionically conductive at operating temperatures. The outside of the ceramic is coated with platinum electrodes: one electrode exposed to the exhaust gas, the second to outside air. Between these electrodes is a zirconia solid electrolyte. The voltage generated depends on the oxygen concentration difference between each electrode. The exhaust platinum electrode is a catalyst at the exhaust gas surface and equilibrates the mixture, consuming, by catalytic reaction, any unreacted oxygen and CO, HC, and H_2, yielding a net amount of oxygen present.

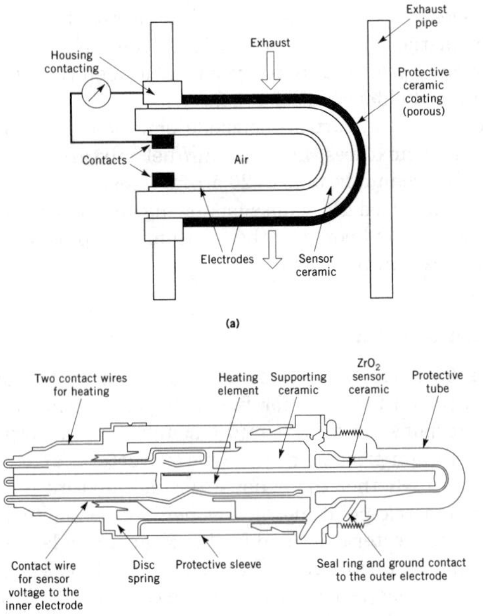

Figure 2. Schematic of lambda sensor (**a**) in exhaust pipe and (**b**) internal components. Courtesy of Robert Bosch.

Computer Controller. The computer controller takes inputs of speed, load, and temperature to assist the engine in cold starting and to select the optimal air/fuel trim adjustment for optimal power, fuel economy, and emissions control. The oxygen sensor closed loop system automatically compensates for changes in fuel content or air density.

Closed loop control has been designed for both carburetors and fuel injection metering systems. The latter were used in almost all 1990 models. Two types of fuel metering exist: a single fuel injector to serve all cylinders, called single-point fuel injection; and fuel injectors for each cylinder, called multipoint fuel injection.

The performance of the catalytic converter is affected by the conditions of air/fuel control provided by the fuel metering system. A slowly responding fuel metering system can dramatically decrease the conversion efficiency of the converter compared to a fast response multipoint fuel injection system.

On-Board Diagnostics. State of California regulations require that vehicle engines and exhaust emission control systems be monitored by an on-board system to assure continued functional performance. The program is called OBD-II, and requires that engine misfire, the catalytic converter, and the evaporative emission control system be monitored. The U.S. EPA is expected to adopt a similar regulation.

One system for measuring catalyst failure is based on two oxygen sensors, one located in the normal control location, the other downstream of the catalyst. The second O_2 sensor indicates relative catalyst performance by measuring the ability to respond to a change in air/fuel mixture. Other techniques using temperatures sensors have also been described.

Oxidation Catalyst. An oxidation catalyst requires air to oxidize unburned hydrocarbons and carbon monoxide. Air is provided with an engine driven air pump or with a pulse air device. Oxidation catalysts were used in 1975 through 1981 models but thereafter declined in popularity. Oxidation catalysts may be used in the future for lean burn engines and two-stroke engines.

The oxidation catalyst (OC) operates according to the same principles described for a TWC catalyst except that the catalyst only oxides HC, CO, and H_2. It does not reduce NO_x emissions because it operates in excess O_2 environments.

Exhaust Gas Recirculation. In one method of NO_x emission control, exhaust gas is fed back into the inlet manifold and mixed with the fuel and inlet air. The resultant mixture upon combustion in the cylinder results in lower peak combustion temperature and lower NO_x formation because the reaction of $N_2 + O_2 \rightarrow NO_x$ is strongly dependent on the combustion flame temperature. The degree of NO_x depression is dependent on the amount of exhaust gas recirculation (EGR). EGR provides a diluent gas having high molecular weight and CO_2 which absorbs heat. Also, EGR affects the flame speed of the mixture, and thus provides a certain antiknock quality to the combustion process. The impact of EGR on engine parameters has been detailed.

EGR can seriously degrade engine performance, especially at idle, under load at low speed, and during cold start. Control of the amount of EGR during these phases can be accomplished by the same electronic computer controller used in the closed loop oxygen sensor TWC system.

Other Emissions Control

Evaporative Emission. Fumes emitted from stored fuel or fuel left in the fuel delivery system are also regulated by U.S. EPA standards. Gasoline consists of a variety of hydrocarbons ranging from high volatility butane (C-4) to lower volatility C-8 to C-10 hydrocarbons. The high volatility HCs are necessary for cold start, and are especially necessary for temperatures below which choking is needed to start the engine. Stored fuel and fuel left in the fuel system evaporates into the atmosphere.

The common method of controlling evaporative emission is an activated charcoal canister that connects to the intake manifold. When the engine is shut off, the valve permits hydrocarbon fumes to be absorbed and stored by the activated charcoal.

Crankcase Emissions. Exhaust gases are also found in the crankcase. Control systems are required to feed these gases back into the inlet manifold so that the hydrocarbons and carbon monoxide contained are consumed in the combustion process. The control devise used is called a positive crankcase ventilation (PCV) valve. This device or one of similar function is required by law.

Alternative Fuels

Under the National Energy Policy Act of 1992 nonpetroleum-based transportation fuels are to be introduced in the United States. Such fuels include natural gas (see GAS, NATURAL), liquefied petroleum gas (qv) (LPG), methanol (qv), ethanol (qv), and hydrogen (qv), although hydrogen fuels are not expected to be a factor until after the year 2000 (see also ALCOHOL FUELS; HYDROGEN ENERGY).

Future Engines and Emission Control Systems

Two engines were under development as of 1993: the two-stroke engine and the lean burn engine. Driving development of these engines were the fact of global warming and the need for fuel economy.

Emission Control Technologies. The California low emission vehicle (LEV) standards have spawned investigations into new technologies and methods for further reducing automobile exhaust emissions.

Technologies being investigated (ca 1993) included improved catalysts that can be located closer to the exhaust manifold (high temperature resistant), reduced thermal mass in the exhaust system, use of alternative fuels, use of air pumps to assist catalyst light off, and engine adjustments that yield higher exhaust temperatures for this period. Three new technologies are undergoing intense development: the electrically heated catalyst, the hydrocarbon trap system, and the exhaust gas igniter.

JOHN J. MOONEY
Engelhard Corporation

Code of Federal Regulations, Title 40, Part 86, Washington, D.C., July 1, 1992.

K. C. Taylor, in J. R. Anderson and J. R. Boudart, eds., *Catalysis Science and Technology*, Berlin, 1984, Chapt. 2, pp. 119–170.

D. D. Eley, H. Pines, and P. B. Wiez, eds., *Advances in Catalysis*, Vol. 33, Academic Press, Inc., New York, 1985.

C. D. Falk and J. J. Mooney, *Three-Way Conversion Catalysts–Effect of Closed Loop Feedback Control and Other Parameters on Catalyst Efficiency*, SAE 800462, Society of Automotive Engineers, Warrendale, Pa., 1980.

EXHAUST CONTROL, INDUSTRIAL

Limits for exhaust emissions from industry, transportation, power generation (qv), and other sources are increasingly legislated (see also EXHAUST CONTROL, AUTOMOTIVE). One of the principal factors driving research and development in the petroleum (qv) and chemical processing industries in the 1990s is control of industrial exhaust releases. Much of the growth of environmental control technology is expected to come from new or improved products that reduce such air pollutants as carbon monoxide (qv), CO, volatile, organic compounds (VOCs), nitrogen oxides (NO_x), or other hazardous air pollutants (see AIR POLLUTION). The mandates set forth in the 1990 amendments to the Clean Air Act (CAAA) push pollution control methodology well beyond what, as of this writing, is in general practice, stimulating research in many areas associated with exhaust system control (see AIR POLLUTION CONTROL METHODS). In all, these amendments set specific limits for 189 air toxics, as well as control limits for VOCs, nitrogen oxides, and the so-called criteria pollutants. An estimated 40,000 facilities, including establishments as diverse as bakeries and chemical plants are affected by the CAAA.

There are 10 potential sources of industrial exhaust pollutants which may be generated in a production facility: *(1)* unreacted raw materials; *(2)* impurities in the reactants; *(3)* undesirable by-products; *(4)* spent auxiliary materials such as catalysts, oils, solvents, etc; *(5)* off-spec product; *(6)* maintenance, ie,wastes and materials; (*(7)* exhausts generated during start-up and shutdown; *(8)* exhausts generated from process upsets and spills; *(9)* exhausts generated from product and waste handling, sampling, storage, and treatment; and *(10)* fugitive sources.

Exhaust streams generally fall into two general categories, intrinsic and extrinsic. The intrinsic wastes represent impurities present in the reactants, by-products, co-products, and residues as well as spent materials used as part of the process, ie, sources *(1)–(5)*. These materials must be removed from the system if the process is to continue to operate safely. Extrinsic wastes are generated during operation of the unit, but are more functional in nature. These are generic to the process industries overall and not necessarily inherent to a specific process configuration, ie, sources *(6)–(10)*. Waste generation may occur as a result of unit upsets, selection of auxiliary equipment, fugitive leaks, process shutdown, sample collection and handling, solvent selection, or waste handling practices (see also WASTES, INDUSTRIAL).

Control Strategy Evaluation

There are two broad strategies for reducing volatile organic compound (VOC) emissions from a production facility: *(1)* altering the design, operation, maintenance, or manufacturing strategy so as to reduce the quantity or toxicity of air emissions produced, or *(2)* installing after-treatment controls to destroy the pollutants in the generated air emission stream.

The most widely used approach to exhaust emission control is the application of add-on control devices. For organic vapors, these devices can be one of two types, combustion or capture. Applicable combustion devices include thermal incinerators (qv), ie, rotary kilns, liquid injection combustors, fixed hearths, and fluidized-bed combustors; catalytic oxidization devices; flares; or boilers/process heaters. Primary applicable capture devices include condensers, absorbers, and absorbers, although such techniques as precipitation and membrane filtration are finding increased application. a comparison of the primary control alternatives is shown in Table 1.

The most desirable of the control alternatives is capture of the emitted materials followed by recycle back into the process. However,

Table 1. Emission Control Technologies

Technology	Reduction effectiveness	Recovery	Waste generation	Advantages	Disadvantages
activated carbon adsorption	90–98%	chemical recovery possible with regeneration	spent carbon or regenerant	good for wide variety of VOCs	carbon replacement, regeneration costs, potential for bed fires
adsorption in wet scrubbers	75 90%+	chemical recovery possible through decanting/ distillation	spent solvent or regenerant	simple operation	not efficient at low concentration
vapor condensation	50–80%	chemical recovery possible through decanting/ treatment	liquid wastes, needs off-gas treatment	simple operation, effective for high VOC concentration	low removals applicability limits to some VOCs, high power costs
thermal oxidation	99%	heat recovery	NO_x generation, CO_2 generation	handles any VOC concentration	high operating costs, capital costs, temperatures, and maintenance
catalytic oxidation	95–99%	heat recovery	spent catalyst regeneration acids and alkalines	simple systems, lower T than thermal economical operation	fouling of catalysts, temperature limits

the removal efficiencies of the capture techniques generally depend strongly on the physical and chemical characteristics of the exhaust gas and the pollutants considered. Combustion devices are the more commonly applied control devices, because these are capable of a high level of removal efficiencies, ie, destruction for a variety of chemical compounds under a range of conditions. Although installation of emission control devices requires capital expenditures, these may generate useful materials and be net consumers or producers of energy. The selection of an emission control technology is affected by nine interrelated parameters: *(1)* temperature, T, of the inlet stream to be treated; *(2)* residence time; *(3)* process exhaust flow rate; *(4)* auxiliary fuel needs; *(5)* optimum energy use; *(6)* primary chemical composition of exhaust stream; *(7)* regulations governing destruction requirements; *(8)* the gas stream's explosive properties or heat of combustion; and *(9)* impurities in the gas stream. Given the many factors involved, an economic analysis is often needed to select the best control option for a given application.

Capture devices are discussed extensively elsewhere (see AIR POLLUTION CONTROL METHODS). Oxidation devices are either thermal units that are heat alone or catalytic units in which the exhaust gas is passed over a catalyst usually at an elevated temperature. These latter speed oxidation and are able to operate at temperatures well below those of thermal systems.

Oxidization Devices

Thermal Oxidation. Thermal oxidation is one of the best known methods for industrial waste gas disposal. Unlike capture methods, eg, carbon adsorption, thermal oxidation is an ultimate disposal method destroying the objectionable combustible compounds in the waste gas rather than collecting them. There is no solvent or adsorbent of which to dispose or regenerate. On the other hand, there is no product to recover. A primary advantage of thermal oxidation is that virtually any gaseous organic stream can be safely and cleanly incinerated, provided proper engineering design is used (see INCINERATORS).

A thermal oxidizer is a chemical reactor in which the reaction is activated by heat and is characterized by a specific rate of reactant consumption. There are at least two chemical reactants, an oxidizing agent and a reducing agent. The rate of reaction is related both to the nature and to the concentration of reactants, and to the conditions of activation, ie, the temperature (activation), turbulence (mixing of reactants), and time of interaction.

Some of the problems associated with thermal oxidizers have been attributed to the necessary coupling of the mixing, the reaction chemistry, and the heat release in the burning zone of the system. These limitations can reportedly be avoided by using a packed-bed flameless thermal oxidizer which is under development.

Catalytic Oxidization. A principal technology for control of exhaust gas pollutants is the catalyzed conversion of these substances into innocuous chemical species, such as water and carbon dioxide. This is typically a thermally activated process commonly called catalytic oxidation, and is a proven method for reducing VOC concentrations to the levels mandated by the CAAA (see CATALYSIS). Catalytic oxidation is also used for treatment of industrial exhausts containing halogenated compounds.

As an exhaust control technology, catalytic oxidation enjoys some significant advantages over thermal oxidation. The former often occurs at temperatures that are less than half those required for the latter, consequently saving fuel and maintenance costs. Lower temperatures allow use of exhaust stream heat exchangers of a low grade stainless steel rather than the expensive high temperature alloy steels. Furthermore, these lower temperatures tend to avoid the emissions problems arising from the thermal oxidation processes.

Critical factors that need to be considered when selecting an oxidation system include *(1)* waste stream heating value and explosive properties. Low heating values resulting from low VOC concentration make catalytic systems more attractive, because low concentrations increase fuel usage in thermal systems; *(2)* waste gas components that might affect catalyst performance. Catalyst formulations have overcome many problems owing to contaminants, and a guard bed can be used in catalytic systems to protect the catalyst; *(3)* the type of fuel available and optimum energy use. Natural gas and No. 2 fuel oil can work well in catalytic systems, although sulfur in the fuel oil may be a problem in some applications. Other fuels should be evaluated on a case-by-case basis; and *(4)* space and weight limitations on the control technology. Catalysts are favored for small, light systems.

There are situations where thermal oxidation may be preferred over catalytic oxidation: for exhaust streams that contain significant amounts of catalyst poisons and/or fouling agents, thermal oxidation may be the only technically feasible control; where extremely high VOC destruction efficiencies of difficult to control VOC species are required, thermal oxidation may attain higher performance; and for relatively rich VOC waste gas streams, ie, having ±20–25% lower explosive limit (LEL), the gas stream's explosive properties and the potential for catalyst overheating may require the addition of dilution air to the waste gas stream.

Catalysts. For VOC oxidation, a catalyst decreases the temperature, or time required for oxidation, and hence also decreases the capital, maintenance, and operating costs of the system.

Catalysts vary both in terms of compositional material and physical structure. The catalyst basically consists of the catalyst itself, which is a finely divided metal; a high surface area carrier; and a support structure (see CATALYSTS–SUPPORTED). Three types of conventional metal catalysts are used for oxidation reactions: single- or mixed-metal oxides, noble (precious) metals, or a combination of the two.

Mechanistic Models. A general theory of the mechanism for the complete heterogeneous catalytic oxidation for low molecular weight vapors at trace concentrations in air does no exist. As with many catalytic reactions, however, certain observations have led to a general hypothesis.

The most cost-effective catalytic oxidation systems require use of a solid catalyst material having a high specific surface area, ie, high surface area per net weight of catalyst. The presence of many small pores necessarily introduces pore transport diffusion resistance as a factor in the overall, or global, kinetics. The overall process consists of: *(1)* transport of reactants from the bulk fluid through the gas film boundary layer to the surface of the catalytic particle; *(2)* transport of reactants into the catalyst particle by diffusion through the catalyst pores; *(3)* chemisorption of at least one reactant on the catalyst surface; *(4)* chemical reaction between chemisorbed species or between a chemisorbed species and a physisorbed or fluid-phase reactant; *(5)* desorption of reaction products from the catalyst surface; *(6)* diffusive transport of products through the catalyst pores to the surface of the catalyst particle; and *(7)* mass transfer of products through the exterior gas film to the bulk fluid.

In principle, any of these steps or some combination can be rate controlling. In practice, temperature plays a primary role in determining the rate-controlling stage. Any comprehensive analysis of actual catalytic oxidation systems of practical interest must include a quantitative understanding of the relative effects of mass transfer (steps, *1,2,6,7*) and surface reaction (steps *3,4,5*). The temperature relationship of these two mechanisms is shown in Figure 1. As a catalyst is heated, conversion of the pollutant is negligible until a critical temperature is reached, then the rate of conversion increases rapidly with rising temperature. This is referred to as the kinetically limited region. Conversion increases in this region because catalytic reaction rates increase with temperature, until the catalyst's normal operating temperature is achieved. Then the conversion rate increases only slightly with further temperature rise in the mass-transfer limited region. At some advanced temperature, the conditions reach a point where thermal oxidation begins to play a role, and the rate of conversion again increases rapidly.

Reaction Rate. The kinetics for a single catalytic reaction can be modeled as $-r_m = k(T)f(C)n$, where $-r_m$ is the rate of the main reaction; $k(T)$ is the rate constant, a function of temperature, $T;f(C)$ is the function of reactant and product concentration, C; and n is the effectiveness factor, which accounts for pore-diffusional resistance.

The volatile organic compounds on the list of hazardous air pollutants under the CAAA have been classified into four main categories: *(1)* pure hydrocarbons (qv), *(2)* halogenated hydrocarbons (see CHLOROCARBONS AND CHLOROHYDROCARBONS), *(3)* nitrogenated hydrocarbons (see CYANIDES), and *(4)* oxygenated hydrocarbons (see ALDEHYDES; ETHERS; KETONES). The compounds in these groups are characterized by oxidation reactions.

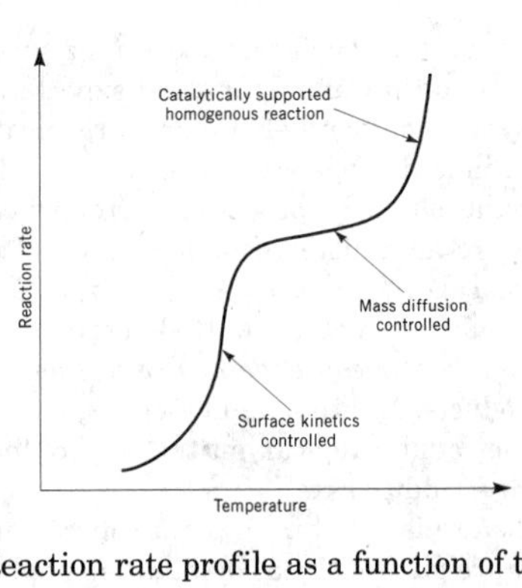

Figure 1. Reaction rate profile as a function of temperature.

Mixture Effects. Care must be taken in determining the oxidation kinetics for a mixture of chemicals. In principle, given one set of conditions and a two-component mixture, the overall conversion of one component A may be controlled by mass transfer to the catalyst surface and the conversion of another component B by surface-reaction kinetics. Of course, the controlling regime (mass transfer or reaction) can change with temperature. Thus for two independent parallel reactions the combined effect of diffusional and reaction rate resistances can have a considerable influence on the relative rate of the two reactions. Additionally, a third, fourth, or nth component can conceivably affect the other components by, for instance, competing more successfully for active surface sites than B while simultaneously influencing the mass transfer of A. Thus even for a simple two- or three-component mixture, interpretation of observed results can be difficult. Extrapolation of mixture behavior from single-component data is ill-advised.

One important consideration of any catalyst oxidation process for a complex mixture in the exhaust stream is the possible formation of hazardous incomplete oxidation products.

Design and Operation. The destruction efficiency of a catalytic oxidation system is determined by the system design. Design and operational characteristics that can affect the destruction efficiency include inlet temperature to the catalyst bed, volume of catalyst, and quantity and type of noble metal or metal oxide used. Catalytic oxidation systems are normally designed for destruction efficiencies that range form 90 to 98%.

Catalyst Inhibition. A number of potential applications for catalytic oxidation of organic materials have resulted in serious odor or eye irritation, or visible emission problems. Some of these failures are a result of fouling of the catalyst surface. Others occur because materials such as halogens in the gas stream interfere with or suppress the activity of the catalyst, or because the substances react with the precious metals, rendering them permanently inactive. Finally, all catalysts eventually deteriorate by aging or thermal processes.

The four basic mechanisms of catalyst decay are fouling or masking, poisoning, thermal degradation through aging or sintering, and loss of catalyst material through formation and escape of vapors. Poisoning and vapor transport are basically chemical phenomena, whereas fouling is mechanical.

Avoiding Catalyst Deactivation. Catalyst deactivation is more easily prevented than cured. Poisoning by impurities may be prevented by removing impurities from the reactants. Carbon deposition and coking may be prevented by minimizing formation of precursors and manipulating mass-transfer regimes so as to minimize the carbon's or coke's effect on activity. Most sintering is irreversible, or reversible only with great difficulty, so it is important to choose reaction, ie, lower temperatures, that do not sinter the catalyst. Additionally, when process upsets that could release inhibitors or cause small fluctuations in the heating value of the oxidizer are highly probable a thermal system is favored over a catalytic one.

Heat- and mass-transfer effects should be avoided because these disguise the intrinsic kinetics. Experiments should be designed to study one deactivation process at a time, and accelerated targets must be representative of the process. Deactivation can be accelerated by using smaller amounts of catalyst, operating at higher temperatures or different pressures, at greater residence times, or at different gas compositions.

Whereas changing catalyst volume or residence time rarely yields complications, changing temperature or pressure could introduce sintering. The properties of the catalyst should be measured both before and after deactivation and inlet and outlet streams should be analyzed by chromatography (qv) or spectrometry.

Catalyst Reactivation. In most processes catalysts inevitably lose activity, and when the activity has declined to a critical level, the catalyst needs to be discarded or regenerated. Regeneration is only possible when the deactivation is reversible by chemical washing or heat treatment or oxidation.

Exhaust Control Technologies

In addition to VOCs, specific industrial exhaust control technologies are available for nitrogen oxides, NO_x, carbon monoxide, CO, halogenated hydrocarbon, and sulfur and sulfur oxides, SO_x.

Nitrogen Oxides. The production of nitrogen oxides can be controlled to some degree by reducing formation in the combustion system. The rate of NO_x formation for any given fuel and combuster design are controlled by the local oxygen concentration, temperature, and time history of the combustion products. Techniques employed to reduce NO_x formation are collectively referred to as combustion controls and U.S. power plants have shown that furnace modifications can be a cost-effective approach to reducing NO_x emissions. Combustion control technologies include operational modifications, such as low excess air, biased firing, and burners-out-of-service, which can achieve 20–30% NO_x reduction; and equipment modifications such as low NO_x burners, overfire air, and reburning, which can achieve 40–60% reduction. As of this writing, approximately 600 boilers having 10,000 MW of capacity use combustion modifications to comply with the New Source Performance Standards (NSPS) for NO_x emissions.

When NO_x destruction efficiencies approaching 90% are required, some form of post-combustion technology applied downstream of the combustion zone is needed to reduce the NO_x formed during the combustion process. Three post-combustion NO_x control technologies are utilized: selective catalytic reduction (SCR); nonselective catalytic reduction (NSCR); and selective noncatalytic reduction (SNCR).

Carbon Monoxide. Carbon monoxide is emitted by gas turbine power plants, reciprocating engines, and coal-fired boilers and heaters. CO can be controlled by a precious-metal oxidation catalyst on a ceramic or metal honeycomb. The catalyst promotes reaction of the gas with oxygen to form CO_2 at efficiencies that can exceed 95%. CO oxidation catalyst technology is broadening to applications requiring better catalyst durability, such as the combustion of heavy oil, coal (qv), municipal solid waste, and wood (qv). Research is underway to help cope with particulates and contaminants, such as flyash and lubricating oil, in gases generated by these fuels.

Halogenated Hydrocarbons. Destruction of halogenated hydrocarbons presents unique challenges to a catalytic oxidation system. The first step in any control strategy for halogenated hydrocarbons is recovery and recycling. However, even upon full implementation of economic recovery steps, significant halocarbon emissions can remain. In other cases, halogenated hydrocarbons are present as impurities in exhaust streams. Impurity sources are often intermittent and dispersed.

The principal advantage of a catalytic oxidation system for halogenated hydrocarbons is in operating cost savings. Catalytically stabilized combusters improve the incineration conditions, but still must employ very high temperatures as compared to VOC combustors.

Uses

Catalytic oxidation of exhaust streams is increasingly used in those industries involved in surface coatings, printing inks, solvent usage, chemical and petroleum processes, engines, cross media transfer, and a number of other industrial and commercial processes.

RONALD L. BERGLUND
The M. W. Kellogg Company

J. C. Summers, J. E. Sawyer, and A. C. Frost "The 1990 Clean Air Act and Catalytic Emission Control Technology for Stationary Sources," in R. G. Silver, J. E. Sawyer, and J. C. Summers, eds., *Catalytic Control of Air Pollution: Mobile and Stationary Sources*," ACS Symposium Series 495, 1992.

M. S. Jennings, N. E. Krohn, and R. S. Berry, *Control of Industrial VOC Emissions by Catalytic Incineration*, Vol. 1, U.S. Environmental Protection Agency, Research Triangle Park, N.C., July 1984.

K. R. Bruns, "Use of Catalysts for VOC Control," presented at the *New England Environmental Expo*, Boston, Mass., Apr. 10–12, 1990.

L. M. Campbell, D. K. Stone, and G. S. Shareef, *Sourcebook: NO_x Control Technology Data*, EPA Report NO. EPA600/S2-91/029., Washington, D.C., Aug. 1991.

EXPECTORANTS, ANTITUSSIVES, AND RELATED AGENTS

Expectorants

Expectorants enhance the production of respiratory tract fluid and thus facilitate the mobilization and discharge of bronchial secretions. Historically, expectorants have been divided into two classes based on specific mechanisms of action. Stimulant expectorants increase respiratory tract secretion by a direct effect on the bronchial secretory cells. Sedative expectorants act by gastric reflex stimulation. Many compounds classed as expectorants have been inadequately studied and the mechanisms of action are not known with certainty.

The clinical effectiveness of expectorants is a topic of significant controversy. The controversy results from a lack of accepted test methods for evaluating expectorants, with a consequent lack of significant objective data to support effectiveness. For most expectorant products, effectiveness is primarily based on a long history of use and a widespread subjective impression. However, the clinical effectiveness of one expectorant, guaifenesin, is supported by objective data. A randomized, double-blind, placebo-controlled study in 40 patients suffering from chronic bronchitis accompanied by productive cough was submitted to the FDA after this expectorant was classified by the FDA's Advisory Review Panel on Over-the-Counter (OTC) Cold, Cough, Allergy, Bronchodilator, and Antiasthmatic Drug Products as Category III, ie, available data insufficient to classify as safe and effective. Statistical data analysis convinced the FDA that guaifenesin loosens and thins sputum and bronchial secretions and makes expectoration easier by increasing sputum volume and reducing sputum viscosity. As a result, guaifenesin is the only expectorant permitted for use in U.S. over-the-counter products.

Guaiacols. Cresote, obtained from the pyrolysis of beechwood, and its active principles guaiacol and cresol have long been used in expectorant mixtures. The compounds are usually classed as direct-acting or stimulant expectorants, but their mechanisms of action have not been well studied.

Guaifenesin (**1**), formerly known as glyceryl guaiacolate, is the synthetic guaiacol derivative that has received the greatest acceptance as an expectorant. This compound is widely used, both in single-entity cough preparations and in combination with other active ingredients. It is the only expectorant permitted for use in U.S. over-the-counter products by the *Code of Federal Regulations* final monograph on expectorant products.

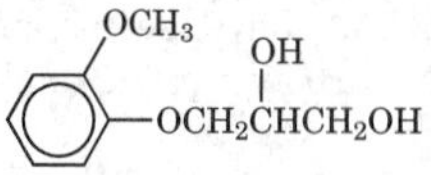

(**1**)

Guacetisal, the acetylsalicylic acid ester of guaiacol, has been shown to retain both antiinflammatory and expectorant activity. It is used in Italy for symptomatic relief of painful respiratory disorders.

Volatile Oils. Limited evidence suggests that these compounds act by direct stimulation of the bronchial secretory cells. Compounds in this category were administered by a number of different routes, including oral, topical, by aerosol, as a cream, and sometimes in lozenges. Volatile oils are not permitted for use in U.S. over-the-counter products by the *Code of Federal Regulations* final monograph on expectorant products. Menthol and camphor, however, are permitted in U.S. over-the-counter topical antitussive products, eg, lozenges, chest rubs, vaporizer additives.

The volatile oils are isolated from plant sources and are terpenoid in structure. They are purified by a combination of physical and chemical processes. Individual components of the oils are often isolated by crystallization or, in some cases, prepared synthetically. Volatile oils include turpentine, oil of lemon, oil of anise, and oil of eucalyptus. Although they are not oils, camphor, menthol, and thymol are the primary components of the volatile oils of camphor tree, mints, and thyme, respectively.

Iodides and Other Inorganic Compounds. Inorganic compounds such as potassium iodide, hydriodic acid, antimony potassium tartrate, and ammonium chloride are thought to act by gastric reflex stimulation.

Miscellaneous Natural Products. Four natural products still used occasionally as expectorants outside the United States are squill, horehound, cocillana, and ipecac.

Mucolytics

Mucolytics reduce the viscosity of tenacious and purulent mucus, thus facilitating removal. The distinction between mucolytics and other classes of expectorants is frequently blurred. Steam, sometimes in conjunction with surfactants or volatile oils, has long been used to decrease viscosity by physical hydration. However, agents that chemically depolymerize certain components of mucus are available. Trypsin and other proteolytic enzymes have shown good clinical activity because of their ability to cleave glycoproteins. Pancreatic dornase, which depolymerizes DNA found in purulent mucus, also has shown clinical utility.

Several mucolytics reduce the viscosity of mucus by cleaving the disulfide bonds that maintain the gel structure, eg, *N*-acetyl-L-cysteine, mesha, bromhexine, and ambroxol. Carbocysteine appears to have a direct action on mucus glycoprotein production.

cis-4-Hydroxymethyl-2-iodomethyl-1,3-dioxolane, a compound structurally similar to one component of iodinated glycerol, was shown to exhibit expectorant activity and mucolytic activity of the same order of magnitude as carbocysteine in rabbits. Decasilate has been shown to possess significant mucolytic activity in rabbits.

Antitussives

Through the centuries, cough remedies have been a popular item and were found in most homes. Today, over 300 prescription and over-the-counter preparations are available, and retail sales of nonprescription cough syrups, expectorants, and cough drops in 1995 are expected to exceed $2 billion.

The Cough Reflex. Coughing is a protective reflex and is one of several important mechanisms for clearing the respiratory tract of excessive secretions and foreign debris, thus ensuring normal gas exchange and minimizing infection. It can be described as occurring in three phases: *(1)* a short deep inspiration of air into the lungs; *(2)* compression of the air by closure of the glottis, and contraction of the thoracic, abdominal, and diaphragmatic muscles; and *(3)* rapid expulsion of the air when the glottis opens.

Coughing can be caused by a variety of irritants, environmental factors, and pathological disorders. The common cold is the most frequent cause of transient coughs. Cigarette smoking is the most common cause of chronic persistent cough.

Theoretically, the act of coughing can be affected directly or indirectly by one or more of the following: *(1)* elimination of the ultimate cause of the cough by treating the responsible pathological condition, or by removing the anatomical or environmental irritant; *(2)* raising the threshold for stimulation of cough peripherally by anesthetizing sensory nerve endings in the respiratory tract; *(3)* interruption of the sensory impulses to the medulla; *(4)* specific depression of the cough center in the medulla; *(5)* interruption of conduction along the motor pathways; *(6)* nonspecific depression of the central nervous system; and *(7)* facilitating bronchial drainage and mucociliary clearance. Most therapeutic agents act by one or more of mechanisms *(2)*, *(4)*, and *(7)*.

Centrally Active Antitussives. Centrally active antitussives depress the medullary cough center, thus raising the threshold for sensory cough impulses. The most well-known compounds in this category are the narcotics. Unfortunately, many of them have the disadvantage of addiction. Molecular modifications of the morphine skeleton and the synthesis of totally new structures have produced more specific drugs without the disadvantage of addiction. Many of the synthetic compounds also possess other useful properties including local anesthetic and antispasmodic activity. Narcotic antitussives include codeine (**2**), ethylmorphine, pholcodine, hydromorphone, hydrocodone, dihydrocodeine, butorphanol, and drotebanol.

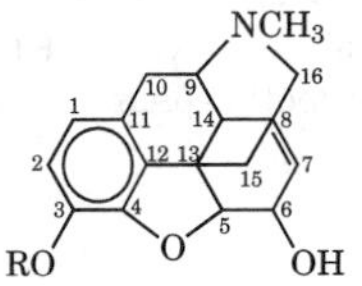

(**2**) R = CH_3

Among the nonopiate narcotics, two compounds, methadone and meperidine, have shown antitussive activity. However, the addiction potential of these two agents limits their clinical use for treating cough. Side effects associated with narcotics include nausea, anorexia, and constipation; most of them also diminish ciliary activity and produce a drying effect on the respiratory tract mucosa.

Nonnarcotic Antitussives. The most centrally active, nonnarcotic antitussive is dextromethorphan (**3**). It is similar to codeine in terms of potency and mechanism of action, ie, it is a direct depressant of the cough center. It is unique in that even though it is structurally related to codeine, it is not addictive. Further, dextromethorphan, the dextrorotatory optical isomer of 3-methoxy-*N*-methylmorphinan, possesses only antitussive activity, whereas the levorotatory isomer, levorphan, possesses both analgesic and antitussive activity. D-3-Methyl-*N*-methylmorphinan has also been synthesized and is reported to be a potent antitussive.

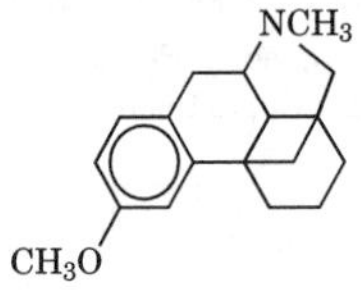

(**3**)

Included in the nonnarcotic class of antitussives are many compounds that do not possess a morphine skeleton and which vary widely from each other with respect to structural features and pharmacologic profiles. They include levopropoxyphene, noscapine, benzonatate, dimethoxanate, pipazethate, chlophedianol, the citrate salt of isoaminile, caramiphen, oxeladin, diphenhydramine, picoperine, oxalamine, zipeprol, ethyl dibunate, diazepam, clonazepam, nifedipine, verapamil, flunarizine, glaucine, fominoben, arbutin, and Δ′-tetrahydrocannabinol.

Except for the addiction liability of some of the narcotic antitussives, side effects for most of the centrally acting compounds are relatively few and mild at therapeutic doses. Qualitative comparisons of both side effects and pharmacological profiles have been summarized for many of the compounds listed above.

Peripherally Active Antitussives. Peripherally active antitussives act by raising the threshold for cough at the sensory nerve endings, or by facilitating bronchial drainage, mucociliary clearance, or both. Agents that act by the first mechanism include local anesthetics that desensitize the mucosa of the respiratory tract, and antispasmodics that relax the smooth muscles of the bronchi (see NEUROREGULATORS). Most of these compounds also have a central antitussive component as a part of their pharmacological profile and have been previously discussed. Expectorants and mucolytics act primarily by the second mechanism.

Economic Aspects

Sales figures for expectorants and antitussives are usually combined under the general headings of cough preparations or cough and cold preparations. Antitussives and expectorants are frequently formulated together for the treatment of cough, or formulated with antihistamines and bronchodilators for the treatment of cold symptoms. At the retail level, sales of nonprescription cough syrups, elixirs, and expectorants at all outlets were $365.2 million in 1991 compared to $304.8 million in 1978. Sales of prescription cough preparations and expectorants were $163.1 million in 1991. Among narcotic antitussives, codeine is by far the leading product. However, sales of codeine-containing cough preparations have declined in recent years, probably because of the increased acceptance of nonnarcotic antitussives such as dextromethorphan.

Health and Safety

Safety and efficacy data on a number of antitussives and expectorants have been reviewed by the FDA's Advisory Review Panel on Over-the-Counter (OTC) Cough, Cold, Allergy, Bronchodilator, and Antiasthmatic Products. The conclusions and recommendations regarding the effectiveness, safety, labeling, and suitability for marketing of over-the-counter preparations have been reported. After review of these recommendations, FDA has issued final monographs for over-the-counter antitussives and for expectorants. LD_{50} data for most of the compounds described have been reported.

DAVID B. PAUL
ROBERT H. DOBBERSTEIN
Sandoz Pharmaceuticals Corporation

A. G. Gilman, L. S. Goodman, T. W. Rall, and F. Murad, eds., *Goodman and Gilman's The Pharmacologic Basis of Therapeutics*, Macmillan Publishing Co., New York, 1985, 1839 pp.

J. E. F. Reynolds and A. B. Prasad, eds., *Martindale The Extra Pharmacopoeia*, 28th ed., The Pharmaceutical Press, London, 1982, 2025 pp.

R. J. Lewis and R. L. Tatken, eds., *Registry of Toxic Effects of Chemical Substances*, Vols. I and II, Department of Health and Human Services, Washington, D.C., 1980.

EXPERT SYSTEMS

At a basic level of description, expert systems are computer programs. However, they are different from most other computer programs in several ways. Functionally, they differ because they can perform problem-solving or decision-making tasks in some well-defined domain at a performance level comparable to human experts. In terms of content, expert systems primarily encode and manipulate symbolic knowledge about some domain or some type of problem-solving task, rather than mathematical equations, algorithms, or numerical data. Because of their emphasis on knowledge rather than on numerical computation, expert systems are often known, more appropriately, as knowledge-based systems. Knowledge-based systems are characterized by their dependence on fairly specialized knowledge, of the kind that humans accumulate through experience, understanding, or insight. By contrast to mathematical models, which describe physical and chemical phenomena, knowledge-based systems are qualitative models that capture human decision-making processes.

Importance of Expert Systems. Advantages of knowledge-based technology include the following: knowledge-based systems can automate complex tasks and do them quickly, even in fairly demanding domains, at a level comparable to their human counterparts; they can handle large amounts of data without suffering from information overload; they are tireless and potentially available round-the-clock; they are consistent, ie, if the inputs are the same, so are the outputs; they provide a readily available, lasting repository of expertise; and they have had proven successes in a variety of tasks, from diagnosis to design to data interpretation, and a variety of domains, from medicine to finance to chemical engineering.

Knowledge-Based Systems in Chemical Engineering. In the chemical engineering domain, knowledge-based technology has been applied mainly to the tasks of diagnosis and design.

Examples of diagnostic applications include Diad-Kit/Boiler, an on-line expert system for performance monitoring and diagnosis of steam boilers, and CatCracker, a knowledge-based system for diagnosing operating problems in fluidized catalytic cracking units. Examples of design applications include BioSep Designer, which can automatically configure a process flow sheet and size the equipment involved in downstream separation of fermentation products; and Still, a tool for automated process and mechanical design of distillation columns.

Apart from design and diagnosis, other types of problem-solving tasks have also received attention in the chemical engineering domain, although not to the same degree. For example, knowledge-based reasoning techniques have been applied to simulation of discrete event processes, hazard and operability studies, intelligent multivariable process control, process operations planning, and scheduling of batch processes.

Technology Overview

From a technology perspective, knowledge-based systems (KBS) represent a new software methodology for solving certain types of problems effectively. It is important to understand what is encoded in knowledge-based systems, and how KBS technology differs from conventional numeric computational techniques.

Representation. From a software viewpoint, knowledge-based technology provides ways to represent knowledge and reason with it. The knowledge to be represented includes facts, descriptions, relationships, and problem-solving knowledge.

Reasoning. Knowledge-based systems need ways to manipulate their representations to produce useful results. Representation and reasoning are closely intertwined; how knowledge is represented influences and constrains the types of reasoning methods that can be brought to bear on the representations. Some fundamental techniques for reasoning are deduction, pattern matching, and search.

Comparison to Conventional Numerical Programs. Fundamentally, knowledge-based systems manipulate symbols that have some meaning in the external world. They typically use qualitative methods rather than quantitative ones. Knowledge-based systems also make use of heuristic strategies for arriving at feasible solutions. There may even be more than one correct solution for the problem. By contrast, conventional numeric programs are usually oriented toward finding exact or optimal solutions.

Human cognitive tasks also have to cope with uncertainty. For example, real world constraints are subject to change without notice, one's knowledge may apply only partially to a particular situation, and the data needed to infer conclusions may be incompletely known. Knowledge-based systems deal with such problems by explicitly representing and reasoning with uncertainty. By contrast, conventional numeric techniques are applied to well-defined problems that can be mathematically characterized. Uncertainty is implicitly dealt with by making simplifying assumptions, or using empirically determined model parameters. The key difference is that the computational process does not deal with the uncertainty; it is the developer of the computational algorithm who does. These remarks are applicable primarily to deterministic numeric models. Stochastic models do explicitly address uncertainty by incorporating random distributions.

The types of data structures and processing techniques used in knowledge-based systems are different from those used in conventional, numerical programs. For example, knowledge-based systems use data structures such as rules and objects and process these using pattern matching, inheritance, and message passing. Often, knowledge-based systems are data-driven, ie, processing is initiated in response to changes in data, not in a predetermined sequence. By contrast, conventional computer programs use data structures that are more suited for representing, storing, and manipulating numbers, eg, arrays, records, fields, and tables. The types of processing techniques used include arithmetic operations, logical comparisons, iterations, and relational calculus. In general, processing tends to be program-driven, ie, a sequence of instructions acts upon data in a predetermined fashion.

Knowledge-based systems differ from conventional programs in the types of applications for which they are used. Knowledge-based technology is most appropriate for implementing tasks that have some cognitive equivalent.

Theory

From the viewpoint of AI and cognitive science, the theoretical question underlying knowledge-based technology is how to model expert problem-solving behavior. The predominant models of representation and reasoning in AI are logic, rules, and objects. More recent approaches to knowledge-based systems include task-specific models.

Logic. The rule-based representation used in many expert systems has its roots in formal logic. Logic programming has been applied successfully to build knowledge-based applications for certain problems. The logic approach supports a declarative style of representation; knowledge about a domain is represented in the form of facts or axioms. The fact base can then be queried about the truth or falsehood of some proposition to be tested. The logic-based system sets up the test proposition as a goal to prove, and uses rules of inference to do so. The most general of the inference rules is a method called resolution theorem proving.

At the implementation level, the Prolog language is probably the most popular software for developing logic-based expert systems. In the chemical engineering domain, the logic-based approach has not been applied extensively. In the United States, rule-based and object-based systems have enjoyed much greater popularity.

Rules. Rules, first pioneered by early applications such as Mycin and R1, are probably the most common form of representation used in knowledge-based systems. The basic idea of rule-based representation is simple. Pieces of knowledge are represented as IF–THEN rules. IF–THEN rules are essentially association pairs, specifying that IF certain preconditions are met, THEN certain fact(s) can be concluded. The preconditions are referred to as the left-hand side (LHS) of the rule, while the conclusions are referred to as the right-hand side (RHS). In simple rule-based systems, both the preconditions and conclusions are variable–value pairs. In general, both the preconditions and conclusions can have conjunctions (AND), disjunctions (OR), and negations (NOT). Conjunctions make a rule more specific, whereas disjunctions make a rule more general. A rule with a disjunction is equivalent to two separate rules with the same conclusion.

Figure 1 shows the architecture of a rule-based system. Fundamentally, there are three components: a rule base, a working memory, and an inference engine. The rule base contains discrete fragments of knowledge, all expressed as IF–THEN associations over variables. The working memory contains facts or assertions that are true at any stage in the computational process, ie, it contains variable–value bindings. The inference engine is a domain and knowledge-independent processing mechanism. The inference engine searches the rule base, matching variable–value bindings in the working memory to preconditions and conclusions of rules in the rule base. When a match occurs, the rule is said to have fired.

Rules may represent either guidelines based on experience, or compact descriptions of events, processes, and behaviors with the details and assumptions omitted. In either case, there is a degree of uncertainty associated with the application of the rule to a given situation. Rule-based systems allow for explicit ways of representing and dealing with uncertainty. This includes the representation of the uncertainty of individual rules, as well as the computation of the uncertainty of a final conclusion based on the uncertainty of individual rules, and uncertainty in the data.

Rule-based systems address the issue of explaining conclusions as well. During the inference engine's processing, two types of explanatory questions can be answered: why and how. Why questions arise when the system makes requests for data in order to validate hypotheses. The system responds to this type of question by referencing the rule which has the data under question. Similarly, a rule-based system responds to the how question by tracing through all the rules that led to the final conclusion, and referencing them.

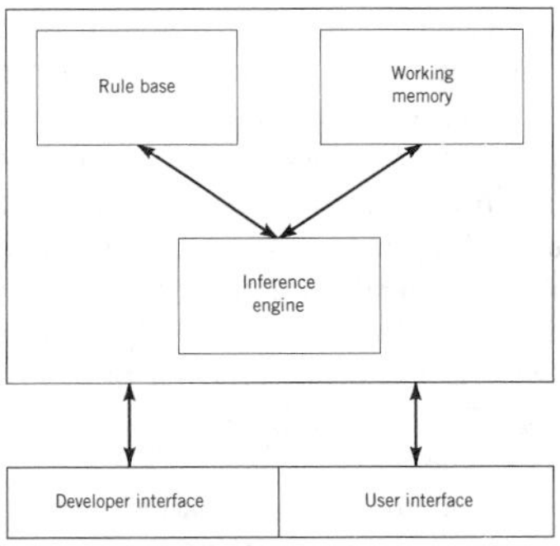

Figure 1. Architecture of a rule-based system.

Shortcomings. Rule representations based solely on variable–value bindings are poor at generalization. Reasoning mechanisms other than IF–THEN inference are hard to encode. Certain types of relationships (eg, taxonomic and causal) essential to capturing knowledge about a domain are outside the scope of a rule-based representation.

Object-Oriented Representation. Object-oriented representation addresses many of the shortcomings of rules, especially when used in combination with a rule-based representation. In contrast to the type of data structures used in traditional programming, eg, variables, records, or linked lists, the basic data structure in object-oriented systems is an object with associated properties. In object-oriented systems, the notion of inheritance is important. Inheritance means that a property of a class is automatically acquired by subclasses and objects of that class.

Another important idea in object-oriented systems is being able to specify how slots can obtain values and what types of values they can hold. This allows specification of data types for slots, initial and default values, and restrictions on value ranges. Also, procedures to execute for obtaining the value can be specified, or even procedures to execute after getting the value. These types of procedures are referred to as daemons.

These discussions address the issue of representation using objects, but what about reasoning? One approach is to couple the object representation with a rule-based system and let the associated inference engine take care of the reasoning. This is the technique used in a number of knowledge-based applications, and it successfully addresses many limitations of pure rule-based systems. However, some of the other restrictions of rule-based systems, namely limited problem-solving capabilities and the inability to explicitly structure the reasoning strategy, still remain. To address these, object-oriented systems offer powerful capabilities for implementing many different reasoning strategies, via encapsulation and message passing. Encapsulation allows procedures and data to be bundled together into objects. Rather than data being passive, and acted upon by procedures, encapsulation allows data to be active, and procedures to work differently depending on the object they are attached to. Daemons are an example of this. The second feature, message passing, allows objects to communicate. The overall reasoning process is accomplished as follows: objects send messages to other objects; the recipient objects respond by executing their associated procedures; results or values are returned to the sending objects. Using these advanced features of object systems requires the use of an integrated programming language such as Lisp or C. These types of capabilities are somewhat on the fringes of mainstream knowledge-based systems technology. They are more applicable in the area of model-based systems.

One of the many advantages of an object-oriented representation is the fact that it allows representation of all the declarative information known about a process or domain in a structured way. Representing the domain knowledge declaratively offers the flexibility to choose whatever reasoning strategy is appropriate for the problem to be solved. In addition, the idea of inheritance greatly facilitates building and maintaining systems using an object-oriented representation. On the downside, object systems do not generally come with built-in problem-solving strategies, uncertainty handling capability, or explanation features. This is up to the system developer to implement using a programming language. For a range of knowledge-based applications, this potential hurdle can be avoided by coupling the object system with a rule base. Many commercially available tools support this hybrid architecture. For other types of problems, the hybrid approach may not be sufficient, and the full power of object-oriented programming may have to be used.

Advanced Topics. Much of the knowledge in mainstream knowledge-based systems is associational in nature. Causal relationships, for example, are encoded as direct associations between symptoms and root causes, suppressing the details of any intermediate causal links. Knowledge in this form is also referred to as compiled knowledge. The advantage of this approach is that at problem-solving time, the knowledge is available to the system in the form needed. No explicit reasoning is required to derive the association. The downside is that if the knowledge-based system runs into a situation where those details become important, then it may fail. This is the brittleness problem.

Model-based reasoning attempts to address the brittleness problem by modeling the behavior of the important entities in the domain. The objective of this modeling effort is to enable the reasoning system to derive any associational knowledge at problem-solving time in a flexible way. Modeling reduces the burden of having to explicitly represent the behaviors of the entities in a compiled form, but requires additional development time. The advantage is that the modeling effort can be leveraged for future applications, eg, if generic models are developed for process equipment, then they can be reused for other processes.

In general, the model-based reasoning approach is best applied, not as a method in itself, but as an add-on to a knowledge-based system. The main reason is that modeling is hard, and problem-solving based solely on fundamental models is computationally complex.

Rules and objects are examples of general problem-solving approaches, ie, the representations and reasoning constructs can be applied to different tasks. The work of mapping the needs of the task to the constructs available is left to the developer of the knowledge-based system. This can be a difficult knowledge engineering problem. Issues of how to appropriately represent the task-level knowledge and how to structure the design rules in the rule base, for example, have to be resolved. Thus, rules and objects can be regarded as fine-grained approaches to representation and reasoning, in the sense that they deal with problem-solving at a low level.

Task-specific problem-solving approaches address these issues of knowledge engineering, representational ease, and representational adequacy. Such approaches are oriented toward characterizing the representation and reasoning needs of different cognitive tasks in a generic way. The advantage is that if a knowledge-based system has to be constructed for such a task, the developer has a ready-made blueprint to work with. The blueprint specifies what representational constructs are required, what reasoning strategies are appropriate, what knowledge is required to build the application, and how to go about acquiring the knowledge. Some researchers in this area have also focused on building tools specific to the tasks, so that even the representation and reasoning methods are available to the developer in a prepackaged form.

Reasoning with time and having to solve problems under time constraints place unique burdens on a problem-solving system. To address these, knowledge-based systems need additional capabilities: problem-solving strategies to deal with the effect of time, increased processing speed, and enhanced integration capabilities, to name a few. Although real-time often means different things to different people, there is general agreement among researchers on the types of issues that real-time systems need to address. These are speed, responsiveness, pre-emption, adaptability, and temporality.

Building Applications

The purpose of this section is to provide some understanding of where and how to appropriately apply knowledge-based technology, and to give an industrial perspective on the process of developing and delivering knowledge-based applications. The literature contains numerous detailed discussions concerning knowledge-based system development.

Technical and Business Issues. In general, the successful application of any technology is dependent on both technical and business issues. For knowledge-based systems, six critical issues are problem identification, user acceptance, measurement of impact, appropriateness, feasibility, and cost.

Approaches to Development. In general, there are two approaches to knowledge-based systems development: the do-it-yourself approach, and the technology approach. In the do-it-yourself approach, the experts, who are the knowledge sources, learn enough about the technology and the tools to construct their own knowledge-based systems. This approach is successful when applied to small, well-defined problems, but education in the technology is a critical prerequisite. The do-it-yourself approach is also well-suited to highly decentralized organizations, where the burden of development and deployment is often on local groups of users.

The technology approach is one in which knowledge-based system professionals interact with experts and end users to develop applications. This is often the only option if the application involves even moderate complexity, and low end expert system shells prove to be inadequate to represent the knowledge or the reasoning. Irrespective of the approach taken, the development process can be roughly broken down into the eight steps: problem selection, problem analysis, application design, knowledge acquisition, prototyping, validation, delivery, and follow-up.

Implementation Tools. Choosing an appropriate tool for knowledge-based application development is not easy. A number of knowledge level criteria have to be considered, including representational capabilities, eg, rules and objects; knowledge organization capabilities, eg, rule sets and taxonomies; reasoning capabilities, eg, backward and forward chaining, rule prioritization; uncertainty handling features, eg, certainty factors and belief values; truth maintenance or assumption maintenance capabilities; and special-purpose features such as temporal reasoning capabilities, which may be needed for real-time applications, or planning and scheduling systems.

In addition, other capabilities relevant to system development, end user interaction, and application integration have to be taken into account: development features, eg, graphic displays of knowledge organization, knowledge-base maintenance features, explanation facilities, system documentation capability, end user interface development tools, availability of an integrated programming language for custom coding, access to external data files, interface to external programs, hardware platforms supported, vendor reliability and service, and price.

In selecting an appropriate tool, two other factors may also be important. First, it may be beneficial to leverage higher initial cost for a more full-featured tool, if future applications are likely to need those capabilities. Second, more than one tool may be necessary to serve the needs of different types of applications, ie, the adage "one size does not fit all" holds well in the tool business. Available tools in the market can be categorized into roughly three classes: low end tools, general-purpose tools of medium complexity, and high end tools and special-purpose tools.

Alternative Technologies

For a number of cognitive or interpretive tasks, there are alternatives to mainstream knowledge-based systems that may be more appropriate, especially if adaptive behavior and learning capability are important to system performance. Two approaches that embody these characteristics are neural networks (nets) and case-based reasoning.

Neural Nets. Neural nets arose out of an alternative approach to solving the problems raised by AI. Instead of modeling high level reasoning processes, neural networks attempt to produce intelligent behavior using the brain as the architectural model. The motivation behind neural network research is the modeling of cognitive activities such as perception, learning, and data interpretation, which symbolic AI approaches do not address adequately. From a practical standpoint, neural nets attempt to address many limitations of knowledge-based systems, including lack of adaptability (self-modification in response to a changing world); lack of robustness (tolerance for missing, bad, or incomplete information); problems of knowledge acquisition (extracting knowledge from experts); problems of storage capacity (amount of knowledge that can be stuffed into a knowledge base); problems of scalability (performance of large systems in comparison to small ones); and problems of speed, especially with increase in the size of the system.

Neural nets consist of many individual computing elements that are interconnected. Each individual computing element, or node, is computationally simple, yet complex behavior can arise out of the multiplicity of the nodes and their interconnectivity. The physical realization of this computing machinery can either be in hardware form, or in the form of a software emulation run on a conventional computer. Instead of symbolic representations, the knowledge inside a neural net is distributed all over the network in the form of connection strengths between computing elements.

In general, neural nets are particularly good at pattern recognition, classification, and data interpretation tasks.

When considering neural network technology as an alternative to knowledge-based systems, there are certain limitations of neural nets that should be recognized. Neural nets, in general, do not handle time well. Their design does not support end-user interaction, eg, asking the user for missing data. They do not have explanation capability. Their performance is usually highly dependent on the amount and quality of training data used.

In the chemical engineering domain, neural nets have been applied to a variety of problems. Examples include diagnosis, process modeling, process control, and data interpretation. Industrial application areas include distillation column operation, fluidized-bed combustion, petroleum refining, and composites manufacture.

Case-Based Reasoning. Case-based reasoning is a relatively new alternative to expert systems. The case-based approach is founded on the observation that humans often solve problems by drawing on specific cases or episodes from their past experience. As a simplification, one might say that the emphasis in case-based reasoning is on memory rather than knowledge. The types of memory addressed include semantic memory (how static facts, and their relation to one another, are stored), episodic memory (how information about temporal events is stored), and conceptual memory (how concepts are stored in episodic form). By modeling memory, case-based reasoning theory attempts to cohesively explain many aspects of cognition, learning, and problem solving, bridging research in both AI and psychology.

Case-based reasoning, as a general model of problem solving, has several advantages: knowledge acquisition, although not eliminated, is simplified, eg, only actual episodic cases have to be extracted from the expert, not generalized rules; problem solving behavior can evolve through repeated application of the case memory, ie, CBR systems can learn from experience; and problem solving behavior can be robust, ie, even if retrieved cases fail to apply exactly to a new situation, they may work after modification.

At the same time, there are also many open research issues for which general solutions are not yet available, including the content and structure of cases, the organization and indexing of cases, the problems associated with search as case memories grow larger, the evolution of indexing schemes with experience, metrics for deciding which cases are applicable to a new situation, mechanisms for modifying cases, and mechanisms for repairing solutions when modified cases fail to work.

At the present time (ca 1995), there are almost no well-known applications of CBR technology in the chemical engineering domain. A few industrial applications in other domains have been reported. The applications cover problem-solving tasks such as scheduling, planning, design, and diagnosis. A few case-based development tools have also emerged. These include CBR-Express (Inference Corporation) and Remind (Trinzic Corporation).

T. S. Ramesh
Mobil Research and Development Corporation

J. P. Ignizio, *Introduction to Expert Systems*, McGraw-Hill Book Co., Inc., New York, 1991.

M. Mavrovouniotis, ed., *Artificial Intelligence Applications in Process Engineering*, Academic Press, Inc., New York, 1990.

A. E. Nisenfeld, ed., *Artificial Intelligence Handbook*, Vols. 1 and 2, Instrument Society of America, Research Triangle Park, N.C., 1989.

W. A. Taylor, *What Every Engineer Should Know About Artificial Intelligence*, MIT Press, Cambridge, Mass., 1988.

EXPLOSIVES AND PROPELLANTS

EXPLOSIVES

Propellants and explosives are chemical compounds or mixtures that rapidly produce large volumes of hot gases when properly initiated. Propellants burn at relatively low rates measured in centimeters per second; explosives detonate at rates of kilometers per second.

Propellants and explosives in large-scale use are based mostly on a relatively small number of well-proven ingredients. Propellants and explosives for military systems are manufactured in the United States primarily in government-owned plants, where they are also loaded into munitions. Composite propellants for large rockets are produced mainly by private industry, as are small arms propellants for sporting weapons.

Explosives and propellants have relatively large amounts of available energy stored compactly and readily deliverable. The power output depends on the rate at which energy is liberated. Propellants are used wherever a readily controllable source of energy is required for periods of time ranging from milliseconds in guns to seconds in rockets. The gases evolved are employed as a working fluid for propelling projectiles and rockets, driving turbines, moving pistons, shearing bolts and wires, operating pumps, and starting engines. Explosives are used wherever very rapid rates of energy application and high pressures are essential. They are employed to produce high intensity shock waves in air, water, rock, and metal; for blasting, cratering, mining, and other civil engineering purposes; for metal welding and forming, cutting, and fragmentation; in shaped charges and many specialty devices requiring high rates of energy transmission; and for initiation of detonation phenomena. The terms burning and deflagration are often used synonymously to describe the gradual consumption of a propellant grain by a flame off the surface and to contrast it with the much more rapid, violent, and destructive phenomena associated with the detonation of explosives.

The development of explosives began in Europe with the formulation and use of black powder in about the middle of the thirteenth century, accelerated in the nineteenth century with the nitration of many compounds to produce high energy explosives, and greatly intensified during World War II. There has been enormous growth in the explosives field in the latter half of the twentieth century made possible in part by modern electronic instrumentation, high speed photography (qv), computers, military and space research and the opportunities in worldwide mining and civil engineering programs. The most common explosive compounds are listed in Table 1.

General Characteristics

Exothermic oxidation–reduction reactions provide the energy released in both propellant burning and explosive detonation. The

Table 1. Explosive Substances

Name	Code	Use
	Inorganic salts	
ammonium nitrate	AN	solid oxidizer
ammonium perchlorate		solid oxidizer
lead azide		primary explosive
ammonium picrate	AP	secondary high explosive
2,4-diamino-1,3,5-trinitrobenzene	DATB	secondary high explosive
diazodinitrophenol	DDNP	primary explosive
ethylene glycol dinitrate	EGDN	liquid explosive
lead styphnate		primary explosive
2,4,6-trinitrotoluene	TNT	secondary high explosive
picric acid	PA	secondary high explosive
	Aliphatic nitrate esters	
mannitol hexanitrate	MN	primary explosive
nitrocellulose	NC	secondary explosive used in propellants
nitroglycerin	NG	liquid secondary explosive ingredient in commercial explosives and propellants
nitromethane	NM	liquid secondary explosive
pentaerythritol tetranitrate	PETN	secondary high explosive used as booster
	Nitramines	
cyclotrimethylene-trinitramine	RDX	secondary high explosive
trinitrophenylmethyl-nitramine	tetryl	secondary explosive used as booster
ethylenedinitramine	EDNA	secondary high explosive
tetrazene		primary explosive
tetranitromethane	TNM	liquid explosive
cyclotetramethylenetetra-nitramine	HMX	secondary high explosive

reactions are either internal oxidation–reductions, as in the decomposition of nitroglycerin and pentaerythritol tetranitrate, or reactions between discrete oxidizers and fuels in heterogeneous mixtures.

An activation energy of 125–250 kJ/mol (30–60 kcal/mol) is usually required to initiate the reaction. Once initiated, the heat evolved is sufficient to cause the reaction to continue and become self-sustaining. Most explosives and propellants are organic compounds or mixtures of compounds that contain carbon, hydrogen, oxygen, and nitrogen. Metallic fuels such as aluminum may be added to increase the heat of reaction. The most common gaseous products of the oxidation–reduction reactions are hydrogen, water, carbon monoxide, carbon dioxide, and nitrogen. Other products depend on the reactants involved.

The specific stimulus that triggers an explosive detonation or propellant burning depends on the material involved and the system environment. In most cases, heat is the ultimate cause of the activation. Gun and rocket propellants are initiated by igniter compositions that produce hot gases and solids at relatively low pressures. Explosives are detonated by high pressure, high temperature shock waves that heat the explosive to rapid reaction temperatures by adiabatic compression. Mechanical impact, frictional forces, electric discharge, and other ultimate sources of heat may also act as initiating stimuli. The minimum quantity of energy required for initiation depends on the chemical characteristics of the material, its physical properties including mass, geometry, density, the degree of confinement, the rate at which the energy is delivered, the environment in which the energy release occurs, and the type of initiating stimulus provided. Explosives and propellants are often ranked in order of their sensitivity and response to a specific stimulus in a given environment.

Propellant Burning. Propellants generally operate at low pressures down to about 3–4 MPa (500 psi) in rockets and up to about 689 MPa (100,000 psi) in high performance guns (see EXPLOSIVES AND PROPELLANTS–PROPELLANTS). The process is characterized by a reaction front that moves in a direction normal to the exposed surface of the grain, proceeding from the outside in laminar layers. The rate of burning depends on the intrinsic rate of decomposition of the propellant formulation and the rate of heat transfer from the hot gases above the propellant surface. One equation that defines the rate of propellant burning is $dx/dt = bP^n$, where x = a coordinate normal to the grain surface, t = time, b = a constant that depends on the propellant temperature and composition, P = pressure at which burning occurs, and n = pressure exponent that depends on the propellant composition and the pressure. Values of n range from nearly 0 to 0.5 for low pressure rocket propellants, and 0.7 to 1.0 for high pressure gun propellant.

Explosive Detonation. Detonations proceed as a result of a reaction front moving in a direction normal to the surface of the explosive. The detonation rate, D, is stable and constant and is primarily governed by the physical and chemical properties of the explosive, the diameter of the charge, its degree of confinement, and particularly its density, p, with which it varies in a linear fashion for most explosives: $D_i = D_o + M(\rho_i - \rho_o)$, where D_i = linear detonation rate, dx/dt, at density ρ_i, D_o = linear detonation rate at density ρ_o, and M = a constant characteristic of the explosive composition. Typical values of D_o at ρ_o = 1.0 g/cm are about 5000–6000 m/s. Values of M are about 3000/4000 m/s.

Deflagration to Detonation Transfer. The same compound or mixture may burn or detonate, depending on the type and intensity of initiation, the degree of confinement, and the physical and geometric characteristics of the material.

Because unwanted detonations of propellants are likely to be catastrophic, the conditions at which deflagration to detonation transformations occur have been intensively studied. A simplified view of the process by which a deflagration is transformed to a detonation suggests that an initial burning of the type associated with propellant combustion occurs. This is followed by convective burning in which the hot gases penetrate the pores of the explosive. The combustion front may then build up into shock-wave pressures that produce a low velocity detonation which is rapidly converted to a full-fledged detonation with its characteristic shock pressure and temperature. Conditions that minimize energy losses and increase the likelihood of buildup of a shock wave tend to enhance the likelihood of a deflagration converting to a detonation.

Pyrotechnic Compositions. Pyrotechnic compositions engage in oxidation–reduction reactions that resemble those of propellants and explosives, but generally produce little or no gas. These are heterogeneous mixtures of a finely powdered metal, metal alloy, or organic fuel and inorganic oxidizers. Such compositions are commonly used for flares, signals, tracers, incendiaries, delays, igniters, heating mixtures, and in devices where the formation of much gas is unacceptable either because the gas pressure causes unwanted changes in the reaction rate or the system is not designed to withstand the pressure without rupturing. The properties of typical pyrotechnic and explosive compositions are compared in Table 2.

Safety Considerations

The catastrophic effects and the increasingly severe legal and economic implications of a disastrous explosion have caused a large effort to be devoted to the safety of explosive operations. Many governmental regulations control the classification, shipping, and handling of explosive materials. The publications in the field of safety have increased greatly, and numerous symposia have been held on the subject. There is an annual explosive safety seminar conducted by the Explosives Safety Board of the United States Department of Defense.

The hazard posed can be limited by maintaining a zone free of people and property around a storage area of explosive material. The minimum radius of the zone depends on the type and quantity of explosive, the extent and type of barricading, and the magnitude of loss

Table 2. Comparison of Pyrotechnic Compositions with High Explosives

Component	Composition, wt %	Heat of reaction, kJ/g[a]	Gas volume,[b]	Relative brisance, % TNT	Ignition temp,[c] °C	Impact test,[d] % TNT
		Pyrotechnic composition				
delay						
barium chromate	90					
boron	10	2.010	13	0	450	12
delay						
barium chromate	60					
zirconium–nickel alloy	26					
potassium perchlorate	14	2.081	12	0	485	23
flare						
sodium nitrate	38					
magnesium	50					
laminac	5	6.134	74	17	640	19
smoke						
zinc	69					
potassium perchlorate	19					
hexachlorobenzene	12	2.579	62	17	475	15
photoflash						
barium nitrate	30					
aluminum	40					
potassium perchlorate	30	8.989	15	15	700	26
		High explosive				
TNT		4.560	710	100	310	100
RDX		5.694	908	140	260	35

[a] To convert J to cal, divide by 4.184.
[b] At STP.
[c] 5-s value.
[d] Pyrotechnic compositions produce only a mild ignition on impact.

that would be encountered if an explosive incident occurred. The maximum distance to which hazardous explosive effects propagate depends on the blast overpressure created, which as a first approximation is a function of the cube root of the explosive weight, W. This is termed the quantity distance and is defined as $D = KW^{\frac{1}{3}}$, where D is the allowable distance from the explosive site and K is a constant that depends on the classification of the explosive, the storage conditions, and the potential effect of a possible explosion on the people, structures, and material within the zone specified. Specific values for K are available.

Highly detailed and systematic techniques have been developed to identify possible failure modes, and to quantify the probability of occurrence and impact on operations, including hazards analysis and risk or reliability assessment.

Environmental Impact. U.S. federal and state environmental legislation to limit, control, and remove pollutants entering the environment has resulted in numerous programs relating to propellant and explosive manufacture, storage, use, and disposal.

Pollutant Reduction. Pollutants from explosives are primarily produced by waste from the explosives manufacture, such as the acids used in nitration (qv). Pollutants may also be produced during incorporation of the explosives in munitions in the use of industrial explosives, and in clean-up and disposal operations. Water-soluble materials may be treated by such methods as ion exchange, reverse osmosis, and biodenitrification. Solid wastes can be incinerated.

Demilitarization and Disposal of Explosive Material. An important consequence of international agreements to greatly reduce the stockpiles of conventional and nuclear munitions is the intensification of a program to develop procedures to destroy, recycle, and/or reclaim explosives, propellants, and pyrotechnic material efficiently and without significant environmental impact.

The procedures commonly used to demilitarize conventional munitions include munitions disassembly, washout or steamout of explosives from projectiles and warheads, incineration of reclaimed explosives, and open burning or detonation. Open burning and detonation of large quantities of energetic compositions are no longer permitted in many areas of the United States. Increasing constraints are being placed even on the uncontrolled destruction of small quantities of these substances.

Performance of Explosives

A wide variety of procedures have been developed to evaluate the performance of explosives. These include experimental methods as well as calculations based on available energy of the explosives and the reactions that take place on initiation. Both experimental and calculational procedures utilize electronic instrumentation and computer codes to provide estimates of performance in the laboratory and the field.

The experimental procedures depend to a large extent on the use to which the explosive is to be put. Comparisons are often made to proven explosives of known performance. The most commonly used tests include those that measure sensitivity, thermal characteristics, physical properties, performance, vulnerability, environmental impact and safety, and producibility. Military requirements place a premium on maximum-energy explosives having very long shelf lives, capable of being used in a wide variety of environment conditions, and often subject to extraordinary stresses of handling and use. By contrast, design of explosives for industrial use emphasizes minimum cost and moderate energy formulations that may be tailored for specific uses. These latter do not require the length of shelf life imposed on military material.

Primary Explosives

Explosives are commonly categorized as primary, secondary, or high explosives. Primary or initiator explosives are the most sensitive to heat, friction, impact, shock, and electrostatic energy. These have been studied in considerable detail because of the almost unique capability, even when present in small quantities, to rapidly transform a low energy stimulus into a high intensity shock wave.

Primary explosives are used to initiate the next element in an explosive chain which consists of explosive charges of increasing mass and decreasing sensitivity. They are arranged in sequence to amplify

Table 3. Properties of Primary Explosives

Property	Mercury fulminate	Lead azide	Silver azide	Normal lead styphnate	DDNP	Tetrazene
molecular weight	285	291	150	468	210	188
color	gray	white	white	tan	yellow	light yellow
crystal density, g/cm^3	4.43	4.38	5.1	3.10	1.63	1.7
crystal form	orthorhombic	orthorhombic monoclinic		cubic	tabular	
melting point, °C	160[a]		252	[a]	157	140–160[a]
heat of formation, kJ/g[b,c]	−0.925	−1.45	−2.07	17.9	4.00	1.13
heat of combustion, kJ/g[b,c]	3.93	2.64	4.34	5.24	13.58	
heat of detonation, kJ/g[b,c]	1.79	1.54	1.90	1.91	3.43	2.75
gas volume, cm^3/g at STP	316	308		368	876	
activation energy, kJ/mol[b]	29.8	172	146	230	230	
collision constant, $\log_{10}$/s	10.8	14.0				
detonation rate, km/s	5.4	5.10	6.8	5.2	6.9	
density, g/cm^3	4.2	4.0	5.1	2.9	1.60	
specific heat, J/(g·K)[b]	0.50	0.46	0.50	0.67		
thermal conductivity, W/(m·K) × 10^{-4}	1837	1256	837			
vacuum stability at 100°C, mL/gas per 40 h at STP	[a]	<1	<1	<1	<1	>5
weight loss at 100°C, %	<1	<1	<1	<1	<5	<5
explosion temperature at 5 s, °C	190–260	345	290	265–280	195	160
effect of prolonged storage	detonates at 80°C	stable	stable	stable	stable	stable
relative impact test value, % TNT	5	11	18	8	15	5
friction pendulum	reacts	reacts	reacts			
static discharge max energy for nonignition, J[b]	0.07	0.01	0.007	0.001	0.25	0.036
relative energy output, % TNT						
lead block	50	40	45	40	110	50
sand test	45	40		25	105	50

[a] Material explodes.
[b] To convert J to cal, divide by 4.184.
[c] Based on liquid H_2O formed.

the input stimulus to an output level of sufficient intensity to maximize the probability of initiating the main charge. Overall energy intensification is about 10 million to one. Primary explosives are used in military detonators, commercial blasting caps, and in stab and percussion electrical primers. Primary explosives may be initiated electrically, by penetration of the element using a firing pin (stab detonator), or by the shock from an exploding wire. The properties of commonly used primary explosives are shown in Table 3. The characteristics of other materials as initiators have also been studied.

Secondary Explosives

Aliphatic Nitrate Esters. Aliphatic nitrate esters, such as glycerol trinitrate (nitroglycerin), ethylene glycol dinitrate (nitroglycol), cellulose nitrate (nitrocellulose), and pentaerythritol tetranitrate (PETN), are among the most powerful explosives available. These nitrate esters are generally less stable than aromatic nitro compounds or nitramines because the former tend to hydrolyze autocatalytically to form nitric and nitrous acids, which further accelerate decomposition. Liquid nitrates, ie, nitroglycerin, are usually less stable than crystalline compounds because of the higher energy state of the liquid phase.

Nitramines. The four most important nitramine explosives are cyclotrimethylenetrinitramine (RDX), cyclotetramethylenetetranitramine (HMX), nitroguanidine (NQ), and 2,4,6-trinitrophenylmethylnitramine (tetryl). Tetryl has been increasingly replaced by HMX and RDX and is no longer used as a booster explosive in the United States. Both RDX and HMX are used as high energy explosives. They are also incorporated in high performance rocket propellants and propellants of reduced sensitivity to stimuli such as fragment impact. Nitroguanidine is employed almost exclusively in gun propellants.

Nitroaromatics. The commonly used nitroaromatic explosives contain three NO_2 groups, generally in the 1, 3, and 5 positions. Aromatics are most often nitrated to the trinitro stage with mixed acid. Further nitration is difficult, and aromatics having four or more nitro groups attached to the ring tend to be relatively unstable. The most extensively used explosive is trinitrotoluene (TNT); however, hexanitrostilbene (HNS), $C_{14}H_6N_6O_{12}$; hexanitroazobenzene (HNAB), $C_{12}H_4N_8O_{12}$; and di- and triaminotrinitrobenzene (DATB and TATB) have also found important application because of low sensitivity to impact, shock, and friction, and their excellent stability at elevated temperatures. Ammonium picrate (AP) has been used in armor-piercing gun projectiles because of its insensitivity to impact and shock.

Applications

Military. The single-component explosives most commonly used for military compositions are TNT, RDX or HMX, nitrocellulose, and nitroglycerin. The last two are used almost exclusively to make propellants. The production volume of TNT far exceeds that of any other explosive. It is used as manufactured, as a base of binary slurries with other high melting explosives, or in ternary systems generally containing a binary mix and aluminum. Other methods of explosive preparation include the pressing of polymeric-based formulations (PBX's) of high energy explosives.

Industrial. In the United States, private corporations operate most of the government-owned plants that make explosives for military use. Explosives and explosive components are also made for industrial use. Ammonium nitrate explosives typified by the water gels, slurries, emulsions, and the ammonium nitrate fuel oil blasting agents (ANFO) and unprocessed ammonium nitrate dominate the sales of industrial explosives and blasting agents in the United States.

Specialized Uses of Explosives. In addition to the conventional use of explosives for mining, civil engineering, and military purposes, an increasing variety of highly specialized applications as well as unusual forms of explosives have been developed to meet unique requirements. Included among the unusual applications are metal forming and metal cladding (see METALLIC COATINGS–EXPLOSIVELY CLAD METALS) or metal welding to molecularly bond layers of different metals, metal cutting, and compaction of metal powders. Among the specialty forms of explosives used for these and many purposes are flexible sheet explosive generally consisting of a polymeric binder and a high performance explosive such as RDX or PETN; flexible or very rigid explosive foams formulated from nitromethane and a polymeric foam producer; fuel-air explosive systems in which a combustible hydrocarbon such as propane is dispersed in air and then ignited to form a high pressure blast wave; plastic bonded explosive (PBXs) containing RDX or HMX and a polymer designed to give very high mechanical strength; and low sensitivity liquid explosives such as hydroxyl ammonium nitrate used as a component in liquid propulsion guns.

More Powerful Explosives. The search for higher energy, yet safe explosives, primarily for military purposes, is concentrated to a large extent on the synthesis of thermally stable molecules having mass densities >1.9 g/mL, the density of HMX, and at least 10% for energy than HMX. Hydrogen-free molecules containing only carbon, nitrogen, and oxygen have been synthesized. The atoms are often bound together in compact strained rings. These small-scale molecules provide additional latent strain energy. Research has also been conducted to find high energy polymeric binders for use with RDX, HMX, and higher energy explosives.

Among the molecules that have received greatest attention are 2,6-bis(picrylamino)-3,5-dinitropyridine (PYX); 3,6-dinitro-*s*-tetrazine; 2,4,6-trinito-*s*-trizene; octanitrocubane (ONC); 1,3,3-trinitroazetidine polynitroadamantane (TNAZ); and 1,4-dinitroglycouril (DINGU).

Insensitive Explosives. The catastrophic propagation of explosive detonations that have been accidentally initiated during peacetime or by military action during wartime, and the high costs of shipping and storing explosive systems have led to a search for low sensitivity explosives and explosive and propellant formulations.

The criteria for insensitive explosives subjected to hazard tests permit no reaction more violent than burning in slow and fast cookoff tests and fragment and bullet tests, no propagation in sympathetic detonation tests, no detonation when struck by a shaped charge jet, no sustained burning when hit by a small fragment, and such special tests as may be required by the use of the explosive. The general approach to this difficult problem of maximizing the energy of the explosive while minimizing sensitivity lies in the use of RDX or HMX (80–90%) thoroughly coated with a plasticized polymer similar to that use in some plastic bonded explosives.

Relatively insensitive explosives of medium energy levels have also been formulated using propellant-type formulations containing ammonium perchlorate, aluminum, and a polybutadiene or polyester binder.

VICTOR LINDNER
U.S. Armament Research,
Development and Engineering Agency

B. T. Federoff, O. E. Sheffield, and S. Kaye, eds., *The Encyclopedia of Explosives and Related Items*, PATR 2700, Vols. 1–10, ARDEC, Dover, N.J.

T. W. Urbanski, *Chemistry and Technology of Explosives*, Vols. 1–3, Pergamon Press, New York, 1967; Vol. 4, 1984.

T. N. Hall and J. R. Holden, *Navy Explosives Handbook, Explosion Effects and Properties, Part 3, Properties of Explosives and Explosive Compositions*, NSWC, White Oak, Md., MP-8116, Oct. 1988.

B. M. Dobratyz, *LLNL Explosives Handbook, Properties of Chemical Explosives and Explosive Simulants*, UCRL 52997, LLNL, University of California, Livermore, Mar. 1981.

PROPELLANTS

Propellants are mixtures of chemical compounds that produce large volumes of high temperature gas at controlled, predetermined rates, and can sustain combustion without requiring atmospheric oxygen for the purpose. Principal applications are in launching projectiles from guns, rockets, and missile systems. Propellant-actuated devices are used to drive turbines, move pistons, operate rocket vanes, start aircraft engines, eject pilots, jettison stores from jet aircraft, pump fluids, shear bolts and wires, and act as sources of heat in special devices. Solid propellants are compact, and have a long storage life, and may be handled and used without any need for taking exceptional precautions.

General Characteristics

Gun Propellants. Solid gun propellants are employed in the form of dense cylindrical or spherical grains, elongated hollow or split sticks, or as sheets of plasticized nitrocellulose. Gun propellants are almost always based on nitrocellulose to provide mechanical strength. These also may contain inert or energetic liquid plasticizers (qv) or a combination of the two to improve physical and processing characteristics, high explosives to increase available energy, stabilizers to prolong storage life, and a small amount of inorganic additives to facilitate handling, improve ignitibility, and decrease muzzle flash. Single-based propellants, used exclusively in guns, derive energy primarily from nitrocellulose. Double-based nitrocellulose propellants contain liquid energetic plasticizers such as nitroglycerin, and are used in rockets as well as guns. Triple-based propellants contain crystalline additives, eg, nitroguanidine, as well as nitrocellulose and energetic additives. Both double- and triple-based propellants are used in guns and rockets. Low sensitivity propellants (LOVA) have also been developed for use in guns; these contain a high energy component, eg, cyclotrimethylene trinitramine (RDX) in a polymeric binder. Gun propellants are made mostly by an extrusion process that produces small grains in large numbers. Nitrocellulose serves on the energetic binder. Typical components of nitrocellulose propellants and their functions are

nitrocellulose	energetic binder
polyglycol diols	nonenergetic binder
nitroglycerin,nitroglycerin, metriol trinitrate, diethylene glycol dinitrate, triethylene glycol dinitrate, dinitrotoluene	plasticizers, energetic
dimethyl, diethyl, or dibutyl phthalates, triacetin	plasticizers, fuels
diphenylamine, diethyl centralite, 2-nitrodiphenylamine, acardite, diethyl diphenylurea	stabilizers
lead salts, eg, lead stannate, lead stearate, lead salicylate	ballistic modifiers in rocket propellants
carbon black	opacifier
lead stearate, graphite, wax	lubricants
potassium sulfate, potassium nitrate, cryolite (potassium aluminum fluoride)	flash reducers in gun propellants
ammonium perchlorate, ammonium nitrate	oxidizers, inorganic
RDX, HMX, nitroguanidine, and other nitramines	oxidizers, organic
lead carbonate, tin	decoppering agents in gun propellants

The percentage compositions of representative nitrocellulose-based gun propellants are shown in Table 1.

Rocket Propellants. Solid rocket propellants are mostly based on chemically cross-linked polymeric elastomers to provide the mechanical properties required in launchings and the environmental

Table 1. Gun Propellant Composition[a], wt %

Component	M1[b]	M2	M5	M6	M8	M9	M10	M15	M17	M26	M30	M31[c]	IMR
nitrocellulose (% *N*)	85.0 (13.15)	77.5 (13.25)	82.0 (13.25)	87.0 (13.15)	52.2 (13.25)	57.8 (13.25)	98.0 (13.15)	20.0 (13.15)	22.0 (13.15)	67.5 (13.15)	28.0 (12.6)	20.0 (12.6)	100.0 (13.15)
nitroglycerin		19.5	15.0		43.0	40.0		19.0	21.5	25.0	22.5	19.0	
nitroguanidine								54.7	54.7		47.7	54.7	
ethyl centralite		0.6	0.6		0.6	0.7		6.0	1.5	6.0	1.5		
diphenylamine	1.0[d]						1.0						0.7[d]
dinitrotoluene	10.0			10.0									8.0[e]
dibutylphthalate	5.0			3.0	3.0							4.5	
potassium nitrate		0.7	0.7		1.2	1.50				0.75			
barium nitrate		1.4	1.4							0.75			
potassium sulfate	1.0[f]			1.0[d]				1.0					1.0[d]
cryolite								0.3	0.3		0.3	0.3	
graphite		0.3	0.3			0.10[d]			0.15[d]				

[a] All compositions are solvent extruded as grains except M8 which is solventless-rolled as sheet.
[b] Also may contain 1.0 wt % basic lead carbonate.
[c] Also contains 1.5 wt % 2-nitrodiphenylamine.
[d] On added basis.
[e] Added as a coating.
[f] If required, on added basis.

conditions experienced in storage, shipment, and handling (see ELASTOMERS, SYNTHETIC). Double- and triple-based nitrocellulose propellants are also employed as rocket propellants.

Polymer-based rocket propellants are generally referred to as composite propellants, and often identified by the elastomer used, eg, urethane propellants or carboxy- (CTPB) or hydroxy- (HTPB) terminated polybutadiene propellants. The cross-linked polymers act as a viscoelastic matrix to provide mechanical strength, and as a fuel to react with the oxidizers present. Ammonium perchlorate and ammonium nitrate are the most common oxidizers used; nitramines such as HMX or RDX may be added to react with the fuels and increase the impulse produced. Many other substances may be added including metallic fuels, plasticizers, stabilizers, catalysts, ballistic modifiers, and bonding agents. Typical components are listed in Table 2.

Nitrocellulose-based rocket propellant grains contain energetic liquid plasticizers such as nitroglycerin, stabilizers, ballistic modifiers, nonenergetic plasticizers, inorganic oxidizer salts, organic explosives, and metallic fuels similar to those used in gun propellants. When these latter components are included, the composition is referred to as a composite-modified double-based propellant (CMDB). Nitrocellulose-based propellants have also been made using isocyanatecurable elastomers which permit a reduction in the amount of nitrocellulose used and an increase in the nitroglycerin contents. Aluminum is most commonly used to increase the impulse of both composite and nitrocellulose-based rocket propellants because of its highly exothermic reaction with the oxidizer.

Liquid Propellants. Liquid propellants have long been used to obtain maximum controllability of rocket performance and, where required, maximum impulse. Three classes of rocket monopropellants exist that differ in their chemical reactions that release energy: *(1)* those consisting of, eg, hydrogen peroxide, H_2O_2; ethylene oxide, C_2H_4O; and nitroethane, $CH_3CH_2NO_2$; that can undergo internal oxidation–reduction reactions; *(2)* those consisting of unstable molecules such as hydrazine, N_2H_2, and acetylene, C_2H_2; and *(3)* those consisting of stable mixtures of two or more compounds that are mutually compatible. Table 3 lists common and experimental liquid rocket bipropellants. Liquid propellants for guns are in the experimental stage, with the most common formulation connecting of a mixture of hydroxyl ammonium nitrate, triethylene ammonium nitrate, and water (20:63:17).

Rocket propellants may produce undesirable smoke-forming combustion products in the exhaust plumes, ie, these products become visible signatures of the location of the source of smoke, and can interfere with optical guidance systems. Smoke formation is caused primarily by particulate matter, such as aluminum oxide, from aluminum in the rocket propellant, or to a lesser extent by compounds of iron, lead, or copper. This is referred to as primary smoke. Water in the propellant combustion products produces the secondary smoke that forms the contrails associated with missile flight.

Rocket propellants are made mostly by a casting process, as distinct from the extrusion process used to make the very much smaller and more numerous gun propellant grains.

Selection Criteria

Energy Considerations. The selection of gun and rocket propellants involves two principal considerations: the total amount of energy required and the mass rate at which the hot gases produced must be delivered to meet system performance requirements. The energy delivered per unit mass depends on the chemical energy of the propellant components, the characteristics of the products of combustion, the chemical equilibria which prevail among the reaction products, and the efficiency with which the system converts thermal to kinetic energy. The rate at which energy is produced depends on the intrinsic burning characteristics of the propellant, its burning surface area, and the operating pressure and temperature of the system. Control of the burning surface area is obtained by using appropriate grain geometries and the required number of grains. Uncontrolled burning can result in intolerably high pressure or, in the worst case, catastrophic detonation.

Performance Calculations. The energy evolved by a propellant may be estimated from the percentage composition, reaction products, the heats of formation of the reactants and the products, the propellant density, and the gases and solids produced. The composition and flame temperatures of the products are determined from the applicable enthalpy–temperature and chemical equilibrium functions of the various molecular species and the operating conditions in the combustion chamber. The most important thermodynamic–thermochemical characteristics of propellant combustion products, in addition to gas volume and flame temperature, are heat capacity, heat capacity ratio, and the covolume of the gases at high pressures.

Mechanical Characteristics. Large rocket grains in particular must have adequate mechanical properties to enable them to withstand the stresses imposed during handling and firing. The mechanical properties depend primarily on the characteristics of the binder, the percentage of solids present, and the particle-size distribution of the solids.

Table 2. Typical Components of Composite Rocket Propellants

Component	Characteristics
	Binders
polysulfides	reactive group (mercaptyl, —SH), is cured by oxidation reactions; low solids loading capacity and relatively low performance; mostly replaced by other binders
polyurethanes, polyethers, polyesters	reactive group (hydroxyl, —OH), is cured with isocyanates; intermediate solids loading capacity and performance
polybutadienes copolymer of butadiene and acrylic acid	reactive group (carboxy, —COOH, or hydroxyl, —OH), is cured with difunctional epoxides or aziridines; intermediate solids loading capacity and better performance than polyurethanes; less than adequate cure stability and mechanical characteristics
terpolymers of butadiene, acrylic acid, and acrylonitrile	superior physical properties and storage stability
carboxy-terminated polybutadiene	cured with difunctional epoxides or aziridines; have good solids loading capacity, high performance, and good physical properties
hydroxy-terminated polybutadiene	cured with diisocyanates; have good solids loading and performance characteristics, and good physical properties and storage stability
	Oxidizers
ammonium perchlorate	most commonly used oxidizer; it has a high density, permits a range of burning rates, but produces smoke in cold or humid atmosphere
ammonium nitrate	used in special cases only, it is hygroscopic and undergoes phase changes, has a low burning rate, and forms smokeless combustion products
high energy explosives (RDX–HMX)	have high energy and density; produce smokeless products; have a limited range of low burning rates
	Fuels
aluminum	most commonly used; has a high density; produces an increase in specific impulse and smoky and erosive products of combustion
metal hydrides	provide high impulse, but generally inadequate stability; produce smoky products and have a low density
	Ballistic modifiers
metal oxides	iron oxide most commonly used
ferrocene derivatives	permit a significant increase in burning rate
other	coolants for low burning rate and various special types of ballistic modifiers
	Modifiers for physical characteristics
plasticizers	improve physical properties at low temperatures and processibility; may vaporize or migrate; can increase energy if nitrated
bonding agents	improve adhesion of binder to solids

Gun Propellants. Although the stresses on individual gun propellant grains are less severe because of the small size, these propellants must withstand much higher weapon pressures and accelerations than rocket propellants. Formulation options are usually more limited for gun propellants than for rocket propellants because the products of combustion must not foul or corrode a gun, should have a low flame temperature, and should exhibit minimum flash and smoke characteristics.

Shelf Life Characteristics. The chemical safe life of all standard propellants is measured in years. Both gun and rocket propellants contain chemical stabilizers that combine with the products of decomposition to prevent autocatalytic breakdown of the propellant composition. The useful service life of gun propellants may be as long as 25 to 50 years. The useful service life of rocket propellants may be significantly less than the chemical safe life if gassing occurs, motor bonding deteriorates, or significant physical changes take place. Generally such effects are produced by high temperature storage or high–low temperature cycling, particularly if moisture is present. Relatively little degradation occurs at ambient temperature conditions.

Table 3. Liquid Rocket Bipropellants

Oxidant	Fuel	Ratio oxidant/fuel	Specific impulse, s[a]
O_2	H_2	4.0	341
O_2	B_2H_6	2.0	344
O_2	N_2H_4	0.90	313
O_2	JP4	2.60	301
F_2	H_2	9.00	410
F_2	JP4	2.40	317
F_2	N_2H_4	2.30	363
IRFNA[b]	C_2H_5OH	2.50	219
IRFNA[b]	UDMH[c]	3.00	288
H_2O_2	C_2H_5OH	4.0	230
H_2O_2	JP4	6.5	233
H_2O_2	N_2H_4		245
N_2O_4	N_2H_4		249
ClF_3	H_2	11.50	318
ClF_3	N_2H_4		292

[a] Calculated values.
[b] IRFNA contains 20–40% lithium nitrate, 55–75% red fuming nitric acid (RFNA), and 4–5% SiO_2; mp = −54°C.
[c] UDMH = unsymmetrical dimethylhydrazine.

The Burning Process

The mass rate of a propellant grain burning at a given pressure and temperature depends on the amount of heat evolved during decomposition and the amount of heat transferred to the burning surfaces of the propellant from the hot gases above it. This rate is also influenced by the tangential velocity of the propellant gases and the radiation from the surroundings. Propellants burn in parallel layers so that the surface recedes in all directions normal to the original surface. The geometry of the grain on completion of burning is similar to its geometry at the start. Propellant burning at high gun pressures proceeds more smoothly and is less subject to erratic behavior than burning at very low pressures because the conditions are appropriate for maximum energy transfer in minimum time.

The composition of the propellant determines the rate of exothermic molecular breakdown at a given temperature and pressure. As the reaction rate increases, the rates of heat production and transfer increase with associated increases in the linear burning rate of the propellant. The heat evolved per gram of propellant in an inert atmosphere is its heat of explosion, Q. It may be readily calculated or experimentally determined in a calorimetric bomb (see THERMOGRAPHY). Values range from ca 2.09 kJ/g (500 cal/g) for cool propellants to ca 6.27 kJ/g (1.3 kcal/g) for maximum energy propellants. The operating pressure of the system is the dominant influence on the burning rate of propellants.

Mechanism of Burning. Much of the information available on the burning process of nitrocellulose propellants is based on the decomposition of nitrate esters and the reaction of oxides of nitrogen with the products of decomposition. The three reaction zones identified are *(1)* the foam zone where molecular bond breakage, primarily the O–N bond in cellulose nitrate–nitroglycerin-type propellants, occurs. Large volatile molecules are produced, such as aldehydes, alcohols, and low molecular weight oxygenated compounds. *(2)* The fizz zone which is above the foam zone and results from partial reaction among the materials ejected from the foam surface. Aldehydes and alcohols are converted to smaller molecules; nitrogen, water, carbon monoxide, carbon dioxide, and nitric oxide are also formed. About half the total heat evolved by the propellant is liberated in the fizz zone. Finally *(3)*, the flame zone, where thermodynamic equilibrium is established. This zone defines the flame temperature of the propellant, which may range from ca 1500 K for cool propellant to 3500 K for very hot ones. Here nitric oxide reacts with the reaction products formed in the fizz zone to produce carbon monoxide, carbon dioxide, hydrogen, water, nitrogen, and a small percentage of other molecules.

Additives. Although the burning rate of nitrocellulose propellants at high gun pressures is not significantly affected by the presence of additives, the addition of 1 to 2% of some metallic salts such as lead acetyl salicylate, lead stearate, and lead stannate to double-based propellants increases their burning rates at much lower rocket pressures.

Composite Propellants. A number of analytical models have been developed to quantitatively define the burning characteristics of composite propellants. The granular diffusion model postulates that the primary reaction zone of ammonium perchlorate propellants lies almost entirely in the gas phase. The Beckstead-Derr-Price model considers both the gas-phase and condensed-phase reactions.

Burning-Rate Equations. The design of propellants for gun or rocket performance requires a knowledge of the exact rate at which the products of combustion are produced under the prevailing conditions of pressure and temperature. Burning-rate equations have been developed to describe the performance of solid propellants based on the assumption that all the exposed propellant surfaces are ignited simultaneously and burn at the same linear rate. For example: $r = a + bP$ (eq. 1) and $r = cP_n$ (eq. 2), where r is the linear burning rate, P is the pressure, n is the pressure exponent, and a, b, and c are constants that vary with temperature. Equation 1 is often used for propellants burning at high gun pressures, whereas equation 2 is associated with low pressure rocket systems.

Burning Control. In order to produce propellant gas at a predetermined rate, a propellant composition is selected having the required burning rate for the operating pressures in the gun or rocket. The geometry of the propellant is then designed so that the necessary burning surface is available to provide the required mass rate of gas evolution. Control of the total burning surface is achieved by establishing the number of grains to be used, the geometrical configuration, and in the case of rocket propellants, the cementing of non-combustible inhibitors on the grain surface or by bonding the exterior of the propellant grain surface to the motor wall.

Uncontrolled Burning. Because propellants contain potentially explosive components, a controllable burning process may change under certain exceptional conditions to uncontrollable burning with consequences comparable to a detonation. The transition from deflagration to detonation in explosives and propellants has been intensively studied, and the available evidence indicates that detonation is most likely to occur when the burning conditions can lead to the initiation and maintenance of a high pressure shock wave. Uncontrolled burning in rocket propellants is most likely to occur when using high energy explosives such as RDX or HMX and energetic plasticizers. It may occur in gun propellants if high loading densities are used and ignition is not rapid; if the grains can be substantially compacted; or if the entire propellant charge is accelerated by localized ignition at the charge base.

Nonconventional Methods of Gun Propulsion

Advanced gun propulsion programs are pursued primarily to obtain projectile velocities considerably greater than the approximately 1.5–2.0 km velocity obtainable using conventional propellants and guns that rely on traditional interior ballistics. Hypervelocities would offer the possibility of achieving extraterrestrial orbits using gun-type systems instead of missiles. The advanced propulsion programs are of three types: those using chemical propellants for accelerating projectiles by unusual methods; those using electrical sources of energy; and those combining these two procedures. Propulsion programs include the traveling charge, the two-stage light gas gun, the Ram accelerator, and electric guns.

VICTOR LINDNER
Armament Research, Development and Engineering Agency

F. N. Kelly, in C. Boyars and K. Klager, eds., *Propellants Manufacture, Hazards and Testing, Advances in Chemistry Series 88*, American Chemical Society, Washington, D.C., 1969, p. 188.

B. T. Federoff and O. E. Sheffield, *The Encyclopedia of Explosives and Related Items*, Vols. 1–10, TR 2700, ARDEC, Dover, N.J.

R. F. Gould, ed., *Propellant Manufacture, Hazards and Testing, Advances in Chemistry Series 88*, American Chemical Society, Washington, D.C., 1969.

EXTRACTION

LIQUID–LIQUID

The physical process of liquid–liquid extraction separates a dissolved component from its solvent by transfer to a second solvent, immiscible with the first but having a higher affinity for the transferred component. The latter is sometimes called the consolute component. Liquid–liquid extraction can purify a consolute component with respect to dissolved components which are not soluble in the second solvent, and often the extract solution contains a higher concentration of the consolute component than the initial solution. In the process of fractional extraction, two or more consolute components can be extracted and also separated if these have different distribution ratios between the two solvents.

The principle of liquid–liquid extraction, and some of the special terminology, are illustrated in Figure 1 which shows a single contacting stage. If equilibrium is fully established after contact, the stage is defined as an ideal or theoretical stage. The two resulting liquid phases are the raffinate from which most of solute C has been removed, and the extract, consisting mainly of solvent B and C.

In the simplest case, the feed solution consists of a solvent A containing a consolute component C, which is brought into contact with a second solvent B. For efficient contact there must be a large interfacial area across which component C can transfer until equilibrium is reached or closely approached. On the laboratory scale this can be achieved in a few minutes simply by hand agitation of the two liquid phases in a stoppered flask or separatory funnel. Under continuous flow conditions it is usually necessary to use mechanical agitation to promote coalescence of the phases. After sufficient time and agitation, the system approaches equilibrium which can be expressed in terms of the extraction factor ϵ for component C:

$$\epsilon = \frac{\text{quantity of C in B-rich phase}}{\text{quantity of C in A-rich phase}} = m\frac{\text{B}}{\text{A}} \tag{1}$$

where B and A refer to the quantities of the two solvents and m is the distribution coefficient.

The component C in the separated extract from the stage contact shown in Figure 1 may be separated from the solvent B by distillation (qv), evaporation (qv), or other means, allowing solvent B to be reused for further extraction. Alternatively, the extract can be subjected to back-extraction (stripping) with solvent A under different conditions, eg, a different temperature; again, the stripped solvent B can be reused for further extraction. Solvent recovery is an important factor in the economics of industrial extraction processes.

Whereas Figure 1 assumes a physical extraction based on different solubilities as expressed by the distribution coefficient, many extractions depend on chemical changes. In such cases the component C in the feed solvent may not itself have any solubility in the extracting solvent B, but can be made to react with an extractant to produce a compound or species which is soluble in B. Many metals can be extracted from aqueous solutions of their salts into organic carrier solvents by using organic extractants which can form organometallic compounds

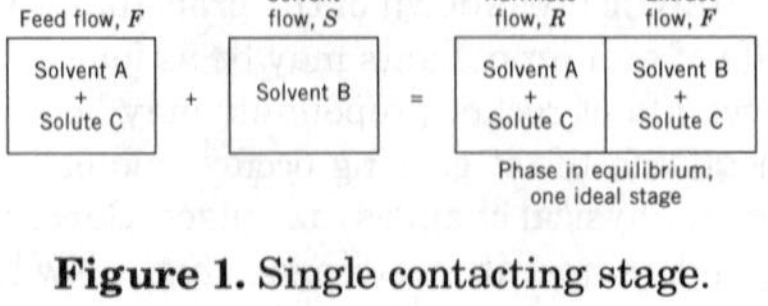

Figure 1. Single contacting stage.

or complexes. Stripping of the metals from the organic to an aqueous phase can be effected by changing a chemical condition such as pH.

Principles

Physical Equilibria and Solvent Selection. In order for two separate liquid phases to exist in equilibrium, there must be a considerable degree of thermodynamically nonideal behavior. If the Gibbs free energy, G, of a mixture of two solutions exceeds the energies of the initial solutions, mixing does not occur and the system remains in two phases. For the binary system containing only components A and B, the condition for the formation of two phases is

$$\frac{d^2G}{dx_A^2} > \qquad (2)$$

The selection of solvents for a given separation depends largely 0 on equilibrium considerations. Other important factors include cost, ease of solvent recovery by distillation (qv) or other means, safety and environmental impact, and physical properties which must permit easy phase dispersion and separation. Solvent selection is therefore a broad-based exercise which is hard to quantify. However a useful quantitative approach has been proposed for comparing simplified equilibrium estimations on the basis of regular solution theory.

Chemical Equilibria. In many cases, mass transfer between two liquid phases is accompanied by a chemical change. The transferring species can dissociate or polymerize depending on the nature of the solvent, or a reaction may occur between the transferring species and an extractant present in one phase.

In addition to the liquid–liquid reaction processes, there are many cases in both analytical and industrial chemistry where the main objective of separation is achieved by extraction using a chemical extractant. The technique of dissociation extraction is very valuable for separating mixtures of weakly acidic or basic organic compounds.

In hydrometallurgical separations, a metal ion in aqueous solution can be selectively converted to an organometallic compound or complex which is soluble in an organic carrier solvent.

Chelating extractants owe effectiveness to the attraction of adjacent groups on the molecule for the metal. Anionic extractants are commonly based on high molecular weight amines. Solvating extractants contain one or more electron donor atoms, usually oxygen, which can supplant or partially supplant the water which is attached to the metal ions.

Interfacial Mass-Transfer Coefficients. Whereas equilibrium relationships are important in determining the ultimate degree of extraction attainable, in practice the rate of extraction is of equal importance. Equilibrium is approached asymptotically with increasing contact time in a batch extraction. In continuous extractors the approach to equilibrium is determined primarily by the residence time, defined as the volume of the phase contact region divided by the volume flow rate of the phases.

The rate of mass transfer depends on the interfacial contact area and on the rate of mass transfer per unit interfacial area, ie, the mass flux. The mass flux very close to the liquid–liquid interface is determined by molecular diffusion in accordance with Fick's first law:

$$N = -D\frac{\partial c}{\partial z} \qquad (3)$$

where N refers to the flux in the z direction, c is the concentration of the consolute component, and D is its molecular diffusivity in the solvent.

Mass-Transfer Coefficients with Chemical Reaction. Chemical reaction can occur in any of the five regions shown in Figure 2, ie, the bulk of each phase, the film in each phase adjacent to the interface, and at the interface itself. Irreversible homogeneous reaction between the consolute component C and a reactant D in phase B can be described as

$$\text{C} + z\text{D} \rightarrow \text{products} \qquad (4)$$

The equations of combined diffusion and reaction, and their solutions, are analogous to those for gas absorption (qv). It has been shown how

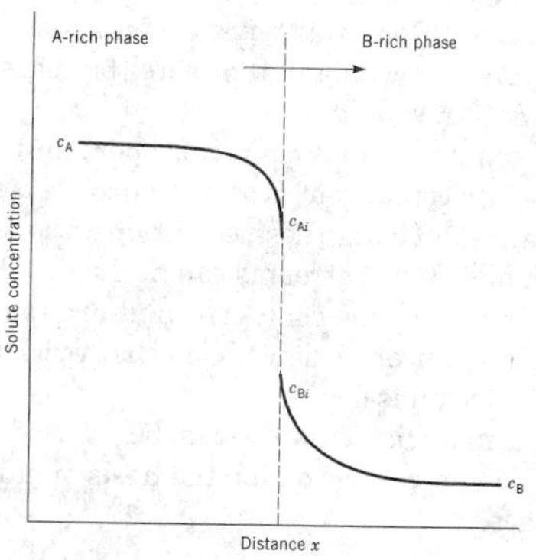

Figure 2. Concentration profiles near an interface where the arrow represents the direction of mass transfer, c_A = concentration of C in A-rich phase, c_B = concentration of C in B-rich phase, and the subscript i denotes the interface.

the concentration profiles and rate-controlling steps change as the rate constant increases. When the reaction is very slow and the B-rich phase is essentially saturated with C, the mass-transfer rate is governed by the kinetics within the bulk of the B-rich phase. This is defined as regime 1. For a slow reaction defined as regime 2, the consolute component C is almost entirely depleted in the bulk of the B-rich phase and the mass transfer of C between the phases controls the rate of the reaction. For a very fast reaction the depletion of C affects the concentration profile in the diffusion film. The steepening of the concentration profile for regime 3 leads to an enhancement in the film mass-transfer coefficient in the B-rich phase. Finally, the case of an instantaneous reaction (regime 4) leads to the formation of a thin reaction zone to which components C and D diffuse in stoichiometric amounts.

Interfacial Contact Area and Approach to Equilibrium. Experimental extraction cells such as the original Lewis stirred cell are often operated with a flat liquid–liquid interface the area of which can easily be measured. In the single-drop apparatus, a regular sequence of drops of known diameter is released through the continuous phase. These units are useful for the direct calculation of the mass flux N and hence the mass-transfer coefficient for a given system. In industrial equipment, however, it is usually necessary to create a dispersion of drops in order to achieve a large specific interfacial area, a, defined as the interfacial contact area per unit volume of two-phase dispersion. Thus the mass-transfer rate obtainable per unit volume is given as

$$(N\cdot a) = K_A a(c_A - c_A^*) \qquad (5)$$

Calculation of Equilibrium Stages. Multistage contacting can be arranged in a concurrent, crosscurrent, or countercurrent manner. The sequence of stages is sometimes referred to as a cascade, referring to the early use of gravity overflow from stage to stage. The countercurrent arrangement represents the best compromise between the objectives of high extract concentration and a high degree of extraction of the solute, for a given solvent-to-feed ratio. For the case of a partially miscible ternary system, the number of ideal stages in a countercurrent cascade can be estimated graphically on a triangular diagram, using the Hunter-Nash method.

Fractional Extraction. Fractional extraction is the separation of two or more consolute components by solvent extraction. Single-solvent fractional extraction has been known for many years, but the range of solvents available is limited because of the requirement that the solvents must be sparingly miscible with each of the feed components.

Dual solvent fractional extraction makes use of the selectivity of two solvents (A and D) with respect to consolute components.

Differential Contacting. Although the equilibrium stage concept has proved extremely useful in describing the performance of mixer-settlers and plate columns having discrete stages, it is not appropriate for spray towers, packed columns, etc, in which no discrete stages can be identified. In such differential types of contactors, equilibrium between phases is never reached and therefore the mass-transfer rate is important in the design procedure.

A differential countercurrent contactor operating with a dilute solution of the consolute component C and immiscible components A and B is shown in Figure 3. Under these conditions, the superficial velocities of the A-rich and B-rich streams can be assumed not to vary significantly with position in the contactor, and are taken to be U_A and U_B, respectively. The concentration of C in the A-rich stream is c_A and that in the B-rich stream is c_B.

A steady-state material balance can be carried out on a small section of length dz and volume dz (on the basis of unit cross-sectional area) in the contactor:

$$U_B dc_B = U_A dc_A = K_A a(c_A - c_A)dz \quad (6)$$

Axial Dispersion. Elementary texts assume that all the fluid in each phase has the same resident time in a countercurrent extractor. In practice, the two phases rarely move countercurrently in plug flow because of axial mixing which arises from the action of turbulent eddies, circulation currents, or the effects of drop wakes. The effect is to flatten the axial concentration profiles within each phase. Axial mixing can lead to a reduction in the effective driving force for mass transfer which in turn reduces the NTU below that expected for the plug flow case. An important feature of the profile is the discontinuity or "jump" in concentration which occurs at entry to the contactor when the liquid in the feed line enters the mixed region of the column.

Two alternative approaches are used in axial mixing calculations. For differential contactors, the axial on model is used:

$$N = -E\frac{\partial c}{\partial z} \quad (7)$$

For contactors in which discrete well-mixed compartments can be identified, for example sieve-plate columns, axial mixing effects are incorporated into the stagewise model by means of the backflow ratio α which is defined as the fraction of the net interstage flow of one phase which is considered to flow in the reverse direction. For a contactor in which there are many compartments, the axial dispersion coefficient and the backflow ratio, α, are interrelated as follows:

$$E = \frac{UH}{\ln((1 + \alpha)/\alpha)} \quad (8)$$

where H is the height of one compartment and U is the superficial velocity. The detailed calculations of concentration profiles and mass-transfer rates with axial mixing require the

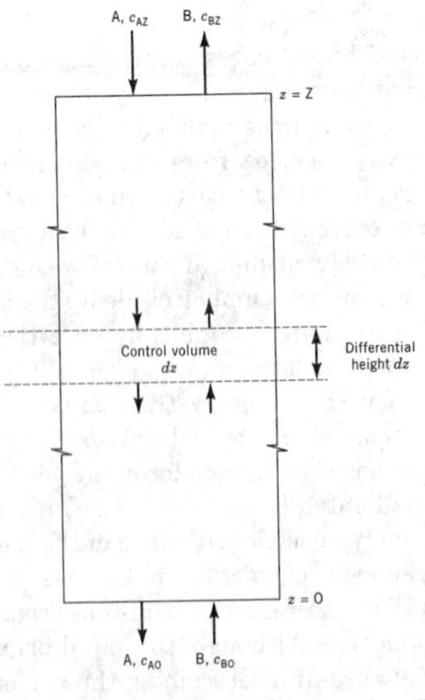

Figure 3. Mass transfer in a differential contactor. Terms are defined in the text.

solution of a fourth-order differential equation (dispersion model) or the equivalent difference equation (backflow model) along with appropriate boundary conditions.

Drop Diameter. In extraction equipment, drops are initially formed at distributor nozzles; in some types of plate column the drops are repeatedly formed at the perforations on each plate. Under such conditions, the diameter is determined primarily by the balance between interfacial forces and buoyancy forces at the orifice or perforation.

In many types of contactors, such as stirred tanks, rotary agitated columns, and pulsed columns, mechanical energy is applied externally in order to reduce the drop size and thereby increase the rate of mass transfer.

Holdup and Flooding. The volume fraction of the dispersed phase, commonly known as the holdup h, can be adjusted in a batch extractor by means of the relative volumes of each liquid phase added. However, in a countercurrent column contactor, the holdup of the dispersed phase is considerably less than this, because the dispersed drops travel quite fast through the continuous phase and therefore have a relatively short residence time in the equipment. The holdup is related to the superficial velocities U of each phase, defined as the flow rate per unit cross section of the contactor, and to a slip velocity U_s:

$$U_s = U_d/h + U_c/(1 - h) \quad (9)$$

As the throughput in a contactor represented by the superficial velocities U_c and U_d is increased, the holdup h increases in a nonlinear fashion. A flooding point is reached at which the countercurrent flow of the two liquid phases cannot be maintained. The flow rates at which flooding occurs depend on system properties, in particular density difference and interfacial tension, and on the equipment design and the amount of agitation supplied.

The nonuniformity of drop dispersions can often be important in extraction. This nonuniformity can lead to axial variation of holdup in a column even though the flow rates and other conditions are held constant.

Coalescence and Phase Separation. Coalescence between adjacent drops and between drops and contactor internals is important for two reasons. It usually plays a part, in combination with breakup, in determining the equilibrium drop size in a dispersion, and it can therefore affect holdup and flooding in a countercurrent extraction column. Secondly, it is an essential step in the disengagement of the phases and the control of entrainment after extraction has been completed.

Membrane Extraction. An extraction technique which uses a thin liquid membrane or film has been introduced. The principal advantages of liquid-membrane extraction are that the inventory of solvent and extractant is extremely small and the specific interfacial area can be increased without the problems which accompany fine drop dispersions.

Supercritical Extraction. The use of a supercritical fluid such as carbon dioxide as extractant is growing in industrial importance, particularly in the food-related industries. The advantages of supercritical fluids (qv) as extractants include favorable solubility and transport properties, and the ability to complete an extraction rapidly at moderate temperature.

Two-Phase Aqueous Extraction. Liquid–liquid extraction usually involves an aqueous phase and an organic phase, but systems having two or more aqueous phases can also be formed from solutions of mutually incompatible polymers such as poly(ethylene glycol) (PEG) or dextran.

Because of the growth in biotechnology, two-phase aqueous extraction is becoming more important industrially. Two-phase aqueous systems have low interfacial tension, low interphase density difference, and high viscosity in comparison with most aqueous–organic systems. Although interfacial contact is very efficient, the separation of the phases after contact can be slow, requiring centrifugation. The performance of a spray column for two-phase aqueous extraction has also been reported.

Equipment and Processing

Laboratory Extractors, Pilot-Scale Testing, and Scale-Up. Several laboratory units are useful in analysis, process control, and process studies. The AKUFVE contactor incorporates a separate mixer and centrifugal separator. It is an efficient instrument for rapid and accurate measurement of partition coefficients, as well as for obtaining reaction kinetic data. Miniature mixer–settler assemblies set up as continuous, bench-scale, multistage, countercurrent, liquid–liquid contactors are particularly useful for the preliminary laboratory work associated with flow-sheet development and optimization because these give a known number of theoretical stages.

Because the factors relating to mass transfer and fluid dynamics of the systems in an extractor are extremely complex, particularly for mixed solvents and feedstocks of commercial interest, pilot-scale testing remains an almost inevitable preliminary to a full-scale contactor design. These tests provide *(1)* total throughput and agitation speed; *(2)* HETS or HTU; *(3)* stage efficiency; *(4)* hydrodynamic conditions, such as droplet dispersion, phase separation, flooding, emulsion layer formation, etc; *(5)* selection of direction of mass transfer; *(6)* solvent-to-feed ratio; *(7)* material of construction and its wetting characteristics; and *(8)* confirmation of the desired separation in cases where equilibrium data are not available.

For design of a large-scale commercial extractor, the pilot-scale extractor should be of the same type as that to be used on the large scale. Reliable scaleup for industrial-scale extractors still depends on correlations based on extensive performance data collected from both pilot-scale and large-scale extractors covering a wide range of liquid systems. Only limited data for a few types of large commercial extractors are available in the literature.

Commercial Extractors. Extractors can be classified according to the methods applied for interdispersing the phases and producing the countercurrent flow pattern. Figure 4 summarizes the classification of the principal types of commercial extractors.

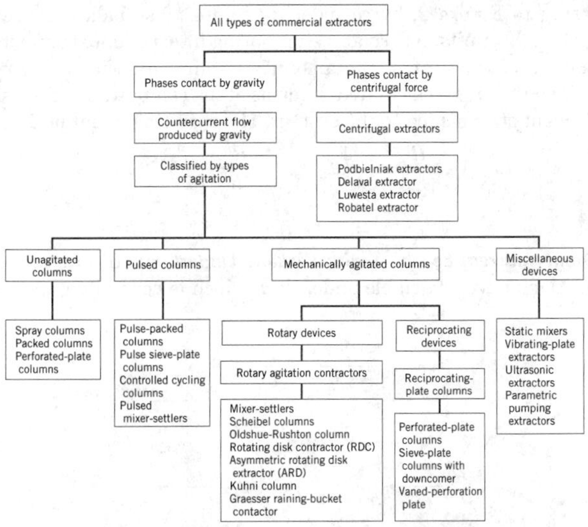

Figure 4. Classification of commercial extractors.

Organic Processes

Petroleum and Petrochemical Processes. The first large-scale application of extraction was the removal of aromatics from kerosene to improve its burning properties. Solvent extraction is also extensively used to meet the growing demand for the high purity aromatics such as benzene, toluene, and xylene (BTX) as feedstocks for the petrochemical industry. Additionally, the separation of aromatics from aliphatics is one of the largest applications of solvent extraction.

Pharmaceutical Processes. The pharmaceutical industry is a principal user of extraction because many pharmaceutical intermediates and products are heat-sensitive and cannot be processed by methods such as distillation. A useful broad review can be found in the literature. Extraction is used in the production of antibiotics, vitamins, sulfa drugs, methaqualone, phenobarbital, antihistamines, cortisone, estrogens and other hormones (qv), and reserpine and alkaloids (qv).

Food Processing. Food processing (qv) makes use of solvent extraction in several ways. Industrial refining of fats and oils using propane is known as the Solexol process. Solvent extraction is used in many protein refining processes, for example the extraction of fish protein from ground fish using *i*-propyl alcohol. Recovery of lactic acid by an extractive fermentation has recently been reported. The applications of extraction in the food industry have been reviewed.

Other Organic Processes. Solvent extraction has found application in the coal-tar industry for many years, as for example in the recovery of phenols from coal-tar distillates by washing with caustic soda solution. Solvent extraction of fatty and resimic acid from tall oil has been reported. Dissociation extraction is used to separate *m*-cresol from *p*-cresol and 2,4-xylenol from 2,5-xylenol. Solvent extraction can play a role in the direct manufacture of chemicals from coal, treatment of industrial effluents, biopolymer extraction, and difficult separations.

Inorganic Processes

Nuclear Fuel Reprocessing. Spent fuel from a nuclear reactor contains ^{238}U, ^{235}U, ^{239}Pu, ^{232}Th, and many other radioactive isotopes (fission products). Reprocessing involves the treatment of the spent fuel to separate plutonium and unconsumed uranium from other isotopes so that these can be recycled or safely stored (see NUCLEAR REACTORS–NUCLEAR FUEL RESERVES).

Copper. The recovery of copper, Cu, from ore leach liquors as a stage in the hydrometallurgical route to the pure metal is one of the largest applications of liquid–liquid extraction.

Nickel and Cobalt. Often present with copper in sulfuric acid leach liquors are nickel and cobalt. In the case of chloride leach liquors, separation of cobalt from nickel is inherently simpler because cobalt, unlike nickel, has a strong tendency to form anionic chloro-complexes. Thus cobalt can be separated by amine extractants, provided the chloride content of the aqueous phase is carefully controlled. A successful example of this approach is the Falconbridge process developed in Norway.

Extraction of Nonmetallic Inorganic Compounds. Phosphoric acid is usually formed from phosphate rock by treatment with sulfuric acid, which forms sparingly soluble calcium sulfate from which the phosphoric acid is readily separated. However, in special circumstances it may be necessary to use hydrochloric acid.

TEH C. LO
T. C. Lo & Associates
MALCOLM H. I. BAIRD
McMaster University

T. C. Lo, M. H. I. Baird, and C. Hanson, eds., *Handbook of Solvent Extraction*, Wiley-Interscience, New York, 1983.

J. D. Thornton, ed., *The Science and Practice of Liquid–Liquid Extraction*, Oxford University Press, Oxford, U.K., 1992.

R. E. Treybal, *Liquid Extraction*, 2nd ed., McGraw-Hill, New York, 1963.

J. C. Godfrey and M. J. Slater, eds., *Liquid–Liquid Extraction Equipment*, John Wiley & Sons, Ltd. Chichester, U.K., 1994.

LIQUID–SOLID

Many substances used in modern processing industries occur in a mixture of components dispersed through a solid material. To separate the desired solute constituent or to remove an unwanted component from the solid phase, the solid is contacted with a liquid phase in the process called liquid–solid extraction, or simply leaching. In leaching, when an undesirable component is removed from a solid with water, the process is called washing.

In the biological and food processing industries many products are extracted from their original structure by liquid–solid extraction (see BIOTECHNOLOGY; FOOD PROCESSING). Sugar (qv) is extracted from sugar beets using hot water; instant coffee (qv) is leached from ground roasted coffee using water; soluble tea (qv) is leached from tea leaves; pharmaceutical components, flavors, and essences are leached from plant roots, leaves, and stems (see FLAVORS AND SPICES; PHARMACEUTICALS); and oil is extracted from peanuts, soybeans, sunflower and cotton (qv) seeds, and halibut livers by solvents such as hexane, acetone, or ether (see SOLVENTS, INDUSTRIAL; SOYBEANS AND OTHER OILSEEDS). These are all examples of liquid–solid extraction.

Large-scale leaching also occurs in the metal processing industries, where useful metals frequently occur mixed with large quantities of unwanted matter, and leaching is used to remove the metals as soluble salts.

Mechanisms of Extraction

If the solute is uniformly distributed through the solid phase the material near the surface dissolves first to leave a porous structure in the solid residue. In order to reach further solute the solvent has to penetrate this outer porous region; the process becomes progressively more difficult and the rate of extraction decreases. If the solute forms a large proportion of the volume of the original particle, its removal can destroy the structure of the particle which may crumble away, and further solute may be easily accessed by solvent. In such cases the extraction rate does not fall as rapidly.

In general, the following steps can occur in an overall liquid–solid extraction process: solvent transfer from the bulk of the solution to the surface of the solid; penetration or diffusion of the solvent into the pores of the solid; dissolution of the solvent into the solute; solute diffusion to the surface of the particle; and solute transfer to the bulk of the solution. Any one of the five basic processes may be responsible for limiting the extraction rate.

The overall extraction process is sometimes subdivided into two general categories according to the main mechanisms responsible for the dissolution stage: *(1)* those operations that occur because of the solubility of the solute in or its miscibility with the solvent, eg, oilseed extraction, and *(2)* extractions where the solvent must react with a constituent of the solid material in order to produce a compound soluble in the solvent, eg, the extraction of metals from metalliferous ores. In the former case the rate of extraction is most likely to be controlled by diffusion phenomena, but in the latter the kinetics of the reaction producing the solute may play a dominant role.

Diffusion and Mass Transfer During Leaching. Rates of extraction from individual particles are difficult to assess because it is impossible to define the shapes of the pores or channels through which mass transfer has to take place. However, the nature of the diffusional process in a porous solid could be illustrated by considering the diffusion of solute through a pore. This is described mathematically by the diffusion equation, the solutions of which indicate that the concentration in the pore would be expected to decrease according to an exponential decay function.

Process Design

In most leaching operations the maintenance of constant fluid flows, pressures, and temperatures are important. These, together with the need to provide a sufficient contact time between the solvent and the solids, usually indicate a need for continuous, multistage, countercurrent processes in which fresh solvent is fed to the final stage while the solids are fed to the first stage. The objective is to be able to operate at steady conditions, and to be able to avoid extraction of undesirable material while preventing loss of solvent for both economic and safety reasons. This is usually achieved through the use of the usual control equipment, and recording instruments provide a useful means of studying plant performance. There are other factors which must be taken into account in the early stages of a design such as the particle size of the solid and the solvent employed.

Equilibrium Relationships and Mass Balances

The solid can be contacted with the solvent in a number of different ways but traditionally that part of the solvent retained by the solid is referred to as the underflow or holdup, whereas the solid-free solute-laden solvent separated from the solid after extraction is called the overflow. The holdup of bound liquor plays a vital role in the estimation of separation performance. In practice both static and dynamic holdup are measured in a process study, other parameters of importance being the relationship of holdup to drainage time and percolation rate. The results of such studies permit conclusions to be drawn about the feasibility of extraction by percolation, the holdup of different bed heights of material prepared for extraction, and the relationship between solute content of the liquor and holdup.

Single-Stage Leaching. A single-stage leaching process is shown in Figure 1. The solution overflow rate is V kg/h; the mass fraction of solute in the overflow solution is x_A; and the liquid in the slurry is flowing at L kg/h, and has a composition y_A. The mass flow of dry inert solids in the slurry is B kg/h.

The material balance equations are, for the total solution:

$$L_0 + V_2 = L_1 + V_1 = M \tag{1}$$

where M is the total input flow rate of solution to the unit; for the solute component A:

$$L_0 y_{A0} + V_2 x_{A2} = L_1 y_{A1} + V_1 x_{A1} = M x_{Am} \tag{2}$$

where $x_1 = y_1 = x_{Am}$; and for the solids:

$$B = L_0 N_0 = L_1 N_1 \tag{3}$$

where N_i is the mass concentration of inert solids in the ith stream, ie, kg of inert solid per kg solution. From these balances the concentration of the discharged solution can be estimated.

Countercurrent Multistage Leaching. Countercurrent extraction offers the most economical use of solvent, permitting high concentrations in the final extract and high recovery from the initial solid but utilizing the least amount of solvent. When the amount of solvent removed with the insoluble solid in the underflow is constant, it is convenient to define the ratio

$$R = \frac{\text{amount of solvent in overflow}}{\text{amount of solven in underflow}} = \frac{V_n}{L_n} \tag{4}$$

If perfect mixing occurs in each stage and the solute is not adsorbed preferentially at the surface of the solid, then the concentration of the solution in the underflow is the same as that in the overflow and

$$R = \frac{\text{amount of solute in overflow}}{\text{amount of solute in underflow}} = \frac{V_n x_{An}}{L_n y_{An}} \tag{5}$$

Referring to Figure 2, by considering solute mass balances over n, $(n - 1)$, ... 2, 1 units in turn and eliminating intermediate solute mass fractions and flow rates, the amount of solute associated with the leached solid may be calculated in terms of the composition of the solid and solvent streams fed to the system. The resulting equation is

$$L_0 y_{A0} = \frac{R^{n+1} - 1}{R - 1} L_n y_{An} - \frac{R^n - 1}{R - 1} V_{n+1} x_{An+1} \tag{6}$$

Countercurrent Leaching with Variable Underflow. In practice most applications have a variable underflow, which is normally greatest at

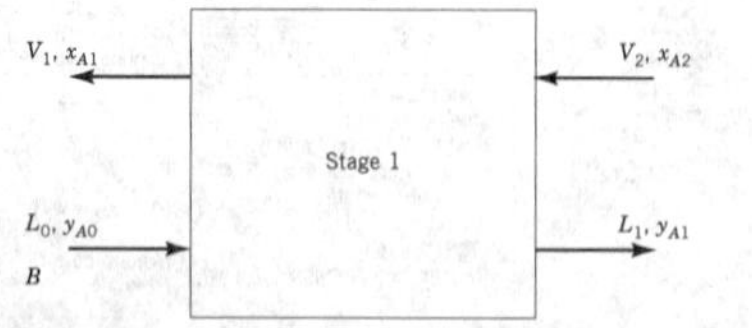

Figure 1. Flow diagram for single-stage leaching.

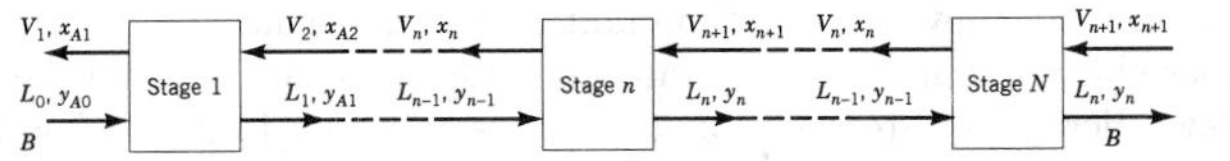

Figure 2. Flow diagram for countercurrent multistage leaching.

that point in the process where the solute concentration in the solvent is highest.

Extractors

The variety of extractors used in liquid–solid extraction is diverse, ranging from batchwise dump or heap leaching for the extraction of low grade ores to continuous countercurrent extractors to extract materials such as oilseeds and sugar beets where problems of solids transport have dominated equipment and development.

Safety and Environmental Considerations

Solvent flammability, the solvent, and dust loading in the atmosphere of the working environment and of the products in the case of edible materials are the main factors that constitute health and safety hazards in extraction plants. General safety and environmental standards must therefore be applied (see PLANT SAFETY) and due recognition taken of the most recently published national regulations relating to acceptable threshold limit values (TLVs) for solvents and dusts.

Disposal of exhausted solids can be easily overlooked at the plant design stage, particularly when these have no intrinsic value; alternative disposal methods might include landfill of inert material or incineration, hydrolysis, or pyrolysis of organic materials. Liquid, solid, and gaseous emissions are all subject to the usual environmental considerations.

RICHARD J. WAKEMAN
University of Exeter

J. M. Coulson, J. F. Richardson, J. R. Backhurst, and J. H. Harker, *Chemical Engineering*, 4th ed., Vol. 2, Pergamon Press, Oxford, U.K., 1991.

A. R. Burkin, *The Chemistry of Hydrometallurgical Processes*, E. & F. N. Spon, London, 1966.

W. L. McCabe and J. C. Smith, *Unit Operations in Chemical Engineering*, 3rd ed., McGraw-Hill Book Co., Inc., London, 1976.

R. H. Perry and D. Green, eds., *Perry's Chemical Engineers Handbook*, 50th ed., McGraw-Hill Book Co., Inc., 1984.

EXTRATERRESTRIAL MATERIALS

Extraterrestrial materials are samples from other bodies in the solar system that can be studied in earth-bound laboratories. Sensitive and ever-improving analytical techniques are used to provide information at levels of detail and sophistication that cannot be matched by telescopic or spacecraft investigations (see ANALYTICAL METHODS). Much of the knowledge of early solar system bodies, processes, environments, and chronology has come from the study of these samples. Extraterrestrial materials that are available for laboratory study include meteoritic materials that fall naturally to the earth, some meteoritic material that has been captured in space, and lunar samples that were recovered by the Apollo and Luna sample-return missions flown to the moon during the years 1969 to 1972. The meteoritic materials in existing collections include samples from asteroids, comets, and the moon, and probably Mars. The comet and asteroid samples are the best preserved solids from the early solar system and are the oldest and most cosmochemically primitive samples available for direct study. Because of their primitive and unfractionated nature, these samples provide the best estimate of the composition of the sun and the solar system as a whole. It has been shown that many meteorites contain preserved interstellar grains, particles older than the Sun that formed around other stars and served as the initial building blocks of the solar system.

Meteorites

Meteorites by definition are extraterrestrial materials that fall from the sky and actually hit the surface of the earth. Meteorites fall randomly to the earth but are not found randomly distributed on the Earth's surface. The highest general concentrations of meteorites occur in Antarctica, where long exposure time and the combined effects of ice movement and sublimation concentrate meteorites on top of blue ice fields. In Antarctica and elsewhere, meteorites are often found in clusters created by the breakup of a larger body during hypervelocity entry into the atmosphere. When a meteor breaks up at high altitude, the resulting fragments impact over an elliptical region several kilometers across the ground, forming a strewn field where sometimes thousands of individual specimens are found. Because of atmospheric breakup, the number of individual meteorite specimens that are collected is much larger than the actual number of meteoroids that produced them. Meteoroids are themselves fragments of bodies that broke up in space, and the actual lineage of meteoritic samples may trace back to a relatively limited number of initial parent bodies.

Types. Most meteorites can be classified into definite groups distinguished by elemental, mineralogical, petrographic, and isotopic composition. The general groups are the chondrites, achondrites, irons, and stony irons. Although fragments of one meteorite class are often found inside another as a result of collisional mixing, in general the bulk properties of meteorites fall into quantified groups without a continuum of compositions between established groups. It is likely that some groups are samples of single asteroids, the apparent source of most meteorites. The majority of asteroids are located in the asteroid belt between Jupiter and Mars, and are believed to be relic solar nebula planetismals that escaped incorporation into planets.

Chondrites. Over 90% of the meteorites that are observed to fall out of the sky are classified as chondrites, samples that are distinguished from terrestrial rocks in many ways. One of the most fundamental is age. Like most meteorites, chondrites have formation ages close to 4.55 Gyr. Chondrites also have basically undifferentiated elemental compositions for most nonvolatile elements and match solar abundances except for moderately volatile elements. The most compositionally primitive chondrites are members for the type 1 carbonaceous (CI) class.

Another unique property of chondrites is the presence of chondrules, objects found in nearly all chondrites except those of the CI class. Chondrules are millimeter-sized, spheroidal bodies composed predominantly of olivine, $(Mg, Fe)_2SiO_4$, pyroxene, and a glass of approximate feldspathic composition. It is believed that these were objects individually orbiting the Sun that formed by rapid melting and cooling of millimeter-sized precursors. The processes must have been highly efficient, as chondrules comprise $> 75\%$ of the mass of many meteorites. It is possible that chondrules were primary building blocks of the earth and the terrestrial planets.

Chondrites are divided into eight subclasses distinguished by elemental, isotopic, and mineralogical composition. A characteristic distinguishing different chondrite groups is the abundance and oxidation state of iron, Fe. Chondrite classes are also distinguished by their abundances of both volatile and refractory elements and by the oxygen isotope compositions. This remarkable distinction, illustrated in the three-isotope plot shown in Figure 1, correlates with compositional and mineralogical classification.

Within each chondrite class there are petrographic grades that relate to alteration processes that occurred within the meteorite parent body. The grades range from 1 to 6, although no class has examples in more than four grades. Grades 3 to 6 represent the effects of thermal metamorphism where the higher number is the more strongly altered. Grades 1 and 2 occur only for the CI and CM chondrites, respectively. CI and CM chondrites have been extensively altered by aqueous alteration in their parent bodies probably as a result of the melting of ice followed by reactions of preexisting phases.

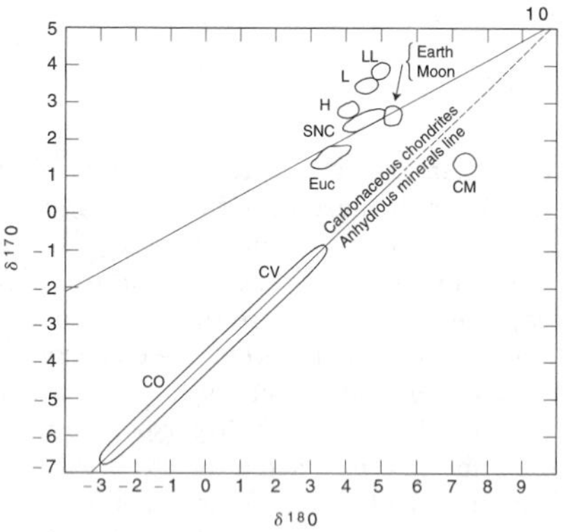

Figure 1. The bulk oxygen isotopic composition of different meteorite classes where (—) is the terrestrial fractionation line. The δ notation refers to the normalized difference between ^{17}O:^{16}O or ^{18}O:^{16}O ratios to those in standard mean ocean water (SMOW) in relative units of parts per thousand. The meteorites formed from materials that were enriched or depleted in ^{16}O and their bulk compositions plot off of the terrestrial line which has a slope of 1/2 owing to mass dependent fractionation. Some of the anhydrous minerals from carbonaceous chondrites fall on a line having a slope of 1. These anomalous compositions may be produced by mixing with an ^{16}O-rich component that has a different nucleosynthetic history than mean solar system material.

Achondrites. The achondrites are differentiated stony meteorites that are apparently derived from parent bodies that were heated to at least partial melting temperatures. Achondrites do not contain chondrules and do not have elemental compositions that match solar abundances for condensable elements. These materials are old but their properties were more determined by planetary processes such as melting and differentiation than by primary nebular processes, such as condensation and accretion. Many of the achondrite subclasses can be combined into three basic groups: HED, SNC, and lunar groups. The HED group comprises the howardite, eucrite, and diogenite subclasses, which together are responsible for more than 6% of all meteorite falls. The SNC achondrites are composed of the Shergotty, Nakhla, and Chassigny subgroups.

Irons. Approximately 4% of meteorite falls are irons. Because they are distinctive rocks and weather relatively slowly, most meteorites that were not seen to fall, but were found accidentally, are irons. Iron meteorites are composed of metallic iron and siderophile elements that fractionated from molten parent bodies. They may have been cores of asteroids or they may have only been localized metal accumulations.

Origin. Typical meteorites have formation ages of 4.55 Gyr and exposure ages of only 10^7 years, during which time they existed as meter-sized bodies unshielded to the effects of cosmic rays. With the exception of the SNC (Martian) and lunar meteorites, it is widely believed that most conventional meteorites are asteroid fragments liberated relatively recently by collisions and transferred to the earth by gravitational perturbations. The principal source location is thought to be the zone 2.5 AU (one AU is equal to the mean distance between the Earth and the sun) from the sun where the orbital frequency about the sun is exactly three times that of Jupiter.

Interplanetary Dust

Interplanetary dust particles (IDPs) are the submillimeter-size regime of the solar system meteoroid inventory ranging in size from tens of nanometers to 1000 km in diameter. These particles are short lived in the interplanetary medium because of the effects of self-collisions and orbital decay caused by the drag component of sunlight (the Poynting-Robertson effect). Most of the IDPs that now reach the earth were liberated from comets and asteroids within the last 10^6 years. Over 40,000 tons of IDPs impinge on the earth annually and cumulatively they are the dominant meteoritic mass input on time scales shorter than 10^7 years. Interplanetary dust in the size range of a few micrometers to a millimeter can be collected in and below the atmosphere, and recovered particles provide an important sampling of asteroids and comets. These samples complement conventional meteorites because they include specimens of objects that for a variety of reasons do not either reach the earth or survive atmospheric entry in greater than cm-sized pieces to become conventional meteorites. Dust provides a broader and less biased sampling of interplanetary materials because small samples of fragile materials can survive atmospheric entry without crushing. Additionally, sunlight pressure effects cause all dust beyond the earth's orbit to evolve toward the sun where collisions with earth are possible.

Collection. IDPs can be collected in space although the high relative velocity makes nondestructive capture difficult. Below 80 km altitude, IDPs have decelerated from cosmic velocity and collection is not a problem; however, particles that are large or enter a very high velocity are modified by heating.

The flux of 10-μm particles is $1/(m^2 \cdot d)$, a value high enough that these IDPs can be collected directly using high altitude aircraft. The spatial density of 10-μm IDPs at 20-km altitude is $10^{-3}/m^3$. Particles >100 μm fall at a rate of only $1/(m^2 \cdot yr)$ and can only effectively be collected after they have fallen to the ground and concentrated in a surface deposit. The small particles are collected primarily using stratospheric aircraft although they have also been recovered by melting pristine Antarctic ice. The larger IDPs have been collected from deep sediments, Greenland ice, and Antarctic ice, and a few other selected terrestrial environments that allow the extraterrestrial to be efficiently isolated and distinguished from terrestrial particles. Many of the $>100 - \mu$m cosmic particles are spherules and their shape assists in making a distinction from other materials.

The most common IDPs are black objects having approximately solar elemental composition except for very volatile elements such as the noble gases. There are particles that deviate strongly from this pattern but they are rare and are usually dominated by a single mineral such as FeS, olivine, or FeNi metal. Most of the particles can be grouped into two classes: one contains hydrated minerals such as serpentine and smectite; the other, ones that are anhydrous.

Origin. Individual IDPs are short-lived and the long-term presence of dust in the solar system implies that there must be sources capable of generating the approximately 10^7 kg/s of new dust required to balance losses by collisions, ejection by radiation pressure, and spiraling into the sun owing to the Poynting-Robertson effect. For sizes >10 μm, it has long been known that comets and asteroids are the principal source. Particles are liberated from asteroids by collisions of both asteroids and asteroid debris. Dust from comets is released when solar heating sublimes the ice matrix in comets. Dust is ejected, and because of the effects of light-pressure drag and ejection velocity, it forms the dust tail that can extend to lengths of over 10^8 km. Most of the comet dust that is collectable on the earth is believed to have been derived from the Kuiper belt comets that reside in a flattened distribution extending from the region of the outer planets to distances of a few 100 AU from the sun. The dust from both comets and asteroids is believed to be samples of early solar system materials that have been relatively well preserved over the age of the solar system inside moderately small bodies.

DONALD E. BROWNLEE
University of Washington

J. F. Kerridge and M. S. Matthews, eds., *Meteorites and the Early Solar System*, University of Arizona Press, Tucson, Ariz., 1988.

R. Hutchinson, *The Search for Our Beginning*, Oxford University Press, New York, 1983.

H. Y. McSween, *Meteorites and Their Parent Planets*, Cambridge University Press, New York, 1987.

F

FANS AND BLOWERS

Fan is the generic term for low pressure air- and gas-moving devices using rotary motion. Fans are subdivided into centrifugal and axial-flow types, depending on the direction of air flow through the impeller. In centrifugal fans, the air is introduced into the center of a revolving wheel or rotor with peripheral blades. Air is drawn through the blades and forced out in centrifugal flow into a scroll or volute housing where a portion of the kinetic energy is converted to pressure or static head. In axial-flow fans the air continues to move directly forward through the fan along the axis of the shaft. Kinetic energy is imparted to the air by the shape and arrangement of the blades. After discharge through the blades, although the general flow direction is still forward, a spiral component of velocity generally has been added to the air. A propeller-type fan is the most common axial-flow fan but more complicated types are in use where the blades resemble vanes in a turbine.

Blower is a term applied to a centrifugal fan generally when it is used to force air through a system under positive pressure. It generally implies a fan developing a reasonably high static pressure of at least 500 Pa (several inches of water). High speed centrifugal blowers (≥ 3600 rpm) are also available in one or more stages to compress air to pressures of 108–150 kPa (1–7 psig). The term blower is also applied to relatively low pressure positive displacement compressors of the rotary lobe, screw, or sliding vane types where the discharge pressure is usually less than 205 kPa (15 psig). Positive displacement blowers are outside the scope of this article.

When a fan is placed at the end of a system so that most of the system pressure drop is on the suction side of the fan, it is commonly called an exhaust fan or an exhauster. This term may also be applied to a ventilating fan where the primary function is to exhaust air from a room or an open hood.

Centrifugal compressors or turbocompressors are high volume centrifugal devices capable of gas compression varying from 105 to >1500 kPa (0.5 to several hundred psig). These generally consist of a number of stages of alternating rotating and stationary turbine blades and turn at very high speeds (see HIGH PRESSURE TECHNOLOGY).

The total pressure produced by a fan can be measured with an impact probe pointed directly upstream. The pressure so measured is a combination of both the static pressure and the kinetic energy pressure equivalent. Static pressure can be measured using a properly designed static wall tap or using the static pressure parts of a pitot tube. The latter represents the true pressure head exclusive of velocity effects. The difference between the total (impact) pressure and the static pressure is the velocity pressure or velocity head. Pressure readings are normally expressed in millimeters or inches of water (1 mm water = 9.807 Pa; 1 in. water = 248.8 Pa) and are referred to atmospheric pressure (101.3 kPa) as the reference base. Thus barometric pressure must be added to obtain absolute pressure. The total pressure rise produced by a fan is the difference in total pressure between the fan outlet and inlet. The fan static pressure is the total pressure rise for the fan reduced by the discharge velocity pressure. Inlet velocity head is assumed to be zero for fan rating purposes.

Centrifugal Fans

Performance Testing. Although fan performance characteristics can be roughly estimated during the early stages of design, fan efficiency losses and slip cannot be estimated accurately from theory alone. Therefore, the exact performance characteristics of a new fan design must be determined by testing. Test conditions such as provision for steady and uniform flow of air approaching the fan inlet must be carefully controlled, because any inlet disturbances can affect performance. For this reason, fan field tests are seldom reliable, and most testing is performed in a laboratory on a test block following procedures set forth in standards for performance including sound.

Types and Characteristics. The four basic fan wheel and blade designs are forward-curved blades, backward-curved blades, straight radial blades, and airfoil design.

Fan Laws and Their Applications

Manufacturers' performance ratings are generally based on atmospheric pressure at sea level, 20°C, and 50% rh. Changes in temperature, gas density, and fan speed affect the performance. For most fan users, the following four laws are adequate.

When fan speed is changed (*1*) the capacity or flow rate varies directly with the speed ratio; (*2*) discharge pressure varies directly with the square of the speed; (*3*) power varies directly with the cube of the speed (at constant inlet density with no change in temperature, absolute pressure, or composition); and (*4*) discharge pressure and power requirements at a constant capacity and fan speed vary directly with gas density, p.

Fan Selection. A fan is selected according to its location in the air-flow system, system performance and control characteristics, cost, efficiency, control stability, flexibility, and noise level.

Duct Connections. Performance curves are measured under ideal laboratory conditions. However, to obtain the same performance curve from a fan in a field installation, the system must approach the characteristics of the test conditions at least in that part of the system close to the fan. Both inlet and outlet duct connections can influence fan performance significantly. These connections can actually change the shape of the fan curve. Therefore, no single correction factor can account for the performance change over its entire range. Although poor outlet connections affect performance, improper inlet connections generally hurt performance more, reducing it the most near free-delivery conditions and the least at peak pressure. Poor performance can result from fan inlet eccentric or spinning flow and discharge ductwork that does not permit development of full fan pressure. Sometimes inlet restrictions starve a fan and limit performance. To obtain rated performance, the air must enter the fan uniformly over the inlet area without rotation or unusual turbulence.

The velocity of air discharging from a fan is not uniform across the discharge outlet but tends to be higher toward the outside of the scroll. The discharge duct evens the velocity distribution into the standard turbulent-flow distribution some distance downstream and converts part of the discharge velocity to static pressure. If a fan is operated without an outlet duct (discharging into a large plenum or the atmosphere), it loses 1–1.5 velocity heads.

Flow Control. In order to control flow, either the system characteristics or the fan characteristics must be changed. Generally, flow control affects the energy input to the fan. Low cost control devices often result in reduced fan efficiency and increased power consumption. Thus if flow reduction is to occur for a long time with powerful fans, more energy-efficient control devices should be considered.

The simplest and cheapest control device is a damper, butterfly valve, or an orifice placed in the duct to throttle the flow and change the system resistance characteristics.

Changing centrifugal fan characteristics usually results in greater energy savings than changing system characteristics. If fan pressure can be reduced together with flow, the most desirable method of energy conservation is to change the speed, because that leaves the efficiency unchanged. If fan capacity is to be changed only infrequently, speeds of belt-driven fans can be adjusted easily with sheave changes. Where frequent speed changes are required, variable-speed motors and drives (electric or hydraulic) are the best but the most expensive.

Inlet-vane control can be used to change the shape of the fan performance curve through imparting spin to the air entering the fan. As more spin is imparted, less energy can be transferred to the air from the blades and static pressure output is reduced.

Motor and Drive. The preferred prime mover for a fan is usually an electric motor. For fans of low to moderate power, V-belt drives are frequently employed. This permits selection of fans that can be operated over a wide range of speeds. Fans requiring powerful motors, 37–

75 kW (50–100 hp) and higher, are generally directly connected to the motor and driven at synchronous speed.

When selecting the motor, power requirements, effect of temperature changes on load, and motor starting current and torque have to be considered.

Other Selection Problems. Additional considerations can arise when fans must handle solids or gases of low density, or must be operated in parallel or series. A complicated flow system involving several fans in parallel, all of which are in series with a common exhaust fan, can lead to surging and vibration unless selected carefully. Maximum tip speed, bearing types, single- and double-inlet fans, and wheel and shaft natural frequency and rigidity must also be considered.

Low Density Gases. A fan may have to operate on low density gas because of temperature, altitude, gas composition (high water vapor content of the gas can be a cause of low density), reduced process pressure, or a combination of such causes. To develop a required pressure, the fan has to operate at a considerably higher speed than it would at atmospheric pressure, and hence it must operate much closer to top wheel speed. Bearing life is shorter, and the fan tends to vibrate more or can be overstressed more easily by a slight wheel unbalance. Abrasion of the blades from dust particles is more severe. Therefore, a sturdier fan is needed for low density gas service.

Mechanical Considerations. The mechanical design of a fan and the various forces that fan parts must withstand are discussed in the literature. The forces result from a combination of fluid, inertial, and vibrational effects.

Tangential forces from air compression act on the blades and are transmitted through the fan hub to the shaft in the form of a resisting torque. Axial thrust may be developed on the fan wheel and shaft because of pressure differences about the wheel and the directional change in momentum of the air at the wheel inlet. The net unbalanced axial thrust must be taken by a thrust bearing that transmits it through the bearing supports to the fan foundation. At maximum fan efficiency, the radial fluid forces acting on the wheel are nearly balanced, but the volute can be correctly designed for only one rating condition. Therefore, as fan operation departs from maximum efficiency, unbalanced radial thrust increases which must be carried by the bearings to the foundation. Centrifugal forces also act on the wheel. If the center of the wheel and shaft rotation does not coincide exactly with the center of the rotating mass, a flexural force produces bending of the shaft, apparent as vibration. In a rotating elastic system, dangerous vibrations are likely to occur at critical speeds. The application of repeated external forces such as flow surging or wheel unbalance excites the elastic structure and causes it to vibrate. If the excitation frequency is close to the natural frequency, resonance can occur with large amplitude vibrations. All of these forces must be carried by the bearings. Thus it is common to use heavier components as fans are called on to operate at higher speeds or higher pressure differentials.

Axial Flow Fans

Axial flow fans, in which the air flow is parallel to the fan axis, are the workhorse fans in many petrochemical and utility industry applications. These are the first choice of air mover whenever large volumes of air at low (most commonly up to 500 Pa (2.0 in. H_2O)) pressures are needed. Axial flow fans range in size from 25 mm diameter (cooling computers) up to 12.3 m in diameter (cooling condenser water in power plant cooling towers). These fans are used in air-cooled heat exchangers for process cooling in many chemical plants in sizes of 1.8–4.3 m (see HEAT-EXCHANGE TECHNOLOGY). Axial fans from 0.6 to 9 m diameter are used in heating, ventilation, and air conditioning (HVAC) applications in homes and office buildings around the world. Most commonly, axial flow fans are used in short ducts called fan rings or cylinders, discharging into the atmosphere. Most large fans in cooling towers have velocity recovery stacks that capture the wasted velocity pressure energy and convert it back into useful work.

Axial fans are classified as propeller, tube-axial, and vane-axial. The choice of fan required is determined by the resistance (static pressure) the fan must work against as well as the volume flow required. Propeller fans usually discharge into a plenum or directly into the atmosphere. Tube-axial fans are usually mounted in ducts as in an air conditioning system. Vane-axial fans are also mounted in ducts but feature a stationary guide vane on the discharge side that straightens the air flow to improve efficiency. Tube-axial fans can work at static pressures up to 623 Pa (2.5 in. H_2O); vane-axial fans can work up to 2000 Pa (8 in. H_2O).

Design Elements. Ideal conditions are obtained in the design of an axial-flow fan when energy transfer from the blade to the gas is uniform along the length of the blade, resulting in uniform pressure generation, minimum losses, and maximum efficiency and stability. Because the blade linear velocity varies with position from tip to hub, attainment of a uniform pressure rise along the blade at different radii requires variation of the blade angle form hub to tip. Hub size is increased for higher pressure designs where it is impractical to generate equal pressures nearer the center of the wheel. Close clearance between blade tips and fan housing is a stringent requirement to prevent backflow losses at the housing wall. Inlet and outlet connections are carefully designed to minimize turbulence.

Capacity Control. Variable air flow fans are needed in the process industry for steam or vapor condensing or other temperature critical duties. These also produce significant power savings. Variable air flow is accomplished by (*1*) variable speed motors (most commonly variable frequency drives (VFDs); (*2*) variable pitch fan hubs; (*3*) two-speed motors; (*4*) selectively turning off fans in multiple fan installations; or (*5*) variable exit louvers or dampers. Of these methods, VFDs and variable pitch fans are the most efficient. Variable louvers, which throttle the airflow, are the least efficient.

Fan Rating. Axial fans have the capability to do work, ie, static pressure capability, based on their diameter, tip speed, number of blades, and width of blades.

Efficiency. Fan efficiency describes a fan's ability to do work and is calculated as total efficiency (Eff) using total pressure (TP) or static efficiency (Eff) using static pressure (SP). Total efficiencies for axial fans range from 55 to 80%; static efficiencies range from about 40 to 65%. When pressure is in pascals and flow rate in m^3/s,

$$\text{Total Eff} = \frac{\text{TP} \times \text{flow rate} \times 10^{-3}}{\text{kW}}$$

$$\text{Static Eff} = \frac{\text{SP} \times \text{flow rate} \times 10^{-3}}{\text{kW}}$$

Performance Curves. Fan manufacturers furnish fan performance curves for each type fan available. These are typically based on 61 m/s (12,000 ft/min) tip speed and 1.20 kg/m^3 (0.75 lb/ft^3) density. To select a fan for a specific duty requires knowledge of the flow, static pressure resistance, and density of the actual operating conditions.

Selection. The fan selection process consists of determining the exact operating point that coincides with the design static pressure, air flow, and density required by the system resistance line.

Application Criteria. The design and construction of axial fans is dictated by size and function. Small fans are usually molded plastic having fixed blades. Most fans larger than one meter in diameter feature hollow fiber glass or extruded or cast aluminum blades that can be adjusted to the proper pitch when the fan is at rest, to provide the required air flow at the design speed. To perform properly, the output of the fan in terms of air flow and static pressure capability must match the system resistance at the design air density. This requirement dictates the fan diameter, rpm, number and types of blades, and blade pitch settings. If a fan has fixed pitch blades, the fan speed must be adjusted. Another increasingly important requirement is fan noise. To meet the maximum allowable noise, fan speed is normally limited. Often fan diameters are determined by space limitations as well as by volume flow requirements.

Noise of Fans

Fan noise is demanding and receiving much attention because of environmental laws. The basic control document is the federal OSHA limitation of 90 dB(A) at an operator's work place for 8-h exposure. There are other limitations on entire plant noise at the boundary of new plants from local ordinances which are typically more severe than the OSHA limitation.

Effect of Vibration. A fan blade is continuously vibrating millions of cycles up and down in operation over a short period of time. Each time a blade tip moves past an obstruction it is loaded and then unloaded. If forced by virtue of tip speed and number of blades to vibrate at its natural frequency, the amplitude is greatly increased and internal stresses result. It is very important when selecting or rating a fan to avoid operation near the natural frequency.

Fatigue from Vibrations. To avoid the possibility of fatigue caused by a resonance, more extensive use of fiber glass blades is being made rather than metallic blades such as aluminum. Fiber glass composite blades, which are often hollow, are not notch sensitive, ie, a small scratch or crack does not spell the disaster it would in a metal blade. Secondly, the lighter mass of a fiber glass blade means less kinetic energy to dissipate in the event of an accident, because the destructive energy in a fan wreck is directly proportional to the mass weight of the blades.

Uses

Fans and blowers are the most widely used mechanical devices for moving air and gases in both large and small volumes. Uses include ventilation, mechanical draft for combustion (including forced- and induced-draft fans and primary- and secondary-air fans), local exhaust for fume and dust containment at hoods and equipment enclosures, forced- and induced-draft cooling for spray towers, cooling towers and ponds, and air-cooled heat exchangers, and conveying of solids (see also HEAT-EXCHANGE TECHNOLOGY). Other applications include air or gas movement in dryers, gas-recirculation fans, air supply for air curtains and air-blast operations, and a great many miscellaneous process industry uses often involving hot and corrosive gases.

ROBERT C. MONROE
Hudson Products Corporation

ASHRAE 1992 Handbook—HVAC Systems and Equipment, American Society of Heating, Refrigerating and Air-Conditioning Engineers, Inc., Atlanta, Ga., 1992.

R. Jorgensen, *Fan Engineering*, 8th ed., Buffalo Forge Co., Buffalo, N.Y., 1983.

Air-Cooled Heat Exchangers for General Refinery Service, API 661, American Petroleum Institute, Washington, D.C., Apr. 1992.

W. C. Osborne, *Fans*, 2nd ed., Pergamon Press, New York, 1977.

FAT REPLACERS

Fat in Foods

A good compilation of the functions of fats in various food products is available. Some functions are quite subtle, eg, fats lend sheen, color, color development, and crystallinity. One of the principal roles is that of texture modification which includes viscosity, tenderness (shortening), control of ice crystals, elasticity, and flakiness, as in puff pastry. Fats also contribute to moisture retention, flavor in cultured dairy products, and heat transfer in deep fried foods. For the new technology of microwave cooking, fats assist in the distribution of the heating patterns of microwave cooking.

It is likely that no one material will replace fats in food; rather, replacement will consist of mixtures, with each ingredient addressing one or more of the roles played by fats in foods.

Fat-Altered Foods. The Food Labeling and Education Act, effective November 1992, defines low fat, low cholesterol, and fat-free foods. Low fat must contain less than 3 grams of fat per serving and per 100 grams of food, reduced fat must be 50% of the usual fat content and reduction must exceed 3 grams per serving, fat-free must have less that 0.5 grams of fat per serving, and percent fat-free must describe foods that fit into the definition of low fat. Foods labeled "lite" must be one-third lower in calories. If more than 50% of the calories are from fat, it must have the fat content reduced by 50%. If lite means other than calories, it must be specified.

With the extensive knowledge available in oil chemistry, development of designer fats and oils is possible. This is of special interest to nutritionists who see the possibilities for structurally designed fats to meet developing knowledge in clinical nutrition and food product development. There is more activity in dairy products than anywhere else in the food industry. Ice milk and frozen yogurt, early leaders in the field, rose rapidly in sales then plummeted. Fat-free ice cream has been marketed, but final results are not yet available. Sales of these products have not cannibalized traditional ice cream. Fat-free ice creams have encountered strong resistance in some segments of the retail trade. Retailers in Maine and New York, states with important dairy producing industries, refuse to sell such products.

One frozen dessert was made with Simplesse, a protein-based fat mimetic that contains no fat. Other dairy product developments include a fat flavor, produced by encapsulating milk fatty acids in maltodextrins; fat-free cottage cheeses; and 2% fat milk, prepared by steam stripping cream with partial fat addback, with a cholesterol level about 60% lower than the starting material.

Activity in the cereal field includes the introduction of a full line of fat-free, cholesterol-free loaf cakes, crunch cakes, and cookies; a light frosting mix and a light pancake mix are also included.

There is available a fat- and cholesterol-free salad dressing, and a fat-reduced, but not fat-free, dressing. There are Federal Standards of Identity for dressings; one labeling requirement is a minimum fat content. The standards and labeling for lowered fat constitute a problem for the regulatory agencies.

Another food product which faces standards problems is processed meat. A great deal of interest has been focused on hamburgers, especially by the fast food restaurants. Carrageenan and hydrolyzed vegetable protein has been used to produce a hamburger with less than 10% fat. Another approach uses frozen soy protein isolate to admix with ground beef to produce hamburger, with an accompanying drop in fat content from 42% down to 24%. Oatrim, a hydrolyzed oat flour, has also been employed as a mixer to produce a low fat hamburger, although at a somewhat elevated price.

The complexity of total replacement has slowed the rate of introduction of new materials, but most ingredient producers introduce a product which replaces one or two aspects of fat functionality and has already been cleared for use in foods by the FDA.

Fat Extenders

Carbohydrates. The materials offered for fat replacement are either carbohydrate or protein and protein-like compounds. Table 1 lists carbohydrate fat-sparing agents.

Starches for fat replacement originate from corn, potato, tapioca, oat, and rice. Starches are comprised chiefly of straight (amylose) and branched (amylopectin) chains of glucose. The ratio for branched to straight chains has an effect on the nature of the resultant starch; this ratio varies with the starch source. Hydrolysis of these starches yields dextrins and maltodextrins, which are also useful in replacing food fats.

Celluloses. Complex carbohydrates including gums, cellulose, methylcellulose, and carboxymethylcellulose also have found application in fat replacement. A good summary of the application of these materials is available.

Microcrystalline cellulose is a nonfibrous form of cellulose obtained by breaking fibrous plant cell walls into fragments, sized from 25 μm to a few tenths of a μm. In commercial preparations 60% of the fragments are below 0.2 μm. In use, microcrystalline cellulose is combined with sodium carboxymethylcellulose (CMC) to produce a colloidal gel. In this combination, CMC serves as a protective colloid, and permits ready dispersion of the powdered mix. When dispersed in a food, the

Table 1. Carbohydrate Fat-Sparing Agents[a]

Trade name	Chemical composition	Producer
Amalean	high amylose corn starch	American Maize
Sta-Slim	modified potato and tapioca starches	Staley Mfg.
Stellar	hydrolyzed corn starch	Staley Mfg.
Paselli SA2	low DE hydrolyzed potato starch[b]	Avebe America
Oatrim	oat maltodextrin	Rhône-Poulenc
Maltrin	corn maltodextrins	Grain Processing
Sta-Dri	waxy maize maltodextrins, DE[b] series	Staley Mfg.
Rice* Complete	hydrolyzed rice solids	Zumbro
Rice* Trin	rice maltodextrins	Zumbro
Rice* Pro	rice protein and maltodextrin	Zumbro
N-Lite B	corn maltodextrin	Nat'l Starch
N-Lite D	modified starch	Nat'l Starch
N-Lite F	mix of starch, milk, and guar gum	Nat'l Starch
N-Lite L	modified starch	Nat'l Starch
N-Lite LP	pre-gelled starch	Nat'l Starch
N-Oil	tapioca dextrin	Nat'l Starch
Slenderlean	modified starch	Nat'l Starch
Lycasin	hydrogenated corn starch hydrolysate	Roquette
Trim Choice	oat maltodextrin	Con Agra
Superbase	mix of rice maltodextrin, starch, xanthan, and whey protein	Excel
Litesse	polydextrose	Pfizer

[a] Many of these products have multiple suppliers.
[b] DE = dextrose equivalent.

mixture sets up a three-dimensional network held together by weak hydrogen bonds. This structure gives the mix its functional properties, ie, emulsion and foam stabilization, high temperature stability, thickening, suspension, and ice crystal control in frozen desserts.

Plant Gums. There are a large number of plant gums that find application as fat replacers. Many are dried plant exudates. Plant gums include polydextrose (Pfizer), plant gums from arabic, guar, locust bean, tamarind, and tara seeds, gum tragacanth, xanthan, derived from the pure culture fermentation of an organism, *Xanthomonas campestris*, Konjac flour, and pectin.

Marine Gums. There are several related gums of marine origin. Carrageenan, from red seaweed, is probably the best known and yields three basic types of gum when extracted, ie, kappa, iota, and lambda. Agar, a gum known for its gel properties, is derived from red algae gathered from in-shore sources, mainly in the Far East, mostly from the Philippines.

Microparticulate Protein Fat Mimetics

There are only a few fat replacement products based on protein. LITA is a corn protein–polysaccharide compound; the role of the polysaccharide is to stabilize the protein (zein). The final product is 87% protein and 5% polysaccharide. The mixture, spray dried after processing, claims to look like cream on rehydration. It is low in viscosity, flavor, and lubricity, and is stable to mild heating.

Simplesse, the best known entry in the protein field, is the subject of several patents. A good scientific description of the product and the process is available. The original product was prepared by treating acidified whey concentrate to simultaneous pasteurization and homogenization. The resultant product is spherical and is in the 1-μm particle size range. The product owes its effectiveness to this particle size, which is below the threshold of particle sensing and serves as a surrogate disperse phase. The effect is that of creaminess in the absence of fat. A later product is based on combined egg white and skim milk, and a further development employs unacidified whey protein concentrate.

The long-standing practice of admixing isolated soy protein with ground meat to lower fat content should not be overlooked. Soy protein has found its best application in hamburger patties, though there are applications in other ground meat products.

Synthetic Fats

Standing in contrast to the above approaches is the strategy of creating new molecules with fat and oil-like physical properties, but a molecular configuration not recognized by the human digestive system; hence they are noncaloric. Olestra, the first such product, achieves its effect by attaching fatty acids to a sucrose, rather than a glycerol backbone. It has most of the qualities for fat replacement including utility in deep fat frying. It has been delayed since the 1970s because of lack of FDA clearance. There is a claim that the potential volume of use is so huge that it constitutes a new development in the approval process.

The size of the fat replacement market has stimulated many attempts to synthesize fat replacer molecules. These efforts, described as attempts to produce "acaloric compounds with fat-like properties, but whose ester bonds have been modified," would include glycerol ethers, pseudofats, and carbohydrate fatty acid esters. Research into this group of compounds indicates that as the number of ester groups increases, there is decreased digestion; ie, at eight ester groups there is virtually no digestive lipolysis and the compound is passed through undigested. The octaester has been shown to lower cholesterol, especially low density cholesterol. Mitsubishi and Unilever are both working on partially esterified sucrose.

Another group of synthetic fat compounds is centered around polycarboxylic esters. Trialkoxycarballylate (TATCA) and trialkoxycitrate (TAC) (Best Foods) have been tried in mayonnaise and margarine. Similar materials have been developed based on malonic acid esters. Work on propoxylated glycerols (Atlantic Refining Company (ARCO)) is still at the bench level. Silicone oils, eg, phenylmethylpolysiloxane and organosilanes, have been investigated as fat replacers, but the controversy over breast implants and the possible side effects of silicones may cast a cloud over this product area.

Caprenin, suggested as a cocoa butter substitute, has useful functional properties with the added value of fewer calories than cocoa butter. It is a triglyceride comprised of two short-chain fatty acids and behenic acid (docosanoic acid), a fully saturated 22-carbon fatty acid. Caprenin is found in high concentration in canola (rapeseed) oil, and can be synthesized from erucic acid, which is also high in earlier strains of canola.

ROY E. MORSE
Consultant
NORMAN SINGER
Consultant

A. M. Thayer, *C & E News*, 9–12, (June 3, 1991).
M. Glicksman, *Food Technol.*, 94–103 (Oct. 1991).
B. Summerkamp and M. Hesser, *Bulking Agents and Fat Substitutes: Analysis of a Dynamic Industry*, HRA, Inc., Prairie Village, Kans., 1989.
D. J. Hamm, *J. Food Sci.* **49**, 419–423 (1984).

FATS AND FATTY OILS

Fats and oils are composed primarily of triglycerides (**1**), esters of glycerol and fatty acids. However, some oils such as sperm whale, jojoba, and orange roughy are largely composed of wax esters. Waxes (qv) are esters of fatty acids with long-chain aliphatic alcohols, sterols, tocopherols, or similar materials.

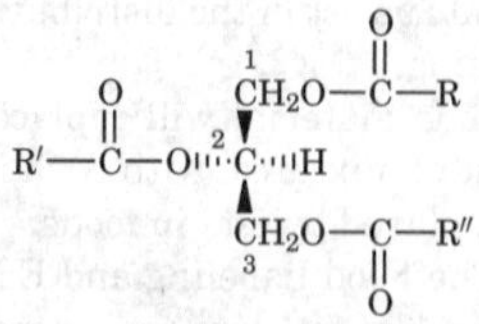

(**1**)

Fatty acids derived from animal and vegetable sources generally contain an even number of carbon atoms since they are biochemically derived by condensation of two carbon units through acetyl or malonyl coenzyme A. However, odd-numbered and branched fatty acid chains are observed in small concentrations in natural triglycerides, particularly ruminant animal fats through propionyl and methylmalonyl coenzyme A, respectively. The glycerol backbone is derived by biospecific reduction of dihydroxyacetone.

Structure (1) shows the stereochemistry of the triglyceride molecule. Positions are numbered by the stereochemical numbering (sn) system. In chemical processes the 1 and 3 positions are not distinguishable. However, for biological systems, the enantiomeric (R or S) form is important.

Fatty acids may be saturated, monounsaturated, or polyunsaturated according to the number of double bonds int he alkyl chain. Naturally occurring double bonds are almost exclusively *cis* (Z) in configuration. The most common fatty acids in animal and vegetable fats and oils are dodecanoic (lauric, 12:0), hexadecanoic (palmitic, 16:0), octadecanoic (stearic, 18:0), 9-*cis*-octadecenoic (oleic, 18:1), 9-*cis*,12-*cis*-octadecadienoic (linoleic, 19:2), and 9-*cis*,-15-*cis*-octadecatrienoic acid (linolenic, 18:3).

Fats and oils are distinguished by their physical state; fats are solid at ambient temperature, whereas oils are liquid. Some edible triglycerides, such as butter, lard, vegetable oils, shortenings, and margarines, have substantial quantities of both liquid and solid components at ambient temperature. Commercial products may be derived from animal carcasses by rendering, or vegetable sources by pressing or solvent extraction.

Composition

Natural fats and oils are composed principally of triglycerides, but other components may be present in minor quantities. These components may have important effects on the nature and quality of the oil or fat.

Free Fatty Acids and Partial Glycerides. After harvest, many crude oil crops contain lipase enzymes that cleave triglycerides into fatty acids and partial glycerides. Elevated free fatty acid concentrations are undesirable because they cause high losses during further processing of the oil. Diglycerides and monoglycerides are formed by hydrolysis.

Free fatty acids are removed by refining, physical refining, or deodorization. Mono- and diglycerides are not removed by alkali refining or bleaching and may have an adverse effect on the quality of the oil.

Phospholipids. Glycerides esterified by fatty acids at the 1,2 positions and a phosphoric acid residue at the 3 position constitute the class called phospholipids. The identity of the moiety (other than glycerol) esterified to the phosphoric group determines the specific phospholipid compound. The three most common phospholipids in commercial oils are phosphatidylcholine or lecithin, phosphatidylethanolamine or cephalin, and phosphatidylinositol. These materials are important constituents of plant and animal membranes. The phospholipid content of oils varies widely. Most oils contain 0.1 to 0.5%. Some phospholipids, such as dipalmitoylphosphatidylcholine ($R = R' =$ palmitic; $R'' =$ choline), form bilayer structures known as vesicles or liposomes. The bilayer structure can microencapsulate solutes and transport them through systems where they would normally be degraded. This property allows their use in drug delivery systems (qv).

Sterols. Sterols are tetracyclic compounds derived biologically from terpenes. They are fat-soluble and therefore are found in small quantities in fats and oils. Cholesterol is a common constituent in animal fats such as lard, tallow, and butterfat.

Tocopherols and Tocotrienols. Algae and plants used as sources of edible oils contain tocopherols, phenolic materials that function as antioxidants (qv). Mammals do not synthesize these compounds, and residues present in their bodies are present because of ingestion. Tocopherols are designated as being vitamin E active. Tocotrienols differ from tocopherols by the presence of three isolated double bonds in the branched alkyl side chain.

Several other naturally occurring antioxidants have been identified in oils. Sesamol occurs as sesamoline, a glycoside, in sesame seed oil. Ferulic acid is found esterified to cycloartenol in rice bran oil and to β-sitosterol in corn oil. Although it does not occur in oils, rosemary extract has also been found to contain powerful phenolic antioxidants.

Carotenoids and Other Pigments. Carotenoids contain conjugated double bonds, a strong chromophore which produces red and yellow coloration in vegetable oils. Carotenoids are tetraterpene hydrocarbons formed by the condensation of eight isoprene units. Another class of compounds, the xanthophylls, is produced by hydroxylation of the carotenoid skeleton. β-Carotene is the best known component of the carotenoids because it is the precursor for vitamin A.

Green coloration, present in many vegetable oils, poses a particular problem in oil extracted from immature or damaged soybeans. Chlorophyll is the compound responsible for this defect.

Processing of Fats and Oils

Fats and oils are derived from animals, plants, or fish by rendering (animal tissues), pressing, or solvent extraction. Crude oils from these processes are often of insufficient quality to be used directly, particularly for edible products. Impurities such as pigments, phosphatide, volatile odorous compounds, and certain metals must be removed by further processing, eg, degumming and dewaxing, refining, bleaching, hydrogenation, randomization/interesterification, physical fractionation, and deodorization.

Sources of Fats and Oils

Fats and oils may by synthesized in enantiomerically pure forms in the laboratory or derived from vegetable sources (mainly from nuts, beans, and seeds), animal depot fats, fish, or marine mammals. Oils obtained from other sources differ markedly in their fatty acid distribution. One variation in composition is the chain length of the fatty acid. Another variation of the fatty acid is the degree of unsaturation.

Castor oil (qv) contains a predominance of ricinoleic acid, which as an unusual structure inasmuch as a double bond is present in the 9 position, whereas a hydroxyl group occurs in the 12 position. The unusual structure of ricinoleic acid affects the solubility and physical properties of castor oil.

Solid fats may show drastically different melting behavior.

Biotechnology is slowly revolutionizing the edible oils industry. Crop breeding, genetic engineering, tissue culture, and mutation selection are avenues being pursued to deliver desirable fatty acid compositions into agronomically favored plants. Oils from microbial sources may offer unique fatty acid compositions. Canola, high oleic sunflower, and high oleic canola oils are recent successes in harnessing the biosynthetic factories.

Separation of a fat or oil from its source material can be accomplished by several different methods. Selection of an extraction process is based on *(1)* obtaining oil substantially undamaged and relatively free of undesirable impurities, *(2)* achieving the highest practical yield, and *(3)* obtaining the maximum economic return on the oil and coproducts. Processes include rendering, mechanical pressing, and solvent extraction.

Physical Properties of Fats and Oils

The physical properties of fats and oils have been reviewed.

Crystallization and Melting Behavior. Pure compounds usually display sharp melting points and impure compounds show broad melting behavior. However, even pure triglycerides show complex melting behavior because of their tendency to pack in several different crystal lattice forms (polymorphism). Triglycerides having three identical fatty acids pack into three distinct polymorphs: *(1)* β, the most stable form shows a triclinic subcell, *(2)* β', a less stable crystal which suggests orthorhombic packing, and *(3)* α, a loosely packed triglyceride which packs hexagonally. Rapid cooling of the triglyceride leads initially to the α' form followed by slow reorganization to β' and β

forms. Mixed glycerides, with more than one type of saturated fatty acid, pack with defects in the structure and chains appear to tilt to correct for these defects. Glycerides with unsaturated fatty acids must pack to accommodate the bend in the alkyl chain caused by the cis double bond. Perhaps the most widely studied fat, cocoa butter, may show as many as seven distinct polymorphic forms.

Some general trends in specific heats have been suggested: *(1)* for solid fats, there is little variation in specific heat for saturated fats and their fatty acids as chain length varies. Specific heat varies directly with the degree of unsaturation. Specific heat of a solid is less than that of the liquid at the same temperature; *(2)* specific heat of liquid fatty acids and glycerides increases with increasing chain length but decreases with increasing unsaturation. For both liquids and solids, specific heat increases with increasing temperature; and *(3)* mixed-acid glycerides have lower specific heats than their corresponding simple glycerides.

Viscosity. Fats, oils, fatty acids, and other fatty acid derivatives show relatively high viscosities compared to other liquids because of the intermolecular interaction of long alkyl chains. Some general trends are that longer chain lengths and lower unsaturation produce higher viscosities. Fatty acids are more viscous than esters because of the greater tendency to form hydrogen bonds. Castor oil is in a class by itself because of its side-chain hydroxyl group, which can form hydrogen bonds. Derivatives of castor oil are consequently useful as specialty lubricants. Fats and oils behave as Newtonian liquids, except at very high shear rates where degradation may begin to occur.

Surface and Interfacial Tension. Commercial oils tend to have lower surface and interfacial tensions because of the presence of polar surface-active components such as monoglycerides, phospholipids, and soaps. Purification of oils on a Florisil column can be used to obtain higher and more consistent values. Monoglycerides and phospholipids can reduce the interfacial tension between an oil and water. Emulsions (qv) may be formed that are relatively stable. Food products such as mayonnaise, margarine, and nonseparating salad dressings are commercial examples of stable emulsions.

Density. The density of liquid oils at 15°C does not vary markedly with changes in composition. Values generally range from 0.912 to 0.964 g/mL. Density increases with decreasing molecular weight and increasing unsaturation.

Smoke, Flash, and Fire Points. These thermal properties may be determined under standard test conditions. These values are profoundly affected by minor constituents in the oil, such as fatty acids, mono- and diglycerides, and residual solvents. These factors are of commercial importance where fats or oils are used at high temperatures such as in lubricants or edible frying fats.

Refractive Index. Refractive index of a fat or oil increases with molecular weight and unsaturation.

Absorption Spectra. Infrared spectra of fats and oils are similar regardless of their composition. The principal absorption seen is the carbonyl stretching peak which is virtually identical for all triglyceride oils. The most common application of infrared spectroscopy is the determination of trans fatty acids occurring in a partially hydrogenated fat.

Solubility Properties. Fats and oils are characterized by virtually complete lack of miscibility with water. However, they are miscible in all proportions with many nonpolar organic solvents.

Chemical Properties

Most triglyceride fats and oils have only two reactive functional groups: the ester linkage joining the fatty acid to the glycerol backbone and double bonds in the alkyl side chain. There is a free hydroxyl group in the side chain of ricinoleic acid found in castor oil and a carbonyl group in the licanic acid side chain of oiticica oil. The double bond influences the reactivity of the adjacent allylic carbon atom, particularly when multiple double bonds are present. Reactions at the carbonyl ester linkage of triglycerides include saponification, alcoholysis, and acidolysis. Reactions of triglycerides at double bonds in the alkyl chain include hydrogenation, isomerization, and oxidation.

Chemical reactions can cause serious quality problems for oils. Hydrolytic and oxidative rancidity can cause oils to become unacceptable to consumers for edible or other uses.

Analytical Methods

Specifications. The quality of individual crude oils is specified by trading rules established by organizations such as the National Soybean Processors Association, National Renderers Association, or National Institute of Oilseed Processors. Standardized tests are defined by the American Oil Chemists Society (AOCS), the Association of Official Analytical Chemists (AOAC), and the American Society for Testing Materials (ASTM). Crude oils must contain minimal amounts of foreign material, protein, volatile or toxic solvents, pesticides, heat-transfer media, moisture, and foreign adulterating fats. They must also not be abused or mishandled which causes them to become oxidized. Crude oils must not show excessive loss on refining which adds to costs and waste disposal problems. Oil processors must also meet specifications for their customers which include measures of oil quality, such as free fatty acid level, color, and peroxide value. Other specifications may relate to functionality in the customer's product, such as melting range or fatty acid composition.

Uses of Fats and Oils

Fats and oils are used in food components and cooking oils, soaps and detergents, drying oils, and manufacture of fatty acids and derivatives.

G. L. HASENHUETTL
Kraft General Foods

Y. H. Hui, ed., *Bailey's Industrial Oil and Fat Products*, 5th Ed., Vols. 1–5, John Wiley & Sons, Inc., New York, 1996.

F. D. Gunstone, J. L. Harwood, and F. B. Padley, eds., *The Lipid Handbook*, 2nd ed., Chapman & Hall, New York, 1994.

M. I. Gurr, and J. L. Harwood, *Lipid Biochemistry: An Introduction*, 4th ed., Chapman & Hall, New York, 1991.

McCutcheon's Handbook, Vol. 1, *Emulsifiers and Detergents*, Vol. 2: *Functional Materials*, McCutcheon's Directories, Glen Rock, N.J., 1996.

FEEDS AND FEED ADDITIVES

NONRUMINANT FEEDS

Feed Ingredients

Both swine and poultry diets are comprised primarily of grains such as corn and grain sorghum, with occasional use of wheat, barley, and other small grains (see WHEAT AND OTHER CEREAL GRAINS). Soybean meal is the primary source of protein in these diets, but animal by-products such as meat and bone meal, poultry by-product meal, feather meal, and fish meal contribute significant amounts of protein and provide some of the minerals required for growth, maintenance, reproduction, and lactation (see SOYBEANS AND OTHER OILSEEDS). Many human and industrial by-products are also used in swine and poultry feeds, eg, dried bakery products, produced from leftover bread and other bakery waste; inedible fats from the processing of vegetable oils for human consumption; large amounts of fats and oils from the restaurant and fast food trade that are produced as by-products of cooking food; and numerous other products that would otherwise go unused. An excellent review of the characteristics of many common feed ingredients is available, as are nutrient composition tables for ingredients most commonly used in animal feeds.

Because of the simplicity of swine and poultry feeds, most feed manufacturers add vitamins (qv) and trace minerals to ensure an adequate supply of essential nutrients. Amino acids (qv) such as methionine, lysine, threonine, and tryptophan, produced by chemical synthesis or by fermentation (qv), are used to fortify swine and poultry diets. The use of these supplements to provide the essential amino acids permits diets with lower total crude protein content.

Virtually all broiler and turkey diets, and much of the swine feeds, are pelleted prior to feeding. Pelleted feeds are consumed in greater quantity than are feeds in a mash or meal form, and generally result in more rapid weight gain and better feed conversion. Proper pelleting of feed also aids in reducing the potential of salmonellas or other bacterial contamination of feeds. Pelleting is accomplished by forcing the feed through a die having many small holes. Pelleting is improved by steaming the feed to gelatinize the starch provided by the grains, by adding low (<2%) levels of fat to the feed, or by addition of various types of pellet binders. There are several types of pellet binders, including bentonite clays, lignosulfonates, and grain starch products, that result in firmer, more durable pellets able to withstand the rigors of mechanical feed handling systems.

Nutrient Requirements of Swine and Poultry

Numerous researchers at state and governmental research institutes have defined the requirements of swine and poultry for virtually all known nutrients. In addition, the nutrient composition of common ingredients has been determined. Through the use of a mathematical technique known as linear programming, aided by the use of high speed computers, poultry and swine nutritionists are able to formulate nutritionally balanced diets for all species of animals. As ingredient prices change as a result of supply and demand, diet composition can be changed almost instantly.

There is no best feed composition because animals thrive on diets composed of many different types of ingredients. Swine and poultry generally adapt readily and rapidly to changes in ingredient composition, as long as the diets provide adequate levels of essential nutrients.

Reference Diets for Chickens. Poultry can be grown on many diverse types of diets. Because of the high percentage of chickens grown under an integrated system, it is sometimes difficult to purchase small quantities of high quality feeds. Persons who sometimes utilize chickens for laboratory or research animals may require information regarding formulas that can be mixed from readily available ingredients.

Feed Additives for Nonruminants

Feed additives are common to swine and poultry feeds for a number of purposes. Antibiotics (qv), used to promote growth and improve feed utilization, are poorly digestible and function primarily by controlling the bacterial flora of the intestinal tract. Antibiotics fed to animals for this purpose generally are not used for human antibiotic therapy. Other antibiotics, used for disease therapy, are generally injected or are absorbed into the body tissues where they are effective against the disease-causing organisms. Other feed additives include antioxidants (qv) to protect feeds against oxidative rancidity, mold inhibitors to prevent development of molds, and anticoccidials.

Official information concerning FDA approval of antibiotics and other drugs is available in the *Code of Federal Regulations*. This document is revised at least once per year and updated in individual issues of the *Federal Register*. An effective and less expensive way to maintain information regarding the status of approved feed additives for animal feeds is to subscribe to the *Feed Additive Compendium*, published yearly and updated monthly.

There are a large number of feed additive products classified as generally recognized as safe (GRAS). Some restrictions are placed on quantity of some products, eg, selenium and ethoxyquin. GRAS products include a wide range of materials, ranging from ammoniated cottonseed meal to xanthan gum. A list of all GRAS substances for animal feeds is available.

The GRAS listing does not include widely used historical products such as grains, sugar, salt, etc. In general, feed ingredients listed in the American Association of Feed Control Officials (AAFCO) official publication are considered in the GRAS category.

A number of products designated GRAS are being scrutinized by the FDA because of advertisements and claims made by producers or manufacturers of these products. Statements that indicate that feeding such products improves animal performance may require substantive data to support such claims in the future.

PARK W. WALDROUP
University of Arkansas

M. S. Ash, *Animal Feeds Compendium, Agricultural Economic Report No. 656*, U.S. Dept. of Agriculture, Economic Research Service, Washington, D.C., 1992.

National Research Council, *Nutrient Requirements of Poultry*, 8th ed., National Academy Press, Washington, D.C., 1984.

National Research Council, *Nutrient Requirements of Swine*, 9th ed., National Academy Press, Washington, D.C., 1988.

Feed Additive Compendium, Miller Publishing Co., Minneapolis, Minn.

PET FOODS

All pet foods sold in the United States are subject to scrutiny by both competitors and feed control officials, including the Food and Drug Administration (FDA), Association of American Feed Control Officials (AAFCO), U.S. Department of Agriculture (USDA), Federal Trade Commission (FTC), American Animal Hospital Association (AAHA), American Veterinary Medical Association (AVMA), and Pet Food Institute (PFI). A European group organized to assure fair trade and free circulation of products through Europe (FEDIAF) also monitors every aspect of U.S. pet foods and follows American trends. More is known about the nutrition of dogs and cats than is known about the nutrition of humans.

Pet foods are different from other animal feeds. Most pet foods are processed in highly sophisticated plants using equipment, sanitation, and quality control exceeding standards observed in many plants producing human-grade foods. Pet foods may be stored for up to a year following manufacture before being consumed. This possible delay in the consumption of pet foods requires more careful ingredient selection, preservation of freshness with antioxidants (qv), packaging and processing to avoid insect infestations and rancidity, and careful storage. Pet foods may contain expensive ingredients to provide desirable promotional and marketing copy.

Pets are fed a wide range of commercial foods, which vary on a dry basis from 15 to 60% protein and 5 to 50% fat. Some pet foods are expensive, nutrient-rich foods that contain twice as much nutrition density as needed. Although small quantities of a high calorie food may be consumed, approximately equivalent nutrition may be obtained by pets consuming larger quantities of a less concentrated food.

Types of Commercial Pet Foods

Pet foods are produced in canned, semimoist, and dry forms. Canned pet foods contain approximately 78 to 82% water and have a strong appeal to both pets and owners. Semimoist foods have moisture contents of 25 to 50%. Dry-type foods contain 10 to 12% moisture and supply about 90% of the nutrition consumed by dogs and 72% of the nutrition eaten by cats.

Therapeutic foods have been developed to meet the needs of pets that have nephritic failure, allergies, thyroid problems, geriatric difficulties, and obesity. Most of these therapeutic diets are dispensed by veterinarians, though some are available in pet food outlets and human-food stores stocking pet foods. Treats are usually snacks that may be nutritionally complete or may provide a tasty morsel as a reward. The number of treat products has escalated rapidly.

Canned and Semimoist Foods. Canned and dry foods are nutritionally comparable on a moisture-free basis. Some canned foods are basically dry foods to which gravy, moisture, and flavor enhancers have been added. Almost all animals tend to prefer moist foods to dry, and canned foods are desirable for geriatric dogs and cats, particularly those having gum and dental deterioration. Canned foods can be gulped by dogs and consumed quickly by cats.

Dry Foods. Dry foods are concentrated sources of nutrition and provide the most economical nutritional value because water in canned foods is expensive. Dry foods tend to scrape the teeth as pets eat, minimizing tartar deposition. When dry food is moistened prior to being consumed, tartar accumulates in a manner comparable to deposits observed with canned foods. Approximately 95 to 98% of dry-type cat and dog foods are made by the extrusion process; the remainder is made by pelleting or baking.

Pet Food Formulation

Weights of adult cats in normal physical condition vary from 2 to 6 kg, which is contrasted with the 1 to 100 kg encountered in adult dogs of different breeds. Dogs have proportionately longer digestive tracts and can digest foods more efficiently than cats. This difference in digestibility helps account for the requirement by cats for higher protein diets.

Animal food ingredients are selected to provide desirable contributions of nutrient availability, digestibility, droppings condition, palatability, processing characteristics, ethical desirability, and economics. Modern commercial pet foods contain about 50 nutrient and nonnutrient additives. Each nutrient is supplied at a near-optimum bioavailable level.

Ingredients used in pet foods are usually high in nutritional quality but generally not desirable as human foods primarily because they do not conform to human taste or processing expectations. By-products such as rendered proteins and fat converted into pet foods may have a derivation unappealing to humans, yet after processing may actually be more free of microorganisms and toxins than foods consumed by humans.

Nutritive Ingredients. Nutrients include amino acids (qv), fats, carbohydrates, fibers, minerals, and vitamins (qv). Some ingredients, such as niacin, supply only niacin, whereas salt provides both essential sodium and chlorine. Meat and bone meal may contain all of the nutrients, but not the correct quantities and ratios needed by dogs and cats, and the minimum required level of some nutrients for some species may be toxic to others.

Proteins. Huge amounts of concentrated proteins, available as oilseed plant by-products from the brewing, distilling, starch, and oil industries, when properly supplemented provide excellent sources of amino acids for pets. The world production of oilseed meals is estimated to be 109×10^6 t. Horses, sheep, cattle, swine, and poultry also use oilseeds efficiently and provide intense competition for the use of these plant proteins. Plant proteins are heated during processing to inactivate enzymes that could otherwise be detrimental.

Soybean products that have been processed to remove a portion or all of the carbohydrates and minerals are used to make textured vegetable proteins which can be formed into various shapes and textures (see SOYBEANS AND OTHER OILSEEDS). Many canned dog foods utilize the textured vegetable protein chunks with added juices, flavor enhancers, vitamins, and minerals to produce canned dog foods that have the appearance of meat chunks. Similarly, those proteins can be combined with uncolored ingredients to imitate marbling and form pet foods with chunk-meat appearance. This processing is commonly used in semimoist pet foods.

Plant proteins from single sources, such as soybean meal, may be abundant in specific amino acids that are deficient in some cereal grains. Thus a combination of soybean meal and corn with their amino acid symbiosis may provide an excellent amino acid profile for dogs. Plant protein mixtures alone do not meet the amino acid needs for cats, because taurine is not generally present in plant proteins.

In the United States, more than 16.3×10^6 kg of human-inedible raw materials are available each year, and the rendering industry is a valuable asset in diverting these into valuable ingredients for use primarily in animal foods. Fish meal production worldwide in 1986 was estimated at 6.23×10^6 t, which with the 125×10^6 t of meat and bone meal plus 6.67×10^6 t of feather meal and poultry by-product meal is the primary source of animal proteins used by the pet food industry.

Milk and egg products are highly desired in pet foods since they supply the highest quality amino acid profiles, with nearly 100% digestibility.

Meat derived from crippled, old, discarded, injured animals, and those that have recently died (designated as 4-D beef), as well as USDA rejected meats, are used in canned pet foods. Fresh meats of human-grade are produced in excess of human needs and are used, including wing-tips, gizzards, livers, necks, backs, and meat still attached to bones.

Fats and Oils. Fats and oils from rendering animal and fish offal and vegetable oilseeds provide nutritional by-products used as a source of energy, unsaturated fatty acids, and palatability enhancement. Fats influence the texture in finished pet foods. The use and price of the various melting point fats is determined by the type and appearance of the desired finished food appearance.

Large quantities of fat are used from the fast food industry; these fats may have dissolved plastics from restaurant wrappers which can restrict spray nozzle orifices as the fats cool during spraying on pet foods (see FATS AND FATTY OILS).

Carbohydrates and Plant Products. The world supply of excess grains provides desirable sources of carbohydrates (qv) and fibers (qv) for animals, including pets. Most grains are relatively low in proteins and, unless processed for starch or alcohol, are generally ground whole and used in animal feeds. Thus the contribution of the accompanying protein, vitamins, minerals, and fibers can be accounted for advantageously during pet food formulation. Corn, wheat, and rice are the most desirable common grains and are used extensively in pet foods.

Fibers and Fiber Sources. Fibers are present in varying amounts in food ingredients and are also added separately (see DIETARY FIBER). Some fibers, including beet pulp, apple pomace, citrus pulp, wheat bran, corn bran, and celluloses are added to improve droppings (feces) form by providing a matrix that absorbs water. Some calorie-controlled foods include fibers, such as peanut hulls, to provide gastrointestinal bulk and reduce food intake.

Nonnutrient Additives. Nonnutritional dietary additives provide antioxidants to preserve freshness, flavor enhancers to stimulate food selection, color to meet the owner's expectations, pellet binders to minimize fine particles, mycostats to minimize mold growth, and ingredient-flow enhancers.

Cat-Specific Additives. Cats are more sensitive to some nutritional deviations than are dogs. A dietary deficiency of arginine is more severe in cats than in dogs. This difference is associated with the higher dietary protein in cat foods.

Taurine. Taurine is a sulfonic amino acid derived from methionine and cystine and functions in many biological systems. Although taurine is plentiful in most mammalian tissues as a free acid, the cat's synthesis of taurine is insufficient to meet its biological needs.

Because heat processing during canning inactivates considerably more taurine or forms an inhibition against taurine uptake by the feline, less taurine is required in dry foods than in canned foods. The Feline Nutrition Expert Subcommittee of the Association of American Feed Control Officials (AAFCO), in the nutrient profiles for complete and balanced cat foods, suggests 0.1% taurine in extruded food and 0.2% in canned foods as a result of the extra loss of taurine during the canning processing.

Feather meal, first hydrolyzed and then oxidized, produces cysteic acid an excellent precursor for taurine in cats. Hydrolyzed feather meal may supplement the taurine provided by other dietary animal proteins and help replace part or all of the synthetic taurine in cat food formulations with considerable cost savings.

Phosphoric Acid. To provide safe levels of dietary magnesium and also prevent feline urinary syndrome (FUS), ingredients such as phosphoric acid, which acidulates the urine, are added at carefully controlled levels to produce an acidic urine of approximately pH 6.5.

Other Additives. Cats cannot convert tryptophan to niacin, or carotene to vitamin A in sufficient amounts to meet their needs. These deviations, as compared with other animals, need not produce problems because added dietary sources of niacin and vitamin A provide the needs of cats.

AAFCO Nutrient Profiles. Pet food products provide package claims of "complete and balanced for specific physiological states" to provide the pet owner with confidence and to assure that pets receive nutritionally desirable foods. Before the promulgation and acceptance of the Association of American Feed Control Official (AAFCO) Nutrient Profiles, a number of references were used for complete and balanced recommendations. The Canine Nutrition Expert (CNE) subcommittee was formed to establish new profiles for complete and balanced dog foods. The AAFCO–CNE nutrient profiles are considered the AAFCO-recognized authority on canine nutrition. The Feline Nutrition Expert (FNE) subcommittee was appointed following the development of AAFCO–CNE recommendations to compile profiles for complete and balanced cat foods. These AAFCO–FNE nutrient protocols for cats were published, and include protocols for adequate testing of cat food products, which are monitored by AAFCO.

Economic Aspects

The annual production of pet foods is approximately 6.35×10^6 t, valued at $8.6 billion. It has been estimated that there are as many as 15,000 different brand labels and package sizes of pet foods, marketed by 3,000 manufacturers. Conservative estimates are closer to 5,000 brands and sizes, with 1,800–1,900 registered in the state of Texas.

JAMES CORBIN
University of Illinois at Urbana-Champaign

National Research Council, *Nutrient Requirements of Cats*, National Academy Press, Washington, D.C., 1986.

National Research Council, *Nutrient Requirements of Dogs*, National Academy Press, Washington, D.C., 1974 and 1985.

Official Publication of Association of American Feed Control Officials, Association of American Feed Control Officials, Georgia Department of Agriculture, Capitol Square, Atlanta, Ga.

RUMINANT FEEDS

Ruminants, which consume plant material grown on land that may be unsuitable for crop farming, need not compete with humans and nonruminant livestock for feed resources. At least one-third of the world's land area is more suitable for grazing than for cultivation. The high fiber content forage produced on this land is largely undigested by monogastrics. Ruminant animals, whose ruminal microflora ferment and digest cellulose (qv), the predominant component of fiber and the earth's most abundant carbohydrate, utilize a large amount of the plant energy produced on this land. Anatomical differences between monogastric and ruminant animals allow the ruminant to be more efficient in digesting cellulose, but generally less efficient in gaining weight and converting feed to gain, because of energetic losses resulting from the fermentation (qv) process.

Feeds

Forages/Roughages. Approximately 75–80% of the feed fed to ruminants during their lifetime production cycle is forage/roughage material. Roughages are made up predominantly of the stem or stalk portion of plants and usually include the seeds and leaves of these plants. These feeds are typically high in fiber, >50% neutral detergent fiber; low in starch, <4%; and moderately low in crude protein, <20%. Roughages are not only a source of nutrition to the ruminant but also help to maintain normal rumen function. Roughages are important to the ruminant animal for a variety of reasons including maintenance-level feeding of herd or flock animals; weaning of young ruminants from milk onto solid feed; preventing metabolic diseases, eg, bovine ketosis and ovine pregnancy toxemia; and maintaining proper fat level, approximately 3.5%, in milk produced by dairy cows. Ruminants consume forages either by grazing or being fed harvested material.

The moisture content at which a plant is harvested usually determines the storage method. Low (15–25%) moisture forages are often stored in some type of bale form in the presence of air. Stave silos, oxygen-limiting silos, concrete bunker silos, and large plastic bags are all methods of storing high (40–75%) moisture forages.

Several different sources of low moisture forage, eg, prairie hay, alfalfa, bromegrass, orchard grass, and blends of hay grown specifically for the purpose of harvesting; and roughage, eg, crop residues such as corn stalks, soybean stubble, or small grains straw, are available.

High Energy Feeds. Concentrated sources of energy are fed to ruminants to allow young ruminant animals to grow more quickly and efficiently. Feedstuffs of this nature are generally high in readily fermentable carbohydrates, ie, they are high starch-containing feedstuffs. Feedstuffs containing high amounts of starch (qv) are often from the seeds of plants such as corn, grain sorghum, oats, and barley.

By-products of agricultural commodities are used as readily fermentable energy-containing feedstuffs. Molasses is a commonly used by-product of the sugar-refining industry. The addition of molasses in feed increases feed acceptability, reduces dustiness of the feed, and improves feed pelleting. Molasses from citrus and wood (qv) processing also is available. Other useful, energy-containing by-products include wheat bran, wheat middlings and shorts, dried citrus pulp, dried beet pulp, dried bakery products, hominy, oat groats, potato meal, whey, corn gluten feed, and rice bran. Since over 450 million metric tons of agricultural by-products and residues exist, a great quantity of ruminant feedstuffs is available that is not used by humans and nonruminant livestock. These by-products, often used as feed because of low cost and availability, may present other problems such as lower energy content than corn and low acceptability.

Besides changing the source of energy, feed processing methods influence the amount of energy available to the ruminant, ie, by influencing the sites of digestion and absorption in the ruminant animal. Fermentation products produced in the rumen and absorbed through the ruminal wall do not contain as much energy as the carbohydrates absorbed through the small intestine. Methods of processing include grinding, rolling, cracking, extruding, steam flaking, heating, wetting, and gelatinizing.

Various sources of lipid have been incorporated into ruminant diets to increase the energy density and provide the large amount of energy needed for slaughter animals to achieve market weight or for dairy cows to produce milk (see MILK AND MILK PRODUCTS). Fats also reduce the dustiness of feeds, increase the feedstuffs' ability to pellet, and improve feed acceptability. The predominant feed source of lipid is from animal origin. Animal fat is typically higher in saturation and is often referred to as grease. Various vegetable sources of lipid also are available.

Lipids present in the diet may become rancid. When fed at high (>4–6%) levels, lipids may decrease diet acceptability, increase handling problems, result in poor pellet quality, cause diarrhea, reduce feed intake, and decrease fiber digestion in the rumen. To alleviate the fiber digestion problem, calcium soaps or prilled free fatty acids have been developed to escape ruminal fermentation.

Supplements

Protein. Although most feedstuffs contain protein, supplemental protein or nitrogen often is needed to meet animal physiological requirements (see PROTEINS). Practical situations in which supplemental protein is required include the feeding of growing/immature animals, lactating females, females in the last trimester of pregnancy, and ruminants grazing rangelands. Soybean meal is the most frequently used source of supplemental protein in the United States. Cottonseed meal is another important protein supplement. Both meals are by-products from oil extraction of the seeds. Raw soybeans also may be used as a supplemental protein source. Various milling, distilling, and brewing by-products are available as supplemental protein sources. Legume forages, such as alfalfa or clover, are considered high quality, readily available protein sources. Animal sources of supplemental protein include meat and bone meal; blood meal, 80% crude protein (CP); fish meal; other marine products; and hydrolyzed feathermeal, 85–90% CP. Additionally, synthetic amino acids are available commercially.

Minerals. Supplementation of macrominerals to ruminants is sometimes necessary. Calcium and phosphorus are the minerals most often supplemented in ruminant diets.

Vitamins. The B-vitamins and vitamin K, $C_{31}H_{46}O_2$, are synthesized by ruminal microorganisms and their supplementation is usually unnecessary. However, supplementation with B-vitamins is beneficial in certain situations. Polioencephalomalacia (PEM), a nervous disorder, is alleviated by intravenous injections of thiamine hydrochloride, $C_{12}H_{18}Cl_2N_4OS$. Niacin supplementation has been shown to partially alleviate subclinical ketosis, increase milk production, and increase average daily weight gain under some conditions.

Vitamins A, D, and E are required by ruminants and, therefore, their supplementation is sometimes necessary.

Additives

Several feed additives for ruminants are available. All additives increase animal growth or efficiency of weight gain, and many provide additional benefits. Additives can be classified into groups based on function.

Ionophores, widely used feed additives that alter ruminal fermentation, interact with the transport of ions across ruminal bacterial cell membranes. The two FDA-approved ionophores for nonlactating ruminants are monensin, $C_{36}H_{62}O_{11}$, and lasalocid, $C_{34}H_{54}O_8$. Zeolites, ion-exchange compounds have been researched to some extent and have been proposed to improve NPN utilization. However, no improvement in NPN utilization was found with lambs fed zeolites. Direct-fed microbials are feed additives composed of microbes and/or ingredients to stimulate microbial growth which allegedly result in a more efficient microbial population.

A problem common to animals consuming a high energy diet or lush, immature legume vegetation is increased susceptibility to bloat. Antifoaming agents available to prevent bloat include silicones, detergents, vegetable oils, animal fats, animal mucins, and liquid paraffins.

Buffers are used to stabilize ruminal pH at 6.0–6.8. Available buffers include bicarbonate, calcium carbonate (limestone), $CaCO_3$, and bentonite, $Al_2O_3 \cdot SiO_2 \cdot H_2O$.

Many ruminal bacteria require one or more branched-chain volatile fatty acids (VFA) for proper growth. Therefore, supplementation of these VFA has been practiced; however, results have been variable.

Increased hormonal levels have resulted in better ruminant animal performance. Estrogen-containing compounds, eg, zearalenone $C_{18}H_{22}O_5$, are available to improve growth and feed efficiency. These estrogens are given as ear implants.

Antibiotics (qv) have been fed at subtherapeutic levels to promote ruminant animal growth. Tylosin, $C_{46}H_{77}NO_{17}$, and avoparcin, $C_{89}H_{101}ClN_9O_{36}$, are examples of antibiotics used as feed additives in ruminant diets.

Propionic acid, $C_3H_6O_2$, and ammonia, NH_3, are additives used to prevent molding of feed. Bentonite, hemicellulose extracts, and lignin sulfonate are used to hold feed pellets together.

Young Animal Feeds

The rumen is not functional at birth and milk is shunted to the abomasum. One to two weeks after birth, the neonate consumes solid food if offered. A calf or lamb that is nursing tends to nibble the mother's feed. An alternative method of raising the neonate is to remove it from its mother at a very young age, < 1week. In this instance, the neonate requires complete dietary supplementation with milk replacer. Sources of milk replacer protein have traditionally included milk protein but may also include soybean proteins, fish protein concentrates, field bean proteins, pea protein concentrates, and yeast protein. Information on the digestibility of some of these protein sources is available.

By approximately eight weeks after birth, the ruminant has developed a fully functional rumen capable of extensive fermentation of feed nutrients.

Several sources of energy are available for the neonate. Lipids are highly digestible, ca 90%, by neonatal calves and continue to increase in digestibility as the calf matures. Lipid sources include milk fat, tallow, and corn oil. Carbohydrates are another source of energy for the young ruminant. Lactose, glucose, and galactose are efficiently digested. Hydrolyzed starch has been used successfully to replace a portion of the energy in diets fed to young ruminants. Protein sources given to calves just starting to consume solid food should contain protein from natural plant sources or from natural plant sources with milk by-products.

Creep feeding often is used in the production of beef cattle. Nursing calves are given access to feedstuffs physically separated from their mothers. One means of creep feeding is to allow the calves access to concentrate feeds, eg, corn, oats, or barley. Another means of creep feeding is to allow calves access to an ungrazed forage stand. This form of creep feeding allows calves to consume a much higher quality forage than is otherwise available.

GREGORY D. SUNVOLD
GEORGE C. FAHEY, JR.
University of Illinois, Urbana

D. C. Church, in D. C. Church, ed., *The Ruminant Animal*, Prentice Hall, Inc., Englewood Cliffs, N. J., 1988, p. 1.

P. J. Van Soest, *Nutritional Ecology of the Ruminant: Ruminant Metabolism, Nutritional Strategies, the Cellulolytic Fermentation and the Chemistry of Forages and Plant Fibers*, Cornell University Press, Ithaca, N.Y., 1987.

P. R. Cheeke, *Applied Animal Nutrition: Feeds and Feeding*, MacMillan Publishing Co., New York, 1991.

R. A. Britton, in M. L. Pinkston, ed., *Proceedings of the 39th Annual Pfizer Research Conference*, New York, 1991, p. 124.

FEEDSTOCKS

COAL CHEMICALS

Coal is used in industry both as a fuel and in much lower volume as a source of chemicals. In this respect it is like petroleum and natural gas, consumption of which also is heavily dominated by fuel use. Much of the recent activity in coal conversion has been focused on production of synthetic fuels, but significant progress has been made on use of coal as a chemical feedstock.

The term *feedstock* in this article refers not only to coal, but also to products and coproducts of coal conversion processes used to meet the raw material needs of the chemical industry.

Coal Carbonization

The thermal degradation of coal in the absence of air is known as pyrolysis or carbonization. This reaction also is referred to as coking because of its large volume use to prepare coke for blast furnaces. The type and quantity of products produced by coal pyrolysis depend on type of coal, rate of heating, and final temperature. In general, rapid heating affords higher liquid yields than slow pyrolysis. Low temperature carbonization or mild gasification is performed at 450–700°C and is used in some areas to prepare smokeless fuels. High temperature carbonization is performed in the range of 900–1100°C and is generally the process used to prepare blast furnace coke. Low temperature carbonization gives fewer gaseous and more liquid products than carbonization at higher temperature.

Most coal-tar chemicals are recovered from coproduct coke ovens. Since the primary product of the ovens is metallurgical coke, production of coal chemicals from this source is highly dependent on the level of activity in the steel industry. In past years most large coke producers operated their own coproduct recovery processes. Because of the decline in the domestic steel industry, the recent trend is for independent refiners to collect crude coal tars and light oils from several producers and then separate the marketable products.

When coal is coked at a temperature of approximately 1000°C, about 70–75% of the product is coke. Nearly 20% of the product is a light gas, mostly methane and hydrogen, that typically is used as fuel to heat the ovens. Coal tars amount to about 4% of the product and light oil or naphtha is about 1%. Ammonia is recovered in an amount equal to about 0.3% of the feed coal. The ammonia is usually converted to ammonium sulfate and sold as a fertilizer. Little or no ammonia is produced in low temperature carbonization.

Many valuable chemicals can be recovered from the volatile fractions produced in coke ovens. For many years coal tar was the primary source for chemicals such as naphthalene, anthracene, and other aromatic and heterocyclic hydrocarbons. Much of the production of these chemicals, especially tar bases such as the pyridines and picolines, is based on synthesis from petroleum feedstocks. Nevertheless, a number of important materials continue to be derived from coal tar.

Approximately 50–55% of the product from a coal-tar refinery is pitch and another 30% is creosote. The remaining 15–20% is the chemical oil, about half of which is naphthalene. Creosote is used as a feedstock for production of carbon black and as a wood preservative.

In 1990, U.S. coke plants consumed 3.61×10^7 t of coal, or 4.4% of the total U.S. consumption of 8.12×10^8 t. Coke production is in a period of decline because of reduced demand for steel and increasing use of technology for direct injection of coal into blast furnaces. The decline in coke production and trend away from recovery of coproducts is reflected in a 70–80% decline in volume of coal-tar chemicals since the 1970s.

In 1990, U.S. production of crude coal tar was 597,000 m^3 (700,000 t) and production of crude light oil was 255,000 m^3 (200,000 t). Crude coal tar and light oil were refined to produce 110,000 m^3 of crude naphthalene (freezing point 76–79°C), 8700 m^3 of crude tar acid oils (tar acid content 5–24%), and 297,000 m^3 of creosote oils. Coal-tar pitch production in 1990 amounted to 600,000 t.

The principal producers of coal-tar chemicals in the United States include AlliedSignal, Aristech, Koppers Industries, and Cooper Creek Chemical.

As the economic value of coproducts has decreased, it has become more difficult to provide capital for environmental controls on air emissions and wastewater streams such as toxic phenolic effluents from chemical recovery operations. Some former coke and manufactured gas sites may require remediation to clean up contaminated soil and groundwater. These difficulties will force the shutdown of some operations and discourage recovery of coproducts in future installations.

Coal Hydrogenation

A new generation of coal liquefaction processes has been under study since the 1980s, differing from the previous versions in that a two-stage approach is employed. A two-stage process has the advantage of allowing optimum conditions to be established in separate reactors for the coal dissolution and upgrading reactions, resulting in higher distillate yield, lower gas yield, and higher quality products. All recent direct liquefaction work in the United States has been devoted to study of catalytic two-stage liquefaction. The Wilsonville, Alabama process development unit has been one of the most active facilities for study of coal liquefaction. In general, oils produced from direct coal liquefaction are lower boiling, have lower hydrogen content, and higher oxygen and nitrogen content than typical petroleum crudes. Also, they differ from petroleum in that they contain mostly condensed cyclic compounds, few paraffins, and no residuum. Virtually all of the work on coal liquid upgrading has concentrated on study of refining conditions for production of gasoline, diesel, jet fuel, and heating oil.

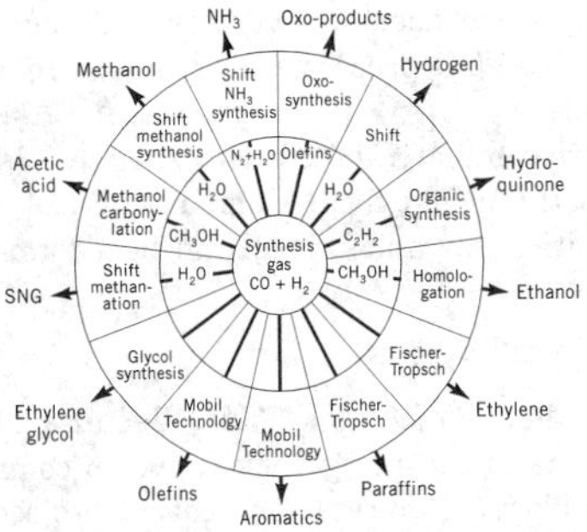

Figure 1. The use of synthesis gas as a chemical feedstock. SNG is substitute natural gas.

Synthesis Gas Generation

Selection of Feedstock. The variety of chemicals available by conversion of synthesis gas is illustrated in Figure 1. Synthesis gas can be produced by steam reforming of natural gas and naphtha, partial oxidation of heavy oil and petroleum coke, and coal gasification. In 1977 the relative ratio of synthesis gas produced by these methods was 87:10:3, respectively. Virtually any hydrocarbon source can be used to generate synthesis gas, and the choice of feedstock for a given project can be made only after careful evaluation of all the alternatives. Most economic studies have shown that the cost of producing liquids from synthesis gas is dominated by the cost of producing the gas. This cost can be as high as 60–75% of the overall cost of the final product. Because of the high cost of producing synthesis gas from coal under economic conditions prevailing in the early 1990s, coal will be the favored feedstock for chemicals only under special circumstances. Some of the factors that influence the choice of feedstock and potentially favor coal are feedstock cost, availability, and C/H ratio.

Gasification Chemistry. Gasification of coal involves many reactions. For many gasifiers, particularly those operating at higher temperatures, these reactions approach equilibrium, and the product mixture is strongly dependent on the feed composition and gasifier temperature and pressure.

Coal Gasifier Designs. After many years of development three general gasifier designs have achieved commercial status. These types are fixed-bed, fluidized-bed, and entrained-bed gasifiers. Each design has advantages and disadvantages and may be more suitable than other designs for a given type of coal or process application (see COAL CONVERSION PROCESSES–GASIFICATION).

Synthesis Gas Chemicals

Synthesis gas chemicals from coal include hydrocarbons via the Fischer-Tropsch Process, ammonia and hydrogen production, methanol production, methanol/higher alcohol mixtures, liquid-phase methanol, dimethyl ether, gasoline range hydrocarbons from the Mobil MTG and MTO Process, aldehydes via oxo synthesis, and acetic acid and anhydrides.

Emerging Processes. There is extensive literature on production of chemicals from synthesis gas. In addition to the foregoing, reports or patents have been published on ethylene glycol, vinyl acetate, aliphatic amines, acrylic acid, and homologation of carboxylic acids. The amidocarbonylation reaction is a new procedure for utilization of synthesis gas to introduce amido and carboxylate groups into olefins or aldehydes. The reaction is a potentially useful technique for industrial production of amino acids and derivatives. Also, biological conversion offers an alternative route for utilization of synthesis gas to make chemicals. Production of ethanol and acetic acid by bioconversion of synthesis gas are areas under active investigation.

Carbide Chemicals

Calcium carbide has been used in steel production to lower sulfur emissions when coke with high sulfur content is used. The principal use of carbide remains hydrolysis for acetylene (C_2H_2) production.

Newer coal-based methods of acetylene manufacture under development include the AVCO process, based on the reaction of coal in a hydrogen plasma.

Companies with commercial facilities using coal as a chemical feedstock include Eastman Chemical Company, Sasol Chemical Industries, Dakota Gasification, and several coal-based ammonia producers.

Planned Coal Chemical Projects

Texaco Gasification Projects in China. Because of large domestic coal reserves, China is becoming very active in commercialization of coal conversion projects. Two coal gasification projects using the Texaco process are currently under construction, and two additional Texaco gasification projects were announced in early 1992.

TVA Urea Project. In Round 4 of DOE's Clean Coal Technology program, the Tennessee Valley Authority (TVA) proposed construction of a 250 MW IGCC power plant that would coproduce urea and electric power from 730,000 metric tons of coal a year. The project was not selected in Round 4, but is being submitted again for consideration in Round 5, which is expected to give chemical projects favorable treatment.

PAUL R. WORSHAM
Eastman Chemical Company

J. Falbe, ed., *Chemical Feedstocks from Coal*, John Wiley & Sons, Inc., New York, 1982.

R. K. Hessley and R. H. Schlosberg, in J. G. Speight, ed., *Fuel Science and Technology Handbook*, Marcel Dekker, Inc., New York, 1990, p. 768.

M. G. Thomas, in B. R. Cooper and W. A. Ellingson, eds., *The Science and Technology of Coal and Coal Utilization*, Plenum Press, New York, 1984, p. 231.

K. R. Payne, ed., *Chemicals from Coal: New Processes*, John Wiley & Sons, Inc., New York, 1987.

PETROCHEMICALS

By definition, petrochemicals are either isolated from or derived from natural gas or petroleum. Other sources of feedstocks such as coal and agricultural products remain very small sources of petrochemicals. The choice of feedstock for a given operator is an economic decision, but may be constrained by hardware limitations on the part of the operator or the availability of indigenous hydrocarbon supplies in a given geographical region. Because the primary uses of natural gas and petroleum are as sources of energy, the costs of petrochemical feedstocks are closely related to the alternative energy values of the various feedstocks. The petrochemical producer must buy feedstocks out of the alternative energy markets. Hence the structure and economics of the energy industry in a given area affect the choice of petrochemical feedstocks. Seasonal effects in the prices of energy products introduce variability in petrochemical feedstock prices and affect the choice of preferred feedstock for a given operator at a point in time. International events such as the OPEC oil embargo of 1973–1974, the Iranian revolution of 1979, or the Middle East war of 1990–1991 have dramatically affected the costs and choice of petrochemical feedstocks for a period of time. Government regulations, such as the Clean Air Act Amendments of 1990, also impact petrochemical feedstock availability, relative cost, and choice of alternative feedstocks.

It is convenient to divide the petrochemical industry into two general sectors: (*1*) olefins and (*2*) aromatics and their respective derivatives. Olefins are straight- or branched-chain unsaturated hydrocarbons, the most important being ethylene (qv), propylene (qv), and butadiene (qv). Aromatics are cyclic unsaturated hydrocarbons, the most important being benzene (qv), toluene (qv), *p*-xylene, and *o*-xylene (see XYLENES AND ETHYLBENZENE) There are two other large-volume petrochemicals that do not fall easily into either of these two categories: ammonia (qv) and methanol (qv). These two products are derived primarily from methane (natural gas) (see HYDROCARBONS, C_1–C_6).

Olefins are produced primarily by thermal cracking of a hydrocarbon feedstock which takes place at low residence time in the presence of steam in the tubes of a furnace. In the United States, natural gas liquids derived from natural gas processing, primarily ethane and propane, have been the dominant feedstock for olefins plants, accounting for about 50 to 70% of ethylene production. Most of the remainder has been based on cracking naphtha or gas oil hydrocarbon streams which are derived from crude oil.

In most of the rest of the world the olefins industry was originally based on naphtha feedstocks. Naphtha is the dominant olefins feedstock in Europe and Asia. In the middle 1980s several large olefins complexes were built outside of the United States based on gas liquids feedstocks, most notable in western Canada, Saudi Arabia, and Scotland. In each case the driving force was the production of natural gas, generally associated with crude oil production, which was in excess of energy demands.

Since the early 1980s olefin plants in the United States were designed to have substantial flexibility to consume a wide range of feedstocks. Most of the flexibility to use various feedstocks is found in plants with associated refineries, where integrated olefins plants can optimize feedstocks using either gas liquids or heavier refinery streams. Companies whose primary business is the production of ethylene derivatives, such as thermoplastics, tend to use ethane and propane feedstocks which minimize by-product streams and maximize ethylene production for their derivative plants. Flexibility allows the operator to pick and choose the most attractive feedstock available at a given point in time.

Aromatics are produced primarily from two sources. The most important in the United States is refinery catalytic reformer operations. The second source of aromatics is pyrolysis gasoline from olefins plants.

Olefin Feedstocks

Olefin Feedstock Selection. The selection of feedstock and severity of the cracking process are economic choices, given that the specific plant has flexibility to accommodate alternative feedstocks. The feedstock prices are influenced primarily by energy markets and secondarily by supply and demand conditions in the olefins feedstock markets.

Simply looking at the feedstock prices or price ratios is insufficient to accurately identify the most attractive feedstock because the values of all of the coproducts and by-products must also be taken into account. This is usually accomplished by calculating the cost to produce ethylene with all other coproduct and by-product yields credited against the cost of ethylene.

Aromatic Feedstocks

The preferred feedstocks for aromatics production are naphthas containing high concentrations of the naphthene precursors to benzene, toluene, and xylenes. The Second source of aromatics feedstock is pyrolysis gasoline from olefins plants.

Other Feedstocks

The only other petrochemical feedstock of significant commercial use is methane (natural gas) which is used primarily to produce ammonia and methanol. Consumption factors are about 28 GJ and 31 GJ per metric ton, respectively (58,300 and 64,700 BTU/lb).

Alternative Feedstocks

Alternative feedstocks for petrochemicals have been the subject of much research and study over the past several decades, but have not yet become economically attractive. Chemical producers are expected to continue to use fossil fuels for energy and feedstock needs for the next 75 years. The most promising sources which have received the most attention include coal, tar sands, oil shale, and biomass. Near-term advances in coal-gasification technology offer the greatest potential to replace oil- and gas-based feedstocks in selected applications (see FEEDSTOCKS–COAL CHEMICALS).

Because oil and gas are not renewable resources, at some point in time alternative feedstocks will become attractive; however, this point appears to be far in the future. Of the alternatives, only biomass is a renewable resource (see FUELS FROM BIOMASS). The only chemical produced from biomass in commercial quantities at the present time is ethanol by fermentation.

DARRYL C. AUBREY
Sacred Heart University Chem Systems Inc.

P. Spitz, *Petrochemicals, The Rise of an Industry*, Wiley-Interscience, 1988.

Synthetic Organic Chemicals Report, U.S. International Trade Commission, Washington, D.C., 1990.

Fertilizer Materials, Current Industrial Reports, U.S. Dept. of Commerce, Bureau of the Census, Washington, D.C.

FERMENTATION

Originally, fermentation described the anaerobic evolution of carbon dioxide from the action of yeast on fruit sugars or malted grains, and the origins of fermentation as an industry are deeply rooted in alcoholic beverage production. Although most traditional fermentation products rely on anaerobic bacterial and yeast metabolism, newer products, eg, antibiotics (qv) and enzymes, tend to rely on aerobic bacteria, fungi, and mold metabolism. Antibiotics dominant the fermentation market. Other products include organic acids, enzymes, baker's yeast, ethanol, amino acids (qv), and steroid hormones (qv). The early 1990s U.S. market was estimated in the $9-10 \times 10^9$ range. The world market is probably three times this figure.

Commercial processes can be classified into five groups: *(1)* the production of biomass, primarily yeast and animal feed additives; *(2)* the production of simple or complex metabolites such as antibiotics, amino acids, vitamins (qv), organic acids etc.; *(3)* the conversion of a substance or groups of substances, often referred to as biotransformation, into another substance, eg, steroids (qv); *(4)* the production of enzymes, eg, amylases, lipases, and chymosin; (see ENZYMES) and *(5)* the production of biotechnologically derived molecules such as insulin and polypeptide hormones that are not normally synthesized by microorganisms. There are numerous advantages offered by fermentation processes. Microbes often accomplish in a single step an economically feasible molecular change that otherwise involves arduous chemical synthesis. Moreover, use of microbial enzymes can often replace the adverse conditions, eg, high temperature, pressures, and drastic pHs, routinely employed in chemical synthesis. Additionally, fermentation often permits the manufacture of purer compounds having greater specificity, especially when a specific chiral enantiomer is preferred. Fermentation may also permit the use of cheaper raw materials.

Fermentation Companies and Products

Commercial fermentation products number well into the hundreds. Examples follow: (*1*) Foods, eg, beer, distilled spirits, cheese, and vinegar. (*2*) Food additives, eg, gluconic acid, citric acid, lactic acid, β-carotene, xanthan gum, colorants, and flavors. (*3*) Enzymes; eg, amylases, lipases, fruit processing enzymes, food processing enzymes, and diagnostic enzymes. (*4*) Amino acids: eg, lysine, tryptophan, glutamate, threonine, arginine, and histidine. (*5*) Vitamins. (*6*) Biotechnology products, eg, human growth hormone. (*7*) Others, eg, steroids, and steroid conversion products and diagnostic reagents. (*8*) Antibiotics. Fermented beverage and food products are the largest commercial processes. There are thousands of breweries and hundreds of distilleries worldwide. Countries where baker's yeast is manufactured for food and feed include: Argentina, Australia, Austria, Belgium, Brazil, Canada, Chile, Colombia, Denmark, Egypt, Finland, France, Germany, Guatemala, Holland, Iran, Japan, Mexico, Peru, Philippines, Republic of Korea, South Africa, Spain, Sweden, Tunisia, Turkey, United Kingdom, United States, Uruguay, Zaire, and Zambia; Food and feed yeast are also produced in Belgium, Finland, France, and the United States. Excluding the alcohol and food industry, the principal companies for which fermentation is a significant business component numbers well over 150. Numerous other companies practice fermentation in some small capacity.

Most of the larger well-established pharmaceutical companies, eg, Smith-Kline-Beecham plc, Pfizer Inc., Eli Lilly & Company, Merck & Company, Novo Labs, Ajinomoto, Miles Inc., Takeda, Zeneca plc, Glaxo, Bristol-Myers-Squibb, and Gist-Brocades have fermentation plants. A selection of newer biotechnology companies utilizing fermentation is listed in Table 1.

Microbiological Aspects

Commercial fermentation processes use a variety of microorganisms. The strain chosen for development and production plays a significant role in determining the process design and engineering of the fermentation plant and downstream processes. Therefore, the isolation, improvement, and preservation of high-yielding strains is of fundamental importance. Traditionally, microorganisms have been selectively cultured, and exposed to a battery of techniques that increase their ability to over-produce the product of interest. The complexity of biochemical pathways and control mechanisms means that for most products, the careful and rigorous selection of microorganisms is a key to obtaining good titers. In this respect, genetic engineering (qv) is playing an ever increasing role and recent success include improvements with *Streptomyces* and fungal cultures. However, for most fermentations process conditions, (eg, growth media, pH, temperature, mixing, and oxygenation, must be manipulated to optimize titers. As an example, continued improvements in penicillin fermentation have resulted in titers of >75 g/L which are ca 15,000 times that of the wild-type. Strain preservation can take many forms. The most widely used techniques are lyophilization, refrigeration, freezing (−5

Table 1. Biotechnology Companies Utilizing Fermentation Technology

Company	Country	Emphasis
Allelix	Canada	therapeutics, biochemicals
Amgen	United States	therapeutics
Bideco	Switzerland	therapeutics
Biocon	Ireland	biochemicals
Biogen	United States	therapeutics, diagnostics
Biotransplant	United States	post-surgery therapeutics
Green Cross	Korea	fine chemicals, biochemicals
British Biotechnology	United Kingdom	therapeutics, diagnostics
Calgene	Australia	agriculture
Cangene	Canada	therapeutics
Cetus (Chiron)	United States	strains, biochemicals, therapeutics
DNAP	United States	artificial snow/agriculture
Dowelanco	United States	crop protection
Envirogen	United States	environmental decontamination
Enzymatix	United Kingdom	biochemicals, enzymes
Genencore	United States	enzymes, biochemicals
Genentech (Roche)	United States	therapeutics
Genzyme	United States	biochemicals, therapeutics
Lucky Biotech	Korea	biochemicals
Mogen	Holland	agriculture
Mycogen	United States	crop protection
Regeneron	United States	therapeutics
Serono	Switzerland	therapeutics
Zeagen (Coors)	United States	vitamins and pigments

to −20°C), low temperature freezing (−70 to −80°C), and storage in liquid nitrogen (−196°C).

Components of Fermentation Processes

Most fermentation processes consist of several distinct operations. First, the culture medium used for inoculum propagation and production fermentation is formulated. Then, sterile medium is prepared along with the fermenters and related equipment. Then, a pure and active seed culture is propagated. Next, the culture is transferred and grown in the production fermenter under optimum conditions. At the end of the fermentation, the product is harvested from the cells or the supernatant. Finally, effluent and cellular debris is treated and appropriate disposal arranged.

All microorganisms require water, sources of carbon, energy, and nitrogen minerals. Certain vitamins and growth factors are often required, and oxygen is required by all aerobic microorganisms (see AERATION). Both chemically defined components (purified carbohydrates and inorganic salts) and nondefined materials (eg, starches, molasses, fish meal, soybean meal, cottonseed flour, casein hydrolysate, yeast products, peanut flour and lard) may be employed. The growth medium is added to the fermenter and heat-sterilized using steam (batch sterilization) or sterilized as it is added to the fermenter (continuous sterilization). Exceptions to the sterilization rule include beer brewing, lactic acid production, vinegar formation, and L-sorbose production. Air for oxygenation is usually filter-sterilized. The sterile medium is inoculated with an actively growing culture and the fermentation is conducted under appropriate conditions of pH, temperature, and aeration. Extra-sterile raw materials are often added to the fermentation in batches or on a continuous basis to maintain production and minimize substrate inhibition. Fermenter contents are kept in a mixed state by agitators or by the air pumped into the fermenter for oxygenation. Foam formation is minimized by defoaming agents, either biological oils or chemicals.

There are essentially three different types of fermentations: continuous, batch, and fed-batch. The majority of industrial processes are fed-batch, ie, during the fermentation, additional sterile nutrients (usually carbohydrates and nitrogen sources) are fed to the batch. Occasionally, specific intermediates are also added. Batch fermentations contain all the nutrients when the fermenter is inoculated and only pH control or foam control agents are added during the course of the fermentation. Beer, wine, and certain vitamins are made by batch fermentation. In a continuous fermentation, nutrients are added continuously and the product is also harvested continuously. Continuous fermentations are rare.

Production fermenters range in size from a few liters to ca 500 m^3 for production of citric acid and other commodity-scale products. Most production fermenters are in the 25–100 m^3 range. Fermentations for products of recombinant technology and many biochemicals are usually carried out on a relatively small scale (1–5 m^3). These fermentations nearly always employ a growth phase, followed by an induction phase during which product is formed. The inducer used depends on the promoter of the genetic construct of the recombinant microorganism. These fermentations are relatively short (12–20 h). A typical antibiotic fermentation may last up to 7 d.

Once a fermentation is complete, the product must be separated from the waste. The product is typically in the fermentation supernatant (extracellular), but there are many exceptions, including baker's yeast; other whole cell products eg, bacitracin, vitamin B_{12} and β-carotene; and products that have to be extracted from broken or lysed cell mass (intracellular), such as some enzymes and most genetic engineering products. Extracellular products are usually isolated by first removing the cell mass by filtration, sedimentation, or centrifugation. Excess water is removed and the product is then crystallized out or extracted by a variety of chemical methods. Intracellular products are collected by breaking open microbial cells, usually by mechanical methods. The isolated material must then be further purified. Of all the products, those produced by genetically engineered microorganisms require the most stringent and elaborate purification procedures. For these drugs, which are typically protein-based, the purification cost often represents more than 90% of the total manufacturing cost.

Economic Aspects

Excluding alcoholic beverages, fermentation products span the spectrum from high value, low volume products, eg, vitamin B_{12}, insulin, gene-splicing enzymes that can command in excess of $1,000/kg to low value, high volume commodity-type products, eg, citric acid, glutamic acid, and aspartic acid, that sell for a few dollars per kilogram. A few fermentation products rank in annual worldwide sales above $50,000,000: many antibiotics, enzymes, organic acids, amino acids, insulin, vitamins B_{12}, and biotin.

Future Developments

Commercialized products in the 1990s such as an enzyme to prevent bread staling from Novo Labs, a fungus-based protein source for humans from ICI and Kelcotrol, and a new water-based food gum from Merck and Company, are expected to buoy the demand for fermentation capacity. Moreover, increased worldwide regulatory acceptance for microbial insecticides is expected and fermentation products such as several pigments, flavors, fragrances, and human therapeutics or components are on the threshold of commercialization. In addition, the application of microbes to reclaim land contaminated with organics, eg, phenol, tar, oil, benzene etc, is becoming more economical compared to traditional methods (see WASTE TREATMENT, HAZARDOUS WASTES). Other markets that hold promise include those for inoculating farm animals with flora that protect against various infections, and the use of microorganisms from harsh environments, eg, sea-bed steam vents, for production of enzymes.

SURJIT S. SENGHA
Lexin Pharmaceutical Corporation

B. Atkinson and F. Mavituna, *Biochemical Engineering and Biotechnology Handbook*, 2nd ed., Macmillan Publishers Ltd., Basingstoke, U.K., 1991.

A. L. Demain and N. A. Solomon, eds., *Manual of Industrial Microbiology and Biotechnology*, American Society for Microbiology, Washington, D.C., 1986.

A. H. Rose, ed., *Economic Microbiology*, Vols. 1–5, Academic Press, Ltd., London, 1977–1981.

R. Gerals, ed., *Prescott and Dunns Industrial Microbiology*, 4th ed., The Avi Publishing Co. Inc., Westport, Conn., 1982.

D. I. C. Wang and co-workers, *Fermentation and Enzyme Technology*, John Wiley & Sons, Inc., New York, 1979.

FERRITES

The term *ferrite* is commonly used generically to describe a class of magnetic oxide compounds which contain iron oxide as a principal component. In metallurgy (qv), however, the term ferrite is often used as a metallographic indication of the α-iron crystalline phase.

Ferrites can be classified according to crystal structure, ie, cubic vs hexagonal, or magnetic behavior, ie, soft vs hard ferrites. A systematic classification as well as some applications are given in Table 1 and Figure 1.

Common Properties of Spinel Ferrites and M-Type Ferrites

The commercial sintered spinel and M-type ferrites have a porosity of 2–15 vol % and a grain size in the range of 1–10 μm. In addition, these materials usually contain up to about 1 wt% of a second phase, eg, $CaO + SiO_2$ on grain boundaries, originating from impurities or sinter aids.

Table 1. Systematic Classification of Ferrites

Crystal chemistry[a]	Formula[b]	Magnetic nature: Soft	Intermediate	Hard	Appearance: Polycrystalline	Single crystalline	Bonded powder	Application
				Cubic				
spinel	$MeFe_2O_4$	X			X	X		recording heads
		X			X			core material for various inductors, transformers, and TV deflection units
	$CoFe_2O_4$, γ-Fe_2O_3		X				X	recording tape
garnet	RFe_5O_{12}		X			X		microwave, magnetooptics, bubble-information storage
perovskite	$R'FeO_3$					X		electroceramic devices
ortho-ferrite[c]	$RFeO_3$		X			X		bubble-information storage
				Hexagonal				
magnetoplumbite	$R'Fe_{12}O_{19}$			X[d]	X		X	permanent magnets
	$R'Fe_{12-x}Me'O_{19}$			X[e]			X	tape for perpendicular recording
W(= MS)	$R'Me_2Fe_{16}O_{27}$			X[d]	X		X	permanent magnets[f]
	$R'Co_2Fe_{16}O_{27}$		X[e]			X		microwaves, shielding[f]
X(= M_2S)	$R'MeFe_{28}O_{46}$			X[d]	X		X	permanent magnets[f]
	$R'CoFe_{28}O_{46}$		X[e]			X		microwaves, shielding[f]
Y(= ST)	$R'_2Me_2Fe_{12}O_{22}$		X[e]		X	X		microwaves, shielding
Z(= MST)	$R'_3Me_2Fe_{24}O_{41}$			X[d]				permanent magnets[f]
	$R'_3Co_2Fe_{24}O_{41}$		X[e]		X			microwaves, shielding

[a] $M = BaFe_{12}O_{19}$; $S = Me_2Fe_4O_8$; $T = Ba_4Fe_8O_7$. See Fig. 1.
[b] Me = Fe^{2+}, Ni^{2+}, Mn^{2+}, Co^{2+}, Zn^{2+} etc; Me0.167 $>^{3+}$, (Ti^{4+} + Co^{2+}), etc; R = Y, Nb, etc; *Rh* >, Sr, Pb, Ca.
[c] These materials are orthorhombic.
[d] Preferred axis (uniaxial anisotropy).
[e] Preferred plane (planar anisotropy).
[f] Potential application.

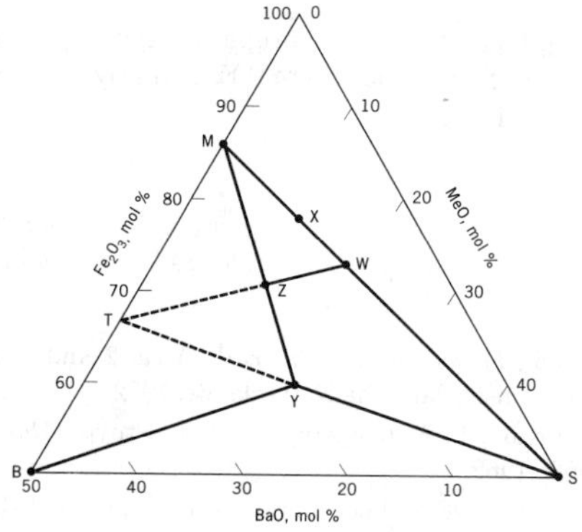

Figure 1. Composition diagram for hexagonal ferrites where $B = BaFe_2O_4$, $M = BaFe_{12}O_{19}$, $S = MeFe_2O_4$, $T = BaFe_4O_7$, $W = BaMe_2Fe_{16}O_{27}$, $X = BaMeFe_{14}O_{23}$, $Y = BaMeFe_6O_{11}$, $Z = Ba_3Me_2Fe_{24}O_{41}$, and Me is as defined in Table 1.

Ferrites are oxides and thus rather inert with respect to water, bases, and organic solvents. However, they may be attacked by acids having sufficiently high strength (pH < 2).

Being ceramic materials, ferrites are also resistant to high temperatures, at least up to the sintering temperature (1200–1400°C). However, noticeable reduction may take place at temperatures > 1100 °C and low oxygen partial pressures.

Ceramic ferrites cannot explode or release poisonous gases, and generally do not contain toxic elements. However, permanent magnets based on Sr-ferrite contain strontium, Sr, which is in principle toxic. In dense (porosity $< 10\%$) materials the Sr is firmly bound; however, in porous (porosity $> 10\%$) materials the second phase may dissolve partially in water or acids giving rise to release of Sr. Even in the latter case the effect is limited.

Crystal Chemistry and Physical Properties

The magnetic properties of ferrites result from the electronic configuration and mutual interactions of the ions present. Thus investigation of the crystal structure is fundamental to the understanding of these materials. Although the specific structures of spinel ferrites and M-type ferrites differ, both classes can be considered to be composed of two sublattices: an anionic lattice having relatively large anions and a cationic lattice containing the smaller cations, which fill interstitial sites.

Spinel Ferrites. In spinel ferrites having the composition AB_2O_4, where A and B are metals, cubic close-packed oxygen ions leave two kinds of interstitial sites for the cations: tetrahedral or A-sties, surrounded by four oxygen ions; and octahedral or B-sites, surrounded by six oxygen ions. A wide variety of transition-metal cations can be fit into these interstitial sites. The most important family of spinels is $Me^{2+}Fe_2^{3+}O_4$, where Me = Mg, Mn, Fe, Co, Ni, Zn, Cu, etc, either singly or in combination. But similar ferrites having less than two Fe-ions per formula unit are also of industrial significance because of the high electrical resistivity.

In the cases of existing unpaired d-electrons, transition-metal ions possess a net magnetic moment. In a spinel these magnetic moments interact through the anions (super exchange), resulting in a situation where the moments of both A-site and B-site ions are aligned, ie, A–A and B–B parallel, but A–B antiparallel, a ferrimagnetic ordering. The net magnetic moment per unit formula can be calculated from the distribution of the cations over these sites.

It is possible to systematically alter the net magnetic moment of ferrites by chemical substitutions. A very important industrial application is the increase of the magnetic moment in mixed MnZn-ferrites and NiZn-ferrites.

The direction of the alignment of magnetic moments within a magnetic domain is related to the axes of the crystal lattice by crystalline electric fields and spin-orbit interaction of transition-metal d-ions.

The net macroscopic B and the resulting μ_i result from two types of magnetization processes. First there is a contribution from the rotation of the magnetization inside each individual magnetic domain, from the preferred direction toward the direction of the applied magnetic field until the sum of the magnetostatic energy (minimal if B lies along H) and the anisotropy energy has reached its minimum value. Secondly, domain walls move. Domains having favorable magnetization directions with respect to the applied magnetic field grow at the expense of others, thus further minimizing the total magnetostatic energy.

Stresses, which can for example be introduced during cooling after ceramic sintering, during machining of sintered products, or simply when product parts are clamped together before use, lead to anisotropy by the magnetostriction effect. Magnetocrystalline anisotropy, magnetostriction, and magnetic permeability depend markedly on chemical composition. These dependencies have been extensively investigated within the ternary diagrams $MnFe_2O_4$–$ZnFe_2O_4$–Fe_3O_4 and $NiFe_2O_4$–$ZnFe_2O_4$–Fe_3O_4.

Properties can also be manipulated by adding specific dopants: Co^{2+} ions are, for example, introduced for extra anisotropy compensation; and Ti^{4+} ions are substituted to form pairs with Fe^{2+} ions and thus to reduce electron hopping. Extensive investigations have been carried out involving the addition of dopants such as CaO (typically 0.1 mol %) and SiO_2 (0.01 mol %) in order to provide ceramic grains having electrically insulating grain boundaries, thus markedly increasing the effective resistivity of the ferrite product.

At high frequencies ferrites exhibit energy losses resulting from various physical mechanisms at different frequencies and appearing as heat dissipation. Hysteresis losses arise from irreversible domain wall jumps. A second important loss contribution comes from eddy currents, induced by alternating magnetic fluxes. A third main loss contribution is from magnetic resonances.

M-Type Ferrites. The magnetism of $BaFe_{12}O_{19}$ comes from the Fe^{3+}-ions, each carrying a magnetic moment of 5 μ_B. These are aligned by either parallel or antiparallel ferromagnetic interaction. Ions of the same crystallographic position are aligned parallel, constituting a magnetic sublattice. The interaction between neighboring ions of different sublattices is a result of superexchange by oxygen. It is the magnetic structure in terms of sublattices and their mutual orientation that governs magnetic behavior, which in turn is described in terms of intrinsic and material properties.

The intrinsic magnetic properties may be subdivided into primary and secondary. The primary properties, such as the saturation magnetization J_s and the magnetocrystalline anisotropy constant K_1, are directly related to the magnetic structure. The secondary properties, such as the anisotropy field strength H_A and the specific domain wall energy (γ_w), are derived from the primary ones. These latter govern the actual magnetic behavior.

The intrinsic properties may be modified by substitution. Ba can be fully replaced by Sr or Pb and partly by Ca (< 40 mol %). CaM, stabilized with 0.03 mol % La_2O_3, is also possible. The intrinsic properties of these M-ferrites vary somewhat and other factors such as sintering behavior and price of raw materials often dictate the commercial viability. Large-scale production is concentrated on BaM and SrM. High quality magnets are generally based on SrM, and somewhat lower priced magnets are based on BaM.

Substitution for Fe^{3+} has a drastic effect on intrinsic magnetic properties. Partial substitution by Al^{3+} or Cr^{3+} decreases J_s without affecting K_1 seriously, resulting in larger H_A and H_c values. Substitution by Ti^{4+} and Co^{2+} causes a considerable decrease in K_1; the uniaxial anisotropy ($K_1 > 0$) may even change into planar anisotropy ($K_1 < 0$). Intermediate magnetic structures are also possible.

Processing

Commercial ferrites are produced by a ceramic process involving powder preparation, shaping, firing, and finishing (see CERAMICS). The powder preparation is usually the classical one involving the mixing of powder raw materials, prefiring, milling, and granulating. The raw materials are oxides or carbonates, the main component always iron oxide. The purity of the raw materials is an important factor with respect to the processing and final quality. Mixing can be done in different ways, depending on the nature and quality of raw materials and of the final product. During prefiring the different compounds react in the solid state to form the final compound or intermediate compounds, losing the volatile substances such as CO_2. Mostly this process is accompanied by homogenization on a local scale. In addition, there is some densification. To limit the effect of densification and to facilitate the handling of the material, it is often granulated before prefiring. To enable shaping and sintering, the prefired material has to be milled down to micrometer-sized particles. The last milling step is generally wet milling to prevent agglomeration effects. During or after milling, binders and lubricants are usually added to facilitate granulating and pressing.

In some cases it may be advantageous to deviate from the classical technology. For example, in wet-chemical preparation better chemical and morphological control may be achieved by starting from salt solutions.

Shaping is often done by dry pressing, which in fact is a simple and effective method to make the variety of shapes needed for electronic applications.

During firing, formation of the proper compound is completed and densification occurs from about 50 to 90% solid by volume, implying a linear shrinkage of 10–25%.

F. X. N. M. KOOLS
D. STOPPELS
Philips Components

E. P. Wohlfarth, ed., *Ferromagnetic Materials*, Vols. 2 and 3, North Holland Publishing Co., Amsterdam, the Netherlands, 1982.

A. Broese van Groenou, P. F. Bongers, and A. L. Stuyts, *Mater. Sci. Eng.* **3**, 317–392 (1968–1969).

E. C. Snelling, *Soft Ferrites*, 2nd ed., Butterworth & Co. Publishers Ltd., Kent, U.K., 1988.

A. Goldman, *Modern Ferrite Technology*, Van Nostrand Reinhold Co., Inc., New York, 1990.

FERROELECTRICS

Polarization which can be induced in nonconducting materials by means of an externally applied electric field $\overline{E}$ is one of the most important parameters in the theory of insulators, which are called dielectrics when their polarizability is under consideration. Experimental investigations have shown that these materials can be divided into linear and nonlinear dielectrics in accordance with their behavior in a realizable range of the electric field. The electric polarization $\overline{P}$ of linear dielectrics depends linearly on the electric field $\overline{E}$, whereas that of nonlinear dielectrics is a nonlinear function of the electric field. The most important materials among nonlinear dielectrics are ferroelectrics which can exhibit a spontaneous polarization $\overline{P}_s$ in the absence of an external electric field and which can split into spontaneously polarized regions known as domains. It is evident that in the ferroelectric the domain states differ in orientation of spontaneous electric polarization, which are in equilibrium thermodynamically, and that the ferroelectric character is established when one domain state can be transformed to another by a suitably directed external electric field. It is the reorientability of the domain state polarizations that distinguishes ferroelectrics as a subgroup of materials from the 10-polar-point symmetry group of pyroelectric crystals.

Properties

It is this high intrinsic dielectric susceptibility response that is the phenomenon most used in the practical application of polycrystalline ceramic ferroelectrics. Ferroelectric ceramics having relative permittivities $\epsilon_{ij}/\epsilon_0 = K_{ij}$ ranging up to 10,000, where ϵ_0 is the dielectric permittivity of vacuum, are widely used in many types of capacitors including the multilayer variety (see ADVANCED CERAMICS; CERAMICS; CERAMICS AS ELECTRICAL MATERIALS).

The piezoelectric and electrostrictive voltage coefficients, d_{ijm} and M_{mnij}, of the ferroelectrics are very large because of the large polarizability. Thus a second principal application of the ferroelectrics uses this high electromechanical coupling for efficient transduction between electrical and mechanical signals in sonic and ultrasonic transducers and filter applications.

Both the spontaneous polarization $\overline{P}_s$ and the remanent polarization $\overline{P}_R$ are strong functions of temperature, particularly near the transition temperature T_c in ferroelectrics; $\Delta\overline{P}_R = \pi\Delta T$, where π is the pyroelectric coefficient. Many ferroelectrics have large pyroelectric coefficients and can be used in thermometry and in bolometry sensing devices of infrared radiation (see INFRARED TECHNOLOGY AND RAMAN SPECTROSCOPY; SENSORS).

Many ferroelectrics are high band gap insulating crystals and have good transparency in both the visible and near-ir spectral regions. In the single-domain state, many ferroelectric crystals also exhibit high optical nonlinearity and this, coupled with the large standing optical anisotropies (birefringences) that are often available, makes the ferroelectrics interesting candidates for phase-matched optical second harmonic generation (SHG).

One area of application utilizes the interaction between the dielectric polarization and the electrical transport processes in ferroelectrics. In single dielectric crystals the effects of the domain polarizations on the drift and retrapping of photogenerated carriers give most interesting photoferroelectric effects. Of more immediate applicability, however, are the large effects of the dielectric changes at the ferroelectric phase transition on the potential barriers at grain boundaries in suitably prepared semiconducting ceramic ferroelectrics. These barium titanate-based compositions show strong positive temperature coefficients of resistivity (PTC effects) and are finding widespread use in temperature and current control for domestic, industrial, and automotive applications.

Materials

Oxygen Octahedra. An important group of ferroelectrics is that known as the perovskites. The perfect perovskite structure is a simple cubic, having the general formula ABO_3, where A is a monovalent or divalent materials such as Na, K, Rb, Ca, Sr, Ba, or Pb, and B is a tetra- or pentavalent cation such as Ti, Sn, Zr, Nb, Ta, or W.

$PbZrO_3$–$PbTiO_3$-Based Materials. Since the mid-1950s, solid solutions of $PbZrO_3$–$PbTiO_3$ (PZT) ceramics having the perovskite structure have gained rising interest because of the superior piezoelectric properties.

Preparation of Ferroelectric Materials

Ceramics. The properties of ferroelectrics, basically determined by composition, are also affected by the microstructure of the densified body which depends on the fabrication method and condition. The ferroelectric ceramic process is comprised of the following steps: *(1)* selection of raw oxide materials, *(2)* preparation of a powder composition, *(3)* shaping, *(4)* densification, and *(5)* finishing.

Powder Preparation. Mixing. The most widely used mixing method is wet ball milling, which is a slow process, but it can be left unattended for the whole procedure.

Calcination. Calcination involves a low (<1000 °C) temperature solid-state chemical reaction of the raw materials to form the desired final composition and structure such as perovskite for $BaTiO_3$ and PZT.

Shaping. The calcined powders must be milled and a binder (usually organic materials) added if necessary for the forming procedure.

Densification. Sintering, hot-pressing, or hot-isostatic-pressing methods may be used to densify the shaped green ferroelectric ceramics to ~95–100% of theoretical value.

Finishing. The densified ferroelectric ceramic bodies usually require machining and metallizing for dimension and surface roughness control and electrical contact.

Thin-Film Ferroelectrics. The trends in integrated circuits (qv) and packaging technologies toward miniaturization have stimulated the development of ferroelectric thin films (see PACKAGING, ELECTRONIC MATERIALS; THIN FILMS). Advances in thin-film growth processes offer the opportunity to utilize the material properties of ferroelectrics such as pyroelectricity, piezoelectricity, and electrooptic activity for useful device applications. The primary impetus of the activity in ferroelectric thin-film research is the large demand for the development of nonvolatile memory devices, also called FERRAMS (ferroelectric random access memories).

Several techniques have been investigated for the preparation of ferroelectric thin films. The thin-film growth processes involving low energy bombardment include magnetron sputtering, ion beam sputtering, excimer laser ablation, electron cyclotron resonance (ECR) plasma-assisted growth, and plasma-enhanced chemical vapor deposition (PECVD) (see PLASMA TECHNOLOGY). Other methods are sol-gel, metal organic decomposition (MOD), thermal and *e*-beam evaporation, flash evaporation, chemical vapor deposition (CVD), metal organic chemical vapor deposition (MOCVD), and molecular beam epitaxy (MBE).

Ferroelectric Ceramic–Polymer Composites. The development of active ceramic–polymer composites was undertaken for underwater hydrophones having hydrostatic piezoelectric coefficients larger than those of the commonly used lead zirconate titanate (PZT) ceramics. It has been demonstrated that certain composite hydrophone materials are two to three orders of magnitude more sensitive than PZT ceramics while satisfying such other requirements as pressure dependency of sensitivity. The idea of composite ferroelectrics has been extended to other applications such as ultrasonic transducers for acoustic imaging, thermistors having both negative and positive temperature coefficients of resistance and active sound absorbers.

Applications

Multilayer Capacitors. Multilayer capacitors (MLC), at greater than 30 billion units per year, outnumber any other ferroelectric device in production.

Piezoelectric and Electrostrictive Device Applications. Devices made from ferroelectric materials utilizing their piezoelectric or electrostrictive properties range from gas igniter to ultrasonic cleaners (or welders).

Table 1. Properties of Relaxor and Normal Ferroelectrics

Property	Normal ferroelectrics	Relaxor ferroelectrics
permittivity temperature dependence $\epsilon = \epsilon\ (T)$	sharp first- or second-order transition above Curie temperature	broad–diffuse phase transition about Curie maxima
permittivity temperature and frequency dependence $\epsilon = \epsilon\ (T, \omega)$	weak frequency dependence	strong frequency dependence
remanent polarization	strong remanent polarization	weak remanent polarization
scattering of light	strong anisotropy (birefringent)	very weak anisotropy (pseudocubic)
diffraction of x-rays	line splitting owing to spontaneous deformation from paraelectric to ferroelectric phase	no x-ray splitting giving a pseudocubic structure

Composite Devices. Composites made of active-phase PZT and polymer-matrix phase are used for the hydrophone and medical imaging devices (see COMPOSITE MATERIALS–POLYMER-MATRIX; IMAGING TECHNOLOGY).

Relaxor Ferroelectrics. The general characteristics distinguishing relaxor ferroelectrics, eg, the $PbMg_{1/3}Nb_{2/3}O_3$ family, from normal ferroelectrics such as $BaTiO_3$, are summarized in Table 1.

Relaxor ferroelectrics have been extensively investigated since the late 1970s because of the ability to generate large electrically induced strains, minimal hysteresis of the strain–electric field response, and minimal thermal strain. These materials also show promise in capacitor applications because of large dielectric permittivities. The field-induced piezoelectric and elastic properties of relaxor ferroelectrics have been investigated for transducer applications, including three-dimensional medical ultrasonic imaging devices.

Polymer Ferroelectrics. Polymer ferroelectrics are used as audio frequency transducers, such as microphones, headphones, and loudspeaker tweeters having excellent frequency response and low distortion because of the low density, lightweight transducer film; ultrasonic transducers for underwater applications, such as hydrophones, and for medical imaging applications; electromechanical transducers for computer and telephone keypads and a variety of other contactless switching applications; and pyroelectric detectors for infrared imaging and intruder detection.

Economic Aspects

Whereas the composite market is growing with the invention of new devices, total unit volume and dollar amounts are small compared to the ferroelectric capacitor and ferroelectric–piezoelectric ceramic markets (see MEDICAL IMAGING TECHNOLOGY). Ferroelectric thin films have not, as of this writing, been commercialized. Demand for PTC ferroelectrics has been decreasing rapidly. Approximately 40% of the U.S. electronic ceramics industry is represented by ferroelectrics. Japanese suppliers, however, generally dominate the electronic ceramic business.

SEI-JOO JANG
The Pennsylvania State University

B. Jaffe, W. R. Cooke, Jr., and H. Jaffe, *Piezoelectric Ceramics*, Academic Press, New York, 1971.

A. J. Moulson and J. M. Herbert, *Electroceramics*, Chapman and Hall, London, 1990.

R. E. Newnham, D. P. Skinner, and L. E. Cross, *Mater. Res. Bull.* **13**, 525 and 599 (1978).

L. E. Cross, *Ferroelectrics* **76**, 241–267 (1987).

FERTILIZERS

Nutrient Requirements of Plants

For satisfactory growth, most plants must have access to at least 22 different chemical elements. Deficiency of any of these essential elements, regardless of how much is required by the plant, can become the limiting factor in retarding healthy plant growth. This important principle, known as the law of the minimum, was one of a number of important nineteenth century contributions of Justus von Liebig. Elements now recognized as absolutely essential to healthy plant growth are given in Table 1.

To agronomists concerned with ensuring an adequate supply of these elements to crops, important considerations are *(1)* the route by which the plant ingests each of the elements, *(2)* the amount of each element required, and *(3)* the chemical forms of the element that are accepted by the plant. Hydrogen and oxygen enter the plant in the form of water, which is absorbed almost entirely through the roots. In healthy plant tissue, up to 95% of the tissue weight consists of water obtained by this route.

Of the dry matter, 44% consists of carbon, all of which enters the plant as gaseous carbon dioxide absorbed from the atmosphere through the leaves. This carbon is transformed to the myriad of organic plant constituents through photosynthesis. In the photosynthesis process, gaseous oxygen is expelled from the leaves into the atmosphere. The green pigment chlorophyll, present in most leaves, is a participant in the photosynthesis process. All of the remaining 21 essential elements present in plants, under usual growing conditions, enter the plant only through root absorption. These are the elements that can be provided by, or supplemented by, the application of fertilizers to the soil. Foliar absorption of at least some of these elements is also possible by spraying solutions onto leaves, but this is a highly specialized practice.

To determine the feasibility of, or need for, fertilization requires knowing *(1)* which of the required elements, if any, are deficient in the soil; *(2)* what chemical forms of the deficient elements are assimilable by the plants and thus suitable as fertilizers; *(3)* what quantity of fertilizer material is required to meet the needs of the crop; and *(4)* whether the crop yield increase resulting from fertilizer application would warrant the cost of the fertilizer production and application.

Plant nutrients usually are categorized as being structural elements, primary or secondary nutrients, or micronutrients. Most micronutrient elements are sufficiently present in native soils or are impurities in nonmicronutrient fertilizers applied to soils. Thus fertilization to provide these micronutrient elements specifically can

Table 1. Chemical Elements Essential to Healthy Growth of Plants

Essential macronutrients	Essential micronutrients	Beneficial	Essentiality not demonstrated
	Metals		
potassium	iron	aluminum	chromium
calcium	copper	strontium	tin
magnesium	manganese	rubidium	nickel
	zinc		
	molybdenum		
	cobalt		
	vanadium		
	sodium		
	gallium		
	Nonmetals		
carbon	boron	selenium	fluorine
hydrogen	silicon		bromine
oxygen	chlorine		
phosphorus	iodine		
nitrogen			
sulfur			

often be omitted. Important exceptions arise in cases of localized soil deficiencies, in special requirements of some crops, or soil depletion resulting from repeated cropping.

The secondary nutrients, calcium, magnesium, and sulfur, are in no way secondary in regard to plant need. They are, however, secondary regarding the need to be furnished through fertilizer application, because these elements are abundant components of soil minerals at most locations. These elements also are incidental components of many fertilizers. Moreover, the widespread agricultural practice of liming by application of pulverized limestone or dolomite to the soil, intended chiefly for control of soil pH, incidentally adds calcium and magnesium (see LIME AND LIMESTONE).

The elements nitrogen, phosphorus, and potassium are primary nutrients not only because healthy plant growth requires them in relative abundance, but also because these are the primary elements that most often must be furnished by fertilizers. None of these elements is a principal component of the usual soil minerals. The principal task of the chemical fertilizer industry worldwide is to furnish agriculture with chemical forms of nitrogen (N), phosphorus (P), and potassium (K) that, when applied to the soil, can be readily assimilated by crop plants. Ability to include secondary nutrients and micronutrients in fertilizers when needed is also a responsibility of the industry, but it is the production of N, P, and K fertilizers that defines the industry.

Nature of Chemical Fertilizers

Chemical Content. Numerous chemical compounds have been shown to be suitable as sources of primary nutrients in fertilizers. A partial listing is given in Table 2. Because the route of these nutrients into the plant is through root absorption from the soil solution, solubility of the compounds in the soil solution is of prime importance. All of the compounds in Table 2 are highly soluble in water, with the exception of dicalcium phosphate. Dicalcium phosphate, nevertheless, is suitably soluble in most soil solutions and is recognized as a highly acceptable fertilizer component. Agronomic response to phosphorus in fertilizers can be suitably predicated by laboratory measurement of solubility in certain neutral or alkaline citrate solution reagents. Solubility in pure water therefore is not a necessity, although a certain degree of water solubility is recommended by most agronomists.

Table 2. Compounds Agronomically Effective in Fertilizers

Compound	Formula	Primary nutrient content, %[a] N	P_2O_5	K_2O
	Nitrogen sources			
ammonia	NH_3	82.2		
ammonium sulfate	$(NH_4)_2SO_4$	21.2		
ammonium nitrate	NH_4NO_3	35.0		
sodium nitrate	$NaNO_3$	16.5		
calcium nitrate	$Ca(NO_3)_2$	17.0		
urea	$CO(NH_2)_2$	46.6		
calcium cyanamide	$CaCN_2$	34.9		
monoammonium phosphate	$(NH_4)H_2PO_4$	12.1	61.7	
diammonium phosphate	$(NH_4)_2HPO_4$	21.2	53.7	
potassium nitrate	KNO_3	13.8		46.6
	Phosphorus sources			
monocalcium phosphate	$Ca(H_2PO_4)_2$		60.6	
dicalcium phosphate	$CaHPO_4$		52.1	
monoammonium phosphate	$(NH_4)H_2PO_4$	12.1	61.7	
diammonium phosphate	$(NH_4)_2HPO_4$	21.2	53.7	
potassium phosphate	K_3PO_4	0.0	33.4	66.5
	Potassium sources			
potassium chloride	KCL			63.1
potassium sulfate	K_2SO_4			54.0
potassium magnesium sulfate	K_2SO_4 2 $MgSO_4$			22.7
potassium nitrate	KNO_3		13.8	46.6
potassium phosphate	K_3PO_4		33.4	66.5

[a] Pure salt basis.

The legal basis for the sale of fertilizers throughout the world is laboratory evaluation of content as available nitrogen, phosphorus, and potassium. By convention, numerical expression of the available nutrient content of a fertilizer is by three successive numbers that represent the percent available of N, P_2O_5, and K_2O, respectively. Thus, for example, a 20–10–5 fertilizer contains available nitrogen in the amount of 20% by weight of N, available phosphorus in amount equivalent to 10% of P_2O_5, and available potassium in amount equivalent to 5% K_2O. The numerical expression of these three numbers is commonly referred to as the analysis or grade of the fertilizer. Accepted procedures for laboratory analysis are fixed by laws that vary somewhat from country to country.

In the United States, the analytical methods approved by most states are ones developed under the auspices of the Association of Official Analytical Chemists (AOAC). Penalties for analytical deviation from guaranteed analyses vary, even from state to state within the United States. The legally accepted analytical procedures, in general, detect the solubility of nitrogen and potassium in water and the solubility of phosphorus in a specified citrate solution.

As is evident from the listing in Table 2, the fertilizer manufacturer has a wide array of compounds from which to choose. Final choices of products and processes therefore rest heavily on other factors such as availability and cost of raw materials, economy of processing, safety of product, economy of handling and shipping, acceptability of physical form and physical behavior of the product, and farmer acceptance.

Chemical fertilizers need not be pure compounds. In fact, impurities in products may beneficially furnish secondary or micronutrients, or the presence of impurities may improve the physical condition of the fertilizer. The presence of impurities in large proportion, however, is disadvantageous because of dilution, which lowers the plant food content (grade) of the product.

Commercial fertilizers are produced as *(1)* single-nutrient materials, which contain only one of the primary nutrient elements; *(2)* binutrient materials, which contain two of the primary nutrients; or *(3)* multinutrient products, which contain all three primary nutrients. The two- and three-component products are frequently referred to as mixed fertilizers. The proportioning of nutrients in mixed fertilizers is variable, depending entirely on farmer preference or on soil analysis to determine relative needs of the soil and the intended crop.

Physical Properties. The physical form and stability of a fertilizer product is of an importance almost equal to that of its chemical content. Commercial fertilizers of importance include not only solids, but also fluids, both solutions and suspensions, and even a gas (anhydrous ammonia).

Factors of importance in regard to the physical properties of solid fertilizers include particle size, particle strength, caking tendency, chemical stability, and hygroscopicity.

Nitrogen Fertilizers

Consumption of nitrogen, in general, is more than double that of either P_2O_5 or K_2O. This reflects both the greater need for nitrogen exhibited by most food crops and the greater economic benefits obtainable from nitrogen fertilizer application. Additionally, there is a tendency for applied nitrogen to be lost by leaching and volatilization, and a consequent requirement for more frequent replenishment than is usual for P_2O_5 and K_2O.

Historic Nitrogen Sources. Nearly all commercial nitrogen fertilizer is today derived from synthetic ammonia. However, prior to the introduction of ammonia synthesis processes in the early 1900s, dependence was entirely on other sources. These sources are still utilized, but their relative importance has diminished. They include crop rotation, natural organics, mineral nitrogen, and by-product ammonia.

Nitrogen Fertilizers from Synthetic Ammonia. Worldwide, the yearly production of synthetic ammonia is about 120×10^6 t, of which about 85% finds use in fertilizers. Essentially all the processes employed for ammonia synthesis are variations of the Haber-Bosch process developed in Germany from 1904–1913. The principal routes by which synthetic ammonia is processed into finished fertilizers are shown in Figure 1. Also included are U.S. consumption data on each of these products for the crop year ended June 30, 1990.

Phosphate Fertilizers

Historic Sources. One fertilization practice, initiated in the early nineteenth century, that was effective in furnishing phosphorus was the application of ground bones to soil. In about 1830, in England, sulfuric acid pretreatment of bones was begun as an effective method of increasing the availability of the phosphorus to growing plants. About 10 years later, similar sulfuric acid treatment of mineral phosphate ore was begun to produce the effective fertilizer product now known as ordinary (or normal) superphosphate.

Phosphate Fertilizers from Mineral Phosphates. Essentially all fertilizer phosphate now is derived from mineral phosphates. Deposits of mineral phosphate are abundant and widely dispersed throughout the world. Nearly all of the mineable deposits are minerals of the apatite group represented by the general formula $Ca_5(F,Cl,OH,0.5CO_3)(PO_4)_3$. As mined, essentially all the ores require beneficiation to reduce the content of clay, silica, or other extraneous material. Beneficiation methods commonly used are washing (gravity separation) and/or flotation (qv) using various flotation agents.

The routes by which mineral phosphates are processed into finished fertilizers are outlined in Figure 2. Notable is the large, steady increase in the importance of monoammonium and diammonium phosphates as finished phosphate fertilizers at the expense of ordinary superphosphate, and to some extent at the expense of triple superphosphate. In the United States about 65% of the total phosphate applied is now in the form of granular ammonium phosphates, and additional amounts of ammonium phosphates are applied as integral parts of granulated mixtures and fluid fertilizers.

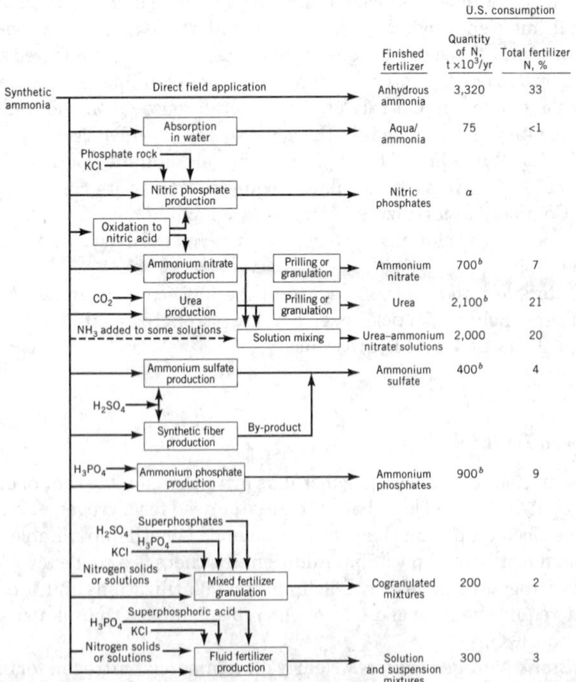

Figure 1. Routes for making synthetic ammonia into fertilizers. The consumption data are for the year ended June 30, 1990. [a]Significant quantities are made and used in foreign countries; [b]includes quantities applied in dry blends. Courtesy of TVA.

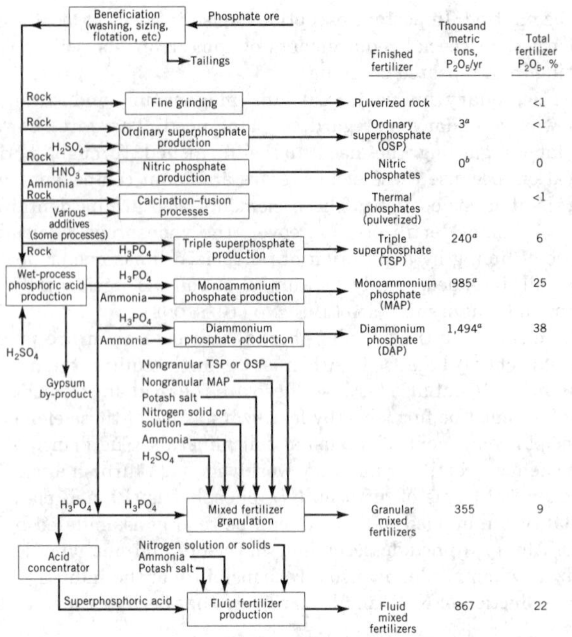

Figure 2. Routes for making finished fertilizer from mineral phosphate. Consumption data are for year ending June 30, 1990. [a]Includes quantities applied in dry blends; [b]significant quantities are made and used in some foreign countries.

Potash Fertilizers

World consumption of potassium salts presently exceeds 28×10^6 t of K_2O equivalent per year. About 93% of that is for fertilizer use (see POTASSIUM COMPOUNDS). The potash industry is essentially a mining and beneficiation industry. The two main fertilizer materials, KCl and K_2SO_4 are produced by beneficiating ores at the mine sites. The upgraded salts then are shipped to distributors and manufacturers of mixed goods.

Some 90% of all potash fertilizer is KCl; the proportion is slightly less in the United States. Most of the remainder of fertilizer potash is sulfate, either as K_2SO_4 or the double salt langbeinite, $K_2SO_4 \cdot 2\,MgSO_4$. A small but increasing amount of KNO_3 is used, mostly as a specialty fertilizer.

For most of the decade prior to 1965, the United States was the world's largest single producer of potash. More than 90% of this production came from mines in New Mexico. As the grades of these deposits lowered and production costs rose, expanding production in Saskatchewan, Canada, increasingly met the growing United States market. In the 1990s, the United States is by far the largest single importer of potash, almost entirely from Canada. Potash mines have long been worked also in Germany and in the former Soviet Union.

Potassium Chloride. The principal ore encountered in the U.S. and Canadian mines is sylvinite, a mechanical mixture of KCl and NaCl. Three beneficiation methods used for producing fertilizer grades of KCl are thermal dissolution, heavy media separation, and flotation. The choice of method depends on factors such as grade and type of ore, local energy sources, amount of clay present, and local fuel and water availability and costs.

Potassium Sulfate. Potassium sulfate is a preferred form of potash for crops that have a low tolerance for chloride. Tobacco and potatoes are two such crops. K_2SO_4 is produced most often from langbeinite by metathetical reaction in aqueous solution.

Potassium Nitrate. Potassium nitrate, known but little used as a fertilizer for many years, may be reclaimed as a by-product of the production of sodium nitrate from natural deposits of caliche in Chile. KNO_3 also has been produced by the double decomposition reaction between sodium nitrate and potassium chloride.

The U.S. domestic commercial potassium nitrate of the 1990s contains 13.9% N, 44.1% K_2O, 0–1.8% Cl, 0.1% acid insoluble, and 0.08% moisture. The material is manufactured by Vicksburg Chemical Company using a process developed by Southwest Potash Division of AMAX Corporation. This process uses highly concentrated nitric acid to catalyze the oxidation of by-product nitrosyl chloride and hydrogen chloride to the more valuable chlorine.

Potassium nitrate is being used increasingly on intensive crops such as tomatoes, potatoes, tobacco, leafy vegetables, citrus, and peaches. The properties that make it particularly desirable for these crops are low salt index, nitrate nitrogen, favorable $N:K_2O$ ratio, negligible Cl^- content, and alkaline residual reaction in the soil.

Potassium Phosphates. Because of the very high analysis of potassium phosphates, eg, KH_2PO_4 is 0–50–33; KPO_3, 0–60–39, and the freedom from chloride and high solubility in liquid fertilizers, methods for economical production of these materials have been sought for many years. These methods have generally been based on the reaction of phosphoric acid or P_2O_5 and KCl, where a variety of potassium phosphates are formed. As of this writing (ca 1993) no proposed process is in large-scale operation.

Mixed Fertilizers

For most crop–soil combinations, optimum fertilization calls for application of more than one of the three primary nutrients and sometimes for application of secondary or micronutrients. Exact requirements often are determined by soil analysis. Mixed fertilizers are provided to supply single fertilizers that contain these nutrients in the required proportions. Thus the farmer needs only to make a single application. There are, however, both economic and agronomic considerations, that sometimes favor multiple applications of single-nutrient fertilizers over application of a single, multinutrient product. One such example is the use of low cost anhydrous ammonia as the nitrogen application along with supplementary application of a primarily phosphate–potash fertilizer. A similar situation, but agronomically dictated, is in the growing of corn. Side dressing using only nitrogen fertilizer at the proper stage of growth often is profitable to corn growers. Phosphate and potash are usually supplied to the corn crop as a separate, initial application of a mixture.

For the year ended June 30, 1990, about 39% of the total primary nutrient used in the United States was applied in mixtures, whereas the remaining 61% was applied by direct application. Because plant nutrient requirements are quite variable, depending on soils, crop varieties, locality, and previous fertilization history, producers of mixed fertilizers are faced with producing many grades (nutrient ratios) of product.

Production of mixed fertilizers involves, in most cases, a bringing together of some combination of the basic nitrogen, phosphate, and potash fertilizers previously described. This production of mixtures, therefore, results in additional expense, over and above that incurred in producing the feed materials. The added convenience must be worth this added expense. Because of the need for a variety of locally tailored grades, mixed fertilizer production plants tend to be regionally located and to serve only local or regional markets.

Nongranular Mixtures. Prior to about 1945, all commercial mixed fertilizers were produced in powdered, ie, nongranular, form. Undesirable characteristics of these nongranular mixed fertilizers were the dustiness and caking tendency, which interfered with handling and application, and low analysis, which resulted in high handling costs. In the United States, production of such nongranular mixtures declined and essentially ceased during the late 1940s and early 1950s, in favor of the production of granulated mixtures. In a few parts of the world, however, production of nongranular mixtures by this method persists.

Granulation of Mixed Fertilizers. A granular fertilizer, in contrast to a pulverized or nongranular one, is a fertilizer in which the individual particles are relatively large and are restricted to closely specified upper and lower size limits. Generally accepted limits in the United States and Canada are 6 Tyler mesh (3.35 mm dia) as the upper limit and 16 Tyler mesh (1.19 mm dia) as the lower limit.

Steam Granulation. The development of granulation methods for mixed fertilizers took somewhat divergent paths in the United States and Europe. U.S. granulation methods usually involve considerable chemical reaction among feed materials in the granulator as a method of developing heat and providing the plasticity required for granulation. With superphosphate as one of the feed materials, reactions during granulation typically result in some loss of P_2O_5 water solubility, but not of citrate solubility which is the basis for fertilizer sale in the United States.

In Europe, where fertilizer sale often is based on water solubility of P_2O_5, granulation methods were developed that largely avoid chemical reaction in the granulator. In such procedures, the feed materials are finely ground fertilizer materials such as ammonium sulfate, ammonium nitrate, urea, superphosphates, ammonium phosphates, and potash salts. Plasticity and agglomeration are promoted by the addition of water and steam while the mass is being subjected to vigorous rolling action in one of several types of granulation equipment. Granulator types used successfully include inclined rotating pans, rotary drums, and pug mills or pin mills. Because steam is essential to provide both heat and moisture, this general type of granulation is commonly referred to as steam granulation. Detailed discussion of the steam granulation process is available.

Steam granulation is practiced in Europe, Australia, and elsewhere, chiefly in small plants in which superphosphate, either ordinary or triple, is a primary ingredient. However, for many of the larger operations, superphosphates have been replaced by ammonium phosphates as the principal P_2O_5 source, and granulation procedures involving chemical reactions are employed in Europe as well as in the United States.

Slurry Granulation. Slurry granulation is granulation in which at least one of the primary feed ingredients to the granulator is a slurry. As developed and employed in the United States, slurry granulation of mixed fertilizer involves preparation first of an ammonium phosphate slurry of closely controlled composition by the reaction of wet-process phosphoric acid with ammonia in a vessel known as a preneutralizer. This slurry then is fed to a granulator along with other solid, liquid, or gaseous, ie, NH_3, fertilizer materials.

As of the end of 1988, there were only 50 mixed fertilizer granulation plants operating in the United States. Some of these employed the slurry granulation process, although there has been considerable conversion to melt granulation using a pipe reactor. The decline in the number of U.S. regional granulation plants from the high of 250 in 1962 is a direct result of competition from the bulk blending method of mixed fertilizer production and from fluid mixed fertilizers.

Melt Granulation. Melt granulation is similar to slurry granulation except that feed of slurry to the granulator is replaced by feed of a hot, concentrated, almost anhydrous melt.

Compaction Granulation. Compaction granulation is based on the principle that many solid materials are semiplastic and deform under high pressure. In compaction granulation of mixed fertilizers a powdered feed mixture containing at least one such plastic material is subjected to high pressure between counter-rotating steel rolls to form a dense, stable sheet of mixture. This sheet then is crushed and sized to yield granule-size chips. These chips, although irregular in shape, are generally acceptable as granular fertilizer because, in comparison to pulverized materials, they exhibit the favorable characteristics of improved flowability, nondustiness, and low caking tendency. Compaction is widely used by the basic potash producers to produce granular size potassium chloride from finer material.

Bulk Blending. Over half of all the mixed fertilizer made and distributed in the United States is in the form of bulk blends. Bulk blends are simple dry mixtures of two or more previously prepared granular size fertilizer products.

Since its introduction in the 1950s, bulk blending has shown continued growth in the United States at the expense of the cogranulation (ammoniation–granulation) methods of mixed fertilizer production.

Plant Design and Operation. There are over 5000 bulk blending plants in the United States, each located in a farming area and serving farmers within a radius that seldom exceeds 40 km. The average annual production per plant is only 5000 tons, and more than half the plants produce only 1000 to 4000 tons. Designs of the plants vary considerably in regard to storage facilities and materials handling, proportioning, and mixing equipment.

The flexibility of the bulk blending system and the close relationship with the farmer allow the bulk blender to provide a number of valuable supplementary services, such as adding herbicides, insecticides, micronutrients, or seeds to the blends; bagging blends; liming; and sampling soil. Consultation services and custom application can also be provided as can sale of anhydrous ammonia or nitrogen solution.

Problems in Bulk Blending. Technical problems are relatively few in the bulk blending process. Some precautions must be taken to protect hygroscopic starting materials, such as ammonium nitrate and urea, from excessive humid exposure. Also, there are several material combinations that must be avoided in blends because of chemical incompatibility. In particular, urea must not be mixed with ammonium nitrate because the combination is extremely hygroscopic and spontaneously becomes wet. Another combination that usually is avoided is urea and superphosphate, because the mixture tends to react and release water of hydration, which causes wetting. Another problem that can become quite serious is particle size incompatibility of some blend ingredients. Rather close matching of particle size is required to avoid segregation susceptibility, whereas rather wide deviations in particle density and shape are not troublesome.

Bulk Blending Outside the United States. In such places as Canada, Brazil, and Australia, which have vast agricultural areas similar to those in the United States, bulk blending is very successful. In Europe, where the farming areas are smaller and agronomic requirements are less varied, cogranulation in relatively large, centrally located plants are preferred. For some small countries and developing countries that are dependent on imported fertilizer materials, bulk blending at the port of entry plus bagging for distribution can be an effective system. Overall, it seems that bulk blending is likely to at least maintain its present large market share in the United States and probably increase its share elsewhere.

Fluid Mixtures. About one-third of all the nitrogen fertilizer used in the United States is applied as the fluid anhydrous ammonia, and an additional 20% is applied in the form of urea–ammonium nitrate solutions. Since the late 1950s, there has also been a steady increase in the amount of mixed fertilizer applied in fluid form. At present, about 20% of all mixed fertilizer applied in the United States is in fluid form, as compared to about 60% as bulk blends and about 20% as cogranulated solids. Of the fluid mixed fertilizers applied, about 60% are true solutions, whereas the remainder are suspensions. Fluid fertilizer plants, like bulk blending ones, are usually small, local enterprises serving local farmers. This is an understandable necessity because of the high water content of fluid fertilizers and the economic infeasibility of transporting products of high water content over long distances. These local plants enjoy some of the same beneficial characteristics as bulk blend plants, in that the products are tailored to local needs and can be delivered and applied in bulk.

Secondary and Micronutrients in Fertilizers

Secondary Nutrients.

Calcium. Soil minerals are a main source of calcium for plants, thus nutrient deficiency of this element in plants is rare. Calcium, in the form of pulverized limestone or dolomite, frequently is applied to acidic soils to counteract the acidity and thus improve crop growth. Such liming incidentally ensures an adequate supply of available calcium for plant nutrition. Although pH correction is important for agriculture, and liming agents often are sold by fertilizer distributors, this function is not one of fertilizer manufacture.

Some commonly used primary nutrient fertilizers are incidentally also rich sources of calcium. Ordinary superphosphate contains monocalcium phosphate and gypsum in amounts equivalent to all of the calcium originally present in the phosphate rock. Triple superphosphate contains soluble monocalcium phosphate equivalent to essentially all the P_2O_5 in the product. Other fertilizers rich in calcium are calcium nitrate, calcium ammonium nitrate, and calcium cyanamide.

Magnesium. Most of the supplemental magnesium is supplied as dolomitic limestone but other important sources are magnesium sulfate (Epsom salts), and potassium–magnesium sulfate (langbeinite), as well as organic chelate compounds.

Sulfur. Sulfur occurs in plants at levels as high as that of phosphorus. It is absorbed by plants as the sulfate anion, SO_4^{2-}, which is a constituent of many fertilizers. As environmental goals of eliminating sulfur oxide emissions from industrial plants are met (see EXHAUST CONTROL, INDUSTRIAL), sulfur deficiency in soils can be expected to increase. All indications are that fertilizers need to contain increasing amounts of sulfur.

Some of the principal forms in which sulfur is intentionally incorporated in fertilizers are as sulfates of calcium, ammonium, potassium, magnesium, and as elemental sulfur.

Granular sulfur-containing fertilizers used in bulk blends include ammonium sulfate, ammonium phosphate–sulfate, potassium sulfate, and potassium–magnesium sulfate.

Micronutrients. Attention to meeting the micronutrient needs of crops has greatly increased as evidenced in an analysis undertaken by TVA and the Soil Science Society in 1972. The micronutrient elements most often found wanting in soil–crop situations are boron, copper, iron, manganese, molybdenum, and zinc. Some of these essential micronutrients can be harmful to plants when used in excess.

The problem in micronutrient fertilization is that of adequately identifying the need, reducing this to a prescription, compounding the fertilizer, and distributing it evenly in the soil. Prescription compounding is being used increasingly as deficiencies are better identified, but a "shotgun" approach is also used in which a multitude of micronutrients are included in small amounts.

Generally, soluble materials are more effective as micronutrient sources than are insoluble ones. For this reason, many soil minerals that contain the micronutrient elements are ineffective sources for plants. Some principal micronutrient sources and uses are summarized below. In this discussion the term frits refers to a fused, pulverized siliceous material manufactured and marketed commercially for incorporation in fertilizers. Chelates refers to metalloorganic complexes specially prepared and marketed as especially soluble, highly assimilable sources of micronutrient elements (see CHELATING AGENTS).

Boron. The principal materials used are borax, sodium pentaborate, sodium tetraborate, partially dehydrated borates, boric acid, and boron frits. Both soil and foliar application are practiced, but soil applications remain effective longer. Boron toxicity is not often observed in field applications (see BORON COMPOUNDS).

Copper. Some 15 copper compounds (qv) have been used as micronutrient fertilizers. These include copper suflates, oxides, chlorides, and cupric ammonium phosphate, and several copper complexes and chelates. Both soil and foliar applications are used.

Iron. As with copper, some dozen or more materials are used as fertilizer iron sources. These include ferrous and ferric oxides and sulfides and ferrous ammonium phosphate, ferrous ammonium sulfate, frits, and chelates. In many instances, organic chelates are more effective than inorganic materials. Recommended application rates range widely according to both type of micronutrient used and crop.

Manganese. Commonly used manganese fertilizer materials are manganous and manganic sulfates, chlorides, carbonates, oxides, frits, and chelates.

Molybdenum. The commonly used molybdenum materials are sodium molybdate, ammonium molybdate, molybdenum trioxide, molybdenum sulfate, and frits.

Zinc. Zinc, one of the most widely needed and used micronutrients, is applied as sulfates (both basic and normal hydrates), carbonate, sulfide, phosphate, oxide, chelates, and other organic materials.

Micronutrients in Granular Fertilizers. In the production of granular fertilizers, it is relatively simple and effective to incorporate micronutrient materials as feeds in the granulation process. A problem with this method, however, is that granulation processes are most efficient and economical when operated continuously to produce large tonnages of the same or similar composition. As a result, granulation processes more often use the shotgun approach.

Micronutrients in Bulk Blends. Prescription formulation of fertilizers containing micronutrients is more practical in the production of bulk blends and fluid fertilizers, because these processes are more adaptable to frequent changes in grades, small-batch preparation, and thus to custom formulation. The problems with formulation and distribution in bulk blends are in achieving even distribution and in avoiding segregation of the micronutrients in storage, transportation, and application. The fundamental concepts for handling these problems are the same as those for avoiding segregation of primary nutrients in bulk blends.

Coating granular fertilizers or bulk blends with powdered micronutrients overcomes the disadvantages of using granular micronutrient additives. When the coating is done well, segregation is not a problem because the micronutrient essentially becomes an integral part of each granule.

Micronutrients in Fluid Fertilizers. In terms of homogeneity and even distribution, fluid fertilizers are probably the best micronutrient carriers. Fluid carriers of micronutrients usually are nitrogen solutions, clear liquid mixtures, or suspensions. Some micronutrients, however, are applied as simple water solutions or suspensions. Foliar micronutrient sprays often contain only the micronutrient material in water solution.

Micronutrient Reactions in Mixtures. The chemical systems that can be created in mixtures of six micronutrients in NPK mixed fertilizers are many and highly complex. Many chemical reactions occur, and the solubility of the reaction products determines the micronutrient solubility. The six principal micronutrients fall into two groups on the basis of the tendency to undergo adverse reactions in carrier fertilizers. Boron and molybdenum undergo few reactions that lower their solubility. Copper, iron, manganese, and zinc react to form more insoluble compounds. Of these adverse reactions those of copper are the least detrimental.

Raw Material Resources for Fertilizers

Resources for Nitrogen Fertilizers. The production of more than 95% of all nitrogen fertilizer begins with the synthesis of ammonia, thus it is the raw materials for ammonia synthesis that are of prime interest. Required feed to the synthesis process (synthesis gas) consists of an approximately 3:1 mixture (by volume) of hydrogen and nitrogen. Technologies for extracting hydrogen (qv) from both hydrocarbons (qv) and water (qv) are available. The extent of reserves of the preferred raw material, which is natural gas, and the cost of extracting hydrogen are increasingly serious problems for the fertilizer industry. Raw materials used to produce hydrogen in the order of decreasing preference are natural gas, straight-run naphthas, fuel oil, coal (qv), and water. The availability of hydrogen for ammonia synthesis, thus, is an inextricable part of the world energy supply–demand situation (see FUEL RESOURCES; GAS, NATURAL).

Resources for Phosphate Fertilizers. Natural mineral deposits are the source of phosphorus for essentially all manufactured phosphate fertilizer (see PHOSPHORUS; PHOSPHORUS COMPOUNDS). Phosphate deposits are numerous throughout the world but size and quality vary widely. Those minerals that contain enough phosphorus to be potential sources for industrial use are called phosphate rock and, sometimes, phosphate ore.

At the present rate of world phosphate rock consumption (150×10^6 t/yr), the total world reserve is sufficient for about 200 years, and the resource would be sufficient for nearly 900 years. At expected increased rates of consumption, the reserves and resources are adequate for at least 150 years and 700 years, respectively. At projected rates of consumption, the high grade reserves in Florida probably will be exhausted by the year 2000. Rock production from the Florida reserve presently constitutes about 80% of all United States production and about one-third of world production. This rate of depletion is causing increased interest in western United States reserves which represent nearly 80% of present U.S. total reserves.

Billions of metric tons of phosphate rock also are present offshore in the oceans, eg, best estimates are that a billion tons of pellets that may contain about 30% P_2O_5 are present in a Baja California–Mexico deposit alone. Other areas in the world that contain large, unevaluated amounts of phosphate include Australia, Alaska, Africa, the Near East, Peru, Colombia, Brazil, the People's Republic of China, Mongolia, and the former Soviet Union.

Resources for Potash Fertilizers. Potassium is the seventh most abundant element in the earth's crust. The raw materials from which potash fertilizer is derived are principally bedded marine evaporite deposits, but other sources include surface and subsurface brines. Both underground and solution mining are used to recover evaporite deposits, and fractional crystallization (qv) is used for the brines.

Of the 67×10^9 t of total estimated reserves and resources in Canada, nearly 5×10^9 t is recoverable by conventional mining methods, and the remainder by solution mining. As of 1974, Canada had about half of the known world reserves and about 90% of known world resources of potassium. The known recoverable potash reserves are sufficient for more than 1000 years at any foreseeable rate of consumption.

Resources of Sulfur. In most of the technologies employed to convert phosphate rock to phosphate fertilizer, sulfur, in the form of sulfuric acid, is vital. Treatment of rock with sulfuric acid is the procedure for producing ordinary superphosphate fertilizer, and treatment of rock using a higher proportion of sulfuric acid is the first step in the production of phosphoric acid, a production intermediate for most other phosphate fertilizers. Over 1.8 t of sulfur is consumed by the world fertilizer industry for each ton of fertilizer phosphorus produced, ie, 0.8 t of sulfur for each ton of P_2O_5. Total 1991 world production of sulfur in all forms was 55.6×10^6 t. The largest proportion of this production (41.7%) was obtained by removal of sulfur compounds from petroleum and natural gas (see SULFUR REMOVAL AND RECOVERY). Deep mining of elemental sulfur deposits by the Frasch hot water process accounted for 16.9% of world production; mining of elemental deposits by other methods accounted for 5.0%. Sulfur was also produced by roasting iron pyrites (17.6%) and as a by-product of the smelting of nonferrous ores (14.0%). The remaining 4.8% was produced from unspecified sources.

Major current sources, ie, elemental deposits, natural gas, petroleum, pyrites, and nonferrous sulfides are expected to last only to the end of the twenty-first century at the world consumption rate of 55.6×10^6 t/yr of the 1990s. However, vast additional resources of sulfur, in the form of gypsum, could provide much further extension but would require high energy consumption for processing.

Resources of Secondary and Micronutrients. *Calcium.* Calcium is the fifth most abundant element in the earth's crust. There is no foreseeable lack of this resource as it is virtually unlimited. Primary sources of calcium are lime materials and gypsum, generally classified as soil amendments (see CALCIUM COMPOUNDS).

Magnesium. Important sources of magnesium are dolomitic limestone, Epsom salts, calcined kieserite, magnesia, potassium–magnesium sulfate, and a few organic sources. Dolomite, seawater, and well and lake brines, the sources of most of the world's magnesium and magnesium compounds, are available in unlimited quantities.

Boron. Virtually all United States boron production and about three-fifths of the world production comes from bedded deposits and lake brines in California. U.S. reserves are adequate to support high production levels. Turkey is the only other boron-producing country of significance.

Copper. World copper reserves are estimated at 408×10^6 t copper; one-fifth being in the United States. Requirements for fertilizer are very small.

Iron. World reserves are placed at 236×10^9 t of ore containing 90×10^9 t of iron; world resources are estimated at 180×10^9 t of iron. Only a small fraction of world production is required for fertilizer use.

Manganese. U.S. resources of manganese are estimated to be on the order of 67×10^6 t Mn. World reserves are about 1.8×10^9 t Mn, and resources 1.45×10^9 t Mn. Only small amounts are required for fertilizer use.

Molybdenum. U.S. reserves and resources are about 3 and 13×10^6 t, respectively. World reserves and resources are about 6 and 23×10^6 t, respectively. The requirement for fertilizer is very small.

Zinc. Zinc deposits in the United States extend from Maine through the Appalachian Mountains, and west through the Mississippi Valley into the Rocky Mountain states. U.S. reserves are estimated to be 27×10^6 t Zn. World reserves and resources are 135 and 110×10^6 t, respectively. The requirements for fertilizer are relatively small.

Environmental Aspects

Fertilizer Use. The worldwide use of fertilizers has an important, positive effect on the environment. Conservative estimates indicate that about 30% of world food production is directly attributable to fertilizer use. Without fertilizer, therefore, at least 40% more virgin land would have to be devoted to agriculture, and 40% more labor and other resources would have to be expended. Even more serious would be the effects of land tillage and cropping without nutrient replenishment. Past experience has shown that, under such a condition, crop yields progressively decrease, the land eventually becomes barren, and forces of wind and water erosion prevail.

Pollution of ground and surface water by fertilizer run-off does occur and is now the subject of much investigation (see GROUNDWATER MONITORING; SOIL STABILIZATION). The problem apparently is serious only under certain conditions of excessive fertilizer application, abnormal soil porosity, or improper choice of fertilizer type. In most applications, prudent use of fertilizer does not contribute to this problem.

Fertilizer Production. The fertilizer industry, like the chemical industry as a whole, is fully subject to stringent national, state, and local antipollution regulations. For the most part, abatement of pollution in the fertilizer industry is handled by employment of standard procedures such as the scrubbing of gaseous effluents and purification treatment for liquid effluents. Problems that are somewhat unique to the industry are nitrous oxide emissions from nitric acid plants and granular fertilizer plants, particulate emissions from ammonium nitrate and urea prilling towers, strip mining of phosphate ore, gypsum disposal from wet-process acid plants, and fluorine emissions from phosphate processing and from gypsum disposal ponds.

Economic Aspects

The economic magnitude of the U.S. fertilizer industry is indicated by the yearly value of products, which approaches \$3 billion. U.S. fertilizer consumption represents only about 13% of the total world consumption; thus, the annual value of worldwide consumption is at least \$22 billion. The world investment for production facilities to produce these quantities of fertilizer is also very high.

GEORGE HOFFMEISTER
Consultant

T. P. Hignett, ed., *Fertilizer Manual*, International Fertilizer Development Center, Muscle Shoals, Ala., 1978.

M. H. McVicar, *Using Commercial Fertilizers*, 3rd ed., The Interstate Printers and Publishers, Inc., 1970.

N. Robinson, in F. E. Khasawneh, E. C. Sample, and E. J. Kamprath, eds., *The Role of Phosphorus in Agriculture*, American Society of Agronomy, Madison, Wis., 1979.

J. M. Potts, ed., *Fluid Fertilizers*, Bulletin Y-185, Tennessee Valley Authority, Muscle Shoals, Ala., 1984.

J. J. Mortvedt and co-eds., *Micronutrients in Agriculture*, Soil Science Society of America, Madison, Wis., 1972.

FIBER OPTICS

As of this writing (1993), regenerative undersea fiber systems are in operation that have capacities of 80,000 voice channels per fiber pair. These systems are limited more by the electronics driving the laser sources than by the properties of the glass fiber used to transmit the signal. This limitation has precipitated the development of a new generation of fiber-optic systems where the glass fiber, once solely a passive component, is becoming an active amplifying component. Rare-earth-doped fibers are being manufactured for use in optical amplifiers. When pumped by light of an appropriate wavelength, an optical signal triggers emission at the signal wavelength, resulting in amplification.

Figure 1 gives a comparison of analogue and digital transmission schemes. The incoming signal (Fig. **1a**) can be transformed directly into an intensity variation of the light beam (Fig. **1b**). A photodetector at the receiver converts this varying intensity into an electrical signal which is then amplified to reproduce the original waveform. Such a signal becomes increasingly degraded and distorted during transmission and amplification. Improved fidelity is provided by digital encoding (Fig. **1c**). In the digital scheme the signal is encoded by flashes of light at regularly timed intervals. The sampling rate must be at least twice that of the highest frequency component for accurate representation of the waveform. A voice signal having a maximum frequency of 4000 Hz must be sampled at a rate of 8000/s. The binary coding of 0s and 1s corresponds to the absence and presence of light. Representation of the height of a voice waveform requires eight bits (a bit is a 0 or a 1). Therefore, to sample a voice wave for one second the digital system requires 64,000 bits (8,000 samples × 8 bits/sample). Although the intensity of the light signal diminishes over distance, as long as it remains above the threshold of the detector the signal can be cleanly regenerated because a pulse is either present or absent. In this manner, noise is eliminated.

Principles of Light Guidance

Light guidance is governed by the structure of the lightguide itself. The refractive index, n, of a material is defined as the ratio of the speed of light in a perfect vacuum to the speed of light through that material. This property is a function of both the composition of the

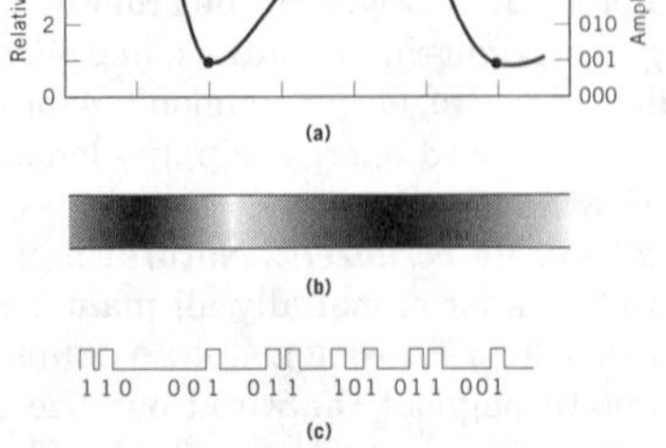

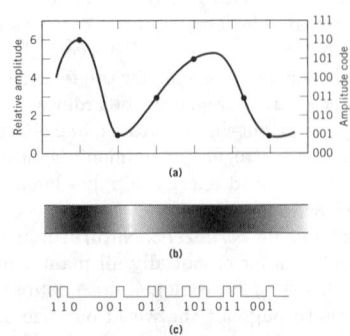

Figure 1. (**a**) A transmission signal and its (**b**) analogue and (**c**) digital encoding.

material and the wavelength of the transmitted light. The higher the refractive index of a material, the more light is retarded, or slowed, in passing through it. At the interface of two materials of different refractive indexes, light is refracted, or bent toward the higher index medium by an angle the sine of which is proportional to the relative indexes of the two media. This property is known as Snell's law. Light that travels in a medium of higher refractive index and impinges on the surface of a medium with a lower refractive index at an angle less than the critical angle, θ_c, is totally reflected. This property of total internal reflection provides a means to transmit light over long distances without radiative losses.

The ability of a waveguide to collect light is determined by the numerical aperture (NA) which defines the maximum angle at which light entering the fiber can be guided. ($NA = \sin\theta_c = (n_1^2 - n_2^2)^{1/2} \simeq n_1(2\Delta)^{1/2}$) where $\Delta = (n_1 - n_2)/n_1$ and typically $\Delta \ll 1$; n_1 is the refractive index of the core; and n_2 is the index of the cladding. Lightguide structures are shown in Figure 2. Whereas Figures **2a** and **2b** are multimode structures having relatively higher Δs and core diameters on the order of 50 μm, Figure **2c** is a single-mode fiber of lower refractive index and a core diameter < 10 μm. The number of modes that can propagate in a fiber is governed by Maxwell's equations for electromagnetic fields, and is related to a dimensionless quantity V called the normalized frequency: $V = (2\pi a/\lambda)NA \simeq (2\pi a n_1/\lambda)(2\Delta)^{1/2}$, where λ is the wavelength of light in vacuum and a is the radius of the fiber core.

Attenuation. The exceptional transparency, or low attenuation, of silica-based glass fibers has made them the predominant choice for optical transmission because of the low level of absorption and scattering of light as it traverses the material. Together these comprise optical attenuation, or loss, measured in dB where $\text{loss(dB)} = 10\log(I_0/I)$, I_0 is the input intensity and I is the output intensity. Values for loss are typically given per kilometer of fiber.

Additional optical attenuation may result from large-scale imperfections or defects of the glass. Fluctuations longer than the wavelength of light, such as diameter variations, may cause Mie scattering. Even in a nearly perfect glass, absorptions from low level cation impurities are detrimental. Similarly, point defects in the anion network can play a role in determining loss; suboxides of germanium can be formed at high temperature in an oxygen-deficient reaction and result in coloration, especially after exposure to radiation. If these extrinsic losses are avoided, as is typical of current production, then fiber attenuation is dominated by Rayleigh scattering and decreases as λ^{-4} until the multiphonon edge is intersected.

Dispersion. The effects of dispersion on the ultimate system performance are as important as the attenuation. Dispersion arises from the variation of the velocity of light with the wavelength of the light. Two types of dispersion which occur are intermodal, found only in multimode fibers, and chromatic, (material a property of the glass and waveguide a funcion of fiber design) which is important for single-mode performance.

Waveguide dispersion depends on how much of the power travels in the cladding of the fiber relative to the amount traveling in the core.

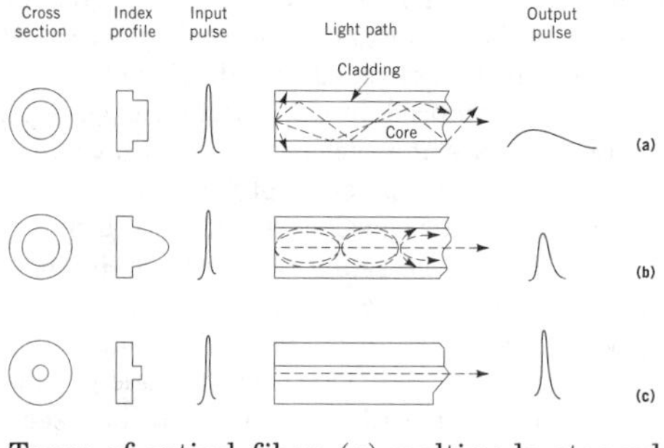

Figure 2. Types of optical fiber: (**a**) multimode stepped index, (**b**) multimode graded index, and (**c**) single-mode stepped index.

Proper design of t'e fiber's core diameter and refractive index profile may be fine-tuned to completely cancel the material dispersion at a given wavelength, or to flatten the total dispersion over a range of wavelengths.

Optical Fiber Fabrication

Viable glass fibers for optical communication are made from glass of an extremely high purity as well as a precise refractive index structure.

Double Crucible. The earliest attempts at producing high purity glass employed the double crucible technique. In spite of the elegance of this technique, contamination occurred during processing, leading to impurity levels in the glass on the order of parts per million rather than the ppb levels necessary.

High silica glasses having lower losses at wavelengths from the visible to the infrared became available through technology using vapor-phase techniques where silicon tetrachloride, $SiCl_4$, is the precursor to silicon dioxide, SiO_2, of near-intrinsic purity. Other compositions could be produced as well using other chloride vapors such as phosphorus oxychloride, $POCl_3$, and germanium tetrachloride, $GeCl_4$, to dope the silica and provide changes in the refractive index of the glass. Two methods evolved. Inside vapor-phase deposition followed from chemical vapor deposition (CVD) processes used in the electronics industry (see ELECTRONIC MATERIALS; INTEGRATED CIRCUITS), and outside vapor-phase techniques such as OVPO and VAD.

Inside Processes. Inside processes such as modified chemical vapor deposition (MCVD) followed CVD techniques. In CVD the concentration of the reactants was kept low to inhibit homogeneous gas-phase reaction in favor of heterogeneous wall reactions which produce a vitreous (amorphous), particle-free deposit on the tube wall. Deposition was continued until a sufficient thickness of glass was produced; then the tube was collapsed to form a solid rod, or preform. This preform was then drawn to a relatively low loss fiber. MCVD, which is widely used commercially, used high concentrations to produce high deposition rates. Small, colloidal-size particles deposit on the wall, are sintered by an oxygen–hydrogen torch traversing the length of the tube to yield glass of required purity and transparency.

One of the areas critical to the MCVD process was understanding the chemistry of the oxidation reactions. It was necessary to control the incorporation of GeO_2 while minimizing OH^- formation. Additionally, understanding the mechanism of particle formation and deposition was critical to further scale-up of the process.

Thermophoresis is a process by which particles in a temperature gradient travel toward the cooler region when bombarded by more energetic particles on the hot side. In the MCVD process the reactants enter the tube, are reacted in the hot zone of the torch, deposit thermophoretically downstream of the torch, and are subsequently sintered to a clear glass as the torch passes over the deposited particulate layer. After repeated passes have been made and the desired structure has been deposited, the direction of the torch is reversed and the tube is collapsed to form a solid preform.

The chemistry of the MCVD process has been studied using a variety of techniques. Infrared (ir) spectroscopy was used to investigate the oxidation reaction of $SiCl_4$ and $GeCl_4$ (see INFRARED TECHNOLOGY AND RAMAN SPECTROSCOPY). It was concluded that at temperatures lower than 1600 K the degree of reaction of the $SiCl_4$, $GeCl_4$, and $POCl_3$ is controlled by reaction kinetics, whereas at higher temperatures the reaction is dominated by thermodynamic equilibria. Rate studies have shown that the residence times typically experienced in the hot zone are sufficient to reach equilibrium above 1700 K.

Incorporation of OH is another critical aspect of the oxidation chemistry. Reduction to the ppb level is necessary for the manufacture of low loss optical fiber. During the deposition phase of MCVD, the chlorine level is between 3 and 10% owing to the oxidation of the chloride reactants. This level of chlorine leads to a reduction in the OH incorporation by a factor of about 4000. During the collapse phase of the process, the chlorine level is significantly reduced and OH incorporation can be high as a result of the presence of hydrogen from either the oxy–hydrogen torch or impurities in the starting gases. A small concentration of chlorine is added to suppress OH incorporation.

The process by which the glass consolidates was found to be viscous sintering controlled by the viscosity of the glass at the consolidation temperature. The driving force is the reduction of surface energy via a decrease in the surface area.

Another process closely related to MCVD is plasma chemical vapor deposition (PCVD) (see PLASMA TECHNOLOGY). This inside process uses the same precursor chemicals as MCVD to form a glass coating inside the tube which is collapsed to form a solid glass preform. The reaction within the tube, however, is quite different. Here the reaction is initiated by a nonisothermal microwave plasma which traverses the inside of the tube. The plasma requires a pressure of a few hundred Pascals and is generated by a microwave cavity which operates at 2.45 GHz. The glass is deposited not as a soot, but as a thin glass layer with efficiencies approaching 100%. In addition to the high efficiency, complex waveguide structures may be formed because many thin layers are produced by rapidly traversing the plasma. This method provides a smoother refractive index variation, which is especially advantageous for multimode fiber preforms.

Outside Processes. The outside vapor deposition (OVD) process was developed by Corning Glass Works. Soot is deposited layer by layer on a rotating mandrel at a temperature such that the soot particles are partially sintered. The precursor chemicals are the same as those used in the MCVD process but are oxidized by a gas–oxygen torch by a similar chemical reactions.

The vertical axial deposition (VAD) process was developed by a consortium of Japanese cable manufacturers and Nippon Telephone and Telegraph (NTT). This process also forms a cylindrical soot form. However, deposition is achieved end-on without use of a mandrel and subsequent formation of a central hole. Both the core and cladding are deposited simultaneously using more than one torch. Whereas the OVD, PCVD, and MCVD processes build a refractive index profile layer by layer, the VAD process uses gaseous constituents in the flame to control the shape and temperature distribution across the face of the growing soot boule.

Although the control in VAD is more difficult, this process has an advantage over OVD, especially for multimode fiber. The thermal mismatch between core and cladding materials caused by the heavy doping necessary to achieve the desired refractive index profile causes cracking in the OVD preforms at the inner surface as the glass cools through the glass transition, T_g. The VAD preforms withstand the stress because they possess no central hole to result in tensile stress. The primary obstacle for VAD was the creation of an optimized refractive index profile to minimize intermodal dispersion. Then it was discovered that the composition could be graded by control of the boule's surface-temperature distribution. This temperature distribution depends on the shape of the boule's growing face. Through understanding the relationship between the temperature and germania concentration, fiber properties equivalent to those formed by OVD and MCVD were achieved.

Fiber Drawing and Strength

Preforms manufactured by MCVD, PCVD, OVD, and VAD all must be drawn into fiber in a similar manner. Standard fibers are drawn to 125 μm in diameter from preforms on the order of 2 to 7.5 cm diameter. Fibers are drawn by holding the preform vertically and lowering it into a furnace. The preform is heated to a temperature at which the glass softens (2200°C) until a gob of glass stretches from the tip of the preform and drops under the force of gravity. A neck-down region is formed at this point, providing the transition between preform and fiber. Fiber is drawn by means of a capstan system, and its diameter is controlled by a diameter monitor that adjusts the draw speed at a fixed furnace temperature. The result is long lengths of uniform fiber. To preserve the intrinsic strength of the pristine glass surface, a protective coating must be applied before the fiber is contacted by the capstan.

Overcladding. Fiber manufacturing and drawing technology have advanced to the point that the optical losses are limited almost entirely by the intrinsic loss of the glass. Initially all of the fiber manufacturing processes (MCVD, PCVD, OVD, and VAD) produced preforms yielding on the order of 10 km of fiber. This situation changed as single-mode fiber usage grew. The proportion of core glass to the total amount of glass in single-mode fiber is much lower than in multimode fiber. Standard single-mode core diameter is ca 8 μm, whereas multimode core diameters may be 50 or 62.5 μm. This led to a method of manufacturing larger preforms by shrinking a second silica tube over a core rod made by vapor deposition or depositing a thick soot layer on same to provide additional cladding. This procedure, known as overcladding, increased the length of fiber drawn from a single preform to more than 100 km, for example, and significantly reduced the cost of producing fiber.

Sol–Gel Processing. Fibers can be designed so that light travels in the inner 30–40 μm of the fiber, which accounts for only about 5% of the fiber mass. Thus, using a core rod, the remaining 95% could be manufactured from less expensive, lower purity materials typically obtained by sol–gel processing (see SOL–GEL TECHNOLOGY).

Defects. The ever-increasing demand for high data rate systems is forcing the search for an even greater understanding of those defects which produce attenuation of only hundredths of a dB/km. Profile control to produce zero dispersion at operating wavelengths is necessary, and environmental effects such as radiation and hydrogen exposure must be minimized. In addition, for reliability, higher strength must be achieved, necessitating the understanding of flaw distributions and growth of flaws (fatigue).

Optical Amplifiers

Throughout the first two decades of their existence optical fibers served a passive role, ie, in the transmission of encoded light signals. In the late 1980s erbiumdoped fiber amplifiers (EDFAs) were introduced, making it possible to amplify a 1.55 μm optical signal without first converting it to an electronic signal. The amplifier consists of a section (tens of meters) of single-mode optical fiber having about 100 ppm of erbium incorporated in the core. This fiber section becomes an amplifier when a continuous source of pump light, usually 0.98 or 1.48 μm, propagates through the fiber. As the optical signal, usually 1.53 to 1.6 μm, travels through the length of fiber containing erbium ions, excited by pump light, amplification occurs by the stimulated emission of photons from the excited state. Noise in the form of broad-band spontaneous emission accompanies this process; however, the signal-to-noise ratio is kept to an acceptable level even when cascading many of these devices for system applications.

A number of means have been developed to produce erbium-doped optical fibers. The compatibility of aluminum and erbium extends to the vapor phase where complex aluminum–erbium chlorides exist at vapor-pressure orders of magnitude higher than $ErCl_3$. The passage of aluminum chloride, Al_2Cl_6, vapor over a heated erbium oxide, Er_2O_3, or erbium chloride, $ErCl_3$, source permit doping in MCVD reactions. In addition, erbium chelates and other organic precursors can be introduced into VAD or OVD flows to produce doped soot. Finally, doping with solutions containing rare-earth ions or sol–gel doping of soot or MCVD substrate tubes prior to the final collapse to a solid preform also provide the means for controlled introduction of the ions.

EDFAs are being introduced into long-distance, particularly undersea, systems which operate at wavelengths near 1.6 μm. In addition to providing an inexpensive means of amplification EDFA also make it possible to amplify numerous wavelengths near 1.6 μm, whereas semiconductor amplifiers suffer crosstalk when amplifying more than one wavelength (see SEMICONDUCTORS). The first transatlantic all-optical system is expected to be in operation in 1995. It should have a potential capacity of greater than 700,000 voice channels per fiber pair.

A commercially attractive fiber amplifier to be used in existing terrestrial networks, operated at wavelengths near 1.3 μm, is being sought. Praseodymium and neodymium in fluoride-based glasses have shown some promise, but such amplifier fibers are not in commercial use as of this writing. Future generations of optical systems are expected to make use of additional active fiber components including fiber grating filters, fiber sensors (qv), and fiber lasers.

Economic Aspects

The market for optical fibers is projected to reach $3.5 billion by 1998. In addition, according to ElectroniCast (San Mateo, Calif.), the total market for passive optical components, optical electronics, connectors, and fiber-optic cable is predicted to increase form $1.76 billion (U.S.) in 1992 to over $4 billion in 1997, and $10 billion by 2002.

SANDRA KOSINSKI
JOHN B. MACCHESNEY
AT&T Bell Laboratories

R. M. Klein, in D. R. Uhlmann and N. J. Kreidl, eds., *Glass: Science and Technology*, Vol. 2., *Processing I*, Academic Press, Inc., Orlando, Fla., 1984, pp. 285–339.

E. J. Friebele and D. L. Griscom, in M. T. Tomazawa and R. H. Doremus, eds., *Treatise on Materials Science and Technology*, Vol. 17, Academic Press, Inc., New York, 1979, pp. 257–351.

E. Desurvire, *Sci. Am.*, 114–121 (Jan. 1992).

FIBERS

SURVEY

This overview of fibers and fiber products introduces the underlying concepts that govern the manufacture and properties of these materials. The field of fibers is an evolving one, with new technologies being developed constantly. With the increasing use of fibers in nontraditional textile applications, such as geotextiles (qv), fiber-reinforced composites, specialty absorption media, and as materials of construction, new fiber types and new processing technologies can be anticipated.

Classification

A fiber may be described as a flexible, macroscopically homogeneous body having a high ratio of length to width and being small in cross section. A significant segment of the world's agricultural activity is concerned with the growth and harvesting of natural fibers, and the production of synthetic fibers is an important activity of the worldwide chemical industry. The textile and paper industries are the prime converters of fibers into end products whose properties can be directly related to the unique combination of properties characteristic of fibers. The paper industry uses almost exclusively a natural cellulosic fiber derived from wood. The textile industry, on the other hand, uses a variety of natural and synthetic fibers in the manufacture of its wide range of products. Textile fibers may be classified according to their origin, as follows:

Naturally occurring fibers
- vegetable: based on cellulose, eg, cotton, linen, hemp, jute, and ramie
- animal: based on proteins, eg, wool, mohair, vicuna, other animal hairs, silk
- mineral: eg, asbestos

Synthetic fibers
- based on natural organic polymers
  - rayon: regenerated cellulose
  - acetate: partially acetylated cellulose derivative
  - triacetate: fully acetylated cellulose derivative
  - azlon: regenerated protein
- based on synthetic organic polymers
  - acrylic: polyacrylonitrile (also modacrylic)
  - aramid: aromatic polyamides
  - nylon: aliphatic polyamides
  - olefin: polyolefins
  - polyester: polyesters of an aromatic dicarboxylic acid and a dihydric alcohol
  - spandex: segmented polyurethane
  - vinyon: poly(vinyl chloride)
  - vinal (or vinylon): poly(vinyl alcohol)
  - carbon/graphite: derived from polyacrylonitrile, rayon, or pitch
  - specialty fibers: poly(phenylene sulfide) and polyetheretherketone
- based on inorganic substances
  - glass, metallic, ceramic

Fibers manufactured from natural organic polymers are either regenerated or derivative; historically they have been designated "man-made" fibers, and were thereby distinguished from fibers based on synthetic organic polymers. A regenerated fiber is one formed when a natural polymer, or its chemical derivative, is dissolved and extruded as a continuous filament, and the chemical nature of the natural polymer is either retained or regenerated after the fiber formation process. A derivative fiber is one formed when a chemical derivative of the natural polymer is prepared, dissolved, and extruded as a continuous filament, and the chemical nature of the derivative is retained after the fiber formation process.

Synthetic fiber manufacture is based on three methods of fiber formation or spinning. In the context of fiber manufacture, spinning refers to the overall process of polymer liquefaction (dissolution or melting), extrusion, and fiber formation. The three principal methods are melt spinning, dry spinning, and wet spinning, although there are many variations and combinations of these basic processes.

In the traditional methods of fiber manufacture the filaments are obtained in continuous form. When several such filaments are combined together and slightly twisted to maintain unity, the product so obtained is called a multifilament yarn. A typical yarn may contain 100 single filaments. Individual filaments, considerably larger in cross section than those used in multifilament yarns, may also be used in certain applications, and these are referred to as monofilaments. Frequently it is desired to obtain fibers in finite lengths for subsequent manufacture into spun yarns by conventional textile spinning operations. In this case thousands of continuous filaments are collected together into a continuous rope of parallelized filaments called a tow. The tow is converted into staple length fiber by simply cutting it into specified lengths.

Fiber Consumption Trends

For centuries the textile industry relied exclusively on the natural fibers, particularly cotton (qv), wool (qv), and silk (qv). With the commercialization of synthetic fibers, starting with those based on natural polymers in the 1930s, and then with those based on synthetic polymers in the 1940s, the textile industry has truly undergone a revolution. The 1990 total world production of textile fibers amounted to approximately 40×10^6 t, distributed among the principal types of fibers. Nearly all regions of the world are involved with fiber production. Polyester and polyamide fibers alone account for 63% of the synthetic fibers produced worldwide in 1990. The textile industry finds itself with an ever-increasing number of fibers from which to manufacture its products. Not only are there many more fiber types for the textile industry to use, but also it is recognized that many advantages are to be gained by blending the various fibers in an almost infinite number of combinations. Many current textile products are blends of natural and synthetic fibers that incorporate the desirable attributes of both.

Types of Fibrous Materials

Fibers are used in the manufacture of a wide range of products that can generically be referred to as fibrous materials. The properties of such materials are dependent on the properties of the fibers themselves and on the geometric arrangement of the component fibers in the structure.

Several types of fibrous materials can be distinguished primarily on the basis of fiber organization. At one extreme are isotropic assemblies where the fibers are arranged in a completely random fashion with no preferred orientation in any of the three principal spatial axes. Isotropic fiber assemblies are quite rare in most end products, and in fact are quite difficult to achieve because of the high length to width ratio of fibers. This high aspect ratio causes fibers to align themselves in a stress field, thereby creating a preferred orientation of the fibers in the processing direction. Thus considerable effort must be exerted to overcome the tendency for preferred orientation when an isotropic fiber assembly is desired. Alternatively, and more commonly, there are various anisotropic fibrous materials with the fibers arranged in well-defined spatial patterns.

Textile Yarns. The high degree of structural anisotropy of textile yarns is reflected in large differences between axial and transverse physical properties. The structural anisotropy of yarns, and of the individual fibers, imparts a unique combination of high strength and low bending rigidity to textile yarns.

Textile yarns are produced from staple (finite length) fibers by a combination of processing steps referred to collectively as yarn spinning. The staple fibers may be either natural fibers, such as cotton or wool, or any of a number of synthetic fibers.

Textile yarns are also produced from continuous filament synthetic fibers. Such yarns are referred to as multifilament yarns, and are characterized by nearly complete filament alignment and parallelization with respect to the yarn axis. Such yarns are quite compact and smooth in appearance. A variety of processes have been developed to introduce bulk and texture in multifilament yarns. These processes are designed to disrupt the high degree of filament alignment and parallelization and to produce yarns with properties generally associated with spun yarns. Schematic representations of typical yarns are given in Figure 1.

Textile Fabrics. Yarns are used principally in the formation of textile fabrics either by weaving or knitting processes.

Nonwovens. The term nonwoven simply suggests a textile material that has been produced by means other than weaving, but these materials really represent a rather unique class of fibrous structure. In nonwovens the fibers are processed directly into a planar sheet-like fabric structure, bypassing the intermediate one-dimensional yarn state, and then either bonded chemically or interlocked mechanically (or both) to achieve a cohesive fabric.

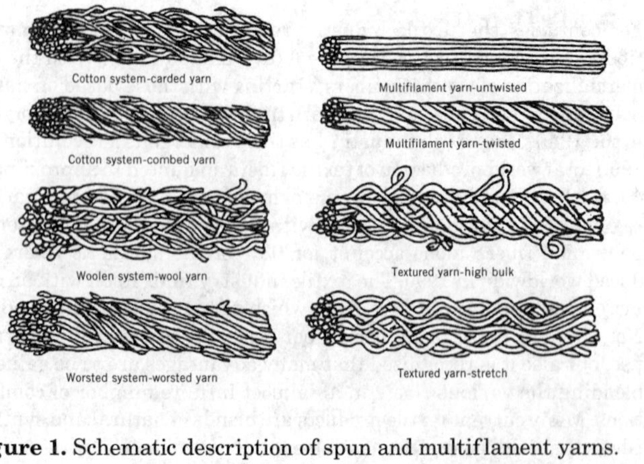

Figure 1. Schematic description of spun and multiflament yarns.

The type of nonwoven material that is produced depends largely on the fiber type used and on the method of manufacture. Typically, air-laid nonwovens are less dense and compact and tend to be softer, more deformable, and somewhat weaker. Wet-laid or paper-like nonwovens are more dense, stronger, more brittle, and less permeable to fluids. It is dangerous, however, to generalize. Nonwovens by either process can be produced to achieve a wide variety of products with a broad range of physical properties.

Fiber Properties

The properties of textile fibers may be conveniently divided into three categories: geometric, physical, and chemical, as shown in Table 1.

Mechanical Behavior. Because textile fibers are used primarily as elements of construction, a sound engineering approach to their mechanical properties is necessary. Since textile fibers are exposed to a variety of chemical environments in the normal course of manufacturing and processing into yarns and fabrics, as well as during their ultimate use by the consumer, it is frequently important to evaluate mechanical properties in relation to a fiber's chemical environment. Not only do such measurements make possible a more probable prediction of fiber behavior under other than standard conditions, but also it may be possible to deduce important information about fiber structure.

The mechanical properties of fibers or of any other material may be described in terms of six factors: strength, elasticity, extensibility, resilience, stiffness, and toughness. Information about these mechanical properties is obtained from stress–strain or load-deformation curves that are graphical records of tensile, compressive, or shearing stresses as a function of deformation. In view of a fiber's geometric shape and dimensions, these curves are usually evaluated under uniaxial tension.

Structure of Fiber-Forming Polymers

With the exception of glass fiber, asbestos (qv), and the specialty metallic and ceramic fibers, textile fibers are a class of solid organic polymers distinguishable from other polymers by their physical properties and characteristic geometric dimensions (see GLASS; REFRACTORY FIBERS). The physical properties of textile fibers, and indeed of all materials, are a reflection of molecular structure and intermolecular organization. The ability of certain polymers to form

Table 1. Classification of Fiber Properties

Geometric	Physical	Chemical
Length	*Density*	*Response to*
average value	linear	acids
distribution	bulk	alkalies
		oxidation
Cross section	*Thermal*	reduction
average value	melting point	heat
distribution	transitions	
shape	conductivity	*Sorption*
		moisture
Crimp	*Optical*	dyes
frequency	birefringence	
form	refractive index	*Swelling*
	luster and color	anisotropy
	Electrical	
	resistivity	
	dielectric constant	
	Surface	
	roughness, friction	
	Mechanical	
	tension, compression,	
	torsion, bending, shear	

fibers can be traced to several structural features at different levels of organization rather than to any one particular molecular property.

Fiber structure is described at three levels of molecular organization, each relating to certain aspects of fiber behavior and properties. First is the organochemical structure, which defines the chemical composition and molecular structure of the repeating unit in the base polymer, and also the nature of the polymeric link. This primary level of molecular structure is directly related to chemical properties, dyeability, moisture sorption, and swelling characteristics, and indirectly related to all physical properties. The chemical structure also determines the magnitude of intermolecular forces, which is important in terms of properties and overall fiber structure. The macromolecular level of structure describes the chain length, chain-length distribution, chain stiffness, molecular size, and molecular shape. The supramolecular organization is the arrangement of the polymer chains in three-dimensional space. The physical properties of fibers are strongly influenced by the organization of polymeric chains into crystalline and noncrystalline domains, and the disposition of these domains with respect to each other. In the case of natural fibers, a further level of structural organization related to the natural growth and development of the fiber must be considered. This morphology is quite complex, with fibrils, microfibrils, and similar structural subunits frequently surrounded by a matrix material in a composite configuration.

All polymeric fibers that are useful in textile applications are semicrystalline, irreversibly oriented polymers, ie, the polymeric chains are partially ordered into regions or domains with near perfect registry so that the laws of x-ray diffraction are obeyed. In other regions of the fiber, the molecular chains or segments of chains are not perfectly ordered and may approach random coil configurations.

All models of fiber structure picture a fiber as a polymeric substance with a high degree of three-dimensional structural regularity, leading directly to the concept of the degree of crystallinity, ie, the fractional crystalline content of a partially crystalline polymeric material. The degree of crystallinity is difficult to evaluate. X-ray diffraction techniques probably provide the best estimates, but unsolved questions remain because the degree of crystallinity is actually a composite value that reflects not only the fractional quantity of ordered material, but also the size or dimensions and the perfection of the crystallites.

The requirement of at least partial crystallinity limits the number of polymers suitable for fiber formation. To a large extent, what is meant by the term fiber-forming polymer is actually crystallizable polymer. The structural characteristics of polymers that allow them to crystallize under appropriate conditions have been summarized as follows. *Regularity*. The polymer chains must be uniform in chemical composition and stereochemical form. *Shape and Interaction*. The shape of the polymer chains must allow close contact or fit to permit effective and strong intermolecular interaction. This is generally achieved by linear macromolecules with no bulky side groups or with side groups that are regularly spaced along the backbone chain. *Repeat length*. The ease of crystallization decreases with increasing length of the polymer repeating unit. *Chain directionality*. Since certain polymer chains have a directionality, the mode of chain packing in a crystallite can take either parallel or antiparallel configurations. *Single-chain conformation*. Crystallization is favored if the chain conformation is compatible with its form in the crystallite. *Chain stiffness*. An optimal stiffness of the polymer is necessary.

Attention must also be focused on the noncrystalline domains. Many important properties of fibers can be directly related to these noncrystalline or amorphous regions. For example, absorption of dyes, moisture, and other penetrants occurs in these regions.

Fiber structure is a dual or a balanced structure. Neither a completely amorphous structure nor a perfectly crystalline structure provides the balance of physical properties required in fibers. The formation and processing of fibers is designed to provide an optimal balance in terms of both structure and properties. Excellent discussions of the structure of fiber-forming polymers and general methods of structure characterization are available.

LUDWIG REBENFELD
TRI/Princeton

B. C. Goswami, J. G. Martindale, and F. L. Scardino, *Textile Yarns: Technology, Structure and Applications*, John Wiley & Sons, Inc., New York, 1977.

J. W. S. Hearle, P. Grosberg, and S. Backer, *Structural Mechanics of Fibers, Yarns, and Fabrics*, John Wiley & Sons, Inc., New York, 1969.

J. W. S. Hearle, *Polymers and Their Properties, Volume 1; Fundamentals of Structure and Mechanics*, Halstead Press, a division of John Wiley & Sons, Inc., New York, 1982.

F. Happey, *Applied Fiber Science*, Vols. 1, 2, and 3, Academic Press, Inc., New York, 1978, 1979.

ACRYLIC

During the 1970s there was rapid growth of acrylic fiber production in Japan, Eastern Europe, and developing countries. By 1981 an estimated overcapacity of approximately 21% had developed. This overcapacity is expected to decrease through the 1990s with continued increases in world production balanced by markets opening in China, Eastern Europe, Russia, and the Americas. Acrylics retain their traditional market, in knitted goods, like sweaters, and men's half-nose, and carpets. In addition, small market shares have developed in many new areas, such as carbon fiber precursors, asbestos replacement fibers, and conductive/metallized fibers.

Physical Properties

Acrylic and modacrylic fibers are sold mainly as staple and tow products with small amounts of continuous filament fiber sold in Europe and Japan. Staple lengths may vary from 25 to 150 mm, depending on the end use. Fiber deniers may vary from 1.3 to 17 dtex (1.2 to 15 den); 3.2 dtex (3.0 den) is the standard form. The appearance of acrylics under microscopical examination may differ from that of modacrylics in two respects. First, the cross sections of acrylics are generally round, bean-shaped, or dogbone-shaped. The modacrylics, on the other hand, vary from irregularly round to ribbon-like. The modacrylics may also contain pigment-like particles of antimony oxide to enhance their flame-retardant properties.

The physical properties of these fibers are compared with those of natural fibers and other synthetic fibers in Table 1.

Chemical Properties

Among the outstanding properties of acrylic fibers is the very strong resistance to sunlight. Acrylic fibers are also resistant to all biological and most chemical agents. In terms of resistance of acrylic fibers to oxidizing agents, Orlon acrylic has been compared to nylon, cotton, and acetate yarns. Acrylic fibers, when heated, discolor and decompose rather than melt, but they have very good color and heat stability at temperatures less than 120°C. The excellent chemical resistance of acrylic fibers may stem from its unique laterally bonded structure.

Flammability

A most important property of acrylic and modacrylic fibers is their flammability and ignition behavior. Fibers used in textiles must not ignite readily when placed in contact with a flame. In this respect acrylic fibers compare favorably to the other natural and synthetic fibers currently on the market. There are, however, significant differences in the burning characteristics of the various fiber types. Cotton and rayon, for example, burn with the formation of a char. Nylon, polyester, olefin, wool, and acrylics, on the other hand, burn and melt simultaneously. More rigorous standards are required for end uses such as carpets, sleepwear, draperies, and bedding. Fibers for these

Table 1. Physical Properties of Staple Fibers

Property	Acrylic	Modacrylic	Nylon-6,6	Polyester	Polyolefin	Cotton	Wool
sp gr	1.14–1.19	1.28–1.37	1.14	1.38	0.90–1.0	1.54	1.28–1.32
tenacity, N/tex[a]							
dry	0.09–0.33	0.13–0.25	0.26–0.64	0.31–0.53	0.31–0.40	0.18–0.44	0.09–0.15
wet	0.14–0.24	0.11–0.23	0.22–0.54	0.31–0.53	0.31–0.40	0.21–0.53	0.07–0.14
loop/knot tenacity	0.09–0.3	0.11–0.19	0.33–0.52	0.11–0.50	0.27–0.35		
breaking elongation, %							
dry	35–55	45–60	16–75	18–60	30–150	<10	25–35
wet	40–60	45–65	18–78	18–60	30–150	25–50	
average modulus, N/tex[a]							
dry	0.44–0.62	0.34	0.88–0.40	0.62–2.75	1.8–2.65		
elastic recovery, %							
2% stretch	99	99–100		67–86		74	99
10% stretch		95	99	57–74	96		
20% stretch							65
electrical resistance	high	high	very high	high	high	low	low
static buildup	moderate	moderate	very high	high	high	low	low
flammability	moderate	low	self-extinguishing	moderate	moderate	spontaneous ignition at 360°C	self-extinguishing
limiting oxygen index	0.18	0.27	0.20	0.21		0.18	0.25
char/melt	melts	melts	melts, drips	melts, drips	melts	chars	chars
resistance to sunlight	excellent	excellent	poor; must be stabilized	good	poor; must be stabilized	fair; degrades	fair; degrades
resistance to chemical attack	excellent	excellent	good	good	excellent	attacked by acids	attacked by alkalies, oxidizing, and reducing agents
abrasion resistance	moderate	moderate	very good	very good	excellent	good	moderate
index of birefringence	0.1		0.6	0.16			0.01
moisture regain, std %	1.5–2.5	1.5–3.5	4–5	0.1–0.2	0	7–8	13–15

[a] To convert N/tex to gf/den, multiply by 11.3.

applications must also be self-extinguishing after removal from the ignition source. The modacrylics, eg, SEF and Kanekalon, melt and self-extinguish. This is generally achieved in acrylonitrile-based fibers by incorporating halogen comonomers such as vinylidene chloride, vinyl chloride, and vinyl bromide.

Fiber Identification

Although visual and microscopical examination, together with simple manual tests, are still the primary methods of identification, there are many new sophisticated instrumental methods available, based on chemical and physical properties. These methods are able to distinguish between closely related fibers which differ only in chemical composition or morphology.

Instrumental Analysis. It is difficult to distinguish between the various acrylics and modacrylics. Elemental analysis may be the most effective method of identification. Specific compositional data can be gained by determining the percentages of C, N, O, H, S, Br, Cl, Na, and K. In addition the levels of many comonomers can be established using ir and uv spectroscopy.

Fiber Characterization

In addition to characterizing the many properties introduced by the choice of monomers and the polymerization process itself, considerable further characterization is required to quantitatively describe the properties imparted by spinning and subsequent downstream processing. These important properties relate to the crystalline order and microstructure of the fibers, and the resultant performance characteristics, such as crimp retention, abrasion resistance, mechanical properties, etc.

Acrylonitrile Polymerization

Except for fibers designed for industrial applications where resistance to chemical attack is of prime importance, all acrylic fibers are made from acrylonitrile combined with at least one other monomer. The comonomers most commonly used are neutral comonomers, such as methyl acrylate and vinyl acetate to increase the solubility of the polymer in spinning solvents, modify the fiber morphology, and improve the rate of diffusion of dyes into the fiber. Sulfonated monomers, such as sodium styrenesulfonate (SSS), sodium methallyl sulfonate (SMAS), and sodium sulfophenyl methallyl ether (SPME) are used to provide dyesites or to provide a hydrophilic component in water reversible crimp bicomponent fibers. Halogenated monomers, usually vinylidene chloride, vinyl bromide, and vinyl chloride, impart flame resistance to fibers used in the home furnishings, awning, and sleepwear markets. Modacrylic compositions are used when the end use requires high flame resistance. Almost all of the modacrylics are flame-resistant fibers with high levels of halogen monomers.

Copolymerization

Homogeneous Copolymerization. Nearly all acrylic fibers are made from acrylonitrile copolymers containing one or more additional monomers that modify the properties of the fiber. Thus, copolymerization kinetics is a key technical area in the acrylic fiber industry. When carried out in a homogeneous solution, the copolymerization of acrylonitrile follows the normal kinetic rate laws of copolymerization. Comprehensive treatments of this general subject have been published. The more specific subject of acrylonitrile copolymerization has been reviewed. The general subject of the reactivity of polymer radicals has been treated in depth.

Heterogeneous Copolymerization. When copolymer is prepared in a homogeneous solution, kinetic expressions can be used to predict copolymer composition. Bulk and dispersion polymerization are somewhat different since the reaction medium is heterogeneous and polymerization occurs simultaneously in separate loci. In bulk polymerization, for example, the monomer swollen polymer particles support polymerization within the particle core as well as on the particle surface. In aqueous dispersion or emulsion polymerization the monomer is actually dispersed in two or three distinct phases: a continuous aqueous phase, a monomer droplet phase, and a phase consisting of polymer particles swollen at the surface with monomer. This affects the ultimate polymer composition because the monomers are partitioned such that the monomer mixture in the aqueous phase is richer in the more water-soluble monomers than the two organic phases.

Polymerization Methods. Acrylonitrile and its comonomers can be polymerized by any of the well-known free-radical methods. Bulk polymerization is the most fundamental of these, but its commercial use is limited by its autocatalytic nature. Aqueous dispersion polymerization is the most common commercial method, whereas solution polymerization is used in cases where the spinning dope can be prepared directly from the polymerization reaction product. Emulsion polymerization is used primarily for modacrylic compositions where a high level of a water-insoluble monomer is used or where the monomer mixture is relatively slow reacting.

Commercial Polymerization Methods. Aqueous media, such as emulsion, suspension, and dispersion polymerization, are by far the most widely used in the acrylic fiber industry. Water acts as a convenient heat-transfer and cooling medium and the polymer is easily recovered by filtration or centrifugation.

Processes using a continuous stirred tank reactor have replaced the semibatch process except where low volume specialty products are made. For startup, the reactor is charged with a certain amount of the reaction medium, usually solvent or pH adjusted water. In more sophisticated processes the start-up period may be minimized by filling the reactor with overflow from a reactor already operating at steady state. The reactor feeds are metered in at a constant rate for the entire course of the production run, which normally continues until equipment maintenance is needed. A steady state is established by taking an overflow stream at the same mass flow rate as the combined feed streams. The main advantage of this process over the semibatch process is that control of molecular weight, dye site level, and polymer composition is greatly improved.

The only other commercial polymerization process used for acrylic fibers is solution polymerization. This type of process can be implemented by feeding the monomers to a continuous mixing tank along with a solvent for the polymer. The overflow stream from this tank is then routed to a form of continuous reactor where the polymerization is carried out in a homogeneous solution. Monomer is removed from the product stream and the resulting polymeric solution is used directly for spinning. An obvious advantage of this process is that considerable cost savings can be achieved by eliminating the filtration, drying, and dope making steps required in the aqueous dispersion process. There are two drawbacks associated with solution polymerization. First, it is difficult to produce dopes of high solids, particularly with organic solvents. Second, most of the effective solvents have very high chain-transfer constants, making it difficult to produce polymer of high molecular weight. Solvents suitable for this type of commercial polymerization are DMF and DMSO.

Because of the highly exothermic nature of acrylonitrile polymerization, bulk processes are not normally used commercially. However, a commercially feasible process for bulk polymerization in a continuous stirred tank reactor has been developed. The heat of reaction is controlled by operating at relatively low conversion levels and supplementing the normal jacket cooling with reflux condensation of unreacted monomer.

The problems of monomer recovery, reaction medium viscosity, and control of reaction heat are effectively dealt with by the process design of Montedison Fibre. This process produces polymer of exceptionally high density. Thus, though the polymer is still swollen with monomer, the medium viscosity remains low because the amount of monomer absorbed in the porous areas of the polymer particles is greatly reduced. The process is carried out in a CSTR with a residence time, Q, such that the product k/d $\times$ Q is greater than or equal to 1. k_d is the initiator decomposition rate constant. This condition controls the autocatalytic nature of the reaction because the catalyst and residence time combination assures that the catalyst is almost totally expended in the reactor.

Solution Spinning

One of the principal problems in early commercialization of acrylic fibers was the lack of a suitable spinning method. The polymer cannot be melt spun, except possibly at high pressure in the presence of water. Solution spinning was the only feasible commercial route. The first real breakthrough occurred at DuPont in 1948 when dimethylformamide (DMF) was used to commercialize Orlon. DMF is still the most important spinning solvent for acrylic fibers, but other effective solvents are in wide use.

Dry Spinning. This is the process first employed commercially by DuPont in 1948. The process is similar to early acetate fiber spinning. The dope is pumped through spinnerettes with 200–900 holes placed at the top of a solvent drying tower. The solvent is removed by circulating an inert gas through the tower at 150–300°C. However, unlike acetate spinning which employs a low boiling solvent (acetone), acrylic solvents are all high boiling and hard to extract completely in the drying tower. Consequently, the fiber from the bottom of the drying tower contains 10–25% solvent. This is removed in a second step by passing the threadline through a hot water bath and possibly by a series of wash rolls. Vapor from the drying tower and wash water from the secondary solvent extraction are routed to a solvent recovery plant where the solvent and wash water are recovered and purified for reuse. This is essential for economic reasons, but environmental protection also requires recovery rather than discharge.

Wet Spinning. Wet spinning differs from dry spinning primarily in the way solvent is removed from the extruded filaments. Instead of evaporating the solvent in a drying tower, the fiber is spun into a liquid bath containing a solvent/nonsolvent mixture called the coagulant. The solvent is almost always the same as the solvent used in the dope and the nonsolvent is usually water.

Tow Processing. After the spinbath step the tow processing is similar for both wet and dry spun yarns. Wet spun yarns, however, may contain 100 to 300% of solvent/nonsolvent mixture per dry pound of fiber whereas dry spun tows generally hold only 10–30% solvent. Therefore, the initial washing steps differ in their details. The key tow processing steps are washing, stretching, finish application, collapse, drying, crimping, and relaxing.

Modification of Properties. *Handle.* For a given fabric construction the denier, compliance, cross-sectional configuration, degree of crimp, moisture absorption, and surface smoothness of the fibers all influence the softness of the final product. Very fine filament deniers are effective where good draping, anticrease properties, and softness of handle are desired. Softness of handle can also be achieved through modifying the fiber cross section. Flattened cross sections reduce the bending modulus. Sheath-core spinning, conjugate spinning, and drawing into modified spinbath compositions also yield modified handle.

Improved Comfort Properties. Wear comfort generally means cotton-like properties. The ability to absorb moisture from the skin and the softness of cotton fabrics are considered to be the two key properties for comfort. The extremely fine denier of cotton fibers accounts for its softness. Both properties can be achieved in acrylic fibers. Improved moisture retention can be achieved by incorporating hydrophilic comonomers that decrease ultimate fiber density, by modifying the fiber spinning process, or by using after-treatments such as modified finishes.

Reduced Pilling. Pilling can be reduced by increasing the likelihood that the pills will break or wear off. Thus the most effective approaches include reducing fiber strength, incorporating defects in the fiber, increasing fiber brittleness, and reducing shear strength.

Improved Hot–Wet Properties. Acrylic fibers tend to lose modulus under hot–wet conditions. Knits and woven fabrics tend to lose their bulk and shape in dyeing and, to a more limited extent, in washing and drying cycles as well as in high humidity weather. Moisture lowers the glass-transition temperature, T_g, of acrylonitrile copolymers and, therefore, crimp is lost when the yarn is exposed to conditions required for dyeing and laundering. A number of polymer and fiber modifications have been devised to overcome this problem, though none has been successful enough to allow acrylics to compete successfully in easy care apparel markets.

Improved Abrasion Resistance. Abrasion resistance is generally improved by reducing the microvoid size and increasing the fiber density. Abrasion-resistant fibers have been produced by incorporating hydrophilic comonomers or comonomers with small molar volumes. Sulfonated monomers, acrylamide derivatives, and *N*-vinylpyrrolidinone are some of the hydrophilic comonomers that can be used to reduce void content; vinylidene chloride, with its relatively small molar volume, is effective in increasing fiber density. The spinning process itself has a significant effect on fiber density and abrasion resistance. Dry spinning, for example, is known to produce a denser fiber structure than conventional wet spinning.

Fiber Whiteness and Thermal Stability. The most effective route to improved whiteness and thermal stability is modification of the polymerization. The polymer must be linear and free of conjugated or unstable chemical structures formed by side reactions during the polymerization process.

Commercial Products

The majority of acrylic fiber production is 3.3–5.6 dtex (3 to 5 den) staple and tow furnished, undyed, in either bright or semidull luster. The principal markets are in apparel and home furnishings. Within the apparel sector these fibers find extensive use in sweaters and in single jersey, double-knit, and warp-knit fabrics for a variety of knitted outerwear garments such as dresses, suits, and children's wear. Other large markets for acrylics in the knit goods area are hand knitting yarns, deep pile fabrics, circular knit, fleece fabrics, half-hose, coarse-cut knitwear, and deep pile fabrics for blankets. Acrylics also hold a strong position in most broadwoven fabric categories including apparel and home furnishings for area rugs, carpets, curtains, and upholstery. Much of the growth in acrylics fibers usage has come from the replacement of wool.

Acrylics and modacrylics are also useful industrial fibers. Fibers low in comonomer content, such as Dolan 10 and DuPont's PAN Type A, have exceptional resistance to chemicals and very good dimensional stability under hot–wet conditions. These fibers are useful in industrial filters, battery separators, asbestos fiber replacement, hospital cubical curtains, office room dividers, uniform fabrics, and carbon fiber precursors. The excellent resistance of acrylic fibers to sunlight also makes them highly suitable for outdoor use. Typical applications include modacrylics, awnings, sandbags, tents, tarpaulins, covers for boats and swimming pools, cabanas, and duck for outdoor furniture.

Besides the standard staple and tow products, acrylic and modacrylic fibers are offered in many forms for specialized applications. Fibers with enhanced properties are in great demand. Yarn bulk is enhanced by using bicomponent or biconstituent fibers. Pilling is reduced by producing fibers that are more brittle, and yarns with exceptionally soft handle are produced by using fine denier fibers or fibers treated with special friction reducing finishes. There have been many efforts to develop premium products based on proprietary technology. Specialty fibers, such as ultrafine denier fibers, acid-dyeable, producer dyed, and pigmented fibers, ion-exchange fibers, metallized and semiconducting fibers, hollow fibers for apparel, coarse denier fibers for wigs, and simulated animal hair, are available. Other examples of specialty applications are the functional fabrics, eg, antimicrobial fabrics. Other functional fabrics are made from moisture-repellent fibers, moisture-absorbent sheath-core fibers, antistatic fibers, and flame-resistant fibers.

Economic Aspects

Because of the rapid capital investment in acrylics that occurred in the early 1970s, there is a large excess capacity. Prices have consequently been soft since 1977. Since that time there has been only minimal investment in plants or equipment and a curtailment in research and development work.

Table 2. Worldwide Synthetic Fiber Capacity and Production, 1991–2001, 10^3 t

Year	Acrylic	Polyester	Nylon	Polypropylene
		Capacity		
1991	2,870	11,200	4,700	3,170
1995	3,360	14,710	5,310	4,000
1998	3,540	15,650	5,410	4,310
2001	3,550	15,980	5,470	4,550
		Production		
1991	2,350	9,170	3,670	2,580
1995	2,620	11,250	4,100	3,230
1998	2,730	13,280	4,390	3,540
2001	2,790	15,050	4,710	3,850

Two U.S. producers remain. Monsanto, best known for its Acrilan acrylic fiber, is by far the largest and most diversified. With new spinning technology, originally introduced as Acrilan II, Monsanto now produces Duraspun, a fiber with enhanced abrasion resistance for the sock market. Other new products utilizing new spinning technology are Softlon and Ultrette, part of Monsanto's HP apparel line. Cytec (formerly American Cyanamid) continues to do well with Creslan acrylic and other premium products like MicroSupreme, a microfiber comparable to Monsanto's Fi-Lana. A third U.S. producer, Mann Industries, Inc. recently withdraw from the acrylic business and sold much of its technology to other producers, like Monsanto. The other U.S. producers are making carbon fiber precursors and carbon fiber. European producers like Courtaulch and Hoescht, are now attempting to sell more fiber in the profitable U.S. market.

Industrial use of acrylic fiber broadwoven goods has increased in recent years. The advantages are uv stability and its wide range of options for dyeing. The primary products are marine fabrics, awnings, and outdoor furniture. The worldwide picture for acrylics is summarized in Table 2.

In general, production is forecast to decline in the most developed countries and in regions that depend heavily on acrylic exports. Production appears to be shifting to the next generation of low cost producers, namely South and East Asia, the Middle East, Africa, and Eastern Europe. Countries such as Taiwan and South Korea are losing their competitive advantage to rising wages and prices.

The principal thrust of Japan's joint research programs is the development of process technologies to reduce conversion costs and development of high value-added products. Energy cost reduction is a prime concern. Opportunities exist in low energy consuming polymerization and spinning technology. High productivity can be achieved by high speed polymerization and spinning, robotization, automation, multi-end spinning, and high speed crimping. Other strategies are withdrawal from unprofitable market sectors and consolidation of production and research and development effort into areas where special cost or property advantages seem apparent. New high volume markets are also being studied. These include acrylics for asbestos replacement, cement reinforcement, and geotextile applications. Products that have limited volume potential but have potential as high value-added premium products are carbon and graphite fiber precursors, high strength fibers, fibers for reverse osmosis, ion-exchange fibers, and functional fibers such as antimicrobial, antistatic, water-repellant, and highly reflectant fibers. Chief among the new products being sought are high strength, high toughness and high modulus fibers, and functional fabrics. Recent developments in man-made fiber technology have been reviewed.

RAYMOND S. KNORR
Monsanto Company

B. G. Frushour and R. S. Knorr, "Acrylic Fibers," in M. Lewin and Eli M. Pearce, eds. *Handbook of Fiber Science and Technology*, Vol. IV, *Fiber Chemistry*, Marcel Dekker, Inc., New York, 1985.

R. W. Moncrieff, *Man-Made Fibers*, 6th ed, John Wiley & Sons, Inc., New York, 1975.

E. Cernia, "Acrylic Fibers," in H. F. Mark, S. M. Atlas, and E. Cernia, eds., *Man-Made Fibers*, Vol. III, Wiley-Interscience, New York, 1968.

R. K. Kennedy, "Modacrylic Fibers," in H. F. Mark, S. M. Atlas, and E. Cernia, eds., *Man-Made Fibers*, Vol. III, Wiley-Interscience, New York, 1968.

J. C. Masson, *Acrylic Fiber Technology and Applications*, Marcel Dekker, Inc., New York, 1995.

CELLULOSE ESTERS

The predominant cellulose ester fiber is cellulose acetate, a partially acetylated cellulose, also called acetate or secondary acetate. It is widely used in textiles because of its attractive economics, bright color, styling versatility, and other favorable aesthetic properties. However, its largest commercial application is as the fibrous material in cigarette filters, where its smoke removal properties and contribution to taste make it the standard for the cigarette industry. Cellulose triacetate fiber, also known as primary cellulose acetate, is an almost completely acetylated cellulose. Although it has fiber properties that are different, and in many ways better than cellulose acetate, it is of lower commercial significance primarily because of environmental considerations in fiber preparation.

Polymer Characteristics

Cellulose triacetate is obtained by the esterification of cellulose (qv) with acetic anhydride (see CELLULOSE ESTERS). Commercial triacetate is not quite the precise chemical entity depicted as (**1**) because acetylation does not quite reach the maximum 3.0 acetyl groups per glucose unit. Secondary cellulose acetate is obtained by hydrolysis of the triacetate to an average degree of substitution (DS) of 2.4 acetyl groups per glucose unit. There is no satisfactory commercial means to acetylate directly to the 2.4 acetyl level and obtain a secondary acetate that has the desired solubility needed for fiber preparation.

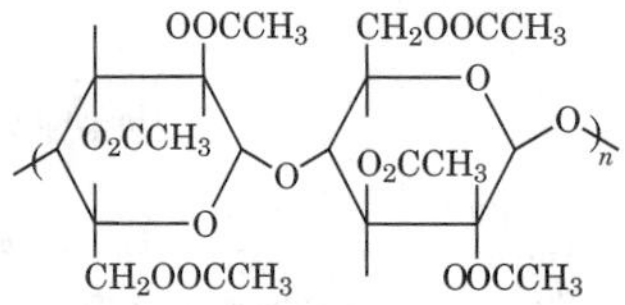

(**1**)

The degree of acetylation is specified by two separate terms: acetyl value (%) and combined acetic acid (%). The ratio of these two values is always 43:60, reflective of the molecular weight ratio of the acetyl group to acetic acid. Commercial cellulose triacetate has a combined acetic acid content of 61.5%, corresponding to 2.92 acetyl groups per glucose unit. Cellulose acetate, with 2.4 acetyl groups per glucose unit, has a combined acetic acid content of approximately 55%.

Fiber Properties

Mechanical Properties. Acetate and triacetate have a tenacity in the range of 0.10–0.12 N/tex (1.1–1.4 gf/den) with a breaking elongation of about 25–30%. Compared to other common textile fibers, acetate and triacetate are relatively weak, eg, 20–25% the tenacity of polyester. This is not necessarily a disadvantage, because fabric construction can be used to obtain the desired fabric performance targets. Pilling, the accumulation of fuzz balls on the fabric with wear, is not a problem as it is with the higher tenacity fibers.

This modulus of elasticity, or Young's modulus, is related to many of the mechanical performance characteristics of textile products. The modulus of elasticity can be affected by drawing, ie, elongating the fiber; environment, ie, wet or dry, temperature; or other procedures. Values for commercial acetate and triacetate fibers are generally in the 2.2–4.0 N/tex (25–45 gf/den) range.

The wet modulus of fibers at various temperatures influences the creasing and mussiness caused by laundering. Acetate, triacetate, and rayon behave quite similarly, with a lower sensitivity than acrylic.

For acetate and triacetate the work of rupture is essentially the same at 0.022 N/tex (0.25 gf/den). This is higher than for cotton (0.010 N/tex = 0.113 gf/den), similar to rayon and wool, but less than for nylon (0.076 N/tex = 0.86 gf/den) and silk (0.072 N/tex = 0.81 gf/den).

The elongation of a stretched fiber is best described as a combination of instantaneous extension and a time-dependent extension or creep. This viscoelastic behavior is common to many textile fibers, including acetate. Conversely, recovery of viscoelastic fibers is typically described as a combination of immediate elastic recovery, delayed recovery, and permanent set or secondary creep. The permanent set is the residual extension that is not recoverable. These three components of recovery for acetate are given in Table 1. The elastic recovery of acetate fibers alone and in blends has also been reported. In textile processing strains of more than 10% are avoided in order to produce a fabric of acceptable dimensional or shape stability.

Absorption and Swelling Behavior. The absorption of moisture by acetate and triacetate fibers generally depends on the relative humidity and whether equilibrium is approached from the dry or wet side. The percentage of moisture regain of commercial fibers (ASTM D1909-68), taken at 65% relative humidity for the absorption cycle, is 6.5 for acetate fiber and 3.5 for triacetate. Heat treatment can lower the moisture regain of triacetate fiber, and values of 2.5–3.2% have been observed.

Percentage of water imbibition is an important property of ease-of-care and quick-drying fabrics. This value is determined by measuring the moisture remaining in a fiber in equilibrium with air at 100% rh while the fiber is being centrifuged at forces up to 1000 g. The average recorded value for acetate is 24%; triacetate not heat-treated, 16%; and heat-treated triacetate, 10%.

Specific Gravity. The values of 1.32 for acetate and 1.30 for triacetate are accepted for fibers of combined acetic acid contents of 55 and 61.5%, respectively.

Refractive Index. The refractive index parallel to the fiber axis (ϵ) is 1.478 for acetate and 1.472 for triacetate. The index perpendicular to the axis (ω) is 1.473 for acetate and 1.471 for triacetate.

Thermal Behavior. Acetate softens and sticks in the 190–205°C range, and fuses at ca 260°C. The apparent shining or glazing temperature is usually lower than the sticking temperature. The sole-plate temperature of an iron should not exceed 170–180°C when used on acetate fabrics. The sticking and glazing temperatures of untreated triacetate fiber are in the same range as those of acetate, whereas those exhibited by heat-treated triacetate fibers are considerably higher. Fabrics made of the latter can be ironed at temperatures as high as 240°C. The melting point of triacetate is ca 300°C.

Acetate and triacetate exhibit moderate changes in mechanical properties as a function of temperature. As the temperature is raised, the tensile modulus of acetate and triacetate fibers is reduced, and

Table 1. Elongation Recovery of Acetate Fibers

Fiber	Immediate elastic recovery, %	Delayed recovery, %	Permanent set, %
acetate multifilament			
at 50% of breaking tenacity	74	26	0
at breaking point	14	16	70
acetate staple yarn			
at 50% of breaking tenacity	58	42	0
at breaking point	12	18	70

the fibers extend more readily under stress. Acetate and triacetate are weakened by prolonged exposure to elevated temperatures in air.

Light Stability. The resistance of textile fibers to sunlight degradation depends on the wavelength of the incident light, relative humidity, and atmospheric fumes. Acetate and triacetate fibers have essentially the same light-absorption characteristics in the visible spectrum; absorption in the uv region is slightly higher. Both fibers, when exposed under glass, behave similarly to cotton and rayon, ie, they are somewhat more resistant than unstabilized pigmented nylon and silk and appreciably less resistant than acrylic and polyester fibers.

Electrical Behavior. Because of their high resistivity both acetate and triacetate yarns readily develop static charges, and an antistatic finish is usually applied to aid in fiber processing. Both yarns have also been used for electrical insulation after lubricants and other finishing agents are removed.

Dyeing Characteristics. Disperse dyes, high melting crystalline compounds with low solubility in the dye bath, are most frequently used for cellulose acetate and triacetate fibers.

Colored acetate and triacetate yarns are produced by incorporating colored pigments (inorganic or organic), soluble dyes, or carbon in the polymer solution before extrusion. Solution-dyed acetate and triacetate yarns are extremely colorfast to washing, dry cleaning, sunlight, perspiration, seawater, and crocking, and usually surpass the performance of vat-dyed yarns. In addition, acetate and triacetate dyed by conventional methods are susceptible to gases or fumes and fade; such fading is absent in solution-spun, pigment-dyed yarn.

Chemical Properties. Under slightly acidic or basic conditions at room temperature, acetate and triacetate fibers are resistant to chlorine bleach at the concentrations normally used in laundering.

Triacetate fiber is significantly more resistant than acetate to alkalies encountered in normal textile operations.

Acetate and triacetate are essentially unaffected by dilute solutions of weak acids, but strong mineral acids cause serious degradation.

Resistance to Microorganisms and Insects. Resistance of triacetate to microorganisms, based on soil-burial tests, is high, approaching that of polyester, acrylic, and nylon fibers.

Manufacture

Cellulose Acetate and Triacetate Polymer. The production of acetate and triacetate polymer is accomplished by the esterification of high purity chemical cellulose, except for special plastic-grade acetates requiring low color and high clarity, where cotton linters are used.

Cellulose Acetate and Triacetate Fibers. Polymer solutions are converted into fibers by extrusion. The dry-extrusion process, also called dry spinning, is primarily used for acetate and triacetate.

The dry-extrusion process consists of four operations; dissolution of the polymer in a volatile solvent; filtration of the solution to remove insoluble matter; extrusion of the solution to form fibers; and lubrication, yarn formation, and packaging.

Anisotropic Solutions

Many cellulosic derivatives form anisotropic, ie, liquid crystalline, solutions, and cellulose acetate and triacetate are no exception. Various cellulose acetate anisotropic solutions have been made using a variety of solvents. The nature of the polymer–solvent interaction determines the concentration at which liquid crystalline behavior is initiated. The better the interaction, the lower the concentration needed to form the anisotropic, birefringent polymer solution. Strong organic acids, eg, trifluoroacetic acid, are most effective and can produce an anisotropic phase with concentrations as low as 28%.

Products

Yarns and Fibers. Many different acetate and triacetate continuous filament yarns, staples, and tows are manufactured. The variable properties are tex (wt in g of a 1000-m filament) or denier (wt in g of a 9000-m filament), cross-sectional shape, and number of filaments. Individual filament fineness (tex per filament or denier per filament, dpf) is usually in the range of 0.2–0.4 tex per filament (2–4 dpf). Common continuous filament yarns have 6.1, 6.7, 8.3, and 16.7 tex (55, 60, 75, and 150 den, respectively). However, different fabric properties can be obtained by varying the filament count (tex per filament or dpf) to reach the total tex (denier).

Yarn Packages. The principal package types used by the textile industry are tubes, cones, and beams.

Staple and Tow. The same extrusion technology that produces continuous filament yarn also produces staple and tow. The principal difference is that spinnerets with more holes are used, and instead of winding the output of each spinneret on an individual package, the filaments from a number of spinnerettes are gathered together into a ribbon-like strand, or tow. A mechanical device uniformly plaits the tow into a carton, from which it can be continuously withdrawn without tangling. Staple is produced by cutting a crimped tow into short (usually 4–5-cm) lengths resembling short, natural fibers.

Fibrillated Fibers. Instead of extruding cellulose acetate into a continuous fiber, discrete, pulp-like agglomerates of fine, individual fibrils, called fibrets or fibrids, can be produced by rapid precipitation with an attenuating coagulation fluid.

Economic Aspects

Cellulose acetate, the second oldest synthetic fiber, is an important factor in the textile and tobacco industries. Acetate belongs to the group of less expensive fibers; triacetate is slightly more expensive. An annual listing of worldwide fiber producers, locations, and fiber types is published by the Fiber Economics Bureau, Inc.

The principal textile applications of both acetate and triacetate fibers are in women's apparel and home-furnishing fabrics. Although the use of acetate fiber for textile applications has generally declined, the total worldwide production of cellulose acetate increased owing to tow for cigarette filters.

The combined annual world acetate production (filament, staple, and tow) peaked in 1980 with 672,000 t, dropped to 574,000 t in 1984, and rose to 713,000 t in 1993. The United States accounted for ca 45% of the world total. Other principal acetate producing countries include the U.K., Japan, Canada, Italy, and the former USSR.

GEORGE A. SERAD
Hoechst Celanese Corporation

G. A. Serad, in J. I. Kroschwitz, ed., *Encyclopedia of Polymer Science and Engineering*, Vol. 3, 2nd ed. Wiley-Interscience, New York, 1985, pp. 200–226.

L. Segal, in N. M. Bikales and L. Segal, eds., *Cellulose and Cellulose Derivatives*, High Polymers Series, Vol. V, Wiley-Interscience, New York, 1971, Chapt XVII-A.

C. J. Maim and G. D. Hiatt, in E. Ott, H. M. Spurlin, and M. W. Graffin, eds., *Cellulose and Cellulose Derivatives*, High Polymers Series, 2nd ed., Vol. V, Wiley-Interscience, New York, 1954, Pt. II.

Data of Fiber Economics Bureau, as published in *Text. Organon*, 112–113, and 115 (June 1986); *Fiber Organon*, 138–139 and 144 (July 1992); and author information.

ELASTOMERIC

Elastomeric fibers can be made from natural or synthetic polymeric materials that provide a product with high elongation, low modulus, and good recovery from stretching. Currently, these fibers are made primarily from polyisoprenes (natural rubber) or segmented polyurethanes and to a lesser extent from segmented polyesters. In the United States the generic designation spandex has been given to a manufactured fiber in which the fiber-forming substance is a long-chain synthetic polymer comprised of at least 85% of a segmented polyurethane; in Europe the equivalent term elastane is commonly used.

Thermoplastic, inelastic fibers, such as nylon and polyester, may be processed to provide spring-like, helical, or zigzag structures. These fibers can exhibit high elongations as the helical or zigzag structure is stretched, but the recovery force is very low. This apparent elasticity results from the geometric form of the filaments as opposed to elastomeric fibers whose elastic properties depend primarily on entropy changes inherent within their polymer structure. Thus processed inelastic fibers must comprise a significant portion of a stretch fabric whereas an elastomeric fiber provides the necessary stretch properties at 5–20% of fabric weight.

Other elastomeric-type fibers include the biconstituents, which usually combine a polyamide or polyester with a segmented polyurethane-based fiber. These two constituents are melt-extruded simultaneously through the same spinneret hole and may be arranged either side by side or in an eccentric sheath–core configuration. As these fibers are drawn, a differential shrinkage of the two components develops to produce a helical fiber configuration with elastic properties. An applied tensile force pulls out the helix and is resisted by the elastomeric component. Kanebo Ltd. has introduced a nylon–spandex sheath–core biconstituent fiber for hosiery with the trade name Sideria.

Nonspandex elastomeric fibers based on segmented polyesters and polyethers are currently being developed that can be melt-spun into threads. Teijin Ltd. produces an elastomeric fiber of this type with the trade name Rexe.

Table 1. Physical Properties of Elastomeric Fibers

Property	Spandex	Extruded rubber	Cut rubber
sizes available[a]	1.1–250 [b]	16–610 tex[b]	2.5–21 μm dia[c]
tenacity, N/tex[d]	0.05–0.13	0.02–0.03	0.01–0.02
elongation, %	400–800	600–700	600–700
modulus[e], N/tex[d]	0.013–0.045	0.004–0.005	0.002–0.004
stability[f]			
uv light	good	fair	fair
ozone	good	poor	poor
NO_x	fair, yellows	poor	poor
active Cl	fair, yellows	poor	poor
body oils	fair	poor	poor
cosmetics	good	fair	fair
dyeability	dyeable	not dyeable	not dyeable
abrasion resistance	very good	poor	poor

[a] Spandex size is usually expressed in denier which is weight in g/9000 m length. However, the SI unit is tex, the weight in g/1000 m. Rubber size is expressed as gauge, which is the reciprocal of diameter or size in inches.
[b] To convert tex to den, multiply by 9.09.
[c] 1,200–10,000 gauge.
[d] To convert N/tex to gf/den, multiply by 11.33.
[e] First cycle stress at 300% elongation.
[f] Both spandex fibers and rubber threads normally contain antioxidants and other stabilizers.

Mechanical Properties

In both rubber thread and spandex fibers, mechanical properties may be varied over a relatively broad range. In rubber, variations are made in the degree of cross-linking or vulcanization by changing the amount of vulcanizing agent, usually sulfur, and the accelerants used. In spandex fibers, many more possibilities for variation are available. By definition spandex fibers contain urethane linkages with the following repeat structure:

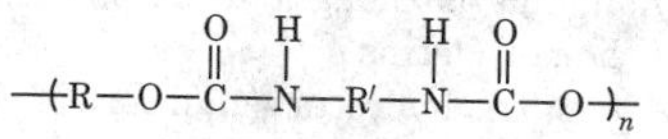

The number of polymers in the classification is obviously very large. Most urethane polymers in current use for the manufacture of spandex fibers are made by the reaction of 1000–4000 molecular weight hydroxy terminated polyethers or polyesters with a diisocyanate at a molar ratio of ca 1:1.4 to 1:2.5, followed by reaction of the resulting isocyanate-terminated prepolymer with one or more diamines to produce a high molecular weight urethane polymer. Small amounts of monofunctional amines may also be included to control polymer molecular weight. Mechanical properties may be affected by changing the particular polyester or polyether glycol, diisocyanate, diamine(s), and monoamine used; they can be further modified by changing the molecular weight of the glycol and by changing the glycol–diisocyanate molar ratio.

The physical characteristics of current commercial rubber and spandex fibers are summarized in Table 1.

Manufacture

Cut Rubber. To produce cut rubber thread, smoked rubber sheet or crepe rubber is milled with vulcanizing agents, stabilizers, and pigments. This milled stock is calendered into sheets 0.3–1.3 mm thickness, depending on the final size of the rubber thread desired. Multiple sheets are layered, heat-treated to vulcanize, then slit into threads for textile uses. Individual threads have either square or rectangular cross-sections.

Extruded Latex Thread. In the manufacture of extruded latex thread, a concentrated (up to ca 50% solids) natural rubber latex is blended with aqueous dispersions of vulcanizing agents, stabilizers, and white pigments. This compounded latex is held under controlled temperature conditions until partial vulcanization occurs. This has the effect of increasing wet strength and thus the processibility of the extruded threads. The matured latex is extruded at constant pressure through precision-bore glass capillaries into a 15–55% acetic acid bath where coagulation into thread form occurs. Threads are removed from the coagulation bath by transfer rollers, washed free of excess acid with water, and conducted through a dryer, after which a silicone oil-based finish is applied and the threads are formed into multiend ribbons. The ribbons are then vulcanized by multiple passes on a conveyer belt through an oven that can increase curing temperature in stages up to about 150°C. After vulcanization the multiend ribbons are packed without support in boxes for shipment to the customer.

Spandex Fibers. Four different processes are currently used to produce spandex fibers commercially: melt extrusion, reaction spinning, solution dry spinning, and solution wet spinning. These processes involve different practical applications of basically similar chemistry. An isocyanate terminated prepolymer is formed by the reaction of a 1000–4000 molecular weight macroglycol with a diisocyanate at a glycol–diisocyanate ratio that may range from 1:1.4 up to about 1:2.5. The soft segment macroglycol can be either a polyether, a polyester, a polycarbonate, hydroxyl-terminated polycaprolactone, or a combination of these. The prepolymer subsequently reacts with either a glycol or diamine(s) at near stoichiometry; a small amount of monofunctional amine may be included to control final polymer molecular weight. If the diol or diamine(s) reaction with the prepolymer is carried out in a solvent, the resulting block copolymer solution may be wet or dry spun into fiber. Alternatively, the prepolymer may be reaction spun by extrusion into a bath containing diamine to form a fiber, or the prepolymer may be permitted to react in bulk with a diol and the resulting polymer melt extruded in fiber form.

Chemical Properties

Stabilization. Both rubber and spandex fibers are subject to oxidative attack by heat, light, atmospheric contaminants such as NO_x, and active chlorine. Both rubber and spandex fibers are likely to contain antioxidants; the spandex fibers may also be stabilized to uv light and to atmospheric contaminants that cause discoloration.

Solvent Resistance. Elastomeric fibers tend to swell in certain organic solvents; rubber fibers swell in hydrocarbon solvents such as hexane. Spandex fibers become highly swollen in chlorinated solvents such as tetrachloroethylene.

Dyeing. Spandex fibers have an affinity for dispersed or acid dyes; rubber fibers normally cannot be dyed.

Economic Aspects

Most process developments have occurred in the United States, Germany, Japan, and Korea. A large proportion of worldwide capacity is controlled by DuPont, either directly or through subsidiaries and joint ventures.

Commercially, elastomeric fibers are almost always used in combination with hard fibers such as nylon, polyester, or cotton. Prices of spandex fibers are highly dependent on thread size; selling price generally increases as fiber tex decreases. Factors that contribute to the relatively high cost of spandex fibers include *(1)* the relatively high cost of raw materials; *(2)* the small size of the spandex market compared to that of hard fibers, which limits scale and thus efficiency of production units; and *(3)* the technical problems associated with stretch fibers that limit productivity rates and conversion efficiencies.

Uses

Elastomeric fibers are used in cut rubber and extruded latex and spandex fibers.

JOHN E. BOLIEK
ARNOLD W. JENSEN
E. I. du Pont de Nemours & Co., Inc.

U.S. Pat. 2,957,852 (Oct. 25, 1960), P. Frankenburg and A. Frazier (to DuPont Co.).

T. Kotani and co-workers, *J. Macromol. Sci-Phys.* **831**, 65 (1992).

U.S. Pat. 3,296,063 (Jan 3, 1967), C. Chandler (to DuPont Co.).

U.S. Pat. 5,028,642 (July 2, 1991), C. W. Goodrich & W. Evans (to DuPont Co.).

OLEFIN

Olefin fibers, also called polyolefin fibers, are defined as manufactured fibers in which the fiber-forming substance is a synthetic polymer of at least 85 wt % ethylene, propylene, or other olefin units. Several olefin polymers are capable of forming fibers, but only polypropylene (PP) and, to a much lesser extent, polyethylene (PE) are of practical importance. Olefin polymers are hydrophobic and resistant to most solvents. These properties impart resistance to staining, but cause the polymers to be essentially undyeable in an unmodified form.

Advances in olefin polymerization provide a wide range of polymer properties to the fiber producer. Inroads into new markets are being made through improvements in stabilization, and new and improved methods of extrusion and production, including multicomponent extrusion and spunbonded and meltblown nonwovens.

Properties

Physical Properties. Table 1 shows that olefin fibers differ from other synthetic fibers in two important respects: *(1)* olefin fibers have very low moisture absorption and thus excellent stain resistance and almost equal wet and dry properties, and *(2)* the low density of olefin fibers allows a much lighter weight product at a specified size or coverage. Thus one kilogram of polypropylene fiber can produce a fabric, carpet, etc, with much more fiber per unit area than a kilogram of most other fibers.

Table 1. Physical Properties of Commercial Fibers

Polymer	Standard tenacity, GPa[a]	Breaking elongation, %	Modulus, GPa[a]	Density, kg/m^3	Moisture regain[b]
olefin	0.16–0.44	20–200	0.24–3.22	910	0.01
polyester	0.37–0.73	13–40	2.1–3.7	1380	0.4
carbon	3.1	1	227	1730	
nylon	0.23–0.60	25–65	0.5–2.4	1130	4–5
rayon	0.25–0.42	8–30	0.8–5.3	1500	11–13
acetate	0.14–0.16	25–45	0.41–0.64	1320	6
acrylic	0.22–0.27	35–55	0.51–1.02	1160	1.5
glass	4.6	5.3–5.7	89	2490	
aramid	2.8	2.5–4.0	113	1440	4.5–7
fluorocarbon	0.18–0.74	5–140	0.18–1.48	2100	
polybenzimidazole	0.33–0.38	25–30	1.14–1.52	1430	15

[a] To convert GPa to psi, multiply by 145,000.
[b] At 21°C and 65% rh.

Tensile Strength. Tensile properties of all polymers are a function of molecular weight, morphology, and testing conditions. Lower temperature and higher strain rate result in higher breaking stresses at longer elongations, consistent with the general viscoelastic behavior of polymeric materials. Similar effects are observed on other fiber tensile properties, such as tenacity or stress at break, energy to rupture, and extension at break. Under the same spinning, processing, and testing conditions, higher molecular weight results in higher tensile strength.

Creep, Stress Relaxation, Elastic Recovery. Olefin fibers exhibit creep, or time-dependent deformation under load, and undergo stress relaxation, or the spontaneous relief of internal stress. High molecular weight and high orientation reduce creep.

Elastic recovery or resilience is the recovery of length upon release of stress after extension or compression. A fiber, fabric, or carpet must possess this property in order to spring back to its original shape after being crushed or wrinkled, Polyolefin fibers have poorer resilience than nylon; this is thought to be partially related to the creep properties of the polyolefins.

Chemical Properties. The hydrocarbon nature of olefin fibers, lacking any polarity, imparts high hydrophobicity and consequently resistance to soiling or staining by polar materials, a property important in carpet and upholstery applications. Unlike the condensation polymer fibers, such as polyester and nylon, olefin fibers are resistant to acids and bases. At room temperature, polyolefins are resistant to most organic solvents, except for some swelling in chlorinated hydrocarbon solvents. At higher temperatures, polyolefins dissolve in aromatic or chlorinated aromatic solvents, and show some solubility in high boiling hydrocarbon solvents. At high temperatures, polyolefins are degraded by strong oxidizing acids.

Thermal and Oxidative Stability. In general, polyolefins undergo thermal transitions at much lower temperatures than condensation polymers; thus, the thermal and oxidative stability of polyolefin fibers are comparatively poor. Preferred stabilizers are highly substituted phenols such as Cyanox 1790 and Irganox 1010, or phosphites such as Ultranox 626 and Irgafos 168 (see ANTIOXIDANTS; HEAT STABILIZERS).

Ultraviolet Degradation. Polyolefins are subject to light-induced degradation; polyethylene is more resistant than polypropylene. Because polyolefins readily form hydroperoxides, the more effective light stabilizers are radical scavengers. Hindered amine light stabilizers (HALS) are favored, especially high molecular weight and polymeric amines that have lower mobility and less tendency to migrate to the surface of the fiber. This migration is commonly called bloom.

Flammability. Most polyolefins can be made fire retardant using a stabilizer, usually a bromine-containing organic compound, and a synergist such as antimony oxide. However, the required loadings are usually too high for fibers to be spun. Fire-retardant polypropylene fibers exhibit reduced light and thermal resistance.

Dyeing Properties. Because of their nonionic chemical nature, olefin fibers are difficult to dye. A broad variety of polymeric dyesites have been blended with polypropylene; nitrogen-containing copolymers are the most favored. In apparel applications where dyeing is important, dyeable blends are expensive and create problems in spinning fine denier fibers. Hence, olefin fibers are usually colored by pigment blending during manufacture, called solution dying in the trade.

Manufacture and Processing

Olefin fibers are manufactured commercially by melt spinning, similar to the methods employed for polyester and polyamide fibers.

Slit-Film Fiber. A substantial volume of olefin fiber is produced by slit-film or film-to-fiber technology. For producing filaments with high linear density, above 0.7 tex (6.6 den), the production economics are more favorable than monofilament spinning. The fibers are used primarily for carpet backing and rope cordage applications.

Bicomponent Fibers. Polypropylene fibers have made substantial inroads into nonwoven markets because they are easily thermal bonded. Further enhancement in thermal bonding is obtained using bicomponent fibers. In these fibers, two incompatible polymers, such as polypropylene and polyethylene, polyester and polyethylene, or polyester and polypropylene, are spun together to give a fiber with a side-by-side or core–sheath arrangement of the two materials. The lower melting polymer can melt and form adhesive bonds to other fibers; the higher melting component causes the fiber to retain some of its textile characteristics. Bicomponent fibers have also provided a route to self-texturing (self-crimping) fibers.

Meltblown, Spunbond, and Spurted Fibers. A variety of directly formed nonwovens exhibiting excellent filtration characteristics are made by meltblown processes, producing very fine, submicrometer filaments. A stream of high velocity hot air is directed on the molten polymer filaments as they are extruded from a spinnerette. This air attenuates, entangles, and transports the fiber to a collection device. Because the fiber cannot be separated and wound for subsequent processing, a nonwoven web is directly formed.

In the spunbond process, the fiber is spun similarly to conventional melt spinning, but the fibers are attenuated by air drag applied at a distance from the spinnerette. This allows a reasonably high level of filament orientation to be developed. The fibers are directly deposited onto a moving conveyor belt as a web of continuous randomly oriented filaments.

Pulp-like olefin fibers are produced by a high pressure spurting process developed by Hercules Inc. and Solvay, Inc. Polypropylene or polyethylene is dissolved in volatile solvents at high temperature and pressure. After the solution is released, the solvent is volatilized, and the polymer expands into a highly fluffed, pulp-like product. Additives are included to modify the surface characteristics of the pulp. Uses include felted fabrics, substitution in whole or in part for wood pulp in papermaking, and replacement of asbestos in reinforcing applications.

High Strength Fibers. The properties of commercial olefin fibers are far inferior to those theoretically attainable. A number of methods, including superdrawing, high pressure extrusion, spinning of liquid crystalline polymers or solutions, gel spinning, and hot drawing produce high strengths, but these methods are tedious and uneconomical for olefin fibers. A high modulus commercial polyethylene fiber with properties approaching those of aramid and graphite fibers is prepared by gel spinning.

Hard-Elastic Fibers. Hard-elastic fibers are prepared by annealing a moderately oriented spun yarn at high temperature under tension. They are prepared from a variety of olefin polymers, acetal copolymers, and polypivalolactone.

Economic Aspects

In the United States, olefin fiber consumption has risen steadily since its introduction in 1961. Olefin fiber is the only synthetic fiber showing market growth in recent years.

Applications

Olefin fibers are used for a variety of purposes from home furnishings to industrial applications. These include carpets, upholstery, drapery, rope, geotextiles (qv), and both disposable and nondisposable nonwovens. Fiber mechanical properties, relative chemical inertness, low moisture absorption, and low density contribute to desirable product properties.

C. J. Wust, Jr.
L. M. Landoll
Hercules Incorporated

M. Ahmed, *Polypropylene Fibers: Science and Technology*, Elsevier Science Publishing Co., Inc., New York, 1982, pp. 344–346.

L. M. Landoll, "Olefin Fibers," in J. I. Kroschwitz, ed., *Encyclopedia of Polymer Science and Engineering*, 2nd ed., Vol. 10, John Wiley & Sons, Inc., New York, 1987, pp. 373–395.

POLYESTER

Properties

The Textile Fiber Product Identification Act (TFPIA) requires that the fiber content of textile articles be labeled. The Federal Trade Commission established and periodically refines the generic fiber definitions. The current definition for a polyester fiber is "A manufactured fiber in which the fiber-forming substance is any long-chain synthetic polymer composed of at least 85% by weight of an ester of a substituted aromatic carboxylic acid, including but not restricted to terephthalate units, and para substituted hydroxybenzoate units."

Poly(ethylene terephthalate), the predominant commercial polyester, has been sold under trademark names including Dacron (DuPont), Terylene (ICI), Fortrel (Wellman), Trevira (Hoechst Celanese), and others. PET is a fiber of great commercial significance, useful in cordage, apparel fabrics, industrial fabrics, conveyor belts, laminated and coated substrates, and numerous other areas.

In the late 1980s, new fully aromatic polyester fibers were introduced for use in composites and structural materials. In general, these materials are thermotropic liquid crystal polymers that are melt-processible to give fibers with tensile properties and temperature resistance considerably higher than conventional polyester textile fibers. Vectran (Hoechst Celanese and Kuraray) is a thermotropic liquid crystal aromatic copolyester fiber composed of *p*-hydroxybenzoic acid and 6-hydroxy-2-naphthoic acid.

Most polyester fiber produced is standard molecular weight (ca 0.6 dL/g intrinsic viscosity), round cross-section PET. However, to engineer specific properties for special uses, many product variants have been developed and commercialized. These variants include using alternative cross sections, controlling polymer molecular weight, modifying polymer composition by using comonomers, and using additives including delusterants, pigments, and optical brighteners.

Changing the cross section of standard PET by the use of specially designed spinneret capillaries can change fabric visual and tactile aesthetics. High molecular weight polymer is used for high strength industrial fibers in tires, ropes, and belts.

Standard polyester fibers contain no reactive dye sites. PET fibers are typically dyed by diffusing dispersed dyestuffs into the amorphous regions in the fibers. Copolyesters from a variety of copolymerizable glycol or diacid comonomers open the fiber structure to achieve deep dyeability. In addition to dyeability, polyesters with a high percentage of comonomer to reduce the melting point have found use as fusible binder fibers in nonwoven fabrics. Specially designed copolymers have also been evaluated for flame-retardant PET fibers.

Physically or chemically modifying the surface of PET fiber is another route to diversified products. Hydrophilicity, moisture absorption, moisture transport, soil release, color depth, tactile aesthetics, and comfort all can be affected by surface modification.

Fine Structural Properties. The performance and properties of PET fibers are significantly impacted by the relative amounts of amorphous and crystalline structures, the orientation of the structures with respect to the fiber axis, and the size distribution of the crystalline regions. Density, mechanical, and thermal properties are significantly affected by the degree of crystallinity.

Mechanical Properties. Polyester fibers are formed by melt spinning generally followed by hot drawing and heat setting to the final fiber form. The molecular orientation and crystalline fine structure developed depend on key process parameters in all fiber formation steps and are critical to the end use application of the fibers.

Molecular orientation and crystallinity generally increase with draw ratio, increasing break tenacity and Young's modulus while decreasing fiber break elongation. Typical properties of continuous filament and staple poly(ethylene terephthalate) fibers are shown in Table 1. Fiber dimensional stability and Young's modulus also are dependent on the heat-setting process. Fibers that are relaxed, or heat-set under no restraint, show a low shrinkage and low initial modulus. Annealed fibers, which are heat-set under tension at constant length, have a low shrinkage and maintain a high initial modulus. Other factors, including polymer molecular weight or the presence of comonomers, can significantly affect the fiber mechanical properties.

Chemical Properties. The hydrolysis of PET is acid- or base-catalyzed and is highly temperature-dependent and relatively rapid at polymer melt temperatures. In general, the hydrolysis and chemical resistance of copolyester materials is less than that for PET and depends on both the type and amount of comonomer.

At room temperature, PET is resistant to organic and moderate strength mineral acids. At elevated temperatures, PET strength loss in moderate strength acids can be appreciable. Strong acids such as concentrated sulfuric acid dissolve and depolymerize PET.

Polyester fibers have good resistance to weakly alkaline chemicals and moderate resistance to strongly alkaline chemicals at room temperature.

Polyester fibers have excellent resistance to soap, detergent, bleach, and other oxidizing agents. PET fibers are generally insoluble in organic solvents, including cleaning fluids, but are soluble in some phenolic compounds, eg, *o*-chlorophenol.

Thermal Properties. The melting point of poly(ethylene terephthalate) is generally reported to be 258–265°C and is generally considered to be independent of molecular weight. Copolymerization with generated or added comonomers results in a decrease in the melting point and a disruption of the crystalline order (lower crystallinity) dependent on the amount of comonomer present.

The glass-transition temperature, T_g, of dry polyester is approximately 70°C and is slightly reduced in water. The glass-transition temperatures of copolyesters are affected by both the amount and chemical nature of the comonomer.

Other Properties. Polyester fibers have good resistance to uv radiation although prolonged exposure weakens the fibers. PET is not affected by insects or microorganisms and can be designed to kill bacteria by the incorporation of antimicrobial agents. The oleophilic surface of PET fibers attracts and holds oils.

Table 1. Mechanical Properties of PET Fibers

	Staple/tow		Continuous filament		
Property	Regular tenacity	High tenacity	POY[a]	Regular tenacity	High tenacity
break tenacity, N/tex[b]	03.–0.5	0.5–0.6	0.2–0.3	0.4–0.5	0.6–0.9
elongation, %	40–60	20–30	110–250	20–40	10–25
elastic recovery, % at 5% elongation	75–80	90		88–93	90
stiffness, N/tex[b]	1–2	5–6	0.2–0.5	1–3	5–7
toughness, N/tex[b]	0.02–0.15	0.02–0.10	0.10–0.20	0.04–0.10	0.04–0.07

[a] POY = partially oriented yarn.
[b] To convert N/tex to gf/den, multiply by 11.33.

Manufacturing and Processing

Terephthalic acid (TA) or dimethyl terephthalate (DMT) reacts with ethylene glycol (2G) to form bis(2-hydroxyethyl) terephthalate (BHET) which is condensation polymerized to PET with the elimination of 2G. Molten polymer is extruded through a die (spinneret) forming filaments that are solidified by air cooling. Combinations of stress, strain, and thermal treatments are applied to the filaments to orient and crystallize the molecular chains. These steps develop the fiber properties required for specific uses. The two general physical forms of PET fibers are continuous filament and cut staple.

Texturing. This is a process applied to continuous filament yarns to introduce loops and bends in the individual filaments. In the 1970s, with the advent of POY, false twist texturing grew explosively, and it accounts for most of the textured yarn produced in the 1990s.

Economic Aspects

PET is based on petroleum and the price of polyester fiber fluctuates with the price of *p*-xylene and ethylene raw materials as well as with the energy costs for production. With the ability to interchange with other fibers, especially cotton in cotton blends, the price of polyester is affected by the price and availability of cotton as well as the supply and demand of polyester.

Safety and Environment Factors

Health and Safety. PET fibers pose no health risk to humans or animals.

Environmental Factors. PET materials are not dangerous to the environment and cannot contaminate surface or ground water. A key environmental advantage for PET material is the ability to recycle.

Applications

Staple. PET staple is widely used in 100% polyester or cotton-blend fabrics for apparel. Along with cotton blends, polyester blends with rayon or wool are also important.

In addition to fabrics, PET staple is used in a wide variety of other applications. High tenacity staple fibers are widely used in sewing thread. Staple PET fibers have been engineered for use in rugs, carpets, and filling products, including furniture, pillows, mattresses, sleeping bags, and stuffed toys. Polyester staple fibers are commonly used in nonwovens for applications in diaper coverstock, filters, linings and interfaces, and disposable towels and wipes.

Filament. Fully drawn flat yarns and partially oriented (POY) continuous filament yarns are available in yarn sizes ranging from about 3.3–33.0 tex (30–300 den) with individual filament linear densities of about 0.055 to 0.55 tex per filament (0.5–5 dpf). The fully drawn hard yarns are used directly in fabric manufacturing operations, whereas POY yarns are primarily used as feedstock for draw texturing. Both textured and hard yarns are used in apparel, sleepwear, outerwear, sportswear, draperies and curtains, and automotive upholstery.

High molecular weight polyester is commonly used to make high strength industrial fibers. These fibers are commonly used in applications requiring high strength and stability, including tire cord, seat belts, industrial belts and hoses, ropes, cords, and sailcloth.

Polyesters are also used in continuous filament spunbonded nonwovens. These spunbonded fabrics are available in a wide range of thicknesses and basis weights and can be used for electrical insulation, coated fabric substrates, disposable apparel for clean rooms, hospitals, and geotextiles (qv).

S. M. HANSEN
P. B. SARGEANT
E. I. du Pont de Nemours & Co., Inc.

H. Luckert and co-workers, in *Ullman's Encyclopedia of Industrial Chemistry*, Vol. A10, 5th ed., VCH Publishers, New York, 1987, p. 511.

A. Ziabicki and H. Kawai, eds., *High Speed Fiber Spinning*, John Wiley & Sons, Inc., New York, 1985.

C. J. Heffelfinger and L. K. Knox, in O. J. Sweeting, ed, *Science & Technology of Polymer Films*, Vol. 1, John Wiley & Sons, Inc., New York, 1971.

S. C. Winchester, D. A. Shiffler, and S. M. Hansen, in J. J. McKetta, ed., *Encyclopedia of Chemical Processing and Design*, Vol. 40, Marcel Dekker, Inc., New York, 1992, p. 152.

POLY(VINYL ALCOHOL)

The principal fiber types that fall under the category of vinyl fibers are fibers that contain at least 85% by weight of vinyl chloride, known generically as vinyon fibers, and those that are composed of at least 50% by weight of vinyl alcohol and are referred to as vinal fibers. The latter are by far larger volume commercial products. Other fibers in this category are based on vinylidene chloride or tetrafluoroethylene (see VINYL POLYMERS–VINYL CHLORIDE POLYMERS; VINYLIDENE CHLORIDE MONOMER AND POLYMERS; FLUORINE COMPOUNDS, ORGANIC–POLYTETRAFLUOROETHYLENE).

Vinal fibers, or poly(vinyl alcohol) fibers, are not made in the United States, but the fiber is produced commercially in Japan, Korea, and China where the generic name vinylon is used. These materials are the subject of this article (see also VINYL POLYMERS–VINYL ALCOHOL POLYMERS).

Manufacture of Fiber

Raw Material. PVA is synthesized from acetylene or ethylene by reaction with acetic acid (and oxygen in the case of ethylene), in the presence of a catalyst such as zinc acetate, to form vinyl acetate which is then polymerized in methanol. The polymer obtained is subjected to methanolysis with sodium hydroxide, whereby PVA precipitates from the methanol solution.

Other Spinning Processes. Of scientific interest only are spinning processes using dimethyl sulfoxide (DMSO) solvent, and a solvent of ethylene glycol and other polyols, as well as dry-jet wet spinning (gel spinning).

Fiber Mechanical Properties

The mechanical properties of PVA fiber vary depending on the conditions of fiber manufacture such as spinning process, drawing process, and acetalization conditions, and the manufacture conditions of raw material PVA. PVA fibers are characterized by high strength, low elongation, and high modulus. In addition to general-purpose types, high strength types with strength of at least 1.47 N/tex (15 gf/dtex) are produced by alkali spinning; material with a yarn strength of nearly 1.77 N/tex (18 gf/dtex) has become available.

Physical and Chemical Properties

Moisture Absorbency. PVA fiber is more hygroscopic than any other synthetic fiber. The hygroscopicity varies depending on how the fiber is processed after spinning, ie, in heat-drawing, heat-treatment, acetalization, and the like.

Dimensional Stability. The wet heat resistance of PVA fiber is indicated by the wet softening temperature (WTS) at which the fiber shrinks to a specified ratio. At one time, the WTS was not more than 95°C for nonacetalized PVA fiber, but improvement of WTS has been achieved by improvement in heat-drawing and -treating techniques.

On the other hand, water-soluble PVA fibers are available on the market. They are stable in cool water but shrink in warm water and dissolve at 40–90°C.

PVA fiber is better in dimensional stability under dry heat than other synthetic fibers.

Thermal Resistance and Flammability. Thermal analysis of PVA filament yarn shows an endothermic curve that starts rising at around 220°C; the endothermic peak (melting point) is 240°C, varying a little depending on manufacture conditions.

With respect to flammability, PVA fiber has a limiting oxygen index (LOI) of 20, the same as that of polyester and polyamide fibers, and does not drip even when burnt, a feature with is highly valued. A flame-retardant-grade PVA fiber is available; it is obtained by mix-spinning with poly(vinyl chloride).

Weather Resistance. It has been shown for over 40 years that PVA fiber has excellent resistance against exposure to sunlight.

Chemical Resistance. PVA exhibits markedly high resistance to organic solvents, oils, salts, and alkali.

Application

PVA has found uses in a variety of industrial fields thanks to its superior mechanical properties of high strength and modulus and low elongation, resistance to chemicals, in particular to alkali, high durability against uv light, etc. It is used in the rubber industry, agricultural materials, fishing materials, sewing thread, nonwoven fabric, paper, and as a reinforcement.

JUN-ICHI HIKASA
Kuraray Co., Ltd.

Ger. Pat. 685,048 (1931); U.S. Pat. 2,072,302 (1932); Brit. Pat. 386161 (1932), W. O. Herrmann and W. Haehnel.

U.S. Pat. 4,440,711 (Apr. 3, 1984), Y. D. Quon, S. Kavesh, and D. S. Drevorsech (to Allied Corp.).

Jpn. Chem. Fiber Assoc. (Oct. 1975).

I. Sakurada, "Polyvinyl Alcohol Fibers," *Int. Fiber Sci. Tech. Ser.* **16**, (1985).

REGENERATED CELLULOSICS

Fibers manufactured from cellulose are either derivative or regenerated; historically they are designated man-made fibers and distinguished from synthetic fibers based on synthetic organic polymers. A derivative fiber is one formed when a chemical derivative of a natural polymer, eg, cellulose, is prepared, dissolved, and extruded as a continuous filament, and the chemical nature of the derivative is retained after the fiber formation process (see FIBERS–CELLULOSE ESTERS). A regenerated fiber is one formed when a natural polymer, or its chemical derivative, is dissolved and extruded as a continuous filament, and the chemical nature of the natural polymer is either retained or regenerated.

The Viscose Process

The main raw material required for the production of viscose is cellulose (qv), a natural polymer of D-glucose (Fig. 1). The repeating monomer unit is a pair of anhydroglucose units (AGU).

Cellulose is the most abundant polymer, an estimated 10^{11} t being produced annually by natural processes. Supplies for the rayon industry can be obtained from many sources, but in practice, the wood-pulping processes used to supply the needs of the paper and board

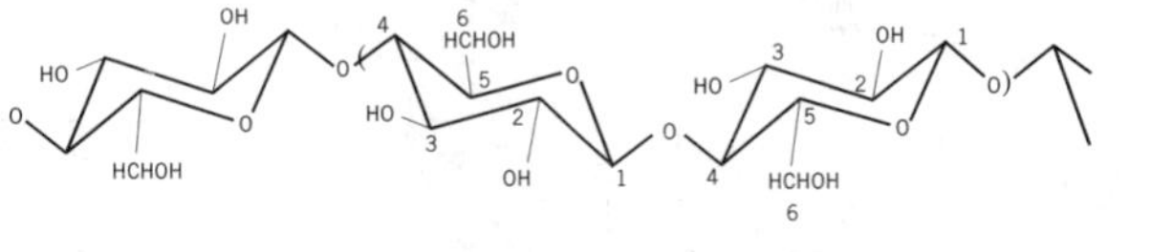

Figure 1. Anhydro-glucose units with 1–4 beta linkages as in cellulose.

industries have been adapted to make the necessary specially pure grade. The trees used to make dissolving pulp are fast-growing hard or soft woods farmed specifically for their high quality pulping.

The final properties of the rayon fiber and the efficiency of its manufacturing process depend crucially on the purity of the pulp used. High tenacity fibers need high purity pulps, and this means pulping to get up to 96% of the most desirable form of cellulose (known as alpha cellulose), and removing most of the unwanted hemicellulose (qv) and lignin. Cellulose molecular weight must be tightly controlled, and levels of foreign matter such as resin, knots, shives, and silica must be very low. Of the two main pulping processes, the sulfite route produces higher yields of lower alpha, more reactive pulp, suitable for regular staple fiber. Prehydrolyzed kraft pulps are preferred for high strength industrial yarn or modal fiber production. The highest purity pulps, up to 99% alpha cellulose, are obtained from cotton fiber. These are no longer used in viscose production but are now the main raw material for cuprammonium rayon.

The flow diagram for the viscose process is given in Figure 2. The sequence of reactions necessary to convert cellulose into its xanthate and dissolve it in soda used to be performed batchwise. Fully continuous processes, or mixtures of batch and continuous process stages, are more appropriate for high volume regular viscose staple production.

Modified Viscose Processes. The need for ever stronger yarns resulted in the first important theme of modified rayon development and culminated, technically if not commercially, in the 0.88 N/tex (10 gf/den) high wet modulus industrial yarn process.

Tire Yarns. A significant strength improvement in processing followed the 1950 Du Pont discovery of monoamine and quaternary ammonium modifiers, which, when added to the viscose, prolonged the life of the zinc cellulose xanthate gel, and enabled even higher stretch levels to be used. Modifiers have proliferated since they were first patented and the list now includes many poly(alkylene oxide) derivatives, polyhydroxypolyamines, and dithiocarbamates.

Fully modified yarns had smooth, all-skin cross sections, a structure made up of numerous small crystallites of cellulose, and filament strengths around 0.4 N/tex (4.5 gf/den). They were generally known as the Super tire yarns. Improved Super yarns (0.44–0.53 N/tex (5–6 gf/den)) were made by mixing modifiers, and one of the best combinations was found to be dimethylamine with poly(oxyethylene)glycol of about 1500 mol wt. Ethoxylated fatty acid amines have now largely replaced dimethylamine because they are easier to handle and cost less.

The strongest fibers were made using formaldehyde additions to the spin bath while using a mixed modified system or using highly xanthated viscoses (50% + CS_2). Unfortunately, problems associated with formaldehyde side reactions made the processes more expensive than first thought, and the inevitable brittleness which results whenever regenerated cellulose is highly oriented restricted the fibers to nontextile markets. The commercial operations were closed down in the late 1960s.

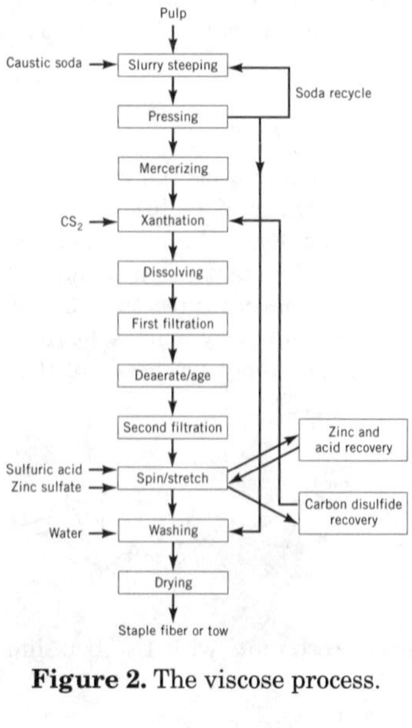

Figure 2. The viscose process.

The formaldehyde approach is still used by Futamura Chemical (Japan). It makes spun-laid viscose nonwovens where the hydroxymethylcellulose xanthate derivative formed from formaldehyde in the spin bath allows the fibers to bond after laying.

High Tenacity Staple Fibers. When stronger staple fibers became marketable, the tire yarn processes were adapted to suit the high productivity staple fiber processes. Improved staple fibers use a variant of the mixed modifier approach to reach 0.26 N/tex (3 gf/den).

The full potential of the mixed modifier tire yarn approach is achieved in the modal or HWM (high wet modulus) staple processes using special viscose-making and spinning systems. Lenzing makes HM333, and BASF (whose viscose operation in the United States has been bought by Lenzing) makes Zantrel. The fibers are most popular in ladies' apparel in the United States for their soft handle and easy dyeing to give rich coloration. They are now being made in finenesses down to 1 dtex (0.9 den) for fine yarns and hydroentangled nonwoven production.

Polynosic Rayons. Another strand of development began in 1952 when Tachikawa patented a method for making strong, high modulus fibers that needed neither zinc nor modifier. The process depends on the fact that minimally aged alk-cell can, after xanthation with an excess of carbon disulfide, be dissolved at low cellulose concentration to give very viscous viscoses containing high molecular weight cellulose. These viscoses have sufficient structure to be spun into cold, very dilute spin baths containing no zinc and low levels of sodium sulfate. The resulting gel-filaments can be stretched up to 300% to give strong, round section, highly ordered xanthate fibers which can then be regenerated.

Bulky Rayons. Permanent chemical crimp can be obtained by creating an asymmetric arrangement of the skin and the core parts of the fiber cross section. Skin cellulose is more highly ordered than core cellulose and shrinks more on drying. If, during filament formation in the spin bath, the skin can be forced to burst open to expose fresh viscose to the acid, a fiber with differing shrinkage potential from side-to-side is made, and crimp should be obtained. Crimp is most important in rayon used for hygienic absorbent products.

Process conditions that favor chemical crimp formation are similar to those used for improved tenacity staple (zinc/modifier route). However, spin bath temperature should be as high as possible (ca 60°C) and the spin-bath acid as low as possible (ca 7%).

Cross-sectional modifications of a more extreme nature than skin-bursting, which nevertheless do not form crimp, have grown in importance since the early 1980s. These yield a permanent bulk increase which can be translated into bulky fabrics without the need for special care.

Inflation had long been known as an intermittent problem of textile yarn manufacture. It was caused by the skin forming too quickly to allow the escape of gases liberated by the regeneration reactions, or as a result of air in the viscose at the jet. All early inflation processes were difficult to control, and after World War II they were neglected until the 1960s. However, their development led to an increased understanding of the inflation process and the identification of conditions which could yield a continuously hollow staple fiber in large-scale production, such as the development of solid Y-shaped and X-shaped multilimbed fibers. Their shape and relative stiffness enable them to absorb more fluid between, as opposed to inside, the fibers. They are therefore as absorbent in use as the inflated versions but do not require the extra process chemicals, and are easier to wash and dry in production and use. They are the most important bulky rayons now in production.

Alloy Rayons. It is possible to produce a wide variety of different effects by adding materials to the viscose dope. The resulting fibers become mixtures or alloys of cellulose and the other material. The two most important types of alloy arise when superabsorbent or flame retardant fibers are made.

American Enka and Avtex both produced superabsorbent alloy rayons by adding sodium polyacrylate, or copolymers of acrylic and methacrylic acids, or sodium carboxymethylcellulose to the viscose. Their use in tompons, the only real market which developed, declined after the Toxic Shock Syndrome outbreak in the early 1980s. Other polymers that have formed the basis of absorbent alloys are starch, sodium alginate, poly(ethylene oxide), poly(vinyl pyrrolidinone), and sodium poly(acrylamido-2-methyl-2-propane sulfonic acid).

Flame retardancy can be obtained by adding flame retardant chemicals to make up about 20% of the fiber weight. Propoxyphosphazine (Ethyl Corporation) retardants were use in Avtex's PFR fiber, and a bis(5,5-dimethyl-2-thiono-1,3,2-dioxaphosphorinanyl) oxide powder (Sandoz) was the basis of later European FR fiber developments. Alloys with inorganic salts such as silicates or aluminates are possible, the salts being converted to fibrous polyacids when the cellulose is burnt off. This latter approach is the basis of the Visil flame-retardant fiber introduced by Kemira Oy Saeteri.

Cuprammonium Rayon

Asahi Chemical Industries (ACI, Japan) is now the leading producer of cuprammonium rayon. Its continuing success with a process which has suffered intense competition from the cheaper viscose and synthetic fibers owes much to ACI's development of high speed spinning technology and efficient copper recovery systems.

Direct Dissolution Processes

The routes to regenerated cellulose fibers already described cope with the difficulties of making a good cellulose solution by going through an easy to dissolve derivative, eg, xanthate, or complex, eg, cuprammonium. The ideal process, one that could dissolve the cellulose directly from ground wood, is still some way off, but since the early 1980s significant progress has been made.

The Finnish viscose producer Kemira Oy Saeteri collaborated with Neste Oy on the development of a carbamate derivative route. This system is based on work that showed that the reaction between cellulose and urea gives a derivative easily dissolved in dilute sodium hydroxide:

$$\text{cell—OH} + NH_2\text{—}\overset{\overset{\displaystyle O}{\|}}{C}\text{—}NH_2 \longrightarrow \text{cell—O—}\overset{\overset{\displaystyle O}{\|}}{C}\text{—}NH_2 + NH_3$$

Neste patented an industrial route to a cellulose carbamate pulp which was stable enough to be shipped into rayon plants for dissolution as if it were xanthate. The carbamate solution could be spun into sulfuric acid or sodium carbonate solutions, to give fibers which when completely regenerated had similar properties to viscose rayon. When incompletely regenerated they were sufficiently self-bonding for use in papermaking. The process was said to be cheaper than the viscose route and to have a lower environmental impact. It has not been commercialized, so no confirmation of its potential is yet available.

Asahi has been applying the steam explosion treatment to dissolving-pulp to make it dissolve directly in sodium hydroxide, and they claim a solution of 5% of steam-exploded cellulose in 9.1% NaOH at 4°C being spun into 20% H_2SO_4 at 5°C. The apparently poor fiber properties (best results being 0.16 N/tex (1.8 gf/den) tenacity dry, with 7.3% extension) probably arise because the fibers were syringe extruded at 8.3 tex/filament (75 den/fil). Asahi feels that this could be the ultimate process for large-scale production of regenerated cellulose fibers.

The Courtaulds Tencel Process. The increasing costs of reducing the environmental impact of the viscose process coupled with the increasing likelihood that the newer cellulose solvents would be capable of yielding a commercially viable fiber process led Courtaulds Research to embark on a systematic search for a new fiber process in the late 1970s.

By 1980, NMMO was shown to be the best solvent, provided well-known difficulties associated with its thermal stability could be avoided by appropriate chemical engineering. Filaments obtained from the first single-hole extrusion experiments had promising properties so Courtaulds committed the resources in 1982 to build the first small pilot plant to test the feasibility of overcoming the solvent handling and recovery problems that had prevented earlier commercial exploitation. Scale-up to 1000 kg/week pilot line was possible in 1984, and in 1988 a 25,000 kg/week semicommercial line was commissioned to allow a thorough test of the engineering and end-use development aspects. By 1997 Courtaulds was projected to have two plants with a combined capacity of 73,000 t/yr.

Comparisons of Tencel with viscose in both laboratory and test markets proved that the fibers were sufficiently different to deserve separate marketing strategies. Tencel is stronger than any other cellulosic, especially when wet; easy to process into yarns and fabrics alone or in blends; easy to blend (unique fiber presentation); easy to spin to fine count yarns; very stable in washing and drying; thermally stable; easy to dye to deep vibrant colors; capable of taking the latest finishing techniques to give unique drape; and comfortable to wear.

The new fiber has physical properties (Table 1) sufficiently different from regular rayon to allow an initial market development strategy that does not erode the position of the traditional viscose fiber. The unique strength, texture, and coloration potential of the fiber enable it to command premium prices in up-market mens' and ladies' outerwear.

Fiber Properties

The bulk properties of regenerated cellulose are the properties of Cellulose II, which is created from Cellulose I by alkaline expansion of the crystal structure (see CELLULOSE). The key textile fiber properties for the most important current varieties of regenerated cellulose are shown in Table 1. Fiber densities vary between 1.53 and 1.50.

A discussion of the fiber properties is complicated by the versatility of cellulose and its conversion routes. Many of the properties can be varied over wide ranges depending on the objectives of the producer.

Thermal Properties. Fibers are not thermoplastic and stable to temperatures below 150°C, with the possible exception of slight yellowing. They ignite at 420°C and have a heat of combustion of 14,732 J/g (3.5 kcal/g)

Moisture Regain. The fibers are all highly hydrophilic with moisture regains at 65% rh ranging from 11 for the polynosics to 13 for regular rayon.

Chemical Properties. The fibers degrade hydrolytically when contacted with hot dilute or cold concentrated mineral acids. Alkalies cause swelling (maximum with 9% NaOH at 25°C) and ultimately disintegration. They are unaffected by most common organic solvents and dry-cleaning agents. They are degraded by strong bleaches such as hypochlorite or peroxide.

Optical Properties. The fibers are birefringent.

Electrical. The electrical properties of the fiber vary with moisture content. The specific resistance of the fibers is around 3×10^6 ohms/cm at 75% rh and 30°C. The dielectric constant (100 kHz) is 5.3 at 65% rh and 3.5 at 0% rh.

Environmental Issues

Rayon is unique among the mass-produced synthetic fibers because it is the only one to use a natural polymer (cellulose) directly. Cellulosic fibers therefore have much to recommend them provided that the processes used to make them have minimal environmental impact.

Liquid Effluents. Recycling of acid, soda, and zinc have long been necessary economically, and the acid–soda reaction produced, sodium sulfate, is extracted and sold into other sectors of the chemical industry.

Gaseous Effluents. Twenty percent of the carbon disulfide used in xanthation is converted into hydrogen sulfide (or equivalents) by the regeneration reactions. Ninety to 95% of this hydrogen sulfide is recoverable by scrubbers that yield sodium hydrogen sulfide for the tanning or pulp industries, or for conversion back to sulfur. Up to 60% of the carbon disulfide is recyclable by condensation from rich streams,

Table 1. Properties of Selected Commercial Rayon Fibers

Property	Cuprammonium	Regular rayon	Improved rayon	Modal	Polynosic	Y-Shaped rayon[a]	Solvent-spun rayon[b]
dry tenacity, cN/tex[c]	15–20	20–24	24–30	34–36	40–65	18–22	40–44
extensibility at break (dry), %	7–23	20–25	20–25	13–15	8–12	17–22	14–16
wet tenacity, cN/tex[c]	9–12	10–15	12–16	19–21	30–40	9–12	34–38
extensibility at break (wet), %	16–43	25–30	25–35	13–15	10–15	23–30	16–18
water imbibition, %	100	90–100	90–100	75–80	55–70	100–110	65–70
cellulose DP[d]	450–550	250–350	250–350	300–500	550–700	250–350	550–600
initial wet modulus[e]	30–50	40–50	40–50	100–120	140–180	35–45	250–270

[a] The Y-shaped rayon data are based on Courtaulds Galaxy fiber.
[b] The solvent-spun rayon data are based on Courtaulds Tencel fiber.
[c] To convert cN/tex to gf/den, divide by 8.82.
[d] DP = degree of polymerization.
[e] The load required to extend the wet fiber by 5% × 20.

but costly carbon-bed absorption from lean streams is necessary to recover the remaining 20 + %. The technology is becoming available to deal with this, but there remains the danger that cost increases resulting from the necessary investments will make the fibers unattractively expensive compared to synthetics based on cheap nonrenewable fossil fuels.

Energy Use. Energy consumption in the xanthate process compares favorably with the synthetics. The methodology of assessing the energy usage of products and processes is currently the subject of much debate, and a standardized approach has yet to emerge. Not surprisingly, most of the published work on fibers was carried out during the last energy crises in the 1973–1981 period.

Fiber Disposal. Cellulosic fibers, like the vegetation from which they arise can become food for microorganisms and higher life forms, ie, they biodegrade.

It is also possible to liberate and use some of the free solar energy that powered the manufacture of sugars and cellulose during photosynthesis. This can be achieved by burning or by anaerobic digestion. If future landfills are lined and operated with moisture addition and leachate recycling, then energy generation and the return of landfill sites to normal use can be accelerated.

Economic Aspects

Since the early 1980s, the viscose-based staple fibers have, like the cuprammonium and viscose filament yarns in the 1970s, ceased to be commodities. They have been repositioned from the low cost textile fibers that were used in a myriad of applications regardless of suitability, to premium priced fashion fibers delivering comfort, texture, and attractive colors in ways hard to achieve with the synthetics. They are still widely used in blends with polyester and cotton to add value, where in the 1980s they would have been added to reduce costs.

Future Possibilities

The cellulose polymer and its conversion routes have already proved to be capable of adaptation to meet a wide range of market demands. The advances being made in getting cellulose into solution with minimal environmental impact augur well for the development of streamlined routes from tree to fiber or fabric.

The progress on understanding the biogenesis of cellulose may yet yield an industrial route from atmospheric carbon dioxide and water, although the simple "plant more forests" option will be more attractive to many. Between these two extremes, in concept and in timing, the production of cellulose from sugar using bacteria, eg, *Acetobacter xylinum*, is becoming feasible. The costs of cellulose obtained in this way are currently very high compared with growing wood, but cost reduction proposals have bene made. These involve either identifying, culturing, or genetically engineering strains of bacteria that produce cellulose microfibers while metabolizing a wider range of foods more efficiently.

C. R. WOODINGS
Courtaulds

A. F. Turbak and co-workers, *A Critical Review of Cellulose Solvent Systems*, ACS Symposium Series 58, ACS, Washington, D.C., 1977.

J. Dyer and G. C. Daul, *The Handbook of Fibre Science and Technology*, Vol. 4, *Fibre Chemistry*, Marcel Dekker Inc., New York, 1985, Chapt. 11.

W. Albrecht and co-workers, in *Man Made Fibre Yearbook*, CTI, Maryland, 1991, pp. 26–44.

VEGETABLE

Vegetable fibers, as the name implies, are derived from plants. The principal chemical component in plants is cellulose, and therefore they are also referred to as cellulosic fibers. The fibers are usually bound by a natural phenolic polymer, lignin, which also is frequently present in the cell wall of the fiber; thus vegetable fibers are also often referred to as lignocellulosic fibers, except for cotton, which does not contain lignin.

Vegetable fibers are classified according to their source in plants as follows: *(1)* the bast or stem fibers, which form the fibrous bundles in the inner bark (phloem or bast) of the plant stems, are often referred to as soft fibers for textile use; *(2)* the leaf fibers, which run lengthwise through the leaves of monocotyledonous plants, are also referred to as hard fibers; and *(3)* the seed-hair fibers, the source of cotton (qv), are the most important vegetable fiber. There are over 250,000 species of higher plants; however, only a very limited number of species have been exploited for commercial uses (less than 0.1%). The commercially important fibers are given in Table 1.

World markets for vegetable fibers have been steadily declining in recent years, mainly as a result of substitution with synthetic materials.

General Properties

Chemical Composition. Chemically, cotton is the purest vegetable fiber, containing over 90% cellulose (qv) with little or no lignin. The other fibers contain 70–75% cellulose, depending on processing. Boiled and bleached flax and degummed ramie may contain over 95% cellulose. Kenaf and jute contain higher contents of lignin, which contributes to their stiffness. Although the cellulose contents are fairly uniform, the other components, eg, hemicelluloses (qv), pectins, extractives, and lignin (qv), vary widely without obvious pattern. These differences may characterize specific fibers.

Fiber Dimensions. Except for the seed-hair fibers, the vegetable fibers of bast or leaf origins are multicelled and are used as strands. However, for papermaking the strands are broken down to the ultimate cells.

Physical Properties. Bast and leaf fibers are stronger (higher tensile strength and modulus of elasticity) but lower in elongation (extensibility) than cotton. Vegetable fibers are stiffer but less tough than

Table 1. Vegetable Fibers of Commercial Interest

Commercial name	Source	Botanical name of plant	Growing area
		Bast or soft fibers	
China jute	Abutilon	*Abutilon theophrasti*	China
flax		*Linum usitatissimum*	north and south temperate zones
hemp		*Cannabis sativa*	all temperate zones
jute		*Corchorus capsularis; C. olitorius*	India
kenaf		*Hibiscus cannabinus*	India, Iran, CIS, South America
ramie		*Boehmeria nivea*	China, Japan, United States
roselle		*Hibiscus sabdariffa*	Brazil, Indonesia (Java)
sunn		*Crotalaria juncea*	India
urena	cadillo	*Urena lobata*	Zaire, Brazil
		Leaf or hard fibers	
abaca		*Musa textilis*	Borneo, Philippines, Sumatra
cantala	Manila maguey	*Agave cantala*	Philippines, Indonesia
caroa		*Neoglaziovia variegata*	Brazil
henequen		*Agave fourcroydes*	Australia, Cuba, Mexico
istle		*Agave* (various species)	Mexico
mauritius		*Furcraea gigantea*	Brazil, Mauritius, Venezuela, tropics
phormium		*Phormium tenax*	Argentina, Chile, New Zealand
pineapple	piña	*Ananas comasus*	Hawaii, Philippines, Indonesia, India, West Indies
sansevieria	bowstring hemp	*Sansevieria* (entire genus)	Africa, Asia, South America
sisal		*Agave sisalana*	Haiti, Java, Mexico, South Africa
		Seed-hair fibers	
coir	coconut husk fiber	*Cocos nucifera*	tropics, India, Mexico
cotton		*Gossypium* sp.	United States, Asia, Africa
kapok		*Ceiba pentandra*	tropics
milkweed floss		*Chorisia* sp.	North America
		Other fibers	
broom root	roots	*Muhlenbergia macroura*	Mexico
broom corn	flower head	*Sorghum vulgare technicum*	United States
crin vegetal	palm leaf segments	*Chamaerops humilis*	North Africa
palmyra palm	palm leaf stem	*Brossus flabellifera*	India
pissava	palm leaf base fibers	*Attalea funifera*	Brazil
raffia	palm leaf segments	*Raphia raffia*	East Africa

synthetic fibers. Kapok and coir are relatively low in strength; kapok is known for its buoyancy.

Among the bast textile fibers, the density is close to 1.5 g/cm^3, or that of cellulose itself, and they are denser than polyester. Moisture regain (absorbency) is highest in jute at 14%, whereas that of polyester is below 1%. The bast fibers are typically low in elongation and recovery from stretch. Ramie fiber has a particularly high fiber length/width ratio.

The microfibrils in vegetable fibers are spiral and parallel to one another in the cell wall. The spiral angles in flax, hemp, ramie, and other bast fibers are lower than cotton, which accounts for the low extensibility of bast fibers.

Processing and Fiber Characteristics

Bast Fibers. Bast fibers occur in the phloem or bark of certain plants. The bast fibers are in the form of bundles or strands that act as reinforcing elements and help the plant to remain erect. The plants are harvested and the strands of bast fibers are released from the rest of the tissue by retting, common for isolation of most bast fibers. Retting involves the biological (bacterial) breakdown of pectins in the plant through immersion of the stalks in ponds or streams for periods of time sufficient to release the desired outer bast fibers from the woody inner core. The retted material is then further processed by breaking, scutching, and hackling.

Leaf (Hard) Fibers. Hard or cordage fibers are found in the fibrovascular systems of the leaves of perennial, monocotyledonous plants growing in Central America, East Africa, Indonesia, Mexico, and the Philippines. They are generally of the *Agave* and *Musa* genera. The leaf elements are harvested by cutting at the base with a sickle-like tool, and bundled for processing by hand or by machine decortication. In the latter case, the leaves are crushed, scraped, and washed. The fibers are generally coarser than the bast fibers and are graded for export according to national rules for fineness, luster, cleanliness, color, and strength.

Seed- and Fruit-Hair Fibers. The seeds and fruits of plants are often attached to hairs or fibers or encased in a husk that may be fibrous. These fibers are cellulosic based and of commercial importance, especially cotton (qv), the most important natural textile fiber.

Economic Aspects

The principal bast and leaf fibers are produced in yields of 2–5%, with some exceptions such as flax (15%) and kapok (17%), on a green plant basis. Vegetable fiber production on the world market has dropped 25–35% since 1970 because of periods of economic recession and synthetic fiber replacements. Imports of vegetable fibers have dropped 70–90% since 1970 (with the exception of flax wastes). These market trends reflect the recessions and substitution with synthetic fibers. Although most vegetable fibers are converted to lower cost commodity products, some of the fibers are converted into the most expensive products in their respective industries, eg, U.S. currency (paper). The value of most of the fibers has dropped since 1981.

Uses

Vegetable fibers have application in a broad range of fibrous products, including textiles and woven goods, cordage and twines, stuffing and upholstery materials, brushes, and paper. The traditional uses for the vegetable fibers have been eroded by substitution with synthetics on the world market. The declining uses include cordage, mats, filling material, brushes, etc. However, the unique properties of the bast fibers have allowed continued use in such specialty papers as bank notes, some writing papers, and cigarette papers.

Recent work by the USDA and Kenaf International (Texas) has demonstrated the potential of both growing and processing kenaf fibers for newsprint and other paper products in the United States. Another promising potential use for vegetable fibers is in the new lignocellulosic-based composites under development in various parts of the industrialized world. Such products are already utilized in the automotive industry for automobile interior door and head liners and as trunk liners.

RAYMOND A. YOUNG
University of Wisconsin, Madison

L. Rebenfeld, in J. I. Kroschwitz, ed., *Encyclopedia of Polymer Science and Engineering*, Vol. 6, 2nd ed., Wiley-Interscience, New York, 1986, p. 647.

W. E. Morton and J. W. S. Hearle, *Physical Properties of Textile Fibers*, 2nd ed., John Wiley & Sons, Inc., New York, 1975, p. 284.

E. E. Nelson, in S. P. Parker, ed., *Encyclopedia of Science and Technology*, Vol. 9, McGrawHill, Inc., New York, 1982, pp. 8–12.

J. G. Cook, *Handbook of Textile Fibers I. Natural Fibers*, 5th ed., Merrow Publishing, Durham, UK, 1984.

FILLERS

A filler is a finely divided solid added to a liquid, semisolid, or solid composition, eg, paint, paper, plastics, or elastomers, to modify the composition's properties and reduce its costs. Fillers can constitute either a major or a minor part of a composition. The structure of filler particles ranges from precise geometrical forms, such as spheres, hexagonal plates, or short fibers, to irregular masses. Fillers are generally used for nondecorative purposes, although they may incidently impart color or opacity to a composition. Lower cost materials possessing modest performance-enhancing properties are discussed here, in contrast to functional fillers or engineered materials designed to achieve very specific results.

Fillers can be classified according to their source, function, composition, or morphology. None of these classification schemes is entirely adequate due to overlap and ambiguity of the categories. Morphological distinctions, used here for the discussion of general filler properties, are either crystalline, eg, fibers, platelets, polyhedrons, and irregular masses; or amorphous, eg, fibers, flakes, solid spheres, hollow spheres, and irregular masses. The compositional scheme used for the compilation of data on specific fillers classifies fibers as either inorganic, eg, carbonates, hydroxides, metals, oxides, silicates, sulfates, and sulfides; or organic, eg, carbon, celluloses, lignins, polymers, proteins, and starches.

Properties

The properties of fillers which influence a given end use are many. The overall value of a filler is a complex function of intrinsic material characteristics, eg, true density, melting point, crystal habit, and chemical composition; and of process-dependent factors, eg, particle-size distribution, surface chemistry, purity, and bulk density. Fillers impart performance or economic value to the compositions of which they are part. These values, often called functional properties, vary according to the nature of the application. A quantification of the functional properties per unit cost in many cases provides a valid criterion for filler comparison and selection. The following are summaries of key filler properties and values: particle morphology, size, and distribution, surface area, surface energy or wettability, acid–base behavior, loading and packing, true density or specific gravity, bulk density, color and brightness, refractive index, free moisture, thermal stability and expansion, and hardness.

Filled Polymer Systems

Fillers play an important role as reinforcement for elastomers. They are used extensively in all subclasses of plastics, ie, general-purpose, specialty, and engineering plastics (qv). Fillers are not, however, a significant factor in fibers (qv).

Elastomers. In the rubber industry the terms filler, reinforcement, and pigment have been used for the same material, derived from different sources. Most rubber technologists distinguish inert fillers from reinforcing fillers. Inert fillers, such as clay, improve the workability of the unvulcanized rubber stock but have little effect on the final properties; reinforcing fillers improve the mechanical properties of the vulcanized rubber. In practice, these materials impart abrasion resistance, tear resistance, tensile strength, and stiffness. The reinforcing action of a given material is primarily dependent on its chemical composition and the type of elastomer in which it is compounded. Table 1 lists typical elastomer fillers and their uses.

Table 1. Elastomer Fillers

Filler	Specific gravity	Compatible elastomer[a]	Uses
alumina	2.7	NR, CR, SRs	hose, mats
asbestos	2.4	NR, SRs	mats, tile
barium sulfate	4.3	NR, CR, SRs	O-rings, belts
carbon blacks[b]	1–2.3		
N110		IR, NR	tires, pads
N220		NR, SBR	tire treads
N550		NR, SRs	extruded goods
N660		BR, NR	tubes
N762		NR, SRs	footwear
N774		NR, SRs	tire carcasses
N990		CR, EPDM	extruded goods
calcium carbonate	2.7–2.9	NR, SRs	footwear, mats
kaolin clay	2.6	NR, SBR, EPM, EPDM	flooring, footwear
mica	2.8	NR, SRs	molded goods
resins	1.2	NBR, CR, NR, SBR	footwear, coatings
silicas			
colloidal	1.3	NR, SBR	sponge
diatomaceous	2.2	NR, IIR	carpet backing
novaculite	2.7	NR, SRs	molded goods
wet process	2	IIR, CR, NBR, NR	hygienic goods
pyrogenic	2.2	silicone rubber	electrical goods
surface treated		NR, SRs	specialty goods
talc	2.8	NR, CR, IIR, EPM	molded goods
natural materials[c]	1.1	NR, SBR, CR	tape, extruded goods
wood and shell flour	0.9–1.6	CR, NR, SRs	footwear

[a] NR, natural rubber; CR, chloroprene; SRs, synthetic rubbers; IR, natural isoprene; SBR, styrene–butadiene rubber; BR, butadiene; EPDM, ethylene–propylene–diene; EPM, ethylene–propylene polymer; IIR, isobutylene–isoprene; NBR, nitrile–butadiene.
[b] Carbon black grades identified by four characters (ASTM D1765-67), ie, cure rate of normal (N) or slow (S), digit classifying typical particle size in nm, and two arbitrarily assigned characters.
[c] For example, rice.

Rubber technologists manipulate the formula so as to keep costs down and to optimize a large number of properties, eg, recovery, rebound, or nerve, retardation (scorch resistance), abrasion resistance, elongation, hardness, modulus of elasticity (stiffness), permanent set, resilience, tear resistance, and tensile strength.

Plastics. In the plastics industry, the term filler refers to particulate materials that are added to plastic resins in relatively large, ie, over 5%, volume loadings. Except in certain specialty or engineering plastics applications, plastics compounders tend to formulate with the objective of optimizing properties at minimum cost rather than maximizing properties at optimum cost. Table 2 lists typical plastic fillers and their uses. Properties affected by fillers include resin viscosity, resin curing, tensile strength, compressive strength, fire resistance, and electrical resistance and conductivity.

Other Filler Systems

Paper. Paper is prepared by depositing cellulose pulp fibers on a continuous wire screen from a dilute suspension in water (see PAPER). Fillers, or loading materials, are finely divided solids that are incorporated into the paper sheet structure by adding them to the pulp slurry prior to its deposition on the wire. Finely divided solids dispersed in water containing an adhesive and then coated on the paper after it has been formed are usually termed coating pigments. Here the term filler is used for both loading materials and coating pigments. Most pa-

Table 2. Plastic Fillers and Their Primary Functions

Filler	Specific gravity	Compatible resins[a]	Primary function
alumina trihydrate	2.42	varied	flame retardance
carbon blacks	1–2.3	varied	optical, electrical, mechanical
calcium carbonate			
mineral	2.60–2.75	PVC, HDPE	multiple, cost
synthetic	2.7	rigid PVC, PP	impact, weathering
kaolin	2.58–2.63	varied	bloom prevention, asbestos
feldspar	2.61	ABS, EVA, SMC	translucency
organics			
wood flour	0.65	varied	reinforcement
starch	1.5	varied	biodegradability
silicas, synthetic			
fumed silica	2.2	FRP, PVC, epoxy	rheology
silica gel		PVC, LDPE	gloss reduction
precipitated silica	1.9–2.2	PVC, PE, EVA	thixotropy
fused silica	2.18	general	electrical
silicas, natural			
crystalline silicas	2.65	general	mechanical, cost
diatomaceous silica	2.65	LDPE films	antiblocking
sphericals			
hollow glass	0.15–0.30	PVC, SMC, BMC	weight reduction
fly ash	0.30–1.0	varied	cost reduction
solid glass	2.5	varied	mechanical
talcs	2.7–2.8	PP, HDPE, TPE, PVC	reinforcement

[a] HDPE, high density polyethylene; PP, polypropylene; EVA, ethylene–vinyl alcohol; SMC, sheet-molding compound; FRP, fiber-reinforced plastic; LDPE, low density polyethylene; PE, polyethylene; BMC, bulk molding compound; TPE, thermoplastic elastomer.

per contains 1–40 wt % of fillers (ca 1993). The optical and mechanical properties of filled paper are superior to those of unfilled paper.

Paint. The liquid phase of paint formulations, usually termed the vehicle, contains volatile and nonvolatile fractions. When paint is applied, the volatile fraction of the vehicle evaporates and the nonvolatile fraction polymerizes to become a film matrix in which prime pigments and fillers are embedded (see PAINT). Prime pigments are coloring and opacifying agents. Fillers in paint are variously referred to as inerts, extender, or supplemental pigments. The primary paint fillers are TiO_2, kaolin clays, calcium carbonate, silica, talc, and zinc oxide. The move toward lower volatile organic carbon (VOC) is changing the nature of the volatile fraction in paints and accelerating the search for new or better fillers. In particular, new viscosity, compatibility, gloss, and applications problems are driving innovations in the market. Properties affected by paint fillers include gloss, hiding power, mechanical properties, and weather resistance.

Economic Aspects

All markets, except paper, have been adversely affected by the downturn in the automotive and construction industries in the early 1990s. Most large-volume fillers are sufficiently diversified so that their growth trends follow GNP. There are some exceptions.

Health and Safety Factors

The principal hazard involved in the handling and use of many fillers is inhalation of airborne particles (dusts) in the respirable size range, ie, 10 μm and below. Filler dusts may be classified as nuisance particulates, fibrogens, and carcinogens.

The American Conference of Governmental and Industrial Hygienists (ACGIH) establishes TLVs for the airborne concentration of many fillers in workroom air. In addition, concern for the toxicity of many metals and their compounds is limiting the use of many fillers, eg, Pb, Co, Cr, and Ba compounds, and possibly the use of certain organometallic surface coatings. Suppliers have information on proper usage and handling of their products. The use of NIOSH–OSHA-approved dust masks or respirators is required when dust concentrations exceed permissable exposure limits.

JAMES S. FALCONE, JR.
West Chester University

W. C. McCrone and co-workers, *Particle Atlas*, Vols. 1–6, Ann Arbor Science, Ann Arbor, Mich. 1973–1978.

T. H. Ferrigno in H. S. Katz and J. V. Milewski, eds., *Handbook of Fillers for Plastics*, Van Nostrand Reinhold, New York, 1987, pp. 8–63.

F. W. Billmeyer, Jr. *Textbook of Polymer Science*, 4th ed., Wiley–Interscience, New York, 1984, p. 349.

FILM AND SHEETING MATERIALS

Film and sheet are defined as flat unsupported sections of a plastic resin whose thickness is very thin in relation to its width and length. Films are generally regarded as being 0.25 mm or less, whereas sheet may range from this thickness to several centimeters thick. Film and sheet may be used alone in their unsupported state or may be combined through lamination, coextrusion, or coating. They may also be used in combination with other materials such as paper, foil, or fabrics.

Film or sheet generally function as supports for other materials, as barriers or covers such as packaging, as insulation, or as materials of construction. The uses depend on the unique combination of properties of the specific resins or plastic materials chosen. When multilayer films or sheets are made, the product properties can be varied to meet almost any need. Further modification of properties can be achieved by use of such additives or modifiers as plasticizers (qv), antistatic agents (qv), fire retardants, slip agents, uv and thermal stabilizers, dyes or pigments (qv), and biodegradable activators.

Properties and Test Methods

Film and sheet materials have an amazing range of properties so that a product may generally be formed or produced to meet the needs of a specific end use. Films as thin as 1.5 μm are produced for capacitor insulation, whereas cast sheet products ranging up to 5.7 cm are used for the construction industry. Films are generally wound on spools or in rolls in widths from 3 mm to several meters. Sheeting materials may be wound in rolls, but more likely are cut and shipped as flat sheet 1–3 meters in width and length. The products may be very stiff and have high tensile strengths, or may be rubbery or flimsy. They may be impermeable to water or gases, or may dissolve or be porous. Some may be crystal clear or opaque, colorless to brilliantly hued. Films may degrade quickly or last indefinitely as the base material for information storage. They may be excellent insulators for use as protective materials, or may be compounded to conduct small currents. Films may be inert to chemical attack or be made easily printable, even degradable. Sheets may be intractable or readily thermoformed into complex shapes. Some films may dissolve in water, whereas others may act as barriers to moisture permeation almost as effectively as metal.

Film and sheeting materials test methods have been standardized by ASTM, DIN, and others.

Tensile properties of importance include the modulus, yields, F^5 (strength at 5% elongation), and ultimate break strength. Tear strength is critical for packaging and other film end uses.

The slip characteristics of film and sheeting are also critical to processing and use. It may influence the hand, but is particularly critical to the free passage of film over rolls and through equipment.

Transparency, color, haze, and gloss are all important elements of film and sheet used for optical end uses. These include packaging, reprographic and photographic uses, glazing, solar, etc.

Moisture and gas barrier properties are of prime interest in packaging applications. These are measured in standard tests or functional tests, which in recent years have been improved to give more meaningful predictive information.

Other important properties that can be measured in the laboratory include sealability, printability, or coating adhesion.

Dimensional stability of films or sheet when exposed to temperature or humidity are important.

Thermoformability is a property required by the many sheet materials used in the thermoforming industry. These properties are unique for the specific forming methods used, and are best determined by actual thermoforming tests on small-scale equipment.

An important property of all film and sheet products is the gauge (or thickness) uniformity. Machine direction uniformity is vital to consistent processibility in web-handling equipment and to controlled economic production. More important, perhaps, is uniformity of transverse gauge. Irregularities in TD gauge, particularly consistent ones, lead to gauge bands, soft spots, honeycomb or chain-like defects, which in turn lead to poor handleability, bad coating or printing, improper tracking, and totally unusable product.

Performance measures such as slittability (ease of slitting the film web into smaller widths), or cuttability (ease of trimming, stamping, or cutting the sheet by shearing) can be determined usually by actual performance testing on processing machines.

Because of the nature of film and sheet products, ie, large area-to-volume ratio, optical defects in the polymer may be readily evident and unacceptable. Thus the presence of degraded polymer, gels, fisheyes, contamination, or improperly dispersed additives, pigments, or colorants may result in a product that is aesthetically displeasing or functionally unacceptable. In addition, such contaminants may cause mechanical or electrical failures under stress. Visual inspection, for optical defects on the moving web, or microscopic inspection are used to measure and define these problems.

Table 1 lists some typical properties or ranges or properties for the more common film and sheet products.

Table 1. Physical and Mechanical Properties[a]

Material	Tensile strength, MPa[b]	Elongation, %	Impact strength, (kN·m)/m[c]	Tear strength, N/mm[d]	Burst strength (Mullen)
ABS	50	25			
cellophane	80	30	31–58	0.8–8	30–50
CA	75	15–55	10.8	1.6–3.9	30–60
CTA	86	10–50		1.6–11.8	50–70
fluorocarbons					
ETFE	52	300		235–350	
FEP	20	300		49	10
PCTFE	52	50–150		1–15.7	23–30
PTFE	20	100–400		3.9–39	
PVF	86	115–250		3.9–39	19–70
ionomer	34	250–400		11.8–49	
nylon-6	225	85–120		6.3–11	
PC	67	40–100		8–10	
PET	210	60–165		20–118	55–80
PI	172	70		3	75
PE					
LDPE	15	200–600	27–42	20–118	10–12
LLDPE	38	400–800	31–50		
MDPE	24	200–500	15–23	20–118	
HDPE	38	10–50	4–12	20–118	
UHMWPE	29	300			
PMMA	59	4–12			
PP[e]	50–275	35–500	19–58	1.2–3.9	
PS[f]	70	3–60		0.8–5.9	16–35
PVC					
rigid	58	25–50		3.9–275	30–40
plasticized	52	3–100		2–275	20

[a] See Table 1 for ASTM test methods.
[b] To convert MPa to psi, multiply by 145.
[c] To convert (kN·m)/m to (kgf·cm)/mil, divide by 3.861.
[d] To convert N/mm to gf/25 μm, multiply by 2.549; to convert to ppi, divide by 0.175.
[e] Biaxial orientation.
[f] Oriented.

Materials

Materials include acrylonitrile–butadiene–styrene, acrylic polymers, cellophane, cellulosics, fluoropolymers, ionomer, polyamide, polycarbonate, polyester, polyimide, polyethylene, polypropylene, polystyrene, vinyl films, polyurethane, specialty films (polysulfones (PSO), poly(phenyl sulfide) (PPS), liquid crystal polyesters (LCP), poly(acrylates), polyetherketones, acetal resins, poly(vinyl butyral) film), and water-soluble films (poly(vinyl alcohol) (PVOH), methylcellulose, poly(ethylene oxide), or starch (qv)).

Manufacture

The processes used commercially for the manufacture of film and sheeting materials are generally similar in basic concept, but variations in equipment or process conditions are used to optimize output for each type of film or sheeting material. The nature of the polymer to be used, its formulation with plasticizers (qv), fillers (qv), flow modifiers, stabilizers, and other modifiers, as well as its molecular weight and distribution are all critical to the processibility and final properties of the product. Most polymers are amenable to one or another of the common manufacturing processes, but some are so intractable they can only be made into film by skiving from large casting of the resin.

The basic methods for forming film or sheeting materials may be classified as follows: melt extrusion, the most important method; calendering; solution casting; and chemical regeneration. Of special note is the use of biaxial orientation as part of the critical manufacturing steps for many film and sheet products.

Economic Aspects

Low density polyethylene films dominate the market in volume, followed by polystyrene and the vinyls. High density polyethylene, poly(ethylene terephthalate), and polypropylene are close in market share and complete the primary products. A number of specialty resins are used to produce 25,000–100,000 t of film or sheet, and then there are a large number of high priced, high performance materials that serve niche markets.

Packaging (qv) represents the largest market area for film and sheeting materials. The largest nonpackaging markets for film and sheet include agricultural (mulching) and construction film, trash liners and bags, pressure-sensitive tape base, insulation board, and diaper liners. These are relatively low cost, high volume uses. There are a large number of premium end uses for film and sheet that demand high performance, long life, or high purity. These uses are filled by higher value films such as polyester, BON, BOPP, polycarbonates, acrylics, fluorocarbons, polyurethanes, polysulfone, polyimides, and many new products. These markets include magnetic tape base, photographic film base, microfilm, drafting and layout bases, color separation films, cartoon cells, imaging films, electrical and electronic insulation or support films, capacitor dielectric, permanent labels, solar control window films, laminates for safety glazing, and high impact glazing.

Sheeting materials are utilized in a wide variety of applications. About 30% of the sheet market is foamed sheet, mostly all polystyrene, used for egg trays, food and drink containers, and insulation. About 10% of the market for solid sheeting is used for glazing, lighting, and outdoor signs. This market has seen growth with the need for high security, impact-resistant glazing for banks, storefronts, ice rinks, and transportation. Polycarbonate, acrylics, and cellulosics dominate these markets. Thermoformed applications account for about 40% of the use for sheet. Vacuum-formed or pressure-formed products include blister and skin packaging, food and drink containers (cups, tubs, trays, and bowls for single use), toys, auto and appliance parts, and luggage. Polystyrene, polypropylene, HDPE, thermoplastic polyester, ABC, and vinyls are used in these thermoforming markets.

K. J. MACKENZIE
Consultant

Chemical Economics Handbook, SRI International, Menlo Park, Calif.

Modern Plastics, McGraw-Hill Book Co., Inc., New York. Particularly the annual market review in the January issues.

Packaging, Cahners Publishing, Newton, Mass. Publishes annual encyclopedia issue.

FILTRATION

Filtration is the separation of two phases, particulate form, ie, solid particles or liquid droplets, and continuous, ie, liquid or gas, from a mixture by passing the mixture through a porous medium. This article discusses the more predominant separation of solids from liquids.

Filtration is often referred to as mechanical separation because the separation is accomplished by physical means. This does not preclude chemical or thermal pretreatment used to enhance filtration. Although some slurries separate well without chemical conditioning, most pulps of a widely varying nature can benefit from pretreatment (see FLOCCULATING AGENTS).

In a filtration system (Fig. 1) the porous filtration medium is housed in a housing, with flow of liquid in and out. A driving force, usually in the form of a static pressure difference, must be applied to achieve flow through the filter medium. It is immaterial from the fundamental point of view how the pressure difference is generated but there are four main types of driving force, ie, gravity, vacuum, pressure, and centrifugal. The two-types of filtration used most often in practice are cake, or surface, filtration (Fig. 2) and deep bed filtration (Fig. 3). This division usually is unambiguous but in some cases, such as cartridge filters, there is no sharp dividing line.

Separation Efficiency. Similarly to other unit operations in chemical engineering, filtration is never complete. Some solids may leave in the liquid stream, and some liquid will be entrained with the separated solids. As emphasis on the separation efficiency of solids or liquid varies with application, the two are usually measured separately. Separation of solids is measured by total or fractional recovery, ie, how much of the incoming solids is collected by the filter. Separation of liquid usually is measured in how much of it has been left in the filtration cake for a surface filter, ie, moisture content, or in the concentrated slurry for a filter-thickener, ie, solids concentration.

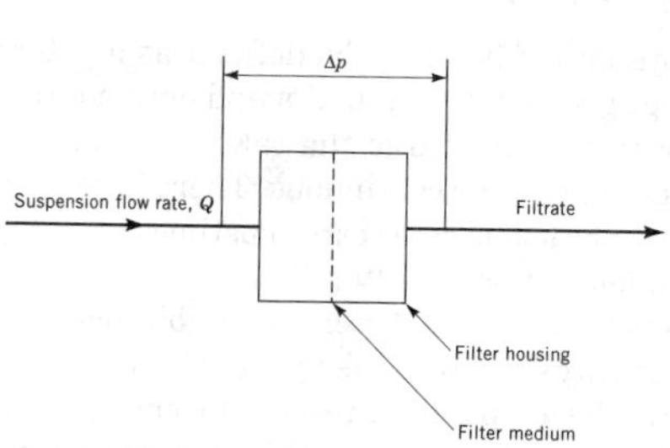

Figure 1. Schematic diagram of a filter.

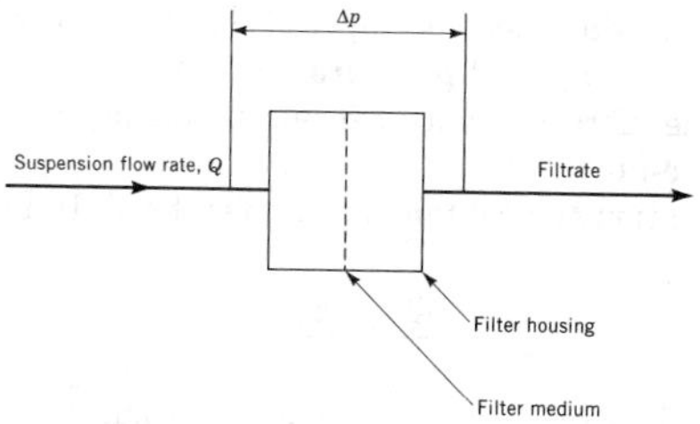

Figure 2. Schematic diagram of a surface filter, ie, the cake filtration mechanism.

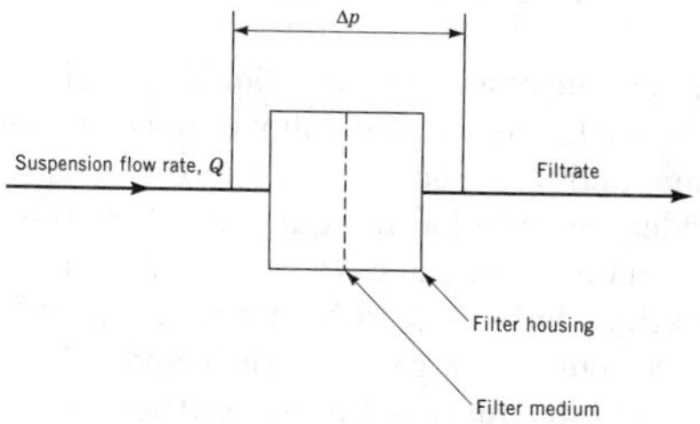

Figure 3. Mechanism of deep bed filtration.

Filtration-Related Processes

Several processes are used to enhance the filtration process itself. They may also be related processes in their own right. They include washing of solids, cake dewatering, pretreatment of suspensions (addition of inert filter aids), mechanical squeezing of cakes, electro-kinetic effects (Table 1), and magnetic separation.

Cake Filtration Theory

It does not matter, from the fundamental point of view, how the pressure drop is generated in the filter. In the case of the centrifugal filters there is an additional phenomenon of the mass forces acting on the liquid within the cake. The conventional filtration theory must be amended to include this effect.

Carman-Kozeny Equation. Flow through packed beds under laminar conditions can be described by the Carman-Kozeny equation in the form

$$\frac{Q}{A} = \frac{\Delta p}{\mu L}\frac{\epsilon^3}{5(1-\epsilon)^2 S_0^2} \tag{1}$$

where Q is volumetric flow rate, A is face area of the bed, L is depth of the bed, Δp is applied pressure drop, ϵ is voidage of the bed (porosity), S_0 is volume specific surface of the bed, and μ is liquid viscosity.

Darcy's Law and the Basic Filtration Equation. Darcy's law combines the constants in the last term of equation 1 into one factor K known as the permeability of the bed, ie,

$$K = \frac{\epsilon^3}{5(1-\epsilon)^2 S_0^2} \tag{2}$$

Table 1. Summary of Electro-Kinetic Effects

Effect	Fluid	Particles	Electric field
electrophoresis	still	moving	applied
sedimentation (or migration) potential	still	moving	measured
electroosmosis	moving	still	applied
streaming potential	moving	still	measured
electroosmotic pressure	still	still	applied

where K is a constant for incompressible solids. For compressible cakes, K depends on applied pressure, approach velocity, and concentration, and therefore presents serious problems in cake filtration testing and scale-up.

Modern filtration theory tends to prefer the Ruth form of Darcy's law, ie,

$$\frac{Q}{A} = \frac{\Delta p}{\mu R} \tag{3}$$

where R is the bed resistance. In cake filtration the bed resistance consists of the medium resistance in series with the resistance of the deposited cake, assuming no penetration of solids into the filtration medium; the general filtration equation is then written as

$$\frac{Q}{A} = \frac{\Delta p}{\mu R + \sigma \mu c V/A} \tag{4}$$

where σ is specific cake resistance, μ is liquid viscosity, c is solids concentration in the feed, V is filtrate volume collected since commencement of filtration, and R is medium resistance.

There is hidden assumption in equation 4 that the volume of the solids and liquid retained in the cake is negligible. This is reasonable at low concentrations but can lead to errors at high solids concentration and moisture contents of cakes. A corrected value of the concentration c can be used in equation 4 to reduce the errors, ie,

$$c(\text{corrected}) = \frac{1}{(1/c) - (1/\rho_s) - [(m-1)/\rho]} \tag{5}$$

where m is the mass ratio of wet to dry filter cake, ρ_s is solids density, and ρ is liquid density. This correction is necessary only at feed concentrations of greater than 200 g/L.

The scale-up of conventional cake filtration uses the basic filtration equation (eq. 4).

Benefits of Prethickening. The feed solids concentration has a profound effect on the performance of any cake filtration equipment. It affects the capacity and the cake resistance, as well as the penetration of the solids into the cloth which influences filtrate clarity and medium resistance. Thicker feeds lead to improved performance of most filters through higher capacity and lower cake resistance.

Prethickening of filter feeds can be done with a variety of equipment such as gravity thickeners, hydrocyclones, or sedimenting centrifuges. Even cake filters can be designed to limit or completely eliminate cake formation and therefore act as thickening filters and be used in this thickening duty.

Pressure Filtration. High pressure drops have a twofold effect, ie, on capacity and on displacement dewatering which often follows.

The most important feature of the pressure filters which use hydraulic pressure to drive the process is that they can generate a pressure drop across the medium of more than 1×10^5 Pa which is the theoretical limit of vacuum filters. While the use of a high pressure drop is often advantageous, leading to higher outputs, drier cakes, or greater clarity of the overflow, this is not necessarily the case. For compressible cakes, an increase in pressure drop leads to a decrease in permeability of the cake and hence to a lower filtration rate relative to a given pressure drop.

Optimization of Cycle Times. In batch filters, one of the important decisions is how much time is allocated to the different operations such as filtration, displacement dewatering, cake washing, and cake discharge, which may involve opening of the pressure vessel. All of this has to happen within a cycle time t_c which itself is not fixed, though some of the times involved may be defined, such as the cake discharge time.

If all of the nonfiltration operations are grouped together into a downtime, t_d, assumed to be fixed and known, an optimum filtration time t_{opt} in relation to t_d can be derived by optimizing the average dry cake production obtained from the cycle. For constant pressure filtration and where the medium resistance R and the specific cake resistance σ are constant, the following equation applies:

$$t_{\text{opt}} = t_d \left[1 + \sqrt{\frac{2\mu R}{\sigma c \Delta p t_d}} \right] \tag{6}$$

where Δp and c are the operating pressure drop and the feed solids concentration, respectively.

When the medium resistance R is small compared with the specific cake resistance σ, the second term in the above equation becomes negligible and the optimum filtration time t_{opt} becomes equal to downtime t_d. For any other case, t_{opt} is always greater than t_d. It follows, therefore, that the filtration time should be at least equal to the sum of the other nonfiltration periods involved in the cycle.

Vacuum Filters

In vacuum filters, the driving force for filtration results from the application of a suction on the filtrate side of the medium. Although the theoretical pressure drop available for vacuum filtration is 100 kPa, in practice it is often limited to 70 or 80 kPa.

In applications where the fraction of fine particles in the solids of the feed slurry is low, a simple and relatively cheap vacuum filter can yield cakes with moisture contents comparable to those discharged by pressure filters. Vacuum filters include the only truly continuous filters built in large sizes that can provide for washing, drying, and other process requirements.

Vacuum filters are available in a variety of types, and are usually classified as either batch operated or continuous. An important distinguishing feature is the position of the filtration area with respect to gravity, ie, horizontal or nonhorizontal filtering surface.

Vacuum filters include the nutsche filter, enclosed agitated vacuum filters, the vacuum leaf filter, the tipping pan filter, horizontal rotating pan filters, horizontal belt vacuum filters, rotary vacuum drum filters, and rotary vacuum disk filters.

Batch Pressure Filters

Excluding variable chamber presses, which rely on mechanical squeezing of the cake and are discussed in a separate section, pressure filters may be grouped into two categories, ie, plate-and-frame filter presses, and pressure vessels containing filter elements. The latter group also includes cartridge filters; these are discussed separately. All of the above pressure filters are suited to handling different types of cake. Pressure vessel filters (leaf-type) handle incompressible or slightly compressible cakes. Filter presses handle both compressible and incompressible cakes, especially with the flexibility potential of membranes. Variable chamber presses cannot be used on incompressible materials. Cylindrical element filters, ie, candle filters, are used for clarification applications, using precoat and often body-feed, resulting in cakes that are slightly compressible. Cartridge filters are for clarification only, with little if any cake formed.

Mechanical Batch Compression Filters. In conventional cake filtration the liquid is expelled from the slurry by fluid pressure in a fixed-volume filtration chamber; in mechanical compression this is achieved by reduction of the volume of the retaining chamber. This compression of either a slurry or a cake, which might have been formed by conventional filtration, offers advantages to industries handling a variety of different materials. Such materials include highly compressible, sponge-like solids; very fine particles such as clays; fibrous pulps; gelatinous mixtures like starch residues or some pharmaceuticals; and flocculated wastewater sludges.

Continuous Pressure Filters

A continuous pressure filter may be defined as a filter that operates at pressure drops greater than 100 kPa and does not require interruption of its operation to discharge the cake; the cake discharge itself, however, does not have to be continuous. There is little or no downtime involved, and the dry solids rates can sometimes be as high as 1750 kg/m²h with continuous pressure filters.

Most continuous pressure filters available (ca 1993) have their roots in vacuum filtration technology. A rotary drum or rotary disk vacuum filter can be adapted to pressure by enclosing it in a pressure cover; however, the disadvantages of this measure are evident. The enclosure is a pressure vessel which is heavy and expensive, the progress of filtration cannot be watched, and the removal of the cake from the vessel is difficult. Other complications of this method are caused by the necessity of arranging for two or more differential pressures between the inside and outside of the filter, which requires a troublesome system of pressure regulating valves.

Despite the disadvantages, the advantages of high throughputs and low moisture contents in the filtration cakes have justified the vigorous development of continuous pressure filters.

Horizontal or vertical vessel filters, especially those with vertical rotating elements, have undergone rapid development with the aim of making truly continuous pressure filters, particularly but not exclusively for the filtration of fine coal. There are basically three categories of continuous pressure filters available, ie, disk filters, drum filters, and belt filters including both hydraulic and compression varieties.

Disk filters include the McGaskell and Gaudfrin disk filters, the KDF filter, and the KHD pressure filter.

Drum filters include the TDF drum filter and the BHS-Fest filter.

Continuous compression filters include belt presses and screw presses.

Thickening Pressure Filters. The most important disadvantage of conventional cake filtration is the declining rate due to the increased pressure drop caused by the growth of the cake on the filter medium. A high flow rate of liquid through the medium can be maintained if little or no cake is allowed to form on the medium. This leads to thickening of the slurry on the upstream part of the medium; filters based on this principle are sometimes called filter thickeners.

The methods of limiting cake growth are classified into five groups, ie, removal of cake by mass forces (gravity or centrifugal), or by electrophoretic forces tangential to or away from the filter medium; mechanical removal of the cake by brushes, liquid jets, or scrapers; dislodging of the cake by intermittent reverse flow; prevention of cake deposition by vibration; and cross-flow filtration by moving the slurry tangentially to the filter medium so that the cake is continuously sheared off. The extent of the commercial exploitation of these principles in the available equipment varies, but cross-flow filtration is exploited most often.

Centrifugal Filters

The driving force for filtration in centrifugal filters is centrifugal forces acting on the fluid. Such filters essentially consist of a rotating basket equipped with a filter medium. Similar to other filters, centrifugal filtration does not require a density difference between the solids and the suspending liquid. If such density difference exists sedimentation takes place in the liquid head above the cake. This may lead to particle size stratification in the cake, with coarser particles being closer to the filter medium and acting as a precoat for the fines to follow.

In centrifuges, in addition to the pressure due to the centrifugal head due to the layer of the liquid on top of the cake, the liquid flowing through the cake is also subjected to centrifugal forces that tend to pull it out of the cake. This makes filtering centrifuges excellent for dewatering applications. From the fundamental point of view, there are two important consequences of these additional dewatering forces. First, Darcy's law and all of the theory based on it is incomplete because it does not take into account the effect of mass forces. Second, pressures below atmospheric can occur in the cake in the same way as in gravity fed deep bed filters. The conventional filtration theory has been modified to make it applicable to centrifugal filters.

Due to good performance and high cost, centrifuges are often referred to as the Rolls-Royces of solid–liquid separation. They have parts rotating at high speeds and require high engineering standards of manufacture, high maintenance costs, and special foundations or suspensions to absorb vibrations. Another feature distinguishing the filtering centrifuges from other cake filters is that the particle size range they are applicable to is generally coarser, from 10 μm to 10 mm. In particular, cake filters that move the cake across the filter medium are restricted to using metal screens, which by their very nature are coarse. No cloth can withstand the abrasion due to the cake forced on the cloth and pushed over its surface. Only the fixed bed, batch-operated centrifuges can use cloth as the filtration medium and be used, therefore, with fine suspensions.

LADISLAV SVAROVSKY
Consultant Engineers and Fine Particle Software

L. Svarovsky, ed., *Solid–Liquid Separation*, 3rd ed., Butterworths, London, 1990.

F. M. Tiller, "Tutorial: Interpretation of Filtration Data, I," *Fluid/Part. Sept. J.* **3**(2) (June 1990).

F. M. Tiller, "Tutorial: Interpretation of Filtration Data, II," *Fluid/Part. Sep. J.* **3**(3) (Sept. 1990).

L. Svarovsky, *Solid–Liquid Separation Processes and Technology*, Vol. 5, *Handbook of Powder Technology*, Elsevier, Amsterdam, the Netherlands, 1985.

FINE ART EXAMINATION AND CONSERVATION

SCIENTIFIC EXAMINATION

The application of scientific techniques to the study of art objects is an interdisciplinary undertaking. The physical scientist is trained to approach a stated problem by analyzing for the identification of measurable variables and devising means to obtain numerical values for these variables. On the other hand, the art historian relies on the trained eye, enabling visual recognition of stylistic characteristics and the more subjective comparison of these with observations about numerous other art works. Communication between these specialists has required mutual efforts. The development of scientific examinations of art objects has had a synergistic relation with the growth of a new profession: that of the art conservator, a specialist having both scientific and artistic training. The conservator consults and collaborates with both scientists and curators, providing appropriate care to objects in the collections to promote long-term preservation.

There are several possible reasons why a scientific study of an art work may be desirable. An obvious one is in cases where the authenticity of an object is doubted on stylistic grounds, but no unanimous opinion exists. The scientist can identify the materials, analyze the chemical composition, and then investigate whether these correspond to what has been found in comparable objects of unquestioned provenance. If the sources for the materials can be characterized, eg, through trace element composition or structure, it may be possible to determine whether the sources involved in the procurement of the materials for comparable objects with known provenance are the same. Comparative examination of the technological processes involved in the manufacture allows for conclusions as to whether the object was made using techniques actually available to the people who supposedly created it. Additionally, dating techniques may lead to the establishment of the date of manufacture.

Although the cases in which scientific analysis is called upon to assist in the authentication process tend to draw the most attention and have, occasionally, spectacular results, these are by no means the most important applications of scientific examination in the study of art works. More fruitful and satisfying are those interdisciplinary projects in which an object or group of objects, of unquestionable provenance, is studied jointly by the historian and the scientist, in order to better evaluate historical context. Also of value is the study of the history of technology that affords insights into the history of the development of civilization.

From the museum professional's point of view probably the most important need is for scientific support in the preservation of art objects and other cultural materials.

Methodologies

Of specific concern in the museum laboratory is the use of nondestructive techniques where the term nondestructive is taken much more literally than in the general analytical field. The term is taken to mean not requiring sample removal at all. The physical integrity of the object is the first and overriding concern in the museum.

Optical Techniques. The most important tool in a museum laboratory is the low power stereomicroscope. This instrument, usually used at magnifications of 3–50×, has enough depth of field to be useful for the study of surface phenomena on many types of objects without the need for removal and preparation of a sample. The information thus obtained can relate to toolmarks and manufacturing techniques, wear patterns, the structure of corrosion, artificial patination techniques, the structure of paint layers, or previous restorations. Any art object coming into a museum laboratory is examined by this microscope (see MICROSCOPY; SURFACE AND INTERFACE ANALYSIS).

Higher magnifications are obtainable using the polarizing research microscope. Samples have to be removed, however, and prepared according to the microscopic technique to be used. The research microscope is also of great importance in the morphological identification of materials used in the making of art objects, such as pigments (qv), fibers (qv), or woods.

The very high powers of magnification afforded by the electron microscope, either scanning electron microscopy (sem) or scanning transmission electron microscopy (stem), are used for identification of items such as wood species, in technological studies of ancient metals or ceramics, and especially in the study of deterioration processes taking place in various types of art objects.

Examinations utilizing uv or ir radiation are frequently used. Ultraviolet light has been in use for a long time in the examination of paintings and other objects, especially for the detection of repairs and restorations. Infrared irradiation is often used in the examination of paintings because, owing to the limited absorption by the organic medium, ir light can penetrate deeply in the paint layers. It is reflected or absorbed in varying degrees by different pigments. Study of the reflected ir image may enable the detection of changes in composition, pentimenti, ie, changes made by the artist to already painted areas, restorations, and, especially important, underdrawings, ie, working drawings applied by the artist on the prepared ground surface. Infrared illumination is also used frequently in the examination of paper artifacts.

Structural Analysis. Some of the optical techniques are also used for structural analysis. Microscopic examinations of metallurgical cross sections or of sections through the paint layers of a painting are indeed structural examinations, as is ir reflectography.

The most well known structural examination technique is probably x-radiography, one of the earliest scientific techniques used in the examination of art objects.

The advantages of three-dimensional imaging and image processing available with computerized tomography (CT) have not gone unnoticed. Whereas CT scan equipment is certainly not one of the standard instruments in the museum laboratory, access to this technology in medical or industrial facilities has resulted in a number of highly informative studies.

In the examination of works of art on paper, the variations in density are often too small to be revealed by conventional x-radiography. Instead, much benefit has been derived from beta-radiography, where the β-radiation is provided by an extended radioisotope source, typically a sheet of plastic in which a certain amount of carbon-14, ^{14}C, is incorporated homogeneously through labeling on the monomer. The paper is placed on the extended source, with a photographic film in close contact on top of it.

A variation on the conventional technique of x-radiography is xeroradiography, where imaging is obtained by electrostatic rather than photographic means.

Neutron activation autoradiography was developed specifically for application in the study of cultural materials, and has been applied very successfully to the examination of paintings.

Chemical Analysis. Virtually any technique for chemical analysis, whether qualitative or quantitative, elemental or molecular, organic or inorganic, has been applied in order to solve a wide variety of questions relating to art objects. In authentication studies, the pigments identified in a painting or the alloy composition determined for a metal object may be compared with data obtained from objects of unquestioned provenance, and thus provide evidence with regard to the authenticity of the object under investigation. In addition to providing reference information, similar measurements, when performed on well provenienced objects, also often lead to inferences with regard to the technology involved in the manufacture of the object, and thus the date and place of introduction of these techniques. Trace-element concentrations of many materials indicate the geographic origins of these materials, and trade and exchange relationships. Identification of deterioration products clarifies questions regarding the decay processes acting on an object, enabling the conservator to design an effective treatment and prevent further deterioration.

Dating. Radiocarbon dating has probably gained the widest general recognition. Until the quite recent development of accelerator-assisted mass spectrometry, application of this technique to dating of art objects, as opposed to archaeological remains, has been limited severely by the large samples needed.

A fundamental assumption in radiocarbon dating is that the formation rate of ^{14}C is constant. This is only acceptable as a first approximation. In reality, this rate undergoes significant variations over both short and long periods. Hence, for more accurate dating, corrections are in order. These are obtained by dating samples from individual annual growth rings from long-lived trees, such as the bristle-cone pine and the Irish oak. Graphs in which the radiocarbon dates for such samples are plotted against their actual age determined by dendrochronological technique are used to apply what are called the bristle-cone pine corrections to the radiocarbon date obtained for an unknown material.

Thermoluminescence dating has, in the relatively short time since its development, had a tremendous impact on the technical study of ceramic materials. Applications of thermoluminescence dating extend well beyond the first human ceramic production. In fact, this technique has been used in geological studies such as the dating of lava flows. A related technique is based on the same principles, only the trap population is measured by means of esr spectrometry. Amino acid racemization dating is applied, in particular, to the dating of shell, bone, or ivory objects. Another relative dating technique for fossil materials is known as nitrogen–fluorine dating.

A few techniques exist that do not provide for direct dating but rather give information as to whether the object is of modern manufacture. One of these is ^{210}Pb dating.

Indirect dating, or dating by inference, is the reaching of certain conclusions regarding the date of manufacture of an object from other information obtained in its study, such as the use of certain materials or working techniques

Types of Objects and Methods Used

Paintings. A typical technical examination of a painting, intended to ascertain its period, condition, and the degree of previous restoration would likely entail most of the following steps.

A close inspection under normal illumination reveals many indications of the condition of the painting and previous repairs. Also, because oil paints become more transparent with age, pentimenti, which originally would have been invisible after the overpainting, can be observed. Raking light illumination is very useful to determine the extent of cracking, distortions of the support, delaminations of the paint layers, etc. This stage of the examination is often done in close cooperation with stylistic experts. Thus, obvious problematic areas can be identified before the other tests are started.

Information obtained through examination with the stereomicroscope at low magnification relates to the characteristics of the craquelure, the pigments used (grain size, morphology), buildup of paint layers (visible in damages and along cracks), technique of the artist (brush work, use of pure and mixed colors, use of glazes, etc), and condition (eg, damages and losses, amount of inpainting or overpainting). Examination under uv illumination principally provides information regarding more recent restorations or changes.

The infrared reflectogram, in the form of an ir photograph, or the image from an ir-sensitive vidicon system, gives more evidence of restorations as well as pentimenti. A variation of this technique is transmitted ir photography, where the light source and the camera are placed on the opposite sides of the painting.

X-radiography reveals evidence of damages and losses in the paint layers and the support. Dimensional changes may also be detected, eg, a cut-down painting on canvas lacks, on one or more sides, the telltale deformations of the canvas caused by the tacking to the stretcher.

Identification of pigments present on paintings of unquestioned attribution provides reference information regarding the use of certain pigments in given periods or schools. Most pigment identifications are done in one of three ways: microscopy and microchemical tests, x-ray diffraction-powder analysis, or energy dispersive x-ray fluorescence spectrometry.

Microscopic examination of cross sections through the paint layers gives definite information regarding the paint-layer sequence in the area from which the sample was taken. This information illustrates the artist's use of underlayers and glazes, superposition of compositional elements, and changes in composition.

The techniques frequently used for identification of the binding media include relatively simple solubility tests, ir absorption spectrophotometry, especially Fourier transform ir (Ftir), specific staining techniques, uv-fluorescence microscopy, thin layer chromatography (tlc), high pressure liquid chromatography (hplc), gas chromatography (gc), and mass spectrometry (ms) (see CHROMATOGRAPHY).

In the case of a panel painting, a small sample of the wooden support can be removed, from which a microscopic specimen can be prepared in order to identify the wood used for the panel.

Metal Objects. Examinations of metal objects generally include the characterization of the metal, determination of the techniques involved in the manufacture, and study of aging phenomena. Of the latter, the state of corrosion is especially important, both in the examination of an object with the purpose of determining its authenticity, as well as in making an assessment of its state if conservation is needed. The layer of corrosion products covering the surface of the metal, the so-called patina, can be studied for its composition and structure (see CORROSION AND CORROSION INHIBITORS). Identification of corrosion products is often performed by means of x-ray diffraction analysis. The structure of the patina is studied using the low power stereomicroscope, where attention is directed to the growth pattern, crystal size, layering, and adherence to the metal.

An important tool in the analysis of the structure of the object is x-radiography. For metal objects, high energy x-rays are needed, obtained from either an industrial radiography instrument or from radioisotope sources (gamma-radiography).

Chemical analysis of the metal can serve various purposes. For the determination of the metal-alloy composition, a variety of techniques has been used. In the past, wet-chemical analysis was often employed, but the significant size of the sample needed was a primary drawback. Nondestructive, energy-dispersive x-ray fluorescence spectrometry is often used when no high precision is needed. For more precise quantitative analyses samples have to be removed from below the surface to be analyzed by means of atomic absorption, spectrographic techniques, etc.

Trace-element analysis of metals can give indications of the geographic provenance of the material. Both emission spectroscopy and activation analysis have bene used for this purpose. Another tool in provenance studies is the measurement of relative abundances of the stable lead isotopes.

Ceramics. Examinations of ceramic objects involve a variety of techniques, depending on the type of information sought. If an assessment of condition and state of repair is made, the most important tools are the low power stereomicroscope, x-radiography, and examination under uv light. For the identification of glass degradation products, a number of chemical analytical techniques can be used, especially x-ray diffraction. X-ray diffraction can also be used to great advantage in the analysis of the mineral composition of a ceramic paste. Not only can the naturally occurring minerals provide clues to the geographic provenance of the clay, but the presence of certain minerals formed at high temperatures can give indications of the firing temperatures.

Elemental chemical analysis provides information regarding the formulation and coloring oxides of glazes and glasses. Energy-dispersive x-ray fluorescence spectrometry is very convenient. However, using this technique the analysis for elements of low atomic numbers is quite difficult, even when vacuum or helium paths are used. The electron-beam microprobe has proven to be an extremely useful tool for this purpose. Emission spectroscopy and activation analysis have also been applied successfully in these studies.

Trace-element analysis, using emission spectroscopy and, especially, activation analysis has been applied in provenance studies on archaeological ceramics with revolutionary results.

Microscopic examination of ceramic paste, both at low magnification and at high power with prepared cross sections, can be used for petrographic study of the mineral composition and also for the determination of techniques involved in the manufacturing process.

In archaeology, the ceramic typologies have always formed the basis for the establishment of chronologies.

Stone Objects. The technical examination of stone objects begins with the use of the low power stereomicroscope. This study yields information regarding toolmarks and, hence, cutting techniques, wear patterns, and wear of toolmark edges.

When a question exists about whether the carving is ancient, an examination under uv illumination can be very helpful.

The first question about a stone object is often which stone was used. X-ray diffraction provides a relatively easy answer for such questions of identification.

The polarizing microscope is one of the most important tools in the study of stone. Petrographic study of a thin section can be extremely helpful in studying the aged surface layer and often yields information that can be used in the assignment of the geographic origin of the materials.

Trace-element analysis provides another approach to these studies, and activation analysis has been applied successfully in provenance studies or, eg, limestone sculpture.

For marble provenance studies, the most successful technique seems to be the measurement, through mass spectrometry, or the abundance ratios of the stable isotopes of carbon and oxygen.

Organic Materials. Museums contain large numbers of objects made out of components from plants or animals, including wood, eg, furniture, carvings; fibers, eg, textiles (qv), paper (qv); fruits, skin, eg, leather (qv), parchment; bone; ivory; etc.

Textiles. Specific questions arising in the study of textiles include the identification of the textile fiber.

For more conclusive identification, a combination of a number of techniques may be necessary: optical microscopy on whole fibers and cross sections, electron microscopy, chemical characterization, and measurements of physical properties.

A second common question relates to the nonfiber components of textiles, which arise from processing and finishing, ie, scouring, weaving, dyeing, glazing. The problem lies in the very low amounts of these materials, even though they have significant effects on properties such as color or hand, ie, the feel of the textile, necessitating the use of highly sensitive techniques. A small sample is removed, and the dyestuff is stripped from the fiber through treatment with acid followed by further extraction in an organic solvent. This solution can then be used to identify the dye by means of spectrophotometric absorption techniques or tlc. Elemental chemical analysis, such as through emission spectroscopy, is used for identification of the mordant.

Dating of textiles is possible by means of radiocarbon dating.

Paper. Microscopic examination can identify the fibers present in the pulp. Inks, watercolor pigments and media, etc, are analyzed similarly to the pigments and media for paintings. However, sample removal tends to be far more disfiguring and hence constitutes an even more restrictive factor. Watermarks are studied with the aid of beta-radiography. Examination in infrared illumination can assist in the reading of documents of which the ink has faded.

CONSERVATION

The conservator is responsible for the physical integrity of the object and the provision of the appropriate care, including such interventions as cleaning, stabilization, consolidation, and restoration, as well as preventive care. This specialist has a thorough training in art history enabling communication on a professional level with the curator, who generally is responsible for the works in the collection and who collaborates with the conservator with regard to considerations of an aesthetic nature.

Ethical Considerations

The practitioners of conservation have formulated a number of ethical professional standards, many of which are basic to the contemporary approach towards the conservation of art objects and other historic materials. Central to all of this is a fundamental respect for the integrity of the object *per se*. Careful consideration should be given to the treatment's potential for causing any immediate or future damage to the object. Any proposed treatment should be thoroughly evaluated as to not only immediate but also long-term possible effects besides the intended beneficial ones.

The arrest of deterioration and the prevention of its recurrence has higher priority than restoration. Thus, identification of the causes of a problem and the design of measures to stabilize and consolidate the object are primary considerations. Removal of the symptoms and restoration of the visual appearance comes only after the physical integrity has been safeguarded.

Inducing any changes to the original parts in order to minimize the visual impact of repairs and restorations is prohibited.

It is imperative that extensive documentation be kept of any treatment, including photographs during all stages of the work. This documentation should then be stored to be available for consultation in the future, when a renewed conservation assessment of the object may be necessary.

Training and Organization

In the United States, a number of academic graduate conservation programs have been established. Candidates for admission to such schools must satisfy a number of stringent demands, which generally include an undergraduate degree in art history or archaeology, extensive undergraduate science coursework, and demonstrable manual skills and dexterity, as evident from a portfolio. After three to four years of training, successful completion is met by a degree or certificate in conservation. An advanced internship with a practicing conservator for another one or two years is generally deemed necessary for the acquisition of the minimum experience needed to be allowed to work independently. Training programs have been established in many countries. An alternative method of training is that of the apprenticeship, an experienced conservator training and sharing knowledge with future colleagues.

Professional organizations for conservators have been established on both the national and international scales. Most conservators are members of the International Institute for the Conservation of Historic and Artistic Works (IIC). Several countries or geographic areas have their own subdivisions of IIC. In the United States, an independent but related organization exists, the American Institute for Conservation of Historic and Artistic Works (AIC).

Several international organizations have been established that can offer conservation advice or even practical help in areas of the world where such is not readily available, eg, the International Centre for the Study of the Preservation and the Conservation of Cultural Properties (ICCROM), based in Rome. The Conservation Information Network (CIN), a collaborative venture of seven institutions in the United States, Canada, and Europe, provides on-line bibliographic and technical information.

Deterioration Mechanisms and Conservation

Metal Objects. *Deterioration.* Apart from physical damage that can result from carelessness, abuse, and vandalism, the main problem with metal objects lies in their vulnerability to corrosion (see CORROSION AND CORROSION CONTROL). The degree of corrosion depends on the nature and age of the object. Corrosion can range from a light tarnish, which may be aesthetically disfiguring on a polished silver or brass artifact, to total mineralization, a condition not uncommon for archaeological material.

Conservation. Because the most common conservation problem with metal objects occurs when corrosion processes form a threat to the safety of the object or disfigure its appearance to an unacceptable degree, many conservation treatments are intended to stabilize the corrosion processes and to remove aesthetically displeasing corrosion crusts. The latter requires a great deal of thought and discussion as to when a corrosion layer ceases to be a desirable patina and becomes unacceptable.

On highly polished surfaces, as on silverware, the slightest tarnish constitutes a disfiguring effect and must be removed and prevented. Often the cleaning is done by careful polishing with a mild abrasive, such as precipitated chalk. On ancient silver, such polishing results in an undesirable shine. Thus to preserve the soft luster of the metal, cleaning is preferably accomplished through chemical or electrochemical means.

Removal of disfiguring corrosion crusts can be accomplished in several ways. One approach is mechanical cleaning, which has the advantage of control in the degree and extent of cleaning. A second possibility is the use of chemical means. Finally, electrolytic reduction can be used with great advantage in those cases where the original surface needs to be recovered.

In order to prevent recurrence of the corrosion, a lacquer can be applied. Alternatively, the environment of the object can be strictly controlled with regard to relative humidity and pollutants.

Bronze disease necessitates immediate action to halt the process and remove the cause.

Some special problems are encountered in the treatment of iron, especially that recovered from underwater sites. Corrosion stimulators, especially chlorides, must be removed. One treatment used in cases where large amounts of objects are recovered is hydrogen reduction at high temperatures. Other reduction treatments are electrolytic.

Stone Objects. *Deterioration.* An important source of damage to stone objects is mechanical in nature. Both breakage and abrasion account for much of the losses on objects made of this relatively fragile material. More difficulties are offered by the processes of a chemical nature which play a role in stone deterioration.

Conservation. The objectives in the treatment of stone objects are primarily cleaning, stabilization, consolidation, repair, and restoration. Cleaning can vary from a light dusting to the removal of stubborn grime and stains with solvents and detergents.

Stabilization involves the removal of the cause of deterioration, which is frequently soluble salts present in the stone.

Consolidation is the introduction into a stone's interior of a substance which adds extra mechanical strength. Such impregnations have been done using both synthetic resins and inorganic compounds.

Ceramics. *Deterioration.* Ceramic objects are fragile, and mechanical damages through breakage and abrasions are the most likely source of destruction. Low fired ceramics can suffer through the rehydration of the body material; this process results in a complete loss of mechanical strength. The presence of soluble salts in porous ceramic bodies has the same disastrous results as in stone.

Conservation. Most conservators use adhesives based on poly(vinyl acetate), acrylic resins, some polyesters, and epoxies. Epoxies and polyesters still have the disadvantage of discoloration, especially noticeable in mended glass. However, there is not much choice in adhesives for glass. Low fired, porous ceramics are subject to loss of structural strength, and become sensitive to moisture. Occasionally, an impregnation with poly(vinyl acetate) or acrylic resins becomes necessary. In archaeological ceramics, salts may have been introduced during burial. These salts must be removed through desalination techniques such as those used for stone.

The layers of decay products and hydrated glass formed on the surface of ancient glass often start to spall off. In such cases, an impregnation with an acrylic resin may be helpful.

Organic Materials. Environmental conditions, especially temperature and relative humidity, cause deterioration of objects made from plant or animal materials.

Another source of damage is light, especially uv light. Photochemically induced deterioration is a principal cause of damage in textiles, via the fading of dyes and decomposition of fibers.

Chemical degradation, whether thermally or photoinduced, primarily results from depolymerization, oxidations, and hydrolysis. A special case is the degradation of cellulose nitrate, which for a long time was used as the base for photographic film.

Modern synthetic polymers are the subject of increasing research by conservation scientists. Not only does their frequent use in conservation treatments require a better understanding of their long term stability, but also many objects, including those in collections of contemporary art and in history and technology museums, are made out of these new materials.

Often when objects are composites of many different materials special problems arise. Conditions that may negatively affect any one component are harmful to the object in its entirety. Then, conditions favorable to the survival of one component may affect the stability of another negatively. A very difficult problem results when one of the components promotes chemical-decay reactions in another of the materials.

Paintings. Deterioration. Paintings are composite objects that have high vulnerability. The various materials are adhered to each other, especially in a laminated structure, to form a source of potential trouble. Any dimensional change in one of the components or between the components as a consequence of changes in environmental conditions results in a strain on the adhesion of the various parts. Strains can lead to failure of the adhesion. This is one of the principal causes of losses in panel paintings, where the dimensional changes in the wooden support cause losses in adhesion between the paint layer and the support.

Conservation. Conservation problems in paintings can be considered according to the stratum in which these occur, ie, in the varnish, the paint layers, or the support.

Upon aging, all natural-resin varnishes exhibit problems resulting from chemical deterioration caused mainly by autoxidation and loss of volatile oils. Treatment can vary from simple dusting to complete removal of the varnish coat. Two frequently used treatment methods are regeneration and replacement.

The most common problem in the paint layers, which can have a wide variety of causes, is loss of adhesion.

Flaking paint is treated by infusion of an adhesive in the areas where needed, followed by resetting the flakes on the substrate; the softening of the paint needed to bend it back is effected through solvent action or heat. Losses can be filled and inpainted.

Failure of a canvas support occurs when, as a result of aging, the fabric becomes hard and brittle and the fibers break. The fabric loses all strength and is unable to function as a support for the paint layers. The common solution, when this has progressed so far as to endanger the safety of the painting, is the lining or relining. This procedure involves the application of a new fabric to the back of the old one.

Wooden panels have a tendency to warp and eventually crack. If the deformation of these supports becomes dangerous to the paint layers, some measures must be taken. In the past, restriction of the movement of the support was attempted through the application of rigid cradles, ie, thick wooden strips running both horizontally and vertically across the back. This was not very successful. Cracks developed between the strips of the cradle and the cleavage between ground and wood often became more aggravated. There is no easy and safe solution to this problem. The best answer lies in prevention through strict climate control of the painting's environment. In cases where the deformation has progressed too far and immediate danger to the painting exists, a transfer can be done in which the painting is again faced on the front, and the original support is tooled down until the groundlayer is exposed. A new support can then be applied to the back. However, these transfers are generally regarded as highly undesirable.

Wall paintings become endangered when the support wall decays. This is often the case with the walls on which the Italian Medieval and Renaissance artists applied their masterpieces. One technique developed for the removal of fresco paintings from the wall, called the strappo technique, involved the attachment of a canvas using a strong glue facing to the front of the painting. Whereas such detachments have undoubtedly saved a number of masterpieces, they do impose significant risks of damage, partial loss, and changes in appearance. Hence, removal of the wall painting from its support is no longer considered to be an acceptable practice except as a measure of last resort, when all efforts to stabilize the work *in situ* have failed.

Works on Paper. Deterioration. Paper is subject to deterioration from several sources. Hydrolytic degradation of cellulose is acid-catalyzed; hence, the acidity present in aged paper from manufacturing techniques, acidic inks, housing material, and degradation processes, leads to discoloration and embrittlement.

A second degradation process is oxidation, often photoinduced especially by exposure to light not filtered for uv.

Attack by molds in high humidity environments can cause weakening of the support by breakdown of sizing, and result in disfiguring stains.

Conservation. In an attempt to save paper, preventive conservation care deserves the highest priority, because it reduces the need for potentially hazardous, complicated, and expensive treatments later. Problems which have a structural impact on long-term stability of paper should be given a higher priority than problems which are merely cosmetic in nature.

In the paper conservation laboratory, washing is probably the most common treatment. The amount of care bestowed on single works of art on paper in museums would be impractical for individual documents or books in archival and library collections, where the number of items exceeds that in art museum collections by several orders of magnitude. Hence, ongoing development and testing of appropriate technologies for mass deacidification is especially of great importance for libraries and archives. Proposed mass deacidification processes involve the use of aqueous or nonaqueous solutions of various magnesium and calcium carbonates, or diethylzinc as a gas-phase reagent.

Brittle, fragile, or torn paper must be reinforced, either by archival quality housing or conservation treatment (such as mending or lining) using appropriate materials and procedures. A great deal of research is underway to investigate the effects of these treatments on the aging of paper.

Textiles. Deterioration. The causes of degradation phenomena in textiles are many and include pollution, bleaches, acids, alkalies, and, of course, wear. The single most important effect, however, is that of photodegradation. Both cellulosic and proteinaceous fibers are highly photosensitive. Light is also the principal cause of damage to both natural and synthetic dyestuffs that fade through photodegradation.

Other causes of worry to the textile conservator are the dangers of mold growth and of insect damage.

Conservation. Washing is a rather popular treatment for the removal of acidic or otherwise unwanted soluble compounds. For textiles that cannot be exposed to aqueous treatment, solvent based dry cleaning may be the preferred mechanism for cleaning.

Wooden Objects. Wooden objects cause several special problems, particularly when recovered from underwater sites. Waterlogged wood, if allowed to dry out, suffers irreparable damage through warping and cracking. It is, therefore, kept underwater until it arrives in the laboratory, where a few treatments are available. Freeze drying has proven to be extremely effective. Another well established treatment is the immersion of the object in a tank with water, in which poly(ethylene glycol) (PEG) is subsequently introduced in a slowly increasing concentration, until impregnation of the wood with PEG has been achieved. Yet another technique involves impregnation with a solution of rosin in acetone.

Furniture is subject to damage by dimensional changes caused by the environment, as well as physical damage through abuse and wear. Insect damage can be severe; because the infestation is in depth, a penetrating treatment must be used. For lesser infestations, an insecticide is sometimes injected into the insect holes; larger infestations, however, require fumigation.

Preventive Conservation

No conservation treatment can completely undo damages to art objects. However, damage can often be prevented. Many deterioration processes are dependent upon environmental conditions.

Temperature and Humidity. Temperature is probably the easiest environmental factor to control. The main concern is that the temperature remains constant to prevent the thermal expansions and contractions that are particularly dangerous to composite objects. Another factor regarding temperature is the inverse relation to relative humidity under conditions of constant absolute humidity, such as exist in closed areas. High extremes in temperature are especially undesirable, as they increase reaction rates. Areas in which objects are exhibited and stored must be accessible; thus a reasonable temperature setting is generally recommended to be about 21°C.

Although the temperature can be controlled with a well-designed air-conditioning system, the small fluctuations which most cycling systems cause may be very harmful. The temperature–time record should be a continuous, flat graph.

More difficulties are encountered with determining humidity settings. Not only is it important that the relative humidity remain constant, but also the desirable value depends on the materials. Many of the larger institutions have sought to solve the problem through the installation of extensive climate control systems, which serve to keep both temperature and humidity at a constant value by means of heating, refrigeration, humidification, and dehumidification. However, the main difficulties are encountered with humidification in extremely dry conditions such as occur during the heating season along the northeastern seaboard.

An alternative to macroclimate systems is the creation of microclimates. The objects are placed within smaller spaces, such as cases, in which an ideal environment is maintained. One possibility is to install equipment to control the climate in individual cases, or groups of cases with similar materials, by mechanical means.

Rigid monitoring of the climatic conditions is an absolute requirement for a successful preventive conservation program. The desirability of a continuous measurement and a written record favors the use of recording thermohydrographs.

Biodeterioration. For objects made out of organic materials, mold and insect attack are the principal causes of damage. Microbiological organisms can also be responsible for serious deterioration of outdoor stone.

Most emphasis is placed on a program of rigorous preventive maintenance. Appropriate climate conditions can help to prevent mold attack, which typically only occurs at elevated relative humidity. Storage furniture which provides an effective barrier for insects, regular inspection of the collections, monitoring of all collection areas with insect traps, and access control measures which minimize the chance of insect entry into the collection areas, are some aspects of an effective pest control management program.

Light. It is imperative that the exposure of photosensitive materials to uv light is kept to a minimum. In many museums, the use of natural daylight for illumination is preferred on aesthetic grounds, as its spectral distribution results in an ideal color rendition. As the uv component does not play a role in the visualization of the object and is the most harmful, it must be completely removed. This can be done through the use of appropriate filters, available either in the form of rigid sheets of plexiglass doped with a strongly uv-absorbing organic material or as flexible sheets, adhered to the glass windows. A variation of the latter, nowadays often used in museums, is a laminated glass with a uv-absorbing interlayer.

When artificial light is used, those light sources which produce a significant amount of uv radiation, such as fluorescent bulbs, should also be provided with a suitable filter. Since visible light also contributes to many photochemical degradation processes. Strict limits on total illumination are imposed for the exhibit of especially sensitive materials.

Pollutants. The problems posed by air pollutants are very serious. Within a museum, measures can be taken to remove harmful substances as efficiently as possible by means of the installation of appropriate filter systems in the ventilation equipment. Proposed specification values for museum climate-control systems require filtering systems having an efficiency for particulate removal in the dioctyl phthalate test of 60–80%. Systems must be able to limit both sulfur dioxide and nitrogen dioxide concentrations to $<10\ \mu g/m^3$, and ozone to $<2\ \mu g/m^3$.

In the preservation of outdoor art objects, the problems caused by air pollution are overwhelming. For metal objects, such as bronze sculpture, a possible solution is the application of a protective barrier layer in the form of a surface coating. Acrylic lacquers, sometimes doped with a corrosion inhibitor such as benzotriazole, and waxes are most frequently used for this purpose. The tremendous cost involved in an effective program of maintenance, necessary in the use of barrier surface layers, is often an insurmountable impediment. For outdoor art objects composed of other materials, eg, stone, there are no efficient protection measures available.

Physical Safety. Preventive conservation also involves ensuring the physical safety of objects. Objects should be guarded against acts of vandalism or damage inflicted by touching them.

Severe damage has sometimes been inflicted through faulty design or manufacture of mounting devices. Moreover, objects are especially subject to serious danger of damage during transit, as for example in travelling exhibits. Travel from one climatic environment to another, as well as lack of climate control in the transportation vehicles, may result in large fluctuations of temperature and humidity. During transport, the object also is in danger of shock, eg, from dropping its crate, and vibrations from the transport vehicle. Measurements of mechanical properties and the use of computer modeling have contributed significantly to identification and quantification of these factors, as well as to the design of packing crates that minimize the risk to the objects.

Conflicting Interests and Shared Responsibilities. The responsibility of collections holding institutions, museums, archives, libraries, etc, toward their collections is not a simple one. On the one hand, there is the obligation to preserve the collections for future generations. On the other hand, there is the mandate for the use of the collections in promoting and facilitating scholarly studies, education, and public enjoyment. Conditions optimal for preservation often severally limit and inhibit access. The study and exhibit value of objects lies to great part in their continued availability for that same purpose. Thus, curators, collections managers, and conservators share a common goal, ie, to make the collection available for the purposes for which it was brought together, while at the same time taking every measure and precaution possible to preserve that use and value of the objects.

LAMBERTUS VAN ZELST
Conservation Analytical Laboratory
Smithsonian Institution

M. Jones, ed., *Fake? The Art of Deception*, British Museum Publications, London, 1989.

S. G. E. Bowman, ed., *Science and the Past*, British Museum Press, London, 1991.

G. Thomson, *The Museum Environment*, Butterworths, London, 1978.

K. Bachmann, ed., *Conservation Concerns: A Guide for Collections and Curators*, Smithsonian Institution Press, Washington, D.C., 1992.

FINE CHEMICALS

PRODUCTION

The requirement for more and more sophisticated organic chemicals has contributed substantially to the emergence of the fine chemicals industry as a distinct entity. Fine chemical manufacturers are backward integrated and production-oriented, and they service the mega enterprises within the chemical industry. The fine chemicals industry has its own characteristics with regard to R&D, production, marketing, and finance.

In the chemical business products may be described as commodities, fine chemicals, or specialties. Various commodities are also known as petrochemicals, basic chemicals, organic chemicals (large-volume), monomers, commodity fibers, and plastics. Advanced intermediates, building blocks, bulk drugs and bulk pesticides, active ingredients, bulk vitamins, and flavor and fragrance chemicals are all fine chemicals. Adhesives (qv), diagnostics, disinfectants, electronic chemicals, food additives (qv), mining chemicals, pesticides, pharmaceuticals (qv), photographic chemicals, specialty polymers, and water treatment chemicals are all specialties. The added value is highest for specialties.

In terms of volume, the borderline between commodities and fine chemicals falls somewhere between about 1,000 and 10,000 t/yr. In terms of unit prices the line typically varies between $2.50/kg, and $10/kg.

Research and Development

Product innovation absorbs considerable resources in the fine chemicals industry, in part because of the shorter life cycles of fine chemicals as compared to commodities. Consequently, research and development (R&D) plays an important role. The main task of R&D in fine chemicals is scaling-up lab processes, as described, eg, in the ORAC data bank or as provided by the customers, so that the processes can be transferred to pilot plants (see PILOT PLANTS) and subsequently to industrial-scale production.

Plant Design

The number of products offered by a fine chemicals manufacturer typically exceeds the number of production trains. Yet, for reasons of economy of scale, the production capacity considerably exceeds the yearly requirement for each product. Furthermore, the product portfolio is regenerated at a fast pace. This set of circumstances leads to the multipurpose plant, as opposed to a dedicated plant. A multipurpose plant is capable of handling several types of chemical reactions and performing a series of unit operations.

A production train in its simplest form consists of a jacketed reaction vessel made from stainless or glass-lined steel and equipped with an agitator, a manhole, and pipe connections. Solid raw materials are charged through the manhole, and liquid and solvents feed through a manifold system installed on top of the respective reactor and connected to one of the inlet nozzles of the reactor head. After completion the reaction mixture, typically a solid–liquid slurry, is discharged through the bottom valve. The solid–liquid separation is performed in the centrifuge which only holds a part of the reaction mixture. The reaction vessel in this simple configuration also serves as a feeding tank for the centrifuge. A new batch can only be initiated after the centrifugation is completed. This module I-type production train can be used, for example, for the formation and precipitation of salts of organic acids, ie, metal carboxylates, from an aqueous solution.

In module II a crystallization vessel, jacketed and connected to cooling water, is added. Thus the salt formation step, which may require heating, is separated from the crystallization (qv), which is completed upon cooling.

In module III, feeding tanks for raw materials and solvents, and a holding tank for the mother liquor, are added.

If the reaction temperature is close or equal to the boiling temperature of the solvent, module IV is used. It differs from module III by the addition of an overhead condenser connected with a vacuum system, a phase separator, and a distillate receiving tank. Also, the jacket for heating/cooling is substituted for a half coil allowing more rapid heat transfer.

In the design of a fine chemicals plant, equally important to the choice and positioning of the equipment is the selection of its size, especially the volume of the reaction vessels. A large plant using large equipment would be expected to be more economical to run than a small one.

Depending primarily on the differing quantities of the fine chemicals to be produced in the same multipurpose unit, the concentration of substances in the reaction mixture, and the reaction time, there is, however, an upper limit for the size of the reaction vessel and the ancillary equipment. Some factors run countercurrent to the economy of scale and point to small-sized equipment. Four of these are *(1)* length of the production campaign; *(2)* in the case of expensive fine chemicals the value of one batch in one piece of equipment becomes very high, sometimes in excess of $1 million and, therefore, the risk of false manipulations becomes excessive; *(3)* the dimensions of existing buildings, tank farms, and the capacity of utilities often determine an upper limit of the equipment size; and *(4)* consumption of solvents to clean the equipment increases with volume, and this means an additional burden for waste treatment facilities.

Apart from determinating the optimum size of equipment, the degree of flexibility is another key plant design parameter. Flexibility means cost; thus only as much flexibility as required by the processes should be built.

Full process control (qv) computerization of multipurpose plants is much more difficult than for single-product continuous units. However, the first computerized fine chemical plants were brought on-line by ICI Organics Division in the early 1970s and there are many others. The additional (ca 10%) cost of computerization has been estimated to give a savings of 30 to 45% in labor costs.

The percentage costs associated with a multipurpose plant equipped with 6.3 m^3 reactors are as follows: 27% for building, including the warehouse, where the building has a foundation; 23% for process equipment, ie, reactors, tanks, feeders, hoppers, heat exchangers, condensors, pumps, including vacuum pumps, centrifuges, suction filters (nutsches), filters, and dryers, sieves, mills; 11% for erection, ie, installation of equipment, piping, insulation, and painting; 5.5% for electrical materials and installation; 12% for process control/instrumentation; 16% for engineering, including profit and general overhead; and 5.5% for various other costs.

Plant Operation

Production Planning. Whereas continuous plants run 24 hours per day, there is more freedom in establishing operating schedules for multipurpose plants. Depending on the work load and the flexibility of the workforce, schedules can also be changed during the course of a year. Common schedules include one shift, ie, 8 hours per day for 5 days per week, and two shifts, ie, 16 hours per day for 5 days per week. In this latter case frequently some minimum activity is maintained during the night, such as supervision of reflux reactions, solvent distillations, or dryers. Three shifts, ie, 24 hours per day, 7 days per week, are also possible. In terms of production cost this last is the most advantageous scheme. Higher salaries for night work is more than offset by lower fixed costs. Also, only part of the workforce has to adhere to this scheme.

Production planning for a fine chemicals company operating multipurpose plants is a demanding task. The goal must be to achieve optimum capacity utilization, important to the profitability of the company. However, conflicting interests of marketing, manufacturing, and controlling have to be aligned.

Quality Control. Because fine chemicals are sold according to specifications, adherence to constant and strict specifications, at risk because of the batchwise production and the use of the same equipment for different products in multipurpose plants, is a necessity for fine chemical companies. For the majority of the fine chemicals, the degree of attention devoted to quality control (qv) is not at the discretion of the individual company. This is particularly the case for fine chemicals used as active ingredients in drugs and foodstuffs (see FINE CHEMICALS–STANDARDS). Standards for drugs are published in the *United States Pharmacopeia* (USP) in the United States and the *European Pharmacopeia* in Europe.

Standards for food-grade chemicals in the United States are published in the *Food Chemicals Codex* (FCC), for laboratory reagents in *Reagent Chemicals—ACS Specifications*, and for electronic-grade chemicals in the *Book of SEMI Standards* (BOSS) by Semiconductor Equipment and Materials International (SEMI). The latter two product categories, with the exception of reagent chemicals used as diagnostics, are not subject to GMP regulations. In these cases most customers expect the producer to comply with a quality standard such as ISO 9001/9002, ie, only released raw materials can be used, the process must be validated, and documentation of each produced batch is kept.

The analytical laboratory is expected to be certified or at least to comply with one of the quality standards for analytical labs set forth by ISO GLP or EN 45,000 ff.

Cost Calculation. The main elements determining production cost are identical for fine chemicals and commodities. A breakdown of production cost is given in Table 1.

Industrial Strategies. Outsourcing of the manufacture of fine chemicals by the life science and other specialty chemical companies has been one of the driving forces for the development of the fine chemicals industry. The advantages of making fine chemicals include maintaining control over the whole supply chain as well as expertise in chemical manufacturing; avoiding the risk of dissipating confidential technology to third parties; occupying idle production capacity; avoiding bad experiences with fine chemicals manufacturers; and cost savings. The advantages to buying are allocating financial, technical, and human resources for the core business, eg, drug or pesticide discovery in the life science industry; avoiding the risk of building a new plant at a time in which the ultimate success of a new specialty is not yet known; liberating production capacity for new fine chemicals by transferring production of older ones to fine chemicals manufacturers; past favorable experience with fine chemicals manufacturers; and the ability to pay as earn, ie, expending no capital in anticipation of future sales.

Assuming more or less comparable production costs for the fine chemical manufacturer and the life science industry, and recognizing that the fine chemicals manufacturer needs to make a profit, it appears *a priori* that buying is more expensive than making. This conclusion, however, is only valid for certain cases because outsourcing constitutes a kind of insurance premium in new product development. If everything proceeds according to plan, buying is more expensive than making; if, however, the new product (specialty) launch is delayed or fails, the life science company fares much better with buying.

Table 1. Cost Calculations for Fine Chemicals Value-Added Structure

Parameter	Cost, %
Variable costs	
raw materials	26
utilities	3
Manufacturing cost	
direct[a]	14
waste pretreatment[b,c]	6
waste treatment	3
allocated plant costs	4
Capital cost	
plant depreciation	11
plant adaptation[d]	5
inventories	6
Corporate overhead and miscellaneous	
R&D	9
general management and miscellaneous[e]	13
Total	*100*

[a] Labor, maintenance, etc.
[b] Recovery of solvents or reagents, extraction and neutralization of mother liquors, precipitation and separation of salts, and scrubbing of gaseous effluents.
[c] The waste pretreatment frequently (and sometimes dramatically) reduces the output of a process, which is not considered in this cost breakdown.
[d] Specific or new equipment.
[e] Administration, marketing, customer service, packaging, transport, insurance, etc.

Economic Aspects

There are about 500 companies worldwide involved in fine chemicals. Most are in Europe, followed by the United States and Japan. Fine chemical companies are generally either small and privately held or divisions of larger companies, such as Eastman Fine Chemicals (United States) and Lonza (Switzerland).

Whereas all larger companies are market-oriented and offer a list of catalog products, smaller ones are more (single) customer focused and offer primarily toll or custom manufacturing services.

There are also companies that concentrate on physical operations in connection with fine chemicals manufacturing. Activities include drug and pesticide formulation, distillation, and milling/sieving/drying.

Total sales of the fine chemical industry were estimated to amount to about $15 billion in 1993, by an in-depth analysis of the product portfolio of representative fine chemicals manufacturers and by performing top-down analyses of the life science industry. However, some consulting firms gave higher ($\sim\$60 \times 10^9$) figures for the size of the fine chemicals business.

Products

If fine chemicals are classified according to applications, the most prominent categories in terms of tonnage volumes are the ones used in the production for agrochemicals, followed by pharmaceutical fine chemicals. Within agrochemicals, triazine herbicides (see HERBICIDES), from the key intermediate cyanuric chloride (see CYANURIC AND ISOCYANURIC ACIDS), are produced in quantities exceeding 100,000 metric tons per annum. Chloroacetanilides, from the key intermediates 2,6-diethylaniline and chloroacetylchloride, and phenoxy herbicides, from L-2-chloropropanoic acid rank between 50,000 and 100,000 t/yr. Also phosgene-derived thiocarbamate and urea herbicides, as well as dithiocarbamate fungicides are very large-volume products (see FUNGICIDES, AGRICULTURAL).

Within pharmaceuticals, the highest volume categories are vitamins (qv), painkillers, and β-lactam antibiotics.

In terms of types of molecular structures, heterocyclic compounds are the most important fine chemical category, especially fine chemicals having an *N*-heterocyclic structure as found in the vitamins B_1, B_2, B_6, H, PP, and folic acid.

From the point of view of application, pharmaceutical fine chemicals constitute the largest part of all fine chemicals, both in terms of number of products and volume of sales. About 40–50% of the total fine chemicals sales comes from pharmaceutical fine chemicals; about 20 to 25% are agrochemicals, and the rest belong to other categories.

The future development of the fine chemicals industry depends mainly on the development of demand. The growth of the fine chemicals business is mainly fostered by the introduction of new pharmaceuticals, agrochemicals, engineering plastics (qv), and other specialties requiring high value organic intermediates.

The strategy of the life science industry is also important. As strict financial management of all operations is becoming more and more imperative, investing in fine chemical manufacturing becomes less of a good risk/benefit ratio, in part because investment decisions for new plants have to be made years before the anticipated launch of a new drug or pesticide.

Future growth of the fine chemicals industry also depends on environmental regulations. All fine chemical companies need to allocate an increasing share of both R&D expenditure and capital investment on searching for clean processes and installing the equipment necessary for the reduction of emissions into the air, purification of wastewater, and pretreatment of solid waste.

PETER POLLAK
Lonza Ltd.

G. Schuch and J. König, *Chem.-Ing. Tech.* **64**(7), 587–593 (1992).

P. Pollak, *Chimica Oggi*, 11–16 (Jan./Feb. 1992); *Eur. Chem. News*, 23–26 (Oct. 5, 1992).

H. H. Szmant, *Organic Building Blocks of the Chemical Industry*, John Wiley & Sons, Inc., New York, 1989, pp. 575–637.

P. Pollak and G. Romeder, *Perform. Chem.*, 36–38 (Feb. 1988); 44–54 (Apr. 1988).

STANDARDS

Fine chemicals are generally considered chemicals that are manufactured to high and well-defined standards of purity, as opposed to heavy chemicals made in large amounts to technical levels of purity.

The fine chemicals standards discussed herein are primarily those originating in the United States. Much discussion has occurred regarding harmonization of the world's standards. It is not yet clear, however, what impact the International Standards Organization Quality Management Standards (ISO 9000) may have on the manufacture and specifications of fine chemicals.

Standards for drugs are established by the United States Pharmacopeial Convention, Inc. (USPC), and have been published in revisions of the *United States Pharmacopeia* (USP), which is now combined with the *National Formulatory* (NF).

Standards for food-grade chemicals in the United States are set by the Committee on Food Chemicals Codex of the National Academy of Sciences (NAS) which publishes them in the *Food Chemicals Codex* (FCC) (see also FOOD ADDITIVES). Standards for laboratory reagents are set by the American Chemical Society (ACS) Committee on Analytical Reagents and are published in *Reagent Chemicals–ACS Specifications*. Standards for electronic-grade chemicals, which have extremely low limits for trace ions, are published annually in *The Book of SEMI Standards* (BOSS) by Semiconductor Equipment and Materials International (SEMI).

The publications detailing standards generally include both specifications and methods of analysis for the substances. The establishment of standards of quality for chemicals of any kind presupposes the ability to set numerical limits on physical properties, allowable impurities, and strength, and to provide the test methods by which conformity to the requirements may be demonstrated.

One of the tasks of greatest importance in providing compendial standards is that of either obtaining, adapting, or developing the test methods for determining compliance with the standards. Such methods must be capable of routine use in many laboratories by different personnel using different equipment.

The USP and the NF are recognized in the Food, Drug, and Cosmetic Act (1938) as establishing legal drug standards that the Food and Drug Administration (FDA) is responsible for enforcing. The FCC is recognized by FDA regulations, on an individual substance basis, as defining food grade. The ACS reagent specifications have no special legal status per se, but are used by the USP–NF, the FCC, SEMI, and in government procurement, and are referenced in FDA regulations and in American Society for Testing and Materials (ASTM) methods.

Federal regulation of drugs, food additives, and reagents used in *in vitro* diagnostic tests has increased markedly (see MEDICAL DIAGNOSTIC REAGENTS). With increasing consumer concern about the safety of fine chemicals, especially those used in foods and drugs, the United States government has become increasingly sensitive to the manner in which standards are set. Freedom of information legislation has confirmed the public's right to know, and this has introduced the objective of due process into the development of standards. The USP–NF and the FCC have mechanisms that make possible public participation in setting the standards with which these agencies are involved. Neither ACS nor SEMI has such a formal mechanism, but individuals from industry, government, and academia serve on the ACS Reagent Committee, and committee meetings of both the ACS and SEMI are open to the interested public upon request.

Chemical and Other Standards Used in Analysis. *National Institute of Standards and Technology (NIST).* The NIST is the source of many of the standards used in chemical and physical analyses in the United States and throughout the world. The standards prepared and distributed by the NIST are used to calibrate measurement systems and to provide a central basis for uniformity and accuracy of measurement.

United States Pharmacopeia. Reference standards are required in many USP and NF tests, and in a few FCC tests. The USP distributes such standards domestically and has authorized international distribution by a number of organizations or companies.

Impact of the Food, Drug, and Cosmetic Act on Fine Chemicals

FDA Quality Standards. In the drug field, specifications and testing methods for antibiotics and biologicals are set by the FDA. Also, specifications and testing methods are prescribed for colorants. Many food-additive petitions are granted with the requirement that certain specifications be met.

Device Legislation. Regulations covering medical devices define reagents used in *in vitro* diagnostic tests as devices (see PROSTHETIC AND BIOMEDICAL DEVICES).

Regulations Concerning Good Manufacturing Practice. Chemicals that are drugs, as defined in the Food, Drug, and Cosmetic Act, are subject to the requirement of the Act that they be made under conditions of Current Good Manufacturing Practice (CGMP). Specific GMP regulations for such chemicals have not been published, but the regulations that have been published for dosage form drugs include many points that should be considered.

The primary thrust of GMP is that it is not enough merely to make chemicals to meet USP or other applicable specifications. The chemicals must be made under clean and sanitary conditions, procedures and processes must be validated and documented, and processing and packaging must be carried out under conditions that preclude mixup and mislabeling. Records must be kept of complaints, and the manufacturer must know enough about the storage properties of the products to specify storage conditions and, if necessary, expiration dates on the label.

A manufacturer of drug chemicals is required to register with the FDA, and is subject to FDA inspection at least once every two years.

SAMUEL M. TUTHILL
NORMAN C. JAMIESON
Mallinckrodt Chemical, Inc.

The U.S. Pharmacopeia 23, (USP 23–NF 18), The U.S. Pharmacopeial Convention, Inc., Rockville, Md., 1990.

Food Chemicals Codex, 4th ed., National Academy of Sciences, National Research Council, Washington, D.C., 1996.

Reagent Chemicals—American Chemical Society Specifications, 8th ed., ACS, Washington, D.C., 1993.

The Book of SEMI Standards, Semiconductor Equipment and Materials International, Mountain View, Calif., 1995.

FIREBRICK. See REFRACTORIES.

FIRE CLAY. See CLAYS; REFRACTORIES.

FIRE PREVENTION AND EXTINCTION. See FLAME RETARDANTS; PLANT SAFETY.

FLAME RETARDANTS

OVERVIEW

Terminology

Some pertinent definitions include *fire retardant* (flame retardant), used to describe polymers in which basic flammability has been reduced by some modification as measured by one of the accepted test methods; *fire-retardant chemical*, used to denote a compound or mixture of compounds that when added to or incorporated chemically into a polymer serves to slow or hinder the ignition or growth of fire, the foregoing effect occurring primarily in the vapor phase; *materials*, single substances of which things are constructed that may be composed of single or blended polymers, may be layered or fiber-reinforced, and might contain a variety of additives; and *products*, consumer items made of one or more materials.

Measuring Fire Performance of Products

Laws have been promulgated to improve the fire performance of everyday fuels. Most of the fire test methods in regulations have been developed by consensus standards organizations in response to a particular fire hazard. The two leading entities are the American Society for Testing and Materials (ASTM) and the National Fire Protection Association (NFPA). Methods are then referenced in the model building codes, such as the Standard Building Code (Southern Building Code), Basic Building Code (Building Code Officials Administration International), and the Uniform Building Code (International Conference of Building Officials), as well as NFPA's National Fire Codes, National Electrical Code, and Life Safety Code. Selected portions of these structures are in turn incorporated into laws by a governmental jurisdiction. In addition, there are a number of voluntary practices. For example, Underwriters Laboratories (UL) allows the use of its endorsement on products that meet their test criteria, and the upholstered furniture industry has adopted voluntary cigarette ignition-resistance standards.

Fire test methods attempt to provide correct information on the fire contribution of a product by exposing a small sample to conditions expected in a fire scenario.

The assessment of the contribution of a product to the fire severity and the resulting hazard to people and property combines appropriate product flammability data, descriptions of the building and occupants, and computer software that includes the dynamics and chemistry of fires.

Methods for Improved Performance

The materials of attention in promoting fire safety are generally organic polymers, both natural, such as wood (qv) and wool (qv), and synthetic, nylon (see POLYAMIDES), vinyl, and rubber (qv). Less fire-prone products generally have either inherently more stable polymeric structures or fire-retardant additives. The former are usually higher priced engineering plastics (qv) which achieve increased stability at elevated temperatures by incorporating stronger (often aromatic) chemical bonds in the backbone of the polymer. Examples are the polyimides, polybenzimidazoles, and polyetherketones. There are also some advanced polymers, such as the polyphosphazenes and the polysiloxanes, which have strong inorganic backbones. Thermally stable pendent groups are also necessary. Strongly bonded polymers may, however, be brittle or difficult to process.

Fire-retardant additives are most often used to improve fire performance of low-to-moderate cost commodity polymers. These additives may be physically blended with or chemically bonded to the host polymer. They generally effect either lower ignition susceptibility or, once ignited, lower flammability. Ignition resistance can be improved solely from the thermal behavior of the additive in the condensed phase. Retardants such as hydrated alumina add to the heat capacity of the product, thus increasing the enthalpy needed to bring the polymer to a temperature at which fracture of the chemical bonds occurs. The endothermic volatilization of bound water can be a significant component of the effectiveness of this family of retardants. Other additives, such as the organophosphates, change polymer decomposition chemistry. These materials can induce the formation of a cross-linked, more stable solid and can also lead to the formation of a surface char layer. This layer both insulates the product from further thermal degradation and impedes the flow of potentially flammable decomposition products from the interior of the product to the gas phase where combustion would occur.

RICHARD G. GANN
National Institute of Standards and Technology

R. G. Gann, R. A. Dipert, and M. J. Drews, in J. I. Kroschwitz, ed., *Encyclopedia of Polymer Science and Engineering*, 2nd ed., John Wiley & Sons, Inc., New York, 1986, pp. 154–210.

J. W. Lyons, *The Chemistry and Uses of Fire Retardants*, Wiley-Interscience, New York, 1970, Chapt. 5.

S. J. Ainsworth, *Chem. Eng. News* **70**, 34 (Aug. 31, 1992).

G. L. Nelson, ed., *Fire and Polymers*, ACS Symposium Series 425, American Chemical Society, Washington, D.C., 1990.

ANTIMONY AND OTHER INORGANIC FLAME RETARDANTS

Flame retardancy can be imparted to plastics by incorporating elements such as bromine, chlorine, antimony, tin, molybdenum, phosphorus, aluminum, and magnesium, either during the manufacture or when the plastics are compounded into some useful product. Phosphorus, bromine, and chlorine are usually incorporated as some organic compound. The other inorganic flame retardants are discussed herein.

Addition of approximately 40% of the halogen flame retardants are needed to obtain a reasonable degree of flame retardancy. This usually adversely affects the properties of the plastic. The efficiency of the halogens is enhanced by the addition of inorganic flame retardants, resulting in the overall reduction of flame-retardant additive package and minimizing the adverse effects of the retardants.

Hydrated metal oxides such as alumina hydrate are usually used alone because these are not synergistic with the halogens. They are useful in applications in which the halogens are excluded or low processing temperatures are used.

Antimony Compounds

Antimony compounds used as flame retardants include antimony trioxide (Sb_2O_3) (Table 1), antimony pentoxide (Sb_2O_5) (Table 2), and sodium antimonate (Na_3SbO_4).

Toxicity. Antimony has been found not to be a carcinogen or to present any undue risk to the environment. However, because antimony compounds also contain minor amounts of arsenic which is a poison and a carcinogen, warning labels are placed on all packages of antimony trioxide.

Mixed Metal Antimony Synergists. Worldwide scarcities of antimony have prompted manufacturers to develop synergists that contain less

Table 1. Physical Properties of Antimony Trioxide

Property	Grade[a]		
	Ultra fine	High tint	Low tint
specific gravity	5.3–5.5	5.3–5.8	5.3–5.8
particle size, μm	0.25–0.45	0.8–1.8	1.9–3.2

[a] All grades are white powders.

Table 2. Properties of Antimony Pentoxide and Sodium Antimonate

Property	Sb_2O_5	Na_3SbO_4
particle size, μm	0.03	1–2
surface area, m^2/gm	50	
specific gravity	4.0	4.8
surface activity	weakly acidic	basic
refractive index, n_D^{20}	1.7	1.75

antimony. Other metals have been found to work in concert with antimony to form a synergist that is as effective as antimony alone. Thermoguard CPA from Elf Atochem NA, which contains zinc in addition to antimony, can be used instead of antimony oxide in flexible poly(vinyl chloride) (PVC) as well as some polyolefin applications. The Oncor and AZ products which contain silicon, zinc, and phosphorus from Anzon Inc. can be used in a similar manner. The mixed metal synergists are 10 to 20% less expensive than antimony trioxide.

Antimony–Halogen Synergism. Antimony synergists are used almost exclusively with either brominated or chlorinated organic flame retardants. These work in concert with one another and provide a highly effective flame-retardant system. Antimony and the halogens react at flame temperatures to form the corresponding trihalide or oxyhalide. The product formed depends on the mole ratios of the reactants and the structure of the organic halogen compound. The active flame-retarding species, ie, the tribromide and the trichloride, are formed directly when the mole ratio of halogen to antimony is at least 3-to-1 and the halogen compound is capable of dehydrohalogenating.

Boron Compounds

In 1990 approximately 4500 metric tons of boron flame retardants were used in the United states to impart flame retardancy to plastics. The most widely used is zinc borate, prepared as an insoluble double salt from water-soluble zinc and boron compounds. Manufacturers and trade names of boron flame retardants are listed include Climax Performance Materials (ZB 467, ZB 223, ZB 113, ZB 237, ZB 325), U.S. Borax (Firebrake ZB), and Buckman Laboratories (Brusan M-11). Other boron compounds used include barium metaborate, boric acid and sodium borate, and ammonium fluoroborate.

Boron Mechanism. Boron functions as a flame retardant in both the condensed and vapor phases. Under flaming conditions boron and halogens form the corresponding trihalide. Because boron trihalides are effective Lewis acids, they promote cross-linking, minimizing decomposition of the polymer into volatile flammable gases. These trihalides are also volatile; thus they vaporize into the flame and release halogen which then functions as a flame inhibitor.

Boron also reacts with hydroxyl-containing polymers such as cellulose. When exposed to a flame the boron and hydroxyl groups form a glassy ester that coats the substrate and reduces polymer degradation. A similar type of action has been observed in the boron–alumina trihydrate system.

Alumina Trihydrate

In 1990, approximately 66,000 metric tons of alumina trihydrate, $Al_2O_3{\cdot}3H_2O$ (Table 3), the most widely used flame retardant, was used to inhibit the flammability of plastics processed at low temperatures. Alumina trihydrate is manufactured from either bauxite ore or recovered aluminum by either the Bayer or sinter processes. Manufacturers of alumina trihydrate include Solem Industries, Aluchem, Alcoa, Custom Grinding Sales, R. J. Marshall, Georgia Marble, and Hitax. Alumina trihydrate is the least expensive and least effective of the flame retarders. It is also limited to plastics that are not processed higher than 220°C.

Table 3. Physical Properties of Alumina Trihydrate

Property	Value
density, g/mL	2.42
refractive index, $n^{20}{}_D$	1.579
average particle size, μm	1–100
Mohs' hardness	2.5–3.5
color	white
water solubility	insoluble

Mechanism. Alumina trihydrate functions as a flame retardant in both the condensed and vapor phases. When activated, it decomposes endothermically, eliminating water.

$$2\ Al(OH)_3 \rightarrow Al_2O_3 + 3\ H_2O$$

In the flame phase the water vapor forms an envelope around the flame, which tends to exclude air and dilute the flammable gases.

Other Inorganic Materials

Other inorganic flame retardants include magnesium hydroxide, molybdenum oxides, and tin.

Applications

Poly(vinyl chloride). PVC is a hard, brittle polymer that is self-extinguishing. In order to make PVC useful and more pliable, plasticizers (qv) are added. More often than not the plasticizers are flammable and make the formulation less flame-resistant. The flame resistance of the poly(vinyl chloride) can be increased by the addition of an inorganic flame-retardant synergist, eg, antimony oxide, mixed metal antimony synergists, zinc borate, molybdenum oxide, zinc stannates, and alumina trihydrate.

Unsaturated Polyesters. There are two approaches used to provide flame retardancy to unsaturated polyesters. These materials can be made flame-resistant by incorporating halogen when made, or by adding some organic halogen compound when cured. In either case a synergist is needed.

Olefin Polymers. The flame resistance of polyethylene can be increased by the addition of either a halogen synergist system or hydrated fillers. Similar flame-retarder packages are used for polypropylene.

IRVING TOUVAL
Touval Associates

M. J. Drew, C. W. Jarves, and G. C. Lickfield, in G. L. Nelson, ed., *Fire and Polymers*, ACS Symposium Series 425, Washington, D.C.

J. W. Hastie, *High Temperature Vapors*, Academic Press, Inc., New York, 1975.

I. Touval, *J. Fire Flam.* **3**, 130 (1972).

HALOGENATED FLAME RETARDANTS

Halogenated flame retardants fall into two general classes, additive and reactive. Additives are mixed into the polymer in common polymer processing equipment. Other ingredients such as stabilizers, pigments (qv), and processing aids are often incorporated at the same time. Reactive flame retardants literally become part of the polymer by either reacting into the polymer backbone or grafting onto it.

Fundamentals of Flammability

In order for a solid to burn it must be volatilized, because combustion is almost exclusively a gas-phase phenomenon. In the case of a polymer, this means that decomposition must occur. The decomposition begins in the solid phase and may continue in the liquid (melt) and gas phases. Decomposition produces low molecular weight chemical compounds that eventually enter the gas phase. Heat from combustion causes further decomposition and volatilization and, therefore, further combustion. Thus the burning of a solid is like a chain reaction. For a compound to function as a flame retardant it must interrupt this cycle in some way. There are several mechanistic descriptions by which flame retardants modify flammability: inert gas dilution, thermal quenching, protective coatings, physical dilution, and chemical interaction.

Flammability Testing

One problem associated with discussing flame retardants is the lack of a clear, uniform definition of flammability. Hence, no clear, uniform definition of decreased flammability exists. The latest American Society for Testing and Materials (ASTM) compilation of fire tests lists over one hundred methods for assessing the flammability of materials. Several of the most common tests used on plastics include ASTM E162-87, ASTM E119-88, MVSS 302, ASTM D2863-87, ASTM E662-83, ASTM E84, UL 723, UL 910, UL 94, UL 790, ASTM E108-90, UL 1715, CAL 133, CAL 117, ASTM E1353-90, ASTM E1354-90, ASTM E1354-90, and ASTM E906.

Table 1. Additive Flame Retardants

Common name	Mol formula	Bromine, %	Specific gravity	Mp, °C°C
	Brominated			
ethylenebisdibromonorbornanedicarboximide	$C_{20}H_{20}Br_4N_2O_4$	45	2.07	294
tetrabromobisphenol A	$C_{15}H_{12}O_2Br_4$	58.4	2.17	180
tris-dibromopropylisocyanurate	$C_{12}H_{15}Br_6N_3O_3$	65.8		106–108
ethylenebistetrabromophthalimide	$C_{18}H_4N_2O_4Br_8$	68	2.66	445
tetrabromobisphenol S-bis(2,3 dibromopropylether)	$C_{18}H_{14}Br_8O_4S$	70.8		52–55
tetrabromocyclooctane	$C_8H_{12}Br_4$	74.7	2.27	73
dibromoethyldibromocyclohexane	$C_8H_{12}Br_4$	74.7	2.38	70–76
tetradecabromodiphenoxybenzene	$C_{18}Br_{14}O_2$	82	3.25	370
decabromodiphenyl oxide	$C_{12}Br_{10}O$	83	3.00	305
	Polymeric and Oligomeric Brominated			
tetrabromobisphenol A carbonate oligomer, phenoxy end capped	$(C_{16}H_{12}O_3Br_4)_n$	52	2.2	210–230
epoxy oligomers of tetrabromobisphenol A	$(C_{18}H_{16}O_3Br_4)_n$	52–54		
poly(dibromostyrene)	$(C_8H_6Br_2)_n$	59	1.9	155–165[a]
brominated polystyrene, low molecular weight	$(C_8H_{5.3}Br_{2.7})_n$	66	2.1	130–140
poly(pentabromobenzylacrylate)	$(C_{10}H_6Br_5O_2)_n$	70	2.05	210

[a] A higher molecular weight version has a mp of 210–230°C.

Table 2. Brominated Reactive Flame Retardants and Intermediates

Compound	Molecular formula	Bromine, %	Specific gravity	Mp, °C
diester/ether diol of tetrabromophthalic anhydride	$C_{15}H_{16}O_7Br_4$	46	1.80	liquid
tetrabromobisphenol A-bis(2-hydroxyethyl ether)	$C_{19}H_{20}O_4Br_4$	51.6	1.8	116
tetrabromobisphenol A	$C_{15}H_{12}O_2Br_4$	58.4	2.17	180
disodium salt of tetrabromophthalate	$C_8O_4Br_4Na_2$	61	2.8	>500
tetrabromodipentaerythritol	$C_9H_{20}O_2Br_4$	63.2	1.98	81.5–82.5
tribromophenyl allyl ether	$C_9H_7Br_3O$	64.6	2.20	74
tribromostyrene	$C_8H_5Br_3$	68		65–67
2,4,6-tribromophenol	$C_6H_3OBr_3$	72.5	2.22–2.55	95.5
tribomoneopentyl alcohol	$C_5H_9OBr_3$	73.6	2.28	62–67
pentabromobenzyl bromide	$C_7H_2Br_6$	84.8		
hexachlorocyclopentadiene	C_5Cl_6	78.0[a]	1.710	11[b]
chlorendic acid	$C_9H_2O_4Cl_6$	55.0[a]		[c]

[a] Chlorine, %.
[b] 239°C bp.
[c] Decomposes to the anhydride.

Flame Retardants

Compounds of chlorine and bromine are the halogen compounds having commercial significance as flame-retardant chemicals. Halogenated flame retardants can be broken down into three classes: brominated aliphatic, chlorinated aliphatic, and brominated aromatic. As a general rule, the thermal stability increases as brominated aliphatic < chlorinated aliphatic < brominated aromatic. The thermal stability of the aliphatic compounds is such that with few exceptions, thermal stabilizers such as a tin compound must be used. Brominated aromatic compounds are much more stable and may be used in thermoplastics at fairly high temperatures without the use of stabilizers and at very high temperatures with stabilizers.

Antimony–Halogen Synergism. Antimony oxide is commonly employed as a fire-retardant supplement for halogen-containing polymer systems as a means of reducing the halogen levels required to obtain a given degree of flame retardancy. This reduction is desirable because the required halogen content may be so high that it affects the physical properties of the final polymer. In many cases, the antimony oxide is used simply to give a more cost-effective system.

Brominated Additive Flame Retardants. Additive flame retardants are those that do not react in the application designated. There are a few compounds that can be used as an additive in one application and as a reactive in another. Tetrabromobisphenol A (TBBPA) is the most notable example. Table 1 lists the properties of most commercially available bromine-containing additive flame retardants.

Chlorinated Additive Flame Retardants. Bis(hexachlorocyclopentadieno)cyclooctane and pentabromochlorocyclohexane are chlorinated compounds used as additive flame retardants.

Oligomeric Flame Retardants. There are several oligomeric flame retardants. The principal advantage claimed for these materials is their resistance to bloom and plate-out. All of the available oligomeric flame retardants are brominated (Table 1).

Reactive Flame Retardants. Table 2 lists the commercially available reactive flame retardants and intermediates.

Economic Aspects

There are a relatively small number of producers of halogenated flame retardants, especially for brominated flame retardants, where three producers account for greater than 80% of world production.

Health and Safety

In general, the acute toxicity of halogenated flame retardants is quite low. Continual use of decabromidiphenyl oxide has been placed in question based on the discovery that under certain laboratory conditions brominated dibenzo-*p*-dioxins are generated.

Research sponsored by the Brominated Flame Retardants Industry Panel regarding the use of brominated flame retardants shows that there is no evidence that the use of decabromodiphenyl oxide leads to any unusual risk. In addition, a study by the National Bureau of Standards (now National Institute of Science and Technology) showed that the use of flame retardants significantly decreased the hazards associated with burning of common materials under realistic fire conditions. Work in Japan confirms this finding.

ALEX PETTIGREW
Ethyl Technical Center

J. Troitzsch, *International Plastics Flammability Handbook*, Hanser Publishers, Munich, Germany, 1990.

M. Lewin, S. M. Atlas, and E. M. Pearce, eds., *Flame Retardant Polymeric Materials*, Plenum Press, New York, 1975.

D. Price, B. Iddon, and B. J. Wakefield, eds., *Bromine Compounds Chemistry and Applications*, Elsevier, Amsterdam, the Netherlands, 1988.

J. A. Barnard and J. N. Bradley, *Flame and Combustion*, Chapman and Hall, London, 1985.

PHOSPHORUS FLAME RETARDANTS

One of the principal classes of flame retardants used in plastics and textiles is that of phosphorus, phosphorus–nitrogen, and phosphorus–halogen compounds (see also FLAME RETARDANTS FOR TEXTILES). Detailed reviews of phosphorus flame retardants have been published (see also PHOSPHORUS COMPOUNDS).

Mechanisms of Action

Condensed-Phase Mechanisms. The mode of action of phosphorus-based flame retardants in cellulosic systems is probably best understood. Cellulose (qv) decomposes by a noncatalyzed route to tarry depolymerization products, notably levoglucosan, which then decomposes to volatile combustible fragments such as alcohols, aldehydes (qv), ketones (qv), and hydrocarbons (qv). However, when catalyzed by acids, the decomposition of cellulose proceeds primarily as an endothermic dehydration of the carbohydrate to water vapor and char. Phosphoric acid is particularly efficaceous in this catalytic role because of its low volatility (see PHOSPHORIC ACIDS AND PHOSPHATES). Also, when strongly heated, phosphoric acid yields polyphosphoric acid which is even more effective in catalyzing the cellulose dehydration reaction. The flame-retardant action is believed to proceed by way of initial phosphorylation of the cellulose.

Vapor-Phase Mechanisms. Phosphorus flame retardants can also exert vapor-phase flame-retardant action. Both physical and chemical vapor-phase mechanisms have been proposed for the flame-retardant action of certain phosphorus compounds, such as triphenyl phosphate.

Interaction with Other Flame Retardants. Some claims have been made for a phosphorus–halogen synergism. A few cases are well established; however, phosphorus–halogen interactions are often merely additive, and in some cases slightly less than additive.

Antagonism between antimony oxide and phosphorus flame retardants has been reported in several polymer systems, and has been explained on the basis of phosphorus interfering with the formation or volatilization of antimony halides, perhaps by forming antimony phosphate.

Commercial Phosphorus-Based Flame Retardants

Many thousands of phosphorus compounds have been described as having flame-retardant utility. The compounds demonstrating commercial utility are much more limited in number. They include inorganic phosphorus compounds (red phosphorus, ammonium phosphates, insoluble ammonium polyphosphate, phosphoric acid-based systems for cellulosics), additive organic phosphorus flame retardants (melamine phosphates and other amine phosphates, trialkyl phosphates, dimethyl methylphosphonate, diethyl ethylphosphonate), halogenated alkyl phosphates and phosphonates (2-chloroethanol phosphate (3:1), 1-chloro-2-propanol phosphate (3:1), 1,3-dichloro-2-propanol phosphate (3:1), bis(2-chloroethyl) 2-chloroethylphosphonate, diphosphates, oligomeric 2-chloroethyl phosphate, 2-chloroethyl 2-bromoethyl 3-bromoneopentyl phosphate, oligomeric cyclic phosphonates, pentaerythritol phosphates, cyclic neopentyl thiophosphoric anhydride, aryl phosphates, phosphine oxides), reactive organic phosphorus compounds (organophosphorus monomers, phosphorus-containing diols and polyols, oligomeric phosphate–phosphonate), reactive organophosphorus compounds in textile finishing (tetrakis(hydroxymethyl)phosphonium salts, dimethyl 3-[(hydroxymethyl)amino]-3-oxopropylphosphonate, and phosphorus-containing polymers (polyester fibers containing phosphorus).

Health, Safety, and Environmental Factors

Toxicology. The structure–toxicity relationships of organophosphorus compounds have been extensively researched and are relatively well understood. The phosphorus-based flame retardants as a class exhibit only moderate-to-low toxicity.

A particular mode of neurotoxicity was discovered for tricresyl phosphate that correlated with the presence of the *o*-cresyl isomer (or certain other specific alkylphenyl isomers) in the triaryl phosphates. The use of low ortho-content cresols has become the accepted practice in industrial production of tricresyl phosphate.

Mutagenic and later carcinogenic properties were found for tris(2,3-dibromopropyl) phosphate, a flame retardant used on polyester fabric in the 1970s. This product is no longer on the market. The chemically somewhat-related tris(dichloroisopropyl) phosphate has been intensively studied and found not to display mutagenic activity. Tris(2-chloroethyl) phosphate appears to be a weak tumor-inducer in a susceptible rodent strain.

There appears to be no documented case of any type of fire retardant contributing to human fire casualties. Most smoke inhalation casualties appear to be caused by carbon monoxide.

Effects on Visible Smoke. Smoke is a main impediment to egress from a burning building. Although some examples are known where specific phosphorus flame retardants increased smoke in small-scale tests, other instances are reported where the presence of the retardant reduced smoke. The effect appears to be a complex function of burning conditions and of other ingredients in the formulation.

Environmental Considerations. The phosphate flame retardants, plasticizers, and functional fluids have come under intense environmental scrutiny. Results published to date on acute toxicity to aquatic algae, invertebrates, and fish indicate substantial differences between the various aryl phosphates. The EPA has summarized this data as well as the apparent need for additional testing.

Tests in pure water, river water, and activated sludge showed that commercial triaryl phosphates and alkyl diphenyl phosphates undergo reasonably facile degradation by hydrolysis and biodegradation. The phosphonates can undergo biodegradation of the carbon-to-phosphorus bond by certain microorganisms.

Economic Aspects

The largest volume use of phosphorus-based flame retardants may be in plasticized vinyl. Other use areas for phosphorus flame retardants are flexible urethane foams, polyester resins and other thermoset resins, adhesives, textiles, polycarbonate–ABS blends, and some other thermoplastics. Development efforts are well advanced to find applications for phosphorus flame retardants, especially ammonium polyphosphate combinations, in polyolefins, and red phosphorus in nylons. Interest is strong in finding phosphorus-based alternatives to those halogen-containing systems which have encountered environmental opposition, especially in Europe.

EDWARD D. WEIL
Polytechnic University

E. D. Weil, in R. E. Engel, ed., *Handbook of Organophosphorus Chemistry*, Marcel Dekker, Inc., New York, 1992, pp. 683–738.

A. Granzow, *Accounts Chem. Res.* **11**(5), 177–183 (1978).

E. D. Weil, "Additivity, Synergism and Antagonism in Flame Retardancy—Recent Developments," paper presented at *3rd Annual BCC Conference on Recent Advances in Flame Retardancy of Polymeric Materials*, Stamford, Conn., May 19–21, 1992.

A. R. Horrocks, *Rev. Prog. Coloration* **16**, 62–101 (1986).

FLAME RETARDANTS FOR TEXTILES

Although the terms *resistant* and *retardant* have similar meanings, flame resistant is normally used when referring to that property of a material which prevents it from burning when an external source of flame is removed; flame retardant is used when referring to the chemicals or chemical treatment applied to a material to impart flame resistance. Flameproof or fireproof, on the other hand, refer to materials totally resistant to flame or fire. No appreciable change in the physical or chemical properties is noted. Asbestos (qv) is an example of a fireproof material.

Another related term is smolder resistance. Smolder resistance implies resistance to ignition by a smoldering source, such as a lit cigarette, place on the surface of a fabric or in the crevice formed between two butting fabrics. A fabric can be smolder resistant and not flame resistant, or vice versa.

Flame Resistance

Factors Affecting Performance. The flame resistance of a textile fiber is affected by the chemical nature of the fiber, its ease of combustion, the fabric weight and construction, the efficiency of the flame retardant, the environment, and laundering conditions.

The weight and construction of the fabric affect its burning rate and ease of ignition. Lightweight, loose-weave fabrics burn much faster than heavier weight fabrics; therefore, a higher weight add-on of fire retardant is needed to impart adequate flame resistance.

Mechanism of Flame Retardants. The burning process of cellulose depends on both a source of ignition and the presence of oxygen. A low temperature degradation of cellulose proceeds by the formation of levoglucosan, which in turn undergoes dehydration and polymerization, leading to tars, flammable gases, liquids, and other solids. The flammable gases thus produced ignite, causing the liquids and tars to volatize to some extent. This produces additional volatile fractions which ignite and produce a carbonized residue that does not burn readily. The process continues until only carbonaceous material remains. After the flame has subsided, the carbonized residue slowly oxidizes and glowing continues until the carbonaceous char is consumed.

In general, cotton treated with an effective flame retardant provides the same decomposition products upon burning as does untreated cotton; however, the amount of tar is greatly reduced, with a corresponding increase in the solid char. Consequently, as decomposition takes place, smaller amounts of flammable gases are available from the tar, and greater amounts of nonflammable gases from the decomposition of the char fraction. Char is essentially carbon. Its oxidation causes afterglow. Phosphorus-containing compounds, in some cases polymers, are particularly effective in inhibiting char oxidation. Numerous studies have been made on burning of untreated and flame-retardant-treated cellulose.

Several theories have been postulated to explain the various types of flame retardants for cotton. These theories include coating, gas, thermal, and dehydration or chemical.

Durability of Retardant Finishes

Fire resistance of a treated cellulosic fabric is reduced when the retardant contains acid groups and the treated fabric is soaked or laundered in water containing calcium, magnesium, or alkali metal ions. Phosphate- and carbonate-based detergents affect durability of fire retardants. Soap-based detergents can result in a substantial loss of fire resistance because of the deposit of fatty acid salts. Phosphorus-based flame retardants are adversely affected by water hardness and laundry bleach, sodium hypochlorite. Exposure to sunlight and weathering can lead to sufficient loss of flame retardant so that the fabric is no longer flame resistant. Similarly, a combination of sunlight followed by laundering or autoclaving can also lead to loss of flame resistance in a cellulosic fabric.

Nondurable Finishes. Flame-retardant finishes that are not durable to laundering and bleaching are, in general, relatively inexpensive and efficient. In some cases, a mixture of two or more salts is more effective than either of the components alone.

The water-soluble flame retardants are most easily applied by impregnating the fabric with a water solution of a retardant, followed by drying. The water-soluble flame retardants used most widely for textiles are listed in Table 1. Less commonly used retardants include sulfamates of urea or other amides and amines; aliphatic amine phosphates, such as triethanolamine phosphate, phosphamic acid (amido phosphoric acid, $H_2PO_3NH_2$), and its salts; and alkylamine bromides, phosphates, and borates.

Semidurable Finishes. Semidurable fire retardants resist removal from 1 to approximately 15 launderings. Such retardants are adequate for applications such as drapes, upholstery, and mattress ticking. If they are sufficiently resistant to sunlight or can be easily protected from actinic degradation, they can also be applied to outdoor textile products. The principal disadvantage of water-soluble

Table 1. Water-Soluble Flame-Retardant Formulations,[a] % Composition

Formulation	Borax	Boric acid	Diammonium phosphate	Sodium phosphate dodecahydrate	Other
	$Na_2B_4O_7 \cdot 10\ H_2O$	H_3BO_3	$(NH_4)_2HPO_4$	$Na_3PO_4 \cdot 12\ H_2O$	
1	70	30			
2	47	20	33		
3		50		50	
4		50	50		
5	50	35		15	
6			25		75[b]
7	15	47			38[c]

[a] 100% Ammonium bromide, NH_4Br, is also used.
[b] Ammonium sulfamate, $NH_4OSO_2NH_2$.
[c] 18% Sodium phosphate, Na_3PO_4, and 20% sodium tungstate dihydrate, $Na_2WO \cdot 2H_2O$.

flame retardants is their lack of durability. This undesirable property can be overcome by precipitating their inorganic oxides on the fabric, eg, $WO_3 \cdot xH_2O$ and $SnO_2 \cdot yH_2O$:

$$2\ Na_2WO_4 + SnCl_4 + (2x + y)H_2O \rightarrow 4\ NaCl + 2\ WO_3 \cdot xH_2O + SnO_2 \cdot yH_2O$$

There are several methods for introducing the insoluble deposits into the fabric structure. The multiple bath method, in which the fabric is first impregnated with a water-soluble salt or salts in one bath and is then passed into a second bath which contains the precipitant, is used most often. Most semidurable retardants used on cotton are based on a combination of phosphorus and nitrogen compounds.

Early Durable Finishes. Early studies to produce durable flame retardants for cellulose were based on treatment with inorganic compounds containing antimony and titanium.

Outdoor Finishes. Excellent fire-resistant fabric has been obtained by treating fabric with a suspension or emulsion of insoluble fire-retardant salts or oxides, eg, antimony(III) oxide, along with a chlorinated organic vehicle such as chlorinated paraffin.

In the 1990s, two types of flame retardants are preferred for outdoor fabrics, ie, a system based on phosphorus and nitrogen such as the precondensate–NH_3 finish and an antimony–bromine system based on decabromodiphenyl oxide and antimony(III) oxide.

FWWMR Finish. The abbreviation for fire, water, weather, and mildew resistance, FWWMR, has been used to describe treatment with a chlorinated organic metal oxide. Plasticizers, coloring pigments, fillers, stabilizers, or fungicides usually are added. However, hand, drape, flexibility, and color of the fabric are more affected by this type of finish than by other flame retardants. Add-ons of up to 60% are required in many cases to obtain adequate flame resistance. Durability of this finish is good and fabric processed properly retains its flame resistance after four to five years of outdoor exposure. This type of finish is suited for very heavy fabrics, eg, tents, tarpaulins, or awnings.

Test Methods

Numerous tests covering flame retardancy and related matters are available. The requirements most often specified for fire resistance of a textile material are that it must pass either Federal Specification Method 5903 or NFPA 701.

Types of Retardants

Fire Retardants for Cellulosics. Phosphorus-containing materials are by far the most important class of compounds used to impart durable flame resistance to cellulose. Flame-retardant finishes containing phosphorus compounds usually also contain nitrogen or bromine or sometimes both.

Flame retardant fabrics and finishes include mesylated and tosylated celluloses, urea–phosphate type, phosphonomethylated ethers, amide-based systems, cyanamide, dialkyl phosphite and related retardants (Pyrovatex CP "New", dialkylphosphonopropionamides, triazines), THPC-based retardants (THPC–amide process, THPC–urea–disodium phosphate, THPOH–amide process, THP–amide process), ammonia–gas-cured flame retardants (THPOH–NH_3 process, precondensate–NH_3 process), and pentamethylphosphorotriamide.

Application techniques include radiation and incorporation of flame retardants in fiber.

Textile-Specific Uses of Flame Retardants

Flame retardants are used in smolder-resistant upholstery fabric, combination flame retardant–durable press performance, flame-retardant treatments for wool, thermoplastic fibers (Tris, decabromodiphenyl oxide–polyacrylate finishes, Antiblaze 19, nylon finishes), polyester–cotton fiber blends (THPOH–ammonia–Tris finish, decabromodiphenyl oxide–polyacrylate finish, THPC–amide–poly(vinyl bromide) finish, THPOH–NH_3 and Fyrol 76, LRC-100 finish, phosphonium salt–urea precondensate), cotton–wool blends, and core-yarn fabric.

Economic Aspects

The identification of Tris as a potential carcinogen dealt a resounding blow to the flame-retardant finishing industry. From 1977 to 1984, several principal suppliers of flame-retardant chemicals either reduced the size of their operations or abandoned the market completely. However, Albright and Wilson Corporation (U.K.) continues to produce THPC–urea precondensate and market it worldwide, and Westex Corporation (Chicago) continues to apply precondensate–NH_3 finish to millions of yards of goods for various end uses. American Cyanamid reentered the market with a precondensate-type flame retardant based on THPS.

The largest commission finishers of fire-resistant textiles in the United states (ca 1993) are Westex and MF&H Textiles, Inc. (Butler, Georgia). Specialized flame-retardant applications to cotton-wrapped polyester, Kevlar, nylon, and glass core yarns are beginning to attract the interest of the industry for special-purpose fabrics.

Health and Safety

Because Tris polyester flame-retardant chemical has been demonstrated to be a potential carcinogen, workers in this field have tested a number of commonly used chemicals for potential mutagenicity. Neither the THPOH–NH_3 finish nor its extracts caused a significant systematic increase in mutations when tested by the Ames mutagenicity test. The Hooker Chemical Co. has reported results of tests conducted by an independent laboratory which indicate no significant mutagenic potential from any of the company's proprietary textile flame retardants. Although Fyrol 76 was reported to be nontoxic, results from its mutagenic screening are not known. Stauffer's substitute for Tris, Fyrol FR2, was accused of mutagenic activity by the Environmental Defense Fund, and has been withdrawn from the market by the company. A study has been made by the National Toxicology Program Study on the carcinogenicity of THPC and THPS which concluded that there is no evidence of carcinogenic activity for either compound in rats or mice.

Regulatory Legislation. In February 1978, the Consumer Products Safety Commission approved changes in the FF-3 and FF-5 standards for children's sleepwear. It eliminated the melt–drip time limit and coverage for sizes below 1 and revised the method of testing the trim. This permits the use of untreated 100% nylon and 100% polyester for children's sleepwear.

TIMOTHY A. CALAMARI, JR.
ROBERT J. HARPER, JR.
United States Department of Agriculture

W. C. Kuryla and A. J. Papa, eds. (1973–1975), *Flame Retardancy of Polymeric Materials*, Vol. 5, Marcel Dekker, Inc., New York, 1979.

J. W. Lyons, *The Chemistry and Uses of Fire Retardants*, Wiley-Interscience, New York, 1970.

W. A. Reeves, G. L. Drake, Jr., and R. M. Perkins, *Fire-Resistant Textiles Handbook*, Technomic Publishing Co., Inc., Westport, Conn., 1974.

Textile Flammability, A Handbook of Regulations, Standards and Test Methods, American Association of Textile Chemists and Colorists, Research Triangle Park, N.C., 1975.

FLAVOR CHARACTERIZATION

Flavor characterization attempts to define what causes flavor and to determine if human response to flavor can be predicted. The ways in which simple flavor active substances, flavorants, produce perceptions are described both in terms of the physiology, ie, transduction, and psychophysics, ie, dose-response relationships, of flavor. Progress has been made in understanding how perceptions of simple flavorants are processed into hedonic behavior, ie, degree of liking, or concept formation, eg, crispy or umami (savory). However, it is unclear how complex mixtures of flavorants are perceived or what behavior they cause. Flavor characterization involves the chemical measurement of individual flavorants and the use of sensory tests to determine their impact on behavior.

Sensory Analysis. Sensory analysis is concerned with the similarities in human flavor perception using methods that are designed to average out certain differences and to detect others. A collection of people (a panel) tastes or smells the same material and reports their perceptions according to previously explained guidelines. Using statistical methods, the similarities, if any, in the panelists' perceptions can be isolated. Sensory analysis requires a large amount of time to design, execute, and analyze, and is therefore expensive. Consequently, manufacturers concerned with flavor are motivated to find less labor intensive instrumental procedures to predict the flavor perceptions of people. Although instrumental methods used for routine quality control are often less expensive than sensory tests, they are indirect and their accuracy must be established using direct sensory methods.

Flavor Perception

Flavor, ie, the human perceptions of flavorants, is generally defined in terms of odor and taste; a third component, texture, also may be included. Odor is a result of stimuli interacting with specialized receptors in the nose. Taste results from the interactions between stimuli and receptor organs on the tongue and in the mouth. There are, however, no single identifiable organs involved with the perception of texture. The presence of pain, the sense of touch, and the detection of sound all contribute to the perception of texture; thus the texture of a food includes all perceptions detected in the mouth that are owing to neither odor nor taste. In the broadest definition, flavor is the combined perception of odor, taste, and texture. Flavor is also used to denote a collection of flavorants that might be added to a food (see FLAVORS AND SPICES).

Psychophysics. Psychophysics, ie, the study of the relationship between sensory perceptions and the stimuli that produce them, has produced some very useful concepts for flavor characterization. One relationship between sensory intensity and stimulus intensity is called Weber's law. For determination of just noticeable difference (JND), a discriminability test is used in which people are asked to discriminate between a stimulus at two different intensities. The JND is the value of this minimally detectable difference. Data from such experiments contain only physical variables, ie, stimulus intensities. The gradual development, by Fechner, Plateau, and Stevens, of psychological variables corresponding to the perceived intensity of sensations led to a more refined form of Weber's law, generally referred to as Steven's law: $\Psi = k\Phi^n$, where Ψ is a measure of the perceived intensity of the sensation and Φ is a measure of the physical intensity of the stimulus which produced the response; k and n are constants. The constant k depends only on the units chosen for the variables Ψ and Φ, ie, the scales. However, the exponent n is characteristic of what is being measured and is independent of stimulus intensity. Two stimuli used for the same function, eg, sweetness, may have different Stevens' law exponents, as is the case with sucrose, $n = 1.5$, and saccharin, $n = 0.8$.

Table 1 lists several different sensory qualities and their corresponding Stevens' law exponents. These values were determined by a process called magnitude estimation, ie, people associate numbers proportional to their perception of the intensity of a sensation.

Table 1. Stevens' Law Exponents for Different Sensory Stimuli[a]

Stimulus	Exponent, n
electric shock	3.5
temperature	1.6
loudness of sound	0.6
brightness of light	0.3
sweetness of sucrose	1.5
bitterness of quinine	0.6
saltiness	1.0
sourness	1.0
odor of n-heptane	0.6

[a] Stevens' law, $\Psi = k\Phi^n$, where k and n are constants.

Flavor Intensity. In most sensory tests, a person is asked to associate a name or a number with his perceptions of a substance he sniffed or tasted. The set from which these names or numbers are chosen is called a scale. The four general types of scales are nominal, ordinal, interval, and ratio. Each has different properties and allowable statistics. The measurement of flavor intensity, unlike the evaluation of quality, requires an ordered scale, the simplest of which is an ordinal scale.

Flavor Description. Typically, a sensory analyst determines if two samples differ, and attempts to explain their differences so that changes can be made. The Arthur D. Little flavor profile (FP), quantitative descriptive analysis (QDA), and spectrum method are three of the most popular methods designed to answer these and more complicated questions. All three methods involve the training of people in the nominal scaling of the flavor qualities present in the food being studied, but they differ in their method for quantitation.

Threshold, Saturation, and Adaptation. Several aspects of flavor perception are not accounted for in the Weber-Stevens' law, eg, threshold, saturation, and adaptation. For every sense there is a minimum detectable stimulus intensity called the threshold. The threshold value is not absolute but is greatly affected by the presence of other stimuli, eg, the threshold for geosmin in both beet juice and fish flesh is ca 50 times higher than it is in water. Saturation is the concentration of a stimulus above which no increase in perception can be detected.

Exposure to a flavor over time always results in a decrease in the perceived intensity. This dynamic effect of flavorants, called adaptation, is a central part of the process by which people experience flavors in foods as well as in sensory tests. Measuring the dynamics of flavor perception is an emerging technology made possible by inexpensive computing. Called time-intensity analysis, these methods are finding wide applications in taste analysis.

Discriminant Sensory Analysis. Discriminant sensory analysis, ie, difference testing, is used to determine if a difference can be detected in the flavor of two or more samples by a panel of subjects. These differences may be quantitative, ie, a magnitude can be assigned to the differences but the nature of the difference is not revealed. These procedures yield much less information about the flavor of a food than descriptive analyses, yet are extremely useful; eg, a manufacturer might want to substitute one component of a food product with another safer or less expensive one without changing the flavor in any way. Several formulations can be attempted until one is found with flavor characteristics that cannot be discriminated from the original or standard sample.

The development of precise and reproducible methods of sensory analysis is prerequisite to the determination of what causes flavor, or the study of flavor chemistry. Knowing what chemical compounds are responsible for flavor allows the development of analytical techniques using chemistry rather than human subjects to characterize flavor. Routine analysis in most food production for the quality control of flavor is rare. Once standards for each flavor quality have been synthesized or isolated, they can also be used to train people to do more rigorous descriptive analyses.

Flavor Chemistry

A persistent idea is that there is a very small number of flavor qualities or characteristics, called primaries, each detected by a different kind of receptor site in the sensory organ. It is thought that each of these primary sites can be excited independently but that some chemicals can react with more than one site producing the perception of several flavor qualities simultaneously. Sweet, sour, salty, bitter, and umami qualities are generally accepted as five of the primaries for taste; sucrose, hydrochloric acid, sodium chloride, quinine, and glutamate, respectively, are compounds that have these primary tastes.

Odor Compounds

The relationship between molecular structure and sensory properties is very unclear for compounds with odor. It seems likely that there is a set of odors that could be called primaries, but a widely accepted list of such primary odor qualities has not been devised. Molecular size and shape have been used to describe the features that distinguish different odor qualities. There are seven odor primaries: ethereal, camphoraceous, musty, floral, minty, pungent, and putrid. Like all theories attempting to relate chemical structure with sensory properties, these ideas have little predictive value.

Identification of Odor Components. The methods used to isolate, concentrate, and identify odor components are not unique to flavor chemistry; the use of sensory analyses to monitor their progress is. Once a particular compound has been identified and a standard unambiguously synthesized, its odor characteristic must be verified by sensory analysis. The complexity of natural products requires the use of separation techniques having the highest available resolution, such as gas and liquid chromatography. In addition to high resolution, the most sensitive analytical techniques such as mass spectroscopy must be used to detect the highly potent trace components.

Since the early 1980s gas chromatography–olfactometry (gco) has emerged from a long history as a simple bioassay for odor in gas chromatography effluents to become a quantitative method for the characterization of odor and aroma. Quantitative gco is exemplified by CharmAnalysis and Aroma Extraction Dilution Analysis in which a series of dilutions are chromatographed separately. As each diluted extract is sniffed eluting from the gas chromatograph, the weaker odors drop below the threshold and cannot be detected. Combination of the data produced in these sessions produces a chromatogram. The taller peaks indicate odors that are well above their thresholds, whereas the smaller peaks represent compounds that may be below their thresholds. The quantitative bioassay data produced in dilution-based gco indicate the relative potency of odorants in a complex mixture; however, there are gco methods based on the perception of intensity instead of the measurement of potency. Most of the earlier methods involved the scaling of perceived intensity. A more developed method of this type, called Osme, is based on the computerized recording of lever position matched with perception of intensity.

TERRY E. ACREE
Cornell University

H. T. Lawless and B. P. Klein, *Sensory Science Theory and Application in Foods, ift Basic Symposium Series*, Marcel Dekker, Inc., New York, 1991, p. 441.

M. O'Mahony, *Sensory Evaluation of Foods*, Marcel Dekker, Inc., New York, 1986, p. 487.

T. E. Acree and R. Teranishi, eds., *Flavor Science: Sensible Principles and Techniques*, ACS Books, Washington, D.C., 1993.

C. T. Ho and C. H. Manley, *Flavor Measurement, ift Basic Symposium Series*, Marcel Dekker, Inc., New York, 1993.

M. Meilgaard, G. V. Civille, and B. T. Carr, *Sensory Evaluation Techniques*, CRC Press, Inc., Boca Raton, Fla., 1991.

FLAVORS AND SPICES

FLAVORS

Flavor is viewed as a division between physical (appearance, texture, and consistency), and chemical sense (smell, taste, and feeling). The Society of Flavor Chemists, Inc. defines flavor as "the sum total of those characteristics of any material taken in the mouth, perceived principally by the senses of taste and smell and also the general senses of pain and tactile receptors in the mouth, as perceived by the brain."

The acceptability of food is determined by its flavor, and a large number of flavoring substances are used. Industrial flavorings are needed for the commercial preparation of foods. However, most of the daily food intake, even in industrialized countries, contains flavor naturally or flavor formed during cooking and preparation for human consumption. Only a minor part of the daily food intake is covered by foods containing added flavorings.

Flavors do several things in food systems. Foremost among these functions is their ability to render food more acceptable and enjoyable. Flavors are often used to create the impression of flavor where little or none exists; to alter the flavor of a product, eg, the flavor of dairy products; to modify, supplement, or enhance an existing flavor, eg, the butter flavor in margarine and the meat or chicken flavor in bouillon. The addition of flavoring is often necessary to compensate for the loss of flavor during the processing of foods, eg, pasteurized foods, concentrated citrus fruit juices, alcoholic beverages, or during the freezing, filtration, pasteurization, and long-term storage of foods.

Food Acceptance. Four features of food are recognized to determine acceptance, ie, flavor, nutritive value, appearance, and mouthfeel. When all four aspects are in proper quantitative proportions, a food finds general acceptance. When all four are interdependent, appearance takes precedence over the others. However, a report by the Food Marketing Institute has shown that consumers placed nutrition second to flavor in importance. A food must have the expected or proper appearance and color before it will be readily consumed.

Taste. Certain basic principles are involved in the physiology of flavor perception. Researchers studying taste generally agree that there are at least five tastes, ie, salty, sour, bitter, sweet, and umami. Umami can be defined to the Japanese as the taste of three broths, ie, Kombu, Shiitake, and Katsuobushi. In English, the narrower definitions of the taste of monosodium L-glutamate (MSG), or the broad but vague concept of savory, meaty, or brothy are used (see FLAVOR CHARACTERIZATION). Tastes are perceived by certain sensory cells or taste buds, contained in the approximately 10,000 papillae located on the tongue.

Taste-active chemicals react with receptors on the surface of sensory cells (taste buds) in the papillae, causing electrical depolarization (drop in the voltage across the sensory cell membrane). The collection of biochemical events that are involved in this process is called transduction. There are several aspects that affect the extent and character of taste and smell; people differ considerably in their sensitivity to and appreciation of smell and taste, and there is a lack of a common language to describe smell and taste experiences.

Odor. The physiology of odor, which is the determining characteristic of flavor, is more complex and less understood than that of taste. It has been claimed that odor is 80% of flavor. A large number of odors are distinguishable, but it is not known how this is accomplished. Olfactory response is only observed when the substance contacts the olfactory membrane, called the olfactory mucosa or olfactory epithelium, which occupies an area of about 2.5 cm^2 in each nostril. Above the nasal passages, the two olfactory clefts are separated by the nasal septum. For a substance to have an odor, it must be capable of reaching the olfactory epithelium high up in the nose, and must come in contact with the olfactory cilia membrane.

The odor of a substance is most logically attributed to its molecular structure. As in taste, its perception is preceded by the process called transduction, in which a chemical reaction with a receptor cell excites a nerve center, giving a sensation. Brain functions such as emotion, attention, cognition, etc mediate these sensations into perceptions. Although odor quality appears to be associated with chemical structures, it has not been possible to predict odor type accurately on this basis.

Whatever the physiology of odor perception may be, the sense of smell is keener than that of taste. If flavors are classed into odors and tastes as is common practice in science, it can be calculated that there are probably more than 10^4 possible sensations of odor and only a few, perhaps five, sensations of taste.

Flavor Materials. Materials for flavoring may be divided into several groups. The most common groupings are either natural or artifi-

Table 1. Pineapple Flavor, Natural and Artificial

Ingredient	Wt %
pineapple juice conc, 60 brix	60.0
pineapple fortifier artificial	1.0
ethyl alcohol, 95%	15.0
water	24.0

cial flavorings. Natural materials include spices and herbs; essential oils and their extractives, concentrates, and isolates; fruit, fruit juices, and fruit essence; animal and vegetable materials and their extracts; and aromatic chemicals isolated by physical means from natural products, eg, citral from lemongrass and linalool from bois de rose.

Artificial materials include aliphatic, aromatic, and terpene compounds that are made synthetically, as opposed to those isolated from natural sources. Natural and artificial flavors are defined as a combination of natural flavors and artificial flavors. It is assumed that whichever portion is in greater amount becomes the first portion of the name (Table 1).

Artificial flavors are defined in the *Code of Federal Regulations* (CFR) as any substance or substances the function of which is to impart flavor, and which are not derived from natural sources. These items include the list of substances found in CFR 21 parts 172.515, 182.60, and the Flavor and Extracts Manufacturing Association's Generally Recognized as Safe (FEMA GRAS) lists.

In 1992 a large number of flavor materials were allowable on the FEMA and the FDA lists (Tables 2 and 3).

In commerce, several classifications of flavoring and compounded flavorings are listed according to composition to allow the user to conform to state and federal food regulations and labeling requirements, as well as to show their proper application. Both supplier and purchaser are subject to the control of the FDA, USDA, and the Bureau of Alcohol, Tobacco, and Firearms (BATF). The latter regulates the alcoholic content of flavors and the tax drawbacks on alcohol, ie, return of a portion of the tax paid on ethyl alcohol used in flavoring.

One class of flavorings, known as true fruit, is composed of fruit juices, their concentrates, and their essences (Table 4). A second group, fruit flavors with other natural flavors (WONF), contains fruit concentrates or extracts that may be fortified with natural essential oils or extractives (isolates), or other naturally occurring plants; (eg, Apple WONF contains apple juice conc, 72 brix (40.0 wt %), apple essence, 150-fold (20.0 wt %), apple fortifier natural (1.0 wt %), ethyl alcohol, 95% (15.0 wt %), and water (24.0 wt %). A third class, artificial fruit flavors, includes fruit concentrates fortified with synthetic materials. It is important to note that other than fruit flavors can also be fortified with synthetic materials, eg, in the making of an artificial maple or meat flavor (Table 5).

Table 2. Chemical Classes Approved for Use in Flavors

	Compounds		
Chemical class	1992	1965	Example
sulfur	152	13	thioester, thiol, mercaptan
nitrogen	99	21	amino acids, pyrazine, ester
acids	67	42	
esters	546	372	methyl, ethyl, allyl, terpene
acetals	28	21	
aldehydes	122	21	terpene
ketones	144	64	terpene, ionone, pyrone
alcohols	143	80	terpene, phenols
ethers	52	51	dioxane, furan, oxide
hydrocarbon	8		terpene
miscellaneous	11		
Total	*1415*	*730*	

Table 3. Natural and Food Ingredients Used in Flavors[a]

Compound	Number of items
Natural flavors[b]	
absolutes	19
botanical extracts	114
botanicals	249
concretes	3
essential oils	144
oleoresins	20
miscellaneous	9
Total	*558*
Food ingredients[c]	
emulsifiers	51
preservatives	49
anticaking agents	12
multipurpose	180
flavor	18
Total	*310*

[a] Information courtesy of Flavor Knowledge Systems, Glenview, Illinois.
[b] FEMA and FDA listings.
[c] FDA listing.

Table 4. True Fruit Apple Flavor

Ingredient	Wt %
apple juice conc, 72 brix[a]	80.0
apple essence[b]	5.0
ethyl alcohol, 95%	15.0

[a] Brix = g of sugar per 100 g liquid.
[b] 150 = fold, ie, one gallon (3.785 L) of concentrated distillate is obtained from 150 gallons of single-fold juice.

Table 5. Pineapple Flavor, Artificial

Ingredient	Wt %
allyl cyclohexane propionate	1.4
allyl caproate	13.0
methyl-*b*-methylthiolpropionate	0.2
geranyl propionate	0.5
ethyl isovalerate	1.0
ethyl butyrate	1.0
γ-nonalactone	0.1
maltol	1.0
vanillin	0.5
2,5-dimethyl-2 (*OH*)-4-(2*H*)-furanone	0.2
orange oil	1.0
ethyl alcohol, 95%	46.0
propylene glycol	34.1

A flavor is composed of two parts, a flavor portion and a diluent portion. The flavor portion is composed of three parts, a character item, a contributory item, and a differential item. A character item is a material whose smell or taste is reminiscent of the named flavor, a chemical additive, or a blend of chemicals that provides the major part of a flavor's sensory identity (more or less characteristic). A contributory item is an additive which, when smelled and/or tasted, helps to create, enhance, or potentiate the named flavor (not characteristic but essential by virtue of its acting together with the latter to produce a definite character). A differential item is an additive or a combination or additives which when smelled or tasted have little if any character reminiscent of the named flavor (neither characteristic or essential).

It is in this area where the greatest number of examples of creativity occur.

The function of the flavor portion is to give the flavor a name and to add character fixation to the flavor. Character fixation is the use of relatively high boiling point solids, at concentrations that are above their threshold values (at the use level), because once they exceed their threshold concentrations, the perception of the flavor does not change.

The diluent portion is the largest portion of a flavor. Its function is to keep the flavor homogeneous and to determine the form of the flavor. The form of a flavor is essentially its physical appearance, ie, liquid, powder, or paste, ie, that which makes the flavor applicable.

One class of flavorings, known as true fruit, is composed of fruit juices, their concentrates, and their essences. A second group, fruit flavor (WONF), contains fruit concentrates or extracts that may be fortified with natural essential oils or extractives (isolates), or other naturally occurring plants.

The apple fortifier is composed of a blend of botanical extractives and natural chemicals (isolates and those derived via natural processes). Pineapple fortifier artificial is composed in some part, if not altogether, of artificial chemicals, botanical extractives, essential oils, etc.

Compounding. In the compounding technique, constituents are selected or rejected because of their odor, taste, and physical chemical properties, eg, boiling point, solubility, and chemical reactivity, as well as the results of flavor tests in water, milk, or an appropriate medium. A compound considered to be characteristic is then combined with other ingredients into a flavor and tasted as a finished flavor in the final product by an applications laboratory.

A flavor is tried at several different levels and in different mediums until the most characteristic one is selected. This is important because the character of the material is known to change quality with concentration and environment.

Specifications. Specifications for many of the essential oils and artificial flavorings are available. Physical specifications encourage standardization and uniformity in basic flavor and perfume materials. Although compliance with specifications does not guarantee the flavor quality standards will be accepted, the specifications fill a need and provide valuable reference for the flavor industry.

The *Food Chemical Codex* defines food-grade quality for the identity and purity of chemicals used in food products. In the United States, the FDA adopts many of the *Food Chemicals Codex* specifications as the legal basis for food-grade quality of flavor and food chemicals.

Flavor Precursors. The characteristic flavors of foods, such as in fruits and vegetables, are considered to result from enzymatic action upon certain more complex components during the normal developmental or ripening process. It has been found that the flavor of fruit can be increased by a process called precursor atmosphere (PA). When apples were stored in a controlled atmosphere containing butyl alcohol, the butyl alcohol levels increased by a factor of two, and the polar products, butyl esters, and some sesquiterpene products increased significantly. The process offers the possibility of compensating for loss of flavor in fruit handling and processing owing to improper transportation conditions or excessive heat.

Another process employed to increase the formation of volatile compounds in fruit is that of bioregulators. When a bioregulator is applied to lemon trees, an increase in both the aldehyde and alcohol fractions of the lemon oil extracted from the fruit of the treated lemon trees was observed.

Enzymes not only produce characteristic and desirable flavor but also cause flavor deterioration (see ENZYMES, INDUSTRIAL). The latter enzyme types must be inactivated in order to stabilize and preserve a food. Freezing depresses enzymatic action. A more complete elimination of enzymatic action is accomplished by pasteurization.

The creation of flavor by enzymes is used in the fermentation process to prepare products such as alcoholic beverages, cheese, pickles, vinegar, bread, and sauerkraut. In some vegetables, such as Cruciferae (mustard) and Alliacae (onion and garlic), the flavor components are released enzymatically when the tissue is crushed or broken. Several essential oils are also created enzymatically. A more complex flavor development occurs in the production of chocolate. The chocolate beans are first fermented to develop fewer complex flavor precursors; upon roasting, these give the chocolate aroma. The flavor development process with vanilla beans also allows for the formation of flavor precursors. The green vanilla beans, which have little aroma or flavor, are scalded, removed, and allowed to perspire, which lowers the moisture content and retards the enzymatic activity. This process results in the formation of the vanilla aroma and flavor, and the dark-colored beans that after drying are the product of commerce.

The use of dry heat, as in roasting, baking, and frying, develops flavor characteristics not found in the unheated product. The distinctive flavor development in breakfast foods, ie, the crust of baked bread, the aroma of roasted coffee, etc, can be directly attributed to the chemical combinations brought about during the heat-treatment operation. These types of flavor are generally characterized by the presence of pyrazines in the product.

Flavor Regulations. The Pure Food and Drug Act of 1906 introduced federal regulations to combat food and drug adulteration and fraud. The law was superseded by the Food Drug, and Cosmetic Act of 1938 and the Amendments of 1954, 1958, 1960, and 1962 (see FOOD ADDITIVES). Flavor regulations differ from country to country, but progress is being made to harmonize regulations in the interest of international trade. There are at least five different ways countries can control and administer flavor regulations: countries may have a positive list of flavor materials, a negative list of flavor materials, a mixed-system having both a positive and negative list, no flavor regulations, or countries may demand prior government approval before a material can be used.

Sensory Evaluation. The type of food and its processing affect flavoring efficiency; therefore, flavor materials must be taste-tested in the food itself. Because there has been a lack of standardization of testing techniques, a committee on sensory evaluation of the Institute of Food Technologists has offered a guide which is designed to help in developing standard procedures. For each type of problem, appropriate taste tests are suggested together with the type of panel, number of samples per test, and analysis of data.

FRANK FISCHETTI, JR.
Craftmaster Flavor Technology, Inc.

H. W. Schultz, H. Day, and L. M. Libby, eds., *The Chemistry and Physiology of Flavor*, Avi Publications Co., Westport, Conn. 1967.

R. W. Moncrieff, *The Chemical Senses*, CRC Press, Cleveland, Ohio, 1967.

T. E. Furia and N. Bellanca, *Fenaroli's Handbook of Flavor Ingredients*, 2nd ed., Vol. 2, CRC Press, Boca Raton, Fla., 1975.

H. Heath, *Source Book of Flavors*, Avi Publications Co., Inc., Westport, Conn., 1981.

SPICES

A spice is any aromatic, pungent, or colored vegetable product used in preparation or cooking to give zest and a piquant or pleasing flavor or color to foods; it is not a food of itself. The vegetable product may be fresh, dried, or ground.

In the 1990s, flavorists, traders, and culinary experts have expanded the term spice to include many vegetable products. Spices can be divided into four general categories. The traditional or so-called true or tropical spices, eg, black and white pepper, cloves, cinnamon, ginger, nutmeg, and mace, may be buds, fruit, bark, roots, or other parts of tropical plants. *Herbs*, eg, sage, rosemary, marjoram, and oregano, are usually the leafy parts of plants that grow in the temperate zones. *Spice seeds*, eg, mustard, celery, caraway, dill, fennel, and anise, grow in both tropical and temperate zones. Lastly, *dehydrated aromatic vegetables*, eg, onion, garlic, parsley, and sweet pepper, are often dried to prevent spoilage.

Most spice products are dried to prevent spoilage. The reduction of water limits mold formation, but only slightly affects the aroma or pungency. The dried product maintains its character and pound for pound is stronger in aroma and flavor than the fresh spice, since a nonessential component has been substantially removed. In areas where a spice is grown, the same product that is dried for storage and shipment is often used fresh for flavoring.

Many spices are processed to produce essential oils, oleoresins, essences, tinctures, extracts, resinoids, etc. These processes separate nonflavor components and further concentrate the aromatic or pungent principles of the spices. Such products allow a wider variety of uses and applications of the vital spice components.

Labeling of Spices for Foods and Beverages. For the labeling of spices used in foods and beverages the United States Food and Drug Administration (FDA) does not permit every item contained in the four general categories to be labeled simply as spice. The FDA does not differentiate between culinary herbs and spices, but it does require that those substances that traditionally have been regarded as foods, eg, bell peppers, onion, garlic, and celery, be labeled separately. Also, spices used to impart color, eg, turmeric, saffron, paprika, or annatto, must be listed separately by name or as spice and coloring (see COLORANTS FOR FOOD, DRUGS, COSMETICS, AND MEDICAL DEVICES).

For the products under its jurisdiction, eg, meat and meat products, the United States Department of Agriculture (USDA) has requirements similar to those of the FDA. However, mustard and spices that impart color must always be listed separately; onion and garlic powder may be listed simply as flavors.

The American Spice Trade Association (ASTA) accepts spice as any dried plant product used primarily for seasoning purposes.

History

In the 1990s the United States is the leading factor in the spice trade. The western hemisphere has become a large producer of important spices, among them aromatic seeds and herbs, eg, mustard, capsicum, dehydrated onion and garlic, basil, tarragon, coriander, cardamom, ginger, and sesame seed. Brazil has become a factor in black pepper production. New York remains the main port of entry for spices (ca 1993) and the United States is the largest user.

Production and Economics of Spices

Spices have become commercial products in over 70 countries of the world and may be produced in almost every country that can grow crops. However, many species of botanicals can be grown only in particular climates or have particular soil requirements. The warm, moist, tropical climates foster the growth of more species than any other areas; the traditional or tropical spices originated in these areas.

The most important considerations in marketing and establishing a crop from a new source are constancy of supply and quality. For some spices, it is difficult to reduce labor costs, as some crops demand individual manual treatment even if grown on dedicated plantations.

Economic Market. The spice trade is controlled by many direct elements and responds slowly to supply and demand fluctuations. Resupply depends on growth to plant maturity, which for certain items, such as black pepper or nutmeg, can be several years. The raw material is directly affected by climate, adverse weather conditions, and control of plant diseases and insect and animal pests. Limited agricultural scientific advances are applied to the cultivation of the spice botanicals, and there are many grades of product and degrees of quality caused by different growing or processing conditions, sometimes by unknown factors as well. Local government control, nationalization, and political unrest also can cause shortages, and business manipulations to alter prices and supply are common.

United States Imports of Spices and Oleoresins. The consumption of spices continues to increase in the United States. The demand for ethnic foods, and the trend toward less salt, glycerides, and fat, has stimulated more spice and condiment use. The United States consumes approximately 25% of the spices produced in the world. In 1995, imports accounted for about 60% of U.S. seasoning needs compared to 80% in the early 1980s.

United States Exports of Spices and Oleoresins. The United States (ca 1996) is the foremost grower of peppermint, spearmint, lemon, lime, and grapefruit products. The mints are processed to essential oils, and the citrus fruit are sold as fresh fruit or processed to frozen concentrates and essential oils; thus they do not qualify as spices.

Many imported raw spices are processed and packaged in the United States and then exported. Because of strict United States food laws, many foreign purchasers prefer spice available from the United States, whether domestic or imported. Thus imported products may end up as a significant part of the export list.

Information regarding U.S. production of oleoresins is not available. It is estimated that there is a decline in domestic production of oleoresins of those spices imported in large volume, such as black pepper, capsicums of all types, and turmeric, since these oleoresins are more frequently produced in the growing areas. However, the manufacture of specialty oleoresins produced from selected imports will continue, and oleoresin production from domestically grown spices is expected to increase.

Market Trends of the 1990s. The United States spice market can be divided into three sectors based on application: industrial, ie, food processing and manufacture; institutional, ie, restaurants, hospitals, schools, and military; and retail. The food manufacturers and institutions account for almost 65% of U.S. spice usage, an increase from about 40% in the 1980s. Retail food outlets make up most of the remainder.

Synthetics. The lack of spice products to satisfy demand and the wide variation in price and availability have caused the manufacture of selected synthetics, chemically identical to the component in the natural spice, to replace the vital components of some spices. However, synthetic organic chemistry is not yet able to manufacture economically the many homologous piperine components in black pepper or those capsaicin amides in red pepper.

Some spice characters can be synthesized, eg, cinnamic aldehyde from petrochemical raw materials and vanillin from lignin.

Other synthetics with cost advantages and large volume productions are *l*-carvone, the primary component in natural spearmint essence; *d*-carvone, the primary component in natural dill and caraway; anethol, in place of anise and fennel spices; and smaller amounts of thymol replacing thyme and disulfide synthetics for onion and garlic. All of these synthetics must be labeled as artificial which may limit their use among consumers.

Specifications, Analysis, and Quality

Spices are natural agricultural products and exhibit a range of variations of many specific characteristics. The most important quality assessment is the subjective physical observation of the whole or ground spice by an expert. The macroscopic and microscopic examination of spice is the criterion for the continued analysis of the product to determine adherance to specifications.

Physical appearance, as well as flavor quality and strength, can be influenced by soil conditions, rainfall, storms, blights, insects, growing and harvesting methods, storage, etc. All of these must be considered to evaluate a particular lot and to harvest, sell, or buy the lot and use it in a food product.

The FDA applies the Federal Food, Drug and Cosmetic Act to the spice industry and its products. The FDA has established a definition for spice which is somewhat general. It states, however, that vegetables such as onions, garlic, and celery are regarded as foods, not spices, even if dried.

The USDA considers most spices generally recognized as safe (GRAS). There are no standards of identity or legal definitions of spices. Spices used in drugs must meet the official standards of the *U.S. Pharmacoepia* in force. Advisory specifications may also be applied in commercial spice trading.

JAMES A. ROGERS, JR.
Consultant

History of Spices, American Spice Trade Association Inc., New York, 1960.

F. Rosengarten, Jr., *The Book of Spices*, Pyramid Books, New York, 1973.

Spice Section, Tropical Products Bulletin, U.S. Dept. of Agriculture, Foreign Agricultural Service, Washington, D.C., Apr. 1996.

U.S. Spice Trade, FTEA 1-93, U.S. Dept. of Agriculture, Foreign Agricultural Service, Washington, D.C., Apr. 1993.

Official Analytical Methods of The American Spice Trade Association, 3rd ed., American Spice Trade Association, Englewood Cliffs, N.J., 1985.

FLOCCULATING AGENTS

Flocculation is defined as the process by which fine particles, suspended in a liquid medium, form stable aggregates called flocs. The degree of flocculation can be defined mathematically as the number of particles in a system before flocculation divided by the number of particles (flocs) after flocculation. Flocculation makes the suspension nonhomogeneous on a macroscopic scale. A complete or partial separation of the solid from the liquid phase can then be made by using a number of different mechanical devices. Flocculating agents are chemical additives which, at relatively low levels compared to the weight of the solid phase, increase the degree of flocculation of a suspension. They act on a molecular level on the surfaces of the particles to reduce repulsive forces and increase attractive forces.

Applications

The principal use of flocculating agents is to aid in making solid–liquid separations. These applications include:

1. Removing small amounts of suspended inorganic or organic particles from surface water prior to its use as drinking water or industrial process water.
2. Concentrating the organic solids in municipal or industrial wastewater to produce a sludge with a minimum volume and water content for incineration or other means of disposal, and a clarified (very low suspended solids) water that can be discharged or recycled. This operation is often called dewatering (qv).
3. Removing suspended inorganic material from waste streams generated in the beneficiation of ores or nonmetallic minerals, to form a concentrated slurry that can be used for reclamation of mined out areas or other uses and a clarified water that can be discharged or recycled.
4. Separating the solid and liquid phases in leaching operations, where a valuable material is contained in the liquid phase, so its recovery is to be maximized.
5. Binding fine cellulose fibers and solid inorganic additives to long cellulose fibers as the paper pulp is being formed into sheets on a paper machine (see PAPERMAKING ADDITIVES).

Chemical Composition

Flocculants can be classified as inorganic or organic. The inorganic group as well as some highly charged cationic organic flocculants are sometimes referred to as coagulants; however, no such distinction is made in this article.

Inorganic Flocculating Agents. The inorganic flocculating agents are water-soluble salts of divalent or trivalent metals. For all practical purposes these metals are aluminum, iron, and calcium. The principal materials currently in use are: aluminum sulfate, aluminum chloride hydroxide, sodium aluminate, ferric chloride, ferric sulfate, ferrous sulfate, calcium hydroxide, and lime.

Organic Flocculants. The organic flocculants are all water-soluble natural or synthetic polymers. Since the 1950s the use of natural products as flocculating agents as steadily declined as more effective synthetics have taken their place. The only natural polymers used to a significant degree as flocculants are starch (qv) and guar gum. Examples of synthetic polymers include acrylamide–acrylic polymers and their derivatives, polyamines and their derivatives, poly(ethylene oxide), and allylamine polymers.

Mechanism of Flocculation

In order to form flocs the individual particles must move and collide. Flocculation can be classified as either orthokinetic or perikinetic. In the first case particle motion results from turbulence in the suspension, and in the latter from Brownian motion. Orthokinetic motion is almost always the case in industrial applications. At very close distances, polar materials are attracted by dipole-induced dipole interactions commonly called van der Waals forces. In most aqueous suspensions, ionization of surface groups gives the particle an overall negative charge. The charged particles in suspension are surrounded by a group of positive ions referred to as the double layer. As particles approach each other the resulting electrostatic repulsion of the double layers prevents flocculation. Increasing the ionic strength of the liquid medium reduces the repulsion until the particles start to aggregate at the critical flocculation concentration. As the charge of these positive ions forming the double layer is increased by adding higher charged ions to the system, the double layer gets nearer to the surface allowing the particles to become closer and be attracted by the van der Waals forces. This is the explanation for the empirically derived Schulze-Hardy rule that the critical flocculation concentration of positive ions for a particular system decreases proportionally with the sixth power of the charge. This mechanism is called double-layer compression and is often cited for the inorganic flocculating agents, such as alum and ferric salts, which add trivalent ions to the system. However, this explanation of the action of aluminum and ferric salts does not take into account the fact that they are present at least partially as polymeric species when added to many systems, and that polymeric precipitates may be formed at the usual concentrations and the pH range that they are used.

The second flocculation mechanism is referred to as the charge patch or electrostatic mechanism. A highly cationic polymer is adsorbed on a negative particle surface in a flat conformation. This promotes flocculation by first reducing the overall negative charge on the particle thus reducing interparticle repulsion. A third mechanism is called bridging. Some individual segments of a very high molecular weight polymer, usually a high molecular weight anionic polyacrylamide, adsorb on a surface. Large segments of the polymer extend into the liquid phase where other segments are adsorbed on other particles, effectively linking the particles together with polymer bridges. In contrast to the first two mechanisms, bridging is strongly affected by molecular weight and the ionic content of the solution.

A fourth mechanism is called sweep flocculation. It is used primarily in very low solids systems such as raw water clarification. Addition of an inorganic salt produces a metal hydroxide precipitate which entrains fine particles of other suspended solids as it settles.

Flocculant Performance and Selection

There is no comprehensive quantitative theory for predicting flocculation behavior that can be used for flocculant selection. This must ultimately be determined experimentally. There are three variables that affect the results obtained in any particular flocculation system. These are the type of flocculant, type of substrate, and type of mechanical treatment of the flocculated substrate. The size and physical properties of the flocs that form, rather than the degree of flocculation, are the key elements in determining the practical effectiveness of a flocculant in any specific application. The effect of mechanical treatment can be viewed in terms of the type of force applied to the flocs. In thickeners and settling basins the flocs are acted on by gravity and by the weight of material added on top of them. In vacuum filters the flocs are subjected to atmospheric pressure. In belt presses and plate-and-frame filters the flocs are subjected to mechanical pressure and in centrifuges they are subject to centrifugal forces. In a flowing system, such as a continuous paper machine, they are subjected to shear and elongational forces on the same scale as the particle size. In addition to the type of force that is applied to the flocs, the kinetics of floc formation also plays an important role in the results obtained in their application.

There are some general principles that can serve as guidelines for initial screening in terms of both flocculant chemistry and molecular weight. In general the large flocs formed by high molecular weight polymers tend to settle faster than smaller ones.

In the case of thickeners, the process of compaction of the flocculated material is important. The flocs settle to the bottom and gradually coalesce under the weight of the material on top of them. As the bed of flocculated material compacts, water is released. Usually the bed is slowly stirred with a rotating rake to release trapped water. The concentrated slurry, called the underflow, is pumped out the bottom. Compaction can often be promoted by mixing coarse material with the substrate because it creates channels for the upward flow of water as it falls through the bed of flocculated material. The amount of compaction is critical in terms of calculating the size of the thickener needed for a particular operation. The process of compaction has been extensively reviewed in the literature.

For most substrates the operating dosage of flocculant necessary to give the settling rate necessary to operate a thickener is well below the maximum amount that can be adsorbed on the substrate. As more and more polymer is added above this operating dosage the flocs can become larger and somewhat sticky. The bed of flocculated material then becomes very viscous. The rake mechanism may become overloaded and the flocculated material may not flow into the underflow pump. The dosage response and the sensitivity to overdosing may affect the selection of flocculating agent.

For filter belt presses and centrifuges, resistance to shear and mechanical pressure is the most important parameter. In general, flocs produced by charge patch neutralization are stronger than those produced by inorganic salts alone.

For vacuum filters, both the rate of filtration and the dryness of the cake may be important. The filter cake can be modeled as a porous solid, and the best flocculants are the ones that can keep the pores open.

Retention aid polymers are used in a very high shear environment, so floc strength and the ability for flocs to reform after being sheared is important. The optimum floc size is a compromise. Larger flocs give better free drainage, but tend to produce an uneven sheet owing to air breakthrough in the suction portions of the paper machine. In some cases the type of floc needed for retention can be seen as similar to that needed for vacuum filtration. Floc size can be controlled by both the type of flocculant and the addition point.

General guidelines concerning the initial selection of flocculant chemistry are *(1)* suspensions of organic materials, such as municipal waste, are usually treated with a cationic flocculant, either inorganic or organic; and *(2)* suspensions of inorganic materials such as clay are usually treated with an anionic polymer or a combination of an anionic polymer with a cationic flocculating agent.

Laboratory Flocculant Testing. The objective of laboratory testing of flocculants is to determine which chemical composition and molecular weight will give the best cost performance. The usual method is to simulate on a laboratory scale the formation of flocs and then subject them to the same or similar types of forces as would be encountered in a full-scale dewatering device.

Operating Parameters and Control

Flocculating agents differ from other materials used in the chemical process industries in that their effect not only depends on the amount added, but also on the concentration of the solution and the point at which it is added. The process streams to which flocculants are added often vary in composition over relatively short time periods. This presents special problems in process control.

Dilution. In many applications, dilution of the flocculant solution before it is mixed with the substrate stream can improve performance. The mechanism probably involves getting a more uniform distribution of the polymer molecules.

Addition Point. The flocculant addition point in a continuous system can also have a significant effect on flocculant performance.

Automatic Control. In some industries, the waste streams can vary in composition over a relatively short time period. When the solids level of a slurry changes, the entire dosage response may change. Automatic systems are available for thickeners that adjust the dosage according to the incoming solids level, overflow turbidity, and streaming current potential.

Analysis

Inorganic flocculants are analyzed by the usual methods for compounds of this type. Residual metal ions in the effluent are measured by spectroscopic techniques such as atomic absorption.

The detection of organic polymers in solution represents a more difficult problem, especially in industrial water and wastewater. In theory, charged polymers react with polymers of the opposite charge in solution and such reactions can be used to titrate the concentration of polymer present. There are a number of techniques using this method.

The molecular weights and molecular weight distributions of lower molecular weight polymeric flocculants are determined by viscosity measurements. High molecular weight acrylamide-based polymers are characterized by light scattering techniques.

Toxicology and Environmental Issues

Based on animal studies and mutagenicity studies, trace amounts of organic polymers do not appear to present a toxicity problem in drinking water. The reaction products with both chlorine and ozone also appear to have low toxicity. The principal concern is the presence of unreacted monomer and other toxic and potentially carcinogenic nonpolymeric organic compounds in commercial polymeric flocculants. The principal compounds are acrylamide in acrylamide based polymers, dimethyldiallyammonium chloride in allylic polymers, and epichlorohydrin and chlorinated propanols in polyamines, as well as the reaction products of these compounds with ozone and chlorine.

Until 1990 the EPA maintained a list of chemicals suitable for potable water treatment in the United States. Since then the entire question of certification and standards has been turned over to a group of organizations headed by the National Sanitation Foundation, which has issued voluntary standards. As of January 1992, standards had been issued for most of the principal inorganic products, but only for three polymers, poly(DADMAC), Epi-DMA (epichlorohydrindimethylamine) polymers and polyacrylamide. Certifications for commercial products meeting specified standards are issued by the National Sanitation Foundation, Underwriter Laboratories, and Safe Water Additives Institute (SWAI).

The same questions about the safety of organic flocculants have been raised in other countries. The most drastic response has occurred in Japan and Switzerland where the use of any synthetic polymers for drinking water treatment is not permitted.

Economic Aspects

The principal trend in the flocculants market is the gradual replacement of low price inorganics, especially alum, with higher priced polymers. The strongest growth is expected for the polymers, which have increased in value in the United States at an annualized rate of 12.5%.

In terms of value, the alum market share is expected to decline. Alum is facing strong competition from polyaluminum chloride both in water treatment and paper, and from iron salts in water treatment.

The polymer market in the United States is dominated by synthetics with natural polymers constituting about one-eighth in monetary terms. Of the synthetic polymers, most are based on acrylamide. A list of producers is as follows; producers in the left-hand column also produce polyamines and polyquaternaries.

Producers of synthetic flocculants based on acrylamide include BASF, Betz Laboratories, Calgon, Cytec Industries, Floerger, Nalco, Polypure, Rohm GmbH, Sankyo Chemical Industries, Sanyo Chemical Industries, Sumitomo Chemical Co., Allied Colloids, Diafloc, Dow, Chemische Fabrik Stockhausen, KEMIRA Oy, Kyoritzu Yuki, Mitsubishi Chemical Industries, Mitsui-Cyanamid, and Toagosei Chemical Industries.

Prices of natural products such as starch, which is produced in many countries, and guar, which is produced mainly in India and Pakistan, are affected by unpredictable factors such as the weather. Toward the end of 1991 prices were rising; however, in the future an oversupply might cause a large drop in prices. Because of the amounts used, starch is usually purchased locally, and pricing fluctuates with local farm prices and conditions.

HOWARD I. HEITNER
Cytec Industries

F. Halverson, in K. J. Hipolit, ed., *Chemical Processing Aids in Papermaking: A Practical Guide*, TAPPI Press, Atlanta, Ga., 1992, pp. 103–127.

J. Gregory, in B. M. Moudgil and P. Somasundaran, eds., *Flocculation, Sedimentation and Consolidation*, American Institute of Chemical Engineers, New York, 1985, pp. 125–138.

FLOTATION

Flotation or froth flotation is a physicochemical property-based separation process. It is widely utilized in the area of mineral processing also known as ore dressing and mineral beneficiation for mineral concentration. In addition to the mining and metallurgical industries, flotation also finds applications in sewage treatment, water purification, bitumen recovery from tar sands, and coal desulfurization as well as environmental cleaning.

Technology

The flotation process is based on the exploitation of wettability differences of particles to be separated. Differences of wettability among solid (mineral) particles can be natural, or can be induced by the use of chemical adsorbates. Because the largest segment of industrial applications is conducted in water, with air, the following discussion is confined mainly to these fluids.

Flotation in the minerals industry applies to a particle size range of about 500 μm (eg, coal cleaning) to 20–10 μm (eg, copper ore concentration); however, 65 mesh (212 μm) to 270 mesh (53 μm) is typical. Water-polluting particulates can be aggregated to flotable sizes.

The flotation process is based on the exploitation of wettability differences of particles to be separated. Differences of wettability among solid (mineral) particles can be natural, or can be induced by the use of chemical adsorbates. Because the largest segment of industrial applications is conducted in water, with air, the following discussion is confined mainly to these fluids.

Flotation in the minerals industry applies to a particle size range of about 500 μm (eg, coal cleaning) to 2–10 μm (eg, copper ore concentration); however, 65 mesh (212 μm) to 270 mesh (53 μm) is typical. Figure 1 summarizes the main steps in mineral processing using froth flotation.

The flotation step is accomplished by the preparation of a pulp, consisting of a solid–liquid slurry that may contain up to 40% solids, to which chemical reagents known as collectors are added in a conditioning tank. The products from the flotation cell are a concentrate and a tailings stream. The concentrate proceeds to the next step for further cleaning or treatment by hydro- or pyrometallurgical methods for the extraction of metals and other valuable compounds, while

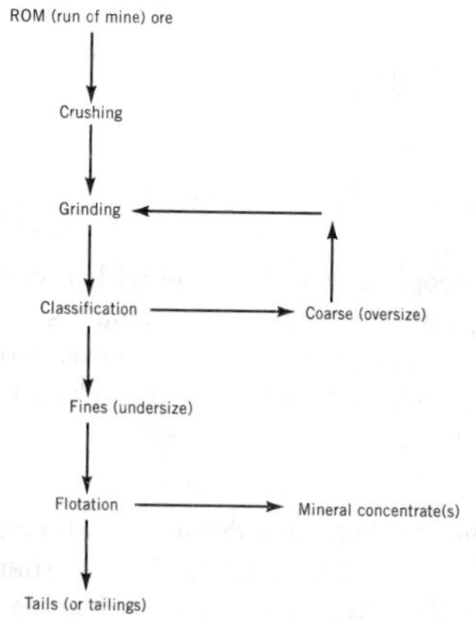

Figure 1. A generic ore beneficiation flow sheet.

the tailings, which are ore components stripped of their valuable mineral content, are collected in lagoons known as tailings ponds. A typical froth flotation process can treat a ROM ore that assays 0.5% to a few percent copper to give a mineral concentrate analyzing 35% copper with a recovery of more than 85% of the copper content of the original ore.

The actual flotation phenomenon occurs in flotation cells usually arranged in batteries and in industrial plants and individual cells can be any size from a few to 30 m^3 in volume.

Process Design and Machinery. Following the field work of geologists and mining engineers and analyses (assays) to establish the grades (concentrations) of values in ores, a mineral concentration flow sheet is established on the basis of a number of preliminary tests. These include studies of comminution properties of the ore, liberation properties of the minerals, and optimization of conditions at which they occur. Reagent testing, choice of flotation conditions, pH, collectors, frothers, and auxiliary reagents follow. The locked cycle test is a design aid that allows the simulation of a full-scale flotation procedure prior to pilot-plant testing.

Flotation cells, also called flotation machines, exist in numerous designs differing in mode of agitation and method of gas introduction and dispersion. The processes that occur in a typical flotation cell consist of agitation, particle–bubble collision and attachment, flotation of particle–bubble aggregates, collection of aggregates in a froth layer at the top of the cell, removal or mineral-laden froth as concentrate, and flow of the nonfloating fraction as tailings slurry.

Interfacial Phenomena

Flotation is a surface chemistry-based process, where numerous phenomena that simultaneously occur at the solid–liquid–air interfacial region determine its outcome. In this context, the variable known as contact angle θ illustrated in Figure 2, is an important correlative parameter. At $\theta = 0°$, the liquid spreads on the solid; in aqueous media in contact with air such a solid is said to be hydrophilic and is wetted by water. Air bubbles do not adhere to hydrophilic solids in water. Conversely, hydrophobic solids are not wetted by water; air bubbles do adhere to them and the value of the contact angle is larger than zero degrees, ie, $\theta > 0°$.

The three interfacial tensions at equilibrium (Fig. 2) conform to Young's equation (eq. 1): where γ represents solid–gas, solid–liquid, and liquid–gas interfacial tensions as indicated by subscripts.

$$\gamma_{sg} - \gamma_{sl} = \gamma_{lg} \cos\theta \qquad (1)$$

Electrical Phenomena at the Solid–Liquid Interface. Solid particles, such as minerals, in contact with water, as in a flotation pulp, undergo an electrical charge rearrangement at the water–solid interface because of hydration and ion dissolution from the lattice, ion adsorption from the aqueous environment, as well as lattice defects and substitutions. Thus an electrical double layer surrounding the particles

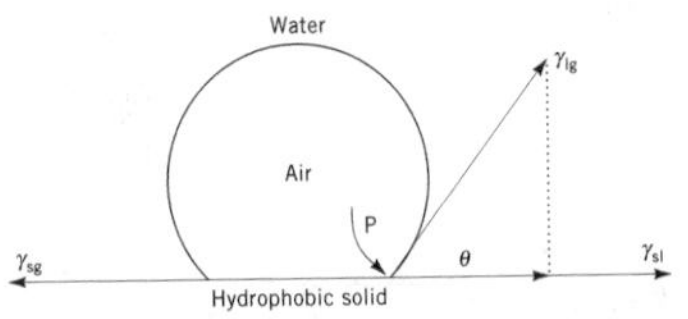

Figure 2. The concept of contact angle with a captive bubble in an aqueous medium, adhering to a hydrophobic solid: P is the three-phase contact point. Here, the vector γ_{lg} passes through P and forms a tangent to the curved surface of the air bubble. The contact angle θ is drawn into the liquid.

is established. The double layer consists of wall charges, then a layer of strongly bonded ions followed by, moving further into the liquid, a layer that consists of compactly packed, solvated ions (Stern layer) after which the diffuse part of the double layer starts.

The chemical composition, stoichiometry, and crystal structures of the solids in contact with water also play important roles in the degree of hydration that occurs at the solid–liquid interface and adsorption phenomena that affect the flotation process.

The zeta potential is essentially the potential that can be measured at the surface of shear that forms if the solid were to be moved relative to the surrounding ionic medium. Techniques for the measurement of the zeta potentials of particles of various sizes are collectively known as electrokinetic potential measurement methods and include microelectrophoresis, streaming potential, sedimentation potential, and electroosmosis. In principle, zeta potential allows the definition of an isoelectric point (IEP) for each mineral (or suspended solid) that defines the conditions at which the measured zeta potential is equal to zero. IEP is also known as point of zero charge (PZC) when the surface carries no net charge. This is the condition at which the net charge on the wall of the particle is electrostatically compensated by an equal and oppositely signed quantity of charge in the diffuse part of the electrical double layer.

Chemicals in the Flotation Process

Flotation reagents are used in the froth flotation process to *(1)* enhance hydrophobicity, *(2)* control selectivity, *(3)* enhance recovery and grade, and *(4)* affect the velocity (kinetics) of the separation process. These chemicals are classified based on utilization: collector, frother, auxiliary reagent, or based on reagent chemistry: polar, nonpolar, and anionic, cationic, nonionic, and amphoteric. The active groups of the reagent molecules are typically carboxylates, xanthates, sulfates or sulfonates, and ammonium ions, as well as alcoholic hydroxyl.

An inherent drawback of classifying flotation reagents according to their function in the flotation process is that what acts as a frother in one flotation system might play both collecting and frothing roles in another.

Interaction of Solids With Flotation Reagents. For flotation to occur with the aid of reagents, such compounds must adsorb at the solid–liquid interface unless the solid to be floated is naturally hydrophobic. In this latter case only depression can be attempted by the use of additional ions or depressants that hinder bubble–particle adhesion. Frothers (typically long-chain alcohols) and/or modifying agents such as hydrocarbon oils can, however, be used to enhance the collection of naturally hydrophobic solids such as MS_2, talc, or plastics.

The following mechanisms of adsorption are responsible for the formation of mineral–reagent bonds: electrostatic interactions, hydrogen bond formation, collectors fitting into lattice cavities, chemical bond formation (chemisorption), crystal field adsorption, and hydrophobic bonding.

Flotation Kinetics

Flotation process kinetics determine the residence time, the average time a given particle stays in the flotation pulp from the instant it enters the cell until it exits. One way to study flotation kinetics is to record flotation recoveries as a function of time under a given set of conditions such as pulp pH, collector concentration, particle size, etc. The data allow the derivation of a simplified expression that describes the rate of the process:

$$(dC/dt = -kC^n) \tag{2}$$

where C = concentration of solids left in the flotation cell, t = time, k = rate constant, and n = order of the process.

Two technologically significant concepts in mineral concentration processes including froth flotation are recovery and grade. Recovery quantifies the percentage of value mineral collected in the froth layer whereas grade represents the chemical analyses of starting materials and products. Recovery and grade vary inversely to one another.

Applications

Applications include sulfide ore flotation, nonsulfide ore flotation, soluble salt flotation, coal flotation, water treatment, ion flotation and foam separation, the gamma flotation process, two-liquid flotation, skin flotation, and piggyback flotation.

Environment, Safety, and Future Developments

New technologies based on the process of froth flotation in areas outside mineral technology are being developed. These include plastics recycling, glass recycling, and recovery of radioactive contaminants or heavy-metals removal from soil; newsprint deinking is an established flotation technology. Similarly, ion and precipitate flotation as well as foam fractionation are areas poised for increased activity due to their potential usefulness in environmental site cleanup operations. In fact, the utilization of the unit operations and unit processes of mineral processing is fundamental to the separations technology on which recycling relies.

Ore flotation processes treat millions of tons of minerals per year, and since these are associated with mining activity they appear to be associated with physical damage to the environment. However, these technologies have a long history of practice and associated environmental control procedures. Specially lined tailings ponds, turbidity, and toxic chemical abatement approaches designed to eliminate environmental damage, as well as revegetation of old mining sites tailings ponds, dams, and dikes, are widespread practices. Furthermore, the flotation process is a technology well established in sewage treatment and water purification, and it can be used for the removal of harmful ions from effluents. Therefore it is safe and the flotation industry is self-regulating.

BAKI YARAR
Colorado School of Mines

A. M. Guadin, *Flotation*, McGraw-Hill Book Co., Inc., New York, 1957.

M. C. Fuerstenau, ed., *Flotation: A.M. Gaudin Memorial Volume*, 2 vols., American Institution of Mining Metallurgical and Petroleum Engineers, Inc., New York, 1976.

J. Leja, *Surface Chemistry of Froth Flotation*, Plenum Press, New York, 1982.

N. L. Weiss, ed., *SME Mineral Processing Handbook*, Vol. 1, AIME Inc., New York, 1985, pp. 5.1–5.110.

FLOW MEASUREMENT

Flow measurement is a broad field covering a spectrum ranging from the minuscule flow rates associated with the pharmaceutical industry to the immense volumes involved in rivers. This measurement is an essential part of the production, distribution, consumption, and disposal of all liquids and gases including fuels, chemicals, foods, and wastes.

Flow Meter Selection

A number of considerations should be evaluated before a flow measurement method can be selected for any application. These considerations can be divided into four general classifications: fluid properties; ambient environment; measurement requirements; and economics.

Flow Calibration Standards

Flow measuring equipment must generally be wet calibrated to attain maximum accuracy, and principal flow meter manufacturers maintain extensive facilities for this purpose. In addition, a number of governments, universities, and large flow meter users maintain flow laboratories. Calibrations are generally performed with water or air using one or more of four basic standards: weigh tanks, volumetric tanks, pipe provers, or master flow meters.

Flow Meter Classifications

Flow meters have traditionally been classified as either electrical or mechanical depending on the nature of the output signal, power requirements, or both. However, improvement in electrical transducer technology has blurred the distinction between these categories. Many flow meters previously classified as mechanical are now used with electrical transducers.

The flow meters discussed herein are divided into two groups, based on the method by which the basic flow signal is generated. The first group consists of meters in which the signal is generated from the energy of the flowing fluid. The second group comprises those flow meters that derive their basic signal from the interaction of the flow and an external stimulus. Meters can be further divided into three subgroups, depending on whether fluid velocity, the volumetric flow rate, or the mass flow rate is measured.

Fluid Energy Activated Flow Meters

Fluid energy activated flow meters include positive-displacement flow meters (eg, reciprocating piston meters, bellows or diaphragm meters, nutating disk meters, rotary impeller vane and gear meters), differential-pressure flow meters (eg, orifice plates, venturi tubes, flow nozzles, elbow meters, wedge meters, pitot tubes, and laminar flow elements), target flow meters, variable-area flow meters, head-area meters (eg, weirs, flumes), cup and vane anemometers, current meters, turbine meters, and oscillatory flow meters (eg, fluid oscillation, vortex precession, vortex shedding).

External Stimulus Flow Meters

External stimulus flow meters are generally electrical in nature. These devices derive their signal from the interaction of the fluid motion with some external stimulus such as a magnetic field, laser energy, an ultrasonic beam, or a radioactive tracer. They include electromagnetic flow meters, momentum flow meters (eg, Coriolis-type flow meters, axial-flow angular-momentum flow meters), ultrasonic flow meters (eg, passive or turbulent noise flow meters, Doppler or frequency-shift flow meters, and transit time flow meters), laser Doppler velocimeters, correlation flow meters (eg, tracer type, cross correlation), and thermal flow meters (eg, hot-wire and hot-film anemometers, differential-temperature thermal flow meters).

THOMAS H. BURGESS
Bailey-Fischer & Porter Company

D. W. Spitzer, *Industrial Flow Measurement*, Instrument Society Of America, Research Triangle Park, N.C., 1984.

Fluid Meters–Their Theory and Application, 6th ed., American Society of Mechanical Engineers, New York, 1971.

D. W. Spitzer, ed., *Flow Measurement*, Instrument Society of America, Research Triangle Park, N.C., 1991.

P. Ackers and co-workers, *Weirs and Flumes*, John Wiley & Sons, Inc., New York, 1978.

FLUIDIZATION

Gas–solids fluidization is the levitation of a bed of solid particles by a gas. Intense solids mixing and good gas–solids contact create an isothermal system having good mass transfer. The gas-fluidized bed is ideal for many chemical reactions, drying, mixing, and heat-transfer applications. Solids can also be fluidized by a liquid or by gas and liquid combined. Liquid and gas–liquid fluidization applications are growing in number, but gas–solids fluidization applications dominate the fluidization field. This article discusses gas–solids fluidization. The basic concepts of a gas-fluidized bed are illustrated in Figure 1.

Fluidized-bed applications in the 1990s may be separated into catalytic reactions, noncatalytic reactions, and physical processes. Examples of fluidized-bed applications include the following:

Chemical Catalytic Processes
- Fluid catalytic cracking (FCC) of heavy petroleum fractions
- Phthalic anhydride
- Acrylonitrile
- Aniline
- Synthesis of polyethylene and polypropylene
- Fischer-Tropsch synthesis
- Oxidation of SO_2 to SO_3
- Chlorination or bromination of methane, ethylene, etc
- Maleic anhydride
- Pyridine

Chemical Noncatalytic Processes
- Roasting of sulfide and sulfate ores (ZnS, pyrites, Cu_2S, $CuCoS_4$, nickel sulfides)
- Calcination (limestones, phosphates, aluminum hydroxide)
- Incineration of waste liquids and solids refuse
- Coking (thermal cracking)
- Combustion of coal and other fuels
- Gasification of coal, peat, wood wastes
- Carbonization of coal (decomposition without oxygen)
- Fluoridation of UO_2 pellets
- Catalyst regeneration
- Hydrogen reduction of ores
- Titanium dioxide

Physical Processes
- Drying (eg, phosphates, coal, PVC, polypropylene, foods)
- Granulation (eg, pharmaceuticals, fertilizers)
- Classification
- Blending
- Coating (eg, polymer coat on metal object)
- High temperature baths
- Airslide conveying
- Absorption (eg, CS_2)
- Filtering of aerosols

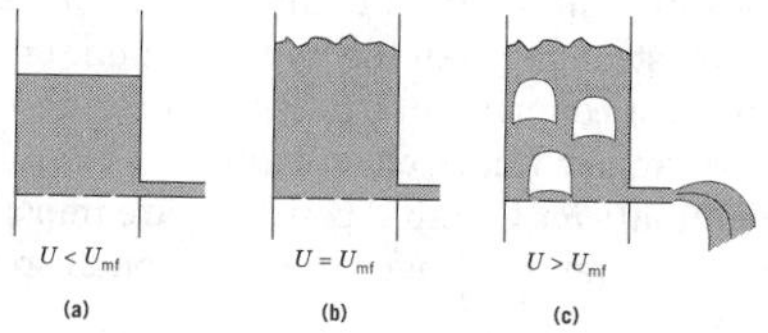

Figure 1. Fluidized-bed behavior where U is the superficial gas velocity and U_{mf} is the minimum fluidization velocity: (**a**) packed bed, (**b**) fluid bed, uniform expansion; and (**c**) bubbling fluid bed.

Medical beds

Quenching, annealing, tempering

Particle Properties

Fluidized-bed design procedures require an understanding of particle properties. The most important properties for fluidization are particle size distribution, particle density, and sphericity.

Particle Size. The solids in a fluidized bed are never identical in size and follow a particle size distribution. An average particle diameter, d_p, is generally used for design. It is necessary to give relatively more emphasis to the low end of the particle-size distribution (fines), which is done by using the surface mean diameter, d_{sv}, to calculate an average particle size:

$$d_{sv} = 1/\Sigma(x_i/d_{pi}) \tag{1}$$

Particle-size distribution is usually plotted on a log-probability scale, which allows for quick evaluation of statistical parameters.

Solid Density. Solids can be characterized by three densities: bulk, skeletal, and particle. The bulk density is the weight of a sample of solids divided by its volume. Skeletal density is the true density of the solid material, and particle density is the weight of a representative solid particle divided by its volume. For nonporous particles, skeletal density equals particle density. Particle density is most useful for calculating hydrodynamic properties.

Sphericity. Sphericity, ψ, is a shape factor defined as the ratio of the surface area of a sphere the volume of which is equal to that of the particle, divided by the actual surface area of the particle.

$$\psi = d_{sv}/d_v \tag{2}$$

Angles of Repose and Internal Friction. The angle of repose is the angle that a pile of solids forms with the horizontal plane. The angle of internal friction is the angle with the horizontal that the flow, no-flow boundary forms when solids are flowing over themselves. This angle is a slight function of the solids flow rate. However, a typical angle of internal friction for a nonsticky material without sharp corners generally exceeds 65°. When designing fluidized-bed internal baffles, the baffles generally are angled at greater than 65° to the horizontal to prevent a zone of stagnant solids forming on top.

Terminal Velocity. A knowledge of terminal velocity is important in fluidized beds because it relates to how long particles are retained in the system. If the operating superficial gas velocity in the fluidized bed far exceeds the terminal velocity of the bed particles, the particles are quickly removed. Large particles will be removed almost immediately when gas velocity exceeds the single particle terminal velocity, whereas small particles will take a long time to entrain because of interactions between neighboring particles.

Minimum Fluidization Velocity. There is a minimum superficial gas velocity required to just fluidize a bed of solids. The minimum fluidization velocity can be estimated from the Zenz plot assuming the voidage at minimum fluidization is 0.5. Alternatively, it can be estimated via a correlation that gives a result equivalent to the plot using a voidage of 0.4. Using both methods defines the range within which a measured value for minimum fluidization velocity falls.

Particle Regimes. Particles are generally classified with respect to how they fluidize in air at ambient conditions on the Geldart diagram (Fig. 2).

Interparticle Forces. Interparticle forces are often neglected in the fluidization literature, although in many cases involving small particles these forces are stronger than the hydrodynamic ones used in most correlations. The most common interparticle forces encountered in gas fluidized beds are van der Waals, electrostatic, and capillary. Interparticle forces predominate for Group C powders, are important for Group A, can can on rare occasions influence Group B and D powders.

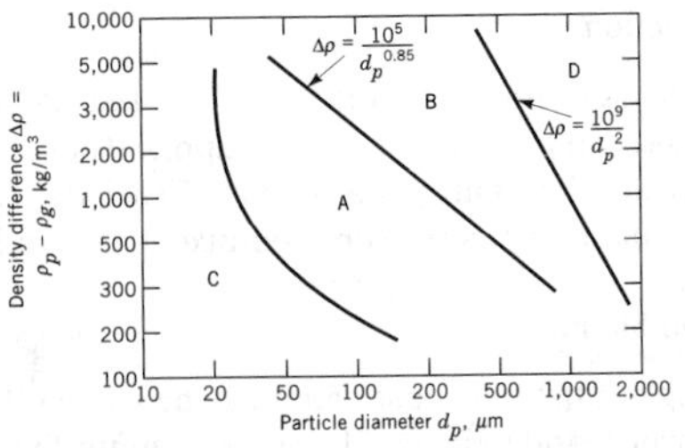

Figure 2. Geldart group particle classification diagram for air at ambient conditions. Group A consists of fine particles; B, coarse particles; C, cohesive, very fine particles; and D, moving and spouted beds.

Fluidization Regimes

The different fluidized-bed regimes are a function of gas velocity. At a low gas velocity, the solids are in a packed-bed or fixed-bed state. As the gas velocity is increased, the drag and buoyancy forces eventually overcome the weight of the particles and interparticle forces, and the particles are completely supported by the gas.

Figure 3 shows how three reactors might appear when operating in the three most common commercial fluidization regimes.

Pressure Drop. The pressure drop across a two-phase suspension is composed of various terms, such as static head, acceleration, and friction losses for both gas and solids. The measurement of pressure drop across the bed is the most common and useful diagnostic technique employed for control of fluidized beds. In a dense fluidized bed, the measured pressure drop per unit length is approximately equal to the weight of the solids per unit volume, or the fluidized apparent density.

Effects of Temperature and Pressure on Minimum Fluidization Velocity. Many basic fluid-bed properties are affected by temperature and pressure. Pressure has little effect on the minimum fluidization velocity, U_{mf}, of fine Group A particles. However, the larger Group B and D particles show a decrease in minimum fluidization velocity with increasing pressure. Increasing temperature increases viscosity and, therefore, reduces U_{mf} for Group A and most Group B particles.

Bubbles and Fluidized Beds. Bubbles, or gas voids, exist in most fluidized beds and their role can be important because of the impact on the rate of exchange of mass or energy between the gas and solids in the bed. Bubbles are formed in fluidized beds from the inherent instability of two-phase systems. Bubbles, which are inherently undesirable, can grow to a large size and cause contact inefficiencies brought on by significant gas bypassing. Bubble size control is achieved by controlling particle-size distribution or by increasing gas velocity.

Solids and Gas Mixing. Solids in an unrestricted fluidized bed can be almost completely backmixed, giving the bed uniform solids properties and a constant temperature throughout. The engine driving the solids mixing and circulation is the drag exerted by the gas on the particles.

Mass Transfer. Mass transfer in a fluidized bed can occur in several ways. Bed-to-surface mass transfer is important in plating applications. Transfer from the solid surface to the gas phase is important in drying, sublimation, and desorption processes. Mass transfer can be the limiting step in a chemical reaction system. In most instances, gas

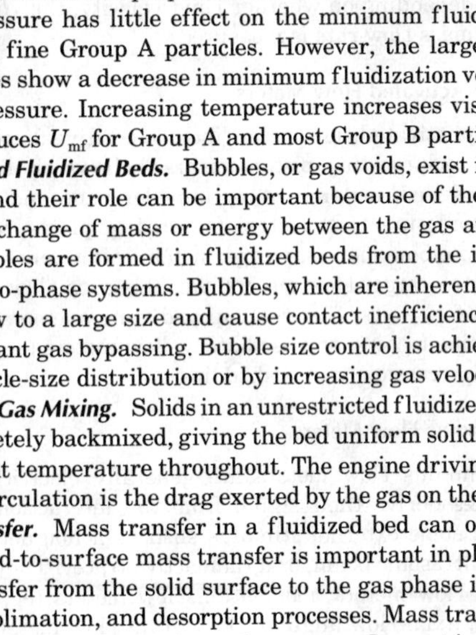

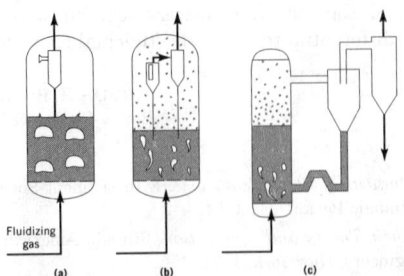

Figure 3. Schematics of commercially used beds, where the shaded area represents the solids: (**a**) vigorously bubbling, (**b**) turbulent, and (**c**) fast fluidized.

from bubbles, gas voids, or the conveying gas reacts with a solid reactant or catalyst.

Heat Transfer. One of the reasons fluidized beds have wide application is the excellent heat-transfer characteristics. Particles entering a fluidized bed rapidly reach the bed temperature, and particles within the bed are isothermal in almost all commercial situations. Gas entering the bed reaches the bed temperature quickly. In addition, heat transfer to surfaces for heating and cooling is excellent.

Distributor Design

Good gas distribution is necessary for the bed to operate properly, and this requires that the pressure drop over the distributor be sufficient to prevent maldistribution arising from pressure fluctuations in the bed. Because gas issues from the distributor at a high velocity, care must also be taken to minimize particle attrition. Many distributor designs are used in fluidized beds. The most common ones are perforated plates, plates with caps, and pipe distributors.

Jet Penetration. At the high gas velocities used in commercial practice, there are jets of gas issuing from distributor holes. It is essential that jets not impinge on any internals, otherwise the internals may be quickly eroded.

Particle Attrition. Distributor jets are a potential source of particle attrition. Particles are swept into the jet, accelerated to a high velocity, and smash into other particles as they leave. To reduce attrition at distributors, a shroud or larger-diameter pipe is often added concentric to the jet hole. The shroud causes a reduction in gas velocity when the jet impinges on the solids while still allowing a high gas velocity, and hence the required pressure drugs, at the orifice.

Bed Internals. Various types of internals that may be found in commercial fluidized beds include solids and gas distributors; cyclones and cyclone diplegs; solids return and withdrawal lines; heat-transfer tubes; supports, hangers, and guides for heat-transfer tubes; baffles; secondary gas-injection nozzles; and pressure, temperature, and sample probes.

Entrainment

Entrainment, or elutriation, is the carryover of particles from a fluidized bed with the exiting gas. When the gas velocity exceeds the terminal velocity of a Group B particle, the particle is usually removed from the bed. For Group A and C powders, the gas drag needs to overcome interparticle forces as well, and gas velocities of many times the single-particle velocity are needed to entrain particles from the bed at high rates. Knowledge of the entrainment rate is important in order to estimate cyclone inlet loading, solids loss rates, and to predict bed particle size changes resulting from the selective loss of fines.

Transport Disengaging Height. The height above the bed at which entrainment becomes essentially constant with height is termed the transport disengaging height (TDH). It is desirable to locate cyclones and vessel outlets above TDH so as to minimize solids loading to the cyclones.

Cyclones. Cyclones are an integral part of most fluidized-bed systems. A cyclone is an inexpensive device having no moving parts that separates solids and gases using centrifugal force. Cyclones are used commercially to remove solids with high efficiency down to about 5 μm.

Circulating Fluidized Beds

Circulating fluidized beds (CFBs) are high velocity fluidized beds operating well above the terminal velocity of all the particles or clusters of particles. A very large cyclone and seal leg return system are needed to recycle solids in order to maintain a bed inventory. There is a gradual transition from turbulent fluidization to a truly circulating, or fast-fluidized bed, as the gas velocity is increased, and the exact transition point is rather arbitrary. The solids are returned to the bed through a conduit called a standpipe. The return of the solids can be controlled by either a mechanical or a nonmechanical valve.

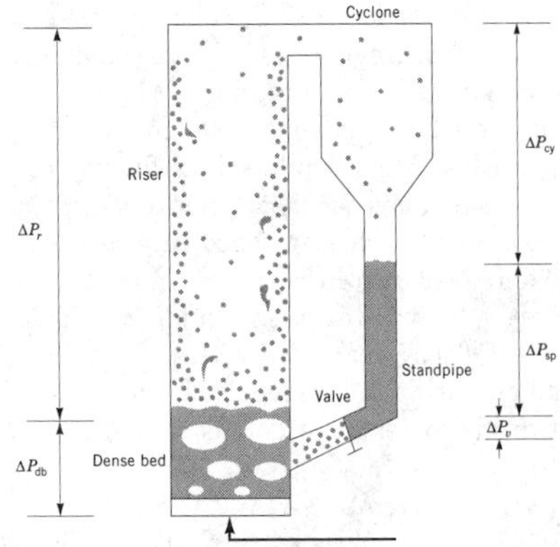

Figure 4. CFB pressure balance, where $\Delta P_{sp} = \Delta P_v + \Delta P_{db} + \Delta P_r + \Delta P_{cy}$. A high gas velocity in a fast bed results in a high solids entrainment rate. Head buildup in the standpipe overcomes head buildup in the dense bed, riser, cyclone, and sealing device (valve), and allows solids to recirculate.

The bed level is not well defined in a circulating fluidized bed, and bed density usually declines with height. Axial density profiles for different CFB operating regimes show that the vessel does not necessarily contain clearly defined bed and freeboard regimes. The solids may occupy only between 5 and 20% of the total bed volume.

Pressure Balance and Standpipes. The pressure balance around the loop of a circulating fluidized bed is illustrated in Figure 4.

Nonmechanical Valves. Nonmechanical valves, which have no moving parts in the solids flow path, are often used to control the flow of Group B solids. Examples of nonmechanical valves are an L-valve, J-valve, loop seal, and reverse seal.

Scale Up of Fluidized Bed

The greatest problems in scaling up fluidized beds have been encountered using Group B solids where bubbles can grow to very large sizes and bubble-size control is more difficult than with Group A solids.

Several attempts have been made to determine proper scaling relationships for Group B fluidized beds. The basic scaling factors are the Reynolds number

$$Re_p = d_p U \rho_g / \mu \tag{3}$$

and the Froude number,

$$\text{Froude number} = \frac{U^2}{g d_p} \tag{4}$$

where density ratio is ρ_p/ρ_g; length ratios, L/d_p and D/d_p; and other factors such as ψ, sphericity, particle size distribution, and bed geometry.

There are some data to suggest that hydrodynamic similarity improves scale up for two-phase systems such as fluidized beds, even though it is not as convincing as single-phase evidence.

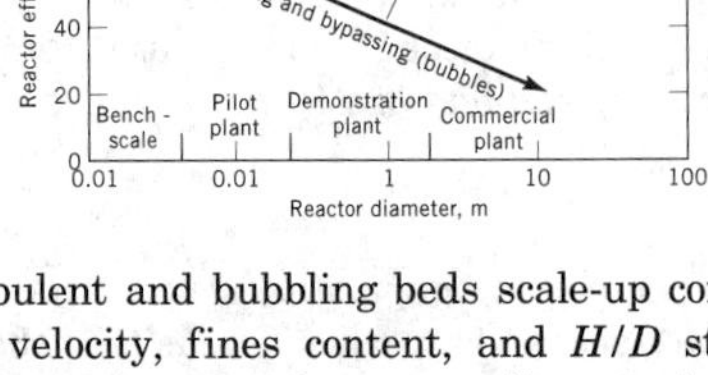

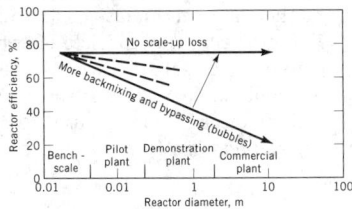

Figure 5. Turbulent and bubbling beds scale-up comparison where increasing gas velocity, fines content, and H/D staging can help maintain reactor efficiency as the reactor diameter increases. A 100% efficiency is equivalent to plug flow.

Group A particles cause fewer scale-up problems because fluidized beds of Group A particles generally are operated in the vigorously bubbling or turbulent fluidization regimes. Also, it is not unusual for a maximum stable bubble–gas void to be on the order of 25 mm or less for these particles. Thus a pilot-plant facility can generally be operated using the same gas void size that a commercial unit would experience. An example of scale-up concerns and ways to avoid them for Group A powders is shown in Figure 5. In this case, the efficiency was maintained at 80% for a turbulent fluidized bed, ie, there was no scale-up loss, but efficiency decreased with scale up for a bubbling bed. Adding fines and operating at a higher gas velocity in a bubbling bed, ie, moving it toward turbulence, can offset scale-up loss.

AMOS A. AVIDAN
Mobil Research and Development Corporation
DESMOND F. KING
Chevron Research and Technology Company
TED M. KNOWLTON
Institute of Gas Technology
Particulate Solids Research Inc.
MEL PELL
E. I. du Pont de Nemours & Co., Inc.

D. Kunij and O. Levenspiel, *Fluidization Engineering*, 2nd ed., Butterworth-Heinemann, Boston, Mass., 1991.

J. F. Davidson, R. Clift, and D. Harrison, eds., *Fluidization*, 2nd ed., Academic Press, London, 1985.

F. A. Zenz and D. F. Othmer, *Fluidization and Fluid Particle Systems*, Reinhold Publishing Corp., New York, 1960.

M. Pell, *Gas Fluidization*, Elsevier, Amsterdam, the Netherlands, 1990.

FLUID MECHANICS

Fluid mechanics is both a descriptive science of the phenomena that occur when fluids flow and a quantitative science showing how these phenomena may be described in mathematical terms. To a practicing chemical technologist, fluid mechanics is an entire body of knowledge, theoretical and empirical, qualitative and quantitative, allowing analysis of the performance of complex plant equipment handling moving fluids.

General Principles

Fluid mechanics became an exact science with the application of the conservation principles. These include the law of the conservation of mass; including the law of conservation of individual species, electric charge, etc; the law of conservation of energy; and the law of conservation of momentum, ie, Newton's second law. Fluid mechanics differs from conventional mechanics in that the former mainly treats bodies that are capable of unlimited deformation. Flow is regarded as primary and elasticity as secondary. Thus instabilities in flow, eg, turbulence, play a large role in fluid mechanics, whereas these are of minor concern in solid mechanics. In some instances the two converge, as in the treatment of extrusion of very viscous or plastic materials in which fluid-like and solid-like behavior are of equal importance.

The most useful mathematical formulation of a fluid flow problem is as a boundary value problem. This consists of two main parts: a set of differential equations to be satisfied within a region of interest and a set of boundary conditions to be satisfied on the surfaces of that region. Sometimes additional conditions are also of interest, eg, when one is investigating the stability of a flow.

Equations of Motion

The starting point for obtaining quantitative descriptions of flow phenomena is Newton's second law, which states that the vector sum of forces acting on a body equals the rate of change of momentum of the body. This force balance can be made in many different ways. It may be applied over a body of finite size or over each infinitesimal portion of the body. It may be utilized in a coordinate system moving with the body (the so-called Lagrangian viewpoint) or in a fixed coordinate system (the Eulerian viewpoint). Described herein is derivation of the equations of motion from the Eulerian viewpoint using the Cartesian coordinate system. The equations in other coordinate systems are described in standard references.

General Equation of Motion. Neglecting relativistic effects, the rate of accumulation of mass within a Cartesian volume element $dx \cdot dy \cdot dz$ must equal the sum of the rates of inflow minus outflow. This is expressed by the equation of continuity:

$$\frac{\partial \rho}{\partial t} + \frac{\partial}{\partial x}\rho u + \frac{\partial}{\partial y}\rho v + \frac{\partial}{\partial z}\rho w = 0 \qquad (1)$$

where ρ is the fluid density; u, v, and w are the x-, y-, and z-components of velocity; and t is the time.

Similarly, a momentum balance can be made on the same infinitesimal volume element. First it is necessary to have a firm idea of the nature of the forces that might be exerted on the element. There may be body forces which, in Newtonian mechanics, are pictured as acting at a distance and include gravitational, electrostatic, and magnetic forces and, if a noninertial frame of reference is used, those forces resulting from accelerations of the frame, eg, centrifugal force. In addition there are forces arising from the continuum stresses exerted on the faces or edges of the element. These stresses arise from intermolecular forces and motions of molecules that express themselves macroscopically as pressure, viscosity, yield stress, etc. These stresses behave mathematically as the components of a symmetrical, second-order tensor. At boundaries between phases, additional forces, which are generalizations of surface tension, must be included. These forces are discussed in standard texts.

In Figure 1, the force balance in Cartesian coordinates for a body not intersected by phase boundaries is

$$\begin{aligned} \rho\frac{Du}{Dt} &= X + \frac{\partial \sigma_x}{\partial x} + \frac{\partial \tau_{yx}}{\partial y} + \frac{\partial \tau_{zx}}{\partial z} \\ \rho\frac{Dv}{Dt} &= Y + \frac{\partial \sigma_y}{\partial y} + \frac{\partial \tau_{zy}}{\partial z} + \frac{\partial \tau_{xy}}{\partial x} \\ \rho\frac{Dw}{Dt} &= Z + \frac{\partial \sigma_z}{\partial z} + \frac{\partial \tau_{xz}}{\partial x} + \frac{\partial \tau_{yz}}{\partial y} \end{aligned} \qquad (2)$$

where Du/Dt, Dv/Dt, and Dw/Dt are the components of acceleration; X,Y, and Z are body forces per unit volume; σ_i is the normal stress on the i-face; and τ_{ij} is the shear stress on the i-face exerted in the j-direction. The operator D/Dt is called the material, total, or substantial derivative and can be expressed as:

$$\frac{D}{Dt} = \frac{\partial}{\partial t} + u\frac{\partial}{\partial x} + v\frac{\partial}{\partial y} + w\frac{\partial}{\partial z} \qquad (3)$$

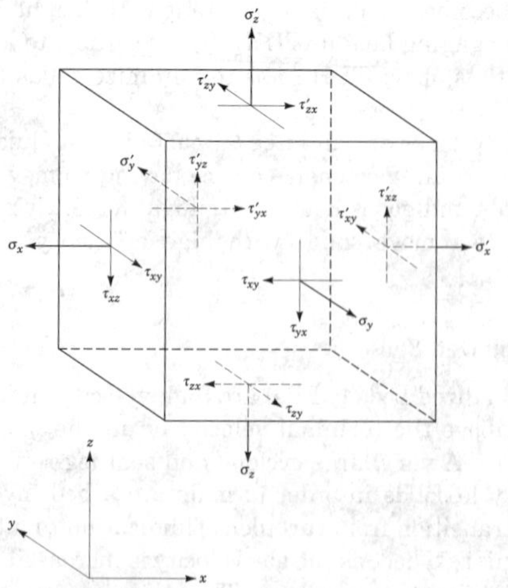

Figure 1. Normal and shear forces on a differential volume $dx \cdot dy \cdot dz$, where $\sigma_i' = \sigma_i + (\frac{\partial \sigma_i}{\partial_i})\, di$ and $\tau_{ij}' = \tau_{ij} + (\frac{\partial \tau_{ij}}{\partial_i})\, di$.

Equation 2 gives the relation between stresses and accelerations obtained from momentum balances. To proceed further requires use of the constitutive equations which codify the material properties through additional relations between the stresses (τ_{xy}, etc) and the rates of strain ($\partial u/\partial x$, etc). The constitutive equation for a given fluid is found empirically or theoretically by use of some theory of material properties. The simplest model is one in which the various stresses are expressed as linear combinations of the rates of strain. When the fluid is homogeneous and isotropic, this relation leads to the Navier-Stokes equations. Fluids that obey these equations are by definition Newtonian. The conditions of homogeneity and isotropy ensure that only two material constants, the shear viscosity, μ, and the dilational viscosity, λ, are needed to describe the fluid. By defining pressure as the negative of the average of the three normal stresses, σ_i, one finds that $\lambda = 2/3\mu$, hence only a single material viscosity μ is required. As defined, the pressure is usually identified with the thermodynamic pressure for purposes such as determining physical properties. Equations 4 and 5 show the constitutive equation so derived, and equation 6 shows the form taken by the equation of motion in the X-direction when these are inserted into the force balance.

$$\begin{aligned} \sigma_x &= -P + 2\mu\frac{\partial u}{\partial x} - \frac{2}{3}\mu\left(\frac{\partial u}{\partial x} + \frac{\partial v}{\partial y} + \frac{\partial w}{\partial z}\right) \\ \sigma_y &= -P + 2\mu\frac{\partial v}{\partial y} - \frac{2}{3}\mu\left(\frac{\partial u}{\partial x} + \frac{\partial v}{\partial y} + \frac{\partial w}{\partial z}\right) \\ \sigma_z &= -P + 2\mu\frac{\partial w}{\partial z} - \frac{2}{3}\mu\left(\frac{\partial u}{\partial x} + \frac{\partial v}{\partial y} + \frac{\partial w}{\partial z}\right) \end{aligned} \tag{4}$$

$$\begin{aligned} \tau_{xy} &= \tau_{yx} = \mu\left(\frac{\partial u}{\partial y} + \frac{\partial v}{\partial x}\right) \\ \tau_{yz} &= \tau_{zy} = \mu\left(\frac{\partial v}{\partial z} + \frac{\partial w}{\partial y}\right) \\ \tau_{zx} &= \tau_{xz} = \mu\left(\frac{\partial w}{\partial x} + \frac{\partial u}{\partial z}\right) \end{aligned} \tag{5}$$

$$\begin{aligned} \rho\frac{Du}{Dt} &= X - \frac{\partial P}{\partial x} + \frac{\partial}{\partial x}\left\{\mu\left[2\frac{\partial u}{\partial x} - \frac{2}{3}\left(\frac{\partial u}{\partial x} + \frac{\partial v}{\partial y} + \frac{\partial w}{\partial z}\right)\right]\right\} \\ &\quad + \frac{\partial}{\partial y}\left[\mu\left(\frac{\partial u}{\partial y} + \frac{\partial v}{\partial x}\right)\right] + \frac{\partial}{\partial z}\left[\mu\left(\frac{\partial w}{\partial x} + \frac{\partial u}{\partial z}\right)\right] \end{aligned} \tag{6}$$

Frictionless Flow. There is actually a constitutive equation that is even simpler than that for the Newtonian fluid. This arises when internal fluid friction is neglected. Although this is not precisely true of a real fluid, with the possible exception of liquid helium-3 at temperatures below 2 K, in many situations the flow calculated under this assumption are very close to those actually observed, at least in a significant portion of the flow field. The importance of the frictionless flow theory lies in the wide variety of available solutions, and in the powerful techniques of calculation available. Notable is Bernoulli's theorem.

Flow Phenomena

Flow Past Bodies. A fluid moving past a surface of a solid exerts a drag force on the solid. This force is usually manifested as a drop in pressure in the fluid. Locally, at the surface, the pressure loss stems from the stresses exerted by the fluid on the surface and the equal and opposite stresses exerted by the surface on the fluid. Both shear stresses and normal stresses can contribute; their relative importance depends on the shape of the body and the relationship of fluid inertia to the viscous stresses, commonly expressed as a dimensionless number called the Reynolds number (Re), $LV\rho/\mu$. The character of the flow affects the drag as well as the heat and mass transfer to the surface. Flows around bodies and their associated pressure changes are important.

Free Flows: Jets, Wakes, and Plumes. When a smoothly flowing stream is disturbed by inserting a solid body or by injecting additional fluid, the disturbance is confined to a narrow region downstream of the initial point. When the disturbance is caused by the injection of a fluid at high velocity, the disturbed region is called a jet. When the velocity is lower than that in the main stream, the disturbance is termed a wake. The term plume is usually applied to a case in which an additional effect, such as buoyancy, plays a significant role and for which the turbulence characteristics of the main stream play the deciding role in its spread.

Jets are used industrially to perform many mixing operations ranging from gradual mixing of tanks to the rapid mixing needed in chemical reactors or flames (see MIXING AND BLENDING). Mixing in turbulent wakes is important in the disposal of wastes at sea and in the dispersion of potential atmospheric pollutants downwind of structures. Most interest in plumes is centered around the behavior of gases emitted from smokestacks into the atmosphere.

Other Flow Phenomena. The flow of easily compressible fluids, ie, gases, exhibits features not evident in the flow of substantially incompressible fluid, ie, liquids. These differences arise because of the ease with which gas velocities can be brought to or beyond the speed of sound and the substantial reversible exchange possible between kinetic energy and internal energy. The Mach number, the ratio of the gas velocity to the local speed of sound, plays a central role in describing such flows.

Flow in Porous Media. Flow of fluids through fixed beds of solids occurs in situations as diverse as oil-field reservoirs, catalyst beds and filters, and absorption (qv) towers. The complex interconnected pore structure of such systems makes it necessary to use simplified models to make practical quantitative predictions. One of the more successful treatments of single-phase pressure drop through such systems employs the results for flow through tubes, using average velocities and tube diameters.

Non-Newtonian Fluids: Die Swell and Melt Fracture. For many fluids the Newtonian constitutive relation involving only a single, constant viscosity is inapplicable. Either stress depends in a more complex way on strain, or variables other than the instantaneous rate of strain must be taken into account. Such fluids are known collectively as non-Newtonian and are usually subdivided further on the basis of behavior in simple shear flow, ie, flow between sliding planes or, to a good approximation, between two mutually rotating cylinders. Figure 2 illustrates the behavior of several of these. The types illustrated are examples of the generalized Newtonian fluid. This simple generalization uses the Navier-Stokes equations but takes viscosity to depend in some way on a single quantity: the rate of strain.

Gas–Liquid Flow. When two or more fluids flow together, a much greater range of phenomena occurs as compared to flow of a single phase. In a conduit many of the technically significant phenomena have to do with the positions assumed by the phase boundaries, and these are governed by the flow conditions rather than by the walls of the conduit. In addition to the densities and viscosities, surface properties can be important. Methods for quantitative calculation for gas–liquid flow are poorly developed in comparison to those for single-phase flow. However, many useful calculations can be made, particularly when the effects of changes in an already existing operation are

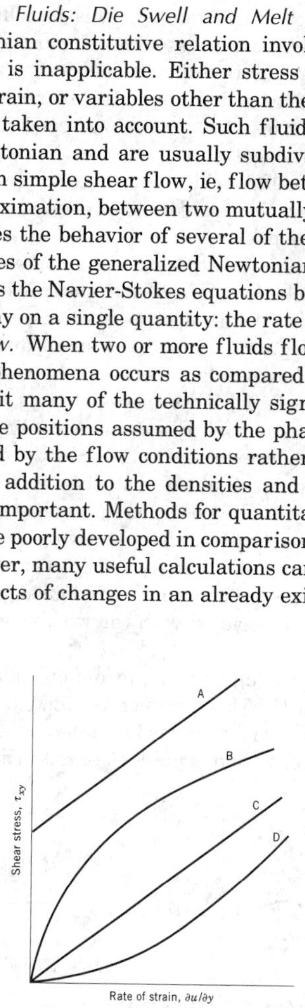

Figure 2. Fluid behavior in simple shear flow where A is Bingham; B, pseudoplastic; C, Newtonian; and D, dilatant.

of interest. All of these calculations depend on an accurate picture of the flow regime, used to denote the significant configurational characteristics of the flow, eg, unidirectional, recirculating, steady, unsteady, liquid-dispersed, gas-dispersed, etc.

Flow Instability. In many flow situations it is found that a mathematically valid solution to the Navier-Stokes equations is closely verified by experiment over some ranges of the variables, but when the variables are changed new flow patterns that are not in keeping with that solution are observed. The change is often rather sudden, eg, the laminar–turbulent transition. Such behavior occurs because solutions to the Navier-Stokes equations are generally not unique. When more than one solution exists, it is possible to observe one flow pattern under one set of circumstances and a different pattern under another. At the present stage of development of fluid mechanics, an experiment must be performed to determine whether a given solution applies. In some cases, however, it is possible, by analyzing the equations of motion, to determine the criteria by which one flow pattern becomes unstable in favor of another. The mathematical technique used most often is linearized stability analysis, which starts from a known solution to the equations and then determines whether a small perturbation superimposed on this solution grows or decays as time passes.

Fluids in Motion. Many of the instabilities associated with fluids in motion are of the shear-flow type. In shear flow the velocity varies principally in a direction perpendicular to the flow direction, eg, pipe flow, boundary layer flow, jet flow, and wake flow. Such flows change from laminar to turbulent when the Reynolds number based on some characteristic length scale exceeds a critical value, which may be different for each flow configuration. In pipe flows and some types of boundary layer flows, the instability is usually referred to as Tollmien-Schlichting instability. This instability occurs in essentially four steps: *(1)* small two-dimensional waves form and are linearly amplified; *(2)* the two-dimensional waves develop into finite three-dimensional waves and are amplified by nonlinear interactions; *(3)* a turbulent spot forms at some localized point in the flow; and *(4)* the turbulent spot propagates until the spot fills the entire flow field with turbulence. At low Reynolds numbers, viscosity damps out the instability. At high Reynolds numbers, however, viscosity provides the destabilizing mechanisms.

In jet flow and wake flow there occurs another type of instability called Kelvin-Helmholtz instability. This instability may be illustrated by considering the development of a small disturbance in the flow situation given in Figure 3. Suppose a small disturbance causes a slight waviness of the boundary between the two flows. The fluid on the convex sides of each flow moves slightly faster and that on the concave sides moves slightly slower. According to the Bernoulli equation, this disturbance decreases the pressure on the convex sides of each flow, and thus the initial disturbance is amplified. Kelvin-Helmholtz instability is also observed in horizontal concurrent flows of stratified immiscible fluids. Raising of a wave at the interface forces the upper fluid to increase in velocity and drop in pressure as it passes the wave. If the pressure decrease is large enough, it overcomes the increase in potential associated with the raising of the wave and the disturbance grows.

An instability involving transition from one laminar flow pattern to another occurs in the flow between concentric, mutually rotating cylinders. When the inner cylinder is rotated at an angular velocity below a critical value, motion is purely circumferential (Couette flow).

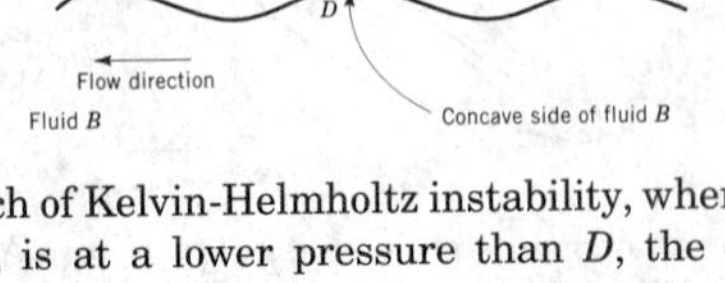

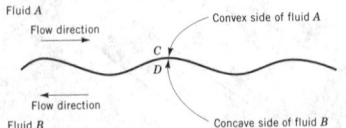

Figure 3. Sketch of Kelvin-Helmholtz instability, where C, the convex side of fluid A, is at a lower pressure than D, the concave side of fluid B.

Above this value, however, centrifugal force destabilizes the flow and a series of laminar, cellular vortices known as Taylor cells is superimposed on the main flow.

When a fluid of low viscosity is used to displace a fluid of higher viscosity from a porous medium or a pipe, the displacement can become unstable by a process known as fingering. Small fingers of the low viscosity fluid, once formed, become regions of lower pressure drop. The displacing fluid flows preferentially into these fingers which grow and ultimately reach the outlet, leaving a portion of the viscous fluid undisplaced.

Fluids at Rest. Fluids at rest may be set into motion by impressing upon them gradients in body or surface forces. Benard instability refers to the formation of convection cells within a fluid as result of the action of a gravitational field on density differences induced by a temperature gradient in the fluid. The onset of instability is described by a critical Rayleigh number (Ra):

$$Ra = \frac{g(\frac{\partial \rho}{\partial T})\Delta T l_b^3}{\mu \alpha} \tag{7}$$

Cellular motions may also arise from gradients in surface tension caused by variations in temperature or concentration. This behavior is commonly referred to as Marangoni instability. Circulation of fluid is promoted by surface tension gradients but inhibited by viscosity, which slows the flow, and by molecular diffusion, which tends to even out the concentration differences. The onset of instability is described by a critical Marangoni number (Ma), an analogue of the Rayleigh number:

$$Ma = \frac{(\frac{\partial \sigma}{\partial c})(\frac{\partial c}{\partial x}) l_M^2}{\mu \mathcal{D}} \tag{8}$$

Surface tension is also responsible for the varicose or Rayleigh breakup of liquid strands into droplets.

Atomization. A gas or liquid may be dispersed into another liquid by the action of shearing or turbulent impact forces that are present in the flow field. The steady-state drop size represents a balance between the fluid forces tending to disrupt the drop and the forces of interfacial tension tending to oppose distortion and breakup. When the flow field is laminar the ability to disperse is strongly affected by the ratio of viscosities of the two phases. Dispersion, in the sense of droplet formation, does not occur when the viscosity of the dispersed phase significantly exceeds that of the dispersing medium. More commonly, atomization occurs under turbulent conditions. The mechanism of atomization and its quantitative description are still incompletely understood.

Secondary Flows. In the flow around a bend, a pattern of secondary flow develops that is superimposed on the main flow so that the streamlines are actually helical. The choice is somewhat arbitrary as to which flow to call main and which to call secondary. Adopting the convention that the main flow is parallel to the tube axis, then at the axis the secondary flow is directed outward toward the section of pipe having the weakest curvature, returning inward along the pipe wall. This type of secondary flow is a consequence of the inertia of the fluid. It is most obvious at low Reynolds numbers but is also significant in turbulent flow.

Solutions to Equations of Motion

Three basic approaches have been used to solve the equations of motion. For relatively simple configurations, direct solution is possible. For complex configurations, numerical methods can be employed. For many practical situations, particularly three-dimensional or one-of-a-kind configurations, scale modeling is employed and the results are interpreted in terms of dimensionless groups. This section outlines the procedures employed and the limitations of these approaches (see COMPUTER-AIDED ENGINEERING (CAE)).

Exact Solutions to the Navier-Stokes Equations. As was true for the inviscid flow equations, exact solutions to the Navier-Stokes equations are limited to fairly simple configurations that allow for considerable simplification both in the equation and in the boundary conditions.

Numerical Solution of the Equation in Motion. Usually an analytical solution to the equation of motion cannot be found; recourse must be had to numerical (computational) methods. The field of computational fluid dynamics (CFD) is moderately mature.

In general, a computational method includes the task of dividing the flow domain of interest into a network of elements with adequate grid resolution, obtaining a set of algebraic equations for the unknown dependent values at the grid intersections or nodes by discretization of the set of pertinent partial differential equations (PDE), and then prescribing an efficient algorithm for solving the set of algebraic equations. The computational results are values of the dependent variables, such as velocity, pressure, temperature, etc, at the grid points which are studied with the help of a graphics display package.

Turbulence Modeling. The time-dependent Navier-Stokes equations are generally considered adequate to represent turbulent flows. Direct numerical solution (DNS) of these equations is limited to low Reynolds numbers. At higher Reynolds numbers, the number of grid points required to resolve small eddies and the small time step size needed to obtain meaningful results make the computation of turbulent flows encountered in engineering practice by DNS outside the capability of present computers. A less computationally intensive technique called large eddy simulation (LES) calculates the three-dimensional time-dependent details of the largest scales of motion using a simple subgrid scale model for the smaller eddies. However, the method is still very computationally intensive. Both DNS and LES are used primarily for studying the physics of turbulence, but LES has the potential of becoming an engineering tool in the near future.

Eddy Viscosity Models. A large number of closure models are based on the Boussinesq concept of eddy viscosity:

$$-\overline{u'v'} = v_t \frac{\partial u}{\partial y} \tag{9}$$

These models are usually categorized according to the number of supplementary partial differential transport equations which must be solved to supply the modeling parameters.

Reynolds Stress Models. Eddy viscosity is a useful concept from a computational perspective, but it has questionable physical basis. Models employing eddy viscosity assume that the turbulence is isotropic. Another limitation is that the turbulent stresses are related only to one velocity scale. In the so-called Reynolds stress models, transport equations for $u'_iu'_j$ are introduced to allow for the individual stresses to develop quite differently. These transport equations are often referred to as second-order closure and are solved simultaneously with the dissipation equation and the Reynolds averaged equations. Thus RSM models require more computer time and memory because of the larger number of PDE involved and the nonlinearity and strong coupling of the set of equations. This extra effort, however, yields more realistic results for flows having rotation, curvature, and strong swirl.

Convergence and Accuracy. Converged numerical solutions can be quite tedious to obtain for complex flows because the nonlinearity of the finite-difference equations usually makes the numerical procedure susceptible to numerical instabilities. These instabilities are often caused by poor estimate of the initial (guessed) field variables and/or rather poor choice of the relaxation parameters. The accuracy of the converged numerical solution depends on the values of the residual sources when the iterative calculations are terminated and the extent by which the finite-difference equations approximate the PDE. In general, smaller grid sizes are needed, particularly in areas where the gradients are steep, to get accurate solutions. A good check for the accuracy is to reduce the grid size until a grid independent solution is obtained.

Complex Geometries. To extend the capabilities of finite-difference and finite-volume methods to handle complex geometries, a boundary fitted coordinate system (BFCS) is now being used and is available in many commercial flow codes.

Modeling. The majority of technological flow problems are not solved by integrating the equations of motion. Instead, most are solved by carrying out laboratory experiments which are then correlated, ie, interpreted, so as to yield useful information about systems that may differ greatly in size and in fluid properties. For the behavior of the experimental model to duplicate that of a system of interest, two criteria must, in principle, be met: the experimental apparatus must be geometrically similar to the system of interest; and certain dimensionless groupings of variables must be duplicated on the two scales. There are two basic methods available for determining the dimensionless groups appropriate to a given situation: dimensional analysis (qv), which can be applied when the equations governing the process are not known; and similarity analysis, which proceeds from the governing equations and offers physical insights into the meanings of the groups.

Integral Forms of Equations of Motion

The solutions of some problems in fluid mechanics require the detailed integration of the differential equations of motion, whereas many others can be solved to a sufficient degree of accuracy by examining only the overall balances of mass, momentum, and energy.

Conservation of Mass. The general equations for the conservation of mass are the scalar equations

$$\underbrace{\frac{\partial}{\partial t}\int\int\int c_i dB}_{\text{accumulation}} + \underbrace{\int\int c_i V \cos\theta dS}_{\text{convection}} + \underbrace{\int\int J_i dS}_{\text{diffusion}} - \underbrace{\int\int\int R_i dB}_{\text{chemical reaction}} = 0 \tag{10}$$

The conservation of mass gives comparatively little useful information until it is combined with the results of the momentum and energy balances.

Conservation of Momentum. The general equation for the conservation of momentum is

$$\underbrace{\frac{\partial}{\partial t}\int\int\int \rho \vec{d} B}_{\text{accumulation}} + \underbrace{\int\int \rho \vec{V} \cos\theta dS}_{\text{convection}} - \underbrace{\int\int\int \vec{d}B}_{\text{creation by body forces}} - \underbrace{\int\int \vec{d} S}_{\text{creation by surface forces}} = 0sB \tag{11}$$

The creation terms embody the changes in momentum arising from external forces in accordance with Newton's second law ($F = ma$). The body forces arise from gravitational, electrostatic, and magnetic fields. The surface forces are the shear and normal forces acting on the fluid; diffusion of momentum, as manifested in viscosity, is included in these terms. In practice the vector equation is usually resolved into its Cartesian components and the normal stresses are set equal to the pressures over those surfaces through which fluid is flowing.

Conservation of Energy. The energy associated with a unit mass of the flowing fluid may be considered as the sum of its potential, kinetic, and internal energies:

$$E = gZ + \frac{V^2}{2} + U \tag{12}$$

The equation describing the conservation of energy is the scalar equation:

$$\underbrace{\frac{\partial}{\partial t}\int\int\int \rho E dB}_{\text{accumulation}} + \underbrace{\int\int \rho EV \cos\theta dS}_{\text{convection}} + \underbrace{\int\int J_E dS}_{\text{nonconvective flux}} = 0 \tag{13}$$

The nonconvective energy flux across the boundary is composed of two terms: a heat flux and a work term. The work term in turn is composed of two terms: useful work delivered outside the fluid, and work done by the fluid inside the control volume B on fluid outside the control volume B, the so-called flow work.

The total energy balance may be regarded as the first law of thermodynamics for flowing fluid. It is an essential tool in analyzing the performance of machines that interconvert heat, work, and internal energy. It contains, however, no explicit terms relating to such matters as dissipation of energy within the system by friction, and there must be recourse to more specific equations such as the mechanical energy balance.

The Mechanical Energy Balance. The mechanical energy equations can be viewed as a definition of a dissipation term ΔF which describes the irreversible conversion into heat of energy available for mechanical work.

Two approaches to this equation have been employed. *(1)* The scalar product is formed between the differential vector equation of motion and the vector velocity and the resulting equation is integrated. This is the most rigorous approach and for laminar flow yields an explicit equation for ΔF in terms of the velocity gradients within the system. *(2)* The overall energy balance is manipulated by asserting that the local irreversible dissipation of energy is measured by the difference:

$$-\delta(\Delta F) = \delta Q - TdS = \delta Q - dU - Pd\left(\frac{1}{P}\right) - \sum \overline{F}_i dn_i \qquad (14)$$

Both approaches yield the same general result, which is

$$\Delta F = -W_M + \int_1^2 \frac{dP}{\rho} + g(Z_1 - Z_2) + \frac{1}{2}\left(\frac{\overline{V}_1^2}{\alpha_1} - \frac{\overline{V}_2^2}{\alpha_2}\right) \qquad (15)$$

The work term W_M is restricted to the mechanical work delivered to the outside via normal and shear forces acting on the boundary. Electrochemical work, ie, by electrolysis of the fluid, is excluded. Evaluation of the integral requires knowledge of the equation of state and the thermodynamic history of the fluid as it passes from S_1 to S_2, neither of which is needed in the overall energy balance.

Flow Measurement

There are dozens of flow meters available for the measurement of fluid flow. The primary measurements used to determine flow include differential pressure, variable area, liquid level, electromagnetic effects, thermal effects, and light scattering. Most of the devices discussed herein are those used commonly in the process industries; a few for the measurement of turbulence are also described.

Measurement by Differential Pressure. The most widely used devices for the flow of fluids are those that utilize a fixed constriction in the path of flow to produce a difference in pressure between the upstream and downstream measuring points: orifice meters, venturi meters, and target meters (Fig. 4).

Measurement by Variable Area. Meters that operate on the principle of variable area incorporate an adjustable constriction in the path of the flow that may be varied so as to maintain a constant differential pressure. The most commonly used device of this type is the rotameter (see Fig. 4).

Measurement by Liquid Level. The flow rate of liquids flowing in open channels is often measured by the use of weirs.

Measurement by Electromagnetic Effects. The magnetic flow meter is a device that measures the potential developed when an electrically conductive flow moves through an imposed magnetic field. The device is useful for the measurement of slurries and other fluid systems where an accumulation of another phase could interfere with flow measurement by other devices.

Measurement by Thermal Effects. When a fine wire heated electrically is exposed to a flowing gas, it is cooled and its resistance is changed. The hot-wire anemometer makes use of this principle to measure both the average velocity and the turbulent fluctuations in the flowing system.

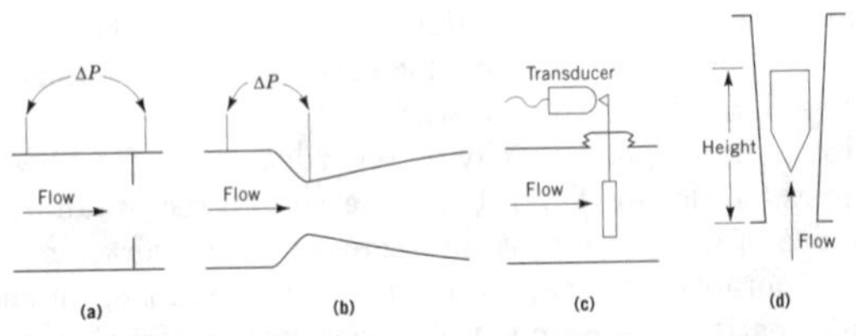

Figure 4. Common flow meters: (**a**) orifice, (**b**) venturi, (**c**) target, and (**d**) rotameter.

Because of its small size and portability, the hot-wire anemometer is ideally suited to measure gas velocities either continuously or on a troubleshooting basis in systems where excess pressure drop cannot be tolerated. Furnaces, smokestacks, electrostatic precipitators, and air ducts are typical areas of application.

Laser Anemometry. When light is scattered from a particle moving at a velocity, V, its frequency is altered by the fraction $(\cos\alpha' - \cos\beta')V/C$, where C is the velocity of light in the fluid and α' and β' are the angles between the particle track and the incident and scattered beams, respectively. This principle is employed in the laser anemometer, which uses a laser beam as the source of monochromatic light, particles in the fluid as scattering centers, and appropriate electronic circuitry to measure the frequency shift (see LASERS). This method has been successfully employed to study turbulent fluctuations and to measure local velocities in stirred vessels.

NOMENCLATURE

B = control volume
d = diameter of drop or bubble
$\mathcal{D}$ = molecular diffusivity
F_i = partial molal free energy of component i
ΔF = irreversible energy dissipation per unit mass of fluid
J_i = nonconvective flux of component i through surface of control volume
L = characteristic length in system, length to achieve specified approach to fully developed flow, thickness of porous bed
l_b, l_M = distance between plates (Benard instability), characteristic length (Marangoni instability)
Ma = Marangoni number, $(\partial\sigma/\partial c)(\partial c/\partial x)l_M^2/\mu\mathcal{D}$
n_i = moles of component i per unit mass of fluid
P = thermodynamic pressure
Q = heat added to a unit mass of fluid
Ra = Rayleigh number, $g(\partial\rho/\partial T)\Delta T l_b^3/\mu\alpha$
Re = Reynolds number, $DV\,\rho/\mu$, $dV\,\rho/\mu$, $LV\,\rho/\mu$, etc
R_i = rate of generation of component i per unit volume
S = entropy, surface area
T, ΔT = temperature, temperature difference
u, v, w = components of velocity in the x, y, z directions
u' = dimensionless reduced velocity
u_i = instantaneous fluid velocity in the i direction
$\overline{u_i' u_j'}$ = Reynolds stresses
U = internal energy
V = velocity
$\overline{V}$ = average velocity, eg, in a pipe
W_M = mechanical work done by a fluid on its surroundings
x, y, z = space coordinates to describe flow; z also indicates height above datum plane
α = thermal diffusivity (Rayleigh number)
μ = shear viscosity
λ = dilational viscosity, relaxation time
ν_t = eddy (or turbulent) viscosity
ρ, ρ_i = density or density of fluid i
$\Delta\rho$ = density difference
σ = interfacial tension
s_i = normal stress exerted in i direction on plane perpendicular to i axis
τ, τ_w = shear stress or shear stress at the wall
τ_{ij} = shear stress exerted in j direction on plane perpendicular to i axis

J. S. SON
Shell Development Company

R. B. Bird, W. E. Stewart, and E. H. Lightfoot, *Transport Phenomena*, John Wiley & Sons, Inc., New York, 1960.

R. S. Brodkey, *The Phenomena of Fluid Motions*, Addison-Wesley Publishing Co., Inc., Reading, Mass., 1967.

H. Schlichting, *Boundary Layer Theory*, McGraw-Hill Book Co., Inc., New York, 1960.

L. K. Spink, *Principles and Practice of Flow Meter Engineering*, The Foxboro Co., Foxboro, Mass., 1976.

FLUORESCENT WHITENING AGENTS

The operation of whitening, ie, bleaching or brightening, is concerned with the preparation of fabrics whose commercial value is dependent on the highest possible whiteness.

With the aid of fluorescent whitening agents (FWAs), also referred to as optical brighteners or fluorescent brightening agents, optical compensation of the yellow cast may be obtained. The yellow cast is produced by the absorption of short-wavelength light (violet-to-blue). With FWAs this lost light is in part replaced; thus a complete white is attained without loss of light. This additional light is produced by the whitener by means of fluorescence. Fluorescent whitening agents absorb the invisible uv portion of the daylight spectrum and convert this energy into the longer-wavelength visible portion of the spectrum, ie, into blue to blue-violet light (Fig. 1).

A fluorescent whitener should be optically colorless on the substrate, and should not absorb in the visible part of the spectrum. In the application of FWAs it is possible to replace the light lost through absorption, thereby attaining a neutral, complete white. Further, through the use of excess whitener, still more uv radiation can be converted into visible light, so that the whitest white is made more sparkling. Since the fluorescent light of a fluorescent whitener is itself colored, ie, blue-to-violet, the use of excess whitener always gives either a blue-to-violet or a bluish green cast.

The toxicological properties of fluorescent whiteners have been summarized. Commercial products investigated thus far have been found to be completely harmless. More than 2000 patents for FWAs exist; there are several hundred commercial products and approximately one hundred producers and distributors.

In 1992, world consumption of FWAs was estimated at 60,000 metric tons. Fifty percent was consumed by the detergent industry, 33% by the paper industry, and 17% by the textile industry. Whitener levels in detergents have stabilized and growth rates track population growth. The paper industry has shown higher growth rates due to a trend to higher whites. The rate of growth has been 4% a year (ca 1992). The textile industry has shown moderate growth (2–4%) largely due to greater usage of cotton fabric. The usage of FWAs in plastics is less than 1% of the total consumption.

Many chemical compounds have been described in the literature as fluorescent, and since the 1950s intensive research has yielded many fluorescent compounds that provide a suitable whitening effect; however, only a small number of these compounds have found practical uses. Collectively these materials are aromatic or heterocyclic compounds; many of them contain condensed ring systems. An important feature of these compounds is the presence of an uninterrupted chain of conjugated double bonds, the number of which is dependent on substituents as well as the planarity of the fluorescent part of the molecule. Almost all of these compounds are derivatives of stilbene

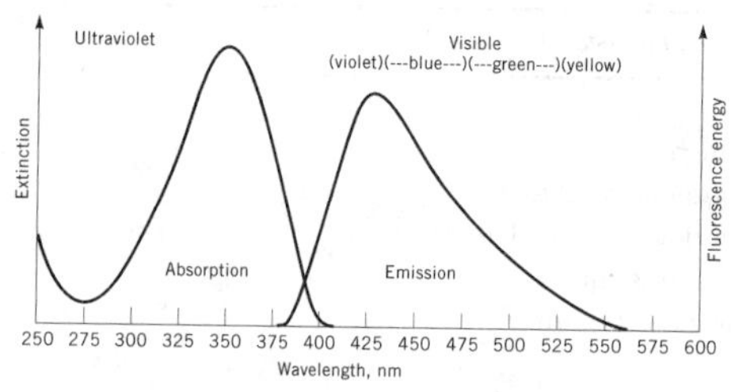

Figure 1. Absorption and emission spectra in solution.

Table 1. Use of Fluorescent Whitening Agents

Industry	Proportion,[a] %	Fundamental structures
textile	20	>30
detergents	45	7[a]
paper	35	
synthetic fibers, plastics	1	9[a]

[a] Values are approximate.

or 4,4′-diaminostilbene; biphenyl; 5-membered heterocycles such as triazoles, oxazoles, imidazoles, etc; or 6-membered heterocycles, eg, coumarins, naphthalimide, *s*-triazine, etc.

Uses

Initially, FWAs were used exclusively in textile finishing; the detergent and paper industries followed thereafter. These products are also used in fiber spinning masses, plastics, and paints. There are more than a thousand known products derived from 200 compounds based on ca 40 fundamental structures. The approximate use distribution of whiteners is shown in Table 1.

Measurement of Whiteness. The Ciba-Geigy Plastic White Scale is effective in the visual assessment of white effects, but the availability of this scale is limited. Most evaluations are carried out by instrumental measurements, utilizing the CIE chromaticity coordinates or the Hunter Uniform Color System (see COLOR).

HAROLD J. MCELHONE, JR.
CIBA-GEIGY Corporation

"Fluorescent Whitening Agents," in F. Coulston and F. Korte, eds., *Environmental Quality and Safety*, Suppl. Vol. 4, Georg Thieme Verlag, Stuttgart, and Academic Press, Inc., New York, 1975.

H. Gold, "Fluorescent Brightening Agents," in K. Venkataraman, ed., *The Chemistry of Synthetic Dyes*, Vol. 5, Academic Press, Inc., New York, 1971.

P. Stensby, *Soap Chem. Special.* **43**(I–V), 41 (Apr.; 84, May; 80, July; 94, Aug.; 96, Sept. (1967)).

J. B. Kramer, in O. Hutzinger, ed., *The Handbook of Environmental Chemistry*, Vol. 3, Part F, Springer-Verlag, Berlin, 1992.

FLUORINE

Fluorine, F_2, is a diatomic molecule existing as a pale yellow gas at ordinary temperatures. Its name is derived from the Latin word *fleure*, meaning to flow, alluding to the well-known fluxing power of the mineral fluorite, CaF_2, which is the most abundant naturally occurring compound of the element. Fluorine has a single naturally occurring isotope, ^{19}F, and has an atomic weight of 18.9984. Fluorine, the most electronegative element and the most reactive nonmetal, is located in the upper right corner of the Periodic Table. It electron configuration is $1s^22s^22p^5$.

The main use of fluorine in the 1990s is in the production of UF_6 for the nuclear power industry. However, it use in the preparation of some specialty products and in the surface treatment of polymers is growing.

Fluorine, which does not occur freely in nature except for trace amounts in radioactive materials, is widely found in combination with other elements, accounting for ca 0.065 wt % of the earth's crust. The most important natural source of fluorine for industrial purposes is the mineral fluorspar, CaF_2, which contains about 49% fluorine.

Physical Properties

Fluorine is a pale yellow gas that condenses to a yellowish orange liquid at −188°C, solidifies to a yellow solid at −220°C, and turns white in a phase transition at −228°C. Fluorine has a strong odor that is easily detectable at concentrations as low as 20 ppb. The odor

Table 1. Physical Properties of Fluorine

Property	Value
melting point, °C	-219.61^a
boiling point, °C	-188.13^a
heat of vaporization, ΔH_{vap}, at -188.44°C and 98.4 kPa, J/mol[b]	6544
heat capacities, J/(mol·K)[b] solid at -223°C	49.338
at -238°C	
density of liquid at bp, kg/m^3	1516^a
density of solid, kg/m^3	1900^c
viscosity, mPa·s(= cP) liquid at -187.96°C	0.257
at -203.96°C gas at 0°C and 101.3 kPa[d]	0.414
	0.0218
thermal conductivity, gas at 0°C and 101.3 kPa, W/(m·K)	0.02477

[a] Generally accepted value. [b] To convert J to cal, divide by 4.184. [c] Mean estimate value. [d] To convert kPa to mm Hg, multiply by 7.5.

resembles that of the other halogens and is comparable to strong ozone (qv). Physical properties are given in Table 1.

Chemical Properties

Fluorine is the most reactive element, combining readily with most organic and inorganic materials at or below room temperature. Many organic and hydrogen containing compounds, in particular, can burn or explode when exposed to pure fluorine. With all elements except helium, neon, and argon, fluorine forms compounds in which it shows a valence of -1. Fluorine reacts directly with the heavier helium-group gases xenon, radon, and krypton to form fluorides (see HELIUM-GROUP GASES, COMPOUNDS). Fluorine is the most electronegative element and thus can oxidize many other elements to their highest oxidation state.

The reactivity of fluorine compounds varies from extremely stable, eg, compounds such as sulfur hexafluoride, nitrogen trifluoride, and the perfluorocarbons (see FLUORINE COMPOUNDS, ORGANIC); to extremely reactive, eg, the halogen fluorides. Another unique property of nonionic metal fluorides is great volatility.

Fluorine is the first member of the halogen family. However, many of its properties are not typical of the other halogens. Fluorine has only one valence state, -1, whereas the other halogens also form compounds in which their valences are $+1$, $+3$, $+5$, or $+7$. Fluorine also has the lowest enthalpy of dissociation relative to the other halogens, which is in part responsible for its greater reactivity. Furthermore, the strength of the bond fluorine forms with other atoms is greater than those formed by the other halogens.

Reactions. *Metals.* At ordinary temperatures, fluorine reacts vigorously with most metals to form fluorides.

Nonmetals. Sulfur reacts with fluorine to yield the remarkable stable sulfur hexafluoride, SF_6.

Silicon and boron burn in fluorine forming silicon tetrafluoride SiF_4, and boron trifluoride, BF_3, respectively. Selenium and tellurium form hexafluorides, whereas phosphorus forms tri- or pentafluorides. Fluorine reacts with the other halogens to form eight interhalogen compounds (see FLUORINE COMPOUNDS, INORGANIC–HALOGENS).

Water. Fluorine reacts with water to form hydrofluoric acid HF, and oxygen difluoride, OF_2.

Oxygen. Oxygen does not react directly with fluorine under ordinary conditions, although in addition to oxygen difluoride, three other oxygen fluorides are known. Dioxygen difluoride, O_2F_2, trioxygen difluoride, O_3F_2, and tetraoxygen difluoride, O_4F_2, are produced in an electric discharge at cryogenic temperatures by controlling the ratio of fluorine to oxygen.

Nitrogen. Nitrogen usually does not react with fluorine under ordinary conditions and is often used as a diluent to moderate fluorinations. However, nitrogen can be made to produce nitrogen trifluoride, NF_3, by radiochemistry, glow discharge, or plasma synthesis.

Noble Gases. Fluorine has the unique ability to react with the heavier noble gases to form binary fluorides.

Hydrogen. The reaction between fluorine and hydrogen is self-igniting and extremely energetic. It occurs spontaneously at ambient temperatures.

Ammonia. Ammonia (qv) reacts with excess fluorine in the vapor phase to produce N_2, NF_3, N_2F_2, HF, and NH_4F.

Organic Compounds. The reaction of pure or undiluted fluorine and organic compounds is usually accompanied by either ignition or a violent explosion of the mixture because of the very high heat of reaction. However, useful commercial scale syntheses using fluorine are undertaken. Volatile compounds may be fluorinated in the gas phase by moderating the reaction using an inert gas such as nitrogen, by reducing reaction temperatures (≤ -78°C), and/or by the presence of finely divided packing materials. Solutions or dispersions of higher boiling materials may be fluorinated in inert solvents such as 1,1,2-trichloro-1,2,2-trifluoroethane or some perfluorocarbon fluids, eg, Fluorinert FC-27 or FC-75 (3M). Efficient removal of the very high reaction heat, which leads to molecular fragmentation and runaway reactions, is the underlying principle in any of the aforementioned approaches.

Polymers. The dilution of fluorine using an inert gas significantly reduces the reactivity, thus allowing controlled reactions to take place with hydrocarbon polymers, even at elevated temperatures. Fluorine may also be used in conjunction with other reactive gases, eg, oxygen and water vapor, to activate polymer surfaces in order to improve chemical bonding and adhesion.

Carbon and Graphite. Fluorine reacts with amorphous forms of carbon, such as wood charcoal, to form carbon tetrafluoride, CF_4, and small amounts of other perfluorocarbons.

Fluorine reacts with high purity carbon or graphite at elevated temperatures under controlled conditions to produce fluorinated carbon, $(CF_x)_n$.

Manufacture

Fluorine is produced by the electrolysis of anhydrous potassium bifluoride, KHF_2 or KF·HF, which contains various concentrations of free HF. The fluoride ion is oxidized at the anode to liberate fluorine gas, and the hydrogen ion is reduced at the cathode to liberate hydrogen.

Commercial Cells. All commercial fluorine installations employ cells having operating currents of $\geq$ 5000 A. The C and E type of the U.S. Atomic Energy Commission (AEC) (now the Dept. of Energy) cell designs predominate in the United States and Canada. The other cell type used in the United States is a proprietary design developed by Allied Chemical Corporation (now AlliedSignal, Inc.). This latter cell has a capacity of 5000 A and is used by AlliedSignal Inc. at its Metropolis, Illinois, plant. Table 2 gives the operating characteristics of a typical commercial size cell (AEC E-type).

Anodes. Fluorine cell anodes are the most important cell component, and their design and materials of construction are key factors in determining productivity and cell live. In the 1990s, anodes are made from petroleum coke and a pitch binder which is calcined at temperatures below that needed to convert the material to graphite. The anode carbon has low electrical resistance, high physical strength, and is resistant to reaction with fluorine.

Table 2. AEC E-Type Cell Operating Characteristics

Characteristic	Value
current, A	6000
operating voltage, V	9–12
cell operating temperature, °C	90–105
hydrogen fluoride in electrolyte, %	40–42
effective anode area, m^2	3.9
anode current density, A/m^2	1500
anodes	32
anode life, A·h	$40–80 \times 10^6$

Other Cell Components. American fluorine manufacturers use Monel or steel cathodes. Polytetrafluoroethylene (PTFE) provides the most satisfactory electrical insulation.

Cells must be fitted with mild steel jackets and/or coils to remove heat during cell operation and to provide heat to maintain the electrolyte molten during shutdown. All commercial cells are totally jacketed.

Heat Transfer. A large portion of cell operating voltage is consumed in ohmic processes which generate heat and are a result of the large separation between anode and cathode and the resistivity of the electrolyte. Approximately 34.8 MJ (33,000 Btu) must be removed per kilogram of fluorine produced from any fluorine cell. This is accomplished by jacketing the cell and/or by using cooling tubes.

Raw Material. The principal raw material for fluorine production is high purity anhydrous hydrofluoric acid. Each kilogram of fluorine generated requires ca 1.1 kg HF.

Process. The generation of fluorine on an industrial scale is a complex operation. The basic raw material, anhydrous hydrogen fluoride, is stored in bulk and charged to a holding tank from which it is continuously fed to the cells. Electrolyte for the cells is prepared by mixing KF·HF with HF to form KF·2HF. The newly charged cells are started up at a low current, which is gradually increased at a conditioning station separate from the cell operating position until full current is obtained at normal voltages. After conditioning, cells are connected in series using ca 12 V provided for each cell by a low voltage, 6000 A d-c rectifier. Hydrogen fluoride content is maintained between 40 and 42% by continuous additions. The electrolyte level must be set and controlled at a certain level below the cell head in order to maintain a seal between the fluorine and hydrogen compartments. The cells are operated at 95–105°C and cooled with water at 75°C.

Equipment. Fluorine can be handled using a variety of materials. System cleanliness and passivation are critical to success. Materials such as nickel, Monel, aluminum, magnesium, copper, brass, stainless steel, and carbon steel are commonly used.

All equipment, lines, and fittings intended for fluorine service must be leaktight, dry, and thoroughly cleansed of all foreign matter before use. After cleaning, the system should be filled with dry nitrogen.

The corrosion resistance of all materials used with fluorine depends on the passivation of the system. This is a pickling operation intended to remove the last traces of foreign matter, and to form a passive fluoride film on the metal surface.

Carbon steel or bronze-body gate valves are commonly used in gaseous fluorine service at low pressure. Plug valves, having Monel bodies and plugs, are recommended for moderate pressure service below 500 kPa (<5 atm). For valve-stem packing PTFE polymer is recommended and it must be maintained leaktight. Valves lubricated or packed with grease or other organics should never be used. Bellows-type valves having Monel or stainless steel bellows are recommended for high pressure service, but not ball valves.

Compressors and blowers for gaseous fluorine service vary in design from multistage centrifugal compressors to diaphragm and piston types. Standard commercial instrumentation and control devices are used in fluorine systems. Pressure is measured using Bourdon-type gauges or pressure transducers. Stainless steel or Monel construction is recommended for parts in contact with fluorine. Standard thermocouples are used for all fluorine temperature-measuring equipment, such as the stainless-steel shielded type, inserted through a threaded compression fitting welded into the line. For high temperature service, nickel-shielded thermocouples should be used.

Dilute mixtures (eg, 10 or 20% F_2 in N_2) are generally less hazardous than pure fluorine, but the same precautions and procedures should be employed.

Economic Aspects

Availability and Shipping. Fluorine gas is packaged and shipped in steel cylinders conforming to U.S. Dept. of Transportation (DOT) specifications 3A1000 and 3AA1000 under a pressure of 2.86 MPa (415 psi). World fluorine producers include Air Products & Chemicals (Allentown, Pa.), AlliedSignal, Inc. (Morristown, N.J.), Asahi Glass Co., Ltd (Tokyo), Ausimont SpA (Milan, Italy), British Nuclear Fuels plc (Preston, U.K.), Cameco Corp. (Port Hope, Ontario, Canada), Central Glass Co., Ltd. (Tokyo), Comurhex (Pechiney group) (Paris), Daikin Industries (Osaka, Japan), Kanto Denka Kogyo Co., Ltd. (Tokyo), and Solvay Fluor und Derivate, GmBH (Hannover, Germany). Mixtures of 10 and 20% fluorine in nitrogen or other inert gases are commercially available in cylinders and tube trailers from Air Products and Chemicals, Inc.

Health and Safety

Fluorine, the most reactive element known, is a dangerous material but may be handled safely using proper precautions. In any situation where an operator may come into contact with low pressure fluorine, safety glasses, a neoprene coat, boots, and clean neoprene gloves should be worn to afford overall body protection. This protection is effective against both fluorine and the hydrofluoric acid which may form from reaction of moisture in the air.

In addition, face shields made of conventional materials or, preferably, transparent, highly fluorinated polymers, should be worn whenever operators approach equipment containing fluorine under pressure. A mask having a self-contained air supply or an air helmet with fresh air supply should always be available. Leaks in high pressure systems usually result in a flame from the reaction of fluorine with the metal. Shields should be provided for valves, pressure-reducing stations, and gauges.

Toxicity. Fluorine is extremely corrosive and irritating to the skin. Inhalation at even low concentrations irritates the respiratory tract; at high concentrations fluorine inhalation may result in severe lung congestion. The American Conference of Governmental Industrial Hygienists (ACGIH) has established the 8-h time-weighted average TLV as 1 ppm or 1.6 mg/m^3, and the short-term exposure limit TLV as 2 ppm or 3.1 mg/m^3.

Burns. Skin burns resulting from contact with pure fluorine gas are comparable to thermal burns and differ considerably from those produced by hydrogen fluoride. Fluorine burns heal much more rapidly than hydrofluoric acid burns.

Disposal. Fluorine can be disposed of by conversion to gaseous perfluorocarbons or fluoride salts. Because of the long atmospheric lifetimes of gaseous perfluorocarbons (see ATMOSPHERIC MODELING), disposal by conversion to fluoride salts is preferred.

Uses

Elemental fluorine is used captively by most manufacturers for the production of various inorganic fluorides. The market for gaseous fluorine is small, but growing. The main use of fluorine is in the manufacture of uranium hexafluoride, UF_6.

Another large use for elemental fluorine is in the production of sulfur hexafluoride, SF_6, a gaseous dielectric for electrical and electronic equipment. An important newer use of fluorine is in the preparation of a polymer surface for adhesives (qv) or coatings (qv).

GEORGE SHIA
AlliedSignal, Inc.

A. J. Rudge, in A. Kuhn, ed., *Industrial Electrochemical Processes*, Elsevier Publishing Co., Amsterdam, the Netherlands, 1971, Chapt. 1.

E. A. Ranken and C. V. Borzileri, "The Safe Handling of Fluorine," *Health and Safety Manual, Suppl. 21.12*, University of California, Lawrence Livermore National Laboratory, Berkeley, Apr. 1987.

J. H. Simons, ed., *Fluorine Chemistry*, Vol. 1, Academic Press, Inc., New York, 1950.

M. Stacey, J. C. Tatlow, and A. G. Sharpe, eds., *Advances in Fluorine Chemistry*, Vol. 2, Butterworths Inc., Washington, D.C., 1961.

FLUORINE COMPOUNDS, INORGANIC

INTRODUCTION

Fluorine (qv), the most electronegative element, is much more reactive than the other elements. Indeed, fluorine reacts with virtually every other element, including the helium-group elements. Because of unique properties, fluorine has been called a superhalogen and several of its compounds called superacids. The term *superacid* is used for systems having higher acidities than anhydrous sulfuric or fluorosulfuric acid.

The basic fluorine-containing minerals are fluorite, commonly called fluorspar, CaF_2; and fluorapatite, commonly called phosphate rock. The reaction of calcium fluoride and sulfuric acid produces hydrogen fluoride. Fluorosilicic acid is produced from fluorapatite as a by-product in the production of phosphoric acid. The boiling point of hydrogen fluoride, 19.54°C, is much higher than that of HCl, −84.9 °C, owing to extensive molecular association via hydrogen bonding in the former. Hydrogen fluoride is the most common reagent for production of fluorine compounds. Elemental fluorine, a pale greenish yellow gas, is produced by electrolysis of anhydrous potassium fluoride–hydrogen fluoride melts.

The fluoride ion is the least polarizable anion. It is small, having a diameter of 0.136 nm, 0.045 nm smaller than the chloride ion. The small size of F^- allows for high coordination numbers and leads to different crystal forms and solubilities, and higher bond energies than are evidenced by the other halides.

A number of elements exhibit the highest oxidation state only because fluorides and oxidation states of +6 and +7 are not uncommon. Fluorine forms very reactive halogen fluorides. Fluorine's special properties lead to many applications. Its complexing properties account for its use as a flux in steelmaking and as an intermediate in aluminum manufacture. The reaction of fluorides with hydroxyapatite, $Ca_5(PO_4)_3OH$, which is found in tooth enamel, to form less soluble and/or more acid-resistant compounds, led to the incorporation of fluorides in drinking water and dentifrices (qv) to reduce dental caries. Many fluorides are volatile and in many cases are the most volatile compounds of an element. This property led to the use of UF_6 for uranium isotope enrichment, critical to the nuclear industry, and the use of metal fluorides in chemical vapor deposition (WF_6, MoF_6, ReF_6), in ion implantation (qv) for semiconductors (qv) BF_3, PF_3, AsF_5, etc), and as unreactive dielectrics (SF_6). Fluorinated steroids, other fluorinated drugs, and anesthetics have medical applications. The stability, lack of reactivity and, therefore, lack of toxicity of some fluorine compounds are also demonstrated by studies reporting survival of animals in an atmosphere of 80% SF_6 and 20% oxygen, and use of perfluorochemicals as short-term blood substitutes because of the ability to efficiently transport oxygen and carbon dioxide. Fluorides including HF, BF_3, SbF_5, PF_5, and several complexes, eg, BF_4^-, PF_6^-, SbF_6^-, and AsF_6^-, are used in many applications in catalysis (qv).

Sources and Applications

The earth's crust consists of 0.09% fluorine. Among the elements fluorine ranks about thirteenth in terrestrial abundance. The ores of most importance are fluorspar, CaF_2; fluorapatite, $Ca_5(PO_4)_3F$; and cryolite, Na_3AlF_6. Fluorspar is the primary commercial source of fluorine. Twenty-six percent of the world's high quality deposits of fluorspar are in North America. Most of that is in Mexico. The majority of the fluorine in the earth's crust is in phosphate rock in the form of fluorapatite which has an average fluorine concentration of 3.5%. Recovery of these fluorine values as by-product fluorosilicic acid from phosphate production has grown steadily, partially because of environmental requirements (see PHOSPHORIC ACID AND PHOSPHATES).

Production of hydrogen fluoride from reaction of CaF_2 with sulfuric acid is the largest user of fluorspar and accounts for approximately 60–65% of total U.S. consumption.

Synthesis

Most inorganic fluorides are prepared by the reaction of hydrofluoric acid with oxides, carbonates, hydroxides, chlorides, or metals. Routes starting with carbonate, hydroxide, or oxide are the most common and the choice is determined by the most economical starting material.

Safety, Toxicity, and Handling

Hazards associated with fluorides are severe. Anhydrous or aqueous hydrogen fluoride is extremely corrosive to skin, eyes, mucous membranes, and lungs; it can cause permanent damage and even death. Detailed information about safety, toxicity, and handling can be obtained from the producers of hydrogen fluoride, eg, Elf Atochem North America, Inc., DuPont, and AlliedSignal. Fluorides susceptible to hydrolysis can generate aqueous hydrogen fluoride. Ingestion of excess fluorides may cause poisoning or damage to bones and/or teeth. Fluorine-containing oxidizers can react with the body in addition to causing burns.

Because hydrogen fluoride is extremely reactive, special materials are necessary for its handling and storage. Anhydrous hydrogen fluoride is produced and stored in mild steel equipment. Teflon or polyethylene are frequently used for aqueous solutions.

The OSHA permissible exposure limit and the American Conference of Governmental Industrial Hygienists (ACGIH) established threshold limit value (TLV) for fluorides is 2.5 mg of fluoride per cubic meter of air. This is the TLV–TWA concentration for a normal 8-h work day and a 40-h work week.

CHARLES B. LINDAHL
TARIQ MAHMOOD
Elf Atochem North America, Inc.

R. J. Gillespie, *Acc. Chem. Res.* **1**, 202 (1968).
R. E. Banks, *J. Fluorine Chem.* **33**, 3–26 (1986).
H. Moissan, *Compt. Rend.* **12**, 1543 (1886).
Threshold Limit Values for Chemical Substances and Physical Agents, 1992–1993, The American Conference of Governmental Industrial Hygienists, Cincinnati, Ohio.

ALUMINUM

Both the binary and complex fluorides of aluminum have played a significant role in the aluminum industry. Aluminum trifluoride, AlF_3, and its trihydrate, $AlF_3 \cdot 3H_2O$, have thus far remained to be the only binary fluorides of industrial interest. The nonahydrate, $AlF_3 \cdot 9H_2O$, and the monohydrate, $AlF_3 \cdot H_2O$, are of only academic curiosity. The monofluoride, AlF, and the difluoride, AlF_2, have been observed as transient species at high temperatures.

Aluminum Trifluoride

Aluminum trifluoride trihydrate, $AlF_3 \cdot 3H_2O$, appears to exist in a soluble metastable α-form as well as a less soluble β-form. Aluminum trifluoride trihydrate is prepared by reacting aluminum hydroxide and aqueous hydrofluoric acid. It can also be advantageously made by a dry process in which dried $Al(OH)_3$ is treated at elevated temperatures with gaseous hydrogen fluoride. High temperature corrosion-resistant alloys such as Monel, Inconel, and titanium are used in the construction of fluidized-bed reactors.

The third process involves careful addition of aluminum hydroxide to fluorosilicic acid which is generated by fertilizer and phosphoric acid-producing plants.

The principal producers of aluminum trifluoride in North America are Alcan, Alcoa, and AlliedSignal. It is also produced in other countries, eg, France, Mexico, Norway, Italy, Tunisia, and Japan.

The principal use of AlF_3 is as a makeup ingredient in the molten cryolite, $Na_3AlF_6 \cdot Al_2O_3$, bath used in aluminum reduction cells in the Hall-Haroult process and in the electrolytic process for refining of aluminum metal in the Hoopes cell.

High Purity Aluminum Trifluoride. High purity anhydrous aluminum trifluoride that is free from oxide impurities can be prepared by reaction of gaseous anhydrous HF and $AlCl_3$ at 100°C, gradually raising the temperature to 400°C. It can also be prepared by the action of elemental fluorine on metal/metal oxide and subsequent sublimation or the decomposition of ammonium fluoroaluminate at 700°C.

Relatively smaller amounts of very high purity AlF_3 are used in ultra low loss optical fiber–fluoride glass compositions, the most common of which is ZBLAN containing zirconium, barium, lanthanum, aluminum, and sodium (see FIBER OPTICS). High purity AlF_3 is also used in the manufacture of aluminum silicate fiber and in ceramics for electrical resistors (see CERAMICS AS ELECTRICAL MATERIALS; REFRACTORY FIBERS). Anhydrous aluminum trifluoride, AlF_3, is a white crystalline solid. Physical properties are listed in Table 1.

Health and Safety. Owing to very low solubility in water and body fluids, AlF_3 is relatively less toxic than many inorganic fluorides. The toxicity values are oral LD_{LO}, 600 mg/kg; subcutaneous, 3000 mg/kg. The ACGIH adopted (1992–1993) TLV for fluorides as F^- is TWA 2.5 mg/m^3. Pyrohydrolysis and strong acidic conditions can be a source of toxicity owing to liberated HF.

Fluoroaluminates

Several fluoroaluminates are known to exist but sodium hexafluoroaluminate, Na_3AlF_6, has dominated industrial applications. More recently potassium tetrafluoroaluminate, $KAlF_4$, has provided a noncorrosive and inexpensive flux in the manufacture of aluminum parts for various applications.

The common structural element in the crystal lattice of fluoroaluminates is the hexafluoroaluminate octahedron, AlF_6^{3-}. The differing structural features of the fluoroaluminates confer distinct physical properties to the species as compared to aluminum trifluoride.

Cryolite. Cryolite constitutes an important raw material for aluminum manufacturing. The natural mineral is accurately depicted as $3\ NaF{\cdot}AlF_3$, but synthetic cryolite is often deficient in sodium fluoride. Physical properties are given in Table 2.

The only commercially viable source of cryolite deposits has been found in the south of Greenland at Ivigtut. For the most part the ore from Ivigtut is a coarse-grained aggregate carrying 10–30% of admixtures, including siderite, quartz, sphalerite, galena, chalcopyrite, and pyrite, in descending order of frequency.

The mineral cryolite is usually white, but may also be black, purple, or violet, and occasionally brownish or reddish. The lustre is vitreous to greasy, sometimes pearly, and the streak is white. The crystals are monoclinic, differing only slightly from orthorhombic symmetry, and have an axial angle of 90°11′.

Synthetic Cryolite. As of this writing, the supply of cryolite is almost entirely met by synthetic material which possesses the same properties and composition with a minor difference in that it is deficient in NaF. Synthetic cryolite also commonly contains oxygen, hydroxyl group, and/or sulfate groups. The NaF deficiency does not interfere for most applications but the presence of moisture leads to the fluorine losses as HF on heating. Because synthetic cryolite is lighter than the natural mineral, losses by dusting are also higher.

Table 1. Physical Properties of Anhydrous Aluminum Trifluoride

Property	Value
mol wt	83.977
mp, °C	1278[a]
transition point, °C	455
density, g/cm^3	3.10
dielectric constant	6

[a] Sublimes.

Table 2. Physical Properties of Cryolite

Property	Value
mol wt	209.94
mp, °C	1012
density, g/cm^3	
monoclinic crystal at 25°C	2.97
solid at 1012°C	2.62
liquid at 1012°C	2.087
hardness, Mohs'	2.5
electrical conductivity, $(\Omega{\cdot}cm)^{-1}$	
solid at 400°C	4.0×10^{-6}
liquid at 1012°C	2.82
viscosity, liquid at 1012°C, mPa·s(= cP)[a]	6.7
solubility in water at 25°C, g/100 g	0.0042

[a] To convert Pa to mm Hg, multiply by 7.

There are several processes available for the manufacture of cryolite. The choice is mainly dictated by the cost and quality of the available sources of soda, alumina, and fluorine. Starting materials include sodium aluminate from Bayer's alumina process; hydrogen fluoride from kiln gases or aqueous hydrofluoric acid; sodium fluoride; ammonium bifluoride, fluorosilicic acid, fluoroboric acid, sodium fluosilicate, and aluminum fluorosilicate; aluminum oxide, aluminum sulfate, aluminum chloride, aluminum hydroxide; and sodium hydroxide, sodium carbonate, sodium chloride, and sodium aluminate.

The manufacture of cryolite is commonly integrated with the production of aluminum hydroxide and aluminum trifluoride. Significant amounts of cryolite are also recovered from waste material in the manufacture of aluminum. In spite of the fact that cryolite is relatively less soluble, its fluoride toxicity by oral routes are reported to be about the same as for soluble fluorides: LD_{50} = 200 mg/kg; for NaF, 180 mg/kg; KF, 245 mg/kg.

The effective dissolution of Al_2O_3 by molten cryolite to provide a conducting bath has spurred the need for its use in manufacture of aluminum. Another use for cryolite is in the production of pure metal by electrolytic refining.

Canada, the United States, and South America are the principal exporters of cryolite and Russia and Europe import cryolite. Primary producers in North America are Alcan, Alcoa, and Reynolds Aluminum.

Potassium Tetrafluoroaluminate. Potassium tetrafluoroaluminate, $KAlF_4$, is a more recent addition to the industrially important fluoroaluminates, mainly because of developments in the automotive industry involving attempts to replace the copper and corrosive solder employed in the manufacture of heat exchangers. Potassium tetrafluoroaluminate in mixtures with other fluoroaluminates, potassium hexafluoroaluminate, K_3AlF_6, and potassium pentafluoroaluminate monohydrate, $K_2AlF_5{\cdot}H_2O$, has emerged as a highly efficient, noncorrosive, and nonhazardous flux for brazing aluminum parts of heat exchangers.

Both $KAlF_4$ and K_3AlF_6 are white solids. The former is less soluble (0.22%) in water than the latter (1.4%). The generally cubic form of $KAlF_4$ inverts to the orthorhombic modification between −23 and 50°C.

An early method of preparation of $KAlF_4$ involved combining aqueous solutions of HF, AlF_3, and KHF_2 in stoichiometric proportions and evaporating the suspension to a dry mixture. The product was subsequently melted and recrystallized. Some of the other conventional technical methods comprise reacting aluminum hydroxide, hydrofluoric acid, and potassium hydroxide followed by separation of the product from the mother liquor; concentrating by evaporation, a suspension obtained by combining stoichiometric amounts of components; and melting together comminuted potassium fluoride and aluminum fluoride at 600°C and grinding the resulting solidified melt.

DAYAL T. MESHRI
Advance Research Chemicals, Inc.

U.S. Pat. 4,983,373 (Jan. 8, 1991), H. P. Withers Jr. and co-workers (to Air Products & Chemicals, Inc.).

U.S. Pat. 4,428,920 (Jan. 31, 1984), H. Willenberg and co-workers (to Kali-Chemie Aktiengesellachaft).

AMMONIUM

Much of the commercial interest in the ammonium fluorides stems from their chemical reactivity as less hazardous substitutes for hydrofluoric acid.

Ammonium Fluoride

Ammonium fluoride is a white, deliquescent, crystalline salt. It tends to lose ammonia gas to revert to the more stable ammonium bifluoride. Its solubility in water is 45.3 g/100 g of H_2O at 25°C and its heat of formation is −466.9 kJ/mol(−116 kcal/mol). Ammonium fluoride is available principally as a laboratory reagent.

Ammonium Bifluoride

Properties. Ammonium bifluoride, NH_4HF_2, is a colorless, orthorhombic crystal. The compound is odorless; however, less than 1% excess HF can cause an acid odor. A number of chemical and physical properties are listed in Table 1.

Ammonium bifluoride dissolves in aqueous solutions to yield the acidic bifluoride ion; the pH of a 5% solution is 3.5. In most cases, NH_4HF_2 solutions react readily with surface oxide coatings on metals; thus NH_4HF_2 is used in pickling solutions (see METAL SURFACE TREATMENTS). Many plastics, such as polyethylene, polypropylene, unplasticized PVC, and carbon brick, are resistant to attack by ammonium bifluoride.

Manufacture. Commercial ammonium bifluoride, which usually contains 1% NH_4F, is made by gas-phase reaction of one mole of anhydrous ammonia and two moles of anhydrous hydrogen fluoride; the melt that forms is flaked on a cooled drum. Production of bifluoride from fluoride by-products from the phosphate industry has had little if any commercial significance.

Precautions in Handling. Ammonium bifluoride, like all soluble fluorides, is toxic if taken internally. Hydrofluoric acid burns may occur if the material comes in contact with moist skin.

Applications. Ammonium bifluoride solubilizes silica and silicates by forming ammonium fluorosilicate, $(NH_4)_2SiF_6$. Inhibited 15% hydrochloric acid containing about 2% ammonium bifluoride has been used to acidize oil wells in siliceous rocks to regenerate oil flow. The use of ammonium bifluoride is important in locations where dissolved silicates foul boiler tubes with scale that cannot be removed using usual cleaning aids. Ammonium bifluoride is also used as an etching agent for silicon wafers.

Rapid frosting of glass is accomplished in a concentrated solution of ammonium bifluoride and hydrofluoric acid with nucleating agents that assure uniform frosts. Ammonium bifluoride is used as a sour or neutralizer for alkalies in commercial laundries and textile plants. Treatment also removes iron stain.

Table 1. Properties of Ammonium Bifluoride, NH_4HF_2

Property	Value
melting point, °C	126.1
boiling point, °C	239.5
solubility in water at 25°C, wt %	41.5
specific gravity	1.50

Ammonium fluorides react with many metal oxides or carbonates at elevated temperatures to form double fluorides. The double fluorides decompose at even higher temperatures to form the metal fluoride and volatile NH_3 and HF. This reaction produces pure salts less likely to be contaminated with oxyfluorides.

JOHN R. PAPCUN
Atotech

Fr. Pat. 1,546,234 (Nov. 15, 1968), (to Farbenfabriken Bayer A.-G.).

R. C. Weast, ed., *Handbook of Chemistry and Physics*, 59th ed., The Chemical Rubber Co., Cleveland, Ohio, 1978.

H. Schutza, M. Eucken, and W. Namesh, *Z. An. All. Chem.* **292**, 293 (1957).

Glass Frosting and Polishing Technical Service Bulletin 667, Atotech USA Inc., Cleveland, Ohio.

ANTIMONY

Antimony Trifluoride

Properties. Antimony trifluoride, SbF_3, is a very hygroscopic, white, crystalline solid, mp = 292°C. It can be sublimed under vacuum. It is very soluble in water, hydrofluoric acid, and polar organic solvents such as alcohols and ketones. Antimony trifluoride is a mild fluorinating reagent.

Preparation. Antimony trifluoride can be readily prepared by dissolving Sb_2O_3 in an excess of anhydrous hydrogen fluoride or in aqueous acid of 40% or higher strength hydrofluoric acid, followed by evaporation of the solution to dryness.

Uses. The market for SbF_3 in the United States is less than 5 t/yr. More recent uses of SbF_3 have been in the manufacture of fluoride glass and fluoride glass optical fiber preform and fluoride optical fiber in the preparation of transparent conductive films.

Antimony Pentafluoride

Properties. Antimony pentafluoride, SbF_5, is a colorless, hygroscopic, very viscous liquid that fumes in air. Its viscosity at 20°C is 460 mPa·s(= cP) which is very close to the value for glycerol.

Preparation. Antimony pentafluoride can be prepared by direct fluorination of SbF_3 or antimony or by reaction of $SbCl_5$ with HF.

Uses. Antimony pentafluoride is a moderate fluorinating reagent and a powerful oxidizer. Antimony pentafluoride is used to saturate double bonds in straight-chain olefins, cycloolefins, aromatic rings, and in the fluorination of halocarbons and CrO_2Cl_2, $MoCl_5$, WCl_6, PCl_3, P_4O_{10}, $SiCl_4$, $TiCl_4$, and SiO_2. Antimony pentafluoride forms intercalation compounds with graphite and fluorinated graphite. These nonstoichiometric substances may have potential use as superconducting materials.

Hexafluoroantimonates

Hexafluoroantimonic acid, $HSbF_6{\cdot}6H_2O$, is prepared by dissolving freshly prepared hydrous antimony pentoxide in hydrofluoric acid or adding the stoichiometric amount of 70% HF to SbF_5. Both of these reactions are exothermic and must be carried out carefully. The superacid systems $HSO_3F{\cdot}SbF_5$ and $HF{\cdot}SbF_5$ (fluoroantimonic acid) are used in radical polymerization and in carbocation chemistry. Anhydrous salts, $MSbF_6$, where M = H, NH_4, and alkali metal, and $M(SbF_6)_2$, where M is an alkaline-earth metal, can be used as photoinitiators for the production of polymers.

Environmental and Safety Aspects

OSHA has a TWA standard on a weight of Sb basis of 0.5 mg/m^3 for antimony in addition to a standard TWA of 2.5 mg/m^3 for fluoride. NIOSH has issued a criteria document on occupational exposure to inorganic fluorides. Antimony pentafluoride is considered by the EPA to be an extremely hazardous substance and releases of 0.45 kg or more reportable quantity (RQ) must be reported. Antimony trifluoride is on the CERCLA list and releasing of 450 kg or more RQ must be reported.

TARIQ MAHMOOD
CHARLES B. LINDAHL
Elf Atochem North America, Inc.

I. G. Ryss, *The Chemistry of Fluorine and Its Inorganic Compounds*, State Publishing House of Scientific, Technical, and Chemical Literature, Moscow, USSR, 1956; English trans., AEC-tr-3927, Office of Technical Services, U.S. Dept. of Commerce, Washington, D.C., 1960, pp. 283–295 (Part I).

G. A. Olah and co-workers, *J. Am. Chem. Soc.* **97**(19), 5477–5481 (1975).

R. J. Gillespie, in V. Gold, ed., *Proton Transfer Reactions*, Chapman and Hall, London, 1975, p. 27.

U.S. Pat. 4,136,102 (Jan. 23, 1979), J. V. Crivello (to General Electric Co.).

ARSENIC

Great care should be exercised in the handling and use of all arsenic compounds (qv) because NIOSH has determined inorganic arsenic to be a carcinogen and OSHA considers inorganic arsenic to be a cancer hazard. The OSHA permissible exposure limit is 10 $\mu g/m^3$, averaged over any 8-h period. The OSHA action level is 5 $\mu g/m^3$, averaged over any 8-h period. The OSHA limits have the force of law and are much lower than the 0.2 mg/m^3 of ACGIH.

Arsenous Fluoride

Arsenous fluoride, AsF_3, is a colorless liquid, mp = −5.95 °C, bp = 57.13 °C at 99 kPa (742.5 mm Hg), and sp gr = 2.67, having a standard enthalpy of formation of −858.1 kJ/mol(−205.1 kcal/mol). Arsenic(III) fluoride can be prepared by fluorination of arsenous oxide using sulfuric acid and calcium fluoride, or using hydrofluoric acid or fluorosulfuric acid; from thermal decomposition of $AsBr_4AsF_6$; from the fluorination of gallium arsenide using F_2 or NF_3; from As_2O_3, CaF_2, and concentrated H_2SO_4; from disproportionation of graphite intercalated compounds of AsF_5; from the reaction of arsenous trichloride with NaF at 300°C in the presence of $ZnCl_2$ or KCl, and from the fluorination of arsenous trichloride with antimony trifluoride or zinc fluoride.

It is used as a fluorinating reagent, in semiconductor doping, to synthesize some hexafluoroarsenate compounds, and in the manufacture of graphite intercalated compounds (see SEMICONDUCTORS). AsF_3 has been used to achieve >8% total area simulated air-mass 1 power conversion efficiencies in Si *p-n* junction solar cells (see SOLAR ENERGY). It is commercially produced, but usage is estimated to be less than 100 kg/yr.

Arsenic Trifluoride Oxide

Arsenic trifluoride oxide, $AsOF_3$, has been reported to be produced by the uv photolysis of O_3 or HOF in the presence of AsF_3.

Arsenic Pentafluoride

Arsenic pentafluoride, AsF_5, melts at −79.8 °C and boils at −52.8 °C. At the boiling point the liquid has a density of 2.33 g/mL.

Arsenic pentafluoride can be prepared by reaction of fluorine and arsenic trifluoride or arsenic; from the reaction of NF_3O and As; from the reaction of $Ca(FSO_3)_2$ and H_3AsO_4; or by reaction of alkali metal or alkaline-earth metal fluorides or fluorosulfonates with H_3AsO_4 and H_2AsO_3F.

It is used as a fluorinating reagent and in syntheses of some hexafluoroarsenate compounds. Arsenic pentafluoride is also used to dope semiconductors; to produce conductive polymers; and in conducting-oriented fibers.

Hexafluoroarsenic Acid and the Hexafluoroarsenates

Hexafluoroarsenic acid can be prepared by the reaction of arsenic acid with hydrofluoric acid or calcium fluorosulfate and with alkali or alkaline-earth metal fluorides or fluorosulfonates. The hexafluoroarsenates can be prepared directly from arsenates and hydrofluoric acid, or by neutralization of $HAsF_6$.

Because of the special stability of the hexafluoroarsenate ion, there are a number of applications of hexafluoroarsenates. For example, onium hexafluoroarsenates have been described as photoinitiators in the hardening of epoxy resins (qv). Lithium hexafluoroarsenate has been used as an electrolyte in lithium batteries (qv). Hexafluoroarsenates, especially alkali and alkaline-earth metal salts or substituted ammonium salts, have been reported to be effective as herbicides (qv).

CHARLES B. LINDAHL
TARIQ MAHMOOD
Elf Atochem North America, Inc.

O. Ruff, *Die Chemie des Fluors*, Springer-Verlag, Berlin, 1920, p. 27.

Ger. Pat. DD248249 A3 (Aug. 5, 1987), P. Wolter, M. Schoenherr, D. Hass.

U.S. Pat. 3,769,387 (Oct. 30, 1973), R. A. Wiesboeck and J. D. Nickerson.

Ger. Offen. 2,618,871 (Nov. 11, 1976) and 2,518,652 (May 2, 1974), J. V. Crivello (to General Electric Co.).

BARIUM

Barium Fluoride

Barium fluoride, BaF_2, is a white crystal or powder. Under the microscope, crystals may be clear and colorless. Reported melting points vary from 1290 to 1355°C. The boiling point is 2260°C, and the density 4.9 g/cm^3. The solubility in water is about 1.6 g/L at 25°C.

High purity BaF_2 can be prepared from the reaction of barium acetate and aqueous HF, by dissolving the impure material in 2–12 *N* HCl and recrystallizing at −40°C, by vacuum distillation of the metal fluoride impurities from a BaF_2 melt, by purification of the aqueous acetate solution by ion exchange followed by fluorination, by solvent extraction using dithiocarbamate and CCl_4, and by solvent extraction using acetonitrile.

Barium fluoride is used commercially in combination with other fluorides for arc welding (qv) electrode fluxes. However, this usage is limited because of the availability of the much less expensive naturally occurring calcium fluoride.

The toxicity of barium fluoride has received only little attention. OSHA has a TWA standard on the basis of Ba of 0.5 mg/m^3 for barium fluoride in addition to a standard TWA on the basis of F or 2.5 mg/m^3.

TARIQ MAHMOOD
CHARLES B. LINDAHL
Elf Atochem North America, Inc.

A. A. Lugina and co-workers *Zh. Neorg. Khim 1981*, **26**(2), 332–336.

Jpn. Pat. 90-144378 (June 4, 1990), K. Kobayashi, K. Fujiura, and S. Takahashi.

EP 90-312689 (Nov. 21, 1990), J. A. Sommers, R. Ginther, and K. Ewing.

A. M. Garbar, A. N. Gulyaikin, G. L. Murskii, I. V. Filimonov, and M. F. Churbanov, *Vysokochist, Veshchestva* (6), 84–85 (1990).

A. M. Garbar, A. V. Loginov, G. L. Murskii, V. I. Rodchenkov, and V. G. Pimenov, *Vysokochist, Veshchestva* (3), 212–213 (1989).

J. Chen and co-workers, *J. Non-Cryst. Solids* **140**(1–3), 293–296 (1992).

BORON

BORON TRIFLUORIDE

Boron trifluoride (trifluoroborane), BF_3, is a colorless gas when dry, but fumes in the presence of moisture, yielding a dense white smoke of irritating, pungent odor. It is widely used as an acid catalyst for many types of organic reactions, especially for the production of polymer and petroleum (qv) products.

Physical Properties.

The physical properties are listed in Table 1. Water and aqueous mineral acids react with BF_3 to yield the hydrates of BF_3 or the hydroxyfluoroboric acids, fluoroboric acid, or boric acid. Solution in aqueous alkali gives the soluble salts of the hydroxyfluoroboric acids, fluoroboric acids, or boric acid. Boron trifluoride, slightly soluble in many organic solvents including saturated hydrocarbons (qv), halogenated hydrocarbons, and aromatic compounds, easily polymerizes unsaturated compounds such as butylenes (qv), styrene (qv), or vinyl esters, as well as easily cleaved cyclic molecules such as tetrahydrofuran. Other molecules containing electron-donating atoms such as O, S, N, P, etc, eg, alcohols, acids, amines, phosphines, and ethers, may BF_3 to produce soluble liquid or solid adducts.

Chemical Properties. Some reactions are listed in Table 2.

Manufacture. Boron trifluoride is prepared by the reaction of a boron-containing material and a fluorine-containing substance in the presence of an acid. The traditional method used borax, fluorspar, and sulfuric acid. In another process, fluorosulfonic acid is treated with boric acid.

Shipment and Handling. The gas is nonflammable and is shipped in DOT 3A and 3AA steel cylinders at a pressure of approximately 12,410 kPa (1800 psi). Boron trifluoride is classified as a poison gas, both domestically and internationally.

Economic Aspects. The sole United states producer of boron trifluoride is AlliedSignal, Inc.

Health and Safety Factors. Boron trifluroide is primarily a pulmonary irritant. The toxicity of the gas to humans has not been reported, but laboratory tests on animals gave results ranging from an increased pneumonitis to death. The TLV is 1 ppm. High concentrations burn the skin similarly to acids such as HBF_4. No chronic effects have been observed in workers exposed to small quantities of the gas at frequent intervals over a period of years.

Uses. Boron trifluoride is an excellent Lewis acid catalyst for numerous types of organic reactions.

FRANCIS EVANS
GANPAT MANI
AlliedSignal, Inc.

H. S. Booth and D. R. Martin, *Boron Trifluoride and Its Derivatives*, John Wiley & Sons, Inc., New York, 1949.

C. A. Wamser, *J. Am. Chem. Soc.* **73**, 409 (1951).

D. R. Martin and J. M. Canon, in G. A. Olah, ed., *Friedel-Crafts and Related Reactions*, Vol. 1, Wiley-Interscience, New York, 1963, pp. 399–567.

A. V. Topchiev, S. V. Zavgorodnii, and Y. M. Paushkin, *Boron Fluoride and Its Compounds as Catalysts in Organic Chemistry*, Pergamon Press, New York, 1959.

Table 1. Physical Properties of Boron Trifluoride

Property	Value
molecular weight	67.8062
melting point, °C	−128.37
boiling point, °C	−99.9
density, critical, d_c, g/cm^3	ca 0.591
entropy, $S_{298.15}$, J/(mol·K)[a]	254.3
Gibbs free energy of formation, $\Delta G_{f298.15}$, kJ/mol[a]	−1119.0

[a] To convert J to cal, divide by 4.184.

Table 2. Reactions of Boron Trifluoride

Reactant	Temperature, °C	Products	Formula
sodium[a]		boron, amorphous, sodium fluoride	NaF
magnesium, molten alloys	no reaction		
calcium	1600	calcium hexaboride	CaB_6
aluminum	1200	aluminum boride (1:12), tetragonal boron	AlB_{12}
	1650[b]	β-rhombohedral boron	
titanium	1600	titanium boride	TiB_2
copper, mercury, chromium, iron	RT or below	no reaction[c]	
sodium nitrate, sodium nitrite	180	sodium fluoroborate, boric oxide	$NaBF_4$

[a] With incandescence. [b] Further reaction. [c] Even when subjected to pressure for a considerable length of time; also no reaction with red-hot iron.

FLUOROBORIC ACID AND FLUOROBORATES

Fluoroboric Acid and the Fluoroborate Ion

Fluoroboric acid, generally formulated as HBF_4, does not exist as a free, pure substance. The acid is stable only as a solvated ion pair, such as $H_3O^+BF_4^-$; the commercially available 48% HBF_4 solution approximates $H_3O^+BF_4^-\cdot 4\ H_2O$. Fluoroboric acid and many transition-metal salts are used in the electroplating (qv) and metal finishing industries. Some of the alkali metal flururoborates are used in fluxes.

Properties. Table 1 lists some of the physical properties of fluoroboric acid. It is a strong acid in water, equal to most mineral acids in strength.

Manufacture, Shipping, and Waste Treatment. Fluoroboric acid (48%) is made commercially by direct reaction of 70% hydrofluoric acid and boric acid, H_3BO_3. The reaction is exothermic and must be controlled by cooling.

Fluoroboric acid and some fluoroborate solutions are shipped as corrosive material, generally in polyethylene-lined steel pails and drums or in rigid nonreturnable polyethylene containers.

Waste treatment of fluoroborate solutions includes a pretreatment with aluminum sulfate to facilitate hydrolysis, and final precipitation of fluoride with lime.

Economic Aspects. In the United States fluoroboric acid is manufactured by Atotech USA, Inc., General Chemical, C.P. Chemical Company, Fidelity Chemical Products, and Chemtech Harstan.

Toxicity. Sodium fluoroborate is absorbed almost completely into the human bloodstream and over a 14-d experiment all of the $NaBF_4$ ingested was found in the urine.

Uses. Printed circuit tin–lead plating is the main use of fluoroboric acid. However, the Alcoa Alzak process for electropolishing aluminum requires substantial quantities of fluoroboric acid.

Main Group

Properties. A summary of the chemical and physical properties of alkali-metal and ammonium fluoroborates is given in Table 2. Chem-

Table 1. Physical Properties of Fluoroboric Acid

Property	Value
heat of formation, kJ/mol[a]	
aqueous, 1 molal, at 25°C	−1527
BF_4^-, gas	−1765 ± 42
entropy of the BF_4^- ion, J/(mol·K)[a]	167
specific gravity, 48% soln	1.37
surface tension, 48% soln at 25°C, mN/m(= dyn/cm)	65.3

[a] To convert J to cal, divide by 4.184.

Table 2. General Properties of Metal Fluoroborates

Compound	Molecular weight	Color	Mp, °C	Density,[a] g/cm^3
$LiBF_4$	93.74	white		
$NaBF_4$	109.79	white	406 dec	2.47
				210[b]
KBF_4	125.92	colorless	530 dec	2.498
$RbBF_4$	172.27		612 dec	2.820
				10[c]
$CsBF_4$	219.71	white	555 dec	3.20
				30[c]
NH_4BF_4	104.84	white	487 dec	1.871[b]
$NaBF_3OH$				2.46

[a] Unless otherwise stated, at 20°C.
[b] At 15°C
[c] At 100°C.

ically these compounds differ from the transition-metal fluoroborates usually separating in anhydrous form. This group is very soluble in water, except for the K, Rb, and Cs salts which are only slightly soluble. Many of the soluble salts crystallize as hydrates.

Manufacture. Fluoroborate salts are prepared commercially by several different combinations of boric acid and 70% hydrofluoric acid with oxides, hydroxides, carbonates, bicarbonates, fluorides, and bifluorides. Glass vessels and equipment should not be used.

Economic Aspects. In the United States the sodium, potassium, ammonium, and magnesium fluoroborates are sold by Advance Research Chemicals, Atotech USA, Inc., and General Chemical. The lithium compound is available from Advance Research Chemicals, Cyprus Foote Mineral, and FMC Lithium Corp. of America. Small amounts of other fluoroborates are sold by Alfa Inorganics, Inc. and Ozark-Mahoning Co.

Uses. Alkali metal and ammonium fluoroborates are used mainly for the high temperature fluxing action required by the metals processing industries. Other uses are as catalysts and in fire-retardant formulations.

Transition-Metal and Other Heavy-Metal Fluoroborates

The physical and chemical properties are less well known for transition metals than for the alkali metal fluoroborates (Table 3).

Economic Aspects. Most fluoroborate solutions listed in Table 4 are manufactured by Atotech USA, Inc., General Chemical, Chemtec/Harstan, C.P. Chemical Company, and Fidelity Chemical Products.

Uses. Metal fluoroborate solutions are used primarily as plating solutions and as catalysts.

Table 3. Properties and Metal Fluoroborates[a]

Compound	Color	Specific gravity	Solubility
$Mn(BF_4)_2 \cdot 6H_2O$	pale pink	1.982	water, ethanol
$Fe(BF_4)_2 \cdot 6H_2O$	pale green	2.038	water, ethanol
$Co(BF_4)_2 \cdot 6H_2O$	red	2.081	water, alcohol
$Ni(BF_4)_2 \cdot 6H_2O$	green	2.136	water, alcohol
$Cu(BF_4)_2 \cdot 6H_2O$	blue	2.175	water, alcohol
$AgBF_4 \cdot H_2O$	colorless		water, less sol in alcohol, sol benzene, sol ether
$Zn(BF_4)_2 \cdot 6H_2O$	white	2.120	water, alcohol
$Cd(BF_4)_2 \cdot 6H_2O$	white	2.292	water, alcohol
$In(BF_4)_3 \cdot xH_2O$	colorless		water
$TlBF_4 \cdot H_2O$	colorless		water
$Sn(BF_4)_2 \cdot xH_2O$	white		water
$Pb(BF_4)_2 \cdot H_2O$	colorless		

[a] Crystalline solids.

Table 4. Commercial Metal Fluoroborate Solutions

Metal cation	Formula	% Metal	Specific gravity, g/cm^3
antimony(II)	$Sb(BF_4)_3$	12.8	1.42
cadmium	$Cd(BF_4)_2$	19.7	1.60
cobalt(II)	$Co(BF_4)_2$	11.8	1.42
copper(II)	$Cu(BF_4)_2$	12.2	1.48
indium	$In(BF_4)_3$	15.3	1.55
iron(II)	$Fe(BF_4)_2$	10.3	1.47
lead(II)	$Pb(BF_4)_2$	28.9	1.75
nickel(II)	$Ni(BF_4)_2$	11.2	1.47
tin(II)	$Sn(BF_4)_2$	20.2	1.61
zinc	$Zn(BF_4)_2$	11.0	1.39

Electroplating. Metal fluoroborate electroplating (qv) baths are employed where speed and quality of deposition are important. High current densities can be used for fast deposition and near 100% anode and cathode efficiencies can be expected. Because the salts are very soluble, highly concentrated solutions can be used without any crystallization. The high conductivity of these solutions reduces the power costs. The metal content of the bath is also easily maintained and the pH is adjusted with HBF_4 or aqueous ammonia. The disadvantages of using fluoroborate baths are treeing, lack of throwing power, and high initial cost. Treeing and throwing power can be controlled by additives; grain size of the deposits can also be changed. As of this writing, metals being plated from fluoroborate baths are Cd, Co, Cu, Fe, In, Ni, Pb, Sb, and Zn. Studies on Fe, Ni, and Co fluoroborate baths describe the compositions and conditions of operation as well as the properties of the coatings. Iron foils electrodeposited from fluoroborate baths and properly annealed have exceptionally high tensile strength.

JOHN R. PAPCUN
Atotech

H. S. Booth and D. R. Martin, *Boron Trifluoride and Its Derivatives*, John Wiley & Sons, Inc., New York, 1949, pp. 87–165.

I. G. Ryss, *The Chemistry of Fluorine and Its Inorganic Compounds*, State Publishing House for Scientific, Technical, and Chemical Literature, Moscow, USSR, 1956; F. Haimson, English trans., *AEC-tr-3927*, U.S. Atomic Energy Commission, Washington, D.C., 1960, pp. 505–579.

Chemical Economics Handbook, Stanford Research Institute, Menlo Park, Calif., 1975, p. 739.5030H.

J. F. Jumer, *Met. Finish.* **56**(8), 44 (1958); **56**(9), 60 (1958).

R. C. Weast, ed., *Handbook of Chemistry and Physics*, Vol. 59, The Chemical Rubber Co., Cleveland, Ohio, 1978.

CALCIUM

Calcium Fluoride

Fluorspar is used directly in the manufacture and finishing of glass (qv), in ceramics (qv) and welding (qv) fluxes, and in the extraction and processing of nonferrous metals.

As of 1993 fluorspar was still the principal source of fluorine for industry. Fluorspar deposits are commonly epigenetic, ie, the elements moved from elsewhere into the country rock. For this reason, fluorine mineral deposits are closely associated with fault zones. In the United States, significant fluorspar deposits occur in the Appalachian Mountains and in the mountainous regions of the West, but the only reported commercial production in 1993 was from the faulted carbonate rocks of Illinois.

Preparation. CaF_2 is manufactured by the interaction of H_2SiF_6 with an aqueous carbonate suspension; by the reaction of $CaSO_4$ with NH_4F; by the reaction of HF with $CaCO_3$ in the presence of NH_4F; by reaction of $CaCO_3$ and NH_4F and $CaCO_3$; and from the thermal decomposition of calcium trifluoroacetate.

Table 1. Physical Properties of Calcium Fluoride

Property	Value
formula weight	78.08
composition, wt %	
Ca	51.33
F	48.67
melting point, °C	1402
boiling point, °C	2513
thermal conductivity, crystal at 25°C, W/(m·K)	10.96
density, solid at 25°C, g/mL	3.181
hardness, Mohs' scale	4
solubility in water, g/L at 25°C	0.146
refractive index at 24°C, 589.3 nm	1.43382
dielectric constant at 30°C	6.64

Worldwide, large deposits of fluorspar are found in China, Mongolia, France, Morocco, Mexico, Spain, South Africa, and countries of the former Soviet Union. The United States imports fluorspar from most of these countries.

Properties. Some of the important physical properties of calcium fluoride are listed in Table 1. Pure calcium fluoride is without color. However, natural fluorite can vary from transparent and colorless to translucent and white, wine-yellow, green, greenish blue, violet-blue, and sometimes blue, deep purple, bluish black, and brown. These color variations are produced by impurities and by radiation damage (color centers).

Mining. Underground mining procedures are used for deep fluorspar deposits, and open-pit mines are used for shallow deposits or where conditions do not support underground mining techniques.

Beneficiation. Most fluorspar ores as mined must be concentrated or beneficiated to remove waste. Metallurgical-grade fluorspar is sometimes produced by hand sorting lumps of high grade ore. In most cases the ore is beneficiated by gravity concentration with fluorspar and the waste minerals, having specific gravity values of >3 and <2.8, respectively.

Economic Aspects. For many years the United States has relied on imports for more than 80% of fluorspar needs.

Grades. Fluorspar is marketed in several grades: metallurgical fluorspar (metspar) is sold as gravel, lump, or briquettes. The minimum acceptable assay is 60% effective calcium fluoride.

Ceramic-grad fluorspar and acid-grade fluorspar have the typical analyses shown in Table 2. Optical-grade calcium fluoride, for special glasses and for growing single crystals, is supplied in purities up to 99.99 CaF_2.

Health and Safety Factors. The low solubility of calcium fluoride reduces the potential problem of fluoride-related toxicity. However, because the solubility of calcium fluoride in stomach acid is higher, continued oral ingestion of calcium fluoride could produce symptoms of fluorosis.

A significant hazard results from contact of calcium fluoride with high concentrations of strong acids because of evolution of toxic concentrations of hydrogen fluoride.

Beneficiation facilities require air and water pollution control systems, including efficient control of dust emissions, treatment of process water, and proper disposal of tailings. Contact with fluorspar may irritate the skin and eyes.

Consumption and Uses. Acid-grade fluorspar, which is >97% calcium fluoride, is used primarily in the production of hydrogen fluoride. Ceramic-grade fluorspar, containing 85 to 95% CaF_2 content, is used in the production of glass and enamel, to make welding rod coatings, and as a flux in the steel industry. Metallurgical-grade fluorspar, containing 60 to 85% or more CaF_2, is used primarily as a fluxing agent by the steel industry.

In the ceramic industry, fluorspar is used as a flux and as an opacifier in the production of flint glass, white or opal glass, and enamels (see ENAMELS, PORCELAIN OR VITREOUS). Fluorspar is used in the manufacture of aluminum, brick, cement, and glass fibers, and is used by the foundry industry.

A small but artistically interesting use of fluorspar is in the production of vases, cups, and other ornamental objects popularly known as Blue John, after the Blue John Mine, Derbyshire, U.K. Optical quality fluorite, sometimes from natural crystals, but more often artificially grown, is important in use as infrared transmission windows and lenses and optical components of high energy laser systems (see INFRARED TECHNOLOGY AND RAMAN SPECTROSCOPY; LASERS).

Table 2. Analyses of Ceramic- and Acid-Grade Fluorspar, wt %[a]

Assay	Ceramic	Acid
CaF_2	90.0–95.5	96.5–97.5
SiO_2	1.2–3.0	1.0
$CaCO_3$	1.5–3.4	1.0–1.5
MgO		0.15
B		0.02
Zn		0.02
Fe_2O_3	0.10	0.10
P_2O_5		0.03
$BaSO_4$		0.2–1.3
R_2O_3[a]	0.15–0.25	0.1–0.3

[a] R_2O_3 is any trivalent metal oxide, eg, Al_2O_3.

TARIQ MAHMOOD
CHARLES B. LINDAHL
Elf Atochem North America, Inc.

D. R. Shawe, ed., *Geology and Resources of Fluorine in the United States*, U.S. Geological Survey Professional Paper 933, Washington, D.C., 1976, pp. 1–5, 18, 19, 82–87.

M. M. Miller, *Fluorspar 1991 Annual Report*, U.S. Dept. of Interior, Bureau of Mines, Washington, D.C.

COBALT

Cobalt Difluoride

Cobalt difluoride, CoF_2, is a pink solid having a magnetic moment of 4.266×10^{-23} J/T (4.6 Bohr magneton) and closely resembling the ferrous (FeF_2) compounds. Physical properties are listed in Table 1. Cobalt(II) fluoride is highly stable.

CoF_2 is manufactured commercially by the action of aqueous or anhydrous hydrogen fluoride (see FLUORINE COMPOUNDS, INORGANIC–HYDROGEN) on cobalt carbonate (see COBALT COMPOUNDS) in a plastic, ie, polyethylene/polypropylene, Teflon, Kynar, rubber, or graphite-lined container to avoid metallic impurities.

Cobalt difluoride, used primarily for the manufacture of cobalt trifluoride, CoF_3, is available from Advance Research Chemical, Inc., Aldrich Chemicals, and PCR in the United States, Fluorochem in the

Table 1. Physical Properties of the Cobalt Fluorides

Parameter	Cobalt difluoride[a]	Cobalt trifluoride
molecular weight	96.93	115.93
melting point, °C	1127	926
solubility, g/100 g[b]		
water	1.36	dec
density, g/cm^3	4.43	3.88

[a] The bp of CoF_2 is 1739°C.
[b] CoF_2 is also soluble in mineral acids.

U.K., and Schuhardt in Germany. CoF_2 is shipped as a corrosive and toxic material in DOT-approved containers.

Cobalt Trifluoride

Cobalt(III) fluoride or cobalt trifluoride, CoF_3, is one of the most important fluorinating reagents. Physical properties may be found in Table 1. CoF_3, a light brown, very hygroscopic compound, is a powerful oxidizing agent and reacts violently with water, evolving oxygen.

Cobalt trifluoride is readily prepared by reaction of fluorine (qv) and $CoCl_2$ at 250°C or CoF_2 at 150–180°C. CoF_3 is used for the replacement of hydrogen with fluorine in halocarbons; for fluorination of xylylalkanes, used in vapor-phase soldering fluxes; formation of dibutyl decalins; fluorination of alkynes; synthesis of unsaturated or partially fluorinated compounds; and conversion of aromatic compounds to perfluorocyclic compounds (see FLUORINE COMPOUNDS, ORGANIC).

CoF_3 is available from Advance Research Chemicals, Inc., Aldrich Chemicals, Aesar, Johnson/Matthey, PCR, Pfaltz & Bauer, Noah Chemicals, and Strem Chemicals of the United States, Fluorochem of the U.K., and Schuhardt of Germany. Demand for cobalt trifluoride varies from 100 to 1500 kg/yr.

Dust masks should be used while handling both the cobalt fluorides and all other cobalt compounds. CoF_3 is shipped as an oxidizer and a corrosive material.

DAYAL T. MESHRI
Advance Research Chemicals, Inc.

R. D. Fowler and co-workers, *Ind. Eng. Chem.* **39**, 292 (1947).

Eur. Pat. Appl EP 281,784 (Sept. 14, 1988), W. Bailey and J. T. Lilack (to Air Products and Chemicals, Inc.).

U.S. Pat. 4,849,553 (July 18, 1989), W. T. Bailey, F. K. Schweighardt, and V. Ayala (to Air Products and Chemicals, Inc.).

Ger. Pat. DD 287,478 (Feb. 28, 1991), W. Radeck and co-workers (to Akademie der Wissenschaften der DDR).

COPPER

Copper(II) Fluorides

Copper(II) forms several stable fluorides, eg, cupric fluoride, CuF_2, cupric fluoride dihydrate, $CuF_2 \cdot 2H_2O$, and copper hydroxyfluoride, CuOHF, all of which are interconvertible. Physical properties of CuF_2 are listed in Table 1. Copper(I) fluoride is believed to be unstable and no evidence for its existence has been found using mass spectrometry.

Manufacture. Several methods of synthesis for anhydrous CuF_2 have been reported, the most convenient and economical of which is the reaction of copper carbonate and anhydrous hydrogen fluoride to form the monohydrate, $CuF_2 \cdot H_2O$, and then fluorination with HF at elevated temperatures.

Uses. Copper(II) fluoride is used as a fluorinating reagent in the fluorination of partially hydrogenated silanes; in superconductors; as a cathode material for high energy density primary and secondary batteries (qv) for the skeletal rearrangements of olefins; low temperature isomerization of pentane and hexane; as a selective herbicide; as a termite repellant; as a fungicide; in the manufacturing of conductive bicomponent fibers for electromagnetic shields; as a catalyst for the removal of nitrogen oxides from flue gases, and for the synthesis of heterocyclic tetraaromatics. The dihydrate is used in the casting of gray iron.

The high purity anhydrous copper(II) fluoride must be stored in a tightly closed or sealed container under an atmosphere of argon. The dihydrate may be stored in polyethylene-lined fiber drums.

Copper(II) fluoride, is available in the United States from Advance Research Chemicals, Aldrich Chemicals, Atomergic, Aesar, Johnson/Matthey, Cerac Corp., and PCR Corp.

Table 1. Physical Properties of CuF_2

Property	Value
molecular weight	101.54
melting point, °C	785 ± 10
boiling point, °C	1676
solubility in water, g/100 g	4.75
density, g/cm^3	4.85

DAYAL T. MESHRI
Advance Research Chemicals, Inc.

Jpn. Kokai Tokkyo Koho, 02, 302,311 (Dec. 14, 1990), I. Harada, M. Aritsuka, and A. Yoshikawa (to Mitsui Tiatsu Chemicals).

Jpn. Kokai Tokkyo Koho 01, 133,921 (Nov. 18, 1987), S. Aoki and co-workers (to Fujikura Ltd.).

Eur. Pat. 286,990 (Apr. 17, 1987), F. W. Dampier and R. M. Mank (to GTE Laboratories, Inc.).

Jpn. Kokai Tokkyo Koho, 63, 49,255 (Mar. 2, 1988), Y. Kawasaki (to Matsushita Electric Industrial Co. Ltd.).

GERMANIUM

Germanium Difluoride

Germanium difluoride is a white solid, mp 110°C, and d_{23} = 3.7 g/cm^3. Germanium difluoride is soluble in ethanol. Germanium difluoride can be prepared by reduction of GeF_4 by metallic germanium, by reaction of stoichiometric amounts of Ge and HF in a sealed vessel at 225°C, by Ge powder and HgF_2, and by GeS and PbF_2. GeF_2 has been used in plasma chemical vapor deposition of amorphous film (see PLASMA TECHNOLOGY; THIN FILMS).

Germanium Tetrafluoride

Germanium tetrafluoride is a gas having a garliclike odor, a reported triple point of −15 °C and 404.1 kPa (4.0 atm), and a vapor pressure near 100 kPa (ca 1 atm) at −36.5 °C. Germanium tetrafluoride can be prepared by thermal decomposition of barium hexafluorogermanate, $BaGeF_6$. GeF_4 is used in ion implantation (qv) in semiconductor chips. Germanium tetrafluoride acts as a Lewis acid to form complexes with many donor molecules. The tetrafluoride is commercially available.

Fluorogermanates

Fluorogermanic acid solutions, H_2GeF_6, are prepared by reaction of germanium dioxide and hydrofluoric acid or by hydrolysis of germanium tetrafluoride. Addition of potassium fluoride, barium chloride, or other salts results in hexafluorogermanates.

TARIQ MAHMOOD
CHARLES B. LINDAHL
Elf Atochem North America, Inc.

N. Barlett and K. C. Yu. *Can. J. Chem.* **39**, 80 (1961).

EP 229707 A1 (July 22, 1987), S. Ishihara, M. Hirooka, J. Hanna, and I. Shimizu.

A. Gottdang and co-workers, *Nucl. Instrum. Methods Phys. Res., Sect. B* **B55**(1–9), 310–313 (1991).

HALOGENS

The halogen fluorides are binary compounds of bromine, chlorine, and iodine with fluorine. Of the eight known compounds, only bromine trifluoride, chlorine trifluoride, and iodine pentafluoride have been of commercial importance.

The halogen fluorides are best prepared by the reaction of fluorine with the corresponding halogen. These compounds are powerful oxidizing agents; chlorine trifluoride approaches the reactivity of fluorine.

The halogen fluorides offer an advantage over fluorine in that the former can be stored as liquids in steel containers and, unlike fluorine, high pressure is not required. Bromine trifluoride is used as an oxidizing agent in cutting tools used in deep oil-well drilling, whereas chlorine trifluoride is used to convert uranium to UF_6 in nuclear fuel processing (see NUCLEAR REACTORS; PETROLEUM).

Except for iodine pentafluoride, the halogen fluorides have no commercial importance as fluorinating agents. Their extreme reactivity and the accompanying energy release of the reaction can be sufficient to disrupt C—C bonds and can result in explosive reactions or fires. In addition, both halogens are generally then introduced into organic compounds, giving rise to a complex mixture of products.

Chemical Properties

Reactions With Metals. All metals react to some extent with the halogen fluorides, although several react only superficially to form an adherent fluoride film of low permeability that serves as protection against further reaction. This protective capacity is lost at elevated temperatures, however. Hence, each metal has a temperature above which it continues to react. Mild steel reacts rapidly above 250°C. Copper and nickel lose the ability to resist reaction above 400 and 750°C, respectively.

Reactions with Nonmetals. Few elements withstand the action of interhalogen compounds at elevated temperatures, and many react violently at or below ambient temperatures. The oxidation of the element proceeds to its highest valence state, whereas the halogen other than fluorine is reduced either to the element or a lower valent interhalogen derivative. In general, reactions of halogens and halogen fluorides yield mixtures.

Reactions With Inorganic Compounds. In an investigation of the reactions of BrF_3 with oxides, little or no reaction was found with the oxides of Be, Mg, Ce, Ca, Fe, Zn, Zr, Cd, Sn, Hg, Th, and the rare earths, whereas the oxides of Mo and Re formed stable oxyfluorides. Manganese dioxide reacted incompletely but $KMnO_4$ released oxygen quantitatively. Complete replacement of oxygen took place with oxides of B, Ti, V, Cr, Cu, Ge, As, Se, Nb, Sb, Te, I, Ta, W, Tl, Pb, Bi, and U at 75°C. Water reacts violently with all halogen fluorides. The hydrolysis process can be moderated by cooling or dilution. Salts of halides other than fluorides react with halogen fluorides to produce the corresponding metal fluoride and release the free higher halogen.

Reactions with Organic Compounds. Most organic compounds react vigorously exhibiting incandescence or even explosively with ClF_3 and BrF_3. For this reason, only the less reactive iodine pentafluoride is used as a fluorinating agent to any extent.

Liquid Halogen Fluorides as Reaction Media. Bromine trifluoride and iodine pentafluoride are highly dimerized and behave as ionizing solvents.

Economic Aspects

U.S. production of bromine trifluoride is several metric tons per year, mostly used in oil-well cutting tools. Air Products and Chemical, Inc. is the only U.S. producer. U.S. chlorine trifluoride production is several metric tons per year. Most of the product is used in nuclear fuel processing. As of 1993, Air Products and Chemicals, Inc. was the only U.S. producer. U.S. production of iodine pentafluoride is several hundred metric tons per year. The two U.S. producers are Air Products and Chemicals, Inc. and AlliedSignal, Inc.

Shipping and Specifications

Bromine trifluoride is commercially available at a minimum purity of 98%. Chlorine trifluoride is commercially available at 99% minimum purity. Iodine pentafluoride is commercially available at a minimum purity of 98%. All three are shipped in steel cylinders and classified as oxidizers and poisons by DOT.

Handling

The halogen fluorides are highly reactive compounds and must be handled with extreme caution. The more reactive compounds, such as bromine trifluoride and chlorine trifluoride, are hypergolic oxidizers and react violently and sometimes explosively with many organic and inorganic materials at room temperature. At elevated temperatures, these cause immediate ignition of most organic substances and many metals.

Materials of Construction. Nickel, Monel, copper, mild steel, 304 stainless steel, and aluminum have been found to be suitable metals of construction for handling halogen fluorides. Equipment should be carefully and completely degreased and passivated with low concentrations of fluorine or the gaseous halogen fluoride before use.

Disposal. Moderate amounts of chlorine trifluoride or other halogen fluorides may be destroyed by burning with a fuel such as natural gas, hydrogen, or propane. The resulting fumes may be vented to water or caustic scrubbers. Alternatively, they can be diluted with an inert gas and scrubbed in a caustic solution.

Toxicity

The time-weighted average (TWA) concentrations for 8-h exposure to bromine trifluoride, bromine pentafluoride, chlorine trifluoride, chlorine pentafluoride, and iodine pentafluoride have been established by ACGIH on a fluoride basis to be 2.5 mg/m^3. No toxicity data have been reported on the other halogen fluorides, but all should be regarded as highly toxic and extremely irritating to all living tissue.

Uses

Chlorine trifluoride is utilized in the processing of nuclear fuels to convert uranium to gaseous uranium hexafluoride. Chlorine trifluoride has also been used as a low temperature etchant for single-crystalline silicon. Bromine trifluoride and chlorine trifluoride are used in oil-well tubing cutters. Iodine pentafluoride is an easily storable liquid source of fluorine having few of the hazards associated with other fluorine sources. It is used as a selective fluorinating agent for organic compounds.

WEBB I. BAILEY
ANDREW J. WOYTEK
Air Products and Chemicals, Inc.

L. S. Boguslavskaya and N. N. Chuvatkin, in L. German and S. Zemskov, eds., *New Fluorinating Agents in Organic Synthesis*, Springer-Verlag, Berlin, 1989, pp. 140–196.

Handling Hazardous Materials, NASA SP-5032, NASA, Washington, D.C., 1965, Chapt. 4.

H. C. Fielding, in R. E. Banks ed., *Organofluorine Chemicals and their Industrial Applications*, Ellis Horwood Publishers, Chichester, U.K., 1979.

R. A. Rhein and M. H. Miles, *Bromine and Chlorine Fluorides: A Review*, Naval Weapons Center technical publication 6811, NWC, China Lake, Calif., 1988.

HYDROGEN

Hydrogen fluoride, HF, is the most important manufactured fluorine compound. It is the largest in terms of volume, and serves as the raw material for most other fluorine-containing chemicals. It is available either in anhydrous form or as an aqueous solution (usually 70%). Anhydrous hydrogen fluoride is a colorless liquid or gas having a boiling point of 19.5°C. It is a corrosive, hazardous material, fuming strongly, which causes severe burns upon contact. Rigorous safety precautions are the standard throughout the industry, and in practice hydrogen fluoride can be handled quite safely.

Properties

Physical Properties. Physical properties of anhydrous hydrogen fluoride are summarized in Table 1. Figure 1 gives the partial pressures of HF and H_2O in aqueous HF solutions. Hydrogen fluoride is unique among the hydrogen halides in that it strongly associates to form polymers in both the liquid and gaseous states. This high degree of association results in highly nonideal physical properties.

Chemical Properties. Hydrogen fluoride, characterized by its stability, has a dissociation energy of 560 kJ (134 kcal), which places HF among the most stable diatomic molecules. Hydrogen fluoride is, however, highly reactive, and it has a special affinity for oxygen compounds, reacting with boric acid to form boron trifluoride and with sulfur trioxide and sulfuric acid to form fluorosulfonic acid. This last reaction demonstrates the dehydrating power of anhydrous hydrogen fluoride. HF belongs to the only class of compounds that readily react with silica and silicates, including glass (qv).

The strong catalytic activity of anhydrous hydrogen fluoride results from the ability to donate a proton, as in the dimerization of isobutylene. Anhydrous hydrogen fluoride is an excellent solvent for ionic fluorides. Whereas hydrogen fluoride is a fairly weak acid as a solute, it is strongly acidic as a solvent.

Table 1. Properties of Anhydrous Hydrogen Fluoride

Property	Value
composition, wt %	
H	5.038
F	94.96
boiling point at 101.3 kPa,[a] °C	19.54
critical pressure, MPa[a]	6.48
critical temperature, °C	188.0
critical density, g/mL	0.29
melting point, °C	−83.55
density, liquid, 25°C, g/mL	0.958
heat of vaporization, 101.3 kPa,[a] kJ/mol[b]	7.493
vapor pressure, 25°C, MPa[a]	122.9
viscosity, liquid, 0°C, mPa·s(= cP)	0.256
refractive index, liquid, 25°C, 589.3 nm	1.1574
dielectric constant, at 0°C	83.6

[a] To convert kPa to psi, multiply by 0.145. [b] To convert J to cal, divide by 4.184.

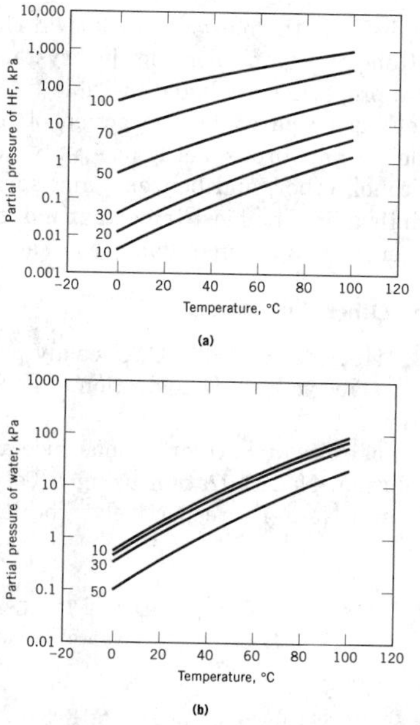

Figure 1. Partial pressures over HF–water solutions where the numbers represent the quantity of HF in solution expressed as wt % (**a**) of HF and (**b**) of H_2O.

Manufacture

Raw Materials. Essentially all hydrogen fluoride manufactured worldwide is made from fluorspar and sulfuric acid, according to the reaction $CaF_2(s) + H_2SO_4 \rightarrow CaSO_4(s) + 2\ HF(g)$. Generally, yields on both fluorspar and sulfuric acid are greater than 90% in commercial plants.

Technology. The key piece of equipment in a hydrogen fluoride manufacturing plant is the reaction furnace. The reaction between calcium fluoride and sulfuric acid is endothermic (1400 kJ/kg of HF) (334.6 kcal/kg), and for good yields, must be carried out at a temperature in the range of 200°C. Most industrial furnaces are horizontal rotating kilns, externally heated by, for example, circulating combustion gas in a jacket. Other heat sources are possible, eg, supplying the sulfuric acid value as SO_3 and steam (qv), which then react and condense, forming sulfuric acid and releasing heat.

In all HF processes, the HF leaves the furnace as a gas, contaminated with small amounts of impurities such as water, sulfuric acid, SO_2, or SiF_4. Various manufacturers utilize different gas handling operations, which generally include scrubbing and cooling.

Some manufacturers recover by-products from the process. Fluosilicic acid, which is used in water fluoridation, can easily be recovered from the plant vent gases. The calcium sulfate discharged from the furnace can also be recovered for use in applications such as cement and road aggregate.

Alternative Processes. Because of the large quantity of phosphate rock reserves available worldwide, recovery of the fluoride values from this raw material source has frequently been studied. Strategies involve recovering the fluoride from wet-process phosphoric acid plants as fluosilicic acid, H_2SiF_6, and then processing this acid to form hydrogen fluoride. Numerous processes have been proposed, but none has been commercialized on a large scale. Other technologies proceeding via intermediates such as NH_4F or KHF_2 are also possible. A more recently developed process involves the reaction of the H_2SiF_6 and phosphate rock, producing a calcium silicon hexafluoride ($CaSiF_6$) intermediate that can be converted to CaF_2 and then to HF by reaction of H_2SO_4. All of the processes produce silica. The quality of the silica varies greatly, and its value as a coproduct has a significant impact on the processes' economics.

Shipping. Hydrogen fluoride is shipped in bulk in tank cars (specification 112S400W) and tank trucks (specification MC312). A small volume of overseas business is shipped in ISO tanks.

Materials of Construction. Acceptability of materials of construction for hydrogen fluoride handling is affected by such variables as temperature, hydrogen fluoride strength, and method of use. Mild steel is generally used for most anhydrous hydrogen fluoride applications, up to 66°C. At higher temperature, Monel, a nickel–copper alloy, is suitable, as is Hastelloy-C, a nickel–molybdenum–chromium alloy.

Aqueous hydrogen fluoride of greater than 60% may be handled in steel up to 38°C, provided velocities are kept low (<0.3 m/s) and iron pickup in the process stream is acceptable. Otherwise, rubber or polytetrafluoroethylene (PTFE) linings are used. For all applications, PTFE or PTFE-lined materials are suitable up to the maximum use temperature of 200°C.

Economic Factors

Production. Global hydrogen fluoride production capacity in 1992 was estimated to be 875,000 metric tons. An additional 204,000 metric tons was used captively for production of aluminum fluoride.

North America accounts for about 38% of the worldwide hydrogen fluoride production and 52% of the captive aluminum fluoride production. Table 2 summarizes North American capacity for hydrogen fluoride as well as the captive capacity for aluminum fluoride production. North American HF production capacity has declined since the

Table 2. Hydrofluoric Acid Capacity in North America

Producer	Market, t/yr	AlF_3 production
Alcan		55,000
Alcoa		45,000
Allied-Signal	142,000	
Du Pont	68,000	
Atochem	22,000	
Quimica Fluor	68,000	
Fluorex	18,000	
Industrias Quimica de Mexico	6,000	16,000
Quimobasicos	6,000	
Total	*330,000*	*116,000*

early 1980s. Production is expected to continue to decline in the short term because of chlorofluorocarbon (CFC) cutbacks, but is expected to rebound later in the 1990s as replacement hydrochlorofluorocarbons and hydrofluorocarbons are introduced to the marketplace.

Fluorspar Supply. Production costs of hydrogen fluoride are heavily dependent on raw materials prices, particularly those of fluorspar, and significant changes have occurred in this area. Identified world fluorspar resources amount to approximately 400 × 10^6 metric tons of fluospar. Of this 400 × 10^6 t, however, only 243 × 10^6 t is considered to be reserves and an additional 93 × 10^6 t is considered reserve base, ie, recoverable at higher market prices.

Based on previous splits in milling operations, about a 60% yield or 146 × 10^6 t of acid-grade spar could be expected. At the production rates of the early 1990s, this would be a 24-yr supply. Additional supplies are expected to be brought into production, however, and no decline in available reserves is expected through the year 2000.

Purity is expected to become a significant concern as reserves are depleted. Higher levels of impurities in the fluorspar may require modifications to HF production technology to produce high quality hydrogen fluoride.

Uses

In the North American HF market, approximately 70% goes into the production of fluorocarbons, 4% to the nuclear industry, 5% to alkylation processes, 5% to steel pickling, and 16% to other markets. This does not include the HF going to aluminum fluoride, the majority of which is produced captively for this purpose.

Health, Safety, and Environmental Aspects

Mild exposure to HF via inhalation can irritate the nose, throat, and respiratory system. The onset of symptoms may be delayed for several hours. Severe exposure via inhalation can cause nose and throat burns, lung inflammation, and pulmonary edema, and can also result in other systemic effects including hypocalcemia (depletion of body calcium levels), which if not promptly treated can be fatal. Permissible air concentrations are OSHA PEL, 3 ppm (2.0 mg/m^3) as F; OSHA STEL, 6 ppm (5.2 mg/m^3) as F; and ACGIH TLV, 3 ppm (2.6 mg/m^3) as F. Ingestion can cause severe mouth, throat, and stomach burns, and may be fatal.

Both liquid HF and the vapor can cause severe skin burns which may not be immediately painful or visible. HF can penetrate skin and attack underlying tissues, and large (over 160 cm^2) burns may cause hypocalcemia and other systemic effects which may be fatal.

Hydrogen fluoride is not a carcinogen. However, HF is highly reactive, and heat or toxic fumes may be evolved. Reaction with certain metals may generate flammable and potentially explosive hydrogen (qv) gas.

The hydrogen fluoride industry has undertaken a significant effort to investigate the behavior of HF releases so as better to define the risks associated with an accidental spill, and to design effective mitigation systems.

ROBERT A. SMITH
AlliedSignal, Inc.

C. L. Yaws and L. S. Adler, *Chem. Eng.*, 119 (Oct. 28, 1974).

T. A. O'Donnell, in J. C. Bailar and co-workers, eds., *Comprehensive Inorganic Chemistry*, Vol. 2, Pergamon Press, Oxford, U.K., 1973, pp. 1038–1054.

Hydrofluoric Acid, Anhydrous—Technical, Properties, Uses, Storage, and Handling, E. I. du Pont de Nemours & Co., Inc., Wilmington, Del., 1984.

Hydrofluoric Acid, Anhydrous, Product Safety Data Sheet, AlliedSignal Inc., Morristown, N.J., 1991.

IRON

Iron(II) Fluoride

Anhydrous iron(II) fluoride, FeF_2, is a white solid, The off-white to buff-colored appearance of the material is attributed to the partial oxidation of Fe^{2+} to Fe^{3+}. FeF_2 is highly stable. It is sparingly soluble in water but the solubility can be increased by the addition of aqueous HF or any strong acid. Physical properties are listed in Table 1. FeF_2 holds great promise in the field of advanced magnets known as the iron–boron–rare-earth-alloy sintered magnets.

Table 1. Physical Properties of Iron Fluorides

Property	FeF_2	FeF_3
mol wt	93.84	112.84
density, g/cm^3	4.09	3.87
mp, °C	1100	1000[a]
bp, °C	1837	

[a] Sublimes.

FeF_2 was first prepared by the action of gaseous hydrogen fluoride over $FeCl_2$ in an iron boat. The only reported industrial application for FeF_2 is its use in rust removal solutions based on oxalic acid. The anhydrous salt is commercially available in 100-g to 5-kg lots from Advance Research Chemicals, Aldrich Chemicals, Cerac, Johnson/Matthey, PCR, and other suppliers in the United States. Toxicity of iron(II) fluoride has not been determined. FeF_2 is shipped as a nonhazardous material in plastic containers.

Iron(III) Fluoride

Iron(III) fluoride, FeF_3, is the most widely known fluoride of iron. It is light greenish (lime green) in color and the crystals have a rhombic structure. Physical properties are listed in Table 1.

Anhydrous FeF_3 is prepared by the action of liquid or gaseous hydrogen fluoride on anhydrous $FeCl_3$ (see IRON COMPOUNDS). FeF_3 is insoluble in alcohol, ether, and benzene, and sparingly soluble in anhydrous HF and water. The most important industrial application of the iron(III) fluoride is in the manufacture of Fe–Co–Nd magnets.

Hydrated Salts and Other Compounds

Hydrated iron(III) fluoride, $FeF_3 \cdot 3H_2O$, is easily prepared from yellow Fe_2O_3 and hydrofluoric acid. Dehydration of $FeF_3 \cdot 3H_2O$ produces oxyfluorides of iron.

In the presence of excess HF, complex ions such as FeF_4^- and FeF_6^- are formed in solution. Neutralization using a base such as NaOH produces $NaFeF_4$ and Na_3FeF_6, respectively. The latter is used as a fluorinating agent.

DAYAL T. MESHRI
Advance Research Chemicals, Inc.

Jpn. Kokai Koho, 63,249,304 (Oct. 17, 1988), A. Kobayashi and T. Sato (to Hitachi Metals Ltd.).

U.S. Pat. 4,828,743 (May 7, 1989), S. Rahfield and B. Newman (to Boyle Midway Household Products Inc.).

LEAD

Lead Difluoride

Lead difluoride, PbF_2, has the highest melting and boiling points among all the dihalides of lead. Two colorless crystalline forms are known. The α-PbF_2 is orthorhombic in structure and is stable at ordinary temperatures. Upon heating to 200°C it transforms to the cubic β-form. Table 1 lists some of the physical properties of PbF_2. PbF_2 is readily prepared by the action of hydrogen fluoride on lead hydroxide, lead carbonate, or α-lead oxide.

Table 1. Physical Properties of Lead Fluorides

Property	PbF_2	PbF_4
mol wt	245.19	283.2
density, g/cm^3	8.24	6.7
melting point, °C	855	600
boiling point, °C	1290	decomposes
solubility, in water, g/100 g	0.0641	[a]

[a] Material hydrolyzes to PbO_2 and HF.

PbF_2 exhibits very good electrical insulating properties and optical transparency. It is thus used in a variety of types of glass (qv). It is also used in printing, photography (qv), brazing, scintillation counters, dielectric interference filters, as a mild fluorinating reagent, as a source material for PbF_4, and as an ingredient in lead–acid batteries (qv).

High purity lead difluoride is available form Advance Research Chemicals, Aldrich Chemicals, Johnson/Matthey, Atomergic, Cerac, and other suppliers in the United States. The U.S. annual consumption varies between 500 to 2500 kg/yr.

Lead Tetrafluoride

Like all the lead tetrahalides, lead tetrafluoride, PbF_4, is very reactive. It is relatively the most stable halide, however. It should be handled in a dry box or under an atmosphere or dry nitrogen. Properties of PbF_4 are given in Table 1.

PbF_4, produced by various routes including the *in situ* species, is a very effective fluorinating agent for olefins. It is an oxidizing agent. It is also used in the preparation of biologically active steroids where the fluorine is added in a cis configuration to the double bond.

Lead fluorides are highly toxic and should be handled with great care. The ACGIH adopted toxicity value for lead compounds as Pb is TWA 0.15 mg/m^3 and for fluorides as F^- 2.5 mg/m^3. PbF_4 is prepared by the action of elemental fluorine on very dry PbF_2 at 280–300°C.

DAYAL T. MESHRI
Advance Research Chemicals, Inc.

A. Bowers and co-workers, *J. Am. Chem. Soc.* **84**, 1050 (1962).

LITHIUM

Lithium Fluoride

Properties. Lithium fluoride, LiF, is a white nonhygroscopic crystalline material that does not form a hydrate. The solubility in water is quite low and chemical reactivity is low, similar to that of calcium fluoride and magnesium fluoride. Several chemical and physical properties of lithium fluoride are listed in Table 1.

Manufacture. Lithium fluoride is manufactured by the reaction of lithium carbonate or lithium hydroxide with dilute hydrofluoric acid.

Uses. Lithium fluoride is used primarily in the ceramic industry to reduce firing temperatures and improve resistance to thermal shock, abrasion, and acid attack.

Table 1. Properties of Lithium Fluoride

Property	Value
melting point, °C	848
boiling point, °C	1681
crystalline form	cubic (NaCl)
a_0, nm	0.401736
density at 20°C, g/cm^3	2.635
standard heat of formation, kJ/mol[a]	−613.0
heat capacity, C_p, J/(mol·K)[a]	42.01
heat of vaporization, kJ/mol[a]	213

[a] To convert kJ to kcal, divide by 4.184.

JOHN R. PAPCUN
Atotech

M. C. Ball and A. A. Norbury, *Physical Data for Inorganic Chemists*, Longman, Inc., New York, 1974.

H. C. Hodge and F. N. Smith, in J. H. Simon, ed., *Fluorine Chemistry*, Vol. 4, Academic Press, New York, 1965, p. 199.

F. M. Cox, *Proc. 2nd Int. Conf. on Luminescence Dosimetry*, Oak Ridge National Laboratory, CONF 680920, Oak Ridge, Tenn., Sept. 1968; F. M. Cox, A. C. Lucas, and B. M. Kaspar, *Health Phys.* **30**, 135 (1976); J. F. Valley, C. Pache, and P. Lerch, *Helv. Phys. Acta* **49**(2), 171 (1976).

G. L. Green, J. B. Hunt, and R. A. Sutula, *U.S. Nat. Tech. Inform. Serv.*, A.D. Rep. 1973, No. 758001.

MAGNESIUM

Magnesium Fluoride

Properties. Magnesium fluoride, MgF_2, is a fine white crystalline powder with low chemical reactivity. This relative inertness makes possible some of its uses, eg, stable permanent films to alter light transmission properties of optical and electronic materials. Chemical and physical properties are listed in Table 1.

Table 1. Chemical and Physical Properties

Property	Value
melting point, °C	1263
boiling point, °C	2227
standard heat of formation, kJ/mol[a]	−112.4
heat of fusion, kJ/mol[a]	58.2
density, g/cm^3	3.127
solubility in water, 25°C, g/100 g of solvent	0.013

[a] To convert kJ to kcal, divide by 4.184.

Manufacture. Magnesium fluoride is manufactured by the reaction of hydrofluoric acid and magnesium oxide or carbonate.

Toxicity. The lethal dose of MgF_2 to guinea pigs by ingestion is 1000 mg/kg.

Uses. Established uses of magnesium fluoride are as fluxes in magnesium metallurgy and in the ceramics industry. A proposed use is the extraction of aluminum from arc-furnace alloys with Fe, Si, Ti, and C. Other uses include optical crystals.

JOHN R. PAPCUN
Atotech

I. G. Ryss, *The Chemistry of Fluorine and Its Inorganic Compounds*, State Publishing House for Scientific and Technical Literature, Moscow, 1956;

Engl. transl. by F. Haimson for the U.S. Atomic Energy Commission, *AEC-tr-3927*, Washington, D.C., 1960, p. 812.

Technical data, The Harshaw Chemical Co., Crystal and Electronics Dept., Solon, Ohio.

H. E. Thayer, *Proc. of the 2nd U. M. Internat. Conference, Peaceful Uses of Atomic Energy, (Geneva)*, **4**, 22 (1958); W. E. Dennis and E. Proudfoot, *U.K. At. Energy R&D* **B**(C) TN-88 (1954).

U.S. Pat. 3,920,802 (Nov. 18, 1975), R. H. Moss, C. F. Swinehart, and W. F. Spicuzza (to Kewanee Oil Co.).

MERCURY

Mercury(I) Fluoride

Mercury(I) fluoride, Hg_2F_2, also known as mercurous fluoride, is a light-sensitive golden yellow material decomposing in water at 15°C. Some of the physical properties are listed in Table 1.

Table 1. Properties of Mercury Fluorides

Property	Hg_2F_2	HgF_2
mol wt	439.22	238.61
density, g/cm^3	8.73	8.95
melting point, °C	>570 dec	645

Several preparatory methods for the manufacture of Hg_2F_2 have been reported. Whereas no commercial applications for Hg_2F_2 have been found, it is available from Advance Research Chemical and Aldrich Chemicals in the United States.

Mercury(II) Fluoride

Mercury(II) fluoride, HgF_2, also known as mercuric fluoride, is a white, hygroscopic solid which turns yellow instantly on exposure to moist air. It must be handled in a dry box or under an atmosphere of dry nitrogen. Some of its physical properties are listed in Table 1. Mercury(II) fluoride is easily prepared by passing pure elemental fluorine over predried $HgCl_2$ at 100–150°C until all the chloride ions have been replaced. Mercury(II) fluoride has been used in the process for manufacture of fluoride glass (qv) for fiber optics (qv) applications and in photochemical selective fluorination of organic substrates. It is available from Advance Research Chemicals, Aldrich Chemicals, Johnson/Matthey, Aesar, Cerac, Strem, and PCR in the United States.

Mercury salts are highly toxic and must be handled carefully. Strict adherence to OSHA/EPA regulations is essential.

DAYAL T. MESHRI
Advance Research Chemicals, Inc.

Jpn. Kokai Tokkyo Koho, JP 63239,137 (Oct. 5, 1988), N. Mitachi, Y. Ooishi, and S. Sakaguchi (to Nippon Telegraph and Telephone Co.).

M. H. Habibi and T. E. Mallouk, *J. Fluorine Chem.* **51**(2), 291–294 (1991).

MOLYBDENUM

Molybdenum Hexafluoride

Molybdenum hexafluoride, MoF_6, is a volatile liquid at room temperature. MoF_6 should therefore be handled in a closed system or in a vacuum line located in a chemical hood. The known physical properties are listed in Table 1.

Molybdenum hexafluoride can be prepared by the action of elemental fluorine on hdyrogen-reduced molybdenum powder (100–300 mesh (ca 149–46 μm)) at 200°C. Molybdenum hexafluoride is used in the manufacture of thin films (qv) for large-scale integrated circuits (qv) commonly known as LSIC systems, in the manufacture of metallized ceramics, and chemical vapor deposition of molybdenum and molybdenum–tungsten alloys.

Table 1. Physical Properties of MoF_6

Property	Value
mol wt	209.93
melting point, °C	17.4
boiling point, °C	35.0
solubility, g/100 g	[a]
density, liquid, g/cm^3	2.544

[a] Hydrolyzes in water.

Molybdenum hexafluoride is classified as a corrosive and poison gas. It is stored and shipped in steel, stainless steel, or Monel cylinders approved by DOT. It is available from Advance Research Chemicals Inc., Aldrich Chemicals, Atomergic, Cerac, Johnson/Matthey, Pfaltz & Bauer, and Strem Chemicals.

Other Molybdenum Fluorides

Three other binary compounds of molybdenum and fluorine are known to exist: molybdenum trifluoride, MoF_3, molybdenum tetrafluoride, MoF_4, and molybdenum pentafluoride, MoF_5. Also known are the two oxyfluorides, molybdenum dioxydifluoride, MoO_2F_2, and molybdenum oxytetrafluoride, $MoOF_4$.

DAYAL T. MESHRI
Advance Research Chemicals, Inc.

Jpn. Kokai Tokkyo Koho JP 61 224,313 (Oct. 6, 1986), S. Tsujiku and coworkers.

Ger. Offen. 3,639,080 (May 21, 1987), Y. S. Liu and C. P. Yakmyshyn.

NICKEL

Nickel Fluoride Tetrahydrate

Nickel fluoride tetrahydrate, $NiF_2{\cdot}4H_2O$, and its anhydrous counterpart, nickel fluoride, NiF_2, are the only known stable binary compounds of nickel and fluorine. The former is a greenish light yellow crystal or powder prepared by the addition of nickel carbonate to 30–50% aqueous HF solution. The tetrahydrate has high solubility in aqueous HF, eg, 13.3 wt % in 30% HF. It is slightly soluble in water and insoluble in alcohol and ether.

Historically, the annual consumption of nickel fluoride was on the order of a few metric tons. Usage is dropping because nickel fluoride is listed in the EPA and TSCA's toxic substance inventory. Small quantities for research and pilot-plant work are available from Advance Research Chemicals, Aldrich Chemicals, Johnson/Matthey, Pfaltz and Bauer, PCR, and Strem Chemicals of the United States, Fluorochem of the United Kingdom, and Morita of Japan.

Nickel Fluoride, Anhydrous

Anhydrous nickel fluoride, a light yellow colored powder, is prepared by the action of anhydrous HF on anhydrous $NiCl_2$, or nickel fluoride tetrahydrate at 300°C.

Nickel fluoride is used in marking ink compositions (see INKS), for fluorescent lamps as a catalyst in transhalogenation of fluoroolefins, in the manufacture of varistors, as a catalyst for hydrofluorination, in the synthesis of XeF_6, and in the preparation of high purity elemental fluorine for research and for chemical lasers (qv). Small quantities are stored and shipped in polyethylene bottles, whereas large amounts are shipped in fiber board drums with polyethylene liners. All nickel compounds are considered as suspected carcinogens and are listed in the EPA and TSCA's toxic substances inventory.

Physical Properties. Anhydrous nickel fluoride has a mol wt of 96.71; mp, 1450°C; bp, 1740°C; solubility in water of 4.0 g/100 g, and density, g/mL, of 4.72.

Other Nickel Fluorides. Nickel trifluoride has been observed.

Nickel Fluoride Complexes

Nickel tetrafluoroborate, $Ni(BF_4)\cdot xH_2O$, can be prepared by dissolving nickel carbonate in tetrafluoroboric acid, HBF_4. Nickel tetrafluoroborate, commercially available as a hydrated solid, and also as a 50% solution, plays an important role in the electroplating (qv) and electronics industries. Its consumption is several hundred metric tons a year. It is available from Advance Research Chemicals, Aldrich Chemicals, Aesar Chemicals, Johnson/Matthey, Harshaw M & T Chemicals, and from various other sources.

The complex hexafluoronickelates, M_2NiF_6 (M = Na, K, Rb, and Cs) and M_3NiF_6 (M = Na), K, Rb, and Cs), are prepared by reaction of elemental fluorine, chlorine trifluoride, or xenon difluoride and a mixture of nickel fluoride and alkali metal fluorides or other metal halides. These hexafluoronickelates can be used as fluorinating reagents, as a source of high purity elemental fluorine, and as high energy solid propellant oxidizers.

DAYAL T. MESHRI
Advance Research Chemicals, Inc.

Jpn. Kokai, 75,139,395 (Nov. 7, 1975), M. Matsuura and co-workers (to Matsushita Electrical Industrial Co.).

B. Zemva and J. Slivnik, *Vestn. Slov. Kem. Drus.* **19**(1–4), 43 (1972).

L. B. Asprey, *J. Fluorine Chem.* **7**(1–3), 359 (1976).

U.S. Pat. 4,711,680 (Dec. 8, 1987), K. O. Christe (to Rockwell Int. Corp.).

NITROGEN

Nitrogen Trifluoride

Physical Properties. Nitrogen trifluoride, NF_3, is a colorless gas. Selected physical properties of NF_3 are given in Table 1.

Chemical Properties. NF_3 can be a potent oxidizer, especially at elevated temperature. Nitrogen trifluoride acts primarily upon the elements as a fluorinating agent, but is not a very active one at lower temperatures. At elevated temperatures, NF_3 pyrolyzes with many of the elements to produce N_2F_4 and the corresponding fluoride.

Hydrogen, hydrides, and hydrocarbons react with NF_3 with the rapid liberation of large amounts of heat. This is the basis for the use of NF_3 in high energy chemical lasers (qv). NF_3 reacts with F_2 and certain Lewis acids under heat or uv light to form the corresponding NF_4^+ salts.

Manufacture and Economics. Nitrogen trifluoride can be formed from a wide variety of chemical reactions. Only two processes have been technically and economically feasible for large-scale production: the electrolysis of molten ammonium acid fluoride; and the direct fluorination of the ammonia in the presence of molten ammonium fluoride.

As a result of the development of electronic applications for NF_3, higher purities of NF_3 have been required, and considerable work has been done to improve the existing manufacturing and purification processes.

Table 1. Physical Properties of Nitrogen Trifluoride

Property	Value
boiling point, °C	−129.0
liquid density at −129°C, g/mL	1.533
heat of vaporization, kJ/mol[a]	11.59
triple point, °C, 0.263 Pa[b]	−206.8
heat of fusion, J/mol[a]	398
heat of transition, kJ/mol[a]	1.513
water solubility, 101.3 kPa,[b] 25°C	1.4×10^{-5} mol NF_3/mol H_2O

[a] To convert kJ to kcal, divide by 4.184. [b] To convert kPa to mm Hg, multiply by 7.5.

Production of NF_3 is less than 100 t/yr in the United States. Air Products and Chemicals, Inc. is the only commercial producer in the United states.

Handling and Toxicity. Nitrogen trifluoride gas is noncorrosive to the common metals at temperatures below 70°C and can be used with steel, stainless steel, and nickel.

Nitrogen trifluoride is a toxic substance and is most hazardous by inhalation. NF_3 induces the production of methemoglobin, which reduces the level of oxygen transfer to the body tissues. At the cessation of NF_3 exposure methemoglobin reverts back to hemoglobin. The OSHA permissible exposure limits is set as a TLV–TWA of 29 mg/kg or 10 ppm.

Environmental impact studies on NF_3 have been performed. Although undiluted NF_3 inhibits seed growth, no effect on plant growth was observed when exposed to 6,000 ppm · min of NF_3 and only minor effects were observed at the 60,000 ppm · min exposure level. Exposure of microbial populations to 25% NF_3 in air for seven hours showed normal growth. NF_3 is not an ozone-depleting gas.

Uses. The principal use of NF_3 is as a fluorine source in the electronics industry. The use of NF_3 as a dry chemical etchant has been reviewed. Another use of NF_3 is as a fluorine source for the hydrogen and deuterium fluoride (HF/DF) high energy chemical lasers.

PHILIP B. HENDERSON
ANDREW J. WOYTEK
Air Products and Chemicals, Inc.

R. E. Anderson, E. M. Vander Wall, and R. K. Schaplowsky, "Nitrogen Trifluoride," *USAF Propellant Handbook*, AFRPL-TR-77-71, Contract F04611-76-C-0058, Aerojet Liquid Rocket Co., Sacramento, Calif., 1977.

Nitrogen Trifluoride: Safety, Applications, and Technical Data Manual, Air Products and Chemicals, Inc., Allentown, Pa., 1992.

U.S. Pat. 4,091,081 (May 23, 1978), A. J. Woytek and J. T. Lileck (to Air Products and Chemicals, Inc.).

J. A. Barkanic and co-workers, *Solid State Technol.* **32**, 172 (1984).

OXYGEN

Oxygen Difluoride

Oxygen difluoride, OF_2, is the most stable binary compound of oxygen and fluorine.

Physical Properties. An extensive tabulation of the physical properties of OF_2 is available. Selected data are mp −224 °C; bp, −145 °C; critical temperature −58 °C; density of liquid, in g/mL from −145 to −153 °C, *t* in K, $d = 2.190 - 0.00523\,t$; heat of formation 31.8 kJ/mol (7.6 kcal/mol); and heat of vaporization 11.1 kJ/mol (2.65 kcal/mol).

Chemical Properties. *Reactions with Metals.* Many common metals react with OF_2, but the reaction stops after a passive metal fluoride coating is formed.

Reactions with Nonmetallic Elements and Inorganic Compounds. Mixtures of OF_2 with carbon, CO, CH_4, H_2, or H_2O vapor explode when ignited with an electrical shock. Elemental B, Si, P, As, Sb, S, Se, and Te react vigorously on slight warming to produce fluorides and oxyfluorides. Oxides such as CrO_3, WO_3, As_2O_3, and CaO react with OF_2 to form fluorides. The corresponding chlorides react with OF_2 to form the respective fluorides and liberate free chlorine in the process.

In aqueous solution, OF_2 oxidizes HCl, HBr, and HI (and their salts), liberating the free halogens. Oxygen difluoride reacts slowly with water and a dilute aqueous base to form oxygen and fluorine Nitric oxide and OF_2 inflame on contact.

Oxygen Difluoride a Source of the OF Radical. The existence of the OF radical was first reported in 1934. The O–F bond length is 0.135789 nm.

Reactions with Organic Compounds. Tetrafluoroethylene and OF_2 react spontaneously to form C_2F_6 and COF_2. Ethylene and OF_2 may react explosively, but under controlled conditions monofluoroethane and 1,2-difluoroethane can be recovered. Benzene is oxidized to quinone and hydroquinone by OF_2. Methanol and ethanol are oxidized at room temperature. Organic amines are extensively degraded by OF_2 at room temperature, but primary aliphatic amines in a fluorocarbon solvent at −42 °C are smoothly oxidized to the corresponding nitroso compounds.

The reaction of OF_2 and various unsaturated fluorocarbons has been examined and it is claimed that OF_2 can be used to chain-extend fluoropolyenes, convert functional perfluorovinyl groups to acyl fluorides and/or epoxide groups, and act as a monomer for an addition-type copolymerization with diolefins.

Preparation. The most satisfactory method of OF_2 generation is probably the fluorination of aqueous NaOH.

Handling and Safety Factors. Oxygen difluoride can be handled easily and safely in glass and in common metals such as stainless steel, copper, aluminum, Monel, and nickel, from cryogenic temperatures to 200°C. At higher temperatures only nickel and Monel are recommended. Oxygen difluoride must be regarded as a highly poisonous gas, somewhat more toxic than fluorine.

Dioxygen Difluoride

Dioxygen difluoride, O_2F_2, prepared by passing a 1:1 mixture of O_2 and F_2 through a high voltage electric discharge tube cooled by liquid nitrogen, has also been prepared by uv irradiation of O_2 and F_2 and by radiolysis of liquid mixtures of O_2 and F_2.

Physical Properties. Because O_2F_2 is unstable, it is difficult to purify. Consequently, some of the reported physical properties are open to question. Selected data are density, in g/mL, from −87 to −156 °C, $d = 2.074 - 0.00291\,t$; heat of formation 19.8 kJ/mol (4.73 kcal/mol); and heat of vaporization 19.2 kJ/mol (4.58 kcal/mol) at −57 °C.

Chemical Properties. The weakest bond in O_2F_2 is the O–F bond and the mechanisms of reaction of O_2F_2 can probably be explained by the formation of F· and ·OOF and not two ·OF radicals. If O_2F_2 is allowed to react quickly with other compounds, simple fluorination usually results. The controlled reactions of O_2F_2, however, yield products that appear to be formed via an ·OOF intermediate.

Uses

Oxygen difluoride is mainly a laboratory chemical. It has been suggested as an oxidizer for rocket applications and has been used for small tests in this area.

Dioxygen difluoride has found some application in the conversion of uranium oxides to UF_6, in fluorination of actinide fluorides and oxyfluorides to AcF_6, and in the recovery of actinides from nuclear wastes (see ACTINIDES AND TRANSACTINIDES; NUCLEAR REACTORS–WASTE MANAGEMENT).

Higher Oxygen Fluorides

Several higher oxygen fluorides, O_3F_2, O_4F_2, O_5F_2, and O_6F_2, and radicals such as $\cdot O_3F$ have been reported. Only ·OF, OF_2, O_2F_2, ·OOF, and O_4F_2, however, have been satisfactorily characterized.

I. J. SOLOMON
IIT Research Institute
JEANE'NE M. SHREEVE
University of Idaho

A. G. Streng, *Chem. Rev.* **63**, 607 (1963).

R. Rhein and G. Cady, *Inorg. Chem.* **3**, 1644 (1964).

J. B. Nielsen and co-workers, *Inorg. Chem.* **29**, 1779 (1990); S. A. Kinkead, L. B. Asprey, and P. G. Eller, *J. Fluorine Chem.* **29**, 459 (1985); Yu. M. Kiselev and co-workers, *Zh. Neorg. Khim.* **33**, 1252 (1988); J. G. Malm, P. G. Eller, and L. B. Asprey, *J. Am. Chem. Soc.* **106**, 2726 (1984).

L. B. Asprey, S. A. Kinkead, and P. G. Eller, *Nucl. Technol.* **73**, 69 (1986).

PHOSPHORUS

The majority of the fluorine in the earth's crust is present in the form of the phosphorus fluoride fluoroapatite, $Ca_5(PO_4)_3F$. During phosphate processing, fluorine values are partially recovered as by-product fluorosilicic acid. The amount of fluorosilicic acid recovered has grown steadily, in part because of environmental requirements.

Physical properties of phosphorus fluorides are given in Table 1.

Phosphorus Pentafluoride

Phosphorus pentafluoride was first prepared in 1876 through fluorination of phosphorus pentachloride using arsenic trifluoride. Other routes to PF_5 have included fluorination of PCl_5 by HF, AgF, benzoyl fluoride, SbF_3, PbF_2, or CaF_2.

Phosphorus pentafluoride is a colorless gas which fumes in contact with moist air and reacts immediately with water to hydrolyze. It behaves as a Lewis acid showing electron-accepting properties. Because it is a strong acceptor, PF_5 is an excellent catalyst, especially in ionic polymerizations. Phosphorus pentafluoride is also used as a source of phosphorus for ion implantation (qv) in semiconductors (qv).

Phosphorus Trifluoride

Phosphorus trifluoride is usually prepared by fluorination of PCl_3 with CaF_2, AsF_3, SbF_3, AgF, PbF_2, ZnF_2, or NaF; reaction of fluorosulfonate, $CaF(FSO_3)$, using molten H_3PO_3; by the reaction of phosphorus oxide and F_2 or NF_3 gas or reaction of PH_3 and NF_3; by the reaction of KHF_2 and PCl_3 or PBr_3. PF_3 is commercially available.

Phosphorus trifluoride is a nearly odorless gas that does not fume in air, and reacts slowly with water but rapidly with base. It may be very toxic, and great care should be taken in handling it. Phosphorus trifluoride acts as a Lewis base.

Phosphorus Oxyfluoride

Phosphorus oxyfluoride is a colorless gas which is susceptible to hydrolysis. It can be prepared by fluorination of phosphorus oxytrichloride using HF, AsF_3, or SbF_3.

Phosphorus Thiofluoride

Phosphorus thiofluoride can be prepared at a low temperature by uv radiation of OCS and PF_3; by the reaction of PF_5 and $(C_2H_5)_4NSH$ in acetonitrile; by the reaction of PF_3 and SF_6 at elevated temperature, or with H_2S; by the reaction of $PSCl_3$ and NaF; and by the high temperature reaction of PF_3 and S.

Fluorophosphoric Acids and the Fluorophosphates

The three primary fluorophosphoric acids, monofluorophosphoric acid, H_2O_3PF, difluorophosphoric acid, HO_2PF_2, and hexafluorophosphoric acid, HPF_6 can be prepared by reaction of phosphoric acid or phosphoric anhydride using varying amounts of HF or phosphorus oxyfluoride and HF or water, or both. All three fluorophosphoric acids are commercially available.

Table 1. Properties of Phosphorus Fluorides

Property	PF_5	POF_3	PF_3	PSF_3
melting point, °C	−91.6	−39.1	−151.5	0.15
boiling point, °C	−84.8	−39.7	−101.8	−0.5
density, liquid, at bp, g/mL			1.6	
critical temperature, °C	>25	73.3	−2.05	−0.73
critical pressure, MPa[a]		4.23	4.33	
heat of fusion, kJ/mol[b]	12.1	14.9		
heat of vaporization, kJ/mol[b]	16.7	23.2	16.5	
heat of formation, $-\Delta H_f$, kJ/mol[b]	1210		946	

[a] To convert MPa to atm, divide by 0.101. [b] To convert kJ to kcal, divide by 4.184.

A number of salts of the monofluoro- and hexafluorophosphoric acids are known and some are commercially important. The salts of difluorophosphoric acid are typically less stable toward hydrolysis and are less well characterized. Sodium monofluorophosphate, the most widely used dentifrice additive for the reduction of tooth decay, is best known (see DENTIFRICES). Several hexafluorophosphates can be prepared by neutralization of the appropriate base using hexafluorophosphoric acid. The monofluorophosphates are usually prepared by other methods because neutralization of the acid usually results in extensive hydrolysis.

The acids are generally shipped and stored in United Nations (UN) 6HA1 heavy plastic drums with steel overpacks. Aluminum is also satisfactory for storage and use of concentrated solutions of the difluoro acid and hexafluoro acid.

Experimentation with test animals and laboratory and plant experience indicate that the fluorophosphoric acids are less toxic and dangerous than hydrogen fluoride. However, they contain, or can hydrolyze to, hydrofluoric acid and must be treated with the same care as hydrofluoric acid.

Fluorophosphate Esters

The esters of monofluorophosphoric acid are of great interest because of their cholinesterase inhibiting activity which causes them to be highly toxic nerve gases and also gives them medical activity (see ENZYME INHIBITORS). The most studied is the bis(1-methylethyl)ester of phosphorofluoridic acid also known as diisopropyl phosphorofluoridate, DFP,

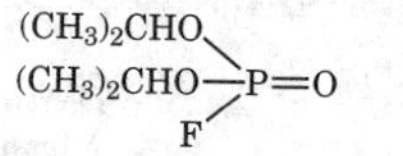

and as the ophthalmic ointment or solution Isoflurophate USP. It is used as a parasympathomimetic agent, and as a miotic in glaucoma and convergent strabismus. Developed during World War II as a nerve gas, it is prepared by reaction of PCl_3 and isopropanol, followed by chlorination and conversion to the desired product using NaF.

CHARLES B. LINDAHL
TARIQ MAHMOOD
Elf Atochem North America, Inc.

R. Schmutzler, in M. Stacey, J. D. Tatlow, and A. G. Sharpe, eds., *Advances in Fluorine Chemistry*, Vol. 5, Butterworth & Co., Inc., Washington, D.C., 1965, pp. 31–287.

G. I. Drozd, *Usp. Khim.* **39**, 3 (1970).

W. Lange, in J. H. Simons, ed., *Fluorine Chemistry*, Vol. I, Academic Press, Inc., New York, 1950, pp. 125–188.

E. L. Muetterties and co-workers, *J. Inorg., Nucl. Chem.* **16**, 52 (1960).

POTASSIUM

Potassium Fluoride

Properties. Anhydrous potassium fluoride is a white hygroscopic salt that forms two hydrates, $KF{\cdot}2H_2O$ and $KF{\cdot}4H_2O$. Chemical and physical properties of KF are summarized in Table 1.

Manufacture. Commercial KF is manufactured from potassium hydroxide and hydrofluoric acid, followed by drying in a spray dryer or flaking from a heated drum.

Toxicology. Ingestion of potassium fluoride may cause vomiting, abdominal pains, and diarrhea.

Uses. Potassium fluoride is used in the manufacture of silver solder fluxes and in fluxes for various metallurgical operations. For many types of replacement of halogens by fluorine in organic compounds, KF is the most frequently used fluoride.

Table 1. Properties of Potassium Fluoride and Potassium Bifluoride

Property	KHF_2	Value
melting point, °C	238.8	857
boiling point, °C		1505
specific gravity at 25°C	2.37	2.48
solubility in H_2O, 25°C, g/100 g	39.2[a]	49.6
heat of fusion, kJ/mol[b]	−920.4	28.2
heat of vaporization, kJ/mol[b]	6.6	173

[a] At 20°C.
[b] To convert kJ to kcal, divide by 4.184.

Potassium Bifluoride

Properties. Other names for potassium bifluoride are potassium hydrogen difluoride and potassium acid fluoride. This white crystalline salt is a soft, waxy solid. Chemical and physical properties are summarized in Table 1.

Manufacture. Potassium bifluoride is produced from potassium hydroxide or potassium carbonate and hydrofluoric acid.

Toxicology and Handling. The TLV for KHF_2 is 2.5 mg/m^3. Potassium bifluoride crystals may break down to a fine white powder that is readily airborne. In this form, the salt is quite irritating to the nasal passages, eyes, and skin.

Uses. A primary use for potassium bifluoride is in the electrolyte for cells in fluorine manufacture.

JOHN R. PAPCUN
Atotech

H. C. Hodge and F. A. Smith, in J. H. Simons, ed., *Fluorine Chemistry*, Vol. 4, Academic Press, Inc., New York, 1965, p. 200.

I. Rajagonal and K. S. Rajams, *Met. Finish.* **76**(4), 43 (1978).

A. R. Basbour, L. F. Belf, and M. W. Bruxton, in M. Stacy and co-eds., *Advances in Fluorine Chemistry*, Vol. 3, Butterworth, Washington, D.C., 1963, pp. 181–250.

F. Naso and L. Ronzini, *J. Chem. Soc. Perkin Trans. 1*, 340 (1974); J. H. Clark and J. M. Miller, *J. Am. Chem. Soc.* **99**, 498 (1977); J. H. Clark, J. Emsley, and O. P. A. Hoyta, *J. Chem. Soc. Perkin Trans. 1*, 1091 (1977).

RHENIUM

Rhenium Hexafluoride

Rhenium hexafluoride, ReF_6, is a pale yellow solid at 0°C, but a liquid at ambient temperature. In the presence of moisture it hydrolyzes rapidly forming HF, ReO_2, and $HReO_4$ (see RHENIUM AND RHENIUM COMPOUNDS). It is not safe to store ReF_6 in a glass trap or glass-lined container. Leaks in the system can initiate hydrolysis and produce HF. The pressure buildup causes the system to burst and an explosion may result.

Properties. Some physical properties of ReF_6 are mol wt, 300.19; mp, 18.5°C, bp, 33.7°C; solubility in HF, 52.5 g/100g; specific gravity, 3.58; and vapor pressure at 20.3°C, 61 kPa. The compound can be handled in dry metal vacuum lines made of copper, nickel, stainless steel, or Monel. It forms a passive fluoride film on the surface which protects these metals from further corrosion.

Rhenium hexafluoride is readily prepared by the direct interaction of purified elemental fluoride over hydrogen-reduced, 300 mesh (ca 48 μm) rhenium powder at 120°C.

Rhenium hexafluoride is used for the deposition of rhenium metal films for electronic, semiconductor, and laser parts, and in chemical vapor deposition (CVD) processes.

Rhenium hexafluoride is a costly (ca \$3000/kg) material and is often used as a small percentage composite with tungsten or molybdenum. The addition of rhenium to tungsten metal improves the ductility and high temperature properties of metal films or parts.

Table 1. Rhenium Oxyfluorides

Compound	$ReOF_5$	ReO_2F_3	ReO_3F	$ReOF_4$	$ReOF_3$
preparative route	$ReO_2 + F_2$	$ReO_2 + F_2$	$KReO_4 + IF_5$	$ReF_6 + M(CO)_x$	$ReF_6 + M(CO)_x$
color	cream	pale yellow	yellow	blue	black
mp, °C	40.8[a]	90	71	107.8[a]	>200

[a] Transition point.

Rhenium hexafluoride (99.5% pure) is commercially available from Advance Research Chemicals, Atomergic, Atochem, Spectra Gases, and Matheson Gas of the United states, Fluorochem of the United Kingdom, and other sources. Because of its high irritating and corrosive nature it is classified as a corrosive, poisonous liquid and is shipped in steel, stainless steel, or Monel cylinders. All precautions must be taken to avoid breathing of vapors or contact with skin.

Other Rhenium Fluoride Compounds

Rhenium heptafluoride, ReF_7, is obtained by the direct interaction of elemental fluorine with hydrogen-reduced rhenium powder. It is a pale yellow solid, mol wt 319.19; mp, 48.3°C; and bp, 73.7°C.

Rhenium pentafluoride, ReF_5, is obtained along with rhenium tetrafluoride, ReF_4, when reduction of ReF_6 is carried out with metal carbonyls (qv). ReF_5 is a greenish yellow solid with mp 48°C.

Rhenium also forms several important oxyfluorides. Properties are summarized in Table 1.

DAYAL T. MESHRI
Advance Research Chemicals, Inc.

G. H. Cady and C. B. Hargreaves, *J. Chem. Soc.*, 1563–1568 (1961).

F. A. Cotton and G. Wilkinson, *Advanced Inorganic Chemistry, A Comprehensive Text*, 3rd ed., Interscience Publishers, Division of John Wiley & Sons, New York, 1972, p. 977.

SILVER

Silver Subfluoride

Pure silver subfluoride, Ag_2F, is a greenish shiny crystalline material. Silver subfluoride is prepared by heating a concentrated solution of silver fluoride with metallic silver powder.

Silver Fluoride

Anhydrous silver fluoride, AgF, is a golden yellow solid in its pure form and is classified as a soft fluorinating agent.

Preparation. Silver fluoride can be prepared by dissolving Ag_2O or Ag_2CO_3 in anhydrous hydrogen fluoride or aqueous hydrofluoric acid, evaporating to dryness, and then treating with methanol or ether.

Properties. Silver fluoride is light sensitive and has a specific gravity of 5.852. It melts at 435°C into a black liquid which boils at 1150°C. Unlike the other halides, it is extremely soluble (182 g/100 g) in water and in anhydrous hydrogen fluoride (83.2 g/100 g at 11.9°C).

Uses. Silver fluoride is used as soft (mild) fluorinating agent for selective fluorination, as a cathode material in batteries (qv), and as an antimicrobial agent. Silver fluoride is commercially available from Advance Research Chemicals, Inc., Aldrich Chemicals, Cerac Corp., Johnson/Matthey, PCR, Atochem, and other sources in the United States.

Silver Difluoride

Silver difluoride, AgF_2, is a black crystalline powder. AgF_2 is prepared by the action of elemental fluorine on AgF or AgCl at 200°C. Silver difluoride should be stored in Teflon, passivated metal containers, or in sealed quartz tubes.

Properties. Silver difluoride melts at 690°C, boils at 700°C, and has a specific gravity of 4.57.

Uses. AgF_2 is a powerful fluorinating agent. Silver difluoride is commercially available from the same sources as those of AgF. In spite of the technical success in laboratory experiments, silver fluorides have found limited use on a large scale mainly because of the high cost of the reagents.

Silver Trifluoride

Silver trifluoride, AgF_3, has been prepared from anhydrous HF solutions of AgF_4^- salts by addition of BF_3, PF_3, or AsF_5. No commercial source is available.

Silver Fluorocomplexes

The silver fluorocomplexes, ie, silver hexafluoroantimonate, $AgSbF_6$; silver hexafluorophosphate, $AgPF_6$; silver tetrafluoroborate, $AgBF_4$; and other salts such as silver trifluoromethane sulfonate, CF_3SO_3Ag, and silver trifluoroacetate, CF_3COOAg, play an important role in the synthesis of organic compounds and have gained potential industrial importance. These complex salts are slightly soluble in anhydrous hydrogen fluoride and very soluble in water. Except for the melting point of CF_3COOAg, not many other physical properties are known. These salts are corrosive and are to be considered toxic because of the presence of Ag^+ ions. Skin contact and inhalation should be avoided. These salts are commercially available. Worldwide consumption of fluorocomplex salts varies between 100 to 300 kg/yr. The most popular salt is $AgBF_4$.

DAYAL T. MESHRI
Advance Research Chemicals, Inc.

D. T. Meshri and W. E. White, *George H. Cady ACS Symposium*, Milwaukee, Wis., June 1970.

Jpn. Kokai 75,131,034 (Oct. 16, 1961), T. M. Saaki (to Japan Storage Battery Co. Ltd.).

SODIUM

Sodium Fluoride

Sodium fluoride, NaF, is a white, free-flowing crystalline powder, mp 992°C, bp 1704°C, with a solubility of 4.2 g/100 g water at 10°C.

Sodium fluoride is normally manufactured by the reaction of hydrofluoric acid and soda ash (sodium carbonate), or caustic soda (sodium hydroxide). The salt is packaged in 45-kg multiwall bags or fiber drums of 45, 170, or 181 kg. Both sodium fluoride and sodium bifluoride are poisonous if taken internally. Dust inhalation and skin or eye contact may cause irritation of the skin, eyes, or respiratory tract, and should be avoided.

Fluoridation of potable water supplies for the prevention of dental caries is one of the principal uses for sodium fluoride. Other uses are as a flux for deoxidizing (degassing) rimmed steel (qv), and in the resmelting of aluminum.

Sodium Bifluoride

Sodium bifluoride (sodium acid fluoride, sodium hydrogen fluoride), $NaHF_2$ or NaF·HF, is a white, free-flowing fine granular material. Its solubility in water is 3.7 g/100g solution at 20°C. The same reactants are used for manufacture as for sodium fluoride. The dried salt is shipped in 45-kg multiwall bags and in 57-, 170-, and 180-kg fiber drums. Sodium bifluoride, by itself or in conjunction with other materials, is a good laundry sour. Leather (qv) bleaching and cleaning of stone and brick building faces are other uses for this material.

WERNER H. MUELLER
Hoescht Celanese Corporation

Fluoridation Engineering Manual, EPA, Office of Water Programs, Washington, D.C., 1972.

S. Budavari, ed., *Merck Index*, 11th ed., Merck & Co., Inc., Rahway, N.J. 1989.

SULFUR

SULFUR FLUORIDES

Sulfur Hexafluoride

Sulfur hexafluoride, SF_6, molecular weight 146.07, is a colorless, odorless, tasteless gas. It is not flammable and not particularly reactive. Its high chemical stability and excellent electrical characteristics have led to widespread use in various kinds of electrical and electronic equipment such as circuit breakers, capacitors, transformers, microwave components, etc (see ELECTRONIC MATERIALS). Sulfur hexafluoride has been commercially available as AccuDri, SF_6 (AlliedSignal, Inc.) since 1948. It is also produced by Air Products and Chemicals in the United States.

Properties. Properties are given in Table 1.

Manufacture and Quality Control. Sulfur hexafluoride is manufactured by combining sulfur vapor and pure elemental fluorine.

Economic Aspects and Shipping. Consumption of SF_6 has increased gradually as dielectric uses have broadened. Sulfur hexafluoride is packaged as a liquefied gas in DOT 3AA 2015 steel cylinders containing 52 kg. Larger quantities are available in tube trailers containing ca 11,000 kg.

Health and Safety Factors. Sulfur hexafluoride is a nonflammable, relatively unreactive gas that has been described as physiologically inert. The current OSHA standard maximum allowable concentration for human exposure in air is 6000 mg/m^3. It should be noted, however, that breakdown products of SF_6, produced by electrical decomposition of the gas, are toxic.

Sulfur Tetrafluoride. Sulfur tetrafluoride, SF_4, molecular weight 108.06, is a highly reactive colorless gas that fumes in moist air and has an irritating odor that resembles sulfur dioxide.

Physical Properties. Selected physical properties are given in Table 2.

Chemical Properties. Sulfur tetrafluoride reacts rapidly with water to give hydrofluoric acid and thionyl fluoride. With alcohols, mixtures of alkyl fluorides and alkyl ethers are obtained. Alcohols bearing electron-withdrawing groups can be converted to the corresponding fluorides in high yield. Sulfur tetrafluoride replaces the carbonyl oxygen with fluorine.

Table 1. Physical Properties[a] of Sulfur Hexafluoride

Property	Value
sublimation point, °C	−63.9
triple point, °C	−50.52
density, solid at −195.2 °C, g/cm^3	2.863
vapor pressure of saturated liquid, MPa[b]	2.3676
heat of formation, kJ/mol[c]	−1221.66
viscosity, liquid, mPa·s(= cP)	0.277
thermal conductivity, liquid, W/(m·K)	0.0583
refractive index, n_D	1.000783
dielectric constant, liquid	1.81

[a] All data refer to 25°C and 101.3 kPa (1 atm), unless otherwise stated. [b] To convert MPa to atm, divide by 0.101. [c] To convert J to cal, divide by 4.184.

Table 2. Physical Properties of Sulfur Tetrafluoride

Property	Value
molecular weight	108.055
melting point, °C	−121.0
boiling point, °C	−38
density, liquid, at −73 °C, g/mL	1.9190
vapor pressure at 25°C, MPa[a]	2.0219

[a] To convert MPa to atm, divide by 0.101.

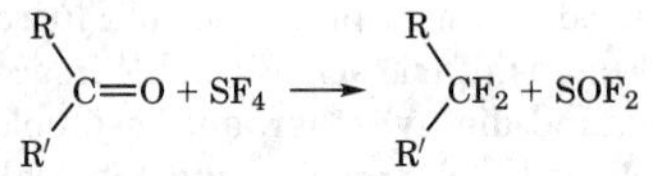

Sulfur tetrafluoride reacts with most inorganic oxides and sulfides to give the corresponding fluorides. An extensive review of SF_4 in organic fluorination is available.

Preparation. For commercial production, SF_4 is made by direct combination of sulfur with elemental fluorine. Commercial applications of SF_4 are limited. It is available from Air Products and Chemicals.

Toxicity. Sulfur tetrafluoride has an inhalation toxicity comparable to phosgene. The current OSHA standard maximum allowable concentration for human exposure in air is 0.4 mg/m^3 (TWA). On exposure to moisture, eg, on the surface of skin, sulfur tetrafluoride liberates hydrofluoric acid and care must be taken to avoid burns.

Other Sulfur Fluorides

Although eight other binary sulfur fluorides have been synthesized and characterized, proof of the existence of several members of this group was dependent on modern instrumental methods of analysis because of extreme instability. SF_5 and S_2F_{10} are stable, however, the latter is noted for its extreme toxicity. All sulfur fluorides other than SF^6 must be considered extremely toxic.

As a group, these materials have no technological utility because of instability, toxicity, and difficulty of preparation. An excellent review of many of these compounds is available. They include sulfur pentafluoride, disulfur decafluoride, thiothionyl fluoride and difluorodisulfane, and difluoromonosulfane and difluorodisulfane difluoride.

FRANCIS E. EVANS
GANPAT MANI
AlliedSignal Inc.

Threshold Limit Values for Chemical Substances and Physical Agents, American Conference of Governmental Industrial Hygienists, Cincinnati, Ohio, 1990–1991.

D. R. James, Technical Note No. 1, *Cooperative Research and Development Agreement (CRADA), Investigation of S_2F_{10} Production and Mitigation in Compressed SF_6-Insulated Power System*, Oak Ridge National Laboratory, Oak Ridge, Tenn., Dec. 28, 1992.

S. D. Barrett, *The A to Z of SF_6, Special Report Electric Light and Power*, TID ed., Dec. 1972.

F. Y. Chu and co-workers, *Conference Record of the IEEE International Symposium on Electrical Insulation*, 1988, pp. 131–134.

FLUOROSULFURIC ACID

Fluorosulfuric acid, HSO_3F, is a colorless-to-light yellow liquid that fumes strongly in moist air and has a sharp odor. It is a strong acid and is employed as a catalyst and chemical reagent in a number of chemical processes, such as alkylation (qv), acylation, polymerization, sulfonation, isomerization, and production of organic fluorosulfates.

Table 1. Physical and Chemical Constants of Fluorosuluric Acid

Property	Value[a]
molecular weight	100.07
boiling point, °C	162.7
freezing point, °C	−88.98
density, g/mL	1.726
viscosity, mPa·s(= cP)	1.56
dielectric constant	ca 120

[a] All values are at 25°C.

Properties. Selected physical properties of fluorosulfuric acid are shown in Table 1. Fluorosulfuric acid is soluble in acetic acid, ethyl acetate, nitrobenzene, and diethyl ether, and insoluble in carbon disulfide, carbon tetrachloride, chloroform, and tetrachloroethane. Many inorganic and organic materials dissolve in fluorosulfuric acid; the physical and chemical properties of such solutions have been extensively investigated.

Preparation and Manufacture. Commercially, fluorosulfuric acid is made by processes utilizing the product as a solvent. Solutions of HF and SO_3 in fluorosulfuric acid are mixed in stoichiometric quantities, or SO_3 and HF are separately introduced into a stream of fluorosulfuric acid to produce essentially pure HSO_3F. Some of the product is then recycled.

Economic Aspects. U.S. manufacturers of fluorosulfuric acid are AlliedSignal and DuPont.

Health and Safety Factors. Fluorosulfuric acid is a strong acid capable of causing severe burns similar to those experienced with sulfuric and hydrofluoric acids. In addition, the fumes of fluorosulfuric acid are extremely irritating, and breathing of the fumes is to be avoided. Material safety data sheets and other literature from manufacturers describe additional precautions in handling large quantities of fluorosulfuric acid.

Derivatives. The nonmetallic inorganic derivatives of fluorosulfuric acid are generally made indirectly, although complex fluorosulfates of the Group 15 (V) elements and of xenon can be made directly, as can the NO^+ and NO_2^+ salts.

FRANCIS E. EVANS
GANPAT MANI
AlliedSignal, Inc.

R. C. Thompson in G. Nickless, ed., *Inorganic Sulphur Chemistry*, Elsevier, Amsterdam, the Netherlands, 1968, pp. 587–606.

A. W. Jache in H. J. Emeleus and A. G. Sharpe, eds., *Advances in Organic Chemistry and Radiochemistry*, Vol. 16, Academic Press, Inc., New York, 1974, pp. 177–200.

TANTALUM

Tantalum Pentafluoride

Tantalum pentafluoride, TaF_5, a white solid with a reported mp of 97°C and a bp of 229°C, is the only known binary fluoride.

There are a number of methods of preparation for TaF_5. For example, tantalum pentafluoride has been produced by the reaction of F_2 or ClF_3 and Ta metal, by contacting Ta_2O_5 with excess HF in the presence of a dehydrating agent and by the reaction of Ta-containing ores and HF-H_2SO_4 followed by extraction with an organic solvent.

TaF_5 has been used as a superacid catalyst for the conversion of CH_4 to gasoline-range hydrocarbons (qv), in the manufacture of fluoride glass and fluoride glass optical fiber preforms, and incorporated in semiconductor devices.

TARIQ MAHMOOD
CHARLES B. LINDAHL
Elf Atochem North America, Inc.

J. H. Canterford and R. Cotton, *Halides of the Second and Third Row Transition Metals*, John Wiley & Sons, Inc., New York, 1968.

I. R. Beattie, K. M. S. Livingston, G. A. Ozin, and D. J. Reynolds, *J. Chem. Soc. A.*, (6), 958–965 (1969).

TIN

Stannous Fluoride

Stannous fluoride, SnF_2, is a white crystalline salt that has mp 215°C, bp 850°C, and is readily soluble in water and hydrogen fluoride. Stannous fluoride is used widely in dentifrices (qv) and other dental preparations because of its anticaries effect. Other uses of SnF_2 are in the synthesis of fluorophosphate glasses having low melting temperatures, in formation of transparent film, and in the preparation of optically active alcohols.

Fluorostannites and Fluorostannates

Complexes of the type SnF_3, where M is NH_4, Na, K, and Cs, have been crystallized from aqueous solutions. From molten mixtures of SnF_2 and Naf, RbF, and CsF, both the $MSnF_3$ (M = Na, K, Rb, and Cs) and the fluorostannate salts, $MSnF_5$ (M = Na, K, Rb, and Cs) have been obtained.

Stannic Fluoride

Stannic fluoride, SnF_4, is a white solid that sublimes at 705°C and hydrolyzes in water to form insoluble stannic acid. It can be prepared by reaction of fluorine and probably ClF_3 or BrF_3 with virtually any tin(II) or tin(IV) compound. Stannic fluoride is used in the manufacture of glass (qv).

Stannous Fluoroborate

Stannous fluoroborate, $Sn(BF_4)_2$, is prepared in electrochemical cells using tin and fluoroboric acid. The main use of stannous fluoroborate is in electroplating (qv).

Hexafluorostannates

The hexafluorostannate anion, SnF_6^{2-}, forms readily and is stable over a wide pH range. Numerous hexafluorostannates have been prepared by dissolving stannates in excess hydrofluoric acid, dissolving stannic acid in excess HF and neutralizing, or by reaction of salts and SnF_4.

Safety, Handling, and Toxicity

Stannous fluoride is used in dentifrices and dental preparations. The OSHA permissible exposure limit and ACGIH established TLV for fluoride is 2.5 mg/m^3 or air.

CHARLES B. LINDAHL
TARIQ MAHMOOD
Elf Atochem North America, Inc.

J. C. Muhler and co-workers, *J. Am. Dent. Assoc.* **50**, 163 (1955).

J. E. Ellingsen, B. Svatun, and G. Roella; *Acta Odontol. Scand.* **38**(4), 219–222 (1980).

G. D. Lukiyanchuk, V. K. Gonsharuk, E. V. Merkulov, and T. I. Usol'tseva, *Fiz. Khim. Stekla*, **18**(2), 141–145 (1992).

TITANIUM

Titanium(III) Fluoride

Titanium trifluoride, TiF_3, is a blue crystalline solid that undergoes oxidation to TiO_2 upon heating in air at 100°C. Titanium trifluoride is prepared by dissolving titanium metal in hydrofluoric acid or by passing anhydrous hydrogen fluoride over titanium trihydrate at 700°C or over heated titanium powder. The ACGIH adopted toxicity values (1992–1993) for TiF_3 is as TWA for fluorides as F^- 2.5 mg/m^3.

This material is available from Advance Research Chemicals, Inc., Aldrich Chemical Co., Inc., Aesar, Johnson/Matthey, Cerac, PCR, and Pfaltz & Bauer in the United States, Fluorochem of the United Kingdom, and Schuchardt of Germany. No commercial applications have been reported.

Titanium(IV) Fluoride

Titanium tetrafluoride, TiF_4, has potential for use in dental hygiene products. It is used in infrared transmitting halide glass. TiF_4 is a colorless, very hygroscopic solid and is classified as a soft fluorinating reagent. TiF_4 melts at temperatures >400 °C. It is soluble in water, alcohol, and pyridine, and has a density of 2.79 g/mL. The most economical and convenient method of preparing is the action of liquid anhydrous HF on commercially available titanium tetrachloride in Teflon or Kynar containers.

Total consumption of TiF_4 in both the United States and Europe is less than 500 kg/yr. TiF_4 is available from Advance Research Chemicals, Inc., Aldrich, Aesar, Johnson/Matthey, Cerac, PCR, and Pfaltz & Bauer of the United States, Fluorochem of the United Kingdom, and Schuchardt of Germany.

Fluorotitanates

Hexafluoroanions of Group 4 (IVB) are octahedral crystals that are quite stable in acidic media. All three hexafluoroacids are known, ie, hexafluorotitanic acid, hexafluorozirconic acid, H_2ZrF_6, and hexafluorohafnic acid, H_2HfF_6. Fluorotitanic acid is used as a metal surface cleaning agent, as a catalyst, and as an aluminum finishing solvent. Fluorotitanates are used in abrasive grinding wheels and for incorporating titanium into aluminum alloys.

The total U.S. consumption of H_2TiF_6 is 20 t/yr. It is packaged in DOT approved polyethylene-lined drums and the salts in polyethylene-lined fiber board drums.

DAYAL T. MESHRI
Advance Research Chemicals, Inc.

L. Skartveit, K. A. Selvig, S. Myklebust, and A. B. Tveit, *Acta Odontol. Scand.* **48**(3), 169–174 (1990).

A. Jha and J. M. Parker, *Phys. Chem. Glasses.* **32**(1), 1–2 (1991).

TUNGSTEN

Tungsten Hexafluoride

Physical Properties. Tungsten(VI) fluoride, WF_6, is a colorless gas that condenses at ca 100 kPa (1 atm) and 17.1°C to a water-white liquid that may be colored owing to metallic impurities. The physical properties of tungsten hexafluoride are given in Table 1.

Chemical Properties. Tungsten hexafluoride is readily hydrolyzed by water to give tungsten trioxide and hydrogen fluoride. It is a strong fluorinating agent and reacts with many metals at room temperature.

Table 1. Physical Properties of Tungsten Hexafluoride

Property	Value
boiling point, °C	17.2
triple point, °C, 55.1 kPa[a]	2.0
liquid density at 15°C, g/mL	3.441
transition point, °C, 32.0 kPa[a]	−8.2
specific heat at 25°C, J/(mol·K)[b]	118.92

[a] To convert kPa to mm Hg, multiply by 7.5. [b] To convert kJ to kcal, divide by 4.184.

Manufacture and Economics. Tungsten hexafluoride is produced commercially by the reaction of tungsten powder and gaseous fluorine at a temperature in excess of 350°C. U.S. production is several metric tons per year. Essentially all of the product is used in CVD. Air Products and Chemicals, Inc. (Allentown, Pa.) and Bandgap Technology Corp. (Broomfield, Colo.) are the only U.S. producers.

Because of the development of electronic applications for WF_6, higher purities of WF_6 have been required. Most metal contaminants and gaseous impurities are removed from WF_6 by distillation. HF, which has a similar vapor pressure to WP_6, must be removed by adsorption.

Handling and Toxicity. Tungsten hexafluoride is irritating and corrosive to the upper and lower airways, eyes, and skin. It is extremely corrosive to the skin, producing burns typical of hydrofluoric acid. The OSHA permissible exposure limits is set as a time-weighted average of 2.5 mg/kg or 0.2 ppm. Monel and nickel are the preferred materials of construction for cylinders and delivery systems.

Uses. The primary use of WF_6 is for blanket and selective deposition of tungsten and tungsten silicide films in the manufacture of VLSI electronic devices.

PHILIP B. HENDERSON
ANDREW J. WOYTEK
Air Products and Chemicals, Inc.

G. H. Cady and G. B. Hargreaves, *J. Chem. Soc.*, 1563 (1961).

H. F. Priest, *Inorg. Synth.* **3**, 181 (1950).

D. A. Bohling and M. George, *Semicond. Int.* **14**, 104 (1991).

J. E. J. Schmitz, *Chemical Vapor Deposition of Tungsten and Tungsten Silicides for VLSI/ULSI Applications*, Noyes Publications, Park Ridge, N.J., 1992.

ZINC

Zinc Fluoride

Anhydrous zinc fluoride, ZnF_2, melts at 872–910°C, has a solubility of only 0.024 g/100 g anhydrous HF at 14.2°C, and can be prepared by slowly drying zinc fluoride tetrahydrate, $ZnF_2 \cdot 4H_2O$, in a current of anhydrous hydrogen fluoride to minimize hydrolysis and formation of the oxide.

Zinc fluoride has been used as a mild fluorinating reagent in replacement of chlorine in halogenated hydrocarbons. It is also used as a catalyst in several applications including cyclization processes. High purity ZnF_2 is used in the synthesis of fluorophosphate glass, fluoride glass, high conducting oxyfluoride glass, as fluoride glass films, in the manufacture of fluoride glass optical fibers, and in the preparation of optical transmitting glass (see GLASS; FIBER OPTICS).

OSHA has a standard time-weighted average (TWA) or 2.5 mg/m^3 based on fluoride. NIOSH has issued a criteria document on occupational exposure to inorganic fluorides.

Zinc Fluoride Tetrahydrate. Zinc fluoride tetrahydrate is prepared by reaction of ZnO and aqueous HF. $ZnF_2 \cdot 4H_2O$ has a water solubility of about 1.6 g/100 mL solution at 25°C.

Fluorozincates. Potassium fluorozincate, $KznF_3$, and sodium fluorozincate, $NaZnF_3$, are used as catalysts in alginate dental impression materials.

PHILIP B. HENDERSON
ANDREW J. WOYTEK
Air Products and Chemicals, Inc.

A. W. Jache and G. H. Cady, *J. Phys. Chem.* **56**, 1106 (1952).

U.S. Pat. 3,728,405 (Sept. 14, 1970), J. Allan (to E. I. du Pont de Nemours & Co., Inc.).

ZIRCONIUM

Zirconium Difluoride

Zirconium difluoride, ZrF_2, prepared by Knudsen cell techniques, is not commercially available.

Zirconium Trifluoride

Zirconium trifluoride, ZrF_3, was first prepared by the fluorination of ZrH_2 using a mixture of H_2 and anhydrous HF at 750°C. This compound is of academic interest rather than of any industrial importance.

Zirconium Tetrafluoride

Zirconium tetrafluoride, ZrF_4, is one of the many important inorganic fluorides that have played a role in the development of heavy-metal fluoride glass (HMFG) technology. Table 1 summarizes some of the physical properties of zirconium tetrafluoride. The anhydrous salt is prepared by several methods, eg, by reacting $ZrCl_4$ with liquid anhydrous HF. The principal application of ZrF_4 has been the manufacture of HMFGs of which the most widely investigated is the system composed of Zr, Ba, La, Al, and Na, also popularly known as the ZBLAN glasses. This system has revolutionized the optics industry because of the significantly superior qualities of these glasses over conventional silica glasses.

Table 1. Properties of ZrF_4

Property	Value
mol wt	167.21
specific gravity	4.54
solubility in water at 25°C	1.388

High purity ZrF_4 is available in the United States from Advance Research Chemicals, Inc., Air Products and Chemicals, Inc., Johnson-Matthey/AESAR group, Aldrich Chemical, and EM Industries, Inc. Consumption of ZrF_4 in the United States is less than 5000 kg/yr.

Fluorozirconic Acid and Fluorozirconates

Hexafluorozirconic acid, H_2ZrF_6, is formed by dissolving freshly prepared oxide, fluoride, or carbonate of zirconium in aqueous HF. This acid is produced commercially in a concentration range of 10 to 47%. The acid can be stored at ambient temperatures in polyethylene or Teflon containers without decomposition for at least two years.

Hexafluorozirconic acid is used in metal finishing and cleaning of metal surfaces, whereas the fluorozirconates are used in the manufacture of abrasive grinding wheels, in aluminum metallurgy, ceramics industry, glass manufacturing, in electrolytic cells, in the preparation of fluxes, and as a fire retardant (see ABRASIVES; METAL SURFACE TREATMENTS).

High purity hexafluorozirconic acid and its salts are produced by Advance Research Chemicals of the United States, and Akita and Moritta of Japan. The technical-grade green-colored material is supplied by Cabot Corp. of the United States.

DAYAL T. MESHRI
Advance Research Chemicals, Inc.

H. P. Withers Jr., V. A. Monk, and G. A. Cooper, *Proceedings of the SPIE International Society of Optical Engineers*, Vol. 1048, 1989, pp. 72–77; U.S. Pat. 4,983,372 (Jan. 8, 1991), H. P. Withers, Jr. and V. A. Monk (to Air Products & Chemicals, Inc.).

FLUORINE COMPOUNDS, ORGANIC

INTRODUCTION

Physical Properties

Substitution of fluorine for hydrogen in an organic compound has a profound influence on the compound's chemical and physical properties. Several factors that are characteristic of fluorine and that underlie the observed effects are the large electronegativity of fluorine, its small size, the low degree of polarizability of the carbon–fluorine bond and the weak intermolecular forces.

The replacement of chlorine by fluorine results in a nearly constant boiling point (bp) drop of approximately 50°C for every chlorine atom that is replaced. A similar boiling point effect with hydrocarbons is apparent, even though the molecular weight of the fluorocarbon is much higher than the corresponding hydrocarbon analogue. An analogous drop in the corresponding fluorocarbon freezing point results in a widened liquid range for applications like lubricating fluids and greases. One other significant property difference, attributed to weak intermolecular forces, can be found in the very low surface tensions of fluorocarbons as compared to hydrocarbons and water.

The low surface tension of highly fluorinated organic compounds is commercially important for their application in surfactants, antisoiling textile treatments, lubricants, and specialty wetting agents.

In contrast, the viscosities of fluorocarbons are higher than those of the corresponding hydrocarbons. This can be explained by the greater stiffness of the fluorocarbon chain arising from the large replusive forces between molecules, and from the greater density imparted by the more massive fluorine atoms (vs hydrogen). The fluorocarbon viscosity drops rapidly with increasing temperature and is accompanied by a simultaneous large decrease in density.

The refractive indexes and dielectric constants for the fluorocarbons are both lower than that for the corresponding hydrocarbon analogue.

Preparation

There are many known ways to introduce fluorine into organic compounds, but hydrogen fluoride, HF, is considered to be the most economical source of fluorine for many commercial applications.

Halogen Exchange. The exchange of another halogen atom in an organic compound for a fluorine atom is the most widely used method of fluorination.

Replacement of Hydrogen. Three methods of substitution of a hydrogen atom by fluorine are *(1)* reaction of a C–H bond with elemental fluorine (direct fluorination, *(2)* reaction of a C–H bond with a high valence state metal fluoride like AgF_2 or CoF_3, and *(3)* electrochemical fluorination in which the reaction occurs at the anode of a cell containing a source of fluoride, usually HF.

Telomer Formation. Fluorinated compounds with active C–Br or C–I bonds can add to fluoroolefins to form addition products in high yield. The olefin most often used is tetrafluoroethylene. Telomerization involves reaction of a telogen, or addition agent like $CBrF_3$, CF_3I, or C_2F_5I, with the olefin to form longer chain addition products called telomers.

Aromatic Ring Fluorination. The formation of an aryl diazonium fluoride salt, followed by decomposition, is a classical reaction (the Schiemann reaction) for aryl fluoride preparation.

Chemical Properties and Applications

Substitution of fluorine into an organic molecule results in enhanced chemical stability. The resulting chemical reactivity of adjacent functional groups is drastically altered due to the large inductive effect of fluorine. These effects become more pronounced as the degree of fluorine substitution is increased, especially on the same carbon atom. This effect demonstrates a maximum in fluorocarbons and their derivatives.

Fluorinated Alkanes. Their lack of reactivity leads to use in certain commercial applications where stability is valued when in contact with highly reactive chemicals.

Fluorinated Olefins. Certain chlorofluorocarbons (CFCs) are used as raw materials to manufacture key fluorinated olefins to support polymer application.

Fluorinated Aromatic Hydrocarbons. Many aromatic fluorocarbon derivatives, eg, hexafluorobenzene, pentafluorotoluene, and perfluoronaphthalene, are examples of compounds that readily undergo nucleophilic ring substitution reactions with loss of one or more fluorine substituents.

Fluorinated Heterocyclic Compounds. The direct action of fluorine on uracil yields the cancer chemotherapy agent, 5-fluorouracil, as one special example of a selective fluorination on a commercial scale.

Fluorinated Acids. Generally, their reactions are similar to organic acids and they find applications, particularly trifluoroacetic acid and its anhydride, as promotors in the preparation of esters and ketones and in nitration reactions.

Fluorinated Biologically Active Compounds. Many biologically active compounds are prepared from fluorobenzene, difluorobenzene, benzotrifluoride, and fluorinated steroids.

Many fluorinated, biologically active agents have been developed and successfully used in the treatment of diseases. The biological property of fluorinated organics has been further extended to applications in the agrochemical and pest management fields. Examples include analgesics, antiviral agents, appetite depressants, tranquilizers, diuretics, inhalation anesthetics, herbicides, insecticides, and fungicides.

Economic Aspects

The CFC commodity application is undergoing significant change owing to environmental pressures. Development of acceptable, alternative fluorinated compounds is extremely expensive. As the largest global supplier, DuPont plans to spend $1 billion by the mid-1990s to develop CFC alternatives. The five-year toxicity testing program alone has been estimated at up to $5 million for each candidate compound tested.

The HCFC and HFC refrigeration alternatives are estimated to be two to five times higher in price, and some of the viable alternatives demonstrate a lower heat-transfer efficiency than the current CFCs. Total production will continue to drop due to conservation in use and elimination of emissive uses along with substitutions in refrigeration applications. The global CFC market is estimated at $4 billion with one-half of that being in the United States. The largest supplier is DuPont with 25% of the global market share.

The Minnesota Mining and Manufacturing Company (3M) manufactures specialty perfluorochemicals using mainly electrochemical fluorination methods at their St. Paul, Minnesota, Decatur, Alabama, and Cordova, Illinois sites. Their capacity is not reported, but is estimated at over 5000 t as fluorinated inert fluids, surfactants, and fire extinguishment chemicals. Asahi, DuPont, and Hoescht all use fluoroolefin telomerization technology at a variety of their sites to manufacture a line of perfluorinated specialty chemicals for stain-resistant treatment and surfactant applications. Globally, these telomer-based fluorochemicals are estimated to be over 5000 t per year with DuPont having one-third of the total.

Aromatic fluorine compounds are varied in kind and in volume. Mallinckrodt Specialty Chemicals division of the Imcera Group claim their continuous process capacity for fluorobenzene is 1200 t per year from their St. Louis, site. EniChem reports a 910 t per year capacity for fluorinated aromatics at their Trissino, Italy site using a novel continuous diazotization process coupled with electrofluorination technology. Hoescht AG has announced plans to double its unspecified capacity at Griesheim, Germany for fluoroaromatics. ICI in the United Kingdom and DuPont at their Deepwater, New Jersey, facility also have fluoroaromatic capabilities. Other smaller suppliers also manufacture fluorinated aromatic compounds for specialty applications, but their capacities are again unreported. Fluoroaromatics are basic intermediate building blocks leading toward the more advanced

aromatic fluorine intermediates (AFI) including fluorinated aniline, quinoline, biphenyl, and phenol compounds. AFI uses are in surfactants, pharmaceuticals, agrochemicals, electronics, and biomedical applications.

Safe Handling Aspects

Existing fluorine compounds cover the range from biologically inert materials like fluorocarbon fluids suitable for potential blood substitutes through biologically active materials like the very highly toxic octafluoroisobutylene.

WILLIAM X. BAJZER
YUNG K. KIM
Dow Corning Corporation

R. E. Banks, *Preparation, Properties and Industrial Applications of Organofluorine Compounds*, Ellis Horwood Ltd., Chichester, U.K., 1982.

J. F. Liebman, A. Greenberg, and W. R. Dolbier, Jr., eds., *Fluorine-Containing Molecules: Structure, Reactivity, Synthesis and Applications*, VCH Publishers, Inc., New York, 1988.

G. A. Olah, R. D. Chambers, and G. K. S. Prakash, eds., *Synthetic Fluorine Chemistry*, John Wiley & Sons, Inc., New York, 1992.

G. Siegemund and co-workers, *Fluorine Compounds, Organic*, in W. Gerhartz and co-workers, eds., *Ullmann's Encyclopedia of Industrial Chemistry*, 5th rev. ed., Vol. A11, VCH Publishers, New York, 1988.

DIRECT FLUORINATION

Before the LaMar process was developed in 1969, the use of direct fluorination was usually considered the classical method of fluorination and other approaches were regarded as modern methods. Now only telomerization reactions using tetrafluorethylene and reactions in hydrogen fluoride-based electrochemical cells are more widely used than direct fluorination on a commercial scale; however, this may change in the future. Direct fluorination not only gives higher yields in most cases but preparation in this manner is applicable to a wider range of organofluorine compounds and classes of compounds inaccessible by these more established technologies. Many compounds are uniquely prepared in the laboratory by direct fluorination, and ton quantities of various fluorocarbon materials are available from 3M Company manufactured by new direct fluorination technology.

Metal Fluoride Method

Fluorination of organic compounds using high valency metallic fluorides may be represented as follows:
Exchange of halogen with fluorine of the metal fluorides, MF_n:

$$\begin{matrix} \diagdown \\ \diagup \end{matrix}C\!-\!X + M^+F^- \longrightarrow \begin{matrix} \diagdown \\ \diagup \end{matrix}C\!-\!F + M^+X^-$$

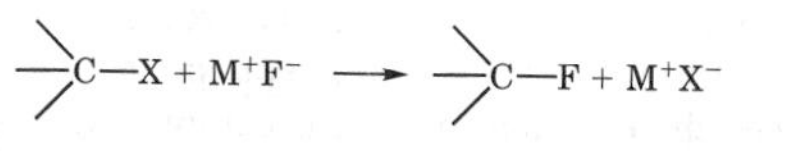

where X = Cl, Br, or I and M = K, Sb, AgHg₂, or Hg.
Replacement of hydrogen with the fluorine of metal fluorides:

$$\begin{matrix} \diagdown \\ \diagup \end{matrix}C\!-\!H + 2\,MF_n \longrightarrow \begin{matrix} \diagdown \\ \diagup \end{matrix}C\!-\!F + HF + 2\,MF_{n-1}$$

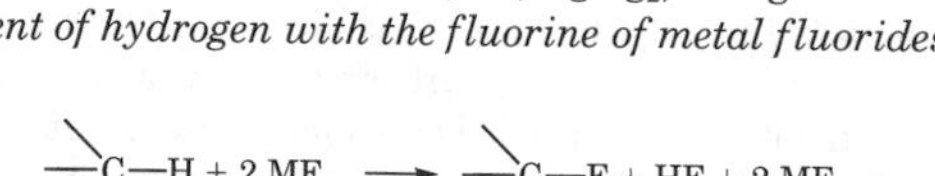

Addition to double bonds:

$$\begin{matrix} \diagdown \\ \diagup \end{matrix}C\!=\!C\begin{matrix} \diagup \\ \diagdown \end{matrix} + 2\,MF_n \longrightarrow F\!-\!\begin{matrix} \diagdown \\ \diagup \end{matrix}C\!-\!C\begin{matrix} \diagup \\ \diagdown \end{matrix}\!-\!F + 2\,MF_{n-1}$$

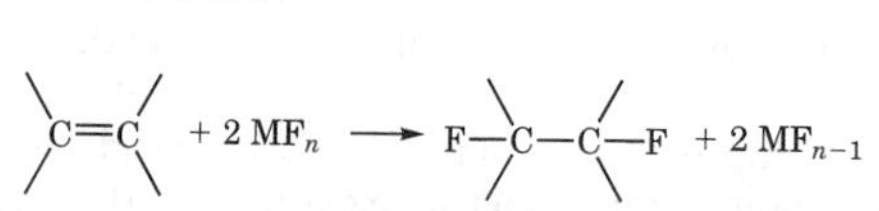

The requirement that organic compounds be vaporized at temperatures averaging 280°C across the bed of cobalt trifluoride or silver difluoride causes serious limitations to the broad applicability of the synthesis of organofluorine compounds using metal fluoride technology. There are at least two companies, Imperial Smelting, Ltd. of Britain and Air Products and Chemicals of Allentown, Pa., still active in this field; the number of organic compounds that can be prepared effectively with this technique numbers approximately 100. Fused-ring aromatic compounds are the most able to survive these harsh fluorination conditions.

Hydrogen Fluoride Electrochemical Cell Methods

Direct fluorination using hydrogen fluoride electrochemical cell methods is mechanistically similar in some regards to direct fluorination with F_2. This method uses an electrolytically activated fluoride ion produced by a Simons' designed hydrogen fluoride electrochemical cell as its primary means of fluorination. The Simons' electrochemical cell fluorination technology is practiced widely by Minnesota Mining & Manufacturing Company (3M) of St. Paul, Minn. In this method, organic precursors are dissolved in liquid hydrogen fluoride and a voltage slightly under the voltage required for generation of elemental fluorine is applied across carbon electrodes.

The principal disadvantage to electrochemical fluorination is the requirement that the organic material be at least somewhat soluble in the polar liquid hydrogen fluoride. Therefore 3M product lines are generally based on perfluoro amines and functionalized materials such as carboxylic acids or sulfonic acids which are soluble in hydrogen fluoride (see FLUORINE COMPOUNDS, ORGANIC–FLUORINATED HIGHER CARBOXYLIC ACIDS; FLUOROETHERS AND FLUORAMINES; PERFLUOROALKANESULFONIC ACIDS).

Fluorocarbons produced by electrochemical fluorination often have small quantities (1–5%) of up to 20 by-products produced by rearrangement. Rearrangement is not characteristic of modern direct fluorination technology using elemental fluorine. By-product formation is a particular disadvantage for applications such as production of fluorocarbon oxygen carriers, fluorocarbon blood, and other biomedical fluorocarbon products where high purity materials are required.

Direct Fluorination Using Elemental Fluorine

Kinetic as well as thermodynamic problems are encountered in fluorination. The rate of reaction must be decelerated so that the energy liberated may be absorbed or carried away without degrading the molecular structure. The most recent advances in direct fluorination are the LaMar process and the Exfluor process, which is practiced commercially by 3M.

Thermochemistry. Thermodynamic considerations are of utmost importance in fluorinations. The limiting parameter to be considered in attempting to develop a satisfactory method for controlling reactions of elemental fluorine is the weakest bond in the reactant compound.

Procedures or conditions that reduce the atomic fluorine concentration or decrease the mobility of hydrocarbon radical intermediates, and/or keep them in the solid state during reaction, are desirable. It is necessary to reduce the reaction rate to the extent that these hydrocarbon radical intermediates have longer lifetimes, permitting the advantages of fluorination in individual steps to be achieved experimentally.

Steric Factors. Initially, most of the collisions of fluorine molecules with saturated or aromatic hydrocarbons occur at a hydrogen site or at a π-bond (unsaturated) site. When collision occurs at the π-bond, the double bond disappears but the single bond remains because the energy released in initiation is insufficient to fracture the carbon–carbon single bond. Once carbon–fluorine bonds have begun to form on the carbon skeleton of either an unsaturated or alkane system, the carbon skeleton is somewhat sterically protected by the sheath of fluorine atoms.

The nonbonding electron clouds of the attached fluorine atoms tend to repel the oncoming fluorine molecules as they approach the carbon skeleton. This reduces the number of effective collisions, making it possible to increase the total number of collisions and still not accelerate the reaction rate as the reaction proceeds toward completion. This protective sheath of fluorine atoms provides the inertness of Teflon and other fluorocarbons. It also explains the fact that greater success in direct fluorination processes has been reported when the hydrocarbon to be fluorinated had already been partially fluorinated by some other process or was prechlorinated.

Kinetic Control. Molecular relaxation processes such as vibrational or rotational relaxations or thermal conduction make it possible to dissipate the energy released during fluorination. Such relaxation processes can minimize the chances that the energy required to break the weakest bond is appropriately localized if the reaction sites are widely distributed over the system.

Reactant molecules are able to withstand more fluorine collisions, as they become more highly fluorinated, without decomposition because some sites are sterically protected, ie, collisions at carbon–fluorine sites are obviously nonreactive. The fluorine concentration may therefore be increased as the reaction proceeds to obtain a practical reaction rate.

Experimental Techniques

The typical fluorination apparatus used in the LaMar process for these reactions is simple in design (Fig. 1). It is essential that the materials of construction be resistant to fluorine.

Aerosol-Based Direct Fluorination. A technology that works on liter and half-liter quantities has been introduced. This new aerosol technique, which functions on principles similar to LaMar direct fluorination, uses fine aerosol particle surfaces rather than copper filings to maintain a high surface area for direct fluorination.

Modern Direct Fluorination. Direct fluorination technology has been scaled up at Exfluor Research Corp. of Austin, Tex. Using direct fluorination, it is possible to produce almost any desired fluorocarbon structure for which there is a hydrocarbon or organic structural precursor. There are two basic approaches to controlling direct fluorination: the LaMar method where the rate of fluorine addition is the limiting factor, and the Lagow-Exfluor method in which the rate of addition of the hydrocarbon is the limiting factor.

Applications

In 1954 the surface fluorination of polyethylene sheets by using a solid CO_2 cooled heat sink was patented. Later patents covered the fluorination of PVC and polyethylene bottles. Studies of surface fluorination of polymer films have been reported. The fluorination of polyethylene powder was described as a fiery intense reaction, which was finally

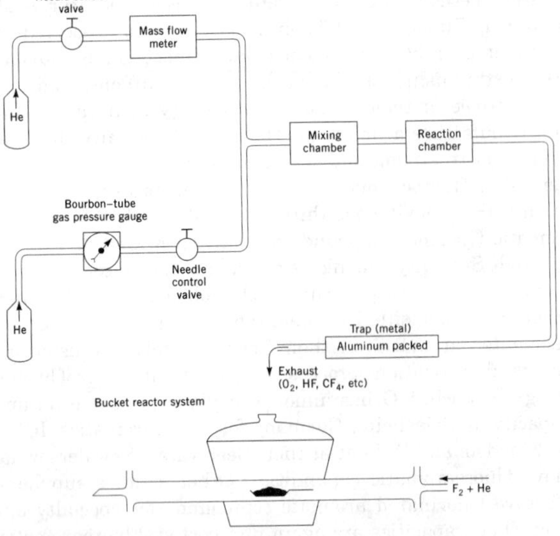

Figure 1. Diagram of typical fluorination apparatus.

controlled by dilution with an inert gas at reduced pressures. Direct fluorination of polymers was achieved in 1970. More recently, surface fluorinations of poly(vinyl fluoride), polycarbonates, polystyrene, and poly(methyl methacrylate), and the surface fluorination of containers have been described. Partially fluorinated poly(ethylene terephthalate) and polyamides such as nylon have excellent soil release properties as well as high wettability. The most advanced direct fluorination technology in the area of single-compound synthesis and synthesis of high performance fluids is currently practiced by 3M Co. of St. Paul, Minn., and by Exfluor Research Corp. of Austin, Tex.

The following companies manufacture organic fluorine compounds by direct fluorination techniques: 3M; Exfluor; Air Products and Chemicals, Inc. (Allentown, Pa.), MarChem, Inc. (Houston, Tex.), Ozark-Mahoning, Inc. (Tulsa, Okla.), and PCR, Inc. (Gainesville, Fla.).

Simple and Complex Organic Molecules. Using modern direct fluorination technology, the synthesis of even the most complex perfluorocarbon structures from hydrocarbon precursors is now possible.

Hydrocarbon Polymers. Direct fluorination can be used for the direct synthesis of fluorocarbon polymers and for producing fluorocarbon coatings on the surfaces of hydrocarbon polymers.

Surface Fluorination of Polymers. Fluorocarbon-coated objects have many practical applications because the chemically adherent surface provides increased thermal stability, resistance to oxidation and corrosive chemicals and solvents, decreased coefficient of friction and thus decreased wear, and decreased permeability to gas flow.

Natural and Synthetic Rubber. Fluorination of natural or synthetic rubber creates a fluorocarbon coating which is very smooth and water-repellent.

Blow-Molded Containers. A surface-fluorination process (Airopak) has been developed by Air Products & Chemicals for the blow-molding industry to produce solvent-resistant polyolefin containers.

RICHARD J. LAGOW
University of Texas at Austin

C. M. Sharts, *J. Chem. Ed.* **45**, 3 (1968).

R. E. Banks, *Fluorocarbons and Their Derivatives*, Oldbourne Press, London, 1964, p. 87.

R. J. Lagow and J. L. Margrave, *Proc. Natl. Acad. Sci.* **67**(4)8A (1970).

U.S. Pat. 5,093,432 (Mar. 3, 1992), T. R. Bierschenk, R. J. Lagow, T. J. Juhlke, and H. Kawa (to Exfluor Research Corp.).

FLUORINATED ALIPHATIC COMPOUNDS

Perfluorocarbons and Hydrofluorocarbons

Properties. Aliphatic PFCs have an unusual combination of physical properties relative to their hydrocarbon counterparts. The volatilities of PFCs are much higher than expected based on their molecular weights. Perfluorocarbons containing up to four carbon atoms boil somewhat higher than the corresponding hydrocarbons; the reverse is true of PFCs with more carbon atoms. Liquid PFCs are two to three times as dense as hydrocarbons with the same carbon skeleton, and aliphatic PFCs have among the lowest dielectric constants, refractive indexes, and surface tensions of any liquids at room temperature. The physical properties of some fluorinated aliphatic compounds are listed in Table 1. Hydrofluorocarbons invariably have higher refractive indexes, dielectric constants, and surface tensions, but lower densities than their PFC counterparts.

Manufacture. The direct fluorination of hydrocarbons with elemental fluorine is extremely exothermic and difficult to control. Special methods including metal packing techniques, jet reactors, and high dilution have been developed to control the reaction, but currently they have limited industrial importance. Poly(carbon monofluoride), $(CF)_x$ is one product that is made commercially by direct fluorination (of graphite). The disadvantages of direct fluorination have been overcome by the use of fluorine carriers, in particular, high valence metal fluorides such as cobalt trifluoride. Aliphatic PFCs are made commercially also by the electrolysis of hydrocarbons in anhydrous HF.

Table 1. Physical Properties of Fluorinated Aliphatic Compounds

Designation	Formula	Mol wt	Bp, °C	Mp, °C	Liquid density, g/mL at °C
PFC-14	CF_4	88.0	−128.1	−183.6	1.613_{-130}
PFC-116	CF_3CF_3	138.02	−78.2	−100.6	1.600_{-80}
CFC-11	CCl_3F	137.36	23.8	−111	1.476_{25}
CFC-12	CCl_2F_2	120.91	−29.8	−158	1.311_{25}
CFC-113	$CClF_2CCl_2F$	187.38	47.6	−35	1.565_{25}
HCFC-22	$CHClF_2$	86.47	−40.8	−160	1.194_{25}
HCFC-123	$CHCl_2CF_3$	152.93	28.7	−107	1.475_{15}
HCFC-141b	CCl_2FCH_3	116.95	32	−103.5	1.250_{10}
HFC-23	CHF_3	70.01	−82.2	−155.2	1.442_{-80}
HFC-134a	CH_2FCF_3	102.03	−26.5	−101	1.21_{25}
HFC-152a	CH_3CHF_2	66.05	−25.8	−117	1.023_{-30}
H-1301	$CBrF_3$	148.92	−57.8	−168	1.538_{25}
H-1211	$CBrClF_2$	165.37	−3.9	−161	1.850_{15}

Health and Safety Factors. Completely fluorinated alkanes are essentially nontoxic. However, some fluorochemicals, especially functionalized derivatives and fluoroolefins, can be lethal. Monofluoroacetic acid and perfluoroisobutylene are notoriously toxic.

Uses. The chemical inertness, thermal stability, low toxicity, and nonflammability of PFCs coupled with their unusual physical properties suggest many useful applications. However, the high cost of raw materials and manufacture has limited commercial production to a few, small-volume products. Carbon tetrafluoride and hexafluoroethane are used for plasma, ion-beam, or sputter etching of semiconductor devices (see ION IMPLANTATION). Hexafluoroethane and octafluoropropane have some applications as dielectric gases, and perfluorocyclobutane is used in minor amounts as a dielectric fluid. Perfluoro-1,3-dimethylcyclohexane is used as an inert, immersion coolant for electronic equipment, and perfluoro-2-methyldecalin is used for pin-hole leak testing of encapsulated electronic devices. Perfluoroperhydrophenanthrene has several diverse applications, ranging from a vapor-phase soldering agent for fabrication of printed circuits to a substitute for internal eye fluid in remedial eye surgery.

Chlorofluorocarbons and Hydrochlorofluorocarbons

Properties. The physical properties of aliphatic fluorine compounds containing chlorine are similar to those of the PFCs or HFCs. They usually have high densities and low boiling points, viscosities, and surface tensions. The irregularity in the boiling points of the fluorinated methanes, however, does not appear in the chlorofluorocarbons. Their boiling points consistently increase with the number of chlorines present.

Manufacture. The most important commercial method for manufacturing CFCs and HCFCs is the successive replacement of chlorine by fluorine using hydrogen fluoride.

Economic Aspects. Trichlorofluoromethane, dichlorodifluoromethane, and trichlorotrifluoroethane account for over 95% of the total production. Between 1986 and 1991 the production of CFCs has decreased dramatically owing to global adherence to the provisions of the Montreal Protocol and eventually will be phased out entirely. Estimates of the distribution by use in 1986 and subsequent reductions in use are shown in Table 2.

Health and Safety Factors. The toxicity of aliphatic CFCs and HCFCs generally decreases as the number of fluorine atoms increases, but there are exceptions as in the case of 141b vs 142b. Also, some derivatives like HCFC-132b can have low acute but high chronic toxicities.

Montreal Protocol. In response to the growing scientific consensus that CFCs and Halons would eventually deplete the ozone layer, the United Nations Environmental Programme (UNEP) began negotiations in 1981 aimed at protecting the ozone layer. In September

Table 2. Worldwide Estimates of CFC Use by Industry, 1991 vs 1986

Application	1986 Total uses, %	Reduction since 1986, %
propellants	28	58
refrigerants	23	7
cleaning	21	41
foam blowing agents	26	35
other uses	2	

of 1987, the Montreal Protocol on Substances that Deplete the Ozone Layer was signed by 24 nations and took force on January 1, 1989. This treaty called for *(1)* limiting production of specified CFCs, including 11, 12, 113, 114, and 115, to 50% of 1986 levels by 1998; *(2)* freezing production of specified Halons 1211, 1301, and 2402 at 1986 levels starting in 1992; and *(3)* convening the signatories yearly to reevaluate and update the Protocol articles in light of recent developments.

In April of 1991, the U.S. National Aeronautics and Space Administration concluded that ozone depletion was occurring even faster than has been estimated, and at the third meeting of the parties to the Montreal Protocol in June of 1991, an earlier phaseout of controlled substances was proposed.

Chlorofluorocarbon Alternatives

Properties. The ideal substitute should have identical or better performance properties than the CFC it replaces, must not harm the ozone layer, and must have a short atmospheric lifetime to ensure a low greenhouse warming potential (GWP). It also must be nontoxic, nonflammable, thermally and chemically stable under normal use conditions, and manufacturable at a reasonable price. The chemical industry has found substitutes that match many but not all of these criteria.

The general strategy has been to incorporate at least one hydrogen atom in the proposed CFC substitute's structure which provides a means for its destruction via hydrogen atom abstraction by tropospheric hydroxyl radicals. The alternatives that have been identified for the various markets are listed in Table 3.

The HCFCs have relatively small but non-zero ozone depletion potentials (ODP) and low global warming potentials (GWP). Recent results indicate even these values may be 15% too high. The HFCs have zero ODPs and low-to-moderate GWPs. Perfluorocarbons also have zero ODPs, but very large GWPs.

Manufacture. The manufacture of CFC alternatives is a far more complex challenge than production of the CFCs themselves. The very design feature which makes the alternatives tropospherically labile, the hydrogen atom substituent, also significantly complicates their manufacture because of potential by-product formation or catalyst inactivation. At least a dozen different routes to HFC-134a have been identified, but a simple, single-step process is very unlikely. A two-step process that has been commercialized first involves reaction of trichloroethylene with HF in the vapor or liquid phase to form HCFC-133a, which is then separated and reacts again with HF to form HFC-134a.

Table 3. Alternatives to CFCs

CFC	Application	Near-term substitute	Long-term substitute
CFC-11	blowing agents and refrigerants	HCFC-123 HCFC-22 HCFC-141b HCFC-142b	HFCs HFC-152a blends
CFC-12	refrigerants	HFC-134a HCFC-22	HFC-134a HFC-152a blends
CFC-113	cleaning agents	blends/azeotropes HCFC-225ca/cb	HFCs
CFC-114	blowing agents and refrigerants	HCFC-124 HCFC-142b blends/azeotropes	HFCs
CFC-115	refrigerants	HFC-125 blends/azeotropes	HFC-125
H-1301	fire extinguishant	HFC-23 HFC-125 PFC-31-10	HFC-23 HFC-227ea
H-1211	fire extinguishant	HCFC-123 HBFC-22B1 HBFC-124B1 PFC-51-14	HFCs

Since HFC-134a likely will be the single largest-volume CFC alternative produced, many manufacturers around the world are in the process of or have plans to commercialize it, each under their own trade name.

Economic Aspects. Manufacturing facilities for CFC alternatives are just now coming on line. The size of the markets for the alternatives is estimated to be quite large (several thousand t/yr), but it will not be as large as the prior markets for CFCs themselves. This is largely because of the higher cost of the alternatives, typically 3–5 times that of the incumbents.

Health and Safety Factors. The toxicity of CFC alternatives is the subject of intense study. Fifteen fluorocarbon producers have formed the Program for Alternative Fluorocarbon Toxicity testing (PAFT) to share the costs associated with determining safe operating and handling procedures for the proposed CFC alternatives.

Hydrofluorocarbons generally are less toxic than HCFCs, with the notable exception of HFC-152, CH_2FCH_2F, which apparently can be metabolically converted to monofluoroacetic acid and is therefore quite toxic.

Fluorocarbons Containing Other Halogens

Properties. The physical and chemical properties of bromo- and iodofluorocarbons are similar to those of the chlorofluorocarbons except for higher densities and generally decreased stability. The stability of these compounds decreases as the ratio of bromine or iodine to fluorine increases. Iodofluorocarbons and most bromofluorocarbons readily lose iodine or bromine radicals under photolytic, thermal, or radical initiation to give the corresponding carbon-centered radical.

Manufacture. Brominated fluoromethanes are prepared industrially by the halogen exchange of tetrabromomethane or by the bromination of CH_2F_2 or CHF_3 at elevated temperatures. Other bromo- or iodofluorocarbons can be prepared by halogenating suitable fluorocarbons, including fluoroolefins, or by halogen exchange of perfluoroiodocarbons. The higher molecular weight perfluoroalkyl iodides are prepared by telomerization of tetrafluoroethylene with lower molecular weight perfluoroalkyl iodides.

Health and Safety Factors. Fluorocarbons containing bromine or iodine are more toxic than the corresponding chlorocompounds. When the ratio of the fluorine to other halogens is high, the toxicity can be quite low, especially for bromofluorocarbons.

Uses. The most important industrial products of this class have been the fire-extinguishing agents $CBrClF_2$ and $CBrF_3$.

BRUCE E. SMART
RICHARD E. FERNANDEZ
E. I. du Pont de Nemours & Co., Inc.

R. E. Banks, B. E. Smart, and J. C. Tatlow, eds., *Organofluorine Chemistry, Principles and Commercial Applications*, Plenum Press, New York, 1994.

FLUOROETHANOLS

Monofluoro Derivative

2-Fluoroethanol (ethylene fluorohydrin, β-fluoroethyl alcohol), FCH_2CH_2OH, is a colorless liquid with an alcohol-like odor; mp, −26.45 °C; bp, 103.55°C; d_4 1.1297. It is miscible with water, stable to distillation, and low in flammability. It is the least acidic of the fluoroethanols, although more acidic than ordinary alcohols. Its most notable difference from the other fluoroethanols is its extreme toxicity. In its chemical reactions, 2-fluoroethanol behaves like a typical alcohol. 2-Fluoroethanol is not currently produced in commercial quantities.

Because of its high toxicity, special procedures should be followed by users of 2-fluoroethanol. Another potential hazard is the formation of the alcohol as a minor by-product in reactions such as those involving boron trifluoride and ethylene oxide. Despite these problems, several potential uses for the alcohol and its derivatives have been reported. The alcohol has been used to control rodent populations and, when labeled with ^{18}F, as a radiodiagnostic agent. Various derivatives have shown promise as herbicides or as agents to control mites and other plant pests.

Difluoro Derivative

2,2-Difluoroethanol, F_2CHCH_2OH, is a colorless liquid with an alcohol-like odor; mp, 28.2°C, bp, 96°C. It is stable to distillation and miscible with water and many organic solvents.

2,2-Difluoroethanol is prepared by the mercuric oxide-catalyzed hydrolysis of 2-bromo-1,1-difluoroethane with carboxylic acid esters and alkali metal hydroxides in water. Its chemical reactions are similar to those of most alcohols. 2,2-Difluoroethyl difluoromethyl ether, made from the alcohol and chlorodifluoromethane in aqueous base, has been investigated as an inhalation anesthetic. Methacrylate esters of the alcohol are useful as a sheathing material for polymers in optical applications. This alcohol has also been reported to be useful as a working fluid in heat pumps.

Trifluoroethanol

2,2,2-Trifluoroethanol, CF_3CH_2OH, is a colorless liquid with an ethanol-like odor; mp, −45 °C; bp, 73.°C, and dielectric constant (25°C), 26.14. It is the most acidic fluoroethanol. It is stable to distillation and miscible with water and many organic solvents. It has the unusual property of dissolving most polyamides, both nylons and polypeptides, at room temperature. Because of its excellent combination of physical and thermodynamic properties, 2,2,2-trifluoroethanol-water mixtures (also known as fluorinols) have application as working fluids in Rankine cycle engines for recovering energy from waste heat sources.

Chemically, 2,2,2-trifluoroethanol behaves as a typical alcohol. Its alkoxides react with bromoethane to give trifluoroethyl ethyl ether, bp 50.3°C. Similarly prepared is bis(trifluoroethyl) ether used as the convulsant drug Flurothyl as a substitute for electric shock therapy. As the trichlorosulfonate ester, trifluoroethanol is used to introduce the trifluoroethyl group into the anxiolytic drug Halazepam. 2,2,2-Trifluoroethanol is also the starting material for the anesthetic Isoflurane (1-chloro-2,2,2-trifluoroethyl difluoromethyl ether) and Desflurane (2-difluoromethoxy-1,1,1,2-tetrafluoroethane).

Trifluoroethanol was first prepared by the catalytic reduction of trifluoroacetic anhydride. The largest producer of trifluoroethanol is Halocarbon Products Corp. Toxicity studies on trifluoroethanol show acute oral LD_{50}, 240 mg/kg; acute dermal LD_{50}, 1680 mg/kg; and acute inhalation $L(ct)_{50}$, 4600 ppm·h.

ARTHUR J. ELLIOT
Halocarbon Products Corporation

F. L. M. Pattison, *Toxic Aliphatic Fluorine Compounds*, Elsevier Publishing Co., New York, 1959, p. 65.

F. L. M. Pattison and co-workers, *J. Org. Chem.* **21**, 739 (1956).

K. N. Campbell, J. O. Knobloch, and B. K. Campbell, *J. Am. Chem. Soc.* **72**, 4380 (1950).

FLUOROETHERS AND FLUOROAMINES

Perfluoroaliphatic ethers and perfluorotertiary amines together with the perfluoroalkanes and cycloalkanes comprise a class of extremely unreactive materials known in the industry as inert fluids. These fluids are colorless, odorless, essentially nontoxic, nonflammable, dense, and extremely nonpolar. In the electronics industry, the lower molecular weight compounds find application in the areas of heat transfer, testing, and vapor-phase soldering. Higher molecular weight polymers and oligomers are used in a variety of applications, including hazardous duty vacuum pump fluids, specialty greases, and various specialty cosmetics and lubricants.

Many perfluoroaliphatic ethers and tertiary amines have been prepared by electrochemical fluorination, direct fluorination using elemental fluorine, or, in a few cases, by fluorination using cobalt trifluoride.

Physical Properties

Perfluorinated compounds boil at much lower temperatures and have lower heats of vaporization than the corresponding hydrocarbon analogues even though they have considerably higher molecular weights (Table 1). Many of the unusual properties of the perfluorinated inert fluids are the result of the extremely low intermolecular interactions.

Thermal Stabilities. The perfluoroethers have thermal stabilities comparable to these of the perfluoroalkanes.

Electrical Properties. The low polarizability of perfluorinated liquids makes them excellent insulators.

Chemical Properties

The inert character of the perfluoroethers and tertiary amines is demonstrated by their lack of basicity or reactivity as compared with their hydrocarbon analogues. Both classes of compounds are nonflammable.

Solvent Properties. In comparison to the more familiar hydrocarbon systems, the solvent properties of the perfluorinated inert liquids are also unusual owing to their nonpolar nature and low intermolecular forces. They are generally very poor solvents for most organic compounds.

Economic Aspects

Information on the production levels of the perfluoroethers and perfluorotertiary amines is not disclosed, but the products are

Table 1. Physical Properties of Some Perfluorinated Liquids

Name	Bp, °C	d_4^{25}	d_D^{24}
perfluoro-4-methylmorpholine	51	1.70	1.267
perfluoro-2-ethyltetrahydrofuran	56	1.69	1.263
perfluorohexane	58	1.68	1.252
Galden HT 70[a]	70	1.73	
perfluorotriethylamine	71	1.73	1.262
perfluoro-4-ethylmorpholine	72	1.74	1.273
perfluoro-4-isopropylmorpholine	95	1.79	1.283
perfluorobutyl ether	102	1.71	1.261
Fluorinert FC-75	103	1.76	1.276
perfluorooctane	103	1.77	1.272
perfluorononane	123	1.80	1.276
perfluorotripropylamine	130	1.82	1.279
perfluorobis(2-butoxyethoxy)methane	178	1.76	
perfluorotributylamine	178	1.86	1.291
perfluoro(diethylamino)ethyl ether	178		
perfluorohexyl ether	181	1.81	1.278
K7 fluid	250	1.82	

[a] A mixture of perfluorinated polyethers marketed by Montefluos.

available commercially and are marketed, for instance, as part of the Fluorinert Electronic Liquids family by 3M Co.

Environmental, Health, and Safety Factors

Over the years animal studies have repeatedly shown that perfluorinated inert fluids are nonirritating to the eyes and skin and practically nontoxic by ingestion, inhalation, or intraperitoneal injection. Perfluorinated ethers and perfluorinated tertiary amines do not contribute to the formation of ground level ozone and are exempt from VOC regulations.

RICHARD M. FLYNN
3M Company

T. Abe and S. Nagase, in R. E. Banks, ed., *Preparation, Properties and Industrial Applications of Organofluorine Compounds*, Ellis Horwood, Chichester, U.K., 1982, p. 19.

W. V. Childs and co-workers, in H. Lund and M. M. Baizer, eds., *Organic Electrochemistry*, 3rd ed., Marcel Dekker, Inc., New York, 1991, p. 1103.

Fluorinert Liquids, technical notebook, 3M Co., St. Paul, Minn., 1987.

PERFLUOROEPOXIDES

Almost all the work on perfluoroepoxides has been with three compounds: tetrafluoroethylene oxide (TFEO), hexafluoropropylene oxide (HFPO), and perfluoroisobutylene oxide (PIBO). Most of this work has dealt with HFPO, the most versatile and by far the most valuable of this class of materials.

Physical Properties

In general, the perfluoroepoxides have boiling points that are quite similar to those of the corresponding fluoroalkenes. Little physical property data concerning these compounds have been published.

Chemical Properties

There are three general reactions of perfluoroepoxides: pyrolyses (thermal reactions), electrophilic reactions, and by far the most important, reactions with nucleophiles and bases.

Preparation

A large number of methods have been used to prepare perfluoroepoxides. All of these methods must contend with the great chemical reactivity of the epoxide product, especially with subsequent ionic and thermal reactions which result in the loss of the desired epoxide. The reaction of perfluoroalkenes with alkaline hydrogen peroxide is a good general method for the preparation of the corresponding epoxides, with the exception of the most reactive of the series, TFEO.

Reaction of perfluoroalkenes and hypochlorites has been shown to be a general synthesis of perfluoroepoxides. This appears to be the method of choice for the preparation of epoxides from internal fluoroalkenes. The direct oxidation of fluoroalkenes is also an excellent general synthesis procedure for the preparation of perfluoroepoxides.

PAUL R. RESNICK
E. I. du Pont de Nemours & Co., Inc.

P. Tarrant and co-workers, *Fluorine Chem. Rev.* **5**, 77 (1971).

H. Millauer, W. Schwertfeger, and G. Siegemund, *Angew. Chem. Int. Ed. Engl.* **24**, 161 (1985).

L. F. Sokolov, P. I. Valov, and S. V. Sokolov, *Usp. Khim.* **53**, 1222 (1984).

Eur. Pat. 64,293 (Dec. 10, 1986), M. Ikeda, M. Miura, and A. Aoshima (to Asahi Kasei Kogyo K.K.).

FLUORINATED ACETIC ACIDS

Fluoroacetic acid, FCH_2COOH, is noted for its high toxicity to animals, including humans. It is sold in the form of its sodium salt as a rodenticide and general mammalian pest control agent. The acid has mp, 33°C; bp, 165°C. The acid is the toxic constituent of a South African plant *Dichapetalum cymosum*, better known as gifblaar. At least 24 other poisonous plant species are known to contain it.

Chemically, fluoroacetic acid behaves like a typical carboxylic acid, although its acidity is higher ($K_a = 2.2 \times 10^{-3}$) than the average. It can be prepared from the commercially available sodium salt by distillation from sulfuric acid.

Sodium Fluoroacetate. Sodium fluoroacetate, FCH_2COONa, known as Compound 1080, is a hygroscopic white solid, mp, 200–202°C. Its solubility at 25°C in g/100 g solvent is water, 111.

Sodium fluoroacetate is usually made by displacing the halogen from an ester of bromo- or chloroacetic acid with potassium fluoride or, in one instance, antimony fluoride, followed by hydrolysis with aqueous sodium hydroxide.

Toxicity. Sodium fluoroacetate is one of the most effective all-purpose rodenticides known. However, it is extremely dangerous to humans, to common household pets, and to farm animals, and should only be used by experienced personnel. The rodent carcasses should be collected and destroyed because they remain poisonous for a long period of time to any animal that eats them.

One characteristic of fluoroacetate toxicity is the wide range in lethal doses for different species ranging from (LD_{50}, mg/kg) 0.06 in dogs, 0.2 in cats, 0.4 in sheep or rabbits, 2–10 in humans, 5 in rats, 7 in mice, to about 400 in toads. The only suggested antidotes for the poisoning are 1,2,3-propanetriol monoacetate, acetamide, and other acetate donors, but these only have an effect if administered before significant amounts of fluoroacetate have been converted to fluorocitrate.

Fluoroacetamide. Fluoroacetamide, FCH_2CONH_2, is a white water-soluble solid having mp 108°C. It has been used as a rodenticide and has been reported to have a better acceptability to rats than sodium fluoroacetate. However, like the latter compound, its misuse has caused deaths to farm animals and pets.

Tull Chemical Co. (Oxford, Ala.) is the only producer of sodium fluoroacetate. It is usually packed in 8-oz (227-g) or 5-kg cans and is almost exclusively exported.

Difluoroacetic Acid

Difluoroacetic acid, $F_2CHCOOH$, is a colorless liquid with a sharp odor; mp, 35°C; bp, 134°C. Difluoroacetic acid undergoes reactions typical of a carboxylic acid. The acid can be synthesized in several different ways. The reaction of tetrafluoroethylene with ammonia to give 2,4,6-tris(difluoromethyl)-s-triazine, followed by its alkaline hydrolysis, has been reported to give the acid in 80% overall yield.

Difluoroacetic acid is much less toxic than fluoroacetic acid (LD_{50} = 180 mg/kg mouse iv).

Trifluoroacetic Acid

Physical Properties. Trifluoroacetic acid, CF_3COOH, is a colorless liquid with a sharp odor resembling that of acetic acid. Its physical properties are shown in Table 1. It is a strong carboxylic acid. It is miscible with water, fluorocarbons, and most common organic solvents.

Chemical Properties. Trifluoroacetic acid undergoes reactions typical of a carboxylic acid.

Preparation. Because of its stability to further oxidation, trifluoroacetic acid can be prepared by the oxidation of compounds containing a trifluoromethyl group bonded to carbon.

Health and Safety Factors. Unlike fluoroacetic acid, trifluoroacetic acid presents no unusual toxicity problems. However, owing to its strong acidity, its vapors can be irritating to tissue, and the liquid acid can cause deep burns if allowed to contact the skin. The acid

Table 1. Physical Properties of Trifluoroacetic Acid

Property	Value
freezing point, °C	−15.36
boiling point, °C	71.8
density at 25°C, g/mL	1.4844
viscosity at 25°C, mPa·s(= cP)	0.813
dielectric constant at 25°, ϵ	42.1

can be safely stored in containers made of glass or common corrosion-resistant alloys and metals such as stainless steel or aluminum.

Economic Aspects. Halocarbon Products Corp. is the largest producer of trifluoroacetic acid.

ARTHUR J. ELLIOTT
Halocarbon Products Corporation

F. L. M. Pattison, *Toxic Aliphatic Fluorine Compounds*, Elsevier Publishing Co., New York, 1959, p. 16.

FLUORINATED HIGHER CARBOXYLIC ACIDS

Perfluorinated carboxylic acids are corrosive liquids or solids. The acids are completely ionized in water. The acids are of commercial significance because of their unusual acid strength, chemical stability, high surface activity, and salt solubility characteristics. The perfluoroalkyl acids with six carbons or less are liquids; the higher analogues are solids (Table 1).

Preparation

There are five methods for the preparation of long-chain perfluorinated carboxylic acids and derivatives: electrochemical fluorination, direct fluorination, telomerization of tetrafluoroethylene, oligomerization of hexafluoropropylene oxide, and photooxidation of tetrafluoroethylene and hexafluoropropylene.

Derivatives

In general, the reactions of the perfluoro acids are similar to those of the hydrocarbon acids. Salts are formed with the ease expected of strong acids. The metal salts are all water soluble and much more soluble in organic solvents than the salts of the corresponding hydrocarbon acids. Esterification takes place readily with primary and secondary alcohols. Acid anhydrides can be prepared by distillation of the acids from phosphorus pentoxide. The amides are readily prepared by the ammonolysis of the acid halides, anhydrides, or esters and can be dehydrated to the corresponding nitriles.

The ammonium salts, $C_nF_{2n+1}COONH_4$, where n equals 7 and larger, are particularly useful as emulsifiers in the polymerization of fluorinated olefin monomers such as tetrafluoroethylene or vinylidene fluoride.

Table 1. Properties of Perfluoroalkylcarboxylic Acids[a]

Acid	Bp, °C	Mp, °C	Density at 20°C, g/mL
perfluoropropanoic	96		1.561
perfluorobutanoic	120	−17.5	1.641
perfluoropentanoic	139		1.713
perfluorohexanoic	157		1.762
perfluorocyclohexane carboxylic[b]	168		1.789
perfluoroheptanoic	175		1.792
perfluorooctanoic	189	52–54	1.792
perfluorodecanoic	218		
perfluorotetradecanoic	270		

[a] Except where noted.
[b] Cyclo-$C_6F_{11}COOH$.

Table 2. Properties of Perfluoralkanedicarboxylic Acids

n	Acid	Mp, °C	Bp, $°C_{kPa}$[a]	Bp of ester,[b] $°C_{kPa}$[a]
1	perfluoromalonic	117		$58–58_{1.2}$
2	perfluorosuccinic	115–116	150_2	173
3	perfluoroglutaric	78–88	$134_{0.4}$	$100_{4.5}$
4	perfluoroadipic	134		$108–110_4$
6	perfluorosuberic	154–158		$156–159^c{}_{3.6}$
8	perfluorosebacic			$102–113_{0.005}$
12	perfluorotetradecanedioic	191		

[a] To convert kPa to mm Hg, multiply by 7.5
[b] Methyl ester unless otherwise noted.
[c] Ethyl ester.

Perfluorodicarboxylic Acids

The lowest members of the series of perfluoroalkanedicarboxylic acids have been prepared and are stable compounds. They have been synthesized by oxidation of the appropriate chlorofluoroolefin as well as by electrochemical fluorination and direct fluorination. Perfluoromalonic acid is an oxidation product of $CH_2{=}CHCF_2CH{=}CH_2$. Perfluorosuccinic acid has been produced by oxidation of the appropriate olefin or by electrochemical fluorination of succinyl chloride or butyrolactone and subsequent hydrolysis. Table 2 lists some typical properties of perfluoroalkanedicarboxylic acids and their esters.

Toxicology and Safety. Because of their strong acidity, the perfluorinated carboxylic acids themselves are corrosive to the skin and eyes.

The salts of the perfluorinated acids are not corrosive, so one is in a better position to discuss toxicity not related to corrosivity. The toxicity of the salts varies depending on the exact structure. The ammonium salt of perfluorooctanoic acid is nonirritating to the skin and moderately irritating to the eyes. Its oral toxicity is rated at moderate; the LD_{50} is 540 mg per kg of body weight. There has been some concern in the past that ammonium perfluorooctanoate was teratogenic. More recent results indicate that it is neither embryotoxic nor teratogenic. Although ammonium perfluorooctanoate was fed to albino rats for two years, no compound-induced carcinogenicity was found in the study. There were statistically significant compound-related benign testicular tumors. Prolonged or repeated exposure can cause liver damage which results in jaundice or tenderness of the upper abdomen. The dust from the ammonium salts of the perfluorinated acids is irritating to breathe, and they should only be handled in a well-ventilated area, or preferably under a hood.

PATRICIA M. SAVU
3M Company

A. M. Lovelace, W. Postelnek, and D. A. Rausch, *Aliphatic Fluorine Compounds*. ACS monograph 138 Reinhold Publishing Co., New York, 1958.

H. C. Fielding, in R. E. Banks, ed., *Organofluorine Chemicals and their Application*, Ellis Howard, 1979, p. 216.

U.S. Pat. 3,132,185 (May 5, 1964), R. E. Parsons (to E. I. du Pont de Nemours and Co., Inc.); U.S. Pat. 3,226,449 (Dec. 28, 1965), W. A. Blanchard and J. C. Rhode (to E. I. du Pont de Nemours and Co., Inc.); U.S. Pat. 3,234,294 (Feb. 8, 1966), R. E. Parsons (to E. I. du Pont de Nemours and Co., Inc.); U.S. Pat. 4,425,199 (Jan. 10, 1984), M. Hamada, J. Ohmura, and F. Muranaka (to Asahi Kasaei Kogyo Kabushi Kaisha).

H. Millauer, W. Schwertfeger, and G. Siegemund, *Ang. Chem. Int. Ed. Eng.* **24**, 161–179 (1985).

PERFLUOROALKANESULFONIC ACIDS

Perfluoroalkanesulfonic acids and their derivatives are of commercial significance because of their unusual acid strength, chemical stability, and the surface activity of the higher members of the series (eight carbons and larger).

Table 1. Boiling Points of Perfluoroalkanesulfonic Acids

Compound	Bp, °C/kPa[a]	Bp, °C[b]
CF_3SO_3H	60/0.4	166
$C_2F_5SO_3H$	81/2.9	175[c]
$C_4F_9SO_3H$	76–84/0.017	200[c]
$C_5F_{11}SO_3H$	110/0.67[d]	212[c,e]
$C_6F_{13}SO_3H$	95/0.47	225[c]
$C_8F_{17}SO_3H$	133/0.8	249
4-CF_3(cyclo-C_6F_{10})SO_3H	120/0.4	241
4-C_2F_5(cyclo-C_6F_{10})SO_3H		254

[a] To convert kPa to mm Hg, multiply by 7.5. [b] At 101.3 kPa = 1 atm. [c] Estimated. [d] The hydrate, $C_5F_{11}SO_3H{\cdot}H_2O$. [e] $C_5F_{11}SO_3H$ anhydrous.

Preparation

The synthetic operations employed when the perfluoroalkanesulfonic acid is derived from electrochemical fluorination, which is the best method of preparation, are shown in equations 1–3.

$$R_hSO_2F + HF \rightarrow R_fSO_2F + H_2 \quad (1)$$

where R_h is an alkyl group and R_f is a perfluoroalkyl group

$$R_fSO_2F + KOH \rightarrow R_fSO_3K + HF \quad (2)$$

$$R_fSO_3K + H_2SO_4 \rightarrow R_fSO_3H + KHSO_4 \quad (3)$$

The boiling points of a series of perfluoroalkanesulfonic acids are listed in Table 1.

Trifluoromethanesulfonic Acid

The first member of the series, CF_3SO_3H, has been extensively studied. Trifluoromethanesulfonic acid is a stable, hydroscopic liquid which fumes in air. Addition of an equimolar amount of water to the acid results in a stable, distillable monohydrate, mp 34°C, bp 96°C at 0.13 kPa (1 mm Hg).

Trifluoromethanesulfonic acid is miscible in all preparations with water and is soluble in many polar organic solvents such as dimethylformamide, dimethylsulfoxide, and acetonitrile.

Alkyl esters of trifluoromethanesulfonic acid, commonly called triflates, have been prepared from the silver salt and an alkyl iodide, or by reaction of the anhydride with an alcohol. Triflates are among the best leaving groups known, so they are commonly employed in anionic displacement reactions.

The metallic salts of trifluoromethanesulfonic acid can be prepared by reaction of the acid with the corresponding hydroxide or carbonate or by reaction of sulfonyl fluoride with the corresponding hydroxide. The lithium salt of trifluoromethanesulfonic acid, CF_3SO_3Li, commonly called lithium triflate, is used as a battery electrolyte in primary lithium batteries because solutions of it exhibit high electrical conductivity, and because of the compound's low toxicity and excellent chemical stability.

Trifluoromethanesulfonic acid anhydride, bp 84°C, is prepared by refluxing the acid over an excess of phosphorous pentoxide. Trifluoromethanesulfonic acid is available from the 3M Company as Fluorochemical Acid FC-24; the lithium salt is available as Fluorochemical Specialties FC-122, FC-123, and FC-124.

Higher Perfluoroalkanesulfonic Acids

These acids show the same general solubilities as trifluoromethanesulfonic acid, but are insoluble in benzene, heptane, carbon tetrachloride, and perfluorinated liquids. All of the higher perfluoroalkanesulfonic acids have been prepared by electrochemical fluorination.

The longer-chain acids and their salts, particularly $C_8F_{17}SO_3H$ and higher are surface-active agents in aqueous media. Generally, derivatives of the longer-chain perfluoroalkanesulfonic acids have a number of unique surface-active properties and have formed a basis for a number of commercial products. Derivatives of *N*-alkyl perfluorooctanesulfonamidoethanol, $C_8F_{17}SO_2N(R)CH_2CH_2OH$, and polymers of *N*-alkyl perfluorooctanesulfonamidoethyl methacrylate, $C_8F_{17}SO_2N(R)CH_2CH_2OCOC(CH_3){=}CH_2$, impart soil, oil, and water repellency to treated fabrics and paper; this forms the basis for 3M's Scotchguard and Scotchban products.

Higher perfluoroalkanesulfonates are slightly more reactive than triflates toward nucleophilic displacements.

Difunctional Perfluoroalkanesulfonic Acids

Alpha, omega-perfluoroalkanedisulfonic acids were first prepared by aqueous alkali permanganate oxidation of the bis-sulfone, $RSO_2(CF_2CF_2)_nSO_2R$. Carbonyl sulfonyl fluorides of the formula $FCO(CF_2)_nSO_2F$ have been prepared by electrochemical fluorination of hydrocarbon sultones. Fluorosulfonyldifluoracetyl fluoride is an important industrial intermediate used in the production of DuPont's Nafion ion-exchange membrane.

PATRICIA SAVU
3M Center

T. Abe and S. Nagase, in R. E. Banks, ed., *Preparation, Properties, and Industrial Applications of Organofluorine Compounds*, Ellis Howard, 1982, p. 37.

R. D. Howells and J. D. McCown, *Chem. Rev.* **77**, 69 (1977).

P. J. Stang, M. Hanack, and L. R. Subramanian, *Synthesis*, 85 (1982).

R. E. Banks, ed., *Preparation, Properties, and Industrial Applications of Organofluorine Compounds*, Ellis Howard, 1982, p. 37.

FLUORINATED AROMATIC COMPOUNDS

Aromatic fluorine compounds have been known for nearly a century, but numerous applications have surfaced only in recent years. The special properties conferred by fluorine justify the higher costs required to produce fluoroaromatics. The unusual physiochemical and biological properties that fluorine imparts to aromatics result from the small size of fluorine (it is bioisosteric with both the hydrogen atom and the hydroxyl group) and from its striking electronic properties, including high electronegativity and the ability to alter polarity of adjacent groups, as well as to donate electrons by resonance. Other significant properties are the enhanced stability of the C–F bond in the absence of activating groups, high lipid solubility, hydrogen-bonding potential (acceptor role), and enzyme inhibition. The carbon–fluorine link in fluoroaromatics has been of considerable value as a label for metabolic, mechanistic, and structural studies.

Fluorine-containing aromatics have been incorporated into drugs (hypnotics, tranquilizers, antiinflammatory agents, analgesics, antibacterials, etc) and into crop protection chemicals (herbicides, insecticides, fungicides). Liquid crystals, positron emission tomography, and imaging systems are newer use areas for fluoroaromatics and fluoroheterocyclics.

For fluorine-free products, the lability of fluorine in fluoronitrobenzenes and other activated molecules permits it to serve as a handle in hair-dye manufacturing operations, high performance polymers such as polyetheretherketone (PEEK), production of drugs such as diuretics, and fiber-reactive dyes. Labile fluorine has also been used in analytical applications and biological diagnostic reagents.

Preparative Methods

Ring-Fluorinated Aromatics and Heterocyclics. Recent advances in synthetic methods include discoveries of new fluorinating agents and modifications of known methods. Some of these efforts were stimulated by objectives to effect selective fluorination of natural or biologically active compounds. The need to prepare ^{18}F-labeled pharmaceuticals for use in positron emission tomography also accelerated the need for improved aromatic fluorination techniques. They include substitutive-aromatic fluorination, diazotization routes, exchange fluorination, saturation–rearomatization, and fluoroaliphatic thermolytic routes.

Side-Chain Fluorinated Aromatics and Heterocyclics. Benzotrifluorides generally are prepared from trichloromethylaromatics with metal fluorides or hydrogen fluoride. Industrial processes feature reaction with hydrogen fluoride under high pressure, atmospheric pressure, or vapor-phase conditions. A potential simplification is the single-step conversion of toluene to benzotrifluoride employing chlorine–hydrogen fluoride (CCl_4 diluent, 460°C).

Benzotrifluorides also are prepared from aromatic carboxylic acids and their derivatives with sulfur tetrafluoride (SF_4).

A significant technical advance features perfluoroalkylation of aromatics (devoid of electron-withdrawing groups) with carbon tetrachloride–hydrogen fluoride to give high selectivity of benzotrifluorides.

The economically attractive oxidative fluorination of side chains in aromatic hydrocarbons with lead dioxide or nickel dioxide in liquid HF stops at the benzal fluoride stage.

Construction of benzotrifluorides from aliphatic feedstocks, or cyclization, represents a new technique with economic potential.

Ring-Fluorinated Benzenes

Fluorobenzene. Fluorobenzene (monofluorobenzene), C_6H_5F, has a molecular weight of 96.1, and is a colorless mobile liquid with a pleasant aromatic odor (Table 1).

Reactions. Fluorobenzene electrophilic substitution reactions are more para directing than are the same chlorobenzene reactions. The Friedel-Crafts ketone synthesis is of commercial importance in upgrading fluorobenzene for drug, polymer, and electronic applications.

The presence of activating groups, eg, *o, p* nitro groups, makes aromatic fluorine reactive in nucleophilic displacement reactions. Numerous applications have been developed based on the lability of fluoronitroaromatics.

4-Bromofluorobenzene can be selectively converted to 4-fluorophenylmagnesium bromide for subsequent incorporation into the silicon-containing fungicide, flusilazole. Boron-containing fluoroaromatics are commercially offered as laboratory reagents.

Enzymatic oxygenation of aryl fluorides without ring opening (biotransformation reaction) provides a new production tool to fluoroaromatic fine chemicals.

Manufacture. Fluorobenzene is produced by diazotization of aniline in anhydrous hydrogen fluoride at 0°C, followed by *in situ* decomposition of benzenediazonium fluoride at 20°C.

Applications. Fluorobenzene is used in crop protection chemicals, drugs, and other medical applications, liquid crystals, and dyes.

Difluorobenzenes. Interest in the commercialization of difluoroaromatics in crop protection chemicals and drugs continues to be strong. Numerous liquid crystals containing the 1,2-difluorobenzene moiety have been synthesized. Table 2 lists physical properties of commercially significant intermediates such as *o*-, *m*-, and *p*-difluorobenzene, 2,4-difluoroaniline and 2,6-difluorobenzonitrile. The LD_{50} values for the three isomeric difluorobenzenes are identical: 55 g/m^3 for 2 h (inhalation, mouse).

Table 1. Physical Properties of Fluorobenzene

Property	Value
melting point, °C	−42.22
boiling point, °C	84.73
density, 25°C, g/mL	1.0183
viscosity at 19.9°C, mPa·s(= cP)	0.585
specific heat, 25°C, J/mol[a]	146.3
solubility in water, 30°C, g/100 g	0.154

[a] To convert J to cal, divide by 4.184.

Table 2. Properties of Fluorinated Aromatic Compounds[a]

Component	Mol wt	Mp, °C	Bp, °C[b]	Refractive index, n^t_D	Specific gravity, d^t_4
	Difluorobenzenes				
$C_6H_4F_2$					
1,2-difluorobenzene	114.09	−34	91–92	1.4452^{20}	1.1496^{25}
1,3-difluorobenzene	114.09	−59.3	82–83	1.4410^{20}	1.1572^{20}
1,4-difluorobenzene	114.09	−13	88–89	1.4421^{20}	1.1716^{20}
	Trifluorobenzenes				
$C_6H_3F_3$					
1,2,3-trifluorobenzene	132.08		94–95	1.4230^{20}	1.280^{20}
1,2,4-trifluorobenzene	132.08		88	1.4230^{20}	1.264^{20}
1.3,5-trifluorobenzene	132.08	−5.5	75.5	1.4140^{20}	1.277^{20}
	Tetrafluorobenzenes				
$C_6H_2F_4$					
1,2,3,4-tetrafluorobenzene	150.08	−42	95	1.4069^{20}	1.422^{25}
1,2,3,5-tetrafluorobenzene	150.08	−48	83	1.4011^{25}	1.393^{20}
1,2,4,5-tetrafluorobenzene	150.08	4	90	1.4045^{20}	1.424^{25}
	Pentafluorobenzene				
C_6HF_5	168.07	−48	85	1.3881^{25}	1.531^{20}
	Fluorotoluenes				
2-fluorotoluene	110.13		113–114	1.4704^{25}	
3-fluorotoluene	110.13		115	1.4691^{20}	
4-fluorotoluene	110.13		116	1.4690^{20}	
2-chloro-6-fluorotoluene	144.58		155	1.5026^{20}	
	Fluoroanilines				
$C_6H_4F(NH_2)$					
2-fluoroaniline	111.12	−29	175	1.5406^{25}	
3-fluoroaniline	111.12		186	1.5445^{25}	
4-fluoroaniline	111.12	−1.9	187	1.5375^{25}	
$C_6H_3F_2(NH_2)$					
2,4-difluoroaniline	129.11	−7.5	169.5	1.5043^{25}	
$C_6H_3(Cl)(F)(NH_2)$					
3-chloro-4-fluoroaniline	145.57	44–47	227–228		
	Fluorobenzonitriles				
2,6-difluorobenzonitrile	139.11	30–32	99[c]	1.4875^{25}	

[a] Colorless unless otherwise noted. [b] At 101.1 kPa = 1 atm unless otherwise noted. [c] At 2.67 kPa = 20 mm Hg.

Hexafluorobenzene. The development of commercial routes to hexafluorobenzene included an intensive study of its derivatives. Particularly noteworthy was the development of high temperature lubricants, heat-transfer fluids, and radiation-resistant polymers.

Hexafluorobenzene. Hexafluorobenzene, C_6F_6, is a colorless liquid with a sweet odor. Physical properties of hexafluorobenzene are given in Table 3.

Manufacture. One commercial process features a three-stage saturation–rearomatization technique using benzene and fluorine gas as raw materials. Principal problems with this method are the complex nature of the process, its dependence on fluorine gas which is costly to produce, and the poor overall utilization of fluorine, because nearly one-half of the input fluorine is removed during the process. An alternative hexafluorobenzene process features exchange fluorination (KF) of hexachlorobenzene in the presence of polar solvents or under solvent-free conditions (450–540°C, autoclave).

Reactions. Hexafluorobenzene is susceptible to attack by nucleophilic agents to give pentafluorophenyl compounds of the general formula C_6F_5X.

Fluorobiphenyls. Fluorobiphenyls are incorporated into the analgesic and antiinflammatory drugs diflunisal and flurbiprofen.

Fluorinated biphenyls have been incorporated into numerous liquid crystal structures.

They can be synthesized by diazotization–fluorination, Gomberg-Bachmann arylation, or Ullmann coupling reactions.

Table 3. Physical Properties of Hexafluorobenzene

Property	Value
mol wt	186.06
melting point, °C	5.10
boiling point, °C	80.261
density, 25°C, g/mL	1.60682
refractive index, n_D^{25}	1.3761

Fluoronaphthalenes and Other Fused-Ring Fluoroaromatics

Few applications for fluoronaphthalenes and related polycyclic structures have materialized. The fused-ring bicyclic, sulindac, a monofluorinated indene-3-acetic acid, is used as an antiinflammatory agent.

2-Fluoronaphthalene is prepared in 54–67% yield from 2-naphthylamine by the Balz-Schiemann reaction or in 51% yield by pyrolysis of indene and chlorofluoromethane at 600°C.

1,4-Difluoronaphthalene is prepared from 4-fluoro-1-naphthylamine by the Balz-Schiemann reaction. It is used in chemical carcinogenesis studies as a synthon for highly condensed difluoro-polycyclic aromatic hydrocarbons.

Octafluoronaphthalene is prepared in 53% yield by defluorination of perfluorodecahydronaphthalene over iron or nickel at 500°C.

Fused-ring polycyclic fluoroaromatics can be made from the corresponding amino fused-ring polycyclic or from preformed fluoroaromatics.

Side-Chain Fluorinated Aromatics

Trifluoromethyl aromatics are used widely in the production of drugs, crop-protection chemicals, germicides, dyes, etc.

General Properties. The trifluoromethyl group is stable under different reaction conditions, eg, the multistep classical transformation of benzotrifluoride to trifluoroacetic acid features successive nitration, reduction, and oxidation.

Benzotrifluoride is stable at 350°C in the presence of iron or copper. The trifluoromethyl group is sensitive to hydrolysis in acidic media. Benzotrifluoride resists ring oxidation. The trifluoromethyl group is inert to numerous reducing agents. Care must be exercised in handling trifluoromethylphenyl organometallics. A compilation of reactive chemical hazards of trifluoromethylphenyl organometallics was published in 1990.

Reactions. Benzotrifluoride undergoes electrophilic substitution reactions, eg, halogenation, nitration, typical of an aromatic containing a strong electron-withdrawing groups. The trifluoromethyl group (sometimes referred to as a pseudohalogen) is meta directing. The strong electron-withdrawing effect of a trifluoromethyl group activates ortho and para halogen toward nucleophilic attack. Such chlorine lability is utilized in the manufacture of crop control chemicals containing trifluoromethyl and nitro groups.

Benzotrifluoride. Benzotrifluoride (α,α,α-trifluorotoluene), $C_6H_5CF_3$ (mol wt, 146.11), is a colorless liquid (Table 4). Benzotrifluoride can be produced by the high pressure reaction of benzotrichloride with anhydrous hydrogen fluoride (AHF).

Table 4. Physical Properties of Benzotrifluoride

Property	Value
mol wt	146.11
melting point, °C	−29.02
boiling point, °C	102.05
density, 25°C, g/mL	1.1814
viscosity at 38°C, mPa·s(= cP)	0.488
surface tension at 20°C, mN/m(= dyn/cm)	23.39
latent heat of vaporization, 102.05°C, J/mol[a]	32.635.2
solubility in water, g/100 g at room temperature	0.045

[a] To convert J to cal, divide by 4.184.

Table 5. Properties of Benzotrifluoride Derivatives

Component	Mol wt	Mp, °C	Bp, $°C_{kPa}$[a]	Refractive index n^t_D	Specific gravity, d^t_4
Aminobenzotrifluoride					
$H_2NC_6H_4CF_3$					
2-aminobenzotrifluoride	116.13	34	$174–175_{100.4}$	1.4800^{25}	1.290^{25}
3-aminobenzotrifluoride	116.13	5–6	187–188 $86_{2.67}$	1.4788^{20}	1.305^{25}
Monochlorobenzotrifluorides					
$CF_3C_6H_4Cl$					
2-chlorobenzotrifluoride	180.56		152.5	1.4550^{20}	1.367^{20}
4-chlorobenzotrifluoride	180.56	−36	139 $29.5_{1.33}$	1.4444^{25}	1.338^{25}
Dichlorobenzotrifluorides					
$CF_3C_6H_3Cl_2$					
2,4-dichlorobenzotrifluoride	215.00	−26	177.5	1.4793^{25}	$1.501^{15.5}$
3,4-dichlorobenzotrifluoride	215.00	−12.4	173.5	1.4736^{25}	1.478^{25}

[a] To convert kPa to mm Hg, multiply by 7.5.

Benzotrifluoride Derivatives. Laboratory recipes for 45 benzotrifluorides were published in 1985. Physical properties and toxicity of commercially significant benzotrifluoride derivatives are listed in Table 5.

Arylfluoroalkyl Ethers

α,α,α-Trifluoromethoxybenzene, $C_6H_5OCF_3$, and other arylfluoroalkyl ethers and thioethers (HCF_2O—, HCF_2CF_2O—, CF_3CH_2O—, and CF_3S—), are assuming greater importance as crop-protection chemicals and pharmaceuticals.

Properties. The trifluoromethoxy (CF_3O—) group in $ArOCF_3$ exhibits unusual stability to strong acids and bases (including organometallic reagents), as well as to strong oxidizing and reducing conditions. The thermal stability is exceptional.

The CF_3O— group exerts predominant para orientation in electrophilic substitution reactions such as nitration, halogenation, acylation, and alkylation.

Trifluoromethoxybenzenes ($ArOCF_3$). Trifluoromethoxybenzene (α,α,α-trifluoroanisole, phenyl trifluoromethyl ether) $C_6H_5OCF_3$ (mol wt 162.11), is a colorless liquid, bp 102°C, mp −50°C, n^{20}_D 1.4060, d^{25}_4 1.226, flash point (closed cup), 12°C. Depending on the ring substituent, trifluoromethoxybenzenes can be made by the sequential chlorination–fluorination of anisole(s).

Trifluoromethylthioaromatics ($ArSCF_3$). Trifluoromethylthioaromatics (aryl trifluoromethyl sulfides) can be made by sequential chlorination–fluorination (SbF_3 or HF) of the corresponding thioanisole.

Difluoromethoxyaromatics ($ArOCHF_2$) and Sulfur Analogues ($ArSCHF_2$). Difluoromethyltion of phenyl (or thiophenols) with chlorodifluoromethane, $CHClF_2$, and aqueous caustic in dioxane gives good yields of aryldifluoromethyl ethers.

Tetrafluoroethoxyaromatics ($ArOCF_2CF_2H$). Tetrafluoroethoxyaromatics are produced by base-calatlyzed addition of tetrafluoroethylene to phenols.

Aryltrifluoroethyl Ethers ($ArOCH_2CF_3$). 2,2,2-Trifluoroethoxybenzenes are obtained from the reaction of activated haloaromatics with sodium 2,2,2-trifluoroethoxide in polar solvents.

Fluorinated Nitrogen Heterocyclics

Ring- or side-chain fluorinated nitrogen heterocyclics have been incorporated into crop-protection chemicals, drugs, and reactive dyestuffs. Key intermediates include fluorinated pyridines, quinolines, pyrimidines, and triazines. Physical properties of some fluorinated nitrogen heterocyclics are listed in Table 6.

Table 6. Properties of Miscellaneous Fluorinated Heterocyclic Compounds

Component	Mol wt	Mp, °C	Bp, °C/kPa[a]	Refractive index, n_D^t	Specific gravity, d_4^t
Fluoropyridines					
C_5H_4FN					
2-fluoropyridine	97.09		$126_{100.4}$	1.4678^{20}	1.1281^{20}
3-fluoropyridine	97.09		$105–107_{100.3}$	1.4700^{20}	1.125^{25}
4-fluoropyridine	97.09	100[b]	$108_{100.0}$	1.4730^{20}	
$C_5H_3F_2N$					
2,4-difluoropyridine	115.08	134–135[b]	104–105		
2,6-difluoropyridine	115.08		$124.5_{99.1}$	1.4349^{25}	1.265^{25}
$C_5H_2F_3N$					
2,4,6-trifluoropyridine	133.07		94–95		
C_5HF_4N					
2,3,5,6-tetrafluoropyridine	151.06		102		
C_5H_5N					
pentafluoropyridine	169.05	−41.5	83.3	1.3856^{20}	
Perfluoroalkylpyridines					
2-trifluoromethylpyridine	147.10		$143_{99.4}$	1.4144^{25}	
3-trifluoromethylpyridine	147.10		113–115	1.4150^{25}	
4-trifluoromethylpyridine	147.10		108–110	1.4144^{25}	
2-chloro-5-trifluoromethylpyridine	181.55	32–34	152_{100}		1.417^{20}
Fluoropyrimidines					
2,4,6-trifluoro-pyrimidine	134.06		$98;60_{24}$	1.4015^{25}	
2,4,5,6-tetrafluoropyrimidine	152.06		89	1.3875^{25}	
5-chloro-2,4,6-trifluoropyrimidine	168.51		114.5	1.4390^{20}	
Fluorotriazines					
2,4,6-trifluoro-1,3,5-triazine	135.05	−38	$72.4_{101.67}$	1.3844^{24}	1.60^{25}
2,4,6-tris(trifluoromethyl)-1,3,5-triazine	285.07	−24.8	95–96	1.3161^{25}	1.593^{25}

[a] To convert kPa to mm Hg, multiply by 7.5.
[b] Mp of HCl salt.

MAX M. BOUDAKIAN
Chemical Consultant

A. E. Pavlath and A. L. Leffler, *Aromatic Fluorine Compounds*, ACS Monograph No. 155, Reinhold, N.Y., 1962.

G. Schiemann and B. Cornils, *Chemie und Technologie Cyclischer Fluorverbindungen*, F. Enke Verlag, Stuttgart, 1969.

I. L. Knunyants and G. G. Yakobson, eds., *Syntheses of Organic Fluorine Compounds*, Springer-Verlag, Berlin, 1985.

R. E. Banks, ed., *Organofluorine Chemicals and Their Industrial Applications*, E. Horwood, Chichester, U.K., 1979.

POLYTETRAFLUOROETHYLENE

Polytetrafluoroethylene (PTFE), more commonly known as Teflon (DuPont), a perfluorinated straight-chain high polymer, has a most unique position in the plastics industry due to its chemical inertness, heat resistance, excellent electrical insulation properties, and low coefficient of friction over a wide temperature range. Polymerization of tetrafluoroethylene (TFE) monomer gives this perfluorinated straight-chain high polymer with the formula–$(CF_2–CF_2)_n$–. The white to translucent solid polymer has an extremely high molecular weight, in the $10^6–10^7$ range, and consequently has a viscosity in the range of 1 to 10 GPa·s ($10^{10}–10^{11}$ P) at 380°C.

Because of its chemical inertness and high molecular weight, PTFE melt does not flow and cannot be fabricated by conventional techniques. The suspension-polymerized PTFE polymer (referred to as granular PTFE) is usually fabricated by modified powder metallurgy techniques. Emulsion-polymerized PTFE behaves entirely differently from granular PTFE. Coagulated dispersions are processed by a cold extrusion process (like processing lead). Stabilized PTFE dispersions, made by emulsion polymerization, are usually processed according to latex processing techniques.

Manufacturers of PTFE include Daikin Kogyo (Polyflon), DuPont (Teflon), Hoechst (Hostaflon), ICI (Fluon), Ausimont (Algoflon and Halon), and the CIS (Fluoroplast). India and The People's Republic of China also manufacture some PTFE products.

Monomer

Preparation. The manufacture of tetrafluoroethylene (TFE) involves the following steps. The pyrolysis is often conducted at a PTFE manufacturing site because of the difficulty of handling TFE.

$$CaF_2 + H_2SO_4 \rightarrow CaSO_4 + 2\ HF$$

$$CH_4 + 3\ Cl_2 \rightarrow CHCl_3 + 3\ HCl$$

$$CHCl_3 + 2\ HF \xrightarrow[\Delta]{SbF_3} CHClF_2 + 2\ HCl$$

$$2\ CHClF_2 \rightarrow CF_2{=}CF_2 + 2\ HCl$$

Properties. Tetrafluoroethylene (mol wt 100.02) is a colorless, tasteless, odorless, nontoxic gas (Table 1).

Uses. Besides polymerizing TFE to various types of high PTFE homopolymer, TFE is copolymerized with hexafluoropropylene, ethylene, perfluorinated ether, isobutylene, propylene, and in some cases it is used as a termonomer.

Manufacture of PTFE

Polytetrafluoroethylene is manufactured and sold in three forms: granular, fine powder, and aqueous dispersion; each requires a different fabrication technique.

Polymerization. In aqueous medium, TFE is polymerized by two different procedures. When little or no dispersing agent is used and vigorous agitation is maintained, a precipitated resin is produced, commonly referred to as granular resin. In another procedure, called aqueous dispersion polymerization, a sufficient dispersing agent is employed and mild agitation produces small colloidal particles dispersed in the aqueous reaction medium; precipitation of the resin particles is avoided.

Granular Resins. Granular PTFE is made by polymerizing TFE alone or in the presence of trace amounts of comonomers.

Table 1. Physical Properties of Tetrafluoroethylene

Property	Value
boiling point at 101.3 kPa[a] °C	−76.3
freezing point, °C	−142.5
dielectric constant at 28°C at 101.3 kPa[a]	1.0017
thermal conductivity at 30°C, mW/(m·K)	15.5

[a] To convert kPa to atm, multiply by 0.01.

Fine Powder Resins. Fine powder resins are made by polymerizing TFE in an aqueous medium with an initiator and emulsifying agents.

Aqueous Dispersions. The dispersion is made by the polymerization process used to produce fine powders of different average particle sizes.

Filled Resins. Fillers such as glass fibers, graphite, asbestos, or powered metals are compounded into all three types of PTFE. Compounding is achieved by intimate mixing.

Properties

The properties described herein are related to the basic structure of polytetrafluoroethylene and are exhibited by both granular and fine powder products. Polytetrafluoroethylene does not dissolve in any common solvent. At ca 342°C, virgin PTFE changes from white crystalline material to almost transparent amorphous gel. Differential thermal analysis indicates that the first melting of virgin polymer is irreversible and that subsequent remeltings occur at 327°C, which is generally reported as the melting point. Virgin PTFE has a crystallinity in the range of 92–98%, which indicates an unbranched chain structure.

Mechanical Properties. Mechanical properties of PTFE depend on processing variables, eg, preforming pressure, sintering temperature and time, cooling rate, void content, and crystallinity. Properties, such as the coefficient of friction, flexibility at low temperatures, and stability at high temperatures, are relatively independent of fabrication. Molding and sintering conditions affect flex life, permeability, stiffness, resiliency, and impact strength. The physical properties of PTFE have been reviewed and compiled (Table 2).

Filled Resins. Filled compositions meet the requirements of an increased variety of mechanical, electrical, and chemical applications. Physical properties of filled granular compounds are shown in Table 3.

Chemical Properties. Vacuum thermal degradation of PTFE results in monomer formation.

Radiation Effects. Polytetrafluoroethylene is attacked by radiation. An increase in stiffness in material irradiated in vacuum indicates cross-linking. Degradation is the result of random scission of the chain.

Absorption, Permeation, and Interactions. Polytetrafluoroethylene is chemically inert to industrial chemicals and solvents even at elevated temperatures and pressures. Absorption of a liquid is usually a matter of the liquid dissolving in the polymer; however, in the case of PTFE, no interaction occurs between the polymer and other substances. Submicroscopic voids between the polymer molecules provide space for the material absorbed; which is indicated by a slight weight increase and sometimes by discoloration. The sorption behavior of perfluorocarbon polymers is typical for nonpolar partially crystalline polymers.

Table 2. Typical Mechanical Properties of Molded and Sintered PTFE Resins

Property	Granular resin	Fine powder	ASTM method
tensile strength at 23°C, MPa[a]	7–28	17.5–24.5	D638-61T
elongation at 23°C, %	100–200	300–600	D628-61T
flexural strength at 23°C, MPa[a]	does not break		D790-61
flexural modulus at 23°C, MPa[a]	350–630	280–630	D747-61T
impact strength, J/m[b]			
21°C	106.7		D256-56
24°C	160		
77°C	>320		
hardness durometer, D	50–65	50–65	D1706-59T
static coefficient of friction with polished steel	0.05–0.08		

[a] To convert MPa to psi, multiply by 145.
[b] To convert J/m to ft·lbf/in., divide by 53.38.

Table 3. Properties of Filled PTFE Compounds

Property	Unfilled	Glass fiber, 15 wt %	Graphite, 15 wt %	Bronze, 60 wt %
specific gravity	2.18	2.21	2.16	3.74
tensile strength, MPa[a]	28	25	21	14
elongation, %	350	300	250	150
thermal conductivity, mW/(m·K)	0.244	0.37	0.45	0.46
hardness, Shore durometer, D	51	54	61	70
wear factor, 1/Pa[b]	5×10^{-14}	28×10^{-17}	100×10^{-17}	12×10^{-17}
coefficient of friction static, 3.4 MPa[a] load	0.08	0.13	0.10	0.10

[a] To convert MPa to psi, multiply by 145. [b] To convert 1/Pa to (in.^{3}·min)/(ft·lbf·h), divide by 2×10^{-7}.

As an excellent barrier resin, PTFE is widely used in the chemical industry. Gases and vapors diffuse through PTFE more slowly than through most other polymers.

Electrical Properties. Polytetrafluoroethylene is an excellent electrical insulator because of its mechanical strength and chemical and thermal stability as well as excellent electrical properties.

Fabrication

Granular Resins. Virgin PTFE melts at about 342°C; viscosity, even at 380°C, is 10 GPa·s (10^{11} P). This eliminates processing by normal thermoplastic techniques, and other fabrication techniques had to be developed: the dry powder is compressed into handleable form by heating above the melting point. This coalesces the particles into a strong homogeneous structure; cooling at a controlled rate achieves the desired degree of crystallinity.

Molding. Many PTFE manufacturers give detailed descriptions of molding equipment and procedures. Round piston molds for the production of solid or hollow cylinders are the most widely used. Automatic molding permits high speed mass production; it is preferable to machining finished material. Isostatic molding allows uniform compression from all directions.

Sintering. Electrical ovens with air circulation and service temperatures up to 400°C are satisfactory for sintering. In free sintering, the cheapest and most widely used process, a preformed mold is placed in an oven with a temperature variation of ±2 °C.

Ram Extrusion. In ram extrusion, a small charge of PTFE powder is preformed by a reciprocating ram and sintered. Subsequent charges are fused into the first charge, and this process continues to form homogeneous long rods.

Fine Powder Resins. Fine powder PTFE resins are extremely sensitive to shear. They must be handled gently to avoid shear, which prevents processing. However, fine powder is suitable for the manufacture of tubing and wire insulation for which compression molding is not suitable. A paste-extrusion process may be applied to the fabrication of tubes with diameters from fractions of a millimeter to about a meter, walls from thicknesses of 100–400 μm, thin rods with up to 50-mm diameters, and cable sheathing.

Dispersion Resins. Polytetrafluoroethylene dispersions in aqueous medium contain 30–60 wt % polymer particles and some surfactant. The type of surfactant and the particle characteristics depend on the application. These dispersions are applied to various substrates by spraying, flow coating, dipping, coagulating, or electrodepositing.

Effects of Fabrication on Physical Properties of Molded Parts. The physical properties are affected by molecular weight, void content, and crystallinity. Molecular weight can be reduced by degradation but not increased during processing. These factors can be controlled during

Table 4. Applications of Polytetrafluoroethylene Resins

Resin grade	Processing	Main uses
	Granular	
agglomerates	molding, preforming, sintering, ram extrusion	gaskets, packing seals, electronic components
coarse	molding, preforming, sintering	tape, molded shapes
finely divided	molding, preforming, sintering	molded sheets, tape, wire wrapping
presintered	ram extrusion	rods and tubes
	Fine powder	
high reduction ratio	paste extrusion	wire coating
medium reduction ratio	paste extrusion	tubing, pipe
low reduction ratio	paste extrusion	thread-sealant tape, pipe liners
	Dispersion	
general-purpose	dip coating	impregnation, coating, packing
coating	dip coating	film coating
stabilized	coagulation	bearings

molding by the choice of resin and fabricating conditions. Void distribution (or size and orientation) also affects properties; however, it is not easily measured. Preforming primarily affects void content, sintering controls molecular weight, and cooling determines crystallinity.

Applications

Consumption of PTFE increases continuously as new applications are being developed. Electrical applications consume half of the PTFE produced; mechanical and chemical applications share equally the other half. Various grades of PTFE and their applications are shown in Table 4.

Economic Aspects

Polytetrafluoroethylene homopolymers are more expensive than most other thermoplastics because of high monomer refining costs. Although fine powder sales have increased in recent years, the sales of granular PTFE are the highest on a worldwide basis. Most of the resin is consumed in the United States, followed by Europe and Japan.

Health and Safety

Exposure to PTFE can arise from ingestion, skin contact, or inhalation. The polymer has no irritating effect to the skin, and test animals fed with the sintered polymer have not shown adverse reactions. Dust generated by grinding the resin also has no effect on test animals. Formation of toxic products is unlikely. Only the heated polymer is a source of a possible health hazard.

Because PTFE resins decompose slowly, they may be heated to a high temperature. The toxicity of the pyrolysis products warrants care where exposure of personnel is likely to occur. Prolonged exposure to thermal decomposition products causes so-called polymer fume fever, a temporary influenza-like condition.

SUBHASH V. GANGAL
E. I. du Pont de Nemours & Co., Inc.

C. A. Sperati, in J. Brandrup and E. H. Immergut, eds., *Polymer Handbook*, 2nd ed., John Wiley & Sons, Inc., New York, 1975 pp. V-29–36.

Mechanical Design Data, Teflon Fluorocarbon Resins, bulletin, E. I. du Pont de Nemours & Co., Inc. Wilmington, Del., Sept. 1964.

Teflon Occupational Health Bull. **17**(2), (1962) (published by Information Service Division, Dept. of National Health and Welfare, Ottawa, Canada).

PERFLUORINATED ETHYLENE–PROPYLENE COPOLYMERS

Perfluorinated ethylene–propylene (FEP) resin is a copolymer of tetrafluoroethylene (TFE) and hexafluoropropylene (HFP); thus its branched structure contains units of $-CF_2-CF_2-$ and units of $-CF_2-CF(CF_3)-$. It retains most of the desirable characteristics of polytetrafluoroethylene (PTFE) but with a melt viscosity low enough for conventional melt processing. The introduction of hexafluoropropylene lowers the melting point of PTFE from 325°C to about 260°C.

As a true thermoplastic, FEP copolymer can be melt-processed by extrusion and compression, injection, and blow molding. Films can be heat-bonded and sealed, vacuum-formed, and laminated to various substrates. Chemical inertness and corrosion resistance make FEP highly suitable for chemical services; its dielectric and insulating properties favor it for electrical and electronic service; and its low frictional properties, mechanical toughness, thermal stability, and nonstick quality make it highly suitable for bearings and seals, high temperature components, and nonstick surfaces.

Properties. The crystallinity of FEP polymer is significantly lower than that of PTFE (70 vs 98%). The structure resembles that of PTFE, except for a random replacement of a fluorine atom by a perfluoromethyl group (CF_3).

Only one first-order transition is observed, the melting point. Increasing the pressure raises mp.

The polymer is thermally stable and can be processed at ca 270°C. Thermal degradation is a function of temperature and time, and the stability is therefore limited.

The primary effect of radiation is the degradation of large molecules to small molecules. Molecular weight reduction can be minimized by excluding oxygen. If FEP is lightly irradiated at elevated temperatures in the absence of oxygen, cross-linking offsets molecular breakdown.

Extensive lists of the physical properties of FEP copolymers are available. Most of the important properties of FEP are similar to those of PTFE; the main difference is the lower continuous service temperature of 204°C of FEP compared to that of 260°C of PTFE. The flexibility at low temperature and the low coefficients of friction and stability at high temperature are relatively independent of fabrication conditions. Unlike PTFE, FEP resins do not exhibit a marked change in volume at room temperature, because they do not have a first-order transition at 19°C. They are useful above −267°C and are highly flexible above −79°C.

Because of excellent electrical properties, FEP is a valuable and versatile electrical insulator. Within the recommended service temperature range, PTFE and FEP have identical properties as electrical insulators. Volume resistivity, which is $>10^{17}$ Ω/cm, remains unchanged even after prolonged soaking in water; surface resistivity is $>10^{15}$ Ω/sq.

The FEP resin is inert to most chemicals and solvents, even at elevated temperatures and pressures. However, it reacts with fluorine, molten alkali metal, and molten sodium hydroxide.

Articles fabricated from FEP are unaffected by weather, and their resistance to extreme heat, cold, and uv irradiation suits them for applications in radar and other electronic components.

Teflon FEP fluorocarbon film transmits more ultraviolet, visible light, and infrared radiation than ordinary window glass. The refractive index of FEP film is 1.341–1.347.

Fabrication

Effects of Fabrication on Product Properties. Extrusion conditions have a significant effect on the quality of the product. Contamination can be the result of corrosion, traces of another resin, or improper handling. Corrosion-resistant Hastelloy C parts should be used in the extruder. Surface roughness is the result of melt fracture or mechanical deformation. Melt fracture can be eliminated by increasing the die opening, die temperature, and the melt temperature and reducing the extrusion rate. Bubbles and discoloration are caused by resin degradation, air entrapment, or condensed moisture. Excessive drawdown, resin degradation, or contamination can result in pinholes, tears, and cone breaks. The blisters are caused by degassing of primary coatings, and loose coating are caused by rapid cooling and long cones.

Economic Aspects

Because of the high cost of hexafluoropropylene, FEP is more expensive than PTFE. During the 1980s, FEP sales increased rapidly because of usage in plenum cable, but because there ar other polymers that can be used in this application, the growth rate for FEP is expected to slow down.

Health and Safety

The safety precautions required in handling TFE–HFP copolymers are the same as those applied to handling PTFE. Large quantities have been processed safely by many different fabricators in a variety of operations. With proper ventilation, the polymer can be processed and used at elevated temperatures without hazard.

Applications

The principal electrical applications include hook-up wire, interconnecting wire, coaxial cable, computer wire , thermocouple wire, plenum cable, and molded electrical parts. Principal chemical applications are lined pipes and fittings, overbraided hose, heat exchangers, and laboratory ware. Mechanical uses include antistick applications, such as conveyor belts and roll covers. A recent development of FEP film for solar collector windows takes advantage of light weight, excellent weatherability, and high solar transmission. Solar collectors made of FEP film are efficient, and installation is easy and inexpensive.

SUBHASH V. GANGAL
E. I. du Pont de Nemours & Co., Inc.

S. V. Gangal, "Tetrafluoroethylene Polymers, Tetrafluoroethylene–Hexafluoropropylene Copolymers," in J. I. Kroschwitz, ed., *Encyclopedia of Polymer Science and Engineering*, 2nd ed., Vol. 16, John Wiley & Sons, Inc., New York, 1989, pp. 601–613.

Electrical/Electronic Design Data for Teflon, E. I. du Pont de Nemours & Co., Inc., Wilmington, Del.

Teflon Fluorocarbon Resins, Mechanical Design Data, 2nd ed., E. I. du Pont de Nemours & Co., Inc., Wilmington, Del., 1965.

Safe Handling Guide, Teflon Fluorocarbon Resins, du Pont Materials for Wire and Cable, bulletin E-85433, E. I. du Pont de Nemours & Co., Inc., Wilmington, Del., 1986.

TETRAFLUOROETHYLENE–ETHYLENE COPOLYMERS

Modified ethylene–tetrafluoroethylene copolymers are the products of real commercial value because they have good tensile strength, moderate stiffness, high flex life, and outstanding impact strength, abrasion resistance, and cut-through resistance. Electrical properties include low dielectric constant, high dielectric strength, excellent resistivity, and low dissipation factor. Thermal and cryogenic performance and chemical resistance are good. These properties, combined with elasticity, make this material an ideal candidate for heat-shrinkable film and tubing. This family of copolymers can be processed by conventional methods such as melt extrusion, injection molding, transfer molding, and rotational molding.

Ethylene and tetrafluoroethylene are copolymerized in aqueous, nonaqueous, or mixed medium with free-radical initiators. The polymer is isolated and converted into extruded cubes, powders, and beads, or a dispersion. This family of products is manufactured by DuPont, Hoechst, Daikin, Asahi Glass, and Ausimont and sold under the trade names of Tefzel, Hostaflon ET, Neoflon EP, Aflon COP, and Halon ET, respectively.

Manufacture

Tetrafluoroethylene–ethylene copolymers have tensile strengths two to three times as high as the tensile strength of polytetrafluoroethylene or of the ethylene homopolymer. Because these copolymers are highly crystalline and fragile at high temperature, they are modified with a third monomer, usually a vinyl monomer free of telegenic activity.

Copolymerization is effected by suspension or emulsion techniques under such conditions that tetrafluoroethylene, but not ethylene, may homopolymerize.

Aqueous emulsion polymerization is carried out using a fluorinated emulsifier, a chain-transfer agent to control molecular weight, and dispersion stabilizers such as manganic acid salts and ammonium oxalate.

Properties

The equimolar copolymer of ethylene and tetrafluoroethylene is isomeric with poly(vinylidene fluoride) but has a higher melting point and a lower dielectric loss. This polymer can be dissolved in certain high boiling esters at temperatures above 230°C.

Transitions. Samples containing 50 mol % tetrafluoroethylene with ca 92% alternation were quenched in ice water or cooled slowly from the melt to minimize or maximize crystallinity, respectively. The dynamic mechanical behavior showed that the α relaxation occurs at 110°C in the quenched sample; in the slowly cooled sample it is shifted to 135°C. The β relaxation appears near −25 °C. The γ-relaxation at −120 °C in the quenched sample is reduced in peak height in the slowly cooled sample and shifted to a slightly higher temperature.

Physical and Mechanical Properties. Modified ethylene–tetrafluoroethylene copolymer has a good combination of mechanical properties, including excellent cut-through and abrasion resistance, high flex life, and exceptional impact strength. As wire insulation, it withstands physical abuse during and after installation.

Modified ETFE is less dense, tougher, and stiffer and exhibits a higher tensile strength and creep resistance than PTFE, PFA, or FEP resins. It is ductile, and displays in various compositions the characteristic of a nonlinear stress–strain relationship.

Thermal Properties. Modified ETFE copolymer has broad operating temperature range up to 150°C for continuous exposure. Cross-linking by radiation improves the high temperature capability further. However, prolonged exposure to higher temperatures gradually impairs the mechanical properties and results in discoloration.

Friction and Bearing Wear of the Glass-Reinforced Copolymer. Glass reinforcement improves the frictional and wear properties of modified ETFE resins (HT-2004). Their frictional and wear characteristics, combined with outstanding creep resistance, indicate suitability for bearing applications.

Electrical Properties. Modified ethylene–tetrafluoroethylene is an excellent dielectric. Its low dielectric constant confers a high corona-ignition voltage. The dielectric constant does not vary with frequency or temperature. Both dielectric strength and resistivity are high. The loss characteristics are minimum; the dissipation factor, although low, increases at higher frequencies.

Chemical Resistance and Hydrolytic Stability. Modified ethylene–tetrafluoroethylene copolymers are resistant to chemicals and solvents that often cause rapid degradation in other plastic materials. Organic compounds and solvents have little effect. Strong oxidizing acids, organic bases, and sulfonic acids at high concentrations and near their boiling points affect ETFE to varying degrees. Physical properties remain stable after long exposure to boiling water.

Water absorption of Tefzel is low (0.029% by weight), which contributes to its outstanding dimensional stability as well as to the stability of mechanical and electrical properties regardless of humidity.

Vacuum Outgassing and Permeability. Under vacuum, modified ethylene–tetrafluoroethylene copolymers give off little gas at elevated temperatures. The following permeability values were determined on Tefzel film (100 μm) at 25°C (1 nmol/m·s·GPa = 0.5 cc·mil/100 in.^{2} d·atm).

Fabrication

Modified ethylene–tetrafluoroethylene copolymers are commercially available in a variety of physical forms and can be fabricated by conventional thermoplastic techniques. Commercial ETFE resins are marketed in melt-extruded cubes, that are sold in 20-kg bags or 150-kg drums.

Melt Processing. Articles are made by injection molding, compression molding, blow molding, transfer molding, rotational molding, extrusion, and coating. Films can be thermoformed and heat sealed. Because of high melt viscosity, ETFE resins are usually processed at high (300–340°C) temperatures.

Forming and Machining. Articles can be formed below the melting point with conventional metal-forming techniques.

Coloring and Decorating. Commercial pigments that are thermally stable at the resin processing temperature may be used.

Assembly. The success of many applications depends on the ability of ETFE fluoropolymer to be economically assembled. Assembly processes include screw assembly, snap-fit and press-fit joints, cold or hot heading, spin welding, and ultrasonic welding.

Potting. Potting of wire insulated with Tefzel has been accomplished with the aid of a coating of a colloidal silica dispersion.

Bonding. Surface treatment, such as chemical etch, corona, or flame treatments, is required for adhesive bonding of Tefzel. Polyester and epoxy compounds are suitable adhesives.

Health and Safety

Large quantities of Tefzel have been processed and used in many demanding service applications. No cases of permanent injury have been attributed to these resins, and only limited instances of temporary irritation to the upper respiratory tract have been reported.

As with other melt-processible fluoropolymers, trace quantities of harmful gases, including hydrogen fluoride, diffuse from the resin even at room temperature. Therefore, the resins should be used in well-ventilated areas.

Bulk quantities of Tefzel fluoropolymer resins should be stored away from flammable materials. In the event of fire, personnel entering the area should have full protection, including acid-resistant clothing and self-contained breathing apparatus with a full facepiece operated in the pressure-demand or other positive-pressure mode.

Applications

Tefzel 200 is a general-purpose, high temperature resin for insulating and jacketing low voltage power wiring for mass transport systems, wiring for chemical plants, and control and instrumentation wiring for utilities. In injection-molded form, it is used for sockets, connectors, and switch components. Because of excellent mechanical properties it provides good service in seal glands, pipe plugs, corrugated tubing, fasteners, and pump vanes. In chemical service, it is used for valve components, laboratory ware, packing, pump impellers, and battery and instrument components.

SUBHASH V. GANGAL
E. I. du Pont de Nemours & Co., Inc.

S. V. Gangal, "Tetrafluoroethylene Polymers, Tetrafluoroethylene–Ethylene Copolymers," in J. I. Kroschwitz, ed., *Encyclopedia of Polymer Science and Engineering*, 2nd ed., Vol. 16, Wiley-Interscience, New York, 1989, pp. 626–642.

Tefzel Fluoropolymer, Design Handbook, E. I. du Pont de Nemours & Co., Inc., Wilmington, Del., 1973.

Tefzel—Properties Handbook, E-31301-3, E. I. du Pont de Nemours & Co., Inc., Wilmington, Del., Dec. 1991.

Tefzel Fluoropolymers, Safe Handling Guide, Bulletin E-85785, E. I. du Pont de Nemours & Co., Inc., Wilmington, Del., May 1986.

TETRAFLUOROETHYLENE–PERFLUOROVINYL ETHER COPOLYMERS

Perfluoroalkoxy (PFA) fluorocarbon resins are designed to meet industry's needs in chemical, electrical, and mechanical applications. These melt processible copolymers contain a fluorocarbon backbone in the main chain and randomly distributed perfluorinated ether side chains:

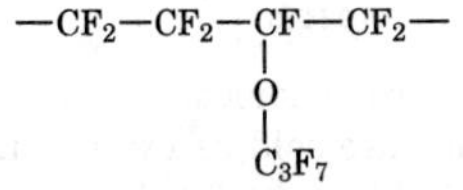

A combination of excellent chemical and mechanical properties at elevated temperatures results in reliable, high performance service to the chemical processing and related industries. Chemical inertness, heat resistance, toughness and flexibility, stress-crack resistance, excellent flex life, antistick characteristics, little moisture absorption, nonflammability, and exceptional dielectric properties are among the characteristics of these resins.

Copolymerization

Tetrafluoroethylene–perfluoropropyl vinyl ether copolymers are made in aqueous or nonaqueous media. In aqueous copolymerizations water-soluble initiators and a perfluorinated emulsifying agent are used. Molecular weight and molecular weight distribution are controlled by a chain-transfer agent. In nonaqueous copolymerization, fluorinated acyl peroxides are used as initiators that are soluble in the medium; a chain-transfer agent may be added for molecular weight control.

Properties

Mechanical Properties. Table 1 shows the physical properties of Teflon PFA. At 20–25°C the mechanical properties of PFA, FEP, and PTFE are similar; differences between PFA and FEP become significant as the temperature is increased. Tests at liquid nitrogen temperature indicate that PFA performs well in cryogenic applications.

Optical Properties and Radiation Effects. Within the range of wavelengths measured (uv, visible, and near-ir radiation), Teflon PFA fluorocarbon film transmits slightly less energy than FEP film. In thin sections, the resin is colorless and transparent; in thicker sections, it becomes translucent. It is highly transparent to ir radiation; uv absorption is low in thin sections. Weather-O-Meter tests indicate unlimited outdoor life. Like other perfluoropolymers, Teflon PFA is not highly resistant to radiation.

Fabrication

Teflon PFA resins are fabricated by the conventional melt-processing techniques used for thermoplastics. Processing equipment is constructed of corrosion-resistant materials and can be operated at 315–425°C. A general-purpose grade, PFA 340, is designed for a variety of molding and extrusion applications, including tubing, shapes,

Table 1. Properties of Teflon PFA

Property	Teflon 340	Teflon 350
nominal melting point, °C	302–306	302–306
specific gravity	10.6	1.8
tensile strength at 23°C, MPa[a]	28	31
creep resistance[b] tensile modulus at 20°C, MPa[a]	270	270
hardness Durometer	D60	D60
water absorption, %	0.03	0.03

[a] To convert MPa to psi, multiply by 145.
[b] Apparent modulus after 10 h: stress = 6.89 MPa at 20°C, 6.89 kPa at 250°C.

and molded components, in addition to insulation for electrical wire and cables.

Health and Safety

Safe practices employed for handling PTFE and FEP resins are adequate for Teflon PFA; adequate ventilation is required for processing above 330–355°C.

Applications and Economic Aspects

The perfluorovinyl ether comonomer used for PFA is expensive, as is PFA. Most PFA grades are sold as extruded, translucent cubes in various colors. Teflon PFA can be fabricated into high temperature electrical insulation and components and materials for mechanical parts requiring long flex life.

SUBHASH V. GANGAL
E. I. du Pont de Nemours & Co., Inc.

S. V. Gangal, in J. I. Kroschwitz, ed., *Encyclopedia of Polymer Science and Engineering*, 2nd ed., Vol. 16, Wiley-Interscience, New York, 1989, pp. 614–626.

Technical Information, No. 11, Processing Guidelines for Du Pont Fluoropolymer Rotocasting Powders of Tefzel and Teflon PFA, E. I. du Pont de Nemours & Co., Inc., Wilmington, Del., 1982.

Teflon PFA Fluorocarbon Resins: Melt Processing of Teflon PFA TE-9705, PIB #1 bulletin, E. I. du Pont de Nemours & Co., Inc., Wilmington, Del., 1973.

Handbook of Properties for Teflon PFA, sales brochure, E46679, E. I. du Pont de Nemours & Co., Inc., Wilmington, Del., Oct. 1987.

POLY(VINYL FLUORIDE)

Homopolymers and copolymers of vinyl fluoride are based on free-radical polymerization of vinyl fluoride and comonomers, usually under high pressure. Poly(vinyl fluoride) homopolymers and copolymers have excellent resistance to sunlight degradation, chemical attack, water absorption, and solvent, and have a high solar energy transmittance rate. These properties have resulted in the utilization of poly(vinyl fluoride) (PVF) film and coating in outdoor and indoor functional and decorative applications. These coatings are used where exceptional high temperature stability, outdoor longevity, stain resistance, adhesion, and release properties are required.

Polymer Properties

Poly(vinyl fluoride) (PVF) is a semicrystalline polymer with a planar, zig-zag configuration. The degree of crystallinity can vary significantly from 20–60% and is thought to be a function of defect structures.

PVF with controlled amounts of head-to-head units varying from 0 to 30% have been prepared. This series of polymers shows melting point distributions ranging from 220°C for purely head-to-tail polymer down to about 160°C for polymer containing 30% head-to-head linkages. This study, however, does not report the extent of branching in these polymers. More recent studies have shown that the extent of branching has a pronounced effect upon the melting temperature of PVF. Change of polymerization temperature from 90 to 40°C produces a change in branch frequency from 1.35 to 0.3%, whereas the frequency of monomer reversals is nearly constant (12.5 ± 1%).

PVF has low solubility in all solvents below about 100°C. Highly polar solvents, such as *N,N*-dimethylacetamide and γ-butyrolactone, can swell PVF below 100°C. PVF is more thermally stable than other vinyl halide polymers. It is transparent to radiation in the uv, visible, and near ir radiation from 350 to 2,500 nm. Radiation between 7,000 and 12,000 nm is absorbed. Expansion to low dose γ-radiation produces cross-links in PVF and actually increases tensile strength and etching resistance, whereas the degree of crystallinity and melting point are reduced.

Fabrication and Processing

PVF is converted to thin films and coatings. Processing of PVF, eg, by melt extrusion, depends on latent solvation of PVF in highly polar solvents and its subsequent coalescence. Poly(vinyl fluoride) can be applied to substrates with solvent-based or water-borne dispersions, or by powder-coating techniques.

Economic Aspects

Poly(vinyl fluoride) is available from DuPont both as a resin and as transparent and pigmented films under the trademark Tedlar PVF film.

Health and Environment

Vinyl fluoride is flammable in air between the limits of 2.6 and 22% by volume. Minimum ignition temperature for VF and air mixtures is 400°C. A small amount, <0.2%, of terpenes is added to VF to prevent spontaneous polymerization. The U.S. Department of Transportation has classified the inhibited VF as a flammable gas.

Toxicity studies, ie, survival and time to incapacitation, of polymers, cellulosics, and airplane interior materials containing PVF films expose mice to pyrolysis products and show PVF thermal degradation products to have relatively low toxicity.

Uses

PVF is weatherable, mechanically strong over a wide temperature range, and inert toward a wide variety of chemicals. It finds wide use as a protective or decorative coating in outdoor and indoor applications.

S. EBNESAJJAD
L. G. SNOW
DuPont Company

D. E. Brasure and S. Ebnesajjad, in J. I. Kroschwitz, ed., *Concise Encyclopedia of Polymer Science and Engineering*, John Wiley & Sons, Inc., New York, 1990, pp. 1273–1275.

D. E. Brasure and S. Ebnesajjad, in J. I. Kroschwitz, ed., *Encyclopedia of Polymer Science and Engineering*, 2nd ed., Vol. 17, John Wiley & Sons, Inc., New York, 1989, pp. 468–491.

K. U. Usmanov and co-workers, *Russ. Chem. Rev.* **46**(5), 462–478 1977; trans. from Usp. Khim. **46**, 878–906 (1977).

POLY(VINYLIDENE FLUORIDE)

Poly(vinylidene fluoride) is the addition polymer of 1,1-difluoroethene, commonly known as vinylidene fluoride and abbreviated VDF or VF_2.

PVDF is a semicrystalline polymer that contains 59.4 wt % fluorine and 3 wt % hydrogen and is commercially polymerized in emulsion or suspension using free-radical initiators. It has the characteristic resistance of fluoropolymers to harsh chemical, thermal, ultraviolet, weathering, and oxidizing or high energy radiation environments. Because of these characteristics it has many applications in wire and cable products, electronic devices, chemical and related processing fields, as a weather-resistant binder for exterior architectural finishes, and in many specialized uses. The polymer is readily melt-processed using conventional molding or extrusion equipment; porous membranes are cast from solutions, and finishes are deposited from dispersions using specific solvents.

Polymer

Polymerization. PVDF is manufactured using radial initiated batch polymerization processes in aqueous emulsion or suspension.

Table 1. Properties of Poly(vinylidene fluoride)

Property	Value
specific gravity	1.75–1.80
water absorption, 24 h at 23°C, %	0.04
refractive index, n_D	1.42
melting peak, T_m, °C	156–180
crystallization peak, T_c, °C	127–146
glass transition, T_g, °C	−40
specific heat, kJ/kg·K[a]	1.26–1.42
thermal conductivity, W/K·m	0.17–0.19
tensile stress at yield, MPa[b]	28–57
tensile stress at break, MPa[b]	31–52
modulus of elasticity in flexure, MPa[b]	1140–2500
limiting oxygen index, %	43

[a] To convert kJ to kcal, divide by 4.184. [b] To convert MPa to psi, multiply by 145.

Table 2. Electrical Properties of Poly(vinylidene fluoride)

Property	Method	Value
volume resistivity, Ω·cm	ASTM D257	$1.5–5 \times 10^{14}$
surface arc resistance, s	ASTM D495	50–60
dielectric strength, kV/mm	ASTM D149	63–67
dielectric constant at 25°C	ASTM D150	
1 kHz		8.15–10.46
10 kHz		8.05–9.90
100 kHz		7.85–9.61
dissipation factor		
1 kHz		0.005–0.026
10 kHz		0.015–0.021
100 kHz		0.039–0.058

Polymer Properties. Typical PVDF design properties are shown in Table 1. Properties of PVDF depend on molecular weight, molecular weight distribution, chain configuration, ie, the sequence in which the monomer units are linked together, including side groups or branching, and crystalline form. Some electrical properties are shown in Table 2.

Economic Aspects

Because of its excellent combination of properties, processibility, and relatively low price compared to other fluoropolymers, PVDF has become the largest-volume fluoropolymer after PTFE.

After 10 years of unabated rapid growth in the plenum wire and cable market, fluoropolymers including PVDF, primarily the flexible VDF/HFP copolymer, are beginning to lose market share to lower priced PVC-alloys. The loss of market share in the plenum market probably will be compensated by growth of PVDF in other fields; thus during the mid-1990s the total volume of PVDF may grow about 5% annually.

Health and Safety Factors

PVDF is a nontoxic resin and may be safely used in articles intended for repeated contact with food.

PVDF is not hazardous under typical processing conditions. If the polymer is accidentally exposed to temperatures exceeding 350°C, thermal decomposition occurs with evolution of toxic hydrogen fluoride (HF).

Some silica-containing additives such as glass and titanium dioxide lower the thermal stability of PVDF and should be used with caution. Processors should consult the resin producer about safe processing practice.

JULIUS E. DOHANY
Consultant

J. E. Dohany and J. S. Humphrey, in J. I. Kroschwitz, ed., *Encyclopedia of Polymer Science and Engineering*, 2nd ed., Vol. 17, p. 532.

J. E. Dohany, in R. B. Seymour and G. S. Kirshenbaum, eds. *High Performance Polymers: Their Origin and Development*, Elsevier Science Publishing Co., New York, 1986, p. 287.

A. J. Lovinger in G. C. Bassett, ed., *Developments in Crystalline Polymers*, Vol. 1, Applied Science Publishers, Ltd., Barking, U.K., 1982, pp. 195–273.

POLYCHLOROTRIFLUOROETHYLENE

Many challenging industrial and military applications have utilized polychlorotrifluoroethylene (PCTFE) since 1937. The unique combination of chemical inertness, radiation resistance, low vapor permeability, electrical insulation properties, and thermal stability of this polymer filled an urgent need for a specialty thermoplastic material for use in the gaseous UF_6 diffusion process for the separation of uranium isotopes.

Properties

The physical properties of PCTFE are primarily determined by a combination of molecular weight and percent crystallinity (45–65%). The high molecular weight thermoplastic has a melt temperature (T_m) of 211–216°C, a glass-transition temperature (T_g) of 71–99°C, and is thermally stable up to 250°C, with a useful temperature range from −240 to 205°C.

The typical mechanical properties that qualify PCTFE as a unique engineering thermoplastic are provided in Table 1. Other unique aspects of PCTFE are low temperature properties, resistance to cold flow due to high compressive strength, and low coefficient of thermal expansion over a wide temperature range.

The high fluorine content contributes to thermal stability and resistance to attack by most chemicals and oxidizing agents; however, PCTFE does swell slightly in halogenated compounds, ethers, esters, and selected aromatic solvents. PCTFE is compatible with liquid oxygen.

Manufacture and Processing

The synthesis of the high molecular weight polymer from chlorotrifluoroethylene has been carried out in bulk, solution, suspension, and emulsion polymerization systems using free-radical initiators, uv, and gamma radiation. Emulsion and suspension polymers are more thermally stable than bulk-produced polymers. Polymerizations can be carried out in glass or stainless steel agitated reactors. Polymer work-up or processing to a dry form is critical.

The lower molecular weight oils, waxes, and greases of PCTFE can be prepared directly by telomerization of the monomer or by pyrolysis of the higher molecular weight polymer.

PCTFE plastics can be processed by the standard thermoplastic fabrication techniques, eg, extrusion, injection, compression, and transfer molding, and they readily machinable.

Table 1. Mechanical Properties of Polychlorotrifluoroethylene

Property	Value
tensile strength, MPa[a]	32–39
compressive strength, MPa[a]	38
modulus of elasticity, MPa[a]	1400
hardness, Shore D	76
deformation under load, at 25°C, 24 h, 7 MPa[a], %	0.3
heat deflection temperature, at 0.46 MPa[a], °C	126

[a] To convert MPa to psi, multiply by 145.

Economic Aspects

Several worldwide commercial manufacturers of PCTFE (AlliedSignal, Daikin Koyoo, and Ucine Kuhlmann) and vinylidene fluoride-modified copolymers offer a variety of products, eg, molding powder, pellets, oils, waxes, and greases. PCTFE thermoplastics are used in high technology, specialty engineering areas where the unique combination of properties and part reliability demands a high performance thermoplastic polymer at a premium price of \$40–100/kg.

Health and Safety Factors

In general, the PCTFE resins have been found to be low in toxicity and irritation potential under normal handling conditions.

Uses

The principal uses of PCTFE plastics remain in the areas of aeronautical and space, electrical/electronics, cryogenic, chemical, and medical instrumentation industries, where its unique properties are required by the application.

G. H. Millet
J. L. Kosmala
3M Company

T. Hashimoto, H. Kawasaki, and H. Kawai, *J. Polymer Sci.* **16**(2) 271–288 (1978).

R. E. Mowers *Cryogenic Properties of Poly(Chlorotrifluoroethylene), Technical Document Report No. RTD-TDR-63-11,* Air Force Contract No. AF04(611)-6354, 1962.

A. W. Myers and co-workers, *Mod. Plast.* **37**, 139 (1960).

J. L. Currie, R. S. Irani, and J. Sanders, *Factors Affecting the Impact Sensitivity of Solid Polymer Materials in Contact with Liquid Oxygen*, ASTM Spec. Tech. Publ. 986, ASTM, Philadelphia, Pa., 1988, pp. 233–247.

R. E. Schawmm, A. F. Clark, and R. P. Reed, *A Compilation and Evaluation of Mechanical, Thermal and Electrical Properties of Selected Polymers*, NBS Report, AEC SAN-70-113, SANL 807 Task 7, SANL Task 6, National Technical Information Service, U.S. Dept. of Commerce, Springfield, Va., Sept. 1973, pp. 335–443.

S. Chandrasekaran, "Chlorotrifluoroethylene Homopolymer" under "Chlorotrifluoroethylene Polymers," in J. I. Kroschwitz, ed., *Encyclopedia of Polymer Science and Engineering*, 2nd ed., Vol. 3, John Wiley & Sons, Inc., New York, 1985, pp. 463–480.

BROMOTRIFLUOROETHYLENE

Bromotrifluoroethylene is a valuable reagent for the synthesis of trifluorovinylic compounds by means of its intermediate organometallic compounds.

Physical Properties

The monomer, bromotrifluoroethylene, $CF_2{=}CFBr$, is a colorless gas; bp −3.0 °C at 101 kPa (754 mm Hg); 58°C at 790 kPa (100 psig); and d_4^{25} 1.86 g/cm^3. Because it is spontaneously flammable in air; its odor is that of its oxidation products, mixed carbonyl halides.

Chemical Properties

Many reactions of bromotrifluoroethylene have been studied. Under basic conditions it adds alcohols such as methanol or ethanol, forming ethers with the general formula $ROCF_2CFBrH$. Similarly, diethylamine adds to it, giving $(C_2H_5)_2NCF_2CFBrH$. Vapor-phase photochemical bromination of bromotrifluoroethylene gives the expected adduct, $CF_2BrCFBr_2$. On the other hand, photochemical chlorination results in only 60% of the expected adduct and 40% scrambled bromo and chloro products.

Another class of reactions that bromotrifluoroethylene undergoes is cycloaddition with acetylenes or olefins.

Manufacture

Bromotrifluoroethylene is prepared from chlorotrifluoroethylene by the following high yield steps:

$$CF_2{=}CFCl + HBr \rightarrow CF_2BrCFClH \xrightarrow{Zn} CF_2{=}CHF + ZnBrCl$$

$$CF_2{=}HF + Br_2 \rightarrow CF_2BrCHFBr \xrightarrow{KOH} CF_2{=}CFBr + KBr + H_2O$$

Bromotrifluoroethylene is manufactured and sold in commercial quantities with a purity of 99.9% by the Halocarbon Products Corp.

Polymers

The olefin can be polymerized in trichlorofluoromethane solution. Prepared either way, the homopolymer is a white powder soluble in acetone and useful as a hard, chemically resistant coating for metal or fabric surfaces. The addition of small amounts of chain-transfer agents to the polymerization mixture gives a lower molecular weight homopolymer that is softer and more soluble. Copolymers of bromotrifluoroethylene with many other monomers such as chlorotrifluoroethylene, tetrafluoroethylene, or trifluoronitrosomethane have been reported. Neither the homopolymer nor the copolymers have any commercial utility.

Telomers. Bromotrifluoroethylene telomers have been prepared using chain-transfer agents such as CF_3SSCF_3, C_2F_5I, CBr_4, of CBr_3F. Commercially available bromotrifluoroethylene telomers have densities of 2.14–2.65 g/cm^3 and viscosities of 2–4000 nm^2/s(= cSt). These fluids are expensive but are made in small volume for the aerospace industry.

Arthur J. Elliott
Halocarbon Products Corporation

J. C. Blazejewski, D. Cantacuzene, and C. Wakselman, *Tetrahedron Lett.*, 2055 (1974).

W. R. Cullen and P. Singh, *Can. J. Chem.* **41**, 2397 (1963).

S. W. Hansen, T. D. Spawn, and D. J. Burton, *J. Fluorine Chem.* **35**, 415 (1987).

P. L. Heinze and D. J. Burton, *J. Org. Chem.* **53**, 2714 (1988).

POLY(FLUOROSILICONES)

The presence of carbon–fluorine bonds in organic polymers is known to characteristically impart polymer stability and solvent resistance. The poly(fluorosilicones) are siloxane polymers with fluorinated organic substituents bonded to silicon. Poly(fluorosilicones) have unique applications resulting from the combination provided by fluorine substitution into a siloxane polymer structure (see Silicon compounds–silicones).

Properties

Fluorosilicone elastomers can be formulated to provide specific durometer (hardness), tear strength, modulus, and solvent resistance properties.

Fluid and Chemical Resistance. Fluorosilicone elastomers and greases are especially suited for applications involving repeated exposure to fuels, oils, hydraulic fluids, and various chemicals. Fluid resistance is excellent to almost all solvents including alcohol–hydrocarbon mixtures currently being evaluated as alternative fuels. Exceptions to this rule are highly polar solvents, such as esters and ketones.

Heat Resistance. Thermal cycling does not lead to embrittlement. Elastomer service temperatures range from −60 to 200°C.

Low Temperature Properties. The ability to retract to 10% of their original extension after a 100% elongation at low temperature is an important test result. Fluorosilicones can typically pass this test down to −59 °C. The brittle point is approximately −68 °C.

Electrical Properties. Like unfluorinated silicone counterparts, fluorosilicone elastomers have inherently good electrical insulating properties. The dielectric properties remain relatively unchanged when the elastomer is exposed to severe environments.

Manufacture

Polymerization. Cyclotrisiloxanes are strained ring compounds. Polymerization is driven by relief of this ring strain. Acid- or base-catalyzed equilibration reactions of cyclotrisiloxane with a measured amount of end-blocking agent lead to fluid polymers with predictable molecular weight distributions.

Compounding. Fluorosilicone gums are compounded generally with fumed or precipitated silica fillers, hydroxy-containing low viscosity silicone oils, and readily available peroxides to produce various rubber products.

Vulcanization. Fluorosilicone elastomers can be peroxide-vulcanized by a free-radical mechanism using vinyl side groups that have been incorporated into the basic polymer structure during the initial polymerization process.

Fabrication. Fluorosilicones can be molded, extruded, or calendered by any of the conventional methods employed in the industry. Compression molding is the most widely used method and is ideal for a great many fabrications at 115–170°C and 5.5–10.3 MPa (800–1500 psi). Injection molding becomes increasingly important for high production operations.

Economic Aspects

Globally, there are but a small number of basic fluorosilicone producers: General Electric Company and Dow Corning Corporation in the United States, ShinEtsu in Japan, and Wacker Chemie in Germany. Growth of U.S. fluorosilicone elastomers was estimated at an average of 7.5% per year for the period 1989 through 1993.

Health and Safety Factors

Information on fluorosilicone polymers is limited to safe handling information available in specific fluorosilicone product brochures. No known chronic health effects have been reported.

Uses

Fluorosilicones are used in surface protection, foam control, fluids (oils) and greases, gels, sealants, and rubber.

WILLIAM X. BAJZER
YUNG K. KIM
Dow Corning Corporation

D. J. Cornelius and C. M. Monroe, *Polym. Eng. Sci.* **25**(8), 467–473 (1985).

M. T. Maxson and K. F. Benditt, SAE Technical Paper Series, Paper No. 88023, *SAEQ. Trans.* **97**(Pt. 2), 1–7 (1989).

M. S. Virant, L. D. Fiedler, T. L. Knapp, and A. W. Norris, SAE Technical Paper Series, Paper No. 910102, *SAEQ. Trans.* **100**(5), 37–48 (1991).

FLUORITE, FLUOROSPAR. See FLUORINE COMPOUNDS, INORGANIC–CALCIUM.

FLUOROALUMINATES, FLUOROBERYLLATES, FLUOROPHOSPHATES, FLUOROSILICATES, AND SIMILAR ENTRIES. See FLUORINE COMPOUNDS, INORGANIC.

FLUOROCHEMICALS. See FLUORINE COMPOUNDS, ORGANIC.

FOAMED PLASTICS

Foamed polymers, otherwise known as cellular polymers or polymeric foams, or expanded plastics have been important to human life since primitive people began to use wood, a cellular form of the polymer cellulose. Cellular polymers have been commercially accepted in a wide variety of applications since the 1940s. The total usage of foamed plastics in the United States has risen from 441×10^3 t in 1967 to a projected 2.8×10^6 t in 1995.

Classification

A cellular plastic has been defined as a plastic the apparent density of which is decreased substantially by the presence of numerous cells disposed throughout its mass. In this article the terms cellular plastic, foamed plastic, expanded plastic, and plastic foam are used interchangeably to denote all two-phase gas–solid systems in which the solid is continuous and composed of a synthetic polymer or rubber.

Theory of the Expansion Process

Foamed plastics can be prepared by a variety of methods. The most important process, by far, consists of expanding a fluid polymer phase to a low density cellular state and then preserving this state. This is the foaming or expanding process. Other methods of producing the cellular state include leaching out solid or liquid materials that have been dispersed in a polymer, sintering small particles, and dispersing small cellular particles in a polymer. The latter processes are relatively straightforward processing techniques but are of minor importance.

The expansion process consists of three steps: creating small discontinuities or cells in a fluid or plastic phase; causing these cells to

Table 1. Methods for Production of Cellular Polymers

Type of polymer	Extrusion	Expandable formulation	Froth foam	Compression mold	Injection mold	Sintering
cellulose acetate[a]	+					
epoxy resin[b]		+	+			
phenolic resin		+				
polyethylene[a]	+	+		+	+	+
polystyrene	+	+			+	+
silicones		+				
urea-formaldehyde resin			+			
urethane polymers[b]		+	+		+	
latex foam rubber			+			
natural rubber	+	+		+		
synthetic elastomers	+	+		+		
poly(vinyl chloride)[a]	+	+	+	+	+	
ebonite				+		
polytetrafluoroethylene						+

[a] Also by leaching. [b] Also by spray.

Table 2. Physical Properties of Commercial Rigid Foamed Plastics

Property	Cellulose acetate	Polystyrene, extruded sheet		PVC		Polyurethane: Polyether		Polyurethane: Isocyanurate, laminate
density, kg/m[3a]	96	96	160	32	64	32	64	32
mechanical properties compressive strength, kPa[b] at 10%	862	290	469	345	1035	138	482	117–206
tensile strength, kPa[b]	1172	2070–3450	4137–6900	551	1207	138	620	248–290
flexural strength, kPa[b]	1014			586	1620	413	1380	
shear strength, kPa[b]	965			241	793	138	413	117
compression modulus, MPa[c]	38–90			13.1	35	2.0	10.3	
flexural modulus, MPa[c]	38			10.3	36	5.5	5.5	
shear modulus, MPa[c]				6.2	21	1.2	3.4	1.7
thermal properties thermal conductivity, W/(m·I)	0.045	0.035	0.035	0.023		0.016–0.025	0.022–0.030	0.019
coefficient of linear expansion, 10^{-5}/°C						5.4	7.2	
max service temperature, °C	177	77–80	80			93	121	149
specific heat, kJ/(kg·K)[d]						ca 0.9	ca 0.9	
electrical properties dielectric constant	1.12	1.27	1.28			1.05	1.1	
dissipation factor	20	0.00011	0.00014			13	18	
moisture resistance water absorption, vol %	4.5							
moisture vapor transmission, g/(m·s·GPa)[e]		86	56	15		35	50	230

[a] To convert kg/m³ to lb/ft³, multiply by 0.0624. [b] To convert kPa to psi, divide by 6.895. [c] To convert MPa to psi, multiply by 145. [d] To convert kJ/(kg·K) to Btu/(lb·F), divide by 4.184. [e] To convert GPa to psi, multiply by 145,000.

grow to a desired volume; and stabilizing this cellular structure by physical or chemical means.

Initiation and Growth of Cells. The initiation or nucleation of cells is the formation of cells of such size that they are capable of growth under the given conditions of foam expansion. The growth of a hole or cell in a fluid medium at equilibrium is controlled by the pressure difference (ΔP) between the inside and the outside of the cell, the surface tension of the fluid phase γ, and the radius r of the cell:

$$\Delta P = 2\gamma/r \tag{1}$$

Stabilization of the Cellular State. The increase in surface area corresponding to the formation of many cells in the plastic phase is accompanied by an increase in the free energy of the system; hence the foamed state is inherently unstable. Methods of stabilizing this foamed state can be classified as chemical, eg, the polymerization of a fluid resin into a three-dimensional thermoset polymer, or physical, eg, the cooling of an expanded thermoplastic polymer to a temperature below its second-order transition temperature or its crystalline melting point to prevent polymer flow.

Manufacturing Processes

A summary of the methods for commercially producing cellular polymers is presented in Table 1.

Expandable Formulations

Physical Stabilization Process. Cellular polystyrene, the outstanding example; poly(vinyl chloride); copolymers of styrene and acrylonitrile (SAN copolymers); and polyethylene can be manufactured by this process.

Chemical Stabilization Processes. This method is more versatile and thus has been used successfully for more materials than the physical stabilization process. Chemical stabilization is more adaptable for condensation polymers than for vinyl polymers because of the fast yet controllable curing reactions and the absence of atmospheric inhibition. Foamed plastics produced by these processes include polyurethane foams, polyisocyanurates, and polyphenols.

Decompression Expansion Processes

Physical Stabilization Process. Cellular polystyrene, cellulose acetate, polyolefins, and poly(vinyl chloride) can be manufactured by this process.

Chemical Stabilization Processes. Cellular rubber and ebonite are produced by chemical stabilization processes.

Dispersion Processes

Frothing. The frothing process for producing cellular polymers is the same process used for making meringue topping for pies. A gas is dispersed in a fluid that has surface properties suitable for producing a foam of transient stability. The foam is then permanently stabilized by chemical reaction. The fluid may be a homogeneous material, a solution, or a heterogeneous material. Foamed plastics produced by frothing include latex foam rubber, urea–formaldehyde resins, and polyurethanes.

Syntactic Cellular Polymers. Syntactic cellular polymer is produced by dispersing rigid, foamed, microscopic particles in a fluid polymer and then stabilizing the system. The particles are generally spheres or microballoons of phenolic resin, urea–formaldehyde resin, glass, or silica. The fluid polymers used are the usual coating resins, eg, epoxy resin, polyesters, and urea–formaldehyde resin.

Properties of Cellular Polymers

Mechanical Properties of Commercial Foamed Plastics. The properties of commercial rigid foamed plastics are presented in Table 2. The properties of commercial flexible foamed plastics are presented in Table 3.

The properties that are achieved in commercial structural foams (density >0.3 g/cm³) are shown in Table 4.

Structural Variables

The properties of a foamed plastic can be related to several variables of composition and geometry often referred to as structural variables. These variables include polymer composition, density, cell structure (ie, cell size, cell geometry, and the fraction of open cells), and gas composition.

Table 3. Physical Properties of Commercial Flexible Foamed Plastics

Property	Expanded NR[a]		Expanded CR[a]	Expanded SBR	Latex foam rubber	PE extruded plank			PE sheet, extruded	Polypropylene, sheet	Polyurethane, Standard cushioning		PVC		Silicon, Sheet
density, kg/m^3 [b]	56	320	192	72	80	35	96	144	43	10	16	24	112	96	160
cell structure[c]	C	C	C	C	O	C	C	C	C		O	O	C	O	O
tensile strength, kPa[d]		206	758	551	103	138	413	690	41		88	118	24	3.4	
tensile elongation, %			500		310	60	60	60	276	138–275	160	205		220	310
rebound resilience, %					73				50			40			
tear strength, $(N/m)^e \times 10^2$						10.5	26	51	26		3.3	4.4			
max service temp., °C	70	70	70	70		82	82	82	82	121					260
thermal conductivity, W/(m·K)	0.036	0.043	0.065	0.030		0.053	0.058	0.058	0.040–0.049	0.039			0.040		0.086

[a] NR = natural rubber; CR = chloroprene rubber. [b] To convert kg/m^3 to lb/ft^3, multiply by 0.0624. [c] C = closed; O = open. [d] To convert kPa to psi, multiply by 0.145. [e] To convert N/m to lbf/in., divide by 1.75.

Table 4. Typical Physical Properties of Commercial Structural Foams

Property	ABS		Nylon[a]	PC[b]	Polyester[c]	HDPE	Polypropylene		High impact polystyrene		Polyurethane	
glass-reinforced	no	yes	yes	no	30%	no	no	20%	no	20%	no	no
density, g/cm^3	0.80	0.85	0.97	0.80	1.10	0.60	0.60	0.73	0.70	0.84	0.40	0.60
tensile strength, kPa[d]	18,600	48,000	101,000	37,900	76,000	8,900	13,800	20,700	12,400	34,500	11,000	23,400
compression strength, kPa[d] at 10% compression	6,900			51,700	76,000	8,900					5,500	19,300
flexural strength, kPa[d]	25,500	82,700	172,000	68,900	137,900	18,800	22,000	41,400	31,000	58,600	22,000	41,400
flexural modulus, GPa[e]	0.86	5.2	5.2	2.1	6.6	0.83	0.83	2.8	1.4	5.2	0.7	1.1
max use temperature, °C	82		203	132	193	110	115					

[a] Nylon-6,6 glass-reinforced. [b] Polycarbonate. [c] Thermoplastic polyester. [d] To convert kPa to psi, divide by 6,895. [e] To convert GPa to psi, multiply by 145,000.

Rigid Cellular Polymers. A separate class of high density, rigid cellular polymers has grown continually since the 1970s to become significant commercially. These are the structural foams with a density >300 kg/m^3. They are treated here as a separate category of rigid foams.

Compressive strength and modulus are widely used as general criteria to characterize the mechanical properties of rigid plastic foams. Rigid cellular polymers generally do not exhibit a definite yield point when compressed but instead show an increased deviation from Hooke's law as the compressive load is increased. Structural variables that affect the compressive strength and modulus of a rigid plastic foam are, in order of decreasing importance: plastic-phase composition, density, cell structure, and plastic state.

The creep characteristic of plastic foams must be considered when they are used in structural applications. Data on the deformation of polystyrene foam under various static loads have been compiled. There are two types of creep in this material: short-term and long-term. The minimum load required to cause long-term creep in molded polystyrene foam varies with density ranging from 50 kPa (7.3 psi) for foam density 16 kg/m^3 (1 lb/ft^3) to 455 kPa (66 psi) at foam density 160 kg/m^3 (10 lb/ft^3).

The successful application of time–temperature superposition for polystyrene foam is particularly significant in that it allows prediction of long-term behavior from short-term measurements. This is of interest in building and construction applications.

Structural Foams. Structural foams are usually produced as fabricated articles in injection molding or extrusion processes. The optimum product and process match differs for each fabricated article, so there are no standard commercial products for one to characterize. Rather there are a number of foams with varying properties. The properties of typical structural foams of different compositions are reported in Table 4. The most important structural variables are again polymer composition, density, and cell size and shape.

Flexible Cellular Polymers. The application of flexible foams has been predominantly in comfort cushioning, packaging, and wearing apparel, resulting in emphasis on a different set of mechanical properties than for rigid foams. The compressive nature of flexible foams (both static and dynamic) is their most significant mechanical property for most uses (Table 3). Other important properties are tensile strength and elongation, tear strength, and compression set. These properties can be related to the same set of structural variables as those for rigid foams.

Other Properties. The thermal, electrical, acoustical, and chemical properties of all cellular polymers are of such a similar nature that the discussions of these properties are not separated into rigid and flexible groups.

Thermal Properties. More information is available relating thermal conductivity to structural variables of cellular polymers than for any other property.

The following separation of the total heat transfer into its component parts, even if not completely rigorous, proves valuable to understanding the total thermal conductivity, k, of foams:

$$k = k_s + k_g + k_r + k_c \qquad (2)$$

where k^s, k_g, k_r, and k_c are the components of thermal conductivity attributable to solid conduction, gaseous conduction, radiation, and convection, respectively.

As a good first approximation, the heat conduction of low density foams through the solid and gas phases can be expressed as the product of the thermal conductivity of each phase times its volume fraction. Most rigid polymers have thermal conductivities of 0.07–0.28 W/(m·K) and the corresponding conduction through the solid phase of a 32 kg/m^3 (2 lbs/ft^3) foam (3 vol %) ranges 0.003–0.009 W/(m·K). In most cellular polymers this value is determined primarily by the density of the foam and the polymer-phase composition. Smaller variations can result from changes in cell structure.

Although conductivity through gases is much lower than that through solids, the amount of heat transferred through the gas phase in a foam is generally the largest contribution to the total heat transfer because the gas phase is the principal part of the total value (ca 97 vol % in a 32 kg/m^3 foam). The thermal conductivities of the halocarbon gases are considerably less than those of oxygen and nitrogen. It has, therefore, proved advantageous to prepare cellular polymers using such gases that measurably lower the k of the polymer foam.

The variation in total thermal conductivity with density has the same general nature for all cellular polymers. The increase in k at low densities is owing to an increased radiant heat transfer; the rise at high densities to an increasing contribution of k_s.

The thermal conductivity of most materials decreases with temperature and can change upon aging under ambient conditions if the gas composition is influenced by such aging. Thermal conductivity of foamed plastics has been shown to vary with thickness. This has been attributed to the boundary effects of the radiant contribution to heat-transfer.

The specific heat of a cellular polymer is simply the sum of the specific heats of each of its components.

The coefficients of linear thermal expansion of polymers are higher than those for most rigid materials at ambient temperatures because of the supercooled-liquid nature of the polymeric state, and this applies to the cellular state as well. When cellular polymers are used as components of large structures, the coefficient of thermal expansion must be considered carefully because of its magnitude compared with those of most nonpolymeric structural materials.

Because the cellular materials, like their parent polymers, gradually decrease in modulus as the temperature rises rather than undergoing a sharp change in properties, it is difficult to precisely define the maximum service temperature of cellular polymers. The upper temperature limit of use for most cellular polymers is governed predominantly by the plastic phase.

Work aimed at developing tests to evaluate the performance of plastic foams in actual fire situations continues. All plastic foams are combustible, some burning more readily than others when exposed to fire. Some additives, when added in small quantities to the polymer, markedly improve the behavior of the foam in the presence of small fire sources.

Plastic foams are advantageous compared to other thermal insulations in several applications where they are exposed to moisture pickup, particularly when subjected to a combination of thermal and moisture gradients.

Electrical Properties. Cellular polymers have two important electrical applications. One takes advantage of the combination of inherent toughness and moisture resistance of polymers along with the decreased dielectric constant and dissipation factor of the foamed state to use cellular polymers as electrical-wire insulation. The other combines the low dissipation factor and the rigidity of plastic foams in the construction of radar domes. Polyurethane foams have been used as high voltage electrical insulation.

Environmental Aging. All cellular polymers are subject to a deterioration of properties under the combined effects of light or heat and oxygen. The response of cellular materials to the action of light and oxygen is governed almost entirely by the composition and state of the polymer phase.

Comfort cushioning is the largest single application of cellular polymers; flexible foams are the principal contributors to this field. However, the rapid growth rate of structural, packaging, and insulation applications has brought their volume over that of flexible foams during the past few years. Table 5 shows U.S. consumption of foamed plastics by resin and market.

Commercial Products and Processes

Flexible Polyurethane. These foams are produced from long-chain, lightly branched polyols reacting with a diisocyanate, usually toluene diisocyanate (TDI), to form an open-celled structure with free air flow during flexure. During manufacture these foams are closely controlled for proper density, ranging from 13 to 80 kg/m^3 (0.8–5 lbs/ft^3), to achieve the desired physical properties and cost.

In flexible polyurethane foams, the primary blowing agent is carbon dioxide, which is formed by the reaction of water and toluene diisocyanate. Softer foams with lower densities require an auxiliary blowing agent such as CFC-11, HCFC-141b, or methylene chloride. Since the load bearing characteristics of the foam are of great importance to the ultimate consumer this property is also closely controlled during manufacture.

Table 5. Market for Cellular Polymers, 10^3 t

Item	1967	1982	1995[a]
	By market		
insulation	58	261	472
flooring	20	98	154
other construction	9	136	288
cushioning	52	195	336
other furniture	40	103	175
packaging	43	177	311
transportation	76	140	238
consumer	44	136	225
bedding	18	57	113
appliances	14	40	61
other	68	225	408
Total	*441*	*1567*	*2781*
	By resin		
flexible urethane	181	511	844
rigid urethane	68	248	449
styrene	125	410	699
vinyl	61	232	413
others	6	165	376
Total	*441*	*1567*	*2781*

[a] Projected as of 1994.

Applications. Carpet underlayment is a substantial market. Most furniture cushioning is made from blocks of slab-produced polyurethane foam in the density range of 16 to 29 kg/m^3 (1.0–1.8 lbs/ft^3). The furniture market for polyurethane foams tends to reflect the current economic trends.

For passenger car seating about 90% is made by the molded foam process. The transportation market has experienced a decline since 1979 due to decreased automotive production and also because U.S. cars have been downsized, resulting in the use of less polyurethane foam per ca.

Consumption of polyurethane foam in bedding reached a maximum in 1978 and has since declined. Textile uses, however, are a relatively stable area and consist of the lamination of polyester foams to textile products, usually by flame lamination or electronic heat sealing techniques.

Rigid Polyurethane. These foams are characterized by closed-celled structure and very high compressive strength. They are produced by using a highly branched, short-chain polyol reacted with an aromatic isocyanate of two or more functionality which is often polymeric. Pour-in-place and free rise rigid polyurethane foams usually have a density in the region of 32.0 kg/m^3 (2.0 lbs/ft^3), although molded rigid foams have densities ranging up to 640 kg/m^3 (40 lbs/ft^3) in structural foams. Insulation effectiveness is one of the outstanding characteristics of rigid polyurethane foams which display thermal conductivities as low as 0.017 W/(m·K).

Process and Equipment. Rigid polyurethane foam processes use the same high or low pressure pumping, metering, and mixing equipment for flexible foams. Subsequent handling of the mixture is determined by the end product desired. Processes include lamination, pour-in-place, molding, bun stock, box foams, and spray.

Applications. The principal use for rigid polyurethane foams is for insulation in various forms utilized by a variety of industries. Packaging constitutes another significant use and is often a foam-in-place operation to protect industrial equipment such as pumps or motors.

Polystyrene. There are five basic types of polystyrene foams produced in a wide range of densities and employed in a wide variety of applications: *(1)* extruded polystyrene board; *(2)* extruded polystyrene sheet; *(3)* expanded bead molding; *(4)* injection molded structural foam; and *(5)* expanded polystyrene loose-fill packaging.

Expanded polystyrene (EPS) beadboard insulation is produced with expandable polystyrene beads. These beads are produced by impregnating with 5 to 8% pentane and sometimes with flame retardants such as hexabromocyclododecane, pentabromomonochlorocyclohexane, or a synergistic mixture of antimony trioxide and dicumyl peroxide during suspension polymerization. The beads are preexpanded by fabricators with steam or vacuum and then allowed to age. The preexpanded beads are fed to the steam heated block molds where further expansion and fusion of beads take place. The molded blocks are then sliced into various sizes needed for specific applications after curing.

Expanded polystyrene bead molding products account for the largest portion of the drinking cup market and are used in fabricating a variety of other products including packaging materials, insulation board, and ice chests. The insulation value, the moisture resistance, and physical properties are inferior to extruded boardstock, but the material cost is much less.

Expanded polystyrene loose-fill packaging materials are produced normally by extrusion process followed by multiple steam expansions to give low density foam shapes that resemble "S", "8", and hollow shells. Expandable polystyrene loose-fill packaging material is also produced by suspension polymerization process with blowing agent incorporated into the polymer during the polymerization. These products are used as dunnage or space filling materials for cushion packaging. They have good shock absorbency, excellent resiliency, and are odorless.

Extruded polystyrene board was first introduced in the early 1940s by Dow Chemical Co. with the tradename Styrofoam. The Styrofoam process consists of the extrusion of a mixture of polystyrene and volatile liquid blowing agent expanded through a die to form boards in various sizes. The continuous boards are then passed through the finishing equipment for further sizing.

In residential sheathing insulation, fiberboard is still the most widely used product, although the use of extruded and molded polystyrene foam and of foilfaced isocyanurate foam is increasing depending on the cost, the amount of insulation required, and compatibility of insulation with other construction systems. In cavity-wall insulation, mineral wool, polyurethane, urea–formaldehyde, and fiber glass are widely used, although fiber glass batt is the most economical insulation for stud-wall construction. In mobile and modular homes, cellular plastics are used widely because of their light weight and more efficient insulation value.

Extruded polystyrene foam sheet is primarily produced in a single-screw tandem extrusion line. Primary application of foam sheet is as a packaging material in items such as disposable dishes and food containers, trays for meat, poultry and produce products, and egg cartons.

Injection molded structural foam is used widely for high density items such as picture frames, furniture, appliances, housewares, utensils, toys, pipes, and fittings. Most of these products are produced by injection molding or profile extrusion methods from impact modified polystyrene. Almost all high density foam products are produced with a chemical blowing agent that releases either nitrogen or carbon dioxide, typically sodium bicarbonate or azodicarbonamides. Medium density products can be produced with either a physical or chemical blowing agent, or a combination of both.

Poly(vinyl chloride). Cellular poly(vinyl chloride) (PVC) foam is available in both flexible and rigid foams. Flexible PVC foams are primarily produced by spread coating and calendering of fluid plastisols by means of a chemical blowing agent or mechanical frothing with air. Flexible PVC foams also are made by the extrusion process. Rigid PVC foams are produced by the extrusion or injection molding processes. Blowing is achieved by a chemical blowing agent or gas injection into the extruder.

Raw Materials. PVC is inherently a hard and brittle material and very sensitive to heat; it thus must be modified with a variety of plasticizers, stabilizers, and other processing aids to form heat-stable flexible or semiflexible products or with lesser amounts of these processing aids for the manufacture of rigid products (see VINYL POLYMERS–VINYL CHLORIDE POLYMERS).

Applications. Furniture and motor vehicle upholstery is the largest market for flexible vinyl foams. Because of better aesthetics (leather-like plastics), comfort, and favorable pricing, they are expected to show good growth in upholstery, carpet backing, resilient floor coverings, outerwear, footwear, luggage, and handbags. The only application for flexible vinyl foams in protective packaging applications is for stretch pallet wraps. These wraps are produced by extrusion.

Rigid vinyl foams in construction markets have grown substantially due to improved techniques to manufacture articles with controlled densities and smooth outer surfaces. Wood molding substitute for door frames and other wood products is an area that has grown. Rigid vinyl foams are also used in the manufacture of pipes and wires as resin extenders and in sidings and windows as the replacement of wood or wood substitutes.

Polyethylene. There are three basic types of polyethylene foams of importance: *(1)* extruded foams from low density polyethylene (LDPE); *(2)* foam products from high density polyethylene (HDPE); and *(3)* cross-linked polyethylene foams. Other polyolefin foams have an insignificant volume as compared to polyethylene foams and most of their uses are as resin extenders.

Extruded low density foam produced from LDPE is a tough, flexible, and resilient closed-celled foam used in a wide variety of applications such as cushion packaging and safety components.

HDPE foam is primarily used as a high density rigid product. Shipping pallets are a rapidly growing market at a projected growth rate of about 26% per year through the mid-1990s. Most of these products are produced by thermoforming sheet and injection molding.

Cross-linked polyethylene foams are produced by either radiation or chemical cross-linking of an extruded expandable sheet containing a chemical blowing agent. These products have finer texture and a softer, more resilient feel than extruded low density polyethylene foams and are used in comfort cushioning and cushion packaging applications.

Kanegafuchi Chemical of Japan has introduced a chemical cross-linking process for producing PE foams by the bead technique similar to EPS. Their Eperan beads have been used to produce molded articles as cushioning materials, sound insulating panels, etc. Asahi-Dow and BASF have also been reported to have developed similar products.

Health and Safety

Flammability. Plastic foams are organic in nature and, therefore, are combustible. They vary in their response to small sources of ignition because of composition and/or additives. All plastic foams should be handled, transported, and used according to manufacturers' recommendations as well as applicable local and national codes and regulations.

Virtually all plastic foams are blown with inert gases (CO_2, N_2, H_2O). Among these blowing agents, hydrocarbons and some of the HCFs and HFCs are flammable and pose a fire hazard in handing at the manufacturing plants.

Atmospheric Emissions. Certain organic compounds are found to be smog generating substances because of their high photochemical reactivity at ambient conditions. Since fully or partially halogenated hydrocarbons are considered to have low reactivity in the lower atmosphere (troposphere), substitution of photochemically reactive compounds for the current blowing agents may reduce ozone depletion in the stratosphere, but has adverse impact on the indoor ambient air quality. Therefore, ozone/oxidant interaction with the total environment needs to be considered in developing environmentally acceptable alternative blowing agents.

Toxicity. The presence of additives or unreacted monomers in certain plastic foams can limit their use where food or human contact is anticipated. Heavy metals can also be found in various additives. The manufacturers' recommendations or existing regulations again should be followed for such applications.

KYUNG W. SUH
The Dow Chemical Company

K. C. Frisch and J. H. Saunders, *Plastic Foams*, Vol. 1, Pts. 1 and 2, Marcel Dekker, Inc., New York, 1972 and 1973.

C. J. Benning, *Plastic Foams*, Vols. 1 and 2, Wiley-Interscience, New York, 1969.

N. C. Hilyard and co-workers, *Mechanics of Cellular Plastics*, Macmillan Publishing Co., Inc., New York, 1982.

D. Klempner and K. C. Frisch eds., *Polymeric Foams*, Hanser Publishers, New York, 1991.

FOAMS

Foam is a nonequilibrium dispersion of gas bubbles in a relatively smaller volume of liquid. An essential ingredient in a liquid-based foam is surface-active molecules. These reside at the interfaces and are responsible for both the tendency of a liquid to foam and the stability of the resulting dispersion of bubbles. Important uses for custom-designed foams vary widely from familiar examples of detergents, cosmetics, and foods, to fire extinguishing, oil recovery, and a host of physical and chemical separation techniques. Unwanted generation of foam, on the other hand, is a common problem affecting the efficiency and speed of a vast number of industrial processes involving the mixing or agitation of multicomponent liquids. In all cases, control of foam rheology and stability is desired. These physical properties, in turn, are determined by both the physical chemistry of their liquid–vapor interfaces and by the structure formed from the collection of gas bubbles.

Physical Chemistry of Interfaces

The chemical composition, physical structure, and key physical properties of a foam, namely its stability and rheology, are all closely interrelated. Since there is a large interfacial area of contact between liquid and vapor inside a foam, the physical chemistry of liquid–vapor interfaces and their modification by surface-active molecules plays a primary role underlying these interrelationships.

For aqueous solutions, the chemical constituents most commonly responsible for foaming are surfactants, ie, surface-active agents. Such molecules find wide use in other settings, and are distinguished by having both hydrophilic and hydrophobic regions.

Reduced Surface Tension. Just as surfactants self-organize in the bulk solution as a result of their hydrophilic and hydrophobic segments, they also preferentially adsorb and organize at the solution–vapor interface. In the case of aqueous surfactant solutions, the hydrophobic tails protrude into the vapor and leave only the hydrophilic head groups in contact with the solution. The favorable energetics of the arrangement can be seen by the reduction in the interfacial free energy per unit area, or surface tension, σ.

Gibbs Elasticity and Marangoni Flows. The reduction of surface tension with increasing surfactant adsorption gives rise to a nonequilibrium effect which can, in some cases, promote foaming. A sudden increase in the interfacial area by mechanical perturbation or thermal fluctuation results in a locally higher surface tension because the number of surfactant molecules per unit area simultaneously decreases. The Gibbs elasticity, E, is often used to quantify the instantaneous change in surface tension σ with area A, ie, $E = d\sigma/d\ln A$. If the film of liquid separating two neighboring bubbles in a foam develops a thin spot, the surface tension gradient in the vicinity of the thin spot will induce a Marangoni flow of liquid toward the direction of higher σ. This flow of liquid toward the thin spot helps heal the fluctuation and thus keeps the neighboring bubbles from coalescing.

Interfacial Forces. Neighboring bubbles in a foam interact through a variety of forces which depend on the composition and thickness of liquid between them, and on the physical chemistry of their liquid–vapor interfaces. For a foam to be relatively stable, the net interaction must be sufficiently repulsive at short distances to maintain a significant layer of liquid in between neighboring bubbles. Interfacial forces include the van der Waals interaction, the electrostatic double layer interaction, and disjoining pressure.

Physical Properties of Foam

Based on the underlying physical chemistry of surfactants at interfaces, important features of foam structure, stability, rheology, and their interrelationships can be considered as ultimately originating in the molecular composition of the base liquid.

Structure. Foam structure is characterized by the "wetness" of the system. Foams with arbitrarily large liquid to gas ratios can be generated by excessive agitation or by intentionally bubbling gas through a fluid. If the liquid content is sufficiently great, the foam consists of well-separated spherical bubbles that rapidly rise upwards displacing the heavier liquid. Such a system is usually called a froth, or bubbly liquid, rather than a foam.

If there are sufficiently strong repulsive interactions, such as from the electric double-layer force, then the gas bubbles at the top of a froth collect together without bursting. Furthermore, their interfaces approach as closely as these repulsive forces allow; typically on the order of 100 nm. Thus bubbles on top of a froth can pack together very closely and still allow most of the liquid to escape downward under the influence of gravity while maintaining their spherical shape. Given sufficient liquid, such a foam can resemble the random close-packed structure formed by hard spheres.

A dry foam, by contrast, is one with so little liquid that the bubbles are severely distorted into approximately polyhedral shapes. Typically this occurs for foams with less than 1% liquid by volume.

A complete characterization of the structure of a foam requires a characterization of the structure of the bubbles that comprise the foam. The total liquid content can be readily found from the mass densities of the foam and the liquid from which it was made. However, a more detailed determination of the bubble structure, including their average size, their shape, their structure and their size distribution is much more difficult, and is typically impeded by the problems in visualizing the interior of a foam. Even in the absence of any intrinsic optical absorption of the liquid, the strong mismatch in the indexes of refraction between the gas and the fluid results in a large scattering of light, usually precluding direct visualization of the interior structure of a foam. As a result, other, less direct, methods have been developed, and must be used, except in exceptional cases where the foam structure has been optimized for visualization.

One optical imaging technique that circumvents the problem of multiple light scattering is to estimate the bubble size distribution from the area individual foam bubbles occupy at a glass surface. Such experiments, and the systematic differences between bulk and surface bubble distributions, have been reviewed. Another technique that also directly measures the bubble size distribution is the use of a Coulter counter, where individual bubbles are drawn through a small tube and counted. This yields a direct measure of the bubble size distribution, but it is invasive and cannot probe the structure of the foam.

One technique that does probe the foam structure directly is cryomicroscopy. The foam is rapidly frozen, and the solid structure is cut open and imaged with an optical or electron microscope. Such methods are widely applicable and provide a direct image of the foam structure; however, they destroy the sample and may also perturb the foam structure in an uncontrolled manner during the freezing.

Stability. Control of foam stability is important in all applications, whether degradation of a custom foam is to be minimized or whether excessive foaming is to be prevented. In all cases, the time evolution of the foam structure provides a natural means of quantifying foam stability. There are three basic mechanisms whereby the structure may change: by the gravitational segregation of liquid and bubbles, by the coalescence of neighboring bubbles via film rupture, and by the diffusion of gas across the liquid between neighboring bubbles.

Any means of characterizing foam structure can be used to study foam evolution, provided that the measurement can be made noninvasively and sufficiently rapidly. One technique that has been applied successfully is the measurement of the change in the pressure head over an evolving foam. Multiple light scattering can also be used to follow the time evolution of a foam. A common engineering technique for determining foam stability entails measuring the amount of foam produced. For defoaming applications, this is often a more important measure of stability than the foam structure.

Rheology. The rheology of foam is striking; it simultaneously shares the hallmark rheological properties of solids, liquids, and gases, and their mechanical response to external forces can be very complex.

One simple rheological model that is often used to describe the behavior of foams is that of a Bingham plastic. This applies for flows over length scales sufficiently large that the foam can be reasonably considered as a continuous medium. The Bingham plastic model combines the properties of a yield stress like that of a solid with the viscous flow of a liquid.

While the Bingham plastic model is an adequate approximate description of foam rheology, it is by no means exact, especially at low strain rates. More detailed models attempt to relate the rheological properties of foams to the structure and behavior of the bubbles.

To determine rheological parameters such as the yield stress and effective viscosity of a foam, commercial rheometers are available; rotational and continuous-flow-tube viscometry are most commonly employed (see RHEOLOGICAL MEASUREMENTS). However, obtaining reproducible results independent of the sample geometry is a difficult goal which arguably has not been achieved in most of the experiments reported in the scientific literature.

Production

Several techniques are available for the generation of special-purpose foam with the desired properties. The simplest method is to disperse compressed gas directly into an aqueous surfactant solution by means of a glass frit. A variation of this method that allows for control of liquid content is to simultaneously pump gas and surfactant solution through a bead pack or steel wool, for example, at fixed rates. Less reproducible mechanical means of foam generation include brute force shaking and blending. For highly reproducible foams composed of small bubbles, such as shaving creams, the aerosol technique is especially suitable.

Applications

Foams have a wide variety of applications that exploit their different physical properties. The low density, or high volume fraction of gas, enable foams to float on top of other fluids and to fill large volumes with relatively little fluid material. These features are of particular importance in their use for fire fighting. The very high internal surface area of foams makes them useful in many separation processes. The unique rheology of foams also results in a wide variety of uses, as a foam can behave as a solid, while still being able to flow once its yield stress is exceeded. Foams are used in food, oil recovery, detergents, textiles, and cosmetics.

Safety, Health, and Environment

Foams play important roles in environmental issues, both beneficial and detrimental.

Natural Waters. Many water systems have a natural tendency to produce foam upon agitation. The presence of pollutants exacerbates this problem. This was particularly severe when detergents contained surfactants that were resistant to biodegradation. Then, water near industrial sites or sewage disposal plants could be covered with a blanket of stable, standing foam. However, surfactant use has switched to biodegradable molecules, which has greatly reduced the incidence of these problems.

Wastewater Treatment. The treatment of wastewater, either from sewage or from industrial processes, typically entails a preliminary filtration to remove the large volumes of solids, and then a slower settling to remove the sand and gravel (see WATER). The water is then treated by an activated sludge process to remove the remaining dissolved solids and organic colloidal particles. Activated sludge is a biomass that assists in the degradation of the organic waste in the water. The process entails a mixing and aeration of the wastewater with the activated sludge, which can lead to problems of foaming. The foams produced can be quite stable, resulting in additional problems for waste disposal. The foams produced in this process differ from those normally encountered in that the foam producing and stabilizing agents are microbial, primarily including *Nocardia, Microthrix parvicella*, and *Rhodococcus*. These foams are more difficult to treat with defoaming agents. Moreover, it is very difficult to predict the degree of foamability of the waste being treated. In other, more specialized wastewater treatments, these problems do not arise, and defoaming agents can be used effectively.

Chlorofluorocarbon Alternatives. There still is no completely satisfactory propellant for use in the aerosol method of foam production. Chlorofluorocarbons, still widely used, are harmful to atmospheric ozone and low molecular weight hydrocarbons, now popular, eg, in producing shaving cream, are explosive and promote the greenhouse effect (see FLUORINE COMPOUNDS, ORGANIC–FLUORINATED ALIPHATIC COMPOUNDS).

DOUGLAS J. DURIAN
UCLA
DAVID A. WEITZ
Exxon Research & Engineering Company

J. H. Aubert, A. M. Kraynik, and P. B. Rand, *Sci. Am.* **254**, 74 (May 1986).

J. J. Bikerman, *Foams*, Springer-Verlag, New York, 1973.

H. C. Cheng and T. E. Natan, in N. P. Cheremisinoff, ed., *Encyclopedia of Fluid Mechanics*, Vol. 3, Gulf Publishing Co., Houston, Tex., 1986, p. 3.

A. M. Kraynik, *Ann. Rev. Fluid Mech.* **20**, 325 (1988).

FOOD ADDITIVES

Definition and Regulatory Considerations

According to the *U.S. Code of Federal Regulations*, food additives may be defined as "substances ... the intended use of which results or may reasonably be expected to result, directly or indirectly, either in their becoming a component of food or otherwise affecting the characteristics of food" (*Title 21, Part 170.3*, Apr. 1, 1990). Canada and the European Community have adopted similar definitions. According to this broad definition, a food additive is synonymous to a food ingredient. In practice, however, the word additive is limited to substances that are used in small quantities.

In the United States, substances permitted in food and beverages are regulated by the U.S. Food and Drug Administration (FDA), an agency of the Department of Health and Human Services. Additives used in meat and poultry (see MEAT PRODUCTS) are regulated by the U.S. Department of Agriculture, and additives for alcoholic beverages are regulated by the Bureau of Alcohol, Tobacco and Firearms of the U.S. Department of Treasury (see BEER; BEVERAGE SPIRITS, DISTILLED; WINE). Premarketing approval is required.

In the United States, additional ramifications may be expected from FDA's announcement of final regulations for new food labeling requirements under the directive of the Nutrition Labeling and Education Act of 1990. Among other things, these regulations limit health claims that can be made on food labels. They also require new information on nutrient content, and limit the use of descriptors such as low and free in association with calories, fat levels, and other food product characteristics.

In Europe, the formation of the European Economic Community has created a requirement to bring food additive approvals of the member nations into alignment, so as to eliminate differences in laws that hinder the movement of foodstuffs among these nations.

Classes of Food Additives

Acidulants. Acidulants, the most versatile and widely used ingredients in the food industry, function well as flavoring agents. They include adipic acid, citric acid, fumaric acid, gluconolactone, lactic acid, malic acid, phosphoric acid, and tartaric acid.

Anticaking Agents. Anticaking agents function by absorbing excess moisture, or by coating particles and making them water repellent. They include calcium silicate, $CaSiO_3$, and calcium and magnesium salts of long-chain fatty acids, such as calcium stearate, $C_{36}H_{70}CaO_4$.

Antifoaming Agents. Polydimethylsiloxane, or silicone, is used at a level of approximately 10 parts per million to control foam in food products.

Antioxidants. Antioxidants work by donating a hydrogen atom to the reactive peroxide radical, ending the chain reaction. Both synthetic and natural antioxidants exist. The most commonly used synthetic antioxidants include butylated hydroxyanisole (BHA), $C_{11}H_{16}O_2$, butylated hydroxytoluene (BHT), $C_{15}H_{24}O$, propyl gallate (PG), $C_{10}H_{12}O_5$, and *tert*-butylhydroquinone (TBHQ), $C_{14}H_{22}O_2$. Although BHT was never removed from the GRAS list, continuing concern over its safety has resulted in decreased usage.

The most popular natural antioxidants on the market are rosemary extracts and tocopherols. Certain compounds, known as chelating agents (qv), react synergistically with many antioxidants. Citric acid and ethylenediaminetetraacetic acid (EDTA), $C_{10}H_{16}N_2O_8$, are the most common chelating agents used.

Another group of compounds called oxygen scavengers retard oxidation by reducing the available molecular oxygen. Products in this group are water soluble and include erythorbic acid, $C_6H_8O_6$, and its salt sodium erythorbate, $C_6H_8O_6Na$, ascorbyl palmitate, $C_{22}H_{38}O_7$, ascorbic acid, $C_6H_8O_6$, glucose oxidase, and sulfites.

Bulking Agents and Bulking Sweeteners. Bulking agents are substances that add bulk to food products while contributing fewer calories than the ingredients they replace.

Polydextrose, a polymer of glucose that contains traces of sorbitol and citric acid, is the most widely used soluble bulking agent in the United States.

Low calorie bulking agents represent an ingredient category having a great deal of potential, and several companies are developing products. The most common are naturally derived polymers of glucose and other sugars (polydextrose falls into this category), enantiomers of natural sugars, or synthetic polymers. Bulking sweeteners provide a bulking effect, along with some of the sweetness and functional properties of sugar. Products that fall into this category include mannitol, $C_6H_{14}O_6$, isomaltitol, some L-sugars, and fructooligosaccharides.

Colorants. According to U.S. regulations, colorants are divided into two classes: certified and exempt. The FD&C certified colors are all water-soluble dyes, but can be transformed into insoluble pigments known as lakes by precipitating the dyes with aluminum, calcium, or magnesium salts on a substrate of aluminum hydroxide.

Exempt colors do not have to undergo formal FDA certification requirements, but are monitored for purity. The colorants exempt from FD&C certification are annatto extract, β-carotene, beet powder, β-apo-8′-carotenol, canthaxanthin, caramel, carmine, carrot oil, cochineal extract, cottonseed flour, ferrous gluconate, fruit juices, grape skin extract, paprika, paprika oleoresin, riboflavin, saffron, titanium dioxide, turmeric, turmeric oleoresin, ultramarine blue, and vegetable juices.

Dietary Fiber. Dietary fiber (qv) is a broad term that encompasses the indigestible carbohydrate and carbohydrate-like components of foods that are found predominantly in plant cell walls (see CARBOHYDRATES). It includes cellulose (qv), lignin (qv), hemicelluloses (qv), pentosans, gums (qv), and pectins.

Emulsifiers. The chemical structures of emulsifiers, or surfactants (qv), enable these materials to reduce the surface tension at the interface of two immiscible surfaces, thus allowing the surfaces to mix and form an emulsion. An emulsifier consists of a polar group, which is attracted to aqueous substances, and a hydrocarbon chain, which is attracted to lipids. Emulsifiers include mono- and diglycerides, lecithin, propylene glycol esters, lactylated esters, sorbitan and sorbitol esters, polysorbates, and sucrose esters.

Enzymes. In the food industry, the largest use of enzymes is in starch processing, cheese production, fruit and vegetable juice processing, baking, and brewing. Commercial enzyme preparations are obtained from animals and plants via extraction, or through cultivation of select microorganisms. Enzymes are divided into six main classes: hydrolases, isomerases, ligases, lyases, oxidoreductases, and transferases.

Fat Replacers. Two classes of fat replacers exist: mimetics, which are compounds that help replace the mouthfeel of fats but cannot substitute for fat on a weight for weight basis; and substitutes, compounds having physical and thermal properties similar to those of fat, that can theoretically replace fat in all applications. Because fats play a complex role in so many food applications, one fat replacer is often not a satisfactory substitute. Thus a systems approach to fat replacement, which relies on a combination of emulsifiers, gums, and thickeners, is often used.

Existing fat mimetics are either carbohydrate-, cellulosic (fiber)-, protein-, or gum-based.

As of this writing, only one fat substitute, caprenin, a triglyceride composed of capric acid, $C_{10}H_{20}O_2$, caprylic acid, $C_8H_{16}O_2$, and behenic acid, $C_{22}H_{44}O_2$, has had any commercial application.

Firming Agents. During thermal processing and freezing, the bonds of pectic substances in plant walls that help to stabilize structure are modified, resulting in an unacceptably soft product. The cell wall structure of fruits and vegetables can be strengthened by adding polyvalent cations that promote the cross-linking of the free-carboxyl groups of pectic substances. Fruits such as tomatoes, berries, and apple slices are commonly firmed by added calcium salts prior to processing. Acidic aluminum salts are added during the preparation of pickles and relishes to provide the same effect.

Flavors. Flavorings are used in the food industry to replace or enhance flavors that are lost during processing, to create flavor combinations that do not exist in nature, and to mask objectionable flavors. Over 6000 flavor ingredients exist. They include essential oils, oleoresins, fruit juices and concentrates, botanical and animal extracts, aroma chemicals, and compounded flavors.

Flavor Enhancers. Flavor enhancers have the ability to enhance flavors at a level below which they contribute any flavor of their own. Worldwide, the most popular flavor enhancers are monosodium L-glutamate (MSG), $NaC_5H_8NO_4$, and the 5′-ribonucleotides: disodium 5′-inosinate (IMP), $C_{10}H_{11}N_4O_8P \cdot 2\,Na$, and disodium 5′-guanylate (GMP), $C_{10}H_{12-}N_5O_8P \cdot 2\,Na$.

Ammonium glycyrrhizinate (AG), $C_{42}H_{65}NO_{16}$, is a flavor enhancer derived from licorice root. Maltol, $C_6H_6O_3$, and ethyl maltol, $C_7H_8O_3$, are used as flavor enhancers in products such as cake mixes, confections, cookies, ice cream, fruit juices, puddings, and beverages.

Flour Bleaching Agents and Bread Improvers. Benzoyl peroxide, $C_{14}H_{10}O_4$, is a bleaching agent that is typically added at the flour mill at a level between 0.015 and 0.075%. This additive oxidizes the carotenoid pigments, resulting in a white flour. Gases that exert an effect on the flour upon immediate contact include chlorine gas, chlorine dioxide, ClO_2, nitrosyl chloride, NOCl, nitrogen oxides, N_xO_x, and nitrogen tetroxide, N_2O_4. Others that exert their effect when the flour is made into dough include potassium bromate, $KBrO_3$, potassium iodate, KIO_3, calcium iodate, $Ca(IO_2)_3$, and calcium peroxide, CaO_2.

Formulation Aids. Formulation acids, which include carriers, binders, fillers (qv), plasticizers (qv), and film-formers, are ingredients used in processing to impart a particular physical state or textural characteristic. Table 1 gives an overview of the formulation aids used in the food industry.

Fumigants. Fumigants are volatile substances used for controlling insects or pests. They include ethylene oxide, C_2H_4O, propylene oxide, C_3H_6O, and methyl bromide, CH_3Br.

Gases. Gases provide three basic functions as food ingredients: preservation, carbonation, and aeration.

Humectants. In certain foods, it is necessary to control the amount of water that enters or exits the product. It is for this purpose that humectants are employed. Polyhydric alcohols (polyols), which include propylene glycol, $C_3H_8O_2$, glycerol, $C_3H_8O_3$, sorbitol, $C_6H_{14}O_6$, and mannitol $C_6H_{14}O_6$, contain numerous hyroxyl groups (see ALCOHOLS, POLYHYDRIC). Their structure makes them hydrophilic and enables them to bind water in foods.

Leavening Agents. Sodium bicarbonate, $NaHCO_3$, is the most commonly used product, but ammonium bicarbonate, NH_4HCO_3, and potassium bicarbonate, $KHCO_3$, are used as well. When used alone, sodium bicarbonate reacts to give products a bitter, soapy flavor. Thus it is always combined with a leavening acid.

Leavening acids are classified according to the rate at which they release carbon dioxide from sodium bicarbonate. Most products use both slow- and fast-acting leavening acids to obtain appropriate volume. The leavening acids most frequently used include potassium acid tartrate, sodium aluminum sulfate, δ-gluconolactone, and ortho- and pyrophosphates.

Lubricants and Release Agents. Ingredients that fall into this category include oils, lecithin, starch, distilled acetylated monoglycerides, and magnesium silicate, $MgSiO_3$.

Nonnutritive Sweeteners. As of this writing there are only three nonnutritive sweeteners approved for used in the United States: aspartame, $C_{14}H_{18}N_2O_5$, saccharin, $C_7H_5NO_3S$, and acesulfame K, $C_4H_5NO_4S \cdot K$.

Nutrients. In the United States, foods are either restored, enriched, or fortified with nutrients. The enrichment program followed in the United States is *(1)* the enrichment of flour, bread, and degerminated and white rice using thiamin, $C_{12}H_{17}N_5O_4S$, riboflavin,

Table 1. Formulation Aids Used in Food Processing

Category	Function	Typical applications
	Carriers	
starches	allow addition of incompatible substances to food product	cheese, dry mixes, flour, flavor compounds
dextrins		
cellulose compounds		
silicas		
	Binders	
starches	hold food together	prepared meat, fish, poultry, chewing gum, confections
salts		
dextrins		
oils		
gums		
	Fillers	
maltodextrin	add bulk to food products	confections, dietary products, chewing gum, cereal mixes
polydextrose		
starches		
	Plasticizers	
oils	maintain soft texture of food products	chewing gum, confections, margarine, cheese products
waxes		
resins		
humectants		
	Film formers	
carnauba wax	increase palatability, preserve gloss, inhibit discoloration, protect food surfaces	confections, snack foods, nuts, fresh and dried fruits and vegetables
paraffin		
sodium caseinate		
mineral oil		

$C_{17}H_{20}N_4NaO_9P$, niacin, $C_6H_5NO_2$, and iron; *(2)* the retention or restoration of thiamin, riboflavin, niacin, and iron in processed food cereals; *(3)* the addition of vitamin D to milk, fluid skimmed milk, and nonfat dry milk; *(4)* the addition of vitamin A, $C_{20}H_{30}O$, to margarine, fluid skimmed milk, and nonfat dry milk; *(5)* the addition of iodine to table salt; and *(6)* the addition of fluoride to areas in which the water supply has a low fluoride content.

Preservatives. Most preservatives do not kill microorganisms present in food. Rather, they prevent further growth and proliferation of anything that is present by either lowering the water activity or increasing the pH of the foods in which they are used. Preservatives include benzoates, sorbates, propionates, organic acids, sulfur dioxide and sulfites, parabens, sodium nitrate and sodium nitrites, and natamycin and nisin.

Processing Aids. Manufacturing aids used to improve the appearance or performance of food products include clarifying agents (flocculants), clouding agents, catalysts, and filter aids.

Solvents. Solvents are generally used to either extract particular compounds, such as an essential oil from a plant, or to carry additives into a food system, such as a flavor into a powdered mix. Common solvents include ethanol, C_2H_6O, glycerine $C_3H_8O_3$, propylene glycol, $C_3H_8O_2$, triethyl citrate, $C_{12}H_{20}O_7$, polyhydric alcohols, carbon dioxide, acetylated monoglycerides, hexane, C_6H_{14}, methylene chloride, CH_2Cl_2, acetone, C_3H_6O, and trichloroethylene, C_2HCl_3.

Stabilizers and Thickeners. Many food products receive their textural properties from a group of compounds known as hydrocolloids. Hydrocolloids fall into two classes: polysaccharides and proteins. They include locust bean gum, guar gum, gum arabic, carrageenan, xanthan gum, cellulose, agar, starch, pectin, alginates, and gelatin.

Market Overview

The U.S. market for food additives reached about $4.3 billion in 1993–1994. The additives and ingredients industry remains quite fragmented. Competitors are drawn from a wide range of industries that include commodity grain and oilseed processors, other agricultural material processors, and bulk, specialty, and fine chemical suppliers. The large majority of participants have under $100 million in sales; only a handful have over $300 million.

Overall, the world market for food additives was approximately $10 billion in 1993–1994. Besides the United States, the other principal markets were the European Community (about $3.4 billion) and Japan (about $2.5 billion). Growth rates are expected to average about 3% per year throughout most of the 1990s. Highest growth rates should continue to be realized in segments where the ingredients and additives are needed for low calorie, low fat, and other nutrition-oriented products. These include nonnutritive sweeteners, fat replacers, bulking agents, thickeners, and dietary fibers. Artificial flavors, colors, and preservatives are expected to experience slow or no growth.

LESLIE J. FRIEDMAN
C. GAIL GREENWALD
Arthur D. Little, Inc.

Code of Federal Regulations, Title 21 Part 170.3, U.S. Government Printing Office, Washington, D.C., Apr. 1, 1990.

R. C. Lindsay, in O. R. Fennema, ed., *Food Chemistry*, 2nd ed., Marcel Dekker, Inc., New York, 1985, pp. 665–666.

FOOD PACKAGING

The principal functions of food packaging are to protect the food contents form physical damage, losses, or deterioration, and to facilitate distribution from processor to consumer. Food packaging also must attractively identify the product and must perform these functions at minimum system cost because the package itself has no intrinsic value to the consumer. As of the 1990s, packagers must actively consider the ultimate disposal of the used package in selecting food packaging. In 1992, food packaging represented about 57% of the United States' more than $70 billion packaging industry.

Packaged Food Classification

Approximately half of all the food products in the United States are fresh or minimally processed. Fresh food products include meats, vegetables, and fruits that are unprocessed except for removal from the original environment and limited trimming and cleaning. Fresh foods are handled to retard deterioration, which is relatively rapid at ambient or higher temperature. Meats are chilled rapidly to below 10°C (50°F) and most vegetables and fruits are generally reduced to below 4.4°C (40°F) by low temperature air, water, or ice. Minimally processed foods include those that have been altered to help retard deteriorative processes. For example, most dairy products must be refrigerated after pasteurization (see MILK AND MILK PRODUCTS); nitrite-cured meats must be kept refrigerated after processing to minimize microbial growth.

Fully processed foods are intended for long-term shelf life at ambient temperature, and include almost all heat processed, dried, etc, foods (see FOOD PROCESSING).

Each category of food has its own protection requirements filled by packaging, and so a variety of packaging is applied.

Paper, Metal, and Glass Packaging

Paper. The largest volume packaging materials in the United States, accounting for 40% of all food packaging, are paper (qv) and paperboard. The packaging properties of paper, its strength, and mechanical properties depend on the treatment of the wood fibers and on the incorporation of fillers (qv) and binding materials at the paper mill. The physicochemical properties of paper and paperboard, such as permeability to vapor and gases, are derived from impregnation, coating, and/or laminating. Materials used to enhance barrier include plastics, such as polyethylene, and resins, such as urea–formaldehyde. Laminated and coated papers often warrant the designation of protective materials (see BARRIER POLYMERS). Many converted papers, however, offer little more than protection from light and mechanical damage.

Paper may be used as flexible packaging material components or as material for construction of rigid containers. Paper packaging include folding paperboard cartons, composite paperboard cans, and corrugated fiberboard shipping containers.

Metal Can Packaging. Nearly 130 billion metal cans are produced annually in the United States, of which 100 billion are two-piece aluminum used for beer and carbonated beverage packaging, but with increasing numbers for still beverage packaging. About 30 billion of the cans are three- and two-piece steel for food and pet food containment. About one-third of all cans are self-manufactured by the packager. The leading merchant can manufacturers are Crown Cork and Seal, American National Can, and Ball.

The primary function of interior can coatings is to prevent interaction between the can and its contents. Exterior can coatings may be used to provide protection against the environment, or as decoration to give product identity as well as protection.

Enamel is applied to steel in the flat before fabrication. Cans manufactured by the draw and ironing operation must be coated internally after fabrication because of the metal deformation with surface disruption that takes place. Types of internal enamel for food containers include oleoresins, vinyl, acrylic, phenolic, and epoxy–phenolic.

More than two-thirds of aluminum cans are recaptured and returned for recycling into more cans. Because of the vast quantities of scrap steel available from automobiles and appliances, recycling of steel cans has been growing at a relatively modest rate.

Glass Packaging. Glass is used for carbonated beverages, beer, and still beverage spackaging. This $4.5 billion industry is declining in importance because of weight, relative fragility, and energy requirements. Glass is recyclable, and so cullet, or crushed reused glass, is a part of every raw material batch (see RECYCLING, GLASS). Glass is virtually chemically inert and impermeable. The principal American glass bottle and jar makers include Owens-Illinois, Vitro, Ball, and American National Can.

Protective coatings capture the original strength of glass containers and delay deterioration. Coatings include hot end treatments in which newly formed hot bottles are subjected to an atmosphere of vaporized metallic compound. This atmosphere reacts with the glass surface, chemically bonding to form a primer that provides permanency to the cold end treatment. The second step of the protective coating, usually an emulsion of polyethylene, is applied after the cooling section of the annealing layer. The second coating imparts lubricity to the container surface. This prevents abrasions or other surface damage from bottle-to-bottle or bottle-to-guide rail contact during normal package handling.

Plastic Packaging

Plastic materials represent less than 10% by weight of all packaging materials. They have a value of over $7 billion including composite flexible packaging; about half is for film and half for bottles, jars, cups, tubs, and trays. The principal materials used are high density polyethylene (HDPE) for bottles, low density polyethylene for film, polypropylene (PP) for film, and polyester (PET) for both bottles and films. Plastic resins are manufactured by petrochemical companies, eg, Union Carbide and Mobil Chemical for low density polyethylene (LDPE), Solvay for high density polyethylene, Himont for polypropylene, and Shell and Eastman for polyester.

Plastic packaging materials are thermoplastic, ie, reversibly fluid at high temperatures and solid at ambient temperatures. These materials may be modified by copolymerization, additives in the blend, alloying, and surface treatment and coating. Properties of principal plastic packaging materials are given in Table 1.

Three-Dimensional Plastic Packaging. Thermoforming is the most common method of fabricating plastic sheet into three-dimensional packaging. Conventional thermoforming of polystyrene and PVC sheet is the most widely used technique for making packages for dairy products and for disposable cups and trays.

Thermoforming may be integrated with filling and sealing on thermoform/fill/seal machines operated in-plant by food packagers. The base web is gripped and moved through heating and forming operations. The open-top pieces are filled and a second web of material is heat-sealed to the flange of the base by heated pressure bars. Cutting may take place during or after sealing.

Extrusion blow molding produces narrow-neck bottles from high and low density polyethylene and other plastics.

Injection stretch blow molding may be performed on a single one-stage machine in sequence or on two independent sequential two-stage machines. PET carbonated beverage bottles are usually produced by injection stretch blow molding. Multilayer injection stretch blow molding has been commercialized for both narrow neck and wide-mouth containers.

Recognition of the merits of hot filling foods into plastic packaging, followed by sealing and cooling, has led to a need for high oxygen-barrier plastic containers capable of resisting temperatures up to 85–90°C for brief periods. Polyester requires physical modification to resist heat, drying to remove water, partial crystallization, and heat stabilization. In heat stabilization, the container is molded and briefly secured in the mold while at elevated temperature rather than be-

Table 1. Properties of Plastic Packaging Materials

Material	Specific gravity	Water-vapor transmission rate, μmol/(m^2·s)	Gas transmission, nmol/(m^2·s)	Use temperature, °C Maximum	Use temperature, °C Minimum
ionomer	0.94–0.96	13–21	1800–3870	71	−73
nylon					
uncoated	1.13–1.14	240–260	21	176–232	−59
saran coated, one side	1.13–1.14	2	4	93	−40
polyester (PET)					
uncoated	1.35–1.39	13	40	204	−62
saran coated, one side	1.4	6	3.2	82[a]	−51
metallized	1.35–1.39	0.3–1.4	0.32	204	−62
polyethylene					
LDPE	0.910–0.925	12	2000–6700	65	−51
HDPE	0.941–0.965	3–6.5	260–2000	121	−51
LLDPE	0.915–0.935	12	2000–6700	76–82	−51
LDPE/12% EVA copolymer	0.94	39	4100–5200	60	−51
polypropylene					
nonoriented	0.88–0.90	5–6.5	670–3300	121	>0
oriented	0.905	3–4	880	121	−51
oriented and metallized	0.905	1	24–80	121	−51
poly(vinyl chloride)	1.21–1.37	>40	40–12000	93	0
poly(vinylidene chloride)	1.64–1.71	0.5–3	0.64–14	ca 82[a]	>−18

[a] Coating softens.

ing chilled immediately. The crystalline structure of the material is thereby altered to resist moderately elevated temperatures. This technique is employed to produce PET bottles for filling with hot liquids.

Some blow-molded bottles are produced in blow-mold/fill/seal operations designed for aseptic packaging. Blow-mold/fill/seal systems are used commercially for beverages and for pharmaceutical packaging.

AARON L. BRODY
Rubbright-Brody, Inc.

M. Bakker, ed., *The Wiley Encyclopedia of Packaging Technology*, John Wiley & Sons, Inc., New York, 1986.

G. L. Robertson, *Food Packaging-Principles and Practice*, Marcel Dekker, Inc., New York, 1993.

F. A. Paine and H. Y. Paine, *A Handbook of Food Packaging*, Leonard Hill Ltd., London, 1983.

F. A. Paine and H. Y. Paine, *Principles of Food Packaging*, Leonard Hill, Ltd., London, 1983.

S. E. M. Selke, *Packaging and the Environment-Alternatives, Trends and Solutions*, Technomic Publishing Co., Inc., Lancaster, Pa., 1990.

S. Sacharow and A. L. Brody, *Packaging: An Introduction*, Harcourt Brace Jovanovich Publications, Inc., Duluth, Minn., 1987.

FOOD PROCESSING

Food processing operations can be grouped into three categories: preparation, assembly, and preservation of foods. Preparation processes are used to convert raw plant or animal tissue into edible ingredients. This may include separation of inedible and hazardous components, extraction or concentration of nutrients, flavors, colors, and other useful components, and removal of water. Assembly processes are used to combine and form ingredients into consumer products. Preservation processes are used to prevent the spoilage of foods. Five sources of food spoilage must be addressed in order to deliver fresh, safe foods and ingredients: microbial contamination, including viruses; enzyme activity from enzymes in the food itself and from external enzymes such as from microbial activity; chemical deterioration such as oxidation and nonenzymatic browning; contamination from animals, insects, and parasites; and losses owing to mechanical damage such as bruising. Preservation processes can be used to extend the shelf life of fresh foods, such as produce, or to manufacture products for long-term storage where shelf lives are measured in years. The processing of foods is regulated by federal food laws that cover good manufacturing practices, nutritional content of foods, and food and ingredient standards (see also FOOD ADDITIVES; FOOD PACKAGING).

Plants and animals are the primary sources of food. The food processing industry devotes considerable research to the selection and improvement of plants and animals for raw materials. Genetic engineering (qv), as well as conventional breeding methods, are being used to improve the yield, color, flavor, texture, nutrient content, and resistance to diseases, insect loss, and climatic stress (see also FEEDS AND FEED ADDITIVES; FERTILIZERS; GROWTH REGULATORS). However, product quality can vary owing to weather, soil, growing practices, harvest methods, and post-harvest handling. Thus food processing unit operations must be designed to accept raw materials having a wide range of qualities. In addition, provision often must be made for profitable use of by-products and waste streams.

Regulations

Food processing operations are usually regulated and mandated by national and international laws, regulations, and standards which define nutritional requirements, certain ingredients, process conditions, and even the composition of some products. Food safety and toxicology regulations include standards for toxic and carcinogenic substances in foods; pathogenic microbes; and physical hazards.

Process Optimization

Food processing operations can be optimized according to the principles used for other chemical processes if the composition, thermophysical properties, and structure of the food is known. However, the complex chemical composition and physical structures of most foods can make process optimization difficult. Moreover, the quality of a processed product may depend more on consumer sensory responses than on measurable chemical or physical attributes. Retention levels of ascorbic acid, $C_6H_8O_6$, or thiamine can often be used as an indicator of process conditions.

Theoretical Basis. Food preservation theory has yielded mathematical models for predicting the heating times and temperatures needed to produce foods free of pathogenic or spoilage microbes. Mild heat treatments used to inactivate viruses, vegetative pathogenic bacteria such as *Salmonella* sp., and certain yeasts (qv) and molds, are referred to as pasteurization operations.

Spore-forming bacteria are among the most heat-resistant organisms known. Research since the early 1920s has been directed toward the development of mathematical models to predict the rate of heat inactivation of *Clostridium botulinum* spores as a function of heating time and temperature, and the composition of the suspending media. Spore germination can be inhibited by antibiotic substances produced by several types of lactic acid producing bacteria. These substances, called bacteriosins, are finding increased use in preventing the growth of gram-positive bacteria.

Two other broad areas of food preservation have been studied with the objective of developing predictive models. Enzyme inactivation by heat has been subjected to mathematical modeling in a manner similar to microbial inactivation. Chemical deterioration mechanisms have been studied to allow the prediction of shelf life, particularly the shelf life of foods susceptible to nonenzymatic browning and lipid oxidation.

Water Activity. The rates of chemical reactions as well as microbial and enzyme activities related to food deterioration have been linked to the activity of water (qv) in food. Water activity, at any selected temperature, can be measured by determining the equilibrium relative humidity surrounding the food. Water activity can be related to the moisture content of the food as measured by standard moisture tests.

Preservation of Foods

Preservation operations to reduce or eliminate food spoilage can be grouped into five categories: heat treatments; storage near or below the freezing point of water; dehydration and control of water activity; chemical preservation; and use of mechanical operations such as washing, peeling, filtration, centrifugation, grinding, ultrahigh hydrostatic pressure, and most importantly, the packaging. Most food preservation technologies use two or more preservation operations because virtually all processed foods are packaged.

Short-Term Storage. Short-term storage operations include refrigeration, heat treatment, and preservatives.

Long-Term Storage. Inactivation of microbes and enzymes in foods and food ingredients is necessary to ensure a long useful packaged shelf life. This can be achieved by using one or more preservation operations such as applying heat; using storage temperatures below −18 °C; drying to water activities below 0.65; and by adding chemical preservatives such as organic acids (acetic or lactic) or table salt.

Thermal Preservation Technology. The heat preservation of foods can be accomplished by various combinations of heating times and temperatures, depending on the number and type of heat-resistant spores present, the composition of the food, and the physical characteristics of the food and package.

The inactivation of heat-resistant spores appears to follow first-order kinetics. Thus if the rate of inactivation of a spore population is known at several temperatures, and the rate of heating of the slowest point in a package can be determined or calculated from heat-transfer principles, then the time needed to sterilize the package can be calculated for any external heating condition.

The establishment of safe thermal processes for preserving food in hermetically sealed containers depends on the slowest heating volume of the containers. Heat-treated foods are called commercially sterile.

Chemical changes in foods resulting from heating, such as the loss of pigments, flavors, and vitamins, can also be approximated by first-order kinetics.

Rapid heating and cooling of liquid foods, such as milk, can be performed in a heat exchanger and is known as high temperature–short time (HTST) processing. HTST processing can yield heat-preserved foods of superior quality because heat-induced flavor, color, and nutrient losses are minimized.

Equipment and processes for thermal preservation depend on the physical form of the food and its pH. Foods having a pH < 4.5 often can be sterilized, for commercial purposes, at or near a temperature of 100°C. Commercial sterility for these products means that the product will not spoil owing to microbial growth as long as the pH remains at or below 4.5 The spores of *Bacillus coagulans* are an important exception. This latter microbe is found in tomato products, and these products are often adjusted to a pH of 4.0 or lower, or given an additional heat treatment.

Acid foods generally require the simplest equipment for heat preservation. The food can be heated to 100°C and filled hot into suitable containers. The containers are sealed, inverted to sterilize the closure, held at the filling temperature for a short time to ensure that the package is thoroughly heated, and then cooled. Tomato sauces, jellies, fruits, fruit juices (qv), and pickles are routinely preserved in this fashion.

Low acid foods have a pH > 4.5, require sterilization at temperatures above 100°C, and thus require treatment in pressure vessels. Heat preservation processes above 100°C can be carried out in batch or continuous heat-exchange equipment.

Freezing Preservation. The rate of loss of color, flavor, texture, and nutrients, the growth of microbes, and the activity of enzymes and other life forms are all functions of temperature. Thus lower storage temperatures prolong the useful life of foods.

Equipment for food freezing is designed to maximize the rate at which foods are cooled to −18 °C to ensure as brief a time as possible in the temperature zone of maximum ice crystal formation. This rapid cooling favors the formation of small ice crystals which minimize the disruption of cells and may reduce the effects of solute concentration damage. Rapid freezing requires equipment that can deliver large temperature differences and/or high heat-transfer rates.

Many formulated foods and certain animal products tolerate freezing and thawing well because their structures can accommodate ice crystallization, movement of water, and related changes in solute concentrations. Starches can be modified for freeze–thaw stability against gel breakdown through several cycles. By contrast, most fruits and vegetables lose significant structural quality on freezing and during storage because their rigid cell structures fail to accommodate to ice crystal formation. Frozen food storage equipment must be designed to minimize temperature fluctuations.

Most frozen foods have a useful storage life of one year at −18 °C. However, foods high in fat such as sausage products may become rancid after two weeks in frozen storage if not protected from oxygen by special packaging and antioxidants.

Food freezing equipment can be classified by the method and medium of heat transfer used. High velocity air is the most common medium used for direct contact freezing of nonpackaged foods. However, rapid freezing requires high air velocities and low operating temperatures. For these and other reasons many foods are frozen in equipment using conduction or liquid heat-transfer methods. Capital and energy savings, reduced moisture loss, and elimination of defrosting are some advantages of these methods.

Liquid heat-transfer media for immersion freezing include solutions of edible salts, sugars, alcohols, and esters. These heat-transfer agents offer high heat-transfer rates, reduced pumping costs, and allow operating at higher refrigerant temperatures.

Conduction freezing between chilled plates is a very cost-effective method of heat removal for products that can be packaged in a geometry to fit between refrigerated plates. Cryogenic freezing equipment uses liquid nitrogen or carbon dioxide snow. These units have the advantage of portability and simplicity and can produce extremely fast freezing rates. The refrigerant can be sprayed directly on the product to ensure rapid heat transfer.

The quality of a frozen food may be determined more by the temperature at which it is stored than by the method or rate of freezing. Storage temperatures may fluctuate as products move from manufacturing through distribution channels to the consumer's home freezer. The useful shelf life of a frozen food may be severely limited by exposure to storage temperatures above −18 °C, even for a few hours.

Dehydration Processing. Dehydration is one of the oldest means of preserving food. Microbes generally do not grow below a minimum water activity, A_w, of 0.65.

Foods dried to water activities in the range of 0.65 to 0.85 are often referred to as intermediate moisture foods. These partially dried foods tend to be soft and to rehydrate easily. The remaining water acts as a plasticizer. Because molds and yeast may be able to grow in these partially dried products, they must be preserved by heat, vacuum, or modified atmosphere packaging, refrigeration, or chemical means.

Foods high in sucrose, protein, or starch (qv) tend to bind water less firmly and must be dried to a low moisture content to obtain microbial stability.

Fresh plant and animal tissue when dried to a water activity much below 0.97 show irreversible disruption of metabolic processes. Products susceptible to oxidation and oxidative rancidity can be treated with antioxidants and vacuum or inert gas packed to minimize exposure to oxygen. Low temperature storage can further reduce the rate of chemical deterioration.

Continuous hot air driers are used to prepare most of the high quality, dried, piece-form fruits and vegetables produced in the United States. Liquids and pastes are commonly dried in spray, drum, or freeze dryers. Particulate foods can be dried in batch or continuous air-fluidized beds or freeze dryers. Many agricultural commodities are sun-dried when weather conditions at harvest provide low humidity, warm temperatures, and good air circulation.

Chemical Preservation. Food additives (qv) can enhance the effectiveness of food preservation by heat, refrigeration, and drying methods. The addition of a food-grade acid to a low acid food to shift the pH to a value below 4.5 allows heat preservation at a temperature of 100°C instead of in the range of 121°C. Antioxidants such as butylated hydroxyanisole (BHA) can be added to potato chips to reduce the need for expensive oxygen-impermeable flexible packaging. Sulfur dioxide is used in wine (qv) and in dry fruit and vegetable products to preserve colors and flavors and prevent nonenzymatic browning.

Food can be preserved by fermentation (qv) using selected strains of yeast, lactic acid-producing bacteria, or molds. The production of ethanol (qv), lactic and other organic acids, and antimicrobial agents in the food, along with the removal of fermentable sugars, can yield a product having an extended shelf life.

Lactic acid-producing bacteria associated with fermented dairy products have been found to produce antibiotic-like compounds called bacteriocins. Concentrations of these natural antibiotics can be added to refrigerated foods in the form of an extract of the fermentation process to help prevent microbial spoilage. Other natural antibiotics are produced by *Penicillium roqueforti*, the mold associated with Roquefort and blue cheese, and by *Propionibacterium* sp., which produce propionic acid and are associated with Swiss-type cheeses.

Ionizing radiation is considered to be a chemical preservation method and applications must be cleared by the Food and Drug Administration for use, not only on a product-by-product basis, but also on a dose basis.

Other Technologies. Several technologies for the preservation of foods using a minimum of heat are being explored. The application of ultrahigh pressure to the preservation of foods has been investigated. Capacitance discharge has also been investigated as a means to pasteurize or commercially sterilize foods which can pass between plates sufficiently close together to allow an electric field of approximately 25,000 V/cm. Very high intensity flashes of visible light can be used to pasteurize fruit juices using a minimum of heating in a manner which appears to be similar to capacitance discharge.

Computer Integrated Manufacturing, Instrumentation, and Controls. Large food processing firms are exploring the use of computer integrated manufacturing. Thermal processing controls have been developed to the point where time and temperature process deviations can be corrected on line. Freezer, dryer, and vacuum evaporator operating conditions can be controlled and optimized using systems already available to the process industry.

An important aspect of food processing, common with other processing industries, is yield of finished product from starting raw materials for any shift and for specific unit operations. Computer-integrated manufacturing can start with the measurement of material flows and build upon this information. Instrumentation for the on-line measurement of specific food qualities of importance to the consumer such as food flavor, aroma, texture, and microbial content are under development. These quality factors are monitored using statistical quality control (qv) procedures using standard sampling plans and control strategies.

DANIEL F. FARKAS
Oregon State University

D. R. Heldman and D. B. Lund, eds., *Handbook of Food Engineering*, Marcel Dekker, Inc., New York, 1992.

N. N. Potter, *Food Science*, 5th ed., Van Nostrand Reinhold, New York, 1995.

D. K. Tressler, W. B. Van Arsdel, and M. J. Copley, eds., *The Freezing Preservation of Foods*, 4th ed., Vols. 1–4, Avi Publishing Co., Westport, Conn., 1979.

Canned Foods, Principles of Thermal Process Control, Acidification, and Container Closure Evaluation, 5th ed., The Food Processors Institute, Washington, D.C., 1988.

FOODS, NONCONVENTIONAL

Nonconventional foods differ from the usual materials of plant and animal origin used for human food or animal feed (see FEEDS AND FEED ADDITIVES; FOOD PROCESSING). These materials can be produced from chemical feedstocks, eg, carbohydrates (qv), hydrocarbons (qv), or other industrial organics, by processes such as microbiological, enzymatic, or chemical synthesis, or from existing natural products, containing carbohydrates, proteins (qv), and fats, by physical, chemical, microbiological, or enzymatic modification.

Single-Cell Protein

Two broad classes of microorganisms are of interest (ca 1993) for single-cell protein (SCP) production, ie, photosynthetic organisms, including algae and certain bacteria: and nonphotosynthetic organisms, including bacteria, actinomycetes, yeasts, molds, and higher fungi. In addition, two different uses of SCP are distinguished, ie, food for humans and feed for animals.

Photosynthetic Organisms. Mass cultivation of algae in ponds or tanks under photosynthetic conditions, using incident sunlight as the energy source and CO_2 as the carbon source, has been investigated in Japan, Taiwan, Mexico, Algeria, India, and in California in the United States. Artificial illumination systems have been used for experimental mass cultivation of algae and in bioregenerative systems for converting CO_2 and human wastes into breathable oxygen and food as part of life-support systems for long-duration space exploration missions.

Centrifugation; flocculation using $Al_2(SO_4)_3$, $Ca(OH)_2$, or cationic polymers; sedimentation; filtration; treatment with ion-exchange resins and drum; sand bed; and sun drying have been investigated for separating, concentrating, and drying algal cells.

Product quality is an important consideration in producing algae for food or feed use. Algal cells must be dried using time–temperature combinations sufficient to destroy pathogenic bacteria and viruses that may be present in ponds, particularly in those culture systems based on sewage.

The nitrogen requirements for algal growth can be met by either ammonium salts, urea, or nitrates.

Key factors influencing growth include temperature; pH; availability of CO_2, nitrogen, phosphorus, and other inorganic nutrients; and availability of sunlight as influenced by latitude, cloud cover, and depth of the culture pond or tank.

Table 1 lists compositional and nutritional data of selected algae. More extensive complications on algae are available.

Nonphotosynthetic Organisms. Nonphotosynthetic microorganisms of interest in SCP production include bacteria, actinomycetes, yeasts, molds, and higher fungi. Carbon and energy sources considered for growing these organisms include carbohydrates such as simple sugars, starches, and cellulose (qv); agricultural, forestry, pulp (qv), paper, and food processing wastes containing these carbohydrates; and hydrocarbons and chemicals derived from them, including alcohols and organic acids.

Commercial-scale operations are conducted in batch, fed-batch, or continuous culture systems. Fermentation vessels include the conventional baffled aerated tank, with or without impeller agitation, and the air-lift tower fermentors in which air is sparged into an annular space between the fermentor wall and internal cylinder. A corrosion-resistant grade of stainless steel (316 L) is usually used for fermentor construction; wood or concrete tanks have been used with agricultural or food wastes.

The FDA regulations provide for the use of dried cells of the yeasts *S. cerevisiae*, *K. marxianus var. fragilis*, and *C. utilis* in foods. Folic acid contents must not exceed 0.04 mg/g. Baker's yeast protein concentrate has been approved by the FDA for use as a functional food additive.

Derived Plant and Animal Products

Leaf Protein Concentrates. Leaf protein concentrates (LPC) are prepared by crushing plant material, extracting the juices, and either using the juice per se or recovering the protein from the juice by heating or chemical precipitation. Dehydrated alfalfa (lucerne) has a long history of use as a source of plant protein for animal feeds. The leaves of alfalfa and other crops are a source of protein that can be extracted to give a concentrated product having increased protein and decreased fiber contents suitable for animal feeding. Plants, other than alfalfa, considered as sources of LPC include pea vines, clover, field beans, mustard, kale, fodder radish, banana leaves, and aquatic plants.

Process development for LPC production dates from the United Kingdom and Hungary from 1920–1940. Table 2 presents some of the processing methods that are used or under development in the 1990s.

Table 3 gives approximate analyses of several LPC products.

LPC products prepared from *Brassica* sp., soybeans, sugar beets, and tobacco have been investigated for their functional properties including nitrogen solubility, fat- and water-binding capacity, emulsification, gelation, and foaming capacity and stability. The 1993 market

Table 1. Composition of Photosynthetically Grown Algae, %[a]

Organism	Nitrogen	Crude[b] protein	Fat	Ash
Chlorella sp.	9.3	58	9	3
Scenedesmus acutus	8.2–10.2	51.4–63.6	11.2–14.3	7.9–16.7
Spirulina sp.	8.8–11.2	55–70	4–7	5–10
Spirulina platensis	8.0	50.0	0.5	11.0

[a] Dry wt basis. [b] Crude protein = %nitrogen × 6.25. Does not accurately reflect true protein content. Algal cells may contain nonprotein nitrogen substances, eg, 4–6% nucleic acids, dry wt basis.

Table 2. Selected Processes for Leaf Protein Concentrate Production, 1993

Process	Description
Rothamsted	pulping in ribbed rollers; precipitating protein at 80°C or at pH 4.0; drying in air below 80°C
Pro-Xan	chopping; ammoniation to pH 8.5; roll or twin-screw pressing; sieve purification; coagulation with steam at 85°C; dewatering; drying
Vepex	mechanical disintegration; multistage pressing; add-back of liquor to press cake; coagulation at 82°C with addition of flocculents; centrifugation; evaporation; drying
Istituto di Industrie Agarie[a]	chopping and screw pressing; centrifugation; coagulation at pH 8.5; treatment with polyelectrolyte; centrifugation; precipitation at pH 4.0; centrifugation; drying

[a] Pisa, Italy.

Table 3. Analysis of Leaf Protein Products, wt %[a]

Leaf protein	Source	Protein[b]	Fat	Crude fiber	Ash
white LPC[c]	alfalfa	88.7	0.6	1.0	0.4
white LPC	quinoa	93.2	0.7	0.9	2.0
Pro-Xan[d]	alfalfa	61.9	8.9	1.7	11.1
Brassica napus[e]		58.5	15.6	2.0	10.0

[a] Dry wt basis. [b] Protein = %nitrogen × 6.25. [c] Nitrogen-free extract = 9.3%, and soluble solids = 0.3% [d] Whole leaf protein concentrate, 16.5% nitrogen-free extract and 7.9% soluble solids. [e] Commonly known as Late Korean rape.

for LPC-type products in the United States was for dried alfalfa meal for animal feed.

Seed-Meal Concentrates and Isolates. Seed-meal protein products include flours, concentrates, and isolates, particularly soy protein products. These can be used as extenders for meat, seafood, poultry, eggs, or cheese (see SOYBEANS AND OTHER OILSEEDS). Detailed information on soybean and other seed-meal production processes is available.

Soybean concentrate production involves the removal of soluble carbohydrates, peptides, phytates, ash, and substances contributing undesirable flavors from defatted flakes after solvent extraction of the oil. Typical concentrate production processes include moist heat treatment to insolubilize proteins, followed by aqueous extraction of soluble constituents; aqueous alcohol extraction; and dilute aqueous acid extraction at pH 4.5.

Vegetable proteins other than that from soy have potential applicability in food products. Functional characteristics of vegetable protein products are important factors in determining their uses in food products. Concentrates or isolates of proteins from cotton (qv) seed, peanuts, rape seed (canola), sunflower, safflower, oats, lupin, okra, and corn germ have been evaluated for functional characteristics, and for utility in protein components of baked products, meat products, and milk-type beverages (see DAIRY SUBSTITUTES).

To modify functional properties, vegetable proteins such as those derived from soybean and other oil seeds can be hydrolyzed by acids or enzymes to yield hydrolyzed vegetable proteins (HVP). Hydrolysis of peptide bonds by acids or proteoyltic enzymes yields lower molecular weight products useful as food flavorings. However, the protein functionalities of these hydrolysates may be reduced over those of untreated protein.

Under FDA regulations, HVP products are permitted as optional ingredients in standardized canned foods such as pears, mushrooms, and tuna, and as a flavoring ingredient in nonstandardized foods. The U.S. Department of Agriculture has cleared HVP as a flavoring ingredient in various meat products.

Fish Protein Concentrates and Isolates. Fish protein concentrates (FPC) and isolates (FPI) are produced for human food use from whole edible species of fish using sanitary processing methods; fish meal and fish solubles are produced for animal feed. FPC raw materials include whole hake, hake-like fish, and herring of the genera *Clupea*, and menhaden and anchovy of the species *Engraulis mordax* without removal of heads, fins, tails, or intestinal contents. FPI raw materials include edible portions of fish body generally recognized as safe for human consumption after removal of heads, fins, tails, bones, scales, viscera, and intestinal contents. In the United States, FDA regulations describe the production processes for preparing FPC and FPI. The FDA regulations also specify that FPC and FPI contain minimum protein contents of 75 and 90%, respectively, a maximum fat content of 0.5%, and a maximum moisture content of 10% by weight. FPC must be free of *Escherichia coli, Salmonella*, and other food pathogens and have a total bacterial plate count of not more than 10,000 per gram. Amino acid profiles of FPC are excellent and compare favorably with whole egg except for tryptophan and lysine.

Economic conditions in the United States have not favored the production of FPC and FPI having desirable functional and nutritional characteristics at prices competitive with those of conventional protein sources.

Numerous seafood analogue products, eg, crab, shrimp, and lobster analogues, have been prepared by modifying the structural and textural properties of fish proteins.

Synthetic Protein Products

Plastein Synthesis. Plasteins are mixtures of high molecular weight proteinaceous peptides. They are synthesized by enzyme-catalyzed growth of peptide chains from lower molecular weight peptides. The process by which plasteins are formed is called the plastein reaction and is the reverse of the proteolytic enzyme hydrolysis of peptide bonds of proteins.

Plasteins are still in the experimental stage of development. Further work is needed on the scale-up of processing conditions for plastein synthesis which would lead to commercially useful products and on the functional utility of plasteins as ingredients in foods.

Synthetic Proteins. Protein-like polypeptides can be synthesized chemically from ammonia, water, and carbon dioxide, or from mixtures of amino acids which are now manufactured by chemical or microbiological synthesis. The possibility exists for protenoids to be nutritionally imbalanced, have mammalian toxicity, and undesirable tastes, and odors.

Product Quality and Safety

The Protein Advisory Group, ad hoc, is the working group of the WHO UN system involving WHO, FAO, and the UN International Children's Emergency Fund (UNICEF). It has developed guidelines for the evaluation of novel sources of protein, eg, single-cell protein; clinical testing of novel sources of protein; human testing of supplementary food mixtures; and nutritional and safety aspects of novel protein sources for animal feed.

In general, nonconventional protein foods must be competitive with conventional plant and animal protein sources on the bases of cost delivered to the consumer, nutritional value to humans or animals, functional value in foods, sensory quality, and social and cultural acceptability.

In the U.S., novel food ingredients or food ingredients produced by novel processes must be cleared by the FDA. In the case of meat and poultry, novel ingredients must also be cleared by the U.S. Department of Agriculture's Food Safety and Inspection Service (FSIS).

JOHN H. LITCHFIELD
Battelle Memorial Institute

J. H. Litchfield in J. L. Marx, ed., *A Revolution in Biotechnology*, Cambridge University Press, Cambridge, U.K., 1989, pp. 71–81.

I. Goldberg and R. Williams, eds., *Biotechnology and Food Ingredients*, Van Nostrand Reinhold Co., Inc., New York, 1991.

L. Telek and H. D. Graham, eds., *Leaf Protein Concentrates*, Avi Publishing Co., Westport, Conn., 1983.

F. H. Steinke, D. H. Waggle, and M. N. Volgarev, eds., *New Protein Foods in Human Health*, CRC Press, Boca Raton, Fla., 1992.

FOOD TOXICANTS, NATURALLY OCCURRING

Toxicants are substances which, upon ingestion, product changes in homeostasis that are threatening to the normal function of the organism. There are substantial differences in the toxicity thresholds of individuals to specific agents. Factors affecting toxicity include body weight, sex, age, general state of health, and the presence of potentiating or inhibitory substances.

Toxic Proteins, Peptides, Amides, and Amino Acids

Nitrogenous compounds are the most frequently implicated natural toxicants in foods. These compounds may be grouped either according to gross manifestations or specific structural characteristics. Accordingly, vitamin-destroying enzymes, hemagglutenins, enzyme inhibitors, and many hepatotoxins, are of protein, peptide, or amino acid composition. Many of the hepatotoxins are also carcinogens.

Enzyme inhibitors (qv) of a protein nature are of significant concern because of widespread occurrence. The most common of these affect the pancreatic enzymes, trypsin and chymotrypsin, and are found in legumes, as well as in egg whites and potatoes.

Many protein inhibitors cause little nutritional difficulty because these compounds are heat labile under ordinary processing and cooking procedures including microwaving, and significant numbers are water soluble. Many are found in highest concentrations in the outer portions of plants, eg, wheat bran; thus normal peeling and milling operations also give some protection. It has been demonstrated that treatment with compounds such as sodium sulfite, ascorbic acid, and cupric sulfate, as well as fermentation with *Rhizopus oligosporus*, such as in the production of tempeh, is also an effective means of reducing trypsin inhibitor activity in both fresh and hardened common beans.

Reports of specific amino acid toxicities from normal eating patterns are rare. Although all amino acids except alanine have been shown to be toxic, the probability of intoxication is very remote. Humans seem able to tolerate all amino acids in excess of 10 times the recommended intake.

Lathyrus sativus (khesari), which constitutes a principal food crop for many in India, has been known for decades to cause neolathyrism. The causative agent, *N*-oxalyl-L-α-diaminopropionic acid (ODAP), produces partial or total loss of control of the lower limbs with associated neurological symptoms. Common methods of preparation, including soaking in lime water and boiling, are effective in destroying this amino acid.

The seeds of legumes may contain hemagglutenins and lectins that may cause destruction of the epithelia of the gastrointestinal tract; interfere with cell mitosis; cause hemorrhage; impede renal, cardiac, and hepatic function; and produce red blood cell agglutination. Many of these compounds are rendered inactive by moist heat, and the toxicity may be further reduced or neutralized by digestive enzymes, making them poorly absorbed. Because lectins reach the colon mostly in an inactive state, they appear to protect humans from colon cancer by causing hypersecretion of intestinal mucus, or by direct toxic effect on tumor cells.

Saponins disrupt red blood cells and may produce diarrhea and vomiting. They may also have a beneficial effect by complexing with cholesterol and thus lowering serum cholesterol levels. In humans, intestinal microflora seem to either destroy saponins or inactivate them in small concentrations.

Acute toxicoses resulting from consumption of toxic mushrooms is infrequent, yet of increasing concern because of the practice of gathering fungi in the wild. The most serious of these toxicoses result from the Amanita family of mushrooms which contains several toxic peptides belonging to the amatoxin and phallotoxin groups.

Both the common cultivated mushroom as well as the Shitake mushroom contain hydrazines that have been shown to be carcinogenic precursors in experimental animals. Hydrazine levels, however, vary considerably as a function of variety, processing, storage, and preparation. One week of refrigerated storage reduces levels significantly, and all hydrazines are lost during canning and/or cooking.

Phytoalexins

Phytoalexins are low molecular weight compounds produced in plants as a defense mechanism against microorganisms. They do, however, exhibit toxicity to humans and other animals in addition to microbes. Coumarins, glycoalkaloids, isocoumarins, isoflavonoids, linear furanocoumarins, stilbenes, and terpenes all fall into the category of phytoalexins. Because phytoalexins are natural components of plants, and because their concentration may increase as a response to production and management stimuli, it is useful to recognize the possible effects of phytoalexins in the human diet.

Linear furanocoumarins are potent photosensitizing agents in celery, parsley, parsnips, limes, and figs. The most commonly reported symptoms include contact dermatitis and photodermatitis, particularly on the hands and forearms.

Enumerable phytoalexins, including furanosesquiterpene, ipomeamarone, eudesmanes, and others, have been isolated from mold-infected sweet potatoes. The clinical symptoms seem to revolve around lung edema. Whereas high concentrations of these chemicals can occur in damaged sweet potatoes, the occurrence is much less (by as much as 20-fold) in nondamaged sweet potatoes. Of possible concern to human health is the fact that blemishes sufficient to result in large increases in concentration of lung-edema toxins are not always easily detected by the naked eye. Additionally, these compounds are heat stable.

Oligosaccharides

Oligosaccharides, specifically the α-galactosides raffinose, stachyose, and verbascose, are widely present in legumes and are indigestible by humans because of a lack of α-galactosidase. As a result, these compounds undergo fermentation in the colon with the concomitant production of CO_2, H_2, and CH_4, commonly referred to as flatulence. Reports have shown germination to be effective in reducing α-galactoside content of cowpeas and other legumes.

Goitrogens are compounds that produce goiter by interfering with thyroxine synthesis in the thyroid gland. Foodborne goitrogens are often characterized by the presence of sulfur and most are thiocyanates or closely related compounds. Because of their widespread occurrence in Cruciferae, goitrogens are among the most common and longest recognized substances of toxic nature in the human food supply.

Oxalates, Phytates, and Other Chelates

Of nutrient chelates in the human diet, oxalates and phytates are the most common. Oxalic acid, found principally in spinach, rhubarb leaves, beet leaves, some fruits, and mushrooms, is a primary chelator of calcium. Oxalate present in pineapple, kiwifruit, and possibly in other foods, occurs as calcium oxalate, CaC_2O_4. This compound is in the form of needle-like crystals, known as raphides, which can produce painful sensations in the mouth when eaten raw. The effects of oxalic acid in the diet may be twofold. First, it forms strong chelates with dietary calcium, rendering the calcium unavailable for absorption and assimilation. Secondly, absorbed oxalic acid causes assimilated Ca to be precipitated as insoluble salts that accumulate in the renal glomeruli and contribute to the formation of renal calculi.

Phytic acid, although restricted to a more narrow range of food products, mainly grains, complexes a broader spectrum of minerals than does oxalic acid. Decreased availability of P is probably the most widely recognized result of excessive intakes of phytic acid, yet Ca, Cu,

Zn, Fe, and Mn are also complexed and rendered unavailable by this compound. High intakes of both calcium and vitamin D help to offset the deleterious effects of oxalates.

Vasoactive and Psychoactive Amines and Alkaloids

Most compounds producing hypertensive episodes are classified as amines and are found in greatest concentration in banana, plantain, tomato, avocado, pineapple, broad beans, and various cheeses. Amines that are vasoactive include dopamine, $C_8H_{11}NO_2$; tyramine; histamine, $C_5H_9N_3$; tryptamine, $C_{10}H_{12}N_2$; noradrenaline, $C_8H_{11}NO_3$; and dihydroxyphenylalanine (DOPA), $C_9H_{11}NO_4$.

Patients receiving monoamine oxidase inhibitors (MAOI) as antidepressant therapy have been especially subject to the hypertensive effects of vasoactive amines. These dietary amines have also been implicated as causative agents in migraine. Other naturally occurring alkaloids (qv) have been recognized for centuries as possessing neurological stimulant and depressant properties.

Caffeine, a xanthine derivative, has been consumed for thousands of years and is present in over 60 plant species including coffee (qv) beans, tea (qv) leaves, cacao beans (see CHOCOLATE AND COCOA), and cola nuts. In addition to naturally occurring sources, caffeine has been used as a food ingredient (flavoring) for over 100 years. Caffeine produces stimulatory effects by facilitating mental and muscular effort and diminishes drowsiness and fatigue. Individual thresholds of toxicity vary considerably, but symptoms such as restlessness, increased respiration, muscular tension and twitching, and tachycardia may imply acute toxicity.

Depressant symptoms, which include burning abdominal pain, decreased excitability, convulsions, nausea, and coma, become the general syndrome for all oral alkaloid poisoning. Myristicin, found in both nutmeg and mace, is a psychoactive agent that may be fatal in infants who consume as little as two whole nutmegs. Its toxicity resemble alcohol intoxications.

Several glycoalkaloids present in food are of toxicological interest. Solanine, found in potatoes, tomatoes, apples, eggplant, and sugar beets, has been responsible for several cases of moderate to severe poisoning. Solanine is a cholinesterase inhibitor (see CHOLINE), and toxic doses are probably ca 200 mg. Market potatoes contain about 1–5 mg of solanine per 100 g fresh weight. The USDA establishes solanine levels of 20 mg/100 g as the limit for safe consumption.

Antinutrients

The presence of antivitamins in certain foods means that merely assuring an adequate intake of a vitamin is no guarantee that a deficiency state cannot exist physiologically. The enzyme thiaminase acts by either specific splitting of the thiamine molecule or nonspecific hydrolysis. Niacin inhibitors, acting through nicotinamide mononucleotide (NMN) depression in erythrocytes, have been studied in corn and millet, and a biotin antagonist, avidin, has long been recognized in raw egg white. Linatine, found in flaxseed, is the only pyridoxine antagonist known and seems to function by the formation of a stable complex. Yeast and pea seedlings contain specific pantothenic acid antagonists, although the structure and mode of action are unexplained. Riboflavin antagonism, found only in the Akee plum of Jamaica, is rather rare, but is of interest because it can be fatal.

The antagonisms that exist between unsaturated fatty acids, and carotene and vitamin E are complicated and largely undefined. Linoleic acid acts as an antivitamin to *dl*-α-tocopherol (vitamin E) by reducing availability through direct intestinal destruction. Various lipoxidases destroy carotenes and vitamin A.

Investigations have focused on the content of polyphenolics, tannins, and related compounds in various foods and the influence on nutrient availability and protein digestibility. It has been established that naturally occurring concentrations of polyphenoloxidase and polyphenols in products such as mushrooms can result in reduced iron bioavailability. Likewise, several studies have focused on decreased protein digestibility caused by the tannins of common beans and rapeseed (canola).

Vitamin Toxicity

Reported cases of vitamin toxicity owing to overdose are usually associated with increased over-the-counter availability of supplemental vitamins and indiscriminate supplementation. Fat-soluble vitamins tend to accumulate in the body with relatively inactive mechanism for excretion and cause greater toxicological difficulties than do water-soluble vitamins.

Infants may be sensitive to doses of vitamin A in the range of 75,000–200,000 IU (90.6–1.5 g). Dangerous doses of vitamin D seem to lie in the range of 1000–3000 IU/kg body wt (25–75 μg/kg body wt). Cases of toxicity of both vitamins E and K have been reported, but under ordinary circumstances these vitamins are considered relatively innocuous.

Of the water-soluble vitamins, intakes of nicotinic acid on the order of 10 to 30 times the recommended daily allowance (RDA) have been shown to cause flushing, headache, nausea, and moderate lowering of serum cholesterol with concurrent increases in serum glucose. Toxic levels of folic acid are ca 20 mg/d in infants, and probably approach 400 mg/d in adults. The body seems able to tolerate very large intakes of ascorbic acid (vitamin C) without ill effect, but levels in excess of 9 g/d have been reported to cause increases in urinary oxalic acid excretion.

Essential Minerals and Heavy Trace Elements

Ingestion of at least 10 times normal levels of essential minerals would be required to approach toxic proportions (see MINERAL NUTRIENTS). The only exceptions occur in cases of plant foods grown on soils unusually high in Mo, Se, and Cu. Levels can reach toxic quantities in these cases, but these are rare occurrences.

Cases involving human toxicity from heavy trace elements, such as Pb, Hg, As, and Cd, are much more common but are almost exclusively traced to accidental contamination rather than true natural occurrences.

Cyanogenic Glycosides

Complex glycosides, which upon hydrolysis yield hydrogen cyanide, are commonly found among plant materials. The toxicity of this class of compounds, found in the bitter almond, pits of stone fruits, sorghum, and lima beans, is directly related to HCN liberation upon digestive hydrolysis.

Nitrates, Nitrites, and Nitrosamines

The carcinogenicity of nitrosamines has created widespread concern over the safety of food products that are significant sources of nitrates and nitrites. Nitrosamines are readily formed by reaction of secondary amines with nitrites at acid pH, conditions which may occur in the gastrointestinal tract.

Nitrates are found in fairly high concentrations in beets, spinach, kale, collards, eggplant, celery, and lettuce. Additionally, nitrates and nitrites are commonly used in the curing solutions of bacon, ham, and other cured meats. In cured meats, nitrates and nitrites control the growth of microorganisms, particularly *Clostridium botulinum*, and also serve as color preservatives.

Although the potentially carcinogenic nitrosamines may be present in foods, particularly cured meats, occurrence is infrequent and at low levels. USDA regulations stipulate that ascorbic acid be added to cured meats at five times the level of nitrates and nitrites to prevent the formation of carcinogenic *N*-nitroso compounds (see FOOD ADDITIVES).

Sodium Chloride

Excessive intake of NaCl contributes to increased fluid retention, and in some individuals there may be a relationship between NaCl intake and hypertension. Both consumers and food processors have reduced use of NaCl.

Toxins

Mycotoxins. The condition produced by the consumption of moldy foods containing toxic material is referred to as mycotoxicosis. Molds and fungi fall into this category and several derive their toxicity from the production of oxalic acid, although the majority of mycotoxins are much more complex.

Mycotoxins find their way into the human diet by way of mold-contaminated cereal and legume crops, meat, and milk products. Corn and peanuts probably represent the most common sources of mycotoxins in the human diet. Many mycotoxins are acutely toxic as well as being potent carcinogens.

Many parasitic fungi have been shown to produce toxins; however, the toxins of *Aspergillus* and *Penicillium* have perhaps the greatest potency against humans.

Seafood Toxins. Virtually scores of fish and shellfish species have been reported to have toxic manifestations. Most of these toxicities have been shown to be microbiological in origin. There are a few, however, that are natural components of seafoods.

Several species of the moray eel (*Gymnothorax*) have caused toxic reactions, especially in Japan. The toxic principle appears to be proteinaceous and is found predominately in the blood but it may occur in the flesh as well. Its exact structure remains somewhat uncertain.

Amnesic shellfish poisoning resulted in four deaths in northeastern Canada in the early 1990s, and domoic acid, the causative agent, was first documented on the Washington and Oregon coasts in 1991. The toxin is produced by a single-celled phytoplankton that constitutes part of the food chain of some shellfish, including Dungeness crab.

Pufferfish toxin, isolated from a dozen or more species, has been identified as having the empirical formula $C_{11}H_{17}N_3O_8$, but the structure is not well-established, nor is it certain that the same structure is universally responsible for poisoning, although this is assumed to be the case. The so-called paralytic shellfish poisoning reported in many areas of the world has a microbiological etiology, and is thus more accurately a contamination rather than a natural toxicosis.

The liver of sharks and other oily fishes sometimes accumulate toxic levels of vitamin A, and cases of acute poisoning have been reported both among Eskimos and the Japanese.

Legislation and Regulatory Considerations

There exists little specific legislation dealing with natural toxicants in foods. The *1958 Food Additives Amendment to the Federal Food, Drug, and Cosmetic Act* stipulates that no substance that has been shown to be carcinogenic to either humans or animals may be added to the food supply. Accordingly, those foods that contain added carcinogens are subject to the Delaney Clause. Maximum tolerances of heavy metals, such as Pb and Hg, have been established by FDA at 0.5 ppm in the food product. For aflatoxins, there is presently a zero tolerance in effect (based on the Delaney Clause), and screening is generally on a qualitative basis. With these exceptions, natural toxicants in food products are generally not treated by specific food legislation. Naturally occurring toxicants have been reviewed in greater detail elsewhere (see TOXICOLOGY).

FRED H. HOSKINS
Washington State University

R. L. Hall and S. L. Taylor, *Food Tech.* **43**(9), 270 (1989).

J. M. Jones, *Food Safety*, Eagan Press, St. Paul, Minn., 1992.

M. R. A. Morgan and D. T. Coxon, *Natural Toxicants in Food*, DCH Publishers, New York, 1987, pp. 221–230.

R. P. Sharma and D. K. Salunke, *Toxicants of Plant Origin*, CRC Press, Boca Raton, Fla., 1989, pp. 179–236.

FORENSIC CHEMISTRY

Forensic science is an applied science having a focus on practical scientific issues that come up during criminal investigations or at trial. Some components are unique to the field because it is conducted within the legal arena.

Physical Evidence

Forensic scientists work with physical evidence, ie, "data presented to a court or jury in proof of the facts in issue and which may include the testimony of witnesses, records, documents or objects." Physical evidence is real or tangible and can literally include almost anything, eg, the transient scent of perfume on the clothing of an assault victim; the metabolite of a drug detected in the urine of an individual in a driving-under-the-influence-of-drugs case; the scene of an explosion; or bullets removed from a murder victim's body.

Examination of physical evidence provides two subtle and different types of conclusion. All members of a class or group have identical characteristics. Types of physical evidence which exhibit class characteristics are paint (qv), glass (qv), fibers (qv), fabric, building material, etc. This type of physical evidence is said to be identified. The best that chemical and physical examinations can ever do is to place items into groups of similarly manufactured items. It is not possible to differentiate one item of evidence as being uniquely distinguishable from another.

Some types of physical evidence, because of the manner in which the material is made, are unique; such evidence can be individualized. Examination can show an item of individualized evidence is unique and comes from one, and only one source. The classic example is fingerprints. Other categories of evidence exhibiting individualization are handwriting, markings on bullets fired from the same gun, broken pieces of glass or plastic which can be physically fit together again, and forensic deoxyribonucleic acid (DNA) evidence.

Physical evidence serves two purposes. In some cases it is used to prove a component or element of a crime. The other purpose for which physical evidence is used is to develop associative evidence in a case. Physical evidence may help to prove a victim or suspect was at a specific location, or that the two came in contact with one another.

Most of the forensic science or crime laboratories located in North America are associated with law enforcement agencies, medical examiner–coroner departments, or prosecutors' offices. There are a large number of independent consultants, also. Laboratories exist at the municipal, county, state, and federal levels of government. There are approximately 300 government-operated forensic science laboratories in the United States.

Forensic science laboratories are generally divided into separate specialty areas. These typically include forensic toxicology, solid-dose drug testing, forensic serology, trace evidence analysis, firearms and tool mark examination, questioned documents examination, and latent fingerprint examination. Laboratories principally employ chemists, biochemists, and biologists at various degree levels.

The bulk of the scientific testing in crime laboratories involves the analysis and characterization of either synthetic or biochemical organic substances or both. Additionally there are a number of evidence categories classified as inorganic.

Forensic Testing

Toxicology. Psychoactive substances, illicit and ethical (licit) drugs and alcohol (ethanol), are the greatest source of physical evidence analyzed in most crime laboratories (see PSYCHOPHARMACOLOGICAL AGENTS). Drug testing falls into two categories: solid dose samples and toxicology (qv) related cases, eg, blood, urine, or tissue specimens in post-mortem cases or cases involving driving under the influence of alcohol or drugs, as well as workplace or employee drug testing.

Blood and urine are most often analyzed for alcohol by headspace gas chromatography (qv) using an internal standard, eg. 1-propanol. Breath alcohol testing is accomplished by a number of techniques.

The oldest reliable procedure involves bubbling a measured volume of deep-lung air containing alcohol through an acidic solution of potassium dichromate, $K_2Cr_2O_7$. Newer instruments rely on infrared spectroscopy to measure the blood alcohol concentration in breath.

Driving under the influence of alcohol cases are complicated because people sometimes consume alcohol with other substances. The most common illicit substances taken with alcohol are marijuana and cocaine. Forensic toxicology laboratories having large caseloads rely on immunoassay (qv) techniques to screen specimens. Immunoassay technology involves the manufacture of antibodies that are specific to particular drugs or to a class of drugs.

There are several immunological techniques in use. Enzyme multiplied immunological technique (EMIT) employs an enzymatic reaction to determine concentration, whereas radioimmunoassay (RIA) uses radioactively tagged reagents such as ^{125}I to measure concentration.

Thin-layer chromatography (tlc) is frequently used. A drawback to tlc, however, is that the technique is not especially sensitive and low levels of drugs may be missed.

Gas chromatography (gc) and gas chromatography-mass spectroscopy (gc/ms) are the most common analytical procedures used in modern forensic toxicology laboratories (see ANALYTICAL METHODS, HYPHENATED INSTRUMENTS). Drugs are separated from their biological matrices, ie, blood, urine, liver, etc, by liquid–liquid or solid-phase extraction (qv) utilizing the solubility of the suspect drug in acid or alkaline aqueous solution relative to the organic solution containing the specimen.

Solid-Dose Narcotics and Dangerous Drugs. Solid-dose drug testing differs from forensic toxicology in that the solid form of the drug is tested, rather than a biological specimen containing the drug and its metabolite. The typical drugs of abuse in North America are heroin; cocaine, ie, free-base, crack, and the HCl salt; marijuana; hashish, a concentrated form of marijuana; amphetamine; methamphetamine; phencyclidine; and LSD.

Trace Evidence. Trace evidence refers to minute, sometimes microscopic material found during the examination of a crime scene or a victim's or suspect's clothing. Trace evidence often helps police investigators develop connections between suspect and victim and the crime scene. The challenge to the forensic scientist is to locate, collect, preserve, and characterize the trace evidence.

Trace evidence in criminal investigations typically consists of hairs; both natural and synthetic fibers (qv), fabrics; glass (qv); plastics; soil; plant material; building material such as cement (qv), paint (qv), stucco, wood (qv), etc, flammable fluid residues, eg, in arson investigations; explosive residues, eg, from bombings, and so on. Perhaps the simplest examination done is the physical match. Other examinations result only in demonstrating class characteristics.

Microscopy (qv) plays a key role in examining trace evidence owing to the small size of the evidence and a desire to use nondestructive testing (qv) techniques whenever possible. Polarizing light microscopy is a method of choice for crystalline materials. Other microscopic procedures involving infrared, visible, and ultraviolet spectroscopy (qv) also are used to examine many types of trace evidence.

More traditional analytical techniques also are used. Capillary column gas chromatography is the method of choice for characterizing flammable fluid residues in arson cases. Scanning electron microscopy (sem) and energy dispersive x-ray analysis (edx) are used frequently in gunshot residue examination and to characterize evidence of an inorganic origin. Pattern recognition examinations are important in footwear and tire impression cases. Lasers (qv) and other high intensity or alternative light sources are useful in crime laboratories to visualize latent fingerprints, seminal fluid stains, obliterated writings, and erasures, and to aid in specialized photographic work. Infrared and ultraviolet light sources are also used to view items of evidence.

Forensic Serology. Blood, often associated with crimes of violence, is powerful physical evidence. Its presence suggests association with the criminal act and blood can be used to associate suspects and locations with the bleeder. Blood is a complex mixture of cellular material, proteins, and enzymes and several tests are available for suspected bloody evidence. A typical test protocol involves *(1)* determining whether blood is present, *(2)* determining if it is human blood, *(3)* typing the blood, and *(4)* when applicable, performing DNA typing.

Many of the chemical tests of the presence of blood rely on the catalytic peroxidase activity of heme. Species origin tests, used to determine whether the specimen is human or from another source, are immunological in nature.

Blood collected as evidence in criminal acts is usually dried and deposited on a variety of substrates. Sample size is usually on the order of a 2 or 3 mm diameter stain. Traditional typing involves ABO blood grouping, and characterizing stable polymorphic proteins or enzymes present in blood by means of electrophoresis (see FRACTIONATION, BLOOD; ELECTROSEPARATIONS, ELECTROPHORESIS).

More recently, the forensic application of DNA testing has dramatically enhanced the ability to determine the source of a blood sample. Two procedures are in forensic use: restriction fragment length polymorphism (RFLP) and polymerase chain reaction (PCR).

BARRY A. J. FISHER
Scientific Services Bureau
Los Angeles County Sheriff's Department

A. Moenssens, F. E. Imbau, and J. E. Starrs, *Scientific Evidence in Criminal Cases*, 3rd ed., Foundation Press, Mineola, N.Y., 1986.

B. A. J. Fisher, *Techniques of Crime Scene Investigation*, 5th ed., CRC Press, Boca Raton, Fla., 1992.

R. Saferstein, ed., *Forensic Science Handbook*. Prentice-Hall, Englewood Cliffs, N.J., 1982 and *Forensic Science Handbook, Vol. II*, 1988.

P. L. Kirk, *Crime Investigation*, 2nd ed., John Wiley & Sons, Inc., New York, 1974.

FORMALDEHYDE

Formaldehyde, $H_2C{=}O$, is the first of the series of aliphatic aldehydes. Annual worldwide production capacity as of this writing (ca 1994) exceeded 15×10^6 t (calculated as 37% solution). Because of its relatively low cost, high purity, and variety of chemical reactions, formaldehyde has become one of the world's most important industrial and research chemicals.

Physical Properties

At ordinary temperatures, pure formaldehyde is a colorless gas with a pungent, suffocating odor. Physical properties are summarized in Table 1.

Formaldehyde is produced and sold as water solutions containing variable amounts of methanol.

Density and refractive index are nearly linear functions of formaldehyde and methanol concentration. The refractive index may

Table 1. Properties of Monomeric Formaldehyde

Property	Value
density, at −80°C, g/cm^3	0.9151
boiling point at 101.3 kPa,[a] °C	−19
melting point, °C	−118
heat of vaporization,[b] ΔH_v at 19°C, kJ/mol[c]	23.3
heat of formation, ΔH°_f at 25°C, kJ/mol[c]	−115.9
heat of combustion, kJ/mol[c]	563.5
heat of solution at 23°C kJ/mol[c]	
in water	62
in methanol	62.8
flammability in air lower/upper limits, mol %	7.0/73

[a] To convert kPa to mm Hg, multiply by 7.5.
[b] At 164 to 251 K, $\Delta H_v = (27,384 + 14.56T - 0.1207T^2)$ J/mol[c].
[c] To convert J to cal, divide by 4.184.

be expressed by a simple approximation for solutions containing 30–50 wt % HCHO and 0–15 wt % CH_3OH: $n_D^{18} = 1.3295 + 0.00125F + 0.0000113M$.

Viscosities have been measured for representative commercial formaldehyde solutions. Over the ranges of 30–50 wt % HCHO, 0–12 wt % CH_3OH, and 25–40°C, viscosity in mPa·s(= cP) may be approximated by viscosity = $1.28 + 0.039F + 0.05M - 0.024t$.

Chemical Properties

Formaldehyde is noted for its reactivity and its versatility as a chemical intermediate. It is used in the form of anhydrous monomer solutions, polymers, and derivatives.

Formaldehyde condenses with itself in an aldol-type reaction to yield lower hydroxy aldehydes, hydroxy ketones, and other hydroxy compounds; the reaction is autocatalytic and is favored by alkaline conditions.

An important synthetic process for forming a new carbon–carbon bond is the acid-catalyzed condensations of formaldehyde with olefins (Prins reaction).

A commercial process based on the Prins reaction is the synthesis of isoprene from isobutylene and formaldehyde through the intermediacy of 4,4-dimethyl-1,3-dioxane.

With acidic catalysts in the liquid phase, formaldehyde and alcohols give formals, eg, dimethoxymethane from methanol.

Monosubstituted acetylenes add formaldehyde in the presence of copper, silver, and mercury acetylide catalysts to give acetylenic alcohols (Reppe reaction).

Primary and secondary amines readily give alkylaminomethanols; the latter condense upon heating or under alkaline conditions to give substituted methyleneamines. With ammonia, the important industrial chemical, hexamine, is produced.

Mono- and dimethylol derivatives are made by reaction of formaldehyde with unsubstituted amides.

Formaldehyde reacts with syn gas (CO,H_2) to produce added value compounds, eg, glycolaldehyde and glycolic acid.

Manufacture

Most of the world's commercial formaldehyde is manufactured from methanol and air either by an oxidation process using a silver catalyst or one using a metal oxide catalyst.

Economic Aspects

The growth rate of U.S. formaldehyde uses has declined since the 1960s.

Since the cost of methanol represents over 60% of formaldehyde's production costs, the formaldehyde price normally reflects the methanol price. Also, freight is a significant cost for formaldehyde since 1–3 kg of water may be shipped with every kg of formaldehyde. The significant price increase in the early 1970s was due to the sudden rise in hydrocarbon prices caused by the Organization of Petroleum Exporting Companies (OPEC) cartel increasing oil prices.

Storage and Handling

As opposed to gaseous, pure formaldehyde, solutions of formaldehyde are unstable. Both formic acid (acidity) and paraformaldehyde (solids) concentrations increase with time and depend on temperature.

Paraformaldehyde solids can be minimized by storing formaldehyde solutions above a minimum temperature for less than a given time period. The addition of methanol as an inhibitor or of another chemical as a stabilizer allows storage at lower temperatures and/or for longer times. Stabilizers for formaldehyde solutions include hydroxypropylmethylcellulose, methyl- and ethylcelluloses, poly(vinyl alcohol)s, or isophthalobisguanamine at concentrations ranging from 10 to 1000 ppm.

Materials of construction preferred for storage vessels are 304-, 316-, and 347-type stainless steels or lined carbon steel.

Health and Safety Factors

Sources of human exposure to formaldehyde are engine exhaust, tobacco smoke, natural gas, fossil fuels, waste incineration, and oil refineries. It is found as a natural component in fruits, vegetables, meats, and fish and is a normal body metabolite. Facilities that manufacture or consume formaldehyde must control workers' exposure in accordance with the following workplace exposure limits in ppm: action level, 0.5; TWA, 0.75; STEL, 2. In other environments such as residences, offices, and schools, levels may reach 0.1 ppm HCHO due to use of particle board and urea–formaldehyde foam insulation in construction.

Formaldehyde causes eye, upper respiratory tract, and skin irritation and is a skin sensitizer. Although sensory irritation, eg, eye irritation, has been reported at concentrations as low as 0.1 ppm in uncontrolled studies, significant eye/nose/throat irritation does not generally occur until concentrations of 1 ppm, based on controlled human chamber studies.

Formaldehyde is classified as a probable human carcinogen by the International Agency for Research on Cancer (IARC) and as a suspected human carcinogen by the American Conference of Governmental Industrial Hygienists (ACGIH).

Uses

Formaldehyde is a basic chemical building block for the production of a wide range of chemicals finding a wide variety of end uses such as wood products, plastics, and coatings, eg, amino and phenolic resins, 1,4-butanediol, polyols, acetal resins, hexamethylenetetramine, slow-release fertilizers, methylenebis(4-phenyl isocyanate), chelating agents, formaldehyde–alcohol solutions, paraformaldehyde, and trioxane and tetraoxane.

H. ROBERT GERBERICH
GEORGE C. SEAMAN
Hoechst Celanese Corporation

J. F. Walker, *Formaldehyde*, 3rd ed., Reinhold Publishing Corp., New York, 1974.

Formaldehyde, Environmental Health Criteria 89, World Health Organization, Geneva, Switzerland, 1989, p. 17.

Y. Kuraishi and K. Yoshikawa, *Chem. Econ. Eng. Rev.* **14**, 31 (June 1982).

C. W. Horner, *Chem. Eng.* **84**, 108 (July 4, 1977).

FORMIC ACID AND DERIVATIVES

FORMIC ACID

Formic acid (methanoic acid) is the first member of the homologous series of alkyl carboxylic acids. It occurs naturally in the defensive secretions of a number of insects, particularly of ants.

Formic acid was a product of modest industrial importance until the 1960s when it became available as a by-product of the production of acetic acid by liquid-phase oxidation of hydrocarbons. Since then, first-intent processes have appeared, and world capacity has climbed to around 330,000 t/yr, making this a medium-volume commodity chemical. Formic acid has a variety of industrial uses, including silage preservation, textile finishing, and as a chemical intermediate.

Physical Properties

Formic acid is a colorless, highly corrosive liquid with an intense and pungent smell. Its basic physical properties are given in Table 1. Formic acid is totally miscible with water, acetone, ether, ethyl acetate, methanol, and ethanol.

Table 1. Physical Properties of Pure Formic Acid

Property	Value
molecular weight	46.026
melting point, °C	8.4
boiling point, °C	100.7
density, g/cm^3 at 20°C	1.220
refractive index n_D^{20}	1.3714
surface tension at 20°C, mN/m(= dyn/cm)	37.67
viscosity at 20°C, mPa·s(= cP)	1.784
dielectric constant at 20°C	57.9

Chemical Properties

Formic acid exhibits many of the typical chemical properties of the aliphatic carboxylic acids, eg, esterification and amidation, but, as is common for the first member of an homologous series, there are distinctive differences in properties between formic acid and its higher homologues. The smaller inductive effect of hydrogen in comparison to an alkyl group leads, for example, to formic acid ($pK_a = 3.74$) being a considerably stronger acid than acetic acid ($pK_a = 4.77$) and in fact the strongest of the simple, unsubstituted carboxylic acids. Formic acid forms esters with primary, secondary, and tertiary alcohols.

Manufacture

Formic acid is currently produced industrially by three main processes: *(1)* acidolysis of formate salts, which are in turn by-products of other processes; *(2)* as a coproduct with acetic acid in the liquid-phase oxidation of hydrocarbons; or *(3)* carbonylation of methanol to methyl formate, followed either by direct hydrolysis of the ester or by the intermediacy of formamide.

Shipping and Storage

Formic acid is commonly shipped in road or rail tankers or drums. For storage of the 85% acid at lower temperatures, containers of stainless steel (ASTM grades 304, 316, or 321), high density polyethylene, polypropylene, or rubber-lined carbon steels can be used. For higher concentrations, Austenitic stainless steels (ASTM 316) are recommended. The DOT hazard classification of formic acid is "corrosive material."

Economic Aspects

Around 60% of the production is based on methyl formate. Of the remainder, about 60% comes from liquid-phase oxidation and 40% from formate salt-based processes. The largest single producer is BASF, which operates a 100,000 t/yr plant at Ludwigshafen in Germany. The only significant U.S. producer of formic acid is Hoechst Celanese, which operates a butane oxidation process.

Health and Safety Factors

The main hazard in normal handling of formic acid is likely to arise from its corrosive effect on the skin and mucous membranes. Suitable protective equipment should be worn when handling the acid, and rubber or PVC gloves, rubber boots, and goggles are needed during bulk handling operations.

DAVID J. DRURY
BP Chemicals Ltd.

H. W. Gibson, *Chem. Rev.* **69**, 673 (1969).

W. Reutemann and H. Kiecka, *Ullmann's Encyclopaedia of Industrial Chemistry*, Vol. A12, VCH, New York, 1989, p. 13.

Formic Acid, Product Technigram, BP Chemicals, Ltd., London, 1991.

N. I. Sax and R. J. Lewis, *Dangerous Properties of Industrial Materials*, 7th ed., Van Nostrand Reinhold Co., Inc., New York, 1989, p. 1766.

FORMAMIDE

Formamide (methanamide), $HCONH_2$, is the first member of the primary amide series and is the only one liquid at room temperature. It is hygroscopic and has a faint odor of ammonia. Formamide is a colorless to pale yellowish liquid, freely miscible with water, lower alcohols and glycols, and lower esters and acetone. It is virtually immiscible in almost all aliphatic and aromatic hydrocarbons, chlorinated hydrocarbons, and ethers. By virtue of its high dielectric constant, close to that of water and unusual for an organic compound, formamide has a high solvent capacity for many heavy-metal salts and for salts of alkali and alkaline-earth metals. It is an important solvent, in particular for resins and plasticizers. As a chemical intermediate, formamide is especially useful in the synthesis of heterocyclic compounds, pharmaceuticals, crop protection agents, pesticides, and for the manufacture of hydrocyanic acid.

In the 1990s, formamide is mainly manufactured either by direct synthesis from carbon monoxide and ammonia, or more importantly in a two-stage process by reaction of methyl formate (from carbon monoxide and methanol) with ammonia.

Properties

Table 1 lists the important physical properties of formamide.

Reactions

As a result of its bifunctionality, formamide is a highly reactive intermediate that is useful in a wide variety of synthetic applications.

Shipment

Formamide is a registered substance, eg, in TSCA, EINECS, and MITI, and can, therefore, be produced in and imported into the United States, EEC, and Japan in compliance with the above-mentioned acts.

Formamide is best shipped in containers made of stainless steel or in drums made of, or coated with, polyethylene.

Economic Aspects

The estimated capacity of formamide was approximately 100,000 t/yr worldwide in 1990. In 1994, there are only three significant producers; BASF in Germany is the leading manufacturer. Most of the formamide produced is utilized directly by the manufacturers.

Table 1. Physical Properties of Formamide

Property	Condition	Value
molecular formula		CH_3NO
molecular weight		45.04
melting point, °C		2.55
boiling point, °C	101.3 kPa[a]	210.5 (decomp.)
density, g/cm^3	20°C	1.1334
dielectric constant	20°C	109 ± 1.5
heat of combustion, kJ/mol[b]		−568.2
heat of vaporization, kJ/mol[b]		64.98
electrical conductivity, S(= Ω^{-1})	25°C	2×10^{-7}
refractive index n_D^{15}	15°C	1.4491
coefficient of expansion, cm^3/K		0.000775
specific heat, kJ/(kg·K)[b]	19°C	2.30
surface tension, mN/m(= dyn/cm)	20°C	58.35
dynamic viscosity, mPa·s(= cP)	15°C	4.320

[a] To convert kPa to mm Hg, multiply by 7.5.
[b] To convert J to cal, divide by 4.184.

Storage

The shelf life is unlimited in sealed containers. The product is neither explosive nor spontaneously flammable in air. It is stable to the effects of light and air below ca 100°C. However, the product is combustible and should accordingly be stored with adequate precautions.

Manufacture and Processing

To prevent contact with formamide, an approved organic vapor respirator, a face shield, goggles, coveralls, and other protective clothing should be worn as necessary. Spilled material must be disposed of in accordance with local, state, and federal regulations.

Health and Safety

Formamide exhibits no particular acute toxicity with oral, dermal, and other applications in rats and other species. Precautions that should be observed when handling formamide include avoidance of prolonged inhalation of vapors or contact of the liquid with skin or eyes. When handling the chemical its teratogenic property has to be taken into account.

Small quantities of spilled formamide can be washed away with plenty of water. Larger amounts should be absorbed appropriately or pumped into containers for proper disposal by incineration or biological degradation in a sewage water treatment plant.

A. Hohn
BASF AG

H. Bipp, in B. Elvers, S. Hawkins, M. Ravenscroft, J. Rounsaville, and G. Schuls, eds., *Ullmann's Encyclopedia of Industrial Chemistry*, 5th ed., Vol. A12, VCH Publishers, Weinheim, Germany, pp. 1–5.

Formamide, BASF Technical Leaflet 1983; BASF data sheet, BASF AG, Operating Division Zwischenprodukte, Ludwigshafen, Germany, 1989.

DIMETHYLFORMAMIDE

N,N-Dimethylformamide (DMF) is a clear, colorless, hygroscopic liquid with a slight amine odor. The solvent properties of DMF are particularly attractive. Because of its high dielectric constant, aprotic nature, wide liquid range, and low volatility, it is frequently used for chemical reactions and other applications requiring high dissolving power. It is often referred to as "the universal solvent."

Physical and Chemical Properties

Table 1 lists some physical properties for DMF.

Manufacture and Shipment

There are two processes used commercially for DMF manufacture. A two-step process involves carbonylation of methanol to methyl formate, and reaction of the formate with dimethylamine. A second process is the direct carbonylation of dimethylamine in the presence of a basic catalyst or a transition metal.

DMF can be purchased in steel drums (DOT 17E, UN1A1, 410 lbs net = 186 kg), tank trucks, and railcars. On October 1, 1993, new regulations in the United States were established for DMF under HM-181; the official shipping name is *N,N*-dimethylformamide (shipping designation UN 2265, Packing Group III, Flammable Liquid).

Table 1. Physical Properties of DMF, $HCON(CH_3)_2$

Property	Value
molecular formula	C_3H_7NO
molecular weight	73.09
boiling point, °C	153.0
freezing point, °C	−60.4
density at 20°C, g/cm^3	0.949
vapor pressure at 20°C, kPa[a]	
refractive index, n_D^{25}	1.4269
dielectric constant, ϵ, 10 kHz, 25°C	36.7
viscosity, 25°C, mPa·s(= cP)	0.802
hydrogen-bonding index[b]	6.4

[a] To convert kPa to mm Hg, multiply by 7.5.
[b] On a DuPont scale.

Economic Aspects

World capacity of DMF is about 267,000 t as of this writing (ca 1994), with plans for an additional 50,000 t announced for India and Indonesia. World production is considerably less than capacity and is probably around 125,000 t. DMF production and export are closely tied to economic conditions and exchange rates, as demonstrated by the dramatic turnaround from 1983 when a substantial portion of U.S. demand was met by imports. Demand in the United States has slowed since the early 1980s because of decreased acrylic fiber production. As with all organic solvents, growing concerns about emissions have caused many users to make more efficient use of DMF. Increasingly, it is being recovered and purified for reuse.

Ideal materials for its handling and storage are nonalloy (carbon) steels, stainless steels, and aluminum. Seals and other soft materials should be made of polytetrafluoroethylene, polyethylene, or high molecular weight polypropylene. Ethylene–propylene rubber O-rings and diaphragms can also be used with DMF. Graphite can be used to lubricate moving parts in contact with DMF. Since DMF is hygroscopic, it should be kept under a blanket of dry nitrogen. High purity DMF required for acrylic fiber production is best stored in aluminum tanks.

Uses

The two largest uses for DMF in the United States have been in pharmaceutical processing and acrylic fiber production. However, in 1990, DuPont ceased U.S. production of acrylic fibers. The remaining U.S. producers use dimethylacetamide, DMAC, in their fiber-spinning processes and acrylic fiber production no longer consumes significant quantities of DMF in the United States. Worldwide demand for DMF in acrylic fiber production has held up better than in the United States.

Another significant application for DMF is as a solvent for depositing polyurethane coatings on leather and artificial leather fabrics. This use, too, has fallen somewhat since the late 1970s, as changing fashions have decreased the demand for "wet look" fabrics, and it now accounts for about 5% of U.S. demand for DMF.

Another use is in various extraction and absorption processes for the purification of acetylene or butadiene and for separation of aliphatic hydrocarbons, which have limited solubility in DMF, from aromatic hydrocarbons. DMF has also been used to recover CO_2 from flue gases.

DMF is used extensively as a solvent, reagent, and catalyst in synthetic organic chemistry. Several comprehensive reviews describe its uses in this area.

Health and Safety Factors

The acute toxicity of DMF is relatively low. Skin absorption is an important route by which DMF can be introduced into the body.

The chemical has been linked to alcohol intolerance among workers, which is manifested by a reddening of the skin upon ingesting alcoholic beverages shortly after exposure to DMF. Chronic exposure to high levels of DMF causes liver damage.

The International Agency for Research on Cancer (IARC) has concluded that evidence associated DMF with cancer in animals is "inadequate," but has classified DMF as "possibly carcinogenic to humans."

The American Conference of Governmental Industrial Hygienists (ACGIH) has recommended that time weighted-average exposures for

DMF not exceed 10 ppm or 30 mg/m^3 (skin designation, 1989 standard) for an eight-hour work day. In the United States, OSHA has accepted the ACGIH limits in setting regulations for worker exposures.

Although it is a versatile and generally stable solvent, DMF must be used with care in some applications. It can react violently with very strong reducing and oxidizing agents.

JOHN A. MARSELLA
Air Products and Chemicals, Inc.

Dimethylformamide. Properties, Uses, Storage, and Handling, Du Pont product brochure H-10902, Oct. 1988.

R. S. Kittila, *Dimethylformamide Chemical Uses*, E. I. du Pont de Nemours & Co., Inc., Wilmington, Del., 1967.

G. L. Kennedy, *CRC Crit. Rev. Toxicol.* **17**, 129–182 (1986). Extensive review of toxicology.

H. Bipp and H. Kieczka in *Ullmann's Encyclopedia of Industrial Chemistry*, 5th ed., Vol. A12, VCH Publishers, New York, 1989, pp. 1–12.

FRACTIONATION, BLOOD

CELL SEPARATION

Whole blood is seldom used in modern blood transfusion. Blood is separated into its components. Transfusion therapy optimizes the use of the blood components, using each for a specific need. Red cell concentrates are used for patients needing oxygen transport, platelets are used for hemostasis, and plasma is used as a volume expander or a source of proteins needed for clotting of the blood.

Blood Component Therapy

Blood is composed of a cellular portion, the formed elements, suspended in plasma. When a test tube with blood is centrifuged, the formed elements are packed onto the bottom of the tube, leaving plasma on top (Fig. 1).

Collection of Blood and Blood Components. Blood can be collected in the form of whole blood donations. In the United States, one unit, ie, 450 mL, or blood is collected from a healthy volunteer blood donor who is allowed to donate blood once every 10 weeks. A unit of blood is typically separated into a red cell fraction, ie, red cell concentrate; a platelet fraction, ie, random donor platelets (RDP); and plasma.

Blood components are also collected through apheresis. In apheresis, advanced blood cell separators are used to collect one or more specific blood components from a donor. The two principal components collected through apheresis are plasma and single-donor platelets (SDP).

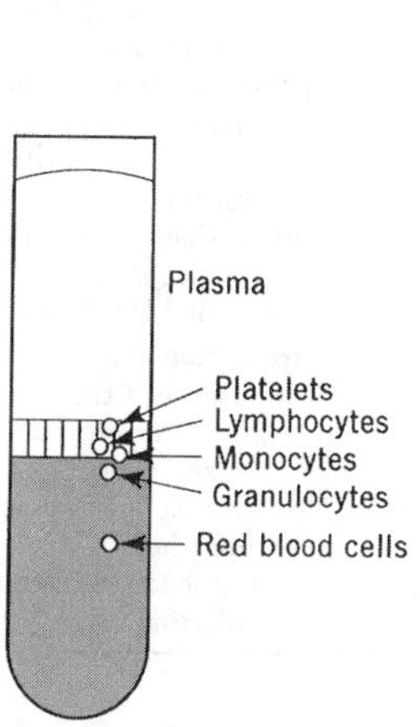

Figure 1. Distribution of component cells by density in a centrifuged sample of human blood.

The objectives of collection, separation, preparation, and storage of blood components are *(1)* to provide a safe blood product through careful screening and testing of the donor and collected product; *(2)* to maintain sterility of the product by adequate cleansing of the venepuncture site(s) and sterile processing methods which are essential to avoid bacterial contamination of the products; *(3)* to maintain viability and function of the components, ie, separation and storage methods need to be optimal for the specific transfused product; and *(4)* to make optimal use of blood components by transfusion of only those components indicated for the malignancy of the patients.

Function and Use of Blood Components

Primary blood components include plasma, red blood cells (erythrocytes), white blood cells (leukocytes), platelets (thrombocytes), and stem cells. Plasma consists of water; dissolved proteins, ie, fibrinogen, albumins, and globulins; coagulation factors; and nutrients. The principal plasma-derived blood products are single-donor plasma (SDP), produced by sedimentation from whole blood donations; fresh frozen plasma (FFP), collected both by apheresis and from whole blood collections; cryoprecipitate, produced by cryoprecipitation of FFP; albumin, collected through apheresis; and coagulation factors, produced by fractionation from FFP and by apheresis.

Centrifugation Methods

Each type of blood cell has its own distribution of mass densities. Most blood cell separators are based on the formation of blood components into layers by density gradient only. Some cell separators, ie, Haemonetics MCS, apply methods based on a combination of mass density and cells size.

Density Gradient Separation. Based on specific density, each cell in a test tube finds its own position (see Fig. 1), ie, red cells at the bottom, then granulocytes, monocytes, lymphocytes, platelets, and plasma on top.

Many cell separation methods are based on the formation of layers by mass-density gradient. The simplest method is based on spinning down a bag of blood and expressing off the different layers. The more complex apheresis machines, eg, Baxter Fenwall CS3000 and COBE Spectra, are based on continuous-flow principles. Other blood cell separators utilize a batch processing method and discontinuous flow, eg, the Haemonetics V50plus and Mobile Collection System.

In some cases, density gradient solutions are used to separate a specific layer of cells. A solution, like Ficoll or Percoll, with a mass density between the density of the cells that are to be collected and the other cells, is added to the blood product.

Countercurrent Separation and Elutriation. The process known as elutriation in cell separation is a refined method for separation of cells having close mass densities. Cells can be separated by making use of differences in the critical velocity of cells. If the mass densities of two cells are identical, but the sizes are different, then the larger particle has a higher critical velocity than the smaller one.

Cell separation techniques that use an inward flow component are referred to as countercurrent separation techniques. The two principal applications of countercurrent flow are found in the Beckman elutriators and the Haemonetics apheresis equipment. The Beckman elutriators are capable of very specific cell separation of small batches of cells. The Haemonetics surge technique can separate platelets and lymphocytes from four liters of donor blood in one hour and forty minutes.

Filtration

Filtration (qv) is applied in blood cell separation to remove leukocytes from red blood cell (RBC) and platelet concentrates. Centrifugational blood cell separators do not reduce white blood cells (WBC) in red cell and platelet products sufficiently to avoid clinical complications such as GvHD and alloimmunization. A post-apheresis filtration step is needed to further reduce the WBC lead.

Filter Design. Modern leukocyte-reduction filters have become highly efficient as a result of careful filter design and advanced biomaterials. General design considerations include high flow rate through the filter, low retention volume, and hydrophilic filter media that does not require priming prior to filtration.

Two types of leuko-reduction filters are applied in blood cell separation: those for filtering red cells and those for platelet filtration. Filters designed for use with RBCs consist of two filtration layers: an upstream screen filter for trapping large particles and microaggregates, and a downstream adsorption filter for leukocyte reduction. The filters for platelet concentrates have only the adsorption filter. The two principal filter designs commercially available (ca 1993) are a relatively flat, large diameter disk-shaped filter, ie, the RC and PL filters of Pall Corporation, and filters having a relatively small cross-sectional area but greater depth, eg, the SepaCell filters of Asahi Corp. The white cell adsorption filter layer is typically of a nonwoven fiber design.

Mechanisms of Leukocyte Adsorption. The exact mechanism of leukocyte adhesion to filter media is not yet fully understood. Multiple mechanisms simultaneously contribute to the adhesion of cells to biomaterials, however, physical and biological mechanisms have been distinguished. Physical mechanisms include barrier phenomenon, surface tension, and electrostatic charge; biological mechanisms include cell activation and cell to cell binding.

Emerging Cell Specific Technologies

A number of cell specific technologies for cell separation are emerging. Cell specific typing through surface antigen marking is possible. The feasibility of separating two subgroups from white blood cell concentrates is being investigated; ie, stem cells, which are marked by the CD34 antigen, and lymphokine activated killer cells (LAK) (CD8). Purified stem cells can be used for cancer therapy (see CHEMOTHERAPEUTICS, ANTICANCER), but also for curing various hematologic genetic diseases through DNA engineering of the purified stem cells (see GENETIC ENGINEERING). The LAK cells are used to treat immune diseases.

Technologies to purify cells from white cell concentrates are in the research stage. Principles used include antibodies covalently bound to a surface, antibody-coated microbeads in a column, magnetic microparticles that have been coated with antibodies, and hollow fibers that have been coated with antibodies.

Regulations, Storage, and Shipment

Blood transfusion is highly regulated worldwide by government institutions, such as the USFDA, and through associations of blood banks, such as the American Association of Blood Banks (AABB). Strict regulations on good manufacturing practices (GMP) have been established to ensure maximum safety of the transfused products. Each blood component has specific storage requirements in terms of optimal temperature, additives, expiration, and storage containers.

For optimal functionality, platelets require a stable and well-balanced pH, gas exchange, ambient temperature, and gentle agitation. Special plastics have been developed for optimal storage of platelets.

THEODOOR HEIN SMIT SIBINGA
Haemonetics

W. H. Dzik, *Transfusion*, 334–339 (1992).

T. S. Kickler and W. Bell, *Transfusion*, 411–414 (1989).

A. Latham, *Vox Sanguinis* **51**, 249–252 (1986).

T. L. Simon, "Apheresis Principles and Practices," in E. C. Rossi, T. L. Simon, and G. S. Moss, eds., *Principles of Transfusion Medicine*, Williams & Wilkins, Baltimore, Md., 1991.

PLASMA FRACTIONATION

Human blood plasma contains over 700 different proteins (qv). Some of these are used in the treatment of illness and injury and form a set of pharmaceutical products that have become essential to modern medicine (Table 1). Preparation of these products is commonly referred to as blood plasma fractionation, an activity often regarded as a branch of medical technology, but which is actually a process industry engaged in the manufacture of specialist biopharmaceutical products derived from a natural biological feedstock (see PHARMACEUTICALS).

In 1993 there were over 100 organizations undertaking plasma fractionation worldwide, having plant capacities ranging from 4 to 1800 m^3/yr. Table 2 lists the six commercial manufacturers in the United States.

Manufacturing and Processing

Plasma fractionation is unusual in pharmaceutical manufacturing because it involves the processing of proteins and the preparation of multiple products from a single feedstock. A wide range of unit operations are utilized to accomplish these tasks. They are listed in Table 3;

Table 1. Pharmaceutical Plasma Derivatives[a]

Product	Clinical application	Molecular weight × 10^3
	Albumin[b]	
human serum albumin	protein and volume replacement	68
plasma protein fraction	volume replacement	68
	Coagulation proteins[c]	
Factor VIII	hemophilia A treatment	300
Factor IX complex	treatment of hemophilia B and other coagulation disorders	57
antiinhibitor coagulant complex[d]	hemophilia A treatment where Factor VIII antibodies are present	
	Inhibitors[c]	
α-1-proteinase inhibitor	emphysema treatment	52
antithrombin III	antithrombin III deficiencies treatment	58
	Immunoglobulins[e]	
immune globulin intravenous (normal)	immunoglobulin (IgG) replacement; treatment of immune disorders	150
immune globulin intravenous	treatment of cytomegalovirus (CMV) infection in immune-suppressed individuals	150
immune serum globulin (normal)	prevention of hepatitis A and rubella infections	150
hepatitis B immune globulin	prevention of hepatitis B infection	150
pertussis immune globulin	prevention of whooping cough infection	150
rabies immune globulin	prevention of rabies infection	150
rho(D) immune globulin	prevention of hemolytic disease of the newborn	150
tetanus immune globulin	treatment or prevention of tetanus infection	150
vaccinia immune globulin	prevention of small-pox infection	150
varicella immune globulin	prevention of chicken-pox infection	150

[a] U.S. Licensed.
[b] Active component is albumin.
[c] Active component is indicated product.
[d] Active component is not known.
[e] Active component is IgG.

Table 2. 1993 Plasma Fractionators

Organization	Capacity, m^3/yr	Location
Alpha Therapeutics	1800	Los Angeles
Cutter Biologicals	1800	West Haven, Conn.
Hyland Therapeutics	1800	Glendale, Calif.
Armour Pharmaceuticals	1300	Collegeville, Pa.
Immuno	450	Rochester, Mich.
Ortho Diagnostic	50	Raritan, N.J.

Table 3. Plasma Fractionation Unit Operations[a]

Unit operation	Method/ technology	FVIII	FIX	IgG(im)	IgG(iv)	PPF	Albumin
	Protein separation[b]						
fractional precipitation[c]	cold-ethanol precipitation		++	+	+	+	+
fractional extraction[d]	from cold-ethanol ppt			+	+	+	+
solid–liquid separation	centrifugation	+	++	++	++	++	++
	depth filtration			++	++	++	++
selective adsorption	depth filtration			+	+	+	+
selective adsorption/ desorption[e]	ion-exchange chromatography	+	+		++		++
	immuno-affinity chromatography	++	++				
	Virus inactivation, in-process						
heat treatment	carbohydrate stabilized	++	++		++		
	fatty acid stabilized					++	++
chemical treatment[f]	solvent–detergent treated	++	++		++		
	Formulation and finishing[g]						
selective adsorption	depth filtration			+	+	+	+
membrane filtration	cross-flow filtration	++	++	++	++	+	+
	dead-end filtration	+	+	+	+	+	+
stabilization	chemical additives	+	++		+	+	+
dispensing	aseptic-dispensing	+	+	+	+	+	+
drying	freeze drying	+	+	++	++		
	Virus inactivation, terminal						
heat treatment	pasteurization					+	+
	dry heating	++	++				

[a] +, method in common use; ++, optional method, depending on procedures used by different manufacturers; im, intramuscular; iv, intravenous. [b] Size exclusion by gel filtration is an optional method for FVIII. [c] Charge reduction (pH, temperature) and other precipitation are common methods of FVIII fractionation. [d] Extraction from other precipitates is an optional method for FVIII fractionation. [e] Affinity chromatography is an optional method for FIX. [f] Optional chemical treatments include potassium thiocyanate for FIX and acid/enzyme treatment for IgG (iv). [g] Selective proteolysis by acid/enzyme treatment is an optional method for IgG (iv).

some are common to a number of products and all must be closely integrated.

Principal Unit Operations. Principal unit operations include protein precipitation, solid–liquid separation, protein adsorption, membrane separations (ultrafiltration/diafiltration, sterile filtration), freeze-drying, and inactivation and removal of viruses.

Process Rationale. The products of plasma fractionation must be both safe and efficaceous, having an active component, protein composition, formulation, stability, and dose form appropriate to the intended clinical application. Processing must address a number of specific issues for each product. Different manufacturers may choose a different set or combination of unit operations for this purpose.

Human plasma is collected from donors either as a plasma donation, from which the red cells and other cellular components have been removed and returned to the donor by a process known as plasmapheresis, or in the form of a whole blood donation. These are referred to as source plasma and recovered plasma, respectively. In both instances the donation is collected into a solution of anticoagulant to prevent the donation from clotting and to maintain the stability of the various constituents. Regulations in place to safeguard the donor specify both the frequency of donation and the volume that can be taken on each occasion.

Following donation, the separated plasma is frozen and transported to a fractionation plant, where it is held in frozen storage before being released for processing. On entering processing, plasma is vulnerable to bacterial contamination and proteolytic degradation. The more labile constituents are particularly at risk. The early process steps aim for a degree of purification, the creation of a stable environment free from bacterial growth, and, where possible, a significant reduction in process volume. These objectives can be met by precipitation processes. Ideally, a range of intermediate products are produced at this stage that are held in storage pending release for further purification. The subdivision of processes in this manner carries a number of advantages including flexibility in scheduling and batch sizing, as well as in maximizing the utilization of limiting or expensive resources.

Factor VIII, immunoglobulin, and albumin are all held as protein precipitates, the first as cryoprecipitate and the others as the Cohn fractions FI + II + III (or FII + III) and FIV_4 + V (or FV), respectively. Similarly, Fractions FIV_1 + FIV_4 can provide an intermediate product for the preparation of antithrombin III and α-1-proteinase inhibitor. This ability to reduce plasma to a number of compact, stable, intermediate products, together with the bacteriacidal properties of cold-ethanol, are the principal reasons these methods are still used industrially.

Direct ion-exchange adsorption is used to recover Factor IX from the FVIII-depleted plasma that remains following cryoprecipitation. Alternatively, Factor IX can be recovered from Cohn Fraction III.

A number of other plasma products are entering into clinical use; growth is expected in at least some of these areas. Fibrinogen, previously withdrawn because of the hepatitis risk, can now be supplied in a virally inactivated form suitable either for infusion or as part of a fibrin sealant kit used for wound healing. Fibrinogen can be recovered from cryoprecipitate, Cohn Fraction I, or from side fractions of Factor VIII processing.

Another by-product of Factor VIII processing having clinical value is von Willebrand factor. It has been recovered from side fractions using ion-exchange and affinity chromatography.

Alpha-1-proteinase inhibitor and antithrombin III are used to treat people with hereditary deficiencies of these proteins. Both can be recovered from Cohn Fraction IV using ion-exchange chromatography and affinity chromatography, respectively. Some manufacturers recover antithrombin III directly from the plasma stream by affinity adsorption.

Economic Aspects

Estimates for a number of economic aspects of plasma fractionation can be made. The world capacity for plasma fractionation exceeded 20,000 t of plasma in 1990 and has increased by about 75% since 1980, with strong growth in the not-for-profit sector.

The clinical use of plasma products varies widely between countries with commercial products being imported in some instances to meet demand. In the United States, the market for plasma products increased from $250,000,000 in 1980 to over $850,000,000 in 1991. This expansion resulted from a 60% increase in the use of albumin, a 70% increase in the use of Factor IX concentrate, the introduction of intravenous immunoglobulin (IgG iv) and a 500% increase in the price of FVIII.

Regulation and Control

The preparation of clinical products from human plasma is regulated as a pharmaceutical manufacturing operation by national authorities who are responsible for giving authorization to distribute a product in their country. This is done by the Food and Drug Administration (FDA) in the U.S. and by the Medicines Control Agency (MCA) in the U.K.

Health, Safety, and Environmental Factors

The possibility that infectious donations of plasma may enter the fractionation process places staff at risk; the transmission of hepatitis B to fractionation workers had been reported before the screening of plasma for hepatitis B infection was introduced. The extensive testing of donations that takes place (ca 1993) reduces this risk substantially. Nevertheless it is assumed that plasma for fractionation may be contaminated with viruses such as hepatitis B, hepatitis C, and HIV, and appropriate precautions should be taken.

Ethanol (qv), the principal bulk reagent in plasma fractionation, is categorized as a highly flammable material with vapor concentration of 3–19% ethanol being explosive at temperatures above the flash point of 13°C. These properties must be considered in the design and specification of equipment and facilities involved in ethanol fractionation. Once the fractionation process has been completed, there are waste solutions containing up to 40% ethanol which require disposal. Some manufacturers recycle this material using distillation (qv), a procedure which must be regulated and controlled to the satisfaction of local customs and excise authorities.

PETER R. FOSTER
Scottish National Blood Transfusion Service

M. H. Stryker, M. J. Bertolini, and Y. L. Hao, *Adv. Biotechnol. Processes* **4**, 275–336 (1985).

J. R. Harris, ed., *Blood Separation and Plasma Fractionation*, John Wiley & Sons, Inc., New York, 1991.

J. S. Finlayson and D. L. Aronson, *Semin. Thromb. Hemost.* **6**, 1–74 (1979); **6**, 85–139 (1980).

R. Madhok, C. D. Forbes, and B. L. Evatt, eds., *Blood, Blood Products and HIV*, 2nd ed., Chapman and Hall, London, 1994.

FRACTURE MECHANICS

Fracture mechanics is a methodology which characterizes the resistance of a material to crack propagation. Provided specific requirements are met, a material property can be measured which describes the performance of the material when a sharp or natural crack is present. This property is called the plane strain fracture toughness and is independent of the specimen geometry used to make the measurement. The measured fracture toughness can then be used in the design of a component or structure to avoid fracture. The concepts of fracture mechanics for brittle crack growth were originally proposed in 1920. However, it was not until World War II that the technology was substantially developed.

Fracture mechanics is now quite well established for metals, and a number of ASTM standards have been defined. For other materials, standardization efforts are underway.

Inherent Flaws

Perhaps the single most important concept in fracture mechanics is the existence of inherent flaws. Any material contains imperfections or defects. Examples are voids at grain boundaries in metals or a dirt particle in a polymer. It is from these defects that cracks eventually begin to grow, because defects cause local stress concentrations.

Linear Elastic Fracture Mechanics

A crack in a body may grow as a result of loads applied in any of the three coordinate directions, leading to different possible modes of failure. The most common is an in-plane opening mode (Mode I). The other two are shear loading in the crack plane (Mode II) and antiplane shear (Mode III), as defined in Figure 1. Only Mode I loading is considered herein.

Crack Tip Stresses. The simplest case for fracture mechanics analysis is a linear elastic material where stress, σ, is proportional to strain, ϵ, giving

$$\sigma = E\epsilon \tag{1}$$

where E is the elastic modulus. In this case the stresses and strains around the tip of a sharp crack, for a given applied remote load, can be shown to vary inversely with the square root of the distance r from the crack tip and go to infinity precisely at the crack tip.

Clearly the stresses and strains are not actually infinite in real materials, but it is found that in relatively brittle materials, where the amount of plastic yielding is small, failure is controlled by K and fracture occurs when a critical value of K is applied. This critical value is the fracture toughness of the material, denoted K_{IC} where the subscript I indicates Mode I loading as shown in Figure 1.

Energy Release Rates. The analysis can equivalently be conducted in terms of the energy dissipated in growing the crack, rather than the stresses. The energy release rate, G, is the rate at which energy is released from the overall system as the crack grows. This is the energy released per unit area of crack surface formed, rather than a time-dependent rate, and so describes the energy available to drive the crack through the material. G can be thought of as the energy which would be available to drive the crack if the crack were to grow at the current applied remote stress level. In fact, the crack does not begin to grow until this available energy is sufficient to overcome the resistance of the material to crack growth. As for K_{IC}, there is a critical energy release rate, G_{IC}, at which the crack begins to grow. The two approaches are complementary and, by considering the rates of change of the stresses and displacements around the crack tip per unit of crack growth, it can be shown that for plane strain conditions:

$$K^2 = EG/(1 - \nu^2) \tag{2}$$

G can be calculated from the change in compliance, ie, the reciprocal of stiffness, of the structure or test specimen.

Crack Tip Yielding and Constraint. In any real material there is some plastic yielding at the crack tip, even though the overall fracture behavior of the material can be characterized using stress intensity concepts which consider the stresses in the elastic region. Figure 2 shows this zone schematically. The plastic zone is larger near the free surfaces of specimen than at the center. This is because the high stresses and strains in the plane of the specimen (the XY plane) lead to a tendency to contract laterally (in the Z direction) at the crack tip. This can occur at the free surfaces, but near the center the unyielded material surrounding the crack tip tends to resist this contraction. A positive stress in the Z direction is generated. Because yield criteria such as those of von Mises are based on the difference between the principal stresses in the three coordinate directions, this Z direction stress acts to restrict the total amount of plastic yielding in the center of the specimen. The degree of constraint and the size of the plastic zone is approximately constant above some minimum distance into the specimen.

Elastic–Plastic Fracture Mechanics

In more ductile materials the assumptions of linear elastic fracture mechanics (LEFM) are not valid because the material yields more at

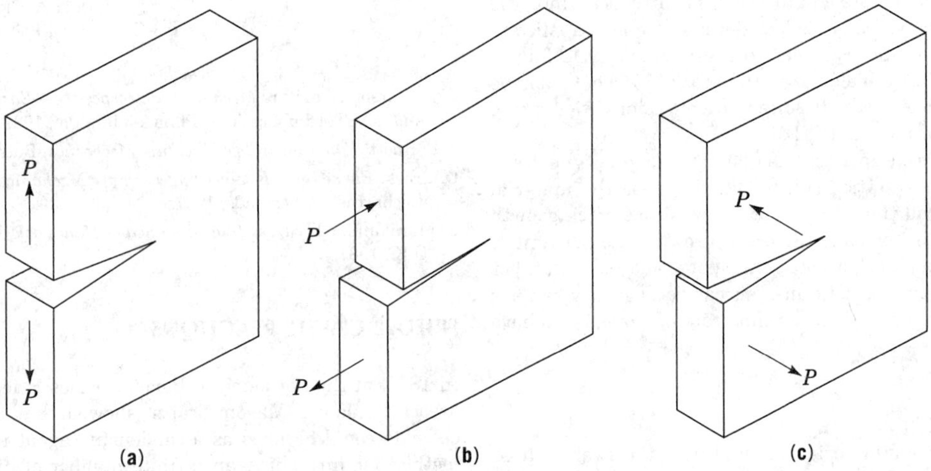

Figure 1. Three modes of fracture where P is load: (**a**) Mode I, (**b**) Mode II, and (**c**) Mode III.

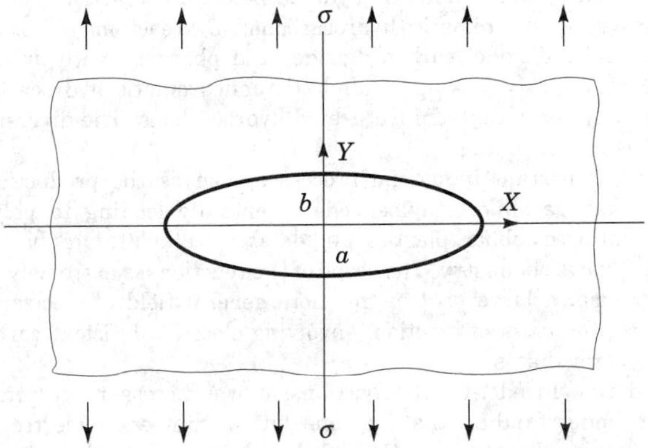

Figure 2. The crack tip plastic zone, defined in the text.

the crack tip, so that the stresses and strains around the tip are no longer dominated by the linear elastic analysis. In order to characterize the fracture behavior of the material it is necessary to look inside the plastic zone at the distribution of stress and strain. If it is assumed that the material deforms in tension according to a power law hardening stress–strain curve,

$$\sigma = \sigma_0 \epsilon^n \quad (3)$$

then it is possible to calculate the variation of stress and strain within the yielded crack tip zone. This stress and strain distribution is known as an HRR field. The magnitude of the product of stress and strain is characterized by a different parameter, the J-integral, and this product varies inversely with the distance from the crack tip, the same as in the linear elastic case. However, the individual variation of stress or strain alone is no longer the simple inverse square root form of the linear elastic case, but a more complex expression dependent on the shape of the stress–strain curve.

Micromechanisms

For a perfectly brittle material the fracture process involves the breaking of bonds at an atomic level to form new surfaces. This was the original concept and was based on studies of the fracture of glass (qv). The energy needed per unit area of new surface formed, γ, was considered. Because the growth of a crack forms two new surfaces, the energy release rate needed to drive the crack in a brittle material is 2γ. However, in most engineering materials there is also a significant amount of plastic energy which must be dissipated in order to grow the crack, which gives a condition for crack growth of the form:

$$G \text{ or } J = (2\gamma + W_p) \quad (4)$$

where W_p is the plastic work that must be done to grow the crack. Fracture mechanics attempts to quantify the amount of energy needed to grow the crack, by treating the behavior on a macroscale continuum basis. The objective is to subtract all nonessential work done in deforming the test specimen to determine the energy specific to the crack growth process alone. Locally at the crack tip various microscale processes are occurring to actually fail the material. There is some process zone, inside the plastic zone adjacent to the blunted crack tip, where material has been deformed so much that it begins to rupture.

Practical Testing Procedures

Specimen Requirements. Fracture toughness tests divide into two distinct groups, K_{IC} (or G_{IC}) measurements for relatively brittle materials and J-integral tests for more ductile materials. The reason for this becomes apparent when the K_{IC} test is considered. Fracture mechanics testing is required to follow certain procedures so that a valid measurement is obtained. Specifically, the initial crack in the test specimen must be sharp and the thickness and depth of the specimen must be sufficient to generate the plane strain constraint discussed earlier.

Fracture Toughness Testing. Some typical fracture toughness test geometries include the single-edge notch (SEN), the center notch (CN), the compact tension (CT), and the three-point bend (TPB). A fracture toughness test, where unstable brittle fracture occurs at a critical applied load, leads to a load-displacement curve having an essentially linear loading curve and a catastrophic drop in load at fracture.

J-Integral Testing. The J-integral test is rather different from the K_{IC} test because a curve of crack growth resistance against crack growth is generated. This requires more than one specimen, with each specimen loaded to different levels to generate different amounts of crack growth.

Fatigue

Fatigue cracking is slow crack growth under repeated cyclic loading or constant static load. For cyclic loading unnotched fatigue data are often presented in the form of S–N curves, where applied stress, S, is plotted against cycles to failure, N. This method leads to widely scattered data because the total lifetime is the sum of the time to initiate a crack from an existing inherent flaw, the size of which is variable between specimens, and the time to grow the crack to final failure of the specimen. This variability is overcome by using fracture mechanics because a deliberate initial crack is introduced into the specimen, which is then cyclically loaded. Because the load applied to each cycle is generally quite low, relative to the single-cycle failure load, the amount of plastic yielding at the crack tip should not be excessive and the problem is well suited to characterization using linear elastic fracture mechanics (LEFM).

Fracture mechanics concepts can also be applied to fatigue crack growth under a constant static load, but in this case the material behavior is nonlinear and time-dependent. Slow, stable crack growth data can be presented in terms of the crack growth rate per unit of time against the applied K or J, if the nonlinearity is not too great. For extensive nonlinearity a viscoelastic analysis can become very complex and a number of schemes based on the time rate of change of J have been proposed.

Impact Loading

The fracture toughness of a material can vary with the strain rate at which the test is conducted. This is why materials such as polymers tend to be more brittle when loaded rapidly. For this reason, fracture mechanics tests are also conducted under impact loading conditions. The three-point bend specimen is probably the most commonly used geometry for impact tests.

It is to be expected that the fracture toughness of a material where the crack is propagating at some speed varies with the speed at which the crack is growing. Because the toughness of a material tends to fall with increasing rate, up to some limit, once a crack begins to propagate in an unstable manner it accelerates to a high (over 100 m/s) speed. Therefore, the calculation of the fracture toughness becomes an extremely complex problem, as the available energy to drive the crack is controlled by dynamic effects. Specialized techniques are needed to calculate the energy release rate during rapid crack propagation and to measure the fracture toughness experimentally.

Environment and Temperature

All of the tests discussed herein could be conducted at different temperatures and in different environments. The effect of a particular environment or temperature can be significant and the experimenter should take care to ensure that the data is collected under the appropriate conditions for the problem being considered. Stress corrosion cracking is the term used for the combined effect of an aggressive environment and an applied stress on the propagation of a crack. The variety of combinations of materials, environments, and potential interactions is so great that any fracture toughness measurement must be performed in the environment in which the component being designed is to be used.

Applications and Design Issues

Fracture toughness is often used purely as a relative comparison of the cracking resistance of different materials. However, it has the potential to also be used for quantitative prediction of actual component or structure performance. A number of rather elaborate design procedures have been developed, one of which is the R6 design code. This design procedure essentially considers the two competing failure processes of gross plastic yielding across the uncracked cross section, leading to failure by plastic collapse of the structure, and failure by crack growth. It attempts to define combinations of these two modes which would lead to failure when neither alone would be sufficient.

The use of fatigue data and crack length measurements to predict the remaining service life of a structure under cyclic loading is possibly the most common application of fracture mechanics for performance prediction. In complex structures the growth of cracks is routinely monitored at intervals, and from data about crack growth rates and the applied loadings at that point in the structure, a decision is made about whether the structure can continue to operate safely until the next scheduled inspection.

BARRY A. CROUCH
E. I. du Pont de Nemours & Co., Inc.

J. M. Barsom, *Fracture Mechanics Retrospective—Early Classic Papers (1913–1965)*, ASTM Publications, Philadelphia, Pa., 1987.

J. F. Knott, *Fundamentals of Fracture Mechanics*, Butterworths, London, 1979.

D. Broek, *Elementary Engineering Fracture Mechanics*, Martinus Nijhoff, Dordrecht, the Netherlands, 1987.

Y. Murakami, ed., *Stress Intensity Factors Handbook*, Pergamon Press, Oxford, U.K., 1987.

FRIEDEL-CRAFTS REACTIONS

In 1877, at the Sorbonne in Paris, Charles Friedel and his American associate, James Mason Crafts, showed that anhydrous aluminum chloride could be used as a condensing agent in a general synthetic method for furnishing an infinite number of hydrocarbons. In work stretching over 14 years, they extended their discoveries of the catalytic effect of aluminum chloride in a variety of organic reactions: *(1)* reactions of alkyl and acyl halides and unsaturated compounds with aromatic and aliphatic hydrocarbons; *(2)* reactions of acid anhydrides with aromatic hydrocarbons; *(3)* reactions of oxygen, sulfur, sulfur dioxide, carbon dioxide, and phosgene with aromatic hydrocarbons; *(4)* cracking of aliphatic and aromatic hydrocarbons; and *(5)* polymerization of unsaturated hydrocarbons. The diversity of reactions is astounding.

Many important industrial processes such as the production of high octane gasoline, ethylbenzene (eventually leading to polystyrene), synthetic rubber, plastics, and detergent alkylates are based on Friedel-Crafts chemistry. The scope of the reactions is extremely wide as they form a large part of the more general field of electrophilic reactions, the class of reactions involving electron deficient carbocationic intermediates.

To define Friedel-Crafts reactions, it was necessary to come to a clear understanding that not one but a number of electrophilic reactions are classified as Friedel-Crafts type. Friedel-Crafts-type reactions are generally considered to be any substitution, isomerization, elimination, cracking, polymerization, or addition reaction that takes place under the catalytic effect of Lewis acid-type acidic halides (with or without cocatalysts) or Brønsted acids. Friedel-Crafts reactions are not limited to the formation of carbon–carbon bonds but also lead to formation or cleavage of carbon–oxygen, carbon–nitrogen, carbon–sulfur, carbon–halogen, carbon–metals, and many other types of bonds. Friedel-Crafts reactions can be divided into two general categories: alkylations and acylations. Within these two broad areas there is considerable diversity.

Catalysts

Friedel-Crafts catalysts are electron acceptors, ie, Lewis acids. The alkylating ability of benzyl chloride was selected to evaluate the

Table 1. Friedel-Crafts Catalyst Activities

Group	Characteristic	Examples
A	very active, high yields but extensive intra- and intermolecular isomerization	$AlCl_3$, $AlBr_3$, AlI_3, $GaCl_3$, $GaCl_2$, $GaBr_3$, GaI_3, $ZrCl_4$, $HfCl_4$, $HfBr_4$, HfI_4, SbF_5, NbF_5, $NbCl_5$, TaF_5, $TaCl_5$, $TaBr_5$, MoF_6, and $MoCl_5$
B	moderately active, high yields without significant side reactions	$InCl_3$, $InBr_3$, $SbCl_5$, WCl_6, $ReCl_5$, $FeCl_3$, $AlCl_3{-}RNO_2$, $AlBr_3{-}RNO_2$, $GaCl_3{-}RNO_2$, $SbF_5{-}RNO_2$, and $ZnCl_2$
C	weak, low yields without side reactions	BCl_3, BBr_3, BI_3, $SnCl_4$, $TiCl_4$, $TiBr_4$, $ReCl_3$, $FeCl_2$, and $PtCl_4$
D	very weak or inactive	many metal, alkaline-earth, and rare-earth element halides

relative catalytic activity of a large number of Lewis acid halides. The results of this study suggest four categories of catalyst activity (Table 1). Catalysts include acid halides (Lewis acids), metal alkyls and alkoxides, protic acids (Brønsted Acids), acidic oxides and sulfides (acidic chalcogenides), acidic cation-exchange resins, and superacids (Brønsted superacids, super Lewis acids, Brønsted-Lewis superacids, solid superacids, and superacidic zeolites).

GEORGE A. OLAH
V. PRAKASH REDDY
G. K. SURYA PRAKASH
University of Southern California

G. A. Olah, ed., *Friedel-Crafts and Related Reactions*, Vols. 1–4, Wiley-Interscience, New York, 1963–1965.

G. A. Olah, *Friedel-Crafts Chemistry*, John Wiley & Sons, Inc., New York, 1973.

R. M. Roberts and A. A. Khalaf, *Friedel-Crafts Alkylation Chemistry*, Mercel Dekker, Inc., New York, 1984.

G. A. Olah, G. K. Prakash, and J. Sommer, *Superacids*, John Wiley & Sons, Inc., New York, 1985, pp. 24–27 and references therein.

FRUIT JUICES

In 1993, frozen concentrated fruit juice for manufacturing was the principal juice product of international commerce. Consumers have demanded more ready-to-serve juice products, especially chilled single-strength juices. Commercial aseptic packaging of single-strength juices and juice drinks worldwide permits packaging in soft plastic packages that can be stored at higher temperatures for convenience and economy for the consumer. Aseptic packaging is also used for more economical storage and transport at higher temperatures of bulk single-strength and concentrated juices.

Raw Materials

In the 1990s, raw materials are selected for suitability for juice production, except for apple juice production which still uses much cull fruit. Variety and maturity are important factors affecting suitability for juice production. When a range of varieties is available, eg, apples, blends are used to achieve uniformity of flavor. Fruit grown in a warmer climate and having a high sugar (qv) content may be mixed with fruit from a cooler region to achieve a desired sugar–acid balance. This sugar–acid ratio is based on °Brix readings, ie, wt % sugar, obtained with a refractometer, and on total acidity obtained by titration.

Juice factories frequently employ field persons to advise growers on the application of sprays to the growing crops so that residues on harvested fruit are within prescribed limits. They also may sample the crop before harvest for analysis, and coordinate harvesting with factory production schedules. Payment for raw materials is frequently based on specifications that are either official government grades or stated market standards. Official graders may be employed to test each load.

Manufacturing

Figure 1 describes many of the steps in the production of fruit juices. citrus have a thick, relatively tough peel that must be kept separate from the juice during extraction. Deciduous fruit are crushed whole and the juice is then separated from the pulp, peel, and seeds, usually in a pressing operation.

Tropical Fruit

Pineapple juice has been the most popular tropical fruit juice consumed in the U.S. for many years. It has long been a by-product of the pineapple canning industry. Juice pressed from the core, and from trimmings of fruit cut into cylinders for canning, is combined with

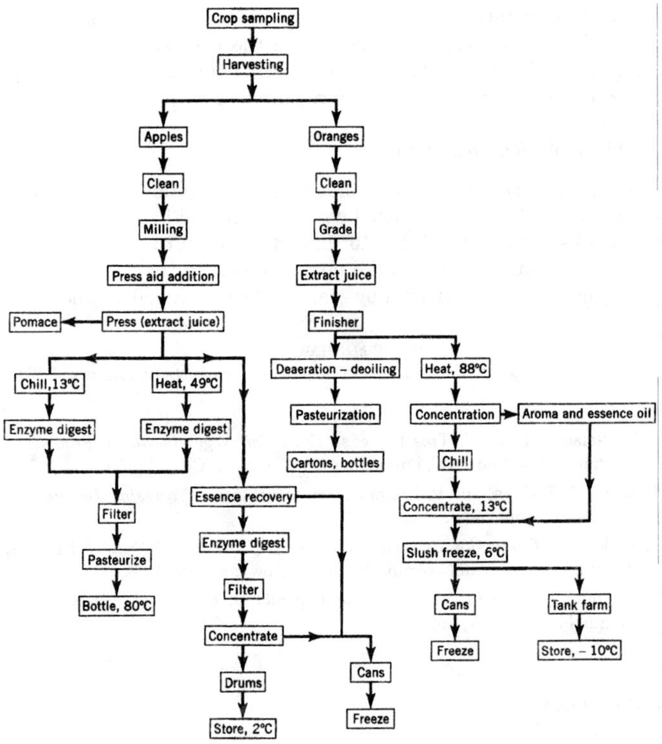

Figure 1. Manufacturing process for citrus (orange) and deciduous (apple) juices.

juice collected from the cutting table to provide most of the pineapple juice sold commercially. This juice contains up to 40% pulp. The pulp content is reduced, usually by finishing with a screw-type finisher followed by centrifugation, before pasteurization and concentration. Pineapple juice can be concentrated in a falling film plate-type evaporator with an essence recovery step prior to concentration. As with other fruit juices, concentrated pineapple juice with added aqueous essence of either 60 °Brix with high pulp content, or 72 °Brix with low pulp content, is the principal product traded internationally. A higher quality pineapple juice made using whole ripe pineapples is being sold at a premium price in the European Economic Community (EEC), accounting for nearly 20% of the pineapple juice consumed.

The production and sale of other tropical fruit juices has more recently received significant attention in Europe and, especially, in North America. Many tropical fruit juices are too pulpy or have harsh or exotic flavors which make 100% juice products unacceptable to most U.S. consumers. They are more acceptable as nectars containing 25–50% juice or as blended fruit drinks where their strong flavors are diluted or modified. Mango and papaya are tropical fruits available in limited supply as concentrated juices or purees. Available single-strength purees include guava, banana, kiwifruit, lulo, soursop, and umbu.

Fruit Juice Drinks

In most markets fruit juice must be 100% juice and contain no additives. Commonly, a juice drink contains 10% fruit juice, which usually is a blend of several fruits. The 1990 Federal Nutrition and Labeling Act requires declaration of juice content so that the consumer can make a more informed choice. With cocktails and juice drinks, added sugars, acids, flavorings, colorings, and nutrients can be used to provide a wide variety of stable products of uniform quality. Because drinks require less juice than 100% juice products, the drinks can be sold at a lower price.

Consumption in the fruit drink/nectar market varies widely from country to country. In the United States juice consumption is 73% vs only 27% for drink/nectar consumption. In Japan the reverse is true with drink/nectar accounting for 73% of the market. In the United Kingdom, 80% of the market is derived from fruit juice, but in Europe as a whole, drink/nectar outsells pure juice by 30%.

World Supply and Consumption

The principal fruit juice and beverage markets are the United States, where 12.1×10^6 L is consumed per year; followed by Canada, 5.6×10^6 L; Western Europe, 1.2×10^6 L; and Japan, 0.76×10^6 L. In all significant markets orange juice predominates. In the United States apple juice is second, followed by grape and then grapefruit juice.

PHILIP E. SHAW
Citrus and Subtropical Products Laboratory, USDA

P. E. Nelson and D. K. Tressler, eds., *Fruit and Vegetable Juice Processing Technology*, 3rd ed., AVI Publishing Co., Westport, Conn., 1980.

S. Nagy, C. S. Chen, and P. E. Shaw, eds., *Fruit Juice Processing Technology*, AgScience, Auburndale, Fla., 1993.

D. Hicks, ed., *Production and Packaging of Non-carbonated Fruit Juices and Fruit Beverages*, Van Nostrand Reinhold, New York, 1990.

D. A. Kimball, *Citrus Processing Quality Control and Technology*, Van Nostrand Reinhold, New York, 1991.

FUEL CELLS

Fuel cells are electrochemical devices that convert the chemical energy of a fuel directly into electrical and thermal energy. In a typical fuel cell, gaseous fuels are fed continuously to the anode (negative electrode) compartment, and an oxidant, eg, oxygen or air, is fed continuously to the cathode (positive electrode) compartment. The electrochemical reactions take place at the electrodes to produce an electric (direct) current. The fuel cell theoretically has the capability of producing electrical energy for as long as the fuel and oxidant are fed to the electrodes. In reality, degradation or malfunction of components limits the practical operating life of fuel cells.

Besides the direct production of electricity, heat is also produced in fuel cells. This heat can be effectively utilized for the generation of additional electricity or for other purposes, depending on the temperature. A practical consideration for fuel cells is compatibility with the available fuels and oxidants. As of this writing the electrochemical reactions involving hydrogen and oxygen (or air) are the only practical ones. The oxygen (qv) is usually derived from air. Hydrogen (qv) is available from several fuel sources, eg, steam-reformed fossil fuels (see COAL; PETROLEUM), coal gasification (see COAL CONVERSION PROCESSES), steam-reformed methanol (qv), etc (see FUELS, SYNTHETIC–GASEOUS FUELS).

General Characteristics

One of the main attractive features of fuel cell systems is the expected high fuel-to-electricity efficiency. This efficiency, which runs from 40–60% based on the lower heating value (LHV) of the fuel, is higher than that of almost all other energy conversion systems. The two primary impediments to the widespread use of fuel cells are high initial cost and short operational lifetime. These two aspects are the focus of research.

Types of Fuel Cells

A variety of fuel cells has been developed for terrestrial and space applications. Fuel cells are usually classified according to the type of electrolyte used in the cells as polymer electrolyte fuel cell (PEFC), alkaline fuel cell (AFC), phosphoric acid fuel cell (PAFC), molten carbonate fuel cell (MCFC), and solid oxide fuel cell (SOFC). These fuel cells are listed in Table 1 in the approximate order of increasing operating temperature, ranging from ~80 °C for PEFCs to ~1000 °C for SOFCs. The physicochemical and thermomechanical properties of materials used for the cell components, ie, electrodes, electrolyte, bipolar separator, current collector, etc, determine the practical operating temperature and useful life of the cells.

Table 1. Fuel Cell Components and Operating Conditions[a,b]

Characteristic	PEFC	AFC[c]	PAFC	MCFC	SOFC[d]
anode	Pt black or Pt/C	Ni	Pt/C	Ni–10% Cr	$Ni–ZrO_2$ cermet
cathode	Pt black or Pt/C	Li-doped NiO	Pt/C	Li-doped NiO	Sr-doped $LaMnO_3$
pressure, MPa[e]	0.1–0.5	~0.4	0.1–1	0.1–1	0.1
temperature, °C	80	260	200	650	1000
electrolyte, wt %	Nafion[f]	85% KOH	100% H_3PO_4	62% Li_2CO_3–38% K_22CO_3[g]	yttria-stabilized ZrO_2

[a] AFC = alkaline fuel cell; MCFC = molten carbonate fuel cell; PAFC = phosphoric acid fuel cell; PEFC = polymer electrolyte fuel cell; and SOFC = solid oxide fuel cell. [b] All cells are bipolar having a filter-press or flat-plate construction, except where otherwise indicated. [c] Used in the *Apollo* program. [d] Tubular cells. [e] To convert MPa to psi, multiply by 145. [f] Fluorinated sulfonic acid, registered trademark of E. I. du Pont de Nemours & Co., Inc. [g] In mol %.

Applications

Fuel cells operating on pure H_2 and O_2 provide a useful power source in remote areas such as in space or under the sea where system weight and volume are important parameters. On the other hand, fuel cell power plants operating on fossil fuels and air offer the potential for environmentally acceptable, highly efficient, and low cost power generation. Thus fuel cells can be considered for terrestrial applications where environmental pollution or noise would be objectionable, and they can be located near the point of use of the electricity such as on an urban site, rather than at a remote location. An analysis, summarized in Table 2, describes the minimum technical requirements for fuel cells for four different types of applications, ie, for buildings, industry, transportation, and utilities. The fuel cell technologies which are most likely to be used in the various applications, and their anticipated capacities are given.

Economic Aspects

From the standpoint of commercialization of fuel cell technologies, there are two challenges: initial cost and reliable life. The initial selling price of the 200-kW PAFC power plant from IFC was about $3500/kW. A competitive price is projected to be about $1500/kW or less for the utility and commercial on-site markets. For transportation applications, cost is also a critical issue. The fuel cell must compete with conventional mass-produced propulsion systems. Furthermore, it is not clear if the manufacturing cost per kilowatt of small fuel cell systems can be lower than the cost of much larger units. The life of a fuel cell stack must be five years minimum for utility applications, and reliable, maintenance-free operation must be achieved over this time period. The projection for the PAFC stack is a five year life, but reliable operation has yet to be demonstrated for this period.

Table 2. Minimum Technical Requirements for Fuel Cell Applications

Characteristic	Buildings	Industry	Transportation	Utility
technology	PEFC/PAFC	SOFC	PEFC	MCFC
efficiency, %	35	40	35	45
system life, yr	2/15	15	2+	20
capacity, kW	5–200	200–2000	5–200	2000+
operating temp., °C	90/194	1000	90	700
heat recovery	important	important		important

Several activities, if successful, would strongly boost the prospects for fuel cell technology. These include the development of *(1)* an active electrocatalyst for the direct electrochemical oxidation of methanol; *(2)* improved electrocatalysts for oxygen reduction; and *(3)* a more CO-tolerant electrocatalyst for hydrogen. A comprehensive assessment of the research needs for advancing fuel cell technologies, conducted in the 1980s, is available.

KIMIO KINOSHITA
ELTON J. CAIRNS
University of California, Berkeley

A. J. Appleby and D. G. Lovering, eds., *Fuel Cells 2, Proceedings of the 2nd Grove Anniversary, London, Sept. 24–29, 1991*, Elsevier, Amsterdam, the Netherlands, 1992. Reprinted from *J. Power Sources* **37** (1992).

A. J. Appleby and F. R. Foulkes, *Fuel Cell Handbook*, Van Nostrand Reinhold, New York, 1989.

N. Q. Minh, *J. Am. Ceram. Soc.* **76**, 563 (1993).

Energy **11**(1/2), (1986).

FUEL RESOURCES

U.S. Energy Production, Consumption, and Availability

Coal production is most significant fuel source followed by natural gas and petroleum. Coal is overwhelmingly the most significant energy source used to generate electricity.

During the period 1970–1990, increased emphasis was placed on renewable energy resources, including wood and wood waste; municipal solid waste and refuse-derived fuel; other sources of biomass and waste, eg, agricultural crop wastes, tire-derived fuels,and selected hazardous wastes burned as fuel substitutes in cement kilns; wind and solar energy; geothermal steam and hot water; and other unconventional energy sources. Estimates of the contribution of these energy sources vary. As of this writing biofuel utilization in the United States runs about 3.7 EJ/yr (3.5×10^{15} Btu/yr) in support of process energy needs for industry, cogeneration facilities, and small standalone power plants, and geothermal energy is about 0.21 EJ/yr (0.2×10^{15} Btu/yr).

Coal Availability and Utilization. There are vast reserves of coal (qv) and lignite in the United States. The total reserve base exceeds 425 billion metric tons equivalent to 11,200 EJ (10.6×10^{18} Btu) and is distributed throughout 32 states. This reserve base has increased by 8.3% since the 1970s despite the high levels of fuel production. Total U.S. recoverable reserves exceed 240 billion metric tons or 6100 EJ (5.8×10^{18} Btu) and are distributed among three geographic areas: the Appalachian, Interior, and Western coal-producing regions. Coal production and consumption in the 1990s reflects the shift toward the use of western, lower sulfur coal.

Environmental considerations also were reflected in coal production and consumption statistics, including regional production patterns and economic sector utilization characteristics. Average coal sulfur content and coal ash content declined, clearly reflecting a trend toward utilization of coal that produces less SO_2 and less flyash to capture.

Oil and Natural Gas Availability and Utilization. U.S. resources and reserves of petroleum (qv) and natural gas, including natural gas liquids (NGL), are limited. Since 1976, the United States has experienced a significant decline in oil reserves. Similarly, from 1976 there was a net reserve loss of dry natural gas.

U.S. distribution of oil and natural gas reserves is centered in Alaska, California, Texas, Oklahoma, Louisiana, and the U.S. outer-continental shelf.

The decrease in petroleum and natural gas reserves has encouraged interest in and discovery and development of unconventional sources of these hydrocarbons. Principal alternatives to conventional petroleum reserves include oil shale and tar sands. The unconventional reserves of natural gas occur principally in the form of recoverable methane from coal beds.

Domestic petroleum, natural gas, and natural gas liquids production has declined at a rate commensurate with the decrease in reserves. Much of the production in the early 1990s is the result of enhanced oil recovery techniques: water flooding, steam flooding, CO_2 injection, and natural gas reinjection.

Whereas the use of petroleum and natural gas is significant in the electricity-generating sector, this usage declined from 1970 to 1990, in part owing to the 1977 Fuel Use Act. The legislation of the 1990s and the growth of independent power producers (IPP) generating electricity for utilities in combined cycle combustion turbine (CCCT) facilities, may mean a reversal in the trend for oil and natural gas utilization for power generation (qv). In any event, total U.S. oil and gas consumption remains high, and these are the fuels of choice for residential, commercial, industrial, and transportation applications.

Other Fuel Availability and Utilization. Nuclear, hydroelectric, and geothermal resources now contribute some 9.8 EJ (9.3×10^{15} Btu) annually to the U.S. economy. Of these energy sources, nuclear power is the dominant force, having over 70% of the total,but nuclear electricity generation may have peaked at 6.5 EJ for political and social reasons. Hydroelectric power generation remains relatively stable. Geothermal energy (qv) has been developed to only a modest extent.

Biomass and waste fuels contributed some 3.7 EJ to the economy. These fuels include wood and wood waste; spent pulping liquor at pulp and paper mills; agricultural materials such as rice hulls, bagasse, cotton gin trash, coffee grounds, and a variety of manures. When wood waste and numerous other forms of biomass are added to municipal solid waste (MSW), refuse-derived fuel (RDF), methane recovered from landfills and sewage treatment plants, and special industrial and municipal wastes such as tire-derived fuel, these together contribute about 5 EJ (4.7×10^{15} Btu) to the U.S. economy. Of these fuels, wood and the biofuels are typically employed in industrial settings either to generate process steam or to cogenerate electricity and process steam.

Other sources of energy worth noting are the extensive wind farms, solar projects, and related emerging unconventional technologies. These renewable resources provide only small quantities of energy to the U.S. economy as of this writing.

Trends in Energy Technology and Future Fuel Consumption. Attention is being paid to increasing the efficiency of fuel utilization as well as to reducing the formation of airborne emissions ranging form particulates NO_x and SO_2 to the management of air toxics such as HCl and trace metals.

Coal. Technologies traditionally deployed for coal utilization include using pulverized coal (PC), cyclone, and stoker-fired boilers. For PC boilers, technologies being deployed or developed include the use of micrometer-sized coal, staged fuel–staged air low NO_x burners, limestone injection multistage burners (LIMB), reburning for NO_x control, and advanced techniques for overfire air management. Cyclone-fired boilers also are capitalizing on reburning technologies and air management techniques. Further, both PC and cyclone-fired boilers are utilizing cofiring techniques, blending nitrogen and sulfur-free biomass fuels, and low sulfur tire-derived fuels with the coal for both cost control and emissions reduction.

Advanced coal utilization technologies include the development of bubbling, circulating, and pressurized fluidized-bed combustion for electricity generation and process energy production.

Petroleum and Natural Gas. The dominant technologies under development for oil and gas include advances in combustion turbine design.

Other Fuels. The emerging technologies for unconventional and renewable fuels somewhat mirror those associated with coal: cofiring of biofuels and coal in PC and cyclone boilers, fluidized-bed combustion systems fed with biofuel or a blend of various solid fuels, combustion air management, and gasification–combustion systems.

World Fuel Reserves, Production, and Consumption

The overwhelming sources of petroleum reserves and supply are in the Middle East. Other significant sources of reserves include Russia, the North Sea, North American countries, and parts of southeast Asia. There are also significant concentrations of coal reserves in Russia

Table 1. Projections of World and U.S. Energy Consumption, 2000–2005, EJ[a]

Energy source	2000 Base	2000 Low	2000 High	2005 Base	2005 Low	2005 High
		World projection[b]				
oil	159	147	170	171	152	190
gas	92	89	94	107	100	112
coal	112	107	116	129	118	136
nuclear	26	25	26	31	27	32
other	35	34	36	42	39	46
Total	*424*	*402*	*442*	*480*	*436*	*516*
		U.S. projection				
oil	40	39	41	42	40	44
gas	25	24	25	26	25	27
coal	21	21	22	22	22	23
nuclear	7	7	7	7	7	8
other	9	9	10	10	10	10
Total	*102*	*100*	*105*	*107*	*104*	*112*

[a] To convert EJ to Btu, multiply by 9.48×10^{14}.
[b] World consumption totals also include the United States.

and China. The dominant coal-producing countries include China and the United States, plus Poland, South Africa, Australia, India, Germany, and the United Kingdom. China is the single largest coal producer and consumer, utilizing over 22 EJ/yr (21×10^{15} Btu/yr) of this solid fossil fuel.

Canada obtains some of its energy from the Athabasca tar sands development (the Great Canadian Oil Sands Project). Oil shale is burned at two 1600-MW power plants in Estonia for electricity generation. World reserves of tar sands total some 6400 EJ (6.1×10^{18} Btu), and world reserves of oil shale total some 20,400 EJ (19.3×10^{18} Btu). Renewable and unconventional energy sources are used more extensively in other parts of the world than in the United States. Biofuels are a significant contributor to certain economies, with proportional contributions as follows: Kenya, 75%; India, 50%; China, 33%; Brazil, 25%; and Scandinavia, 10%. Peat is a significant source of energy for Russia, Finland, and Ireland.

Given the world trends in fuels availability and consumption, projections of energy production and consumption have been made, as shown in Table 1.

DAVID A. TILLMAN
JEFFREY B. WARSHAUER
DAVID E. PRINZING
Enserch

D. A. Tillman, *The Combustion of Solid Fuels and Wastes*, Academic Press, Inc., San Diego, Calif., 1991.

The U.S. Coal Industry, 1970–1990: Two Decades of Change, Energy Information Agency, Washington, D.C., 1992.

EIA's Annual Energy Review 1992, Energy Information Agency, Washington, D.C., 1993.

FUELS FROM BIOMASS

Projections indicate that by the year 2000, the biomass energy contribution will increase to about 4.2×10^{15} kJ/yr(1.9×10^6 BOE/d), ie, over 4% of total U.S. primary energy consumption. Land- and water-based vegetation, organic wastes, and photosynthetic organisms are categorized as biomass and are nonfossil, renewable carbon resources from which energy, eg, heat, steam, and electric power, and solid, liquid, and gaseous fuels, ie, biofuels, can be produced and utilized as fossil fuel substitutes.

The capture of solar energy as fixed carbon in biomass via photosynthesis is the initial step in the growth of biomass. It is depicted by the equation

$$CO_2 + H_2O + \text{light} \xrightarrow{\text{chlorophyll}} (CH_2O) + O_2$$

The primary features of biomass-to-energy technology as a source of synthetic fuels are illustrated in Figure 1.

Distribution of Carbon. Estimation of the amount of biomass carbon on the earth's surface is a problem in global statistical analysis. The results of one such study are summarized in Table 1.

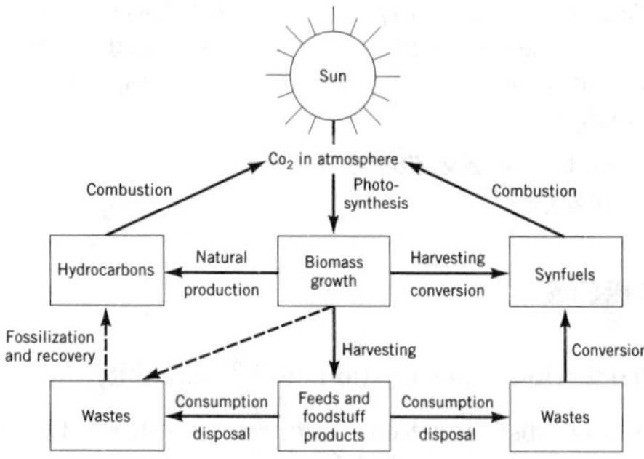

Figure 1. Biomass-to-energy technology.

Table 1. Estimate of Net Photosynthetic Production of Dry Biomass Carbon and Standing Biomass Carbon for World Biosphere[a]

Ecosystem	Area, 10^6 km^2 [b]	Mean net carbon production $t/(hm^2 \cdot yr)$[b]	Mean net carbon production 10^9 t/yr[b]	Standing biomass carbon t/hm^2	Standing biomass carbon 10^9 t
tropical rain forest	17.0	9.90	16.83	202.5	344
boreal forest	12.0	3.60	4.32	90.0	108
tropical seasonal forest	7.5	7.20	5.40	157.5	118
temperate deciduous forest	7.0	5.40	3.78	135.0	95
temperate evergreen forest	5.0	5.85	2.93	157.5	79
Total	*48.5*		*33.26*		*744*
extreme desert-rock, sand, ice	24.0	0.01	0.02	0.1	0.2
desert and semidesert scrub	18.0	0.41	0.74	3.2	5.8
savanna	15.0	4.05	6.08	18.0	27.0
cultivated land	14.0	2.93	4.10	4.5	6.3
temperate grassland	9.0	2.70	2.43	7.2	6.5
woodland and shrubland	8.5	3.15	2.68	27.0	23.0
tundra and alpine	8.0	0.63	0.50	2.7	2.2
swamp and marsh	2.0	13.50	2.70	67.5	14.0
lake and stream	2.0	1.80	0.36	0.1	0.02
Total	*100.5*		*19.61*		*85*
Total continental	*149.0*		*52.87*		*829*
open ocean	332.0	0.56	18.59	0.1	3.3
continental shelf	36.6	1.62	4.31	0.004	0.1
estuaries excluding marsh	1.4	6.75	0.95	4.5	0.6
algae beds and reefs	0.6	11.25	0.68	9.0	0.5
upwelling zones	0.4	2.25	0.09	0.9	0.04
Total marine	*361.0*		*24.62*		*4.5*
Grand total	*510.0*		*77.49*		*833.5*

[a] Dry biomass is assumed to contain 45% carbon.
[b] $1\ km^2 = 1 \times 10^6\ m^2$ (0.3861 sq. mi); to convert $t/(hm^2 \cdot yr)$ to short ton/(acre · yr), divide by 2.24.

Human activity, particularly in the developing world, continues to make it more difficult to sustain the world's biomass growth areas. It has been estimated that tropical forests are disappearing at a rate of tens of thousands of hm^2 per year.

The remaining carbon transport mechanisms on earth are primarily physical mechanisms, such as the solution of carbonate sediments in the sea and the release of dissolved carbon dioxide to the atmosphere by the hydrosphere. The great bulk of carbon, however, is contained in the lithosphere as carbonates in rock. These carbon deposits consist of lithospheric sediments and atmospheric and hydrospheric carbon dioxide. Together, these carbon sources comprise 99.9% of the total carbon estimated to exist on the earth. Fossil fuel deposits are only about 0.05% of the total, and the nonfossil energy-containing deposits make up the remainder, about 0.02%.

Biomass carbon is thus a very small, but important, fraction of the total carbon inventory on earth. It has served as a primary energy source for the industrialized nations of the world; it continues to do so for developing countries. Biomass carbon may again become a dominant source of energy products throughout the world because of fossil fuel depletion and environmental problems. The utilization of biomass carbon as a primary energy source does not add any new carbon dioxide to the atmosphere; it is simply recycled between the surface of the earth and the air over a period of time that is extremely short compared to the recycling time of fossil-derived carbon dioxide.

Energy Potential

The energy content of standing biomass carbon, ie, the above-ground biomass reservoir that in theory could be harvested and used as an energy resource (Table 1), is about 110 times the world's annual energy consumption. Biomass should therefore be considered as a raw material for conversion to large supplies of renewable substitute fossil fuels. Under controlled conditions, dedicated biomass crops could be grown specifically for energy applications.

A realistic assessment of biomass as an energy resource is made by calculating average surface areas needed to produce sufficient biomass at different annual yields to meet certain percentages of fuel demand for a particular country. These required areas are then compared with available surface areas.

Although relatively large areas are required, the use of land-or freshwater-based biomass for energy applications is still practical. It is possible that biomass for both energy and foodstuffs, or energy and forest products applications, can be grown simultaneously or sequentially in ways that would benefit both. Relatively small portions of the bordering oceans also might supply needed biomass growth areas, ie, marine plants would be grown and harvested.

Waste biomass is another large source of renewable carbon supply. It consists of a wide range of materials and includes municipal solid wastes (MSW), municipal biosolids, industrial wastes, animal manures, agricultural crop and forestry residues, landscaping and tree clippings and trash, and dead biomass that results from nature's life cycles.

To assess the potential availability and impact of energy from wastes on energy demand, the energy contents and availabilities of different types of wastes generated must be considered.

Several studies estimate the potential of available virgin and waste biomass as energy resources (Table 2).

U.S. Market Penetrations. U.S. consumption of biomass energy in 2000 is projected to be about 50% greater than the consumption of biomass energy in 1990.

A projection of biomass energy consumption in the United States for the years 2000, 2010, 2020, and 2030 is shown in Table 3 by end use sector. This analysis is based on a National Premiums Scenario which assumes that specific market incentives are applied to all new renewable energy technology deployment. The scenario depends on the enactment of federal legislation equivalent to a fossil fuel consumption tax.

The market penetration of synthetic fuels from biomass and wastes in the United States depends on several basic factors, eg, demand, price, performance, competitive feedstock uses, government incentives, whether established fuel is replaced by a chemically identical fuel or a different product, and cost and availability of other fuels such as oil and natural gas.

U.S. capacity for producing biofuels manufactured by biological or thermal conversion of biomass must be dramatically increased to approach the potential contributions based on biomass availability.

Global Market Penetration. Biomass energy is a significant source of energy in the developing regions of the world. It is a small energy resource in the industrialized areas relative to fossil fuels. The markets for biomass energy as replacements and substitutes for fossil fuels are large and have only been developed to a limited extent. As fossil fuels are either phased out because of environmental issues or become less available because of depletion, biomass energy is expected to acquire an increasingly larger share of the organic fuels market.

Table 2. Potential U.S. Biomass Energy Available in 2000, EJ[a],[b]

Energy source	Estimated recoverable	Theoretical maximum
wood and wood wastes	11.0	26.4
municipal solid wastes		
combustion	1.9	2.1
landfill methane	0.2	1.1
herbaceous biomass and agricultural residues	1.1	15.8
aquatic biomass	0.8	8.1
industrial solid wastes	0.2	2.2
biosolids methane	0.1	0.2
manure methane	0.05	0.9
miscellaneous wastes	0.05	1.1
Total	*15.4*	*57.9*

[a] 1 EJ = 0.9488×10^{15} Btu.
[b] Gross heating value of biomass or methane. Conversion of biomass or methane to another biofuel requires that the process conversion efficiency be used to reduce the potential energy available. These figures do not include additional biomass from dedicated energy plantations.

Table 3. Projected Biomass Energy Consumption in the United States from 2000 to 2030, EJ[a]

End use sector	2000	2010	2020	2030
industry[b]	2.85	3.53	4.00	4.48
electricity[c]	3.18	4.41	4.95	5.48
buildings[d]	1.05	1.53	1.90	2.28
liquid fuels[e]	0.33	1.00	1.58	2.95
Total	*7.41*	*10.47*	*12.43*	*15.19*

[a] 1 EJ = 0.9488×10^{15} Btu. Assumes market incentives of 2 ¢/kWh on fossil fuel-based electricity generation, \$2.00/10^6 Btu on direct coal and petroleum consumption, and \$1.00/10^6 Btu on direct natural gas consumption.
[b] Combustion of wood and wood wastes.
[c] Electric power derived from present (ca 1992) technology via the combustion of wood and wood wastes, MSW, agricultural wastes, landfill and digester gas, and advanced digestion and turbine technology.
[d] Biomass combustion in wood stoves.
[e] Ethanol from grains, and ethanol, methanol, and gasoline from energy crops.

Chemical Characteristics of Biomass

The chemical characteristics of biomass vary over a broad range because of the many different types of species

Biomass Conversion

Various processes can be used to produce energy or gaseous, liquid, and solid fuels from biomass and wastes. In addition, chemicals can be produced by a wide range of processing techniques.

Land-based feeds, ie, trees, plants, and grasses can be converted to produce energy thermal, steam, electric products, solid fuels char,

Table 4. Biomass Feedstock Characteristics that Affect Suitability of Conversion Process

Feedstock characteristic	Process type			
	Physical	Thermal	Biological	Chemical
water content	+	+	+	+
energy content	+	+		+
chemical composition		+	+	+
bulk component analysis	+	+	+	+
size distribution	+	+	+	+
noncombustibles	+	+	+	
biodegradability			+	
carbon reactivity		+		+
organism content/type			+	

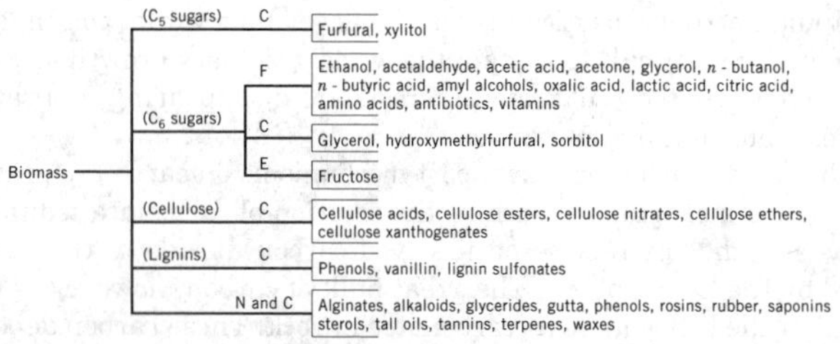

Figure 2. Primary biomass-derived chemicals. Dominant processing methods are chemical (C), fermentation (F), enzymic (E), and natural (N) processes; products in parentheses represent intermediates.

combustibles, gaseous fuels methane (SNG), hydrogen, low and medium thermal-value gas, light hydrocarbons.

Water-based feeds, ie, single-cell algae, multicell algae, and water plants can be converted to produce liquid fuels such as methanol, ethanol, higher hydrocarbons, oils and chemicals.

Table 4 lists the important feed characteristics to be examined when developing a successful conversion process for a specific biomass feedstock. The need to meet environmental regulations can affect processing costs.

The primary types of conversion processes for biomass can be divided into four groups, ie, physical, biological–biochemical, thermal, and chemical.

Physical Processes. Physical processes include particle size reduction, separation, drying, and fabrication.

Biological–Biochemical Processes. Biomass can be subjected to fermentation conditions to form a variety of products. Two of the most common fermentation processes, anaerobic digestion and alcoholic fermentation, yield methane and ethanol, respectively. Biochemical processes include those that occur naturally within the biomass and biophotolysis.

Thermal Processes. Thermal processes for the production of energy and fuels from biomass and wastes usually involve irreversible chemical reactions, heat, and the transfer of chemical energy from reactants to products. The two largest classes of thermal processes are combustion and pyrolysis. A third class of processes can be described either as a combination of combustion and pyrolysis reactions, such as biomass gasification, or as a thermochemical process in which conversion of the feed is facilitated by a reactant such as water or hydrogen.

Another approach to the production of energy products from biomass and wastes is based on hydrogenation. Hydrogen, which can be either generated from the feed or the conversion products, or obtained from an independent source, reacts directly with the feed organics or intermediate process streams at elevated pressures and temperatures to yield substitute fuels.

Chemical Processes. In the production of chemicals from biomass, wood is still the raw material of choice for the manufacture of certain chemicals, although many of them cannot compete with fossil-based products. However, specialty chemicals are often manufactured from nonwoody biomass because they occur naturally in certain plant species or can easily be derived from these plants. Figure 2 lists some of the more important primary biomass-derived chemicals, the principal intermediates, and the dominant processing methods used.

Biomass Production

The manufacture of synfuels and energy products from virgin biomass requires that suitable quantities be grown, harvested, and transported to the conversion plant site. Many variables must be considered when selecting the proper species or mixture of species for operation of a system: growth cycle; fertilization; insolation; temperature; precipitation; propagation and planting procedures; soil and water needs; harvesting methods; disease resistance; growth area competition from biomass for food, feed, and fiber; growth area availability; possibilities for simultaneous or sequential growth of biomass for synfuels and foodstuff or other applications; and nutrient depletion. At least 250,000 botanical species, of which only about 300 are cash crops, are known in the world. A relatively small number are, and will be, used as biomass feedstocks for the manufacture of synfuels and energy products.

In the ideal case, biomass chosen for energy applications should be high yield, low cash-value species that have short growth cycles and grow well in the area and climate chosen for the biomass energy system. Fertilization requirements should be low and possibly nil if the species selected fix ambient nitrogen, thereby minimizing the amount of external nutrients that must be supplied to the growth areas. After harvesting, growth should commence again without the need for replanting. Surprisingly, several biomass species meet many of these idealized characteristics and appear to be quite suitable for energy applications. There are a number of important factors that relate to biomass production for energy applications.

Photosynthesis. The basic biochemical pathways in ambient carbon dioxide fixation involve decomposition of water to form oxygen, protons, and electrons; transport of these electrons to a higher energy level via Photosystems I, II, and several electron transfer agents; concomitant generation of reduced nicotinamide adenine dinucleotide (NADPH) and adenosine triphosphate (ATP); and reductive assimilation of carbon dioxide to carbohydrate.

The maximum efficiency with which photosynthesis can occur has been estimated by several methods. The upper limit has been projected to range from about 8 to 15%, depending on the assumptions made; ie, the maximum amount of solar energy trapped as chemical energy in the biomass is 8 to 15% of the energy of the incident solar radiation.

The biochemical pathways involved in the conversion of carbon dioxide to carbohydrate play an important role in understanding the molecular events of biomass growth. Three different biochemical energy transfer pathways occur during carbon dioxide fixation. One pathway, the Calvin three-carbon cycle, involves an initial three-carbon intermediate of phosphoglyceric acid (PGA). This cycle, often referred to as the reductive pentose phosphate cycle, is used by autotrophic photochemolithotropic bacteria, algae, and green plants. Plant biomass species that use the Calvin cycle are called C_3 plants; it is common in many fruits, legumes, grains, and vegetables. Typical C_3 biomass species are peas, sugar beet, spinach, alfalfa, Chlorella, Eucalyptus, potato, soybean, tobacco, oats, barley, wheat, tall fescue, sunflower, rice, and cotton.

The second pathway is called the C_4 cycle because the carbon dioxide is initially fixed as the four-carbon dicarboxylic acids, malic or aspartic acids. C_4 biomass often occurs in areas of high insolation, hot daytime temperatures, and seasonal dry periods. Typical C_4 biomass includes important crops such as corn, sugarcane, and sorghum, and forage species and tropical grasses such as Bermuda Grass.

The third pathway is called crassulacean acid metabolism (CAM). CAM refers to the capacity of chloroplast-containing tissues to fix carbon dioxide in the dark via phosphoenolpyruvate carboxylase leading to the synthesis of free malic acid. Biomass species in the CAM category are typically adapted to arid environments, have low photosynthesis rates, and have high water usage efficiencies. Examples are cactus plants and the succulents, such as pineapple. Relatively few CAM plants have been exploited commercially.

Significant differences in net photosynthetic assimilation of carbon dioxide are apparent between C_3, C_4, and CAM biomass species. One of the principal reasons for the generally lower yields of C_3 biomass is its higher rate of photorespiration; if the photorespiration rate could be reduced, the net yield of biomass would increase. Considerable research is in progress to achieve this rate reduction by chemical and genetic methods, but as yet, only limited yield improvements have been made.

The specific carbon dioxide-fixing mechanism used by a plant will affect the efficiency of photosynthesis, so from an energy utilization standpoint, it is desirable to choose plants that exhibit high photosynthesis rates to maximize the yields of biomass in the shortest possible time. There are numerous factors that affect the efficiency of photosynthesis other than the carbon dioxide-fixing mechanism, eg, insolation; amounts of available water, nutrients, and carbon dioxide; temperature; and transmission, reflection, and biochemical energy losses within or near the plant. For lower plants such as the green algae, many of these parameters are under human control. For conventional biomass growth subjected to the natural elements, it is not feasible to control all of them.

Land Availability. The availability of sufficient land suitable for production of land-based biomass can be estimated for the United States by several techniques. One method relies on the land capabilities classification scheme developed by the U.S. Department of Agriculture, in which land is divided into eight classes. Classes I to III are suited for cultivation of many kinds of crops; Class IV is suited only for limited production; and Classes V to VIII are useful only for permanent vegetation such as grasses and trees.

There is ample opportunity to produce biomass for energy applications on nonfederal land not used for foodstuffs production. Large areas of land in Classes V to VIII not suited for cultivation and sizable areas in Classes I to IV not being used for crop production also would appear to be available for biomass energy applications; land used for crop production could be considered for simultaneous or sequential growth of biomass for foodstuffs and energy. Portions of federally owned lands, which are not included in the survey, might also be dedicated to biomass energy applications.

Water Availability. The production of marine biomass in the ocean, even on the largest scale envisaged for energy applications, would require only a very small fraction of the available ocean areas. The U.S. Navy has estimated that a square area 753 km on each edge off the coast of California may be sufficient to produce enough giant brown kelp for conversion to methane to supply all of the nation's natural gas needs. This large area is very small when compared with the total area of the Pacific Ocean. Also, the benefits to other marine life from a large kelp plantation have been well documented. Any conflicts that might arise would be concerned primarily with ocean traffic.

Freshwater biomass in theory can be grown on the 20 million hm^2 of fresh water in the United States. However, several difficulties mitigate against large-scale freshwater biomass energy systems. About 80% of the fresh water in the United States is located in the northern states, while several of the freshwater biomass species considered for energy applications require a warm climate such as that found in Gulf states. The freshwater areas suitable for biomass production in the southern states, however, are much smaller than those in the North, and the density of usage is higher in southern inland waters.

Land-Based Biomass. Much effort to evaluate land-based biomass energy applications has been expended. This work aims at selecting high yield biomass species, characterizing physical and chemical properties, defining growth requirements, and rating energy use potential. Several species have been proposed specifically for energy usage, while others have been recommended for multiple uses, one of which is as an energy resource. Most land-based biomass plantations operated for energy production or synfuel manufacture also will yield products for nonenergy markets. Land-based biomass for energy production can be divided into forest biomass, grasses, and cultivated plants.

Table 5. Annual Biomass Yields on Fertile Sites[a]

Dry organics, t/(hm^2·yr)	Climate	Ecosystem type	Remarks
1	arid	desert	better yield if hot and irrigated
2		ocean phytoplankton	
2	temperate	lake phytoplankton	little influence by humans
3		coastal phytoplankton	probably higher in some polluted estuaries
6	temperate	polluted lake phytoplankton	in agricultural and sewage runoffs
6	temperate	freshwater submerged macrophytes	
12	temperate	deciduous forest	
17	tropical	freshwater submerged macrophytes	
20	temperate	terrestrial herbs	possibly higher yields if grazed
22	temperate	agriculture, annuals	
28	temperate	coniferous forests	
29	temperate	marine submerged macrophytes	
30	temperate	agriculture, perennials	
30		salt marsh	
30	tropical	agriculture, annuals	including perennials in continental climates
35	tropical	marine submerged macrophytes	including coral reefs
38	temperate	reedswamp	
40	subtropical	cultivated algae	better yield if CO_2 supplied
50	tropical	rain forest	
75	tropical	agriculture, perennials, reedswamp	

[a] Average net values.

Water-Based Biomass. The average net annual productivities of dry organic matter on good growth sites for land- and water-based biomass are shown in Table 5. With the exception of phytoplankton, which generally has lower net productivities, aquatic biomass seems to exhibit higher net organic yields than land biomass. Water-based biomass considered to be the most suitable for energy applications include the unicellular and multicellular algae and water plants.

Systems Analysis

The overall design of an integrated biomass-to-synfuel system is very important to its successful operation. The system is large and requires coordination of many different operations, such as planting, growing, harvesting, transporting, and converting biomass to gaseous and liquid synfuels. The detailed design of a biomass-to-synfuel system depends on several parameters, such as the type, size, number, and location of the biomass growth and processing areas. In the ideal case, synfuel production plants are located in or near the biomass growth areas to minimize cost of transporting the harvested biomass to the plant. All nonfuel effluents are recycled to the growth areas. This type of synfuel plantation, if developed, would be equivalent to an isolated system with inputs of solar radiation, air, carbon dioxide, and minimal water, and one output, synfuel. The nutrients are kept within the ideal system so that the addition of external fertilizers and chemicals is not necessary. Also, environmental and disposal problems are minimized. Various modifications of this idealized design can be conceived for large-scale usage.

Economics. The practical value of biomass energy ultimately depends on the end-user costs of salable energy and biofuels. Consequently, many economic analyses have been performed on biomass production, conversion, and integrated biofuels systems. Conflicts abound when attempts are made to compare results developed by two or more groups for the same biofuels because methodologies are not the same. Comparative analyses, especially for hypothetical processes conducted by an individual or group of individuals working together, should be more indicative of the economic performance and ranking of groups of biofuels systems.

Several important generalizations can be made. The first is that fossil fuel prices are primary competition for biomass energy. In the context of biomass energy costs, dry, woody, and fibrous biomass species have an energy content of approximately 20 MJ/kg (8600 Btu/lb) or 20 GJ/t (17.2 MBtu/short ton). If such types of biomass were available at delivered costs of $1.00/GJ ($1.054/MBtu), or $20.00/dry t ($18.14/dry short ton), biomass on a strict energy content basis without conversion to biofuels would cost less than most of the delivered fossil fuels. The U.S. Department of Energy has set cost goals of delivered biomass energy crops at $1.90–2.13/GJ ($2.00–2.25/MBtu) and $0.18/L ($0.67/gal) for fuel ethanol from biomass without subsidies in 2000, $0.22 to $0.26/L ($0.85 to $1.00/gal) by the year 2007 for biocrude-derived gasoline, $3.32/GJ ($3.50/MBtu) for methane from the anaerobic digestion of biomass by the year 2000, and 4.5 cents/kWh for electricity from biomass by the late 1990s.

It is essential to recognize several other factors, in addition to the cost of virgin biomass and its conversion to biofuels, when considering whether the costs of biomass energy are competitive with the costs of other energy resources and fuels. Some potential biomass energy feedstocks have negative values; ie, waste biomass of several types such as municipal biosolids, municipal solid wastes, and certain industrial and commercial wastes must be disposed of at additional cost by environmentally acceptable methods. Many generators of these wastes will pay a service company for removing and disposing of the wastes, and many of the generators will undertake the task on their own. These kinds of feedstocks often provide an additional economic benefit and revenue stream that can help support commercial use of biomass energy.

Another factor is the potential economic benefit that may be realized due to possible future environmental regulations from utilizing both waste and virgin biomass as energy resources.

Energetics. The net energy production efficiency of an integrated biomass energy system is extremely important to its development and practical use. The ultimate goal is to design and operate environmentally acceptable systems to produce new supplies of salable energy whether they be low heat value gas, substitute natural gas, substitute gasolines or diesel fuels, methanol, ethanol, hydrogen, or electric power from biomass at the lowest possible cost and energy consumption. It is necessary to quantify how much energy is expended and how much salable energy is produced in each fully integrated system. An energy budget similar to an economic budget should be prepared because the capital, operating, and salable energy cost projections and the conversion process efficiency are insufficient to choose and design the best systems. These values do not necessarily correlate with net energy production. Also, the capital energy investment consumed during construction of the system should be recovered during its operation. Comparative analyses of similar systems for production of synthetic liquid and gaseous fuels from the same feedstock or of different systems that yield the same fuels from different biomass should be performed by consideration of the economics and the net energetics.

To permit comparison, net energy analyses must be clearly specified as to all details. On the whole, net energy production in a modern corn-to-ethanol plant would appear to be borderline if petroleum fuels comprise a significant part of the nonfeed energy inputs. However, to improve net energy production, these fuels can be replaced by fuels generated within the system or by renewable fuels, and credit can be taken for the by-products. Another route is to use more of the corn plant or the stillage as feedstock to the fermentation plant.

Commercial Use of Biomass Energy in the United States

With two possible exceptions, ie, a few small tree plantations, and fuel ethanol from corn, sugarcane, and sugar beet, there is no biomass species grown in the United States specifically for conversion to biofuels. However, conversion processes in commercial use span the basic technologies of combustion, gasification, and liquefaction. These include the combustion of wood, wood wastes, forestry residues, and agricultural residues such as rice husks and bagasse for power production; combustion of MSW and RDF for simultaneous disposal and energy recovery; biological gasification of animal manures in farm- and feedlot-scale anaerobic digestion systems for simultaneous waste disposal and production of biogas as well as upgraded solids for feed, fertilizer, and animal bedding; biological gasification of municipal wastewater biosolids by anaerobic digestion for simultaneous waste disposal and biogas production; biological gasification of MSW in sanitary landfills and recovery of biogas for fuel and power production, which also mitigates environmental and safety problems caused by gas migration in the landfill; thermal gasification of biomass for LHV gas production for on-site use; rapid pyrolysis of wood wastes for fuel oil and chemicals; and alcoholic fermentation of starchy and sugar crops for fuel ethanol for use as a fuel extender, octane enhancer, and oxygenate in motor gasoline blends.

Inventories of commercial usage in the United States are available

Biomass Production. There is no biomass species grown and harvested in the United States specifically for conversion to biofuels, with the possible exceptions of feedstocks for fuel ethanol and a few tree plantations. However, the advances in biomass growth technologies developed in the United States for agricultural crops, trees, and aquatic species, and that are commercial and being improved further through research, are available for growth of biomass energy crops. Multicropping designs and multiple-use crops will be the most likely candidates for biomass energy when conditions warrant commercial plantations.

Economic and Legislative Impacts

Biofuels usage has slowly increased since the mid-1980s because of environmental problems, eg, MSW disposal; favorable legislation, eg, tax incentives and PURPA; and combinations of both, eg, the required use of oxygenated transportation fuels. Although environmental problems continue to increase, many tax incentives for alternative renewable energy resources have been reduced or eliminated. Commercialization of biomass energy is driven by waste disposal, alternative fuels and environmental issues, and the available tax incentives.

The Energy Policy Act of 1992 (H.R. 776) has liberalized the rules concerning biofuels and provides tax incentives for increased usage. Many states also have gasohol fuel tax exemptions in place, and some have enacted legislation that requires use of oxygenated fuels under certain conditions. Most of these laws impact favorably on biofuels usage.

Many energy analysts believe that it is only a matter of time before petroleum prices, the economic parameter that influences almost all other energy prices, begin to return to market prices of at least $3.97/m^3 ($25/bbl). It is widely believed that the gas bubble, which provided excess gas deliverability in the 1980s, will decline in the 1990s. Thus, energy prices are expected to rise again under any scenario. If petroleum prices stabilize at, or continue to increase to, levels over $3.97/m^3 ($25/bbl) it is expected that this, along with environmental issues, will provide the market forces that will increase biomass energy usage.

Research

A large variety of biomass feedstock developments and advanced conversion processes for the production of energy, fuels, and chemicals are in the research stage in the United states. The research is aimed at reducing the cost of biomass and increasing the efficiency of production of the final products.

Feedstock Development. Most of the research in process in the United States in the early 1990s on the selection of suitable biomass species for energy applications was limited to laboratory studies and small-scale test plots. Many of the research programs on feedstock development were started in the 1970s or early 1980s.

Considerable research has been conducted to screen and select nonwoody herbaceous plants such as switchgrass as candidates for biomass plants in the continental United States; other research has concentrated on cash crops such as sugarcane; and still other research has emphasized tropical grasses. Overall research on the development of herbaceous energy crops shows that a broad range of plant species may ultimately be prime energy crops.

Research to develop trees such as hybrid poplar as energy crops via short-rotation intensive culture made significant progress in the 1980s. Projections indicate that yields of organic matter can be substantially increased by coppicing techniques and genetic improvements.

Aquatic biomass, particularly micro- and macroalgae, are more efficient at converting incident solar radiation to chemical energy than are most other biomass species. For this reason, and the fact that most aquatic plants do not have commercial markets, research was performed in the late 1970s and 1980s to evaluate several species as energy crops. The overall goals of the research have generally been directed either to biomass production, often with simultaneous waste treatment, for subsequent conversion to fuels by fermentation, or to species that contain valuable products. The aquatics studied and their main applications are microalgae for liquid fuels, the macrophyte water hyacinth for wastewater treatment and conversion to methane, and marine macroalgae for specialty chemicals or conversion to methane.

Conversion. The direct combustion of biomass for heat, steam, and power has been, and is expected to continue to be, the principal end use of biomass energy. Conventional biomass-fired technology uses a variety of combustion equipment designs that are usually capable of burning a wet, nonhomogeneous fuel with large variations in moisture content and particle size. Even though the burning of biomass is one of the oldest energy-producing methods used, research continues to make significant advances in the art and science of biomass combustion. Recent U.S. legislation concerned with air quality and waste biomass disposal has had a significant impact on the direction of ongoing research to develop advanced biomass combustion systems.

A large amount of research was performed in the 1970s and 1980s on the anaerobic digestion of biomass to develop biological gasification processes capable of producing methane. Research in the early 1990s has addressed several potentially beneficial methods of improving the process, eg, two-phase digestion in which the acetogenic and methanogenic phases are physically separated.

Some of the other research studies have addressed topics such as high solids biomass digestion, utilization of superthermophilic organisms, advanced reactor designs, landfill gas enhancement, and microbiology of the mixed cultures involved in methane fermentation.

Extensive research and pilot studies have been carried out since 1970 to develop thermochemical processes for biomass conversion to energy and fuels. Basic studies on the effects of various operating conditions and reactor configurations have been performed in the laboratory and at the process development unit (PDU) and pilot scales on steam, steam-air, air-blown, and oxygen-blown gasification, and on hydrogasification. Other research has also been done on the rapid pyrolysis of biomass which, in addition to gaseous products, yields coproduct liquids and solids.

Figure 3 outlines most of the biomass liquefaction methods under development.

Limited research has continued on the utilization of seed and vegetable oils as motor fuels, particularly as substitute diesel fuels and diesel fuel extenders. Work has focused on studies of the yields and properties of the methyl and ethyl esters (biodiesel) formed by transesterifying the oilseed and vegetable oil triglycerides, the performance

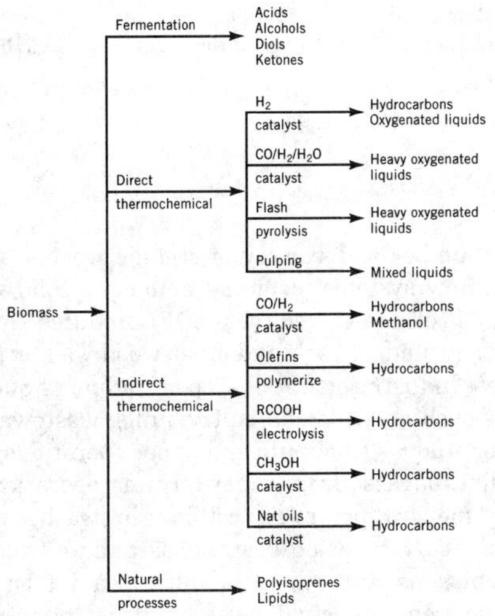

Figure 3. Biomass liquefaction routes under development (ca 1992).

of neat oils and oil–diesel fuel blends as fuels for compression ignition engines, improvement of the transesterification process and the fuel characteristics of the resulting esters as diesel fuels, upgrading vegetable oils to gasolines and diesel fuels by hydrocracking processes, and field tests of the liquid fuels made from seed and vegetable oils in trucks and buses.

Most of the research done since 1970 on the direct thermochemical liquefaction of biomass has been concentrated on the use of various pyrolytic techniques for the production of liquid fuels and fuel components. Some of the techniques investigated are entrained-flow pyrolysis, vacuum pyrolysis, rapid and flash pyrolysis, ultrafast pyrolysis in vortex reactors, fluid-bed pyrolysis, low temperature pyrolysis at long reaction times, and updraft fixed-bed pyrolysis. Other research has been done to develop low cost, upgrading methods to convert the complex mixtures formed on pyrolysis of biomass to high quality transportation fuels, and to study liquefaction at high pressures via solvolysis, steam–water treatment, catalytic hydrotreatment, and noncatalytic and catalytic treatment in aqueous systems.

Prospects. Despite the slow development of renewable biomass as a primary source of energy, the large research effort in progress on feedstock production and conversion is expected to lead to greater commercialization of advanced energy and organic chemical processes based on biomass.

Growing environmental concerns, federal and state environmental regulations, and stabilization of crude oil prices over $25/bbl are expected to be the driving forces behind increased usage of biomass energy. In the long term, carbon taxes applied to fossil fuel usage, especially for vehicles and utility power plants, and oil depletion are expected to provide very strong incentives to convert to renewable biomass energy resources for both mobile and stationary applications.

DONALD L. KLASS
Entech International, Inc.

D. L. Klass, ed., *Energy from Biomass and Wastes*, Vols. 4–16, IGT, Chicago, 1980–1993.

D. L. Klass and G. H. Emert, eds., *Fuels from Biomass and Wastes*, Ann Arbor Science Publishers, Inc., Ann Arbor, Mich., 1981, 592 pp.

First Biomass Conference of the Americas: Energy, Environment, Agriculture, and Industry, NREL/CP-200-5768, DE930/0050, Proceedings Vols. I–III, National Renewable Energy Laboratory, Golden, Colo., 1993, 1942 pp.; D. L. Klass, "Highlights of the First Biomass Conference of the Americas," *SERBEP Update*, Oct. 1993, 6 pp.

Second Biomass Conference of the Americas: Energy, Environment, Agriculture and Industry, NREL/CP-200-8098, Proceedings, National Renewable Energy Laboratory, Golden, Colo., 1995, 1741 pp.; D. L. Klass, "Highlights of the Second Biomass Conference of the Americas," *SERBEP Update*, Oct. 1995, 12 pp.

FUELS FROM WASTE

A significant number and variety of organic wastes are combusted in energy recovery systems including municipal solid waste (MSW), various forms of refuse-derived fuel (RDF) produced from MSW, and municipal sewage sludge; bark and other wood wastes from sawmills and other forest industry operations; spent pulping liquor from chemical pulp mills such as kraft and sulfite mills; wastewater treatment solids (WTS) or sludges from pulp and paper operations; agribusiness wastes including bagasse from sugar-refining operations, rice hulls, orchard and vineyard prunings, cotton gin trash, and a host of other food and fiber-producing operations; manure from feedlots and dairy cattle, chickens, and other agricultural animals; methane-rich gases generated from municipal waste landfills; industrial trash and specific wastes such as demolition debris, broken pallets, unrecyclable paper wastes, and related materials; off-gases from pulp mills and chemical manufacturers; incinerable hazardous wastes generated regularly as a function of production processes, eg, spent solvents, or found on Superfund sites targeted for clean-up; and a broad range of other specific specialty wastes. The practice of incinerating these materials has become increasingly prevalent (ca 1990) in order to accomplish disposal in a cost-effective, environmentally sensitive manner. The combustion of such wastes already contributes some 5 EJ (5×10^{15} Btu) to the U.S. economy and over 15 EJ ($>14 \times 10^{15}$ Btu) to the economies of the industrialized world. Combustion of such wastes reduces the volume of material which must be disposed of in a landfill, reduces the airborne emissions resulting from plant operations and landfill operations, and permits some economic benefit through energy recovery.

The technologies used to combust wastes depend on the form and location of components to be burned. Typically solid wastes are burned, alone or in combination, and both with and without supplementary fossil fuels. Solid wastes can be burned in mass-burn or pile-burning systems such as hearth furnaces, spreader–stokers, ashing and slagging rotary kilns, or fluidized beds. The choice of combustion technology depends on the degree of waste preparation which is practical; the availability of existing combustion systems, eg, a spreader–stoker for hog fuel utilization, adapted to the cofiring of hog fuel and WTS; and the type of energy recovery contemplated. Energy recovery from the solid wastes can be accomplished in the form of medium or high pressure steam, eg, 4.5–8.6 MPa (44–85 atm) (672–783 K), suitable for cogeneration or condensing power generation purposes; low pressure steam, eg, 314–1030 kPa (3.1–10.2 atm), saturated, suitable for process purposes; or the direct production of process heat in the form of heated air or hot combustion products. Energy recovery from gaseous wastes can be accomplished through electricity generation from gas-fired boilers, combustion turbines, or internal combustion engines. Alternatively, these gaseous fuels can be used to generate process heat in conventional fashion.

Critical concerns associated with energy generation from wastes include fuel composition characteristics; combustion characteristics; formation and control of airborne emissions including both criteria pollutants and air toxics, eg, trace metals; and the characteristics of bottom and flyash generated from waste combustion.

Fuel Characteristics of Organic Wastes

Fuel characteristics of organic wastes include physical characteristics such as state, specific gravity, bulk density, porosity, and void volume, and related thermal properties; traditional chemical analyses such as proximate and ultimate analyses, including chlorine; calorific content; elemental analyses of the ash, including trace metal contents, base–acid, slagging, and fouling ratios of the various ash products; and certain chemical structural analyses such as aromaticity. These characteristics are governed by the sources of waste-based fuels. They determine the performance of materials in fuel preparation systems such as particle size reduction and drying systems, and also govern the combustion characteristics of the various wastes being burned.

Sources of Waste-Based Fuels. The general architecture of waste-based fuels is a function of waste origination. MSW characteristics are governed by the product composition of the waste stream, as shown in Table 1. The composition of RDF is governed by the processing technologies used to generate the fuel. The composition of MSW and RDF ultimately is a function not only of the general composition of the waste stream and the RDF production technology, but also of community and industrial recycling programs.

The basic architecture of wood-waste fuels is governed by sawmill or plywood mill configuration, and the consequent blend of bark, trim ends, sawdust, planer shavings, and related residuals. The characteristics of pulp-mill wastes, eg, bark, WTS, and spent pulping liquor, also are determined by the production processes. The characteristics of wastes from food processing, eg, bagasse, rice hulls, peach pits, cotton gin trash, etc, are governed by the basic product manufacturing technology and its efficiency of separation.

Physical Properties. The greatest degree of definition exists for wood and related biofuels. The least degree of definition exists for MSW, related RDF products, and the broad array of hazardous wastes. Table 2 compares the physical property data of some representative combustible wastes with that of the traditional fossil fuel bituminous coal.

Specific gravity is the most critical of the characteristics in Table 2. It is governed by ash content of the material, is the primary determinant of bulk density along with particle size and shape, and is related to specific heat and other thermal properties.

Table 1. Product Composition for Municipal Solid Waste, Wt %

Product	1990[a]	2000[b]
paper and paperboard	38.3	41.0
yard waste	17.0	15.3
food waste	7.7	6.8
plastics	8.3	9.8
wood	3.7	3.8
textiles	2.2	2.2
rubber and leather	2.5	2.4
glass	8.8	7.6
metals	9.4	9.0
miscellaneous	2.1	2.1

[a] Approximate.
[b] Estimated.

Table 2. Physical Properties of Waste-Based Fuels

Fuel	Specific gravity[a]	Bulk density,[b] kg/m^3	Moisture content, wt %
municipal waste		160–320	25–35
waste paper	1.2–1.4	80–160	15–25
waste wood	0.37–0.65	100–320	5–15[c]
			40–65[d]
bagasse			50–55
rice hulls			7–10
orchard and vineyard prunings	0.45–0.55		20–40
bituminous coal	1.12–1.35	672–1393	3.5–5.0

[a] Ovendry.
[b] To convert kg/m^3 to lb/ft^3, divide by 16.01.
[c] Dry waste.
[d] Wet waste.

Chemical Composition. The analysis of solid fuels contains the bases for calculating reactivities, ie, volatile:fixed carbon ratios, volatile carbon:total carbon ratios, hydrogen:carbon and oxygen:carbon ratios, and aromaticity, which is estimated from the chemical components of the waste stream.

Combustion of Solid Waste-Based Fuels

It is useful to examine the combustion process applied to solid wastes as fuels and sources of energy. All solid wastes are quite variable in composition, moisture content, and heating value. Consequently, they typically are burned in systems such as grate-fired furnaces or fluidized-bed boilers where significant fuel variability can be tolerated.

Combustion characteristics of consequence include the overall mechanism of solid waste combustion, factors governing rates of waste fuels combustion, temperatures associated with waste oxidation, and pollution-formation mechanisms.

Applications of Fuels From Waste

Because fuels from combustible organic wastes have long been economic in specific industries such as pulp mills, sawmills, sugar mills or factories, and other biomass processing operations, and because municipal waste-to-energy is becoming increasingly cost effective, these systems are continuing to be installed. The typical industrial system is used either to generate process steam or to generate both electricity and steam in a cogeneration application.

Since the early 1980s, there have been several stand-alone power plants built to fire biomass wastes including such materials as wood waste, rice hulls, and vineyard prunings; these facilities typically generate 20–50 MW_e for sale to electric utilities. MSW and RDF are typically consumed in condensing power plants generating 15–50 MW_e while reducing the volume of solids to be landfilled.

There has been increased interest in firing wood waste as a supplement to coal in either pulverized coal (PC) or cyclone boilers at 1–5% of heat input. This application has been demonstrated by such electric utilities as Santee-Cooper, Tennessee Valley Authority, Georgia Power, Delmarva, and Northern States Power.

DAVID A. TILLMAN
Enserch Environmental

D. A. Tillman, *The Combustion of Solid Fuels and Wastes*, Academic Press, San Diego, Calif., 1991.

W. R. Seeker, W. S. Lanier, and M. P. Heap, *Municipal Waste Combustion Study: Combustion Control of MSW Combustors to Minimize Emissions of Trace Organics*, EER Corporation, Irvine, Calif., 1987.

A. M. Ujihara and M. Gough, *Managing Ash From Municipal Waste Incinerators*, Resources for the Future, Washington, D.C., 1989.

National Incinerator Testing and Evaluation Program: Mass Burning Incinerator Technology, Vol. II, Lavalin, Inc. Quebec City, Quebec, Canada, 1987.

FUELS, SYNTHETIC

GASEOUS FUELS

Whereas coal continues to be a principal source of substitute natural gas, a more recently recognized source is petroleum (qv).

Gas from Coal

Coal can be converted to gas by several routes, but often a particular process is a combination of options chosen on the basis of the product desired, ie, low, medium, or high heat-value gas. In a very general sense, coal gas is the term applied to the mixture of gaseous constituents that are produced during the thermal decomposition of coal at temperatures in excess of 500°C (>930 °F), often in the absence of oxygen (air). A solid residue (coke, char), tars, and other liquids are also produced in the process.

The tars and other liquids (liquor) are removed by condensation leaving principally hydrogen, carbon monoxide, and carbon dioxide in the gas phase. This gaseous product also contains low boiling hydrocarbons, sulfur-containing gases, and nitrogen-containing gases including ammonia and hydrogen cyanide. The solid residue is then treated under a variety of conditions to produce other fuels which vary from a purified char to different types of gaseous mixtures. The amounts of gas, coke, tar, and other liquid products vary according to the method used for the carbonization (especially the retort configuration), and process temperature, as well as the nature (rank) of the coal.

Typical analyses and heat content of common fuel gases vary (Table 1) and depend on the source as well as the method of production.

In the United States, so-called second-generation coal gasification processes came into being as a result of the recognized need to develop reliable, domestic energy sources to replace the rapidly diminishing supply of conventional fuels. More recently, the biological conversion of coal and synthesis gas (carbon monoxide–hydrogen mixtures) into liquid fuels by methanogenic bacteria has received some attention.

Carbonization. Next to combustion, carbonization represents one of the largest uses of coal. Carbonization is essentially a process for the reproduction of a carbonaceous residue by thermal decomposition, accompanied by simultaneous removal of distillate, of organic substances. This process may also be referred to as destructive distillation. It has been applied to a whole range of organic materials, more particularly to natural products such as wood, sugar, and vegetable matter to produce charcoal. However, in the present context, coal usually yields coke, which is physically dissimilar from charcoal and appears with the more familiar honeycomb-type structure.

Coal carbonization processes are generally defined according to process operating temperature. Terms are defined in Table 2.

Gasification. The gasification of coal is essentially the conversion of coal by any one of a variety of processes to produce combustible gases. Primary gasification is the thermal decomposition of coal to produce mixtures containing various proportions of carbon monoxide, carbon dioxide, hydrogen, water, methane, hydrogen sulfide, and nitrogen, as well as products such as tar, oils, and phenols. A solid char product may also be produced, and often represents the bulk of the weight of the original coal.

Table 1. Analyses of Fuel Gases

Type of fuel gas	Gas composition, vol %							Heat value,[a] MJ/m^3
	CO	CO_2	H_2	N_2	O_2	CH_4	Illuminants	
blast-furnace	27.5	10.0	3.0	58.0	1.0	0.5		3.8
producer (bituminous)	27.0	4.5	14.0	50.9	0.6	3.0		5.6
blue-water	42.8	3.0	49.9	3.3	0.5	0.5		11.5
carburetted water	33.4	3.9	34.6	7.9	0.9	10.4	8.9[b]	20.0
retort coal	8.6	1.5	52.5	3.5	0.3	31.4	2.2[c]	21.5
coke-oven	6.3	1.8	53.0	3.4	0.2	1.6	3.7[d]	21.9
natural								
mid-continent		0.8		3.2		96.0		36.1
Pennsylvania				1.1		67.6	31.3[e]	46.0

[a] To convert MJ/m^3 to Btu/ft^3, multiply by 26.86.
[b] 6.7 vol % C_2H_4 plus 2.2 vol % C_6H_6.
[c] 1.1 vol % each of C_2H_4 and C_6H_6.
[d] 2.7 vol % C_2H_4 plus 1.0 vol % C_6H_6.
[e] 31.3 vol % C_2H_6.

Table 2. Coal Carbonization Methods

Carbonization process	Final temperature, °C	Products	Processes
low temperature	500–700	reactive coke and high tar yield	rexco (700°C) made in cylindrical vertical retorts; coalite (650°C) made in vertical tubes
medium temperature	700–900	reactive coke with high gas yield, or domestic briquettes	town gas and gas coke (obsolete); phurnacite, low volatile steam coal, pitch-bound briquettes carbonized at 800°C
high temperature	900–1050	hard, unreactive coal for metallurgical use	foundry coke (900°C); blast-furnace coke (950–1050°C)

Secondary gasification involves gasification of the char from the primary gasifier, usually by reaction of the hot char and water vapor to produce carbon monoxide and hydrogen.

The gaseous product from a gasifier generally contains large amounts of carbon monoxide and hydrogen, plus lesser amounts of other gases and may be of low, medium, or high heat value depending on the defined use.

The importance of coal gasification as a means of producing fuel gas(es) for industrial use cannot be underplayed. But coal gasification systems also have undesirable features. A range of undesirable products are also produced which must be removed before the products are used to provide fuel and/or to generate electric power.

Chemistry. Coal gasification involves the themal decomposition of coal and the reaction of the carbon in the coal, and other pyrolysis products with oxygen, water, and hydrogen to produce fuel gases such as methane by internal hydrogen shifts $C_{coal} + H_{coal} \rightarrow CH_4$, or through the agency of added (external) hydrogen $C_{coal} + 2\ H_2 \rightarrow CH_4$, although the reactions are more numerous and more complex.

Process Parameters. The most notable effects in gasifiers are those of pressure and coal character. Some initial processing of the coal feedstock may be required. The type and degree of pretreatment is a function of the process and/or the type of coal.

Depending on the type of coal being processed and the analysis of the gas product desired, some or all of the following processing steps may be required: *(1)* pretreatment of the coal (if caking is a problem); *(2)* primary gasification of the coal; *(3)* secondary gasification of the carbonaceous residue from the primary gasifier; *(4)* removal of carbon dioxide, hydrogen sulfide, and other acid gases; *(5)* shift conversion for adjustment of the carbon monoxide–hydrogen mole ratio to the desired ratio; and *(6)* catalytic methanation of the carbon monoxide–hydrogen mixture to form methane. If high heat-value gas is desired, all of these processing steps are required because coal gasifiers do not yield methane in the concentrations required.

There are three fundamental reactor types for gasification processes: *(1)* a gasifier reactor, *(2)* a devolatilizer, and *(3)* a hydrogasifier. Gasification processes can be divided into four categories based on reactor (bed) configuration: *(1)* fixed bed, *(2)* moving bed, *(3)* fluidized bed, and *(4)* entrained bed.

Combustion. Coal combustion, not being in the strictest sense a process for the generation of gaseous synfuels, is nevertheless an important use of coal as a source of gaseous fuels.

There are two principal methods of coal combustion: fixed-bed combustion and combustion in suspension.

A significant issue in combustors in the mid-1990s is the performance of the process in an environmentally acceptable manner through the use of either low sulfur coal or post-combustion clean-up of the flue gases. Thus there is a marked trend to more efficient methods of coal combustion and, in fact, a combustion system that is able to accept coal without the necessity of a post-combustion treatment or without emitting objectionable amounts of sulfur oxides, nitrogen oxides, and particulates is very desirable.

The parameters of rank and moisture content are regarded as determining factors in combustibility as it relates to both heating value and ease of reaction as well as to the generation of pollutants.

Chemistry. In direct combustion coal is burned to convert the chemical energy of the coal into thermal energy, ie, the carbon and hydrogen in the coal are oxidized into carbon dioxide and water.

The complex nature of coal as a molecular entity has resulted in the chemical explanations of coal combustion being confined to the carbon in the system. The hydrogen and other elements have received much less attention but the system is extremely complex and the heteroatoms, eg, nitrogen, oxygen, and sulfur, exert an influence on the combustion. It is this latter that influences environmental aspects.

Combustion Systems. Combustion systems vary in nature depending on the nature of the feedstock and the air needed for the combustion process. However, the two principal types of coal-burning systems are usually referred to as layer and chambered. The former refers to fixed beds; the latter is more specifically for pulverized fuel.

Gas from Other Fossil Fuels

As of this writing natural gas is a plentiful resource, and there has been a marked tendency not to use other fossil fuels as SNG sources. However, petroleum and oil shale (qv) have been the subject of extensive research efforts.

Petroleum. Thermal cracking (pyrolysis) of petroleum or fractions thereof was an important method for producing gas in the years following its use for increasing the heat content of water gas. Many water gas sets operations were converted into oil-gasification units. Some of these have been used for base-load city gas supply, but most find use for peak-load situations in the winter.

A second group of refining operations which contribute to gas production are the catalytic cracking processes, such as fluid-bed catalytic cracking, and other variants, in which heavy gas oils are converted into gas, naphthas, fuel oil, and coke.

As in the case of coal, synthetic natural gas can be produced from heavy oil by partially oxidizing the oil to a mixture of carbon monoxide and hydrogen, $2\ CH_{petroleum} + O_2 \rightarrow 2\ CO + H_2$, which is methanated catalytically to produce methane of any required purity.

When relatively light feedstocks, eg, naphthas having ca 180°C end boiling point and limited aromatic content, are available, high nickel content catalysts can be used to simultaneously conduct a variety of near-autothermic reactions. This results in the essentially complete conversions of the feedstocks to methane: $CH_3(CH_2)_3CH_3 + 2\ H_2O \rightarrow 4\ CH_4 + CO_2$.

Oil Shale. Oil shale is a sedimentary rock that contains organic matter, referred to as kerogen, and another natural resource of some consequence that could be exploited as a source of synthetic natural gas. However, as of this writing, oil shale has found little use as a source of substitute natural gas.

Biomass. Biomass is simply defined for these purposes as any organic waste material, such as agricultural residues, animal manure, forestry residues, municipal waste, and sewage, which originated from a living organism.

Biomass is another material that can produce a mixture of carbonaceous solid and liquid products as well as gas: $C_{organic} \rightarrow C_{coke/char/carbon}$ + liquids + gases.

Biomass resources are variable, but it has been estimated that they represent substantial amounts (up to 20×10^6 mJ(20×10^{15} Btu)) of energy, ie, ca 19% of the annual energy consumption in the United States.

Gas Treating

The reducing conditions in gasification reactors effect the conversion of the sulfur and nitrogen in the feed coal to hydrogen sulfide, H_2S, and ammonia, NH_3. Some carbonyl sulfide, COS, carbon disulfide, CS_2, mercaptans, RSH, and hydrogen cyanide, HCN, are also formed in the gasifier. These compounds, along with carbon dioxide, are removed simultaneously, either selectively or nonselectively, from the gas stream in the clean-up stages of the process using commercially available physical or chemical solvents and scrubbing agents.

JAMES G. SPEIGHT
Western Research Institute

J. G. Speight, ed., *Fuel Science and Technology Handbook*, Marcel Dekker, Inc., New York, 1990.

J. G. Speight, *The Chemistry and Technology of Coal*, 2nd ed., Marcel Dekker, Inc., New York, 1994.

J. G. Speight, *The Chemistry and Technology of Petroleum*, 2nd ed., Marcel Dekker, Inc., New York, 1991.

J. G. Speight, *Gas Processing: Environmental Aspects and Methods*, Butterworth Heinemann, Oxford, U.K., 1993.

LIQUID FUELS

The creation of liquids to be used as fuels from sources other than natural crude petroleum broadly defines synthetic liquid fuels. Synthetic liquid fuels have characteristics approaching those of the liquid fuels in commerce, specifically gasoline, kerosene, jet fuel, and fuel oil. For much of the twentieth century, the synthetic fuels emphasis was on liquid products derived from coal upgrading or by extraction or hydrogenation of organic matter in coke liquids, coal tars, tar sands, or bitumen deposits. More recently, however, much of the direction involving synthetic fuels technology has changed. There are two reasons.

The potential of natural gas, which typically has 85–95% methane, has been recognized as a plentiful and clean alternative feedstock to crude oil. Estimates place worldwide natural gas reserves at ca1×10^{14} $m^3(3.5 \times 10^{15}$ $ft^3)$ corresponding to the energy equivalent of ca1×10^{11} $m^3(637 \times 10^9$ bbl) of oil. As of this writing, the rate of discovery of proven natural gas reserves is increasing faster than the rate of natural gas production. Many of the large natural gas deposits are located in areas where abundant crude oil resources lie such as in the Middle East and Russia. However, huge reserves of natural gas are also found in many other regions of the world, providing oil-deficient countries access to a plentiful energy source. The gas is frequently located in remote areas far from centers of consumption, and pipeline costs can account for as much as one-third of the total natural gas cost. Thus tremendous strategic and economic incentives exist for on-site gas conversion to liquids.

In general, the proven technology to upgrade methane is via steam reforming to produce synthesis gas, $CO + H_2$. Such a gas mixture is clean and when converted to liquids produces fuels substantially free of heteroatoms such as sulfur and nitrogen. Two commercial units utilizing the synthesis gas from natural gas technology in combination with novel downstream conversion processes have been commercialized.

The direct methane conversion technology which has received the most research attention involves the oxidative coupling of methane to produce higher hydrocarbons such as ethylene. These olefinic products may be upgraded to liquid fuels via catalytic oligomerization processes.

A second trend in synthetic fuels is increased attention to oxygenates as alternative fuels as a result of the growing environmental concern about burning fossil-based fuels. The environmental impact of the oxygenates, such as methanol, ethanol, and methyl *tert*-butyl ether (MTBE) is still under debate, but these alternative liquid fuels are gaining new prominence.

Despite reduced prominence, coal technology is well positioned to provide synthetic fuels for the future. World petroleum and natural gas production are expected ultimately to level off and then decline. Coal gasification to synthesis gas is utilized to synthesize liquid fuels in much the same manner as natural gas steam reforming technology. Although as of this writing world activity in coal liquefaction technology is minimal, the extensive development and detailed demonstration of processes for converting coal to liquid fuels should serve as solid foundation for the synthetic fuel needs of the future.

Coal, tar, and heavy oil fuel reserves are widely distributed throughout the world. In the Western hemisphere, Canada has large tar sand, bitumen (very heavy crude oil), and coal deposits. The United States has very large reserves of coal and shale. Coal comprises ca 85% of the U.S. recoverable fossil energy reserves. Venezuela has an enormous bitumen deposit and Brazil has significant oil shale reserves. Coal is also found in Brazil, Colombia, Mexico, and Peru. Worldwide, the total resource base of these reserves is immense and may constitute >90% of the hydrocarbon resources in place.

Indirect Liquefaction/Conversion to Liquid Fuels

Indirect liquefaction of coal and conversion of natural gas to synthetic liquid fuels is defined by technology that involves an intermediate step to generate synthesis gas, $CO + H_2$. The main reactions involved in the generation of synthesis gas are the coal gasification reactions:

Combustion

$$C + O_2 \rightleftharpoons CO_2 \ \Delta H_{298K} = -394 \text{ kJ/mol}(-94.2 \text{ kcal/mol}) \quad (1)$$

Gasification

$$C + 1/2\ O_2 \rightleftharpoons CO \ \Delta H_{298K} = -111 \text{ kJ/mol}(-26.5 \text{ kcal/mol}) \quad (2)$$

$$C + H_2O(g) \rightleftharpoons CO + H_2 \ \Delta H_{298K} = +131 \text{ kJ/mol}(31.3 \text{ kcal/mol}) \quad (3)$$

$$C + CO_2 \rightleftharpoons 2\ CO \ \Delta H_{298K} = +172 \text{ kJ/mol}(41.1 \text{ kcal/mol}) \quad (4)$$

Water gas shift

$$CO + H_2O(g) \rightleftharpoons CO_2 + H_2 \ \Delta H_{298K} = -41 \text{ kJ/mol}(-9.8 \text{ kcal/mol}) \quad (5)$$

the methane steam reforming reactions:

Partial oxidation

$$CH_4 + 1/2\ O_2 \rightleftharpoons CO + 2\ H_2 \ \Delta H_{298K} = -36 \text{ kJ/mol}(-8.6 \text{ kcal/mol}) \quad (6)$$

Reforming

$$CH_4 + H_2O(g) \rightleftharpoons CO + 3\ H_2 \Delta H_{298K} = 206 \text{ kJ/mol}(49.2 \text{ kcal/mol}) \quad (7)$$

and the water gas shift reaction (eq. 5), used to increase the H_2/CO ratio of the product synthesis gas.

Coal Upgrading via Fischer-Tropsch. Industrial operation of the Fischer-Tropsch synthesis involved five steps: (*1*) synthesis gas manufacture; (*2*) gas purification by removal of water and dust, and hydrogen sulfide and organic sulfur compounds; (*3*) synthesis of hydrocarbons; (*4*) condensation of liquid products and recovery of gasoline from product gas; and (*5*) fractionation of synthetic products. Only the synthesis reactor and its method of operation were unique to the process.

Fischer-Tropsch liquid obtained using cobalt catalysts is roughly equivalent to a very paraffinic natural petroleum oil but is not so complex a mixture. Straight-chain, saturated aliphatic molecules predominate but monoolefins may be present in an appreciable concentration.

Natural Gas Upgrading via Fischer-Tropsch. A study of the economics of Fischer synthesis led to the conclusion that the large-scale production of gasoline from natural gas offered hope for commercial utility. In the Hydrocol process (Hydrocarbon Research, Inc.) natural gas was treated with high purity oxygen to produce the synthesis gas which was converted in fluidized beds of iron catalysts.

The Shell middle distillate synthesis (SMDS) process developed by Shell Oil Company, uses remote natural gas as the feedstock. This two-step process involves Fischer-Tropsch synthesis of paraffinic wax called the heavy paraffin synthesis (HPS). The wax is subsequently hydrocracked and hydroisomerized to yield a middle distillate boiling range product in the heavy paraffin conversion (HPC). In the HPS stage, wax is maximized by using a proprietary catalyst having high selectivity toward heavier products and by the use of a tubular, fixed-bed Arge-type reactor. The HPC stage employs a commercial hydrocracking catalyst in a trickle flow reactor. The HPC step allows for production of narrow range hydrocarbons not possible with conventional Fischer-Tropsch technology.

Shell's two-step SMDS technology allows for process flexibility and varied product slates. The liquid product obtained consists of naphtha, kerosene, and gas oil in ratios from 15:25:60 to 25:50:25, depending on process conditions. The products manufactured are predominantly paraffinic, free from sulfur, nitrogen, and other impurities, and have excellent combustion properties.

Liquid Fuels via Methanol Synthesis and Conversion. Methanol is produced catalytically from synthesis gas. By-products such as ethers, formates, and higher hydrocarbons are formed in side reactions and are found in the crude methanol product. Whereas for many years methanol was produced from coal, after World War II low cost natural gas and light petroleum fractions replaced coal as the feedstock.

The most significant development in synthetic fuels technology since the discovery of the Fischer-Tropsch process is the Mobil methanol-to-gasoline (MTG) process. Methanol is efficiently transformed into C_2–C_{10} hydrocarbons in a reaction catalyzed by the synthetic zeolite ZSM-5. The reaction sequence can be summarized as

$$n/2[2\ CH_3OH \rightleftarrows CH_3OCH_3 + H_2O] \rightarrow C_nH_{2n} \rightarrow n\ [CH_2] \quad (8)$$

where $[CH_2]$ represents an average paraffin–aromatic mixture. Two versions of the MTG process, one using a fixed bed, the other a fluid bed, have been developed.

Table 1 contains typical gasoline quality data. MTG gasoline typically contains 60 vol % saturates, ie, paraffins and naphthenes; 10 vol % olefins; and 30 vol % aromatics. Sulfur and nitrogen levels in the gasoline are virtually nil.

Because the MTG process produces primarily gasoline, a variation of that process has been developed which allows for production of gasoline and distillate fuel. This process integrates two known technologies, methanol-to-olefins (MTO) and Mobil olefins-to-gasoline-and-distillate (MOGD). The combined process produces gasoline and distillate in various proportions and, if needed, olefinic by-products.

The gasoline product from the integrated MTO/MOGD process is predominately olefinic and aromatic. The gasoline quality (ca 89 octane) is comparable to FCC gasoline. After hydrofinishing, the distillate product is mostly isoparaffinic and has high cetane index, low pour point, and negligible sulfur content. MOGD diesel fuel has somewhat lower density than typical conventional fuels (0.8 vs 0.86). Low aromatics levels contribute to a stable jet fuel with very little smoke emission during combustion. MOGD diesel and jet fuels meet or exceed all conventional specifications.

Table 1. MTG Gasoline Quality

Parameter	Average	Range
density at 15°C, kg/m^3	730	728–733
Reid vapor pressure, kPa[a]	86.2	83.4–91.0
octane number		
research	92.2	92.0–92.5
motor	82.6	82.2–83.0
durene content, wt %	2.0	1.74–2.29
induction period, min	325	260–370
distillation at 70°C, % evaporation	31.5	29.5–34.5
distillation end point, °C	204.5	196–209

[a] To convert kPa to psia, multiply by 0.145.

Direct Conversion of Natural Gas to Liquid Fuels

Direct upgrading routes which have been extensively studied include direct partial oxidation to oxygenates, oxidative coupling to higher hydrocarbons, and pyrolysis to higher hydrocarbons. Owing to the inert nature of methane, the technology is limited by the yields of desired products which in turn affects the process economics. Plants to produce acetylene from methane by high temperature pyrolysis routes have been commercialized.

Generally, the most developed processes involve oxidative coupling of methane to higher hydrocarbons. Oxidative coupling converts methane to ethane and ethylene by

$$2\ CH_4 + 1/2\ O_2 \rightarrow H_3CCH_3 + H_2O \quad (9)$$

$$H_3CCH_3 + 1/2\ O_2 \rightarrow H_2C{=}CH_2 + H_2O \quad (10)$$

The process can be operated in two modes: co-fed and redox. The co-fed mode employs addition of O_2 to the methane/natural gas feed and subsequent conversion over a metal oxide catalyst. The redox mode requires the oxidant to be from the lattice oxygen of a reducible metal oxide in the reactor bed. After methane oxidation has consumed nearly all the lattice oxygen, the reduced metal oxide is reoxidized using an air stream. Both methods have processing advantages and disadvantages. In all cases, however, the process is run to maximize production of the more desired ethylene product. The ethylene is then converted to liquid fuels via olefin oligomerization processes.

Some emerging technologies in this area are the ARCO gas-to-gasoline process and the OXCO process.

Oxygenate Fuels

Alcohols and ethers, especially methanol, ethanol, and methyl *tert*-butyl ether (MTBE), have been widely used separately or in blends with gasolines (reformulated gasoline) and other hydrocarbons to fuel internal combustion engines. Fuel properties of key oxygenates are presented in Table 2. These compounds, as a class, may be considered to be partially oxidized, ie, each has a mole of oxidized hydrogen. They differ from the hydrocarbons that make up gasoline principally in lower heating values and in higher vaporization heat requirements. This constitutes a serious disadvantage to the substitution of oxygenates, especially lower alcohols, for motor gasoline.

Table 2. Fuel Oxygenates Properties

Oxygenate	Blending octane, 1/2 (RON + MON)[a]	Heat of combustion, MJ/L[b]	Specific gravity	Boiling point, °C
methanol	101	18.0	0.79	64.6
ethanol	101	21.3	0.79	78.5
2-propanol	106	26.4	0.79	82.4
2-butanol	99	28.3	0.80	99.5
tert-butyl alcohol	100	28.1	0.80	82.6
MTBE[c]	108	30.2	0.75	55.4
ETBE[d]	111	32.5	0.74	72.8
TAME[e]	102	31.2	0.77	86.3
gasoline	87	34.8	0.74	

[a] RON = research octane number; MON = motor octane number.
[b] To convert MJ/L to Btu/gal, multiply by 3589.
[c] MTBE = methyl *tert*-butyl ether.
[d] ETBE = ethyl *tert*-butyl ether.
[e] TAME = *tert* – amyl methyl ether.

Other properties which greatly influence the potential of oxygenates as fuels include octane performance, solubility in gasoline, effect on gasoline vapor pressure, sensitivity to water, and evaporative/exhaust emissions. Oxygenate fuels tests are often debated because the tests employed were developed for conventional gasolines.

The addition of small percentages of oxygenates to gasoline can produce large gains in octane. Thus, as blending components in gasoline, oxygenates improve octane quality. As neat fuels for spark-ignition engines, octane values for oxygenates are not useful in determining knock-limited compression ratios for vehicles because of the lean carburetor settings relative to gasoline. Neither do these values represent the octane performance of oxygenates when blended with gasoline.

In part because neat alcohols are insufficiently volatile to enable a cold engine to start, even at moderate temperatures, the use of neat alcohols for automotive motor fuel is problematic. Manufacturers have, however, reported that alcohol-powered cars, after being started and warmed up, can have the same or better driveability as gasoline cars. As for gasoline vehicles, port fuel injector fouling has occurred in some methanol vehicles and has affected driveability and emissions. Other problems related to high alcohol content gasoline in conventional engines include vapor lock and corrosion. Flexible-fuel vehicles (FFV), which can operate on either neat methanol or gasoline, or mixtures thereof, are being evaluated.

Gasoline blends containing oxygenates change the emissions characteristics of a motor vehicle designed for gasoline. Oxygenates and oxygenate-blends approved for use by the U.S. government are expected to have desirable emissions features as automotive fuels, and governmental environmental mandates and regulations have necessitated increased examination and implementation of oxygenates as fuels. As of this writing, however, no process can produce alcohols or ethers at equivalent or lower cost per volume than gasolines derived from natural petroleum.

Direct Liquefaction of Coal

Direct liquefaction, the production of liquids from feed coal in a single processing scheme without a synthesis gas intermediate step, includes two routes for the upgrading of coal: hydrogenation and pyrolysis. In hydrogenation, the conversion of coal to liquids having higher hydrogen-to-carbon ratio involves the addition of hydrogen. Generally, the additional hydrogen required is added either from molecular hydrogen or from a hydrogen-donor solvent such as tetralin. Processes classified under pyrolysis are those which produce liquids by removal of carbon. This occurs when coal is thermally processed under inert or reducing atmospheric conditions.

Hydrogenation processes include the H-coal process, the solvent-refined coal process, the Exxon donor solvent-coal liquefaction process, and staged coal liquefaction at Wilsonville, Alabama. Coal pyrolysis processes include the COED process, the Occidental Petroleum coal conversion process, the TOSCOAL process, and coal carbonization.

Other Processes

Shale Oil. In the United States, shale oil, or oil derivable from oil shale, represents the largest potential source of liquid hydrocarbons that can be readily processed to fuel liquids similar to those derived from natural petroleum. Some countries produce liquid fuels from oil shale. There is no such industry in the United States. Petroleum supply and price stability has since severely curtailed shale oil development. In addition, complex environmental issues further prohibit demonstration of commercial designs.

Economic Aspects of Synthetic Fuels

As of this writing (ca 1994), processes for production of synthetic liquid fuels by upgrading natural gas, coal, or heavy oil are generally not directly competitive with crude oil upgrading. Nevertheless, synthetic fuels technology is projected to play a primary role in providing liquid fuels once crude oil depletion is of concern.

SCOTT HAN
CLARENCE D. CHANG
Mobil Research and Development Corporation

F. Mako and W. A. Samuel, in R. A. Meyers, ed., *Handbook of Synfuels Technology*, McGraw-Hill, Inc., New York, 1984, pp. 2-5–2-43.

H. D. Schindler, *Final Technical Report on DOE Contract No. D-AC01-87ER30110*, Vol. 2, Department of Energy, Washington, D.C., 1989.

M. Crow and co-workers, *Synthetic Fuel Technology Development in the United States—A Retrospective Assessment*, Praeger Publishing, New York, 1988.

American Petroleum Institute, *Alcohols and Ethers-A Technical Assessment of Their Application as Fuels and Fuel Components*, API Publication 4261, American Petroleum Institute, Washington, D.C., July 1988.

FUNGICIDES, AGRICULTURAL

Pathogenic fungi cause a substantial reduction in expected crop yields; further losses can result during storage of harvested crops. For those fungi that can seriously affect economically important plants (Table 1), means have been sought to control these infections by crop rotation and husbandry, genetic manipulation of the plant species, and external treatment of plants using agricultural fungicides.

Agricultural fungicide application accounts for about 20% of all pesticide use. Agricultural fungicides can be applied to the soil to control fungi that are resident there, to the seed or foliage of the plant to be protected, or to harvested produce to prevent storage losses. Those applied to the soil are in many instances nonselective, volatile soil sterilants, such as formaldehyde (qv), which kill all soil organisms, including fungi. Soil and crop storage fungicides, which represent only a very small fraction of the fungicides used, are covered elsewhere. Seed and foliar-applied agricultural fungicides, listed in Table 2, are discussed herein.

Because of the wide diversity of chemical structures encountered, fungicides are classified herein as being nonsystemic or systemic. The nonsystemic fungicides have a protectant mode of action and must be applied to the surface of a plant generally before infection takes place. These do not translocate from the site of application. The systemic fungicides can penetrate the seed or plant and are then redistributed within to unsprayed parts or subsequent new growth, rendering protection from fungal attack or eradicating a fungus already present.

Nonsystemic Fungicides

From 20 to 25 nonsystemic fungicides are utilized in agriculture, although use is declining. These are some of the oldest known fungicides and cover a wide range of chemistry from simple inorganic salts to highly complex organic structures. Selective accumulation by spores plays a dominant role in the toxicity of many of these compounds. The majority are regarded as general cell poisons and can be used only when they are not able to penetrate host plant tissue in appreciable amounts. The fungal pathogen is controlled before it infects the plant so that the resulting efficacy is primarily achieved through protecting the plant rather than curing the disease. The mode of action, ie, biochemical basis for activity, of most known nonsystemic fungicides is generally nonspecific, and inhibition at multiple sites results ultimately in interference with energy producing or transferring processes, which disrupts fungal respiration and membranes. Nonsystemic fungicides include sulfur, copper, mercury, tin, thiocarbamate and thiurame derivatives, phthalimides and some trichloromethylthiocarboximides, aromatic hydrocarbons, and dicarboximides.

Site-Specific Systemic Fungicides

In general, the systemic fungicidal treatment of crop plants is only possible using inhibitors of fungal-specific targets, and there has

Table 1. Important Diseases of Crop Plants

Fungal class	Pathogen: Scientific name	Pathogen: Common name
Phycomycetes subclass oomycetes	*Phytophthora infestans*	potato late blight
	Plasmopara viticola	downy mildew of grape
	Pseudoperonospora cubensis	cucumber downy mildew
	Pythium spp.	damping off diseases
Ascomycetes	*Erysiphe graminis*	powdery mildew of wheat/barley
	Gaeuman-nomy-ces graminis	take-all of oats and wheat
	Podosphaera leucotricha	apple powdery mildew
	Pyrenophora teres	net blotch of barley
	Pyricularia oryzae	rice blast
	Rhynchosporium secalis	leaf scald of barley
		brown rot of pome fruit
	Sclerotinia spp.	
		leaf spot of brassicas and legumes
	Sphaerotheca fuliginea	cucurbit powdery mildew
	Uncinula necator	grape powdery mildew
	Venturia inaequalis	scab of apple
	Mycosphaerella fijiensis	sigatoka disease of bananas
Basidiomycetes	*Puccinia* spp.	leaf rusts of wheat and oats
		black scurf of potato
	Rhizoctonia spp.	sheath blight of rice
		sharp eyespot of wheat
	Tilletia spp.	bunts of wheat
	Uromyces spp.	bean rusts
	Ustilago spp.	smuts of wheat, barley, oat, and maize
Deuteromycetes		early blight of potato
	Alternaria spp.	tobacco brown spot
		leaf spot of brassicas
	Botrytis spp.	grey mold of grape and other crops
		leaf spot of sugarbeet
	Cercospora spp.	
		brown eyespot of coffee
		wilts, broad range of hosts
	Fusarium spp.	
		ear blight of wheat
		root and foot rot of wheat
	Helminthosporium spp.	
		leaf spot of maize
	Pseudocercosporella herpotrichoides	eyespot of wheat
	Septoria nodorum	glume blotch of wheat
	Septoria tritici	wheat leaf blotch

been considerable progress in developing agricultural fungicides having high levels of fungal specificity. Elucidation of the biochemical mechanisms of action of compounds has led to the discovery of some novel compounds. Many of the fungicides introduced since the 1970s have been systemic fungicides which inhibit fungal growth at various stages of fungal development. These fungicides are often active at very low levels compared with nonsystemics and tend to exhibit a much narrower activity spectrum as a consequence of their action against a specific biochemical target. Precise biochemical targets have been defined for many of the different classes of fungicide chemistries. Some have a biochemical target site in common. The selectivity of systemic fungicides can be attributed to differences in a number of factors. These include uptake and accumulation in the fungal cell, inherent differences at the target site, differences in metabolism of the fungicide by the plant or fungi, and the degree of importance of the target system to the survival of the fungus.

Table 2. Alphabetical List of Seed and Fungicides

Common name	Trademark	Common name	Trademark
anilazine	Dyrene	fuberidazole	Voronit
benalaxyl	Galben	furalaxyl	Fongarid
benomyl	Benelate	imazalil	Fungaflor
blasticidin S	Bla-S	imibenconazole	Manage
bupirimate	Nimrod	iprobenphos	Kitazin P
buthiobate	Denmert	iprodione	Rovral
captafol	Difolatan	isoprothiolane	Fuji-one
captan	Orthocide	kasugamycin	Kasumin
carbendazim	Bavistin	mancozeb	Dithane M-45
carboxin	Vitavox	maneb	Dithane M-22
chinomethionat	Morestan	mepronil	Basitac
chloroneb	Demosan	metalaxyl	Ridomil
chlorothalonil	Bravo	methfuroxam	Trivax
chlozolinate	Serinal	metsulfovax	Provax
cymoxanil	Curzate	myclobutanil	Systhane
cyproconazole	Alto	nabam	Parzate
dichlofluanid	Euparen	nuarimol	Trimidal
dichlone	Phygon	ofurace	Oturanic
dicloran	Allisan	oxadixyl	Sandofan
diclomezine	Monguard	oxycarboxin	Plantvax
dimethirimol	Milcurb	polyoxin B	Polyoxin AL
dinocap	Karathane	polyoxin D	Polyoxin Z
dithianon	Delan	prochloraz	Sportak
dodemorph	Milan	procymidone	Sumisclex
dodine	Cyprex	propiconazole	Tilt
ediphenphos	Hinosan	pyroquilon	Funorene
ethirimol	Milcap	quintozene	Botrilex
etridazole	Terrazole	tebuconazole	Folicur
fenarimol	Rubigan	tetraconazole	Eminent
fenfuram	Pano-ram	thiabendazole	Mertect
fenpiclonil	Beret	thiophanate methyl	Topsin M
fenpropidin	Patrol	thiram	Tersan
fenpropimorph	Corbel	triadimefon	Bayleton
fentin acetate	Brestan	triarimol	Trimidal
fentin hydroxide	Du-ter	tricyclazole	Beam
ferbam	Fermate	tridemorph	Calixin
flusilazole	Nustar	triforine	Cela W524
flutriafol	Impact	vinclozolin	Ronilan
flutolanl	Moncut	zineb	Dithane Z-78
folpet	Phaltan	ziram	Milbam, Zerlate
fosetyl-Al	Aliette		

Site-specific systemic fungicides include mitochondrial respiration inhibitors (carboxin, oxycarboxin, flutolanil, fenfuram, mepronil, methfuroxam, and metsulfovax); microtubulin polymerization inhibitors (thiabendazole, fuberidazole, carbendazim, benomyl, and thiophanate methyl); inhibitors of sterol biosynthesis (C-14 demethylation inhibtors [piperazines, pyridines, pyrimidines, and azoles]); Δ^{14}-reduction and Δ^{8}-Δ^{7}-isomerization inhibitors (the morpholines); RNA biosynthesis inhibitors (phenylamides and hydroxypyrimidines); phospholipid biosynthesis inhibitors (ediphenphos and iprobenphos); melanin biosynthesis inhibitors (tricyclazole, pyroquilon, and the experimental compound PP389); fungal protein biosynthesis inhibitors (blasticidin S and kasugamycin); and cell wall biosynthesis inhibitors (polyoxin B and polyoxin D).

Resistance to Fungicides

Well over 100 plant pathogens have become resistant to various fungicides under field conditions. Failure of the acyl alanines, benzimidazoles, thiophanates, carboxanilides, dicarboximides, hydroxypyrimidines, some organophosphates, and most of the antibiotics has occurred. In other cases, a moderate decrease in sensitivity without a rapid loss of disease control has been observed as in the case of sterol biosynthesis inhibitors (triazoles, pyrimidines, and imidazoles) and organophosphates. The most effective approach is to use fungicides having different modes of action in combination, either as mixtures or in alternation, possibly utilizing both specific site and multisite inhibitors. Because of resistance problems great importance is attached to chemistries that inhibit novel fungal enzyme targets.

Economic Aspects

Within a particular market segment the pricing of fungicide products from the various manufacturers is extremely uniform and tends to be dictated at least in part by the cost of established products that have stood the test of time balanced against the needs of the grower to demonstrate a clear cost-benefit advantage from their use.

Newer fungicides, in order to retain cost-effectiveness, need to be very highly active, which also serves to achieve efficacy in the field at low dose rates thus keeping environmental pollution problems as small as possible. More in-depth knowledge of fungal biochemistry and the molecular events involved in host/pathogen interactions should facilitate the identification of novel fungal targets for use in a biorational approach to fungicide discovery, through the application of computer-aided molecular design (CAMD) approaches to the molecular modeling (qv) of the target to design new fungicides. Recombinant DNA technologies are expected to play an escalating role in the validation of such biorational targets.

BARRY A. DREIKORN
W. JOHN OWEN
Dow Elanco

H. Buchenauer, in G. Haug and H. Hoffmann, eds., *Chemistry of Plant Protection*, Vol. 6, Springer-Verlag, Berlin, 1990.

C. R. Worthing and R. J. Hance, eds., *The Pesticide Manual*, 9th ed., The British Crop Protection Council, Farnham, U.K., 1991.

W. Köller, ed., *Target Sites of Fungicide Action*, CRC Press, Boca Raton, Fla., 1992.

I. Denholm, A. L. Devonshire, and D. W. Holloman, eds., *Resistance 91: Achievements and Developments in Combating Pesticide Resistance*, Elsevier Applied Science, London, 1992.

FURAN DERIVATIVES

Furan (**1**) is a 5-membered heterocyclic, oxygen-containing, unsaturated ring compound. From a chemical perspective it is the basic ring structure found in a whole class of industrially significant products. The furan nucleus is also found in a large number of biologically active materials. Compounds containing the furan ring (as well as the tetrahydrofuran ring) are usually referred to as furans. From a manufacturing standpoint, however, furfural (**2**) is the feedstock from which all of the commercial furan derivatives are derived.

Furan is produced from furfural commercially by decarbonylation; loss of carbon monoxide from furfural gives furan directly. Tetrahydrofuran (**3**) is the saturated analogue containing no double bonds.

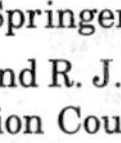

furan (**1**)

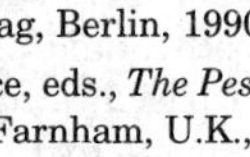

furfural (**2**)

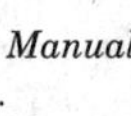

tetrahydrofuran (**3**)

Furfural is derived from biomass by a process in which the hemicellulose fraction is broken down into monomeric 5-carbon sugar units, which then are dehydrated to form furfural.

This article primarily discusses simple furans in which the nucleus occurs as a free monocycle. Sometimes polychlorinated isobenzofurans are referred to simply as *furans*. These isobenzofurans, comprised of chlorine-containing fused rings, are not simple furans. This presentation is generally limited to compounds (including resins or polymers) derived from furfural.

Tetrahydrofuran (**3**) is produced commercially from furfural by decarbonylation followed by hydrogenation.

The furan nucleus is a cyclic, dienic ether with some aromaticity. It is the least aromatic of the common 5-membered heterocycles.

The balance between aromatic and aliphatic reactivity is affected by the type of substituents on the ring. Furan functions as a diene in the Diels-Alder reaction. With maleic anhydride, furan readily forms 7-oxabicyclo [2.2.1]hept-5-ene-2,3-dicarboxylic anhydride (**4**).

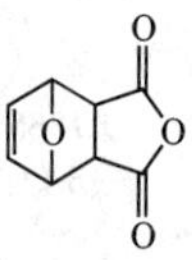

(**4**)

Radicals derived from furan are named similarly to analogous radicals in the benzene series. Typical radicals are 2 (or α)-furyl, 2-furfuryl, 2-furoyl, and 2-furfurylidene.

Furfural

Furfural can be classified as a reactive solvent. It resinifies in the presence of strong acid. Furfural is an excellent solvent for many organic materials, especially resins and polymers. On catalyzation and curing of such a solution, a hard rigid matrix results, which does not soften on heating and is not affected by most solvents and corrosive chemicals.

Furfural is formed by a series of reactions when biomass materials containing hemicellulose are treated with acid at an elevated temperature. When treated with aqueous acid, the hemicellulose is depolymerized to give primarily xylose, which under the reaction conditions loses three molecules of water and cyclizes to give furfural.

Physical Properties. When freshly distilled, furfural is a colorless liquid with a pungent, aromatic odor reminiscent of almonds. Furfural is miscible with most of the common organic solvents, but only slightly miscible with saturated aliphatic hydrocarbons. Inorganic compounds, generally, are quite insoluble in furfural.

Important physical properties of furfural, as well as similar properties for furfuryl alcohol, tetrahydrofurfuryl alcohol and furan are given in Table 1.

Chemical Properties. The chemical properties of furfural are generally characteristic of aromatic aldehydes but with some differences attributable to the furan ring. Furfural resinifies in the presence of acid and heat. Open chain compounds are formed from furfural under strong oxidizing conditions.

Furfural is very thermally stable in the absence of oxygen. At temperatures as high as 230°C, exposure for many hours is required to produce detectable changes in the physical properties of furfural, with the exception of color. However, at a temperature above 250°C, in a closed system, furfural will spontaneously and exothermically decompose to furan and carbon monoxide with a substantial increase in pressure.

Furfural can be reduced to 2-furan-methanol, referred to herein as furfuryl alcohol, or converted to furan by decarbonylation over selected catalysts.

Acetals are readily formed with alcohols and cyclic acetals with 1,2 and 1,3-diols.

Just as most other aldehydes do, furfural condenses with compounds possessing active methylene groups.

Furfural is a resin former under the influence of strong acid. It will self-resinify as well as form copolymer resins with furfuryl alcohol, phenolic compounds, or convertible resins of these.

Manufacture. Furfural is produced from annually renewable agricultural sources such as nonfood residues of food crops and wood wastes. The pentosan polysaccharides, xylan and arabinan, commonly known as hemicellulose, are the principal precursors of furfural and are always found together with lignin and cellulose in plant materials. Theoretically, all pentosan-containing substances are potentially

Table 1. Physical Properties of Furan Derivatives

	Furfural	Furfuryl alcohol	Furan	Tetrahydrofurfuryl alcohol
General properties				
molecular weight	96.09	98.10	68.08	102.13
boiling point at 101.3 kPa (1 atm), °C	161.7	170	31.36	178
freezing point, °C	−36.5		−85.6	<−80
refractive index, n_D				
20°C	1.5261	1.4868	1.4214	1.4250
density, d_4, at 20°C	1.1598	1.1285	0.9378	1.0511
vapor density (air = 1)	3.3	3.4	2.36	3.5
critical pressure, P_c, MPa[a]	5.502		5.32	
critical temperature, T_c, °C	397		214	
solubility, wt %, in water				
	20°C	8.3	∞	∞
25°C			1	
alcohol; ether	∞	∞	∞	∞
Fluid properties				
viscosity, mPa·s(= cP)				
20°C			0.38	6.24
25°C	1.49	4.62		
surface tension, mN/m (= dyn/cm)				
25°C		ca 38		37
29.9°C	40.7			
Electrical properties				
dielectric constant				
20°C	41.9			
23°C				13.6

[a] To convert MPa to atm, divide by 0.101.

usable for the production of furfural. Only a relatively few, however, are commercially significant.

Furfural is commercially produced in batch or continuous digesters where the pentosans are first hydrolyzed to pentoses (primarily xylose), which are then subsequently cyclodehydrated to furfural.

In all processes, raw material is charged to the digester and heated with high pressure steam. Enough excess steam is used to drive the furfural out of the reaction zone as vapor. The condensed reactor vapors are fed to a stripping column from which an enriched furfural–water distillate mixture is taken overhead and condensed. The liquid passes into a decanter where it separates into two layers. The furfural-rich lower layer containing about 6% water is processed further to obtain the furfural of commerce, and the water-rich layer containing about 8% furfural is recycled back to the stripper column as reflux.

Uses. Furfural is primarily a chemical feedstock for a number of monomeric compounds and resins.

Hydrogenation to furfuryl alcohol is the largest use. Some of the furfuryl alcohol is further hydrogenated to produce tetrahydrofurfuryl alcohol. The next major product is furan, produced by decarbonylation. Furan is a chemical intermediate; most of it is hydrogenated to tetrahydrofuran, which in turn is polymerized to produce polytetramethylene ether glycol (PTMEG). The principal direct application of furfural is as a selective solvent, and it also used as a reactive solvent in certain applications.

Furfural has been used as a component in many resin applications, most of them thermosetting.

Furfuryl Alcohol

Physical Properties. Furfuryl alcohol (2-furanmethanol) is a liquid, colorless, primary alcohol with a mild odor. On exposure to air, it gradually darkens in color. Furfuryl alcohol is completely miscible with water, alcohol, ether, acetone, and ethyl acetate, and most other organic solvents with the exception of paraffinic hydrocarbons. It is an excellent, highly polar solvent, and dissolves many resins. The physical constants of furfuryl alcohol are listed in Table 1.

Chemical Properties. Furfuryl alcohol undergoes the typical reactions of a primary alcohol such as oxidation, esterification, and etherification. Although stable to strong alkali, furfuryl alcohol is very sensitive to acid, thus imposing limitations on the conditions used in many of the typical alcohol reactions. It is a more reactive solvent than furfural and easily resinifies under the influence of acid and heat.

Under acidic conditions, furfuryl alcohol polymerizes to black polymers, which eventually become crosslinked and insoluble in the reaction medium. Copolymer resins are formed with phenolic compounds, formaldehyde and/or other aldehydes. Ethoxylation and propoxylation of furfuryl alcohol provide useful ether alcohols.

The chemistry of furfuryl alcohol polymerization involves reactions that give linear chains or oligomers containing essentially two repeating units (**5, 6**) with (**6**) predominating.

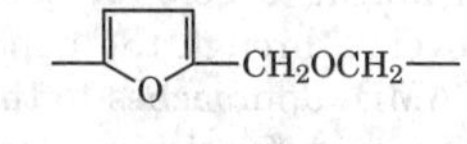

(5)

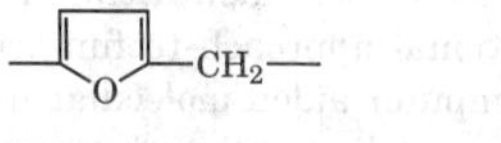

(6)

Development of color has been shown to be due to conjugated sequences along linear oligomeric chains (**7**).

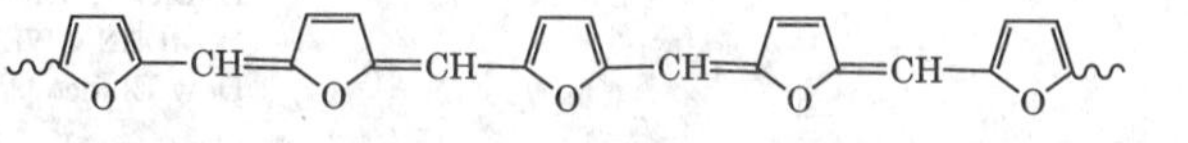

(7)

Manufacture. Furfuryl alcohol has been manufactured on an industrial scale by employing both liquid-phase and vapor-phase hydrogenation of furfural. Copper based catalysts are preferred because they are selective and do not promote hydrogenation of the ring.

Uses. Furfuryl alcohol is widely used as a monomer in manufacturing furfuryl alcohol resins, and as a reactive solvent in a variety of synthetic resins and applications. The final cross-linked products display outstanding chemical, thermal, and mechanical properties. Many commercial resins of various compositions and properties have been prepared by polymerization of furfuryl alcohol and other co-reactants such as furfural, formaldehyde, glyoxal, resorcinol, phenolic compounds and urea.

A number of applications of furfuryl alcohol are based on its reactive solvent properties. When a resin solution is cured, a hard, rigid, thermoset matrix results, often with outstanding properties. For example, furfuryl alcohol is a reactive solvent for phenolic resins in the manufacture of refractories for ladles holding molten steel.

The industrial value of furfuryl alcohol is a consequence of its low viscosity, high reactivity, and the outstanding chemical, mechanical, and thermal properties of its polymers, corrosion resistance, nonburning, low smoke emission, and excellent char formation.

Furan

Physical Properties. Furan, a colorless liquid with a strong ethereal odor, is low-boiling and highly flammable. It is miscible with most common organic solvents but only very slightly soluble in water. The physical properties of furan are listed in Table 1.

Chemical Properties. Furan is a heat-stable compound, although at 670°C in the absence of catalyst, or at 360°C in the presence of nickel, it decomposes to form a mixture consisting mainly of carbon monoxide, hydrogen, and hydrocarbons. Substitution and addition reactions can be effected under controlled conditions, with reaction occurring first in the 2- and 5-positions.

Strong dienophiles add to furan. Although both endo and exo isomers are formed initially, the former rapidly isomerize to the latter in solution, even at room temperature.

Catalytic hydrogenation of furan to tetrahydrofuran is accomplished in either liquid or vapor phase.

Manufacture. Furan is produced commercially by decarbonylation of furfural in the presence of a noble metal catalyst.

Uses. Furan is utilized as a chemical building block in the production of other industrial chemicals for use as pharmaceuticals, herbicides, stabilizers, and fine chemicals. Furan is readily hydrogenated, hence it is a source of commercial tetrahydrofuran (THF).

Washing and cleaning agents containing salts of maleic acid–furan copolymers form complexes with alkaline earth ions. These cleaning compositions do not contain phosphorus or nitrogen and find use in metal, foodstuff, and machine dishwashing products.

Tetrahydrofurfuryl Alcohol

Physical Properties. Tetrahydrofurfuryl alcohol (2-tetrahydrofuran-methanol) (**8**) is a colorless, high boiling liquid with a mild, pleasant odor. It is completely miscible with water and common organic solvents. Tetrahydrofurfuryl alcohol is an excellent solvent, moderately hydrogen-bonded, essentially nontoxic, biodegradable, and has a low photochemical oxidation potential. Most applications make use of its high solvency. Selected physical properties of tetrahydrofurfuryl alcohol are listed in Table 1.

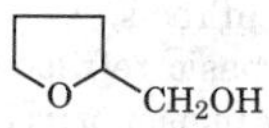

(**8**)

Chemical Properties. Without inhibitors, tetrahydrofurfuryl alcohol is susceptible to autoxidation. In the absence of air, however, no observable changes occur even after several years storage. In the presence of air, if a stabilizer such as Naugard is added, tetrahydrofurfuryl alcohol remains colorless after protracted periods of storage.

The reactions of tetrahydrofurfuryl alcohol are characteristic of its structure, involving primary alcohol and cyclic ether functional groups. As a primary alcohol, it undergoes normal displacement or condensation reactions affording new functional groups, (eg, halides, esters, alkoxylates, ethers, glycidyl ethers, cyanoethyl ethers, amines, etc). As a cyclic ether, it is typically unreactive, but the ring can be forced to open by hydrolysis or hydrogenolysis to give a variety of open-chain compounds.

All the common monobasic and dibasic esters of tetrahydrofurfuryl alcohol have been prepared by conventional techniques. Tetrahydrofurfuryl acrylate and methacrylate, specialty monomers, have been produced by carbonylation (nickel carbonyl and acetylene) of the alcohol as well as by direct esterification and ester interchange.

Manufacture. Tetrahydrofurfuryl alcohol is produced commercially by the vapor-phase catalytic hydrogenation of furfuryl alcohol. Liquid-phase reduction is also possible.

Uses. Tetrahydrofurfuryl alcohol is of interest in chemical and related industries where low toxicity and minimal environmental impact are important. For many years tetrahydrofurfuryl alcohol has been used as a specialty organic solvent. The fastest growing applications are in formulations for cleaners and paint strippers, often as a replacement for chlorinated solvents. Other major applications include formulations for crop sprays, water-based paints, and the dyeing and finishing of textiles and leathers. Tetrahydrofurfuryl alcohol also finds application as an intermediate in pharmaceutical applications.

Other Furan Derivatives

Other furan compounds, best derived from furfural, are of interest although commercial volumes are considerably less than those of furfural, furfuryl alcohol, furan, or tetrahydrofurfuryl alcohol. Applications include solvents, resin intermediates, synthetic rubber modifiers, therapeutic uses, as well as general chemical intermediates.

Compounds containing the furan ring are generally excellent solvents. Some are miscible with both water and with hexane. Presence of the ether oxygen adds polarity as well as the potential for hydrogen bonding. Ring-substituted derivatives of furfural and furfuryl alcohol are reactive solvents, similar to the parent compounds.

Compounds containing the furan or tetrahydrofuran ring are biologically active and are present in a number of pharmaceutical products.

Derivatives of furan and tetrahydrofurfuryl alcohol are used in the polymerization of synthetic rubber to control stereoregularity and other properties.

Health and Safety

As with all chemical compounds, the Material Safety Data Sheet (MSDS) for each of the specific furan derivatives should be reviewed before starting to work with these materials. Additional information on toxic effects of most of these compounds can be found in RTECS (*Registry of Toxic Effects of Chemicals*), HSDB (*Hazardous Substances Data Bank* from the National Library of Medicine), and standard works on toxicology. Toxicology studies are taking place on a continuing basis with many chemicals, including furan derivatives. New data may change the perspective on toxicity of these chemicals.

Precautions should be taken when working with these compounds because furfural, furfuryl alcohol, and furan are moderately toxic, tetrahydrofurfuryl alcohol is less so. Since regulations change from time to time, up-to-date exposure limit recommendations from OSHA (Occupational Safety and Health Agency) or ACGIH (American Council of Governmental Industrial Hygienists) need to be consulted and followed.

R. H. KOTTKE
Great Lakes Chemical Corporation

A. P. Dunlop and F. N. Peters, *The Furans*, ACS Monograph 119, Reinhold Publishing Corp., New York, 1953.

M. V. Sargent and T. M. Cresp "Furans" in D. Barton and W. D. Ollis, eds., *Comprehensive Organic Chemistry; The Synthesis and Reaction of Organic Compounds*, Vol. 4, Pergamon Press Ltd., Oxford, U.K., 1979.

A. Gandini "Furan Polymers" in J. I. Kroschwitz, ed., *Encyclopedia of Polymer Science and Technology*, 2nd ed., Vol. 7, John Wiley & Sons, Inc., New York, 1987.

F. M. Dean, *Advances in Heterocyclic Chemistry*, **30**, 168; **31**, 237 (1982).

FURNACES, ELECTRIC

INTRODUCTION

The term *electric furnace* applies to all furnaces that use electrical energy as their sole source of heat. Electric furnaces are used mainly for heating solid materials to desired temperatures below their melting points for subsequent processing, or melting materials for subsequent casting into desired shapes, ie, electric heating furnaces or electric melting furnaces.

Classification is by the manner in which the electrical energy is converted into heat. Thus three distinct types of widely used industrial furnaces can be distinguished: electric resistance furnaces, electric arc furnaces, and electric induction furnaces. The conversion of electrical energy into heat in each type of furnace is schematically illustrated in Figure 1.

Economic Aspects

Electric furnaces often are selected in preference to fuel-fired furnaces because the former are characterized by a significantly lower total operating costs. The increasing importance of environmental considerations tends to favor selection of electric furnaces. In electric furnaces (particularly resistance and induction furnaces), air pollution control is required only where pollutants emanate from the charge, but cost is much lower in the absence of a large hot gas volume.

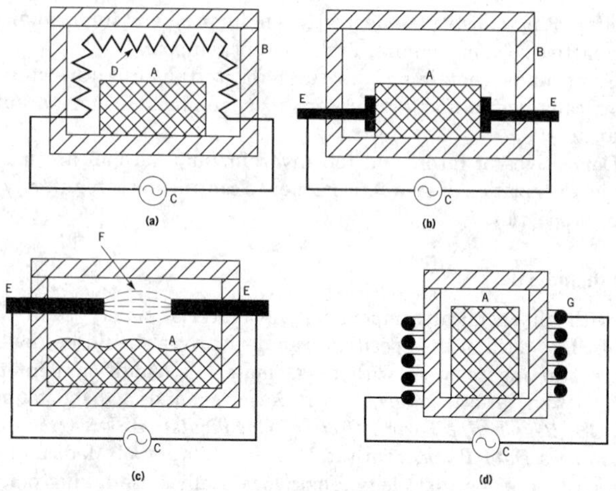

Figure 1. Main types of electric furnaces: **(a)** resistance furnace, indirect heat (resistor furnace); **(b)** resistance furnace, direct heat; **(c)** arc furnace; **(d)** induction furnace. A, charge to be heated or melted; B, refractory furnace lining; C, electric power supply; D, resistors; E, electrodes; F, electric arc; G, induction coil.

Electric furnaces have a much higher efficiency and therefore release considerably less heat into their surroundings, thereby minimizing the need to cool the work area.

RICHARD A. SOMMER
Consultant

W. Trinks, *Industrial Furnaces, Fuels, Furnace Types and Furnace Equipment—Their Selection and Influence Upon Furnace Operation*, 4th ed., Vol. II, John Wiley & Sons, Inc., New York, 1967, 358 pp., emphasis placed on heating furnaces (fuel-fired and electric) rather than melting furnaces.

K. Kegel, ed., *Elektrowaerme, Theorie und Praxis*, Union Internationale d'Electrothermie, Verlag W. Girardet, Essen, Germany, 1974, 902 pp., a complete and up-to-date reference for all electric furnaces.

A. Muhlbaur, "Electrical Energy for Heating Processes for the Future," *Electrowarme Int.* **49**(B3), B58 (1991), a review of various electric heating and melting industrial applications.

ARC FURNACES

Arc furnaces used in electric melting, smelting, and electrochemical operations are of two basic designs: the indirect and the direct arc. The arc of the indirect-arc furnace is maintained between two electrodes and radiates heat to the charge. The arcs of the direct-arc furnace are maintained between the charge and the electrodes, making the charge a part of the electrical power circuit. Not only is heat radiated to the charge, but the charge is heated directly by the arc and the current passing through the charge.

Indirect-Arc Furnaces

Indirect-arc furnaces have been used primarily in foundries for melting copper, copper alloys, and other nonferrous metals having a low melting point. They have also been used for producing molten iron and, occasionally, molten steel. The typical indirect-arc furnace is a single-phase furnace utilizing two horizontally mounted graphite electrodes, each of which project into an end of a refractory-lined horizontally mounted cylindrical steel shell.

Although rocking of the furnace to intermittently cover and hence protect up to 90% of the refractory, as well as improved refractories, has done much to make the indirect-arc furnace more viable, these furnaces are becoming less common, primarily due to high operating costs as a result of erosion of the refractory by the intense arc radiation.

Direct-Arc Furnaces

Open-Arc Furnaces. Most of the open-arc furnaces are used in melting and refining operations for steel and iron. Although most furnaces have three electrodes and operate utilizing three-phase a-c power to be compatible with power transmission systems, d-c furnaces are becoming more common. Open-arc furnaces are also used in melting operations for nonferrous metals (particularly copper), slag, refractories, and other less volatile materials.

A standard melting furnace consists of a refractory-lined steel shell with water-cooled upper sidewalls and roof (the lower portion is refractory to contain the molten metal); graphite electrodes; electrical equipment, bus bars, and flexible conductors to energize the electrode(s), equipment to regulate the position of the electrodes and thereby control the energy input; a means to access the inside of the furnace through a door; a method to tilt the furnace to empty it; and a means to allow the furnace to be recharged.

The refractory linings for the hearth and lower walls of furnaces designed for melting ferrous materials may be acidic, basic, or neutral. Silica has been widely used in the past, and is still being used in a number of iron and steel foundries. Alumina, a neutral refractory, is normally used for furnace roofs and in the walls for iron foundries, but basic brick can also be used in roofs.

Magnesite or dolomite, basic refractories, are used primarily in furnaces where the sulfur or phosphorus content of the metal, or both, must be reduced.

Almost all the electrodes used in open-arc furnaces are prefabricated and are made of regular or dense graphite.

The voltage chosen for open-arc furnaces must be high enough to compensate for the voltage drops caused by the resistance and inductance of the primary and secondary electrical circuits and still have the required power input available to sustain the arcs. In the smaller furnaces, the voltage must be high enough to penetrate any thin oxide coatings on the scrap. Also, it must provide a sufficient area of meltdown; otherwise, the electrodes bore a small hole through the scrap, melting insufficient metal to cover the hearth resulting in high consumption of the bottom refractories. The highest phase-to-phase no-load voltage for a 200 kVA production furnace usually is 200 V, and 1000 V for a 120,000 kVA furnace is not uncommon. Lower voltages are also available for the operator to use during a furnace refining cycle; the lowest voltage is approximately one-third of the highest voltage. However, high productivity operations generally do not make use of the lower voltage taps.

Other types of open-arc furnaces include the vacuum-arc furnace and the plasma-arc furnace.

D-C Arc Furnace. With the advent of more economical thyristor-controlled d-c power supplies, as well as limitations imposed by power companies on arc furnace-generated flicker, d-c furnaces have become more common, particularly in countries with weak power grids such as Japan. These furnaces are nearly identical to their counterparts, except they typically have a single electrode passing through the roof and a means to collect the current through a furnace bottom electrode.

Submerged-Arc Furnace. Furnaces used for smelting and for certain electrochemical operations are similar in general design to the open-arc furnace in that they are usually three-phase, have three vertical electrode columns and a shell to contain the charge, but direct current may also be utilized. They are used in the production of phosphorus, calcium carbide, ferroalloys, silicon, other metals and compounds, and numerous types of high temperature refractories.

Arc-Resistance Furnace. The arc-resistance furnace is similar to the submerged-arc furnace except the electrodes of the former are most often in direct contact with material, usually slag or a nonmetallic material, but they may also arc to the slag layer. Even when the electrode is in contact with the melt there are still minute arcs between the bottom and sides of the electrode, because it is not wetted by the slag, and the majority of the heat is developed in the melt in the immediate vicinity of the electrode tip. The furnace interior may be filled with a burden of unmelted charge above the melt, as in a submerged-arc furnace, or may contain a bare molten bath. The primary difference between the arc-resistance and submerged-arc furnace is that the former exhibits ohmic conductance.

Graphite electrodes are used primarily in smaller furnaces or in sealed furnaces. Prebaked carbon electrodes, made in diameters of <152 cm or 76 by 61 cm rectangular, are used primarily in smelting furnaces where the process requires them. However, self-baking electrodes are preferred because of their lower cost.

Health and Safety

All new furnace installations require pollution control equipment. This normally consists of off-gas afterburning (sometimes with energy recovery), and dust collection equipment, typically a baghouse.

For arc furnace worker safety, high power electrical systems require proper design and precautions, and handling of molten materials requires a minimum of fire-retardant clothing and often dust masks. Water must be prevented from coming in contact with the melt. Furthermore, since open-arc furnace noise levels commonly exceed 100 dBA, hearing protection is a necessity. Noise is normally not a problem with smelting furnaces.

J. KEVIN COTCHEN
MAN GHH Corporation

J. W. Wild, *ISS-AIME Electric Furnace Proceedings*, Vol. 34, I&SS, St. Louis, Mo., 1976, pp. 301–303.

AIME Electric Furnace Steelmaking, Vol. 1, Wiley-Interscience, Inc., New York, 1962, pp. 153–174.

C. R. Taylor, ed., *Electric Furnace Steelmaking*, Iron and Steel Society, Warrendale, Pa., 1985, pp. 63–70.

C. R. Davis and B. H. Baker, *ISS-AIME Electric Furnace Proceedings*, Vol. 33, Houston, Tex. 1975, pp. 136–144.

INDUCTION FURNACES

Induction furnaces utilize the phenomena of electromagnetic induction to produce an electric current in the load or workpiece. This current is a result of a varying magnetic field created by an alternating current in a coil that typically surrounds the workpiece. Power to heat the load results from the passage of the electric current through the resistance of the load. Physical contact between the electric system and the material to be heated is not essential and is usually avoided. Nonconducting materials cannot be heated directly by induction fields.

The efficiency of an induction furnace installation is determined by the ratio of the load useful power, P_n, to the input power, P_o, drawn from the utility. Losses that must be considered include those in the power converter (transformer, capacitors, frequency converter, etc), transmission lines, coil electrical losses, and thermal loss from the furnace.

Induction Heating

Design. The coil of an induction heater typically encircles the load. The current intensity within the load is greatest at the surface and diminishes to zero at the center. This crowding of the current close to the surface is known as skin effect.

Power Supplies and Controls. Induction heating furnace loads rarely can be connected directly to the user's electric power distribution system. If the load is to operate at the supply frequency, a transformer is used to provide the proper load voltage as well as isolation from the supply system. Adjustment of the load voltage can be achieved by means of a tapped transformer or by use of a solid-state switch. The low power factor of an induction load can be corrected by installing a capacitor bank in the primary or secondary circuit.

Economics. Induction heating equipment installations can require significant investment in electric power components as well as the work handling equipment made necessary by the process. These costs can be offset by savings in plant space, reduction in metal loss, precise control of product temperature, and reduced in-process inventory.

Applications. A unique capability of induction heating is apparent in its ability to heat the surface of a part to a high temperature while the interior remains at room temperature.

Induction heating is used to heat steel reactor vessels in the chemical process industry. Zone control of process temperature can be obtained by proper design of coil structure. Applications requiring high process temperatures generally not achievable by other means are possible when induction heating of a graphite susceptor is combined with the use of low conductivity high temperature insulation such as flake carbon interposed between the coil and the susceptor.

Induction Melting

Induction melting applications almost always contain the liquid metal charge within a hearth formed by a suitable refractory material. It is possible to design the hearth to satisfy a wide variety of application requirements ranging from a few kilograms to hundreds of tons of metal and for operation in normal or hostile environments.

Coreless Induction Furnaces

Coreless furnaces derive their name from the fact that the coil encircles the metal charge but the coil does not encircle a magnetic core.

Frequency Selection. When establishing the specifications for a coreless induction furnace, the material to be melted, the quantity of metal to be poured for each batch, and the quantity to be produced per hour must be considered simultaneously. Graphs have been developed that combine these factors with practical experience to indicate possible solutions for a specific requirement.

Operation. Small- and medium-sized coreless induction furnaces powered from high frequency power supplies can be started with a charge of metal pieces at room temperature, usually scrap material of appropriate alloy. As the metal charge begins to melt, a molten pool is established and the charge compacts, allowing additional charge to be added. Alloy additions and temperature adjustments complete the melting cycle. Large induction melting furnaces are most often started with an initial charge of molten metal.

Channel Induction Furnaces

The term channel induction furnace is applied to those in which the energy for the process is produced in a channel of molten metal that forms the secondary circuit of an iron core transformer. The primary circuit consists of a copper coil which also encircles the core. This arrangement is quite similar to that used in a utility transformer. Metal is heated within the loop by the passage of electric current and circulates to the hearth above to overcome the thermal losses of the furnace and provide power to melt additional metal as it is added.

Inductor. The channel inductor assembly consists of a steel box or case that contains the inductor refractory and the inductor core and coil assembly. The channel is formed within the refractory. Inductor power ratings range from 25 kilowatts for low temperature metals to 5000 kilowatts for molten iron.

Hearth. The hearth of a channel induction furnace must be designed to satisfy restraints that are imposed by the operating inductor, ie, the inductor channels must be full of metal when power is required, and it is also necessary to provide a sufficient level of metal above the channels to overcome the inward electromagnetic pressure on the metal in the channel when power is applied.

The refractory used to construct the hearth can be in the form of bricks or preformed shapes, or it may be monolithic. Often a furnace design utilizes all three.

Operation. Channel furnaces can be used for melting or holding metal. In either case, the inductor and the hearth refractory are preheated to avoid thermal shock as the liquid metal is introduced at startup. Once the inductor channel has been flooded, it is rarely emptied until the inductor is taken out of service. Inductor life can vary from six months to a number of years depending on the metal alloy and the size and power rating of the inductor. Channel melting furnaces are often designed so that a large portion of their total capacity can be discharged by tilting or rotating the furnace. Dry or preheated metal is added to the furnace at the melt rate of the furnace.

Applications. Small- and medium-sized foundries producing iron castings for automotive and other similar applications often utilize channel melting furnaces. They are also used in brass and other nonferrous melting and holding applications.

RICHARD A. SOMMER
Consultant

American Foundrymen's Society, Inc., *Refractories Manual*, 2nd ed., Des Plaines, Ill., 1989.

American Society for Metals, *Metals Handbook, Heat Treating*, Vol. 4, 9th ed., Metals Park, Ohio, 1991.

S. L. Semiatin and D. E. Stutz, *Induction Heat Treatment of Steel*, American Society for Metals, Metals Park, Ohio, 1986.

W. Trinks, *Industrial Furnaces*, Vol. 1, 4th ed., John Wiley & Sons, Inc., New York, 1951.

RESISTANCE FURNACES

The most widely used and best known resistance furnaces are indirect-heat resistance furnaces or electric resistor furnaces. They are categorized by a combination of four factors: batch or continuous; protective atmosphere or air atmosphere; method of heat transfer; and operating temperature. The primary method of heat transfer in an electric furnace is usually a function of the operating temperature range. The three methods of heat transfer are radiation, convection, and conduction. Radiation and convection apply to all of the furnaces described. Conductive heat transfer is limited to special types of furnaces.

Operating temperature ranges are classified as low, medium, and high; there is no standard or precise definition of these ranges. Generally, a low temperature furnace operates below 760°C, medium temperature ranges from 760–1150°C, and furnaces operating above 1150°C are high temperature furnaces. There is often indiscriminate use of the words furnace and oven. The term oven should be used when temperatures are below 760°C, and the word furnace applied for higher temperatures. The term furnace is used here regardless of operating temperature.

Furnace Atmospheres

Electric furnaces can operate either with air in the interior of the furnace or with a protective atmosphere; the choice is dictated by the process requirements of the work. The furnace must be designed for the atmosphere to be used, because the combination of temperature and atmosphere are significant factors in selecting internal materials used in the furnace construction; this applies particularly to the selection of heating element (resistor) material. It is feasible and common to design an electric furnace that can operate in both air and protective atmospheres although shortened element life generally results from frequent alternating between reducing atmospheres and oxidizing atmospheres. There are exceptions to this rule as some resistor materials must be periodically oxidized, if used in a reducing atmosphere. Other resistor materials are limited to a particular atmosphere.

Low Temperature Convection Furnaces

Low temperature convection furnaces are designed to transfer the heat from the heating elements by forced convection. Convection is normally used in furnaces operating below 760°C because it is the most effective means of heat transfer that can maintain good uniformity of temperature on various workload configurations. Convection furnaces also are used (in this range of temperatures) where it is important that no part of the work load exceed the controlled temperature. This is accomplished by shielding the work load from any view of the heating elements and by controlling the temperature of the air or atmosphere, which carries the heat from the heating elements to the work, at the desired maximum temperature.

Radiation Furnaces

Low Temperature Radiation Furnaces. These are of the infrared heater type. Heat transfer is by direct radiation from a high temperature heating element. Control of the heat is obtained by controlling the time of exposure to the heat radiation. This type of furnace is normally used for such applications as drying of paint films. Heating elements are nickel–chrome resistance wire which is wound on ceramic supports or contained in sheaths.

Medium Temperature Radiation Furnaces. The temperature range is generally 760–1150°C. Most of the heat is transferred directly to the work by radiation from the heating elements and by radiation to the furnace refractory which reradiates the energy to the work. Heating elements may be located in the side-walls, roof, or floor of the furnace. Location of heating elements must be selected to assure uniform temperatures in the work zone.

High Temperature Radiation Furnaces. These furnaces are similar in construction to medium temperature radiation furnaces, but operate above 1150°C. The insulation system must be designed to withstand the high temperatures, and internal structural parts become critical.

Elements commonly used as resistors are either silicon carbide, carbon, or high temperature metals, eg, molybdenum and tungsten. The latter impose stringent limitations on the atmosphere that must be maintained around the heating elements to prevent rapid element failure, or the furnace should be designed to allow easy, periodic replacement.

Vacuum Radiation Furnaces. Vacuum furnaces are used where the work can be satisfactorily processed only in a vacuum or in a protective atmosphere. Most vacuum furnaces use molybdenum heating elements. Because all heat transfer is by radiation, metal radiation shields are used to reduce heat transfer to the furnace casing.

Conduction Furnaces

Conduction furnaces utilize a liquid at the operating temperature to transfer the heat from the heating elements to the work being processed. Conduction furnaces are of three general types. One has a pot or crucible with suitable exterior insulation. Sheathed resistance elements are inside the pot, which contains molten lead or another low melting metal. The molten metal can be the conductive medium that transfers heat to the work immersed in it, or the molten metal may be the work. Such furnaces are often used to supply molten-type metal, lead, zinc, etc.

The salt-bath furnace is another type of conduction furnace. A molten salt not only provides the medium for conductive heat transfer, the salt is the heating resistor.

The third type of conduction furnace is a fluidized bed. In this design the product to be heated is submerged in sand, which is supported by a high porosity plate. Heated air (or atmosphere) is recirculated through the porous plate and sand, which gives a high heat-transfer efficiency to the product.

Direct-Heat Electric-Resistance Furnaces

Direct-heat electric furnaces use the material to be heated as the resistor, and the furnace consists of an insulated enclosure to retain the heat, a power source of suitable voltage, and means of attaching the power leads to the work. This type of furnace has several limitations that have prevented widespread use. Since the work is the resistor, it must have a uniform cross section between power connection points, and the material must be homogeneous. Varying sections or nonuniformities in the material can produce hot or cold spots in proportion to the change in electrical resistance. Also, a given furnace must be designed for work in which each piece to be heated has about the same resistance and power requirements.

There are large-scale operations using direct-heat resistance furnaces. These are mainly used in melting bulk materials where the liquid material serves as a uniform resistor. The most common application for this type of direct-heat electric resistance furnace is the melting of glass and arc furnaces for the melting of steel.

Applications

Electric furnaces are used for annealing, brazing, carburizing, galvanizing, forging, hardening, melting, sintering, enameling, and tempering metals, most notably aluminum, copper, iron and steel, and magnesium alloys.

ROBERT R. WALTON
Wellman Furnaces, Inc.

W. Trinks and M. Mawhinney, *Industrial Furnaces, Principles of Design and Operation*, 5th ed., Vol. 1, John Wiley & Sons, Inc., New York, 1961, 486 pp.

W. Trinks and M. Mawhinney, *Industrial Furnaces, Fuels, Furnace Types and Furnace Equipment—Their Selection and Influence Upon Furnace Operation*, 4th ed., Vol. 2, John Wiley & Sons, Inc., New York, 1967, 358 pp.

V. Paschkis and J. Persson, *Industrial Electric Furnaces and Appliances*, 2nd ed., Interscience Publishers, a division of John Wiley & Sons, Inc., New York, 1960.

FURNACES, FUEL-FIRED

A furnace is a device (enclosure) for generating controlled heat with the objective of performing work. In fossil-fuel furnaces, the work application may be direct (eg, rotary kilns) or indirect (eg, plants for electric power generation). The furnace chamber is either cooled (waterwall enclosure) or not cooled (refractory lining). In this article, furnaces related to metallurgy such as blast furnaces are excluded because they are covered under associated topics.

Fuels

Fuel-fired furnaces primarily utilize carbonaceous or hydrocarbon fuels. Since the purpose of a furnace is to generate heat for some useful application, flame temperature and heat transfer are important aspects of furnace design. Heat transfer is impacted by the flame emissivity. A high emissivity means strong radiation to the walls.

The carbon–hydrogen ratio of fuels is a variable used widely in fuel technology to estimate emissivity.

Power-Plant Furnaces

In 1991, over two-thirds of the electric power consumed in the United States was generated by fossil energy. The bulk of electric power generation comes from coal-fired power-plant furnaces supplying steam to turbogenerators. In terms of megawatts supplied, the coal-fired power plant is, therefore, the foremost component of importance in the energy supply system. Power-plant furnaces are of waterwall type and are generally designed for steam pressures in the range of 12.4–24.1 MPa (1800–3500 psi); the latter value is referred to as supercritical, ie, > 22.1 MPa abs (3208 psia).

Industrial Furnaces

Generally speaking, industrial furnaces are an order of magnitude smaller than power-plant furnaces since the applications are usually on an individual basis (hospital complex, chemical plant, paper mill, etc) rather than feeding power to a regional electric grid. Like the power-plant furnace, the function of the industrial furnace usually is to generate steam, generally for a chemical process, mechanical power, or heating application, rather than electric power generation. There are also many fired heaters that utilize the hot exhaust gases directly for heating, drying, roasting, calcining, etc. Industrial boilers include package boilers, paper mill (recovery) furnaces, large industrial furnaces (Waterwall), and refractory-wall furnaces.

Analysis

The rising demand for fuel efficiency and performance guarantees are imposing increasing requirements for analytical complexity. Analytical emphasis has shifted toward a heat-transfer-oriented view of predicting furnace performance: *(1)* wall absorption rate and gas temperature profiles, and *(2)* pollutant formation.

The analytical mechanisms for predicting the corresponding pollutant formation associated with fossil-fuel-fired furnaces lag the thermal performance prediction capability by a fair margin. The most firmly established mechanism at this time is the prediction of thermal NO_x formation. The chemical kinetics of pollutant formation is, in fact, a subject of research.

Fluidized-Bed Combustion

New furnace concepts in evolutionary stages include fluidized-bed furnaces, coal gasification furnaces, and MHD furnaces. Of these technologies, fluidized-bed combustion has reached commercial-scale operations.

CARL R. BOZZUTO
ABB Power Plant Laboratories

J. G. Singer, *Combustion Engineering—A Reference Book on Combustion, A Reference Book on Fuel Burning and Steam Generation*, Combustion Engineering, Inc., Windsor, Conn., 1991.

L. Douglas Smoot, ed., *Coal Science and Technology, Fundamentals of Combustion*, Vol. 20, Elsevier, New York, 1993.

M. Bashar and T. S. Czarnecki, "Design and Operation of a Lignite-Fired CFB Boiler Plant," *Proceedings of the Tenth International Conference on Fluidized Bed Combustion*, San Francisco, May 1–4, 1989.

K. A. Bueters and W. W. Habelt, *NO_x Emissions from Tangentially Fired Utility Boilers*, AIChE Symposium Series, Vol. 71, American Institute of Chemical Engineers, New York, 1975, Parts I–II.

FUSION ENERGY

As far as is known, nuclear fusion, which drives the stars, including the sun, is the primary source of energy in the universe. It occurs when the nuclei of lighter elements, such as hydrogen, are fused together at extremely high temperatures and pressure to form heavier elements, such as helium. Whereas practical methods for harnessing fusion reactions and realizing the potential of this energy source have been sought since the 1950s, achieving the benefits of power from fusion has proved to be a difficult, long-term challenge.

Fusion is widely held to be the ultimate resource for the world's long-term energy needs. The fuel reserves for fusion are virtually limitless and available to all countries. Fusion fuels can be extracted from water. Additionally, fusion promises to be an energy source which is potentially safe and environmentally benign. Radiological and proliferation hazards are much smaller than for fission power plants. The atmospheric impact is negligible compared to fossil fuels, and adverse impacts on the Earth's ecological and geophysical processes are smaller than for large-scale renewable energy sources. The economics and costs of fusion power plants are still being studied, but appear comparable to those for other medium- and long-term energy sources. The tantalizing promise of affordable essentially unlimited supplies of clean, safe energy, free of political boundaries, has motivated a worldwide research effort to develop this energy resource.

In order to effect a fusion reaction between two atomic nuclei, it is necessary that these nuclei be brought together closely enough to experience an attractive nuclear force. All nuclei are positively charged and repel one another via Coulomb's law, the electrostatic law of the repulsion of like charges. This electrostatic barrier can be overcome by imparting sufficient kinetic energy to the reacting species so that the nuclei can approach closely enough together that quantum mechanical tunneling can occur. The repulsive forces increase rapidly with the magnitude of the nuclear charge; therefore, nuclear fusion research has concentrated on the lightest elements and the isotopes having the lowest atomic numbers.

The reactions of deuterium, tritium, and helium-3, ^{3}He, having nuclear charges of 1, 1, and 2, respectively, are the easiest to initiate. These have the highest fusion reaction probabilities and the lowest reactant energies.

Plasma Conditions Required for Net Energy Release

The most promising approach to attaining significant reaction rates is to heat the reacting species to a high temperature, thereby imparting large kinetic energies to the nuclei in the form of thermal motions. By doing so, the particles, eg, deuterons and tritons, may scatter among themselves many times before undergoing fusion reactions, without losing significant energy from the system. At any given temperature, a system of particles in thermal equilibrium is characterized by a Maxwellian distribution of kinetic energies. The particles at the high energy end of this distribution account for most of the fusion reactions in fusion experiments.

The fusion fuel, when undergoing thermonuclear reactions, exists as an ionized gas called a plasma. In physics, the plasma state usually means a high temperature gas of net electrical neutrality consisting of free electrons and ions exhibiting collective behavior. The collection of charged particles exhibits characteristics of an electrically conducting fluid that can interact with electromagnetic fields. As such, its physical behavior is much more complex than that of an ordinary gas, and plasma confinement can be disrupted or reduced by many different kinds of plasma instabilities and other loss mechanisms.

In a plasma undergoing fusion reactions, the reactivity, and thus the fusion-power output rate, increases with increasing temperature. However, over a wide range of temperatures, as the temperature of the plasma is raised, the radiation losses are also increased, primarily because of bremsstrahlung, or continuum, ie, braking radiation from the electrons. For any fusion-fuel system there exists a unique temperature at which the fusion power production is precisely balanced by the radiation losses. This temperature is called the ideal ignition temperature, and equals about 50 million K (5 keV) for a D–T plasma (1 keV = 11.6×10^6 K). For a D–D plasma, this temperature is considerably higher, about 400×10^6 K (40 keV), a fact which considerably increases the difficulty of using pure deuterium fuel. Furthermore, a fusion system must be operated above the ideal ignition temperature for net power production, typically by a factor of 2–5.

Besides having to satisfy a minimum temperature requirement, the plasma must be sufficiently dense and contained for a long enough time to yield net power. If the plasma burns above the ideal ignition temperature for some time period, τ, the fusion energy released must at least equal the energy required to heat the plasma to that temperature plus the energy radiated during that period. It can be shown that this condition is met by requiring that the product of the plasma density, n, and confinement time, τ, exceed a characteristic value which depends only on the temperature. The minimum value of the product $n\tau$ represents the least stringent condition for the plasma to be a net producer of fusion energy. For D–T plasmas, this minimum occurs at a temperature of about 100×10^6 K, for which $n\tau \sim 10^{20}$ s/m^3. This minimum $n\tau$ product is called the Lawson criterion. For D–D, the minimum $n\tau$ product is about 10^{22} s/m^3 at a higher temperature, again indicating that a pure deuterium system requires a higher quality of confinement. A commonly used measure of the quality of plasma confinement is given by the triple product of the plasma density, n, ion temperature, T_i, and energy confinement time, τ, usually expressed in units of keV·s/m^3. A primary goal of fusion research is to achieve $n\tau T_i$ values of $\sim 10^{22}$ keV·s/m^3, as required for a D–T reactor. Experiments as of this writing (1993) have reached a value of 1.1×10^{21} keV·s/m^3 in the JT-60 tokamak in Japan.

Plasmas at fusion temperatures cannot be kept in ordinary containers because the energetic ions and electrons would rapidly collide with the walls and dissipate their energy. A significant loss mechanism results from enhanced radiation by the electrons in the presence of impurity ions sputtered off the container walls by the plasma. Therefore, some method must be found to contain the plasma at elevated temperature without using material containers.

Once a fusion reaction has begun in a confined plasma, it is planned to sustain it by using the hot, charged-particle reaction products, eg, alpha particles in the case of D–T fusion, to heat other, colder fuel particles to the reaction temperature. If no additional external heat input is required to sustain the reaction, the plasma is said to have reached the ignition condition. Achieving ignition is another primary goal of fusion research.

Paths to Fusion Power

Two diverse technical approaches to fusion power, magnetic confinement fusion, also known as magnetic fusion energy (MFE) and inertial confinement fusion, also known as inertial fusion energy (IFE) are being pursued worldwide. These form the basis of a large number of fusion research programs. Magnetic confinement techniques, studied since the 1950s, are based on the principle that charged particles such as electrons and ions, ie, deuterons and tritons, tend to be bound to magnetic lines of force. Thus the essence of the magnetic confinement approach is to trap a hot plasma in a suitably chosen magnetic field configuration for a long enough time to achieve a net energy release, which typically requires an energy confinement time of about one second. In the alternative IFE approach, fusion conditions are achieved by heating and compressing small amounts of fuel ions, contained in capsules, to the ignition condition by means of tightly focused energetic beams of charged particles or photons. In this case the confinement time can be much shorter, typically less than a millionth of a second.

Tokamak. The design concept that has come the closest by far to achieving energy breakeven conditions is the tokamak. Invented in the 1950s by the Russian physicists Andrei Sakharov and Igor E. Tamm, the tokamak derives its name from the Russian acronym for toroidal magnetic chamber. Technical progress in tokamaks was dramatic in the late 1980s and early 1990s. Central ion temperatures of 400×10^6 K have been reached, and energy confinement times have increased from 0.02 to about 1.4 seconds for strongly heated plasmas. The result has been $n\tau T_i$ triple products, of about 10^{21} keV·s/m^3, compared to the value of $\sim 10^{22}$ keV·s/m^3 required for a steady state D–T reactor.

Future Developments and Applications

The goal of fusion development is central station electrical power generation. Using the D–T fuel cycle, power would be extracted from the thermalization of the neutron kinetic energy deposited in the blanket. Pulsed systems such as inertial fusion require storage techniques to provide a continuous output of electrical power. In some cases, this storage medium may be simply the thermal blanket surrounding the reaction chamber. In MFE, significant technological challenges include the development of large superconducting magnets, efficient current drive systems, and adequate diverter plates and plasma facing components to handle the high particle and radiation heat loads. Provisions must also be made for the replacement and maintenance of components by remote handling techniques.

Fusion energy research is also the primary avenue for the development of plasma physics as a scientific discipline. The technologies and the science of plasmas developed en route to fusion power are already important in other applications and fields of science.

WILLIAM R. ELLIS
Raytheon Engineers & Constructors

H. A. Bethe, *Phys. Rev.* **55**, 103 (1939).

S. Glasstone and R. H. Lovberg, *Controlled Thermonuclear Reactions*, D. Van Nostrand, New York, 1960.

R. A. Gross, *Fusion Energy*, John Wiley & Sons, Inc., New York, 1984.

L. Spitzer, Jr., *Physics and Fully Ionized Gases*, 2nd rev. ed., John Wiley & Sons, Inc., New York, 1962.

G

GALLIUM AND GALLIUM COMPOUNDS

Gallium is a scarce but not a rare element. It is found most commonly in association with its immediate neighbors in the Periodic Table, ie, zinc, germanium, and aluminum. The concentration of gallium in the earth's crust, 10–20 g/t (10–20 ppm), is comparable to that of lead and arsenic.

There is an abundance of aluminum ores and alumina plants, and less importantly to gallium production zinc ores and plants; these are the main sources of gallium.

Properties

Physical Properties. Gallium, at. wt 69.717, has two stable isotopes, ^{69}Ga, 60.4%, and ^{71}Ga, 39.6%, and twelve unstable isotopes, from mass 63 through 76. The radius of the atom is 0.138 nm, and of the ions Ga^{3+} and Ga^{+}, is 0.062 nm and 0.133 nm, respectively. Solid gallium has a metallic, slightly bluish appearance.

The physical properties of gallium, especially its thermal properties, are exceptional. It has a low mp and vaporizes above 2200°C, ie, it has the longest liquid interval of all the elements. Also, it is easily supercooled. However, it expands during solidification by 3.2%, a property shared by only two other elements, germanium and bismuth. Its crystal structure is unusual for a metal. Gallium crystallizes in the orthorhombic system, and it is very anisotropic. This latter property is attributed to the existence of Ga–Ga covalent bonds along its [001] axis. Principal physical properties of the normal form are listed in Table 1.

Because gallium embrittles aluminum, the DOT Office of Hazardous Materials has classified gallium in Hazard Class HM-181 and has placed restrictions on air shipment. IATA (international air transport regulation) classifies gallium as corrosive.

Chemical Properties. In accordance with its normal potential, gallium is chemically similar to zinc and is somewhat less reactive than aluminum.

Extraction

A minor amount of gallium is extracted as a by-product from the zinc industry. The gallium content of sphalerites generally is concentrated in the residues of zinc distillation and in the iron mud resulting from purification of zinc sulfate solutions. Gallium is extracted from these streams by acidic solutions and gallium salts are recovered by liquid–liquid extraction.

Recovery from Bayer Liquor. The significant amount of primary gallium is recovered from the alumina industry. The main source is the sodium aluminate liquor from Bayer-process plants that produce large quantities of alumina. Several methods have been developed to recover gallium from Bayer liquor, eg, carbonation, electrolysis, chemical reduction, liquid–liquid extraction, and ion-exchange resins. Presently, gallium is mainly recovered by liquid–liquid extraction with 8-hydroxyquinoline derivatives.

Purification. Extraction from aluminum or zinc ores produces crude gallium metal or concentrates. These concentrates are transformed to sodium gallate, gallium chloride, or gallium sulfate solutions which are purified, then electrolyzed. Gallium is deposited as a liquid. The purification of the gallium salt solutions is carried out by solvent extraction and/or by ion exchange.

Ultrahigh (>99.99999% = 7.N) purity metallic gallium is achieved by a combination of several operations such as filtration, electrochemical refining, heating under vacuum, and fractional crystallization.

Recycling. A large part of the wastes from the gallium arsenide, GaAs, industry is recovered for both economical and environmental reasons. Several processes are effective and are being used to recover both the gallium either as a metal, a salt, or a hydroxide for recycling, and the arsenic in some form for recycling or disposal. Thermal decomposition of gallium arsenide waste is one method which competes with the use of hydrometallurgical routes in caustic soda media.

Production

Total worldwide gallium production capacity excluding the CIS, for which data are not available, is estimated to be at 250 t/yr.

The gallium either comes from mining sources or is recycled from scrap. Scrap-recycling capacity is taking a larger place each year. In the United States, the main processors are Eagle-Picher Industries and Recapture Metals; in Japan, Rasa Industries, Mitsubishi Metals, and Sumitomo Metal Mining; and in Europe, Rhône-Poulenc.

Alloys and Intermetallic Compounds

Alloys. Gallium has complete miscibility in the liquid state with aluminum, indium, tin, and zinc. No compounds are formed. Systems obtained when gallium is in the presence of bismuth, cadmium, germanium, mercury, lead, silicon, or thallium present miscibility gaps. No intermetallic compounds are formed.

Intermetallic Compounds. Numerous intermetallic gallium–transition element compounds have been reported. Gallium also forms numerous compounds with lanthanide and yttrium, as well as actinides.

Compounds Other Than Intermetallic

Compounds include gallium hydrides; gallium halides; gallium oxyhalides; gallium halogenates; gallium sulfohalides; gallium oxides; gallates of numerous metallic elements; gallium chalcogenides; gallium compounds with nitrogen, phosphorus, arsenic, and antimony; and carbon compounds of gallium, trialkyl gallium among others.

Toxicology

The toxicity of metallic gallium or gallium salts is very low. The corrosive, poisonous, or irritating nature of some gallium compounds is attributable to the anions or radicals with which it is associated. Gallium metal-organics, such as $Ga(CH_3)_3$, react vigorously with air, and can be explosive.

Uses

Gallium Solar Neutrino Experiment. Among the few potential experiments for the detection of the overwhelming majority of solar neutrinos, which are low energy pp-neutrinos, is the radiochemical gallium solar neutrino experiment.

Electronic and Magnet Applications. Gallium is used in production of diodes, transistor amplifiers, integrated circuits, photovoltaics, lasers, photodetectors, optical IC, and magnets of the type FeNdB.

Medical Uses of Gallium. Gallium can be used to detect such diseases as Hodgkin's disease, lymphomas, and interstitial lung disease. In dental applications gallium alloys are nonstaining and used in the fabrication and repair of dental prostheses.

Catalysis Application. Gallium is used in catalysts for aromatization in the petroleum industry.

Table 1. Physical Properties of Normal Gallium

Property	Value
melting point, °C	29.7714
boiling point, °C	ca 2200
density at mp, g/cm^3	
solid	5.904
liquid	6.095
vapor pressure at 1198 K, Pa[a]	0.14

[a] To convert Pa to mm Hg, divide by 133.3.

Economic Aspects

Despite very strong growth in the market for many gallium-containing devices used in electronics, achieving a balance in the gallium industry has been difficult economically. Because of a significant over-capacity in the gallium industry, prices have been weak and are decreasing.

JEAN LOUIS SABOT
HUBERT LAUVRAY
Rhône-Poulenc

K. Wade and A. J. Banister, *The Chemistry of Al, Ga, In, and Tl*, Pergamon Press, New York, 1974.

A. J. Downs, ed., *Chemistry of Aluminium, Gallium, Indium, & Thalium*, Routledge, Chapman, and Hall, London, June 1993.

D. F. Ferry, *Gallium, Arsenide Technology*, McMillan Co., New York, 1985.

M. J. Howes and D. V. Morgan, *Gallium Arsenide Materials, Devices, and Details*, John Wiley & Sons, Ltd., Chichester, U.K., 1985.

GAS, NATURAL

Natural gas is a mixture of naturally occurring hydrocarbon and non-hydrocarbon gases found in porous geologic formations beneath the earth's surface. Methane is a principal constituent and the mixture may contain higher hydrocarbons such as ethane, propane, butane, and pentane. Gases such as carbon dioxide, nitrogen, hydrogen sulfide, various mercaptans, and water vapor along with trace amounts of other inorganic and organic compounds can also be present. Natural gas is found in a variety of geological formations including sandstones, shales, and coals.

Discussions of natural gas can involve the following definitions:

Associated gas: free natural gas in immediate contact, but not in solution, with crude oil in the reservoir

Dissolved gas: natural gas in solution in crude oil in the reservoir

Dry gas: gas where the water content has been reduced by a dehydration process or gas containing little or no hydrocarbons commercially recoverable as liquid product

Liquefield natural gas (LNG): natural gas that has been liquefied by reducing its temperature to 111 K at atmospheric pressure; it remains a liquid at 191 K and 4.64 MPa (673 psig)

Natural gas liquids (NGL): a liquid hydrocarbon mixture which is gaseous at reservoir temperatures and pressures, but recoverable by condensation or absorption

Nonassociated gas: free natural gas not in contact with, nor dissolved in, crude oil in the reservoir

Sour gas: gas found in its natural state containing compounds of sulfur at concentrations exceeding levels for practical use because of corrosivity and toxicity

Sweet gas: gas found in its natural state containing such small amounts of sulfur compounds that it can be used without purification with no deleterious effect on piping or equipment, and without the potential for health hazards

Wet gas: unprocessed or partially processed natural gas produced from strata containing condensible hydrocarbons

Gas Reserves and Production

Data compiled for 1992 placed the world's estimated proved natural gas reserves at approximately 1.24×10^{14} m^3(4.38×10^{15} ft^3). The worldwide natural gas reserves have continued to increase as the demand for gas has increased and exploration efforts have expanded. In 1992, the principal political/geographical entities of the United States, the Confederation of Independent States (CIS), and the Oil Producing and Exporting Countries (OPEC) held 3.9, 40.0, and 40.0% of the world natural gas reserves, respectively.

Natural gas production on a worldwide basis has also continued to increase. Worldwide natural gas production increased at a rate of approximately 3.7%/yr for the period 1986–1990. By 1990, the annual production level had reached an energy equivalent of 73.1×10^{18} J(1.75×10^{16} kcal) or approximately 2.1×10^{12} m^3(7.42×10^{13} ft^3) and provided 22.9% of the total energy used. Government-owned companies dominate the lists of international organizations holding and producing the available natural gas reserves. Several of the international companies are based in the United States. The reserve holdings and the production data for these organizations reflect their activities throughout the world.

Properties

The composition of natural gas at the wellhead depends on the characteristics of the reservoir and is highly variable with respect to both the constituents present and the concentrations of these constituents.

The physical properties of the principal constituents of natural gas are listed in Table 1. These gases are odorless, but for safety reasons, natural gas is odorized before distribution to provide a distinct odor to warn users of possible gas leaks in equipment. Sulfur-containing compounds such as organic mercaptans, aliphatic sulfides, and cyclic sulfur compounds are effective odorants at low concentrations and are added to natural gas at levels ranging from 4 to 24 mg/m^3.

The pressure–volume–temperature (PVT) behavior of many natural gas mixtures can be represented over wide ranges of temperatures and pressures by the relationship, $PV = ZnRT$ where P is the absolute pressure; V, the volume; Z, the compressibility factor for the mixture; n, the number of moles of gas; R, the universal gas constant; and T, the temperature in Kelvin.

Processing

Natural gas obtained at the wellhead usually undergoes some type of treatment or processing prior to its use for safety, economic, or system and material compatibility reasons.

Dehydration. Produced gas is usually saturated with water vapor at the wellhead temperature and pressure. Generally, these water-vapor levels are reduced to concentrations no greater than 112 mg/m^3 (7 lbs/10^6 ft^3) gas to prevent condensation during transmission in high pressure pipelines and to reduce the possibility of corrosion. Usually

Table 1. Physical Constants of Natural Gas Constituents

Compound	Formula	Mol wt	Boiling point, K[a]	Critical pressure, kPa[b]	Critical temperature, K
methane	CH_4	16.043	111.64	4595	190.56
ethane	C_2H_6	30.070	184.55	4871	305.34
propane	C_3H_8	44.097	231.08	4247	369.86
2-butane	C_4H_{10}	58.123	261.37	3640	407.86
n-butane	C_4H_{10}	58.123	272.65	3796	425.17
2-pentane	C_5H_{12}	72.150	301.00	3381	460.44
n-pentane	C_5H_{12}	72.150	309.23	3369	469.71
n-hexane	C_6H_{14}	86.177	341.89	3012	507.38
n-heptane	C_7H_{16}	100.204	371.58	2736	540.21
n-octane	C_8H_{18}	114.231	398.83	2487	568.83
n-decane	$C_{10}H_{22}$	142.285	447.32	2104	617.60
nitrogen	N_2	28.013	77.35	3400	126.21
oxygen	O_2	31.999	90.20	5043	154.59
carbon dioxide	CO_2	44.010	194.68[c]	7384	304.22
hydrogen sulfide	H_2S	34.076	212.88	8963	373.41
water	H_2O	18.015	373.16	22055	647.14
air		28.963	78.83	3771	132.43

[a] At atmospheric pressure, 101.3 kPa (1 atm).
[b] To convert kPa to psi, multiply by 0.145.
[c] Denotes sublimation temperature.

the process selected for dehydration involves either liquid or solid desiccants. Dehydration may also be accomplished by expansion refrigeration which utilizes the Joule-Thompson effect.

Natural Gas Liquids. Natural gases containing high concentrations of the higher hydrocarbons are processed both to reduce the potential for condensation of these higher molecular-weight compounds during transmission and subsequent use, and to recover the natural gas liquid (NGL) products which can be marketed in both the fuel and petrochemical feedstock market.

Natural gas liquids are recovered from natural gas using condensation processes; absorption processes employing hydrocarbon liquids similar to gasoline or kerosene as the absorber oil; or solid-bed adsorption processes using adsorbents such as silica, molecular sieves, or activated charcoal.

Acid Gas Constituents. There are more than 30 processes available for removing the acid gas constituents such as hydrogen sulfide, carbon dioxide, and other organic sulfur compounds, ie, carbonyl sulfide, organic mercaptans, and disulfides. Because of the toxicity of hydrogen sulfide, requirements for removal are severe.

Both batch processes and continuous processes are used. Batch processes are used when the daily production of sulfur is small and of the order of 10 kg. When the daily sulfur production is higher, of the order of 45 kg, continuous processes are usually more economical. Using batch processes, regeneration of the absorbant or adsorbant is carried out in the primary reactor. Using continuous processes, absorption of the acid gases occurs in one vessel and acid gas recovery and solvent regeneration occur in a separate reactor.

Iron sponge is the oldest and most widely used batch process for removing sulfur compounds from natural gas.

There are numerous chemical and physical solvents available for use in continuous acid gas removal processes. The chemical absorbants include aqueous solutions of organic amines such as monoethanolamine, diethanolamine, triethanolamine, diglycolamine, or methyldiethanolamine. Adsorption systems employing molecular sieves are available for feed gases having low acid gas concentrations. Another process option is based on the use of polymeric, semipermeable membranes.

Nitrogen. The separation of nitrogen from natural gas relies on the differences between the boiling points of nitrogen (77.4 K) and methane (91.7 K) and involves the cryogenic distillations of a feed stream that has been preconditioned to very low levels of carbon dioxide, water vapor, and other constituents that would form solids at the low processing temperatures.

Specifications

Whereas there is no universally accepted specification for marketed natural gas, standards addressed in the United States are listed in Table 2.

Table 2. Natural Gas Pipeline Specifications[a]

Characteristic	Specification	Test method[b]
water content, mg/m^3	64–112	ASTM (1986) D1142
hydrogen sulfide, mg/m^3	5.7	GPA (1968) Std. 2265
gross heating value,[c] MJ/m^3	35.4	GPA (1986) Std. 2172
hydrocarbon dew point at 5.5 MPa,[d] K	264.9	ASTM (1986) D1142
mercaptan content, mg/m^3	4.6	GPA (1968) Std. 2265
total sulfur, mg/m^3	23–114	ASTM (1980) D1072
carbon dioxide, mol %	1–3	GPA (1990) Std. 2261
oxygen, mol %	0–0.4	GPA (1990) Std. 2261

[a] Gas must be commercially free of sand, dust, gums, and free liquid. Delivery temperature, 322.16 K; delivery pressure, 4.83 MPa.
[b] ASTM = American Society for Testing Materials; GPA = Gas Processors' Association.
[c] To convert MJ/m^3 to Btu/ft^3, multiply by 26.86.
[d] To convert MPa to psi, multiply by 145.

Transmission and Storage

As exploration and production activities have expanded both the natural gas resource base and the worldwide proven reserves, long-distance gas transmission pipelines have been constructed to link these resources to the industrialized areas and population centers of the world. The availability of high tensile-strength steel pipe and the development of techniques to construct, weld, and lay large diameter high pressure pipelines make it possible to economically transport natural gas to the marketplace. These transmission systems, coupled with localized, lower pressure distribution networks, bring gas to large segments of the world. Natural gas production and transmission systems are complemented by underground storage systems.

Liquefied natural gas (LNG) also plays a large role in both the transportation and storage of natural gas. At a pressure of 101.3 kPa (1 atm), methane can be liquefied by reducing the temperature to about −161 °C. When in the liquid form, methane occupies approximately 1/600 of the space occupied by gaseous methane at normal temperature and pressure.

Uses

Fuel. Natural gas is used as primary fuel and source of heat energy throughout the industrialized countries for a broad range of residential, commercial, and industrial applications.

Chemical Use. Both natural gas and natural gas liquids are used as feedstocks in the chemical industry. The largest chemical use of methane is through its reactions with steam to produce mixtures of carbon monoxide and hydrogen.

Ethylene is produced by steam-cracking the ethane and propane fractions obtained from natural gas, and the butane fraction can be catalytically dehydrogenated to yield 1,3-butadiene, a compound used in the preparation of many polymers. The *n*-butane fraction can also be used as a feedstock in the manufacture of methyl tertiary butyl ether (MTBE).

Production

Natural gas is produced from reservoirs containing both oil and gas (associated gas) and from nonassociated reservoirs holding only gas. These reservoirs may be relatively shallow and require wells drilled to depths of a few hundred meters. However, production is also being realized from reservoirs located at substantial depths requiring wells drilled to depths in excess of 6100 m. Production takes place both at onshore installations and on offshore platforms which service wells drilled to provide access to reservoirs located below the floor of the ocean.

Economic Aspects

Economically, natural gas represents an attractive energy option in those industrialized areas of the world where business infrastructures and delivery systems have been established to provide reliable service.

Outlook. New technologies are continually being developed which will expand the opportunities for using natural gas. The outlook for natural gas depends on the ability of the international natural gas community to continue to demonstrate that the resource base and the economic competitiveness of natural gas are adequate to justify its use for short-term and long-term applications.

KERMIT E. WOODCOCK
Consultant
MYRON GOTTLIEB
Gas Research Institute

1991 Gas Facts, The American Gas Association, Arlington, Va., 1991.

Oil & Gas Journal Data Book—1993 Edition, PennWell Publishing Co., Tulsa, Okla., 1993.

F. S. Manning and R. E. Thompson, *Oil Field Processing of Petroleum*, Vol. 1, PennWell Publishing Co., Tulsa, Okla., 1991.

R. N. Maddox, *Gas Conditioning and Processing*, Vol. 4, Campbell Petroleum Series, Norman, Okla., 1985.

GASOLINE AND OTHER MOTOR FUELS

Gasoline and other motor fuels comprise the largest single use of energy in the United States, and in 1988 accounted for 21% of all energy usage. The cost of this energy has been and is expected to continue to be a primary factor in the national economy. Moreover, the fraction of resources from which these fuels come that is provided by foreign sources is a matter of political concern. The fraction of total crude oil produced domestically shrunk from 73% in 1970 to 49% in 1991 and is predicted to continue to drop. In the 1970s, two Organization of Petroleum Exporting Countries (OPEC) embargoes resulted in rapid increases in the price of crude oil and therefore motor fuels. These increases triggered programs designed to develop alternative sources of fuels such as coal, oil shale, and natural gas. In the 1990s, as a result of lower price volatility and improved energy efficiencies, the inflation adjusted cost of driving is about one-half of what it was in the 1960s. Alternative fuels are more important for the potential to reduce emissions and improve air quality than for securing energy self-sufficiency.

General Aspects of Manufacture of Motor Fuels

All motor fuel in the United States is manufactured by private companies. Many of these are vertically integrated. That is, the same company finds the crude oil or buys it from a producing government, refines it into finished products, and then sells to independent retailers who specialize in that company's blended products or sells at company operated service stations. There are also a significant number of companies that participate in only some aspects of the business cycle such as refining or marketing.

Four groups are involved in the production or use of motor fuels in the United States: *(1)* manufacturers of the vehicles; *(2)* manufacturers and/or marketers of the fuels; *(3)* purchasers and users of fuels and vehicles; and *(4)* federal, state, and local regulatory agencies.

All four groups, or stakeholders, must work together to guarantee that fuels and vehicles are well matched. The American Society of Testing and Materials (ASTM) was founded in 1902 to promote just such a need for all products. ASTM Committee D-2 provides a forum for regulators, vehicle manufacturers, fuel producers, and consumers to develop and recommend nonbinding standards for petroleum products.

Although ASTM specifies certain quality levels, there are a number of factors that contribute to other quality levels in the marketplace. At times, government regulations are more restrictive than ASTM specifications, especially with respect to environmental issues. Secondly, competitive forces may encourage companies to provide fuel quality that is better than that defined by ASTM. Thirdly, ASTM specifications do not have the force of law, and certain companies may decide to exceed or not meet their recommended values. In response to this last factor, some states have adopted ASTM fuel quality specifications as state regulations, thus forcing a minimum quality level in the field.

GASOLINE

Gasoline demand is largely determined by the growth in the number of cars, the kilometers of paved roads available for driving, population, and economic growth. A primary factor in moderating the demand for gasoline has been the dramatic improvement in automotive fuel economy since the mid-1970s.

Economic predictions for future gasoline demand are somewhat divided. Some economists argue that the number of cars have reached saturation levels, that the population is aging, and that future cars should have better fuel economy. These factors suggest that gasoline demand should decline over the next 15–20 years. Others believe that vehicle numbers and use should continue to increase and that fuel economy is not expected to improve. This scenario predicts gasoline demand growth. Projections of future gasoline demand range from an increase of 1.1% per year to a decrease of 0.4% per year over the time frame of 1990–2010. Gasoline demand in other industrialized regions follows a pattern similar to that in the United States.

Requirements of Good Gasoline

To satisfy high performance automotive engines, gasoline must meet exacting specifications, some of which are varied according to location and based on temperatures or altitudes. The fuel must evaporate easily and burn completely when the spark plug fires in each cylinder. Early detonation of the fuel in the cylinder can cause destructive engine knock. The fuel must be chemically stable. It should not form gums or other polymeric deposit precursors. There should be no particulate contaminants or entrained water. Contaminants must be prevented from the point of manufacture in the refinery all the way through the distribution system until the fuel is metered from the vehicle tank into the engine.

Octane. Octane is probably the single most recognized measure of gasoline quality. The octane value of gasolines are posted on service station dispensers, and most drivers recognize a fuel in which the octane is too low. Broadly speaking, octane is a measure of the combustion characteristics of gasoline. Low octane gasoline has a tendency to preignite, causing rapid energy release and pressure fluctuations in the cylinder which result in a loud metallic noise commonly called knock. In addition to producing an objectionable sound, knock reduces the amount of useful work that can be extracted from the engine. Under extreme conditions of prolonged knock, overheating and even engine damage can occur. The damage is typically caused by catastrophic melting of piston crowns or head gaskets.

Chemical Factors. Because knock is caused by chemical reactions in the engine, it is reasonable to assume that chemical structure plays an important role in determining the resistance of a particular compound to knock. The chemical factors affecting knock are

Change in structure	Knocking tendency
longer paraffin chains	+
isomerizing normal paraffins	−
aromatizing normal paraffins	−
alkylating aromatics	−
saturating aromatic rings	+

Knocking may also be reduced by the use of alkyllead compounds, which are effective radical traps. Lead compounds are now banned in the United States.

Vehicle Factors. Because knock is a chemical reaction, it is sensitive to temperature and reaction time. Engine operating and design factors which affect the tendency to product knocking are presented below.

Increased knock	Decreased knock
higher compression ratio	increased turbulence
advanced spark schedule	exhaust gas recycle
higher coolant temperature	cooled air charge
turbocharging or supercharging	high altitudes
combustion chamber deposits	high humidity

Measuring Octane. Two different values need to be considered when discussing octane measurements. One is the knocking tendency of the fuel, called the fuel octane number. The other is the knocking tendency of the vehicle, called octane number requirement.

The octane number of a fuel is determined in a single cylinder cooperative fuel research (CFR) engine by comparing its knocking tendency to various primary reference fuels (PRF) mixtures of *n*-heptane and *i*-octane. Its measured octane is equal to the octane of the PRF which has the same knocking intensity. Knock intensity is controlled to an average value by varying the compression ratio of the engine. In practice, the exact value of a fuel's octane number is determined to the nearest 0.1 octane number by interpolation from two PRFs that are no more than two octane numbers apart. The CFR engine is operated at two conditions to simulate on-road driving conditions. The less severe condition measures research octane number (RON) and the more severe condition measures motor octane number (MON).

The octane number requirement (ONR) of a car is the octane number which causes barely audible, ie, trace knock when driven by a trained rater. The Coordinating Research Council (CRC), a research organization funded jointly by the American Petroleum Institute (API) and the American Automobile Manufacturers Association (AAMA), has defined test procedures for measuring ONR. Each year, CRC members measure ONR of more than 100 cars and publish the results.

Volatility. The properties of a gasoline which control its ability to evaporate are critical to good operation of a vehicle. In an Otto cycle engine, the fuel must be in the vapor state for combustion to take place. The volatility or vaporization characteristics of a gasoline are defined by three ASTM tests: Reid vapor pressure (RVP), the distillation curve, and the vapor/liquid ratio (V/L) at a given temperature, ASTM D323, D86, and D2533, respectively.

Startability. In order to achieve combustion in an Otto cycle engine, the air/fuel ratio in the combustion chamber must be near the stoichiometric ratio. When the engine is first started, the vehicle is designed to meter extra fuel and less air to the engine so that there is adequate vapor in the engine to support combustion. The ability of a fuel to achieve good starting can be correlated with RVP and a measure of the front end of the distillation curve, either E70 or T_{10}. Usually, minimum levels of RVP such as 60 kPa (0.6 bar) and E70 minimum of 10% (or a T_{10} maximum of 60°C) are satisfactory for good startability in most winter locations. At higher temperatures, lower RVP and front-end volatilities are adequate to provide good starting characteristics.

Vapor Lock. At the other end of the spectrum from starting is vapor lock, a problem of too much volatility. Vapor lock occurs when too much of the fuel evaporates and either starves the engine for fuel or provides too much fuel to the engine. It occurs on days that are warmer than usual and when the car has reached full operating temperatures.

Vehicle manufacturers minimize these problems by keeping the fuel system cool and under positive pressure. Fuel manufacturers minimize the problems by seasonal volatility blending.

Warm-Up. Warm-up refers to that period of operation beginning immediately after the car has started and continuing until the engine has reached normal operating temperatures, usually after 10 minutes or so of operation. During this period, the vehicle designer wants to get the vehicle equivalence ratio to stoichiometric as soon as possible to minimize emissions. On the other hand, if the mixture is leaned out too soon, the car experiences poor driveability during the warm-up period. From the fuel's perspective, the middle of the distillation curve plays the largest role in achieving good warm-up performance.

The most common expression for controlling driveability is known as the driveability index (DI), which has the form: $DI = 1.5\ T_{10} + 3\ T_{50} + T_{90}$. It is generally felt that fuels which have values of DI below 570 when T is in °C (1200 when T is in °F) provide good warm-up driveability performance.

Icing. At temperatures within 5°C of freezing and under conditions of high humidity, ice can form in the intake system of vehicles with carburetors or throttle body fuel injectors. In the extreme, ice can clog the carburetor jets and stall the car completely. After the car is fully warmed up, there is generally enough heat in the intake system to prevent any ice buildup. The vehicle manufacturer minimizes icing problems by rapid heating of the intake system, which also helps lower emissions, and through the use of multipoint fuel injectors.

On the fuel side icing tendencies can be reduced by proper volatility blending and, if necessary, by the use of additives, such as isopropyl alcohol, which depresses the freezing point of water, or a surfactant, which prevents ice crystals from sticking to and building up on metal surfaces.

Back-End Volatility. The portion of the gasoline that boils above 150°C is referred to as the back end. Generally, as the engine heats up this material evaporates. However, if there are too many back ends in the gasoline, then not all may boil off and the performance of the lubricant may be degraded. Very heavy molecules, such as those having more than 12 carbon atoms, may contribute to combustion chamber deposits. Condensed ring aromatics are particularly effective contributors to these deposits.

Vehicle volatility requirements are a strong function of ambient temperatures. ASTM has defined five volatility classes based on expected minimum and maximum daily temperatures. Each month, each state is assigned a volatility class, depending on its temperature history.

Cleanliness. Good gasoline must be both chemically and physically clean. Chemical cleanliness or stability means that it does not contain nor react under conditions of storage and use to form unwanted by-products such as gums, sludge, and deposits. Chemical cleanliness is assured by controlling the hydrocarbon composition and by appropriate additives. Physical cleanliness means that there are no undissolved solids or large amounts of free water in the gasoline. Stability is measured by a number of ASTM tests: D381 (Existent Gum), D525 (Oxidative Stability), and D873 (Potential Gum).

Manufacture

Distillation. Petroleum refining begins with the distillation of crude oil into a number of different fractions. In many cases, two distillations are carried out, one at atmospheric pressure and one under vacuum. Table 1 shows typical boiling ranges for the various crude oil fractions and typical yields from Arab Light, a common crude oil.

Catalytic Cracking. As of this writing (ca 1994), over 50% of the gasoline in the United States is obtained by catalytic cracking which uses a fluidized bed of powdered or small diameter catalysts that are continuously regenerated in an adjacent vessel called a regenerator. The principal class of reactions in the fluidized-bed catalytic cracking (FCC) process converts high boiling, low octane normal paraffins to lower boiling, higher octane olefins, naphthenes (cycloparaffins), and aromatics. FCC naphtha is almost always fractionated into two or three streams. Typical properties are shown in Table 2.

Thermal Cracking. Certain cracking conversion processes are carried out without catalysts. Heavy residuum streams in the refinery can be cracked thermally to produce coke and a mixture of lighter products. If naphtha is produced, it may require extensive treating before it can be used directly in gasoline or used as feed to other processes. Heavy distillate may also be thermally cracked using high (~ 800 °C) temperature steam. This process, steam cracking, is used to generate olefins for use in chemicals plants, but also generates material in the naphtha range which, if the quality is appropriate, may be used in gasoline.

Reforming. Catalytic reforming is a process to increase the octane of gasoline components. The feed to a reforming process is naphtha (usually virgin naphtha) boiling in the 80–210°C range. The catalysts

Table 1. Properties of Crude Oil Fractions[a]

Fraction	Boiling range, °C	Yield, %
gas	<0	<1
virgin naphtha		
light	0–100	18
heavy	100–200	
gas oil/kerosene	200–400	33
residue	>400	48

[a] From Arab Light crude.

Table 2. Properties of FCC Naphtha

Property	FCC naphtha		
	Light	Intermediate	Heavy
boiling range, °C	<105	105–160	160–220
RON	91	88	91
MON	79	77	81
aromatics, vol %	10	35	65
olefins, vol %	60	20	15

are platinum on alumina, normally with small amounts of other metals such as rhenium.

Depending on the catalysts and operating conditions, the following types of reactions occur to a greater or lesser extent: *(1)* heavy paraffins lose hydrogen and form aromatic rings; *(2)* cycloparaffins lose hydrogen to form corresponding aromatics; *(3)* straight-chain paraffins rearrange to form isomers; and *(4)* heavy paraffins are hydrocracked to form lighter paraffins.

Reformers generate highly aromatic, high octane product streams, and a great deal of hydrogen.

Some of the negative aspects of reformate are production of benzene, polynuclear or multiring aromatics (PNAs), and light gas (C_1–C_4).

Alkylation. Alkylation is the chemical combination of two light hydrocarbon molecules to form a heavier one and involves the reaction of butenes in the presence of a strong acid catalyst such as sulfuric or hydrofluoric acid. The product is a heavier multibranched isoparaffin. Propene and the various pentenes may also be used, to produce C_7 or C_9 isoparaffins, respectively.

Isomerization. Isomerization is a catalytic process which converts normal paraffins to isoparaffins. The feed is usually light virgin naphtha and the catalyst, platinum on an alumina or zeolite base. Octanes may be increased by over 30 numbers when normal pentane and normal hexane are isomerized.

Hydrogen Processing. Hydrogen is probably the most valuable refinery chemical in terms of its ability to improve the quality of refinery streams. It can be used to remove unwanted species such as sulfur and nitrogen.

Blending Agents. Blending agents are components of gasoline that are used at levels up to 20% and which are not natural components of crude oil. As of this writing, all blending agents are oxygenated compounds such as ethers and alcohols and are used for one or more of a variety of reasons, including to increase the octane of the fuel, to reduce vehicle emissions, and/or to use renewable resources to reduce dependency on imported crude oil.

Additives. Gasoline additives are used to improve the performance of the fuel either because the hydrocarbon components themselves contain some deficiency or because it is more effective to add a small amount of additive than to change the composition of the gasoline. Lead antiknock additives were a good case in point. Octane can be increased either by processing steps such as reforming, which requires significant investment and increased operating costs, or by using additives which is considerably less expensive. Additives are added in parts per million levels to distinguish them from blending agents which are added in the percents. They include dyes, antioxidants, metal deactivators, corrosion inhibitors, antiicing additives, detergent additives, carburetor detergents, fuel injector detergents, intake valve detergents, and demulsifiers.

Blending and Distribution

When blending gasoline from its components, refinery operators must balance a number of factors in the most economical way. First, the components must be used at the same rate at which these are produced or else the refinery either runs out of material or drowns in excess components. Secondly, each gasoline fuel grade must be produced to the specifications set by marketers and regulators. The specifications should not be exceeded in a way which increases manufacturing cost. Finally, blend targets must take into account the fact that the gasoline might not be sold immediately and may travel in pipelines, barges, or tankers and then sit in a distribution terminal. The time between production in the refinery and sale to the customer can be as long as one month.

Most refineries accomplish the difficult task of blending through the use of sophisticated linear programming algorithms. The linear programs are run in a few different time scales ranging from weekly to yearly to help refineries plan their seasonal operations. In the winter, gasoline demand is down and heating oil demand is up; in the summer, gasoline demand is at its peak.

Many gasoline properties, especially octane and RVP, do not blend linearly. Proper prediction of the octane of refinery blends is important because octane has traditionally been one of the most expensive gasoline properties, and raising pool octane often entails significant investment and increased operating costs. Also, it is possible to meet targets for the different grades by properly choosing blend stocks to take advantage of octane bonuses available from the nonlinear blending characteristics.

Gasoline blends are shipped from the refinery to a storage terminal. From the storage terminal, the gasoline is shipped by tank truck to individual service stations. The trucks, which have 40,000 L capacity, have 4–5 compartments so that they can deliver different grades at the same time. Service station tanks have a capacity of 12,000–15,000 L and are buried underground. Submerged turbine pumps transfer the gasoline from the tanks to dispensers at the dispensing islands. Tanks have been made of a number of materials, although the most popular is reinforced fiber glass. As a result of environmental concerns about gasoline leakage from underground tanks, the tanks have double-wall construction and have leak detectors between the two walls.

Fuel Economy

Fuel economy, typically expressed as distance driven per volume of fuel consumed, ie, in km/L (mi/gal), is measured over two driving cycles specified by the Federal Test Procedure. The statutory fuel economy standards are based on a harmonic average of the city and highway tests assuming that 45% of distance is accumulated under highway conditions and 55% is accumulated under city driving conditions.

Fuel economy is measured using a carbon balance method calculation. The carbon content of the exhaust is calculated by adding up the carbon monoxide, carbon dioxide, and unburned hydrocarbons concentrations. Then using the percent carbon in the fuel, a volumetric fuel economy is calculated. If the heating value of the fuel is known, an energy specific fuel economy in units such as km/MJ can be calculated as well.

Emissions

As a result of atmospheric pollution levels that exceed the National Ambient Air Quality Standards (NAAQS) in many parts of the United States, both the federal government and the State of California have implemented standards for exhaust and evaporative emissions from new vehicles. The first of these standards went into effect in 1968 and mandated that the vapors from the vehicle crankcase be routed back through the engine and burned. Since then, the standards have continued to grow stricter, as shown in Table 3. California has mandated that starting in 1998 a certain percentage of new vehicles sales must be zero emissions vehicles (ZEV). These ZEV vehicles are envisioned to be electric battery vehicles.

The emissions standards have forced changes in vehicle hardware. Many of these changes in hardware have resulted in changes in fuel as well. For example, starting in 1975, vehicle manufacturers installed noble metal catalysts in order to meet federal standards, and because the catalysts could not tolerate lead, all fuel manufacturers were mandated to sell at least one grade of unleaded gasoline. As demand for

Table 3. Federal Light-Duty Exhaust Emission Standards

Year	Emissions, g/km[a] HC[b]	CO	NO_x	Automotive hardware changes
precontrol	9.4	56	3.9	
1970	2.5	21		lean combustion
1972	1.8	17		
1973			1.9	exhaust gas recycle
1975	0.94	9.4		oxidation catalysts
1977			1.2	
1980	0.25	4.4		
1981		2.1	0.6	three-way catalysts, oxygen sensors
1994[c]	0.16[d]	2.1	0.25	
2003[e]	0.078[d]	1.1	0.12	

[a] To convert g/km to g/mi, multiply by 1.609.
[b] HC = hydrocarbons.
[c] Standards phase in over three years.
[d] Nonmethane hydrocarbons.
[e] If necessary and technically feasible.

Table 4. California Phase 2 Gasoline Composition

Property	Limit per liter	Values for averagers Average	Cap
RVP, kPa[a]	48		48
sulfur, ppm	40	30	80
aromatics, vol %	25	22	30
olefins, vol %	6	4	10
T_{90}, °C	149	143	166
T_{50}, °C	99	93	104
oxygen, wt %	1.8–2.2		2.7
benzene, vol %	1.0	0.8	1.2

[a] To convert kPa to psi, multiply by 0.145.

unleaded gasoline grew, new processing capacity was required and different blends developed.

In addition to setting standards for exhaust emissions, the government set standards for evaporative emissions. These refer to hydrocarbons that escape from the vehicle when fuel evaporates, either while the car is operating (running losses), while it is sitting and not being operated (diurnal emissions), or immediately after operation (hot soak). In order to control evaporative emissions, auto manufacturers have installed canisters of activated charcoal in their vehicles since 1972.

Although the charcoal canisters are about 95% effective, fuel volatility still impacts the mass of vapors that break through the canister. Therefore, EPA mandated that starting in the summer of 1992, RVP levels be reduced below the levels specified in ASTM D4814. Class C regions, generally the northern part of the country, are limited to a maximum RVP of 62 kPa (9.0 psi) vs an ASTM limit of 79 kPa (11 psi), and the southern Class B regions are limited to a maximum RVP of 54 kPa (7.8 psi) vs 69 kPa (9.0 psi) for ASTM.

The Clean Air Act Amendments of 1990 introduced a new concept in emission reduction: reducing exhaust emissions by controlling the composition of the fuel. This law mandated gasoline marketers to change gasoline composition so that emissions from existing vehicles would be reduced. Reduction targets of 15% and at least 20% were set for 1995 and 2000, respectively. Reductions are to be measured against 1990 vehicles and industry average gasoline. The reductions are for hydrocarbons (summer ozone season only), and for air toxics (year round). Air toxics are defined as the sum of the emissions of benzene, 1,3-butadiene, formaldehyde, acetaldehyde, and polycyclic organic material. Reformulated gasoline (RFG) may not result in any increase in NO_x emissions. Additionally, RFG must contain no more than 1% benzene and at least 2% oxygen.

Compliance with the RFG regulations is measured by using a formula to calculate the emissions from a given fuel. This formula, developed by EPA, predicts average exhaust and evaporative emissions as a function of gasoline chemical and physical parameters.

In the winter, the Clean Air Act mandates that gasoline in all areas which exceed the NAAQS for CO must contain at least 2.7% oxygen. This is based on the assumption that adding oxygen to the fuel reduces CO emissions.

The state of California has taken a different conceptual approach to reducing emissions through control of gasoline composition. Instead of defining a performance target, ie, 25% reduction, the State has defined composition targets which are aimed at achieving emissions reductions. These targets, shown in Table 4, took effect in 1996.

Gasoline composition may also be regulated in Europe. A tripartite initiative is being carried out among the European Commission, the oil industry, and the automotive industry. Based on an analysis of the required improvements in air quality, new regulations are to be written that control vehicle emissions and fuel composition into the twenty-first century.

DIESEL FUEL

As a fuel for internal combustion engines, diesel fuel ranks second only to gasoline. The volume of diesel fuel used relative to gasoline is expected to grow somewhat in the United States, although not as much as predicted in the 1980s. Diesel cars, thought at one time to be very promising, have encountered significant customer resistance in the United States. In other countries, diesel engines have captured a much larger share of the passenger car market.

Combustion in Diesel Engines

Unlike the spark-ignited gasoline engine, the diesel engine, first used by Rudolf Diesel in the 1890s to burn finely powdered coal dust, employs compression ignition. Liquid fuel was employed soon after.

There are two categories of diesel engines: direct injection (DI) and indirect injection (IDI). In DI engines, the fuel is injected directly into the combustion chamber. In IDI engines, there is a small prechamber into which the fuel is injected. The fuel starts to ignite in the prechamber and the hot burning gases are forced out into the main combustion chamber through a small passage. IDI engines may operate at higher speeds and use lower pressure injector systems which tend to be less expensive. They are used mainly on passenger cars although progress has also been made in producing small, high speed DI engines for passenger car use.

Requirements for Good Diesel Fuel

Diesel fuel is used in a wide variety of vehicular engines ranging from small IDI powered passenger cars to large trucks and construction equipment. There are actually three grades of diesel fuel defined in ASTM D975, the specification for diesel fuels. The first is Grade 1-D, suitable for high speed engines which operate under widely varying conditions of speed and load. Grade 1-D also has excellent low temperature properties. Grade 2-D is a general-purpose diesel suitable for use either in automotive or nonautomotive applications. It can be used in high speed engines involving relatively high loads and uniform speeds. Grade 4-D is much more viscous and is used in low and medium speed engines having sustained loads at substantially constant speed. Most cars and trucks use 2-D, a general-purpose grade. 1-D, a more volatile, lower density, lower aromatic fuel, is used in cold weather and in municipal buses.

Ignition Quality. The ability of diesel fuel to burn with the proper characteristics is described by its cetane number, a measure of ignition delay. Excessively long ignition delays (low cetane number) cause rough engine operation, misfiring, incomplete combustion, and poor startability.

The procedure for measuring the cetane number of diesel fuel (ASTM D613) is similar to that used for measuring gasoline octane number. Cetane (n-hexadecane), $C_{16}H_{34}$, is defined as having a cetane number of 100; α-methyl-naphthalene, $C_{11}H_{10}$, is defined as having a cetane number of 0. 2,2,4,4,6,8,8-Heptamethylnonane (HMN), $C_{16}H_{34}$, which can be produced in high purity, is used as the low reference fuel and has a cetane number of 15. Blends of cetane and HMN represent intermediate ignition qualities according to the formula: cetane number = %cetane + 0.15(%HMN).

The cetane number of a fuel depends on its hydrocarbon composition. In general, normal paraffins have high cetane numbers, isoparaffins and aromatics have low cetane numbers, and olefins and cycloparaffins fall somewhere in between. Diesel fuels marketed in the United States have cetane numbers ranging between 35 and 65. Most manufacturers specify a minimum cetane number of 40–45.

Cetane number is difficult to measure experimentally. Therefore, various correlation equations have been developed to predict cetane number from fuel properties.

Cold Temperature Properties. Diesel fuel must be able to be pumped and to flow through all filters and injectors at the lowest temperature that may be encountered in use. When the temperature is lowered, wax molecules in the fuel start to crystallize. This temperature is known as the wax appearance point or cloud point. These temperatures, which are generally the same, are measured by ASTM D2500 and D3117, respectively. If the temperature is lowered still further, the fuel gels and does not flow. This is the pour point and is measured by ASTM D97. These tests measure the ability of a fuel to operate in a diesel engine. Generally, the cloud point of a fuel is 4–6°C above the pour point, although fuels having differences of 11°C are not uncommon. The true operability temperature is somewhere in between the two; cloud point is too high and pour point is too low. Many engine manufacturers recommend fuels having pour points of 6°C below the lowest temperature at which the engine is expected to operate. Additives may be used to extend downward the operating range of diesel fuel.

Volatility. Volatile light fractions in diesel fuel help to provide easy engine starting but are generally low in cetane number and energy content. Heavy fractions, which have good cetane and energy content, can contribute to deposit formation and hard starting if present in too high concentrations. Desirable quality characteristics are obtained by careful blending of refinery streams.

Viscosity. For optimum performance of diesel engine injector pumps, the fuel should have the proper viscosity. Too low viscosity results in excessive injector wear and leakage. Viscosity that is too high may cause poor atomization of the fuel upon injection into the cylinders.

Density. The greater the density of diesel fuel, the greater its heat content per unit volume and therefore the greater its power or fuel economy. Because diesel fuel is purchased on a volume basis, density is often stipulated in purchase specifications and measured on delivery.

Flash Point. Specifications for flash point vary with grade; the lowest value is 38°C for grade 1-D. Controlling flash point is important in order to prevent the vapor space in storage and vehicle tanks from being in the explosive range. Setting the flash point at 38°C protects most storage vessels from exploding.

Carbon Residue. The tendency of a diesel fuel to form carbon deposits in an engine can be roughly predicted by one of two carbon residue tests: the Ramsbottom Coking Method (ASTM D524) or the Conradson Carbon Test (ASTM D189). For use in high speed diesel engines operating over a range of loads and speeds, ASTM specifications call for no more than 0.15% Ramsbottom carbon residue.

Sulfur. Sulfur in diesel fuel should be kept below set limits for both environmental and operational reasons. Operationally, high levels of sulfur can lead to high levels of corrosion and engine wear owing to emission of SO_3 that can react with condensed water during start-up to form sulfuric acids. Diesel fuel may contain high concentrations of sulfur (up to 5000 ppm), especially compared to gasoline (330 ppm average). The relationships between fuel sulfur content and total exhaust particulates has been well documented. Through negotiations between the Engine Manufacturers Association and API, it was agreed that sulfur content of highway diesel could be reduced, and EPA specified a maximum level of 0.05 wt % starting in October 1993.

Ash Content. The fuel injectors of diesel engines are designed to very close tolerances and are sensitive to any abrasive material in the fuel. Therefore, the maximum permissible ash content of the fuel is specified. The permissible amount of ash is between 0.01 and 0.1 wt %, depending on the grade of diesel.

Aromatics Content. Aromatic compounds have very poor ignition quality and, although they are not specifically limited in ASTM D975, there are practical limitations to using high aromatic levels in highway diesel fuel. The federal government began effectively limiting aromatic content to below 40% starting in October 1993 by specifying a minimum cetane index of 40. California limited aromatic levels below 10% beginning in the same time period, also because of emissions concerns.

Stability. Diesel fuel can undergo unwanted oxidation reactions, leading to insoluble gums and also to highly colored by-products. Stability is measured using ASTM D2274.

Diesel Fuel Manufacture

The biggest factors in determining how diesel fuel is blended in a given refinery are the availability of high cetane stocks. Because straight-run distillates contain the greatest amount of normal paraffins and cycloparaffins and the least amounts of branched paraffins and aromatics, these are the preferred stocks for diesel blending. Cracked stocks, which are relatively rich in aromatics, are less desirable form the standpoint of ignition quality. However, these have high energy density and good cold temperature properties. In the United States, where a high level of cracking is necessary to meet gasoline demand, the large supply of cracked fractions and the relatively small supply of straight-run distillates make substantial use of cracked stocks economically necessary. This has been made possible through the use of cetane improvers to improve cetane and through the use of hydrogenation to improve stability.

Other additives include corrosion inhibitors, detergent additives, cold flow improvers, and oxygenates.

Diesel Environmental Regulations

Emission standards have been set for heavy-duty vehicles in much the same manner as they have been set for gasoline engines. Because heavy-duty vehicles are primarily diesels, the focus is on diesel engine emissions. Standards have been written in units of grams per brake-horsepower-hour (g/bhph) = g/kW·h × 1.34, which normalize the emissions according to the total energy output of an engine over the specified driving cycle.

Diesel manufacturers have found it difficult to meet the stringent emissions targets. Development of exhaust treatment devices to reduce particulates and meet NO_x standards has been underway. These devices either trap or catalytically oxidize the particles or both.

California has taken a slightly different approach from that of the federal government. Starting in October, 1993, all diesel fuel in California must contain no more than 10% aromatics, 500 ppm sulfur, and meet all other ASTM specifications. Alternative formulations are possible if these are shown to have equivalent NO_x emissions to a base reference fuel. In addition to the specifications that apply to the commercial fuel, other aspects of the reference fuel composition are tightly controlled. The fuel must also have a minimum cetane number of 48 without cetane improvers.

ALTERNATIVE FUELS

Alternative fuels fall into two general categories. The first class consists of fuels that are made from sources other than crude oil but that have properties the same as or similar to conventional motor fuels. In this category are fuels made from coal and shale. In the second category are fuels that are different from gasoline and diesel fuel and which require redesigned or modified engines. These include methanol, compressed natural gas (CNG), and liquefied petroleum gas (LPG). Use of alternative fuels has been promoted for two reasons: potential emissions benefits and reduced dependence on imported petroleum.

ALBERT M. HOCHHAUSER
Exxon Research and Engineering Company

J. B. Heywood, *Internal Combustion Engine Fundamentals*, McGraw-Hill Book Co., Inc., New York, 1988.

K. Owen and T. Coley, *Automotive Fuels Handbook*, Society of Automotive Engineers, Warrendale, Pa., 1990.

L. M. Gibbs, *SAE Trans. J. Fuels Lubricants* **99**(4), 618–638 (1990).

G. T. Kalghatgi, *SAE Trans. J. Fuels Lubricants* **99**(4), 639–667 (1990).

R. Tupa and C. J. Dorer, *SAE Trans.* **95**(6), 340–374 (1986).

GASTROINTESTINAL AGENTS

Antipeptic Ulcer Therapy

The primary aim in antiulcer therapy is either prevention of acid secretion from the gastric parietal cells or neutralization of the acid before it comes into contact with the ulcerated areas of the gastrointestinal tract. Long established therapy of duodenal ulcers, prior to the introduction of cimetidine, $C_{10}H_{16}N_6S$, included diet, antacids, and anticholinergics. Of these, only antacids have been shown conclusively to be effective. The antacids can produce healing, but only when given in large and frequent doses. Antacids, used mainly as adjunctive therapy and generally available without a prescription, are indicated for short-term therapy of esophageal reflux (heartburn) and gastric upset. Antacids include aluminum hydroxide gel, precipitated calcium carbonate, magnesia and alumina suspension, magnesium oxide, magnesium trisilicate, magaldrate, and sodium bicarbonate.

The therapy for duodenal ulcers changed drastically upon the introduction of cimetidine in the United States in 1977. Cimetidine was the first of a series of agents which work by antagonism of histamine responses at the site responsible for acid secretion. This receptor is different from the H_1 receptor targeted for classical antihistaminic agents for treatment of allergies. Since the approval of cimetidine, three additional H_2 blockers have been approved for acute therapy of duodenal ulcers, ie, famotidine, $C_8H_{15}N_7O_2S_3$; nizatidine, $C_{12}H_{21}N_5O_2S_2$; and ranitidine; $C_{13}H_{22}N_4O_3S$.

A new class of agents, which specifically prevent the secretion of hydrochloric acid by the gastric parietal cells, acts by selective inhibition of the H^+ proton pump. The compound marketed in the United States (ca 1992), omeprazole, $C_{17}H_{19}N_3O_3S$, is a potent antisecretory agent and is used only for acute therapy of ulcerative disease and esophagitis.

Drugs acting by mechanisms not related to inhibition of gastric acid secretion have been introduced into therapy in the United States. The two compounds falling into this category are sucralfate, which works by an as yet to be determined mechanism, and misoprostil, a prostanoid derivative indicated only for prevention of gastrointestinal damage induced by nonsteroidal antiinflammatory agents such as aspirin. Therapy also has been changed by the acceptance that *Helicobacter pylori*, an organism able to withstand the acid media of the stomach, is associated with gastritis and ulcerative disease. No specific agents have been approved for eradication of *Helicobacter*, but combinations of several antibiotics are effective and some bismuth-containing compounds may act by their effect against the organism.

The market for antiulcer agents is large and is comprised of both prescription and over the counter (OTC) products. The estimated prescription market is over $3 billion annually in the United States, whereas the more difficult to estimate OTC market is in the range of $500 million annually. Several pharmaceutical companies are attempting to obtain approval for OTC use of prescription-only agents, and patent protection for several of the histamine antagonists runs out in the mid-1990s.

Laxatives

Laxatives facilitate the passage and elimination of feces. Laxatives have traditionally been classified as bulking agents; contact, ie, stimulant; saline, ie, osmotic; and emollients, ie, lubricant. Emollient or lubricant laxatives such as mineral oil are not discussed herein.

Bulk laxatives include carboxymethylcellulose sodium, polycarbophil, and psyllium hydrophilic mucciloid.

Contact laxatives include bisacodyl, phenolphthalein, cascara sagrada, castor oil, docusate calcium, and docusate sodium.

Saline or osmotic laxatives include lactulose, magnesium citrate solution, magnesium sulfate, sodium phosphate, and polyethylene glycol electrolyte preparation.

Antidiarrheal Therapy

Commonly used antidiarrheals work by one of two mechanisms: effects on net intestinal secretion, or a decrease in intestinal propulsive motility. Narcotic analgesics are constipating and are antidiarrheal owing to the effect on intestinal propulsion. This is not a smooth muscle relaxing activity but results from the increase in nonpropulsive phasic smooth muscle contractions, and is apparently the mechanism of such commonly used compounds as codeine sulfate, diphenoxylate, difenoxin, and loperamide. Some opiates also have been found to have effects on intestinal secretion. Bismuth subsalicylate (Pepto-Bismol) is effective and works by several mechanisms. Sulfasalazine and mesalamine are useful in therapy of inflammatory bowel disease.

Antiemetics

These drugs may act at a variety of physiological locations, eg, blockade of dopamine and effects at the chemoreceptor trigger zone, vestibular apparatus, or serotonin (5-HT_3) receptors. Antiemetics include benzquinamide, cyclizine hydrochloride, dimenhydrinate, dronabinol, meclizine hydrochloride, ondansetron, and prochlorperazine maleate.

Gastric Prokinetics

Gastric prokinetics are a new class of agent that can augment gastric emptying. In most cases these also increase the barrier pressure of the lower esophageal sphincter and increase acid clearance making them useful for therapy of esophageal reflux (heartburn) and esophagitis. They also may be effective in increasing gastric emptying in diabetic gastroparesis and other diseases in which there is a decrease in the ability for the stomach to empty. Cisapride (Propulsid) has been accepted by the FDA for the treatment of gastroesophageal reflux. It also has been shown to have significant effect on gastric emptying and may be useful in diabetic gastroparesis and other disease in which delayed gastric emptying is a problem. The prevalent hypothesis of its activity is that it has agonistic activity on the serotonin (5-HT_4) receptors on the nervous network in the gastrointestinal tract. Another prokinetic is metoclopramide.

HENRY I. JACOBY
Discovery Research Consultants

D. R. Bennett, ed. *Drug Evaluations Annual 1993*, American Medical Association, Chicago, Ill.

USP XXII, 22th rev., 1990, The U.S. Pharmacopoeial Convention, Rockville, Md.

Merck Index, 11th ed., Merck, Rahway, N.J., 1989.

C. A. Fleeger, ed., *USAN and USP Dictionary of Drug Names*, U.S. Pharmacopeial Convention Inc., Rockville, Md.

GELATIN

Gelatin is a protein obtained by partial hydrolysis of collagen, the chief protein component in skin, bones, hides, and white connective tissues of the animal body. Type A gelatin is produced by acid processing of collagenous raw material; type B is produced by alkaline or lime processing. Because it is obtained from collagen by a controlled partial hydrolysis and does not exist in nature, gelatin is classified as a derived protein.

Uses of gelatin are based on its combination of properties; reversible gel-to-sol transition of aqueous solution; viscosity of warm aqueous solutions; ability to act as a protective colloid; water permeability; and insolubility in cold water, but complete solubility in hot water. It is also nutritious. These properties are utilized in the food, pharmaceutical, and photographic industries. In addition, gelatin forms strong, uniform, clear, moderately flexible coatings which readily swell and absorb water and are ideal for the manufacture of photographic films and pharmaceutical capsules.

Chemical Composition and Structure

Gelatin is not a single chemical substance. The main constituents of gelatin are large and complex polypeptide molecules of the same amino acid composition as the parent collagen, covering a broad molecular weight distribution range. In the parent collagen, the 18 different amino acids are arranged in ordered, long chains, each having ~95,000 mol wt. These chains are arranged in a rod-like, triple-helix structure consisting of two identical chains, called α_1, and one slightly different chain called α_2. These chains are partially separated and broken, ie, hydrolyzed, in the gelatin manufacturing process. Different grades of gelatin have average molecular weight ranging from ~20,000 to 250,000.

Analysis shows the presence of amino acids from 0.2% tyrosine to 30.5% glycine. The five most common amino acids are glycine, 26.4–30.5%; proline, 14.8–18%; hydroxyproline, 13.3–14.5%; glutamic acid, 11.1–11.7%; and alanine, 8.6–11.3%. The remaining amino acids in decreasing order are arginine, aspartic acid, lysine, serine, leucine, valine, phenylalanine, threonine, isoleucine, hydroxylysine, histidine, methionine, and tyrosine.

Stability. Dry gelatin stored in airtight containers at room temperature has a shelf life of many years. However, it decomposes above 100°C. Aqueous solutions or gels of gelatin are highly susceptible to microbial growth and breakdown by proteolytic enzymes. Stability is a function of pH and electrolytes and decreases with increasing temperature because of hydrolysis.

Physical and Chemical Properties

Commercial gelatin is produced in mesh sizes ranging from coarse granules to fine powder. In Europe, gelatin is also produced in thin sheets for use in cooking. It is a vitreous, brittle solid, faintly yellow in color. Dry commercial gelatin contains about 9–13% moisture and is essentially tasteless and odorless with specific gravity between 1.3 and 1.4. Most physical and chemical properties of gelatin are measured on aqueous solutions and are functions of the source of collagen, method of manufacture, conditions during extraction and concentration, thermal history, pH, and chemical nature of impurities or additives.

Economic Aspects

Of the gelatin produced in the United States, 55% is acid processed, ie, type A. The U.S. food industry consumes about 20,000 t/yr, with an annual growth rate of 0.5%; the pharmaceutical industry consumes about 10,000 t/yr; and the photographic industry about 7,000 t/yr. In the United States, the pharmaceutical gelatin market is expected to grow on the average of 2.5% per year. The photographic gelatin market has been stable or growing slightly. Color paper and x-ray products use over 55% of the photographic gelatin in the United States, with graphic arts and instant films using an additional 30%.

THOMAS R. KEENAN
Kind & Knox Gelatine, Inc.

A. G. Ward and A. Courts, eds., *The Science and Technology of Gelatin*, Academic Press, Inc., New York, 1977.

K. Ridgway, ed., *Hard Capsules Development and Technology*, The Pharmaceutical Press, London, 1987.

J. Photogr. Sci. **40**,(5,6), 122–251 (1992).

S. J. Band, ed., "Photographic Gelatin," *Proceedings of the Fifth RPS Symposium*, Oxford, U.K., 1985, The Imaging Science and Technology Group of the Royal Photographic Society, 1987.

GEMSTONES

GEMSTONE MATERIALS

There are three types of gemstone materials as defined by the U.S. Federal Trade Commission: *(1)* natural gemstones are found in nature and at most are enhanced; *(2)* imitation or simulated, fake, faux, etc, material resembles the natural material in appearance only and is frequently only colored glass or even plastic; and *(3)* synthetic material is the exact duplicate of the natural material, having the same chemical composition, optical properties, etc, as the natural, but made in the laboratory. Moreover, the word gem cannot be used for synthetic gemstone material. The synthetic equivalent of a natural material may be used as an imitation of another, eg, synthetic cubic zirconia is widely used as a diamond imitation.

Synthetic gemstone materials often have multiple uses. Synthetic ruby and colorless sapphire are used for watch bearings, unscratchable watch crystals, and bar-code reader windows. Synthetic quartz oscillators are used for precision timekeeping, citizen's band radio (CB) crystals, and filters. Synthetic ruby, emerald, and garnets are used for masers and lasers.

In the gemstone jewelry market, synthetics provide a less expensive alternative to natural gemstones, but of a better quality than that available in costume jewelry. In general, a synthetic should be available for no more than 10% of the cost of equivalent-quality natural gemstone to be commercially viable. Synthetics are frequently divided into three groups: *(1)* luxury synthetics, involving slow and difficult growth processes, produced in small quantities for a price-restricted market; *(2)* intermediates; and *(3)* low cost synthetics, produced on a large scale.

Properties

The important properties are those of importance in natural gemstones. First is hardness, H. A value of 7 or greater on Mohs' scale is desirable to avoid scratches from the quartz (H = 7) sand present in dust. Next is color or a total lack of color, as in diamond and its simulants. A high refractive index (RI) permits the return by total internal reflection of most of the light falling onto a well-cut gemstone, giving brilliance, and a high dispersion (DISP) spreads the internally reflected light into spectral colors, resulting in fire.

Several gemstone species occur in various colors, depending on the presence of impurities or irradiation-induced color centers. Any material can have its color modified by the addition of various impurities: synthetic ruby, sapphires, and spinel are produced commercially in over 100 colors.

Manufacture

The most frequently used techniques for the commercial manufacture of synthetic gemstone materials are summarized in Table 1. Only rarely used for synthetics are such alternative growth techniques as the Bridgman technique of solidification in a crucible and the float zone technique, both involving growth from the melt.

Table 1. Techniques for Commercial Gemstone Material Synthesis

Technique	Material
	Crystal growth from the melt
Verneuil (flame fusion)	ruby, sapphires, and stars; spinel; rutile; strontium titanate
Czochralski (pulling)	ruby and sapphire; alexandrite; garnets: YAG and GGG
float zone	ruby and sapphire; alexandrite
skull melting	cubic zirconia
	Crystal growth from solution
flux	alexandrite; emerald; ruby and sapphire; spinel
hydrothermal	colorless, amethyst, citrine, and smoky quartz; emerald; ruby and sapphire
high pressure	diamond;[a] jadeite[a]
	Other techniques
complex chemical	opal

[a] Grown for other purposes or experimental production.

Materials

Alexandrite. Alexandrite, which is a colorless chrysoberyl, $BeAl_2O_4$, when pure, has a color change derived from Cr.

Beryl. Beryl, $Be_3Al_2Si_6O_{18}$, is called aquamarine when pale green or blue from the presence of Fe, emerald when dark green from Cr or at times V, and morganite or red beryl when pink or red, respectively, from Mn. Only synthetic emerald is in commercial production.

Corundum. Crystalline Al_2O_3, corundum, is called ruby when colored red by about 1% Cr, and sapphire for colorless and other colors particularly when blue from charge transfer between about 0.01% each of Fe^{2+} and Ti^{4+}.

Cubic Zirconia. As of this writing (ca 1994), cubic zirconia, ZrO_2, is the best diamond imitation available. This material can also be made in almost any color.

Diamond. The synthesis by a high pressure process of single-crystal diamond large enough for gemstone use was revealed by the General Electric Company in 1971. The yellow color (containing N) is grown much more easily than colorless (pure) and blue (B). None of these is likely to be viable for use in jewelry in the near future.

Garnets. Both YAG, yttrium aluminum garnet, $Y_3Al_5O_{12}$, and GGG, gadolinium gallium garnet, $Gd_3Ga_5O_{12}$, have the garnet structure and were used at one time as diamond imitations. These have been supplanted by cubic zirconia.

Opal. Opal is the only commercial synthetic gemstone material that is not a single crystal. It consists of a three-dimensional diffraction grating of geometrically aligned spheres of $SiO_2 \cdot xH_2O$, where x is usually <10%.

Quartz. When colorless, quartz is also known as rock crystal; irradiation of this produces smoky quartz. The name citrine is used when quartz is colored by Fe, and irradiation of this can produce purple-colored amethyst under certain circumstances.

Rutile. Rutile, a form of TiO_2, was at one time used as a rather poor diamond imitation. Related is strontium titanate, $SrTiO_3$, now more properly called synthetic tausonite.

Spinel. Colorless (pure), blue (Co), and other colored synthetic spinels made by the Verneuil process are widely seen in class rings and in other jewelry uses, where the blue is often mislabeled as synthetic sapphire.

Other Synthetic Materials. Many other natural gemstone materials have been duplicated in the laboratory on an experimental basis, often only in small sizes. Examples include tourmaline, topaz, and zircon. Of some potential is synthetic jadeite, one of the two forms of jade.

Discredited Synthetics. There are several materials that have in the past been considered to be synthetics, but were found on closer examination not to deserve such a designation, being merely imitations. Examples include imitation coral, lapis lazuli, and turquoise, all made by ceramic processes.

KURT NASSAU
Nassau Consultants

Guides for the Jewelry Industry, U.S. Federal Trade Commission, Washington, D.C., Feb. 27, 1979 (under revision in 1993).

K. Nassau, *Gems Made by Man*, Gemological Institute of America, Santa Monica, Calif., 1980.

C. S. Hurlbut, Jr. and R. C. Kammerling, *Gemology*, 2nd ed., John Wiley & Sons, Inc., New York, 1991.

K. Nassau, *The Physics and Chemistry of Color*, John Wiley & Sons, Inc., New York, 1983.

GEMSTONE TREATMENT

Color and clarity are two of the attributes that give gemstones used in jewelry value. Gemstones deficient in either color or clarity can be enhanced. Almost worthless material can at times be converted into valuable-appearing gemstones. An estimated two-thirds of all colored gemstones used in jewelry have been treated. Accordingly, the identification of the use of treatments and the disclosure of enhancements to the purchaser are important.

Some treatments are practiced so widely that untreated material is essentially unknown in the jewelry trade. The heating of pale Fe-containing chalcedony to produce red-brown carnelian is one of these.

The stability of a particular treatment is also important. The enhancement should survive during normal wear or display conditions.

Heat Treatments

The most commonly seen of the gemstones that have been enhanced by heat treatment are listed in Table 1. Parameters for specifying the conditions for heat treatment of a gemstone material include the maximum temperature reached and the time for which the maximum temperature is sustained; the rate of heating to temperature, the rate of

Table 1. Gemstones Enhanced by Heating

Material	Change[a]	Product	Use[b]
amber	clarified, sun-spangled	amber	F
amber	reconstructed, aged	amber	R
beryl	green to blue	aquamarine	W
chalcedony	pale to red-brown or red	carnelian, agate, tiger's eye, etc	W
corundum	develop, intensify, or lighten blue	blue sapphire	W
corundum	develop or intensify yellow	yellow sapphire	W
corundum (ruby)	remove off-shades	ruby	F
corundum (ruby, sapphires)	remove silk, remove or develop asterism	starting material	W
corundum	diffuse in color or asterism	ruby, sapphires	R
diamond	change color after irradiation	starting material	R
quartz	amethyst to yellow citrine	starting material	W
quartz	crackled and dyed	various colors	R
zircon	brown to colorless or blue	starting material	W
zoisite	brown to deep purple-blue	starting material	W

[a] All product colors listed are stable.
[b] Prevalence of treatment occurring in product: R = rare to occasional; F = frequent; W = widespread or near-total.

Table 2. Rays and Particles Commonly Used for the Irradiation of Gemstones

Irradiation type	Average energy, eV	Coloration uniformity	Induced radioactivity	Localized heating
x-rays	1×10^4	poor	none	none
γ-rays from Co-60 or Cs-137	1×10^6	good	none	none
neutron beam	1×10^6	good	strong	none
electron beam	1×10^6	poor	none	strong
electron beam	2×10^7	good	some	strong

Table 3. Gemstones Enhanced by Irradiation

Material	Change or product	Comments[a]	Use[b]
corundum	colorless to yellow	S,U,R	R
diamond	near colorless to black, blue, green, yellow, or red	S,R	F
pearl	darken to black	S	F
quartz	colorless to smoky	S,R	W
quartz	amethyst to amethyst–citrine	S,R	F
spodumene	pink kunzite to deep green	U,R	R
topaz	colorless or pale to blue	S,R	W
topaz	colorless or pale to brown	S,U,R	R
tourmaline	colorless or pale to red or multicolor	S,R	F

[a] S = stable; U = unstable, may fade; R = can be reversed by another treatment.
[b] Prevalence of treatment occurring in product: R = rare or occasional; F = frequent; W = widespread to near-total.

cooling down from temperature, and any holding stages while heating and cooling; the chemistry and pressure of the atmosphere; and any material in contact with the gemstone. Exact conditions for heat treatments vary widely according to the natural materials used.

Irradiation Treatments

The process of irradiation involves the exposure of a specimen to one of a variety of radiations. A summary is given in Table 2.

When radiation interacts with matter, a displacement of the outermost electrons in atoms occurs. This displacement can lead to the formation of color centers or to valence state changes. The most commonly seen gemstones enhanced by irradiation are summarized in Table 3. When properly performed, there is no significant residual radioactivity.

Other Treatments

Other treatments fall into three groups: impregnations, surface modifications and composite gemstones.

Identification of Treated Gems

A trained gemologist, taught by the Gemological Institute of America of Santa Monica, California, and New York, the Gemmological Association of Great Britain of London, or elsewhere, is needed for identification of treated gems. This topic is also discussed in textbooks. In some materials the induced change is the exact equivalent of a process that also occurs naturally, so that such treatments cannot be identified.

KURT NASSAU
Nassau Consultants

K. Nassau, *Gemstone Enhancement*, 2nd ed., Butterworths, Boston, Mass., 1994.

R. T. Liddicoat, Jr., *Handbook of Gem Identification*, 12th ed., Gemological Institute of America, Santa Monica, Calif., 1989.

C. S. Hurlbut, Jr. and R. C. Kammerling, *Gemology*, 2nd ed., John Wiley & Sons, Inc., New York, 1991.

B. W. Anderson and E. A. Jobbins, *Gem Testing*, 10th ed., Butterworths, London, 1990.

GENETIC ENGINEERING

PROCEDURES

The contemporary meaning of genetic engineering implies a use of the techniques of molecular biology, especially recombinant deoxyribonucleic acid (DNA) techniques, rather than breeding in the formation of new genotypes. Recombinant DNA molecules are composed of two parts: first, a vector where the function is to provide the biochemical functions necessary for replication of the recombinant DNA molecule, and secondly, the passenger DNA which is joined to the vector and is replicated passively under control of the vector. Recombinant DNA technology allows the construction *in vitro* of DNA molecules that are not found in nature and the subsequent introduction into organisms, resulting in new genotypes and phenotypes of the recipient. Particularly in plant and animal science, a gene may be introduced into an organism by recombinant DNA technology and the line with the desired properties further manipulated by breeding.

Analysis of DNA Information

Molecular biology is an information-based science. In this context information can be defined as the negative logarithm of the probability of a system occupying a particular state, given the total number of states available to it. In a DNA sequence there are four possible states at each position, corresponding to the four nucleic acid bases, adenine (A), $C_5H_5N_5$; guanine (G), $C_5H_5N_5O$; thymine (T), $C_5H_6N_2O_2$; and cytosine (C), $C_4H_5N_3O$. A DNA of chain length n therefore has 4^n possible arrangements. Because there are so many potential sequences of a DNA molecule, and because DNA molecules of the same base composition can have similar biochemical properties but very different sequences, standard biochemical techniques cannot address the most biologically important property of DNA, its information content. Genetic engineering techniques allow the analysis and manipulation of genetic information based on its nucleotide sequence.

Sequence-Dependent Cleavage of DNA by Restriction Enzymes. Bacteria in nature are constantly exposed to exogenous DNA, primarily from bacteriophage (viruses) in the environment. Probably as a defense system, many bacteria contain a two-part DNA restriction and modification system. Restriction enzymes are of several types; the most useful for cloning, the Type II restriction enzymes, recognize specific sequences, usually 4–8 base pairs (bp) in length, and cut DNA molecules within these sequences. In nature, restriction enzymes serve as a sort of immune mechanism. Invading viruses are inactivated by restriction enzyme digestion of their DNA.

Location of Specific Sequences to DNA Restriction Fragments. A second technique that is universally applied to DNAs large and small is that of southern blotting. In these experiments, DNA fragments separated by gel electrophoresis are denatured *in situ* and transferred by capillary action to a nitrocellulose or nylon membrane, thereby making a contact print of the DNA in the gel. The single-stranded DNA fragments are bound irreversibly to the filter which is then immersed in a solution containing a single-stranded nucleic acid probe. The probe forms double helical base-paired hybrid regions with filter-bound DNA of complementary sequence. Solution conditions of hybridization can be set up to distinguish exact from inexact matches. The hybridized probe is detected by directly exposing the filter to x-ray film, if the probe is radioactive, or by enzymatic staining, if the probe is labeled by chemical modification. In either case the contact print of the gel shows the mobilities of the DNA fragments complementary to the probe. The specificity of Watson-Crick base pairing in DNA allows single fragments to be detected in the midst of a large excess of noncomplementary DNA. This specificity is the basis for, among other techniques, genetic fingerprinting of individual human DNAs and the use of species-specific gene probes for detection of bacterial species.

Restriction Sites as Genetic Markers. In the early 1980s workers recognized that the presence of a restriction enzyme site is a chromosomal marker that can be assayed by southern blotting of genomic DNA. Thus, if the pattern of a southern blot is different for the DNA of different individuals in the population, and if one or more patterns cosegregate with a mutant gene causing a disease, the restriction pattern is a surrogate diagnostic marker for the disease. This phenomenon, termed restriction fragment length polymorphism (RFLP), can be used to predict an inheritance pattern in the absence of any other information about the disease other than its pattern of heredity.

Gene Isolation by Recombinant DNA Techniques

Workers in the early 1970s recognized that restriction enzymes provided tools not only for DNA mapping but also for construction of new DNA species not found in nature. A collection of recombinant DNA species consisting of many passenger sequences joined to identical vector molecules is called a library. Individual recombinant DNAs are isolated from single clones of the library for detailed analysis and manipulation.

Plasmid DNAs. Plasmids are nucleic acid molecules capable of intracellular extrachromosomal replication. Ultimately a plasmid is defined by its mode of DNA replication. DNA replication is initiated at a single, characteristic sequence, termed the origin. The origin sequence determines the copy number of the plasmid relative to the host chromosome and the host enzymes that are involved in plasmid replication.

Plasmids can be introduced into cells by several methods. The most common method is transformation, where the recipient cells are made competent to receive DNA by washing with a solution of Ca^{2+} or other inorganic ions. Then the naked DNA is added directly; a fraction of the cells take up the DNA and replicate it. Some but not all plasmids also transfer by conjugation, a sexual process where the DNA is donated from one cell to another after physical contact.

Most plasmids are topologically closed circles of DNA. They can be separated from the bulk of the chromosomal DNA by virtue of their resistance to denaturation in alkaline solution.

Plasmid Vectors for Facile Introduction of Passenger DNA and Selection of Recombinants. Three parts of a vector are key to its utility. The origin sequence allows the replication of plasmid DNA in high copy number relative to the chromosome. A gene encoding resistance to an antibiotic allows growth of plasmid-containing cells in the media. The third region of the plasmid allows the introduction of passenger DNA.

Construction of a Recombinant Plasmid by Joining Vector and Passenger DNA. The unique restriction sites in the plasmid vector DNA provide sites in the molecule for insertion of restriction-digested DNA fragments.

Identification of the Desired Passenger Sequence in Plasmid Cloning Experiments. The objective of recombinant DNA construction is to obtain a clone of a single DNA sequence. If more than a single restriction fragment is ligated into different vector molecules, the result is a library of clones, all of which have the same vector sequence but with different passengers. Libraries are often described by the source of the passenger DNA. Genomic DNA libraries contain the total chromosomal DNA inserted into a vector. Copy DNA (cDNA) libraries contain passenger DNA derived by copying messenger RNA (mRNA) into DNA using the enzyme RNA-dependent DNA polymerase. This process is known as reverse transcription.

The number of independent sequences in a DNA population is defined as its complexity. In a cloning experiment, the complexity of a library reflects the complexity of the passenger DNA population.

Given a library of sufficient complexity, it is then necessary to find the clone of interest against the background of recombinant clones containing other passenger sequences. Several strategies are employed. The simplest method is to use direct phenotypic selection of DNA capable of genetic complementation of a mutant in the host. Alternatively, genes encoding antibiotic resistance have been identified by direct phenotypic selection. The most common screening method uses a radioactively labeled probe to hybridize to DNA from the recombinant bacteria. In these experiments, DNA from bacterial colonies is denatured *in situ* for hybridization by a variation of southern blotting.

Vectors for Cloning Larger Fragments of DNA. Plasmid DNAs used in molecular cloning have a practical limit in the amount of DNA that can be inserted into them. When complex libraries are needed, for example to isolate a mammalian gene, other cloning strategies are needed. These strategies are based on the replication of the bacteriophage lambda.

Isolation of DNA for Phage Cloning. Because lambda-derived cloning vectors accept only a narrow size range of DNA inserts, a library constructed from completely restriction-digested DNA is unlikely to be representative of the total passenger DNA population. In order to construct representative libraries the passenger DNA population is partially digested using a frequently cutting, ie, 4 bp recognition sequence, restriction enzyme under conditions where the average size of the products is close to 20 kb. The passenger DNA fragments are then separated by agarose gel electrophoresis or by sucrose density gradient centrifugation to eliminate those smaller and larger fragments in the digestion products.

Screening of Recombinant Phage by DNA Hybridization or Antibody Recognition. In screening a recombinant phage library, the phage are plated at high density so that plaques are nearly contiguous. Then the plate is blotted with nitrocellulose or nylon filter paper. Dipping the filter into alkaline solution lyses the phage particles and denatures the DNA, making it ready for hybridization with a DNA probe. The phage from the hybridizing region of the plate are diluted to a low density, plated, and rescreened. After two or three screenings, a clonal population of recombinant phage are present. The passenger DNA can be analyzed by southern blotting, restriction mapping, and sequencing.

Cosmid Vectors. Whereas the amount of information required for growth of a phage is large, that required for replication and selection of a plasmid is much less. Addition of the cos sequences governing phage DNA packaging to a plasmid (hence the name cosmid) allows recombinant molecules of the appropriate size to be packaged into infectious particles.

Yeast Artificial Chromosomes for Insertion of Passenger DNA Molecules Larger than 10^5 bp. The sizes of the passenger DNA molecules inserted into cosmid and phage vectors are limited by the requirement for packaging of the recombinant DNA into the phage head. In addition, recombinant plasmids in *E. coli* are often unstable if these are larger than 50–100 kb. A system for constructing libraries from complex DNAs has been developed that overcomes many of these limitations. Reasoning that individual eukaryotic chromosomes are extremely large ($>10^6-10^7$ bp in most organisms) DNA molecules, a vector that provides the sequences necessary for faithful replication and segregation of an individual yeast chromosome to the recombinant was constructed. Passenger DNA inserted into these vectors is replicated by yeast as an extra, linear chromosome in the nucleus. In addition to a cloning site and sequences allowing the preparation of the vector from *E. coli*, these yeast artificial chromosome (YAC) vectors contain sequences required for chromosomal propagations.

A YAC library can represent all the sequences of the human genome (3×10^9 bp) with a 99% probability of finding an individual sequence in as few as 50,000 clones. Mapping and sequencing of complex genomes, as in the Human Genome Project, is expected to use YAC libraries preferentially.

Clones Linked Together to Form Larger Maps. Many eukaryotic genes are larger than the carrying capacity of a single phage or cosmid vector. Genomic DNA from eukaryotes contains noncoding sequences termed introns within the coding sequence; therefore, the DNA required to encode a protein can be much longer than that predicted from the size of the corresponding mRNA.

As a result of these considerations it is usually necessary to identify clones in the library that are linked to the first clone obtained by screening. The library is rescreened using a restriction fragment from one end of the passenger DNA segment in the first clone as a probe. The process, termed chromosome walking, may be reiterated several times until the contiguous inserts, termed *contigs*, cover the entire chromosomal region of interest. Complete contig maps have been determined for single chromosomes of several common laboratory organisms as a guide to genomic sequencing.

Experimental Protocols. In many cases, optimized reaction buffers, nucleotides, and enzymes are packaged in kits along with detailed protocols by the manufacturers. Beyond the information provided by reagent manufacturers, laboratories need a collection of experimental protocols. One such collection is distributed on a subscription basis and updated quarterly.

Analysis of DNA Sequences

After a desired clone is obtained and mapped with restriction enzymes, further analysis usually depends on the determination of its nucleotide sequence. The nucleotide sequence of a new gene often provides clues to its function and the structure of the gene product. Additionally, the DNA sequence of a gene provides a guidepost for further manipulation of the sequence, for example, leading to the production of a recombinant protein in bacteria.

The sequence of a gene predicts the sequence of the protein it encodes. The relationship between nucleotide sequence of a DNA or its mRNA and the amino acid sequence of the protein it encodes is given by the genetic code. Several strategies are available for identifying protein-coding regions. The frequency at which synonomous codons are used varies in different organisms. Sequences are searched using a database of codon frequency in the organism of interest to identify the most likely coding regions. In addition, a DNA sequence can be translated in all three reading frames. A database of known protein coding sequences is then searched using the predicted amino acid sequences as a query. Statistically significant homologies can provide a clue to the structure and function of the protein(s) encoded by the cloned DNA.

Determination of DNA Sequence Information. Almost all DNA sequence is determined by enzymatic methods which exploit the properties of the enzyme DNA polymerase.

Computer Analysis of DNA Sequence Information

The amount of information from a single DNA sequencing project can be staggering. Therefore, it is almost always necessary to analyze these data by computer methods. A number of commercial systems are available for analysis of DNA sequence information, operating on a variety of platforms, ranging from personal computers through workstations and supercomputers, depending on the intensity of the task. The field is not fully developed and research is ongoing in algorithm development, database manipulation, and network applications, among others.

Assembly and Analysis of the Results from a Sequencing Project. The size of a gene almost always exceeds the data available from a single DNA sequencing experiment. It is necessary to identify contiguous regions of sequence information and assemble these into completed projects. This can be done by relatively simple string matching, followed by highlighting points where two sets of overlapping information disagree. Resolution of discrepancies is then a matter of the investigator's judgment.

Uses of Sequence Information

DNA sequence information is the starting point for other applications, including the expression of a gene product, the search for related sequences in biological samples, *in vitro* mutagenesis of the sequence, and structure–function studies of gene expression.

Specific Amplification of Related Sequences by the Polymerase Chain Reaction. If the sequence of a gene is known, primers that are unique to the gene can be synthesized. An oligonucleotide longer than 15–18 residues is likely to be unique even in a complex genome. Such an oligonucleotide, if hybridized to a single-stranded DNA, can be used to prime DNA polymerase so that the DNA is replicated. If two primers are made, each complementary to one strand of a gene, then each strand of the DNA located between them can be specifically replicated by DNA polymerase. The newly replicated DNA strands can be separated by heating and can then serve as template for another round of primed synthesis, leading to another doubling in the concentration of the original amplified sequence. Because the concentration of the DNA of interest doubles with each cycle, at least a 50,000-fold increase in its concentration is achievable within a few hours. This polymerase chain reaction (PCR) is used in a variety of experimental manipulations and diagnostic procedures. Sequences less than a few hundred base pairs long are most efficiently amplified by PCR but even this relatively limited information can be fruitful.

Expression of Genes in a Heterologous Host

In many cases it is possible to synthesize the product of a gene in a different organism, eg, bacteria, yeast, or higher eukaryote. Recombinant DNAs directing the synthesis of the gene product must contain information specifying a number of biochemical processes.

Replication of the Recombinant DNA. In bacteria, replication of the recombinant DNA is provided by origin sequences, derived usually from plasmids indigenous to the host.

Selection of Recombinants. Selection of recombinants is provided either by a gene specifying antibiotic resistance or the ability to allow growth of recombinants in the absence of a particular nutrient.

Transcription of the Foreign Gene. Promoters are sequences preceding the start of transcription that direct RNA polymerase action. In general, these are specific to an organism and must be supplied, eg, when expressing a mammalian DNA sequence in bacteria.

Translation of the Foreign Gene. The translation of a mRNA into a protein is governed by the presence of appropriate initiation sequences that specify binding of the mRNA to the ribosome.

Stability and Purification of the Recombinant Protein. There are no hard and fast rules specifying, eg, whether a recombinant protein is available in a soluble state in the cell. In some cases, the expression system must be engineered by *in vitro* mutagenesis to optimize overall yield of the protein.

Mutagenesis of Cloned DNA

Mutational analysis of a cloned gene is often essential for identifying structure–function relationships in its expression or in the protein encoded by the cloned gene. Alternatively, expression of a recombinant protein is often dependent on the codon usage optimal for the host. A number of techniques are available for mutagenesis. Randomized treatment of DNA using a chemical mutagen continues to be useful. In addition, a short synthetic DNA can be made having specific or random mutations introduced during synthesis. This altered information can then be incorporated into the cloned gene. A few of the more general approaches are described herein. Other methods include linker-scanning mutagenesis and mutagenic PCR.

Protein Pharmaceutical Products

Development of recombinant proteins for pharmaceutical use has grown exponentially since 1982, when recombinant human insulin received approval from the United States Food and Drug Administration. Paralleling the development of small-molecule pharmaceuticals, previously approved proteins are being tested for new applications. Table 1 lists recombinant proteins approved by FDA as of August 1991. At that time, twice as many applications were pending and 10 times as many were in clinical trials.

Regulatory and Safety Issues

Safety regulations have been modified to recognize that no new hazards are created by recombinant DNA research, eg, DNA introduction does not make a pathogen out of a nonpathogen.

Table 1. Approved Biotechnology Drugs, 1991

Product	Company	Indication
human insulin	Lilly	diabetes (1982)
murine monoclonal antibodies	Ortho	reversal of acute kidney transplant rejection (1985)
interferons-alpha	Schering-Plough; Hoffmann-LaRoche; Interferon Sciences	hairy cell leukemia (1986); genital warts (1988, 1989); AIDS-related Kaposi's sarcoma (1988); non-A/non-B hepatitis (1991)
somatotropin	Genentech; Eli Lilly	human growth hormone deficiency in children (1985, 1987)
tissue plasminogen activator, alteplase	Genentech	acute myocardial infarction (1987); acute pulmonary embolism (1990)
erythropoietin	Amgen	anemia associated with chronic renal failure (1989)
hepatitis B vaccine	Merck	hepatitis B prevention (1989); interferon gamma (Genentech); management of chronic granulomatous disease (1990)
colony stimulating factors	Immunex; Hoechst-Roussel; Amgen	bone marrow transplantation (1991); treatment of chemotherapy-induced neutropenia (1991)

FRANCIS J. SCHMIDT
University of Missouri, Columbia

F. M. Ausubel and co-workers, eds., *Current Protocols in Molecular Biology*, Wiley-Interscience, New York.

J. Sambrook, E. F. Fritsch, and T. Maniatis, *Molecular Cloning: A Laboratory Manual*, Cold Spring Harbor Laboratory Press, Cold Spring Harbor, N.Y., 1989.

B. D. Davis, ed., *The Genetic Revolution*, Johns Hopkins University Press, Baltimore, Md., 1991.

ANIMALS

Transgenic Animals

A transgenic animal is an animal that has a modified gene inserted into its DNA. This modified or foreign gene is called a transgene. Transgenic animals are produced that either over or under express specific proteins within certain cells. This leads to animals having unique characteristics.

Agricultural uses of this technology include insertion of genes into farm animals for improved milk production, growth rate, and disease resistance. Biomedical uses of this technology include the development of lines of transgenic laboratory animals as experimental models for human diseases and the production of farm animals that produce recombinant pharmaceutical proteins in their milk. The production and use of transgenic laboratory and farm animals represents an evolving technology of engineering animal species for specific roles in science and agriculture.

Methods of Gene Transfer in Animals. Transfer of foreign genes into animals is done at an early stage of embryonic development (one cell to blastocyst stage) prior to implantation or placentation. Embryos at this stage of development can be grown outside the uterus of the mother (*in vitro* culture) in specialized medium containing nutrients that support their growth. For best results, micromanipulation and gene transfer are performed on one-cell embryos because integration of the transgene into the DNA of a one-cell embryo theoretically assures that all the cells of the adult animal carry the foreign gene. In practice, gene transfer into one-cell embryos does not always result in an adult that has the transgene in every cell. This is because the transgene may not actually integrate into the embryonic DNA until several cleavages of the embryo have occurred. This results in an adult animal that is genetically mosaic, harboring mixtures of cells having different sites of transgene integration or populations of transgenic and nontransgenic cells. Gene transfer into later stage embryos, ie, two-cell, four-cell, or greater, is less desirable because the different cells would not be expected to integrate the transgene equivalently. Therefore, genetic mosaicism of the adult is more likely than if single-cell embryos are used. Three methods of gene transfer are used to insert foreign DNA into the early embryo. They are microinjection, retroviral infection, and embryonic stem cells.

Expression of Foreign DNA in Transgenic Animals. Animals that carry a transgene do not have equivalent levels of expression of the foreign protein. In some animals the transgene can be detected in the DNA but no foreign protein is detected in the body. This suggests that the gene is nonfunctional, ie, not expressing mRNA and not producing protein. Other animals carry the transgene and express the foreign protein at much too high levels. This can cause abnormal development and pathological conditions.

Regulation of expression of the protein coding portion of the transgene should be under the control of the promotor region of the DNA construct. However, other factors, including the flanking DNA sequences at the site of transgene insertion, can influence the level of expression of the foreign DNA. Poor reproducibility of expression of foreign genes in transgenic animals may be partially corrected by dominant control regions. These DNA sequences can be inserted into DNA constructs to provide more consistent expression of proteins in transgenic animals.

Production of Multiple Animals From Single Embryos

One goal of animal agriculture is to exploit the superior genetics of animals. A technology that accomplishes this goal is artificial insemination. An average female can be mated to a superior male to produce above average progeny. Artificial insemination is so widely used in the dairy industry that very few bulls breed nearly all the dairy cows in the United States. It is much more difficult to exploit the genetics of superior female cattle because they produce few gametes (ova). Therefore, it is necessary to develop methods to mass produce identical copies of embryos that are collected from a superior female animal. Two methods are available to do this: embryo splitting and embryo cloning. These are illustrated in Figure 1.

Nucleic Acid Analysis of Individual Animals

Genes associated with milk production in dairy cattle and growth rate in meat-producing animals, ie, cattle, swine, and poultry, are of great economic importance. Most of these traits are believed to be quantitatively inherited. In other words, several (possibly 25 to 100) different genes are inherited and the additive effect of these genes determine the ultimate phenotype, ie, level of milk production or growth rate. The location of a gene that contributed to quantitative inheritance is known as a quantitative trait loci (QTL). If the genetic value of a gene is known, then animals at or before birth can be tested for their complement of genetic material. The genetic value of an animal, based on the additive inheritance of numerous genes, can then be calculated, and superior animals can be selected at an early age before time and money are invested. For this reason, finding the location of the genes that control production, as well as the alleles, ie, DNA sequences, for superior production, is a significant strategy for engineering improved animal species through biotechnology.

Two possibilities exist. First, although relatively few genes controlling quantitative traits have been discovered or completely understood, multiple alleles are found for some of these genes. Therefore, alleles associated with superior production can be identified and animals having the superior allele selected. Unfortunately the vast majority of genes that control production have not yet been discovered.

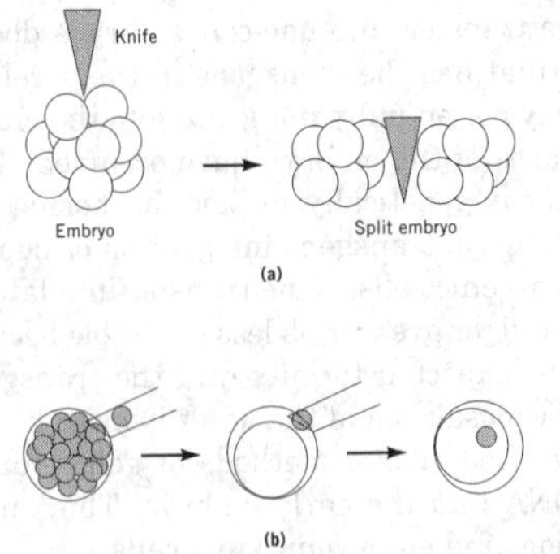

Figure 1. Methods for production of identical cattle. (**a**) Embryo splitting. A microdissection knife is used to cut an early embryo into two identical halves. (**b**) Embryo cloning. A cell from a morula stage embryo is removed and then inserted into an oocyte that has had its nucleus removed. Following fusion of the two cells the newly produced one-cell embryo is identical to the original morula. The procedure can be repeated to give multiple copies of the original morula.

Therefore, a second method of genetic selection can be performed without knowledge of the specific genes that control quantitative traits. For this method, genetic markers, polymorphic gene sequences with unknown function, located close to QTL, are used to select superior farm animals. The use of these methods, ie, either using markers located within known QTL or using those markers located near undiscovered QTL to find superior animals is known as marker assisted selection.

Genetic Analysis of Embryos

It is possible to select superior embryos during their initial development. An embryo is allowed to develop *in vitro* to the blastocyst stage. Several cells are dissected form the embryo for genetic analysis. This procedure is not harmful to the embryo. The DNA within these cells is analyzed by PCR. In the polymerase chain reaction (PCR), DNA is denatured into single strands and specific primers are allowed to anneal to the denatured DNA. The sequence of DNA between the primers is then filled in by a DNA polymerase enzyme. This process is repeated over and over until the sequence between the two primers is amplified nearly one billion-fold. Each genetic test involves different PCR primers and a different PCR reaction. At the end of the genetic analysis, only embryos with the desired genotype are removed from *in vitro* culture and transferred to recipient females. Embryos of the undesirable genotypes can be frozen indefinitely in liquid nitrogen or discarded. This process is used to test for and select embryo sex, milk protein genes and somatotropin genotype.

MATTHEW LUCY
University of Missouri, Columbia
ROBERT COLLIER
Monsanto Company

V. G. Pursel and co-workers, *Science* **244**, 1281 (1989).
J. M. Robl and N. L. First, *J. Reprod. Fertil. Suppl.* **33**, 101.
R. S. Prather, T. T. Stumpf, and L. F. Rickords, *Anim. Biotech.* **3**, 67 (1992).
M. R. Dentine, *Anim. Biotech.* **3**, 81 (1992).

MICROBES

Gene Technologies

The techniques involved in isolation and manipulation of genes from several microbes have become routine. Genes are cloned into plasmid or phage-based vectors and introduced into the microbe of choice. The number of vectors that have been engineered to be versatile is on the increase for several systems. Some vectors are commercially available for gene manipulations from several vendors. In addition, construction of gene libraries and screening are also performed by small biotechnology companies. Most of the reagents needed for gene manipulation such as oligonucleotides, peptides, and antibodies are commercially available. In addition, kits are also accessible for some of the routine tasks such as plasmid isolation, deoxyribonucleic acid (DNA) sequencing, etc.

Gene Transfer Methods. When bacteria are exposed to an electric field a number of physical and biochemical changes occur. The bacterial membrane becomes polarized at low electric field. The nature of the membrane disturbance is not clearly understood but bacteria, yeast, and fungi are capable of DNA uptake. This method, called electroporation, has been used to transform a variety of bacterial and yeast strains that are recalcitrant to other methods.

Gene Alteration and Amplification. A divergent set of technologies converged in the mid-1980s. These included efficient methods of oligonucleotides synthesis and the use of these oligonucleotides as primer *in vitro* to amplify DNA. A combination of these technologies resulted in two powerful methodologies, ie, site-directed mutagenesis and polymerase chain reaction (PCR) amplification.

Gene Expression

The methods involved in the production of proteins in microbes are those of gene expression. Several plasmids for expression of proteins having affinity tails at the C- or N-terminus of the protein have been developed. These tails are useful in the isolation of recombinant proteins. Most of these vectors are commercially available along with the reagents that are necessary for protein purification. A majority of recombinant proteins that have been attempted have been produced in *E. coli*. In most cases these recombinant proteins formed aggregates resulting in the formation of inclusion bodies. These inclusion bodies must be denatured and refolded to obtain active protein, and the affinity tails are useful in the purification of the protein. Some of the methods described herein involve identification of functional domains in proteins.

Display of Peptides and Proteins on Phages. The use of phages to display peptides and proteins is an extremely versatile method allowing rapid screening of large random libraries in a relatively short time for a specific function, eg, binding of a hormone to a receptor.

Gene Expression Systems. One of the potentials of genetic engineering of microbes is production of large amounts of recombinant proteins. This is not a trivial task. Each protein is unique and the stability of the protein varies depending on the host. Thus it is not feasible to have a single omnipotent microbial host for the production of all recombinant proteins. Rather, several microbial hosts have to be studied. Expression vectors have to be tailored to the microbe of choice. Hosts include *Escherichia coli, Erwinia, Pseudomonas, Bacillus* sp., *Clostridium, Streptomyces, Rhodococcus*, lactic acid bacteria, yeasts, and filamentous fungi.

Commercial Products from Genetically Engineered Microbes

Genetically engineered microbes are used to produce several commercial products. The number of products is likely to increase exponentially throughout the 1990s. They include therapeutics, bulk enzymes, antibiotics, chemicals, and crop protection products.

Future Prospects. One of the advances in the recombinant DNA technology is the rapidity with which most of the experiments can be performed. There has also been constant improvement in the quality of reagents. Automation in DNA sequencing should also have an impact on characterization of genes. Isolation of genes for a variety of enzymes having useful properties from thermophilic bacteria and methanogens should evolve. As rapid progress is made in this arena microbes are expected to be tailored to produce useful commercial products.

VASANTHA NAGARAJAN
E. I. du Pont de Nemours & Co., Inc.

M. A. Innis and co-workers, *PCR Protocols: Guide to Methods and Applications*, Academic Press, Inc., New York, 1990.

D. V. Goeddel, *Methods Enzymol.* **185** (1990).

M. Kobayashi, T. Nagasawas, and H. Yamada, *Trends. Biotechnol.* **10**, 402–408 (1992).

W. W. M. Adams, *Ann. Rev. Microbiol.* **47**, 627–658 (1993).

PLANTS

Several discoveries in the 1980s and 1990s permitted the transition of plant molecular biology from a fledgling science to commercial reality. These discoveries ranged from the identification of biologically important genes to the development of methods to introduce new genes into plants and regulate gene expression. The former process is commonly referred to as transformation. Nearly five dozen plant species have been transformed and the list of plant species subject to transformation include principal field crops such as corn, cotton, rape, rice, soybean, and wheat. In addition, several horticultural species such as tomato, potato, petunia, chrysanthemum, apple, walnut, melons, etc, have been subject to transformation. More than 500 field tests have been conducted and transgenic plants such as transgenic tomato, soybean, corn, rape, potato, petunia, melons, and cucumbers are in the advanced stages of commercial development and regulatory process.

Four methods have been extensively investigated for the introduction of transferred deoxyribonucleic acid (T-DNA) into plants. These include agrobacterium mediated T-DNA transfer, direct uptake of DNA by protoplasts, particle acceleration techniques such as electrostatic discharge or biolistics gun technology, and DNA uptake into partially digested immature embryos. By far the most commonly used method for gene introduction into dicotyledonous plants is the agrobacterium technology.

Expression of genes that have been introduced into plants is regulated by promoters, although the extent of regulation of gene activity by the promoter is influenced at least to some extent by the insertion site of the gene within the chromosome. The choice of promoters is dictated by the tissue and developmental specificity required for gene expression. By far the most commonly used promoter for constitutive gene expression in both mono- and dicotyledonous plants is the Cauliflower mosaic virus (CaMV) 35S promoter.

In order to determine which plants cells have been transformed, selectable marker genes are introduced during transformation. These marker genes permit selective growth of transgenic cells on the medium used for tissue propagation whereas the nontransgenic cells are killed.

Two specific applications of plant biotechnology include engineering tolerance to the widely used herbicide glyphosate, $C_3H_8NO_5P$, and increasing starch biosynthesis in plants. The first application deals with a trait which directly impacts the farmer during the production phase of agriculture; the second application deals with a trait that impacts the consumer of agricultural products. These traits may be referred to as agronomic and quality traits, respectively.

A number of other agronomic and quality traits are being investigated. These include insect, virus, disease, and nematode resistance, fertilizer-use efficiency, ripening control, fruit firmness, etc. Of these traits the most advanced agronomic trait for bioengineering is insect resistance. Insect resistant cotton and corn have been obtained by introduction and expression of a *Bacillus thurigiensis* kurastaki gene (BtK gene). Insect-resistant potato has been obtained by expression of a BtT gene which encodes a protein, selectively toxic to the Colorado potato beetle, a principal pest of potato. Virus resistance, conferred by expression of the viral coat protein (CP) gene in transgenic plants, has also received considerable attention. Products such as potato, squash, melons, etc, based on this technology are in advance stages of development and commercialization.

Both tomato fruit ripening and fruit firmness are among the advanced quality traits that are being investigated. A variety of approaches, based on inhibition of ethylene production are being pursued for enhancement of shelf life of tomato. For enhancing fruit firmness, cell wall hydrolytic enzymes such as polygalacturonidase and pectin methylesterase are being investigated.

JANICE W. EDWARDS
GANESH M. KISHORE
DAVID M. STARK
Monsanto Company

R. B. Horsch and co-workers, *Science* **223**, 496 (1984).

F. J. Perlak and D. A. Fischhoff, *Advances in Engineered Pesticides*, Marcel Dekker Inc., New York, 1993, pp. 199–211.

S. R. Padgette and co-workers, *J. Biol. Chem.* **266**, 22364 (1991).

D. M. Stark and co-workers, *Science* **258**, 287 (1992).

GEOTEXTILES

A geotextile, a subset of geosynthetics, is defined as any permeable textile material used with foundation soil, rock, earth, or any other geotechnical engineering-related material, as an integral part of a synthetic project, structure, or system.

The varied uses of geotextiles include drainage, dissimilar materials separation, erosion control, environmental protection, highway pavement rehabilitation, as component materials for geocomposites used in a variety of applications. Within each of these categories of use there exist numerous types of installations using the geotextiles.

Manufacturing Process

Polymers used in the manufacturing process in decreasing order of use include polypropylene, polyester, polyamide (nylon), polyethylene, and others to a much lesser extent. The fibers used in manufacturing the geotextiles are made by melt-spinning, ie, melting the polymer material and forcing it through a spinneret. Hardening of the fiber filaments is accomplished for the geotextile materials mostly through a process of simultaneous stretching and cooling of the fibers. The stretching of the fibers produces a more orderly arrangement of the molecules in the fibers, resulting in increased strength of the fiber. Monofilament fibers are further processed to form the various types of yarns used in the manufacturing of geotextiles. Multifilament yarn is formed by several monofilaments being twisted together to form the yarn. Staple fibers are formed from a rope-like bundle of monofilaments being crimped and cut into 2.5- to 10-cm (1- to 4-in.) lengths. Staple yarns are formed from these staple fibers by twisting into longer fibers (yarns). Another type of material used in geotextiles is called slit film. A continuous flat sheet of the polymer is cut into fibers by slitting with knives, or through the use of air jets. The resulting ribbons are slit-film fibers.

Geotextiles may be woven, nonwoven, or knitted. All types, woven, nonwoven, or knitted, are susceptible to degradation owing to the effects of ultraviolet light and water. Thus stabilizing agents are added to the base polymeric material to lessen the effects of exposure to ultraviolet light and water.

Woven and nonwoven geotextiles are the most common types presently in use in geotechnical engineering. A woven geotextile is formed through the conventional textile weaving process resulting in a screen-like or mesh material with various sizes of mesh openings, depending on the tightness of the weave.

Nonwoven Geotextiles. There are two basic manufacturing processes by which a nonwoven geotextile is produced: a one-stage or continuous process where the fibers are spun and bonded in one continuous process, and a two-stage process where fibers are laid down and then bonded. The most common types of bonding processes are the spun-bonded process, melt- or heat-bonded, resin-bonded, and needle-punched.

Physical Properties

The polymer used, the manufacturing process, and additives used in producing the geotextile determine the appearance and physical properties of the product. Properties may be classified as index properties or design and performance properties. Index properties provide a means of material differentiation, quality control in the manufacturing process, and quality assurance for the specifying agency. Design or performance properties define how the geotextile performs under specific installation conditions.

Economic Aspects

Geotextiles are a relatively new concept for solving problems in geotechnical engineering. They have gained wide use as not only an economical solution to these problems, but in many instances as the only viable solution to a complex engineering problem.

The installed price of geotextiles ranges from less than $1.20/m^2 to over $12/m^2 ($1–10/yd^2), depending on the geotextile and geographic location. The savings compared to what in the past were considered to be conventional solutions to geotechnical problems can range anywhere from a few thousand dollars per project to well over a million dollars, depending on the magnitude of the project.

L. DAVID SUITS
New York State Department of Transportation

R. M. Koerner, *Designing with Geosynthetics*, Prentice-Hall, Inc., Englewood Cliffs, N.J., 1986.

B. R. Christopher and R. D. Holtz, *Geotextile Design and Construction Guidelines*, Participant Notebook, rev. ed., National Highway Institute, Washington, D.C., 1992.

B. R. Christopher and R. D. Holtz, *Geotextile Engineering Manual*, Course Text, National Highway Institute, Washington, D.C., 1984.

A Design Primer: Geotextiles and Related Materials, Industrial Fabric Association International, St. Paul, Minn., 1990.

GEOTHERMAL ENERGY

Heat emanating from within the earth is one source of geothermal energy. This vast repository of energy is generated from the decay of natural radioisotopes and heat from the molten core of the earth.

Natural sources of geothermal fluids for heating and bathing have been utilized since prehistoric times. In the 1800s and 1900s applications of hydrothermal resources expanded widely to include space and district heating, agriculture, aquaculture, industrial processing, and, most recently, electric power generation (qv). Historically this energy was utilized by diverting surface hot water or steam sources. As technology progressed, wells were drilled to tap geothermal fluids more efficiently, and improvements in drilling technology have enabled access to and recovery of deeper and hotter fluids. In addition, development of techniques to extract geothermal heat from rock in which no natural mobile fluids exist is under way to bring this energy to the surface.

The useful applications of hydrothermal resources depend on the temperature of the extracted fluid. Figure 1 shows the distribution of thermal energy use in the United States as a function of temperature. It is clear that relatively low temperature fluids can be effectively applied for purposes such as greenhouse heating, fish farming, and especially space heating. Waters at higher temperatures can be used for a variety of industrial processes. All direct uses of geothermal energy require that the point of application be essentially co-located with the source of the hot water. Transportation of hot fluid over more than a few kilometers is impractical. The energy can, however, be converted to electricity. It is then possible to apply

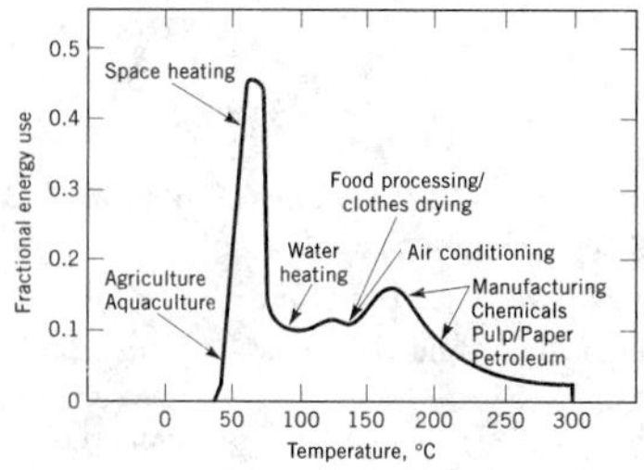

Figure 1. Thermal energy use vs temperature. Electricity generation is practical from thermal energy sources hotter than 150°C.

the power generated by hydrothermal energy in a variety of ways and at distant locations. The efficiency of electrical generation is directly related to the thermal quality of the resource. Using the most advanced power generating equipment, it may be economical to generate electricity from geothermal waters at temperatures as low as 150°C.

To successfully compete with the multitude of energy sources available, geothermal energy must be available and retrievable in both a convenient and an economical manner. As of this writing, these conditions have only been met using high grade geothermal resources in the form of hot water and steam and these particular hydrothermal resources are limited. Most of the world's accessible geothermal energy is found in rock that is hot but dry. Although research and development have demonstrated success in extracting thermal energy on a limited basis, the vast hot dry rock resource has not yet been shown to be an economically feasible source of energy on a scale large enough for practical use.

Geothermal Energy Resources

Type. Figure 2 is a generalized view of a cross-section of the earth indicating the various types of geothermal resources.

Magnitude. Whereas the total amount of energy stored within the earth is extremely large, only a very small fraction of that energy is accessible, in part because drilling and energy extraction costs escalate rapidly with depth. Commercial drilling can reach depths of about 9000 m, so all of the thermal energy in the earth to that depth can be considered part of the geothermal resource base. This resource base has been estimated to be on the order of 1×10^7 EJ (1×10^7 quads where 1 quad $= 10^{15}$ BTU) in the United States alone. One quad is equivalent to 3×10^7 m^3(182×10^6 bbl) of oil. The amount of geothermal energy estimated to exist according to resource type is

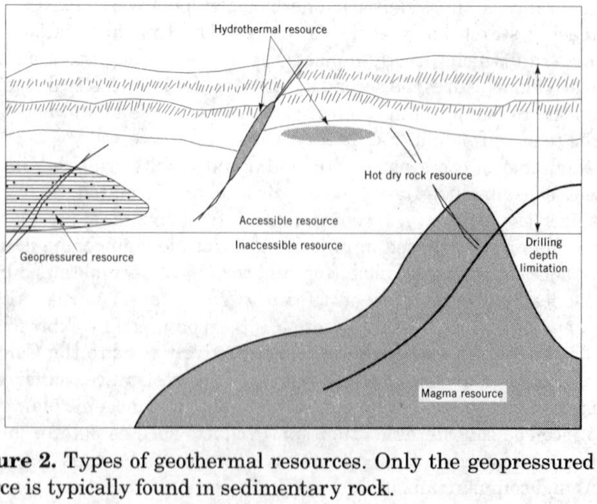

Figure 2. Types of geothermal resources. Only the geopressured resource is typically found in sedimentary rock.

Resource type	Accessible energy, EJ
hydrothermal	130,000
geopressured	540,000
hot dry rock	10,000,000
magma	500,000
Total	*11,170,000*

The total world consumption of energy in all forms is only about 300 EJ (300 quads); thus the earth's heat has the potential to supply all energy needs for the foreseeable future. Economic and environmental considerations, however, may preclude the utilization of all but a small part of this potential resource. Only a miniscule fraction of this energy supply has been tapped.

Hydrothermal Resources

Hydrothermal resources are characterized by the presence of heat relatively close to the earth's surface coincident with a trapped body of water to absorb that heat and provide a mechanism for its transfer to the surface. Hydrothermal resources occur, in general, throughout the world, in regions where continental plates meet, and the upwelling of magma and the earth's crust lead to rock temperatures that are higher than the worldwide average. Within these areas, the places where water is trapped in geologic formations are scattered and difficult to predict. Most of these areas have been identified by surface exposure of the fluid resource.

Temperatures of hydrothermal reservoirs vary widely, from aquifers that are only slightly warmer than the ambient surface temperature to those that are 300°C and hotter. The lower temperature resources are much more common. The value of a resource for thermal applications increases directly with its temperature, and in regions having hotter water more extensive use of geothermal resources has been implemented. Resources in remote areas often go unused unless hot enough to be employed in generating electricity.

Drilling and Field Development. The techniques for drilling hydrothermal wells have been adapted from those in use in the oil and gas industry (see GAS, NATURAL; PETROLEUM). Rotary drilling rigs are normally employed along with conventional drilling equipment such as steel casing, drilling lubricants, and casing cements.

Worldwide Hydrothermal Development. Electricity generation from hydrothermal resources is rapidly increasing in a number of developing countries. The relatively simple engineering, straightforward components, and ease of repair of hydrothermal plants make these plants ideal for application in nations with unsophisticated economies.

Economics. The cheapest electricity comes from The Geysers Geothermal Field in California. Electricity costs from hot water resources using flashed steam and binary plants, respectively, are progressively higher. Environmental concerns may add to the capital cost of a hydrothermal plant. Hydrogen sulfide abatement systems can run several million dollars. The exact costs vary with the composition of the gas to be treated. Liquid-dominated sources which are high is dissolved solids incur added capital and operating costs to pay for the collection and disposal of the spent brine and any precipitated solid residues.

The development of hydrothermal resources was given a strong push by a variety of governmental regulations instituted in response to the oil crisis of the mid-1970s. Perhaps the most significant law was the Public Utilities Regulatory Policies Act (PURPA), passed in 1978. The growth of the geothermal energy industry during the 1980s can be attributed in large part to such government policies.

Whereas there are significant known hydrothermal reserves and an estimated large amount of undiscovered hydrothermal energy, the future growth of the industry is tied closely to energy prices and environmental regulations. Hydrothermal energy utilized for the production of electricity is expected to remain exclusively a western U.S. resource.

Geopressured Resources

The Resource. Geopressured resources consist of highly overpressured mixtures of hydrocarbons, predominantly methane, and water, in sedimentary formations. The potentially useful energy in geopressured resources exists as three components: fossil chemical from the methane, heat from the water, and mechanical from the high pressure of the fluid. Geopressured resources are generally found very deep in the earth at levels of 3600 to 6000 m or more.

The wellbore can occur when the structure of the formation sand is disturbed by the turbulent flow, greatly reducing the energy production capacity of the well. The high salinity of geopressured water leads to a spent fluid disposal problem. The most common solution is to pump the saline water down a nearby well into a geological formation at a shallower depth than the geopressured resource. The formation of calcium carbonate scale creates significant operational difficulties in utilizing highly saline geopressured fluids. Scale inhibitors and the requirement for frequent removal of accumulated scale from piping and equipment can both add substantially to the maintenance cost of geopressured facilities. In one proposed power plant design, the mechanical power is first utilized in a pressure-reduction turbine, then the hydrocarbon and aqueous fluids are separated, and the water is fed to the heat exchanger of a binary power plant. The gas is used to produce electricity through conventional technology or sold directly to off-site users. Most of the work on developing techniques to exploit geopressured resources has been carried out under the auspices of the U.S. Department of Energy (DOE).

Economics. The cost of energy from geopressured resources is not competitive. The high costs can be related to a number of factors. The multiple energy forms, each effectively requiring its own generating plant, result in higher capital costs. Additionally, the salinity of the fluid leads to problems in corrosion and scaling as well as in disposal. Finally, the depth at which the resource exists means that drilling costs are high. Commercial utilization of geopressured resources for electricity production is not expected in the foreseeable future.

Direct Uses of Geopressured Fluids. Many of the use typical of hydrothermal energy, such as greenhouse, fishfarm, and space heating, have been proposed for geopressured resources, but none have been commercially developed.

Hot Dry Rock

The Resource. To utilize HDR resources, a practical means of accessing the hot rock and transporting its energy to the surface must be developed. In effect, an underground heat exchanger must be created to transfer the thermal energy of the rock to a mobile fluid. Because of the low thermal conductivity of hard rock, the surface area of the heat exchanger would have to be extremely large.

The Technology. The basic technique for extracting energy from HDR was conceived and patented in the early 1970s. It is based on drilling and hydraulic fracturing technologies developed in the petroleum and geothermal industries. The first step in constructing a heat mine is to drill a well into sufficiently hot and impervious rock, with the exact depth of the well to be determined by local heat-flow and other thermal conditions. Wells drilled for HDR applications are similar to many aspects to hydrothermal wells except that these wells are deeper and sometimes penetrate into a much greater depth of hard, crystalline rock. For this reason, specialized drilling and logging equipment which can withstand extended exposure to high temperatures is required. After the well has been completed, a segment of the bottom portion of the well is blocked off using a packer which provides pressure isolation. Water under high pressure is pumped through the packer and forced into joints in the surrounding rock body to form a reservoir consisting of a relatively small amount of water in a very large volume of rock.

To complete the system, a second well is drilled into the reservoir at some distance from the first, using a seismic map of the reservoir as a guide. In operation, water pumped down one well heats as it flows through the joints in the reservoir rock and returns to the surface through the second well, where its thermal energy is extracted using binary technology. The water can then be recycled.

Issues related to operation of a HDR geothermal energy system are the efficiency of energy extraction from the rock in the reservoir region; the impedance to flow as the water traverses the reservoir body; and the water losses resulting from leakage from the reservoir.

Economics. The costs of developing HDR resources are closely tied to the depth at which sufficiently hot rock is found. In the eastern U.S., it is generally necessary to drill much deeper to reach hot rock. Because drilling is the most expensive single factor in HDR developments, the first HDR electric plants are expected to be built in the western United States.

A number of studies have been conducted to assess the economics of producing electricity from HDR. The studies indicate that electric power from high grade HDR resources could be competitive in the 1990s, whereas that derived from medium grade resources is likely to be only marginally so. Because direct heating applications do not require temperatures as high as those needed for electricity production and do not suffer the losses inherent in conversion of thermal to electric energy, even lower grade HDR resources may prove to be competitive in site-specific space or industrial heating applications. No commercial HDR facility has been built as of this writing.

Geochemistry and Environmental Aspects. The environmental characteristics of HDR make it potentially the cleanest geothermal energy source and thus place it among the most promising of all energy resources. When operated as a closed loop, no significant amounts of air, water, or terrestrial pollutants are produced. Because the active reservoir is located thousands of feet below the water table, there is no danger of ground or surface water contamination. Finally, when a plant is decommissioned at the end of its useful life, the underground system could be permanently shut in by techniques already well known and proven in the oil, gas, and geothermal industries.

HDR Outside the U.S. The extraction of geothermal energy from HDR is being evaluated at a number of locations around the world. All work is based on the same general technical approach as that employed in the U.S. Exploratory HDR work has been done in Japan. An HDR has begun in Sweden, Switzerland, and Russia, and more extensive work is being done in England, France, and Germany.

Magma

The Resource. The core of the earth is generally believed to consist of molten rock known as magma. The energy content of the core is essentially boundless, but it is unreachable from a practical standpoint because it lies many kilometers below the surface. The technology to drill to those depths does not exist. In volcanically active regions, however, magma intrusions can be found relatively close to the surface in some localities. In these areas magma is a potentially useful geothermal resource, but relatively little is known about intrusive magma chambers.

The U.S. Geological Survey estimates that the total amount of magma energy existing within 10 km of the surface is on the order of 50,000–500,000 EJ in molten or partially molten magma. It has been estimated that only 2 km^3 of magma could provide enough energy to operate a 1000-MW electric power plant for 30 years.

Work on the extraction of useful energy from magma has been limited primarily to paperpaper and laboratory studies aimed at understanding the formation, extent, cooling, and other facets of magmatic bodies. Field drilling has been limited.

Economics. A cost estimate of power from magma has been developed for the California Energy Commission estimating that a 50-MW electricity-producing facility could produce power for 5.6 cents/kWh. Utilization of magma energy will take a concerted effort in resource identification and verification, drilling technology, and materials development.

DAVID DUCHANE
Los Alamos National Laboratory

R. DiPippo, *Geothermal Energy as a Source of Electricity*, U.S. Department of Energy, Washington, D.C., 1980.

J. Kestin, ed., *Sourcebook on the Production of Electricity from Geothermal Energy*, U.S. Department of Energy, Washington, D.C., 1980.

L. M. Edwards, G. V. Chilingar, H. H. Rieke III, and W. H. Fertl, eds., *Handbook of Geothermal Energy*, Gulf Publishing Co., Houston, Tex., 1982.

H. C. H. Armstead and J. W. Tester, *Heat Mining*, E. & F. N. Spon, London, 1987.

GERNANIOL. See TERPENOIDS.

GERANIUM OIL. See OILS, ESSENTIAL; PERFUMES.

GERMANIUM AND GERMANIUM COMPOUNDS

Germanium, Ge, at. no. 32, having electronic configuration [Ar] $3d^{10}4s^24p^2$, is a semiconducting metalloid element found in Group 14 (IVA), Period 4 of the Periodic Table. Although it looks like a metal, it has a diamond cubic crystal structure and is fragile like glass. Its electrical resistivity is about midway between that of metallic conductors and good electrical insulators. The first significant use was in solid-state electronics, and, using germanium, the transistor was invented. The entire modern field of semiconductors owes it development to the early successful use of germanium. Whereas germanium is still used in the field of electronics, its use in infrared optics surpassed electronic applications in the 1970s. Germanium has also found widespread use in the fields of γ-ray spectroscopy, catalysis, fiber optics, and photovoltaic cells.

Occurrence

The crust of the earth is estimated to contain Ge at a concentration of 1–7 g/t. Germanium, which usually occurs widely dispersed in minerals such as sphalerite, rarely occurs in concentrated form. Most germanium production has been from zinc smelters and copper smelters.

Germanium also occurs in significant concentrations in many coals around the world. When coal is burned in power-generating or coking plants, the germanium tends to concentrate in the fly ash or flue dust produced. Production from coal in the 1990s is reported in the former USSR and in China.

Properties

The physical, thermal, and electronic properties of germanium metal are shown in Table 1.

Chemical Properties

Germanium Metal. Germanium is quite stable in air up to 400°C, where slow oxidation begins. The metal resists concentrated hydrochloric acid, concentrated hydrofluoric acid, and concentrated sodium hydroxide solutions, even at their boiling points. Nitric acid attacks germanium at all temperatures more readily than does sulfuric acid. Germanium reacts readily with mixtures of nitric and hydrofluoric acids and with molten alkalies, and more slowly with aqua regia. In compounds, germanium can have a valence of either 2 or 4.

Germanium halides include germanium tetrachloride, $GeCl_4$, germanium tetrabromide, $GeBr_4$, germanium tetraiodide, GeI_4, and germanium tetrafluoride, GeF_4.

Germanium oxides include germanium dioxide, GeO_2, and germanium monoxide, GeO. Germanates include sodium heptagermanate, $Na_3HGe_7O_{16}\cdot 4\ H_2O$. Germanides include magnesium germanide, Mg_2Ge. Germanes include germane, GeH_4. Organogermanium compounds include spirogermanium (2-aza-8-germaspiro-[4,5]-decane-2-propanamine-8,8-diethyl-*N*,*N*-dimethyl dihydrochloride), $C_{17}H_{36}$-

Table 1. Properties of Germanium

Parameter	Value
atomic weight	72.59
density at 25°C, g/cm^3	5.323
Mohs' hardness	6.3
melting point, °C	937.4
boiling point, °C	2830
heat capacity at 25°C, J/(kg·K)	322
heat of combustion, J/g	7380
vapor pressure at 2080°C, kPa	1.33
coefficient of linear expansion, 10^{-6}/K	
at 100 K	2.3
at 300 K	6.0
thermal conductivity, W/(m·K)	
at 100 K	232
at 300 K	59.9
intrinsic resistivity at 25°C, Ω·cm	53
intrinsic conductivity type	N (negative)
band gap, direct at 25°C, minimum eV	0.67

$GeN_2 \cdot 2HCl$, and carboxyethyl germanium sesquioxide (3,3′-germanoic anhydride dipropanoic acid), $C_6H_{10}Ge_2O_7$.

Alloys. Many Ge alloys have been prepared and studied.

Manufacturing and Processing

Ore Processing. No mineral is mined solely for its germanium content. Almost all of the Ge recovered worldwide is a by-product of other mining. In the United States, zinc concentrates have been roasted and then sintered for zinc recovery. The sinter fume is chemically leached, and the germanium is selectively precipitated from the leach solution by fractional neutralization and sent to the germanium refinery.

Purification. Regardless of the source of Ge, all Ge concentrates are purified by similar techniques. The ease with which concentrated germanium oxides and germanates react with concentrated hydrochloric acid and the low boiling point (83.1°C) of the resulting $GeCl_4$ make chlorination a standard refining step. An oxidizing agent is often added to the primary distillation or to the subsequent fractionation, or both, to suppress the volatility of arsenic. Other purification steps are used to separate objectionable impurities present. The fractionation is usually done in glass or quartz because most subsequent uses of $GeCl_4$ require metallic impurity levels of no more than about 1 mg/kg (1ppm).

Economic Aspects

World reserves of germanium have been estimated at 4000 metric tons, but it is impossible to discuss reserves without considering price. In many applications, the cost of germanium is but a small part of the overall cost of the product. Thus substantially higher germanium prices would have little impact on such uses but would expand germanium reserves significantly.

Because germanium is almost always recovered as a by-product, its price and availability over the long range are subject to supply and demand considerations for its host products, usually zinc and copper, as well as for itself. This is not the case over the short term (6–12 months) because producers often recover germanium from stockpiles of smelter residues that can last for many months. Therefore, short-term pricing is largely controlled only by demand.

An annual review of germanium is published by the U.S. Bureau of Mines and a broader survey is published every five years.

Toxicology

Germanium compounds generally have a low order of toxicity. Only germane, GeH_4, is considered toxic, having a maximum time-weighted average 8-h exposure limit of only 0.2 ppm.

JACK H. ADAMS
DENNIS THOMAS
Eagle-Picher Industries, Inc.

F. Glocking, *The Chemistry of Germanium*, Academic Press, Inc., London, 1969.

M. Lesbre, P. Mazerolles, and J. Satgé, *The Organic Compounds of Germanium*, John Wiley & Sons, Ltd., London, 1971.

The Economics of Germanium, 6th ed., Roskill Information Services, Ltd., London, 1990.

V. A. Nazarenko, *Analytical Chemistry of Germanium*, trans. N. Mandel, John Wiley & Sons, Inc., New York, 1974.

GLASS

Common usage of the term *glass* follows the definition of G. W. Morey: "Glass is an inorganic substance in a condition which is continuous with, and analogous to, the liquid state of that substance, but which, as the result of a reversible change in viscosity during cooling, has attained so high a degree of viscosity as to be, for all practical purposes, rigid" (*The Properties of Glass*, 2nd ed., Reinhold, 1954, p. 28). Both organic and inorganic materials may form glasses if their structure is noncrystalline, ie, if they lack long-range order. This includes some plastics, metals, and organic liquids. In principle, rapid cooling could prevent crystallization of any substance if the final temperature is sufficiently low to prevent structural rearrangement. Thus glasses are formed primarily for kinetic reasons.

Glasses can be prepared by methods other than cooling from a liquid state, including solution evaporation, sintering of gels, reaction sputtering, vapor deposition, neutron bombardment, and shock-wave vitrification. These techniques suggest that the purely kinetic explanation of the glassy state is subject to question and that the previous definitions need modification.

Structure

The basic structural unit of silicate glasses is the silicon–oxygen tetrahedron in which a silicon atom is tetrahedrally coordinated to four surrounding oxygen atoms. Oxygens shared between two tetrahedra are called bridging oxygens. In pure vitreous silica, SiO_2, virtually all oxygens are bridging. Glass-forming systems other than silica have been examined. Both nmr and x-ray diffraction results led to the suggestion that the boroxyl ring is the structural unit of vitreous B_2O_3.

The random-network theory of glass was formulated in 1932. It proposes that atoms present in glass form a three-dimensional connected structure without periodic order and with energy content comparable to that of the corresponding crystalline material. According to this theory, the coordination number of an atom determines its role in a glass structure, and the following four rules should be fulfilled for an oxide to form glass: (*1*) each oxygen atom must be linked to no more than two cations; (*2*) the number of oxygen atoms around any one cation must be small, ie, three or four; (*3*) the oxygen polyhedra must share corners, not edges or faces, to form a three-dimensional network; and (*4*) at least three corners must be shared. For one-component glasses, each polyhedron shares corners with at least three other polyhedra in such a way that the network is continuous in three dimensions. In multicomponent glasses, additional cations are distributed throughout holes in the network.

Although x-ray structural work strongly supports the random network theory, there are aspects of the random-network theory that are often criticized. It is possible, for example, to form glasses when no three-dimensional network is possible. Furthermore, modifying cations have been shown to occur at regular interatomic distances ranging over several coordination shells.

Composition

Conditions favorable for glass formation may be deduced from either geometric or bond strength considerations. On the basis of the rules discussed above, the following oxides should be glass formers: B_2O_3, SiO_2, GeO_2, P_2O_5, As_2O_5, P_2O_3, As_2O_3, Sb_2O_3, V_2O_5, Sb_2O_5, Nb_2O_5, and Ta_2O_5. In fact, they are all so used. The only fluoride that fulfills the rules of glass formation is BeF_2, which readily forms a glass.

Glass formers generally have cation–oxygen bond strengths greater than 335 kJ/mol (80 kcal/mol). In multiple-component systems, oxides with lower bond strengths do not become part of the network and are called modifiers. Oxides with energies of ca 335 kJ/mol may or may not become part of the network and are referred to as intermediates. The dissociation energies used to predict glass formation are calculated, taking into account the coordination number of the cation. In multiple-component glasses, the terms formers, modifiers, and intermediates are frequently used to define the role of the individual oxides. However, an element such as lead may be either a modifier or intermediate, depending on its coordination and the glass system considered.

Glass formation of individual oxides can be predicted from the melting point, and individual bond energies can be normalized by dividing by the melting point of the oxide. This ratio is relevant because the melting point is related to the amount of thermal energy available to rupture bonds. If the bond energy is large and the melting point low, glass formation is favored.

Other correlations of glass formation and properties have been offered. For example: (*1*) cation valence should be three or greater; (*2*) glass formation should increase with decreasing cation size; and (*3*) the Pauling electronegativity should be between 1.5 and 2.1. Using these criteria, four types of oxides are described: (*1*) strong glass formers such as Si, B, Ge, As, and P; (*2*) intermediate formers that require rapid cooling, such as Sb, V, W, Mo, and Te; (*3*) oxides that form glasses in binary mixtures with nonglass formers, such as Al, Ga, Ti, Ta, Nb, and Bi; and (*4*) oxides that do not form glasses.

Glass composition work starts with the application of structural and bonding rules of glass formation. Numerous ternary systems and their glass-forming regions have been investigated. There are three types of ternaries: Type A, single former and two modifiers; Type B, two formers and one modifier; and Type C, three glass formers.

Single-Phase Glasses

Single-phase glasses include vitreous silica, multicomponent silicate systems, soda–lime glasses, borosilicate glasses, aluminosilicate glasses, lead glasses, borate glasses, phosphate glasses, chalcogenide glasses, halide glasses, and metallic glasses.

Glass Ceramics and Phase-Separated Glasses

Glass is a good medium for controlled crystallization, and has become the basis for a number of unique crystalline materials known as glass-ceramics (qv) (Table 1). The separation of a single glass into multiple glassy phases can make the article cloudy and adversely affect its chemical durability. Controlled phase separation, however, can produce opaque, white opal glass or, after a leaching step, lead to new materials such as porous glass or 96% silica glass. The colloidal suspension of multiple phases in transparent glass produces precise colors for products such as optical filters. Furthermore, photosensitive and photochromic glasses change their optical transmission and color, sometimes reversibly, upon stimulus by a combination of light and heat treatment. All these transformations generally depend on the phenomena of diffusion, nucleation, and growth.

Table 1. Commercial Glass-Ceramics

Commercial designation	Principal crystalline phases	Properties	Application
Schott Zerodur	β-quartz solid solution, SiO_2	zero expansion	electric range tops, telescope mirrors
Corning 9632[a]	β-quartz	low expansion, high strength, thermal stability, chemical durability	electric range tops (transparent)
Corning 8603	lithium metasilicate, $Li_2O{\cdot}SiO_2$; lithium disilicate $Li_2O{\cdot}2SiO_2$	photochemically machinable	fluid amplifiers
Corning 9608	β-spodumene solid solution, $Li_2O{\cdot}Al_2O_3{\cdot}(SiO_2)_{4-10}$	low expansion, high chemical durability	cooking utensils
Corning 9617[a]	β-spodumene solid solution; anatase, TiO_2	low expansion, high strength, thermal stability, chemical durability	electric range tops (opaque)
Corning 0330[b]	β-spodumene solid solution; rutile TiO_2	mechanical, chemical, thermal stability	exterior, interior cladding, laboratory bench tops
Corning 9455	β-spodumene solid solution; mullite, $3Al_2O_3{\cdot}2SiO_2$	low expansion, high thermal and mechanical stability	heat exchangers
Neoceram (Japan)	β-spodumene	low expansion	cooking ware
Corning 9606	cordierite, $2MgO{\cdot}2Al_2O_3{\cdot}5SiO_2$; spinel, $MgO{\cdot}Al_2O_3$; MgO-stuffed β-quartz; quartz, SiO_2	low expansion, high transparency to radar	missile radomes
Corning 9625[c]	α-quartz solid solution, SiO_2; spinel, $MgO{\cdot}Al_2O_3$; enstatite, $MgO{\cdot}SiO_2$	very high strength	classified
Corning 9658	fluorphlogopite solid solution, $KMg_3AlSi_3O_{10}F_2$; mullite, $3Al_2O_3{\cdot}2SiO_2$	machinable, high dielectric strength, thermal stability	precision dielectric components, insulators, high vacuum components
High K (Corning)	$(Ba,Sr,Pb)Nb_2O_6$	high dielectric constant	capacitors

[a] Surface strengthening by differential crystallization.
[b] Surface strengthening by $Na^+ \Leftrightarrow Li^+$ ion exchange.
[c] Surface strengthening by 2 $Li^+ \Leftrightarrow Mg^{2+}$ ion exchange.

Properties

Rheological. The viscosity of a glass determines its melting, forming, and annealing procedures as well as the limitations of its use at high temperature. Viscosity is ordinarily measured between 10^{13} and 10 Pa·s (10^{14} and 100 P); at room temperature it is greater than 10^{19} Pa·s (10^{20} P). Properties are shown in Table 2. The effect of modifiers on the viscosity of glass at high temperature depends on their polarizability or ionic field strength. Low field-strength modifiers decrease the viscosity of silica more than high field-strength ones. At low temperature, the effects of a modifier on viscosity are largely controlled by its coordination number. Modifiers with higher coordination numbers tend to increase low temperature viscosity as a result of packing restraints.

Glass is usually melted and fined at viscosities between 5 and 50 Pa·s (50–500 P) but forming and final viscosity requirements vary greatly.

Viscosities of glass are compared qualitatively. A hard glass has a high softening point and a soft glass has a lower softening point. Long

Table 2. Properties of Glass Fibers

Property	Value for				
	E-glass	T-glass	C-glass	SF-glass	S-glass
thermal expansion × $10^7/°C$[a]	60	80	72	75	34
strain point,[b] °C	507		435		760
annealing point,[b] °C	657		585		810
softening point,[b] °C	846	715	752	675	970
density, g/cm^3	2.60	2.54	2.61	2.57	
Young's modulus, GPa	74		70		87
dielectric constant, 1 MHz	6.4	7.3	7.8	8.3	
refractive index,[c] 1 MHz	1.548	1.541	1.549	1.537	1.523

[a] At 0–300°C.
[b] These data subject to normal manufacturing variations.
[c] Refractive index may be either the sodium yellow line (589.3 nm) or the helium line (587.6 nm). Values at these wavelengths do not vary in the first three places beyond the decimal point.

and short refers to the temperature difference between the softening and strain point of glass. A long glass has a large temperature difference between its softening and strain point, ie, it solidifies (sets up) slower than a short glass as temperature decreases.

Thermal. The thermal expansion of a glass determines the range of materials to which it can be safely sealed. It also affects the ability of the glass to survive thermal shock or cycling. The usefulness of glass as a heat exchanger or a thermal barrier, and its ease of melting and forming, depend on its heat-transfer properties and emissivity. The upper use temperature is a function of all of these properties.

Stresses caused by steady-state thermal gradients may or may not cause failure, depending on the degree of constraint imposed by some parts of the item upon others or by the external mounting. Thus, under minimum constraint and maximum uniformity of gradient through the thickness, very large temperature differences can be tolerated.

Heat Transmission. At room temperature, the thermal conductivity of glasses ranges from 0.67 to 1.21 W/(m·K), with the most common compositions near the upper end of the range. Thermal conductivity of glass-ceramics ranges from 1.7 to 3.8 W/(m·K). Above 500°C, conductivity increases rapidly because of radiation within the glass.

Liquidus Temperature. The liquidus temperature determines the susceptibility of a glass to devitrification and therefore influences its forming limitations and often its heat-treating requirements.

Gas Permeation. Gas permeation through glass is of crucial importance to any high vacuum system. With the advent of high intensity tungsten–halogen lamps, the lamp envelopes must contain gas under pressure for a long time at high temperatures. The permeation rate varies with temperature, glass type and composition, and gas. Helium is the most mobile of the gases and silica is the most permeable glass. The addition of a modifier blocks the openings in the glass network and slows the permeation rate.

Mechanical. Because of its amorphous structure, glass is brittle, reasonably abrasion resistant, and nearly perfectly elastic as long as its temperature is low enough to prevent viscous flow. Among other things, glasses do not contain crystallographic planes which slip relative to each other and therefore permit the material to deform plastically when stress is applied. Several steps are commonly taken to make the glass more break-resistant. The elimination of surface flaws, eg, microcracks or bruise checks, either by careful handling during and after forming, annealing, etc, or by acid etching the surface to remove these flaws, prevents a crack from starting. Secondary treatments, eg, ion exchange, thermal tempering, glazing, or differential crystallization, provide a compressive stress on the surface of the glass that must be overcome before the applied tensile stress causes breakage.

Theoretical calculation places the intrinsic strength of glass as high as 35 GPa (5×10^6 psi); the strength of flame-polished silica rods is 14 MPa (2.0×10^6 psi). However, the practical strength of glass is a small fraction of these figures because of stress concentrations introduced by surface imperfections.

Glasses, like other brittle materials, deform elastically until they break in direct proportion to the applied stress. The Young's modulus E is the constant of proportionality between the applied stress and the resulting strain. It is about 70 GPa (10^7 psi) [(0.07 MPa stress per μm/m strain = (0.07 MPa·m)/μm)] for a typical glass. A high modulus glass is one in which the atoms are closer together or the bonds are stronger and therefore more resistant to rupture.

On the Mohs' scale of scratch hardness, glass lies between apatite, 5, and quartz, 7.

Electrical. Glasses are used in the electrical and electronic industries as insulators, lamp envelopes, cathode ray tubes, and encapsulators and protectors for microcircuit components, etc. Besides their ability to seal to metals and other glasses and to hold a vacuum and resist chemical attack, their electrical properties can be tailored to meet a wide range of needs. Generally, a glass has a high electrical resistivity, a high resistance to dielectric breakdown, and a low power factor and dielectric loss.

The properties of some glasses and glass-ceramics make them more suitable for this application than others. The relative suitability of a glass of glass-ceramic material is expressed by a factor of merit which is proportional to the material's strength and inversely proportional to its Young's modulus, expansion coefficient, and rate of microwave energy absorption.

Chemical Stability. The resistance of glass to chemical corrosion is frequently the reason for its use. However, the durability of a glass varies from highly soluble to highly durable, depending on its composition and the solvent considered.

Optical. Optical glasses are usually described in terms of their refractive index at the sodium D line (589.3 nm), and their v value (or Abbe number) which is a measure of the dispersion or variation of index with wavelength, ie, $v = (n_D - 1)/(n_F - n_C)$, in which n_F is the refractive index at the hydrogen F line (486.1 nm) and n_C is the refractive index at the hydrogen C line (656.3 nm). These data have been incorporated into a six-digit numbering system used to identify optical glasses. Glasses with index n_D less than 1.60 and a v value of 55 or above are defined as crown glasses; those with a v value below about 50 are defined as flint glasses. Glasses with n_D greater than 1.60 are defined as crown glasses if v is 50 or above. Crowns are usually alkali silicate glasses and flints are lead alkali silicates. Phosphate and borate-based glasses are used to obtain high index or low dispersion glasses or both.

Reflectance is dependent on the condition of the glass surface. It can be specular, as in polished or precision molded surfaces, or diffuse for ground or irregularly etched surfaces.

The spectral transmission of glass is determined by reflection at the glass surfaces and the optical absorption within the glass. Overall transmission of a flat sample at a particular wavelength is equal to $(1 - R)^2 e^{-\beta t}$, where β is the absorption coefficient, t the thickness of glass, and R the air—glass reflection coefficient. Transmission is a function of wavelength.

Transmission may be controlled by the types of glass used, eg, silicate, phosphate, borate, etc, or by the control or addition of coloring additives, melting atmosphere, melting temperatures, and cooling schedules. In glasses containing suspended particles, transmission is diffuse.

The stress optical coefficient, B, or Brewster's constant, is a measure of this proportionality, and varies greatly with glass composition. Values of B range from about −1 for extra dense flints containing 80% PbO to ca 3.5 for 96% silica glass. For soda–lime glass, B is about 2.5.

Interaction of glass with low energy visible and uv radiation may result in alteration of the electronic states. These changes may cause coloration or luminescence effects, eg, photochromic or photosensitive glass. When changes in color are produced by sunlight, the effect is frequently referred to as solarization. Effects produced by ions, gamma rays, and x-rays are relevant to glasses used as electronic tube envelopes, radiation-shielding windows, and dosimeters. The stability

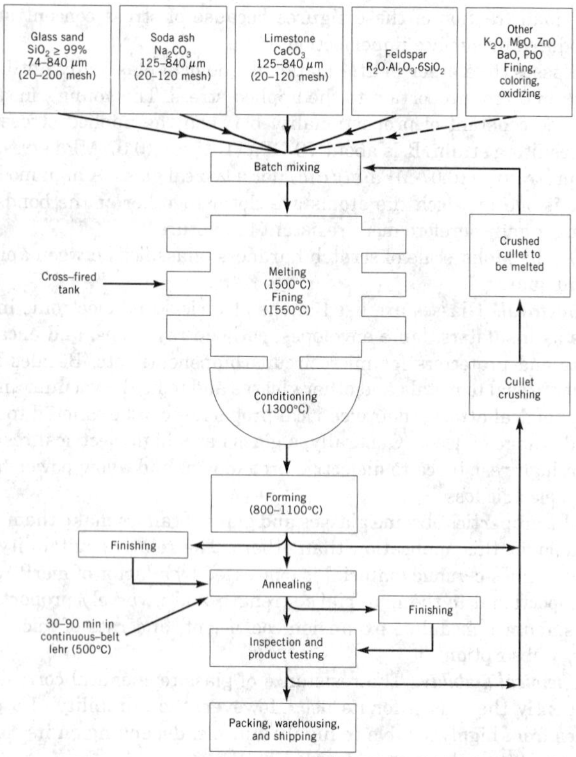

Figure 1. Glass manufacture. Temperatures are for common soda–lime glass. Other glasses may require appreciably different temperatures.

of glass in neutron fields depends on the relative absence of boron in the glass composition.

Manufacture and Processing

Most glass articles are manufactured by a process in which raw materials are converted at high temperatures to a homogeneous melt that is then formed into the articles. The flow diagram in Figure 1 summarizes the details of glass container manufacturing. The vapor deposition of SiO_2 from a flame fed with $SiCl_4$ and oxygen is the basis for manufacturing high purity glass used for blanks which are redrawn into optical fibers. Fused silica items that cannot be formed from viscous melts of SiO_2 or quartz are prepared by vapor deposition.

Forming. Molten glass is either molded, drawn, rolled, or quenched, depending on desired shape and use. Bottles, dishes, optical lenses, television picture tubes, etc, are formed by blowing, pressing, casting, and/or spinning the glass against a mold to cool it and to set its final shape. Window glass, tubing, rods, and fiber are formed by drawing the glass in air (or on a bath of molten tin as in the float process) until it sets up and can be cut to length. Art glass is usually hand-formed by blowing and shaping it while soft. Glass that is intended to be crushed into powder, called frit, is quenched between water-cooled rollers or ladled or poured directly into water (dry-gauged) and then dried.

Secondary Operations. In secondary forming, a piece of preformed glass is reheated and reworked into the finished product.

Glass can be cut by five methods. (*1*) Flame cut-off is accomplished with pinpointed flames which heat the glass until it is soft enough to separate. (*2*) Sawing is done with band, wire, or circular blades used in conjunction with a loose or bonded abrasive. (*3*) To score-break glass, the piece is scored with a tool such as a file, a diamond, or a steel or carbide wheel, and then bent to apply tension to the scored area. (*4*) Score-thermal crack-off is done by applying heat opposite a mechanically produced score on the glass surface. (*5*) Localized heating with pinpointed burners followed by rapid chilling, usually with water, accomplishes a thermal crack-off.

Glass is drilled with carbide or bonded-diamond drills under a suitable coolant such as water or kerosene.

Glass is ground with sand, garnet, corundum, silicon carbide, boron carbide, or diamond.

This process is similar to grinding, but the polishing compound, usually cerium oxide, zirconium oxide, or ferric oxide, is finer. The polishing tool may be plastic, cellulose, felt, or pitch.

Treatment of glass surface with a chemical may alter its strength, appearance, or durability.

Quality control is governed by the uses for which a glass is designated. Generally, each manufacturer has strength and abrasion standards; a notable exception is ophthalmic lenses which the FDA specifies must withstand a 15.88 mm-dia steel ball dropped from 1.27 m.

Methods of cleaning glass are closely associated with the types of glass and its particular use. Aqueous solvents frequently used range form chromic–sulfuric acid mixtures for cleaning borosilicate laboratory ware to common detergent solutions.

Other Methods. Special applications require nonstandard methods of manufacture. Vapor deposition is principally used in the manufacture of high quality glass such as optical fibers and optical mirrors.

Bulk-fused silica is commercially produced by a variety of techniques, including vapor deposition.

Glass Films. Glass films are formed by both reactive and nonreactive deposition methods. Examples of nonreactive deposition methods are evaporating, sputtering, and ion-implantation or ion-plating. glass films are used in the semiconductor industry because of their dielectric properties, and they are used for encapsulating integrated circuits and other electronic devices because they provide a reasonable hermetic seal.

Reactive sputtering is a process in which highly active metal atoms react with a gas before their deposition on cold substrate.

Thermal oxidation is one of the oldest and most commonly used methods of forming a primary passiviting film of SiO_2 on silicon. A glass layer can also be made by anodic oxidation on a metal or semiconductor surface, such as silicon, by making it the anode in an electrolytic cell, immersing this anode in a suitable electrolyte (often aqueous), and passing a current through it.

Economic Aspects

Glass manufacture is classified according to the product into flat, container, fiber, or specialty glass.

Flat Glass. In the United States the main producers of flat glass are PPG Industries, Libbey-Owens-Ford (LOF), Guardian Industries, Ford Motor Company, and AFG Industries. Growth of this industry segment depends on the construction and automotive markets, 57 and 25% of the market, respectively. The float process produces more than 85% of all flat glass. The remainder is thin sheet for picture glass or rolled and patterned glass. A small but growing market is that for thin sheet glass for liquid crystal displays. Production of flat glass is cyclical but increases in volume by approximately 2–3% per year. Since 1990, the value of exports has exceeded that of imports; the balance is strongly influenced by the variations in current exchange rates.

Container Glass. Statistics for food, beverage, drug and cosmetic, and household and industrial containers are compiled and published by the Glass Packaging Institute (GPI). About 75% of the containers are narrow neck. Approximately 85% of container glass is clear; the remainder is mostly amber. Other tints make up a small percentage of the total. In the United States, principal producers are Owens-Illinois, Anchor, and Kerr glass companies.

Fiber Glass. Fiber glass is classified as either textile or wool. Many companies produce textile fibers for draperies, upholstery, tires, reinforced plastics, paper, and tape, including Manville/Schuller, Owens-Corning Fiberglass, PPG Industries, and Nicofibers. Glass wool is primarily used for building insulation, industrial equipment and pipe insulation. Growth of the insulation market has been faster than the textile area because of the demand for additional insulation in both new and existing buildings.

Specialty Glass. The pressed-and-blown or hollow-ware industry is comprised of over one hundred companies in the United States, including Corning Glass Works, Owens-Illinois, General Electric, and Anchor Hocking. The wide variety of products is divided into categories of pressed-and-blown glass for table, kitchen, art, and novelty applications and products of purchased glass. The latter consists of items for scientific, technical, and industrial uses such as electrical and electronic products, laboratory glassware, optical and ophthalmic glass, etc.

DAVID C. BOYD
PAUL S. DANIELSON
DAVID A. THOMPSON
Corning Incorporated

G. W. Morey, *The Properties of Glass*, 2nd ed., Reinhold Publishing Corp., New York, 1954, p. 28.

R. H. Doremus, *Glass Science*, John Wiley & Sons, Inc., New York, 1973.

E. B. Shand, *Glass Engineering Handbook*, 2nd ed., McGraw-Hill, Book Co., Inc., New York, 1958.

F. V. Tooley, ed., *The Handbook of Glass Manufacture*, Ashlee Publishing Co., New York, 1984.

GLASS-CERAMICS

Glass-ceramics are polycrystalline materials formed by the controlled crystallization of glass. Most commercial glass-ceramic products are formed by highly automated glass-forming processes and converted to a crystalline product by the proper heat treatment. Glass-ceramics can also be prepared via powder processing methods in which glass frits are sintered and crystallized. The range of potential glass-ceramic compositions is therefore extremely broad, requiring only the ability to form a glass and control its crystallization.

Glass-ceramics can provide significant advantages over conventional glass or ceramic materials, by combining the ease and flexibility of forming and inspection of glass with improved and often unique physical properties in the glass-ceramic. They possess highly uniform microstructures, with crystal sizes on the order of 10 micrometers or less; this homogeneity ensures that their physical properties are highly reproducible.

Unlike conventional ceramic materials, glass-ceramics are fully densified with zero porosity. They generally are at least 50% crystalline by volume and often are greater than 90% crystalline.

More than $500 million in glass-ceramic products are sold yearly worldwide. These range from transparent, zero-expansion materials with excellent optical properties and thermal shock resistance to jade-like highly crystalline materials with excellent strength and toughness. The highest volume is in cookware and tableware, architectural cladding, and stovetops and stove windows. Glass-ceramics are also referred to as Pyrocerams, vitrocerams, devitrocerams, sitalls, slagceramics, melt-formed ceramics, and devitrifying frits.

Design

There are three key variables in the design of a glass-ceramic. The glass composition controls much of the glass workability as well as its nucleation and crystallization properties. The glass-ceramic phase assemblage is responsible for many of its physical and chemical properties, such as thermal expansion, hardness, and electrical characteristics. The crystalline microstructure is the key to many mechanical and optical properties, such as strength, toughness,and transparency.

Properties

Thermal Properties. Many commercial glass-ceramics have capitalized on their superior thermal properties, particularly low or zero thermal expansion coupled with high thermal stability and thermal shock resistance.

Mechanical Properties. Because of the nature of the crystalline microstructure, strength, elasticity, toughness (resistance to fracture propagation), and abrasion resistance are higher in glass-ceramics than in glass.

Optical Properties. Glass-ceramics may be either opaque or transparent. When the crystals are much smaller than the wavelength of light as in some mullite and spinel glass-ceramics, or when the crystals have low birefringence and the index of refraction is closely matched, as in some Mg-stuffed β-quartz glass-ceramics, excellent transparency can be achieved.

Chemical Properties. Generally, highly siliceous glass-ceramics with low alkali residual glasses, such as glass-ceramics based on β-quartz and β-spodumene, have excellent chemical durability and corrosion resistance similar to that obtained in borosilicate glasses.

Electrical Properties. In general, glass-ceramics have such high resistivities that they are used as insulators.

Glass-Ceramic Families

All commercial as well as most experimental glass-ceramics are based on silicate bulk glass compositions. Glass-ceramics can be further classified by the composition of their primary crystalline phases, which may consist of silicates, oxides, phosphates, or borates. The principal commercial glass-ceramics are based on silicate crystals.

New and Potential Applications.

Glass-ceramics will find numerous applications in the rapidly growing field of optoelectronics. Glass-ceramics will also play a role in the burgeoning field of information storage and display. Finally, glass-ceramics will play a key role in the growing arsenal of advanced materials, both alone and in combinations with other materials.

LINDA R. PINCKNEY
Corning Incorporated

G. H. Beall, *Ann. Rev. Mater. Sci.* **22**, 91–119 (1992).

P. W. McMillan, *Glass-Ceramics*, 2nd ed., Academic Press, Inc., New York, 1979.

Z. Strnad, *Glass-Ceramic Materials, Glass Science and Technology*, Vol. 8, Elsevier, Amsterdam, 1986.

GLASSES, ORGANIC–INORGANIC HYBRIDS

The search for new high performance materials has spurred the development of composites combining high modulus/high thermal stability inorganic glasses and low modulus/low thermal stability polymeric glasses. Research has resulted in a novel class of amorphous polymer–glass composites referred to as organic–inorganic hybrids or inorganic–organic hybrids, depending on the component with the highest volume fraction. These materials are synthesized in a variety of ways but ultimately exhibit near-molecular-level mixing of the matrix and the filler. Hence the term hybrid. Typically, this high degree of mixing results in transparent materials which exhibit significant increases in thermomechanical properties owing to extensive interaction between the polymeric and inorganic phases. However, the relatively high volume fraction of polymer included in these materials normally limits their service temperatures to well below 400°C.

Organic–Inorganic Glass Systems

Many polymers and oligomers have been utilized in the synthesis of these hybrids, including poly(alkenes, acrylates, ethers, esters, amides, imides, and dienes). Some novel inorganic precursors have been used, such as clay. The use of metal alkoxides including methoxides, ethoxides, isopropoxides, as well as phenyl and other organically substituted derivatives of silicon, titanium, zirconium, and aluminum, have been reported. The research concerning the sol–gel processing of silica and its gel-to-glass transition far outweighs that done on the other alkoxides.

Organic–inorganic glass systems include those based on polydimethyl siloxane, poly(tetramethylene oxide), nitrile rubber, nylon-6, poly(ethylene oxide), poly(ethyloxazoline), poly(2-methyl-2-oxazoline), poly(vinyl acetate), poly(*p*-phenylene vinylene), poly(*N*-vinylpyrrolidinone), polymethacrylates [poly(methyl methacrylate), poly(*n*-butyl methacrylate)], poly(arylene ether) ketone, polyimide, hydrogenated polybutadiene, and sodium poly(4-styrene sulfonate).

Applications

There are several areas of application for organic–inorganic hybrids: microelectronic usage, abrasion-resistant coatings, thermal–oxidative-resistant coatings, nonlinear optical devices, hydrogels, biosensors, and cross-linking agents.

A. B. Brennan
T. M. Miller
University of Florida, Gainesville

A. B. Brennan, B. Wang, D. E. Rodrigues, and G. L. Wilkes, *J. Inorgan. Organomet. Polym.* **1**, 167–187 (1991).

Y. Chujo and T. Saegusa, *Advances in Polymer Science* **100**, 11–29 (1992).

C. J. T. Landry and co-workers, *Polymer* **33**, 1496–1506 (1992).

B. M. Novak and M. W. Ellsworth, *Mater. Sci. Eng.* **A162**, 257–264 (1993).

GLASSY METALS

Metallic glasses do not occur naturally but are produced by various techniques, the oldest of which is by cooling molten metals so rapidly that the atoms do not get to form regular crystalline structures. The atoms are frozen in random or nonrepeating atomic patterns similar to those found in organic glasses. By metallic it is meant that the amorphous material is composed primarily, but not necessarily exclusively, of metallic elements that exhibit the properties of metals, such as electrical or magnetic behavior.

Primarily through modification of the chemistries, glasses have been produced that can truly be considered bulk, even exceeding 1 cm in all dimensions. Warm extrusion and consolidation of metallic glass powders produce blocks of glassy metals that can later be machined into components. Besides rapid solidification into thin ribbons or flakes, there exists a wide range of techniques to produce metallic glass. Methods available include chemical means, mechanical alloying, vaporization, and solid-state reactions.

Interest is maintained in these materials because of the combination of mechanical, corrosion, electric, and magnetic properties. However, it is their ferro-magnetic properties that lead to the principal application of glassy metals. The soft magnetic properties and remarkably low coercivity offer tremendous opportunities for this application.

A limitation of metallic glass is that it exists in metastable form, which means that it tends to crystallize if heated with sufficient thermal energy to allow the kinetics of crystallization, ie, both nucleation and growth, to occur. If glassy metal alloys were all intrinsically unstable, however, they would be much less promising as an engineering material. Understanding the solid-state structure, the unusual mechanical properties, the liquid-like electrical properties, and the ferromagnetic properties of metallic glasses has been the focus of research and development efforts since the late 1950s.

Examples of the various categories of metallic glasses are given in Table 1.

Table 1. Metallic Glasses

Alloy	Useful properties/applications
$Fe_{78}B_{13}Si_9$	Metglas 2605S2,[a] good magnetic properties
$Fe_{80}B_{20}$	Metglas 2605[a]
$Pd_{80}Si_{20}$	easy to form glass, thick samples produced
$Pt_{60}Ni_{15}P_{25}$	
$Cu_{84.3}P_{15.7}$ eutectic	brazing foil
$Al_{85}Ni_5Fe_2Gd_8$	high strength low density thick ribbons
$Mg_{80}Cu_{15}Sn_5$	low density cast 4-mm dia rods
$Co_{83}Gd_{17}$	sputtered sample can support magnetic bubbles
$Mo_{64}Re_{16}P_{10}B_{10}$	superconducting below 8.7 K
$W_{60}Ir_{20}B_{20}$	crystallization temperature above 1200 K
$Zr_{41.2}Ti_{13.8}Cu_{12.5}Ni_{10}Be_{22.5}$	14-mm rod produced

[a] Produced by AlliedSignal.

Formability

To form an amorphous, ie, glassy metal alloy from the liquid state means that the crystallization step must be avoided during solidification. This can be understood by considering a time–temperature–transformation (TTT) diagram. Nucleating phases require an incubation time to assemble atoms through a statistical process into the correct crystal structure which is capable of surmounting an activation barrier ΔG^*. Incubation times can vary from fractions of a second to many seconds. the shape of the TTT curve is in the form of a *C* because of competing phenomena. As temperature is lowered, the free energy available to nucleate and grow a crystalline phase increases but the kinetic ability to do so through atomic diffusion decreases, resulting in a nose at T'. The glass-forming ability of a material is determined by the kinetics of this process, followed by the initial stages of crystal growth. If the liquid alloy is cooled from above the melting temperature to a temperature below the nose of the TTT curve at T' within a time less than the time where crystallization begins, the alloy's liquid-like structure becomes frozen in when the temperature drops below the glass-forming temperature, T_g, whereby an amorphous solid is obtained. Once the alloy is below T_g, diffusion processes are such that growth of a crystalline embryo is essentially halted, and the liquid structure is preserved.

A reduced glass temperature can be defined as $T_{rg} = T_g/T_l$, which represents a measure of glass-forming ability. The higher the T_g and the lower the liquidus temperature, T_l, the easier it is to supercool the metal melt to a glassy state. Conventional theory predicts T_{rg} = 0.65–0.7 for good glass formers.

Processing

Traditionally, production of metallic glasses requires rapid heat removal from the material which normally involves a combination of a cooling process that has a high heat-transfer coefficient at the interface of the liquid and quenching medium, and a thin cross section in at least one-dimension. Besides rapid cooling, a variety of techniques are available to produce metallic glasses. Processes not dependent on rapid solidification include plastic deformation, mechanical alloying, and diffusional transformations.

Splat quenching or gun techniques involve rapid solidification through atomizing molten metal by blowing it out a tube. The resulting liquid vapor is quenched by impingement upon a metal substrate having a high thermal coefficient. Melt-spinning can produce large quantities of very uniform ribbons, filaments, or tapes.

Lasers can be used to obtain very fast quench rates up to 10^8 K/s.

Metallic glass powders can be made in various sizes through atomization and comminution processes.

Ion implantation has a large (10^{14} K/s) effective quench rate. This surface treatment technique allows a wide variety of atomic species to be introduced into the surface. Sputtering and evaporation methods are other very slow approaches to making amorphous films, atom by atom.

Investigations have been carried out on the formation of amorphous phase through solid-state reactions, typically by the use of mechanical alloying and interdiffusion between thin films.

Crystallization

Metallic glasses are metastable. A combination of thermal energy and time leads to crystallization. At room temperature, this may require a very long time but at moderate temperatures of 373 K or more (depending on the alloy) devitrification can occur in minutes. Usually the crystallization temperature is given as the temperature at which crystallization begins as the alloy is heated at a constant rate. Crystallization leads to an increase in density of about 1%. In general, T_x, the crystallization temperature, is somewhere between 0.4 and 0.65 T_m where T_m is the melting temperature.

Crystallization, by definition, implies that the initial structure be a glass, followed by the nucleation and growth of a crystalline phase, be it the equilibrium one or a metastable phase. The process is a first-order transformation and involves atomic diffusion, or at least atomic shuffles. Types of crystallization reactions that occur include polymorphous crystallization, which is a composition invariant transformation such as that in Fe–B, and eutectic crystallization, T_e, in FeNiPB glass, where fine lamellae of iron–nickel austenite and metastable $(FeNi)_3$ PB phases grow cooperatively.

Crystallization need not always be deleterious to properties such as strength. In Al-based glasses, partially crystallized material actually increased the fracture strength by 30%.

Mechanical Properties

Of the various physical properties, it is the mechanical properties that make metallic glasses so unique when compared to their crystalline counterparts. A metallic glass obtains its mechanical strength in quite a different way from crystalline alloys. The disordered atomic structure increases the resistance to flow in metallic glasses so that these materials approach their theoretical strength. An attractive feature is that metallic glasses are equally strong in all directions because of the random order of their atomic structure.

The ductility of glassy metals varies according to the kind of stress applied. Owing to the large intrinsic ductility, metallic glasses can be plastically deformed into useful shapes at no loss of mechanical strength.

The fracture of metallic glasses occurs by the formation of shear bands having a 45° orientation relative to the tensile axis. Typical veining on the fracture surface is a characteristic of almost all metallic glasses.

The deleterious embrittlement of a glassy metal during annealing is also accompanied by a change in fracture mode. Almost all metallic glasses containing Fe, Co, and Al show this behavior.

At low applied stress levels crystalline alloys display fatigue properties superior to metallic glasses. At large loads approaching the fracture strength, metallic glasses can survive many more cycles than comparable strength crystalline alloys. This is because the metallic glass deforms elastically up to the high stress levels, whereas crystalline alloys generate localized regions of plastic deformation which lead to nucleation of fatigue cracks.

Properties

Chemical. Along with magnetic and mechanical behavior, high corrosion resistance is one of the most desirable properties of metallic glasses.

Magnetic. More experimental research and development work has been done on magnetic effects in metallic glasses than on any other property (Table 2). It is their soft magnetism and microstructural homogeneity that has led to the most significant applications.

Electrical. Unlike their crystalline counterparts, amorphous metals generally have high electrical resistivity not only at room temperature, but also, because of a very small temperature coefficient, near absolute zero. Certain metallic glasses, eg, $La_{80}Au_{20}$, do show superconductivity.

Table 2. Magnetic Properties

Alloy	Coercive field, H_c, A/m[a]	Curie temp, θ_c,°C
$Fe_{80}B_{20}$[b]	3.1	374
$Fe_{80}P_{16}B_1C_3$[c]	4.0	292
$Fe_3Co_{72}P_{16}B_6Al_3$	1.2	260
$Fe_{96.8}Si_{3.2}$[d]	20–40	730

[a] To convert A/m to oersted, multiply by 0.0126. [b] Metglas 2605. [c] Metglas 2615. [d] Material is crystalline.

Uses

The magnetic properties of glassy metals provide the only commercial use in bulk quantities, although brazing foils provide another niche for metallic glasses. Metallic glasses have yet to find their way into commercial products in structural applications, in spite of the great strength of some glasses. This is related to their shearing instability and size limitations of the glassy product which requires that it employ a fabrication step to obtain bulk material. This situation is changing, however; currently, major breakthroughs are being made in producing bulk metallic glasses by conventional techniques.

GARY J. SHIFLET
University of Virginia

A. L. Greer, *Science*, **267**, 1947–1953 (Mar. 31, 1995).

GLYCEROL

Glycerol, propane-1,2,3-triol, glycerin (USP), a trihydric alcohol, is a clear, water-white, viscous, sweet-tasting hygroscopic liquid at ordinary room temperatures above its melting point. Glycerol occurs naturally in combined form as glycerides in all animal and vegetable fats and oils, and is recovered as a by-product when these oils are saponified in the process of manufacturing soap, when the fats are split in the production of fatty acids, or when fats are esterified with methanol in the production of methyl esters. Since 1949 it has also been produced commercially by synthesis from propylene. The latter currently accounts for ca 30% of United States production.

The uses of glycerol number in the thousands, with large amounts going into the manufacture of drugs, cosmetics, toothpastes, urethane foam, synthetic resins, and ester gums. Tobacco processing and foods also consume large amounts either as glycerol or glycerides.

Properties

Physical properties of glycerol are shown in Table 1. Glycerol is completely soluble in water and alcohol, slightly soluble in diethyl ether, ethyl acetate, and dioxane, and insoluble in hydrocarbons.

Glycerol, the simplest trihydric alcohol, forms esters, ethers, halides, amines, aldehydes, and such unsaturated compounds as acrolein. As an alcohol, glycerol also has the ability to form salts such as sodium glyceroxide.

Grades. Several grades of refined glycerol are marketed; specifications vary depending on the consumer and the intended use. USP-grade glycerol is water-white and suitable for use in foods, pharmaceuticals, and cosmetics, or where the highest quality is demanded. Food Kosher glycerin meets all USP requirements and is produced synthetically or from 100% vegetable glycerides.

Economic Aspects

Commercial production and consumption of glycerol has generally been considered a fair barometer of industrial activity, as it enters

Table 1. Physical Properties of Glycerol

Property	Value
mp, °C	18.17
bp at 101.3 kPa,[a] °C	290
sp gr, 25/25°C, 100% glycerol in air	1.2620
n_D^{20}	1.47399
surface tension at 20°C, mN/m(= dyn/cm)	63.4
viscosity at 20°C, mPa·s(= cP)	1499

[a] To convert kPa to mm Hg, multiply by 7.5.

into such a large number of industrial processes. It generally tends to rise in periods of prosperity and fall in recession times.

Handling and Storage

Most crude glycerol is shipped to refiner in standard tank cars or tank wagons. Imported crude arrives in bulk, in vessels equipped with tanks for such shipment or in drums. Refined glycerol of a CP or USP grade is shipped mainly in bulk in tank cars or tank wagons.

Health and Safety Factors

Glycerol, since 1959, is generally recognized as safe (GRAS) as a miscellaneous or general-purpose food additive under the CFR, and it is permitted in certain food packaging materials.

The aquatic toxicity (TLm96) for glycerol is > 1000 mg/L, which is defined by NIOSH as an insignificant hazard.

Derivatives

Glycerol derivatives include acetals, amines, esters, and ethers. Of these the esters are the most widely employed. Alkyd resins are esters of glycerol and phthalic anhydride. Glyceryl trinitrate (nitroglycerin) is used in explosives and as a heart stimulant. Included among the esters also are the ester gums (rosin acid ester of glycerol), mono- and diglycerides (glycerol esterified with fatty acids or glycerol trans-esterified with oils), used as emulsifiers and in shortenings. The salts of glycerophosphoric acid are used medicinally.

Acetins. The acetins are the mono-, di-, and triacetates of glycerol that form when glycerol is heated with acetic acid.

LOWEN R. MORRISON
Procter & Gamble

Physical Properties of Glycerin and Its Solutions, Glycerin Producers' Association, New York, 1975.

E. Woollatt, *The Manufacture of Soaps, Other Detergents and Glycerin*; John Wiley & Sons, Inc., New York, 1985, pp. 296–357.

SDA Glycerin and Oleochemicals Statistics Report, The Soap and Detergent Association, New York, 1992.

GLYCOLS

ETHYLENE GLYCOL AND OLIGOMERS

Glycols are diols, compounds containing two hydroxyl groups attached to separate carbon atoms. In an aliphatic chain, ethylene glycol, the adduct of water and ethylene oxide, is the simplest glycol and is the principal topic of this article. Diethylene, triethylene, and tetraethylene glycols are oligomers of ethylene glycol. Polyglycols are higher molecular weight adducts of ethylene oxide and are distinguished by intervening ether linkages in the hydrocarbon chain. These polyglycols are commercially important; their properties are significantly affected by molecular weight. They are water soluble, hygroscopic, and undergo reactions common to the lower weight glycols.

The uses for ethylene glycol are numerous. Some of the applications are polyester resins for fiber, PET containers, and film applications; all-weather automotive antifreeze and coolants, defrosting and deicing aircraft; heat-transfer solutions for coolants for gas compressors, heating, ventilating, and air-conditioning systems; water-based formulations such as adhesives, latex paints, and asphalt emulsions; manufacture of capacitors; and unsaturated polyester resins. The oligomers also have excellent water solubility but are less hygroscopic and have somewhat different solvent properties. The number of repeating ether linkages controls the influence of the hydroxyl groups on the physical properties of a particular glycol.

Physical Properties

Ethylene glycol and its lower polyglycols are colorless, odorless, high boiling, hygroscopic liquids completely miscible with water and many organic liquids. Physical properties of ethylene glycols are listed in Table 1. Ethylene glycols markedly reduce the freezing point of water.

Chemical Properties

The hydroxyl groups on glycols undergo the usual alcohol chemistry, giving a wide variety of possible derivatives. Hydroxyls can be converted to aldehydes, alkyl halides, amides, amines, azides, carboxylic acids, ethers, mercaptans, nitrate esters, nitriles, nitrite esters, organic esters, peroxides, phosphate esters, and sulfate esters. The largest commercial use of ethylene glycol is its reaction with dicarboxylic acids to form linear polyesters.

Manufacture

In 1937 the first commercial application of the Lefort direct ethylene oxidation to ethylene oxide followed by hydrolysis of ethylene oxide became, and remains in the 1990s, the main commercial source of ethylene glycol production.

Economic Aspects

During the 1980s through the early 1990s glycol prices varied widely owing to capacity variations and supply and demand imbalances. The current North American ethylene glycol capacity estimate (ca 1994) is 3.5×10^6 t. Although accurate figures are not available, worldwide ethylene glycol nameplate capacity is estimated at 8.8×10^6 metric tons.

Health, Safety, and Environmental Factors

Laboratory biochemical oxygen demand (BOD) tests using unacclimated biomass show that ethylene glycol is readily biodegraded in a system which attempts to simulate the dilute biological conditions of a river and lake. Ethylene glycol can be treated effectively in conventional wastewater treatment plants and does not persist in the environment under expected conditions.

None of the glycols are highly irritating to the mucous membranes or skin. A splash of these neat materials in the eye may produce marked irritation, but permanent damage should not be expected.

Table 1. Properties of Glycols

Property	Ethylene glycol	Diethylene glycol	Triethylene glycol	Tetraethylene glycol
mol wt	62.07	106.12	150.17	194.23
sp gr, 20/20°C	1.1155	1.1185	1.1255	1.1247
bp at 101.3 kPa,[a] °C	197.6	245.8	288	dec
mp, °C	−13.0	−6.5	−4.3	−4.1
viscosity at 20°C, mPa·s(= cP)	20.9	36	49	61.9

[a] To convert kPa to mm Hg, multiply by 7.5.

Ethylene glycol is not recommended for use as an ingredient in food or beverages, or where there is a significant contact with food or potable water.

There is no evidence of genetic toxicity for ethylene, diethylene, and triethylene glycols from a battery of *in vitro* tests. Tetraethylene glycol is not believed to be mutagenic.

Derivatives

In addition to oligomers ethylene glycol derivative classes include monoethers, diethers, esters, acetals, and ketals as well as numerous other organic and organometallic molecules. These derivatives can be of ethylene glycol, diethylene glycol, or higher glycols and are commonly made with either the parent glycol or with sequential addition of ethylene oxide to a glycol alcohol, or carboxylic acid forming the required number of ethylene glycol subunits.

Ethylene glycol monoethers are commercially manufactured by reaction of an alcohol with ethylene oxide.

Glycol monoethers are widely used in cleaning formulations. They are used as jet fuel additives to inhibit icing in fuel systems and as solvents and cosolvents for conventional solvent-based lacquer, enamel, and wood stain industrial coating systems as well as cosolvents for waterborne industrial coating systems.

Ethylene glycol diethers are made by derivatizing both glycol hydroxyl groups. Strong solvating and stability properties allow numerous applications for glycol diethers including adhesives and coatings, ink formulations, cleaning compounds, batteries, electronics, polymer solvents, polymer plasticizers, bold refining, and gas purification.

Cyclic polyethers, or crown ethers, are cyclic structures containing ethylene glycol units as $\leftarrow CH_2CH_2O \rightarrow_n$ with $n > 2$ and generally between 4 and 8. Crown ethers greatly improve the solubility of salts in organic solvents and have a wide range of applications in organic reactions, especially as phase-transfer reagents.

Glycols can be used in the manufacture of poly(ethylene glycol) (PEG). PEG find applications in agriculture products, ceramics, chemical intermediates, coatings and adhesives, cosmetics and toiletries, electronics, foods and feeds, household products, lubricants, metal processing, mining, paper, petroleum, pharmaceuticals, photography, plastics and resins, printing, rubber chemicals, textiles and leather, and wood products.

Ethylene glycol (as well as the higher glycols) can be esterified with traditional reagents such as acids, acid chlorides, acid anhydrides, and via transesterification with other esters. Low molecular weight glycol esters are good solvents for cellulose esters and printing inks, and are employed in industrial extraction processes and the protective coating industry. Fatty acid esters, together with other surfactants, are good emulsifying, stabilizing, dispersing, wetting, foaming, and suspending agents.

Ethylene glycol in the presence of an acid catalyst readily reacts with aldehydes and ketones to form cyclic acetals and ketals. Cyclic acetals and ketals are used as protecting groups for reaction-sensitive aldehydes and ketones in natural product synthesis and pharmaceuticals.

Diethylene Glycol. Physical properties of diethylene glycol are listed in Table 1. Manufacture of unsaturated polyester resins and polyols for polyurethanes consumes 45% of the diethylene glycol. Approximately 14% is blended into antifreeze. It is also converted to morpholine used in natural gas dehydration and in such applications as plasticizers for paper, fiber finishes, compatibilizers for dye and printing ink components, latex paint antifreeze, and lubricants in a number of applications.

Triethylene Glycol. Physical properties of triethylene glycol are listed in Table 1.

Triethylene glycol is an efficient hygroscopicity agent with low volatility, and about 45% is used as a liquid drying agent for natural gas. It is also used as a solvent and as a vinyl plasticizer.

Tetraethylene Glycol. Physical properties of tetraethylene glycol are listed in Table 1.

Tetraethylene glycol is used in moisturizing and plasticizing cork, adhesives, and other substances. It may be used directly as a plasticizer or modified by esterification with fatty acids to produce plasticizers. Tetraethylene glycol also has found application in the separation of aromatic hydrocarbons form nonaromatic hydrocarbons (BTX extraction).

M. W. Forkner
J. H. Robson
W. M. Snellings
Union Carbide Corporation

G. O. Curme and F. Johnston, *Glycols*, ACE Monograph No. 114, Reinhold Publishing Corp., New York, 1952, Chapt. 2.

J. March, *Advanced Organic Chemistry*, 4th ed., John Wiley & Sons, Inc., New York, 1992.

NTP Technical Report on the Toxicology and Carcinogenesis Studies of Ethylene Glycol (CAS No. 107-21-1) in B6C3F1 Mice (Feed Studies), NIH Publication No. 91-3144, U.S. Department of Health and Human Services, Washington, D.C., Public Health Service, 1991.

PROPYLENE GLYCOLS

The propylene glycol family of chemical compounds consists of monopropylene glycol (PG), dipropylene glycol (DPG), and tripropylene glycol (TPG). These chemicals are manufactured as copoducts and are used commercially in a large variety of applications. They are available as highly purified products which meet well-defined manufacturing and sales specifications. All commercial production is via the hydrolysis of propylene oxide.

The propylene glycols are clear, viscous, colorless liquids that have very little odor, a slightly bittersweet taste, and low vapor pressures. The most important member of the family is monopropylene glycol, also known as 1,2-propylene glycol, 1,2-dihydroxypropane, 1,2-propanediol, methylene glycol, and methyl glycol. All of the glycols are totally miscible with water.

Propylene glycol, when produced according to the U.S. Food and Drug Administration good manufacturing practice guidelines at a registered facility, meets the requirements of the U.S. Food, Drug, and Cosmetic Act as amended under Food Additive Regulation CFR Title 21, Parts 170–199. It is listed in the regulation as a direct additive for specified foods and is classified as generally recognized as safe (GRAS). In addition, it meets the requirements of the *Food Chemicals Codex* and the specifications of the *U.S. Pharmacopeia XXIII*. Because of its low human toxicity and desirable formulation properties it has been an important ingredient for years in food, cosmetic, and pharmaceutical products.

Chemistry

Monopropylene glycol (1,2-propanediol) is a difunctional alcohol with both a primary and a secondary hydroxyl.

1,2-Propylene glycol undergoes most of the typical alcohol reactions, such as reaction with a free acid, acyl halide, or acid anhydride to form an ester; reaction with alkali metal hydroxide to form metal salts; and reaction with aldehydes or ketones to form acetals and ketals. The most important commercial application of propylene glycol is in the manufacture of polyesters by reaction with a dibasic or polybasic acid. Polyethers are also products of commercial importance.

Stereochemical and Structural Isomers

Propylene glycol, dipropylene glycol, and tripropylene glycol all have several isomeric forms. Propylene glycol has one asymmetric carbon and thus there are two enantiomers: (*R*)-1,2-propanediol and (*S*)-1,2-propanediol. 1,3-Propanediol is a structural isomer. Dipropylene glycol exists in three structural forms and since each structural isomer has two asymmetric carbons there are four possible stereochemical isomers per structure or a total of twelve isomers. These twelve consist of four enantiomer pairs and two meso- compounds. Tripropylene glycol has four structural isomers and each structural isomer has three

Table 1. Properties of Glycols

Physical properties	Propylene glycol	Dipropylene glycol	Tripropylene glycol
formula	$C_3H_8O_2$	$C_6H_{14}O_3$	$C_9H_{20}O_4$
molecular weight	76.1	134.2	192.3
boiling point at 101.3 kPa,[a] °C	187.4	232.2[b]	265.1[b]
vapor pressure, kPa,[a] 25°C	0.017	0.0021	0.0003
density at 25°C, g/mL	1.032	1.022	1.019
viscosity at 25°C, mPa·s(= cP)	48.6	75.0	57.2

[a] To convert kPa to mm Hg, multiply by 7.5.
[b] Varies with isomer distribution.

asymmetric carbons so each structural isomer has eight possible stereochemical isomers or a total of 32 isomers of tripropylene glycol.

Physical and Chemical Properties

Table 1 lists various physical and chemical properties and constants for the propylene glycols.

Consumption and Use

Consumption of propylene glycol follows an erratic pattern in the United States which reflects domestic economic conditions. Principal uses of propylene glycol in the United States are in unsaturated polyester resins; cosmetics, pharmaceuticals, foods; pet food; tobacco humectant; functional fluids; paints and coatings; liquid detergents; and others.

Dipropylene Glycol. Dipropylene glycol is similar to the other glycols in general properties, and its fields of use are comparable. The greater solvency of dipropylene glycol for castor oil indicates its usefulness as a component of hydraulic brake fluid formulations; its affinity for certain other oils has likewise led to its use in cutting oils, textile lubricants, and industrial soaps. It is also used as a reactive intermediate in manufacturing polyester resins, plasticizers, and urethanes.

Tripropylene Glycol. Tripropylene glycol is an excellent solvent in many applications where other glycols fail to give satisfactory results. Its ability to solubilize printing ink resins is especially marked.

Toxicology

All of the propylene glycols display a low acute oral toxicity in laboratory rats. Propylene glycols produce a negligible degree of irritation upon eye or skin contact.

Inhalation of the vapors of any of the propylene glycols appears to present no significant hazard in ordinary applications and this is reflected in the fact that OSHA has not found it necessary to establish a permissible exposure level in the workplace. However, in 1985 the American Industrial Hygiene Association reviewed human experience and animal data and established a Workplace Environmental Exposure Level (WEEL) guideline for propylene glycol at 50 ppm total vapor and aerosol averaged over an eight-hour period.

Environmental Considerations

The propylene glycols vary in biodegradability, but it is expected that they will exhibit moderate to high biodegradability in a natural environment.

All of the propylene glycols are considered to be practically nontoxic to fish on an acute basis ($LC_{50} > 100$ mg/L) and practically nontoxic to aquatic invertebrates, also on an acute basis.

ALTON E. MARTIN
FRANK H. MURPHY
The Dow Chemical Company

J. W. Lawrie, *Glycerol and the Glycols*, Monograph Series, American Chemical Society, 1928, New York, p. 388.

I. Mellan, *Polyhydric Alcohols*, Spartan Books, Washington, D.C., 1962, p. 46.

G. O. Curme and F. Johnston, *Glycols*, Reinhold Publishing Corp., New York, 1953.

A Guide to Glycols, The Dow Chemical Co., Midland, Mich., 1992.

OTHER GLYCOLS

Glycols such as neopentyl glycol, 2,2,4-trimethyl-1,3-pentanediol, 1,4-cyclohexanedimethanol, and hydroxypivalyl hydroxypivalate are used in the synthesis of polyesters and urethane foams. Their physical properties are shown in Table 1.

Neopentyl Glycol

Neopentyl glycol, or 2,2-dimethyl-1,3-propanediol is a white crystalline solid at room temperature, soluble in water, alcohols, ethers, ketones, and toluene but relatively insoluble in alkanes.

Chemical Properties. Neopentyl glycol can undergo typical glycol reactions such as esterification, etherification, condensation, and oxidation:

Manufacture. Commercial preparation of neopentyl glycol can be via an alkali-catalyzed condensation of isobutyraldehyde with 2 moles of formaldehyde (crossed Cannizzaro reaction).

Derivatives. A number of derivatives of neopentyl glycol have been prepared; some show promise for commercial applications. They include organophosphorus derivatives, acetals and ketals, cyclopropane derivatives, diamine, simple esters, and polyesters.

2,2,4-Trimethyl-1,3-Pentanediol

2,2,4-Trimethyl-1,3-pentanediol is a white, crystalline solid. Trimethylpentanediol is soluble in most alcohols, other glycols, aromatic hydrocarbons, and ketones, but it has only negligible solubility in water and aliphatic hydrocarbons.

Chemical Properties. Trimethylpentanediol, with a primary and a secondary hydroxyl group, enters into reactions characteristic of other glycols.

Manufacture and Processing. 2,2,4-Trimethyl-1,3-pentanediol can be produced by hydrogenation of the aldehyde trimer resulting from the aldol condensation of isobutyraldehyde.

Table 1. Physical Properties of Several Glycols

Properties	Neopentyl glycol	2,2,4-Trimethyl-1,3-pentanediol	1,4-Cyclohexane-dimethanol[a]	Hydroxypivalyl hydroxypivalate
molecular formula	$C_5H_{12}O_2$	$C_8H_{18}O_2$	$C_8H_{16}O_2$	$C_{10}H_{20}O_4$
mol wt	104.2	146.2	144.2	204.3
melting range, °C	124–130	46–55	45–50[b]	46–50
boiling point, °C, at 101.3 kPa[c]	212	236	286	290
boiling range, °C		215–235		
density, g/cm³				
at 20°C	1.06		1.02	1.02
at 15°C		0.937		
heat of combustion, kJ/mol[d]	−3100	−5050	−4849[e]	

[a] Mixture of isomers, cis/trans ratio (wt %) =~32/68.
[b] Mp of cis isomer = 41 °C; mp of trans isomer = 70 °C.
[c] To convert kPa to mm Hg, multiply by 7.5.
[d] To convert kJ to kcal, divide by 4.184.
[e] Paar bomb.

Derivatives. Derivatives include simple polyesters, and esters, ketals.

1,4-Cyclohexanedimethanol

1,4-Cyclohexanedimethanol, 1,4-dimethylolcyclohexane, or 1,4-bis(hydroxymethyl) cyclohexane, is a white, waxy solid. The commercial product consists of a mixture of cis and trans isomers.

1,4-Cyclohexanedimethanol is miscible with water and low molecular weight alcohols and appreciably soluble in acetone. It has only negligible solubility in hydrocarbons and diethyl ether.

Chemical Properties. The chemistry of 1,4-cyclohexanedimethanol is characteristic of general glycol reactions; however, its two primary hydroxyl groups give very rapid reaction rates, especially in polyester synthesis.

Manufacture. The manufacture of 1,4-cyclohexanedimethanol can be accomplished by the catalytic reduction under pressure of dimethyl terephthalate in a methanol solution.

Derivatives. Derivatives include mixed phosphonate esters, diesters, and polyesters.

Hydroxypivalyl Hydroxypivalate

Hydroxypivalyl hydroxypivalate or 3-hydroxy-2,2-dimethylpropyl 3-hydroxy-2,2-dimethylpropionate is a white crystalline solid at room temperature.

Hydroxypivalyl hydroxypivalate is soluble in most alcohols, ester solvents, ketones, and aromatic hydrocarbons. It is partially soluble in water.

Chemical Properties. Both hydroxy groups on hydroxypivalyl hydroxypivalate are primary, which results in rapid reactions with acids during esterification. The absence of hydrogens on the carbon atom beta to the hydroxyls is a feature this glycol shares with neopentyl glycol, resulting in excellent weatherability.

Manufacture. Hydroxypivalyl hydroxypivalate may be produced by the esterification of hydoxypivalic acid with neopentyl glycol or by the intermolecular oxidation–reduction (Tishchenko reaction) of hydroxypivaldehyde using an aluminum alkoxide catalyst.

Derivatives. Polyester resins are useful for preparation of coatings exhibiting a combination of hydrolytic stability, excellent weather resistance, and good flexibility.

T. E. PARSONS
Eastman Chemical Company

Publication No. N-261A, N-307E, N-392B, and N-330B, Eastman Chemical Co., Kingsport, Tenn.

GOLD AND GOLD COMPOUNDS

Properties

Gold, Au, atomic number 79, is a third row transition metal in Group 11 (IB) of the Periodic Table. It occurs naturally as a single stable isotope of mass 197, ^{197}Au, which is also formed via the decay of ^{197}Pt (half-life 20 min) formed in the irradiation of platinum with slow neutrons. Selected properties are shown in Table 1. Gold is characterized by high density, high electrical and thermal conductivities, and high ductility. At least 26 unstable gold isotopes have been made; the most frequently used is ^{198}Au, which has a half-life of 2.7 d.

Gold is the most noble of the noble metals. Other than in the atomic state, the metal does not react with oxygen, sulfur, or selenium at any temperature. It does, however, react with tellurium at elevated (ca 475°C) temperatures to produce gold ditelluride, $AuTe_2$, which is also found in the naturally occurring mineral, calaverite, AuTe. Gold reacts with the halogens, particularly in the presence of moisture.

Gold reacts with various oxidizing agents at ambient temperatures provided a good ligand is present to lower the redox potential below that of water. Thus, gold is not attacked by most acids under ordinary conditions and is stable in basic media. Gold does, however, dissolve readily in 3:1 hydrochloric–nitric acid (aqua regia) to form $HAuCl_4$ and in alkaline cyanide solutions in the presence of air or hydrogen peroxide to form $(Au(CN)_2)^-$. These reactions are important to the extraction and refining of the metal.

Table 1. Gold Properties

Property	Value
atomic weight	196.9665
melting point, K	1337.59
boiling point, K	3081
density[a] at 273 K, g/cm^3	19.32
Brinell hardness (10/500/90), annealed at 1013 K, kgf/mm^2	25
tensile strength, annealed at 573 K, MPa[b]	123.6–137.3
elongation, annealed at 573 K, %	39–45
compressibility at 300 K, Pa^{-1}[c]	6.01×10^{-12}
heat of fusion, J/mol[d]	1.268×10^4
vapor pressure at 1000 K, Pa[c]	5.5×10^{-8}
specific heat at 298 K, J/(g·K)[d]	1.288×10^{-1}
thermal conductivity at 273 K, W/(m·K)	311.4
thermal expansion at 273–373 K, K^{-1}	1.416×10^{-7}
electrical resistivity at 273 K, Ω·cm	2.05×10^{-6}

[a] The commercially accepted value has been given. Measured values and density calculations from x-ray data show some variations.
[b] To convert MPa to psi, multiply by 145.
[c] To convert Pa to mm Hg, divide by 133.3.
[d] To convert J to cal, divide by 4.184.

At high temperature, attack by concentrated sulfuric and nitric acids is slow and is negligible for phosphoric acid. Gold is very resistant to fused alkalies and to most fused salts except peroxides. Gold readily amalgamates with mercury. Gold is very corrosion and tarnish resistant and imparts corrosion resistance to most of the commonly used gold alloys.

Extraction and Refining

Placer mining is the oldest form of gold mining. The technique is still in use in places where appropriate alluvial or marine deposits exist, such as Alaska, but requires large quantities of water. At present, most gold is obtained either by deep mining, most notably in South Africa, or by open pit mining such as in the United States.

Mined gold ore is milled sufficiently to allow separation of the gold; recoveries usually are 92–96% of the ore's gold content. After size reduction, various pretreatment and recovery processes may be used such as gravity concentration, flotation, roasting, chlorination, and cyanidation, the choice depending on the metallurgical characteristics of the ore.

Refractory or difficult to treat ores, ie, those containing excessive amounts of sulfur as pyrites or pyrrhotites, arsenic, tellurides, or carbonaceous material, are pretreated prior to extraction. Pretreatment may include fine grinding, roasting, or pressure oxidation (autoclaving), as well as biological preoxidation.

The impure gold concentrates from any of the primary recovery processes are melted under oxidizing conditions to remove most of the copper and other base metals, leaving gold plus silver. If the silver content is very low, the gold can be recovered by the Wohlwill electrolysis process in a chloride solution.

Treatment of impure gold is largely via the Miller process, in which chlorine is bubbled through the molten metal and converts the base metals to chlorides, which volatilize. Silver is converted to the chloride, which is molten and can be poured. If platinum-group metals are present, the chlorine process is unsuitable.

In the refining of gold, associated silver and the platinum-group metals are difficult to remove. Gold-silver alloys, named doré metal, that contain moderate amounts of gold can be treated by electrolysis in a nitrate solution at room temperature. Silver also can be removed from doré metal by treatment using hot sulfuric acid.

In refining precious metal scrap and some concentrates, the gold is converted to $HAuCl_4$ by treatment with aqua regia. After heating to remove nitrogen oxides, gold is precipitated from solution by reduction with sulfur dioxide or ferrous sulfate.

A number of special problems arise in the treatment of precious metal wastes of various types, eg, fabrication scrap and dusts (sweeps), and in treating scrap comprising gold-clad base metals. The scraps may contain copper, nickel, zinc, iron, tin, and lead, as well as gold, silver, or platinum-group metals. In refining, zinc can be fumed away as zinc oxide, and iron, lead, and some tin may be removed as slag. The precious metals remain with copper and most of the nickel. This product can be used as the anode in a sulfate solution and most of the copper and nickel removed, the precious metals remaining as an anode slime or mud.

Sweeps and related materials containing nonmetallic particles can be treated by adding appropriate fluxes to produce a low melting slag.

Production

Gold is widely distributed and the average content in the earth's crust is estimated to be 3.5 ppb. The gold content of ocean water varies considerably with location. The average value is of the order of 10 ppt, which is well below the concentration (3 ppm) required for economic recovery.

From primary lodes in rocks, gold has been released by weathering in conjunction with erosion by flowing streams in the form of fine grains or nuggets or as residual or stream placers which have often become buried by substantial layers of rock, as in South Africa.

In the United States, about 90% of gold production originates from ores and placer deposits. The remainder is recovered primarily as a by-product of the refining of base metals, chiefly copper. The principal gold producing states are Nevada (60%) and California (10%) followed by Montana, Utah, South Dakota, Washington, Colorado, Alaska, Idaho, Arizona, and New Mexico.

Economic Aspects

The price of gold has declined substantially from its historic high of U.S. $27,328/kg ($850/troy oz) reached in January 1980. The price of gold reflects gold's status as both a commodity an investment vehicle. In real terms, ie, in constant 1991 currency, the average price of gold in the early 1990s differs little from average levels since the beginning of this century.

Specifications

Dentistry. Most casting alloys meet the composition and properties criteria of specification no. 5 of the American Dental Association which prescribes four types of alloy systems constituted of gold–silver–copper with addition of platinum, palladium, and zinc. Composition ranges are specified, as are mechanical properties and minimum fusion temperatures. Wrought alloys for plates also may include the same constituents. Similarly, specification no. 7 prescribes nickel and two types of alloys for dental wires with the same alloy constituents.

Analytical and Test Methods

The method used to determine gold depends on the state of the sample to be analyzed and on its expected gold content and interfering impurity levels. Test methods include gravimetric methods, trimetric methods, spectrophotometric methods, and emission spectrography, atomic absorption, and neutron activation.

Health and Safety Factors

Chrysotherapy, therapy with gold compounds, along with some immunosuppressive and antimalarial drugs, remains as one of the few treatments capable of slowing or halting the damage caused by rheumatoid arthritis.

Radioactive ^{198}Au, prepared by irradiating natural gold in a nuclear reactor, has been used in radiotherapy either as grains which can be implanted in cancerous tissue such as of the prostate or nasopharynx or infused in colloidal form as in the treatment of bladder cancer. Colloidal ^{198}Au also has been employed in radiosynovectomy.

The labeling of various biological macromolecules with gold using such soluble species as $[Au(CN)_2]^-$, $[AuCl_4]^-$, and $[AuI_4]^-$, eliminates phasing problems in the determination of three-dimensional structures by x-ray diffraction because the high x-ray scattering power of Au permits its ready location in electron-density maps. Conjugates of colloidal gold with various proteins have been used as an aid to visualization in immunoelectron microscopy.

Uses

Gold Metal. Besides its use for monetary reserves, gold is used in the private sector principally for investment and fabrication. By far, the largest commercial use is jewelry. In the electronics industry, gold is used as fine wires or thin film coatings and frequently in the form of alloys to economize on gold consumption and to impart properties such as hardness.

In dentistry, gold is used for a variety of restorations.

Alloys. Except for white golds, the carat golds used in jewelry are alloys of gold, silver, and copper. Frequently these are modified by other metals, mainly zinc, as well as cadmium, nickel, platinum, or palladium.

The silver-rich phases in the gold–silver–copper system are white but lack tarnish resistance. Therefore, two other alloy systems are used primarily as white golds: gold–nickel–copper (often together with zinc) and gold–palladium alloys which usually also contain other metals such as silver, nickel, platinum, and zinc.

Gold and its alloys have many technological applications, in most cases in the form of thin films produced by electro- or chemical deposition.

A multitude of alloys has been developed for applications in the electronics, aerospace, and nuclear power industries.

Electrodeposition. Electrodeposition is the most widely practiced technique for producing gold or gold alloy coatings for jewelry and for decorative and industrial purposes.

Electroless Plating. Surface coatings that are obtained from baths comparable to those used for electrodeposition but that are applied without current are produced by electroless plating. The process operates either by electrochemical displacement (atomic exchange) or by chemical reduction.

Deposition from Liquid Gold. Liquid gold refers either to organic suspensions or emulsions containing finely divided gold powder or to solutions of organogold compounds in organic solvents. Liquid golds are formulated as paints or inks which are applied by painting, brushing, screen printing, etc, and that yield well-bonded gold films after subsequent drying and firing.

Gold and gold alloy films prepared form liquid gold are used widely for decorative and industrial purposes, particularly in electronics, and for optical and heat-concentrating and heat-shielding applications.

Catalysis. Although the literature contains numerous references to catalysis by gold and gold compounds, particularly gold alloys or bimetallic clusters, practical applications in this area remain extremely meager.

Derivatives

Gold Compounds. The chemistry of nonmetallic gold is predominantly that of Au(I) and Au(III) compounds and complexes. In the former, coordination number two and linear stereochemistry are most common. The majority of known Au(III) compounds are four coordinate and have square planar configurations. In both of these common oxidation states, gold preferably bonds to large polarizable ligands and, therefore, is termed a class b metal or soft acid.

Common gold compounds include halides (eg, gold(III) chloride, gold(III) iodide), cyanides, oxides and hydroxides (eg, gold(III) hydroxide, $Au(OH)_3$), and sulfides (eg, gold(I) sulfide, Au_2S).

Organogold Compounds. Both alkyl and aryl complexes of Au(I) and Au(III) as well as olefin and acetylene complexes have been prepared and studied.

Cluster Compounds. More recently, an increasing amount of interest has developed in gold-containing bimetallic cluster compounds which permit investigation at the molecular level of the metal–metal interactions thought to occur in bimetallic catalysts or alloys. Most often, these compounds are prepared from organophosphine stabilized Au(I) compounds such as halides or alkyls and generally contain one to three gold phosphine fragments bonded to one or more transition metal atoms, most often as carbonyl species.

J. G. COHN
ERIC W. STERN
Engelhard Corporation

R. J. Puddephatt, *The Chemistry of Gold*, Elsevier Scientific Pub. Co., Amsterdam, 1978.

W. S. Rapson and T. Groenewald, *Gold Usage*, Academic Press, Inc., New York, 1978.

Annual Reports on Gold, Gold Fields Mineral Services, Ltd., London.

Gold, Annual Reports; Mineral Commodity Summaries; Minerals Yearbooks; Mineral Industry Surveys; Mineral Trade Notes, U.S. Dept. of Interior, Bureau of Mines, Washington, D.C.

GRIGNARD REACTIONS

The term *Grignard reaction* refers to both the preparation of a class of organomagnesium halide compounds and their subsequent reaction with a wide variety of organic and inorganic substrates.

The general sequence of the reactions is now embodied in the following generic forms, where RX = an organic halide (most typically a chloride or bromide, although fluorides can be induced to react); S = a coordinating solvent (such as an ether or an amine); and AZ = a substrate with an electronegative group, Z:

$$RX + Mg + n\,S \rightarrow RMgX{\cdot}S_n$$

$$RMgX{\cdot}S_n + AZ \rightarrow RAZMgX{\cdot}S_n$$

$$RAZMgX{\cdot}S_n \rightarrow RA + ZMgX{\cdot}S_n$$

The heterolysis of AZ is dependent on the substrate and does not always occur. The final isolation of the product usually involves a hydrolysis step.

The development of improved industrial procedures, including the substitution of tetrahydrofuran (THF) for diethyl ether and the demonstration that the less reactive, but significantly less expensive, vinyl and aryl chlorides could be successfully used, has greatly expanded the commercial possibilities of this reaction. In the flavor, fragrance, pharmaceutical, and fine chemical industries, its use can generally be regarded as routine. Tens of thousands of metric tons of Grignard reagents are produced annually for captive use or merchant sale.

The great value of the Grignard reaction to the synthetic chemist is its general applicability as a building block for an impressive range of structures and functional groups. The Grignard reagent can act both as a prototypical carbon nucleophile that can undergo addition and substitution reactions and as a strong base that can deprotonate acidic substrates, resulting in the conjugate base or in some cases elimination reactions. Grignard reagents react with most functional groups containing polar multiple bonds (eg, ketones, nitriles, sulfones, and imines), highly strained rings (epoxides), acidic hydrogens (eg, alkynes), and certain highly polar single bonds (eg, carbon–halogen and metal–halogen).

Preparation of Grignard Reagents

A Grignard reagent is prepared by first adding magnesium and a partial charge of solvent to the reactor, followed by the addition of RX, in the remaining solvent, to the reaction flask.

Solvent Preparation. The most critical aspect of the solvent is that is must be dry (less than 0.02 wt % of H_2O) and free of O_2.

Other considerations for the solvent are the solubility of the Grignard reagent and the temperatures required for initiation and adventitious reactions of the Grignard with the solvent. Based on these three considerations, the best general solvent for the preparation of a Grignard reagent is THF. However, other solvents that are commonly used are diethyl ether, methyl *t*-butyl ether, di-*n*-butyl ether, glycol diethers, toluene, dioxane (R_2Mg), and hexane.

Magnesium Preparation. A surface coating resulting from the oxidation or hydration of the metal surface is the principal problem encountered for the magnesium reaction component. Fortunately, there are dozens of methods to remove the inert coating, thus activating the magnesium. For industrial use, the best method is using freshly chipped Mg turnings with a small quantity of the desired Grignard added to the reactor before addition of RX.

The Organohalogen Component. Just as for Mg and the solvent, the organic halide must be dry (less than 0.02 wt % of H_2O) and free of O_2. The relative reactivity of the halogens is reflected in the rate of disappearance of Mg, which follows the general order I > Br > Cl >> F. Unfortunately, the rate of disappearance Mg of does not always correlate with the formation of active Grignard. Typically, the more reactive the RX is, the higher the probability of forming a homocoupled product. Therefore, when choosing X, the rate of reactivity, product selectivity, and cost must be taken into account.

Other Methods. There are several common alternative methods for making Grignard reagents. Metal-exchange reactions are straightforward and MgR_2 can easily be prepared by this route.

Hydromagnesation reactions allow for the economical preparation of a Grignard reagent from an olefin.

Industrial Manufacturing Process

In spite of its industrial use for many years, the commercial-scale production of Grignard reagents has not been extensively described. The only practically important method is the batch method described by Grignard in 1900, namely formation of the Grignard reagent, reaction with a substrate, followed by hydrolysis of the reaction mixture.

The equipment can usually be constructed of carbon steel except for the hydrolysis vessel, which is usually glass-lined to avoid corrosion by aqueous acids. All vessels must be supplied with an inert gas (nitrogen or argon) for purging and blanketing and are vented to release offgases. It is imperative that the reaction vessel be protected with a rupture disk.

Analysis of Grignard Reagents

There are three potential problems that may occur during Grignard reagent preparation: oxidation by O_2, hydrolysis by H_2O, or homocoupling during the addition of alkyl or aryl halide. All three of these reactions decrease the active Grignard reagent while maintaining the same equivalents of base. Consequently, the concentration of a Grignard reagent should not be assumed, based on the reactants. The disadvantages of not analyzing the Grignard reagent are improper stoichiometry, potentially deleterious side reactions, highly exothermic quenching processes, phase splits, waste disposal, and cost problems. The analytical technique must be able to differentiate between active Grignard and total basicity. Many methods are available to measure the active Grignard, ranging from titration to electrophilic quenching followed by gc analysis.

Economic Aspects

The Grignard reaction has been commercially important for a number of years, and for certain industrial processes it remains the favored (or only) practical route to construct various element-to-carbon bonds.

There are five components to the cost of using a Grignard reagent: *(1)* magnesium metal, *(2)* the halide, *(3)* the solvent, *(4)* the substrate, and *(5)* disposal of the by-products. Prices for tetrahydrofuran and diethyl ether, the two most commonly used solvents, have increased. The cost of the halide depends on its structure, but as a general rule the order of cost is chloride < bromide < iodide.

Health and Safety Factors

Fire Hazards. The hazards associated with the manufacture, transport, and use of Grignard reagents are related to the flammability of the solvents employed and the exothermic reactions involved in their preparation and use.

Toxicology. Because of their high reactivity, there is little meaningful information on the health hazards of Grignard reagents per se. Rather, consideration needs to be given to the reagents employed, including the solvents and the products (or by-products) of the reaction. Some starting materials, such as organic halides (notably methyl bromide and vinyl chloride), are particularly toxic.

Regulatory Considerations. Commercial use of a Grignard reagent in the United States requires that it appear on the Environmental Protection Agency (EPA) list of Chemical Substances in Commerce. A corresponding registration exists for the European community and for Japan.

Because they are classified as flammable liquids, Grignard reagents in the United States must be packaged in drums or other suitable containers bearing a red U.S. Department of Transportation label.

Reactions and Applications of Grignard Reagents

Reactions and applications of Grignard reagents include asymmetric syntheses using Grignard reagents, Grignard reactions with inorganic chlorides, Grignard reagents as bases, metal-assisted modified Grignard reactions, intramolecular Grignard reactions, Grignards as methacrylate polymerization catalysts, and Grignard reagents as supports for the Ziegler-Natta process.

GARY S. SILVERMAN
Elf Atochem North America
PHILIP E. RAKITA
Elf Atochem Japan

G. S. Silverman and P. E. Rakita, eds., *The Grignard Reagent Handbook*, Marcel Dekker, Inc., New York, 1996.

M. Kharasch and O. Reinmuth, *Grignard Reactions of Nonmetallic Substances*, Prentice-Hall, Inc., New York, 1954.

C. Raston and G. Salem, *Chem. Met.-Carbon Bond* **4**, 159 (1987).

M. Okubo and K. Matsuo *Rev. Heteroatom Chem.* **10**, 213 (1994).

GROUNDWATER MONITORING

Groundwater monitoring is used to analyze the impact of a variety of surface and subsurface activities, including seawater intrusion, application of agricultural products such as herbicides, pesticides, and fertilizers, residential septic systems, and industrial waste ponds. Another focus of groundwater monitoring has been contamination associated with waste landfills and ruptured underground petroleum storage tanks.

The design of a groundwater monitoring strategy requires a basic understanding of groundwater flow systems. The majority of groundwater flow occurs in formations known as aquifers. At least two types of data can be retrieved using groundwater wells, ie, groundwater pressure and groundwater quality. A monitoring well allows measurement of these properties at a specific point in an aquifer. Monitoring wells come in a variety of sizes and materials, but each is simply a pipe extending from the ground surface to a point in theaquifer at which the pressure or contaminant is to be assessed. Monitoring wells are functional only in the saturated zone of the subsurface. Within the unsaturated soil zone, tensiometers, soil moisture blocks, and psychrometers have been used to assess fluid pressures. Fluid samples are retrieved using suction cup lysimeters for subsequent quality analysis.

Groundwater Pressure and Energy

The energy state of soil water can be defined with respect to the Bernoulli equation, neglecting thermal and osmotic energy as

$$E = z + P/\gamma + v^2/2g \tag{1}$$

where E is the energy per unit weight, P the pressure, γ the specific weight, z the elevation, and v the average velocity. The three energy terms represented by the right-hand side of the equation are potential energy, pressure energy, and kinetic energy, respectively. In most groundwater applications, the kinetic energy term is much less significant than the other two and is neglected. Thermal gradients cause moisture to migrate toward colder regions. However, thermal energy has been neglected in the present formulation and the equation cannot be used to simulate problems where there is a significant temperature gradient present.

When the energy terms are expressed as energy per unit weight, the term head is often used. Therefore, the total head, h, is equal to the elevation head, z, plus the pressure head, P/γ:

$$h = z + \frac{P}{\gamma} \tag{2}$$

Calculation of Groundwater Flow

The framework for the solution of porous media flow problems was established by the experiments of Henri Darcy in the 1800s. The relationship between fluid volumetric flow rate, Q, hydraulic gradient, and cross-sectional area, A, of flow is given by the Darcy formula:

$$Q = KA\frac{h_1 - h_2}{\Delta l} \tag{3}$$

The constant K, which maintains the equality, has been termed the hydraulic conductivity, permeability, or simply conductivity. The permeability is generally accepted to be a constant for a saturated soil, except for very small gradients. Here h_z represents the hydraulic head at location z, whereas Δl is the hydraulic length between points 1 and 2. A is an area perpendicular to the discharge vector.

This form of Darcy's law is applicable only to saturated flow. There are distinctions between the state of soil water in the saturated and unsaturated regions. These distinctions lead to an alternative form of Darcy's law for the case of unsaturated flow.

The vertical component of flow can be determined if a well is screened at two different elevations. Frequently, nested wells are used instead of a single well and multiple screenings to determine the vertical component of flow. Nested wells must be situated close enough to one another so horizontal gradients do not become a factor. Nested wells can also be used to analyze multilayer aquifer flow. There are many situations involving interaquifer transport owing to leaky boundaries between the aquifers. The primary case of interest involves the vertical transport of fluid across a horizontal semipermeable boundary between two or more aquifers.

Monitoring Well Design for Contaminant Transport Studies

Monitoring wells are installed by first completing a soil boring to the approximate depth of groundwater measurements. Drilling methods for the borehole include auger, mud rotary, cable tool, jetted wells, and driven wells. During the drilling, a boring log is prepared that records details of the subsurface materials encountered as the depth progresses. A well casing is installed in the borehole with a well screen at or near the bottom of the borehole. In the vicinity of the well screen, a filter pack of natural, ie, typically sand or pea gravel, or synthetic materials is used to preclude clogging of the well screen. Often, a secondary filter pack consisting of finer materials is placed above the primary filter. Above this is the virtually impermeable bentonite seal. A neat cement grout above this seal extends to the ground surface.

A variety of techniques can be used to retrieve the groundwater sample once the well is in place. Pumps, bailers, and syringes are among the devices used to draw the sample to the surface.

Design of a groundwater monitoring program minimally includes consideration of materials, location, indicator parameters, and timing.

Data analysis is aided by a variety of statistical techniques to assess significance, highlight trends, and form mathematical models of any correlations developed.

CAROL J. MILLER
Wayne State University

D. M. Nielsen, ed., *Practical Handbook of Ground-Water Monitoring*, Lewis Publishers, Inc., Chelsea, Mich., 1991.

C. W. Fetter, *Applied Hydrogeology*, 2nd ed., Macmillan Publishers, New York, 1988.

M. Barcelona, A. Wehrmann, J. Kelly, and W. Pettyjohn, *Contamination of Ground Water: Prevention, Assessment, Restoration*, Pollution Technology Review No. 184, Noyes Data Corporation, Park Ridge, N.J., 1990.

J. Devinny, L. Everett, J. Lu, and R. Stollan, *Subsurface Migration of Hazardous Wastes*, Van Nostrand Reinhold, New York, 1990.

GROWTH REGULATORS

ANIMAL

The growth of animals can be defined as an increase in mass of whole body, tissue(s), organ(s), or cell(s) with time. Improved understanding of the control of metabolic aspects of growth has provided the opportunity to regulate animal growth. Improvement of rate and efficiency of growth benefits the producer. Improvement in composition of meat animals benefits the producer through more efficient gain and greater value, and benefits the processor through less labor requirement for trimming and removal of fat. The consumer benefits by receiving a quality, desirable food at a cost reflective of efficient production.

Four general classes (ca 1993) of growth regulators are approved by the Food and Drug Administration (FDA) for use in food-producing animals in the United States. These include naturally occurring and synthetic estrogens and androgens, ie, anabolic steroids; ionophores; antibiotics; and bovine somatotropin. Compounds in the first class, anabolic steroids, act as metabolism modifiers to alter nutrient partitioning toward greater rates of protein synthesis and deposition, thereby increasing the weight at which 25 to 30% lipid content in the body or carcass is achieved. Ionophores have highly selective antibiotic activity and appear to enhance feed conversion efficiency through effects on ruminal microbes. Antibiotics, administered at subtherapeutic doses, enhance growth through improving feed conversion efficiency and/or growth rate, with no consistent effect on body or carcass composition.

Two other classes of growth regulators, ie, somatotropin or somatotropin secretogogues, and select synthetic phenethanolamines, have been investigated for the ability to alter growth; no compound in either class has yet been approved by the FDA for use in animals raised for meat. In 1993, the FDA approved administration of recombinant bovine somatotropin for increasing milk production in dairy cows. Administration of native or recombinant somatotropin (ST) to growing pigs, cattle, and lambs dramatically enhances rate, efficiency, and composition of gain. Likewise, experimental dietary administration of select synthetic phenethanolamines, most of which are β-adrenergic agonists, also has produced striking changes in rates of skeletal muscle and adipose tissue growth and accretion in growing cattle, lambs, pigs, and poultry.

Somatotropin, the β-adrenergic agonists, and the anabolic steroids are considered metabolism modifiers because these compounds alter protein, lipid, carbohydrate, mineral metabolism, or combinations of these; and they partition nutrient use toward greater rates of protein deposition, ie, muscle growth, and lesser rates of lipid accretion.

Anabolic Steroids

Several anabolic steroid implants have been approved for use in beef cattle in the United States, but only one, zeranol, is approved for use in lambs. Anabolic steroids are not used for growth regulation in swine or poultry.

Commercial products approved by the Food and Drug Administration include the naturally occurring hormone estradiol (Compudose); the natural hormone progesterone, used in combination with estradiol or estradiol benzoate, ie, Steer-oid, Synovex-S, and Synovex-C for calves; the fungal metabolite zeranol (Ralgro) which has estrogenic properties; the synthetic progestin melengestrol acetate (MGA); testosterone propionate in combination with estradiol benzoate, ie, Synovex-H or Heifer-oid; and a synthetic testosterone analogue, trenbolone acetate (TBA) which is used alone, ie, Finaplix, or in combination with estradiol, ie, Revalor.

Economics. Estimates of anabolic steroids in growing cattle indicate that savings associated with reduced feed costs are approximately \$50.00 per animal. Increased value of the carcass resulting from the increased amount of saleable lean meat produced is estimated to range from \$15.00 to \$30.00 per animal.

Withdrawal from anabolic steroid treatment is not required before slaughter because residue levels in edible tissues are negligible, and are significantly lower than other sources of estradiol such as the normal endogenous production in humans and the phytoestrogens consumed in plant food sources.

Ionophores

An ionophore may be defined as an organic substance that binds a polar compound and acts as an ion-transfer agent to facilitate movement of monovalent, eg, sodium and potassium, and divalent, eg, calcium, ions through cell membranes. The change in electrical charge in membranes influences the transport of nutrients and metabolites across the cell membrane, but the exact mechanism by which ionophores improve growth performance in growing ruminants is not known.

Monensin and other ionophores are being fed to over 90% of feedlot cattle grown for beef to enhance efficiency of gain; improvements of 5–10% are common. Ionophores also are used as anticoccidial drugs in poultry production and have similar, but lesser, effects in ruminants.

Doses range from 6 to 33 ppm in the diet, but very little if any ionophore can be measured in the circulation after feeding. Tissue concentrations are also very low.

Antibiotics

Antibiotics approved for use as growth enhancers in livestock and poultry include bacitracins, bambermycins, lincomycin, penicillin, streptomycin, tetracyclines, tiamulin, tylosin, and virginiamycin.

Chemically synthesized antimicrobials used in animal and poultry feeds include arsenicals, eg, arsanilic acid, sodium arsanilate, and roxarsone; sulfa drugs, eg, sulfadimethoxine, sulfamethazine, and sulfathiazole; carbadox; and nitrofurans, eg, furazolidone and nitrofurazone.

Growth Hormone-Releasing Factor

Exogenous administration of the naturally occurring growth hormone-releasing factor (GRF(1-44NH_2)) stimulates ST secretion and increases circulating concentrations of ST in growing pigs, cattle, and sheep. Administration of GRF is presumed to act through the same mechanisms involved in ST mediation of metabolism and tissue growth.

Health and Safety Information

The U.S. Food and Drug Administration's Center for Veterinary Medicine thoroughly evaluates the proposed use of any compound, natural or synthetic, used in food-producing animals for human food safety, safety to the animal of intended use, and safety to the environment. When a compound receives approval by the FDA, the efficacy and safety have been extensively investigated, and necessary labeling, handling, use, and withdrawal time requirements, if any, are determined. The Food Safety and Inspection Service (FSIS) of the USDA is responsible for ensuring that USDA-inspected meat and poultry products are safe, wholesome, and free of adulterating residues.

DONALD H. BEERMANN
Cornell University

D. L. Hancock, J. F. Wagner, and D. B. Anderson, *Growth Regulation in Farm Animals, Advances in Meat Research*, Vol. 7, Elsevier Science Publishers Ltd., Essex, U.K., 1991, pp. 255–297.

F. N. Owens, J. Zorrilla-Rios, and P. Dubeski, *ibid.*, pp. 321–342.

R. D. Boyd and D. E. Bauman, *Animal Growth Regulation*, Plenum Publishing Corp., New York, 1989, pp. 257–293.

D. H. Beermann, *The Endocrinology of Growth, Development, and Metabolism in Vertebrates*, Academic Press, Inc., San Diego, Calif., 1993, pp. 345–366.

PLANT

The arrival of new plant growth regulators on the market in the early 1990s, especially synthetic ones, is in a static state. Many growth regulators continue to be used experimentally, but the transition to approved usage is being delayed for several reasons. On the financial side, a number of mergers, buy-outs, and other dispositions of chemical companies has led to a decrease in the number of commercial compounds available. Moreover, the cost of registration has prompted some producers to withdraw chemicals from the market. Some older plant growth regulators have undergone several trade name changes, adding confusion to the field.

The ideal plant growth regulator should leave no harmful persistent residue in a finished product or crop and the paradigm compounds are ones that have high specific activity, are target specific, and are environmentally biodegradable.

Natural Products

The most acceptable growth regulators appear to be those compounds that already occur in nature (Table 1) and elicit certain desirable responses in economic crops.

Synthetics

Commercially available synthetic plant growth regulators include alar B, amidochlor, ancymidol, butralin, chlormequat chloride, chlorpropham, 3-CPA, 4-CPA, 2,4-dichlorophenoxyacetic acid, dimethipin, dormex, etacelasil, ethoxyquin, flumetralin, flurprimidol, folcysteine, inabenfide, maleic hydrazide, mefluidide, mepiquat chloride, merphos, morphactin, naphthalene acetic acid, naphthalene acetamide, Off-Shoot-O, paclobutrazol, *N*-(phenylmethyl)-1-*H*-purine-6-amine, *N*-phenylthalamic acid, and sevin.

Experimental synthetic plant growth regulators are AC 94377, BAS 111, CPPU, CN-11-3183, Razor, DCPA, Dacthal Rid, Cimectacarb, CGA 163935, Figaron, HOE 074 784, PPG 1721, and AC 310449.

Table 1. Natural Plant Growth Regulators

Product	Trade name	LD_{50}, g/kg
Available natural products		
brassinolide		rat[a]
24-epibrassinolide		rat, 1[a]
cytokinins, mixed[b]	Cytogen	rabbit, 10
	Trigger	rat, 5
n-decanol	Off-Shoot-T, Royaltac, Sucker Plucker, Antak	rat, 12.8
dikegulac	Atrimmec, Atrinal	mouse, 19.5; rat, 31
ethylene	Cerone, Prep, Ethrel,	rat, 4.22
ethephon	Chipcor, Florel	
gibberellins		
GA_3	Berelex, Gib-Tabs, Gib-Sol, Pro-Gibb	mouse, 15
GA_4 and GA_7	Pro-Gib 47, Regulex	no toxicity
lactic acid	Propel	4.94
Available natural product derivatives		
benzylamine purine	Promalin (as a mixture with GA_4 and GA_7)	mouse, 1.69
indole-3-*n*-butyric acid	IBA, Hormodin, Rhizopon (AA), Jiffy Grow	mouse, 100
N-(phosphonomethyl)-glycine	Roundup, Glyphosate, Polado	rat, 3.9

[a] Oral.
[b] Kinetin (3); 6-(4-hydroxy-1,3-dimethylbut-*trans*-2-enylamino)-9-β-D-ribofuranosylpurine.

HORACE G. CUTLER
USDA, ARS

H. G. Cutler and B. A. Schneider, *Plant Growth Regulator Handbook of The Plant Growth Regulator Society of America*, 3rd ed., Boyce Thompson Institute, New York, 1990.

W. T. Thompson, *Agricultural Chemicals, Book III—Miscellaneous Agricultural Chemicals, 1991–92 Revision*, Thompson Publications, Fresno, Calif., 1991.

L. G. Nickell, *Plant Growth Regulators*, Springer-Verlag, New York 1982, p. 173.

R. M. Sacher, "Strategies to Discover Plant Growth Regulators for Agronomic Crops," in *Chemical Manipulation of Crop Growth and Development*, 1982, p. 167.

GUMS

The term *gums* denotes a groups of industrially useful polysaccharides or their derivatives that hydrate in hot or cold water to form viscous solutions, dispersions, or gels (Table 1).

Gums are used in industry because their aqueous solutions or dispersions possess suspending and stabilizing properties. In addition, gums may produce gels or act as emulsifiers, adhesives, flocculants, binders, film formers, lubricants, or friction reducers, depending on the shape and chemical nature of the particular gum. Considerable research has been carried out to relate the structure and shape (conformation) of some gums to their solution properties.

Economic Aspects

Gums fall into a category of specialty chemicals called thickeners and stabilizers. This market is dominated by starch, starch derivatives, and cellulosics. Although the gums only represent approximately 5% of the sales by weight, they represent approximately 25% in dollars. Estimates for the sales and prices of the principal gums are shown in Table 2.

Table 1. Classification of Gums

Algal	Botanical	Microbial
agar	*seed gums*	dextran
algin	guar gum	xanthan gum
carrageenan	locust bean gum	gellan gum
	plant exudates/extracts	welan gum
	gum arabic	rhamsan gum
	gum ghatti	
	gum tragacanth	
	karaya gum	
	pectin	

Table 2. Estimates of Markets for Gums

Gum	Sales, 10^3 t U.S.	World	Price, \$/kg
	Thickeners		
gum arabic	4.5	18	5.5–7.7
gum tragacanth	0.23		33–88
guar	36	73	0.88–1.32
locust bean gum	2.3		07.7–10.5
xanthan	9.1	23	12.4–15.4[a] 9.9
	Gelling agents		
agar	0.45		28.6–33
alginate	3.2	18	13.2–15.4
carrageenan	3.2–5		8.8–17.6[b]
pectin	2.3–3.2	16–20.5	11–15.4

[a] Industrial.
[b] Food.

Algal (Seaweed) Gums

Agar. This gum is extracted from certain marine algae belonging to the class *Rhodophyceae*, red seaweed, which abound off the coasts of Japan, Mexico, Portugal, and Denmark. Important species include *Gelidium cartilagineum* and *Gracilaria confervoides*. Agar is an alternating copolymer of 3-linked β-D-galactopyranose and 4-linked 3,6-α-L-galactopyranose units.

Limited data are available for the types or grades of commercial agar, which is usually in the form of chopped shreds, sheets, flakes, granules, or powder. The official specifications for agar are provided in the USP and the *Food Chemicals Codex*.

Harvesting or collection of red seaweed is carried out by hand. After collection, the seaweed is dried and bleached in the sun prior to baling. Commercial extraction procedures involve washing, chemical extraction, filtration, gelation, freezing, bleaching, washing, drying, and milling.

Agar is insoluble in cold water, but is soluble in boiling water. On cooling to about 35°C, a firm gel forms at concentrations greater than 0.5%. Agar is used as a stabilizer in a variety of food applications but the principal use is in gelled media for the culture of yeasts, molds, and bacteria.

Algin. Algin occurs in all members of the class *Phaeophyceae*, brown seaweed, as a structural component of the cell walls in the form of the insoluble mixed calcium, magnesium, sodium, and potassium salt of alginic acid.

Alginates available for industrial use include the sodium, potassium, ammonium, and mixed sodium—calcium salts of alginic acid and propylene glycol alginate.

The ability of alginates to form edible gels by reaction with calcium salts is an important property. In practice, alginate gels are obtained using three principal methods, namely diffusion setting, internal setting, or setting by cooling. All species of brown algae contain algin; however, most of the algin produced commercially is isolated only from a few species. In the United States, algin is extracted from the giant kelp, *Macrocystis pyrifera*. In Europe, algin is extracted from *Ascophyllum nodosum, Laminaria hyperborea*, and *Laminaria digitata*.

The commercial processes for the production of algin involve extraction with alkali, followed by a series of washing, drying, and milling steps.

Propylene glycol alginate is the only organic derivative of algin that is widely used in industrial applications, principally the food industry. This product is prepared by the reaction of partially neutralized alginic acid with propylene oxide.

The toxicological properties of alginates have been extensively investigated, and it has been established that alginates are safe to use in foods. Alginates are used in a variety of food and industrial applications as gelling and stabilizing agents.

Carrageenan. The term *carrageenan* is the generic description for a complex mixture of sulfated polysaccharides that are extracted from certain genera and species of red seaweeds, including Chondrus crispus, Eucheuma, and Gigartina stellata. Red algae are abundant on the northeast coat of the North American continent. *Eucheuma* species grow in tropical waters extending form the Philippines to the east coast of Africa.

The seaweed is harvested by raking and hand-gathering. The collected seaweed is dried mechanically in many areas and shipped to the processing plants. Carrageenan is extracted from the seaweed with hot water at slightly alkaline pH. The aqueous extract is filtered and recovered by alcohol precipitation, dried, and milled.

The Food and Drug Administration lists carrageenan as GRAS, an approved food additive. The major applications are as a stabilizer and gelling agent in foods.

Botanical Gums

The botanical gums represent a family of polysaccharides obtained from a wide variety of plant sources. They are subdivided into exudate gums, seed gums, and gums obtained by extraction of plant tissue.

The properties of a botanical gum are determined by its source, the climate, season of harvest, and extraction and purification procedures. The considerable viscosity variation observed among gums from different sources determines, in part, their uses.

Plant Exudates. Most plant families include species that exude gums, and those that produce copious quantities represent a ready supply of gums. The plants are usually shrubs or low growing trees. Collection is by hand and labor costs represent a large proportion of the cost of these gums. Raw gum prices have remained low and steady because of the low labor costs in the producing countries of the Middle East and North Africa.

Natural gums are exuded in a variety of shapes characteristic of the species of origin. These shapes include the globular shape of gum arabic and the flakes or thread-like ribbons of gum tragacanth.

The quality of individual gums is mainly determined by color and taste or odor. Many gums are colorless when secreted, but darken on aging. Most gums are usually tasteless unless contaminated by the bitter flavors of tannins which precludes their use in foods.

Although many plant gum exudates are known, only gum arabic, ghatti, karaya, and tragacanth have wide industrial use. Gum arabic is a dried exudate from a species of the acacia tree, Acacia senegal, found in various tropical and semitropical areas of the world. Gum karaya or sterculia gum is the dried exudate of the Sterculia urens tree, which is now cultivated in India, the primary producing area. Gum tragacanth is an exudate from several species of tree, of the genus Astralagus, found in the dry, mountainous regions of Iran, Syria, and Turkey. Gum ghatti is an exudate from Anogeissus latifolia, a tree that is found in India and Sri Lanka.

Pectin. Pectin is a generic term for a group of polysaccharides, mainly partially methoxylated polygalacturonic acids, which are located in the cell walls of all plant tissues. The main commercial sources of pectin are citrus peel and apple pomace. The pectin is extracted, the extract purified, and the pectin precipitated; increased extraction times lead to the production of low methoxyl pectins. The main use for pectin is as a gelling agent for jams and jellies.

Seed Gums. The two major gums in this category are locust bean gum and guar gum. Other seed gums, including tamarind, psyllium, quince, and larch gum, have only limited use in specific applications.

Locust bean gum is produced by milling the seeds from the leguminous evergreen plant, Ceratonia siliqua, or carob tree, which is widely growth in the Mediterranean area.

Guar gum is derived from the seed of the guar plant, *Cyamopsis tetragonolobus*, a pod-bearing nitrogen-fixing legume grown extensively in Pakistan, India, and on a commercial scale since 1946, in the southwestern United States. During processing, the seed coat is removed by heating and milling. The endosperm, comprising approximately 40% of the seed, is then separated from the germ by various milling processes. The final milled endosperm is commercial guar gum. Many derivatives of guar, including cationic, carboxymethyl, carboxymethylhydroxypropyl and oxidized guar, have been prepared. One in particular, hydroxypropyl guar gum, is of industrial importance.

Structurally, both locust bean gum and guar gum comprise a straight chain of D-mannose with D-galactose sidechains. Locust bean gum is used as a stabilizer in several food applications; particularly dairy applications. Guar gum, which has lower cost, is used in a wide range of both food and industrial applications.

Microbial Gums

Although several microbial gums have been produced in commercial quantities, only xanthan, gellan, welan and rhamsan gums have significant current application.

Xanthan Gum. The polysaccharide, xanthan gum is produced commercially by culturing Xanthomonas campestris purely under aerobic conditions. When the fermentation is complete, the gum is recovered from the fermentation broth by precipitation with isopropyl alcohol, and dried, milled, tested, and packed. Xanthan gum is a cream-colored powder that dissolves in either hot or cold water to produce solutions with high viscosity at low concentration. These solutions exhibit pseudoplasticity, ie, the viscosity decreases as the shear rate increases. Xanthan gum has excellent stability to salts, and in acids and bases. The most unusual property of xanthan gum is the reactivity with the galactomannans, guar gum and locust bean gum.

The FDA issued a food additive order in 1969 that allowed the use of xanthan gum in food products without specific quantity limitations. Xanthan gum is used as a thickener and stabilizer in numerous food, industrial, and oil field applications.

Gellan Gum. Gellan gum is the generic name for the extracellular polysaccharide produced by the bacterium, Pseudomonas elodea (ATCC 31461). Proprietary to Kelco Division of Merck & Co., Inc., gellan gum is manufactured in an aerobic, submerged fermentation. Gellan gum has been permitted for use in foods in Japan since 1988, and in 1992 received general food approval in the United States. It is used as a stabilizer, thickener, and gelling agent in a variety of food applications.

Welan Gum. This gum is produced by a carefully controlled aerobic fermentation using an Alcaligenes strain (ATCC 31555). Welan has similar properties to xanthan gum except that it has improved thermal stability and compatibility with calcium at alkaline pH. It is used as a drilling fluid viscosifier and in specialized cement and concrete applications.

Rhamsan Gum. Rhamsan gum is produced by Alcaligenes strain (ATCC 31961). Solutions of rhamsan have high viscosity at low shear rates and low gum concentrations, making it useful for several industrial applications.

JOHN K. BAIRD
Kelco Division of Merck & Company, Inc.

R. L. Whistler, *Industrial Gums*, 3rd ed., Academic Press, Inc., New York, 1992.

P. A. Sandford and J. K. Baird, in G. O. Aspinall, ed., *The Polysaccharides*, Vol. 2, Academic Press, Inc., New York, 1983.

J. K. Baird and D. J. Pettitt, in I. Goldberg and R. Williams, eds., *Biotechnology and Food Ingredients*, Von Nostrand Reinhold, New York, 1992, Chapt. 9.

H

HAFNIUM AND HAFNIUM COMPOUNDS

Hafnium, Hf, is in Group 4 (IVB) of the Periodic Table, as are the lighter elements zirconium and titanium. Hafnium is a heavy gray-white metallic element never found free in nature. It is always found associated with the more plentiful zirconium. The two elements are almost identical in chemical behavior.

Hafnium is obtained as a by-product of the production of hafnium-free nuclear-grade zirconium. Hafnium's primary use is as a minor strengthening agent in high temperature nickel-base superalloys. Additionally, hafnium is used as a neutron-absorber material, primarily in the form of control rods in nuclear reactors.

Properties

Physical Properties. Hafnium is a hard, heavy, somewhat ductile metal having an appearance slightly darker than that of stainless steel. Physical properties of hafnium are summarized in Table 1. These data are for commercially pure hafnium which may contain from 0.2 to 0.3% zirconium. Although a number of radioactive isotopes have been artificially produced, naturally occurring hafnium consists of six stable isotopes.

Chemical Properties. Hafnium's aqueous chemistry is characterized by a high degree of hydrolysis, the formation of polymeric species, a very slow approach to true equilibrium, and the multitude of complex ions that can be formed. Partially reduced di- and trihalides have been produced by reducing anhydrous hafnium tetrahalides with hafnium metal.

Hafnium is a highly reactive metal. The reaction with air at room temperature is self-limited by the adherent, highly impervious oxide film which is formed. This film provides oxidation stability at room temperature and resistance to corrosion by aqueous solutions of mineral acids, salts, or caustics.

Occurrence and Mining

The primary commercial source is zircon (zirconium orthosilicate). Zircon sand is found in heavy mineral sand layers of ancient ocean beaches. Principal zircon sand producing countries are Australia, South Africa, the United States, and Russia. Zircon is always a coproduct form the mining of rutile and ilmenite mineral sands to supply the titanium oxide pigment industry. Baddeleyite, a naturally occurring zirconium oxide, is available form South Africa and Russia.

Most of the heavy mineral sands operations in the world are similar. Typically the quartz sand overburden is bulldozed away to reach the heavy mineral sand layer, which usually has 2 to 8% heavy minerals. The excavation is flooded and the heavy mineral sands layer is mined by a floating dredge with a cutter-head-suction. The sand slurry is pumped to a wet-mill concentrator mounted on a barge behind the dredge. Wet concentration using screens, cones, spirals, and sluices removes roots, coarse sand, slimes, quartz, and other light minerals. The tailings are returned to the back end of the excavation. Rehabilitation of worked-out areas is about a 10-year project which includes replacing the overburden and topsoil to pre-existing levels and contours, and reestablishing the natural vegetation, usually from company-owned nurseries.

Table 1. Physical Properties of Hafnium

Property	Value
atomic weight	178.49
density, at 298 K, kg/m^3	13.31×10^3
melting point, K	2504
boiling point, K	4903
specific heat, at 298 K, J/(kg·K)[a]	144
thermal conductivity at 273 K, W/(m·K)	23.3

[a] To convert J to cal, divide by 4.184.

Manufacture

Decomposition of Zircon. Zircon sand is inert and refractory. Therefore the first extractive step is to convert the zirconium and hafnium portions into active forms amenable to the subsequent processing scheme. For the production of hafnium, this is done in the United States and France by carbochlorination. In the Ukraine, fluorosilicate fusion is used. Caustic fusion is the usual starting procedure for the production of zirconium chemicals which does not involve hafnium separation.

Separation of Hafnium. Many methods have been proposed for the separation of hafnium and zirconium; three different industrial methods are in use: liquid–liquid extraction, molten salt distillation, and fluorozirconate crystallization.

Reduction. Hafnium metal is obtained by reducing hafnium tetrachloride with liquid magnesium (Kroll process) or by electrolysis of hafnium tetrachloride in a molten chloride–fluoride salt.

Refining. Kroll-process hafnium sponge and electrowon hafnium do not meet the performance requirements for the two principal uses of hafnium metal. Further purification is accomplished by the van Arkel-de Boer, ie, iodide bar, process and by electron beam melting.

Economic Aspects

Total hafnium available worldwide from nuclear zirconium production is estimated to be 130 metric tons annually. The annual usage, in all forms, is about 85 t. The balance is held in inventory in stable intermediate form such as oxide by the producers Teledyne Wah Chang (Albany, Oregon) and Western Zirconium in the United States; Cezus in France; Prinieprovsky Chemical Plant in Ukraine; and Chepetsky Mechanical Plant in Russia (crystal bar). Demand for hafnium has not shown significant growth since the late 1980s, nor has pricing changed.

Health and Safety

High surface area forms of hafnium metal such as foil, fine powder, and sponge are very easily ignited, and fine machining chips can be pyrophoric. Most hafnium compounds require no special safety precautions because hafnium is nontoxic under normal exposure. Acidic

Table 2. Physical Properties of Some Hafnium Compounds

Property	HfB_2	HfC	HfO_2	HfN	HfS_2	$HfSe_2$	$HfSi_2$	HfF_4	$HfCl_4$	$HfBr_4$	HfI_4
melting point, °C	3370	3830	2810	3330			1750	>968	432[a]	424[a]	449[a]
specific gravity, measured, g/cm^3	10.5	12.2	9.68		6.03	7.46	7.2			4.90	
color	gray	gray	white	gold	purple-brown	dark brown		white	white	white	yellow-orange
hardness,[a] kgf/mm^2	2900[c]	2300[d]	1050[e]	1640[f]			930[f]				

[a] At 3.34 MPa (33 atm). [b] 1 kgf/mm^2 = 9.8 MPa. [c] Vickers' hardness. [d] Knoop hardness. [e] Diamond pyramid hardness (DPH), 2 kg. [f] Microhardness, 50 gf/mm^2 = 490 kPa.

compounds such as hafnium tetrachloride hydrolyze easily to form strongly acidic solutions and to release hydrogen chloride fumes, and these compounds must be handled properly.

Hafnium Compounds

Most hafnium compounds have been of slight commercial interest aside from intermediates in the production of hafnium metal. However, hafnium oxide, hafnium carbide, and hafnium nitride are quite refractory and have received considerable study as the most refractory compounds of the Group 4 (IVB) elements. Physical properties of some of the hafnium compounds are shown in Table 2.

RALPH H. NIELSEN
Teledyne Wah Chang Corporation

K. L. Komarek, ed., *Hafnium: Physico-Chemical Properties of Its Compounds and Alloys*, International Atomic Energy Agency, Vienna, 1981.

E. M. Sherwood and I. E. Campbell, in D. E. Thomas and E. T. Hayes, eds., *The Metallurgy of Hafnium*, U.S. Government Printing Office, Washington, D.C., 1960.

D. J. Cardin, M. F. Lappert, and C. L. Raston, *Chemistry of Organo-Zirconium and Hafnium Compounds*, Halsted Press, Division of John Wiley & Sons, Inc., New York, 1986.

J. Wang, H. P. Li, and R. Stevens, *J. Mater. Sci.* **27**, 5397–5430 (1992).

HAIR PREPARATIONS

Hair products are normally cosmetics and are thus subject to all laws and regulations that control the labeling and claims of all cosmetic products. There are, however, several significant variations to this premise, ie, hair colorants, professional use only products, and products that make drug claims.

Shampoos

The largest segment of the hair care market is the shampoo category. With development of mass marketing, the number of shampoos available has reached immense proportions; this has been aided by the availability of synthetic detergents. Synthetic detergents, which have replaced soap-based products used in shampoos, allow for greater formulating flexibility and control, and meet new product standards. In the 1990s a shampoo must not only cleanse and have tolerance to hard water, but it should also be able to provide different performance attributes.

Properties. The primary purpose of a shampoo is to clean the hair and scalp. In its cleansing process the typical shampoo must be able to remove the various soils found on the hair and scalp, ie, natural oily exudates and scales; conditioners and setting products that may be applied; and airborne soils that accumulate on the hair and scalp. The shampoo should leave hair soft, lustrous, and in a manageable condition without leaving a harsh, dry, raspy feel. A good performing shampoo should not overclean by removing all the oils from the hair.

There are a number of factors in formulating an acceptable shampoo product. In addition to cleansing, the shampoo should have good lathering properties. Further, an acceptable shampoo should be safe for repeated use, nontoxic, nonirritating, adequately preserved, chemically and physically stable, and have a pleasant fragrance.

Product Forms. Shampoos consist of an aqueous solution or dispersion of one or more cleansing additives, together with other ingredients to enhance performance and consumer acceptability, ie, foam enhancers, preservatives, colors, fragrances, and pearling agents in the case of opaque shampoos. Other additives found in shampoo compositions include thickeners, conditioners, antidandruff agents, sequestrants, and buffering agents.

Shampoos have been prepared in various forms, and have included systems that are thick and thin, clear and opaque, pourable liquids, solids, gels, pastes, powders, flakes, and aerosol types. In many cases, shampoos have been prepared and directed for various hair types, eg, normal, dry, damaged, and color treated.

Synthetic Detergents. Synthetic detergents are now (ca 1993) the backbone of most shampoo products. These surfactants are classified according to the electrical properties of their hydrophyllic groups in aqueous solutions; ie, designated as anionic for those negatively charged, nonionic for those with no charge, cationic for those with a positively charged hydrophyll, and amphoteric for those having both positive and negative ionic features.

The anionics are used primarily in shampoo preparations because of their superior foaming and cleansing properties. The amphoterics are low foaming detergents and are generally regarded as low cleansers, but they are very mild and often are found in baby-type shampoos. Cationic surfactants are poor foamers and have low detergency; however, they are substantive to hair and, as a consequence, are used primarily in systems for hair conditioning purposes. Nonionic detergents are poor foamers and are used primarily in shampoos for property modifications such as conditioning, solubilizing, viscosity, etc.

Anionic surfactants include primary alkyl sulfates (lauryl sulfate), alkyl ether sulfates, alkyl sulfosuccinate half esters, fatty acid–sarcosine condensates, fatty acid–peptide condensates, alkyl monoglyceride monosulfates, acyl isethionates, alpha-olefin sulfonates, and alkyl sulfoacetates.

Nonionic surfactants include alkanolamides, amine oxides, and ethoxylated nonionics.

Most amphoteric surfactants are derivatives of imidazoline or betaine. Sodium lauroamphoacetate has been recommended for use in non-eye stinging shampoos. Combinations of amphoterics with cationics have provided the basis for conditioning shampoos.

Shampoo Additives. Additives include thickeners (alkanolamides, inorganic salts, derivatives of cellulose, natural gums, carboxyvinyl polymers, polyvinyl alcohols, magnesium aluminum silicates, glycol stearates, and fatty alcohols), opacifiers (glycol mono- and diesters, higher fatty alcohols, stearate soaps, and latex copolymer emulsions), conditioners (lanolin and its derivatives, fatty amine oxides, cationic polymers, cationic guar gums, fatty amines and alcohols, alkanolamides, quaternary ammonium compounds, glycerin, sugar, sorbitol, fatty glycerides, esters, silicones, beer, egg, honey, milk, and herb extracts), and preservatives (methyl and propyl parabens, DMDM hydantoin, quaternium-15, phenoxyethanol, imidazolidinyl urea, and a mixture of methylchloroisothiazolinone and methylisothiazolinone.

Baby Shampoos. These shampoos, specifically marketed for small children, feature a non-eye stinging quality. The majority of the products in this category are based on amphoteric detergent systems.

Medicated Dandruff Shampoos. Shampoo additives used to treat dandruff include antimicrobial additives, eg, quaternary ammonium salts; keratolytic agents, eg, salicylic acid and sulfur; heavy metals, eg, cadmium sulfide; resorcinol; and many others. Selenium sulfide, coal tar, or zinc pyrithione are the most used active antidandruff shampoo additives.

Two-in-One Shampoos. These shampoos are combination cleansing and conditioning products. They are based on conventional anionic detergents to provide desired physical shampoo properties combined with conditioning additives. Most common conditioning additives used are silicones, quaternaries, cationic guar gums, and polymers.

Manufacture, Evaluation, and Safety. The manufacture of shampoos is a relatively simple operation requiring a suitable stainless steel kettle, with provisions for heating and cooling and equipped with an appropriately sized mixer.

Laboratory methods are valuable to help assess such factors as foaming, cleansing, rinsing, and wet and dry combing effects in the development of a shampoo. These can be determined under standardized, controlled conditions. More critical evaluations can be made through half-head salon comparisons to competitive products and by panel tests under actual use conditions.

Shampoos generally do not represent a hazard with regard to skin and eye safety; once used, shampoos are almost immediately rinsed and have little contact time on sensitive areas. To assure this safety, provisions to test the finished product for skin and eye irritation should be made.

Hair Conditioners

The term hair conditioner can be applied to such products as rinses, hair dressings and intensive treatments. Conditioners are used to provide different effects to the hair; primarily ease of combing, sheen, and soft feel.

Hair Rinses. These products generally are designed to be used in conjunction with shampoos to provide special benefits to hair, eg, wet and dry combing ease, antistatic effects, shine, manageability, and detangling. The active ingredients in most creme rinses are quaternary ammonium compounds such as steartrimonium chloride and cetrimonium chloride. Other additives useful in after-rinse hair conditioners include certain fatty amines, amine oxides, and cationic polymers. After-shampoo conditioners also are used to improve the finish of hair with respect to manageability, body, texture, etc. Additives used to obtain these effects include protein additives, silicones, and lanolin and its derivatives. Most rinses are opaque products, although clear versions can be found. These products range in consistency from pourable liquids to thick creams.

Hairdressings. Products associated with final grooming effects to hair are termed hairdressings. They are used to impart not only a holding effect to hair, but also provide an added benefit of giving hair a natural, healthy, lustrous appearance. Hairdressings can be found as liquid or cream emulsions, gels, or as hydroalcoholic preparations.

Historically, the main constituent of brilliantines is an oil, usually a mineral oil type, in a rather high concentration. In certain brilliantine compositions, vegetable and animal oils are used as substitutes for mineral oil. In these systems, because of their potential for rancidity, antioxidants must be included. Other alternatives to mineral oils that have found utility in brilliantines are the polyethylene glycols. Use of these materials offers the advantage of chemical stability to rancidity. Other additives found in brilliantines to improve their aesthetics include colorants, fragrance, medicated additives, lanolin, and fatty acid esters.

Solid brilliantines and pomades may be considered heavy-duty type hairdressings. Their main component is petrolatum with additions of various waxes to obtain different consistency ranges.

Hairdressing products have been prepared by dilution of various mostly synthetic oils with alcohol. Additives found in these products include ethoxylated and propoxylated glycols, ethoxylated ethers, various lanolin derivatives, and ethoxylated and propoxylated diols and triols. In addition, alcoholic hair tonics may contain quaternary conditioners, keratolytic agents, hair setting resins, colorants, and fragrance.

Emulsified hair dressings have been formulated both in liquid and cream forms. They are either water-in-oil or oil-in-water emulsions. Mineral oil is commonly used in water-in-oil hairdressing emulsions. Emulsifiers include magnesium, zinc, or aluminum stearate, beeswax, borax, sorbitan sesquioleate, ethoxylated fatty alcohols, ethyoxylated lanolin alcohols, polyglyceryl esters, and acetlyated ether esters. Other emulsifiers include absorption bases which allow use of as much as 80% water in preparing these product forms.

Clear gel hairdressings come in two forms, ie, microemulsions or setting gels. Microemulsions are systems containing mineral oil which can be blended with emollients, conditioning additives, lanolins, and protein compounds. Setting gels are formed through use of high molecular weight polymers such as the methylcellulose ethers and carboxyvinyl polymers.

Fixatives

Fixatives are liquid products used to achieve a desired hairstyle and to temporarily hold the style in place. These products can be grouped into two classes, ie, styling products and finishing sprays. Styling products are used primarily on wet hair to make combing easier and to give the hair some tack so that the style remains in place as the hair is dried. Examples of this type of fixative include styling mousses, gels, lotions, and spray gels. Finishing sprays, eg, pump and aerosol hair sprays, are applied to the hair after the style is dried and set.

When formulating a hair fixative, the balance between the two principal benefits, hold and styling ease, must be selected. Hold is characterized by the stiffness of the polymer film and its ability to remain stiff when exposed to high humidity. Styling ease is characterized by the product's ability to decrease surface friction of the hair during the combing, drying, and styling process. These benefits tend to be inversely proportional.

Styling Products. The setting or holding ingredient in styling products is the film-forming polymer. Amounts of total polymer vary from product to product depending on the performance desired; levels can be found from 1 to 7%, with the majority formulated from 3–4%. Holding polymers include polyvinylpyrrolidinone (PVP), polyvinylpyrrolidinone/vinyl acetate copolymer (PVP/VA), polyquaternium-11, and polyquaternium-4. Other ingredients include natural gums, conditioning agents (polyquaternium materials, quaternary ammonium salts, silicone compounds, and natural oils), solvents (deionized water, ethanol, propylene or butylene glycol, sorbitol, and ethoxylated nonionic surfactants), gelling agents (Carbomer 940, hydroxyethylcellulose and hydroxypropyl methylcellulose), foaming agents/propellants (nonionic surfactants, hydrocarbon propellants), and preservatives.

The manufacturing of styling products is relatively simple. Generally, a tank with simple agitation is sufficient for low viscosity products, ie, mousse, spray gels, and sculpting lotions. Styling gels are not as easy to manufacture. The gelling materials used are hygroscopic and tend to clump when added to water. Mousses pose little manufacturing problem, but because they are aerosolized they must be filled with special equipment.

Gels are usually sold in a low density polyethylene tube or in a bottle with a pump. Spray gels, heat-activated sprays, scrunching sprays, etc, are generally in high density polyethylene bottles with pump spray devices. Mousses are generally packed in aluminum cans because they are pressurized and the high water content tends to corrode tin-plated steel aerosol cans.

Finishing Sprays. The primary setting agent in finishing sprays is the film-forming polymer. Polymer concentrations range form 2–7% in aerosols and up to 7% in pump versions.

Vinyl acetate (VA)/crotonates copolymer was the first polymer used in fixatives to contain carboxylic acid groups which, depending on neutralization percent, could produce variations in film properties; eg, stiffness, humidity resistance, resiliency, tack, and removability by shampoo. It has largely been replaced in hair sprays by newer polymers.

VA/crotonates/vinyl neodecanoate copolymer is the most used polymer in aerosol hair sprays (ca 1993). Like its precursor above, it has free carboxylic acid groups which can be neutralized to give various film properties. Recommended neutralizing agents include aminomethyl propanol, ammonium hydroxide, and dimethyl stearamine. Recommended percent neutralization is 90%, but products can be found in the 80–110% range. Other polymers include ethyl and butyl esters of poly(vinyl methyl ether)/maleic anhydride (PVM/MA) copolymer, and octylacrylamide/acrylates/butylaminoethyl methacrylate copolymer. Other ingredients include plasticizers and other film-modifying additives (polymer neutralizer, glycols, phthalates, fatty alcohols, silicone derivatives, lanolin derivatives, various oils, various fatty esters, and fragrance), solvent systems (ethanol, water, dimethyl ether), propellants (propane, isobutane, butane, dimethyl ether, and hydrofluorocarbon 152A).

Finishing sprays are easily prepared as simple solutions of the polymers, neutralizers, plasticizers, fragrance, etc, in ethanol.

Flame extension, flashbacks, and flashpoints must be determined for aerosol sprays, and the shipping cases must be properly labeled according to U.S. Department of Transportation (DOT) standards. In both pump and aerosol products, spray rates, spray patterns, and particle size distribution have to be optimized.

Aerosol finishing sprays generally are packaged in tin-plated steel containers consisting of a dome, body, and base which may or may not be lined for protection against corrosion. For products containing higher water contents, aluminum is generally used for the package. Nonaerosol sprays are typically packaged in high density polyethylene.

Health and Safety Factors. Finishing spray products generally have high alcohol contents which create a flammability hazard. Deliberate inhalation of aerosols poses a potential health hazard to the consumer that could be fatal. Additionally, spraying a high alcohol content finishing spray into the eyes can cause severe irritation. Appropriate warning must be displayed on the package.

Environmental Regulation. In 1978, federal regulation banned the use of chlorofluorocarbons in hair sprays. This forced a dramatic change in the technology of aerosol hair sprays, requiring new formulations and new dispensing parts to accommodate hydrocarbon propellants. In the early 1990s, California and New York enacted strict limits on allowable VOC content.

The primary VOCs in hair sprays are lower order alcohols and propellants. In nonaerosols, with typical VOC levels of 88–92%, a drop to 80% can be accomplished with total water addition of about 15%, which causes an increase in solution viscosity and a heavier, wetter spray. In aerosols with typical VOC levels of 93–98%, a much bigger decrease is needed. This in turn causes difficulties in spraying and may require the use of dimethyl ether as a component of the propellant system. Polymer suppliers are working to develop new materials which may be adaptable to these high water content formulas with the overall goal of approaching the performance standards of anhydrous fixative sprays.

Coloring Preparations

Among the desired properties of a good hair dye, toxicological safety is of primary importance. Modern hair colorants can be divided into temporary, semipermanent, and permanent systems. These categories are characterized by the durability of the color imparted to the hair, the type of dye employed, and the method of application.

Temporary Hair Colorants. Temporary hair colorants use large dye molecules that deposit on the surface of the hair without penetrating the cuticle. Dyes used for this class of hair color are shown in Table 1.

Semipermanent Hair Colorants. This system uses so-called direct dyes which penetrate into the cortex but slowly diffuse out again when the hair is washed. Nitro and anthraquinone dyes are used mainly, and azobenzenes less frequently. A partial list of semipermanent dyes includes HC Yellow No. 2, HC Yellow No. 4, Disperse Blue 3, Disperse Violet 1, HC Red No. 1, and HC Orange No. 1. More extensive data are available.

The naturally derived dyes used by the Egyptians and other ancient civilizations are actually examples of semipermanent dyes. The best known dye of this kind comes from the Henna plant and is still in use after thousands of years. The extract of the Henna plant contains lawsone (2-hydroxy-1,4-naphthoquinone). This dye produces a reddish color on hair, which is best used to produce a warming effect on brown hair.

Permanent Hair Colorants. Permanent colorants produce hair coloration that is formed inside the hair by hydrogen peroxide-induced coupling reactions of colorless dye precursors.

Color-forming reactions are accomplished by primary intermediates, secondary intermediates, and oxidants. Primary intermediates include the so-called para dyes, *p*-phenylenediamine, *p*-toluenediamine, *p*-aminodiphenylamine, and *p*-aminophenol, which form a quinone monoimine or diimine upon oxidation. The secondary intermediates, also known as couplers or modifiers, couple with the quinone imines to produce dyes. Secondary intermediates include *m*-diamines, *m*-aminophenols, polyhydroxyphenols, and naphthols. Some of the more important oxidation dye colors are given in Figure 1. An extensive listing is available.

Also present but not essential in permanent hair colorants are nitro dyes which dye hair without oxidation. These dyes, nitro derivatives of aminophenols and benzenediamines, impart yellow, orange, or red tones.

Attempts to broaden the range of materials available as dye precursors have been made. Oxidative dyes based on pyridine derivatives produce less sensitization than those based on benzene derivatives; however, they lack tinctorial power, lightfastness, and availability. Derivatives of tetraamino-pyrimidine are claimed to act as primary intermediates to give intense shades with good fastness and excellent toxicological properties.

Oxidative hair dye products usually are formulated at a pH of 9.5–10.5. Ammonia, buffered with oleic acid, is the most commonly used alkalizing agent.

The colorant formulation is mixed just before application with an oxidizing agent, ie, developer. Hydrogen peroxide is the preferred developer, usually at a concentration of 6%.

Several products have appeared on the market that are positioned as being more gentle than the usual permanent hair color products. They differ from other permanent hair dye products in that they employ an alkalizing agent other than ammonia to obtain a pH about one unit lower than conventional products. Usually they use lower concentrations of peroxide. These products are positioned as longer lasting semipermanent hair color products.

One development (ca 1993) in hair coloring involves the formation of pigments within the hair that are very similar to natural melanin. Thus either catalytic or air oxidation of 5,6-dihydroxyindole can be effectively used to permanently dye hair within a short time. The formed color can, if required, be further modulated with dilute H_2O_2 or can be even totally removed from hair by this oxidant.

Metallic dyes are among the older hair color materials known. Commercial products are based on a 1% solution of lead acetate in an aqueous, slightly acidic, alcoholic medium. Precipitated sulfur appears to be essential.

The purpose of bleaching is to lighten or altogether decolorize the natural pigment with minimal damage to the hair itself. An ammoniacal solution is added just before use to activate the hydrogen peroxide. The alkaline solution can be formulated into a shampoo vehicle with oleate soaps or ethyoxylated fatty alcohols. Bleaches of the simple ammoniacal peroxide type give limited lightening, which can be in-

Table 1. Temporary Dye Colors

FDA Designation	CI name	Type
Ext. D&C Violet No. 2	Acid Violet 43	anthraquinone
D&C Red No. 33	Acid Red 33	azo
D&C Brown No. 1	Acid Orange 24	disazo
D&C Green No. 5	Acid Green 25	anthraquinone
Ext. D&C Yellow No. 7	Acid Yellow 1	nitro
D&C Red No. 22	Acid Red 87	xanthene
FD&C Blue No. 1	Acid Blue 9	triphenylmethane
FD&C Green No. 3	Food Green 3	triphenylmethane

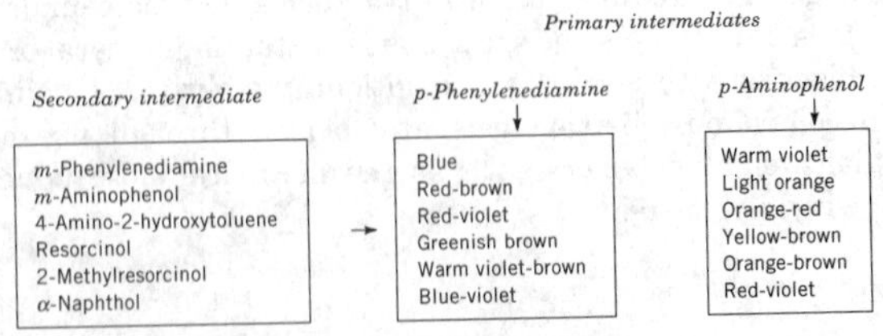

Figure 1. Quinine imines from primary intermediates couple with secondary intermediates to form various colors.

creased with bleach accelerators or boosters, including one or more per salts such as ammonium, potassium, or sodium persulfate or their combinations.

Hair Coloring Regulation Issues. In the United States the classification of color additives is complex. Under the Federal Food, Drug and Cosmetic Act, all cosmetic colors must be the subject of an approved color additive petition to the Food and Drug Administration, but there is an exception for coal-tar colorants used to color hair. For any other cosmetic use than coloring hair the FDA can require a certification on each manufactured batch of colorant to assure conformance with the approved specifications. Many of the approved color additives are restricted in their potential use. These restrictions can be found in the color additive regulations in the *Code of Federal Regulations* at 21 CFR 73 and 74.

Permanent Waving Preparations

Chemistry of Hair Waving. A particular geometry innate to each individual hair is the result of processes of keratinization and follicular extrusion that transform viscous mixtures of polypeptide chains into strong, resilient, and rigid keratin fiber. Waving entails softening keratin, molding it to a desired shape, and finally annealing the newly imparted configuration. The underlying mechanism of waving is thus essentially molecular and involves a manipulation of physicochemical interactions that stabilize the keratin structure. Covalent cystine cross-links are broken by treatment with a reducing agent. This softens the hair so that it can be molded into a new shape. The cross-links are then reformed by treatment with an oxidizing agent.

In the typical waving procedure, freshly shampooed and still damp hair is wetted with waving lotion and wound onto plastic curlers with the help of porous end papers or sponges. Hair is then left to process, rinsed thoroughly after 10–20 min, and neutralized while still on rods. After neutralization, hair is unwound, rinsed again, and either freely dried or set in the desired style.

Waving Lotions. The reagent most frequently used for the reduction of hair is thioglycolic acid. Conventional waving lotions contain 0.5–0.8 *M* thioglycolic acid adjusted to and maintained at pH 9.1–9.5. The neutralizing base is ammonia, alkanolamines, or both.

Alkali sulfites have gained a stronghold in the hair waving market, focusing on the soft wave and casual styles for which the expectations of the waving performance are less rigorous. Reduced danger of over processing and lack of odor are clear benefits.

Waving products have appeared on the market formulated in the neutral pH range, ie, so-called acid waves. They are based either on the thioglycolic acid or its glyceryl esters. The waving performance of these products is mediocre.

In Asia, particularly Japan, the use of cysteine as a waving agent is widespread. As a waving agent cysteine is a poor performer and, in most formulations, thioglycolic acid is added to improve waving efficacy.

Waving lotions frequently are formulated with a number of additives with the intention of enhancing the efficacy and the aesthetics of the process. Thus surfactants of the nonionic type are used to improve the wetting of hair and penetration, a hydrogen bond breaking agent such as urea is added to intensify the swelling of hair, ammonium sulfate is used to decrease swelling, and latex emulsions and polyacrylates are employed as opacifiers. Conditioning materials used include mineral oil, lanolin, and hydrolyzed protein; the addition of cationic polymers has been patented.

Neutralizing Lotion. The principal active ingredient of cold wave neutralizers is usually an oxidizing agent. The most popular is hydrogen peroxide. Aqueous solutions of sodium bromate occasionally are used. Wetting and foaming additive agents occasionally are used to improve spreading and retention of the neutralizer in the hair. Acids such as citric acid and tartaric acid are suggested for the deswelling of hair and thus improvement in its overall condition. Conditioning agents such as stearalkonium chloride are frequently employed to assure smooth texture, easy combing, and control of flyaway. Conditioning additives based on polymeric silicones have been patented.

Evaluation. Two parallel approaches are used in the industry to assess the efficacy of waving formulations. These are full- and half-head tests against established products, and laboratory evaluation of hair tresses processed according to waving instructions.

Health and Safety. The dermal toxicology of alkaline solutions of thioglycolic acid has been reviewed extensively. The reagent has been found harmless to normal skin when used under conditions adopted for cold waving. Hand protection is recommended for the professional hairdressers who routinely handle these products.

Manufacturing. The highly reactive nature of the active components of the permanent waving products requires rigorous control at every production state.

Hair Straightening Preparations

Temporary Hair Straightening. The most frequently used technique in this category is hot combing. The function of the pressing oil is to act simultaneously as a protective heat-transfer agent between the comb and the hair and as a lubricant to reduce the drag of the comb. Pressing oils are usually based on petrolatum and mineral oil mixed with some wax and a perfume.

Permanent Hair Straightening. The basic technical premise underlying permanent hair straightening is similar to that adhered to in waving. It thus is not surprising that many hair straightening compositions are just thickened versions of permanent waving products. Alkaline thioglycolate (6–8%) is formulated into a thick oil-in-water (o/w) emulsion or cream using generous concentrations of cetyl alcohol and stearyl alcohol and high molecular weight polyethylene glycol, together with a fatty alcohol sulfate as emulsifier. Hair straightening compositions based on mixtures of ammonium bisulfite and urea have been introduced and have found some application in the Caucasian hair market. An important class of permanent straighteners in use is that based on alkali as an active ingredient. Sodium hydroxide, potassium hydroxide, or a sodium carbonate guanidine combination is used at concentrations of 1.5–3% in a heavy cream base.

Professional Use Products

Many products in the hair care and hair color categories are distributed solely for professional use by cosmetologists, beauticians, and hairdressers in their places of business. The Fair Packaging and Labeling Act does not apply to products used in professional establishments. Specifically, this means that these products are not required to have an identity statement or a list of ingredients.

Economic Aspects

Retail sales of hair preparations have more than doubled from $\$2 \times 10^9$ in 1978 to ca$\$4.2 \times 10^9$ in 1991. While price increases over this 13-year period were clearly a factor, a variety of novel and functional products have been introduced into this market.

Regulations

Definitions. Cosmetic products in the United States are regulated by FDA under the authority of two different laws, ie, the Federal Food, Drug and Cosmetics Act and the Fair Packaging and Labeling Act. Each of these Acts imposes slightly different conditions and labeling requirements for the products under their jurisdiction.

Drug Products. Although most hair care products are cosmetics and are regulated as such, some products also can be drugs. If the product is intended to treat or prevent a disease condition or to affect the structure or function of the body, the product is a drug. Therefore, any product that makes a representation that it can control dandruff, treat psoriasis or seborrheic dermatitis, grow hair, prevent baldness, or other similar claims, is considered a drug product. Products that make both drug and cosmetic claims are considered drugs and must be in compliance with both the drug and cosmetic regulations.

STANLEY POHL
JOSEPH VARCO
PAUL WALLACE
LESZEK J. WOLFRAM
Clairol, Inc.

J. F. Corbett and K. Ventakamaran, eds., *Chemistry of Synthetic Dyes*, Vol. 5, Academic Press, Inc., New York, 1971.

M. S. Balsam and E. Sagarin, eds., *Cosmetics, Science and Technology*, Wiley-Interscience, New York, 1972.

C. Zviak, ed., *The Science of Hair Care*, Marcel Dekker, New York, 1986.

HARDNESS

Hardness is a measure of a material's resistance to deformation. In this article hardness is taken to be the measure of a material's resistance to indentation by a tool or indenter harder than itself. This seems a relatively simple concept until mathematical analysis is attempted; the elastic, plastic, and elastic recovery properties of a material are involved, making the relationship quite complex. Further complications are introduced by variations in elastic modulus and frictional coefficients.

As a consequence, although the precise analysis of the indentation process continues, numerous practical applications of indentation hardness are in use and others are being developed. The impetus to this development is that whatever the numerical value of indentation hardness, it is clearly related to many other material properties of greater interest to engineers such as strength, wear resistance, and machinability. The relationship to the strength properties of materials is the most important. The indentation hardness test provides at once a simple, rapid, and essentially nondestructive means of testing a material and discovering its strength.

A hardness indentation causes both elastic and plastic deformations which activate certain strengthening mechanisms in metals. Dislocations created by the deformation result in strain hardening of metals. Thus the indentation hardness test, which is a measure of resistance to deformation, is affected by the rate of strain hardening.

Anisotropy in metals and composite materials is common as a result of manufacturing history. Anisotropic materials often display significantly different results when tested along different planes. This applies to indentation hardness tests as well as any other test.

Many types of hardness tests have been devised. The most common in use are the static indentation tests, eg, Brinell, Rockwell, and Vickers. Dynamic hardness tests involve the elastic response or rebound of a dropped indenter, eg, Scleroscope (Table 1).

Table 1. Hardness Tests Described by ASTM Standards

Common name	Title	ASTM number
Brinell	Brinell Hardness of Metallic Materials	E10
Rockwell	Rockwell Hardness and Rockwell Superficial Hardness of Metallic Materials	E18
Vickers DPH	Test Method for Vickers Hardness of Metallic Materials	E92
Knoop/DPH	Test Method Microhardness of Materials	E384
Scleroscope	Recommended Practice for Scleroscopic Hardness Testing of Metallic Materials	E448
International Rubber	Test Method for Rubber Property International Hardness	D1415
Durometer	Test Method for Rubber Property Durometer Hardness	D2240
Barcol	Test Method for Indentation Hardness of Rigid Plastics via Barcol Impresser	D2583
Portable	Test Method for Indentation Hardness of Metal using Portable Hardness Testers	E110
Webster	Webster Hardness Gauge	B647

Although indentation hardness tests are usually classified as nondestructive, they do in fact leave a permanent indentation on the surface of the workpiece.

Hardness Conversions

Despite variations in hardness test procedures and the variations in physical properties of the materials tested, hardness conversions from one test to another are possible. This approximate relationship is only consistent within a single-material system, eg, iron, steel, or aluminum. Conversion of hardness data to some measure of strength is also possible and has been done for several common materials.

RONALD D. CROOKS
Consultant

L. Small, *Hardness Testing*, American Society for Metals International, 1987.

A. R. Fee and co-workers, *Mechanical Testing*, Vol. 8, American Society for Metals International, 1985, pp. 71–113.

H. Scott and co-workers, *Metals Handbook*, American Society for Metals International, pp. 93–104.

V. E. Lysaght and A. DeBellis, *Hardness Testing Handbook*, American Chain and Cable Co., 1969.

HAZARD ANALYSIS AND RISK ASSESSMENT

The purpose of hazard analysis and risk assessment in the chemical process industry is to *(1)* characterize the hazards associated with a chemical facility; *(2)* determine how these hazards can result in an accident; and *(3)* determine the risk, ie, the probability and the consequence of these hazards. The complete procedure is shown in Figure 1.

Hazard Identification Procedures

Methods for performing hazard analysis and risk assessment include safety review, checklists, Dow Fire and Explosion Index, what-if analysis, hazard and operability analysis (HAZOP), failure modes and effects analysis (FMEA), fault tree analysis, and event tree analysis.

Scenario Identification

An important part of hazard analysis and risk assessment is the identification of the scenario, or design basis by which hazards result in accidents. Hazards are constantly present in any chemical facility. It is the scenario, or sequence of initiating and propagating events, which makes the hazard result in an accident.

It is not practicable to perform detailed studies on all possible scenarios; thus many studies focus on identifying the worst practicable scenario and the worst potential scenario. The worst practicable scenario considers scenarios which have a reasonable chance for occurrence. The worst potential scenario is a scenario leading to the largest catastrophe.

Source Modeling and Consequence Modeling

Once the scenario has been identified, a source model is used to determine the quantitative effect of an accident. This includes either the release rate of material, if it is a continuous release, or the total amount of material released, if it is an instantaneous release.

Once the source modeling is complete, the quantitative result is used in a consequence analysis to determine the impact of the release. This typically includes dispersion modeling to describe the movement of materials through the air, or a fire and explosion model to describe the consequences of a fire or explosion. Other consequence models are available to describe the spread of material through rivers and lakes, groundwater, and other media.

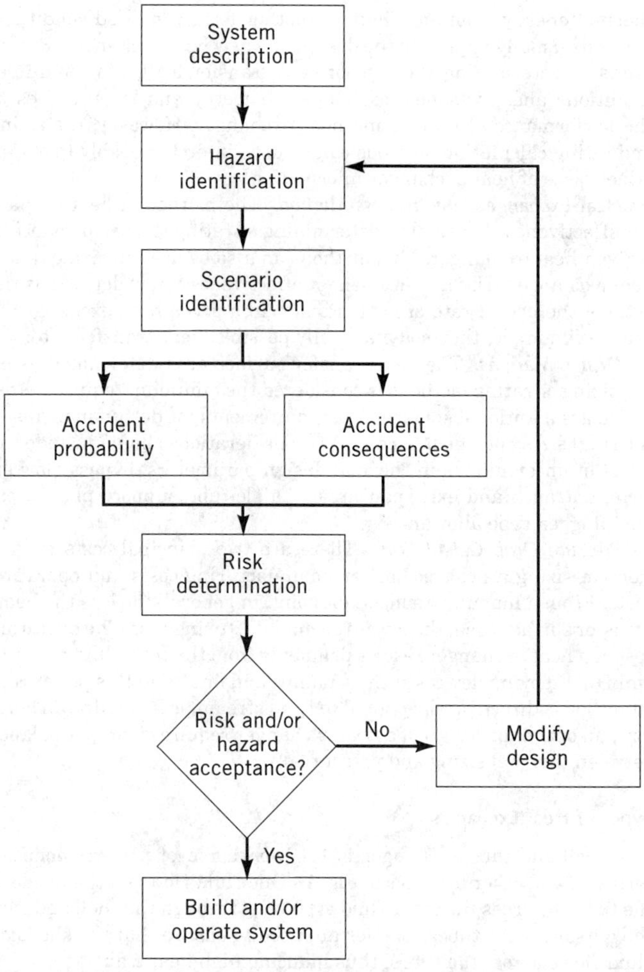

Figure 1. Flow chart representing the complete hazard identification and risk assessment procedure.

Probability

In order to complete an assessment of risk, a probability must be determined. The easiest method for representing failure probability of a device is an exponential distribution.

$$R(t) = e^{-\mu t} \quad (1)$$

where $R(t)$ is the reliability, μ is the failure rate in faults per time, and t is the time.

Once the reliability is defined, the failure probability, $P(t)$, follows.

$$P(t) = 1 - R(t) = 1 - e^{-\mu t} \quad (2)$$

A considerable assumption in the exponential distribution is the assumption of a constant failure rate. Real devices demonstrate a failure rate curve more like that shown in Figure 2. For a new device, the failure rate is initially high owing to manufacturing defects, material defects, etc. This period is called infant mortality. Following this is a period of relatively constant failure rate. This is the period during which the exponential distribution is most applicable. Finally, as the device ages, the failure rate eventually increases.

The next step is to develop a method to determine the overall reliability and failure probability for systems constructed of a variety of individual components. This requires an understanding of how components are linked. Components are linked either in series or in parallel. For series linkages, overall failure results from the failure of any of the components. For parallel linkages, all of the components must fail.

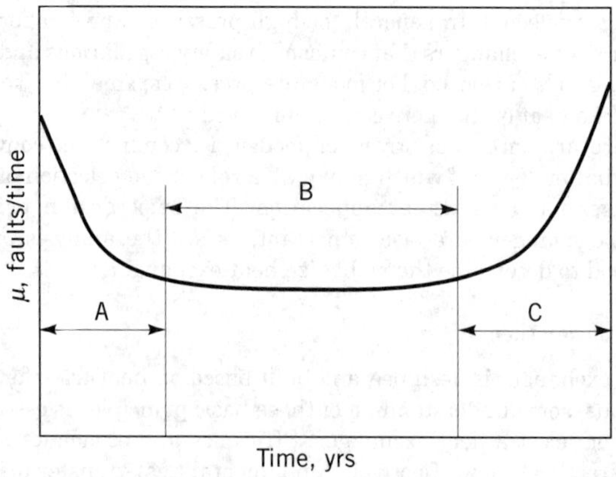

Figure 2. Failure rate curve for real components. A, infant mortality; B, period of approximately constant μ; and C, old age.

The computational technique for linkages is the following: for series linkages, the reliabilities of the individual components are multiplied together; for parallel linkages the failure probabilities are multiplied together.

The numbers computed using this approach are only as good as the failure rate data for the specific equipment. Frequently, failure rate data are difficult to acquire.

Hazard Acceptance and Inherent Safety

The remaining step in the hazard identification and risk assessment procedure shown in Figure 1 is to decide on risk acceptance. For this step, few resources are available and analysts are left basically by themselves.

A more recent concept which could have significant impact on future designs is that of inherent safety. This basic principle states that what is not there cannot be blown up or leak into the environment. Thus, the idea is to avoid the hazard in the first place.

Hazard avoidance is performed by three techniques. First, there is substitution. This means substituting a less hazardous material for the material in use. The second method for inherent safety is attenuation, ie, operating the process at lower temperatures and pressures. The last inherent safety technique is intensification. This means using much smaller inventories of hazardous raw and intermediate materials, and reducing process hold-up and inventories.

DANIEL A. CROWL
Michigan Technological University

Guidelines for Hazards Evaluation Procedures: Second Edition with Worked Examples, American Institute of Chemical Engineers, Center for Chemical Process Safety, New York, 1992.

D. A. Crowl and J. F. Louvar, *Chemical Process Safety: Fundamentals with Applications*, Prentice Hall, Englewood Cliffs, N.J., 1990.

Guidelines for Chemical Process Quantitative Risk Analysis, American Institute of Chemical Engineers, Center for Chemical Process Safety, New York, 1989.

F. P. Lees, *Loss Prevention in the Process Industries*, Butterworths, London, 1986.

HEAT-EXCHANGE TECHNOLOGY

HEAT TRANSFER

In order to select a proper heat exchanger for a given application, various factors such as pressure, temperature, size, fouling factor, and the use of toxic or corrosive fluids must be considered. These pressure and temperature requirements mainly dictate the type of heat exchanger selected. In general, for high pressures and temperatures, tubular heat exchangers that conform to safety regulations and manufacturing codes are used. For moderate pressures, small but very efficient plate heat exchangers can be employed.

There are three heat-transfer modes, ie, conduction, convection, and radiation, each of which may play a role in the selection of a heat exchanger for a particular application. The basic design principles of heat exchangers are also important, as are the analysis methods employed to determine the right size heat exchanger.

Heat-Transfer Theory

A heat exchanger is designed and built based on heat-transfer principles; thus, some understanding of these basic principles is essential to design or select a heat exchanger. Efficiency and economics may depend directly on how effectively fundamental heat-transfer principles are applied in the design of the heat exchanger.

Conduction Heat Transfer. When there is a temperature difference in a body, there is an energy transfer from the high temperature region to the low temperature region, a phenomenon called an energy transfer by conduction. Although conduction occurs in liquids and gases, the contribution to heat transfer is relatively small as compared to convection or radiation for these cases. In a solid such as a metal tube wall or a flat wall made of multicomponent materials, however, conduction is the dominant heat-transfer mode. In most conventional heat exchangers, heat transfer occurs between two fluids separated by solid walls, which are either a tube wall in tubular heat exchangers or a plane wall in plate heat exchangers.

Fourier's Law of Heat Conduction. The heat-transfer rate, Q, per unit area, A, in units of W/m^2 ($Btu/(ft^2 \cdot h)$) transferred by conduction is directly proportional to the normal temperature gradient:

$$\frac{Q}{A} \sim \frac{dT}{dx} \tag{1}$$

or in equation form the heat-transfer rate Q becomes,

$$Q = -kA\frac{dT}{dx} \tag{2}$$

where the proportionality constant, k, is called the thermal conductivity of the material. The minus sign, required in equation 2 to ensure that the direction of the heat transfer is positive when the temperature gradient is negative, is necessary because thermal energy flows in the direction of decreasing temperature.

Convection Heat Transfer. Convective heat transfer occurs when heat is transferred from a solid surface to a moving fluid owing to the temperature difference between the solid and fluid. Convective heat transfer depends on several factors, such as temperature difference between solid and fluid, fluid velocity, fluid thermal conductivity, turbulence level of the moving fluid, surface roughness of the solid surface, etc. Owing to the complex nature of convective heat transfer, experimental tests are often needed to determine the convective heat-transfer performance of a given system. Such experimental data are often presented in the form of dimensionless correlations.

Convective heat transfer is classified as forced convection and natural (or free) convection. The former results from the forced flow of fluid caused by an external means such as a pump, fan, blower, agitator, mixer, etc. In the natural convection, flow is caused by density difference resulting from a temperature gradient within the fluid.

Newton's Cooling Law of Heat Convection. The heat-transfer rate per unit area by convection is directly proportional to the temperature difference between the solid and the fluid which, using a proportionality constant called the heat-transfer coefficient, h, becomes

$$Q = hA(T_{\text{fluid}} - T_{\text{solid}}) \tag{3}$$

Basic Thermal Design Methods for Heat Exchangers

The basic heat-transfer principles of sizing and rating heat exchangers are important to design. Sizing refers to determining the amount of heat-transfer surface area required to transfer a specified quantity of thermal energy from one fluid to another for given fluid conditions and thus usually applies to the design of a new heat exchanger. Rating refers to determining the rate of heat transfer for given fluid-inlet conditions and given heat-exchanger geometry and thus applies to the performance of an existing heat exchanger. However, the sizing and rating calculation methods can be used interchangeably to obtain either piece of heat exchanger information.

Heat-Exchanger Effectiveness Method. The method of heat-exchanger effectiveness is useful in determining or rating the performance of a given heat exchanger. This method can also be used in sizing a new heat exchanger. The heat-exchanger effectiveness, ϵ, is defined as the ratio of the actual rate of heat transfer in a given heat exchanger to the maximum, ie, thermodynamically possible, heat-transfer rate.

Design Margins. The heat-transfer surface area determined using the sizing or rating methods is considered the minimum required area. There are additional surface-area requirements for design margins in the final sizing of a heat exchanger. Considerations should include the effect of uncertainties in thermal design parameters, bypass flow effects, entrance and exit span areas, baffle-tube support plate area, and plugged tube allowance.

Pressure Drop Calculations. There are two principal costs to consider in sizing a heat exchanger: manufacturing costs and operating costs. From a manufacturing standpoint, in general, the less the heat-transfer surface area, the lower the manufacturing cost. The operating cost of a heat exchanger results primarily from the cost of the power to run fluid-moving devices such as pumps and fans, and this power consumption is directly proportional to fluid stream pressure drop. Therefore, an optimum design of a heat exchanger requires a proper balance between thermal sizing and pressure drop.

Types of Heat Exchangers

The shell-and-tube exchanger is the workhorse of power, chemical, refining, and other industries (Fig. 1). One fluid flows on the inside of the tubes whereas the other fluid is flowing through the shell and over the outside of the tubes. Baffles are used to ensure that the shellside fluid flows across the tubes, thus inducing high heat transfer.

Plate heat exchangers, which are used as an alternative to shell-and-tube heat exchangers in relatively low temperature and pressure applications involving liquids and two-phase flows, have some important advantages over shell-and-tube exchangers. Plate heat exchangers include plate–frame heat exchangers, plate–fin heat exchangers, and plate–coil heat exchangers. The advantages of using plate heat exchangers are less surface for heat transfer is required, resulting in weight, volume, and cost advantage over shell-and-tube and other noncompact heat exchangers; thermal rating of plate heat

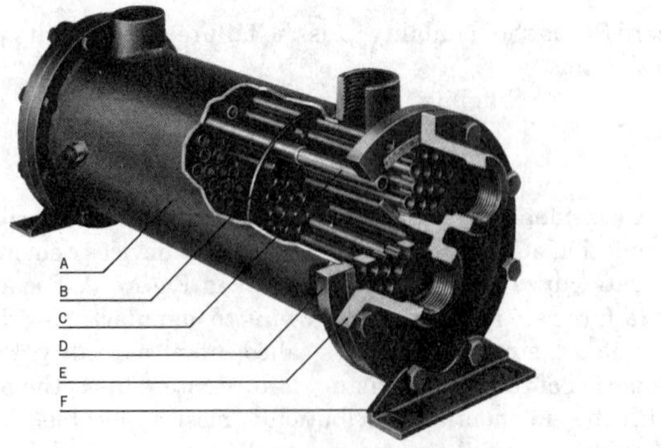

Figure 1. Shell-and-tube heat exchanger: A, shell of high strength; B, tube sheet; C, tubes (normally small diameter tubes are seamless, but large diameter tubes (>1 in.) are welded tubes); D, bonnets; E, baffles to assure more efficient circulation by providing minimum clearance between tubes and tube holes as well as baffles and shells; and F, mounting brackets. Courtesy of Basco.

exchangers can readily be increased or decreased by varying the number of plates, which is important if substantial changes in load occur; and the increased effectiveness of plate heat exchangers reduces the required cooling flow rate, resulting in savings relative to piping, pumps, valves, and operating cost. In spite of these advantages, however, the plate heat exchangers are rarely used, even in relatively low temperature and pressure applications. This may be because of the widespread familiarity with shell-and-tube exchangers and the large number of manufacturers of shell-and-tube exchangers. Plate heat exchangers are not normally used in nuclear applications for safety considerations. Note that ASME Boiler and Pressure Vessel codes do not recognize the plate heat exchangers.

The principal disadvantage of plate–frame heat exchangers is the large number of surfaces that must be sealed by gaskets.

Cold-plate heat exchangers for electronics cooling applications operate at heat-flux levels typically on the order of 2–10 W/cm^2. The electronics industry has targeted heat-flux capacities of up to 25 W/cm^2 for the next generation cold-plate for advanced applications. Achieving this level of cooling requires development of methods of providing high surface-density cooling within coolant passages in the cold plate. Potential scenarios that might provide high heat flux cooling may include high density ribbed or finned surfaces or impinging jet cooling of the cold-plate primary surface.

Use of Heat Exchangers

Heat exchangers are used whenever energy has to be transferred, and the proper design and use of heat exchangers are vitally important for efficient operation of an industrial system, for energy conservation, and ultimately for the protection of the environment. Despite decades of continuous research and development, there are numerous design and operating problems originating from a lack of understanding of basic flows and heat-transfer phenomena such as flow distribution in manifolds, flow-induced vibration in two-phase flows, heat-transfer enhancement, fouling, etc.

In order to help companies and organizations overcome problems associated with heat exchangers, Heat Transfer Research Inc. (HTRI) was established in the United States in 1962, and the Heat Transfer and Fluid Flow Service (HTFS) was established in the United Kingdom in 1967. These organizations provide results of heat-transfer and fluid-flow research, design methods, supporting computer programs, and proprietary equipment testing. In addition, the American Society of Mechanical Engineers (ASME), American Society for Testing and Materials (ASTM), and Tubular Exchanger Manufacturers Associations Inc. (TEMA) provide various safety and design codes and technical services. More recent reference books come with software disks, and analyses or design calculations of various energy systems can be conducted.

Fundamental issues involved in the use of various heat exchangers have been summarized in a thermal science workshop sponsored by the National Science Foundation. There are a number of areas that require different types of heat exchangers. Some of the emerging technologies where heat exchangers are expected to play a critical role are electronic cooling, micro and macro gravity applications, ozone depletion, global warming and other environmental issues, biotechnology, high temperature superconductors, and ultrahigh temperature waste-heat recovery.

Header Design

Headers, ie, manifolds and tanks, are the chambers or transition ducts at each end of the heat-exchanger core on each fluid side for distributing fluid to the core at the inlet and collecting fluid at the exit. These may be classified broadly as normal, turning, and oblique flow headers. Poor design of headers reduces heat-transfer performance significantly and may also increase pressure drop substantially owing to flow maldistribution, flow separation, and jet effects. Thus header design is an important problem for all heat exchangers where fluid from the inlet pipe is distributed to the exchanger core via manifolds and tanks. If novel heat-exchanger applications are contemplated, the header volume must be a very small fraction of the total exchanger volume, particularly for highly compact heat-exchanger applications.

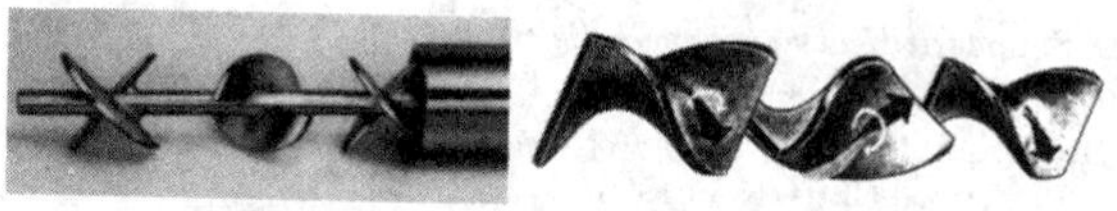
(a) (b)

Figure 2. Static mixers which provide a continuous mixing and processing unit with a nonmoving part. These static mixers can be easily installed in new and existing pipelines. (**a**) Courtesy of Ross; (**b**) courtesy of Chemineer.

No design theory and modeling is available to obtain uniform flow for normal headers, ie, diffusers having downstream flow resistance and turning headers, with or without vanes. Only very limited design information is available for oblique flow headers. Manifolds in a heat exchanger can be further classified into four types: dividing, combining, parallel, and reverse flow manifolds. Parallel and reverse flow manifolds are those which combine dividing and combining flow manifolds. In a parallel flow manifold, the flow directions in dividing and combining flow headers are the same; in a reverse flow manifold, the flow directions are opposite. The objective of the manifold design is to obtain a uniform flow distribution in the heat-exchanger core, with the manifold occupying the smallest fraction of volume of the total heat exchanger.

Several investigators have conducted analytical and numerical studies on dividing and combining flow manifolds. Friction was shown always to increase the flow imbalance in a combining flow manifold and friction might either increase or decrease the flow imbalance in a dividing flow manifold depending on the area ratio. The larger the cross-sectional area of dividing and combining flow manifolds, the better the flow distribution has been reported to be.

The flow distribution is a direct consequence of the static pressure difference between dividing and combining flow headers. There are two factors controlling the pressure variations in manifold headers: friction and momentum. In a combing flow header, these two factors lower the pressure along the header in the flow direction. However, in a dividing flow header, these two factors work in opposite directions. The friction effect lowers the pressure along the header, whereas the momentum effect increases the pressure. The flow velocity decreases in the flow direction owing to fluid loss into channels, creating momentum deficiency along the dividing flow header and thus increasing pressure. Furthermore, the pressure increases near the end of the dividing flow header due to the conversion of kinetic energy to stagnation pressure.

Performance Enhancement in Heat Exchangers

Static Mixer. To enhance the performance of conventional shell-and-tube heat exchangers, one can use static mixer elements inside tubes as shown in Figure 2. Process fluid is continuously mixed, thus producing performance enhancement.

Advanced Heat-Transfer Fluid. A conventional heat-exchanger system requires a high volumetric flow rate, resulting in the consumption of a large amount of pumping power. The use of an advanced heat-transfer fluid has been proposed to increase the convective heat-transfer coefficient by increasing the effective thermal capacity of working fluids, a technique that would permit the use of a smaller volumetric flow rate and smaller heat exchangers.

YOUNG I. CHO
Drexel University
S. M. CHO
Foster Wheeler Energy Corporation

H. Martin, *Heat Exchangers*, Hemisphere Publishing Corp., New York, 1992.

E. A. D. Saunders, *Heat Exchangers: Selection, Design, and Construction*, John Wiley & Sons, Inc., New York, 1988.

B. K. Hodge, *Analysis and Design of Energy Systems*, 2nd ed., Prentice-Hall, Inc., Englewood Cliffs, N.J., 1990.

E. Choi, Y. I. Cho, and H. G. Lorsch, *Int. J. Heat Mass Trans.* **37**, 207–215 (1994).

HEAT-TRANSFER MEDIA OTHER THAN WATER

At temperatures below 0°C or above 200°C, heat-transfer media other than water often are the more optimum choice. Glycol/water, typically ethylene glycol or propylene glycol solutions, are widely used for liquid-phase secondary cooling and heating applications. Using appropriate inhibitors, the glycol-based fluids can be used over a temperature range of −50 to 175°C. Glycol/water solutions intended for heat-transfer systems are manufactured by Union Carbide Corp. and Dow Chemical.

High Level Heat-Transfer Media

The heat-transfer medium must exhibit sufficient thermal stability at the service temperature and high enough flash and fire points to permit safe operation. Most high level heat-transfer fluids are used at temperatures above the flash and fire points under proper protection from flames and arcs, but are not used above their autoignition temperatures. Other factors that must be considered include ease of reprocessing fluid that experiences thermal degradation and ease of monitoring the system to detect the presence of decomposition products or contaminants. Compatibility with process fluids and the ability to resist damage from nuclear radiation also may be important.

Several generalizations can be made concerning thermal stability and degradation of organic heat-transfer media. *(1)* Aromatic materials exhibit thermal stabilities that generally are superior to aliphatic compounds. *(2)* The recommended maximum operating temperature for commercially available products is a rough measure of relative thermal stability. *(3)* Fluid degradation should produce a minimum of volatile materials *(4)* Degradation should not produce reactive or corrosive materials. *(5)* Oxidation stability may be an important factor if air is present at high temperatures. Low insoluble sludge formation is an advantage.

Vapor-Phase and Liquid-Phase Operation

When establishing whether liquid-phase or vapor-phase systems are better, it is necessary to consider the overall process and economics, the thermal tolerance of the process, and the required equipment. In many cases, the costs for the two systems do not differ significantly. In vapor-phase systems, heat is transferred at the saturation temperature of the vapor, which affords uniform and precisely controlled temperatures. In liquid-phase systems, the temperature of the fluid necessarily changes as heat is transferred, therefore, temperatures are not uniform even if large circulation rates are employed for the heat-transfer fluid. In systems having multiple heat users, a combination of both vapor and liquid phase may be preferred. For small, compact systems, natural-convection vapor-phase systems generally are preferred. Electrically heated packaged units are commonly used for small liquid-phase systems. For large systems, heat losses from fluid piping may be greater for vapor-phase systems. Larger vapor-phase systems frequently require forced circulation condensate return when there are several users at different temperature levels.

Heat-Transfer Fluids

Heat-transfer fluids are listed in Table 1.

Gases. The common permanent gases can be used as heat-transfer media and are the only substances capable of spanning the entire range of temperatures required in industrial applications. Gas systems for heat transfer are characterized by low rates of heat transfer, large volumetric flow rates, and high pumping costs. However, most of these disadvantages can be offset by operating at moderate pressures, eg, 2 mPa (20 atm), and by using extended (finned) surfaces for heat transfer. Commonly used gases include air, flue gases, nitrogen, carbon dioxide, hydrogen, helium, and argon. Superheated steam is also frequently used as a heat-transfer fluid.

Table 1. Commercially Available Heat-Transfer Fluids

Fluid	Chemical composition	Temp. range, °C Min	Max	AIT,[a] °C
	Petroleum oils			
Mobiltherm 603	paraffinic oil	40	290	350
Caloria HT 43	paraffinic oil	40	315	354
Thermia Oil C	paraffinic oil	40	290	
Calflo FG	paraffinic oil	40	260	285
Calflo AF	paraffinic oil	40	290	343
Calflo HTF	paraffinic oil	40	325	355
Multitherm PG-1	mineral oil	65	315	366
Multitherm IG-2	paraffinic oil	65	815	371
Multitherm 503	1-decene dimer	−50	260	324
Paratherm NF	mineral oil	65	315	366
Paratherm HE	paraffinic oil	65	315	371
Therminol HFP	paraffinic oil	4	300	385
Therminol Experiment	mineral oil	−15	315	324
	Synthetic fluids			
Tetralin	hydronaphthalene	40	310	384
UCON HTF-500	polyalkylene glycol	40	260	415
Dowtherm A	diphenyl/diphenyl oxide	40	400	621
Dowtherm G	aryl ethers	0	370	584
Dowtherm LF	alkylated aromatic	−40	340	467
Dowtherm J	alkylated aromatic	−70	315	420
Dowtherm Q	alkylated aromatic	0	330	411
Dowtherm HT	hydrogenated terphenyls	10	340	350
Therminol 55	alkylated aromatic	−20	290	357
Therminol 59	alkylated aromatic	−45	315	404
Therminol 60	polyaromatic mixture	−50	315	446
Therminol 66	hydrogenated terphenyls	0	340	374
Therminol 75	alkyl polyphenyls	160	400	538
Therminol LT	alkylated aromatic	−70	315	429
Therminol D-12	synthetic hydrocarbon	−45	260	277
Therminol VP-1	diphenyl/diphenyl oxide	40	400	621
Marlotherm S	dibenzylbenzenes	−15	350	500
Marlotherm L	benzyl toluenes	−50	350	500
Thermalane L	synthetic paraffin	−45	260	332
Thermalane 600	synthetic paraffin	−15	300	377
Thermalane 800	synthetic paraffin	−15	325	377
Syltherm 800	dimethylsiloxane polymer	−40	400	385
Syltherm XLT	polydimethylsiloxane	−70	260	350
Hitec Salt	nitrates and nitrites	150	540	

[a] AIT = autoignition temperature.

Liquid Metals. Liquid metals are used as heat-transfer media at temperature levels as high as can be contained by suitable materials of construction. High rates of heat transfer are achieved with liquid metals; thus, they are suitable for operation requiring high heat flux or low temperature differences. The most commonly used liquid metal is sodium–potassium eutectic.

Comparison of Heat-Transfer Fluids. A dozen fluids may fulfill the operating requirements of a specific applications. Final fluid selection should be based on safety of the fluid during service, heat-transfer rate, operating pressure drop, and system cost.

Low Level Heat-Transfer Media

Refrigeration is required to cool to temperatures lower than those attainable using cooling water or ambient air. Low temperature processing also requires a suitable low temperature fluid. Several fluids are used for both of these types of service. There are several types of refrigeration systems, each of which requires a suitable working fluid or refrigerant. Refrigerants absorb heat not wanted or needed and reject it elsewhere.

Gas-Cycle Systems. The prevailing gas that is used in closed gas-cycle refrigeration is air.

Steam-Jet Systems. Low pressure water vapor can be compressed by high pressure steam in a steam jet. In this way, a vacuum can be created over water with resultant evaporation and cooling; water, therefore, serves as a refrigerant.

Absorption Systems. Absorption refrigeration cycles employ a secondary fluid, the absorbent, to absorb the primary fluid, refrigerant vapor, which has been vaporized in the evaporator. Only two refrigerant–absorbent pairs have found extensive commercial use: ammonia–water and water–lithium bromide.

Mechanical Compression Systems. Several types of fluids are used as refrigerants in mechanical compression system: ammonia, halocarbon compounds, hydrocarbons, carbon dioxide, sulfur dioxide, and cryogenic fluids. A wide temperature range therefore is afforded.

Secondary Coolants. In many refrigeration applications, heat is transferred to a secondary coolant which is in turn cooled by the refrigerant. The secondary fluid may be any liquid that transfers heat without a change in its state. Secondary coolants frequently are called brines because such fluids originally were mixtures of salts and water. Common refrigeration brines are water solutions of calcium chloride or sodium chloride. These brines must be inhibited against corrosion. Organic fluids also are mixed with water to serve as secondary coolants. The most commonly used fluid is ethylene glycol. Others include propylene glycol, methanol, ethanol, glycerol, and 2-propanol. These solutions must also be inhibited against corrosion. Because of the relatively high viscosities of brines, many of the common refrigerants are sometimes employed as secondary coolants. Some of the synthetic heat-transfer fluids such as Dowtherm J, Therminol LF, and Syltherm XLT, are often used as the secondary fluid of choice because these offer excellent heat-transfer properties and viscosity remains low (ca10 mPa·s(= cP)) at −70°C.

Heat Pumps. Heat pumps involve the application of external power to pump heat form a lower temperature to a higher temperature. Heat pumps are frequently used for space heating and are simply refrigeration cycles operated in reverse.

Thermal Engine Cycles. Thermal engine cycles operating with organic refrigerants are employed to recover energy from waste heat streams at temperatures below 150°C. Recovery of such heat is justified only when recovery cannot be effected through process-oriented heat utilization. Typical systems employ the Rankine cycle to product electrical or shaft power. The most frequently used refrigerants are halocarbons and hydrocarbons.

PAUL E. MINTON
Union Carbide Corporation

R. L. Green, A. H. Larsen, and A. C. Pauls, "Get Fluent About Heat Transfer Fluids," *Chem. Eng.* (Feb. 1989).

J. Tria, R. L. Green, and A. H. Larsen, "Heat Transfer Fluid Properties: Their Role In Temperature Control," *Process Heating*, 27–32 (Nov.–Dec. 1994).

Equipment For Systems Using Dowtherm Heat Transfer Media, Dow Chemical Co., Midland, Mich.

Design and Operational Considerations for High Temperature Organic Heat Transfer Systems, Dow Chemical Co., Midland, Mich.

HEAT PIPES

Heat pipes are used to perform several important heat-transfer roles in the chemical and closely allied industries. Examples include heat recovery, the isothermalizing of processes, and spot cooling in the molding of plastics. In its simplest form the heat pipe possesses the property of extremely high thermal conductance, often several hundred times that of metals. As a result, the heat pipe can produce nearly isothermal conditions making an almost ideal heat-transfer element. In another form the heat pipe can provide positive, rapid, and precise control of temperature under conditions that vary with respect to time.

The heat pipe is self-contained, has no mechanical moving parts, and requires no external power other than the heat that flows through it.

Principles of Operation

The heat pipe achieves its high performance through the process of vapor-state heat transfer. A volatile liquid employed as the heat-transfer medium absorbs its latent heat of vaporization in the evaporator (input) area and releases it at the condenser (output) area. The highest possible latent heat of vaporization is desirable to achieve maximum heat-transfer and temperature uniformity with minimum vapor mass flow.

The unique aspect of the heat pipe lies in the means of returning the condensed working fluid from the condenser to the evaporator. Condensate return is accomplished by means of a specially designed wick. The surface tension of the liquid is the active force that produces wick pumping, which is a familiar process in lamp wicks and sponges. Using proper design, a substantial flow rate can be sustained against the pressure head of the counterflowing vapor or even against a slight gravitational head. In those applications where the heat source is below the heat sink, the condensate returns by gravity, ie, without the wick.

The heat pipe consists, then, of the following components: a closed, evacuated chamber (evacuation is required to establish a contaminant-free system and to prevent air or other gases from interfering with the desired vapor flow), a wick structure of appropriate design, and a thermodynamic working fluid having a substantial vapor pressure at the desired operating temperature. A schematic drawing of an elemental heat pipe is shown in Figure 1. The following basic condition must be satisfied for proper operation:

$$\Delta P_c > \Delta P_l + \Delta P_v + \Delta P_g \tag{1}$$

that is, for liquid return, the pressure difference owing to capillarity, ΔP_c, must exceed the sum of the opposing evaporator-to-condenser pressure differential in the vapor, ΔP_v, plus the pressure differential in the liquid caused by gravity, ΔP_g, plus that caused by frictional losses, ΔP_l, in the liquid. Under this condition, there is liquid flow toward the evaporator, and heat can be transferred. The pressure difference in the vapor is a direct function of the mass flow rate and an inverse function of the cross-sectional area of the vapor space. The mass flow rate is related directly to the transferred power and inversely to the latent heat of vaporization. The gravitational head is the elevation of the evaporator with respect to the condenser. It can be either positive or negative, depending on whether it aids or opposes the desired flow in the wick.

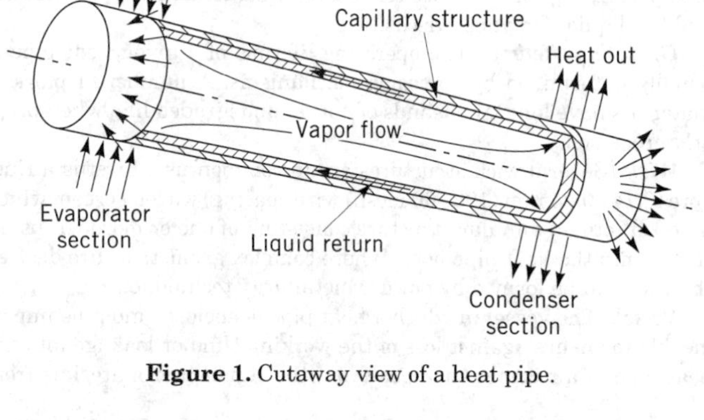

Figure 1. Cutaway view of a heat pipe.

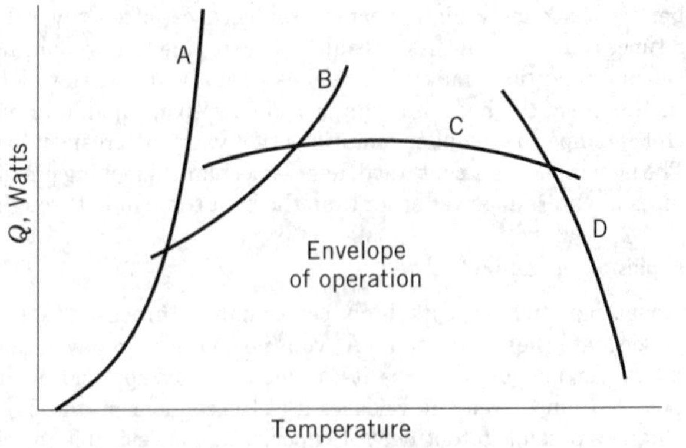

Figure 2. The throughput, Q, vs operating temperature showing the envelope of heat pipe operating limits.

Design Features

The heat pipe has properties of interest to equipment designers. One is the tendency to assume a nearly isothermal condition while carrying useful quantities of thermal power. A second property, closely related to the first, is the ability of the heat pipe to effect heat-flux transformation. The third characteristic of interest grows directly from the first, ie, the high thermal conductance of the heat pipe can make possible the physical separation of the heat source and the heat consumer (heat sink). The fourth characteristic, temperature flattening of nonuniform inputs or outputs, makes use of all three of the preceding properties.

Operational Limits

Although there are several limits which apply to heat pipe operation, these generally lend themselves to specific design solutions or occur at sufficiently high levels of performance to permit a wide latitude of practical applications. The envelope of these limits is shown generically in Figure 2. Curve A represents limits associated with vapor flow, ie, either insufficient working fluid vapor pressure is available to transport vapor along the length of the heat pipe (viscous limit) or the vapor flow has reached the sonic velocity (sonic limit). Curve B represents the entrainment limit which occurs when friction with the outgoing vapor prevents the returning liquid from reaching the evaporator. Curve C describes a wicking limit and occurs when the capillary pressure developed in the wick can no longer support the total pressure drop in the fluid flow path (includes liquid, vapor, and gravitational effects). Curve D is the boiling limit which occurs when vapor is generated within the capillary structure in an uncontrolled manner, much the same as film boiling.

Selection of Materials

Working Fluid. Qualitatively, for high power throughput under typical operating conditions, it is advantageous to have a high latent heat of vaporization, high surface tension, high liquid and vapor densities, and low liquid and vapor viscosities.

Operating Lifetime. The operating lifetime of a given heat pipe is usually determined by corrosion mechanisms. A number of pairs of materials have long (thousands of hours) undegraded life, when properly processed.

Wick. Several wick structures are in common use. First is a fine-pore (0.14–0.25 mm (100–60 mesh) wire spacing) woven screen, which is rolled into an annular structure consisting of one or more wraps inserted into the heat pipe bore. Where complex geometries are desired the wick can be formed by powder metallurgy techniques.

Vessel. The vessel in which a heat pipe is enclosed must be impermeable to ensure against loss of the working fluid or leakage into the heat pipe of air combustion gases or other undesired materials from the external environment. The vessel, as well as the wick, must be compatible with the working fluid. Where possible, the wick and vessel are made of the same material to avoid the formation of galvanic corrosion cells in which the working fluid can serve as the electrolyte.

WALTER B. BIENERT
Dynatherm Corporation
DONALD M. ERNST
Thermacore, Inc.
G. YALE EASTMAN
DTX Corporation

G. Y. Eastman, *Sci. Am.* **218**(5), 38 (1968).

P. D. Dunn and D. A. Reay, *Heat Pipes*, 4th ed., Pergamon Press, Inc., New York, 1993. *Heat Pipe Design Handbook*, Dynatherm Corp., Cockeysville, Md., 1972.

NETWORK SYNTHESIS

Synthesis of heat-exchange (energy) networks has shown large economic rewards and the most promise for future universal use.

The high cost of energy mandates carefully planned heat-exchange networks for economically and environmentally viable plants. The process of developing a good heat-exchange or energy network is most easily viewed as a multiple-tier optimization problem. Minimum lifetime plant cost is the objective. The process includes minimizing capital and operating costs plus an emphasis on developing a robust design that includes the necessary flexibility and operability characteristics.

Design Scheme

The process design engineer must understand and have a plan for using the principles, design rules, and techniques for energy, capital, and operability trade-offs. Iteration with the systematic design of other plant subsystems and between the tiers is used where appropriate; however, any significant change in results from a tier usually requires a completely different solution from the following tiers.

Once the network has been synthesized, traditional design techniques are used to set stream flows through parallel units and analyze individual heat exchangers, heaters, and coolers. Optimization of these individual units is also considered at this point. The optimization of heat exchangers involves such things as baffle and tube sheet layout as well as pressure drop.

Problem Specification. The problem of heat-exchange network synthesis can be described as follows. A set of cold streams, $i = 1$ to M, initially at supply temperature T_{si} is to be heated to target temperature T_{ti}. Simultaneously, a set of hot streams, $j = 1$ to N, initially at supply temperature T_{sj} is to be cooled to target temperature T_{tj}. The best network of heat exchangers, heaters, and coolers to accomplish the required temperature change is desired. Appropriate placement of letdown turbines and similar energy conversion devices is also needed. Best usually means most economic for the capital cost and utility costs available.

Representation. The impetus for heat-exchange network design was the development of an adequate means to represent the problem. The temperature and heat (enthalpy) relationship for any process stream can be represented on a temperature–enthalpy diagram. Figure 1 shows four different streams on such a diagram.

An important feature of the temperature–enthalpy representation is that individual streams can be lumped together in an ensemble of single hot and single cold composite streams. The composite representation helps structure the solution to a given heat-exchange network problem. If one composite curve is slid toward the other, the curves usually approach each other at a single pinch point. This is the point where the curves would first touch when slid horizontally. The pinch is analogous to the minimum temperature of approach in the design of an individual heat exchanger. The importance of the temperature pinch point to network synthesis cannot be overstated.

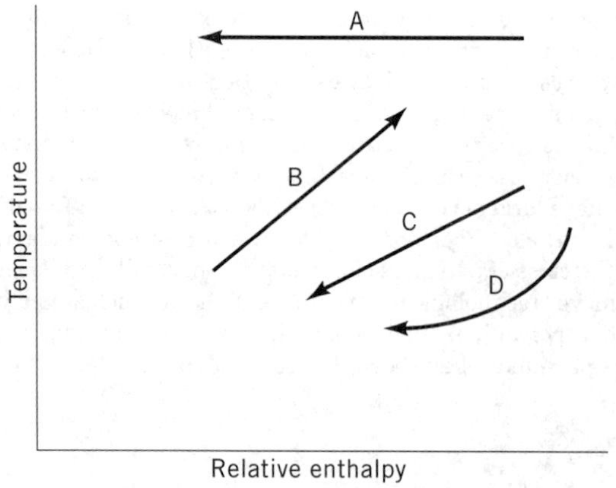

Figure 1. Temperature–enthalpy representation of stream where A represents a pure component that is condensing, eg, steam; B and C represent streams having constant heat capacity, C_p, that are to be heated or cooled, respectively; and D represents a multicomponent mixture that changes phase as it is cooled.

Alternative representations of stream temperature and energy have been proposed. Perhaps the best known is the heat-content diagram, which represents each stream as an area on a graph where the vertical scale is temperature, and the horizontal is heat capacity times flow rate.

In addition to a thermodynamic representation of streams, a means of representing the matching of streams for heat transfer is needed. The flowsheet representation has traditionally been used.

Idealization. For the purpose of network synthesis, the overall heat-transfer coefficient is usually idealized as a constant value. This independence of the heat-transfer coefficient makes possible the iterations necessary to solve the network problem.

Limiting Network Conditions

Heat-exchange network design has been made easier by the development of various limiting conditions. A detailed understanding of any model requires a clear view of the model at limiting conditions. A variety of limits have been developed and exploited in heat-exchange network design.

Feasibility (Heat-Transfer Pinch). A network must be in heat balance, but much more can be said about required heating and cooling utilities. The concept of composite streams permits quick determination of utility requirements. If one composite stream is slid horizontally, it will approach the other. The limiting condition of maximum energy recovery (MER_+) occurs when the composite streams just touch (zero-temperature driving force) or when the need for either a hot or cold utility is eliminated. If the streams touch, any possible network having limiting utility rates requires infinite heat-exchange area. On the other hand, if only a heating or cooling utility is required, the limiting case might represent a practical solution. Furthermore, the limiting utility rates for any given temperature of approach at the pinch, ΔT_p, can be found from the temperature–enthalpy diagram.

Minimum Area. The limit of minimum network area has been presented in the literature. If idealized double-pipe exchangers are used, a heat-exchange network having minimum area can quickly be developed for any ΔT_p. In the limiting case, where all heat-transfer coefficients are assumed to be equal, the area for this network can easily be obtained from the composite streams by

$$A_{\Delta T_p} = \frac{1}{U}\int_{T_{\text{lowest}}}^{T_{\text{highest}}} \frac{\partial Q}{T_j - T_i}$$

where A is the minimum possible network area at the given ΔT_p, U is the heat-transfer coefficient, and Q is the heat transferred. The vertical distance between the two composite stream curves, $T_j - T_i$, is the driving force at each point along the composite streams. This integration is similar to that used for the design of individual heat exchangers. The minimum-area value is not of great interest in itself; however, it does provide a limiting value much like the Carnot cycle for heat engines.

Minimum Number of Exchangers. The fewest number of matches or exchangers that are required in a network can be developed as a limit. The number needed, $E_{\min}$, is generally one less than the total number of streams, S (process and utility), involved in the network:

$$E_{\min} = S_{\text{process}} + S_{\text{utilities}} - 1$$

When this equation holds, each stream match (exchanger) must provide that one of the two streams involved reaches its target temperature. Such a network is called acyclic. In an acyclic network, it is not possible to trace a closed path along stream lines from exchanger to exchanger and return to the starting point without retracing some of the path.

There are exceptions to this simple equation that occur infrequently but nevertheless must be considered. A more complete relationship for the number of exchangers, E, in a network is obtained by applying Euler's network relation from graph theory:

$$E = S_{\text{process}} + S_{\text{utilities}} - P + L$$

where P is the number of independent heat loads that can be identified. This means P is the number of possible subproblems, each of which are in heat balance. In practice this is almost always one, but if it exceeds one, the minimum number of exchangers is reduced. Cyclic paths in a network are called heat-load loops.

Unfortunately, the minimum number of exchangers is not the same as the number of shells required if conventional shell-and-tube heat exchangers are used. If ΔT_p is large, the difference is not significant. However, as more energy recovery is obtained, ΔT_p must be reduced and exchangers near the pinch may require several shells in series to maintain an acceptable F factor. As for the limit of minimum area, the number of exchangers is not of paramount concern. Rather, the number of exchanger shells and the area distribution among the shells are of primary importance.

Synthesis Algorithms

The synthesis of heat-exchange networks takes place after the feasibility of developing a network has been established for the selected ΔT_p and other constraints. The primary objective is to obtain the lowest cost structure for the utility rates selected. Like many engineering optimization problems, there is not a single distinct solution but rather a set of nearly equal cost solutions. This set of solutions has been called juxtaoptimum. The selection of the best solution for a given application can be highly dependent on the difficult-to-quantify factors: operability, controllability, and flexibility. Ideally, synthesis methods would lead to all members of the juxtaoptimum set without wasting time developing poor networks. The designer could then pick a design based on these secondary factors. Synthesis methods include combinatorial, heuristic, inventive, and evolutionary.

Network Optimization

Process calculations for traditional unit-operations equipment can be divided into two types: design and performance. Sometimes the performance calculation is called a simulation. The design calculation is used to roughly size or specify the equipment. Following the design guideline, a particular piece of equipment is chosen. It is then necessary to calculate the performance of the selected item in the service of interest.

The detailed network calculations are time-consuming, and computer-aided methods are virtually required to evaluate the alternatives in the optimization process. The number of variables to consider is large. Parameters that can be varied include exchanger areas, pressure drops, and detailed mechanical aspects of the individual exchangers. These include such items as baffle cut and spacing, tube diameter, gauge, and pitch. If the network under consideration is in an existing plant, installed exchangers may be selected for use. Network operation (start-up, turndown, and shutdown) should also be of concern at this point.

A flow-sheeting program is needed to effectively handle all of these detailed design calculations. Moreover, a program capable of optimizing the selected variables is desirable.

Design Practice

Many of the calculations necessary for heat-exchange network design can be done by hand and need to be understood by every process engineer. In fact, a clear understanding of the methodology involved can be most easily accomplished using a sheet of graph paper with development of some of the curves and networks discussed. Emphasis needs to be placed on developing an overall understanding of process-energy use rather than simply adding heat exchangers to recover energy.

Computational resources make it attractive for the average process designer to use a computer program to determine and evaluate alternative network configurations. Commercial heat-exchange network programs are available and these programs typically take the raw stream data available to the process engineer for the plant and calculate the limiting network conditions. Alternative networks based on synthesis algorithms are then determined. Some of the programs even do a more detailed design of individual exchangers and subsequent optimization of the network. Following the accurate determination of flows and other stream conditions, several of the available simulators prepare the input necessary for traditional process simulation programs. However, none of these programs relieves the process engineer of the need to be the architect of the energy system.

Future Developments

Energy Networks. Heat-exchange networks are concerned with just a portion of the total energy in a process plant. Ideally, a process plant should be examined for its total energy consumption. Other plant energy systems are under consideration and should eventually be included in this type of analysis. This would include not only process thermal energy and shaft energy, but pumping requirements and electrical power as well.

Separation Networks. The area of separation networks is important in chemical-process plants. Many separations are conducted by distillation, and there has been considerable interest in the process of designing sequences of distillation columns. Columns are separation-process-run by energy, and there is an obvious tie-in with heat exchange. Some progress has been made in the analysis of energy input to and withdrawal from a distillation column. The techniques for separation-process synthesis must be integrated with those for energy network synthesis.

EDWARD HOHMANN
California State Polytechnic University

D. Boland and B. Linnhoff, *Chem. Eng.*, 22 (Apr. 1979).

V. Cena, C. Mustacchi, and F. Natali, *Chem. Eng. Sci.* **32**, 1227 (1977).

K. C. Hohmann and F. J. Lockhart, "Optimum Heat Exchanger Network Synthesis," *AICHE 82nd National Meeting*, Atlantic City, N.J., 1976.

E. C. Hohmann and D. B. Nash, "A Simplified Approach to Heat Exchanger Network Analysis," *85th National AIChE Meeting*, Philadelphia, Pa., June 1978.

HEAT-RESISTANT POLYMERS

The intensive interest in heat-resistant polymers results from new technological processes requiring higher use temperatures for development. Thermally stable or high performance polymers dictate high melting (softening) temperatures, resistance to oxidative degradation at elevated temperature, resistance to other (nonoxidative) thermolytic processes, and stability to radiation and chemical reagents.

A number of thermally stable polymers have been synthesized, but in general the types of structures that impart thermal resistance also result in poor processing characteristics. Attempts to overcome this problem have largely been concentrated on the incorporation of flexible groups into the backbone or the attachment of stable pendent groups. Among the class of polymers claimed to be thermally stable only a few have achieved technological importance, some of which are polyamides, polyimides, polyquinoxalines, polyquinolines, and polybenzimidazoles. Of these, polyimides have been the most widely explored.

Polyimides

In addition to high temperature property retention, polyimides (PI) also exhibit chemical resistance and relative ease of synthesis and use. Polyimides are available as films, fibers, enamels or varnishes, adhesives, matrix resins for composites, and molding powders. They are used in numerous commercial and military aircraft as structural composites. Work continues on these materials, including the more recent electronic applications.

Synthesis and Properties. Several methods have been suggested to synthesize polyimides. The predominant one involves a two-step condensation reaction between aromatic diamines and aromatic dianhydrides in polar aprotic solvents.

The polymers are stable up to 550°C (10% weight loss by tga) in N_2 atmosphere with the glass-transition temperatures, T_g, ranging from 281–344°C. The inherent viscosities of the polymers in H_2SO_4 are up to 0.32 dL/g. Also prepared were aromatic PIs containing triphenylamine units. These polymers show a 10% weight loss at 520°C in air, with T_gs in the range 287–331°C. Synthesis of some bismaleimides has been reported from epoxy resins which show thermal stability up to 370°C. Thermotropic poly(ester–imides) have been obtained from trimellitic acid with phosphonate or phosphate groups in the main chain. These polymers are stable in the range 410–425°C (5% weight loss) in air.

Structurally different PIs have been synthesized with alternating rigid (pyromellitimide) and semiflexible (polymethylene) units along the chain backbone. Copoly(amide–imides) comprise an important class of copolyimides that have been developed into a commercial product.

Aliphatic–aromatic poly(amide–imides) based on *N,N'*bis(carboxyalkyl) benzophenone-3,3′,4,4′-tetracarboxylic diimides have shown a 10% weight loss at 400°C.

Aromatic copoly(amide–imide)s with *s*-triazine rings in the repeating unit of the backbone are also possible from a diacyl chloride reacting with preformed imide groups and diamines containing *s*-triazine rings. Poly(phenylquinoxaline–amide–imides) are thermally stable up to 430°C and are soluble in polar organic solvents.

Polyetherimide synthesis has been achieved by reaction of a dianhydride containing an ether linkage with a diamine, reaction of a diamine containing an ether linkage with a dianhydride, or nucleophilic displacement of halo or nitro groups of a bisimide by bisphenol dianion. Such PIs exhibit good thermal stability and melt processibility.

A large variety of bisimides and polymers containing maleimide and citraconimide end groups have also been reported.

A cross-linked and crystalline copoly(ester–imide) containing an alkene function was made by reaction of an unsaturated diacid chloride containing a cyclic imido group with ethylene glycol at low temperatures.

Terpoly(amide–imide–urethanes) have been synthesized in yields up to 50–75% by the reaction of 4-carboxy-*N*-(*p*-hydroxyphenyl)phthalimide with diisocyanates in N-methyl-2-pyrrolidinone containing 5% lithium chloride.

Polypyromellitimide films based on cyclotriphosphazene and bisaspartimide-derived diamines have shown thermal stabilities up to 800°C.

Some novel copolyimides containing metal phthalocyanines are possible by treating copper, cobalt, nickel, and zinc phthalocyaninotetramines with pyromellitic dianhydride (PMDA) and tetracarboxylic dianhydride (BTDA).

Silicon-containing PIs, useful as insulation and protective materials, demonstrate adhesion to fibers, fabrics, glass, quartz, and carbon. The synthetic method used is the reaction of the silicon-containing dianhydride with diamines.

Three important linear aromatic PIs, namely LARC-TPI, LARC-160, and LARC-13 were developed by researchers at NASA-Langley Research Center.

Polyoxadiazoles

Poly(1,3,4-oxadiazole) (POD) is a widely used isomer of the oxadiazole family of thermally stable polymers.

Synthesis and Properties. Polyoxadiazoles containing aromatic moieties with aliphatic linkages/groups have been widely explored in the literature. The aromatic moieties increase the rigidity of the polymer; the presence of aliphatic groups makes the chain more flexible and processible.

One series of POD has been prepared from the corresponding dicarboxylic acid/acid chlorides and hydrazine sulfate in polyphosphoric acid (PPA). Aromatic PODs containing amide and imide groups have been synthesized by the solution polycondensation method. Carbazole-containing PODs have been obtained by cyclodehydration (in the presence of $POCl_3$) of polyhydrazides of the corresponding dicarboxylic acids. Research activities in the area of PODs containing aromatic groups have been centered around the production of highly processible, soluble, and thermally stable polymers. In this particular class of PODs, the imide- and phenylene-containing backbones have been widely explored.

Fully aromatic, thermally (up to 250°C) and hydrolytically resistant films of PODs have been realized from polyhydrazides. Films of these polymers are useful as seawater desalination membranes.

Polyoxadiazole–imides containing hexafluoroisopropylidene (HFIP) groups are soluble in common solvents and still retain good mechanical and thermal properties. Thermally stable POD films containing pyridine rings have potential application as reverse osmosis membranes. A general method for the preparation of copolyoxadiazoles, ie, poly(aryl ether oxadiazoles), has been developed where the generation of an aryl ether linkage is the polymer-forming reaction. Synthetic methods are based on either an oxadiazole-activated or an hydrazide-activated halodisplacement with phenoxides.

The synthesis of phenoxaphosphine-containing PODs by the cyclodehydration of polyhydrazides obtained from 2,8-dichloroformyl-10-phenylphenoxaphosphine-10-oxide and aliphatic and aromatic dihydrazides has been described.

Polyquinoxalines

Polyquinoxalines (PQ) have proven to be one of the better heat-resistant polymers with regard to both stability and potential application. The aromatic backbones are derived from the condensation of a tetramine with a bis-glyoxal.

Polyphenylquinoxalines (PPQ) are easier to make than the polyquinoxalines and offer superior solubility, processibility, and thermooxidative stability. The PPQs exhibit excellent high temperature adhesive, composite, and film properties. However, to increase the use temperature of PPQs, acetylene groups have been placed on the backbone and subsequently thermally cured.

Synthesis and Properties. A number of monomers have been used to prepare PQs and PPQs, including aromatic bis(*o*-diamines) and tetramines, aromatic bis(α-dicarbonyl) monomers (bisglyoxals), bis(phenyl-α-diketones) and α-ketones, bis(phenyl-α-diketones) containing amide, imide, and ester groups between the α-diketones. Significant problems encountered are that the tetraamines are carcinogenic, difficult to purify, and have poor stability, and the bisglyoxals require an arduous synthesis.

Polyquinolines

Polyquinolines are some of the most versatile thermally stable polymers.

Synthesis and Properties. Polyquinolines are formed by the step-growth polymerization of *o*-aminophenyl (aryl) ketone monomers and ketone monomers with alpha hydrogens (mostly acetophenone derivatives). Both AA–BB and AB-type polyquinolines are known, as well as a number of copolymers. The wide variety of ketomethylene and amino ketone monomers that could be synthesized, and the ability of the quinoline-forming reaction to generate high molar mass polymers under relatively mild conditions, allow the synthesis of a series of polyquinolines with a wide structural variety. Thus polyquinolines with a range of chain stiffness from a semirigid chain to rod-like macromolecules have been synthesized. Polyquinolines are most often prepared by solution polymerization of bis(*o*-amino aryl ketone) and bis(ketomethylene) monomers, where R = H or C_6H_5, in *m*-cresol with di-*m*-cresyl phosphate at 135–140°C for a period of 24–48 h. Polyquinolines have also been obtained by a post-polymerization thermal treatment of poly(enamino nitriles). In an effort to increase the processibility of polyquinolines, fluoromethylene groups have been successfully incorporated into the chain in place of Ar in the bis(ketomethylene) moiety.

Hexafluoroisopropylidene (HFIP)-Containing Polymers

Much attention has been paid to the synthesis of fluorine-containing condensation polymers because of their unique properties and different classes of polymers including polyethers, polyesters, polycarbonates, polyamides, polyurethanes, polyimides, polybenzimidazoles, and epoxy prepolymers containing pendent or backbone-incorporated bis-trifluoromethyl groups have been developed. These polymers exhibit promise as film formers, gas separation membranes, seals, soluble polymers, coatings, adhesives, and in other high temperature applications. Such polymers show increased solubility, glass-transition temperature, flame resistance, thermal stability, oxidation and environmental stability, decreased color, crystallinity, dielectric constant, and water absorption.

Hexafluoroisopropoxy (HFIP-O) Group-Containing Polymers

Several classes of polymers containing the HFIP-O group have been reported. These polymers show promise as film formers, gas separation membranes, coatings, seals, and other high temperature applications due to the properties imparted by this function, similar in many ways to the HFIP group.

Synthesis and Properties. Several polymers containing HFIP-O groups have been investigated, the most common being epoxies and polyurethanes. Numerous avenues to produce these materials have been explored. The synthesis of two new fluorinated bicyclic monomers and the use of these monomers to prepare fluorinated epoxies with improved physical properties and a reduced surface energy have been reported. The monomers have been polymerized with the diglycidyl ether of bisphenol A, and the thermal and mechanical properties of the resin have been characterized.

PATRICK E. CASSIDY
TEJRAJ M. AMINABHAVI
V. SREENIVASULU REDDY
Southwest Texas State University

H. R. Kricheldorf and R. Huner, *J. Polym. Sci., Polym. Chem. Ed.* **30**, 337 (1992).

I. K. Varma and C. K. Geetha, *J. Appl. Poly. Sci.* **22**, 411 (1978).

P. M. Hergenrother, *J. Marcromol. Sci. Rev. Macromol. Chem.* **6**, 1 (1971).

P. E. Cassidy, T. M. Aminabhavi, and J. M. Farley, *J. Macromol. Sci., Rev. Macromol. Chem. Phys.* **C29**, 365 (1989).

P. E. Cassidy and co-workers, *European Polymer J.* **31**, 353 (1995).

HEAT STABILIZERS

Heat stabilizers protect polymers from the chemical degrading effects of heat or uv irradiation. These additives include a wide variety of chemical substances, ranging from purely organic chemicals to metallic soaps to complex organometallic compounds. By far the most common polymer requiring the use of heat stabilizers is poly(vinyl chloride) (PVC). However, copolymers of PVC, chlorinated poly(vinyl chloride) (CPVC), poly(vinylidene chloride) (PVDC), and chlorinated polyethylene (CPE), also benefit from this technology. The discussion centers on heat stabilizers for PVC because this polymer is the most important class of halogenated polymers requiring these chemical additives.

In normal operations, PVC resin is intimately mixed with the desired ingredients under high intensity shear mixing conditions to result in a homogeneous dry powder compound. The heat stabilizers can be either liquids or powders and are added early in the blending cycle to afford stabilizing action during this operation. Preheating the resin to about the glass-transition temperature facilitates the adsorption of the liquid additives giving the final compound better powder flow properties and decreasing the bulk density. Post-compounding operations, eg, extrusion pelletizing, can increase the overall heat history of the polymer, thus necessitating slightly higher levels of heat stabilizers to compensate for this.

Function of Stabilizers

PVC degradation proceeds by both free-radical and ionic reactions, although the latter appears to be the more important route. Lewis acid catalysts, such as zinc chloride or hydrogen chloride, can greatly accelerate the rate of dehydrochlorination of the polymer. Heat stabilizers serve several distinct functions during PVC processing: absorption of hydrogen chloride, replacement of labile chlorines, prevention of autoxidation, and disruption of polyunsaturated sequences. An ancillary function of many heat stabilizers is provision of uv stability, leading to good weathering properties for the final articles.

Stabilizer Test Methods

In general, heat stabilizers are tested as a component in a complete formulation where each ingredient has a measured effect on the overall performance. Stabilizer performance is generally evaluated by visually inspecting the color of the test pieces as a function of heating and processing time. Static oven aging, dynamic two-roll milling, and torque rheometry are three of the most common tests used to evaluate heat stabilizers.

Classes of Heat Stabilizers

Organotin Compounds. Organotin-based heat stabilizers are the most efficient and universally used PVC stabilizers. These are all derivatives of tetravalent tin, and all have either one or two alkyl groups covalently bonded directly to the tin atom. The commercially important alkyltins are the methyltin, *n*-butyltin, and *n*-octyltin species.

Alkyltin Intermediates. For the most part, organotin stabilizers are produced commercially from the respective alkyltin chloride intermediates. There are several processes used to manufacture these intermediates. The desired ratio of monoalkyltin trichloride to dialkyltin dichloride is generally achieved by a redistribution reaction involving a second-step reaction with stannic chloride (tin(IV) chloride).

Stabilizer Synthesis. The selected alkyltin chloride intermediate reacts with either a carboxylic acid or a mercaptan in the presence of an appropriate base, such as sodium hydroxide, to yield the alkyltin carboxylate or alkyltin mercaptide heat stabilizer. Alternatively, the alkyltin chloride can react with the base to yield the alkyltin oxide, which may or may not be isolated, for subsequent condensation with the selected carboxylic acid or mercaptan.

Costabilizers. In most cases the alkyltin stabilizers are particularly efficient heat stabilizers for PVC without the addition of costabilizers. Mercaptans are quite effective costabilizers for some of the alkyltin mercaptides, particularly those based on mercaptoethyl ester technology. Combinations of mercaptan and alkyltin mercaptide are currently the most efficient stabilizers for PVC extrusion processes.

The various lubricants formulated into PVC to improve the processing can also enhance the performance of the stabilizer.

Commercial Stabilizers. In general, the alkyltin mercaptides exhibit the highest overall heat stability together with imparting excellent rheological properties to the polymer. The alkyltin carboxylates, on the other hand, are unsurpassed for imparting excellent weathering properties, but generally give poor rheological characteristics. Table 1 lists the commercially important alkyltin stabilizer compounds.

Economics. The pricing of stabilizers is generally based on the PVC processing application, the type of PVC used, and the other microingredients present in the formulation.

Health and Safety Aspects. Many of the alkyltin stabilizers are considered safe to use in almost every conceivable end use for PVC. Particularly, the U.S. FDA, German BGA, and Japanese JHPA have sanctioned the use of mixtures of dimethyltin and monomethyltin isooctyl thioglycolate, mixtures of di-*n*-octyltin and mono-*n*-octyltin isooctyl thioglycolate, and poly(di-*n*-octyltin maleate) as the primary heat stabilizers for PVC used for food packaging purposes. These same *n*-octyltin products are also approved for use in pharmaceutical applications such as pill containers and PVC tubings. Since the migration into water of the alkyltin mercaptide-based products from PVC is found to be extremely low, most of these stabilizers are suitable for pipes carrying drinking water according to NSF International, a private industry supported regulatory agency complying with all current U.S. EPA guidelines.

Studies demonstrate that the levels of volatile tin compounds in the air during PVC processing operations are significantly below the TLV of 0.1 mg/m^3 for tin compounds established for the U.S. workplace.

Mixed Metal Stabilizers. The second most widely used class of stabilizers are the mixed metal combinations. These products predominate in the flexible PVC applications in the United States: however, they find competition from the lead-based products in Europe. The only noteworthy flexible PVC application where the mixed metal products do not dominate is for electrical wire and cable coatings where the lead products are preferred, although alternative mixed metal stabilizers are continually being sought to replace the leads in this application as well.

The commercially important alkali and alkaline-earth metals used in these stabilizer systems are based on the salts and soaps of calcium, zinc, magnesium, barium, and cadmium. Other organic compounds,

Table 1. Commercially Important Alkyltin Compounds

Name	Structure
poly(dibutyltin maleate)	$\text{+}(C_4H_9)_2SnOOCCH{=}CHCOO\text{+}_n$
poly(dioctyltin maleate)	$\text{+}(C_8H_{17})_2SnOOCCH{=}CHCOO\text{+}_n$
dibutyltin bis(butyl maleate)	$(C_4H_9)_2Sn(OOCCH{=}CHCOOC_4H_9)_2$
dimethyltin bis(2-ethylhexyl thioglycolate)	$(CH_3)_2Sn(SCH_2COOC_8H_{17})_2$
dibutyltin bis(2-ethylhexyl thioglycolate)	$(C_4H_9)_2Sn(SCH_2COOC_8H_{17})_2$
dioctyltin bis(2-ethylhexyl thioglycolate)	$(C_8H_{17})_2Sn(SCH_2COOC_8H_{17})_2$
dibutyltin sulfide	$[(C_4H_9)_2SnS]_3$
methyltin tris(2-ethylhexyl thioglycolate)	$CH_3Sn(SCH_2COOC_8H_{17})_3$
butyltin tris(2-ethylhexyl thioglycolate)	$C_4H_9Sn(SCH_2COOC_8H_{17})_3$
octyltin tris(2-ethylhexyl thioglycolate)	$C_8H_{17}Sn(SCH_2COOC_8H_{17})_3$
methyltin tris(2-mercaptoethyl oleate)	$CH_3Sn(SCH_2CH_2OCOC_{17}H_{33})_3$
methyltin (2-mercaptoethyl oleate)sulfide	$CH_3Sn(SCH_2CH_2OCOC_{17}H_{33})(S)$

such as phosphites, epoxides, polyols, and β-diketones, can also be added to enhance the performance further.

The most popular commercial products are combinations including calcium–zinc, barium–calcium–zinc, barium–zinc, and barium–cadmium. Modern calcium–zinc, barium–zinc, and barium–calcium–zinc mixtures are touted as effective replacements for many of these barium–cadmium formulations. The safety of these newer products are unquestioned and certain calcium–zinc mixtures are widely used to stabilize PVC food packaging, mineral water bottles, and pharmaceutical containers throughout the world. In many applications, particularly in plasticized PVC, the mixed metal products effectively offer the right combination of processibility, heat and light stability, low odor, and nonsulfur staining characteristics to be the best choice of stabilizer.

Mixed Metal Stabilizer Synthesis. The mixed metal salts and soaps are generally prepared by reaction of commercially available metal oxides or hydroxides with the desired C_8–C_{18} carboxylic acids.

Commercial Stabilizers. There is a great variety of commercial formulations utilizing the mixture of the alkali and alkaline-earth metal salts and soaps. In many cases, products are custom formulated to meet the needs of a particular application or customer. The acidic ligands used in these products vary widely and have dramatic effects on the physical properties of the PVC formulations. The choice of ligands can affect the heat stability, rheology, lubricity, plate-out tendency, clarity,heat sealability, and electrical and mechanical properties of the final products. No single representative formulation can cover the variety of PVC applications where these stabilizers are used.

Typically, solid stabilizers utilize natural saturated fatty acid ligands with chain lengths of C_8–C_{18}. To complete the package, the solid products also contain other solid additives, such as polyols, antioxidants, and lubricants. Liquid stabilizers can make use of metal soaps of oleic acid, tall oil acids, 2-ethyl-hexanoic acid, octylphenol, and nonylphenol.

Costabilizers. There are several commercially used costabilizers and one or more are generally formulated into the mixed metal stabilizer system. Some of these additives can also be added by the PVC compounder or processor in addition to the stabilizer package to further enhance the desired performance characteristics. The epoxy compounds and phenolic antioxidants are among the most commonly used costabilizers with the mixed metal stabilizers. Other costabilizers include polyols, phosphites, β-diketones, specialty amines, and hydrotalcite.

Economics. As with the alkyltin stabilizers, the market pricing of mixed metal stabilizers tend to be directed by the particular application. The basic metal salts and soaps tend to be less costly than the alkyltin stabilizers.

Lead Stabilizers. The myriad of toxicological and ecotoxicological problems surrounding the use of any lead chemicals has restricted lead stabilizers to uses in flexible PVC wire and cable coatings in the United States. In Europe and Asia, the lead stabilizers predominate for wire and cable uses and are also widely used to stabilize PVC pipe and weatherable building profiles. These are solid products and are supplied as powders, flakes, or strands, usually in special packaging to control dusting. The commonly used commercial lead-based PVC stabilizers rely on one or more lead(II) oxide groups bound to the primary bivalent lead salt.

Environmental agencies such as The World Health Organization (WHO) and the U.S. EPA are continually lowering recommended human exposures to lead compounds. Lead products have a high refractive index and can only be used in opaque applications. Overbased lead salts have a high degree of reactivity and tend to interact, many times disfavorably, with other ingredients in the formulation. They especially react with almost any source of sulfur to form black lead sulfide, the so-called lead stain phenomenon. In some applications, they remain highly effective PVC heat stabilizers.

Stabilization Mechanism. Traditionally, lead salts were thought to perform only as acid scavengers during PVC stabilization; it is likely that this activity leads to good long-term stability. Bivalent lead compounds can readily form complexes, and recently workers have proposed that these products also displace labile chlorines on the polymer in a fashion similar to the mixed metal stabilizers.

Table 2. Principal Lead Stabilizers

Stabilizer	Formula	PbO, %	Specific gravity
tribasic lead sulfate	$PbSO_4 \cdot 3PbO \cdot H_2O$	89	6.9
dibasic lead phosphite	$PbHPO_3 \cdot 2PbO \cdot 1/2H_2O$	90	6.1
dibasic lead phthalate	$C_4H_4(COO)_2Pb \cdot 2PbO$	80	4.2
basic lead carbonate	$2PbCO_3 \cdot Pb(OH)_2$	87	6.7
dibasic lead stearate	$Pb(OOCC_{17}H_{35})_2 \cdot 2PbO$	55	2.0
lead stearate	$Pb(OOCC_{17}H_{35})_2$	29	1.4

Lead Stabilizer Synthesis. Most commercial stabilizers are produced by reaction of a water slurry of lead oxide with the appropriate acid while heating, yielding a solid product with a particle size of about 1 μm. Most often, the lead product is treated with a coating agent to reduce dusting and improve dispersability in the PVC. The stabilizer is then filtered, dried, and packaged. During the drying and coating step, other coadditives such as pigments, lubricants, and fillers can be blended into the mixture to make a total package formulation.

Commercial Stabilizers. There are six lead salts and soaps that typically are used in the commercial PVC stabilizers. The lead stearate soaps are often combined with the lead salts to provide lubrication and added stabilizer activity. The key to the high activity of these stabilizers is the very high lead content. Table 2 describes six commonly used lead stabilizers.

Economics. The lead-based stabilizers tend to be priced relatively low. The lead phthalates tend toward the higher end of this range, whereas the pipe one-pack products fall into the low end.

Health and Safety Aspects. Worldwide, there is continuing pressure by environmental and human toxicologists to reduce the use of heavy metals such as lead in every application. In the United States, the EPA and Occupational Safety and Health Administration (OSHA) provide regulations over the producers and users of lead stabilizers. The permissible exposure level (PEL) of workers is regulated at 50 $\mu g/m^3$ of airborne lead per 8-h workday. Further, workers are not allowed to be exposed to greater than 30 $\mu g/m^3$ for more than 30 days per year.

Antimony Mercaptide Stabilizers. Antimony tris(laurylmercaptide) and antimony tris(isooctyl thioglycolate) are typical of this class of heat stabilizers. These compounds were used mainly in rigid PVC applications, particularly pipes, competing with the alkyltin mercaptides. Their use greatly diminished during the late 1980s and early 1990s, particularly because questions raising doubts about the toxicological safety of antimony compounds have arisen. The performance of the antimony products has also reduced their uses in many processes. For example, they are less compatible with PVC, which leads to a cloudy appearance in clear applications; they react with sulfur sources to form Sb_2S_3, an orange-colored by-product; and they detract from the weatherability of PVC formulations. Further, because the cost of tin metal decreased significantly during the 1980s, antimony pricing has continued to rise, making these products less economically attractive than they once were.

Commercial Stabilizers. The performance of the antimony stabilizers is significantly enhanced by adding polyhydroxybenzene compounds, eg, catechol, to the PVC.

KEITH A. MESCH
Morton International, Inc.

L. I. Nass, in L. I. Nass, ed., *Encyclopedia of PVC*, Vol. 1, Marcel Dekker, New York, 1976.

E. D. Owen, in E. D. Owen, ed., *Degradation and Stabilization of PVC*, Elsevier, London, 1984, Chapt. 5.

J. Edenbaum, ed., *Plastics Additives and Modifiers Handbook*, Van Nostrand Reinhold, New York, 1992, Sect. II.

H. Andreas, in R. Gächter and H. Müller, eds., *Plastics Additives Handbook*, Hanser, Munich, 1983, Chapt. 4.

HELIUM GROUP

GASES

The helium-group gases are helium, He; neon, Ne; argon, Ar; krypton, Kr; xenon, Xe; and radon, Rn. These are all members of Group 18 (VIIIA) of the Periodic Table and are characterized by completely filled valence electron shells. Historically, they have been called the rare, noble, or inert gases. But although comparatively rare, krypton, xenon, and radon are not completely inert; all three form stable molecules with highly electronegative elements such as F, Cl, and O. Although inert enough, helium and argon are not rare; both are bulk items of commerce. At some future time, however, when helium-bearing natural gases have been depleted and the atmosphere becomes its only source, helium is expected to return to being a truly rare gas.

Occurrence

On the earth, the only practical sources of the stable helium-group gases are the atmosphere and certain helium-bearing natural gases. Faint concentrations of helium and argon are occluded in some minerals. The neon, krypton, xenon, and argon were likely part of the original mass that condensed to form the earth. However, the earth's gravitational field is inadequate to prevent helium's escape from the atmosphere. Helium is being formed continuously on earth by α-decay of heavier elements such as uranium and thorium, the α-particle being simply a fully ionized helium atom. Thus, the atmospheric helium concentration represents a dynamic equilibrium between the gain of helium diffusing from the earth's crust and the loss of helium into space. The total terrestrial inventory of helium is estimated to be 4.9×10^{14} m^3, where the volume is measured throughout at 101.3 kPa (1 atm) absolute pressure and 15°C.

Argon-40 is created by the decay of potassium-40. The various isotopes of radon, all having short half-lives, are formed by the radioactive decay of radium, actinium, and thorium. Krypton and xenon are products of uranium and plutonium fission, and appreciable quantities of both are evolved during the reprocessing of spent fuel elements from nuclear reactors.

The principal source of helium is certain natural gas fields. In the United States, recovery of helium is economical only for helium-rich gases containing more than about 0.3 vol % helium. Most of the United States helium resources are located in the midcontinent and Rocky Mountain regions, and about 89% of the known United States supply is in the Hugoton field in Kansas, Oklahoma, and Texas; the Keyes field in Oklahoma; the Panhandle and Cliffside fields in Texas; and the Riley Ridge area in Wyoming.

Resources and Conservation. The availability of the helium-group gases from the atmosphere is unlikely to change. There are no environmental sinks for these practically inert materials, and quantities removed from the atmosphere are eventually returned.

Helium continues to be a strategic material, but its importance has shifted from a simple lifting gas to a unique medium essential to high technology. The cost of separation from fuel gases and of long-term storage is great, and the probability of discovering new sources is uncertain at best. Yet the future cost of satisfying all helium requirements by extraction from the atmosphere is assuredly even higher.

Physical Properties

Pure Elements. All of the helium-group elements are colorless, odorless, and tasteless gases at ambient temperature and atmospheric pressure. Chemically, they are nearly inert. A few stable chemical compounds are formed by radon, xenon, and krypton, but none has been reported for neon and helium. The helium-group elements are monoatomic and are considered to have perfect spherical symmetry.

Some of the physical properties of helium-group elements are summarized in Table 1.

Table 1. Physical Properties of the Helium-Group Elements

Property	^{3}He	^{4}He	Ne	Ar	Kr	Xe	Rn
at. no.	2	2	10	18	36	54	86
at. wt.	3.0160	4.0026	20.183	39.948	83.80	131.30	222
normal bp, K	3.1905	4.224	27.102	87.28	119.79	165.2	211
density, kg/m^3, 0°C[a]	0.1347	0.17850	0.9000	1.7838	3.7493	5.8971	9.73
viscosity, 25°C, Pa·s							
17.2[b]	19.85	31.73	22.64	25.3	23.1	23.3	21,39
soly. in water, 20°C, mL/kg[c]	8.61	10.5	33.6	59.4	108.1	230	21

[a] Gas at 101.3 kPa. To convert kPa to psi, multiply by 0.145.
[b] Estimated.
[c] mL (101.32 kPa, 0°C) dissolved per kg water having partial pressure of pertinent gas of 101.32 kPa.

Quantum Mechanical Effects. The very light gases show significant deviations from the classical law of corresponding states, especially at cryogenic temperatures. This anomalous behavior is caused by quantum mechanical effects that become increasingly significant with decreasing atomic weight. The liquid and solid phases of the two helium isotopes exhibit physical characteristics found in no other substances.

The helium isotopes have the largest zero-point motion of any substance. This is manifested in several unique characteristics of condensed helium. The heliums are the only known substances that do not freeze under their own vapor pressure. All other materials have a triple point, that unique temperature and pressure at which the solid, liquid, and vapor phases coexist. Neither helium-3 nor helium-4 has a triple point. Under moderate pressure, both remain liquid to absolute zero. Even at the lowest temperatures, a substantial pressure is required to solidify helium, and then the solid formed is one of the softest, most compressible known. The melting curves of both helium isotopes show a minimum.

Liquid Helium-4. Because the helium-4 atom contains an even number of fermions, it is a boson. When saturated liquid helium is cooled below 2.175 K, it undergoes what is generally recognized as the manifestation of a Bose-Einstein condensation. The liquid displays a striking and unique change of properties; it becomes a superfluid.

In many ways, helium II behaves as a liquid having a vanishingly small viscosity and a very high thermal conductivity.

Another unique phenomenon exhibited by liquid helium II is the Rollin film. All surfaces below the lambda point temperature that are connected to a helium II bath are covered with a very thin (several hundredths μm) mobile film of helium II.

Several oscillatory phenomena occur in helium II. Ordinary sound waves are propagated in a normal manner in helium II. To distinguish ordinary sound from the other wave-like phenomena, it is called first sound. In first sound, the superfluid and normal components oscillate exactly in phase, thus producing the density wave associated with normal sound. When the superfluid and normal components oscillate exactly out of phase, there is no periodic change in bulk liquid density, but there is a periodic change in the concentration of the superfluid component. This concentration difference is observable as a periodic change in temperature. Second sound, then, is a propagation of heat pulses that has all of the characteristics associated with wave propagation such as reflection, refraction, and interference. Third sound involves a simultaneous thermal and density wave propagation in the Rollin film in which free oscillation of the normal fluid component is inhibited by viscous wall effects. Fourth sound is the transmission of correlated thermal and density waves through densely packed beds of fine powders which effectively immobilize the normal fluid component.

Helium-3 has not one but three superfluid phases, and these have properties quite different from superfluid helium-4. The phases are magnetic, and many of the physical properties are anisotropic. A principal distinguishing feature of the three superfluid phases is their distinctive nuclear magnetic resonance characteristics.

A number of mixtures of the helium-group elements have been studied and their physical properties are found to show little deviation from ideal solution models.

The fundamental differences in the quantum mechanical character of the two helium isotopes created much interest in the properties of mixtures. Several reviews are available. Mixtures of isotopes of a single element usually behave quite ideally, but in the case of ^3He–^4He solutions, nonideality reaches the point of forming two immiscible liquid phases.

Below about 0.5 K, the interaction between ^3He and ^4He in the superfluid liquid phases becomes very small, and in many ways the ^4He component behaves as a mechanical vacuum to the diffusional motion of ^3He atoms.

Production

Helium is separated from helium-bearing natural gases usually, but not always, in the process of removing nitrogen to improve the fuel value of the gas. Thus, in a sense, the principal part of commercial helium is as a by-product. Argon, neon, krypton, and xenon, as well as small quantities of helium, are obtained from by-product streams that are concentrated during the separation of air to produce oxygen and nitrogen. Radon is collected as a daughter product of the fission of radium, and helium-3 is the daughter product of tritium.

Commercial Distribution

Commercially pure (≥ 99.997%) helium is shipped directly from helium-purification plants located near the natural-gas supply to bulk users and secondary distribution points throughout the world. Commercially pure argon is produced at many large air-separation plants and is transported to bulk users up to several hundred kilometers away by truck, by railcar, and occasionally by dedicated gas pipeline. Normally, only crude grades of neon, krypton, and xenon are produced at air-separation plants. These are shipped to a central purification facility from which the pure materials, as well as smaller quantities and special grades of helium and argon, are then distributed. Radon is not distributed commercially.

Because of liquid helium's uniquely low temperature and small heat of vaporization, containers for its storage and transportation must be exceedingly well insulated.

Economic Aspects

Since 1987, the market for U.S.-produced helium has grown at an average annual rate of 9%. Exports of helium started in the early 1960s, and became a significant fraction of total production only after the development of equipment and techniques for transoceanic shipment of bulk liquid helium. Because argon is a by-product of air separation, its production is ca 1% that of air feed.

Uses

The primary domestic uses of helium are magnetic resonance imaging, cryogenics, welding, and pressurizing and purging. Minor uses included controlled atmospheres, breathing mixture, leak detection, and other uses such as lifting gas, heat transfer, etc.

The main uses for argon are in metallurgical applications and in electric lamps. Neon, krypton, and xenon, because of high costs, are limited to specialized uses in research, instrumentation, and electric lamps. There are no significant technical uses for radon.

Nuclear Reactors. Programs for the development of fusion power provide applications for the helium-group gases. The magnetic plasma confinement approaches require massive quantities of liquid helium to cool the large superconducting magnets and cryopumps. In the laser-ignited, inertial confinement approaches, xenon discharge lamps are a common means for energizing the large pulsed lasers. To improve energy yields, liquid helium is used to freeze the tritium fuel into a uniform, thin solid film on the inner surface of the glass microballoon targets.

In particle-track detectors, neon and helium are the common filling gases for spark chambers and streamer chambers, and liquid helium, neon, neon–hydrogen mixtures, and xenon have been used in bubble chambers. Helium-3 ions are used as bombarding particles in activation analysis.

SHUEN-CHENG HWANG
WILLIAM R. WELTMER, JR.
The BOC Group, Inc.

R. E. Stanley and A. A. Moghissi, eds., *Noble Gases*, CONF 730915 (Library of Congress 75-27055), ERDA TIC, 1973.

K. D. Timmerhaus and co-eds., *Advances in Cryogenic Engineering*, Vols. 1–38, Plenum Press, New York, 1960–1992.

International Cryogenic Engineering Conferences, Vols. 1–14, IPC Science and Technology Press and Butterworth, London, 1967–1993.

B. Elvers, S. Hawkins, and G. Schulz, eds., *Ullmann's Encyclopedia of Industrial Chemistry*, Vol. A17, VCH Publishers, Weinheim, Germany, 1991.

COMPOUNDS

The heavy gases, krypton, xenon, and radon, had been shown to react with fluorine and other powerful oxidants to form a number of stable products. As of this writing no stable compounds of the lighter gases, helium, neon, and argon, have been found. A complete survey of helium-group (noble-gas) compounds may be found in the literature. Some of the properties of the fluorides, oxides, and oxofluorides are given in Table 1. Prior to the discovery of these compounds, hydrates and other clathrate compounds of the helium-group gases were known. A discussion of clathrates may also be found in the literature.

Xenon Compounds

Xenon compounds include halides (xenon difluoride, XeF_2, xenon tetrafluoride, XeF_4, and xenon hexafluoride, XeF_6), oxides (xenon trioxide, XeO_3, and xenon tetroxide, XeO_4), oxofluorides (xenon oxide tetrafluoride, $XeOF_4$, xenon oxide difluoride, $XeOF_2$, xenon trioxide difluoride, XeO_3F_2, and xenon dioxide tetrafluoride, XeO_2F_4), xenates ($MHXeO_4 \cdot 1.5H_2O$ where M = Na, Rb, Cs), and perxenates ($Na_4XeO_6 \cdot 6H_2O$, $K_4XeO_6 \cdot 9H_2O$, $Li_4XeO_6 \cdot 2H_2O$, and $Ba_2XeO_6 \cdot 1.5H_2O$), complex salts and molecular adducts, and derivatives in which xenon is bonded to polyatomic groups through oxygen, nitrogen, and carbon.

Krypton Compounds

Krypton compounds include krypton difluoride, KrF_2, complex salts ($KrF^+SbF_6^-$; $KrF^+Sb_2F_{11}^-$; $Kr_2F_3^+SbF_6^-$; $KrF^+AsF_6^-$;$Kr_2F_3^+AsF_6^-$; $KrF^+Nb_2F_{11}^-$; $KrF^+TaF_6^-$; $KrF^+Ta_2F_{11}^-$; $KrF^+PtF_6^-$; and $KrF^+AuF_6^-$), krypton bonded to oxygen ($Kr(OTeF_5)_2$), and complex salts in which krypton is bonded to nitrogen ($RC{\equiv}N{-}KrF^+AsF_6^-$ where R = H, CF_3, C_2F_5, n-C_3F_7).

Radon Compounds

Radon compounds include radon fluoride and complex salts ($RnF^+SbF_6^-$, $RnF^+Sb_2F_{11}^-$, and $RnF^+BiF_6^-$).

Table 1. Physical and Thermodynamic Properties of Helium-Group Gas Fluorides, Oxofluorides, and Oxides

Compound	Melting point, °C	Color	ΔH_{sub}, kJ/mol[a]	ΔH_f, kJ/mol[a]	ΔG_f, kJ/mol[a]	Density, g/cm^3
KrF_2	dec 25[b]	colorless	41	60.2		3.22
XeF_2	129.03[c]	colorless	55.71	−162.8	−86.08	4.32
XeF_4	117.10[c]	colorless	60.92	−267.1	−145.5	4.04
XeF_6	49.48	colorless[d] or yellow-green[e,f]	59.12[g]	−338.2[g]	−169.0[g]	3.56[g,h]
$XeOF_2$	dec ca 0[i]	yellow				
$XeOF_4$	−46.2	colorless		−25[j]		3.11[e,k]
XeO_2F_2	30.8[i]	colorless		234[j]		4.10
XeO_3	dec ca 25[i]	white	100[j]	402		4.55
XeO_4	dec <0[i]	yellow solid at −196°C		642		

[a] To convert J to cal, divide by 4.184. [b] Decomposes at ~10% h^{-1} [c] Triple point. [d] Solid. [e] Liquid. [f] Vapor. [g] Phase I. [h] Phase II has a density of 3.71 g/cm^3; Phase III, 3.82 g/cm^3; and Phase IV, 3.73 g/cm^3. [i] Explosive. [j] Estimated. [k] Liquid at 22.5°C.

Prospects for Argon Compounds

Experimental evidence for ArF^+ in the gas phase has been obtained, leading to $D_o(ArF^+) \geq 1.655$ eV and confirming the instability of HeF^+ and NeF^+ in their electronic ground states.

Methods of Preparation of Binary Fluorides

All helium-group gas chemistry originates with the binary fluorides. The amounts of the binary xenon fluorides suitable for synthetic work are generally prepared by heating mixtures of xenon and fluorine to 250–400°C in nickel or Monel vessels. Although all three fluorides coexist in equilibrium, suitable adjustments of temperature, pressure, and xenon:fluorine ratio can be made to yield primarily difluoride, tetrafluoride, or hexafluoride.

Krypton difluoride cannot be synthesized by the standard high pressure–high temperature means used to prepare xenon fluorides because of the low thermal stability of KrF_2. There are three low temperature methods which have proven practical for the preparation of gram and greater amounts of KrF_2. Radon fluoride is most conveniently prepared by reaction of radon gas with a liquid halogen fluoride (ClF, ClF_3, ClF_5, BrF_3, or IF_7) at room temperature.

Uses

Stable noble-gas compounds have no industrial uses as of this writing but are frequently utilized in laboratories as fluorinating and oxidizing agents. A particularly important use for unstable noble-gas halides is as the gain medium in excimer lasers, which are increasingly being employed as high power sources of tunable laser light in the ultraviolet and visible spectral regions.

GARY J. SCHROBILGEN
J. MARC WHALEN
McMaster University

N. Bartlett and F. O. Sladky, in J. C. Bailar, Jr., H. J. Emeléus, R. Nyholm, and A. F. Trotman-Dickenson, eds., *Comprehensive Inorganic Chemistry*, Vol. 1, Pergamon Press, New York, 1973, pp. 213–330.

H. Selig and J. H. Holloway, in F. L. Boschke, ed., *Topics in Current Chemistry*, Vol. 124, Springer-Verlag, Berlin, 1984, pp. 33–90.

B. Žemva, *Croat. Chem. Acta* **61**, 163 (1988).

G. J. Schrobilgen, in R. D. Chambers, G. A. Olah, and G. K. S. Prakash, eds., *Synthetic Fluorine Chemistry*, John Wiley & Sons, Inc., New York, 1992, Chapt. 1, pp. 1–30.

HEMATOLOGY. See AUTOMATED INSTRUMENTATION, HEMATOLOGY.

HEMICELLULOSE

Hemicellulose is the least utilized component of the biomass triad comprising cellulose, lignin, and hemicellulose.

Pure hemicellulose components are seldom extracted directly from their source. Extracts are a mixture of polysaccharides, lignin, and lignin—hemicellulose complexes (by chemical linkages and possibly physical interactions) characteristic of their origin and the solvent employed. Hemicellulose has a lower degree of polymerization (DP) than cellulose (about 200 vs more than 10,000) and its lower limits have not been clearly defined. The extract may contain two or more polymers of similar composition but different structures (polydiversity) or of different distributions and amounts of branching or bonding in otherwise similar molecules (polydispersity). If a single polymer is present, it may exhibit a spectrum of molecular weights (polymolecularity) which may exhibit a Gaussian or biased distribution. A pure hemicellulose component is one where polydiversity has been avoided and a degree of heterogeneity has been attained compatible with end use application.

The most common hemicellulose in angiosperms is composed of D-xylose arranged in a linear manner. D-Mannose is derived from a glucomannan which is the most common hemicellulose in most gymnosperms. Both contain other sugars and exist in a variety of configurations and molecular weights. This article concentrates primarily on the components of tracheids and fibers of arborescent plants.

The common hemicellulose components of arborescent plants are listed in Table 1.

Isolation and Analysis

Techniques for the isolation of hemicellulose depend on the intended end use and whether it occurs in soluble waste material or is part of a solid matrix. Isolation is more difficult from solids as diminution of particle size and removal of undesired encrustants such as lignin is necessary to increase accessibility and destroy lignin hemicellulose bonds. Delignification techniques, except for those using ethanolamine, employ oxidants. Peroxides, peroxyacetic acid, and chlorine dioxide have been used but the most common reagents are chlorine and acidified sodium chlorite.

Delignification extracts varying amounts of hemicellulose. Low reaction temperatures and (where possible) high salt concentrations minimize losses and concomitant chemical degradations such as oxidation and the effects of pH. Pectic substances and easily soluble arabinans, arabinogalactans, galactoglucomannans, xylans, and compression wood galactans are found in waste chlorite liquors. Carbonyl groups (excepting carboxyls) are frequently reduced with a suitable reagent before alkaline extractions are attempted to minimize β-elimination reactions. The use of an inert atmosphere during alkaline extraction prevents oxidation by oxygen.

Table 1. Hemicellulose Components of Arborescent Plants

Gymnosperms	Dicotyledons	Monocotyledon (bamboo)
arabino-(4-*O*-methyl-glucurono)xylan	*O*-acetyl-(4-*O*-methyl-glucurono)xylan	arabino-(4-*O*-methyl-glucurono)xylan
O-acetylgalacto-glucomannan (0.1:1:3)	glucomannan	heteroxylans
O-acetylgalacto-glucomannan (1:1:3)	arabinogalactan	D-glucans
arabinogalactan	pectic substances	arabinogalactans
pectic substances	tension wood components	pectic substances
compression wood components		

Extraction of hemicellulose is a complex process that alters or degrades hemicellulose in some manner. Alkaline reagents that break hydrogen bonds are the most effective solvents but they de-esterify and initiate β-elimination reactions. Polar solvents such as DMSO and dimethylformamide are more specific and are used to extract partially acetylated polymers from milled wood or holocellulose.

The separation of the polysaccharide components utilizes their different solubilities, polar groups, extents of branching, molecular weights, and molecular flexibilities and may be accomplished batchwise or with easily automated column techniques such as column or high performance liquid chromatography. These procedures have been summarized in several reviews.

The increasing sophistication of analytical techniques coupled with suitable fractionation procedures has made the heterogeneity of hemicellulose components increasingly apparent. These techniques common to polymer chemistry include gas chromatography–mass spectroscopy, and proton and ^{13}C-nuclear magnetic resonance spectroscopy. Molecular studies employ viscometry, osmometry, x-ray techniques, light scattering, and chromatographic and centrifugal techniques, as well as the use of optical rotatory dispersion and circular dichroism.

Pulping

The complex behavior of hemicellulose during pulping has been reviewed. When hemicellulose and lignin dissolve with the help of chemical transformations, fresh cellulosic surfaces are created and competition for deposition in these spaces arises between the dissolved components. Hemicellulose degradations also occur and are related to pH (acid hydrolysis, β-elimination reactions, redox reactions, etc), and pyrolytic effects to an extent dependent upon the time, temperature, and liquor composition of the cook. These reactions are rendered more complex because the cell wall controls the diffusion of the reactants and products into and out of the fiber so that hemicellulose may not be able to react or diffuse out of the fiber before the cook is completed. Under pulping conditions, the formation and cleavage of lignin-hemicellulose bonds and possible carbohydrate–carbohydrate bonds occurs complicating the nature of the product. The extent to which these competing reactions is accomplished is reflected in product composition and end use quality and is the subject of much empirical research.

Those pulps with about 15% hemicellulose are usually used for paper manufacture, whereas those with 5% or less are used where a high cellulose content is required. The proportion of glucomannan is slightly greater in sulfite pulps, whereas the quantity of xylan is somewhat greater in kraft and soda pulps. These proportions can be altered slightly by changes in the cooking schedule.

Suitable pretreatment of wood before pulping alters the behavior of hemicellulose significantly. Saponification of the acetyl groups of softwood before sulfite cooking results in glucomannan retention in the final product. Those treatments that limit the peeling reaction during alkaline pulping processes (reductions with $NaBH_4$, H_2S, or oxidations with chlorite, polysulfide, anthraquinone, etc), can result in polysaccharide retention. The pretreatment of wood with mineral acid or liberated acids of wood at elevated temperatures (ie, 170°C for 30 min) diminishes the DP of hemicellulose components sufficiently that they will be consumed mostly during a subsequent alkaline cook. The resulting pulp behaves more like cotton cellulose in many industrial applications.

The Effect of Hemicellulose in Commercial Products

Hemicellulose components have an effect on the properties of products in which they are present. In the case of viscose manufacture much of the hemicellulose which remains in the product has only a marginal effect on strength properties and brightness and no effect on heat stability. It does contribute to swelling in yarn, and that remaining in the spent viscose liquor is harmful to filter presses. Increased clogging of spinnerets, low color index, and decreased yarn strength can also result when resin, cations, and hemicellulose are together in the steeping liquor.

When uronic acid and L-arabinofuranosyl branches are lost as a result of processing, the poor solubility of xylan acetate in organic solvents contributes adversely to cellulose acetate processing. The haze density (opaqueness) of solutions and solid products increases as hemicellulose components become less branched. Glucomannan in an acetate product contributes less haze than xylans but more than arabinoxylans. It is responsible for the false viscosity of cellulose acetate solutions derived from wood pulp. Glucomannan also causes filtration difficulties during processing, but the effects of xylans are unpredictable.

An understanding of the effect of hemicellulose on paper products is less clear because the formation of paper webs largely depends on their structure, which is influenced by many factors besides hemicellulose. Hemicellulose and related gums and mucilages help maintain a random dispersion of fibers in the furnish which results in more uniform and mechanically stronger paper webs. It also increases the rate at which pulp fibers respond to mechanical action (beating) and increases fiber bonding. The strong interfiber bonds formed after drying can alter paper structure and lead to losses in tearing strength and decreases in opacity if they are too extensive.

Applications

Hemicellulose and hemicellulose-like polysaccharides are beneficial components of foodstuffs because of their interactions between water and water-insoluble components. Endogenous polysaccharides are responsible for processing characteristics as well as texture and mouthfeel. Other properties such as gel formation, swelling of dough, fermentation, and optical properties are achieved using suitable treatments with enzymes and chemicals.

Besides inherent usefulness as a naturally occurring component of some manufactured products, hemicellulose can be utilized either as a polymer or as the source of chemical intermediates. The former use is complicated since the mixture of polysaccharides and lignin in extracts requires special treatment if one component is to be isolated. As a result, the naturally occurring gums and mucilages are at a competitive advantage in the marketplace.

Larch arabinogalactan is easy to isolate and requires limited purification for many uses. The mixture of sugars, oligosaccharides, degraded hemicellulose, and lignin found in the liquors and condensates from Asplund-like and prehydrolysis pulping processes is used for binding and extending animal fodder. The sugars and oligosaccharides in waste sulfite liquor can be used for furfural production and the growth of yeast. Pectin is used in the food industry, and apart from specialty uses finds limited application in the paper industry.

Derivatives of hemicellulose components have properties similar to the cellulosic equivalents but modified by the effects of their lower molecular weight, more extensive branching, labile constituents, and more heterogeneous nature. Acetates, ethers, carboxymethylxylan, and xylan–poly(sodium acrylate) have been prepared.

NORMAN S. THOMPSON
Consultant

D. Clayton and co-workers, "Chemistry of Alkaline Pulping," in *Pulp and Paper Manufacture*, 3rd ed., Vol. 5, Alkaline Pulping, The Joint Textbook Committee of the Paper Industry, TAPPI, CPPA, Technology Park, Atlanta, Ga., 1989.

K. Shimizu, in N. S. Hon and N. Shiraishi, eds., *Wood and Cellulosic Chemistry*, Marcel Dekker, Inc., New York, 1991, Chapt. 5.

R. L. Whistler and C.-C. Chen, in M. Lewin and I. S. Goldstein, eds., *Wood Structure and Composition*, Vol. 11, International Fiber Science and Technology Series, Marcel Dekker, Inc., New York, 1991, Chapt. 7.

N. S. Thompson, in I. S. Goldstein ed., *Organic Chemicals from Biomass*, CRC press, Boca Raton, Fla., 1981.

HEPARIN. See Blood, coagulants and anticoagulants.

HERBICIDES

Herbicide Classes and Databases

Herbicides can be classified as selective and nonselective. Selective herbicides, like 2,4-D (2,4-dichlorophenoxyacetic acid), metolachlor, and EPTC, are more effective against some types of plants than others, eg, broadleaved plants vs grasses. Glyphosate is representative of the nonselective herbicides used for total vegetable control.

The classes of herbicidally active toxophores are limited in number. Arbitrary classification by toxophore reveals eight generic herbicide groupings, ie, triazines, amides (haloacetanilides), carbamates, toluidines (dinitroanilines), ureas, plant growth hormones (phenoxy acids), diphenyl ethers, and miscellaneous unrelated compounds. Classification of commercial herbicides by chemical structure yields 10 related groupings with subgroups, ie, phenoxy alkanoic acids; bipyridiniums; benzonitriles with phthalic compounds; dinitroanilines; acid amides; carbamates; thiocarbamates; heterocyclic nitrogen compounds including triazines, pyridines, pyridazinones, sulfonylureas, and imidazoles; substituted ureas; and miscellaneous groupings that include halogenated aliphatic carboxylic acids, inorganics and organometallics, and derivatives of biologically important amino acids.

Herbicides are also sometimes classified according to mode of action, selectivity, registered uses, and toxicity. The ever-increasing importance of herbicides and other pesticides and agrochemicals to a wide range of users, regulators, and researchers has led to the development of multiple and extensive computer databases. The primary database resources contain collected information relevant to herbicides, and numerous resource publications are available to those needing information on the various aspects of herbicides.

Database	Type of information
Agribusiness	agricultural chemicals and finance; agribusiness companies, product development history, and government policies
Agricola	National Agricultural Library database; general coverage of U.S. agriculture
Agrochemicals handbook	active components of agrochemicals
BIOSIS previews (Biological Abstracts)	research literature in biological, biomedical, and life sciences
Chemical Abstracts search	chemical literature and applications
CAB Abstracts	general agricultural and biological information, including weed science, from U.S. sources
Claims/U.S. patent abstracts	patents issued by the U.S. Patent Office; chemical patent records from 1950 to present
Derwent world patents index	patent data from 30 patent-issuing authorities around the world; agricultural chemical patents from 1965 to present
EMMI	U.S. EPA Environmental Monitoring Methods Index, including regulatory lists, analytical methods, detection and regulatory limits
Enviroline	coverage of worldwide environmental information
Pollution abstracts	references to environmental literature including pollutant sources and control
Toxline	National Library of Medicine toxicological information; references to pesticides, herbicides, environmental pollution, carcinogenic chemicals, food contamination, toxicological analyses
Toxnet	National Library of Medicine Toxicology Data Network, including Hazardous Substances Data Bank (HSDB), Registry of Toxic Effects on Chemical Substances (RTECS), Toxic Chemical Release Inventory (TRI), plus several other toxicological/carcinogenesis-related files

Herbicide Development

Examination of the various classified listings of herbicides provides insight into the processes and approaches that lead to the discovery of new pesticides. The four principal development approaches are random screening, imitative chemistry, testing natural products, and biorational development.

Quantitative Structure–Activity Relationship Design. Increasing economic pressures toward more, better, and cheaper pesticides have led to the development and application of the Quantitative Structure–Activity Relationship (QSAR) paradigm and related experimental design principles for pesticides. Theoretically, quantitative determination of the relationships between chemical structure and biological and environmental properties of a molecule should permit the design of a novel molecule with exactly those properties considered ideal for the intended application.

Modes of Herbicide Action

Modes of herbicide action include Photosystem I inhibitors [acifluorfen ($C_{14}H_{17}ClF_3NO_5$), nitrofen ($C_{12}H_7Cl_2NO_3$), oxyfluorofen ($C_{15}H_{11}ClF_3NO_4$)], electron transport between Photosystem I and Photosystem II inhibitors, Photosystem II inhibitors [atrazine ($C_8H_{14}ClN_5$), metribuzin ($C_8H_{14}N_4O_5$), diuron ($C_9H_{10}Cl_2N_2O$), bromacil ($C_9H_{13}BrN_2O_2$), ioxynil ($C_7H_3I_2NO$), dinoseb ($C_{10}H_{12}N_2O_5$), bromoxynil ($C_7H_3Br_2NO$), dinitrocresol ($C_7H_6N_2O_5$)], bleaching herbicides [fluridone ($C_{19}H_{14}F_3NO$), flurochloridone ($C_{12}H_{10}Cl_2F_3NO$), flurtamone, S3442 ($C_{17}H_{19}NO_2$), diflufenican ($C_{19}H_{11}F_5N_2O_2$), difunon ($C_{14}H_{12}N_2O_2$), norflurazon ($C_{12}H_9ClF_3N_3O$), amitrole ($C_2H_4N_4$), fluometuron ($C_{10}H_{11}F_3N_2O$), fomesafen ($C_{15}H_{10}ClF_3N_2O_6S$)], chlorophyll biosynthesis inhibitors [oxadiazon ($C_{15}H_{18}Cl_2N_2O_3$), DTP, MK-616 ($C_{14}H_{12}ClNO_2$), clethodim ($C_{17}H_{26}ClNO_3S$), sethoxydim ($C_{17}H_{29}NO_3S$), haloxyfop, methyl ($C_{16}H_{13}ClF_3NO_4$), tralkoxydim, fenoxaprop, ethyl ($C_{18}H_{16}ClNO_5$), fluazifop, butyl ($C_{19}H_{20}F_3NO_4$), alachlor ($C_{14}H_{20}ClNO_2$), metolachlor ($C_{15}H_{22}ClNO_2$), diclofop, methyl ($C_{16}H_{14}Cl_2O_4$), CDEC ($C_8H_{14}ClNS_2$), diallate ($C_{10}H_{17}Cl_2NOS$), EPTC ($C_9H_{19}NOS$), triallate ($C_{10}H_{16}Cl_3NOS$), metflurazon ($C_{13}H_{11}ClF_3N_3O$)], lipid and wax synthesis inhibitors, radical damage to antioxidative systems and cellular components inducers [paraquat ($C_{12}H_{14}N_2$), diquatop ($C_{12}H_{12}N_2$), tridiphane ($C_{10}H_7Cl_5O$)], herbicidal inhibition of enzymes [MAA (CH_5AsO_3), MSMA ($CH_5AsO_3 \cdot Na$), DSMA ($CH_5AsO_3 \cdot 2Na$), AMA, cacodylic acid ($C_2H_7AsO_2$), glufosinate ($C_5H_{12}NO_4P$), ammonium glufosinate ($C_5H_{12}NO_4P \cdot H_3N$)], amino acid and nucleotide biosynthesis inhibitors [phaseolotoxin ($C_{12}H_{33}N_8O_9PS$), glyphosate ($C_3H_8NO_5P$), rhizobitoxine ($C_7H_{14}N_2O_4$), chlorsulfuron ($C_{12}H_{12}ClN_5O_4S$), chlorimuron, ethyl ($C_{15}H_{15}ClN_4O_6S$), sulfometuron ($C_{15}H_{16}N_4O_5S$), bensulfuron, methyl ($C_{16}H_{18}N_4O_7S$), imazaquin ($C_{17}H_{17}N_3O_3$), imazapyr ($C_{13}H_{15}N_3O_3$), imazethapyr ($C_{15}H_{19}N_3O_3$), imazamethabenz ($C_6H_{20}N_2O_3$)], cell division inhibitors [trifluralin ($C_{13}H_{16}F_3N_3O_4$), oryzalin ($C_{12}H_{18}N_4O_6S$), pendimethalin ($C_{13}H_{19}N_3O_4$), nitralin, dinitramine ($C_{11}H_{13}F_3N_4O_4$), asulam ($C_8H_{10}N_2O_4S$), propham ($C_{10}H_{13}NO_2$), chloropropham ($C_{10}H_{12}ClNO_2$), barban ($C_{11}H_9Cl_2NO_2$), butylate ($C_{11}H_{23}NOS$), cycloate ($C_{11}H_{21}NOS$), propachlor ($C_{11}H_{14}ClNO$), DCPA ($C_9H_9Cl_2NO$), pronamide ($C_{12}H_{11}Cl_2NO$), bensulfide ($C_{11}H_{24}NO_4PS_3$), cinmethylin ($C_{18}H_{26}O_2$)], and plant growth regulator synthesis and function inhibitors [(naphtalene acetic acid ($C_{12}H_{10}O_2$), indolebutyric acid ($C_{12}H_{13}NO_2$), 2,4-D ($C_8H_6Cl_2O_3$), 2,4,5-T ($C_8H_5Cl_3O_3$), MCPA ($C_9H_9ClO_3$), dicamba ($C_8H_6Cl_2O_3$), chloramben ($C_7H_5Cl_2NO_2$), picloram ($C_6H_3Cl_3N_2O_2$), naptalam ($C_{18}H_{13}NO_3$), TIBA ($C_7H_3I_3O_2$), diclofop ($C_{15}H_{12}Cl_2O_4$), ethephon ($C_2H_{16}ClO_3P$), tetcyclacis ($C_{13}H_{12}ClN_5$),

AMO-1618 ($C_{19}H_{31}N_2O_2 \cdot Cl$), chlormequat chloride ($C_5H_{13}ClN \cdot Cl$), mepiquat chloride ($C_7H_{16}N \cdot Cl$), ancymidol ($C_{15}H_{16}N_2O_2$), uniconazole ($C_{15}H_{18}ClN_3O$), paclobutrazol ($C_{15}H_{20}ClN_3O$), BAS 11100W ($C_{16}H_{23}N_3O_2$)].

Environmental Fate of Herbicides

Herbicide Fates in Plants. Beyond modes of action and structure–activity relationships, developers of new herbicides must also consider uptake by plants, translocation within the plant, and possible deactivation of herbicides by contact with soil. Some of these problematic factors can be addressed as part of the QSAR studies and during the screening process. Considerable attention is also being paid to the use of safeners which protect the crop from herbicides that specifically target the weeds usually associated with that crop. Environmental protection and pesticide regulation concerns are the driving forces in the current efforts toward minimizing application rates, optimizing delivery through improved formulations and application equipment, and increasing target specificity. These research and development efforts include other important and related areas of interest to chemists, eg, the fate and detection of herbicides in the soil and ground and surface water.

Factors Affecting Environmental Fate. The fate of herbicides in the environment is influenced by many chemical, biological, and physical factors. The principal transport and dissipation pathways include sorption to organic and mineral soil and sediment constituents; transport to groundwater in the solution phase by mass flow and/or diffusion; transport to surface water in either the solution or sorbed phases; loss to the atmosphere through volatilization, with redeposition at a later time and location; transformation or mineralization by biological, chemical, or photochemical processes; and uptake by plant or animal species. These processes do not operate as isolated systems, but occur simultaneously and involve significant interaction and feedback. Although the environmental fates of most herbicides are controlled primarily by one or two of the outlined processes, all of these factors influence the fate to some extent.

Measurement of Environmental Fate. Continued concern is expressed over the potential contamination of surface and groundwaters by agricultural chemicals. Herbicides have received much of this attention, due to their widespread use and the large total volume applied. However, this perceived threat to groundwater resources appears to be largely unfounded. A survey of private wells and public water well supplies in the United States has revealed that $< 1\%$ contain herbicides at levels that would affect human or animal health. In addition, contaminated sources can usually be attributed to point rather than nonpoint sources. Nonpoint sources are generally treated by modifications in agricultural management practices. Typical modifications would include the use of alternative herbicide formulations, the splitting of the herbicide application in time, or the installation of vegetative buffer strips to trap runoff.

A re-evaluation of the water quality problem has revealed that surface water resources, rather than groundwater resources, are at higher risk of contamination from agricultural chemicals.

The public health implications of drinking water contamination by herbicides are unclear. The levels that have been detected in groundwater are generally in the part per billion (ppb) or part per trillion (ppt) range and are below estimated acute toxicity levels. However, the long-term health effects of this exposure are generally unknown. Several studies have demonstrated that the mortality from some types of cancer is significantly higher in rural residents of many corn belt states. The U.S. Environmental Protection Agency (EPA) developed (ca 1993) a classification scheme in an attempt to further evaluate the carcinogenic potential of herbicides and pesticides. In this system, chemicals are placed in one of five groups, A–E, according to their carcinogenic potential, ranging from definite (A) human carcinogens to no evidence of carcinogenicity for humans (E). The principal difference between these groups is the amount of accumulated evidence demonstrating carcinogenic potential.

This classification scheme is used in part in the determination and calculation of health advisory (HA) drinking water levels or carcinogenic risk estimates. The majority of herbicides in use in the United States (ca 1993) for which HAs have been issued fall into Group D, with a smaller percentage falling into Group C. This would indicate that there are insufficient data to classify the carcinogenic potential of many herbicides. The lack of data does indicate however, that further testing will be required before the carcinogenic potential of many herbicides is known. Based on available HAs and the U.S. EPA classification scheme, acifluorfen, alachlor, amitrole, haloxyfop–methyl, lactofen, and oxadiazon have been listed as B2 carcinogens. Further information on carcinogenic risk assessment is available.

Since 1984, dramatic technical advances have been made in the analysis of trace organic chemicals in the environment. Indeed, these advances have been largely responsible for the increased public and governmental awareness of the wide distribution of herbicides in the environment. The ability to detect herbicides at ppb and ppt levels has resulted in the discovery of trace herbicide residues in many unexpected and unwanted areas. The realization that herbicides are being transported throughout the environment, albeit at extremely low levels, has caused much public and governmental concern. However, the public health implications remain unclear.

Traditionally, herbicides have been analyzed by gas chromatography (gc) or spectrophotometric methods. The method of choice when accuracy and sensitivity are of the utmost importance is gc, especially when combined with mass spectrometry. However, several other methods are used for routine monitoring or screening purposes. High pressure liquid chromatography (hplc) provides detection limits that nearly rival gc and require significantly less sample preparation and cleanup. Advances in the 1980s have made thin-layer chromatography (tlc) a valuable tool in herbicide analysis. Another analytical tool that has received much attention and shows great promise for routine analysis is enzyme immunoassay (eia). This technique offers the advantages of a low cost analysis, few interferences, high specificity and sensitivity, and a minimal amount of sample preparation.

A mobility ranking based on soil thin-layer chromatography (stlc) is used to classify the herbicide leaching potential of various herbicides. The rankings range from I (immobile) to V (very mobile) with intermediate categories of II (low mobility), III (intermediate), and IV (mobile). This method is widely used and has been accepted for submission of leaching data for herbicide registration purposes by the U.S. EPA. A comprehensive search of the STORET water quality database, maintained by the U.S. EPA Office of Water, is used to evaluate the potential water quality implications of various herbicides.

Herbicide Groups

Herbicides can be grouped according to common structural features. Sometimes the assignment is arbitrary when there are a multitude of functional groups, eg, acifluorfen which is a diphenyl ether (phenoxy compound) as well as a trifluoromethyl compound.

Phenoxyalkanoics. The phenoxyalkanoic herbicide grouping is composed of two subgroups, the phenoxyacetic acids and the phenoxypropionic acids. They are widely used for foliar control of broadleaf weeds. The more heavily functionalized phenoxypropionic acid herbicides are relatively new herbicides compared to the phenoxyacetic acids and are used primarily for selective control of grassy weeds in broadleaf crops.

Considerable concern has been raised over the carcinogenic potential of the phenoxyacetic acid herbicide 2,4-D. However, the World Health Organization (WHO) has evaluated the environmental health aspects of this chemical and concluded that 2,4-D posed an insignificant threat to the environment. They did indicate, however, that only limited data on toxicology in humans are available. An HA has been issued for the phenoxyacetic acid herbicide MCPA. It was found in 4 of 18 SW samples analyzed and in none of 118 GW samples, and has been placed in group D for carcinogenic potential. EPA has published two gc methods for the analysis of the phenoxyalkanoic herbicides.

Table 1. Environmental Health Advisories for Herbicides

Herbicide	Health advisories[a] SW	Health advisories[a] GW	Mobility[b]	Carcino-genic potential group[c]	Analytical methods[d]
Bipyridinium compounds					
diquatop			immobile		hplc
paraquat		0/843	immobile	E	hplc
Benzonitrile, acetic acid, and phthalic compounds					
chloramben	13/34	1/566	very mobile	D	gc[e]
DCPA	386/1995	12/982		D	gc[e]
dicamba	262/806	2/230	very mobile	D	gc[e]
dichlobenil			low		
endothall	0/3	0/604		D	gc
naptalam					uv
Dinitroaniline and derivatives					
benefin					gc
dinitramine					gc
dinoseb	1/89	0/1270		D	
fluchloralin					gc
oryzalin					uv
pendimethalin					gc
trifluralin	172/2047	1/507	immobile	C	ir
Acid amides					
alachlor					gc
bensulide			immobile		hplc
diphenamide	0/3	0/678	intermediate	D	gc
metolachlor	2091/4161	13/596		C	gc
napropamide					
pronamide	20/391			C	gc
propachlor	34/1690	2/99	intermediate	D	gc
propanil			low		
Phenyl carbamates					
chloropropham			low		
karbutilate					hplc
propham	1/392	0/583	intermediate	D	hplc
Thiocarbamates					
asulam					uv
butylate	91/836	2/152		D	gc, glc
EPTC					hplc
thiobencarb			relatively immobile		gc, glc
triallate					gc, glc
vernolate					hplc
Triazines					
ametryn	2/1190	24/560	intermediate	D	general[f]
atrazine	4123/10,942	343/5208	intermediate	C	gc
cyanazine	1708/5297	21/1821	intermediate	D	ir
hexazinone			relatively immobile	D	gc
metribuzin	938/4651	0/416		D	general[f]
prometon	386/1419	36/746	intermediate	D	gc
prometryn			low		
propazine	33/1097	15/906	intermediate	C	general[f]
simazine	922/5873	202/2654	intermediate	C	gc
terbutryn					general[f]
Pyridines					
clopyralid			minimal[g]		
fluroxypyr			varied		
picloram	420/744	3/64	mobile[g]	D	general[h]
triclopyr			intermediate		hplc
Pyridazinones					
norflurazon			low		
pyrazon					uv
Sulfonylureas					
chlorimuron, ethyl			mobile		
chlorsulfuron			intermediate		hplc, gc
metsulfuron, methyl			to very mobile		gc
sulfometuron			mobile to very mobile		
Imidazole compounds					
buthidazole					eia
imaza-methabenz					eia
imazapyr					eia
imazaquin			mobile to very mobile[i]		eia
imazethapyr			immobile to mobile[i]		eia
Other heterocyclic nitrogen derivatives					
amitrole			mobile		vis
bentazon			very mobile	D	hplc
isoxaben			immobile		
Ureas and uracils					
bromacil	0/3	0/841	mobile	C	glc
chloroxuron			immobile		glc
diuron	0/25	0/1337	low	D	ir
fluometuron	0/14	0/156	intermediate	D	uv
linuron					uv
tebuthiuron			intermediate to very mobile	D	uv
terbacil				E	uv
Aliphatic-carboxylic					
dalapon	0/14	0/14	very mobile	D	ir
TCA			very mobile		
Inorganics and metal organics					
AMS				D	titration
Miscellaneous trifluromethyl compounds					
acifluorfen				B_2	hplc
fluridone					gc
lactofen					hplc
Amino acid analogues					
glufosinate,			intermediate		
glyphosate	0/6	0/98	immobile to low mobility	D	hplc
Other miscellaneous compounds					
cinmethylin					gc
ethofumesate					gc
tridiphane					

[a] SW = surface water; GW = ground water. Positive results/number of tests.
[b] Mobility ranking based on soil thin-layer chromatography (stlc).
[c] Group A, human carcinogen; Group B, probable human carcinogen; Group C, possible human carcinogen; Group D, not classifiable; Group E, no evidence of carcinogenicity for humans.
[d] gc = gas chromatography; hplc = high pressure liquid chromatography; ir = infrared spectroscopy; uv = ultraviolet spectroscopy; glc = gas-liquid chromatography; eia = enzyme immunoassay; vis = visible spectroscopy.
[e] Gc for chlorinated pesticides can be used.
[f] General draft method for nitrogen- and phosphorus-containing pesticides.[g]
[g] Mobility has been reported to be mobile and minimal in different studies.
[h] General draft method for determination of chlorinated acids in water.
[i] Mobility is a function of soil pH.

Bipyridiniums. The bipyridinium herbicides (Table 1), paraquat and diquat, are nonselective contact herbicides and crop desiccants.

Diquat is also used as a general aquatic herbicide. Paraquat and diquat are much more toxic than most herbicides, and ingestion of sufficient quantities can result in death if prompt medical treatment is not obtained.

Benzonitrile, Acetic Acid, and Phthalic Compounds. Benzonitrile herbicides (Table 1) are generally used for pre-emergence and post-emergence control of broadleaf weeds. Dichlobenil also controls grass weeds and dichlobenil, endothall, and fenac are used as aquatic herbicides. Most benzonitriles are selective in their control. Benzonitrile herbicides are acidic in nature, thus their environmental fate is influenced by changes in soil pH. Sorption of these herbicides is expected to increase with decreasing pH.

Dinitroanilines and Derivatives. Dinitroaniline herbicides are used principally for the selective, pre-emergence control of annual grasses and broadleaved weeds.

Acid Amides. The principal use of acid amide herbicides is the selective control of seedling grass and certain broadleaved weeds. The majority of acid amide herbicides are applied pre-emergence or pre-plant incorporated, except for propanil which is applied post-emergence.

Phenylcarbamates. Phenylcarbamate herbicides represent one of two subgroups of carbamate herbicides, the phenylcarbamates and the thiocarbamates. The carbamate herbicides are used, in general, for the selective pre-emergence control of grass and broadleaved weeds. Exceptions would include barban, desmedipham, and phenmedipham, which are applied post-emergence.

Thiocarbamates. Thiocarbamate herbicides are nonionic. Diallate and triallate were strongly sorbed to both cation- and anion-exchange resins but minimally to kaolinite or montmorillonite. This behavior suggests a physical, rather than ionic mechanism of attraction.

Triazines. Triazine herbicides are one of several herbicide groups that are heterocyclic nitrogen derivatives. Triazine herbicides include the chloro-, methylthio-, and methoxytriazines. They are used for the selective pre-emergence control and early post-emergence control of seedling grass and broadleaved weeds in cropland. In addition, some of the triazines, particularly atrazine, prometon, and simazine, are used for the nonselective control of vegetation in noncropland. Simazine may be used for selective control of aquatic weeds.

Pyridines and Pyridazinones. Pyridine herbicides are auxin-type herbicides generally used for selective control of broadleaved weeds in cropland, rangelands, and noncroplands. The pyridazinones are used primarily for the selective pre- and post-emergence control of seedling grass and broadleaved weeds in cotton and sugarbeets.

Sulfonylureas. Sulfonylurea herbicides are a relatively new class of herbicides generally used for selective pre- and post-emergence control of broadleaved weeds in croplands. Sulfometuron–methyl is used for broad-spectrum selective or nonselective weed control in noncroplands.

Imidazoles. Imidazole herbicides are generally used for selective pre- and post-emergence control of grass and broadleaved weeds in croplands. Buthidazole and imazapyr are used for broad-spectrum, nonselective weed control in noncroplands.

Other Heterocyclic Nitrogen Derivative Herbicides. The herbicides in this group are heterocyclic nitrogen derivatives that do not readily fall into one of the previously discussed groups. They have a wide range of uses and properties. Most of these herbicides are used for selective, pre- and/or post-emergence weed control. Amitrole is used for post-emergence, nonselective weed control in noncroplands and also as an aquatic herbicide.

Ureas and Uracils. Urea herbicides are generally used for selective pre-emergence and early post-emergence control of seedling grass and broadleaved weeds. Uracil herbicides are generally used for selective control of annual and perennial weed control in certain crops and for general weed control in noncrop areas. Bromacil, linuron, and tebuthiuron are used for the nonselective control of weeds in noncropland.

Aliphatic–Carboxylics. These are used primarily for the selective control of annual and perennial grass weeds in cropland and noncropland. Dalapon is also used as a selective aquatic herbicides.

Metal Organics and Inorganics. The metal organic herbicides are arsenicals used for the selective, post-emergence control of grass and broadleaved weeds in cropland and noncroplands.

Miscellaneous Trifluoromethyl Compounds. The herbicides in this group are used for a wide variety of weed-control purposes. Acifluorfen, lactofen, and oxyfluorfen are used for selective, pre-, and post-emergence weed control in croplands. Fluorochloridone is used for selective, pre-emergence weed control in cropland, and fluridone, fomesafen, and mefluidide are used for post-emergence control. Fluridone is also used as an aquatic herbicide.

Amino Acid Analogues. Amino acid analogue herbicides also control a large variety of weeds. Glyphosate and glufosinate are used for the broad-spectrum, nonselective control of grass and broadleaved weeds. Diethatyl is used for selective, pre-emergence control of grass and broadleaved weeds. Flamprop is used to control the growth of wild oats in wheat.

Miscellaneous Other Herbicides. The herbicides in this group are not readily included in any of the preceding groups. Acrolein (2-propenal) is used as a contact, aquatic herbicide. Sethoxydim, clethodim, and tridiphane are used for selective, post-emergence weed control. Cinmethylin and clomazone are used for selective pre-emergence control and ethofumesate for selective pre- and post-emergence weed control.

Economic Aspects

During the period from 1979 through 1991, the estimated U.S. total annual volume of herbicide usage increased somewhat from 254 million kg active ingredient (AI) to 285 million kg. Peak herbicide usage of 306 million kg occurred in 1984. During the years between 1979 and 1992, agricultural uses accounted for 76 to 81% of the herbicide applied in the United States. Combined U.S. government, industrial, and commercial herbicide usage during those years ranged from 14% in 1979, to 19% in 1986, and 17% in 1991. Home gardens and lawns received the remaining 4 to 5%.

Based on 1990–1991 estimates, the most used herbicides in the United States are, in descending order of usage, atrazine, alachlor, metolachlor, 2,4-D, trifluralin, cyanazine, EPTC, metham–sodium, glyphosate, and butylate.

Although the ratios have varied from year to year since 1979, the selective herbicides used in corn production have accounted for approximately 21% of herbicide use on a per crop basis. Herbicide use in soybean and cotton production combined account for ca 23% of the selective herbicide market. Graminicides, which selectively kill grasses, constitute 40% of the total market, leaving a market share of approximately 16% for the nonselective herbicides.

Innovative Weed Management Agents

Adoption by the agricultural community requires that an innovative weed management agent must be an effective control of the target species, be cost-effective, and be practical to employ. It must not interfere with crop production practices such as crop rotation or the use of other pesticides. Additionally, new weed-control agents cannot pose a significant threat to human health or the environment. Considerable costs are incurred in the development, registration, production, and marketing of weed control agents. These costs require that an herbicide have sufficient long-term market viability and market niche potential to justify these costs in time and money. The need for safe and effective methods of crop production in an environment that contains competitive weeds is becoming increasingly critical.

Weed Management Strategies. Managers of agroecosystems are being encouraged to manage weed populations at levels that are below their economic optimum thresholds, rather than attempting to eliminate or control all noncrop plants, regardless of their actual impact. Decisions concerning management of weed populations should be governed by both agroecological principles and site-specific considerations in the context of an overall integrated pest management program. However, the practical implementation of integrated pest management (IPM) programs can be difficult.

Nonchemical or traditional practices, such as weed seed removal, optimal crop seeding rates, crop selection, enhanced crop competitiveness, crop rotation, and mechanical weed control are all important components of an effective weed management program. In the context of modern intensive chemical herbicide application, nonchemical practices may represent an innovative approach to weed management and should receive careful consideration.

Natural Products and Allelopathic Compounds as Herbicides. There is growing concern that compounds that do no occur in nature may produce unanticipated health and environmental problems. However, plants, fungi, marine organisms, and certain bacteria produce a vast array of organic compounds, and many of these natural products exhibit biological activity. In nature, these compounds are produced in minute quantities and present interesting chemical problems in detection, identification, quantification, and production of active and stable analogues. Although these compounds appear to be ecologically safe in naturally occurring amounts, the large quantities required for agricultural applications may cause environmental problems similar to those associated with chemical herbicides.

Investigations of natural product chemistries have aided in the development of bialaphos, cinmethylin, picloram, glufosinate, and other important herbicides. Additional compounds may be found through investigations of natural products that cause plants and other organisms to undergo rapid physiological change, such as plant hormones and phytotoxins. Many plant hormones and phytotoxins are also produced by microorganisms. Additionally, microorganisms have been reported to contain novel natural products that could provide basic structural templates for the development of new herbicides.

Plant Pathogens and Insects as Control Agents. Concerns about accumulations of chemical control agents in the environmental and food resources have also increased interest in microbial weed control agents. Controlling weeds with carefully screened plant pathogens offers several benefits, including a high degree of specificity for a given target weed, low potential for negative human health and environmental impact, inability to accumulate in the food chain, and other advantages. The high degree of host specificity may limit the market size for some biological control agents, but these bio-control agents can be combined with chemical herbicides and other pathogens to increase the spectrum of weeds controlled. The marketing of biological control agents may also be constrained by slow expression of phytotoxicity, pathogen dependence on optimum environmental conditions, potential resistance of the weed towards the pathogen, and lack of formulation stability under field conditions and during preuse storage. These constraints can be addressed by genetic manipulation of selected pathogenic strains to produce more effective control agents and by the investigation of the mechanisms of disease resistance in plants.

There are two principal approaches to the biological control of weeds. The first approach is referred to as classical or inoculative biological weed control. The intent of classical biological weed control approaches is the management of introduced weed populations by introduction of host-specific pathogens from the weed's native range, thus moderating the growth of weed populations by the reestablishment of an old association between host and pathogen populations in the expanded range.

An additional approach to biological weed control is referred to as the inundative or augment approach to biological weed management. This approach utilizes pathogenic propagules formulated as a weed control agent, eg, mycoherbicides. The mass-inoculation of pathogenic propagules in an effective formulation can enhance the dissemination and survival of the pathogens, overwhelm target weed resistance, and produce results similar to those achieved with chemical herbicides. Mycoherbicides often contain native pathogens that are active against native weeds and are thus highly selective against the target weed species.

Research concerning plant pathogen control agents has resulted in two commercially available mycoherbicides. The mycoherbicide Collego is a formulated product consisting of propagules of the fungus *Colletotrichum gloeosporioides*, and Devine is a formulated product containing the sexual spores of the oomycete, *Phytophthora palmivora*. Devine is used to control stranglevine (*Morrenia odorata*) in citrus, and Collego is used in northern jointvetch (*Aeschynomene virginica*) control in rice and soybean.

Control of Weed Seeds. If agents that control weed seed germination could be applied prior to planting, interference from weeds would be prevented until reintroduction of weed propagules. Additionally, if a very large portion of the weed seed bank could be stimulated to germinate prior to planting, weeds could be controlled by a single cultivation or application of nonselective herbicide.

Development of agents that stimulate weed seed germination and/or attack weed seeds would have a profound impact on weed management. However, herbicide development programs in the early 1990s do not focus on identifying agents that are effective on weed propagules. A systematic search for compounds that render weed seeds nonviable or cause them to germinate simultaneously could provide important new weed management tools.

JUDITH M. BRADOW
CHRISTOPHER P. DIONIGI
RICHARD M. JOHNSON
SUHAD WOJKOWSKI
U.S. Department of Agriculture

Herbicide Handbook, 7th ed., Weed Science Society of America, Champaign, Ill., 1994.

W. Draber and T. Fujita, eds., *Rational Approaches to Structure, Activity, and Ecotoxicology of Agrochemicals*, CRC Press, Boca Raton, Fla., 1992.

R. Grover and A. J. Cessna, eds., *Environmental Chemistry of Herbicides*, Vols. 1 and 2, CRC Press, Boca Raton, Fla., 1991.

U.S. EPA, *Manual of Chemical Methods for Pesticides and Devices*, 2nd ed., Assoc. Off. Anal. Chem., Arlington, Va., 1992.

Whitehead, R. ed., *U.K. Pesticide Guide* 1995, CAB International, Wallingford, Oxon, U.K.

HEXANES. See HYDROCARBONS.

HIGH PERFORMANCE FIBERS

High performance fibers are generally characterized by remarkable unit tensile strength and resistance to heat, flame, and chemical agents that normally degrade conventional fibers. Applications include uses in the aerospace, biomedical, civil engineering, construction, protective apparel, geotextiles, and electronic areas.

Preparation and Properties

The principal classes of high performance fibers are derived from rigid-rod polymers, gel spun fibers, modified carbon fibers, synthetic vitreous fibers, and poly(phenylene sulfide) fibers.

Rigid-Rod Polymers. Rigid-rod polymers are often liquid crystalline polymers classified as lyotropic, such as the aramid Kevlar, or thermotropic liquid crystalline polymers, such as Vectran.

Liquid Crystallinity. The liquid crystalline state is characterized by orientationally ordered molecules. The molecules are characteristically rod- or lathe-shaped and can exist in three principal structural arrangements: nematic, cholesteric, and smectic.

Industrial Lyotropic Liquid Crystalline Polymers (LCPs). In the 1970s, researchers at DuPont reported that the processing of extended chain all para-aromatic polyamides from liquid crystalline solutions produced ultrahigh strength, ultrahigh modulus fibers. The greatly increased order and the long relaxation times in the liquid crystalline state compared to conventional systems led to fibers with highly oriented domains of polymer molecules. The most common lyotropic aramid fiber is poly(*p*-phenylene terephthalamide) (PPT) which is marketed as Kevlar by DuPont. Aramid fiber is available from Akzo under the trade name Twaron. These fibers are used in body armor, cables, and composites for sports and space applications.

In 1985, Teijin Ltd. introduced Technora fiber, previously known as HM-50, into the high performance fiber market. Technora is based on the 1:1 copolyterephthalamide of 3,4′-diaminodiphenyl ether and *p*-phenyl-enediamine. Technora is a wholly aromatic copolyamide of PPT, modified with a crankshaft-shaped comonomer, which results in the formation of isotropic solutions that then become anisotropic during the shear alignment during spinning. The polymer is synthesized by the low temperature polymerization of *p*-phenylenediamine, 3,4′-diaminophenyl ether, and terephthaloyl chloride in an amide solvent containing a small amount of an alkali salt.

Heterocyclic Rigid-Rod Polymers. PBZ, a family of *p*-phenylene–heterocyclic rigid-rod and extended chain polymers includes poly(*p*-phenylene-2,6-benzobisthiazole) (*trans*-PBZT) and poly(*p*-phenylene-2,6-benzobisoxazole) (*cis*-PBO). PBO is undergoing commercial development.

Nomex fiber was commercialized for applications requiring unusually high thermal and flame resistance. It retains useful properties at temperatures as high as 370°C. Nomex has low flammability and has been found to be self-extinguishing when removed from the flame. An outstanding characteristic is low smoke generation on burning. MPD-1 fibers may be obtained by the polymerization of isophthaloyl chloride and *m*-phenylenediamine in dimethylacetamide with 5% lithium chloride. Fibers are dry spun directly from solution.

Poly(2,2′-(*m*-phenylene)-5,5′-bisbenzimidazole) is a textile fiber marketed by Hoechst Celanese which does not form liquid crystalline solutions, owing to its bent meta backbone monomeric component. PBI has excellent resistance to high temperature and chemicals. PBI is being marketed as a replacement for asbestos and as a high temperature filtration fabric with excellent textile apparel properties. Typical properties of stabilized PBI are a tenacity of 0.27 N/tex (3.1 gf/den), a fiber breaking elongation of 30%, an initial modulus of 3.9 N/tex (45 gf/den), a density of 1.43 g/cm^3, and a moisture regain of 15% (at 21°C and 65% relative humidity).

Solution dyeing of PBI is necessary because the glass-transition temperature (T_g) of PBI is greater than 400°C, and as a result dye molecules only slowly diffuse into the PBI fiber structure.

Industrial Thermotropic LCPs. Vectran, poly(6-hydroxy-2-naphthoic acid-*co*-4-hydroxybenzoic acid), is currently the only thermotropic fiber which is commercially available. Vectran is synthesized by the melt acidolysis of *p*-acetoxybenzoic acid and 6-acetoxy-2-naphthoic acid. Vectran HS fibers are reported to have typical tensile strength and modulus values of 2 N/tex (23 gf/den) and 46 N/tex (525 gf/den), respectively. The melting point and density are reported to be 330°C and 1.4 g/cm^3. The fibers have excellent chemical resistance except for their resistance to alkali.

Gel Spun Fibers. In the mid-1970s it was discovered at the Dutch States Mines Co. (DSM) that through an ingenious new method of gel spinning ultrahigh molecular weight polyethylene it was possible to produce fibers having twice the tenacity of Kevlar, which was then considered to be the strongest known fiber. These high performance polyethylene fibers (HPPE) produced by the DSM subsidiary company, Stamicarbon, are called Dyneema and those produced by the Allied Signal Corp. in the United States are sold under the trade name of Spectra 1000. The commercial products have somewhat lower strengths than the laboratory fibers but still are in the high 2.6 N/tex (30 gf/den) range.

Properties. Fiber property comparisons for the different products are given in Table 1. The attributes of HPPE fibers include high strength; high abrasion resistance; high uv stability as compared to other synthetics; high resistance to acids, alkali, organic chemicals, and solvents; and low density. Disadvantages are a low melting point of about 150°C, which means performance is limited to no more than 120°C; difficult processing; and poor surface adhesion properties.

Table 1. Properties of Commercial HPPE Fibers

Fiber	Tenacity, N/tex[a]	Initial modulus, N/tex[a]	Elongation at break, %
Dyneema	1.01–3.57	57–128	3–7
Spectra 1000	3.4–3.57	162–171	3–7

[a] To convert N/tex to gf/den, multiply by 11.33.

Elongatable Carbonaceous Fiber. It is difficult to weave or knit regular carbon fiber. To overcome this drawback, an exciting new modification of carbon fiber technology was developed; by using less stringent carbonizing conditions and only partially carbonizing the precursor fibers, improved textile fiber properties have been achieved.

Properties. Unlike regular carbon fibers, these new products do not conduct electricity, but do exhibit good textile processing properties and possess exceptional ignition-resistant, flame-retardant, and even fire blocking properties.

Vitreous Fibers. Man-made vitreous fibers (MMVF) comprise a number of glass and specialty glass fibers and also refractory ceramic fibers. The vitreous state in glass is somewhat analogous to the amorphous state in polymers. However, unlike organic polymers, it is not desirable to achieve the crystalline state in glass. Glasses are produced from glass-forming compounds such as SiO_2, P_2O_5, etc, which are mixed with other intermediate oxides such as Al_2O_3, TiO_2, or ZnO, and modifiers or fluxes like MgO, Li_2O, BaO, CaO, Na_2O, and K_2O.

A wide range of glass compositions is available to suit many textile fiber needs; the three most common glass compositions are referred to as E, S, and AR glasses. AR glass is a special glass with higher contents of Zr_2O designed to resist the calcium hydroxide in the cementitious products where it is used. S glass is a magnesium–aluminum–silicate cross-linked glass used where high mechanical strength or higher application temperatures are desired. E glass is a member of the calcium–aluminum–silicate family containing less than 2% alkali (see composition in ASTM specification D578-89a) and is the predominant glass used to make textile and continuous filament fibers.

Manufacture. Vitreous fibers are produced by several processes including the continuous drawing process, the rotary process, and flame attenuation.

Properties. Glass fibers made from various compositions have softening points in the range 650–970°C. Fiber length and diameter distributions are significant factors in determining thermal and acoustical insulation properties. Slag wool and rock wool fibers are prepared from the slag from pig iron blast furnaces. Seventy percent of the slag wool in the United States is used for ceiling tiles.

Refractory Ceramic Fibers (RCF). These MMVF materials constitute only about 1% of the vitreous fiber market but have exceptional high temperature performance characteristics. They are produced by using high percentages of Al_2O_3 about 50/50 with SiO_2 as is or modified with other oxides like ZrO_2 or by using Kaolin clay which has similar high amounts of Al_2O_3. Specially prepared ceramic fibers are used to protect space vehicles on re-entry and can withstand temperatures above 1250°C.

Sulfar Fibers. Ryton fibers are high performance products developed by Phillips Petroleum Co. by reaction of *p*-dichlorobenzene with sodium sulfide in the presence of a polar solvent. The U.S. Federal Trade Commission granted the fiber the new generic name of Sulfar. The fiber has excellent chemical and high temperature performance properties.

Properties. As prepared, the polymer is not soluble in any known solvents below 200°C and has limited solubility in selected aromatics, halogenated aromatics, and heterocyclic liquids above this temperature. The properties of Ryton staple fibers are in the range of most textile fibers and not in the range of the high tenacity or high modulus fibers such as the aramids. The density of the fiber is 1.37 g/cm^3

Table 2. Classification of High Performance Fibers and High Technology Textiles by Property

Property	Fiber types	Applications
high tenacity and high modulus	aramids, gel spun polyethylene, polyarylate	tires, antiballistic, ropes, optical cables
resistant to heat and flame	aramids, PPS, PEEK, PBI, polyimides, EDF	protective clothing for various applications
resistance to chemical agents	PPS, fluorocarbon, polyolefins	filters, geotextiles, marine applications
microtex and hollow fibers	most synthetics and regenerated fibers	filtration, leisure, insulation, biomedical
intricate shapes and porosities	most synthetics and regenerated fibers	fashion, fragrances, antimicrobial, fiber optics, specialty wipes

which is about the same as polyester. However, its melting temperature of 285°C is intermediate between most common melt spun fibers (230–260°C) and Vectran thermotropic fiber (330°C). The main advantage of Ryton is not any particular excelling property, but rather the ability to retain its standard properties under adverse conditions.

Applications

One review indicates that high technology textile uses will account for 50% of all worldwide fiber consumption by the year 2000 compared to 10–15% in 1990. In some instances various technologies and concepts are combined or refined to produce a textile product for the desired application(s). Thus sophistication and enhancement of properties may be introduced at the fiber, yarn, and/or fabric levels.

Structure/Property Classification. The relationship between structure and properties of textile or fibrous substrates and their applications is one method of classifying nontraditional or high technology textiles (Table 2). At the fabric or product level, the classes may be described as coated and laminated fabrics, composites and fiber-reinforced materials, three-dimensional fabric structures, and fabrics containing polymers or structural features that impart multifunctional properties or allow the fibrous substrate to act as an intelligent material.

Classification by Types of Application. Another way to classify high performance fibers and high technology textile materials or products is by types of applications. A scheme of 10 main categories has been adopted. They include transportation, manufacturing, agriculture and forestry, civil engineering and construction, fishery and marine, protective clothing, sports and leisure, biomedical and health care, defense and aerospace, and energy use and conservation.

MALCOLM POLK
Georgia Institute of Technology
TYRONE L. VIGO
U.S. Department of Agriculture
ALBIN F. TURBAK
Consultant

A. K. Dinghra and H. G. Lauterbach, in J. Kroschwitz, ed., *Encyclopedia of Polymer Science and Engineering*, Vol. 6, John Wiley & Sons, Inc., New York, 1986, p. 756.

M. Lewin and J. Preston, eds., *High Technology Fibers, Part A*, Marcel Dekker, Inc., New York, 1985; *Part B*, 1989; *Part C*, 1993.

T. L. Vigo and A. F. Turbak, eds., *High-Tech Fibrous Materials: Composites, Biomedical Materials, Protective Clothing, and Geotextiles*, American Chemical Society, Washington, D.C., 1991, 398 pp.

T. L. Vigo and B. J. Kinzig, eds., *Composite Applications: The Role of Matrix, Fiber, and Interface*, VCH Publishers, New York, 1992, 407 pp.

HIGH PRESSURE TECHNOLOGY

High Pressure in the Chemical Industry

The use of high pressure in the chemical industry may be traced from efforts to liquify the so-called permanent gases in the mid-nineteenth century, to the synthesis of ammonia, methanol, and urea and the hydrogenation of coal to produce hydrocarbon products. The discovery of polyethylene in 1930 led to the development of continuous processes for the production of the polymer, some of which operate at pressures as high as 350 MPa (51,000 psi). The design of equipment for continuous chemical processes operating at pressures above about 20 MPa (2900 psi) is discussed herein.

Other Industrial Applications. Apart from mechanical applications in which hydraulic pressure is used to supply power or to generate liquid jets for mining minerals or cutting materials, many of the specialized applications such as isostatic compaction are batch operations. Very high dynamic pressures produced in materials by shock waves generated by exploding charges adjacent to the material are used for welding, forming, and metal cutting operations.

Design of Thick-Walled Cylinders

Elastic Behavior. Early in the twentieth century it was considered prudent to design pressure vessels so that they operated within the elastic range of the material of construction at all times. To that end designers arranged for the working pressure to be less than the yield pressure of the cylindrical shell, ie, the pressure to cause initial yielding of the bore of the shell. The ratio of the yield pressure to the working pressure was regarded as the factor of safety with respect to yielding. This philosophy continued with the introduction of the pressure vessel codes in which the factor of safety with respect to yielding is expressed as the ratio of the yield strength in tension of the material of construction to the maximum permitted design stress generated by the internal pressure.

The state of stress at radius r in the wall of a cylinder of inner radius r_i and outer radius r_o (radius ratio $k = r_o/r_i$) subjected to an internal pressure P_i has been shown to be equivalent to a simple shear stress, τ, which varies across the wall thickness in accordance with equation 1 together with a superimposed uniform tensile stress.

$$\tau = \frac{k^2}{(k^2 - 1)} \left(\frac{r_i}{r} P_i 2\right) \tag{1}$$

The maximum value of the shear stress occurs at the bore and is given by

$$\tau = \frac{k^2}{(k^2 - 1)} P_i \tag{2}$$

If it is assumed that the uniform tensile stress has no significant effect on yield, then the yield pressure, P_y, of a cylinder subjected solely to an internal pressure may be calculated from

$$P_y = \tau_y \frac{(k^2 - 1)}{k^2} \tag{3}$$

where τ_y is the yield strength of the material of construction in torsion. Equation 3 predicts the correct yield pressure only if the material is isotropic, and the cylinder free from residual stress prior to the application of pressure and sufficiently long for there to be no end effects.

Criteria of Elastic Failure. If, as is usually the case, the design needs to be based on tensile data, then a criterion of elastic failure has to be invoked, introducing some uncertainty in the calculated yield pressure. Of the criteria which have been formulated the two most important for ductile materials are the maximum shear stress criterion and the shear strain energy criterion. According to the former, $\tau_y = \sigma_y/2$ and according to the latter, favored for the gun steels used in the construction of high pressure vessels, $\tau_y = \sigma_y/\sqrt{3}$, where σ_y is the yield strength of the material in tension. The yield strength of steels which can be forged and uniformly heat treated yet retain sufficient ductility to resist fast fracture is such that the working pressure of a vessel having a bore of 350 mm is restricted to about 200 MPa (29,000 psi). To achieve a higher yield pressure and hence working pressure, steels having a greater yield stress would be required. Alternatively, the inherent strength of the steel may be used more effectively by prestressing cylinders to ensure a more uniform stress distribution under load.

Partially Plastic Cylinders. As the internal pressure is increased above the yield pressure the inner layers of the cylinder are stressed plastically, whereas the outer ones remain elastic. Analysis of the stresses and strains in a partially plastic thick-walled cylinder made of a material which work-hardens is very complicated. However, if it is assumed that the material yields at a constant value of the yield shear stress, that the elastic–plastic boundary is cylindrical and concentric with the bore of the cylinder, and that the axial stress is the mean of the tangential and radial stresses, then it may be shown that the internal pressure, P_e, needed to take the boundary to any radius r_e such that $r_i < r_e < r_o$ is given by

$$P_e - \tau_y \left\{ 1 - \left(\frac{r_e}{r_o}\right)^2 + \ln\left(\frac{r_e}{r_i}\right)^2 \right\} \qquad (4)$$

If yield and subsequent plastic flow of the material occurs in accordance with the maximum shear stress criterion, then $\sigma_y/2$ may be substituted for τ_y in equation 4 and in equation 5 below. On the other hand, if the shear strain energy criterion is thought to be applicable, it may be assumed as a first approximation that the corresponding value is $\sigma_y/\sqrt{3}$.

Collapse and Bursting Pressures. If the pressure is sufficiently large to push the plastic–elastic boundary to the outer surface of the cylinder so that the fibers at that surface yield, then the wall is unrestrained and the cylinder is said to collapse. With an ideal material which does not work-harden, the collapse pressure, P_c, sometimes called the full plastic flow pressure, would be the bursting pressure of the cylinder. It is given by equation 4 when $r_e = r_o$; thus,

$$P_c = 2\tau_y \ln k \qquad (5)$$

In reality the bursting pressure is a little higher than the collapse pressure, as most high strength steels exhibit some strain hardening.

Pre-Stressed Cylinders. Apart from using a steel of higher tensile strength, nothing can be done to affect the bursting pressure of a cylinder of given radius ratio. On the other hand, the pressure to cause initial yielding of the cylinder can be raised by developing compressive stresses at the bore prior to the application of pressure, using a technique known as pre-stressing. Procedures such as compound shrinkage, tape winding and autofrettage used to pre-stress vessels and pressure-containing components all have their origin in the design of ordnance.

Effect of Temperature. Both high and low temperatures affect the mechanical properties of metals. In general, the ductile properties, and in particular the toughness and impact strength, of most low alloy steels, decrease sharply as the temperature is reduced, and care must be taken with choice of materials if low temperature embrittlement is to be avoided. On the other hand, the yield and tensile strength of steels decrease as the temperature increases, and allowance must be made for it in estimating the static strength of a thick-walled cylinder at temperatures above ambient. Above about 350°C, creep starts to become an important factor with Ni–Cr–Mo steels of the type used for high pressure applications, and the stresses in the wall of the vessel and its deformation are no longer independent of time. In addition to the effect of temperature on mechanical properties, temperature gradients generated in the walls of vessels as a result of applied heat or of heat liberated by exothermic reactions proceeding within the vessel cause thermal stresses which may need to be considered when estimating the stresses in a thick-walled cylinder subjected to both internal pressure and heat flux.

Static Design Criteria. At one time ICI designed reactors for the manufacture of low density polyethylene (LPDE) with a factor of safety of 2.5 with respect to bursting pressure at room temperature, using steel forgings which had a transverse impact strength > 34 J as measured in an Izod test. The toughness requirement ensured that the steel had sufficient ductility to resist fast fracture. These criteria were modified by ICI in light of its experience as well as by other companies, but at present (ca 1996) the design of vessels for LDPE service is based almost exclusively on the estimated bursting or collapse pressure of the vessel together with the fracture toughness properties of the material of construction. The equivalent fracture toughness to an impact strength of 34 J is approximately 120 MPa$\sqrt{\text{m}}$. The ASME Subcommittee on Pressure Vessels (Section VIII) proposes to use these criteria in the new division of the pressure vessel code dealing with the design of vessels at high pressure.

Manufacture of Vessels and Tubes. The choice of steel for early forged pressure vessel bodies and closures was influenced by ordnance practice. Vessels designed by ICI for the production of LDPE during the 1950s were made of 2.5% Ni–Cr–Mo steel, a gun steel which could be heat-treated to achieve a uniform ultimate tensile strength (UTS) of about 965 MPa throughout the wall thickness while retaining adequate ductility and impact strength. Later, as the required reactor volume increased, 3.5% Ni–Cr–Mo–V steel having a UTS of 112 MPa was used. In the United States, similar gun steels, 2300 series nickel steels and Ni–Cr–Mo–V steels, are widely used for forged pressure vessels. Probably the largest vessels currently in use for the manufacture of LDPE are about 1.25 m external diameter and 10 m long.

Tubes for the early LDPE plants up to 3 m long having a bore of 12–25 mm were made from bored bar in 12–14% chromium steel to BS 970 EN56C by a number of cold reducing operations. Subsequently, tubes were made in material equivalent to EN56C by several companies in the United States, using a cold pilgering process.

In the late 1950s, the Timken Company began to produce tubes 10 m long having a bore of 25–50 mm in AISI 4300 series steels. The process starts by piercing a billet, which is then hot rolled or hot Assel elongated to reduce the outside diameter to the required size. This hot worked tubing may then be finished, heat treated and sold as hot worked tubing or it may become the hollow for subsequent cold reduction by cold pilgering. After cold pilgering, the tube is finished, heat treated and sold as cold reduced or "Rotorolled" tubing. A fine bore finish can be achieved by honing.

Effect of Fatigue. The maximum operating pressure and throughput of the LDPE processes developed in the 1960s and 1970s was governed more by problems associated with the design of the reciprocating compressors than with design of the vessels and piping.

In most LDPE plants a primary compressor raises the pressure of ethylene to about 25–30 MPa and a secondary compressor to 150–315 MPa. The major problems in the development of secondary compressors for higher operating pressures and throughputs were, firstly, the restriction of ethylene leakage past the plunger or piston to an acceptable level and secondly, the avoidance of fatigue failure of high pressure components such as cylinders and valves. The stress in the cylinder of a reciprocating compressor fluctuates cyclically between suction and discharge pressures and a compressor running at 5 Hz will complete 10^7 cycles in about 23 days of continuous operation; hence the design of the pressure-containing parts must be based on high cycle fatigue data.

In addition to the problems associated with the design of reciprocating compressors, some LDPE processes use pulsed tubular reactors which are subjected to severe fatigue conditions. These reactors may be 1000 m long and are constructed by joining together lengths of high pressure tubing ~ 10 m long with a bore diameter in the range of 25–75 mm and a radius ratio of about 2.5. The tubes are fitted with water-cooled jackets to remove the heat of polymerization but the resulting polymer tends to stick on the bore surface of the reactor and reduces heat transfer. To increase the flow velocity through the reactor and slough the polymer from the wall, the pressure at the exit of the reactor is rapidly reduced from its maximum value of ~300 MPa to ~200 MPa every few minutes. The depth and frequency of the so-called bump cycle varies widely from plant to plant and may give rise to fatigue failures unless care is taken in the design of the tubes.

Nature of Fatigue Failure. When the pressure in a thick-walled cylinder of ductile material is increased to cause bursting, the fracture nearly always take a spiral course across the wall thickness, preceded by considerable plastic deformation. On the other hand, a fatigue crack generated by a pulsating internal pressure in a plain cylinder initiates and propagates from inclusions or defects intersecting the bore surface and usually spreads outward across the wall so that its shape in the axial–radial plane is approximately semicircular. Provided the cylinder is made of ductile material, the crack usually continues to propagate under the cyclic loading conditions until it intersects the outside surface and leakage ensues. If the material lacks toughness, propagation of the crack may be interrupted part way through the wall by the intervention of fast fracture.

The experimentally measured fatigue strength of cylinders of different radius ratios subjected to repeated internal pressure cannot be related to the fatigue strength of the material in other forms of loading, eg, reversed bending. The limiting maximum shear stress which can be endured indefinitely at the well finished bore of a plain cylinder made of most of the materials tested is about 1/3 of the UTS for those materials whose tensile strength is less than 1000 MPa. If the cylinder contains a cross-bore or has a sharp change of bore diameter, the fatigue strength will be reduced by an amount dependent on the stress concentration associated with the design feature. Every effort should be made to reduce the stress concentration by design changes such as the removal of sharp corners. For maximum fatigue strength the component should be fabricated from steel free from inclusions and dissolved gases and to this end vacuum arc remelted steel is often used.

In the life range of 10^6 to 10^7 cycles, the fatigue strength of thick-walled cylinders depends not only on the maximum range of bore shear stress generated by the cyclic pressure, but also on the associated mean shear stress. Compound construction or autofrettage increases the fatigue strength of cylinders as a result of the reduction in the mean shear stress at the bore of the cylinder brought about by shrinkage or overstrain. Both techniques are employed for increasing the fatigue strength of components such as packing cups and suction and discharge valves used in the construction of the cylinders of secondary compressors for LDPE plants. Whereas compound shrinkage is confined to components which are of simple shape, autofrettage can be used for components of more complicated shape.

Any surface treatment which sets up compressive stresses in the bore layers of a cylinder will increase its endurance to high cycle fatigue, and peening, nitriding, honing, etc, are widely used for increasing the fatigue strength of compressor components.

Because the limiting maximum shear stress that can be endured by a plain thick-walled cylinder indefinitely is about 1/3 of the UTS of the material of construction, it might be thought that increasing the tensile strength would be a good way to increase the fatigue strength. Tests on plain and cross-bored cylinders in various states of hardness show that the fatigue limit is raised as the UTS increases, but at much higher pressures and shorter lives, the higher strength cylinders survive for fewer cycles than those of lower strength. Furthermore, if the tensile strength is increased too much the crack may not reach the outer surface before fast fracture intervenes.

An important application of linear elastic fracture mechanics is the estimation of the critical crack or defect size which will cause fast fracture to occur from a knowledge of the fracture toughness of the material. Standard procedures for fracture toughness testing of materials give reproducible values of minimum fracture toughness, but lack of accepted stress intensity factors for internally pressurized components has, until recently, limited this application. Another application is the estimation of the rate of growth of fatigue cracks so that inspection frequencies may be established to ensure that cracks are not allowed to reach critical size.

Mechanical Joints

Pressure Vessels. Removable closures for pressure vessels usually consist of three elements: a cover to the opening in the vessel, a coupling device holding the cover in position against the internal pressure, and a sealing ring or gasket between the cover and the vessel.

Generally it is desirable to have as few openings as possible in the walls of a vessel and to this end it is customary to accommodate ancillary equipment, such as stirrer glands, pipe connections, and small fittings for measuring instruments in the more lightly stressed end cover.

Usually this component has to withstand not only the stress produced by the internal pressure acting on the cover, but that produced by the initial tightening of the seal. Coupling devices include bolted flanges, screwed plugs, and split collars engaging with buttress-shaped grooves machined on the vessel and end cover. The high stress concentration at the root of the first active thread in a screwed plug may initiate fatigue cracking and this aspect of closure design has received considerable attention in recent years.

The sealing ring or gasket is nearly always made of a softer material than the vessel and end cover so that when it is tightened to make an initial pressure-tight seal, it deforms and follows the irregularities in the mating surfaces closely. Thus, deformation is almost entirely confined to the sealing ring, which may be replaced as necessary. Gaskets or sealing rings are usually of the self-sealing type; they include the lens, cone, delta, wave, and O-rings.

An alternative method of making a pressure-tight seal between a vessel and its cover is to develop such a high compressive stress between the two components that they yield along a narrow, axisymmetrical, circular band of contact. By this means the asperities on the mating surfaces are smoothed out and what is usually described as a metal-to-metal seal is formed. This technique makes it difficult to refurbish the mating surfaces of large vessels, and it is almost entirely confined to fittings used for connecting small-bore pipelines and to sealing adjacent packing cups or valve components in reciprocating compressors.

Tube Connections. Tubes having a wide range of bore sizes are required to operate at pressures in the region of 150–300 MPa (22,000–44,000 psi). Cone and collar type union connectors are nearly always used to connect small bore tubes at pressures up to about 400 MPa. The ends of tubes having a bore greater than about 12 mm are usually threaded and coupled together with loose screwed flanges. Cone rings, lens rings or metal lip sealing rings may be used to make the seal between the ends of adjacent lengths of tube.

Safety and Testing

The safe operation of high pressure plant necessitates suitable in service inspection to ensure that the equipment remains within acceptable design limits. Nondestructive testing plays an important role and dye penetrant, magnetic particle, eddy current, and ultrasonic techniques are widely used to detect flaws or fatigue cracks in high pressure components.

K. E. Bett
University of London

W. R. D. Manning and S. Labrow, *High Pressure Engineering*, Leonard Hill, London, 1971.

B. Crossland, K. E. Bett, H. Ford, and A. K. Gardner, *Proc. Inst. Mech. Engrs.* **200**(A4), 237 (1983).

HIGH PURITY GASES

High purity industrial gases are routinely delivered in large quantities having purities exceeding 99.999% (>5 nines pure). There are many applications for gases where purity even higher than 99.999% is required.

There is no universally accepted definition of what purity levels correspond to high purity. However, gases having total impurities specified <1 ppm on a molar or volume basis must be manufactured and handled differently from regular gases if that specification is to be maintained. A good working definition of high purity is gases having certain individual impurities held to levels <0.1 ppm.

Depending on a volume, high purity gases can be delivered using either bulk systems, where a plant-wide distribution system is integrated with central gas storage facilities, or cylinders, where a short local distribution system is supplied from a single high pressure cylinder.

Gases used in the manufacture of semiconductor materials fall into three principal areas: the inert gases, used to shield the manufacturing processes and prevent impurities from entering; the source gases, used to supply the molecules and atoms that stay behind and contribute to the final product; and the reactive gases, used to modify the electronic materials without actually contributing atoms or molecules.

Production and Purification

The separations processes used for manufacturing high purity gases are generally the same as those used for making lower purity products. Purification by distillation and adsorption are often used. Chemical conversion processes, where the impurities are converted into more easily separable forms through a selective chemical reaction, are often employed in point-of-use purifiers.

Bulk Gases. The bulk gases are usually characterized by high volume flow requirements in the manufacturing process. Historically these have consisted of nitrogen, oxygen, argon, hydrogen, and to a lesser extent, helium.

Nitrogen. Because of numerous applications in semiconductor manufacturing, high purity nitrogen is produced both at high volumes and at some of the lowest impurity levels seen for any of the high purity gases.

Distillation. All high purity nitrogen is manufactured from air using multistage cryogenic distillation.

Chemical Conversion. In both on-site and merchant air separation plants, special provisions must be made to remove certain impurities. The main impurity of this type is carbon monoxide, CO. The most common approach for CO removal from the feed air to the air separation unit is chemical conversion to CO_2 using an oxidation catalyst in the prepurification unit. The CO_2 is removed by a prepurification unit in the air separation unit.

At throughputs below 500 nm^3/h, a wide variety of inert gas purification processes based on chemical conversion can be used to produce high purity nitrogen. Typically, the impure nitrogen is passed through a bed of reagents, where the conversion reactions occur, causing the impurities to remain behind in the bed of reagents. These processes require that the bulk of the oxygen in the feed be removed by some other method.

Oxygen. High purity oxygen for use in semiconductor device manufacture is produced in relatively small quantities compared to nitrogen. There are two different purification processes in general use for manufacturing the gas: distillation and chemical conversion plus adsorption.

Distillation. As for nitrogen, all high purity oxygen is derived from air through the air separation process using cryogenic distillation. Generally, air separation units that manufacture commercial purity oxygen also remove nitrogen and other light impurities to levels low enough for high purity applications.

Chemical Conversion and Adsorption. Where additional distillation is not practical, hydrocarbons and heavy noble gases can also be removed by combining chemical conversion with adsorption. Commercial purity oxygen is passed through a high temperature bed of oxidation catalyst; where hydrocarbon impurities are oxidized to CO_2 and H_2O. The CO_2 and H_2O products from catalytic oxidation of hydrocarbon impurities are removed using a temperature swing adsorption (TSA) process. The adsorbent is typically one of the molecular sieves.

Argon. High purity argon has many applications as an inert gas during the manufacture of semiconductor devices. In these applications, nitrogen is a reactive impurity which, in addition to O_2, H_2O, CO_2, CO, and all hydrocarbons, must be removed from the argon to low levels.

Distillation. Conventional purity argon is separated from air using a combination of distillation and chemical conversion. High purity argon is made the same way.

Chemical Conversion. Except for control of nitrogen impurity levels, the same chemical conversion methods used for nitrogen purification at low flow rates can also be used for argon purification.

Hydrogen and Helium. Whereas hydrogen and helium are very different chemically, these gases have low boiling points and are normally liquefied during manufacture. Because the boiling points are so low, even very small amounts of trace impurities tend to freeze and form solid deposits. To prevent formation of these deposits, trace impurities must be removed prior to liquefaction. Similar methods are used to purify hydrogen and helium prior to liquefaction.

Hydrogen. High purity hydrogen is usually delivered and stored as a cryogenic liquid and vaporized when needed. This vaporized liquid seldom needs any further processing to meet high purity specifications. High pressure hydrogen gas can be delivered and further purified on-site to meet high purity specifications. This is accomplished using combinations of chemical conversion, cryogenic adsorption, and palladium membrane processes.

Helium. High purity helium is usually not required in large quantities and is therefore not commonly delivered as a cryogenic liquid. Instead, high pressure cylinders are filled from a liquid helium source by the gas supplier and then transported to the customer. When high purity helium is required, the high pressure gaseous helium is processed through an on-site purifier.

Because helium is not chemically reactive, the same chemical conversion processes used for purification of nitrogen and argon are also applied to helium purification.

Specialty Gases. The specialty gases are generally more reactive than the bulk gases.

Purification of specialty gases can be divided into two areas: purification done by the gas supplier on a bulk scale prior to filling the cylinder or other delivery container, and purification carried out by the consumer on a point-of-use scale generally just prior to use.

Bulk Purification. Many specialty gases originate as by-products or low purity intermediate chemicals produced during the course of manufacturing something else. The purification processes tend to utilize standard methods and are done on a continuous basis.

Distillation. Processes which utilize either simple liquid vapor flash processes or multistage distillation are often used for purification of bulk specialty gases.

Adsorption and Chemical Conversion. In some cases, removal of moisture or oxygen added by small amounts of air contamination is all that is necessary to make a gas suitable for high purity applications. With a limited objective, it is usually most effective to use an adsorption process and may be designed either with or without the capability for repeated generation.

Chemical conversion processes can also be used for moisture and oxygen removal.

Delivery and Control

Once a gas has been purified, it must be brought to its intended point of use without being degraded by the addition of excessive contamination.

Delivery methods for high purity gases can be divided both according to chemical reactivity with respect to the containment system and, to a less significant degree, according to volume throughput requirements. Many highly flammable gases such as H_2 and SiH_4 are still inert with respect to the containment systems. Even though special provisions must be made because of flammability, the technology used to deliver flammable gases is similar to that employed for inert gases.

Bulk Gases. Attaining high purity gases where they are used requires a suitable gas distribution system. To achieve a high purity distribution system, there must be an absence of dead zones, external leakage, outgassing, and particulate contamination.

Specialty Gases. The purity of specialty gases depends on the systems and procedures adopted by the distributors for bulk gas supply and cylinder preparation, filling, and delivery. Most of the precautions taken into consideration in the bulk gases delivery system are also applied for specialty gases to eliminate recontamination.

Analysis and Certification

Ensuring that the purification and delivery processes are working properly is essential to successful applications of high purity gases.

Particulates. Separation of particulate impurities is an important part of the process for manufacturing high purity gases and the most common approach is through filtration technology.

Description of Analytical Methods. Procedures for analyzing both gas and particulate impurities in high purity bulk gas products are available. Typical gaseous impurities include oxygen, moisture, carbon monoxide, carbon dioxide, total hydrocarbons (THC), argon, and nitrogen analyzed to the low ppb level. Particle impurities are analyzed to the 0.1-μm level.

Gaseous Impurities. Instrument calibration [excluding the atmospheric pressure ionization mass spectrometer (apims)] typically consists of two steps: zeroing and spanning. Zeroing is accomplished by allowing the analyzer to sample a gas, the contaminant level of which is below the lower detection limit of the analyzer. This is called a zero gas. A typical method for generating zero gas is to take the actual sample gas and run it through a gas purifier.

Spanning, accomplished using a sample gas containing a known volume concentration of impurity, is performed at levels that are the same order of magnitude as the required detection. The actual span concentration is selected so that the majority of expected measurements fall at or within its value.

The calibration procedures for an atmospheric pressure ionization mass spectrometer (apims) involves the generation of separate calibration curves for each of the monitored impurities.

Particulate Impurities. Particle counters require factory calibration every two years. In addition, the background signal associated with both the instrument and its sampling system must be quantified so that it may be subtracted from sample measurements. To accomplish this, an absolute filter (<0.01 μm rating) is employed. The absolute filter removes all particles entering the sampling system, so any particle registered by the counter can be directly attributed to the sampling system or instrument noise.

Sampling and Analysis Guidelines. As a general safety consideration, all gases should be vented to an external area and whenever possible, inert gases should be used as the test gas for piping systems.

Applications of High Purity Gases

The applications of high purity gases are primary in the semiconductor industries. In addition to the microelectronics industry, other applications for high purity oxygen include fiber optics manufacturing, production of pharmaceuticals, and usage as calibration media in research and development laboratories, and in the pollution control field. Applications for high purity hydrogen include oxidation processing and epitaxial growth for both silicon and gallium arsenide.

High purity argon is used in the high technology fields of electronics, fiber optics, research and development, powder metal spraying, and hot isostatic pressing.

Other applications of high purity specialty gases include hydrogen bromide for etching single-crystal silicon, polysilicon, and aluminum; nitrogen trifluoride as a fluorine source for *in situ* cleaning processes for chemical deposition equipment, semiconductor etching and deposition, and high energy chemical lasers; and sulfur hexafluoride as a key etching material in certain semiconductor manufacturing processes.

WALTER H. WHITLOCK
EDWARD F. EZELL
SHUEN-CHENG HWANG
The BOC Group, Inc.

R. DiNapoli and A. M. Sass, in J. J. McKetta, ed., *Encyclopedia of Chemical Processing and Design*, Vol. 31, Marcel Dekker, Inc., New York, 1990, p. 236.

W. H. Whitlock, "The Ultra-High Purity Challenge", in *Separation of Gases, Proceeding of the Fifth BOC Priestley Conference*, Birmingham, U.K., Sept. 19–21, 1989, Royal Society of Chemistry, 1990.

HIGH TEMPERATURE ALLOYS

High temperature alloys are those combinations of metals that are used specifically for their heat resisting properties. Physical properties such as melting temperatures, elastic modulus, density, and thermal conductivity of the elemental metals that serve as the bases for most high temperature alloys are listed in Table 1. Contrary to expectations, melting point is not the primary indicator of adequate high temperature strength. For example, nickel, which has the lowest melting point of any element in Table 1, is the choice for the most severe high temperature structural applications in air. There are several reasons for this, including the lack of an allotropic phase transformation in nickel below its melting point, its high tolerance for alloying elements without causing a phase change from the close packed face-centered cubic (fcc) crystal structure, and the ability to produce a very stable precipitate, γ'-(Ni_3Al), that is the primary source of high temperature strengthening in nickel-base superalloys.

Density is a particularly important characteristic of alloys used in rotating machinery, because centrifugal stresses increase with density. Alloys which contain the heavier elements, ie, molybdenum, tantalum, or tungsten, have correspondingly high densities.

Thermal expansion coefficients are important for mating components, as well as in the development of thermal stresses, which vary directly with the expansion coefficients. In general, thermal expansion coefficients are inversely proportional to melting point.

Mechanical Behavior

Creep Rupture. Metals and their alloys lose appreciable strength at elevated temperatures. For most materials, the ultimate tensile and yield strengths fall off regularly as the temperature increases. The exceptions are some intermetallics, eg, nickel aluminide(3:1), Ni_3Al, and alloys containing a large volume fraction of Ni_3Al, eg, single-crystal alloys PWA 1480. Not only is temperature important, but time also influences mechanical behavior at elevated temperatures. Stresses well below the yield strength can cause gradual deformation and eventual

Table 1. Physical Properties of High Temperature Metals

Metal	Melting point, °C	Density, g/cm^3	Thermal expansion coefficient at RT, 10^6/°C	Thermal conductivity at RT, W/(m·K)[a]
Co	1495	8.85	13.8	69.0
Ni	1453	8.90	13.3	92
Fe	1537	7.87	11.76	75
Cr	1890	7.2	6.2	66.9
Nb	2468	8.6	7.1	52.3
W	3410	19.3	4.5	20.1
Ta	2996	16.6	6.6	54
V	1900	6.1	9.7	31
Mo	2610	10.22	5.4	146

[a] To convert W/(m·K) to cal/(cm·s·°C), multiply by 2.39×10^{-3}.

fracture if sustained long enough. For this reason, the standard tensile test is inadequate for providing design data at elevated temperatures, and creep-rupture tests must be conducted, in which time-dependent deformation and fracture are determined from periodic measurements under a fixed stress or load.

Relaxation. Relaxation also is associated with creep at high temperatures. If a specimen is stretched or compressed and is then held over a period of time at a high temperature with its ends in fixed positions, the stresses within the specimen gradually diminish. After sufficient time has elapsed, the tensile or compressive stresses may relax to only a fraction of the original values. Creep and relaxation are thus complementary processes, and an alloy having high creep strength generally resists relaxation.

Relaxation is an important example of a creep phenomenon encountered in practice. Bolts, studs, flanges, and springs of all kinds are subject to relaxation when used at high temperatures. Bolts can become loose so that bolted joints develop leaks after operation at elevated temperatures.

Fatigue. Engineering components often experience repeated cycles of load or deflection during their service lives. Under repetitive loading most metallic materials fracture at stresses well below their ultimate tensile strengths, by a process known as fatigue. The actual lifetime of the part depends on service conditions, eg, magnitude of stress or strain, temperature, environment, surface condition of the part, as well as on the microstructure.

Resistance to fatigue fracture is an important consideration in selecting materials for many high temperature applications, most notably in rotating machinery such as gas or steam turbines. Generally, two classifications of fatigue behavior are made, depending on the parameter being reversed, stress, or strain.

High temperature materials which exhibit the greatest resistance to high cycle fatigue on a strength basis, ie, fatigue limit/tensile strength vs N_f, are composite materials and dispersion strengthened alloys. Precipitation hardened alloys, on the other hand, usually demonstrate very poor fatigue resistance relative to the tensile strengths, perhaps because the precipitates become unstable as a result of repeated cycling.

The rate of crack growth is often a more useful parameter than is fatigue life. Fracture mechanics (qv) techniques have been widely applied to the crack growth behavior of high temperature alloys.

Summary of Strengthening Methods. In practice, few alloys are strengthened by only one or merely a few of the mechanisms described, as shown in Table 2.

Table 2. Strengthening Mechanisms in High Temperature Alloys

Alloys	Primary strengthening	Secondary strengthening
superalloys		
Ni-base	precipitation of γ'	solid soln
Co-base	solid soln	precipitation of carbides
bcc refractory metals	cold working	solid soln, precipitation
directionally solidified eutectics	composite	solid soln, precipitation
dispersion-strengthened alloys	dispersion	solid soln, grain size
intermetallics	long-range order	solid soln, precipitation

Surface Stability

Oxidation. Immense progress in technology has imposed ever-increasing demands on the mechanical and chemical properties, in particular the oxidation and scaling resistance, of metallic materials.

The scale morphology is dependent on the conditions of reaction, the time of oxidation, the composition of the corrosive medium, and the type and composition of the particular alloy involved. Complex alloys may form two or more layers differing in either composition or microstructure or both. In order to maintain good oxidation resistance at least one of the layers must be compact and preferably be a slow growing oxide.

The oxidation of most modern alloys is dependent on the formation of a compact protective film of a slow growing chemically stable oxide such as chromium(III) oxide, C_2O_3, alumina, Al_2O_3, or silica, SiO_2. The oxidation behavior of multicomponent γ'-strengthened alloys can be estimated by considering the Ni, C_2, and Al, content of the alloy.

Hot Corrosion. Hot corrosion is an accelerated form of oxidation that arises from the presence not only of an oxidizing gas, but also of a molten salt on the component surface. The molten salt interacts with the protective oxide so as to render the oxide nonprotective. Most commonly, hot corrosion is associated with the condensation of a thin molten film of sodium sulfate, Na_2SO_4, on superalloys commonly used in components for gas turbines, particularly first-stage turbine blades and vanes.

The deposition of molten Na_2SO_4 in gas turbines is believed to be related to the reaction between the residual sulfur in fuel and sodium, which may be contained either in the fuel or the intake air. The sodium in the air is normally present as an aerosol of sea salt.

For most alloys, the corrosion rate displays a maximum at 850–900°C, and decreases very rapidly at temperatures up to 1000°C, again strongly suggesting that a molten salt is necessary in order to initiate hot corrosion.

It is generally conceded that the chromium content is the most important factor in hot corrosion resistance. For this reason cobalt alloys, which generally contain 20% or more chromium, display better hot corrosion resistance than nickel alloys, which typically contain 8–15% chromium.

Coatings. It is common practice to apply some type of protective coating to extend the surface stability of superalloy or refractory metal components. Superalloy coatings provide an aluminum reservoir for growth of protective Al_2O_3 scales and inhibit further oxidation. Coating microstructures can be varied from the relatively brittle high aluminum content intermetallic matrix phase, eg, CoAl or NiAl, to the relatively ductile low aluminum content metal plus intermetallic-type structures, eg, Co + CoAl. During cyclic oxidation and hot corrosion, excessive spallation occurs. Coatings have been developed, particularly Co–25%Cr–2%Al–0.1%Y and Ni–15%Cr–6%Al–0.1%Y, that exhibit excellent resistance to thermal cycling. These overlay coating compositions are based on the knowledge that aluminum is required to form protective Al_2O_3, and chromium is necessary to enhance Al_2O_3 formation and to improve hot corrosion resistance further. Excellent resistance to spallation is generally attributed to the addition of yttrium. A compact adhesive $Al_2O_3/CoAl_2O_4$ scale is formed having a low parabolic growth rate, thereby protecting the underlying base alloy. Both $CoAl_2O_4$ and $NiAl_2O_4$ spinels grow rather slowly when compared with most oxides. Such coatings degrade after a period of time, because in addition to attack of the protective oxide by spallation, erosion, and chemical means, inward diffusion of aluminum and chromium occurs. Therefore, although coatings can improve the oxidation resistance, degradation can occur and oxidation of the alloy ensues. NiCoAlY compositions offer superior elevated temperature oxidation resistance and diffusional stability on nickel-base superalloys. Additions of cobalt to NiCrAlY enhance hot corrosion resistance and improve coating ductility. CoCrAlY coatings provide superior protection in a hot corrosion environment.

In the case of refractory metals, coatings generally are silicides, applied by pack cementation or slurry processes. Typical silicide compositions are Si–20Cr–20Fe for niobium alloys and $MoSi_2$ for molybdenum alloys.

Specific Alloy Systems

Plain Carbon and Low Alloy Steels. For the purposes herein plain carbon and low alloy steels include those containing up to 10% chromium and 1.5% molybdenum, plus small amounts of other alloying elements. These steels are generally cheaper and easier to fabricate than the more highly alloyed steels, and are the most widely used class of alloys within their serviceable temperature range.

Of the common alloying elements in steel (qv), molybdenum is the most effective in increasing creep–rupture strength, and the carbon–molybdenum steels generally have more than twice the creep–rupture strength of plain carbon steel at the same temperature. The most commonly used steels for high temperature service contain from 0.5 to 1.5% molybdenum.

Chromium is the most effective addition to improve the resistance of steels to corrosion and oxidation at elevated temperatures, and the chromium–molybdenum steels are an important class of alloys for use in steam (qv) power plants, petroleum (qv) refineries, and chemical-process equipment. The chromium content in these steels varies from 0.5 to 10%. As a group, the low carbon chromium–molybdenum steels have similar creep–rupture strengths, regardless of the chromium content, but corrosion and oxidation resistance increase progressively with chromium content. Most of the chromium–molybdenum steels are used in the annealed or in the normalized and tempered condition; some of the modified grades have better properties in the quench and tempered condition.

Stainless Steels. Steels containing 11% and more of chromium are classed as stainless steels. The prime characteristics are corrosion and oxidation resistance, which increase as the chromium content is increased. Three groups of wrought stainless steels, series 200, 300, and 400, have composition limits that have been standardized by the American Iron and Steel Institute (AISI).

Although the stainless steels usually are specified in the wrought condition, a number of iron–chromium, iron–chromium–nickel, and nickel or cobalt-base alloys are produced as castings. Castings are classified as heat resistant when utilized in applications at 650°C or higher. Examples of Fe–Ni–Cr heat resisting castings are HC (equivalent to wrought AISI type 446), HH (AISI type 309), and HD (AISI type 327). These alloys are used in metallurgical furnaces, oil-refinery furnaces, power plant equipment, gas turbines, and in the manufacture of glass and synthetic rubber. Iron–chromium castings containing 10–30% Cr are useful chiefly for oxidation resistance, whereas iron alloys containing more than 18% Cr and more than 7% Ni have superior strength and ductility. Other iron-base alloys containing more than 10% Cr and more than 25% Ni are used in both reducing and oxidizing atmospheres.

The 12%-chromium ferritic superalloys are a group of proprietary steels that are essentially modifications of AISI 403 stainless steel. Examples are Crucible 422, Lapelloy (AISI 619), and Jessop-H46. The modifications include adding up to several percent of molybdenum and/or tungsten to stiffen the matrix, and up to 0.5% of niobium and vanadium to improve the dispersion and stability of the carbides. Up to 2% nickel, copper, and aluminum also may be present in these steels. The modified steels have a substantially greater creep–rupture strength than the standard AISI 403 stainless steel, and about the same level of corrosion and oxidation resistance. Typical applications include high temperature bolts, blades for jet-engine compressors and for high temperature steam turbines, compressor and turbine disks for jet engines, boiler, superheater, and reheater tubes and valve parts. These steels are available in most wrought forms.

The precipitation-hardening stainless steels are proprietary grades hardened by both the martensitic transformation and precipitation hardening. These contain higher amounts of chromium (16–17%) and nickel (4–7%) than the 12% chromium ferritic alloys. These steels are normally used at lower temperatures than the 12% chromium ferritic superalloys.

The highly alloyed austenitic stainless steels are proprietary modifications of the standard AISI 316 stainless steel. These have higher creep-rupture strengths than the standard steels, yet retain the good corrosion resistance and forming characteristics of the standard austenitic stainless steels.

Nickel-Base Superalloys. The nickel-base superalloys are the most complex in composition and microstructures and, in most respects, the most successful high temperature alloys. The earliest superalloys were wrought, ie fabricated to final size by a mechanical working operation. Later alloys have incorporated higher aluminum plus titanium contents, as well as molybdenum for solid-solution strengthening (Nimonics 115 and 120).

Alloys developed by processing through the investment casting process had higher strength and design flexibility, which led to many further advances through air cooling. The cast alloys tended to contain less chromium, which was replaced by molybdenum, tungsten, and tantalum, while retaining high volume fractions (to 60%) of γ'.

Apart from γ' and solid-solution strengthening, many alloys benefit from the presence of carbides, carbonitrides, and borides.

An undesirable feature of the most highly alloyed superalloys is the tendency to develop unwanted phases such as σ and μ. Sigma (σ) phase, a platelike intermetallic compound of two or more transition metals, eg, Cr_xFe_y or $(CrMo)_x(NiCo)_y$ where x and y can vary from 1 to 7, may precipitate from alloys containing a high refractory metal content, eg, IN-100. There is a critical temperature range, centered around 800°C, for the precipitation of σ, and precipitation leads to a decrease in rupture properties. Low temperature ductility also is adversely affected. The recognition that σ and other topologically close packed (tcp) phases are electron compounds where precipitation from solution can be predicted by knowledge of the average electron vacancy number of the alloy matrix was the basis of the Phacomp system for predicting safe alloy compositions. In brief, the total concentration of elements having high electron vacancy number (Cr, Mo, W, Mn) must be limited to avoid σ-phase precipitation, either during alloy processing or in service. Computer programs derived from the principles of electron vacancy numbers and phase stability are used in engineering specifications for superalloys utilized in aircraft engines and industrial gas turbines.

The temperature capability of nickel-base alloys has been improved markedly by processing techniques such as vacuum arc melting.

Cleanliness is critical in modern superalloys because of the role of inclusions in initiating fatigue cracks and fracture. Several refining processes to produce cleaner superalloys have been introduced, including the use of ceramic-foam filters in conjunction with vacuum induction melting, the reintroduction of electroslag remelting, and the development of electron-beam cold-hearth refining. Directional solidification (DS), in which heat withdrawal is made to occur parallel to the ingot axis, has been introduced to produce large columnar grains parallel to that axis. The technique, which resembles the Bridgeman method for growing single crystals, produces increased ductility at intermediate temperatures (760°C), improved rupture strength in thin sections, and improved low cycle fatigue life. The best elevated temperature properties of nickel-base alloys are obtained by an adaptation of the DS process to produce single crystals.

The coarse grains developed by conventional casting processes usually are deleterious to fatigue life. For parts such as turbine disks that are life-limited by fatigue rather than creep, fine grains are produced by powder metallurgical techniques.

Some of the superalloys can be welded by arc melting processes, as well as by resistance and electron-beam techniques. Alloys having low contents are readily weldable.

Binary Fe–Ni alloys as well as several alloys of the type Fe–Ni–X, where X = Cr or Co, are utilized for their low thermal expansion coefficients over a limited temperature range. Other elements also may be added to provide altered mechanical or physical properties. Common trade names include Invar (64%Fe–36%Ni), Elinvar (52%Fe–36%Ni–12%Cr) and super Invar (63%Fe–32%Ni–5%Co). These alloys, which have many commercial applications, are typically used at low (25–500°C) temperatures.

The latest class of ODS alloys to be developed, based on Ni, Cr, and Al, relies on an Al_2O_3 protective scale for dynamic oxidation resistance; Y_2O_3 is the dispersoid in each of these alloys.

The inadequate low temperature strength displayed by many dispersion strengthened alloys led to attempted combinations of dispersion, and γ'-hardening through the mechanical alloying technique, ie, simultaneous melding of all constituents: master alloy, solutes, and ox-

Table 3. Commercially Available Refractory Metal Alloys

Alloy designation	Nominal compositions, wt %
unalloyed niobium	Nb–0.030O–0.01C–0.03N
Nb–1Zr	Nb–1Zr
WC-103	Nb–10Hf–1Ti
FS-85	Nb–27Ta–10W–1Zr
SCB-291	Nb–10W–10Ta
B-88	Nb–28W–2Hf–0.07C
WC-129Y	Nb–10W–10Hf–0.24Y
D-43	Nb–10W–1Zr–0.1C
unalloyed tantalum	Ta–0.015O–0.01C–0.01N
Ta–10W	Ta–10W
T-111	Ta–8W–2Hf
T-222	Ta–10W–2.5Hf–0.01C
Astar 811C	Ta–8W–1Re–1Hf–0.025C
unalloyed molybdenum	Mo–0.04C–0.003O–0.001N
Mo-TZM	Mo–0.5Ti–0.1Zr–0.03C
Mo–42Re	Mo–42Re
Mo–50Re	Mo–50Re
unalloyed tungsten	W–0.01C–0.006O–0.005N
W–3Re	W–3Re
W–5Re	W–5Re
W–25Re	W–25Re
W–0.3Hf–0.025C	W–0.3Hf–0.025C
W–4Re–0.3Hf–0.025C	W–4Re–0.3Hf–0.025C
W–24Re–0.3Hf–0.025C	W–24Re–0.3Hf–0.025C

ide dispersoids, in a special high energy ball mill. At low temperatures γ'-hardening is achieved and above 1000°C strength is retained to a greater extent than for conventional alloys.

The principal applications of nickel-base superalloys are in gas turbines, where they are utilized as blades, disks, and sheet metal parts.

Iron–Nickel Base Superalloys

Iron–nickel base superalloys were developed primarily from the stainless steels. In the United States, these alloys included 19-9 DL, 16-25-6, and A-286. Later, higher nickel contents were employed to take advantage of the superior oxidation resistance of nickel and the beneficial effects of γ'-forming elements. All iron–nickel base superalloys rely on solid solution hardening to some extent.

Because iron–nickel alloys tend to contain large amounts of ferrite stabilizers such as chromium and molybdenum, the minimum nickel content required to maintain a fcc matrix is about 25 wt %. High iron contents lower cost, increase fabricability, and tend to raise the melting point, at the expense of poorer oxidation resistance than nickel-base alloys. Chromium is added for surface protection and solid-solution strengthening of gamma. Molybdenum also is added for solid-solution strengthening, but is present also in carbides and γ'. Small quantities of boron or zirconium are added to improve workability and stress–rupture properties, and carbon is useful as a deoxidant and to provide MC carbides to help refine grain size during hot working. Finally, ductilizing effects may be realized with small addition of magnesium, calcium, and certain rare-earth elements. Iron–nickel alloys are used extensively in aircraft gas turbines and in the space shuttle main engine.

Cobalt-Base Superalloys. Cobalt-base superalloys are used principally where operating metal temperatures range from 650 to 1000°C and stresses are relatively low. Strengthened primarily by carbide precipitation and solid-solution effects, these alloys are widely used as forgings and castings for nozzle vanes in gas turbine engines, because of good thermal shock and hot corrosion resistance, and in sheet metal assemblies, such as combustion chamber liners, tail pipes, and afterburners. The cobalt alloys generally are inferior in strength to the strongest cast nickel-base superalloys. Cobalt-base alloys generally rely on chromium for high temperature corrosion resistance, and most contain at least 20–25% Cr to form protective Cr_2O_3.

Refractory Metals and Their Alloys. Many elements which could be called refractory are found in the Periodic Table, but those which have received the most attention for potential structural applications are the bcc metals, tantalum, molybdenum, vanadium, niobium, and tungsten, all of which melt at or above about 2000°C, (see Table 1). Commercially available alloys are as shown in Table 3.

Since about 1950, arc melting of refractory metals in vacuum or inert atmospheres by the consumable electrode technique has been used commercially to produce large ingots and billets. Extrusion, forging, and sheet rolling technologies have advanced rapidly so that many of the refractory metal alloys are now available in various mill forms. Electron-beam melting, plasma-arc spraying, fused-salt electroplating, and vapor deposition are among the specialized techniques being used to produce and fabricate the refractory metals.

In many high temperature applications in the electrical and electronics industry, the refractory metals are protected by a vacuum or an inert gas, so that oxidation is not a problem. However, for most other high temperature applications, poor oxidation resistance has limited use. The oxides of the refractory metals, rather than existing as tight, protective barriers, suffer from porosity at moderately elevated temperatures, volatility at higher temperature, and spalling of the oxide scales away from the substrate, especially at corners and edges.

In addition to oxidation itself, gas diffusion into the base metal can be more damaging than the actual loss of metal from the surface.

Efforts in the 1960s concentrated on developing oxidation-resistant coatings. The most successful coatings are disilicides of the base metal, 2–5 mm thick, usually applied by a high temperature pack cementation process. Aluminide coatings also have been applied successfully. Other coating systems have been based on noble metals and Ni–Cr alloys.

Joining is another problem. Fusion welding in inert atmospheres often develops recrystallized structures in the heat-affected zones, so that welded parts lose strength and are embrittled. Special joining techniques being used to help overcome these deficiencies include electron-beam and solid-phase welding, and the development of special brazing materials.

New Materials and Processes. New materials and processes include aligned eutectics, oxide and fiber-reinforced superalloys, intermetallic compounds and other ordered phases including titanium aluminides, nickel aluminides, and iron aluminides.

NORMAN S. STOLOFF
Rensselaer Polytechnic Institute

A. K. Vasudevan and J. J. Petrovic, *Mat. Sci. and Eng.* **155A**, 1–17 (1992).

J. L. Smialek and G. H. Meier, in C. T. Sims, N. S. Stoloff, and W. C. Hagel, eds., *Superalloys II*, John Wiley & Sons, Inc., New York, 1987, pp. 293–320.

G. E. Wasielewski and R. A. Rapp, in C. T. Sims and W. C. Hagel, eds., *The Superalloys*, John Wiley & Sons, Inc., New York, 1972, pp. 287–316.

G. H. Gessinger, *Powder Metallurgy of Superalloys*, Butterworths, London, 1984, p. 282.

R. W. Buckman, Jr., in J. L. Walter, M. R. Jackson, and C. T. Sims, eds., *Alloying*, ASM, Materials Park, Ohio, 1988, pp. 419–445.

HIGH TEMPERATURE COMPOSITES. See COMPOSITE MATERIALS.

HISTAMINE AND HISTAMINE ANTAGONISTS

The history of histamine, $C_5H_9N_3$, and the development of antihistamines have been reviewed.

Histamine Synthesis, Metabolism, and Distribution

The synthesis and disposition of histamine is well described both in allergy textbooks and in review articles.

Synthesis. Histamine, 2-(4-imidazolyl)ethylamine, is formed by decarboxylation of histidine by the enzyme L-histidine decarboxylase. Most histamine is stored preformed in cytoplasmic granules of mast cells and basophils. Histamine release is mainly caused by cross-linking of immunoglobulin E on the mast cell surface by antigens.

Histamine Release. Histamine release is mainly caused by cross-linking of immunoglobulin E on the mast cell surface by antigens. Basophil degranulation is caused mainly by histamine-releasing factors produced by inflammatory cells, such as neutrophils, platelets, and eosinophils. After its release, histamine diffuses rapidly into the blood stream and surrounding tissues.

Metabolism. Metabolism of histamine occurs via two principal enzymatic pathways. Most (50 to 70%) histamine is metabolized to *N*-methylhistamine by *N*-methyltransferase, and some is metabolized further by monoamine oxidase to *N*-methylimidazoleacetic acid and excreted in the urine. The remaining 30 to 40% of histamine is metabolized to imidazoleacetic acid by diamine oxidase, also called histaminase. Only 2 to 3% of histamine is excreted unchanged in the urine.

Histamine in the Brain. There is evidence that histamine functions as a neurotransmitter or a neuromodulator in the brain. In the brain, histamine is related to functions such as the regulation of neuroendocrine and cardiovascular systems, thermoregulation, the circadian rhythm of sleep-wakefulness, behavior, vestibular function, cerebral vascular regulation, and antinociception and analgesia.

Histamine in the Cardiovascular System. Histamine is present in sympathetic nerves and has a distribution within the heart that parallels that of norepinephrine. A physiological role for cardiac histamine as a modulator of sympathetic responses is highly plausible.

Histamine Receptors

The actions of histamine are mediated through at least three distinct receptors.

The H_1 Receptor and its Ligands. The H_1 receptor mediates most of the important histamine effects in allergic diseases. These include smooth muscle contraction, increased vascular permeability, pruritus, prostaglandin generation, decreased atrioventricular node conduction time with resultant tachycardia, activation of vagal reflexes, and increased cyclic guanosine monophosphate (cGMP) production.

In the histamine molecule there are two principal structural elements: an imidazole moiety and an ethylamine side chain. Only the N_π;-position is absolutely necessary for H_1 agonism. The imidazole ring can be replaced, eg, 2-pyridylethylamine, 2-thiazolylethylamine, or substituted at the 2-position. 2-Methylhistamine is often used as a selective H_1 agonist; however, larger substituents are not allowed unless a phenyl ring is used. 2-Phenylhistamine analogues appear to be very selective H_1 receptor agonists.

The classical H_1 receptor antagonists are reversible, competitive, dose-dependent inhibitors of the action of histamine on H_1 receptors. Histamine H_1 antagonists are usually divided into two classes: the first-generation or classical H_1 antagonists and the second-generation H_1 antagonists. The main distinction between the first- and second-generation drugs is the absence of sedative and anticholinergic side effects in the latter.

The classical histamine H_1 receptor antagonists are structurally very similar, all being substituted ethylamines. The classical H_1 receptor antagonists can be subdivided into six classes: aminoalkylethers (diphenhydramine, $C_{17}H_{21}NO$), ethylenediamines (tripelennamine, $C_{16}H_{18}N_3$), alkylamines (chlorpheniramine, $C_{16}H_{19}ClN_2$), piperazines (hydroxyzine, $C_{21}H_{27}ClN_2O_2$), phenothiazines (promethazine, $C_{17}H_{20}N_2S$), piperidines (cyproheptadine, $C_{21}H_{21}N$).

Several antihistamines have been derived from classical H_1-receptor antagonists, but do not penetrate into the brain. They include acrivastine ($C_{22}H_{24}N_2O_2$), cetirizine ($C_{21}H_{25}ClN_2O_3$), ebastine ($C_{32}H_{39}NO_2$), epinastine ($C_{16}H_{15}N_3$), loratadine ($C_{22}H_{23}N_2O_2Cl$), pibaxizine ($C_{24}H_{29}NO_4$), and terfenadine ($C_{32}H_{41}NO_2$).

Some of the second-generation H_1 antagonists have nonclassical structures, eg, astemizole ($C_{28}H_{31}FN_4O$), azelastine ($C_{22}H_{24}ClN_3O$), emedastine ($C_{17}H_{26}N_4O$), levocabastine ($C_{26}H_{29}FN_2O_2$), and mizolastine ($C_{24}H_{25}FN_6O$).

The H_2 Receptor and its Ligands. The H_2 receptor mediates effects, through an increase in cyclic adenosine monophosphate (cAMP), such as gastric acid secretion; relaxation of airway smooth muscle and of pulmonary vessels; increased lower airway mucus secretion; esophageal contraction; inhibition of basophil, but not mast cell histamine release; inhibition of neutrophil activation; and induction of suppressor T cells. There is no evidence that the H_2 receptor causes significant modulation of lung function in the healthy human subject or in the asthmatic.

Combined H_1/H_2 receptor stimulation by histamine is responsible for vasodilation-related symptoms, such as hypotension, flushing, and headache, as well as for tachycardia stimulated indirectly through vasodilation and catecholamine secretion.

Structural requirements of histamine as an H_2 agonist are considered to be the protonated side-chain nitrogen atom and the ability of the imidazole amidine system to undergo a tautomeric shift. 4-Methylhistamine is often used as a selective H_2 agonist. Larger substituents are not allowed. Methylation of the amine group is allowed, but leads to nonselective analogues. H_2 agonists are divided in three chemical classes, ie, analogues of histamine, dimaprit, and impromidine.

The prototype examples of H_2 antagonists are cimetidine ($C_{10}H_{16}N_6S$), famotidine ($C_8H_{15}N_7O_2S_3$), and ranitidine ($C_{13}H_{22}N_4O_3S$).

The H_3 Receptor and its Ligands. The H_3 receptor has been reported to modulate the release of a variety of neurotransmitters and can be regarded as a general regulatory mechanism.

In contrast to the development of selective agonists for the H_1 and H_2 receptor, potent agonists for the H_3 receptor can be obtained by simple modification of the histamine molecule. Histamine itself is already a rather potent agonist of the H_3 receptor, although it is of course not very selective.

H_3 receptor antagonists include betahistine ($C_8H_{12}N_2$), clobenpropit ($C_{14}H_{17}N_4SCl$), and thioperamide ($C_{15}H_{24}N_4S$).

Uses of Histamine Receptor Ligands

H_1 Antihistamine Treatment in Allergic Diseases. H_1-receptor antagonists are used for the symptomatic treatment of several allergic diseases where histamine release form mast cells is induced via immunological or nonimmunological mechanisms. H_1-receptor antagonists are used mainly in allergic seasonal or perennial rhinoconjunctivitis and urticaria.

Clinical Efficacy and Side-Effects of H_1 Antihistamines. It is evident from the mechanism of action of antihistamines and the etiology of allergic diseases that antihistamines in no sense achieve a cure of the patient's allergy. After the administration of a therapeutic dose, a temporal blockade of the effects of histamine is obtained. Whereas classical antihistamines needed at least twice daily administration, for most of the more recently introduced agents administration once daily is sufficient.

Nevertheless, although the nonsedating H_1 antihistamines have substantially improved the acceptability and clinical efficacy of this class of compounds, these do not provide complete relief; eye disease responds less well than nasal disease, of the rhinitis symptoms nasal congestion responds poorly, breakthrough symptoms occur at high pollen counts, and only some 70% of patients report excellent to good treatment responses. Considerable research therefore still continues in the H_1 antihistamine field. New antihistamines are continually being introduced.

The classical H_1 antihistamines are not very specific, and several compounds have some degree of anticholinergic activity. Anticholinergic effects can present side effects such as dry mouth, blurred vision, and urine retention, whereas the anticholinergic action of some antihistamines is probably the reason for effectiveness in motion sickness. Interactions in the brain with noradrenergic, serotonergic, and dopaminergic uptake systems may play a role in behavioral effects of H_1 antihistamines.

At therapeutic doses, the classical H_1-receptor antagonists generally produce sedation. This usually unwanted effect is probably caused by the H_1-receptor blockade in the CNS.

The second-generation H_1 antihistamines generally present few side effects and, in particular, are considered not to cause sedation, mainly because of reduced ability to penetrate the CNS. However, terfenadine and astemizole have been associated with prolongation of the QT-interval in the electrocardiogram (ECG) and ventricular arrhythmias, generally at higher than therapeutic plasma levels.

Azelastine and levocabastine have been developed for topical application. In contrast to earlier reports of sensitization with older antihistamines locally applied to the skin, sensitization has not been reported with local application to the nose or eyes.

H_2 Agonists and Antagonists. H_2-antagonists inhibit histamine-induced gastric acid secretion. They are used widely in the treatment of peptic ulcer disease and esophageal reflux.

H_3 Agonists and Antagonists. No clear therapeutic indications have been reported for H_3 receptor ligands, yet with their use insights in the role of histamine H_3 receptors in various (patho)physiological processes have been obtained. Interesting options for therapeutic application of H_3 agonists could be in asthmatic diseases, gastrointestinal disorders, and in the regulation of sleep/wakefulness patterns.

Economic Aspects

The sales of antagonists of H_1 receptors, used in the treatment of allergic diseases, represent 1% of the overall pharmaceutical market, ie, \$1.7 billion (U.S.). Sales of H_2 antagonists, used mainly in peptic ulcer disease and esophageal reflux, represent 3.5% of the world market, ie \$6 billion (U.S.). H_3 agonists or antagonists have not yet found a clear indication.

MONIQUE M.-L. JANSSENS
Janssen Research Foundation
HENDRIK TIMMERMAN
ROBERT LEURS
Leiden/Amsterdam Center for Drug Research

E. Middleton, Jr. and co-workers, eds., *Allergy, Principles and Practice,* 3rd ed., The C. V. Mosby Co., St. Louis, Mo., 1988.

B. Uvnäs, ed., *Histamine and Histamine Antagonists,* Springer-Verlag, Berlin, 1991.

R. Leurs, H. Van der Goot, and H. Timmerman, *Adv. Drug Res.* **20**, 217 (1991).

HOLLOW-FIBER MEMBRANES

A hollow-fiber membrane is a capillary having an inside diameter of >25 μm and an outside diameter <1 mm and whose wall functions as a semipermeable membrane. The fibers can be employed singly or grouped into a bundle which may contain tens of thousands of fibers and up to several million fibers as in reverse osmosis (Fig. 1). In most cases, hollow fibers are used as cylindrical membranes that permit selective exchange of materials across their walls. However, they can also be used as containers to effect the controlled release of a specific material, or as reactors to chemically modify a permeate as it diffuses through a chemically activated hollow-fiber wall, eg, loaded with immobilized enzyme.

The excellent mass-transfer properties conferred by the hollow-fiber configuration led to numerous applications. Commercial applications have been established in the medical field in gas separations and pervaporation; other applications are in various stages of development.

Hollow-fiber membranes may be divided into two categories: open hollow fibers where a gas or liquid permeates across the fiber wall, while flow of the lumen medium gas or liquid is not restricted, and loaded fibers where the lumen is filled with an immobilized solid, liquid, or gas. The open hollow fiber has two basic geometries: the first is a loop of fiber or a closed bundle contained in a pressurized vessel. Gas

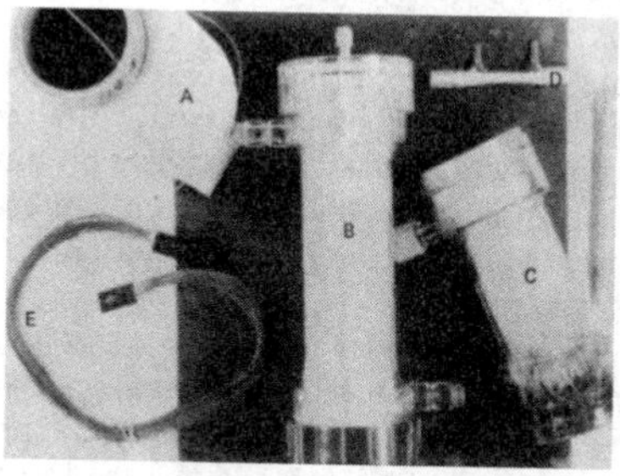

Figure 1. A, hollow-fiber spool; B, hollow-fiber cartridge employed in hemodialysis; C, cartridge identical to item B demonstrating high packing density; D, hollow-fiber assembly employed for tissue cell growth; E, hollow-fiber bundle potted at its ends to be inserted into a cartridge or employed in a situation that requires mechanical flexibility.

or liquid passes through the small diameter fiber wall and exits via the open fiber ends. In the second type, fibers are open at both ends. The feed fluid can be circulated on the inside or outside of the relatively large diameter fibers. These so-called large capillary (spaghetti) fibers are used in microfiltration, ultrafiltration, pervaporation, and some low pressure (<1035 kPa = 10 atm) gas applications.

In open fibers the fiber wall may be a permselective membrane, and uses include dialysis, ultrafiltration, reverse osmosis, Donnan exchange (dialysis), osmotic pumping, pervaporation, gaseous separation, and stream filtration. Alternatively, the fiber wall may act as a catalytic reactor and immobilization of catalyst and enzyme in the wall entity may occur. Loaded fibers are used as sorbents, and in ion exchange and controlled release. Special uses of hollow fibers include tissue-culture growth, heat exchangers, and others.

Hollow fibers offer three primary advantages over flat-sheet or tubular membranes. First, hollow fibers exhibit higher productivity per unit volume; second, they are self-supporting; and third, high recovery in individual units can be tolerated.

Properties

Morphology. The desired fiber-wall morphology frequently dictates the spinning method. The basic morphologies are isotropic, dense, or porous; and asymmetric (anisotropic), having a tight surface (interior or exterior) extending from a highly porous wall structure. The tight surface can be a dense, selective skin, permitting only diffusive transport, or a porous skin, allowing viscous flow of the permeate as in conventional ultrafiltration or reverse osmosis. Membrane-separation technology is achieved by use of these basic morphologies.

Mechanical Considerations and Fiber Dimensions. The hollow fiber is self-supporting, and is actually a thick wall cylinder. The ratio of outside to inside diameter in some reverse-osmosis applications is about 2 to 1 thus providing the strength to withstand high operating pressures, commercially up to 10,000 kPa (96 atm), without collapsing. A hollow fiber that is exposed to external pressure would exhibit a collapse pressure P_c that depends on the inner and outer fiber radii (IR, OR) and the Young's modulus E and Poisson ratio v of the material. The approximate relationship is given by the expression

$$P_c = \frac{2E}{(1 - v^2)} [(OR - IR)/(OR + IR)]^3 \tag{1}$$

When the operation of the hollow-fiber membrane is to be reversed, and permeation from the bore to outer zone is required, circumferential stress and pressure drop along the fiber capillary (bore) must be considered in the design of the fiber unit.

Spinning

In preparation of permselective hollow-fiber membranes, morphology must be controlled to obtain desired mechanical and transport properties. Fiber fabrication is performed without a casting surface. Therefore, in the moving unsupported thread line, the nascent hollow-fiber membrane must establish mechanical integrity in a very short time.

There are three conventional synthetic fiber spinning methods that can be applied to the production of hollow-fiber membranes: melt spinning, solution (wet) spinning, and a combination of these first two methods, dry-jet wet-spinning.

Spinnerettes. In all methods, a tubular cross section is formed by delivering the spinning dope through an extrusion orifice. Four schemes of spinnerette nozzle cross sections are *(1)* the segmented-arc design; *(2)* the plug-in-orifice design; *(3)* the tube-in-orifice jet design; *(4)* the multiannular design.

Macrovoids. Hollow-fiber membranes that are solution-spun by the foregoing methods can exhibit large voids in conical, droplet, or lobe configurations. These voids may extend through the entire fiber cross section. The voids, in general, result from fast coagulation of a spinning solution that is relatively low in either polymer concentration or viscosity. The use of a less severe quenching medium, on the other hand, yields a macrovoid-free hollow fiber.

The presence of macrovoids in hollow-fiber membranes is a serious drawback since it increases the fragility of the fiber and limits its ability to withstand hydraulic pressures. Such fibers have lower elongation and tensile strength.

Fiber Treatment

Treatment In-line. The coagulated fiber on the moving threadline may be subjected to cooling (for melt spinning, or for dry-jet wet spinning conducted at high temperatures), washing to remove trace solvents and dope additives, swelling with diluted solvent and/or plasticizers, stretching between godets, and heat-treating (annealing) to consolidate its morphology and impose transport properties (such as closing the skin pores of an asymmetric hollow-fiber membrane for reverse-osmosis applications). In the continuous processing of hollow fibers, these steps add little to costs but are fundamental to achieving the desired functionality of the product.

Post-Treatment of Hollow Fibers. End use of the hollow-fiber membrane dictates the type of post-treatment, if any. There are three main categories: fibers that are spun, fibers that will be chemically or physically modified, and fibers that will serve as a porous matrix for support of another (active) polymer deposited (or entrapped) upon (or within) its walls. There is no theoretical impediment to the inclusion of all conventional treatments in the spinning line: photochemical cross-linking, fluorination, and antiplasticizers have been successful.

Fiber Modification. Chemical modification of the fiber is usually a separate operation. The largest such commercial processing is the deacetylation of cellulose acetate hollow fibers, which converts them into regenerated cellulose hollow fibers employed in hemodialysis.

Composite Hollow-Fiber Membrane. Composite membranes consist of highly porous substrates, having minimum resistance to the permeates, which support ultrathin semipermeable membranes. The production scheme for a composite hollow-fiber membrane, consisting of polysulfone coated by polyethyleneimine (PEI) that is cross-linked *in situ* (on the exterior surface of the fiber), is shown in Figure 2.

Although these composite fibers were developed for reverse osmosis their acceptance in the desalination industry has been limited due to insufficient selectivity and oxidative stability. The concept, however, is extremely viable; composite membrane flat films made from interfacial polymerization have gained wide industry approval. Hollow fibers using this technique to give equivalent properties and life, yet to be developed, should be market tested during the 1990s.

Interpenetrated Wall Matrix. Ion-exchange hollow fibers can be produced by polymerizing an ionic monomer within the porous wall matrix of a hollow fiber. Requirements of such a fabrication are *(1)* the monomers should not dissolve or plasticize the polymer from which the fibers are made; *(2)* the heat generated during the polymerization and contraction prior to the formation of new interpenetrating polymer should be minimized; and *(3)* the polymerization should not occur

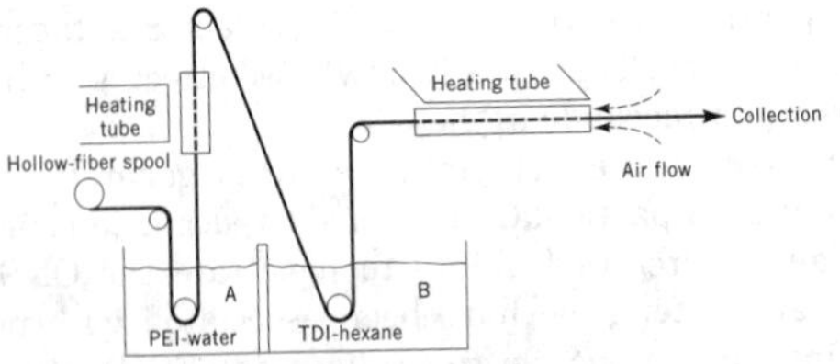

Figure 2. Composite hollow-fiber production scheme (PEI = polyethyleneimine; TDI = toluene 2,4-diisocyanate). Anisotropic (porous skin) polysulfone hollow fiber is rolled into bath A and is lifted vertically (to avoid droplet formation) into a heating tube. The fiber is then passed through bath B and is annealed in a ventilated heating tube (110°C).

within the lumen (and hence cause plugging of the fiber). One drawback of such fibers is brittleness.

Fiber Handling and Unit Assembly. Most hollow fibers can be collected on spools by winding machines analogous to those used in the textile industry. Individual or multifilaments can be crosswound, or may be wound in a simple parallel arrangement (for highly plasticized, or large ID fibers, where cross-winding intersections may weaken the structure). Subsequent handling of the filament depends on the intended use of the hollow fiber.

Assembling and potting (cementing together) of hollow-fiber bundles, as shown in Figure 1, require great care and precision technology. The potting agent must be compatible with the function assigned to the fiber, as well as with the fiber material. Another factor important in the selection of a potting agent is its surface tension (ability to wet fibers yet not excessively wick). Commonly employed potting agents include epoxy resins, polyurethanes, and silicone rubbers.

In general there are three main types of hollow-fiber flow configurations. In the most common, for reverse osmosis and ultrafiltration, the feed enters outside the fiber; permeate is inside the fibers and flow is countercurrent. In the second, for large diameter fibers, where the feed has a high loading of particulates, the feed is through the fiber bore; permeate is outside the fiber; flow is usually countercurrent. In cross flow, shell side feed is prevalent as with microfiltration. Gas permeation uses all three flow patterns.

Materials

The components employed in spinning-dope formulations must be consistent in every batch preparation, because numerous parameters are involved in the spinning process. Thus stringent criteria are imposed on the selection of components to be used in each spinning operation. The components are rigorously tested for purity, molecular weight, molecular weight distribution, chemical composition, viscoelastic properties, and other specific parameters that might influence hollow-fiber production and final membrane properties. This often requires close cooperation between the producers of the polymer and the hollow fiber manufacturers. Materials include cellulose, cellulose ester (cellulose acetate and cellulose triacetate), polysulfone, poly(methyl methacrylate), polyamide, other nitrogen-containing polymers (polybenzimidazole, polyacrylonitrile (PAN)), and glass and inorganic hollow-fiber membranes.

Sorbent Fibers

Filled Fibers. Interest in the encapsulation of specific active materials (eg, activated charcoal, enzymes, drugs) led to the development of encapsulation spinning, usually employing a wet- or dry wet-spinning process.

The rationale for the development of such fibers is demonstrated by their application in the medical field, notably hemoperfusion, where cartridges loaded with activated charcoal-filled hollow fiber contact blood. Low molecular weight body wastes diffuse through the fiber walls and are absorbed in the fiber core. In such processes, the blood does not contact the active sorbent directly, but faces the nontoxic, blood compatible membrane. Other uses include waste industrial applications as general as chromates and phosphates and as specific as radioactive/nuclear materials.

Hollow Fiber with Sorbent Walls. A cellulose sorbent and dialyzing membrane hollow fiber was reported in 1977 by Enka Glanzstoff AG. This hollow fiber, with an inside diameter of about 300 μm, has a double-layer wall. The inner wall consists of Cuprophan cellulose and is very thin, approximately 8 μm. The outer wall, which is ca 40-μm thick, consists mainly of sorbent substance bonded by cellulose. The advantage of such a fiber is that it combines the principles of hemodialysis with those of hemoperfusion. Two such fibers have been made: one with activated carbon in the fiber wall, and one with aluminum oxide, which is a phosphate binder.

Future Prospects

Hollow-fiber membranes are subjected to extensive studies for gaseous separation (eg, CO_2, H_2, O_2, N_2, H_2S, CO, CH_4), where the capillary configuration has an advantage over the spiral-wound flat film and plate-and-frame devices. Another significant area of development and commercialization is pervaporation. These membranes are dense, rather than porous structures. Generally asymmetric composite constructions are employed with the ultrathin membranes on an open support.

Considerable research and development effort is being placed on a chlorine-resistant membrane that will maintain permeability and selectivity over considerable time periods (years).

When pressurized liquid is used to separate micrometer-size particles from fluids, the process is called microfiltration. Generally particle sizes are from 0.02 to 10 μm. Thus compared to ultrafiltration and reverse osmosis, fluxes and pore sizes are large, osmotic pressure low, and pressures moderate. Two types of microfiltration processes exist, crossflow and deadend. Commercially, the former is growing at the expense of the latter.

Sorbent fibers were developed in the late 1970s, in particular by California Institute of Technology and Gulf South Research Institute. The concept of encapsulation within a hollow fiber, gas, liquid, suspended solid, catalyst, or others, has potential.

IRVING MOCH, JR.
E. I. du Pont de Nemours & Co., Inc.

S. Loeb and S. Sourirajan, *Adv. Chem. Ser.* **38**, 117 (1962).

J. Scott, ed., *Hollow Fibers Manufacture and Applications, Chemical Technology Review No. 194*, Noyes Data Corp., Park Ridge, N.J., 1981.

S. Torrey, ed., *Membrane and Ultrafiltration Technology, Developments Since 1981*, Noyes Data Corp., Park Ridge, N.J., 1984.

R. W. Baker and co-workers, *Membrane Separation Systems, Recent Developments and Future Directions*, Noyes Data Corp., Park Ridge, N.J., 1991.

HOLMIUM. See LANTHANIDES.

HOLOGRAPHY

Holography involves an image recording technique whereby the complete wave information (optical or other wave phenomena) emanating from a three-dimensional scene is captured in a suitable material and a reconstruction step where the information is replayed to reconstruct the true three-dimensional character of the recorded scene. The three-dimensional character of scenes reconstructed with a hologram is illustrated by photographs of the reconstruction from a single hologram taken from different viewing angles as shown in Figure 1. As implied by the name hologram, which literally means "whole record," a hologram contains the complete intensity and phase information associated with a given scene.

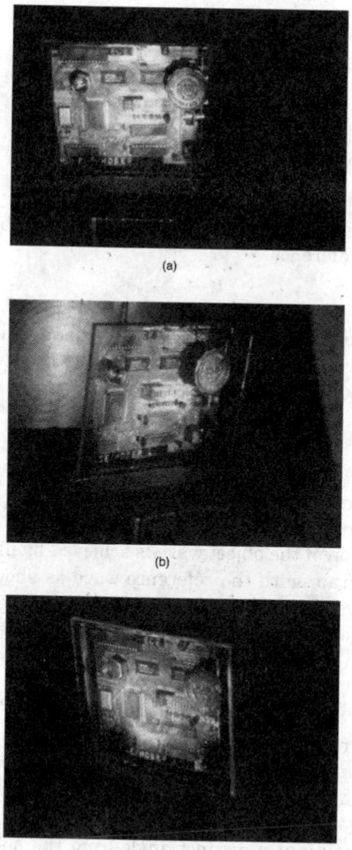

Figure 1. Reconstructions from a hologram. (**a**) Normal viewing angle; (**b**) viewing from left side; (**c**) viewing from right side. Courtesy of Mr. Tae Jin Kim.

In optical holography, the phase information carried by an object wave to be recorded is transformed into a complex interference pattern by combining it with a mutually coherent reference beam. The resulting nonuniform intensity distribution is then recorded with either standard photographic film or other specialized materials for holography. This indirect method of capturing both the intensity and phase information is necessary because the extremely high frequencies of light make direct sensing and recording impossible.

With the advent of new materials including polymers and photorefractive media, which require less handling complexity, relatively recent applications involving the storage and manipulation of information have been introduced. Significant advances in optoelectronics such as the introduction of compact semiconductor and solid-state lasers have also helped to fuel such new applications.

The Holographic Principle

Holography involves the recording of the mutual interference pattern due to two mutually coherent optical fields. A generic holographic recording experiment is shown in Figure 2: an expanded and collimated laser beam is split into two paths, with one falling directly on the holographic material and the other scattering off an object to be collected on the same materials. The first is called the reference and the second the object. The intensity distribution falling on the film is given by equation 1, where S is the object wave amplitude and R represents the plane wave reference.

$$I \propto |S + R|^2 = |S|^2 + |R|^2 + SR^* + RS^* \qquad (1)$$

Film and other holographic recording media respond to the overall exposure or total optical energy per unit area deposited during the exposure time τ given by $\epsilon = \tau I$. An assumed linear relationship between

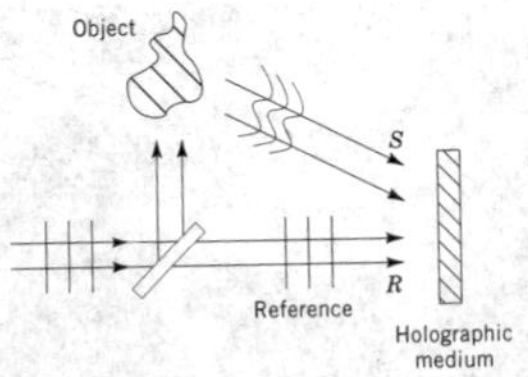

Figure 2. Holographic recording system. S is the object wave and R the plane wave reference.

the exposure and the transmittance of the film after exposure and development is given by equation 2, where t_0 is the average transmittance of the film and κ is a constant that depends on the material and processing.

$$t = t_0 + \kappa\tau(|S|^2 + |R|^2 + SR^* + RS^*) \quad (2)$$

The last two terms in the transmittance expression enable the holographic reconstruction process.

Reconstruction of the object wave is achieved by illumination of the developed hologram with the reference wave as shown in Figure 3**a**. The diffracted wave amplitude from the hologram is given by equation 3, where the first term represents the attenuated reference wave after passage through the hologram.

$$A_{\mathrm{diff}} \propto tR = [t_0 + \kappa\tau(|S|^2 + |R|^2)]R + \kappa\tau|R|^2S + \kappa\tau R^2S^* \quad (3)$$

The second term represents a virtual image of the original object signal which can be viewed by an observer looking at the hologram along the original object signal direction. The last term represents an unfocused wave which carries the complex conjugate of the signal amplitude emerging at a distinct angle from the hologram as shown in Figure 3**a**. This term can be brought to a focus by reconstructing the hologram with a plane wave traveling in exactly the opposite direction as the original reference wave as shown in Figure 3**b**. If $R = e^{ikx}$ represents the original plane wave reference, then $R^* = e^{-ikx}$ represents a wave traveling in the opposite case. In such a case, the diffracted wave amplitude is given by equation 4, where the last term represents a real image formed precisely at the location of the original object.

$$A_{\mathrm{diff}} \propto tR^* = [t_0 + \kappa\tau(|S|^2 + |R|^2)]R^* + \kappa\tau R^{*2}S + \kappa\tau|R|^2S^* \quad (4)$$

Hologram Varieties

The rich variety in the types of holograms stems from the specifics of how the interference patterns are recorded. Although a more complete classification of the various hologram types is possible, the most important classification parameters are the material perturbation (amplitude or phase), material thickness (thin or thick), and the recording format (transmission or reflection).

Material Perturbation. Exposure of a suitable holographic material to recording waves must induce perturbations of its optical transmission properties to record holograms. The perturbations may include a complex set of physical effects, but in most cases they can be lumped into one of two categories: *(1)* absorption perturbations and *(2)* phase perturbations. A unit amplitude plane wave at wavelength λ passing through a homogenous material of thickness T, refractive index n, and absorption constant α emerges with amplitude A, where the first multiplicative term describes the effect of material absorption and the second represents the imposed change in phase of the wave.

$$A = e^{-\alpha T} e^{i\frac{2\pi}{\lambda}nT} \quad (5)$$

Holographic information can be recorded by spatially modulating either (or a combination of) the absorption or phase (index or thickness change).

Materials that respond to incident exposure by absorption perturbations include photographic film and photochromic glasses/crystals. Phase holograms can be recorded in a large variety of materials, the most popular of which are dichromated gelatin, photopolymers, thermoplastic materials, and photorefractive crystals.

Material Thickness. Holograms recorded in a material whose physical thickness is small when compared with the grating spacing is considered a thin hologram; the effect on an incident optical wave is characterized by a spatially varying transmittance function. The output of such a hologram can be determined by multiplying the input field with the transmittance of the hologram and analyzing the resultant diffraction using the principles of coherent optics theory. The most prominent features of thin gratings are lack of strong angular selection in reconstruction and limited diffraction efficiency. A quantity that is often used to delineate the boundary between thin and thick holograms (thick implies that angular/wavelength selectivity effects are exhibited) is

$$A = \frac{2\pi\lambda T}{n_0\Lambda_g^2} \quad (6)$$

where n_0 is the average refractive index of the hologram, Λ_g is the grating spacing, and T is the physical thickness of the hologram. Values of $Q < 1$ imply a thin hologram and $Q > 1$ imply a thick hologram.

Computer-Generated Holograms. Instead of physically recording an object wave by mixing it with a reference on a suitable recording medium, holograms can be prepared in the following two steps. First, the desired object wave distribution can be calculated across the plane where the holographic plate would have been positioned. Next, a transparency is produced using computer-controlled photolithography, which can be as simple as a laser printer with proper photographic demagnification, that yields a reconstruction of the object when properly illuminated.

Display Holograms

Display holography is by far the most familiar and well-developed holographic application to date with several museums dedicated to holographic art. These holograms range from extremely simple ornaments to large color portrait holograms with strikingly real appearance. The one practical feature of all display holograms is that they be observable with white light so that potentially dangerous laser beams need not be used. Display holograms include the reflection hologram, the achromatic hologram, the rainbow hologram, true color holograms, and composite holograms.

Holographic Applications

Holographic applications include holographic interferometry, holographic optical elements, holographic data storage, and holographic pattern recognition.

JOHN H. HONG
Rockwell International Science Center

P. Hariharan, *Optical Holography: Principles, Techniques, and Applications*, Cambridge University Press, Cambridge, 1984.

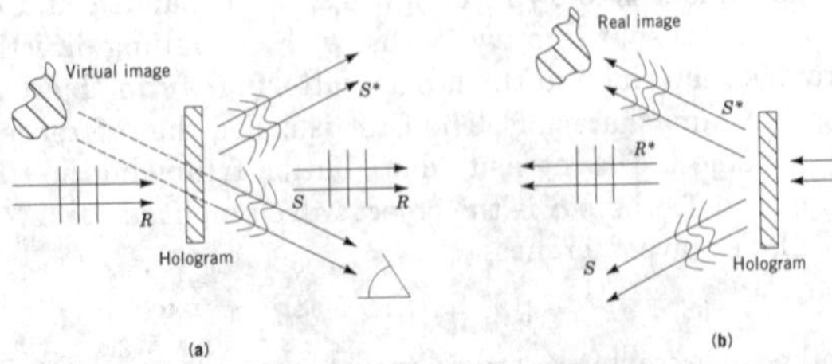

Figure 3. Holographic reconstruction. (**a**) Reconstruction of virtual image of object; (**b**) reconstruction of real image of object.

R. J. Collier, C. B. Burkhardt, and L. H. Lin, *Optical Holography*, Academic Press, Inc., New York, 1971.

N. Abramson, *The Making and Evaluation of Holograms*, Academic Press, Inc., London, 1980.

J. W. Goodman, *Fourier Optics*, McGraw-Hill Book Co., Inc., San Francisco, 1968.

HORMONES

SURVEY

Vertebrate Hormones

The term *hormone* is used to denote a chemical substance, released from a cell into the extracellular fluid in low quantities, which acts on a target cell to produce a response. Hormones are classified on the basis of chemical structure; most hormones are polypeptides, steroids, or derived from single amino acids (Table 1).

Polypeptide hormones are synthesized as part of a larger precursor molecule or prohormone. Cleavage of the prohormone by specific cellular enzymes, ie, peptidases, produces the secreted form of the hormone.

Steroids are synthetic products of cholesterol. The chemical structure of a steroid hormone is determined by sequential enzymatic processing of the cholesterol molecule. Steroid products differ among steroid-secreting glands because of differences in enzyme processing.

Amino acid-derived hormones include the catecholamines, epinephrine and norepinephrine, and the thyroid hormones, thyroxine and triiodothyronine. Catecholamines are synthesized from the amino acid tyrosine by a series of enzymatic reactions that include hydroxylations, decarboxylations, and methylations. Thyroid hormones also are derived from tyrosine; iodination of the tyrosine residues on a large protein backbone results in the production of active hormone.

Mechanisms of Action. Biologically effective concentrations of hormones range between 10^{-7} to 10^{-12} M. A fundamental defining characteristic of a hormone is that it binds to a stereospecific cellular receptor to activate a response. Binding to a receptor protein activates an intracellular transduction process that mediates the hormone action. Hydrophilic molecules, eg, polypeptide hormones and catecholamines, bind to membrane receptors. Common intracellular transducers for these hormones include cyclic AMP, calcium, and phosphatidyl inositides. Termination of the hormone action occurs after metabolism of the hormone or the hormone-receptor complex.

Effects and Secretion. An endocrine gland is defined classically as a tissue consisting of hormone-secreting cells that synthesize products for release into the blood. The pattern of release is reflected by changes in plasma hormone concentration. Analysis of hormone concentration is performed using radioimmunoassay based on competition of a hormone for binding to a specific antibody, or bioassay based on measurement of a specific biological response. Most hormones are secreted episodically, showing minute-to-minute changes in the circulation. Also, rhythms of hormone secretion are common, having periodicities that vary from hours to days.

Endocrine Pathology. The health of an organism is dependent on the maintenance of hormone concentrations within a normal physiological range. Pathology occurs when hormone concentrations are higher or lower than normal for extended periods. Excess hormone concentrations can result from over-production by endocrine cells.

Endocrinologists use hormones or hormone analogues to treat endocrine disease. Potent analogues of some hormones have been synthesized that antagonize the action of the natural hormone. Their use is effective in treating endocrine problems related to excess or inappropriate hormone production. To offset endocrine gland removal or reduced function, replacement therapy is performed. In most cases, synthetic hormones are administered. The availability of protein hormones for therapeutic or experimental use has been increased greatly through the application of genetic engineering techniques. The deoxyribonucleic acid (DNA) sequence of the human gene coding for a

Table 1. Hormones in Vertebrates

Tissue of origin/ hormone	Chemical nature	Site of action	Effect
		Adrenal	
adrenal cortex			
aldosterone	steroid	kidney	electrolyte and water metabolism
cortisol	steroid	multiple tissues	protein, lipid, and carbohydrate metabolism; cardiovascular stability; immune responses
corticosterone			
adrenal medulla			
epinephrine	catecholamine	cardiac muscle	increase heart rate
		skeletal muscle	vasodilation
		skin and kidney	vasoconstriction
		liver	vasodilation and glycogenolysis
		adipose tissue	lipolysis
		intestinal smooth muscle	relaxation
norepinephrine	catecholamine	cardiac muscle	increase heart rate
		skeletal muscle, skin, and kidney	vasoconstriction
		liver	vasoconstriction and glycogenolysis
		adipose tissue	lipolysis
		intestinal smooth muscle	relaxation
Leu-enkephalin, Met-enkephalin	polypeptide	multiple tissues	endogenous opiates
		Cardiovascular tissue	
endothelial cells			
endothelin	polypeptide	vascular smooth muscle	vasoconstriction
heart			
atrial natriuretic hormone (ANH)	protein	kidney	increased sodium excretion and decreased renin secretion
		vascular smooth muscle	vasodilation
		adrenal cortex	decreased aldosterone secretion
		Gastrointestinal (GI) tract	
gastrin	polypeptide	stomach	increased acid secretion
secretin	polypeptide	pancreas	increased water and biocarbonate secretion
cholecystokinin (CCK)	polypeptide	gallbladder	contraction
		pancreas	increased enzyme secretion
motilin	polypeptide	gastrointestinal tract	smooth muscle contraction
neurotensin	polypeptide	stomach	decreased acid secretion and emptying
		pancreas	increased bicarbonate secretion
		intestine	increased motor activity

Table 1. Hormones in Vertebrates *(continued)*

Tissue of origin/ hormone	Chemical nature	Site of action	Effect
peptide tyrosine tyrosine (PYY)	polypeptide	stomach	decreased acid secretion and emptying
somatostatin	polypeptide	GI tract	decreased GI hormone secretion; decreased adsorption of nutrients; decreased gastric emptying and gall bladder contraction
		Gonadal tissue	
corpus luteum			
progesterone	steroid	uterus	proliferation and vascularization of the endometrium; preparation for ovum implantation and maintenance of pregnancy
		mammary glands	alveolar development
relaxin	protein	uterus	cervical softening
ovary			
estrone	steroid	uterus	endometrial proliferation
estradiol	steroid	ovary	increased cell division and follicle growth
		mammary glands	duct development; development of secondary sex characteristics
inhibin	protein	pituitary	inhibits FSH
activin	protein	pituitary	stimulates FSH
testis			
testosterone	steroid	accessory sex organs	maturation and normal function; development of secondary sex characteristics
inhibin	protein	pituitary	inhibits FSH
activin	protein	pituitary	stimulates FSH
		Hypothalamic/brain hormones	
corticotropin-releasing hormone (CRH)	polypeptide	pituitary	release of ACTH and β-endorphin
dopamine	catecholamine	pituitary	inhibition of prolactin
gonadotropin-releasing hormone (GnRH)	polypeptide	pituitary	release of LH and FSH
growth hormone-releasing hormone (GRH)	polypeptide	pituitary	release of growth hormone
somatostatin	polypeptide	pituitary	inhibition of growth hormone and TSH
thyrotropin-releasing hormone (TRH)	tripeptide	pituitary	release of TSH and prolactin
vasopressin	polypeptide	pituitary	release of ACTH and β-endorphin
		Pituitary	

Table 1. Hormones in Vertebrates *(continued)*

Tissue of origin/ hormone	Chemical nature	Site of action	Effect
anterior pituitary (adenohypophysis)			
adrenocorticotropic hormone (ACTH)	polypeptide	adrenal	secretion of adrenocortical steroids
β-endorphin	polypeptide	multiple tissues	endogenous opiate
follicle-stimulating hormone (FSH)	glycoprotein	gonads	steroid and peptide secretion
γ-melanocyte-stimulating hormone (γ-MSH)	polypeptide	adrenal	potentiates response to ACTH
growth hormone	protein	multiple tissues	growth of bone and muscle; metabolism of carbohydrate and lipid; anabolic effect on mineral metabolism
luteinizing hormone	glycoprotein	gonads	steroid secretion
		ovary	ovulation
		testes	development of interstitial tissue
prolactin	protein	mammary gland	proliferation; milk secretion
		corpus luteum	development and functional activity
thyroid-stimulating hormone (TSH)	glycoprotein	thyroid	growth and secretion
intermediate pituitary (pars intermedia)			
β-endorphin	polypeptide	multiple tissues	endogenous opiate
β-lipotropin	polypeptide	adipose tissue	lipolysis
α-melanocyte-stimulating hormone (α-MSH)	polypeptide	melanophores	skin pigmentation
γ-melanocyte-stimulating hormone (γ-MSH)	polypeptide	adrenal	potentiates responses to ACTH
posterior pituitary (neurohypophysis)			
oxytocin	polypeptide	uterus	contraction
		mammary gland	milk ejection
vasopressin	polypeptide	kidney	water reabsorption
		vascular smooth muscle	vasoconstriction
		liver	glycogenolysis
		Thyroid	
parathyroid			
parathyroid hormone	polypeptide	kidney, bone, and GI tract	mobilization of calcium and phosphorus
pineal			

Table 1. Hormones in Vertebrates *(continued)*

Tissue of origin/ hormone	Chemical nature	Site of action	Effect
melatonin	indole	skin	inhibits pigmentation
		reproductive tissue	regulates function
thyroid			
thyroxine	amino acid derived	multiple tissues	increases metabolic rate and oxygen consumption
triiodothyronine			
calcitonin	polypeptide	bone and kidney	mobilization of calcium and phosphorus
Various organs			
kidney			
erythropoeitin	glycoprotein	bone marrow	erythrocyte production
renin		adrenal	aldosterone secretion
angiotensin II (AII)	polypeptide	vascular smooth muscle	vasoconstriction
		brain	drinking
pancreas			
glucagon	polypeptide	liver	glycogenolysis and gluconeogenesis
insulin	polypeptide	multiple tissues	carbohydrate utilization
		adipose tissue	lipogenesis
pancreatic peptide	polypeptide		function unknown
somatostatin	polypeptide		inhibition of all pancreatic hormones
placenta			
chorionic gonadotropin	glycoprotein	corpus luteum	maintenance of function
placental lactogen	polypeptide	material tissues	insulin-like effects
relaxin	protein		cervical softening

specific protein hormone is cloned. When inserted into rapidly replicating bacteria, enormous quantities of mammalian hormones can be produced. This approach has generated highly purified recombinant human hormones, including insulin and growth hormone.

The field of endocrinology has as its primary focus the study of the structure and function of endocrine glands and their secretory products. However, findings (ca 1994) have fostered a broader scope for endocrinology. Hormone secretion from cells that exist outside the classic endocrine glands can occur. The ability to synthesize and secrete hormones has been demonstrated for nonglandular tissue, including neurons, ie, cells identified in the central and peripheral nervous systems, and leukocytes. To identify cells that synthesize hormones, molecular biological approaches have been used to measure the messenger ribonucleic acid (mRNA) that encodes for a specific hormone or for an enzyme required for hormone synthesis.

Transport in the blood is no longer a requisite for a hormonal response. Responses can occur after release of hormones into the interstitial fluid with binding to receptors in nearby cells, called paracrine control, or binding to receptors on the cell that released the hormone, called autocrine control. A class of hormones shown to be synthesized by the tissue in which they act or to act in the local cellular environment are the prostaglandins (qv). These ubiquitous compounds are derived from arachidonic acid which is stored in the cell membranes as part of phospholipids. Prostaglandins bind to specific cellular receptors and act as important modulators of cell activity in many tissues.

A hormone can have multiple biological effects that are conferred by binding to receptors on specific target cells. A broader view of what constitutes a hormone is exemplified by the discovery that identical peptides are present as secretory products in neurons of the peripheral and central nervous system, as well as glandular cells in the gut, pancreas, and other tissues. In addition to affecting local cell activity in the nervous system and in nonneural tissue, many of these peptides also have been shown to act after secretion into the circulation.

Nontraditional Hormones. Novel hormones identified in cardiovascular tissue have profound effects on maintenance of blood pressure and blood volume in mammals. Atrial natriuretic hormone (ANH) is a polypeptide hormone secreted from the atria of the heart. When the cardiac atrium is stretched by increased blood volume, secretion of ANH is stimulated; ANH in turn increases salt and water excretion and reduces blood pressure. Endothelin is a polypeptide hormone secreted by endothelial cells throughout the vasculature. Although endothelin is released into the circulation, it acts locally in a paracrine fashion to constrict adjacent vascular smooth muscle and increase blood pressure.

Two protein hormones, inhibin and activin, have been identified in gonadal tissue. Inhibin has been isolated from ovarian follicular fluid and found to inhibit pituitary secretion of FSH. Activin is secreted by the ovary and the testes into the circulation. In addition, both inhibin and activin have intragonadal autocrine and paracrine effects that influence gonadal steroidogenesis.

Erythroid differentiation factor (EDF) is a protein isolated originally from the culture fluid of a human leukemia cell line; it induces the proliferation and differentiation of hematopoietic progenitor cells.

Insect and Plant Hormones

Insect Hormones. Insects and crustaceans must shed their exoskeleton in order to grow. This molting process is called ecdysis and is initiated by a steroid hormone called ecdysone, which is secreted by the Y-organ located at the base of the antennae. The corpus cardiaca of the brain produces prothoracotropic hormone, which acts on the prothoracic gland to secrete ecdysone. Another hormone, juvenile hormone, secreted by the corpora allata in insects, prevents development from the larval to the adult stage; when juvenile hormone concentrations decrease, metamorphosis is initiated.

Plant Hormones. Plant hormones are organic substances, active in small ($<1\ \mu$M) amounts, which are formed in one part of a plant and usually translocated to other sites to induce specific biochemical or morphological responses. Auxins derived from the amino acid tryptophan induce elongation in shoot cells; the principal naive auxin in higher plants is indole-3-acetic acid. Another class of hormones, the gibberellins, are four-ring structures that occur as 19-carbon or 20-carbon, mono-, di-, or tricarboxylic acids. Originally isolated from fungi, gibberellins are natural products of higher plants and an intact ring structure is essential for their activity in stimulating cell division and cell elongation. Cytokinins are isopentenyl adenine derivatives that promote cell division. Growth-inhibiting hormones are important for inducing periods of plant dormancy; abscisic acid is a natural growth-inhibiting hormone. Plants also synthesize brassinosteroids that contain the steroid nucleus, are active in concentrations lower than those of other plant hormones, ie, pM, and act alone or synergise with other plant hormones to affect plant growth.

Pherohormones

Pherohormones, or pheromones, are interorganismal hormones that transmit information between members of a species. Insect behavior is affected by different pheromones used as sex attractants or as chemical markers of food sources. The chemical structure of pheromones varies. These may be small molecules that rapidly disperse and serve to signal alarm, eg, 4-methyl-3-heptanone in ants, or larger, less volatile compounds that persist for longer periods and function as sex attractants, eg, 3,13-octadecadien-1-ol. There is evidence for olfactory communication in a variety of mammalian species, including humans. In male and female rodents, different volatile constituents in urine induce aggressive behavior, delay or accelerate puberty, or attract the opposite gender. Originally isolated from porcine testicles, 16-androstenes are C-19 steroids that are volatile, have a pronounced odor, and have been implicated as sex attractants in some mammals.

WILLIAM C. ENGELAND
University of Minnesota

K. L. Becker, *Principles and Practice of Endocrinology and Metabolism*, J. B. Lippincott Co., Philadelphia, Pa., 1990.

H. Laufer and R. G. H. Downer, *Endocrinology of Selected Invertebrate Types*, Alan R. Liss, Inc., New York, 1988.

T. C. Moore, *Biochemistry and Physiology of Plant Hormones*, 2nd ed., Springer-Verlag, New York, 1989.

G. D. Prestwich and G. J. Blomquist, *Pheromone Biochemistry*, Academic Press, Inc., Orlando, Fla., 1987.

HORMONES, ADRENAL-CORTICAL

Adrenal-cortical hormones comprise an important drug class. Though not without adverse effects, these compounds are the drugs of choice in the treatment of afflictions ranging from the moderate skin rash to severe acute inflammatory disorders.

The adrenal cortex releases both mineralocorticoids (from the zona glomerulosa) and glucocorticoids (from the zona fasciculata/reticularis). In addition, some androgenic and estrogenic steroids are synthesized by the adrenal gland. Once released by the adrenal cortex, the primary endogenous function of glucocorticosteroids is in influencing carbohydrate and protein metabolism. Mineralocorticoids regulate sodium reabsorption in the collecting tubules of the kidney. As with the mineralocorticoids, the generation of glucocorticosteroids is intricately balanced; where the production and regulatory process malfunctions, the result is either an excess (eg, Cushing's syndrome) or a deficiency (eg, Addison's disease) in glucocorticoid levels.

Corticosteroids and their metabolites were early recognized as possessing powerful antiinflammatory and immunomodulatory properties. Even prior to 1950, reports of the antiarthritic properties of cortisone (**1**) by Hench and co-workers indicated the potential for these compounds to reduce the suffering of patients with inflammatory diseases. This awareness, combined with the first synthesis of naturally occurring glucocorticoids (11-desoxycorticosterone), led not only to the massive increase in research in the area of steroid synthesis and physiology, but to a Nobel prize in 1950 for early steroid pioneers Hench, Reichstein, and Kendall.

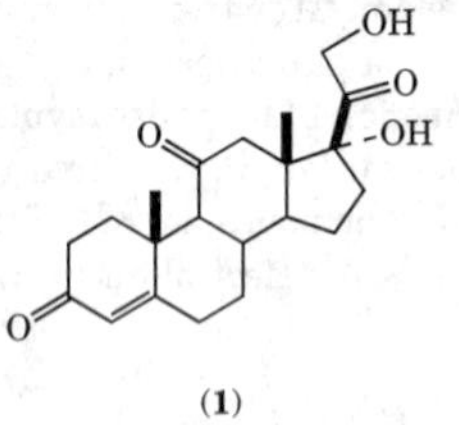

(**1**)

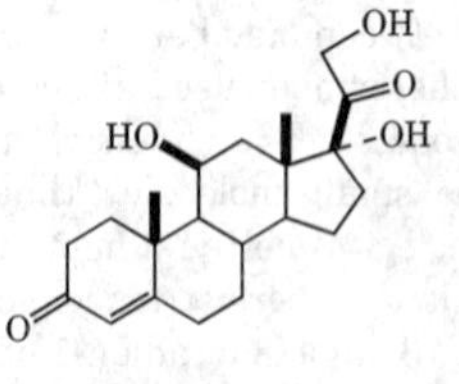

(**2**)

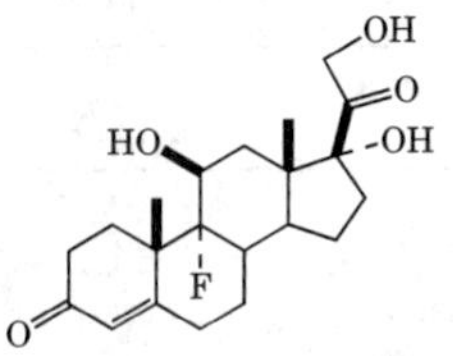

(**3**)

The synthesis of 9α-fluorocortisol (**3**) opened the way for development of many more highly active antiinflammatory agents, and indeed some of today's most active antiinflammatory agents bear some resemblance to this 9α-fluorinated steroid.

All natural adrenocorticoids are derivatives of the planar ring system 5α-pregnane (**4**). Substituents lying above the plane of the rings are assigned a β-configuration, indicated by a dotted line. Angular methyl groups at C-10 and C-13 have the β-configuration and are often shown simply by solid bonds. Tertiary hydrogen atoms at C-8, C-9, C-14, and C-17 are usually omitted unless their stereochemistry differs from that shown in (**4**).

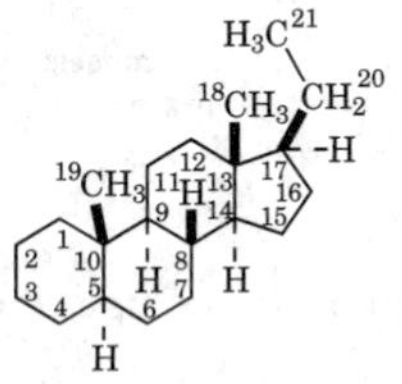

(**4**)

Clinical Use of Adrenal-Cortical Steroids

The predominant clinical use of corticosteroids is a result of their associated antiinflammatory properties. These are commonly used as topicals for the suppression of symptoms, including inflammation, occurring in a particular disease state; these compounds are rarely considered curative in their usage. Many other disease states do, however, respond well symptomatically to treatment with corticosteroid therapy.

Mineralocorticoid therapy is a less common, though still important aspect of the medicinal use of adrenal-cortical steroids. Very important applications include use in treating adrenocortical insufficiency or in other adrenocoritcoid replacement therapies, and in the management of salt-losing forms of congenital adrenogenital syndrome. Aldosterone, owing to its short half-life, is not used therapeutically; desoxycorticosterone acetate (a natural aldosterone precursor) and fludrocortisone are the only clinically significant mineralocorticoidal agents.

Adrenocortical Biosynthesis and Metabolism

Corticosteroids are biosynthesized from cholesterol and released as needed by the adrenal cortex; they are not stored. Cholesterol undergoes a series of irreversible oxidations during which carbons 22 through 27 are cleaved, resulting in pregnenolone. Reversible isomerization of Δ^5 to Δ^4 results in progesterone, the key intermediate in both mineralocorticoid and glucocorticoid biosynthesis. Cortisol is the product of the 11β-, 17α-, and 21-oxidations (flavoprotein, cytochrome P-450-mediated) of progesterone; aldosterone is the result of oxidations at the 11β-, 18-, and 21-positions. Glucocorticosteroid biosynthesis is outlined in Figure 1, and key steps in the mineralocorticoid biosynthesis are highlighted in Figure 2.

The major metabolic transformations of the adrenal cortical hormones generally follow the metabolism of cortisol.

Mechanism of Action of Antiinflammatory Steroids

Receptor Structures for Glucocorticoids and Mineralocorticoids. The effects of adrenal-cortical steroids are thought to result from their interaction with intracellular receptors, and a great deal of attention has

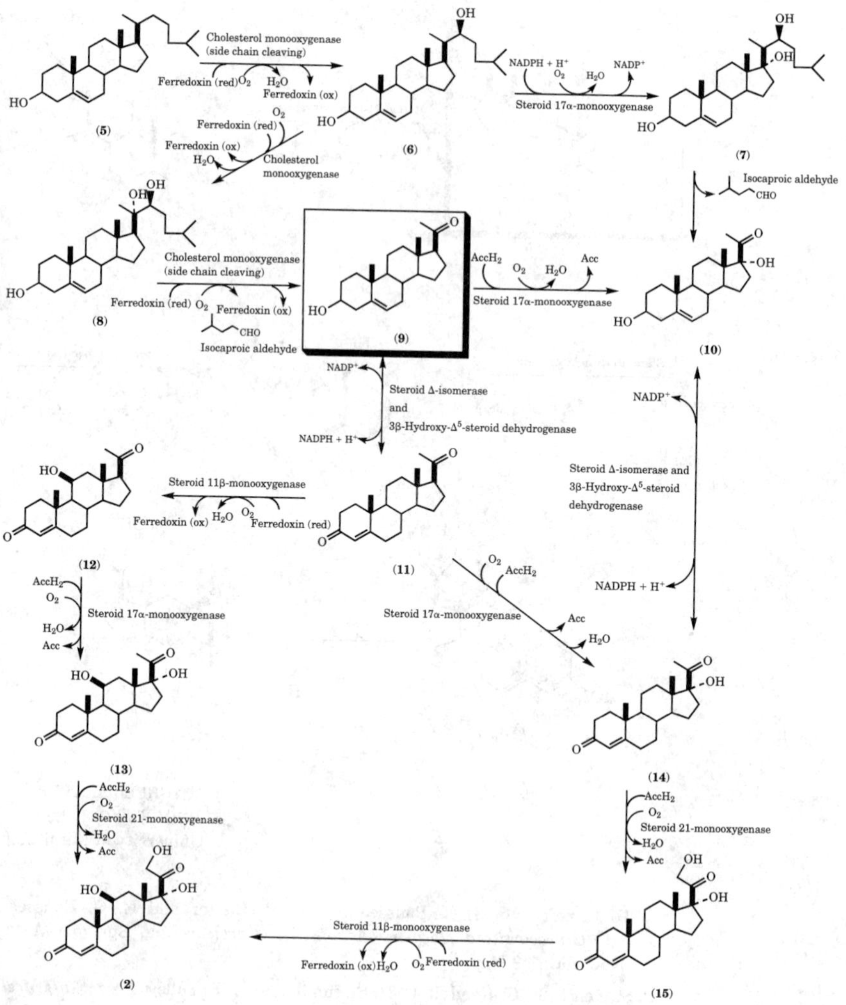

Figure 1. Biosynthetic pathways for formation of cortisol from cholesterol.

focused on determining the structure and function of the glucocorticoid receptor (GR) as well as the mineralocorticoid receptor (MR). The primary function of the MR seems to be nearly identical to that of the GR, with the primary differences being the restricted expression of the MR (limited mostly to tissues in the kidney, colon, salivary and sweat glands, and hippocampus), and the different proteins encoded by the activated DNA-receptor complex. The highly homologous GR and MR proteins are also very similar in action.

Mechanism of Action. The process by which adrenocortial steroids impart their action is based upon the action of the steroid on a receptor (MR or GR). One result of this process is the lag between optimum pharmacologic activity and peak blood concentrations. The stages of GR goes through in becoming active can be divided into five general steps, and each step is mediated by the glucocorticoid receptor (GR): (*1*) subcellular localization, (*2*) association with heat shock proteins (hsp), (*3*) hormone binding, (*4*) dimerization and DNA binding, and (*5*) transactivation. The use of glucocorticoids leads to antiinflammatory effects by first controlling gene expression, which subsequently leads to the synthesis and/or suppression of inflammation regulatory proteins.

Glucocorticoids have been shown to inhibit gene transcription of proteins involved in the inflammatory process, including the key inflammation mediators called cytokines (IL-1, IL3-6, IL8, GM-CSF, TNFα). Steroids have been also shown to suppress the formation of cytokine receptors.

Mineralocorticoids follow a mechanistic route similar to that of glucocorticoids, though differing in the proteins expressed. The activated MR-DNA complex promotes the expression of aldosterone-induced proteins (AIPs), which then act to increase Na^+ conductance of the luminal membrane and concurrently increase Na^+ pump activity of the basolateral membrane.

Synthesis of Glucocorticoids

Hydrocortisone and Prednisolone. A major difficulty in the manufacture of corticosteroids was the lack of an abundant raw material containing an 11-oxygenated function. Microorganisms were discovered that are capable of introducing an 11β-hydroxyl group into a steroid. *Rhizopus arrhizus* transformed progesterone (11), which was readily available from diosgenin or from the soybean sterol, stigmasterol, into 11α-hydroxyprogesterone; even better results were obtained using *Rhizopus nigricans*. Other organisms include *Corynebacterium simplex* which converts cortisone (**1**) and cortisol (**2**) into their 1-dehydro derivatives, prednisone (**20**) and prednisolone (**21**), respectively. These steroids surpassed their parent hormones in antirheumatic and antiallergic activity and produced lower mineralocorticoid activity and other side effects. Nearly all corticoids on the market other than cortisone are 1-dehydro steroids.

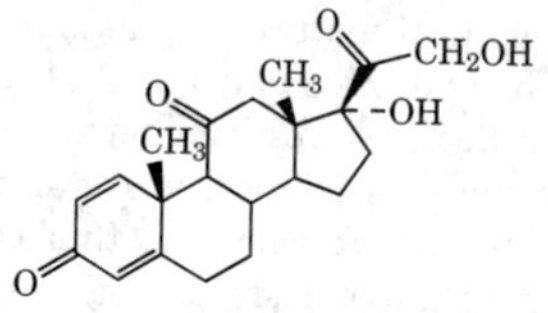

(20)

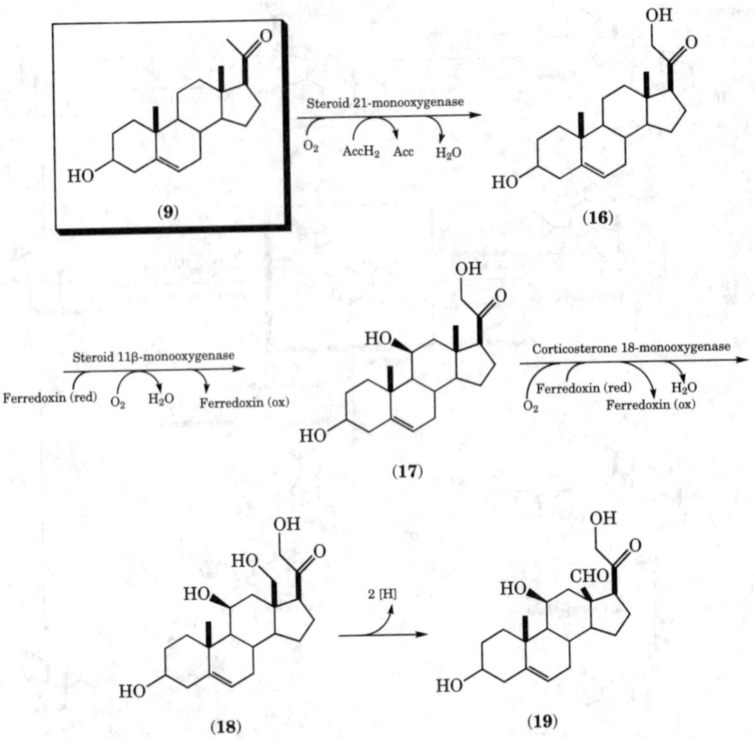

Figure 2. Key steps in the mineralocorticoid biosynthesis.

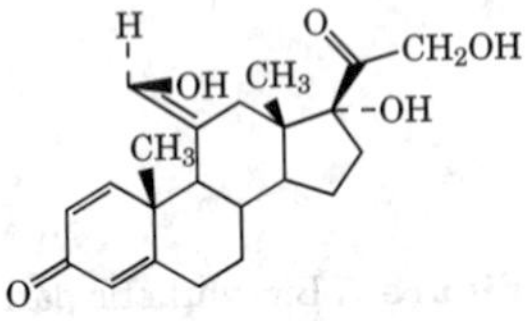

(21)

The Mineralocorticoids

Studies have shown that aldosterone from adrenal extracts was intensely active in the survival and sodium retention assays in the adrenalectomized rat.

Hyperaldosteronism is accompanied by elevation of blood pressure, and can be treated with an aldosterone antagonist, eg, spironolactone, which is synthesized from dehydroepiandrosterone (DHA). The market for mineralocorticoids is extremely modest as their main value lies in the treatment of Addison's disease.

Adrenal-Cortical Steroid Antagonists

Antagonists of glucocorticoid and mineralocorticoid activity have found increased use clinically in the treatment of hypertension, Cushing's disease, and heightened intraocular pressure.

Some general structural features that lead to both observed differences in steroidal conformation (via x-ray structures) and to noticeable effects in bioactivity (RBAs, agonism vs antagonism, magnitude of activity, etc) are the following: (*1*) Unsaturation of the A-ring: for the A-ring diazoles, all the GR antagonist known have the 4-ene-3-one structure. As this is shared with the most potent glucocorticoids, it is thought that this structural feature increases binding affinity to the steroid receptor (*2*) The hydroxy groups at C-11, C-17, and C-21, seem to be individually or collectively responsible for agonistic activity.

Spironolactone is the most clinically useful steroidal aldosterone antagonist; and unlike GR antagonists, this compound is utilized much more frequently than aldosterone agonists. Interfering with Na^+ reabsorption and K^+ secretion in the late distal segment, this compound is predominantly used with other diuretics.

MITCHELL A. AVERY
JOHN R. WOOLFREY
University of Mississippi

J. D. Baxter and G. G. Rousseau, in J. D. Baxter and G. G. Rousseau, ed., *Glucocorticoid Hormone Action: An Overview*, Vol., Springer-Verlag, New York, 1979, pp. 1–24.

P. S. Hench, E. C. Kendall, C. H. Slocumb, and H. F. Polley, *Proc. Staff Meet., Mayo Clinic* **24**, 181 (1949).

J. Fried and E. F. Sabo, *J. Am. Chem. Soc.* **75**, 2273 (1953).

J. Wepierre and J.-P. Marty, *Trends Pharmacol. Sci.*, 23 (1979).

ANTERIOR PITUITARY HORMONES

The hormones of the anterior pituitary gland play a significant role in the maintenance of normal health and body function. This master gland produces hormones involved in the regulation of somatic growth, metabolic rate, carbohydrate and lipid metabolism, lactation, reproduction, and response to stress. Eleven anterior pituitary hormones have been extensively characterized at the protein and genomic levels. These hormones fall into three classic categories, ie, pro-opiomelanocortin-related (POMC-related) hormones, proteins structurally related to prolactin and growth hormone (PRL/GH-related), and glycoproteins. Structural similarities exist within each biochemical category, and the structural data presented represent listings in the Protein Identification Resource (PIR) database, compiled by the National Biomedical Research Foundation, Washington, D.C. Table 1 provides summary information on the principal biologic roles of each of these hormones. The National Hormone and Pituitary Program at the NIH and the USDA Animal Hormone Program provide anterior pituitary hormones free of charge for nonprofit research purposes. All principal hormones of the anterior pituitary gland are available commercially as material purified from anterior pituitary gland extracts, synthetic peptides, or proteins produced by the expression of recombinant DNA. Excellent listings of commercial sources of anterior pituitary hormones are available.

Table 1. Hormones of the Anterior Pituitary and Their Functions

Hormone	Number of amino acids[a]	Principal function
	Pro-opiomelanocortin-derived peptides	
adrenocorticotropin	39	stimulates cortex of adrenal gland to produce glucocorticoids
β-endorphin	31	functions as neurotransmitter; exerts opiate-like analgesia
lipotropin		mobilizes fat; precursor for β-endorphin
β-LPH	91	
γ-LPH	58	
melanotropin		pigmentation
α-MSH	13	
β-MSH	22	
	Prolactin/growth hormone-related peptides	
prolactin	199	supports lactation
growth hormone	191	stimulates body growth; anabolism
	Glycoproteins	
follicle-stimulating hormone	210	supports maturation of ovarian follicles and sperm
luteinizing hormone	204	induces ovulation; maintains testicular function
thyroid-stimulating hormone	211	stimulates thyroid hormone production and thyroid gland growth

[a] From human amino acid sequence (PIR).

Pro-Opiomelanocortin-Derived Hormones

A single parent gene codes for the 267 amino acid glycoprotein which contains the sequences for adrenocorticotropic hormone (ACTH), melanocyte-stimulating hormone (MSH), endorphin, and lipotropin (LPH). This precursor has been named pro-opiomelanocortin. (POMC), and the biosynthesis and structure have been reviewed.

Prolactin/Growth Hormone-Related Hormones

Prolactin and growth hormone (GH) are structurally related peptides approximately 200 amino acids in length. The best documented function of prolactin is to stimulate milk synthesis and secretion. Growth hormone is a potent metabolic regulator and stimulator of body growth.

Glycoprotein Hormones

There are three pituitary glycoprotein hormones, ie, luteinizing hormone (LH), follicle-stimulating hormone (FSH), and thyroid-stimulating hormone (TSH). Luteinizing hormone and FSH control significant aspects of reproduction. These hormones are referred to as gonadotropins, owing to their trophic and stimulatory effects on gonadal tissues. Thyroid-stimulating hormone controls the function of the thyroid gland.

Names are necessary to report factually on available data; however, the U.S. Dept. of Agriculture neither guarantees nor warrants the standard of the product, and the use of the name by the U.S. Dept. of Agriculture implies no approval of the product to the exclusion of others that may also be suitable.

ROBERT L. MATTERI
U.S. Department of Agriculture

R. J. Ryan and co-workers, *Recent Prog. Hormone Res.* **43**, 383–429 (1987).

M. J. Soares, T. N. Faria, K. F. Roby, and S. Deb, *Endocrine Rev.* **12**, 402–423 (1991).

W. J. De Koning, G. A. Walsh, A. S. Wrynn, and D. R. Headon, *Biotechnology* **12**, 988–992 (1994).

J. S. Strobl and M. J. Thomas, *Pharm. Rev.* **46**, 1–34 (1994).

ANTERIOR PITUITARY-LIKE HORMONES

Hormones with structural and functional similarities to the hormones of the anterior pituitary gland fall into two main categories, ie, proteins of the prolactin/growth hormone (PRL/GH) family, and glycoproteins. These hormones are produced in the placenta by chorionic tissue, and so are normally found in the female during pregnancy.

Prolactin/Growth Hormone Family

Chorionic Somatomammotropin. Chorionic somatomammotropin is detectable in maternal serum by immunoassay at about two weeks of pregnancy. Its secretion increases thereafter to very high levels, 20 μg/mL serum, prior to delivery. Human CS behaves much like prolactin in a number of assay systems. Accordingly, hCS is thought to contribute to the development of mammary tissue in preparation for postnatal nursing. Human CS is thought to contribute to the commonly observed resistance to insulin in the mother, which serves to divert carbohydrate energy sources to the developing fetus. Human CS is available for research purposes from a variety of sources, but does not have clinical applications.

Placental Growth Hormone. The secretion of GH-V becomes elevated at about three weeks of pregnancy and increases to approximately 15 ng/mL near term. The physiological role of GH-V is uncertain. GH-V is a potent growth-stimulator but possesses considerably less lactogenic activity than GH-N. There are no clinical applications (ca 1993) for GH-V.

Placental Lactogens. The placentae of rodents and ruminants produce prolactin-related molecules called placental lactogens. These may directly regulate fetal growth. The presence of bPL receptors in bovine endometrium suggests that bPL could regulate the function of maternal tissues which support pregnancy, fetal growth, and development.

There are no commercial uses for ruminant PLs. There may be potential applications as performance stimulators in domestic animal production. Milk yield and feed intake are stimulated by the administration of bPL. The development of recombinant bPL for use in food animals is being investigated by Monsanto Co.

Prolactin-Like Proteins. A number of prolactin-like proteins (PLPs), which are distinct form the PLs, have been identified in ruminants and rodents. The functional roles of PLPs remain to be determined.

Glycoprotein Family

Human Chorionic Gonadotropin. Human CG (hCG) is produced by syncitiotrophoblast cells of the placenta. The secretion of hCG from chorionic cells begins about 10 days after fertilization. The detection of this early rise in hCG in urine forms the basis for a variety of nonprescription home pregnancy tests. Maximum production of CG occurs at approximately 70 days of gestation and declines rapidly, reaching relatively low levels in the serum by the second trimester. The secretion of hCG early in pregnancy prolongs the lifespan of the corpus luteum, an ovarian structure which secretes the steroid hormones progesterone and estradiol, needed to maintain pregnancy. After the first two weeks, the placenta is able to produce sufficient steroid hormones to maintain pregnancy.

Human CG has considerable commercial value in clinical fertility control due to its LH-like bioactivity and natural resistance to biological degradation. The injection of hCG is used to induce ovulation following treatment with high doses of follicle-stimulating hormone (FSH) to stimulate the maturation of multiple ova. Significant producers of hCG for clinical usage are Serono (Italy) and Organon (Sweden). Human CG is available for research purposes from many suppliers.

Equine Chorionic Gonadotropin. Equine CG (eCG) is produced by trophoblast-derived structures in the endometrium known as endometrial cups. The original name of this hormone was pregnant mare serum gonadotropin (PMSG). Equine CG is secreted at high levels during the first trimester of the mare's pregnancy, ie, days 40–130. The high concentration of serum eCG is conducive to hormone extraction and purification. The prolonged biological half-life and FSH bioactivity make eCG extremely valuable commercially. This hormone is used to induce superovulation for commercial and research purposes. In the United States, Food and Drug Administration licensing has been issued only for the use of eCG in swine, as PG-600 (Intervet, Holland). Equine CG is readily available for research purposes from many suppliers.

Names are necessary to report factually on available data; however, the U.S. Dept. of Agriculture neither guarantees nor warrants the standard of the product, and the use of the name by the U.S. Dept. of Agriculture implies no approval of the product to the exclusion of others that may also be suitable.

ROBERT L. MATTERI
U.S. Department of Agriculture

R. J. Ryan and co-workers, *Recent Prog. Hormone Res.* **43**, 383–429 (1987).

G. Baumann, *Endocrine Rev.* **12**, 424–449 (1991).

I. A. Forsyth, *Exp. Clin. Endocrinol.* **102**, 244–251 (1994).

W. J. De Koning, G. A., Walsh, A. S. Wrynn, and D. R. Headon, *Biotechnology* **12**, 988–992 (1994).

POSTERIOR PITUITARY HORMONES

The posterior lobe of the pituitary, ie, the neurohypophysis, is under direct nervous control, unlike most other endocrine organs. The hormones stored in this gland are formed in hypothalamic nerve cells but pass through nerve stalks into the posterior pituitary. The biological activities of pituitary extracts result from peptide hormones, often found in association with higher molecular weight proteins.

The principal hormones of the human posterior pituitary include the two nonapeptides, oxytocin and arginine vasopressin (antidiuretic hormone, ADH). Many other hormones, including opioid peptides, cholecystokinin (CCK), and gastrointestinal peptides, also have been located in mammalian neurohypophysis, but are usually found in much lower concentrations. Studies have demonstrated that oxytocin and vasopressin are synthesized in other human organs, both centrally and peripherally, and there is considerable evidence for their role as neurotransmitters.

Oxytocin, although found in large quantities in both males and females, primarily functions in contracting the uterus during childbirth and releasing milk from mammary tissue. The fowl-pressor effect, ie, elevation of chicken blood pressure following iv administration of peptide, is a third action. It is fundamentally pharmacological, but an assay based on the effect often has been useful for rapid screening of potential active analogues. Arginine vasopressin (AVP) promotes water readsorption by the kidney, ie, the antidiuretic effect, and contraction of the smooth muscle of vascular tissue, ie, the pressor effect. Additional biological activities based on the behavioral properties of both hormones represent areas of intense research interest.

General findings support the concept that neurohypophyseal hormones are relatively flexible peptides which have various binding and active elements that can, in theory, be modified to form either superagonists or hormone inhibitors.

Synthesis of Posterior Pituitary Hormones and Their Analogues

Agonists. Both oxytocin and the vasopressins have been synthesized by numerous routes using both rapid-solution and solid-phase methods. In general, large structural variations, which alter overall geometries, prove destructive to activity. Subtle variations, however, have led to compounds with retained and even markedly enhanced hormonal activities; several of these are summarized in Tables 1 and 2.

Antagonists. Another goal of structure–function studies of peptide hormones is the design of antagonists or hormone inhibitors that may have potential clinical usage. Modifications of the N-terminus of oxytocin and the vasopressins have yielded structures having significant

Table 1. Synthetic Analogues of Oxytocin With High Potency or Selectivity

Peptide	Biological activities[a]			
	Oxytocic	Milk ejection	Antidiuretic	O/A ratio[b,c]
oxytocin	520	474	4	130
1-deaminooxytocin	803	541	19	42
[Thr-4]-oxytocin	923	543	0.9	1026
[Thr-4,Gly-7]-oxytocin	166	802	0.002	83,000

[a] Biological activities are in USP units/mg. [b] O/A ratio = oxytocic/antidiuretic ratio. [c] In the absence of Mg^{2+}.

Table 2. Synthetic Analogues of Vasopressin With High Potency or Selectivity

Peptide	Biological activities[a]		
	Antidiuretic	Vasopressor	A/P ratio[b]
arginine vasopressin	503	487	1
1-deamino[D-Arg]-8-vasopressin	1200	0.39	3000
1-deamino-Val-4, Arg-8	1230	antagonist	infinite

[a] Biological activities are in USP units/mg.
[b] A/P ratio = antidiuretic/pressor activity ratio.

in vitro inhibitory activities. As shown in Table 3, synthetic inhibitors of oxytocin contain variations in the 1 and 2 positions.

Other potential oxytocin antagonists are being developed using leads from naturally occurring, nonpeptide structures, such as an extract from *Streptomyces*. A selective nonpeptide vasopressin V_1 receptor antagonist also has been found.

Degradation. Both oxytocin and vasopressin have short plasma half-lives in humans.

Neurophysins, Hypothalamic Hormones, and Hormone Processing

The neurophysins represent a group of medium-sized (9000–10,000 mol wt) proteins found in the posterior pituitary that act as carriers of oxytocin and the vasopressins. Although the pattern is species-variant, the two principal forms found in higher mammals are neurophysin I (NP-I) and II (NP-II); these appear to be associated with vasopressin and oxytocin, respectively. According to the concept of one neurophysin–one hormone, the carrier proteins and their respective hormones are synthesized as linear peptide chains on human chromosome 20 and then cleaved and cyclized to vield NP-I and vasopressin and NP-II and oxytocin.

Table 3. Synthetic Analogues of Oxytocin and Vasopressin Having Antagonist Activities

Analogue[b]	Activity, pA_2[a]	
	Antioxytocic	Antipressor[c]
Oxytocin		
[Mcpr-1]oxytocin	7.61	weak
[penicillamine-1]oxytocin	6.86	0
[deamino,cyclo(Glu-4, Lys-8)]-oxytocin	8.74	6.3
[Mcpr-1,Tyr(CH_3)-2,Orn-8]-oxytocin	8.52	7.96
Vasopressin		
[Mcpr-1,Tyr(CH_3)-2]AVP	8.13	8.62
[Mcpr-1,Tyr(CH_2CH_3)-2,Val-4]AVP	7.88	8.16
[Mcpr-1,Phe-2,Ile-4,Ala-9]AVP		7.71

[a] pA_2 = -log A_2, where A_2 refers to concentration required to cause 50% inhibition of agonist response. [b] Mcpr = 3 − mercapto − 3,3 − cyclopentamethylene propionic acid. [c] Anti-antidiuretic activity for [Mcpr − 1, Tyr(CH_2CH_3) − 2, Val − 4]AVP = 7.57; for [Mcpr − 1, Phe − 2, Ile − 4, Ala − 9]AVP = 8.38.

In humans, the hypothalamic-derived protein and the hormone noncovalent complexes are packaged in neurosecretory granules, then migrate along axons at a rate of 1–4 mm/h until they reach the posterior pituitary where they are stored prior to release into the bloodstream by exocytosis. Considerable evidence suggests that posterior pituitary hormones function as neurotransmitters, vasopressin acts on the anterior pituitary to release adrenocorticotropic hormone (ACTH) as well as on traditional target tissues such as kidneys. Both hormones promote other important central nervous system (CNS) effects.

Oxytocin and Vasopressin Receptors. The actions of oxytocin and vasopressin are mediated through their interactions with receptors. Different receptor types as well as different second messenger responses help explain their diverse activities in spite of the hormones' structural similarities. Thus oxytocin has at least one separate receptor and vasopressin has been shown to have two principal receptor types, V_1 and V_2. Subclasses of these receptors have been demonstrated, and species differences further complicate experimental analysis.

The ultimate goal of structure–function studies is complete understanding of the hormone's interaction with its receptor(s). Evidence continues to support an oxytocin model in which residues 2 and 5, ie, Tyr and Asn, are vital for interaction with the uterine smooth muscle receptor.

V_1 receptors, found in vascular smooth muscles and in the liver, function by means of a calcium-dependent pathway. V_2 receptors, found in the kidney, modulate the antidiuretic response of vasopressin analogues through a cyclic adenosine monophosphate (AMP)-dependent pathway.

Uses

Oxytocin has been used widely to induce labor, although its suitability has been questioned; prostaglandin E_2 may be a practical alternative.

Hyposecretion by the hypothalamic supraoptic nuclei or injury to the posterior pituitary can give rise to diabetes insipidus; a gene deletion in the rat also has been shown to lead to diabetes insipidus. The symptoms of the disease can be controlled by chronic administration of antidiuretic hormone (ADH). Vasopressin infusion also has been used in the management of gastrointestinal hemorrhage. A vasopressin analogue, desamino-8-D-arginine vasopressin, has proven useful in hemophilia as a hemostatic agent, reducing the need for blood and plasma products.

Vasopressin and Memory. Several hormones, eg, ACTH, vasopressin, and the catecholamines, may be involved in memory retention or consolidation.

ARNO F. SPATOLA
University of Louisville

S. Yoshida and L. Share, eds., *Recent Progress in Posterior Pituitary Hormones*, Elsevier Amsterdam, the Netherlands, 1988.

D. Ganter and D. Pfaff, eds., *Neurobiology of Oxytocin*, Springer-Verlag, Berlin, 1986.

S. Melmed and R. J. Robbins, eds., *Molecular and Clinical Advances in Pituitary Disorders*, Blackwell Scientific Publications, St. Louis, Mo., 1991.

D. M. Gash and G. Boer, eds., *Vasopressin: Principles and Properties*, Plenum, New York, 1987.

HUMAN GROWTH HORMONE

Human growth hormone (hGH), also known as somatotropin, is a protein hormone produced and secreted by the somatotropic cells of the anterior pituitary. Secretion is regulated by a releasing factor, ie, the growth hormone-releasing hormone, and by an inhibitory factor, somatostatin. Human growth hormone plays a key role in somatic growth through its effects on the metabolism of proteins, carbohydrates, and lipids. Human growth hormone exerts its biological effects either through direct action of the hormone at the target tissue or indirectly through the action of a second class of peptide hormones, the , also known as insulin-like growth factors, which are produced primarily in the liver in response to hGH binding to specific receptors there.

Two well-known pathological conditions are the result of an excess or a deficiency of this hormone. The condition in which the body produces an excess of hGH is known as acromegaly or giantism. The condition in which too little is produced is dwarfism.

Chemical and Physical Properties

Human growth hormone is a single polypeptide chain of 191 amino acids having two disulfide bonds, one between Cys-53 and Cys-165, forming a large loop in the molecule, and the other between Cys-182 and Cys-189, forming a small loop near the C-terminus. Molecular mass is 22,125; the empirical formula is $C_{990}H_{1529}N_{262}O_{300}S_7$.

Purified hGH is a white amorphous powder in its lyophilized form. It is readily soluble (concentrations > 10 mg/mL) in dilute aqueous buffers at pH values above 7.2.

hGH Derivatives. Several derivatives of hGH are known. These derivatives include naturally occurring derivatives, variants, and metabolic products, degradation products primarily of biosynthetic hGH, and engineered derivatives of hGH produced through genetic methods. They include methionyl hGH, 20K hGH, acetylated hGH, desamido hGH, sulfoxide hGH, size isomers, proteolytically cleaved two-chain forms, truncated forms, forms bound to binding proteins, and various derivatives resulting from point mutations.

Biological Properties

Human growth hormone is a very complex molecule biologically. Several diverse biological activities such as anabolic, insulin-like, diabetogenic, and lactogenic activities have been ascribed to hGH, which also appears to promote water and salt retention. An in-depth discussion of these activities may be found in several excellent reviews available in the literature.

Uses

Human growth hormone, used as a human pharmaceutical, is approved in the United States for treatment of growth failure owing to hGH deficiency (a condition known as pituitary dwarfism), for replacement therapy in growth-deficient adults, and for chronic renal insufficiency in children. A decision for use in treating Turner's syndrome is expected from the FDA in the very near future. Other potential indications being evaluated include the treatment of burns and wounds, cachexia, osteoporosis, constitutional growth delay, aging, malnutrition, and obesity.

Manufacture and Processing

Since 1985, manufacture of hGH has been almost exclusively by recombinant DNA technology.

Recombinant hGH. Introduction of recombinant DNA technology meant an unlimited supply of hGH could be produced in a number of different systems. *Escherichia coli* (*E. coli*) was the first heterologous host used and originally produced Met-hGH, authentic hGH with the initiating methionine attached. This was followed by production in *E. coli* of precursors of hGH that could be cleaved *in vitro* to yield natural sequence hGH.

Economic Aspects

Human growth hormone is one of the largest selling therapeutic proteins produced by recombinant DNA technology. Upon approval of additional indications, the sales of hGH are expected to increase even more. hGH preparations marketed in the United States include Humatrope (Eli Lilly and Co.), Protropin and Nutropin (Genentech, Inc.), and Genotropin (Kabi).

Metabolism and Disposition

The pharmacokinetics of hGH have been evaluated in animals and humans. After intravenous administration, the elimination of hGH is described by first-order kinetics with a serum half-life of 12–30 min in both animals and humans. Traditionally, intramuscular (im) injection has been the method of choice for delivery of hGH. In general, no significant differences have been observed in the pharmacokinetics or biological activities of recombinant natural sequence hGH, recombinant *N*-methionyl-hGH, or pituitary-derived material in humans. The principal organs involved in the peripheral clearance of hGH from the plasma are the kidney and liver.

Information regarding the metabolic fate of hGH in humans and animals is fragmentary. Application of sensitive and accurate methods of protein analysis such as electrospray mass spectrometry should help resolve some of the deficiencies in the information available.

GERALD W. BECKER
WARREN C. MACKELLAR
RALPH M. RIGGIN
VICTOR J. WROBLEWSKI
Lilly Research Laboratories

C. H. Li and H. Papkoff, *Science* **124**, 1293 (1956).

D. V. Goeddel and co-workers, *Nature* **281**, 544 (1979).

G. Baumann, M. W. Stolar, and T. A. Buchanan, *Endocrinology* **119**, 149 (1985).

V. J. Wroblewski, M. Masnyk, and G. W. Becker, *Endocrinology* **129**, 465 (1991).

BRAIN OLIGOPEPTIDES

Neuropeptides represent the most numerous of the classes of neurotransmitters and neuromodulators (see NEUROREGULATORS). There are at least 50 oligopeptides known to exist in the brain, and the total list of oligopeptide neurotransmitter candidates may eventually be significantly greater. The clinical impact of these peptides on brain function, however, has been limited because most peptides are not metabolically stable, may not cross the blood brain barrier, or there may be no specific antagonists available to test the normal functioning of a peptide system. The development of novel analogues which satisfy many of these concerns represents a significant potential. Central nervous system (CNS) disorders involving Alzheimer's disease, appetite control, depression, and analgesia are all potentially affected by drugs acting at specific neuropeptide systems. The cloning and sequencing of the receptors for these peptides and modeling of their binding sites should be important components in the design of such drugs.

Many oligopeptides identified in the central nervous system exert both gastrointestinal and behavioral effects and also modify pituitary function. Within the brain, almost every possible behavioral action has been associated with some type of neuropeptide. Moreover, many neuropeptides have been associated with several CNS disease states. Opioid peptides and substance P have been associated with pain states, neurotensin has been associated with schizophrenia and the actions of psychotropic drugs, and corticotropin-releasing factor (CRF) has been associated with clinical depression. All neuropeptides are distributed in discrete areas of the brain and many are strategically located to mediate many of these effects. One important aspect of neuropeptide localization is colocalization, ie, the finding that many neuropeptides are stored and released along with other peptide and nonpeptide neurotransmitters. For this reason, neuropeptides have often been termed neuromodulators because they may act to modulate the effects of other neurotransmitters.

Many neurotransmitters are remarkably potent, both *in vivo* as well as in binding to their respective receptors. Most neuropeptide receptors belong to the superfamily of G-protein-coupled receptors. Thus the immediate biological response in the cell to the binding of the peptide to its receptor is the production of an intracellular second messenger system, usually either cyclic adenosine monophosphate (cAMP) or phosphoinositide turnover.

Analysis. Neuropeptides are assayed either by bioassays, including receptor binding assays.

Synthesis and Biosynthesis

Synthesis. In contrast to pituitary hormones, which usually can be obtained in pure form only after extraction from animal tissues, brain oligopeptides are readily available because of their small size. The synthetic replica represents the most economical and readily accessible source for the oligopeptides. Two techniques are available for laboratory synthesis of oligopeptides, ie, solution chemistry and solid-phase peptide synthesis (SPPS).

The use of SPPS allows synthesis of both native oligopeptides and important synthetic analogues. Many of these fragments or substituted analogues have been found to be equal to, or more potent than, the parent molecule and to exhibit long-acting and antagonistic activities.

Biosynthesis. The biosynthesis of neuropeptides is much more complex and involves the multistep process of transcription of specific mRNA from specific genes, formation of a high molecular weight protein product by translation, post-translational processing of the protein precursor to allow for proper packaging within the cell, and final enzymatic cleavage to produce the active peptide product.

Individual Brain Oligopeptides

Table 1 lists neuropeptides, along with their common acronyms and molecular weights.

Tachykinins. All members of this family contain the conserved carboxy-terminal sequence -Phe-X-Gly-Leu-Met-NH_2, where X is an aromatic, ie, Phe or Tyr, or branched aliphatic, eg, Val or Ile, amino acid. In general, this C-terminal sequence is crucial for tachykinin activity.

The preprotachykinin A gene encodes both substance P and substance K, while the preprotachykinin B gene encodes neuromedin K. In the brain, one of the most commonly discussed roles of substance P is the mediation of pain information, where this peptide acts as one of the primary neurotransmitters in the sensory afferent fibers which conduct pain information.

There are a number of analogues of substance P which are potent antagonists at substance P receptors; these include [D-Arg^1, D-Phe^5, D-$Trp^{7,9}$, Leu^{11}]-substance P, [Arg^4, Gly^5, D-Trp^7, $Asp(OC_4H_9)^{10}{}_2$]-substance P, and [Arg^6, D-$Trp^{7,9}$, CH_3-Phe^8]-substance P 6-11.

Endorphins, Enkephalins, and Dynorphins. In 1994 it is clear that three separate opioid peptide families exist, each presenting different gene products, ie, enkephalins, β-endorphin, and dynorphin. All of these peptides share the common sequence Tyr-Gly-Gly-Phe-X, where X is either Leu or Met.

These three peptide groups exhibit different localization patterns in the brain. β-Endorphin cell bodies are mostly localized to the arcuate nucleus of the hypothalamus. In contrast to β-endorphin, enkephalins are found in cell bodies in many different brain areas. Like enkephalin, dynorphin peptides also are widely distributed throughout the brain, although their presence in the supraoptic and paraventricular nuclei of the hypothalamus are particularly well known.

Three separate genes encode the opioid peptides. Enkephalin is derived from preproenkephalin A. β-Endorphin is one of the many products of POMC. Three different dynorphin peptides are derived from the third opioid gene, preproenkephalin B, or preprodynorphin.

All opioid peptides elicit a number of morphinomimetic activities following intracerebroventricular injection. β-Endorphin also stimulates prolactin and growth hormone secretion *in vivo*.

Hundreds of enkephalin analogues have been synthesized in an effort to find a nonaddictive opioid.

The opioid peptides are unique from a pharmacological point of view because there historically exists a number of specific nonpeptide antagonists, typified by naloxone. Nevertheless, there are several peptide antagonists available for these receptors, particularly for delta-opioid receptors.

Corticotropin-Releasing Factor (CRF). As shown in Table 1, CRF is one of the largest of the brain oligopeptides and activity appears to

Table 1. Names and Properties of Selected Brain Oligopeptides

Peptide[a]	Acronym	Species	Molecular weight
adrenocorticotrophic hormone	ACTH	human	4538
angiotensin			
angiotensin I		human	1296
angiotensin II		human	1046
bombesin			1619
bradykinin			1060
calcitonin gene-related peptide	CGRP	human	3787
cholecystokinin	CCK		
CCK-33		human	3950
CCK-8			1149
dynorphins			
dynorphin-A	DYN A	porcine	2146
dynorphin-B	DYN B	porcine	1570
α-neo-endorphin	α-NEO	porcine	1228
endorphins	END		
β-endorphin	β-END	human	3463
α-endorphin			1745
enkephalins			
Leu-enkephalin	L		555
Met-enkephalin	M		573
galanin		human	3156
gastrin		human	2097
hypothalamic-releasing factors			
corticotropin-releasing factor	CRF[b]	human	4754
growth hormone-releasing factor	GRF[c]	human	5037
luteinizing hormone-releasing hormone	LHRH[d]		1182
somatostatin	SRIF[e]		1637
thyrotropin-releasing factor	TRF[f]		362
melanocyte-stimulating hormone	MSH		
α-MSH			1664
β-MSH		human	2659
γ-MSH			1570
neurotensin			1672
neuropeptide Y	NPY	human	4269
peptide YY		human	4307
secretin		human	3038
tachykinins			
substance K	SK		1133
neurokinin B[g]			1210
substance P	SP		1347
vasoactive intestinal peptide	VIP	human	3324
vasopressin	ADH		1083

[a] Alternative nomenclature for hypothalamic releasing factors are indicated.
[b] Corticotropin-releasing factor (CRF) = corticoliberin. [c] Growth hormone-releasing factor (GRF) = growth hormone-releasing hormone (GHRH) = somatoliberin. [d] Luteinizing hormone-releasing hormone (LHRH) = gonadotropin-releasing hormone (GnRH) = gonadoliberin-luteinizing hormone-releasing factor (LRF) = luliberin. [e] Somatostatin (SS) = somatotropin-releasing inhibiting factor (SRIF). [f] Thyrotropin-releasing factor (TRF) = thyrotropin-releasing hormone (TRH) = thyroliberin. [g] Neurokinin B = neuromedin K.

reside in several areas of its sequence, although much of its potency lies in the C-terminal 27 amino acids. CRF is localized in several brain areas. The best studied structure is the parvocellular region of the periventricular nucleus of the hypothalamus, where fibers project to the median eminence. Other areas containing CRF include amygdala, lateral hypothalamus, central gray area, dorsal tegmentum, locus coereleus, parabrachial nucleus, dorsal vagal complex, and inferior olive. CRF is colocalized with a number of other peptides, including enkephalin and vasopressin.

CRF is derived from a precursor of 196 amino acids. The principal function of CRF is to regulate the release of POMC-derived peptides, especially ACTH. Because the role of ACTH is to release glucocorticoids, CRF is a significant regulator of pituitary–adrenal function. CRF also regulates several aspects of the immune system. Perhaps most interestingly, from the point of view of brain function, is the suggestion that CRF plays a principal role in the etiology of CNS depression. In this view, CRF causes many of the known signs of clinical depression, including decreased appetite, decreased libido, and insomnia. Such findings suggest that specific CRF antagonists may be valuable tools in the treatment of depression.

There is a definite lack of practical synthetic CRF analogues (ca 1994). The only CRF antagonist available is α-helical CRF 9-41.

Somatostatin. Somatostatin is somatotropin-releasing inhibiting factor (SRIF or SS). Somatostatin is widely distributed throughout the brain and the gastrointestinal tract; it also is found in mammalian plasma, in the rat retina, and in the human adrenal medulla. In the brain, the highest levels are found in the amygdala and hypothalamus, followed by several areas of frontal cortex.

Somatostatin is derived from a precursor containing 116 amino acids. Somatostatin exerts some neurotropic actions, eg, as a tranquilizer and as a spontaneous motor activity depressor. It also lengthens barbiturate anesthesia time and induces sedation and hypothermia.

Substitution of Trp^8 by D-Trp^8 increased the potency of somatostatin. Most other substitutions, however, are deleterious to biological activity.

Hypothalamic-Releasing Peptides. The highest concentration of gonadoliberin (LRF) is found in the hypothalamus; however, it also has been reported to be present in extrahypothalamic central nervous regions, blood, urine, and placenta. LRF acutely stimulates lutenizing hormone (LH) and follicle-stimulating hormone (FSH) secretion. Additionally, there is evidence that LRF can act on the central nervous system to modulate sexual behavior. Potent and long-acting analogues of LRF have been designed.

Thyroliberin's greatest hypothalamic concentration is found in the mammalian median eminence. The principal function of TRF is to stimulate thyroid-stimulating hormone (TSH) release, but it also releases prolactin and growth hormone under specific conditions. TRF has been reported to alleviate depressive symptoms, to reverse the duration of anesthesia and hypothermia induced by a number of substances, and to increase spontaneous motor activity. It is derived from a precursor that contains five separate copies of this decapeptide, flanked as usual by double basic residues. Several hundred analogues to TRF have been synthesized.

Neurotensin. Immunoreactive neurotensin is present in mammalian gut and is distributed throughout the central nervous system; its highest concentration is in the hypothalamus and in the substantia gelatinosa of the spinal cord. Its overall brain distribution is not unlike that of enkephalin.

The many pharmacological actions of neurotensin include hypotension, increased vascular permeability, hyperglycemia, increased intestinal motility, and inhibition of gastric acid secretion. In the brain, it produces analgesia at remarkably low doses. The smallest sequence possessing most of the neurotensin spectrum of activities and its high potency is the hexapeptide C-terminus. [D-Trp^{11}]-Neurotensin acts like a neurotensin antagonist.

Calcitonin Gene-Related Peptide (CGRP). CGRP has a wide distribution in the nervous system. CGRP is derived from a precursor structurally related to the calcitonin precursor.

CGRP promotes release and inhibits degradation of substance P, thus increasing its effects. Like substance P, CGRP is a vasodilator. Finally, CGRP may have trophic actions on motoneurons. Relatively few CGRP analogues are available (ca 1994). One interesting analogue is CGRP 8-37, which appears to be an antagonist at CGRP receptors and an agonist at calcitonin receptors.

Cholecystokinin. Cholecystokinin (CCK) is a 3-amino acid peptide that shares the same carboxy-terminal pentapeptide sequence with gastrin. There are relatively high quantities of CCK in the brain.

CCK has been detected in two principal forms, ie, the traditional 33-amino acid peptide, and an octapeptide CCK-8.

In addition to its actions on the gastrointestinal tract, CCK in the brain has been associated with control of eating behavior. There is also colocalization of CCK with dopamine, which correlates with biochemical studies suggesting that CCK is a neuromodulator of dopamine function.

The CCK system shares one property with the opioid system, ie, the existence of selective nonpeptide antagonists. Selective, potent peptide antagonists for CCK, eg, Cl-988 and PD 134308, have been developed that may be useful as anxiolytics and as drugs which increase the analgesic effect of morphine but at the same time prevent morphine tolerance.

STEVEN R. CHILDERS
Wake Forest University

D. R. Lynch and S. H. Snyder, *Ann. Rev. Biochem.* **55**, 773 (1986).

T. Hokfelt, *Neuron* **7**, 867 (1991).

S. B. H. Kent, *Ann. Rev. Biochem.* **57**, 957 (1988).

R. A. Houghten, C. Pinilla, S. E. Blondelle, J. R. Appel, C. T. Dooley and J. H. Cuervo, *Nature* **354**, 84 (1991).

SEX HORMONES

Progestins are a class of steroids named for their progestational effects, which are essential for the initiation and continuation of pregnancy. The primary progestin in humans is progesterone, a hormone and intermediate in the synthetic pathways of estrogens, androgens, and corticosteroids. Progesterone is synthesized from pregnenolone, $C_{21}H_{32}O_2$, which is derived from the side-chain cleavage of cholesterol.

The primary sources of progesterone in women are the corpora lutea of the ovary and the placenta. Other naturally occurring progestins such as pregnanediol (5α- or 5β-pregnane-3α,20α-diol and 20-dihydroprogesterone (20α- or 20β-dihydroprogesterone) can result from progesterone metabolism in steroid-responsive tissues. These progestins have much weaker biological activity than progesterone and their physiological significance remains unclear. Progesterone has been used as a therapeutic agent, but is rapidly metabolized with little or no oral activity and a short duration of action. In humans, it is metabolized primarily by the liver into the biologically inactive pregnanediol. All of the progestational agents in clinical use (ca 1994) are synthetic steroidal progestins.

In addition to newer progestin agonists, potent progestin antagonists are available and may eventually replace progestin agonists for many therapeutic uses. Once given only orally or by injection, steroids can now be administered in sustained release formulations such as depot injection, vaginal silicone ring, or subcutaneous silicone rods.

Synthesis of Steroidal Agonists

Progestins are derivatives of a planar tetracyclic structure and fall into one of four structural classes: 5α-pregnanes, 5α-androstanes, estranes, and gonanes.

Three methods of obtaining steroids have been developed for the commercial preparation of progestins, ie, chemical degradation of steroids isolated from plants, microbial degradation of steroids isolated from plants, and total chemical synthesis. Advantages of using naturally occurring steroids as raw materials include availability of large quantities and presence of the required tetracyclic ring structure possessing the correct stereochemical orientation. Industrial processes (ca 1994) for the synthesis of progestins make use of all three methods. Most progestins are prepared from a few steroidal precursors.

17-Ethynyl Steroids. Although 17-ethynyl steroids do not represent a separate skeletal class, they are significant in the development of an orally active progestin. Progress toward orally active progestins has been spurred by the observation that ethynyl groups, introduced at the 17α-position of estradiol and testosterone, result in orally active compounds. Ethynylestradiol is the estrogenic component of most oral contraceptives (ca 1994).

Table 1. Physical Properties of Progestins

Common name	Molecular formula	Mol wt	Melting point, °C
allylestrenol	$C_{21}H_{32}O$	300.483	79.5–80
chlormadinone acetate	$C_{23}H_{29}ClO_4$	404.932	211–212
cyproterone acetate	$C_{24}H_{29}ClO_4$	416.943	200–201
desogestrel	$C_{22}H_{30}O$	310.478	109–110
ethynodiol diacetate	$C_{24}H_{32}O_4$	384.514	129–132, 126–127
gestodene	$C_{21}H_{26}O_2$	310.435	198
hydroxyprogesterone	$C_{21}H_{30}O_3$	330.466	221
hydroxyprogesterone acetate	$C_{23}H_{32}O_4$	372.503	246.5
hydroxyprogesterone caproate	$C_{27}H_{40}O_4$	428.611	119–121
levonorgestrel	$C_{21}H_{28}O_2$	312.451	238–242
lynestrenol	$C_{20}H_{28}O$	284.441	162–164
medroxyprogesterone acetate	$C_{24}H_{34}O_4$	386.530	205–209
megestrol acetate	$C_{24}H_{32}O_4$	384.514	214–216
norgestrel	$C_{21}H_{28}O_2$	312.451	205–207
norethindrone	$C_{20}H_{26}O_2$	298.424	203–204
norethindrone acetate	$C_{22}H_{29}O_3$	340.461	161–163
norgestimate	$C_{23}H_{31}NO_3$	369.25	214–218
progesterone	$C_{21}H_{30}O_2$	314.476	121–122, 127–131
promegestone	$C_{22}H_{30}O_2$	326.478	152

Agonists and Antiprogestins

Steroidal Agonists. Many of the progestins were originally prepared from optically active natural products and contain one enantiomer. In general, steroidal progestins are stable, white to off-white crystalline solids. Because of possible instability of the A-ring, it has been recommended that many of them be stored protected from light. Table 1 lists data for a variety of steroidal progestin agonists. A list of progestins used in the United States and Europe is given in Table 2.

Steroidal Antiprogestins. A significant breakthrough in the progestin antagonist area was the discovery of the 11β-substituted progestin antagonists. The first of these compounds to be used clinically was RU 38486, ie, RU 486 or mifepristone. Modifications in the structure of RU 486 have been made in an effort to improve upon the antiprogestational activity of RU 486 while minimizing antiglucocorticoid activity.

The methodology used in the preparation of RU 486 and other 11β steroids is conjugate addition of a cuprate reagent to the α,β-unsaturated epoxide which provides the 11β-substituted steroid stereospecifically. Subsequent steps lead to the synthesis of RU 486 and its analogues.

Nonsteroidal Agonists and Antagonists. There are no nonsteroidal progestin agonists or antagonists in clinical use. Nonsteroidal compounds having affinity for the progesterone receptor or activity typical of such compounds have been reported. A series of acylanilides, which are structurally related to the antiandrogen flutamide, has been reported to possess potent progestational activity and no antiandrogenic activity.

Structure–Activity Studies

An understanding of the ligand–receptor interaction is crucial in the effort to develop more potent and more selective ligands. Ideally, both the structure of the receptor and the activity of a variety of ligands are known, allowing conclusions to be drawn about structure–activity relationships (SAR). Although the amino acid sequence of the progesterone receptor is known, the three-dimensional structure remains unsolved. Techniques designed to identify which amino acids are important for ligand binding have been developed, including site-directed mutagenesis, affinity labeling, and reactions of specific amino acids with chemical reagents. Another approach has been to use binding data and structures of ligands to develop a three-dimensional map of the receptor.

Table 2. Progestins Marketed in the United States and Europe for Contraceptive and Noncontraceptive Uses

	Primary use[a]			Primary use[a]	
Progestin	C	NC	Progestin	C	NC
algestone acetophenide	+		lynestrenol	+	+
allylestrenol		+	medrogestone		+
chlormadinone	+	+	medroxyprogesterone acetate	+	+
chlormadinone acetate	+	+	megestrol acetate		+
cyproterone acetate		+	milbolerone		+
demegestone	+		nomegestrol acetate		+
desogestrel	+		norethindrone	+	+
dydrogesterone		+	norethindrone acetate	+	+
ethisterone		+	norethindrone enanthate	+	+
ethynodiol diacetate	+		norethynodrel		+
gestodene	+		norgestimate	+	
gestrinone		+	norgestrel	+	+
gestronol hexanoate		+	norgestrienone	+	
hydroxyprogesterone hexanoate		+	progesterone	+	+
levonorgestrel	+	+	promegestrone		+

[a] C = contraceptive; NC = noncontraceptive.

The prediction of protein structure through computational techniques has been applied to steroid hormone receptors. Hydrophobic cluster analysis (HCA) has been used to develop a model of the hormone binding domain of steroid receptors. Using HCA as a first approximation, the hormone-binding domains are proposed to be similar to a cleaved portion of human α_1-antitrypsin, the crystal structure of which is known. Another analysis has predicted that the hormone-binding domains of steroid receptors are similar to the substilisin-like serine proteases. Both models await further validation through comparison with biological data and, hopefully, the eventual solution of the three-dimensional structure of the progestin receptor.

Modes of Structural Studies. Determining the three-dimensional structure of relatively small steroids has been accomplished through x-ray crystallography and nuclear magnetic resonance (nmr) spectroscopy. If the structure is unknown or theoretical, or the compound is difficult to crystallize, molecular modeling techniques may be employed to predict the preferred conformation.

Quantitative Structure–Activity Relationships. Many quantitative structure–activity relationship (QSAR) studies of progestins have appeared in the literature and an extensive review of this work is available. QSAR studies attempt to correlate electronic, steric, and/or hydrophobic properties to progestational activity or receptor binding affinity.

Examination by ^{13}C nmr of eight analogues of norethindrone, where structures varied at the C-11, C-18, and $\Delta^{15,16}$ position, suggests a correlation between the ^{13}C-resonance of C-17 and the relative binding affinity (RBA) to the progesterone receptor and to the progestin/androgen RBA ratios. A nonlinear relationship has been found between the biological activity and lipophilicity of esters of norethindrone and levonorgestrel. The relationship between progestational potency and both calculated lipophilicity and steric effects of 13β-substituents has been explored by measuring the activity of the 13β-ethyl and 13β-acetyl pregnanes on uterine endometrial proliferation *in vivo*. Binding energies of progesterone analogues, calculated from their binding affinities for the rabbit uterine progesterone receptor, have also been used to draw conclusions about receptor–ligand interactions. A successful QSAR has been obtained for affinity to the progesterone receptor for 55 progesterone derivatives using the minimal steric, ie, topological, method.

Two correspondence analysis methods have been described. Using the minimum spanning tree method, many test compounds have been found to separate into four main branches corresponding to favorable androgen receptor (AR) binding, progesterone receptor (PR) binding, glucocorticoid receptor (GR) binding, and a nonspecific branch with AR/PR and GR binding. A related method, that of correspondence factoral analysis, also has been developed. This method allows representation of compounds and biological variables on the same two-dimensional graph. The results agree with those found using the minimum spanning tree method with the same compounds showing the highest specificities for the tree receptors.

Mechanism of Action of Progestins and Antiprogestins

Like all steroid hormones, progestins exert their biological effects by altering the rate at which certain genes are expressed in hormone-responsive cells. Steroids bind in a competitive and reversible fashion to hormone-specific receptor proteins which act as transcription factors regulating ribonucleic acid (RNA) synthesis within the cell nucleus. For the progestin receptor, the binding of steroid is an absolute requirement for the receptor to interact with deoxyribonucleic acid (DNA) and initiate gene transcription in intact cells.

The binding of hormone to the receptor triggers a number of changes in the receptor that allow the hormone-receptor complex to bind to DNA and stimulate transcription. The occurrence of this ligand-induced transcriptional activity is termed receptor activation. The binding of progestin to the receptor induces the phosphorylation of serine residues in the receptor, the release of heat shock proteins normally associated with the hormone-free receptor, and the dimerization of activated receptors. The binding of progestin antagonists induces changes in receptor structure and phosphorylation different from those due to progestin agonist binding. A monoclonal antibody directed to the carboxyl terminus of the human progestin receptor can differentiate between agonist- and antagonist-receptor complexes. It is these ligand-induced changes in receptor structure which are believed to be the basis for the biological effects of progestin agonists and antagonists.

Classifications of Progestin Antagonists. Progestin antagonists can be classified based on their ability to form hormone-receptor complexes capable of binding to progesterone response elements on the nuclear DNA. Type I antagonists bind to the receptor but do not induce the receptor activation necessary for DNA binding. Type II antagonists, such as RU 486, bind to the receptor and induce the binding of receptor to DNA. Type I antagonists may be viewed as pure antagonists because, owing to their blockade of receptor-DNA binding, no gene transcription is possible. However, type II antagonists may induce transcription of some genes. Type I antagonists include ZK 98299; type II antagonists such as RU 486 include ZK 112993, ORG 31710, and ORG 31806.

Regulation of Progestin Receptors. The biological response of a tissue to a steroid hormone depends on the number of steroid receptors in the tissue. This makes regulation of receptor concentration an issue in determining the effectiveness of a steroid. Progestin receptors can be regulated by a number of hormones. Estrogens increase the levels of the progestin receptor in most target tissues and cells. In contrast, progestins decrease the number of progestin receptors, increasing the rate of receptor degradation and decreasing the rate of receptor synthesis. Whereas estrogens and progestins are the primary regulators of progestin receptor concentrations, other factors and hormones may also play a role. Insulin, insulin-like growth factor, and epidermal growth factor can increase progestin-receptor concentrations in cultured breast and uterine cells. Luteinizing hormone and follicle stimulating hormone, the pituitary hormones responsible for ovulation, induce progestin receptors in cultured ovarian granulosa cells.

Assay Methods for Evaluating Progestin Agonists and Antagonists. Assay methods include progestin receptor binding assays, cell-based assays measuring cell proliferation and gene transcription, and *in vivo* assays measuring progestational effects at the tissue and whole animal levels.

Effects of Progestins on Various Target Tissues. Progestins exert a variety of direct and indirect effects on all tissues and organs within the reproductive system. Organs that contain progestin receptors, and thus can respond directly to progestins, include the uterus, cervix, vagina, breast, ovary, brain, and pituitary gland. Receptors have also been identified in rat thymus, human lymphocytes, and bone cells. Several factors are responsible for progestin effects including the dose and potency of the drug itself, the duration of treatment, the route of administration, and the ability of the drug to affect progestin-receptor number and estrogen levels.

The uterus is exceptionally sensitive to the effects of estrogens and progestins. Progestins serve to suppress the stimulatory effects of estrogen on uterine growth, an effect termed antiestrogenic. Progestins also act on the uterine myometrium, reducing the frequency of contractions. In nonpregnant women, menstrual bleeding is the most noticeable effect of progestins on the lower reproductive tract. Elsewhere within the lower reproductive tract, progestins have other effects. The quantity of cervical mucus is lessened and the mucus is more viscous under the influence of progestins. The motility of the oviduct is affected by progestins which can act to delay the transport of ova from the ovary to the uterus.

The breast is another tissue sensitive to estrogens and progestins. In concert with estrogens, progestins stimulate mammary gland growth and regulate milk production.

The effects of progestins are not limited to the reproductive system. Progestins exert direct actions on bone osteoblast cells, acting as a bone-forming hormone. Within the brain, progestins modulate neuroendocrine function, mood, and behavior via intracellular nuclear receptors.

Progestins have indirect effects on other nonreproductive systems and organs, effects not necessarily involving progestin receptors. Some synthetic progestins, especially medroxyprogesterone acetate (MPA) and other acetoxyprogesterone derivatives, have sufficient glucocorticoid-like activity to lower adrenocorticotrophin hormone (ACTH) and cortisol levels via a negative feedback effect. Depending on dose and potency, synthetic progestins can impair glucose tolerance and increasing insulin resistance.

There has been great interest in understanding the impact of synthetic progestins on the cardiovascular system and the risk of coronary heart disease in women. Coronary heart disease has been linked with high plasma levels of total cholesterol and low density lipoprotein cholesterol (LDL) and low plasma levels of high density lipoprotein cholesterol (HDL). Estrogens tend to increase plasma levels of cardiac-protective HDL and decrease LDL levels, whereas progestins have opposite effects. At clinically used doses, norgestrel and levonorgestrel have a greater effect on HDL and LDL levels than estrane derivatives such as norethindrone. The newer gonane-based progestins, eg, desogestrel, gestodene, and norgestimate, have minimal or even favorable effects on the LDL/HDL ratio.

Pharmacokinetics and Metabolism of Therapeutic Progestins. Norethindrone is rapidly and completely absorbed after oral administration. There is a significant loss of norethindrone owing to a first-pass effect. Approximately 36% of an oral dose is lost as a result of metabolism within the intestinal wall and liver. In women, 60.8% of norethindrone in the blood is bound to albumin, 35.5% to sex hormone binding globulin (SHBG), and 3.7% is free. The half-life for elimination has been measured in a number of studies in which the values ranged from 3.4 to 13.4 hours, thereby giving an average of approximately 8 hours.

Levonorgestrel is 100% orally absorbed with no first-pass effect. Peak plasma levels are reached between 0.5 and 2 hours. Plasma levels of levonorgestrel decline in a biphasic fashion; the initial phase has a half-life of 50 to 180 minutes followed by a second phase with a half-life of 10 to 26 hours. In plasma, approximately 98% of levonorgestrel circulates bound, ~50% to albumin, 48% to SHBG, and 2% free.

Norgestimate is completely and rapidly absorbed after oral administration with peak plasma levels reached within two hours. Over the course of two weeks following administration of ^{14}C-norgestimate, 35–49% is excreted in the urine and 16–49% in the feces.

Gestodene is completely absorbed after oral administration with no first-pass metabolism. The half-life of elimination is approximately 20 hours. Repeated daily dosing results in steady-state serum levels after about 10 days. These levels are approximately four to five times those following a single dose, due to an increase in serum SHBG. The vast majority of gestodene in the serum is protein-bound. At steady-state conditions, about 75% of gestodene is bound to SHBG and 24% to albumin. In contrast, gestodene is 50% SHBG-bound and 48% albumin-bound after a single dose. Regardless of the frequency of administration, less than 1–2% of circulating gestodene is free and biologically active.

Desogestrel is rapidly and nearly completely converted to the biologically active metabolite, 3-ketodesogestrel. The oral administration of equal doses of desogestrel or 3-ketodesogestrel produces nearly identical serum concentrations of 3-ketodesogestrel. The affinity of 3-ketodesogestrel for the progestin receptor in human uterine myometrium is approximately six times greater than that of progesterone and 57 times greater than that of desogestrel. 3-Ketodesogestrel binds extensively to albumin and SHBG; only approximately 2.5% of 3-ketodesogestrel is free in circulation.

Accurate pharmacokinetic and metabolism studies on MPA have been difficult because the radioimmunoassays employed cannot differentiate between medroxyprogesterone acetate (MPA) and its metabolites. Values of the mean elimination half-life of MPA were similar, being 33.8 and 39.7 hours when measured by hplc and radioimmunoassay, respectively. Approximately 94% of MPA in the blood is bound to albumin. When taken orally, MPA is rapidly absorbed with little or no first-pass metabolism. Peak serum levels are reached after three hours. Steady state occurs after three days of daily administration.

After oral administration, peak plasma levels of mifepristone (RU 486) are reached in one hour and over 95% is bound to plasma proteins. The plasma half-life of RU 486 is approximately 24 h.

JOSEPH W. GUNNET
LISA A. DIXON
The R. W. Johnson Pharmaceutical Research Institute

R. A. Hill, D. N. Kirk, H. L. J. Makin, and G. M. Murphy, eds., *Dictionary of Steroids*, Chapman and Hall, London, 1991.

J. W. Goldzieher and K. Fotherby, eds., *Pharmacology of the Contraceptive Steroids*, Raven Press, New York, 1994.

T. Ojasoo, J.-C. Doré, J.-P. Mornon, and J.-P. Raynaud, in M. Bohl and W. M. Duax, eds., *Molecular Structure and Biological Activity of Steroids*, CRC Press, Boca Raton, 1992, Fla., 1992, p. 157.

M. K. Agarwal, ed., *Antihormones in Health and Disease*, S. Karger, Basel, Switzerland, 1991.

ESTROGENS AND ANTIESTROGENS

Estrogens are a group of naturally occurring steroid sex hormones which are characterized by their ability to induce estrus in the female mammal. They are derivatives of the planar tetracyclic structure estra-1,3,5(10)-trien-3-ol, and the three principal estrogens in humans are estrone (E_1), estradiol (E_2), and estriol (E_3). The two synthetic steroidal estrogens which have attained the greatest degree of therapeutic use are ethinyl estradiol (EE) and its 3-methyl ether, mestranol. In contrast to the naturally occurring estrone derivatives, these acetylenic analogues are orally active and are the main estrogenic components of combination oral contraceptives and certain estrogen replacement products.

Diethylstilbestrol (DES), which was first synthesized in the 1930s, is the most widely studied nonsteroidal estrogen and has been extensively reviewed. It is an extremely potent estrogen, possessing four times the oral potency of estradiol, but carcinogenicity problems have limited its use. For the purposes of this article, antiestrogens are compounds that counteract the biological activity of estrogens at the receptor level.

Chemistry

Steroidal and Nonsteroidal Estrogens. Modification of the basic steroid skeleton and the nature of the functional groups in the B, C, and D rings while maintaining the phenolic A-ring has continued to be a primary approach in the development of new estrogens with unique biological profiles. A series of patents from Schering AG has described the synthesis of various 14–17 α- and β-ethano-bridged estratriene compounds. All are reported to be potent estrogens. A series of 17-halomethylene estratrienes have also been reported by Schering AG to be potent estrogens.

11 β-Nitrate Esters. It was discovered in the late 1930s that introduction of a 17 α-alkynyl group in estradiol gave orally active estrogens such as EE (ethinyl estradiol), which have been widely employed with synthetic progestins (pharmaceutical agents which have effects similar to progesterone) in oral contraceptives. The 17-alkynyl group is more resistant to metabolism in the liver than are the naturally occurring estrogens, and attention has refocused on the search for new orally active super estrogens which could be used at lower doses and theoretically reduce the metabolic burden. As was the case with the hormone antagonists, introduction of functionality at the 11-position of the estrane skeleton has resulted in the generation of orally active potent estrogen agonists. Agonists are drugs that stimulate activity at cell receptors normally stimulated by naturally occurring substances. Various estrane derivatives have been converted with ceric ammonium nitrate selectively and efficiently to the corresponding $9\alpha,11\beta$-hydroxy nitrate esters which were then deoxygenated at C-9 with triethylsilane—boron trifluoride etherate to yield the desired 11β-nitrate esters; standard transformations then gave the 7α-methyl target compounds such as (**1**).

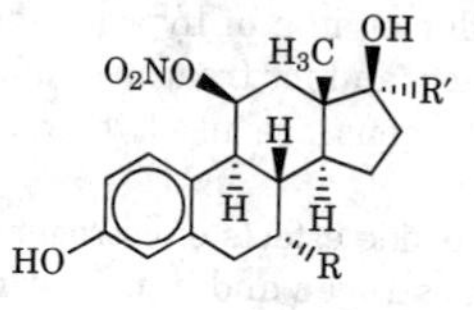

(**1**)

where R′ = R = H for CDB-3280 R′ = C≡CH; R = H for CDB-3294 R′ = H; R = CH_3 for CDB-1357 R′ = C≡CH; R = CH_3 for CDB-3322

C-11 functionalized estrone derivatives are synthetically available by the following process: treatment of 7α-methylestrone acetate (**2**) with four equivalents of ceric ammonium nitrate in 90% acetic acid provides the 9α-hydroxy-11β-nitrate ester in good yield. Subsequent deoxygenation of the C-9 benzylic position of the nitrate ester with retention of configuration is effected utilizing triethylsilane and boron trifluoride etherate to produce compound (**3**). Reduction with $NaBH_4$ affords the key target (**1**). A nearly identical approach has been used to prepare (**3**) which was converted to (**1**) and also ethinylated at C-17 with acetylene and potassium *t*-butoxide to give CDB-3322.

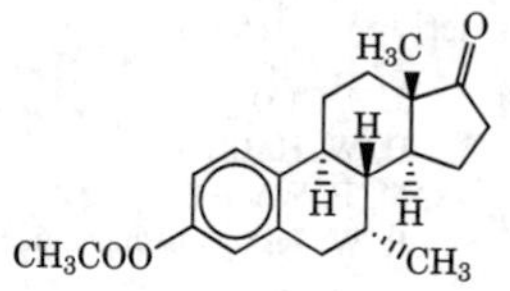

(**2**)

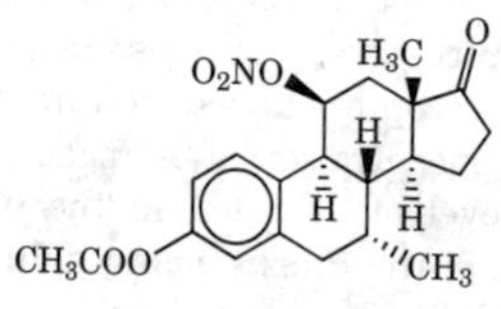

(**3**)

Steroidal Antiestrogens. The balance of estrogenic and antiestrogenic activity expressed by the nonsteroidal antiestrogens varies widely across species, target organs, cells, and genes, depending on which indicator of response is measured. Tamoxifen, Nolvadex (ICI 46474), is a synthetic nonsteroidal antiestrogen which has been used for the control of hormone-sensitive breast cancer. However, tamoxifen is a partial estrogen agonist and as such has estrogen-like stimulatory activity on the uterus, vagina, mammary glands, and the pituitary–ovarian axis in animals. Attempts to synthesize nonsteroidal antiestrogens devoid of partial estrogen agonist activity have met with limited success. The emphasis has returned to the synthesis of steroidal antiestrogens. The ICI compounds, ICI 164384, and ICI 182780 are examples of "pure" antagonists.

Based on the potent antagonist activity of the above ICI compounds and the observations that halogenation of the 16α-position of the D-ring in the steroid nucleus often leads to compounds with increased affinity for the estrogen receptor, a series of 7α-undecanamide-substituted 17β-estradiols with 16-halogen substituents were synthesized. These 16-halo-7-alkylamide antiestrogens, characterized by EM-139 and EM-170 demonstrated potent and pure antagonistic activity *in vivo* in screens where tamoxifen exhibited estrogenic activity and was only a weak partial antiestrogen. The synthesis of EM-139 and related 16α-halosteroids has been carried out utilizing an enol acetate as a key intermediate.

Related 7-substituted 19-norsteroids have appeared in the patent literature as potential antiestrogens including a $14\alpha,17\alpha$-ethanoestratriene (Schering AG) and a 7-dimethylaminoethoxyphenyl analogue (Roussel-Uclaf).

The 11-position of estradiol analogues has been a fruitful site of exploration in the development of hormone antagonists, eg, the antiprogestin RU 486. The Roussel group has also uncovered novel antiestrogens by investigating various substituents at the 11β-position, eg, a bis-dimethylaminophenyl compound and a phenoxyoctanamide.

Nonsteroidal Antiestrogens. The first generation of nonsteroidal antiestrogens all demonstrated some degree of agonist activity. The search for compounds with improved specificity has continued and include triarylethylenes (TAEs) (toremifene, 4-iodotamoxifen, 4-bromotaxomifen, nitromifene), phenylhydrazones (compound A-007 (DEKK-TEC), chromenes (centchroman), benzopyran derivatives (CDRI-85/287), benzofuran ring analogues (morpholino derivatives, 1,3,8-trichloro-6-methyldibenzofuran (MCDF)), benzothiophenes (LY 117018 and LY 156758), carbocyclic triarylethylenes (dihydronaphthalene derivatives, benzo[*a*]fluorenes), and diphenylmethanes and diphenylethanes (CGS-20267, D-18954).

Modeling and Crystallographic Studies of Estrogen Agonists and Antagonists

There have been a number of studies that have attempted to describe the mechanism of action of estrogen agonists and antagonists at the molecular level based on binding studies, molecular modeling strategies, conformational analyses, and similar strategies. This is an area where limited progress has been achieved owing primarily to the lack of potent specific antagonists which totally block the action of endogenous and exogenous steroids.

Pharmacology

Mechanism of Estrogen Action. Estrogen molecules are lipophilic and diffuse through the plasma membrane of all cells. The steroid ligands encounter their specific receptors only within target cells. In cells lacking estrogen receptors, estrogens are not readily retained and exit the cell. Estrogen receptors appear in target tissues before ovary maturation, and the concentration of estrogen receptor in the uterus correlates with the level of estrogen in the blood. The number of estrogen-binding sites in the uterus changes during the estrual cycle from a minimum of 1000 sites per cell during estrus (ovulation), increasing to 3500 sites per cell and reaching a maximum at proestrus, which is immediately prior to the next cycle, of 5000 sites per cell.

The estrogen receptors are large protein molecules that are mainly localized in the cell nucleus. The receptor contains three principal domains: an estrogen-binding site at the C-terminal region, a DNA-binding domain in the middle of the protein molecule which is capable of binding to specific regions within the target genes called estrogen response elements, and a modulating domain at the N-terminus. Recent evidence showed that the estrogen receptor is bound to specific estrogen response elements of a variety of genes with or without estrogen.

Estrogen exerts hormonal effects by first binding with high affinity to the receptor to form an estrogen receptor complex. The hormone binding then induces conformational changes in the steroid-binding domain and other domains of the receptor, a process which is referred to as receptor activation or transformation. The transformed or activated estrogen receptor complex alters the interaction with target genes, leading to an increase of the affinity to DNA and other nuclear components. As a result, estrogen induces changes in the activity of transcription machinery associated with the target genes, which in turn modulates the expression of target genes and further regulates cell function, growth, or differentiation.

Out of the trillions of cells in the human body, only special cell types or tissues elicit biochemical or physiologic responses to estrogen. These target cells possess estrogen receptors, although the receptor concentration varies between cell types. The responses to estrogen of target cells are also divergent. The mechanism for the cell- or tissue-specificity of estrogen response is still under active investigation. However, several elements are considered to modulate the nature of cell- or tissue-specific effects of estrogen following estrogen receptor binding, eg, specific target genes and the associated gene network in each type of cell; cell-specific regulating factors for transcription of the genes, such as regulatory sequences, transcriptional factors, and local chromatin conformation of the genes and nuclear matrix; and the relative response of the cell to other synergistic molecules of estrogen action.

The presence of estrogen-receptor and estrogen-target genes containing estrogen response elements as well as appropriate transcription machinery determine the cell or tissue's competence to respond to estrogen. Factors such as the cell- or tissue-specific array of target genes, the associated transcription regulatory factors, the chromatin structure, and nuclear matrix surrounding the genes allow different target cells to elicit diverse cell-specific responses to a certain ligand. The molecular features of different ligands affect the stability of ligand-receptor binding and the interaction of ligand-receptor complex with DNA and transcription machinery which in turn governs the nature of induced estrogen responses, ie, active or latent, strong or weak, being agonist in one tissue and being antagonist in another tissue.

Biosynthesis. Natural estrogens are produced by steroidogenesis in various tissues. The ovary is the primary source of the hormone in nonpregnant women. Estradiol is the most potent and primary product of the ovary, although the organ also produces estrone.

Metabolism and Distribution. Estrogens are readily absorbed through the gastrointestinal (GI) tract. During this process the unconjugated estrogens are converted primarily to estrone. Therefore, estrogens, if taken orally, cause increased serum estrone levels. Both endogenous and exogenous estrogens are metabolized similarly. Maximal serum estrogen levels after oral ingestion are reached in 4–6 h. Inactivation of estrogen in the body is carried out mainly in the liver.

The absorption of estrogen is efficient through other modes of delivery (transdermal, vaginal, nasal, or intramuscular). Various vehicles by which estrogen is administered give different rates of absorption. Circulating estrogens are tightly conjugated with sex hormone-binding globulin and weakly bound to albumin.

Pharmacological Effects. Three principal natural estrogens (E_1, E_2, and E_3) are produced by the ovary and play important roles in the development and support of female reproduction. The normal menstrual cycle is modulated by a variety of ovarian steroids and peptides, among which estrogen is a primary regulator.

Estrogens stimulate cellular proliferation, induce RNA and protein synthesis of uterine endometrium and the fibrous connective tissue framework for ovaries, and increase the size of the cells. This effect leads to the growth and regeneration of the endometrial layer and spiral arterioles, and increase in the number and size of endometrial glands. Under the influence of estrogen, vaginal mucosa becomes thicker, as cervical mucus becomes thinner.

Breast development is initiated by estrogens with both ductal and stromal growth, resulting in breast enlargement. Estrogens also promote body hair and female distribution of fat in the breasts, buttocks, and thighs. Estrogens modulate bone growth in a biphasic manner. The positive effects of estrogens on salt and water retention usually cause edema and decrease of bowel motility. Estrogens also stimulate the synthesis and secretion of prolactin in pituitary lactotrophic cells.

Estrogens have an influence on hepatic metabolism. The production of sex hormone-binding globulin, thyroxine-binding globulin, blood-clotting factors (VII to X) and plasminogen in the liver is stimulated by estrogens. Estrogens promote the production of high density lipoprotein (HDL), especially HDL2 and its apolipoproteins A1 and A2, in liver. Contrarily, estrogens inhibit the hepatic formation of low density lipoprotein (LDL). These effects vary with different types and doses of estrogens and the route of administration.

Therapeutic Uses

Estrogens, along with progestins, are used to suppress ovulation for fertility control in the form of contraceptives (see CONTRACEPTIVES). Estrogens are applied to treat gonadal failure, to induce and maintain secondary sexual characteristics due to inadequate production of ovarian steroids in hypogonadal individuals. Estrogens are also indicated for hormone replacement therapy in postmenopausal women, for the preparation of the endometrium of hypogonadal women before donor egg and embryo transfer, and for treatment of breast cancer.

Chemotherapy. Estrogens are used to treat prostate cancer and breast cancer.

Adverse Effects. The side effects of estrogens vary according to the type and dosage of the estrogen and if progestins are co-administered. The most common side effect is nausea. Breast enlargement, enlargement of endometrial tissue, and intermenstrual bleeding are other common side effects. Low dose estrogens, such as multiphasic oral contraceptives which contain 30–35 μg of estrogen and new versions of progestogen, significantly reduce intermenstrual bleeding.

Epidemiological studies of breast cancer have suggested that in the presence of other cocarcinogens such as virus, chemicals, and radiation, estrogen could induce breast cancer. The causation of breast cancer also include genetic factors. Clinical, biological, and epidemiological data indicate that exogenous estrogens are linked to endometrial cancer. A series of case-control and follow-up studies have linked "high dose" oral contraceptive use to cervical intraepithelial neoplasia and frankly invasive cervical carcinoma. High dosages of oral estrogens have been reported to increase the risk for jaundice, cholestatic hepatitis, gallstones, and hepatic vein blood clots. Estrogens promote the development of hepatic neoplasms associated with increased hepatic cell regenerative activity.

J. Z. GUO
D. W. HAHN
M. P. WACHTER
R. W. Johnson Pharmaceutical Research Institute

A. E. Wakeling and J. Bowler, *J. Steroid Biochem.* **30**, 141 (1988).

J. Gorski and co-workers, *Biol. Reprod.* **48**, 8 (1993).

J. H. Clark, W. T. Schrader, and B. W. O'Malley, in J. D. Wilson and D. W. Foster, eds., *Textbook of Endocrinology*, 8th ed., W. B. Saunders, Philadelphia, Pa., 1991.

D. W. Hahn, A. Phillips, and J. L. McGuire, in *Aktuelle Aspekte der Hormonalen Kontrazeption*, P. J. Keller, ed., Karger, New York, 1991.

HYDANTOIN AND ITS DERIVATIVES

Hydantoin is an accepted name for 2,4-imidazolidinedione. This ring system rarely occurs in nature, although some natural products with hydantoin substructures are known. A huge number of derivatives have been prepared.

Physical Properties

Hydantoins are crystalline solids with high melting points, particularly those compounds in which nitrogen is unsubstituted, because this allows intermolecular association by hydrogen bonds. Hydantoins are weak acids which dissociate at the imidic N − 3—H atom because this allows more efficient delocalization of the negative charge than ionization at N-1.

Several structure–acidity relationships have been established for hydantoin derivatives. Thus, ionization is known to be unaffected by alkyl substituents at N-1 and at C-5. However, aryl and other electron withdrawing groups can considerably enhance the acidity of hydantoins. Introduction of an arylmethylene side chain at C-5 increases the acidity of the N-1 hydrogen, making it measurable. This is due to delocalization of the negative charge at N-1 into the C-5 substituent.

Solvent variation can greatly affect the acidity of hydantoins. Water provides a better stabilization for the hydantoin anion and hence an increased acidity when compared to DMSO. 2-Thiohydantoin (pK_a 8.5) is a slightly stronger acid than hydantoin (pK_a 9.0). 4-Thiohydantoins appear to be weaker acids.

Spectral Properties. Hydantoin derivatives show weak absorption in the uv-visible region, unless a part of the molecule other than the imidazolidinedione ring behaves as a chromophore; however, pK_a values have been determined by spectrophotometry in favorable cases. Absorption of uv by thiohydantoins is more intense. Several pK_a values of thiohydantoins have been determined by uv-visible spectrophotometry.

Chemical Properties Hydantoins can react with electrophiles at both nitrogen atoms and at C-5. The electrophilic carbonyl groups can be attacked by nucleophiles, leading to hydrolysis of the ring or to partial or total reduction of the carbonyl system. Other reactions are possible, including photochemical cleavage of the ring.

Synthesis

Synthesis From α-Amino Acids and Related Compounds. Addition of cyanates, isocyanates, and urea derivatives to α-amino acids yields hydantoin precursors. This method is called the Read synthesis, and can be considered as the reverse of hydantoin hydrolysis. Thus the reaction of α-amino acids with alkaline cyanates affords hydantoic acids, which cyclize to hydantoins in an acidic medium.

In a modification of the original method, Read replaced α-amino acids with α-amino nitriles. Chlorosulfonyl isocyanate is an excellent alternative to alkaline cyanates in the preparation of hydantoins from sterically hindered or labile amino nitriles.

Substitution of alkaline cyanates by isocyanates allows the preparation of 3-substituted hydantoins, both from amino acids and amino nitriles.

A variety of α-amino acid derivatives, including the acids themselves, halides, esters, and amides can be transformed into hydantoins by condensation with urea. α-Hydroxy acids and their nitriles give a similar reaction.

Synthesis From Aldehydes and Ketones. Treatment of aldehydes and ketones with potassium cyanide and ammonium carbonate gives hydantoins in a one-pot procedure (Bucherer-Bergs reaction) that proceeds through a complex mechanism. Some derivatives, like oximes, semicarbazones, thiosemicarbazones, and others, are also suitable starting materials.

Synthesis From Thiohydantoins. A modification of the Bucherer-Bergs reaction consisting of treatment of an aldehyde or ketone with carbon disulfide, ammonium chloride, and sodium cyanide affords 2,4-dithiohydantoins. 4-Thiohydantoins are available from reaction of amino nitriles with carbon oxysulfide. Both thiocompounds can be transformed into hydantoins.

Health and Safety Factors (Toxicology)

The acute toxicity of hydantoin derivatives seems to be low. Most studies on long-term toxicity of hydantoins deal with phenytoin, due to the wide use of this compound as an anticonvulsant. Long-term toxic effects of phenytoin include folate deficiency due to impaired folate absorption, hypocalcemia and osteomalacia, alterations of carbohydrate metabolism, gingival hyperplasia, and teratogenic effects. These are grouped under the term fetal hydantoin syndrome, which is associated with continued use of hydantoins during the early stages of pregnancy, and consists of mild retardation of physical and mental indexes, dysmorphic faces and, occasionally, cleft palate, cleft lip, and cardiac defects. Formation of cyanide by degradation of hydantoin derivatives used as antiseptics for water treatment has been described, and this fact might have toxicological relevance.

Applications

Halogenated Hydantoins. Halogenation has been achieved by use of a variety of halogenating reagents. These derivatives are employed as reagents in synthesis and analysis and also as disinfectants and biocides in water treatment.

N-Methylolhydantoins. 1,3-Bis(hydroxymethyl)-5,5-dimethylhydantoin is used extensively as a preservative in cosmetic and industrial applications, and carries EPA registration for the industrial segment. 1-Hydroxymethyl-5,5-dimethylhydantoin is used as an odorless donor of formaldehyde for adhesive applications.

Epoxy Resins. Urethane and ester-extended hydantoin epoxy resins cured with several compounds seem to have better properties than the previous ones.

5-Substituted Hydantoins. 5-Methylhydantoin has been elected from several structures as a formaldehyde scavenger for color photosensitive materials and water-thinned inks and coatings. Although the hydantoin ring itself does not present any medicinal activity, many 5,5-disubstituted hydantoins have shown interesting biological properties, and some of them are used in medicine, particularly as anticonvulsants. Several types of hydantoin derivatives find use as herbicides.

CARMEN AVENDAÑO
J. CARLOS MENÉNDEZ
Universidad Complutense

E. Ware, *Chem. Rev.* **46**, 403 (1950).

E. S. Schipper and A. R. Day, in R. C. Elderfield, ed., *Heterocyclic Compounds*, Vol. 5, John Wiley & Sons, Inc., New York, 1957.

C. Avendaño and G. G. Trigo, *Adv. Heterocycl. Chem.* **38**, 177 (1985).

J. Philip, I. J. Holcomb, and S. A. Furasi, in K. Florey, ed., *Analytical Profiles of Drug Substances*, Vol. 13, Academic Press, Orlando, Fla., 1984.

HYDRAULIC FLUIDS

The moving parts of many industrial machines are actuated by fluid that is under pressure. A system used to apply the fluid can consist of a reservoir, a motor-driven pump, control valves, a fluid motor, and piping to connect these units, eg, a hydraulic system. Generally, petroleum lubricating oils, and sometimes water are used as the pressure-transmitting or hydraulic fluids. Lubricating oil is not only suitable for pressure transmission and controlled flow, but it also minimizes friction and wear of moving parts and protects ferrous surfaces from rusting.

Hydraulic actuation is based on Pascal's discovery that pressure which has developed in a fluid acts equally and in all directions throughout the fluid and behaves as a hydraulic lever or force multiplier.

Viscosity Classification

The viscosity classification for hydraulic fluids and other industrial liquid lubricants is defined by ASTM D2422 (ISO STD 3448), which establishes 18 viscosity grades in the range of 2–1500 mm^2/s (= cS) covering approximately the range from kerosene to cylinder oils. Classification is based on the principle that the midpoint kinematic viscosity of each grade should be about 50% higher than that of the preceding one.

Different fluids have different rates of change in viscosity with temperature. The viscosity index (VI), a method of applying a numerical value to this rate of change, is based on a comparison with the relative rates of change of two arbitrarily selected types of oils that differ widely in this characteristic. A high VI indicates a relatively low rate of change of viscosity with temperature; a low VI indicates a relatively high rate of change of viscosity with temperature.

Types

Antiwear premium hydraulic fluids represent the largest volume of hydraulic fluids used. Shortly after their introduction in 1960, a second product group was formulated, characterized by the same antiwear characteristics but having lower pour points and higher viscosity indexes. These were formulated for use in mobile and marine applications subject to temperature extremes.

The largest volume of hydraulic fluids are mineral oils containing additives to meet specific requirements. These fluids comprise over 80% of the world demand [ca 3.6×10^9 L (944×10^9 gal)]. In contrast world demand for fire-resistant fluids is only about 5% of the total industrial fluid market. Fire-resistant fluids are classified as high water-base fluids, water-in-oil emulsions, glycols, and phosphate esters. Polyolesters having shear-stable mist suppressant also meet some fire-resistant tests.

Synthetic Fluids

The starting materials for synthetic lubricants are synthetic base stocks, often manufactured from petroleum, made by synthesizing compounds which have adequate viscosity for use as lubricants. The primary performance features of synthetic lubricants are outstanding flow characteristics at extremely low temperatures and stability at extremely high temperatures. Synthesized hydrocarbons, organic esters, polyglycols, and phosphate esters account for over 90% of the volume of synthetic lubricant bases in use. Other synthetic lubricating fluids include a number of materials that generally are used in low volumes, eg, silicones, silicate esters, and halogenated fluids.

Additives

The additives most commonly used in hydraulic fluids include pour-point depressants, viscosity index improvers, defoamers, oxidation inhibitors, rust and corrosion inhibitors, and antiwear compounds.

Properties

Hydraulic fluid functions include transmitting pressure and energy; sealing close-clearance parts against leakage; minimizing wear and friction in bearings and between sliding surfaces in pumps, valves, cylinders, etc; removing heat; flushing away dirt, wear particles, etc; and protecting surfaces against rusting. The hydraulic-fluid properties that are used to characterize a suitable product and ASTM test designations are

Property	ASTM test designation
specific gravity	D1298
pour point	D97
flash point	D92
kinematic viscosity	D445
viscosity index	D2270
color, ASTM	D1500
acid number	D664 or 974
rust inhibition	D665
foaming characteristics	D892
oxidation stability	D943
hydrolytic stability	D2619
lubricity testing	
four-ball method	D2266
vane pump wear test	D2882
FZG method	D5182
emulsion characteristics	D1401
water content	D1744

Environmental Aspects

Developments in hydraulic fluids are driven by environmental concerns including disposal of waste, waste minimization, biotoxicity, effects on human health, and the ecology.

Used oil disposal trends include waste minimization such as by reclaiming used fluid on site, as well as recycling of mineral oil lubricants instead of disposing by incineration.

Human and environmental welfare for lubricants and their use is addressed in Material Safety Data Sheets (MSDS). These MSDS address toxicology and health concerns based on the components in the lubricant as well as indicating the proper response in case of a spill. Environmental hazards of the lubricant are covered on European and Japanese MSDS.

Changes in fluid compositions include the reduction and removal of zinc from hydraulic fluids.

Vegetable and seed oils as well as some synthetic basestocks present a new class of biodegradable basestocks. Aquatic toxicity criteria toward fish is found to be acceptable for this class of fluids as measured by EPA 560/6-82-002 and OECD 203:1-12. Biodegradable hydraulic fluids are typically made from canola oil (rapeseed oil) or sunflower oil and contain performance additives for antiwear, demulsibility, etc.

Economic Aspects

Hydraulic fluids are the second largest use of lubricants for automotive and industrial markets. Estimates for 1992 are that 1.089×10^9 L (81×10^6 gal) of hydraulic fluids were sold out of 8.9×10^9 L (2.3×10^9 gal) of total industrial lubricating fluids. The world market is shown in Table 1. Most hydraulic fluids were mineral oil-based products. The remainder represented principally fire-resistant hydraulic fluids and synthetic-based lubricants.

Uses

Hydraulic actuation is applied to machine tools, presses, draw benches, jacks, and elevators as well as to die-casting, plastic-molding,

Table 1. Geographical Marketing of Hydraulic Fluids

Geographical area	Sales, %: Hydraulic fluids	Sales, %: Industrial lubes[a]
North America	29	26
Western Europe	22	27
Central and Eastern Europe	26	20
Far and Middle East	23	27

[a] South America and Africa also have about 4% of the industrial lubricant market.

welding, coal-mining, and tube-reducing machines. Hydraulic loading is used for pressure, sugar-mill, and paper-machine press rolls, and calender stacks.

Hydrostatic Transmissions. The most recent use of hydraulic power has been in hydrostatic transmissions which are used in many self-propelled harvesting machines and garden tractors and in large tractors and construction machines. Applications in trucks for highway operation also are being developed.

Electrorheological Fluids. Electrorheological fluids are a newer category of hydraulic fluids being actively pursued for use in shock absorbers. An electric field causes the fluid to thicken.

Fire-Resistant Hydraulic Fluids. Fire-resistant hydraulic fluids are used where the fluid could spray or drip form a break or leak onto a source of ignition, eg, a pot of molten metal or a gas flame. Conditions such as these exist in die-casting machines or in presses located near furnaces.

Synthetic Lubricants. Some of the primary applications for synthetic lubricants include the following.

Field of service	Synthetic fluids used
industrial	
circulating oils	polyglycols, synthetic hydrocarbon fluid (SHF), organic esters
gear lubricants	polyglycols, SHF
hydraulic fluids (fire-resistant)	phosphate esters, polyglycols
compressor oils	polyglycols, organic esters, SHF
gas turbine oils	SHF, organic esters
greases	SHF
automotive	
passenger car engine oils	SHF, organic esters
commercial engine oils	SHF, organic esters
gear lubricants	SHF
brake fluids	polyglycols
aviation	
gas turbines	organic esters
hydraulic fluids	SHF, phosphate esters, silicones
greases	silicones, organic esters, SHF

ALAN D. DENNISTON
Unocal Corporation

J. G. Wills, *Lubrication Fundamentals*, Marcel Dekker, Inc., New York, 1980.

Lubrication **78**(4) (1992).

1992 Report on U.S. Lubricating Oil Sales, National Petroleum Refiners Association, Washington, D.C., 1993.

E. R. Booser, *CRC Handbook of Lubrication*, Vol. 1, Boca Raton, Fla., 1983.

HYDRAULIC SEPARATION. See MINERAL RECOVERY AND PROCESSING.

HYDRAZINE AND ITS DERIVATIVES

Hydrazine (diamide), N_2H_4, a colorless liquid having an ammoniacal odor, is thesimplest diamine and unique in its class because of the N—N bond.

Hydrazine and its simple methyl and dimethyl derivatives haveendothermic heats of formation and high heats of combustion. Hencethese compounds are used as rocket fuels. Other derivatives are used as gas generators and explosives. Hydrazine, a base slightly weaker than ammonia, forms a series of useful salts. As a strong reducing agent, hydrazine is used for corrosion control in boilers and hot-water heating systems; also for metal plating, reduction of noble-metal catalysts, and hydrogenation of unsaturated bonds in organic compounds. Hydrazine is also an oxidizing agent under suitable conditions. Having two active nucleophilic nitrogens and four replaceable hydrogens, hydrazine is the starting material for many derivatives, among them foaming agents for plastics, antioxidants, polymers, polymer cross-linkers and chain-extenders, as well as fungicides, herbicides, plant-growth regulators, and pharmaceuticals. Hydrazine is also a good ligand; numerous complexes have been studied. Many heterocyclics are based on hydrazine, where the rings contain from one to four nitrogen atoms as well as other heteroatoms.

The many advantageous properties of hydrazine assure continued-commercial utility. Hydrazine is available in anhydrous form as well as aqueous solutions, typically 35, 51.2, 54.4, and 64 wt % N_2H_4 (54.7, 80, 85, and 100% hydrazine hydrate).

Physical Properties

Anhydrous hydrazine is a colorless, hygroscopic liquid having a musty ammoniacal odor. It fumes in air owing to the absorption of water and perhaps also of carbon dioxide, forming carbazic acid, $CH_4N_2O_2$. Hydrazine is miscible with water, alcohol, amines, and liquid ammonia, but has only limited solubility in other solvents. Its physical properties are more like the isoelectronic hydroxylamine or hydrogen peroxide rather than ethane, owing to hydrogen bonding, which is exemplified in relatively high melting (2°C) and boiling (113.5°C) points, as well as an abnormally high (39.079 kJ/mol (9.340 kcal/mol)) heat of vaporization as compared to 14.64 kJ/mol (3.50 kcal/mol) for the isoelectronic ethane.

Chemical Properties

Thermal Decomposition. Hydrazine is a high energy compound having a high positive heat of formation; however, elevated (>200°C) temperatures are needed before appreciable decomposition occurs. The decomposition temperature is lowered significantly by many catalysts, particularly copper, cobalt, molybdenum, ruthenium, iridium, and their oxides. Iron oxides (rust) also catalyze decomposition.

Acid–Base Reactions. Anhydrous hydrazine undergoes self-ionization to a slight extent, forming the hydrazinium, $N_2H_5^+$, and the hydrazine, $N_2H_3^-$, ions:

$$2\ N_2H_4 \rightleftarrows N_2H_5^+ + N_2H_3^-\ K_i = 10^{-25}$$

Hydrazinium salts, $N_2H_5^+X^-$, are acids in anhydrous hydrazine, metallic hydrazides, $M^+N_2H_3^-$, are bases.

Hydrazine as Nucleophile. Reaction of hydrazine and carbon dioxide or carbon disulfide gives, respectively, hydrazinecarboxylic acid, $NH_2NHCOOH$, and hydrazinecarbodithioic acid, $NH_2NHCSSH$, in the form of the hydrazinium salts. These compounds are useful starting materials for further synthesis.

Reductions. Hydrazine is a very strong reducing agent. In the presence of oxygen and peroxides, it yields primarily nitrogen and water with more or less ammonia and hydrazoic acid. It is used in metal reductions, hydrogenations, carbonyl reductions, catalytic hydrogenations, diazene reductions, aldehyde syntheses, and olefin syntheses.

Alkylhydrazines. Mono- and higher substituted alkyl hydrazines can be made by alkylation of hydrazine using alkyl halides.

Substituted alkyl hydrazines are prepared from suitable alkylating agents. Epoxides yield hydroxyalkylhydrazines; aziridines give β-aminoalkylhydrazines; sultones yield ω-sulfoalkylhydrazines; and acrylonitrile, β-cyanoethylhydrazine.

A general synthesis for arylhydrazines is via diazotization of aromatic amines, followed by reduction of the resulting diazonium salt.

Hydrazides and Related Compounds. Substitution of the hydroxyl group in carboxylic acids with a hydrazino moiety gives carboxylic acid hydrazides. In this formal sense, a number of related compounds fall within this product class although they are not necessarily prepared this way. Some of the more common of these compounds include thiohydrazides, sulfonylhydrazides, semicarbazide, thiosemicarbazide, carbohydrazide, thiocarbohydrazide, amidrazones, hydrazidines, aminoguanidine, diaminoguanidine, and triaminoguanidine. Carboxylic acid hydrazides are prepared from aqueous hydrazine and the carboxylic acid, ester, amide, anhydride, or halide. The reaction usually goes poorly with the free acid.

Hydrazones and Azines. Depending on reaction conditions, hydrazines react with aldehydes and ketones to give hydrazones, azines, and diaziridines. Hydrazones are formed from mono- and *N,N*-disubstituted hydrazines. Many of these compounds are highly colored and have found use as dyes and photographic chemicals. Several pharmaceuticals and pesticides are members of this class.

Heterocyclics. One of the most characteristic and useful properties of hydrazine and its derivatives is the ability to form heterocyclic compounds. Numerous pharmaceuticals, pesticides, explosives, and dyes are based on these rings.

Manufacture

The commercially feasible processes involve partial oxidation of ammonia (or urea) using hypochlorite or hydrogen peroxide. Most hydrazine is produced by some variation of the Raschig process,which is based on the oxidation of ammonia using alkaline hypochlorite. Ketazine processes are modifications in which the oxidation is carried out in the presence of a ketone such as acetone or butanone. A process developed by Produits Chimiques Ugine Kuhlmann (PCUK) and practiced by Elf Atochem (France) and Mitsubishi Gas (Japan) involves the oxidation of ammonia by hydrogen peroxide in the presence of butanone and another component that apparently functions as an oxygen-transfer agent. The oxidation of benzophenone imine has received much attention but is not commercial.

Economic Aspects

The estimated world production capacity for hydrazine solutions is 44,100 t on a N_2H_4 basis. About 60% is made by the hypoclorite–ketazine process, 25% by the peroxide–ketazine route, and the remainder by the Raschig and urea processes. In addition there is anhydrous hydrazine capacity for propellant applications.

World demand for hydrazine solutions is about 31,000 t N_2H_4, excluding Eastern Europe, Russia, and mainland China. The demand is nearly equally divided between captive use and merchant business.

Hydrazine is a mature product likely to grow worldwide in stepwith the gross national product. However, annual growth in chemical blowing agents in southeast Asia may be higher, perhaps >5%. In water treatment, the growth might be somewhat less, perhaps 2–3%. A significant new application for hydrazine, in the manufacture of sodium azide for automobile air bags, could require an additional 3000 t/yr of N_2H_4.

Shipment and Specifications

Shipment of hydrazine solutions is regulated in the United States by the Department of Transportation (DOT) which classifies all aqueous solutions between 64.4 and 37% N_2H_4 as "Corrosive" materials with a subsidiary risk of "Poison". Hydrazine has been identified by both the Environmental Protection Agency and the DOT as a hazardous material and has been assigned a reportable quantity (RQ) of 0.450 kg (1 lb) if spilled.

Handling and Storage

Hydrazine is a base, a reducing agent, and a high energy compound; it is also volatile and toxic. These properties determine its proper handling, storage, use, and disposal. Inadvertent contact with acids or oxidizing agents must be avoided because extremely exothermic reactions and evolution of gases may result. Hydrazine is not sensitive to shock or friction. In the absence of decomposition catalysts, liquid anhydrous hydrazine has been heated to over 200°C without appreciable decomposition. Hydrazine fires are effectively combated using water because hydrazine ismiscible with water in all proportions.

The broad explosive range of hydrazine vapor is a concern. An inert gas blanket helps avoid formation of explosive mixtures. Hydrazine may ignite wood, rags, paper, or other common organic materials. Thus these should not be used near hydrazine. Use protective clothing to avoid body contact and provide adequate ventilation to reduce inhalation danger.

Materials of Construction. Materials generally considered satisfactory for all N_2H_4 concentrations, including anhydrous, are 304L and 347 stainless steels having less than 1.0 wt % molybdenum, a catalyst for the decomposition of hydrazine. For concentrations less than 10%, cold-rolled steel is satisfactory. Among the nonmetallic materials, poly(tetrafluoroethylene), polyethylene, and polypropylene are suitable; PVC is not recommended. Ethylene–propylene–diene monomer (EPDM) rubber and polyketones and polyphenylene sulfides are reportedly suitable for use with anhydrous hydrazine.

Disposal. Spills and wastewater containing hydrazines must be contained and treated. Proper disposal methods for the hydrazines make use of their reductive properties. Fuel-grade hydrazines may be burned, but aqueous solutions less than 50% may require supplementary fuel. Chemical destruction of dilute (preferably 5% or less) solutions can be achieved with various oxidants such as NaOCl, $Ca(OCl)_2$, H_2O_2, and acidified permanganate; however MMH and UDMH, may form mutagenic nitrosamines. A method is described for treating contaminated wastewater in which N_2H_4, MMH, UDMH, and *N*-nitrosodimethylamine are effectively decomposed at a controlled pH of ~5 by a uv-induced chlorination. Ozonation of these three hydrazines yields a variety of products, including methanol, formaldehyde monomethylhydazone, and formaldehydedimethylhydrazone; and tetramethyl tetrazene from the oxidative coupling of two molecules of UDMH. Methyldiazene was also found as an oxidation product of MMH. The rates of decomposition from aqueous solutions are greatest at pH 9.1 in the presence of uv light. Blowdown from boilers treated with hydrazine for corrosion control is effectively treated by neutralization with lime and chlorination. A vapor suppressant foam system (ASE95 polyacrylic/MSAR combination) has been evaluated for covering hydrazine fuel spills to minimize the release of toxic fumes.

Toxicology

Hydrazine is toxic and readily absorbed by oral, dermal, or inhalation routes of exposure. Contact with hydrazine irritates the skin, eyes, and respiratory tract. Liquid splashed into the eyes may cause permanent damage to the cornea. At high doses it can cause convulsions, but even low doses may result in central nervous system depression. Death from acute exposure results from convulsions, respiratory arrest, and cardiovascular collapse. Repeated exposure may affect the lungs, liver, and kidneys. Evidence is limited as to the effect of hydrazine on reproduction and/or development; however, animal studies demonstrate that onlydoses that produce toxicity in pregnant rats result in embryotoxicity.

The TLV is set at 0.1 ppm (hydrazine); 0.2 ppm (MMH); and 0.5ppm (UDMH). The International Agency for Research on Cancer (IARC) classifies hydrazine as a 2B or possible human carcinogen. The American Conference of Governmental Industrial Hygienists (ACGIH) classifies hydrazine as an A2 or suspect human carcinogen.

Uses

The principal applications of hydrazine solutions include chemical blowing agents, 40%; agricultural pesticides, 25%; and water treatment, 20%. The remaining 15% finds use in a variety of fields including pharmaceuticals, explosives, polymers and polymer additives, antioxidants, metal reductants, hydrogenation of organic groups, photography, xerography, and dyes.

HENRY W. SCHIESSL
Olin Corporation

E. Schmidt, *Hydrazine and Its Derivatives*, John Wiley & Sons, Inc., New York, 1984. Contains an exhaustive bibliography.

P. A. S. Smith, *Derivatives of Hydrazine and Other Hydronitrogens Having N-N Bonds*, The Benjamin/Cummings Publishing Co., Inc., Reading, Mass., 1983.

P. A. S. Smith, *The Chemistry of Open-Chain Nitrogen Compounds*, Vols. 1 and 2, W. A. Benjamin, Inc., Menlo Park, Calif., 1965–1966.

Houben-Weyl, *Methoden der Organischen Chemie*, 4th ed., Band X/2, Stickstoff Verbindungen 1, Teil 2, Georg Thieme Verlag, Stuttgart, Germany, 1967.

HYDRIDES

Hydrides are compounds that contain hydrogen in a reduced or electron-rich state. Hydrides may be either simple binary compounds or complex ones. In the former, the negative hydrogen is bonded ionically or covalently to a metal, or is present as a solid solution in the metal lattice. In the latter, which comprise a large group of chemical compounds, complex hydridic anions such as BH_4^-, AlH_4^-, and derivatives of these, exist.

Commercial applications of hydrides have become important and some of these compounds have become industrial chemicals manufactured and used on a large scale.

Simple (Binary) Hydrides

Ionic Hydrides. The ionic or saline hydrides contain metal cations and negatively charged hydrogen ions. They crystallize in the cubic lattice similar to the corresponding metal halide, and when pure, are white solids. When dissolved in molten salts or hydroxides and electrolyzed, hydrogen gas is liberated at the anode. Their densities are greater than those of the parent metal, and their formation is exothermic. All are strong bases.

Physical properties of the alkali metal hydrides are given in Table 1. Sodium hydride finds commercial usage in organic synthesis in condensation and alkylation reactions.

Table 2 gives thermochemical data of alkaline-earth metal hydrides. All form orthorhombic crystals. Calcium hydride is a convenient portable source of hydrogen gas, which results from its reaction with water.

Covalent Hydrides. Table 3 gives some properties of these compounds. Transition-metal hydrides, ie, interstitial metal hydrides, have metallic properties, conduct electricity, and are less dense than the parent metal. These hydrides are much harder and more brittle than the parent metal, and most have catalytic activity. They include titanium hydride, TiH_2, zirconium hydride, ZrH_2, rare earth hydrides (lanthanum dihydride, lanthanum trihydride, cerium hydride, CeH_2), group 5 (VB) hydrides (Table 4), and hydrogen storage alloys (FeTi, $LaNi_5$, Mg_2TiH_6, and $MgTi_2H_6$).

Table 1. Physical Properties of Alkali Metal Hydrides

Hydride	Mp, °C	$\Delta H_{(298)}$, kJ/mol[a]	$\Delta F_{(298)}$, kJ/mol[a]	S, J/(mol·K)[a]	Lattice energy, kJ/mol[a]	Density, g/cm³
LiH	688	−90.7	−70	25	916	0.77
NaH	420 dec	−56.5	−37.7	48	791	1.36
KH	dec	−57.9	−37.3	61	720	1.43
RbH	300 dec					2.60
CsH	dec					3.4

[a] To convert J to cal, divide by 4.184.

Table 2. Physical Properties of Alkaline-Earth Metal Hydrides

Hydride	$\Delta H_{(298)}$, kJ/mol[a]	$\Delta F_{(298)}$, kJ/mol[a]	S, J/(mol·K)[a]	Density, g/cm³
CaH_2	−186.3	−147.4	42	1.90
SrH_2	−180.5	−138.6	54	3.27
BaH_2	−171.2	−132.3	67	4.16

[a] To convert J to cal, divide by 4.184.

Table 3. Properties of Covalent Hydrides

Hydride	Formula	Mp, °C	Bp, °C	Density[a] g/cm³	Density[a] g/L
beryllium hydride[b]	BeH_2	125 dec	220[c]		
magnesium hydride	MgH_2	280 dec		1.45	
aluminum hydride	AlH_3				(−185)
silane[d]	SiH_4	−185[e]	−119.9	0.68[f]	1.44[g] (20)
germane	GeH_4	−165	−90	1.523[f] (−142)	3.43[g] (0)
stannane	SnH_4	−150	52		
arsine	AsH_3	−116.9[e]	−62	1.604[f] (64)	2.695[g]

[a] Temperature in °C is given in parentheses. [b] ΔH_{298} = 19.3 kJ/mol (4.6 kcal/mol). [c] Begins to dissociate. [d] ΔH_{298} = 30.55 kJ/mol (7.30 kcal/mol). [e] Freezing point. [f] Liquid. [g] Gas at atmospheric pressure.

Table 4. Group 5 (VB) Hydrides

Compound	Formula	Density, g/cm³
vanadium hydride	VH	5.4
niobium hydride	NbH	6.6
niobium dihydride[a]	NbH_2	
tantalum hydride	TaH	15.1

[a] Decomposes slowly.

Complex Hydrides

The complex hydrides are a large group of compounds in which hydrogen is combined in fixed proportions with two other constituents, generally metallic elements. These compounds have the general formula $M(M'H_4)_n$, where n is the valence of M, and M′ is a trivalent Group 3 (IIIA) element such as boron, aluminum, or gallium. The most important complex hydrides are listed in Table 5.

Borohydrides. The alkali metal borohydrides are the most important complex hydrides. They are ionic, white, crystalline, high melting solids that are sensitive to moisture but not to oxygen. They include lithium borohydride, $LiBH_4$, and sodium borohydride, $NaBH_4$.

Complete hydrolysis of $NaBH_4$ produces 2.37 L hydrogen (STP) per gram of borohydride; similarly, addition of acid to a cold aqueous solution liberates the theoretical amount of hydrogen. The inorganic reductions of $NaBH_4$ are numerous and varied. Sodium borohydride reacts with boron halides to form diborane, B_2H_6, which is more conveniently handled as the monomer BH_3 complexed with an ether, sulfide, or amine. Sodium borohydride is used extensively for the reduction of organic compounds. Sodium borohydride is manufactured from sodium hydride and trimethyl borate in a mineral oil medium at about 275°C. Sodium borohydride is classified as a flammable solid. It is available as powder, caplets, and granules and as a 12% solution in caustic soda. The principal uses of $NaBH_4$ are in synthesis of pharmaceuticals and fine organic chemicals; removal of trace impurities from bulk organic chemicals; wood-pulp bleaching, clay leaching, and vat-dye reductions; and removal and recovery of trace metals from plant effluents.

Potassium borohydride was formerly used in color reversal development of photographic film and was preferred over sodium borohydride because of its much lower hygroscopicity. Because other borohydrides are made from sodium borohydride, they are correspondingly more expensive. Generally their reducing properties are not sufficiently different to warrant the added cost. Zinc borohydride, $Zn(BH_4)_2$, however, has found many applications in stereoselective reductions.

Borohydride Derivatives. Modification of the BH_4^- anion has provided derivatives of widely differing reducing properties. Alkoxyborohydrides, such as sodium trimethyoxyborohydride, $NaBH(OCH_3)_3$, exhibit enhanced reducing power but are less selective and more sensitive to decomposition by water. Sodium cyanoborohydride, $NaBH_3CN$,

Table 5. Complex Hydrides

Formula	Density, g/cm^3	Mp, °C
$LiBH_4$	0.66	278
$NaBH_4$	1.074	505
KBH_4	1.177	585
$Be(BH_4)_2$	0.702	123 dec
$Mg(BH_4)_2$		320 dec
$Ca(BH_4)_2$		260 dec
$Zn(BH_4)_2$		>50 dec
$Al(BH_4)_3$	0.549	−64.5[a]
$Zr(BH_4)_4$	1.13	28.7
$Th(BH_4)_4$	2.59	204 dec
$U(BH_4)_4$	2.67	100 dec
$(CH_3)_4NBH_4$	0.84	>310
$(C_2H_5)_4NBH_4$	0.926	225 dec
$(C_4H_9)_4NBH_4$		>300
$(C_8H_{17})_3CH_3NBH_4$	0.9	ca 30
$C_{16}H_{33}(CH_3)_3NBH_4$	0.9	ca 160
$NaBH_3CN$	1.20	240 dec
$NaBH(OCH_3)_3$	1.24	230 dec
$LiAlH_4$	0.917	190 dec
$NaAlH_4$	1.28	178
$Mg(AlH_4)_2$		140 dec
$Ca(AlH_4)_2$		>230 dec
$LiAlH(OCH_3)_3$		
$LiAlH(OC_2H_5)_3$		
$LiAlH(OC_4H_9)_3$	1.03	>400
$NaAlH_2(OC_2H_4OCH_3)_2$	1.122	205 dec
$NaAlH_2(C_2H_5)_2$		85

[a] Bp, 44.5°C.

on the other hand, shows weakened reducing properties and is unique among the complex hydrides because it is stable in acidic aqueous solutions to a pH of about 3.

Sodium or tetramethylammonium triacetoxyborohydride has become the reagent of choice for diastereoselective reduction of β-hydroxyketones to antidiols. Trialkylborohydrides, eg, alkali metal tri-*sec*-butylborohydrides, show outstanding stereoselectivity in ketone reductions.

Aluminohydrides. In general, the aluminohydrides are more active and powerful reducing agents than the corresponding borohydrides. They decompose vigorously with water. Reaction also occurs with alcohols, although more moderately, providing a route to substituted derivatives.

Freshly prepared lithium aluminum hydride is a white crystalline solid that tends to become gray during storage, although very little loss in purity occurs. Although lithium aluminum hydride is best known as a nucleophilic reagent for organic reductions, it converts many metal halides to the corresponding hydride, eg, Ge, As, Sn, Sb, and Si. Commercial manufacture of $LiAlH_4$ uses the original synthetic method, ie, addition of a diethyl ether solution of aluminum chloride to a slurry of lithium hydride.

Sodium aluminum hydride can be prepared from NaH, but direct synthesis from the elements is more economical.

Aluminohydride Derivatives. The few known derivatives of the aluminohydrides are principally alkoxy substitutions, including the trimethoxy, $LiAlH(OCH_3)_3$, triethoxy. $LiAlH(OC_2H_5)_3$, and tri-*t*-butoxy aluminohydrides, $LiAlH(O\text{-}t\text{-}C_4H_9)_3$.

Health and Safety Factors

In general, hydrides react exothermically with water, resulting in the generation of hydrogen. This hydrolysis reaction is accelerated by acids or heat and, in some instances, by catalysts. Because the flammable gas hydrogen is formed, a potential fire hazard may result unless adequate ventilation is provided. Ingestion of hydrides must be avoided because hydrolysis to form hydrogen could result in gas embolism.

Another aspect of the hydrolysis of hydrides is the alkalinity that results, especially from alkali metal and alkaline-earth hydrides. This alkalinity can cause chemical burns in skin and other tissues. Hydrolysis considerations obviously demand that hydrides be kept away from contact with acids.

Although there is little toxicity information published on hydrides, a threshold limit value (TLV) for lithium hydride in air of 25 μg/m^3 has been established. More extensive data are available for sodium borohydride in the powder and solution forms. The acute oral LD_{50} of $NaBH_4$ is 50–100 mg/kg for $NaBH_4$ and 500–1000 mg/kg for the solution. The acute dermal LD_{50} (on dry skin) is 4–8 g/kg for $NaBH_4$ and 100–500 mg/kg for the solution. The reaction or decomposition by-product sodium metaborate is slightly toxic orally (LD_{50} is 2000–4000 mg/kg) and nontoxic dermally.

EDWARD A. SULLIVAN
Morton International

E. R. H. Walker, *Chem. Soc. Rev.* **5**, 23 (1976).

R. M. Adams and A. R. Siedle, *Boron, Metallo-Boron Compounds and Boranes*, Wiley-Interscience, New York, 1964, Chapt. 6, pp. 373–506.

W. M. Mueller, J. P. Blackledge, and G. G. Libovitz, *Metal Hydrides*, Academic Press, Inc., New York 1968, Chapt. 12, pp. 546–674.

B. D. James and M. G. H. Wallbridge, *Progrin Inorganic Chemistry*, Vol. 11, Wiley-Interscience, New York, 1970, pp. 99–231.

HYDROBORATION

Hydroboration is the addition of a boron–hydrogen bond across a double or triple carbon–carbon bond to give an organoborane:

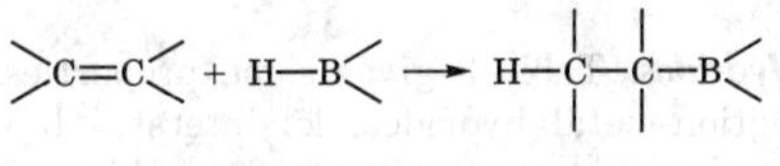

The boron atom in organoboranes can be replaced with other elements, usually with high stereoselectivity; many functional groups are tolerated. Consequently, organoboranes are among the most versatile synthetic intermediates, and their role in organic synthesis is constantly increasing.

One of the newer and more fruitful developments in this area is asymmetric hydroboration and homologation giving chiral organoboranes, which can be transformed into chiral carbon compounds of high optical purity. Other new directions focus on catalytic hydroboration, asymmetric allylboration, cross-coupling reactions, and applications in biomedical research.

The Hydroboration Reaction

Diborane, the first hydroborating agent studied, reacts sluggishly with olefins in the gas phase. In the presence of weak Lewis bases, eg, ethers and sulfides, it undergoes rapid reaction at room temperature or even below 0°C. The catalytic effect of these compounds on the hydroboration reaction is attributed to the formation of monomeric borane complexes from the borane dimer. Stronger complexes formed by simple tertiary amines react with olefins at elevated temperatures.

Mono-, di-, and trialkylboranes may be obtained from olefins and the trifunctional borane molecule.

Mechanism. The characteristic features of hydroboration were originally accounted for in terms of a simple four-center transition-state model (Fig. **1a**) serving as a useful working hypothesis. Models based on orbital symmetry considerations were also advanced (Fig. **1b**).

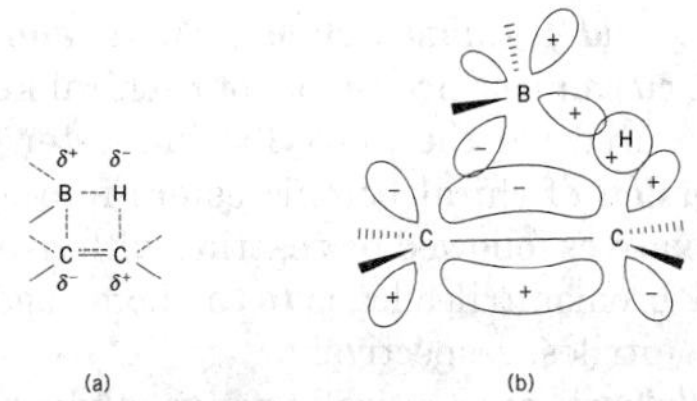

Figure 1. Simple four-center transition-state models of hydroboration.

Hydroborating Agents

Mono- and dialkylboranes obtained by controlled hydroboration of hindered olefins and by other methods can serve as valuable hydroborating agents for more reactive olefins. Heterosubstituted boranes are also available and used for this purpose. These borane derivatives show differences in reactivity and selectivity.

Borane Complexes. Borane solutions in tetrahydrofuran are commercially available or can be prepared by absorbing gaseous diborane in tetrahydrofuran. Diborane can be conveniently generated from the reaction of sodium borohydride with boron trifluoride etherate. Although borane–THF is a useful reagent, it must be stabilized with small amounts of sodium borohydride for longer storage at 0°C. Borane–dimethyl sulfide complex (BMS) is free of these inconveniences. Its disadvantage is the unpleasant smell of dimethyl sulfide, which is volatile and water insoluble. Borane–triethylamine complex is used when slow liberation of borane at elevated temperatures is advantageous. However, borane–amine adducts with certain hindered tertiary amines hydroborate olefins at room temperature.

Monosubstituted Boranes. Only a few monoalkylboranes are directly available by hydroboration. Tertiary hexylborane, 2,3-dimethyl-2-butylborane (thexylborane, Thx BH_2), easily prepared from 2,3-dimethyl-2-butene, is the best studied.

Monoisopinocampheylborane, $IpcBH_2$ is an important asymmetric hydroborating agent. It is prepared from α-pinene either directly or, better, by indirect methods.

A number of less hindered monoalkylboranes are available by indirect methods. Monohalogenoboranes are conveniently prepared from borane–dimethyl sulfide and boron trihalides (BX_3, where X = Cl, Br, I) by redistribution reaction. The products are liquids, soluble in various solvents and stable over prolonged periods.

Disubstituted Boranes. Even slight differences in steric or electronic effects of substituents may have an effect on the hydroboration reaction course. These effects are well demonstrated in disubstituted boranes, and consequently a range of synthetically useful reagents has been developed. Primary dialkylboranes react readily with most alkenes at ambient temperatures and dihydroborate terminal acetylenes.

In contrast to simple unhindered dialkylboranes, borinanes and borepanes do not redistribute readily. These boraheterocyclic reagents can be prepared by hydroboration of the corresponding dienes with borane, 9-BBN, or monochloroborane, followed by thermal isomerization or reduction, respectively. Dicyclohexylborane, Chx_2BH is prepared in quantitative yield by the hydroboration of cyclohexene with borane.

9-Borabicyclo [3.3.1] nonane, 9-BBN, is the most versatile hydroborating agent among dialkylboranes. It is commercially available or can be conveniently prepared by the hydroboration of 1,5-cyclooctadiene with borane, followed by thermal isomerization of the mixture of isomeric bicyclic boranes initially formed. A very hindered hydroborating agent is dimesitylborane.

Dihalogenoboranes are conveniently prepared by the redistribution of borane–dimethyl sulfide with boron trihalide–dimethyl sulfide complexes.

Several oxygen- and sulfur-substituted boranes have been reported. 1,3,2-Benzodioxaborole (catecholborane, CB) is the one best studied. It is commercially available or can be prepared by the reaction of catechol with borane·THF, or by other procedures.

The hydroboration of alkenes with catecholborane under the influence of Wilkinson's catalyst proceeds at room temperature, whereas heating is necessary in the absence of catalyst. This result provided the impetus for studies of catalytic hydroboration. The studies focused on catecholborane, rhodium, iridium, and lanthanide salts catalysts.

Alkylchloro- and alkylbromoboranes are valuable reagents for the synthesis of di- and trialkylboranes having different alkyl groups. Thexylchloroborane, ThxBHCl, is a very useful reagent. It can be prepared by the reaction of monochloroborane with 2,3-dimethyl-2-butene, or from thexylborane and hydrogen chloride.

Among chiral dialkylboranes, diisopinocampheylborane is the most important and best-studied asymmetric hydroborating agent. The most convenient synthesis, providing product of essentially 100% ee, involves the hydroboration of α-pinene with borane–dimethyl sulfide in tetrahydrofuran.

Selectivity

Chemoselectivity. Double and triple carbon–carbon bonds are more reactive than most other functionalities toward borane and substituted boranes. Consequently, many functional groups are tolerated in the hydroboration reaction. Using a suitably chosen hydroborating agent, only aldehydes, ketones, and carboxylic acids must be protected. However, it may not be necessary to protect ketones in the catalytic hydroboration for isolated functionalities. Hydroboration of allylic and vinylic derivatives leads to α-, β-, or γ-substituted organoboranes prone to eliminations or rearrangements. In some cases such transformations may be synthetically useful.

Conjugated ketones are either reduced to allylic alcohols or undergo 1,4-addition to give enolboranes.

Directive Effects. Hydroboration of olefins involves predominant placement of the boron atom at the less hindered site of the double bond. The direction of addition is governed by polarization of the boron–hydrogen bond and by combination of steric and electronic effects of substituents at the double bond.

Functional groups influence the regioselectivity of hydroboration by inductive, mesomeric, and steric effects, their magnitude depending on the proximity of the double bond and the functional group.

Stereoselectivity. The addition of a boron–hydrogen bond across the double bond proceeds cleanly in a cis fashion leading to simple diastereoselection for suitably substituted double bonds. Double bonds are approached by the hydroborating agent from the less sterically hindered face. The thermodynamically less stable addition products may result, as has been demonstrated for β-pinene and camphene. Borane discriminates well between faces differing significantly in steric hindrance. When the difference is small, low selectivity results. Bulky, sterically demanding hydroborating agents show higher stereoselectivity.

Reactions of Organoboranes

Organoboranes available by hydroboration and also by other methods are versatile synthetic intermediates. The organic groups attached to boron can be transferred, usually with high stereoselectivity, to hydrogen, oxygen, nitrogen, halogens, sulfur, selenium, metal atoms, and carbon. Consequently, carbon–hetero atom and carbon–carbon bonds can be constructed. The combination of hydroboration and functionalization corresponds overall to anti-Markovnikov addition of the elements of HX to a double bond.

$$RCH{=}CH_2 \xrightarrow{HB\langle} RCH_2CH_2B\langle \longrightarrow RCH_2CH_2X$$

Replacement of Boron by Hydrogen or a Hetero Atom. These reactions include protonolysis, oxidation, halogenolysis, replacement of boron by nitrogen, replacement of boron by sulfur and selenium, and mercuration.

Carbon-Carbon Bond Formation. These reactions include coupling of organic groups attached to boron, cross-coupling of organoboranes with organic halides, organoborate rearrangements, single-carbon insertion reactions (carbonylation, cyanidation, carbanion coordination), reactions with acyl carbanion equivalents, α-alkylation of carbonyl compounds and derivatives, α-bromination transfer, addition to carbonyl compounds, β-alkylation of carbonyl compounds, and reactions of boron-stabilized carbanions.

Thermal Isomerization of Organoboranes. Trialkylboranes undergo isomerization under the action of heat, generally at temperatures above 100°C, the boron atom moving to the least hindered site of the alkyl group.

Concerted Reactions of Organoboranes. Allylic organoboranes react via cyclic transition states with aldehydes, ketones, alkynes, allenes, and electron-rich or strained alkenes. Bicyclic structures, which can be further transformed into boraadamantanes, are obtained from triallyl- or tricrotylborane and alkynes. The addition proceeds in three discrete steps and the intermediates can be isolated. Unsaturated organoboranes react in the Diels-Alder reaction.

Polymerization. Hydroboration of α,ω-dienes with monoalkylboranes gives reactive organoboron polymers which can be transformed into polymeric alcohols or polyketones by carbonylation, cyanidation, or the DCME reaction followed by oxidation.

Synthesis of Isotopically Labeled Compounds. Organoborane reactions have been applied for the synthesis of isotopically labeled compounds important in chemical and biological research and in modern medical imaging techniques, such as positron emission tomography (pet) and magnetic resonance imaging (mri). Organoboranes tolerate a wide range of physiologically active functionalities and hence are well-suited intermediates in radiopharmaceutical pathways.

Asymmetric Synthesis Via Chiral Organoboranes

Asymmetric induction in the hydroboration reaction may result from the chirality present in the olefin (asymmetric substrate), in the reagent (asymmetric hydroboration), or in the catalyst (catalytic asymmetric hydroboration).

Synthesis of Chiral Alcohols, Ketones, Halides, Deuterated Hydrocarbons, and Amines. The chiral organoboranes produced by asymmetric hydroboration can be transformed into the heterosubstituted chiral products applying methodologies developed for achiral organoboron compounds. Thus the organoboranes obtained from alkenes are oxidized to the corresponding chiral alcohols. Among other examples are the alcohols derived from heterocyclic olefins, dienes, functionalized olefins, and deuterium- or tritium-labeled chiral alcohols.

Synthesis of Chiral Alkanes, Alkenes, and Alkynes. An efficient general synthesis of α-chiral (Z)- and (E)-alkenes in high enantiomeric purity is based on the hydroboration of alkynes and 1-bromoalkynes, respectively, with enantiomerically pure IpcR*BH readily available by the hydroboration of prochiral alkenes with monoisopinocampheylborane, followed by crystallization.

It is also possible to prepare α-chiral acetylenes and alkanes by this method. In a shorter synthesis of α-chiral alkynes, a prochiral disubstituted (Z-alkene is hydroborated with diisopinocampheylborane and the trialkylborane produced is treated with alkynyllithium followed by iodine.

Synthesis of α-Chiral and Homologated Aldehydes, Acids, and β-Chiral Alcohols. A general approach to these compounds is based on the reaction of dichloromethyllithium with boronic esters. Rearrangement of the complex followed by reduction with potassium triisopropoxyborohydride provides the homologated boronic ester, which can be oxidized to the corresponding alcohol or transformed into the homologated aldehyde by reaction with methoxy(phenylthio) methyllithium. β-Chiral alcohols not available in high optical purity by asymmetric hydroboration of terminal alkenes are readily prepared by this method.

Synthesis of α- and β-Chiral Ketones, Esters, and Nitriles. Chiral boronic esters are convenient precursors of α-chiral ketones (R*COR′), which can be prepared via the dialkylborinic ester or dialkylthexyl route. The conversion of chiral boronic esters into optically pure B-alkyl-9-BBN derivatives followed by reaction with α-bromoketones, α-bromoesters, or α-bromonitriles leads to the homologated β-chiral ketones, esters, and nitriles, respectively.

Asymmetric Allylboration. Optically active allylic boronates and dialkylboranes transfer the allylic group to aldehydes enantioselectively. The asymmetric allyl- and crotylboration of aldehydes has emerged as an effective alternative to the aldol methodology in reactions involving acyclic stereoselection.

Homologation of Boronic Esters. A convenient general method of enantioselective carbon–carbon bond formation, not involving hydroboration, is based on the homologation reaction of boronic esters derived from optically active 1,2-diols, eg, 2,3-pinanediol.

Enolboration. The aldol reaction is one of the most powerful methodologies for the formation of carbon–carbon bonds in a steredefined manner. Boron enolates are important intermediates for this transformation, since transition states of boron-mediated aldol reactions appear tightly organized, transmitting well the spatial arrangement to the aldol product.

MAREK ZAIDLEWICZ
Nicolaus Copernicus University

H. C. Brown, in *Organic Synthesis via Boranes*, John Wiley & Sons, Inc., New York, 1975.

A. Pelter, K. Smith, and H. C. Brown, in *Borane Reagents*, Academic Press, London, 1988.

K. Smith, in M. Schlosser, ed., *Organometallics in Synthesis*, John Wiley, & Sons, Ltd., Chichester, U.K., 1994, p. 461.

M. Zaidlewicz, in *Houben-Weyl. Stereoselective Synthesis*, Part D, G. Thieme, Stuttgart, Germany, 1995.

HYDROCARBON OXIDATION

This article provides a brief introduction to the homogeneous free-radical oxidations of paraffinic and alkylaromatic hydrocarbons.

Kinetics of Chain Reactions

One characteristic of chain reactions is that frequently some initiating process is required. In hydrocarbon oxidations radicals must be introduced and to be self-sustained, some source of radicals must be produced in a chain-branching step. Moreover, new radicals must be supplied at a rate sufficient to replace those lost by chain termination. In hydrocarbon oxidation, this usually involves the hydroperoxide cycle.

The predominant radicals in hydrocarbon oxidation are usually alkylperoxy radicals, particularly when sufficient oxygen is present to scavenge alkyl radicals. These radicals are weak hydrogen abstractors and they build up to relatively high concentrations, becoming the dominant species in the radical flux. Depending on their structures, they can participate in various reactions other than chain propagation. They are usually the main participants in bimolecular radical reactions. Other radicals are less involved in bimolecular radical reactions because their high reactivities cause them to be present at very low concentrations.

Product Sequences in Hydrocarbon Oxidation

Under conditions of low rates and high chain lengths, hydroperoxides may be the principal precursors for all other products. At high rates, however, significant quantities of chain-termination products (alcohols, ketones, and aldehydes) can be produced directly from peroxy radicals without going through hydroperoxides. In any event, there is a sense in which these products may be thought of as primary.

All components of the reaction mixture, whatever their source, are subject to the same kind of radical attacks as the starting substrate(s). Any free-radical oxidation is inevitably a cooxidation of substrate(s) and products. The yields of final products are determined by two factors: (*1*) how much is produced in the reaction sequence, and (*2*) how much product survives the reaction environment. By kinetic correlations and radiotracer techniques, it is possible to estimate these relationships and develop a mathematical model of the system.

Primary and secondary alcohols oxidize rapidly (frequently ca 6–10 times as fast as the parent hydrocarbon), and probably efficiently, to the corresponding carbonyl compounds (see Fig. 1). Carbonyl compounds can be primary (from radicals or hydroperoxides) or secondary (from alcohols). Thus the picture emerges of hydrocarbon oxidations occurring through complicated series-sequential pathways as in Figure 1, where clearly other reactions could be going on as well. All possible pathways are pursued to some extent; traffic along any pathway is a function of energy requirements and relative concentrations.

Carbonyl intermediates are also susceptible to further oxidation. Aldehydes can oxidize very rapidly to acids; peracids are likely intermediates.

Two important inefficient routes for aldehyde oxidation are the decarbonylation of the intermediate acyl radical, which is especially important at higher temperature (it is the source of much of the carbon monoxide produced in hydrocarbon oxidations), and a bimolecular radical reaction in which acyloxy radicals are generated; they are unstable and decarboxylate readily, providing much of the carbon dioxide produced in hydrocarbon oxidations.

Ketones oxidize about as readily as the parent hydrocarbons or even a bit faster.

Acids are usually the end products of ketone oxidations but vicinal diketones and hydroperoxyketones are apparent intermediates. Acids are readily produced from vicinal diketones, perhaps through anhydrides (via, eg, a Bayer-Villiger reaction).

The carboxyl function appears to deactivate the group to which it is attached and acetic acid is particularly resistant to further oxidation. It can be produced in reasonable efficiencies even when most other products are further oxidized to a significant extent. Beginning with propionic acid all the higher saturated acids are significantly less resistant to oxidation than acetic.

Esters are also formed in hydrocarbon oxidations. The clear implication is that most esters come from esterification reactions of free acids and alcohols produced by the oxidation.

In addition to production of simple monofunctional products in hydrocarbon oxidation there are many complex, multifunctional products that are produced by less well-understood mechanisms. There are also important influences of reactor and reaction types (plug-flow or batch, back-mixed, vapor-phase, liquid-phase, catalysts, etc).

Efficiency of Intermediate Formation. The variation of the efficiency of a primary intermediate with conversion of the feed hydrocarbon can be estimated. Ratios of the propagation rate constants (k_2/k_1) and reactor type (batch or plug-flow vs back-mixed) are important parameters. Even materials which are rather resistant to oxidation ($k_2/k_1 = \sim 0.1$) are consumed to a noticeable degree at high conversions. The use of plug-flow or batch reactors can offer a measurable improvement in efficiencies in comparison with back-mixed reactors. Intermediates that cooxidize about as readily as the feed hydrocarbon (eg, ketones with similar structure) can be produced in perhaps reasonable efficiencies but, except at very low conversions, are subject to considerable loss through oxidation. They may be suitable coproducts if they are also precursors to more oxidation-resistant desirable materials. Intermediates which oxidize relatively rapidly ($k_2/k_1 = 5-50$; eg, alcohols and aldehydes) are difficult to produce in appreciable amounts, even in batch or plug-flow reactors.

The maximum yield of a primary intermediate, as well as the efficiency (and conversion) at maximum yield, can also be estimated.

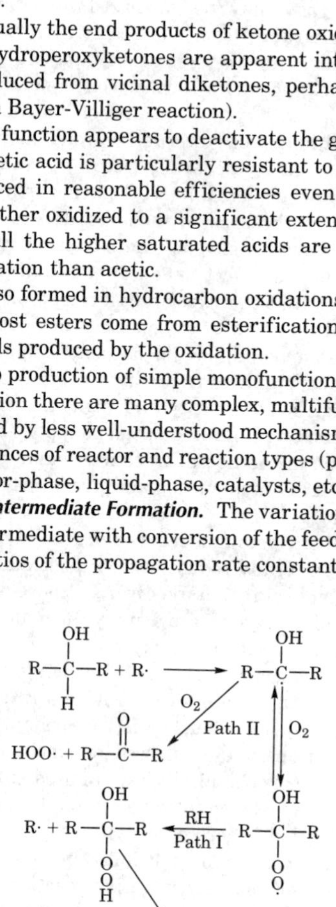

Figure 1. Production of carbonyl compounds from alcohols by various oxidation routes.

Vapor-Phase Oxidation

Above about 250°C, the vapor-phase oxidation (VPO) of many organic substances becomes self-sustaining. Such oxidations are characterized by a lengthy induction period. During this period, peroxides accumulate until they can provide a source of new radicals to sustain a chain reaction. Once a critical threshold peroxide concentration is reached, the reaction accelerates very rapidly.

The NTC Phenomenon. VPO reactions of typical alkanes may be considered conveniently in three temperature regions. Under some circumstances, particularly at pressures not greatly exceeding atmospheric, a curious and fundamentally important phenomenon known as the negative temperature coefficient (NTC) region is observed between the low and intermediate temperature ranges. In the NTC zone, increasing temperature actually results in lower reaction rates.

Notwithstanding some problems and conflicts, there is widespread agreement that the NTC phenomenon may well be related to the reversibility of the following equation:

$$R\cdot + O_2 \rightleftarrows ROO\cdot \qquad (1)$$

This reversal can lead to production of HOO· radicals and olefins. HOO· radical is an effective inhibitor which may well be responsible for the NTC effect.

Cool Flames. An intriguing phenomenon known as "cool" flames or oscillations appears to be intimately associated with NTC relationships. A cool flame occurs in static systems at certain compositions of hydrocarbon and oxygen mixtures over certain ranges of temperature and pressure.

When the necessary conditions are met, a cool flame seems to arise when heat generation during the low temperature oxidation exceeds heat losses. This leads to increasing temperature and increasing rates because of a higher radical generation rate by the low temperature chain-branching agent, ROOH. At a critical temperature, a cool flame appears. This causes the temperature to rise into the NTC region. Provided the temperature does not rise enough to permit the intermediate temperature chain-branching agent, HOOH, to become effective in providing new radicals, the flame can be quenched and the temperature drops. When the temperature returns to the low temperature reaction region, if sufficient amounts of the reactants remain, the whole process can proceed through another cycle. If, on the other hand, the temperature rise is great enough to bring about significant reaction of the intermediate temperature chain-branching agent, two-stage ignition occurs, ie, the cool flame is followed, after ca 1 s by a hot flame (explosion or rapid combustion). The relationships that give rise to cool flames in static reactors are also associated with oscillations in flow reactors.

Even though there is some challenge, the reversibility of reaction 1 as the root cause of the NTC and cool flame phenomena is widely accepted. It is also generally accepted that, as a consequence of reversibility, there is a temperature-dependent switchover from a low temperature chain-branching agent, ROOH, to an intermediate temperature chain-branching agent, HOOH. However, there are some instances where NTCs and cool flames are reported, but the generation of HOO· by the postulated reactions is not possible. The oxidations of acetone and methane are notable examples. These cases apparently require the presence of some other intermediate that can trigger the switchover to HOO· radicals. This role is often assigned to formaldehyde.

Effect of Pressure. As pressure is increased, the production of olefins is suppressed and the NTC region disappears. The reaction rate also increases significantly and, therefore, essentially complete oxygen conversion can be attained at lower temperatures. The product distribution shifts toward oxygenated materials that retain the carbon skeleton of the parent hydrocarbon.

Molecular Structure

Molecular structure strongly influences the reaction rate in low temperature VPO but has less effect in the intermediate and high temperature regions. At low temperature and low pressure, relative rates vary from 1.0 for pentane to 1380 for decane. This variation is much greater than the variation in the ease of abstractability of the hydrogen atoms.

The similarity of oxidation rates of different hydrocarbons in the higher temperature regions is probably related to the predominance of alkyl radical cracking reactions under these conditions. The products of such reactions would be similar for most common hydrocarbons.

Methane. As our most abundant hydrocarbon, methane offers an attractive source of raw material for organic chemicals. Successful commercial processes of the 1990s are all based on the intermediate conversion to synthesis gas. An alternative one-step oxidation is potentially very attractive on the basis of simplicity and greater energy efficiency. However, such processes are not yet commercially viable.

At ordinary pressures, the rate of homogeneous oxidation of methane is not very high (<500°C). Reaction 1 is highly reversed, ie, the ratio of $CH_3\cdot$ to $CH_3OO\cdot$ is very high. The radicals present in highest concentration tend to be the principal participants in radical–radical reactions. The generation of intermediate ethane from methyl radicals has been found to be a significant path in the oxidation of methane.

Although there are no new methane VPO competitive processes, current technology may be useful for the production of impure methanol in remote areas for use as a hydrate inhibitor in natural gas pipelines.

Ethane. Ethane VPO occurs at lower temperatures than methane oxidation but requires higher temperatures than the higher hydrocarbons. This is a transition case with mixed characteristics. Low temperature VPO, cool flames, oscillations, and a NTC region do occur. At low temperatures and pressures, the main products are formaldehyde, acetaldehyde ($HCHO : CH_3CHO$ = ca 5), and carbon monoxide. These products arise mainly through ethylperoxy and ethoxy radicals.

As the temperature is raised, ethylene and hydrogen peroxide become important products.

Increasing pressure increases yields of methanol and ethanol, increases the $C_2H_5OH{:}CH_3OH$ ratio, and reduces yields of acetaldehyde and formaldehyde.

Propane. The VPO of propane is the classic case. The low temperature oxidation (beginning at ca 300°C) readily produces oxygenated products. A prominent NTC region is encountered on raising the temperature and cool flames and oscillations are extensively reported as complicated functions of composition, pressure, and temperature. There can be a marked induction period.

A process to produce propylene by VPO of propane was patented in the former USSR in 1987. Similar processes have the potential to coproduce hydrogen peroxide. Yields of hydrogen peroxide as high as 1 mol/mol propylene produced have been reported with 60–70% propylene selectivity.

Butane. The VPO of butane is, in most respects, quite similar to the VPO of propane. However, at this carbon chain length an important reaction known as back-biting first becomes significant.

Isobutane. The VPO of isobutane is similar to the VPO of other low molecular weight paraffins; however, among the significant differences, conjugate olefin can be a main product even in the low temperature region.

Higher Hydrocarbons. The VPO of higher hydrocarbons is similar to that of the lower members of the series with two significant additional complications: (*1*) the back-biting reactions of alkylperoxy radicals, particularly at positions 2 or 3 carbons removed from the peroxy position, and (*2*) above the NTC region, radical fragmentation.

More complete understanding of VPO of hydrocarbons, especially in the fuel range, is vital for continued development of modern technology. The quenching phenomena associated with NTC regions and cool flames have been recognized as producing significant emissions of unburned and partially oxidized fuel. A significant correlation between cool flame phenomena and octane number has been reported. The use of a flow reactor under cool flame conditions has been suggested as a rapid method for octane number determination.

Liquid-Phase Oxidation

Although there are many similarities between VPO and liquid-phase oxidation (LPO), there are sharp distinctions. First, of course, in LPO with a gaseous oxidant (such as air), it is necessary that the oxidant (oxygen) be transported from the vapor phase into the liquid phase before LPO can occur. Moreover, it is possible to have a homogeneous catalyst dissolved in the liquid phase. If the solvent and/or products are volatile, they may be vaporized in significant amounts into any vent gas issuing from the reactor. Thus each individual component may have its own residence time in the system. Under these conditions, the hydrocarbon conversion no longer has a simple relationship with the exposure of individual components to oxidation conditions and alternative relationships are needed for kinetic studies. LPOs generally occur at lower temperatures than VPOs. LPOs do not exhibit NTC phenomena; oscillations are known to occur, but these seem to be related to catalyst valence shift cycles and involve oxygen mass-transfer rate limitations. Another rate reduction effect is encountered if the temperature increases enough so that the vapor pressure approaches the imposed system pressure. In this case, oxygen is swept out of solution and the reaction stops; a boiling liquid does not support LPO. Generally, olefins are not significant products of the LPO of saturated materials. The high heat capacity and high heat of vaporization of the solvent, combined with limited oxygen supply, tend to damp temperature excursions in the liquid. Of course, runaway VPO reactions are possible if the LPO is quenched and unconverted oxygen collects in the vapor space.

The lower temperatures and reduced degree of oxygen starvation in LPO (vs VPO) generally reduce carbon monoxide production from acyl radicals (a major source) markedly. As a consequence, acids, from further oxidation of aldehydes, are usually the main products.

Mass Transfer. The transfer of oxygen from the vapor to the liquid phase is a critical part of any LPO involving a gaseous oxidant. A simplified view of this process can provide a rationalization of the observed system behavior when air is the oxidant. The typical air-sparged LPO system can be considered to consist of two zones. The zone near the sparger is chemically rate limited since the liquid contains enough dissolved oxygen to scavenge alkyl radicals. Under these conditions, the reaction is zero order with respect to oxygen; the kinetics are thus determined by the initiating and chain-branching, chain-terminating, and chain-propagating reactions. The second zone is mass transfer rate limited. The rate can be first-order with respect to oxygen.

Pressure. Within limits, pressure may have little effect in air-sparged LPO reactors. The optimum pressure is likely to be determined by the permissible maximum gas holdup and/or the desirable maximum vapor load in the vent gas.

Reactor Configuration. The horizontal cross-sectional area of a reactor is a critical parameter with respect to oxygen mass-transfer effects in LPO since it influences the degree of interaction of the two types of zones. Reactions with high intrinsic rates, such as aldehyde oxidations, are largely mass-transfer rate-limited under common operating conditions. Such reactions can be conducted effectively in reactors with small horizontal cross sections. Slower reactions, however, may require larger horizontal cross sections for stable operation.

Catalysts and Promoters. The function of catalysts in LPO is not well understood. Perhaps they are not really catalysts in the classical sense because they do not necessarily speed up the reaction. They do seem to be able to alter relative rates and thereby affect product distributions, and they can shorten induction periods. The basic function in shortening induction periods appears to be the decomposition of peroxides to generate radicals.

Increasing efforts to heterogenize homogenous catalysts for LPO are apparent. Significant advantages in product recovery, catalyst use, and catalyst recovery are recognized. In some instances, however, the active catalyst is reported to be materials dissolved from the solid catalyst.

Propane. Propane is difficult to oxidize in LPO because of its volatility and lack or reactivity. It can, however, be oxidized with a suitable solvent and sufficiently high pressures and temperatures. The principal products are acetone and isopropyl alcohol.

Butane. Butane LPO has been a significant source for the commercial production of acetic acid and acetic anhydride for many years. Methanol carbonylation is now the dominant process for acetic acid production, but butane LPO in established plants remains competitive.

Isobutane. Isobutane can be oxidized noncatalytically to give predominantly *t*-butyl hydroperoxide (TBHP). A significant outlet for TBHP is the molybdenum-complex catalyzed production of propylene oxide, a process developed by Oxirane. The *tert*-butyl alcohol coproduct is used mostly to make methyl *tert*-butyl ether, a gasoline additive.

Higher Paraffins. The LPO of paraffins to produce synthetic fatty acids (SFAs) has been practiced extensively in the former USSR and Eastern bloc countries. Elsewhere, varying degrees of interest in the process have been evident in Germany, Japan, China, and a number of other countries. Although the basic mechanisms are similar, several difficulties are encountered in higher paraffin LPO that are not so evident in the LPO of, eg, butane. For one thing, the desired product SFAs have virtually the same susceptibility to further oxidation, on a per carbon atom basis, as the feed hydrocarbon, contrasting sharply with acetic acid from butane LPO. An even more significant limitation is that a higher paraffin, once attacked, is highly susceptible to the back-biting reaction before being converted to nonradical products. The typical SFA process uses a manganese catalyst with a potassium promoter (for solubilization) in a batch reactor.

Cyclohexane. The LPO of cyclohexane supplies much of the raw materials needed for nylon-6 and nylon-6,6 production. Cyclohexanol (A) and cyclohexanone (K) may be produced selectively by using a low conversion process with multiple stages. A one-step LPO of cyclohexane directly to adipic acid has received a lot of attention but has not been implemented on a large scale.

Alkylaromatics. The aromatic ring is fairly inert toward attack by oxygen-centered radicals. Aromatic acids consisting of carboxyl groups substituted on aromatic rings are good candidates for production by LPO of alkylaromatics since their k_2/k_1 ratios are low. Terephthalic acid (TPA) and dimethyl terephthalate (DMT) are the outstanding examples of high volume chemicals made in this way; efficiencies are ca 95%.

The aromatic core or framework of many aromatic compounds is relatively resistant to alkylperoxy radicals and inert under the usual autoxidation conditions. Consequently, even somewhat exotic aromatic acids are resistant to further oxidation; this makes it possible to consider alkylaromatic LPO as a selective means of producing fine chemicals. Such products may include multifunctional aromatic acids, acids with fused rings, acids with rings linked by carbon–carbon bonds, or through ether, carbonyl, or other linkages. The products may even be phenolic if the phenolic hydroxyl is first esterified.

Synthetic phenol capacity in the U.S. was reported to be ca 1.6×10^6 t/yr in 1989, almost completely based on the cumene process. Some synthetic phenol is made from toluene by a process developed by The Dow Chemical Company. Toluene is oxidized to benzoic acid in a conventional LPO process. Liquid-phase oxidative decarboxylation with a copper-containing catalyst gives phenol in high yield.

Dihydroxyarenes can be produced from the corresponding diisopropylarenes in a manner similar to the production of phenol from cumene.

An oxirane process utilizes ethylbenzene to make the hydroperoxide, which then is used to make propylene oxide. The hydroperoxide-producing reaction is similar to the first step of cumene LPO except that it is slower. In the epoxidation step, α-phenylethyl alcohol is the coproduct. It is dehydrated to styrene.

CHARLES C. HOBBS
Consultant, sponsored by Hoechst Celanese Corporation

S. Al-Makaika, *Atmos. Oxid. Antioxid.* 45–82 (1993).

C. Walling, *Free Radicals in Solution*, John Wiley & Sons, Inc., New York, 1957.

S. W. Benson and P. S. Nangia, *Acc. Chem. Res.* **12**(7), 223–228 (1979).

R. A. Sheldon and J. K. Kochi, *Metal-Catalyzed Oxidations of Organic Compounds*, Academic Press, Inc., New York, 1981.

HYDROCARBON RESINS

Hydrocarbon resin is a broad term that is usually used to describe a low molecular weight thermoplastic polymer synthesized via the thermal or catalytic polymerization of coal-tar fractions, cracked petroleum distillates, terpenes, or pure olefinic monomers. These resins are used extensively as modifiers in the hot melt and pressure sensitive adhesive industries. They are also used in numerous other applications such as sealants, printing inks, paints, plastics, road marking, carpet backing, flooring, and oil field applications. They are rarely used alone.

Typical hydrocarbon resins range in appearance from hard, brittle solids to viscous liquids. They may come in flakes, pellets, drums, or in molten form. Depending on application requirements, many resins are available as solutions in organic solvents or oils. Anionic, cationic, or nonionic emulsion forms are also manufactured. Hydrocarbon resins typically have a number average molecular weight (M_n) of less than 2000. The colors of these resins range from water-white to dark brown. Water-white resins usually are produced from the Lewis acid polymerization of pure olefinic monomers or by the hydrogenation of catalytically or thermally produced precursors. Colors are determined on the Gardner and Saybolt scales.

Physical Properties

Most hydrocarbon resins are composed of a mixture of monomers and are rather difficult to fully characterize on a molecular level. The characteristics of resins are typically defined by physical properties such as softening point, color, molecular weight, melt viscosity, and solubility parameter. These properties predict performance characteristics and are essential in designing resins for specific applications. Actual characterization techniques used to define the broad molecular properties of hydrocarbon resins are Fourier transform infrared spectroscopy (ftir), nuclear magnetic resonance spectroscopy (nmr), and differential scanning calorimetry (dsc).

Polymerization

Most commercial hydrocarbon resins produced from olefinic feedstocks are synthesized via carbocationic polymerization. Very similar catalyst systems are used in the synthesis of coumarone–indene, petroleum, terpene, and pure monomer-based resins.

Typically, Lewis acids are used in the synthesis of hydrocarbon resins. Examples are $AlCl_3$, BF_3, $(CH_3CH_2)_2AlCl$, $CH_3CH_2AlCl_2$, and complexes of these acids. Friedel-Crafts (Lewis) acids have been shown to be much more effective in the initiation of cationic polymerization when in the presence of a cocatalyst such as water, alkyl halides, and protic acids. Virtually all feedstocks used in the synthesis of hydrocarbon resins contain at least traces of water, which serves as a cocatalyst.

Economic Aspects

The projected annual growth rate for hydrocarbon resins through the early to mid-1990s has been approximated at 4–8% based on various sources. Production figures may be gained from the U.S. International Trade Commission reports or from various consulting firms, which specialize in monitoring hydrocarbon resin and related industries. Consumption and growth predictions with respect to specific end uses may also be gained from these sources.

R. DERRIC LOWERY
Exxon Chemical Company

U.S. Pat. 4,078,132 (Mar. 7, 1978), A. Lepert (to Exxon Research and Engineering Co.).

U.S. Pat. 3,968,088 (July 6, 1976), H. Asai and A. Wada (to Nippon Zeon Co., Ltd.).

U.S. Pat. 4,629,766 (Dec. 16, 1986), B. Bossaert, A. Malatesta, and J. Mourand (to Exxon Research and Engineering).

U.S. Pat. 4,048,095 (Sept. 13, 1977), E. R. Ruckel and R. T. Wojcik (to Arizona Chemical Co.).

HYDROCARBONS

SURVEY

Hydrocarbons, compounds of carbon and hydrogen, are structurally classified as aromatic and aliphatic; the latter includes alkanes (paraffins), alkenes (olefins), alkynes (acetylenes), and cycloparaffins. Crude petroleum oils, which span a range of molecular weights of these compounds, excluding the very reactive olefins, have been classified according to their content as paraffinic, cycloparaffinic (naphthenic) or aromatic.

Hydrocarbons are important sources for energy and chemicals and are directly related to the gross national product. The United States has led the world in developing refining and petrochemical processes for hydrocarbons from crude oil and natural gas. Hydrocarbons from crude oil have become the energy sources of the industrial world, largely replacing wood and even displacing coal. However, in the U.S., crude oil production peaked at 1.3×10^6 t/d(9.6×10^6 bbl/d) (conversion factors vary depending on oil source) in 1970, causing increased reliance on foreign oil sources. Since the crude oil embargo in 1973, a number of alternative energy sources have been investigated to reduce the U.S. international trade deficit. The fossil-fuel era may turn out to have been a brief interlude between the wood-burning era of the nineteenth century and the renewable energy sources era of the twenty-first century.

Hydrocarbon resources can be classified as organic materials, which are either mobile such as crude oil or natural gas, or immobile materials including coal, lignite, oil shales, and tar sands. Most hydrocarbon resources occur as immobile organic materials which have a low hydrogen-to-carbon ratio. However, most hydrocarbon products in demand have a H:C higher than 1.0.

Products	Molar H:C ratio
natural gas	4.0
LPG	2.5
gasoline	2.1
fuel oil	
light	1.8
heavy	1.3
coal	0.8

Immobile hydrocarbon sources require refining processes involving hydrogenation. Additional hydrogen is also required to eliminate sources of sulfur and nitrogen oxides that would be emitted to the environment. Resources can be classified as mostly consumed, proven but still in the ground, and yet to be discovered. A reasonable estimate for the proven reserves for crude oil is 140×10^9 t(1.0×10^{12} bb). Since 1950, the dominance of reserves has been in the eastern hemisphere and in offshore fields. Proved world gas reserves are nearly 4×10^{12} trillion metric feet, with 31% in the Middle East. Another factor to be considered is the fraction of crude that can be obtained from a reservoir. The average primary recovery from a reservoir was about 25–30% of the crude oil in place in 1993.

Hydrocarbons as Energy Sources

Hydrocarbons from petroleum are still the principal energy source for the U.S. About 60% of the world's energy is supplied by gas and oil and about 27% from coal.

A significant obstacle to increased gas use is the lack of sufficient transportation and distribution systems. Environmental concerns have encouraged reliance on natural gas as a cleaner burning fuel. Combustion of natural gas emits about half the CO_2 that coal generates at equivalent heat output. However, low oil prices have caused the number of operating drilling rigs in the United States to drop to well below the peak in the 1980s, cutting production of gas.

The U.S. refines about 25% of the world's crude oil, and because of its declining oil reserves, must import additional crude oil.

Natural gas imports have grown more slowly because imports from overseas require governmental licenses and cryogenic liquefaction plants are very expensive. Natural gas imports are chiefly by pipeline from Canada.

Gas and oil are the principal energy sources even though the U.S. has large reserves of coal. Although the use of coal and lignite is being encouraged as an energy source, economic and environmental considerations have kept petroleum consumption high. The use of compressed natural gas (CNG) is expected to grow in response to the Clean Air Act of 1990. Reliance on foreign imports has remained high, increasing since the collapse of oil prices in 1986.

In 1990, U.S. energy consumption by end user sector was 35.8% residential and commercial, 37% industrial, and 27.2% transportation. The breakdown of consumption by source was 41.2% petroleum, 23.8% natural gas, 23.5% coal, 7.6% nuclear, and 3.9% hydroelectric and other.

Hydrocarbons as Chemical Intermediates

Because of the time lag involved in collecting information on U.S. chemicals production and sales literature figures are always 2–3 years out of date. In 1991, total production of primary chemical products (ie, C_2–C_5 olefins and paraffins, C_6–C_8 aromatics, plus miscellaneous other compounds used as intermediates in synthesis of other chemicals) was 5.41×10^7 kg. Only about half that amount was actually sold on the open market; the rest was used internally as feedstock for other chemicals. Total U.S. primary chemical sales in 1991 amounted to 2.76×10^7 kg for a total value of \$$9.63 \times 10^9$.

Raw Materials. Petroleum and its lighter congener, natural gas, are the predominant sources of hydrocarbon raw materials, accounting for over 95% of all such materials.

Coal is used mainly to produce synthesis gas, a mixture of CO and hydrogen. Much of the production of synthesis gas is unreported, as it typically is never isolated. For economic reasons, few chemicals are made from synthesis gas in the U.S. With the exception of politically dictated operations, Fischer-Tropsch production of the hydrocarbons is much more expensive than production and refining of petroleum. The cost of Fischer-Tropsch hydrocarbons from coal has historically been $10–15 per barrel higher than petroleum.

There are some chemicals that can be made economically from coal or coal-derived substances. Methanol and CO are used to make acetic anhydride and acetic acid. Methanol itself can be made from synthesis gas over a copper–zinc catalyst.

Though there has been much discussion about using biomass as a renewable resource for hydrocarbon production, few chemicals are currently made in significant quantities from biological feedstocks. The most important is ethanol, used as an oxygenated additive to reformulated gasoline, which is meant to burn more cleanly than normal gasoline. That use is made economically feasible only by significant government subsidies in the form of tax exemptions.

Synthesis Gas Chemicals. Hydrocarbons are used to generate synthesis gas, a mixture of carbon monoxide and hydrogen, for conversion to other chemicals. The primary chemical made from synthesis gas is methanol, though acetic acid and acetic anhydride are also made by this route. Carbon monoxide is produced by partial oxidation of hydrocarbons or by the catalytic steam reforming of natural gas. About 96% of synthesis gas is made by steam reforming, followed by the water gas shift reaction to give the desired H_2/CO ratio.

Aliphatic Chemicals. The primary aliphatic hydrocarbons used in chemical manufacture are ethylene, propylene, butadiene, acetylene, and *n*-paraffins.

Cyclic Hydrocarbons. The cyclic hydrocarbon intermediates are derived principally from petroleum and natural gas, though small amounts are derived from coal. Most cyclic intermediates are used in the manufacture of more advanced synthetic organic chemicals and finished products such as dyes, medicinal chemicals, elastomers, pesticides, and plastics and resins.

Hydrocarbons as End Use Chemicals

Hydrocarbons are used in lubricants, agriculture/food, surfactants, coatings, and polymers.

Regulatory Issues

Regulatory issues are increasingly driving the petroleum industry, taking an ever-growing share of capital. New environmental regulations govern every aspect of operation from drilling to refining. The number of regulations has increased dramatically since the 1970s. They include the 1990 Clean Air Act, the Water Quality Act of 1987 and Clean Water Act of 1977, the Resource Conservation and Recovery Act (RCRA) of 1976, as amended in a more comprehensive form in 1984, the Comprehensive Environmental Response, Compensation, and Liability Act (CERCLA/Superfund) amended by SARA in 1986, the Toxic Substances Control Act (TSCA) of 1976, the Occupational Safety and Health Act (OSHA), as amended in 1990, Hazardous Waste Operations and Emergency Response regulations, and the Pollution Prevention Act of 1990.

DAVID E. MEARS
Unocal
ALAN D. EASTMAN
Phillips Petroleum

M. W. Ball, D. Ball, and O. S. Turner, *The Fascinating Oil Business*, Bobbs Merrill Co., Inc., Indianapolis, Ind., 1965.

I. I. Nesterov and F. K. Salmanov, in R. F. Meyer, ed., *The Future Supply of Nature-Made Petroleum and Gas*, Pergamon Press, New York, 1977, p. 185.

World Energy Outlook, Chevron Corp., Richmond, Calif., Apr. 1990.

Environmental Statutes, 1993 Edition, Government Institutes, Inc., Rockville, Md., Feb. 1993.

ACETYLENE

Acetylene, C_2H_2, is a highly reactive, commercially important hydrocarbon. It is used in metalworking (cutting and welding) and in chemical manufacture. Chemical usage has been shrinking due to the development of alternative routes to the same products based on cheaper raw materials. The reactivity of acetylene is related to its triple bond between carbon atoms and, as a consequence, its high positive free energy of formation. Because of its explosive nature, acetylene is generally used as it is produced without shipping or storage.

Physical Properties

The physical properties of acetylene have been reviewed in detail. The triple point is at −80.55°C and 128 kPa (1.26 atm). The temperature of the solid under its vapor at 101 kPa (1 atm) is −83.8°C. The vapor pressure of the liquid at 20°C is 4406 kPa (43.5 atm). The critical temperature and pressure are 35.2°C and 6190 kPa (61.1 atm). The density of the gas at 20°C and 101 kPa is 1.0896 g/L. The specific heats of the gas, C_p and C_v (at 20°C and 101 kPa) are 43.91 and 35.45 J/mol·°C (10.49 and 8.47 cal/mol, °C), respectively. The heat of formation ΔH_f at 0°C is 227.1 kJ/mol (54.3 kcal/mol). The solubility in water at 20°C is 16.6 g/L at 1520 kPa (15.0 atm) and 1.23 g/L at 101 kPa. Acetylene forms a hydrate of approximate stoichiometry $C_2H_2 \cdot 6\ H_2O$. The solubility in organic solvents is much greater than in water, 237 g/L in acetone, for example, at 15°C and 1520 kPa (15.0 atm) total pressure.

Chemical Properties

Acetylene is highly reactive due to its triple bond and high positive free energy of formation. Extensive reviews of acetylene chemistry are available. Important reactions involving acetylene are hydrogen replacements, additions to the triple bond, and additions by acetylene to other unsaturated systems. Moreover, acetylene undergoes polymerization and cyclization reactions. Hydrogenation, halogenation, hydrohalogenation, hydration, and vinylation are important addition reactions. In the ethynylation reaction, acetylene adds to a carbonyl group. The formation of a metal acetylide is an example of hydrogen replacement.

Copper acetylides form under a variety of conditions. Cuprous acetylide formation from Ilosvay's solution is the basis for the qualitative detection of trace amounts of acetylene. The method can be used to determine acetylene in parts per billion concentrations. The acetylides of copper, silver, and mercury are explosive, the explosiveness depending on the formation conditions. Materials containing copper or silver should not be used in an acetylene system.

Handling of Acetylene

The design of equipment for the handling and use of acetylene must take into consideration the possibility of acetylene decompositions. The design parameters must consider various factors, namely pressure, temperature, source of ignitions, and ultimate pressures which may result from a decomposition.

Acetylene in Cylinders. Acetylene cylinders are constructed to stabilize acetylene and, thereby, safely avoid the hazard of a detonation. The basic feature of an acetylene cylinder that is different from all other cylinders is that it is entirely filled with a monolithic porous mass. It is this monolithic mass that stabilizes the acetylene and permits its safe shipment. After the cylinder has been manufactured, a specified quantity of solvent is added, usually acetone. Acetone dissolves many times its own volume of acetylene and its purpose is to increase the amount of acetylene that may be safely charged and shipped. The acetylene is, therefore, not a free gas but is in solution.

Acetylene cylinders are fitted with safety devices to release the acetylene in the event of fire. Cylinders manufactured in the United States are equipped with safety devices which contain a fusible metal that melts at 100°C.

The manufacture and shipping of acetylene cylinders in the United States are in compliance with the specifications and regulations of the Department of Transportation. The specifications are verified by the Bureau of Explosives of the American Association of Railroads. DOT-8 and DOT-8AL specify the requirements for manufacture and testing of steel shell, porous mass, and quantity of acetone (therefore, acetylene) which may be charged.

Commercially pure acetylene can decompose explosively (principally into carbon and hydrogen) under certain conditions of pressure and container size. It can be ignited, ie, a self-propagating decomposition flame can be established, by contact with a hot body, by an electrostatic spark, or by compression (shock) heating.

The flammability range for acetylene–air at atmospheric pressure is ca 2.5–80% acetylene in tubes wider than 50 mm. The range narrows to ca 8–10% as the diameter is reduced to 0.8 mm. Ignition temperatures as low as 300°C have been reported for 30–75% acetylene mixtures with air and for 70–90% mixtures with oxygen.

Manufacture

From Calcium Carbide. Acetylene is generated by the chemical reaction between calcium carbide and water with the release of 134 kJ/mol (900 Btu/lb of pure calcium carbide).

$$CaC_2 + 2\ H_2O \rightarrow Ca(OH)_2 + C_2H_2$$

Because of the exothermic reaction and the evolution of gas, the most important safety considerations in the design of acetylene generators are the avoidance of excessively high temperatures and high pressures.

Most carbide acetylene processes are wet processes from which hydrated lime, $Ca(OH)_2$, is a by-product. It is marketed for industrial wastewater treatment, neutralization of spent pickling acids, as a soil conditioner in road construction, and in the production of sand–lime bricks.

Basically two reaction schemes are used for acetylene generation. The one most widely used in the United States is carbide-to-water generation. Standards for the design and construction of acetylene-generating equipment using this technique have been developed over the years by the acetylene industry. Underwriters Laboratories, Inc. have generally accepted design criteria for acetylene generating equipment.

Although there are numerous variations in the design of commercially available carbide-to-water acetylene generators, all consist of a water vessel or reaction chamber, a carbide feed mechanism, and a carbide storage container that empties into the feed mechanism.

Water-to-carbide generation has found only limited acceptance in the United States and Canada, but has been used frequently in Europe for small-scale generation.

Purification of Carbide Acetylene. The purity of carbide acetylene depends largely on the quality of carbide employed and, to a much lesser degree, on the type of generator and its operation.

The maximum amount of impurities in U.S. Grade B acetylene (Carbide Generated Acetylene) is 2% on a dry basis. This gas meets commercial requirements for acetylene used in cutting and welding. Production of U.S. Grade A acetylene used in sensitive chemical reactions requires further purification to reduce impurities to 0.5%. There are four main impurities: phosphine, ammonia, hydrogen sulfide, and organic sulfides. The purification involves oxidation of phosphine to phosphoric acid, the neutralization and absorption of ammonia, and the oxidation of hydrogen sulfide and organic sulfur compounds. Many processes are employed depending on the type and amount of impurities and the end use of the gas. These wet or dry processes range from simply passing the gas over purifying media to multistep chemical treatments.

RICHARD E. GANNON
Textron Defense Systems
ROBERT M. MANYIK
Union Carbide Corporation
C. M. DIETZ
H. B. SARGENT
R. O. THRIBOLET
R. P. SCHAFFER
Consultants

S. A. Miller, *Acetylene—Its Properties, Manufacture and Uses*, Vol. 1, Academic Press, Inc., New York, 1965.

Acetylene Transmission for Chemical Synthesis, Pamphlet G 1.3, Compressed Gas Association, Arlington, Va., 1984.

R. J. Tedeschi, *Acetylene-based Chemicals from Coal and Other Natural Resources*, Marcel Dekker, Inc., New York, 1982.

S. A. Miller, *Acetylene—Its Properties, Manufacture and Uses*, Vol. 2, Academic Press Inc., New York, 1966.

C_1–C_6

METHANE, ETHANE, AND PROPANE

Physical Properties

Methane, ethane, and propane are the first three members of the alkane hydrocarbon series having the composition, C_nH_{2n+2}. Selected properties of these alkanes are summarized in Table 1.

Manufacturing and Processing

The main commercial source of methane, ethane, and propane is natural gas, which is found in many areas of the world in porous reservoirs; they are associated either with crude oil (associated gas) or in gas reservoirs in which no oil is present (nonassociated gas). These gases are basic raw materials for the organic chemical industry as well as sources of energy. The composition of natural gas varies widely but the principal hydrocarbon usually is methane.

Relatively small amounts of methane, ethane, and propane are produced as by-products from petroleum processes, but these usually are consumed as process or chemical feedstock fuel within the refineries. Some propane is recovered and marketed as liquefied petroleum gas (LPG).

Table 1. Selected of Methane, Ethane, and Propane

Property	Methane	Ethane	Propane
mol formula	CH_4	C_2H_6	C_3H_8
mol wt	16.04	30.07	44.09
mp, K	90.7	90.4	85.5
bp, K	111	185	231
explosivity limits, vol %	5.3–14.0	3.0–12.5	2.3–9.5
autoignition temp, K	811	788	741
flash point, K	85	138	169
heat of combustion, kJ/mol[a]	882.0	1541.4	2202.0
vapor pressure at 273 K, MPa[b]		2.379	0.475
specific heat at 293 K, J/(mol·K)[a]	37.53	54.13	73.63
density at 293 K, kg/m[3c]	0.722	1.353	1.984
hazards[d]			

[a] To convert J to cal, divide by 4.184. [b] To convert MPa to atm, divide by 0.101. [c] To convert kg/m^3 to lb/ft^3, divide by 16.0. [d] Fire explosion, asphyxiation; no significant toxic effects.

There are, however, a variety of other sources of methane that have been considered for fuel supply. For example, methane present in coal deposits and formed during mining operations can form explosive mixtures known as fire damp. In Western Europe, some methane has been recovered by suction from bore holes drilled in coal beds and the U.S. Bureau of Mines has tested the economic practicality of such a system. Removal of methane prior to mining the coal would reduce explosion hazards associated with coal removal. As much as 11.3×10^9 m^3 (400 trillion (10^{12}) cubic feet or 400 TCF) of methane might be recoverable from U.S. coal beds.

Methane also is commonly produced by the decomposition of organic matter by a variety of bacterial processes, and the gas is used as a fuel in sewage plants. Methane also is called marsh gas because it is produced during the decay of vegetation in stagnant water.

There has been considerable research into the production of substitute natural gas (SNG) from fractions of crude oil, coal, or biomass.

Production and Shipment

World natural gas reserves and production are shown in Table 2. The deposits of natural gas are extensive and provide sources of feedstock and fuel.

The large-scale use of natural gas requires a sophisticated and extensive pipeline system. In many underdeveloped areas, large quantities of natural gas are being flared because they must be produced with crude oil. However, the opportunity for utilizing the streams or for bringing the gas to industrial markets is being developed. Several large-scale ammonia plants have been built in developing countries (Pakistan, Saudi Arabia, Iran, etc). In some cases, pipeline delivery is feasible, namely from Algeria to France and from Libya to Italy. A third possibility is liquefaction of the methane and shipment in specially designed refrigerated tanker ships.

Large-scale recovery of natural gas liquid (NGL) occurs in relatively few countries. This recovery is almost always associated with the production of ethylene by thermal cracking. Some propane also is used for cracking, but most of it is used as LPG, which usually contains butanes as well. Propane and ethane also are produced in significant amounts as by-products, along with methane, in various refinery processes, eg, catalytic cracking, crude distillation, etc. They either are burned as refinery fuel or are processed to produce LPG and/or cracking feedstock for ethylene production.

Uses

Methane. The largest use of methane is for synthesis gas, a mixture of hydrogen and carbon monoxide. Synthesis gas, in turn, is the primary feed for the production of ammonia and methanol. Synthesis gas is produced by steam reforming of methane over a nickel catalyst. Methane is also used for the production of several halogenated products, principally the chloromethanes. Due to environmental pressures, this outlet for methane is decreasing rapidly.

Ethane and Propane. The most important commercial use of ethane and propane is in the production of ethylene by way of high temperature (ca 1000 K) thermal cracking. Ethane has been investigated as a feedstock for production of vinyl chloride, at scales up to a large pilot plant, but nearly all vinyl chloride is still produced from ethylene. Propane's largest use outside of steam cracking is as fuel, since propane is the chief constituent of NGL.

Table 2. World Natural Gas Production, 1993

Country	Production, 10^9 m^{3a}	Share, %
CIS (former USSR)	761.0	35.0
United States	544.5	25.0
Canada	156.5	7.2
the Netherlands	85.5	3.9
United Kingdom	63.2	2.9
Indonesia	53.3	2.4
Algeria	50.6	2.3
Mexico	37.1	1.7
Saudi Arabia	32.1	1.5
Iran	31.4	1.3
Norway	24.8	1.1
United Arab Emirates	24.3	1.1
Australia	24.0	1.1
next seven	135.2	6.1
all others	156.3	7.2
Total	*2179.8*	*100.0*

[a] To convert m^3 to ft^3, multiply by 35.3.

BUTANES

Butanes are naturally occurring alkane hydrocarbons that are produced primarily in association with natural gas processing and certain refinery operations such as catalytic cracking and catalytic reforming. The term butanes includes the two structural isomers, *n*-butane, $CH_3CH_2CH_2CH_3$, and isobutane, $(CH_3)_2CHCH_3$ (2-methylpropane).

Properties

The properties of butane and isobutane have been summarized in Table 3. The alkanes have low reactivities as compared to other hydrocarbons. Much alkane chemistry involves free-radical chain reactions that occur under vigorous conditions, eg, combustion and pyrolysis. Isobutane exhibits a different chemical behavior than *n*-butane, owing in part to the presence of a tertiary carbon atom and to the stability of the associated free radical.

Reactions of n-Butane. The most important industrial reactions of *n*-butane are vapor-phase oxidation to form maleic anhydride, thermal cracking to produce ethylene, liquid-phase oxidation to produce acetic acid and oxygenated by-products, and isomerization to form isobutane.

Reactions of Isobutane. The addition of isobutane to various C_3–C_4 alkenes is used in the production of high quality gasoline blending stock.

Table 3. Properties of Butane

Property	*n*-Butane	Isobutane
molecular weight	58.124	58.124
normal fp in air at 101.3 kPa,[a] K	134.79	113.55
normal bp at 101.3 kPa,[a] K	272.65	261.43
flammability limits at 293.15 K and 101.3 kPa,[a] vol %		
in air		
lower	1.8	1.8
upper	8.4	8.4
autoignition temp in air at 101.3 kPa,[a] K	693	693
flash point, K	199	190
heat of combustion, kJ/mol[b]		
gross[c]		
gas	2880	2866
liquid	2853	2847
heat of vaporization at normal bp, kJ/mol[b]	22.39	21.30
vapor pressure at 310.93 K, kPa[a]	356	498
thermal conductivity at 101.3 kPa,[a] W/(m·K)		
density, kg/m^3 gas, at 101.3 kPa[a]	2.5379	2.5285
stoichiometric combustion flame temp, K, in air	2243	2246
maximum flame speed in air, m/s	0.37	0.36

[a] To convert kPa to atm, divide by 101.3.
[b] To convert J to cal, divide by 4.184.
[c] Real gas at 101.3 kPa[a] and 288.7 K; liquid at saturation pressure and 298.15 K.

Shipment

Butanes are shipped by pipeline, rail car, sea tanker, barge, tank truck, and metal bottle throughout the world. All U.S. container shipments must meet Department of Transportation regulations. Domestic water shipments are regulated by the U.S. Coast Guard.

Health and Safety

n-Butane and isobutane are colorless, flammable, and nontoxic gases. They are simple asphyxiants, irritants, and anesthetics at high concentrations. Because they are heavier than air, they should not be used near sparking motors or other nonexplosion-proof equipment. Contact of the liquid form of the hydrocarbons with the skin can cause frostbite. Both butane and isobutane form solid hydrates with water at low temperatures. Hydrate formation in liquefied light petroleum product pipelines and certain processing equipment can lead to pluggage and associated safety problems.

Uses

Butanes are used as gasoline blending components, liquefied gas fuel, and in the manufacture of chemicals. *n*-Butane and small amounts of isobutane are blended directly into motor fuel to control the fuel's volatility.

Pentanes

There are three isomeric pentanes, ie, saturated aliphatic hydrocarbons of molecular formula C_5H_{12}. They are commonly called *n*-pentane, Isopentane (2-methylbutane), and neopentane (2,2-dimethylpropane).

Properties

Each isomer has its individual set of physical and chemical properties; however, these properties are similar (Table 4). The fundamental chemical reactions for pentanes are sulfonation to form sulfonic acids, chlorination to form chlorides, nitration to form nitropentanes, oxidation to form various compounds, and cracking to form free radicals. Many of these reactions are used to produce intermediates for the manufacture of industrial chemicals. Generally the reactivity increases from a primary to a secondary to a tertiary hydrogen.

Occurrence and Recovery

Pentanes occur chiefly in straight-run gasoline, natural gasoline, and in certain refinery streams. Appreciable quantities of pentanes are produced in catalytic cracking, while smaller amounts come from hydrocracking and catalytic reforming.

Table 4. Properties of Pentanes

Property	*n*-Pentane	Isopentane	Neopentane
molecular weight	72.151	72.151	72.151
normal freezing point, K	143.429	113.250	256.57
normal bp, K	309.224	301.002	282.653
water solubility at 25°C, g C_5H_{12}/100 kg H_2O	9.9	13.2	
spontaneous ignition temp in air, K	557.0	700.0	729.0
flash point, K	233.0	213.0	198.0
heat of combustion, kJ/mol[a] at 298 K			
liquid	3245	3239	3230
gas	3272	3264	3253
heat of vaporization, kJ/mol[a]	25.77	24.69	22.75
dielectric constant	1.843	1.843	1.801
ASTM octane number			
research	61.8	93.0	85.5
motor	63.2	89.7	80.2

[a] To convert J to cal, divide by 4.184.

Most pentanes are still blended into motor fuel, though increasingly strict vapor pressure regulations may end this practice in the United States by the year 2000.

Health and Safety

Pentanes are only slightly toxic. Because of their high volatilities and consequently, their low flash points, they are highly flammable. Pentanes are classified as nonreactive, ie, they do not react with fire-fighting agents. Pentanes are classified as simple asphyxiants and anesthetics.

The ICC classifies all three pentanes as flammable liquids and requires that they be affixed with a red label for shipping. Because of their high vapor pressure, *n*-and isopentane are transported in heavy-walled drums and neopentane is transported in cylinders.

Uses

The main use for pentanes has been in motor fuel, though regulations limiting fuel vapor pressure are decreasing the amount of pentanes, particularly isopentane, present in gasoline during warm parts of the year.

Isopentane can be alkylated with light olefins to give gasoline material; however, the resulting alkylate is lower quality (research octane = 78–80) than that produced from isobutane (research octane = 90–98). Some outlet has to be found for the increasing amount of pentane displaced from gasoline by vapor pressure regulation, and it is likely that much of that pentane will find its way into alkylation streams.

Some isopentane is dehydrogenated to isoamylene and converted, by processes analogous to those which produce methyl *t*-butyl ether (MTBE) to *t*-amyl methyl ether (TAME), which is used as a fuel octane enhancer like MTBE.

HEXANES

Hexane refers to the straight-chain hydrocarbon, C_6H_{14}; branched hydrocarbons of the same formula are isohexanes. Hexanes include the branched compounds, 2-methylpentane, 3-methylpentane, 2,2-dimethylbutane, 2,3-dimethylbutane, and the straight-chain compound, *n*-hexane. Commercial hexane is a narrow-boiling mixture of these compounds with methylcyclopentane, cyclohexane, and benzene; minor amounts of C_5 and C_7 hydrocarbons also may be present. Hydrocarbons in commercial hexane are found chiefly in straight-run gasoline which is produced from crude oil and natural gas liquids. Smaller volumes occur in certain petroleum refinery streams.

Properties

Properties of the principal hydrocarbons found in commercial hexane are shown in Table 5. The flash point of *n*-hexane is −21.7 °C and the autoignition temperature is 225°C. The explosive limits of hexane vapor in air are 1.1–7.5%. Above 2°C the equilibrium mixture of hexane and air above the liquid is too rich to fall within these limits.

Manufacture

Commercial hexanes are manufactured by two-tower distillation of a suitable charge stock, eg, straight-run gasolines that have been distilled from crude oil or natural gas liquids that have been stripped from natural gas. Highly pure *n*-hexane can be produced by adsorption on molecular sieves.

Health and Safety

Hexane is classified as a flammable liquid by the ICC, and normal handling precautions for this type of material should be observed.

n-Hexane can be grouped with the general anesthetics in the class of central nervous system depressants. Hexane vapors are mildly irritating to mucous membranes. Exposure to concentrations in excess

Table 5. Properties of Hydrocarbons Found in Commercial Hexanes

Hydrocarbon	Freezing point, °C	Normal bp, °C	Liquid density, kg/m^3 at 20°C	Liquid refractive index, n_D^{20}
2-methylbutane	−159.900	27.852	619.67	1.35373
n-pentane	−129.730	36.065	626.20	1.35748
cyclopentane	−93.866	49.262	745.38	1.40645
2,2-dimethylbutane	−99.870	49.741	649.16	1.36876
2,3-dimethylbutane	−128.538	57.988	661.64	1.37495
2-methylpentane	−153.660	60.271	653.15	1.37145
3-methylpentane		63.282	664.31	1.37652
n-hexane	−95.322	68.736	659.33	1.37486
methylcyclopentane	−142.455	71.812	748.64	1.40970
benzene	5.533	80.100	879.01	1.50112
cyclohexane	6.554	80.738	778.55	1.42623
2,2-dimethylpentane	−123.811	79.197	673.85	1.38215
2,4-dimethylpentane	−119.242	80.500	672.70	1.38145
1,1-dimethylcyclopentane	−69.795	87.846	754.48	1.41356

of 1% hexane may cause dizziness, unconsciousness, prostration, and death. Prolonged skin contact with hexane results in irritation and dermatitis. Direct contact with lung tissue can result in chemical pneumonitis, pulmonary edema, and hemorrhage.

Uses

Other than fuel, the largest volume appliction for hexane is in extraction of oil from seeds, eg, soybeans, cottonseed, safflower seed, peanuts, rapeseed, etc. Hexane is also a desirable solvent and reaction medium in the manufacture of polyolefins, synthetic rubbers, and some pharmaceuticals.

Cyclohexane

Cyclohexane, C_6H_{12}, is a clear, essentially water-insoluble, noncorrosive liquid that has a pungent odor. It is easily vaporized, readily flammable, and less toxic than benzene. Structurally, it is a cycloparaffin.

Properties

Properties of cyclohexane are given in Table 6.

Stereochemistry. Cyclohexane can exist in two molecular conformations: the chair and boat forms. The predominant stereochemistry of cyclohexane has no influence in its use as a raw material for nylon manufacture or as a solvent.

Reactions. The most important commercial reaction of cyclohexane is its oxidation (in liquid phase) with air in the presence of soluble cobalt catalyst or boric acid to produce cyclohexanol and cyclohexanone. Cyclohexanol is dehydrogenated with zinc or copper catalysts to cyclohexanone which is used to manufacture caprolactam.

Table 6. Properties of Cyclohexane

Property	Value
mol wt	84.156
fp, °C	6.554
bp, °C	80.738
flammability limits (in air), vol %	1.3–8.4
flash point (closed up), °C	−17
heat of transition, kJ/kg[a]	80.08
heat of fusion, kJ/kg[a]	31.807
heat of vaporization at 25°C, kJ/kg[a]	392.50
vapor pressure, kPa[b]	
at 30°C	16.212
at 80°C	99.095

[a] To convert J to cal, divide by 4.184. [b] To convert kPa to atm, divide by 101.3.

Occurrence

Cyclohexane is present in all crude oils in concentrations of 0.1–1.0%.

Manufacture and Shipment

Essentially all high purity cyclohexane is made by hydrogenation of benzene. A small amount of cyclohexane of lower purity is produced by fractional distillation from crude oil and from catalytic reformer effluent.

Economic Aspects

In a well-designed multistage hydrogenation unit, operating costs are small as a result of recovery of the heat of hydrogenation between reactor stages by steam generation or integration with other process units, or by more efficient one- and two-stage processes. Consequently, the principal costs in cyclohexane manufacture are maintenance expenses, interest and return charges on the plant and working capital, and the cost of benzene and high purity hydrogen. The price of cyclohexane is dependent on the price of benzene. Virtually all cyclohexane goes to the production of nylon. The U.S. accounts for about a third of the world's consumption of cyclohexane, or 3.785×10^6 m^3/yr (~1 billion gal/yr).

Health and Safety

The threshold limit value (TLV) for cyclohexane is 300 ppm (1050 mg/m^3). With prolonged exposure at 300 ppm and greater, cyclohexane may cause irritation to eyes, mucous membranes, and skin. At high concentrations, it is an anesthetic and narcosis may occur. Because of its relatively low chemical reactivity, toxicological research has not been concentrated on cyclohexane.

Uses

Almost all of the cyclohexane that is produced in concentrated form is used as a raw material in the first step of nylon-6 and nylon-6,6 manufacture. Cyclohexane also is an excellent solvent for cellulose ethers, resins, waxes, fats, oils, bitumen, and rubber.

ALAN D. EASTMAN
Phillips Petroleum
DAVID MEARS
Unocal Corporation

Synthetic Organic Chemicals, United States Production and Sales, 1991, U.S. International Trade Commission Publication 2607, Washington, D.C., Feb. 1993.

Gas Processors Suppliers Association Engineering Data Book, 9th ed., Gas Processors Suppliers Association, Tulsa, Okla., 1972, 1977.

W. Braker and A. L. Mossman, *Matheson Gas Data Book,* 5th ed., Matheson Gas Products, East Rutherford, N.J., 1971.

HYDROCHLORIC ACID. See HYDROGEN CHLORIDE.

HYDROCOLLOIDS. See GUMS.

HYDROGEL. See CONTACT LENSES.

HYDROGEN

Hydrogen, the lightest element, has three isotopes: hydrogen, H, at wt 1.0078; deuterium, D, at wt 2.0141; and tritium, T, at wt 3.0161.

Hydrogen is very abundant, being one of the atoms composing water; deuterium and tritium occur naturally on earth, but at very low levels. Tritium, a radioactive low energy beta-emitter with a half-life of 12.26 yr, is useful as a tracer in hydrogen reactions.

Whereas hydrogen atoms exist under certain conditions, the normal state of pure hydrogen is the hydrogen molecule, H_2, which is the lightest of all gases. Molecular hydrogen is a product of many reactions, but is present at only low levels (0.1 ppm) in the earth's atmosphere. The hydrogen molecule exists in two forms, designated *ortho*-hydrogen and *para*-hydrogen, depending on the nuclear spins of the atoms. Many physical and thermodynamic properties of H_2 depend on the nuclear spin orientation, but the chemical properties of the two forms are the same.

Hydrogen is a very stable molecule having a bond strength of 436 kJ/mol (104 kcal/mol), and is not particularly reactive under normal conditions. However, at elevated temperatures and with the aid of catalysts, H_2 undergoes many reactions. Hydrogen forms compounds with almost every other element, often by direct reaction of the elements. The explanation for its ability to form compounds with such chemically dissimilar elements as alkali metals, halogens, transition metals, and carbon lies in the intermediate electronegativity of the hydrogen atom.

Hydrogen is one of the most important industrial commodities. It is used in the production of ammonia, urea, methanol and higher alcohols, and hydrochloric acid; as a reducing agent; and to desulfurize or hydrogenate various petroleum and edible oils. Hydrogen, produced as a by-product, is used in a multitude of industrial processes as a fuel, and liquid hydrogen is an important cryogenic fluid. Almost all commercial hydrogen is produced by reaction of water and hydrocarbons, the steam reforming reaction, or by partial oxidation of hydrocarbons. Electrolysis is practiced to a limited extent.

Hydrogen is seen by many as having a central role in the future energy equation. This role is envisioned as one of an energy carrier. Hydrogen would be produced from primary renewable energy sources, eg, solar energy, by various water splitting techniques. The hydrogen produced would be shipped to various locations and used as a fuel or chemical commodity. Upon combustion, hydrogen returns to water, accompanied by virtually no pollution and no greenhouse gas production, in contrast to what occurs when hydrocarbons are burned.

Physical and Thermodynamic Properties

Tables 1, 2, and 3 outline many of the physical and thermodynamic properties of para- and normal hydrogen in the solid, liquid, and gaseous states, respectively.

Bonding of Hydrogen to Other Atoms. The hydrogen atom can either lose the 1*s* valence electron when bonding to other atoms, to form the H^+ ion, or conversely, it can gain an electron in the valence shell to form the hydride ion, H^-. Most hydrogen compounds are formed through covalent bonding of hydrogen to the other atoms.

Reactions of Synthesis Gas. The main hydrogen manufacturing processes produce synthesis gas, a mixture of H_2 and CO. Synthesis gas can have a variety of H_2-to-CO ratios, and the water gas shift reaction is used to reduce the CO level and produce additional hydrogen, or to adjust the H_2-to-CO ratio to one more beneficial to subsequent processing. Synthesis gas is used mainly to produce ammonia and methanol.

Other Reactions of Hydrogen. Sulfur, nitrogen, and oxygen are heteroatoms, which are abundant in many fuel sources such as petroleum, coal, and oil shale. These elements are considered pollutants and detriments to the refining process. Hydrogen is used to reduce the levels of these contaminants. Hydrogen reacts with a number of metal oxides at elevated temperatures to produce the metal and water.

Reactions of Atomic Hydrogen. Atomic hydrogen is a very strong reducing agent and a highly reactive radical that can be produced by various means.

Absorption of Hydrogen in Metals. Many metals and alloys absorb hydrogen in large amounts. The absorption is largely reversible for palladium and for some other metals and alloys. Hydrogen diffuses and absorbs in many metals, with detrimental effects. Hydrogen ex-

Table 1. Physical and Thermodynamic Properties of Solid Hydrogen

Property	Hydrogen, *para*-	Hydrogen, Normal
mp, K (triple point)	13.803	13.947
vapor pressure at mp, kPa[a]	7.04	7.20
density at mp, (mol/cm^3) × 10^3	42.91	43.01
heat of fusion at mp, J/mol[b]	117.5	117.1
heat of sublimation at mp, J/mol[b]	1023.0	1028.4
thermal conductivity at mp, mW/(cm·K)	9.0	9.0
dielectric constant at mp	1.286	1.287

[a] To convert kPa to mm Hg, multiply by 7.5. [b] To convert J to cal, divide by 4.184.

Table 2. Physical and Thermodynamic Properties of Liquid Hydrogen

Property	Hydrogen, *para*-	Hydrogen, Normal
mp, K (triple point)	13.803	13.947
normal bp, K	20.268	20.380
density at bp, mol/cm^3	0.03511	0.03520
density at mp, mol/cm^3	0.038207	0.03830
compressibility factor at mp, Z = PV/RT	0.001606	0.001621
coefficient of volume expansion, $(-\partial V/V\partial T)_p$ at triple point, K^{-1}	0.0102	0.0102
heat of vaporization at triple point, J/mol[a]	905.5	911.3
viscosity at triple point, mPa·s(= cP)	0.026	0.0256
thermal conductivity at triple point, mW/(cm·K)	0.73	0.73
dielectric constant at triple point	1.252	1.253

[a] To convert J to cal, divide by 4.184.

Table 3. Physical and Thermodynamic Properties of Gaseous Hydrogen[a]

Property	Hydrogen, *para*-	Hydrogen, Normal
density at 0°C, (mol/cm^3) × 10^3	0.05459	0.04460
compressibility factor, Z = PV/RT, at 0°C	1.0005	1.00042
adiabatic compressibility, $(\partial V/V\partial P)_s$, at 300 K, MPa^{-1b}	7.12	7.03
coefficient of volume expansion, $(\partial V/V\partial P)_p$, at 300 K, K^{-1}	0.00333	0.00333
C_p at 0°C, J/(mol·K[c]	30.35	28.59
C_v at 0°C, J/(mol·K)[c]	21.87	20.30
enthalpy at 0°C, J/mol[c,d]	7656.6	7749.2
internal energy at 0°C, J/mol[c,d]	5384.5	5477.1
entropy at 0°C, J/(mol·K)[c,d]	127.77	139.59
velocity of sound at 0°C, m/s	1246	1246
viscosity at 0°C, mPa·s(= cP)	0.00839	0.00839
thermal conductivity at 0°C, mW/(cm·K)	1.841	1.740
dielectric constant at 0°C	1.00027	1.000271
isothermal compressibility $1/V(\partial V\partial P)_T$, at 300 K, MPa^{-1b}	−9.86	−9.86
self-diffusion coefficient at 0°C, cm^2/s		1.285
gas diffusivity in water at 25°C, cm^2/s		4.8 × 10^{-5}
Lennard-Jones parameters collision diameter, σ m × 10^{10}		2.928
interaction parameter, ε/k, K		37.00
heat of dissociation at 298.16 K, kJ/mol[c]	435.935	435.881

[a] All values at 101.3 kPa (1 atm). [b] To convert MPa to atm, divide by 0.101. [c] To convert J to cal, divide by 4.184. [d] Base point (zero values) for enthalpy, internal energy, and entropy are 0 K for the ideal gas at 101.3 kPa (1 atm) pressure.

posure, under certain conditions, can seriously weaken and embrittle steel and other metals.

Manufacture

The principal commercial processes specific for the manufacture of hydrogen are steam reforming, partial oxidation, coal gasification, and water electrolysis. However, these are not of equal economic importance. In the U.S., the bulk of the industrial hydrogen is manufactured by steam reforming of natural gas. Relatively small quantities of hydrogen are produced by steam reforming of naphtha, partial oxidation of oil, coal gasification, or water electrolysis. Worldwide, hydrogen as a raw material for the chemical industry is derived as follows: 77% from natural gas/petroleum, 18% from coal, 4% by water electrolysis, and 1% by other means. Significant quantities of hydrogen, especially to satisfy refinery H_2 demand, are produced as by-product $-H_2$.

Hydrogen Purification

A wide range and a number of purification steps are required to make available hydrogen/synthesis gas having the desired purity that depends on use. Technology is available in many forms and combinations for specific hydrogen purification requirements. Methods include physical and chemical treatments (solvent scrubbing); low temperature (cryogenic) systems; adsorption on solids, such as active carbon, metal oxides, and molecular sieves, and various membrane systems. Composition of the raw gas and the amount of impurities that can be tolerated in the product determine the selection of the most suitable process.

Environmental Considerations

Short-term environmental concerns associated with hydrogen production are minimized by the use of steam reforming of natural gas and by the recovery of hydrogen as a by-product. The methane steam reforming process is one of the most environmentally acceptable. The environmental concerns become greater for the partial oxidation of heavier fuels and coal gasification technologies. Although the technology exists for effective environmental controls for these latter processes, the controls place a heavy economic burden on the overall system.

Coal feedstocks present the most serious environmental problems because of potential particulate emissions from coal-handling and processing facilities. Additionally, disposal of ash and slag solids removed from the gasification step must be in an environmentally safe manner. Some gasification produces significant amounts of liquid by-products such as tars, phenols, and naphthas which must be either recovered or incinerated. Some coals contain significant amounts of sulfur which must be stripped out of the raw syngas as hydrogen and carbonyl sulfides, necessitating further processing in a sulfur recovery unit. Condensate streams from gasifiers may also contain hydrogen cyanide and soluble metals in addition to ammonia which further complicates disposal.

Partial oxidation of heavy liquid hydrocarbons requires somewhat simpler environmental controls. The principal source of particulates is carbon, or soot, formed by the high temperature of the oxidation step. The soot is scrubbed from the raw synthesis gas and either recycled back to the gasifier, or recovered as solid pelletized fuel. Sulfur and condensate treatment is similar in principle to that required for coal gasification, although the amounts of potential pollutants generated are usually less.

Shipment and Storage

Whereas the safe storage of hydrogen has been practiced for many years, as of this writing almost all hydrogen is used near the production site.

Hydrogen Gas. Hydrogen is supplied by many vendors as a high pressure gas in steel cylinders. The use of hydrides as a means of storing hydrogen is not yet (ca 1994) of commercial importance. Hydride storage has been used in demonstrations, eg, to power automobiles. Hydride formulations and properties are available.

Storage in Microcapsules. Storage in microcapsules is under development and has no commercial significance as of this writing. Glass microcapsules behave very much like metal hydrides. Future storage methods may involve existing underground formations that previously held natural gas.

Economic Aspects

The United States consumes about 1.2 EJ (1.1×10^{15} Btu(1.1 quad)) of hydrogen annually. Most U.S. hydrogen production, estimated at over 6.5×10^{10} m^3/yr(2.3×10^{12} ft^3/yr), is used captively in the production of ammonia and methanol as well as in refinery operations. Sales or merchant use may total about 2.0×10^9 m^3, ca 3%, of production and is principally divided between two U.S. producers, Praxair and Air Products. Additional merchant hydrogen capacity is located in Canada. Merchant hydrogen uses are divided among chemicals (83%), electronics (5%), metals (5%), government (4%), float glass (1%), and foods (1%). The balance is miscellaneous uses.

Liquid Hydrogen. The use of liquid hydrogen is a well established technology because of its use in the space program. Liquid hydrogen is, however, more difficult to produce and maintain than liquid natural gas. Refrigeration costs are high owing to the low (bp = 20.4 K) liquefication temperature. There are a number of special problems associated with liquid hydrogen. Examples are the need to precool the gas to the inversion temperature before the hydrogen can cool on expansion to liquefy, and the exothermic ortho-to-para conversion after liquefaction.

Large-scale use of liquid hydrogen has led to the construction of large insulated storage tanks such as the 1893 m^3 (500,000 gal) liquid hydrogen storage sphere erected at the then Atomic Energy Commission (now DOE) test site in Nevada in 1963. Liquid hydrogen has been transported by rail in tank cars of 36 and 107 m^3 (9,500 and 28,000 gal) capacity. These latter have a special Linde cryogenic insulation and operate under less than 133 mPa (1 μm Hg) absolute pressure with a heat-transfer coefficient of 0.1163 W/(m^2·K).

The ICC classifies hydrogen as a flammable gas and requires that it carry a red label. Data on storage is available. The production and handling of flammable gases and liquefied flammable gases is regulated by OSHA.

Health and Safety Factors

Hydrogen gas is not considered toxic but it can cause suffocation by the exclusion of air. The main danger in the use of liquid and gaseous hydrogen lies in its extreme flammability in oxygen or air.

Mandatory regulations governing the distribution of liquid or gaseous hydrogen are available, as are guidelines on the safe use of liquid hydrogen and gaseous hydrogen. Other reports concerning hydrogen safety may be found.

T. A. CZUPPON
S. A. KNEZ
D. S. NEWSOME
The M. W. Kellogg Company

A. G. Sharpe, *Inorganic Chemistry*, 3rd ed., Longman Scientific and Technical, Burnt Hill, Essex, U.K. (co-published in U.S. by John Wiley & Sons, Inc. New York), 1992, p. 211.

K. M. Mackay and M. F. A. Dove in J. C. Bailar, H. J. Emeleus, R. Nyholm, and A. F. Trotman-Dickenson, eds., *Comprehensive Inorganic Chemistry*, Vol. 1, Pergamon Press, New York, 1973, p. 93.

R. D. McCarty, *Hydrogen Technological Survey—Thermophysical Properties*, NASA SP-3089, U.S. Government Printing Office, Washington, D.C., 1975, pp. 518–519.

R. D. McCarty, J. Hord, H. M. Roder, *Selected Properties of Hydrogen (Engineering Design Data)*. U.S. Dept. of Commerce, National Bureau of Standards, Washington, D.C., 1981, pp. 6–291.

HYDROGEN BROMIDE. See BROMINE COMPOUNDS.

HYDROGEN CHLORIDE

History and Occurrence

Hydrogen chloride, HCl, exists in solid, liquid, and gaseous states and is very soluble in water. It is found naturally in gases evolved from volcanoes, particularly those in Mexico and South America. Its formation is attributed to the high temperature reaction of water with the salts found in seawater. The original atmosphere of the earth is considered to have contained water, carbon dioxide, and hydrogen chloride in the ratio of 20:3:1. Hydrogen chloride was also detected in the atmosphere of the planet Venus. The dissociation of HCl is considered the source of chlorine detected in the spectra of distant stars.

Hydrochloric acid is also present in the digestive system of most mammals. The gastric mucosa lining the human stomach produces about 1.5 L/d of gastric juices, containing an acid concentration in the range of 0.05 to 0.1 *N*. A deficiency of hydrochloric acid impairs the digestive process, particularly of carbohydrates and proteins, and excess acid causes gastric ulcers.

Physical and Thermodynamic Properties

Anhydrous Hydrogen Chloride. Anhydrous hydrogen chloride is a colorless gas that condenses to a colorless liquid and freezes to a white crystalline solid. The physical and thermodynamic properties of HCl are summarized in Table 1 for selected temperatures and pressures. The high thermal stability of hydrogen chloride is a consequence of the large enthalpy of its formation.

Hydrogen Chloride–Water system. Hydrogen chloride is highly soluble in water and this aqueous solution does not obey Henry's law at all concentrations.

Hydrogen chloride and water form four hydrates. The dihydrate is formed when a saturated solution is cooled at atmospheric pressure. The monohydrate has a melting point of −15.35 °C; the trihydrate has a melting point of −24.9 °C; the hexahydrate is very unstable and has a melting point of −70 °C. Addition of hydrogen chloride to pure water lowers the freezing point until a eutectic temperature of about −85 °C is reached at 25% HCl. Hydrogen chloride and water form constant boiling mixtures. Hydrogen chloride is completely ionized in aqueous solutions at all but the highest concentrations. The viscosity of hydrochloric acid solutions, η, increases slightly with increasing concentration. The specific heat of aqueous solutions of hydrogen chloride decreases with acid concentration. The electrical conductivity of aqueous hydrogen chloride increases with temperature.

Hydrogen Chloride–Water–Inorganic Compound Systems. Salting out metal chlorides from aqueous solutions by the common ion effect upon addition of HCl is utilized in many practical applications. The properties of the $FeCl_2 \cdot HCl \cdot H_2O$ system are important to the steel-pickling industry. Other metal chlorides that are salted out by the addition of hydrogen chloride to aqueous solutions include those of magnesium, strontium, and barium.

Metal chlorides which are not readily salted out by hydrochloric acid can require high concentrations of HCl for precipitation. This property is used to recover hydrogen chloride from azeotropic mixtures. The solubility of chlorine in hydrochloric acid is an important factor in the purification of by-product hydrochloric acid.

Hydrogen Chloride–Organic Compound Systems. The solubility of hydrogen chloride in many solvents follows Henry's law. Notable exceptions are HCl in polyhydroxy compounds such as ethylene glycol, which have characteristics similar to those of water.

Chemical Properties

Reactions of Anhydrous Hydrogen Chloride. Hydrogen chloride reacts with inorganic compounds by either heterolytic or homolytic fission of the H–Cl bond. However, anhydrous HCl has high kinetic barriers to either type of fission and hence, this material is relatively inert.

Anhydrous HCl protonates the Group 5 (V) hydrides. The heavier transition-metal oxides require a higher reaction temperature, and the primary reaction product is usually the corresponding oxychlorides. Thermodynamic considerations for the reaction

$$M + nHCl \rightarrow MCl_n + n/2H_2$$

indicate that most metals should react with HCl. However, this reaction is kinetically slow at all but elevated temperatures.

Hydrogen chloride and oxygen react in the gaseous state to liberate chlorine. Anhydrous HCl forms addition compounds at lower temperatures with halogen acids such as HBr and HI, and also with HCN. Hydrogen chloride reacts with sulfur trioxide, yielding liquid chlorosulfuric acid.

Reaction with Organic Compounds. Hydrogen chloride adds to carbon–carbon double and triple bonds in a variety of organic compounds. Acetylene and hydrogen chloride historically were used to make chloroprene. The olefin reaction is used to make ethyl chloride from ethylene and to make 1,1-dichloroethane from vinyl chloride.

Lower alcohols such as methanol can be converted to the corresponding alkyl chlorides by carrying out the reaction

$$ROH + HCl \rightarrow RCl + H_2O$$

using either a liquid or a solid catalyst.

The introduction of the chloromethyl group to both aliphatic and aromatic compounds is carried out by reaction of paraformaldehyde and hydrogen chloride. This method is used for synthesizing methyl chloromethyl ether, benzyl chloride, and chloromethyl acetate.

Table 1. Physical and Thermodynamic Properties of Anhydrous Hydrogen Chloride

Property	Value
melting point, °C	−114.22
boiling point, °C	−85.05
heat of vaporization at −85.05 °C, kJ/mol[a]	16.1421
triple point, °C	−114.25
critical pressure, P_c, MPa[b]	8.316
critical volume, V_c, L/mol	0.069
critical density, g/L	424
compressibility coefficient	0.00787
heat capacity, C_p, J/(mol · k)[a]	
liquid at 163.16 K	60.378
solid at 147.16 K	48.98
surface tension at 118.16 K, mN/cm(= dyn/cm)	23
viscosity, mPa·s(= cP)	
liquid at 118.16 K	0.405
vapor at 273.06 K	0.0131
thermal conductivity, mW/(m·K)	
liquid at 118.16 K	335
vapor at 273.16 K	13.4
density, g/cm^3 liquid at 118.16 K, solid	1.045
rhombic at 81 K	1.507
cubic at 98.36 K	1.48
refractive index	
liquid at 283.16 K	1.254
gas at 273.16 K	1.0004456
dielectric constant	
liquid at 158.94 K	14.2
gas at 298.16 K	1.0046
electrical conductivity, $(\Omega \cdot m)^{-1}$	
at 158.94 K	1.7×10^{-7}
at 185.56 K	3.5×10^{-7}

[a] To convert J to cal, divide by 4.184. [b] To convert MPa to atm, divide by 0.101.

Hydrochloric Acid. Most metals and alloys react with aqueous hydrochloric acid via

$$M + n\, H_3O^+ \rightarrow M^{n+} + n\, H_2O + n/_2\, H_2$$

This is essentially a corrosion reaction involving anodic metal dissolution where the conjugate reaction is the hydrogen evolution process.

Oxides and hydroxides react with HCl to form a salt and water as in a simple acid–base reaction. However, reactions with low solubility or insoluble oxides and hydroxides is complex and the rate is dependent on many factors similar to those for reactions with metals.

HCl can be electrolyzed to produce H_2 and chlorine. Many organic reactions are catalyzed by acids such as HCl. Typical examples of the use of HCl in these processes include conversion of lignocellulose to hexose and pentose, sucrose to inverted sugar, esterification of aromatic acids, transformation of acetaminochlorobenzene to chloroanilides, and inversion of methone.

Manufacturing and Processing

Hydrogen chloride is produced by the direct reaction of hydrogen and chlorine, by reaction of metal chlorides and acids, and as a by-product from many chemical manufacturing processes such as chlorinated hydrocarbons.

Hydrogen Chloride Produced from Incineration of Waste Organics. Environmental regulations regarding the disposal of chlorine-containing organic wastes has motivated the development of technologies for burning or pyrolyzing the waste organics and recovering the chlorine values as hydrogen chloride. Several catalytic and noncatalytic processes have been developed to treat these wastes to produce hydrogen chloride.

Hydrogen Chloride from Hydrochloric Acid Solutions. Gaseous hydrogen chloride is obtained by partially stripping concentrated hydrochloric acid using an absorber–desorber system.

Purification. Gaseous HCl from all the manufacturing processes described invariably contains moisture, and sometimes organic species. H_2SO_4 drying can be used to remove small amounts of water, reducing the residual water content to less than 0.02%. If the water content is high, it can be removed as concentrated hydrochloric acid by cooling the gas mixture before drying with sulfuric acid. Addition of chlorosulfuric acid to this stream reduces the water content to less than 10 ppm. This mixture also removes unsaturated organics such as ethylene and vinyl chloride and certain organic compounds such as monochloroacetic acid.

Chlorine can be removed by either activated carbon adsorption or by reaction with olefins such as ethylene over-activated carbon at temperatures of 30–200°C. Addition of liquid high boiling paraffins can reduce the chlorine content in the HCl gas to less than 0.01%. Solid absorbents generally remove the organics from HCl^-. Crude HCl recovered from production of chlorofluorocarbons by hydrofluorination of chlorocarbons contains unique impurities which can be easily removed.

Use of air or purified HCl gas as stripper is practiced to remove volatile dissolved organics and chlorine from aqueous HCl.

Materials of Construction, Storage, and Handling

Gaseous Hydrogen Chloride. Cast iron, mild steel, and steel alloys are resistant to attack by dry, pure HCl at ambient conditions and can be used at temperatures up to the dissociation temperature of HCl.

Aqueous Hydrochloric Acid. Tantalum and zirconium exhibit the highest corrosion resistance to HCl. However, the corrosion resistance of zironium is severely impaired by the presence of ferric or cupric chlorides. Tantalum–molybdenum alloys containing more than 50% tantalum are reported to have excellent corrosion resistance. Common plastics and elastomers show excellent resistance to hydrochloric acid within the temperature limits of the materials. Carbon and graphite rendered impervious with 10–15% phenolic, epoxy, or furan resin are among the most important materials for hydrochloric acid service up to 170°C. The most important applications of these materials for hydrochloric acid service are heat exchangers and centrifugal pumps. Glass and ceramic-coated equipment is widely used for handling hydrochloric acid.

Production and Economic Aspects

Over 90% of the HCl produced in the United States originates as a coproduct form various chlorination processes; direct generation of HCl from H_2 and Cl_2 accounts for only about 8% of the total production.

As capital costs increase, operating costs decrease. Data indicate the direct route from H_2 and Cl_2 to be the most economic among all the technologies.

Health and Safety Factors

Hydrogen chloride in air is an irritant, severely affecting the eye and the respiratory tract. The vapor in the air, normally absorbed by the upper respiratory mucous membranes, is lethal at concentrations of over 0.1% in air, when exposed for a few minutes. The maximum allowable concentration under normal working conditions has been set at 5 ppm.

Hydrogen chloride in air can also be a phytotoxicant. Tomatoes, sugar beets, and fruit trees of the prunus family are sensitive to HCl in air. Exposure of concentrated hydrochloric acid to the skin can cause chemical burns or dermatitis.

Storage and Handling

All Department of Transportation (DOT), Environmental Protection Agency (EPA), and Occupational Safety and Health Act (OSHA) rules and regulations should be reviewed prior to handling hydrochloric acid and all the regulations must be followed. All employees handling HCl must be trained to ensure that they are familiar with the appropriate materials safety data sheets and applicable regulations.

The U.S. Department of Transportation classifies HCl as a corrosive material and requires that it be transported in DOT-approved delivery vessels.

Uses

Hydrogen chloride and the aqueous solution, muriatic acid, find application in many industries. In general, anhydrous HCl is consumed for its chlorine value, whereas aqueous hydrochloric acid is often utilized as a nonoxidizing acid. The latter is used in metal cleaning operations, chemical manufacturing, petroleum well activation, and in the production of food and synthetic rubber.

Most of the HCl produced is consumed captively, ie, at the site of production, either in integrated operations such as ethylenedichloride–vinyl chloride monomer (EDC/VCM) plants and chlorinated methane plants or in separate HCl consuming operations at the same location.

Anhydrous Hydrogen Chloride. In the U.S., all ethylene dichloride (EDC) is produced from ethylene, either by chlorination or oxychlorination (oxyhydrochlorination).

Most of the HCl consumed in the manufacture of methyl chloride from methanol is a recycled product.

Several methods are available for generating chlorine from HCl. These include electrolysis of metallic chloride solutions, electrolysis of hydrochloric acid, oxidation of hydrogen chloride to chlorine with nitric acid, and oxidation of hydrogen chloride to chlorine using oxygen in the presence of catalysts (Deacon process and the modified Deacon process).

Perchloroethylene (PCE) and trichloroethylene (TCE) can be produced either separately or as a mixture in varying proportions by reaction of C_2-chlorinated hydrocarbons.

Most ethyl chloride is produced by the hydrochlorination of ethylene using anhydrous HCl.

Other uses for anhydrous HCl include use in cottonseed delinting and disinfecting, as a catalyst promoter for petroleum isomerization, in the production of agricultural chemicals, and in the preparation of hydrochloride salts in the pharmaceutical industry.

Aqueous Hydrochloric Acid. The largest captive use of aqueous HCl is for brine acidification prior to electrolysis in chlorine/caustic cells and the largest merchant markets for HCl are steel pickling and oil-well acidizing.

MOHAMED W. M. HISHAM
TILAK V. BOMMARAJU
Occidental Chemical Corporation

S. Austin and A. Glowacki, *Ullmann's Encyclopedia of Industrial Chemistry*, 5th ed., Vol. A13, VCH, Weinheim, Germany, 1989, p. 283.
"Ethylene Dichloride," *Chemical Economics Handbook*, SRI International, Menlo Park, Calif., June 1992.
J. E. Buice, R. L. Bowlin, K. W. Mall, and J. A. Wilkinson, *Encyclopedia of Chemical Processing and Design*, Vol. 26, Marcel Dekker, Inc., New York, 1987, p. 396.
Hydrochloric Acid, Chemical Products Synopsis, Mansville Chemical Products Corp., Asbury, N. J., Jan. 1993.

HYDROGEN ENERGY

Fossil fuels provided about 80% of total world energy demand in 1990, and if oil and natural gas usage continues at the mid-1990s rate, known reserved are not expected to last into the year 2030. Coal reserves are much larger, but of all the fossil fuels coal is considered to be the least environmentally benign.

The burning of fossil fuels emits various pollutants into the atmosphere, ie, carbon dioxide; nitric and nitrous oxides, NO_x; as well as sulfur dioxide from sulfur-containing fuels such as coal; and products of incomplete combustion including carbon monoxide, unburned hydrocarbons, and particulates. Hydrogen, however, when combusted, produces mainly water.

Hydrogen, the most abundant element in the universe, is not normally found in its unreacted state in nature; it is almost always necessary to synthetically produce the hydrogen to be utilized. Hydrogen is made by splitting water into hydrogen and oxygen, a process that utilizes energy; therefore hydrogen is not usually considered a source of energy, but rather a form of stored energy. The energy to produce hydrogen from water can come from a wide variety of sources including solar energy, wind, hydroelectric, nuclear, natural gas reformation, or coal gasification. The hydrogen produced can then be stored, transported, and utilized in conventional energy consumption devices such as automobiles and gas appliances.

Hydrogen Fuel

Properties. When compared to other common fuels on a per unit mass basis, hydrogen is found to have a considerably higher heating value than many of the hydrocarbon fuels in common usage. This makes hydrogen especially attractive for use where fuel weight is of considerable importance, such as in aviation or space applications. Although hydrogen offers a much higher heating value per unit weight, it is also substantially more voluminous than hydrocarbon fuels (Table 1). The energy required to ignite hydrogen is an order of magnitude less than that required to ignite hydrocarbon fuels, but that energy must be supplied at a higher temperature than that required for most hydrocarbon fuels.

Production. Proposed methods for hydrogen production include natural gas reformation, coal gasification, electrolysis, and biological production of hydrogen.

Transporting. Small volumes of gaseous hydrogen are typically transported in high pressure containers via truck, train, or barge. The preferred method for transporting large volumes of hydrogen efficiently is through an underground pipeline system similar to existing natural gas networks.

Economic Aspects

Although hydrogen can be produced utilizing water and renewable energy, as of this writing the majority of hydrogen is synthesized from fossil fuels, eg, by methane reformation and coal gasification processes. As natural gas and petroleum reserves become depleted, a shift from a liquid fossil fuel base to cleaner burning fuels is expected.

The use and effective costs of various energy alternatives are shown in Table 2.

Table 1. Combustion Properties for Common Hydrocarbon Fuels

Fuel	Quantity of fuel[a]/GJ[b]	Flammability limit in air, vol % gas: Lower	Higher	Max flame speed, cm/s	Spontaneous ignition temp, °C	Ignition energy, mJ[b]
hydrogen, m^3	98	4	75	265	571	0.02
methane, m^3	29	4	16	34	632	0.47
propane, m^3	11	2	11	40	504	0.31
butane, m^3	9	2	10	37	431	0.76
methanol, kg	47	6	50	49	470	
gasoline, kg	22	0.8	6	34	447	1.35
diesel (heptane), kg	22	1	8	40	247	0.70
coal, kg						
anthracite	30					
bituminous	36					
lignite	58					

[a] Lower heat value fuel. [b] To convert J to cal, divide by 4.184.

Hydrogen Storage

Numerous methods of storing hydrogen are used including compressed gaseous hydrogen, liquid hydrogen, and metal hydrides. Most commercially viable hydrogen energy applications employ one or more of these storage technologies. Other hydrogen storage methods have been proposed, including solid hydrogen (slush) and microspheres, but as yet that technology has not proven feasible.

Hydrogen Energy Applications

Future hydrogen energy applications include the hydrogen internal combustion engine (water induction and direct cylinder induction), hydrogen in aviation, the hydrogen homestead, including a hydrogen range, and a hydrogen water heater and furnace, and the hydrogen fuel cell.

Table 2. Projected Synthetic and Fossil Fuel Prices, \$/GJ[a,b]

Fuel	Use cost	Effective cost[c]
Gaseous fuel		
hydrogen		
coal	12.08	20.86 (10.17)
hydro	15.01	12.31 (6.00)
solar	18.94	15.53 (7.58)
synthetic natural gas	11.45	24.81
natural gas	7.04	
Liquid fuel		
hydrogen		
coal	14.23	25.36
hydro	17.89	14.67
solar	22.80	18.70
synthetic gasoline	19.85	34.97
synthetic jet fuel	15.65	30.77
fuel oil	9.14	
gasoline	11.23	21.40
jet fuel	7.42	17.59

[a] To convert J to cal, divide by 4.184. [b] Estimates for the year 2000 in 1990 \$U.S. [c] In internal combustion engine, unless otherwise noted (fuel cells).

ROGER E. BILLINGS
International Academy of Science

R. E. Billings, *The Hydrogen World View*, International Academy of Science, Independence, Mo., 1991.

R. E. Billings, M. Sanchez, P. A. Cherry, and D. B. Eyre, "LaserCell Prototype Vehicle," *Project Hydrogen '91 Conference*, International Academy of Science, Independence, Mo., 1991.

J. O.'M. Bockris, T. N. Veziroglu, and D. Smith, *Solar Hydrogen Energy: The Power to Save the Earth*, MacDonald Optima, London, 1991.

T. N. Veziroglu and R. E. Billings, eds., *Project Hydrogen '91 Conference*, American Academy of Science, Independence, Mo., 1991.

HYDROGEN FLUORIDE. See FLUORINE COMPOUNDS, INORGANIC.

HYDROGEN IODINE. See IODINE COMPOUNDS.

HYDROGEN-ION ACTIVITY

Hydrogen ions are involved in a wide variety of natural and industrial reactions, and the equilibrium positions as well as the rates of these reactions are therefore dependent on hydrogen-ion concentration. The hydrogen ion is more correctly termed hydronium ion. The unhydrated proton does not exist in aqueous solution but rather is bound to several molecules of water. This ion, sometimes represented as $H(H_2O)_n^+$, is usually written simply as H^+. More important is the distinction between the hydrogen ion concentration and its activity. The hydrogen ion concentration, or total acidity, is obtained by titration and corresponds to the total concentration of hydrogen ions available in a solution, ie, free, unbound hydrogen ions as well as hydrogen ions associated with weak acids. The hydrogen ion activity refers to the effective concentration of unbound hydrogen ions, ie, the form which affects physicochemical reaction rates and equilibria. The effective concentration of hydrogen ion in solution is expressed in terms of pH, which is the negative logarithm of the hydrogen-ion activity, a_{H^+}

$$\mathrm{pH} = -\log_{10} a_{H^+} \qquad (1)$$

The relationship between activity, a, and concentration, c, is

$$a = \gamma c \qquad (2)$$

where the activity coefficient γ is a function of the ionic strength of the solution and approaches unity as the ionic strength decreases; ie, the difference between the activity and the concentration of free hydrogen ions diminishes as the solution becomes more dilute. The pH of a solution may have little relationship to the titratable acidity of a solution that contains weak acids or buffering substances; the pH of a solution indicates only the free hydrogen-ion activity. If total acid concentration is to be determined, an acid–base titration must be performed.

pH Determination

Two methods are used to measure pH: electrometric and chemical indicator. The most common is electrometric and uses the commercial pH meter with a glass electrode. This procedure is based on the measurement of the difference between the pH of an unknown or test solution and that of a standard solution.

More recently, two different types of nonglass pH electrodes have been described which have shown excellent pH-response behavior. In the neutral-carrier, ion-selective electrode type of potentiometric sensor, synthetic organic ionophores, selective for hydrogen ions, are immobilized in polymeric membranes. Another type of pH sensor is based on an integrated ion-selective electrode and insulated-gate field-effect transistor. These sensors, usually termed ion-selective field-effect transistors (ISFETs), are based on the modulation of the transistor source-drain current by a potential (or charge) applied to the transistor gate region.

The second method for measuring pH, the optical indicator method, has more limited applications. The success of this procedure depends on matching the color that is produced by the addition of a suitable indicator dye to a portion of the unknown solution with the color produced by adding the same quantity of the same dye to a series of standard solutions of known pH. The indicator dyes can also be immobilized onto paper strips (eg, litmus paper) or, more recently, have been placed onto the distal end of fiber-optic probes which, when combined with photometric readout, provide more quantitative indicator-dye pH determinations.

Accuracy and Interpretation of Measured pH Values. To define the pH scale and permit the calibration of pH measurement systems, a series of reference buffer solutions have been certified by the U.S. National Institute of Standards and Technology (NIST). The acidity function which is the experimental basis for the assignment of pH, is reproducible within about 0.003 pH unit from 10 to 40°C. However, errors in the standard potential of the cell, in the composition of the buffer materials, and in the preparation of the solutions may raise the uncertainty to 0.005 pH unit. The accuracy of the practical scale may be further reduced to 0.008–0.01 pH unit as a result of variations in the liquid-junction potential.

Sources of Error. Several common causes of measurement problems are electrode interferences and/or fouling of the pH sensor, sample matrix effects, reference electrode instability, and improper calibration of the measurement system.

pH Measurement Systems

Glass Electrodes. The glass electrode is the hydrogen-ion sensor in most pH-measurement systems. The pH-responsive surface of the glass electrode consists of a thin membrane formed from a special glass that, after suitable conditioning, develops a surface potential that is an accurate index of the acidity of the solution in which the electrode is immersed. To permit changes in the potential of the active surface of the glass membrane to be measured, an inner reference electrode of constant potential is placed in the internal compartment of the glass membrane. The inner cell commonly consists of a silver–silver chloride electrode or calomel electrode in a buffered chloride solution. Immersion electrodes are the most common glass electrodes. Miniature and microelectrodes are also used widely, particularly in physiological studies. Capillary electrodes permit the use of small samples and provide protection from exposure to air during the measurements. The composition of the glass has a profound effect on the electrical resistance, the chemical durability of the pH-sensitive surface, and the accuracy of the pH response in alkaline solutions.

Reference Electrodes and Liquid Junctions. The electrical circuit of the pH cell is completed through a salt bridge that usually consists of a concentrated solution of potassium chloride. The solution makes contact at one end with the test solution and at the other with a reference electrode of constant potential. The liquid junction is formed at the area of contact between the salt bridge and the test solution.

The commercially used reference electrode–salt bridge combination usually is of the immersion type. Some provision is made to allow a slow leakage of the bridge solution out of the tip of the electrode to establish the liquid junction with the test solution.

Combination electrodes have increased in use and are a consolidation of the glass and reference electrodes in a single probe, usually in a concentric arrangement, with the reference electrode compartment surrounding the pH sensor. The advantages of combination electrodes include the convenience of using a single probe and the ability to measure small volumes of sample solution or in restricted-access containers.

Theoretical considerations favor liquid junctions by which cylindrical symmetry and a steady state of ionic diffusion are achieved.

Samples that contain suspended matter are among the most difficult types from which to obtain accurate pH readings because of the so-called suspension effect, ie, the suspended particles produce abnormal liquid-junction potentials at the reference electrode. Internal consistency is achieved by pH measurement using carefully prescribed measurement protocols, as has been used in the determination of soil pH.

Another effect that may result in spurious pH readings is caused by streaming potentials. Presumably, these are attributable to changes in the reference electrode liquid junction that are caused by variations in the flow rate of the sample solution. This problem may be avoided by maintaining constant flow and geometry characteristics and calibrating the system under operating conditions that are identical to those of the sample measurement.

pH Instrumentation. The pH meter is an electronic voltmeter that provides a direct conversion of voltage differences to differences of pH at the measurement temperature.

Because of the very large resistance of the glass membrane in a conventional pH electrode, an input amplifier of high impedance (usually 10^{12}–10^{14} Ω) is required to avoid errors in the pH (or mV) readings.

In addition, most devices provide operator control of settings for temperature and/or response slope, isopotential point, zero or standardization, and function (pH, mV, or monovalent–bivalent cation–anion). Microprocessors are incorporated in advanced-design meters to facilitate calibration, calculation of measurement parameters, and automatic temperature compensation.

Temperature Effects. The emf, E, of a pH cell may be written

$$E = E_g^{o\prime} - k\text{pH} \tag{3}$$

where k is the Nernst factor (2.303 RT)/F, and $E^{o\prime}{}_{;g}$ includes the liquid-junction potential and the half-cell emf on the reference side of the glass membrane. Changes of temperature alter the scale slope because k is proportional to T. The scale position also is changed because the standard potential is temperature dependent: $E^{o\prime}{}_{;g}$ is usually a quadratic function of the temperature.

The objective of temperature compensation in a pH meter is to nullify changes in emf from any source except changes in the true pH of the test solution. Nearly all pH meters provide automatic or manual adjustment for the change of k with T.

Nonaqueous Solvents

The activity of the hydrogen ion is affected by the properties of the solvent in which it is measured. Scales of pH only apply to the medium, ie, the solvent or mixed solvents, eg, water–alcohol, for which the scales are developed. The comparison of the pH values of a buffer in aqueous solution to one in a nonaqueous solvent has neither direct quantitative nor thermodynamic significance. Consequently, operational pH scales must be developed for the individual solvent systems.

Other difficulties of measuring pH in nonaqueous solvents are the complications that result from dehydration of the glass pH membrane, increased sample resistance, and large liquid-junction potentials. These effects are complex and highly dependent on the type of solvent or mixture used.

Indicator pH Measurements

The indicator method is especially convenient when the pH of a well-buffered colorless solution must be measured at room temperature with an accuracy no greater than 0.5 pH unit. Under optimum conditions an accuracy of 0.2 pH unit is obtainable.

Because they are weak acids or bases, the indicators may affect the pH of the sample, especially in the case of a poorly buffered solution. Variations in the ionic strength or solvent composition, or both, also can produce large uncertainties in pH measurements, presumably caused by changes in the equilibria of the indicator species. Specific chemical reactions also may occur between solutes in the sample and the indicator species to produce appreciable pH errors.

Industrial Process Control

The pH meters and electrodes for process control do not differ materially from those used for measurements in the laboratory, but the emphasis in industrial applications is on rugged construction to withstand mechanical stresses and extremes in ambient conditions.

The pH meter usually is coupled to a data recording device and often to a pneumatic or electric controller. The controller governs the addition of reagent so that the pH of the process stream is maintained at the desired level.

RICHARD A. DURST
Cornell University
ROGER G. BATES
University of Florida

R. G. Bates, *Determination of pH, Theory and Practice*, 2nd ed., Wiley-Interscience, New York, 1973.

Y. C. Wu, W. F. Koch, and R. A. Durst, *Standardization of pH Measurements*, National Bureau of Standards Special Publication 260-53, U.S. Government Printing Office, Washington, D.C., 1988.

G. Eisenman, ed., *Glass Electrodes for Hydrogen and Other Cations*, Marcel Dekker, New York, 1967.

H. Galster, *pH Measurement: Fundamentals, Methods, Applications, Instrumentation*, VCH, New York, 1991.

HYDROGEN PEROXIDE

Hydrogen peroxide, H_2O_2, mol wt 34.016, is a strong oxidizing agent commercially available in aqueous solution over a wide range of concentrations. It is a weakly acidic, nearly colorless clear liquid that is miscible with water in all proportions. The atoms are covalently bound in a nonpolar H—O—O—H structure having association (hydrogen bonding) somewhat less than that found in water. Now manufactured primarily in large, strategically located anthrahydroquinone autoxidation processes, its many uses include bleaching wood pulp and textiles, preparing other peroxygen compounds, and serving as a nonpolluting oxidizing agent.

Physical Properties

Properties of pure hydrogen peroxide are listed in Table 1. In aqueous solution the hydrogen bonds (association) between water and H_2O_2 molecules are appreciably more stable than those between molecules of the individual species. This increase in attraction forces is evidenced from many properties such as heat of mixing, vapor pressure, viscosity, dielectric constant, etc. Physical constants have been determined or calculated for aqueous H_2O_2 solutions, the only form in which hydrogen peroxide is commercially available (Table 2).

Chemical Properties

Hydrogen peroxide is a weak acid, having a $pK_a = 11.75$.

$$H_2O_2 + H_2O \rightarrow HO_2^- + H_3O^+ \tag{1}$$

Table 1. Properties of Hydrogen Peroxide

Property	Value
mp,[a] °C	−0.41
bp, °C	150.2
density at 25°C, g/mL	1.4425
viscosity at 20°C, mPa·s(= cP)	1.245
surface tension at 20°C, mN/m(= dyn/cm)	80.4
specific conductance at 25°C, $(\Omega\cdot cm)^{-1}$	4×10^{-7}
heat of fusion, J/g[b]	367.52
specific heat at 25°C, J/(g·K)[b]	2.628
heat of vaporization at 25°C, kJ/g[b]	1.517
dissociation constant[c] at 20°C	1.78×10^{-12}
heat of dissociation, kJ/mol[b]	34.3

[a] Tends to supercool. [b] To convert J to cal, divide by 4.184. [c] At zero ionic strength.

Table 2. Physical Properties of Aqueous Hydrogen Peroxide

Liquid, wt % H_2O_2	Freezing point, °C	Boiling point,[a] °C	Vapor,[a] wt % H_2O_2	Density at 25°C, g/mL	ΔH_{vap} at 25°C kJ/g[b]
10	−6.4	101.7	0.9	1.0324	2.357
30	−25.7	106.2	4.2	1.1081	2.192
50	−52.2	113.8	13.0	1.1914	2.017
70	−40.3	125.5	33.4	1.2839	1.832
90	−11.5	141.3	75.0	1.3867	1.627

[a] At 101.3 kPa (1 atm). [b] To convert J to cal, divide by 4.184.

Free-Radical Formation. Hydrogen peroxide can form free radicals by homolytic cleavage of either an O—H or the O—O bond.

Decomposition. The decomposition of hydrogen peroxide may be homogeneous or heterogeneous and can occur in the vapor or the condensed phase. Although there is considerable evidence that the decomposition occurs as a chain reaction involving free radicals, the products of the decomposition are water and oxygen gas. Decomposition of hydrogen peroxide must be controlled at all times, in part because of the economic impact, but more importantly because the resultant simultaneous generation of oxygen and heat may cause serious safety problems.

Stabilization. Pure hydrogen peroxide solutions are relatively stable and can be stored for extended periods in clean passive containers. Commercial solutions, however, invariably contain or may be exposed to varying amounts of catalytic impurities and must therefore contain reagents which deactivate these impurities, either by adsorption or through formation of complexes.

Molecular Addition. Oxyacid salts, metal peroxides, nitrogen compounds, and others form crystalline peroxyhydrates in the presence of hydrogen peroxide.

Substitution. A variety of peroxygen compounds can be formed through substitution reactions of hydrogen peroxide with organic reagents. Inorganic peroxygen compounds can be prepared through similar reactions with inorganic reagents.

Oxidation. Hydrogen peroxide is a strong oxidant. Hydrogen peroxide oxidizes a wide variety of organic and inorganic compounds, ranging from iodide ions to the various color bodies of unknown structure in cellulosic fibers.

Reduction. Hydrogen peroxide reduces stronger oxidizing agents such as chlorine, sodium hypochloride, potassium permanganate, and ceric sulfate.

Manufacture

Hydrogen peroxide is composed of equal molar amounts of hydrogen and oxygen and can be formed directly by catalytically combining the gaseous elements. It can also be formed from compounds that contain the peroxy group; from water and oxygen by thermal, photochemical, electrochemical or similar processes; and by the uncatalyzed reaction of molecular oxygen with appropriate hydrogen-containing species. It has been manufactured commercially by processes based on the reaction of barium peroxide or sodium peroxide with an acid, the electrolysis of sulfuric acid and related compounds, the autoxidation of 2-alkylanthrahydroquinones, isopropyl alcohol, and hydrazobenzene, and more recently by the Huron-Dow process through the cathodic reduction of oxygen in an electrolytic cell using dilute sodium hydroxide as the electrolyte. By far, the majority of hydrogen peroxide produced since 1957 has been based on the autoxidation of 2-alkylanthrahydroquinones.

Purification and Concentration. The crude product from any hydrogen peroxide process can be used as such, but commercial grades are further purified, concentrated, and stabilized.

Procedures include solvent extraction followed by optional air stripping to remove residual solvent and treatment with synthetic resins, polyethylene, waxes, carbon, and aluminum and magnesium hydroxides and alumina. Active ion-exchange resins have been used to remove both metallic and acidic impurities. More recent patented methods for purifying crude hydrogen peroxide include further contact with an aromatic gasoline in static mixers, followed by a series of coalescing steps to effect phase separation, passing the solution through columns packed with halogen-containing porous styrene-divinylbenzene copolymer resin, and passing the solution through an anion-exchange resin which has been pretreated with various chelating agents to remove metal ions and organics.

Concentration of hydrogen peroxide prepared by the autoxidation processes can be carried out safely and conveniently by distillation at reduced pressure.

Economic Aspects

U.S. production of hydrogen peroxide is shown in Table 3. The driving influence on H_2O_2 capacity and production has come from the pulp and paper sector, as this industry converts from chlorine bleaching and the attendant dioxin and chloroform formation to the environmentally more benign peroxide chemistry.

Health and Safety Factors

Hydrogen peroxide, especially in high concentrations, is a high energy material and a strong oxidant. The Comprehensive Environmental Response, Compensation, and Liability Act (CERCLA) reportable spill quantity for greater than 52% hydrogen peroxide is 1 lb (0.45 kg). It is considered an acute, reactive, and pressure hazard under Superfund and Reauthorization Amendments (SARA) Title III. However, it can be handled safely if proper personal protective equipment is worn and the proper precautions are observed. Some generally applicable control measures and precautions for handling hydrogen peroxide include the use of adequate ventilation to keep airborne concentrations below exposure limits, 8 h TWA, 1.4 mg/m^3, use of coverall chemical splash goggles in combination with a full-length face shield if spraying is a potential occurrence, use of a NIOSH/MSHA-approved respirator if airborne concentration can exceed exposure limits, and use of neoprene or other impervious and compatible gloves. Other clothing items such as impervious aprons, pants, jackets, hoods, boots, and totally encapsulating chemical suits with breathable air supply should be available for use as necessary.

Health and Physiological Effects. Hydrogen peroxide is irritating to the skin, eyes, and mucous membranes. However, low concentrations (3–6%) are used in medicinal and cosmetic applications.

Decomposition and Explosive Hazards. The principal hazards associated with hydrogen peroxide include (*1*) decomposition of H_2O_2 with unrelieved pressure buildup; (*2*) spontaneous combustion of mixtures of H_2O_2 and readily oxidizable material; (*3*) inadvertent admission of incompatible materials into a tank containing H_2O_2 or vice versa; (*4*)

Table 3. U.S. Production of Hydrogen Peroxide[a,b]

Year	Production, t × 10^3
1960	26.0
1970	55.7
1980	105.8
1982	98.5
1984	126.3
1986	138.0
1988	161.0
1990	216.6
1992	271.8
1994	360[c]
1996	470[c]
1998	580[c]
2000	700[c]

[a] Courtesy U.S. Dept. of Commerce, Bureau of the Census. [b] 100% basis. [c] Estimated.

decomposition of H_2O_2 to form an oxygen-rich vapor phase; (*5*) deflagration, detonation of a condensed-phase mixture of H_2O_2 and organics initiated by shock or thermal effects; and (*6*) explosive reaction of H_2O_2 vapor.

Uses

Bleaching. The largest single use for hydrogen peroxide in the United States and North America is wood pulp bleaching, but consumption for the manufacture of chemicals, environmental applications, and for bleaching cotton, wool, and other textiles is significant.

Environmental concerns have led the pulp and paper industry to turn to alkaline solutions of hydrogen peroxide as a replacement for chlorine and, in some cases, for hypochlorite and chlorine dioxide in bleaching applications.

Environmental Applications. Hydrogen peroxide is an ecologically desirable pollution-control agent because it yields only water or oxygen on decomposition. It has been used in increasingly greater amounts to convert domestic and industrial effluents to an environmentally compatible state. Hydrogen peroxide or a peroxycarboxylic acid made from H_2O_2 is used in the manufacture of a number of organic and inorganic chemicals.

Derivative Formation. Hydrogen peroxide is an important reagent in the manufacture of organic peroxides, including *tert*-butyl hydroperoxide, benzoyl peroxide, peroxyacetic acid, esters such as *tert*-butyl peroxyacetate, and ketone derivatives such as methyl ethyl ketone peroxide. These are used as polymerization catalysts, cross-linking agents, and oxidants.

Mining. Hydrogen peroxide, in combination with various carbonates or bicarbonates, is used as an oxidant for the in-place solution mining of low grade uranium ores.

Propellant. The catalytic decomposition of 70% hydrogen peroxide or greater proceeds rapidly and with sufficient heat release that the products are oxygen and steam. The thrust developed form this reaction can be used to propel torpedoes and other small missiles.

WAYNE T. HESS
E. I. du Pont de Nemours & Co., Inc.

W. C. Schumb, C. N. Satterfield, and R. L. Wentworth, *Hydrogen Peroxide*, Reinhold Publishing Corp., New York, 1955.

H. Pistor, in K. Winnacker and L. Kuchler, eds., *Chemische Technologie, Band 1 Anorganische Technolgie 1*, Hanser, Munich, 1969 (industrial processes).

J. G. Wallace, *Hydrogen Peroxide in Organic Chemistry*, E. I. du Pont de Nemours & Co., Inc., Wilmington, Del., 1962.

R. Powell, *Hydrogen Peroxide Manufacture*, Chemical Process Review No. 20, Noyes Development Corporation, Park Ridge, N.J., 1968.

HYDROGEN SULFIDE. See SULFUR COMPOUNDS.

HYDROMETALLURGY. See METALLURGY, EXTRACTIVE.

HYDROQUINONE, RESORCINOL, AND CATECHOL

Hydroquinone, resorcinol, and catechol (or pyrocatechol) are represented by structures (**1**), (**2**), and (**3**), respectively.

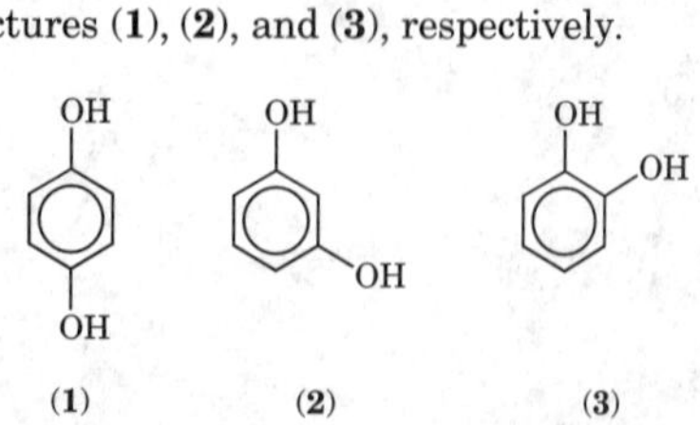

(**1**) (**2**) (**3**)

Manufacture and Processing

Dihydroxybenzenes (dihydric phenols) are industrially prepared according to four different routes: (*1*) oxidation of aniline (selective access to hydroquinone); (*2*) alkali fusion of *m*-benzenedisulfonic acid (selective access to resorcinol); (*3*) oxidation of *p*- or *m*-diisopropylbenzene (selective access to hydroquinone or resorcinol); and (*4*) hydroxylation of phenol by hydrogen peroxide (simultaneous access to hydroquinone and catechol). The main breakthrough of the 1980s was the discovery of titanium silicalite (TS-1), a synthetic zeolite from the ZSM family containing no aluminum in which some titanium atoms replace silicon atoms in the cristalline system, as catalyst of phenol hydroxilation by hydrogen peroxide. A process based on this catalyst is operative on a 10,000 t/yr scale.

World production capacities according to countries and process types are presented in Table 1.

Miscellaneous Preparations. Many laboratory innovations have been run. Processes include those starting from cyclohexene, starting from phenol, starting from chlorophenols, starting from bisphenol A, starting from benzene, as well as pyrolysis of vegetals and biochemical routes.

Synthesis of Derivatives

Hydroquinone resorcinol and catechol are important industrial intermediates, and there has been significant research and development of processes for manufacturing their derivatives.

Catechol Derivatives. An elegant synthesis of trimethoxybenzaldehyde starting from guaiacol and formaldehyde (**2**) has been developed. A new synthesis of carbofuran has been described via 2,3-dihydro-2,2-dimethyl-7-hydroxybenzofuran, starting from *o*-methallyloxyphenol in the presence of trivalent aluminum derivatives.

Hydroquinone Derivatives. Thermotropic polymers are polymers in which the main chain contains a mesomorphic state between the disordered liquid and the crystalline state. Many important industrial developments have been devoted to this class of polymers. Melting point decrease of polyesters containing only rigid units in the principal chain can be obtained in many ways, but in particular by introducing substituents or functional groups allowing internal rotations of the rigid units to occur. Therefore, development of hydroquinone containing substituents such as halogens, eg, chlorohydroquinone, alkyls, or phenyl was initiated.

The selective monochlorination of hydroquinone by $SOCl_2$ or a combination of HCl and H_2O_2 has been studied in various solvents. The synthesis of chloranil has been improved. The old processes start-

Table 1. Manufacturing Processes and World Production Capacities for Dihydroxybenzenes

Dihydroxybenzene	World capacity,[a] t	Process	Location
hydroquinone	~45,000–50,000	aniline oxidation	ex Comecon[b], PRC
		phenol hydroxylation	France, U.S., Italy, Japan
		p-diisopropylbenzene hydroperoxidation	U.S., Japan
resorcinol	30,000–35,000	benzenedisulfonic acid alkali-fusion	Japan, U.S., Italy, Germany, U.K., Puerto Rico
		m-diisopropylbenzene hydroperoxidation	Japan
catechol	>25,000	phenol hydroxylation	France, U.S., Italy, Japan
		coal-tar distillation	U.K., ex Comecon[b]

[a] Estimated for 1994. [b] Comecon = Council for Mutual Economic Assistance (Communist-bloc nations).

ing from phenol or 2,4,6-trichlorophenol have been replaced by new ones involving hydroquinone chlorination and thence avoiding traces of pentachlorophenol.

Resorcinol Derivatives. Aminophenols are important intermediates for the syntheses of dyes or active molecules for agrochemistry and pharmacy. Syntheses have been described involving resorcinol reacting with amines.

2,4-Dihydroxybenzophenones are used for the syntheses of dyes, polymers, and medicines. They are prepared by the condensation of resorcinol with benzoic acids.

The synthesis of 2,4-dihydroxyacetophenone by acylation reactions of resorcinol has been extensively studied. The reaction is performed using acetic anhydride, acetyl chloride, or acetic acid. The condensation of an aldehyde with resorcinol gives rise to calix arene.

Resorcinol carboxylation with carbon dioxide leads to a mixture of 2,4-dihydroxybenzoic acid and 2,6-dihydroxybenzoic acid. The condensation of resorcinol with chloroform under basic conditions, in the presence of cyclodextrins, leads exclusively to 2,4-dihydroxybenzaldehyde. Finally, the synthesis of 1,3-bis(2-hydroxyethoxy)benzene has been described with ethylene glycol carbonate in basic medium, in the presence of phosphines.

Uses and Specification

Hydroquinone is used in a broad range of applications such as photographic developers, polymerization inhibitors, rubber antioxidants, food antioxidants, synthesis intermediates, and water treatment. Catechol is mainly used for synthesis in food, pharmaceutical, or agrochemical ingredients. A specific application of *tert*-butylcatechol is as a polymerization inhibitor. The main uses of resorcinol are in the manufacture of rubber and wood adhesives.

Hydroquinone is available in photographic, inhibitor, and technical grade. Manufacturer's specifications are given in Table 2.

Catechol is available in standard and extra pure grade. Manufacturer's specifications for catechol and resorcinol are given in Table 3.

Health and Safety

Dihydroxybenzenes (DHBs) are slightly more acutely toxic than phenol. Contact with dihydroxybenzene through oral, dermal, or respiratory routes can induce significant systemic exposure. Skin or eye effects have been demonstrated during chronic or accidental professional exposure.

In humans, cases of dermatitis have been described after contact with DHBs. Combined exposure to hydroquinone and quinone airborne concentrations causes eye irritation, sensitivity to light, injury of the corneal epithelium, and visual disturbances. Cases with an appreciable loss of vision have occurred. Long-term exposure causes staining due to irritation or allergy of the conjunctiva and cornea and also opacities. Resorcinol and catechol are also irritants for eyes.

Acute intoxication with DHBs occurs mainly by the oral route; symptoms are close to those induced by phenol poisoning including nausea, vomiting, diarrhea, tachypnea, pulmonary edema, and CNS excitation with possibility of seizures followed by CNS depression. Convulsions are more frequent with catechol as well as hypotension due to peripheral vasoconstriction. Hypotension and hepatitis seem more frequent with hydroquinone and resorcinol. Methemoglobinemia and hepatic injury may be noted within a few days after intoxication by DHBs.

All operations producing dust require the usual measures to prevent dust in the atmosphere exceeding the allowable daily concentration. If this is not feasible, personal protection devices should be used. Especially when hydroquinone is present as a powder, adequate eye protection should be provided.

Table 2. Manufacture's Specifications for Hydroquinone[a]

Specification	Photographic	Inhibitor	Technical
appearance	crystals	crystals	crystals
color	white	white to light tan	light tan
melting point, °C	171.0–174.0	171.0–174.0	170.0 min
assay, wt % (min)	99.5	99	99
heavy metal content, eg, Pb % (max)	0.001		
ash content, wt % (max)	0.05	0.05	
iron content, wt % (max)	0.001		

[a] Data courtesy of Rhône-Poulenc.

Table 3. Manufacture's Specifications for Catechol and Resorcinol[a]

Specification	Catechol		Resorcinol	
grade	standard	extra pure	technical	USP XX
appearance	flakes	white crystals	flakes[b]	crystals[c]
melting point, °C	102	102		109.1
assay, wt (min)	98	99.5	99	99
ash content, wt % (max)			0.005	0.05

[a] Data courtesy of Rhône-Poulenc. [b] White or slightly colored. [c] White or nearly white.

LÉON KRUMENACKER
MICHEL COSTANTINI
P. PONTAL
J. SENTENAC
Rhône-Poulenc Recherches

Fr. Pat. 2,523,575 (Mar. 19, 1982), A. Esposito, C. Neri, and F. Buonomo (to Anic).

Fr. Pat. 2,486,523 (July 11, 1980), K. Formanek, D. Michelet, and D. Petre (to Rhône-Poulenc).

U.S. Pat. 4,324,731 (Apr. 13, 1982), D. Michelet and S. Veracini (to Rhône-Poulenc).

HYDROTHERMAL PROCESSING

Hydrothermal processing encompasses a broad set of technologies which share a range of operating temperatures and pressures and require the use of engineered pressure systems. Table 1 compares the distinguishing features of these technologies, all of which utilize water heated to temperatures above its boiling point (nominally 100°C). These diverse technologies have common needs for equipment and plant design. Many of the designs can be translated from one application to another.

Process Equipment

Hydrothermal Synthesis Systems. In consideration of scale-up of a hydrothermal process for high performance materials, several criteria must be considered. First, the mode of operation, which can be either continuous, semicontinuous, or batch, must be determined. Factors to consider are the operating conditions, the manufacturing demand, the composition of the product mix (single or multiple products), the amount of waste that can be tolerated, and the materials of construction requirements.

Conventional reactors used for conducting hydrothermal reactions, leaching or precipitation, are frequently called autoclaves. Vessels can be vertical, horizontal, or spherical. Steam agitated vessels are fabricated from welded stainless steel cylinders with spherical heads. Horizontally oriented vessels are divided into several mixed sections. The body can be lined with rubber, lead, or alloy steel.

Tubular continuous autoclaves were invented in the early 1930s and have been used extensively in Germany for leaching bauxite. A commercial design based on semicontinuous operation was developed for manufacture of silicate powders.

Table 1. Features of Hydrothermal Processes

Process	Temp., °C	Pressure, MPa[a]	Conditions	Application
pressure leaching/ precipitation	100–200	<5	reducing or inert	treatment of ores to recover metals
hydrothermal synthesis	100–350	<20	oxidizing or inert	prpn of fine particle oxides
hydrothermal crystallization	350–450	<200	inert	prpn of large crystals and gemstones
wet oxidation	150–350	<20	oxidizing	destruction of organic sludges
supercritical water oxidation	400–600	22–30	oxidizing	destruction of hazardous and toxic waste

[a] To convert MPa to psi, multiply by 145.

In order to facilitate pressure letdown and heat recovery, an antipressure or receiving vessel is used.

Wet Oxidation Reactor Design. Several types of reactor designs have been employed for wet oxidation processes. Zimpro, the largest manufacturer of wet oxidation systems, typically uses a tower reactor system. A horizontal, stirred tank reactor system, known as the Wetox process, was initially developed by Barber-Colman, and is also offered by Zimpro.

Wetox uses a single-reactor vessel that is baffled to stimulate multiple stages. The design allows for higher destruction efficiency at lower power input and reduced temperature.

Another wet oxidation system is designed to be operated in a deep well, up to 1500 m below the ground surface. Called a vertical tube reactor, it may be suitable for destruction of municipal sludge and toxic wastes in some geological locations having large (>380 L/min) waste flows.

Other above-ground continuous flow systems have been designed and operated for SCWO processes.

WILLIAM J. DAWSON
Chemical Materials International
PIETER KRIJGSMAN
CEC Company

W. J. Dawson and M. K. Han, "Development and Scale-Up of Hydrothermal Processes for Synthesis of High Performance Materials," *Proceedings of the Milton E. Wadsworth IV International Symposium on Hydrometallurgy*, Aug. 1–5, 1993.

R. A. Laudise, *Chem. Eng. News* (Sept. 28, 1987).

R. M. Barrer, *Hydrothermal Chemistry of Zeolites*, Academic Press, Inc., New York, 1982.

W. J. Dawson, *Cer. Bull.* **67**,(10), 1673–1678 (1988).

HYDROXYBENZALDEHYDES

Hydroxybenzaldehydes are organic compounds of the general formula:

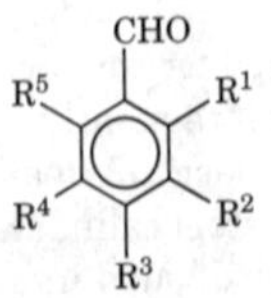

where R^1 through R^5 = H or OH but at least one R group is OH. All of the isomeric mono-, di-, and trihydroxybenzaldehydes have been isolated. The higher polyhydroxybenzaldehydes are unknown. This article deals primarily with *p*-hydroxybenzaldehyde and salicylaldehyde which together represent more than 99% of the hydroxybenzaldehydes market.

Of the two commercially important monohydroxybenzaldehydes the ortho isomer (salicylaldehyde) is the more important one. Salicylaldehyde (salicylic aldehyde, salicylal) is a colorless, oily liquid, the only hydroxybenzaldehyde liquid at room temperature with a pungent irritating odor. It occurs naturally in beer, oils of spirea, bird cherries and cassia, and in coffee, grape, tea, and tomato. Salicylaldehyde and its derivatives are utilized as ingredients in agricultural chemicals, electroplating, perfumes, petroleum chemicals, polymers, and fibers.

p-Hydroxybenzaldehyde (4-formylphenol) is a colorless to faint tan solid, with a slight, agreeable, aromatic odor. It occurs naturally in some plants in small amounts.

Physical Properties

Physical constants for salicylaldehyde and *p*-hydroxybenzaldehyde are listed in Table 1. Spectral data have been published.

The location of the hydroxyl and aldehyde groups ortho to one another in salicylaldehyde permits intramolecular hydrogen bonding, and this results in the lower melting point and boiling point and the higher acid dissociation constant observed relative to *p*-hydroxybenzaldehyde.

Chemical Properties

The effect of the aldehyde group on the phenolic hydroxyl group is primarily an increase in its acidity: both 2-hydroxy- and 4-hydroxybenzaldehydes are stronger acids than phenol ($pK_a(H_2O, 20\ °C) = 9.89$). The aldehyde group, however, has little effect on the reaction of the hydroxyl group. The deactivating effect of the phenolic hydroxyl on the aldehyde group is more pronounced, but the hydroxybenzaldehydes still undergo most of the normal aldehyde reactions.

Reactions of the Hydroxyl Group. The hydroxyl proton of hydroxybenzaldehydes is acidic and reacts with alkalies to form salts.

Reactions of the Aromatic Ring. The aromatic ring of hydroxybenzaldehydes participates in several typical aromatic electrophilic reactions, including halogenation, sulfonation and diazonium coupling.

Reactions of the Aldehyde Group. Reactions of the aldehyde group include oxidation, the Canizzaro reaction, reduction, reactions with amines and amides, aldol reactions, and the Perkin reaction.

Manufacture

The main processes for the manufacture of hydroxybenzaldehydes are based on phenol. The most widely used process is the saligenin process. Although 4-hydroxybenzaldehyde can be made by the saligenin route, it has been made historically by the Reimer-Tiemann process, which also produces salicylaldehyde. Other routes for hydroxybenzaldehydes are the electrolytic or catalytic reduction of hydroxybenzoic acids and the electrolytic or catalytic oxidation of cresols.

Table 1. Physical Data for *o*- and *p*-Hydroxybenzaldehydes

Property	Value: Salicylaldehyde	Value: *p*-Hydroxybenzaldehyde
mp, °C	−7	117
bp, °C (kPa)[a]	197(100)	310(100)
	80 (1.6)	170 (1.3)
density, g/cm^3	1.167 (20°C)	1.143 (117°C)
solubility, g/100 g water	1.7 (86°C)	1.3 (30.5°C)
acid dissociation constant pK_a, H_2O, 25°C	8.14	7.6

[a] To convert kPa to mm Hg, multiply by 7.5.

Economic Aspects

Rhône-Poulenc (RP), producing both in Europe and the U.S., is the only producer of salicylaldehyde worldwide, for merchant sales. A large portion of it is used captively in the manufacture of coumarin. The remainder is available for the merchant market.

Worldwide capacity figures for salicylaldehyde are not published; however, the estimated capacity is approximately 4000–6000 t/yr. The supply–demand picture for salicylaldehyde has been well balanced in the 1990s, as RP has expanded capacity to meet the growing market need. Chuo Kasein (Japan), various Chinese companies, and Hoechst (France) are producers of *p*-hydroxybenzaldehyde.

Health and Safety Factors

Salicylaldehyde has a moderate acute oral toxicity. *p*-Hydroxybenzaldehyde has a low acute oral toxicity. Neither material is likely to present a problem from ingestion incidental to its handling and industrial use. It should be recognized, however, that serious effects may result if substantial amounts are swallowed.

Tests performed on rabbits indicate that neither material is absorbed through the skin in toxic amounts. Skin contact with *p*-hydroxybenzaldehyde is essentially nonirritating; however, contact with salicylaldehyde is capable of causing a severe burn, especially in case of prolonged or repeated contact. *p*-Hydroxybenzaldehyde is slightly irritating to the eyes and can cause slight transient irritation and slight transient corneal injury. Salicylaldehyde is appreciably irritating to the eyes and may cause pain, irritation, and some corneal injury.

CHRISTIAN MALIVERNEY
MICHEL MULHAUSER
Rhône-Poulenc
Recherches

Flavor and Extract Manufacturers' Association of the United States (FEMAUS), *Gov. Rep. Annouce. (U.S.)* **85**(6), 48 (1985).

For the general chemical properties, see S. Patai, ed., *The Chemistry of the Carbonyl Group*, Vol. 1, Wiley-Interscience, New York, 1966; J. Zabicky, Vol. 2, ed., 1970.

HYDROXYCARBOXYLIC ACIDS

Lactic Acid

Lactic acid (2-hydroxypropanoic acid), $CH_3CHOHCOOH$, is the most widely occurring hydroxycarboxylic acid and thus is the principal topic of this article. Lactic acid is a naturally occurring organic acid that can be produced by fermentation or chemical synthesis. It is present in many foods both naturally or as a product of *in situ* microbial fermentation, as in sauerkraut, yogurt, buttermilk, sourdough breads, and many other fermented foods. Lactic acid is also a principal metabolic intermediate in most living organisms, from anaerobic prokaryotes to humans.

Two significant producers of lactic acids are CCA Biochem by of the Netherlands, with subsidiaries in Brazil and Spain, and Sterling Chemicals, Inc. in Texas City, Tex. CCA uses carbohydrate feedstocks and fermentation technology, and Sterling uses a chemical technology. Lactic acid has been considered a relatively mature fine chemical in that only its use in new applications, eg, as a monomer in plastics or as an intermediate in the synthesis of high volume oxygenated chemicals, would cause a significant increase in its anticipated demand.

Physical Properties. Pure, anhydrous lactic acid is a white, crystalline solid with a low melting point. However, it is difficult to prepare the pure anhydrous form of lactic acid; generally, it is available as a dilute or concentrated aqueous solution. The properties of lactic acid and its derivatives have been reviewed. A few important physical and thermodynamic properties from this reference are summarized in Table 1.

Table 1. Physical and Thermodynamic Properties of Lactic Acid

Property	Value
density, g/mL at 20°C	1.2243
viscosity,[a] mPa·s(= *cP*)	36.9
heat of solution, L(+) at 25°C, kJ/mol[b]	7.79
heat of fusion, kJ/mol[b]	
racemic	11.33
L(+)	16.86
heat of combustion, MJ/mol[b]	
racemic	−1.355
L(+)	−1.343

[a] 88.6 wt % solution at 25°C. [b] To convert J to cal, divide by 4.184.

Lactic acid is also the simplest hydroxy acid that is optically active. L(+)-Lactic acid (**1**) occurs naturally in blood and in many fermentation products. The chemically produced lactic acid is a racemic mixture and some fermentations also produce the racemic mixture or an enantiomeric excess of D(−)-lactic acid (**2**).

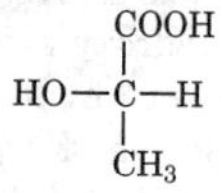
(**1**)

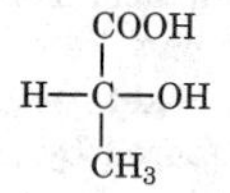
(**2**)

Many of the physical properties are not affected by the optical composition, with the important exception of the melting point of the crystalline acid, which is estimated to be 52.7–52.8°C for either optically pure isomer, whereas the reported melting point of the racemic mixture ranges from 17 to 33°C.

Chemical Properties. Its two functional groups permit a wide variety of chemical reactions for lactic acid. The primary classes of these reactions are oxidation, reduction, condensation, and substitution at the alcohol group.

Economic Aspects. The incentive for economical production of lactic acid is coming from the development of new, large-volume uses of lactic acid, particularly as feedstocks for biodegradable polymers and oxygenated chemicals. The advent and deployment of highly efficient, membrane-based separation processes and chemical and catalytic conversion technologies, together with commercial interest by several agriprocessing and chemical companies, will lead to the production of low cost lactic acid, which in turn will result in new opportunities for large-scale use of lactic acid.

Specifications, Quality Control, and Analytical Methods. Lactic acid is generally sold under four general product categories: synthetic, fermentation, heat-stable fermentation, and technical.

Lactic acid is generally recognized as safe (GRAS) for multipurpose food use. Lactate salts such as calcium and sodium lactates and esters such as ethyl lactate used in pharmaceutical preparations are also considered safe and nontoxic.

Uses. Currently, the principal use of lactic acid is in food and food-related applications, which in the United States accounts for approximately 85% of the demand. The rest (~15%) of the uses are for nonfood industrial applications.

Hydroxyacetic Acid

Hydroxyacetic acid (glycolic acid), $HOCH_2COOH$, is the first and simplest member of the family of hydroxycarboxylic acids. It occurs naturally as the chief acidic constituent of sugar-cane juice and also occurs in sugar beets and unripe grape juice. It is widely used as a cleaning agent for a variety of industrial applications, and also as a specialty chemical and biodegradable copolymer feedstock.

Properties. Glycolic acid is a colorless, translucent solid; mp = 10°C; bp = 112°C; *d* at 25°C = 1.26 g/mL; K_a at 25°C = 1.5×10^{-4};

pH at 25°C = 0.5; heat of combustion = 697.1 kJ/mol (166.6 kcal/mol); heat of solution = −11.55 kJ/mol; and flash point >300°C.

Glycolic acid is soluble in water, methanol, ethanol, acetone, acetic acid, and ethyl acetate. It is slightly soluble in ethyl ether and sparingly soluble in hydrocarbon solvents.

Reactions. Because it contains both a carboxyl and a primary hydroxyl group, glycolic acid can react as an acid or an alcohol or both. Thus some of the important reactions it can undergo are esterification, amidation, salt formation, and complexation with metal ions, which lead to many of its uses. As a fairly strong acid it can liberate gases (often toxic) when it reacts with the corresponding salts.

Manufacture, Processing, and Economic Aspects. Hydroxyacetic acid is produced commercially in the U.S. by the reaction of formaldehyde with carbon monoxide and water.

Other Hydroxy Acids

Apart from lactic and hydroxyacetic acids, other α- and β-hydroxy acids have been small-volume specialty products produced in a variety of methods for specialized uses.

Preparation. The general preparation of α-hydroxy acids is by the hydrolysis of an α-halo acid or by the acid hydrolysis of the cyanohydrins of an aldehyde or a ketone. β-Hydroxy acids may be made by catalytic reduction of β-keto esters followed by hydrolysis. β-Hydroxy acids can also be prepared by the Reformatsky reaction. γ-Hydroxy acids are seldom obtained in the free state because of the ease with which they form monomeric inner esters, which form stable five-membered rings. Thus the lactones of these acids are the common chemical forms and among these lactones γ-butyrolactone is one of the larger volume specialty chemicals derived from dehydrogenation of 1,4-butanediol.

Reactions and Uses. The common reactions that α-hydroxy acids undergo such as self- or bimolecular esterification to oligomers or cyclic esters, hydrogenation, oxidation, etc, have been discussed in connection with lactic and hydroxyacetic acid. A reaction that is of value for the synthesis of higher aldehydes is decarbonylation under boiling sulfuric acid with loss of water.

β-Hydroxy acids lose water, especially in the presence of an acid catalyst, to give α,β-unsaturated acids, and frequently β,γ-unsaturated acids. γ-Hydroxybutyric acid and its derivatives, particularly its sodium salt, have been studied and used as anesthetics, tranquilizers, sedatives, and hypnotics in surgery and general obstetrics.

Certain bacterial species produce polymers of γ-hydroxybutyric acid and other hydroxyalkanoic acids as storage polymers. These are biodegradable polymers with some desirable properties for manufacture of biodegradable packaging materials, and considerable effort is being devoted by ICI Ltd. and others to the development of bacterial fermentation processes to produce these polymers at a high molecular weight.

γ-Butyrolactone undergoes amination reactions with methylamine or ammonia to produce *N*-methyl-2-pyrrolidinone (NMP) or 2-pyrrolidinone (PDO) respectively, both of which are commercially important derivatives.

Other multifunctional hydroxycarboxylic acids are mevalonic and aldonic acids which can be prepared for specialized uses as aldol reaction products (mevalonic acid) and mild oxidation of aldoses (aldonic acids).

RATHIN DATTA
Consultant

C. H. Holten, A. Muller, D. Rehbinder, *Lactic Acid*, International Research Association, Verlag Chemie, Copenhagen, Denmark, 1971.

Lactic Acid and Lactates, product bulletin, Purac Inc., Arlington Heights, Ill., 1989.

R. Datta and co-workers, *FEMS Microbiol. Revs.* **16**, 221–231 (1995).

Glycolic.(Hydroxyacetic) Acid: Properties, Uses, Storage, and Handling; Glycolide S.G.: Properties, Uses, Storage, and Handling, bulletins, Du Pont Chemicals, Wilmington, Del., 1992.

HYDROXY DICARBOXYLIC ACIDS

Many natural and synthetic organic compounds are hydroxy dicarboxylic acids. This article discusses mainly malic and tartaric acids; thiomalic acid is included because of its structural similarity to malic acid.

Malic Acid

Malic acid (hydroxysuccinic acid, hydroxybutanedioic acid, or 1-hydroxy-1,2-ethanedicarboxylic acid), $C_4H_6O_5$, is a white, crystalline material. The levorotatory isomer, *S*(−)-malic acid (L-malic acid), is a natural constituent and common metabolite of plants and animals. The racemic compound, *R,S*-malic acid (DL-malic acid), is a widely used food acidulant. This material is also used in some industrial applications as a sequestrant and as a buffer for pH control. *R*(+)-Malic acid (D-malic acid) is available only as a laboratory chemical. Following the introduction of a modern, continuous manufacturing process in the early 1960s, malic acid gradually became a large-volume industrial organic acid.

Physical Properties. Malic acid crystallizes from aqueous solutions as white, translucent, anhydrous crystals. The S(−) isomer melts at 100–103°C and the R(+) isomer at 98–99°C. On heating, D,L-malic acid decomposes at ca 180°C, by forming fumaric acid and maleic anhydride. Under normal conditions, malic acid is stable; under conditions of high humidity, it is hygroscopic. Malic acid is a relatively strong acid. Its dissociation constants are given in Table 1.

Chemical Properties. Because of its chiral center, malic acid is optically active.

Reactions. Malic acid undergoes many of the characteristic reactions of dibasic acids, monohydric alcohols, and α-hydroxycarboxylic acids.

Manufacture. In the United States, Canada, and Europe, only the synthetic *R,S*-malic acid is produced commercially, whereas both the S and R,S forms are produced in Japan.

Aqueous fumaric acid is converted to levorotatory malic acid by the intracellular enzyme, fumarase, which is produced by various microorganisms.

The commercial synthesis of *R,S*-malic acid involves hydration of maleic acid or fumaric acid at elevated temperature and pressure.

Energy And Environmental Considerations. The energy requirements to produce malic acid via conventional processes are fairly moderate. Malic acid production generates low levels of solid, airborne, and liquid waste. Solid waste is primarily nontoxic malic acid salts resulting form regenerating carbon cells and ion-exchange resins. Airborne emissions are primarily particulates. A 1% malic acid solution is readily biodegradable, with BOD of 5300 mg/L.

Table 1. Physical Properties of *R,S*-Malic Acid

Property	Value
mol wt	134.09
melting point, °C	ca 130
d_4^{20}	1.601
dissociation constant	
K_1	4×10^{-4}
K_2	9×10^{-6}
viscosity (50% aqueous solution at 25°C), mPa·s(s = cP)	6.5
solubility in nonaqueous solvents, % wt/wt	
ethanol	45.5
acetone	17.8
methanol	82.7

Shipping and Storage. Malic acid is shipped in 50-lb, 100-lb, and 25-kg, multiwall paper bags or 100-lb (45.5 kg) fiber drums. Malic acid can be stored in dry form without difficulty, although conditions of high humidity and elevated temperatures should be avoided to prevent caking.

Economic Aspects. Malic acid is manufactured in over 10 countries. The production is primarily used for food (26.6%) and beverages (54.7%); however, some industrial applications (18.7%) exist, eg, coatings, polymers, and resins. (Historical patterns of use in the United States have been stable and are as noted in parentheses).

Health and Safety. The U.S. FDA has affirmed *R,S*- and *S*(−)-malic acid as substances that are generally recognized as safe (GRAS) as flavor enhancers, flavoring agents and adjuvants, and as pH control agents. *R,S*-and *S*(−)-malic acid may not be used in baby foods. Malic acid is also cleared to correct natural acid deficiencies in juice or wine.

Uses. *R,S*-Malic acid is utilized in a variety of food and beverage and some industrial applications because of its unique combination of properties. These include having unusual taste-blending characteristics, flavor-fixing qualities, the ability to retain sour taste longer, high water solubility, and chelating and buffering properties. Malic acid is also a reactive intermediate in chemical synthesis.

Thiomalic Acid

Thiomalic acid (mercaptosuccinic acid), $C_4H_6O_4S$, mol wt = 150.2, is a sulfur analogue of malic acid. The properties of the crystalline, solid thiomalic acids are given in Table 2. The racemic acid has the following acid dissociation constants at 25°C: $pK_{a1} = 3.30$; $pK_{a2} = 4.94$.

R,S-Thiomalic acid can be prepared from bromosuccinic acid by reaction with K_2S. The two enantiomers can be obtained from the corresponding optically active potassium bromosuccinates.

Thiomalic acid is a skin sensitizer and an antidote in heavy-metal poisoning. Traditionally, it was a component of cold permanent hair-waving solutions and of rust-removing and corrosion-inhibiting compositions. Sodium aurothiomalate (Myochrisin) and other gold thiomalate complexes have antiarthritic properties. The well-known insecticide, malathion, is the thiomalate *S*-ethyl ester of *O,O*-dimethylphosphonodithioic acid.

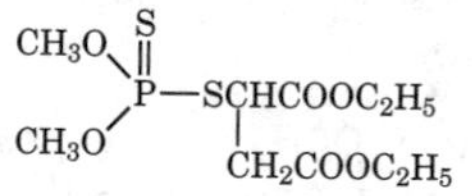

Tartaric Acid

Tartaric acid (2,3-dihydroxybutanedioic acid, 2,3-dihydroxysuccinic acid), $C_4H_6O_6$, is a dihydroxy dicarboxylic acid with two chiral centers. It exists as the dextro- and levorotatory acid: the meso form (which is inactive owing to internal compensation), and the racemic mixture (which is commonly known as racemic acid). The commercial product in the United States is the natural, dextrorotatory form, (*R-R**, *R**)-tartaric acid (L(+)-tartaric acid). This enantiomer occurs in grapes as its acid potassium salt (cream of tartar). In the fermentation of wine, this salt forms deposits in the vats.

Physical Properties. When crystallized from aqueous solutions above 5°C, natural (*R-R**,*R**;)-tartaric acid is obtained in the anhydrous form. Below 5°C, tartaric acid forms a monohydrate which is unstable at room temperature. Some of the physical properties of (*R-R**,*R**;)-tartaric acid are listed in Table 3.

Table 2. Properties of Thiomalic Acids

Acid	Mp, °C	Solubility		$[\alpha]_D^{17}$[a]
		Water	Ethanol	
R,S-	151	very sol	very sol	
R	154	sol	sol	+64.4°
S	152–153	sol	slightly sol	−64.8°

[a] 5% acid in ethanol.

Table 3. Physical Properties of (*R-R,*R**;)-Tartaric Acid**

Property	Value
mol wt	150.086
mp, °C (anhydrous)	169–170
d^{20}, g/cm^3	1.76
heat of solution, kJ/mol[a]	−13.8

[a] To convert kJ to kcal, divide by 4.184.

The solubility of (*R-R**,*R**;)-tartaric acid in water varies from 115g/100g H_2O at 0°C to 343g/100g H_2O at 100°C. One hundred grams of absolute ethanol dissolves 20.4 g of tartaric acid at 18°C, and 100 g of ethyl ether dissolves 0.3 g at 18°C.

Chemical Properties. The notation used by *Chemical Abstracts* to reflect the configuration of tartaric acid is as follows: (*R-R**,*R**;)-tartaric acid, (*S-R**,*R**;)-tartaric acid, and *meso*-tartaric acid. Racemic acid is an equimolar mixture of the two optically active enantiomers and, hence, like the meso acid, is optically inactive.

When free (*R-R**,*R**;)-tartaric acid is heated above its melting point, amorphous anhydrides are formed which, on boiling with water, regenerate the acid. Further heating causes simultaneous formation of pyruvic acid, $CH_3COCOOH$; pyrotartaric acid, $HOOCCH_2CH(CH_3)COOH$; and, finally, a black, charred residue. In common with other hydroxy organic acids, tartaric acid complexes many metal ions.

Occurrence. (*R-R**,*R**;)-Tartaric acid occurs in the juice of the grape and in a few other fruits and plants. It is not as widely distributed as citric acid or *S*(−)-malic acid. The only commercial source is the residues from the wine industry. The racemic acid is not a primary product of plant processes but is formed readily from the dextrorotatory acid by heating alone or with strong alkali or strong acid. *meso*-Tartaric acid is not found in nature. It is obtained from the other isomers by prolonged boiling with caustic alkali.

Manufacture. The chemical reactions involved in tartaric acid production are formation of calcium tartrate from crude potassium acid tartrate, formation of tartaric acid from calcium tartrate, formation of Rochelle salt from argols, and formation of cream of tartar from tartaric acid and Rochelle salt (RS) liquors.

Economic Aspects. The estimated total worldwide market for tartaric acid is 58,000 t and potassium bitartrate (acid basis) is 20,000 t.

Health and Safety. The FDA affirmed (*R-R**,*R**;)-tartaric acid as a generally- recognized-as-safe (GRAS) food substance.

Uses. Tartaric acid is used in carbonated beverages, wine making, and other foods. It is also used to produce emulsifiers, in the manufacture of pharmaceuticals, and in many industrial uses.

Salts. Rochelle salt is used in the silvering of mirrors. Its properties of piezoelectricity make it valuable in electric oscillators. Medicinally, it is an ingredient of mild saline cathartic preparations, eg, compound effervescing powder. In food, it can be used as an emulsifying agent in the manufacture of process cheese. Cream of tartar is used in baking powder and in prepared baking mixes.

GARY T. BLAIR
JEFFREY J. DEFRATIES
Haarmann & Reimer Corporation

Chem. Mktg. Rep., 35 (Jan. 25, 1993).

Food Chemicals Codex, 3rd ed., 3rd Suppl., National Academy of Sciences, National Research Council, Washington, D.C., 1992.

Code of Federal Regulations, 21 CFR 184.1069, Office of the Federal Register, U.S. Government Printing Office, Washington, D.C., 1993.

H. W. Ockerman, *Source Book for Food Scientists*, The Avi Publishing Co., Inc., Westport, Conn., 1978, p. 276.

HYGROMETRY. See DRYING.

HYPNOTICS, SEDATIVES, ANTICONVULSANTS, AND ANXIOLYTICS

The clinical effects observed with the variety of therapeutic agents that are used as hypnotics, sedatives, anxiolytics, and anticonvulsants reflect a spectrum of related activities at the cellular level characterized by a series of generalized responses within the central nervous system (CNS) that result in alterations in the dynamic state of brain function. Considerable advances have been made in the elucidation of the molecular events surrounding the activities of these agents and the γ-aminobutyric acid (GABA)/benzodiazepine (BZ) receptor complex. Serotonin, ion channel(s), and purinergic recognition sites have been implicated in the mechanism of action of a variety of anxiolytics, hypnotics, sedatives, and anticonvulsants, but the molecular targets for many of these agents, both in terms of efficacy and side effect liabilities, remain unknown.

Hypnotic agents depress the CNS and induce sleep when given at appropriate doses. Such agents typically act either by enhancing inhibitory neurotransmitter actions, eg, GABA, in the CNS or by inhibiting the actions of excitatory neurotransmitters, eg, glutamate. At lower doses, this class of compound can be sedating and can also have a calming or anxiolytic action. At higher doses, the traditional hypnotic agents, but not the BZs, can produce coma, a degree of anesthesia, and eventually death.

Anticonvulsants or antiepileptics are agents that prevent epileptic seizures or modulate the convulsant episodes elicited by seizure activity. Certain of these agents, eg, the BZs, are also hypnotics, anxiolytics, and sedatives, reinforcing the possibility of a common focus of action at the molecular level.

Anxiolytics are compounds that act primarily to relieve the symptoms of anxiety although such agents can also be used as anticonvulsants, sedatives, hypnotics, and anesthetic agents. The principal class of anxiolytics, the BZs, shows dependence liability whereas newer agents such as buspirone and ritanserine produce antianxiety effects via central serotoninergic systems.

Hypnotics and Sedatives

As a therapeutic class, hypnotics are nonselective CNS depressants that elicit drowsiness and a natural sleep state from which the individual can be aroused. The effects of hypnotics are generally dose-dependent.

Alcohols. Ethanol, C_2H_5OH is a potent sleep inducing agent that in the form of wine, liquor, and beer has been in recreational use since prehistory. Ethanol is a sedative, anxiolytic, and hypnotic agent. It can prove fatal in excess both acutely, owing to CNS depressive actions, and chronically as the result of progressive alcohol-induced tissue atrophy, most notably cirrhosis of the liver and neurological damage. Whereas ethanol is typically viewed as a CNS stimulant, this perceived effect results from a depression of tonic inhibitory neurotransmission processes. Chlorinated alcohols such as ethchlorvynol, C_7H_9ClO, and chloral hydrate, $C_2H_3Cl_3O_2$, are also useful hypnotics. The latter compound also possesses anticonvulsant and muscle relaxant activities. Both ethanol derivatives are associated with nausea, vomiting, dizziness at high doses and a typical hangover effect reflecting a residual depression of the CNS. Chloral hydrate irritates the gastrointestinal tract and can cause ataxia, nightmares, and allergic reactions. Ethchlorvynol has many properties in common with chloral hydrate and in addition has hypotensive actions.

Barbiturates. The barbiturates represent a seminal class of hypnotic agents that have been largely replaced in clinical use by the benzodiazepines (BZs). As a class, barbiturates have a broad spectrum of CNS depressant activity from mild sedation to general anesthesia. Representative compounds include amobarbital, $C_{11}H_{18}N_2O_3$, pentobarbital, $C_{11}H_{18}N_2O_3$, phenobarbital, $C_{12}H_{12}N_2O_3$, and the 5-allyl-5-(1-methylbutyl) analogue, secobarbital, $C_{12}H_{17}N_2O_3$.

Benzodiazepines. Many of the 1,4-BZs are routinely used as hypnotics. Examples include chlordiazepoxide, $C_{21}H_{14}Cl_2N_3O$, clorazepate, $C_{16}H_{13}ClN_2O_4$, diazepam, $C_{16}H_{13}ClN_2O$, flurazepam, $C_{21}H_{23}ClFN_3O$, lorazepam, $C_{15}H_{10}Cl_2N_2O$, triazolam, $C_{17}H_{12}Cl_2N_4$, estazolam, $C_{16}H_{11}ClN_4$, quazepam, $C_{17}H_{11}ClF_4N_2S$, and temazepam, $C_{16}H_{13}ClN_2O_2$. Unlike the barbiturates, the BZs are not general CNS depressants and are generally considered to be the safest available as hypnotics. As a class, BZs are also muscle relaxants and interact with alcohol. These latter properties, in addition to dependence liability and amnesia, limit their usefulness in the clinical setting.

Miscellaneous Agents. Compounds having sedative-hypnotic properties include the phenothiazines, methotrimeprazine, $C_{19}H_{24}N_2OS$ and promethazine, $C_{17}H_{20}N_{20}S$, methyprylon, $C_{10}H_{17}NO_2$, glutethemide, $C_{13}H_{15}NO_2$, methaqualone, $C_{16}H_{14}N_2O$, meprobamate, $C_9H_{18}N_2O_4$, carbromal, $C_7H_{13}BrN_2O_2$, bromoisovalum, $C_6H_{11}BrN_2O_2$, ethinamate, $C_9H_{13}NO_2$, etomidate, $C_{14}H_{16}N_2O_2$, and paraldehyde, $C_6H_{12}O_3$. Of these, glutethimide is addicting and shows little advantage in use as compared to the BZs or barbiturates whereas meprobamate is the most clinically useful.

A newer class of hypnotic at the preclinical stage as of this writing (ca 1995) are the neurosteroids, also known as the epalons. These also interact with the $GABA_A$/BZ receptor complex, have shown interesting activity in preclinical models, although their degrees of side effect liability appears the same as that of the BZs.

Melatonin, $C_{13}H_{16}N_2O_2$, and related compounds have marked effects on circadian rhythm. Novel ligands for melatonin receptors such as $C_{17}H_{16}N_2O_2$, have potential use in the treatment of the sleep disorders associated with jet lag. Such agents may also be useful in the treatment of seasonal affective disorder (SAD), the depression associated with the winter months. Histamine, adenosine and neuropeptides such as corticotropin-like intermediate lobe peptide (CLIP) and vasoactive intestinal polypeptide (VIP) also have sedative–hypnotic activities.

BZ analogues that can act as antagonists or inverse agonists of the classical anxiolytic agonist, diazepam, have also been discovered as the result of the intensive effort following the discovery of the GABA/BZ receptor complex. Flumazenil, $C_{15}H_{14}FN_3O_3$, an imidazoBZ, is in clinical use as an analeptic. It is used in reversing the sedation associated with the BZs, specifically in terms of outpatient anesthesia.

Anticonvulsants

Barbiturates. Phenytoin, $C_{15}H_{12}N_2O_2$, is used in the treatment of all types of seizure except absence and is the drug of choice in status epilepticus but is not very useful, in treating myoclonic or atonic seizures. Phenytoin exerts its anticonvulsant effects without causing a general depression of the CNS. The barbiturate phenobarbital is a long-acting anticonvulsant used against generalized and partial seizures but it is ineffective in absence seizures. Primidone, $C_{12}H_{14}N_2O_2$, is an analogue of phenobarbital that is used for the treatment of generalized tonic–clonic seizures.

Succinimides. Ethosuximide, $C_7H_{11}NO_2$, and the related succinimide, methsuximide, $C_{12}H_{13}NO_2$, are used in absence seizure treatment. Like the other anticonvulsants discussed, the mechanism of action of the succinimides is unclear.

Benzodiazepines. Several BZs have anticonvulsant activity and are used for the treatment of epilepsy producing their anticonvulsant actions via interactions with the $GABA_A$/BZ receptor complex to enhance inhibitory GABAergic transmission.

Miscellaneous. The iminostilbene, carbamazepine, $C_{15}H_{12}N_2$-O, is a congener of the tricyclic antidepressant, imipramine. Initially used in the treatment of trigeminal neuralgia, it is effective against all types of epilepsy except absence. Predominant usage is in generalized tonic, tonic–clonic, and partial seizures. Oxcarbazepine, $C_{15}H_{12}N_2O_2$, is better tolerated than carbamazepine.

Unlike the heterocyclic anticonvulsants, valproate, $C_8H_{16}O_2$, is a branched chain carboxylic acid. The compound has minimal sedative and CNS side effects and is indicated for use in a wide variety of seizure types including tonic–clonic and absence seizures.

The oxazolidinedione, trimethadione, $C_6H_9NO_3$, at one time the drug of choice for the treatment of absence seizures, has been replaced by ethosuximide and valproate.

Acetazolamide, $C_4H_6N_4O_3S_2$, is a sulfonamide having antiepileptic activity. Paraldehyde has been used in the treatment of status epilepticus but has been replaced by the BZs.

Progabide, $C_{17}H_{16}ClFN_2O_2$, is a GABA receptor agonist that interacts with both $GABA_A$ and $GABA_B$ receptor subtypes and issued for the treatment of complex partial seizures, generalized tonic–clonic, atonic, and myoclonic seizures as both mono- and cotherapy. Gabapentin, $C_9H_{17}NO_2$, an analogue of GABA, has anticonvulsant activity.

Lamotrigine, $C_9H_7Cl_2N_5$, was approved for use as an anticonvulsant in 1992 and is thought to act by inhibiting the release of the excitatory neurotransmitter, glutamate. Felbamate, $C_{11}H_{14}N_2O_4$, is a newer anticonvulsant approved by the FDA at the end of 1992 but its use was significantly proscribed due to significant side effect liabilities. Stiripentol, $C_{14}H_{18}O_3$ is an older agent that may produce its anticonvulsant actions via a direct or indirect GABAergic mechanism.

Zonisamide, $C_8H_8N_2O_3S$, is a broad-spectrum anticonvulsant that was approved in Japan in 1989 and as of this writing is in Phase III clinical trials in the United States. Remacemide, $C_{17}H_{20}N_2O$, is a prodrug of an *N*-methyl-D-aspartate antagonist having anticonvulsant activity in preclinical animal models that is being evaluated clinically for complex partial seizures. Topiramate, $C_{12}H_{21}NO_8S$, is another potential anticonvulsant agent currently in Phase III clinical trials. The purine, BW A78U, $C_{14}H_{164}N_6$, and mioflazine, $C_{29}H_{30}Cl_2F_2N_4O_2$, an inhibitor of adenosine transport, are also active in animal models of epilepsy and are noteworthy given the potential role of the neuromodulator adenosine, as an endogenous anticonvulsant also.

Anxiolytics

Ethanol is the most widely used antianxiety agent.

Benzodiazepines. The breakthrough in antianxiety therapy occurred in 1958 when the BZ, chloradiazepoxide, was discovered. Other BZ anxiolytic agents include clorazepate, diazepam, lorazepam, alprazolam, $C_{17}H_{13}ClN_4$, oxazepam, $C_{15}H_{11}ClN_2O_2$, halazepam, $C_{17}H_{12}ClF_3N_2O$, and prazepam, $C_{19}H_{17}ClN_2O$.

Nonbenzodiazepine Benzodiazepine Receptor Ligands. The triazolopyridazine, CL 218872, $C_{13}H_9F_3N_4$, and the pyrazolopyridine, CGS 9896, $C_{16}H_{10}ClN_3O$, are the prototypic agents of this type although neither proceeded beyond early clinical trials.

Serotonin Receptor Anxiolytics. The arylpiperazine buspirone, $C_{21}H_{31}N_5O_2$, was the first of a series of partial agonists active at the $5HT_{1A}$ receptor that include ipsapirone, $C_{19}H_{23}N_5O_3S$ and tandospirone, $C_{21}H_{29}N_5O_2$. Buspirone is especially effective in treating anxiety in the aged population.

Newer experimental approaches to anxiety therapy include ligands interacting with the ligand-gated ion channels that are selectively activated by nicotine, $C_{10}H_{14}N_2$, the well-known active ingredient of cigarettes which has anxiolytic actions. Cholecystokinin B receptor ligands, specifically the dipeptoid, CI-988, $C_{35}N_{42}N_4O_6$, have demonstrated anxiolytic activity in preclinical models but are not being pursued clinically.

A newer, highly experimental approach to anxiety therapy is the use of antisense oligonucleotides to the anxiogenic peptide, NPY.

Economic Aspects

Sales for hypnotic–sedative agents in the U.S. for 1994 were estimated to be $150 million. U.S. sales of antianxiety agents including both the BZs and buspirone-like agents for 1993 were estimated at $1.1 billion.

MICHAEL WILLIAMS
MARK W. HOLLADAY
Abbott Laboratories

A. G. Romero and R. B. McCall, *Ann. Rep. Med. Chem.* **27**, 21–30 (1992).

Merck Manual, 16th ed., Merck, Sharp and Dohme Research Laboratories, Rahway, N.J., 1992, pp. 1436–1444.

R. Levy and co-workers, eds., *Antiepileptic Drugs*, 4th ed., Raven Press, New York, 1995.

HYPOGLYCEMIC AGENTS.

See INSULIN AND OTHER ANTIDIABETIC AGENTS.

I

ICE. See REFRIGERATION; WATER.

ICE CREAM. See MILK PRODUCTS.

IDITOL. See SUGAR ALCOHOLS.

ILMENITE. See TITANIUM COMPOUNDS.

IMAGING TECHNOLOGY

Imaging systems, consisting of specialty chemicals and techniques, are used to produce copies or photographic representations of macroscopic entities that can be seen by the human eye. Moreover, imaging systems are utilized to produce representations of what is outside the range of human vision.

An alphabetized list of *Encyclopedia* articles that are directly related to the various imaging technologies follows.

COLOR PHOTOGRAPHY	PHOTOCONDUCTIVE POLYMERS
COLOR PHOTOGRAPHY, INSTANT	PHOTOGRAPHY
ELECTRICALLY CONDUCTIVE POLYMERS	PRINTING PROCESSES
ELECTROPHOTOGRAPHY	RADIOACTIVE TRACERS
FINE ART EXAMINATION AND CONSERVATION	RADIOPAQUES
INFORMATION STORAGE MATERIALS	SURFACE AND INTERFACE ANALYSIS
LITHOGAPHIC RESISTS	THERMOGRAPHY
MEDICAL IMAGING TECHNOLOGY	
NONDESTRUCTIVE EVALUATION	X-RAY TECHNOLOGY
PHOTOCHEMICAL TECHNOLOGY	

IMINES, CYCLIC

Ethyleneimine (aziridine, azacyclopropane) is the smallest cyclic imine consisting of a three-membered *N*-heterocyclic ring ($n = 2$):

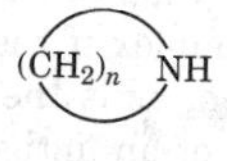

This article describes ethyleneimine and the most important aziridine derivatives. Unsubstituted ethyleneimine is industrially the most important representative of the aziridine class. The BASF group is by far the largest manufacturer of ethyleneimine and has production plants in Germany and the U.S. Another important producer is the Nippon Shokubai Co. Ltd. of Japan.

Physical Properties

Ethyleneimine (EI) and its two most important derivatives, 2-methylaziridine (propyleneimine), and 1-(2-hydroxyethyl)aziridine (HEA) are colorless liquids. They are miscible in all proportions with water and the majority of organic solvents. Ethyleneimine is not miscible with concentrated aqueous NaOH solutions (>17% by weight). Ethyleneimine has an odor similar to ammonia. The physical properties of ethyleneimine and the derivatives mentioned are given in Table 1.

Table 1. Physical Properties of Ethyleneimine and Derivatives

Property	EI	PI[a]	HEA[b]
solidification point, °C	−74	−65	
boiling point, °C	57	66	156
density, g/mL	0.837[c]	0.8017[d]	1.088
refractive index n_D	1.4130[c]	1.4084[d]	1.453[d]
flashpoint, °C	−13	−10	67
viscosity at 25°C, mPa·s(= cP)	0.418	0.491	
dielectric constant at 25°C	18.3		

[a] Propyleneimine. [b] Hydroxyethylaziridine. [c] At 20°C. [d] At 25°C.

Chemical Properties

Ethyleneimine is the only C_2H_5N isomer stable at room temperature provided CO_2 is excluded from the air. Unexpectedly, ethyleneimine has the highest calculated relative heat of formation of the C_2H_5N isomers. Ethyleneimine shows a lower basicity than noncyclic aliphatic amines such as dimethylamine.

Reactions

Depending on the experimental conditions used, the basicity or the ring strain can be the driving force in reactions involving ethyleneimine. With catalysis by Brönsted or Lewis acids, the aziridine ring can be opened by a large number of nucleophiles to give β-substituted ethylamines. In the absence of strong nucleophiles and at elevated temperatures, preparation of polyethyleneimines from aziridines is possible by acid-catalyzed reaction of the aziridine with itself. On the other hand, ethyleneimine and other aziridines substituted only on carbon show the typical reactions of a secondary amine, such as addition onto unsaturated systems, complex formation with metals, and reaction with halogen compounds. At low temperatures and alkaline pH the *N*-substituted aziridines are generally formed in these reactions. High temperatures and catalysis by acids or nucleophiles promote secondary reactions with opening of the three-membered ring, and these can be used for synthesis of heterocyclic compounds.

Nucleophilic Ring Opening. Opening of the ethyleneimine ring with acid catalysis can generally be accomplished by the formation of an intermediate aziridinium salt, with subsequent nucleophilic substitution on the carbon atom which loses the amino group. In the following, R represents a Lewis acid, usually H^+; A^- = the nucleophile.

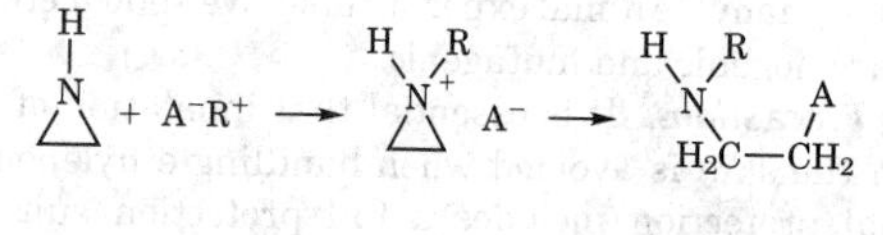

Electrophilic Reactions on the Aziridine Nitrogen. The generalized reaction of aziridines with an electrophile (R^+) is as follows.

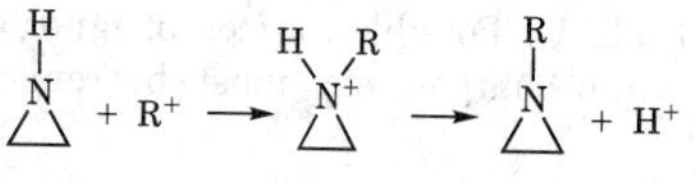

Reactions with Transition-Metal Compounds. The numerous published products of reactions of transition-metal compounds with aziridines can be divided into complexes in which the aziridine ring is intact, compounds formed by reaction of aziridine with the ligands of a complex, and complexes in which the aziridine molecule is fragmented (imido complexes).

Other reactions include reductive ring opening, oxidative ring opening, thermal and photochemical reactions, and polymerization.

Preparation

The Wenker process has been carried out by BASF and various other companies since the end of the 1960s. In this process the hemisulfate of monoethanolamine, a nonvolatile, crystalline substance, is used for the alkaline cyclization. The reaction can be carried out under pressure.

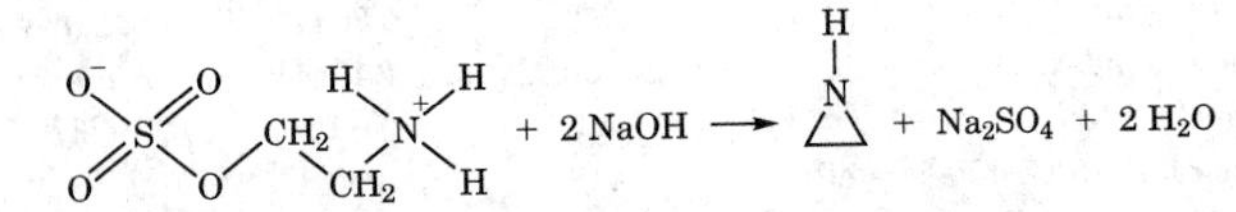

A production plant for salt-free ethyleneimine synthesis by catalytic dehydration of monoethanolamine in the gas phase has started operation at the Japanese company Nippon Shokubai.

Economic Aspects

Because of its toxicity, ethyleneimine monomer is not sold by the BASF group or Nippon Shokubai, currently the only large producers. Ethylenimine is used onsite for further reaction to produce polymers and intermediates. The BASF Corporation has started production of ethyleneimine in the U.S. in Freeport, Texas. The current world ethyleneimine production capacity is more than 12,000 t/yr. In the U.S. alone, the consumption of polyethyleneimines approximately doubled between 1983 and 1990. One of the largest markets worldwide is the paper industry, which uses ethyleneimine polymers and their derivatives as process chemicals for paper making. Large amounts of ethyleneimine-based polymers are also used as oil field chemicals and flocculating (aggregating) agents. In addition, ethyleneimine derivatives can be used for a large number of special applications, such as enzyme immobilization, textile finishing, membranes, etc.

Storage

Resistant materials for the storage and handling of ethyleneimine are low carbon steel, V2A and V4A chrome nickel steel, and (strengthened) glass and enamel. Copper, silver, and golds, containing these metals must be avoided.

Health and Safety Factors

Toxicology. Ethyleneimine is highly toxic on inhalation, on contact with the skin, and if swallowed.

The odor threshold for detection of ethyleneimine is 2 ppm. The maximum permissible concentration of ethyleneimine in the air at the place of work is 0.5 ppm (as specified in statutory regulations in the U.S. and in Germany). Animal experiments have shown ethyleneimine to be both carcinogenic and mutagenic.

Handling Precautions. It is essential that inhalation of vapors and contact with the skin is avoided when handling ethyleneimine. Suitable personal protection includes a full protection suit, preferably made of butyl rubber, and a breathing and face mask (hood with independent air supply). Ethyleneimine and other low molecular weight aziridines are very highly flammable and can form explosive mixtures with air (Table 1). Possible sources of ignition (open flames, electric sparks, static charges, etc) must be removed when using ethyleneimine.

G. SCHERR
U. STEUERLE
R. FIKENTSCHER
BASF AG

O. C. Dermer and G. E. Ham, *Ethylenimine and Other Aziridines*, Academic Press, Inc., New York, 1969.

E. J. Goethals, in K. J. Ivin and T. Saegusa, eds., *Ring-Opening Polymerization*, Vol. 2, Elsevier Applied Science Publishers, New York, 1984.

D. A. Tomalia and G. R. Killat, in J. I. Kroschwitz, ed., *Encyclopedia of Polymer Science and Engineering*, Vol. 1, John Wiley & Sons, Inc., New York, 1985.

D. Horn and F. Linhart, in J. C. Roberts, ed., *Paper Chemistry*, Blackie, London, 1991.

IMMUNOASSAY

Immunoassay is a method that identifies and quantifies unknown analytes using antibody–antigen reactions. Techniques are based in immunochemistry, analytical chemistry, and biochemistry, with a history of development paralleling advances in microbiology and immunology (see also IMMUNOTHERAPEUTIC AGENTS).

The success of immunoassay in clinical diagnostics, and the generic nature of immunoassay technology has resulted in the application of the method in other areas. By the late 1980s, commercial immunoassay products and systems were available for detection and diagnosis in environmental, food, and chemical processing applications. Whereas the application of immunoassays in these areas is small in relation to clinical diagnostic immunoassays, nonclinical applications of immunoassays hold the potential for significant growth by the year 2000 (see AUTOMATED INSTRUMENTATION, CLINICAL CHEMISTRY).

The Antibody–Antigen Reaction

Immunoassays are based on the binding and complexing of an antigen to an antibody, and the use of some physical or chemical means to measure and quantify the antigen–antibody complex. The antibody–antigen reaction is a typical reversible bimolecular reaction having rate constants for the forward and backward reactions which are dependent on concentration of the antigen (Ag) and antibody (Ab), affinity for the antigen as defined by the association constant of the antibody for its antigen, temperature, pH, and other environmental conditions. This reaction is represented by an equation common to reversible receptor-ligand assays:

$$\mathrm{Ab} + \mathrm{Ag} \rightleftarrows \mathrm{AbAg} \tag{1}$$

and the equilibrium constant for the reaction is determined by the mass action equation:

$$K = \frac{[AbAg]}{[Ab][Ag]} \tag{2}$$

where $[Ab]$ is the concentration of antibody sites for antigen, and $[Ag]$ is the concentration of free antigen. The association, or affinity, constant for the antibody–antigen reaction is then further defined as the equilibrium constant at half-saturation of the antibody with antigen. Because at half-saturation AbAg and Ab are equal, these cancel in the above equation and the association constant, K_a, is equal to the reciprocal of the free antigen concentration:

$$K_a = \frac{1}{[Ag]} \tag{3}$$

Thus, if the antibody has a high affinity for the antigen, it has a high association constant. Typical association constants range from 10^6 to 10^{10} L/mol, and as high as 10^{13} L/mol for some monoclonal antibodies.

The definition of an association constant for an antibody–antigen reaction can become more complex if the antibody–antigen reaction involves a multivalent antigen, as is the case when a polyclonal antiserum is used for detection of an antigen. This type of multivalent binding is termed avidity and is defined by the equation:

$$x\mathrm{Ab} + y\mathrm{Ag} \rightarrow \mathrm{Ab}_x\mathrm{Ag}_y \tag{4}$$

Definition of the association (or avidity) constant for such multivalent antibody–antigen reactions must consider not only the heterogeneity of the antibodies and the antigen determinant site(s), but also an apparent additive effect of binding two antigen molecules to a single antibody. Such effects lead to a multiplying of the individual association constants and an apparent large increase in the total association–avidity constant. This multiplication of avidity through multivalent binding has been exploited to increase the sensitivity of many immunoassays. A more detailed discussion of antigen–antibody reaction kinetics may be found in the literature.

Basic Technology

The principal approach to immunoassay is illustrated in Figure 1, which shows a basic sandwich immunoassay. In this type of assay, an

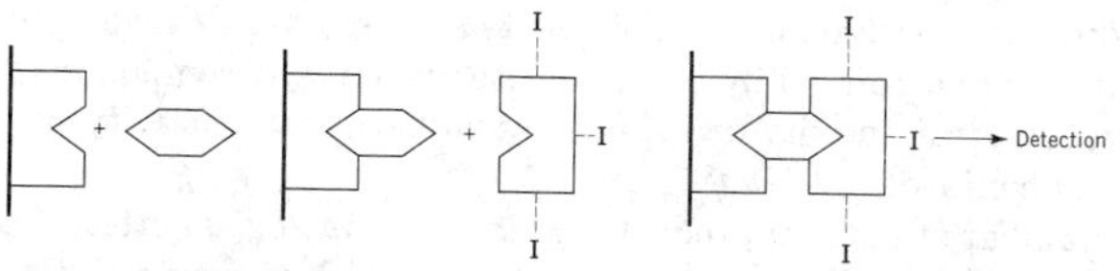

Figure 1. A principal approach to immunoassay, the sandwich immunoassay, where the thick line represents the solid matrix, the antibody, the antigen, and I an indicator molecule such as an enzyme, fluorophore, or radioisotope.

antibody to the analyte to be measured is immobilized onto a solid surface, such as a bead or a plastic (microtiter) plate. The test sample suspected of containing the analyte is mixed with the antibody beads or placed in the plastic plate, resulting in the formation of the antibody–analyte complex. A second antibody which carries an indicator reagent is then added to the mixture. This indicator may be a radioisotope, for RIA; an enzyme, for EIA; or a fluorophore, for fluorescence immunoassay (FIA). The antibody-indicator binds to the first antibody–analyte complex, free second antibody-indicator is washed away, and the two-antibody–analyte complex is quantified using a method compatible with the indicator reagent, such as quantifying radioactivity or enzyme-mediated color formation.

In fact, most RIAs and many nonisotopic immunoassays use a competitive binding format. In this approach, the analyte in the sample to be measured competes with a known amount of added analyte that has been labeled with an indicator that binds to the immobilized antibody. After reaction, the free analyte–analyte-indicator solution is washed away from the solid phase. The analyte-indicator on the solid phase or remaining in the wash solution is then used to quantify the amount of analyte present in the sample as measured against a control assay using only an analyte-indicator.

There are many variations on these two basic approaches for immunoassays.

Immunoassay Design. The basic reagent and design requirements of an immunoassay are antibody, antigen, conjugates of either or both, and a means for separating bound and unbound reagents.

There are many possible means for quantification of the antigen–antibody reaction. Immunoassays may be classified according to the technology used for detection and quantification of the analyte being detected, eg, turbidimetric agglutination immunoassays, radioimmunoassay, enzyme immunoassay, fluorescence immunoassay, and chemiluminescent immunoassay.

Comparison of Methodologies. A heterogenous immunoassay is a multistep assay requiring the sequential addition of reagents with washing steps between reagent additions. Most immunoassay kits and many commercial immunoassay analyzers are based on heterogenous EIA or FIA.

During the 1980s, a number of homogenous immunoassays were developed and commercialized. Homogenous immunoassays occur in one vessel, requiring no separation of components prior to quantification. The advantages of homogenous immunoassays are simple formats and rapid data output producing user-friendly and cost-effective products. Technical challenges to consider, however, are the necessity to remove or minimize background interference from the reagents and nonspecific binding reactions.

Monoclonal vs Polyclonal Antibodies. A continuing question facing the developer of an immunoassay is whether to use monoclonal (MAb) or polyclonal (PAb) antibodies in the assay. Polyclonal antibodies are the natural mixture of antibodies resulting from the immune response to an antigen. A family of antibodies results, each binding specifically to a different antigenic determinant (or part of a determinant) on the same antigen.

Whereas such diversity in the immune response may have evolved as a protective means to the host animal, PAbs present problems to the immunoassay developer looking for high antibody specificity and low total protein.

In 1975, the first successful production of MAbs was reported. By fusing normal antibody-producing cells with a B-cell tumor (myeloma), hybridoma cell lines resulted which produced antibodies having a specificity to only one determinant on an antigen; ie, all the antibodies produced from the cell line are identical.

The singularity of MAbs and the ease of mass production appeared to be the answer to rapid development of highly specific immunoassays. Whereas MAbs appear to be the choice for use in immunoassays, a majority of immunoassay developers and suppliers use polyclonal antibodies. The primary reason for this choice lies in the investment of time and costs required to fuse, clone, and screen thousands of hybridomas to discover those producing MAbs having the high avidity required for an assay. In addition, MAb-producing hybridoma cells can be extremely unstable, losing antibody production capabilities or simply dying out in a few passages (generations).

The question of whether to use MAbs or PAbs in an assay is a matter of assay requirements (specificity and sensitivity) and economics and cannot be answered on technical merit alone.

Immunoassay–DNA Probe Hybrid Assays

Nucleic acid (deoxyribonucleic acid (DNA) and ribonucleic acid (RNA)) probes utilize labeled, ie, radioactive, enzymatic, or fluorescent, fragments of DNA or RNA (the probe) to detect complimentary DNA or RNA sequences in a sample. Because the probe is tailored for one specific nucleic acid, these assays are highly specific and very sensitive.

As the result of high specificity and sensitivity, nucleic acid probes are in direct competition with immunoassay for the analytes of some types of clinical analytes, such as infectious disease testing. Assays are being developed, however, that combine both probe and immunoassay technology. In such hybrid probe–immunoassays, the immunoassay portion detects and amplifies the specific binding of the probe to a nucleic acid. Either the probe *per se* or probe labeled with a specific compound is detected by the antibody, which in turn is labeled with an enzyme or fluorophore that serves as the basis for detection.

Hybrid probe–immunoassays are expected to find a specific niche in clinical analysis, especially as a means to adapt probe assays to existing immunoanalyzers which are locked into a specific enzyme or fluorescence detection technology. Commercialization of the first of these assays is expected by the year 2000.

Immuno(bio)sensors

Immunoassay technology is also being applied in the development of antibody-based biosensors, or immunosensors. A biosensor is an electronic detection device containing a biological molecule, such as an antibody, enzyme, or receptor, as its basic detection element. The ideal biosensor employs a homogenous (one-step, nonprep) format, real-time detection (results in less than one minute) in a cost-effective, portable, user-friendly design. Immunosensors have been designed which use both direct and indirect immunoassay technology to detect specific analytes within a minute or less in a variety of matrices. Indirect immunosensors may employ EIA, FIA, or CLIA principles whereby enzyme-, fluorophore- or chemiluminescent-labeled or antibody, respectively. Measurements may be based on the perturbation of an electrical field by the antibody–antigen binding event; changes in light scattering fluorescence or chemiluminescence on an optical fiber; or changes in the weight of an antibody–antigen complex as compared to the weight of antibody alone on a piezoelectric crystal. Immunosensors promise to become principal players in chemical, diagnostic, and environmental analyses by the latter 1990s.

RICHARD F. TAYLOR
Arthur D. Little, Inc.

J. Clausen, *Immunochemical Techniques for the Identification and Estimation of Macromolecules*, 3rd ed., Elsevier, Amsterdam, the Netherlands, 1988.

A. Kawamura, Jr., ed., *Fluorescent Antibody Techniques and Their Applications*, University of Tokyo Press, Baltimore, 1977.

E. T. Maggio, ed., *Enzyme Immunoassay*, CRC Press, Boca Raton, Fla., 1981.

R. F. Taylor, ed., *Protein Immobilization: Fundamentals and Applications*, Marcel Dekker, Inc., New York, 1991.

IMMUNOSUPPRESSANTS.

See ANALGESICS, ANTIPYRETICS, AND ANTI-INFLAMMATORY AGENTS; CHEMOTHERAPEUTICS, ANTICANCER; IMMUNO THERAPEUTIC AGENTS.

IMMUNOTHERAPEUTIC AGENTS

The immune system is the primary mechanism of defense against invasive disease for human beings. Advances in immunology during the last part of the twentieth century have continued at a rapid rate and cytokines and immune cells having specific markers continue to be defined. A number of natural and synthetic immunotherapeutic agents have been discovered that can modulate components of the normal or aberrant immune system, through stimulation or suppression. However, most of these substances also have inherent adverse side effects.

Immunotherapy for Various Disease States

Immunodeficiencies. Whereas elimination and neutralization of the antigen by avoidance and isolation is feasible for minor cases of allergies, patients with immunodeficiency are at greater health risk and require increased amount of care to maintain optimal health. Antibiotic therapy is an important modality for the treatment of immunodeficient patients and continuous prophylactic antibiotic treatment is often beneficial, especially when there is a potential for overwhelming infection, other forms of immune therapy have proven insufficient, or there is a high risk for a specific infection.

In passive immunotherapy immune globulin (Ig) is an effective replacement in most forms of antibody deficiency. Plasma is rarely indicated in the 1990s because of the risk of disease, particularly AIDS transmission. Because plasma contains many factors in addition to immunoglobulins (Igs), plasma is, however, of particular value in patients with protein-losing enteropathy, complement deficiencies, and refractory diarrhea.

Problems associated with active immunotherapy for a faltering immune system involve identifying the requirements for appropriate immunotherapeutic agents so that both efficacy and safety can be ensured. Immunological disturbances in patients can lead to pathogenesis of many diverse problems, such as senescence, primary and acquired immunodeficiency syndromes, acute and chronic infections, autoimmune diseases, and cancer. Immunoenhancing agents may be useful where a certain degree of immune capacity is present, but are of limited value in treating cellular or phagocytic immunodeficiencies.

Antiviral agents are used in attempts to combat the devastating effect of HIV on the immune system. As of this writing there are three principal approaches to the treatment of AIDS: (*1*) use of anti-HIV agents to destroy the virus or control its growth; the National Cancer Institute (NCI) encourages submission of synthetic and characterized natural products for anti-HIV screening; (*2*) immunotherapy to restore impaired immune functions; and (*3*) treatment of specific opportunistic infections or tumors.

The majority of antiviral agents, whether purines, pyrimidinones, amantadines, or others, were designed primarily to interfere with viral replication. Usually antiviral chemotherapy results in a wide variety of side effects and a narrow separation between efficacy and toxicity. Some antiviral agents, eg, inosine pranohex, $C_{52}H_{78}N_{10}O_{17}$, are also immunomodulators. Plants and microorganisms produce unique and diverse chemical structures, some of which act as immunomodulators. Of specimens used in traditional medicine, approximately 450 plant species have shown antiviral activity out of 4000 plants screened. Several tannins exhibit strong inhibition of tumor promotion experimentally.

Whereas over 200 plant constituents are reported to have antiviral activity, as determined by *in vitro* methods, only 31 compounds have shown antiviral activity *in vivo*. Immunotherapeutic activity has not been determined.

A number of natural products exhibit immunological effects. Echinacea extracts contain a large number of constituents and the immunostimulatory activity of such extracts has been demonstrated. The biological activity of these extracts has warranted development of fermentation methods for large-scale production.

Cianidanol (Ci) or cathechin, $C_{15}H_{14}O_6$, a flavonoid found primarily in higher woody plants, has been shown to have both some specific and nonspecific effects on the immune system. Cianidanol exerts an immunoenhancing effect on the function of various peripheral mononuclear blood cells. The T-cell activation by Ci at low doses stimulates the spontaneous and mitogen-induced proliferation and Ig secretion of human peripheral blood mononuclear cells.

Ling Zhi-8 (LZ-8) is an immunomodulatory protein isolated from the mycelial extract of *Ganoderma lucidium*, that has been purified and shown to stimulate mouse spleen and human peripheral blood lymphocytes.

Polysaccharides obtained from different sources have been shown to have immunostimulant and antitumor activity. One example is a glucan, isolated from yeast, a branched β-1,3-polyglucopyranose originally present in the yeast cell wall as a component of zymosan. It is known to be a broad-spectrum enhancer of host defense mechanisms. Immunopharmacological studies of this glucan demonstrated antitumor effects; prevention of carcinogenesis; increase in host resistance to bacterial, viral, fungal, and parasitic infections; and an increase in phagocytic and proliferative activity of the reticuloendothelial system.

Synthetic immunomodulators that have been developed are listed in Table 1. These compounds have been shown to modulate the immune system. Some also have antitumor activity.

Rheumatoid Arthritis. Steroids used for RA treatment generally influence the synthesis and response to IL-1. Glucocorticoids, eg, prednisone, can affect virtually every aspect, phase, and cell type involved in immunologic and inflammatory reactions. Some rheumatologists are now using immunosuppressive drugs such as methotrexate in early stages of RA with significant success. Antimalarials, gold compounds, penicillamine, and sulfasalazine are all used in antirheumatics. Traditional antirheumatic drugs having immunological activity are listed in Table 2. Immunosuppressive drugs that are increasingly used for treatment of severe and active RA are given in Table 3.

A new generation of antiinflammatory agents having immunosuppressive activity has been developed. The appearance of preclinical and clinical reports suggest that these are near entry to the pharmaceutical market. For example, tenidap has been demonstrated to

Table 1. Synthetic Immunomodulators

Compound	Molecular formula
amiprilose	$C_{14}H_{27}NO_6$
bucillamine	$C_7H_{13}NO_3S_2$
ditiocarb sodium	$C_5H_{11}NS_2Na$
inosine pranobex	$C_{10}H_{12}N_4O_5 3C_9H_9NO_3 3C_5H_{13}NO$
muramyl dipeptide	$C_{19}H_{32}N_4O_{11}$
muroctasin	$C_{43}H_{78}N_6O_{13}$
platonin	$C_{38}H_{61}N_3S_3 2I$
procodazole	$C_{10}H_{10}N_2O_2$
tenidap	$C_{14}H_9ClN_2O_3S$
tetramisole	$C_{11}H_{12}N_2S$
tilorone	$C_{25}H_{34}N_2O_3$
thymopentin	$C_{30}H_{49}N_9O_9$
tolfenamic acid	$C_{14}H_{12}ClNO_2$
ubenimex	$C_{16}H_{24}N_2O_4$

Table 2. Traditional Antirheumatic Drugs Having Immunological Activity

Compound	Molecular formula
prednisone	$C_{21}H_{26}O_5$
tenidap[a]	$C_{14}H_9ClN_2O_3S$
SKF 86002	$C_{16}H_{12}FN_3S$
E-5110	$C_{20}H_{29}NO_3$
auranofin	$C_{20}H_{34}AuO_9PS$
aurothioglucose	$C_6H_{11}AuO_5S$
gold sodium thiomalate	$C_4H_5AuO_4S$
chloroquine	$C_{18}H_{26}ClN_3$
mefloquine	$C_{17}H_{16}F_6N_2O$
tetrandrine	$C_{38}H_{42}N_2O_6$
general IL-1 inhibitor	
D-penicillamine	$C_5H_{11}NO_2S$
sulfasalazine	$C_{18}H_{14}N_4O_5S$

[a] This drug is considered an immunomodulator.

Table 3. Immunosuppressive Drugs

Compound	Molecular formula
Synthetic products	
methotrexate	$C_{20}H_{22}N_8O_5$
azathioprine	$C_9H_7N_7O_2S$
cyclophosphamide	$C_7H_{15}Cl_2N_2O_2P$
mycophenolate	$C_{23}H_{31}NO_7$
brequinar	$C_{23}H_{15}F_2NO_2$
deoxyspergualin	
Natural products	
cyclosporin A	$C_{62}H_{111}N_{11}O_{12}$
sirolimus	$C_{51}H_{79}NO_{13}$
FK 506	$C_{44}H_{69}NO_{12}$
taxol	$C_{47}H_{51}NO_{14}$
mizoribine	$C_9H_{13}N_3O_6$

inhibit IL-1 production from human peripheral blood monocytes in culture.

Cancer. Modern cancer therapy has been primarily dependent upon surgery, radiotherapy, chemotherapy, and hormonal therapy (see CHEMOTHERAPEUTICS, ANTICANCER; HORMONES; RADIOPHARMACEUTICALS). Chemotherapeutic agents may be able to retard the rate of growth, but are unable to eradicate the entire population of neoplastic cells without significant destruction of normal host tissue. This serious side effect limits general use. More recently, the immunotherapeutic approach to cancer has involved modification and exploitation of the cellular and molecular mechanisms in host defense, regulation of tissue proliferation, tissue differentiation, and tissue survival. The results have been more than encouraging.

Natural products that are immune stimulants, eg, *Bacillus calmette guerin* (BCG), the purified protein derivative (PPD) of tuberculin, and crude lymphokine preparations, have been used to attempt regression of numerous primary or metastatic skin malignancies. This modality has also been used to treat basal cell carcinoma, mycosis fungoids, lymphangiosarcoma, reticulum cell sarcoma, and breast cancer. However, this local response appears to be selective for tumor cells and is relatively sparing of the normal adjacent tissues.

In the early 1990s immunologic treatment modalities for primary or metastatic malignant melanoma include interferons as single agents, or in combination with other cytotoxic drugs, monoclonal antibodies, and melanoma vaccines.

The concept that a weak or suppressed host immune response may be overridden by active immunization appears to be valid. Vaccines of autologous tumor extracts have been used to treat patients with malignant melanoma. Interesting and significant progress has been made in the treatment of neoplasia using interferons and interleukin-2.

Cytokines, eg, interferons, interleukins, tumor necrosis factor (TNF), and certain growth factors, could have antitumor activity directly, or may modulate cellular mechanisms of antitumor activity. Cytokines may be used to influence the proliferation and differentiation of T-cells, B-cells, macrophage–monocyte, myeloid, or other hematopoietic cells. Alternatively, the induction of interferon release may represent an important approach for synthetic–medicinal chemistry, to search for effective antiinflammatory and antifibrotic agents. Inducers of interferon release may also be useful for lepromatous leprosy and chronic granulomatous disease. The potential cytokine and cytokine-related therapeutic approaches to treatment of disease are summarized in Table 4.

The conjugation of monoclonal antibodies (MoAbs) to radioisotopes, chemotherapeutic agents, and protein toxins has also been given consideration. Large amounts of human MoAbs can be produced by biotechnological means.

The combination of different therapeutic modalities, surgery, chemotherapy, or radiation, as well as a combination of different chemotherapeutic and biotechnology-engineered agents, have been used in the treatment of a number of neoplasms and led to higher rates of response. Investigations using biological response modifiers have produced very encouraging clinical benefits, as seen in therapy of breast cancer, myeloma, non-Hodgkin's lymphomas, hairy cell leukemia, essential thrombocythemia, and renal cell carcinoma.

Transplantation. Advances in surgical techniques and the availability of selective immunosuppressants, along with careful patient selection and proper post-surgery management, are factors in making

Table 4. Cytokine and Cytokine-Related Therapeutic Approaches to Disease

Cytokine/antibody[a]	Disease target	Species tested
IL-1	radiation/cytotoxic injury bacterial infection	rodent
TNF-α	autoimmune lupus nephritis	rodent
TNF-β	tumor destruction	rodent
INFs	antiinflammatory immunoregulation	rodent and human
INF-α, β, γ	tumor destruction, tumor and lymphocyte-induced angiogenesis	human rodent
INF-γ	rheumatoid arthritis	human
	lepromatous leprosy	human
	chronic granulomatous disease	human
IFN inducers		
poly I:C	fibrosis; transplantation	rodent
tilorone	adjuvant arthritis	rodent
	DTH granuloma	rodent
IL − 2 + LAK cell or tumor infiltrating lymphocyte	tumor destruction	rodent and human
GM-CSF, G-CSF, M-CSF, multi-CSF	cytotoxic injury; bone marrow transplantation; myelodysplastic syndromes; AIDS neutropenia	rodent and human
CSF-1 (M-CSF)	tumor destruction	rodent
basic FGF (bovine)	cartilage repair	rabbit
GM-CSF Ab + IL − 3Ab	cerebral marlaria	rodent
IL-4 Ab	allergy; parasitic infection	rodent

[a] Ab = antibody; IL = interleukin; TNF = tumor necrosis factor; INF = interferon; LAK = lymphocyte-activated killer; CSF = colony stimulating factors; and FGF = fibroblast growth factor.

transplantation the treatment of choice for organ failure. But the use of transplants is still limited because of the acute immune rejection phenomenon (host vs graft reaction, or HVGR) which may destroy the transplanted tissue within days to months after the surgery. Safe and effective immunotherapeutic agents, used to control and regulate the HVGR, are crucial to the developments of better transplantation methods. The primary goal is to achieve selective suppression of the recipient's immune response to the foreign antigens in the graft, ie, to attain specific immunological tolerance only to specific antigens.

Although there is no reliable method as of this writing for induction of Ag-specific unresponsiveness, some degree of tolerance has been observed by use of nonspecific immunosuppressive therapy. Nonspecific immunosuppressive therapy in an adult patient is usually through cyclosporin, started intravenously at the time of transplantation, and given orally once feeding is tolerated. Typically, methylprednisone is started also at the time of transplantation, then reduced to a maintenance dose. Azathioprine may also be used in conjunction with the prednisone to achieve adequate immunosuppression.

A number of fungal immunosuppressives have been isolated from fermentation broths and demonstrated to have immunotherapeutic efficacy. Other than cyclosporin, two fungal metabolites, sirolimus, previously known as rapamycin, and FK-506 are in various stages of development.

STEWART WONG
Jing Xing Health and Safety Resources, Inc.

R. Berkow and A. J. Flether, eds., *The Merck Manual of Diagnosis and Therapy*, Merck & Co., Inc. Rahway, N.J., 1992.

C. J. Dunn, in E. S. Kimball, ed., *Cytokines and Inflammation*, CRC Press, Inc., Boca Raton, Fla., 1991.

I. G. Otterness, T. J. Carty, and L. Loose, in A. J. Lewis and N. R. Ackerman, eds., *Therapeutic Approaches to Inflammatory Diseases*, Elsevier Science Publishing Co., Inc., New York, 1991.

INCENDIARIES. See CHEMICALS IN WAR; PYROTECHNICS.

INCINERATORS

Municipalities and industries are encouraged to reduce waste generation. Nevertheless, even under maximum use of source reduction and recycling (qv), significant quantities of waste continue to be generated (see WASTES, INDUSTRIAL). As of this writing, high temperature-incineration is the preferred technology for managing these wastes. Properly designed incinerators have the capability to destroy nearly 100% of all types of liquid organic wastes and an estimated 60% of solid wastes. Incineration is extremely limited in the United States. Roughly 60% of total U.S. wastes generated annually are classified as hazardous waste and less than 0.5% is incinerated. About 15% of the municipal solid waste generated is disposed of in incineration systems and effectively all of the medical waste.

High ($300–1900/t) capital investment and operating costs help discourage use of incineration for hazardous wastes. These costs for incinerators are well above those for alternative treatment methods such as biological treatment, $60–770/t; landfill in drums, $260–740/t or in bulk, $90–150/t; or deep well injection, $20–140/t. A primary contributor to the high operating cost of incinerators is the need for auxiliary fuel, particularly for the disposal of liquid wastes having high water content or solid wastes having low heating value. In addition, gas scrubbers can consume large quantities of chemicals, especially if chemical addition is not carefully controlled. Moreover, because incineration systems are typically complex, highly skilled operators are required to ensure efficient and reliable operation.

U.S. Regulations Impacting Design and Operation of Incinerators

U.S. regulations governing the design and operation of incinerators include the Resource Conservation and Recovery Act (RCRA), the Toxic Substances Control Act (TSCA), and the Clean Air Act Amendments of 1990 (CAA). Many states are authorized to regulate hazardous waste and incinerator programs, and state regulations are generally more stringent than federal.

Solid Waste Incineration

Polymeric or carbonaceous solids are degraded by high temperature. In the presence of oxygen, any carbon, hydrogen, and sulfur are oxidized to CO_2, H_2O, and SO_2, respectively. The rate of incineration increases rapidly with temperature. A range of 700–760°C is generally required for combustion, and most general-purpose incinerators operate between 760 and 1100°C. For an incinerator to operate without auxiliary fuel or air preheating, the waste feed or refuse must contain less than 50% moisture or 60% ash, and have more than 25% combustibles.

Atmospheric Conditions. In addition to complete combustion, wastes may be destroyed by treatment at high temperatures either without oxygen (pyrolysis), using limited oxygen (partial combustion), or in reactive atmospheres (gasification), such as those containing steam, hydrogen, or carbon dioxide.

Refuse Benefaction. It is extremely difficult to burn and recover useful energy from unsorted municipal waste because of its heterogeneity in size, shape, chemical composition, and heating value. However, preparation of the waste before thermal treatment facilitates burning. Such pretreatment contributes to the front-end costs but reduces furnace costs. The waste is upgraded by separation of the nonorganic fraction and drying, shredding, and densifying solids. These fuels thus prepared are referred to as refuse-derived fuels (RDF) (see FUELS FROM WASTE)

Incinerator Types. Incinerator types include moving-grate incinerators, multichamber incinerators, nonconventional incinerators (suspension-fired units, slagging incineration, starved-air incinerators, vortex incinerators, multihearth furnace, and fluidized-bed incinerators), and rotary kiln incinerators.

Factors Affecting Destruction of Solid Wastes. The analysis of the evolution and/or destruction of hydrocarbons during the incineration of solid hazardous wastes involves heat transfer, mass transfer, and reaction kinetics (see HEAT-EXCHANGE TECHNOLOGY, HEAT TRANSFER). Figure 1 is a generalized flow chart for the processes experienced by solids during incineration.

Vapor Pressures and Adsorption Isotherms. The key variables affecting the rate of destruction of solid wastes are temperature, time, and gas–solid contacting. The effect of temperature on hydrocarbon vaporization rates is readily understood in terms of its effect on liquid and adsorbed hydrocarbon vapor pressures. For liquids, the Clausius-Clapeyron equation yields

$$p = A_1 \exp\left(\frac{-\Delta H_{vap}}{RT}\right) \qquad (1)$$

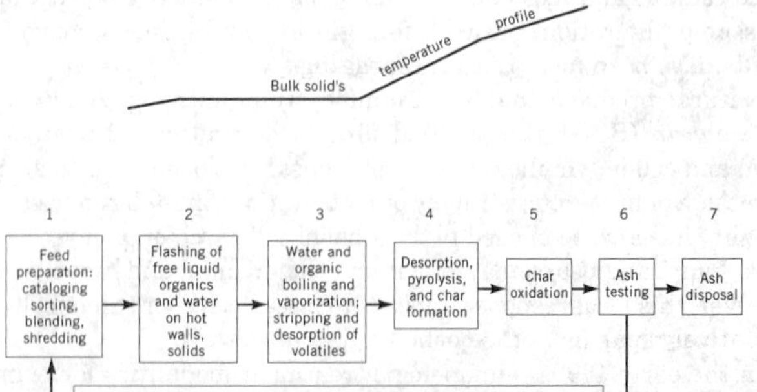

Figure 1. Generalized process flow chart for the thermal treatment of solid wastes. To a certain extent, steps 2, 3, 4, and 5 always proceed in parallel because of mixing limitations, nonhomogeneities in the waste, and unevenness in its heating.

where A_1 is a constant, p is the partial pressure of the hydrocarbon, T is temperature, R the ideal gas constant, and ΔH_{vap} is the heat of vaporization which is assumed to be independent of temperature. Heats of vaporization for liquid hydrocarbons are typically 30–40 kJ/mol (7.2–9.6 kcal/mol).

For adsorbed hydrocarbons, the adsorption–desorption process can be thought of as a reaction and the adsorption isotherm as a description of the reaction at equilibrium.

Desorption Rates From Wet and Dry Solids. The presence of water complicates the desorption process in several ways, the most obvious being a large thermal effect associated with its high heat of vaporization. Adsorbed moisture can also dramatically affect hydrocarbon adsorption isotherms by competing for adsorption sites, thus decreasing the number of available sites, and affecting the affinity of the remaining sites for hydrocarbons. In addition, the vaporization of adsorbed and condensed moisture creates a stream of water vapor which strips hydrocarbons from porous solids.

Particle Size and Desorption Rates. Intraparticle mass transfer controlled the rate of desorption when single particles were involved and interparticle mass transfer controlled in a bed of particles in a rotary kiln. These results apply to full-scale kilns. As particle size is increased, intraparticle resistances to heat and mass transfer eventually begin to dominate.

In general, the desorptive behavior of contaminated soils and solids is so variable that the required thermal treatment conditions are difficult to specify without experimental measurements. Experiments are most easily performed in bench- and pilot-scale facilities. Full-scale behavior can then be predicted using mathematical models of heat transfer, mass transfer, and chemical kinetics.

Heat Transfer in Rotary Kilns. Heat transfer in rotary kilns occurs by conduction, convection, and radiation. In a highly simplified model, the treatment of radiation can be explained by applying a one-dimensional furnace approximation. The gas is assumed to be in plug flow; the absorptivity, α_g, and emissivity, ϵ_g, of the gas are assumed equal ($\alpha_g = \epsilon_g$); and the presence of water in the solids is taken into account. Energy balances are performed on both the gas and solid streams. Parallel or countercurrent kilns can be specified.

Mass Transfer and Kinetics in Rotary Kilns. The rates of mass transfer of gases and vapors to and from the solids in any thermal treatment process are critical to determining how long the waste must be treated. Oxygen must be transferred to the solids. However, mass transfer occurs in the context of a number of other processes as well. The complexity of the processes and the parallel nature of steps 2, 3, 4, and 5 of Figure 1, require that the parameters necessary for modeling the system be determined empirically.

Software for performing practical calculations related to hazardous waste incineration (HWI) is available, and performs three types of calculations: (*1*) thermochemical calculations such as material and energy balances relating to incinerator temperature, excess air, and feed heating value; (*2*) stoichiometric calculations to give exhaust gas compositions and flow rates; and (*3*) preliminary incinerator design calculations. Required input data for the design calculations include either the combustion gas velocity or the length-to-diameter ratio of the incinerator. The volumetric heat release rate for the facility is also required.

Pollutant Emissions from Solid Waste Incinerators. Oxides of nitrogen (NO and NO_2) and sulfur (primarily SO_2) are emitted from most combustion systems including hazardous waste incinerators. The two principal mechanisms by which NO_x is formed are summarized in Figure 2. The thermal NO_x pathway is important in any high temperature process containing N_2 and O_2. The fuel NO pathway is also important if the fuel or waste contains nitrogen.

Metals entering a solid waste incinerator can leave the system with the bottom ash, the captured fly ash, or the exhaust gases. The fly ash is typically enriched in the heavy metals.

The emissions of polychlorinated dibenzo-*p*-dioxins (PCDD) and polychlorinated dibenzo-furans (PCDF) from incinerators are of interest to the public, scientists, and engineers. The U.S. EPA classifies 2,3,7,8-tetrachlorodibenzo-*p*-dioxin (2,3,7,8-TCDD) as the most potent carcinogenic compound it has evaluated. It is also listed as the agency's most potent reproductive toxin.

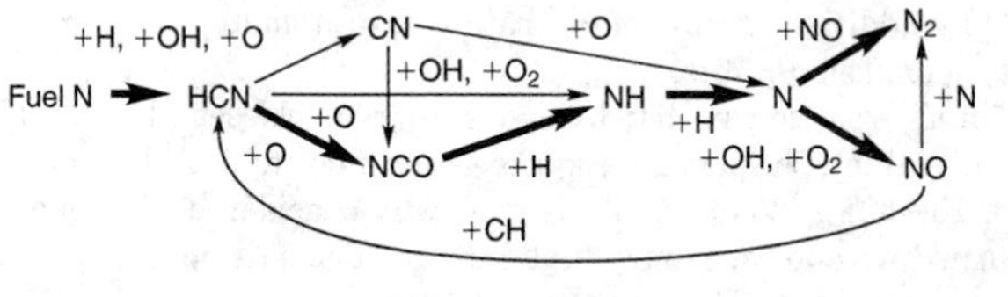

Figure 2. Schematic showing reaction pathways by which fuel nitrogen, N, is converted to NO and N_2. The bold lines indicate the key pathways. Thermal NO is formed from N_2: $N_2 + O \rightarrow NO + N$.

The proposed mechanism by which chlorinated dioxins and furans form has shifted from one of incomplete destruction of the waste to one of low temperature, downstream formation on fly ash particles. Two mechanisms are proposed, a *de novo* synthesis, in which PCDD and PCDF are formed from organic carbon sources and Cl in the presence of metal catalysts, and a more direct synthesis from chlorinated organic precursors, again involving heterogeneous catalysis. Bench-scale tests suggest that the optimum temperature for PCDD and PCDF formation in the presence of fly ash is roughly 300°C.

Chlorine may be formed by the Deacon reaction at temperatures below about 900°C. Significant reductions in PCDD and PCDF emissions have been shown with the upstream injection of $Ca(OH)_2$ at about 800°C.

Liquid Waste Incineration

Incinerators. Vertical furnaces are normally used for wastes containing high salt concentrations. Investment is typically higher than for furnaces of horizontal orientation as burners and controls are located in an elevated position, installation of furnace refractory is more difficult, and additional structural steel to support the furnace is required.

For given combustion air, waste, and auxiliary fuel feed rates to the incinerator, furnace residence time decreases as furnace pressure decreases.

Regulations require that the incinerator furnace be at normal operating conditions, including furnace temperature, before hazardous wastes are injected. This requires auxiliary fuel burners for furnace preheating. In addition, the burners provide heat when the wastes burned are of low heating value. Auxiliary burners are sized for conditions where liquid wastes are injected without the addition of high heating value wastes.

The furnace is constructed with a steel shell lined with high temperature refractory (see REFRACTORIES). Refractory type and thickness are determined by the particular need. Where combustion products include corrosive gases such as sulfur dioxide or hydrogen chloride, furnace shell temperatures are maintained above about 150–180°C to prevent condensation and corrosion on the inside carbon steel surfaces. Where corrosive gases are not present, insulation is sized to maintain a shell temperature below 60°C to protect personnel.

Three types of refractory are used. Castable refractory, similar to concrete, is placed in the shell using forms and poured in place or blown in. Plastic refractory is prepared in a stiff consistency and is either hammered or rammed in place. Plastic refractories are typically used for repairs. Fire brick is the most commonly used refractory. It is bonded in place using thin mortar joints.

Quench systems are used to cool hot furnace gases from 980–1200°C to 120–150°C. This allows less expensive materials of construction such as fiber glass reinforced plastic (FRP) (see GLASSES, ORGANIC–INORGANIC HYBRIDS) to be used downstream of the quench and in gas cleaning equipment, and reduces the volume of gas flow, resulting in smaller equipment. Water or air quenching systems are typically used.

Control systems are used to regulate the addition of liquid waste feed, auxiliary fuel, and combustion air flows to the incinerator furnace. In addition, scrubber operation is automated to help ensure meeting emission limits.

Liquid wastes can be divided into two classes: low and high heating value. The former requires auxiliary fuel. A heating value above 16 MJ/kg (7000 Btu/lb) is generally considered high enough to be burned without auxiliary fuel, but this depends on the specifics of excess air and desired flame temperature.

The steps in waste destruction are (*1*) heatup of the waste to its boiling point, (*2*) vaporization of the waste droplets, (*3*) heatup of any waste residue to combustion temperature, and (*4*) destruction by combustion reaction.

The factors which govern the efficiency of waste destruction include atomization, ie, mean drop size, and size distribution, temperature, residence time, O_2 concentration, and flow patterns.

Pollutant Emissions from Liquid Waste Incinerators. Wastes considered for incineration are usually organic in nature, so that the vast majority of the waste ends up as CO_2, H_2O, and N_2. Although CO_2 is of increasing concern because it is a greenhouse gas (see ATMOSPHERIC MODELING), emission issues have generally been related to the combustion products of sulfur, halogens, and metallic components of the waste. Carbon monoxide, which is not thermodynamically stable under normal incineration conditions, is also regulated. If CO is found in incinerator off-gas it is evidence of poor fuel–oxidant mixing or insufficient effective residence time, possibly owing to cold spots, oxidant starved zones, or bypassing.

Sulfur generally becomes SO_2, although some smaller amounts are possibly converted to SO_3, depending on temperature. Chlorine mostly results in HCl, but some Cl_2 and atomic Cl forms as well.

HCl can be absorbed into water to make a concentrated (usually 20 wt %) HCl solution, whereas SO_2 and Cl_2 must be scrubbed using a basic reagent (usually caustic or lime) to be effectively removed.

Oxides of nitrogen, NO_x, can also form. These are generally at low levels and too low an oxidation state to consider water scrubbing. A basic reagent picks up the NO_2, but not the lower oxidation states; the principal oxide is usually NO, not NO_2. Generally, control of NO_x is achieved by control of the combustion process to minimize NO_x, ie, avoidance of high temperatures in combination with high oxidant concentrations, and if abatement is required, various approaches specific to NO_x have been employed. Examples are NH_3 injection and catalytic abatement.

Alkali metals form basic oxides that are very reactive toward acidic species such as the acid gases, silicates, and aluminates. These form stable salts with acid gases if the off-gas contains such gases. Sodium, the most common of these metals, prefers to form chlorides ahead of sulfates. Sodium carbonate only forms in the absence of halides and sulfur oxides, SO_x. There usually is too little NO_x present to form nitrates (see SODIUM COMPOUNDS).

Alkali metal halides can be volatile at incineration temperatures. Rapid quenching of volatile salts results in the formation of a submicrometer aerosol which must be removed or else exhaust stack opacity is likely to exceed allowed limits. Sulfates have low volatility and should end up in the ash. Alkaline earths also form basic oxides. Calcium is the most common and sulfates are formed ahead of halides. Calcium carbonate is not stable at incineration temperatures.

Transition metals are more likely to form oxides. Silica and alumina form complexes with the basic oxides, eg, alkali, metals, alkaline earths, and some transition-metal oxidation states, in the ash.

The best approach toward estimating the chemistry of most contaminant species is to assume chemical equilibrium. Computer programs and databases (qv) for calculating chemical equilibria are widely available. Care must be taken that all species of concern are in the database referenced by the program being used, and if necessary, important species must be added in order to get the complete picture.

The equilibrium approach should not be used for species that are highly sensitive to variations in residence time, oxidant concentration, or temperature, or for species which clearly do not reach equilibrium. There are at least three classes of compounds that cannot be estimated well by assuming equilibrium: CO, products of incomplete combustion (PICs), and NO_x.

Particulate Pollutant Control Equipment. Particulate pollutant control equipment includes venturi scrubbers, electrostatic precipitators, and baghouses.

Incinerators for Vapors and Gases

Undesirable combustible gases and vapors can be destroyed by heating to the autoignition temperature in the presence of sufficient oxygen to ensure complete oxidation to CO_2 and H_2O. Gas incinerators are applied to streams that are high energy, eg, pentane, or are too dilute to support combustion by themselves. The gas composition is limited typically to 25% or less of the lower explosive limit. Gases that are sufficiently concentrated to support combustion are sometimes burned in flares, waste-heat boilers, in conjunction with other fuels in boilers and kilns, or used as process fuel. Occasionally, such gases may be burned in specially designed furnaces incorporating heat or material recovery, eg, chlorine.

Catalytic Incinerators. Catalytic incinerators, often used to remove hydrocarbons from exhaust gas streams, are more compact than direct-flame incinerators, operate at lower temperatures, often require little fuel, and produce little or no NO_x from atmospheric fixation. However, the catalytic bed must be preheated and carefully temperature controlled. Thus these are generally unsuited to intermittent and highly variable gas flows.

Direct-Flame Incinerators. In direct-flame incineration, the waste gases are heated in a fuel-fired refractory-lined chamber to the autoignition temperature where oxidation occurs with or without a visible flame. A fuel flame aids mixing and ignition. Excess oxygen is required, because incomplete oxidation produces aldehydes, organic acids, carbon monoxide, carbon soot, and other undesirable materials.

Flares. Flares are used for burning concentrated gases that support combustion such as hydrocarbon blowdown gases, tank venting, and emergency releases. These are located well above other structures. Pilot flames must be arranged to ensure ignition of all combustibles, especially when the flare handles only emergency releases. Design details to be considered are (*1*) pilot flames that stay lit and can be relit even in very high winds or heavy rains, (*2*) flare height and location to protect both personnel and the surroundings, (*3*) protection against flashback to the process, (*4*) production of explosive mixtures in the flare pipe from intermittent flows, and (*5*) entrained combustible liquid removal from the flare gas to prevent burning liquid droplets falling from the flare.

For environmental reasons, burning should be smokeless. Long-chain and unsaturated hydrocarbons crack in the flame producing soot. Steam injection helps to produce clean burning by eliminating carbon through the water gas reaction.

R. BERTUM DIEMER, JR.
THOMAS D. ELLIS
E. I. du Pont de Nemours & Co., Inc.
GEOFFREY D. SILCOX
JOANN S. LIGHTY
DAVID W. PERSHING
University of Utah

C. R. Dempsey and E. T. Oppelt, *J. Air Waste Manage. Assoc.* **43**, 25–73 (1993).

W. P. Linak and J. O. L. Wendt, *Prog. Energ. Combust. Sci.* **19**, 145–185 (1993).

J. M. Beer and N. A. Chigier, *Combustion Aerodynamics*, Halsted Press, New York, 1972, pp. 125–126.

INCLUSION COMPOUNDS

Notwithstanding the immense number and great variety of inclusion compounds, all of them may be classified into three main categories being either a complex, a cavitate, or a clathrate according to the

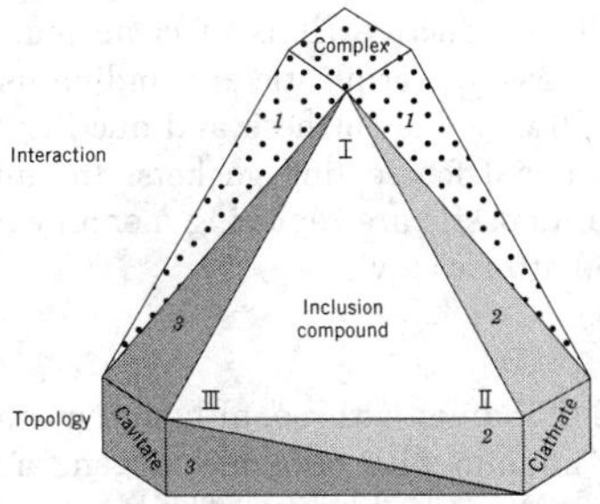

Figure 1. Classification/nomenclature of host–guest type inclusion compounds, definitions and relations: (*1*) coordinative interaction, (*2*) lattice barrier interaction, (*3*) monomolecular shielding interaction; (I) coordination-type inclusion compound (inclusion complex), (II) lattice-type inclusion compound (multimolecular/extramolecular inclusion compound, clathrate), (III) cavitate-type inclusion compound (monomolecular/intramolecular inclusion compound).

criteria given in Figure 1. Typical examples for each class of inclusion compounds are the crown complexes, the calix-cavitates, and the hydroquinone clathrates, but in many of the recently known inclusion situations there are borderline cases treated as complex-clathrate hybrids (coordinatatoclathrates or clathratocomplexes depending on the dominant inclusion character.) By way of contrast, the description addition compound (adduct) may be used to the best advantage if a cavity does not exist either at the host molecule or in the lattice buildup. Inclusion compound, therefore, is the generic term of choice which refers to the presence of any not precisely defined cavity. In a more detailed topological characterization, there are two-dimensional open intercalates (layer- or sandwich-type inclusions), one-dimensional open channel inclusions (tubulates), and totally enclosed cage inclusions (cryptates).

Intramolecular Cavity Inclusions: Cavitates

Cavitates include crown macroring inclusion compounds (coronates), cryptates, podates, cyclophane host inclusion compounds, calixarene inclusion compounds, cyclodextrin and amylose inclusion compounds, cucurbituril inclusion compounds, molecular cleft inclusion compounds, and anionic guest inclusion compounds.

Extramolecular Cavity Inclusions: Lattice-Type Inclusion Compounds Clathrates

These compounds include Hofmann- and Werner-type inclusion compounds, inclusion compounds of urea, thiourea and selenourea, inclusion compounds of gossypol, inclusion compounds of phenolic hosts, inclusion compounds of deoxycholic acid (choleic acids), inclusion compounds of macrocyclic and oligocyclic lattice hosts and recently designed organic host lattices.

Preparation and Characterization of Inclusion Compounds

There are several ways to prepare inclusion compounds. In solution, they may simply be formed by dissolving together host and guest in a common solvent. Inclusion formation in solution applies only for intramolecular cavity inclusions and complexes. Crystalline inclusion compounds may be prepared by crystallization from the guest solvent or by cocrystallization of host and guest from an inert solvent. Solid inclusion compounds are also formed by direct exposure of the host to the vapor or liquid guest or, sometimes, by grinding solid host and guest together. Moreover, replacement of an included guest has been demonstrated in particular cases.

Appropriate guest molecules are those that have a suitable size and shape to accommodate the host cavity and that complement the host cavity chemically.

Stabilities of inclusion compounds span a wide range. Some are very stable at ambient conditions and require heating to considerable temperatures or treatment under high vacuum to cause decomposition. Others are only stable when in contact with mother liquor or excess guest solvent from which the inclusion compound was grown. A simple yet informative way for estimation of inclusion stabilities is to relate the decomposition point of the inclusion compound to the usual boiling point of the respective guest liquid.

Uses

Inclusion compounds open up a wide area of applications. An important aspect in this connection is the specific microenvironment created by the host enclosure of the guest which exerts an influence on the physical spectroscopic, chemical, and other properties of the guest.

Retardation and Control. This influence may manifest itself in a reduced volatility, and therefore, lower possible storage and handling problems of a compound when included; toxic and hazardous substances become safer.

Shielding and Stabilization. Inclusion compounds may be used as sources and reservoirs of unstable species. The inner phases of inclusion compounds uniquely constrain guest movements, provide a medium for reactions, and shelter molecules that self-destruct in the bulk phase or transform and react under atmospheric conditions.

Solubilization and Activation. Compounds included in a host take solubility properties of the host shell, and thus, become more soluble when trapped in polar or apolar media, depending on the nature of the host. This leads to important uses in chemical synthesis known as the phase-transfer principle.

Organized Media Effects. Another general reason for using host-guest inclusion chemistry in synthesis is controlled selectivity and artificial enzyme mimicry.

Sensing. Crown compounds modified by responsible chromogenic groups (chromoionophores) proved valuable tools for measuring metal ions and even enantiomeric guest concentrations in solution. Ion selective electrodes based on crown compounds and podands as the sensitive component have broad analytical applications from industrial wastewater control to clinical bedside monitoring of blood.

EDWIN WEBER
Technische Universität Bergakademie Freiberg

J. L. Atwood, J. E. D. Davies, and D. D. MacNicol, eds., *Inclusion Compounds*, Vols. 1–3, Academic Press, Inc., London, 1984; Vols. 4–5, Oxford University Press, Oxford, U.K., 1991.

E. Weber, ed., *Molecular Inclusion and Molecular Recognition—Clathrates I and II*, Springer, Berlin-Heidelberg, 1987 and 1988.

F. Vögtle, *Supramolecular Chemistry—An Introduction*, John Wiley & Sons, Ltd., Chichester, U.K., 1991.

J.-M. Lehn, *Supramolecular Chemistry—Concepts and Perspectives*, VCH Verlagsgesellschaft, Weinheim, Germany, 1995.

INDANTHRENE, INDANTHRENE DYES. See DYES, ANTHRAQUINONE.

INDENE. See HYDROCARBON RESINS.

INDICATORS. See HYDROGEN-ION ACTIVITY.

INDIGOID DYES. See DYES, NATURAL.

INDIUM AND INDIUM COMPOUNDS

Indium

Indium, an element of Group 3 (IIIA) of the Periodic Table, occurs between gallium and thallium. It is a soft, lustrous, silver-white metal, highly malleable and ductile.

Table 1. Physical Properties of Indium

Property	Value
atomic weight	114.82
atomic number	49
melting point, °C	156.61
boiling point, °C	2080
specific heat at 25°C, kJ/(kg·K)[a]	0.233
coefficient of linear expansion, 0–100°C, $\times 10^6$°C^{-1}	24.8
electrical resistivity, Ω·m at 3.38 K	superconducting
density, kg/m^3 at 20°C	7.300
thermal conductivity at 0°C, W/(m·K)	83.7

[a] To convert J to cal, divide by 4.184.

The abundance of indium in the earth's crust is probably about 0.1 ppm, similar to that of silver. It is found in trace amounts in many minerals, particularly in the sulfide ores of zinc and to a lesser extent in association with sulfides of copper, tin, and lead. Indium follows zinc through flotation concentration, and commercial recovery of the metal is achieved by treating residues, flue dusts, slags, and metallic intermediates in zinc smelting and associated lead (qv) and copper (qv) smelting (see METALLURGY, EXTRACTIVE; ZINC AND ZINC ALLOYS).

Properties. Table 1 lists many of the physical, thermal, mechanical, and electrical properties of indium. The highly plastic nature of indium, which is its most notable feature, results from deformation from mechanical twinning. Indium retains this plasticity at cryogenic temperatures. Indium does not work-harden, can endure considerable deformation through compression, cold-welds easily, and has a distinctive cry on bending as does tin.

Economic Aspects. Production of indium had been reported from Belgium, Canada, China, France, Germany, Italy, Japan, the Netherlands, Peru, the U.K., and the U.S., as well as countries in the CIS (the former Soviet Union).

Production reported from Belgium, the Netherlands, and the U.K. is thought to be from imported concentrates, residues, and scraps. Production in the CIS and the People's Republic of China is difficult to confirm. Many producers do not report indium production. Research and product development program indicate that a strong growth rate should continue for some time. Increase in indium production are expected to be easily accomplished.

Safety and Handling. Physiologically indium is a nonessential element (see MINERAL NUTRIENTS). It is classified as toxic, but there have been no reported cases of systemic effects in human exposure to indium. The threshold limit value of the ACGIH is 0.1 mg/m^3. The primary toxic effects of ionic indium are on the kidneys, but there may be effects on the respiratory system.

Alloys. Indium alloys with a wide range of metals, and many binary and ternary systems have been studied extensively. Indium generally increases the strength, corrosion resistance, and hardness of a system to which it is added. Low melting point alloys of indium with lead, tin, bismuth, and cadmium having melting points as low as 47°C are used in surgical casts, patternmaking, lens blocking, turbine blade machining, fire door safety links, and sprinkler heads (see HIGH TEMPERATURE ALLOYS). The alloy of ln 15%–Ag 85%–Cd 5% is used in control rods for nuclear rectors (qv). Tin–indium, lead–tin–indium, and lead–silver–indium solder alloys have melting points in the range of 100–300°C.

Uses. Indium's first commercial use was in the production of dental alloys (see DENTAL MATERIALS), but its first significant use was in the production of bearing for heavy-duty and high speed service (see BEARING MATERIALS). The solder and alloy market, including low melting or fusible alloys, is a principal user of indium (see SOLDER AND BRAZING ALLOYS). Applications for electroplated indium coating include indium bump bonding for silicon semiconductor die attachment to packaging substrates and miscellaneous applications where the physical or chemical properties of indium metal are desired as a plated deposit. Electrically conductive films of clear indium oxide (doped with tin oxide) on glass (qv) or plastic are finding use in many applications (see THIN FILMS). The softness and ductility of indium make it an excellent material for sealing gaskets. Indium chemicals and electroplated metal deposits are replacing mercury (qv) in the manufacture of alkaline batteries (qv).

Indium Compounds

The usual valence of indium is three, although monovalent and bivalent compounds of indium with oxygen, halogen, and Group 15 (VA) and 16 (VIA) elements are well known. The lower valence compounds tend to disproportionate into trivalent compounds and indium metal; the trivalent compounds are stable.

Indium compounds include halides (indium trichloride, $InCl_3$, indium dichloride, $InCl_2$, indium monochloride, InCl), oxides (indium troxide, In_2O_3, indium suboxide, In_2O, indium monoxide, InO), sulfates (trisulfate, $In_2(SO_4)_3$, indium acid sulfate crystal, $In(HSO_4)_2$), sulfides (In_2S_3), other salts (indium nitrate trihydrate, $In(NO_3)_3\ 3H_2O$, indium phosphate, $InPO_4$), organic compounds (trimethyl indium, $In(CH_3)_3$, triethyl, $In(C_2H_5)_3$, triphenyl, $In(C_6H_5)_3$), and intermetallic and semiconducting compounds (carbon-free indium-based fullerenes, indium antimonide, indium arsenide, and indium phosphide, InP).

JAMES A. SLATTERY
Indium Corporation of America

World Minor Metals Survey, 2nd ed., Metal Bulletin, London, 1981.

J. M. Ramaradhya, R. C. Bell, S. Brownlow, and C. J. Mitchell, in S. D. Snell and L. S. Ettre, eds., *Encyclopedia of Industrial Chemical Analysis*, Vol. 19, John Wiley & Sons, Inc., New York, 1971, p. 518.

C. E. T. White and H. Okamoto, eds., *Phase Diagrams of Indium Alloys and their Engineering Applications*, ASM International, Materials Park, Ohio, 1992.

S. C. Sevov and T. C. Corbett, *Science* **262**, 880–883 (1993).

L. G. Stevens and C. E. T. White, "Indium and Bismuth," in *Metals Handbook*, Vol. 2, 10th ed., ASM International, Materials Park, Ohio, 1990.

INDOLE

Indole is a heteroaromatic compound consisting of a fused benzene and pyrrole ring, specifically benzo[*b*]pyrrole. The indole ring is incorporated into the structure of the amino acid tryptophan and occurs in proteins and in a wide variety of plant and animal metabolites.

Properties

Indole is a colorless solid, mp 53–53°C, which is reasily soluble in most organic solvents but sparingly soluble in water. Indole has a musty odor which is very persistent and its derivatives have some applications in the formulation of fragrances.

The industrial source of indole has been isolation from coal-tar distillate. Several patents for the manufacture of indole have been issued with aniline and ethylene glycol, aniline and ethylene oxide, 2-ethylaniline, and *N*-ethylanline as the starting materials.

Reactivity

Indole is a heterocyclic analogue of naphthalene. The basic reactivity patterns of indole can be understood as resulting from the fusion of an electron-rich protein pyrrole ring with a benzene ring.

Reactions include electrophilic aromatic substitution (eg, halogenation, nitration, *C*-acylation, and alkylation), *N*-alkylation, arylation, lithiation and subsequent transformations, and oxidation.

Syntheses

Although there are a wide variety of indole ring syntheses, most of the more useful examples fall within a small number of groups. Indole syntheses usually start with an aromatic compound, either monosubstituted or ortho-disubstituted.

Processes include the Fischer indole synthesis from arylhydrazones and related sigmatropic syntheses, reductive cyclizations of nitro compounds, the Madelung synthesis from anilides and related base-catalyzed condensations, and transition-metal catalyzed cyclizations.

Biologically Active Indole Derivatives

Synthetic Derivatives of Indoles as Pharmaceuticals. Thousands of indole derivatives have been prepared and evaluated as potential pharmaceuticals. Of those which have been put into use perhaps the most important are the nonsteroidal antiinflammatory agent, indomethacin, and the β-adrenergic blocker, pindolol.

Naturally Occurring Compounds. Many derivatives of indole are found in plants and animals where they are derived from the amino acid tryptophan. Several of these have important biological function or activity. Serotonin functions as a neurotransmitter and vasoconstrictor. Melatonin production is controlled by the circadian cycle and its physiological level influences daily and seasonal rhythms in humans and other species. Indole-3-acetic acid is a plant growth stimulant used in several horticultural applications.

The largest single class of naturally occurring indoles are the plant alkaloids. These occur with a wide range of structural diversity and are typically derived from tryptophan and terpenoid structural units. Several of these compounds are pharmacologically significant. Reserpine acts as a tranquilizer and hypotensive agent. The dimeric vinca aklaloids, vincristine and vinblastine, are used in the treatment of Hodgkin's disease, leukemia, and other forms of cancer. Derivatives of the ergot alkaloid lysergic acid are used in the treatment of migraine and the diethylamide is lysergic acid diethylamide.

Toxic Indole Derivatives. There are several documented cases where indole derivatives, both natural and of synthetic origin, have been linked to pathological effects in humans. 3-Methylindole, which is produced by bacterial fermentation in cattle, can lead to pulmonary edema. The pyridoindoles Trp-P-1 and Trp-P-2 are genotoxic substances which originate from pyrolysis of tryptophan and have been identified in foods cooked at excessively high temperatures. 4-Chloro-6-methoxyindole, which can be extracted from fava beans, yields a potent mutagen on interaction with nitrate ion.

RICHARD J. SUNDBERG
University of Virginia

C. W. Bird and G. W. H. Cheeseman, eds., *Comprehensive Heterocyclic Chemistry*, Vol. 4, Pergamon Press, Oxford, 1984, Chapts. 3.04, 3.05, and 3.06.

W. J. Houlihan, eds., *The Chemistry of Heterocyclic Compounds*, Vol. 25, Parts 1, 2, and 3, Wiley-Interscience, New York, 1972.

R. J. Sundberg, *The Chemistry of Indoles*, Academic Press, Inc., New York, 1970.

INDOPHENOL. See SULFUR DYES.

INDULINES. See AZINE DYES.

INDUSTRIAL ANTIMICROBIAL AGENTS

Industrial antimicrobial agents are chemicals used to prevent the adverse consequences of microbiological activity in processes and products. Some are unique to this segment and others are drawn from the antimicrobial agents used in medicine, agriculture, and sanitary applications. Industrial antimicrobials are selected where process or strictly physical conditions, such as irradiation or heat, are impractical or ineffective in controlling microbiological activity.

Chemical suppliers include basic manufacturers of active ingredients, formulators, and distribution or service industries. The relative importance of each sector depends greatly upon the industry being supplied. In many instances, the vendor may supply a number of performance chemicals (eg, corrosion control agents or stabilizers) in addition to the antimicrobial agent.

Microbiology

Industrial antimicrobial agents are targeted against bacteria, mold, fungi, yeasts, and algae. Sometimes the mere presence of organisms is a problem but more frequently microbes consume desireable compounds and produce harmful by-products.

Organisms causing problems in industrial situations are by nature self-sufficient in synthesizing required biochemicals from the most basic molecules. The type and availability of nutrient may be the most significant determinant in the microbiological ecology of the products and processes that require the use of an industrial antimicrobial agent.

Antimicrobial Activity

Microbiologists have developed the following hierarchy to categorize the expected level of performance of antimicrobial treatments:

Category	Result of treatment
sterilant	completely kills all life forms
sporicidal	kills spores
disinfectant	kills all infectious bacteria
cidal	kills all organisms of type (eg, tuberculocidal)
sanitizer	reduces number of organisms to safe level
antiseptic	prevents infection
static	prevents growth of organism (eg, fungistatic)

Measures of Activity. The potency of an antimicrobial to kill or render the organisms inactive is measured by the minimum inhibitory concentration (MIC). The lowest concentration inhibiting growth is the MIC for that organism (the lower the number, the more potent the antimicrobial). The MIC is of little value to the users of antimicrobials because it is not a true predictor of performance in any particular application, and more application-specific tests have been developed.

Applications

Practically, the broad general area of industrial antimicrobial concerns is divided into process and preservative disciplines. Figure 1 show the key application areas of process and companion-preservative uses.

Regulation of Antimicrobial Agents

The key trend affecting this industry is the regulation of products and their uses by the U.S. EPA. This regulation has curtailed the introduction of new ingredients, selectively limited the markets for a number of products, and eliminated products by suspension or cancellation.

Process uses	**Preservative uses**
Swimming pool sanitizers	Jet fuel
Cooling water	Paper
Metalworking	Paint film
Pigment slurries	Caulks
Petroleum recovery	Adhesives
Pulp	Leather
Paint (in-can)	Wood
Latex	Textile
Adhesive	Plastics
Hide processing	Cosmetics
Sapstain	
Laundry sanitizers	

Figure 1. Applications of industrial antimicrobial agents.

Several antimicrobials have been banned or severely restricted by the EPA based on documented or suspected toxicity or environmental problems. Others have been discontinued in the face of testing costs required by the EPA reregistration program mandated by the Federal Insecticide, Fungicide, and Rodenticide Act (FIFRA) of 1988. Some of the significant products that have become obsolete are 2,4,5-trichlorophenol, sodium 2,4,5-trichlorophenate, hexachlorophene, 2,2′-methylenebis(4-chlorophenol), phenylmercuric acetate, and 3,4,5-tribromosalicylanilide.

Toxicology. Industrial antimicrobial agents are regulated in the United States as pesticides under FIFRA. Thus the industrial antimicrobial regulation is strongly influenced by agricultural pesticide considerations. The hallmark of the EPA allowance is registration, which is the culmination of a process that begins with the submission of toxicology, chemistry, and environmental data showing that the product can be used without "adverse effects on man and the environment." The most often quoted toxicology data is the acute oral LD_{50}. This number is a crude measurement of the potential hazard to humans (and other mammals) from ingestion.

Reregistration. In addition to its authority to control the availability of new active ingredients, formulations, and the way in which the formulas are used, EPA is reapproving the use of older products through a process known as reregistration.

Manufacturers and Suppliers

Manufacturers and suppliers of industrial antimicrobial active ingredients and formulations include the following companies:

ANGUS Chemical Co.
Northbrook, Ill.
BASF
Parsippany, N.J.
Calgon Corp.
Moon Run, Pa.
Dearborn Division, W. R. Grace
Lake Zurich, Ill.
Hüls America
Piscataway, N.J.
Lonza Inc.
Fairlawn, N.J.
Mooney Chemical, Inc.
Cleveland, Ohio
Osmose Wood Preserving Co.
Buffalo, N.Y.
Rohm and Haas Co.
Philadelphia, Pa.
Troy Chemical Corp.
Newark, N.J.
Zeneca (formerly ICI)
Wilmington, Del.
Atochem North America
Philadelphia, Pa.
Betz Laboratories
Trevose, Pa.
CSI
Charlotte, N.C.
Great Lakes Chemicals
West Lafayette, Ind.
Mentor, Ohio
Miles, Inc.
Pittsburgh, Pa.
Morton International, Inc.
Danvers, Mass.
Petrolite Corp.
St. Louis, Mo.
Stepan Co.
Northfield, Ill.
Union Carbide Chemicals
Danbury, Conn.
Baker Performance Chemicals
Houston, Tex.
Buckman Laboratories
Memphis, Tenn.
Dow Chemical USA
Midland, Mich.
Hickson Corp.
Conley, Ga.
Lehn & Fink Products Co.
Montvale, N.J.
Monstanto Co.
St. Louis, Mo.
Olin Corp.
Cheshire, Conn.
Givaudan Rorer
Clifton, N.J.
Sherex
Dublin, Ohio
Vinings Industries, Inc.
Atlanta, Ga.
ISK Biotech
Nako Chemical Co.
Naperville, Ill.

Antimicrobial End Use

The duration of required antimicrobial action can range from less than an hour per treatment (eg, swimming pools) to years (wood poles and pilings) and retreatment may be very practical or problematic. The chemistry of the mechanism of action (MOA) is not always known, but the successful active ingredient reflects a balance between the reactivity and durability appropriate to the market. Many active ingredients are not equally active against both bacteria and fungi. Bacteria are most likely to be the major problem when liquid water is present and fungi are more significant when intermittently wetted. Because of the very practical differences presented by the chemistry of the treatment site and the microbial ecology, a diverse range of active ingredient chemistries enjoy niche positions.

An abbreviated list showing the more significant active ingredient chemistry by antimicrobial end use is shown in Figure 2.

Process uses	
Swimming pool sanitizers	Chlorinated isocyanurates/hypochlorites
Cooling water	Bromine/isothiazoline compounds
Metalworking	Organo nitrogen e.g. triazine compounds
Pigment slurries	Benzisothiazalone
Petroleum recovery	Acrolein/fatty amines
Pulp	Methylene bis isothiocyanate/hexachloro dimethylsufone
Paint (in can)	Triazine/aminoalcohol
Latex	Dibromodicyanobutane
Adhesive	Formaldehyde
Hide processing	Phenyl phenol
Sapstain	Quarternary ammonium and iodocarbamate
Laundry sanitizers	Quarternary ammonium
Disinfection	Phenols/gluteraldehyde
Equipment sanitizers	Iodophors/hydrogen peroxide/quaternaries

Preservative uses	
Marine antifouling	Copper
Jet Fuel	Organoboron
Paint film	Benzimidizole/chlorophthalonitrile/folpet
Caulks	Zine omadine
Leather	Nitrophenol
Wood	Chrome-copper-arsenate compounds
Textile	Copper quinolates/naphthenates
Plastics	Oxybisphenoxarsine
Cosmetics	Methyl isothiazolinone/parabens

Figure 2. Commonly used antimicrobial agents.

Classes of Antimicrobial Compounds

Classes include quaternary ammonium compounds, phenolics [sodium orthophenyl phenate, orthophenyl and its sodium salt, *p*-chloro-*m*-xylenol, and 2-(2′,4′-dichlorophenoxyl)-5-chlorophenol(2,4,4′-trichloro 2′-phenoxyphenol)], chlorine and bromine oxidizing compounds (*p*-tolydiiodomethyl sulfone and 3-iodo-2-propynylbutyl carbamate), organometallics [10,10′-oxybisphenoxarsine (OBPA), organotin compounds, and copper quinolinolate (oxine copper)], organosulfur compounds, heterocyclics (Captan and Folpet), and other nitrogen compounds.

Miscellaneous, New, and Developmental Antimicrobial Agents. Acrolein (qv) is a unique chemical used for petroleum production. Biobor has become the antimicrobial addition of choice for aviation fuels. Chlorophthalonil (tetrachloroisophthalnitrile) is a significant agricultural fungicide, in addition to being one of the most important latex paint film preservatives (producer, ISK).

Table 1. Antimicrobials

Chemical	Name
methyl-3,5,7-triaz-1-azonia-tricyclodecane chloride	BusanR 1024
1-(hydroxymethyl)5,5-dimethyl hydantoin	Glyco Serve
2-bromo-2-nitropropanediol	Bronopol
methanol, [[2-(dihydro-5-methyl-3 (2*H*)-oxazoly)-1-methylethoxy]ethoxy] methoxyl	CosanR 145
alkyl(61% C_{12}, 23% C_{14}, 11% C_{16}, 5% C_{18}) dimethyl benzyl ammonium chloride	

Table 2. Developmental Industrial Antimicrobials

Chemical name	Structure	Producer
2(*n*-octyl)-4,5-dichloro-isothiozal-3-one[a]	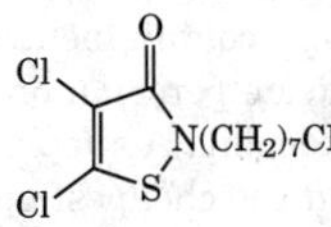	Rohm and Haas
decylthioethylamine[b]	$CH_3(CH_2)_9SCH_2CH_2NH_2$	Dow
decylthioethylamine-hydrochloride[b]	$CH_3(CH_2)_9SCH_2CH_2NH_3^+Cl^-$	Dow
N-hydroxymethyl-3,5-dimethylpyrazole	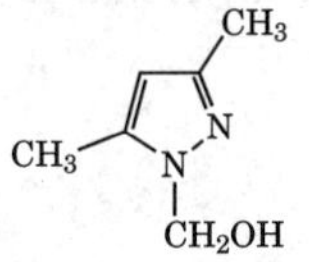	Buckman

[a] Trade name, Sea Nine; used as a marine antifoulant. [b] Trade name, XV-40304. OIL; used for cooling water.

As noted earlier, the most significant trend in the industry has been the decline in the number of available active ingredients due to EPA regulation. Table 1 shows the new active ingredients, together with the year of their obtaining an EPA registration.

Table 2 shows some of the developmental products that have EPA applications pending and may be available in the near future.

THOMAS MCENTEE
Morton International, Inc.

S. S. Block, ed., in *Disinfection, Sterilization, and Preservation*, Lea and Febiger, Philadelphia, Pa., 1991, pp. 20–24.

H. Rossmoore, *ASTM Standards on Materials and Environmental Microbiology*, American Society for Testing and Materials, Philadelphia, Pa., 1987.

J. R. Back, B. R. Friedfeld, and A. A. Boccone, *Biocides—United States*, C. K. Kline, Fairfield, N.J. 1984, p. 29.

C. G. Hollis and co-workers, in H. W. Rossmoore, ed., *Biodeterioration and Biodegradation*, Vol. 8, Elsevier Applied Science, London, 1991.

INDUSTRIAL HYGIENE

Industrial hygiene is devoted to the anticipation, recognition, evaluation, and control of environmental factors or stresses arising in or from the workplace that may cause sickness, impaired health and well-being, or significant discomfort and inefficiency among workers or among the citizens of the community. It is a profession practiced by over 11,000 industrial hygienists in the United States and many more worldwide. U.S. industrial hygienists are typically members of the American Industrial Hygiene Association (AIHA), which is the largest industrial hygiene organization, the American Conference of Governmental Industrial Hygienists (ACGIH), and the American Academy of Industrial Hygiene (AAIH). Many are certified industrial hygienists (CIH) as a result of meeting the requirements of the American Board of Industrial Hygiene (ABIH). Outside the United States, industrial (also called occupational) hygienists are members of such professional associations as the British Occupational Hygiene Society (BOHS) and the International Occupational Hygiene Association (IOHA).

Industrial hygienists work closely with members of several other professions concerned with workplace health and safety, eg, occupational medicine, occupational health nursing, and safety engineering. All of these groups are involved in the implementation of the laws that regulate workplace health and safety. In the United States the principal law is the Occupational Safety and Health Act (OSHA) enforced by the U.S. Department of Labor (U.S. DOL). Similar laws are in place in almost every country in the world and are proposed by such international organizations as the World Health Organization (WHO) and the International Labor Organization (ILO).

A partial list of the hazards or conditions arising from the workplace (see also PLANT SAFETY) and with which industrial hygienists are concerned includes the following.

Chemical
- carcinogens
- acute poisons
- reproductive hazards
- corrosives
- irritants
- pneumoconiosis producing ducts
- neurotoxins
- nephro (kidney) toxins

Physical
- noise
- heat
- cold
- ionizing radiation
- microwave radiation
- extremely low frequency (ELF) radiation
- vibration
- magnetic fields
- ultraviolet radiation
- infrared radiation
- laser radiation

Ergonomic
- repetitive strain injury (RSI)
- carpal tunnel syndrome
- back injury
- lifting hazards
- visual display units
- human/machine interaction

Industrial hygienists must be able to anticipate what workplace materials or events may give rise to any of these hazards, to recognize the hazards that occur, to evaluate a hazard to determine the degree of risk it presents, and to control hazards so as to reduce risk. The industrial hygienist's job begins when a new chemical or process is conceived. Based on data from animal experiments and/or human epidemiology relating to a substance or an analogous chemical it is possible to estimate the toxicity of the substance. Whenever possible, it is best to avoid using potentially dangerous chemicals. Similarly, potentially hazardous processes that produce excessive noise, heat, or other stress-related situations should be anticipated and avoided. However, the industrial hygienist can usually devise ways to use potentially dangerous chemicals safely.

Recognition of Potential Hazards

The process of recognition of potential hazards is based on extensive knowledge of what kinds of hazards may occur in any industry, process, or job activity. Chemical hazard sources include the following.

Fugitive Emissions. Fugitive emissions or leaks occur wherever there are breaks in a barrier that maintains containment.

Process Operations. There are a few actions the operators may need to take which can involve contact with process materials. Sampling of process streams is one such task.

Material Handling. Some material-handling steps are difficult to accomplish with total containment. Solids handling is often done by open means both because the hazard is perceived to be less and because it is more difficult to design totally closed solids handling systems.

Maintenance. Open system maintenance can add to exposure by disturbing and dispersing deposits of materials in equipment. Most maintenance (qv) is done while the plant is in operation. Thus the maintenance workers are in close proximity to operating equipment for long periods of time. Maintenance that exposes workers to health risks include welding, painting, sandblasting, insulation, chemical cleaning, and catalyst handling.

Waste Handling. Housekeeping procedures in general can have a significant impact on employee exposure, and certain waste handling procedures can result in very serious exposure if proper precautions are not taken.

Air cleaning systems are often used to remove dust or vapors from plant or process exhaust streams. It is necessary to enter the air cleaner periodically for inspection or repair. Dust deposits inside the equipment are likely to be stirred up and inhaled by unprotected workers.

Wastewater treatment facilities may receive chemical process wastes and spills. These wastes may volatize on emerging from a closed sewer system into open waste treatment tanks particularly if hot streams have heated the tank. These releases can occur without warning and result in unexpected employee exposure.

Hazard Evaluation

The evaluation phase of industrial hygiene is the process of making measurements on some set of samples which permits a conclusion about the degrees of hazard. Before conducting an evaluation, it is necessary to make a number of choices of what and where to sample, when to sample, how long to sample, how many samples to take, what sampling and analytical methods to use, what exposure criteria to use in the analysis of the data, and how to report the results. These choices as a whole constitute the evaluation plan. The object is to find if one or more workers have an unacceptable probability of being exposed in excess of some established limit.

Decision Process. In many cases, the decision regarding the need for exposure reduction measures is obvious and no formal statistical procedure is necessary. However, as exposure criteria are lowered, and control becomes more difficult, close calls become more common, and a logical decision-making process is needed. A typical process is shown in Figure 1.

Generic Exposure Assessment

In the U.S., the Occupational Safety and Health Administration is in the process of developing a generic exposure assessment standard which would apply to chemicals having permissible exposure limits (PELs) listed in Part 1910 of the *Code of Federal Regulations*. Exceptions are the few chemicals for which there are other detailed regulations. This standard is to prescribe how to determine if exposure measurement is required, the frequency of measurement when required, how measurement may be discontinued, what records are to be kept, and how to notify employees.

Other Agents

Evaluations of occupational exposure to physical agents such as noise, radiation or heat, biological agents, and multiple chemical agents are similar to the process for single chemical substances, but have some key differences.

Control

The evaluation phase should be planned to yield the data needed to draw accurate conclusions about control needs. The need for certainty depends on how difficult it is to achieve control.

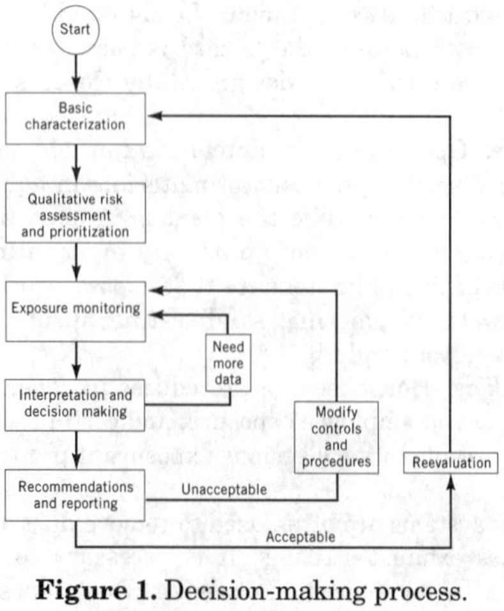

Figure 1. Decision-making process.

Although the evaluation phase comes chronologically between the recognition and control phases, the control options play a considerable role in the extent or intensity of the evaluation phase.

Options. Traditional control options for overexposure are material substitution, process change, containment, enclosure, isolation, source reduction, ventilation, provide personal protection, change work practices, and improve housekeeping. A simple way of looking at selection of control options is to find the cheapest option that results in the desired amount of exposure reduction. It is not actually that simple, however, because the various options differ in ways other than cost and degree of control. Some of the other factors to consider in selection of control options are operability, reliability, and acceptability.

JEREMIAH LYNCH
Exxon Chemical Company

American Conference of Governmental Industrial Hygienists, *Threshold Limit Values for Chemical Substances and Physical Agents in the Work Environment with Intended Changes for 1993–1994*, ACGIH, 1991, P.O. Box 1937, Cincinnati, Ohio, 45201.

W. A. Burgess, *Recognition of Health Hazards in Industry*, Wiley-Interscience, New York, 1981.

R. J. Lewis, *Sax's Dangerous Properties of Industrial Materials*, Van Nostrand Reinhold, New York, 1992.

S. Lipton and J. Lynch, *Handbook of Health Hazard Control in the Chemical Process Industry*, John Wiley & Sons, Inc., New York, 1994.

INFORMATION RETRIEVAL

The literature of chemistry and associated fields has increased enormously since 1980. Establishment of subspecialties and newly defined disciplines as well as increased research output have led to an explosion of journals, books, and on-line databases, all of which attempt to capture, record, and disseminate this plethora of knowledge. Tertiary reference tools in chemistry and technology (eg, *Kirk-Othmer*, 4th ed.) help track the primary literature. Excellent references that discuss basic chemical information tools are *The Literature Matrix of Chemistry, Chemical Information Sources*, and *How to Find Chemical Information*.

Retrieval of chemical information will continue to be an issue of accessing what is available in the fastest, most cost-effective manner. Changes in the ways information is located and retrieved will be driven by technological advances in computer hardware, development of software, and progress in telecommunications. The resources available through Internet are increasing daily. Content of electronic databases has remained basically the same; what has changed are the tools to access the information. Publishers continue to explore the possibilities of electronic media. Ease of information access, for example, through the use of a natural language interface, or the ability to query multilingual databases using a single language, is another emerging issue. The Special Interest Group on Information Retrieval of the Association of Computing Machinery (SIGIR ACM) meets annually to address these and other information issues. Proceedings of their meetings provide an overview of advances in information technology and access. As technology becomes more complex, issues of ownership, copyright protection, reuse of retrieved information, and access costs will need to be examined and resolved (see COPYRIGHTS AND TRADEMARKS).

Libraries and information centers are rapidly moving from purchase and ownership of print and on-line resources to rapid access to these resources; a change in philosophy from "just in case" to "just in time." Libraries are making hard decisions about what to purchase and what to exclude because of the magnitude and cost of a complete collection of available information. These factors have forced severe cutbacks in what is actually bought as well as increases in cooperative purchasing and loan agreements between libraries on the local, regional, national, and even international levels. Increased computer power and technology have made geographic boundaries and limitations obsolete. Documents, such as journal articles, can be obtained quickly from other libraries and commercial document delivery vendors. Paper copy remains the preferred format for delivery of such documents. Documents can be delivered by regularly scheduled mail, by overnight delivery, or by facsimile transmission.

Although electronic publishing of journals is in its infancy, more and more full text journals are becoming available on-line, allowing printing of a document from the user's computer. The limitation of ASCII format (text only, no graphics) is being addressed by vendors. The American Association for the Advancement of Science's publication *Online Journal of Current Clinical Trials*, which debuted in July 1992, supports text and nontext and publishes a paper within days of acceptance. The Research Libraries Group has produced ARIEL, a software package that allows image scanning of a document transmitted through Internet. ARIEL provides images and text of greater resolution than fax and uses standard personal computer (PC) hardware.

Location of and access to chemical and technical information other than journal articles is available through computerized information networks. Electronic bulletin board systems (BBS) provide a telecommunications tool to anyone who has a computer and a modem. Questions can be posted and read by thousands of bulletin board users worldwide, and files and software are easily transferred from virtually anywhere to one's computer.

Networks. The rise in popularity and use of Internet has dramatically changed the way information is disseminated. Internet is a worldwide link of thousands of separately administered computer networks of many sizes and types. Each of these networks is connected to as many as tens of thousands of computers; the total number of individual Internet users is in the millions.

The three basic Internet applications of remote login, electronic mail, and file transfer are building blocks of more sophisticated applications that offer increased functionality and ease of network use. Tools such as Gopher, Wide Area Information Servers (WAIS), and World Wide Web (WWW) go beyond the three basic Internet functions to make information on the network easier to locate and use. Detailed descriptions of these tools are available. This trend toward more powerful, user-friendly networked information resource access systems should continue as Internet grows and matures.

Copyright. Any discussion of emerging technologies in storage and retrieval of information leads to a question of copyright compliance. Copyright in the electronic and computer age, using the internal and external networks and document delivery mechanisms outlined above, is a complex and unresolved issue. One of the primary reasons for using document delivery services, other than for obtaining information that is not locally available, is copyright compliance. Document suppliers generally handle payment of any required copyright fees. In the electronic and computer age, establishing who owns a particular piece of information can be difficult, as can determining what can legally be done with information once it is obtained.

Budgeting. Changes in the storage and retrieval of chemical information require that libraries and information centers now consider not only what should be purchased but also what monies should be allocated for the purchase of information in nonprint formats such as CD-ROMs (compact disk read-only memory) and on-line databases. Coupled with this is budgeting for the cost of hardware and software to enable the rapid and cost-effective delivery of needed information. The geometric increase in sources, both printed and on-line, has increased the role of the information specialist as an expert in the delivery of chemical information. Retrieval from increasingly diverse and complex sources becomes the paramount issue for searchers of chemical literature in the 1990s.

On-Line Database Resources

The on-line information industry has grown dramatically since 1972 when Dialog Information Services, Inc. (Dialog) offered the first publicly available commercial databases (qv). Databases covering virtually all important subject areas were developed, and significant publications became available via thousands of bibliographic, abstract, textual, directory, and numeric databases. For electronic databases to be made available publicly, development was required in three primary technologies: computers, communications, and databases themselves.

Database Producers. Producers of databases, also known as database publishers or information providers, determine the content of the databases, produce them, and typically lease or license them to private organizations or database vendors. Database producers may be categorized as government, not-for-profit, commercial/industrial, and mixed.

Primary database vendors that offer or are developing fixed price options include Mead Data Central, Dow Jones News/Retrieval and DataTimes, NewsNet, Dialog, and OCLC. The environment of the 1990s is one in which database vendors and producers are competing for survival, and end users are leading the way to what could represent significant growth in the on-line database industry. Vendors include BRS Online Products, Cambridge Crystallographic Data Centre, Chemical Information Systems, Data-Star, DIALOG Information Retrieval Service, ESA-IRS, MDL Information Systems, Inc, Mead Data Central, MEDLARS, ORBIT ONline Service, Questel, and STN International.

Chemical Information and Search Methods and Services

Chemical information is reported and recorded in many forms, and a wide variety of databases have evolved to collect the various types of information. Bibliographic, business, structure, numeric, spectra, and reaction databases currently are available.

Bibliographic–Technical. These include Agrochemical Handbook, Analytical Abstracts (AA), APILIT, APIPAT, CA File, CA Registry File, CA Search CAB ABSTRACTS, Ceramic Abstracts, Chemical Engineering and Biotechnology Abstract (CEBA), Chemical Journals of the American Chemical Society (CJACS), Chemical Journals of the Royal Society of Chemistry, Chemical Safety NewsBase (CSNB), Chinese Patent Abstracts in English Database, CLAIMS, COMPENDEX PLUS, CORROSION, Current Biotechnology Abstracts (CBA), Current Patents (Evaluation/Fast-Alert), Dissertation Abstracts Online, EMBASE, Energy Science & Technology, EPAT, European Directory of Agrochemical Products, GenBand, INPADOC, INSPEC, JAPIO, Rapra Abstracts, SciSearch, Thomas Register Online, U.S. Patents Fulltext, World Patent Index (WPI), World Surface Coatings, and World Textiles.

Business–Industrial. These include ABIINFORM, BIOBUSINESS BUSINESSWIRE, CENDATA, Chemical Industry Notes (CIN), Commerce Business Daily, Conference Papers Index, Dialog Journal Name Finder, Dialog Product Name Finder, DISCLOSURE DATABASE, Dun's Market Identifiers, ERIC, Federal Register, Food Science and Technology Abstracts, Harvard Business Review, Health Periodicals Database, Health Planning and Administration, ICC International Business Research, International Pharmaceutical Abstracts (IPA), INVESTEXT, MANAGEMENT CONTENTS, Marquis Who's Who (MWW), MATERIALS BUSINESS FILE, NTIS Bibliographic Database, Nursing and Allied Health, PHARMACEUTICAL NEWS INDEX (PNI), Pharmaceutical Business News, Pharmaprojects, Pollution Abstracts, PR Newswire, PTS News, PTS PROMPT, SEC ONline, Textile Technology Digest, Textline, THOMAS REGISTER ONLINE, TOXLINE, Trade and Industry, and World Textiles.

Structure. Structure searching involves matching a query compound against a machine-readable file of chemical structures. Structure searching determines if a compound is present in a file and retrieves it along with any associated information. Chemical structure files are compiled as novel chemicals, and compounds are registered and given unique identifiers, eg, the CAS Registry Number, which is assigned sequentially to each new structure entering the system.

Substructure searching involves retrieval of all the compounds in a file containing some specified portion of a chemical structure, irrespective of the rest of the molecule in which the query substructure occurs.

Databases include the Beilstein File, the Gmelin File, the Registry File, the Description, Acquisition, Retrieval, and Correlation File, and the Structure and Nomenclature Search System.

Numeric. Researchers routinely use reported numeric measurements and data in their work. Numeric databases include the Beilstein Handbook of Organic Chemistry, the Gmelin Handbook of Inorganic and Organometallic Chemistry, property data networks [the Materials Property Data Network Inc. (MPD) and Chemical Property Data Network (CPDN)], and TDS NUMERICA.

Cambridge. The Cambridge Structural Database is an integrated system of programs for searching, retrieving, and analyzing data on more than 96,000 organic and organometallic structures, which were determined by x-ray and neutron diffraction. About 15,000 compounds a year are being added to the database.

Spectra. The ability to consult collections of standard spectra is crucial in the analysis of unknown compounds. A long history of data collection efforts has been aimed at these applications. Among the best known of the published handbooks are the Sadtler Spectral Data Sheets, which include ir, Raman, and nmr spectra. On-line sources include the Chemical Information System, SpecInfo, and The Canadian Scientific Numeric Database Service (CAN/SND).

Reactions. CASREACT File is a chemical reaction database containing over 118,500 records with reaction information derived from documents covered in the Organic Section of *Chemical Abstracts*. The file is available from STN.

Integrated Systems. Until recently, each of the numerous databases and sources of information available to chemists and technologists had to be searched individually, and selected results either printed for file storage or downloaded to an in-house or private computer system for easy future access.

Molecular Design Limited (MDL) has marketed an Integrated Scientific Information System (ISIS), which provides the capability to query multiple systems, including binary, text, proprietry, and relational databases across global networks, thereby providing transparent desktop access to multiple autonomous data sources.

Patent Information and Search Methods/Services

Patents are unique as primary source documents because of their stylized format, specialized language, presence of legally significant claims, descriptive drawings, and frequent disclosure of chemical compositions as generic (Markush) structures. A comprehensive review of electronic databases that contain patent information, including legal aspects, is available. Databases include bibliographic databases [World Patents Index, U.S. Patents, CLAIMS, APIPAT, EPAT, JAPIO, INPADOC, and PHARM.

Full-Text Databases. Two vendor-provided full-text patent databases are LEXPAT, produced by Mead Data Central, and PATFULL, produced by Dialog Information Services.

Chemical Substructure Databases. Several patent databases are searchable by chemical substructure. These are designed to give higher relevance of retrieval when searching chemical compounds than the bibliographic or full-text databases. They include MPHARM, WPIM (World Patents Index Markush), and MARPAT.

CD-ROM Databases. Since about 1989, CD-ROM format bibliographic and image patent databases have become available as current awareness, reference, or image storage and retrieval tools. The databases are designed for stand-alone PC or local network use, and in some cases they may be alternatives to retrieving patent information on-line. U.S. patent information resources available on CD-ROM include APS (Automated Patent Searching), CASSIS, FullText, OG/PLUS, Patent-Images, and PatentView.

The European Patent office ESPACE series of CD-ROM products are ESPACE-EP, ESPACE-FIRST, ESPACE-UK, ESPACE-WORLD, and ESPACE-ACCESS.

MARKUSH TOPFRAG. Derwent's TOPFRAG family of products is PC-based software that automates the selection of search codes and strategies.

Environmental and Safety Information and Search Methods and Services

There are public and private databases. Public databases are produced by the government and private enterprise and are commercially available through database vendors, such as STN, DIALOG, BRS, ORBIT, and NLM, and through various universities. Private databases are produced by government agencies, corporations, or other organizations for in-house use by their employees or others affiliated with them. The information in these databases may be made available on a need-to-know basis to individuals or corporations in the public or private sectors. Knowledge of the existence of private databases is usually obtained by personal contact within an organization or thorough disclosure in published literature. Private industrial databases are not usually accessible by the public, although information contained in them on hazardous materials must be reported to the EPA under provisions of TSCA (Toxic Substances Control Act) Section 8e.

Public Databases. The most comprehensive list of publicly available databases is the two-volume *Gale Directory of Databases*.

Private (EPA) Databases. The U.S. EPA maintains a list of approximately 600 current information systems, as well as some of the models and databases used within the organization. The list is published in *Information Systems Inventory* (ISI) which is updated yearly and maintained by the Information Management and Services Division of the Office of Information Resources Management.

On-Line Search Aids

MACCS-II. The Molecular Access System is a chemical information management system from Molecular Design Limited (MDL), San Leandro, California. It offers menu-driven graphical input for building, maintaining, and accessing chemical structures and any associated data, eg, chemical and physical properties, biological activity, toxicity data, pricing, safety, and supplier information.

MACCS-II enables direct interface with other database management systems, such as the Relational Database Management System (RDBMS) and Oracle, so that databases that contain text and numeric data, for which special interfaces are normally needed, can be constructed.

Optical Disk-Based Information and Document Image Systems

Optical-based storage technology has joined paper, microfilm, and electronic/magnetic technologies as another medium for the storage, retrieval, and management of information (see INFORMATION STORAGE MATERIALS). Optical media differ from magnetic media in that the information is encoded and read by means of laser optics. Information stored on optical disk may be either in a searchable text format (ASCII) or in a format containing only bit-mapped images, usually obtained as output from a scanner. Through the scanning and digitization process, pages that consist of printed text, graphics, photographs, drawings, handwriting, tables, etc, are converted to their binary representation and are stored as bit-mapped images on the optical media. The information stored on optical disk often has counterparts in other formats, such as printed publications or on-line databases or files. Advantages of document imaging systems for complementing, enhancing, or replacing traditional paper- or microfilm-based systems include increased storage capacity, ease of access via automated retrieval, simultaneous searching and viewing at multiple workstations, speed of access and delivery resulting in productivity gains, improved customer service, document security, document integrity via preservation and elimination of lost or misfiled documents, and networking and integration capabilities for these systems. Among the types of optical media that have been developed, the two most common for information storage and retrieval systems are both optical disk-based systems, namely CD-ROM and WORM (write once read many).

Several projects such as CORE (Chemistry On-Line Retrieval Experiment) at Cornell University, Project Mercury at Carnegie-Mellon University, Right-Pages at AT&T, and Red Sage at the University of California, San Francisco are in progress and illustrate the issues arising from implementation of on-line information systems that combine text and image.

Private Bibliographic and Text Databases

Personal computers have introduced new ways to handle private bibliographic and text files. The most important factors to consider to achieve satisfactory results in building a bibliographic or text database are the type of information to be stored and the needs of the user. Types of information include correspondence, research results and documentation, meeting notes, and bibliographic references. Needs of the user to be considered should include the potential number of users of the database, restrictions for the access and display of the information because of privacy or proprietary reasons, and the retrieval mechanisms (eg, by keyword, authority list, controlled vocabulary, author, title, date, or other document or information attributes). In addition, criteria for selecting and encoding information for the database need to be established.

The type of hardware or computer system to be used and the potential size of the database should also be considered. Another factor frequently overlooked in private database creation is commitment to the support and maintenance of the database. Support involves training users, solving software and hardware problems, and upgrading the software when new features become available or are needed by the end user. Maintenance of the database include adding new information, deleting information no longer wanted, and correcting information in the database when errors are detected. Software packages available for building databases range from generic personal computer database management systems, which require customizing to software designed specifically for bibliographic files. Most of the software is for Macintosh or IBM compatible PCs. Some examples are PROCITE, NOTEBOOK II, ENDNOTE, LIBRARY MASTER, PERSONAL FILE SYSTEM, ASKSAM, BASISPLUS, and PERSONAL LIBRARIAN.

CYNTHIA S. BARCELON-YANG
EVELYN L. BROWNLEE
EMMETT D. CALHOUN
BRUNO A. CAPUTO
CHARLES C. CUMBO
JOSEPH P. DANISZEWSKI
DOUGLAS A. ECKEL
KENNETH H. GLASPEY
DARLYN C. GREEN-KOCHER
MARIANNE B. GRUBER
MARGARET M. ISSELMANN
THOMAS C. JOHNS
ALICIA P. KING
DAVID M. KRENTZ
FLUORENCE H. KVALNES
LURAY M. MINKIEWICZ
BEHROOZ NAZER
ANGELA K. G. PARSONS
CAROL R. PERROTTO
RITA D. RATLIFF
JEANETTE C. SIKES
AMIE H. WEBSTER
Du Pont Company

H. Skolnik, *The Literature Matrix of Chemistry*, John Wiley & Sons, Inc., New York, 1982, p. vi.

G. Wiggins, *Chemical Information Sources*, McGraw-Hill Book Co., Inc., New York, 1991.

R. Maizell, *How to Find Chemical Information*, John Wiley & Sons, Inc., New York, 1987.

INFORMATION STORAGE MATERIALS

OPTICAL

A most important element in computer technology is data storage. Progress in microelectronics, therefore, is directly linked to progress in data storage, that is the ability to store large amounts of information in the smallest possible space, irreversibly or preferably reversibly.

For data storage two types of memory are distinguished: main memory with moderate capacity (currently 16×10^6 bit in each element) and extremely short access time ($\leq 10^{-7}$ s), and mass memory with very high capacity ($>10^7$ bit in each element) and moderate access time ($\geq 10^{-12}$ s).

The most important mass memories use magnetic media in the form of magnetic tapes or disks (floppy disk and hard disk). Laser addressed optical mass memories are of increasing commercial importance.

Polymers are only marginally important in main memories of semiconductor technology, except for polymeric resist films used for chip production. For optical mass memories, however, they are important or even indispensable, being used as substrate material (in WORM, EOD) or for both substrate material and the memory layer (in CD-ROM). Peripheral uses of polymers in the manufacturing process of optical storage media are, eg, as binder for dye-in-polymer layers or as surfacing layers, protective overcoatings, uv-resist films, photopolymerization lacquers for replication, etc.

Classification

In general, the commercially used optical data storage media deposit the information on disks or cards (two-dimensional data deposition, Table 1). Data storage systems, which store data in three or more dimensions, are being developed.

Depending on the method of data read-out, respectively read-in/read-out, two systems are distinguished: mechanooptical systems with usually disk-shaped media (optical disks), and purely optical systems with card-shaped media without moving parts (optical memory cards).

Planar Data Deposition (2-D Storage)

Data Storage With Prerecorded Information. CD-ROM disks are nearly identical to the well-known compact disk digital audio (CD-DA; short CD). The information on a CD-ROM is stamped in the form of clearly defined pits on the disk surface during the disk's manufacture, using injection molding or injection stamping techniques. A metal stamper transfers the digital information to the disk's surface.

Writable, Nonerasable Data Storage. In many applications, data storage systems are required which enable the user to write data and text as well as in some cases digitized graphics and pictures. They should allow fast access to the stored information at all times. The information itself, however, many not be changed, erased, or overwritten. Examples for these applications are mainly office files,

Table 1. Methods of Two-Dimensional Data Storage on Disks

Symbol	Typical properties	Examples	Application of polymers
CD-ROM[a]	not erasable or rewritable	technology identical with audio compact disk (CD-DA)	substrate and information layer from polycarbonate (PC)
WORM[b]	writable not erasable	polymeric or glassy substrates with metal or alloy layer, dye-in polymer film	substrate $\theta \leq 51/4$ in.[c]: PC $\theta > 51/4$ in.: glass, partly PC
EOD[d]	erasable rewritable	polymeric or glassy substrates with magnetooptical recording layer (MOR); phase change recording layer (PCR); photochromic dyestuff	substrate $\theta \leq 51/4$ in.: PC, partly glass $\theta > 51/4$ in.: not on the market

[a] Compact disk–read only memory. [b] Write once, read many times. [c] 5 1/4 in. is a standard disk size. [d] Erasable optical disk.

especially those which require mandatory storage, protection against manipulation, and forgery-proof documents, eg, expense statements, payment and salary files, bank statements, employee files, or production instructions. In general, these storage systems could substitute for filing of documents on paper or archiving on microfiche.

The conditions of these target applications are fulfilled by WORM-disks. A WORM disk generally consists of two polycarbonate disks with 130 mm diameter in sandwich construction. A breakthrough in WORM applications is the introduction of a WORM disk in CD format (so-called CD-WORM or CD-R; R = recordable), offering the possibility of using CD-WORM disks in customary, new generation CD players (multimedia operation).

WORM disks must fulfill the following requirements: high storage capacity ($>2 \times 10$ bit/cm^2), short access time $<10^{-1}$ s), read back with high signal-to-noise ratio $>$ 47 dB, long shelf life of the information ($>$20 years), low storage costs ($\leq 10^{-7}$ \$/bit), low error rate (BER = bit error rate $\leq 10^{-12}$ bit, after error correction), and high reliability (MTBF = mean time between failure $>$2000 h).

Writing Techniques. WORM disks differ depending on their data writing techniques, which can be divided into three classes: ablative writing, bubble forming, and phase change.

Dyes for WORM-Disks. Regarding their memory layer, dye-in-polymer systems show advantages over metal layers in their higher stability, lower toxicity, lower heat conductivity, lower melting and sublimation temperature, and simpler manufacturing technique (substrate coating by sublimation or spincoating).

The following requirements need to be fulfilled by dyes or dye-in-polymer systems as active components in WORM-disks: high absorption capability at the wavelength of the write laser (wavelength 780–840 nm is low writing energy); defined threshold to avoid destructive reading; low heat conductivity parallel to the disk surface to yield focused pits, ie, high storage density, low toxicity, good solubility in solvents (eg, pentanol, hexanol) which do not attack the disk material (generally polycarbonate); and good film forming, ie, low cost manufacturing technique.

The dyes can be classified in four groups: methine dyes, naphthalocyanine derivates, naphthoquinone derivatives, and metal complexes.

Erasable, Rewritable Optical Data Storage. Erasable optical disk (EOD) systems are challenging classic magnetic media in some areas of application, primarily magnetic tape and the hard disk, but mostly optical media complement magnetic media.

High demands are made on erasable, rewritable optical data storage: high storage density ($\geq 2 \times 10^7$ bit/cm^2), short access time ($\geq$ 30 ms), reading with high signal-to-noise ratio ($\geq$47 dB), high data transfer rate ($\geq$0.7 MByte/s), high number of read/write cycles ($\geq 10^7$), long guaranteed shelf life of data ($\geq$10 years), low susceptibility to dirt and disturbances, trouble-free removable media, low cost per bit ($\leq 10^{-7}$ \$/bit), low bit error rate (BER $\leq 10^{-12}$, after correction), and high reliability (MTBF $\geq 10^5$ h).

Magnetooptical Recording. In a simplified way, a magnetooptical recording (MO-R) system can regarded as a CD recorder using polarizing optics, a laser of controllable intensity, and a magnetic coil. A disadvantage of MO storage technology in comparison to magnetic hard disks is the longer average access time: typically 30 ms, as compared to 10 ms. This is caused by the mechanical inertia of MO write/read heads.

As storage medium a (pregrooved) disk containing a hard magnetic layer is used where the information is stored in magnetic domain patterns.

In the thermomagnetic write process, a focused laser beam with a typical power of 10 mW locally heats the magnetic layer to its Curie temperature, where the spontaneous magnetization and the coercivity vanish. During cooling, the magnetization in the heated spot can be reversed by a small magnetic field (demagnetizing field of the layer itself and externally applied field), creating a magnetic domain.

The read process utilizes the polar magnetooptical Kerr effect: the polarization plane of the reflected light is rotated clock- or counterclockwise by a perpendicular magnetization (up or down) in the film according to the domain pattern. The minimum usable domain size is determined by the opticall resolution during read-out.

Table 2. Properties of MO Materials

Material[a]	Microstructure	Deposition temp, °C	H_c,[b] kA/m	θ_K[c]	SNR[d] dB
α-GdTbFe	no grains	RT	>400	0.3	58
α-TbFeCo	no grains	RT	>400	0.3–0.4	61
Co-ferrite	0.1–0.5 μm	500	>250	10/μm	35[e]
Ba-ferrite	40 nm	620	>100	1/μm	50[f]
Co/Pt	10–30 nm	RT	>100	0.2	55

[a] Chemical stability of α-GdTbFe and α-TbFeCo is poor; it is good for other materials.
[b] H_c = coercive force. To convert kA/m to kOe, divide by 79.58.
[c] θ_K = Kerr rotations in degrees.
[d] Signal-to-noise ratio measured at conditions: wavelength (λ) = 800 nm, carrier frequency (f) = 1 MHz, linear velocity of the disk (v) = 5 m/s, bandwidth (BW) = 30 kHz, unless otherwise noted.
[e] f = 0.5 MHz, v = 2 m/s, BW unknown.
[f] f = 1.5 MHz, v = 2 m/s, BW unknown.

Two write methods are in use: magnetic field modulation (MFM) and laser modulation (LM).

Materials for MO Media. Amorphous alloys of rare-earth elements (eg, Gd, Tb, Dy) and transition metals (Fe, Co) are the materials of choice for today's MO media. Other material classes principally suited for MO media are Co/Pt multilayers, garnets and ferrites. The quality of disks is mainly characterized by the signal-to-noise ratio (SNR). The features of MO media classes relevant for recording are summarized in Table 2.

Exchange-Coupled RE-TM Layers. It is difficult to obtain RE-TM thin films that exhibit all the desired magnetooptic and micromagnetic properties. In most cases, the optimization of one property adversely affects another property. New and interesting possibilities in this regard utilize sandwich structures of two or more RE-TM films (exchange-coupled layers, ECLs) with different magnetic properties. One layer (the storage layer) can, eg, be optimized toward high coercivity for storage, and the other one toward high Kerr rotation for read-out.

The most spectacular applications of ECLs are the possibility of direct overwrite (DOW) with laser modulation and of magnetically induced superresolution.

Stability of Domains. In bilayers used for DOW or superresolution, the bias layer is homogeneously magnetized in the ground state after a complete write or read cycle, whereas the storage layer contains magnetic domains leading to magnetic interface walls between the two layers. For the domains to be stable, the wall energy density has to be lower than the coercive (area) density $2 \cdot H_c M_s \cdot d$ of both layers (thickness, d). In order to achieve stable storage in exchange-coupled layers, different stack design and material compositions have been used.

Interface Walls and Domain Formation. The existence of a macroscopically large interface wall between two magnetic layers gives rise to new effects in the domain formation process that are not possible in single layers.

Layer Stacks and Protective Layers. The layer stack of an MO disk consists mainly of an MO layer, a dielectric antireflection layer, and a metallic reflection layer (Fig. 1). The dielectric layers have to fulfill several functions. They must provide a barrier against oxygen and moisture, an antireflection layer for coupling in of the laser light, and heat insulation of the recording layer (in a quadrilayer configuration).

MO Media Summary. When compared to magnetic recording on hard disks, the advantage of MO data storage is the removability of the disks and the high storage capacity (especially on multiplatter (juke-box) systems) whereas the access time has not yet been reached.

Phase Change Recording (PC-R). Erasable and rewritable optical data storage disks in PC (phase change) technology unite the advantages of optical storage (high storage density, exchangeable media, robustness) with direct overwrite. On a PC medium the information is

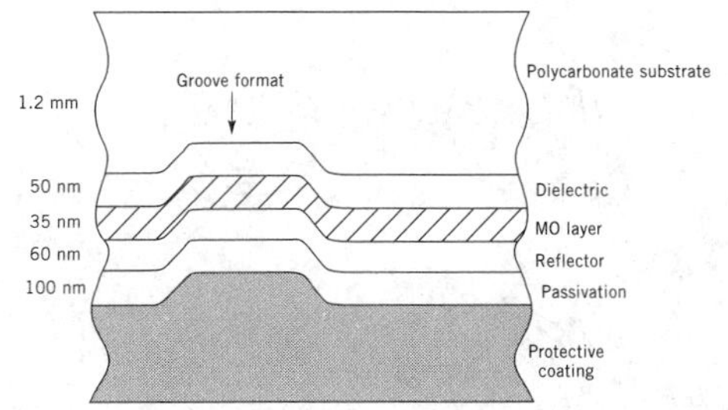

Figure 1. Cross section of an MO disk (trilayer configuration). Examples of the various layers are dielectric, AlN, Si_3N_4; MO layer, TbFeCo; reflector, Al; passivation, AlN.

stored as amorphous spots in a thin crystalline film. The PC technology is based on fast, reversible transformations between crystalline and amorphous phases of certain alloys. The data rate is limited by the crystallization time.

Materials for PC Media. Crystalline alloys of elements from the third to the sixth main group are preferred, eg, GeSbTe, InSeSb.

Projected Reversible Optical Recording

Possible techniques and materials include reversible data storage in dye polymer layers, reversible generation of depressions, reversible bubble generation, reversible deformation, materials with reversible coloring, metallic alloys with optically controllable coloring, photochromic organic dyes, LC side-chain polymers with dyes, and biopolymers for reversible data storage.

Card-Shaped Optical Data Storage. Rather than storing exclusively alphanumeric data like the current magnetic devices or chip-cards, optoelectronic memory cards offer the possibility of storing pictures, graphics, drafts, fingerprints, possibly even sounds, as well as alphanumeric data. Such a credit card-sized storage medium is called optical memory card (OMC).

Data Deposition in Three or More Dimensions

Holographic Information Storage. Data storage technology aims to concentrate data in the smallest possible space, that is, to achieve maximum spatial storage density. This can be realized by three-dimensional data deposition as a bit oriented format or as an interference pattern (hologram). Therefore, holographic storage techniques attract great interest.

The highest information density is achieved by storing information in superimposed volume-phase holograms, so-called thick holograms. The thickness of the storage material must be greater than the the wavelength by a factor of >100, so that several thousand holograms can be stored by angle-selective superimpostion in the same volume element. With each new hologram the incidence angle is altered by a fraction of a degree. Of special interest is 3-D optical data storage by two-photon excitation.

Materials. For holographic information storage, materials are required which alter their index of refraction locally by spotwise illumination with light. Suitable are photorefractive inorganic crystals, eg, $LiNbO_3$, $BaTiO_3$, $LiTaO_3$, and $Bi_{12}SiO_{20}$. Also suitable are photorefractive ferroelectric polymers like poly(vinylidene fluoride-*co*-trifluorethylene) (PVDF/TFE). Preferably transparent polymers are used which contain approximately 10% of monomeric material (so-called photopolymers, photothermoplasts). These polymers additionally contain different initiators, photoinitiators, and photosensitizers.

Photochemical Hole Burning

Aside from holography, photon-gated spectral hole burning (photochemical hole burning (PHB), also named persistent spectral hole burning (PSHB)) is another possibility for achieving multidimensional information storage. Besides the three spatial dimensions, the parameters of frequency and electrical field strength can be used to store information. Another possible advantage of the PHB method is its potential multiplexing characteristics; by synchronous recording and read-out, exceptional data transfer rates can be achieved.

A great disadvantage of PHB is the necessity to operate at very low temperatures (<20 K). Therefore, this recording technique currently has no practical significance.

Materials. Beside inorganic materials (eg, barium chloride/fluoride crystals, doped with 0.05% samarium), transparent thermoplasts, are preferred for the PHB technique, doped with small amounts of suitable organic pigments such as phthalocyanines, 9-aminoacridine, 1,4-dihydroxyanthraquinone, 2,3-dihydroporphyrin, and tetraphenylporphyrin derivates.

Substrate Materials for Optical Memories

High demands are placed on the substrate material of disk-shaped optical data storage devices regarding the optical, physical, chemical, mechanical, and thermal properties. In addition to these physical parameters, they have to meet special requirements regarding optical purity of the material, processing characteristics, and especially in mass production, economic characteristics (costs, processing). The question of recyclability must also be tackled.

The birefringence of substrate materials for optical data storage devices requires special attention, especially in the case of EOD(MOR) disks. Birefringence has no importance for glass substrates (glass does not exhibit any significant birefringence) and is only a subordinate factor for polymeric protective layers of aluminum substrates because of their reflective read/write technique.

In the case of polymeric substrate materials, high birefringence not only makes exact focusing of the laser beam for reading and writing more difficult, but also effects an unwanted rotation of the plane of polarization during reading, which can severely impair the signal-to-noise ratio, especially using EOD(MOR) disks. Therefore, polymers with low anisotropy of their molecular optical polarizability are preferred.

Nonpolymeric Substrate Materials

MO disks have to meet particularly high demands in terms of low birefringence; for WORM and EOD(PCR) disks a higher resistance to heat softening is wanted and for all optical storage disks with diameters exceeding 5.25 in. (>130 mm), increased requirements exist regarding smallest warp (thermally or by moisture absorption), distortion, and creep (at very high rotation speeds). These increased demands can be met by glass. Disadvantages of glass are high cost, high weight, and low impact strength.

Polymeric Substrate Materials

Polycarbonates. Currently, all audio CDs (CD-AD), all CD-ROM, and the biggest fraction of substrate disks for WORM and EOD worldwide are manufactured from a modified bisphenol A–polycarbonate (BPA–PC) (**1**).

For substrates of WORM and EOD(PCR) disks the industry in the future wants polymers that have a markedly improved resistance to heat softening compared to BPA-PC and, if possible, a lower water absorption and lower birefringence, but otherwise maintain the good characteristics in toughness, production, and cost. This goal is being approached in different ways: further modification of BPA-PC, newly developed polymers, improvement of the processing characteristics of uv-curable cross-linked polymers, and development of special copolymers and polymer blends, eg, PPE/PS.

Poly(methyl methacrylate). PMMA offers distinct advantages over BPA-PC with respect to significantly lower birefringence, higher modulus, and lower costs, but has not been successful as a material for audio CDs and CD-ROM as well as a substrate material for WORM and EOD disks because of its high water absorption (which makes it prone to warp) and its unsuitability for metallizing, and less so because of its low resistance to heat softening.

Cyclic Polyolefins (CPO) and Cycloolefin Copolymers (COC). Japanese and European companies are developing amorphous cyclic polyolefins as substrate materials for optical data storage. The materials are based on dicyclopentadiene and/or tetracyclododecene where R = H, alkyl, or $COOCH_3$.

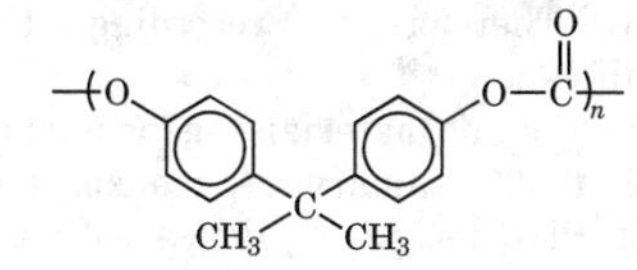

(1)

The principal advantages of CPO over the current substrate materials based on BPA–PC are very low water absorption (depending on type less than 0.1%), resistance to polar solvents and to acids and bases, low birefringence (comparable to that of PMMA), and short vacuum time (for sputtering ferrimagnetic layers). Disadvantages include high proneness to warp, low impact resistance, significantly higher melt viscosity (longer cycle time in disk production), unsatisfactory metallizability, and high expense.

Cyclic polyolefins (CPO) and, especially cycloolefin copolymers (COC) and blends of cycloolefin copolymers with suitable engineering plastics have the potential to be interesting materials for substrate disks for optical data storage.

Cross-Linked Polymers. Special, uv-curable epoxy resins (qv) for substrate disks for optical data storage (Sumitomo Bakelite, Toshiba) excel by means of their very low birefringence (<5 nm/mm) and high Young's modulus. Resistance to heat softening and water absorption are similar to BPA–PC, but impact resistance is as low as that of PMMA. The decisive disadvantage for industrial mass production of substrate disks is the extremely long curing time (eg, 5 h at 100°C); thus efforts to reduce curing time have been numerous.

Other Polymers. These polymers have not gained commercial importance: polystyrene (PS), poly(vinyl chloride) (PVC), cellulose acetobutyrate (CAB), bis(diallylpolycarbonate) (BDPC), poly(ethylene terephthalate) (PET), styrene–acrylonitrile copolymers (SAN), poly(vinyl acetate) (PVAC), and for substrates with high resistance to heat softening, polysulfones (PSU) and polyimides (PI).

Comparison of Substrate Materials

The most important properties of polymeric substrate materials based on BPA–PC, PMMA, and CPO (three different products) are compared in Table 3.

Table 3. Comparison of Characteristic Properties of Substrate Materials

Properties	BPA-PC[a]	PMMA[b]	CPO[c]	CPO[c]	CPO[c]
light transmission, %	>90	>90	>90	>90	92
birefringence, nm/mm	<40	<20	<25	<20	<20
T_g, °C	145	105	140	171	150
tensile stress at yield, MPa[d]	62	73	64	42	59
elongation at break, %	>50	5	10	16	3
impact strength, notched kJ/m²[e]	20–30	2	4	3	
water absorption (100% rh), %	<0.35	2.1	<0.01	0.2	0.01
melt flow index (280°C), g/10 min	57		15		
vacuumizing time,[f] min	78	200	14		

[a] CD-modified BPA-polycarbonate (CD 2005, Bayer AG).
[b] Poly(methyl methacrylate).
[c] Cyclopolyolefins.
[d] To convert MPa to psi, multiply by 145.
[e] To convert kJ/m² to ftlbf/in.², divide by 2.10.
[f] Mitsui method.

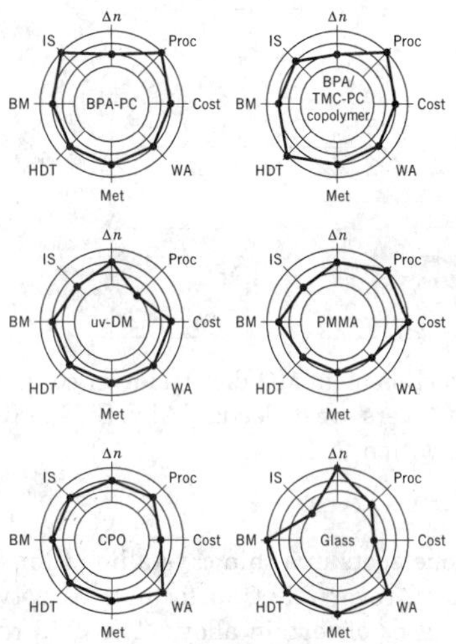

Figure 2. Qualitative comparison of substrate materials for optical disks: Δn = birefringence; IS = impact strength; BM = bending modulus; HDT = heat distortion temperature; Met = metallizability; WA = water absorption; Proc = processibility. The materials are bisphenol A–polycarbonate (BPA-PC), copolymer (20:80) of BPA-PC and trimethylcyclohexane–polycarbonate (TMC-PC), poly(methyl methacrylate) (PMMA), uv-curable cross-linked polymer (uv-DM), cyclic polyolefins (CPO), and, for comparison, glass.

In spite of the decision of commercial suppliers of digital optical storage disks, it is of interest to compare the most important properties of glass and polymers and rate them according to their characteristics relevant to optical storage disks. For a direct visual comparison, Figure 2 depicts the most important properties relevant to data storage in spider charts. The innermost circle symbolizes the most unfavorable case of each property and the outermost circle the most favorable case.

GUENTHER KAEMPF
University of Aachen
DIETER MERGEL
University of Essen

M. Emmelius, G. Pawlowski, and H. W. Vollmann, *Angew. Chem.* **101**, 1475 (1989).

D. J. Gravesteijn, C. Steenbergen and J. van der Veen, *Proc. SPIE Int. Soc. Opt. Eng.* **420**, 327 (1983).

K. L. Mittal, *Proc. Amer. Chem. Soc. Symp. on Polymers in Information Storage Technology*, Los Angeles, (Sept. 25–30, 1988), Plenum Press, New York and London, 1989.

R. Gamino and T. Suzuki, eds. *Magneto-Optical Recording Materials*, IEEE press, Dec. 1995.

M. Chen and A. Rubin, *Proc. Soc. Photo Opt. Instrum. Eng.* **1078**, 150 (1989).

H. Yamazaki, Y. Sugiyama, R. Chiba, and S. Yagi, *Adv. Mater.* **5**, 214 (1993).

Ulstrastructure Processing of Advanced Materials, D. R. Uhlmann and D. R. Ulrich, eds., John Wiley & Sons, Inc., New York, 1992, p. 475

F. J. A. M. Greidanus and W. B. Zeper, *MRS Bulletin* **15**(4), 31 (1990).

D. Mergel and co-workers, *SPIE* **1274**, 270 (1990).

R. M. Pisipati, H. Schmid, and G. Kaempf, *MRS Bull.* **15**(4), 46 (1990).

G. Kaempf and co-workers, *Kunststoffe* **82**, 385 (1992); *Kunstst. / German Plastics* **82**, 9 (1992); *J. Polym. Adv. Technol.* **3**, 169 (1992).

Hoechst High Chem Mag. **14**, 48 (1993).

Mod. Plastics Int., 11 (July 1993).

MAGNETIC

At present most information (~95%) is still stored on paper, 3% on microfiche, and the remaining 2% by magnetic–optical–magnetooptical and semiconductor storage devices. Nevertheless, magnetic recording represents a multibillion dollar industry and is still a growing market.

In this article the main focus is magnetic recording (MR) and rather less on magnetooptic recording (MOR) (see INFORMATION STORAGE MATERIALS, OPTICAL). Both methods are used for professional as well as for consumer applications. The various applications of magnetic recording include audio, video, or data recording. Each application has its own type of media in the form of tape, floppy, or hard disk. At present MOR media are only available on hard disks. The trends in magnetic recording technology are continually increasing recording densities and storage capacity with a decreasing price per bit.

Progress to Higher Densities

Three key developments have led to increased density: (*1*) in 1970 a ferrite head was used together with a particulate medium having a coercivity of 28 kA/m (352 Oe) with a head-medium spacing of 430 nm, followed in 1980 by (*2*) a configuration of a plated medium (H_c = 56 kA/m), a thin-film head, and a spacing of 200 nm. In 1990 (*3*) a magnetoresistive read head combined with an inductive write head were introduced together with a sputtered medium having a H_c = 120 kA/m and a head/medium distance of 100 nm. Not only media are being developed with thin-film technologies but also film heads. The small spacing for high densities has introduced an additional field of research on protective layers and tribology.

Magnetic Properties of Recording Materials

The relation between the three important values for a magnetic material is shown in equation 1:

$$Bm = v_0(H + M) \qquad (1)$$

where B is the magnetic induction or flux density generated by a magnetic field H. The magnetic induction also consists of a contribution from the magnetization M. The B in free space is $\mu_0 H$ wheras the contribution from the M of the material is $\mu_0 M$. The vector sum of both is thus $B = u_0(H + M)$ in which μ_0 is the permeability in free space ($\mu_0 = 4 \times 10^{-7}$ hy/m). The relative permeability in a material is defined as μ_r and given by $\mu_r = \mu/\mu_0$. A classification for the various types of magnetic materials is given by their susceptibility (χ = M/H) or permeability (μ = B/H). The most important materials for recording can be divided into ferromagnetic materials like Fe, Co, Ni (χ = 50 to 10^4), and ferrimagnetic materials, eg, γ-Fe_2O_3, ferrites. The media used for recording are termed magnetically hard; in comparison with permanent magnet materials it would be better to define them as semihard. The magnetic head materials have the properties of soft magnetic materials. Consequently their properties are very different.

Intrinsic and Extrinsic Properties. The materials Fe, Co, and Ni and their alloys and oxides are mostly used for recording applications materials. Their magnetic properties are described by intrinsic and extrinsic parameters. The intrinsic properties (saturation magnetization, M_s, magnetocrystalline anisotropy, K, Curie temperature, T_c, and magnetostriction, m_s) are determined by the type and number of atoms, their arrangement in the crystal structure and their temperature. The extrinsic properties (remanent magnetization M_r, coercivity H_c, and permeability μ) can also be influenced by the size and shape of the magnetic material and its (magnetic) history. Consequently, in the case of thin-film media, microstructure and morphology play a key role in determining extrinsic properties.

Figure 1 shows the two hysteresis loops for a medium and a head material. The coercivity, H_c, the saturation magnetization, M_s or induction, B_s, remanent magnetization, M_s or induction, B_r, and the permeability, μ, differ for the two materials.

The hysteresis loop, in general, supplies information about the magnetic properties such as H_c, M_s, M_r, preferred direction of the magnetization or anisotropy, and it can even give some idea about the magnetization reversal process involved. In general, recording media requirements are high coercivity, high remanent magnetization, high squareness ($S = M_r/M_s$) of the hysteresis loop, and low noise.

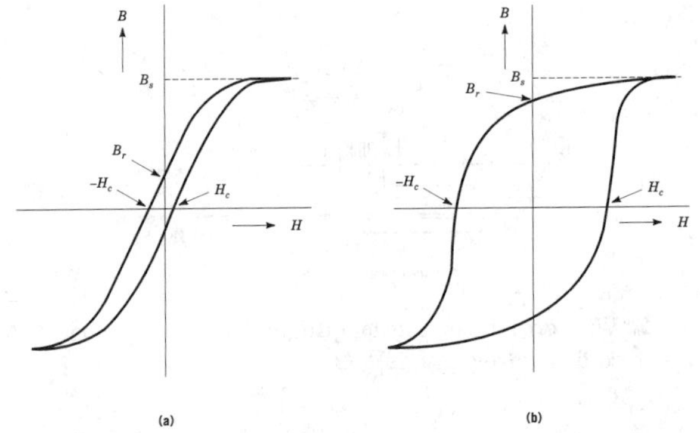

Figure 1. Basic hysteresis properties for (**a**) a recording head and for (**b**) a magnetic medium. See text.

The properties for head materials can be summarized as large saturation magnetization for producing a large gap field, high permeability at all frequencies in order to ensure high efficiency, small coercivity with low hysteresis loss, low magnetostriction for obtaining low-medium contact noise, and small but not zero magnetic anisotropy to suppress the domain noise. To ensure good reliability and a long operating time, the head materials must exhibit a good thermal stability and a high resistance to wear and corrosion. The choice of materials and preparation technologies are the tools for tailoring head and medium properties.

Switching-Field Distribution. Both M_r and H_c have a strong relation with the recording process. M_r determines the maximum output signal of a recording medium and hence the signal-to-noise ratio. H_c ascertains how easily data can be recorded and erased or changed, but it also determines the maximum head field.

The slope of the hysteresis loop in H_c is also an important parameter. From this slope, the parameter S^* can also be derived. The S^* is defined in relation to the slope of the loop at H. In the case of longitudinal recording experimental data have shown that there is a connection between S^* and recording parameters. Although S^* is normally used as a switching field distribution (sfd) parameter, it is not always suitable. The sfd can be seen as a distribtution function of the number of units reversing at a certain field. For a particulate medium without collective behavior, this function is closely related to the particle size distribution, as differently sized and shaped particles reverse at different fields. Of course the shape, orientation, and interaction between particles influence the sfd as well. Media with a high H_c and a small sfd are more suitable for high density recording because the distribution of the switching fields is very small.

Magnetic Recording

There are two modes of magnetic recording dependent on the direction of the magnetization (magnetic anisotropy) namely longitudinal (LMR) and perpendicular magnetic recording (PMR). In the former the magnetic anisotropy lies in the plane of the medium, and in the case of PMR the anisotropy is directed parallel to the medium normal. Magnetooptic recording (MOR) also requires a perpendicular magnetic anisotropy.

Modern recording technologies are based on digital signal processing, even for audio and video recording. Digital technologies provide a much lower signal-to-noise ratio, comfort of error detection, and correction and integration of large-scale integration (LSI) circuits technology. The input data may be either analogue (audio, video; using an a-d converter) or digital (computer data). Examples are the minidisc and dcc (digital compact cassette) for audio application and the HDTV digital VTR.

Longitudinal Recording. The principles of LMR are given in Figure 2. High density recording depends on shortening the recording

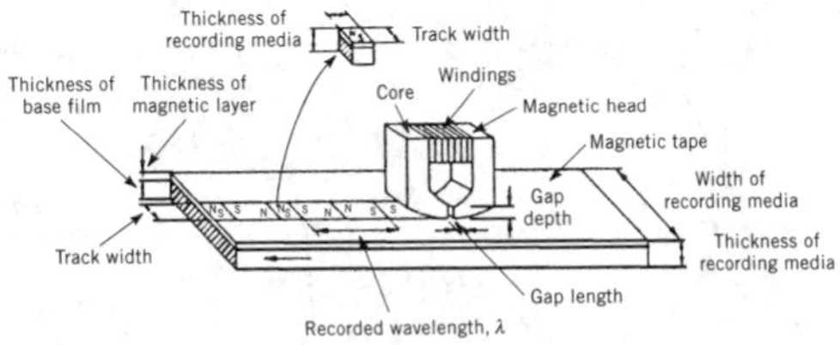

Figure 2. Written bits in the longitudinal recording mode (LMR). Shortest recorded wavelength is $\lambda_m/2$.

wavelengths and also narrowing the track width. The basic recording principles are described in detail in the literature.

Medium Noise. In magnetic recording systems three types of noise must be considered: medium, head, and electronic noise. Medium noise is the most important factor influencing the performance of the recording system. Local structural and chemical variations in the medium cause distortions in the magnetization pattern resulting in noise in the signal. The packing density (ratio between a magnetic and nonmagnetic material) is the key factor for this noise. In general smaller particles have a statistical reduction of noise. This rule is valid if the medium does not have exchange coupling and small effects of magnetostatic interaction.

Perpendicular Recording. In the case of perpendicular recording the easy axis of the magnetization is perpendicular to the medium. The most important reason for developing the perpendicular mode of recording is that in the high density area the transition length becomes more and more important.

Demagnetization. The internal microstructure as well as the macroscopic shape can influence the internal magnetizing field. In the case of PMR and MOR the anisotropy must be perpendicular, which means that the highest demagnetization field is directed opposite to the magnetization of the sample. This means that the anisotropy energy in the perpendicular direction should be larger than the demagnetizing energy for the perpendicular direction of the magnetization.

Magnetooptic Recording. This mode of recording combines the advantages of optical and magnetic recording. The number of commercial systems is growing. The minidisk is a good example for using MO-recording technologies for consumer applications. Reversed domains are formed with a focused laser spot (diameter $<1\ \mu$m) in a perpendicular magnetically saturated disk. Writing is based on the thermomagnetic writing process which means that the applied field, heating the material above its Curie temperature.

Pulsed diode laser beam has been used for writing at a wavelength of 800 nm and 10 mW/100 ns. The disk is rotated at a frequency of 37 Hz. The external applied field perpendicular to the medium is lower than the H_c of the medium at room temperature in order to limit the switching to the heated spot only. The same laser as for the write operation is used for reading but at a lower power level. Magnetooptical read-out is based on the fact that a change in polarization direction occurs when linearly polarized light is reflected from a magnetic surface (polar Kerr effect).

Erasing of the information can be realized by heating the film with a laser beam while reversing the applied field. This is one of the limitations of MO because it slows down the data rate by a factor of ~2.

Magnetization Reversal Mechanisms

The reversal of the magnetization is a basic principle of magnetic recording. Magnetization in a material can be reversed by applying a field, and finally the whole material will be saturated in a direction parallel to the field. The two different states of + and − magnetization is the basic idea for digital information storage. The mode of magnetization reversal depends on the material and its size and shape. The two principal methods for reversing the magnetization are rotation and domain-wall motion. Both modes can be examined in terms of energy considerations. The coherent rotation mechanism and the incoherent rotation mechanism only occur in single-domain particles. A particle is single domain below certain dimensions. Above a critical radius a domain wall can exist and the reversal takes place by domain-wall motion.

Switching of a Single-Domain Particle. If an isolated ferromagnetic particle is considered, then the magnetization reversal depends on the dimension of this particle. For recording the so-called single-domain particles dispersed in the medium are most important to deal with.

Beside single-domain particles, multidomain particles also exist. The distribution in domains lowers the magnetostatic energy but increases the exchange energy caused by the domain walls. The reversal in such particles mainly takes place by domain-wall motion. This kind of reversal mechanism influences the coercivity.

Reversal Mechanism in Thin Films. The ideal magnetic structure for magnetic recording medium consisting of a polycrystalline microstructure is that a crystallite reversed its magnetization by rotation and not by domain-wall motion. In other words, for high density recording the crystallites should act as independent single-domain particles consequently without exchange coupling, but depending on the distance still with magnetostatic coupling. In practice the thin films possess a wide distribution of grain size, and not all crystallites are completely separated from each other. This influences the reversal behavior.

There are two main models from the microstructural point of view, namely the particulate and the continuous microstructural model. In the first one, the crystals that are formed during film deposition are believed to interact only through magnetostatic interaction. No exchange force acts over the column boundaries due to physical separation. In the continuous model the reversal mechanism is thought to take place by Bloch walls as in stripe domains, hindered by the column boundaries which can increase the coercivity of the medium.

Hysteresis Loop and Reversal Process in Thin-Film Media. Materials used as storage media should have a nonequilibrium behavior which can be designated as a memory. For a magnetic recording medium this behavior is represented by hysteresis loops. The transition between two states of +M and −M represent the presence of information. Due to the Weiss domain theory the atomic moments in a ferromagnetic material are ordered. The difference between the demagnetized and the magnetized state is due to the dimensions and the number of domains having opposite directions of magnetization.

In the transition from the demagnetized state (H = O; M = O) to saturation (H = large; M = M), small domains (aligned favorably with the field) grow in the direction of the field (wall motion). Increasing the field another reversal mechanism is relevant, namely the rotation of the magnetization into the easy axis. At very high fields the moments lying in the direction of the easy axis, which is close to the applied field, are coherently rotated in the direction of the field. The final state of the material is a single domain (if the applied field is sufficiently high). In this description the magnetization is reversed mainly by domian-wall motion which means movement and bowing of the wall.

Preparation of Thin-Film Media

The general properties for media are a sufficient magnetization, M, for reading by the head with an acceptable S/N and an acceptable field strength to create a magnetization reversal directly related to the coercivity H_c. The latter parameter should not be too high for successful writing by the head field but it must be large enough to protect the medium against an unwelcome reduction of the signal during storage by demagnetization. For high density recording a significant potential for changing the signal during the required storing time is the self-demagnetizing field originated in the material itself and is proportional to the medium magnetization. Consequently the H_c must become higher for a more strongly magnetizable media and that is also the case if the recording density increases.

Based on the preparation technology and morphology two different types of recording media can be prepared, namely, particulate-coated media and thin-film media. The first consists of discrete magnetic particles dispersed in organic resins and the second is created on the substrate (tape, floppy, hard disk) by depositing a continuous layer of a magnetic metal, alloy, or oxide. In general the essential design parts of such thin-film media consists of a substrate made of glass, aluminum, polyester, etc; a transition (intermediate or seed layer) between the substrate and the magnetic (recording) layer; and a covering layer. They all consist of different materials, chemical compositions, microstructures, and thicknesses.

Thin-Film Media Preparation Technologies. A thin film can be defined as an area (volume) on top of a carrier (substrate) with properties differing from it. The interface between substrate and thin film has a great influence on the properties of the layer. The interface is determined by the properties of the substrate, the material(s) used for the thin film, and the method of deposition. During thin-layer processes the environment can be a liquid, gas, or vacuum. Such layers can be deposited by electro- or by electroless plating, chemical vapor deposition, and physical vapor deposition methods.

Using the above-mentioned technologies, multilayers consisting of a few monolayers of ferromagnetic material alternating with a nonferromagnetic material can also be prepared. This technology is used for preparing media for magnetooptic recording and thin-film heads based on the magnetoresistance principle. Most of the media (ca 1993) are prepared by physical vapor deposition technologies such as evaporation and sputtering.

Multilayer Technologies. During the 1970s and 1980s enormous advances in thin-film preparation processing have been made in the field of so-called artificial structuring of materials; semiconductors, metals, and insulators have been prepared in various sizes and geometrics. Several classes of layered structures have been made using metal compounds. Generally, the total multilayer thickness is in the tens of micrometer range wheras the individual layers can be varied from one to tens of nm. Layered structures can be deposited by sequential deposition of two or more materials. After preparing a multilayered (metal) film, by alternating deposition of two elements, a periodicity alloy along the film normal should appear if the following conditions are satisfied: the layer thicknesses are determined on an atomic scale, a layered structure is formed, and the interdiffusion is sufficiently suppressed.

The choice of materials for metallic systems is still expanding and at present various examples of combinations with different atomic radii are being prepared. Here multilayered techniques also show possibilities for new material syntheses. In contrast to materials prepared by chemical procedures, superlattices are made far from equilibrium. Most of the stacked layers have more or less sharply defined boundaries and some have a noncrystalline structure in the individual layers or one of the layers is noncrystalline.

All these possible combinations strongly affect the electronic and physical properties of materials, especially when the dimensions of the layer structure become comparable with the characteristic lengths relevant to the properties in the particular materials. The application of multilayer structures in recording technologies plays an important role. Multilayer configurations are used in thin-film heads (magnetoresistance type) for magnetic recording as well as in media for magnetooptic recording. Crucial aspects important for these types of layers are the sharpness of the interface and the flatness. These aspects are strongly related to the method of preparation, the type of materials, and the substrates used. Generally speaking the layer growth mechanism plays the key role in the final interface structures.

Microstructure and Morphology of Thin-Film Media

The process parameters (flux rate, substrate temperature, etc), type of material (desorption, dissociation, and diffusion-energy terms), and the substrate properties influence the growth process. Depending on the process, film materials, and substrate behavior, all types of layer structures can be grown (amorphous, polycrystalline, and single crystal).

In most cases the final properties of the deposited layers differ principally from materials made by the standard metallurgical methods. The substrate temperature is the most important process parameter for explaining the morphology of evaporated films. This result is modified for the sputtering process and is extended with a second parameter, namely the sputter gas pressure. In this work the influence of the surface roughness is also considerable.

Generally, deposited films have higher defect densities than those of bulk materials. The defects in polycrystalline thin films are grain boundaries, column boundaries, voids, dislocations, and interior gas bubbles. Defects are mostly responsible for the low temperature and interdiffusion processes. In the case of polycrystalline films the grain boundary is the most important detail. Epitaxial growth is a very special form of nucleation and growth and has a unique orientation relation with the substrate. Single-crystalline films can be prepared by the correct choice of the substrate material and deposition parameters. Epitaxial growth is also the base for textural growth in the case of polycrystalline films.

Columnar Structure and Grain Size. Most of the deposited films reported in the literature have a so-called columnar structure. Depending on the substrate materials the columnar diameter can increase with the layer thickness or it is constant through the layer thickness. Further, the columnar diameter depends on the argon gas pressure during sputtering, the substrate temperature, and the bias voltage on the substrate. Grain size or columnar size are strongly influenced by the microstructure of the underlayer/substrate.

Crystal Structure and Texture. Many of the Co–X–Y thin-film media where X(= Cr, P) and Y(= Ta, Pt, Ni) do have a hexagonal close packed (hcp) structure with the texture axis (c-axis) parallel or perpendicular to the substrate surface depending on the properties of the Cr underlayer. It is also possible for the films to have fcc phases. The texture of a polycrystalline material can be simply defined as the crystallographic preferential orientation.

Chemical Inhomogenities or Compositional Separation. Compositional separation at the grain boundaries influences the magnetic interactions of the individual grains. Deposition parameters such as the temperature, substrate material, and the use of a seed layer play an important role. There are, in principle, two driving forces for obtaining the compositional separation, namely the temperature and deposition geometry.

Co Binary and Ternary Alloyed Thin Films. Most of the thin-film media for longitudinal and perpendicular recording consist of Co–X–Y binary or ternary alloys. In most cases Co–Cr is used for perpendicular recording while for the high density longitudinal media Co–Cr–X is used (X = Pt, Ta, Ni). For the latter it is essential to deposit this alloy on a Cr underlayer in order to obtain the necessary inplane orientation. A second element combined with Co has important consequences for the Curie temperature (T_c) of the alloy, at which the spontaneous magnetization disappears. Adding Cr to Co has two important effects; reduction of the T_c and M_s. For recording applications these values should be optimized. The T_c must not be too close to room temperature, because then the magnetic behavior becomes too sensitive for the temperature variations. M_s should have a certain value because otherwise the information cannot be read by the head.

Another favorable influence on the film morphology is the reduction of the column dimensions and the appearance of the compositional separation.

Surface Properties. In the case of very high density recording the surface becomes more and more important. Therefore, analyses of the chemical and structural properties (eg, surface topology) in relation to the magnetic properties are necessary. One conclusion is obvious, namely that for films with different surface and bulk hystereses, the magnetization cannot be homogenous throughout the film during all stages of the hysteresis curve. Therefore, this aspect should be studied in more detail because as film media are becoming still thinner, the surface volume ratio will be more important.

Microstructure and Magnetic Properties of Thin-Film Media

Magnetic Structure. An important characteristic of a medium is its magnetic structure, the magnetic unit (intrinsic domain structure or written bit) in the magnetizable layer which has, in principle, two oppositely stable directions parallel to the anisotropy axis. The switching of the magnetic units can be achieved by a sufficient applied field. Study of the magnetic structures and their switching behavior, etc, can be carried out by several techniques. The study of the M–H loop gives information about the macroscopic behavior of the media. Increasing density requires more knowledge about the micromagnetic behavior. More insight can be obtained by computer simulations. New experimental methods are available and being developed for collecting more information about the mesomagnetic (an area between macro and micro) properties of the media like the methods for observing the magnetic domains, domain walls, written bits, and stray fields using the Bitter-colloid sem method, magnetooptic Kerr (MO-Kerr) observations, Lorentz tem observations, electron holography, scanning electron microscope polarization analysis (sempa), anomalous hall measurements in combination with photolithography, and the magnetic force microscope (mfm).

Magnetization of Deposited Alloys. Frequently the relation between the magnetization and composition is presented by the Slater-Pauling curve which gives the relation between the saturation magnetization (M_s) and homogenous bulk Co–X. In general, most papers report that the M_s of sputtered and evaporated films, deposited at higher substrate temperatures, is found to be larger than that for bulk alloys having the same average chemical composition. Although in literature various origins have been proposed, the most likely explanation is the phase separation.

Coercivity of Thin-Film Media. The coercivity in a magnetic material is an important parameter for applications but it is difficult to understand its physical background. For thin-film recording media, values of more than 250 kA/m have been reported. First of all the coercivity is an extrinsic parameter and is strongly influenced by the microstructural properties of the layer such as crystal size and shape, composition, and texture. These properties are directly related to the preparation conditions. Material choice and chemical inhomogeneties are responsible for the M_s of a material and this is also an influencing parameter of the final H_c. In crystalline material, the crystalline anisotropy field plays an important role. It is difficult to discriminate between all these parameters and to understand the coercivity origin in the different thin-film materials in detail.

Thin-Film Media for Various Types of Recording

Thin-film media can be made by various technologies, eg, sputtered deposited Co–Cr–X films for longitudinal applications, laminated media for hard disk application, metal evaporated tape, and multilayers for possible applications in magnetooptic recording.

J. C. LODDER
University of Twente

C. D. Mee and E. D. Daniel, *Magnetic Recording*, Vols. 1–3, McGraw-Hill Book Co., Inc., 1987–1988; and updated ed., *Magnetic Recording Handbook, Technology and Application*, 1990.

K. H. J. Buschow and co-workers, eds., *High Density Digital Recording*, Kluwer Academic Publishers, Series E: Applied Sciences, Vol. 229, NATO ASI series, Dordrecht, the Netherlands, 1993.

T. C. Arnoldussen and co-workers, eds., *Noise in Digital Magnetic Recording*, World Scientific Publishing Co., Singapore, 1992.

H. N. Bertram, *Theory of Magnetic Recording*, Cambridge University Press, Cambridge, U.K., 1994.

INFRARED DETECTORS. See INFRARED TECHNOLOGY AND RAMAN SPECTROSCOPY; PHOTODETECTORS.

INFRARED TECHNOLOGY AND RAMAN SPECTROSCOPY

INFRARED TECHNOLOGY

The infrared (ir) region of the electromagnetic spectrum has been used for a great variety of purposes, from radiant heating to fiber optic communications. The applications most relevant to chemical technology are spectrometry and radiometry. Spectrometry is a molecular analysis method based on measuring the wavelength dependent interaction of radiation with matter. Radiometry measures the amount of radiation, thereby remotely monitoring thermal factors. The ir region of the electromagnetic spectrum lies between the visible and microwave regions, and is generally taken to be from 0.75 to 1000 μm (13,333 to 10 cm^{-1}). This wide expanse is further divided by spectroscopists into the near infrared, 0.75–2.5 μm (13,333–4,000cm^{-1}); the mid-infrared, 2.5–25 μm (4000–400 cm^{-1}); and the far infrared, 25–1000 μm (400–10 cm^{-1}). These divisions roughly correspond to regions having different kinds of spectral features and involving different kinds of equipment. For radiometry the ir spectrum is divided up differently. The short-wave infrared (swir) extends from 1 to 3 μm, the mid-wave infrared (mwir) from 3 to 8 μm, and the long-wave infrared (lwir) from 8 to 13 μm.

Equipment

Sources. Although broad-band sources are based on thermal emission, few produce a true blackbody emission spectrum. Blackbody sources are used for the calibration of radiometric equipment and standards for determining absolute emissivities or reflectivities.

The optimum source for spectrometry depends on the spectral region. A quartz halogen lamp having a tungsten filament is excellent for the near infrared region. Because there are no suitable envelope materials for longer wavelengths, mid-infrared sources must function exposed to air, which reduces attainable source temperatures and intensities. The three common types of mid-infrared sources are globars, Nernst glowers, and nichrome coils. The mid-infrared sources can be used for the high energy end of the far infrared region, but any constant-emissivity thermal source hot enough to be useful at far infrared wavelengths has the majority of its output in the mid-infrared region. The high pressure mercury arc in a quartz envelope is the preferred source for wavelengths beyond 100 μm.

Lasers (qv) are the principal monochromatic sources in the infrared region. The most ubiquitous is the diode laser. Most commercial diode lasers are based on the Group 13–15 (III–V) compound gallium arsenide, GaAs, in which some of the gallium, arsenic, or both have been replaced by other Group 13 (III) or Group 15 (V) elements to produce the desired laser wavelength. There are a number of ion-doped glass and crystal lasers in the near infrared. Neodymium ion, at 1.06 μm, is the most common.

Dye lasers produce laser action from organic dyes that are dissolved in a liquid solvent and driven by a flashlamp or another laser. These are principally used in the visible spectrum, but dyes are available out to 1.32 μm.

Infrared Materials and Optics. There are many metal halides and chalcogenides that are transparent over large portions of the infrared spectrum. Glass (qv), however, is limited to wavelengths shorter than 2.7 μm because of a strong absorption at 2.8 μm. The transmission of other oxides ends partway through the mid-infrared region. The alkali halides are the most common mid-infrared window materials.

At far infrared wavelengths polyethylene and poly(ethylene terephthalate) (PET or Mylar) are transparent and can be used for windows. Lenses are less commonly used in the ir region than in the visible and uv ranges, because ir lens materials tend to be fragile, expensive, or reflective.

The reflectance of metals is generally higher in the infrared range than in the visible. Gold is the most widely used reflector when high reflectance is important. The reflectance of aluminum dips at 0.82 μm, but is above 97% for wavelengths longer than 1.5 μm, making aluminum an economical choice. Aluminum oxidizes, so it is often given a protective coating. Magnesium fluoride, MgF_2, and SiO are common protective films.

Colored absorption filters, which are commonly used in the visible spectrum, are used as long-wave pass filters in the near infrared region, where their component materials can transmit. Many semiconductors have a sharp transmission onset and can be used as long-wave pass filters in the near and mid-infrared regions. The most commonly used infrared filter is the multilayer interference filter, which consists of a series of thin layers built up on a transparent substrate.

Dichroic polarizers, the most common type in the visible spectrum, contain linear dye molecules oriented in one direction so that they preferentially absorb light of one polarization. Dichroic filters can produce a beam of good polarization purity, and are inexpensive, but they can only be used for low power applications. The simplest type of infrared polarizer, often referred to as a stack-of-plates polarizer, consists of a series of transparent, high refractive index plates positioned in the beam at Brewster's angle. Stack-of-plates polarizers can handle high powers, but are sensitive to orientation. Interference filters are essentially stacks of plates, so they can be used as polarizers. Wire grid polarizers are compact and insensitive to orientation, but expensive.

Optical Fibers. Transmission within an optical fiber relies on the total internal reflection of the electromagnetic wave as it travels in the fiber core. Silica-based glass fiber, used in the telecommunications industry, is the most mature of the fiber technologies. The fiber is strong and inexpensive. Silica fibers can transmit well from 0.25 to 2 μm, but different formulations are used for the uv and ir ends of that range. For somewhat longer wavelengths, fluoride glasses are the preferred materials, and the fluorozirconates are the best developed of these. The fluoride glasses are not as strong as the silica glasses and are moisture sensitive. Chalcogenide fibers are required for even longer wavelengths. Most chalcogenide fibers are nonstoichiometric mixtures of sulfur, selenium, or tellurium with one Group 15 (V) element or one Group 14 (IV) element, or both. Chalcogenide fibers are not as strong as fluoride fibers, but are not moisture sensitive.

Detectors. Modern infrared detectors fall into two classes, thermal and quantum (photon). Thermal detectors have sensitivities that are independent of wavelength, but respond slowly because the detector must change temperature. Quantum detectors are generally faster and more sensitive, but have a sensitivity that rises smoothly with increasing wavelength up to a long-wavelength limit beyond which it drops rapidly.

Only a few types of thermal detectors are in common use. Thermocouples and themopiles (groups of thermocouples connected in series) use the thermal emf of a junction to measure radiation. Their principal use in in energy measurement, such as laser power meters. The pyroelectric detector is by far the most commonly used type of thermal detector.

In quantum detectors, electrons are directly excited to higher energy levels by the absorption of photons. All quantum detectors have a maximum wavelength (minimum wavenumber) beyond which they do not function, which is defined by the size of the energy gap the electrons must jump. Photoemissive detectors, such as phototubes and photomultipliers, are limited to wavelengths of roughly 1 μm or shorter. Most other infrared quantum detectors use photosensitive semiconductors as sensors (qv). They can be either photoconductive, in which photon absorption changes electrical resistance, or photovoltaic (photodiodes), in which photon absorption causes a voltage.

Radiometers and Thermal Imagers. A radiometer is essentially a calibrated detector and its supporting electronics used to measure the radiance of a source. Numerous accessories can be added to this basic description. An infrared thermometer is a radiometer that incorporates a filter to limit the measurement to a narrow wavelength range.

Thermal imagers are used for radiometry, spectroradiometry, and thermal imaging. The conceptually simplest imager uses a staring array in which the whole image is focused on the array simultaneously, but most systems scan the image over the array in some manner.

Grating Spectrometers. Spectrometers can be classified as spectrophotometers and spectroradiometers. Spectrophotometers are intended for testing a sample by measuring its interaction with radiation, and have a built-in source. Spectroradiometers analyze the emission from a test object; thus the sample is the source. Infrared grating spectrometers work like their uv/visible counterparts, but they have been largely displaced by Fourier-transform spectrometers.

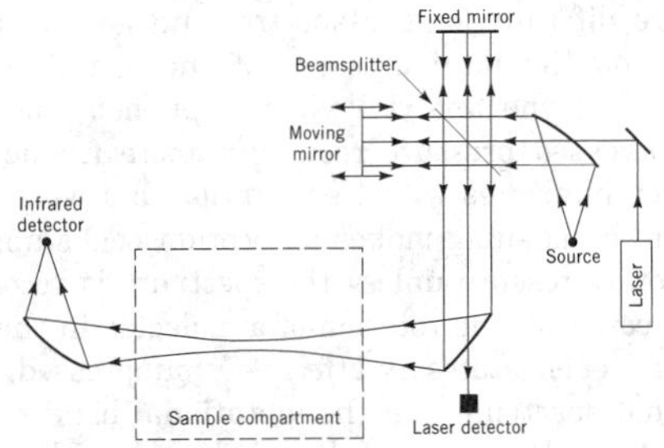

Figure 1. Schematic of the optical layout of a Fourier-transform spectrometer.

Fourier-Transform Infrared Spectrometers. Most Fourier-transform infrared (ftir) spectrometers are based on an interferometer designed by Michelson in 1881. Figure 1 shows the basic structure. Interference at the beamsplitter as the mirror moves modulates each wavelength at a different frequency, allowing the instrument to measure the intensity at each wavelength separately.

Interference Filter Spectrometers. Spectrometers based on bandpass interference filters can be used wherever the higher resolution afforded by grating and interferometer-based instruments is not needed, so these are particularly popular for near infrared applications. Filter spectrometers are generally smaller and more rugged than grating and interferometer instruments, and are particularly useful as industrial and portable instruments. The simplest filter instrument uses a series of bandpass filters mounted on a wheel in front of a detector.

Acousto-Optic Filters. The newest type of spectrometer to become commercially available is the acousto-optic tunable filter (AOTF). An AOTF is a solid-state, electronically tunable bandpass filter based on the diffraction of optical waves by acoustic waves in an optically anisotropic crystal. AOTFs are small, inexpensive, and tune or scan rapidly, but have low resolution.

Spectrometry

The range of photon energies [160–0.12 kJ/mol (38–0.03 kcal/mol)] within the infrared region corresponds to the energies of vibrational and rotational transitions of individual molecules, of electronic transitions in many semiconductors, and of vibrational transitions in crystalline lattices.

Normal Modes, Group Frequencies, and Band Shapes. The number of allowed vibrational modes in a molecule is 3N-6, or 3N-5 for a planar molecule, where N is the number of atoms in the molecule. The allowed vibrational modes involve groups of bonds or all bonds in a molecule, and are called normal modes. The normal modes depend on the symmetry of the molecule and can be identified using the rules of group theory.

Predicting the normal modes of most molecules is very difficult. Certain chemical bonds and functional groups usually produce an infrared peak near the same wavenumber location regardless of the molecule in which they occur. These characteristic band location ranges are called group frequencies, and are strong diagnostic tools for identifying an unknown sample.

Nothing affects the appearance of the infrared spectrum of a substance more than its physical state. A gaseous molecule is free to rotate, so it has a purely rotational far infrared spectrum and mid-infrared vibrational transitions consisting of many sharp rotational lines. Although such densely structured spectra potentially provide the greatest amount of data, these are often more difficult to use analytically than condensed-phase spectra. Group frequency correlations are difficult for gas spectra, and gas-phase linewidths are dependent on the total pressure of the sample (not just the partial pressure of the analyte) via the phenomenon of pressure broadening. Increased pressure results in more frequent molecular collisions, which interferes with free rotation and widens absorption peaks. Pressure broadening makes absorption peaks appear to grow in size with total pressure unless the spectrum is recorded at high resolution. By contrast, the rotation of a molecule in the liquid (neat or solution) and solid states is effectively suppressed. There is no purely rotational spectrum and the vibrational band shapes have a symmetric, statistical appearance that is largely independent of the molecule involved. To a good approximation, the band shape, $A_{\tilde{\nu}}$, fits the Lorentz function:

$$A_{\tilde{\nu}} = \frac{A_{\text{peak}}\gamma^2}{\gamma^2 + (\tilde{\nu} - \tilde{\nu}_0)^2}$$

where A_{peak} and $\tilde{\nu}_0$ are the absorbance and wave number, respectively, at the peak (center) of the band, $\tilde{\nu}$ is the wave number and γ is the half width of the band at half height. Liquid and solid band positions are usually shifted slightly downward from vapor positions.

Transmission, Absorption, and Beer's Law. The majority of infrared spectrometry is still done by the classic method of transmission spectrometry; the intensity of an infrared beam passing completely through a sample is measured. The standard description of how much radiation passes through the sample is that of Beer's law (or the Bouguer-Beer-Lambert law):

$$I = I_0 10^{-acl}$$

where l_0 and l are the intensities of the infrared beam entering and exiting the sample, respectively, a and c are the absorption coefficient (or extinction coefficient or absorptivity) and the concentration of the absorbing species, and l is the path length of the beam in the sample.

Near-Infrared Spectrometry. A renaissance in near infrared spectrometry started in the late 1960s when the near infrared region began to be used for the analysis of moisture, oil, and protein levels in grain and grain products. The strength of near infrared analysis is its ability to handle condensed-phase samples with little or no sample preparation. The near infrared region is good both for samples that are hard to thin or dilute, like agricultural products, and for process-line monitoring, where analysis time must be kept to a minimum.

Sampling Methods. A wide variety of methods exists for the investigation of samples by infrared spectrometry. The choice depends on the nature of the sample, the kind of information desired, and the time available. Transmission, emission, and photoacoustic absorption are applicable to both gases and condensed phases, but the reflection methods (eg., diffuse, specular, and attenuated total reflection) require a sample surface, so they apply only to liquids and solids.

Data Analysis. The computerization of spectrometers and the concommitant digitization of spectra have caused an explosive increase in the use of advanced spectrum analysis techniques. Data analysis in infrared spectrometry is a very active research area and software producers are constantly releasing more sophisticated algorithms. Each instrument maker has adopted an independent format for spectrum files, which has created difficulties in tranferring data. The Joint Committee on Atomic and Molecular Physical Data has developed a universal format for infrared spectrum files called JCAMP-DX. Most instrument makers incorporate in their software a routine for translating their spectrum files to JCAMP-DX format.

The classic method of quantifying using Beer's law is still the preferred method where applicable. A number of multivariate analysis methods have been developed to make use of data from many wave numbers (even the complete spectrum) when a single wave number is inadequate. Most of the methods use a set of calibration spectra (a training set) to generate a model of how the spectral data are related to parameters of the calibration set.

Hyphenated Techniques. Hyphenated techniques are methods in which two or more analytical technologies have been joined together in a single process. Most commonly, the combination is a separation technique with an instrumental analysis technique, in this case, infrared spectrometry, used as a detector. Development of hyphenated infrared techniques has centered on combining a chromatographic method with FTIR. Commercial instruments are available that combine FTIR with gas chromatography (GC-FTIR), high performance liquid chromatography (HPLC-FTIR), and supercritical fluid chromatography (SFC-FTIR). These combine the power of chromatography to handle mixtures with the identification ability of infrared analysis.

Applications. The most ubiquitous use of infrared spectrometry is chemical identification. It has long been an important tool for studying newly synthesized compounds in the research lab, but industrial identification uses cover an even wider range. In many industries ir spectrometry is used to assay feedstocks (qv). The polymer and coating industries use ir spectrometry extensively, measuring everything from composition and cure level, to weathering effects and the structure of multilayer films. The hyphenated techniques are used to analyze complex mixtures, especially for fragrances, flavors and foodstuffs, and for environmental remediation and monitoring.

Radiometry

Radiance, the flow of energy per unit area per unit solid angle through a surface normal to the direction of flow, is the basic radiometric parameter because all other quantities can be determined from it by integration, and because it is constant along any line radiating out from a source. Most radiometry applications depend on detecting temperature changes induced in surfaces. Radiometry can detect flow blockages in pipes, liquid levels in tanks, flaws in internal structures, mechanical stresses in surfaces, and electrical stresses in circuits.

ROGER W. JONES
AMES Laboratory, Iowa State University

G. W. Chantry, *Long-Wave Optics*, Vols. 1 and 2, Academic Press, Inc., Orlando, Fla., 1984.

W. L. Wolfe and G. J. Zissis, eds., *Infrared Handbook*, rev. ed., Environmental Research Institute of Michigan, Ann Arbor, Mich., 1985.

P. R. Griffiths and J. A. de Haseth, *Fourier Transform Infrared Spectrometry*, John Wiley & Sons, Inc., New York, 1986.

P. Coleman, ed., *Practical Sampling Techniques for Infrared Analysis*, CRC Press, Boca Raton, Fla., 1993.

RAMAN SPECTROSCOPY

Photons can interact weakly with a molecule, resulting in the excitation of a molecular vibration, and the subsequent scattering of a new photon at a slightly longer wavelength (lower energy) than the incident photon. This effect is called Raman scattering, and its measurement, Raman spectroscopy. The exciting light is usually a visible or near-infrared (nir) laser.

Although a handful of industrial laboratories have employed Raman spectroscopy for many years, interest has blossomed since the early 1980s. Both the solution of several vexing experimental problems and the development of sophisticated user-friendly instrumentation have contributed to this growth. Specialist reviews of Raman work in most primary areas of application exist, including reviews for inorganic materials, semiconductors (qv) and superconductors, polymers, process control (qv), remote sensing (usually through a fiber-optic probe), and biological systems. Moreover, comprehensive reviews of the most recently published material in the field appear biennially.

As of the 1990s using compact lasers, which provide intense monochromatic light, efficient spectrographs, and sensitive charge-coupled device (CCD) detectors under computer control, Raman spectra can sometimes be recorded in times as short as a few ms and are routinely recorded in a few seconds.

Theory

Every molecule generates its own characteristic vibrational Raman spectrum, which can be used for qualitative identification. Raman scattering arises from interaction between the electromagnetic field of a photon and the electric field of the electron cloud of a molecule. The strength of the scattering depends on the magnitude of the change in deformation of the electron cloud in the external field, caused by a molecular vibration. The magnitude is given by the (spatial) derivative of the molecular polarizability. The fundamental selection rule in Raman spectroscopy is that the vibration must cause a change in the polarizability of the molecule.

Raman and infrared bandwidths, expressed in wave numbers, are usually the same. Both spectroscopies have similar strengths and weaknesses for structure elucidation.

The intensity of Raman-scattered light increases as the fourth power of the frequency of the incident light. Thus spectra are stronger if excited with uv lasers, but for practical reasons most Raman spectroscopy is performed using green or nir lasers.

For most purposes only the Stokes-shifted Raman spectrum, which results from molecules in the ground electronic and vibrational states being excited, is measured and reported. Anti-Stokes spectra arise from molecules in vibrational excited states returning to the ground state. The relative intensities of the Stokes and anti-Stokes bands are proportional to the relative populations of the ground and excited vibrational states. These proportions are temperature-dependent and follow a Boltzmann distribution.

Raman scattering is very inefficient compared to either the Rayleigh scattering or fluorescence. Other tools to increase detection of Raman signal include modern spectrographs which have high throughput, and CCD detectors which have high quantum efficiency and very low dark current.

Sample Preparation

Ease of Sample Preparation. A significant practical advantage of Raman spectroscopy is the ease of sample handling and preparation. Spectra of solids, liquids, and gases can often be obtained without any sample preparation.

Fluorescence Interference. The historical drawback to widespread use of Raman spectroscopy has been the strong fluorescence background exhibited by many materials, even those which are nominally nonfluorescent. This fluorescence often arises from an impurity in the sample, but may be intrinsic to the material being studied. Several methods have proved useful in reducing this background. One of the simplest is sample purification. Another method, called photobleaching, works on robust solids but may cause photodecomposition in many materials. The simplest solution to the fluorescence problem is excitation in the near infrared (750 nm–1.06 μm).

Intensity Enhancement Mechanisms

Resonance Raman Spectroscopy. If the excitation wavelength is chosen to correspond to an absorption maximum of the species being studied, a 10^2–10^4 enhancement of the Raman scatter of the chromophore is observed. This effect is called resonance enhancement or resonance Raman (RR) spectroscopy. There are several mechanisms to explain this phenomenon, the most common of which is Franck-Condon enhancement. RR spectroscopy has been an important biochemical tool, and it may have industrial uses in some areas of pigment chemistry.

Surface-Enhanced Raman Spectroscopy. A second technique for increased sensitivity uses the strong enhancement of the electric field of a light wave at certain rough metal surfaces. This surface-enhanced Raman scattering (sers) results in a 10^3–10^6 increase in signal of molecules in contact with the surface.

Instrumentation.

In a typical Raman experiment the sample is illuminated using a laser and the scattered light is collected through low *f*-number optics and focused into a spectrograph or interferometer.

Modern Raman instruments utilize either a dispersive element (grating) or an interferometer. In most cases the spectrum is analyzed by a grating spectrograph fitted with a CCD array detector.

Lasers having wavelengths ranging from the deep uv to the near infrared have been used in Raman spectroscopy. In industrial laboratories, the most common laser is the Nd:YAG operating at 1.06 μm. Increasingly, diode lasers or other lasers operating in the 750–785-nm region are encountered.

The slow-scan CCD, also called the scientific CCD, or in the spectoscopy literature simply CCD, is the detector of choice for most applications of Raman spectroscopy.

Dispersive Spectrographs. The preferred instrument is a single-stage spectrograph preceded by a Rayleigh line rejection filter. Holographic (Bragg diffraction) filters are most commonly employed, because these offer high (70–80%) throughput and can operate as close as 50 cm^{-1} from the exciting line.

Fourier-Transform Raman Spectroscopy. The first and most popular form of near-infrared Raman spectroscopy employs modified Fourier-transform infrared (ftir) spectrometers, which allow use of Nd:YAG 1064-nm excitation.

Raman Microspectroscopy. Raman spectra of small solids or small regions of solids can be obtained at a spatial resolution of about 1 μm using a Raman microprobe. A widespread application is in the characterization of materials.

Fiber-Optic Probes. Fiber-optic probes provide remote sampling capabilities to Raman instrumentation, are stable, and give reproducible signals. Their historical niche has been in environmental monitoring. More recently these probes have been used in chemical process control and related areas such as incoming materials inspection.

KENNETH A. CHRISTENSEN
ELIZABETH A. TODD
MICHAEL D. MORRIS
University of Michigan

J. G. Grasselli and B. J. Bulkin, eds., *Analytical Raman Spectroscopy*, John Wiley & Sons, Inc., New York, 1991.

B. Schrader, A. Hoffmann, and S. Keller, *Spectrochim. Acta, Part A* **47A**, 1135–1148 (1991).

D. E. Battey, J. B. Slater, R. Wludyka, H. Owen, D. M. Pallister, and M. D. Morris, *Appl. Spectrosc.* **47**, 1913–1919 (1993).

D. B. Chase and J. F. Rabolt, eds., *Fourier Transform Raman Spectroscopy*, Academic Press, New York, 1994.

INITIATORS

FREE-RADICAL INITIATORS

Free-radical initiators are chemical substances that, under certain conditions, initiate chemical reactions by producing free radicals. Initiators contain one or more labile bonds that cleave homolytically when sufficient energy is supplied to the molecule. The energy must be greater than the bond dissociation energy (BDE) of the labile bond. Radicals are reactive chemical species possessing a free (unbonded or unpaired) electron. Radicals may also be positively or negatively charged species carrying a free electron (ion radicals). Initiator-derived radicals are very reactive chemical intermediates and generally have short lifetimes, ie, half-life times less than 10^{-3} seconds.

The principal commercial initiators used to generate radicals are peroxides and azo compounds. Lesser amounts of carbon–carbon initiators and photoinitiators, and high energy ionizing radiation are also employed commercially to generate radicals.

There are three general processes for supplying the energy necessary to generate radicals from initiators: thermal processes, microwave or ultraviolet (uv) radiation processes, and electron transfer (redox) processes. Radicals can also be produced in high energy radiation processes. Once formed, radicals undergo two basic types of reactions: propagation reactions and termination reactions.

Radicals are employed widely in the polymer industry, where their chain-propagating behavior transforms vinyl monomers into polymers and copolymers. The mechanism of addition polymerization involves all three types of reactions discussed above, ie, initiation, propagation by addition to carbon–carbon double bonds, and termination.

Two other important commercial uses of initiators are polymer cross-linking and polymer degradation.

Structure–Reactivity Relationships. Much has been written about the structure–reactivity of radicals. No single unifying concept has satisfactorily explained all radical reactions reported in the literature. A longstanding correlation of structure and reactivity involves comparisons of the energies required to homolytically break covalent bonds to hydrogen. It is assumed that this energy, the hydrogen bond dissociation energy (BDE), reflects the stability and the reactivity of the radical coproduced with the hydrogen atom. However, this assumption should really be limited to radical reactivity and selectivity in hydrogen atom abstraction reactions, and can be particularly misleading for reactions with polar transition states, in which radicals can behave either as nucleophiles or electrophiles. Nevertheless, the correlation of radical reactivity with BDE is quite useful. Table 1 shows some general BDE values for the formation of various carbon and oxygen radicals from various precursors. According to the theory, the higher the BDE, the higher the reactivity and the lower the stability of the radical formed by removal of a hydrogen atom.

The choice of an initiator for a given radical process depends on the reaction conditions and reactivity of the initiator. These two factors must be balanced so that the desired reaction is achieved.

Activation Parameters. Thermal processes are commonly used to break labile initiator bonds in order to form radicals. The amount of thermal energy necessary varies with the environment, but absolute temperature, T, is usually the dominant factor. The energy barrier, the minimum amount of energy that must be supplied, is called activation energy, E_a. A third important factor, known as the frequency factor, A, is a measure of bond motion freedom (translational, rotational, and vibrational) in the activated complex or transition state. E_a and A are known as the activation parameters and, along with T, are related to the decomposition rate, k_d, by the equations:

$$k_d = Ae^{(-E_a/RT)} \text{or} \ln k_d = \ln (A) - E_a/RT$$

Half-Life. Once these activation parameters have been determined for a initiator, half-life times at a given temperature, ie, the time required for 50% decomposition at a selected temperature, and half-life temperatures for a given period, ie, the temperature required for 50% decomposition of an initiator over a given time, can be calculated. Half-life data are useful for comparing the activity of one initiator with another when the half-life data are determined in the same solvent and at the same concentration and, preferably, when the initiators are of the same class.

Commercial initiators are primarily organic and inorganic peroxides, aliphatic azo compounds, certain organic compounds with labile carbon–carbon bonds, and photoinitiators.

Table 1. Bond Dissociation Energies

Precursor	BDE, kJ/mol[a]
$(R)_3C$—H	381
$(R)_2CH$—H	406
RCH_2—H	418
CH_3—H	439
RO—H	439
RCOO—H	444
C_6H_5—H	469
HO—H	498

[a] To convert kJ/mol to kcal/mol, divide by 4.184.

Organic Peroxides

Organic peroxides are compounds possessing one or more oxygen–oxygen bonds. They have the general structure ROOR′ or ROOH, and decompose thermally by the initial cleavage of the oxygen–oxygen bond to produce two radicals:

$$\text{ROOR}' \rightarrow \text{RO}\cdot + \cdot\text{OR}'$$

Following radical generation, the radicals produced (RO· and R′O·) can initiate the desired reaction. However, when the radicals are generated in commercial applications, they are surrounded by a solvent, monomer, or polymer "cage." When the cage is solvent, the radical must diffuse out of this cage to react with the desired substrate. When the cage is monomer, the radical can react with the cage wall or diffuse out of the cage. When the cage is polymer, reaction with the polymer can occur in the cage. Unfortunately, other reactions can occur within the cage and can adversely affect efficiency of radical generation and radical reactivity. If the solvent reacts with the initiator radical, then solvent radicals may participate in the desired reaction.

Two secondary propagating reactions often accompany the initial peroxide decomposition: radical-induced decompositions and β-scission reactions. Both reactions affect the reactivity and efficiency of the initiation process.

Approximately 100 different organic peroxide initiators, in well over 300 formulations, are commercially produced throughout the world, primarily for the polymer and resin industries.

The eight classes of organic peroxides that are produced commercially for use as initiators are listed in Table 2. Included are the 10-h half-life temperature (ie, the temperature at which 50% of the peroxide decomposes in 10 h) ranges for the members of each peroxide class.

Inorganic Peroxides

Inorganic peroxide–redox systems have been employed for initiating emulsion homo- and copolymerizations of vinyl monomers. These

Table 2. Commercial Organic Peroxide Classes

Organic peroxide class	Structure[a]	10-h $t_{1/2}$[b,c], °C
diacyl peroxides	R—C(=O)—OO—C(=O)—R	21–75
dialkyl peroxydicarbonates	RO—C(=O)—OO—C(=O)—OR	49–51[d]
tert-alkyl peroxyesters	R—C(=O)—OO—*t*-R	38–107
OO-tert-alkyl *O*-alkyl monoperoxycarbonates	RO—C(=O)—OO—*t*-R	99–100
di(*tert*-alkylperoxy)ketals	R′(R)C(OO—*t*-R)$_2$	92–110
di-*tert*-alkyl peroxides	*t*-R—OO—*t*-R	115–128
tert-alkyl hydroperoxides	*t*-R—OO—H	[e]
ketone peroxides	HOO—C(R′)(R)—(OO—C(R′)(R))$_x$—OOH + other structures	[e]

[a] $x = 0$ or 1. [b] Temperature at which $t_{1/2} = 10$ h. [c] In benzene, unless otherwise noted. [d] In trichloroethylene (TCE). [e] Not applicable.

Table 3. Commercial Azo Initiators

Name	Structure
2,2′-azobis[4-methoxy-2,4-dimethyl]-pentanenitrile	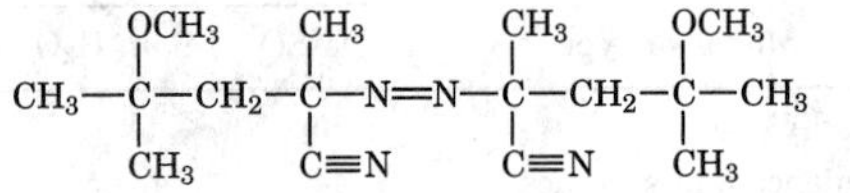
2,2′-azobis[2,4-dimethyl]-pentanenitrile	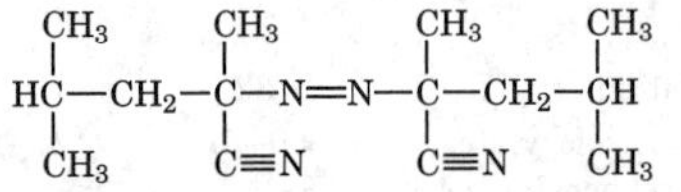
2,2′-azobis[iso-butyronitrile]	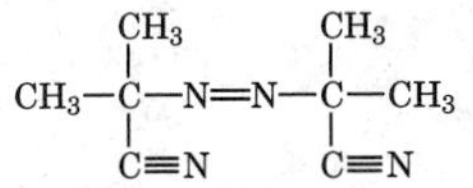
2,2′-azobis[2 methyl-butyronitrile]	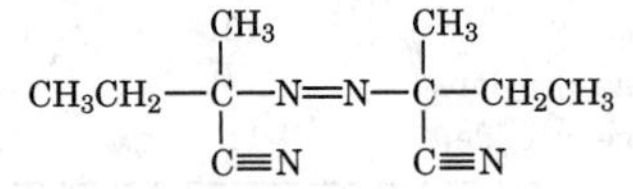
1,1′-azobis[cyclo-hexanecarbonitrile]	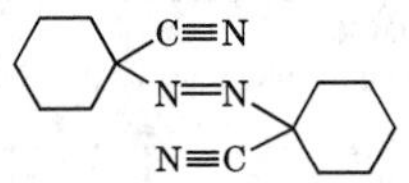
4,4′-azobis[4-cyanova-leric acid]	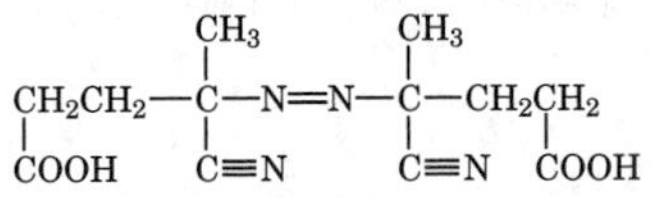
dimethyl-2,2′-azobis-[2-methylpropionate]	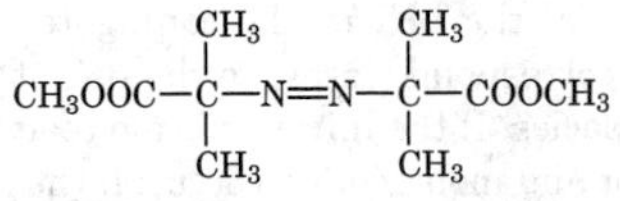
azobis[2-acetoxy-2-propane]	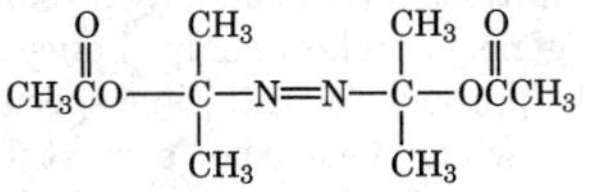
2,2′-azobis[2-amidinopro-pane]dihydrochloride	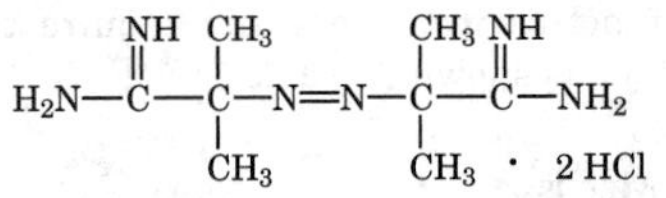

systems include hydrogen peroxide–ferrous sulfate, hydrogen peroxide–dodecyl mercaptan, potassium peroxydisulfate–sodium bisulfate, and potassium peroxydisulfate–dodecylmercaptan. Potassium peroxydisulfate, KSO (or the corresponding sodium or ammonium salt), is an inorganic peroxide that is used widely in emulsion polymerization (eg, latexes, rubbers, etc), usually in combination with a reducing agent.

When handling and using peroxide initiators, care should be exercised because they are thermally sensitive and decompose (sometimes violently) when exposed to excessive temperatures, especially where they are in their pure or highly concentrated states. However, they are useful as initiators because of their thermal instability. What may be a safe temperature for one peroxide can be an unsafe temperature for another, since peroxide initiators encompass a wide activity range. Because some peroxides are shock- or friction-sensitive in the pure state, they are generally desensitized by formulating them into solutions, pastes, or powders with inert diluents. All manufacturers' literature should be carefully scrutinized and the peroxide safety literature should be reviewed before handling and using specific peroxide initiator compositions.

Azo Compounds

Generally, the commercially available azo initiators are of the symmetrical azonitrile type:

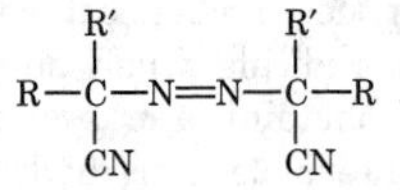

The symmetrical azonitriles are solids with limited solubilities in common solvents. Some commercial aliphatic azo compounds and their 10-h half-life temperatures are listed in Table 3.

Care should be exercised in handling and using azo initiators in their pure and highly concentrated states, because they are thermally sensitive and can decompose rapidly when overheated. Although azonitriles are generally less sensitive to contaminants, the same cautions that apply to peroxides also should be applied to handling and using azo initiators. The manufacturers' safety literature should be read carefully. The potential toxicity hazards of decomposition products must be considered when using azonitriles.

Carbon–Carbon Initiators

Carbon–carbon initiators are hexasubstituted ethanes that undergo carbon–carbon bond scission when heated to produce radicals. The thermal stabilities of the hexasubstituted ethanes decrease rapidly as the sizes of the alkyl groups increase. The 10-h half-life temperature range of this class of initiators is very broad, extending from about 100°C to well above 600°C. An extensive compilation of half-life data on carbon–carbon initiators has been published. The commercially available carbon–carbon initiators are tetrasubstituted 1,2-diphenylethanes which undergo homolyses to generate low energy, *tert*-aralkyl radical pairs. Three carbon–carbon initiators are currently available commercially, 2,3-dimethyl-2,3-diphenylbutane (**1**), 3,4-dimethyl-3,4-diphenylhexane (**2**), and 1,1,2,2-tetraphenyl-1,2-bis(trimethylsiloxy)ethane (**3**).

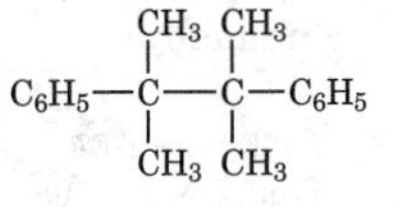

(**1**)

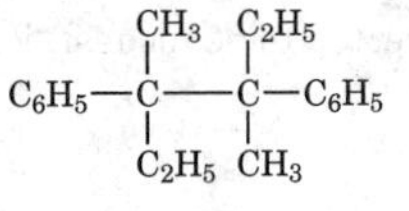

(**2**)

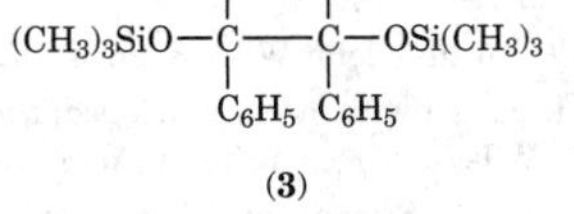

(**3**)

Other Radical Generating Systems

There are many chemical methods for generating radicals reported in the literature that do not involve conventional initiators. Most of these radical-generating systems cannot broadly compete with the use of conventional initiators in industrial polymer applications owing to cost or efficiency considerations. However, some systems may be well-suited for initiating specific radical reactions or polymerizations, eg, grafting of monomers to cellulose using ceric ion.

Initiation Through Radiation and Photoinitiators

High energy ionizing radiation sources (eg, x-rays, γ-rays, α-particles, β-particles, fast neutrons, and accelator-generated electrons) can generate radical sites on organic substrates. If the substrate is a vinyl monomer, radical polymerization can occur. If the substrate consists of a polymer and a vinyl monomer, then polymer cross-linking, degradation, grafting of the monomer to the polymer, and homopolymerization of the monomer can all occur. Radical polymerizations of vinyl monomers with ionized plasma gases have been reviewed.

Initiation of radical reactions with uv radiation is widely used in industrial processes. In contrast to high energy radiation processes where the energy of the radiation alone is sufficient to initiate reactions, initiation by uv irradiation usually requires the presence of a photoinitiator, ie, a chemical compound or compounds that generate initiating radicals when subjected to uv radiation. There are two types of photoinitiator systems: those that produce initiator radicals by intermolecular hydrogen abstraction [benzophenone, 4-phenylbenzophenone, xanthone, thioxanthone, 2-chlorothioxanthone, 4,4′-bis(*N,N′*-dimethylamino)benzophenone (Michler's ketone), benzil, 9,10-phenanthraquinone, and 9,10-anthraquinone] and those that produce initiator radicals by photocleavage [α,α-dimethyl-α-hydroxyacetophenone, (1-hydroxycyclohexyl)-phenylmethanone, benzoin ethers (methyl, ethyl, isobutyl), α,α-dimethoxy-α-phenylacetophenone, α,α-diethoxyacetophenone, 1-phenyl-1,2-propanedione, 2-(*O*-benzoyl)oxime, diphenyl(2,4,6-trimethylbenzoyl)phosphine oxide, and α-dimethylamino-α-ethyl-α-benzyl-3,5-dimethyl-4-morpholinoacetophenone].

Economic Aspects

The principal worldwide producers of organic peroxide initiators (and their trade names) include Elf Athochem (Luperco, Luperox, Lupersol, Lucidol (U.S.), Luchem, Alperox, Decanox, Peroximon, and Retilox), Akzo-Nobel (Trigonox, Perkadox, Cadox, Cadet, Laurox, Liladox, Kenodox, Lucidol (Europe), Butanox, and Cyclonox), Aztec (Aztec), Peroxid-Chemie (Interox), Witco (Esperox, Esperal, USP, Quickset, and Hi Point), Nippon Oil & Fats Company (Nyper, Perbutyl, Percumyl, Perhexa, Permek, and Peroyl), Norac (Superox), Hercules (DiCup, VulCup), and Sanken Kako (Sanperox). The principal worldwide producers of organic azo initiators are DuPont (Vazo), Elf Atochem (Ficel), and Wako. The worldwide market for organic azo initiators is small, being only about 10% of the market for organic peroxide initiators. Ciba Geigy is a significant supplier of photoinitators (Darocur, Irgacure). The market for these initiators has been reviewed. Because most of the consumption of organic peroxides and azo initiators is in the developed countries, market growth in the 1990s is expected to be modest, ie, 2–3% annually.

José Sanchez
Terry N. Myers
Elf Atochem North America, Inc.

D. Swern, ed., *Organic Peroxides*, Vols. I, II, and III, Wiley-Interscience, New York, 1970, 1971, and 1972.

W. Ando, ed., *Organic Peroxides*, John Wiley & Sons, Inc., New York, 1992.

C. S. Sheppard, in J. I. Kroschwitz, ed., *Encyclopedia of Polymer Science and Engineering*, Vol. 2, Wiley-Interscience, New York, 1985, pp. 143–157.

S. P. Pappas, ed., *UV Curing: Science and Technology*, Technology Marketing Corporation, Stamford, Conn., 1978, Chapt. 1.

ANIONIC INITIATORS

In anionic polymerization, the reactive propagating intermediate generated by the initiation reaction is an anion, ie, a species which carries a formal negative charge, with a corresponding positively charged counterion. In living anionic polymerization, the kinetic steps of chain termination and chain transfer are absent. This unique aspect of many anionic polymerizations provides a methodology for preparing polymers with control of the significant variables affecting polymer properties including molecular weight, molecular weight distribution, block copolymer composition, and microstructure, as well as molecular architecture (linear, branched, and cyclic macromolecules). An important consideration for preparation of polymers with well-defined structures and low degrees of compositional heterogeneity is the choice of a suitable initiator.

In general, an appropriate initiator is a species which has approximately the same structure and reactivity as the propagating anionic species, ie, the pK_a of the conjugate acid of the propagating anion should correspond closely to the pK_a of the conjugate acid of the initiating species. If the initiator is too reactive, side reactions between the initiator and monomer can occur; if the initiator is not reactive enough, then the initiation reaction may be slow or inefficient.

Table 1. Relationships Between Monomer Reactivity, Carbanion pK_a Stability, and Suitable Initiators

Monomer type	pK_a[a] In DMSO[b]	pK_a[a] In H_2O	Initiators[c]
ethylene	56		RLi
dienes and styrenes	44		NH_2^-, RLi, RMt
	43		aromatic radical anions,[d] cumyl K,Mt,
acrylonitrile	32		RMgX
alkyl methacrylates, alkyl acrylates	30–31	27–28	fluorenyl−, $RArC^{-2}$, ketyl radical anions[e]
vinyl ketones	26	19	
oxiranes	29–32	16–18	RO−
thiiranes	17	12–13	
nitroalkenes	17	10–14	
siloxanes		10–14	RO−,OH−
β-lactones	12	4–5	RCOO−
alkyl cyanoacrylates	12.8		HCO^{3-}, H_2O
vinylidene cyanide	11	11	

[a] pK_a of the conjugate acid of the anionic propagating intermediate.
[b] pK_a values in DMSO.
[c] Mt refers generally to alkali metals (Li, Na, K, Rb, Cs).
[d] For example, naphthalene radical anion 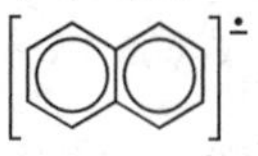 with counterion (Li+, Na+, K+).
[e] $Ar_2\dot{C}O$.

The general relationship between monomer structural type, pK_a and appropriate initiating species is shown in Table 1. Those monomers which form the least stable anions, ie, which have the largest values of pK_a for the corresponding conjugate acids, are the least reactive monomers and require the use of the most reactive initiators as shown in Table 1.

Alkali Metals

The use of alkali metals for anionic polymerization of diene monomers is primarily of historical interest. The electron-transfer mechanism of the anionic polymerization of styrenes and 1,3-dienes initiated by alkali metals has been described in detail; the dimerization of radical anion intermediates is the important step.

Aromatic Radical Anions

Many aromatic hydrocarbons react with alkali metals in polar aprotic solvents to form stable solutions of the corresponding radical anions. These solutions can be analyzed by uv-visible spectroscopy and stored for further use.

Sodium naphthalene and other aromatic radical anions react with monomers such as styrene by reversible electron transfer to form the corresponding monomer radical anions which rapidly dimerize.

Monomers which can be polymerized with aromatic radical anions include styrenes, dienes, epoxides, and cyclosiloxanes. Aromatic radical anions which are too stable do not efficiently initiate polymerization of less reactive monomers; thus the anthracene radical anion cannot initiate styrene polymerization.

Alkyllithium Compounds

Anionic polymerization of vinyl monomers can be effected with a variety of organometallic compounds; alkyllithium compounds are the most useful class. A variety of simple alkyllithium compounds are available commercially. Most simple alkyllithium compounds are soluble in hydrocarbon solvents such as hexane and cyclohexane and they can be prepared by reaction of the corresponding alkyl chlorides with lithium metal.

Simple alkyllithium compounds are aggregated in solution, in the solid state, and even in the gas phase. The important differences between the various alkyllithium compounds are their degrees of aggregation in solution and their relative reactivity as initiators for anionic polymerization of styrene and diene monomers.

The kinetics of initiation reactions of alkyllithium compounds often exhibit fractional kinetic order dependence on the total concentration of initiator, consistent with initiation by the unassociated form of the alkyllithium.

The use of aliphatic solvents causes profound changes in the observed kinetic behavior for the alkyllithium initiation reactions with styrene, butadiene, and isoprene, ie, the inverse correspondence between the reaction order dependence for alkyllithium and degree of organolithium aggregation is generally not observed. Also, initial rates of initiation in aliphatic solvents are several orders of magnitude less than those observed, under equivalent conditions, in aromatic solvents. Furthermore, pronounced induction periods are observed in aliphatic hydrocarbon solvents.

The relative reactivities of alkyllithiums as polymerization initiators are intimately linked to their degree of association. In the following the average degree of association in hydrocarbon solution, where known, is indicated in brackets after the alkyllithium. For styrene polymerization, the relative reactivity of alkyllithium initiators is menthyllithium $> sec\text{-}C_4H_9Li > i\text{-}C_3H_7Li > i\text{-}C_4H_9Li > n\text{-}C_4H_9Li > t\text{-}C_4H_9Li$. For diene polymerization, menthyllithium $> sec\text{-}C_4H_9Li > i\text{-}C_3H_7Li > t\text{-}C_4H_9Li > i\text{-}C_4H_9Li > n\text{-}C_4H_9Li$.

Alkyllithium compounds are primarily used as initiators for polymerizations of styrenes and dienes.

Quantitative Analysis of Alkyllithium Initiator Solutions. The amount of carbon-bound lithium is calculated from the difference between the total amount of base determined by acid titration and the amount of base remaining after the solution reacts with either benzyl chloride, allyl chloride, or ethylene dibromide.

Copolymerization Initiators. The copolymerization of styrene and dienes in hydrocarbon solution with alkyllithium initiators produces a tapered block copolymer structure because of the large differences in monomer reactivity ratios for styrene ($r_s < 0.1$) and dienes ($r_d > 10$). In order to obtain random copolymers of styrene and dienes, it is necessary to either add small amounts of a Lewis base such as tetrahydrofuran or an alkali metal alkoxide (MtOR, where Mt = Na, K, Rb, or Cs).

Difunctional Initiators

These initiators are of considerable interest for the preparation of triblock copolymers, telecheclic polymers, and macrocyclic polymers.

Aromatic radical anions, such as lithium naphthalene or sodium naphthalene, are efficient difunctional initiators. However, the necessity of using polar solvents for their formation and use limits their utility for diene polymerization.

The methodology for preparation of hydrocarbon-soluble, dilithium initiators is generally based on the reaction of an aromatic divinyl precursor with two moles of butyllithium.

Although a plethora of divinyl aromatic compounds have been investigated as precursors for hydrocarbon-soluble dilithium initiators, the only system which has been demonstrated to produce a hydrocarbon-soluble dilithium initiator is based on 1,3-bis (1-phenylethenyl)benzene.

Functionalized Initiators

The use of alkyllithium initiators which contain functional groups provides a versatile method for the preparation of end functionalized polymers and macromonomers. For a living anionic polymerization, each functionalized initiator molecule produces one macromolecule with the functional group from the initiator residue at one chain end and the active carbanionic propagating species at the other chain end.

Other Initiators

Other initiators include cumyl potassium, 1,1-diphenylmethylcarbanions, fluorenyl carbanions, enolate initiators, and alkoxide-type initiators.

Health and Safety Factors

Hydrocarbon solutions of alkyllithium compounds are air and moisture sensitive and should be either handled in an inert atmosphere or by using syringes using recommended procedures for handling air-sensitive compounds. Alkyllithium reagents react with acidic compounds that contain reactive hydrogens such as water, alcohols, phenols, acids, and even primary and secondary amines. The reaction of butyllithium with water produces butane and lithium hydroxide, which can lead to spontaneous ignition in the presence of oxygen. Contact of alkyllithium solutions with air does not generally lead to spontaneous ignition; however, if large surface areas are formed, for example in a spill, spontaneous ignition can occur. Carbon dioxide fire extinguishers must not be used because carbon dioxide reacts exothermically with alkyllithium compounds. It is prudent to have an all-purpose fire extinguisher available when working with these organometallic compounds. Suitable fire-extinguishing chemicals include powdered limestone and powders containing sodium chloride and sodium bicarbonate.

RODERIC P. QUIRK
University of Akron
VICTOR M. MONROY
General Tire, Inc.

M. Morton, *Anionic Polymerization: Principles and Practice*, Academic Press, Inc., New York, 1982.

M. Szwarc, *Carbanions, Living Polymers and Electron Transfer Processes*, Wiley-Interscience, New York, 1968.

S. Bywater, in J. I. Kroschwitz, ed., *Encyclopedia of Polymer Science and Engineering*, Vol. 2, 2nd ed., John Wiley & Sons, Inc., New York, 1985.

M. Fontanille, in G. C. Eastmond and co-eds., *Comprehensive Polymer Science*, Vol. 3, *Chain Polymerization I*, Pergamon Press, Elmsford, N.Y., 1989.

CATIONIC INITIATORS

Cationic polymerization may be induced by a variety of physical (high energy radiation, direct or indirect uv radiation, electroinitiation) and chemical methods (protic acids, Friedel-Crafts acids, stable cation salts, cation donor in conjunction with a Friedel-Crafts acid). The most important initiating system is the cation donor (initiator)/Friedel-Crafts acid (coinitiator) system, which has found many applications. Butyl rubber, a copolymer of isobutylene and isoprene containing 0.5–2.5% isoprene to make vulcanization possible, is the most important commercial polymer made by cationic polymerization. Another important commercial application of cationic polymerization is the manufacture of polybutenes, low molecular weight copolymers of isobutylene, and a smaller amount of other butenes used in adhesives, sealants, lubricants, viscosity improvers, etc.

Unless one is working with superdried systems or in the presence of proton traps, adventitious water is always present as a proton source. Polymerization rates, monomer conversions, and to some extent polymer molecular weights are dependent on the amount of protic impurities; therefore, well-established drying methods should be followed to obtain reproducible results.

In place of a proton source, ie, a Brønsted acid, a cation source such as an alkyl halide, ester, or ether can be used in conjunction with a Friedel-Crafts acid. Initiation with the ether-based initiating systems in most cases involves the halide derivative which arises upon fast halidation by the Friedel-Crafts acid, MX_n. The efficiency of the initiator/coinitiator system depends greatly on the monomer in question. As a general rule, the stability (reactivity) of the initiating cation should be close to that of the propagating chain end. Since initiation involves two subsequent events, ie, ion generation and cationation, species on the two extremes are less active or may be completely inactive, because they form ionic species very slowly and/or in extremely low concentration, or would form ions in high concentration that are, however, too stable to cationate the monomer.

The activity of an initiating system is also affected by the nature of the Friedel-Crafts acid. The following Friedel-Crafts acidity scale can be established: $BF_3 < AlCl_3 < TiCl_4 < BCl_3 < SbF_5 < SbCl_5 < BBr_3$. The advantage of the $TiCl_4$ and the aluminum-based systems is their relative insensivity toward solvent polarity. The activity of the BCl_3- or BBr_3-based system is greatly solvent-dependent, ie, sufficient activity only occurs in polar solvent.

Solvent polarity and temperature also influence the results. The dielectric constant and polarizability, however, are of little predictive value for the selection of solvents relative to polymerization rates and behavior. Evidently every system has to be examined independently. In cationic polymerization of vinyl monomers, chain transfer is the most significant chain-breaking process. The activation energy of chain transfer is higher than that of propagation; consequently, the molecular weight of the polymer increases with decreasing temperature.

Initiation by a carbocation source provides control of the head-group (controlled initiation) when used in conjunction with a Friedel-Crafts acid (eg, $(C_2H_5)_3Al$, $(CH_3)_3Al$, $(C_2H_5)_2AlCl$, BCl_3 for isobutylene, or I_2 and zinc halides for vinyl ethers) where chain transfer to monomer is absent or negligible, or in the presence of a proton trap to abort chain transfer to monomer. That is, initiation from tertiary, allylic, and benzylic halides gives rise to macromolecules carrying tertiary, allylic, and benzylic head-groups. Initiation by halogens results in head-groups carrying the halogen. Controlled initiation, however, is achieved only when polymer formation from adventitious protic impurities is also absent or negligible.

A special case of controlled initiation is the inifer method. The word inifer (from *ini*tiator trans*fer* agents) describes compounds that function simultaneously as initiators and as chain transfer agents. The inifer technique provided the first carbocationic route toward the synthesis of telechelic (α,ω functional) polymers.

Although it was long believed that most Friedel-Crafts acids, particularly halides of boron, titanium, and tin, require an additional cation source to initiate polymerization, recent results show that in many systems Friedel-Crafts acids alone are able to initiate cationic polymerization. The mechanism of initiation appears to be halometalation, as originally suggested by Sigwalt and Olah.

Many initiating systems used in the cationic polymerization of vinyl monomers can also be used to initiate ring-opening polymerization of cyclic monomers such as cyclic ethers, acetals, lactams, lactones, and siloxanes. Polymerization of cyclic monomers may involve different type of ionic as well as covalent growing species. Under certain conditions, termination processes may be absent. The polymerization of cyclic monomers, however, is almost always complicated by inter- and intramolecular chain transfer to polymer. The later results in cyclic oligomer formation. The extent of cyclic oligomer formation can be minimized in the polymerization of epoxides by the recently discovered activated monomer mechanism. Cyclic ether and acetal polymerizations are also important commercially. Polymerization of tetrahydrofuran is used to produce polyether diol, and polyoxymethylene, an excellent engineering plastic, is obtained by the ring-opening polymerization of trioxane with a small amount of cyclic ether or acetal comonomer to prevent depolymerization.

Recently a variety of initiating systems have been described that allow not only controlled initiation but also controlled propagation in the polymerization of vinyl monomers. In these living polymerization systems, chain braking (chain transfer and irreversible termination) is absent. The key to these living polymerizations is the high stability of the growing end, where the nucleophilic counteranion interacts strongly with the cationic active site. Living polymerizations have also been reported with initiating systems, forming nonnucleophilic counteranions in the presence of added Lewis bases (electron donors) and, in the presence of common ion salts, shifting the ionic dissociation equilibrium toward the nondissociated species. With these systems, rapid advances have been made toward the synthesis of well-defined materials with controlled architecture, molecular weight, molecular weight distributions and end-functionalities by cationic polymerization.

Since the discovery of living cationic systems, cationic polymerization has progressed to a new stage, where the synthesis of designed materials is now possible.

RUDOLF FAUST
University of Massachusetts, Lowell

J. P. Kennedy and E. Marechal, *Carbocationic Polymerization*, Wiley-Interscience, New York, 1982.

J. P. Kennedy and B. Ivan, *Designed Polymers by Carbocationic Macromolecular Engineering*, Hanser Publishers, Munich, Germany, 1991.

INKS

Writing inks differ from printing inks in that the latter are generally applied to a substrate by means of a printing press. Printing inks as supplied to the graphic arts industry are used in much greater volume by far as compared to writing inks. This article is divided into a discussion of printing inks, followed by some miscellaneous categories of ink, including ink for ball-point pens, with which the greatest amount of ink writing is done. The number of printing-ink manufacturing establishments in the United States is approximately 450. This includes some 50 captive ink plants.

Printing Inks

Printing ink is a mixture of coloring matter dispersed or dissolved in a vehicle or carrier, which forms a fluid or paste which can be printed on a substrate and dried. The colorants used are generally pigments, toners, dyes or combinations of these materials, which are selected to provide the desired color contrast with the background on which the ink is printed. The vehicle used acts as a carrier for the colorant to the substrate. Printing inks are applied in thin films on many substrates such as paper, paper board, metal sheets and metallic foil, plastic films and molded plastic articles, textiles, and glass. Printing inks can be designed to have decorative, protective, or communicative functions. In some cases, combinations of these functions are achieved.

There are four principal classes of printing ink, which vary considerably in physical appearance, composition, method of application, and drying mechanism. These also fall into two general types of consistency or viscosity, paste and liquid. The classes are letter press and lithographic (litho) inks, which are called paste inks, and flexographic (flexo) and rotogravure (gravure) inks, which are called liquid inks.

The four key properties of inks are drying, rheology (or flow), color, and end use properties. Use properties are those considerations that determine how printed substrates function throughout all processing and usage from the time of printing throughout the useful life of the printed product.

Other Ingredients. Other ingredients in inks, besides pigments and dyes, include driers, waxes, antioxidants, and miscellaneous additives (lubricants, surfactants, thickeners, gellants, defoamers, and preservatives).

Letterpress and Litho Newsprint Inks

The U.S. news ink industry, with a sales potential of over $350 million annually, represents a dynamic, ever-changing segment of the graphic arts market. A gamut of products available commercially meet the needs of technical specifications, environmental trends, and quality requirements. From the three printing modes currently available, the web offset lithographic process and water-based flexo continue to grow at the expense of letterpress printing. The change in the market share is governed by the growing demand for print quality, which the letterpress process cannot deliver.

The printing of newspapers is conducted at very high speeds, often reaching 3000 feet per minute. Inks dry by absorption of liquid into the porosity of the substrate. Evaporation of water in a flexo publication ink can accelerate the drying process.

Web Offset. This is, by far, the largest type of newspaper printing. Three different press ink feed configurations used by the industry, ie, open fountain, injector, and keyless, require inks with specific rheological characteristics. Web offset lithographic printing uses planographic, aluminum-based printing plates, fountain solution, and an ink formulated to accept and properly emulsify water.

Web offset inks are highly pigmented, to yield the desired print density at a thin printing film (1 μm). The viscosity of offset inks is relatively high, but varies with the press configuration. Inks of two distinct chemistries are used in this process: a traditional type based on mineral oil, and a newer one containing soya bean oil. The resins employed are selected from low cost nonfunctional hydrocarbon-type resins, more complex hydrocarbon resins modified with rosin and/or phenolics, and gilsonite (for black only).

Recently developed low rub blacks offer smudge-resistant print. Their share of the market is growing rapidly.

A variety of additives are used to control the properties of wetting and dispersion of pigments, flow, lithography, and rub-off of inks. These additives belong to classes of materials such as surfactants, bentonite clays, alkyds, functional resins, polymers, etc.

Letterpress. This is the oldest printing process still in use. Inks in the printing process are transferred directly from a raised area to a substrate. The printing plates contain a thick layer of photopolymer deposited over a plastic or aluminum base.

Basic raw materials for letterpress inks, such as mineral oils, soya bean oil, resins, and pigments, are essentially the same as those used in web offset inks. Inks are tinctorially weaker, relatively fluid, and their low and high shear viscosities are low.

Flexo. Printing is conducted with a printing plate similar to letterpress. However, the chemistry of the photopolymer is somewhat different in order to make the plate water-insensitive. A high quality of print has been demonstrated by several newspapers utilizing this printing process. The printed matter is virtually smudge-free.

Typical inks are water-based, with acrylic emulsion resins as the main binder. Inks of this type occasionally use natural products such as starches, lignins, and lignin derivatives. Hence, ecologically, this process is more desirable. Press ready inks are very fluid and of low viscosity. Inks contain a variety of additives for the elimination of foaming, dispersion of pigments, rheological modifiers, slip agents, etc.

Web Heat-Set Publication and Commercial Inks. Almost all heat-set inks are now printed on web offset presses, and are based on vehicles containing synthetic resins and/or some natural resins. These are dissolved in hydrocarbon solvent fractions which are specially fractionated for use in the ink industry. They dry in less than one second by means of solvent evaporation in a heatset oven.

Sheet-Fed Offset Inks. Inks for these presses are based on vehicles containing phenolic-modified, maleic-modified, or unmodified resins dissolved in solvents. Some inks also contain alkyds, which may be modified with other polymers, such as urethanes, styrene, and the like. On coated stocks, many sheet-fed inks quick set to a tack-free state by precipitation and solvent separation and then dry fully by oxidation. The most commonly used oils are linseed, soya, and tall oils. Special acrylic resins have been developed for use in quickset inks, and offer nonskinning properties and excellent press stability.

Duplicator and Business Form Inks. The inks can contain drying oil alkyds along with hydrocarbon resins and high boiling (200–370°C) hydrocarbon solvents. Business form inks closely resemble the lithographic heatset or quickset inks. Business forms have also been printed by the ink jet method. These inks are usually based on water, glycols, and dyes.

Folding-Carton Inks. The majority of folding-carton inks are based on various quickset vehicles, as described above. However, when maximum gloss, good rub, and product resistance are required, they contain mainly oleoresinous vehicles. They dry by oxidation to form tough, glossy films. Ultraviolet light-cured inks are also used in litho printing of high quality folding cartons.

Metal Container Inks. Ink vehicles for metal containers that are printed on special flat sheet-fed litho presses are based mainly on blends of oleoresinous varnishes containing alkyds, polyesters, and melamine resins. These inks dry during a 5–15 minute cure at 150–250°C in long gas-fired ovens. Polymerization, oxidation, and cross-linking reactions accounts for their drying and hardening.

The principal method of decorating cans is printing in the round. Ink vehicles for letterset printing of two-piece aluminum or steel containers are mainly based on special polyester vehicles used in conjunction with melamine cross-linkers.

Plastics. Vehicles in offset inks for plastics (polyethylene, polystyrene, vinyl) are based on hard drying oleoresinous varnishes which sometimes are diluted with hydrocarbon solvents. Uv inks are widely used for decoration of these preformed plastic containers.

Manufacture. Paste inks are produced in two ways: *(1)* by mixing predispersed (preground) or flushed pigment concentrates with vehicles, solvents, oils, and compounds, and filtering, or *(2)* by mixing dry pigments or resin-coated pigments with vehicles and compounds and then dispersing them with various types of ink mills. The more fluid inks (news, flexo, or gravure) usually are delivered in tank trucks directly to the printer. Ink vehicles are usually produced in separate resin/varnish plants.

Control of inks is done by examining their color strength, hue, tack, rheology, drying rate, stability, and product resistance. Weather-Ometers, Fade-Ometers, glossmeters, printability testers, colorimeters, spectrophotometers, viscometers, rub testers, and gas chromatographs are employed to check production batches or to pretest new submissions or raw materials. Proofing presses and sometimes pilot presses are utilized by ink manufacturers to control production and test new formulations.

Flexographic and Rotogravure Inks Flexo and gravure inks are both known as liquid inks because of their low viscosity. The inks for both systems have basic components in common with inks for other printing processes. Vehicles disperse and carry the pigment, and also contribute most to the end use properties. Colorants provide color. Solvents dissolve resins in the vehicle and determine drying rate. Additives modify ink properties to overcome deficiencies.

The vehicle is composed of resins, solvent, and additives. Solvents are required for two reasons. The first requirement a solvent must satisfy is to dissolve the resin; this results in a low viscosity ink suitable for printing. Secondly, the solvent must evaporate quickly and completely from the printed film.

Both flexo and gravure inks are delivered in the form of a virgin ink concentrate, which retards the speed of pigment settling and reduces shipping costs. Solvent is used press-side to reduce the ink to a correct printing viscosity.

Additives are used to provide a specific property. For example, a wax provides rub resistance in the printed film or a surfactant reduces foam generation in the fountain.

Manufacture. Manufacturing processes consist of two general operations, vehicle preparation and pigment dispersion.

Rotogravure Inks. Because there are no rubber or plastic components in contact with the solvents contained in gravure ink formulations, it is permissible to use solvents such as ketones and aromatic hydrocarbons which cannot be tolerated in flexo inks. This provides the gravure ink formulator with much greater latitude in regard to binder selection. In other respects the compositions generally are similar.

There are 10 gravure ink types categorized by the binders or solvents used: A, aliphatic hydrocarbon; B, aromatic hydrocarbon; C, nitrocellulose; D, polyamide resins; E, SS nitrocellulose; M, polystyrene; T, chlorinated rubber; V, vinyls; W, water-based; and X, miscellaneous.

Ketones and esters are required for C-type inks. The usual solvent for D-type inks are mixtures of an alcohol, such as ethyl alcohol or isopropyl alcohol, with either aliphatic or aromatic hydrocarbons. The alcohols, proprietary denatured ethyl alcohol and isopropyl alcohol, are commonly used for E-type inks. Aromatic hydrocarbon solvents are used for M-type inks. T-type inks are also reduced with aromatic hydrocarbons. Ketones are commonly used solvents for V-type inks.

Approximately 50% of all flexographic inks use water as their primary solvent and diluent. They contain vehicles based on either acrylic emulsions, or hydrosols or an alkali-soluble rosin ester having a high acid number.

The main advantages of water inks include environmental desirability, excellent press stability, printing quality, heat resistance, absence of fire hazard, and the convenience and economy of water for reduction and wash-up.

The majority of Type A and Type B inks are used for gift wraps, newspaper supplements, catalogues, advertising inserts, and similar publication work. Inks in the Type C group are used for printing on foil, paper, cellophane, paperboard, coated and uncoated paper, glassine, acetate, metallized paper, and some specialized fabrics. Type D inks have excellent adhesion to many plastic films. They are used in foil, paper, and paperboard as well as on a variety of films. Type E inks are often used on paper and paperboard, some grades of cellophane, shellac or nitrocellulose primed foil pouch stock glassine, and many specialty coated papers and boards.

Water inks are primarily used in packaging gravure on board and paper. Publication gravure printers are actively testing Type W inks for various publication applications. Type V inks are used for printing vinyl films and Saran.

Lamination Inks. This class of ink is a specialized group. In addition to conforming to the constraints described for flexo and gravure inks, these inks must not interfere with the bond formed when two or more films, eg, polypropylene and polyethylene, are joined with the use of an adhesive in order to obtain a structure that provides resistance properties not found in a single film. In addition to polyamide, lamination inks ordinarily contain modifiers such as polyketone resin, plasticizer, and wax to impart specific properties such as block resistance and increased bond strength.

Miscellaneous Inks

Screen Process Inks. Screen-process inks are dispersions of pigments in vehicles which are, for the most part, solutions of resins in solvents of the boiling range of VM & P naphtha. Drying of solvent-based inks is usually by evaporation, but in some cases it is a combination of oxidation and evaporation. Various types of binders are used such as rosin esters, phenolics, cellulose derivatives, vinyls, and oleoresinous varnishes, depending on the film properties desired. Uv inks are also widely used for screen-process printing. After premixing, the ink is ground on a three-roll or media mill. The resulting ink should be short and soft so as not to drag on the squeegee and to release the substrate cleanly after the print is made.

Stamp-Pad Inks. Because it is desirable that the total ink soak into the stock, dyes are used rather than pigments. The vehicles used are usually glycols.

Ball-Point Inks. These inks are medium-viscosity semi-Newtonian fluids of high tinctorial strength which must be slow drying and free of particles so that they continue to feed to the paper without clogging. Drying on the paper is accomplished by rapid penetration and some evaporation. These properties are obtained by strong dye solutions and pigment dispersions in vehicles containing oleic acid and castor oil or a sulfonamide plasticizer.

Water-Based Writing Inks. These consist of very fine pigment dispersions in aqueous media containing small amounts of glycol or glycerol and a dispersing aid. They dry mainly by evaporation and quick wetting of cellulosic fibers in paper substrates.

Engraving Inks. Owing to the thick film that can be deposited (ca 25–150 μm), high strength formulations are not required, but the body of the ink is quite short so as to wipe cleanly from the plate. Drying is by a combination of oxidation or polymerization and by evaporation of solvent. The pigment, including a large percentage of colorless extender pigment, is dispersed on a three-roll mill in a vehicle composed of heat-bodied drying oil or oleoresinous vehicle, sometimes in combination with a resin-solvent type vehicle. Web engraving presses using heat or electron beam curing have been developed. They use appropriate polymerizing vehicles.

Electrostatic Inks. The electrostatic ink, also called an electrostatic toner, is a powder composed of pigment dispersed in a resin. The particles must have the proper electrical properties, particle-size range, and be free-flowing. After the image is deposited on the substrate, it is heat- or solvent-fused to a continuous film.

Decal Inks. The inks must dry completely on the surface by oxidation or solvent evaporation. The formulation of decalcomania inks is governed by the particular printing process employed in printing the transfer paper. Decalcomanias for ceramics require pigments that may be heated to high temperatures. Further, most decalcomanias should use pigments that are fast to light because many are subsequently transferred to outdoor signs or to store windows. Vehicles consist of oleoresinous varnishes containing metallic driers or are resin–solvent types.

Hot-Transfer Inks. One type of hot-transfer ink is made with heat-fusible resins and waxes to be transferred to cloth.

Ink-Jet Printing. The inks formulated for jet printing must be very fluid, stable, and free of any particles that could cause clogging of the jet nozzles, and be capable of depositing and adhering to a substrate with a minimum of character fogging. They are generally formulated with soluble dye colorants in a suitable aqueous or solvent-based vehicle.

Environmental Considerations

General environmental concerns such as use of renewable resources rather than crude-oil-based chemistries, biodegradeable inks and coatings, the pressure to recycle waste materials back into the raw material supply, etc, impact printing ink technology. Technologies developed in response to environmental concerns have been emerging and will continue to emerge as new issues supplant existing ones.

U.S. regulations encompass both federal, state, and local guidelines. In addition, there are numerous voluntary industry guidelines affecting ink making.

R. W. BASSEMIR
A. BEAN
O. WASILEWSKI
D. KLINE
W. HILLIS
C. SU
I. R. STEEL
W. E. RUSTERHOLZ
Sun Chemical Corporation

J. Fetsko, ed., *Raw Material Data Handbooks*, Vols. 1–4, National Printing Ink Manufacturers Association, Hasbrouck Heights, N.J., 1983.

Printing Ink Handbook, National Association of Printing Ink Manufacturers, Hasbrouck Heights, N.J., 1988.

R. Leach, ed., *The Printing Ink Manual*, Van Nostrand Reinhold, Berkshire, U.K., 1988.

J. Fetsko, ed., *Relationship of Ink/Water Interactions to Printability of Lithographic Printing Inks*, Pt. 1, National Printing Ink Manufacturers Association, Hasbrouck Heights, N.J., 1986.

J. Fetsko, ed., *Relationship of Ink/Water Interactions to Printability of Lithographic Printing Inks*, Pt. 2, National Printing Ink Manufacturers Association, Hasbrouck Heights, N.J., 1988.

INORGANIC HIGH POLYMERS

The most commercially successful inorganic polymers to date are the polysiloxanes, owing to their unique high temperature stability, low temperature flexibility, and a number of other advantageous properties such as low surface energy and room-temperature vulcanizability.

Because of increasing technological needs in the area of high performance materials there has been a growing interest in the synthesis and development of new inorganic polymers. The polyphosphazenes and the polysilanes have shown the most promise in this area during the last two decades. In addition to these two, other novel inorganic polymer systems have also recently been developed.

Polyphosphazenes

The polyphosphazenes, sometimes also referred to as polyphosphonitriles, are the most chemically versatile inorganic polymers known to date. Based on their method of synthesis, two different types of phosphazene polymers are now in existence and are undergoing parallel development. The first type, bearing substituents on phosphorus bonded mostly via phosphorus-oxygen and phosphorus–nitrogen linkages, was developed in the mid-1960s as soluble, hydrolytically stable polymers. Polymers of the second type, with substituents linked via direct phosphorus–carbon bonds were first reported in the early 1980s and since then they have also seen significant development.

Synthesis. The synthesis of poly(dichlorophosphazene), the parent polymer to over 300 macromolecules of types (**1**) and (**2**), is carried out via controlled, ring-opening polymerization of the corresponding cyclic trimer, $(N = PCl_2)_3$.

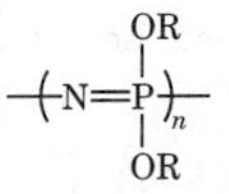

(**1**)

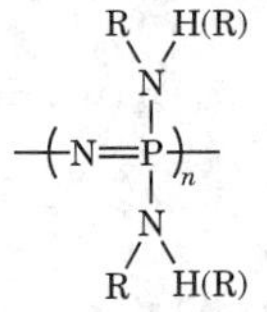

(**2**)

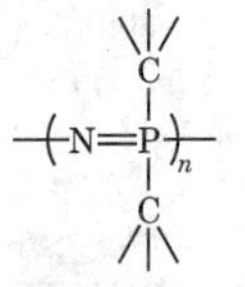

(**3**)

Properties. One of the characteristic properties of the polyphosphazene backbone is high chain flexibility which allows mobility of the chains even at quite low temperatures.

The thermal stability of polymers of types (**1**) and (**2**) is dependent on the nature of the substituents on phosphorous. Polymers with methoxy and ethoxy substituents undergo skeletal changes and degradation above about 100°C, but arloxy and fluoroalkoxy substituents provide higher thermal stability. Most of the P–N- and P–O-substituted polymers either depolymerize via ring-chain equilibration or undergo cross-linking reactions at temperatures much above 150–175°C.

Phosphazene polymers are inherently good electrical insulators unless side-group structures allow ionic conduction in the presence of salts. Polyphosphazenes also exhibit excellent visible and uv-radiation transparency when chromophoric substituents are absent.

Another valuable characteristic of many phospahzene polymers is their flame-retardant behavior and low smoke generation on combustion.

A remarkable feature of phosphazene polymers of types (**1**) and (**2**) is that appropriate substituents (which are readily attached) can be used as toggle switches to turn several properties, such as hydrolytic stability and electrical conductivity, on and off.

Applications. The P–O- and P–N-substituted polymers have so far shown the greatest commercial promise. The fluoroelastomers possess good rubber properties with the added advantages of being nonburning, hydrophobic, and solvent- and fuel-resistant. In addition to these, because of flexibility down to about −60 °C, these polymers have been used in seals, gaskets, and hoses in army tanks, in aviation fuel lines and tanks, as well as in cold-climate oil pipeline applications. These polymers have also found application in various types of shock mounts for vibration dampening.

The aryloxyphosphazene polymers, on the other hand, have been used primarily in wire and cable coatings and jackets and as fire-resistant, low smoke, closed-cell foams and sound-barrier sheets.

Biomedical Applications. In the area of biomedical polymers and materials, two types of applications have been envisioned and explored. The first is the use of polyphosphazenes as bioinert materials for implantation in the body either as housing for medical devices or as structural materials for heart valves, artificial blood vessels, and catheters.

The second type of biomedical application utilizes the versatile chemistry of polyphosphazenes to generate bioactive polymers. Two approaches have been developed: one is to tie or physically entrap biologically active molecules using the phosphazene backbone as the carrier or encapsulant. The other is to attach bioactive molecules to a hydrolyzable (degradable) phosphazene backbone that releases the active species on breakdown of the backbone to harmless species that can be metabolized or directly excreted.

Two crucial aspects of the design of bioactive polyphosphazenes have been carefully developed. One involves the hydrophilicity or hydrophobicity of the polymer, and the other is the stability of the polymer or tactical substituent linkages that allow release of the active agent or ensure its potency to be retained in the bound form.

Polymers Bearing Metal Complexes. A large number of polymers with side groups containing metal complexes have been reported. The complexes are linked to the phosphazene backbone primarily through a ligand on a substituent, although linkages through the skeletal nitrogen or through direct metal-phosphorus bonding with the skeletal phosphorus atoms have also been utilized.

Solid Electrolyte Applications. Among other potentially useful polymers synthesized by the versatile macromolecular substitution process are polymers based on oligoether substituents or heterocyclic substituents that have been under intense investigation for solid electrolyte battery applications. The most promising of these is poly [bis(methoxyethoxyethoxy)phosphazene (MEEP).

Polymers with Alkyl and Aryl Substituents on P

Even though partially alkyl- and aryl-substituted polyphosphazenes are accessible via the ring-opening polymerization followed by the macromolecular substitution route, polymers in which all substituents are attached through direct phosphorus–carbon bonds are not yet accessible by this method.

Synthesis. The first fully alkyl/aryl-substituted polymers were reported in 1980 via a condensation–polymerization route. In addition to providing fully alkyl/aryl-substituted polyphosphazenes, the versatility of the process has allowed the preparation of various functionalized polymers and copolymers.

Properties. The condensation–polymerization reaction yields alkyl- and aryl-substituted polymers with average molecular weights in the range 40,000 to 250,000 (M_n ranges from 20,000 to 100,000). In general, the polymers are soluble in chlorinated solvents such as CH_2Cl_2 and $CHCl_3$. Polymers with phenyl substituents are also soluble in tetrahydrofuran.

The P–N backbone remains quite flexible with small, unbranched alkyl substituents on phosphorus.

Alkyl- and aryl-substituted polyphosphazenes exhibit onset of decomposition at between 350 and 400°C.

Applications. Polymers with small alkyl substituents are ideal candidates for elastomer formulation because of quite low temperature flexibility, hydrolytic and chemical stability, and high temperature stability. In light of the biocompatibility of polysiloxanes and P–O- and P–N-substituted polyphosphazenes, poly(alkyl/arylphosphazenes) are also likely to be biocompatible polymers. A third potential application is in the area of solid-state batteries.

Phosphazenes Containing Skeletal Carbon, Sulfur, and Metal Atoms

The first phosphazene polymers containing carbon, sulfur, and even metal atoms in the backbone have been reported. These were all prepared by the ring-opening polymerization of partially or fully chloro-substituted (or fluoro-substituted) trimers containing one hetero atom substituting for a ring-phosphorous atom in a cyclotriphosphazene-type ring.

An example of polyphosphazene incorporating metal atoms is (**4**), where M = M_0 or W.

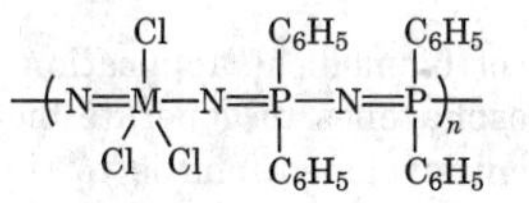

(4)

Poly(alkyl/aryloxothiazenes)

The synthesis of a new class of inorganic polymers (**5**) with a backbone consisting of alternating sulfur(VI) and nitrogen atoms, and with variable aklyl or aryl substituents as well as a fixed oxygen substituent on sulfur, has recently been accomplished. These polymers are structurally analogous to poly(alkyl/arylphosphazenes).

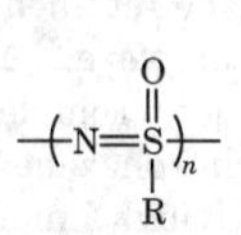

(5)

Synthesis and Properties. The synthesis of (**5**) follows a straightforward route based on readily accessible starting materials and on some novel reactions in organo–inorganic sulfur chemistry, as well as on polycondensation chemistry analogous to that utilized in the preparation of poly(alkyl/arylphosphazenes). The polymers exhibit some interesting characteristics that appear to be related to their unusual repeat unit.

Polysilanes

The polysilanes (**6**) are a unique class of polymers that exhibit σ-conjugation along the backbone.

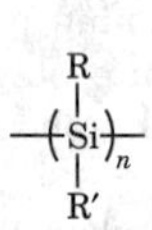

(6)

Synthesis of Polysilanes. The most commonly utilized method is based on the Wurtz-type alkali metal coupling of dichlorosilanes. Other synthesis methods include dehydrogenative coupling, ring-opening polymerization, polymerization of masked disilenes, electrochemical synthesis, and polymer modification.

Properties. Most unsymmetrically substituted dialkyl and alkyl/aryl homopolymers as well as copolymers are soluble in solvents such as tetrahydrofuran or toluene. The longer Si–Si bond length, compared with the C–C bond length, allows quite a bit of flexibility in the backbone such that glass-transition temperatures as low as −76 °C (for poly(n-hexylmethylsilane)), have been observed. On the other hand, as expected, aryl substitution brings about significant increases in T_g. Thus, polysilanes cover the range from rubbery elastomers to brittle solids. Polysilanes are chemically inert to air and water at ordinary temperature, but their reactivity increases in solvent.

Electronic Properties. What distinguishes polysilanes from virtually all other polymers is their backbone σ- conjugation. This lead to strong electronic absorption in the near-uv from a σ–σ transition.

The polysilanes are normally electrical insulators, but on doping with AsF_5 or SbF_5 they exhibit electrical conductivity up to the levels of good semiconductors (qv).

Polysilanes absorb electromagnetic energy and undergo chain scission. This is an extremely important property of these polymers in terms of applications. Photochemistry is exhibited both in solution and in the solid state.

Applications. Polysilanes are used in the manufacture of β-silicon carbide, microlithography, xerography, and photoinitiation.

Polygermanes

Soluble and well-characterized polygermane homopolymers, $(R_2Ge)_n$, and their copolymers with polysilanes have been prepared by the alkali metal coupling of diorgano-substituted dihalogermanes, via electrochemical methods, and by transition-metal catalyzed routes, as with the synthesis of polysilanes.

The polygermanes exhibit many of the same electronic properties as polysilanes, including near-uv photoabsorption, thermochromism, photobleaching, as well as nonlinear optical activity, and have seen a fair amount of theoretical and experimental investigation. However, despite similarities with polysilanes, polygermanes appear to be unlikely candidates for commercial exploitation.

AROOP K. ROY
Dow Cornina Corporation

J. E. Mark, H. R. Allcock, and R. West, *Inorganic Polymers*, Prentice-Hall, Inc., Englewood Cliffs, N.J., 1992.

C. W. R. Wade and co-workers, in C. E. Carraher, J. E. Sheats, and C. U. Pittman, eds., *Organometallic Polymers*, Academic Press, Inc., New York, 1978, pp. 289–300.

H. R. Allcock, in *Inorganic Polymers*, Prentice-Hall, Inc., Englewood Cliffs, N.J., 1992, pp. 95–118.

INSECT CONTROL TECHNOLOGY

In the United States there are more than 10,000 species of insects, mites, and ticks that cause losses to agriculture, but only about 600 species require annual applied control measures.

Role of Chemicals in Insect Control. Increasing use of integrated pest management (IPM) practices and the introduction of the pyrethroids, which are effective at about one-tenth the application rate of the older insecticides, resulted in decreased insecticide use and by 1982 the farm use on primary crops was estimated at 32,000 t, comprised of organochlorines, 6%; organophosphates, 67%; carbamates, 18%; and pyrethroids, 4%. Corn became the most heavily treated crop with 42% of the total, followed by cotton, 24%, and soybean, 16%. In agriculture, the average benefit/cost ratio from insecticide use ranges from \$3 to \$5 return for every \$1 invested by the farmer(s).

The value of insecticides in controlling human and animal diseases spread by insects has been dramatic. It has been shown that between 1942 and 1952, the use of DDT in public health measures to control the mosquito vectors of malaria and the human body louse vector of typhus saved five million lives and prevented 100 million illnesses. Insecticides have provided the means to control such important human diseases as filariasis transmitted by *Culex* mosquitoes and onchocerciasis transmitted by *Simulium* blackflies.

Integrated Pest Management. Although employment of chemicals for insect pest control is essential to modern society, the extensive and injudicious use of chemical insecticides since 1946 has resulted in many problems including *(1)* widespread insect resistance, *(2)* emergence of resurgent and secondary pests whose regulating natural enemies are adversely affected, *(3)* hazards to human health, *(4)* ubiquitous environmental pollution by persistent lipophilic organochlorines, and *(5)* exponentially increasing costs of new insecticides. Most of these unintended consequences of chemical pest control relate to a pervasive eradication philosophy resulting from the euphoria about the effectiveness of successive generations of organochlorine, organophosphorus, carbamate, and pyrethroid insecticides.

The primary goals of IPM are *(1)* to determine how the life system of the pest needs to be modified to reduce the numbers to tolerable levels, ie, below the economic threshold; *(2)* to apply biological knowledge and current technology to achieve the desired modification, ie, applied ecology; and *(3)* to devise procedures for pest control compatible with economic and environmental control aspects, ie, economic and social acceptance.

IPM practices rely heavily on protection and conservation of natural enemies, parasites, predators, and diseases that regulate or balance populations of insect pests. IPM programs are based on two important parameters: the economic injury level defined as that population density of a pest that causes enough injury to justify the cost of remedial treatment, and the economic threshold, defined as that pest density at which control measures should be applied to prevent an increasing insect population from reaching the economic injury level. Whenever applied, IPM practices have consistently resulted in decreases in insecticide applications of 50 to 90% over conventional spray programs.

Insecticide management is concerned with the safe, efficient, and economical handling of insecticides during manufacture, utilization, and disposal. The essential components are selection of the proper insecticide for the IPM program, selection of the mode, timing, and dosage of application, consideration of the problems of resistance and resurgence, the possible effects of insecticide residues on food crops and in the environment, and the impact of these on humans, domestic animals, and wildlife.

Insecticides

Inorganic Stomach Poisons. The fundamental biochemical lesion produced by arsenicals is the result of reaction between As^{3+} and the sulfhydryl groups of key respiratory enzymes such pyruvate and α-ketoglutarate dehydrogenases.

The fluoride ion inhibits enzymes, such as enolase, which require Mg as a prosthetic group, by precipitating a complex magnesium fluorophosphate; thus it prevents phosphate transfer in oxidative metabolism.

Inorganic substances include Borax, $Na_2B_4O_7 \cdot 10H_2O$; sodium tetraborate, $Na_2B_4O_7$; boric acid, H_3BO_3; white phosphorus; silicic acid, SiO_2 or H_2SiO_3; and sulfur and its compounds.

Contact Poisons of Plant Origin. Nicotinoids include nicotine extracted from *Nicotiana tabaccum* and the synthetic imidocloprid, 1-[(6-chloro-3-pyridinyl)methyl]-*N*-nitro-2-imidazolidinimine (bp 137–144°C, vp 0.2 μPa at 20°C).

Nicotine, anabasine, and imidocloprid affect the ganglia of the insect central nervous system, facilitating transsynaptic conduction at low concentrations and blocking conduction at higher levels. Nicotine is used as a contact insecticide for aphids attacking fruits, vegetables, and ornamentals, and as a fumigant for greenhouse plants and poultry mites.

The use of rotenone-bearing roots from *Derris* and *Lonchocarpus* spp. as insecticides in the United States was developed as a result of federal laws against residues of lead, arsenic, and fluorine upon edible produce. Insects poisoned with rotenone exhibit a steady decline in oxygen consumption. Poisoning inhibits the mitochondrial oxidation of Krebs cycle intermediates, which is catalyzed by NAD. Rotenone-containing insecticides have been used as dusts of ground roots, dispersible powders, and emulsive extracts. Their principal uses have been for application to edible produce just prior to harvest and for the control of animal ectoparasites and cattle grubs.

Cevadine, $C_{32}H_{49}O_9N$, and veratridine, $C_{36}H_{51}O_{11}N$, both esters of cevine, are the alkaloids responsible for the insecticidal action. These alkaloids are highly poisonous to mammals and may cause irritation of the eyes and respiratory tract and violent sneezing. Sabadilla is used as a dust or wettable powder of the ground seeds for the control of plant-feeding Hemiptera, and with sugar as a toxic bait for thrips.

This compound from *Ryania speciosa* is effective as both a contact and a stomach poison. The material has been used in the control of the European corn borer and codling moth and is formulated as a wettable powder of ground stems or as a methanolic extract. Ryanodine uncouples the ATP–ADP actomyosin cycle of striated muscle.

Pyrethrum from *Chrysanthemum cinerariacfolium* is an especially valuable insecticide because it produces a rapid paralysis or knockdown of flying insects, and because its toxic components are rapidly inactivated upon exposure to light. Pyrethrum products are used as household insecticides, livestock insecticides, grain protectants, and to control insect pests on edible produce just prior to harvest. The principal use is in aerosol sprays at 0.04 to 0.25% active ingredient (AI) together with 5–10 times this amount of a synergist such as piperonyl butoxide.

Synergists containing the methylenedioxyphenyl moiety, although essentially nontoxic themselves, can activate the pyrethrins up to 30-fold by reacting with the Fe atom of the cytochrome microsomal oxidases to block detoxication. Such synergists are also highly effective when used with rotenone, ryania, and carbamates.

Synthetic Pyrethroid Insecticides. These synthetic pyrethroids have become one of the most important classes of insecticides, with world annual production estimated at 6000 t. They include pyrethroids from chrysanthemic acid [allethrin (*d* 1.005–1.015, vp 16 mPa at 30°C); resmethrin, (5-benzyl-3-furanyl)methyl (±)-*cis,trans*-chrysanthemate (mp 43°C); bioresmethrin; phenothrin, (3-phenoxyphenyl)methyl(±)-*cis,trans*-chrysanthemate; empenthrin, (*E*)-(*RS*)-1-ethynyl-2-methylpent-2-enyl (1*RS*)-*cis,trans*-chrysanthemate; prallethrin, (*RS*)-2-methyl-4-oxo-3-prop-2-ynylcyclopent-2-enyl(1*RS*)-*cis,*-*trans*-chrysanthemate; and tetramethrin, (1,3,4,5,6,7-hexahydro-1,3-dioxo-2*H*-isoindole-2-yl)-methyl(±)-*cis,trans*-chrysanthemate (mp 65–80°C)], pyrethroids with modified chrysanthemate esters [permethrin, (3-phenoxyphenyl)methyl-3-(2,2-dichloroethenyl)-(±)-*cis,trans*-2,2-dimethylcyclopropaneocarboxylate; cyfluthrin, (*R,S*)-α-cyano-(4-fluoro-3-phenoxyphenyl)methyl (1*R,S*)-*cis,trans*-3-(2,2-dichloroethenyl)-2,2-dimethylcyclopropanecarboxylate (mp 60°C, vp <1 mPa at 20°C); deltamethrin, (*S*)-α-cyano-3-phenoxyphenyl)methyl (+)-*cis*-(1*R*,3*R*) 3-(2,2-dibromoethenyl)-2,2-dimethylcyclopropanecarboxylate (mp 98–101°C, *d* 1.108, vp 2 μPa at 25°C); cyhalothrin, (*RS*)-α-cyano-(3-phenoxyphenyl)methyl (±)-*cis*- (1*R,S*) 3-(2-chloro-3,3,3-trifluoropropenyl)-2,2-dimethylcyclopropanecarboxylate (mp 49°C, vp 0.2 μPa at 25°C); bifenthrin, [2-methyl-(1,1′-biphenyl)-3-yl]methyl (*Z*)-(1*R, S*)-*cis*-3-(2-chloro-3, 3, 3-trifluoropropenyl)- 2, 2-dimethylcyclopropanecarboxylate (mp 68–70°C, vp 0.034 mPa at 25°C); tefluthrin, 2, 3, 5, 6-tetrafluoro-4-methylphenylmethyl 2-(±)-(1*RS*, 3*RS*)-*cis*-(2-chloro-3, 3, 3-trifluoropropenyl)-2, 2-dimethylcyclopropanecarboxylate (mp 44°C, vp 80 mPa at 20°C); fenpropathrin, α-cyano-(3-phenoxyphenyl)methyl 2,2,3,3-tetramethylcyclopropanecarboxylate (vp 0.73 mPa at 20°C); bioethanomethrin; Kadethrin, 5-(benzyl-3-furanylmethyl)-3-[dihydro-2-oxo-3-(2*H*)-thienylidene)-methyl]-*cis*-(1-*R*,3*S*)-2,2-dimethylcyclopropanecarboxylate (mp 31°C, vp <0.1 mPa at 20°C); tralomethrin, (*S*)-α-cyano-(3-phenoxyphenyl)methyl (1*R*, 3*S*)-2, 2-dimethyl-3-[(*RS*)-1,2,2,2-tetrabromoethyl)]cyclopropanecarboxylate (bp 138–148°C, *d* 1.70, vp 17 mPa at 25°C); and acrinathrin, (*S*)-α-cyano-(3-phenoxyphenyl)methyl (*Z*)-(1*R*-*cis*)-2,2-dimethyl-3-2-[2,2,2-trifluoromethyl)ethoxycarbonyl]vinyl cyclopropanecarboxylate (mp 82°C, vp 0.39 μPa at 25°C], pyrethroid esters of benzene acetate [flucythrinate, (*R,S*)-α-cyano-(3-phenoxyphenyl)methyl (±)-(2*S*)-α-isopropyl-4-difluoromethoxyphenylacetate (*d* 1.189, vp 1.2 μPa at 20°C); and fluvalinate, (*R,S*)-α-cyano-(3-phenoxyphenyl)methyl *N*-[2-chloro-4-(trifluoromethyl)phenyl] DL-valine (*d* 1.29, vp <13 μPa at 25°C)].

The pyrethroids readily penetrate the insect cuticle, and their action is characterized by excitation, incoordination, and paralysis, and results in rapid knockdown. The biochemical lesion is at the sodium channels of the nerve axon.

The pyrethroids are extremely toxic to fish, LC_{50}1 to 10 μg/L. They are also very hazardous to bees and other beneficial insects. Whereas the natural pyrethrins and resmethrin have only a one- to two-day persistence, permethrin, decamethrin, and fenvalerate may persist as foliage residues for two to four weeks and as soil residues for up to 1–2 months. Mammalian toxicity ranges from very safe insecticides such as the natural pyrethrins, allethrin, resmethrin, tetramethrin, and cyfluthrin, to highly toxic compounds such as deltamethrin, flucythrinate, cyhalothrin, and tefluthrin. These compounds are irritating to the skin and nasal membranes, and the lack of a specific antidote is a disadvantage.

Organochlorine Insecticides. The success of DDT as an insecticide stimulated research and discovery in the organochlorine area. Lindane, toxaphene, chlordane, heptachlor, aldrin, dieldrin, and endrin were developed during the decade after World War II. These organochlorine insecticides dominated the market during the period of 1945–1965.

Methoxychlor, 1,1,1-trichloro-2,2-bis-(4-methoxyphenyl)ethane (mp 89°C), is favored for general environmental use, and perthane, 1,1-dichloro-2,2-bis-(4-ethylphenyl)ethane (mp 60–61°C), has been used as a household insecticide.

DDT and its analogues specifically affect the peripheral sense organs of insects and produce violent trains of afferent impulses that result in hyperactivity, convulsions, and paralysis. Death results from metabolic exhaustion and the production of an endogenous neurotoxin. The specific biochemical lesion is at the sodium channels of the nerve axon.

The unusual stability of DDT and its high lipid/H_2O partitioning ($>1 \times 10^6$) has resulted in many environmental problems, eg, soil persistence with a half-life of 2.5–10 yr, bioaccumulation from water to fish at levels $> 1,000,000$, transport through food chains, and ubiquitous tissue storage in humans and animals. These properties are shared by its primary breakdown product, DDE, which is even more environmentally recalcitrant.

Lindane is γ-hexachlorocyclohexane (configuration aaaeee). Cyclodienes are polychlorinated cyclic hydrocarbons with endomethylene-bridged structures, prepared by the Diels-Alder diene reaction, eg, heptachlor, 1,4,5,6,7,8,8-heptachloro-3*a*,4,7,7*a*-tetrahydro-4,7-,methano-1*H*-indene (mp 95°C, vp 0.04 Pa at 25°C); aldrin; dieldrin, 1,2,3,4,10,10-hexachloro-1,4,4a,5,8, 8*a*-hexahydro-6,7-epoxy-1,4-*endo,exo*-5,8-dimethanonaphthalene (mp 176°C, vp 0.4 mPa at 20°C); endrin, 1,2,3,4,10,10-hexachloro-1,4,4a,5,8,8*a*-hexahydro-6,7-epoxy-1,4-*endo,endo*-5,8-dimethanonaphthalene (mp 245 dec, vp 0.022 mPa at 25°C); endosulfan, 6,7,8,9,10,10-hexachloro-1,5,5a,6,9,9*a*-hexahydro-6,9-methano-2,4,3-benzodioxatheipin-3-oxide (mp 70–100°C, vp 1.3 mPa at 30°C); mirex, 1,2,3,4,5,5,6,7,8,9,10,10-dodecachloro-octahydro-1, 3, 4-methano-2*H*-cyclobuta-[*c, d*]-pentalene (mp 485°C); and chlordecone, decachloro-5-oxo-pentacyclo-$[5.3.0.0^{2,6}O^{3,9},O^{4,8}]$-decane (mp 349°C dec).

The cyclodienes, like lindane and toxaphene, affect the nerve axon, producing hyperactivity, convulsions, prostration, and death.

The high lipophilicity of the cyclodienes and the prolonged persistence of dieldrin and heptachlor epoxide (soil half-lives 2–10 yr) have resulted in severe environmental contamination. As a result registrations for aldrin, endrin, dieldrin, heptachlor, mirex, and chlordecone have been cancelled in the United States. The compounds are bioaccumulated from water to fish up to 100,000- to 300,000-fold and are ubiquitous in human fat and milk.

Organophosphorus Insecticides. Presently there are more than 100 commercially available organophosphorous (OP) insecticides with a total world use of the order of 2×10^5 t annually.

It is estimated that more than 25×10^6 different potentially toxic OP esters can be made using Schrader's classic formula for effective phosphorylating agents, where R and R′ are short-chain alkyl, alkoxy, alkylthio, or alkyl-amino groups, and X is a displaceable moiety with a high energy P-bond such as F or acyl anhydride, and the pentavalent phosphorus atom is bonded to oxygen or sulfur.

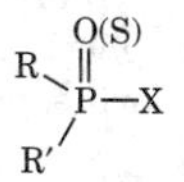

Organophosphorus insecticides are available with very short residual action, eg, tetraethyl pyrophosphate and mevinphos, or with prolonged residual activity, eg, diazinon and azinphos–methyl. The organophosphorus insecticides represent the most versatile class of chemicals employed for insect control.

They include phosphoric acid and phosphorothioic acid anhydrides [tetraethyl pyrophosphate (bp 104–110°C at Pa, *d* 1.185, vp 6.1 mPa at 30°C); and sulfotepp, *O,O,O,′O′*-tetraethyl dithiopyrophosphate (bp 110–113°C at 29 Pa, *d* 1.196, vp 22 mPa at 20°C)], aliphatic phosphorothioate esters [demeton, *O,O*-diethyl *O*-(2-ethylthio)ethyl phosphorothionate (bp 123°C at 0.13 kPa, *d* 1.119); demeton–methyl, a mixture of *O,O*-dimethyl *O*-(2-ethylthio)ethyl phosphorothioate (demeton–methyl O) (bp 74°C at 20 Pa) and *O,O*-dimethyl *S*-(2-ethylthio)ethyl phosphorothioate (demeton–methyl S) (bp 89°C at 20 Pa); oxydemetonmethyl, *O,O*-dimethyl *S*-(2-ethylsulfinyl)-ethyl phosphorothioate; disulfton, *O,O*-diethyl *S*-(2-ethylthio)ethyl phosphorodithioate (bp 62°C at 1.3 Pa, *d* 1.144, vp 24 mPa at 20°C); phorate, *O,O*-diethyl *S*-(ethylthio)methyl phosphorodithioate (bp 118–120°C/0.1 kPa, *d* 1.167, vp 0.11 Pa at 20°C); formothion, *O,O*-dimethyl *S*-(*N*-formyl-*N*-methylcarbamoyl)-methyl phosphorodithioate (mp 25°C, *d* 1.36, vp 0.11 mPa at 20°C); vamidothion, *O,O*-dimethyl-*S*-[2-(*N*-methyl-1-methylcarbamoyl)-ethylthio]-ethyl phosphorothioate (mp 46°C); phenthoate, *O,O*-dimethyl *S*-(α-carboethoxy)benzyl phosphorodithioate (mp 17–18°C, *d* 1,226, vp 5 mPa at 40°C); carbophenothion, *O,O*-diethyl *S*-(4-chlorophenylthio)methyl phosphorodithioate (bp 82°C/1.3 Pa, *d* 1.271, vp 1.1 mPa at 25°C); ethion, *O,O,O,′O′*-tetraethyl *S,S′*-methylene diphosphorodithioate (bp 165°C/53 Pa, vp 0.2 mPa at 20°C); and malathion, *O,O*-dimethyl *S*-(1,2-dicarbethoxy)ethyl phosprodithioate (bp 156–157 C at 93 Pa, *d* 1.23, vp 5.2 mPa at 30°C)], phenyl phosphorothiate esters [parathion, *O,O*-diethyl *O*-(4-nitrophenyl) phosphorothioate (bp 375°C, *d* 1.265, vp 5.3 mPa at 27°C); methyl parathion, *O,O*-dimethyl *O*-(4-nitrophenyl) phosporothioate (mp 34°C, *d* 1.358, vp 1.3 mPa at 23°C); fenitrothion, *O,O*-dimethyl *O*(3-methyl-4-nitrophenyl) phosphorthioate (bp 140–145°C at 13 Pa, *d* 1.3227, vp 0.15 mPa at 20°C); fensulfothion, *O,O*-diethyl *O*-(4-methylsulfinylphenyl) phosphorothioate (bp 138–141°C at 1.3 Pa); ronnel or fenchlorphos, *O,O*-dimethyl *O*-(2,4,5-trichlorophenyl) phosphorothioate (mp 41°C, vp 0.11 Pa at 25°C); cyanophos, *O,O*-diethyl *O*-(4-cyanophenyl) phosphorothioate (mp 14°C, *d* 1.255, vp 0.5 mPa at 20°C); and temephos, *O,O,O′,O′*-tetramethyl *O,O′*-(thiodi-4,1-phenylene)-bisphosphorothioate (mp 30°C, *d* 1.32)], phenyl phosphorodithioate esters [eg, sulprofos, *O*-ethyl, *S*-propyl *O*-[(4-methylthio)phenyl] phosphorodithioate (bp 155°C/13 Pa, *d* 1.20, vp 0.1 mPa at 20°C)], phosphonothioate esters of phenols [fonofos, *O*-ethyl, *S*-phenyl ethylphosphonodithioate (bp 130°C at 13.3 Pa, *d* 1.16 vp 28 mPa at 25°C); EPN, *O*-ethyl *O*-(4-nitrophenyl) phenylphosphonothioate (mp 41°C, *d* 1.268, vp 126 μPa at 25°C); and leptophos, *O*-methyl *O*-(4-bromo-2,5-dichlorophenyl) phenylphosphonothioate (mp 71°C)], vinyl phosphates [dichlorvos, *O,O*-dimethyl *O*-(2,2-dichlorovinyl) phosphate, $(CH_3O)_2P(O)OCH{=}CCl_2$ (bp 140°C at 27 kPa, *d* 1.314, vp 1.6 Pa at 20°C); naled, *O,O*-dimethyl *O*-(1,2-dibromo-2,2-dichloroethyl)phosphate, $(CH_3O_2)P(O)OCHBrCBrCl_2$ (mp 27°C, bp 110°C at 66.7 Pa, *d* 1.96, vp 0.26 Pa at 20°C); trichlorfon, *O,O*-dimethyl-1-hydroxy-2,2,2-trichloroethyl phosphonate, $(CH_3O)_2P(O)CH(OH)CCl_3$ (mp 83°C, *d* 1.73, vp 1 mPa at 20°C); mevinphos, *O,O*-dimethyl *O*-(2-methoxycarbonyl-1-methyl-vinyl) phosphate, $(CH_3O_2)P(O)C(CH_3){=}CHC(O)OCH_3$ (bp 106–107°C at 0.13 kPa, *d* 1.25, vp 0.38 Pa at 21°C); phosphamidon, *O,O*-dimethyl *O*-(2-chloro-2-diethylcarbamoyl-1-methylvinyl)phosphate, $(CH_3O)_2P(O)OC(CH_3){=}C(Cl)C(O)N(C_2H_5)_2$ (bp 160°C at 0.2 kPa, *d* 1.21, vp 3.2 mPa at 20°C); dicrotophos, *cis-O,O*-dimethyl *O*-(2-dimethylcarbamoyl-1-methylvinyl) phosphate, $(CH_3O)_2$ $P(O)C(CH_3){=}CHC(O)N(CH_3)_2$ (mp 54°C, *d* 1.33, vp 0.9 mPa at

20°C), monocrotophos, *cis-O,O*-dimethyl *O*-(2-methylcarbamoyl)-1-methylvinyl phosphate, $(CH_3O_2)P(O)C(CH_3){=}CHC(O)NHCH_3$ (mp 54°C, *d* 1.33, vp 0.9 mPa at 20°C), crotoxyphos, *O,O*-dimethyl-*O*-(1-methyl-2-(1-phenylcarbethoxy)vinyl phosphate (bp 135°C at 4 Pa, *d* 1.19, vp 1.8 mPa at 20°C); chlorfenvinphos, *O,O*-diethyl-*O*-[2-chloro-1-(2,4-dichlorophenyl)-vinyl] phosphate (bp 168–170°C at 67 Pa, *d* 1.36, vp 0.97 mPa at 25°C); tetrachlorvinphos or stirifos, *O,O*-dimethyl *O*-[2-chloro-1-(2,4,5-trichlorophenyl)vinyl] phosphate (mp 97–98°C, bp 5 μPa at 20°C); and propetamphos, *O*-[2-(isopropoxycarbonyl)-1-methylvinyl]-O-methyl, *N*-ethyl phosphoramidothioate (bp 87–89°C at 0.7 Pa, *d* 1.129, vp 1.9 mPa at 20°C)], phosphorothioate esters of heterocyclic enols [chloropyrifos, *O,O*-diethyl *O*-(3,5,6-trichloro-2-pyridinyl) phosphorothioate (mp 42–43°C, vp 2.5 mPa at 25°C); fospirate, dimethyl 3,5,6-trichloro-2-pyridinyl phosphate; etrimfos, *O,O*-dimethyl *O*-(6-ethoxy-2-ethyl-4-pyrimidinyl) phosphorothioate (*d* 1.195, vp 8.4 mPa at 20°C); diazinon, *O,O*-diethyl *O*-(2-isopropyl-4-methyl)-6-pyrimidinyl) thiophosphate (bp 83–84°C at 0.3 Pa, *d* 1.116, vp 18.2 mPa at 20°C); pirimiphos–methyl or pyrimithate, *O,O*-dimethyl *O*-(2-diethylamino-6-methyl-4-pyrimidinyl) phosphorothioate (*d* 1.157, vp 15 mPa at 30°C); isazophos, *O,O*-diethyl *O*-(5-chloro-1-isopropyl-(1*H*)-1,2,3-triazol-3-yl) phosphorothioate (bp 100°C at 1.3 Pa, *d* 1.22, vp 4.3 mPa at 20°C); and triazaphos, *O,O*-diethyl *O*-(1-phenyl-1*H*-1,2,4-triazol-3-yl) phosphorothioate (*d* 1.247, vp 0.39 mPa at 30°C)], phosphorothioate esters of S-methyl heterocycles [azinphos–methyl, *O,O*-dimethyl *S*-4-oxobenzo-[*d*]-(1,2,3-triazin-3-yl-methyl) phosphorodithioate (mp 73–74°C, *d* 1.44, vp $<$ 1 mPa at 20°C); azinphos–ethyl, the corresponding *O,O*-diethyl ester (mp 53°C, *d* 1.284, vp 0.029 mPa at 20°C); phosmet, *O,O*-dimethyl *S*-(*N*-phthalimidomethyl) phosphorodithioate (mp 72°C, vp 0.13 Pa at 50°C); dialifor, *O,O*-diethyl *S*-(2-chloro-1-phthalimidoethyl) phosphorodithioate (mp 67–69°C, vp 133 mPa at 35°C); azamethiphos, *O,O*-dimethyl *S*-[6-chloro-2-oxooxazole-[4,5-*b*]-pyridin-3-(2*H*)-3-yl)-methyl] phosphorodithioate (mp 89–90°C, vp 5 μPa at 20°C); methidathion, *O,O*-dimethyl *S*-[(5-methoxy-2-oxo-1,3-4-thiadiazol-3-(2*H*)-yl)-methyl] phosphorodithioate (mp 39–40°C, vp 0.19 Pa at 20°C); menazon, *O,O*-dimethyl *S*-(4,6-diamino-1,3,5-triazin-2-yl)-methyl phosphorodithioate (mp 160–162°C, vp 0.13 mPa at 25°C); and endothion, *O,O*-dimethyl *S*-[5-methoxy-4-oxo-4*H*-pyran-2-yl)-methyl] phosphorothioate (mp 96°C)], and miscellaneous organophosphorus esters [methoamidophos, *O,S*-dimethylphosphoramidothioate (mp 39–41°C, *d* 1.3, vp 39 mPa at 30°C); acephate, *O,S*-dimethyl *N*-acetyl phosphoramidothioate (mp 72–80°C, vp 0.23 mPa at 24°C); phoxim, *O,O*-diethyl *O*-(cyanobenzilideneamino) phosphorothioate (mp 5°C, bp 102°C at 1.3 Pa, *d* 1.176, vp 16 mPa at 20°C); ethoprophos or ethoprop, *O*-ethyl, *S,S*-dipropyl phosphorothioate (bp 86–91°C at 27 Pa, *d* 1.094, vp 45 mPa at 26°C); cruformate, *O*-methyl *O*-[2-chloro-4-(*t*-butyl)phenyl] *N*-methylphosphoramidate (mp 61°C); isofenphos, *O*-ethyl, *O*-[2-(isopropoxycarbonyl)phenyl]-*N*-isopropyl phosphoramidothioate (bp 120°C at 1.3 Pa, *d* 1.13, vp 0.52 mPa at 20°C); and heptenophos, *O,O*-dimethyl *O*-(7-chlorobicyclo-[3.2.0]-hepta-2,6,dien-6-yl) phosphate (bp 94°C at 0.13 Pa, vp 97 mPa at 20°C)].

The organophosphorus insecticides owe their biological activities to the capacity of the central P atom to phosphorylate the esteratic site of the enzyme acetylcholinesterase (AChE), which is an essential constituent of the nervous system of insects as well as higher animals.

The reactivity of the individual O—P insecticides is determined by the magnitude of the electrophilic character of the phosphorus atom, the strength of the bond P—X, and the steric effects of the substituents.

The organophosphorus compounds, with their tremendous diversity of structures, provide opportunities for the development of true physiological selectivity. Such selectivity is important not only in providing safety for the user of pesticides and for plant and animal possessions but in providing ecological selectivity for wildlife and for the valuable beneficial insects, ie, honeybees, pollinators, parasites, and predators. Extraordinary selectivity has been accomplished with the parathion-type of insecticide by incorporating Cl or CH_3 groups in the meta position of the aryl ring. These groups interact sterically with acetylcholinesterase (AChE), increasing the affinity for the insect enzyme and decreasing it with the mammalian enzyme.

Organophosphorus insecticides are intrinsically reactive and readily degrade by oxidation and hydrolysis and in living organisms. Therefore, their use does not present serious problems of biomagnification and food-chain transfer. Soil persistence is low. However, the organophosphorus insecticides are general biocides that are toxic to nearly all animal organisms. The organophosphorus insecticides are highly toxic to bees and to beneficial parasites and predators. Great care should be taken in handling, applying, and storing these insecticides as the majority of insecticide poisonings throughout the world result from the use of the highly toxic OP compounds. It is essential to wear protective clothing including mask, goggles, and rubber gloves when handling these materials.

Carbamate Insecticides. These are structurally derivatives of the unique plant alkaloid physostigmine. The carbamates may be considered synthetic derivatives of the synaptic neurotransmitter acetylcholine, with very low turnover numbers. The *N,N*-dimethylcarbamates of heterocyclic enols and the *N*-methylcarbamates of a variety of substituted phenols with a wide range of insecticidal activity were described in 1954. The latter are the most widely used carbamate insecticides, and the *N*-methylcarbamates of oximes have subsequently been found to be effective systemic insecticides.

They include carbaryl, 1-naphthyl *N*-methylcarbamate (mp 142°C, *d* 1.232 g/cm^3, vp 0.67 Pa at 20°C); carbofuran, 2,3-dihydro-2,2-dimethyl-7-benzofuranyl *N*-methyl-carbamate (mp 150–152°C); propoxur, 2-isopropoxyphenyl *N*-methylcarbamate (mp 91°C, vp 0.84 mPa); dioxacarb, 2-(1,3-dioxolan-2-yl)-phenyl *N*-methylcarbamate (mp 114–115°C, vp 50 μPa at 20°C); bendiocarb, 2,2-dimethyl-1,3-benzodioxol-4-ol *N*-methylcarbamate (mp 129–130°C, vp 0.7 mPa at 25°C); mexacarbate, 4-dimethylamino-3,5-dimethylphenyl *N*-methyl-carbamate (mp 85°C); isoprocarb, 2-isopropylphenyl *N*-methylcarbamate (mp 93°C, vp 0.38 mPa at 20°C); trimethacarb, 3,4,5-trimethylphenyl *N*-methylcarbamate (mp 117–119°C, vp 6.8 mPa at 25°C); ethiofencarb, 2-[2′-(methylthio)methyl]phenyl *N*-methyl-carbamate (*d* 1.147, vp 13 mPa at 30°C); dimetilan, 2-(*N,N*-dimethylcarbamoyl)-3-methylpyrazol-5-yl *N,N*-dimethylcarbamate (mp 68°C); pyrimicarb, 2-(dimethylamino)-5,6-dimethyl-4-pyrimidinyl *N,N*-dimethylcarbamate (mp 90°C, vp 4 mPa at 30°C); aldicarb, *O*-(methylcarbamoyl)2-methyl-2-methylthiopropionaldehydeoxime (mp 99–100°C, vp 13 mPa at 20°C); methomyl, *S*-methyl, *N*-[(methylcarbamoyl)oxy]thioacetimidate (mp 78–79°C, vp 7 mPa at 20°C); oxamyl, *S*-methyl-1-1(dimethylcarbamoyl)-*N*-[(methylcarbamoyl)oxy]thioformimidate; and thiofanox, 3,3-dimethyl-1-(methylthio)-2-butanone *O*-[(methylamino)carbonyl]oxime (mp 57°C, vp 22.6 mPa at 25°C].

All of the insecticidal carbamates are cholinergic. The insecticides are strong carbamylating inhibitors of acetylcholinesterase and may also have a direct action on the acetylcholine receptors because of their pronounced structural resemblance to acetylcholine.

The *N*-methylcarbamates generally are biodegradable and of low soil persistence. Certain carbamates are highly toxic to birds. Fish toxicity of carbamates is generally low, but these compounds are extremely toxic to bees. In cases of human poisoning, atropine is a specific antidote.

Insect Growth Regulators. These compounds, unlike most conventional insecticides, interfere with biochemical processes that are unique to arthropods; eg, molting, ecdysis, and formation of the chitinous exoskeleton. Therefore, they are selective insecticides with very low mammalian toxicity. They include juvenoids (Table 1).

Chitin Synthesis Inhibitors. These are insect growth regulators that prevent the formation of the insect chitinous exoskeleton and thus produce a critical biochemical lesion during hatching, ecdysis, or pupation. These complex biochemical and physiological processes are unique to arthropods; therefore, the benzoyl phenyl ureas are highly specific insecticides. They include diflubenzuron, 1-(4-chlorophenyl)-3-(2,6-difluorobenzoyl) urea (mp 239°C, vp 0.033

Table 1. Insecticidal Effectiveness of Some Juvenoids

	LD$_{50}$ values				
Compound	*Tenebrio molitor*, μg/pupa	*Galleria melonella*, μg/pupa	*Aedes aegypti*, ppm	*Musca domestica*, μg/pupa	*Heliothis virescens*, ppm
neotenin	0.70	0.060	0.15	>100	24
epiphenonane	0.0024	0.037	0.057	54	2.2
hydroprene	0.25	0.040	0.0078	18	0.30
methoprene	0.0040	5.7	0.00017	0.0035	0.77
fenoxycarb		0.00034	0.000022	0.040	0.018

mPa at 50°C); teflubenzuron, 1-(2,4-difluoro-3,5-dichlorophenyl)-3-(2,6-difluorobenzoyl) urea (mp 222°C, vp 0.8 mPa at 20°C); and azadirachtin and related limonoid triterpenoids.

Acaricides

Chemicals that are especially effective in controlling the mites and ticks of the order Acarina are acaricides. A number of acaricidal chemicals have come into widespread use that have almost specific toxicity to the mites but are inactive against insects. In general, these acaricides are highly stable compounds with comparatively prolonged residual action and low mammalian toxicity. These acaricides exhibit a considerable degree of specificity for various species of acarina and are most useful for the phytophagous Tetranychidae and Eriophyidae. Other acaricides that repel or kill mites and ticks that attack humans and animals are described.

Acaricides include chlorfenethol, 1,1-bis(*p*-chlorophenyl)ethanol (mp 70°C); chlorobenzilate, ethyl *p,p′*-dichlorobenzilate; dicofol, 1,1-bis(*p*-chlorophenyl)-2,2,2-trichloroethanol; tetradifon, 2,4,5,4′-tetrachlorodiphenyl sulfone (mp 148°C); sulphenone, *p*-chlorophenyl phenyl sulfone (mp 98°C); ovex, *p*-chlorophenyl, *p*-chlorobenzenesulfonate (mp 86.5°C); propargite, 2-(*p*-*tert*-butlyphenoxy)cyclohexyl 2-propynyl sulfite; cyhexatin, tricyclohexylhydroxystannane (mp 195°C); and dienochlor, bis-(pentachloro-2, 4-cyclopentadien-1-yl) (mp 122°C).

Miscellaneous Insecticides

Formamidines. These are competitive agonists of octopamine or 1-(*p*-hydroxyphenyl)-2-aminoethanol, an insect neurotransmitter. They include chlordimeform, *N′*-(4-chloro-2-methylphenyl)-*N,N*-dimethylformamidine (mp 35°C, *d* 1.105, vp 0.045 mPa at 20°C), and amitraz, *N′*-2,4-dimethylphenyl-*N*-(*N′*-2,4-dimethylphenyl)-imino *N*-methylmethanimidamide (mp 86°C).

Avermectins and Ivermectin. The avermectins are pentacyclic lactones isolated from fermentation products of *Streptomyces avermitilis*, and ivermectin is a semisynthetic chemical, 22,23-dihydroavermectin B_1. These insecticides appear to function as agonists for the neuroinhibitory transmitter γ-aminobutyric acid (GABA).

Petroleum Oils. Petroleum oil sprays are used as insecticides for dormant sprays in the control of scale insects, mites, and insect eggs; summer foliage sprays for aphids, mealybugs, mites thrips, psyllids, whiteflies, and scale insects; livestock sprays for the control of lice, fleas, and mites; and mosquito larvicides. They also are used as carriers for contact insecticides to increase their effectiveness.

Fumigants. Fumigants are chemicals that are distributed through space as gases and therefore, at a given temperature and pressure, must exist in the gaseous state in sufficient concentration to be lethal to the insect pest. This physical requirement greatly limits the number of insecticides that may be usefully employed as fumigants. Compounds boiling at about room temperature, eg, hydrogen cyanide, methyl bromide, and ethylene oxide, are the most useful general fumigants. For soil fumigation, however, the slower the release of vapors from substances, eg, ethylene dibromide and β,β-dichlorodiethyl ether, that boil as high as 180°C, has proven effective. Other organic toxicants of relatively high vapor pressure, eg, naphthalene and *p*-dichlorobenzene, sublime readily enough to have special uses as fumigants; and contact insecticides, eg, lindane, dichlorvos, and mevinphos, may kill insects by vapor action under certain circumstances. The fumigant action of the less easily vaporizable insecticides is enhanced by the use of atomization, volatilization by heat, or burning in pyrotechnic mixtures. The true insecticidal fumigants are described in Table 2.

The fumigant is applied to an enclosure that is as gastight as possible.

Vacuum fumigation of packaged foods, tobacco, spices, pharmaceuticals, and other commodities is carried out in steel chambers or vaults at a pressure of 2–23 kPa (15–175 mm Hg).

Citrus and deciduous fruit trees have been fumigated for the control of scale insects for many years by hydrogen cyanide introduced under relatively gastight tents of 240–270 g/m^2 (7–8 oz/yd^2) army duck or nylon impregnated with vinyl chloride–vinyl acetate copolymer. Tent fumigation under plastic wrapping is used for termite control.

The fumigant may be injected directly ahead of the plow or applied by more elaborate soil injectors. Small plots are treated by pouring the fumigant into holes punched in the soil at regular intervals and covering immediately. More volatile soil fumigants, eg, methyl bromide, are applied under plastic sheeting that is sealed at the edges with earth.

The fumigants generally are highly reactive compounds that interact with vital biochemical processes within the target pest, usually by a bimolecular process, eg, alkylation. Their reactivity and high vapor pressure make it difficult to prevent widespread contamination of the surrounding air, water, and soil. Soil fumigation practiced with highly persistent, lipophilic compounds (see Table 2) at high dosages of 90–450 kg/ha (80–400 lb/acre) has contaminated the treated food commodity and water in shallow wells used for domestic supply. Fumigants, such as carbon tetrachloride, ethylene oxide, β,β-dichlorodiethyl ether, ethylene dibromide, and 1,2-dibromo-3-chloropropane (DBCP) are chemical carcinogens; the latter two compounds have produced azoospermia in workers exposed in both factory production and agricultural operations. For these reasons, all U.S. usage of DBCP was canceled in 1979.

Microbial Control. Insects are attacked by a multitude of pathogens. More than 450 viruses, 100 bacteria, 460 fungi, 250 protozoa, and 20 rickettsia are recognized as natural enemies. A number of these are adaptable for mechanical dissemination as microbial insecticides for the innoculation of insect populations, soils, fields, orchards, or forests with spores, virus suspensions, or microbial toxins.

Insect pathogens are used in three different ways: *(1)* to maximize the spread of naturally occurring epizootics, *(2)* by introducing them into pest populations as permanent mortality factors, and *(3)* by application as microbial insecticides consisting of suspensions of spores, virus particles, or microbial toxins. Microbial insecticides have great potential usefulness in IPM programs. However, they presently comprise only about 1% of the world insecticide market, but their usage growth rate is presently about 10 times greater than that of conventional synthetic insecticides. Commercial microbial insecticides are listed in Table 3. Bacterial products are presently the most widely used because of ease of production.

Insect Resistance to Insecticides

Many insect species have developed races that are sufficiently resistant to the action of specific insecticides so as to necessitate changes in control practices. This resistance, which has been described as accelerated microevolution, results from the selection of naturally occurring mutants that have acquired a resistant allele (R) that confers some degree of immunity to the insecticide through biochemical, physiological, or behavioristic factors.

Resistance Management. IPM practices that reduce the frequency of exposure to specific insecticides, incorporate source reduction, and incorporate a variety of suppressive measures such as biological and cultural controls and host plant resistance offer the most practical principles for resistance management.

Table 2. Properties of Fumigants

Name	Formula	Bp, °C	Specific gravity: Liquid d_4^{20},[a] g/mL	Specific gravity: Gas (air = 1)	Vp[a,b] kPa	Soly,[a] g/100 mL H_2O	Flammability in air,[c] vol %	TLV safe limit,[d] ppm	Uses
acrylonitrile	$CH_2{=}CHCN$	78	0.797	1.8	11		3	20	mills, commodities
carbon disulfide	CS_2	46.3	1.263	2.6	41.9	0.22	1	10	household
chloropicrin	Cl_3CNO_2	112	1.651	5.7	2.7	0.19	nf	0.1	soil, grain
1,1-dichloro-1-nitroethane	$CH_3CCl_2NO_2$	124	1.415		2.25	0.25		10	grain, stored products
1,2-dichloropropane	$CH_2ClCHClCH_3$	95.4	1.159		28	0.27		50	soil
trans-1,3-dichloropropene	$ClCH{=}CHCH_2Cl$	111	1.224		2.47	0.28	f	1	soil
ethylene chlorobromide	$ClCH_2CH_2Br$	107	1.689		5.3	0.7^{30}	nf		soil, commodities
ethylene dichloride	$ClCH_2CH_2Cl$	83.5	1.257	3.4	10	0.87	6		grain, soil household
ethyl formate	$HCOOC_2H_5$	54	0.917			10		100	dried fruits
ethylene oxide	$(CH_2)_2O$	10.7	0.887^{7}	1.5	146	∞	3	1	packaged foods
hydrogen cyanide	HCN	26	0.688	0.9	84.0	∞	6	10	general
methyl bromide	CH_3Br	4.5	1.732^{0}	3.3	189.3	1.34^{25}	13.5	5	general
methyl formate	$HCOOCH_3$	32	0.974		83.2	30		100	dried fruits
naphthalene	$C_{10}H_8$	218 (mp 80)		4.4	0.01	0.003	f	10	fabric pests, greenhouse
p-dichlorobenzene	$C_6H_4Cl_2$	173.4 (mp 53)		5.1	0.1	0.008^{25}	nf	75	fabric pests
phosphine	PH_3	−87.4	0.746^{-90}	1.2			2	0.3	grain
sulfuryl fluoride	SO_2F_2	−55.2	1.342^{25}	3.5	1601.3	0.075^{25}	nf	5	structural pests
trichloroacetonitrile	Cl_3CCN	85	1.44^{25}						grain
trichloroethylene	$ClCH{=}CCl_2$	86.7	1.470^{15}	4.5	9.7	insol	nf	50	grain

[a] At 20°C unless otherwise indicated by superscript, °C. [b] To convert kPa to mm Hg, multiply by 7.5. [c] nf = nonflammable; f = flammable. [d] Threshold limit value, time-weighted average.

Biochemistry of Resistance. Studies of specific gene regulation of the physiological processes causing insecticide resistance have identified a number of increasingly generalized biochemical mechanisms. Metabolic resistance is a result of enhanced detoxication of the insecticide by enzymatic processes such as glutathion transferases, esterases, epoxide hydrolases, and microsomal oxidases.

Insecticide Formulation

Insecticides are commonly formulated as dusts, baits, water dispersions, emulsions, and solutions. The preparation and use of such formulations involves accessory agents such as dust carriers, solvents, emulsifiers, wetting and dispersing agents, stickers, and deodorants or masking agents.

Application. The three general methods of applying insecticides are spraying, with water or oil as the principal carrier; dusting, with a fine dry powder as the carrier; and fumigation, where the insecticide is applied as a gas.

Genetic Control. Manipulation of the mechanisms of inheritance of the insect pest populations has occurred most successfully through mass release of sterilized males, but a variety of other techniques have been studied, including the environmental use of chemosterilants and the mass introduction of deleterious mutations, eg, conditional lethals and chromosomal translocations.

Repellents and Attractants. Repellants are substances that protect animals, plants, or products from insect attack by making food or living conditions unattractive or offensive. These substances, which may not be poisonous or only mildly toxic, are rarely, if ever, effective against all kinds of insects. Such chemicals can sometimes be employed to advantage where it is impossible to use an insecticide and may afford a greater or lesser degree of protection to manufactured products, growing plants, or the bodies of animals and humans. Among the many examples are the following. *(1)* Repellants against crawling insects. Examples are the creosote lines uses as barriers to the migration of chinch bugs; trichlorobenzene and other chemicals used to protect buildings from termites; heavy oils at the base of poultry roosts as a barrier to poultry mites; and certain chemical bands about tree trunks. *(2)* Repellants against the feeding of insects. These include the application of bordeaux, lime, and similar washes to plants to ward off leafhoppers and some chewing insects; mosquito repellants (ie, *N,N*-diethyl *m*-toluamide (DEET), dimethyl phthalate, and 2-ethyl-1,3-hexanediol) and fly sprays to lessen the attacks of blood-sucking flies and mosquitoes; the application of sulfur to the body to keep chiggers from attacking; the use of smoke and smudges to repel biting flies; the chemical treatment of logs to keep beetle borers from destroying log cabins and other rustic work; and moth balls, oil of cedar, and mothproofing treatments to protect materials from attack by clothes moths and carpet beetles. *(3)* Repellants against the egg laying of insects. Examples are the use of pine-tar oil and diphenylamine to keep screwworm flies from laying eggs about wounds of animals.

The use of specific insect attractants is an important component of modern IPM technology and promotes essentially nonpolluting and more economical pest control practices. These chemicals may have attractant, arrestant, and phagostimulant properties. They are used *(1)* to monitor insect populations in relation to the economic threshold, and to time control operations using a variety of inexpensive sticky

Table 3. Microbial Insecticides in Commercial Use or Development

Product	Pests controlled
	Bacteria
Bacillus popilliae	Japanese beetle larvae in soil
Bacillus thuringiensis aizawa	diamond back larvae, wax moth
Bacillus thuringiensis israelensis	mosquito and black fly larvae
Bacillus thuringiensis kurstaki	caterpillars
Bacillus thuriengensis tenebrionis	Colorado potato beetle
Bacillus sphaericus	mosquito larvae
	Fungi
Beauveria bassiana	European corn borer (China)
Hirsutella thompsoni	citrus red mite
Metarhizum anisopliae	sugar cane spittle bug (Brazil)
Verticillium lecanii	greenhouse aphids, whiteflies
	Protozoa
Nosema locustae	grasshoppers
	Viruses[a]
alfalfa looper NPV	alfalfa looper
beet armyworm NPV	beet armyworm
codling moth NPV	codling moth larvae (California, Europe)
gypsy moth NPV	gypsy moth larvae
heliothis NPV	cotton bollworms, tobacco budworm
pine sawfly NPV	pine sawfly larvae
tussock moth NPV	Douglas fir tussock moth larvae

[a] NPV = nuclear polyhedrosis virus.

traps of flat, cylindrical, and delta design; *(2)* to detect incipient infestations of exotic pests; *(3)* for removal trapping as with the Japanese beetle, bark beetles, and tsetse flies; *(4)* to lure insects to toxic baits; and *(5)* with insect female sex pheromones applied at relatively high dosages, such as for the gypsy moth, codling moth, pink bollworm, and grape berry moth, to cause mating disruption. The semichemicals used are categorized as *(1)* sex pheromones involved in intraspecific chemical communication; *(2)* aggregation pheromones promoting mass attack on host plants; and *(3)* plant kairomones involved in interspecific chemical communication leading to host plant selection. Attractants for more than 250 species of insect pests are available commercially.

Insecticide Residues in Foods

Approximately 70% of all insecticide use is in agriculture, and applications are generally made directly to raw agricultural commodities to protect plants and animals from insect attacks. With the exception of microbial insecticides, nearly all of these uses of insecticides result in residues of the various chemicals and their degradation products, which may be present in detectable amounts, ppb to ppm in food, despite weathering during crop maturation and attenuation during food processing. With increasing use of pesticides and the proliferation of plant protection chemicals, the nature and magnitude of such persisting residues have assumed great significance in public health and in the economics of international commerce in food products.

Residue Persistence Curves. Residue persistence curves define the required interval after application to attenuate to a toxicologically safe level, ie, the maximum residue limit. This defines the safe interval before harvest as expressed on the pesticide label directions.

Toxicology of Insecticide Residues. Risk assessment from the chronic ingestion of insecticide residues is made from the results of lifetime feeding studies at several dosage ranges with mice, rats, and dogs. The observed no adverse effect level (NAOEL) in mg of pesticide ingested per kg of body weight is evaluated by considerations of the test animals' general health, food intake, weight gain, gross histopathology, blood chemistry, and enzyme activity. The acceptable daily intake (ADI) for humans exposed to pesticide residues in the diet is determined from such laboratory animal investigations by incorporating a safety factor to accommodate for inter- and intraspecific variations. This safety factor is set at 10-fold where valid human exposure data are available, at 100-fold where there are valid laboratory data but no human data, and at 1000-fold where adequate chronic exposure data are available. ADI values are established and periodically reviewed by joint committees of the Food and Agricultural Organization (FAO) and WHO of the United Nations.

ROBERT L. METCALF
University of Illinois, Urbana-Champaign

R. L. Metcalf and R. A. Metcalf, *Destructive and Useful Insects*, 5th ed., McGraw-Hill Book Co., Inc., New York, 1993.

R. L. Metcalf and W. H. Luckmann, eds., *Introduction to Insect Pest Management*, 3rd ed., John Wiley & Sons, Inc., New York, 1994.

Agrochemicals Handbook, Royal Society Chemistry, Cambridge, U.K., 1991.

K. H. Büchel, ed., *Chemistry of Pesticides*, Wiley-Interscience, New York, 1983.

INSTRUMENTATION AND CONTROL. See PROCESS CONTROL.

INSULATION, ACOUSTIC

Acoustic insulation may be defined as a material or construction that reduces the passage or transmission of sound into or out of a medium such as air, water, or a solid structure. The term acoustic insulation covers a broad range of materials and mechanisms for the control of sound: sound-absorbing materials that reduce the reflections of impinging sound; sound-blocking materials that reduce sound transmission from one location to another; vibration isolating materials and devices that reduce transmission of vibrations from a vibrating source to potential sound-radiating structures; and vibration damping materials that reduce vibrations and sound radiation in and from materials and structures.

Sound Absorption

When a sound wave strikes a material, a fraction of its energy is reflected and a fraction is dissipated, or absorbed, by the material. The fraction of sound energy absorbed by a material is designated by its sound-absorption coefficient (α). The sound-absorption coefficient of a given material is between zero and one; if it is zero all the impinging energy is reflected and none absorbed; if it is one all the energy is absorbed and none reflected.

Units. The unit of sound absorption is the metric sabin, which is equivalent to one square meter of "perfect" absorption, eg, one square meter of a material with $\alpha = 1.0$. The English unit of sound absorption is the sabin, which is equivalent to one square foot of perfect absorption. The number of metric sabins of absorption provided by an area of material is calculated by multiplying its area by its sound-absorption coefficient. The sound absorption of materials is frequency dependent; most materials absorb more or less sound at some frequencies than at others.

Test Methods. Two basic types of test methods are commonly used to measure sound-absorption in test laboratories: the reverberation room method and the impedance tube method.

Materials. Most sound-absorbing materials are fibrous or porous and are easily penetrated by sound waves. Air particles excited by sound energy move rapidly to and fro within the material and rub against the fibers or porous material. The frictional forces developed dissipate some of the sound energy by converting it into heat. The fibrous materials most often used for sound-absorbing purposes are composed of either glass fibers or mineral fibers.

Foamed plastic acoustical materials are manufactured by two different processes. Both processes involve combining reactants that simultaneously produce a polymer, typically polyurethane, and generate a gas. Bubbles of gas expand the reacting mass and eventually form contiguous polyhedrons. If the contact planes between the polyhedrons rupture and establish openings between the cells, allowing air to penetrate, the material will have useful sound-absorbing properties.

Other fibrous and porous materials used for sound-absorbing treatments include wood, cellulose, and metal fibers; foamed gypsum or Portland cement combined with other materials; and sintered metals.

Resonant Sound Absorbers. Two other types of sound-absorbing treatments, resonant panel absorbers and resonant cavity absorbers (Helmholtz resonators), are used in special applications, usually to absorb low frequency sounds in a narrow range of frequencies. Resonant panel absorbers consist of thin plywood or other membrane-like materials installed over a sealed airspace. Resonant cavity absorbers consist of a volume of air with a restricted aperture of the sound field.

Uses. Sound-absorbing materials are frequently used to reduce reverberation, or the persistence of sound in a space after generation of the sound ceases; to prevent echoes from distant surfaces; to reduce focused reflections from curved surfaces; and to prevent the buildup of sound by multiple reflections within rooms and other enclosures. Sound-absorbing materials also are used to reduce reflections of noise from one location to another.

Products. There is a large number of commercially available sound-absorbing products for use on ceilings, walls, and for other special applications. Sound absorption coefficients and NRC values for some sound-absorbing products and treatments are indicated in Table 1.

Sound Isolation

When a sound wave comes in contact with a solid structure, such as a wall between two spaces, some of the sound energy is transmitted from the vibrating air particles into the structure, causing it to vibrate. The vibrating structure, in turn, transmits some of its vibrational energy into the immediately adjacent air particles on the opposite side, thereby radiating sound to this adjacent space.

The inherent ability of a material to block sound transmission is called its sound transmission loss (TL), typically reported in one-third octave frequency bands. The actual reduction of sound transmitted from one space to another is the noise reduction (NR) of the intervening construction.

Units and Rating Procedures. The unit of sound pressure level is the decibel (dB), defined as follows where L_p is the sound pressure level, p is the measured sound pressure, and p_{ref} is the reference sound pressure of 20 μPa. TL and NR also are expressed in decibels.

$$L_p = 10\log_{10}(p/p_{ref})$$

A single number rating of sound transmission loss, taking into account all frequency bands, is the sound transmission class (STC), defined in ASTM E413-87. Two similar single-number ratings are used for field measurements of sound isolation: noise isolation class (NIC) and field sound-transmission class (FSTC).

Table 1. Sound-Absorption Coefficients (α) for Some Sound-Absorbing Treatments

Treatment	Octave band center frequency, Hz 125	250	500	1000	2000	4000	NRC
1.9-cm thick mineral fiber acoustic tiles[a]							
low absorption	0.40	0.30	0.54	0.78	0.67	0.48	0.55
high absorption	0.67	0.62	0.66	0.88	0.99	0.99	0.80
2.5-cm nubby fiber glass ceiling panels	0.70	0.95	0.75	0.99	0.99	0.99	0.90
fabric-wrapped fiber glass panels							
2.5-cm thick	0.07	0.37	0.73	0.97	0.99	0.99	0.75
5.1-cm thick	0.23	0.81	0.99	0.99	0.99	0.99	0.95
heavy velour draperies[b]	0.15	0.35	0.55	0.70	0.70	0.70	0.60
thin carpet on concrete	0.02	0.05	0.10	0.15	0.25	0.50	0.15
heavy carpet on pad on concrete	0.05	0.10	0.30	0.50	0.70	0.80	0.40
7.6-cm acoustical steel deck[c]	0.73	0.99	0.99	0.89	0.52	0.31	0.85
2.5-cm sprayed cellulose fiber	0.08	0.29	0.75	0.98	0.93	0.76	0.75

[a] Mounting E-400. [b] 50% fullness spaced 15 cm from hard surface. [c] Ribbed deck with ribs filled with fiber glass and sides perforated.

These procedures are used only for rating airborne sound isolation. A related procedure is used for rating the effectiveness of floor/ceiling constructions in reducing impact noise transmission, such as footsteps, from upper floors to rooms below. The noise produced by a standard "tapping machine" is measured in the room below, and is rated in a similar manner to the STC rating procedure. This procedure is described in ASTM E492-90. The result is a single-number rating called the impact isolation class (IIC).

Test Methods. The laboratory test method for determining the sound-transmission loss performance of constructions is defined in ASTM E90-90.

$$TL = L_1 - L_2 + 10\log S - 10\log A_2$$

The purpose of noise reduction measurements in buildings is to determine the overall sound-isolating performance of the construction. Random noise is introduced into a source room. The space-averaged noise is then measured in the source room and the receiving room in one-third octave or octave frequency bands. The noise reduction is determined by subtracting the measured sound pressure levels in the receiving room from those in the source room. The noise isolation class (NIC) is determined using the STC rating procedure. Measurement of field sound-transmission loss is similar to noise reduction, except that it is used to determine the sound-isolating performance of a single element of the construction and can be compared directly to laboratory measurements. The STC rating procedure is used to determine the field sound-transmission class (FSTC). The field test method is defined in ASTM E336-90.

Materials. All common building materials provide some degree of sound isolation when used to separate adjacent spaces. The sound-isolating performance depends on a number of factors including mass, stiffness, size, and complexity of construction. In general, materials used for sound-isolating purposes must be impervious to air penetration.

Uses. Music buildings, studios, performance facilities, and other acoustically critical spaces usually require special sound-isolating constructions to provide high degrees of sound isolation from exterior noises and between rooms. Increases in road and air traffic mean higher levels of environmental noise in many areas, and soundproofing of residential properties and schools in the vicinity of highways and airports is being carried out in many locations in the United States. Within buildings, sound-isolating constructions also are used to enclose noise-producing air-handling units and other mechanical equipment.

Sound-Isolating Constructions. Although some materials are used alone in single-layer constructions for sound-isolating purposes, most sound-isolating construction contain two or more parts, frequently separated by an airspace, ie, double wall constructions. Table 2 provides STC ratings for some typical constructions.

Impact-Isolating Constructions. Adequate impact sound isolation is difficult to achieve when hard materials, such as terrazzo, quarry tile, vinyl tile, hardwood, etc, are used on floors on multistory buildings. Complex constructions incorporating resiliently supported floors and/or ceilings are required to reduce impact noise transmission when hard flooring materials are used. Carpeting can significantly reduce impact noise transmission, especially when installed over resilient padding.

Vibration Isolation

Units. The performance of a vibration isolator is characterized by its transmissibility, defined as the ratio of the force transmitted to the supporting side of the isolator compared with the driving force acting on the vibrating side of the isolator:

$$\text{transmissibility} = \text{output force/input force}$$

Table 2. STC Ratings of Some Common Building Constructions

Construction	STC
wood studs with gypsum board	35
double row of wood studs with gypsum board and fiber glass blankets	56
9-cm steel studs with gypsum board and fiber glass blankets	47
29-cm lightweight concrete block	47
29-cm poured concrete	58

The transmissibility of an isolator varies with frequency and is a function of the natural frequency (f_n) of the isolator and its internal damping.

Another measure of vibration isolation is isolation efficiency, which is one minus transmissibility and is usually defined as the percent of force transmitted through the isolator. Thus an isolator with a transmissibility of 0.75 has an isolation efficiency of 25%. A third measure of vibration isolation is insertion loss, which is the difference between the transmitted vibration with the isolators in place and with no isolators.

Materials. Materials commonly used in vibration isolators include steel in the form of springs, and elastomers in the form of cubes or pads.

Uses. Vibration isolators are used to reduce transmission of vibration into building structures from ventilating fans and other machinery; to isolate electron microscopes and other vibration sensitive equipment from sources of vibration; to reduce transmission of vibrations from motors into vehicle frames and bodies; to reduce transmission from rapid transit and railroad steel rails into bridges and other supporting structures; and to minimize vibration generated by motors and fans in various appliances.

Products. Commercially available vibration isolators include single and multiple coil springs with mounting bases and connectors for HVAC fans and other equipment. Spring hangers, neoprene hangers, and combinations of the two also are available for suspending vibrating equipment. Ribbed or waffled neoprene isolators typically are used to isolate equipment such as electrical transformers, which produce vibrational energy only at higher frequencies. At the other end of the spectrum are pneumatic isolators (air springs) consisting of inflated air bladders of neoprene or rubber with one or more tuned air chambers. They are used to isolate very low frequency vibrations.

Vibration Damping

Vibration damping is a process that reduces the vibrational energy in a system by converting some of the energy into heat. All materials and systems have some inherent damping, just as all materials absorb some sound, although in both cases the amounts can be very small. Damping is a highly complex phenomenon and there are many damping mechanisms, including interface friction, fluid viscosity, turbulence, acoustic radiation, eddy currents, and magnetic and mechanical hysteresis. Mechanical hysteresis is the only damping process that depends on internal friction within a material and is, therefore, also known as material damping.

Units. Two measures that are commonly used to define material damping are the loss factor (η) and amplification at resonance (Q), both of which are dimensionless: $\eta = D/2\pi W = 1/Q$ where D is the energy dissipated per cycle of vibration and W is the average total energy of the vibrating system. These relationships become more complex and less useful for highly damped systems.

Materials. The loss factors for viscoelastic materials are orders of magnitude higher than for rigid materials; as a result, these materials are widely used for damping treatments. The maximum loss factors for these materials at room temperatures range from about 0.2 to about 5.0.

Uses. The purpose of increasing the damping may be to reduce the vibration of the element at its resonant frequency, or it may be to attenuate flexural wave propagation along an extended structure, thereby increasing the sound-transmission loss of the structure.

Products. Damping treatments are available from many manufacturers in sheet form, as tapes adhering to a surface, and in bulk form for spraying or troweling onto a surface. Laminated glass, having a damping plastic interlayer, is available from many glass suppliers.

PARKER W. HIRTLE
CAROL E. PARSSINEN
Acentech Incorporated

Catalog of STC and IIC Ratings for Wall and Floor/Ceiling Assemblies, Office of Noise Control, California Department of Health Services, Berkeley, Calif., 1984.

R. S. Jones, *Noise and Vibration Control in Buildings*, McGraw-Hill Book Co., Inc., New York, 1984.

D. A. Bies and C. H. Hansen, *Engineering Noise Control Theory and Practice*, Unwin Hyman Ltd., London, 1988.

C. M. Harris, ed., *Handbook of Acoustical Measurement and Noise Control*, 3rd ed., McGraw-Hill Book Co., Inc., New York, 1991.

INSULATION, ELECTRIC

It has been predicted that by the year 2000 fiber optic cables will become the largest segment of the insulated wire industry, mostly at the expense of the telephone and electronic communication wires. These cables do not conduct electricity, but rather use light as the vehicle for communicating data.

Each segment of the insulated wire and cable industry has its own set of standards, and cables are built to conform to specifications provided by a large variety of technical associations, such as the The Institute of Electrical & Electronic Engineers (IEEE), The Insulated Cable Engineers Association, (ICEA), National Electrical Manufacturers Association (NEMA), Underwriters Laboratories (UL), Rural Electrification Administration of the U.S. Department of Agriculture (REA), Association of Edison Illumination Companies (AEIC), Military Specifications of the Department of Defense (MIL), American Society for Testing and Materials (ASTM), National Electrical Code (NEC), etc.

Designs and Materials

Data Communication Wires. Electronic cables such as data communication wires employ three basic designs: coaxial, twisted pair, and fiber optics. Cables are available in a variety of constructions and materials, in order to meet the requirements of industry specifications and the physical environment.

Building Wires. These wires conduct electricity at relatively low voltages (eg, 110 V and 220 V). Typically they contain a metallic conductor (copper or aluminum) that is insulated with polymeric compounds based on polyethylene or PVC which are applied over a conductor using an extruder.

Magent Wires. These wires are used principally in the electrical and electronics industries for coils, inductors, transformers, armatures, solenoids, etc.

Specialty Wires. Several categories of specialty wires employ special designs and materials, custom made to fit particular applications and/or specifications.

Appliance wires require a higher temperature rating (105°C or higher). Therefore, the insulation is made of fluorinated thermoplastics, such as polytetrafluoroethylene (PTFE) or fluorinated ethylene–propylene (FEP).

Instrumentation wires contain multiple pairs of conductors, each insulated with flame-retardant PVC and with an overall flame-retardant PVC jacket.

Depending on the specific application, a variety of polymers can be considered for military applications and aerospace wires including PVC, polyamides, PTFE, etc.

Railroad/transit cables are single and multiconductor cables, rated 300 to 2000 V. Their insulation can be based on ethylene–propylene rubber (EPR) and is specially compounded to be flame retardant; the jacket can also be flame retardant with low smoke emission during fire.

Control and signal cables are made up of fine copper wire strands of plain electrolytic copper wire with PVC or EPR-based insulation and an outer jacket of special PVC or ethylene copolymers.

Electric submersible oil well pump cables are rated up to 5 kV. Insulations can be based on polypropylene for low temperature wells or on ethylene–propylene rubber which is compounded with special ingredients in order to resist the environments of high temperature wells.

Power Cables. These high voltage cables have the most complicated designs.

Properties and Test Specifications

The most important electrical properties of insulation are dielectric strength, insulation resistance, dielectric constant, and power factor. Corona resistance, although not strictly an electrical property, is usually considered also.

Electrical and Water Treeing

Treeing is an electrical prebreakdown phenomenon. This type of damage progresses through a dielectric section under electrical stress so that, if visible, its path looks something like a tree. Treeing can occur and progress slowly by periodic partial discharge, it may occur slowly in the presence of moisture without partial discharge, or it may happen rapidly as the result of an impulse voltage. Although generally associated with a-c or impulse voltages, treeing has been observed with high d-c voltage stresses in wet experimental conditions. Treeing may or may not be followed by complete electrical breakdown of the dielectric section in which it occurs. In solid organic dielectrics it is the most likely mechanism of electrical failures which do not occur catastrophically, but rather appear to be the result of a more lengthly process.

Generally, trees occur under the relatively high voltages associated with power cables. Trees can be classified in three classes: electrical, water, and electrochemical.

Test Methods for Electrical and Water Treeing. In order to test resistance against electrical treeing, the concept of the standard defect is used in the needle test and modifications thereof.

The Association of Edison Illumination Companies (AEIC) has approved an accelerated cable life test in which typical underground distribution power cables can be statistically compared based on their resistance to water treeing (number of days to fail). The comparison can be made by varying the type of insulation and/or other cable layers in an environment that contains hot water (90°C) under 8 V/μm (200 V/mil) voltage stresses (four times the typical power cables operating voltages).

Physical, Mechanical, and Environmental Tests

Typical standard tests performed on insulation and/or jacket compounds measure tensile strength, ultimate elongation, modulus, set, tear, heat distortion, heat shock, cold bend and low temperature brittleness, abrasion resistance, and shear resistance. Depending on the environment in which the cable operates, the following tests may be done: resistance to oil or other chemicals, including water absorption; air aging resistance, measured at various temperatures either as percent retention of the sample initial physicals or the time to become brittle; oxygen and ozone resistance; radiation resistance when used in nuclear stations; flame resistance, measured as oxygen index or vertical or horizontal flame tests; smoke tests, using various equipment; and flame and smoke emission for the wires used indoors in the plenum areas are determined by the UL910 test.

Materials Used in Insulated Wires and Cables

The most widely used insulation compounds are based on PE, PP, silicone rubber, EPR, PVC, and fluoroplastics.

Magnet Wires. Magnet wires can be classified as coated wires, coated wires with fibrous wrappings, and wires with impregnated fibrous wrappings; the last two categories are older technologies. Examples of thermoplastic coatings are fluoropolymers, eg, Teflon or polyamides, eg, nylon. Thermosetting coatings are more resistant to cut-through and have superior resistance to heat and solvents. The silicones, polyamides, and fluorocarbons are best suited for very high temperatures applications, the polyurethanes for ease of removal, and epoxies for solvent and chemical resistance. Several other polymers are also used to coat the magnet wires.

Power Cables. The materials mostly used to produce power cables are ethylene copolymers loaded with conductive carbon black for semiconductive shielding layers, polyethylene or ethylene–propylene rubber-based compounds as insulations, and either thermoplastic materials (eg, polyethylene, PVC) or thermosetting (based on chlorinated polyethylene (CPE), chlorosulfonated polyethylene (CSPE), chloroprene, etc) for jackets.

Cross-linked polyethylene (XPLE) and ethylene–propylene rubber (EPR), both themosets, are the primary extruded dielectrics used in medium and high voltage power cables.

Power cable Conductor Shields provide a smooth, continuous, conductive, and isopotential interface between the conductor and insulation. There are two design approaches: most shields are either semiconductive shields that use large amounts of carbon black mixed in polymeric-based formulations, or stress-relieving shields that are based on materials with high dielectric constant. Brand names for the latter are Permashield or Emission Shield.

The insulation shield is a layer applied over the insulation. It plays much the same role as the conductor shield in protecting the insulation from the damaging effects of ionization at the outside of the insulation surface; therefore, it too must always remain in intimate contact with the insulation and be free of voids and defects at the interface. As an integral component of cable grounding, the insulation shield must be a resistive shield, providing a uniform ground around the insulation during field service; it also contributes to the grounding of the cable during switching surges, short circuits, or lightning strikes.

Besides the metallic coverings (based on aluminum, copper and copper alloys, lead, steel, and zinc), the most popular jacketing materials are based on polymeric materials that can be either thermoplastic (with limited high temperature use) or thermosetting.

ARMAND MOSCOVICI
The Kerite Company

Du Pont Co. guide, *How to Specify, Bid and Install Plenum Cable*, Wilmington, Del., 1986, pp. 6–16.

Kerite Co., *Kerite Cable-Dependable Power from Source to Load*, Seymour, Conn.

R. T. Vanderbilt Co., *Rubber Handbook*, Norwalk, Conn., 1991, pp. 701–715.

Union Carbide Corp., *Kabelitems Wire and Cables No. 150, 152, Treeing Update*, Danbury, Conn.

INSULATION, THERMAL

Insulation

A variety of cellular plastics exist for use as thermal insulation as basic materials and products, or as thermal insulation systems in combination with other materials. Polystyrenes, polyisocyanurates (which include polyurethanes), and phenolics are most commonly available for general use; however, there is increasing use of other types, including polyethylenes, polyimides, melamines, and poly(vinyl chlorides) for specific applications.

In the 1990s, however, primary applications include use in refrigerators and freezers, where cellular plastics account for over 90% of total insulation; in the building envelope, ie, from foundation to roof, 70%; and in walls, sheathings, and basements, ca 40%. Other uses include pipelines, refrigerated transportation, chemical processing, road and runway beds, and cryogenic applications.

The newer open-cell foams, based on polyimides, polybenzimidazoles, polypyrones, polyureas, polyphenylquinoxalines, and phenolic resins, produce less smoke, are more fire resistant and can be used at higher temperatures. These materials are more expensive and used only for special applications including aircraft and marine vessels. Rigid poly(vinyl chloride) (PVC) foams are available in small quantities mainly for use in composite panels and piping applications.

As a result of the energy crisis of the early 1970s the use of increased insulation for energy conservation became more economically attractive. All national codes and standards specify the need for levels of insulation in the building envelope (foundation to roof) dependent on the climatic region, and building applications in 1990s account for over 80% of the total volume of cellular plastic materials used for insulation purposes.

Function of Thermal Insulation

Three basic mechanisms of heat transmission occur in thermal insulation: radiation (electromagnetic waves), conduction (atomic or molecular collisions), and convection (fluid motion). The function of thermal insulation is to minimize and control these modes. This is accomplished primarily by introducing low emittance–high reflectance barriers to attenuate radiation; incorporating a large number of small, low density, low thermal conductivity elements to minimize solid conduction and convection; and including a high density, low thermal conductivity gas, or evacuation of encapsulated systems to minimize gas conduction and convection.

Although thermal performance is a principal property of thermal insulation suitability for temperature and environmental conditions; compressive, flexure, shear and tensile strengths; resistance to moisture absorption; dimensional stability; shock and vibration resistance; chemical, environmental, and erosion resistance; space limitations; fire resistance; health effects; availability and ease of application and economics are also considerations.

Cellular Plastics as Insulation

A low ($\leq$0.4 W/(m·K)) thermal conductivity polymer, fabricated into low density foam consisting of a multitude of tiny closed cells, provides good thermal performance. Cellular plastic thermal insulation have been developed in various types, eg, open-cell, closed-cell (the most energy efficient), and closed-cell containing gases with a thermal conductivity approximately one-half that of air. Cellular plastic thermal insulation can be used in the 4-350 K temperature range.

The drawbacks of cellular materials include limited temperature of applications, poor flammability characteristics without the addition of fire retardants, possible health hazards, uncertain dimensional stability, thermal aging and degradation, friability, and embrittlement due to the effects of uv light.

Properties

Table 1 lists and compares significant properties of insulation, which include themal conductivity, fire resistance, and minimal production of toxic gases, primarily during combustion. Other important criteria include water-vapor permeability, resistance to water absorption, and dimensional stability over prolonged periods of submission to extreme environments.

Thermal Conductivity and Aging. Thermal performance is governed by gas conduction and radiation. For closed-cell extruded polystyrene, polyisocyanurate, and phenolic foams containing high molecular weight and other low thermal conductivity gaseous blowing agents (at 24°C, $\lambda < 0.01$W/(m·K)), the initial values of λ as blown, are between 0.013 and 0.02 W/(m·K). These values can be maintained only if the aging process cannot occur, ie, air cannot diffuse into the cells or the blowing agent cannot diffuse out or partially dissolve into the polymer matrix. To accomplish this, the materials must be contained with impermeable, thick membranes such as metal sheet or hole-free foils well adhered to the surfaces of the cellular polymer.

Blowing Agents and Accelerated Aging Testing. The Montreal Protocol, an international agreement whereby CFCs were to be phased out gradually, was developed in 1987. It has subsequently been revised to ensure CFC elimination by the year 2000 at the latest. In addition, there was a nonbinding declaration of intent that the hydrochlorofluorocarbons (HCFCs), the most widely touted substitutes for CFCs, also be phased out no later than 2020.

The search for alternative blowing agents has necessitated a significant change in testing requirements to provide some assessment of aged value. Accepted accelerated testing procedures normally use >25 – mm thick specimens and time exposure periods of 180 days at 24°C or 90 days at 60°C to indicate some amount of aging.

An alternative method known as slicing and scaling has been developed. In this, the rate of diffusion is determined on thin specimens

Table 1. Typical Properties of Cellular Plastic Materials Used as Thermal Insulation

Property	ASTM method	Polyisocyanurate	XEPS[a]	EPS[b]	Polyimide	Polyethylene	Phenolic
density, kg/m^3	C591	30–40	30–48	12–30	8–12	21–32	45–60
closed-cell content, %	> 90	>90	>90	<10	>90	>90	
water-vapor permeability[c]	C355	2–3	0.4–0.15	1–4	high	0.02	< 1
water absorption, vol %	C272	2–5	0.15	2–4		<1	<2
thermal expansion × 10^{-6}/°C	E228	30–40	30–40	30–45	30–40	30–50	20–40
heat capacity, J/(kg·K)[d]	C351	1500	1200–1300	1200–1300			2000
thermal conductivity, W/(m·K)	C518	0.026[e] 0.020[f]	0.028	0.038–0.033	0.043	0.035	0.018
fire resistance	E136	combustible	combustible	combustible	combustible	combustible	combustible
flame spread[c]	E84	25–50	5–15	10–25	12	<25	20–25
smoke development[d]	E84	155–500 55–200	10–40	125	7	<50	5–15
toxicity		toxic gases when burned	CO when burned	CO when burned	CO when burned	CO when burned	CO when burned
dimensional stability, vol %	D2126	0–12	<2	<1			<1
upper temp limit, °C		120	75	75	260	180	150

[a] Extruded. [b] Molded. [c] Arbitrary (qualitative) units given in ASTM test. [d] To convert J to cal, divide by 4.184. [e] Aged unfaced. [f] Impermeable skins.

(6–10 mm thick) and a scaling factor S used to relate the results to thick specimens.

Thickness. Although radiation between parallel surfaces is independent of distance, the measurement of where radiation is significant requires the introduction of an additional variable, thickness. The thickness effect is observed in materials of low density at ambient temperatures and in materials of higher density at elevated temperatures.

Mean Temperature. Thermal performance is highly dependent on mean temperature.

Moisture. Absorbed and retained moisture, especially as ice, has a significant effect on the structural and thermal properties of insulation materials. The freezing of moisture and rupturing of cells result in permanent reduction of thermal and structural performance.

Mechanical Properties and Structural Performance. As a result of the manufacturing process, some cellular plastics have an elongated cell shape and thus exhibit anisotropy in mechanical, thermal, and expansion properties. Efforts are underway to develop manufacturing techniques that reduce such anisotropy and its effects. In general, higher strengths occur for the parallel-to-rise direction than in the perpendicular-to-rise orientation.

Flame Resistance. Fire characteristics of thermal insulations for building applications are generally reported in the form of qualitative or semiquantitative results from ASTM E84 or similar tunnel tests. Similar larger scale tests are used for aircraft and marine applications. New large-scale tests have been developed; the results can be taken to represent actual performance more closely.

Health and Safety Factors

The long-term effects of CFCs and HCFCs leaking into the environment have been discussed. Combustion where all cellular plastics can evolve smoke containing carbon monoxide and in certain cases cyanide and other toxic gases for various constituents involved in their manufacture is also a consideration.

Urea–formaldehyde use has been greatly restricted because of free formaldehyde emissions which can cause eye irritation and in some cases serious illness. Some attempts at developing formaldehyde-free urea-based materials are ongoing.

RONALD P. TYE
Consultant

C. J. Benning, *Plastic Foams*, Vol. 1, *The Physics and Chemistry of Product Performance and Process Technology*, John Wiley & Sons, Inc., New York, 1969.

An Assessment of Thermal Insulation Materials and Systems for Building Applications, DOE Report, BNL-50862 UC-95d, U.S. Dept. of Energy, Washington, D.C., 1978; R. P. Tye and D. L. McElroy, eds., *ASTM STP 718, Thermal Insulation Performance*, American Society for Testing and Materials, Philadelphia, Pa., 1980, pp. 9–26.

W. C. Turner and J. F. Malloy, *Thermal Insulation Handbook*, R. E. Krieger Publishing Co., Inc., Melbourne, Fla., 1981, pp. 191–275.

W. R. Strzepek, in E. C. Guyer and D. L. Brownell, eds., *Handbook of Applied Thermal Design*, McGraw-Hill Book Co., New York, 1989, pp. 3–30 to 3–41.

F. Sherwood-Rowland, *Chlorofluorocarbons and Depletion of Stratospheric Ozone, Improved Thermal Insulation–Problems and Perspectives*, D. A. Brandreth, ed., Technomic Publishing Co., Inc., Lancaster, Pa., 1991, pp. 5–25.

INSULIN AND OTHER ANTIDIABETIC AGENTS

Therapy of Diabetes

The goals of diabetes therapy include elimination of the clinical symptoms of hyperglycemia, prevention of long-term sequelae, and restoration of a sense of well-being. Therapeutic regimens are generally tailored to the individual. Patients having insulin-dependent diabetes mellitus (IDDM) require treatment with insulin. For some patients having noninsulin-dependent diabetes mellitus (NIDDM), careful attention to diet and exercise alone may have a profound impact on the disease. Some patients having NIDDM are best managed with insulin as well, whereas other are best treated using oral blood glucose lowering agents. In virtually all cases the requirement for insulin or an oral agent is reduced by proper attention to exercise and diet.

Insulin

Insulin is a peptide hormone produced in the islets of Langerhans within the pancreas, which acts as the principal regulator of glucose homeostasis. Under normal physiological conditions the β-cells within prancreatic islets secrete insulin into the bloodstream following nutrient ingestion. The insulin is then carried in the blood to targeted tissues, all of which have insulin receptors on their cellular surfaces. Tissues of insulin action include the liver, muscle, and fat. Insulin binding to the insulin receptors on these cells induces a series of intracellular events which culminate in the increased cellular uptake of circulating glucose and other nutrients as well as their storage as glycogen and fat, and increases in gene expression and protein synthesis.

Preparation of Insulins. Until the early 1980s insulin for therapeutic purposes was produced almost exclusively by extraction from beef and pork pancreases. However, these insulin differ from human insulin by one to three amino acids. Because patients treated with purified insulins still develop antibodies to insulin, enzymatic and biosynthetic methods are now used in the preparation of therapeutic insulin identical in sequence to human insulin.

Therapeutic Insulin Preparations. Insulin preparations for therapeutic use differ in time of onset, duration of action, purity, and species of origin. The concentration of all nonprescription insulins available in the United States is U100 (100 units/mL). A unit is the amount of insulin required to reduce the blood glucose of a fasting rabbit to 2.5 mmol/L (45 mg/dL). There are generally 24–30 units/mg of purified insulin. Insulin must be given by hypodemic injection or infusion pump because the hormone is destroyed in the gastrointestinal tract. Insulin preparation may be divided into rapid-, intermediate-, and long-acting, depending on the rapidity of onset and duration of action (Table 1).

Table 1. Characteristics of Insulin Preparations

	Action profile, h[a]			
Composition[b]	Onset	Peak	Duration	Insulin species[c]
		Short-acting		
insulin solution				
unbuffered	0.5	2–5	6–8	H,P,B/P
phosphate buffer	0.5	2–5	6–8	H
		Intermediate-acting		
protamine zinc suspension, phosphate buffer	1–2	4–12	18–26	H,P,B,B/P
amorphous and crystalline suspension, acetate buffer	1–3	6–15	18–26	H,P,B,B/P
NPH 70%, regular 30%	0.5	2–12	24	H,P
NPH 50%, regular 50%	0.5	2–12	24	H
		Long-acting		
crystalline suspension, acetate buffer	4–6	8–30	24–36	H,B

[a] Times are averages and can vary markedly between patients, insulin species, injection site, etc.
[b] For preparation methods, see text.
[c] H = human insulin; P = porcine insulin; B = bovine insulin; B/P = bovine/porcine mixture.

Table 2. Sulfonylureas Used as Oral Hypoglycemic Agents

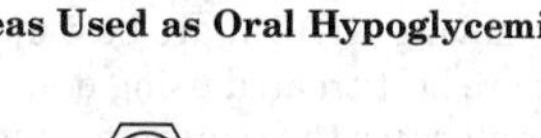

Compound	Molecular formula	R	R′
		First generation	
tolbutamide	$C_{12}H_{18}N_2O_3S$	$H_3C—$	$—(CH_{23}CH_3)$
chlorpropamide	$C_{10}H_{13}ClN_2O_3S$	Cl—	$—(CH_{22}CH_3)$
tolazamide	$C_{14}H_{21}N_3O_3S$	$H_3C—$	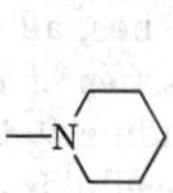
acetohexamide	$C_{15}H_{20}N_2O_4S$	$H_3CCO—$	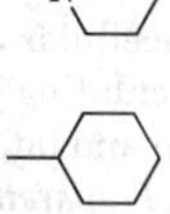
		Second generation	
glyburide (glibenclamide)	$C_{23}H_{28}ClN_3O_5S$	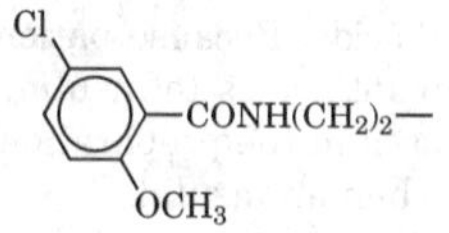	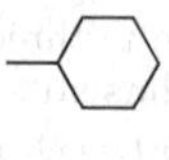
glipizide	$C_{21}H_{27}N_5O_4S$	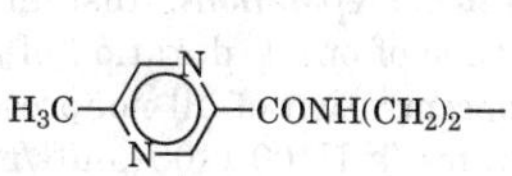	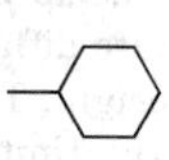
gliclazide	$C_{15}H_{21}N_3O_3S$	$H_3C—$	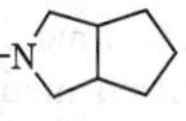

Oral Hypoglycemic Agents

Three classes of oral therapeutic agent are available for treating patients with diabetes mellitus (NIDDM): the arylsulfonylureas (known simply as sulfonylureas), biguanides, and α-glycosidase inhibitors (Table 2). The sulfonylureas appear to act by increasing insulin secretion from pancreas and potentrate its effects on target tissues. The mechanism of action of biguanides is unclear, although it does not increase insulin secretion. The α-glycosidan inhibitor, impair digestion of complex carbohydrates.

STEVEN E. SHOELSON
Joslin Diabetes Center, Harvard Medical School

C. R. Kahn and G. C. Weir, eds., *Joslin's Diabetes Mellitus*, 13th ed., Lea & Febiger, Philadelphia, Pa., 1994.

The Diabetes Control and Complications Trial Research Group, *New Engl. J. Med.* **329**, 977 (1993).

INTEGRATED CIRCUITS

The fundamental cornerstone of the development of the electronics age has been the concurrent decrease in cost associated with designing and manufacturing an increasingly sophisticated and miniaturized unit of integrated circuitry. Many devices are connected (or integrated) into a complete integrated circuit, or chip. Thousands of chips are formed simultaneously on a wafer. In general, a number of wafers are processed at the same time, although some equipment designs operate on one wafer at a time (single-wafer processing).

The nomenclature of integrated circuits has changed as the complexity of ICs has increased: small-scale integration (SSI) has evolved to medium-scale integration (MSI), to large-scale integration (LSI), and to the mature very-large scale integration (VLSI), which has 10^5 or more devices per chip. The next generation of ICs are classified as ultra-large scale integration (ULSI).

A new generation of IC technology has developed roughly every three years. The Rule of Two holds that approximately for every two generations (six years), the device feature size decreases by two, and other properties such as logic gate speed, chip area, power dissipation, and maximum input/output (I/O) pins increase by two. Forecasts predict an ambitious scaling device dimensions of 0.18 μm by 2001. Moreover, the number of chips per wafer is increasing, as well as the size of the chips and wafers. Silicon wafers are now (ca 1995) fabricated in sizes up to 200 mm in diameter, and undergoing development of 300–400 mm sizes.

Trends in the industry include transferring more of the functions that are found on the supporting printed circuit boards onto the wafers themselves, to reduce the amount of chip packaging and required interconnections. This development is known as wafer-scale integration (WSI), where the goal is to design a complete computer on a wafer. These developments have been supported by concurrent advances in computer-aided design (CAD) tools, both in software and hardware, which have been used to develop integrated CAD design systems that perform IC layout design, simulation, and testing.

Silicon, Si, technology predominates in the semiconductor industry. Gallium arsenide, GaAs, is considered a possible substitute for silicon substrates, based on its potential for high speed applications where it can operate at high (1.9 GHz) frequencies using low power consumptions and high sensitivity. One reason that GaAs technology has not fulfilled its promise is that silicon technology has dramatically improved in the interim, particularly with improvements in speed, and has reduced the cost-effectiveness of pursuing GaAs development. Expense has limited usage of GaAs to microwave devices, primarily for military use. However, nonmilitary applications for GaAs devices have been growing, particularly in wireless products such as cellular phones.

Basics of Silicon Technology

The property that allows silicon to function in a number of capacities is electronic configuration. Silicon has four electrons in its outer shell. Its crystalline structure allows other elements to reside next to silicon and share electron orbitals with silicon, altering its electrical properties. These other elements are called dopants, and are introduced into the silicon structure though doping processes. The most common dopant is boron, which has three electrons in its outer shell. Silicon doped with the electron-deficient boron has an overall positive charge and is called a *p*-type silicon. A second common dopant is phosphorus, P, having five electrons in its outer shell. Silicon doped with P has an overall negative charge, and is called *n*-type silicon. Arsenic, As, is another commonly used dopant.

The fabrication of an integrated circuit involves the sequential formation of alternating layers of insulators, semiconductors, and conductors on a silicon wafer. These layers are assembled to form transistor devices that are interconnected to produce particular electrical functions. These layers can be formed by deposition of new material, oxidation of material present on the surface, implantation of additional constituents into surface features, or epitaxial growth of silicon. In order to interconnect the layers, isolate devices from each other, and form integrated circuitry, these layers must be selectively patterned. The patterning is accomplished by photolithography and etching processes.

There are two kinds of integrated circuits (ICs): analogue, or linear ICs, and digital, or logic ICs. Analogue ICs produce, amplify, or respond to various voltages, and are used for any kinds of amplifiers, timers, oscillators, etc. Digital ICs respond to or produce signals that have only two voltage levels. These are used for microprocessors, memories, and microcomputers. It is possible to combine digital and analogue devices on one chip.

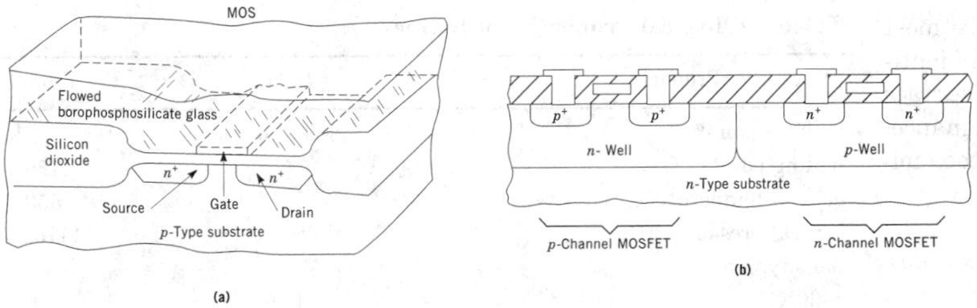

Figure 1. Cross sections of electronics devices used in ICs. (**a**) NMOS transistor; (**b**) a twin-tub CMOS device on an *n*-type substrate. Courtesy of Custom VLSI Microelectronics.

Digital IC families are further divided by design and function. The principal IC technologies include *p*- and *n*-channel metal-oxide semiconductors (PMOS and NMOS, respectively), complementary metal-oxide semiconductors (CMOS), and bipolar and integrated-injection-logic (I^2L) devices. Of these, CMOS designs are by far the most popular, having an estimated 73% of the worldwide market in 1994.

There are several resasons for the widespread use and development of CMOS devices, including low power density, relatively good noise immunity and soft error protection, design simplicity, and the capability to include lower power analogue and digital circuitry on the same chip. The most attractive feature has been the ability to scale CMOS technology to smaller dimensions.

Typical CMOS devices use both NMOS and PMOS transistors to form logic devices. A simple NMOS transistor is shown in Figure 1**a**. A common CMOS design combines NMOS and PMOS constructions in twin-well (twin-tub) structures, as shown in Figure 1**b**.

Crystal Growth and Wafer Preparation

The single-crystal silicon that is used in IC technology starts with a polycrystalline material called electronic-grade silicon (EGS). The polycrystalline EGS is converted to single-crystal silicon via the Czokralski (CZ) crystal growing process, based on the solidification of silicon atoms from the liquid phase at a moving interface. Volume production of 200-mm diameter crystals is standard.

Fabrication Processes

Process Integration. The fabrication of a VLSI integrated circuit involves sequential processing steps. Step interrelationships must be considered in designing a process sequence. The registration of one layer to another layer already present is repeated frequently with resist patterning followed by etching or implantation.

Processing Facilities. Two factors essential to successful IC processing are the quality of the process water and the ambient air. Use of ultrapure filtration and clean-room facilities addresses these concerns. In addition, three areas of facilities management are undergoing intense development to support the production of submicrometer VLSI and ULSI designs. The first is the area of particle control, where the thrust is to reduce even further the typical allowance of 0.05 particles/cm^2 that are less than 0.35 μm in size. There are two approaches to reduce the particle count: prevent particle generation, and prevent particles from depositing on wafers.

The minienvironment approach to contamination control has been increasing in use. Another approach is using integrated processing, where consecutive processes are linked in a controlled environment.

A second area of development that has impacted facility design is the trend to single-wafer processing, allowing enhanced control in processing individual wafers. This should carry greater importance as wafer size goes beyond 200-mm diameter to 300–400 mm.

A third facet of facilities development receiving much attention is the concept of total cost of ownership in evaluating process technology and equipment.

Analytical Techniques

The physical techniques used in IC analysis all employ some type of primary analytical beam to irradiate a substrate and interact with the substrate's physical or chemical properties, producing a secondary effect that is measured and interpreted. The three most commonly used analytical beams are electron, ion, and photon x-ray beams. Each combination of primary irradiation and secondary effect defines a spe-

Table 1. Analytical Techniques Used in VLSI Technology

Energy range, keV	Secondary signal	Acronym	Technique	Application
		Electron beam		
0.020–0.200	electron	leed	low energy diffraction	surface structure
0.300–30	electron	sem	scanning electron microscope	surface morphology
1–30	x-ray	emp	electron microprobe	surface region composition
500–10	electron	aes	Auger spectroscopy	surface layer composition
100–400	electron	tem	transmission electron microscopy	high resolution structure
100–400	electron, x-ray	stem	scanning TEM	imaging, x-ray analysis
100–400	electron	eels	electron energy loss spectroscopy	local small area composition
		Ion beam		
0.5–2.0	ion	iss	ion scattering spectrometry	surface composition
1–15	ion	sims	secondary ion mass spectrometry	trace composition vs depth
1–15	atoms	snms	secondary neutral mass spectrometry	trace composition vs depth
≥1	x-ray	pixe	particle induced x-ray emission	trace composition
5–20	electron	sim	scanning ion microscope	surface characterization
>1000	ion	rbs	Rutherford backscattering	composition vs depth
		Photon beam		
>1	x-ray	xrf	x-ray fluorescence	composition (μm depth)
>1	x-ray	xrd	x-ray diffraction	crystal structure
>1	electron	esca.xps	x-ray photoelectron spectroscopy	surface composition
laser	ion		laser microprobe	composition of irradiated area
laser	light	lem	laser emission microprobe	trace elements (semiquantative)
		Neutron beam		
reactor	gamma	naa	neutron activation analysis	bulk (trace) composition

cific analytical technique. The IC substrate properties that are most frequently analyzed include size, elemental and compositional identification, topology, morphology, lateral and depth resolution of surface features or implantation profiles, and film thickness and conformance. A summary of commonly used analytical techniques for VLSI technology can be found in Table 1.

IRENE SAWCHYN, PH.D.
Consultant

C. Hu, *Semiconductor Int.*, 105–114 (June 1994).

P. Singer, *Semiconductor Int.*, 56–59 (Apr. 1994).

S. M. Sze, ed., *VLSI Technology*, 2nd ed., McGraw-Hill Book Co., Inc., New York, 1988.

H. W. Fry and co-workers, *Solid State Technol.*, 31–40 (Mar. 1994).

INTEGRATED PEST MANAGEMENT SYSTEMS. See INSECT CONTROL TECHNOLOGY; REPELLENTS.

IODINE AND IODINE COMPOUNDS

Iodine

Iodine, I, atomic number 53, atomic weight 126.9044, is a nonmetallic element belonging to the halogen family in Group 7 (VIIA) of the Periodic Table. The only stable isotope has a mass number of 127. There are 22 other iodine isotopes having masses between 117 and 139; 14 of these isotopes yield significant radiation.

Iodine is a bluish black, crystalline solid having a metallic luster. It is obtained in shiny flakes or prills that can be easily crushed to powder. Iodine crystallizes in rhomboidal plates belonging to the triclinic system.

Occurrence in Nature. Iodine is, indeed, one of the scarcest of the nonmetallic elements in the total composition of the earth. Although not abundant in quantity, iodine is distributed in rocks, soils, waters, plants, animal tissues, and foodstuffs. Excepting the possible occurrence of elemental iodine vapor in the air near certain iodine-rich springs, iodine never occurs free in nature. It is always found combined with other elements.

Wherever it occurs, the quantities of iodine are generally exceedingly small, and very sophisticated chemical methods are required to detect them. Only a few substances characteristically contain iodine in relatively large quantities. These are seaweeds, sponges, and corals; the underground waters from certain deep oil-well boring and mineral springs; and, most impressive of all, the vast natural deposits of sodium nitrate, caliche, ore found in the northern part of Chile.

Physical Properties. The cell constants at 18°C are given in Table 1, along with other physical properties. Iodine is only slightly soluble in water and no hydrates form upon dissolution. Iodine dissolves in many organic solvents, and the color of the resulting solutions varies with the nature of the solvent. Iodine vapor is characterized by the familiar violet color and by its unusually high specific gravity, approximately nine times that of air.

Chemical Properties. The electron configuration of the iodine atom is $[Kr]4d^{10}5s^{2}5p^{5}$ and its ground state is $2p^{0}_{3/2}$. Principal oxidation states are −1, +1, +3, +5, and +7, but the oxide IO_2 where iodine has an oxidation state of +4 is also known. Iodine forms thermodynamically stable compounds in all these oxidation states, except the +4. Iodine is the heaviest of all the common halogens and the least electronegative. It is usually less violent in its reactions than the other members of the halogen family. Iodine presents mild oxidizing properties in acidic solutions.

Iodine forms compounds with all the elements except sulfur, selenium, and the noble gases. It reacts only indirectly with carbon, nitrogen, oxygen, and some noble metals such as platinum.

Table 1. Physical Properties of Iodine

Properties	Solid	Liquid
melting point[a]	113.6	
boiling point,[b] °C		185
critical temperature, °C		553
critical pressure, kPa[c]		11753.7
density, g/mL		
60°C	4.89	
120°C		3.960
entropy at 25°C, J/(mol·K)[d,e]	116.81	
specific heat at 25–113.6°C, J/(g·K)[d]	$0.1582 + 1.9628 \times 10^{-4}\ T^{f}$	
viscosity at 116°C, mPa(= cP)		2.268
vapor pressure at 25°C, kPa	0.04133[a]	
thermal conductivity at 24.4°C, W/(m·K)	0.4581	
electrical resistivity at 25°C, Ω·cm	5.85×10^{6}	
dielectric constant at 23°C	10.3	
refractive index, n_D	3.34	

[a] For the solid, between 0 and 113.6°C, $\log p_{kPa} = -(3410.71/T) - 0.3523 \log T - 1.301 \times 10^{-3}T + 14.3140$ where T is in Kelvin. [b] For the liquid, between 113.6 and 186°C, $\log p_{kPa} = -(2300.24/T) + 10.025$, where T is in Kelvin. [c] To convert kPa to atm, divide by 101. [d] To convert J to cal, divide by 4.184. [e] Entropy = 62.25 J/(mol·K). [f] T is in Kelvin.

The chemistry of aqueous iodine has been extensively studied because of the role of iodine as a disinfectant. The system is very complex, owing to the number of oxidation states available to iodine under ambient conditions.

Manufacture and Processing. The production processes are based on the raw materials containing iodine: seaweeds, mineral deposits, and oil-well or natural gas brines.

The earliest successful manufacture of iodine started in 1817 using certain varieties of seaweeds. The seaweed was dried, burned, and the ash lixiviated to obtain iodine and potassium and sodium salts.

The only iodine obtained from minerals has been a by-product of the processing of nitrate ores in Chile.

There are two ways for producing iodine form caliche iodates: first, from solutions containing more equivalent iodine than its solubility as elemental iodine in the same solution of about 0.4 g/L at 25°C; and second, from more diluted equivalent iodine solutions.

About 65% of the iodine consumed in the world comes from brines processed in Japan, the United States, and the former Soviet Union. The predominant production process for iodine from brines is the blow-out process, which was first issued in Japan. Iodine is present in brines as iodide, and its concentration varies from about 10 to 150 ppm. The recovery process can be divided into brine clean-up, iodide oxidation to iodine followed by air blowing out and recovery, and iodine finishing.

Materials of Construction. High silicon iron, Stellite 6, Hastelloy C, and stainless steels types 304, 309, 316, and 317, have low corrosion rates when immersed in an aqueous solution containing 5% I_2 and 7.5 Kl at 25°C. Hastelloy B gives promising results when exposed to liquid and gaseous iodine up to 400°C; molybdenum seems capable of withstanding even higher temperatures.

Among nonmetallic materials, glass, chemical stoneware, enameled steel, acid-proof brick, carbon, graphite, and wood are resistant to iodine and its solutions under suitable conditions, but carbon and graphite may be subject to attack.

Economic Aspects. Iodine traditionally has had boom and bust cycles, but the oversupply scenario appearing by the end of 1990 was more acute than previous lows.

Health and Safety Factors and Regulations. Iodine is much safer to handle at ordinary temperatures than the other halogens because iodine is a solid and its vapor pressure is only 1 kPa (7.5 mm Hg) at 25°C, compared to 28.7 kPa (215 mm Hg) for bromine and 700 kPa

(6.91 atm) for chlorine. When handling properly packed containers, usual work clothes are sufficient. In the handling of solid, unpacked iodine, rubber gloves, rubber apron, and safety goggles are recommended. Respirators or masks are also recommended.

The U.S. Occupational Safety and Health Administration (OSHA) has set a ceiling level for iodine of 0.1 ppm in air. The American Conference of Government and Industrial Hygenists (ACGIH) established 0.1 ppm as the TLV (TWA) for iodine. The maximum allowable concentration in air (MAK value) is also 0.1 ppm.

Empty containers may be destroyed in an incinerator or decontaminated by washing with a dilute thiosulfate or sulfite solution. Bulk wastes should be treated by controlled iodine recovery processes.

Iodine can affect the body if inhaled, if it comes in contact with the eyes or skin, or if it is swallowed. It may enter the body through the skin. Chronic absorption of iodine causes iodism characterized by insomnia, inflammation of the eyes and nose, bronchitis, tremor, diarrhea, and weight loss.

Iodine is not combusible by itself, but can react very vigorously with reducing materials.

Inorganic Iodine Compounds

Inorganic iodine compounds include iodides [alkali metal iodides (KI, mol wt 166.02, mp 686°C); sodium iodide (NaI, mol wt 149.92, mp 662°C)], hydrogen iodide (HI, mol wt 127.93, mp −50.9°C, bp −35.1°C), iodates [potassium iodate (KIO_3, mol wt 214.02); sodium iodate ($NaIO_3$, mol wt 197.90); calcium iodate monohydrate ($Ca(IO_3)_2$ H_2O, mol wt 407.90); iodic acid (HIO_3, mol wt 179.93)], and iodine halides and polyhalides [RbI_3; $KICl_4$; KIBrCl; $N(CH_3)_4I_9$; $N(C_{25}H)_4IBr_2$; iodine monochloride (ICl, mol wt 162.38); and iodine trichloride (ICl_3, mol wt 233.39)].

Organic Iodine Compounds

The organic iodine compounds have lower heats of formation and greater reactivities than their chloro and bromo analogues. As in the case of the inorganic iodides, their indexes of refraction and specific gravities are higher than those of the corresponding chloro and bromo derivatives.

The aliphatic iodine derivatives are usually prepared by reaction of an alcohol with hydrogen iodide or phosphorus triiodide; by reaction of iodine, an alcohol, and red phosphorus; addition of iodine monochloride, monobromide, or iodine to an olefin; replacement reaction by heating the chlorine or bromine compound with an alkali iodide in a suitable solvent; and the reaction of triphenyl phosphite with methyl iodide and an alcohol. The aromatic iodine derivatives are prepared by reacting iodine and the aromatic system with oxidizing agents such as nitric acid, fuming sulfuric acid, or mercuric oxide. They include methane derivatives [methyl iodide (CH_3I, mol wt 141.95); methylene iodide (CH_2I_2, mol wt 267.87, mp 6.0°C, and bp 181°C); and iodoform (CHI_3, mol wt 393.78, mp about 120°C)], and other iodo compounds [thymol iodide ($C_2OH_{24}I_2O_2$, mol wt 550.23); ethyl iodide (C_2H_5I, bp of 72.2°C); and iodobenzene (C_6H_5I, mol wt 204.02, mp −30 C, bp 188–189°C)].

Uses

Iodine has a wide range of uses in the chemical and allied industries. A high percentage of the initial use of iodine lies in the production of intermediates, which are frequently marketed as such. The breakdown of world iodine consumption for 1992 suggested that 35% was used to manufacture pharmaceuticals, 30% for inorganic salts, 15% as sanitizers (iodophors), 12% was used to produce other organic derivatives, 5% for agricultural chemicals, and 3% for catalysts. The expected annual increase in iodine consumption is about 5% in the area of pharmaceuticals and sanitizers, and 2% for the inorganic salts. An increase of 1–2% is estimated for the other areas. Iodine is also used in photography and dyes, inks, and colorants.

ARMIN LAUTERBACH
GUSTAVO OBER
SQM Iodine

R. F. Rolsten, *Iodide Metals and Metal Iodides*, John Wiley & Sons, Inc., New York, 1961.

The Economics of Iodine, 6th ed., Roskill Information Services, Ltd., London, Sept., 1994.

Etude Bibliographique sur l'Iode et les Derives de l'Iode, SQM Iodine Corp. Internal Document, SQM Iodine Corp., Norfolk, Va., Apr. 1993.

IODINE VALUE. See CARBOXYLIC ACIDS; FATS AND FATTY OILS.

IODOACETIC ACID. See ACETIC ACID.

IODOFLUORO HYDROCARBONS. See FLUORINE COMPOUNDS, ORGANIC.

ION EXCHANGE

Ion exchange is a process in which cations or anions in a liquid are exchanged with cations or anions on a solid sorbent. Cations are interchanged with other cations, anions are exchanged with other anions, and electroneutrality is maintained in both the liquid and solid phases. The process is reversible, which allows extended use of the sorbent resin before replacement is necessary.

Many naturally occurring inorganic and organic materials have ion-exchange properties. This article places emphasis on the styrenic and acrylic resins that are made as small beads.

The primary application for ion exchange is the softening and deionization of water. The remaining applications include waste treatment, catalysis, purification of chemicals, plating, hydrometallurgy, food processing, and pharmaceutical uses. Because ion-exchange resins are insoluble polymeric acids and bases, these resins are also useful in removing acids and bases from gaseous streams via the neutralization of functional groups.

Weak and strong acid-type resins are for removal of cations and are called cation exchangers. Weak and strong base resins remove anions and are called anion exchangers. In addition to these four resin types, there are specialty resins used in applications where higher specificity for certain ions under challenging conditions is a critical factor.

Continuous columnar operation of ion-exchange systems is preferred over batch operation. Each column must be taken off-stream periodically to remove the adsorbed ions and restore the resin to the ionic form required for the adsorption step. In this sense, a columnar ion-exchange operation is not continuous. Continuous operation has been approached in a number of designs by using muliple columns in parallel and staging regeneration, or by moving resin, or vessels containing resin, in a direction opposite to the flow of liquid. Some of these approaches have been abandoned; others are increasing in popularity.

Manufacture of Resins

The production of ion-exchange resins is a multiple step process. It begins with the polymerization of monomers to form solid intermediate copolymers that are insoluble in both water and solvents. The copolymers are functionalized during additional steps in different reactors from those used for copolymer production. Conversion to another ionic form may be required after functionalization is completed. Excess water is removed by vacuum filtration prior to packaging. Packaging in fiber drums is common. Alternative containers include metal drums, bulk boxes and bags, and smaller plastic, paper, and burlap bags. Polyethylene liners are used as a barrier between containers and water-containing resin.

Manufacture of ion-exchange resins has traditionally been a batch process. Significant progress was made more recently in the development of a continuous process for the manufacture of copolymer beads.

Copolymerization. The chemistry of the resin matrix, the type and degree of porosity, the particle size, and the particle size distribution are established in the copolymerization step. Formulations and operating procedures must be strictly followed. Reaction vessels must be well designed. Mistakes made during copolymerization are rarely corrected during functionalization.

Functionalization. Copolymers do not have the ability to exchange ions. Such properties are imparted by chemically bonding acidic or basic functional groups to the aromatic rings of styrenic copolymers, or by modifying the carboxyl groups of the acrylic copolymers. There does not appear to be a continuous functionalization process on a commercial scale.

Physical and Chemical Properties

Ion-exchange resins are used repeatedly in a cyclic manner over many years, and deterioration of both physical and chemical properties can be anticipated. Comparison of the properties of used resin with those of new resin is helpful to learning more about the nature and cause of deterioration. Corrective action frequently extends the life of the resin. Comparison of properties must always be made with the resin in the same ionic form.

Particle Shape and Size. With few exceptions, resins are supplied as small, round beads having a diameter between 0.3 and 1.2 mm. Some resins are reduced to a smaller size by grinding to satisfy specific requirements in applications for electric power generation and pharmaceuticals.

Density and Specific Gravity. Density generally pertains to the bulk, or pack-out, weight of wet resin per unit volume. The density is characteristic of the resin and is dependent on the copolymer structure, the degree of cross-linking, the nature of the functional groups, and the ionic form of those groups. A change in density after extended use is a signal that chemical degradation has occurred. The density of most cation exchangers is in the 800–900 g/L range, whereas most anion exchangers are in the 640–740 g/L range.

The specific gravity generally refers to the value determined for wet resin when using a pycnometer. Values range from about 1.04 to about 1.25. Cation exchangers have a greater specific gravity than anion exchangers.

Porosity. The structure of ion-exchange resins is either microporous or macroporous. Microporous resins are more commonly referred to as gel or gellular-type resins.

Capacity. Capacity is a measure of the quantity of ions, acid, or base removed (adsorbed) by an ion-exchange material. The quantity removed is directly correlated with the number of functional groups. Operating capacity, also called the working capacity or column capacity, is a measure of the quantity of ions, acids, or bases adsorbed, or exchanged, under the conditions existing during batch or columnar operation.

Selectivity. A significant exchange of ions does not occur unless the functional group of the resin has a greater selectivity for ions in solution than for ions occupying the functional group, or unless there is a mass action effect, as in regeneration. Selectivity coefficients have been reported in numerous publications for both cations and anions.

The need to know selectivity coefficients precisely is rarely necessary in industrial applications. However, knowledge of relative differences is important when deciding if the reaction is favorable or not.

Selectivity differences increase as the degree of cross-linking of a resin increases, but these differences are relatively minor. Structural composition of the functional groups has a much greater effect on the magnitude of selectivity differences.

Kinetics. The degree to which an ion-exchange reaction is completed depends on a number of factors which include contact time, ionic concentration, degree of cross-linking, and temperature.

Moisture and Water Content. Each resin has a characteristic water content dependent on the resin matrix, the structure of the functional groups, and the ionic form of those groups. Resins are packaged by weight and sold by volume. The dewatering operation prior to packaging is a critical step since removal of too little is costly to the buyer. Analyzing for water content is important to both the seller and user. The quantity of water contained by the resin is recorded on a percentage basis and determined by two methodologies. In each procedure, a small (ca 15 g) sample is removed from a larger composite sample collected during pack-out. In one procedure, the sample is accurately weighed before and after placing in a 105°C oven for at least 8 h. This procedure yields the moisture content typical of resin contained in the shipping containers. In the other procedure, a similar sample is soaked in water, then filtered under vacuum in a Buchner funnel prior to weighing before and after oven drying. The moisture content reported is a pseudoequilibrium value typical of the specific resin and its ionic form.

Swelling and Shrinking. Ion-exchange resins shrink or swell reversibly as they are converted from one ionic form to another. The degree of change is dependent on the resin matrix, the functional group, and the ions adsorbed by the functional groups.

The degree of swelling and shrinking is important for design of ion-exchange columns, especially for the location of the distributors used to disperse incoming fluids, and collect outgoing ones, evenly over the cross-sectional area of the resin bed.

Hydraulic Properties. Both the resistance to liquid flow through a resin bed and the degree to which a resin bed expands during a backwashing step are important design factors for ion-exchanging systems. These characteristics are also critical to those using the resins because movements of resins not only signal the existence of a problem but give indications as to the nature of the problem. Pressure drop and hydraulic information for new resins are available from the resin manufacturer and the supplier of equipment.

Factors which have the greatest impact on pressure drop are the depth of the bed, flow rate, viscosity, temperature, and particle size.

Backwashing is the upward flow of water through a bed of resin at a flow rate sufficient to fluidize the resin, but not so great that resin is carried out of the column with the exiting water. Resins are backwashed to remove dirt and resin fragments, to classify resin particles by size, and to relieve any packing that may have occurred with previous use.

Oxidants, such as dissolved chlorine in water supplies, react with synthetic ion exchangers to cause a loss of capacity, physical weakening of the resin, and partial solubilization of the resin. Anion-exchange resins are most prone to loss of functionality as the oxidant attacks and severs the linkage between nitrogen and carbon on the polymeric structure. In addition to this form of degradation, some of the functional groups of strong base anion exchangers are converted to weak base groups. The overall effect is a loss of both strong base and total capacity with an increase in weak base capacity. Loss of functional groups with cation-exchange resins by oxidative attack is uncommon. The rate of oxidative attack is enhanced by the presence of metals such as iron and copper which serve as catalysts, by higher temperatures, and by higher concentrations of oxidants. Aside from low concentrations of oxidants found in most water supplies, the processing of chemical streams with much higher levels of oxidizing chemicals is practiced occasionally on an industrial basis. The potential dangers are generally recognized.

Ion-exchange resins should not be used at temperatures above those recommended by the manufacturer. Exceptions are made when frequent replacement of resin is an economic advantage over the operating and capital cost of cooling and reheating the process stream. Functional groups are lost from both cation- and anion-exchange resins when the temperature limit is exceeded.

Excessive pressure drop across the resin bed causes fragmentation of the beads. Resins shrink and swell as they are alternately put through adsorption and regeneration cycles. The larger the volume change and the shorter the time involved, the greater the potential for physical damage to the resin particles. The appearance of cracks is the first sign of physical deterioration. Fragmentation into smaller irregularly shaped particles is a sign of further deterioration.

Resins should always be protected freezing, although that may not always be possible. Generally, a few freeze–thaw cycles do not result in visual damage (cracking or fragmentation). Nevertheless, some weakening of the physical structure occurs because fragmentation is apparent if cycling continues.

Cation and anion exchangers lose weight and capacity, cross-linking is reduced, and water-soluble components are released if the radiation tolerance limits have been exceeded. The effects of gamma radiation have been studied more than other types of radiation.

Equipment

Ion-exchange systems in process applications may be batch, semicontinuous, or continuous. Batch opertions are not common but, where used, involve a kettle with mechanical agitation. Injecting with air or an inert gas is an alternative. A screened siphon or drain valve is required to prevent resin from leaving with the product stream.

Semicontinuous and continuous systems are, with few exceptions, practiced in columns. Most columnar systems are semicontinuous since flow of the stream being processed must be interrupted for regeneration. Columnar installations almost always involve the process stream flowing down through a resin bed. Those that are upflow use a flow rate that either partially fluidizes the bed, or forms a packed bed against an upper porous barrier or distributor for process streams.

Systems

Ion-exchange systems vary from simple one-column units, as used in water softening, to numerous arrays of cation and anion exchangers which are dependent upon the application, quality of effluent required, and design parameters.

Cyclic Operation

Resins are seldom used once and discarded. Whether the system is run batchwise or in columns, the resin must be periodically removed from service and regenerated. An exception is the use of a resin as a catalyst in organic reactions. Each cycle consists of two principal steps, rinse and backwash. Failure to use good practices results in poor cyclic performance.

Shipping

Shipping resins in a water wet condition is standard practice. Removal of water by evaporative methods is expensive and not necessary for the majority of application since they take place in aqueous systems. Dry resins (almost always strong acid cation exchangers) are required in several catalytic applications in the chemical industry.

Economic Aspects

Commercial producers of synthetic spherical organic ion-exchange resins include Bayer, Chemolimpex, Dow, Mitsubishi Kasei, Ostion, Purolite, Röhm and Haas, and Sybron.

Replacement sales have become an ever-increasing percentage of total sales in the mature ion-exchange industry. Economic downturns such as that of the early 1990s affect resin purchases; plans for new installations are abandoned or delayed. Greater effort is made to forestall purchase of replacement resin by directing old resin to a cleaner condition. In addition, quality improvement efforts by resin manufacturers have yielded products having greater physical durability, thereby lessening resin losses caused by physical attrition. Competitive technologies such as reverse osmosis have had an increasing impact on lowering the volume of resin required for old as well as new installations. Additionally, the once large installed volume of resin in the uranium industry has disappeared as environmental concerns over nuclear reactors have brought a halt to construction and operation of plants of this type, and as the cold war ended. On the positive side, a substantial growth of installed resin has occurred in processing corn sweetener for use in the beverage industry. Significant growth has taken place in catalytic applications, especially for the production of methyl *tert*-butyl ether, a gasoline octane enhancer used in place of tetraethyllead.

Sales of cation-exchange resins have routinely been slightly more than twice the volume of anion exchangers, although deionization require more anion exchanger than cation. One of the reasons for the opposite ratio is the large volume of cation exchangers used in home and industrial plants for softening water. Another is the chromatographic separation of fructose and glucose in the corn sweetener industry, which requires large volumes of cation-exchange resin. A similar process for recovery of sucrose from beet sugar molasses added to the demand for cation exchangers.

Health and Safety

Ion-exchange resins are not considered hazardous. However, cation exchangers when in the hydrogen form, and anion exchangers when in the hydroxide form, yield acidic and basic solutions, respectively, when in contact with neutral salt solutions. The corrosive potential should not be overlooked, and skin sensitivity has been reported occasionally, especially when gloves are not used when handling resin. Resins which have been used to remove toxic substances may slowly release these materials if the toxic substances are still attached to the resin.

CHARLES DICKERT
Consultant

F. C. Nachod and J. Schubert, eds., *Ion Exchange Technology*, Academic Press, Inc., New York, 1956.

R. Kunin, *Ion Exchange Resins*, Robert E. Krieger Publishing Co., Huntington, N.J., 1972.

K. Dorfner, *Ion Exchangers*, Ann Arbor Science Publishers, Ann Arbor, Mich., 1973.

C. Calmon, *React. Polym.* **4**(2), 131–146 (1986).

ION IMPLANTATION

Modern technology depends on materials with precisely controlled properties. Ion beams are a favored method (and in integrated circuit technology, the prime method) to achieve controlled modification of surfaces and near-surface regions. In every integrated circuit production line there are ion implantation systems. In addition to integrated circuit technology, ion beams are used to modify the mechanical, tribological, and chemical properties of metals, intermetallics, and ceramics without altering their bulk properties.

Ion implantation of materials results from the introduction of atoms into the surface layer of a solid substrate by bombardment of the solid with ions in the eV to MeV energy range. Several ballistic-like atomic processes occur during ion implantation. The ballistic interactions of an energetic ion with a solid are shown schematically in Figure 1. The figure shows sputtering events at the surface, single-ion/single-atom recoil events, the development of a collision cascade involving a large number of displaced atoms, and the final position of the incident ion. The solid-state aspects of ion implanted materials are particularly broad because of the range of physical properties that are sensitive to the presence of trace amounts of foreign atoms. Mechanical, chemical, electrical, optical, magnetic, and superconducting properties are all affected and may even be dominated by the presence of such foreign atoms. The use of energetic ions affords the possibility of introducing a wide range of atomic species independent of thermodynamic factors, thereby making it possible to obtain impurity concentrations and distributions of particular

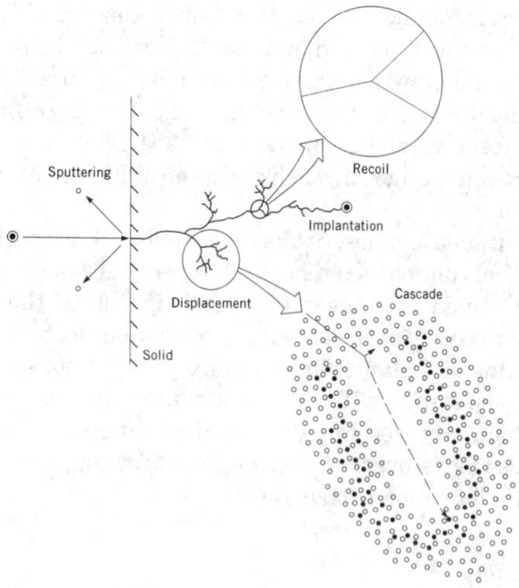

Figure 1. The ballistic interactions of an energetic ion with a solid. Depicted are sputtering events at the surface, single-ion/single-atom recoil events, the development of a collision cascade involving a large number of displaced atoms, and the final position of the incident ion. ○ = normal atom; • = interstitial atom; ⊙ = incident ion.

interest. In many cases, these distributions would not be otherwise attainable.

The implantation system shown in Figure 2**a** illustrates a conventional ion implantation system in widespread use within the semiconductor industry. This directed beam system uses a magnet for mass analysis of the ion beam, electrostatic steering to manipulate the beam, and a magnet to select specific ions from the beam extracted from the ion source.

One ion implantation system which does not use mass analysis and is capable of extremely high ion currents is the broad beam system (Fig. 2**b**). Broad beam ion sources typically employs grids at the front end of the source to obtain electrostatic acceleration of ions. Like conventional ion implantation systems, broad beam systems are also referred to as directed beams.

The plasma source implantation system (Fig. 2**c**) does not use the extraction and acceleration scheme found in traditional ion implanters, but rather the sample to be implanted is placed inside a plasma. This ion implantation scheme evolved from work on controlled fusion devices. The sample is repetitively pulsed at high

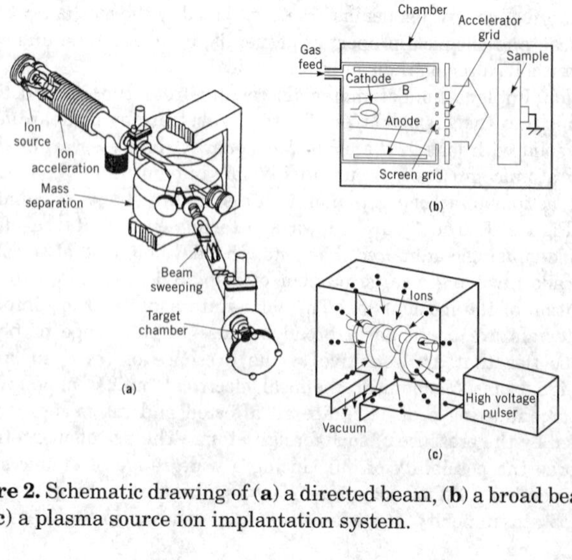

Figure 2. Schematic drawing of (**a**) a directed beam, (**b**) a broad beam, and (**c**) a plasma source ion implantation system.

negative voltages (around 100 kV) to envelope the surface with a flux of energetic plasma ions. Because the plasma surrounds the sample, and because the ions are accelerated normal to the sample surface, plasma-source implantation occurs over the entire surface, thereby eliminating the need to manipulate nonplanar samples in front of the ion beam. In this article, ion implantation systems that implant all surfaces simultaneously are referred to as omnidirectional systems.

Ion implantation (outside the traditional semiconductor applications) for the controlled modification of surface sensitive properties has had two principal thrusts: (*1*) as a metallurgical tool for studying basic mechanisms in areas such as aqueous corrosion, high temperature oxidation, and metallurgical phenomena (eg, impurity trapping); and (*2*) as a means of beneficially modifying the mechanical or chemical properties of materials. Optical/electrical properties, the traditional industrial application of ion implantation, such as the refractive index, reflectance, conductivity, and magnetic properties, can be modified. Chemical properties affected by ion implantation are relevant to the fields of electrochemistry (corrosion), catalysis, and oxidation resistance. The fastest growing research application of ion implantation modifies the mechanical and tribological properties, eg, hardness, modulus, friction, wear resistance, and fatigue resistance, of a material surface.

Some of the advantages of ion implantation in comparison to other surface treatments (such as coatings) are (*1*) surface properties can be optimized independently of the bulk properties; (*2*) the process is not limited by thermodynamic constraints, so solid solubility limits can be exceeded by several orders of magnitude, alloy compositions are not limited by diffusion, and metastable compounds can be produced; (*3*) the process modifies existing surfaces, so there are no interfaces to degrade mechanical properties and original dimensions are retained; (*4*) low process temperatures avoid thermally related degradation in surface finish and bulk mechanical properties; and (*5*) the process is highly controllable and reproducible.

Ion implantation processes also have limitations. An intrinsic basic limitation of directed beam ion implantation is that it is a line-of-sight process; it is not feasible to apply it to samples having complicated geometrics. Secondly, the range of ions in solids is generally low, which leads to shallow penetration and a thin modified layer. Finally, ion implantation as a surface modification tool is generally unfamiliar to most users of other surface modification processes.

These limitations can be addressed in a number of ways. First, plasma source ion implantation techniques have the ability to treat complicated geometries and are presently being evaluated for commercial applications.

The shallow penetration of ion implantation would in itself make it appear useless as a technique for engineering applications; however, there are several situations involving both physical and chemical properties in which the effect of the implanted ion persists to depths far greater than the initial implantation range. The thickness of the modified zone can be extended by combining ion implantation with a deposition technique or if deposition occurs spontaneously during the ion implantation process. In addition, ion implantation at elevated temperatures, but below temperatures at which degradation of mechanical properties could occur, has been shown to increase the penetration depths substantially.

Ion–Solid Interactions

Ion Stopping. Ion–solid interactions are the foundation that underlies the broad application of ion implantation to the modification of materials. The principal features governing the successful exploitation of ion implantation are the range distribution of the energetic ions, the amount and nature of the lattice disorder that is created, and the location of the energetic ions in the crystal lattice. At high dose levels, used to incorporate greater than 5–10 atomic % of implanted species to modify the composition of the target, other phenomena become important: sputtering, ion-induced phase formation, and transformations.

Nuclear collisions can involve large discrete energy losses and significant angular deflection of the trajectory of the ion. In nuclear stopping, the average energy loss results from elastic collisions with target atoms. This process is responsible for the production of disorder by the displacement of atoms from their lattice position. Electronic collisions occur continuously and involve much smaller energy losses per collision, negligible deflection of the ion trajectory, and negligible lattice disorder. Electronic stopping is an inelastic process and results from energy transferred from the ion to the target electrons. Typical units for the energy loss rate are eV/nm or keV/μm.

A proper understanding of the mechanisms of energy loss is important not only in controlling the depth profile of implanted dopant atoms, but also in determining the nature of the lattice disorder produced during ion implantation or ion irradiation of the solid.

Range. For the energy regime normally used in heavy ion implantation (tens to hundreds of keV) the nuclear contribution to the stopping process normally dominates and this is reflected in the particular trajectories as the ion comes to rest within the solid.

In range theory the range distribution is regarded as a transport problem describing the slowing down of energetic ions in matter. Two general methods for obtaining range quantities, one using Monte Carlo or molecular dynamics simulations (more accurate) and the other employing analytical methods (less accurate), have been developed.

The actual distance or range, R, an ion traverses, the ion's projected range, R_p, measured along the vector of the ion's incident trajectory, and the straggle (or deviation) in the ion's projected range, ΔR_p, can be estimated using

$$R(\text{nm}) = \frac{6E(\text{keV})}{\rho(\text{g/m}^3)} \frac{M_2}{Z_2} \frac{M_1 + M_2}{M_1} \frac{\left(Z_1^{2/3} + Z_2^{2/3}\right)^{1/2}}{Z_1} \quad (1)$$

$$R_p \simeq \frac{R}{1 + M_{2/3}M_1} \quad (2)$$

$$\Delta R_p \simeq R_p/2.5 \quad (3)$$

where E is the incident particle energy, ρ is the mass density of the target, M_1 and Z_1 are the mass and atomic number of the incident particle, and M_2 and Z_2 are the mass and atomic number of the target atom.

Implanted Species Concentration. The peak atomic density N_p in the ion implantation distribution is estimated using equation 4 where N_p is in units of atoms/cm^3, ϕ_i is the ion dose in units of atoms/cm^2, and ΔR_p is in units of cm.

$$N_p = \frac{0.4\phi_i}{\Delta R_p} \quad (4)$$

To obtain the peak atomic concentration C_p resulting from this peak number of implanted ions requires knowing N, the atomic density of the substrate. The general relation for the concentration of the implanted species at the peak of the distribution is given by equation 5:

$$C_p = \frac{N_p}{N_p + N} \quad (5)$$

Channeling. The crystal orientation influence of ion penetration is called channeling or the channeling effect. When an ion trajectory is aligned along atomic rows, the positive atomic potentials of the line of atoms steer the positively charged ion within the open space, or channels, between the atomic rows. These channeled ions do not make close-impact collisions with the lattice atoms and have a much lower rate of energy loss and hence a greater range than those of nonchanneled ions. The depth distribution of channeled ions is difficult to characterize under routine implantation conditions. The channeling distribution depends on surface preparation, substrate temperature, beam alignment, and disorder introduced during the implantation process itself. The channeling effect requires that the incident ions be aligned within a critical angle of the crystal axes or planes. The critical angle depends on the ion energy, ion species, and substrate, but is typically less than 5°.

Radiation Damage. It has been known for many years that bombardment of a crystal with energetic (keV to MeV) heavy ions produces regions of lattice disorder. An implanted ion entering a solid with an initial kinetic energy of 100 keV comes to rest in the time scale of about 10^{-13} due to both electronic and nuclear collisions. As an ion slows down and comes to rest in a crystal, it makes a number of collisions with the lattice atoms. In these collisions, sufficient energy may be transferred from the ion to displace an atom from its lattice site. Lattice atoms which are displaced by an incident ion are called primary knock-ons. This process continues and creates a cascade of atomic collisions which is collectively referred to as the collision, or displacement, cascade. The disorder can be directly observed by techniques sensitive to lattice structure, such as electron-transmission microscopy, MeV-particle channeling, and electron diffraction.

Radiation Enhanced Diffusion. Ion irradiation is quite efficient in forming vacancy–interstitial pairs. The atomic displacements resulting from energetic recoiling atoms can be highly concentrated into small localized regions containing a large concentration of defects well in excess of the equilibrium value. If the defects are produced at temperatures where they are mobile, and can in part anneal out, the balance between the rate of formation vs the rate of annihilation leads to a steady-state concentration of defects. Since the atomic diffusivity is proportional to the defect concentration, an excess concentration of defects leads to an enhancement in the diffusional process.

Sputtering. The erosion of a surface by energetic particle bombardment is called sputtering. In this process surface atoms are removed by collisions between the incoming particles and the atoms in the near surface layers of a solid. Sputtering sets the limit of the maximum concentration of atoms that can be implanted and retained in a target material.

Other Processes Utilizing Ion Beams. Materials under ion irradiation undergo significant atomic rearrangement. The most obvious example of this phenomenon is the atomic intermixing and alloying that can occur at the interface separating two different materials during ion irradiation. This process is known as ion beam mixing (IBM). A related process uses ions to bombard material as it is being deposited onto a substrate. This process is called ion beam assisted deposition (IBAD) or ion assisted deposition (IAD).

Ion Implantation Applications

Ion implantation is used to reduce wear and friction; to increase fatigue resistance, to improve aqueous corrosion resistance; to modify fracture toughness, hardness, and microstructure of ceramics; to modify the conductivity, hardness, and oxidation and chemical resistance of polymers; to prepare controlled, reproducible, and unique catalysts, in metastable compound formation; and to deposit coatings such as diamondlike carbon.

KEVIN C. WALTER
MICHAEL NASTASI
Los Alamos National Laboratory

Dearnaley, J. H. Freeman, R. S. Nelson, and J. Stephen, *Ion Implantation*, North Holland, Amsterdam, the Netherlands, 1973.

M. Nastasi, J. W. Mayer, and J. K. Hirvonen, *Ion Solid Interactions: Fundamentals and Applications*, Cambridge University Press, Cambridge, in press, 1998.

J. R. Conrad and K. Sridharan, eds., "Proceedings of the 1st International Workshop on Plasma Based Ion Implantation," *J. Vac. Sci. Tech.* **B12**, 807 (1994).

J. F. Ziegler, ed., *Ion Implantation Technology*, North-Holland, Amsterdam, the Netherlands, 1992.

IONOMERS

The generic term *ionomer* was introduced by DuPont in 1964 in conjunction with the commercialization of the new Surlyn resins to denote a thermoplastic polymer containing both covalent and ionic bonds, and having properties influenced to substantial effect by the ionic bonding. Since that time, the meaning has been expanded to include many compositions such as the glass ionomers used in dentistry which cannot be melt processed. In the interest of clarity and consistency, it is proposed that the term ionomer be reserved for polymers having melt viscosities suitable for conventional melt processing methods. Descriptions such as ion-containing or ion-linked are appropriate for highly viscous or true thermoset materials.

Despite the broad scope of the field and the unusual property combinations obtainable, commercial exploitation has been confined mainly to the original family based on ethylene copolymers. Within certain industries, such as flexible packaging, the word ionomer is understood to mean a copolymer of ethylene with methacrylic or acrylic acid, partly neutralized with sodium or zinc.

Ethylene-Based Ionomers

Physical Properties. The semicrystalline, ethylene-based ionomers of commerce are flexible, transparent polymers notable for high strength and elasticity in both solid and molten states. The ionic bonding is completely reversible and has a strong influence on properties, even at temperatures well above the melting point.

Mechanical Properties. Table 1 shows the general range of mechanical properties available in commercial Surlyn ionomers. The substitution of acrylic acid for methacrylic acid has only minor effects on properties.

The issue of mechanical property changes over time has been addressed and a structural model has been developed. A correlation was established between stiffness and the size of an endotherm (T_i), normally seen in dsc scans of ionomers at about 50°C. This endotherm increases in size with increasing neutralization. The T_i endotherm disappears completely when the dsc measurement is repeated immediately, but then gradually reappears during room-temperature storage.

In addition to time-related effects, the solid-state physical properties are also affected by adsorbed water, which functions as a plasticizer. Water pickup is affected by the nature of the cation, with sodium ionomers absorbing about 10 times the level of the zinc equivalent under the same conditions.

Crystallinity of Ionomers. Ionomers are much less hazy than the ethylene acid copolymers from which they are derived. Studies with optical and electron microscopes have shown that this is due to suppression of the spherulitic structure by the metal ions. Surprisingly, x-ray diffraction has shown that polyethylene crystallinity is present in the ionomers. A typical level of crystallinity is 30%.

Rheological Properties. The melt viscosity of an acid copolymer increases dramatically as the fraction of neutralization is increased.

Softening is apparent over a wide range, while the melt is strong and elastic. This gradual melting is beneficial in heat-sealing applications.

Infrared Spectra of Ionomers. Infrared absorption data, first published in 1964, show that partial neutralization of ethylene–methacrylic acid introduced new absorption bands at 1480–1670 cm^{-1} for the ionized carboxylate group while the 1698 − cm^{-1} band of the free acid carboxyl diminishes in size. In addition to providing information on structural features, the numerous absorption bands are significant in applications technology, providing rapid warmup of film and sheet under infrared radiation.

Table 1. Mechanical Properties of Surlyn Ionomers

Property	Range
stiffness, MPa[a]	90–400
yield point, MPa[a]	8–20
tensile strength, MPa[a]	23–40
elongation at break, %	280–500
Shore D	54–70
brittleness temperature, °C	−100 to −140

[a] To convert MPa to psi, multiply by 145.

Solubility of Ionomers. Ionic bonding with metal ions decreases solubility in organic solvents. At high neutralization levels with alkali metal ions, many ionomers spontaneously form colloidal suspensions in water when stirred vigorously at 100–150°C under pressure. These provide convenient methods for applying thin coatings of ionomers to paper and other substrates.

Electrical Properties. Due to the comparatively low content of polar groups, most commercial ionomers are very good insulating resins.

Permeability. Acid copolymers are less permeable to natural oils than conventional homopolymers, and this difference increases greatly when they are neutralized.

In the area of gas permeability, the low crystallinity of a typical ionomer (~ 30%) results in relatively high permeability to oxygen. For packaging of fresh meat this is advantageous, but in other packaging areas, combination with a barrier layer may be required.

Manufacture and Processing. Most commercial processes involve copolymerization of ethylene with the acid comonomer followed by partial neutralization, using appropriate metal compounds.

Many methods for the conversion of acid copolymers to ionomers have been described by DuPont. The chemistry involved is simple when cations such as sodium or potassium are involved, but conditions must be controlled to obtain uniform products. Solutions of sodium hydroxide or methoxide can be fed to the acid copolymer melt, using a high shear device such as a two-roll mill to achieve uniformity. All volatile by-products are easily removed during the conversion, which is run at about 150°C.

Economic Aspects. Worldwide production is of the order of 110,000 t.

Health and Safety Factors

During processing at elevated temperatures, normal precautions are needed to prevent accidental burns. Surlyn ionomers have U.S. Food and Drug Administration clearance for food contact.

Uses

Flexible packaging is the largest commercial application area for ethylene ionomers. The unusual resilience and roughness of ionomers have resulted in sporting goods applications, including golf ball covers and bowling pin coatings. Ionomers are easily foamed due to high melt strength, and the foams are durable, leading to uses in construction, skilifts, and softball cores.

Noncommercial Ethylene-Based Ionomers

Noncommercial ethylene-based ionomers include amine-linked and complexed ionomers and ethylene–dicarboxylic acid copolymers.

Ionomers Not Based on Ethylene

Ionomers not based on ethylene include styrene-based ionomers, EPDM-derived ionomers, butadiene–methacrylic acid ionomers, telechelic ionomers, pentenamer ionomers, bitumen ionomers, and polyoxymethylene ionomers.

RICHARD W. REES
E. I. du Pont de Nemours & Co., Inc.

W. J. MacKnight and T. R. Earnest, *J. Macromol. Rev.* **16**, 41 (1981).

R. W. Rees, in K. C. Frisch, ed., *Polyelectrolytes*, Technomic Publishing Co., Inc., Westport, Conn., 1976, pp. 177–197.

N. L. Zutty, J. A. Faucher, and S. Bonotto, in N. M. Bikales, ed., *Encyclopedia of Polymer Science and Technology*, Vol. 6, Interscience Publishers, a Division of John Wiley & Sons, Inc., New York, 1967, p. 420.

R. W. Rees, in J. I. Kroschwitz, ed., *Encyclopedia of Polymer Science and Engineering*, 2nd ed., Vol. 4, Wiley-Interscience, New York, 1986, pp. 395–417.

IONOPHORES. See ANTIBIOTICS, POLYETHERS; ANTIBIOTICS, PEPTIDES; CHELATING AGENTS.

IRON

Iron, Fe, from the Latin *ferrum,* atomic number 26, is the fourth most abundant element in the world's crust, outranked only by aluminum, silicon, and oxygen. It is the world's least expensive and most useful metal. Although gold, silver, copper, brass, and bronze were in common use before iron, it was not until humans discovered how to extract iron from its ores that civilization developed rapidly.

Pure iron is a silvery white, relatively soft metal and is rarely used commercially. Typical properties are listed in Table 1.

Iron is alloyed with other elements for commercial applications. The most important alloying element is carbon. Small amounts of carbon alloyed with iron lower the melting point. The distinction between steels and other irons is based on properties and defined by the iron–carbon phase diagram. Steel is generally classified as those iron–carbon alloys (0–2% C) which have a high melting point and can be hot rolled. Iron with carbon up to about 2% can be heated to a temperature at which only one phase (gamma iron) exists. Gamma iron is face-centered cubic (fcc) in structure, and therefore is plastic, or malleable, which allows hot rolling. Cast irons are those which contain sufficient quantities of the eutectic (about 2–5% C) to make the metal too brittle to hot roll; thus the requirement that it be cast. Pig iron from the blast furnace is liquid iron saturated with carbon (>4.3% C) depending on the temperature corresponding to the liquidus line.

Iron is indispensable in the human body. The average adult body contains 3 grams of iron. About 65% is found in hemoglobin, which carries oxygen from the lungs to the various parts of the body. Iron is also needed for the proper functioning of cells, muscles, and other tissues.

Iron Ores

Minerals. Iron-bearing minerals are numerous and are present in most soils and rocks. However, only a few minerals are important sources of iron and thus are called ores. Hematite is the most plentiful iron mineral mined, followed by magnetite, goethite, siderite, ilmenite, and pyrite.

Sources. Iron ore deposits were formed by many different processes, eg, weathering, sedimentation, hydrothermal, and chemical.

Table 1. Properties of Iron

Property	Value
melting point, °C	1537
boiling point, °C	3000
crystal structure[a]	bcc
density,[b] g/cm^3	7.87
thermal conductivity at 0°C, W/(m·K)	79
electrical resistivity at 20°C, $\mu\Omega$·cm	9.71
tensile strength, MPa[c]	240–280
yield strength, MPa[c]	70–140
Young's modulus of elasticity, GPa[c]	195
Poisson's ratio	0.3
thermal expansion from 0–300°C, K^{-1}	12.6×10^{-6}
specific heat at 100°C, J/(g·K)[d]	0.50

[a] Room temperature. [b] Hot rolled. [c] To convert MPa to psi, multiply by 145. [d] To convert J to cal, divide by 4.184.

Iron ores occur in igneous, metamorphic, and sedimentary deposits. Normally, as-mined iron ore contains 25 to 68% iron. The main iron ore deposits in the U.S. lie near Lake Superior in Minnesota (Mesabi range) and Michigan (Marquette range).

Canada's chief deposits occur along the borders between Quebec and Newfoundland in an area called the Labrador Trough, and in an area north of Lake Superior. Most of the deposits are similar to those found in Minnesota and Michigan.

Other countries that have large iron ore deposits include Brazil (Carajas and Quadrilatero Ferrifero deposits), Australia (Pilbara deposits), Ukraine (Krivi Rog deposit), Russia (Kursk deposit), Venezuela (Cerro Bolivar deposit), India (Bihar-Orissa, Hospet, Kudremukh, and Goa deposits), South Africa (Sishen and Thabazimbi deposits), and Sweden (Kiruna, Svappavaara, and Malmberget deposits).

Annual world iron ore production has hovered around 9 to 9.75 × 10^8 t since the mid-1980s. International trade of iron ore peaked in 1980 at 4.24 × 10^8 t; otherwise it has remained fairly steady in the range of 3.6 to 4 × 10^8 t/yr since the mid-1980s. The main exporting countries are Brazil, Australia, India, Canada, South Africa, Russia, Ukraine, and Venezuela.

Beneficiation. Iron ore coming from the mine must be properly sized. A gyratory crusher is normally used for primary crushing down to approximately 300 mm. Secondary crushing down to 25 mm can be done in a cone crusher. Fine grinding can be done by rod mills followed by either ball or pebble mills. In some cases, autogenous grinding can be used to replace the cone crusher and rod mills.

Iron ores of different characteristics and compositions can be blended to a more uniform composition. This can be accomplished during handling operations involved in transporting ore to its point of use, or through special blending facilities, such as stacking and reclaiming. Sand and clay can be removed from iron ore by washing in a log-washer or classifier, followed by screening. Low intensity magnetic separators are used to upgrade iron ores containing magnetite. High intensity magnetic separators are used to upgrade iron ores containing hematite or ilmenite.

Agglomeration. Iron ore concentrates are often too fine to be used directly in ironmaking processes; therefore they must be agglomerated. The agglomerating methods typically used in the iron ore industry are pelletizing, sintering, and, to a limited extent, briquetting and nodulizing.

Ironmaking Processes

Ironmaking refers to those processes which reduce iron oxides to iron. By the nature of the processes, the iron produced usually contains carbon and/or other impurities which are removed in downstream processing. There are three principal categories of ironmaking processes, in order of commercial importance: blast furnace, direct reduction, and direct smelting.

Blast Furnace. The blast furnace is the predominant method for making iron (Fig. 1). In essence, the blast furnace is a large, countercurrent, chemical reactor in the form of a vertical shaft which is circular in cross section. Iron ore, coke, and fluxes constitute the burden which is charged continually into the top. Pressures in the shaft are controlled to 100–300 kPa (1–3 atms) gauge. Preheated air (hot blast) is blown in through water-cooled nozzles (tuyeres) around the circumference of the furnace near the bottom. The oxygen in the air reacts with the coke to form hot reducing gases (mostly carbon monoxide) which ascend through the burden and (*1*) provide heat for melting; (*2*) react with the iron ore to reduce it to iron; and (*3*) heat the ore, coke, and fluxes to reaction temperatures. Nitrogen in the hot blast is heated by the coke combustion, and aids in heat transfer to the burden. The gases leaving the top of the furnace (top gas) are cleaned, cooled, and used as fuel to preheat the air for the hot blast.

Molten iron (hot metal or pig iron) and slag (molten oxides) are produced and accumulate in the bottom of the furnace. The hot metal and slag are drained semicontinuously through a taphole (tapping, or casting) into a trough. The hot metal is separated from the slag by a weir/

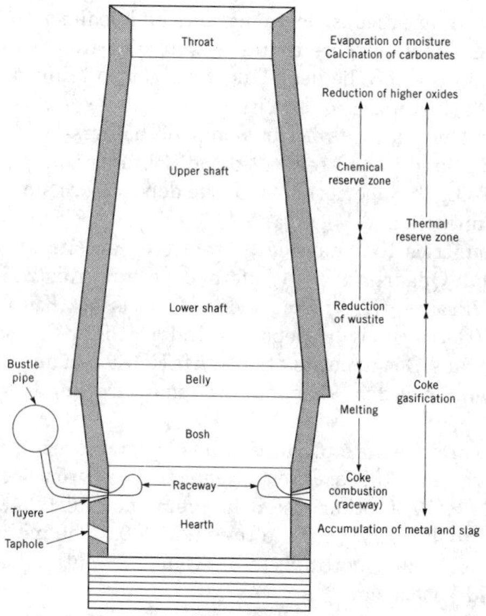

Figure 1. Schematic of a blast furnace.

dam arrangement at the end of the trough, then flows through runners to a refractory-lined rail car. The hot metal is then transported to a nearby site for further processing. About 99% of all pig iron produced in the United States is used for steelmaking. The remainder is cast into pigs for remelting or used directly for iron castings.

Direct Reduction. Direct reduction processes are distinguished from other ironmaking processes in that iron oxide is converted to metallic iron without melting.

Direct Smelting. Direct smelting processes use coal directly instead of coke. Several processes are under development which effectively divide the functions of the blast furnace into two separate but connected unit operations. First, the iron ore is prereduced in a shaft furnace or a fluidized bed, depending on the process and the type of ore used. Second, the prereduced ore is charged into a molten bath into which coal and oxygen or air are also introduced. The gases leaving the smelter are used to perform the reduction in the prereduction vessel. The COREX process is the only one of the newer ironmaking processes operating on a commercial scale.

Production and Economics

The proportion of world pig iron produced in the United States has decreased dramatically since 1950. Also notable is the widening gap between pig iron and steel production, indicating the increasing use of recycled iron or scrap and alternative iron sources such as direct reduced iron (DRI) and hot briquetted iron (HBI). The increased demand for scrap is reflected in higher scrap iron prices which in turn have spurred growth in direct reduction processes.

Economic conditions have made the massive capital requirements of large furnaces too great to manage. As a result, there are two principal thrusts in ironmaking development. First, progress continues to be made in increasing existing blast furnace productivity and in decreasing coke rates. Coal injection to replace coke units has assumed a prominent role. Injection of oxygen and other reductants besides coal are expected to be used more extensively. Increased additions of scrap, DRI, and HBI are expected to play a significant role in efforts to boost productivity and decrease coke rates. Second, development efforts in direct reduction and direct smelting processes have also increased.

Globally, iron production is expected to increase in developing countries as local steel industries grow to supply the increasing demand for steel products. Iron production in already developed countries is expected to stabilize or possibly decline as the opportunities for export diminish.

Environmental Issues

Tremendous progress was made in the 1980s and 1990s in response to environmental issues, especially in the area of emissions from ironmaking facilities. Dust is controlled by protecting open piles, watering roadways, covering conveyor belts and transfer points, controlling fumes through improved casthouse practices, and air cleaning systems ducted to baghouse or other filtration systems. Sulfur dioxide emissions are controlled using off-gas and stack cleaning systems.

Mining practices have been altered to include reclamation of areas where open pit mining has occurred. Safety practices throughout the ironmaking processes are continually being upgraded through training, improved operating practices, and installation of sophisticated detectors and automatic shutdown systems.

Perhaps the biggest environmental challenge for ironmaking processes into the twenty-first century involves responding to the concerns about global warming. Ironmaking processes require the use of carbon-based reductants, and ultimately result in the emission of carbon dioxide.

Cast-Iron Production

Cast irons are normally produced by melting iron or steel scrap along with pig iron. The carbon and silicon levels are adjusted to obtain the desired properties. Melting is done in cupolas, electric induction furnaces, or electric arc furnaces.

Health and Safety

Iron presents minimal health risks. Skin contact should not result in any adverse health effect. Excessive inhalation of dust may be irritating to the respiratory tract. Dust may also cause mechanical irritation on eye contact. Iron dust does present a moderate fire and explosion hazard when exposed to heat and flame.

J. A. LEPINSKI
PT Perkasa Indobaja
JEFFREY C. MYERS
Midrex Direct Reduction Corporation

W. T. Lankford, Jr. and co-workers, *The Making, Shaping, and Treating of Steel,* 10th ed., Association of Iron and Steel Engineers, Pittsburgh, Pa., 1985.

G. S. Brady and H. R. Clauser, *Materials Handbook,* 11th ed., McGraw-Hill Book Co., Inc., New York, 1977.

F. D. DeVaney, "Iron Ore," in *SME Mineral Processing Handbook,* Society of Mining Engineers of AIME, New York, 1985.

J. G. Peacey and W. G. Davenport, *The Iron Blast Furnace,* Pergamon Press, London, 1979.

IRON BY DIRECT REDUCTION

Direct reduction (DR) is the process of converting iron ore (iron oxide) into metallic iron without melting. The metallic iron product, known as direct reduced iron (DRI), is used as a high quality feed material in steelmaking.

In 1994 world production of DRI was 27 million metric tons, and is expected to reach 35 million metric tons annually by the year 2000. The driving force behind this rapid increase in production is the demand for DRI as a high purity supplement to ferrous scrap in electric arc furnace steelmaking.

Physical Properties

DRI can be produced in pellet, lump, or briquette form. When produced in pellets or lumps, DRI retains the shape and form of the iron oxide material fed to the DR process. The removal of oxygen from the iron oxide during direct reduction leaves voids, giving the DRI a spongy appearance when viewed through a microscope. Thus, DRI in

Table 1. Physical Characteristics of DRI[a]

Parameter	Pellets/lump	HBI
density, t/m^3		
bulk	1.6–1.9	2.4–2.8
apparent	3.5	5.0–5.5
porosity, %	50	15
saturated water absorption, wt %	12–15	2–3
nominal size, mm	4–20	30 × 50 × 110

[a] Produced in the MIDREX Direct Reduction Process.

these forms tends to have lower apparent density, greater porosity, and more specific surface area than iron ore. In the hot briquetted form it is known as hot briquetted iron (HBI). Typical physical properties of DRI forms are shown in Table 1.

Chemical Properties

DRI retains the chemical purity of the iron ore from which it is produced; therefore, it tends to be very low in residual elements such as copper, chrome, tin, nickel, and molybdenum.

Production

The reduction of iron ore is accomplished by a series of reactions that are the same as those occurring in the blast furnace stack. These include reduction by CO, H_2, and, in some cases solid carbon, through successive oxidation states to metallic iron, ie, hematite, Fe_2O_3, is reduced to magnetite, Fe_3O_4, which is in turn reduced to wustite, FeO, and then to metallic iron, Fe.

Direct Reduction Processes

Five principal processes produced 95.6% of the total DRI. Natural gas-based direct reduction accounts for 92.5% of worldwide production and coal-based direct reduction accounts for the other 7.5%. The five processes are the MIDREX process, the HYL I process, the HYL III process, the SL/RN process, and the FIOR process.

Other DR Processes. The other DR processes, eg, the CODIR, DRC, ACCAR, and Dav Steel processes, make up 4.4% of worldwide production and mostly consist of coal-based, rotary-kiln processes. All of these are similar to the SL/RN process. In addition, one small coal-based, shaft-furnace plant based on the Kinglor-Metor process is operating.

DR Processes Under Development. The 1990s have seen continuous evolution of direct reduction technology. Short-term development work is focusing on direct reduction processes that can use lower cost iron oxide fines as a feed material. Some examples of these processes include FASTMET, Iron Carbide, CIRCOFER, and an improved version of the FIOR process called FINMET.

Long-term development work is focusing on direct smelting technologies. Multimillion dollar development programs are underway in the United States (AISI Direct Ironmaking process), Australia (HIsmelt process), and Japan (DIOS process). A direct smelting process, called the COREX process, already is in commercial operation in South Africa.

Handling, Shipping, and Storing

In handling, shipping, and storing DRI, care should be taken to avoid oxidation. Millions of tons of DRI in pellet and lump form have been shipped by barge, ocean vessel, truck and rail. The key to avoiding oxidation is simply to keep the material cool and dry. The chemical reactions involved have been well documented. In general, oxidation of DRI takes place in two forms: reoxidation and corrosion.

In comparison, HBI is almost twice as dense as DRI, and thus does not absorb as much water and is much more resistant to reoxidation and corrosion. Several methods of passivating DRI to make it more resistant to reoxidation and corrosion have been developed, but none has been as effective as hot briquetting. Guidelines for offshore shipping of pellet/lump DRI and HBI have been prepared by the International Maritime Organization.

Uses

Over 95% of the world's DRI production is consumed in electric arc furnace steelmaking. The remaining 5% is split among blast furnaces, oxygen steelmaking, foundries, and ladle metallurgy facilities.

J. A. LEPINSKI
PT Perkasa Indobaja

L. von Bogdandy and J. Engell, *The Reduction of Iron Ores, Scientific Basis and Technology*, Springer-Verlag, Berlin, 1971.

R. L. Stephenson, ed., *Direct Reduced Iron—Technology and Economics of Production and Use*, ISS/AIME, Warrendale, Pa, 1980.

Direct Reduction of Iron Ore: A Bibliographical Survey, The Metals Society, London, 1979.

A. Chatterjee, *Beyond the Blast Furnace*, CRC Press, Boca Raton, Fla., 1994.

IRON COMPOUNDS

Iron is the most abundant transition metal in the earth's crust and with the exception of aluminum is more abundant than any other metallic element. It is the lightest element of Group 8 (VIIIB) of the Periodic Table and is the first metallic element in the Table that fails to attain an oxidation state equal to the number of electrons in the valence shell, ie, no compound of Fe(VIII) is known.

The standard aqueous reduction potentials for iron are

$$Fe^{2+} + 2\,e^- = Fe \; E° = -0.44V$$

$$Fe^{3+} + e^- = Fe^{2+} \; E° = +0.77V$$

Iron metal reacts readily with most nonmetals and dissolves in dilute acids to afford the iron(II) cation. Dissolution does not occur in chromic acid, concentrated nitric acids, or hydrogen peroxide, H_2O_2, because the metal is protected by formation of a passivating oxide film, which can be removed mechanically or by acids of coordinating anions such as HCl. The facile interconversion of iron(II) and iron(III) and the ability of the coordination environment to fine tune the redox potential of the couple is reflected in the large variety of functions that iron performs in biological systems.

Salts of iron(II) are known for almost all of the common anions. The exceptions, including NO_2, result from redox incompatibilities. Many of the salts are hydrates and are subject to either efflorescence or hydration.

Salts and Simple Coordination Compounds

These include acetates [anhydrous iron(II) acetate, $Fe(C_2H_3O_2)_2$; and iron(III) acetate, $Fe(C_2H_3O_2)_3$], carbonates [iron(II) carbonate, $FeCO_3$; and iron(II) hydrogen carbonate, $Fe(HCO_3)_2$], citrates [iron citrate; iron(II) citrate; iron(III) citrate; and iron(III) ammonium citrate], cyanides [the hexacyano complexes, ferrocyanide hexakiscyanoferrate-(4−), $(Fe(CN)_6)^{4-}$; tetrapotassium hexakiscyanoferrate trihydrate, $K_4[Fe(CN)_6]{\cdot}3\,H_2O$; tetrasodium hexakiscyanoferrate decahydrate, $Na_4[Fe(CN)_6]{\cdot}10\,H_2O$; tetraammonium hexakiscyanoferrate, $(NH_4)_4[Fe(CN)_6]$; dibarium hexakiscyanoferrate, $Ba_2[Fe(CN)_6]$; dicalcium hexakiscyanoferrate, $Ca_2[Fe(CN)_6]$; dilead hexacyanokisferrate, $Pb_2[Fe(CN)_6]$; tripotassium hexakiscyanoferrate, $K_3[Fe(CN)_6]$; pentacyano complexes; and prussian blue], formates [iron(II) formate dihydrate, $Fe(HCO_2)_2{\cdot}2\,H_2O$; and iron(III) formate, $Fe(HCO_2)_3$], fumarates [iron(II) fumarate, $Fe(C_4H_2O_4)$], halides [iron(II) fluoride, FeF_2; iron(II) fluoride tetrahydrate, $FeF_2{\cdot}4H_2O$; iron(III) fluoride, FeF_3; iron(III) fluoride trihydrate, $FeF_3{\cdot}3H_2O$; iron(II) chloride, $FeCl_2$; iron(II) chloride tetrahydrate, $FeCl_2{\cdot}4H_2O$; iron(III) chloride,

$FeCl_3$; iron(III) chloride hexahydrate, $FeCl_3{\cdot}6H_2O$; iron(II) bromide, $FeBr_2$; iron(III) bromide, $FeBr_3$; iron(II) iodide, FeI_2, gluconates [iron(II) gluconate dihydrate, $Fe[HOCH_2(CHOH)_4CO_2]_2{\cdot}2H_2O$], nitrates [iron(II) nitrate hexahydrate, $Fe(NO_3)_2{\cdot}6H_2O$; iron(III) nitrate nonahydrate, $Fe(NO_3)_2{\cdot}9H_2O$; and iron(III) nitrate hexahydrate, $Fe(NO_3)_3{\cdot}6H_2O$, oxides and hydroxides [iron(II) oxide, FeO; iron(III) oxide, Fe_2O_3; triiron tetroxide or iron(II,III) oxide, Fe_3O_4; iron(II) hydroxide, $Fe(OH)_2$; and iron(III) hydroxide, $Fe(OH)_3$], ferrites, garnets, and ferrates [sodium ferrite, $NaFeO_2$; calcium ferrite; barium ferrite; zinc ferrite; potassium ferrate; barium ferrate; calcium ferrate; and sodium ferrate], perchlorates [iron(II) percholate hexahydrate, $Fe(ClO_4)_2{\cdot}6H_2O$], sulfates [iron(II) sulfate heptahydrate, $FeSO_4{\cdot}7H_2O$; iron(II) sulfate tetrahydrate, $FeSO_4{\cdot}4H_2O$; iron(II) ammonium sulfate or Mohr's salt, $FeSO_4{\cdot}(NH_4)_2SO_4{\cdot}6H_2O$; iron(III) sulfate, $Fe_2(SO_4)_3$], sulfides [iron(II) sulfide, FeS; iron disulfide, FeS_2], chelate compounds [diketones, ie, bis(2,4-pentanedionato)iron(II), $Fe(C_5H_7O_2)_2$, and $Fe(acac)_3$; ethylenediaaminetetraacetic acid, ie, iron(II) ethylenediaminetetraacetic acid, $Fe(EDTA)^-$ or *N,N′*-1,2-ethanediylbis[*N*-(carboxymethyl)glycinato]ferrate(2−); iron(III) ethylenediaminetetraacetic acid, $Fe(EDTA)^{2-}$ or *N,N′*-1,2-ethanediylbis[*N*-(carboxymethyl)glycinato]ferrate(1−); macrocycles, eg, iron(II) phthalocyanine tetrasodium phthalocyaninetetrasulfonatoferrate; oxalates, ie, iron(II) oxalate dihydrate, $FeC_2O_4{\cdot}2H_2O$; tris(ethanedioato)ferrate(3−), $[Fe(C_2O_4)_3]^{3-}$; iron(III) oxalate hexahydrate, $Fe_2(C_2O_4)_3{\cdot}6H_2O$], and polypyridyl ligands [2,2′-bipyridine (bipy); 1,10-phenanthroline (phen); and 2,2′:6′,2″-terpyridine (terpy)].

Organometallic Compounds

Organometallic compounds include carbonyls, [iron pentacarbonyl, $Fe(CO)_5$; diiron nonacarbonyl, $Fe_2(CO)_9$; and triiron dodecacarbonyl, $Fe_3(CO)_{12}$] and metallocenes [bis(cyclopentadienyl)iron or ferrocene, $Fe(C_5H_5)_2$; and bis(cyclopentadienyldicarbonyliron), $Fe(CO)_2(C_5H_5)]_2$].

Compounds of Biochemical Relevance

Iron is perhaps the most important of the transition elements that play a role in biochemistry. It is an essential element for all organisms. The functions of iron-containing metalloproteins include electron transfer, dioxygen transport and storage, activation of dioxygen and hydrogen peroxide with concurrent oxidation of substrates, dismutation of superoxide and peroxide, activation and production of dihydrogen, reduction and rearrangement of substrates, and phosphate hydrolysis, among others. Because of the near total insolubility of iron under physiological conditions, iron metalloproteins and chelate compounds function in the solubilization, uptake, transport, and storage of iron.

Iron-containing proteins are classified as either heme proteins or nonheme iron proteins. The former contain iron that is coordinated to a porphyrin ligand.

Iron Sulfur Compounds. Many molecular compounds are known in which iron is tetrahedrally coordinated by a combination of thiolate and sulfide donors. Of the 10 or more structurally characterized classes of Fe—S compounds, the four shown in Figure 1 are known to occur in proteins.

(μ-Oxo)bis(μ-carboxylato) Diiron Complexes. Several nonheme iron proteins of widely varying functions contain a binuclear iron site as a common structural feature. The proteins include hemerythrin, the O_2-transport protein of marine invertebrates; ribonucleotide reductase, an enzyme which catalyzes the deoxygenation of ribonucleoside diphosphates to deoxyribonucleosides; methane monooxygenase, an enzyme which catalyzes the oxidation of methane to methanol; and purple acid phosphatases, which catalyze the dephosphorylation of phosphoproteins and nucleotides. The site contains two antiferromagnetically coupled iron atoms that are coupled by a bridging oxo or hydroxo group and two bridging carboxylate groups and is recognizable as a portion of the basic ferric acetate structure. The enzymes differ in the nature of the terminal ligands to each iron.

Siderophores. Iron is not readily available at physiological pH because it is present as the insoluble hydrated iron(III) oxide, which has

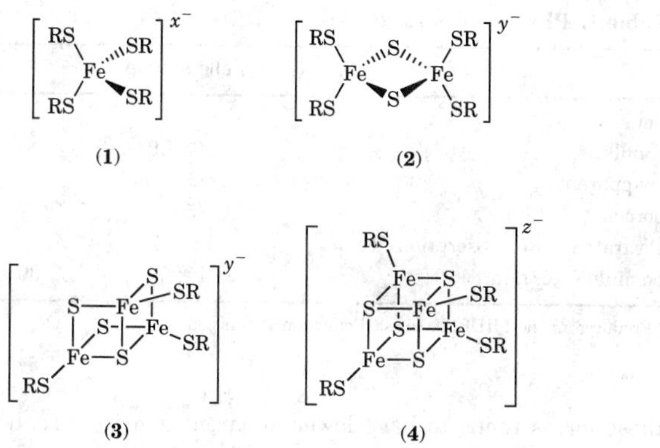

Figure 1. Four important classes of iron compounds where *x*, *y*, and *z* represent 1 or 2, 2 or 3, and 1, 2, or 3, respectively. Structure (**2**) is a 2Fe–2S center; (**3**), a 3Fe–4S; and (**4**) a 4Fe–4S.

$K_{sp} \sim 10^{-39}$. Bacteria synthesize chelating agents to facilitate the solubilization of iron from the environment, transport into the organism, and release of iron. Most contain negatively charged oxygen-donor groups which preferentially complex iron(III) and afford octahedral, high spin complexes called siderophores. The two principal classes of donor groups employed are catecholates (**5**) and hydroxamates (**6**).

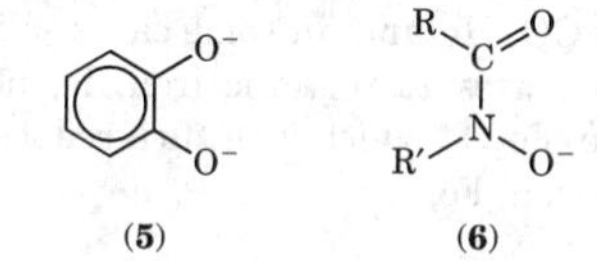

Economic Aspects

Suppliers include Aldrich, Alfa/Aesar, Cerac, Fisher, and Pressure Chemical, among others.

Health and Safety

Most iron salts and compounds may be safely handled following common safe laboratory practices. Some compounds are irritants. A more serious threat is ingestion of massive quantities of iron salts which results in diarrhea, hemorrhage, liver damage, heart damage, and shock. A lethal dose is 200–250 mg/kg of body weight. The majority of the victims of iron poisoning are children under five years of age.

Two compounds associated with particular industrial risks are iron(III) oxide, Fe_2O_3, and iron pentacarbonyl, $Fe(CO)_5$. Chronic inhalation of iron(III) oxide leads to siderosis. Adequate ventilation and mechanical filter respirators should be provided to those exposed to the oxide. Iron pentacarbonyl is volatile and highly toxic.

ALAN M. STOLZENBERG
West Virginia University

N. N. Greenwood and A. Earnshaw, *Chemistry of the Elements*, Pergamon Press, Oxford, U.K., 1984.

F. A. Cotton and G. Wilkinson, *Advanced Inorganic Chemistry*, 5th ed., John Wiley & Sons, Inc., New York, 1988.

G. Wilkinson, R. D. Gillard, and J. A. McCleverty, eds., *Comprehensive Coordination Chemistry*, Pergamon Press, Oxford, U.K., 1987, Vols. 1,2,4, and 6.

Gmelins Handbuch der Anorganischen Chemie, Springer-Verlag, Berlin, System Number 59.

ISOCYANTES, ORGANIC

Isocyanates are derivates of isocyanic acid, HN=C=O, in which alkyl or aryl groups, as well as a host of other substrates, are directly linked to the NCO moiety via the nitrogen atom. Structurally, isocyanates (imides of carbonic acid) are isomeric to cyanates, ROC≡N (nitriles of carbonic acid), and nitrile oxides, RC≡N → O (derivates of carboxylic acid).

Isocyanates are liquids or solids which are highly reactive. The basis for the high reactivity of the isocyanates is the low electron density of the central carbon.

Industrially, isocyantes have become large-volume raw materials for addition polymers, such as polyurethanes, polyureas, and polyisocyanurates. By varying the reactants (isocyanates, polyols, polyamines, and others) for polymer formation, a myriad of products have been developed, ranging from flexible and rigid insulation foams to the high modulus automotive exterior parts to high quality coatings and abrasion-resistant elastomers unmatched by any other polymeric material. The most significant mono-, di-, and oligomeric isocyanates, which constitute over 90% of global isocyanate production, are aromatic isocyanates [toluene 2,4-diisocyanate (TDI), toluene 2,6-diisocyanate (TDI), 4,4′-methylene diphenyl diisocyanate (MDI), 2,4′-methylene diphenyl diisocyanate, polymeric methylene diphenyl diisocyanate (PMDI), *p*-phenylene diisocyanate (PDI), naphthalene-1,5-diisocyanate (NDI)], aliphatic isocyanates [1,6-hexamethylene diisocyanate (HDI), isophorone diisocyanate (IPDI), 4,4′-dicyclohexylmethane diisocyanate (H_{12}MDI), 1,4-cyclohexane diisocyanate (CHDI), bis(isocyanatomethyl)cyclohexane (H_6XDI,DDI), tetramethylxylylene diisocyanate (TMXDI)] and monoisocyanates [methyl isocyanate (MIC), *n*-butyl isocyanate (BIC), phenyl isocyanate (PIC), 3-chlorophenyl isocyanate, 3,4-dichlorophenyl isocyanate, *p*-toluenesulfonyl isocyanate.

Synthetic Methods

Preparation from Amines. The most common method of preparing isocyanates, even on a commercial scale, involves the reaction of phosgene and aromatic or aliphatic amine precursors.

Preparation from Nitrene Intermediates. A convenient, small-scale method for the conversion of carboxylic acid derivatives into isocyanates involves electron sextet rearrangements, such as the ones described by Hofmann and Curtius.

Nonphosgene Preparation. The term nonphosgene route is primarily used in conjunction with the conversion of amines (or the corresponding nitro precursor) to isocyanates via the use of carboxylation agents.

Chemical Properties

Addition Reactions. Isocyanates undergo addition reactions with a wide variety of substrates. Preferred addition occurs across the C═N bond of the NCO moiety.

Insertion Reactions. Isocyanates also may undergo insertion reactions with C—H bonds.

Cycloaddition Reactions. Isocyanates undergo cycloadditions across the carbon–nitrogen double bond with a variety of unsaturated substrates. Addition across the C═O bond is less common.

Oligomerization and Polymerization Reactions. One special feature of isocyanates is their propensity to dimerize and trimerize. Aromatic isocyanates, especially, are known to undergo these reactions in the absense of a catalyst. The dimerization product bears a strong dependency on both the reactivity and structure of the starting isocyanate.

Commercial Manufacturing Processes

Aromatic Isocyanates. A variety of methods are described in the literature for the synthesis of aromatic isocyanates. Only the phosgenation of amines or amine salts is used on a commercial scale. Much process refinement has occurred to minimize the formation of disubstituted ureas arising by the reaction of the generated isocyanate with the amine starting material.

Aliphatic Isocyanates. Conventional aliphatic isocyanates have historically been manufactured using the hydrogen chloride salt slurry approach. Exceptions to this are the longer chain aliphatics which, due to the increased solubility, have reaction rates conducive to the free amine process. An alternative approach, generally referred to as a two-phase phosgenation, has gained wide scale acceptance for the production of aliphatic isocyanates.

Low boiling isocyanates, such as methyl isocyanate, are difficult to prepare via conventional phosgenation due to the fact that the *N*-alkyl carbamoyl chlorides are volatile below their decomposition point. A convenient method for the synthesis of these low boiling materials consists of the reaction of *N,N′*-dimethylurea with toluene diisocyanate to yield an aliphatic–aromatic urea which is pyrolyzed to yield the desired isocyanate. Alternatively, an appropriate aliphatic–aromatic urea can be prepared by the reaction of diphenylcarbamoyl chloride with methylamine.

Specialty Isocyanates. Acyl isocyanates, extensively used in synthetic applications, cannot be directly synthesized from amides and phosgene. Reactions of acid halides with cyanates have been suggested. However, the dominant commercial process utilizes the reaction of carboxamides with oxalyl chloride. Cyclic intermediates have been observed in these reactions which generally give a high yield of the desired products.

Of the many other methods leading to isocyanates, only a few are practical enough in regard to availability of starting materials to be of general applicability. One of the more promising approaches utilizes olefinic substrates which add isocyanic acid in Markovnikov fashion to form alkyl isocyanates. One approach uses the slow addition of the olefin to an excess of solvent and isocyanic acid in the presence of a catalytic amount of inorganic acid. Another approach involves the formation of the dichloro intermediate. The dichloro compound reacts at low temperatures with an excess of isocyanic acid in the presence of a Lewis acid. Pyrolysis approaches can also be used to prepare substituted isocyanates which cannot be prepared using other methods.

Carbodiimide Formation. Carbodiimide formation has commercial significance in the manufacture of liquid MDI. Heating of MDI in the presence of catalytic amounts of phosphine oxides or alkyl phosphates leads to partial conversion of isocyanate into carbodiimide.

Health and Safety Factors

Isocyanates are classified as dangerous substances (EEC Guidelines). They are generally labeled toxic and should be handled with care. Exposure hazards increase substantially when handling vapors or mists. Isocyanate vapors or mists may be irritating to the nose, throat, and lungs. Sensitization may result from excessive exposure.

Repeated or prolonged skin contact may cause irritation, blistering, dermatitis, or skin sensitization. Contact with the eye has been reported to cause irritations in testing with rabbits. For these reasons, isocyanates must be handled in well-ventilated areas. Respirators should be worn whenever the possibility of vapor exposure exists. Chemical goggles should be worn when handling isocyanates. In the event of direct skin contact, use a safety shower immediately, removing all clothing while washing. In all cases, call a physician immediately.

The most overlooked hazard and contaminant is water. Water reacts with isocyanates at room temperature to yield both ureas and large quantities of carbon dioxide. The presence of water or moisture can produce a sufficient amount of CO_2 to overpressurize and rupture containers. For these reasons, the use of dry nitrogen atmospheres is recommended during handling.

Also, the presence of strong bases, even in trace amounts, can promote the formation of isocyanurates or carbodiimides.

Temperature control is important in the handling and storage of isocyanates. Storage at inappropriate temperatures can cause product discoloration, viscosity increases, and dimerization.

Most commercial isocyanates have a high flash point and are classified as Class IIIB combustible liquids. These materials, however, burn in the presence of an existing fire or heat source in the presence of oxygen. In the event of an isocyanate fire, use a carbon dioxide or dry chemical extinguisher. For fires covering large areas, use of a protein foam or water spray is recommended. Personnel engaged in fighting isocyanate fires must be protected against nitrogen dioxide vapors and isocyanate fumes. Firefighters should wear approved positive pressure, self-contained breathing apparatus, and fire-resistant clothing.

Economic Aspects and Applications

Since 1971, the overall demand for isocyanates has increased at a compounded rate of 12%. Although this level will not likely be sustained in the future due to the maturation of key application markets, it is probable that additional growth will occur through the year 2000. This trend will likely include a shift in emphasis from TDI to MDI and polymeric MDI-based materials. New growth opportunities in the construction industry, structural applications, and growth in the automotive industry exist. Third-world markets are also anticipated to provide growth opportunities.

Globally, BASF, Bayer (Miles in North America), Dow, and ICI historically have been the leading producers of aromatic isocyanates. In North America, Olin is a principal supplier of TDI and aliphatic isocyanates. Rhône-Poulenc and Hoechst are principal suppliers in Europe.

Aromatic Isocyanates. In North America, aromatic isocyanates are heavily used as monomers for addition and condensation polymers. The principal applications include both flexible and rigid polyurethane foam and noncellular applications, such as coatings, adhesives, elastomers, and fibers.

Aliphatic Isocyanates. Aliphatic diisocyantes have traditionally commanded a premium price because the aliphatic amine precursors are more expensive than aromatic diamines. They are most commonly used in applications which support the added cost or where the long-term performance of aromatic isocyanates is unacceptable. Monofunctional aliphatic isocyanates, such as methyl and *n*-butyl isocyanate, are used as intermediates in the production of carbamate-based and urea-based insecticides and fungicides.

A number of markets have been established for light-stable, aliphatic diisocyanates in the United States. The largest market is in high performance coatings. The largest coating market is in automotive refinishes. Other coating include uv-cured coating for vinyl tile and sheet flooring, electronic circuit boards, powder coatings, and paints. Hydrogenated MDI ($H_{12}MDI$), *m*-xylylene diisocyanate (XDI), and isophorone diisocyanate are currently used in many of these coating applications.

Aliphatic isocyanates have a small but growing market application in thermoplastic polyurethanes (TPU). Medical applications include wound dressings, catheters, implant devices, and blood bags. A security glass system using light stable TPU as an inner layer is under evaluation for shatteroroof automotive windshield applications.

Developments in aliphatic isocyanates include the synthesis of polymeric aliphatic isocyanates and masked or blocked diisocyanates for applications in which volatility or reactivity are of concern.

Specialty Isocyanates. Specialty isocyanates are organic isocyanates having the isocyanate function attached to a carbonyl group or to elements other than carbon. *p*-Toluenesulfonyl isocyanate is used as a drying agent for organic solvents. Arenesulfonyl diisocyanates, such as *m*-phenylenedisulfonyl diisocyanate, are used as monomers for base-soluble polymers. Arenesufonyl monoisocyanates are used as intermediates for pharmaceuticals and herbicides.

REINHARD H. RICHTER
RALPH D. PRIESTER, JR.
Dow Chemical

A. A. A. Sayigh, H. Ulrich, and W. J. Farrissey, in J. K. Stilles and T. W. Campbell, eds., *Condensation Monomers*, John Wiley & Sons, Inc., New York, 1972, pp. 369–476.

R. Richter and H. Ulrich, in S. Patai, ed., *The Chemistry of Cyanates and their Thio Derivatives*, John Wiley & Sons, Inc., New York, 1977, p. 619.

J. K. Rassmussen and A. Hassner, *Chem. Rev.* **76**, 389 (1976).

H. Ulrich, in W. F. Gum, W. Riese, and H. Ulrich, eds., *Reaction Polymers*, Hanser, New York, 1992, p. 358.

ISOCYANURIC COMPOUNDS.

See CYANURIC AND ISOCYANURIC ACIDS.

ISOPRENE

In the 1990s isoprene is used almost exclusively as a monomer for polymerization.

The isoprene unit exists extensively in nature. It is found in terpenes, camphors, diterpenes (eg, abietic acid), vitamins A and K, chlorophyll, and other compounds isolated from animal and plant materials. The correct structural formula for isoprene was first proposed in 1884.

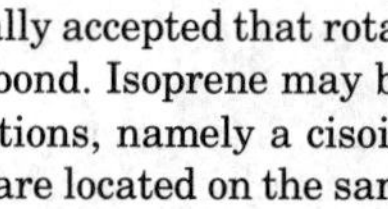

Properties

Isoprene (2-methyl-1,3-butadiene) is a colorless, volatile liquid that is soluble in most hydrocarbons but is practically insoluble in water. Typical properties of isoprene are listed in Table 1.

Conformation. The exact conformation of the isoprene molecule is still in doubt. It is generally accepted that rotation is restricted around the central C—C single bond. Isoprene may be considered as an equilibrium of two conformations, namely a cisoid (*s-cis*) conformation in which both vinyl groups are located on the same side of the C—C bond, and a transoid (*s-trans*) one with the vinyl groups located on the opposite sides of the bond.

Reactions

Isoprene is highly reactive both as a diene and through its allylic hydrogens, and its reactions are similar to those of butadiene. Apart from polymerization, the most widely investigated isoprene reactions are the formation of six-membered rings by the Diels-Alder reaction.

Free-Radical Reactions. Free radicals attack isoprene, and two competing mechanisms, at the double bond or involving C—H bonds, are postulated:

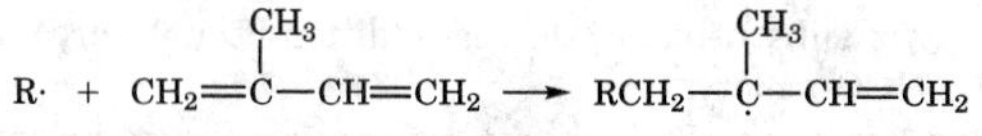

$$R\cdot + R'H \xrightarrow{\text{solvent}} RH + R'$$

Halogens and Halogenated Compounds. The chlorination of isoprene in CCl_4 at −5 to −10°C, using an equimolar ratio of chlorine to isoprene, gives us a mixture of 44% of 1,4-dichloro-2-methyl-2-butene

Table 1. Properties of Isoprene

Property	Value
mol wt	68.11
density of liquid, gm/cm³ at 25°C	0.6759
freezing point, °C	−145.95
bp at 101.3 kPa,[a] °C	34.067
n_D^{30}	1.41524
flash point, °C	−48

[a] To convert kPa to atm, divide by 101.3.

and 14% of 3,4-dichloro-2-methyl-1-butene as addition products, along with 42% of the substitution product, 2-chloromethyl-1,3-butadiene.

Bromination of isoprene using Br_2 at −5°C in chloroform yields only *trans*-1,4-dibromo-2-methyl-2-butene.

The reaction of dihalocarbenes with isoprene yields exclusively the 1,2- (or 3,4-) addition product, eg, dichlorocarbene Cl_2C: and isoprene react to give 1,1-dichloro-2-methyl-2-vinylcyclopropane.

Isoprene reacts with α-chloroalkyl ethers in the presence of $ZnCl_2$ in diethyl ether from 0–10°C.

Hydrocarbons. The reaction of isoprene with toluene, ethylbenzene, or isopropylbenzene is catalyzed by sodium or potassium. The products are chiefly monopentenylated in the side chain, and no information can be obtained on whether the addition is 1,4- or 1,2-, because under these conditions the double bond migrates.

Other Compounds. Primary and secondary amines add 1,4- to isoprene.

Polymerization. Isoprene polymerization can proceed by either 1,4- or 1,2- (vinyl) addition.

Of the many catalysts that polymerize isoprene, four have attained commercial importance. One is a coordination catalyst based on an aluminum alkyl and a vanadium salt which produces *trans*-1,4-polyisoprene. A second is a lithium alkyl which produces 90% *cis*-1,4-polyisoprene. Very high (99%) *cis*-1,4-polyisoprene is produced with coordination catalysts consisting of a combination of titanium tetrachloride, $TiCl_4$, plus a trialkylaluminum, R_3Al, or a combination of $TiCl_4$ with an alane (aluminum hydride derivative).

The polymerization of isoprene by alkali metal and organometallic compounds (other than organolithium) is a heterogeneous reaction both in bulk and hydrocarbon solvents.

Production

The largest capacity has been and remains in the CIS (former USSR) region. Several plants around the world have been shut down, and the trend appears to continue downward. On the other hand the use of isoprene in block copolymers has grown rapidly. This growth has tended to offset some of the decline of *cis*-1,4-polyisoprene.

The principal route for production of isoprene monomer outside of the CIS is recovery from ethylene by-product C_5 streams.

Synthesis. Because of the limited availability of by-product isoprene much effort has been devoted to synthesis of isoprene. Most routes tend to have marginal selectivity and require large amounts of energy. The choice of which route is preferable depends on availability and cost of raw materials and cost of energy. Several synthetic routes have been practiced commercially including propylene dimer, dehydrogenation of tertiary amylenes, isobutylene–formaldehyde, isopentane dehydrogenation and acetone–acetylene.

Health and Safety

Isoprene is not known to present serious toxicological hazards in handling; however, as is the case with many other chemicals, studies concerning the safety of isoprene are ongoing. In humans, a one minute inhalation of 0.16 mg isoprene per liter air is mildly irritating to the mucous membranes of the eyes, nose, and upper respiratory passages. It was proposed that the limit of isoprene concentration on industrial sites be set at 0.04 mg/L air; it was also recommended that the maximum concentration of isoprene in water be set at 0.005 mg/L.

Isoprene is classified by the ICC as a flammable liquid requiring a red label. Because of the potential hazards on its exposure to oxygen, isoprene should be stored in an inert atmosphere (nitrogen) in the presence of at least 50 ppm of *t*-butylcatechol.

Economic Aspects and Applications

Isoprene pricing tends to vary considerably due to a fairly thin commercial market. Because isoprene raw materials are primarily petroleum-based and synthesis or recovery is energy intensive, most pricing is indexed to petroleum and energy. For large-scale applications, monomer production is in tandem with application production. Generally isoprene availability is less than butadiene, and the price is higher. Isoprene is used where the unique properties of the products can command a premium over butadiene. Almost all isoprene produced is used for the preparation of polymers or copolymers. *cis*-Polyisoprene is the largest application, with SIS block polymers being a rapidly growing secondary application. Butyl rubber is a significant third application.

HUGH M. LYBARGER
The Goodyear Tire and Rubber Company

R. Adams, ed., *Organic Reactions*, Vol. 4, John Wiley & Sons, Inc., New York, 1948, pp. 60–173.

W. M. Saltman and E. Schoenberg, in J. R. Elliot, ed., *Macromolecular Syntheses*, Vol. 2, John Wiley & Sons, Inc., New York, 1966, p. 50.

Isoprene, Report No. 28, Stanford Research Institute, Menlo Park, Calif., 1967.

ITACONIC ACID AND DERIVATIVES

Itaconic acid (methylenebutanedioic acid, methylenesuccinic acid) is a crystalline, high melting acid (mp = 167–168°C) produced commercially by fermentation of carbohydrates. Itaconic acid is produced in the broth form citric acid.

Physical and Chemical Properties

Itaconic acid is isomeric with citraconic and mesaconic acids. Under acidic, neutral, or mildly basic conditions and at moderate temperatures, itaconic acid is stable. At elevated temperatures or under strongly basic conditions, the isomers are interconvertible. Itaconic acid, anhydride, and mono- and diesters undergo vinyl polymerization.

Uses

Emulsion stability, flow properties of the formulated coating, and adhesion to substrates are improved in certain polymeric coatings by the acid.

U.S. Pat. 3,165,582 (Dec. 22, 1964), M. Batti (to Miles Laboratories, Inc.).

H. W. Ashton and J. R. Partington, *Trans. Faraday Soc.* **30**, 598 (1934).

R. C. Wilhoit and I. Lei, *J. Chem. Eng. Data* **10**(2), 166 (1965).

R. W. Rees and D. J. Vaughan, *Polym. Prepr. Am. Chem. Soc. Div. Polym. Chem.* **6**, 296 (1965).

J

J ACID. See DYES AND DYE INTERMEDIATES.

JASMIN, JASMINE. See OILS, ESSENTIAL; PERFUMES.

JUNIPER. See OILS, ESSENTIAL.

JUTE. See FIBERS, VEGETABLE.

K

KETENES, KETENE DIMERS, AND RELATED SUBSTANCES

Ketenes are oxo compounds with cumulated carbonyl and carbon–carbon double bonds of the general structure $R_1R_2{=}C{=}O$, where R_1 and R_2 may be any combination of hydrogen, alkyl, aryl, acyl, halogen, and many other functional groups. Ketenes with $R_1 = H, R_2 \neq H$, are sometimes called aldoketenes, those with $R_1, R_2 \neq H$, ketoketenes. The S- and N-analogues of ketenes are called thioketenes ($R_1R_2C{=}C{=}S$) and ketenimines ($R_1R_2C{-}C{=}NR$), respectively.

The parent substance, ketene itself, is the only ketene to be manufactured in very large industrial quantities. Its principal applications are for the manufacture of acetic anhydride and diketene. The latter is an important organic intermediate used as the source of acetoacetic esters, amides, and anilides, which are widely used in the preparation of fine chemicals, pigments, drugs, and agrochemicals. Dimeric long-chain alkylketenes (C_{12}–C_{20}) are used in industrial quantities as paper sizing agents.

The chemistry of ketenes is dominated by their high reactivity: most of them are not stable under normal conditions, and many exist only as transient species. Nucleophilic attack at the *sp*-carbon, [2 + 2] cycloadditions, and ketene insertion into single bonds are the most important and widely used reactions of such compounds.

Monomeric Ketenes

Physical Properties. Ketenes range in their properties from colorless gases such as ketene and methylketene to deep colored liquids such as diphenylketene and carbon subsulfide. Other important physical properties of the parent compound ketene are as follows. Density is 0.65 g/mL at −60°C, whereas vapor density, compared to theoretical, air = 1, is 1.45. Free energy of formation $\Delta G_f^\circ = -49.6 \pm 1.6$ kJ/mol and enthalpy of formation $\Delta H_f^\circ = -47.7$ kJ/mol(−11.4 kcal/mol). The dipole moment is 4.7×10^{-30} C·m (1.41 D).

Manufacture. Ketenes can be considered the internal anhydrides of the corresponding carboxylic acids, and as such can be made by removing a molecule of water from these acids, either directly or indirectly. Numerous methods to convert a carboxylic acid or derivative to the corresponding ketene have been described.

Commercially and industrially most important, ketene itself, $H_2C{=}C{=}O$, is produced by pyrolysis of acetic acid.

Thioketenes can be prepared in several ways, from carboxylic acid chlorides by thionation with phosphorus pentasulfide P_2S_5, from ketene dithioacetals by β-elimination, from 1,2,3-thiadiazoles with flash pyrolysis, and from alkynyl sulfides (thioacetylenes).

Ketenimines are usually prepared from carboxylic acid derivatives such as amides and imino chlorides via elimination and from nitriles via alkylation with alkyl halides under strong basic conditions.

Shipping and Storage. Most ketenes are extremely reactive and unstable so they cannot be stored or transported. Some have been isolated only in solution, or have not been isolated at all, but are used *in situ*. Ketene itself is stable for some hours at −80 °C, but dimerizes within minutes at 0°C. All reactions with ketene on an industrial scale have been performed either directly in the ketene manufacturing plant or by transporting diketene and cracking it back to ketene immediately next to the reaction vessel.

Economic Aspects. Due to the physical nature and instability of these materials, all ketene production is used captively and production figures are not readily available. The economic aspects of the products made from ketenes will be addressed later.

Health and Safety Factors. Ketene itself is a highly poisonous gas, strongly irritating to the eye, the respiratory tract, and the skin. Different, sometimes conflicting values for its toxicity are found in the literature, mainly due to the difficulty in maintaining accurately and measuring low levels of the unstable ketene over hours. Its toxicity is estimated to be of the same order of magnitude as that of phosgene and like the latter it can cause latent damage of the respiratory tract which may become acute only several hours after exposure (pulmonary edema). Repeated or high exposure may cause permanent lung damage.

The OSHA PEL and NIOSH REL (recommended exposure limit)/10 h-TWA exposure limit for ketene is 0.5 ppm (0.9 mg/m^3). No carcinogenic effects of ketene have been reported. Ketene is listed in the EPA TSCA chemical inventory (1990) and in the EPA TSCA Test Submission (TSCATS) Data Base and the 1992 NHOS Hazard Code 41840.

Practically nothing is known about the toxicity of higher ketenes, thioketenes, and ketenimines, but it is prudent to consider them at least as toxic and hazardous as ketene itself. Ketenes and related compounds are highly reactive with a wide variety of substances. They can polymerize violently or even explosively, especially in the presence of bases or strong acids.

Uses. The main use of ketene is in the manufacture of acetic anhydride. The second most important use of ketene is in the production of diketene by controlled dimerization. Diketene has wide utility in the manufacture of pharmaceutical and agricultural chemicals, dyes, pigments and other fine chemicals.

Chlorine adds to ketene to form chloroacetyl chloride. Chloroacetyl chloride (CAC) is used in large volume in the manufacture of the pre-emergence herbicides alachlor and butachlor. Significant volumes of CAC are also used in pharmaceutical manufacture, such as anesthetics of the lidocaine type, and in the production of the tear gas chloroacetophenone.

Of industrial significance is ethyl 4,4,4-trifluoroacetoacetate for the production of herbicides (eg, Monsanto's Dimension) and antimalarial agents such as Roche's Mefloquin, as well as ethyl 4-chloro-chloroacetoacetate for the production of pharmaceuticals. Another principal use of ketene is in production of sorbic acid. Ketene has also been used on a large scale for C-acetylation in the synthesis of the carbapenem antibiotic thienamycin.

Dimeric Ketenes

Physical Properties. Dimeric ketenes are colorless to dark brown liquids or crystalline solids with a broad range of melting and boiling points. Table 1 lists examples of dimeric ketenes and thioketenes.

Chemical Properties. Diketene is a reactive and versatile compound which can undergo reactions with a large variety of compounds. These reactions have been reviewed comprehensively. In most reactions diketene appears to react as acetylketene or one of its tautomeric forms. Diketene usually reacts either at the carbonyl group (nucleophilic attack), or at the olefinic bond (electrophilic attack), either

Table 1. Dimeric Ketenes

Name	Physical properties	Mp or bp,[a] °C
diketene	colorless liquid	mp −7.5, bp 127 (101.3 kPa), bp 69–71 (13.3 kPa), bp 38.5 (3.1 kPa)
hexadecylketene dimer	colorless crystals	mp 64 (88), also 81 (89)
dimethylketene dimer	lachrimatory liquid[b]	bp 170 (97.6 kPa) (90), bp 83–85 (5.3 kPa)
octadecylketene dimer	crystals	mp 80 (91)
tetradecylketene dimer	crystals	mp 57–58 (89)
dimethylketene dimer	white crystals, sublimes at 95°C	mp 108–111 (92)
dimethylthioketene dimer	crystals	mp 123.5–125 (93)
cyclobutane-1,3-dione	white crystals	mp 119–120 dec (94)
dispiro(5.1.5.1)tetra-decane-7,14-dione	crystals	mp 164–165 (95)

[a] At 101.3 kPa unless otherwise stated; to convert kPa to torr, multiply by 7.5.
[b] Density = 1.88 g/mL; n_D = 1.4381.

process resulting almost always in an exothermic reaction and in the opening of the strained diketene ring. The strain energy is = 94.2 kJ/mol (22.5 kcal/mol). The so-formed 1,3-dicarbonyl compounds can react further if other functional groups are present, often forming heterocyclic compounds. Acetoacetylations and the formation of five- and six-membered heterocyclic rings are the very heart of diketene chemistry.

It is, however, possible to perform reactions such as hydrogenation, halogenation, polymerization, and [2 + 2] cycloadditions with the exocyclic double bond of diketene without opening the β-lactone ring.

Shipping and Storage. Because of its extreme reactivity and hazardous properties, diketene is now generally consumed at the site of production.

Economic Aspects. All diketene production is used captively and, therefore production figures can only be estimated by the volume of derivatives output on the merchant market. World production of diketene is probably close to 100,000 metric tons, approximately 20% of that production in the U.S. The world production of the alkylketene dimers is believed to be around 15,000 metric tons, equally split between the United States and the rest of the world.

Health and Safety Factors

Good ventilation, eye and skin protection, and an approved organic vapor respirator should be used when handling diketene.

Diketene is a strongly irritating, powerfully lachrimatory, poisonous liquid, but is considerably less toxic than ketene. The eye and respiratory tract are especially endangered, as diketene quickly damages the tissue of the cornea and lung. Exposure causes a burning sensation in eyes, nose, and throat, as well as respiration difficulties and coughing. At higher levels, loss of consciousness and death can occur. Absorption of liquid diketene by the skin is possible, with local itching and severe burning of the skin. Ingestion causes irritation if the gastrointestinal tract. It is not listed as a carcinogen by International Agency for Research on Cancer (IARC), the National Toxicology Program (U.S.), OSHA, and ACGIH. The greatest hazard is violent exothermic polymerization with quick pressure build-up and rupture of the vessel.

Uses. As the most reactive and economical source of the acetoacetyl moiety, diketene is used as a valuable synthetic intermediate in the manufacture of acetoacetic acid derivatives and heterocyclic compounds which are used as intermediates in the manufacture of dyestuffs, agrochemicals, pharmaceuticals, and polymers.

Acetoacetic Acid Derivatives

The most important use of diketene is for the preparation of derivatives of acetoacetic acid, such as acetoacetate esters, acetoacetamides, and chloroacetoacetates, which have found many uses in life sciences, dyestuffs, adhesives, and coatings.

Physical Properties. Acetoacetic esters are high boiling liquids with pleasant odors. Lower *N*-alkylamides are water-soluble liquids; acetoacetamide and acetoacetarylides are solids. 4-Chloroacetoacetates are high boiling lachrymatory liquids.

Chemical Properties. The acetoacetyl moiety is highly functionalized and can undergo many transformations. At the active methylene group, condensation reactions with other carbonyl compounds, halogenation, alkylation, and nitrosation can occur. At the ester or amide group, decarboxylation and transesterification can occur. At the ketone carbonyl group, reduction and addition of nucleophiles can occur. Combination of such reactions gives access to a broad spectrum of different types of compounds.

Manufacture and Uses. Acetoacetic esters are generally made from diketene and the corresponding alcohol as a solvent in the presence of a catalyst. Methyl acetoacetate (MAA) and ethyl acetoacetate (EAA) are the most widely used esters.

Shipping and Storage

MAA and EAA are stable liquids, and are shipped in nonreturnable 208-L (55-gal) polyethylene-lined drums.

Economic Aspects

Total U.S. annual production of MAA and EAA combined is estimated to be 6000–7000 metric tons. There are only two U.S. producers of these esters at this time, Tennessee Eastman Co. in Kingsport, Tennessee, and Lonza Inc. in Bayport, Texas.

Total U.S. annual production of all arylides combined is estimated to be 12,000 to 13,000 metric tons. The largest volume arylide is AAA (acetoacetanilide) for Pigment Yellow 12 as well as for carboxin.

CLAUDIO ABAECHERLI
Lonza AG
RAIMUND J. MILLER
Lonza, Inc.

H. Kropf and E. Schaumann, eds., in *Houben Weyl Methoden der Organischen Chemie*, Vol. E 15, 4th ed., parts 2 and 3, Thieme Verlag, Stuttgart, Germany, 1993, pp. 1598–3146.

S. Patai, ed., *The Chemistry of Ketenes, Allenes and Related Compounds*, parts 1 and 2, John Wiley & Sons, Inc., New York, 1980.

P. W. Raynolds, in V. H. Agreda and J. R. Zoeller, eds., *Acetic Acid and its Derivatives*, Marcel Dekker, Inc., New York, 1993, p. 161.

R. J. Clemens, *Chem. Rev.* **86**, 241–318 (1986).

KETONES

Ketones are a class of organic compounds that contain one or more carbonyl groups bound to two aliphatic, aromatic, or alicyclic substituents, and are represented by the general formula

$$R-\overset{\overset{\displaystyle O}{\|}}{C}-R'$$

Ketones are named by selecting as the parent compound the longest carbon chain that contains the carbonyl group, and by replacing the terminal "-e" of the parent compound by "-one." The parent chain is numbered in the direction which gives the carbonyl group the lowest number. $CH_3COCH_2CH_3$ thus becomes 2-butanone. In naming complex carbonyl structures containing more than one functional group, the carbonyl group takes precedence over alkene, hydroxyl, and most other groups.

Ketones are an important class of industrial chemicals that have found widespread use as solvents and chemical intermediates. Acetone is the simplest and most important ketone and finds ubiquitous use as a solvent. Higher members of the aliphatic methyl ketone series (eg, methyl ethyl ketone, methyl isobutyl ketone, and methyl amyl ketone) are also industrially significant solvents. Cyclohexanone is the most important cyclic ketone and is primarily used in the manufacture of γ-caprolactam for nylon-6. Other ketones find application in fields as diverse as fragrance formulation and metals extraction.

Physical Properties

The lower molecular weight aliphatic ketones and cycloaliphatic ketones are stable, colorless liquids and generally have a pleasant, slightly aromatic odor. They are relatively volatile with boiling points slightly above those of corresponding symmetrical ketones. The members of the series up to C_5 are fairly soluble in water and are excellent solvents for nitrocellulose, vinyl resin lacquers, cellulose ethers and esters, and various natural and synthetic gums and resins.

In contrast, aromatic ketones are high boiling, colorless liquids that generally have a fragrant odor and are almost insoluble in water. They

Table 1. Physical Properties of Ketones

Systematic name (common name)	Mol wt	Fp, °C	Bp at 101.3 kPa,[a] °C	Refractive index, n^{20}_D	Sp gr 20/20, °C	Viscosity at 20°C, mPa·s(= cP)	Surface tension, mN/m (= dyn/cm) at 20°C	H_{vap} at at 101.3 kPa,[a] kJ/mol[b]	Liquid specific heat capacity at (*T*)°C, J/(kg·K)[b]	Flash point, open cup, °C (closed)	Soly at 20°C, wt % In water	Soly at 20°C, wt % Water in
					Methyl alkyl ketones							
2-propanone (acetone)	58.08	−94.7	56.1	1.3590	0.7905	0.33	24.0	29.53	2224 (30)	−16(−18)	complete	complete
2-butanone (methyl ethyl ketone)	72.10	−85.9	79.57	1.3780	0.8062	0.41	24.6	31.64	2203 (20)	−6(−6)	26.8	11.8
2-pentanone (methyl propyl ketone)	86.13	−77.8	102.4	1.3902	0.8076	0.51 (28.3°C)	23.2	33.39		(7)	4.3	3.3
3-methyl-2-butanone (methyl isopropyl ketone)	86.13	−92	94.2	1.3882	0.8044	0.43 (25°C)	24.6 (25°C)	30.63		(6)	6.53	
4-methyl-2-pentanone (methyl isobutyl ketone)	100.16	−84.0	116.2	1.3957	0.8020	0.61	23.6	35.60	1920 (20)	23 (16)	1.6	1.9
2-hexanone (methyl *n*-butyl ketone)	100.16	−55.8	127.5	1.4007	0.8125	0.62	25.4	36.05	2228 (25)	(35)	1.75	3.7
3-methyl-2-pentanone (methyl *sec*-butyl ketone)	100.16	−83	117.4	1.4001	0.8142			35.12			2.26	
3,3-dimethyl-2-butanone (pinacolone)	100.16	−50	106.4	1.3986	0.8070			33.5			2.0	1.8
2-heptanone (methyl amyl ketone)	114.18	−35	151.5	1.4087	0.8166	0.77	26.1	39.25		47 (49)	0.43	1.45
5-methyl-2-hexanone (methyl isoamyl ketone)	114.18	−73.9	144.9	1.4069	0.8127	0.77	25.3 (25°C)			41 (35)	0.54	1.28
2-octanone (methyl hexyl ketone)	128.22	−20.5	173.3	1.4153	0.8197	0.95 (25°C)	26.6 (25.5)	40.88		(62)		
4-hydroxy-4-methyl-2-pentanone (diacetone alcohol)	116.16	−44.2	169.2	1.4226	0.9406	3.2	31	41.6	1883	61 (47)	complete	complete
					Dialkyl ketones							
3-pentanone (diethyl ketone)	86.3	−39.4	101.8	1.3923	0.8155	0.47	24.7	33.69	2215 (25)	(13)	3.4	2.6
2,4-dimethyl-3-pentanone (diisopropyl ketone)	114.19	−69	125.0	1.399								
2,6-dimethyl-4-heptanone (diisobutyl ketone)	142.24	−46	169.4	1.4172	0.8076	1.02	22.2	39.31		49 (49)	0.05	0.75
3-hexanone (ethyl propyl ketone)	100.16		123.2	1.4003	0.8174		25.04	35.66		35	1.57	
3-heptanone (butyl ethyl ketone)	114.19	−39	147.3	1.4088	0.8197	0.84	25.7	36.59		41 (46)	0.43	0.78
3-octanone (ethyl amyl ketone)	128.22	−46	167–168	1.4150	0.8220							
2,6,8-trimethyl-4-nonanone (isobutyl heptyl ketone)	184.32	−75	218.2	1.4257	0.8180	1.9		44.56		90 (88)	< 0.01	0.2

Table 1. Physical Properties of Ketones *(continued)*

Systematic name (common name)	Mol wt	Fp, °C	Bp at 101.3 kPa,[a] °C	Refractive index, n^{20}_{D}	Sp gr 20/20, °C	Viscosity at 20°C, mPa·s(= cP)	Surface tension, mN/m (= dyn/cm) at 20°C	H_{vap} at at 101.3 kPa,[a] kJ/mol[b]	Liquid specific heat capacity at (*T*)°C, J/(kg·K)[b]	Flash point, open cup, °C (closed)	Soly at 20°C, wt %	
											In water	Water in
					Unsaturated ketones							
3-buten-2-one (methyl vinyl ketone)	70.09	−6	81.4	1.4130								
3-methyl-2-buten-2-one (methyl isopropenyl ketone)	84.12	−54	98	1.4236	0.855							
4-methyl-3-penten-2-one (mesityl oxide)	98.15	−53	129.5	1.4414	0.8521	0.6	28.4	43.1	2176 (20)	29 (31)	3.1	3.4
4-methyl-4-penten-2-one (isomesityl oxide)	98.15		121.5	1.4458	0.8548							
3,5,5-trimethyl-2-cyclohexen-1-one (isophorone)	138.21	−8.1	215.3	1.4775	0.9229	2.6	32	43.4	1799 (20)	104 (85)	4.3 (25°C)	1.2 (25°C)
3,5,5 trimethyl-3-cyclohexen-1-one (β-isophorone)	138.21		181–191		0.89						0.03	
					Diketones							
2,3-butanedione (diacetyl)	86.09	−2.5	90.2	1.3938	0.9843			34.3				
2,3-pentanedione	100.12	−52	111				31	35.4	1983 (20)			
2,4-pentanedione (acetylacetone)	100.11	−23.5	140.4	1.4510	0.9753	0.58		36.55	1956.2			
2,5-hexandione	114.15	−5.4	192.3	1.4256	0.9734	1.6			(15)			
					Cyclic ketones							
cyclopentanone (adipic ketone)	84.12	−50.6	130.8	1.4359	0.9512	1.2	33.35	36.53			29	14
cyclohexanone (pimelic ketone)	98.15	−31.1	155.7	1.4510	0.9482	2.21	35.2	37.62	2039.8 (30.8°C)	46 (43)	2.5	8.0
cycloheptanone	112.17	−21	179	1.4611			26.4			72 (62.5)	0.3	1.4
3,3,5-trimethylcyclohexanone	140.22	−10	188.8	1.4455	0.888	2.54						
					Aromatic ketones							
acetophenone (methyl phenyl ketone)	120.15	19–20	201.7	1.5342	1.0296	0.93		45.69	93 (82)		0.55	1.65
benzophenone (diphenyl ketone)	182.22	48–49.5	305									
1-phenyl-2-propanone (phenylacetone)	134.18	−15		1.5158								
propiophenone (phenyl ethyl ketone)	134.17	18.2	218	1.5265	1.012		37.4	45.44		96 (85)	0.01 (25°C)	

[a] To convert kPa to mm Hg, multiply by 7.5. [b] To convert J to cal, divide by 4.184.

are useful as intermediates in chemical manufacture. Functionalized and cyclic ketones are also good solvents. Ring size and the type and location of functional groups affect odor, color, and reactivity of these ketones.

The physical properties of some common ketones are listed in Table 1. Ketones are commonly separated by fractional distillation, and vapor–liquid equilibria and vapor pressure data are readily available for common ketones. A number of other temperature dependent physical properties for acetone, methyl ethyl ketone, methyl isobutyl ketone, and diethyl ketone have been published.

Chemical Properties

The constituent carbonyl group makes many of the reactions and methods of preparation for ketones similar to those of aldehydes. Ketones, however, generally undergo 1,2-addition reactions across the carbonyl group less readily than aldehydes because of steric hindrance around the carbonyl group. Similarly, the relative reactivity among ketones is influenced by the polarity and electrophilic nature of the substituents in the vicinity of the carbonyl group (eg, hydrogens alpha to the carbonyl group). The chemical properties of diketones, and cyclic and unsaturated ketones such as 2,4-pentanedione, cyclohexanone, and mesityl oxide, respectively, are enhanced, thereby increasing their utility as chemical intermediates.

Reduction. Most ketones are readily reduced to the corresponding secondary alcohol by a variety of hydrogenation processes.

Oxidation. Ketones are oxidized with powerful oxidizing agents such as chromic or nitric acid.

Condensation. Depending on the nature of the hydrocarbon groups attached to the carbonyl, ketones can either undergo self-condensation, or condense with other activated reagents, in the presence of base. Although ketonic carbonyl groups are less reactive than aldehydic carbonyls in the presence of basic catalysts, this is not the case with acid catalysts.

Preparation of Amines. Amines are prepared by heating aliphatic, aromatic, or cyclic ketones with ammonium formate at 165–190°C (Leuckart reaction).

Thermal Stability. The saturated C_4—C_{12} ketones are thermally stable up to pyrolysis temperatures (500–700°C). At these high temperatures, decomposition can be controlled to produce useful ketene derivatives.

Health and Safety Factors

Ketones are flammable substances that do not exhibit a known high degree of chronic toxicity. Low molecular weight (C_3-C_{12}) saturated aliphatic ketones, which represent the bulk of industrially important ketones, may be classified among the solvents of comparatively low toxicity hazard. The eight-hour threshold limit value is generally above 100 ppm, although the odor threshold is in the range 5–25 ppm. High vapor concentrations of these volatile ketones induce anesthesia, however the vapors are so irritating to the eyes and mucous membranes of the respiratory system that the atmosphere generally becomes intolerable before toxic concentrations are achieved. Many ketones are also powerful drying and degreasing agents and prolonged skin contact can cause dermatitis.

The C_3-C_{12} ketones are all highly flammable liquids. The toxicity of unsaturated ketones and diketones is significantly greater. The eight-hour threshold limit value for these materials is ≤50 ppm.

Environmental Aspects

Most industrially important ketones are volatile organic compounds (VOC) which are subject to air pollution control regulations.

The impact of the regulations is to require users and producers of VOC ketones to limit release by either reformulating to new solvent systems, to install environmental control systems which recover and recycle solvents, or reduce emissions with carbon absorption beds or incineration equipment. The use of some individual ketones will decline further, but the overall short-term use of ketones is forecast to remain stable.

John Braithwaite
Union Carbide Chemicals and Plastics Company, Inc.

S. R. Sandler and W. Karo, *Organic Functional Group Preparations*, Vol. 1, 2nd ed, Academic Press, Inc., New York, 1983.

S. Takaoka, *Acetone, Methyl Ethyl Ketone, and Methyl Isobutyl Ketone*, Process Economics Program, Report No. 77, SRI, Menlo Park, Calif., May 1972.

K. Schmitt, *Chem. Ind. (Dusseldorf)* **18**, 4, 204 (1966).

KINETIC MEASUREMENTS

Kinetic measurements are studies of the rates at which chemical reactions occur. Generally, these studies involve preparing a chemical system using reagent concentrations different from the equilibrium values and then monitoring the concentration changes as the system approaches equilibrium, although other, less direct strategies are sometimes exploited. Chemical kinetic data are used in materials science, biochemistry and molecular biology, earth and atmospheric science, and many branches of engineering. Related concepts appear in nuclear physics, but presuppositions and methods are different there.

Kinetic information is aquired for two different purposes. First, data are needed for specific modeling applications that extend beyond chemical theory. These are essential in the design of practical industrial processes and are also used to interpret natural phenomena such as the observed depletion of stratospheric ozone. Compilations of measured rate constants are published in the United States by the National Institute of Standards and Technology (NIST). Second, kinetic measurements are undertaken to elucidate basic mechanisms of chemical change, simply to understand the physical world. The ultimate goal is control of reactions, but the immediate significance lies in the patterns of kinetic behavior and the interpretation in terms of microscopic models.

Explaining chemical change by postulating mechanisms in terms of macroscopic concentrations is expected to continue for the forseeable future. For a fundamental understanding of very simple reactions, however, traditional kinetics is being challenged by theoretical and experimental methods that focus directly on the behavior of individual atoms.

Macroscopic Behavior and the Rate Law

Chemical Equations. Chemical changes are discussed with the aid of the equations used to treat equilibrium, ie, the reaction of reactants A, B, C, and so on, to produce products P, Q, and so forth.

The essential information implied by the chemical equation is the stoichiometry at the macroscopic level, ie, if a moles of A react, then b moles of B do also; p moles of P formed, etc.

A kinetic study typically prepares some set of initial concentrations not at equilibrium and describes the subsequent evolution of each. A basic assumption is that each component evolves according to some differential equation where t represents time.

$$d[A]/dt = f([A],[B],\cdots,[P],\cdots,\text{ other conditions}) \tag{1}$$

In general, the differential equation could be very complicated, eg, the concentrations may be functions of spatial coordinates as well as time. Experimental measurements are arranged to ensure that simplified equations apply.

The Well-Stirred Mixture. A key assumption of most kinetic measurements is that of a well-mixed solution of reactants. Then any component can be characterized by a single time-dependent concentration, applicable to the entire system.

In particularly simple cases, which occur frequently, one may assume a dependence on powers of reactants and ignore products

$$d[A]/dt = -h[A]^x[B]^y[C]^z \tag{2}$$

Experimental Verification of a Rate Law

It is possible to prepare a system having an initial concentration for each component, and then measure a finite, but small, change in the concentration of one component, $\Delta[A]$ for example, over a known interval of time, Δt. The experimental velocity $\Delta[A]/\Delta t$ and the concentrations can be substituted into a proposed rate law, like equation 2, along with postulated values for the exponents $x,y,$ ... to determine an observed constant k_{obs}. If this process is repeated for a reasonable range of concentrations, and the postulated rate law having the same exponents always yields the same k_{obs}, then it is asserted that the rate law has been verified for those concentration ranges and the rate constant determined. This approach is a reasonable strategy for an initial survey of a totally unknown system; but it is wasteful, in that it extracts very little data from each set of initial conditions. More often, the integrated form of the rate law is fit to multiple concentration measurements recorded at different times for each set of initial conditions.

Flooding and Pseudo-First-Order Conditions. Flooding is an experimental strategy that simplifies both measurement and analysis. For an example, consider a reaction that is independent of product concentrations and has three reagents. If a large excess of $[B_i]$ and $[C_i]$ are used, and the disappearance of a lesser amount of $[A]$ is measured, the rate law can be integrated with the assumption that all concentrations are constant excepts $[A]$. Consequently, simple expressions are derived for the time variation of $[A]$. Under flooding conditions and using equation 2, if x happens to be 1, the time-dependent concentration of A exhibits an exponential decrease from its initial value $[A_i]$ to its final equilibrium value, or endpoint, $[A\infty]$:

$$[A(t)] - [A_\infty] = ([A_i] - [A_\infty])\exp(-k_{\text{obs}}t) \tag{3}$$

The conditions chosen make the reaction appear to be first-order overall, although the reaction is really not first-order overall, unless y and z happen to be zero. The pseudo-first-order rate constant k_{obs} is related to the k in the originally postulated rate law by

$$k_{\text{obs}} = k[B_i]^y [C_i]^z \tag{4}$$

If x is not 1, equation 3 is replaced by a different, but still simple, integrated rate law.

The Initial Conditions. One of two very different strategies are used in kinetic measurements to produce the initial, nonequilibrium concentrations of reactants. Either the separate reagents are mixed or a system previously at equilibrium is perturbed.

Mixing known quantities of reagents to produce desired initial concentrations can be carried out with either a continuous flow or a stopped flow apparatus. Engineering details become important for fast reactions, especially those occurring in less than one millisecond.

Perturbation methods can be divided further into two categories. One uses a flash of light (flash photolysis) or other radiation (radiolysis) to create a homogeneous distribution of a desired reagent from some precursor molecule. This method is capable of very fast time resolution; recent advances in lasers allow the study of processes occurring in 10^{-14} s. The other perturbation method does not affect reagents but instead changes some intensive thermodynamic property, such as temperature, pressure, or an electric field. Concentrations are monitored as the system adjusts to the requirements of the new equilibrium.

Indirect and Novel Methods. A direct measurement is a record of changing concentrations as a function of time. In principle, the same information is available as a Fourier transform in the frequency domain. Since the latter part of the nineteenth century, it has been possible to measure the absorption of electromagnetic radiation as a function of frequency (or wavelength) and interpret lineshapes to yield kinetic information on the picosecond time scale. In gases, line widths increase at high pressure owing to collision broadening. Dissociation or ionization may also determine a linewidth. More subtle effects occur in liquids. More recently, chemical reactivity on slower time scales has been measured by lineshape analysis in Mossbauer spectroscopy and magnetic resonance.

Dramatic progress is being made in extending kinetic analyses to microscopic samples, such as single biological cells. One strategy infuses precursor molecules into the correct part of a cell and then uses flash photolysis to start a reaction that is monitored continuously by fluorescence microscopy. Within such small volumes, there may be only thousands of molecules of a given type and their number may change due to statistical fluctuations, even in a system at equilibrium. Such fluctuations obey the same laws of chemical kinetics as any other perturbation; and a study of fluctuations can determine kinetic parameters, although this is far from being a routine technique. Pioneering experiments are even being carried out for kinetic studies on single atoms, not in cells, but isolated in a magnetic or optical trap.

Rapid instrumental measurements have largely, but by no means completely, supplanted an earlier tradition that relied on a more chemical strategy of measuring concentrations of reactants or intermediates by intercepting these with a scavenger that reacted quickly to form a stable product that could be quantified later. Such trapping methods can be construed as being indirect in the sense that they rely on one chemical reaction being faster or slower than another.

Experimental Variation of Chemical Rates with Temperature and Pressure

The experimentally measured dependence of the rates of chemical reactions on thermodynamic conditions is accounted for by assigning temperature and pressure dependence to rate constants.

Microscopic Models in Kinetics

Mechanism is a technical term, referring to a relatively detailed, microscopic description of a chemical transformation, which, nevertheless, still falls far short of a complete dynamical description at the atomic level. A mechanism for a reaction is sufficient to predict the macroscopic rate law of the reaction. This deductive process is valid only in one direction, ie, an unlimited number of mechanisms are consistent with any measured rate law. A successful kinetic study postulates a mechanism, derives the rate law, and demonstrates that the rate law is sufficient to explain experimental data over some range of conditions. New data may be discovered later that prove inconsistent with the assumed rate law and require that a new mechanism be postulated. Mechanisms state, in particular, what molecules actually react in an elementary step and what products these produce. An overall chemical equation may involve a variety of intermediates, and the mechanism specifies those intermediates.

DOUGLAS MAGDE
University of California at San Diego

S. Claesson, ed., *Fast Reactions and Primary Processes in Chemical Kinetics*, 5th ed., Wiley-Interscience, New York, 1967.

K. Kustin, ed., *Fast Reactions, Methods in Enzymology*, Vol. 16, Academic Press, Inc., New York, 1969.

H. Strehlow, *Rapid Reactions in Solution*, VCH, Weinheim, Germany, 1992.

J. W. Moore and R. G. Pearson, *Kinetics and Mechanism*, 3rd ed., Wiley-Interscience, New York, 1981.

KRYPTON. See HELIUM GROUP, GASES.

L

LABELS. See FLAME RETARDANTS; INDUSTIRAL HYGIENE.

LABORATORY INFORMATION MANAGEMENT SYSTEMS

A laboratory information management system (LIMS) is a computer or computer network used to automate the acquisition and management of raw analytical data. In its simplest form, it tracks samples and test results through analytical laboratories and provides summaries of the status of these samples and tests. In its most advanced form, the system is interfaced to the laboratory's instrumentation and communication network to allow automation of data gathering, compilation, and reporting.

The approximately 40 vendors selling LIMS have been able to design the systems with enough flexibility to meet the needs of most laboratories. Commercial systems can usually be delivered quickly and the laboratory benefits from a large user base, compatible accessories, and future enhancements.

The benefits of LIMS depend largely on the needs of the laboratory and the type of system installed. In general, a LIMS improves the management of the laboratory by providing more accurate and timely information regarding work load and work-load distribution. The productivity is increased through the automation of many clerical and routine tasks associated with sample identification, tracking, and transcription of results. The quality of the data can be improved through automated data acquisition, reduction in transcription errors, and automatic enforcement of validation analysis and standardization procedures. Finally, the system can be used to meet regulatory agency compliance requirements.

Suppliers of LIMS include Banyon Systems, Inc., Beckman Instruments Inc., Challenger Group, Inc., Chesapeake Software Inc., Cirrus Technology, Digital Equipment Corporation, DSP Development Corporation, Hewlett-Packard, IBM, Keithley Instruments, Laboratory Data Systems Inc., Laboratory MicroSystems Inc., LabWare Ltd., Northwest Analytical Inc., Novell Inc., PE Nelson Div., Radian Corporation, Statistical Graphics Corporation, 3Com Corporation, Varian Associates Inc., and VG Instruments.

The various functions of a LIMS can be summarized as the analytical or managerial tasks, as follows.

Analytical level tasks

- Sample number generation
- Bar-code label generation
- Sample log-in
- Verification of data format entered into the computer
- Worksheet generation
- Construction and checking of calibrated curves
- Direct data acquisition from chromatographs
- Data collection for analytical instruments
- Entry of instrumental readings
- Manual results entry
- Interpretation of calibrated curves and quality control samples
- Interpretation and acceptance of sample data
- Routine automatic calculations
- Plotting routines for visualization of analytical data

Managerial level tasks

- Acknowledgement of sample receipt
- Backlog investigation
- Sample and status tracking
- Database searches
- Numbers of samples assayed
- Tests utilized
- Numbers of samples analyzed per instrument
- Cost per assay
- Customer charges
- Results collation and presentation
- Report generation
- Scheduling and rescheduling of work
- Archiving and retrieval of data
- Workload status and the justification of equipment
- Regulatory agency compliance
- Audit trail for all database transactions
- Security: class or hierarchy
- Instrument records and calibration where appropriate

Automation

The development and widespread use of computers and microprocessors in control laboratory instruments has made it possible to fully automate a laboratory, including interfacing instruments directly to a LIMS. In the fully automated laboratory, a sample is logged into a LIMS, then transferred to a laboratory where it is prepared for analysis by a robot, which then transfers it to an autosampler or analyzer. Once analyzed, the data is transferred through a communications link to a device which could convert the raw data into information that a customer needs.

Quality Management

The business activity of the organization dictates quality requirements for the LIMS. Security and regulatory requirements for LIMS data define the level of effort expended to validate a LIMS and the data being stored. In addition, the quality of the hardware and software used to implement the LIMS both play a role in determining overall system quality.

A LIMS system must be validated to ensure data security and integrity and to ensure it complies with applicable government regulations. The validation process includes first developing a strategy and setting specific objectives for validation. Modular testing of the system should by employed, and a detailed test protocol should be developed for each LIMS module. Documentation is critical throughout the validation period. The test strategies and procedures, as well as flow charts for the LIMS source code, must be documented. Training documentation for LIMS users and administrators should be included in a complete validation package.

The hardware and software used to implement LIMS systems must be validated. Computers and networks need to be examined for potential impact of component failure on LIMS data. Security concerns regarding control of access to LIMS information must be addressed. Software, operating systems, and database management systems used in the implementation of LIMS systems must be validated to protect against data corruption and loss.

Costs and Benefits

An excellent tutuorial on evaluating the costs and benefits of LIMS has been presented by R. R. Stein. The tutorial explains how to perform a cost–benefit analysis using a pragmatic approach to the economics involved. To assist in the analysis, a detailed list of specific LIMS costs and benefits from the tutorial is included. The size of the system, based on the number of samples and analyses per year, can be used to approximate the cost. Although the tutorial is a good starting point for accurately assessing the costs of a proposed system, the task of measuring the value of the benefits can be much more subjective in nature.

Benefits can be classified as tangible and intangible. Tangible benefits are benefits that are easily assigned a monetary value. These include items such as reductions in the costs of calculating and reporting data, improved capacity through better access to management of data in making assignments, and being able to identify and report sample status more quickly. Intangible benefits might include better overall service to customers and a general perception of better laboratory management through the use of state-of-the-art management techniques. Another significant benefit of a LIMS is the improvement of the overall quality of the laboratory.

Selection

Before deciding on a LIMS product, a complete set of specifications for required functions of the LIMS should be written.

When researching the requirements for a LIMS, answers to the following questions, courtesy of The Royal Society of Chemistry, should be sought. Is the laboratory to be used for a single technique or multiple techniques, for a single product line or product type, or for all types of samples? Would a redesign of experimental procedures produce an increase in efficiency? Are there management requirements, including secretarial assistance? Are computing resources required, including a laboratory management data system, local data collection and storage of instruments results, data storage and filing requirements at laboratory level, use of personal workstations, introduction of local area networks (LANs), mainframe interfaces, and telecommunication facilities? Are there reporting, archiving, and database requirements? Are personnel records to be included in the computerized laboratory management scheme? What overall response and speed of the system is necessary for the turnaround expected? What is to be done about the education of users, a most important requirement that is easily overlooked in the planning stages?

The search for a vendor can begin once the specification is complete. The level to which a vendor's products conform to established requirements is the most important selection criterion for a LIMS. Financial stability of the vendor's organization is another consideration, because ongoing technical support, from the vendor is vital to LIMS implementation and enhancements. The technology used in a vendor's products should be evaluated for robustness and longevity.

Database Management

A database management system (DBMS) is used by most LIMS systems for storing data. Examples of commercially available DBMS are DB2, DBASE, Informix, INGRES, ORACLE, and RDB. All of these DBMS conform to the "relational" model developed by Codd.

Impacts of New Technology

Computer hardware costs have decreased dramatically. As a result, systems have become more affordable. Higher performance of new technology allows more functional capacity to be provided in a smaller, less costly machine.

Advances in network operating systems (NOS) provide database server and independent, simultaneous (distributed) data processing capabilities necessary to support LIMS functions throughout a network of computers.

Another significant impact of new technology is the evolution of the client/server computing model into commercially viable systems. This model incorporates a more powerful computer for data storage and retrieval (the server), connected to client workstations via a network.

A significant concept of the client/server model is to extend the scope of the application to function in an enterprise-wide (possibly worldwide) network of interconnected LANs.

PC workstations have become powerful, are simpler to use, and are generally ubiquitous within both laboratory and office environments.

Databases are becoming more standardized, thereby allowing a greater number of supplemental functions to be added to a LIMS and foreign systems, eg, project accounting, to be more easily integrated.

For these reasons, the desktop and client/server models are expected to increase in percentage of LIMS software offerings and installed base in the future.

Instrumentation advances have increased the power and quality of the fundamental analytical techniques used in conjunction with LIMS. Unfortunately, these advances come at a price of increasing complexity and volume of information.

Data acquisition has evolved, but standards are still lacking. This makes data acquisition the most difficult and time-consuming aspect of the overall LIMS implementation. The time savings which result from automated data capture result in its generally being undertaken, to some degree, despite the difficulties.

Work is being done to create uniform standards for exchange of information between analytical instrumentation and external (host) computers, but the diversity and the competitive nature of the instrumentation marketplace tend to impede these efforts, leading to an environment of constant change and a need for new and rewritten programs to communicate between LIMS and the automated instruments.

MICHAEL G. BARRETT
KENNETH O. MACFADDEN
JEAN A. TAYLOR
W. R. Grace & Co.-Connecticut

R. D. McDowall, ed., *Laboratory Information Management Systems: Concepts, Integration and Implementation*, Sigma Press, New York, 1987, p. 12.

A. Braithwaite, *Anal. Proc.* **24**, 126 (Apr. 27, 1987).

R. D. McDowall, *Lab. Info. Mgt.* **17**, 270 (1992).

R. R. Stein, *Lab. Info. Mgmt.* **13**, 15–36 (1991).

LAMINATED MATERIALS, GLASS

A laminate is an orderly layering and bonding of relatively thin materials. A commonly laminated material is glass. Most commonly, two pieces of float or sheet glass are bonded with poly(vinyl butyral) (PVB) to produce a highly transparent safety glass, eg, an automotive windshield. This combining of transparent abrasion-resistant glass and resilient plastic achieves the durability and safety demanded of such products. Other materials that may be incorporated in laminated glass are colorants, electrically conducting films or wires, and rigid plastics. The value of the laminate is the utilization of the desirable properties from each of the constituents.

Laminated glass is not a true composite material. The glass needs the safety net effect of the interlayer if impacted, and the interlayer needs the durability and rigidity of the glass for useful service other than during impacts. Exceptions where laminated glass more truly fits the definition of a composite are when it is used for noise attenuation or bullet resistance. In these applications, the alternate layering of rigid and soft materials achieves results beyond those produced by either alone.

Properties

Laminated materials frequently have limits on properties below those found in one of the components. Laminated glass with a PVB interlayer has a maximum service temperature not exceeding 70°C, far below that of solid glass.

Glass-PVB laminates become more rigid with a decrease in temperature, and below −7°C approach the performance of solid glass. At temperatures above 38°C these laminates are less rigid and provide improved penetration resistance. Some applications utilize heat-strengthened or tempered glass for additional strength.

Most laminated glass applications are concerned with impact strength, and minimum performance levels are required by specification. Aircraft laminates may utilize electrical resistance heating as deicing for vision enhancement.

Automotive and architectural laminates of PVB develop maximum impact strength near 20°C. This balance is obtained by the plasticizer-to-resin ratio and the molecular weight of the resins. The optical properties of laminated glass are required to be equal to solid glass, because most applications are in vision areas.

Production and Shipment

Chemical attack, particularly from moisture and alkaline conditions, is prevented by use of acidic packing materials and open, ventilated packages. Good crate design and proper handling throughout shipment avoids mechanical damage. Glass-to-glass contact is never permitted. Long-term storage must be in well-ventilated areas, never in sealed containers. In the case of trans-ocean shipping, however, sealed containers are often used with a desiccant added to prevent moisture attack of the glass.

Glass products are shipped and stored in a vertical plane, and during transportation they are placed so that each plate has an edge in the direction of travel.

Economic Aspects

Architectural products currently represent 5–10% of the laminated glass volume. Increased consumer safety awareness and security needs are expanding the flat laminate market. Safety codes (eg, 16CFR 1201 and ANSI Z97.1) specify laminated glass as one means of meeting their requirements. In hurricane-prone areas laminated glass is being specified increasingly by local authorities.

Laminated windshields, as opposed to tempered glass windshields, are gaining in market share outside of North America. The trend toward laminated windshields is expected to continue and nonlaminated windshields will likely be obselete by the year 2000.

Uses

Uses of laminated glass include penetration-resistant windshields, special laminated windshields, aircraft windshields, and architectural products.

R. TERRELL NICHOLS
Ford Motor Company
ROBERT M. SOWERS
Consultant, Ford Motor Company

Safety Code for Safety Glazing Materials for Glazing Motor Vehicles Operating on Land Highways, Z26.1-1973, American National Standards Institute, New York.

Architectural Saflex for Sound Control, Tech. Bulletin No. 6295, Monsanto Polymers and Petrochemicals, St. Louis, Mo., 1972.

FGMA Glazing Manual, Flat Glass Marketing Assoc., Topeka, Kans., 1974.

R. N. Pierce and W. R. Blackstone, *Impact Capability of Safety Glazing Materials, PB195040*, Southwest Research Institute, San Antonio, Tex., 1970; contains detailed descriptions of test equipment, methods, and results for all types of glazings.

LAMINATED MATERIALS, PLASTIC

Laminates are a special form of composite material or reinforced plastic because the continuous reinforcing ply of fibrous material imparts significant strength in the x–y plane.

The reinforcing ply of laminates may be a woven fabric scrim, a nonwoven web of polymer monofilaments, or a mat of fibers. One of the most common reinforcements in use is also one of the oldest, ordinary cellulose fiber paper.

Resins

The commonly used resins in the manufacture of decorative and industrial laminates are thermosetting materials. Thermosets are polymers that form cross-linked networks during processing. The types of thermosets commonly used in laminates are phenolics, amino resins (melamines), polyesters, and epoxies.

Reinforcements

The reinforcing ply in a laminate may make up half or more of its total weight. Therefore, properties of a laminate are strongly dependent on the ply. These piles are specified by basis weight in grams per square meter and may range from as low as 15 g/m^2 for a lightweight overlay sheet to as much as 300 g/m^2 or more for a strong filler sheet.

The most commonly used reinforcement for high pressure decorative and industrial laminates is paper. For use in decorative laminates, the surface paper must be highly refined pure cellulose technically called alpha cellulose.

Other reinforcements that may be used in the substrate layers of decorative laminates and throughout the structure of industrial laminates are woven fabrics of glass or canvas and nonwoven fabrics of various polymeric monofilaments such as polyester, nylon, or carbon fibers.

The reinforcing ply acts as the carrier for the plastic resin during intermediate processing steps known as saturation and B-staging. It is this ply that together with the resin makes a laminate a composite material, and the layering of these piles that makes the final product a laminate.

Manufacture

Treating. Treating is the term used in the laminate industry for the application of the plastic resin to the reinforcing ply or carrier web that eventually forms a ply in the composite laminated product. Typical means to apply the resin are reverse roll coaters, dip and scrape, or dip and squeeze operations.

Collation. Collation is the process by which the individual laminate plies are assembled prior to curing in the press. The buildup of the laminate determines the final properties of the product.

Press Curing. The laminate as an article of manufacture is prepared in a flat-bed press. Modern high pressure presses may be as large as 2 m × 5 m, and low pressure presses are as long as 7.5 m. Normally, high pressure presses have multiple openings or daylights, sometimes 20 or more. Low pressure presses have only a single daylight. Continuous presses are also in use for the manufacture of low pressure products. Some continuous high pressure presses are being built, but they are not in wide usage. During the press operation, which is actually a form of compression molding, the resin-treated laminate plies are heated under pressure and the resins cured.

Finishing and Fabrication. Because laminates are normally pressure cured in flat-bed presses and plies overextend the plates, laminates have rough or uneven edges when removed from the press. These edges are sawed off and the back of the laminate is often sanded to improve the strength of subsequent bonding to various substrates.

Common grades of laminates tend to be thin materials ranging from 0.5–1.5 mm in thickness, therefore for most applications they must be supported. In the manufacture of furniture, cabinetry, and countertops the laminates are bonded to particle board or plywood. Because the laminates consist largely of cellulosic paper, their dimensional stability is similar to wood, particularly to particle board.

In small pieces or as inserts, laminates may be used unsupported because they are quite stiff and strong. An important fabrication operation for laminates is post-forming. This is an operation in which a laminate is heated and bent.

Properties and Grades

Aesthetic properties are of greatest concern in decorative laminates. These include gloss, appearance, cleanability, wear resistance, stain resistance, and other surface properties. Physical properties are of most importance for industrial laminates. These include strength, electrical and thermal properties, expansion coefficient, and punchability. The definitions of the laminate grades in these standards follow.

Decorative laminates include general-purpose type, post-forming type, cabinet liner type, backer type, specific-purpose type, high wear type, and fire-rated type.

Industrial laminates include paper-base grades, fabric-based grades, asbestos-based grade, glass-based grades, nylon cloth grade with phenolic resin binder, flame-resistant grades, and composite-based laminates.

Health and Environmental Concerns

Key resins used in the manufacture of laminates are made with formaldehyde. Plant atmospheres as well as individual operators are monitored to be certain they are exposed to levels of formaldehyde that are below OSHA guidelines of 0.75 ppm.

In the final product, the formaldehyde has completely reacted to form a very inert thermoset resin. Spontaneous emission of formaldehyde from high pressure laminates is measured at approximately the accepted background level of 0.035 ppm. Melamine surfaced laminates are tested and approved for food service equipment by the National Sanitation Foundation.

In fires, melamine–phenolic laminates ignite slowly at high temperatures and burn slowly producing smoke that has about the same toxicity as wood smoke.

Disposal of laminate scrap resulting from edge trim, sanding dust, and fabrication trim presents other problems. The scrap within the manufacturing plant can be ground up and used as part of the fuel source in boilers with the proper permits and controls. Stack effluents from treating operations can be burned off in boiler feedstock or captured in charcoal filter beds. Fabrication shop scrap goes to landfills where it has not generally been a problem because of its density and inertness.

RONALD J. KEELING
Formica Corporation

T. S. Carswell, *Phenoplasts: Their Structure, Properties and Chemical Technology*, Interscience Publishers, Inc., New York, 1947.

The Chemistry of Melamine Crystals, Melamine Chemicals, Inc., Donaldson, La., 1990.

J. Casey, ed., *Pulp and Paper Chemistry and Chemical Technology*, Vol. I, 3rd ed., John Wiley & Sons, Inc., New York, 1980.

High Pressure Decorative Laminates, Standards Publication No. LD3-1991, National Electrical Manufacturers Association (NEMA), Washington, D.C., 1991.

U.S. Pat. 5,215,695 (June 1, 1993), C. Bortoluzzi and R. Bogana (to Abet Laminati SpA).

LAMINATED WOOD-BASED COMPOSITES.

See WOOD-BASED COMPOSITES AND LAMINATES.

LANTHANIDES

Lanthanides is the name given collectively to the fifteen elements, also called the 4*f* elements, ranging from lanthanum, La, atomic number 57, to lutetium, Lu, atomic number 71. The rare earths comprise lanthanides, yttrium, Y, atomic number 39, and scandium, Sc, atomic number 21. The most abundant member of the rare earths is cerium, Ce, atomic number 58.

Occurrence

The lanthanides, distributed widely in low concentrations throughout the earth's crust, are found as mixtures in many massive rock formations, eg, basalts, granites, gneisses, shales, and silicate rocks, where they are present in quantities of 10–300 ppm. Lanthanides also occur in some 160 discrete minerals, most of them rare, but ones in which the rare-earth (RE) content, expressed as oxide, can be as high as 60% rare-earth oxide (REO).

Usually lanthanides are divided into several subgroups: the light lanthanides, from La to Nd, medium lanthanides, from Sm to Dy, and heavy lanthanides, Ho to Lu. Alternatively, nomenclature such as ceric RE, from La to Nd, and yttric RE, from Sm to Lu plus Y, is used.

The relative abundance of certain rare earths in rocks is a powerful tool for the study of the formation of rock deposits. Relative abundances of the rare earths in the earth's crust are, in the U.S., 6,471 t $\times 10^3$; Australia, 754; India, 1,939; South Africa, 987; and China, 36,000.

The rare earths are not so rare. Cerium, the most abundant of the rare-earth elements, is roughly as abundant as tin; thulium, the least abundant, is more common than cadmium or silver. Over 200 rare-earth-containing minerals have been identified. Allanite, apatite, bastnaesite, brannerite, cerite, euxenite, fergusonite, fluocerite, gadolinite, monazite, pyrochlore, samarskite, xenotime, and zircon are considered the principal rare-earth-containing minerals, but only monazite and bastnaesite are processed on a large industrial scale. China has the largest share of the world's reserves of rare earths (more than 75%). The United States, India, South Africa, and Australia also have significant reserves.

Properties

In the sixth row of the Periodic Table, the binding energies of the 4*f* and 5*d* subshells lie close together. Of most direct interest, however, is the electronic configuration of the ions, which are generally trivalent and have configuration $[Xe]4f^n$, the 4*f* subshell being progressively filled from lanthanum to lutetium.

Physical Properties. The rare earths form alloys with most metals. The main physical properties of rare earth metals are given in Tables 1 and 2.

Chemical Properties. Yttrium and the lanthanides are typical hard acids, and bind preferably with hard bases such as oxygen-based ligands. Nevertheless they also bind with soft bases, typically sulfur and nitrogen-based ligands in the absence of hard base ligands. The chemical properties of the trivalent lanthanides vary little from one lanthanide to another in a given compound or solution.

Mining

A limited number of rare-earth minerals are mined for large-scale rare-earth production: monazite, bastnaesite, loparite, xenotime. In addition, since the 1980s rare-earth-containing clays called ionic ore are mined in China.

Processing

Industrial Digestion. The commercial digestion process for monazite uses caustic soda. The phosphate content of the ore is recovered as marketable trisodium phosphate and the rare earths as RE hydroxide.

The commercial 60% of REO concentrate of bastnaesite can be upgraded to 90% REO by leaching with hydrochloric acid then calcining. Other processes include digestion with hydrochloric or sulfuric acids and recovery of RE chlorides or sulphates solutions, respectively.

There are two methods used at various plants in Russia for loparite concentrate processing. The chlorination technique is carried out using gaseous chlorine at 800°C in the presence of carbon. The volatile chlorides are then separated from the calcium–sodium–rare-earth fused chloride, and the resultant cake dissolved in water. Alternatively, sulfuric acid digestion may be carried out using 85% sulfuric acid at 150–200°C in the presence of ammonium sulfate. The ensuing product is leached with water, while the double sulfates of the rare earths remain in the residue.

Ion-Adsorption Deposits. Ion-adsorption clay deposits result from prolonged *in situ* weathering of REO-rich host rocks, most commonly granitic or volcanic rocks, where erosion has been limited to a low extent. The critical requirements for the formation of such deposits are met in southern China. A distinctive feature of the RE distribution pattern in the Chinese ion-adsorption ores is cerium deficiency. Although the reserves of REO in ion-adsorption ores are presently

Table 1. Properties of the Lanthanides[a]

Parameter	Lanthanum	Cerium	Praseodymium	Neodymium	Promethium	Samarium	Europium	Gadolinium	Terbium	Dysprosium	Holmium	Erbium	Thulium	Ytterbium	Lutetium
at no.	57	58	59	60	61	62	63	64	65	66	67	68	69	70	71
at wt	138.91	140.12	140.907	144.24	145	150.35	151.96	157.95	158.9254	162.50	164.930	167.26	168.934	173.04	174.97
mp, °C	918	798	931	1021	1042	1074	822	1313	1365	1412	1474	1529	1545	819	1663
bp, °C	3464	3433	3520	3074	~3000	1794	1429	3273	3230	2567	2700	2868	1950	1196	3402
density, g/cm^3	6.1453	6.770	677.3	7.007		7.520	5.234	7.9004	8.2294	8.5500	8.7947	9.066	9.3208	6.9654	9.8404
heat of fusion, kJ/mol[b]	6.201	5.179	6.912	7.134		8.623	9.221	10.05	10.80	10.782	16.874	19.90	16.84	7.657	18.65
heat of sublimation, at 25°C, kJ/mol[b]	431.0	422.6	355.6	327.6	~348	206.7	144.7	397.5	288.7	290.4	300.8	317.10	232.2	152.1	427.6
conduction electrons	3	3, 3.1	3	3	3	3	2	3	3	3	3	3	3	2	3
crystal structure	hcp	dhcp	dhcp	dhcp		rhomb	bcc	hcp	hcp	hcp	hcp	hcp	hcp	fcc	hcp
radius of atom, nm	0.1879	0.1824	0.1828	0.1821	0.1811	0.1804	0.20418	0.18013	0.17833	0.17743	0.17661	0.17566	0.17462	0.19392	0.17349
Curie point, °C								292.7	220	86	19	18	32		
Néel point, °C		ca 13				15	90		230	178	133	84	56		
valence in aq sol	3	3,4	3	3	3	3	3,2	3	3	3	3	3	3	3	3
color of oxide[c]	white	off-white[d]	black[e]	blue		cream	white, greenish tinge	white	brown[f]	yellowish white	yellowish white	pink	white, greenish tint	white	white
color of aq sol[c]	colorless	colorless[d]	green	rose	yellow	colorless	colorless	colorless	colorless	yellow tint	yellow	pink	white, greenish tint	colorless	colorless
ionic radius, nm	0.1061	0.1034	0.1013	0.0995	0.0979	0.0964	0.0950	0.0938	0.0923	0.0908	0.0894	0.0881	0.0870	0.0858	0.850

[a] All elements silvery in color. [b] To convert J to cal, divide by 4.184. [c] RE_2O_3 unless otherwise noted. [d] CeO_2. [e] Pr_6O_{11}. [f] Tb_4O_7.

Table 2. Lanthanide and Yttrium Distribution in Mineral Sources, wt %[a]

	Bastnaesite	Loparite	Monazite			Xenotime
Rare earth	California	Russia	E. Australia[b]	W. Australia[b]	India[c]	Malaysia
lanthanum	32.0	27.8	20.2	23.9	23.0	0.50
cerium	49.0	57.1	45.3	46.1	46.0	5.00
praseodymium	4.40	3.7	5.40	5.05	5.50	0.70
neodymium	13.5	8.7	18.3	17.4	20.0	2.20
samarium	0.50	0.91	4.60	2.53	4.00	1.90
europium	0.10	0.13	0.10	0.05		0.20
gadolinium	0.30	0.21	2.00	1.49		4.00
terbium	0.01	0.07	0.20	0.04		1.00
dysprosium	0.03	0.09	1.15	0.69		8.70
holmium	0.01	0.03	0.05	0.05		2.10
erbium	0.01	0.07	0.40	0.21		5.40
thulium	0.02	0.07	trace	0.01		0.90
ytterbium	0.01	0.29	0.20	0.12		6.20
lutetium	0.01	0.05	trace	0.04		0.40
yttrium	0.10	0.14	2.10	2.41		60.8

[a] On a basis of 100% REO. [b] Australian monazite usually contains 4–8% thorium and 0.1–0.3% uranium. [c] Indian monazite contains 8–10% thorium.

estimated to be only 1,000,000 t, these deposits are very important because of the rare-earth distribution compared with conventional ores (monazite, bastnaesite).

Separation Processes. The product of ore digestion contains the rare earths in the same ratio as that in which they were originally present in the ore, with few exceptions, because of the similarity in chemical properties. The various processes for separating individual rare earth from naturally occurring rare-earth mixtures essentially utilize small differences in acidity resulting from the decrease in ionic radius from lanthanum to lutetium. The acidity differences influence the solubilities of salts, the hydrolysis of cations, and the formation of complex species so as to allow separation by fractional crystallization, fractional precipitation, ion exchange, and solvent extraction. In addition, the existence of tetravalent and divalent species for cerium and europium, respectively, is useful because the chemical behavior of these ions is markedly different from that of the trivalent species.

Separation processes include selective oxidation, selective reduction, fractional precipitation, fractional crystallization, ion exchange, and liquid–liquid extraction.

Metal Production

The deposition of RE metals from aqueous solutions does not work because of the highly electropositive nature of the REE. Therefore, industrial production of RE metals is carried out by fused salt electrolysis or metallothermic reduction.

Toxicity

The lanthanides are considered only slightly toxic in the Hodge-Sterner classification system and are safely handled with ordinary care. Inhalation of rare-earth vapors or dust should be avoided, and the skin washed thoroughly if it comes into contact with any dust or solution.

Uses

Applications Linked to Chemical and Structural Properties. Lanthanides are used in catalysis, the glass industry, and ceramics. One of the largest and fastest developing applications of cerium oxide is in automotive post-combustion catalysis. Owing to redox properties, CeO_2 acts as an oxygen reservoir to ensure the buffering effect necessary to control the composition of the exhaust gas, particularly to allow the oxidation of CO and hydrocarbons when the medium is globally reducing. Besides and owing to its high thermal stability at the elevated (>800°C) temperatures in the catalytic muffler, cerium oxide acts as a thermal stabilizer of alumina and precious metal particles, the two other components of the catalyst, and avoids sintering that would make them ineffective. Special grades of cerium oxides are developed for this application.

Applications Linked to Physical Properties. Applications involving physical properties use high purity (99.99%) lanthanides and exploit the elements' specific electronic configuration.

Rare-earth ions are used to give characteristic color to glass or ceramics, eg, praseodymium green, neodymium purple, or erbium pink. Other optical applications involve the use of cerium(IV) as an antibrowning agent for glass, in TV face plates for example, or of lanthanum as a component (40% by weight) of high index borate glasses for microscopes, telescopes, and camera lenses.

Applications of rare-earth luminescence developed in the early 1960s, as these elements became available in industrial quantities at a high level of purity. Intense and quasimonochromatic emissions obtained from rare-earth activators diluted in appropriate matrices have been used in many applications since that time. Color television was among the first to use rare-earth-based phosphors.

Rare-earth-based phosphors are also widely used in tricomponent fluorescent lighting, where white light is obtained from the combination of primary monochromatic emissions at 450, 550, and 610 nm.

There is also use of rare-earth-based phosphors in x-ray intensifying screens used in medical radiography.

In order to raise Curie temperatures, rare-earth metals are alloyed with elements having higher ordering temperatures such as the transition metals iron, cobalt, or nickel. High energy magnets can thus be made. Such performances allowed intense magnetic energies in small volumes to be obtained owing to samarium–cobalt magnets, and thus a miniaturization of devices like stepping motors or, more spectacularly, of the small headphones that permitted the appearance of the Walkman in the early 1980s. Even more powerful are the neodymium–iron–boron magnets that appeared in the mid-1980s. These magnets are in the course of a rapid development, eg, in voice coil motors and the automotive industry. There is a promising forecast for several industrial applications.

A relatively new field for the application of rare-earth-transition-element alloys is magnetooptical recording.

Future Applications. The use of gadolinium complexes as contrast agents in magnetic resonance imaging (mri) is growing. Lanthanum–nickel-based alloys are good candidates for rechargeable batteries technologies. Furthermore, cerium additives are becoming important for control of pollution (soots, NO_x), from diesel engines. Perovskite-mixed oxides such as $LaTO_3$, T = Co, Ni, Cu...) are being developed for oxidation catalysis and electrodes or interconnects for solid oxide fuel cells. Cerium sulfide is to be used as an alternative to cadmium sulfoselenide reds in the pigment industry.

Economic Aspects

During the period 1990–2000, the expected average annual growth in overall rare-earth compounds demand is from 3–6% in tonnage. In 2000, the expected world consumption might be roughly 56,000 t, and Asia is expected to account for 50% of the total consumption, the largest market being China (25%). During the 1990s the usage of nonseparated rare earths is expected to increase less than usage of separated materials. The latter should double from 8000 t in 1990 to 16,000 t in 2000. In the market of separated rare earths, magnetism is expected to drive the demand to 8000 t in 2000 as compared to 2000 t in 1990. The other markets including phosphors, metallurgy, catalysts, glass, and ceramics are expected to increase at approximately 5% growth annually.

JEAN-LOUIS SABOT
PATRICK MAESTRO
Rhône-Poulenc Recherches

Gmelin Handbuch der Anorganische Chemie, System No. 39, Rare Earth Metals, 8th ed., Springer-Verlag, Berlin.

F. H. Spedding and A. H. Daane, *The Rare Earths*, John Wiley & Sons, Inc., New York, 1961.

K. A. Gschneider, Jr. and L. Eyring eds., *Handbook on the Physics and Chemistry of Rare Earths*, Vols. 1–22, Elsevier, Amsterdam, the Netherlands.

"Industrial Applications of Rare Earth Elements," *ACS Symp. Ser.*, 164, 1981.

LANTHANUM. See LANTHANIDES.

LASER DYES. See LASERS.

LASERS

Lasers are sources of light, a form of electromagnetic radiation which propagates at a velocity of 3×10^{10} cm/s and is characterized by an oscillating electric field.

The many types of lasers produce light at different wavelengths in the visible, infrared, and ultraviolet regions of the spectrum. Light from lasers has many properties different from those of light from conventional light sources. Laser light can be highly monochromatic, well-collimated, coherent, and in some cases can have extremely high power. These unusual properties lead to a wide variety of applications for lasers in science, engineering, and industry. The term laser is an acronym constructed from light amplification by stimulated emission of radiation.

Lasers as of this writing are familiar tools for applications such as alignment, measurement, as instrumental sources, and in industrial material processing. Use in conjunction with optical fibers has radically changed telecommunications. Lasers also form the basis of many consumer products such as supermarket scanners, compact disk players, and laser printers. They are also used in medical applications.

In the 1990s, significant developments in laser technology continue. These include semiconductor laser technology and semiconductor laser-pumped solid-state lasers. These developments should lead to smaller, more efficient laser devices. Future applications include laser-assisted thermonuclear fusion (see FUSION ENERGY) and laser-assisted separation of isotopes. Among the most important chemical applications are new spectroscopic techniques (see SPECTROSCOPY, OPTICAL), the monitoring of transients in chemical reactions (see KINETIC MEASUREMENTS), and state-selective chemistry, in which the course of a chemical reaction is controlled by selectively exciting certain molecular states. These applications have all been demonstrated on a research scale.

Fundamentals of Lasers

Laser light is produced from transitions between atomic or molecular energy levels. Generation of light requires two energy levels, E_1 and E_2, separated by the photon energy E_p of the light that is to be produced.

$$E_p = h\nu = hc/\lambda = E_2 - E_1 \qquad (1)$$

where h is Planck's constant, c is the velocity of light, ν is the laser frequency, λ the wavelength, and E_1 and E_2 are, respectively, the lower and higher energy levels. The relevant energy levels may be those of atoms or molecules in a gas, as in the case of the helium–neon laser, or of ions embedded in a solid host material, as in a ruby laser, or they may be energy levels that belong to a crystalline lattice as a whole, as in the aluminum gallium arsenide, AlGaAs, semiconductor laser.

The interaction between the light and the energy levels to produce laser operation relies on the phenomenon of stimulated emission. For stimulated emission to occur an atom or molecule must be in one of its excited levels and have a vacant energy level of lower energy. Then incoming light of the proper frequency can trigger a transition from the upper level to the lower level. The photon energy of the incident light must equal the energy difference between the two levels. The light can then stimulate the atomic or molecular system to make the transition. At the same time, the energy stored in the atomic or molecular system is emitted as light, so that there is an increase in the light intensity. The emitted light has the same photon energy as the incident light, travels in the same direction as the incident light, and remains in phase with it. It is this last property which gives rise to many of the other important properties of lasers, including coherence and directionality.

The three requirements for a laser are a material that possesses an appropriate set of energy levels (the active medium), some means for excitation or pumping the atoms or molecules to excited upper energy levels while at the same time leaving lower lying energy levels empty, and some means of resonant feedback to allow the light to pass back and forth through the active medium. During these passes, the light is amplified by the stimulated emission process and increases in intensity.

Laser Types

There are many types of lasers, having a wide variety of methods of construction and based on many different classes of materials. The properties of some commercially available lasers are summarized in Table 1. Typical available characteristics are given.

Other Lasers. There are two other types of lasers which as of this writing are not at the same stage of maturity as those in Table 1. These include chemical lasers, which produce a population inversion by a chemical reaction that leaves the product in an excited state, and

Table 1. Common Commercial Lasers

Laser	Wavelength, μm	Operation	Output, W
	Gas lasers		
He–Ne	0.6328	continuous (CW)	0.0005–0.035
CO_2	10.6	CW	to 25,000
		pulsed	500–1,000[a]
Ar	0.4880, 0.5145 and other lines	CW	to 25
Kr	0.6471 and other lines	CW	to 16
He–Cd	0.442, 0.325	CW	to 0.150
KrF	0.249	pulsed	to 4[b]
	Solid-state lasers		
$Y_3Al_5O_{12}$:Nd (YAG)	1.06, 0.532	CW	to 1800
		pulsed	10[a]
Al_2O_3:Cr	0.6943	pulsed	to 400[b,c]
Al_2O_3:Ti	tunable 0.67–1.05	CW	3.5
		pulsed	
	Semiconductor laser		
AlGaAs	0.8–0.95	CW	to 20
		pulsed	0.050[a]
InGaAsP	1.3, 1.55	CW	to 0.150
		pulsed	
AlInGaP	0.63–0.69	CW	to 0.25
		pulsed	
	Liquid laser		
	tunable, 0.25–1	CW	to 2[d]
		pulsed	

[a] Most common wattage. [b] Units are J/pulse. To convert J to cal, divide by 4.184. [c] Most common value is 10 J/pulse (2.4 cal/pulse). [d] At selected wavelengths.

the free-electron laser (FEL), which directly converts the kinetic energy of a relativistic electron beam into light.

Tunable Lasers. Tunability is an important feature for many spectroscopic and chemical applications. The leading tunable laser in the near ultraviolet, and visible and near infrared regions has been the dye laser, offering reasonably high power, narrow linewidth and a broad tuning range. It has been employed for many studies of molecular structure and chemical reactions. Tunable solid-state lasers, such as titanium-doped sapphire, Ti:sapphire, which offer a broad tuning range without the necessity to change dye materials, have begun to compete strongly in the visible and near-infrared regions and have displaced dye lasers for some applications. Using frequency doubling, these cover most of the range from 0.35 to 1.1 μm.

The free-electron laser (FEL) offers the ultimate in tunability, in principle being unlimited in its tuning range. A FEL represents a very large investment, however.

Laser Safety

The beam from a laser can inflict damage on various parts of the human body. In addition, there are other hazards associated with the use of lasers. Therefore, a well-conceived and well-organized safety program is required for the use of lasers, particularly those of high power.

Effects of Laser Radiation. The structure of the body most easily damaged by laser light is the retina, the photosensitive surface at the back of the eyeball. High power lasers also can cause serious skin burns and damage to the cornea.

Associated Hazards. There are other possible hazards associated with the use of lasers outside of the hazards produced by the optical beam. The most serious is the possibility of electrical shock associated with the high voltage electrical supplies. Such shocks are potentially lethal. In addition, the poisonous or corrosive substances used either in the laser itself, such as dye materials and solvents, or in equipment used in association with lasers, such as modulators, present serious hazards.

Safety Standards. Protection from laser beams involves not allowing laser radiation at a level higher than a maximum permissible exposure level to strike the human body. Maximum permissible exposure levels for both eyes and skin have been defined. One of the most common safety measures is the use of protective eyewear.

One of the most significant laser safety standards is that developed by the Z-136 committee of the American National Standards Institute (ANSI). Although it is voluntary, many organizations use the ANSI standard. It contains a number of items including a recommendation for maximum permissible levels of exposure to laser radiation for various wavelengths, exposure durations, and different parts of the body; separation of lasers into four different classes according to the level of hazard they present; and recommendation of safety practices for lasers in each of the classes.

The U.S. Food and Drug Administration (FDA) adopted a legally binding standard, which took the form of a performance standard for laser products. The standard provides a classification scheme for lasers similar to the ANSI classification. All lasers sold after August 2, 1976 must comply with its provisions.

Several state governments have also passed laws regulating and controlling lasers. Provisions of the laws vary greatly from state to state.

Nonlinear Optics

The electromagnetic field of a light beam produces an electrical polarization vector in the material through which it passes. In ordinary optics, which may be termed linear optics, the polarization vector is proportional to the electric field vector E. However, the polarization can be expanded in an infinite series:

$$P = \chi E(1 + a_2E + a_3E^2 + a_4E^3 + \cdots) \qquad (2)$$

where P is the polarization, E is the electric field, χ is the linear polarizability, and the a_i are constants. The equation is a simplified scalar representation of a tensor equation. The nonlinear coefficients a_i are very small compared to unity. For reasonably small values of electric field, only the first term in the equation is important and in the prelaser era polarization was approximated to be linearly proportional to electric field. When high power lasers became available, the electric fields became much higher and the product a_iE^{i-1} could become large enough to be observable. The second term in equation 2 is of the form χa_2E^2. If E has a sinusoidal variation of the form $E_0\cos\omega t$, the resulting components of polarization are of the form

$$a_2\chi E_0^2 \cos^2 \omega t = 0.5a_2\chi E_0^2(1 + \cos 2\omega t) \qquad (3)$$

The second term on the right-hand side of equation 3, a component oscillating at frequency 2ω, represents the second harmonic of the incident beam. This component of the polarization vector can radiate light at the frequency 2ω.

As the incident radiation at frequency ω and the second harmonic radiation of frequency 2ω propagate through the material, the intensity of the second harmonic radiation builds. However, because of dispersion, the two waves eventually get out of phase and the intensity of the second harmonic radiation decreases. This limitation is avoided by use of a technique called phase matching.

Only certain types of crystalline materials can exhibit second harmonic generation. Because of symmetry considerations, the coefficient a_2 must be identically equal to zero in any material having a center of symmetry. Thus the only candidates for second harmonic generation are materials that lack a center of symmetry. Some common materials which are used in nonlinear optics include barium sodium niobate, $Ba_2NaNb_5O_{15}$; lithium niobate, $LiNbO_3$; potassium titanyl phosphate, $KTiOPO_4$; beta-barium borate, β-BaB_2O_4; and lithium triborate, LiB_3O_5.

JOHN F. READY
Honeywell Technology Center

J. F. Ready, *Industrial Applications of Lasers,* Academic Press, New York, 1978.
A. E. Siegman, *Lasers,* University Science Books, Mill Valley, Calif., 1986.
D. L. Andrews, *Lasers in Chemistry,* Springer-Verlag, Berlin, 1990.
R. W. Boyd, *Nonlinear Optics,* Academic Press, San Diego, Calif., 1991.

LATEX TECHNOLOGY

Latex technology encompasses colloidal and polymer chemistry in the preparation, processing, and conversion of natural and synthetic latices into useful products.

Worldwide concern over the AIDS epidemic has sharply increased the demand for latex used in the preparation of rubber gloves and similar dipped goods. Estimates are for a 5–7% annual growth rate in this market.

Many synthetic latices exist. They contain butadiene and styrene copolymers (elastomeric), styrene–butadiene copolymers (resinous), butadiene and acrylonitrile, chloroprene copolymers, methacrylate and acrylate ester copolymers, vinyl acelate copolymers, vinyl and vinylidene chloride copolymers, ethylene copolymers, fluorinated copolymers, acrylamide copolymers, styrene–acrolein copolymers, and pyrrole and pyrrole copolymers. Many of these latices also have carboxylated versions.

Traditional applications for latices are adhesives, binders for fibers and particulate matter, protective and decorative coatings (qv), dipped goods, foam, paper coatings, backings for carpet and upholstery, modifiers for bitumens and concrete, thread, and textile modifiers. More recent applications include biomedical applications as protein immobilizers, visual detectors in immunoassays (qv), as release agents, in electronic applications as photoresists for circuit boards, in batteries (qv), conductive paint, copy machines, and as key components in molecular electronic devices.

By 1935 emulsion polymerization became the method of choice in making synthetic rubber because of its many advantages: (*1*) the reaction mass viscosity remains low throughout polymerization, providing for improved heat transfer, agitation, and product handling; (*2*) the sensible heat of the water in the emulsion balances the heat of reaction generated by free-radical polymerization; and (*3*) the rate of reaction is rapid, while producing very high molecular weight.

Synthetic Latex Manufacture

Kinetics and Mechanisms. In 1945 the first recognized qualitative theory of emulsion polymerization was presented. This mechanism for classic emulsion preparation was quantified and the polymerization separated into three stages—stage I: particle nucleation; stage II: growth in polymer particles saturated with monomer; and stage III: growth in polymer particles with a decreasing monomer concentration.

Basic Components. The principal components in emulsion polymerization are deionized water, monomer, initiator, emulsifier, buffer, and chain-transfer agent. A typical formula consists of 20–60% monomer, 2–10 wt % emulsifier on monomer, 0.1–1.0 wt % initiator on monomer, 0.1–1.0 wt % chain-transfer agent on monomer, small amounts of various buffers and bacteria control agents, and the balance deionized water.

Process. Commercial processes manufacturing latex can be divided into batch, semibatch, and continuous methods.

Latex Properties

The observable properties of a latex, ie, stability, rheology, film properties, interfacial reactivity, and substrate adhesion, are determined by the colloidal and polymeric properties of the latex particles. Important polymer properties include molecular weight distribution, monomer sequence distribution, glass-transition temperature, crystallinity, degrees of cross-linking, and free monomer. Methods for analyzing each of these properties exist, depending on the end use of the product. An overview of the various polymer colloid characterization methods is available.

Improving Properties Through Compounding. The potential value of most polymers can be realized only after proper compounding. Materials used to enhance polymer properties or reduce polymer cost include antioxidants, cross-linking reagents, accelerators, fillers, plasticizers, adhesion promoters, pigments, etc.

CHESTER H. GELBERT
MICHAEL C. GRADY
E. I. du Pont de Nemours & Co., Inc.

E. Daniels, E. D. Sudol, and M. S. El-Aasser, eds., *Polymer Latexes: Preparation, Characterization and Applications,* ACS Symposium Series, Vol. 492, American Chemical Society, Washington, D.C., 1992.

J. R. Richards, J. P. Congalidis, and R. G. Gilbert, *J. Applied Polym. Sci.* **37,** 2727–2756 (1989).

C. H. Gelbert and H. E. Berkheimer, *Paper C in Educational Symposium No. 18 on Latex Technology,* Rubber Division of the American Chemical Society, Montreal, Canada, 1987.

LATEX. See LATEX TECHNOLOGY; ELASTOMERS, SYNTHETIC; RUBBER, NATURAL.

LEAD

Lead, Pb, is an essential commodity in the modern industrial world, ranking fifth in tonnage consumed after iron, copper, aluminum, and zinc. Slightly over half of the lead produced in the world now comes from recycled sources.

Lead has unique properties: low melting point, ease of casting, high density, softness, malleability, low strength, ease of fabrication, acid resistance, electrochemical reaction with sulfuric acid, and chemical stability in air, water, and earth. The principal uses of lead and its compounds, in descending order of importance, are storage batteries, pigments, ammunition, solders, plumbing, cable covering, bearings, and caulking. In addition, lead is used to attenuate sound waves, atomic radiation, and mechanical vibration. In most of these applications lead is not used in its pure state, but rather as an alloy.

Occurrence and Ores

The occurrence of concentrated and easily accessible lead ore deposits is unexpectedly high, and these are widely distributed through the world. The most important ore mineral is galena, PbS (87% Pb), followed by anglesite, $PbSO_4$ (68% Pb), and cerrusite, $PbCO_3$ (77.5% Pb). The latter two minerals result from the natural weathering of galena.

Most (88%) lead mined in the United States comes from five mines in Missouri. The rest comes from 11 mines in Colorado, Idaho, Montana, Alaska, Washington, and Nevada.

Lead and zinc materials are so intimately mixed in many deposits that they are mined together and then separated. Silver minerals are frequently found in association with galena.

The concentration of lead in ore bodies of commercial interest generally ranges from 2 to 6%; the average is 2.5%. Improvements in ore-dressing techniques have made possible the exploitation of deposits having lead contents less than 2%.

The world reserves of lead are estimated at 71×10^6 t and scattered around the world. Over one-third (25×10^6 t) of this total is located in North America.

Physical Properties

Physical properties of lead are listed in Table 1.

Chemical Properties

Lead forms a series of compounds corresponding to the oxidation states of +2 and +4. The +2 state is the most common. Compounds of lead(IV) are regarded as covalent, those of lead(II) as primarily ionic. Lead is amphoteric, forming plumbous (Pb(II)) and plumbic (Pb(IV)) salts as well as plumbites and plumbates, respectively. Lead is one of the most stable of fabricated materials because of excellent corrosion resistance to air, water, and soil.

Acid Oxidation. Reactions of lead with acid and alkalies are varied. Nitric acid, the best solvent for lead, forms lead nitrate; acetic

Table 1. Physical Properties of Lead

Property	Value
at. wt	207.2
melting point, °C	327.4
boiling point, °C	1770
specific gravity at 20°C	11.35
specific heat, J/(kg·K)[a]	130
vapor pressure, kPa[b] at 980°C	0.133
1420°C	13.33
1600°C	53.3
thermal conductivity[c] at 28°C, W/(m·K)	34.7
electrical resistivity at 20°C, $\mu\Omega$/cm	20.65
specific conductance[d] at 0°C, $(\Omega\cdot cm)^{-1}$	5.05×10^4
viscosity at 440°C, mPa·s(= cP)	2.12
magnetic susceptibility at 20°C, m^3/kg[e]	-0.29×10^{-6}
hardness, Mohs'	1.5
Young's modulus, GPa[f]	16.5
tensile strength, common lead, at 20°C, kPa[f]	14,000

[a] To convert J to cal, divide by 4.184. [b] To convert kPa to mm Hg, multiply by 7.5. [c] Thermal conductivity relative to Ag = 100 is 8.2. [d] Electrical conductivity relative to Cu = 100 is 7.8. [e] To convert m^3/kg to emu/g, multiply by 79.3. [f] To convert GPa to psi, multiply by 145,000.

acid forms soluble lead acetate in the presence of oxygen; sulfuric acid forms insoluble lead sulfate.

There are three common oxides of lead: lead oxide, PbO, also known as litharge; lead dioxide, PbO_2; and lead tetroxide, Pb_3O_6. Lead oxide is used in batteries, glass, and ceramics. Lead dioxide is used in the manufacture of dyes, rubber substitutes, and pyrotechnics.

Processing

Lead is usually processed from ore to refined metal in four stages. These are ore dressing, smelting, drossing, and refining.

Secondary Lead. The emphasis in technological development for the lead industry in the 1990s is on secondary or recycled lead. Recovery from scrap is an important source for the lead demands of the United States and the rest of the world. In the United States, over 70% of the lead requirements are satisfied by recycled lead products. The ratio of secondary to primary lead increases with increasing lead consumption for batteries. Well-organized collecting channels for spent batteries are required for a stable future for lead.

The principal types of scrap are battery plates and paste, drosses, skimmings, and industrial scrap such as solders, babbits, cable sheathing, etc. Some of this material is reclaimed by kettle melting and refining. However, most scrap is a combination of metallic lead and its alloying constituents mixed with compounds of these metals, usually oxides and sulfates. Therefore, recovery as metals requires reduction and refining procedures.

Economic Aspects

The principal U.S. lead producers, ASARCO Inc. and The Doe Run Co., account for 75% of domestic mine production and 100% of primary lead production. Both companies employ sintering/blast furnace operations at their smelters and pyrometallurgical methods in their refineries.

The future use of lead may be decided by the resolution of an environmental paradox. Some markets for lead are being phased out because of environmental concerns, eg, the use of tetraethyllead as a gasoline additive. However, a 1990 State of California law and similar laws in nine eastern U.S. states require that 2% of new cars meet zero-emission standards in 1998. By 2003, this requirement rises to 10% of new vehicles. Zero emission vehicles are generally accepted to mean electric, ie, battery powered cars, and there is a considerable research effort to bring suitable electric vehicles to market by 1998.

Many battery systems for powering electric vehicles; are being investigated. However, the lead–acid battery, despite environmental concerns, is by far the most mature and accepted. If lead–acid battery technology is adopted, the demand for lead is expected to increase strongly. The established world resources of 71×10^6 t can meet the demand for electric vehicles for a long time. Without usage for electric vehicles, the slowly increasing trend evidenced in the use of lead in the United States in the 1990s is likely to go into decline because of environmental regulations limiting the applications of lead in uses other than batteries.

Health and Safety Factors

Exposure to excessive amounts of lead over a long period of time (chronic exposure) increases the risk of developing certain diseases. The parts of the body which may be affected include the circulatory system, nervous system, digestive system, reproductive system, and kidneys. These effects include anemia, muscular weakness, kidney damage, and reproductive effects, such as reduced fertility in both men and women, and damage to the fetus of exposed pregnant women.

Because lead may be ingested and inhaled, and because particle size and chemical composition affect its absorption, it is important that the concentration of lead in the blood be determined on a periodic basis for any person occupationally exposed to lead. Measures to control exposure to lead include the provision of proper ventilation, application of proper work practices, following rules of good hygiene, and wearing respirators and protective clothing. Detailed regulations by OSHA and other United States agencies governing occupational exposure to lead may be found in Part 1910 of Title 29 of the *Code of Regulations.*

Environmental Standards. Lead in the environment is regulated in the United States because of its potential occupational impact, as well as concern about the impact lead may have on the cognitive and physical development of young children. Standards have been set for lead in air, water, and other environmental media.

Preventive Measures. The intake uptake biokinetic model (IUBK) projects the impact of lead in the environment on blood lead. This model assumes conservatively high levels of intake and cannot account for chemical speciation; thus, over-predictions of blood lead levels often occur. Nonetheless, because of the allegations of the impact of blood lead and neurobehavioral development, blood lead levels in children are being reduced administratively to below 10 μg/dL. In order to do so, soil leads are being reduced to a level of between 500–1000 ppm where remediation is required.

The different forms of lead have different bioavailability and this ultimately impacts cleanup levels.

MICHAEL KING
VENKOBA RAMACHANDRAN
ASARCO Inc.

P. Crowson, *Minerals Handbook, 1992–93, Statistics and Analyses of the World's Minerals Industry,* Stockton Press, New York, p. 130.

T. S. Mackey and R. D. Prengaman, eds., *Lead–Zinc '90, AIME Symposium,* Anaheim, Calif., Feb. 1990.

Technical data, American Bureau of Metal Statistics Inc., U.S. Bureau of Mines, Washington, D.C.

R. L. Amistadi, "Whither Lead," paper presented at *Independent Battery Manufacturers Association,* Chicago, Ill., Oct. 20, 1993.

LEAD ALLOYS

Lead alloys, which have low melting points, can be cast into many shapes by using a variety of molding materials and casting processes. About 50% of lead is used pure lead, lead oxides, or lead chemicals; the remainder is used in the form of lead alloys. Lead and its alloys are generally melted, handled, and refined in cast-iron, cast-steel, welded steel, or spun-steel melting kettles without fear of contamination by iron.

Lead is ductile and malleable, and can be fabricated into various shapes by rolling, extruding, forging, spinning, and hammering. The low tensile strength and very low creep strength of lead make it unsuitable for use without the addition of alloying elements. The principal alloying elements used to strengthen lead are antimony, calcium, tin, copper, tellurium, arsenic, and silver. Minor alloying elements are selenium, sulfur, bismuth, cadmium, indium, aluminum, and strontium.

Lead–Antimony Alloys

Properties. Lead–antimony alloys are the most widely used lead alloys. The mechanical properties of lead–antimony alloys containing arsenic, tin, and copper are shown in Table 1. Most lead–antimony alloys of commercial importance contain 11 wt % or less antimony. Lead–antimony alloys are used in lead–acid batteries, ammunition, cable sheathing, anodes, and wrought products such as pipe and sheet.

Lead–Antimony–Tin Alloys. Lead–antimony–tin alloys are used for printing, bearings, solders, slush castings, and specialty castings. These alloys have low melting points, high hardness, and excellent high temperature strength and fluidity. Excellent antifriction properties and good hardness make lead–antimony–tin alloys suitable for journal bearings.

Lead–Calcium Alloys

Lead–calcium alloys are replacing lead–antimony alloys for many applications. Most U.S. original equipment automotive batteries are

Table 1. Mechanical Properties of Lead–Antimony Alloys[a]

Antimony content, wt %	Yield strength after aging, MPa[b,c]		Tensile strength, MPa[b]	Elongation, %
11	68.9	74.4	75.9	5
6	55.2	71.0	73.8	8
3	34.0	55.2	65.5	10
2	24.1	37.9	46.9	15
1	13.8	19.3	37.9	20
0	3.5	3.5	11.7	55

[a] After 30 days, unless otherwise noted. [b] To convert MPa to psi, multiply by 145. [c] Column on left is after one day.

constructed of lead–calcium grids, whereas most U.S. replacement batteries utilize lead–calcium alloys for the negative grid and lead–antimony alloys for the positive grid. Lead–calcium is used worldwide for standby power, submarine, and specialty sealed batteries. Lead–calcium alloy batteries do not require addition of water and can therefore be sealed. Lead–calcium alloys are used for electrowinning anodes, cable sheathing and sleeving, specialty boat keels, and lead alloy tapes. The main use of binary lead–calcium alloys is for the grids in large, stationary standby power batteries.

Lead–calcium alloys can be protected against loss of calcium by addition of aluminum. Tin additions to lead–calcium and lead–calcium–aluminum alloys enhances the mechanical and electrochemical properties. A principal use for lead–calcium–tin alloys is lead anodes for electrowinning.

Lead–Copper Alloys

Copper is an alloying element as well as an impurity in lead. High copper–lead alloys generally contain 60–70% copper. The mechanical properties of sand-cast copper–lead alloys are shown in Table 2. The high copper alloys are difficult to cast and are susceptible to extensive segregation. Cast lead–copper (60–70 wt %) alloys are used as bearing and bushings for high temperature service. More recently, the cast alloys have been replaced by sintered copper powder products infiltrated with lead to produce more uniform distribution of the lead.

Only lead alloys containing copper below 0.08% have practical applications. Lead sheet, pipe, cable sheathing, wire, and fabricated products are produced from lead–copper alloys having copper contents near the eutectic composition. Lead–copper alloys are specified because of superior mechanical properties, creep resistance, corrosion resistance, and high temperature stability compared to pure lead.

Lead–Silver Alloys

Lead–silver alloys show significant age hardening when quenched from elevated temperature. Small additions of silver to lead produce high resistance to recrystallization and grain growth. The principal uses for lead–silver alloys are as anodes and high temperature solders.

Table 2. Mechanical Properties of Rolled Lead Alloys

Property	Lead alloy components, wt %					
	Cu 0.06	Ca 0.06	Ca 0.04, Sn 0.50	Ca 0.065, Sn 0.70	Ca 0.06, Sn 1.30	Sb 6.0
tensile strength, MPa[a]	17.4	32.8	48.8	62.8	69.6	30.6
yield strength, MPa[a]	9.0	25.1	46.0	59.2	66.2	19.5
elongation, %	55	35	15	10	10	35
time to failure,[b] h		7	850	3,000	>10,000	1.5

[a] To convert MPa to psi, multiply by 145. [b] At 20.9 MPa.

Lead–Tellurium Alloys

Tellurium is often used in lead alloys when high mechanical strength at minimal alloy content is required. It is used for pipes and sheets, shielding for nuclear reactors, and cable sheathing.

Lead–Tin Alloys

Lead alloys with tin in all proportions, providing a series of alloys that have wide application in many industries. The principal use of lead–tin alloys is as solders for sealing and joining metals. Lead–tin alloys are used for corrosion-resistant coatings on steel and copper.

Other Alloys

Low Melting Alloys. Lead alloys having large amounts of bismuth, tin, cadmium, and indium that melt at relatively low (10–183°C) temperatures are known as fusible or low melting alloys. The specifications of many of these alloys are listed in ASTM B774-87.

These alloys are used as fuses, sprinkler system alloys, foundry pattern alloys, molds, dies, punches, cores, and mandrels where the low melting alloy is often melted out of a mold. The alloys are also used as solders, for the replication of human body parts, and as filler for tube bending. Lead–indium alloys are often used to join metals to glass.

Reactive Lead Alloys. Strontium–lead alloys over-age rapidly, resulting in significant loss of mechanical strength within several days. The addition of 1 wt % or more of tin is required to resist over-aging. The high cost of lead–strontium alloys compared to lead–calcium alloys has restricted usage. Lead–strontium–tin alloys are used as anodes for copper electrowinning.

Lead–lithium and lead–lithium–tin alloys have been proposed as alloys for lead–acid battery grids because of very rapid aging and very high mechanical properties. These alloys are, however, susceptible to grain boundary corrosion. Lead–lithium alloys containing strontium, barium, and calcium have been used for bearings. The addition of aluminum to prevent oxidation permits casting at high temperatures without oxidation.

Occupational Health and Safety

Because of the toxicity of lead, special care must be taken when working with lead alloys. Lead and its inorganic compounds are neurotoxins which may produce peripheral neuropathy. For an overview of the effects of lead exposure, see Occupational Exposure to Lead, Appendix A (29 CRF 1910.1025).

Despite the benefits of lead and lead alloys, the use of these materials is declining rapidly, owing primarily to environmental health and safety factors.

Other Uses for Lead Alloys

Lead alloys are also used for antifriction, sound attenuation, radiation shielding, and corrosion resistance.

Economic Aspects

Worldwide, about 50% of the lead consumed is used for lead–acid batteries. In the United States, about 97% of the lead used in lead–acid batteries is recycled, making lead the most recycled of any metal in the world. The lead usage throughout the world is about 50% pure lead, primarily for active materials of lead–acid batteries, gasoline additives, pigments, and glasses and glazes. The remaining material is consumed in the form of lead alloys.

The principal manufacturers of lead and lead alloys in North America are Tonolli, NOVA, Cominco, and Noranda in Canada; Doe Run, ASARCO, RSR Corp., Exide, Schuylkill, Sanders, GNB, and East Penn in the U.S.; and Penoles in Mexico. In Europe, the principal manufacturers are Union Meniere in Belgium; Metallgellschaft in Germany; Metalleurop in Germany–France; Brittania Refined Metals and H. J. Enthoven in the U.K.; Nuova SAMIM in Italy; and

Boliden in Sweden. In Asia and Australia the principals are MIM and Pasminco in Australia; Mitsui, Mitsubishi, and Toho Zinc in Japan; and Korea Zinc in Korea.

R. DAVID PRENGAMAN
RSR Corporation

W. Hofmann, *Lead and Lead Alloys,* Springer-Verlag, New York, 1970.

A. Worchester and J. O'Reilly, in *Metals Handbook,* Vol. 2, 10th ed., ASM International, Metals Park, Ohio, 1990, p. 543.

LEAD COMPOUNDS

LEAD SALTS

In general, the chemistry of inorganic lead compounds is similar to that of the alkaline-earth elements. Thus the carbonate, nitrate, and sulfate of lead are isomorphous with the corresponding compounds of calcium, barium, and strontium. In addition, many inorganic lead compounds possess two or more crystalline forms having different properties.

The carbonates, sulfates, nitrates, and halides of lead (except the yellow iodide) are colorless. Bivalent lead forms a soluble nitrate, chlorate, and acetate; a slightly soluble chloride; and an insoluble sulfate, carbonate, chromate, phosphate, molybdate, and sulfide. Highly crystalline basic lead salts of both anyhydrous and hydrated types are readily formed.

All lead-containing compounds are produced from pig lead through a series of suitable steps, except for the small amount of lead in leaded zinc oxide, for which high grade lead ore is used. Most lead compounds are prepared directly or indirectly from lead monoxide, PbO, commonly known as litharge. In general, lead compounds may be formed by one or more of three methods: (*1*) reaction between a slurry of litharge, or a similar lead compound such as the hydroxide or carbonate, and the desired acid; (*2*) reaction between the solution of a lead salt and the desired acid, or solution thereof in the case of an inorganic acid, or soluble salt of the acid (these reactions are facilitated by the fact that the desired lead compound usually is relatively insoluble, thus forming as a precipitate); and (*3*) fusion or calcination of litharge and the desired oxide, such as B_2O, SnO_2, ZrO_2, and TiO_2, resulting in lead borate, lead stannate, lead zirconate, ZrO, and lead titanate. This method is particularly applicable with the oxides of the elements in Groups 14 (IVA), 12 (VA), and 16 (VIA) of the Periodic Table.

Most uses of lead in chemical compounds other than in storage batteries are dissipative. The greater part of the lead used in other forms is recoverable.

Halides

Halides (Table 1) include lead difluoride, PbF_2, lead dichloride, $PbCl_2$, lead dibromide, $PbBr_2$, and lead diiodide, PbI_2.

Oxides

Lead forms two simple oxides, PbO and PbO_2, where it is divalent and tetravalent, respectively. Lead also forms a mixed oxide, Pb_3O_4, and a black oxide which normally comprises 55–85% lead monoxide, the remainder being finely divided metallic lead. The largest market for lead chemicals is the use of lead oxides in lead–acid storage battery electrodes. The ceramics industry is the next largest consumer of lead oxides, for use in glasses, glazes, and vitreous enamels for metal coating and glass decoration, followed by the rubber industries. Some physical properties of lead oxides are given in Table 2.

Lead Hydroxide. Lead hydroxide, $Pb(OH)_2$, mol wt 241.23, starts to dehydrate at about 130°C, and decomposes to lead monoxide at 145°C. Lead hydroxide is prepared by adding alkali to a solution of lead nitrate or by electrolysis of an alkaline solution with a lead anode.

Sulfide and Telluride

Lead Sulfide. Lead sulfide (galena, lead glance), PbS, mol wt 239.25, mp 114°C, d = 7.57–7.59 g/cm³, is metallic black and crystallizes in the cubic system.

Lead Telluride. Lead telluride, PbTe, forms white cubic crystals, mol wt 334.79, sp gr 8.16, and has a hardness of 3 on the Mohs' scale. It is very slightly soluble in water, melts at 917°C, and is prepared by melting lead and tellerium together.

Sulfates

Lead forms a normal and an acid sulfate and several basic sulfates. Basic and normal lead sulfates are fundamental components in the operation of lead–sulfuric acid storage batteries. Basic lead sulfates also are used as pigments and heat stabilizers in vinyl and certain other plastics. Sulfates (Table 3) include lead sulfate, $PbSO_4$, monobasic lead sulfate, $PbO \cdot PbSO_4$, and tetrabasic lead sulfate, $4\ PbO \cdot PbSO_4$.

Lead Nitrate

Lead nitrate, $Pb(NO_3)_2$, mol wt 331.23, sp gr 4.53, forms cubic or monoclinic colorless crystals. Lead nitrate is used in many industrial processes, ranging from ore processing to pyrotechnics to photothermography.

Phosphite

In commercial applications of poly(vinyl chloride) polymers where weathering resistance, thermal stability, and electrical insulating properties are required, a stabilizer system based on dibasic lead

Table 1. Physical Properties of Lead Halides

Property	$PbBr_2$	$PbCl_2$	PbF_2	PbI_2
mol wt	376.04	278.1	245.21	461.05
mp, °C	373	501	855	402
bp, °C	916	950	1290	954
d, g/cm³	6.66	5.85	8.24	6.16
soly at 0°C, g in 100 mL H_2O	0.455	0.673		0.044

Table 2. Physical Properties of Lead Oxides

Property	PbO	PbO_2	Pb_2O_3	Pb_3O_4
mol wt	223.21	239.21	462.42	685.63
mp, °C	897[a]			830[b]
dec, °C	1472[c]	290	370	500
d, g/cm³				
α	9.53	9.375		9.1
β	9.6			

[a] Begins to sublime before melting. [b] When decomposition is prevented by oxygen pressure. [c] Bp.

Table 3. Physical Properties of Lead Sulfates

Property	$PbSO_4$	$PbO \cdot PbSO_4$
mol wt	303.25	526.44
mp, °C	1170[a]	977
d, g/cm³, soly, g/100 mL H_2O	6.2	6.92
at 25°C	4.25×10^{-3}	4.4×10^{-3}[b]
40°C	5.6×10^{-3}	
crystal structure	orthorhombic, monoclinic	monoclinic

[a] Decomposes above 900°C. [b] At 0°C.

phosphite provides a unique balance of properties. Its plasticizer reactivity is in the same range as dibasic lead phthalate, its electrical properties are superior, and it is the only stabilizer known that can provide the required electrical properties and weathering resistance in the absence of carbon pigmentation.

Dibasic Lead Phosphite. Dibasic lead phosphite, $2PbO{\cdot}PbHPO_3{\cdot}1/2H_2O$, is a white crystalline powder, mol wt 742.63, sp gr 6.9, refractive index 2.25 and lead oxide content 90.2%.

Lead Azide

Lead azide, $Pb(N_3)_2$, mol wt 291.23, crystallizes as colorless needles. It is a sensitive detonating agent, exploding at 350°C.

Lead Antimonate

Lead antimonate (Naples yellow), $Pb_3(SbO_4)_2$, mol wt 993.07, *d*-6.58 g/cm^3, is an orange–yellow powder that is insoluble in water and dilute acids, but very slightly soluble in hydrochloric acid.

Acetates

Acetates (Table 4) include anhydrous lead acetate (plumbous acetate), Pb $(C_2H_3O_2)_2$; basic lead acetate (lead subacetate), $2\,Pb(OH)_2{\cdot}Pb(C_2H_3O_2)_2$; lead acetate trihydrate (plumbous acetate trihydrate), $Pb(C_2H_3O_2)_2{\cdot}3\,H_2O$; and lead tetraacetate (plumbic acetate), $Pb(C_2H_3O_2)_4$.

Lead Benzoate

Lead benzoate monohydrate, $Pb(C_6H_5CO_2){\cdot}H_2O$, mol wt 467.43, is a white crystalline powder that loses its water of hydration when heated to 100°C.

Carbonates

Lead Carbonate. Lead carbonate, $PbCO_3$, mol wt 267.22, d = 6.6 g/cm^3, forms colorless orthorhombic crystals; it decomposes at about 315°C.

Basic Lead Carbonate. Basic lead carbonate (white lead), $2PbCO_3{\cdot}Pb(OH)_2$, mol wt 775.67, d = 6.145 g/cm^3, forms white hexagonal crystals; it decomposes when heated to 400°C.

Phthalates

Two commercial forms of lead phthalates, both dibasic, are widely used as heat stabilizers in poly(vinyl chloride) (PVC) polymers and copolymers.

Dibasic Lead Phthlate. Dibasic lead phthalate, $2PbO\ Pb(O_2C)_2C_6H_4{\cdot}1/2H_2O$, is a white crystalline powder, mol wt 826.87, sp gr 4.6, and lead oxide content 79.8% PbO.

Silicates

Lead forms acid, basic, and normal, or metasilicates. Commercial lead silicates (frits) are made to specific $PbO{:}SiO_2$ ratios for the glass and ceramics industries, the rubber industry as vulcanizing agent, and the plastics industry as a heat and light stabilizer. These are supplied as granular or pulverized fusion products. Some physical properties are given in Table 5.

Table 4. Physical Properties of Lead Acetates

Property	Anhydrous	Basic	Trihydrate	Tetraacetate
mol wt	325.28	807.69	379.33	443.77
mp, °C	280	75 (200 dec)	75 (200 dec)	175
d, g/cm^3	3.25		2.55	2.228
refractive index, n_D			1.567[a]	
soly, g/100 mL H_2O				
at 15°C	44.3[b]	6.25	45.61	
100°C	221[c]	25	200	

[a] Along the β-axis. [b] At 20°C. [c] At 50°C.

Table 5. Physical Properties of Lead Silicates

Property	Monosilicate	Bisilicate	Tribasic silicate
mol wt	294.85	343.37	729.63
mp, °C	700–784	788–816	705–733
d, g/cm^3	6.50–6.65	4.60–4.65	7.52
refractive index, n_D	2.00–2.02	1.72–1.74	2.20–2.24

Borate

Lead borate monohydrate (lead metaborate), $Pb(BO_2)_2{\cdot}H_2O$, mol wt 310.82, d = 5.6 g/cm^3 (anhydrous), is a white crystalline powder.

Lead Titanate

Lead titanate (lead metatitanate), $PbTiO_3$, mol wt 302.09, d = 7.52 g/cm^3, forms yellow tetragonal crystals below 490°C an cubic crystals above 490°C.

Lead Zirconate

Lead zircoate, $PbZrO_3$, mol wt 346.41, has two colorless crystal structures.

Health and Safety Factors

Lead is a cumulative poison. Lead poisoning is one of the commonest of occupational diseases. However, the presence of lead-bearing materials in an industrial facility does not necessarily result in exposure on the part of the worker. The lead must be in such a form, and so distributed, as to gain entrance into the body or tissues of the worker. Lead is absorbed into the human body by inhalation of the dust or ingestion of lead-contaminated products.

The toxicity of the various lead compounds may depend on several factors: (*1*) the solubility of the compound in body fluids, (*2*) the fineness of the particles of the compound, and (*3*) conditions under which the compound is being used. Of the various lead compounds, the carbonate, the monoxide and the sulfate are considered to be more toxic than metallic lead or other inorganic lead compounds. Organic lead compounds are rapidly absorbed by the respiratory and gastrointestinal systems and through the skin. Lead and its various compounds are suspected carcinogens and are considered reproductive toxins.

OSHA regulations limit exposure to inorganic lead compounds of an employee, without a respirator to 50 μg/m^3 of air as a time-weighted average (TWA) in an eight-hour shift. This standard, 29 CFR 1910.1025, went into effect on March 1, 1979. The object of the lead standard is to prevent absorption of harmful quantities of lead. The standard is intended to protect workers from not only acute (short-term) effects, but also from chronic (long-term) effects. The standard establishes an action level of 30 μg/m^3 air TWA based on an 8-hr work day. The action level initiates several requirements of the standard, such as exposure monitoring, medical surveillance, employee training and education, and record keeping. The Permissible Exposure Limit (PEL) set by the standard is 50 μg/m^3 air TWA in an eight-hour shift. The PEL initiates several additional requirements of the standard such as respiratory protection, protective clothing, and engineering controls to maintain employee exposures to below the PEL.

In most cases, proper workplace precautions will provide adequate worker protection. All workers handling inorganic lead compounds should avoid creating dust; avoid inhaling or ingesting dust; wear appropriate protective equipment and clothing; and wash thoroughly after handling and before eating, drinking, using tobacco products or applying cosmetics. Adequate care and attention paid to safe handling practices can effectively minimize or eliminate any health risks associated with the handling, storage, use, and disposal of lead compounds. The regional OSHA office should be consulted for the latest rules and regulations.

OSHA regulations apply only to occupationally exposed employees. The EPA has set standards for ambient (outdoor) lead in air levels. The U.S. National Ambient Air Quality Standard for lead is 1.5 $\mu g/m^3$ of air based on measurements collected every six days and averaged over a calender quarter. The installation and proper maintenance of exhaust filtration systems enables most plants to comply with the EPA limits for lead.

DODD S. CARR
Consultant
WILLIAM C. SPANGENBERG
KEVIN CHRONLEY
Hammond Lead Products, Inc.

D. Greninger, V. Kollonitsch, and C. H. Kline, *Lead Chemicals,* International Lead Zinc Research Organization, Inc., New York, 1975.

J. W. Mellor, *Inorganic and Theoretical Chemistry,* Vol. III, Longmans, Green & Co., New York, 1930, pp. 636–888.

A. T. Wells, *Structural Inorganic Chemistry,* 3rd ed., Clarendon Press, Oxford, U.K., 1962, particularly pp. 475–479, 902–903.

INDUSTRIAL TOXICOLOGY

Lead (qv), lead alloys (qv), and lead compounds have been used for thousands of years for a number of purposes. The toxic effects of these materials have been known or suspected for almost as long. Serious efforts intended to reduce occupational and population exposure to lead have been made only since the late 1960s. Although these efforts, largely in the form of regulatory actions, voluntary actions, and increased public awareness, have been highly successful, concern about lead as a significant public health problem has increased. Epidemiological and experimental evidence regarding adverse health effects at successively lower levels of lead exposure have led to downward revision of criteria for acceptable blood lead concentrations. The U.S. Environmental Protection Agency (EPA) has designated 10 $\mu g/dL$ as a target level for regulatory development and enforcement/clean-up purposes. As of this writing, the EPA considers lead to be a significant and widespread health hazard in the United States.

This article has been reviewed by the Office of Pollution Prevention and Toxics (U.S. EPA) and approved for publication. Approval does not signify that the contents necessarily reflect the views and policies of the Agency, nor does mention of commercial products constitute endorsement or recommendation for use.

STEPHEN C. DEVITO
JOSEPH BREEN
U.S. Environmental Protection Agency

J. J. Breen and C. R. Stroup, eds., *Lead Poisoning: Exposure, Abatement, Regulation,* Lewis Publishers/CRC Press, Boca Raton, Fla., 1995.

H. L. Needleman, ed., *Human Lead Exposure,* CRC Press, Inc., Boca Raton, Fla., 1992.

P. J. Landrigan, J. R. Froines, and K. R. Mahaffey, *Top. Environ. Health* **7,** 421–451 (1985).

R. A. Goyer, in M. O. Amdur, J. Doull, and C. D. Klaassen, eds., *Casarett and Doull's Toxicology: The Basic Science of Poisons,* 4th ed., Pergamon Press, Inc., New York, 1991, pp. 639–646.

LEAD POISONING. See LEAD COMPOUNDS.

LEAD STORAGE BATTERIES. See BATTERIES, SECONDARY CELLS (LEAD–ACID).

LEATHER

The temporary preservation of leather is an important practical problem. When an animal is killed and skinned, bacterial degradation starts immediately, but temporary preservation by drying or salting does not result in a usable leather. Treatment to prevent decay is not the same as tanning, which is a slow process and in modern leather production involves a series of chemically interdependent steps. Tanning not only preserves the hide or skin, but also makes the leather resistant to cracking from flexing. The desired properties of leather depend on use. From the same hide, leathers may be obtained for garments, shoes, or mechanical applications. There are however, limitations to the possible use of a particular hide or skin. For example, sheep skin and most other fur skins have very little value for shoes. The heavy skin of alligator would not be satisfactory for gloves.

Hides and Skins

The structures of hides and skins are dependent on the needs of the animal and its environment. The functions of an animal's skin include protection from predators and infection, and maintenance of body temperature. The relative importance of these functions depends on the animal. Methods by which the skin accomplishes these functions is the same for most mammals.

The hide, as it is removed from an animal, has body fat and a thin membrane separating the hide from the fat and flesh of the body of the animal. The area near the inside of the hide is made up of the heaviest fibers of the hide. The fibers are the ultimate in nonwoven structure that give leather its remarkable strength and flexibility. Above the heavy fibers there are glands and hair follicles.

The fiber structure is very fine near the surface of the skin and this fine structure imparts a silky feel to the leather. The value of the skin or hide is then dependent in part on this smoothness.

Sheep and fur animals are protected primarily by their wool or hair. The fiber structure of the skin is very fine and has less strength than calfskin or that of other nonfur mammals. Goats, the animal of choice in areas of harsh climates and limited food supply, are particularly suited to warm and sometimes arid climates.

Chemical Composition. From the point of view of leathermaking, hides consist of four broad classes of proteins: collagen, elastin, albumen, and keratin. The fats are triglycerides and mixed esters. The hides as received in a tannery contain water and a curing agent. Surface dirt is usually about 2–5 wt %. Cattle hides have 5–15% fats depending on the breed and source. The balance of the hide is protein.

Leathermaking occurs in three broad steps: (*1*) removal of all materials that are not a part of the final product; (*2*) rendering the remaining hide substance biorefractive, ie, tanning; and (*3*) treating the stabilized tanned material to impart the characteristics desired in the final product. In all steps the solubility and reactivity of the components form the basis of the treatments employed.

Collagen, the principal protein of the hide, is the material that is made into leather. The reactivity of the collagen toward tanning agents and the dyeability as well as the strength, flexibility, and durability of the collagen when tanned are all important to making leather a material of choice for utility and fashion.

Leather Manufacture

The manufacture of leather follows the same general steps for a great variety of leathers. The largest category of hides tanned is cattle hides. Of those types used in the case of cattle hides, chrome tanning of unhaired hides is by far the dominant system in use throughout the world. The tanning of other types of hides and skins requires variations in the systems used for cattle hides.

Curing. The temporary preservation of hides or skins is known as curing. Curing includes air drying and salt curing. Several methods have received considerable research attention as alternatives to salt curing. These include use of sodium bisulfite as a disinfectant to allow preservation with or without decreased salt in a brine cure; use

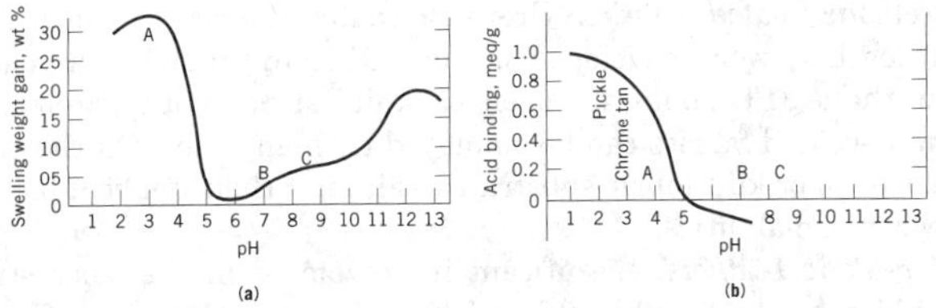

Figure 1. The pH dependence, where A, B, and C represent regions corresponding to the pK_as of glutamic and aspartic acids, lysine, and argenine, respectively, of (**a**) protein swelling, and (**b**) protein acid-binding capacity.

of disinfectants such as quaternary amine for temporary preservation in direct shipping to the tannery from the packing plant; preservation of hides by radiation sterilization; and substitution of materials such as potassium chloride for sodium chloride. These methods have met with only limited commercial success.

Preproduction Handling. Preproduction handling includes soaking, unhairing, trimming, and fleshing. During the unhairing procedure a pH of 12–13 is obtained. The hide swells to two times its original thickness. This swelling has a beneficial opening of the fibers to permit better tannage (Fig. **1a**)

Splitting. In most modern large tanneries that make upholstery leather, and in some that make shoe uppers leather, the hides are split in the lime condition. In splitting the hides are cut to the desired thickness with a horizontal belt knife. The trimmed split is tanned separately.

Deliming and Bating. The limed hides have a pH around 12. Because chrome tanning is done at pH 2–4, the lime must be removed for pH adjustment. In addition, the undesirable materials in the hide, ie, both natural and the degradation products from the unhairing, must be removed.

Pickling. Bated hides or skins are at a near-neutral pH and thus are immediately processed, because under these conditions the protein is subject to bacterial degradation. Pickling is the term used for acidification of the hides. Proteins are amphoteric and at neutral pH hides are flacid and have little acid or base bound. For chrome tanning, the desired pH is about 2.0 (see Fig. **1b**). Once pickled, the hides can be drained and stored indefinitely. The usual procedure in the manufacture of chrome-tanned leathers is the use of a continuous bate, pickle, and tan method.

Chrome Tanning. Chrome tanning has the advantages of light color, speed of processing, low costs, and great stability of the resulting leather. Tanning refers to a specific reaction of the tanning chemicals combining with the hide or skin to stabilize the protein and make it resistant to bacterial degradation. Chrome tanning is done in a drum similar to that used for deliming. The salt solution from the pickle is present and the solution is at about pH 2.0. The chrome-tanning material is usually a basic chromium sulfate. The general formula for the most common commercial chrome-tanning product is $2\ Cr(OH)SO_4 \cdot Na_2SO_4$.

Retanning, Coloring, and Fatliquoring. Chrome-tanned leather is a light blue in color. The fibers are only stabilized against microbial action and do not have the feel of leather. If the leather were dried at this point only a stiff, unattractive product would result. The characteristics of desired leather result from the retanning, coloring, and fatliquoring. Fatliquoring is the term applied to the oiling of leather.

The mechanism of the tannage is accepted to be largely one of replacement of the bound water molecules by the phenolic groups of the tannin and subsequent formation of hydrogen bonds with the peptide bonds of the protein. The effect of this bonding is to make the leather almost completely biorefractive.

Staking. Staking is a mild flexing of the leather to bring it to the desired final softness.

Buffing. The buffing step consists of a light sandpapering of the grain.

Finishing. The finishing of leather has changed greatly because of the development of resin systems in the coating industry. Leather finishes must minimally be abrasion resistant and flexible, and must adhere to the leather. Formation of a tough water-resistant film is also desirable.

Grading, Measuring, and Shipping. The grading is done on the basis of hide defects, shape of the skin, manufacturing defects, or any other factors of importance to the specifications of the sale. The leather in the grade to be shipped is measured for area.

Chemistry of Tanning

Chrome Tanning. Modern chrome-tanning methods are well controlled and employ an extensive knowledge of the chemistry of the system. The most common chromium-tanning material used is basic chromium sulfate, $Cr(OH)SO_4$. Chromium sulfate is described as being 33% basic and in solution gives a pH of ca 2.5.

Ionization of the basic chromium salt results in the formation of complex ions. The basic complex structure penetrates the hide at pH $<$ 3.0. There is a fixation of the chromium to the hide protein primarily by reaction of the chromium and the carboxyl groups of the hide.

The chromium can be stabilized in a limited way to prevent surface fixation by addition of formate ions. This stabilization can then be reversed in the neutralization to a pH of about 4.0 and tannage becomes complete. This simple formate addition has decreased the time of chrome tanning by about 50% and has greatly increased the consistent quality of the leather produced.

Vegetable Tanning. The demand for vegetable tanning has decreased but remains a principal factor in the production of heavy leathers. Vegetable tannins have the general structure of polyphenolic compounds. There are two general classes of vegetable tannins: the hydrolyzable and the condensed. The hydrolyzable tannins are derivatives of pyrogallol. The condensed tannins are derivatives of catechol.

Drying. The retanned leather is stretched to increase the area for the best yield and to produce a flat leather surface. The leather is then dried. There are several drying (qv) systems used which depend on the type and thickness of the leather. Thin garment leathers, made soft by retanning and fatliquoring, may be dried by hanging in a dry loft. Soft leathers can be reworked by mechanical means to the softness desired when dry. For shoe uppers the leather should be held in a stretched condition. Two or three types of drying are commonly in use in the industry: toggle drying, paste drying, and vacuum drying.

Dyeing of Leather

Leather dyeing for any desired color can be done using the dyes developed for the textile industry. The penetration of the dye depends on the pH of the leather and the tannage. Chrome-tanned leather has chromium bonded to the leather fibers. This chromium can act as a mordant for acid dyes, resulting in fast colors and intense shading at the surface of the leather.

Environmental Aspects

The processing of hides and skins into leather results in a large quantity of waste materials. The hide in the salt-cured condition contains salt in a crystalline form, water as salt solution, and as hide liquid components, flesh, blood, manure, and surface dirt from the animal.

Pollution control within the industry has been largely concentrated in the areas of waste treatment techniques. Separate treatment of concentrated waste streams has been most successful and cost effective. These treatments include air oxidation of sulfide wastes from the unhairing process, high exhaustion processing of the trivalent chromium which is then recycled or precipitated, primary waste treatment for high suspended solids, coprecipitation and secondary waste treatment to lower the biological oxygen demand (BOD), and use of the solid wastes for fertilizers or other animal by-products. Moreover, a shift away from highly volatile solvents in the finish drying process to a low solvent system has been effected.

THOMAS C. THORSTENSEN
TSG

F. O'Flaherty, W. T. Roddy, and R. M. Lollar, *The Chemistry and Technology of Leather,* ACS Monograph Series No. 134, Reinhold Publishing Co., New York, 1956–1965.

T. C. Thorstensen, *Practical Leather Technology,* 4th ed., Kreiger Publishing Co., Malabar, Fla., 1983.

K. Bienkiewicz, *Physical Chemistry of Leather Making,* Kreiger Publishing Co., Malabar, Fla., 1983.

G. D. McLaughlin and E. R. Theis, *The Chemistry of Leather Manufacture,* Reinhold Publishing Co., New York, 1945.

LEATHER-LIKE MATERIALS

During the third quarter of the twentieth century, with improved nonwoven fabrics, man-made leathers finally succeeded in simulating leather to such an extent that they are nearly identical in appearance, physical properties, and structure. These leathers have enjoyed success in all leather-use areas. With the technology of microfibers, they continue to evolve both in quality and quantity.

Types of Leather-Like Materials

Leather-like materials now important in the market are of three main classes: (*1*) vinyl-coated fabrics, (*2*) urethane-coated (synthetic) fabrics, and (*3*) man-made leathers. To appreciate their leather-replacement capabilities it is necessary to know the structure of natural leather.

Leather. As a result, leather is made up of interlaced bundles of collagen fibers. Such a unique hierarchical structure gives leather several advantages: (*1*) transformability into any desired shape, (*2*) resistance to penetration of wind, water, and other materials, (*3*) breathability (water vapor and air permeability, and water absorption), (*4*) flexibility, and (*5*) processibility into finished forms having a grain or suede surface.

Vinyl-Coated Fabrics. The construction of vinyl-coated fabrics varies according to its application. The material is durable but stiff and heavy. Incorporating an expanded foam structure into the coating layer reduces the weight, and replacing the woven substrate fabric with a soft knit fabric improves flexibility.

Vinyl-coated fabrics exhibit high density, extremely low water vapor and air permeability, cold touch, poor flex endurance, and plasticizer migration. However, they have good scratch resistance and colorability and are inexpensive.

Urethane-Coated Fabrics. Urethane-coated fabrics, termed synthetic leather, were developed in the 1960s by applying a coat of polyurethane (PU) onto a woven or knit fabric. With poromerics, urethane-coated fabrics can be employed for many uses. On the other hand, woven or knit fabric substrates still limit their application, due to low conformability.

Man-Made Leathers. Significant improvement in the fiber structure of leather is finally achieved by using microfibers as fine as 0.001–0.0001 tex (0.01–0.001 den). With this microfiber, a man-made grain leather Sofrina (Kuraray Co., Ltd.) with a thin surface layer, and a man-made suede Suedemark (Kuraray Co., Ltd.) with a fine nap were first developed for clothing, and have expanded their uses. Ultrasuede (Toray Industries, Inc.) also uses microfibers with a rather thick fineness of 0.01 tex (0.1 den). Contemporary (1995) man-made leathers employ microfibers of not more than 0.03 tex (0.3 den) to obtain excellent properties and appearance resembling leather.

Physical and Chemical Properties

The properties of leather-like materials depend on the polymer used for substrate and coating layer. Feel, hand, and resistance to grain break are affected by the construction. Physical properties of leather and leather-like materials are shown in Table 1.

Manufacture and Processing

Vinyl-Coated Fabrics. Manufacturing methods for vinyl-coated fabrics now available are calendering and extrusion for thick layer, and paste coating for thin layer. Both solid and foam vinyl-coating layers are used. The processes may be followed by heat treatment and pressing with engraved rolls to produce the desired grain surface.

Urethane-Coated Fabrics. Manufacturing methods for urethane-coated fabrics are the dry system and the wet system.

Man-Made Leathers. These materials contain a nonwoven fabric which is impregnated with a polyurethane to improve flexibility, processibility, and conformability.

A manufacturing method proceeds by the following steps: (*1*) A special fiber is made by melt spinning and cut into 25–60-mm length. (*2*) The fibers are carded and cross-lapped to form a batt. (*3*) The batt is needle-punched with a barbed needle and entangled to improve physical strength; the batt is then subjected to sizing and pressing to become a nonwoven fabric with adequate thickness. (*4*) The nonwoven fabric is impregnated with polyurethane and immersed in a DMF—water bath for coagulation to create a substrate with porous polyurethane

Table 1. Physical Properties of Leather and Leather-Like Materials

Product	Substrate/fiber[a]	Coating layer	Weight, g/m^2	Thickness, mm	Density, g/mL	Tensile strength, N/mm^{2b,c}	Elongation at break, %[c]	Tear strength, kN/m[c,d]	Water-vapor permeability,[e] mg/(cm·h)	Flex endurance,[e] 10^3 cycles
					Man-made leather					
Clarino[f]	nonwoven/porous	thick foam	580	1.50	0.39	8 (5)	45 (85)	30 (33)	2.6	>1000
	nonwoven/micro	thin solid	550	1.30	0.42	15 (13)	80 (115)	69 (68)	3.8	>1000
Sofrina[f]	nonwoven/micro	thin solid	220	0.50	0.44	12 (14)	115 (120)	100 (80)	5.0	>1000
Suedemark[f]	nonwoven/micro	suede	120	0.35	0.34	9 (7)	70 (106)	43 (43)	12.4	>1000
					Urethane-coated fabrics					
solid-type	woven/regular	thin solid	393	0.90	0.44	12 (7)	20 (28)	45 (66)	1.5	300
foamed-type	woven/regular	thick foam	404	1.13	0.36	11 (8)	7 (29)	30 (22)	1.5	400
					Vinyl-coated fabrics					
solid-type	knit/regular	thick solid	912	0.91	1.00	9 (7)	33 (229)	31 (35)	0.3	50
foamed-type	knit/regular	thick foam	839	1.03	0.81	6 (4)	31 (200)	24 (27)	0.4	150
					Leather					
carf	grain	450	0.80	0.56	11 (10)	55 (75)	25 (25)	10.1	>1000	
side	grain	1000	1.50	0.67	23 (13)	70 (96)	73 (93)	7.5	500	

[a] Product is nylon in man-made leathers; cotton in urethane- and vinyl-coated fabrics. [b] To convert N/mm^2 to kgf/cm^2, multiply by 10.2; to psi, multiply by 145. [c] Numbers in parentheses are crosswise. [d] To convert kN/m to ppi, multiply by 5.71. [e] At 20°C. [f] Registered trademark of Kuraray Co., Ltd.

structure. For the two-layer substrate, a polyurethane coating process must come between impregnation and coagulation. (*5*) The substrate is prefinished by coating or sanding for grain or suede surface, followed by, as required, embossing, dyeing, and other finishings. Most manufacturing methods now available are similar to this but with some modifications.

Economic Aspects and Application

The production of man-made leather has increased rapidly due to its high quality. Up to 90% is produced in the Far East, and approximately 50% is exported to the U.S. and European countries.

The big three applications of vinyl-coated fabrics are (*1*) automotive (36%), (*2*) bags (17%), and (*3*) interiors (10%). Those of urethane-coated fabrics are (*1*) clothing (18%), (*2*) shoes (18%), and (*3*) accessories of shoes (11%). Those of man-made leathers are (*1*) shoes (46%), (*2*) clothing (13%), and (*3*) bags (10%).

KATSUMI HIOKI
Kuraray Company, Ltd.

Y. Saito and A. Kubotsu, in Sen-I Gakkai ed., *Zusetsu Sen-I no Keitai)* (*A Diagram of Fiber Structure*), Asakura-shoten Co., Ltd., Tokyo, 1983, p. 254.

K. Nagoshi, in J. I. Kroschwitz, ed., *Encyclopedia of Polymer Science and Engineering,* Vol. 8, 2nd ed., John Wiley & Sons, Inc., New York, 1987, pp. 677–697.

T. Yasui and co-workers, *Polyester,* The Textile Institute, Manchester, U.K., 1993, pp. 210, 211.

LECITHIN

Lecithin and other phospholipids are of universal occurrence in living organisms. They are constituents of biological membranes and are involved in permeability, oxidative phosphorylation, phagocytosis, and chemical and electrical excitation.

Lecithin is not only used in the strict scientific sense to describe pure phosphatidylcholine (Fig. 1), but also to describe crude phospholipid mixtures containing phosphatidylcholine (PC), phosphatidylethanolamine (PE), phosphatidylinositol (PI), other phospholipids, and a variety of other compounds such as fatty acids, triglycerides, sterols, carbohydrates, and glycolipids. Commercial lecithin is currently available in more than 40 different formulations varying from crude oily extracts from natural sources to purified and synthetic phospholipids. Many of these products are defined according to the stage of the purification process from which they are obtained and fall into three broad categories (Table 1) varying in their constituents both qualitatively and quantitatively.

Industrial lecithins from a variety of sources are utilized. The main sources include vegetable oils (eg, soybean, cottonseed, corn, sunflower, rapeseed) and animal tissues (egg and bovine brain). However, egg lecithin and in particular soy lecithin are by far the most important in terms of quantities produced, so much so that the term soy lecithin and commercial lecithin are often used synonymously.

Physical Properties

Commercial crude lecithin is a brown to light yellow fatty substance with a liquid to plastic consistency. Its density is 0.97 g/mL (liquid) and 0.5 g/mL (granule). The color is dependent on its origin, process conditions, and whether it is unbleached, bleached, or filtered. Its consistency is determined chiefly by its oil, free fatty acid, and moisture content. Properly refined lecithin has practically no odor and has a bland taste. It is soluble in aliphatic and aromatic hydrocarbons, including the halogenated hydrocarbons; however, it is only partially soluble in aliphatic alcohols. Pure phosphatidylcholine is soluble in ethanol.

Commercial lecithin is soluble in mineral oils and fatty acids but is practically insoluble in cold vegetable and animal oils. It is insoluble but infinitely dispersible in water. Commercial lecithin is a wetting and emulsifying agent. Lecithin is one of the very few natural and edible surface-active agents of this type that is soluble or dispersible in oil.

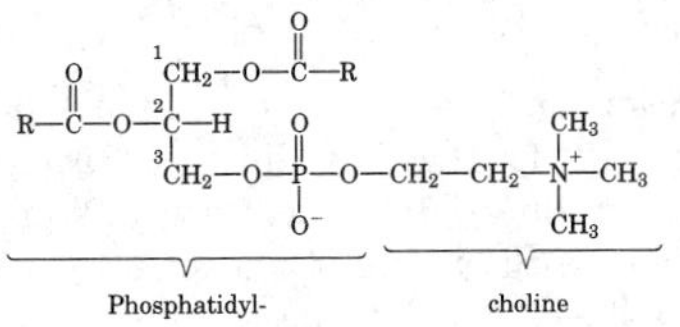

(1)

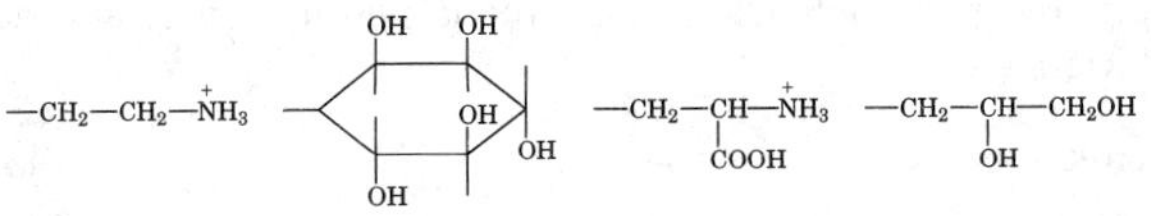

(2) **(3)** **(4)** **(5)**

Figure 1. Chemical structure of phosphatidylcholine (PC) (**1**) and other related phospholipids. $R{-}\overset{\overset{O}{\|}}{C}{-}O{-}$ represents fatty acid residues. The choline fragment may be replaced by other moieties such as ethanolamine (**2**) to give phosphatidylethanolamine (PE), inositol (**3**) to give phosphatidylinositol (PI), serine (**4**), or glycerol (**5**). If H replaces choline, the compound is phosphatidic acid. The corresponding IUPAC-IUB names are (**1**), 1,2-diacyl-*sn*-glycero(3)phosphocholine; (**2**), 1,2-diacyl-*sn*-glycero(3)phosphoethanolamine; (**3**), 1,2-diacyl-**sn**-glycero(3)phosphoinositol; (**4**), 1,2-diacyl-**sn**-glycero(3)phospho-L-serine; and (**5**), 1,2-diacyl-*sn*-glycero(3)phospho(3)-*sn*-glycerol.

Table 1. Categories of Commercial Lecithin

Natural	Refined	Modified
Plastic	*Deoiled*	*Physically*
unbleached		custom-blended
bleached		natural and refined
doubled-bleached		
Fluid	*Fractionated*	*Chemically*
unbleached	alcohol-soluble	
bleached	alcohol-insoluble	
double-bleached		*Enzymatically*

Chemical Properties

In general, the presence of fatty acid groups in the phospholipid molecule permits reactions such as saponification, hydrolysis, hydrogenation, halogenation, sulfonation, phosphorylation, elaidinization, and ozonization.

Manufacture and Processing

Crude soy lecithin is obtained as a by-product during the degumming process of soy oil. Only a minor proportion of the total lecithin that is potentially available in the vegetable processing industry is produced.

Purification Processes.

Separation of neutral and polar lipids, so-called deoiling, is the most important fractionation process in lecithin technology. A classic solvent for the deoiling is acetone.

Due to the possible environmental problems with acetone, new technologies are being developed for the production of deoiled lecithins like an ethanol-based extraction and fractionation or a process involving treatment of lipid mixtures with supercritical gases or supercritical gas mixtures.

Commercial Grades

There are six common grades of lecithin available including (*1*) clarified lecithins, (*2*) fluidized lecithins; (*3*) compounded lecithins; (*4*) hydroxylated lecithins; (*5*) deoiled lecithins and (*6*) fractionated lecithins. Fractions with different phosphatidylcholine content are commercially available. Besides these common commercial grades, more special products are available, eg, enzymatically modified lecithin and phospholipids, semisynthetic phospholipids, and acetylated lecithins.

Economic Aspects

The total commercial lecithin potential if all vegetable oils were degummed worldwide would be 552,000 t. Although soybean, sunflower, and rape lecithins are available in the market, the principal commercial interest is only in soybean lecithin. The annual worldwide production is 130,000 t.

Health and Safety Factors

The phospholipids are biodegradable, but their presence in streams and water resources, especially in the form of soap stock, is undesirable. Fatty acid recovery from phospholipids is less than with neutral oils because of the lower fatty acid content. There are no known health hazards involved in the production of commercial lecithin from crude vegetable oils because the phospholipids are nonvolatile and are a nonirritating food material.

Uses

The worldwide uses of lecithin break down as follows: margarine, 25–30%; baking/chocolate and ice cream, 25–30%; technical products, 10–20%; cosmetics, 3–5%; and pharmaceuticals, 3%.

Cosmetics and Soaps. One to five percent lecithin moisturizes, emulsifies, stabilizes, conditions, and softens when used in products such as skin creams and lotions, shampoos and hair treatment, and liquid and bar soaps. Since the introduction of Capture in 1986, liposomes produced from phospholipids are commercially available worldwide.

Pharmaceuticals. Lecithin and especially purified phosphatidylcholine can act as excipients in pharmaceutical (drug) formulation to enhance and control the bioavailability of the active component. Moreover, phosphalidylcholine can be utilized as a diedelic source, as it involved in the cholesterol metabolism and the metabolism of fats in the liver; also, it can be utilized as a precursor of brain acetylcholine, as neurotransmitter.

ARMIN WENDEL
Rhône-Poulenc Rorer

I. Hanin and G. B. Ansell, eds., "Lecithin: Technological, Biological and Therapeutic Aspects," *Advances in Behavioral Biology,* Vol. 33, Plenum Press, New York, 1987.

I. Hanin and G. Pepeu, eds., *Phospholipids: Biochemical, Pharmaceutical, and Analytical Considerations,* Plenum Press, New York, 1990.

B. F. Szuhaj and G. R. List, eds., *Lecithins,* American Oil Chemist's Society, Champaign, Ill., 1985.

B. F. Szuhaj, ed., *Lecithins: Sources, Manufacture and Uses,* American Oil Chemist's Society, Champaign, Ill., 1989.

LEUKOTRIENES. See ANTIASTHMATIC AGENTS; PROSTAGLANDINS.

LEVULOSE (FRUCTOSE). See CARBOHYDRATES; SUGAR.

LICENSING

A license is an agreement between two or more parties which conveys certain intellectual property rights (eg, patent, copyright, trademark, or know-how) from the licensor (the holder or owner of the intellectual property rights). For the sake of simplicity, herein a licensor (the entity which is licensing-out) and licensee (the entity which is licensing-in) are assumed to be companies; however, an individual, the government, or a nonprofit entity such as a university may also be a licensor or licensee.

Reasons for Licensing

In any contractual arrangement, each party has a particular set of reasons to enter into the agreement. In a licensing arrangement, the licensor holds rights to certain intellectual property that may be of value; the licensee would like access to those property rights in order to make, use, or sell products covered by such intellectual property rights.

Overall Perspective. For both parties, some of the factors which affect the decision to license-out or license-in within a strategic business plan might include (*1*) an assessment of competing products, (*2*) an assessment of alternative approaches, (*3*) an assessment of the real and perceived features, advantages, and benefits of the technology, (*4*) projections of market potential, and (*5*) the life span of the product. In addition, the question should be asked, "Is the technology evolutionary?" in which case a potential licensee would have to decide whether or not the technology is worth the investment, or is it "revolutionary," in which case the potential licensee might not want to "miss the revolution." In all cases, both the licensee and licensor must see a benefit from the license agreement in order for the business relationship to be successful.

Components of a License Agreement

Every license agreement is a negotiated contract between parties, and therefore there are no hard-and-fast rules as to what is required in a license agreement.

Diligence. Although the financial terms of a license agreement generally receive the most attention, "diligence" provision, ie, those provisions which specify the obligations of the licensee to develop the technology diligently, are often more important.

Financial Terms. Although most licenses involve a financial transaction, it should be noted that *royalty-free licenses* can be and often are granted, allowing a licensee to practice an invention or obtain rights under copyrights without making compensation.

A *royalty-bearing license* presumes a financial exchange between the parties, ie, the licensor grants certain intellectual property rights to the licensee for monetary consideration.

License issue fees, also known as upfront payments, are often but not always required from licensees as a gesture of good faith and serious intention by the licensee to develop the technology.

Annual payments are often viewed as a financial diligence provision or as a fee for maintaining the license.

Earned royalties are typically a percentage of the net selling price of the licensed product, although they can be a set amount of money per sale, regardless of the sales price.

Exclusivity and Nonexclusivity. Intellectual property can be licensed exclusively or nonexclusively, with many variations in between.

Field of use restrictions are common in license agreements. For example, a licensor can limit the licensed field of use for the same compound to research reagents, diagnostic products, or therapeutic products. In this instance, a licensor can grant at least three exclusive field of use licenses for one technology.

Other Terms and Conditions. The license agreement should always have a specified term and include provisions for termination. Other standard provisions can be included to address such issues as reporting, infringement, indemnity/warranties, governing law, assignment, or notices.

Option Agreement and Right of First Refusal

An *option agreement* is a contract between two parties in which one party acquires the right (option) to acquire a license to specified intellectual property, usually within a given time period. An agreement which grants one party, the potential licensee, the right of first refusal is often interpreted to mean that the party has the right to match any other offer for the technology.

Types of Intellectual Property Licenses

Types of intellectual property licenses include the patent license, the copyright license, trademark licenses, and tangible property licenses and material transfer agreements.

Valuing Technology

The most commonly asked questions about licenses revolve around pricing and valuing the technology, eg, what is a reasonable royalty? There is no easy answer because many factors must be taken into consideration. In the end, the value of the technology depends on what the licensee is willing to pay and what the licensor is willing to accept for the license. Licenses are usually a negotiated and unique agreement between two parties.

The value of the technology to be licensed depends on its potential technological and market competitive position; its features, advantages, and benefits; and whether the technology is evolutionary or revolutionary.

Every industry has its own sense of what a reasonable royalty might be. Whereas earned royalties are usually based on net sales, a very rough rule of thumb is that an earned royalty reflects approximately 25% of profits earned on the particular technology.

KATHARINE KU
Stanford University

B. Cronin, "Licensing Patents for Maximum Profits", *Int. J. Tech. Mgmt.* **4**(4,5), 411–420 (1989).

R. Goldscheider, "The Art of Licensing Out", *Technology Management: Law, Tactics, Forms,* Clark Boardman Callaghan, New York, 1988, Chapt. 6.

J. D. Major, "Some Practical Intellectual Property Aspects of Technology Transfer", *Int. J. Tech. Mgmt.* **3**(1,2), 43–49 (1988).

A. F. Millman, "Licensing Technology", *Mgmt. Decis.* **21**(3), 3–16 (1983).

LIGHT-EMITTING DIODES. See LIGHT GENERATION.

LIGHT GENERATION

LIGHT-EMITTING DIODES

One of the most fundamental optoelectronic devices is the light-emitting diode (LED). The LED consists of a p–n junction which emits light (ultraviolet, visible, or infrared radiation) in response to a forward current passing through the diode. The wavelength (color) of light emitted is characteristic of the materials employed in the active (light-emitting) region of the device. Such materials typically consist of compound semiconductors from Groups 13 (III) and 15 (V) of the Periodic Table (III–V semiconductors), such as GaAs, GaP, GaAsP, AlGaAs, InGaAsP, AlGaInP, and InGaN. In addition, IV–IV (Groups 14–14) and II–VI (Groups 12–16) semiconductors may also be employed to form LEDs. The exact material choice is dictated by the desired wavelength of emission, performance, and cost of the device.

Materials and Device Physics

Semiconductors. The basic material employed in LEDs is the semiconductor, a solid which possesses a conductivity intermediate between that of a conductor and an insulator. Unlike conductors, semiconductors and insulators possess an energy gap, E_g, between two energy bands, the conduction band (CB) and the valence band (VB). In a semiconductor, this energy gap is typically 0.3–6.0 eV.

Virtually all LEDs are formed from crystalline semiconductors wherein the atoms are arranged in a regular structure similar to that of diamond. This zinc blende crystal structure consists of two interpenetrating face-centered cubic (fcc) lattices, each consisting of different atoms. A few LED materials, eg, SiC and InGaN, are prepared in a different structure, the wurtzite crystal structure. This structure consists of two hexagonal closed-packed lattices composed of different atoms, each lattice interpenetrating the other.

In a semiconductor, the energy gap is generally much larger than thermal energy at room temperature (~25 meV), resulting in limited conductivity. However, the conductivity can be varied over several orders of magnitude by introducing impurities into the crystal structure. These impurities can either contain an excess or deficiency of valence electrons required for bonding to the host crystal. Such impurities are referred to as dopants and are classified as n-type donors (excess electrons) or p-type acceptors (deficient electrons or excess holes). The energy levels of these dopants generally lie within the energy gap of the semiconductor relatively close (<40 meV) to the energy band edges. Consequently, in n-type material sufficient thermal energy exists at room temperature to ionize electrons from the donor states to the conduction band creating mobile electrons. Similarly, in p-type material electrons from the valence band are excited to acceptor states within the gap resulting in the creation of mobile holes in the valence band. These mobile species (electrons in n-type material and holes in p-type material) are referred to as majority carriers.

p–n Junction Diode. A p–n junction diode is formed when two adjacent areas of a single-crystal semiconductor are doped using opposite type impurities. A forward bias, positive voltage on the p-material, nearly equivalent to the band gap energy of the semiconductor, facilitates the injection of carriers, and consequently current flow through the diode.

Radiative Recombination. After the carriers are injected across the junction, the carriers no longer constitute the majority of the mobile charge carriers, and thus are referred to as minority carriers. These minority carriers come into thermal equilibrium near the band edges on the opposing sides of the junction wherein they recombine with a majority carrier. The recombination process can either be radiative, ie, generating light, or nonradiative, ie, generating heat. Radiative transitions typically occur from/to the band edges or nearby impurity states. The wavelength, of the photon emitted in this process is given by:

$$\Delta E = \frac{hc}{\lambda} \tag{1}$$

where ΔE is the difference in the energy levels involved in the transistion (approximately the band gap energy), h is Planck's constant, and c is the speed of light.

Direct and Indirect Energy Gap. The radiative recombination rate is dramatically affected by the nature of the energy gap, E_g, of the semiconductor. The energy gap is defined as the difference in energy between the minimum of the conduction band and the maximum of the valence band in momentum, k, space. For almost all semiconductors, the maximum of the valence band occurs where holes have zero momentum, $k = 0$. Direct semiconductors possess a conduction band minimum at the same location, $k = 0$, Γ point, where electrons also have zero momentum. Thus radiative transitions that occur in direct semiconductors satisfy the law of conservation of momentum.

In an indirect semiconductor, the minimum of the conduction band occurs at a location, X or L point, where electrons have nonzero momentum, $\neq 0$. Consequently, a single recombination event cannot satisfy the law of conservation of momentum, and thus is forbidden. Radiative recombination can occur upon the additional interaction of a phonon (quantized lattice vibration). This is a two-step process with a much lower probability of occuring than a single recombination event. As a result, indirect semiconductors process a much lower internal quantum efficiency than direct semiconductors.

Isoelectronic Centers. The internal quantum efficiency of some indirect materials may be drastically improved by the addition of isoelectronic impurities to the semiconductor lattice which serve as radiative center, ie, N or (Zn, O) in GaP. However, the recombination processes

Table 1. Commercial LED Structures

Emission λ, nm	Active layer material[a]	Structure[b]	Window layer material	Substrate[c]	Lattice matched	Growth technique[d]
450–510	$In_xGa_{1-x}N(x \le 0.2)$	DH	GaN	Al_2O_3 (TS)	no	MOCVD
480	SiC, 6 H[e]	homo	SiC	SiC, 6 H (TS)	yes	MOCVD
555	GaP[e]	homo	GaP	Gap (TS)	yes	LPE
565	GaP:N[e]	homo	GaP	GaP (TS)	yes[f]	LPE
570–630	$GaAs_xP_{1-x}$: N($x \le 0.35$)	homo	GaAsP	GaP (TS)	no	VPE
560–640	$(Al_xGa_{1-x})_{0.5}In_{0.5}P(x \le 0.6)$	DH	AlGaAs or GaP	GaAs (AS)	yes	MOCVD[g]
560–640	$(Al_xGa_{1-x})_{0.5}In_{0.5}P(x \le 0.6)$	DH	GaP	GaP (TS)	yes	MOCVD[g]
650	$GaAs_{0.6}P_{0.4}$	homo	GaAsp	GaAs (AS)	no	VPE
650–880	$Al_xGa_{1-x}As(x \le 0.45)$	SH	AlGaAs	GaAs (AS)	yes	LPE
650–800	$Al_xGa_{1-x}As(x \le 0.45)$	DH	AlGaAs	AlGaAs (TS)	yes	LPE
700	GaP:(Zn,O)[e]	DH	GaP	GaP (TS)	no	LPE
880	AlGaAs:Si	homo	AlGaAs	AlGaAs (TS)	yes	LPE
820–880	$Al_xGa_{1-x}As(x \le 0.1)$	DH	AlGaAs	GaAs (AS)[h]	yes	LPE or MOCVD
900	GaAs:Zn	homo	GaAs	GaAs (AS)	yes	horizontal bridgeman
940	GaAs:Si	homo	GaAs	GaAs (TS)	yes	LPE
1300–1550	InGaAsP	DH	InP	InP (TS)	yes	LPE or MOCVD

[a] All bandgaps are direct unless otherwise noted. [b] DH = double heterostructure, homo = homostructure, and SH = single heterostructure. [c] AS = absorbing substrate, TS = trasparent substrate. [d] LPE = liquid-phase epitaxy, MOCVD = metalorganic chemical vapor disposition, VPE = vapor-phase epitaxy. [e] Band gap is indirect. [f] Some small lattice mismatch occurs as a result of high nitrogen doping level. [g] GaP window grown by VPE. [h] GaAs substrate may be locally removed to form small transparent emission region.

in this case still require more than one event, making such materials generally less efficient than those with direct band gaps.

Active Layer Structures. The structure of the LED active (light-emitting) layer may also strongly influence the efficiency of the device. A variety of layered structures are typically employed in LEDs, including homostructures single-heterostructures and double-heterostructures. The latter results in the most efficient and highest speed LEDs by virtue of the ability to confine both electrons and holes in the active layer.

Light Extraction. Light-emitting diode performance is limited by the ability to extract the photons generated within the high refractive index semiconductor chip (typically $n_1 \sim 3.5$) into the outside world ($n_2 = 1$ for free space or $n_2 \sim 1.6$ for epoxy typically utilized in encapsulation). The large difference in refractive indexes results in a critical angle, θ_c, for internal reflection given by Snell's law:

$$\theta_c = \sin^{-1}(n_2/n_1)$$

of 17° and 27° for LED chips in air and epoxy, respectively.

The extraction efficiency of a LED chip is most strongly influenced by the structure of the chip, eg, the Kness of window layers, and optical transparency of the substrate. As a result, the extraction efficiency (percentage of photons generated within the LED which escape to the outside world) typically ranges from ~4 to 30% for absorbing substrate devices with thin window layers and transparent substrate devices, respectively.

Device Design and Fabrication

Although the LED is one of the most basic optoelectronic devices, a variety of complex and interacting material and structural considerations in designing these devices exists. These include the choice of materials for emission wavelength of the LED as well as the geometry and fabrication methods of the device. The principal structural properties of commercially available LEDs are summarized in Table 1.

Device Performance

The emission wavelength of a LED is determined primarily by the energy gap of the semiconductor material employed in the active layer. Radiative transitions in LEDs occur over a range of energies resulting in a finite width of spectral emission (typically 15–100 nm). Typical emission spectra for various visible LEDs are shown in Figure 1.

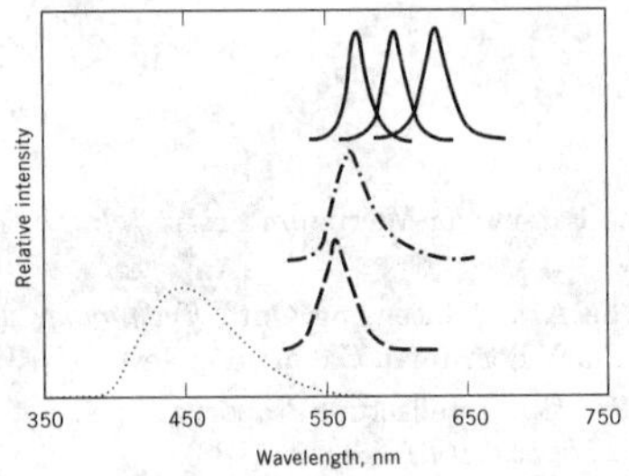

Figure 1. Light-emission spectra of various LEDs: emission from (—) AlGaInP LEDs results from direct gap recombination, whereas that of (----) GaP is from indirect mechanisms; (–·–) represents GaP with an N-isolectronic trap, GaP:N, and (· · ·) represents InGaN:Zn where the broadening in the spectrum results from recombination originating from deep levels within the energy gap.

Performance is described in a variety of ways depending on the desired application of the LED. The power efficiency, P_E, of a LED is the ratio of the output power to the power input to the device, and can be expressed as

$$P_E = C_{\mathrm{ex}}\eta_{\mathrm{int}}\left(\frac{hc}{\lambda}\right)\left(\frac{1}{V}\right)$$

where C_{ex} is the extraction efficiency, η_{int} is the internal quantum efficiency, and V is the applied voltage to the device. The applied voltage is given by the sum of the junction voltage, approximately equivalent to the built-in potential, and the voltage drop given by the dynamic resistance of the diode. Typically, this resistance is dynamic and on the order of 1–3 Ω for communication LEDs and may be as high as 10 Ω for visible/display LEDs.

For LEDs utilized in visible/display applications, the human eye serves as the detector of radiation. Thus a key measure of performance is luminous efficiency which is weighted to the eye sensitivity (CIE) curve.

One of the main advantages of LEDs over other light sources is reliability. Radiative recombination is a natural process which does not necessarily damage the crystal. Consequently, LEDs typically exhibit degradation rates in the range of 5–20% during the first 1000 h of operation and do not reach half-brightness for more than 100,000 h.

Visible/Display Emitters

The material employed for an LED has the greatest effect on the cost and performance of the device. Light-emitting diodes based on GaAsP, the first visible LEDs introduced, are among the highest volume, lowest cost devices available.

The first LEDs to utilize the N-isoelectronic trap were based on GaP:N. These devices, grown by LPE, emit in the yellow-green portion of the spectrum and have efficiencies of 2.5 lm/W. The improved efficiency relative to GaAsP:N results from a much lower dislocation density facilitated by lattice matched growth. The higher efficiency makes these devices useful for a variety of applications, including indicators and moderate performance lighting/display applications.

The AlGaAs materials system is used to produce high performance visible LEDs. This system possesses a direct energy gap in the red to near-ir spectral regime. The system is lattice matched to GaAs, and using LPE growth techniques can be employed to form heterostructure devices in high volumes. The direct band gap facilitates the fabrication of LEDs having typical luminous efficiencies ranging from 2–10 lm/W. The highest performance is obtained for a double heterostructure TS structure. These structures were the first to exceed red-filtered 60-W incandescent sources (3–4 lm/W) in luminous efficiency, opening new markets for LEDs. These high performance LEDs are relatively complex and costly to produce, and thus are only used in applications where high efficiencies are required.

The direct-indirect transition in AlGaAs limits the useful wavelength of these high brightness devices to the 650-nm band, ie, the red. Innovations in lattice matched $(Al_xGa_{1-x})_{0.5}In_{0.5}P/GaAs$ LEDs grown by MOCVD have resulted in emitters which have extended the useful high brightness spectral regime into the orange, yellow, and green. More recently. TS $(Al_xGa_{1-x})_{0.5}In_{0.5}P/GaP$ structures have been realized using the technique of compound semiconductor wafer bonding, doubling the efficiency of these devices in the green to red spectral regime. Performance exceeds 50 lm/W at $\lambda \sim 607$ nm

Another advance in LED technology has been the commercial introduction of blue and deep green emitters, and thus the introduction in 1994 of blue emitters based on the InGaN system was significant. Unlike other III–V LEDs, relatively high internal quantum efficiencies can be achieved even in the presence of highly lattice mismatched growth, ie, high dislocation densities. Such mismatched growth is required because as of this writing (ca 1995) no lattice matched substrate is available for the nitride system. However, high brightness blue, blue-green, and green AlGaN/GaN/InGaN double heterostructure devices grown by MOCVD on sapphire, Al_2O_3, substrates are commercially available.

Progress in the area of II–VI blue and blue-green light-emitter development led to the first injection blue laser diode. This device employed a CdZnSe/ZnSe/ZnSSe quantum well heterostructure having N as the *p*-type dopant. These advances have led to the demonstration of very high efficiency (8 lm/W) ZnSeTe/ZnSe blue-green (510 nm) double heterostructure TS LEDs grown on zinc selenide, ZnSe. Although these devices exhibit excellent initial performance, they are plagued by reliability problems.

Communication Emitters

Figure 2 shows the typical commercial performance of LEDs used for optical data communication. Both free-space emission and fiber-coupled devices are shown, the latter exhibiting speeds of <10 ns. Typically there exists a trade-off between speed and power in these devices, however performance has been plotted as a function of wavelength for purposes of clarity.

Markets and Applications

Light-emitting diodes are the most commercially important compound semiconductor devices in terms of both dollar and volume sales. They are becoming increasingly employed in more advanced applications.

Light-emitting diodes are utilized in a variety of markets. Generally, visible/display LEDs can be grouped into two principal classifications: low end indicators and high brightness emitters. The low end indicators are typically utilized in such markets as consumer, industrial/instrumentation, automotive interiors, and indoor signs. However, more complex and battery-driven applications which require high brightness devices are emerging. In addition, LEDs are being employed to backlight liquid crystal displays (LCDs), further requiring the use of high performance devices. Indicators are also used in automotive interior applications on dashboards within cars. In another application, arrays of LEDs are used in electrophotography and employed to form write bars and erase bars for printing information.

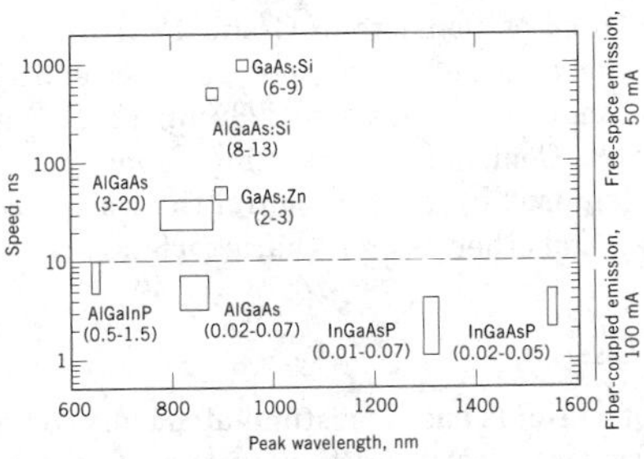

Figure 2. Speed (90–10% fall time) vs peak wavelength for commercial communication LED emitters. Output power levels in mW are given in parentheses. Generally, these exists a tradeoff between speed and output power in these devices. The data for fiber-coupled devices are based on emission into a 50/125-μm fiber, except for AlGaInP which is for 1-mm plastic optical fiber coupling.

The developments in the red AlGaAs, red-orange-yellow-green AlGaInP, and blue-green InGaN systems are on the threshold of invoking a revolution in high performance lighting, competing in markets previously reserved for filtered incandescents.

Light-emitting diodes are used in communications to transmit data optically. These LEDs can generally be divided into two categories: free-space emitters and fiber-coupled devices. More conventional free-space emitters are employed for remote controls; optocouplers, wherein an optical emitter/detector pair provides electrical isolation; and sensors, eg, bar-code wands, slot interrupters for measuring the presence of an object, etc. An emerging application is the ir wireless market, wherein much higher data rates can be achieved using properly designed TS AlGaAs devices.

Much higher data rates can be obtained for small emission area fiber-coupled devices. The selection of the device/fiber depends on the desired application. Plastic optical fiber-coupled devices emitting at 650 nm are available for short-haul low cost data links employed for industrial controls, medical instrumentation, and computer–peripheral communication. Glass optical fiber-coupled emitters are employed to achieve higher data rates and longer link lengths and typically operate at 850, 1300, and 1550 nm.

FRED KISH
Hewlett-Packard Company

A. A. Bergh and P. J. Dean, *Light-Emitting Diodes,* Clarendon, Oxford, U.K., 1976.

M. G. Craford, in L. E. Tannas, Jr., ed., *Flat Panel Displays and CRTs,* Van Nostrand Reinhold, New York, 1985, pp. 289–331.

K. Gillessen and W. Schairer, *Light-Emitting Diodes—An Introduction,* Prentice Hall, Englewood Cliffs, N.J., 1987.

R. H. Saul, T. P. Lee, and C. A. Burrus, in W. T. Tsang, ed., *Semiconductors and Semimetals,* Vol. 22, Academic Press, Inc., Orlando, Fla., 1985.

SEMICONDUCTOR LASERS

Most of the diode lasers emitting in the red and near-infrared (nir) parts of the electromagnetic spectrum are based on compound semiconductors composed of elements of Group 13 (III) and 15 (V) of the Periodic Table. Some of the newer green and blue emitting lasers use compound semiconductors composed of Group 12 (IIB) and 16 (VI) of the Periodic Table. Compounds containing as many as four different elements are often used in order to control the laser wavelength, electrical properties, and other desired characteristics.

Semiconductor Laser

The operation of a laser is based on stimulated emission. A photon generated in the forward biased junction, in the process of spontaneous emission, can be absorbed in the semiconductor or cause an additional transition by stimulating an electron to make a transition to a lower energy state. In this latter process another photon is emitted. The rate of stimulated emission is governed by a detailed balance between absorption and spontaneous and stimulated emission rates. That is, stimulated emission occurs when the probability of a photon causing a transition of an electron from the conduction to the valence band, with the probability emission of another photon, is greater than for the upward transition of an electron from the valence to conduction band upon absorption of the photon. These rates are commonly described in terms of Einstein's theory of A and B coefficients. For semiconductors, a simple condition describing the carrier density necessary for stimulated emission has been derived. Lasing can start when the density of electrons injected into the conduction band exceeds the hole density in the valence band. This is a condition of population inversion which occurs when the separation of the quasi-Fermi levels for the holes, F_p, and the electrons, F_e, is greater than the energy of the emitted photon,

$$F_e - F_p > hv$$

and the photon energy hv must be at least equal to the band gap energy. Thus in semiconductor lasers stimulated emission occurs between distributions of states in the conduction and valence bands. In most other lasers (qv), such as gas or glass lasers, this transition occurs between discrete energy levels.

A semiconductor laser requires a means of generating spatially localized high concentrations of minority carriers, a medium to provide the gain, and a means of providing some feedback to the stimulated emission. The medium is the semiconductor structure arranged in a way which helps to confine the carriers and light. Light is generated in this structure by means of a p–n junction which injects electrons from the valence to the conduction band and thus provides the population inversion. This is followed by recombination of electrons with holes and emission of light. Further recombination can be stimulated (stimulated emission) by light already present in the medium. This optical feedback is carefully arranged by forming a cavity which has two mirrors parallel to each other. Light generated within the cavity is then partially reflected back into the crystal. Such mirrors can be formed in most compound semiconductors lasers by cleaving two ends of a waveguide.

Heterostructures

A more effective carrier confinement is offered by a double heterostructure in which a thin layer of a low band gap material (the active layer) is sandwiched between larger band gap layers. The physical junction between two materials of different band gaps, and chemical compositions, is called a heterointerface. A schematic representation of the band diagram of such a structure is shown in Figure 1.

In addition to the carrier confinement, the active region is crucial in providing light confinement and waveguiding because the index of refraction of this low band gap layer is larger than the indexes of the surrounding layers. Such light confinement is absent in homostructure or single heterostructure lasers.

Lasing occurs whenever the gain arising from stimulated emission exceeds the cavity losses.

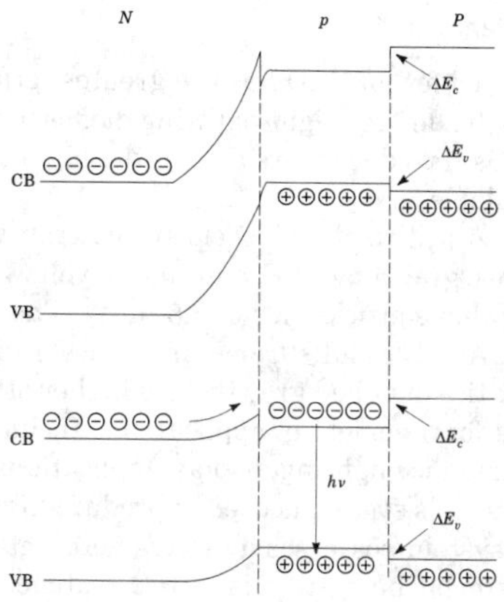

Figure 1. Schematic cross section and the band diagram of a double heterostructure showing the band-edge discontinuities, ΔE_c and ΔE_v, used to confine carriers to the smaller band gap active layer. (**a**) Without and (**b**) with forward bias. See text.

Quantum Wells

Epitaxial crystal growth methods such as molecular beam epitaxy (MBE) and metalorganic chemical vapor deposition (MOCVD) have advanced to the point that active regions of essentially arbitrary thicknesses can be prepared (see THIN FILMS, FILM FORMATION TECHNIQUES). Most semiconductors used for lasers are cubic crystals where the lattice constant, the dimension of the cube, is equal to two atomic plane distances. When the thickness of this layer is reduced to dimensions on the order of 0.01 μm, between 20 and 30 atomic plane distances, quantum mechanics is needed for an accurate description of the confined carrier energies. Such layers are called quantum wells and the lasers containing such layers in their active regions are known as quantum well lasers.

The uncertainty principle, according to which either the position of a confined microscopic particle or its momentum, but not both, can be precisely measured, requires an increase in the carrier energy. In quantum wells having abrupt barriers (square wells) the carrier energy increases in inverse proportion to its effective mass (the mass of a carrier in a semiconductor not being the same as that of the free carrier) and the square of the well width.

The two-dimensional carrier confinement in the wells formed by the conduction and valence band discontinuities changes many basic semiconductor parameters. The parameter important in the laser is the density of states in the conduction and valence bonds. The density of states is greatly reduced in quantum well lasers. This makes it easier to achieve population inversion and thus results in a corresponding reduction in the threshold carrier density. In the quantum well lasers, carriers are confined to the wells which occupy only a small fraction of the active layer volume. The internal loss owing to absorption induced by the high carrier density is very low, as little as $\pi_i^{-}2\ \mathrm{cm}^{-1}$. The output efficiency of such lasers shows almost no dependence on the cavity length, a feature useful in the preparation of high power lasers.

The incorporation of quantum wells into the material has other subtle consequences. The valence band in direct bulk semiconductors is degenerate at the maximum energy point (valence band maximum). At this point there are two types of holes of the same energy, the heavy and light ones. The effective mass of the light hole is similar to that of the electron, whereas that of the heavy hole is about 10 times larger than that of the electron. In quantum wells the two hole bands become separated in energy, and it is said that the degeneracy is lifted. The energy shift owing to quantum well confinement is larger for the light holes. The lasing transition is more likely to occur between the confined electron and heavy hole states.

All the layers of conventional heterostructure lasers, that is, lasers based on quantum wells, must precisely replicate the lattice structure of the substrate. These layers are fairly thick, on the order of 0.1 μm, and any lattice mismatch large enough to alter electronic properties invariably results in generation of defects. Strain can be introduced in quantum wells, with the thickness on the order of 0.01 μm or less, without generating defects.

Materials

Several compound semiconductor systems permit the growth of high quality, thin-layered crystal structures that have large and abrupt changes in the band gap and the index of refraction required in heterostructure lasers. The layer-to-layer changes result from changes in composition. Several of the Group 13–15 (III–V) material systems used for the preparation of laser structures are shown in the band gap-lattice constant plots of Figure 2. All of these materials have cubic structures known as zinc blende. The special nature of the binary semiconductor compounds such as gallium arsenide, GaAs, or indium phosphide, InP, arises from their availability as the substrate materials needed for the growth of lasers.

Two of the materials systems shown in Figure 2 are of particular importance. These are the ternary compounds formed from the Group 13 (III) elements such as Al and Ga in combination with As and quaternary compounds formed from Ga and In in combination with As and P.

An even wider range of wavelength, toward the infrared, can be covered with quantum well lasers. In the Al_xGa_{1-x} As system, compressively strained wells of Ga_xIn_{1-x} As are used. This ternary system is indicated in Figure 2 by the line joining GaAs and InAs.

The usual acceptor and donor dopants for Al_xGa_{1-x} As compounds are from Groups 2 and 12 (II), 14 (IV), and 16 (VI) of the Periodic Table. Group 2 and 12 elements are acceptors and Group 16 elements are donors. Depending on the growth conditions and the growth method, Si and Ge (Group 14) can be either donors or acceptor, ie, be amphoteric, which is of special interest in light-emitting diodes. Another principal material system is $Ga_xIn_{1-x}As_{1-y}P_y$ grown on InP.

Quantum well lasers in this system typically use ternary $In_{0.53}Ga_{0.47}$ As wells and binary InP barriers. All quaternary lasers, ie, lasers in which both the wells and barriers are formed by quarternary compounds, are also being developed. These structures can be lattice matched or strained.

The materials discussed yield lasers operating in the infrared and near visible spectral ranges. Many applications of lasers, such as printing or high density memories, require as short a wavelength as possible. The III–V system most suitable for short wavelength visible operation is the $(Al_xGa_{1-x})In_{1-y}P$ system.

More exotic Group 13–15 (III–V) materials such as gallium nitride are being investigated for shorter lasing wavelengths. One of the prime candidates is the InGaN/GaN system.

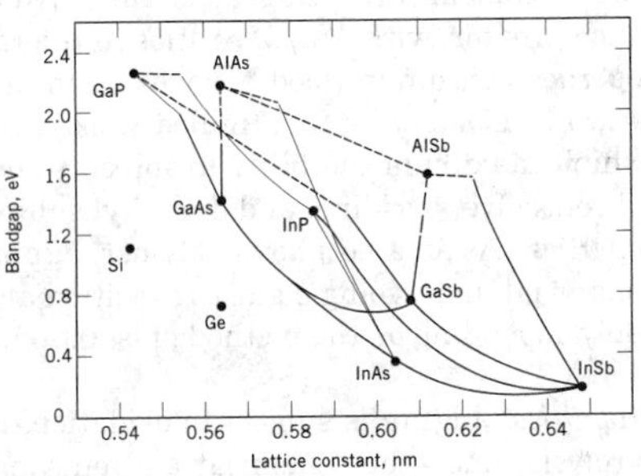

Figure 2. Band gap versus lattice constant for Group 13–15 (III–V) semiconductors where (—) denotes gap and (----) indirect. Lines joining the binary semiconductors indicate possible compositions of ternary compounds. The quaternary compound $Ga_xIn_{1-x}As_{1-y}P_y$ may have any band gap and lattice constant that lie in the region of the plane bound by InP, GaP, GaAs, and InAs. Only the compositions where $x = 2.2\ y$ are lattice matched to InP.

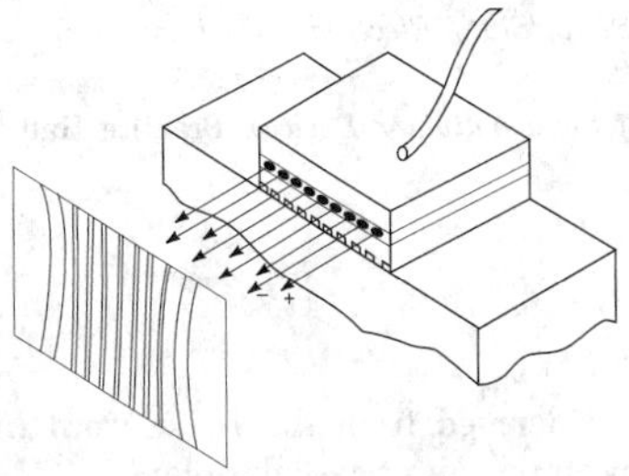

Figure 3. High power array of phase coupled GaAs/AlGaAs lasers mounted p-side down on a thermal heat sink. The $\pi/2$ shift of the neighboring lasers is indicated by the + and − signs. The output pattern consists of two dominant peaks, each associated with the lasers of the same phase, and much weaker peaks in between.

Injection lasers based on Group 2 or 12–16 (II–VI) compounds have also been demonstrated. These large band gap compounds make green lasers a reality and promise an extension of the lasing wavelength into the blue.

Laser Structures

Al_xGa_{1-x} As lasers show outstanding performance in two areas: high power lasers and vertical cavity surface emitting lasers. High power lasers are based on phase coupled arrays of the type illustrated in Figure 3. The individual lasers are formed simply by providing a narrow stripe contact, 5–10 μm in width. Proton implantation, which turns GaAs highly resistive, is used to confine the current to a narrow region close to the stripe (see ION IMPLANTATION). These lasers have no lateral index step to guide the light and are said to be gain guided. Lasing occurs in the areas under and very near the stripe in which the carrier density necessary for threshold is reached.

The large emission aperture and tight current confinement in laser arrays are important for high power operation. In GaAs lasers the output power is limited by catastrophic facet damage, the damage caused by a high light flux at the facet. It must be realized that the submicrometer active layer thickness of double heterostructure and the narrow stripe width result in radiative power fluxes on the order of MW/cm^2. Facet damage is known to occur at flux of approximately 10 MW/cm^2. The large aperture of laser arrays is very beneficial in this respect.

GaAs-based lasers also excel in new applications directed toward optical computing and optical data processing. Such applications require two-dimensional arrays of lasers having a very low threshold current, high beam quality, and high speed of operation. These requirements are well served by vertical cavity surface emitting lasers (VCSELs).

Lasers fabricated from alloys of InGaAsP operate in the 1.3–1.55 μm wavelength range and are used in fiber optic communications.

Most of the lasers discussed operate in a small number of discrete longitudinal modes, the Fabry-Perot modes. The individual modes are very narrow, much less than 0.01 nm, but are separated by spectral distances of ca 1.0 nm. Thus the overall width of the laser spectrum may exceed 4–5 nm. This is unacceptable in communication systems operating at 1.55 μm, the minimum loss region of optical fibers, over distances on the order of 100 km and data rates as high as 10 gigabites per second (Gb/s).

HENRYK TEMKIN
Colorado State University

A. S. Grove, *Physics and Technology of Semiconductor Devices,* John Wiley & Sons, Inc., New York, 1967.

S. M. Sze, *Physics of Semiconductor Devices,* John Wiley & Sons, Inc., New York, 1981.

B. G. Streetman, *Solid State Electronic Devices,* Prentice Hall, Englewood Cliffs, N.J., 1990.

M. Shur, *Physics of Semiconductor Devices,* Prentice Hall, Englewood Cliffs, N.J., 1990.

LIGNIN

The word lignin is derived from the Latin word *lignum,* meaning wood. It is a main component of vascular plants.

According to a widely accepted concept, lignin may be defined as an amorphous, polyphenolic material arising from enzymatic dehydrogenative polymerization of three phenylpropanoid monomers, namely, coniferyl alcohol (**1**), sinapyl alcohol (**2**), and *p*-coumaryl alcohol (**3**).

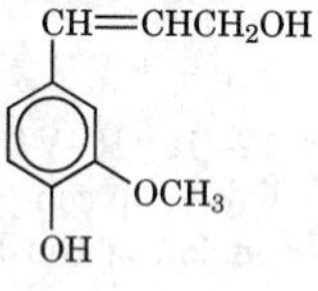

(**1**)

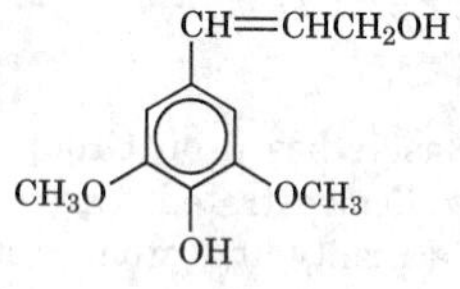

(**2**)

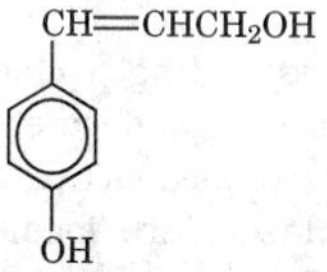

(**3**)

The biosynthesis process, which consists essentially of radical coupling reactions, sometimes followed by the addition of water, of primary, secondary, and phenolic hydroxyl groups to quinonemethide intermediates, leads to the formation of a three-dimensional polymer which lacks the regular and ordered repeating units found in other natural polymers such as cellulose and proteins.

Normal softwood lignins are usually referred to as guaiacyl lignins because the structural elements are derived principally from coniferyl alcohol (more than 90%), with the remainder consisting mainly of *p*-coumaryl alcohol-type units. Normal hardwood lignins, termed guaiacyl–syringyl lignins, are composed of coniferyl alcohol and sinapyl alcohol-type units in varying ratios.

The distribution of lignin in individual cells of lignified wood has been well examined. The lignin concentration is rather uniform across the secondary wall, but there is a significant increase in lignin concentration at the boundary of the middle lamella and primary wall region. This pattern of lignin distribution, with the highest concentration in the interfiber region and a lower, uniform concentration in the bulk of the cell walls, is typical for most wood cells. Thus lignin serves the dual purpose of binding and stiffening wood fibers through its distribution between and in the cell walls.

Lignin performs multiple functions that are essential to the life of the plant. By decreasing the permeation of water across the cell wall in the conducting xylem tissues, lignin plays an important role in the internal transport of water, nutrients, and metabolites. It imparts rigidity to the cell walls and acts as a binder between wood cells, creating a composite material that is outstandingly resistant to compression, impact, and bending. It also imparts resistance to biological degradation.

In commercial chemical pulping of wood, the reverse process in nature is performed to isolate fibers for papermaking. In the process, wood is delignified by chemically degrading and/or sulfonating the lignin to water-soluble fragments. The industrial lignins thus obtained are used in many applications.

Structure and Reactions

The structural building blocks of lignin are linked by carbon–carbon and ether bonds. Units that are trifunctionally linked to adjacent units represent branching sites which give rise to the network structure characteristic of lignin.

Of the functional groups attached to the basic phenylpropanoid skeleton, those having the greatest impact on reactivity of the lignin include phenolic hydroxyl, benzylic hydroxyl, and carbonyl groups. The frequency of these groups may vary according to the morphological location of lignin, wood species, and method of isolation.

Reactions include electrophilic substitution, conversion of aromatic rings to nonaromatic cyclic structures, conversion of cyclic to acyclic structures, ring coupling and condensation reactions, cleavage of ether bonds, cleavage of carbon–carbon bonds, substitution reactions on side chains, and formation and elimination of multiple bond functionalities.

Analytical Methods

Detection of Lignin. The characteristic color-forming response of lignified tissue and some lignin preparations on treatment with certain organic and inorganic reagents was recognized in the early nineteenth century. More than 150 color reactions have now been proposed for the detection of lignin. Reagents used in these reactions may be classified into aliphatic, phenolic, and heterocyclic compounds, aromatic amines, and inorganic chemicals. Among the important reactions are the Wiesner and Mäule color reactions.

Determination of Lignin Content. Lignin content in plants (wood) is determined by direct or indirect methods. The direct method includes measurement of acid-insoluble (ie, Klason) lignin after digesting wood with 72% sulfuric acid to solubilize carbohydrates.

Characterization of Lignin. Lignin is characterized in the solid state by Fourier transform infrared spectroscopy (ftir), uv microscopy, interference microscopy, cross polarization/magic angle spinning nuclear magnetic resonance spectroscopy (cp/mas nmr), photoacoustic spectroscopy, Raman spectroscopy, pyrolysis–gas chromatography–mass spectroscopy, and thermal analysis. In solution, lignins are characterized by spectral methods.

Functional Group Analysis. The total hydroxyl content of lignin is determined by acetylation with an acetic anhydride–pyridine reagent, followed by saponification of the acetate, and further followed by titration of the resulting acetic acid with a standard 0.05 *N* sodium hydroxide solution.

Properties

Molecular Weight and Polydispersity. Because it is not possible to isolate lignin from wood without degradation, the true molecular weight of lignin in wood is not known. The weight-average molecular weight, $\overline{M}_w$, of softwood milled wood lignin is estimated to be 20,000; lower values have been reported for hardwoods.

The polydispersity of softwood milled wood lignin, as measured by $\overline{M}_w/\overline{M}_n = 2.5$, is high compared with that of cellulose and its derivatives. Other lignins show different polydispersity as demonstrated by high pressure size exclusion chromatograms. The polydispersity of lignosulfates is much greater, with $\overline{M}_w/\overline{M}_n$ ratios in the range of 6–8.

Solution Properties. Lignin in wood behaves as an insoluble, three-dimensional network. Isolated lignins (milled wood, kraft, or organosolv lignins) exhibit maximum solubility in solvents such as dioxane, acetone, methyl cellosolve, pyridine, and dimethyl sulfoxide.

Thermal Properties. As an amorphous polymer, lignin behaves as a thermoplastic material undergoing a glass transition at temperatures which vary widely depending on the method of isolation, sorbed water, and heat treatment.

Chemical Properties. Lignin is subject to oxidation, reduction, discoloration, hydrolysis, and other chemical and enzymatic reactions. Many are briefly described elsewhere. Key to these reactions is the ability of the phenolic hydroxyl groups of lignin to participate in the formation of reactive intermediates.

Industrial Lignins

Industrial lignins are by-products of the pulp and paper industry. Lignosulfonate, derived from sulfite pulping of wood, and kraft lignin,

Table 1. European and American Lignin Manufacturers

Producer	Country	Annual capacity, t/yr
Borregaard LignoTech	Norway	160,000
LignoTech Sweden	Sweden	50,000
LignoTech Deutschland	Germany	50,000
LignoTech Iberica	Spain	30,000
LignoTech Finland	Finland	50,000
Avebene	France	80,000
Attisholz	Switzerland	100,000
Georgia Pacific	United States	190,000
LignoTech USA	United States	65,000
Westvaco	United States	45,000
others		290,000
Total		*1,110,000*

derived from kraft pulping, are the principal commercially available lignin types. Organosolv lignins derived from the alcohol pulping of wood are also reported to be available commercially, but their quantities are limited.

The production capacity of lignin in the Western world is estimated to be ca 1.1×10^6 t/yr. Although the production of lignosulfonates has been declining, kraft lignin production has increased. Of the companies listed in Table 1, LignoTech Sweden and Westvaco produce kraft lignins. The rest produce lignosulfonates.

Lignosulfates. Relatively large-volume uses of lignosulfonates include animal feed pellet binders, water-reducing agents in concrete admixtures, dispersants for gypsum board manufacture, thinners/fluid loss control agents for drilling muds, dispersants/grinding acids for cement manufacture, and in dust control applications, particularly road dust abatement.

Lignin technology has advanced significantly, and increased research and development efforts have resulted in specialty uses in several key market areas, eg, dye dispersants, pesticide dispersants, carbon black dispersant, water treatment/industrial cleaning applications, and complexing agent for micronutrients.

Kraft Lignins. In many applications, the base lignin must be modified (ie, through sulfonation or oxidation) prior to use. Once modified, kraft lignins can be used in most of the same applications in which lignosulfonates are used.

The lignins produced in these processes have potential application in wood adhesives, as flame retardants, as slow-release agents for agricultural and pharmaceutical products, as surfactants, as antioxidants, as asphalt extenders, and as a raw material source for lignin-derived chemicals.

STEPHEN Y. LIN
STUART E. LEBO, JR.
LignoTech USA, Inc.

K. V. Sarkanen and H. L. Hergert, in K. Sarkanen and C. Ludwig, eds., *Lignins: Occurrence, Formation, Structure, and Reactions,* Wiley-Interscience, New York.

J. Nakano and G. Meshitsuka, in S. Y. Lin and C. W. Dence, eds., *Methods in Lignin Chemistry,* Springer-Verlag, Berlin, 1992.

S. Y. Lin and I. S. Lin, in *Ullmann's Encyclopedia of Industrial Chemistry,* 5th ed., Vol. 15, VCH, Weinheim, Germany, 1990.

LIGNITE AND BROWN COAL

Lignite and brown coal are common names for coals having properties intermediate between peat and bituminous coal as a result of limited coalification (see COAL). In general, brown coal designates a geologically younger, ie, less coalified, material than the firmer, fibrous lignite. In many English-speaking countries, the consolidated coals are termed *lignite*. In many English-speaking countries, the consolidated coals are termed lignite, and unconsolidated coals are termed brown coal. In Australia, and in Germany and a number of other European countries, the generic term brown coal is used for the whole class, including some coals that are included in the ASTM classification as subbituminous. Lignite signifies the firmer, fibrous, woody variety. Herein lignite is used as the comprehensive term.

Selection of coal for a particular use requires a knowledge of composition greater than that supplied from the ASTM classification. Progress is being made toward classifying all kinds of coal, including lignite, by correlating properties with composition and other qualities.

Lignite is less valuable than coals of higher rank, primarily because its much higher (30–70% as mined) water content and high chemically combined oxygen content result in a relatively low heating value (LHV). In the past, the expense of shipping limited the market largely to the vicinity of the mine. However, in the United States the low sulfur content of lignite has made long distance shipments economically feasible, in order to limit sulfur oxide emissions at electric power generation plants. The increasing worldwide demand for energy together with desire for national self-sufficiency has increased the importance of low heating value coals.

Geology

Lignite was deposited relatively recently (ca $2.5–60 \times 10^6$ yr ago), mainly during the Tertiary era. U.S. deposits include those in the Dakotas, Alaska, Montana, and Wyoming. The Miocene period provided the brown coal deposits that are up to 300 m thick in the Latrobe Valley of Victory in Australia.

Composition, Properties, and Analysis

Macroscopic Appearance. Lignitic coals vary from brown to dull black when moist, although the color may appear considerably lighter when the coal is dried. Breakage is easiest for the unconsolidated coals. Strength and toughness increase as coalification increases.

Physicochemical Structure. Water-filled pores and capillaries of differing diameters permeate the organic gel material that makes up as-mined lignite.

Properties. The apparent density of lignite is 0.8–1.35 g/cm^3, which is lower than values given for higher ranking coals. Therefore, greater volume is required for storage, transportation, and lignite reactors than is needed for an equivalent weight of more mature coals. Lignite generally has lower elasticity and greater plasticity than more mature coals. The tar yield is usually higher for lignite than for more mature coals. Tar yields are important in determining selection for carbonization and for liquid fuel production by pyrolysis.

Oxidation. The high reactivity of lignites with oxygen requires special care during mining, transportation, and storage to avoid spontaneous combustion from heat generation.

Resources and Production

The importance of a coal deposit depends on the amount that is economically recoverable by conventional mining techniques. The world total recoverable reserves of lignitic coals were 3.28×10^{11} metric tons at the end of 1990, of which ca 47% was economically recoverable as of 1992. These estimates of reserves change as geological survey data improve and as the resources are developed.

The extent of lignite production is generally not proportional either to total resources or to known economic reserves. Lack of energy alternatives is a strong motive to developments in lignite production.

Main Deposits and Production Areas. The eastern European reserves of lignitic coals provide the primary solid fuel for the eastern part of Germany, the former Czechoslovakia, Hungary, the former Yugoslavia, and Bulgaria. The importance of lignite as an energy source is great enough in Germany to permit long-range planning that includes removal and relocation of towns or villages situated on deposits in order to permit more complete recovery of the lignite resource. Hard coal is more important in most of the western European countries, with the exception of Austria and Italy.

In the U.S., lignite deposits are located in the northern Great Plains and in the Gulf states. Subbituminous coal is found along the Rocky Mountains. The lignite deposits of North Dakota and Montana extend into Canada as far as Saskatchewan. Canadian deposits are also located in Alberta, Yukon, the Northwest Territories, Ontario, and Manitoba.

Production. The mining or winning of lignitic coal typically involves deposits near the surface. The open-cast, open-cut, or strip-mining techniques employed involve mobile equipment built to provide a range of capacities to over 200,000 m^3/d. The rate of production can be increased rapidly, and the amount of labor per ton of coal mined is less than for underground mining. The quality of the coal, ratio of overburden thickness to seam thickness, stratigraphy, and distance to location of consumption are important in determining the cost to the consumer.

Concern about spontaneous ignition has led some operators to try to match the mining and consumption rates, so that there is little if any reserve, as in minemouth power generation stations. When the coal must be stockpiled, careful stacking minimizes oxygen reaction and overheating. To limit drying, spraying with cold water is useful.

For short distances from the mine, transportation is by truck or conveyer belt. Rail transportation is generally used for greater distances. Slurry pipelines are being considered as an alternative. Drying can be accomplished by evaporative, hydrothermal, or other thermal processes.

Health and Safety Factors

The principal hazards involve the tendency of the coal toward spontaneous combustion as the coal dries, especially at the exposed seam.

Economic Aspects

The price of lignite per mined ton or per heat unit is lower than that for higher rank coals. The market for all coals is primarily as boiler fuel for electric power production. Prices are generally established by contracts between utility and supplier before mining begins. Because of its sulfur content, lignite is becoming more important.

Uses

Most of the world's coal supply is used for combustion to generate steam for electric power production. This is especially true of lignitic coals. Other uses for lignite, such as briquetting, for domestic and industrial fuels; carbonization, to provide coke and liquid by-products; gasification, to provide gaseous fuels; chemical feedstocks, for making fertilizers and other liquid fuels; and direct liquefaction are being developed.

KARL S. VORRES
Argonne National Laboratory

R. A. Durie, ed., *The Science of Victorian Brown Coal: Structure, Properties and Consequences for Utilisation,* Butterworth Heinemann, Oxford, 1991. An excellent reference not only for Victorian Brown Coal, but for lignitic coals of the world.

The World Energy Council issues Conference reports on reserves, resources and production at six-year intervals. More limited reports are issued at two-year intervals.

Symposia on the Technology and Use of Lignite have been held in conjunction with the University of North Dakota Energy and Environmental Research Center and the preceding organizations.

LIME AND LIMESTONE

The elements calcium and magnesium, which are distributed very widely in the earth's crust, most commonly occur in carbonate forms of rock, generally classified as limestone. Although vast strata of this ubiquitous rock are buried so deeply as to be inaccessible, great tonnages of this stone are extracted for commercial use. Limestone, literally one of the most basic raw materials of industry and construction, occurs in varying degrees in nearly every country.

Limestone may be classified as to origin, chemical composition, texture of stone, and geological formation. Chemically it is composed primarily of calcium carbonate, $CaCO_3$, and secondarily of magnesium carbonate, $MgCO_3$, with varying percentages of impurities. Although these carbonates occur in many other rocks, ores, and soils, in its broadest definition limestone is distinguished by a content of more than 50% total carbonate. Limestone's most important chemical characteristic is that when subjected to high temperature it decomposes chemically into lime, calcium oxide, CaO, with decarbonation occurring through the expulsion of carbon dioxide gas. This primary product, known as quicklime, can then be hydrated, or slaked, into hydrated lime, calcium hydroxide, $CA(OH)_2$, ie, the water is chemically combined with the calcium oxide in an equimolecular ratio.

Definitions

In addition to showing varying degrees of chemical purity, limestone assumes a number of widely divergent physical forms, including marble, travertine, chalk, calcareous marl, coral, shell, oolites, stalagmites, and stalactites. All these materials are essentially carbonate rocks of the same approximate chemical composition as conventional limestone.

Limestone is generally classified into the following types: (*1*) high calcium, in which the carbonate content is essentially calcium carbonate having no more and usually less than 5% magnesium carbonate; (*2*) magnesian, which contains both calcium and magnesium carbonates, and has a magnesium carbonate content of 5–20%; and (*3*) dolomitic, which contains >20% but not more than 45.6% $MgCO_3$, the exact amount contained in a true, pure, equimolecular dolomite. The balance is $CaCO_3$.

The carbonate minerals that comprise limestone are calcite (calcium carbonate), which is easily the most abundant mineral type; aragonite (calcium carbonate); dolomite (double carbonate of calcium and magnesium); and magnesite (magnesium carbonate). Individual limestone types are further described by many common names.

Geology

Geographical Occurrence. Significant deposits in the U.K. are chiefly confined to the Devonian, lower Carboniferous, Jurassic, and Cretaceous systems. In the United States, the geological distribution is broader. Quantities of stone occur in the older Cambrian, Ordovician, and Silurian systems, as well as in the previously mentioned groups.

The deposits of limestone in the United States, if no limitation as to quality is placed, are widespread, occurring in nearly every state, usually in tremendous amounts. However, deposits of high purity limestone for making lime are more limited in distribution.

Detailed information concerning the location and analysis of limestone deposits in the United States can be obtained from the various state geological surveys, the U.S. Bureau of Mines, and the U.S. Geological Survey. Descriptive summaries of the limestone deposits in the various states have been published.

Impurities. Alumina in combination with silica is present in limestone chiefly as clay, though other aluminum silicates in the form of feldspar and mica may be found. Siliceous matter other than clay may occur in the free state as sand, quartz, and chert. Chemical and metallurgical limestones should contain less than 1% alumina and 2% silica. Iron compounds in limestone are seldom injurious to a lime product unless a very pure lime is required.

Properties of Limes and Limestones

Color. The purest forms of calcite and magnesite are white, often with an opaque cast, but most conventional limestone, even relatively pure types, are gray or tan. Quicklime is usually white of varying intensity, depending on chemical purity.

Odor. Except for highly carbonaceous species, most limestones are odorless.

Texture. All limestones are crystalline, but there is tremendous variance in the size, uniformity, and arrangement of their crystal lattices. The crystals of the minerals calcite, magnesite, and dolomite are rhombohedral; those of aragonite are orthorhombic. The crystals of chalk and of most quick and hydrated limes are so minute that these products appear amorphous, but high powered microscopy proves them to be cryptocrystalline. Hydrated lime is invariably a white, fluffy powder of micrometer and submicrometer particle size. Commercial quicklime is used in lump, pebble, ground, and pulverized forms.

Hardness. Pure calcite is standardized on Mohs' scale at 3; aragonite is harder, 3.5–4. Dolomite limestone is generally harder than high calcium.

Strength. The compressive strength of limestone varies tremendously, having values from 8.3 to 196 MPa (1,200–28,400 psi).

Luminescence. Limestone possesses only limited luminescent qualities.

Thermal Properties. Because all limestone is converted to an oxide before fusion or melting occurs, the only melting point applicable is that of quicklime. These values are 2570°C for CaO and 2800°C for MgO. Boiling point values for CaO are 2850°C and for MgO 3600°C.

Solubility. High calcium limestone is only very faintly soluble in water, whereas hydrated lime is slightly soluble. The solubility of hydrate on a CaO basis is 1.330 g/L (or 0.13%) of saturated solution at 10°C. Contrary to that of limestone, the solubility of hydrate is in inverse proportion to temperature, decreasing with higher temperatures.

Plasticity. An innate characteristic of a lime putty of paste-like consistency is its plasticity or its ability to be molded under pressure and to retain its altered shape without deformation. This rheological property is important for structural uses of lime in masonry mortar and plaster.

Stability. All calcitic and dolomitic limestones are extremely stable compounds, decomposing only in fairly concentrated strong acids or at calcining temperatures of 898°C for high calcium and about 725°C for dolomitic stones at 101.3 kPa (1 atm). Lime solutions develop a high pH of slightly under 12.5 at 25°C, whereas limestones attain a pH of 8–9.

Chemical Reactions. Chemical reactions include neutralization and causticization. The manufacture of Portland cement is predicated on the high temperature reaction of lime, CaO, with silica and alumina to form tricalcium silicate and aluminate.

Limestone Production. Because more than 99% of U.S. limestone is sold or used as crushed and broken stone, rather than dimension-stone, most of the description of limestone's extraction and processing herein focuses on the former (Fig. 1). Most stone is obtained by open-pit quarrying methods, although there are many large underground mines operating, with more expected in the future.

The quarry operation consists of overburden removal (down to the top of the limestone), followed by drilling and blasting. The shot rock is then loaded into quarry trucks and taken to the primary crusher. Following crushing the stone is conveyed to the secondary crusher, then screened into various sizes before being stockpiled. Where mines are involved, all steps are identical, except that overburden removal is obviated.

Lime Manufacture

Most lime plants worldwide produce their own kiln feed from a contiguous quarry or mine, and are thus integrated lime producers. However, several unintegrated lime plants located on the Great Lakes obtain kiln feed by boat from large commercial quarries in northern Michigan. Most of these plants, among the largest in U.S. lime production, are situated in the Chicago and Detroit areas and in northern Ohio.

Theory of Calcination. The reversible reaction involved in the calcination and recarbonation of lime–limestone is one of the simplest and most fundamental of all chemical reactions. In practice, lime burning can be quite complex, however, and many empirical modifications are often necessary for efficient performance.

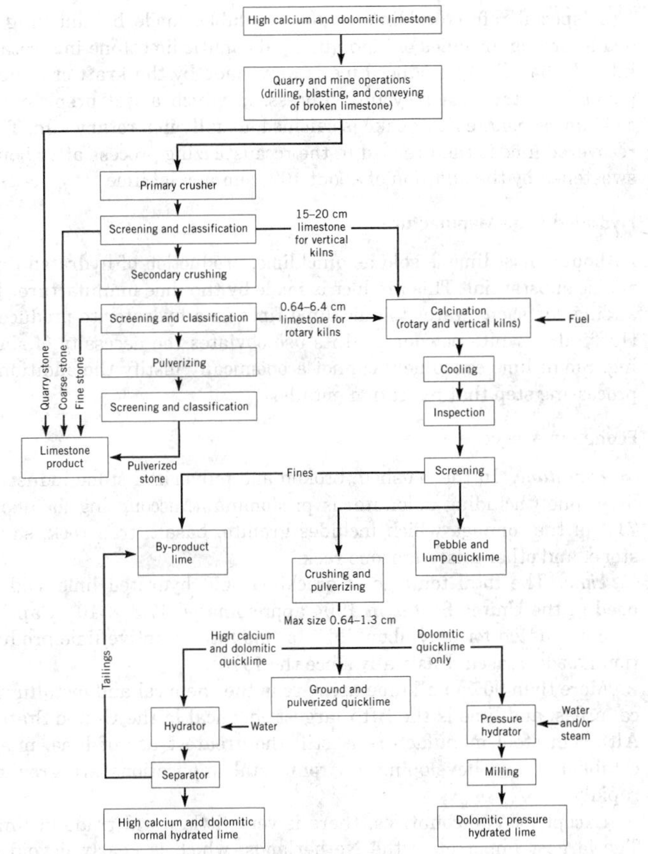

Figure 1. Simplified flow sheet for lime and limestone products.

There are three essential factors in the thermal decomposition of limestone: (*1*) the stone must be heated to the dissociation temperature of the carbonates; (*2*) this minimum temperature (but in practice a higher temperature) must be maintained for a certain duration (ca.2.8 × 10^6 Btu/t); and (*3*) the carbon dioxide evolved must be removed rapidly. However, actual energy consumption in lime burning varies from 3.5 to 8.0 Btu/t, depending principally on the kiln type. In the U.S., about 90% of commercial lime capacity and 60% of captive lime is calcined in rotary kilns. The rotary is not nearly as preeminent in most other countries.

Outside the U.S., particularly in developing countries, the vertical kiln is the most commonly used. One reason for the decline in use in the U.S. is the energy crisis of the 1970s, when supplies of natural gas and fuel oil, the principal fuel for vertical kilns, became stringent and prices escalated rapidly. The vertical kiln has made a slight comeback in the U.S., however, through the introduction of the Maerz parallel flow regenerative kiln.

Two other kiln types used in the U.S. are the FluoSolids and the Calcimatic. With the former the limestone feed granules (65–68 mesh) are calcined on a fluidized bed in a vertical type kiln, producing a very reactive quicklime. The Calcimatic consists of a circular traveling hearth on which the pebble limestone feed is calcined by the time the hearth completes one revolution.

Calcination Products. Quicklime is available in a variety of sizes, depending primarily on the kiln type and the specific use. Rotary kilns produce a pebble lime of 3.81–5.08 cm top size, whereas vertical kilns produce lump lime ranging from 6-in. top size down to 5.08 cm. However, smaller sizes are available, including ground and pulverized. In addition, pelletized lime is also produced by compressing quicklime fines into 2.54-cm pellets.

A special refractory lime is also available, made by sintering or dead-burning granules of high quality dolomitic limestone in a rotary kiln at 1650°C. By-product lime is also made by the kraft or sulfate paper industry in a recycling process, in which a wet prepipicated calcium carbonate filter cake is calcined, usually in a rotary kiln. This recovered lime is then reused in the recausticizing process after being sweetened by the addition of about 10% commercial lime.

Hydrated Lime Manufacture

Although most lime is sold as quicklime, production of hydrated lime is also substantial. This product is made by the lime manufacturer by adding sufficient water to quicklime fines in a hydrator to produce a fluffy, dry, white powder, and its use obviates the necessity of slaking. Small lime consumers cannot economically justify the additional processing step that hydration entails.

Economic Aspects

Limestone. In the crushed, broken and pulverized stone industry, limestone (including dolomite) is predominant, accounting for about 71% of the tonnage which includes granite, basalt, trap rock, sandstone, and other miscellaneous rock.

Lime. The total tonnage of quicklime and hydrated lime sold or used in the United States in 1994 approximated 17.3×10^6. Captive lime accounted for only about 10% of this total. Captive lime production has decreased drastically since the 1970s.

More than 90% of all lime uses are in the chemical and metallurgical areas, and lime is the fifth largest chemical in the United States. Although steel manufacture is still the greatest use of lime, many established and developing environmental applications are growing rapidly.

Except for two countries, there is very little world trade in lime. The largest importer is the Netherlands, which is nearly devoid of limestone and thus imports about 10^6 t annually from Belgium and Germany. The other net importer of consequence is the U.S., which imports ca 150,000 t/yr or less than 1% of U.S. production. About 85% of the U.S. imports are from Canada; the balance is from Mexico.

Uses

Limestone. Limestone is used in construction, building materials, and mineral feed.

Lime. Lime, the versatile chemical, is used in an almost countless number of ways. Among the principal uses are included steel manufacture; environmental uses, eg, water, sewage, industrial and hazardous waste treatment, and scrubbing of power plant stack gases for control of acid rain; gold, copper, and uranium mining; production of precipitated calcium carbonate for use in alkaline sizing for paper manufacture; recausticization at paper mills; chemical manufacturing; construction, eg, for mortar and plaster, lime–soil stabilization for road and airfield construction, and as an antistripping agent in asphalt paving mixes; and agriculture.

K. A. GUTSCHICK
National Lime Association

National Stone Association, Washington, D.C., proceedings of conventions on operating problems and discussions, 1946–1993.

National Lime Association, Arlington, Va., proceedings of conventions and operating meetings.

R. S. Boynton, *Chemistry and Technology of Lime and Limestone,* John Wiley & Sons, Inc., New York, 1966; *Ibid.,* 2nd rev. ed., 1979.

Chemical Lime Facts, 6th ed., National Lime Association, Washington, D.C., 1992.

LIQUEFIED PETROLEUM GAS

Liquefied petroleum gas (LPG) is a subcategory of a versatile class of petroleum products known as natural gas liquids (NGLs) that are produced along with and extracted from natural gas. LPG is also produced from the refining of crude oil. Although LPG is commercially defined as propane, butane, and butane–propane mixtures, commercial availability is primarily limited to propane. There are two grades of specification propane, propane HD-5 or special-duty propane and commercial propane. The primary difference in the two grades is that the propylene content of propane HD-5 is restricted to a maximum of 5 vol %. Propylene is found only in refinery-produced propane. The principal uses of LPGs are as fuels and feedstocks for the production of petrochemicals such as ethylene.

Other natural gas liquids include natural gasoline, which is composed of the pentanes and heavier components of the natural gas stream, and ethane. Most recently ethane has become the principal product of natural gas processing plants.

Properties

Physical properties of the principal components of LPG are summarized in Table 1.

Manufacture and Processing

LPG is recovered from natural gas principally by one of four extraction methods: turboexpander, absorption (qv), compression, and adsorption (qv). Selection of the process is dependent on the gas composition and the degree of recovery of ethane and LPG, particularly from large volumes of lean natural gas.

Purification. The LPG generally requires treatment for removal of hydrogen sulfide, H_2S, organic sulfur compounds, and water in order to meet specifications. Several methods are used: amine treatment, caustic treatment, coalescing, solid-bed dehydration, molecular sieve treatment, solid-bed caustic treatment, and fractionation.

Production and Shipment

Historically, about two-thirds of the LPG produced in the U.S. came from natural gas processing and one-third was produced from refinery operations. The progress of LPG utilization has been closely related to progress in transportation and storage of this fuel. Large volumes of LPG usually are transported by high pressure pipelines. This use of pipelines is increasing rapidly. Large quantities of LPG are transported in railroad tank cars which have an average capacity of 113.5 m^3/car (3×10^4 gal/car), although this use is decreasing.

Economic Aspects

The production and consumption of LPG in the U.S. increased dramatically from its early beginnings in the 1930s until the international energy crises of the 1970s when rising prices and regulatory restraints resulted in reduced domestic production. However, total consumption, including imports, resumed a modest growth characteristic after that time.

Table 1. Physical Properties of LPG Components

Component	Mol formula	Bp, 101.3 kPa,[a] °C	Vapor pressure, 37.8°C, kPa[b]	Liquid density, g/L[a]
ethane	C_2H_6	−88.6		354.9
propane	C_3H_8	−42.1	1310	506.0
isobutane	C_4H_{10}	−11.8	498	561.5
n-butane	C_4H_{10}	−0.5	356	583.0
1-butene	C_4H_8	−6.3	435	599.6
cis-2-butene	C_4H_8	3.7	314	625.4
trans-2-butene	C_4H_8	0.9	343	608.2
n-pentane	C_5H_{12}	36.0	107	629.2

[a] At saturation pressure. [b] To convert kPa to psi, multiply by 0.145.

Storage

Large volumes of LPG are stored to meet peak demand during cold seasons. LPGs are both volatile and flammable and must be stored and handled in special equipment. Standards for storing and handling LPG are published by the National Fire Protection Association and API. Four main types of storage are high pressure storage above ground, low pressure refrigerated storage above ground, frozen earth storage, and underground cavern storage.

R. RAY TAYLOR
Phillips Petroleum Company

LP-Gas Market Facts, National LP-Gas Association, Oak Brook, Ill., 1977.

Liquefied Petroleum Gas Specifications and Test Methods, Gas Processors Association, GPA Publication 2140-92, Tulsa, Okla.

ASTM Standard D1835-91, American Society for Testing and Materials, Philadelphia, Pa., 1992.

Storage and Handling of Liquefied Petroleum Gases, National Fire Protection Association, NFPA 58, Boston, Mass., 1989.

LIQUID CRYSTALLINE MATERIALS

Liquid crystals represent a state of matter with physical properties normally associated with both solids and liquids. Liquid crystals are fluid in that the molecules are free to diffuse about, endowing the substance with the flow properties of a fluid. As the molecules diffuse, however, a small degree of long-range orientational and sometimes positional order is maintained, causing the substance to be anisotropic as is typical of solids. Therefore, liquid crystals are anisotropic fluids and thus a fourth phase of matter. There are many liquid crystal phases, each exhibiting different forms of orientational and positional order, but in most cases these phases are thermodynamically stable for temperature ranges between the solid and isotropic liquid phases. Liquid crystallinity is also referred to as mesomorphism.

Many thousands of organic substances, some rigid-rod polymers, and other macromolecules exhibit liquid crystallinity. The general common molecular feature is either an elongated or flattened, somewhat inflexible molecular framework, which is usually depicted as either a cigar- or disk-shaped entity. The orientational and positional order in a liquid-crystal phase is only partial, with the intermolecular forces striking a very delicate balance involving both attractive and repulsive interactions. As a result, liquid crystals are extraordinarily sensitive to external perturbations, eg, temperature, pressure, electric and magnetic fields, shearing stress, or foreign vapors. For this reason, liquid crystals are used to design practical devices to either monitor ambient changes of various kinds or to transduce an environmental fluctuation into a useful electrical or optical output.

Besides being used in the scientific study of cooperative phenomena and complex fluid phases, liquid crystalline phenomena have received a good deal of attention due to the possibility of practical applications. Liquid crystals are widely used in electrooptic displays. Other applications include radiation and pressure sensors, optical switches and shutters, and thermography. The liquid crystalline structures formed by amphiphilic molecules form the basis for emulsions and are studied thoroughly by researchers in the food, drug, and oil industries. Polymers that form an anisotropic fluid phase are important in the fabrication of lightweight, ultrahigh strength, and temperature-resistant fibers, and are beginning to be used in electrooptic displays. Liquid crystals also appear to play an important role in the structure and biochemical function of living tissue, where the characteristic combination of order and flow mobility is particularly suited to life processes. Certain disease states, eg, atherosclerosis, sickle cell anemia, or cancer, may be associated with physical changes in the liquid-crystalline order within biological structures.

Orientational and Positional Order in Fluids

Solids of mesogenic (liquid-crystal forming) molecules melt to form fluids in which some of the long-range molecular order is retained. At the simplest level, the elongation or flattening of the mesogenic molecules prevents the immediate dissolution of the parent, solid-state order. The loss of positional order of the centers of mass of the molecules in the parent solid may be either partial or complete upon melting, but some degree of orientational order is always retained. The fluid retains many solid-like properties, which are finally eliminated when the substance passes into the normal, isotropic liquid phase at a higher temperature (a second melting point). Solid-like features return if the substance is cooled from the isotropic state; this intermediate state is usually thermodynamically reversible, but in some cases it only forms upon cooling. Partial dissolution of solid-state order also may occur in certain substances by the use of solvents. In this case the molecules are either orientationally ordered in the solvent (some macromolecules), or form aggregates, in which the molecules exhibit long-range positional and/or orientational order. Liquid crystals that are established solely by the adjustment of temperature are referred to as thermotropics, whereas those that form through the addition of a solvent are called lyotropics.

Orientational Distribution Function and Order Parameter. In a liquid crystal a snapshot of the molecules at any one time reveals that they are not randonly oriented. There is a preferred direction for alignment of the long molecular axes. This preferred direction is called the director, and it can be used to define an orientational distribution function, $f(\theta)$, where $f(\theta)\sin\theta d\theta$ is proportional to the fraction of molecules with their long axes within the solid angle $\sin\theta d\theta$.

It is useful to describe the amount of orientational order with a single quantity. X-ray, uv, optical, ir, and magnetic resonance techniques are used to measure the order parameter in liquid crystals.

Positional Distribution Function and Order Parameter. In addition to orientational order, some liquid crystals possess positional order in that a snapshot at any time reveals that there are parallel planes which possess a higher density of molecular centers than the spaces between these planes. If the normal to these planes is defined as the z-axis, then a positional distribution function, $g(z)$, can be defined, where $g(z)dz$ is proportional to the fraction of molecular centers between z and $z + dz$. Since $g(z)$ is periodic, it can be represented as a Fourier series (a sum of a sinusoidal function with a periodicity equal to the distance between the planes and its harmonics). To represent the amount of positional order, the coefficient in front of the fundamental term is used as the order parameter. The more the molecules tend to form layers, the greater the coefficient in front of the fundamental sinusoidal term and the greater the order parameter for positional order.

Bond Orientational Order. In some cases, although the lattice of points of high density of molecular centers parallel to the planes are not correlated from layer to layer, the two principal directions of the lattice are the same for all layers. In these materials, the interactions between the planes do not prevent the planes from translating relative to each other, but do prevent them from rotating relative to each other.

Thermotropic Liquid Crystals

Thermotropic liquid crystals result from the melting of mesogenic solids due to an increase in temperature. Both pure substances and mixtures form thermotropic liquid crystals. In order for a mixture to be a thermotropic liquid crystal, the different components must be completely miscible. Examples include nematic liquid crystals (*p*-methoxybenzylidene-*p′*-*n*-butylaniline (MBBA); *p*-azoxyanisole (PAA); *p*-*n*-hexyl-*p′*-cyanobiphenyl; di-4-methoxyphenyl-*trans*-1,4-cyclohexane-dicarboxylate; and *p*-quinquephenyl), cholesteric liquid crystals ((−)-2-methylbutyl 4-(4′-methoxybenzylideneamino)cinnamate), and smectic liquid crystals (ethyl 4-(4′-phenylbenzylideneamino)benzoate; ethyl 4-(4′-ethoxybenzylideneamino)cinnamate; *p*-*n*-octyloxybenzoic acid; 4-(4′-n-octadecyloxy-3′-nitrophenyl) benzoic acid; diethyl *p*-terphenyl-*p,p″*-carboxylate; 2-(*p*-pentylphenyl-

5-(*p*-pentyloxyphenyl)-pyrimidine; and 4-ethyl-4′-butyloxybenzylodeneaniline). Much more is known about calamitic (rod-like) liquid crystals then discotic (disk-like) liquid crystals, since the latter were discovered only recently.

Nematic. In a nematic liquid crystal, the long axes of the molecules remain substantially parallel, but the positions of the centers of mass are randomly distributed. Therefore, there is orientational order and a nonzero orientational order parameter, but there is no positional order.

If the molecules of a liquid crystal are optically active (chiral), then the nematic phase is not formed. Instead of the director being locally constant as is the case for nematics, the director rotates in helical fashion throughout the sample. Within any plane perpendicular to the helical axis the order is nematic-like. In other words, as in a nematic there is only orientational order in chiral nematic liquid crystals, and no positional order.

Smectic. Smectic liquid crystals are distinguished from nematics by the presence of some positional order (a tendency to form layers) in addition to orientational order. The direction of preferred orientational order is perpendicular to the layers in a smectic *A* liquid crystal and at an angle with the layer normal in a smectic *C* liquid crystal.

In much the same way as a chiral compound forms the chiral nematic phase instead of the nematic phase, a compound with a chiral center forms a chiral smectic *C* phase rather than a smectic *C* phase. In a chiral smectic *C* liquid crystal, the angle the director is tilted away from the normal to the layers is constant, but the direction of the tilt rotates around the layer normal in going from one layer to the next.

Frustrated Phases. Chiral molecules normally form chiral phases, but in some cases this is done in an interesting way. For example, it is not unusual for a chiral molecule to form a smectic *A* phase, which is not chiral. If the molecule is highly chiral, however, twist is sometimes introduced into the smectic *A* phase by an array of grain boundaries which are perpendicular to the smectic *A* layers and parallel to the director. In one compound the normal to the layers is rotated by roughly 17° on either side of a grain boundary and the grain boundaries are separated by about 24 nm, giving this twist grain boundary (TGB) phase a pitch of a little more than 500 nm. In a sense the frustration of an achiral phase of chiral molecules has been relieved by the introduction of these twist grain boundaries.

Discotic Phases. Molecules which are disk-shaped rather than elongated also form thermotropic liquid-crystal phases. Usually these molecules have aromatic cores and six lateral substituents, although the predominance of six lateral substituents is solely historical; molecules with four lateral substituents also can form liquid-crystal phases. Although the flatness of these molecules creates a steric effect promoting alignment of the normal to the disks, the fact that disordered side chains are also necessary for the formation of these phases (as is often the case for liquid crystallinity in elongated molecules) should not be ignored. The most simple discotic phase is the nematic phase, in which the normal to the disks are preferentially aligned along a single direction (director). If the molecules are chiral or if a chiral dopant is added to a discotic liquid crystal, a chiral nematic discotic phase can form.

Metallomesogens. It is also possible to synthesize compounds based on metal atoms which possess liquid-crystal phases. The series based on dithiolene complexes (**1**), where M = Ni, Pd, or Pt, contains a number of compounds which show the liquid-crystal phases typical of rod-like molecules.

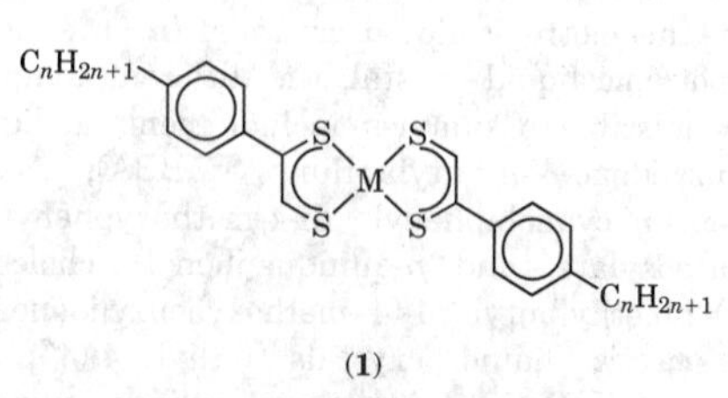

(**1**)

Disk-shaped molecules based on a metal atom possess discotic liquid-crystal phases.

Lyotropic Liquid Crystals Some molecules in a solvent form phases with orientational and/or positional order. In these systems, the transition from one phase to another can occur due to a change of concentration, so they are given the name lyotropic liquid crystals. Of course temperature can also cause phase transitions in these systems, so this aspect of thermotropic liquid crystals is shared by lyotropics. The real distinctiveness of lyotropic liquid crystals is the fact that at least two very different species of molecules must be present for these structures to form.

Amphiphilic Molecules. In just about all cases of lyotropic liquid crystals, the important component of the system is a molecule with two very different parts, one that is hydrophobic and one that is hydrophilic. These molecules are called amphiphilic because when possible they migrate to the interface between a polar and nonpolar liquid.

Even more interesting phenomena occur when amphiphilic compounds are put into water–oil mixtures. If the oil concentration is low, the amphiphilic molecules form micelles and the oil collects inside the micelles. As the oil concentration is increased, the micelles continue to swell with oil until it is safe to say that the system is really composed of volumes of water and volumes of oil separated by a single amphiphilic layer. This type of system is called an emulsion, and thus amphiphiles can serve as emulsifiers.

When a highly polar liquid, a slightly polar liquid, and an amphiphile are mixed together at the right temperature and in the right concentrations, the micelles which form are not spherical. Within this vary narrow concentration range, the micelles are rod-shaped for one part of this range and disk-shaped for another part. In either case the micelles themselves orient their symmetry axes (the long axis for the rod-shaped micelles and the short axis for the disk-shaped micelles) just like a thermotropic liquid crystal.

Polymorphism

A liquid crystal compound in more cases than not takes on more than one type of mesomorphic structure as the conditions of temperature or solvent are changed. In thermotropic liquid crystals, transitions between various phases occur at definite temperatures and are usually accompanied by a latent heat.

An exception to the rule that lowering the temperature causes transitions to phases with increased order sometimes occurs for polar compounds which form the smectic A_d phase (a layered structure formed by molecular dimers). Decreasing the temperature causes a transition from nematic to smectic A_d, but a further lowering of the temperature produces a transition back to the nematic phase (called the reentrant nematic phase). Electric or magnetic fields also may induce mesomorphic phase transitions.

Synthesis

Just because a molecule is long, narrow, and meets the requirement of geometric anisotropy does not ensure that it will have a liquid crystal phase. The particular phase structure that occurs in a compound, ie, smectic, nematic, or chiral nematic, not only depends on the molecular shape but is intimately connected with the strength and position of the polar or polarizable groups within the molecule, the overall polarizability of the molecule, and the presence of chiral centers.

Molecular interactions that lead to attraction include dipole–dipole interactions, dipole-induced dipole interactions, dispersion forces, and hydrogen bonding.

In order for dipole–dipole and dipole-induced dipole interactions to be effective, the molecule must contain polar groups and/or be highly polarizable. Ease of electronic distortion is favored by the presence of aromatic groups and double or triple bonds. These groups frequently are found in the molecular structure of liquid crystal compounds. The most common nematogenic and smectogenic molecules are of the type shown in Table 1. In general, if the X link is rigid, a liquid crystal phase is favored.

The importance of unsaturation is illustrated by the fact that 2,4-nonadienoic acid forms a liquid-crystal phase, whereas the *n*-aliphatic

Table 1. Some Central Linkages Found in Liquid Crystalline Compounds

$$R_1 - \bigcirc - X - \bigcirc - R_2$$

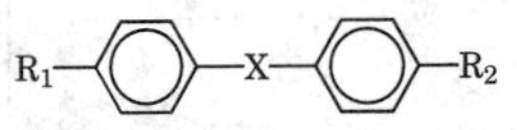

X	Series name
—CH=N—	Schiff bases
—N=N—	diazo compounds
—N=N— ↓ O	azoxy compounds
—CH=N— ↓ O	nitrones
—CH=CH—	stilbenes
—C≡C—	tolans
—OC— $=$ O	esters
— (nothing)	biphenyls

carboxylic acids do not. The two double bonds enhance the polarizability of the molecule and bring intermolecular attractions to a level that is suitable for mesophase formation. The overall linearity of the molecule must not be sacrificed in potential liquid-crystal candidates. Bulky, even if highly polarizable, functional groups or atoms that are attached anywhere but on the end of a rod-shaped molecule are usually less favorable for liquid-crystal formation.

In the case of carboxylic acids, hydrogen bonding can induce liquid-crystal phases by lengthening the molecular unit through dimerization:

$$R-C \begin{matrix} O \cdots H-O \\ O-H \cdots O \end{matrix} C-R$$

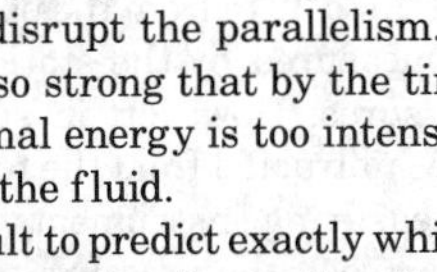

On the other hand, hydrogen bonding may lead to nonlinear molecular associations that disrupt the parallelism. Hydrogen bonding associations may also be so strong that by the time the solid reaches its melting point the thermal energy is too intense to permit substantial order to remain within the fluid.

Although it is difficult to predict exactly which type of liquid-crystal phase will be formed by a molecule meeting the general requirements, rough trends can be recognized. The presence of functional groups that lead to strong lateral interactions, eg, dipoles operating across the long molecular axis, favor the layered smectic structure. When these structural elements are not present but the molecule is otherwise suitable for mesomorphism, ie, is long and narrow, the nematic phase is likely. Longer terminal groups favor the smectic phase over the nematic phase. An asymmetric center on the molecule causes the chiral nematic and chiral smectic *C* phases in place of the nematic and smectic *C* phases.

Goals in liquid crystal synthesis include the design of room temperature thermotropics which are stable, colorless liquid crystalline over a wide range of temperature, and operate at low voltage and power levels.

A good deal of synthesis effort has been devoted to chiral liquid crystals, especially those with chiral smectic *C* phases. The chiral smectic *C* phase is ferroelectric, which gives it properties quite useful for applications. Perhaps the most important property of these phases is that a lateral dipole can produce a spontaneous polarization.

Polymer Liquid Crystals

Both polymer melts and polymer solutions sometimes form phases with orientational and positional order. Thermotropic polymer liquid crystals possess at least one liquid crystal phase between the glass-

transition temperature and the transition temperature to the isotropic liquid. Lyotropic polymer liquid crystals possess at least one liquid crystal phase for certain ranges of concentration and temperature.

Polymer Melts. When a rigid, polarizable monomer forms wither a mainchain polymer with flexible segments in between or a side-chain polymer with flexible segments between the rigid segments and the flexible main chain, liquid-crystal phases are usually stable.

Examples of polymers which form anisotropic polymer melts include petroleum pitches, polyesters, polyethers, polyphosphazines, α-poly-p-xylylene, and polysiloxanes. Synthesis goals include the incorporation of a liquid crystal-like entity into the main chain of the polymer to increase the strength and thermal stability of the materials that are formed from the liquid-crystal precursor, the locking in of liquid crystalline properties of the fluid into the solid phase, and the production of extended chain polymers that are soluble in organic solvents rather than sulfuric acid.

Polymer Solutions. Perhaps the most extensively studied macromolecular liquid crystals are the synthetic polypeptides, such as poly(γ-benzyl L-glutamate) (PBLG). PBLG is a homopolymer of the L-enantiomorph of a single amino acid with the following repeat unit.

$$\begin{matrix} & O \\ & \| \\ \left(NHCHC \right)_n \\ | \\ (CH_2)_2 \\ | \\ COOCH_2 - \bigcirc \end{matrix}$$

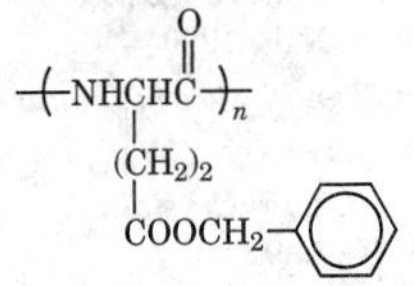

PBLG adopts the α-helical conformation in a number of solvents as a result of intramolecular hydrogen bonding and favorable stacking of the pendent side chains. Thus the polymer assumes an extended, relatively rigid geometry and may become ordered spontaneously at sufficiently high concentrations. The formation of this lyotropic liquid crystal phase occurs at a critical volume fraction of polymer ϕ^* which is inversely proportional to the length-to-diameter ratio of the macromolecule.

A variety of aromatic and extended-chain polyamides that spontaneously form a mesophase in concentrated solutions also have been synthesized.

The polyamides are soluble in high strength sulfuric acid or in mixtures of hexamethylphosphoramide, N,N-dimethylacetamide, and LiCl. The liquid-crystal phase is optically anisotropic and the texture is nematic. The nematic texture can be transformed to a chiral nematic texture by adding chiral species as a dopant or incorporating a chiral unit in the main chain as a copolymer.

Applications. The polyamides have important applications. The very high degree of polymer orientation that is achieved when liquid crystalline solutions are extruded imparts exceptionally high strengths and moduli to polyamide fibers and films. DuPont markets such polymers, eg, Kevlar, and Monsanto has a similar product, eg, X-500, which consists of polyamide and hydrazide-type polymers (see HIGH PERFORMANCE FIBERS; POLYAMIDES, FIBERS). Liquid-crystal polymers are also used in electrooptic displays.

Liquid Crystals in Biological Systems

Many biological systems exhibit the properties of liquid crystals. Considerable concentrations of liquid crystalline compounds have been found in many parts of the body, often as sterol or lipid derivatives. A liquid crystal phase has been implicated in at least two degenerative diseases, atherosclerosis and sickle cell anemia. Living tissue, such as muscle, tendon, ovary, adrenal cortex, and nerve, show the optical birefringence properties that are characteristic of liquid crystals. The liquid crystal state has been identified in many pathological tissues, particularly in areas of large lipid deposits. Massive deposits of liquid crystalline cholesterol derivatives have been found in the kidneys, liver, brain, spleen, marrow, and aorta walls. Certain living sperms possess a liquid crystalline state.

Cell Membrane. The fluid mosaic model of the cell membrane is one in which the phospholipids provide the basic order and integrity of the cell through amphiphilic interaction with the aqueous environment.

Microfilaments and Microtubules. There are two important classes of fibers found in the cytoplasm of many plant and animal cells that are characterized by nematic-like organization. These are the microfilaments and microtubules which play a central role in the determination of cell shape, either as the dynamic element in the contractile mechanism or as the basic cytoskeleton.

Liquid Crystalline Structures. In certain cellular organelles, deoxyribonucleic acid (DNA) occurs in a concentrated form. Striking similarities between the optical properties derived from the underlying supramolecular organization of the concentrated DNA phases and those observed in chiral nematic textures have been described. Concentrated aqueous solutions of nucleic acids exhibit a chiral nematic texture *in vitro.*

PETER J. COLLINGS
Swarthmore College

B. Bahadur, *Liquid Crystals: Applications and Uses,* Vols. 1–3, World Scientific, Singapore, 1990–1992.

P. J. Collings, *Liquid Crystals: Nature's Delicate Phase of Matter,* Princeton University Press, Princeton, N.J., 1990.

P. G. de Gennes and J. Prost, *The Physics of Liquid Crystals,* Clarendon Press, Oxford, U.K., 1993.

A. M. White and A. H. Windle, *Liquid Crystalline Polymers,* Cambridge University Press, Cambridge, U.K., 1992.

LIQUID LEVEL MEASUREMENT

The four process control parameters are temperature, pressure, flow, and level. Modern process level detection systems are varied and ubiquitous; in modern chemical plants there are thousands of processes requiring liquid level indication and liquid level control. From accumulators to wet wells, the need for level devices is based on the need for plant efficiency, safety, quality control, and data logging. Unfortunately, no single level measurement technology works reliably on all chemical plant applications. This fact has spawned a broad selection of level indication and control device technologies, each of which operates successfully on specific applications.

Measurement vs Control

Level devices can be divided into two broad groups: those that indicate level and those that provide means to control level. Indication devices include sight glass gauges, dip stick indicators, and magnetic liquid level indicators.

Floats. Float level switches are suitable for clean liquid applications, primarily for alarm function. A float follows level change, moving a stem and magnetic attraction sleeve within a nonmagnetic enclosing tube.

Displacers. Displacer level switches are suitable for clean and dirty fluids and are used principally to control sump pumps where shifting specific gravity, turbulent surface, and foam are common problems. Displacer(s) are suspended from a range spring connected to a stem and attraction sleeve. With change in level the spring senses the change in buoyancy, causing the stem and attraction sleeve to move within a nonmagnetic enclosing tube.

Buoyancy. Buoyancy level controllers are used to control a process having continuous flow through the vessel. The primary application is in reactors, feedwater heaters, deaerators, and similar processes having boiling and turbulent conditions. A hollow cylinder (displacer) is suspended from a range spring or torque tube. Change in level causes a change in buoyancy which creates a force–balance shift which is transmitted through a metallic seal into the control housing.

Conductivity. Conductivity level switches are generally limited to applications with low pressure, conductive fluids (high pressure models are available). They are alarm or pump control devices. Metal rods are inserted into the vessel with low level a-c voltage applied.

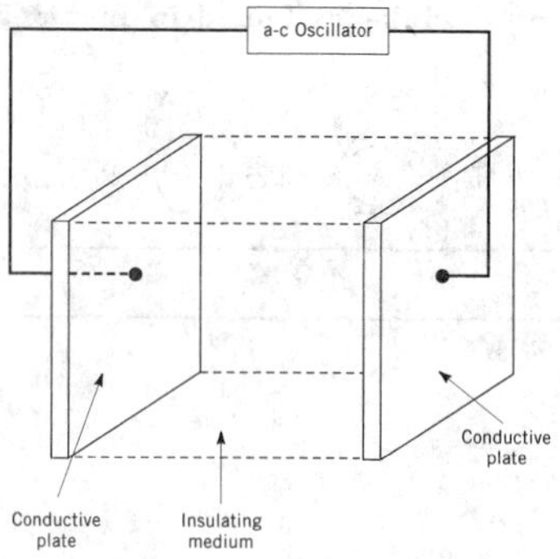

Figure 1. Basic capacitor.

Capacitance. Capacitance-based measurement devices, like conductivity devices, utilize the electrical properties of the medium to derive its measurement. Unlike conductivity devices, capacitance can be used to measure either conductive or nonconductive media (dielectric value: nonconductive $< 10 >$ conductive). Capacitance is developed when an oscillator impresses a high frequency a-c signal across two conductive plates separated by an insulating material, or dielectric (Fig. 1).

There are two general weaknesses associated with capacitance systems. First, because it is dependent on a process medium with a stable dielectric, variations in the dielectric can cause instability in the system. Second, buildup of conductive media on the probe can cause significant output error.

Static Pressure. The static pressure system is based on the fact that the static pressure exerted by a liquid is directly proportional to the height of the liquid above the point of measurement regardless of the volume in the tank, provided that the specific gravity remains constant.

When a pressure gauge is used to measure liquid level in a vessel under atmospheric pressure, the pressure tap is located at the approximate minimum level line of the vessel. When the pressure gauge cannot be located at the minimum tank level, a diaphragm box is used. An air purge or bubbler system are used for corrosive liquids or slurries where the gauge can be located up to 30 m from the point of measurement.

Liquid level can be measured by the static pressure method also at nonatmospheric pressures. However, in such cases the pressure above the liquid must be subtracted from the total head measurement. Differential pressure measuring instruments that measure only the difference in pressure between the pressure tap at the bottom of the tank and the pressure in the vapor space are used for this purpose.

Differential Pressure. Differential pressure transmitters designed for liquid level measurements use solid-state electronics and have a two-wire 4–20 mA d-c output. The Series 1151 differential pressure transmitter manufactured by Rosemount (Minneapolis, Minnesota) uses a capacitance sensor in which capacitor plates are located on both sides of a stretched metal-sensing diaphragm. Foxboro's Model 823 transmitter uses a taut wire stretched between a measuring diaphragm and a restraining element. The Honeywell ST 3000 transmitter contains a solid-state sensing element.

All these devices are filled with silicon oil and have low gradient, corrosion-resistant barrier diaphragms on both the high and low pressure sides of the sensor.

Ultrasonic. Ultrasonic level devices are based on measuring the propagation of inaudible sound waves through air, liquids, or metals at a frequency range of 20 kHz to 4 mHz. The use of ultrasonic energy is different in on/off switches and in transmitters. Switches act on the attenuation of the acoustic signal in the gap between two crystals, while transmitters measure the time of flight of the ultrasonic pulse.

Microwave. Microwave devices utilize high frequency energy to make their measurement. The implementation of microwave energy, as with ultrasonic, is different between on/off (presence or absence) switches and transmitters (continuous measurement).

Fiber Optic. Fiber optic level switches are normally limited to free-flowing, noncoating fluids at low temperatures and pressures.

Thermal Dispersion. Thermal dispersion level switches are used on applications where multiple shifts in liquid characteristics are present. The unit is responsive only to a change in the thermal conductivity of the liquid and ignores shifts in specific gravity, dielectric, density, temperature, and pressure. Units are used for alarm signal; however, pump control may be obtained using two units with a latching relay.

Magnetostrictive. When a ferromagnetic material is subjected to a magnetic field, it expands or contracts in a predictable fashion. This phenomenon is the basis for magnetostrictive measurement. The liquid level gauge consists of three primary parts: a ferromagnetic waveguide protected by a solid outer rod, an electronics assembly that determines the product level based on the waveguide behavior, and a float containing a set of magnets that ride the outside of the gauge's outer rod.

The thermal dispersion technique can also be utilized as a continuous level monitor providing an analogue output of the level in the vessel. This is accomplished utilizing an insertion probe with a separate electronics section.

Radiation. Nuclear radiation level switches and level transmitters are primarily used where process contamination is not allowed, process media prohibits use of other technologies, or where high temperatures prohibit use of other devices. The chief advantage of the nuclear unit is that all elements are completely external to the vessel.

Phase Tracking. The principle of phase tracking uses a high frequency transmission line as a sensor. The sensor is comprised of two parallel conductors and hangs vertically in the tank.

Servo Gauge. Servo gauges are high accuracy, electromechanical devices that are used on inventory control applications where accountability is mandated for custody tranfer of liquids. The area of large API, field-erected, bulk storage vessels is where these devices originally found a niche. Servo gauges use a displacer as a primary element.

Economic Aspects

Following is a list of suppliers for level sensing technologies: sight glasses (John C. Ernst Co., Penberthy Inc.), dip sticks (B&K, Inc. Bagby Gage Pole Co.), magnetic liquid level indicators (Champ Tech, K-Tek, Magnetrol Int.), floats and displacers (Magnetrol Int., SOR, Inc.), buoyancy (Fisher Controls. Int., Inc., Magnetrol Int.), conductivity (B&W Controls, Warrick Controls, Yarway) capacitance (Bindicator, Drexelbrook, Endress & Hauser, Magnetrol Int., Princo, Robertshaw Controls) pressure/differential pressure (Foxboro, Rosemount, Smar, Honeywell, Moore Technologies), thermal dispersion (FCI, Kurz, Magnetrol Int., Sierra), magnetostrictive (MTS-Temposonics, Magnetek, Petrovend, Magnetrol Int.), phase tracking (CTI-Celtek), servo gauge (Enraf-Nonius, L&J, Whessoe-Varec, Magnetrol Int.), radiation (KayRay-Sensall, Ohmart, Ronan, Texas Nuclear) ultrasonic (Bestobell-Mobray, Endress & Hauser, Magnetrol Int., Milltronics, Sensall), microwave (Endress & Hauser, Krohne, Magnetrol Int.), microwave (Saab, TN-Canonbear, Vega, Magnetrol Int.), fiber optic (Besta, Genelco (Bindicator), Honeywell Microswitch).

JERRY BOISVERT
BOYCE CARSELLA, JR.
DAN DEVER, JR.
TED WILLIAMS
Magnetrol International, Inc.

R. C. Whitehead, *Liquid Level Measurement,* Vol. 12, 2nd ed., pp. 481–499.

V. N. Lawford, *Instrumentation Technol.* **21**(12), 30 (1974).

W. W. Schoop, *Instrumentation Control Systems* **46**(5), 73 (1973).

T. S. Imsland, *Instrumentation Control Systems* **42**(5), 120 (1969).

LITERATURE ON CHEMICAL TECHNOLOGY.

See INFORMATION RETRIEVAL; PATENTS LITERATURE.

LITHIUM AND LITHIUM COMPOUNDS

Lithium, Li, an element with unique physical and chemical properties, is useful in a wide range of applications. The estimated increase in future demand has led to the development of lower cost resources as well as additional plant openings.

Many of the properties of lithium are similar to those of magnesium and other of the alkaline-earth metals. This is in accord with the diagonal relationship principle of the Periodic Table. Resemblance to magnesium includes the high solubility of the halides (except the fluoride) in both water and polar organic solvents and the high solubility of the alkyls in hydrocarbons; the low aqueous solubility of the carbonate, phosphate, fluoride, and oxalate; the thermal instability of the carbonate and nitrate; the formation of the carbide and nitride by direct combination; and the reaction with oxygen to form the normal oxide.

There has been significant growth in lithium and lithium compounds usage since 1960. Applications of lithium metal are mainly in batteries (qv) and in the manufacture of lithium derivatives such as the hydride (see HYDRIDES), amide, nitride, and organolithium compounds.

Geochemistry

Lithium is widely distributed in nature. Trace amounts are present in many minerals, in most rocks and soils, and in many natural waters. The lithium content of the earth's crust is estimated to be from 20 to 70 ppm by weight. Ocean water contains about 0.18 ppm Li, whereas many natural brines have several hundred ppm, and a few brines contain more than 1000 ppm Li. Lithium is contained mostly in the accessory minerals, especially in clays. In igneous rocks, the lithium that is present generally is concentrated in the dark ferromagnesian minerals such as biotite, amphiboles, and pyroxenes.

Lithium-bearing minerals occur mainly in granitic pegmatites, which are coarse-grained igneous rocks composed largely of quartz, feldspar, and mica. Only spodumene, $LiAlSi_2O_6$, and petalite, $LiAlSi_4O_{10}$, are important lithium sources from minerals. Lithium also is present in some sedimentary deposits or in brines associated with granitic pegmatites.

Recovery

Recovery from Ores and Clays. The preferred method of extraction of lithium from spodumene ore is the sulfuric acid process, used on ore concentrates of 5–6% Li_2O, representing 62–74% pure spodumene. Methods suitable for extraction from spodumene also can be used for petalite, because the latter mineral converts to β-spodumene-SiO_2 solid solution on heating to a high temperature. Most other processes that have been described for the extraction of lithium can be classified as being either alkaline or ion exchange. Limestone–gypsum roasting and selective chlorination have been demonstrated to be applicable to extracting lithium from clays containing hectorite.

Recovery from Brines

Natural lithium brines are predominately chloride brines varying widely in composition. The economical recovery of lithium from such sources depends not only on the lithium content but on the concentration of interfering ions, especially calcium and magnesium. If the magnesium content is low, its removal by lime precipitation is feasible. Location and availability of solar evaporation (qv) are also important factors.

Brines from the Andes Mountains of South America, especially the Salar de Atacama in Chile containing 0.15% lithium, are the largest commercial source of lithium from brine processing. Lithium is also produced from brines at Clayton Valley, Nevada, where lithium is only present at 0.03% in brines pumped from wells of 90–210 m.

The lithium-bearing oil-field waters of southern Arkansas and eastern Texas contain high concentrations of calcium chloride. Lithium is normally recovered from brines by precipitation of lithium carbonate after solar evaporation and precipitation of other salts, but recovery by ion-exchange technology has been proposed for selected brines.

Lithium Metal

Properties. Lithium, an alkali metal, has a silvery luster, an atomic number of 3, an atomic weight of 6.941, mp 180. 5°C, bp 1336°C, and a $1s^2\ 2s^1$ electronic configuration. It is the first metallic member of Group 1 (IA) in the Periodic Table. Two stable isotopes are present in natural lithium: ^7Li having an abundance of 92.4 at. % and ^6Li, 7.6 at. %. Lithium, density = 0.531 g/cm^3 at 20°C, is the lightest of all solid elements. In general, the properties of lithium are similar to those of the other alkali metals, eg, ease of oxidation to form a univalent ion, the strongly basic property of the hydroxide, etc. In the alkali metal group, lithium has the highest melting point, boiling point, and heat capacity, and the smallest ionic radius, ie, 60 pm. The ionic radius and the resulting high ionic charge density largely account for the unusual properties and effects of lithium such as the powerful fluxing action of Li_2O, in ceramic compositions.

Thin films (qv) of lithium metal are opaque to visible light but are transparent to uv radiation. Lithium is the hardest of all the alkali metals and has a Mohs' scale hardness of 0.6. Its ductility is about the same as that of lead.

Manufacture. An electrolytic process devised in 1893 resembles the one generally used for lithium production. Molten salt electrolysis from a lithium chloride-potassium chloride mixture is performed using graphitized carbon rod as the anode and a carbon steel cell body as the cathode. Modern U.S. installations employ a 55 wt LiCl–45 wt % KCl electrolyte at about 460°C. Two grades are produced by electrolysis. The essential difference is in sodium concentration.

Health and Safety. Lithium metal, UN No. 1415, is classified by the United States Department of Transportation as "Dangerous When Wet." The required shipping label which shows this classification identifies the key hazards: emission of flammable gases on reaction with water, corrositivity to eyes and skin, and solid flammability. Liquid lithium is easily ignited in air and, once it has begun to burn, requires special techniques to extinguish. Under airtight conditions, lithium can be stored indefinitely.

Economic Aspects. Lithim metal is available commercially in ingots, special shapes, shot, and dispersions.

Inorganic Lithium Compounds

The unique properties of the lithium ion result in part from its small size and correspondingly high charge density. The ^6Li isotope which has a large neutron capture radius finds uses in thermonuclear devices, neutron shielding, and tritium production. All of these are especially important for the fusion reactors proposed for future power generation (see FUSION ENERGY). Lithium is also used for treating affective mood disorders and has been proposed for other medical applications (see PSYCHOPHARMACOLOGICAL AGENTS). The size and charge density of this lightest metal ion are expected to lead to additional uses in emerging technologies.

Principal uses of inorganic lithium compounds include glass and ceramic components, aluminum reduction, grease preparation, dehumidification compounds, and sanitizers.

Inorganic lithium compounds include lithium acetate $LiCH_3CO_2$; lithium amide, $LiNH_2$; lithium benzoate, $LiC_7H_5O_2$; lithium borate, $LiBO_2{\cdot}2H_2O$; lithium carbonate, Li_2CO_3; lithium halides; lithium fluoride, LiF; lithium chloride LiCl; lithium bromide, LiBr, and lithium iodide LiI; lithium hydride, LiH; lithium-hydroxide $LiOH{\cdot}H_2O$; lithium hyprochlorite, LiOCl; lithium niobate $LiNbO_3$; lithium nitrate, $LiNO_3$; lithium nitride, Li_3N; lithium oxide, Li_2O; lithium perchlorate, $LiClO_4$; lithium peroxide, Li_2O_2; lithium phosphate, Li_3PO_4; lithium sulfate, Li_2SO_4); lithium silicate, Li_2SiO_3; and other lithium salts.

Organolithium Compounds

Organolithium compounds are organometallic compounds in which the lithium is bonded directly to carbon. Because of the substantial covalent character in these bonds, many of the compounds exist as liquids or as low melting solids and are soluble in organic solvents, such as ethers and liquid hydrocarbons. These compounds are reactive to oxygen and moisture and may ignite spontaneously in the pure state or in concentrated solutions on exposure to air. Organolithium compounds are useful in many Grignard-type reactions employed in the synthesis of pharmaceutical and agricultural products (see GRIGNARD REACTIONS). These compounds are also used as initiators of the stereospecific polymerization of conjugated dienes and vinylaromatic compounds to produce rubbery polymers and plastics.

Organolithium compounds include *n*-butyllithium, $CH_3CH_2CH_2CH_2Li$; *sec*-butyllithium, $CH_3CH_2CH(Li)CH_3$; *tert*-butyllithium, $(CH_3)_3CLi$; hexyllithium, $CH_3CH_2CH_2CH_2CH_2CH_2Li$; methyllithium, CH_3Li; lithium acetylide, $LiC{\equiv}CH{\cdot}H_2NCH_2CH_2NH_2$; and phenyllithium, C_6H_5Li.

Handling and Toxicity of Lithium Compounds

Lithium ion is commonly ingested at dosages of 0.5 g/d of lithium carbonate for treatment of bipolar disorders. However, ingestion of higher concentrations (5 g/d of LiCl) can be fatal. As of this writing, lithium ion has not been related to industrial disease. However, lithium hydroxide, either directly or formed by hydrolysis of other salts, can cause caustic burns, and skin contact with lithium halides can result in skin dehydration. Organolithium compounds are often pyrophoric and require special handling.

CONRAD W. KAMIENSKI
Consultant
DANIEL P. MCDONALD
Catawba Valley Community College
MARSHALL W. STARK
FMC Corporation

R. D. Crozier, *8th Industrial Minerals International Congress,* Boston, 1988, Metal Bulletin, PLC, London, 1988, p. 59.

G. E. Foltz, in J. J. McKelta and W. A. Cunningham, eds., *Encyclopedia of Chemical Processing and Design,* Vol. 28, Marcel Dekker Inc., New York, 1988, pp. 324–344.

R. O. Bach, *Lithium: Current Applications in Science, Medicine and Technology,* John Wiley & Sons, New York, 1985, pp. 337–407.

NIH Publication No. 93-3476, U.S. Dept. of Health and Human Services, Washington, D.C., Jan. 1993.

LITHOGRAPHY. See LITHOGARPHIC RESISTS.

LITHOGRAPHIC RESISTS

In modern lithography, replication is effected by first coating the substrate with a photosensitive polymer film termed a *lithographic resist,* and then exposing the film to a pattern of light. Photochemically induced changes in areas of the film exposed to light alter solubility to such extent that either exposed or unexposed areas can be selectively dissolved to reveal portions of the substrate surface. If the exposed areas of resist become insoluble, then the resist is termed negative-acting; if solubility of the exposed regions increases then the resist is termed positive-acting.

In certain applications, the resist film becomes a permanent, functional component of the device being constructed, typically serving as an electrical insulator or protective encapsulant. In such cases the patterned film often is treated to enhance chemical resistance and mechanical properties. Most commonly, however, the resist pattern is a temporary template, and is removed or stripped after the image has been transferred to the substrate. Numerous methods of image transfer have been practiced, including plating and deposition (*additive* processing), chemical etching or physical milling to erode the exposed substrate (*subtractive* processing) or bombardment with energetic ions which become implanted in the substrate surface.

Lithographic resists enable a host of technological advances that have had far-ranging impacts. For example, the rapid progress in microelectronics technology in large part stems from refinement of the lithographic techniques used to fabricate computer integrated circuits.

Adaptation of materials and processes originally devised for semiconductor manufacture has allowed fabrication of sensors, complex optical and micromechanical assemblies, and devices for medical diagnostics using lithographic resists.

Essential Attributes of Lithographic Resists

Regardless of the specific application, all resists must display certain fundamental functional properties: (*1*) The resist composition must form *uniform, defect-free films* on the substrate of interest. (*2*) The coated film must display adequate *adhesion* to the substrate after coating, and through the develop and image transfer steps. (*3*) The resist must have suitable *radiation sensitivity.* (*4*) The resist must provide *high fidelity reproduction* of the mask image on the substrate. (*5*) The patterned resist must provide an effective *protective barrier* during image transfer. (*6*) After image transfer, the patterned resist must be readily and completely *removable* without substrate damage.

Historical Development of Resist Materials

Most modern lithographic resists are evolved from materials first developed for the printing industry. Increasingly specialized resists and processing techniques have been introduced as applications of lithographic technology have grown in scope and sophistication.

Dichromated Resists. The first compositions widely used as photoresists combine a photosensitive dichromate salt (usually ammonium dichromate) with a water-soluble polymer of biologic origin such as gelatin, egg albumin (proteins), or gum arabic (a starch).

Poly(vinyl cinnamate) Resists. Dichromated resists exhibit numerous shortcomings.

In the 1950s resist systems with substantially improved processing characteristics were developed. The first commercially available member of this class, KPR, introduced in 1953 by Eastman Kodak, is a cross-linking system based on the photodimerization of poly(vinyl cinnamate) chains. Pendant carbon–carbon double bonds on adjacent polymer strands undergo photocyclization to form a crosslinked, insoluble network in exposed areas.

Resists Based on Bis-Azide Cross-Linking Photochemistry. Negative-acting resist compositions that combine an unsaturated hydrocarbon polymer derived from polyisoprene with an organic aromatic bis-azide were introduced in the 1960s. These resists exhibited photosensitivity and adhesion significantly improved over available poly(vinyl cinnamate) systems.

Dry-Film Resists Based on Radical Photopolymerization. Photoinitiated polymerization (PIP) requires that special measures must be taken to apply the chemistry in lithographic applications. The attractive aspect of PIP is that each initiator species produced by photolysis launches a cascade of chemical events, effectively forming multiple chemical bonds for each photon absorbed. The gain that results constitutes a form of "chemical amplification" analogous to that observed in silver halide photography, and illustrates a path for achieving very high photosensitivities.

In 1968, E. I. du Pont de Nemours & Company introduced an innovative PIP resist system (Riston), supplied in the form of a multilayered "dry film" structure. The polymerizable layer is sandwiched between a polyolefin carrier sheet and a transparent polyester cover sheet. There are a number of commercial resist products marketed in dry-film format.

Resist Materials for Imaging with Ionizing Radiation. Although most lithographic processing uses light in the visible to uv wavelength region (known as photolithography), a number of specialized, low volume applications employ high energy radiation and have unique resist requirements as a result. These include very high resolution lithography techniques using beams of electrons or x-rays to induce chemical changes in the resist film. Electron beam or e-beam lithography is used extensively for the fabrication of high resolution photomask patterns, and for direct writing or circuit patterns for low volume, custom microelectronic applications.

In e-beam lithography, a finely focused beam of high energy electrons is scanned in a pattern across a resist-coated substrate. Collisions of impinging (primary) electrons and secondary electrons with atoms of the resist film produced ionic and radical species that undergo further reaction, modifying properties of the surrounding matrix. In x-ray lithography, initial absorption of a photon produces an energetic photoelectron whose collisions with atoms of the resist lead to the same intermediates produced by e-beam radiation.

Modern Resists for Microlithography

Most of the photoresists used in the manufacture of integrated circuits have employed diazonaphthoquinone–novolac resist products. Newer commercial resist systems designed for deep-ultraviolet applications are being used to build leading edge semiconductor products; forecasts suggest that use of such advanced materials will grow several-fold.

Positive-Tone Photoresists Based on Dissolution Inhibition by Diazonaphthoquinones. The intrinsic limitations of previous resist systems led the semiconductor industry to shift to a class of imaging materials based on diazonaphthoquinone (DNQ) photosensitizers. Both the chemistry and the imaging mechanism of these resists differ in fundamental ways from those used previously. The DNQ acts as a dissolution inhibitor for the matrix resin, a low molecular weight condensation product of formaldehyde and cresol isomers known as novolac. The phenolic structure renders the novolac polymer weakly acidic, and readily soluble in aqueous alkaline solutions. In admixture with an appropriate DNQ the polymer's dissolution rate is sharply decreased. Photolysis causes the DNQ to undergo a multistep reaction sequence, ultimately forming a base-soluble carboxylic acid which does not inhibit film dissolution. Immersion of a patternwise-exposed film of the resist in an aqueous solution of hydroxide ion leads to rapid dissolution of the exposed areas and only very slow dissolution of unexposed regions. In contrast with crosslinking resists, the film solubility is controlled by *chemical* and *polarity* differences rather than molecular size.

DNQ–novolac resists exhibit several important practical attributes. First, DNQ photochemistry is not inhibited by oxygen, so unlike free-radical based systems the resist can be exposed in noncontact modes with unimpaired imaging properties. Second, films of the resist dissolve in aqueous base by a surface-limited etching reaction, with no evidence of swelling. The developing solvent is nonflammable and water-based.

DNQ Synthesis and Properties. DNQ photosensitizers are synthesized by base-catalyzed condensation of a diazonaphthoquinone sulfonyl chloride with a mono- or polyhydroxy species to produce a sulfonate ester. Since the structures of the photolabile diazonaphthoquinone group and the photoinert ballast can be readily and independently changed using this route, many DNQ variants have been prepared.

Novolac Synthesis and Properties. Novolac resins are condensation products of phenolic monomers (typically cresols or other alkylated phenols) and formaldehyde, formed under acid catalysis.

Structural Effects. Novolacs used in DNQ-based resists often are prepared from a mixture of cresol isomers, in particular a mixture of *para-* and *meta-*cresols, rather than a single component. The polymerization chemistry is sufficiently general and flexible such that a range of phenolic homologues and derivatives have been incorporated into novolacs with the goal of improving lithographic properties.

Deep-Ultraviolet Chemically Amplified Resists based on Acid Catalysis. In any optical imaging system, the size of the smallest element that can be accurately resolved is related to the wavelength of exposing light: the smaller the wavelength, the finer the feature that can be resolved. In the microlithographic arena, improved resolution is achieved by incrementally shifting the exposure wavelength to smaller values as refinements in optics, tooling and process technology permit. The microelectronics industry is in transition from exposure tools designed to use monochromatic light at $\lambda = 436$ and 365 nm (two strong emission lines in the spectrum of mercury arc lamps) to more advanced exposure tools that use light at 248 nm (the output wavelength of a krypton fluoride excimer laser), the deep-ultraviolet or duv region. This wavelength shift has profound implications from the viewpoint of resist design. DNQ-novolac resists are impractical for duv photolithography for two reasons.

If such a film is exposed at 248 nm, the strong nonbleaching absorbance at the wavelength (due in part to the novolac matrix polymer, and in part, to the DNQ and its photoproduct) sharply attenuates the beam as it passes through the film. The result is highly nonuniform photolysis of photosensitizer in the resist film, with insufficient conversion near the substrate interface.

Second, the overall brightness of light sources available for duv exposure is much less than that of mercury arc lamps used for 365 nm wavelength. Though KrF excimer lasers are generally regarded as powerful sources of uv light, the spectral output line is relatively broad. Limitations of available lens materials make correction for chromatic aberration difficult, so the source output beam must span only a very narrow wavelength range. Introduction of optical line narrowing elements into the beam leads to a large overall attenuation. Roughly speaking, a photoresist designed to be used with duv exposure tooling now at hand must be at least tenfold more efficient in its utilization of absorbed photons than a DNQ–novolac resist. Since the quantum yield for DNQ–novolac systems is on the order of ~0.3, a 10× improvement in efficiency is a considerable challenge.

One potential approach extends the idea of chemical amplification introduced in our preceding description of dry-film resists. In 1982, it was recognized that if a photosensitizer producing an acidic product is photolyzed in a polymer matrix containing acid-labile groups, the acid will serve as a spatially localized catalyst for the formation or cleavage of chemical bonds.

Since the original proof of concept, and a later demonstration of its practical use in semiconductor manufacturing, applications and extensions of this concept have proliferated.

Photoacid Generators. Practical applications of photoacid generators (PAGs) have been actively pursued since onium salts were first reported to serve as photopolymerization catalysts in the 1970s. Onium salts are particularly useful as photoinitiators since they generate both Bronsted acids and free radicals and consequently can simultaneously initiate both cationic and radical cures. Owing to the high quantum efficiency for acid production, these ionic PAG's are well-suited for chemically amplified photoresist applications. Onium salts are one of several classes of photoacid generators. A variety of other PAG compounds (both ionic and nonionic) have been developed as researchers have sought to optimize the following key functional properties:

1. Quantum yield of acid generation
2. Photoacid characteristics

   Acid strength, pK_a

   Acid volatility

   Acid diffusion length
3. Wavelength response
4. Solubility
5. Thermal stability
6. Manufacturing costs
7. Toxicity

Ionic Photoacid Generators. Ease of synthesis, high thermal stability and good quantum yield have made sulfonium and iodonium salts the most widely used onium salts.

Nonionic Photoacid Generators. Such materials include a variety of structural types which may undergo several different photochemical rearrangements. For example, irradiation of 2,6-dinitrobenzyl ester results in an intramolecular rearrangement (the well known *ortho*-nitrobenzyl rearrangement) ultimately producing toluenesulfonic acid. In this case, no radical flux is produced upon photolysis, minimizing the possibility of crosslinking side reactions that can lead to resist scumming or even negative-tone behavior in positive resist systems.

To achieve the best overall resist performance, the optimum PAG for a given resist system, whether ionic or nonionic, must balance the functional properties listed earlier in this section. The development of new photoacid generators, and the characterization of their functional properties, are considered key to the design of resists with increased levels of performance.

Acid-Catalyzed Chemistry. Acid-catalyzed reactions form the basis for essentially all chemically amplified resist systems for microlithography applications. These reactions can be generally classified as either cross-linking (photopolymerization) or deprotection reactions. The latter are used to unmask acidic functionality such as phenolic or pendant carboxylic acid groups, and thus, lend themselves to positive tone resist applications. Acid-catalyzed polymer cross-linking and photopolymerization reactions, on the other hand, find application in negative tone resist systems.

Extending the Chemically Amplified Resist Concept

Microlithography as used in modern semiconductor manufacturing has advanced largely through evolutionary refinement of exposure tool optics and mechanics and resist materials. The switch to duv-CA resists and excimer-laser-based exposure tools in many respects represents a departure from this evolutionary process, and constitutes a considerable technical and financial challenge to the electronics industry.

Thin-Film Imaging Resists. The use of shorter exposure wavelengths and other refinements in exposure tool optics improve resolution, however, this is usually accompanied by an increased difficulty in maintaining the image pattern in focus through the thickness of the resist film; that is, there is a trade-off between exposure tool resolution and its depth-of-focus. A resist material that could be used as a very thin film (at a thickness of ca 200 nm or less), yet still would act as a useful mask throughout subsequent processing steps, would go far in circumventing the optics trade-off. Thin-film imaging (TFI) resists incorporating silicon, when coupled with resist development using oxygen-reactive ion etching (O_2–RIE), are examples of such materials.

Alternative Exposure Technologies. *Optical Enhancements.* The term wavefront engineering has been coined to describe a set of optical techniques intended to extend exposure tool resolution beyond the classical Rayleigh limit. Phase-shift mask lithography is best suited to the fabrication of devices comprised of regular arrays such as electronic memory devices. The complex patterns found in logic and processor circuits present a greater challenge in mask fabrication.

Electron-Beam and Ion-Beam Lithography. The serial nature of electron-beam lithography, where only one feature at a time is exposed, severely limits its utility as a full-scale manufacturing technology. Research is focused on increasing throughput. One such scheme is termed cell projection and makes use of a specially shaped beam to write one or more device cells in a single exposure.

Ion-beam lithography is quite similar conceptually to e-beam lithography. The larger effective mass of the accelerated ion (typically a proton) and its lower penetration depth into the resist greatly reduces scattering in the resist film compared to e-beam exposure. In consequence the ion-beam analogue has a better intrinsic resolution.

Extreme Uv (Euv) Lithography. This term denotes a projection lithography system designed to use light of wavelength near 13 nm (the soft x-ray region), produced by irradiation of a metal target with a pulsed uv laser. At this wavelength, high resolution imaging can be achieved while maintaining acceptable depth of focus if imaging optics with a low numerical aperture are used.

Proximity X-Ray. The practice wherein improved resolution is achieved by decreasing the wavelength of the imaging radiation reaches its limit with x-ray lithography. A significant investment has been made in the use of synchrotrons as light sources for lithography, using emission in the spectral range of 1–2 nm. The high transparency of organic films at this wavelength, combined with a large effective depth of focus, enable the fabrication of resist structure with high aspect ratios.

Even though synchrotrons can provide very high radiation flux, chemically amplified (CA) resists can be advantageously applied. Conventional aqueous-developing resists such as the DNQ–novolac family have been used for very high resolution x-ray lithography, but the radiation sensitivity of such systems is unsuited for manufacturing-scale throughput. The superior radiation sensitivity of CA resists both improves system utilization, and reduces mask radiation damage during exposure.

W. D. HINSBERG
G. M. WALLRAFF
R. D. ALLEN
IBM Research Division

C. G. Willson, L. Thompson and M. Bowden, eds., *Introduction to Microlithography,* 2nd ed., American Chemical Society, Washington, D.C., 1994.

W. Moreau, *Semiconductor Lithography,* Plenum Press, New York, 1988.

E. Reichmanis and A. Novembre, *Ann. Rev. Mater. Sci.* **23,** 11 (1993).

J. Shaw, M. Hatzakis, E. Babich, J. Paraszczak, D. Witman, and K. Stewart, *J. Vac. Sci. Tech. B* **7**(6), 1709 (1989).

LUBRICATION AND LUBRICANTS

The primary purpose of lubrication is separation of moving surfaces to minimize friction and wear. Several distinct regimes are commonly employed to describe the fundamental principles of lubrication. These range from dry sliding to complete separation of two moving surfaces by a fluid lubricant, with an intermediate range involving partial separation in boundary or mixed lubrication. When elastic surface deflections exert a strong influence on the nature of lubrication of a concentrated contact, as in a ball or roller bearing, a regime of elastohydrodynamic lubrication is encountered with its distinctive characteristics.

Petroleum Lubricants

Petroleum (qv) products dominate lubricant production with a 98% share of the market for lubricating oils and greases. While lower cost leads to first consideration of these petroleum lubricants, production of various synthetic lubricants covered later has been expanding to take advantage of special properties such as stability at extreme temperatures, chemical inertness, fire resistance, low toxicity, and environmental compatibility.

Petroleum oils generally range from low viscosity, with molecular weights as low as 250, to very viscous lubricants, with molecular weights up to about 1000. Physical properties and performance characteristics depend heavily on the relative distribution of paraffinic, aromatic, and alicyclic (naphthenic) components. For a given molecular size, paraffins have relatively low viscosity, low density, and higher freezing temperatures. Aromatics have higher viscosity, higher density, and darker color, and undergo rapid change in viscosity with temperature. Alicyclic oils are characterized by low pour point, low oxidation stability, and other properties intermediate to those of the paraffins and aromatics.

Almost all premium lubricants are so-called paraffinic oils composed primarily of both paraffinic and alicyclic structures, with only a minor portion of aromatics. When stabilized with an oxidation inhibitor and fortified with other appropriate additives, these paraffinic-alicyclic compositions provide nonsludging oils that are satisfactory for almost any type of service.

The first step in producing a lubricating oil involves distillation of the crude petroleum. Subsequent refining steps remove undesirable aromatics and the minor portion of sulfur, nitrogen, and oxygen compounds.

Low temperature filtration is a common final refining step to remove paraffin wax in order to lower the pour point of the oil. Finished lubricating oils are then made by blending these refined stocks to the desired viscosity, followed by introducing additives needed to provide the required performance. Table 1 lists properties of typical commercial petroleum oils. Methods for measuring these properties are available from ASTM.

Viscosity

The viscosity of an oil is its stiffness or internal friction. The general ISO international viscosity classification system for industrial oils is given in ASTM D2422 (American National Standard Z11.232). For high speed machines, ISO viscosity-grade 32 turbine and hydraulic oils are a common choice. ISO grades 68 and 100 are applied for more load capacity in slower speed machines where power loss and temperature rise are less of a question.

Oil viscosity decreases with increasing temperature in the general pattern shown in Fig. 1. The great increase in viscosity with high pressure provides the dramatic load capacity in elastohydrodynamic contacts in rolling bearings, gears, and cams at pressures ranging up to 2000 to 3000 MPa (300,000 to 450,000 psi).

Generalized pressure–temperature–viscosity relations have been developed from the extensive data for petroleum and synthetic oils.

Additives

With chemical additives being used in almost all lubricants, their worldwide production has grown to be a $5 billion segment of the chemical industry. Typical volume percentages applied in commercial petroleum lubricants, with lubricants for internal combustion engines accounting for about 72% of the market volume, include automotive and diesel engine oils, ie, straight, single SAE grade, 12%, or multigrade, 20%; automotive gear and transmission oils, 12%; hydraulic and turbine oils, 0.75%; and greases 4%.

Comprehensive reviews of additive practices are available in the literature and in extensive patent coverage. The common types of additives are discussed in approximate order of the frequency of their use.

Oxidation inhibitors. Zinc dialkyl dithiophosphates are the primary oxidation inhibitors in combining these functions with antiwear properties in automotive oils and high pressure hydraulic fluids. Their production volume is followed by aromatic amines, sulfurized olefins, and phenols.

Rust Inhibitors. For mild conditions with a small amount of water present in a large quantity of circulating oil, long-chain amines, alkyl succinic acids, and other mildly polar organic acids find use. For more severe conditions in shipping and storage of machinery, and in outdoor weather, more strongly adherent sodium and calcium sulfonates, organic phosphates, and polyhydric alcohols are used.

Antiwear and Extreme Pressure Agents. Zinc dialkyl dithiophosphates are the most widely used antiwear agents.

Friction Modifiers. The primary products used are fatty acids with 12–18 carbon atoms and fatty alcohols, or esters of fatty acids such as the glycerides of rapeseed and lard oil.

Detergents and Dispersants. Detergents are metal salts of organic acids used primarily in crankcase lubricants.

Pour-Point Depressants. The pour-point of a low viscosity paraffinic oil may be lowered by as much as 30–40°C by adding 1.0% or less of

Table 1. Representative Petroleum Lubricating Oils

Type	Viscosity, mm²/s(= cSt) 40°C	100°C	Flash point, °C	Pour point, °C	Sp gr, at 15°C	Viscosity index	Common additives[a]	Uses
automobile (SAE)								
10W	28	4.9	204	−28	0.878	106	R,O,D,VI,P,	automobile, truck,
20W	48	7.0	218	−24	0.884	103	W,F,M	and marine
30	93	10.8	228	−20	0.890	100		reciprocating
40	134	13.7	238	−16	0.895	97		engines
50	204	17.8	250	−10	0.901	94		
10W-30	62	10.3	208	−36	0.880	155		
20W-40	138	15.3	246	−21	0.897	114		railroad diesels
15W-40	108	15.0	218	−27	0.885	145		diesels
gear (SAE)								
80W-90	144	14.0	192	−22	0.900	93	EP,O,R,P,F	automotive and
85W-140	416	27.5	210	−14	0.907	91	industrial	
								gear units
automatic transmission	38	7.0	188	−40	0.867	140	R,O,W,F,VI,	automotive
							P,M	hydraulic systems
turbine								
light	31	5.4	206	−10	0.863	107	R,O	steam turbines,
medium	64	8.7	220	−6	0.876	105		electric motors,
heavy	79	9.9	230	−6	0.879	103		industrial circulating systems
hydraulic fluids								
light	30	5.3	206	−24	0.868	99	R,O,W	machine tool
medium	43	6.5	210	−23	0.871	98		hydraulic
heavy	64	8.4	216	−22	0.875	97	?	systems
extra low temp	14	5.1	96	−62	0.859	370	R,O,W,VI,P	aircraft hydraulic systems

[a] R, rust inhibitor; O, oxidation inhibitor; D, detergent–dispersant; VI, viscosity-index improver; P, pour-point depressant; W, antiwear; EP, extreme pressure; F, antifoam; and M, friction modifier.

polymethacrylates, polymers formed by Friedel-Crafts condensation of wax with alkylnaphthalene or phenols, or styrene esters.

Viscosity (Viscosity-Index) Improvers. Oils of high viscosity index (VI) can be attained by adding a few percent of a linear polymer similar to those used for pour-point depressants. The most common are polyisobutylenes, polymethacrylates, and polyalkylstyrenes.

Foam Inhibitors. Methyl silicone polymers of 300–1000 mm²/s (= cSt) at 40°C are effective additives at only 3–150 ppm for defoaming oils.

Synthetic Oils

Although synthetic fluids find applications which employ their unique individual characteristics, total production of synthetics represent only on the order of 2% of the lubricant market. Poly(α-olefin)s, esters, polyglycols, and polybutenes represent the types of primary commercial interest. The use of poly(α-olefin)s and esters would expand rapidly if synthetic oils were adopted for factory fill of automotive engines.

Properties and uses of representative synthetics appear in Table 2. In addition to considering their physical properties, selection is needed of appropriate paints, seals, hoses, plastics, and electrical insulation to avoid problems with the pronounced solvency and plasticizing action of many of these synthetic oils.

Greases

A grease is a lubricating oil that is thickened with a gelling agent, eg, a soap (qv). For design simplicity, decreased sealing requirements, and less need for maintenance, greases are almost universally given first consideration as lubricants for ball and roller bearings in electric motors, household appliances, automotive wheel bearings, machine tools, aircraft accessories, and railroad apparatus. Greases are also used for lubrication of small gear drives and for many slow speed sliding applications.

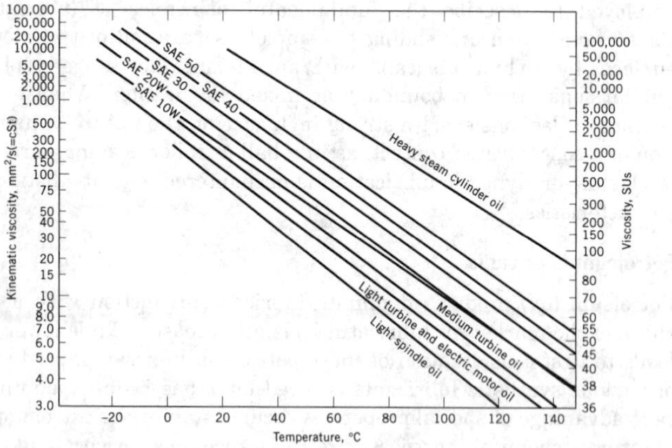

Figure 1. Variation of viscosity with temperature for selected petroleum oils.

Oils in Greases. Essentially the same type of oil is used in compounding a grease as would normally be selected for oil lubrication. Petroleum oils are used in about 99% of the grease produced.

Thickeners. Common gelling agents are the fatty acid soaps of lithium, calcium, sodium, and aluminum in concentrations of 6–

Table 2. Properties of Representative Synthetic Oils

Type	Viscosity, mm^2/s(= cSt) 100 °C	40°C	−54°C	Pour point, °C	Flash point, °C
synthetic hydrocarbons					
Mobil 1, 5W-30[a]	11	58		−54	221
SHC 824[a]	6.0	32		−54	249
SHC 629[a]	19	141		−54	238
organic esters					
MIL-L-7808	3.2	13	12,700	−62	232
MIL-L-23699	5.0	24	65,000	−56	260
MIL-L-6085	3.2	12	10,000	−68	232
Synesstic 68[b]	7.5	65		−34	266
polyglycols					
LB-300-X[c]	11	60		−40	254
50-HB-2000[c]	70	398		−32	226
phosphates					
tricresyl phosphate	4.3	31		−26	240
Fyrquel 150[d]	4.3	29		−24	236
Fyrquel 220[d]	5.0	44		−18	236
Skydrol 500B-4[e]	3.8	11	3,100	−65	182
silicones					
SF-96 (50)[f]	16	37	460	−54	316
SF-95 (1000)[f]	270	650	7,000	−48	316
F-50	16	49	2,500	−74	288
polyphenyl ether					
OS-124[e]	13	373		4	288
silicate					
Coolanol 45[e]	3.9	12	2,400	−68	188
fluorochemical					
Halocarbon 27[g]	3.7	30		−18	none
Krytox 103[h]	5.2	30		−45	none

[a] Mobil Oil Corp. [b] Exxon Corp. [c] Union Carbide Chemicals Co. [d] Akzo Chemicals. [e] Monsanto Co. [f] General Electric Co. [g] Halocarbon Products Corp. [h] DuPont Co.

25 wt %. Finely divided clay particles of the bentonite and hectorite types are also used as grease thickeners after being coated with an organic material such as quaternary ammonium compounds. Several other nonsoap powders for high temperature greases include silica gel, graphite, and polyurea powders.

Gelling action of these thickening agents varies. Oil is believed to be held in the grease structure by a combination of capillary forces, adsorption on the gel-forming molecules, and physical entrapment within fibrous interlacing crystallites in the case of fatty acid soaps. The wide variation in characteristics of petroleum greases using various thickener types is indicated in Table 3.

Additives. Chemical additives similar to those used in lubricating oils also are added to grease to improve oxidation resistance, rust protection, and extreme pressure properties.

Synthetic Grease. Synthetics are commonly employed only when their higher cost is justified by extreme temperatures or by need for special properties which cannot be achieved with petroleum greases. Severe temperature and operating requirements have led to a broad range of synthetic greases for military use.

Mechanical Properties. Greases vary in consistency from soap-thickened oils that are fluid at room temperature to hard brick-type greases that are cut with a knife.

Grade 2 greases are the most commonly used. They generally are sufficiently stiff to avoid mechanical churning which would break down their gel structures, and are adequately soft and oily to provide the lubrication needs of most bearings. Softer greases (down to Grade 000) are used where greater feeding is necessary, as with multiple row roller bearings and various gear mechanisms. Stiffer greases

Table 3. Typical Characteristics of Petroleum Greases

Base	Texture	Dropping point, °C	Continuous use, °C	Water resistant	Mechanical stability
		Soap			
Al	smooth and stringy	90	65	yes	poor
Ba	buttery or fibrous	200+	120	yes	good
Ca	smooth and buttery	100	80	yes	fair
Li	buttery to stringy	200	120	yes	good to poor
Na	buttery or fibrous	200	120	no	good to poor
Sr	buttery or fibrous	200	120	yes	good
complex	smooth and buttery	200+	120	yes	good
		Nonsoap			
modified clay	smooth	260+	140	yes	fair
silica gel	smooth	260+	140	some	poor
carbon black	smooth	260+	140	yes	good
polyurea	smooth	260+	140	yes	good

of Grade 3 consistency are used for prepacked ball bearings where the grease is held by the bearing seals in close proximity with the ball complement. Hard brick greases are applied as blocks that are inserted directly in the sleeve-bearing box, eg, in a paper mill.

Solid-Film Lubricants

These provide thin films of a solid, or a combination of solids, interposed between two moving surfaces to reduce friction and wear. They are coming into more general use for high temperatures, vacuum, nuclear radiation, aerospace, and other environments that prohibit use of oils and greases.

The wide range of solid lubricants can generally be classified as either inorganic compounds or organic polymers, both commonly used in a bonded coating on a matching substrate, plus chemical conversion coatings and metal films. Since solid-film lubricants often suffer from poor wear resistance and inability to self-heal any breaks in the film, search continues for improved compositions.

Inorganic Compounds. The most important inorganic materials are layer–lattice solids in which the bonding between atoms in an individual layer is by strong covalent or ionic forces and those between layers are relatively weak van der Waal's forces. Because of their high melting points, high thermal stabilities, low evaporation rates, good radiation resistance, and effective friction lowering ability, molybdenum disulfide, MoS_2, and graphite are the preferred choices in this group.

Organic Polymers. Self-lubricating polymers are used primarily in three ways: as thin films, as self-lubricating materials or as binders for lamellar solids. Polytetrafluoroethylene (PTFE) is outstanding in this group. Other polymers finding self-lubricating use are fluorinated ethylene–propylene copolymer (FEP), perfluoroalkoxy resin (PFA), ethylene–chlorotrifluoroethylene alternating copolymer (ECTFE), and poly(vinylidene fluoride) (PVDF).

Bonded Solid-Film Lubricants. Although a thin film of solid lubricant that is burnished onto a wearing surface often is useful for break-in operations, over 95% are resin bonded for improved life and performance.

Substrate Properties. Higher hardness of the substrate lowers friction. Wear rate of the film also is generally lower. Phosphate undercoats on steel considerably improve wear life of bonded coatings by providing a porous surface which holds reserve lubricant. The same is true for surfaces that are vapor- or sandblasted prior to application of the solid-film lubricant. Optimum surface roughness usually is 0.05–0.5 um.

Chemical Conversion Coatings. These involve inorganic surface compounds developed by chemical or electrochemical action. One of the best known treatments for steel is phosphating to coat the surface with a layer of mixed zinc, iron, and manganese phosphates. Other films are anodized oxide coatings on aluminum, oxalate on copper alloys, and various sulfides, chlorides, and fluorides.

Metal Films. In many respects, soft metals such as gallium, indium, thallium, lead, tin, gold, and silver are ideal solid lubricants. They have low shear strength, can be bonded strongly to substrate metal as continuous films, have good lubricity, and have high thermal conductivity. Metal films can be applied by electroplating or by vacuum processes, eg, evaporation, sputtering, and ion plating.

Metalworking Lubrication

The purpose of metalworking fluids is both to remove heat from the tool and workpiece and to minimize friction and wear by providing good lubricity. Most metal forming employs petroleum or synthetic oil fortified with additives to provide as much lubricating film support as possible for the high stresses involved at the workpiece contact. Polyol esters are finding broadening use in rolling steel, and poly(alkylene glycol)s and polybutenes are used as dispersions in solvents for cold rolling aluminum foil.

Extreme Ambient Conditions

Gas Lubrication. Despite severe limitations, gas lubrication of bearings has received intensive consideration for its resistance to radiation, for high speeds, temperature extremes, and use of the working fluid (gas) in a machine as its lubricant. A primary limitation is, however, the very low viscosity of gases.

Gases that have been used for bearing lubrication include air, hydrogen, helium, nitrogen, oxygen, uranium hexafluoride, carbon dioxide, and argon.

Liquid Metals. If operating temperatures rise above 250–300°C, where many organic fluids decompose and water exerts high vapor pressure, liquid metals have found some use, eg, mercury for limited application in turbines; sodium, especially its low melting eutectic with 23 wt % potassium, as a hydraulic fluid and coolant in nuclear reactors; and potassium, rubidium, cesium, and gallium in some special uses.

Cryogenic Bearing Lubrication. Cryogenic fluids, such as liquid oxygen, hydrogen, or nitrogen are used as lubricants in liquid rocket propulsion systems, turbine expanders in liquefaction and refrigeration, and pumps to transfer large quantities of liquefied gases.

Nuclear Radiation Effects. Degree of damage suffered by a lubricant depends primarily on the total radioactive energy absorbed, whether it is from neutron bombardment or from gamma radiation. The first change observed with petroleum oils (at about 10^4 gray dosage) is evolution of hydrogen and light hydrocarbon gas as fragments from the original molecule. Unsaturation results in decreased oxidation stability, cross-linking, polymerization, or scission. The general range of tolerance limit of 1 to 4×10^6 Gy($1-4 \times 10^8$ rads) for petroleum oils tends to be somewhat higher than for synthetic oils.

Conventional greases consisting of petroleum oils thickened with lithium, sodium, calcium, or other soaps suffer significant breakdown of the soap gel structure at doses above about 10^5–10^6 Gy (10^7–10^8 rad).

Lubrication with Glass. Softening glass is used as a lubricant for extrusion, forming, and other hot working processes with steel and nickel-base alloys up to about 1000°C, for extrusion and forming titanium and zirconium alloys, and less frequently for extruding copper alloys. Principal types of glasses used are pure fused silica, silica–soda–lime, borosilicates, and aluminosilicates.

Production

Total yearly production of lubricants in the United States has been fairly stable since the 1960s. The production peak of 11.2×10^6 m^3(70.7×10^6 bbl) in 1974 gradually declined to 8.9×10^6 m^3(55.9×10^6 bbl) in 1991, which is about 30% of worldwide production. Automotive lubricants make up about 56% of U.S. production, industrial lubricants 38%, and greases 2%. Future growth rate of the market is expected typically to be 1–3% per year.

Environmental and Health Factors (Toxicology)

Conservation, health, safety, and environmental pollution concerns have led to the creation of wide-reaching legislation. Regulations generally prohibit disposal of lubricants in streams, chemical dumps, or other environmental channels. Over half of disposed lubricants are burned as fuel, usually mixed with virgin residual and distillate fuels.

Waste aqueous metalworking fluids may be successfully treated by conventional means for removal of tramp oil, surfactants, and other chemical agents to provide suitable effluent water quality. Considerable effort is underway to improve and expand recycling of lubricating oils by the following procedures.

Reclamation. This involves simple separation of contaminants by gravity settling of water and dirt, centrifuging, filtering, and membrane techniques.

Reprocessing. The simplest operation involves flash distillation in an evaporator at about 100–200°C in partial vacuum to remove water and low boiling contaminants, eg, gasoline and solvents. This is followed by treatment with fuller's earth or other activated clay for removing oxidation products and most additives to produce a purified, light-colored oil which, with suitable additives, is satisfactory for use as fuel, metalworking base stocks, noncritical lubricants, and concrete form oil.

Rerefining. The technology currently attracting most attention for producing original quality lubricating oil depends on distillation in thin-film evaporators (TFE).

Food Processing. To ensure safe processing of edible food and beverage products, two federal agencies control use of food-grade lubricants: the U.S. Department of Agriculture (USDA) regulates meat and poultry plants, whereas the U.S. Food and Drug Administration (FDA) monitors other food as well as drug manufacturers (see FOOD PROCESSING).

For severe requirements where lubricants contact food on a regular basis, the FDA publishes a list of authorized ingredients in the *Codes of Federal Regulations.* These are included in three classes: (*1*) white mineral oils (21 CFR 172.878) used, for example, as release agents in bakery products, confections, dehydrated fruits and vegetables, and egg whites; (*2*) petrolatums (21 CFR 172.880) used in applications similar to white mineral oils; and (*3*) technical white oils (21 CFR 178.3620) used in processing aluminum foil for food packaging (qv), in manufacture of animal feed and fiber bags, and on food machinery.

E. R. BOOSER
Consultant

E. R. Booser, ed., *Handbook of Lubrication,* Vols. 1, 2, 3, CRC Press, Boca Raton, Fla., 1983, 1984, 1994.

D. Klamann, *Lubricants and Related Products,* Verlag Chemie, Deerfield Beach, Fla., 1984.

Friction, Lubrication, and Wear Technology, ASM Handbook, Vol. 18, ASM International, Metals Park, Ohio, 1992.

B. J. Hamrock, *Fundamentals of Fluid Film Lubrication,* NASA Reference Publication 1255, U.S. Government Printing Office, Washington, D.C., 1991.

LUMINESCENT MATERIALS

CHEMILUMINESCENCE

Chemiluminescence is the emission of light from chemical reactions at ordinary temperatures. Chemiluminescent reactions produce a reaction intermediate or product in an electronically excited state, and radiative decay of the excited state is the source of the light. When the excited state is a singlet, the radiative process is identical to fluorescence; when the excited state is a triplet, phosphorescent emission results. Electronically excited states can emit ultraviolet (uv) or infrared (ir) radiation as well as visible light, and the definition of chemiluminescence is no longer restricted to visible light emission. Moreover, the formation of electronically excited reaction products can be detected by their photochemical reactions, even when radiation is negligible. Thus chemiluminescence is a special case of the more general process of chemiexcitation. It is observed in liquid-, gas-, and solid-phase reactions.

Mechanism

The mechanism of chemiluminescence is still being studied and most mechanistic interpretations should be regarded as tentative. Nevertheless, most chemiluminescent reactions can be classified into (*1*) peroxide decomposition, including bioluminescence and peroxyoxalate chemiluminescence; (*2*) singlet oxygen chemiluminescence; and (*3*) ion radical or electron-transfer chemiluminescence, which includes electrochemiluminescence.

Energy Requirement. Visible light has an energy content of 167 kJ/ein (40 kcal/ein) (red) to 293 kJ/ein (70 kcal/ein) (blue), and an excited state radiating visible light must have that same energy with respect to its ground state. The excitation energy requirement is met by the sum of reaction enthalpy and activation energy.

The Chemiluminescent Pathway. Theory regarding the crossing of ground- to excited-state potential energy surfaces is incomplete; several potential criteria related to efficient chemiexcitation have been considered. First, since the energy released by a reaction can evolve as either vibrational or electronic excitation energy, small or rigid product molecules, which have relatively few vibrational degrees of freedom, should favor electronic excitation. Most likely, the conversion of substantial chemical energy to low energy vibrational excited states is a "forbidden" process analogous to the low probability of the transfer of excitation energy to vibrational energy when the energy gap between available electronic and vibrational quantum states is large. Thus a large energy release combined with a paucity of vibrational modes should favor electronic excitation and chemiluminescence.

Second, excited-state molecular geometry is often different from ground-state geometry. A reaction producing, eg, a bent carbonyl group, may favor chemiluminescence because the carbonyl excited state configuration is unfavorable compared to the planar ground state. Electronic excitation would then be preferred because it requires less molecular motion in the transition state. Orbital symmetry conservation and spin-orbit coupling may also be factors.

Third, singlet excitation, which is required for efficient chemiluminescence, may be favored over triplet excitation when the developing excited state is $\pi \rightarrow \pi^*$ rather than $n \rightarrow \pi^*$. Finally, electron transfer between an anion radical-cation radical pair can produce a neutral excited-state–ground-state product pair, and it has been suggested that reactions of certain peroxides with electron-rich fluorescers can produce an ion-radical pair comprising the fluorescer cation radical and a carbonyl anion radical derived from the peroxide. Electron transfer within the solvent cage then provides the electronically excited fluorescer. Alternatively, it has been suggested that electron-rich fluorescers form charge-transfer complexes with such peroxides, and that reversal of charge during peroxide decomposition is related to fluorescer excitation.

Liquid-Phase Chemiluminescence

Peroxide Decomposition. In many chemiluminescent reactions of peroxides, two carbonyl groups are formed simultaneously by decomposition of an intermediate such as compound (**1**):

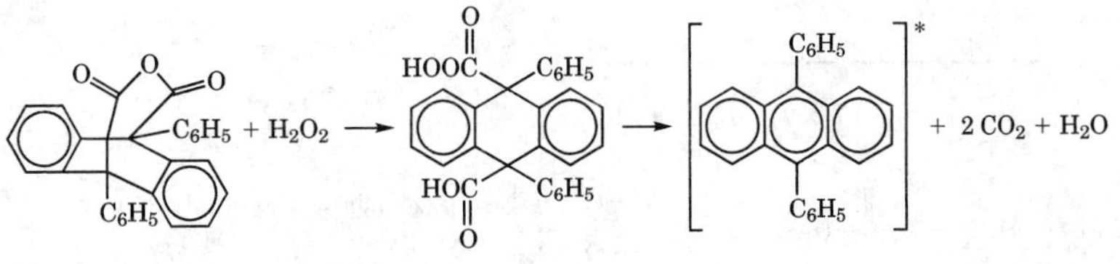

(**1**) (**2**)

In such reactions the substantial heat of the simultaneous (concerted) formation of the carbonyl groups produced meets the energy requirement. Substances that provide this reaction include 1,2-dioxetanes, α-peroxylactones (1,2-dioxetanones), peroxyoxalate, luminol (phthalhydrazide), and organometallics.

1,2-Dioxetanes. Simple dioxetanes (**3**) decompose thermally near or below room temperature to generate excited states of carbonyl products.

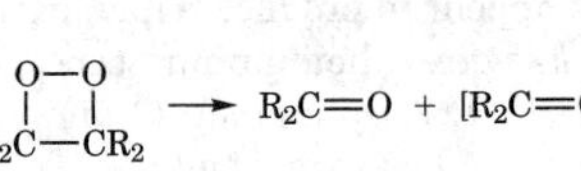

(**3**)

Excitation appears to be general for this reaction but yields of excited products vary substantially with the substituent R. The highest yield reported is from tetramethyl-1,2-dioxetane (TMD) where the yield of triplet acetone is 50% of total acetone formed. Most other dioxetanes investigated provide lower triplet yields, but some provide higher yields of excited singlets.

A range of adamantyl 1,2-dioxetanes have been synthesized that have enzyme (EH) cleavable groups (**4**, where X = 7 − $OOCCH_3$, 6-$OOCCH_3$, or 7-OPO_3Na_2. Enzyme catalyzed decomposition produces a metastable phenoxide intermediate (eg, **5**) which decomposes to produce light; the emitting moiety is the methyl 3-oxybenzoate anion (**6**). 5-Substituents (eg, Cl, Br, HO) on the adamantyl ring increase the rate of light emission from the phenoxide intermediate most likely, in part, due to a hyperconjugation effect.

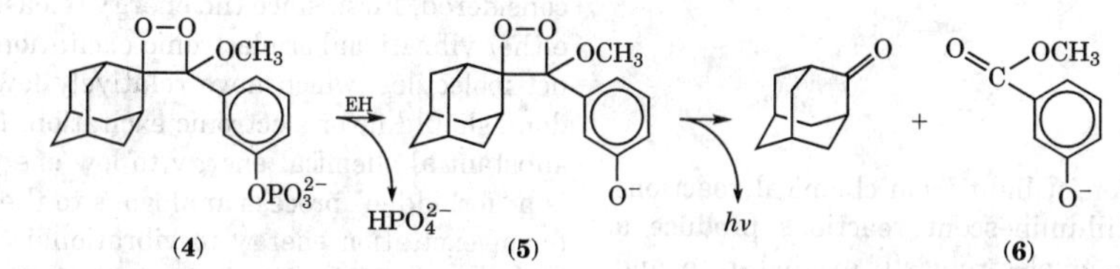

(4) (5) (6)

In addition to ready thermal decomposition, 1,2-dioxetanes are also rapidly decomposed by transition metals, amines, and electron-donor olefins. However, these catalytic reactions are not chemiluminescent as determined by the temperature drop kinetic method.

Dioxetane decomposition has also been proposed to account for chemiluminescence from other reactions, including gas-phase reactions of singlet oxygen with ethylene and vinyl ethers.

α-Peroxylactones (1,2-Dioxetanones). Alkyl-substituted 1,2-dioxetanones are prepared using low temperature techniques. The α-hydroperoxy acids can be prepared in high yield and cyclized to the dioxetanone with dicyclohexylcarbodiimide in carbon tetrachloride at low temperatures.

Dioxetanones decompose near or below room temperature to aldehydes or ketones. The decomposition reactions are weakly chemiluminescent (Qcca10^{-7} ein/mol) because the products are poorly fluorescent. However, addition of 10^{-3}M rubrene provides a Qcca10^{-3} ein/mol, and a Qc on the order of 3–7% was calculated at rubrene concentrations above 10^{-2} M after correcting for yield loss factors.

Long before 1,2-dioxetanones were isolated, they were proposed as key intermediates in bioluminescence. This idea led to the discovery of a number of new chemiluminescent reactions.

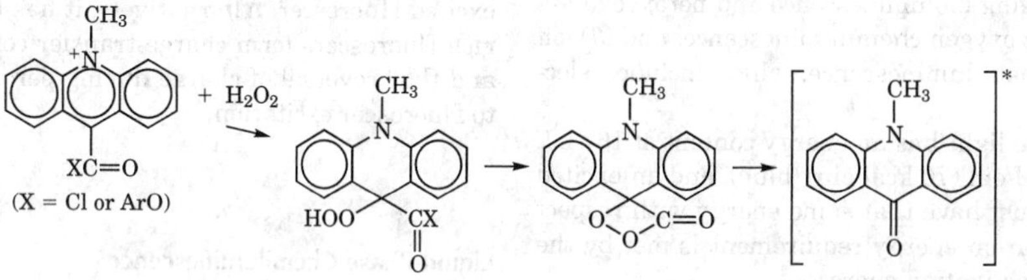

(7) (8) (9)

Peroxyoxalate. The chemical activation of a fluorescer by the reactions of hydrogen peroxide, a catalyst, and an oxalate ester has been the object of several mechanism studies. It was first proposed in 1967 that peroxyoxalate (**10**) was converted to dioxetanedione (**11**), a highly unstable intermediate which served as the chemical activator of the fluorescer (flr).

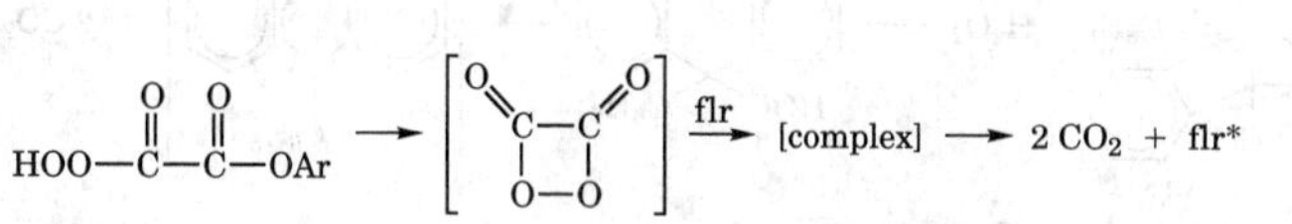

(10) (11)

flr* → flr + $h\nu$

Subsequent studies suggested that the nature of the chemical activation process was a one-electron oxidation of the fluorescer by (**11**) followed by decomposition of the dioxetanedione radical anion to a carbon dioxide radical anion. Back electron transfer to the radical cation of the fluorescer produced the excited state which emitted the luminescence characteristic of the fluorescent state of the emitter.

Peroxyoxalate chemiluminescence is the most efficient nonenzymatic chemiluminescent reaction known. Quantum efficiencies as high as 22–27% have been reported for oxalate esters prepared from 2,4,6-trichlorophenol, 2,4-dinitrophenol, and 3-trifluoromethyl-4-nitrophenol (6,76,77) with the fluorescers rubrene or 5,12-bis(phenylethynyl)naphthacene.

Most peroxyoxalate chemiluminescent reactions are catalyzed by bases and the reaction rate, chemiluminescent intensity, and chemiluminescent lifetime can be varied by selection of the base and its concentration. Weak bases such as sodium salicylate or imidazole are generally preferred. Alternatively, weak acids and certain salts have been found to extend the lifetimes of inherently rapid reactions which occur with highly reactive esters, such as bis(2,4-dinitrophenyl) oxalate.

Peroxyoxalate chemistry has been used to carry out photochemical reactions but does not appear to produce triplet excited states (91).

Luminol (Phthalhydrazide). Chemiluminescence from luminol (3-aminophthalhydrazide), (**12**) isoluminol (4-aminophthalhydrazide), and analogues has been studied extensively.

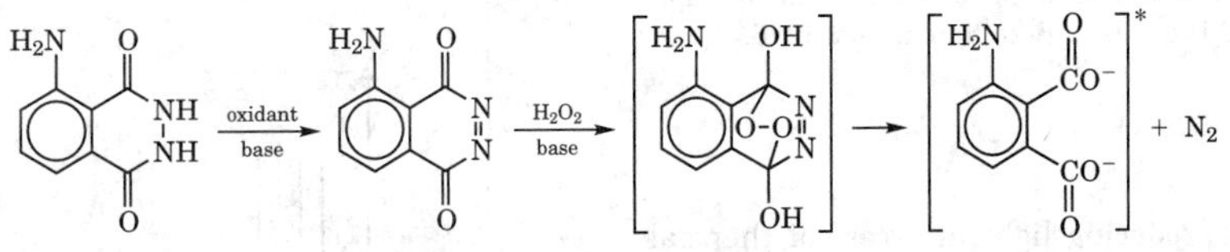

(12) (13) (14) (15)

Reaction takes place in aqueous solution with hydrogen peroxide and catalysts such as Cu(II), Cr(III), Co(II), ferricyanide, hemin, or peroxidase. Chemiluminescent reaction also takes place with oxygen and a strong base in a dipolar aprotic solvent such as dimethyl sulfoxide. Under both conditions Qc is about 1% (light emission, 375–500 nm).

The mechanism appears to follow the equation above. Dianion (**15**) has been shown to be the emitting fluorescer, and reaction of luminol (**12**) with oxygen-18 in KOH–dimethylsulfoxide produced one labeled oxygen in each carboxylate group. A kinetic study of the reaction of (**12**) with aqueous alkaline persulfate indicated a one-step, two-electron oxidation of the mono anion of structure (**12**) to the azoquinone (**13**), and the presence of structure (**13**) has been demonstrated during the chemiluminescent reaction. Compound (**13**) and several analogues have been synthesized and have been shown to be chemiluminescent under luminol conditions. A charge-transfer mechanism has been proposed.

Singlet Oxygen. The electronically excited singlet state of oxygen can be produced by passing ground-state (triplet) oxygen through a microwave discharge, by reaction of hydrogen peroxide with hypochlorite ion, by energy transfer from triplet excited states formed by irradiation to ground-state oxygen, and by low temperature thermal decomposition of the triphenyl phosphite–ozone complex. Chemiluminescence from $^1\Delta g$ oxygen can be strong at high concentrations, but addition of 5×10^{-4} M violanthrone increases the intensity 100-fold.

Electron-Transfer Chemiluminescence. Electron-transfer reactions appear to be inherently capable of producing excited products when sufficient energy is released. This ability may be related to the speed of electron transfer, which is fast relative to atomic motion, so that vibrational excitation is inhibited.

Electrochemiluminescence is somewhat complicated in that three processes can produce light, depending on the energy released by the electron-transfer process and the excitation energy of the aromatic hydrocarbon. In each case a charge-transfer complex between the oppositely charged radicals is probably formed. If sufficient energy is available, the complex can dissociate to one ground-state molecule and one excited singlet molecule, and luminescence is relatively efficient. If only enough energy is available for triplet excitation, a triplet excited state results that can produce excited singlets by triplet–triplet annihilation. If insufficient energy is released even for triplet excitation, luminescence can still be produced by excimer (excited dimer) emission from the complex itself. In the first two cases, the luminescence spectrum matches the normal fluorescence spectrum of the hydrocarbon, whereas in the latter case typical, red-shifted, broad-band excimer emission results. Excitation energy transfer from an excimer produced by electrochemiluminescence to a europium chelate has been reported to produce narrow band europium emission.

Gas-Phase Chemiluminescence

Gas-phase chemiluminescence is illustrated by the classic sodium-chlorine cool flame:

$$Na + Cl_2 \rightarrow NaCl + Cl\cdot$$

$$Cl\cdot + Na_2 \rightarrow NaCl + [Na]^*$$

Intense sodium D-line emission results from excited sodium atoms produced in a highly exothermic step. Many gas-phase reactions of the alkali metals are chemiluminescent, in part because their low ionization potentials favor electron transfer to produce intermediate charge-transfer complexes such as $[Cl^- \cdot Na_2^+]$. There appears to be an analogy with solution-phase electron-transfer chemiluminescence in such reactions.

Excitation energy can be provided by kinetic (translational) and vibrational energies as well as from reaction enthalpy as demonstrated by molecular beam experiments.

Solid-Phase Chemiluminescence

Siloxene. Siloxene is fluorescent and red chemiluminescence results from oxidation with ceric sulfate, chromic acid, potassium permanganate, nitric acid, and several other strong oxidants.

Bioluminescence

Bioluminescence is characteristic of numerous marine and a few land organisms (~666 genera from 13 phyla), extending from single-cell microorganisms such as bacteria and dinoflagellates, to marine vertebrates, such as the hatchet fish. Certain fish, such as the flashlight fish which has a light organ under its eyes, use photobacteria symbiotically to generate light. Marine bioluminescence includes sponges, worms, crustaceans, corals, snails, squids, clams, shrimp, and jellyfish. Bioluminescent land species include fungi, centipedes, millipedes, worms, beetles, and fireflies.

Bioluminescence functions in mating (fireflies, the Bahama fireworm), in the search for prey (angler fish, *Photinus* fireflies), camouflage (hatchet fish, squid), schooling (euphausiid shrimp), and to aid deep water fish (flashlight fish, *Photoblepharon*) to see in the dark ocean depths.

The chemistry of bioluminescence is complex and in general the reactions involve oxygen, a luciferin and a luciferase enzyme (eg, firefly luciferase) or a photoprotein (eg, apoaequorin).

Applications of Chemical Light

Applications include marking and illumination and analytical applications (eg, flow injection analysis and high performance liquid chromatography, direct metal analyses, titration indicators, hydrogen peroxide analysis, clinical analysis, immunoassay, nucleic acid assays, bacteria and biomass determination, oxidation analyses, and air pollution analyses).

IRENA BRONSTEIN
Tropix, Inc.
LARRY J. KRICKA
University of Pennsylvania
RICHARD S. GIVENS
University of Kansas

A. K. Campbell, *Chemiluminescence,* Horwood, Chichester, U.K., 1988.

A. K. Campbell, L. J. Kricka, and P. E. Stanley, eds., *Bioluminescence and Chemiluminescence: Fundamentals and Applied Aspects,* John Wiley & Sons, Ltd., Chichester, U.K., 1994.

K.-D. Gundermann and F. McCapra, *Chemiluminescence in Organic Chemistry,* Springer-Verlag, Berlin, 1987.

J. W. Birks, ed., *Chemiluminescence and Photochemical Reaction Detection in Chromatography,* VCH Publishers, Inc., New York, 1989.

K. van Dyke, ed., *Bioluminescence and Chemiluminescence: Instruments and Applications,* Vols. I and II, CRC Press, Inc., Boca Raton, Fla., 1985.

PHOSPHORS

Luminescence is the process of producing light in excess of thermal radiation following an excitation. A solid material exhibiting luminescence is called a phosphor. Phosphors are usually fine inorganic compound powders of a high degree of purity and a median particle size of 3–15 micrometers but may be large single crystals, used as scintillators, or glasses or thin films. Phosphors may be excited by high energy invisible uv radiation (photoluminescence), x-rays (radioluminescence), high energy electrons (cathodoluminescence), a strong electric field (electroluminescence), or in some cases infrared radiation (up-conversion), chemical reactions (chemiluminescence), or even stress (triboluminescence).

Because phosphors convert the exciting energy to visible radiation, they have many everyday applications; phosphors are responsible for the light generated by fluorescent lamps, televisions, computer terminals, etc.

Phosphors usually contain activator ions in addition to the host material. These ions are deliberately added in the proper proportion during the synthesis. The activators and their surrounding ions form the active optical centers. Table 1 lists some commonly used activator ions.

The optical properties of a phosphor are measured on relatively thick plaques of the phosphor powder. An important optical property for the application of the phosphor is its emission spectrum, the variation in the intensity of the emitted light versus wavelength. Fluorescent lamps must have phosphors which produce white light of high luminous efficiency and with good color rendering properties. Because individual activator centers generally emit in a relatively narrow region of the spectrum producing a colored light, more than one activator or phosphor must be used. Similarly colored televisions employ three phosphors in separate closely spaced dots; one dot contains a phosphor which emits in the blue, one in the green, and one in the red region of the spectrum. In other applications, such as x-ray screens, it is desirable to have an emission spectrum concentrated near the peak in the sensitivity of the receptor, such as the x-ray photographic film. The reflectance spectrum is a graph of the percentage of radiation reflected and absorbed by the powder plaque versus wavelength.

Theory of Luminescence

The Configuration Coordinate Model. To illustrate how the luminescent center in a phosphor works, a configurational coordinate diagram is used in which the potential energy of the luminescent or activator center is plotted on the vertical axis and the value of a single parameter describing an effective displacement of the ions surrounding the activator, Q, is plotted on the horizontal axis (Fig. 1).

Table 1. Common Activator Ions

Type	Important examples	Color range	Others
$s^2 \rightarrow sp$	Sb^{3+}	blue-green	Tl^+, Ga^+
broad band	Sn^{2+}	visible	Bi^{3+}, In^+
$d \rightarrow f$	Eu^{2+}	blue-green	
broad (50 nm)	Ce^{3+}	uv-green	
$O \rightarrow M$	WO_4^{2-}	460–520 nm	MoO_4^{2-}
very broad (100 nm)	VO_4^{3-}	480–580 nm	NbO_4^{3-}
$d_t \rightarrow d_e$	Mn^{2+}	510–580 nm	Mn^{4+}, Fe^{3+}
broad and narrow		green-orange	Cr^{3+}, Ni^{2+}
$f \rightarrow f$	Eu^{3+}	red	Pr^{3+}, Nd^{3+}
narrow	Tb^{3+}	green	Tm3+, Dy^{3+} Er^{3+}, Ho^{3+}

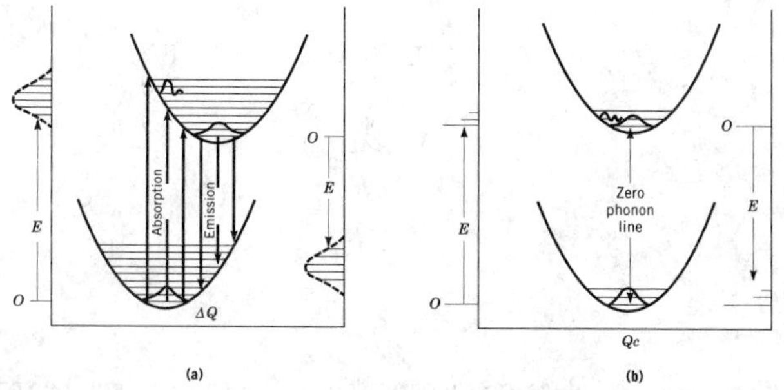

Figure 1. General configurational–coordinate diagrams for (**a**) broad-band absorbers and emitters, and (**b**) narrow-band or line emitters. The ordinate represents the total energy of the activator center and the abscissa is a generalized coordinate representing the configuration of ions surrounding the activator.

Nonradiative Decay. To have technical importance, a luminescent material should have a high efficiency for conversion of the excitation to visible light. Photoluminescent phosphors for use in fluorescent lamps usually have a quantum efficiency of greater than 0.75. All the exciting quanta would be reemitted as visible light if there were no nonradiative losses.

Energy Transfer. In addition to either emitting a photon or decaying nonradiatively to the ground state, an excited sensitizer ion may also transfer energy to another center either radiatively or nonradiatively.

Lamp Phosphors In fluorescent lamps, phosphors are coated on the inside of the lamp tube using a slurry containing the powder and a liquid which is either poured down through the tube, up-flushed, or in some cases the tubes are filled and then drained. Because of concerns over having volatile organic solvents in the air, the liquid medium containing the powder is usually water with an added agent, a thickener, to increase the viscosity of the suspension, such as poly(methacrylic acid) or poly(propylene oxide). Other additives are included, such as dispersants, in order to improve the dispersion of the powder, defoamers (qv), and sometimes powder adherence additives, such as fumed alumina, Alon, or boric oxide.

Lamp phosphors include the calcium halophosphate phosphors, deluxe phosphors, and triphosphors [the red-emitting triphosphor, the green-emitting phosphor ($CeMgAl_{11}O_{19} : Tb^{3+}$, $LaPO_4 : Ce^{3+}$, Tb^{3+}, $GdMgB_5O_{10} : Ce^{3+}$, Tb^{3+}), and the blue-emitting triphosphor components ($BaMg_2Al_{16}O_{27} : Eu^{2+} Sr_{5-x-y}Ba_xCa_y(PO_4)_3Cl : Eu^{2+}$)].

X-Ray Excited Phosphors

X-ray intensifying screens make use of phosphors that convert the high energy x-ray photons to visible radiation which sensitizes a photographic film. In order to be useful as an x-ray phosphor the material must have high x-ray absorption, high density, and the activator must emit efficiently in the blue or green spectral region to match the sensitivity of the film. Conventional screens have used $CaWO_4$ as a broad-band emitter in the uv–blue region of the spectrum.

Divalent europium-activated BaFCl was the first rare-earth-activated x-ray phosphor. Another x-ray phosphor is LaOBr activated with Tm^{3+}.

Scintillators are phosphor materials made in the form of single crystals or optically transparent polycrystalline ceramic or glass rods. These serve as detectors in computer-aided tomography (CAT) and other applications. Other rare-earth phosphors, such as $Gd_2O_2S : Pr^{3+}$, have also been used in CAT applications.

Fuji Corp. commercialized an x-ray photostimulable storage phosphor screen around 1985. In this device the bombardment of the phosphor screen by high energy x-rays generates free electrons and holes which are subsequently trapped. The stored energy can later be released by either thermal or optical stimulation. The stimulation releases the trapped charge carriers which then combine, transferring the recombination energy to a luminescent center, typically Eu^{2+}, which decays radiatively. The intensity of luminescence is proportional to the x-ray dosage. At present the x-ray storage phosphor used in nearly all commercial systems is $BaFBr:Eu^{2+}$.

Phosphors for Cathode Ray Tubes

In colored cathode ray tubes (CRTs), such as those used in televisions and computer terminals, three electron gun beams are focused on three different sets of phosphor dots on the front face of the tube. The dots are produced by using a complicated photolithography process. The phosphor dots are produced by settling the three different phosphors, each of which emits one of the primary saturated colors, red, green, or blue. Each phosphor is deposited separately and the three dots in each set are closely spaced so that the three primary colors are not resolved at normal viewing distances. Instead the viewer has the impression that there is only one color, the color achieved when the three primary colors are added together.

In the U.S. red-emitting are used for color television application by two phosphors: $Y_2O_3Eu^{3+}$ and more recently $Y_2O_2SEu^{3+}$, which has a high efficiency and a nearly ideal spectrum. For the green-emitting component, the U.S. green phosphor (Zn, Cd)S:Cu, Al is used. The blue-emitting component of most television screens and computer terminals is another sulfide, ZnS:Ag,Al.

Light-Emitting Diodes and Electroluminescence

A phosphor which generates light directly when an applied electric field is impressed across it is most desirable for flat panel displays. There are two ways this can be done with present materials. The first is to use a light-emitting diode (LED). These are single crystals usually of GaP doped with trace amounts of nitrogen. The second way to directly convert electric energy into light is with an electroluminescent phosphor. By far the best electroluminescent phosphor is $ZnS:Mn^{2+}$.

ALOK M. SRIVASTAVA
THOMAS F. SOULES
General Electric Company

K. H. Butler, *Fluorescent Lamp Phosphors,* Pennsylvania State University Press, University Park, Pa., 1980.

T. E. Peters, R. G. Pappalardo, and R. B. Hunt Jr., *Lamp Phosphors in Advances in Solid State, Phosphors, Solid State Luminescence,* Academic Press, Inc., New York, 1993.

G. Blasse and B. C. Grabmaier, Luminescent Materials, Springer-Verlag, New York, 1994.

FLUORESCENT PIGMENTS (DAYLIGHT)

Daylight-fluorescent pigments require no artificially generated energy. Daylight, or an equivalent white light, can excite these unique materials not only to reflect colored light selectively, but to give off an extra glow of fluorescent light, often with high efficiency and surprising brilliance. These pigments can also be excited with both short- and long-wave ultraviolet light. The use of a black light markedly increases the brilliance of the pigments, which makes them useful as tracers in many different applications.

Fluorescent pigments are comprised of dyed organic polymers. These polymers are clear and colorless and are formulated to be a solvent for the fluorescent dyestuff. There are many different chemical types of polymers and dyestuffs. In this article, the term dye applies to any organic substance that exhibits strong absorption of light in the visible or even ultraviolet region of the spectrum without regard to any affinity for textile fibers, paper, or other substrates (see DYES, APPLICATION AND EVALUATION).

A fluorescent substance is one that absorbs radiant energy of certain wavelengths and, after a fleeting instant, gives off part of the absorbed energy as quanta of longer wavelengths. In contrast to ordinary colors in which the absorbed energy degrades entirely to heat, light emitted from a fluorescent color adds to the light returned by simple reflection to give the extra glow characteristic of a daylight fluorescent material. This fluorescence phenomenon can lead to reflectance values greater than 100% in a specific part of the spectrum.

Availability

The primary manufacturers of daylight-fluorescent pigment at the present time are Dane and Co. (London); Day-Glo Color Corp. (Cleveland, Ohio); Nippon Keiko Kagaku Co. Ltd. (Tokyo); Nippon Shokubai (Osaka); Lawter Chemical Corp. (Skokie, Illinois); Radiant Color, Division of Magruder (Elizabeth, New Jersey); Sinloihi Co., Ltd. (Kamakura, Japan); and U.K. Seung (Busan, Korea). Smaller regional manufacturers are located in China, India, Russia, and Brazil.

Theory of Fluorescence

Structure. Virtually all important dyes contain aromatic rings in their structures. According to the theory, groups called chromophores have to be present on benzenoid rings in order for compounds to have appreciable light absorption or color. Certain basic groups, so-called auxochromes, are also necessary to bring out or intensify the color.

Chromogens. Organic dyes can be divided into four classes, depending on the type of chromogen or unsaturated system present: (*1*) $n \rightarrow \pi^*$ chromogens, (*2*) donor–acceptor chromogens, (*3*) cyanine-type chromogens, and (*4*) acyclic and cyclic polyene chromogens. Almost all strongly fluorescent dyes fall into classes (*2*) and (*3*), whereas only a few have cyclic polyene chromogens of groups (*4*). The chromogens of class (*1*) are detrimental to fluorescence.

Rigidity and Fluorescence. The more rigid the molecule, the less likely that low energy vibrations are reradiated as heat instead of light. The effect of forming a more rigid structure in fluorescent dyes of the rhodamine series has been clearly demonstrated with the remarkable dye designated Rhodamine 101.

Energy Levels and Light Absorption. A dye molecule of about 50 atoms would have ca 150 normal vibrations of the molecular skeleton. Figure 1 shows the typical transition between various energy states that the π-electrons of a dye molecule can undergo. The singlet ground state of the π-electrons in the molecule is designated S_0 and represents the lowest electronic energy level possible for the molecule. The molecule can be excited to higher electronic states such as S_1 or S_2 with an associated set of vibrational energy levels represented by a series of lines above the particular electronic level.

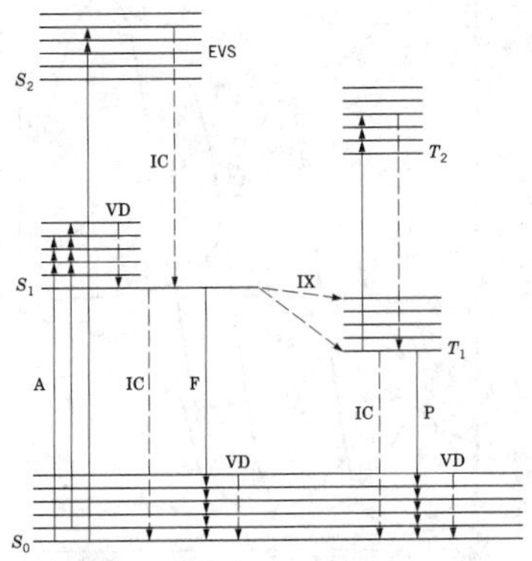

Figure 1. Schematic energy-level diagram for a dye molecule. Electronic states: S_0 = ground singlet state; S_1 = first excited singlet state; S_2 = second excited singlet state; T_1 = first excited triplet state; T_2 = second excited triplet state; EVS = excited vibrational states. Transitions: A = absorption to excited states; VD = vibrational deactivation; IC = internal conversion; F = fluorescence; IX = intersystem crossing; and P = phosphorescence

Figure 1 shows fluorescence from the excited S_1 state. After absorption (A) and vibrational deactivation (VD) occur, the lowest or nearly lowest level of the singlet excited state S_1 is reached. If the molecule is fluorescent with a high quantum efficiency, fluorescent emission of a quantum of light generally occurs, indicated by fluorescence (F).

It can be seen from Figure 1 that transitions other than fluorescence can take place from the S_1 state. The molecule can lose electronic energy by internal conversion, passing through a higher vibrational level of the S_0 state, before undergoing vibrational deactivation by surrounding molecules. The electronic excitation energy can also pass by intersystem crossing (IX) to one of the levels of the first excited triplet state, characterized by two unpaired electrons with parallel spins.

The triplet state has the relatively long lifetime of 10^{-4} s or more. When emission occurs from the triplet state, it is called phosphorescence (P). Inorganic materials can exhibit phosphorescence after a delay as long as several hours but with dyestuffs in resins any delay is quite short.

When measuring the lightfastness of fluorescent materials it is important to keep in mind several criteria. In general, the greater the concentration of the pigment the greater the lightfastness, the thicker the specimen the greater the lightfastness, and in some cases additives such as light stabilizers and antioxidants can have beneficial effects. Such additives are most beneficial when used in an overcoat to protect the pigment. Additives which have a detrimental effect are metals such as iron and zinc, and oxidizing chemicals. Some classes of fluorescent dyes and pigments are significantly more lightfast than others. Fluorescent dyes exhibit concentration quenching such that lightfastness reaches a maximum at a certain concentration, but begins to decrease above this concentration.

Color Formation

Spectral-Energy-Ratio Curves. Figure 2 shows the spectral-energy-ratio curves of three daylight-fluorescent pigments in pigment draw-downs, ie, paint samples drawn down with a bar on special panels, and a curve for a nonfluorescent ink. The lower left part of each of the first three curves is essentially the same as the transmittance or reflectance spectrum of the dye. With a strongly fluorescent substance most of the absorbed energy is stored in the S_1 excited state and is largely given off as fluorescent light of longer wavelengths covering a considerable range.

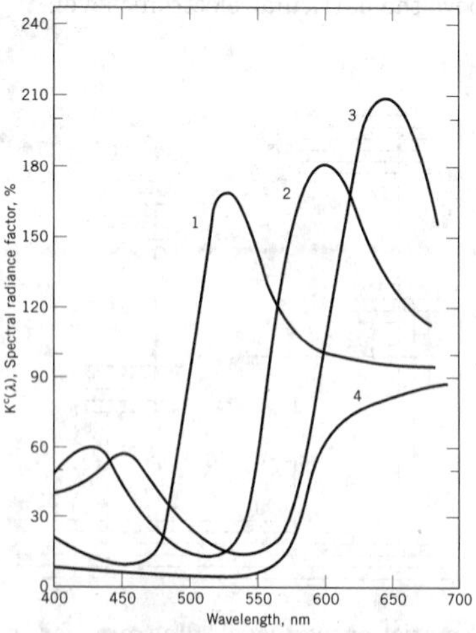

Figure 2. Curves 1, 2, and 3 show the spectral radiance factor for equivalent coatings of separate toluenesulfonamide–melamine–formaldehyde Day-Glo pigments containing 0.5% of a dye, either Alberta Yellow, Rhodamine F5G, or Rhodamine B Extra. Curve 4 is for a bright nonfluorescent red-orange printing ink. The illuminant was Source C. A magnesium oxide-coated block was used as a comparison white.

Effect of Two or More Dyes—Cascading. A most remarkable effect in daylight fluorescence, the transfer of energy from one fluorescent dye to another, can be used to produce colors more brilliant for their particular spectral regions than one dye alone could produce. For example, a yellow dye will absorb blue light and reemit yellow light, which an orange dye will absorb and reradiate as orange light.

Methods of Manufacturing

Methods of manufacturing include bulk pigment polymerization, and emulsion suspension polymerization, in both aqueous and nonaqueous media. Pigments can be either insoluble or soluble in the application system.

Economic Aspects

The market price of fluorescent pigments varies from ca $9/kg for certain grades of material that might find application in the textile and paper coating industries, to over $20/kg for special, high technology products with applications ranging from flexo and felt tip markers to plastics. Growth in the primary market segments such as packaging, safety, signage, toys, etc, approximate GNP growth in most of the world's regions, and new markets that are opening have seen substantially better growth. The total world marketplace can vary widely from year to year, being affected by textile fashions.

Application Properties and Uses

Fluorescent colors are remarkable for their extremely high visibility and their ability to attract attention, and applications utilizing these properties have gained the greatest acceptance. Advertising offers one of the main uses of fluorescent colors. Another large field for fluorescent color is for safety uses. Fluorescent color also is used in the optical-sensing field and in the coding and tracing of documents and other items.

The brilliance of daylight-fluorescent color finds use in most of the color consuming markets because of advances in use technologies. Markets served are injection molded toys, blow molded bottles, high speed sheet-fed and web-offset printing, gravure and flexo printing (water and solvent inks), industrial paint to tempra colors, felt tip pens, paper coating, textile printing and dyeing, plus a variety of specialty applications.

Commercial Properties of Fluorescent Pigments and Colorants

The largest percentage of commercial fluorescent pigments are made by bulk polymerization and are mechanically ground. The sulfonamide melamine–formaldehyde pigments generally have a density of 1.3–1.4 g/mL and average particle sizes from 2.5 to 6 μm. Melting points of the thermoplastic types range from 110–140°C; the thermosets do not melt but soften in the range of 150–170°C. These pigments decompose at about 200°C.

The other mechanically ground pigments are of the ester, amide, and other condensation-type chemistries. The average particle sizes are similar to the formaldehyde types except for toners and some plastic pigments which are coarsely ground. Melting points range from 70 to 170°C depending on the chemistry; however, the density tends to be less than the formaldehyde, in the 1.15 to 1.25 g/mL range. Decomposition of these resins tends to occur between 250 and 300°C but color degradation starts at about 200°C. Special pigments for high temperature applications such as Radiant K and Day-Glo ZQ have been developed which have better color retention properties at elevated temperature.

Plastics. Most manufacturers of fluorescent pigments offer special products for coloring thermoplastic molding resins. Products suitable for this type of use are Day-Glo ZQ Series, Radiant's K-600 and K-700 pigments; and Lawter's TC Series.

Paint. Fluorescent pigments in various types of paint offer an effective way to impart fluorescence. Because of the inferior lightfastness of fluorescent products in thin layers, the paint is generally applied in a 75–150-μm thick layer to optimize the resistance to exterior fading. For maximum color effect and durability, fluorescent paints should be applied over a high grade white substrate and overcoated with a clear uv-absorbing coating that virtually doubles the life of the color effect. The most commonly used paint systems are alkyd enamels or acrylic lacquers. For these paint systems, Day-Glo A or D, and AX Series, Radiant R-103-G, R-105, and R-203-G Series, and Lawter B-3500 Series are recommended.

Gravure Ink. Pigments suitable for this type of application, depending on the nature of the solvent used, are Day-Glo A and AX Series pigments, the Radiant R-105 and R-106 Series, and Dane A and E Series pigments. These products are useful for A-type gravure where aliphatic and small amounts of aromatic solvents are used.

In C- and T-type gravure systems where oxygenated and aromatic solvents are used, the Radiant P-1700 Series and Day-Glo GT as well as other thermoset pigments are recommended.

Flexographic Inks. Fluorescent toners such as the Radiant GF, Lawter HVT, and Day-Glo HM and HMS Series toners are used in solvent flexographic ink formulations. For water-based Flexo Radiant Aquabest and Day-Glo SPL series are used.

Screen Inks. Lawter B-3500 Series, Day-Glo A and AX Series pigments, and Radiant R-105 and R-106 Series are recommended.

Lithographic and Letterpress Printing Inks. Fluorescent pigments of fine enough particle size for litho and letterpress printing inks could not be obtained by standard techniques when starting with dry pigment. However, manufacturing spherical fluorescent particles *in situ* in a paste-ink vehicle made acceptable printing properties possible. Typical products include Day-Glo Starfire Series printing-ink bases, Radiant VF and Visiprint Series printing-ink bases, and Dane OLC Series bases.

Vinyl Products. For vinyl plastisol, organosol products and calendering, Day-Glo Color Corp. offers T, D, VC, and AX-Series pigments, Lawter Chemical offers the B-3500 and G-3000 Series, and Radiant Color offers P-1600 and R-203-G Series. In addition, Day-Glo offers VC Series for vinyl calendering where nonformaldehyde products are needed.

Health and Safety Factors

Good safety practices are recommended when handling fluorescent pigments, including a respirator and dust collecting equipment. The pigments present no unusual fire or explosion hazards. They clean up easily with detergent and water or solvents appropriate for the coating vehicle system.

STEVEN G. STREITEL
Day-Glo Color Corporation

R. Donaldson, "Spectrophotometry of Fluorescent Pigments," *Brit. J. Appl. Phys.* **5**(6), 120 (1954).

K. Venkataraman, *The Chemistry of Synthetic Dyes,* Vol. 3, Academic Press, Inc., New York, 1970, pp. 169–221.

F. Forster, *Fluoreszenz Organischer Verbindungen,* Vandenhoeck and Ruprecht, Gottingen, 1951.

Colour Index, 3rd ed., American Association of Textile Chemists and Colorists, Triangle Park, N.C., 1971–1976.

LUTETIUM. See LANTHANIDES.

M

MACHINING METHODS, ELECTROCHEMICAL

Electrochemical machining (ECM) is an electrolytic process. Metal removal is achieved by electrochemical dissolution of an anodically polarized workpiece which is one part of an electrolytic cell. As of the 1990s, ECM is employed in many ways, eg, by automotive, offshore petroleum, and medical engineering industries, as well as by aerospace firms, which are its principal user.

Theoretical Background

Electrolysis. ECM is similar to electropolishing in that ECM also is an anodic dissolution process. The rates of metal removal offered by the polishing process are, however, considerably less than those needed in metal machining practice. Two observations relevant to ECM can be made. Because the anode metal dissolves electrochemically, the rate of dissolution (or machining) depends, by Faraday's laws of electrolysis, only on the atomic weight A and valency z of the anode material, the current I which is passed, and the time t for which the current passes. Because only hydrogen gas is evolved at the cathode, the shape of that electrode remains unaltered during the electrolysis. This feature is perhaps the most relevant in the use of ECM as a metal-shaping process.

Characteristics of ECM. By use of Faraday's laws if m is the mass of metal dissolved, and because $m = \eta \rho_a$, where η is the corresponding volume and ρ_a the density of the anode metal, the volumetric removal rate of anodic metal $\dot{\eta}$ is given by

$$\dot{\eta} = \frac{AI}{zF\rho_a} \qquad (1)$$

where F, the Faraday constant, equals 96,487 C.

In ECM, electrolytes serve as conductors of electricity and Ohm's law also applies to this type of conductor. The resistance of electrolytes may amount to hundreds of ohms.

Accumulation within the small machining gap of the metallic and gaseous products of the electrolysis is undesirable. If growth were left uncontrolled, eventually a short circuit would occur between the two electrodes. To avoid this crisis, the electrolyte is pumped through the interelectrode gap so that the products of the electrolysis are carried away. The forced movement of the electrolyte is also essential in diminishing the effects both of electrical heating of the electrolyte, resulting from the passage of current and hydrogen gas, which respectively increase and decrease the effective conductivity.

The main advantages of ECM are that the rate of metal machining does not depend on the hardness of the material, complicated shapes can be machined on hard metals, and there is no tool wear.

Electrochemical Machining

Machine Components. Industrial electrochemical machines work on the principles outlined. Particular attention has to be paid to the stability of the electrochemical machine tool frame, and to the machining table which should also be stable and firm. The electrolyte has to be filtered carefully to remove the products of machining and often has to be heated in its reservoir to a fixed temperature, for instance 30°C, before entering the machining apparatus. This procedure is used to provide constant operating conditions.

Rates of Machining. Faraday's laws, embodied in equation 2, can be employed to calculate the rates at which metals can be electrochemically machined:

$$m = \frac{AIt}{zF} \qquad (2)$$

where m is the mass of metal electrochemically machined by a current I, in amperes, passed for a time t in seconds. The quantity A/zF, called the electrochemical equivalent of the anode–metal, corresponds to the atomic weight of the dissolving ions over the valency times the Faraday's constant.

Many factors other than current influence the rate of machining. These involve electrolyte type, rate of electrolyte flow, and other process conditions. If the rates of electrolyte flow are kept too low, the current efficiency of even the most easily electrochemically machined metal is reduced. Insufficient flow does not allow the products of machining to be so readily flushed from the machining gap.

Surface Finish. Besides influencing the rate of metal removal, electrolytes also affect the quality of surface finish obtained in ECM. Depending on the metal being machined, some electrolytes leave an etched finish.

In many applications, a polish is desirable on machined components. The production of an electrochemically polished surface is usually associated with the random removal of atoms from the anode workpiece, the surface of which has become covered with an oxide film. These conditions are determined by the particular metal–electrolyte combination being used. The mechanisms controlling high current density electropolishing in ECM are not completely understood.

Occasionally, metals that have undergone ECM have a pitted surface, the remaining area being polished or matte. Pitting normally stems from gas evolution at the anode electrode; the gas bubbles rupture the oxide film, causing localized pitting.

Process variables also play a significant part in determination of surface finish. For example, the higher the current density, generally the smoother the finish on the workpiece surface.

The distribution of the electric current lines leads to rounding of edges; thus, very sharp corners cannot be produced by ECM. Pulsed ECM (PECM) may be a promising way to improve dimensional accuracy control and also to simplify tool design.

Accuracy and Dimensional Control. Electrolyte selection plays an important role. Sodium chloride, for example, yields much less accurate components than sodium nitrate.

Shaping. Most metal-shaping operations in ECM utilize the same inherent feature of the process whereby one electrode, generally the cathode tool, is driven toward the other at a constant rate when a fixed voltage is applied between them. Under these conditions, the gap width between the tool and the workpiece becomes constant. The rate of forward movement between the tool and the workpiece becomes constant. The rate of forward movement of the tool is matched by the rate of recession of the workpiece surface resulting from electrochemical dissolution.

An inherent feature of ECM, whereby an equilibrium gap width is obtained, is used widely in ECM for reproducing the shape of the cathode tool on the workpiece.

Applications

ECM is used in smoothing of rough surfaces, hole drilling, full-form shaping, electrochemical grinding, and electrochemical arc, or discharge, machining.

Economic Aspects

Computer-controlled equipment and sensors (qv) are available for electrochemical machining systems. However, in the 1990s practical ECM systems are often favored because the amount of control and/or monitoring of the process is far less than that which was required in the 1960s and 1970s. Thus machines are used successfully in which electrical spark detection is eliminated and machining products control, eg, pH monitoring, is nonexistent. The measures in most industrial countries to protect the environment, however, is expected to lead to increased control of electrochemical machining products (normally called sludge), gas generation, and disposal of spent electrolyte solutions.

J. A. McGeough
X. K. Chen
University of Edinburgh

G. Bellows, *Non-Traditional Machining Guide 26 Newcomers for Production,* Metcut Research Associates Inc., Cincinnati, Ohio, 1976, pp. 28–29.

J. A. McGeough, *Principles of Electrochemical Machining,* Chapman and Hall, London, 1974.

J. A. McGeough, *Advanced Methods of Machining,* Chapman and Hall, London, 1988.

J. A. McGeough and A. DeSilva, in *Advanced Manufacturing Processes, Systems, and Technologies—Transactions of the Institution of Mechanical Engineers,* London, 1996, pp. 3–156.

MACROLIDE ANTIBIOTICS. See ANTIBIOTICS, MACROLIDES.

MAGNESIUM AND MAGNESIUM ALLOYS

Magnesium, atomic number 12, is in Group 2 (IIA) of the Periodic Table between beryllium and calcium. It has an electronic configuration of $1s^2 2s^2 2p^6 3s^2$ and a valence of two. The element occurs as three isotopes with mass numbers 24, 25, and 26.

Magnesium occurs widely in nature in the minerals dolomite, magnesite, olivine, brucite, and carnallite, and in the form of magnesium chloride in seawater, underground natural brines, and salt deposits (see also CHEMICALS FROM BRINE; MAGNESIUM COMPOUNDS; OCEAN RAW MATERIALS). Metallic magnesium is produced by electrolysis of molten magnesium chloride or thermal reduction of magnesium oxide.

Elemental magnesium is silvery white. Having a specific gravity of 1.74, it is the lightest structural metal. For engineering applications, it is alloyed with one or more elements, ie, aluminum, manganese, rare-earth metals, lithium, silver, thorium, zinc, and zirconium, to produce alloys having very high strength-to-weight ratios.

In contrast to predictions of eventual exhaustion of high grade domestic ores of many common metals, seawater is a virtually unlimited source of magnesium.

Properties

Table 1 gives some of the physical properties of 99.9% pure magnesium. Magnesium is high in the electrochemical series, having a standard potential of −2.4 V. Like most metals, it is resistant to atmospheric and chemical attack because of a stable protective film, ie, oxide, carbonate, sulfate, fluoride, and others.

Magnesium and water react at acidic pH levels, at elevated temperatures, and in the presence of salts and certain contaminants.

The ability of magnesium metal to reduce oxides of other metals can be exploited to produce metals such as zirconium, titanium, and uranium (see ZIRCONIUM AND ZIRCONIUM COMPOUNDS; TITANIUM AND TITANIUM ALLOYS; URANIUM AND URANIUM COMPOUNDS).

Manufacturing

Magnesium metal can be manufactured by electrolytic and metallothermic reduction. The method of choice depends on several variables, including raw material availability, location, and integration into other chemical facilities. Producers and corresponding capacities are shown in Table 2 (see also ELECTROCHEMICAL PROCESSING, INORGANIC).

Recycling. Substantial quantities of magnesium are recycled annually in the U.S., as well as in many other countries. The largest single recycling (qv) effort in magnesium is in the area of aluminum beverage cans. Because these cans contain around 2% magnesium by weight, this represents approximately 25,000 metric tons of magnesium per year. Most of the remaining recycled magnesium comes from die castings and from scrap generated in the die casting process. This is estimated to be 9,000–11,000 metric tons annually worldwide and the quantity is expected to grow as the volume of die castings expands. Metal coming out of the recycling industry ends up mainly in magnesium alloys or metal going to steel desulfurization.

Table 1. Properties of Magnesium

Properties	Value
atomic weight	24.31
melting point, °C	650
boiling point, °C	1103
density at 20°C, g/cm^3	1.738
electrical resistivity at 0°C, $\Omega\cdot m \times 10^{18}$	4.10
heat of fusion, 20°C, kJ/kg[a]	386
heat of vaporization, 20°C, kJ/kg[a]	5272
specific heat at 20°C, J/(kg·K)[a]	1025
surface tension, 20°C, mN/m(= dyn/cm)	563
viscosity at melting, mPa·s(= cP)	1.25

[a] To convert J to cal, divide by 4.184.

Table 2. Primary Magnesium Producers and Capacity

Producer	Process	Nominal capacity, $t \times 10^3$
Dow Magnesium	electrolytic–seawater	70
Magcorp	electrolytic–brine	34
Northwest Alloys	Magnetherm	40
Timminco	Pidgeon	12
Norsk Hydro, Canada	electrolytic–magnesite	44
Total North America		*200*
Norsk Hydro, Norway	electrolytic–seawater, brine	55
Pechiney, France	Magnetherm	17
Total Europe		*72*
CIS countries	electrolytic–carnalite	53
Brasmag, Brazil	resistance	9
China	electrolytic–magnesite and Pidgeon	30
India	Pidgeon	3
Total other		*95*
Total		*367*

Economic Aspects

The largest growth segment of the magnesium market is in automotive die castings where the lighter weight of magnesium offers significant weight and fuel savings advantages. The ability to cast intricate shapes in magnesium also allows the metal to serve as replacement for some steel fabricated parts on a cost competitive basis.

Health and Safety Factors

Magnesium articles or parts are difficult to ignite because of good thermal conductivity and high (> 450 °C) ignition temperatures. However, magnesium can be a fire hazard in the form of dust, flakes, or ribbon when exposed to flame or oxidizing agents. Magnesium is essential to most plant and animal life (see MINERAL NUTRIENTS). Dietary deficiency, rather than toxicity, is the more significant problem.

Uses

Magnesium is employed in a wide variety of applications, based on its chemical, electrochemical, physical, and mechanical properties. Its uses include nonstructural applications (aluminum alloying; hot metal desulfurization; ductile iron; and chemical, electrochemical, and metal reduction), and structural applications (die castings, gravity castings, and wrought products).

Magnesium Alloys. Magnesium alloys are most commonly designated by a system established by ASTM which covers both chemical compositions and tempers.

Table 3. Chemical Compositions and Physical Properties of Magnesium Cast and Wrought Alloys

Alloy			Nominal composition, %[a]					Physical properties	
ASTM	UNS	Temper	Al	Mn	RE[b]	Zn	Other	Density at 20°C, g/cm^3	Mp[c] °C
			Sand and permanent-mold castings						
AM100A	M10100	-T6	10.0	0.2				1.81	465
AZ63A	M11630	-F	6.0	0.2		3.0		1.82	455
		-T4							
		-T5							
		-T6							
AZ81A	M11810	-T4	7.6	0.2		0.7		1.80	510
AZ91C,E	M11914,-18	-F	8.7	0.2		0.7		1.80	470
		-T4							
		-T6							
AZ92A	M11920	-F	9.0	0.2		2.0		1.83	445
		-T4							
		-T5							
		-T6							
EZ33A	M12330	-T5			3.0	2.7	0.7 Zr	1.80	545
QE22A	M18220	-T6			2.2		2.5 Ag	1.82	550
WE43A	M18430	-T6			3.0		4.0 Y	1.84	543
WE54A	M18410	-T6			3.5		5.2 Y	1.85	549
ZE41A	M16410	-T5			1.2	4.2		1.84	510
ZE63A	M16630	-T6			2.6	5.7	0.7 Zr	1.87	515
ZK51A	M16510	-T5				4.6	0.7 Zr	1.81	550
ZK61A	M16610	-T6				6.0	0.8 Zr	1.83	520
			Die castings						
AM50A	M10500	-F	5.0	0.4				1.78	543
AE42X1		-F	4.0	0.3	2.0			1.79	565
AM60A,B	M10600,-02	-F	6.0	0.2				1.79	541
AS41A,B	M10410,-12	-F	4.2	0.3			1.0 Si	1.77	566
AZ91B,D	M11912,-16	-F	9.0	0.2		0.6		1.80	470
			Sheet and plate						
AZ31B	M11311	-F	3.0		1.0		1.77	565	
AZ31B,C	M11311,-12	-H24	3.0	0.3		1.0		1.77	565
		-H26	3.0	0.3		1.0		1.77	565
		-O	3.0	0.3		1.0		1.77	565
AZ61A	M11610	-F	6.5	0.2		1.0		1.80	510
		-O	6.5	0.2		1.0		1.80	510
			Extruded bars, rods, solid and hollow shapes, and tubes						
AZ80A	M11800	-F	8.5		0.5		1.80	490	
		-T5	8.5		0.5		1.80	4.90	
ZK60A	M16600	-F			5.7	0.55 Zr	1.83	520	
		-T5			5.7	0.55 Zr	1.83	520	

[a] Balance Mg. [b] Rare earths. [c] The solidus temperature (lower limit of alloy melting range).

Composition and Properties of Selected Alloys. Table 3 shows the chemical compositions and physical properties of the magnesium alloys used most commonly in cast and wrought form. Typical mechanical properties at 20–25°C of selected magnesium alloys in various cast forms are given in Table 4.

Heat Treatment. Heat treatment improves the properties of magnesium castings.

Metallography. Most commercial magnesium alloys are either of the solid solution or hypoeutectic type, where intermediary phases are second constituents.

Fabrication. Magnesium alloys are fabricated by common methods, including melting followed by casting, rolling, extrusion, and forging. Total energy required for the manufacture of magnesium sheet or extrusions has been estimated at 75 MJ/kg (32,090 Btu/lb) and 83 MJ/kg (35,980 Btu/lb), respectively. Further fabrication includes forming, joining, and machining after which standard assembly methods are used.

Magnesium alloys are produced from molten magnesium directly from magnesium cells or by remelting magnesium pigs or ingots in oil- or gas-fired steel pots or electric-induction furnaces.

Machining. Magnesium is the easiest of all structural metals to machine. Because of this machinability, it is sometimes used in applications where a large number of machining operations are required.

Some of the advantages of excellent machinability include reduced machining time, resulting in higher productivity for the machine tools and thus lower capital investment; greatly increased tool life; an excellent surface finish with a single large cut; well-broken chips which minimize handling costs; and less tool buildup.

Dry machining is strongly encouraged owing to the fact that the value of turnings or chips produced are significantly higher as a result of ease with which these can be recycled through remelting or through use as the desulfurization reagent.

Coolants or cutting fluids containing animal or vegetable oil must be avoided. The carboxylic acid functions present can undergo reaction with the magnesium on standing.

Corrosion and Finishing. With few exceptions, magnesium exhibits good resistance to corrosion at ambient temperatures unless there is significant water content in the environment in combination with certain contaminants.

Table 4. Mechanical Properties of Magnesium Casting Alloys[a]

Alloy							
ASTM	UNS	Temper	Tensile strength, MPa[b]	Compressive yield strength, MPa[b]	Bearing strength, MPa[b]	Shear strength, MPa[b]	Hardness, Brinell[c,d]
			Sand and permanent-mold castings				
AM100A	M10100	-F	152	83		124	53
		-T4	276	90	476	140	52
		-T5	152	110			58
		-T61	276	131	560	145	69
AZ63A	M11630	-F	200	97	415	125	50
		-T4	276	97	410	124	55
		-T5	200	97	455	130	55
		-T6	276	131	475	138	73
AZ81A	M11810	-T4	276	83	400	165	55
AZ91C,E	M11914,-18	-F	165	97	415		60
		-T4	276	90	415	150	55
		-T6	276	131	460	165	70
AZ92A	M11920	-F	172	97	345	125	65
		-T4	276	97	470	140	63
		-T5	172	117	345	140	69
		-T6	276	152	540	180	81
EZ33A	M12330	-T5	159	110	310	135	50
QE22A	M18220	-T6	276	207			78
WE43A	M18430	-T6	252	187		162	85
WE54A	M18410	-T6	275	171		150	85
ZE41A	M16410	-T5	207	138	485	150	62
ZE63A	M16630	-T6	276				
ZK51A	M16510	-T5	276	165	485	150	65
ZK61A	M16610	-T6	276				70
			Die castings				
AM50A	M10500	-F	200				58
AM60A,B	M10600,-02	-F	220				62
AS41A,B	M10410,-12	-F	210				
AZ91B,D	M11912,-16	-F	230	160		140	63

[a] Properties determined on separately cast test bars using 0.2% offset method.
[b] To convert MPa to psi, multiply by 145.
[c] See HARDNESS.
[d] 500-kg load, 10-mm ball.

CLIFFORD B. WILSON
KEN G. CLAUS
MATTHEW R. EARLAM
JAMES E. HILLIS
The Dow Chemical Company

Kh. L. Strelets, *Electrolytic Production of Magnesium,* TT76-50003, U.S. Dept. of Commerce, Technical Information Service, Springfield, Va., translated by J. Schmorak, Keter Publishing House Jerusalem Ltd., 1977, p. 1.

G. J. Kipouros and D. R. Sadoway, *Advances in Molten Salt Chemistry,* Vol. 6, Elsevier, Amsterdam, 1987.

E. F. Emley, *Principles of Magnesium Technology,* Pergamon Press, New York, 1966, p. 120.

R. S. Busk, *Magnesium Product Design,* Marcel Dekker, Inc., New York, 1987; *Annual Book of ASTM Standards,* ASTM B80, ASTM B107, ASTM B90, American Society of Testing and Materials, Philadelphia, Pa., 1992.

MAGNESIUM COMPOUNDS

There are hundreds of magnesium compounds known, varying from the chlorophyll molecule to asbestos. Only those compounds produced commercially are discussed herein.

Magnesium Acetate

Anhydrous magnesium acetate, a white, crystalline, deliquescent solid, occurs in two forms: α-$Mg(C_2H_3O_2)_2$ formed by the reaction of MgO and concentrated acetic acid (13–33%) in boiling ethyl acetate, and β-$Mg(C_2H_3O_2)_2$, which is formed using 5–6% acetic acid. Of commercial interest is magnesium acetate tetrahydrate, $Mg(C_2H_3O_2)_2 \cdot 4H_2O$, a colorless to white crystalline solid obtained from aqueous solution. Physical properties of magnesium acetate and its hydrates are given in Table 1.

The solubility of magnesium acetate is shown in Figure 1. Aqueous solutions of magnesium acetate are characterized by very high viscosities that are generally attributed to acetate association.

Uses. The largest use for magnesium acetate is in the production of rayon fiber, which is used for cigarette filter tow (see FIBERS, REGENERATED CELLULOSICS).

Magnesium acetate also has uses as a dye fixative in textile printing, as a deodorant, a disinfectant, an antiseptic in medicine, and as a reagent chemical.

Handling and Safety. Magnesium acetate is hygroscopic and should be stored in a cool, dry place. Personal protective equipment to be used when handling magnesium acetate includes chemical safety goggles, chemical resistant gloves, and a NIOSH/MSHA-approved respirator.

Magnesium Alkyls

Magnesium alkyl compounds RMg, RMgR, or RMgR′, along with other compounds are useful as polymerization catalysts.

Properties. Magnesium alkyls are white, crystalline, pyrophoric solids that react vigorously with water, alcohols, and other compounds containing an active hydrogen. Magnesium alkyls are soluble in ether

Table 1. Physical Properties of Magnesium Acetates

Property	α-$Mg(C_2H_3O_2)_2$	β-$Mg(C_2H_3O_2)_2$	$Mg(C_2H_3O_2)_2 \cdot H_2O$	$Mg(C_2H_3O_2)_2 \cdot 4H_2O$	β – Mg-$(C_2H_3O_2)_2 \cdot 4H_2O$
mol wt	142.40	142.40	160.38	214.46	214.46
crystal system	orthorhombic	triclinic	orthorhombic	monoclinic	monoclinic
lattice constants, nm			Pmcn		
a	1.127	1.034	1.175	0.8550	1.296
b	1.501	1.295	1.753	1.1995	0.7647
c	1.100	0.7726	0.6662	0.4807	1.017
angle, degree					
α	112.02				
β		94.53		95.37	113.84
γ		95.80			
density, g/cm^3	1.507	1.502		1.454	
mp, °C	323 dec			80	

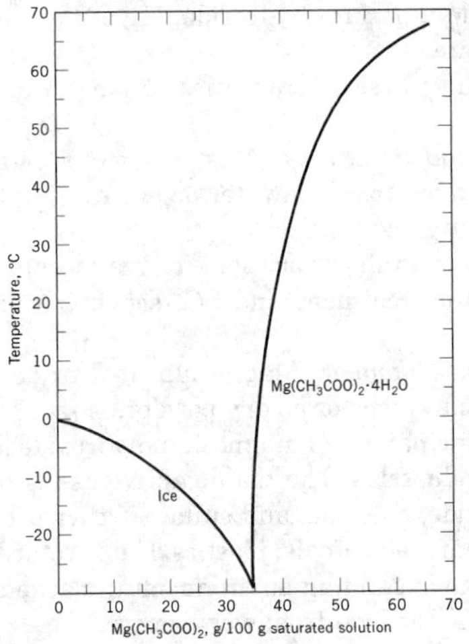

Figure 1. The $Mg(CH_3COO)_2 \cdot H_2O$ system.

solutions but insoluble in benzene and some alkane solutions, and decompose at 170–200°C.

Preparations. Magnesium alkyls may be prepared from a Grignard reagent according to the following disproportionation reaction:

$$2\ RMgX + \text{dioxane} \rightarrow R_2Mg + MgX_2$$

Reactions. The most noted magnesium alkyl reactions involve the solvated Grignard reagents. The more common reactions involving unsolvated magnesium alkyl are: oxidation, hydrolysis, peroxide addition, polymer formation, alkane formation and magnesium hydride formation.

Economic Aspects. Annual worldwide production of magnesium alkyls is <300 t.

Magnesium Bromide

Occurrence. Magnesium bromide, $MgBr_2$, is found in seawater, some mineral springs, natural brines, inland seas and lakes such as the Dead Sea and the Great Salt Lake, and salt deposits such as the Stassfurt deposits.

Properties. Physical properties of anhydrous magnesium bromide and the hexahydrate are shown in Table 2. The solubility of magnesium bromide is 101 g/100 mL of water at 20°C; the solubility of the hexahydrate is 160 g/100 mL of 95% ethanol at 20°C. Magnesium bromide is soluble is alcohols and forms addition compounds with numerous organic substances such as alcohols.

Table 2. Physical Properties of Magnesium Bromide and Magnesium Bromide Hexahydrate

Property	$MgBr_2$	$MgBr_2 \cdot 6H_2O$
mol wt	184.13	292.22
crystal system	hexagonal	monoclinic
density, calculated, g/cm^3	3.855	2.076
mp, °C	711	174.2

Uses. Magnesium bromide is used in medicine as a sedative in treatment of nervous disorders, in electrolyte paste for magnesium dry cells, and as a reagent in organic synthesis reactions.

Magnesium Carbonate

Occurrence. Chemical reactions in the system $MgO \cdot CO_2 \cdot H_2O$ result in a series of normal carbonates. In its natural form magnesite is a member of the calcite group of rhombohedral carbonates. It is the most common species of the naturally occurring magnesium carbonate minerals.

Properties. The physical properties of the normal magnesium carbonates are given in Table 3.

Production. Naturally occurring magnesite is widely distributed throughout the earth's crust and is used as a starting raw material for the production of magnesia, MgO, and other magnesium compounds.

Benefication. The purity of the mined ore can be increased by several processes of mineral benefication (see MINERALS RECOVERY AND PROCESSING). High quality magnesium carbonate is produced by means of froth flotation (qv).

Manufacture. Synthetic forms of magnesium carbonate and basic magnesium carbonates can be produced by the carbonation of magnesium hydroxide slurries.

Hydrated and basic magnesium carbonates can also be produced from calcined dolomite, $CaO \cdot MgO$, or half calcined dolomite, $MgO \cdot CaCO_3$, by means of slaking, carbonation, filtration, and decarbonation.

Table 3. Physical Properties of Magnesium Carbonates

Parameter	Magnesite	Barringtonite	Nesquehonite	Lansfordite
mol wt	84.32	120.35	138.37	174.4
crystal system	hexagonal	triclinic	monoclinic	monoclinic
density, calc, g/cm^3	3.009	2.825	1.837	1.730
hardness, Mohs'	3.5–5.0		2.5	2.5
melting point, °C	402–480[a]			

[a] Material decrepitates.

The principal producers of synthetic normal magnesium carbonates and basic magnesium carbonates are J.T. Baker Inc., Philipsburg, New Jersey; GTE Corp., Sylvannia Chemicals Division, Towanda, Pennsylvania; Mallinkrodt Specialty Chemicals, St. Louis, Missouri; Morton Specialty Chemicals, Manistee, Michigan; and Ube Chemical Industries, Tokyo.

Uses. Natural magnesites are used in the pollution control industry as acid neutralizing agents and in gaseous SO_2 scrubbers (see AIR POLLUTION CONTROL METHODS). The only processing that the natural magnesite receives for these applications is crushing and screening to a uniform size.

Precipitated magnesium carbonate and basic magnesium carbonates are calcined to produce magnesia having surface areas of ca 200 m^2/g. These high surface area magnesias are used as thickening agents in the sheet molding of rubber and as scorch retarders in chloroprene rubbers (see RUBBER, COMPOUNDING). Precipitated carbonates are useful as extenders in paint (qv) and pigments (qv).

USP-grade anhydrous magnesium carbonate is used as a flavor impression intensification vehicle in the processed food industry (see FLAVORS AND SPICES). Basic magnesium carbonates are used as free flowing agents in the manufacture of table salt, as a bulking agent in powder and tablet pharmaceutical formulations, as an antacid, and in a variety of personal care products (see PHARMACEUTICALS).

Health and Safety Factors. Magnesium carbonates and its basic hydrated forms have minimal toxicological effects when encountered at normal exposure levels. However, in response to the possible adverse effects of long-term exposure to magnesium carbonate dust, the ACGIH has established a TLV–TWA of 10 mg/m^3 for magnesite dust that contains no asbestos (qv) fibers and less than 1% free silica.

Magnesium Chloride

Properties. Magnesium chloride, $MgCl_2$, is one of the primary constituents of seawater and occurs in most natural brines and salt deposits formed from the evaporation of seawater. Magnesium chloride, one of the most commercially important magnesium compounds, is available in the anhydrous and hexahydrate forms. The physical properties of these compounds are given in Table 4. Magnesium chloride forms hydrates containing 8 and 12 molecules of water of hydration. Magnesium chloride forms double salts with potassium and ammonium chlorides.

Preparation and Manufacture. Magnesium chloride can be produced in large quantities from *(1)* carnallite or the end brines of the potash industry (see POTASSIUM COMPOUNDS); *(2)* magnesium hydroxide precipitated from seawater; *(3)* by chlorination of magnesium oxide from various sources in the presence of carbon or carbonaceous materials; and *(4)* as a by-product in the manufacture of titanium (see TITANIUM AND TITANIUM ALLOYS).

Economic Aspects. Great Salt Lake Minerals & Chemical Corp. (Utah) is the sole producer of dry magnesium chloride in the U.S.

Uses. Anhydrous magnesium chloride is used mainly as a raw material in the production of magnesium metal. Another important use of magnesium chloride is in the preparation of oxychloride cements, 5 $Mg(OH)_2 \cdot MgCl_2 \cdot 8H_2O$, for flooring (nonsparking), wall plaster compositions, fire-resistant panels, fireproofing of steel beams, and grinding wheels.

Anhydrous magnesium chloride, along with magnesium bromide and magnesium idodide, is used in a process for producing organometallic compositions such as alkyllithium compounds used as reagents in the preparation of pharmaceuticals (qv) and special chemicals.

Table 4. Physical Properties of Magnesium Chloride

Property	$MgCl_2$	$MgCl_2 \cdot 6H_2O$
mol wt	95.22	203.31
crystal system	hexagonal	monoclinic
density, calculated, g/cm^3	2.333	1.585
mp, °C	708	116–118 dec

Table 5. Physical Properties of Magnesium Hydroxide

Property	Value
mol wt	58.32
crystal system	hexagonal
density, g/cm^3, $Mg(OH)_2$	2.36
hardness, Mohs'	2.5
melting point, $Mg(OH)_2$, °C	dec 350[a]
solubility, at 25°C,[b] mg/L	11.7

[a] Begins to lose H_2O. [b] Only fair agreement between data of various authors.

Magnesium Hydroxide

Occurrence. Magnesium hydroxide, $Mg(OH)_2$, occurs naturally as the mineral brucite.

Properties. The physical properties of magnesium hydroxide are listed in Table 5.

Manufacture and Processing. Most commercial-grade magnesium hydroxide is obtained from seawater or brine using lime or dolomitic lime (see LIME AND LIMESTONE).

Uses. Environmental applications of magnesium hydroxide, primarily wastewater treatment and SO_x scrubbing of flue gases, are increasing.

Production and Shipment. Magnesium hydroxide is produced and shipped in aqueous slurry or as dry powder.

Uses. The principal use of magnesium hydroxide is in the pulp (qv) and paper (qv) industries. The main captive use is in the production of magnesium oxide, chloride, and sulfate. Other uses include ceramics, chemicals, pharmaceuticals, plastics, flame retardants/smoke suppressants, and the expanding environmental markets for wastewater treatment and SO_x removal from waste gases.

Economic Aspects. The quantity of magnesium hydroxide shipped and used in the U.S. decreased between 1990 and 1993.

Health and Safety Factors. Magnesium hydroxide is not absorbed by the skin. Dry magnesium hydroxide may irritate the eyes, skin, nasal passages, and respiratory tract. $Mg(OH)_2$ powder is classified by OSHA as a nuisance dust. ACGIH categorizes the powder form as particulates not otherwise classified. Exposure limits are as follows: ACGIH 10 mg/m^3, OSHA 5 mg/m^3 (respirable), and 15 mg/m^3 (total). Magnesium hydroxide is reported in the EPA TSCA inventory.

Magnesium Iodide

Properties. Magnesium iodide can exist as two deliquescent and heat-sensitive compounds: the octahydrate, $MgI_2 \cdot 8H_2O$, and the hexahydrate, $MgI_2 \cdot 6H_2O$. The physical properties of both compounds are shown in Table 6. Magnesium iodide is soluble in alcohols and many other organic solvents, and forms numerous addition compounds with alcohols, ethers, aldehydes, esters, and amines.

Uses. Magnesium iodide is used in the deoxygenation of oxiranes into olefins and iodine. Anhydrous MgI_2 is used in a process for producing organometallic and organobimetallic compositions, which are important in the preparation of pharmaceutical and special chemicals.

Table 6. Physical Properties of Magnesium Iodide and Hydrates

Property	MgI_2	$MgI_2 \cdot 6H_2O$	$MgI_2 \cdot 8H_2O$
mol wt	278.12	386.21	422.24
crystal system	hexagonal	monoclinic	orthorhombic
density, calc, g/cm^3	4.496	2.353	2.098
mp, °C		637 dec	43.5

Table 7. Physical Properties of Magnesium Nitrates

Property	$Mg(NO_3)_2$	$Mg(NO_3)_2 \cdot 6H_2O$
mol wt	148.32	256.38
crystal system		monoclinic
mp, °C		89

Magnesium Nitrate

Anhydrous magnesium nitrate, $Mg(NO_3)_2$, is very difficult to isolate. The commercial product is the deliquescent hexahydrate, $Mg(NO_3)_2 \cdot 6H_2O$. Properties are given in Table 7. Magnesium nitrate is prepared by dissolving magnesium oxide, hydroxide, or carbonate in nitric acid, followed by evaporation and crystallization at room temperature. Most magnesium nitrate is manufactured and used on site in other processes.

Uses. A soluble form of magnesium nitrate is used as a fertilizer in states such as Florida, where drainage through the porous, sandy soil depletes the magnesium (see FERTILIZERS). Magnesium nitrate is also used as a prilling aid in the manufacture of ammonium nitrate. Another use for magnesium nitrate is as an alternative to sulfuric acid in the purification of nitric acid.

Handling and Safety. Magnesium nitrate should be stored in a cool, dry place because it is hygroscopic. Magnesium nitrate is an acute skin, eye and respiratory irritant which can be absorbed into the body via inhalation and ingestion. Personal protection to be used when handling magnesium nitrate includes chemical safety gogles, chemical resistant gloves, and a NIOSH/MSHA approved respirator. Magnesium nitrate is a strong oxidizer and is incompatible with strong reducing agents and strong acids.

Magnesium Oxide

The principal commercial forms of magnesia are dead-burned magnesia (periclase), caustic-calcined (light-burned) magnesia, hard-burned magnesia, and calcined dolomite. These materials are usually formed by the thermal decomposition or chemical reaction of various magnesium compounds including magnesite ore, magnesium hydroxide, magnesium chloride, and synthetic magnesium carbonate. Physical properties of periclase are given in Table 8.

Properties. The properties of magnesia produced by thermal decomposition are determined by the calcination time, temperature, the nature of the magnesium-containing precursor, and other chemical compounds in the process. Increasing calcination time and temperature increases the crystallite size of the magnesia, simultaneously decreasing the surface area and reactivity of the product.

Uses. Dead-burned magnesia is used extensively for refractory applications in the form of basic granular refractories and brick. Hard-burned magnesias may be used in a variety of applications such as ceramics (qv), animal feed supplements, acid neutralization, wastewater treatment, leather (qv) tanning, magnesium phosphate cements, magnesium compound manufacturing, fertilizer, or as a raw material for fused magnesia. A patented process has introduced this material as a cation adsorbent for metals removal in wastewater treatment.

Table 8. Physical Properties of Periclase

Property	Value
mol wt	40.304
crystal form	fcc
density,[a] g/cm^3	3.581
hardness, Mohs'	5.5–6.0
melting point, °C	2827 ± 30
thermal conductivity at 100°C, J/(s·cm·°C)[b]	0.360
electrical resistivity at 27°C, Ω·cm	1.3×10^{15}
specific heat at 27°C, kJ/(kg·K)[b]	0.92885

[a] Determined by x-ray. [b] To convert J to cal, divide by 4.184.

The 1990 U.S. domestic shipments of caustic-calcined (light-burned) and specified magnesias were 36% for animal feeds and fertilizers; 19% for chemical processing; 18% for metallurgical uses such as refractories, electrical, water treatment, stack gas scrubbing, and foundry; 17% for manufacturing of rayon, fuel additives, rubber, pulp and paper, and uranium processing; 3% for construction including oxychloride and oxysulfate cement, general construction, and insulation; 3% for pharmaceuticals and nutrition, ie, medicinal and pharmaceutical usage, and use in sugar and candy; and 4% was unspecified.

Safety. Magnesium oxide (fume) has a permissible exposure limit (PEL) (8 hours, TWA), of 10 mg/m^3 total dust and 5 mg/m^3 respirable fraction. Tumorigenic data (intravenous in hamsters) show a TD_{LO} of 480 mg/kg after 30 weeks of intermittent dosing, and toxicity effects data show a TC_{LO} of 400 mg/m^3 for inhalation in humans. Magnesium oxide is compatible with most chemicals.

Magnesium Peroxide

Industrial production of magnesium peroxide, MgO_2, involves the reaction of magnesium oxide and hydrogen peroxide (qv). A product containing not more than 50% MgO_2 is obtained.

Uses. Magnesium peroxide is used mainly in medicine for treating hyperacidity in the gastric intestinal tract, and in the treatment of metabolic diseases such as diabetes and ketonuria. It is also used in the preparation of toothpaste and antiseptic ointments.

Magnesium Phosphate

An aqueous solution of monoammonium phosphate reacts with MgO to form ammonium magnesium phosphate, $NH_4 \cdot MgPO_4 \cdot 6H_2O$. Properties are given in Table 9. Magnesium phosphate is used in investment castings and magnesium phosphate cements.

Magnesium Sulfate

Magnesium sulfate, $MgSO_4$, is found widely in nature as either a double salt or as a hydrate.

Properties. Physical properties of anhydrous magnesium sulfate, kieserite, and epsomite, are listed in Table 10.

Manufacture and Processing. Anhydrous $MgSO_4$ can be prepared only by dehydration of a hydrate. Aqueous solutions of $MgSO_4$ can be prepared by dissolving MgO, $Mg(OH)_2$, or $MgCO_3$ in sulfuric acid; or absorbing SO_2 using a $Mg(OH)_2$ slurry to form the soluble bisulfite, $Mg(HSO_3)_2$, followed by air oxidation to SO_4^{-2}. Technical-grade

Table 9. Physical Properties of Magnesium Phosphates

Property	Farringtonite	Dittmarite	Struvite
mol wt	262.85	155.33	245.40
crystal system	monoclinic	orthorhombic	orthorhombic
density, calc, g/cm^3	2.76	2.19	1.706
hardness, Mohs'		2	2
melting point, °C	1184		decrepitates

Table 10. Physical Properties of Magnesium Sulfates and Magnesium Sulfate Hydrates

Property	$MgSO_4$	Kieserite	Epsomite
mol wt	120.37	138.38	246.48
crystal system	orthorhombic	monoclinic	orthorhombic
density, g/cm^3			
calculated	2.908	2.571	1.678
observed	2.93		1.677

Table 11. Properties of Magnesium Sulfite Tri- and Hexahydrates

Property	$MgSO_3 \cdot 3H_2O$	$MgSO_3 \cdot 6H_2O$
mol wt	158.42	212.47
crystal system	orthorhombic	hexagonal
calculated density, g/cm^3	2.117	1.723
mp, °C		200 dec

Table 12. Physical Properties of Magnesium Vanadates

Property	$Mg_{1.9}V_3O_8$	MgV_3O_8	$Mg_2V_2O_7$	$Mg_3V_2O_8$
mol wt	327.01	305.12	262.50	302.79
crystal system	monoclinic	orthorhombic	triclinic	orthorhombic
density, g/cm$_3$				
calculated	3.41	3.42	3.26	3.473
observed	3.37	3.39	3.1	

Epsom salt is prepared by dissolving MgO, $Mg(OH)_2$, or $MgCO_3$ in sulfuric acid. To prepare a USP-grade Epsom salt, higher purity MgO or $Mg(OH)_2$ is used.

Economic Aspects. Epsom salt is usually shipped in bulk or in 45-kg bags. Magnesium sulfate solution can be shipped in bulk, in either totes or drums.

Uses. Magnesium sulfate is used primarily in the chemical and pharmaceutical industries. Langbeinite is mined in the Carlsbad region of New Mexico. The International Minerals and Chemical Corp. (Germany) uses langbeinite to produce $MgCl_2$ by crystallizing out and decomposing carnallite, $KCl \cdot MgCl_2 \cdot 6H_2O$. Because it also contains potassium sulfate, langbeinite is used widely as a fertilizer ingredient.

Magnesium Sulfite

Preparation and Properties. The white hexahydrate $MgSO_3 \cdot 6H_2O$, is prepared by adding an excess of sulfur dioxide, SO_2, to a suspension of magnesium hydroxide, $Mg(OH)_2$, or basic magnesium carbonate, $5MgO \cdot 4\ CO_2 \cdot 5H_2O$. The properties of the tri- and hexahydrate are listed in Table 11. Uses include flue gas desulfurization and wood pulping.

Magnesium Sulfonates

Magnesium sulfonates are detergents containing magnesium carbonate or magnesium complexes as the metallic portion, and an oil-soluble magnesium-based substrate, dispersed as a colloid in petroleum oil.

Uses. Principal uses of magnesium sulfonates are as additives to engine oils, automatic transmission fluids, gear oils and industrial oils (see HYDRAULIC FLUIDS). North American producers of magnesium sulfonates include Lubrizol, Witco, and Amoco. In Europe magnesium salicyclates are an alternative detergent manufactured by Shell Chemical.

Magnesium Vanadates

Several forms of magnesium vanadates have been characterized. Some physical properties are summarized in Table 12 (see also VANADIUM AND VANADIUM ALLOYS).

Fuels from areas having natural deposits of vanadium such as Venezuela may contain significant amounts of this metal, which results in deposition of vanadium compounds in the boiler as the fuels are burned (see PETROLEUM).

The recovery of vanadium from slags is of commercial interest because of the depletion of easily accessible ores and the comparatively low concentrations (ranging from less than 100 ppm to 500 ppm) of vanadium in natural deposits.

Magnesium vanadates, like vanadium compounds in general, are known irritants of the respiratory tract and conjunctiva. The threshold limit value (TLV) for vanadium compounds in air recommended by the National Institute of Occupational Safety and Health is 0.05 mg/m^3, based on a typical 8-h workday and 40-h workweek. Chronic inhalation can lead to lung diseases such as bronchitis, bronchopneumonia, and lobar pneumonia.

L. C. JACKSON
S. P. LEVINGS
M. L. MANIOCHA
C. A. MINTMIER
A. REYES GIBSON
P. E. SCHEERER
D. M. SMITH
M. T. WAJER
M. D. WALTER
J. T. WITKOWSKI
Martin Marietta Magnesia Specialties, Inc.

R. C. Weast, ed., *Handbook of Chemistry and Physics,* 70th ed., CRC Press Inc., Boca Raton, Fla., 1989.

R. Boynton, *Chemistry and Technology of Lime and Limestone,* John Wiley & Sons, Inc., New York, 1980.

J. Wicken and L. Duncan, "Magnesite and Related Minerals," in *Industrial Minerals and Rocks,* 5th ed., American Institute of Mining, Metallurgical and Petroleum Engineers, New York, 1983.

D. Kramer, *Annual Report—Magnesium and Magnesium Compounds—1993,* U.S. Department of the Interior, Bureau of Mines, U.S. Government Printing Office, Washington, D.C., 1993, p. 19.

MAGNETIC MATERIALS

BULK

All materials that are magnetized by, ie, exhibit a response in, a magnetic field are magnetic materials and are classified according to the nature of the response, eg, as ferromagnetic or ferrimagnetic, the latter typified by the ferrites. Most commercially important magnetic materials are ferromagnets and ferrimagnets (see also FERRITES).

Soft Magnetic Materials

Soft magnetic materials are characterized by high permeability and low coercivity. There are six principal groups of commercially important soft magnetic materials: iron and low carbon steels, iron–silicon alloys, iron–aluminum and iron–aluminum–silicon alloys, nickel–iron alloys, iron–cobalt alloys, and ferrites. In addition, iron–boron-based amorphous soft magnetic alloys are commercially available. Table 1 summarizes the properties of some of these materials. Table 2 summarizes properties of some ferrites. Properties of amorphous soft magnetic alloys are listed in Table 3.

Uses of Soft Magnetic Materials. Because of low coercivity and high magnetic permeability, iron and low carbon steels tend to be used in static applications. Low carbon steels and the lower grade Fe–Si alloys are used in small motors and generators. The higher grade Fe–Si alloys have traditionally been used in power distribution transformers and large rotating machinery, but certain economical amorphous iron–metalloid alloys, because of their lower resistivity, are increasingly being used in the manufacture of distribution transformers by General Electric, Westinghouse, and Osaka. Ni–Fe alloys, used widely in high quality relays, electronic transformers, converters, and inverters in the electronics industry, have much higher permeability and much lower resistivity than Fe–Si alloys. Soft ferrites (oxides) are suitable for high frequency applications. The Co–Fe alloys are used because of their higher saturation polarization (flux density) electrical resistivity and Curie temperature compared to the iron–nickel alloys, but have the disadvantage of poorer workability and higher cost. Thus, they are used in special applications.

Table 1. Magnetic Properties of Fully Annealed Iron and Iron Alloys

Iron and alloys	B_s, T[b]	d, g/cm^3	Resistivity, $\mu\Omega\cdot$cm	H_c (B_m = 1 T),[b] A/cm[a]	Permeability, A/cm[a] H = 0.8	Permeability, A/cm[a] H = 8	Core loss (1.5 T,[b] 60 Hz), W/kg[c] 0.35 mm	Core loss 0.46 mm	Core loss 0.64 mm
magnetic ingot iron									
cast	2.15	7.85	10.7	0.68	3,500	1,500			
0.2-cm sheet	2.15	7.85	10.7	0.88	1,800	1,575			13.20
electromagnet iron, 0.2-cm sheet	2.15	7.85	12.0	0.81	2,750	1,575			
hydrogen-annealed iron	2.15	7.85	10.1	0.04	14,000	1,580			
low carbon steel, decarburized	2.14	7.85	12.5	0.70	2,000	1,530	8.10	9.2	11.44
cold-rolled									
M36 Si–Fe	2.04	7.75	41.0	0.36	7,400	1,485		3.85	4.73
M22 Si–Fe	1.98	7.65	49.0	0.31	8,100	1,450		3.63	4.29
M6 (110)[001] 3.2% Si–Fe	2.03	7.65	48.0	0.06	16,000	1,820	1.45		

[a] To convert A/cm to Oe, divide by 0.7958. [b] To convert T to G, multiply by 10^4. [c] At thickness shown.

Table 2. Characteristics of Ferrites

Property	MnZn ferrites									NiZn ferrites		
code	H5A	H5B	H5C2	H5E	H6F	H6H3	H6K	H7C1	H7C2	K5	K6A	K8
practical frequency, MHz	<0.2	<0.1	< 0.1	<0.01	0.2–2.0	0.01–0.8	0.01–0.3	<0.3	<0.2	<8	1–50	<200
initial permeability, μ_0	3,300	5,000	10,000	18,000	800	1,300	2,200	2,500	3,900	290	25	16
relative loss factor, tan $\delta/\mu_i \times 10^6$, at (kHz)	<2.5 (10)	<6.5 (10)	<7.0 (10)		<17 (1,000)	<1.2 (100)	<3.5 (100)			<28 (1,000)	<150 (10,000)	<250 (100,000)
temperature coefficient of $\mu_i \times 10^6$ from −30 to 20°C, $(\mu_2 - \mu_1)/\mu_1^2(T_2 - T_1)$	−0.5 to 2.0	−0.5 to 2.0	−0.5 to 1.5	−0.5 to 2.0		0.3 to 2.0	0.4 to 1.2			−4.0 to 2.0		
Curie temperature, °C	>130	>130	>120	>115	>200	>200	>130	>230	>200	>280	>450	>500
saturation flux density, T[a]	0.41	0.42	0.40	0.44	0.40	0.47	0.39	0.51	0.48	0.33	0.30	0.27
disaccommodation factor, $D \times 10^6$ (from 1–10 min), $(\mu_1 - \mu_2)/\mu_1^2 \log(t_2/t_1)$ where t = time	<3	<3	<1	<1	<12	<5	<2			<30	<20	
resistivity, $\Omega\cdot$m	1	1	0.15	0.05	4	25	8	10	2	20 × 10^5	2.5 × 10^5	1.0 × 10^5
applications			transformers			inductors			power supplies		inductors	

[a] To convert T to G, multiply by 10^4.

Table 3. Properties of Amorphous Magnetic Alloys

Alloy	Composition	Saturation induction, B_s, T[a]	Coercive force H_c, A/m	Magnetostriction, $\lambda_s \times 10^{-6}$
		Iron-based		
Metglas 2605SC	$Fe_{81}B_{13.5}Si_{3.5}C_2$	1.61	3.2	30
Metglas 2605S-2	$Fe_{78}B_{13}Si_9$	1.56	2.4	27
Metglas 2605CO	$Fe_{67}Co_{18}B_{14}Si_1$	1.80	4.0	35
Metglas 2605S-3	$Fe_{79}B_{16}Si_5$	1.58	8.0	27
		Iron–nickel-based		
Metglas 2826MB	$Fe_{40}Ni_{38}Mo_4B_{18}$	0.88	1.2	12
		Cobalt-based		
Metglas 2705M	$Co_{67}Ni_3Fe_4Mo_2B_{12}Si_{12}$	0.72	0.4	0.5

[a] To convert T to G, multiply by 10^4.

Soft magnetic ferrites are oxides and they are electrical insulators. Because of their exceptionally higher resistivities, ferrites are particularly suitable for high frequency applications, of about 100,000 cycles (10 kHz).

Hard Magnetic Materials

Hard or permanent magnetic materials are characterized by high coercivity and high energy product. The important commercial hard magnetic materials are hard ferrites such as ferroxdure (Table 4), rare-earth (R)-cobalt alloys, and the ternary alloys based on $Nd_2Fe_{14}B$. The last exhibits the highest coercivities and energy products. The use of Alnico and the binary R–Co alloys has continually decreased because of the high cost of cobalt. These are being replaced by the ternary NdFeB materials, including $Nd_2Fe_{14}B$. The progress in energy products for hard magnetic materials is shown in Figure 1.

Uses. Hard ferrites are used widely in electromechanical devices, eg, generators, relays, motors, and magnetos; electronic applications, eg, loudspeakers, traveling-wave tubes, and telephone ringers and receivers; antitheft tags, holding devices such as door closers, seals, and latches; and are perennial favorites in various toy designs. Loudspeakers are the largest use of permanent magnets (ca 50%). Strontum ferrites exhibit higher coercivities and are increasingly being produced.

The commercial development of magnets based on $Nd_2Fe_{14}B$ boride proceeded rapidly, and they are now being used in many diverse ap-

Table 4. Magnetic Properties of Commercial Permanent Magnet Materials

Material	T_C, °C	$(BH)_{max}$, kJ/m^{3a}	B_r, T^b	H_c, kA/m^c
Ferroxdure ($SrFe_{12}O_{19}$)	450	36	0.42	250
Alnico 9	850	72	1.05	120
$SmCo_5$	724	144	0.87	600
$Sm(Co_{0.68}Cu_{0.10}Fe_{0.21}Zr_{0.01})_{7.4}$	800	240	1.10	510
$Nd_2Fe_{14}B$	312	290	1.23	880

[a] To convert kJ/m³ to G·Oe, multiply by 12.57×10^4. [b] To convert T to G, multiply by 1×10^4. [c] To convert kA/m to Oe, divide by 7.958×10^{-2}.

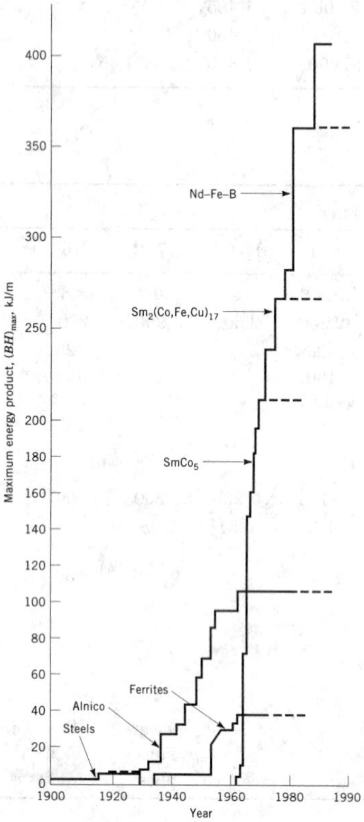

Figure 1. Progress in energy product for hard magnetic materials. To convert J to cal, divide by 4.184.

plications, for example, servo devices for machine tools, for over 30 d-c motors for fully equipped automobiles (windshield wipers, cooling fans, window and antenna lift motors, etc), magnetic resonance imaging (mri), computer disk drives, and medical device applications. Their largest use is in positioning motors for computer hard disk drives. New designs of electrical machines are now taking place.

Economic Aspects

The manufacturers of permanent magnets include Hitachi which produces Alnico Grades 5–9, in cast form in various shapes, of Grades 2 and 5 in sintered form as well as the rare-earth–cobalt (Hicorex); IG Technologies Inc., which produces Alnico, cast grades (Hyflux) 5, 8, 9 in various shapes, sintered, grades 2, 5, 8, cunife magnets, ceramic magnets (Indox) grades 1 and 5, and the rare-earth–cobalt Incor; Crucible Magnetics, which produces Alnico, cast grades 5, 7, and 8, ceramic magnets (Ferrimag), and the rare-earth–cobalt Crucore; Arnold Engineering, which produces Alnico, cast grades 5 and 8, sintered grades 2, 5, 8; GM, Delco Remy Division which produces Nd–Fe–B, limited to several shapes only (Magnaquench), isotropic and anisotropic, many grades, under the name of Permag, in the form of disks, rectangles, and squares, supplied by the Magnetic Materials Division of the Dexter Corp.; Sumitomo, which produces Nd–Fe–B and sintered magnets; 3M Co., which produces magnetic oxides, ferrites rubber bonded to form flexible permanent magnet tape with or without adhesive (Plastiform); and AlliedSignal, which produces the amphorous magnetic alloys known as Metglas.

JACK WERNICK

E. P. Wolfarth, ed., *Ferromagnetic Materials—A Handbook on the Properties of Magnetically Ordered Substances,* Vols. 1 and 2, Elsevier, New York 1980; E. P. Wolfarth and K. H.-J. Buschow, eds., Vols. 3 and 4, Elsevier, 1988.

C. W. Chen, *Magnetism and Metallurgy of Soft Magnetic Materials,* North-Holland, New York, 1977.

F. E. Luborsky, ed., *Amorphous Metallic Alloys,* Butterworths, London, 1983.

J. F. Herbst, "Permanent Magnets," *Am. Sci.,* 251 (May–June 1993).

THIN FILMS AND PARTICLES

The largest use of magnetic films and particles, in the form of tapes and disks for recording and retention of audio, visual, and digital information, is in memory and storage technologies (see INFORMATION STORAGE MATERIALS, MAGNETIC). Price per bit of information, including the cost of the peripheral electronics, and performance, as denoted by access time, generally are used to characterize the various memory technologies. Power, modular capacity, reliability, nonvolatility, etc, are also factors describing the efficacy of memories.

Magnetic Properties and Structure

The static or low frequency magnetic properties pertinent to thin-film materials generally are utilized to characterize magnetic materials. As a first approximation, these properties serve to suggest utility for device applications. Saturation magnetization M_s and Curie temperature T_C are intrinsic (structure insensitive) properties and are equal to the bulk values when thick films are made properly. For very thin highly paramagnetic films, such as those of platinum, sandwiched between ferromagnetic, eg, Co, or antiferromagnetic, eg, Cr, thin films, in the form of multilayered structures being developed for recording heads, a magnetization can be induced in the normally paramagnetic material (see MAGNETIC MATERIALS, BULK). The surface area-to-volume ratio of the individual layers is so large that the atomic moments at the interfaces play an important role.

Fabrication

Fabrication methods include thermal evaporation, sputtering, magnetron sputtering, pulsed laser evaporation, molecular beam epitaxy, chemical vapor deposition, electrolytic and electroless deposition, and growth from solution.

Materials

Magnetic storage materials for storage of audio and video information as well as of digital data are in the form of tape and disks. There are two states of remanent magnetization for recording: longitudinal, in which the magnetization is in the plane of the recording medium; and perpendicular, in which the magnetization is normal to the plane. For particulate media in which the acicular submicronic particles are single domain and embedded in plastic, the magnetization is confined in the direction of the long dimension.

Multilayer materials exhibiting high magnetization and permeability are undergoing considerable research and development for advanced recording heads. The discovery of giant magnetoresistance in multilayered nano-thick magnetic materials is expected to become important for advanced read heads.

Particulate Materials. There are three principal classes of particulate magnetic materials: γ-ferric oxide, γ-Fe_2O_3, and its modifications; chromium dioxide, CrO_2; and iron. A comparison of the remanent magnetization, B_r, and coercivity, H_c, for several material systems is shown in Table 1.

Table 1. Magnetic Properties of Common Magnetic Recording Media

Material	B_r, T[a]	H_c, kA/m[b]
γ-Fe_2O_3	0.11	26
Fe_2O_3–Fe_3O_4	0.15	37
Co–γ-Fe_2O_3	0.15	52
CrO_2	0.15	45
$BaFe_{12}O_{19}$	0.12	64
Fe	0.30	120

[a] To convert T to G, multiply by 10^4. [b] To convert kA/m to Oe, divide by 7.958×10^{-2}.

Recording Heads. Materials that are suitable for read/write recording heads for tapes and disks are characterized by high saturation flux density, low remanent induction to avoid erasure of information when the writing current ceases, and low hysteresis and low eddy-current loss, particularly for high data rates or high frequency operation. In addition, because of the small air gap between the head and recording medium, the head material should be abrasion resistant. Dust particles and the magnetic attraction between head and tape can lead to abrasion.

For general-purpose audio recording, laminated Ni–Fe alloys exhibit the required high saturation and low remanence and eddy-current losses; moreover, abrasion is low. Head wear is improved by use of precipitation hardened material. The spinel structure oxides, manganese–zinc and nickel–zinc ferrites, exhibit good abrasion resistance and high frequency characteristics and in some cases are the preferred material despite their relatively low saturation.

For high quality audio and video recording where the recording medium is CrO_2 or Co impregnated γ-Fe_2O_3, sputtered Sendust alloy films (9.6 wt % Si, 5.4 wt % Al, balance Fe; see also Table 2) and ferrites are used as head materials.

Thin-Film Magnetic Metallic Media. Advanced magnetic recording media are in the form of thin films. The metallic media are typically sputtered films having carbon overcoats for protection. Cobalt-based alloys have been developed for use as longitudinal, ie, c-axis of the crystalline Co-alloy parallel to the plane of the substrate, magnetic recording media (see COBALT AND COBALT ALLOYS). Magnetic disks are presently fabricated alloys on NiP coated aluminum alloy disk substrates.

Magnetooptic Materials. The application of magnetooptic effects to optical memory systems, such as for laser beam writing and magnetooptic read, has been the subject of much research.

Memory systems based on laser writing and reading through the interaction of electromagnetic radiation, either through reflection utilizing the Kerr effect or by transmission utilizing the Faraday effect, have begun to appear in the marketplace.

The magnetic storage media being employed are ternary amorphous alloys (Table 2) composed of the rare-earth elements gadolinium, Gd, and terbium, Tb, with Fe and Co for use in the near infrared.

Table 2. Magnetic Metals Investigated For Thin-Film Devices

Material	Composition, wt %	Application[a]
iron		MR
iron–nickel alloys	Permalloys	MR MD, T, RH
cobalt–nickel	82 Co, 18 Ni	
cobalt–phosphorus	98 Co, 2 P	R
cobalt–nickel–phosphorus	75 Co, 23 Ni, 2 P	R
iron–nickel–chromium	76 Fe, 12 Ni, 12 Cr 74 Fe, 8 Ni, 18 Cr	
Vicalloy II	13 V, 35 Fe, 52 Co	R
Cunife I	60 Cu, 20 Ni, 20 Fe	R
Cunife II	50 Cu, 20 Ni, 27.5 Fe, 2.5 Co	R
Cunico I	50 Cu, 21 Ni, 29 Co	R
Cunico II	35 Cu, 24 Ni, 41 Co	R
manganese bismuth (1:1)		TH
manganese aluminum germanide		TH
manganese gallium germanide		TH
Sendust alloy	85 Fe, 9.6 Si, 5.4 Al	TS, 12-μm films, RH, MRM
RCo(Fe) amorphous alloys[b]	variable	MRM
Co–Fe–Cr–P–C–B amorphous alloys	variable	SMB, H_c < 8 A/m[c]
Co–Cr	18–22 Cr	PR, LR
Co–Cr–Ta/Cr, Co–Pt–Cr/Cr		LR
Pt/Co or Pd/Co multilayers	ultrathin alternating layers	MRM
Pt/Fe epitaxial multilayers	3 nm Pt/2.3 nm Fe	MRM, PR
CoTa–Zr amorphous multilayers on Al_2O_3 separators	variable	HFRH
FeTaN multilayers on Al_2O_3	0.5 μm alloy/0.1 μm Al_2O_3	RH/HDTV
Co/Ni multilayers	variable	MRM, PR
Co/Au multilayers	variable	PR
$Co_{1-x}Pt_x$multilayers	$x = 0.45 - 0.9$	PA, MRM

[a] MR = magnetic recording; MD = magnetoresistive detectors; T = transducers; RH = recording heads; R = recording; TH = thermomagnetic or Curie-point writing; TS = tetrode sputtering; MRM = magnetooptic recording media or magnetooptical recording; SMB = soft magnetic behavior; PR = perpendicular recording; LR = longitudinal recording; HFRH = high frequency recording heads; RH/HDTV = recording heads for high definition television; and PA = perpendicular anisotropy. [b] Where R = Gd or Tb. [c] 8 A/m = 0.10 Oe.

Table 3. Magnetic Superlattices Exhibiting Giant Magnetoresistance

Material system	Multilayer information	Result[a,b]
Fe/Cr	var ((001) Fe/Cr(001))$_n$, 0.9–9.0-nm single-crystal layer thickness on single-crystal GaAs	resistivity lowered by factor of 2 at 4.2 K, switching fields on order of 1 T req'd
Co/Cu	var	strong antiparallel coupling through nonferromagnetic Cu; high (>800 kA/m) fields req'd to change antiferromagnetic spin structure into ferromagnetic
Co–Fe/Cu	1.0 nm Co_9Fe/1.0 nm Cu ion-beam sputtering on MgO(110) substrates	Co–Fe/Cu grew having inplane uniaxial anisotropy, easy axis parallel to cube direction in the MgO(110) plane; saturation field (240 kA/m) at RT for GMR = 45%
Ni,Fe/Cu	$Ni_{81}Fe_{19}$/Cu/Co	GMR enhanced by presence of thin Cu layer; magnetoresistance of >17%for field changes of ±8 kA/m at RT
Ni,Fe/Cu/Co	[$Ni_{80}Fe_{20}$/Cu/Co/Cu] r-f diode sputtering on Si(100) single-crystal wafers at RT	resistance changes ≤70% within a few amperes/m
NiFe/Cu/NiFe/FeMn		resistance changes of 3–4% in fields of 40–800 A/m
Co/Ag	Co(15 nm)/Ag(6.0 nm) electon-beam evaporation on top of a 5.0-nm Cr buffer layer on Si(111) substrates	interface roughness important in understanding connection between GMR and antiferromagnetic coupling

[a] To convert T to G, multiply by 10^4. [b] To convert kA/m to Oe, divide by 7.958×10^{-2}.

These materials are compatible with GaAs-based lasers. These alloys are ferrimagnetic.

The rare-earth (R) garnets, $R_3Fe_5O_{12}$, which are ferrimagnetic, are being investigated for magnetooptic recording.

Amorphous single-domain CoTaZr cores having Al_2O_3 interlayers where the CoTaZr thickness is from 0.23–0.9 μm, depending on the number of layers, and Al_2O_3 is 0.01-μm thick, were evaluated for use as thin-film heads. This material combination is attractive for low noise heads operating at frequencies up to 40 MHz.

Magnetic Superlattices. The discovery in the late 1980s of giant magnetoresistance (GMR) in antiferromagnetically coupled Fe/Cr superlattices stimulated great interest (see Table 3). Properties of metallic superlattices consisting of thin alternate single-crystal layers of different magnetic materials as well as alternate layers of magnetic and nonmagnetic materials were examined.

Magnetoresistive recording heads offer much more sensitivity than inductive heads and there is strong evidence as of this writing (ca 1994) that such heads will be used exclusively by the year 2000. The trend in the development of head materials is toward thin-film media. Although Permalloy films ($Ni_{18}Fe_{19}$) are used for magnetoresistive sensors, the change in resistance is only about 2.5%. Higher magnetoresistive materials are needed.

Magnetic Fluids. Magnetic fluids are stable colloidal suspensions of ferromagnetic particles, such as Fe_3O_4, and of subdomain size (ca 10 nm) in aqueous or organic bases. The fluid behaves as a homogeneous Newtonian liquid and reacts to a magnetic field. These materials are used in bearings, rotary-shaft seals, and feedthroughs (see BEARING MATERIALS).

JACK WERNICK

M. H. Kryder, "Data Storage in 2000: Trends in Data Storage Technologies," *IEEE Trans. Magn.* **25**(6), 4358 (1989).

M. P. Sharrock, "Particulate Magnetic Recording Media: A Review," *IEEE Trans. Magn.* **25**(6), 4374 (1989).

J. M. E. Harper, "Ion Beam Techniques in Thin Film Deposition," *Solid State Technol.*, 129 (Apr. 1987).

B. Heirich and J. A. C. Bland, eds., *Ultrathin Magnetic Structures*, Springer, Berlin, 1994.

MAGNETIC PROPERTIES. See MAGNETIC MATERIALS.

MAGNETIC SPIN RESONANCE

Magnetic spin resonance techniques are among the most powerful methods available for determining primary structure, conformation, and local dynamic properties of molecules in liquid, solid, and even gas phases of organic and inorganic molecules. These methods measure the interactions between matter and external radio frequency (rf) fields in the presence of a static external magnetic field. Under suitable conditions the measurements also may be sufficiently quantitative for analytical applications. Although much of the theory is similar, spin resonance experiments are subdivided into nuclear magnetic (nmr) and electron spin (esr) resonance methods. As of this writing almost all experiments are done using the pulse or Fourier transform (ft) mode of operation and that mode is discussed herein.

Theoretical Background

An unpaired electron or an atomic nucleus, unless it has even integral values for both atomic mass and atomic number, has a nonzero angular momentum and measurable magnetic moment.

A given nucleus can exist in any of $2nI + 1$ states described by a nuclear spin quantum number, m_i, ranging from $+I$ (lowest energy) to $-I$ (highest energy) in unit steps. When placed in an external static magnetic field, B_0, the difference in energy, ΔE, between adjacent states is given by:

$$\Delta E = h\omega_0/2\pi = h\gamma B_0/2\pi = h\nu$$

where h is Planck's constant (6.626×10^{-34} J·s), γ is the magnetogyric ratio, and $\omega_0/2\pi = \nu$ is the Larmor frequency in Hertz. The value of γ for an electron is about 660 times larger than for a hydrogen atom.

The precise value of both the energy and Larmor frequency is proportional to the strength of the static external field. The relative population, N_β/N_α, of the two energy levels is described by the Boltzmann relation,

$$N_\beta/N_\alpha = \exp(-\Delta E/kT)$$

leading to population differences on the order of 5 parts in 10^5 for ^1H at 300 K and 7.05 T(70.5 kG) $= B_0$.

Nuclear Magnetic Resonance. The interaction of a nucleus with B_0 is usually described using vector notation and models as in Figure 1 where the bulk magnetization, M, and the static field B_0 are initially parallel to z. A radio frequency pulse is applied in the xy plane for a duration of t μs, tipping the magnetization by θ radians where $\theta = \gamma B_1 t$. The duration of the 90° pulse required to completely shift magnetization into the xy plane is thus inversely proportional to B_1. The change in orientation of individual spins with respect to the external field is referred to as resonance. Because the net magnetization is now inclined with respect to B_0 in response to the rf source, it precesses about the z axis at the Larmor frequency. The concept of a rotating frame, in which a new set of axes, x' and y', rotate about z at the same frequency as the precessing nucleus, ie, γB_0, is conceptually convenient. Because the rotating frame is the standard reference system for describing magnetic resonance experiments, the prime designations for that system usually are omitted in most theoretical discussions. The application of an rf field, B_1, is considered to be along x' and this leads to a tipping of the magnetization vector in the $y'z'$ plane, as shown in Figure 1. The magnitude of the tipping is determined by the pulse width or duration, ie, the length of time, typically in μs, that B_1 is applied.

The precessing magnetization induces a current in the detector coil which surrounds the sample and is centered on the laboratory y-axis. The time dependent fluctuation in this current decays in intensity as the spin distribution returns to its equilibrium distribution. The response to a single pulse is referred to as a free induction decay (FID) or transient and is converted to a digital form by an analogue-to-digital converter (ADC) for storage in a computer file.

In ft/nmr the detected signal is the sum of the individual decays for all of the nuclei. To convert this information back into a frequency-dependent form in which spectra are generally viewed, it is necessary to apply a Fourier transformation. Mathematically this is expressed as

$$f(\omega) = \int f(t)\exp(i\omega t)dt$$

where $f(\omega)$ is the frequency domain spectrum and $f(t)$ is the time-dependent fluctuation of the amplitude in the composite nmr signal.

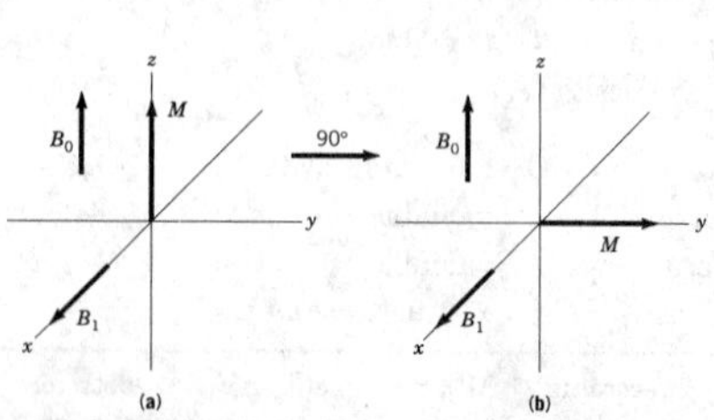

Figure 1. Interaction of nucleus (electron) with static magnetic field, B_0, where the bulk magnetization, M, is (**a**) parallel to B_0 and to the z-axis, and (**b**), upon application of a 90° radio frequency pulse along x, M perpendicular to B_0 and to the z-axis. See text.

Because $f(t)$ is real, $f(\omega)$ has both real, absorption spectrum, and imaginary, dispersion mode components.

The time constants associated with the restoration of magnetic equilibrium after application of a pulse are referred to as relaxation times and their reciprocals as relaxation rates. Relaxation may occur by any of several different mechanisms involving either longitudinal, T_1, or transverse, T_2, processes.

Because the energy differences between adjacent levels are very small in magnetic resonance experiments, the transition probabilities are also very small in the absence of local field fluctuations. Dipole–dipole interactions are the dominant mechanism for creating local field fluctuations and these decrease in intensity with the sixth power of the internuclear distance, so that only the closest nuclei have an appreciable effect on the relaxation rate for a given atom. The strength of the interaction also depends on the square of the product of the gyromagnetic ratios of the interacting nuclei. Other mechanisms, such as quadrupolar relaxation, may become important if either of the interacting nuclei is paramagnetic or the quadrupolar moment is significant.

At a minimum, knowledge of relaxation times is important in determining the rate at which the pulse experiment can be repeated and successive transients added to the FID. The delay between acquisitions may be tuned for either minimum acquisition time, ie, short delays, or for uniform sensitivity by setting the delay to at least five times the longest T_1 in the system.

One additional phenomenon which plays an increasingly important role in nmr-based structure determination is the nuclear Overhauser effect (NOE). The NOE originates in relaxations arising from dipole–dipole interactions through space and decreases in intensity with r^6_{IS}, where r_{IS} is the internuclear separation of two mutually relaxing spins.

The observation of an NOE between two nuclei implies that the nuclei are close together in space, typically separated by less than 0.45 nm. The failure to observe an NOE does not, however, mean that the nuclei are far apart, because the relaxation rate also depends on τ_c.

Resolution is an important consideration in any experiment involving a number of closely spaced maxima. The digital resolution (DR) achievable in an nmr spectrum is defined by the relation

$$\mathrm{DR} = 2\cdot(\text{sweep width})/(\text{number of data points})$$

The acquisition time during which the spectrum is acquired is the reciprocal of DR.

Electron Spin Resonance. Electron spin resonance (esr) also known as electron paramagnetic resonance (epr) is a second magnetic resonance technique which finds particular applications in the study of free radicals, paramagnetic species, and other molecules containing an unpaired electron. The electron has a spin of 1/2 and, because of its low mass, a gyromagnetic ratio of 1760 rad/(T·s). Field strengths required for esr are much lower, typically 0.34 T (3.4 kG) and the frequencies higher (9–35 GHz) than for nmr. Data are usually reported in terms of intensity, signal strength, versus energy where the energy for a transition is usually expressed as

$$\Delta E = g\beta H$$

where g is the dimensionless splitting constant equal to 2.0023 for a free electron, β is the Bohr magnetron, μ_{B} ($1\mu_B = 9.2732 \times 10^{-24}$ J/T), and H is the external field strength. Experimental data are usually reported as first derivative spectra rather than in the absorption mode typical of nmr.

Equipment

Nuclear Magnetic Resonance. In 1994 there were three principal vendors of nmr instrumentation in the U.S., Bruker Instruments (Billerica, Mass.), JEOL USA, Inc. (Peabody, Mass.), and Varian Associates (Palo Alto, Calif.). Details of instrumentation are best obtained directly from manufacturers. A schematic illustrating the principal components of a ft/nmr spectrometer is shown in Figure 2.

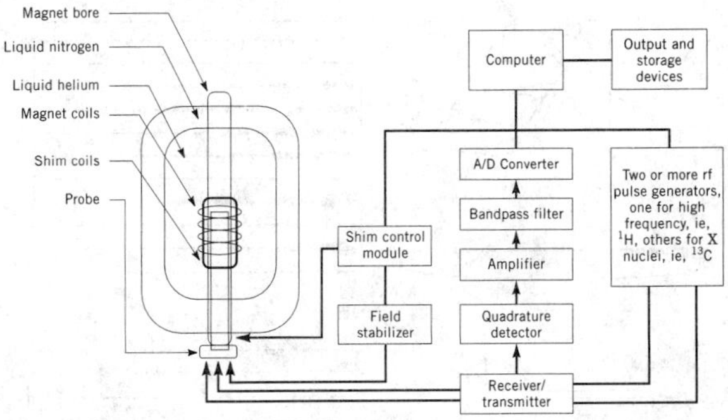

Figure 2. A block diagram schematic representation of a Fourier transform nmr spectrometer, ie, a superconducting magnetic resonance system.

Electron Spin Resonance. Instrumentation for esr differs from nmr instrumentation principally as a consequence of the larger gyromagnetic ratio in this experiment. As a result, the external field is typically 0.35 T (3.5 kG). At this field strength the frequency associated with an unpaired electron is about 9.5 GHz (X-band). Systems operating at frequencies up to 34 GHz (Q-band) are also available.

Health and Safety

Safety considerations for magnetic resonance (mr) experiments have received little attention except for the problems associated with the use of electronic devices such as pacemakers in the magnetic field. However, in a 1990 study of reproductive health involving more than 1900 women working in clinical mr facilities in the United States no substantial differences were reported between the group of women directly involved with mr equipment (280 individuals) and other working women (894 individuals). Conclusions are restricted to exposure to the static external field.

Nuclei of Nuclear Magnetic Resonance

Although nmr can be applied to many different nuclei, the overwhelming interest, especially for the organic chemist, continues to lie with ^1H and ^{13}C nmr. The hydrogen experiment is simpler because ^1H has a natural abundance of more than 99%, a large magnetogyric ratio, and a spin of 1/2. A three-stage model is commonly used to describe nmr experiments. Stages involve preparation, evolution, and detection periods.

Multidimensional nmr

Multidimensional nmr underwent explosive growth in the 1980s and early 1990s and impinges on all aspects of modern nmr. In a typical 2-D experiment, a set of related FIDs are acquired using the same basic pulse sequence while systematically varying the evolution time. The experiment is illustrated schematically in Figure 3 which shows the steps in obtaining a correlated spectroscopy (COSY) spectrum.

There are many instrumental considerations in multidimensional spectroscopy. These are important in defining the appropriate digital resolution in F_1 and F_2. Unlike 1-D nmr, the data size in 2-D experiments can become quite large very rapidly if the wrong parameters are used.

Time constraints are an important factor in selecting nmr experiments. There are four parameters that affect the amount of instrument time required for an experiment. A preparation delay of 1–3 times T_1 should be used.

Window functions for 2-D nmr are necessary because of the poor digital resolution in both dimensions.

High Resolution Solid-State nmr

The ability to acquire high resolution nmr data from solid phases is critical to many applications where, for example, test samples are in-

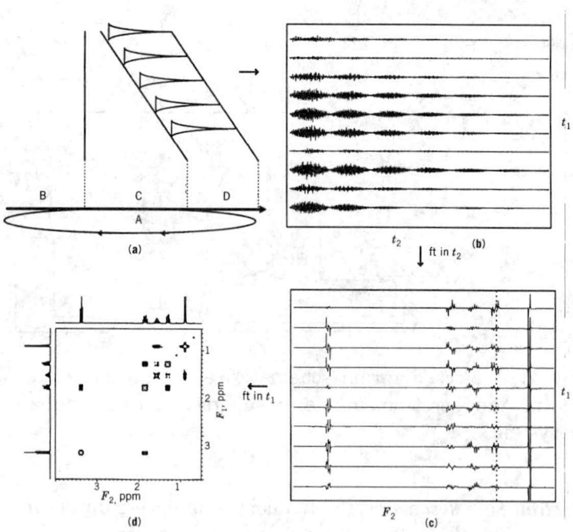

Figure 3. A 2-D nmr experiment of 2-methyl-5-bromopentane [626-88-0], $C_6H_{13}Br$, where t_1 and t_2 correspond to evolution and acquisition time, respectively. (**a**) A set of pulsing experiments differing in evolution times leading to (**b**) a set of related FIDs to which a Fourier transform in t_2 is applied. Each FID leads to (**c**) a set of spectra; and a second Fourier transform along t_1 results in (**d**) the COSY spectrum. In (**a**), A indicates the progression of time through B, preparation, C, evolution and mixing, and D, acquisition. See text.

soluble, exploration of a nascent structure is desired, or samples are cross-linked. However, the use of liquid-phase nmr experiments using solid samples leads to broad lines having little discernible structural detail. Three principal problems in obtaining nmrs of solids are chemical shift anisotropy (CSA), a high degree of dipolar interactions, and very long (10^2–10^3 s) T_1 values typical of rigid molecules.

As originally proposed in 1962, magic-angle spinning (MAS), where the sample is rapidly rotated about an axis inclined to B_0 by 54.74, the magic angle, produces considerable line sharpening.

A further reduction in the dipolar interaction term is achieved by dipolar decoupling. One difficulty with this method is that the optimal mixing time for different carbon environments is not the same; as a result, the integrated intensities are even less quantitative than in liquid-phase ^{13}C nmr. A compromise value is selected by varying the mixing time systematically and looking for a value that provides an adequate signal for all resonances. A second difficulty is that the benefit derived from cross-polarization decreases with increasing proton–X nucleus distance so that it really only affects atoms having covalently bound hydrogen. Thus, quaternary carbon atoms and other X nuclei such as ^{31}P are insensitive to CP.

One final technical improvement in solid-state nmr is the use of combined rotational and multiple pulse spectroscopy (CRAMPS), a technique which also requires a special probe and permits the acquisition of high resolution ^1H and X nucleus nmr from solids.

Applications of solid-state nmr include measuring degrees of crystallinity, estimates of domain sizes and compatibility in mixed systems from relaxation time studies in the rotating frame, preferred orientation in liquid crystalline domains, as well as the opportunity to characterize samples for which suitable solvents are not available. This method is a primary tool in the study of high polymers, zeolites (see MOLECULAR SIEVES), and other insoluble materials.

Electron Spin Resonance

Esr is also a powerful technique both alone or in combination with nmr in the endor method. Precise measurements of the hyperfine couplings and line-shape lead to detailed information concerning the structure and conformation of molecules containing unpaired electrons. It is perhaps the most powerful tool for studying structure in free-radical containing species. A second area of current importance is the study of radiation-induced changes in chemicals.

DAVID J. KIEMLE
WILLIAM T. WINTER
State University of New York, Syracuse

A. E. Derome, *Modern NMR Techniques for Chemical Research,* Pergamon Press, Oxford, U.K., 1987.

J. K. M. Sanders and B. K. Hunter, *Modern NMR Spectroscopy: A Guide for Chemists,* 2nd ed., Oxford University Press, Oxford, U.K., 1993.

L. Kevan and M. K. Bowman, eds., *Modern Pulsed and Continuous-Wave Electron Spin Resonance,* John Wiley & Sons, Inc., New York, 1990.

R. N. Bracewell, *The Fourier Transform and Its Applications,* 2nd rev. ed., McGraw-Hill Book Co., Inc., New York, 1986.

MAGNETIC TAPE. See INFORMATION STORAGE MATERIALS.

MAGNETOHYDRODYNAMICS

Magnetohydrodynamic (MHD) power generation is a method of generating electric power by passing an electrically conducting fluid through a magnetic field (see also POWER GENERATION). By means of the interaction of the conducting fluid with the magnetic field, the MHD generator transforms the internal energy of the conducting fluid into electric power in much the same way as does the interaction of a solid conductor with a magnetic field in a conventional turbogenerator. In principle, the working fluid can be any electrically conducting fluid, such as salt water, liquid metal, or hot ionized gas. For central station power generation applications, the most suitable working fluid is a hot ionized gas. MHD energy conversion systems can be classified into two types. Closed cycle typically operates using a clean gas which is recycled; open cycle generally operates using combustion products which are then discarded.

In its simplest form the MHD generator consists of a duct through which the gas flows, driven by an applied pressure gradient, and a magnet, in which the duct is located. The generator operates in a Brayton cycle similar to that of a turbine. Because the MHD process requires no rotating machinery or moving mechanical parts, the MHD generator can operate at much higher temperature and hence, higher efficiency, than is possible for other power generation technologies. The system of most interest for central station power generation is the open cycle system using electrically conducting coal (qv) combustion products as the working fluid. Coal-burning central station MHD power plants promise to generate power at up to 50% greater efficiency and at a lower cost of electricity than can be achieved using conventional coal-burning power plants of the early to mid-1990s (see COAL CONVERSION PROCESSES).

In addition to potential advantages in efficiency, MHD power generation offers significant potential for reduced environmental intrusion. Effective control of pollutants is inherent in the basic design and operation of MHD power plants.

Key Elements

The principles of MHD power generation are shown in Figure 1. An electrically conducting fluid flows with velocity $\vec{u}$, through a magnetic field $\vec{B}$, to induce an electric field $\vec{E}$ which is orthogonal to both the flow direction and the magnetic field direction. If the flow is contained in a duct, as shown, the two walls perpendicular to the electric field are at different potentials. If these two walls are electrically conducting and connected through an external resistance or load, the field causes a current to flow through the load, thus generating power. The conducting walls, through which current is extracted from the duct, are the electrode walls. The wall which emits electrons is designated

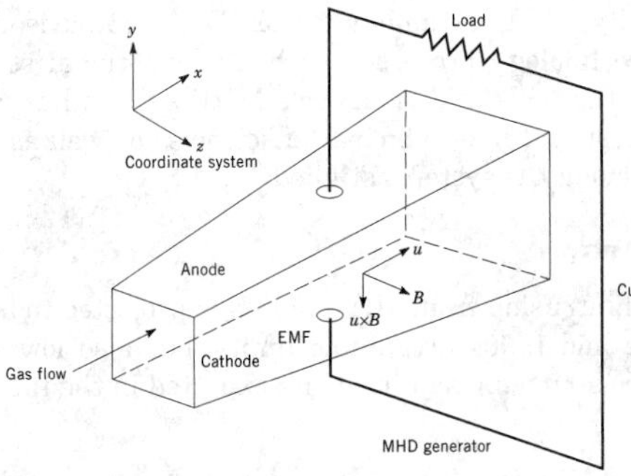

Figure 1. Principles of MHD. See text.

the cathode; the other wall is the anode. The two walls separating the electrode walls are the insulator walls.

For central station power generation the open cycle system using electrically conducting coal combustion products as the working fluid is employed. The fuel typically is pulverized coal burned directly in the MHD combustor, although in some plant designs cleaner fuels made from coal by gasification or by beneficiation have been considered (see FUELS, SYNTHETIC).

In application to electric utility power generation, MHD is combined with steam (qv) power generation, as shown in Figure 2. The MHD generator is used as a topping unit to the steam bottoming plant. From a thermodynamic point of view, the system is a combined cycle. The MHD generator operates in a Brayton cycle, similar to a gas turbine; the steam plant operates in a conventional Rankine cycle.

The MHD topping cycle components are the magnet, the coal combustor, nozzle, MHD channel, associated power conditioning equipment and the diffuser. The magnetic field required in a commercial plant is typically 4.5–6 T ($4.5–6 \times 10^4$ G); hence, the magnet is superconducting, as a conventional magnet would require impractical amounts of electric power.

The steam bottoming plant of the combined MHD–steam power plant consists basically of a heat recovery and seed recovery system (HRSR) and a turbine–generator for additional power production. The HRSR is essentially a heat recovery boiler and oxidant preheater which is fired by the exhaust gases from the MHD channel. In addition to generating steam, the HRSR system must also perform the functions of NO_x and SO_x control, slag tapping, seed recovery, and particulate removal.

Chemical Regeneration. In most MHD system designs the gas exiting the topping cycle exhausts either into a radiant boiler and is used to raise steam, or it exhausts into a direct-fired air heater and is used to preheat the primary combustion air. An alternative use of the exhaust gas is for chemical regeneration, in which the exhaust gases are used to process the fuel from its as-received form into a more beneficial one. Chemical regeneration has been proposed for use with natural gas and oil as well as with coal (see GAS, NATURAL; PETROLEUM).

MHD Fundamentals

The basic principles of MHD power generation are shown in Figure 1. The working fluid is typically a slightly ionized gas, ie, mostly neutral atoms or molecules with a small (<0.1%) fraction of ions and electrons. The working fluid flows with velocity u, in the axial, x, direction, through the MHD channel, which is located in a magnetic field of strength B, in the z direction. An electric field, uB, is induced in the transverse ($-y$) direction, as shown. Because there exists also an internal electric field, E, owing to the load current, the electric field, E', seen by the moving gas is

$$E' = E - uB \tag{1}$$

and the induced current per unit area, j_y, is

$$j_y = \sigma E' = \sigma(E - uB) \tag{2}$$

where σ is the electrical conductivity of the gas. If a load coefficient K is defined, such that

$$K \equiv E/uB \tag{3}$$

the magnitude, j, of the current density can be expressed as

$$j = \sigma uB(1 - K) \tag{4}$$

and the generator power per unit volume, P, is then

$$P = jE = \sigma u^2B^2K(1 - K) \tag{5}$$

Assuming that the current in the gas is carried mostly by electrons, the induced electric field uB causes transverse electron motion (electron drift), which, being itself orthogonal to the magnetic field, induces an axial electric field, known as the Hall field, and an axial body force, F, given by

$$F = jB = \sigma uB^2(1 - K) \tag{6}$$

This force acts in the $-x$ direction and retards the flow. The rate at which the gas does work in pushing itself against this force is

$$Fu = jBu = \sigma u^2B^2(1 - K) \tag{7}$$

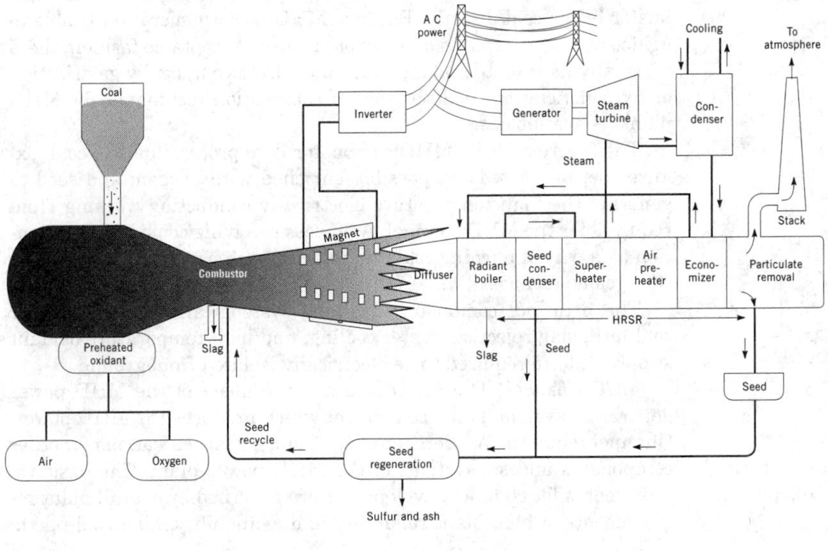

Figure 2. MHD/steam power plant.

The ratio of electrical power output (eq. 5) to the power required to push the gas (eq. 7) can be defined as the efficiency of the process, η_e:

$$\eta_e = jE/juB = K \quad (8)$$

The difference between the push power and the electric power output is

$$juB - jE = j^2/\sigma \quad (9)$$

The term j^2/σ is the rate of dissipation of energy per unit volume by joule heating. This occurs within the working fluid, and so represents a departure from thermodynamic irreversibility rather than an energy loss. This simplified discussion has neglected the effects of axial current flow, ie, Hall current, induced by the axial field.

MHD Generator Geometries. The basic requirement for any generator geometry is that the gas velocity have a component that is not parallel to the magnetic field. For efficiency, the gas velocity should be orthogonal to the magnetic field direction. Development work has focused mainly on the linear generator. There are four basic variations of the linear MHD channel (Fig. 3) which differ primarily in their method of electrical loading.

Flow and Performance Calculations. Electrodynamic equations are useful when local gas conditions (u, σ, B) are known. In order to describe the behavior of the flow as a whole, however, it is necessary to combine these equations with the appropriate flow conservation and state equations. These last are the mass, momentum, and energy conservation equations, an equation of state for the working fluid, an expression for the electrical conductivity, and the generalized Ohm's law.

Electrical Conductivity. In order to conduct electricity, the working fluid must contain charged particles, ie, it must be partially ionized. Some of the gas atoms or molecules must be stripped of one or more of their electrons. The energy required to accomplish this, called the ionization potential, is measured in electron volts. In MHD flows of interest, the required energy is supplied by heating the gas. Thus the ionization process is referred to as thermal ionization.

Efficiency and Economic Factors

Because the MHD generator has no moving mechanical parts, it can operate at a much higher combustion temperature than other power generating systems, allowing the combined MHD–steam cycle to achieve higher thermal efficiency than other systems. The high efficiencies together with competitive capital costs yield very attractive cost of electricity (COE) estimates for MHD. A comparison of about 20 advanced technology processes with a conventional steam plant concluded that the coal-fired open-cycle MHD system has potentially one of the highest coal pile-to-busbar efficiencies as well as one of the lowest COEs among the systems studied.

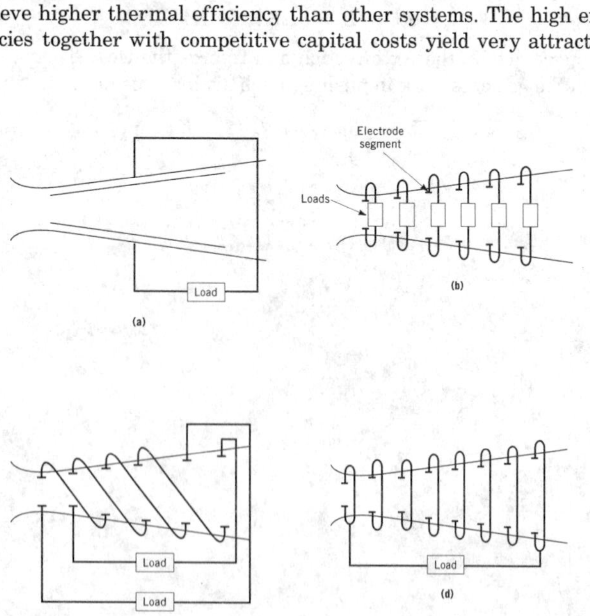

Figure 3. Linear MHD generator configurations: (**a**), two-terminal Faraday or continuous electrode; (**b**), segmented Faraday; (**c**), diagonal connections; and (**d**), Hall geometry.

Environmental Factors

Environmental intrusion from MHD plants is projected to be not only well below the mid-1990s acceptable limits, but also low enough to satisfy the more stringent requirements expected in the future.

Power Plants

A complete MHD power plant must integrate the MHD generator and the bottoming plant to maximize plant efficiency and minimize cost of electricity. Net plant efficiency is maximized by simultaneous optimization of net MHD power generation, ie, MHD generator power minus the cycle compressor power and oxygen plant (if used) compressor power, and of waste heat utilization in the steam bottoming plant. Compromises between performance and cost, particularly of the magnet but also of the oxygen plant and other ancillary equipment, must be made.

Plant Economics. A power plant is evaluated economically in terms of capital costs ($/kW) and levelized costs of electricity (mills/kWh = $\$10^{-3}$/kWh). Important factors are the escalation of costs with time and the cost over time of the capital to build the plant.

Components and Subsystems

High Temperature Air Preheaters. Combustion air–oxidant preheating for open cycle generators is accomplished in one of two ways. One way is to use the heat energy of the MHD generator exhaust gas directly; in this case, the preheater is classified as directly fired and is located in the MHD generator exhaust as part of the bottoming plant (Fig. 2). The other way of preheating combustion air is to use a separate heat source using clean fuel. This type of preheater is classified as separately fired. Directly fired preheat offers the potential of higher cycle efficiencies than can be achieved with separately fired preheat, at the same oxidizer temperature. However, because of the severe difficulties associated with designing a directly fired preheater capable of operating with the seed and ash-laden gases flowing from the generator, first-generation commercial plants are to use separately fired preheat.

More advanced MHD power plants of the future are expected to use preheat temperatures of up to 2000 K. These temperatures, to be achieved by direct firing, require the use of high temperature regenerative heat exchangers. Regenerative heat exchangers of both the fixed-bed and moving-bed types have been considered for MHD use.

Combustor. In the majority of MHD plant designs the MHD combustor burns coal directly. Because MHD power generation is able to utilize pulverized coal in an environmentally acceptable fashion, there is usually no need to make cleaner fuels from coal, eg, by gasification or by beneficiation. A discussion of combustion techniques for MHD plants is available.

The function of the MHD combustor is to process fuel, ie, coal; oxidizer, ie, preheated air, possibly enriched with oxygen; and seed to generate the high temperature electrically conducting working fluid required for the MHD channel. A process receiving considerable attention as a way of burning coal and rejecting ash as slag is that used in the cyclone furnace.

The principal combustor ancillary systems are the systems for coal feed, slag rejection, water cooling, and high temperature oxidant supply. All are required to be electrically isolated from ground.

MHD Channel. The MHD channel, the heart of the MHD power generation system, is the component which produces the MHD power. Channel requirements determine the principal specifications for other components and subsystems of the MHD power plant. The basic requirements for channel development are governed by overall plant requirements of high plant reliability and availability, high coal-pile to bus-bar efficiency, and low cost of electricity. To satisfy these plant requirements, three primary MHD channel design criteria can be identified: (*1*) duration or operating time between maintenance periods; (*2*) fraction of thermal energy input extracted from the gas as electric power output (enthalpy extraction ratio); and (*3*) isentropic efficiency, the ratio of the actual enthalpy change of the gas flowing through the channel to the enthalpy change of an isentropic flow at the same pressure ratio.

From a construction and fabrication point of view, the channel must provide a secure means of containing the working fluid from the combustor and a means of conducting current from the working fluid to the external load, and have adequate durability to satisfy overall power system requirements. Issues related to durability have dominated the development of channel construction methods, particularly of those surfaces which face the hot conducting gas. These surfaces consist of electrodes, which are the current-carrying elements, and insulators, which separate the electrode walls. Durability issues and the resulting designs of gas-side surfaces for coal-fired channels differ from those for clean fuel-fired channels.

Development of durable electrode walls, one of the most critical issues for MHD generators, has proceeded in two basic directions: ceramic electrodes operating at very high surface temperature s ($\geq$2000 K) for use in channels operating with clean fuels such as natural gas, and cooled metal electrodes with surface temperatures in the range 500–800 K for channels operating with slag or ash-laden flows.

Because of the unavailability of electrically insulating materials which can withstand the harsh environment inside coal-fired channels, the insulator walls of the channel are typically made of metal elements which are insulated from each other to prevent any net flow of current. Like electrode walls, insulator walls are designed to operate with a slag coating.

The main objectives of channel mechanical and thermal design are to maintain structural and sealing integrity, to provide adequate cooling of gas-side surface elements, and to use efficiently the magnet bore volume, ie, to maximize the ratio of channel flow cross-section area to the magnet bore cross-section area. This last requirement affects not only the channel mechanical design but also the packaging of channel electrical wires, cooling hoses, and manifolds. In broad terms, MHD channels built to date have fallen into one of three types of construction categories: plastic box construction; window frame construction; or reinforced window frame construction.

Electrical Loading and Control. The function of the channel loading system is to extract from the channel the power generated in each plasma element with minimal losses. This means, that the load circuit impedance must match as closely as possible the channel impedance at all axis locations along the channel, which is achieved by use of multiple power take-off points. Ultimately, power from the separate take-offs must be consolidated into a single terminal pair at the transmission grid, by means of appropriate circuitry. An inverter is necessary between the channel and the transmission grid in order to convert the relatively low voltage (20–40 kV) dc output of an MHD generator to ac at transmission line voltages (200–400 kV). In principle, the power consolidation function can be combined with the inversion function by use of common circuitry; in practice, it is simpler and less costly to separate these functions.

Magnet. The magnetic field for utility scale MHD generators is provided by a superconducting magnet system (SCMS) for economic reasons, as the cost of electricity for a conventional magnet of the required size is prohibitive (see MAGNETIC MATERIALS). The SCMS consists of three principal subsystems: the main magnet and cryostat subsystem, the cryogenic refrigeration system, and the power supply and protection subsystem. Of these, the magnet subsystem is the most critical, having the majority of the design choices and requiring the bulk of the engineering and manufacturing effort.

The magnet is required to provide a field of the required magnitude and, in the case of linear channels, axial profile. Linear MHD systems require a sharp magnetic field reduction at the channel ends. The magnet is wound from a composite of niobium–titanium and copper or aluminum. The Nb–Ti is the superconductor and the copper or aluminum serves to stabilize the conductor, by providing capacity for heat absorption from the joule heating which occurs in the event that the conductor undergoes a transition from its superconducting state into a normally conducting state. Liquid helium cooling to about 4 K is necessary to maintain the superconducting state.

For power plants, magnetic fields of 4.5 to 6 T (4.5–6 $\times$ 10^4 G) are required, over warm bore volumes having typical dimensions of 3–4 m dia and 15–20 m long. Stored energies in such magnets are 2000 MJ (480,000 kcal) or greater.

Some technology for large superconducting magnets has been developed, mainly for bubble chamber and fusion reactor applications, and magnets having stored energy up to 500 MJ (100,000 kcal) have been built. Winding and fabrication techniques for very large saddle-shaped coils need further development.

Studies of MHD superconducting magnets have been performed at the Plasma Fusion Center at the Massachusetts Institute of Technology. These include studies of materials concerned with the properties of highly stressed structural members operating at temperatures near absolute zero, studies of superconductor configurations and winding techniques, studies of shipping and on-site assembly methods to establish the degree of modularity required, and studies of scaling factors and costs. Similar work was done as part of the Italian national superconducting magnet program as well as in India.

Heat Recovery and Seed Recovery System. Although much technology developed for conventional steam plants is applicable to heat recovery and seed recovery (HRSR) design, the HRSR has several differences arising from MHD-specific requirements. First, the MHD diffuser, which has no counterpart in a conventional steam plant, is included as part of the steam generation system. The diffuser experiences high (30–50 W/cm^2) heat transfer rates. Thus, it is necessary to allow for thermal expansion of the order of 10 cm in both the horizontal and vertical directions at the connection between the diffuser and the radiant furnace section of the HRSR.

Secondly, inlet conditions are more severe in the MHD HRSR because the hot gas entering the HRSR is at considerably higher temperature and enthalpy than that entering a conventional boiler.

Third, design constraints are imposed by the requirement for controlled cooling rates for NO_x reduction. And finally, the MHD HRSR conditions are more hazardous to the furnace materials because of the more corrosive environment created by the hotter combustion gases, bearing potassium, which is not found in conventional plants.

Plant Control, Part Load Performance, and Availability. Conventional power plant control practices are all applicable, with some modifications, to the operation of MHD-steam power plants. These can provide for attractive load following characteristics, plant stability, and safe operation. Special control actions to safeguard plant equipment during emergencies and abnormal operating conditions such as sudden loss of MHD generator load are part of the overall plant control strategy.

Analyses have shown, that relatively high plant efficiency can be maintained at part load, by reduction of fuel input, mass flow, and MHD combustor pressure. Reliability, availability, and maintainability are critical factors in the use of MHD for electric utility power generation. The duration and lifetime requirement of the principal components and subsystems of an MHD–steam power plant have been assessed in relation to the requirements of acceptable overall plant reliability and availability. Duration and reliability of the MHD generator channel are critical factors.

Development Programs

The U.S. national MHD development program was funded by the U.S. Department of Energy (DOE), through the Pittsburgh Energy Technology Center. The program objectives were to establish, through proof of concept (POC) testing, an engineering database so that the risks and benefits of the technology can be evaluated by the private sector before proceeding with commercial demonstration. The POC

Table 1. National MHD Programs

Country	Organization	Facility	Thermal input, MW	Magnetic field, T[a]
Australia	Univ. of Sydney, School of Elect. Eng.	disk generator	2	2.7[b]
China	Shanghai Power Equipment Res. Inst.		25	4.5[c]
		SMS	5	
	Institute of Elect. Eng.		25	1.9
India	Bharat Heavy Electricals	pilot plant	5	2[d]
Italy	Industrial MHD Consortium/ Ansaldo	MDA	1	1.4
	CNR/Ansaldo			
	Industrial MHD Consortium			
Japan	Tokyo Inst. of Tech	FUJI-1	2–3	
	Hokkaido Univ.	MDX-1	5	2.5
	Kyoto Univ.			
	Toyohashi Univ. Tech.			2
Romania	ICPET	G-MHD-03	1–2.5	
Russia	Inst. High Temp	U-25G	25	3.5
		U-02	5	1.5–2[e]
	Krzhyzhanovski Power Inst.	M-25	25	
	Inst. of Energy Saving Problems, Ukranian SSR Academy of Sciences	K-1	15	1.8

[a] To convert T to G, multiply by 10^4. [b] Cumulative; clean fuel was used. Test duration = 36 h. [c] Continuous generation. Test duration = 1000 h; at SMS facility, 60 h. [d] LPG was used as fuel. Test duration ≤ 150 h. [e] Test duration = 100 s of h.

program was aimed at performance and lifetime of principal components and subsystems. The national program contained four major elements: conceptual design of MHD repowered plant, integrated topping cycle, integrated bottoming cycle, and seed regeneration.

International MHD Programs. Data on a number of current or previous coal-fired MHD power generation programs conducted in other countries are summarized in Table 1.

ROBERT KESSLER
Textron Systems Division

R. J. Rosa, *Magnetohydrodynamic Energy Conversion,* McGraw-Hill Book Co., Inc., New York, 1968.

M. Petrick and B. Ya. Shumyatsky, eds., *Open Cycle MHD Electrical Power Generation,* Argonne National Laboratory, Argonne, Ill., 1978.

J. B. Heywood and G. J. Womack, eds., *Open Cycle MHD Power Generation,* Pergamon Press, London, 1969.

Published proceedings of: *Symposium on Engineering Aspects of Magnetohydrodynamics,* annually since 1961; *International Conference on Magnetohydrodynamic Electrical Power Generation,* every 3–4 years since 1962.

MAINTENANCE

One of the goals of good management is to operate with a predictive and preventive maintenance program that prevents unscheduled utility service interruptions and machine breakdowns. A reasonable cost of such a program contributes to the profitability of manufacturing plants and results in reduction of operating costs and increased productivity. This all can be accomplished by organizing, managing, and controlling maintenance.

Maintenance Considerations during Facility Construction or Renovation

Maintenance managers should be on board during the design state to ensure the maintainability of a facility. A maintenance manager's case is simple: the original cost of new construction is soon forgotten, but facility maintenance and operational costs continue for the life of a facility.

The maintenance manager oversees important issues during the design and construction phase. Some areas requiring special attention include checking the design from an operational point of view; checking machinery and equipment for operation and maintenance; checking that all equipment is safely installed and can be maintained; checking the marking of supply lines (air, water, gases, chemicals, etc); identifying the electrical circuitry and checking panel labels; reviewing manufacturers' suggestions for spare parts and tools; reviewing test results, ie, strength of materials, pipeline integrity, etc; reviewing balance reports for HVAC work, ie, air, water, electric load, etc; reviewing fire sprinkler alarm and water test; participating in housekeeping and sanitary support during construction; and checking utilities, including electric power, water, drainage, air supply, that will supply new areas without overloading the individual systems. Also, maintenance workers should inspect roof penetrations and the installation of equipment on roofs during construction.

Standardization of equipment and material is another area that lends itself for partnership between the design team and maintenance. When an existing facility is renovated or a new section/building is added, maintenance can help ensure that materials, equipment, and parts are used that match those that were used in the existing facility. Setting such standards helps to cut down inventory and reduce operation and maintenance costs as well.

Maintenance costs influence the bottom line of every balance sheet. As a result, maintenance managers can improve the entire facility by helping make sure that the planning and design of new construction and renovation projects takes place with maintenance in mind.

Developing a Simple Predictive and Preventive Maintenance Program

A PPM program is needed to avoid equipment failures, utility outages, and production interruptions. From a cost savings angle it is extremely important to do preventive maintenance in order to avoid breakdowns. Periodic inspections and a good lubrication program uncover conditions that could lead to breakdowns. When problems are found early, they can be taken care of without work interruption and costly repairs.

A more sophisticated program is a predictive maintenance system. Maintenance gets done when it is really needed. Special tools are required to pick the right intervals. One simple way is to perform preventive maintenance as suggested by the equipment manufacturer. When machinery is operated intermittently, a very cost-effective way is to install elapsed operating time indicators and to do maintenance only after the equipment has operated for a predetermined number of hours. Vibration analysis is another predictive maintenance tool. The instrument measures the machine vibration at different frequencies as measured by the RPM of the shaft for bearings, fans, or other equipment. Vibration spikes indicate an imminent machine problem.

PM programs vary considerably and depend very much on how much a facility can spend and how skilled the technicians are that operate the systems. More expensive and sophisticated are computerized maintenance management systems (CMMS). These systems often include programs that are needed by a facility as well as manufacturing management. Technological developments have made it possible for CMMS to integrate facility, manufacturing, safety, security, and many other functions of a facility. Facility managers can order from an extended menu that includes machinery maintenance, work orders, projects, labor and material cost tracking, scheduling control, stores issues, inventory control, purchasing activity, drawings, reports, analysis, etc.

The keeping of good records is a significant tool in providing good facility maintenance. Especially in chemical-process plants, the quality of record keeping supports efficient operation.

In-house departments must not be allowed to become too large. Departments that try to tool up to do everything in-house often lose cost effectiveness. Having an in-house maintenance department should never rule out the use of contracted maintenance or repair services.

Contracted services should be used for specialized work like elevator service, emergency generator testing, uninterruptible power supply (UPS) service, and maintenance.

Complete Maintenance Contracts. Such complete service contracts can help especially smaller facilities that can only provide an on-site service administrator.

Supplemental Maintenance Contracts. The contractor supplements an existing in-house maintenance program. This is used to relieve peak loading of the in-house maintenance department which may happen during vacation periods, seasonal work, when breakdowns occur, or for large rearrangement jobs.

Manpower Labor Contracts. Some facilities prefer a strict manpower contract that supplies skilled or unskilled labor as required to fill in and supplement an existing workforce.

Specialty Skills Contracts. These contracts might handle work such as janitorial, window washing, painting, gardening, security guards, parking lot sweeping, weed and pest control, instrumentation maintenance and calibration, and other specialized maintenance.

Motivating the Maintenance Workers

Management must motivate employees by creating a professional environment that includes enriching the work itself. Maintenance workers enjoy working in a professional atmosphere, with good shops and tools, readily available job information and material, and proper training so that jobs can be done right the first time. Enriching jobs means challenging employees, encouraging their initiative and creativity, and empowering them to improve productivity and quality performance. On the job training and allowing greater responsibilities provide maintenance workers with an atmosphere of growth in their profession.

Organizing Safety

It is the maintenance manager's responsibility to safeguard the facility and to avoid accidents that cause injuries and losses. The best way to accomplish this is to make every employee aware that safety first must become part of their daily work habits. According to the National Safety Council, the first step in setting up a safety program is to make policy crystal clear to every maintenance worker. Once this policy has been established and is enforced by upper management, maintenance managers can organize a good safety program that should include use of safety factors.

Supervisors are responsible to see that every worker is adequate on the job. Physically, mentally, and emotionally inadequate workers are accident-prone. Personal hazards result from lack of knowledge, conflict of motives, physical and mental factors.

Mock OSHA Inspection. Maintenance can learn a lot about how the Occupational Safety and Health Administration (OSHA) trains their inspectors and what is emphasized in an OSHA inspection. Some of the training of OSHA inspectors follows a program involving the recognition of potential hazards, avoidance of these hazards, and prevention of accidents (RAP).

Use of Outside Contractors. When maintenance uses an outside contractor or a subcontractor, it is important that maintenance managers understand the legal and working relationship between the owner and the contractor. In all negotiations the maintenance manager represents the owner. The prime consideration must always be the safety of their workers, prevention of even minor accidents, and prevention of loss of material and property.

Tools and Equipment. Adequate tools and equipment must be provided for each part of the job and arrangements made for proper care of them.

Including Environmental Care and Appreciation

The responsibility of maintenance managers must include strict compliance to environmental regulations. Pollution prevention is a continuous responsibility of the maintenance management and can only be achieved by educating every worker in the maintenance department and watching over outside contractors. Keeping up with the state-of-the-art of environmental regulation compliance is as important as keeping up with the knowledge of operating a facility. Making sure that every company is a good neighbor by preventing pollutants from the plants entering the environment is as important as the safeguarding of employees from hazardous exposure while at work.

Developing an Energy Conservation Program

Although there are different opinions of how to combat the U.S. national energy shortage, it is agreed that in order to survive as much energy must be conserved as possible. Industrial energy consumption accounts for about half of the total energy used in the United States. Because the cost of fuel, electric power, and water has increased steadily and dramatically since the 1980s, it is of even greater importance to develop an ongoing conservation program. Plant engineers and maintenance managers must realize that such a program saves money and reduces manufacturing cost.

When a new facility has been designed for energy conservation and when an older facility has been remodeled to make it more energy efficient, it becomes the responsibility of the maintenance management to provide a preventive and ongoing maintenance program that assures a continuous energy-conservation effort.

Creating a Partnership with Suppliers and Equipment Manufacturers

In order to be successful any maintenance department needs strong internal customer–supplier partnerships. Such partnerships require opportunity for mutual benefits, predictable performance by each partner, and communication across links.

This concept not only includes the many service contracts that maintenance departments negotiate, but also the daily supplies as well as capital equipment and installation contracts.

Cooperation with Other Service Groups of the Facility

Maintenance departments provide support to other functions of manufacturing management. Some of the corporate staff activities that are of concern to the plant maintenance organization are industrial relations, finance, material control, services, and manufacturing engineering.

Controlling Cost and Assuring Budget Responsibility

A well-managed maintenance program results in a reduction of operating costs and a controlled budget that makes use of goals that were established to ensure the maximum return on the capital invested. This must be done despite rising costs, OSHA regulations, environmental protection, and the energy shortage. To be effective, every maintenance manager always must have the profits of the company in mind. As labor and material costs increase, it is management's duty to devise systems, procedures, and controls that cause a reduction in unproductive time, waste of material, and unnecessary expenses.

ERIC M. BERGTRAUN
National Semiconductor Corporation

The Association for Facilities Engineering, Cincinnati, Ohio, publishes an information services catalog which features *Facility Management Library Reprint Services.*

G. H. Magee, *Facilities Maintenance Management,* R. S. Means, Kingston, Mass., 1988.

R. K. Mobley, *An Introduction to Predictive Maintenance,* Van Nostrand Reinhold, New York, 1990.

B. C. Langley, *Plant Maintenance,* Prentice Hall, New York, 1986.

MALEIC ANHYDRIDE, MALEIC ACID, AND FUMARIC ACID

Maleic anhydride (**1**), maleic acid (**2**), and fumaric acid (**3**) are multifunctional chemical intermediates that find applications in nearly every field of industrial chemistry. Each molecule contains two acid carbonyl groups and a double bond in the α, β position. Maleic anhydride and maleic acid are important raw materials used in the manufacture of phthalic-type alkyd and polyester resins, surface coatings, lubricant additives, plasticizers (qv), copolymers (qv), and agricultural chemicals (see ALKYD RESINS; POLYMERS, UNSATURATED; LUBRICATION AND LUBRICANTS). Both chemicals derive their common names from naturally occurring malic acid.

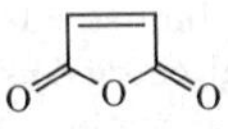

(**1**)

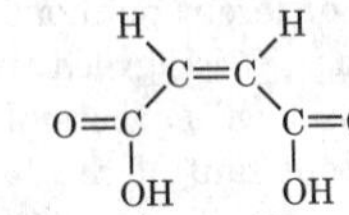

(**2**)

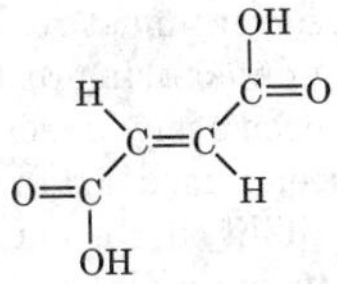

(**3**)

Fumaric acid occurs naturally in many plants and is named after *Fumaria officinalis,* a climbing annual plant, from which it was first isolated. It is used as a food acidulant and as a raw material in the manufacture of unsaturated polyester resins, quick-setting inks, furniture lacquers, paper sizing chemicals, and aspartic acid.

Physical Properties

Physical constants for maleic anhydride, maleic acid, and fumaric acid are given in Table 1. From single crystal x-ray diffraction data, maleic anhydride is a nearly planar molecule with the ring oxygen atom lying 0.003 nm out of the molecular plane.

Maleic and fumaric acids have physical properties that differ due to the cis and trans configurations about the double bond.

Chemical Properties.

Extensive descriptions of the chemistry of maleic anhydride and its derivatives are available in the literature. The broad industrial applications for this chemistry derive from the reactivity of the double bond in conjugation with the two carbonyl oxygens. Reactions include acid chloride formation, acylation, alkylation, amidation, concerted nonpolar reactions, decomposition and decarboxylation, electrophilic addition, esterification, free-radical reactions, grignard-type reactions, halogenation, hydration and dehydration, hydroformylation, isomerization, ligation to metal atoms, nucleophilic addition, oxidation, polymerization, reduction, and sulfonation.

Manufacture

Process Technology Evolution. Growth in the worldwide maleic anhydride industry is exclusively in the butane-to-maleic anhydride route, often at the expense of benzene-based production. Table 2 shows 1995 and estimated 1996, 1997, and 2000 worldwide maleic production capacity broken down in categories of benzene, butane, and phthalic anhydride coproduct. As can be seen from this table, butane routes are expected to grow at the expense of benzene-based processes.

Recovery and Purification. All processes for the recovery and refining of maleic anhydride must deal with the efficient separation of maleic anhydride from the large amount of water produced in the reaction process. Recovery systems can be separated into two general categories: aqueous- and nonaqueous-based absorption systems. Solvent-based systems have a higher recovery of maleic anhydride and are more energy efficient than water-based systems.

Fumaric Acid. Fumaric acid for commerce is derived from maleic acid through catalytic isomerization. Purified maleic anhydride is the main source of maleic acid. High purity fumaric acid is produced through crystallization of the aqueous mixture, washing, and drying. Decolorizing and crystallization techniques are used to treat impure maleic solutions.

Shipment

Molten maleic anhydride is shipped in tank rail cars, tank trucks, and isotanks (for overseas shipments). Solid form maleic anhydride is produced from molten maleic anhydride as briquettes or pastilles weighing 0.5 to 20 g. Fumaric acid is shipped in solid form, the particle size varying based upon the specification.

Economic Aspects

The switch of feedstock from benzene to butane was completed in the U.S. in 1985, being driven by the lower unit cost and lower usage of butane in addition to the environmental pressures on the use of benzene. Worldwide, the switch to butane was continuing with 58% of the total world maleic anhydride capacity based on butane feedstock in 1992. This capacity percentage for butane increased from only 6% in 1978. In 1995, 31% of the total world maleic anhydride capacity was based on benzene feedstock and 2% was derived from other sources, primarily phthalic anhydride by-product streams.

Another characteristic of the maleic anhydride supply picture has been the emergence of newer capacity, predominantly in Asia, which has resulted in a significant oversupply to that area of the world. The historically largest and second largest maleic anhydride markets are North America and Western Europe, respectively. Although the names of the producers in both of these areas of the world have changed rather dramatically between the late 1970s and the early 1990s, the total capacity in these world areas has only declined a small amount. Beginning in the late 1980s, small commercial plants based

Table 1. Physical Properties of Maleic Anhydride, Maleic Acid, and Fumaric Acid

Property	Maleic anhydride	Maleic acid	Fumaric acid
formula	$C_4H_2O_3$	$C_4H_4O_4$	$C_4H_4O_4$
formula weight	98.06	116.07	116.07
mp, °C	52.85	138–139	287
bp, °C	202	ca 138 dec	290
sp gr, at 20/20°C, solid	1.48	1.590	1.635
molar volume		81	79

Table 2. Projected World Maleic Anhydride Capacity by Feedstock

	1995		1996		1997		2000	
Feedstock	10^3 t/yr	%	10^3 t/yr	%	10^3 t/yr	%	10^3 t/yr	%
butane	662.2	67.3	712.2	68.6	756.2	71.2	912.7	75.8
benzene	301.0	30.6	301.0	29.0	282.0	26.5	261.5	21.7
phthalic anhydride coproduct	20.5	2.1	24.5	2.4	24.5	2.3	29.5	2.5
Total	*983.7*	*100.0*	*1037.7*	*100.0*	*1062.7*	*100.0*	*1203.7*	*100.0*

on butane fluidized-bed technologies were built in Asia. As of the end of 1992, fixed-bed butane-based technology accounted for 74% of the butane-based maleic anhydride facilities worldwide with butane-based fluidized-bed technology accounting for the remainder. The butane-based transport-bed process announced by DuPont is scheduled to start up in Spain in 1996.

Health and Safety Factors (Toxicology)

Maleic Anhydride. The ACGIH threshold limit value in air for maleic anhydride is 0.25 ppm and the OSHA permissible exposure level (PEL) is also 0.25 ppm. Maleic anhydride is a corrosive irritant to eyes, skin, and mucous membranes. Pulmonary edema (collection of fluid in the lungs) can result from airborne exposure. Maleic anhydride is combustible when exposed to heat or flame and can react vigorously on contact with oxidizers.

Maleic Acid. Maleic acid is produced by the hydration of maleic anhydride. The hazards of its use are analogous to those of maleic anhydride. It is a skin and severe eye irritant. It is combustible when exposed to heat or flame and can react vigorously with oxidizing agents.

Uses

Maleic anhydride itself has few, if any, consumer uses but its derivatives are of significant commercial interest. The majority of the maleic anhydride produced is used in unsaturated polyester resin (see POLYESTERS, UNSATURATED). Unsaturated polyester resin is then used in both glass-reinforced applications and unreinforced applications.

Fumaric acid and malic acid are produced from maleic anhydride. The primary use for fumaric acid is in the manufacture of paper sizing products (see PAPERMAKING ADDITIVES). Fumaric acid is also used to acidify food as is malic acid. Malic acid is a particularly desirable acidulant in certain beverage selections, specifically those sweetened with the artificial sweetener aspartame.

Lubrication oil additives represent another important market segment for maleic anhydride derivatives. Maleic anhydride is used in a multitude of applications in which a vinyl copolymer is produced by the copolymerization of maleic anhydride with other molecules having a vinyl functionality. The use of maleic anhydride in the manufacture of agricultural chemicals has declined in the U.S. since the early 1980s.

There are numerous further applications for which maleic anhydride serves as a raw material. These applications prove the versatility of this molecule. The popular artificial sweetener aspartame is a dipeptide with one aminoacid (L-aspartic acid) which is produced from maleic anhydride as the starting material. An important future use for maleic anhydride is believed to be the production of products in the 1,4-butanediol-γ-butyrolactone–tetrahydrofuran family. This technology can be used to produce the product mix of the three molecules as needed by the producer.

TIMOTHY R. FELTHOUSE
JOSEPH C. BURNETT
SCOTT F. MITCHELL
MICHAEL J. MUMMEY
Huntsman Corporation

B. C. Trivedi and B. M. Culbertson, *Maleic Anhydride,* Plenum Press, New York, 1982; ~500 pp. on polymers and their applications.

B. M. Culbertson, "Maleic and Fumaric Polymers," in J. I. Kroschwitz, ed., *Encyclopedia of Polymer Science and Engineering,* 2nd ed., Vol. 9, Wiley-Interscience, New York, 1987, pp. 225–294.

S. D. Cooley and J. D. Powers, "Maleic Acid and Anhydride," in *Encyclopedia of Chemical Processing and Design,* Vol. 29, Marcel Dekker, Inc., New York, 1988, pp. 35–55.

K. Lohbeck and co-workers, "Maleic and Fumaric Acids," in *Ullmann's Encyclopedia of Industrial Chemistry,* Vol. A16, 5th ed., VCH, Weinheim, Germany, 1990, pp. 53–62.

MALONIC ACID AND DERIVATIVES

Malonic Acid

Physical Properties. Malonic acid, HOOC—CH_2—COOH, was discovered and isolated in 1858 as a product of malic acid oxidation. The physical properties of malonic acid are listed in Table 1.

Reactions. Malonic acid is a useful tool for synthesizing α-unsaturated carboxylic acids because of its ability to undergo decarboxylation and condensation with aldehydes or ketones at the methylene group. Cinnamic acids are formed from the reaction of malonic acid and benzaldehyde derivatives. If aliphatic aldehydes are used acrylic acids result. Similarly this facile decarboxylation combined with the condensation with an activated double bond yields α-substituted acetic acid derivatives. Reactions of the carboxylic acid groups include monoesterification, diesterification, or conversion with thiols.

Preparation. The industrial production of malonic acid is much less important than that of the malonates. Malonic acid is usually produced by acid saponification of malonates. Further methods which have been recently investigated are the ozonolysis of cyclopentadiene, the air oxidation of 1,3-propanediol, or the use of microorganisms for converting nitriles into acids.

Economic Aspects. Malonic acid is produced by Juzen and Tateyama in Japan as well as Lonza Ltd. in Switzerland and Riedel-De Haen Ltd. in Germany.

Health and Safety Factors (Toxicology). Due to its acidity malonic acid is classified as a mild irritant (skin irritation, rabbits).

Meldrum's Acid. Meldrum's acid is commercially used for the production of monoesters of malonic acid and beta-keto acids.

Malonates

Physical Properties. Industrially, the most important esters are dimethyl malonate, CH_3OCO—CH_2—$COOCH_3$, and diethyl malonate, C_2H_5 – OCO—CH_2—$COOC_2H_5$, whose physical properties are summarized in Table 2. Both are sparingly soluble in water (1 g/50 mL for the diethyl ester) and miscible in all proportions with ether and alcohol.

Reactions. The chemical properties of malonates are highlighted by the acidity of the methylene group (pK_a ~ 13) to such an extent that a proton can be easily detached by a strong base, usually alkoxides.

Manufacture. The predominant manufacturing processes are the hydrogen cyanide process and carbon monoxide process.

Table 1. Properties of Malonic Acid[a]

Property	Value
mol wt	104.06
melting point, °C	135 dec
solubility	
in water at 20°C	139 g/100 mL
in pyridine at 15°C	15 g/100 g

[a] Also called propanedioic acid or methanedicarboxylic acid.

Table 2. Physical Properties of Dimethyl and Diethyl Malonate

Property	Dimethyl malonate[a]	Diethyl malonate[b]
mol wt	132.12	160.17
mp, °C	−62	−50
bp, °C[c]	181.4	199
d_4^{20}, g/mL	1.1544	1.0551
refractive index at 20°C	1.4140	1.4143

[a] Also called propanedioic acid dimethyl ester. [b] Also called propanedioic acid diethyl ester. [c] At 101 kPa = 1 atm.

Economic Aspects. Dimethyl and diethyl malonates are produced via the carbon monoxide process at Hüls (Germany), Juzen (Japan), and Korean Fertilizers (S. Korea); they are produced via the hydrogen cyanide process at Lonza (Switzerland) and Tateyama (Japan). Total capacity is estimated to be about 12,000 t/yr. Furthermore, producers are also reported in the People's Republic of China and in Romania.

Health and Safety Factors. Dimethyl malonate and diethyl malonate do not present any specific danger of health hazard if handled with the usual precautions. Nevertheless, inhalation and skin contact should be avoided.

Diisopropyl Malonate. This dialkyl malonate has gained industrial importance for the synthesis of the fungicide isoprothiolane through condensation with carbon disulfide and ethylene dichloride. Diisopropyl malonate is produced by Mitsubishi Chemical (Japan) using the carbon monoxide process.

Cyanoacetic Acid and Cyanoacetates

Physical Properties. The physical properties of cyanoacetic acid $N{\equiv}C{-}CH_2COOH$ are summarized in Table 3. The industrially most important esters are methyl cyanoacetate and ethyl cyanoacetate. Both esters are miscible with alcohol and ether and immiscible with water.

Reactions. The chemical properties of cyanoacetates are quite similar to those of the malonates.

Manufacture. Cyanoacetic acid and cyanoacetates are industrially produced by the same route as the malonates using a modified hydrogen cyanide process, starting from a sodium chloroacetate solution via a sodium cyanoacetate solution.

Economic Aspects. In order to avoid the extraction and evaporation steps, most of the cyanoacetic acid derivatives are made directly from solution; therefore, only a small portion of the acid produced is traded. Cyanoacetic acid is produced by Boehringer-Ingelheim and Knoll (Germany), Juzen (Japan), as well as Hüls (U.S.).

Methyl cyanoacetate and ethyl cyanoacetate are produced by Lonza (Switzerland) and Huls (U.S.), as well as Juzen and Tateyama (Japan). The total production capacity is estimated to be in the range of 10,000 metric tons per year.

Health and Safety Factors. Handling of cyanoacetic acid and cyanoacetates do not present any specific danger or health hazard if handled with the usual precautions.

Uses. In many cases cyanoacetic acid, cyanoacetates, or cyanocetamide can be used alternatively. The traded cyanoacetic acid is mainly intended for the synthesis of the cough remedy dextromethorphan (see EXPECTORANTS, ANTITUSSIVES, AND RELATED AGENTS) and of the fungicide cymoxanil (see FUNGICIDES, AGRICULTURAL). Otherwise cyanoacetic acid is directly converted as a solution with 1,3-dimethylurea into 2-cyano-*N,N'*-dimethylcarbamoyl acetamide which is further upgraded into the diuretics theophylline and caffeine.

The largest application of methyl and ethyl cyanoacetate is the production of the cyanoacrylate adhesives (qv) widely used within the car and electronic industries.

Malononitrile

Physical Properties. The physical properties of malononitrile $N{\equiv}C{-}CH_2C{\equiv}N$ are listed in Table 4.

Reactions. As in the case of malonates and cyanoacetates, the chemical properties of malononitrile are determined by two reactive centers, namely the methylene group and the two cyano functions. A peculiar reaction of malononitrile is the base-catalyzed dimerization leading to 2-amino-1,1,3-tricyanopropene.

Table 3. Physical Properties of Cyanoacetic Acid, Methyl Cyanoacetate, and Ethyl Cyanoacetate

Property	Cyanoacetic acid	Methyl cyanoacetate	Ethyl cyanoacetate
mol wt	85.06	99.09	113.12
mp, °C	66	−22	−22
bp °C/kPa[a]	108°/1.5 kPa	203°/101 kPa	206°/101 kPa

[a] To convert kPa to mm Hg, multiply by 7.5.

Table 4. Physical Properties of Malononitrile[a]

Property	Value
mol wt	66.06
mp, °C	31
bp at 76 kPa[b], °C	218–219
d_4^{35}, g/mL	1.0494
solubility in water, g/mL	0.133

[a] Also called propanedinitrile and dicyanomethane. [b] To convert kPa to mm Hg, multiply by 7.5.

Manufacture. Malononitrile can be produced batchwise by elimination of water from cyanoacetamide with phosphorous pentachloride. Most of it is now produced continuously starting from cyanogen chloride and acetonitrile in a high temperature gas phase reaction.

Removal of maleic and fumaric acids from the crude malononitrile by fractional distillation is impractical because the boiling points differ only slightly. The impurities are therefore converted into high boiling compounds in a conventional reactor by means of a Diels-Alder reaction with a 1,3-diene.

Economic Aspects. Malononitrile is produced by Lonza Ltd. (Switzerland) using the cyanogen chloride process.

Health and Safety Factors. Malononitrile is usually available as a solidified melt in plastic-lined drums. Remelting has to be done carefully because spontaneous decomposition can occur at elevated temperatures, particularly above 100°C, in the presence of impurities such as alkalies, ammonium, and zinc salts. Occupational exposure to malononitrile mainly occurs by inhalation of vapors and absorption through the skin. Malononitrile has a recommended workplace exposure limit of 8 mg/m^3.

Uses. Malononitrile is extensively used in the life sciences industry. The most important products are vitamin B_1 (thiamine) and bensulfuron-methyl, a sulfonyl urea herbicide. Most other product uses fall under the *N*-containing heterocycles.

PETER POLLAK
GÉRARD ROMEDER
Lonza Ltd.

A. J. Fatiadi, *Synthesis* **3**, 165 (1978).
F. Freeman, *Synthesis* **12**, 925 (1981).
Ger. Offen. 2,329,251 (Dec. 13, 1973), H. Marketz (to Lonza Ltd.).
Ger. Offen. 2,741,383 (Feb. 2, 1979), E. Catalucci (to Lonza AG).

MALTS AND MALTING

Malting is essentially the same process as occurs when seeds fall to the ground or are planted, are moistened by water (qv), and germinate. The terms malt and malting can apply to any germinated grain; however, nearly all commercial malting involves barley. Because the brewing process and finished beer characteristics are a function of malt properties, malting is considered to be a part of the brewing process.

Manufacturing and Processing

Raw Materials. Two principal types of malting-grade barley are in use, ie, six-row and two-row. The main growing areas for barley are North Dakota, Montana, eastern South Dakota, and western Minnesota; six-row barley is predominant. Increasingly significant areas are California, Oregon, Washington, Idaho, and Colorado, where predominantly two-row barley is produced. Less than one-half of the barley grown in the U.S. is processed by the malt industry; the remainder is used as animal feed, and ca 80% of the barley used by the malting industry is the six-row variety (see FEEDS AND FEED ADDITIVES, PET FOODS).

Barley varieties recommended by the American Malting Barley Association, Inc. (Milwaukee, Wisc.) are used for producing brewers' malt. Anheuser-Busch supplements these varieties with some of their own malting barley varieties, whereas Adolph Coors primarily uses their own barley variety (Moravian III). The main malting-grade varieties for two-row barley are Harrington, Moravian III, and B1202 and, for six-row barley, Robust, Excel, Morex, and Azure.

Most of the malting barleys in the world are two-row varieties; these are characterized by larger berries, lower protein content, lower enzyme activity, and higher extract than the predominant six-row varieties used in the United States.

Processing. The U.S. malting process consists of three basic steps: steeping, germination, and kilning (Fig. 1). Foreign malting plants generally are smaller and cover a wide spectrum of configurations; eg, conventional compartments, large drums, semicontinuous (Wanderhaufen) units, towers, continuous (Domalt) units, fleximalt (combined germination and kilning), circular units, and many others. Of special interest is the French development (Nordon et Cie, Nancy, France) of a large flat-bottomed steeping unit which is provided with a reversible turner that can be used to both load and unload barley, and that is easily adapted to aerobic steeping.

New Technology. *Barley Breeding.* The barley breeding programs involve conventional crossbreeding techniques and have resulted in barley varieties of better yield, disease resistance, and malt quality (National Barley and Malt Laboratory, Madison, Wisc.; American Malting Barley Association, Milwaukee, Wisc.; Brewing and Malting Barley Research Institute, Winnipeg, Canada).

A substantical breakthrough in improving malting quality through the use of mutagenic techniques is the development of experimental proanthocyanidin-free varieties which can yield beer that is colloidally stable and thus does not require stabilization in the brewery.

Growth Regulators. Perhaps the most significant scientific contribution to malting technology has been the use of gibberellic acid and an increased understanding of its role in the malt modification process.

Nitrosamines and Kilning. In order to avoid the formation of nitrosamines, which are attributed to the reaction of nitrogen oxides and amines that are present in barley, either indirect heating of kiln air or using low Nox burners (Maxon Corp., Muncie, Ind.) on direct-fired kilns are being applied (see *N*-NITROSAMINES). The introduction of small amounts of sulfur dioxide during the early stages of kilning also reduces formation of nitrosamines.

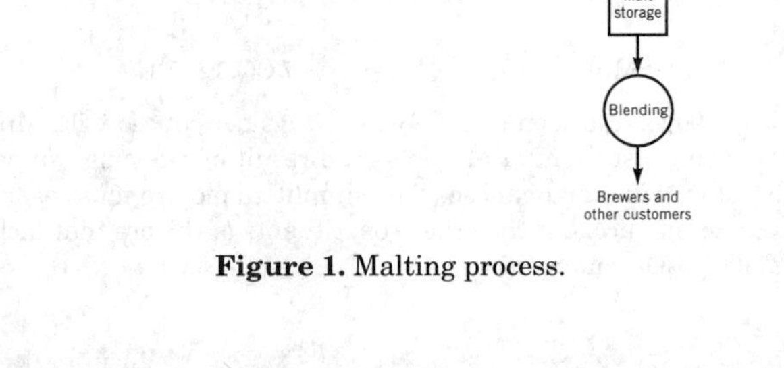

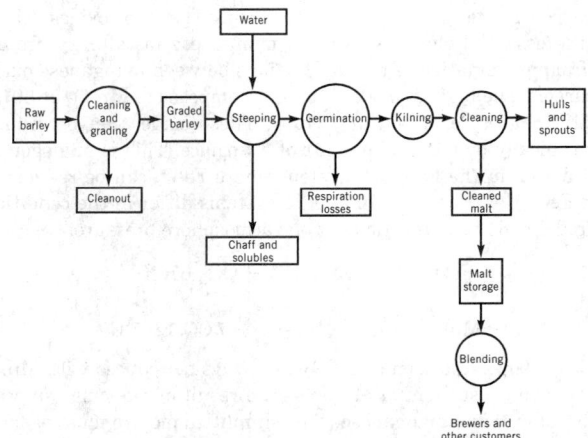

Figure 1. Malting process.

Specialty Malts and Malt Substitutes. Specialty malts, malt substitutes, and alternative brewing technology have the potential to lower brewers' malt usage in favor of lower cost grain bills and processes, but no significant trends among large U.S. brewers to dramatically lower malt usage have been detected.

Economic Aspects

Malt Production and Producers. Because approximately 95% of malt manufactured is used to make beer, malt production follows trends in beer production. Distillers and food malts account for approximately 5% of the U.S. and world malt population.

World usage of malt is slightly higher than that in the U.S. because U.S. brewers use higher quantities of corn and rice adjuncts than foreign brewers. Malt usage in the U.S. has also been declining for many years because of the popularity of lower calorie and lower alcohol beers, which require less malt to make.

Unless changes in malt composition, brewing technology, or beer products develop, it is probable that the unit usage will not decrease substantially below the current level. However, continued growth in lower calorie beer, cost pressures to substitute lower cost adjuncts for malt in the brewing process, or changes in high gravity and other brewing technology could reduce usage.

U.S. and Canadian maltsters are shown in Table 1, along with a range of estimated annual capacities. Since the 1970s, the number of malting companies in the United States has decreased due to mergers, acquisitions, and the closing of smaller malting companies and individual malthouses.

Commercial information on the U.S. malting industry can be obtained from the Beer Institute (Washington D.C.). Data on barley can be obtained from the America Malting Barley Association, Inc. (Milwaukee, Wisconsin). Canadian statistics are available from the

Table 1. U.S. and Canadian Maltsters

Company	Plant location(s)	Est. malt capacity, 10^3 t/yr
Adolph Coors Co.	Golden, Colo.	150–300
Anheuser-Busch Co., Inc.	Idaho Falls, Idaho Manitowoc, Wis. Moosehead, Minn.	400–600
Breiss Malting Co.	Chilton, Wis.	<15
Canada Malting Co., Ltd.	Calgary, Alberta Montreal, Quebec Thunder Bay, Ontario	400–600
Dominion Malting Co.[a]	Winnipeg, Manitoba	5–150
Fleischmann Kurth Malting Co.[b]	Chicago, Ill. Manitowoc, Wis. Milwaukee, Wis. Red Wing, Minn.	300–400
Froedtert Malting Corp.[c]	Milwaukee, Wis. Winona, Minn.	300–400
Great Western Malting Co.[d]	Los Angeles, Calif. Pocatello, Idaho Vancouver, Wash.	300–400
Ladish Malting Co.[e]	Jefferson Junction, Wis. Spiritwood, N.D.	400–600
Miller Brewing Co.	Waterloo, Wis.	<30
Minnesota Malting Co.	Cannon Falls, Minn.	50–150
Prairie Malt Ltd.[f]	Biggar, Saskatchewan	50–150
Rahr Malting Co.	Shakopee, Minn.	300–400
Schreier Malting Co.	Sheboygan, Wis.	50–150
Stroh Brewery Co.	St. Paul, Minn.	<30
Westcan Malting Ltd.[g]	Alix, Alberta	50–150

[a] Partly owned by ADM, Inc. [b] Owned by ADM, Inc. [c] Owned by Grand Malteries Modern (France). [d] Owned by Canada Malting Co. [e] Owned by Cargill, Inc. [f] Majority owned by Schreier. [g] Partly owned by Rahr.

Canadian Grain Commission and Brewing and Malting Barley Research Institute (Winnipeg, Canada).

Malthouses being built in the United States are being designed for large batch sizes (150–200 t malt) and are highly automated to reduce labor and utility costs. However, unless the beer market begins to grow or new uses for malt are found, it is likely that few new malthouses will be built.

Health and Safety Factors. Dust-control systems and good housekeeping are employed in the barley and malt elevators, steephouse, and other processing areas to eliminate or minimize the potential of dust explosions and inhalation. Although low levels of sulfur dioxide are employed (0.1–2.0 kg S/(h·t malt)), the potential of toxic sulfur dioxide concentrations resulting from process or operator error does exist. Fumigants and various cleaning agents are used routinely in the malting industry in accordance with safe operating practices.

Specialty Products and By-Products

A wide variety of special malts are produced which impart different flavor characteristics to beers. These malts are made from green (malt that has not been dried) or finished malts by roasting at elevated temperatures or by adjusting temperature profiles during kilning. A partial list of specialty malts includes standard malts, ie, standard brewers, lager, ale, Vienna, and wheat; caramelized malts, ie, Munich, caramel, and dextrine; and roasted products, ie, amber, chocolate, black, and roasted barley.

Only two companies produce specialty malts in roasters or specialty kilns in North America: Breiss Malting Co. (Chilton, Wisconsin) and Extractos y Maltas (Mexico City). Specialty malts represent less than 2% of malt sold in North America.

Malt syrups, which are extracts of conventional or specialty malts, are produced by three companies in the United States: Breiss Malting Co., Malt Products Corp., and Crompton & Knowles Corp. Malt extracts are used in a variety of food applications and by microbrewers and home brewers. The main by-products from the malting industry are malt sprouts, cleanout material, and small-kerner barley.

JAMES A. DONCHECK
Bio-Technical Resources, L.P.

A. W. MacGregor and R. S. Bhatty, eds., *Barley: Chemistry and Technology,* American Association of Cereal Chemists, Inc., St. Paul, Minn., 1993.

D. E. Briggs and co-workers, *Malting and Brewing Science,* Vol. 1, 2nd ed., Chapman and Hall, New York, 1981.

E. B. Adamic, *The Practical Brewer,* Master Brewers Association of the Americas, Madison, Wis., 1977, pp. 21–39.

MANGANESE AND MANGANESE ALLOYS

Manganese, atomic number 25, atomic weight 54.94, belongs to Group 7 (VII) in the Periodic Table. Its isotopes are ^{51}Mn, ^{52}Mn, ^{54}Mn, ^{55}Mn, and ^{56}Mn, but ^{55}Mn is the only stable one. Manganese, a gray metal resembling iron, is hard and brittle and of little use alone. Its principal use in the metallic form is as an alloying element and cleansing agent for steel, cast iron, and nonferrous metals (see METAL SURFACE TREATMENTS). Manganese is essential to the steel (qv) industry where it is used mostly as a ferroalloy. After iron (qv), aluminum (see ALUMINUM AND ALUMINUM ALLOYS), and copper (qv), manganese ranks along with zinc as the next most used metal (see also MANGANESE COMPOUNDS).

Properties

Table 1 lists properties of manganese.

Table 1. Properties of Manganese

Property	Value
melting point, °C	1244
boiling point, °C	2060
density at 20°C, g/cm^3	7.4
specific heat at 25.2°C, J/g	0.48

Minerals

Manganese, which occurs in many minerals widely distributed in the earth's crust, constitutes about 0.1% of the earth's crust and is the twelfth most abundant element. The principal sources of commercial grades of manganese ore for the world are found in Australia, Brazil, Gabon, the Republic of South Africa, and Ukraine. The chief minerals of manganese are pyrolusite, romanechite, manganite, and hausmannite. There is also wad, which is not a definite mineral but is a term used to describe an earthy manganese-bearing amorphous material of high moisture content.

Ores

In general, only ores containing at least 35% manganese are classified as manganese ores. Ores having 10–35% Mn are known as ferruginous manganese ores, and ores containing 5–10% manganese are known as manganiferrous ores. Ores containing less than 5% manganese with the balance mostly iron are classified as iron ores. No manganese ores of commercial value are to be found in the United States.

Deep-Sea Manganese Nodules. A potentially important future source of manganese is the deep-sea nodules found over wide areas of ocean bottom (see OCEAN RAW MATERIALS). At depths of 4–6 km, billions of metric tons of nodules are scattered over the ocean floor in concentrations of up to 100,000 t/km^2.

Impurities. Impurities usually found in manganese ore may be classified into metal oxides, eg, iron, zinc, and copper; gangue; volatile matter such as water, carbon dioxide, and organic matter; and other nonmetallics. In practice, ores are blended to provide the most economical mixture consistent with the specifications for ferromanganese.

The particle size of manganese ores is an important consideration for the smelting furnace. In general, the ore size for the furnace charge is −75 mm with a limit to the amount of fines (−6 mm) allowed. Neither electric furnaces nor blast furnaces operate satisfactorily when excessive amounts of fines are in the charge.

Most of the principal producers of ferromanganese operate sintering plants as the means of agglomeration of manganese ore fines.

Alloy Processing

Smelting. The greatest application of manganese is in ferrous metallurgy (qv). Manganese alloys are employed as refining agents as well as alloying additions.

Process Chemistry. Manganese is combined with oxygen in its ores, and carbon is the most economical reducing agent for oxides. Therefore, the essential characteristics of manganese metallurgy are evident from examination of the interactions between manganese oxides and carbon. The highest oxide, MnO_2, decomposes to Mn_2O_3 at 507°C, and Mn_2O_3 goes to Mn_3O_4 at 1240°C. These oxides can be reduced exothermically by CO in the shaft of a furnace. This is analogous to the situation in the Fe–O–C system where Fe_2O_3 can be reduced by CO to FeO. However, the Mn and Fe systems differ in the conditions required for the final reactions at one atmosphere pressure:

$$FeO + C \rightarrow Fe + CO + CO_2 \quad 670°C$$

$$7\,MnO + 10\,C \rightarrow Mn_7C_3 + 7\,CO \quad 1267°C$$

From the above, it is seen that Fe will always be reduced with Mn. Fe is a common constituent of Mn ores. The result of reducing Mn ore is an alloy of iron and manganese. This simplified picture must be modified to take into account the other constituents of the ore that include FeO, CaO, Al_2O_3, and SiO_2.

Slag composition influences the amount of manganese recovered as metal. Basic slags, those containing CaO, increase the activity of MnO and therefore promote manganese reduction. Such slags also suppress the activity of SiO_2, thus limiting the introduction of silicon into the alloy.

Because of the limitations imposed by the presence of gangue and the relatively high volatility of manganese, the products obtained when reducing Mn ore with carbon are a carbon saturated alloy and a slag containing manganese. Under suitable conditions the manganese content of the slag can be sufficiently low to justify discarding the slag. Otherwise, the slag is smelted carbothermically, using SiO_2 additions and higher temperature, producing an alloy of silicon and manganese, which is low in carbon, and a discardable slag of low manganese content. Manganese alloys containing lower carbon are produced by refining carbon saturated manganese or silicon manganese alloys.

High Carbon Ferromanganese. Ferromanganese, also known as high carbon ferromanganese, or in the United States as standard ferromanganese, is the largest tonnage manganese alloy used in the steel industry. Of the overall average usage of manganese in steel, 75% is supplied as ferromanganese. The countries in the world that produce ferromanganese and silicomanganese are: Argentina, Australia, Belgium, Brazil, Bulgaria, Canada, Chile, China, Croatia, Czechoslovakia, France, Georgia, Germany, India, Italy, Japan, Korea (North), Korea (South), Mexico, Norway, Peru, the Philippines, Ukraine, and the U.S.

Ferromanganese is produced in blast furnaces and electric smelting furnaces. Economics usually determine which smelting process is chosen for ferromanganese. Both methods require about the same amount of coke for reduction to metal, but in the case of the blast furnace, the thermal energy required for the smelting process is supplied by the combustion of additional coke, which in most countries is a more expensive form of energy than electricity.

Capital requirements for a new facility generally favor the electric furnace process. However, in some countries having integrated steel industries, the availability of excess ironmaking blast furnaces, metallurgical coke, and relatively high cost of electricity, the blast furnace is an attractive choice.

Figure 1 shows a typical electric furnace for producing ferromanganese. Ore and reducing agent are introduced to the top of the furnace. Electric energy is conducted to the charge through carbon electrodes. Slag and metal are withdrawn from the hearth.

The *high manganese slag practice* is used by most plants where high grade manganese ores are smelted and silicomanganese is also produced. Manganese content of this slag ranges from 30–42%.

The *discard slag practice* is followed if the ore is of such low quality that a high degree of manganese extraction is required to achieve the alloy grade, or if the ore contains basic oxides, eg, CaO and MgO, which leads naturally to low manganese slags. Manganese content of the slag from this practice ranges from 10 to 20% and manganese recovery in alloy ranges between 85 and 90%.

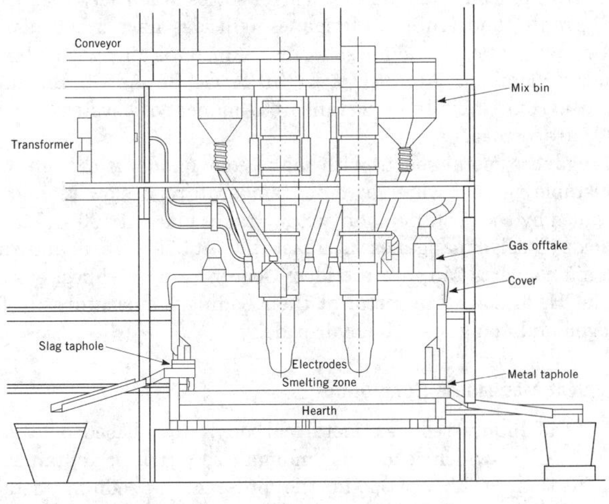

Figure 1. Covered ferromanganese furnace.

Silicomanganese. Silicomanganese is an alloy of manganese and iron containing 12.5 to 18.5% silicon which provides the steel industry with a convenient source of the two most important alloying and deoxidizing elements (Mn,Si) consumed in the production of steel. The production process is similar to that of electric furnace smelting of ferromanganese. It differs in the furnace charge which contains large amounts of quartz (SiO_2) and, if required to adjust slag composition, limestone or dolomite. Smelting temperature is higher and the off-gas is predominantly carbon monoxide. Although silicomanganese can be produced directly from ore, frequently slag from high carbon ferromanganese and drosses from refined ferromanganese operations are used as sources of Mn.

A low carbon grade of silicomanganese containing 28–32% Si and <0.06% C is usually made by a two-stage process. In the first step, low iron silicomanganese containing 16 to 18% silicon and about 2% carbon is made by smelting quartz and ferromanganese slag which is depleted in iron. Subsequently, in a separate furnace process, a mixture containing the crushed low iron silicomanganese, quartz, and coal (qv) or coke is smelted in a slagless process where the quartz is reduced to silicon that displaces the carbon in the remelted silicomanganese. This product is mainly used as the reducing agent in a silicothermic process to produce the low carbon grade of refined ferromanganese.

Refined Ferromanganese. Refined ferromanganese refers to alloys that are not carbon saturated and range from less than 0.10 to 1.50% maximum carbon. Medium carbon grades are used in special grades of steels where in final additions carbon control is important. The low carbon grades are used mainly in the production of certain grades of stainless steels.

Originally CaO and Mn ore mixtures were reacted with silicon in silicomanganese or low carbon silicomanganese to produce medium and low carbon ferromanganese alloys.

In 1971, oxygen refining of high carbon ferromanganese was introduced as a method for producing medium carbon ferromanganese. This manganese oxygen refining (MOR) process is similar to that used by the steel industry to produce steel in the basic oxygen furnace (BOF). The low carbon grade of ferromanganese must still be made by the silicon reduction method.

Manganese Nitride. Manganese nitride is an addition agent for steel where both manganese and nitrogen are required in certain grades of steel. Briquettes of comminuted medium carbon ferromanganese are nitrided in an annealing-type furnace in a nitrogen atmosphere.

Electrolytic Processes

Electrolytic processes include electrolysis of aqueous solutions, which yields high purity Mn, and fused-salt electrolysis.

Uses of Metallic Manganese

Manganese is essential to the production of steel. It is found in varying amounts in all steels and cast iron. About two-thirds of the manganese used in steel is as an alloying element to enhance the hardenability, strength, and other mechanical properties of steel. The remaining third of the manganese is used in steel to combine with the residual sulfur in iron to prevent hot shortness during the rolling process for steel and to improve the deoxidizing effect of aluminum and silicon in molten steel. The bulk of steel production is of the multipurpose low carbon steels containing from 0.15% to 0.8% manganese. In cast iron, manganese neutralizes the sulfur in the iron and adds strength to the final casting. Electrolytic Mn is employed as an alloying element in the nonferrous industry.

Economic Aspects

Outlook. Because the world iron and steel industry consumes about 94% of the manganese mined, the market for manganese is largely dependent on the world steel market. The marked decrease in manganese ore production for the five-year period of 1988–1992 is a reflection of the worldwide decline in the volume of steel production during that same period. Improvements in steelmaking processes that use ferromanganese alloy additions more efficiently along with changes in steel grades that require less manganese have also lessened manganese consumption.

Health and Safety Factors

Health and Environment. Manganese in trace amounts is an essential element for both plants and animals and is among the trace elements least toxic to mammals, including humans. Exposure to abnormally high concentrations of manganese, particularly in the form of dust and fumes, is, however, known to have resulted in adverse effects to humans (see MINERAL NUTRIENTS). Two kinds of disease owing to manganese are known in humans: manganic pneumonia and manganism.

Airborne manganese concentrations in the U.S. range from 0.02 to 0.57 $\mu g/m^3$ in urban areas and 0.0017–0.047 $\mu g/m^3$ in nonurban areas. The ACGIH recommends a TLV of 5 mg/m^3.

Plant Safety. Of the many ferroalloy products produced in electric furnaces, ferromanganese has the greatest potential for furnace eruptions or the more serious furnace explosions.

Most of the serious eruptions of manganese furnaces can be traced to a set of conditions that cause bridging or hang-up of the charge materials so that the normal downward movement through the furnace is disrupted or retarded. Safe operation of ferromanganese furnaces requires careful control of raw material particle size, oxygen content of the ore blend, and charge stoichiometry.

Most modern furnaces are equipped with computers that log raw material usage and other operating data. Many of the larger furnaces have computer systems that are programmed to continually monitor operating data and to automatically make adjustments to obtain optimum performance. The benefits of computer controlled electric smelting furnaces include safer and more efficient operation, as well as increased productivity and lower costs (see PROCESS CONTROL).

LOUIS R. MATRICARDI
JAMES DOWNING
Consultants

B. Wellbeloved, P. M. Craven, and J. Wauzby, in *Ullmann's Encyclopedia of Industrial Chemistry,* Vol. A16, Wiley-VCH, Weinheim, Germany, 1986.

G. Volkert and co-workers, in *Metallurgie der Ferrolegierungen,* Springer, New York, 1972.

MANNITOL. See SUGAR ALCOHOLS.

MANUFACTURED GAS. See FUELS, SYNTHETIC–GASEOUS FUELS.

MANGANESE COMPOUNDS

Manganese is the twelfth most abundant element in the earth's crust and is the fourth most used metal following iron, aluminum, and copper (see also MANGANESE AND MANGANESE ALLOYS). The most used manganese compound of the 1990s, outside of the manganese–iron alloy, ferromanganese, used in steelmaking (see STEEL), is manganese dioxide. Manganese, at. no. 25, belongs to the first transition series and is the principal member of Group 7 (VIIA). It has nine isotopes.

Ground-state electronic configuration is $1s^22s^22p^63s^23p^63d^54s^2$. Manganese compounds are known to exist in oxidation states ranging from −3 to +7. Both the lower and higher oxidation states are stabilized by complex formation. In its lower valence, manganese resembles its first row neighbors chromium and especially iron in the Periodic Table. Commercially the most important valances are Mn^{2+}, Mn^{4+}, or Mn^{7+}.

As the oxidation state of manganese increases, the basicity declines: eg, from MnO to Mn_2O_7. Oxyanions are more readily formed in the higher valence states. Another characteristic of higher valence-state manganese chemistry is the abundance of disproportionation reactions.

$$2\ Mn^{3+} \rightarrow Mn^{2+} + Mn^{4+}$$

$$2\ Mn^{5+} \rightarrow Mn^{4+} + Mn^{6+}$$

$$3\ Mn^{6+} \rightarrow Mn^{4+} + 2\ Mn^{7+}$$

There are approximately 250 known manganese minerals. The primary ores, which typically have a Mn content >35%, usually occur as oxides or hydrated oxides, or to a lesser extent as silicates or carbonates. The manganese-containing minerals of economic significance are pyrolusite, braunite, nsutite, manganite, psilomelane, cryptomelane, hausmannite, jacobsite, bixbyite, rhodonite, rhodochrosite, bementite, todorokite, and ramsdellite. Battery-grade manganese dioxide ores are composed predominantly of nsutite, cryptomelane, and todorokite.

The world's supply of commercial manganese ore comes from Australia, Brazil, Gabon, and the Republic of South Africa.

Deep-sea manganese nodules represent a significant potential mineral resource. Whereas the principal constituent of these deposits is manganese, the primary interest has come from the associated metals that the nodules can also contain (see OCEAN RAW MATERIALS). Recovery is considered an economic potential in the northwestern equatorial Pacific, and to a lesser degree in the southern and western Pacific and Indian Oceans. The U.S. manganese production from domestic ore stopped in 1970. The U.S. depends on imports for its manganese needs and maintains sizable stockpiles for emergencies.

Whereas hydrogen does not react with manganese to form a hydride, hydrogen is soluble to some extent in manganese metal. Exposure of manganese to oxygen leads to the ready formation of manganese oxides, especially at higher temperatures. Nitrogen above 740°C forms solid solutions, as well as several nitrides; such as MnN, Mn_6N_5, Mn_3N_2, Mn_2N, and Mn_4N. Manganese nitrides are used in steelmaking as nitrogen-containing intermediate alloys (see NITRIDES).

Carbon reacts with molten manganese, forming various carbides including $Mn_{23}C_6$, Mn_3C, Mn_7C_3, Mn_2C_7, and $Mn_{15}C_4$. Manganese tricarbide reacts with water to yield about 75% H_2, 12–15% CH_4, and 6–8% ethylene. It is an important factor in a fuel–alloy process designed to produce liquid hydrocarbons (qv). In steel and other ferrous alloys, manganese carbides (qv) achieve the desired mechanical properties. With silicon, manganese forms a series of silicides, eg, Mn_3Si, Mn_5Si_3, MnSi, and $MnSi_{1.7}$. Manganese silicides have excellent heat-resisting properties. Manganese forms compounds only with a limited number of metals, ie, Au, Be, Zn, Al, In, Ti, Ge, Sn, As, Sb, Bi, Ni, and Pd. In the commercially important iron–manganese system, no compounds are formed.

Manganese metal reacts with many compounds. Although Mn is fairly stable against water at room temperature, a slow reaction accompanied by the evolution of hydrogen takes place at 100°C. Most dilute acids dissolve manganese at a fast rate. At 350–875°C, anhydrous ammonia converts Mn into nitrides. Concentrated alkalies, eg, KOH and NaOH, dissolve Mn metal at their boiling temperatures to form hydrogen and manganese(II) hydroxide.

Low Valent Manganese Compounds

A family of highly reduced metal carbonyls (qv) based on the anion $Mn(CO)_4^{3-}$ which contains manganese(−III), is obtained by the reduction of $Mn_2(CO)_{10}$ in the presence of sodium metal in hexamethylphosphoramide.

The manganese(0) compound, $Mn_2(CO)_{10}$, has yellow crystals, mp 154–155°C, and sublimes *in vacuo*. The metal–metal bond is 29.3 pm and has an estimated strength of 142 kJ/mol (34 kcal/mol). This compound is obtained from the monovalent methylcyclopentadienylmanganese tricarbonyl (MMT), $C_9H_7Mn(CO)_3$, by reduction with sodium in diglyme in the presence of CO under pressure. Manganese carbonyl is the parent compound for a large family of manganese carbonyl compounds. Compounds having manganese in the univalent positive state exist only as complexes.

Divalent Manganese

Divalent manganese compounds are stable in acidic solutions but are readily oxidized under alkaline conditions. Most soluble forms of manganese that occur in nature are of the divalent state. Manganese(II) compounds are characteristically pink to colorless, with the exception of MnO and MnS which are green, and $Mn(OH)_2$, which is white. The physical properties of selected manganese(II) compounds are given in Table 1.

Trivalent Manganese

The Mn^{3+} ion is so unstable that it scarcely exists in aqueous solution. In the solid phase, the most stable forms of Mn(III) are manganese sesquioxide, Mn_2O_3, and its hydrate $Mn_2O_3 \cdot nH_2O$, and manganese oxide, Mn_3O_4, which is thermally the most stable manganese oxide. Physical properties of manganese(III) compounds are given in Table 2.

Compounds of trivalent manganese can be made either by oxidation of corresponding manganous compounds or by reduction of the more highly oxidized compounds. The color of Mn(III) compounds in the solid state can vary from red to green. The corresponding aqueous solutions mostly have a reddish purple, almost permanganate-like appearance.

Manganese(III) Oxides. The sesquioxide, Mn_2O_3, exists in dimorphic forms. The α-Mn_2O_3 exists in nature as the mineral bixbyite. Synthetic α-Mn_2O_3 is prepared by the thermal decomposition of the nitrate, dioxide, carbonate, oxalate, or chloride in air in the temperature range of 500–800°C.

Table 1. Physical Properties of Manganese(II) Compounds

Compound	Density,[a] g/cm³	Mp, °C	Bp, °C	Solubility
manganese acetate tetrahydrate	1.589			sl sol H_2O, sol ethanol, methanol
manganese borate				insol H_2O, ethanol, sol dil acids
manganese carbonate[b]	3.125	dec > 200		sol prod H_2O: 8.8×10^{-11} sol in dil acids
manganese chloride	2.977_{25}	650	1190	v sol H_2O, sol pyridine, ethanol, insol ether
manganese hydroxide[c]	3.26_{25}	dec 140		sol acid, sol base at higher temp
manganese nitrate hexahydrate	1.82	25.8	129.4	v sol, H_2O, sol ethanol
manganese(II) oxide[d]	5.37_{23}	1945		insol H_2O
manganese sulfate	3.25	700	dec 850	sol 52 g/100 g H_2O, sl sol methanol, insol ether
manganese dihydrogen phosphate dihydrate		$-H_2O$, 100	sol H_2O, insol ethanol, deliquescent	

[a] Temp. in °C of readings given as subscript. [b] Also known as rhodochrosite. [c] Also known as pyrochroite. [d] Also known as manganosite.

Table 2. Physical Properties of Manganese(III) Compounds

Compound	Density, g/m³	Mp, °C	Solubility
trimanganese tetraoxide[a] α-phase[b]	4.84	1560	insol H_2O
manganese(III) acetate dihydrate			dec H_2O
manganese(III) acetylacetonate		172	insol H_2O, sol org solv
manganese(III) fluoride	3.54	dec (stable to 600)	dec H_2O
α-manganese(III) oxide	4.89_{25}	871–887 dec	insol H_2O
γ-manganese(III) oxide, hydrated	4.2–4.4	250 dec to Mn_2O_3	insol H_2O, disproportionates in dilute acids

[a] Mixed Mn(II), Mn(III) valent compound. [b] Also known as hausmannite.

The mixed valent oxide Mn_3O_4 occurs in nature as the mineral hausmannite. Mn_3O_4 is the most stable of the manganese oxides, and is formed when any of the other oxides, or hydroxides, are heated in air above 940–1000°C.

As seen in some iron oxides, both γ-Mn_2O_3 and Mn_3O_4 have pseudo-spinel structure and tetragonal symmetry. Both oxides can be represented by the general formula $Mn^{II}Mn^{III}{}_2O_4$ where the Mn(II) ions occupy tetrahedral sites and the Mn(III) ions occupy the octahedral sites of the spinel. Thus, potassium maganate(III), $KMnO_2$, is prepared by heating Mn_2O_3 in the presence of KOH. Needle-shaped gray crystals that are readily oxidized by O_2 to higher alkali manganites are formed.

Tetravalent Manganese

By far the most significant manganese(IV) compound is the dioxide MnO_2, found in nature as pyrolusite, a black mineral. There is also a hydrated form approximating $MnO_2 \cdot 2H_2O$, which is formed by precipitation from solutions. The dioxide is a reasonably good conductor and is insoluble in water. This lack of aqueous solubility is responsible for much of its stability, because the Mn(IV) ion is unstable in solution. Physical properties of manganese(IV) compounds are given in Table 3.

In acid solution, MnO_2 is an oxidizing agent, and is used as such in industry.

Most of the simple halides, MnX_4, are unknown except for blue manganese tetrafluoride. Manganese(IV) is amphoteric, appearing as the cation in salts and as an anion in compounds known as manganites, M_2MNO_3, where M is monovalent.

The principal complexes of Mn^{4+} are of the type K_2MnX_6, where X may be fluoride, chloride, cyanide, or iodate, and as a group these materials are readily hydrolyzed. A triperoxymanganate(IV), $K_2H_2MnO(O_2)_2$, is said to be formed when $KMnO_4$ in 30% KOH is treated with H_2O_2 at −18°C.

Water purification chemistry depends on the oxidation of Mn(II) by a suitable oxidant, eg, O_2 at pH > 9, or at lower pH potassium

Table 3. Physical Properties of Manganese(IV) Compounds

Compound	Density, g/cm³	Mp, °C	Solubility
pentamanganese octaoxide[a]	4.85_{20}	550 dec to a Mn_2O_3	insol H_2O
β-manganese(IV) oxide	5.0_{26}	535	insol H_2O
potassium manganate(IV)	3.071_{25}	1100	dec H_2O, disproportionates

[a] Mixed valent Mn(II), Mn(IV) compound.

permanganate or ozone, resulting in the formation of hydrous manganese dioxide. The colloidal properties of hydrous manganese oxide have been studied and exploited in the treatment of potable water.

Manganese Oxides. Manganese(IV) dioxide rarely corresponds to the expected stoichiometric composition of MnO_2, but is more realistically represented by the formula $MnO_{1.7-2.0}$, because it invariably contains varying percentages of lower valent manganese. It also exists in a number of different crystal forms, in various states of hydration, and with a variety of contents of foreign ions. Examples of natural and synthetic manganese dioxides are listed in Table 4.

The term *γ-manganese dioxide* is applied to a series of hydrated manganese dioxides of moderate crystallinity that are suitable for battery purposes.

The electrochemically active phases of manganese dioxide, ie, gamma and rho, typically contain approximately 4% by weight chemically bonded or structural water. The chemical and electrochemical reactivity of manganese dioxide has been shown to result from the presence of cation vacancies in the manganese dioxide crystal lattice.

Synthetic active manganese dioxides, prepared by the reduction of permanganate or the pyrolysis of lower valent manganese salts, have been used as mild, selective, heterogeneous oxidation reagents.

Manganese dioxide, in combination with other metal oxides, forms a series of active catalysts that participate in a variety of environmentally important oxidation and decomposition reactions. The manganese-based catalysts for these applications exhibit a long life and high catalytic activity. At moderately elevated temperatures, manganese dioxide catalysts are used for the complete oxidative degradation of many organic compounds. These catalysts are particularly effective for oxygenated compounds such as alcohols, acetates, and ketones.

Manufacture of manganese metal or compound requires the manganese dioxide of the natural ores to be reduced to lower oxides, principally MnO or other Mn(II) salts. This reduction is usually carried out at 600–900°C by roasting finely powdered MnO_2 mixed with ground coal or heavy oil. Alternatively, the presence of gaseous reductants such as carbon monoxide or hydrogen is employed.

Synthetic Manganese Dioxides. Chemical manganese dioxide (CMD) can be prepared by various methods including the thermal decomposition of manganese salts such as $MnCO_3$ or $Mn(NO_3)_2$ under oxidizing conditions. CMD can also result from the reduction of higher valent manganese compounds, eg, those containing the MnO_4^- ion.

Electrolytic Manganese Dioxide. The anodic oxidation of an Mn(II) salt to manganese dioxide dates back to 1830, but the usefulness of electrolytically prepared manganese dioxide for battery purposes was not recognized until 1918. Initial use of electrolytic manganese dioxide (EMD) for battery use was in Japan where usage continues.

The properties of EMD are summarized in Table 5. EMD is strictly a nonstoichiometric manganese dioxide containing 2–5% lower valent manganese oxides and 3–5% chemically bound water.

EMD is prepared from the electrolysis of acidified manganese sulfate solution and can be summarized as follows:

Anode $$Mn^{2+} + 2\,H_2O \rightarrow MnO_2 + 4\,H^+ + 2e^-$$

Cathode $$2\,H^+ + 2e^- \rightarrow H_2$$

Table 4. Manganese Dioxide Crystal Phases

Phase	Example mineral	Description
α-MnO_2	hollandite group	tunnel structure of corner-shared double chains of (MnO_6) octahedra (2 × 2 channels)
β-MnO_2	pyrolusite	single chains of edge-shared (MnO_6) octahedra (1 × 1 channels)
γ-MnO_2	nsutite	regions of single and double chains of edge-shared (MnO_6) octahedra (both 2 × 1 and 1 × 1 channels)
		layers of edge-shared (MnO_6) octahedra
δ-MnO_2	ramsdellite	double chains of edge-shared (MnO_6) octahedra (2 × 1 channels)

Table 5. Properties of Electrolytic Manganese Dioxide

Property	Typical ranges
density, g/cm^3	4.2–4.5
bulk density, g/cm^3	1.7–2.5
particle size, μm	<74
surface area, m^2/g	30–60
MnO_2, wt %	92 (min)

Overall $$Mn^{2+} + 2\,H_2O \rightarrow MnO_2 + 2\,H^+ + H_2$$

Manganese(V) Compounds

Manganese(V) appears to exist only as the oxyanion MnO_4^{3-} and is generally referred to as manganate(V); occasionally the term hypomanganate is used. Selected manganese(V) compounds and their physical properties are given in Table 6. The most important manganese(V) compound is K_3MnO_4, a key intermediate in the manufacture of potassium permanganate.

Manganese(VI) Compounds

The hexavalent state of manganese is represented by a few alkali metal and alkaline-metal salts of manganic acid, H_2MnO_4, which is known only through its sodium, potassium, rubidium, cesium, barium, and strontium salts. Properties of a few of these salts are given in Table 7.

Potassium manganate(VI), precursor of potassium permanganate, is made commercially in either a one- or two-state fusion reaction, or by anodic oxidation of manganese metal or ferromanganese in KOH.

Alkali manganate(VI) salts are used as oxidants in synthetic organic reactions and their reactions have been observed to be similar to permanganate, except that manganate(VI) exhibits lower reactivity. Additionally, solid $BaMnO_4$ in methylene chloride has been reported to achieve high yields for the oxidation of diols to dialdehydes.

Manganese(VII) Compounds

Permanganic acid $HMnO_4$, is conveniently prepared in the laboratory from barium permanganate and sulfuric acid, or by anodic oxidation of ferromanganese in a divided cell using H_2SO_4 as the electrolyte. Physical properties of Manganese(VII) Compounds are given in Table 8.

Table 6. Physical Properties of Manganese(V) Compounds

Compound	Mp, °C	Solubility
barium manganate(V)[a]	dec 960	insol H_2O
lithium manganate(V)	dec >125	sol 3% LiOH at 0°C
potassium manganate(V)[b]	dec 800–1100	v sol H_2O dec, hygroscopic, sol 40% KOH at −15°C
rubidium manganate(V)		
sodium manganate(V)	dec 1250	v sol H_2O, dec, hygroscopic
strontium manganate(V)–strontium hydroxide		insol H_2O

[a] Density = 5.25 g/m^3. [b] Density = 2.78 g/m^3.

Table 7. Physical Properties of Manganese(VI) Compounds

Compound	Density, g/m^3	Mp, °C	Solubility
barium manganate(VI)	4.85	dec 1150	insol H_2O sol product 2.46×10^{-10}
potassium manganate(VI)	2.80_{23}	dec 190	sol H_2O dec, sol KOH
sodium manganate(VI)		dec 300	sol H_2O dec

Table 8. Physical Properties of Manganese(VII) Compounds

Compound	Density, g/m^3	Mp, °C	Solubility
potassium manganate(VI), manganate(VII) double salt			sol H_2O dec
manganese heptoxide	2.396_{20}	5.9	v sol H_2O, hygroscopic
ammonium permanganate	2.22_{25}	dec >70	8 g/100 g H_2O at 15°C
barium permanganate	3.77	dec 95–100	72.4 g/100 g H_2O at 25°C
calcium permanganate tetrahydrate	2.49	dec 130–140	388 g/100 g H_2O at 25°C, deliquescent
cesium permanganate	3.579	dec 320	0.23 g/100 g H_2O at 20°C
lithium permanganate	2.06_{25}	dec 190	71 g/100 g H_2O at 16°C
magnesium permanganate hexahydrate	2.18	dec 130	v sol H_2O, sol CH_3OH, pyridine, glac acetic acid
potassium permanganate	2.703_{20}	dec 200–300	sol H_2O, acetic acid, trifluoroacetic acid, acetic anhydride, acetone, pyriding, benzonitrile, sulfolane
rubidium permanganate	3.23_{25}	dec 250	1.1 g/100 g H_2O at 19°C
silver permanganate	4.27	110 dec	0.92 g/100 g H_2O at 20°C
sodium permanganate	1.972	36.0	v sol H_2O, deliquescent
zinc permanganate hexahydrate	2.45	dec 90–105	v sol H_2O, deliquescent

Manufacture of Potassium Permanganate. Potassium permanganate may be manufactured by the one-step electrolytic conversion of ferromanganese to permanganate, or by a two-step process involving the thermal oxidation of manganese(IV) dioxide of a naturally occurring ore into potassium manganate(VI), followed by electrolytic oxidation to permanganate.

Oxidation Reactions. Potassium permanganate is a versatile oxidizing agent characterized by a high standard electrode potential that can be used under a wide range of reaction conditions. The permanganate ion can participate in a reaction in any of three distinct redox couples, depending on the nature of the reducing agent and the pH of the system. Typically permanganate oxidation reactions are conducted in an aqueous environment, or in organic cosolvents, which exhibit some degree of stability toward the oxidant. Solvents include acetone, acetic acid, acetic anhydride, *t*-butanol, ethanol, pyridine, and trifluoroacetic acid. Permanganate oxidizes hydrogen, carbon monoxide, and hydrogen peroxide under a variety of pH conditions, and the halides under acidic conditions.

Economic Factors

Manganese-containing ores, concentrates, nodules, or synthetic materials are classified based on manganese content as metallurgical-grade, 38–55 wt % Mn, or chemical- or battery-grade, 44–54 wt % Mn. World manganese mine production capacity of manganese (Mn > 35 wt%) and manganiferrous (Mn from 5–35 wt %) ores is on the order of 10×10^6 t/yr based on manganese content.

Total world capacity for electrolytic manganese dioxide (EMD) is estimated to be in the area of 194,500 t/yr, and annual capacity of chemical manganese dioxide (CMD) is estimated to be in the range of 40,000 t.

The U.S. consumption of manganese is distributed among three industries: iron and steelmaking, where 88% of the Mn is consumed; the manufacture of batteries, where 7% is used; and chemical usage, which accounts for the remaining 5%. Purchases of manganese ore are made on the basis of user requirements, individual specifications, and availability.

Health and Safety. Manganese appears to be an essential trace element for all living organisms (see MINERAL NUTRIENTS). Its concentration in organisms primarily depends on the species. Plants contain between 1–700 mg/kg, ocean fish between 0.3–4.6 mg/kg, and muscles of mammals 0.2–3 mg/kg.

Environmentally, manganese-bearing particulate matter is usually removed from air using dust collecting devices such as electrostatic precipitators, filter systems, cyclones, or wet scrubbers (see AIR POLLUTION CONTROL METHODS). In the case of liquids, soluble manganese can be removed from liquid effluents through precipitation as a hydrous oxide by adjustment of the pH to >8.3 using $Ca(OH)_2$, plus the application of an oxidizing agent such as O_2, Cl_2, ClO_2, NaClO, or $KMnO_4$. Aeration is also effective, provided the pH is raised to above 9.4. The final disposal of the resulting manganese dioxide-containing sludges depends on local conditions and regulations. These sludges are usually deposited in landfills.

Inhalation of particulate manganese compounds, such as manganese dioxide, can lead to an inflammatory response in the lungs of both humans and animals. This response is characteristic of all inhalable particulate matter, however, suggesting that the manganese compound is not specifically responsible. General population exposure to manganese compounds in the air in nonurban areas is about 5 ng/m^3; in urban areas, 33 ng/m^3; and in source dominated areas, 135 ng/m^3. In the soil, manganese is estimated to be in the 40–900 mg/kg range. The maximum reported is 7000 mg/kg. The lowest observed adverse effect level (LOAEL) reported for Mn by inhalation is 0.14 mg/m^3. This results in a chronic exposure by inhalation minimal risk level (MRL) of 0.3 μg/m^3. The primary exposure of the general population to manganese compounds is from diet.

There is conclusive evidence from human studies that inhalation exposure to high levels of manganese compounds can lead to a disabling syndrome of neurological effects termed manganism. It has only been documented in workers exposed to high levels of manganese dust or fumes in mines or foundries, typically following several years of exposure.

Human and animal studies indicate that inorganic manganese compounds have a very low acute toxicity by any route of exposure. Regulations and guidelines have been established in many countries for manganese and its compound. Potassium permanganate under RCRA definition meets the criteria of an ignitable waste, and if discarded is considered a hazardous waste.

KENNETH PISARCZYK
Carus Chemical Company

D. B. Wellbeloved, P. M. Craven, and J. W. Waudby, in *Ullmann's Encyclopedia of Industrial Chemistry,* Vol. A16, VCH, Weinheim, Germany, 1990, p. 80.

R. D. W. Kemmit, in J. C. Bailar, H. J. Emeleus, R. Nyholm, and A. F. Trotman-Dickenson, eds., *Comprehensive Inorganic Chemistry,* Vol. 3, Pergamon Press Ltd., Oxford, U.K., 1973, p. 816.

D. Arndt, *Manganese Compounds as Oxidizing Agents in Organic Chemistry,* Open Court Publishing Co., LaSalle, Ill., 1981, pp. 169–177.

Gmelin Handbook of Inorganic Chemistry, 8th ed., Manganese Part C 2, System Nymber 56, Springer-Verlag, New York, 1975, p. 177.

MARKET AND MARKETING RESEARCH

Market research is a long established technique used to secure data for management to use in its decision making. Market research may be short or long term. Some market analysts use the following time frames: short term, up to 18 months; intermediate term, 18 months to 5 years; long term, 5 to 10 years. In general, short-term market research is synonymous with sales analysis and is used to assist the sales manager in setting goals, measuring performance, and giving the production department operating targets.

Intermediate or long-term market research has as its objectives the quantifying of markets for a particular chemical in terms of tonnages, growth potentials, general location of markets, competitive factors, and the impact of existing or potential government regulations on the market.

Marketing research, as compared to market research, is more directly concerned with identifying existing or potential users of a product, their present sources of supply, the nature and duration of any contracts that exist between producer and buyer, competitors' strategies in product development and pricing, requirements for facilities and personnel to compete successfully, and the status of competition from producers in other countries. Also, government regulations involving production of chemicals, their transportation, and disposal of wastes and by-products have a marked influence on the profitability of most chemical process industry operations.

Market research studies usually originate in the sales or marketing groups of a company. As a general rule, the sales analysis or short-term-type study is done by in-house personnel, often on a continuous basis. Field sales personnel are often used to assist the market research group in securing data. Long-term market research studies may originate in sales or marketing groups if the company already produces the product. If a new product is involved, the study may originate in the research and development group or at the corporate planning level. Marketing research studies usually originate in the higher levels of management, eg, general manager or vice president. This is especially true if the proposed study is for a product new to the company.

Selection of an in-house group or consulting firm to do a market or marketing research study does not follow any set pattern. In the 1990s there are at least 100 well-known and capable consultants or consulting firms in the United States.

Sales Analysis

It is axiomatic that sales analysis depends on detailed records of sales of a specific chemical to a specific company. Paramount to the success of such studies is the existence of data recorded on a systematic and continuous basis. It follows that these studies are done best by an in-house staff on products already produced by the company. However, on occasion, a product new to the company can be studied by the in-house group with the assistance of their field sales force.

Use of Results. The results are most useful in production planning, particularly if grade differences appear to be in the offing, and in assuring that adequate supplies are available for sales.

Market Research

Market research in the chemical process industries differs sharply from consumer market research primarily in the so-called universe. In industry studies, the universe is quite small compared to the consumer market.

Methodology. Practitioners of chemical market research develop individual styles and techniques. However, four elements are essential to every useful study: defining the problem, data gathering, analysis of data, and presentation of findings.

Costs. There are two cost elements in doing marketing research studies: professional charges and out-of-pocket expenses. The actual cost of any study is entirely dependent upon the number of interviews and the type of interviews.

Use of Results. Market researchers are occasionally disappointed in the use made of their reports. They cite instances where action contrary to their recommendations is taken, often with discouraging results; or where no action is taken, and another company successfully takes advantage of the opportunity. It is good practice for a market research manager to follow up a report and try to determine if management is using it in making decisions.

Market and Marketing Research Organizations

The Chemical Management and Resources Association has about 1000 members. The current CMRA defines its purposes to be "to promote the growth and development of marketing management, business development, business intelligence and planning in the chemical or allied process industries through industrial marketing or business/market research; to provide continuing education and foster the development of those so engaged; to contribute and make available to the public, information in the field of chemical and industrial marketing management, and business research; to cooperate with government officials in furthering the national welfare, and to carry out such activities recognized as law for such organizations."

The Federation of European Marketing Research Association (FEMRA) has about 500 members of which about 300 are ECMRA members.

Methodology. The methodology previously outlined for market research studies is applicable to marketing research studies. However, many more elements must be considered, especially in the realm of strategy factors.

Marketing research groups in some chemical companies also conduct studies beyond their normal activities. These include assistance in the market development phases of new product introductions, searches for unfilled needs in products or services which their company may be able to meet, and searches for new uses for existing products.

New product development programs present another type of challenge to the researcher. Often the researcher has no guidelines for evaluating the new product and must formulate a unique plan for developing enough information to construct a matrix that would show the risks and rewards of the project.

Methodology for marketing research studies differs most from market research studies when the researcher evaluates the competitive forces at work and formulates a marketing strategy.

Costs. Because much more personal contact work is required, the cost of marketing research studies is significantly higher than the cost of market research studies. Also, the advisability of using the most senior personnel raises the cost.

Analysis of Data. Again, the basic techniques outlined earlier for market research studies apply to marketing research studies. However, in the realm of competitive forces and the formulation of strategies, the pragmatic judgment of the experienced researcher is essential. In most cases, the researcher does not have hard data to draw on. Instead, to some degree, a series of mental images of the principal competitors has been formulated indicating their probable response to new developments or competitors in a given market.

Then the researcher decides on a strategy for the company. The classic strategies are well known: acquisition, internal development, licensing, and joint ventures.

Marketing Strategy Factors. Of the elements mentioned earlier as factors in determining strategy, several deserve more detailed discussion: pricing, distribution channels, applications research, technical service, and concessions to customers. It is useful to divide the products of the chemical industry into two broad groups: commodity and specialty. Management skills that are successful for commodity chemicals may fail for specialty chemicals and vice versa.

Pricing. Chemical pricing has always been a complex subject, but rapidly escalating raw materials costs, costs of meeting government regulations, inflationary pressures, existence of competition on a world basis, and excessively high costs of capital give management more problems than ever.

Many analysts favor ROS as a benchmark for comparison because it is up to date and simple and because it is increasingly difficult to determine a true ROI based on what profits might be on plants built under inflation, with expensive capital and construction costs.

Under an inflationary economy, the newest plant is usually the highest cost producer unless it features a unique process with significantly higher product yields and/or lower production costs. Recently, the new producer's list price is that of the marketplace, rather than lower as in the past. Pricing of specialty chemicals and specialty chemical systems is based on value to the customer. The pricing of a new chemical that will compete against other chemicals does involve the usual cost elements that set the price. However, it has been shown that an empirical approach may be of value.

Distribution Channels. Most commodity chemicals are primarily sold by the producer to a relatively small number of very large users. However, producers of commodity chemicals also utilize distributors to reach small volume users. A researcher formulating a strategy for a particular company must determine what portion of the planned output will be sold by its sales force and whether distributors or manufacturers' representatives might be beneficial.

Applications Research. Specialty chemical producers devote a larger share of their time and costs to applications research than do producers of most commodity chemicals. The most successful specialty chemical producers have been those companies that are able to respond quickly to customer needs and problems under the conditions found in the customer's plant.

Commodity chemical producers have varying records of performance in applications research. It is usually high on the priority list when the product is still evolving. A researcher planning a strategy must determine if a commitment to applications research is required. If so, the cost of facilities and personnel and the time required to assemble these must be calculated and included in the overall cost of entry.

Technical Service. A researcher planning strategy must determine whether commitment to a technical service facility and personnel is required. If it is, the cost of this commitment must be determined and included in the overall cost of entering the product field.

Concessions to Customers. A researcher formulating a strategy for a client company must take into account any special situations that may exist between a seller and a buyer for a given product.

Presentation of Results. Because the marketing research study is usually more complex and more detailed, a series of reports or presentations may occur, including some or all of the following: overview oral report to top management; overview but more detailed oral report to individual departments or divisions; brief written reports for top management, highly visual in nature; brief written reports for division heads; and a complete written report for reference.

Use of Results. Since a marketing research study is often part of a total feasibility study, the results are usually evaluated by management and a decision is made as to the corporate position. It is incumbent on marketing research managers or their superiors to etermine if the recommendations they made will be considered.

Reports in purchasing research usually differ from conventional market research reports. In many companies, a purchase profile report is prepared. It shows concisely the existing vendor capacities for the raw material, planned expansions or new producers, demands for other uses, and demand within the analyst's company. In some cases, the report includes a world supply and demand balance. A key objective of a purchase profile report is to make the buyer as well informed as the marketing manager of the seller. If this is achieved, the buyer can often secure a beneficial purchase contract.

Company practices differ in who does purchasing research and how it is done. Several patterns are evident. The chemical buyer is responsible for preparing the purchase profiles, possibly with in-house library assistance. Market research analysts are assigned to the purchasing department and prepare some or all of the profiles needed. Outside consultants are used to prepare some of the purchase profiles or as a check on internal procedures and conclusions.

Competitive Intelligence

A few market and marketing analysts in the chemical and chemical process industries are doing competitive intelligence work. This function has its own professional organization, the Society of Competitive Intelligence Professionals, and it has about 2000 members, largely from nonchemical businesses. Competitive intelligence work requires a constant monitoring of announced and rumored developments. It is used to seek out emerging technologies that may impact on some operation of a company and affect its competitive standing.

EDWARD TARNELL
Colin A. Houston & Associates, Inc.

W. E. Cox, Jr. and L. V. Dominquez, *Ind. Market. Manage.* **8,** 81 (1979).

Today's Chemist at Work, 21 (Jan. 1993).

Chem. Eng., 145 (Feb. 1993).

D. D. Lee, *Industrial Marketing Research, Techniques and Practices,* Technomic Publishing Co., Westport, Conn., 1978.

MASS SPECTROMETRY

In its simplest form, a mass spectrometer is an instrument that measures the mass-to-charge ratios *m/z* of ions formed when a sample is ionized by one of a number of different ionization methods. If some of the sample molecules are singly ionized and reach the ion detector without fragmenting, then the *m/z* ratio of these ions gives a direct measurement of the molecular weight.

Ideally, a mass spectrum contains a molecular ion, corresponding to the molecular mass of the analyte, as well as structurally significant fragment ions which allow either the direct determination of structure or a comparison to libraries of spectra of known compounds. Mass spectrometry (ms) is unique in its ability to determine directly the molecular mass of a sample.

If the sample to be analyzed is a mixture, it is not always easy to distinguish between molecular ions and fragment ions. Fragment ions can only be used for structure determination, however, if ascribed to a particular molecular ion. Thus separation of the molecular species in a mixture before mass spectral analysis is very important and chromatographic techniques are often employed (see CHROMATOGRAPHY).

An important alternative to chromatographic separation of a mixture is the use of tandem mass spectrometry, designated mass spectrometry/mass spectrometry (ms/ms). Ms/ms and chromatography are frequently used together to analyze a complex sample.

In the first mass spectrometers, the sample was heated to give a vapor which was then ionized by electron ionization (EI). Some compounds do not give molecular ions by EI and the development of chemical ionization (CI), an attempt to obtain molecular weight information for such compounds, followed.

For thermally labile materials, soft ionization techniques which do not require direct heating of the sample have been developed, such as fast atom bombardment (FAB), field desorption (FD), atmospheric pressure ionization (API) and matrix-assisted laser desorption (MALDI).

Instrumentation

A mass spectrometer consists of four basic parts: a sample inlet system, an ion source, a means of separating ions according to the mass-to-charge ratios, ie, a mass analyzer, and an ion detection system. Additionally, modern instruments are usually supplied with a data system for instrument control, data acquisition, and data processing. Inlet systems are often interfaces for gas chromatography or high performance liquid chromatography (hplc) which may also ionize the sample, eg, thermospray (TSP). Mass analyzers are most often one of the following quadrupoles, magnetic sectors, time of flight, ion traps or fourier transform ion cyclotron resonance (fticr).

Three important parameters for mass spectrometers are mass resolution, mass range, and sensitivity.

Mass Spectrometers for Tandem Mass Spectrometry. To acquire ms/ms spectra the ion of interest is isolated from other sample ions and activated to cause it to fragment. The resulting product ions are mass analyzed to give the ms/ms spectrum. This is most easily achieved using two mass spectrometers in tandem and having a high pressure gas cell between them.

The most popular ms/ms spectrometer is the triple quadrupole; the middle quadrupole is the collision chamber and is not used for mass analysis, and for high resolution ms/ms, 4-sector instruments, which are two magnetic sector mass spectrometers linked by a high pressure gas cell, are used.

Data Systems. A very important part of a mass spectrometer is the computer system used to acquire and process the mass spectral data. Networking ms systems to each other and to laboratory information management system (qv) (LIMS) is becoming more important. The American Society of Mass Spectrometry (ASMS) has overseen the development of NetCDF software that is a standard format for ms data. The use of workstations allows rapid searching of spectral libraries and databases (qv) that usually contain over 10^5 entries. True multitasking, ie, simultaneous data acquisition and data processing on a single processor, is also possible.

Commercial Mass Spectrometers

A shift toward low cost easy to use instruments has occurred, and the major analytical instrument makers have purchased small manufacturers to add a particular technique to their instrument range.

Applications

Biotechnology. There has been a tremendous growth in the application of mass spectrometry in biotechnology (qv) since the 1980s. New ionization methods have steadily expanded the range of biomolecules which can be analyzed by mass spectrometry, and ms/ms data can be used to sequence subpicamole quantities of small peptides.

Environmental. The high sensitivity and specificity of mass spectrometry when coupled with high resolution chromatography make ms an ideal method for use in environmental analyses. A number of standard ms methods exist for target compound analysis. These methods have been written by government agencies such as the U.S. Environmental Protection Agency (EPA).

Environmental applications, such as dioxin analysis, have led to the development of very sensitive sector instruments that, when operating at 10,000 resolution, can quantitatively detect dioxins at femtogram levels in environmental extracts. This level of performance gives overall method sensitivities in the ppm range.

Oil Analysis. Characterization of oil samples is difficult because these are very complex. To simplify oil sample spectra, data is obtained using low energy (<10 eV) electron ionization. The geographic origin of an oil sample can be deduced from the types and abundances of steranes it contains. Sterane distributions have been determined by using ms/ms to monitor fragmentation of the steranes molecular ion, which produces the characteristic fragment ion at *m/z* 217 and/or by high resolution monitoring of *m/z* 217.

Polymers. Mass spectrometric analysis of polymers is problematic because of the wide mass range and difficulties in ionization. One useful ionization method is field desorption (FD). One promising development is the application of MALDI to polymer analysis. Like FD, MALDI does not give fragment ions, and molecular mass distributions for polystyrene samples up to PS70000 have been reported, albeit with low mass resolution.

Inorganic Applications. The three most widely used mass spectrometric techniques in inorganic analysis are secondary ion mass spectrometry (sims), inductively coupled plasma mass spectrometry (icpms), and glow discharge mass spectrometry (gdms). The technique of secondary ion mass spectrometry (sims) is used to map ion abundances across a sample as a function of depth and has found heavy usage in the semiconductor industry. Inductively coupled plasma mass spectrometry (icpms) offers rapid, simultaneous multielement analysis with ppb detection limits for some elements. Glow discharge mass spectrometry (gdms) allows the direct determination of the elemental composition of solid samples.

COLIN MOORE
Uniroyal Chemical Company

J. R. Chapman, *Practical Organic Mass Spectrometry,* 2nd ed., John Wiley & Sons, Inc., Chichester, UK, 1993.

F. C. Walls and co-workers, *Biological Mass Spectrometry,* Elsevier, Amsterdam, the Netherlands, 1990.

F. W. McLafferty, ed., *Tandem Mass Spectrometry,* John Wiley & Sons, Inc., New York, 1983.

D. Colodner, V. Salters, and D. C. Duckworth, *Anal. Chem.* **66,** 1079A (1994).

MATCHES

The word match is of uncertain origin. In common parlance, a match is a short, slender, elongated piece of wood or cardboard, suitably impregnated and tipped to permit, through pyrochemical action between dry solids with a binder, the creation of a small transient flame. The word match also is used for fuse lines which after ignition on one end serve as fire-transfer agents in fireworks and for explosives (qv). Such items belong in the field of pyrotechnics (qv).

Mechanism of Fire Production

The essential chemical reaction takes place on contact of potassium chlorate and red phosphorus, which by itself is one of the most unpredictably hazardous dry reactions in pyrochemistry (see CHLORINE OXYGEN ACID AND SALTS–CHLORIC ACID AND CHLORATES; PHOSPHORUS). In the match head, and separately in the striker, each of two materials is embedded in a matrix of glue so that, on striking under mild friction, a few particles of both materials come harmlessly in contact and react with formation of well-contained sparks. The modifying materials in the match head function as sensitizers (sulfur or rosin), burning-rate modifiers (potassium dichromate or lead thiosulfate), and ash-formers (diatomaceous earth, powdered glass, etc); the latter serve to hold the glowing residue safely together by a sintering process. The glue, starch, and paraffin in the stem below the head act as flame-forming fuels and the neutralizers account for the practically indefinite storage stability of well-made matches. In the striker, the glass powder controls proper bite and sensitivity. The binder is insolubilized to prevent staining of clothing caused by rain or perspiration.

The SAW match is similar to the safety match except that it is richer in fuel, and gives a billowing somewhat wind-resistant flame. The phosphorus sulfide in the tip provides the ignitability on any solid surface, and a little of the same material in the base bulb adds to wind resistance, but otherwise the base is underbalanced in active materials to prevent self-ignition from rubbing during transportation.

Manufacture

The low price of book matches is mainly the result of high speed, mechanized production methods.

The most common size is a book of 20 or 28 (30) matches, and the 40-match size offers additional advertising area. "Ten-strike" matches are included in military food packages and are sometimes used for advertising purposes.

Wooden matches can be made by a veneering method. The alternative method consists in cutting round splints from selected blocks of white pine by means of rows of cutting dies each resembling a large darning needle of which the eye is the cutter. The splints in both types of operation are forced into holes in cast-iron plates and are thus transported through the various dipping operations.

A third type of commercial matches popular in South America is the wax vestas with a center of cotton threads or of a rolled and compressed thin and tough paper surrounded by and impregnated with wax; each match is a miniature candle of long (ca 1 min) burning time.

Two processes precede the affixing of the heads for wooden matches. The first one is glow-proofing of the splint by impregnation with ammonium phosphate or a mixture of it with boric acid. This suppresses continuation of glowing of the carbonized splint after discard and also prevents the burned part with the still-hot tip from falling off and singeing clothing (see also FLAME RETARDANTS). The second impregnation is the soaking up of paraffin wax into the stem for a certain length to assure flame forming and fire transfer to the wood.

Nonstandard and Military Matches. Specialties that occasionally appear on the market are actually fireworks items, made laboriously and at relatively high cost by hand-dipping with limited mechanization. Such matches produce a colored flame, give off perfume or fumigating vapors, or furnish a persistent glow or flame for the purpose of burning in a strong draft. In order to do these things effectively, an enlarged elongated bulb is necessary.

An interesting variation of the regular match is the pull match. The tip part of the match is enclosed in a strip of corrugated paper glued to a flat cardboard (such as a box of cigarettes) and the inside of the corrugated board is covered with striking material. On pulling the match fast enough out of the corrugation, the tip passes and engages the striker and becomes lit.

A curious item is the repeatedly ignitable match. It resembles a tiny pencil, the center part being a safety match composition which is surrounded by a cool-burning chemical mixture whose essential ingredient is nearly always metaldehyde.

The principle of the safety match is also used in the pull-wire fuse lighter used to start a fuse train for the ignition of fireworks items or more frequently for blasting work. This is a reversed pull match whereby the striker material is coated on a pull wire, and the match head material is within a small metal cup in a cardboard tube. Pulling the coated wire vigorously out of the device ignites the match mixture in the tube for fire transfer to the tubular fuse train.

During World War II, the Quartermaster Corps of the United States requested development of a SAW match that would withstand at least six hours submersion in water. Although no match is strikable after prolonged exposure to extremely high humidity, it is possible to prevent infiltration of moisture temporarily, and especially attack by liquid water, by coating the match head and part of the stem with nitrocellulose lacquer. Large numbers of matches protected in this manner were made during World War II.

Formulations. Formulations are by no means a closely guarded secret, mainly since they are only a starting point on the way to producing a satisfactory match and striker adapted to the specific conditions of manufacture.

European matches, mostly of brown or black tips, are basically identical with U.S. matches in their formulations, except that they contain in addition red iron oxide or manganese dioxide of pigment grade in the match heads. Match materials, testing methods, and related matters have been reviewed.

Economic Aspects

The rising popularity of disposable butane lighters and especially the increasing volumes of very inexpensive imported lighters have resulted in a tremendous decline in the use of matches.

The cost and selling price for matches increase considerably with higher quality cover paper, elaborateness of printed messages on and inside the cover (and sometimes even on the splint), and size of the order. In any case, the customer receives exactly the same high quality matches and striking strip.

Book matches are an important medium of advertising since they represent a truly utilitarian item which is more often given away than sold. Advertisers seeking national distribution pay for the message on many millions of books without entering otherwise in the sale of the matches (resale advertising match).

Toxicity and Other Safety Aspects

Potassium chlorate is the only active material that can be extracted in more than traces from a match head and only 9 mg are contained in one head. This, even multiplied by the content of a whole book, is far below any toxic amount for even a small child. No poisonous properties whatsoever can be imputed to the striking strip. SAW matches are similarly harmless but, because of their easy flammability, they should be entirely kept out of a household with smaller children. The same warning may apply to all wooden matches.

Sometimes a match ignites promptly but only a weak and unsatisfactory flame follows. This is the result of prolonged exposure of the matches to a temperature above 54°C in storage. The defect is caused by gradual dissipation of the paraffin wax throughout the splint and is evidenced by the disappearance of the line demarcation, which is clearly visible in book matches.

MARK C. BEAN
D. D. Bean & Sons Company

I. Kowarsky, "Matches," in F. D. Snell and L. S. Ettre, eds., *Encyclopedia of Industrial Chemical Analysis,* Vol. 15, John Wiley & Sons, Inc., New York, 1972.

C. A. Finch, in B. Elvers, S. Hawkins, and G. Schulz, eds., *Ullmann's Encyclopedia of Industrial Chemistry,* Vol. A16, VCH Verlagsgesellschaft GmbH, Weinheim, Germany, 1990, pp. 163–169.

Fed. Spec. EE-M-101 J, 1978.

H. Ellern, *Military and Civilian Pyrotechnics,* Chemical Publishing Co., Inc., New York, 1968.

MATERIALS RELIABILITY

Reliability is a parameter of design like a system's performance or load ratings and is concerned with the length of failure-free operation. Reliability as it relates to products or equipment can be measured in various ways. Since it is a design parameter, it has to be addressed early in the design cycle.

Terminology

Reliability. The reliability of a system is defined as the probability that the system will perform its intended function satisfactorily for a specified interval of time when operating under stated environmental conditions. It has to be realized that supposedly identical products fail at different times; thus, reliability can be quantified only as a probability. For any product there is some underlying function that describes this success pattern.

System Effectiveness. A system is designed to perform some intended function in a prescribed fashion. This overall capability is termed system effectiveness.

From the standpoint of a military product, system effectiveness is the probability that the system meets successfully an operational demand within a given time when operating under specified conditions. From the standpoint of commercial products, system effectiveness is more difficult to define, but basically means customer satisfaction.

Maintainability is important to the customer during the life of the system. For improved maintainability in equipment design the following concepts should be considered: ease of accessibility for repair, captive hardware with quick attach/detach fasteners, color coding, avoiding the need for specialty repair tools, built-in diagnostics, modularity, and standardization.

Design Reliability

Since reliability and maintainability are essentially design parameters, improvements are most easily and economically accomplished early in the design cycle. Useful techniques for design reliability improvement are design review, failure mode and effects analysis, and life-cycle cost.

System Reliability Models

Static reliability models are used in preliminary analyses to determine necessary reliability levels for subsystems and components. A

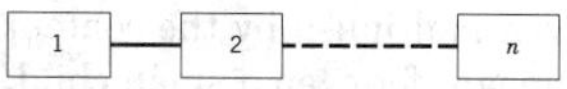

Figure 1. Series block diagram.

subsystem is a particular low level grouping of components. Some trial and error is usually necessary to obtain reasonable groupings for any particular system. Early identification of potential system weaknesses facilitates corrective action.

A reliability block diagram can be developed for the system from the definition of adequate performance. The block diagram represents the effect of subsystem or component failure on system performance. In this preliminary analysis, each subsystem is assumed to be either a success or failure. A reliability value is assigned to each subsystem where the application and a specified time period are given. The reliability values for each subsystem and the functional block diagram are the basis for the analysis.

Series Systems. The series configuration is the most commonly encountered in practice. In a series system, all subsystems must operate successfully for the system to be successful. The reliability block diagram is given in Figure 1.The system reliability is

$$R_s = \prod_{i=1}^{n} R_i$$

where R_i is the reliability for the ith subsystem, and R_s is system reliability.

Parallel Systems. A parallel (or redundant) system is not considered to be in a failed state unless all subsystems have failed. The system reliability is calculated as

$$R_s = 1 - \prod_{i=1}^{n} (1 - R_i)$$

System reliability is improved by providing alternative means for performing the same task.

Systems can have both parallel and series subsystems. Reliability is calculated by successively reducing the system using the basic series or parallel formulas.

Some systems cannot be represented by a simple combination of series and parallel subsystems. The systems are more complex in nature and the concept of coherent systems must be used in a more general and powerful treatment.

Reliability Measures

The reliability function $R(t)$ is defined as

$$R(t) = P(t > t) = 1 - F(t)$$

Life Expectancy of Devices. The expected or average life of devices is defined as

$$E(t) = \int_{-\infty}^{\infty} uf(u)du$$

where $f(t)$ is the probability density function (PDF) for the time-to-failure random variable **t**.. The expected life also can be found from

$$E(t) = \int_{0}^{\infty} R(t)dt, t \geqslant 0$$

The expected life is sometimes used as an indicator of system reliability; however, it can be a false indication and should be used with caution. In most test situations the chance of surviving the expected life is not 50% and depends on the underlying failure pattern.

Failure Rate and Hazard Function. The failure rate is defined as the rate at which failures occur in a given time interval. Considering the time interval $[t_1, t_2]$, the failure rate is given by

$$\frac{R(t_1) - R(t_2)}{(t_2 - t_1)R(t_1)}$$

and this is the rate of failure for those surviving at the beginning of the interval. This formula can be used to calculate failure rate from empirical life-test data.

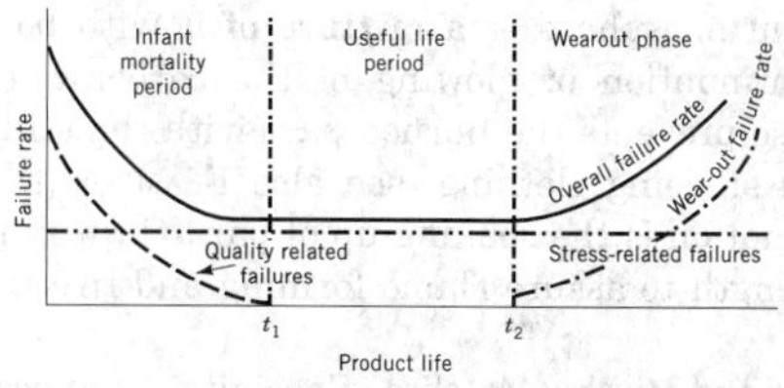

Figure 2. Failure rate vs product life.

The hazard function is defined as the limit of the failure rate as the interval of time approaches zero. The resulting hazard function $h(t)$ is defined by

$$h(t) = \frac{f(t)}{R(t)}$$

The hazard function can be interpreted as the instantaneous failure rate. The quantity $h(t)\Delta t$ for small Δt represents the probability of failure in the interval Δt, given that the device was surviving at the beginning of the interval.

The failure rate changes over the lifetime of a population of devices. An example of a failure-rate vs product-life curve is shown in Figure 2 where only three basic causes of failure are present.

Hazard function, PDF, and reliability function are related for any theoretical failure distribution. The relationships are

$$f(t) = h(t) \exp\left[-\int_0^t h(u)du\right]$$

and

$$R(t) = \exp\left[-\int_0^t h(u)du\right]$$

Conditional Failure Probability. The concept of conditional probability of failure is useful to predict the chances of survival for a device that has been in operation for a period of time and is not in a failed state. Such information is helpful for maintenance planning.

If a device has a reliability function $R(t)$ and has been successfully operating for a period of time T, the conditional reliability function is given by

$$R(t|t > T) = \frac{R(t)}{R(T)}, t > T$$

Exponential Distribution

The exponential distribution has proved to be a reasonable failure model for electronic equipment. However, like any failure model, it has limitations which should be well understood.

Basic Statistical Properties. The PDF for an exponentially distributed random variable **t** is given by

$$f(t, \lambda) = \lambda e^{-\lambda t}, t \geqq 0$$

where λ is the failure-rate parameter. The quantity $\tau = 1/\lambda$ is the mean or expected life, also expressed as *MTBF*.

The reliability function is given by

$$R(t) = e^{-\lambda t}, t \geqq 0$$

or

$$R(t) = e^{-t/\theta}, t \geqq 0$$

whereas the hazard function is

$$h(t) = \lambda = \frac{1}{\theta}$$

The hazard function is a constant which means that this model would be applicable during the midlife of the product when the failure rate is relatively stable. It would not be applicable during the wearout phase or during the infant mortality (early failure) period.

On complex systems, which are repaired as they fail and placed back in service, the time between system failures can be reasonably well modeled by the exponential distribution.

Point Estimation. The estimator for the mean life parameter θ is given by

$$\hat{\theta} = \frac{T}{r}$$

where T is total accumulated test time considering both failed and unfailed (or suspended) items; and r is total number of failures. The reliability function is then estimated by

$$\hat{R}(t) = e^{-t/\hat{\theta}}, t \geqq 0$$

Confidence-Interval Estimates. Confidence-interval estimates for the expected life or reliability can be obtained easily in the case of the exponential using failure-censored (Type II) and time-censored (Type I) life testing. It is possible to specify a test as either time- or failure-truncated, whichever occurs first.

The Nonzero Minimum-Life Case. In many situations, no failures are observed during an initial period of time. For example, when testing engine bearings for fatigue life no failures are expected for a long initial period. Some corrosion processes also have this characteristic. In the following it is assumed that the failure pattern can be reasonably well approximated by an exponential distribution.

The PDF for the two-parameter exponential distribution is given by

$$f(t, \theta, \delta) = \frac{1}{\theta} e^{-(t-\delta)/\theta}, t \geqq \delta \geqq 0, \theta > 0$$

The reliability function is

$$R(t) = e^{-(t-\delta)/\theta}, \geqq \delta \geqq 0$$

The expected life is $(\delta + \tau)$. The quantity δ is referred to as the minimum life parameter.

The Weibull Distribution

The Weibull distribution is a more versatile failure model than the exponential one. It is a popular model and widely used to estimate product reliability because it can be analyzed graphically with Weibull probability paper.

Basic Statistical Properties. The reliability function for the three-parameter Weibull distribution is given by

$$R(t) = \exp\left[-\left(\frac{t-\delta}{\theta-\delta}\right)^{\beta}\right], t \geqq \delta \geqq 0, \beta > 0, \theta > \delta$$

where δ is minimum life, θ is characteristic life, and β is Weibull slope.

The two-parameter Weibull has a minimum life of zero and the reliability function is

$$R(t) = e^{-(t/\theta)^{\beta}}, t \geqq 0$$

The hazard function for the two-parameter Weibull is

$$h(t) = \frac{\beta}{\theta^{\beta}}, t \geqq 0t^{\beta-1}$$

This hazard function decreases with $\beta < 1$, increases with $\beta > 1$, and remains constant for $\beta = 1$. The value of β can give some indication of wearout or infant mortality.

Parameter Estimation. Weibull parameters can be estimated using the usual statistical procedures; however, a computer is needed to readily solve the equations. Graphical estimation can be made on Weibull paper without the aid of a computer, although the results cannot be expected to be as accurate and consistent.

Binomial Distribution

To determine in the laboratory if a component survives in use, a test bogey is frequently established based on past experience. The test bogey is correlated with the particular test used to duplicate (or simulate) field conditions. The bogey can be stated in cycles, hours, revolutions, stress reversals, etc. A number of components are placed on test and each component either survives or fails. The reliability for this situation is estimated.

The failure model is the binomial distribution given by

$$p(y) = \binom{n}{y} R^y(1-R)^{n-y}, y = 0, 1, 2 \cdots n$$

where R is the product reliability; n, the total number of products placed on test; and y, the number of products surviving the test. Furthermore

$$\binom{n}{y} = \frac{n!}{y!(n-y)!}$$

The quantity $p(y)$ is the probability that exactly y out of n components survive the test where the component reliability is R.

Success Testing. Acceptance life tests are sometimes planned with no failures allowed. This gives the smallest sample size necessary to demonstrate a reliability at a given confidence level. The reliability is demonstrated relative to the test employed and the testing period.

LEONARD LAMBERSON
Western Michigan University

C. E. Ebeling, *An Introduction to Reliability and Maintainability Engineering,* The McGraw-Hill Companies, Inc., New York, 1997.

F. Jensen and N. E. Petersen, *Burn-In,* John Wiley & Sons, Inc., New York, 1982.

L. M. Leemis, *Reliability Probabilistic Models and Statistical Methods,* Prentice-Hall, Englewood Cliffs, N.J. 1995.

E. E. Lewis, *Introduction to Reliability Engineering,* John Wiley & Sons, Inc., New York, 1987.

MATERIALS STANDARDS AND SPECIFICATIONS

A standard is a document, definition, or reference artifact intended for general use by as large a body as possible; a specification, which involves similar technical content and similar format, usually is limited in both its intended applicability and its users.

Standardization minimizes disadvantageous diversity, assures acceptability of products, and facilitates technical communication. There are many attributes of materials that are subject to standardization, eg, composition, physical properties, dimensions, finish, and processing. Implicit to the realization of standards is the availability of test methods and appropriate calibration techniques. Apart from physical or artifactual standards, written or paper standards also must be considered, ie, their generation, promulgation, and interrelationships.

The International Organization for Standardization (ISO) defines a standard as the result of the standardization process: "the process of formulating and applying rules for an orderly approach to a specific activity for the benefit and with the cooperation of all concerned and in particular for the promotion of optimum overall economy taking due account of functional conditions and safety requirements." Standardization involves concepts of units of measurement, terminology and symbolic representation, and attributes of the physical artifact, ie, quality, variety, and interchangeability. A specification, however, is defined as "a document intended primarily for use in procurement which clearly and accurately describes the essential technical requirements for items, materials, or services including the procedures by which it will be determined that the requirements have been met." The ISO defines a specification as "a concise statement of a set of requirements to be satisfied by a product, a material or a process indicating, whenever appropriate, the procedure by means of which it may be determined whether the requirements given are satisfied. Notes—(*1*) A specification may be a standard, a part of a standard, or independent of a standard. (*2*) As far as practicable, it is desired that the requirements are expressed numerically in terms of appropriate units, together with their limits." A specification may also be viewed as the technical aspects of the legal contract between the purchaser of the material, product, or service and the vendor of the same and defines what each may expect of the other.

Standards

Objectives and Types. The objectives of standardization are economy of production by way of economies of scale in output, optimization of varieties in input material, and improved managerial control; assurance of quality; improvement of interchangeability; facilitation of technical communication; enhancement of innovation and technological progress; and promotion of the safety of persons, goods, and the environment.

Types of standards include physical or artifactual standards, paper or documentary standards, regulatory standards, voluntary standards, product standards, public and private standards, and consensus standards.

Generation, Administration, and Implementation. The development of a good standard is a lengthy and involved process, whether for a private organization, a nation, or an international body. The generic aspects of the development of a standard are shown in Figure 1.

Standard Reference Materials. An important development in the United States, relative to standardization in the chemical field, is the establishment by NIST of standard reference materials (SRMs), originally called standard samples. The objective of this program is to provide materials that may be used to calibrate measurement systems and to provide a central basis for uniformity and accuracy of measurement. SRMs are well-characterized, homogeneous, stable materials or simple artifacts with specific properties that have been measured and certified by NIST.

Standard reference materials provide a necessary but insufficient means for achieving accuracy and measurement compatibility on a national or international scale. Good test methods, good laboratory practices, well-qualified personnel, and proper intralaboratory and interlaboratory quality assurance procedures are equally important.

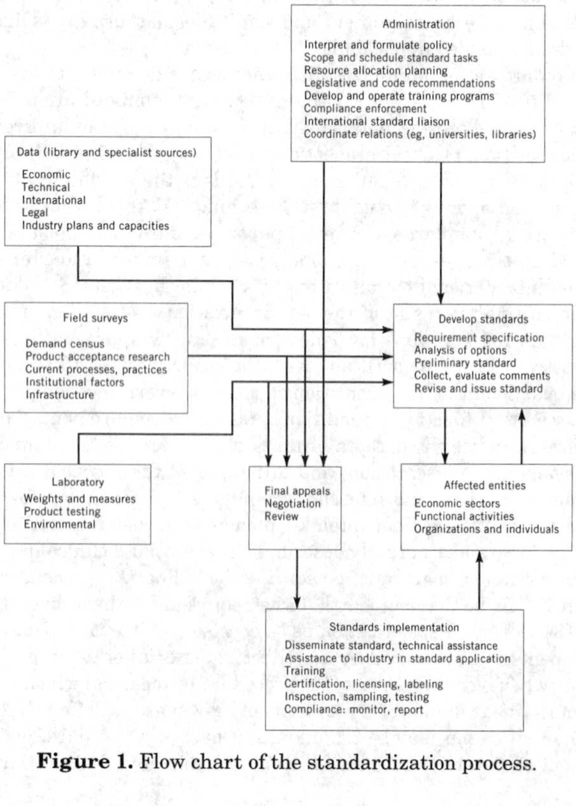

Figure 1. Flow chart of the standardization process.

Analytical standards imply the existence of a reference material and a recommended test method. Analytical standards other than for fine chemicals and for the NIST series of SRMs have been reviewed. Another sphere of activity in analytical standards is the geochemical reference standards maintained by the U.S. Geological Survey and by analogous groups in France, Canada, Japan, South Africa, and Germany.

Chronological standards are needed for an extremely diverse range of fields, eg, astrophysics, anthropology, archaeology, geology, oceanography, and art. The techniques employed for dating materials include dendrochronology, thermoluminescence, obsidian hydration, varve deposition, paleomagnetic reversal, fission tracks, racemization of amino acids, and a variety of techniques related to the presence or decay of radioactive species, eg, ^{14}C, ^{10}Be, ^{18}O, and various decay products of the U and Th series.

Standard Reference Data. The term, standard reference data, implies a data set or collection that has passed some screening and evaluation by a competent body and warrants the body's imprimatur and promotion. Such a data set may be generated expressly for this purpose by especially careful measurements made on a standard reference material or other well-characterized material, eg, the series of standard x-ray diffraction patterns generated by NIST. In some cases, a reference data set may not represent a specific set of real experimental observations but a recommended, consistent set of stated reliability that is synthesized from limited, fragmentary, and conflicting literature data by review, analysis, adjustment, and interpolation.

Standards for Nondestructive Evaluation. Nondestructive evaluation (NDE) standards are important in materials engineering for evaluating the structure, properties, and integrity of materials and fabricated products. Such standards apply to test methods, artifactual standards for test calibration, and comparative graphical or pictorial references. These standards may be used as inspection guides, to define terms describing defects, to describe and recommend test methods, for qualification and certification of individuals and laboratories working in the NDE field, and to specify materials and apparatus used in NDE testing.

Traceability. Measurements are traceable to designated standards if scientifically rigorous evidence is produced on a continuing basis to show that the measurement process is producing data for which the total measurement uncertainty is quantified relative to national or other designated standards through an unbroken chain of comparisons. The intent of traceability is to assure an accuracy level sufficient for the need of the product or service. Traceability also is used by materials engineers for the identification of the origin of a material. This attribution often is necessary where knowledge of composition, structure, or processing history is inadequate to assure the properties required in service.

Basic Standards for Chemical Technology. There are many numerical values that are standards in chemical technology. They include atomic weight (carbon-12, ^{12}C), temperature (the Celsius scale), pressure [one atm. = 0.101325 MPa(14.695 psi)], length (one meter), mass (kilogram), time (the second), standard cell potential, concentration (the mole), and energy (the joule).

Specifications

Objectives and Types. A specification establishes assurance of the fitness of a material, product, process, or service for use. Within a company, the specification is the means by which engineering conveys to purchasing what requirements it has for the material to be supplied to manufacturing. Material specification records provide information regarding a proven successful material that can be used in a new product. Such records also are useful in the rebuilding of components and as defense evidence in a liability suit.

Content. Although formats of materials specifications may vary according to the need, the principal elements are title, statement of scope, requirements, quality assurance provisions, applicable reference documents, preparations for delivery, notes, and definitions.

Strategy and Implementation. The most effective specification is that which accomplishes the desired result with the fewest requirements. Properties and performance should be emphasized rather than how the objectives are to be achieved. Wherever possible, tests should be easy to perform and highly correlatable with service performance. Tests that indicate service life are especially useful. Standard test references, eg, ASTM methods, are the most desirable.

Effective specification control often can be established other than through requirements placed on the end use material, ie, the specification may bear on the raw materials, the process used to produce the material, or ancillary materials used in its processing. Related but supplementary techniques are approved vendor lists, accredited testing laboratories, and preproduction acceptance tests.

Economic Aspects

A proper assessment of the costs and benefits associated with standardization depends on having suitable baseline data with which to make a comparison. Several surveys have shown typical dollar returns for the investment in standardization in the range of 5:1–8:1 with occasional claims made for a ratio as high as 50:1.

Savings include reduced costs of materials and parts procurement; savings in production and drafting practice; reduction in engineering time, eg, design, testing, quality control, and documentation; and reduction in maintenance, field service, and in-warranty repairs.

The ideal specification regards only those properties required to assure satisfactory performance in the intended application and properties that are quantitative and measurable in a defined test. A particularly effective approach is to recognize within a specification or related set of specifications the different levels of quality or reliability required in different applications.

From the customer's point of view, there is an optimal level of standardization. Increased standardization lowers costs but restricts choice. Furthermore, if a single minimal performance product standard is rigorously invoked in an industry, competition in a free market ultimately may lead the manufacturer of a superior product to save costs by lowering his product quality to the level of the standard, thus denying other values to the customer.

Legal Aspects

The increasing incidence of class action suits over faulty performance, the trend toward personal accountability and liability, and the increasing role of consumerism have all affected standardization. Improvement in the technical quality of standards, the involvement of all of the possible stakeholders in standards creation, and endorsement by larger standardizing bodies help to minimize the legal exposure of the individual engineer or company. A particular embodiment of these attitudes is the certification label, ie, a symbol or mark on the product indicating that it has been produced according to the standards of a particular organization.

Antitrust laws sometimes have been invoked in opposition to the collaborative activities of individual companies or private associations, eg, ASTM, in the development of specifications and standards. Although such activities should not constitute restraint of trade, they must be conducted so that the charge can be refuted. Therefore all features of due process proceedings must be observed. Actions aimed at strengthening the voluntary standards system have begun. A recommended national standards policy has been generated by an advisory committee that was initiated by, but is independent of, the ANSI. The Federal Office of Management and Budget has issued a circular establishing a uniform policy for federal participation and the use of voluntary standards.

Education

Seminars, workshops, and short courses sponsored by professional societies and trade associations provide the needed training in materials standards and specifications.

Trends and Outlook

International Standards. International trade is increasing rapidly in volume, in complexity, and in its significance to individual national economies. Thus the move toward more extensive adoption of international standards as well as cross-referencing of equivalent national specifications is understandable. The 1990s U.S.-designed car may be equipped with a German engine and French tires, and be built in part from Japanese steel and Dutch plastics. This composite implies a need for materials standardization accepted on an international level.

Quality and Reliability. Increased requirements for quality and reliability in all products, especially for those of high dollar value and in components of highly integrated technological systems, has led to the formation and broad adoption of the ISO 9000 series of international quality standards for products and services. ISO 9000, which has now been adopted by over 50 countries, is actually a series of five integrated standards developed during the 1980s to provide uniform, worldwide quality assurance requirements. ISO 9000 is the road map to the series and also defines key terms; ISO 9001 relates to design and servicing; ISO 9002 to production and installation; ISO 9003 to final inspection and testing; and finally ISO 9004 provides guidance on implementing these standards.

Environmental Protection and Safety. Increasing concern over the environment and safety issues has led to new standards for exposure of organisms to materials, noise, and electromagnetic radiation (see ENVIRONMENTAL IMPACT; INDUSTRIAL HYGIENE; TOXICOLOGY). The decreasing availability of natural resources forces industry to make use of leaner ores and apply materials that are in short supply more frugally (see MINERALS RECOVERY AND PROCESSING). This usage is expected to result in new analytical standards and compositional specifications. The use of specifications in coping with problems of residual and additive elements in both virgin and recycled materials has been reviewed (see RECYCLING).

Computerization. The computerization of all aspects of industry and commerce, from management to engineering and manufacturing, and from purchasing to sales, has made it vital to standardize the ways materials information is incorporated into machine-readable systems (see COMPUTER-AIDED DESIGN AND MANUFACTURING (CAD/CAM); COMPUTER-AIDED ENGINEERING (CAE)). More than a dozen standards in this area have been developed by ASTM's Committee E49, which has also prepared a guide to the building of materials databases (qv). A particularly important issue is standardization to facilitate the exchange of digital information.

Another standardization matter relative to computerization of materials information is that of terminology (see NOMENCLATURE). Full terminological standardization is not expected to be realized until the twenty-first century, but the hazards of lack of such standardization are exacerbated in computerized systems.

Nonlaboratory Environments. Environments deviating significantly from that of the laboratory (eg, space, ocean, human body, nuclear reactor), yet in which all the usual engineering functions must be performed, also pose problems and opportunities for material standards. Sensors (qv) must measure the attributes of these environments, construction materials must withstand the exposure regimes, performance criteria must be specified, and quality of standards and specifications must be higher.

Units. The SI system of units and conversion factors (qv) has been formally adopted worldwide, with the exception of Brunei, Burma, Yemen, and the U.S. The growing but limited participation of the U.S. in the metrication movement is evident by the passage of the Metric Acts of 1866 and 1975 and the subsequent establishment of the American National Metric Council (private) and the U.S. Metric Board (public) to plan, coordinate, monitor, and encourage the conversion process.

Sources of Information

There are many hundreds of standards-making bodies in the U.S. These comprise branches of state and federal government, trade associations, professional and technical societies, consumer groups, and institutions in the safety and insurance fields. The products of their efforts are heterogeneous, reflecting parochial concerns and different ways of standards development. However, by evolution, blending, and accreditation by higher level bodies, many standards originally developed for private purposes eventually become de facto, if not official, national standards. Individuals seeking access to standards and specifications are referred to the directories listed in *Standards Activities of Organizations in the United States,* by S.J. Chumas, (NBS SP 681, NBS, Gaithersburg, Md., 1984); *Standards and Specifications—Information Sources,* by E.G. Struglia, (Gale Research, Detroit, Mich., 1973), and Technical Data, Technical Indexes, Ltd. (Bracknall, U.K., updated monthly).

JACK H. WESTBROOK
Brookline Technologies

N. E. Promisel and co-workers, *Materials and Process Specifications and Standards,* NMAB Report 33, Washington, D.C., 1977.

NIST Standard Reference Materials Catalog, Superintendent of Documents, U.S. Government Printing Office, Washington, D.C.

The Voluntary Standards System of the United States of America—An Appraisal by the American Society for Testing and Materials, ASTM, Philadelphia, Pa., 1975.

P. L. Ricci and L. Perry, *Standards: A Resource and Guide for Identification and Acquisition,* Stirz, Minneapolis, Minn., 1991.

MEAT PRODUCTS

Meat is not only a flavorful product, but it also provides protein and essential minerals and vitamins, especially B vitamins. Meat consumption varies with social, economic, political, and geographical differences on a worldwide basis.

In the U.S., red meat production remains steady while poultry production is growing more rapidly. Beef consumption reached its highest point in 1976 (40.4 kg) and subsequently decreased to 32.8 kg in 1980 and to 28.9 kg in 1991. Initially, this decrease was a result of consumers turning to pork as an alternative. After declining steadily through the 1980s, per capita pork consumption turned upward again in 1987. However, the rise was short-lived as consumption decreased again in 1990. The real winners in the shift away from beef have been poultry products. The shift in consumption patterns is due to the fact that consumers are becoming more health-conscious and some media or popular press articles have labeled red meat as bad for health and longevity.

Income is also an important factor affecting demand for meat. Demand generally increases with higher income, but consumption tends to level off and may even decline at the highest incomes. Increasing incomes also change the types of meat demanded. Lower incomes may lower meat consumption or bring about a switch to lower priced meats. A major trend has been for consumers to eat more food away from home. Markets for fast foods such as chicken and hamburgers have increased.

Health and Safety Concerns

Fat Intake. Consumers have been warned that a diet high in fat increases the probability of chronic health problems and diseases, including coronary heart disease (CHD).

The American Dietetic Association, the American Heart Association, and the National Heart, Lung and Blood Institute recommend 142–198 g (5–7 oz) of lean, trimmed meat daily. It was also pointed out that trimmed meat, especially red meat, provides large amounts of essential nutrients such as iron, zinc, vitamin B_{12}, and balanced protein. The idea that the risk of CHD and cancer can be greatly reduced by avoiding a meat-centered diet have prompted some consumer groups to demand healthy meat products. In response, meat producers began to produce leaner beef with the use of growth hormones, and meat processors developed various types of low fat meat products.

Growth Promotants. Livestock can be exposed to many chemicals used to promote growth, improve feed utilization, or enhance meat acceptability. Since the early 1980s, bovine somatotropin (BST) and porcine somatotropin (PST) have been extensively studied. Somatotropin is a growth hormone that occurs naturally in animals (see HORMONES–ANTERIOR PITUITARY HORMONES). The safety of beef for human consumption from cattle treated with BST was determined in 1984 by the Food and Drug Administration (FDA). However, not everyone accepts the FDA findings. Some groups or individuals have argued that more testing is needed. The use of BST has been approved in the dairy industry, but the use of PST in the pork industry has not been approved by FDA for commercial use in the United States. Beta-adrenergic agonists that are known to promote growth, such as clenbuterol and cimaterol, improve the growth rate and feed conversion of sheep and poultry. Effects on swine are varied; definitive data on cattle are not yet available. β-Estradiol and zeranol are available compounds that occur naturally and are very effective repartitioning agents, enhancing rates of protein and lean tissue production whenever present at effective levels in cattle depositing fat. Trenbolone acetate, another example of growth promotant, is a member of the group of chemical structures of naturally occurring and synthetic sex steroids used in commercial anabolic steroid implants for growing beef cattle.

Antibiotics. The use of antibiotics in livestock production has caused serious public concern that the hazardous antibiotic residues in meat are contributing to health problems in humans. Some scientists and consumer groups support the notion that continuous feeding of penicillin, tetracycline, and other antibiotics to livestock for disease prevention may result in development of antibiotic-resistant strains of bacteria and subsequently contribute to human illness.

Pathogens. Pathogenic and spoilage microorganisms can be transferred to the meat during post-slaughter processing, storage, and handling. During slaughtering, many pathogens that may be present in the intestinal contents of the animals can contaminate the carcass and subsequently the processing tables and other equipment. Sufficient application of heat during cooking, however, destroys pathogenic and meat spoilage microorganisms and produces meat products that are commercially stable at ambient or refrigeration temperature. In addition, the heat treatment must be sufficiently severe to not only destroy the contaminating bacteria but also certain bacterial spores or toxins.

Trichinosis. Trichinosis is caused by parasitic nematode *Trichinella spiralis* that localizes in the muscles of pigs and some game animals (see ANTIPARASITIC AGENTS). People become infected by eating undercooked meat, most commonly pork or game such as bear. For many years, hotels, restaurants, institutional food suppliers, and consumers cooked pork to 82°C to ensure the destruction of *T. spiralis.* Other methods including freezing (−30 °C for at least 16 h), irradiation (19 to 750 krads), and curing (combined with up to 3.5% salt) have also been used for the destruction of *T. spiralis.*

Meat Processing Ingredients

Meat processing ingredients include meat, salt, water, phosphate, nitrite, extenders, seasonings, curing accelerators, and starter cultures.

Meat Processing Procedures and Machinery

Mechanical Tenderization. Sophisticated advances have been made in improving meat tenderness. Mechanical tenderization involves the application of blades, knives, pins, or needles to meat via mechanical pressure.

Cured Meats. The term meat curing means the addition of salt, nitrite, and/or nitrate, sugar, and other ingredients for the purpose of preserving and flavoring meat.

Sectioned and Formed Products. The meats that are utilized to produce sectioned and formed products may be entire muscles, very coarsely ground meat, or flaked meat. Meat particles can be produced by using a flaking machine that is capable of varying the flake size from very fine to coarsely flaked materials. The mechanical energy that must be applied to the various size of meat pieces and other ingredients to extract myofibrillar proteins can be provided by a mixer, tumbler, or massager.

Minced Products. The products that are included in this class are sausages of the fresh, fermented, dried, and cooked varieties. The meat ingredients can be either ground in a mincer or chopped in a bowl chopper.

Finely Chopped Products. The manufacture of finely comminuted processed meat products is dependent on the formation of a functional protein matrix within the product. The ability of the protein to successfully entrap moisture and fat is affected by many factors. These factors include the water holding capacity of the meat as well as the levels of meat, water, fat, salt, and nonmeat additives in the formulation. The production of finely chopped sausage requires additional particle size reduction with more time in a bowl chopper or passage through an emulsion mill. This type of consistency is often desired for the finely chopped sausages and loaves.

Fermented Products. Fermented meat products such as semidried and dried sausages are generally recognized as safe, if critical points during processing are controlled properly.

Hazard Analysis Critical Control Point

The hazard analysis critical control point (HACCP) concept is a systematic approach to the identification, assessment, prevention, and control of hazards. The system offers a rational approach to the control of microbiological, chemical, environmental, and physical hazards in foods, avoids the many weaknesses inherent in the inspectional quality control approach, and circumvents the shortcomings of reliance on microbiological testing. The food industry and government regulatory agencies are placing greater emphasis on the HACCP system to provide greater assurance of food safety.

HACCP Principles. The National Advisory committee on Microbiological Criteria for Foods established seven principles for the HACCP system: conduct hazard analysis and risk assessment; determine critical control points; establish specifications for each ccp; monitor each ccp; establish corrective action; establish a recordkeeping system; establish verification procedures.

Fat Reduction in Meat Products

The amount of fat, especially saturated fat and cholesterol in meat products, is of concern to a growing number of health-conscious consumers. The introduction of low fat beef in food service establishments as well as closer trimming of retail beef cuts and leaner ground beef in supermarkets across the U.S. demonstrates the meat industry's response to consumer desires for lower fat consumption. In order to be labeled as low fat, a meat product must contain no more than 10% fat. The palatability of ground beef, however, is directly related to the fat content. The overall acceptability of ground beef products is maximized at a fat content of approximately 20%. As the fat content of ground beef decreases, there is a significant decrease in product juiciness and tenderness.

Ingredient Additions and Substitutions. Processed meat products have the greatest opportunity for fat reduction for modification because their composition can be altered by reformulation with a fat replacement (see FAT REPLACERS), eg, added water, protein-based substitutes, carbohydrate-based substitutes, functional blends, and noncaloric synthetic fat substitutes.

Nutritional Labeling

The USDA's Food Safety and Inspection Service (FSIS) regulates the labeling of meat and poultry products, while FDA has responsibility over all other food labeling. The FDA regulations implement the Nutrition Labeling and Education Act of 1990.

Nutritional Labeling Content. As part of its efforts to harmonize labeling requirements with the FDA proposal, the FSIS mandates that nutrition information include the same 15 declarations required by FDA as well as allowing certain optional disclosures. The mandatory disclosures include calories, calories from total fat, total fat to nearest one-half gram, saturated fat to nearest one-half gram, cholesterol in milligrams, total carbohydrates in grams excluding fiber, complex carbohydrates in grams, sugars in grams including sugar alcohols, dietary fiber in grams, protein in grams, sodium in milligrams, vitamin A as a percentage of reference daily intake (RDI), vitamin C as a percentage of RDI, calcium as a percentage of RDI, and iron as a percentage of RDI. If the particular product contains insignificant amounts of eight nutrients, the abbreviated format should include calories, total fat, total carbohydrates, protein, and sodium. The optional disclosures include calories from saturated fat and unsaturated fat, unsaturated fat to nearest 0.5 gram (this is mandatory if fatty acid and/or cholesterol claims are made), polyunsaturated and/or monounsaturated fat to the nearest 0.5 gram, declaration of sugar alcohols in grams, insoluble and soluble fiber, potassium in milligrams, and thiamin, riboflavin, niacin, and other vitamins or minerals (if a claim regarding these nutrients is made).

Service Size. The label presentation should allow the consumers to understand the nutrition contents of individual meat products, compare nutrition contents across product categories, and choose among relevant food alternatives. The establishment of serving sizes has been the most controversial aspect of the nutritional labeling either for the consumers or manufacturers, because there are wide varieties of product sizes on the market, and it is almost impossible to standardize these sizes. In addition, there is also considerable confusion on the definitions of serving and portion. Currently, FDA and USDA's FSIS continue to cooperate and the goal is to establish standards that could be used by food manufacturers to determine label serving sizes and whether a claim such as low sodium meets criteria for the claim.

Nutritional Labeling Descriptors. In order to avoid confusion, descriptive terms must be accompanied by definitions which adequately explain the terms. In the case of nutrition-related claims, analytical sampling offers a means of assuring the accuracy of the stated claims. The USDA's FSIS has proposed a list of descriptors relevant for meat and poultry products.

GLENN R. SCHMIDT
S. RAHARJO
Colorado State University

"Pathogen Reduction, Hazard Analysis, and Critical Control Point (HACCP) Systems," *Fed. Reg.* **61**(144), 38805–38989 (1996).

H. D. Hafs and R. G. Zimbelman, eds., *Low-Fat Meats: Design Strategies and Human Implications,* Academic Press, San Diego, Calif., 1994, 328 pp.

J. R. Romans and co-workers, *The Meat We Eat,* 13th ed., Interstate Publishers, Inc., Danville, Ill., 1994, 1193 pp.

A. M. Pearson and T. R. Dutson, eds., *Advances in Meat Research: Meat and Health,* Vol. 6, Elsevier Applied Science, New York, 1990, 554 pp.

MECHANICAL TESTING. See MATERIALS RELIABILITY.

MEDICAL DIAGNOSTIC REAGENTS

Purified enzymes are widely used in medical diagnostic reagents in the measurement of analytes in urine, plasma, serum, or whole blood. Enzymes are very specific catalysts that can be derived from plants and animals, although microbial fermentation (qv) is the most popular production method. Enzymes are used extensively in diagnostics, immunodiagnostics, and biosensors (qv) to measure or amplify signals of many specific metabolites. Purified enzymes are expensive. This is the main reason for the increasing utilization of reusable immobilized enzymes in clinical analyses (see ENZYME APPLICATIONS; IMMUNOASSAY).

The main development in medical diagnostic reagents since the 1960s has been the steady growth of dry (solid-phase) chemistry systems. Dry chemistry systems have made substantial gains over wet clinical analysis in the number of tests performed in hospitals, laboratories, and homes because of ease, reliability, and accuracy.

Wet chemistry methods for analysis of body analytes, eg, blood glucose or cholesterol, require equipment and trained analysts (see AUTOMATED INSTRUMENTATION). In contrast, dry chemistry systems can be used at home.

Dry chemistry systems are useful not only to diabetics, but also to patients having other medical problems. These systems are also used in animal diagnosis, food, fermentation, agriculture, and environmental and industrial monitoring.

Enzyme-Catalyzed Reactions in Solution

Measurement Considerations. A prototype enzyme-catalyzed reaction where one substrate (S) produces only one product (P) may be described by

$$E + S \underset{k_{-1}}{\overset{k_1}{\rightleftharpoons}} ES \overset{k_2}{\rightarrow} E + P$$

where E is enzyme, ES is the enzyme–substrate complex, and k_i represents the reaction rate constants. The reaction can be followed by monitoring the loss of substrate or the formation of product. A graph of the concentration of substrate, or product, vs time gives an exponential curve. In the equilibrium method (end point) used for S determination, data are collected when the concentration of S or P are time-independent. Methods where data are obtained from the early linear part of the curve are known as kinetic methods (see KINETIC MEASUREMENTS).

Glucose, urea (qv) and cholesterol (see STEROIDS) are the substrates most frequently measured, although there are many more substrates or metabolites that are determined in clinical laboratories using enzymes.

Indicators. There are certain compounds that are suitable as indicators for sensitive and specific clinical analysis. Nicotinamide adenine dinucleotide (NAD) occurs in oxidized (NAD^+) and reduced (NADH) forms. Nicotinamide adenine dinucleotide phosphate (NADP) also has two states, $NADP^+$ and NADPH. NADH has a very high uv–vis absorption at 339 nm, extinction coefficient = $6300(M{\cdot}cm)^{-1}$, but NAD^+ does not. Similarly, NADPH absorbs light very strongly whereas $NADP^+$ does not. In the enzymatic assays of cholesterol, glucose, and urea, oxygen is used and H_2O_2 is formed. The H_2O_2 generated reacts with a chromogen in the presence of the enzyme peroxidase to produce a color change.

Measurement of Analytes. Biochemical reactions used in the measurement of selected analytes are commercially available as prepackaged kits of reagents. Measurement of the reactions given plus many other analytes can be made. Kits are available for measurement of cholesterol, citrate, creatinine, galactose, glucose, lactate, triglycerides and blood urea nitrogen.

Assay of Enzymes. In body fluids, enzyme levels are measured to help in diagnosis and for monitoring treatment of disease. Enzyme levels are determined by the kinetic methods. The assays are set up so that the enzyme concentration is rate-limiting. The continuous flow analyzers, introduced in the early 1960s, solved the problem of the high workload of clinical laboratories. In this method, reaction velocity is measured rapidly; the change in absorbance may be very small, but within the capability of advanced kinetic analyzers.

Enzymes, measured in clinical laboratories, for which kits are available include γ-glutamyl transferase (GGT), alanine transferase (ALT), aldolase, α-amylase, aspartate aminotransferase, creatine kinase and its isoenzymes, galactose-1-phosphate uridyl transferase, lipase, malate dehydrogenase, 5′-nucleotidase, phosphohexose isomerase, and pyruvate kinase.

Immobilized Enzymes in Diagnostic Reagents. The use of immobilized, instead of soluble, enzymes for measurement of analytes has received considerable attention, especially for clinical analyses. Use of immobilized enzymes offers the advantages of greater accuracy, stability, and convenience. Only a few methods utilizing immobilized enzymes have become commercially available.

Dry Chemistry

Background. Enzymes are essential in user-friendly diagnostic dry chemistry systems. Dry chemistry test kits are available in thin strips that are usually disposable. They may be either film-coated or impregnated. The most basic diagnostic strip consists of a paper or plastic base, polymeric binder, and reactive chemistry components consisting of enzymes, surfactants, buffers, and indicators. Diagnostic coatings or impregnation must incorporate all reagents necessary for the reaction. The coating can be either single or multilayer in design. A list of analytes, enzymes, drugs, and electrolytes assayed by dry chemistry diagnostic test kits follows:

Analytes
- glucose
- urea
- urate
- cholesterol (total)
- triglycerides
- bilirubin (total)
- ammonium ions
- creatinine
- calcium
- hemoglobin
- HDL cholesterol
- magnesium(II)
- phosphate (inorganic)
- albumin
- protein in cerebrospinal fluid

Enzymes
- alkaline phosphate (ALP)
- lactate dehydrogenase (LDH)
- creatine kinase
- MB isoenzyme (CK–MB)
- lipase
- amylase (total)

Drugs
- phenobarbitone
- phenytoin
- theophylline
- carbamazepine

Electrolytes
- sodium ion
- chloride
- carbon dioxide

Dry chemistry systems are widely used in physician's offices and hospital laboratories, and by millions of patients in their own homes worldwide. These systems are used for routine urinalysis, blood chemistry determinations, and immunological and microbiological testing. The main advantage of this technology is elimination of the need for reagent preparation and many other manual steps common to liquid reagent systems. This yields greater consistency and reliability of test results. Furthermore, dry chemistry systems have longer shelf stability and hence there is a reduced waste of reagents. Each test unit contains all the reagents and reactants necessary to perform assays.

Molded Dry Chemistry. In general, most enzymes are very fragile and sensitive to pH, solvent, and elevated temperatures. The catalytic activity of most enzymes is reduced dramatically as the temperature is increased. Typical properties of diagnostic enzymes are given in Table 1. The presence of ionic salts and other chemicals can considerably influence enzyme stability. To keep or sustain enzymatic activity, the redox centers must remain intact. The bulk of the enzyme, polymeric in composition, is an insulator; thus, altering it does not reduce the enzyme's catalytic activity. It has been suggested that molding of strips using reaction injection molding (RIM) may lead to useful chemistries, including biosensors, in the future.

Table 1. Properties of Diagnostic Enzymes

Parameter	Cholesterol oxidase (CO)	Cholesterol esterase (CE)	Glucose oxidase (GOD)	Peroxidase (POD)
source	*Streptomyces*	*Pseudomonas*	*Aspergillus*	horseradish
EC	1.1.3.6	1.1.13	1.1.3.4	1.11.1.7
mol wt	34,000	300,000	153,000	40,000
inhibitor	Hg^{2+}, Ag^+	Hg^{2+}, Ag^+	Hg^{2+}, Ag^+, Cu^{2+}	CN^-, S^{2-}
pH, optimum	6.5–7.0	7.0–9.0	5.0	6.0–6.5
temp, optimum, °C	45–50	40	30–40	45
pH stability[a]	5.0–10.0	5.0–9.0	4.0–6.0[b]	5.0–10.0

[a] At 25°C for 20 h, unless otherwise indicated. [b] At 40°C for 1 h.

Application of Diagnostic Technology in Monitoring Diabetes. Very frequent measurements of blood glucose to manage diabetes are one of the most important applications of diagnostic reagents. The blood glucose monitoring market (ca 1995) totaled about $\$7.5 \times 10^8$ in the United States and was expected to grow at a rate of 10% annually.

In the early 1960s, a promising approach to glucose monitoring was developed in the form of an enzyme electrode that used oxidation of glucose by the enzyme GOD. This approach has been incorporated into a few clinical analyzers for blood glucose determination.

Much work has been done on exploration and development of redox polymers that can rapidly and efficiently shuttle electrons. Extensive work has been done on osmium-containing polymers. The most stable and reproducible redox polymer of this kind is a poly(4-vinyl pyridine) (PVP) to which $Os(bpy)_2Cl_2$, where bpy $= 2,2'$ − bipyridine, has been attached to 1/16th of the pendant pyridine groups. The resultant redox polymer is water insoluble and biologically compatible by partial quaternization of the remaining pyridine groups using 2-bromoethyl amine. The newly introduced quaternized amine groups can react with a water-soluble epoxy, eg, polyethylene glycol diglycidyl ether, and GOD to produce a cross-linked biosensor coating film.

Flexible polymer chains have also been used for relays to provide communication between GOD's redox centers and the electrode. These ferrocene-modified siloxane polymers are stable and nondiffusing. Biosensors based on these redox polymers gave good response and superior stability. Commercial electrochemical microbiosensors, eg, Exactech (Medisense) and a silicon-based 6+ system (*i*-Stat) have appeared in the marketplace. These newer technologies should certainly impact rapid blood chemistry determinations by the year 2000.

The next generation of amperometric enzyme electrodes may well be based on immobilization techniques that are compatible with microelectronic mass-production processes and are easy to miniaturize. Integration of enzymes and mediators simultaneously should improve the electron-transfer pathway from the active site of the enzyme to the electrode.

Functionalized conducting monomers can be deposited on electrode surfaces aiming for covalent attachment or entrapment of sensor components. There is a pressing need for an implantable glucose sensor for optimal control of blood glucose concentration in diabetics. Two novel technologies have been used in the fabrication of a miniature electroenzyme glucose sensor for implantation in the subcutaneous tissues of humans with diabetes. An electrodeposition technique has been developed to electrically attract GOD and albumin onto the surface of the working electrode. The resultant enzyme–albumin layer was cross-linked by glutaraldehyde. A biocompatible polyethylene glycol–polyurethane copolymer has also been developed to serve as the outer membrane of the sensor to provide differential permeability of oxygen relative to glucose, in order to avoid oxygen deficit encountered in physiologic tissues.

ARTHUR M. USMANI
Bridgestone/Firestone, Inc.

H. U. Bergmeyer, ed., *Methods of Enzymatic Analysis,* 3rd ed., Academic Press, New York, 1983.

C. G. Guilbault, *Handbook of Enzymatic Methods of Analysis,* Marcel Dekker, Inc., New York, 1976.

A. M. Usmani, in A. M. Usmani and N. Akmal eds., *Diagnostic Biosensor Polymers,* ACS Symposium Series 556, ACS Books, Washington, D.C., 1994.

J. E. Kennamer, A. D. Burke, and A. M. Usmani, in C. G. Gebelein and C. E. Carraher eds., *Biotechnology and Bioactive Polymers,* Plenum Publishing Corp., New York, 1992.

MEDICAL IMAGING TECHNOLOGY

Medical imaging is the application of nonsurgical techniques to produce images of internal organs and tissues.

The five principal imaging technologies involve optical, x-ray, ultrasound, radio frequency (rf), or nuclear techniques. Additionally, medical imaging relies heavily on hundreds of ancillary chemical, computer, detector, electronic, film, and magnetic technologies developed in the latter twentieth century. The discussion herein includes basic imaging principles, and endoscopic, x-ray, ultrasound, magnetic resonance, and nuclear imaging as found in hospitals.

Basic Imaging Principles

An image is a matrix of picture elements (pixels) representing the magnitude of the imaged quantity in a given location. Images may be produced by absorption, emission, or reflection of energy by body tissue.

The signal in an image is defined as the intensity of the energy arising from the imaged tissue. The contrast between two tissues in an image is the difference between the signals of the two tissues. The signal-to-noise ratio (SNR) of a tissue in a medical image is the ratio of the signal intensity of that tissue to the noise level in the image. The SNR is not the best indicator of image quality. Rather, the contrast-to-noise ratio (CNR) between adjacent tissues is the factor which determines the utility of an image, provided sufficient signal exists.

Medical images are annotated with the conventional medical nomenclature for the directions of left, right, superior (toward the head), inferior (toward the feet), anterior (toward the front), and posterior (toward the back) of the body. There are three standard planes: one which is perpendicular to the long axis of the body and divides the body into superior and anterior parts is referred to as an axial plane, one which divides the body into left and right halves is called a sagittal plane, and one dividing the anterior from the posterior is referred to as a coronal plane.

Information from an imaging session may be presented as a projection, tomographic, or volume image.

Endoscopy

Theory and Equipment. Endoscopic imaging involves the production of a true color picture of the inside of the human body using lenses and either hollow pipes, a fiber optic bundle, or a small charge-coupled device (CCD) video camera. All three use a large field-of-view, sometimes referred to as a fish eye, lens to allow a 180° field of view.

Applications. Endoscopy finds applications in a number of investigations of the inside of the human body. The largest use of endoscopic techniques is in the examination of the gastrointestinal tract. Each of the endoscopic imaging procedures is relatively risk free and painless when performed by competent and well-trained individuals using a local anesthetic.

X-Ray Imaging

X-ray medical imaging is the most mature and widely used of the diagnostic imaging modalities (see X-RAY TECHNOLOGY).

Theory and Equipment. The field of x-ray medical imaging can be divided into plane film and CT imaging. Plane film imaging produces projection images of an object placed between a source and a detector which in most cases is a sheet of photographic film (see PHOTOGRAPHY). CT imaging produces tomographic images of a transaxial slice through the body. CT utilizes a source of x-rays and an electronic detector which converts the x-radiation into an electrical signal. Many references on x-ray medical imaging are available.

The continuous and discrete emissions from an x-ray tube cover a broad range of frequencies. It is necessary to image using a narrow band of x-rays because the attenuation of x-rays by body tissues is frequency dependent. Narrowing the band of x-rays emanating from a source is accomplished by sending the beam through filters composed of aluminum, copper, or zirconium. This process is referred to as hardening of the beam.

X-ray vacuum tubes contain a resistively heated tungsten cathode. In some tubes this heat is dissipated by water or oil cooling the anode.

Other tube designs incorporate a rotating anode which spreads the heat out over a larger mass of metal. The x-ray beam is typically pulsed rather than being a continuous wave (CW). X-rays interact with matter in three ways: photoelectric absorption, Compton scattering, and pair production. All three interactions occur when x-rays are absorbed by the human body. The first two dominate, however, owing to the lower energy of the x-rays.

Two general types of detectors are used in x-ray medical imaging: scintillation and gas ionization. Scintillation detectors are used for both conventional projection and computerized tomographic imaging. Ionization detectors have been used only in CT applications. All detectors used in detection of x-ray radiation must be linear and have a maximum efficiency at the wavelength of the x-ray photon to be detected.

Intensifier screens are typically used on both sides of the film, resulting in an overall sensitivity approximately 50 times greater than for the film alone. One drawback associated with intensifying screens is a blurring of the image.

Computerized tomography (CT) imaging is based on obtaining a series of one-dimensional (1-D) projection x-ray images which encompass 180° of projection angles with respect to the imaged object. Each 1-D projection represents the absorption of x-ray radiation along the line from the source to the detector. The 1-D projection images are back-projected using computer programs to produce an image of the internal contents of the original object. The mathematics that describe the signal-generating process in CT imaging are called the Radon and inverse Radon transforms.

Fourth generation detectors utilize a stationary 360° ring of detectors and a rotating x-ray source having a 30° spread. Experimental scanners are being developed with more sources and detectors which allow dynamic x-ray CT imaging of a beating heart. A 3-D image of an object may be obtained by moving either the patient or the source/detector gantry axially with respect to each other. To minimize the x-ray exposure to the patient, two-dimensional (2-D) detector arrays are being used which also eliminate the need for axial motion of the object.

Applications. Applications of x-ray imaging span the entire discipline of medicine. Some of the more common applications are angiography, mammography, and GI, muscular skeletal, neuro, and dental imaging. Muscular skeletal imaging primarily utilizes plane film x-ray imaging. Other nonplane film procedures for muscular skeletal imaging are bone density measurements utilizing dual energy CT scans.

Safety. X-rays are classified as ionizing radiation. These photons possess sufficient energy to ionize molecules, leading to bond breakage and the formation of free radicals. The most significant safety concern centers on radiation dose obtained in x-ray imaging sessions. In addition to the short-term effects of the radiation, there also is concern about the long-term effects.

An ancillary concern arises from the use of contrast agents, eg, gadolinium complexes during CT scans, barium for GI images, and iodine complexes during angiography. The information gained from a medical imaging procedure must always be balanced against health risks of the imaging procedure.

All x-ray equipment must be periodically inspected and the output monitored and calibrated to minimize the chance of accidental overexposure. Another concern involves radiation accumulation by medical personnel operating x-ray equipment. Lead aprons and film badges are used to minimize exposure and to monitor accumulated dose, respectively.

Ultrasound Imaging

Theory. Ultrasound medical imaging is performed by sending a pulse of ultrasound energy into the body and listening for reflections or echoes. For ultrasound technology to be useful in medicine, a method is needed for creating an image from the reflected ultrasound energy. In ultrasound medical imaging the distance between an ultrasound source and a tissue boundary is best determined using a focused beam of ultrasound waves. This 1-D image is not, however, very useful for clinical purposes. A 2-D image is necessary to provide relevant information.

Ultrasound images are typically tomographic images with a slice thickness of 1–2 mm and a field of view of 20–30 cm. A tomographic ultrasound image is generated by sending a series of ultrasound pulses into the portion of anatomy being imaged. Each ultrasound pulse is sent out at a different angle from the source so as to sweep through the anatomy to be imaged in a manner similar to radar sweeping across the sky for airplanes.

The source and detector of ultrasound in an ultrasound medical imager is called a transducer. The transducer is a piezoelectric crystal which physically changes its dimensions when a potential is applied across the crystal.

Typical piezoelectric materials are ceramic crystals and copolymers, such as poly(vinylidene fluoride-*co*-trifluoroethylene), $(—CH_2—Cl_2—)_n—(—CF_2—CFH—)_m$.

The required sweeping of the ultrasound beam across the imaged plane may be accomplished by one of three methods. The transducer may be physically moved through a series of angles to obtain the image, the transducer may be pointed at an ultrasound mirror that rotates through the desired angles, or a linear array of transducers may be employed.

The resolution in an ultrasound image is, among other things, related to the duration of the ultrasound pulse, ie, the shorter the pulse the better the resolution. Imaging may not be performed when the pulse duration is longer than the time to receive an echo. The shorter the ultrasound pulse the more difficult it is to discern it from noise, and the poorer the SNR of the image. As the pulse duration is decreased, the power of the ultrasound pulse is typically increased to compensate for the poorer SNR.

Another factor affecting the SNR in an ultrasound image is interference between reflected signals from small scatters in the tissues.

Contrast in an ultrasound image is related to differences in propagation constants for the tissues. A boundary between two tissues having a large difference in propagation constant reflects large amounts of ultrasound. Ultrasound contrast agents are substances that are introduced into a tissue to change the propagation constant and hence reflect more ultrasound energy. Typical ultrasound contrast agents are lipid-stabilized microbubbles having a diameter of 1–5 micrometers.

Applications. Ultrasound imaging is used for imaging of soft tissues. Its primary advantages are low cost and safety compared to other medical imaging modalities. Ultrasound imaging finds its greatest applications in obstetrics and gynecology for studying the uterus and a fetus, in cardiology for studying the function of the heart, and for imaging of the abdomen.

Safety. High power ultrasound waves can cause local heating and transient cavitation in water. Cavitation is a process in which microscopic gas bubbles expand and collapse as a consequence of the ultrasound wave. The rapid collapse can be adiabatic, causing the energy to be transferred to bond-breaking processes that create free radicals and give rise to the health concern. The rule of thumb in ultrasound medical imaging is to utilize a power level that is as low as reasonably possible.

Magnetic Resonance Imaging

Theory and Equipment. MRI is based on the principles of nmr. This fact makes MRI very interesting to chemists. The less knowledgeable reader is directed to the entry on magnetic spectroscopies (see MAGNETIC SPIN RESONANCE) or one of the bibliographic entries for a more detailed description.

Magnetic resonance imaging (MRI) is a tomographic imaging modality. The basis of MRI states that the resonance frequency of a nucleus is proportional to the magnetic field it is experiencing. If a spatially varying magnetic field is set up across a sample, the nuclei within the sample resonate at a frequency related to their positions.

This simple concept of a 1-D image can be expanded to a 2-D image employing back-projection technology similar to that used in CT imaging. If a series of 1-D images, or projections of the signal in a sample, are recorded for linear 1-D magnetic field gradients applied along several different trajectories in a plane, the spectra can be transformed into a 2-D image using an inverse Radon transform or a back-projection algorithm. This procedure is seldom used. Instead, Fourier-based imaging techniques are used in most MRI.

The contrast between any two tissues may be maximized by prudent choice of the imaging parameters. Clinicians have adopted nomenclature for the various types of images produced as a consequence of the choice of imaging parameters.

The clinician may also change the contrast in an image using a chemical contrast agent. A contrast agent is typically a paramagnetic substance that is introduced into the body and has an affinity for certain tissue types.

Typical MRI contrast agents contain gadolinium. The gadolinium is chelated with a ligand such as ethylenediaminetetraacetic acid (EDTA), diethylenetriaminepentaacetic acid (DTPA), or tetraazacyclododecanetetraacetic acid (DOTA) to lower its toxicity.

Applications. Magnetic resonance imaging finds its greatest use in neuro imaging. MRI has excellent soft tissue specificity and can, therefore, be used to identify many types of lesions in the brain and spinal cord. The utility of MRI in providing structural information about these areas has surpassed that of CT. In addition to the structural information, MRI can also provide functional information. Previously, functional imaging required the use of positron emission tomography (PET).

The second largest application of MRI is in muscular skeletal imaging of joints such as the knee, shoulder, hips, and wrist. Torn ligaments and rips in the cartilage between the tibia and femur are readily seen using MRI.

Another imaging procedure in which MRI is challenging traditional x-ray procedures is magnetic resonance angiography (MRA). Unlike x-ray-based angiography, MRA does not require the injection of contrast agents into the blood stream. MR angiography images flowing blood as opposed to a contrast agent in a blood vessel. As a consequence MRA can detect locations having poor flow which appear normal on an x-ray angiogram because of the presence of contrast agent in the static blood. Magnetic resonance angiography may be performed by one of two techniques: time-of-flight and phase-contrast angiography. Both techniques are routinely used to image flowing blood.

Safety. Because of the relatively young age of MRI there is concern regarding its safety. Users are trying to err on the side of caution. The principal safety concerns are related to the static magnetic field B_0, changing magnetic fields dB_0/dt, tissue heating from r-f power deposition, and acoustic noise. The United States Food and Drug Administration guidelines on static magnetic field limits B_0 to less than 2 T (2×10^4 G). The greatest concern about the health effects of strong magnetic fields are those effects caused by ferromagnetic objects being pulled into the imager while a patient is inside, or the torques created on a ferromagnetic object which might be in the patient's body.

The dB/dt is limited to 6 T/s out of concern that larger values could cause nerve stimulation. The r-f exposure is limited to a specific absorption rate (SAR) of 0.4 W/kg for the whole body, 0.32 W/kg averaged over the head, and less than 8.0 W/kg spatial peak in any one gram of tissue. These numbers are designed to limit the temperature rise to less than 1°C and localized temperature of no greater than 38°C head, 39°C trunk, and 40°C in the extremities.

Magnetic resonance imagers produce a loud knocking sound when the magnetic field gradients are turned on or off. The acoustic noise levels can be high in the bore of the magnet. Patients are usually given ear plugs which can decrease the sound of the knocking by upward of 26 dB.

Nuclear Medicine Imaging

Nuclear medicine imaging involves the use of exogenous radioactive materials to image the body.

Theory and Equipment. The basic principle behind nuclear medical imaging is that a radiopharmaceutical can be introduced into the body which emits radiation detectable outside of the body. Radiopharmaceuticals are biologically active and have a short half-life ($T_{1/2}$). The detectable radiation is typically a γ-ray photon. The radiopharmaceutical must be introduced in sufficient concentration to produce detectable signals outside of the body, but not large enough to be lethal. The more common radioactive nuclei used in radiopharmaceuticals are listed in Table 1. With the exception of xenon, these nuclei are typically bonded to other atoms or complexed with chelates to form the radiopharmaceutical.

Table 1. Radioactive Nuclei Used in Radiopharmaceuticals

Nucleus	Radioactive decay product	γ-Ray energy, keV	$T_{1/2}$	Production[a]
^{201}Tl	γ	70	73 h	CPB
^{133}Xe	γ	81	5.27 d	fission
^{131}I	γ	364	8.05 d	fission
^{123}I	γ	159	13 h	CPB
^{111}In	γ	171, 245	67.9 h	CPB
^{99}Tc^{m}	γ	140.5	6.03 h	^{99}Mo decay
^{82}Rb	β^+	511	1.2 min	^{82}Sr decay
^{67}Ga	γ	93, 184, 300	78.3 h	CPB
^{18}F	β^+	511	110 min	CPB
^{15}O	β^+	511	2 min	CPB
^{13}N	β^+	511	10 min	CPB
^{11}C	β^+	511	20.5 min	CPB

[a] CPB = charged-particle bombardment.

The radiation emitted by the radiopharmaceutical is most often detected using scintillation detectors of NaI(TI).

Gamma radiation may be detected and processed to produce a 2-D planar or a CT emission image.

The scheme used to detect the two 511-keV γ-rays from a β^+ emitter incorporates principles of coincidence detection. The signals from two detectors pointing toward each other along a straight line are processed by circuitry which only produces output when signals are instantaneously detected from each detector. Lead collimators are placed in front of each detector to minimize scattered and random coincidence.

Applications. Brain and central nervous system imaging are common applications of nuclear imaging. Cardiac nuclear imaging using ^{99}Tc^{m}-red blood cells can measure the fraction of blood pumped by the heart during each beat.

Safety. The principal concerns regarding nuclear medical imaging are those associated with the radiopharmaceuticals. Much research has gone into the selection of radiopharmaceuticals exhibiting minimal toxicities, rapid elimination from the body, and short half-life. In addition to health concerns specific to the patient, attention must be paid to minimizing accidental exposure or ingestion of radiopharmaceuticals by the clinical personnel suppressing the imaging procedure.

JOSEPH P. HORNAK
Rochester Institute of Technology

J. T. Bushberg, J. A. Seibert, E. M. Leidholdt Jr., and J. M. Boone, *The Essential Physics of Medical Imaging,* Williams and Wilkins, Baltimore, Md., 1994.

Z. H. Cho, J. P. Jones, and M. Singh, *Fundamentals of Medical Imaging,* John Wiley & Sons, Inc., New York, 1993.

S. Webb, ed., *The Physics of Medical Imaging,* IOP Publishing, Philadelphia, Pa., 1988.

J. P. Hornak and L. M. Fletcher, in E. R. Dougherty, ed., *Digital Image Processing Methods,* Marcel Dekker, New York, 1994.

D. D. Stark and W. G. Bradley, eds., *Magnetic Resonance Imaging,* Mosby, Lanham, Md., 1988.

MELAMINE FORMALDEHYDE RESINS. See AMINO RESINS AND PLASTICS.

MEMBRANE TECHNOLOGY

Membranes have gained an important place in chemical technology and are being used increasingly in a broad range of applications. The key property that is exploited in every application is the ability of a membrane to control the permeation of a chemical species in contact with it. In packaging applications, the goal is usually to prevent permeation completely. In controlled drug delivery applications, the goal is to moderate the permeation rate of a drug from a reservoir to the body. In separation applications, the goal is to allow one component of a mixture to permeate the membrane freely, while hindering permeation of other components.

The market for membranes in industrial separation processes, such as gas separation, reverse osmosis, ultrafiltration, microfiltration, and electrodialysis, is estimated to exceed one billion dollars. Currently, the artificial kidney (hemodialysis) and blood oxygenation markets represent another one billion dollars. All of these applications are growing. Membranes are expected to play a critical role in the next generation of biomedical devices, such as the artificial pancreas and liver. A doubling in the size of the total industry between 1995 and 2005 is likely.

Types of Membrane

In essence, a membrane is a discrete, thin interface that moderates the permeation of chemical species in contact with it. The principal types of membrane are shown schematically in Figure 1.

Preparation of Membranes and Membrane Modules

Because membranes applicable to diverse separation problems are often made by the same general techniques, classification by end use application or preparation method is difficult. The first part of this section is, therefore, organized by membrane structure.

Dense Symmetrical Membranes. These membranes are used on a large scale in packaging applications (see FILM AND SHEETING MATERIALS; PACKAGING). They are also used widely in the laboratory to characterize membrane separation properties. However, it is difficult to make mechanically strong and defect-free symmetrical membranes thinner than 20 μm, so the flux is low, and these membranes are rarely used in separation processes. For laboratory work, the membranes are prepared by solution casting or by melt pressing. Melt forming is commonly used to make dense films for packaging applications, either by extrusion as a sheet from a die or as blown film.

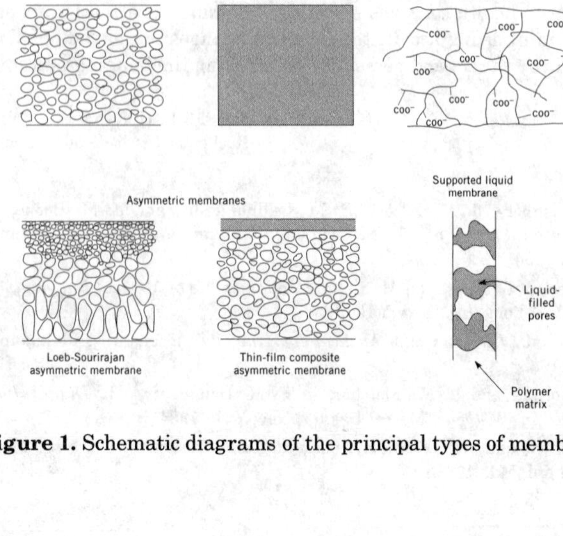

Figure 1. Schematic diagrams of the principal types of membrane.

Microporous Symmetrical Membranes. These membranes, used widely in microfiltration, typically contain pores in the range of 0.1–10 μm diameter. Microporous membranes are generally characterized by the average pore diameter, d, the membrane porosity, ϵ (the fraction of the total membrane volume that is porous), and the tortuosity of the membrane, τ (a term reflecting the length of the average pore through the membrane compared to the membrane thickness). The most important types of microporous membrane are those formed by one of the solution–precipitation techniques; about half of all microporous membranes are made in this way. The remainder is made by various proprietary techniques, the more important of which include irradiation, expanded film, and template leaching.

Asymmetric Membranes. In industrial applications other than microfiltration, symmetrical membranes have been displaced almost completely by asymmetric membranes, which have much higher fluxes. Asymmetric membranes have a thin, permselective layer supported on a more open porous substrate.

Phase inversion, also known as solution precipitation or polymer precipitation, is the most important asymmetric membrane preparation method. Techniques include polymer precipitation by cooling, polymer precipitation by solvent evaporation, polymer precipitation by imbibition of water vapor, and polymer precipitation by immersion in a nonsolvent bath.

A method of making asymmetric membranes involving interfacial polymerization was developed in the 1970s. This technique was used to produce reverse osmosis membranes with dramatically improved salt rejections and water fluxes compared to those prepared by the Loeb-Sourirajan process.

Another important type of composite membrane is formed by solution casting a thin (0.5–2.0 μm) film on a suitable microporous film.

A number of companies have developed ceramic membranes for ultrafiltration and microfiltration applications. Ceramic membranes have the advantages of being extremely chemically inert and stable at high temperatures, conditions under which polymer films fail. Ceramic membranes can be made by three processes: sintering, leaching, and sol–gel techniques.

Liquid Membranes. Although still being explored in a number of laboratories, the more recent development of much more selective conventional polymer membranes has diminished interest in processes using liquid membranes.

Hollow-Fiber Membranes. Most of the techniques listed in the foregoing were developed originally to produce flat-sheet membranes, but the majority can be adapted to produce membranes in the form of thin tubes or filters. Formation of membranes into hollow fibers has a number of advantages, one of the most important of which is the ability to form compact modules with very high surface areas. This advantage is offset, however, by the generally lower fluxes of hollow-fiber membranes compared to flat-sheet membranes made from the same materials and the poor flow distribution, which leads to stagnant areas in the module and membrane fouling.

Hollow fibers are usually on the order of 25 μm to 2 mm in diameter. They can be made with a homogeneous dense structure, or preferably with a microporous structure having a dense permselective layer on the outside or inside surface. The dense surface layer can be integral, or separately coated onto a support fiber. The fibers are packed into bundles and potted into tubes to form a membrane module. More than a kilometer of fibers may be required to form a membrane module with a surface area of one square meter. A module can have no breaks or defects, requiring very high reproducibility and stringent quality control standards.

Hollow-fiber fabrication methods can be divided into two classes. The most common is solution spinning. Solution spinning allows fibers with the asymmetric Loeb-Sourirajan structure to be made. An alternative technique is melt spinning. Melt-spun fibers are usually relatively dense and have lower fluxes than solution-spun fibers, but

Table 1. Characteristics of Module Designs

Property	Hollow-fine fibers	Capillary fibers	Spiral-wound	Plate and frame	Tubular
manufacturing cost, \$/m^2	5–20	20–100	30–100	100–200	50–200
resistance to fouling	very poor	good	moderate	good	very good
parasitic pressure drop	high	moderate	moderate	low	low
suitability for high pressure operation	yes	no	yes	can be done with difficulty	can be done with difficulty
limitation to specific types of membrane	yes	yes	no	no	no

Table 2. Membrane Separation Technologies

Uses	Processes	Status
developed technologies	microfiltration	well-established unit processes; no significant breakthroughs seem imminent
	ultrafiltration	
	reverse osmosis	
	electrodialysis	
developing technologies	gas separation	a number of plants have been installed; market size and number of applications served is expanding rapidly
	pervaporation	
to-be-developed technologies	facilitated transport	significant problems remain to be solved before industrial systems will be installed

because the fiber can be stretched after it leaves the die, very fine fibers can be made.

Membrane Modules. A useful membrane process requires the development of a membrane module containing large surface areas of membrane. The development of the technology to produce low cost membrane modules was one of the breakthroughs that led to the commercialization of membrane processes in the 1960s and 1970s. The earliest designs were based on simple filtration technology and consisted of flat sheets of membrane held in a type of filter press: these are called plate-and-frame modules. Systems containing a number of membrane tubes were developed at about the same time. Both of these systems are still used, but because of their relatively high cost they have been largely displaced by two other designs: the spiral-wound module and the hollow-fiber module.

The choice of the appropriate membrane module for a particular membrane separation balances a number of factors. The principal factors that enter into this decision are listed in Table 1.

Membrane Applications

The principal use of membranes in the chemical processing industry is in various separation processes. These can be classified into technologies that are developed, developing, or to-be-developed, as shown in Table 2. Membranes, or rather films, are also used widely as packaging materials. Membranes are also used in various biomedical applications, for example, in controlled-release technology and in artificial organs, such as the artificial kidney, lung, pancreas, etc (see CONTROLLED-RELEASE TECHNOLOGY).

RICHARD W. BAKER
Membrane Technology & Research, Inc.

R. W. Baker and co-workers, *Membrane Separation Systems,* Noyes Data Corp., Park Ridge, N.J., 1991.

M. Mulder, *Basic Principles of Membrane Technology,* Kluwer Academic Publishers, Dordrecht, the Netherlands, 1991.

W. S. W. Ho and K. K. Sirkar, eds., *Membrane Handbook,* Chapman & Hall, New York, 1992.

MEMORY-ENHANCING DRUGS

Potential memory-enhancing drugs are discussed herein predominantly from the standpoint of treatments that intervene in one or more processes associated with the development of dementia.

As of 1994, no drugs are available to address the etiology of neuronal loss and consequent memory impairment. There are, however, a number of drugs used throughout the world that enhance cerebral metabolism or that palliate cognitive dysfunction through modulation of neurotransmitter systems. Whereas there is considerable controversy surrounding the clinical efficacy of these agents, cognition enhancers are sold worldwide and comprise an annual market estimated to be between \$1 and \$2 billion. Widespread usage results largely from availability and the absence of alternative therapy.

The compounds used to palliate the mnemonic and cognitive decline associated with dementia include cerebral vasodilators and the so-called nootropic agents. These materials enhance cerebral metabolism. Agents which enhance neurotransmitter function are in most cases cholinergic.

Cerebral Metabolism Enhancers

Whereas the majority of agents being evaluated for treatment of dementia have activities associated with specific neuronal systems, cerebral metabolism enhancers have undefined or varied mechanisms. Hydergine (**1**), vinpocetine (**2**), and nimodipine (**3**) initially had been thought to exert their activity through cerebral vasodilation (Fig. 1). However, these are used to treat patients with dementia or other age-related symptoms of compromised cognitive function based on other mechanisms and without a clear understanding of the reasons for beneficial actions. The other agent in this group, acetyl-L-carnitine (**4**) is thought to exert its beneficial effects by its positive influence on energy metabolism in the mitochondria, as well as on cholinergic activity.

Nootropics

The term *nootropic* has been used to describe a class of compounds defined by the ability of its members to facilitate learning. The compounds are most effective in animals that have had their cognitive abilities compromised in some way. The molecular mechanism underlying the cognitive-enhancing effects of this class of molecules is unknown, although interaction with the excitatory amino acid network, muscarinic M-1 receptors, or enzymes such as prolylendopeptidase have been suggested. Piracetam is the classic representative of the group, and many other acetams such as aniracetam, oxiracetam, pramiracetam are undergoing or have undergone clinical evaluation. The nootropics are the largest class of compounds being considered for patients with AD or compromised memory and cognition function.

The mechanism of action of nootropic agents has been proposed to be their ability to facilitate information acquisition, consolidation, and retrieval. No one particular effect has been observed with any consistency for these agents.

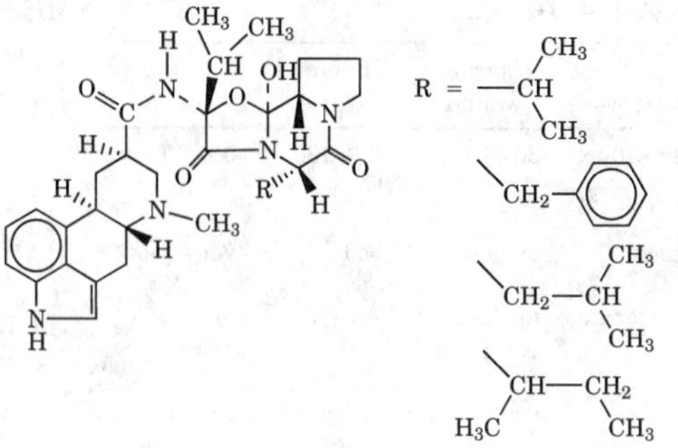

(1)

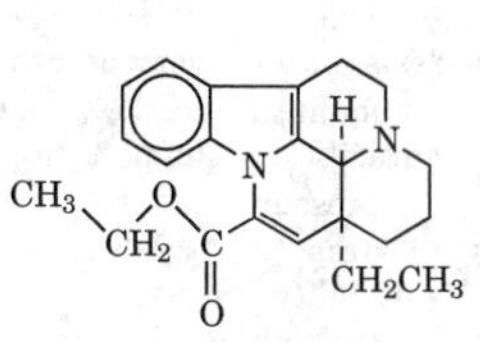

(2)

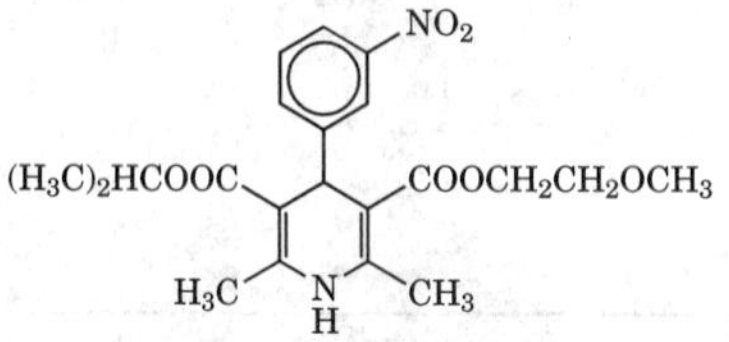

(3)

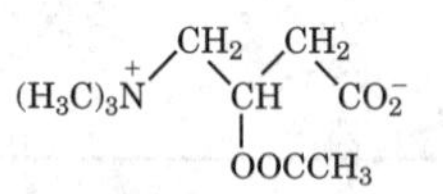

(4)

Figure 1. Structures of cerebral metabolism enhancers.

Cholinomimetics

One of the earliest identified and most consistent neurochemical changes observed in AD is the profound loss of neocortical cholinergic innervation. This loss correlates with the degree of dementia. The role of cholinergic dysfunction in memory impairment and symptoms of dementia is well supported, although this represents only one factor of this disease.

Although controversy exists over the cholinergic involvement in AD dementia, as of 1993 the only AD therapy approved by the U.S. FDA was the cholinesterase inhibitor, tacrine, $C_{13}H_{14}N_2$, sold as Cognex (Warner-Lambert).

Several cholinergic strategies, other than cholinesterase inhibition, have been employed with the intention of ameliorating the symptoms of AD. These include precursor loading acetylcholine release enhancement, and direct activation of both muscarinic and nicotinic receptors.

Acetylcholine Precursors. Early efforts to treat dementia using cholinomimetics focused on choline supplement therapy. α-Glycerylphosphorylcholine (α-GFC) and cytidine-5-diphosphatecholine (CDP-choline) are two more recently studied choline-delivering agents. The former has been reported to increase ACh production and release, and to reverse scopolamine-induced behavioral deficits in rats, as well as to reverse behavioral deficits in old and excitotoxin-lesioned rats. The latter has been shown to be effective in improving behavioral performance in compromised animals.

Acetylcholinesterase Inhibitors. The greatest activity in the area of cholinomimetic treatments for AD has been in the development of agents that retard the degradation of acetylcholine (ACh) through the blockade of acetylcholinesterase (AChE) activity.

Physostigmine, an alkaloid, has been the most extensively studied AChE inhibitor. Through its reactive carbamoyl group, physostigmine acylates the catalytic site of AChE, thereby inhibiting the enzyme. Side effects of physostigmine include gastrointestinal disturbances as well as cardiovascular effects. Other shortcomings of physostigmine are a short half-life and variable bioavailability.

SDZ ENA 713 is long-acting carbamate-containing molecule being investigated for AChE inhibition and AD therapy. The advantage claimed for this compound over physostigmine and tacrine is the CNS specificity of SDZ ENA 713 relative to the other AChE inhibitors. This selectivity may serve to reduce peripheral side effects while maintaining clinical efficacy.

The aminoacridines, tacrine and its 1-hydroxy metabolite, velnacrine, are reversible inhibitors of AChE. Serious hepatotoxicity of tacrine has been documented. More recent data suggest, however, that this toxicity can be reduced by carefully monitoring serum alanine aminotransferase levels. The side effects of tacrine also include gastrointestinal disturbances and emesis, and alternative AChE therapies are being advanced. Velnacrine's limited efficacy and side-effect profile, which includes drug-related hematological changes, caused it to be dropped from further development.

Three structurally unrelated AChE inhibitors being pursued for AD treatment are huperzine A, E2020, and galanthamine. Metrifonate is itself not an AChE inhibitor, but is nonenzymatically converted into an active irreversible inhibitor of the enzyme. The compound is relatively specific for AChE over butyrylcholinesterase and the irreversible nature of its inhibition gives rise to an extended duration of action. Some clinical experience has been gained through its use to treat schistosomiasis and it is undergoing clinical evaluation for AD.

Receptor Agonists

Muscarinic Receptor Agonists. Acetylcholine indirect agonists such as the AChE inhibitors and ACh-releasing agents only have value in treating dementia if enough of the cholinergic arbor in the hippocampus and cortex of affected individuals remains functional. As the cholinergic innervation declines, as is the case upon progression of AD, these therapies lose efficacy. However, there is evidence that post-synaptic receptors actually are preserved in AD. Thus direct muscarinic agonists should remain effective even as the presynaptic cholinergic terminals decline in number.

The identification of multiple subtypes of muscarinic receptors has stimulated a search for subtype specific muscarinic agonists which may limit side effects while increasing efficacy.

Five distinct muscarinic receptors have been identified, designated m1 to m5. CI-979 is a balanced muscarinic agonist having equal affinities for cloned m1 and m2 receptors. It increases central muscarinic tone. CI-979 is well tolerated in humans up to a dose of 1 mg. Whereas balanced muscarinic agents having acceptable therapeutic indexes may be of clinical value, more hope is held for subtype specific agents.

Nicotinic Receptor Agonists. The gastrointestinal and cardiovascular side effects of nicotine limit its therapeutic value. Thus efforts to discover brain-specific nicotinic agonists for AD treatment led to ABT 418. This compound was shown to be 3–10 times more potent than nicotine in enhancing performance of laboratory animals in paradigms designed to measure learning and memory. In contrast, ABT 418 was less potent than nicotine in producing emesis. ABT 418 is being evaluated in human clinical trials.

Acetylcholine Release Modulators

An alternative approach to stimulate cholinergic function is to enhance the release of acetylcholine (ACh). Compounds such as the aminopyridines increase the release of neurotransmitters. The mechanism by which these compounds modulate the release of acetylcholine is likely the blockade of potassium channels. However, these agents increase both basal (release in the absence of a stimulus) and stimulus-evoked release. 4-Aminopyridine was evaluated in a pilot study for its effects in AD and found to be mildly effective.

Unlike the aminopyridines, linopirdine (AVIVA) enhances evoked and not basal release of acetylcholine. Like 4-aminopyridine, linopirdine enhances the release of several neurotransmitters. Because the functions of multiple neurotransmitter systems are decreased in dementias like AD, the property of compounds such as linopirdine to enhance the release of several neurotransmitters offers an advantage over AD therapies aimed at stimulating the cholinergic system alone.

Another compound that affects parameters relating to several neurotransmitter systems is HP749 which is in clinical trials for the treatment of the dementia associated with Alzheimer's disease. This compound has several effects in *in vitro* neurochemical assays, including monoamine reuptake blockade, enhancement of NE release, and inhibition of α_2-adrenergic and muscarinic receptor binding.

ROBERT ZACZEK
ROBERT J. CHORVAT
The DuPont Merck Research Laboratories

B. S. Meldrum, in I. F. C. Rose, ed., *Metabolic Disorders of the Nervous System,* Pitman, London, 1983, pp. 175–187.

S. B. Dunnett and H. C. Fibiger, in A. C. Cuello, ed., *Cholinergic Function and Dysfunction,* Vol. 98, Elsevier Science Publishers BV, Amsterdam, The Netherlands, 1993, p. 413.

L. J. Thal, in R. Becker and E. Giacobini, eds., *Physostigmine in Alzheimer's Disease,* Birkhauser, Boston, Mass., 1991.

L. L. Iversen and co-workers, in R. Becker and E. Giacobini, eds., *Cholinergic Basis for Alzheimer Therapy,* Birkhauser, Boston, Mass., 1991, p. 297.

MERCURY

Mercury, Hg, atomic number 80, atomic weight 200.59, also called quicksilver, is a heavy, odorless metal belonging to Group 12 (IIB) of the Periodic Table. Elemental mercury is a liquid at room temperature, having a characteristic bright, silvery appearance. The symbol Hg is taken from the Latin word *hydrargyrum,* meaning liquid silver. Unlike the other two Group 12 elements, mercury exhibits two valences, mercurous, Hg^+, and mercuric, Hg^{2+}.

Applications of mercury include use in batteries (qv), chlorine and caustic soda manufacture (see ALKALI AND CHLORINE PRODUCTS), pigments (see PIGMENTS, INORGANIC), light switches, electric lighting, thermostats, dental repair (see DENTAL MATERIALS), and preservative formulations for paints (qv). As of the end of the twentieth century, however, increased awareness of and concern for mercury toxicity has resulted in both voluntary and regulatory reduction of mercury usage (see also MERCURY COMPOUNDS).

Occurrence

Mercury metal is widely distributed in nature, usually in quite low concentrations. The terrestrial abundance is on the order of 50 parts per billion (ppb), except in mercuriferous belts and anthropogenically contaminated areas. The most important mineral of mercury is cinnabar, found in rocks near recent volcanic activity of hot spring areas and in mineral veins or fractures as impregnations.

Mercury ore deposits occur in faulted and fractured rocks, such as limestone, calcareous shales, sandstones, serpentine, chert, andesite, basalt, and rhyolite. Deposits are mostly epithermal in character.

Table 1. Physical and Chemical Properties of Mercury

Property	Value
melting point, °C	−38.87
boiling point, °C	356.9
triple point, °C	−38.84168
bp rise with pressure, °C/kPa[a]	0.5595
compressibility (volume) at 20°C, MPa^{-1}[b]	39.5×10^{-6}
conductivity, thermal, $W/(cm^2 \cdot K)$	0.092
critical density, g/cm^3	3.56
critical pressure, MPa[b]	74.2
critical temperature, °C	1677
density, g/cm^3	
at melting point	14.43
0°C	13.595
volume expansion coefficient at 20°C, $°C^{-1}$	182×10^{-6}
magnetic moment, ^{199}Hg, J/T[c]	4.63×10^{-24}
refractive index at 20°C	1.6–1.9
soly in water, $\mu g/L$	20–30
surface tension temp coefficient, mN/(m·°C) (dyn/(cm·°C))	−0.19
viscosity at 20°C, mPa·s(= cP)	1.55

[a] To convert kPa to mm Hg, multiply by 7.5. [b] To convert MPa to atm, divide by 0.101. [c] Value is equivalent to 0.4993 μ_B.

Properties

The physical and chemical properties of mercury are given in Table 1.

The specific heat varies with temperature. For solid mercury it increases, but in liquid mercury it drops such that the specific heat at 210°C is the same as that at −75°C. The vapor pressure *P,* of mercury, also behaves irregularly.

At ordinary temperatures, mercury is stable and does not react with air, ammonia (qv), carbon dioxide (qv), nitrous oxide, or oxygen (qv). It combines readily with the halogens and sulfur, but is little affected by hydrochloric acid, and is attacked only by concentrated sulfuric acid. Both dilute and concentrated nitric acid dissolve mercury, forming mercurous salts when the mercury is in excess or no heat is used, and mercuric salts when excess acid is present or heat is used. Mercury reacts with hydrogen sulfide in the air and thus should always be covered.

The only metals having good or excellent resistance to corrosion by amalgamation with mercury are vanadium, iron, niobium, molybdenum, cesium, tantalum, and tungsten.

Production and Shipment

Primary Production. Mercury ore is mined by both surface and underground methods. The latter furnishes about 90% of the world's production. Mercury is recovered also as a by-product in the mining and processing of precious and base metals (see MINERALS RECOVERY AND PROCESSING).

Mercury is also produced by working mine dumps and tailing piles, particularly those accumulated during turn-of-the-century mining operations. The average grade of mercury ore mined from large mines throughout the world in the latter part of the twentieth century has ranged from 4 to 20 kg/t. Total recovery has approached 95%. The average grade of ore has generally declined as a result of the practice of mining the richest parts of ore bodies to realize a higher profit and because prices have generally increased, allowing lower grade ores to be exploited. Since 1990, the only mine production of mercury in the United States has come as a by-product of gold mining operations (see GOLD AND GOLD COMPOUNDS).

Secondary Production. Smaller quantities of mercury are produced each year from industrial scrap and waste materials. Secondary production of mercury in the U.S. has increased steadily in the 1990s as a result of the rising costs of hazardous waste disposal and increased restrictions on mercury disposal.

Shipping. Prime virgin-grade mercury is packaged in wrought iron or steel flasks containing 34.5 kg of the metal. Mercury of greater purity, produced by multiple distillation or other means, may be marketed in flasks but is usually packaged in small glass or plastic containers.

Processing

Primary. Mercury metal is produced from its ores by standard methods throughout the world. The ore is heated in retorts or furnaces to liberate the metal as vapor which is cooled in a condensing system to form mercury metal. Retorts are inexpensive installations for batch-treating concentrates and soot. Other recovery methods have been used.

Secondary. Scrap material, industrial and municipal wastes, and sludges containing mercury are treated in much the same manner as ores to recover mercury.

Economic Aspects

Mercury toxicity from environmental pollution and occupational exposure has become an area of concern throughout the world. Since the 1980s much effort has been devoted both in terms of legislation and voluntary action, to limiting the production and use of mercury, and to finding technologies designed to supplant the need for mercury. As a result, overall demand for mercury has decreased throughout the industrialized world.

Grades, Specifications, and Quality Control

The commercial-grade (99.9%) purity marketed as prime virgin-grade mercury has a clean bright appearance and contains less than 1 ppm of dissolved base metals. Prime virgin-grade mercury of lower purity is brought up to specification by filtration, redistillation, or electrolytic processing.

Triple-distilled mercury is of highest purity, commanding premium prices. It is produced from primary and secondary mercury by numerous methods, including mechanical filtering, chemical and air oxidation of impurities, drying (qv), electrolysis, and most commonly multiple distillation.

The purity of mercury can be estimated by its appearance. Because mercury has such a high specific gravity, almost all impurities, including amalgams, are lighter and float on the surface, causing the bright, mirror-like surface to become dull and black.

Mercury in the Environment

Releases. Annual global releases of mercury into the atmosphere from natural sources have been estimated to be on the order of 2700–6000 t; annual anthropogenic releases of mercury into the atmosphere have been reported from 630–6800 t.

Global anthropogenic releases of mercury into the biosphere are approximately 11,000 metric tons per year. The release of mercury into aquatic ecosystems results from electricity production, manufacturing processes, domestic wastewater and sewage sludge, mining, and fallout of atmospheric mercury (see METALLURGY–EXTRACTIVE METALLURGY; WASTES, INDUSTRIAL). The release of mercury into soils results from mining and smelting activities, coal fly ash and bottom ash, wood wastes, commercial wastes, sewage sludge, urban refuse, fallout of atmospheric mercury, and other sources.

Regulations. In order to decrease the amount of anthropogenic release of mercury in the United States, the EPA has limited both use and disposal of mercury. In 1992, the EPA banned land disposal of high mercury content wastes generated from the electrolytic production of chlorine-caustic soda, accompanied by a one-year variance owing to a lack of available waste treatment facilities in the U.S. A thermal treatment process meeting EPA standards for these wastes was developed by 1993. The use of mercury and mercury compounds as biocides in agricultural products and paints has also been banned by the EPA.

The State of New Jersey has passed a law restricting the sale and disposal of batteries (qv) containing mercury. California and Minnesota have placed restrictions on the disposal of fluorescent light tubes, which contain from 40–50 mg of mercury per tube, depending on size. After batteries, fluorescent lamps are the second largest contributor of mercury in solid waste streams in the U.S. As a result of increased waste restrictions on the use and disposal of mercury, by the year 2000 mercury in municipal solid waste streams was expected to be about 160 t.

Health and Safety Factors

Exposure. The exposure of humans and animals to mercury from the general environment occurs mainly by inhalation and ingestion of terrestrial and aquatic food chain items. Fish generally rank the highest (10–300 ng/g) in food chain concentrations of mercury.

In occupational settings mercury exposure results predominantly from direct dermal contact or inhalation of mercury vapors. Chlor–alkali plants are one of the principal sources for occupational exposure to mercury. Mining and refining of mercury contribute to exposure, and the processing of cinnabar can result in high exposures to the skin and lungs, producing poisoning in a relatively short time.

Dentists and dental assistants can be exposed to significant quantities of mercury during the preparation and use of mercury amalgam (see DENTAL MATERIALS). People who have mercury amalgam dental restorations may be exposed to 1–2 μg/d of mercury as a result of its dissolution in saliva. Exposure to mercury from amalgam restorations, however, is considered unlikely to pose a health risk to the patient.

Recommended safety measures which minimize occupational exposure to mercury include the use of efficient respirators, adequate ventilation and air-exhaust systems, employee warning signs and messages, training in accident emergency procedures, immediate and thorough cleanup of spills, airtight storage of mercury-containing wastes, frequent monitoring of mercury levels in the work area, and coverall-type work clothes.

Toxicity. The toxic effects of mercury and mercury compounds (qv) are well known, and several detailed discussions on mercury toxicity are available. Toxicity to the central nervous system is more prominent after exposure to mercury vapor than to divalent mercury.

Short-term exposure to mercury vapor may produce symptoms within several hours. These symptoms include weakness, chills, metallic taste, nausea, vomiting, diarrhea, labored breathing, cough, and a feeling of tightness in the chest. Pulmonary toxicity may progress to an interstitial inflammation of the lung with severe compromise of respiratory function. Recovery, although usually complete, may be complicated by residual interstitial growth of excess fibrous tissue.

Chronic exposure to mercury vapor produces an insidious form of toxicity that is manifested by neurological effects and is referred to as the asthenic vegetative syndrome.

The biochemical basis for the toxicity of mercury and mercury compounds results from its ability to form covalent bonds with sulfur. Even in low concentrations divalent mercury is capable of inactivating enzymes containing sulfhydryl (—SH) groups, causing interference with cellular metabolism and function.

The affinity of mercury for sulfhydryl groups provides the basis for treatment of mercury poisoning using chelating agents (qv) such as dimercaprol (for high level exposures or symptomatic patients), or penicillamine (for low level exposures or asymptomatic patients).

STEPHEN C. DEVITO
U.S. Environmental Protection Agency

Characterization of Products Containing Mercury in Municipal Solid Waste in the United States, 1970–2000, OSW No. EPA530-R-92-013 (NTIS No. PB92-162 569), U.S. Environmental Protection Agency, Washington, D.C., 1992.

R. G. Smith, in R. Hartung and B. D. Dinman, eds., *Environmental Mercury Contamination,* Ann Arbor Science Publishers, Inc., Ann Arbor, Mich., 1972, pp. 97–136.

A. M. Fan, in L. Fishbein, A. Furst, and M. A. Mehlman, eds., *Advances in Modern Environmental Toxicology,* Vol. XI. *Genotoxic and Carcinogenic Metals: Environmental and Occupational Occurrence and Exposure,* Princeton Scientific Publishing Co., Inc., Princeton, N.J., 1987, pp. 185–210.

R. A. Goyer, in M. O. Amdur, J. Doull, and C. D. Klaassen, eds., *Casarett and Doull's Toxicology, The Basic Science of Poisons,* 4th ed., Pergamon Press, Inc., New York, 1991, pp. 646–651.

MERCURY COMPOUNDS

Mercury salts exist in two oxidation states: mercurous, Hg+, and mercuric, Hg2+. The former exist as double salts.

Many mercury compounds are labile and easily decomposed by light, heat, and reducing agents. This innate lack of stability in mercury compounds makes the recovery of mercury from various wastes that accumulate with the production of compounds of economic and commercial importance relatively easy (see RECYCLING).

The toxic nature of mercury and its compounds has caused concern over environmental pollution, and governmental agencies have imposed severe restrictions on release of mercury compounds to waterways and the air (see MERCURY). Methods of precipitation and agglomeration of mercurial wastes from process water have been developed. These methods generally depend on the formation of relatively insoluble compounds such as mercury sulfides, oxides, and thiocarbamates. Metallic mercury is invariably formed as a by-product. The use of coprecipitants, which adsorb mercury on their surfaces facilitating removal, is frequent.

The covalent character of mercury compounds and the corresponding ability to complex with various organic compounds explains the unusually wide solubility characteristics. Mercury compounds are soluble in alcohols, ethyl ether, benzene, and other organic solvents. Moreover, small amounts of chemicals such as amines, ammonia (qv), and ammonium acetate can have a profound solubilizing effect (see COORDINATION COMPOUNDS). The solubility of mercury and a wide variety of mercury salts and complexes in water and aqueous electrolyte solutions has been well outlined.

Owing to legal restrictions, use of mercury compounds has declined. The most important areas of mercury compound usage as of the mid-1990s are as preservatives and fungicides in coating compositions, as catalysts, as intermediates in the formation of other compounds, and as a component of compositions used as semiconductors (see COATINGS; CATALYSIS; FUNGICIDES, AGRICULTURAL; SEMICONDUCTORS). Some pharmaceutical uses remain, eg, in ophthalmic preparations and antiseptics (see DISINFECTANTS AND ANTISEPTICS).

All mercury compounds should be stored in amber bottles or otherwise protected from light. In manufacture, glass-lined equipment is preferred, although stainless steel may be used. Stainless steel may cause some discoloration at high temperatures if concentrated acetic acid is used.

Mercury Salts

Mercury salts include mercuric acetate, $Hg(C_2H_3O_2)_2$; mercuric carbonate, $HgCO_3{\cdot}3HgO$; mercuric cyanide, $Hg(CN)_2$; mercuric oxycyanide or basic mercuric cyanide, $Hg(CN)_2{\cdot}HgO$; mercuric fulminate, $Hg(ONC)_2$; mercury fluorides; mercurous chloride, Hg_2Cl_2; mercuric chloride, $HgCl_2$; mercurous bromide, Hg_2Br_2; mercuric bromide, $HgBr_2$; mercurous iodide, Hg_2I_2; mercuric iodide, HgI_2; complex halides; mercurous nitrate, $Hg_2N_2O_6$ or $Hg_2(NO_3)_2$; mercuric nitrate, $Hg(NO_3)_2$; mercuric oxide, HgO; mercurous sulfate, Hg_2SO_4; mercuric sulfate, $HgSO_4$; mercuric sulfide, HgS, and mercury telluride.

Organomercury Compounds

Organomercury compounds include phenylmercuric acetate (PMA), $HgC_8H_8O_2$, 3-chloro-2-methoxypropylmercuric acetate, $ClCH_2C(OCH_3)HCH_2HgOOCCH_3$, and alkyl mercuric compounds, RHgX.

Miscellaneous Compounds of Pharmaceutical Interest

These include antiseptics [ammoniated mercury, $Hg(NH_2)Cl$; *o*-(chloromercuri)phenol (mercarbrolide); merbromin (disodium 2,7-dibromo-4-hydroxymercurifluorescein, commonly called mercurochrome); nitromersol; and mercurophen], antisyphilitics (mercuric salicylate and mercuric succinimide), and diuretics [chlormeodrin (methoxy(urea)propylmercuric chloride)].

Health and Safety Factors

Toxicity. Inorganic mercury compounds, aryl mercury compounds, and alkoxy mercurials are generally considered to be quite similar in their toxicity. Alkyl mercury compounds are considered to be substantially more toxic and hazardous. Mercury and its compounds can be absorbed by ingestion, absorption through the skin, or by inhalation of the vapor. The metal itself, however, rarely produces any harmful effects when ingested.

Alkyl mercury compounds in the blood stream are found mainly in the blood cells, and only to a small extent in the plasma. This is probably the result of the greater stability of the alkyl mercuric compounds, as well as their peculiar solubility characteristics. Alkyl mercury compounds affect the central nervous system and accumulate in the brain.

Environmental Factors. The control, recovery, and disposal of mercury-bearing waste products are as important to the mercurials industry as the manufacturing process.

Safety. The maximum acceptable concentration (MAC) for mercury in all forms except alkyl compounds is 0.05 mg Hg/m^3 air. For alkyl mercury compounds the TLV is set at 0.01 mg Hg/m^3 air.

Suitable ventilating equipment, consisting mainly of carbon absorbers which effectively absorb mercury vapor from recirculated air, must be employed to maintain standards below the value permitted in the occupational environment. When the possibility of higher exposures exists, small disposable masks utilizing a mercury vapor absorbent may be employed.

The control of mercury in the effluent derived from the manufacturing processes used in the preparation of inorganic and organic mercurials is mandated by law in the United States. The concentrations and the total amounts vary with the industry and the location, but generally it is required that the effluent contain not more than 0.01 mg Hg/L.

MILTON NOWAK
WILLIAM SINGER
Troy Chemical Corporation

Makarova and Nesmeyanov, *Methods of Elements—Organic Chemistry,* Vol. 4, North-Holland Publishing Co., Amsterdam, the Netherlands, 1967.

W. E. Machemer, *Devel. Ind. Microbiol.* **20,** 25–39 (1979).

M. Nowak, *Am. Paint J.* (June 14, 1971).

L. G. Hepler and G. Olofsson, *Chem. Rev.* **75,** 585 (1975).

METAL ANODES

In electrolytic processes, the anode is the positive terminal through which electrons pass from the electrolyte. Most materials used in metal anode fabrication are characteristically expensive; use has, however, been justified by enhanced performance and reduced operating cost. An additional consideration that has had increasing influence on selection of the appropriate anode is concern for the environment (see ELECTROCHEMICAL PROCESSING).

Industrial metal anodes can generally be classified in one of two groups. The first group, chlorine-generating anodes, find application primarily in the manufacture of chlorine and caustic, sodium chlorate, and sodium hypochlorite (see ALKALI AND CHLORINE PRODUCTS; CHLORINE OXYGEN ACIDS AND SALTS). The second group, consisting of the oxygen-evolving anodes, do not generate a saleable product directly, but rather facilitate the desired cathodic reaction. Commercial uses include high speed electrogalvanizing of steel (qv), electrowinning of base metals, plating operations, cathodic protection, electrophoretic painting, copper (qv) foil treatment, and, more recently, the primary production of copper foil itself (see ELECTROPLATING; MACHINING METHODS, ELECTROCHEMICAL).

Coating Structure and Morphology

The crystal structure of metal anode coatings has been investigated using x-ray diffraction studies, x-ray fluorescence analysis, and microprobe studies in conjunction with scanning electron micrographs (see MICROSCOPY; X-RAY TECHNOLOGY). However, the role of the titanium metal substrate, or that of the oxides formed when the substrate is anodized or heated in air, is not completely clear for coatings that contain titanium in their solution. Most coatings, even those not containing titanium in the formulation, exhibit traces of titanium dioxide, present either in rutile or anatase form.

Ruthenium–Titanium Oxides. The x-ray diffraction studies of ruthenium–titanium oxide coatings show that the coating components are present as the metal dioxides, each in the rutile form as well as in solid solution with each other.

It appears that the titanium metal substrate on which the coating is deposited plays an important role in the structure and morphology of the coating. The surface layer of rutile titanium dioxide normally found on oxidized titanium metal apparently acts as a seed to initiate growth of the rutile form of the oxide, rather than the anatase form. Interfacial layers of titanium suboxides, known to be electrically conductive, also act to effect a gradual transition from pure metal to pure rutile oxides. Scanning electron micrographs of ruthenium–titanium oxide coatings show a characteristic microcracked surface.

Iridium Oxide. Iridium dioxide coatings, typically used in combination with valve metal oxides, are quite similar in structure to those of ruthenium dioxide coatings.

Platinum–Iridium. There are two distinct forms of 70/30-wt % platinum–iridium coatings. The first, prepared as prescribed in British patents, consists of platinum and iridium metal. The surface morphology of a platinum–iridium metal coating is cracked, but not in the regular networked pattern typical of the DSA oxide materials. The second form consists of Pt metal but the iridium is present as iridium dioxide. Iridium metal may or may not be present, depending on the baking temperature.

Spinel Cobalt Oxides. The cobalt mixed oxide, Co_3O_4, containing Co(II) and Co(III) ions, has the spinel structure. In these coatings containing zirconium, a separate, partially crystalline phase of zirconium dioxide occurs. The coating has large microscopic pores, providing a high surface area which appears to be related to the presence of zirconium. Spinel oxides applied without the presence of zirconium are dense and closely packed coatings.

Operating Performance of Coatings

The key yardstick of performance for a coated metal anode is the period of time, measured in ampere-hours, that the coating operates before it reaches an unacceptable voltage, as measured by the single electrode potential (SEP). Factors influencing the escalation include accumulated average lifetime, operating current density, electrolyte conditions, exposure to oxygen in the anolyte compartment, and cell design.

Chlorine Anodes. In chlorine manufacture, anode operating life is limited by coating wear. The wear rate varies according to cell type. The longest anode coating life has been achieved in diaphragm chlorine cells.

Because mercury (qv) chlorine cells operate at higher current densities and because the mercury cell anode can be adjusted during operation to minimize the anode-to-cathode gap, anode coating life in these cells is much shorter. Because of limited commercial experience with anode coatings in membrane cells, commercial lifetimes have yet to be defined. Expected lifetime is 7–12 years.

Metal anode coatings commercially used for manufacture of sodium chlorate include not only the ruthenium oxide coatings, but also platinum–iridium coatings. Whereas ruthenium oxide coatings might be preferred for a longer performance life and higher resistance to process upsets, platinum–iridium coatings generally operate at higher efficiencies during the first months of operation.

Oxygen-Evolving Anodes. In the case of oxygen-evolving anodes employing iridium dioxide-based coatings, the most significant commercial experience has been in the electrogalvanizing industry where coated titanium anodes have supplanted lead and zinc anodes, and thus lead contamination of the product and waste streams.

In contrast to the coating wear limitation on anode life experienced in chlorine cells, passivation of the substrate beneath the coating is typically the limiting factor for oxygen-evolving anodes. As a result, technology has been introduced to either maintain or modify the titanium surface to increase coating adhesion and significantly improve lifetimes.

Structure Design

Each electrolytic application demands a unique approach to anode structure design and fabrication. Factors such as current distribution, gas release, ability to maintain structural tolerances, electrical resistance, and the practicality of recoating must be taken into account. The most commercially accepted design for diaphragm chlorine cells is that of the expandable anode (Fig. 1).

In mercury chlorine cells, it has been found that cells operate at lower voltages when fitted with anode structures comprised of triangular rods or of vertical blades.

The dimensionally stable characteristic of the metal anode made the development of the membrane chlorine cell possible. These cells are typically arranged in an electrolyzer assembly which does not allow for anode-to-cathode gap adjustment after assembly. Also, very close tolerances are required. The latitude that titanium affords the cell designer has made a wide variety of monopolar and bipolar membrane cell designs possible.

When used in radial cells such as for the production of copper foil and in some electrogalvanizing operations, the anode must be curved to meet the shape dictated by the cathodic drum. The anode must also achieve exact tolerances to assure that a constant anode-to-cathode gap is created and maintained.

Manufacturing Technology

Manufacturing techniques for metal-coated anodes have been developed to a high level of sophistication. By the mid-1990s commercial

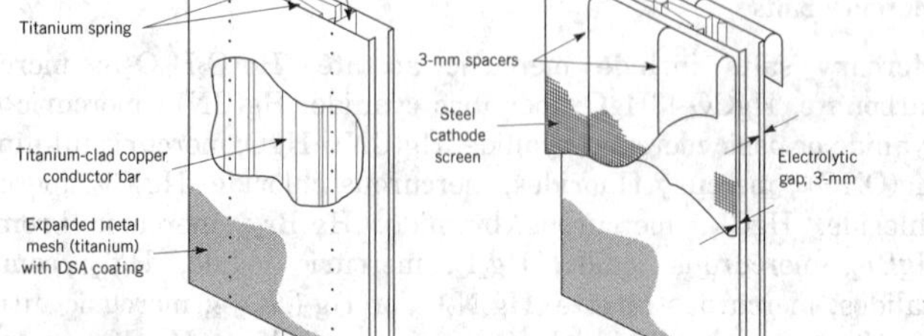

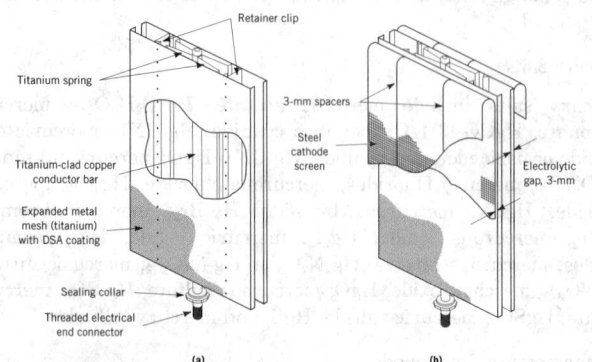

Figure 1. Expandable anode for diaphragm chlorine cells: (**a**) clamped and (**b**) expanded.

plants were producing or recoating in excess of 50,000 anodes annually. This scale of operation requires continuous coating processes (qv) the use of electrostatic spray application, robotics, and strict process control (qv). The high cost of the coating itself demands high utilization efficiency. Quality requirements include consistent coating distribution, strong adherence, and a good surface appearance.

Environmental and Safety Factors

The primary environmental concern for the coating plant is actually the residual material on the anode structures being returned for recoating. Therefore the anode user must enact effective cleaning procedures prior to shipment. Overall, the DSA (Electrode Corp.) has made chlorine manufacture cleaner, more consistent, simpler, and therefore safer.

Economic Aspects

For the most part, coated metal anodes are made available by long-term lease rather than through outright sale. There has been a single-source supply protected by patents in most geographic areas. As the cost-saving and environmental value of this technology continues to increase, upward expectation in respect to prices is created. However, the appearance of new anode producers, together with expiration of some early patents, has created a competitive situation in some applications. This competition can be expected in turn to create a counterbalancing pressure to lower prices.

THOMAS A. LIEDERBACH
Electrode Corporation

U.S. Pat. 3,632,498 (Jan. 4, 1972), H. B. Beer (to Chemnor Aktiengesellschaft).

U.S. Pat. 3,711,385 (Jan. 16, 1973), H. B. Beer (to Chemnor Corp.).

K. J. O'Leary and T. J. Navin, "Morphology of Dimensionally Stable Anodes," paper presented at the *Chlorine Bicentennial Symposium,* San Francisco, Calif., May 1974.

K. L. Hardee, L. K. Mitchell, and R. C. Carlson, in *Proceedings of the AESF Sixth Continuous Steel Strip Coating Symposium,* May 1990.

METAL FIBERS. See COMPOSITE MATERIALS; FIBERS, ACRYLIC

METALLIC COATINGS

SURVEY

Metallic coatings provide an inexpensive way to modify or control the properties of a base material. Coatings (qv) may be functional or decorative, permanent or temporary, sacrificial, or noble. The metallic coating may be continuous, or it may be patterned into discontinuous functional or decorative areas. The base material may be metallic, nonmetallic inorganic, or organic such as plastic, paper, or fiber. The common criterion for the purposes herein is that the metallic coating is functionally bonded to the base material. Galvanized zinc items, gold clad jewelry, electroplated materials, and semiconductor chips are among the most common types of materials having metallic coatings (see ELECTROPLATING; ELECTRONICS, COATINGS; INTEGRATED CIRCUITS; SEMICONDUCTORS).

Metallic coatings are most often selected for protective function. Decorative ability is a common secondary function. Most coatings applied by hot dipping, such as galvanizing or aluminizing, use a film of a more chemically reactive metal over a less reactive material such as iron alloy. These are sacrificial coatings because the zinc or aluminum slowly dissolves instead of the underlying steel.

Decorative coatings must be sufficiently corrosion resistant to maintain an attractive appearance during the anticipated life of the composite product. Composites of multiple metal layers or mixtures of metallic and nonmetallic coatings are becoming increasingly common as material properties are selectively tailored. Another important function of metallic coatings is to provide wear resistance. The most technically demanding types of coatings are those grown in precisely defined multiple layers.

Plating and galvanizing are the most common methods of applying metallic coatings, but many other processes, such as hot dipping, cementation, thermal spraying, sputtering, and chemical vapor disposition, have been developed (see also COATING PROCESSES). All metallizing techniques are potential sources of pollution, a subject of increasing ecological, economic, legislative, and public health interest.

There are a plethora of commercially useful methods for applying a metallic coating. Many more techniques have been demonstrated in the laboratory. Each method has different critical parameters, ie, maximum and minimum coating thickness producible, substrate temperature, bulk or imagewise deposition, coating adhesion value, type and cost of coating equipment, labor requirement, scrap rate, rework capability, and safety and waste disposal aspects. For convenience, metallic coating methods may be divided into classes defined by the general way in which the metal is applied, eg, liquid-phase, gas-phase, and vacuum-phase metallizing or by direct physical or thermal bonding.

The two largest-volume processes, in terms of amount of metal used and amount of surface area coated, are the liquid-phase metallizing processes hot dip coating and electroplating (qv).

Many of the newer coating methods give metallized coatings that are quite distinct from simple single or multiple coatings of metals or alloys. Many of these newer coatings, which cannot be made except by one or a small family of processes, are defined as much by process method as physical and chemical structure. Examples of these coatings include ion-plated surfaces, laser heat-treated surfaces, modifications of existing surfaces to predetermined depths by selective additions of atoms or heat; and Teflon–electroless nickel composite coatings.

Liquid-Phase Metallizing Techniques

Hot Dip Galvanizing. The largest single type of metallic coating process in terms of amount of metal used is hot dip galvanizing. Half of all galvanized zinc is used on steel coils, one-third is employed for coating of parts after fabrication, and the remainder for wires and tubes. The other uses, in approximate order, are electroplating, zinc-filled paints, zinc spray processes, zinc foils having conductive adhesive, and mechanical zinc plating.

In hot galvanizing, zinc is applied to iron (qv) and steel parts by immersing the parts into a bath of molten zinc. Whereas in principle almost any metal could be coated with molten zinc, this coating serves no worthwhile purpose on most metals. The combination of zinc and ferrous materials are almost uniquely suited to each other. Aluminum and cadmium are the only other similar combinations. Zinc provides iron parts with better corrosion protection by developing a coating of zinc and zinc compounds on the base metal surface.

Hot Dipping Using Other Metals. Processes include coating with aluminum, coating with 55% aluminum–zinc alloy, hot dip tin coating of steel and cast iron, hot dip terne coating, and solder coatings.

Electroplating. Many developments in electroplating (qv) have been the direct result of increased coating functionality and economy, and others of the increasing need for environmental and legislative compliance. The field of electroplating has expanded rapidly.

The most common electroplated metal is zinc, followed by nickel, copper, and chromium. Many other metals and metaloids can be electroplated, including manganese, iron, cobalt, gallium, germanium, arsenic, selenium, ruthenium, rhodium, palladium, silver, cadmium, indium, tin, lead, bismuth, mercury, antimony, gold, iridium, and platinum. Types of electroplating include fused salt plating, plating from nonaqueous solvents, and brush plating.

Chromium metal seems to be biologically harmless owing to inertness. Trivalent chromium Cr(III), one of the essential mineral nutrients (qv), has not been shown to be harmful. Hexavalent chromium, Cr(VI), found in chromic acid and used in electroplating baths, is very toxic as well as a suspected carcinogen. Two basic types of chromium are plated: hard chromium up to 100 μm or more, and decorative chromium at 0.25–1.0 μm. No replacement coating has been found for thick hard chromium deposits for wear resistance and parts salvage, although electroless nickel can partially substitute for chromium plating baths. Many decorative applications are being converted to Cr(III). Products from the newest Cr(III) baths are essentially equivalent to those from the older decorative Cr(VI) baths, although some cosmetic color differences exist which have prevented complete conversion to the Cr(III) baths.

Cyanide solutions formerly were ubiquitous in electroplating shops. The best cleaners contained sodium cyanide. The removal of cyanides (qv) from all plating baths has been a general goal since the latter part of the 1980s. As of this writing (ca 1995), cyanide-containing cleaners are rare and are only used for special purposes. Several noncyanide copper strike baths have been introduced.

The greatest tonnage decrease in cyanide plating has occurred for zinc plating. As of this writing, less than one-third of all zinc is plated from cyanide baths. The remainder is plated from alkaline noncyanide, zinc potassium chloride, and zinc ammonium chloride baths.

Chlorinated and fluorinated cleaners have been widely used as degreasers during surface preparation prior to plating. Recognition of the role of many of these solvents as either global warming gases or ozone depletion agents has led to prohibitions against their continued use (see AIR POLLUTION; ATMOSPHERIC MODELING). Many new and improved cleaning systems are being developed, such as alcohol-based cleaners, emulsion cleaners, and cleaners based on natural bioproducts such as limonene. During this process a massive reformulation of many aqueous cleaners has also occurred.

Cadmium usage, illegal in most of Europe, is being discouraged elsewhere. The U.S. military has cadmium specifications for electronic, fastener, and marine equipment, which requires only cadmium. Tin is being substituted for tin–lead as a metallic etch resist during printed circuit board production.

Electroless Plating. The metallizing process known as electroless plating (qv) is mainly used for deposition of copper on plastics and for nickel–phosphorus alloy on plastics and metals. Formaldehyde (qv) is the most common copper reducing agent, giving pure coatings. Sodium hypophosphite is the agent mainly used for electroless nickel. An unusual nickel–phosphorus alloy or solid solution is formed. Unlike electroplating, electroless plating can be used on almost any substrate, metallic or nonmetallic. Often electroless plating is used as the first coating to make glass (qv), ceramic, or plastic conductive, followed by conventional electrolytic plating.

Immersion Plating. A simplified aqueous metal deposition process which does not use electric current, immersion plating works only when a metal of higher electromotive force, such as copper, is deposited on a metal of lower electromotive force, such as iron or aluminum. The coatings are typically very thin and porous. One important application is immersion plating of copper sulfate bath onto steel wire, to use as a drawing agent. Very thin immersion gold deposits are used for inexpensive jewelry.

Miscellaneous Techniques. Lasers (qv) have been used for both electroless and electrolytic plating; selective dissolution has been used from ancient times to give the appearance of a thin plated coating of precious metal; and mercury layers plated onto the surface of analytical electrodes serve as liquid metal coatings.

Gas-Phase Metallizing Techniques

Metal or Thermal Spray Coatings. These specific processes include plasma arc spray, flame spray, laser spray, and electric arc spray, depending on the energy input source. Rods, wires, and powders are used as coating material sources. Thermal spray metal and ceramic coatings have diverse properties suitable for numerous applications, including corrosion resistance, eg, zinc and aluminum, especially against oxidation or salt water corrosion; high temperature oxidation, eg, nickel, cobalt, and chromium alloys; electrical conductivity, eg, radio frequency interface (RFI) shielding by zinc or tin on nonconductors; electrical resistance, eg, insulating layers in induction heating coils and high temperature strain gauges; wear resistance, eg, chromium–nickel–boron alloys and carbide-containing coatings; catalytic surfaces; nuclear moderators; and dimensional buildup for salvaging worn metal parts.

A great advantage of zinc arc spray is that it can be applied to almost any plastic. Zinc arc spray, also suitable for prototypes and small lots of materials, is less suited for very small parts and parts having blind holes or complex interior surfaces, or where warpage is a problem.

Zinc arc spraying or flame spray equipment is hardly more hazardous than a welding torch, and only safety goggles and gloves are required. Safety aspects emphasize reduction of noise and vapor inhalation.

Thermal spray processes can be used to give coatings of chromium carbide or nickel chromium for erosion resistance, copper nickel indium for fretting resistance, tungsten carbide cobalt for wear and abrasion resistance, and even aluminum silicon polyester mixtures for abradability.

Carburizing and Nitriding. Several commonly used metallurgical surface treatments are applied by gas-phase reactions in a reducing atmosphere for carburizing, and in a nitrogen atmosphere for nitriding. These treatments are used to increase the surface hardness of ferrous alloys by diffusion of carbon and nitrogen at high temperature.

Pack Diffusion. Pack diffusion or cementation processes are similar to pack carburizing, and are used to coat iron, nickel, cobalt, and copper with chromium, boron, zinc (Sheradizing), aluminum, silicon, titanium, molybdenum, and other metals.

Chromizing and Related Diffusion Processes. Chromizing is similar to aluminizing. A thin corrosion and wear resistant coating is applied to low cost steels such as mild steel, or to a nickel-based alloy. In the related boronizing process, a thin boron alloy is produced for extreme hardness, wear, and corrosion resistance. Siliconizing is yet another process used especially for coating of the refractory metals Ti, Nb, Ta, Cr, Mo, and W (see REFRACTORIES).

Metals. Aircraft and space vehicles, turbine generators, and other such applications require high strength at high temperatures along with excellent oxidation resistance. Superalloys, ie, complex nickel and cobalt-based alloys, and refractory metals, eg, niobium, tungsten, molybdenum, tantalum, and their alloys, are used for applications at temperatures above 1000°C. In many case the coatings must be resistant both to oxidation and to hot corrosion by sulfidation from sulfur-bearing gases.

Two types of coatings have been used for superalloys: diffusion coatings, in which a layer of nickel, cobalt, platinum, or palladium aluminide, ie, NiAl, CoAl, PtAl, or PdAl, is formed on the surface by diffusion; and overlay coatings, in which a complex coating material such as nickel–cobalt–chromium–aluminum–yttrium, NiCoCrAlY, is applied to the surface. Pack cementation is the most widely used process for applying diffusion coatings to superalloys.

Vacuum-Phase Metallizing Techniques

Vacuum-phase metallizing techniques all depend on the use of a vacuum as part of the metallizing process (see VACUUM TECHNOLOGY). The tonnage of metal deposited by these techniques is insignificant compared to hot dip galvanizing or plating, and the total surface area metallized is much less than that done by plating. However, the economic added value of vacuum metallizing probably exceeds that of either plating or galvanizing because of the extremely rapid growth and deep market penetration of semiconductor devices. Projections call for at least a 10% compounded annual growth rate into the twenty-first century. The state of the art in vacuum metallizing is in semiconductors (qv) processing.

Thermal Evaporation. Thermal evaporation is done in a high vacuum to minimize chemical side reactions of the evaporated active metal. Thermal evaporation is inexpensive and efficient. It is used for low cost items such as aluminized plastic sheet and decorative Christmas tinsel, second surface coating of transparent plastic auto parts, and metallization of glass and ceramics (qv). Aluminum is the predominant metal used. Color effects such as gold or brass are achieved by applying a dyed translucent organic protective over the aluminum.

Sputtering or Glow Discharge. Sputtering can be done using both conductive and nonconductive items. A low pressure atmosphere of argon is used in the deposition vessel. This method gives excellent coating adhesion and more consistent coatings than simple vacuum deposition. It can also be applied to a wider range of materials, including complex metal alloys and oxides. The equipment and operating costs are higher than vacuum deposition, and deposition rates are lower. Reactive sputtering is a variation in which two or more deposition sources are used. This process can be used to give compounds such as silicon carbide on the surface.

Chemical Vapor Deposition. This process is distinct from simple thermal evaporation because chemical vapor deposition (CVD) depends on a chemical reaction at the surface of the part and in that way is analogous to electroless plating. This process uses a gas of one or more chemical species, which react at a heated substrate to form an appropriate film. The film can be metallic, nonmetallic, single element, or compound. Among the coatings available are copper and aluminum conductors, diamond and tantalum oxide dielectrics, lead zirconium titanate piezoelectrics, and bismuth strontium calcium cuprate superconductors (see also ELECTRONICS, COATINGS; THIN FILMS–FILM FORMATION TECHNIQUES). Some of the largest applications are for coating TiC and TiN on cutting tools.

Ion Implantation. Also known as ion plating, ion implantation (qv) is a high vacuum process for modifying the surface properties of any material. This is a surface modification technique rather than a surface coating technique. Semiconductors are commonly modified using this technique, because ion implantation can be done using total or imagewise scanning of the beam over the surface. Nitrogen, chromium, and phosphorus can be used to harden metals and increase the corrosion resistance. Metastable alloys can be produced that are difficult to make by other processes.

Laser Hardening and Modification. Lasers are used to surface harden ductile steels and improve the toughness to a depth of 0.35 mm or more. Lasers can also be used to bond solid or powder coatings to a surface. Typical coatings are nickel or titanium carbide on iron, and nickel, cobalt, manganese, and titanium carbide, TiC, on aluminum. Use of lasers with other specialized coating methods is common.

Metallizing by Direct Physical or Thermal Bonding

Direct bonding techniques are among the oldest types of metallizing, and the most versatile. Many methods depend on heat or pressure and an adhesive layer to glue the coating to the substrate. Methods for metallizing on a metallic surface often depend on removal or displacement of a preexisting surface oxide layer. Many metals form intermetallic alloys or self-diffuse into one another even at room temperature, but the surfaces in contact must be clean and oxide-free. Processes include lamination, mechanical plating, slurry coatings, and roll bonding or strip roll welding.

Environmental Concerns

Each type of metallic coating process has some sort of hazard, whether it is thermal energy, the reactivity of molten salt or metal baths, particulates in the air from spray processes, poisonous gases from pack cementation and diffusion, or electrical hazards associated with arc spray or ion implantation. Vacuum or inert gas operations can produce flammable dusts or powders when opened for cleaning. Most of the hazards are confined to the operator and immediate environs in the operating plant. OSHA is the primary regulator of these hazards in the United States, although many local and state agencies, especially fire departments, also regulate coatings plants. Adequate training, documentation, and protective equipment are the minimum requirements. The principal regulatory burden falls on wastes and discharges which leave the plant.

Table 1. EPA Pretreatment Standards for Aqueous Discharge[a,b]

	Existing sources, PSES[c]		New source, PSNS[c]	
Material, mg/L	1 Day	30 Days	1 Day	30 Days
cadmium	0.69	0.26	0.11	0.07
chromium, total	2.77	1.71	2.77	1.71
copper	3.38	2.07	3.38	2.07
lead	0.69	0.43	0.69	0.43
nickel	3.98	2.38	3.98	2.38
silver	0.43	0.24	0.43	0.24
zinc cyanide	2.61	1.48	2.61	1.48
Total	*1.2*	*0.65*	*1.2*	*0.65*
Treatable	*0.86*	*0.32*	*0.86*	*0.32*
Total toxic organics	*2.13*		*2.13*	

[a] Captive manufacturers performing metal finishing, including electroplating, discharging to POTWs. [b] pH equals 6–10. [c] Maximum value. PSES = pretreatment standards for existing sources; PSNS = pretreatment standards for new sources.

The U.S. EPA regards metallizers such as platers, surface finishers, and printed circuit board producers as among the most important source polluters for metals. Much production of electroplated items has, however, shifted from the United States to less environmentally stringent countries.

The surviving U.S. plants have embraced all types of waste treatment processes (see WASTE TREATMENT, HAZARDOUS WASTE; WASTES, INDUSTRIAL). The most desired pollution prevention processes are those which reduce the total amount of waste discharged. Treatment and disposal are less strongly emphasized options. Zero wastewater discharge facilities and water recycling processes are becoming more common.

Discharge limits vary between localities and among plants. Table 1 shows federal EPA maximum discharge limits for a number of metals for a new metal-finishing installation.

Newer federal limits on metals contents of sewage sludge combined with laws on fuller treatment of the sewage sludge and its allowed disposal methods have affected limits. The metallic content of sludges from municipal waste treatment facilities is becoming of great concern.

Air pollution (qv) is recognized as a significant problem for coating facilities.

Many types of waste treatment and waste minimization processes are in common use in the metallization industry. Air scrubbers are commonly required, even for general acid fume removal from plating shops. Additionally, many newer technologies have been adapted for use in metallizing operations. These include air stripping, antimisting agents, biological destruction, carbon absorption, countercurrent rinsing, crystallization (qv), distillation (qv), Donnan dialysis, electrodialysis, electrowinning, evaporation (qv), filtration (qv), fluocculation, hydrolysis, incineration (see INCINERATORS), ion exchange (qv), metallic replacement, neutralization, oxidation, pH adjustment (see HYDROGEN-ION ACTIVITY), photolysis, precipitation, process modification, reduction, reverse osmosis (qv), salt splitting, sedimentation (qv), solidification, and spray rinsing.

GERALD A. KRULIK
Applied Electroless Concepts, Inc.
NENAD V. MANDICH
HBM Engineering Company

Metal Finishing Guidebook and Directory, Vol. 92, No. 1A, Elsevier Publishing, New York, 1944; collected vols., Publications, Inc., Hackensack, N.J., updated yearly.

L. J. Durney, ed., *Electroplating Engineering Handbook,* 4th ed., Van Nostrand Reinhold Co., Inc., New York, 1984.

W. H. Safranek, *The Properties of Electrodeposited Metals and Alloys,* 2nd ed., American Electroplaters and Surface Finishers Society, Orlando, Fla., 1986.

Plating and Surface Finishing, Semiconductor International, and *Metal Finishing* provide some of the best information on innovative metallizing methods.

EXPLOSIVELY CLAD METALS

Explosive cladding, or explosion bonding and explosion welding, is a method wherein the controlled energy of a detonating explosive is used to create a metallurgical bond between two or more similar or dissimilar metals. No intermediate filler metal, eg, a brazing compound or soldering alloy, is needed to promote bonding and no external heat is applied. Diffusion does not occur during bonding.

The explosive cladding process provides several advantages over other metal-bonding processes:

(1) A metallurgical, high quality bond can be formed between similar metals and between dissimilar metals that are incompatible for fusion or diffusion joining. Brittle, intermetallic compounds, which form in an undesirable continuous layer at the interface during bonding by conventional methods, are minimized, isolated, and surrounded by ductile metal in explosion cladding. Examples of these systems are titanium–steel, tantalum–steel, aluminum–steel, titanium–aluminum, and copper–aluminum.

(2) Explosive cladding can be achieved over areas that are limited only by the size of the available cladding plate and by the magnitude of the explosion that can be tolerated.

(3) Metals having tenacious surface films that make roll bonding difficult, eg, stainless steel/Cr–Mo steels, can be explosion clad.

(4) Metals having widely differing melting points, eg, aluminum (660°C) and tantalum (2996°C), can be clad.

(5) Metals having widely different properties, eg, copper or maraging steel, can be bonded readily.

(6) Large clad-to-backer ratios can be achieved by explosion cladding.

(7) The thickness of the stationary or backing plate in explosion cladding is essentially unlimited. Backers >0.5 − m thick and weighing 50 t have been clad commercially.

(8) High quality, wrought metals are clad without altering chemical composition.

(9) Different types of backers can be clad; clads can be bonded to forged members, as well as to rolled plate.

(10) Clads can be bonded to rolled plate that is strand-cast, annealed, normalized, or quench-tempered.

(11) Multilayered composite sheets and plates can be bonded in a single explosion, and cladding of both sides of a backing metal can be achieved simultaneously.

(12) Nonplanar metal objects can be clad.

(13) The majority of explosion-clad metals are less expensive than the solid metals that could be used instead of the clad systems.

Limitations of the explosive bonding process are as follows:

(1) There are both inherent hazards in storing and handling explosives and undesirable noise and blast effects from the explosion.

(2) Obtaining explosives with the proper energy and detonation velocity is difficult.

(3) Metals to be explosively bonded must be somewhat ductile and resistant to impact.

(4) For metal systems in which one or more of the metals to be explosively clad has a high initial yield strength or a high strain-hardening rate, a high quality bonded interface may be difficult to achieve. This problem increases when there is a large density difference between the metals.

(5) Geometries suited to explosive bonding require straight-line egression of the high velocity jet emanating from between the metals during bonding.

(6) Thin backers must be supported, thus adding to manufacturing cost.

(7) The preparation and assembly of clads is not amenable to automated production techniques and each assembly requires considerable labor.

Theory and Principles

To obtain a metallurgical bond between the two metals, the atoms of each metal must be brought sufficiently close so that their normal forces of interatomic attraction produce a bond. The surfaces of metals and alloys must not be covered with films of oxides, nitrides, or adsorbed gases; when such films are present, metal surfaces do not bond satisfactorily (see METAL SURFACE TREATMENTS).

Explosive bonding is a cold pressure-welding process in which the contaminant surface films are plastically jetted from the parent metals as a result of the high pressure collision of the two metals. The metal plates, which are cleaned of any surface films by the jet action, are joined at an internal point by the high pressure that is obtained near the collision point.

Parallel and Angle Cladding. The arrangements shown in Figures 1 and 2 illustrate the operating principles of explosion cladding. Angle cladding (Fig. 1) is limited to cladding for relatively small pieces. The arrangement shown in Figure 2 is by far the simplest and most widely used.

Jetting. A layer of explosive is placed in contact with one surface of the prime metal plate which is maintained at a constant distance from and parallel to the backer plate, as shown in Figure 2**a.** The jet, which moves in the direction of detonation, is observed between the deflected prime metal and the backer metal.

Typically, jet formation is a function of plate collision angle, collision-point velocity, cladding-plate velocity, pressure at the colli-

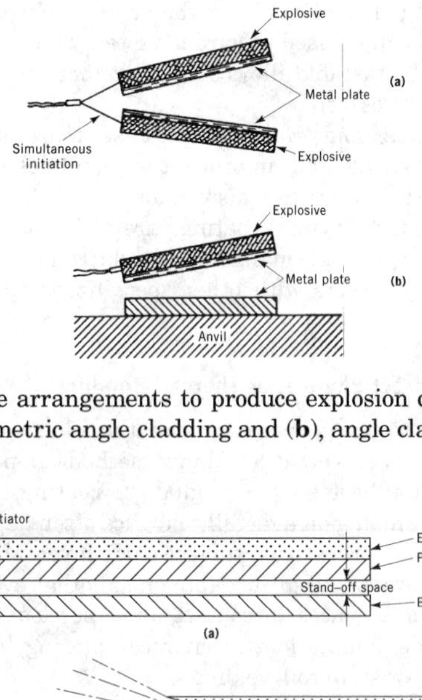

Figure 1. Angle arrangements to produce explosion clads, where (**a**) represents symmetric angle cladding and (**b**), angle cladding.

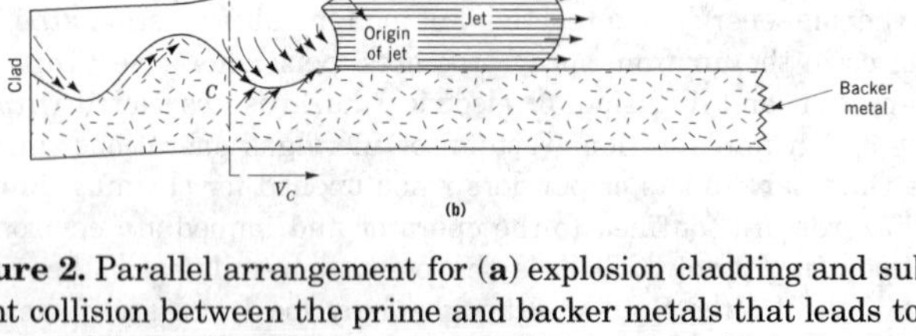

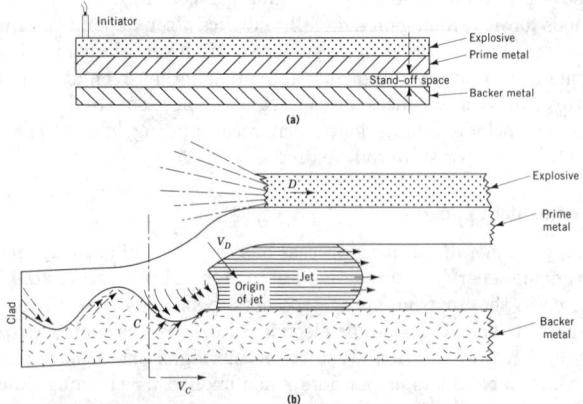

Figure 2. Parallel arrangement for (**a**) explosion cladding and subsequent collision between the prime and backer metals that leads to (**b**) jetting and formation of the wavy bond zone, where V_D is the detonation velocity and V_C, the collosion velocity.

	Zirconium	Magnesium	Stellite	Platinum	Gold	Silver	Niobium	Tantalum	Hastelloy	Titanium	Nickel alloys	Copper alloys	Aluminum	Stainless steels	Alloy steels	Carbon steels
Carbon steels	●	●			●	●	●	●	●	●	●	●	●	●	●	●
Alloy steels	●	●	●					●	●	●	●	●	●	●	●	
Stainless steels			●		●	●	●	●		●	●	●	●	●		
Aluminum		●				●	●	●		●	●	●	●			
Copper alloys						●	●	●		●	●	●				
Nickel alloys		●		●	●			●		●	●					
Titanium	●	●				●	●	●		●						
Hastelloy									●							
Tantalum					●		●	●								
Niobium				●			●									
Silver						●										
Gold																
Platinum				●												
Stellite 6B																
Magnesium		●														
Zirconium	●															

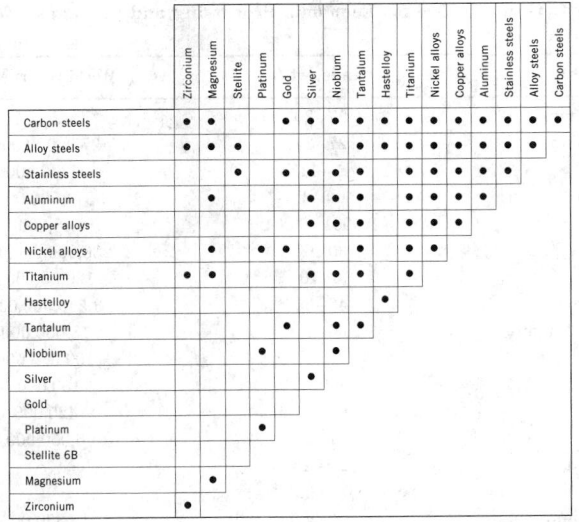

Figure 3. Commercially available explosion-clad metal combinations.

sion point, and the physical and mechanical properties of the plates being bonded.

Nature of the Bond. Extensive metallurgical testing has determined that the best clad properties are obtained when the bond zone is wavy. It is therefore preferable that commercial, explosively bonded metals exhibit a wavy bond–zone interface.

The industrially useful combinations of explosively clad metals that are available in commercial sizes are listed in Figure 3.

Processing

Explosives. The pressure, P, generated by the detonating explosive that propels the prime plate is directly proportional to its density, ρ, and the square of the detonation velocity, V^2_D:

$$P = \frac{1}{4}\rho V_D^2$$

The detonation velocity is controlled by adjusting the packing density or the amount of added inert material.

The types of explosives that have been used include both high (4500–7600 m/s) and low to medium (1500–4500 m/s) velocity materials.

High velocity	Low–medium velocity
trinitrotoluene (TNT)	ammonium nitrate
cyclotrimethylenetrinitramine (RDX)	ammonium nitrate prills sensitized with fuel oil
pentaerythritol tetranitrate (PETN)	ammonium perchlorate
composition B	amatol
composition C_4	amatol and sodatol diluted with rock salt to 30–35%
plasticized PETN-based rolled sheet and extruded cord	dynamites
primacord	nitroguanidine
	diluted PETN

In commercial practice, powdered explosives on an ammonium nitrate basis are used in most cases.

Metal Preparation. The cladding faces usually are surface ground, using an abrasive machine, and then are degreased with a solvent to ensure consistent bond strength.

Assembly, Stand-Off. The air gap present in parallel explosion cladding can be maintained by metallic supports that are tack-welded to the prime and backer plates or by small and light inserts that are placed between the prime and backer. A moderating layer of buffer, eg, polyethylene sheet, water, rubber, paints, and pressure-sensitive tapes, may be placed between the explosive and prime metal surface to attentuate the explosive pressure or to protect the metal surface from explosion effects.

Facilities. The problems of noise, air blast, and air pollution (qv) are inherent in explosion cladding, and clad-composite size is restricted by these problems. Thus the cladding facilities should be in areas that are remote from population centers.

Safety Aspects

All explosive material should be handled and used following approved safety procedures in compliance with applicable federal, state, and local laws, regulations, and ordinances. The Bureau of Alcohol, Tobacco, and Firearms (BATF), the Hazardous Materials Regulation Board (HMRB) of the Department of Transportation (DOT), the Occupational Safety and Health Agency (OSHA), and the Environmental Protection Agency (EPA) in Washington, D.C., have federal jurisdiction on the sale, transport, storage, and use of explosives. Many states and local counties have special explosive requirements.

Uses

Applications such as chemical-process vessels and transition joints represent approximately 90% of the industrial use of explosion cladding. Other applications include conversion-rolling billets, nonplanar specialty products, tube welding and plugging, and refractory metals and alloys.

Production and Markets

Explosion-bonded metals are produced by several manufacturers in the United States, Europe, and Japan. Total world markets for explosion-clad metals are estimated to fluctuate between 30×10^6 to 60×10^6 annually.

OSWALD R. BERGMANN
E. I. du Pont de Nemours & Co., Inc.

R. S. Rinehart and J. Pearson, *Explosive Working of Metals,* MacMillan, New York, 1963.

G. R. Cowan and A. H. Holtzman, *J. Appl. Phys.* **34**(Pt. 1), 928 (1962).

U.S. Pat. 3,397,444 (Aug. 20, 1968), O. R. Bergmann, G. R. Cowan, and A. H. Holtzman (to E. I. du Pont de Nemours & Co., Inc.).

O. R. Bergmann, G. R. Cowan, and A. H. Holtzman, *Trans. Met. Soc. AIME* **236,** 646 (1966).

METALLURGY

SURVEY

Going into the twenty-first century, numerous advanced materials utilized in contemporary society depend on specialty metals often requiring novel processing techniques as well as a detailed knowledge of chemistry and atomic structure (see ABLATIVE MATERIALS; GLASSY METALS; HIGH TEMPERATURE ALLOYS; SHAPE-MEMORY ALLOYS).

Early sources define metallurgy as the process of extracting metal from ores. For many metals, the primary source materials as of the 1990s are still crude metalliferous ores. For some metals however, recycled materials contribute significantly to total metal production. For example, in the United States the recycling (qv) rate of all-aluminum used beverage cans is over 50%. For an energy-intensive metal such as aluminum, this represents a substantial energy saving. Recycled aluminum requires only 5% of the energy needed to make aluminum from bauxite ore (see ALUMINUM AND ALUMINUM ALLOYS; RECYCLING, METALS–NONFERROUS METALS).

Metallurgy includes not only the treatment of crude ore and scrap, but also the processing of intermediates, ie, concentrates, and wastes, such as slags, tailings, etc, for contained metal values. The various areas and subdisciplines comprising metallurgy may be summarized as follows:

Extractive metallurgy			
Mineral processing	Chemical metallurgy	Process metallurgy	Physical metallurgy
comminution	hydrometallurgy	alloying	structure–property relationships
classification	pyrometallurgy	casting	failure analysis
flotation	electrometallurgy	deformation processes	corrosion
	corrosion	heat treatment	
		powder metallurgy	
		nuclear metallurgy	

Definitions

The field of metallurgy has a unique and frequently very specialized vocabulary. Understanding this language helps to clarify certain concepts and processing steps. Definitions of key terms follow.

To *concentrate* is to take an action to intensify in strength or purity by the removal of valueless or unneeded constituents.

Electrometallurgy covers the various electrical processes for the working of metals.

Flotation is the method of mineral separation in which a froth created in water by a variety of reagents floats some finely crushed minerals.

Gangue consists of the undesired minerals associated with ore, mostly nonmetallic.

Hydrometallurgy refers to the treatment of ores, concentrates, and other metal-bearing materials by wet processes.

Leaching is the extracting of a soluble metallic compound from an ore by selectively dissolving it in a suitable solvent.

A *mineral* is an inorganic substance occurring in nature and having a definite chemical composition or a characteristic range of chemical composition, and distinctive physical properties or molecular structure.

Mineral dressing is the physical or chemical concentration of raw ore into a product from which a metal can be recovered at a profit.

Ore is a mineral or aggregate of minerals from which a valuable constituent, especially a metal, can be profitably extracted.

Pyrometallurgy is metallurgy involved in winning and refining metals where heat is used, as in roasting and smelting.

Roasting is the heating of solids, frequently to promote a reaction with a gaseous constituent in the furnace atmosphere.

Smelting is any metallurgical operation in which metal is separated by fusion from those impurities with which it may be chemically combined or physically mixed.

Ore

The metal content of an ore is typically called the ore grade and is usually expressed as weight percent for most metals. For precious metals, however, grade is usually expressed in g/t (oz/short ton). Because the definition of ore is established by economic considerations, there is no upper limit to grade, ie, the richer, the better. There is frequently a lower limit or cutoff grade, however, based on process efficiency and economics. Table 1 shows the average grade of various metalliferous ores that can be processed economically. Also shown is an estimate of the world total reserve base for each metal. For many metals, ore grade depletion has been a serious problem.

Table 1. Grade of Ore for Economic Processing and Estimated World Reserves

Ore	Grade, wt %	World reserves, t
aluminum	27–29	
chromium	27–34	6,778,000,000[a]
cobalt	1–11	8,340,000
copper	0.5–2	574,000,000
gold	0.0001–0.001	48,600
iron	30–60	230,000,000,000
lead	5–10	120,000,000
manganese	45–55	3,540,000,000
molybdenum	0.6–1.8	13,000,000
nickel	1.5–3	109,000,000
platinum	0.001	100,000[b]
silver	0.04–0.08	420,000
tin	1–5	6,050,000
titanium	2.5–25	
uranium	0.1–0.9	
vanadium	1.6–4.5	16,330,000
zinc	10–30	295,000,000

[a] As chromite. [b] Reserves of the platinum-group metals (qv), ie Pt, Pd, Rh, Ru, Ir, and Os, together equal 1×10^8 t.

Modern heap leaching practice has also made possible the treatment of extremely lean (ca 1 g/t) gold ores, which had been considered uneconomic as late as the early 1980s. In addition, changing technology and process innovations have contributed to the extraction of nickel from low grade lateritic ores and to the potential recovery of aluminum from nonbauxitic aluminum resources. Some metals are produced primarily as a by-product from the mining and refining of other metals.

Product Specifications

Impurities in crude metal can occur as other metals or nonmetals, either dissolved or in some occluded form. Normally, impurities are detrimental, making the metal less useful and less valuable. On the other hand, impurities may have commercial value. For example, gold, silver, platinum, and palladium, associated with copper, each has value.

Refined metals, as traded on the open market, vary considerably in composition. However, there are strict specifications for certain impurity elements for a number of metals.

Economic Aspects

Metal price has a very significant influence on production patterns. Price drives metal exploration and development of new resources, and is the main factor in determining materials substitution. High metal prices stimulate exploration activities and new mine development, which in turn increase supply. High metal prices also encourage the search for substitutes, decreasing demand.

Metals present at relatively high concentrations, in the earth's crust, such as iron and aluminum, are the least expensive; rare metals such as gold and platinum are the most valuable. Metal demand also has an important influence on price.

Approximately three-quarters of the elements in the Periodic Table are metals. The winning, refining, and fabrication of these metals for commercial use together represent the complex and diverse field of metallurgy.

BRENT HISKEY
University of Arizona

G. Agricola, *De Re Metallica* (Basel, 1556), H. C. Hoover and L. L. Hoover, trans., Dover Publications, Inc., New York, 1950.

P. W. Thrush, *A Dictionary of Mining, Mineral, and Related Terms,* U.S. Bureau of Mines, Washington, D.C., 1968.

ASTM Standards, Vol 2.01 and Vol 2.04, American Society for Testing and Materials, Philadelphia, Pa., 1993.

Metal Prices in the United States through 1991, U.S. Bureau of Mines, Washington, D.C., 1991.

EXTRACTIVE METALLURGY

Extractive metallurgy deals with the extraction of metals from naturally occurring compounds and the subsequent refinement to a purity suitable for commercial use. These operations, known as winning and refining of metals, follow mining and beneficiation of the ore. They precede the fabricating processes. Selection and design of the extractive processes depend on the raw materials available, and conditions for the refining steps are related to the ultimate use of the metal. Thus mining, extraction, and fabrication are closely interrelated (see also MINERALS RECOVERY AND PROCESSING).

The scientific basis of extractive metallurgy is inorganic physical chemistry, mainly chemical thermodynamics and kinetics (see THERMODYNAMICS). Extractive metallurgy is usually divided into three principal areas: (*1*) pyrometallurgy, which consists of high temperature processes to carry out smelting and refining reactions; (*2*) hydrometallurgy, which is characterized by the use of aqueous solutions and inorganic solvents to achieve the desired reactions; and (*3*) electrometallurgy, in which electrical energy is used to extract and refine metals by electrolytic processes. Electrometallurgy may be carried out either at high temperature or in aqueous solution. An integrated metallurgical flow sheet may include pyrometallurgical, hydrometallurgical, and/or electrometallurgical steps.

Pyrometallurgy

The essential operations of an extractive metallurgy flow sheet are the decomposition of a metallic compound to yield the metal, followed by the physical separation of the reduced metal from the residue. This is usually achieved by a simple reduction, or by controlled oxidation of the nonmetal and simultaneous reduction, or by controlled oxidation of the nonmetal and simultaneous reduction of the metal. This may be accomplished by the matte smelting and converting processes.

In a simple pyrometallurgical reduction, the reducing agent, R, combines with the nonmetal, X, in the metallic compound, MX, according to a substitution reaction of the following type:

$$MX + R \rightarrow M^0 + RX$$

The selection of a particular type of reduction depends on technical feasibility and the economics of the process as well as on physicochemical considerations. In particular, the reducing agent should be inexpensive relative to the value of the metal to be reduced. The product of the reaction, RX, should be easily separated from the metal, easily contained, and safely recycled or disposed.

The preparation, reduction, and refining operations are very much interdependent, and for a given metal must be considered as parts of a single flow sheet. To illustrate the principles of extractive metallurgy, however, it is convenient to discuss the various operations separately.

Preparatory Processes. Processes include drying and calcination, roasting of sulfides, chlorination, and sintering and pelletizing.

Drying and calcination are usually carried out in various types of kilns such as rotary kilns, shaft furnaces, and rotary hearths. The induration step in a pelletizing plant requires similar equipment. The Dwight-Loyd continuous sintering machine is commonly used. It consists of a series of grates mounted as an endless traveling belt. The older roasting furnace is the multiple hearth roaster consisting of 8–12 hearths enclosed in a cylindrical shell. Fluid-bed roasters are well suited for gas–solid reactions. Usage in metallurgical operations started in 1950 and developed rapidly thereafter.

Reduction to Liquid Metal. Reduction to liquid metal is the most common metal reduction process. It is preferred for metal of moderate melting point and low vapor pressure. The common furnaces for production of liquid metal are the bast furnace, the reverberatory furnace, the converter, the flash smelting furnace, and the electric arc furnace (see FURNACES, ELECTRIC).

The reduction of iron oxides by carbon in the iron (qv) blast furnace is the most important of all extractive processes, and the cornerstone of all industrial economies.

The traditional method of lead (qv) smelting is reduction of an oxide in a blast furnace similar in principle to that used for iron, but very different in design and operation.

Many nonferrous metals can be extracted by reduction smelting, eg, copper, tin, nickel, cobalt, silver, antimony, and bismuth. Blast furnaces are sometimes used for the smelting of copper or tin, but flash and reverberatory furnaces are more common for metals other than lead.

Most primary copper is produced by matte smelting, an operation yielding a molten sulfide of copper and iron, called matte, which is further oxidized in the converting step to yield metallic copper. A similar process is used for the extraction of nickel (see NICKEL AND NICKEL ALLOYS).

Reduction to Gaseous Metal. Volatile metals can be reduced as easily and completely separated from the residue before being condensed to a liquid or a solid product in a container physically separated from the reduction reactor. Reduction to gaseous metal is possible for zinc, mercury, cadmium, and the alkalai and alkaline-earth metals, but industrial practice is significant only for zinc, mercury, magnesium, and calcium.

Reduction to Solid Metal. Metals having very high melting points cannot be reduced in the liquid state. Because the separation of a solid metallic product from a residue is usually difficult, the raw material must be purified before reduction.

Very reactive metals, eg, titanium or zirconium, which in the liquid state react with all refractory materials available to contain them, also require reduction to solid metal.

Less reactive metals can also be produced by reduction below their melting point. Several processes, referred to as direct reduction, have been developed for the reduction of iron oxides to solid iron. The reducing agents used commercially are hydrogen, carbon monoxide, natural gas, carbon, and carbonaceous fuels. The metallic product is impure (90–96 wt % Fe), usually charged to a steel-making furnace. The advantage of the direct reduction is that it can be operated on a small scale and the capital investment is much lower than that of a blast-furnace plant (see IRON BY DIRECT REDUCTION).

Refining Processes. All the reduction processes yield an impure metal containing some of the minor elements present in the concentrate, or some elements introduced during the smelting process. These impurities must be removed from the crude metal in order to meet specifications for use. Refining operations may be classified according to the kind of phase involved in the process, ie, separation of vapor from a liquid or solid, separation of a solid from a liquid, or transfer between two liquid phases. In addition, they may be characterized by whether or not they involve oxidation–reduction reactions. They include volatilization, precipitation, and slag refining (eg, steelmaking).

Hydrometallurgy

The treatment of ores by dissolution is a fairly simple operation which imitates natural leaching processes. Hydrometallurgical processes are preferred when the pyrometallurgical route is impossible or impractical. If the metal to be extracted is more reactive than the impurities to be removed, or if the grade of the ore is very low and cannot be upgraded by physical beneficiation, hydrometallurgy becomes important.

The continuing decrease in grade, the increasingly complex nature of available ores, and the need for high purity metallic products has favored the development of hydrometallurgy which has found applications for advanced materials, including ceramics (qv) composites (see COMPOSITE MATERIALS), and nanostructure materials. Concern for air pollution (qv) caused by pyrometallurgical plants and the cost of preventive devices are incentives to consider hydrometallurgy in cases where water pollution can be controlled.

The three main steps are leaching or dissolution of the metals in a suitable aqueous solvent, purification or removal of impurities and/or concentration of the solution, and recovery or precipitation of the reduced metal or its compound from solution. In exceptional cases such as the treatment of seawater or brines for metal values, the first step has been taken by nature (see CHEMICALS FROM BRINE; OCEAN RAW MATERIALS). Most hydrometallurgical processes, however, involve several steps for the purification and/or concentration of the solution before the recovery of a pure product becomes feasible. In addition, steps are required for the physical separation of the solid phases from the liquid, ie, washing, clarification, thickening, filtering, drying (qv), etc. Last but not least, the solvent is usually too valuable to be discarded, and must be regenerated and recycled (see SOLVENT RECOVERY, CONDENSATION). Hydrometallurgical plants operate most efficiently in a closed circuit which limits water pollution.

Hydrometallurgical Flow Sheets. The various hydrometallurgy operations can be combined in many ways to design processes appropriate for specific metals.

Electrometallurgy

The use of electricity for winning and refining metals could not have been developed without an understanding of the basic principles of electrochemistry and the availability of cheap industrial electric power. Electrometallurgy is a more powerful tool than the other extractive processes. It supplies energy to the system in a way that enables a reaction to proceed against its chemical affinity. Chemically reactive metals are more easily recovered by electrometallurgy than by chemical reduction, and can be obtained in high purity. Ellectrometallurgy, however, has practical limitations. First among these is expense. Second, the electrochemical process occurs at an electrolyte–metal interface so that the output of an electrochemical reactor is directly proportional to the area of that interface. Improving the energy efficiency and increasing the electrode area per unit volume of reactor remain a challenge.

Electrowinning from Aqueous Solutions. Electrowinning is the recovery of a metal by electrochemical reduction of one of its compounds dissolved in a suitable electrolyte. Various types of solutions can be used, but sulfuric acid and sulfate solutions are preferred because these are less corrosive than others and the reagents are fairly inexpensive. From an electrochemical viewpoint, the high mobility of the hydrogen ion leads to high conductivity and low ohmic losses, and the sulfate ion is electrochemically inert under normal conditions.

The generalized flow sheet of an aqueous electrowinning process consists of at least three main steps. (*1*) The metal is put into solution by leaching of a calcine, ie, a roasted concentrate of sulfide ore, or by the direct leaching of low grade ores containing oxidized minerals or weathered sulfides. (*2*) The pregnant solution is purified to remove metallic impurities more noble than the metal to be electrowon, and impurities that could reduce the current efficiency. Sometimes, the pregnant solution is treated by solvent extraction to produce a more concentrated electrolyte. (*3*) The purified solution is fed to the electrolysis tanks where the metal is plated on a cathode and oxygen is evolved at an inert anode, usually made of lead or one of its alloys. The sulfuric acid is regenerated by the anodic process and is recycled to leaching.

Electrolysis of Molten Salts. Metals more active than zinc and manganese cannot be recovered by electrodeposition from aqueous solutions. Most of them, however, can be electrowon from a molten electrolyte. A compound of the metal to be electrodeposited is dissolved in a mixture of salts of more active metals in order to achieve a low melting point, a suitable viscosity to density, and a high conductivity. Reduction of the metal as a separate liquid phase having a density different from that of the electrolyte is a preferable means of obtaining electrodeposits. The cell electrolyte cannot be recycled through leaching and purification as in aqueous electrowinning, and the feed must be carefully purified and dehydrated.

Electrorefining. Electrolytic refining is a purification process in which an impure metal anode is dissolved electrochemically in a solution of a salt of the metal to be refined, and then recovered as a pure cathodic deposit. Electrorefining is a more efficient purification process than other chemical methods because of its selectivity. In particular, for metals such as copper, silver, gold, and lead, which exhibit little irreversibility, the operating electrode potential is close to the reversible potential, and a sharp separation can be accomplished, both at the anode where more noble metals do not dissolve and the cathode where more active metals do not deposit. The power consumption of this process is very small, and electrorefining is much less sensitive to the cost of electric power than other electrometallurgical processes.

PAUL DUBY
Columbia University

C. Bodsworth, *The Extraction and Refining of Metals,* CRC Press, Inc., Boca Raton, Fl., 1994.

J. D. Gilchrist, *Extractive Metallurgy,* 3rd ed., Pergamon Press Ltd., Oxford, U.K., 1989.

F. Habashi, *Principles of Extractive Metallurgy,* Gordon and Breach, New York, Vol. 1, "General Principles," 1969; Vol. 2, "Hydrometallurgy," 1970; Vol. 3, "Pyrometallurgy," 1986.

E. Jackson, *Hydrometallurgical Extraction and Reclamation,* Ellis Horwood Ltd., Chichester, U.K., 1986.

POWDER METALLURGY

Powder metallurgy (P/M) is both an ancient art and a modern technology on the cutting edge of new materials and manufacturing processes. Powder metallurgy, a worldwide industry, uses metal powders as the raw material from which to manufacture all kinds of products, but chiefly precision metal shapes and parts. These parts perform in many consumer and industrial products such as automobile engines and transmissions, aircraft engines, aerospace hardware, washing machines, power tools, riding lawn mowers, computer disk drives, and surgical implements.

Metal powders are consolidated into shapes through a number of different production processes (Fig. 1). The most widely used powder metallurgy process covers three basic steps for producing conventional density parts: mixing, compacting, and sintering.

Advantages

The P/M process is cost effective in producing simple or complex parts at, or very close to, final dimensions at hourly production rates that range from a few hundred to several thousand parts. As a result, only minor or no machining is required. P/M parts also may be sized for closer dimensional control and/or coined for both higher density and strength. Both ferrous and nonferrous parts can be oil impregnated to function as self-lubricating bearings.

Powder metallurgy competes with such conventional metal-forming processes as casting, stamping, screw-machining, forging, and permanent mold casting.

Characteristics

Individual Particles. The precise determination of particle size, usually referred to as the particle diameter, can actually be made only for spherical particles (see SIZE MEASUREMENT OF PARTICLES). Metal powder particles are produced in a variety of shapes. The desired shape usually depends to a large extent on the method of fabrication.

The density of a metal powder particle is not necessarily identical to the density of the material from which it is produced because of the particles' internal porosity. Any reaction between two powder particles starts on the surface. The amount of surface area compared to the volume of the particle is, therefore, an important factor in powder technology. The particle—surface configuration, whether it is smooth

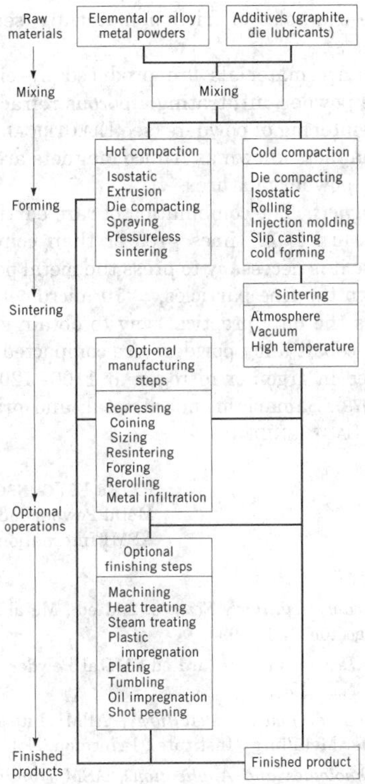

Figure 1. The P/M process.

or contains sharp angles, is another. The particle surface area depends strongly on the method of production.

Powder particles may consist of a single crystal or many crystal grains of various sizes. The microstructure, ie, the crystal grain size, shape, and orientation, depend also on the method of powder fabrication. However, in many cases a correlation exists between particle size and grain size.

The possibility of reducing surface oxide layers in a reducing gas atmosphere (Cu, Fe, Ni, W, Mo, etc) depends on the type of metal. For a given thickness of the oxide layer, the amount of oxide in a powder changes with the particle size.

Particle activity determines the type and rate of the reaction of a powder particle with its environment.

Powder Mass. A mass of powder consists of a large quantity of particles. The most important properties of a good molding-grade powder are flow rate, particle size, and size distribution, apparent density, green strength, compressibility, and dimensional stability during sintering. A powder must flow well in order to fill all parts of the die cavity evenly and move through the automatic equipment (see POWDERS, HANDLING).

Manufacture

The manufacture of metal in powder form is a complex and highly engineered operation. It is dominated by the variables of the powder, namely those that are closely connected with an individual powder particle, those that refer to the mass of particles which form the powder, and those that refer to the voids in the particles themselves. The primary methods for the manufacture of metal powders are atomization, the reduction of metal oxides, and electrolytic deposition.

Other methods include mechanical comminution and condensation of metal vapors followed by deposition on cooler surfaces, which yields metal powders as does decomposition of metal hydrides. Reaction of a metal halide and molten magnesium, known as the Kroll process, is used for titanium and zirconium. This results in a sponge-like product.

Using rapid solidification technology (RST), molten metal is quench cast at a cooling rate up to 10^6°C/s as a continuous ribbon. This ribbon is subsequently pulverized to an amorphous powder. RST powders include aluminum alloys, nickel-based superalloys, and nanoscale powders. RST conditions can also exist in powder atomization.

Other methods of metal powder manufacture are also employed for specific metals. Selective corrosion of carbide-rich grain boundaries in stainless steel, a process called intergranular corrosion, also yields a powder.

Processing

Consolidation. Metal powders are consolidated by heat or by pressure followed by heat, or by heating during the application of pressure. Consolidation produces a coherent mass of definitive size and shape for further working, heat treating, or use as is.

The characteristics of a pressed compact are influenced by the characteristics of the powder: rate and manner of pressure application, maximum pressure applied and for what period of time, shape of die cavity, temperature during compaction, additives such as lubricants and alloy agents, and die material and surface condition. The effect of various compaction variables on the pressed compact are shown in Figure 2.

Consolidation Techniques. Uniaxial pressing of metal powders in a die of specific dimensions and configuration is the most frequently used technique for the consolidation of powders in the manufacture of P/M products. Other consolidation techniques include cold isostatic pressing; hot isostatic pressing; powder rolling; P/M forging; spray forming; warm compacting; slip casting; vibratory consolidation; hot pressing; extrusion, swaging, or rolling; and metal injection molding.

Compacting Lubricants. The surface area of most moldable metal powders is in the range of 500–700 cm^2/g. Finer powders can have a surface area as high as 1500 cm^2/g. A very large number of individual particles is involved. For example, 1 cm^3 uniformly filled with 2-μm spherical particles having a surface area of 1200 cm^2/g contains ca 1.2×10^9 particles. Because of this large surface area, a considerable amount of friction has to be overcome during powder consolidation. To one degree or another, friction is present in all consolidation methods.

Dry lubricants are usually added to the powder in order to decrease the friction effects. The more common lubricants include zinc stearate, lithium stearate, calcium stearate, stearic acid, paraffin, graphite, and molybdenum disulfide. Lubricants are generally added to the powder in a dry state in amounts of 0.25–1.0 wt % of the metal powder. Some lubricants are added by drying and screening a slurry of powder and lubricant. In some instances, lubricants are applied in liquid form to the die wall.

Lubricants protect die and punch surfaces from wear and burn-out of the compact during sintering without objectionable effects or

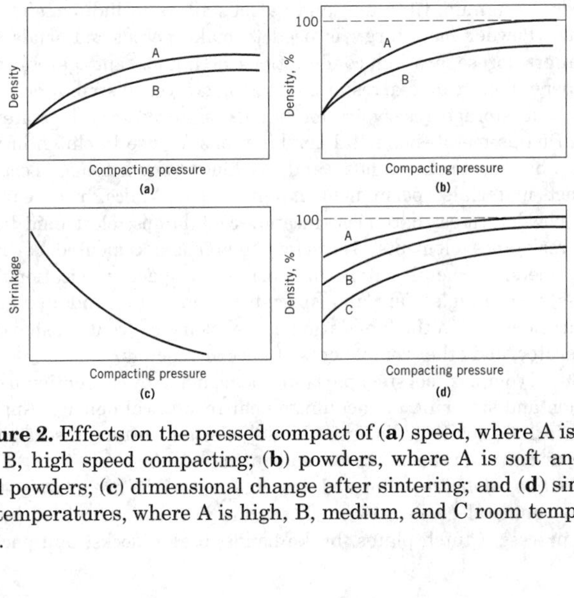

Figure 2. Effects on the pressed compact of **(a)** speed, where A is low and B, high speed compacting; **(b)** powders, where A is soft and B, hard powders; **(c)** dimensional change after sintering; and **(d)** sintering temperatures, where A is high, B, medium, and C room temperature.

residues. They must have small particle size, and overcome the main share of friction generated between tool surfaces and powder particles during compaction and ejection. They must mix easily with the powder, and must not excessively impede powder flow (see LUBRICATION AND LUBRICANTS).

Sintering. Basically a solid-state process, sintering transforms compacted mechanical bonds between the powder particle into metallurgical bonds.

Liquid-phase sintering refers to the sintering of a powder mixture of two or more components, of which at least one has a melting temperature lower than the others. The sintering temperature is then selected in such a manner that a liquid phase is formed in which the solid powder particles of the other components rearrange. A high density powder compact is the result.

In infiltration, the pores of a sintered solid are filled with a liquid metal or alloy. The most common application is the infiltration of a porous sintered steel matrix with copper, to give a copper-infiltrated P/M part. The aims of infiltration are to obtain higher strength and a pore-free structure for plating, machining, and sealing of pressure-tight applications. The liquid and the solid must not react to form a solid compound or alloy having a specific volume as great as or greater than their combined preinfiltration specific volumes. The porosity in the porous matrix should be interconnected. Ideally, the matrix material should be insoluble in the liquid infiltrant. Simple liquid-phase sintering does not give pore-free structures; infiltration does, provided the pores are interconnected.

Post-Sintering Treatments. The sintering process concludes the powder metallurgy processes of production and consolidation. However, some P/M parts may require a number of further operations, such as working treatments, sizing, repressing, coining, heat treatments, and finishing.

Health and Safety

Metal powders possess an immensely high ratio of specific surface area to volume. This characteristic contributes to several potentially hazardous properties such as pyrophoricity, explosiveness, and toxicity. The problems associated with the fine particles can be minimized or eliminated with proper handling and good housekeeping procedures, such as storage in appropriate containers, processing in sparkproof equipment, avoiding exposure to open flames, and minimizing airborne particulates (see POWDERS, HANDLING).

Applications

P/M structural parts can be made from a range of materials including iron, steels, low and high brass, bronze, nickel and nickel-base alloys, copper, aluminum, titanium, and various alloys including refractory metals. Powder metallurgy is used to make porous materials such as filters for separating combinations of liquids and gases, surge dampeners and flame arrestors, metering devices, distribution manifolds, and storage reservoirs for liquids; sintered friction materials classified as metal–nonmetal combinations for use in clutch brakes, brake blocks, brake bands, and packing compositions; electrical contact materials; permanent magnets (eg, Alnico, rare earths); iron powder cores; batteries; incandescent lamps; electronic tubes; and resistance elements; refractory materials; cemented carbides; in cermets; in metal-matrix composites; in space applications (eg, heat-shield shingles, air and space-borne mirrors); and in nuclear applications (eg, in the fabrication of fuel elements, control, shielding, moderator, and other components of nuclear reactors).

Small complex tool steel parts are being made by conventional compaction and sintering in vacuum to near theoretical density. Applications include spade drills, knife blades, slotting cutters, insert blades for gear cutters, reamer blades, and cutting tool inserts.

Friction Materials. Sintered friction materials are classified as metal-nonmetal combinations. These are best manufactured by the P/M process. Clutch plates, brake bands, brake blocks, and packing compositions are examples of friction materials (see BRAKE LININGS AND CLUTCH FACINGS).

Electrical contact materials are produced by either slicing rod made from metal powder, infiltrating a porous refractory skeleton, or compaction and sintering of powders (see ELECTRICAL CONNECTORS).

Permanent magnets known as Alnico magnets are made by pressing and sintering powder mixtures.

Rare-Earth Magnets. The combination of rare earths and cobalt exhibits magnetic flux 4–19 times greater than conventional Alnico magnets. Because it is necessary to press the metal powders in a magnetic field to align the fine particles, ≤ 10 micrometers in size, powder metallurgy is the only practical way to obtain such superalloys. The rare-earth—cobalt alloy powders are compacted in a mechanical press and sintered in argon or nitrogen at 1100–1200°C (see COBALT AND COBALT ALLOYS). Samarium, mischmetal, and praseodymium are mostly used (see LANTHANIDES).

PETER K. JOHNSON
Metal Powder Industries Federation
APMI International

R. M. German, *Powder Metallurgy Science,* 2nd ed., Metal Powder Industries Federation, Princeton, N.J., 1994.

Powder Metallurgy Design Manual, 3rd ed., Metal Powder Industries Federation, Princeton, N.J., 1998.

International Journal of Powder Metallurgy, APMI International (formerly American Powder Metallurgy Institute), Princeton, N.J.

Powder Metal Technologies and Applications, ASM Handbook, Vol. 7, ASM International, Materials Park, Ohio, 1998.

METAL-MATRIX COMPOSITES

A composite material consisting of two or more physically and/or chemically distinct, suitably arranged or distributed phases, generally having characteristics different from those of any components in isolation. Usually one component acts as a matrix in which the reinforcing phase is distributed. When the continuous phase or matrix is a metal, the composite is a metal-matrix composite (MMC). The reinforcement can be in the form of particles, whiskers, short fibers, or continuous fibers (see COMPOSITE MATERIALS).

There are three kinds of metal-matrix composites distinguished by type of reinforcement: particle-reinforced MMCs, short fiber-or whisker-reinforced MMCs, and continuous fiber- or sheet-reinforced MMCs. Table 1 provides examples of some important reinforcements used in metal-matrix composites as well as their aspect (length/diameter) ratios and diameters.

Particle or discontinuously reinforced MMCs have become important because they are inexpensive compared to continuous fiber-reinforced composites and they have relatively isotropic properties compared to the fiber-reinforced composites. Figures **1a** and **b** show typical microstructures of continuous alumina fiber/Mg and silicon carbide particle/Al composites, respectively.

Processing

There are several important fabrication processes for metal-matrix composites.

Table 1. Typical Reinforcements Used in Metal-Matrix Composites

Type	Aspect ratio	Diameter, μm	Examples
particle	~1–4	1–25	SiC, Al_2O_3, BN, B_4C
short fiber or whisker	~10–1000	0.1–25	SiC, Al_2O_3, $Al_2O_3 + SiO_2$, C
continuous fiber	>1000	3–150	SiC, Al_2O_3, C, B, W

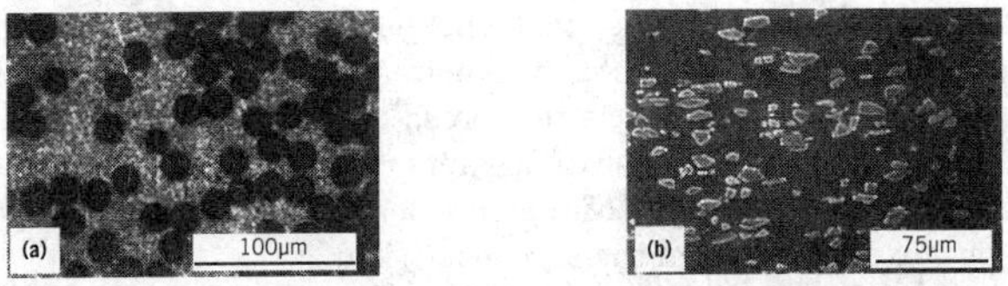

Figure 1. Typical microstructures of some metal-matrix composites: (**a**) continuous alumina fiber/Mg and (**b**) silicon carbide particle/Al composites.

Liquid-State Processes. *Casting* or *liquid infiltration* involves infiltration of a fiber bundle by liquid metal. It is not easy to make MMCs by simple liquid-phase infiltration, mainly because of difficulties with wetting ceramic reinforcement by the molten metal. One liquid infiltration process involving particulate reinforcement, called the Duralcan process, has become quite successful.

Squeeze casting or *pressure infiltration* involves forcing the liquid metal into a fibrous preform. Pressure is applied until the solidification is complete. By forcing the molten metal through small pores of fibrous preform, this method obviates the requirement of good wettability of the reinforcement by the molten metal. The process is conducted in the controlled environment of a pressure vessel and rather high fiber volume fractions; complex shaped structures are obtainable. Although commonly aluminum matrix composites are made by this technique, alumina fiber-reinforced intermetallic matrix composites, eg, TiAl, Ni_3Al, and Fe_3Al matrix materials, have been prepared by pressure casting.

Solid-State Processes. *Diffusion bonding* is a common solid-state welding technique for joining similar or dissimilar metals. Interdiffusion of atoms from clean metal surfaces in contact at an elevated temperature leads to welding. The principal advantages of this technique are the ability to process a wide variety of matrix metals and control of fiber orientation and volume fraction. Among the disadvantages are processing times of several hours, expensive high processing temperatures and pressures, and only objects of limited size can be produced. There are many variants of the basic diffusion bonding process, however, all of them involve a simultaneous application of pressure and high temperature.

Deformation processing of metal–metal composites involves mechanical processing (swaging, extrusion, drawing, or rolling) of a ductile two-phase material. The two phases co-deform, causing the minor phase to elongate and become fibrous in nature within the matrix. These materials are sometimes referred to as *in situ* composites.

Powder processing methods involving cold pressing and sintering, or hot pressing can be used to fabricate MMCs, primarily particle- or whisker-reinforced MMCs. The matrix and the reinforcement powders are blended to produce a homogeneous distribution. Most of the information on this critical step is considered proprietary. The final step is hot pressing, uniaxial or isostatic, to produce a fully dense composite. The hot pressing temperature can be either below or above the matrix alloy solidus.

Deposition techniques for metal-matrix composite fabrication involve coating individual fibers in a tow with the matrix material needed to form the composite followed by diffusion bonding to form a consolidated composite plate or structural shape. The main disadvantage of using deposition techniques is that they are time consuming. Advantages include: (*1*) The degree of interfacial bonding is easily controllable; interfacial diffusion barriers and compliant coatings can be formed on the fiber prior to matrix deposition or graded interfaces can be formed. (*2*) Filament-wound thin monolayer tapes can be produced that are easier to handle and mold into structural shapes than other precursor forms; unidirectional or angle-plied composites can be easily fabricated in this way.

Several deposition techniques are available: immersion, plating, electroplating, spray deposition, chemical vapor deposition (CVD), and physical vapor deposition (PVD) (see THIN FILMS).

In Situ Processes. In these techniques, the reinforcement phase is formed *in situ*. The composite material is produced in one step from an appropriate starting alloy, thus avoiding the difficulties inherent in combining the separate components as done in a typical composite processing. Controlled unidirectional solidification of a eutectic alloy is a classic example of *in situ* processing. Unidirectional solidification of a eutectic alloy can result in one phase being distributed in the form of fibers, and ribbon in the other. Distribution fineness of the reinforcement phase can be controlled by simply controlling the solidification rate. The solidification rate in practice, however, is limited to a range of 1–5 cm/h because of the need to maintain a stable growth front which requires a high temperature gradient. The XD process (Martin Marietta) is another *in situ* process which uses an exothermic reaction between two components to produce a third component.

Spray-Forming of Particulate MMCs. Another process for making particle-reinforced MMCs involves the use of spray techniques that have been used for some time to produce monolithic alloys. One particular example of this, a co-spray process, uses a spray gun to atomize a molten aluminum alloy matrix, into which heated (for drying) silicon carbide particles are injected. The process is totally computer controlled and is quite fast, but it should be noted that it is essentially a liquid metallurgy process. The formation of deleterious reaction products is avoided because the time of flight is extremely short. An advantage of the process is the flexibility it affords in making different types of composites, eg, *in situ* laminates can be made using two sprayers or by selective reinforcement. This process is quite expensive, however, mainly because of the capital equipment.

Interfaces in Metal-Matrix Composites

The interface region in a composite is important in determining the ultimate properties of the composite. At the interface a discontinuity occurs in one or more material parameters such as elastic moduli, thermodynamic parameters such as chemical potential, and the coefficient of thermal expansion. The importance of the interface region in composites stems from two main reasons: the interface occupies a large area in composites, and in general, the reinforcement and the matrix form a system that is not in thermodynamic equilibrium.

Properties

Modulus. Unidirectionally reinforced continuous fiber-reinforced metal-matrix composites show a linear increase in the longitudinal Young's modulus as a function of the fiber volume fraction. Particle reinforcement also results in an increase in the modulus of the composite; the increase, however, is much less than that predicted by the rule of mixtures.

Strength. Prediction of MMC strength is more complicated than the prediction of modulus. Consider an aligned fiber-reinforced metal-matrix composite under a load P_c in the direction of the fibers. This load is distributed between the fiber and the matrix:

$$P_c = P_m V_m + P_f V_f$$

where P_m and P_f are loads on the matrix and fiber, respectively. This equation can be converted to the following rule of mixtures relationship under conditions of isostrain, ie, the strain in the fiber, matrix, and composite is the same:

$$\sigma_c = \sigma_f V_f + \sigma_m V_m$$

where σ is the stress, V is the volume fraction, and *c,f,* and *m* denote composite, fiber, and matrix, respectively. In general, ceramic reinforcements have a coefficient of thermal expansion greater than that of most metallic matrices, thermal stresses are generated in both the components, the fiber and the matrix. The reinforcement also acts as a constraint on the matrix flow, thus increasing its flow stress.

A series of events can take place in response to the thermal stresses: (*1*) plastic deformation of the ductile metal matrix (slip, twinning, cavitation, grain boundary sliding, and/or migration); (*2*) cracking and failure of the brittle fiber; (3) and adverse reaction at the interface; and (4) failure of the fiber–matrix interface.

Toughness. Toughness can be regarded as a measure of energy absorbed in the process of fracture or more specifically as the resistance to crack propagation, K_{Ic}. The toughness of MMCs depends on the matrix alloy composition and microstructure; the reinforcement type, size, and orientation; and processing insofar as it affects microstructural variables, eg, distribution of reinforcement, porosity, segregation, etc.

Thermal Stresses and Properties. In general, ceramic reinforcements (fibers, whiskers, or particles) have a coefficient of thermal expansion greater than that of most metallic matrices. This means that when the composite is subjected to a temperature change, thermal stresses are generated in both components.

Aging. Frequently the metal-matrix alloy used in a MMC has precipitation hardening characteristics, ie, such an alloy can be hardened by suitable heat treatment called aging treatment. It has been shown that the microstructure of the metallic matrix is modified by the presence of ceramic reinforcement and consequently the standard aging treatment for, eg, an unreinforced aluminum alloy, is not valid. The particle- or whisker-type reinforcements such as SiC, B_4C, Al_2O_3, etc, are unaffected by the aging process. These reinforcements, however, can affect the precipitation behavior of the matrix quite significantly. In particular, a higher dislocation density in the matrix metal or alloy than that in the unreinforced metal or alloy is produced.

Fatigue. This is the phenomenon of mechanical property degradation leading to failure of a material or a component under cyclic loading. Many high volume applications of composite materials involve cycling loading situations, eg, automobile components. Application of conventional approaches, such as the stress vs cycle (S-N) curves or the application of linear elastic fracture mechanics (LEFM) to fatigue of composites, is not straightforward. The main reasons for this are the inherent heterogeneity and anisotropic nature of the composites which result in damage mechanisms in composites being very different from those encountered in conventional, homogeneous, or monolithic material. Novel approaches, such as that epitomized by the measurement of stiffness reduction as a function of cycles, are being used to analyze the fatigue behavior of MMCs. Because many applications of composite MMCs involve temperature changes, it is important that thermal fatigue characteristics of composites be evaluated in addition to their mechanical fatigue characteristics.

Creep. The phenomenon of creep refers to time-dependent deformation. In practice, at least for most metals and ceramics, the creep behavior becomes important at high temperatures and thus sets a limit on the maximum application temperature. In general, this limit increases with the melting point of a material. An approximate limit can be estimated to lie at about half of the Kelvin melting temperature. The basic governing equation of steady-state creep can be written as follows:

$$\epsilon = A(\sigma/G)^n \exp(-\Delta Q/kT)$$

where ϵ is the steady-state creep strain rate, ς is the applied strain, n is an exponent, G is the shear modulus, ΔQ is the activation energy for creep, k is the Boltzmann's constant, and T is temperature in kelvin. In general, the creep resistance of metal is improved by the incorporation of ceramic reinforcements. The modeling of creep behavior of MMCs is complicated because in the temperature regime where the metal matrix may be creeping, the ceramic reinforcement is likely to be deforming elastically.

Applications

In aerospace applications, low density coupled with other desirable features, such as tailored thermal expansion and conductivity, high stiffness and strength, etc, are the main drivers. Performance rather than cost is an important item. Inasmuch as continuous fiber-reinforced MMCs deliver superior performance to particle-reinforced composites, the former are frequently used in aerospace applications. In nonaerospace applications, cost and performance are important, ie, an optimum combination of these items is required. It is thus understandable that particle-reinforced MMCs are increasingly finding applications in nonaerospace applications.

Reduction in the weight of a component is a significant driving force for any application in the aerospace field.

An important application of MMCs in the automotive area is in diesel piston crowns. This application involves incorporation of short fibers of alumina or alumina–in the crown of the piston. The replacement of nickel cast iron by aluminum matrix composite results in a lighter, more abrasion resistant, and cheaper product. Another application in the automotive sector involves the use of carbon fiber and alumina particles in an aluminum matrix for use as cylinder liners.

Particulate metal-matrix composites, especially with light metal-matrix composites such as aluminum and magnesium, also find applications in automotive and sporting goods.

Niobium–titanium superconductors are used in magnetic resonance imaging (MRI) techniques for medical diagnostics. The superconducting solenoid made from Nb-Ti/Cu composite wire is immersed in a liquid helium cryogenic Dewar flask. Nb-Ti/Cu superconducting composites are also used in various high energy physics applications, such as particle accelerators.

Metal-matrix composites can be tailored to have optimal thermal and physical properties to meet requirements of electronic packaging systems, eg, cores, substrates, carriers, and housings. A controlled thermal expansion space truss, ie, one having a high precision dimensional tolerance in space environment, was developed from a carbon fiber (pitch-based)/Al composite. Continuous boron fiber-reinforced aluminum composites made by diffusion bonding have been used as heat sinks in chip carrier multilayer boards.

A C/Al composite is useful in heat-transfer applications where weight reduction is an important consideration, eg, in high density, high speed integrated circuit packages for computers and in base plates for electronic equipment. Another possible use of this composite is to dissipate heat from the leading edges of wings in high speed airplanes.

K. K. CHAWLA
University of Alabama at Birmingham

K. K. Chawla, *Composite Materials: Science & Engineering,* 2nd ed., Springer-Verlag, New York, 1998.

M. Taya and R. J. Arsenault, *Metal Matrix Composites,* Pergamon Press, Oxford, U.K., 1990.

S. Suresh, A. Needleman, and A. Mortensen, eds., *Metal Matrix Composites,* Butterworth-Heinemann, Boston, Mass., 1993.

T. W. Clyne and P. J. Withers, *An Introduction to Metal Matrix Composites,* Cambridge University Press, Cambridge, U.K., 1993.

METAL PLATING. See ELECTROPLATING; METALLIC COATINGS.

METAL SURFACE TREATMENTS

CASE HARDENING

The performance of many metallic components can be markedly improved by developing a surface region which is harder than that of its underlying region. Processes to achieve this are called case hardening. The case is the surface region which is hardened, and the region under the case is called the core (Fig. **1a**). The original surface region is an intrinsic part of the case, unlike treatments such as electroplating (qv) where material is deposited on the original surface.

There are many characteristics of hard cases that make their development desirable. One is wear resistance. Usually, the process is designed to develop high compressive residual stresses in the surface,

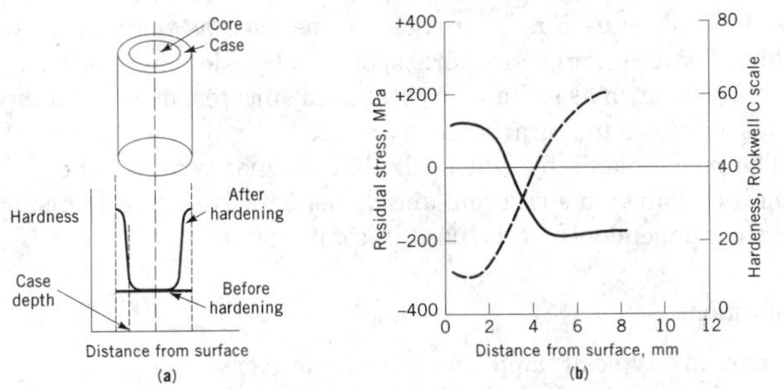

Figure 1. (**a**) Schematic diagram showing definition of case hardening and of case depth where (—) is the diameter; (**b**) residual stress across the radius of a case hardened steel, showing the high compressive residual stress (– – –) at the surface induced by induction heating, as well as the microhardness (—) of the surface. To convert MPa to psi, multiply by 145.

which counteract tensile stresses induced by the loading condition during use of the component (Fig. **1b**).

Failure can be reduced by the presence of compressive residual stresses. and the development of these stresses is an important purpose of case hardening (Fig. **1b**). The relationship between the surface treatment and fatigue life is given in Table 1 for three types of steel (qv), the most common metallic material that is case hardened, and the material emphasized herein. The three main characteristics to control in case hardening are case depth, case microstructure, and case hardness and strength. The methods and steels chosen for case hardening are closely related to those factors.

Conventional Hardening of Steels

Iron–Carbon Phase Diagram. The hardening of steels begins with the formation of the high temperature phase austenite, denoted γ, which is iron with a fcc crystal structure.

Decomposition of Austenite. When steel is cooled sufficiently rapidly from the austenite region to low (eg, 25°C) temperature, the austenite decomposes into a nonequilibrium phase. This phase, called martensite, is body-centered tetragonal. It is the hardest form of steel, and its formation is critical in hardening.

Tempering. Although martensite is hard and strong, it is also brittle, and therefore must be heat treated to improve toughness and ductility. This heat treatment is called tempering, and the structure, carbide particles in ferrite, is called tempered martensite. The longer the tempering time and the higher the tempering temperature, the coarser the structure. This reduces the hardness and strength but increases ductility and toughness. Thus the desired mechanical properties are obtained by proper choice of the tempering temperature and time.

Case Hardening by Heat Treatment

No Chemical Change. By proper choice of the quenching process and the steel, a heat-treated component can be produced that is hard in the surface region, ie, all martensite, and softer in the center portion. However, for many applications the desired properties are obtained if the hardened region (case) is relatively shallow, eg, 0.2 cm. It is difficult to achieve this by conventional methods. Instead, the component is heat treated to achieve desired core properties, the austenite range, which is subsequently cooled rapidly to form martensite. This is case hardening without any chemical change in the surface region.

Processes include case hardening by flame heating, by laser heating, and by electron beam heating.

Chemical Change. In case hardening by carburizing, carbon atoms are deposited by chemical reaction on the surface of the steel while at high temperature in the austenite region. The atoms then move by diffusion into the steel, but are continuously replaced at the surface by more carbon atoms. The rate of deposition exceeds that of diffusion, so that a carbon gradient builds up with time. The steel is then quenched to convert this surface region to martensite, forming a hard surface with a favorable compressive residual stress.

The nitrogen atom, similar in size to carbon, is quite soluble in the high temperature austenite phase of iron. Thus nitrogen can be used to form a hard iron–nitrogen martensite. Nitrogen can also be added simultaneously with carbon to austenite to form hard martensite. However, the nitriding temperature is usually well below that of the formation of austenite (<723°C), and because martensite forms only from austenite, hardening by nitrogen addition occurs by another mechanism. The nitrogen added to the steel surface reacts with iron or specific alloying elements to form a fine dispersion of very hard nitrogen compounds, ie, nitrides (qv), and the nitrides produce a very hard case.

A problem in nitriding is that a layer of nitrides may form on the surface which does not have desirable properties. This is called the white layer or compound layer, and may have to be machined or ground off after nitriding.

Molecular nitrogen, N_2, is stable and relatively inert. It does not decompose to atomic nitrogen to form a nitrogen case. Instead, a gaseous compound containing nitrogen must be used. The common carrier of nitrogen in gas nitriding is ammonia (qv).

Nitrogen can be produced for deposition on the steel surface by ionization of gaseous nitrogen to form an ionic plasma. The plasma gives off visible radiation, the color of which depends on the type of gas being ionized. This process is sometimes called glow discharge nitriding (see PLASMA TECHNOLOGY).

The nitrogen ions are neutralized at the surface, forming atomic nitrogen, which then diffuses into the surface and reacts with iron and alloying elements to form the hard nitrides. A distinct advantage of this method is that the plasma generally deposits nitrogen uniformly on the surface, so that less uneven deposition occurs than is sometimes found with nitriding by flowing hydrogen–ammonia gas.

Because nitriding is carried out in the ferrite and carbide region of the phase diagram, at lower temperatures than in carburizing, the diffusion rate of nitrogen is relatively low and hence the time to attain a suitable case is longer than in carburizing.

An advantage of nitriding is that the core properties can be set by prior heat treatment, such as by quenching and tempering, then the surface nitrided.

Nitrided components possess outstanding fatigue properties.

Case Hardening by Surface Deformation. When a metallic material is plastically deformed at sufficiently low temperature, eg, room temperature for most metals and alloys, it becomes harder. Thus one method to produce a hard case on a metallic component is to plastically deform the surface region. This can be accomplished by a number of methods, such as by forcing a hardened rounded point onto the surface as it is moved. A common method is impinge upon the surface fine hard particles such as hardened steel spheres (shot) at high velocity. This process is called shot peening.

Case Hardening by Ion Implantation

One method of changing the chemistry of the surface of a component without heat treatment or chemical reaction is ion implantation (qv).

Table 1. Improvement in Material Life as a Result of Case Hardening

Steel	Surface hardness, HRC[a]	Method of hardening	Number of shafts tested	Cycles to failure × 10^6
4140	36–42	through-hardened	20	>0.4
4320	40–46	carburized to 1.0–1.3 mm	6	ca 0.8
1137	42–48	induction hardened to 3.0-mm min effective depth and 40 HRC[a]	5	>1.1

[a] HRC = Rockwell C hardness (see HARDNESS).

Not only can elements such as carbon and nitrogen be added, but larger elements such as titanium can be also. A gas of the implant species is produced by vaporization, if necessary, and ionized in some manner, such as by heating or forming a plasma. The ions are removed from the region of formation by a negative electrostatic field, then they enter an accelerator. The ions exit the accelerator with a velocity sufficient to penetrate the surface of the target. The ion beam is focused and its direction controlled by a magnetic or electrostatic lens system. Hence the location of implantation on the surface can be controlled.

The ions not only are implanted in the surface, but cause considerable lattice damage displacing host atoms. An amorphous layer may be formed and the structure is not an equilibrium one. Thus the solubility of the implanted ions may greatly exceed the solubility limit. All of these effects combine to produce a hard case.

The complexity of the apparatus needed for ion implantation makes this method of case hardening of limited application, eg, low loads. Further, the case depth is considerably lower than that produced by carburizing or nitriding.

Ion implantation is being used to form a thin hard case on materials other than steels. Titanium alloys have been successfully implanted with nitrogen. The process has been applied to ceramics to modify the surface region.

Other Methods of Case Hardening

There are a multitude of methods available for case hardening, and the technical details of implementation may be complicated. They include pack carburizing, drip carburizing, vacuum carburizing, plasma carburizing, liquid carburizing, carbonitriding, liquid nitriding, ferritic nitriding, plasma nitrocarburizing, austenitic nitrocarburizing, and boriding or boronizing.

CHARLIE R. BROOKS
University of Tennessee

ASM Handbook, Vol. 4, ASM International, Materials Park, Ohio, 1991; excellent and detailed articles of most case hardening methods.

C. R. Brooks, *Principles of the Surface Treatment of Steels.* Technomic Publishing Co., Lancaster, Pa., 1992.

W. L. Grube and S. Verhoff, in R. Kossowsky, ed., *Surface Modification Engineering,* Vol. II, CRC Press, Boca Raton, Fla., 1989, p. 107.

S. L. Semiatin, D. E. Sutuz, and I. L. Harry, *Induction Heat Treatment of Steel,* American Society for Metals, Metals Park, Ohio, 1986.

CLEANING

Cleaning, the removal of unwanted matter, is the beginning of the treatment cycle for metal. The unwanted matter may be carbon smut, welding flux, ink, oxidation products, oil, fingerprints, or other material. Cleaners may be classified as solvent-based or aqueous. Within the aqueous class there are many subclasses, the most important of which are the alkaline cleaners. There are also a variety of ways to apply cleaners. As of the mid-1990s, solvent-based cleaner usage was declining.

Alkaline Cleaners

Alkaline cleaners are used most commonly for metal surfaces. These are typically composed of a blend of alkaline salt builders, such as sodium phosphates, sodium silicates, sodium hydroxide, or sodium carbonate. In addition, they almost always contain detergents, ie, surfactants (qv), and, optionally, wetting agents, coupling agents, chelating agents (qv), and solubilizers. Alkaline cleaners are usually applied in the range of 38–93°C.

The composition of the builders in an alkaline cleaner is dependent on the metal substrate from which the soil is to be removed. For steel (qv) or stainless steel aggressive, ie, high pH, alkaline salts such as sodium or potassium hydroxide can be used as the main alkaline builder. For aluminum, zinc, brass, or tin plate, less aggressive (lower pH) builders such as sodium or potassium silicates, mono- and diphosphates, borates, and bicarbonates are used.

The mechanisms by which alkaline cleaner removes the soil are saponification, emulsification, and dispersion. These mechanisms can operate independently or be used in combination.

Applications

Cleaners are typically applied either by immersion or by spray. Variations of immersion include electrocleaning, ultrasonic cleaning, and barrel cleaning. Variations of spray cleaning include steam cleaning and power washing. Additionally, solvent cleaners can be applied by vapor degreasing, and both solvent and aqueous cleaners can be applied by mechanical methods.

Economic Aspects

Usage of solvent-based cleaners is decreasing; usage of aqueous-based alkaline, acid, or neutral cleaners is increasing. There are over 60 suppliers of metal cleaners in the United States. One of the leading suppliers of aqueous-based cleaners is the Henkel Corp., Surface Technologies Group. Cleaners, are available under the trademarks Parco, P3, Alumiprep, Ridoline, Metalprep, and Prep-N-Cote, to name just a few.

A trend toward liquid cleaners was evident as of 1994 because of convenience features such as automatic additions of the cleaner by chemical feed pump. Safety features such as minimized heat generation upon blending with water to make the desired concentration are also important.

DONALD P. MURPHY
Henkel Corporation

S. Spring, *Industrial Cleaning,* Prism Press, Melbourne, Australia, 1974.

D. Murphy, *Metals Handbook,* 9th ed., Vol. 5, ASM, Metals Park, Ohio, 1982, pp. 22–25.

D. Murphy and G. Tupper, *Users' Guide to Powder Coating,* Association for Finishing Processes of the Society of Manufacturing Engineers, Dearborn, Mich., 1985, pp. 33–58.

A. Pollack and P. Westphall, *An Introduction to Metal Degreasing and Cleaning,* R. Draper Ltd., Teddington, U.K., 1963.

CHEMICAL AND ELECTROCHEMICAL CONVERSION TREATMENTS

Phosphating

Reactive metal surfaces can be chemically treated and covered with inert, amorphous, or crystalline coatings which grow on the base metal. Phosphates represent the most important area of the conversion coatings. These coatings (qv) are applied as preparation for painting, temporary corrosion protection, lubricant carrier in cold forming, friction improver for stamping and drawing, and as insulation on electrical steels (see also METALLIC COATINGS; METAL TREATMENTS).

The Iron Phosphating Process. Dissolution and oxidation of iron(II) to iron(III) and coating development result in the formation of an amorphous coating which contains iron phosphate, $FePO_4$, as the principal coating constituent. In addition, some iron oxide, Fe_2O_3, which forms from the rearrangement of ferric hydroxide, and some tertiary iron phosphate, called vivianite, $Fe_3(PO_4)_2$, are present. The oxidative condition needed for coating formation is provided by accelerators such as chlorate, nitrate, permanganate, and air entrapped during spraying.

Iron phosphating has been the process of choice for applications where cost considerations override maximum performance needs. Advantages are low chemical cost, low equipment cost, good paint bonding, easy control, and minimum sludge. The disadvantages, thin coatings and poorer corrosion resistance than for zinc phosphating, limit the application of iron phosphating to particular industries. Nevertheless, iron phosphates are the most widely applied conversion coatings when surface preparation for paint application is needed.

The Zinc Phosphating Process. The zinc phosphating reaction involves acid attack on the substrate metal at microanodes and deposition of phosphate crystals at microcathodes. Liberation of hydrogen and the formation of phosphate sludge also occur.

Zinc phosphate coatings form the basis for paint adhesion in a variety of industries. These are used when long-term quality is of concern in applications such as for automotive parts and vehicles, coil-coated products, and appliances.

Testing of Painted Products. The enhancement of paint adhesion is one of the principal functions of conversion coating. A group of tests based on product deformation is used to test the painted product. The appliance and coil-coating industries use the mandrel bend, the cross-hatch adhesion test, and the direct and reverse impact tests. Adhesion after a water soak is judged using a cross-hatch test performed on the exposed surface. Several accelerated corrosion tests are also employed to evaluate the effectiveness of the phosphate coating in the performance of painted products (see CORROSION AND CORROSION CONTROL).

Other Phosphate Coatings. Phosphate coatings are also used as surface treatments for wire drawing, tube drawing, and cold extrusion.

Chromating

Chromating or chromatizing has been widely practiced in the metal industry since the late 1940s to improve corrosion resistance and performance of subsequently applied organic finishes. Commonly, chromates are used to treat wrought alloys, cast alloys, and coatings comprising aluminum, zinc, magnesium, or cadmium. The aerospace, transportation, architecture, appliance, marine, and electronics industries, among others, utilize chromatic coatings extensively (see also COATINGS, MARINE; ELECTRONICS, COATINGS).

Chromate conversion coatings are thin, amorphous or crystalline adherent surface layers on the metal surface. The two classes of chromate coatings are chromium phosphates (green chromates) and chromium chromates (gold chromates).

Anodizing

Whereas many metals can be anodized, aluminum is by far the most widely anodized metal. The anodizing process is comprised of several pre- and post-treatment steps. The anodizing step consists of placing the part to be anodized in a tank where a controlled direct current charge can be applied for a predetermined length of time. At the anode, ie, the part, the aluminum is oxidized to aluminum oxide. The hydrogen ion migrates to the cathode, where it is reduced, forming gas bubbles that are given off to the atmosphere. As this process continues, a uniform porous oxide film is formed on the part. The resulting material having the anodic film can be used for structural purposes, such as aircraft parts, or for decorative purposes, such as windows or picture frames.

After anodizing, the aluminum part can be colored or sealed. The two-step coloring method, sometimes called electrolyte coloring, is the most popular coloring method for architectural purposes. The use of dyes for coloring is becoming more popular because of the almost infinite range of colors that can be produced. Moreover, dyes do not need to be electrically deposited.

As a final step, anodized parts must be sealed to ensure corrosion resistance of the anodic coating. Sealing involves plugging the anodic pores completely so contaminants cannot reach the base material. A variety of sealing methods are used by anodizers (see SEALANTS).

Environmental Issues

There are three aspects of conversion coating application which impact the environment. First, high operating temperatures for cleaners and phosphating baths make these processes energy intensive.

Secondly, because conversion coatings rely on using such metals as zinc, nickel, and chromium, the possibility of discharge of these metals is likely. Some manufacturing locations already fall under legislation prohibiting the use of these metals. Substitutes for Ni-containing phosphates have been developed, but zinc is still part of these systems. Chromium has been eliminated from post-treatment solutions and chromium-containing coatings on aluminum and zinc are being replaced with nonchromium coatings for some applications without any sacrifice of performance.

Lastly, sludge generation is an expected by-product of the conversion coating reaction. The disposal of all sludges is expected to become problematic in the future. General restrictions on the disposal of sludge generated when aluminum is treated in a conversion coating system have been in place since the early 1980s. No economically feasible way to recycle conversion-coating sludge is available, although this problem occupies research efforts of companies worldwide.

MICHAEL PETSCHEL
ROBERT HART
Parker Amchem

W. Rausch, *Die Phosphatierung von Metallen,* Eugene G. Lenz, Verlag, Germany, 1974.

G. Lorin, *Phosphating of Metals,* Finishing Publications Ltd., Middlesex, U.K., 1974.

K. Woods and S. Spring, *Metal Finish.* (Sept. 1980).

"Preparing Steel for Organic Coatings", *Product Finishing Directory,* Gardiner Publications, Inc., Cincinnati, Ohio, 1993.

PICKLING

Pickling is a term used to describe metal-cleaning operations designed to remove oxides from metal surfaces. These oxide films may be the result of in-process operations or simply environmental corrosion. Among the terms commonly used to describe these oxide layers are scale, rust, smut, white corrosion, and black or blue oxide. Although in some cases the oxide films may be removed using alkaline solutions of various compositions, pickling solutions are predominantly acidic, and most often very strongly acidic.

Whereas no single pickle formula is generally effective on all metal alloys, sulfuric acid is generally the most versatile of all the acids. The mechanism by which the acid solutions attack and remove the oxide film varies with the metal, the metal oxides to be removed, and the acid used. In some cases oxides are removed by the acid penetrating film imperfections. Acids dissolve the base metal as well as the oxide layer. Pickle inhibitors minimize or prevent the acid from attacking the base metal, yet allow effective removal of the oxides.

The speed of the pickle reaction is dependent on the concentration and temperature of the pickle, the degree of agitation of either the metal part or the pickle solution, the alloy being pickled, and the acid used. Pickling solutions may be applied by either spray or immersion techniques. Because of the noxious fumes emitted, there must be adequate ventilation.

Alkaline Deoxidizers

In certain applications, and particularly when hydrogen embrittlement caused by acid etching must be prevented, highly caustic alkaline solutions together with complexing agents, eg, gluconates, citrates, and EDTA, can be used to derust or remove light scale on steel alloys.

Economic Aspects

The widespread use of chemical etchants to remove oxides and certain other materials from metal surfaces stems primarily from the very low cost of the pickling chemicals.

Health and Safety

Acids such as sulfuric, hydrochloric, nitric, and especially hydrofluoric as well as strong alkalies such as caustic soda and caustic potash are extremely corrosive to animal and vegetable tissue. Extreme caution must be taken to prevent skin contact, inhalation, or ingestion. Violent reactions may occur when dissolving or diluting many of these chemicals with water.

The Material Safety Data Sheet (MSDS) for each chemical or proprietary blend used in a process must be thoroughly read and understood before the process is put into practice.

KENNETH HACIAS
Henkel Corporation

Metals Handbook, 9th ed., Vol. 5, ASM International, Materials Park, Ohio.

K. Hacias, *Wire J. Int.,* 86–89 (Jan. 1995).

M. Murphy, ed., *Metals Finishing, Guidebook and Directory,* Vol. 90, Metal Finishing, Hackensack, N.J., 1992.

METAL TREATMENTS

Operations performed on metals previously consolidated by processes such as melting and casting are referred to as metal treatments. Most of these treatments are mechanical and/or thermal. Mechanical treatments involve changes in shape by forming or machining. Metal treatments such as joining and coating of metals are not discussed herein (see METALLIC COATINGS; WELDING).

Mechanical Forming

Forming processes and techniques that are available for a particular alloy depend on its workability, which is the ability to be plastically deformed.

Workability Testing. Workability tests measure the amount of deformation that can be tolerated without fracture, or the development of an instability such as buckling or necking. In most workability tests a specimen is deformed to failure at a constant load rate or strain rate in tension, compression, torsion, shear, or bending. The most common technique is a tensile test at a constant strain rate where the load and elongation are measured continuously.

Plastic Deformation. In plastic deformation, crystallographic planes slip past each other. Slip is facilitated by the unique atomic structure of metals, which consists of an electron cloud surrounding positive nuclei. This structure permits shifting of atomic position without separation of atomic planes and resultant fracture. The stress required to slip an atomic plane past an adjacent plane is extremely high if the entire plane moves at the same time. Therefore, the plane moves locally, which gives rise to line defects called dislocations. These dislocations explain strain hardening and many other phenomena.

Hot Working. Plastic deformation at temperatures sufficiently high that strain hardening does not result is termed hot working. The temperature range for successful hot working depends on composition and other factors such as grain size, previous cold working, reduction, and strain rate. The lack of strain-hardening results from sufficient thermal energy for recrystallization, which refers to the formation of new grains. Hot working permits forming of relatively brittle materials that cannot readily be cold worked. Other advantages are grain refinement, reduction of segregation, healing of defects, such as porosity, and dispersion of inclusions.

Cold Working. Cold working involves plastic deformation well below the recrystallization temperature. Required stresses for cold working are greater than for hot working and the amount of strain without heat treatment is limited. Advantages are close dimension control, good surface finish, and increased low temperature strength because of strain hardening. Grain refinement can be achieved by annealing, which entails heating after cold working to temperatures at which recrystallization occurs. The effect of cold working on tensile properties and grain structure and subsequent annealing are shown in Figure 1.

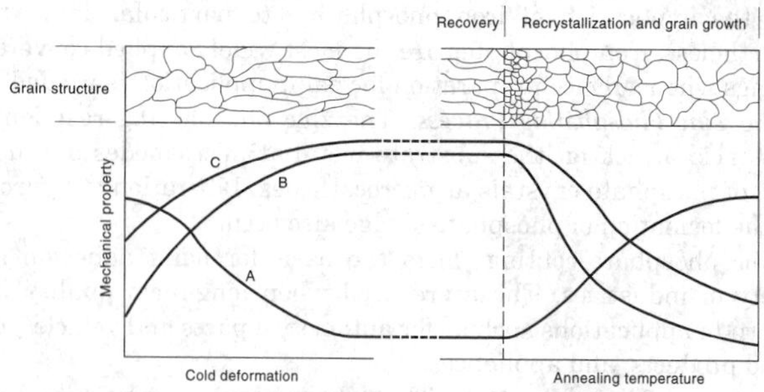

Figure 1. Variation of tensile properties and grain structure with cold working and annealing: A, elongation; B, yield stress; and C, ultimate tensile stress.

Primary Forming Processes. Primary forming operations are usually hot-working operations directed toward converting cast ingots into wrought blooms, billets, bars, or slabs. The large grains typical of cast structures are refined, porosity is reduced, segregation is reduced, inclusions are more favorably distributed, and a shape desirable for subsequent operations is produced. The principal operations used for ingot breakdown are forging, extruding, and rolling. Extrusion differs from forging and rolling in that more deformation occurs in one pass. Forging and rolling include many passes and some reheating. In addition, intermediate conditioning is sometimes necessary.

Secondary Forming Processes. The objective of secondary forming processes, either cold- or hot-working, is to form a shape. Such processes include rolling, open- and closed-die forging, upset forging, extruding, roll forging, ring rolling, deep drawing, spinning, bending, stretching, stamping, drawing, and high velocity forming.

Thermal Treatments

Annealing. In annealing, a cold-worked material is heated to soften it and improve its ductility. The three stages of annealing are recovery, recrystallization, and grain growth (see Fig. 1). Recovery occurs at relatively low temperature and may result in some softening caused mainly by the arrangement of dislocations into a more favorable distribution. Recrystallization is the formation of new grains with a relatively low dislocation density and little internal strain, which replaces strained grains with high dislocation densities. At increasing temperature, the newly formed grains exhibit grain growth. Prolonged exposure at a given temperature also tends to promote grain growth.

Precipitation Hardening. In precipitation hardening, also called age hardening, fine particles are precipitated from a supersaturated solid solution. These particles impede the movement of dislocations, thereby making the alloy stronger and less ductile. In order for an alloy to exhibit precipitation hardening, it must exhibit partial solid solubility and decreasing solid solubility with decreasing temperatures.

An example of the many alloy systems satisfying these requirements is the aluminum—copper system. At about 500–600°C, an alloy containing 4.5% Cu consists only of alpha, a solid solution of Cu in Al. Below 500°C, the phase θ ($CuAl_2$) exists in addition to alpha. The objective of precipitation hardening is to distribute the second phase as fine particles which are effective in blocking dislocation motion.

Precipitation hardening consists of solutioning, quenching, and aging. Solutioning entails heating above the solvus temperature in order to form a homogeneous solid solution and quenching to room temperature retains a maximum amount of alloying element (Cu) in solid solution. The cooling rate required varies considerably with different alloys. In aging, the alloy is heated below the solvus to permit precipitation of fine particles of a second phase θ ($CuAl_2$). The solvus represents the boundary on a phase diagram between the solid-solution region and a region consisting of a second phase in addition to the solid solution.

Heat Treatment of Steel. Steels are alloys having up to about 2% carbon in iron plus other alloying elements. The vast application of steels is mainly owing to their ability to be heat treated to produce a wide spectrum of properties. This occurs because of a crystallographic or allotropic transformation which takes place upon quenching. This transformation and its role in heat treatment can be explained by the crystal structure of iron and by the appropriate phase diagram for steels (see STEEL).

Machining

The term machinability is used to indicate the ease or difficulty with which a material can be machined to the size, shape, and desired surface finish. Machining parameters that affect machinability include feed, speed, depth of cut, cutting fluid, cutting-tool material, and cutting-tool geometry (see TOOL MATERIALS). The relative ease of machining materials also depends on the particular machining operation and corresponding material removal characteristics. Machinability index and machinability rating are used as qualitative measures of machinability under specified conditions. Machinability ratings have been based on one or more of the following criteria: tool life, cutting speed, and power consumption.

Physical characteristics of metals have a significant impact on machinability. These include microstructural features such as grain size, mechanical properties such as tensile properties, and physical properties such as thermal conductivity.

Surface Treatments

In some metal-forming operations such as rod and wire drawing, various surface treatments are applied to the workpiece. These include descaling, cleaning the application of lubricant carriers, and the use of lubricants (see LUBRICATION AND LUBRICANTS). Descaling can be mechanical or chemical (pickling). Lubricant carriers are applied by dipping the workpiece in hot solution or slurries such as lime or phosphate coatings. (see also METAL SURFACE TREATMENTS).

LAURENCE A. JACKMAN
CARLOS N. RUIZ
Teledyne Allvac

L. E. Doyle and co-workers, *Manufacturing Processes and Materials for Engineers,* Prentice-Hall, Inc., Englewood Cliffs, N.J., 1985.

A. G. Guy, *Introduction to Materials Science,* McGraw-Hill Book Co., Inc., New York, 1972.

W. J. Patton, *Materials in Industry,* 3rd ed., Prentice-Hall, Inc., Englewood Cliffs, N.J., 1986.

K. Lang, ed., *Handbook of Metal Forming,* McGraw-Hill Book Co., Inc., New York 1985, pp. 26.18–26.23.

METAL VAPOR SYNTHESIS. See LIGHT GENERATION; THIN FILMS

METHACROLEIN. See ACROLEIN AND DERIVATIVES.

METHACRYLIC ACID AND DERIVATIVES

Methacrylic acid (MAA) was first prepared in 1865 by the hydrolysis of ethyl methacrylate, which was in turn obtained by dehydrating ethyl α-hydroxyisobutyrate. The polymerizability of methacrylic acid was first noted in 1880 when a white powder was obtained in a distillation of methacrylic acid. The acetone cyanohydrin process for the synthesis of methyl methacrylate via the formation of methacrylamide sulfate, which was patented in 1934, still forms the basis for the bulk of the methyl methacrylate (MMA) currently produced.

Table 1. Selected Properties of Methacrylates

Compound	Mol wt	Mp, °C	Viscosity, mPa(= cP)	Flash point, °C	Autoignition temperature, °C
methyl[a]	100.11	−48	0.53	9[b]	435
ethyl[a]	114.14	−17	0.92	16[c]	393
butyl[a]	142.19	−50	0.92	49[d]	294
lauryl[e]	262	−22		110	277
2-dimethyl-aminoethyl[f]	157.2	−30	1.1	75	
2-hydroxyethyl[g]	130.14	−12		66	
2-hydroxypropyl[g]	144.17	−89	7.1	98	
glycidyl	142.1	<−60	5	76	

[a] Heat of polymerization = 57.5 kJ/mol (13.7 kcal/mol). [b] Lower explosion limit (LEL) = 2.1%; upper explosion limit (UEL) = 12.5%z. [c] LEL = 1.8%; UEL to saturation. [d] LEL = 2.0%; UEL = 8%. [e] Made from a mixture of higher alcohols, predominantly C-12. [f] pK_a = 8.4. [g] Heat of polymerization =~ 50 kJ/mol (12 kcal/mol).

Physical Properties

Selected physical properties of various methacrylates are given in Table 1.

Reactions

Methacrylic acid and its ester derivatives are α,β-unsaturated carbonyl compounds and exhibit the reactivity typical of this class of compounds, ie, Michael and Michael-type conjugate addition reactions and a variety of cycloaddition and related reactions. Although less reactive than the corresponding acrylates as the result of the electron-donating effect and the steric hindrance of the α-methyl group, methacrylates readily undergo a wide variety of reactions and are valuable intermediates in many synthetic procedures.

Polymerization

The vast majority of commercial applications of methacrylic acid and its esters stem from their facile free-radical polymerizability (see INITIATORS, FREE-RADICAL INITIATORS). Solution, suspension, emulsion, and bulk polymerizations have been used to advantage. Although of much less commercial importance, anionic polymerizations of methacrylates have also been extensively studied. Strictly anhydrous reaction conditions at low temperatures are required to yield high molecular weight polymers in anionic polymerization. Side reactions of the propagating anion at the ester carbonyl are difficult to avoid and lead to polymer branching and inactivation. Polymerization of methacrylates is also possible via what is known as group-transfer polymerization.

Higher Alkyl and Functional Methacrylates

Most large-scale industrial methacrylate processes are designed to produce methyl methacrylate or methacrylic acid. In some instances, simple alkyl alcohols, eg, ethanol, butanol, and isobutyl alcohol, may be substituted for methanol to yield the higher alkyl methacrylates. In practice, these higher alkyl methacrylates are usually prepared from methacrylic acid by direct esterification or transesterification of methyl methacrylate with the desired alcohol.

Hydroxy functional methacrylates are accessible by the reaction of methacrylic acid and ethylene oxide or propylene oxide in the presence of chromium, iron, or ion-exchange catalysts.

Manufacture and Processing

The basic feedstock for the manufacture of methyl methacrylate and methacrylic acid is, ultimately, natural gas or crude oil. It is convenient to categorize the various manufacturing routes in terms of the specific hydrocarbon raw material used. Propylene (C-3) routes require the addition of one carbon atom. Ethylene (C-2)-based routes require the addition of two carbon atoms to create the four-carbon methacrylate backbone. The commercial viability of a process is determined by the aggregate of raw material cost and utilization (process yield), operating costs (energy), waste disposal costs, environmental impact, and plant capital investment.

Research is currently directed toward development of novel technologies that may present economic advantages with respect to the conventional acetone cyanohydrin (ACH) route. Mitsubishi Gas Chemical Co. has developed and patented a modified acetone cyanohydrin-based route that does not use sulfuric acid and therefore presents the opportunity for reduced waste costs. A novel C-3 route based on the palladium-catalyzed carbonylation of methylacetylene has been developed by Shell Oil Co. There have been significant improvements in catalysts and resulting yields for key transformations in many routes since the 1980s.

MMA from Acetone Cyanohydrin (ACH). The process for conversion of acetone cyanohydrin–H_2SO_4 to methyl methacrylate through methacrylamide sulfate has been practiced commercially since 1937 and is based on technology patented by ICI in 1934. Acetone cyanohydrin is prepared via base-catalyzed reaction of acetone and hydrogen cyanide. Acetone and hydrogen cyanide are obtained as by-products from the commercial production of phenol and acrylonitrile, respectively. Hydrogen cyanide is also manufactured directly by catalytic ammoxidation of methane. Sulfuric acid is used in excess and serves as both reactant and solvent in the reaction with acetone cyanohydrin to form methacrylamide sulfate through an α-sulfatoamide intermediate.

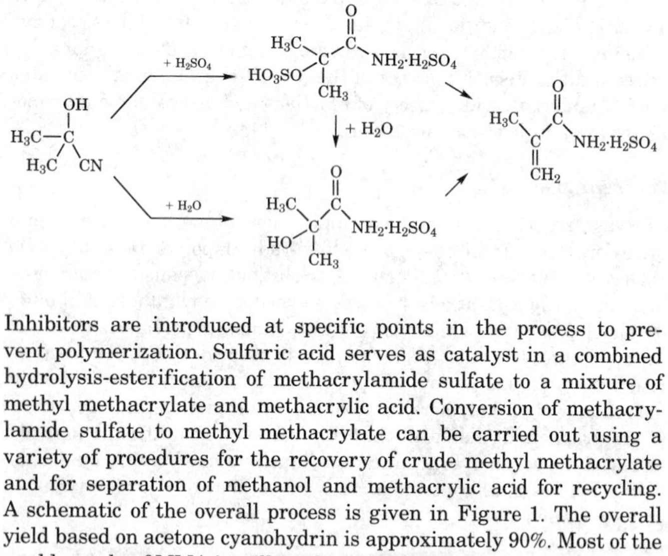

Inhibitors are introduced at specific points in the process to prevent polymerization. Sulfuric acid serves as catalyst in a combined hydrolysis-esterification of methacrylamide sulfate to a mixture of methyl methacrylate and methacrylic acid. Conversion of methacrylamide sulfate to methyl methacrylate can be carried out using a variety of procedures for the recovery of crude methyl methacrylate and for separation of methanol and methacrylic acid for recycling. A schematic of the overall process is given in Figure 1. The overall yield based on acetone cyanohydrin is approximately 90%. Most of the world supply of MMA is still produced by this process.

Ethylene-Based (C-2) Routes. MMA and MAA can be produced from ethylene as a feedstock via propanol, propionic acid, or methyl propionate as intermediates. Propanal may be prepared by hydroformylation of ethylene over cobalt or rhodium catalysts. The propanal then reacts in the liquid phase with formaldehyde in the presence of a secondary amine and, optionally, a carboxylic acid. The reaction presumably proceeds via a Mannich base intermediate which is cracked to yield methacrolein. Alternatively, a gas-phase, crossed aldol reaction with formaldehyde catalyzed by molecular sieves (qv) may be used to form methacrolein. The methacrolein is then oxidized to methacrylic acid.

Isobutylene-Based (C-4) Routes. Isobutylene or *tert*-butyl alcohol can be converted to methacrylic acid in a two-stage, gas-phase oxidation process via methacrolein as an intermediate. The alcohol and isobutylene may be used interchangeably in the processes since *tert*-butyl alcohol readily dehydrates to yield isobutylene under the reaction conditions in the initial oxidation. Variations of this process have been commercialized.

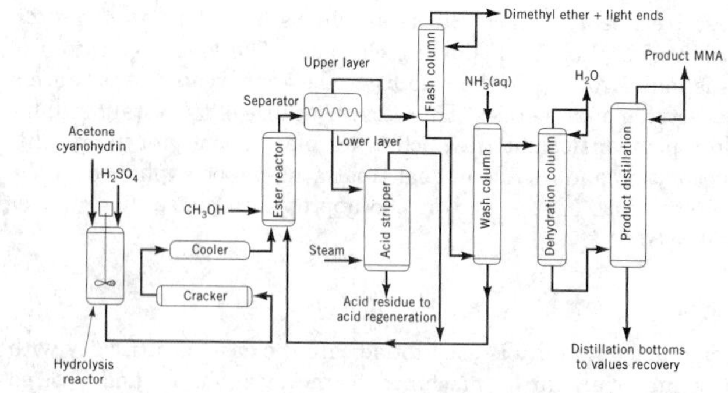

Figure 1. MMA from acetone cyanohydrin via methacrylamide sulfate.

Methacrylonitrile Process. MAA and MMA may also be prepared via the ammoxidation of isobutylene to give methacrylonitrile as the key intermediate. A mixture of isobutylene, ammonia, and air are passed over a complex mixed metal oxide catalyst at elevated temperatures to give a 70–80% yield of methacrylonitrile. Relatively modest yields are obtained in the ammoxidation reaction and the generation of a considerable acid waste stream combine to make this process economically less desirable than the ACH or C-4 oxidation to methacrolein processes.

Uses

Methacrylic acid and methacrylic esters are used in a wide variety of polymers with a broad spectrum of applications. Poly(methacrylic acid) or its neutralized salts are used as additives for detergent builders and rheology modifiers. Methacrylic esters, the most important of which is methyl methacrylate, yield hard, tough polymers in contrast to the softer acrylates. Copolymerization of methacrylic esters with acrylates allows the preparation of hard but flexible polymers for use in paints, polishes, and many other coatings. Methacrylate polymers are prized for their clarity, colorability, color compatibility, weatherability, and ultraviolet light stability, which allows them to be used for both indoor and outdoor applications. The principal end uses of methacrylates are in acrylic sheet and molding resins which find commercial application in signs, displays, glazing compounds, lighting fixtures, building panels, automotive components, plumbing fixtures, and appliances. They are also used as impact modifiers in poly(vinyl chloride) (PVC) siding, film, sheet, and plastic bottle manufacture.

Storage and Handling

Polymerizations of methacrylic acid and derivatives are very energetic (MAA, 66.1 kJ/mol; MMA, 57.5 kJ/mol = 13.7 kcal/mol). The potential for the rapid evolution of heat and generation of pressure presents an explosion hazard if the materials are stored in closed or poorly vented containers. To prolong usable shelf-life, commercially available methacrylic monomers are inhibited with the methyl ether of hydroquinone (MEHQ). Other commonly used inhibitors are alkylphenols and hydroquinone. Once inhibited, methacrylic acid or its esters may be handled as flammable materials. To avoid photoinitiation of polymerization, all methacrylates should be stored with minimal exposure to light.

Most unwanted polymerization events of methacrylic monomers occur because of overheating, leading to inhibitor depletion or oxygen depletion, and in turn to inhibitor inactivation. Care should be taken to avoid stagnant areas in transfer lines or pump heads where polymerization may begin and which may then act to seed polymerization of the bulk material.

The relatively high freezing point of methacrylic acid (15°C) is a problem because the inhibitor tends to partition into the liquid phase upon freezing. Thawing of the material tends to create localized pools of uninhibited methacrylic acid which are extremely susceptible to polymerization. Care should be taken to limit thawing temperatures to less than 40°C and to ensure good mixing of the thawed material. For most polymer applications the removal of the inhibitors from the monomer is unnecessary.

Health and Safety Factors

Methacrylates are slightly to severely irritating to skin and eyes and are considered potential skin sensitizers. Several lifetime exposure studies by a variety of routes (oral capsule, drinking water, inhalation) in rodents have shown it to be noncarcinogenic. Methyl methacrylate also is nonteratogenic (did not produce birth defects) after inhalation exposures. Headaches, vomiting, and drowsiness are symptomatic of overexposure to vapors. In the workplace, the pungent odor and irritant nature of the methacrylate monomers serves as a warning property and tends to keep exposures low.

Appropriate protective clothing and equipment should be worn to minimize exposure to methacrylate liquids and vapors. The working area should be adequately ventilated to limit vapors.

ANDREW W. GROSS
JOHN C. DOBSON
Rohm and Haas Company

R. H. Yocum and E. B. Nyquist, eds., *Functional Monomers*, Marcel Dekker Inc., New York, 1973.

M. Salkind, E. H. Riddle, and R. W. Keefer, *Ind. Eng. Chem.* **51**, 1232 (1959).

METHACRYLIC POLYMERS

The nature of the R group in methacrylic acid ester monomers having the generic formula $CH_2{=}C\ (CH_3)COOR$ generally determines the properties of the corresponding polymers. Methacrylates differ from acrylates in that the α-hydrogen of the acrylate is replaced by a methyl group (see ACRYLIC ESTER POLYMERS). This methyl group imparts stability, hardness, and stiffness to methacrylic polymers. The methacrylate monomers are extremely versatile building blocks. They are moderate-to-high boiling liquids that readily polymerize or copolymerize with a variety of other monomers. All of the methacrylates copolymerize with each other and with the acrylate monomers to form polymers having a wide range of hardness; thus polymers that are designed to fit specific application requirements can be tailored readily.

The uniqueness of methyl methacrylate as a plastic component accounts for its industrial use, and it far exceeds the combined volume of all of the other methacrylates. In addition to plastics, the various methacrylate polymers also find application in sizable markets as diverse as lubricating oil additives, surface coatings (qv), impregnates, adhesives (qv), binders, sealers (see SEALANTS), and floor polishes.

Physical Properties

The nature of the alkyl group from the esterifying alcohol, the molecular weight, and the tacticity determine the physical and chemical properties of methacrylate ester polymers. The physical properties of amorphous methacrylic polymers evidence a principal change in the glass-transition region. Chemical reactivity, mechanical and dielectric relaxation, viscous flow, load bearing capacity, hardness (qv), tack, heat capacity, refractive index, thermal expansivity, creep, and diffusion differ markedly below and above the transition region.

The properties of methacrylic polymers are also affected by molecular weight. Typically, mechanical properties increase as the molecular weight increases. However, beyonds some critical value, about 100,000, the increase is slight and levels off asymptotically.

Mechanical Properties. Methacrylates are harder polymers of higher tensile strength and lower elongation than their acrylate counterparts because substitution of the methyl group for the α-hydrogen on the main chain restricts the freedom of rotation and motion of the polymer backbone.

At room temperature, the first member of the linear aliphatic methacrylate series, poly(methyl methacrylate), is a hard, fairly rigid material which can be sawed, carved, or worked on a lathe. When heated above its T_g, on poly(methyl methacrylate) is a tough, pliable, extensible material that is easily bent or formed into complex shapes, and can be molded or extruded.

Optical Properties. Poly(methyl methacrylate) transmits light in the range of 360–1000 nm almost perfectly (92% compared to the theoretical 92.3%). The wavelength of visible light falls approximately between 400 and 700 nm. At a thickness of 2.54 cm or less, poly(methyl methacrylate) absorbs virtually no visible light. Beyond 2800 nm, essentially all infrared radiation is absorbed. Commercial grades of poly(methyl methacrylate) often contain uv radiation absorbers that block light in the 290–350 nm range. The absorber thus screens the user from sunburn and protects the polymer against long-term degradation from light. Poly(methyl methacrylate)'s transparency to x-rays and radiation has been found to be about the same as that of human flesh or water. Sheets of poly(methyl methacrylate) are opaque to alpha particles, and for thicknesses above 6.35 mm (0.250 in.), the polymer is essentially opaque to beta radiation; poly(methyl methacrylate) is used as a transparent neutron stopper. Most formulations of colorless sheet have high transmittance to standard broadcast and television waves as well as to most radar bands.

Many items, such as magnifiers, reducers, camera lenses, prisms, and especially complex reflex lenses widely used in automotive taillights, are made from poly(methyl methacrylate).

Electrical Properties. The surface resistivity of poly(methyl methacrylate) is higher than that of most plastic materials. Weathering and moisture affect poly(methyl methacrylate) only to a minor degree. High resistance and nontracking characteristics have resulted in its use in high voltage applications, and its excellent weather resistance has promoted the use of poly(methyl methacrylates) for outdoor electrical applications.

Chemical Properties

Methacrylate polymers have a greater resistance to both acidic and alkaline hydrolysis than do acrylate polymers; both are far more stable than poly(vinyl acetate) and vinyl acetate copolymers. There is a marked difference in the chemical reactivity among the noncrystallizable and crystallizable forms of poly(methyl methacrylate) relative to alkaline and acidic hydrolysis. Conventional (ie, free-radical), bulk-polymerized, and syndiotactic polymers hydrolyze relatively slowly compared with the isotactic type. Polymer configuration is unchanged by hydrolysis.

The chemical resistance of poly(methyl methacrylate) may be summarized as follows. PMMA is not affected by most inorganic solutions, mineral oils, animal oils, low concentrations of alcohols; paraffins, olefins, amines, alkyl monohalides; and aliphatic hydrocarbons and higher esters, ie, >10 carbon atoms. However, PMMA is attacked by lower esters, eg, ethyl acetate, isopropyl acetate; aromatic hydrocarbons, eg, benzene, toluene, xylene; phenols, eg, cresol, carbolic acid; aryl halides, eg, chlorobenzene, bromobenzene; aliphatic acids, eg, butyric acid, acetic acid; alkyl polyhalides, eg, ethylene dichloride, methylene chloride; high concentrations of alcohols, eg, methanol, ethanol; 2-propanol; and high concentrations of alkalies and oxidizing agents.

The chemical resistance and excellent light stability of poly(methyl methacrylate) compared to two other transparent plastics is illustrated in Table 1.

Manufacture and Processing

Free-radical polymerization processes are used to produce virtually all commercial methacrylic polymers. Usually free-radical initiators (qv) such as azo compounds or peroxides are used to initiate the polymerizations. Photochemical and radiation-initiated polymerizations are

Table 1. Relative Outdoor Stability of Poly(methyl methacrylate)

Material	Light transmittance, %	After exposure,[a] %	Haze, %	After exposure,[a] %
poly(methyl methacrylate)	92	92	1	2
polycarbonate	85	82	3	19
cellulose acetate butyrate	89	68	3	70

[a] Three-yr outdoor.

also well known. At a constant temperature, the initial rate of the bulk or solution radical polymerization of methacrylic monomers is first-order with respect to monomer concentration, and one-half order with respect to the initiator concentration. Methacrylate polymerizations are markedly inhibited by oxygen; therefore considerable care is taken to exclude air during the polymerization stages of manufacturing.

A substantial fraction of commercially prepared methacrylic polymers are copolymers. Monomeric acrylic or methacrylic esters are often copolymerized with one another and possibly several other monomers. Copolymerization greatly increases the range of available polymer properties. The all-acrylic polymers tend to be soft and tacky; the all-methacrylic polymers tend to be hard and brittle. By judicious adjustment of the amount of each type of monomer, polymers can be prepared at essentially any desired hardness or flexibility. Small amounts of specially functionalized monomers are often copolymerized with methacrylic monomers to modify or improve the properties of the polymer directly or by providing sites for further reactions.

Bulk Polymerization. This is the method of choice for the manufacture of poly(methyl methacrylate) sheets, rods, and tubes, and molding and extrusion compounds. Three bulk polymerization processes are commercially important for the production of methacrylate polymers: batch cell casting, continuous casting, and continuous bulk polymerization. Approximately half the worldwide production of bulk polymerized methacrylates is in the form of molding and extrusion compounds, a quarter is in the form of cell cast sheets, and a quarter is in the form of continuous cast sheets.

Solution Polymerization. The solution polymerization of methacrylic monomers to form solution polymers or copolymers is an important commercial process for the preparation of polymers for use as coatings, adhesives, impregnates, and laminates. Typically the polymerization is done batchwise by adding monomer to an organic solvent in the presence of a soluble peroxide or azo initiator.

Emulsion Polymerization. The principal markets for aqueous dispersion polymers made by emulsion polymerization of methacrylic esters are the paint (qv), paper (qv), textile, floor polish, and leather (qv) industries where they are used principally as coatings or binders. Copolymers of methyl methacrylate with either ethyl acrylate or butyl acrylate are most common.

Suspension Polymerization. This method yields polymethacrylates in the form of tiny beads, which are primarily used as molding powders and ion-exchange resins. Most suspension polymers prepared as molding powders are poly(methyl methacrylate); copolymers containing up to 20% acrylate for reduced brittleness and improved processibility are also common. Suspension polymers of poly(methyl methacrylate) copolymerized with an amino or acid functional monomer, and with a di- or trivinyl monomer for cross-linking, are useful as ion-exchange resins.

Graft Polymerization. Graft copolymers are prepared by attaching one polymer as a branch to the chain of another polymer of different composition. This is usually accomplished by generating radical sites on the first polymer onto which monomer of the second polymer is grafted. The grafting may be accomplished in bulk, solution, or dispersion systems. The presence of distinct, but chemically bonded segments of two polymers often confers interesting and useful properties. Commercially, the most important methacrylate graft copolymers are the MABS and MBS polymers.

The MABS copolymers are prepared by dissolving or dispersing polybutadiene rubber in a methyl methacrylate—acrylonitrile—styrene monomer mixture. MBS polymers are prepared by grafting methyl methacrylate and styrene onto a styrene—butadiene rubber in an emulsion process. The product is a two-phase polymer useful as an impact modifier for rigid poly(vinyl chloride).

Ionic Polymerization. The anionic polymerization of methacrylic monomers to stereoregular or block copolymers is well known. These polymerizations are conducted in organic solvents, primarily using organometallic compounds as initiators. This technology is of minor commercial significance, but is of interest for the preparation of polymers of narrow molecular weight distribution and controlled molecular architecture. Methacrylate monomers do not generally polymerize by a cationic mechanism.

Health and Safety Factors

In general, methacrylate polymers are considered nontoxic. Various methacrylate polymers are used in food packaging (qv) and handling, in dentures and dental fillings (see DENTAL MATERIALS), and as medicine dispensers and contact lenses. However, care must be exercised because additives or residual monomers present in various types of polymers can display toxicity.

During manufacture, considerable care is exercised to reduce the potential for violent polymerizations, and to reduce exposure to flammable and potentially toxic monomers and solvents.

Dust explosions ignited by static discharge are a recognized hazard encountered in the handling of poly(methyl methacrylate) powders or in the fabrication of poly(methyl methacrylate) plastic sheet. Methacrylic solution polymers are treated as flammable mixtures; latex polymers are nonflammable.

Uses

The principal U.S. market for methacrylate resins is for glazing and skylights. Other significant markets include consumer products, transportation signs and lighting fixtures, plumbing (spas, tubs, showers, sinks, etc), and panels and siding. The resins are also used in medicine, optics, and as oil additives.

RONALD W. NOVAK
PATRICIA M. LESKO
Rohm and Haas Company

J. Brandrup and E. H. Immergut, *Polymer Handbook,* 3rd ed., Wiley-Interscience, New York, 1989.

M. Harrington, in I. I. Rubin, ed., *Handbook of Plastic Materials and Technology,* John Wiley & Sons, Inc., New York, 1990.

P. M. Lesko and P. R. Sperry, in P. A. Lovell and M. S. El-Aasser, eds., *Emulsion Polymerization and Emulsion Polymers,* John Wiley & Sons, Ltd., Chichester, U.K., 1997.

METHANOL

Methanol (methyl alcohol), CH_3OH, is a colorless liquid at ambient temperatures with a mild, characteristic alcohol odor. Originally called wood alcohol since it was obtained from the destructive distillation of wood, today commercial methanol is sometimes referred to as synthetic methanol because it is produced from synthesis gas, a mixture of hydrogen and carbon oxides, generated by a variety of sources.

Methanol has been used as a solvent and as a feedstock for bulk organic chemicals (primarily formaldehyde), with modest growth potential. However, after 1990, demand for methanol as a feedstock for methyl *tert*-butyl ether (MTBE) accelerated when the latter became a significant oxygenated component in motor fuels.

Table 1. Physical Properties of Methanol

Property	Value
boiling point, °C	64.70
heat of formation (liquid) at 25°C, kJ/mol[a]	−239.03
heat of fusion, J/g[a]	103
heat of vaporization at boiling point, J/g[a]	1129
heat of combustion (gross) at 25°C, J/g[a]	22,662
flammable limits in air, vol %	
lower	6.0
upper	36
autoignition temperature, °C	464
flash point, closed cup, °C	11
surface tension at 25°C, mN/m(= dyn/cm)	22.1
specific heat of vapor at 25°C, J/(g·K)[a]	1.370
specific heat of liquid at 25°C, J/(g·K)[a]	2.533
vapor pressure at 25°C, kPa[b]	16.96
solubility in water	miscible
density at 25°C, g/mL	0.7866
liquid viscosity at 25°C, mPa·s(= cP)	0.541
thermal conductivity at 25°C, W/(m·K)	0.202

[a] To convert J to cal, divide by 4.184. [b] To convert kPa to mm Hg, multiply by 7.5.

Physical Properties

Important physical properties of methanol are given in Table 1.

Chemical Reactions

Methanol undergoes reactions that are typical of alcohols as a chemical class. Dehydrogenation and oxidative dehydrogenation to formaldehyde over silver or molybdenum oxide catalysts are of particular industrial importance.

Manufacture and Processing

Synthetic methanol production first utilized a zinc–chromium oxide catalyst. The activity of this catalyst required that it be operated at 25–35 MPa (250–350 atm) and 320–450°C. This high pressure process suffered from high capital and compression energy costs, compounded by poor catalyst selectivity.

The high pressure process was rendered obsolete in the mid-1960s when a more active copper-zinc-alumina catalyst that could operate at 5–10 MPa (50–100 atm) and 210–270°C with higher selectivity and stability was developed.

The energy consumption (lower heating value of the feedstock plus fuel) of the low pressure process has successively improved from over 38.3 GJ/t when it was first introduced to 29.0–30.3 GJ/t by the mid-1990s. Natural gas-based reforming plants have advanced to the point where the scope for still further gains in efficiency is small and the gains costly to obtain.

The synthesis reactions are as follows.

$$CO + 2\ H_2 \leftrightarrows CH_3OH \quad (1)$$

$$CO_2 + 3\ H_2 \leftrightarrows CH_3OH + H_2O \quad (2)$$

Subtracting reaction 2 from reaction 1 gives the familiar water gas shift reaction (eq. 3).

$$CO + H_2O \leftrightarrows CO_2 + H_2 \quad (3)$$

Synthesis Gas Generation Routes. Any hydrocarbon that can be converted into a synthesis gas by either reforming with steam (eq. 4) or gasification with oxygen (eq. 5) is a potential feedstock for methanol.

$$C_nH_m + nH_2O \leftrightarrows nCO + (n + m/2\ H_2) \quad (4)$$

$$C_nH_m + (n/2\ O_2 \rightarrow nCO + (m/2\ H_2)) \quad (5)$$

Steam reforming of natural gas accounts for at least 80% of the world's methanol capacity. A steam reformer is essentially a process furnace in which the endothermic heat of reaction is provided by firing across tubes filled with a nickel-based catalyst through which the reactants flow. Several mechanical variants are available (see AMMONIA).

Steam reforming of naphtha is very similar to natural gas reforming, except that naphtha must be vaporized prior to the desulfurization step. Presently there is no significant methanol production from this feedstock owing to the high relative cost of naphtha.

Combined reforming splits the total reforming duty between a conventional fired reformer and a downstream catalytic secondary reformer.

The primary reformer can be designed to operate at higher pressures and lower temperatures, since the residual methane will be further reduced to low levels by reaction with oxygen at the higher operating temperature of the secondary reformer, leading to a decrease in compression requirements and a reduction in the overall energy consumption of the plant. The combined reforming concept has also been employed to use high pressure primary reformers salvaged from old ammonia plants to provide synthesis gas for methanol plants.

Gas-heated reforming is an extension of the combined reforming concept where the primary reformer is replaced by a heat-transfer device in which heat for the primary reforming reaction is recovered from the secondary reformer effluent. Various mechanical designs have been proposed which are variants of a shell-and-tube heat exchanger.

Synthesis gas can be generated from coal and petroleum fractions by a variety of processes (see COAL CONVERSION PROCESSES, GASIFICATION; HYDROGEN). The high cost of coal handling and preparation and treatment of effluents, compounded by continuing low prices for crude oil and natural gas, has precluded significant exploitation of coal as a feedstock for methanol.

Methanol Synthesis. All commercial methanol processes employ a synthesis loop, and Figure 1 shows a typical example as part of the overall process flow sheet. This configuration overcomes equilibrium conversion limitations at typical catalyst operating conditions. A recycle system that gives high overall conversions is feasible because product methanol and water can be removed from the loop by condensation.

The makeup synthesis gas is compressed, mixed with recycled gas, and preheated against the converter effluent gas before entering the converter. The converter effluent is first used to heat saturator water or boiler feedwater before being returned to the loop interchanger and then on to a cooler which condenses the crude methanol–water mixture. Noncondensable gases are disengaged in a catchpot for recycle. A purge is taken from this recycle to remove excess hydrogen, methane, and other inerts. The crude methanol mixture is Purified in a conventional distillation train of one to three towers.

The feature that is most useful in distinguishing commercial methanol processes from one another is the type of reactor used. The four basic types in use are the quench, multiple adiabatic, tube-cooled, and steam-raising converters. There are a variety of proprietary reactor designs commercially available from licensors, all of which are either one of these four types or a combination of two among them.

Health and Safety

Methanol is not classified as carcinogenic, but can be acutely toxic if ingested; 100–250 mL may be fatal or result in blindness. The principal physiological effect is acidosis resulting from oxidation of methanol to formic acid. Methanol is a general irritant to the skin and mucous membranes. Methanol vapor can cause eye and respiratory tract irritation, nausea, headaches, and dizziness.

Methanol does not pose an undue toxicity hazard if handled in well-ventilated areas, and is rated as a slight health hazard by the National Fire Protection Association (NFPA).

Storage and Handling

Methanol is stable under normal storage conditions. Methanol is not subject to hazardous polymerization reactions, but can react violently

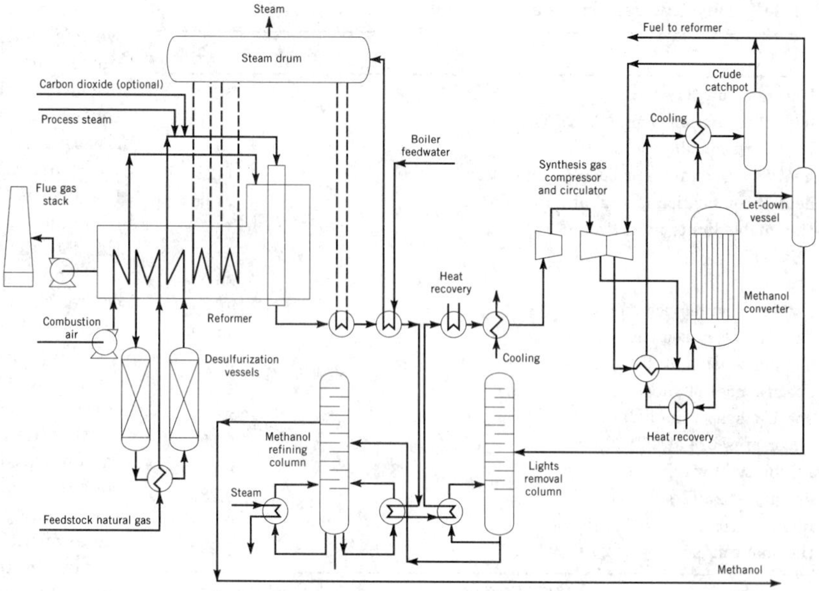

Figure 1. Methanol synthesis process flow sheet.

with strong oxidizing agents. The greatest hazard involved in handling methanol is the danger of fire or explosion. The NFPA classifies methanol as a serious fire hazard.

Normal precautions for storing and handling flammable liquids should be observed, such as diking and fire fighting provisions. Methanol is shipped overland by tank cars, trucks, and drums. The same safety and materials considerations apply to each of these types of containers, with appropriate labeling required by local and Federal authorities. Methanol may be off-loaded either by pumps or by pressurizing it with an inert gas, but never with air.

Personnel involved in the handling of methanol require eye and skin protection from the irritating properties of methanol in the event of a spill.

Uses

The principal use of methanol has traditionally been in the production of formaldehyde, where typically around 40% of the world methanol market is consumed. In the United States, an increasing role for methanol has been found in the oxygenated fuels market from the use of MTBE. Another significant use of methanol is in the production of acetic acid; other uses include the production of solvents and chemical intermediates.

The most recent uses for methanol can be found in the agricultural sector. Test studies are being carried out where methanol is sprayed directly onto crops to improve plant growth. Methanol can be used as a carbon source for the production of single-cell protein (SCP) for use as an animal feed supplement. Denitrification of wastewater in treatment plants offers another potential use for methanol. There are a few such plants in the world; however, this use is not expected to grow appreciably, as there are more proven methods for nitrogen removal commercially available.

ALAN ENGLISH
JERRY ROVNER
John Brown
SIMON DAVIES
Davy Process Technology

1993 World Methanol Conference, Atlanta, Ga.

Y. Kobayashi and H. Nakamura, "MRF Reactor, Commercially Proven Performance and Enhancement for Large Scale Methanol Plant," *AIChE 1993 Spring Meeting,* Houston, Tex.

S. Narula, "Economic Prospects for Methane Conversion Technologies", *AIChE 1989 Spring Meeting,* Houston, Tex.

E. Supp and A. T. Weschler, "Conversion of Ammonia Plants to Methanol Production using Lurgi's Combined Reforming Technology", *AIChE 1992 Spring Meeting,* New Orleans.

MICA

Mica is a generic term that applies to a wide range of hydrous aluminum silicate minerals characterized by sheet or plant-like structure, and possessing to varying degrees, depending on composition and weathering, flexibility, elasticity, hardness (qv), and the ability to be split into thin (1 μm) sheets. All micas form flat six-sided monochromic crystals, and possess cleavage parallel to the basal plane.

Mica exists in nature in a wide variety of compositions. Muscovite and phlogopite are the only natural micas of commercial importance. Vermiculite, although not considered a true mica by most mineralogists, is a micaceous mineral formed from the weathering of phlogopite or biotite and is also of commercial importance. Fluorophlogopite, $K_2Mg_6(Al_2Si_6O_{20})F_4$, is a synthetic mica made from pure chemical oxides.

Mica has been classified into three groups: *(1)* the mica group proper, *(2)* the clintonite or brittle micas group, and *(3)* the chlorite group. Supplementary to these are the vermiculites, which are hydrated compounds that result from the alteration of any one of the micas, but usually biotite. All minerals in these groups belong to the monoclinic crystal system, and all show plane angles of 60 and 120° on the basal section. The crystals usually form in hexagonal or rhombohedral-shaped scales, prisms, or plates. The basic structural unit of mica is a layer composed of two silicon tetrahedral sheets with a central octahedral sheet.

Muscovite is dioctahedral, having a theoretical composition of 11.8% K_2O, 45.2% SiO_2, 38.5% Al_2O_3, and 4.5% H_2O. Muscovite mica formed as a primary mineral in pegmatites and granodiorite differs in physical properties compared to muscovite mica formed by secondary alteration (mica schist). The main differences are in flexibility and ability to be delaminated. Primary muscovite is not as brittle and delaminates much easier than muscovite formed as a secondary mineral. Mineralogical properties of the principal natural micas are shown in Table 1.

Table 1. Mineralogical Properties of Micas[a]

Properties	Muscovite	Biotite	Lepidolite	Phlogopite
sp gr	2.76–3.0	2.7–3.1	2.8–3.3	2.8–2.9
luster	vitreous–pearly	splendent, sometimes submetallic	pearly	pearly
crystal	rhombic or hexagonal	pseudo-rhombohedral	hexagonal	hexagonal
colors	gray, brown, pale green, violet, yellow, dark olive-green, ruby	green, black, yellow	rose red, violet-gray, lilac, yellowish, grayish white, white	yellow-white, gray–green, pearly, brown, black

[a] Optical signs of micas are negative, crystal system is monoclinic, streak is colorless.

Mining

Flake Mica. Flake mica is mined from weathered and hard rock pegmatites, granodiorite, and schist and gneiss by conventional open-pit methods. In soft, residual material, dozers, shovels, scrapers, and front-end loaders are used to mine the ore. Often kaolin, quartz, and feldspar are recovered along with the mica (see also CLAYS; SILICON COMPOUNDS).

Hard rock mining of these ore bodies requires drilling and blasting with ammonium nitrate and dynamite. After blasting, the ore is reduced in size with a drop ball and then loaded on trucks for transportation to the processing plant. Mica, quartz, and feldspar concentrates are separated, recovered, and sold from the hard rock ore.

Sheet Mica. Pockets of mica crystals are found in pegmatite stills and dikes or grandonite ore bodies. Sheet mica is mined by both underground and open-pit mining procedures. Underground mining is accomplished by driving a shaft, formed with tungsten carbide-tipped air drills, hoists, and explosives (see EXPLOSIVES AND PROPELLANTS; TOOL MATERIALS). After blasting, the mica is placed in boxes or bags for transporting to the trimming shed where it is graded, split, and cut to various specified sizes for sale.

Sheet mica is no longer mined in the U.S. Most sheet mica is mined in India.

Beneficiation Processes

Flake or Scrap Mica. In the early to mid-1900s, flake or scrap mica was mainly processed by a jigging procedure which consists of hydraulically washing a pile of bulldozed ore across a series of roll crushers and Trommel screens gaped at different size openings.

The grade of mica produced by jigging is very poor, usually about 75% concentrate, and recovery of available mica low (50%). Specifications on mica have become more stringent, therefore a more efficient processing method has been devised that provides higher quality mica, as well as more efficient recovery.

Because of improved mica processing operations, low cost earthen waste impoundment ponds have been built to store solid waste and thereby provide for a relatively cheap means of meeting new federal and state environmental laws. There are several methods of preparing ore for beneficiation after it arrives at the plant site (Fig. 1).

Flake mica is also produced as a by-product from processing feldspar ore (hard granodiorite) from mica schist which normally contains from 30–60% recrystallized muscovite mica along with quartz and iron minerals. The quartz is usually not suitable for glass sand or high purity material, however.

Sheet Mica. The preparation of sheet mica for feedstock for various punching and machining operations involves cobbing mica blocks or books to remove dirt, rock, and defective mica, trimming and splitting into sizes and thicknesses suitable for punching and milling to desired shapes, and grading the finished mica sheets according to size and quality. The waste mica resulting from cobbing and trimming (scrap mica) is often mixed with flake mica for processing by dry or wet ground procedures.

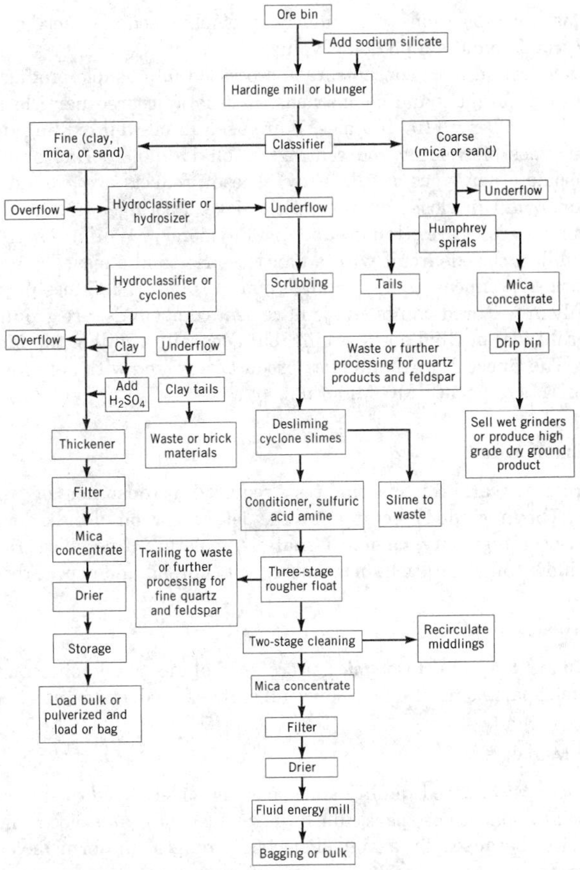

Figure 1. Flow sheet for the acid circuit processing and recovery of mica from weathered granodiorite ore. An alkaline—cationic circuit may be used by inserting a second conditioner containing lignin sulfonate, adjusting the pH to 8.0, and adding NaOH and DRL (distilled tall oil) fatty acid to the first conditioner.

The grade determines whether the mica can be used in high technology electronic instruments, eg, computer-aided tomography (CAT) scan (see MEDICAL IMAGING TECHNOLOGY), or in low technology devices, eg, a toaster. Many types of insulators, as well as the base for electronic circuits, are formed from the high quality sheets of mica by a punch pressing operation.

Procedures for Production

The general pieces of equipment used in grinding flake mica or mica concentrate into saleable mica products are hammer mills of various types, fluid energy mills, Chaser or Muller mills for wet grinding, and Raymond or Williams high side roller mills. Another method is being developed, called a Duncan mill (J. M. Huber, Inc.), that is similar in many respects to an attrition mill. All of these mills are used in conjunction with sieves, and all but some types of hammer mills incorporate air classifiers as a part of the circuit.

Ground Mica. This constitutes by far the largest commercial use for mica. It is largely produced from the beneficiation of weathered and unweathered pegmatites, granodiorite, and metamorphic schists, although some higher grades are produced from trimmings of sheet mica or Type A (low quality) mica blocks.

Wet ground mica products account for approximately 15% of the total mica market. Exact sizing of mica products coupled with surface treatment procedures have led to a greater use for wet ground mica in plastic compositions, particularly automobile bodies. These quality products demand a high dollar value.

Dry ground mica concentrate is processed into usable products by several different grinding methods. Relatively coarse particle sizes (1.651–0.147 mm (10–100 mesh)) are used in oil-well drilling muds, some types of welding rod coatings, asphalt (qv), roofing shingles, and some other types of fillers (qv). These products are ground on a hammer mill in closed circuit with a sieve. Roofing micas produced from mica schist are often ground in a Raymond or Williams high side roll mill in closed circuit with an air classifier and a sieve. The finer particle-size micas $\leq$0.147−0.044 mm (−100 to − 325 mesh), used mainly in textured paints and joint cement compounds, are ground on several types of fluid energy mills, but generally a mill of the Majacs type. The finest dry ground mica product is ground with superheated steam (Micronized, KMG Minerals).

Testing of Mica

There are several conventional tests required by consumers of ground mica. They include screening and the determination of bulk density, true specific gravity, chemical analysis, moisture, free silica, refraction index, oil absorption, brightness, grit content, and aspect ratio.

By-Products of Mica

The main by-products of mica processing plants are kaolin, quartz, and feldspar. Some plants produce all of these products for sale.

Mica Market

Sheet Mica. Good quality sheet mica is widely used for many industrial applications, particularly in the electrical and electronic industries, because of its high dielectric strength, uniform dielectric constant, low power loss (high power factor), high electrical resistivity, and low temperature coefficient. Mica also resists temperatures of 600–900°C, and can be easily machined into strong parts of different sizes and shapes.

Built-Up Mica. When the primary property needed for a particular application is insulation, built-up mica made by binding layered mica splittings together serves as a substitute for the more expensive sheet mica. The principal uses for built-up mica are segment plate, molding plate, flexible plate, heater plate, and tape.

Wet Ground Mica. Wet ground mica is used because of its unique properties, ie, luster, slip and sheen, and high aspect ratio. It is used in wallpaper and coated paper, nacreous pigments, as a coating for rubber, in outdoor house paint, and in aluminum paints. Mica is used in all types of sealers for porous surfaces, such as wallboard masonry, and concrete blocks, to reduce penetration and improve holdout (see SEALANTS), and as a filler in plastics to improve its electrical and thermal resistance and its insulating qualities.

Dry Ground Mica. Dry ground mica produced by hammer milling and screening is used in oil-well drilling, coatings for roofing shingles, roofing felt, and for some types of welding rod flux (see BUILDING MATERIALS, SURVEY).

The largest use for fine, dry ground mica is in the manufacture of wallboard joint cements. Ground mica that is essentially $\leq$0.147 mm-100 mesh and ~70% passing a 0.044 mm (325 mesh) Tyler sieve is used in the joint compound mixture as a filler and extender. These compounds are used to fill joints between panels of gypsum plaster board (see CALCIUM COMPOUNDS). Mica contributes to making a nonabsorbing smooth surface that reduces shrinkage and eliminates cracks. It is also used in the finished coating on ceilings and to prepare thermal insulation and acoustical qualities of ceiling tile and prefabricated concrete.

Fine particle-size dry ground mica is also used as an extender and filler in certain texture and traffic paints. Mica particles are stronger than iron and not brittle like other inerts. It is an antifriction, antifouling, antisettling, anticorrosive, antitarnish, and antisiege agent. It is a superior reinforcing pigment that acts as a sealer over porous surfaces and reduces penetration and flushing (see SEALANTS); moreover, it improves the moisture resistance of protective coatings and adhesion to all types of surfaces.

Micronized mica is a trade name (KMG Minerals, formerly English Mica Co.) for a very fine particle-size dry ground product, usually ground with superheated steam in a special fluid energy mill and used as a replacement for wet ground mica in certain types of paints. Micronized mica, preferably calcined, is also used in cosmetic applications, ie, nail varnishes, lipsticks, eyeshadows, and barrier cream, because is has the advantages of high ultraviolet light stability, excellent lubricity, skin adhesion, and compressibility (see COSMETICS). Some of these micas are coated with oxides like titanium and iron.

Environmental and Health Regulations

Mica mining is subjected to local, state, and federal laws. The Mining, Safety and Health Administration (MSHA) regularly monitors mica mining operations for safety violations.

Both state air and water environmental departments together with the U.S. EPA regulate and oversee air and water quality associated with mica mining operations. Most states have land management departments that regulate dam safety, erosion, sedimentation, and reclamation. The mica mines must control erosion and sedimentation and restore the mined out areas. This is accomplished either by backfilling or contouring and seeding operations, or in cases where this is impractical or undesirable, lakes for water-related recreation may be built. The Corps of Engineers have jurisdiction over laws governing wetlands.

Health regulations are supervised by county and state health departments. There are no known health problems caused by the mica crystal, however, most industrial mica products contain some free silica particles that can cause silicosis and some states require employees who work in mica plants to receive an annual x-ray.

JAMES T. TANNER
North Carolina State University

M. L. Rajgarhia, *Ground Mica,* Mica Manufacturing Co., Private Ltd., Calcutta, India, 1987, p. 30; *British Standards,* British Standards Institute, London.

R. W. Grim, *Clay Mineralogy,* McGraw-Hill Book Co., Inc., New York, 1968, 596 pp.

L. L. Davis, *Minerals Yearbook,* U.S. Bureau of Mines, Washington, D.C., 1991–1993, p. 4, 5, 7–9.

J. B. Preston, "Mica," *Pigment Handbook,* John Wiley & Sons, Inc., New York, 1971, 30 pp.

MICROBIAL POLYSACCHARIDES

Produced by virtually all microbes, carbohydrate polymers serve as intracellular energy stores and cell wall components, among other roles (see CARBOHYDRATES). They are perhaps most apparent when present as extracellular capsules, sheaths, or slime secretions. Extracellular polysaccharides may confer a survival advantage to microbes by protecting against desiccation, acting as buffers against environmental changes, preventing invasion by bacteriophages, and by helping cells adhere to surfaces and to one another. Technological exploitation of microbial polysaccharides has been limited mostly to those produced extracellularly in substantial quantities. Applications of these exopolysaccharides can be based on the unique chemical functionalities present, or on the bulk physical properties of the biopolymer. The physical properties of polymers, including polysaccharides, result from their chemical structures.

Microbial polysaccharides may be categorized into groups based on the types of monomer units present. Two of the most important types of microbial polysaccharides are neutral homopolysaccharides and anionically charged heteropolysaccharides. Other groups also exist, such as charged homopolysaccharides, but are of limited occurrence and not commercially significant as of this writing.

Polysaccharides can be classified by structure, biological origin, or mode of biosynthesis. Most polysaccharides are synthesized by an elaborate sequence of steps catalyzed by cytosolic and membrane-bound enzymes.

Alginates

The term *alginate* refers to the salt forms of alginic acid, a copolymer of β-D-mannopyranosyluronic acid (**1**) and α-L-gulopyranosyluronic acid (**2**) residues linked 1 → 4.

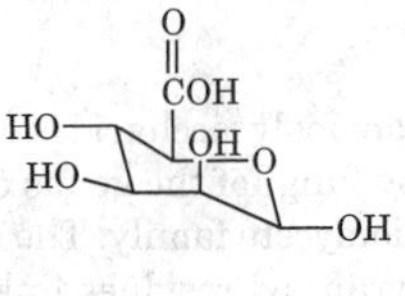

(**1**)

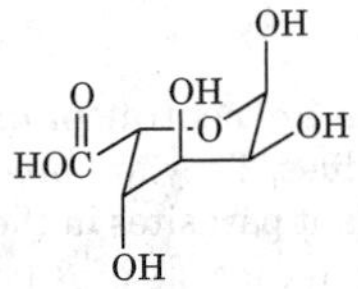

(**2**)

For many years, alginates were derived solely from marine algae, hence the origin of the term algin. Later it was demonstrated that the common nitrogen-fixing soil bacterium *Azotobacter vinelandii* also produces extracellular capsules of alginate. The phenomenon is known to be common among strains of *A. vinelandii,* as well as *A. chroococcum* and *A. beijerinckii.* Many other bacteria have been found to produce extracellular alginate, most notably strains of the opportunistic human pathogen *Pseudomonas aeruginosa,* the nonpathogenic *P. mendocina,* and numerous plant pathogenic species of *Pseudomonas.*

Bacterial alginates contain *O*-acetyl residues linked to the sugar units. The sequences and proportions of the constituent units can vary, and much research has focused on understanding the factors determining these structural variations.

Much of the potential for bacterially produced alginate has not yet been realized, owing mainly to the difficulty in competing with algal-derived products.

Bacterial Cellulose

Although cellulose is usually thought of as a plant-derived polysaccharide, there does exist one well-known example of cellulose (qv) production by a bacterium. This bacterium, *Acetobacter xylinum,* produces a tough, membranous pellicle (cellulose) in liquid cultures.

Production and Utilization. Although bacterial cellulose has been known since the late 1800s, there had been little commercial interest for many years, owing to the abundance and low cost of plant-derived cellulose. However, the ability of bacterial cellulose to form tough, uniform membranes has suggested applications, for example, in ultrafiltration membranes, speaker diaphragms for personal stereo headphones, nonwoven fabrics, coatings (qv), and suspending agents.

Dextrans

Dextran is a term that has traditionally been applied to any extracellular bacterial α-D-glucan synthesized from sucrose in which α(1 → 6) linkages predo minate. Dextrans have been more strictly defined as D-glucans containing chains of D-glucopyranosyl residues consecutively α(1 → 6)-linked, with various degrees of branching through α(1 → 2), α(1 → 3), or α(1 → 4) linkages. A number of lactic acid bacteria produce dextrans, the most notable being *Leuconostoc mesenteroides* and certain *Streptococcus* species.

Biosynthesis. Unlike most microbial exopolysaccharides, dextrans are enzymatically synthesized directly from sucrose. The enzymes, classified as glycosyltransferases, are known generically as dextransucrase.

Derivatives. Derivatives of dextran may be classified into two categories: those in which the dextran chain has been covalently modified, and those in which the structure is unchanged, but the molecular weight has been lowered, either by alteration of biosynthetic conditions or by depolymerization of high molecular weight dextran. Although clinical dextrans are still used in some applications, artificial polymers such as poly(vinylpyrrolidone) are replacing dextran as blood plasma extenders, because dextran tends to elicit an immune response in sensitive individuals.

Many covalently modified derivatives of dextran have been described. Of these, the most important are dextran sulfate and cross-linked dextran.

Production and Utilization. Dextran sulfate displays anticoagulant properties, and has been investigated as a substitute for heparin (see BLOOD, COAGULANTS AND ANTICOAGULANTS). It has been shown that dextran sulfate can inhibit HIV binding to human T-lymphocytes, and is being studied for its potential in the treatment of AIDS and other viruses. Dextran, chemically cross-linked with epichlorohydrin (Sephadex), is useful in gel-filtration chromatography (qv). Derivatives such as diethylaminoethyl–Sephadex and carboxymethyl–Sephadex are used in ion-exchange chromatography.

Cross-linked dextran known as dextranomer (Debrisan), which is similar to Sephadex, has been used in treating wounds. Fluids and small molecules are absorbed into the gel particles, and proteins and cellular material are excluded. Complexes of colloidal iron with dextran, known as iron–dextran, are used in treating iron deficiency anemia. This use is limited mainly to animals, especially pigs, because iron–dextran has been listed as a suspected carcinogen. The ability of dextran to form stable complexes with metals is one of its more useful properties. Large markets for dextran include use in the manufacture of photographic and x-ray films and in aluminum manufacturing, where dextran solutions are sometimes used in the recovery of aluminum from bauxite ores. Dextran has been used as a binder in tobacco products, and its use in shaving creams and other cosmetics (qv) has also been suggested. The U.S. FDA status of dextran as a food additive is not clear. Dextran is produced commercially by fermentation of sucrose with *L. mesenteroides* B-512F.

Although dextran is manufactured using traditional fermentation (qv) methods, there are advantages of using cell-free enzyme preparations to synthesize dextran. These advantages include better control over the synthetic process and greater ease of purification of the end product. Immobilized dextransucrase is especially well suited for the production of low molecular weight, low viscosity oligosaccharides and clinical-sized dextrans.

Emulsan and Liposan

Microorganisms that degrade hydrocarbons and utilize them as carbon sources usually possess some way of rendering the hydrophobic hydrocarbons water soluble, generally by secreting some type of surfactant or emulsifying agent (see SURFACTANTS). Two of the best known compounds in this category are emulsan and liposan. Emulsan, first isolated from cultures of the oil-degrading bacterium *Acinetobacter calcoaceticus* strain RAG-1, consists mainly of a heteropolysaccharide with amino sugars substituted by *O*-esterification with long-chain acyl groups. Another microbial polysaccharide-based emulsifier is liposan, produced by the yeast *Candida lipolytica* when grown on hydrocarbons.

Applications. Proposed uses of emulsan have included scouring crude oil from tankers and other containers, cleaning oil-handling equipment, dispersing oil-soluble pigments (see DISPERSANTS), enhanced oil recovery, and other emulsion-stabilizing applications.

Gellan

Gellan was successfully introduced as a replacement for agar in the early 1980s. Synthesized by the bacterium *Pseudomonas elodea* (*Sphingomonas elodea, Auromonas elodea*), this anionic heteropolysaccharide consists of D-glucopyranosyl (Glc*p*), L-rhamnopyranosyl (Rha*p*), and D-glucopyranosyluronic acid (Glc*p*A) residues linked in repeating units:

$$[\rightarrow 3)\text{-}\beta\text{-D-Glc}p\text{-}(1 \rightarrow 4)\text{-}\beta\text{-D-Glc}p\text{A-}(1 \rightarrow 4)\text{-}\beta\text{-D-Glc}p\text{-}(1 \rightarrow 4)\text{-}\alpha\text{-L-Rha}p\text{-}(1 \rightarrow]_n$$

$$\begin{array}{c} 3 \\ \uparrow \\ 1 \\ \alpha\text{-L-Rha}p \text{ or } \alpha\text{-L-Man}p \end{array}$$

Production and Utilization. Unlike many newly discovered microbial polysaccharides, gellan has become a commercial success in a relatively short period. It was first marketed as a replacement for agar in microbiological applications. The advantages of gellan over agar include higher purity, better clarity, and the ability to obtain strong gels at lower polysaccharide concentrations. Gellan is especially useful in marine microbiology, where agar-degrading microbes are often encountered. Gellan has also found biotechnological applications in plant tissue culture and in cell immobilization. Other uses include as a food additive in icings, frostings, jams, jellies, and fillings, where it can be used as a replacement for agar and carrageenans, and as a general food additive for stabilizing, thickening, and gelling.

Gellan is made by fermentation in a medium containing glucose, salts, and a nitrogen source.

Welan

Welan is produced by an *Alcaligenes* species (ATCC-31555) by aerobic fermentation. The polymer is structurally similar to gellan, sharing the same backbone sequence. It has an additional side group of an α-L-rhamnopyranosyl or an α-L-mannopyranosyl (Man*p*) unit linked ($1 \rightarrow 3$) to a β-D-glucopyranosyl unit in the backbone of the polymer:

Applications. The high heat tolerance and good salt compatibility of welan gum indicate its potential for use as an additive in several aspects of oil and natural gas recovery. Welan also has suspension properties superior to xanthan gum, which is desirable in oil-field drilling operations and hydraulic fracturing projects. It is compatible with ethylene glycol, and a welan–ethylene glycol composition that forms a viscous material is useful in the formulation of insulating materials.

Levan

Levan has been the subject of numerous studies, most of which have focused on structure and biosynthesis. The term levulan first appeared in 1881, and was used to describe a gum consisting of fructose (levulose) units that had been formed by microbial action on molasses. It was named by analogy with the dextrose-containing gum known as dextran.

Levans are water-soluble, nongelling, and generally of lower viscosity than most other gums and polysaccharides, despite their often high molecular weights. The average molecular weight can vary considerably, and depends on the conditions of biosynthesis, usually falling in the range between 10^6 and 10^7. The reasons for the variation can be best explained in terms of the mechanism of enzymatic biosynthesis. The molecular sizes of levan chains depend on the presence or absence of other carbohydrates in the synthetic reaction mixtures and on the length of time the enzyme acts on the levan after all of the sucrose is consumed. The source and purity of the enzyme, reaction temperature, and other factors may also play a role.

Production and Utilization. Although many uses have been proposed for levan, it is not being manufactured or used for any applications as of this writing (ca 1998).

Pullulan

Pullulan is a water-soluble extracellular α-D-glucan elaborated by the fungus *Aureobasidium pullulans* (formerly *Pullularia pullulans*). It is a linear polymer of maltotriose units linked from the reducing end of one trisaccharidic unit to the nonreducing end of the next trisaccharidic unit by $\alpha(1 \rightarrow 6)$ linkages:

$$[\rightarrow 6]\alpha\text{-D-Glc}p(1 \rightarrow 4)\alpha\text{-D-Glc}p(1 \rightarrow 4)\alpha\text{-D-Glc}p(1 \rightarrow)_n$$

Pullulan is generally produced in liquid fermentations, and its accumulation causes a marked increase in the viscosity of the medium. However, as cultures age, the viscosity usually drops off.

Production and Utilization. Pullulan can be used in foods as a viscosifier and low calorie partial replacement for starch, and as a binder. Its unique film-forming and oxygen-barrier properties make it especially useful in protective and adhesive edible coatings (qv). Other applications have been suggested in degradable films and fibers, paper coatings and binders, cosmetics, pharmaceutical tablet coatings, and even in soluble contact lenses (qv) which contain slow-release bioactive medicines. Pullulan fractions of narrow molecular weight ranges are available for use as standards in gel-permeation chromatography.

Scleroglucan

Scleroglucans are neutral, branched homopolysaccharides composed of glucose residues. They are produced by fungi of the genus *Sclerotium,* which are plant parasites in the Basidiomycete family. The main chain of scleroglucan consists of β-D-glucopyranosyl residues linked ($1 \rightarrow 3$) with every third sugar bearing a single D-glycopyranyl residue linked $\beta(1 \rightarrow 6)$. The polysaccharide is insoluble in 2-propanol, which is used to isolate and concentrate the material.

Production and Utilization. The most important potential use for scleroglucan is as a mobility control agent for enhanced oil recovery. Many of its rheological properties, such as the production of highly viscous solutions at low concentrations, and excellent long-term stability at elevated temperature and salt concentrations, are similar to those of xanthan gum.

Curdlan

Curdlan is a neutral $\beta(1 \rightarrow 3)$-linked D-glucan produced by several bacteria, primarily *Alcaligenes faecalis* var. *myxogenes,* as well as by *Agrobacterium radiobacter, Rhizobium meliloti,* and *R. trifolii.*

Applications. Several food uses have been proposed for curdlan including jellies, jams, noodles, and tofu. Its gelling properties make it useful for the preparation of instant puddings and multiple layer puddings. It may be useful as a stabilizing agent in frozen desserts such as ice creams. Curdlan can be added to bind water, add stability, and improve the body and gloss of food products.

Succinoglycan

Succinoglycan is an acidic extracellular polysaccharide produced by several bacteria, including *Alcaligenes faecalis, Agrobacterium radiobacter, Rhizobium meliloti,* and *R. trifolii.* The polymer consists of a repeating octasaccharide, as shown for succinoglycan from *Alcaligenes faecalis* var. *myxogenes.* Succinoglycan has potential use in foods and industrial processes as a thickening agent. Succinoglycan has potential use in foods and industrial processes as a thickening agent.

Xanthan Gum

Xanthan gum is an anionic heteropolysaccharide produced by several species of bacteria in the genus *Xanthomonas; X. campestris* NRRL B-1459 produces the biopolymer with the most desirable physical properties and is used for commercial production of xanthan gum (see GUMS). It is composed of repeating units consisting of a main chain of D-glucopyranosyl residues with trisaccharide side chains made up of D-mannopyranosyl and D-glucopyranosyluronic acid residues.

$$[\rightarrow 4)\text{-}\beta\text{-D-Glc}p\text{-D-Glc}p\text{-}(1 \rightarrow 4)\text{-}\beta\text{-D-Glc}p\text{-}(1 \rightarrow]_n$$

$$\beta\text{-D-Man}p(1 \rightarrow 4)\text{-}\beta\text{-D-Glc}p\text{A}$$

$$\text{-}(1 \rightarrow 4)\text{-}\alpha\text{-D-Man}p$$

Approximately half of the terminal D-mannosyl residues have pyruvate present as 4,6-*O*-(1-carboxyethylidene) substituents.

Xanthan gum has several desirable physical properties that explain the wide application range developed for this polysaccharide. The viscosity of xanthan gum solutions is highly pseudoplastic. Relatively low concentrations of the biopolymer produce highly viscous solutions that maintain viscosity over wide ranges of temperature and pH.

Production and Utilization. The nutritional requirements of *X. campestris* have been studied in order to optimize the production of xanthan gum. Fermentations for the industrial production of xanthan gum are done at 28°C, and utilize glucose concentrations from 1–5%. Higher glucose concentrations do not result in higher levels of gum biosynthesis. Saccharides such as sucrose, starch, and maltodextrins can also be used for gum production.

The unusual rheological properties of xanthan gum have led to its use in a wide variety of food and industrial applications. Large quantities of xanthan gum are used by the oil and natural gas industry in several aspects of hydrocarbon production (see GAS, NATURAL; HYDROCARBONS; PETROLEUM).

Other industrial uses for xanthan gum include thickening textile and carpet printing pastes, suspending pigments in ceramic glazes to improve glaze dispersion, and ink and clay coating formulations in the printing and paper (qv) industries, respectively. Agrochemical producers blend herbicides (qv) and insecticides (see INSECT CONTROL TECHNOLOGY) with xanthan in order to improve application to plants.

Other Microbial Polysaccharides

There are a number of other polysaccharides from fungi and bacteria which have actual or proposed practical applications. These include polysaccharides whose usefulness derives from particular structural features which are important not because of their viscosity or gelling ability, but because they elicit a particular biological response.

Some polysaccharides elicit a general type of immune response or specific response. These immunomodulator polysaccharides often fall into two categories: sulfated polysaccharides and β-D-glucans. Some general examples include curdlan sulfate, which exhibits anti-HIV activity, and sulfated yeast glucan, which enhances immune resistance against bacterial, viral, and fungal infections, and also shows antitumor activity. The β-D-glucans, are not only microbial in origin, but can also be found in plants, fungi, and macroalgae.

The USDA neither guarantees nor warrants the standard of the product, and the use of the name by USDA implies no approval of the product to the exclusion of others that may also be suitable.

GREGORY L. COTE
JEFFREY A. AHLGREN
U.S. Department of Agriculture

R. L. Whistler and J. N. BeMiller, eds., *Industrial Gums, Polysaccharides and their Derivatives,* Academic Press, Inc., San Diego, Calif., 1993.

V. J. Morris, *Agro Food Indus. Hi Tech,* **3,** 3–8 (1992).

V. Crescenzi, *Biotechnol. Prog.* **11,** 251–259 (1995).

I. W. Sutherland, *Trends Biotech.* **16,** 41–46 (1998).

MICROBIAL TRANSFORMATIONS

Microorganisms are of considerable economic importance in the manufacture of antibiotics (qv), alkaloids (qv), vitamins (qv), amino acids (qv), industrial solvents (see SOLVENTS, INDUSTRIAL), organic acids, nucleosides, nucleotides, fermented beverages (see BEER; WINE), and fermented foods (see also FERMENTATION). Microbes also catalyze simple and chemically well-defined reactions involving a variety of other compounds. Some reactions can be carried out more economically by microbial means than by strictly chemical manipulation. Microorganisms have therefore been used in processes that yield a number of important products, eg, L-ascorbic acid, steroid hormones, 6-aminopenicillanic acid, various L-amino acids, L-ephedrine, D-fructose, vinegar (qv), and malt (see BEVERAGE SPIRITS, DISTILLED; MALTS AND MALTING).

Microorganisms and their enzymes have been used to functionalize nonactivated carbon atoms, to introduce centers of chirality into optically inactive substrates, and to carry out optical resolutions of racemic mixtures. Their utility results from the ability of the microbes to elaborate both constitutive and inducible enzymes that possess broad substrate specificities and also remarkable regio- and stereospecificities.

Reactions

Oxidation. Both mono- and polynuclear aromatic hydrocarbons can be oxidized by different microorganisms. For example, *p*-cymene is converted to cumic acid and *p*-xylene to *p*-toluic acid. A high (98%) yield process has been developed for the production of salicylic acid from naphthalene using *Pseudomonas aeruginosa.* Microorganisms are used to construct chiral synthons, synthetic chemical intermediates, from various substituted benzenes. The technology of accumulating *cis*-dihydrodiols (**1**) produced by the action of dioxygenases on benzenoid substrates is available. Mutant *Pseudomonas putida* strains accumulate the chiral diols that permit the highly enantiomerically controlled syntheses of agents such as conduritols (polyols), pinitols, and others.

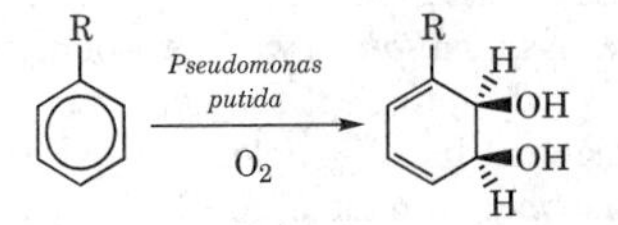

(**1**)

Hydroxylation, dehydrogenation, and β-oxidation reactions occur commonly with microorganisms. When cortisone and hydrocortisone were identified in 1949 as potent antiinflammatory agents and no adequate synthesis existed to meet the sharply increased demand for these compounds, hydroxylation of steroid intermediates at C-11 became crucial for large-scale production. The problem of introducing functionality at that site was solved upon discovery that progesterone (**2**) is oxidized to 11-α-hydroxyprogesterone (**3**) by *Rhizopus arrhizus* and also *Aspergillus niger.*

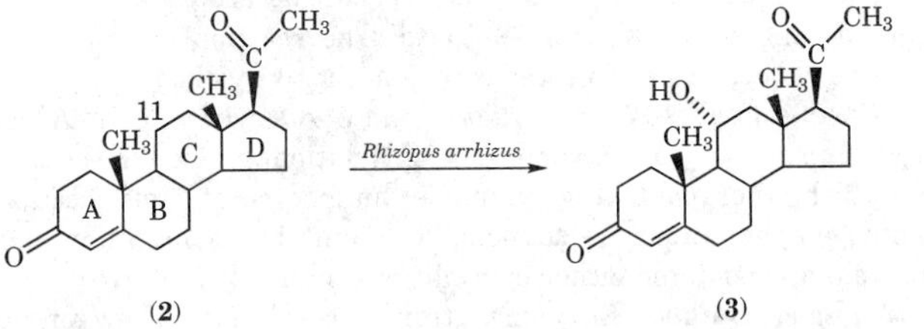

(**2**) (**3**)

Reduction. Stereospecific microbial reductions of ketonic substrates are often dependent on the size and nature of substituents flanking the ketone functional group to be reduced. The reduction of racemic decalone and hexahydroindanone derivatives and of related di- and tricyclic ketones by the fungus *Curvularia falcata* is highly stereospecific, giving alcohols of S-absolute configuration when ketones are flanked by large and small groups. Models of these reactions may help to predict the stereochemistry of optically active alcohols obtainable from ketone substrates.

Hydrolysis. Hydrolysis is one of the most widely used microbiological reactions. Hydrolytic enzymes are generally stable and require no co-factors for catalysis. Hydrolysis of many esters, glycosides, epoxides, lactones, β-lactams, nitriles, and amides has been described. Acetylated steroids have been hydrolyzed with varying degrees of selectivity by numerous organisms. (−)-14-Acetoxycodeine has been converted to (−)-14-hydroxycodeine and atrophine to tropine. The sugar moiety has been removed from cardiac glycosides and saponins. L-Amino acids have been produced from their optically inactive forms.

Carbon–Carbon Bond Formation. Asymmetric microbial acyloin condensation was discovered in 1921 and utilized in 1934 in the stereospecific synthesis of (1*R*, 2*S*)-ephedrine.

In this thiamine pyrophosphate-mediated process, benzaldehyde, added to fermenting yeast, reacts with acetaldehyde (qv), generated from glucose by the biocatalyst, to yield (*R*)-1-phenyl-1-hydroxy-2-propanone; the enzymatically induced chiral center helps in the asymmetric reductive (chemical) condensation with methylamine to yield (1*R*,2*S*)-ephedrine. Substituted benzaldehyde derivatives react in the same manner.

Amination and Hydration. Optically active products which correspond to the naturally occurring L-isomers have been obtained by the asymmetric addition of ammonia or water to fumaric acid. Thus aspartase-producing bacteria have been used in the manufacture of L-aspartic acid.

Other L-amino acids have been obtained using bacterial enzymes. Thus, by the addition of NH_4^+ and pyruvate to the reaction mixtures, L-tyrosine has been produced from phenol, L-DOPA from catechol, L-tryptophan from indole, and 5-hydroxy-L-tryptophan from 5-hydroxyindole. By hydration of DL-mixtures, L-isomers of a few other amino acids have been generated in 95–100% yields. Amination is also involved in the production of 5′-guanosine monophosphate (5′-GMP), a compound used as a food seasoning, by means of two mutants of *Brevibacterium ammoniagenes.* The first mutant ferments glucose to xanthosine-5′-monophosphate (5′-XMP), which is converted to 5′-GMP by the second mutant, either by sequential operation or by mixed cultures of both mutants.

Deamination, Transamination. Two kinds of deamination that have been observed are hydrolytic, eg, the conversion of L-tyrosine to 4-hydroxyphenyllactic acid in 90% yield, and oxidative, eg, isoguanine to xanthine and formycin A to formycin B.

Dehydration. Dehydration of hydroxy fatty acids is quite common. Other compounds undergo the same reaction, eg, elymoclavine to agroclavine, chanoclavine, and other compounds; and *cis*-terpin hydrate to α-terpineol.

N- and O-Demethylation. Microbial *N*- and *O*-demethylation reactions occur with high regiospecifities for multifunctional natural products. The biocatalytic approach is preferable to chemical methodologies, which require drastic conditions and are nonspecific.

Decarboxylation. Decarboxylation of linear and aromatic carboxylic acids and of amino acids is common and of practical interest. L-Lysine can be synthesized by stereospecific decarboxylation of *meso*- (but not DL) $\alpha\alpha'$-diaminopimelic acid. The reaction is catalyzed by *Bacillus sphaericus* and proceeds in quantitative yields.

N-Acetylation, O-Phosphorylation, and O-Adenylylation. *N*-Acetylation, *O*-phosphorylation, and *O*-adenylylation provide mechanisms by which therapeutically valuable aminocyclitol antibiotics, eg, kanamycin, gentamicin, sisomicin, streptomycin, neomycin, or spectinomycin are rendered either partially or completely inactive.

Transglycosylation. Enzymatic transglycosylations allow preparation of various oligosaccharides that cannot be made readily by strictly chemical means. The reaction can be expressed as

$$\underset{\text{donor}}{\text{R—O—R}'} + \underset{\text{acceptor}}{\text{R}''\text{OH}} \rightleftharpoons \underset{\text{product}}{\text{R—O—R}''} + \underset{\text{by-product}}{\text{R}'\text{OH}}$$

Isomerization. Isomerization of the double bonds of steroids is well known. Isomerizations of 5-androstene and 5-pregnene have been investigated but the products are not of practical value. Isomerization is of considerable importance in the manufacture of high fructose syrup, which is used as a food sweetener.

Methodology

The selection of organisms that carry out biotransformation reactions is of paramount importance. However, because only a few systematic examinations of the action of microorganisms on specific classes of organic compounds have been made, there is no assured way to select the organism that performs the desired reaction with the substrate of interest. In many cases the suitable microorganisms have been uncovered by screening randomly selected cultures isolated from various natural sources, eg, soil, decomposing organic material, or spontaneous fermentations. Others have been found by the enrichment method, which utilizes the substrate in question as the only source of carbon and nitrogen for growth and energy.

To conduct biotransformation reactions, the selected organism usually is grown in flasks with or without aeration and shaking until a sufficient amount of cell biomass has been generated (see AERATION; CELL CULTURE TECHNOLOGY). The substrate chemical to be transformed is then added to the cells, the incubation is continued, and the progress of the transformation monitored by suitable chromatographic, spectroscopic, or biological methods. When the maximum transformation has been obtained, the reaction is terminated and the product isolated and identified. There are many variations of this procedure.

A common means of increasing the rate or extent of biotransformation reactions is to expose the microorganisms to a mutagenic agent (such as x-rays, ultraviolet irradiation, nitrosoguanidine, 5-bromouracil, acridine, half-mustards, or novobiocin) which acts by causing DNA base-pair deletions, inversions, transitions, transversions, or deletion of extrachromosomal DNA elements (plasmids). Mutant strains are subsequently screened for enhanced expression of the desired enzyme. However, these types of mutations are characteristically random and multiple and often result in impaired growth rates and stability of the biocatalyst. By genetic engineering, new strains have been prepared for a variety of transformations either by expression of plasmid-borne genes, by site-specific insertions, or by mutations of the chromosome.

OLDRICH K. SEBEK
The Upjohn Company
JOHN P. N. ROSAZZA
University of Iowa

J. B. Jones, C. J. Sih, and D. Perlman, eds., *Applications of Biochemical Systems in Organic Chemistry,* John Wiley & Sons, Inc., New York, 1976, part 1.

H. G. Davies, R. H. Green, D. R. Kelly, and S. M. Roberts, *Biotransformations in Preparative Organic Chemistry,* Academic Press Inc., San Diego, Calif., 1989; K. Faber, *Biotransformations in Organic Chemistry,* Springer-Verlag, New York, 1992; S. M. Roberts, K. Wiggins, and G. Casy, *Preparative Biotransformations: Whole Cells and Isolated Enzymes in Organic Synthesis,* John Wiley & Sons, Inc., New York, 1993; A. R. Battersby, *Enzymes in Organic Synthesis,* Ciba Foundation Symposium 111, Pitman, London, 1985.

C. T. Goodhue, J. P. Rosazza, and G. P. Peruzzotti, in A. L. Demain and N. A. Solomon, eds., *ASM Manual of Industrial Microbiology and Biotechnology,* American Society for Microbiology, Washington, D.C., 1985, pp. 97–121.

K. Kieslich, *Microbial Transformations of Non-Steroid Cyclic Compounds,* G. Thieme, Publishers, Stuttgart, Germany, 1976.

MICROCHEMISTRY. See ANALYTICAL METHODS.

MICROELECTRONICS. See NANOTECHNOLOGY; SEMICONDUCTORS; X-RAY TECHNOLOGY.

MICROBIAL AND VIRAL FILTRATION

Filtration (qv) is the separation of particles from a fluid (liquid or gas) by passage of that fluid through a permeable medium. Sterile filtration ensures complete removal of viable organisms. Advances in membrane technology (qv) have resulted in the availability of filtrative devices for the removal of viruses in addition to bacteria. The filtration process is inherently nondestructive.

Filtration for Bacterial Removal

Mechanisms of Filter Retention. In general, filtrative processes operate via three mechanisms: direct interception, inertial impaction, and diffusional interception. Whereas these mechanisms operate concomitantly, the relative importance and role of each may vary.

Direct interception refers to a sieve-type mechanism in which contaminants larger than the filter pore size are directly trapped by the filter. This type of particle arrest is independent of filtration conditions.

Inertial impaction involves the removal of contaminants smaller than the pore size. Particles are impacted on the filter through inertia. Because the differential densities of the particles and the fluids are very small, inertial impaction plays a relatively small role in liquid filtration, but can play a major role in gas filtration.

Diffusional interception or Brownian motion, ie, the movement of particles resulting from molecular collisions, increases the probability of particles impacting the filter surface. Diffusional interception also plays a minor role in liquid filtration.

Types of Filters. In general, there are two types of filters used for microbial removal: depth and membrane. The first type removes microorganisms and particles mainly through retention by entrapment or impaction and adherence. These rely on filter matrix depth to achieve particulate contaminant retention. The primary mechanism of bacterial cell retention by membrane filters is the sieving effect, due to the highly stable, uniform pore matrix, so that trapped cells are not released. Membrane filtration is the method of choice for pharmaceutical and biological applications where absolute microbe retention is required. Membrane filters used for sterile filtration applications are typically constructed from polymers.

Membrane Filter Ratings. Filters are rated based on the ability to remove particles of a specific size from a fluid. There is, however, no standard on which method is to be used to specify performance. In general, the absolute rating, or cutoff point of a filter refers to the diameter of the largest particle, normally expressed in micrometers which can pass through the filter.

Filter Selection. Considerations include the characteristics of the fluid to be filtered, the level of bioburden present, specifications on effluent quality, the volume of product to be filtered, flow rate, and temperature.

The feed stream must be compatible with the membrane selected. The composition of the feed as well as pH and operating temperature must be considered. For sterile processes, the biological safety of the membrane filter or filter cartridge must be demonstrated by the performance of the USP Class VI (121 C) Plastics Test for Biological Reactivity.

The criteria to be met by the effluent or filtrate must be clearly defined. For aseptic processes a typical requirement is sterilization through an 0.2 μm-rated sterilizing-grade filter.

Flow rate, measured in units of volume per unit time, is dependent on pressure, and resistance. The flow rate achievable through a filtration system is directly related to the applied differential pressure and inversely related to the resistance to flow.

All components of a system contribute a resistance to flow which results in pressure drop. Pressure drops or losses in a system can be caused by piping, connections, valves, and filling heads, as well as by the filter and its assembly.

The temperature of filtration may affect the viscosity of the fluid, the corrosion rate of the housing, and filter medium compatibility. Elevated temperatures tend to accelerate corrosion and may weaken the gaskets and seals of filter housings.

Sterile Filtration of Liquids. The only true test of a sterilizing-grade filter is its microbial retention capability. By FDA definition, sterilizing-grade 0.2 μm-rated filters refer to filters which can remove more than 10^7 cfu/cm^2 of *B. diminuta* and yield sterile effluent. The challenge method for conducting a liquid bacterial challenge is detailed by ASTM F838-83. The bacterial challenge, *B. diminuta* at a minimum concentration of 1×10^7/cm^2 of filter area, is passed through the filter suspended in a sterile carrier fluid under standard test conditions. Bacterial concentrations are determined in the input as well as in the effluent. The entire effluent from the test filter, as well as aliquots of dilutions of effluent, are passed through an analysis membrane. Post-challenge recoveries done using analysis membranes allow for assay of the entire effluent so that even a single microorganism in the effluent is detected.

Factors that could potentially affect microbial retention include filter type, eg, structure, base polymer, surface modification chemistry, pore size distribution, and thickness; fluid components, eg, formulation, surfactants, and additives; sterilization conditions, eg, temperature, pressure, and time; fluid properties, eg, pH, viscosity, osmolarity, and ionic strength; and process conditions, eg, temperature, pressure differential, flow rate, and time.

The ratio of the difference between the numbers of challenge microorganisms recovered upstream and downstream of the test filter to the average total challenge received by the filter provides an indication of the removal efficiency of the filter, ie,

$$\text{removal efficiency, \%} = \frac{\text{Average total challenge} - \text{average total recovery}}{\text{average total challenge}} \times 100$$

Integrity Testing. A bacterial challenge is a destructive test and precludes subsequent use in a filtration operation. Therefore, filter manufacturers provide validation documentation for a filter with correlation of microbial removal to other nondestructive physical integrity tests. In the preparation of parenterals, the filter assembly is subjected to a nondestructive integrity test both prior to and after completion of the filtration operation.

Validation Considerations. It is necessary to validate filter performance, because the efficiency of the given filter is dependent on the physical, eg, viscosity and temperature, as well as the chemical, eg, presence of surfactants, composition of the suspending fluid. Microbial retention is required to be demonstrated under simulated pharmaceutical conditions in order to document the performance claims of the filter.

Sterilization Considerations. In sterile filtration processes the downstream side of the filter must be sterilized and must remain sterile during the entire process. Presterilized (gamma-irradiated) filters may be available or alternatively, filters may be sterilized by the user. The most common method of sterilization is by steam under pressure. The sterilization process must be validated to ensure that sterile conditions are met for a given system.

Sterile Filtration of Gases. Primary applications for sterile gas filtration are the sterilization of fermentor inlet air, fermentor vent gas, vents on water for injection tanks, and vacuum break filters during lyophilization. Typically, the membrane in gas filtration applications is a hydrophobic membrane, although there are applications in which the liquid (condensate) in the system is well controlled and hydrophilic membranes may be used. The effluent for gas filtration applications is typically filtered at the 0.2-μm level.

Verification of the microbial retention efficiency of the membrane filters may be undertaken using either liquid or aerosol challenge tests. A liquid challenge is performed using a protocol similar to that described for liquid filtration.

Aerosol challenges may be conducted using a test setup essentially comprising three components: the nebulizer, mixing chamber, and a sampling system. The challenge microorganism is aerosolized using a nebulizer. The aerosol is then mixed with compressed dry air to ensure that the monodispersed microbial challenge to the filter is delivered as a dry aerosol, rather than as microdroplets. Sampling may be done using a vacuum switch device that alternates between the upstream and downstream impingers, and a split-stream liquid impingement method. Following the challenge, the buffer from the impingers located upstream and downstream of the test filters are assayed using standard microbiological methods.

Filters for use in sterile gas filtration must conform to standards similar to those mandated for sterile liquid filtration.

Filtration for Virus Removal

General Principles. Advances in filtration technology have resulted in the availability of filtration devices for applications involving removal of viruses. The virological safety of biologicals and biopharmaceuticals is a key consideration in their manufacture. Much of the concern regarding viral contaminants in therapeutic agents centers around blood and blood products as well as biopharmaceuticals which have a blood or tissue component to their production. Viruses represent a diverse group which include enveloped and nonenveloped viruses, and ribonucleic acid (RNA) and deoxyribonucleic acid (DNA) viruses of various sizes.

Methods used to ensure virological safety are briefly classified as either virus inactivation or virus removal methods. The former includes chemical inactivation, pasteurization, uv inactivation, and solvent–detergent and ion-exchange (qv) chromatography (qv). There are limitations to the application of these methods.

The most desirable mechanism for the removal of viral particles using filtration is size exclusion. However, as in the case of bacterial removal by filtration, other mechanisms may also influence virus removal. These factors can include viral adsorption to the filter surface by electrostatic interactions; changes in pore size characteristics during filtration owing to deposits of material on the membrane surface; and the filtration conditions. The removal of viruses by particle size minimizes many of the variables affecting the level of retention and can be a predictable means of sterilization. However, concomitant with the requirement for adequate virus removal is the inherent necessity for retention of product concentration and/or activity following filtration processes.

Virological Safety Considerations. Historically, there have been several unfortunate iatrogenic accidents involving virus dissemination via administration of vaccines and blood products. Transmission of Hepatitis A, Hepatitis B, Hepatitis C, and HIV via administration of blood-derived products has been reported.

Minimization of risk of inadvertent exposure to real and theoretical viral contaminants is achieved by incorporation of multiple barriers to virus transmission in an integrated manner. Approaches that have been recommended are prevention of access of virus by screening of raw materials/precursors used; monitoring production, ie, adventitious virus testing; and a general evaluation of the manufacturing process, ie, process validation of viral clearance.

Safety is thus the result of multiple barriers operating in concert. These methods represent the only feasible approach in the face of theoretical risks, which cannot be adequately characterized by classical technology.

Configurations of Virus Filtration Systems. The two principal membrane filtration systems for the removal of viral particles from fluids are single-pass or direct flow filtration and cross- or tangential flow filtration. In the first, the entire volume is filtered through the membrane filter. Typically, the membrane is either a flat sheet cut into disks or a pleated sheet assembled as a cartridge filter. The pore size of these types of membrane filters is generally between 50–100 nm.

In the cross-flow mode, fluid is passed across the membrane surface while a portion of the flow is diverted through the filter (permeate). A portion of flow is returned to the central reservoir as retentate. In this process the volume of fluid in the retentate continually decreases as more of the initial volume is collected as permeate. Viral particles are concentrated in the retentate. Typically the filtration systems utilizing cross-flow are either in the tangential-type system where fluid passes between two flat sheets of membrane material or consist of hollow-fiber filters where the fluid passes through the middle of hollow tubes (see HOLLOW-FIBER MEMBRANES).

Methods to Detect and Quantitate Viral Agents in Fluids. In order to assess the effectiveness of membrane filtration, the ability to quantitate the amount of virus present pre- and post-filtration is critical. The method of choice for filter challenge studies is the plaque assay which utilizes the formation of plaques, localized areas in the cell monolayer where cell death caused by viral infection in the cell has occurred on the cell monolayer. Each plaque represents the presence of a single infectious virus. Virus quantity in a sample can be determined by serial dilution until the number of plaques can be accurately counted.

Effectiveness of Membrane Filtration. *Microfiltration.* Various membrane filters have been used to remove viral agents from fluids. In some cases, membranes which have pores larger than the viral particle can be used if the filtration is conducted under conditions which allow for the adsorption of the viral particle to the membrane matrix. These are typically single-pass systems having pore sizes of 0.10–0.22 μm. By removal standards, these filters remove viruses at a rate on the low end of the desired titer reduction and the removal efficiency varies with differences in fluid chemistry and surface chemistry of viral agents.

Ultrafiltration. Ultrafilters have also been examined for viral removal by size exclusion utilizing tangential flow and hollow-fiber membrane systems. The titer reduction varies depending on virus size and membrane filter pore size distribution. For example, removal of poliovirus by a 30,000 molecular weight polysulfone ultrafilter removes $>10^4$ particles of poliovirus in water of various qualities.

Integrity Testing. As in the case of bacterial removal, it is necessary to carry out an integrity test on the filter. Ideally, the integrity test should be performed both pre- and post-use. This is possible when a nondestructive integrity test method is used. Integrity tests used for virus removal filters include a forward-flow test similar to the test done on bacterial removal filters. In general, the integrity test results must correlate with the virus removal claims, as specified by the filter manufacturer.

Validation Considerations. Mechanisms other then size exclusion may be operative in the removal of viruses from biological fluids. Thus virus removal must be validated within the parameters set forth for the production process and using membrane material representative of the product line of the filter.

The validation study for filtrative virus removal essentially involves challenging (spiking) the product using high titers of infectious virus under conditions that simulate process parameters and quantitating virus in pre- and post-treatment samples. Pre-purification treatments and post-purification modification reactions must also be validated. The choice of virus for validation studies is not as clearly defined as in bacterial filtration where there is an industry-accepted standard. Thus, validation studies should be conducted using a panel of viruses that includes known contaminants which may represent identifiable and theoretical risks to product contamination, for example, HIV in the case of blood products.

Improvements in membrane technology, validation of membrane integrity, and methods to extend filter usage should further improve the performance of membrane filters in removal of viral particles.

HAZEL ARANHA-CREADO
K. OSHIMA
Pall Corporation

F. W. Bowman, M. P. Calhoun, and M. White, *J. Pharm. Sci.* **56,** 222 (1967).

M. Osumi, N. Yamada, and M. Toya, *J. Pharm. Sci. Technol.* **50,** 30 (1996).

Guidelines on Sterile Drug Products Produced by Asceptic Processing, Center for Drugs and Biologics and Office of Regulatory Affairs, U.S. FDA, Washington, D.C., June 1987.

"Indirect Food Additives Subpart B: Substances for Use as Basic Components of Single and Repeat Use Food Contact Surfaces," *Code of Federal Regulations,* Title 21, Part 177, U.S. Government Printing Office, Washington, D.C., 1994.

MICROEMULSIONS

There is no official or universally accepted definition of what constitutes a "microemulsion."

However, the concept of microemulsions holds a central role within the field of surfactant technology. Perhaps the most fundamental fact captured by the term is that, contrary to a popular saying, oil and water can mix (see SURFACTANTS).

Definition of a Microemulsion

The term *micro*emulsion implies a system which (like an emulsion) contains droplets of oil or water, but in which the droplets are too small to scatter light (see EMULSIONS).

A microemulsion is a true, thermodynamically stable, liquid solution that contains water, oil, and at least one amphiphile. Typically the oil is a mineral oil or hydrocarbon, but it may be almost any nonpolar compound. The oil and the amphiphile may be single, pure components; or (as in most commercial formulations), the oil, amphiphile, or both may contain an indefinitely large number of compounds.

Microemulsions and Phase Diagrams

The existence or nonexistence of a microemulsion depends not only on the presence of certain classes of compounds (ie, components), but also on the concentrations of these components. The number of phases present and their compositions, when presented in graphical form, constitute a phase diagram. Moreover, as specified by Gibbs' phase rule, amphiphile–oil–water–phase diagrams form characteristic patterns that change in qualitatively similar ways when the temperature, pressure, concentrations, or molecular structures of the components are changed. Thus, phase diagrams offer not only another way to define microemulsions, but also a rigorous way to clarify differences in terminology and usage.

Figure 1 illustrates the phase diagram of an amphiphile–oil–water system such as $C_4H_9OC_2H_4OH$ ("4E1")–decane–water or $C_6H_{12}(OC_2H_4)_2OH$ ("C6E2")–tetradecane–water. For a real surfactant, such as C12E4, the diagram would be more complicated, because of the occurrence of liquid crystalline phases. Samples whose compositions fall within the tietriangle of Figure 1 form three liquid phases, of compositions, *T, M,* and *B* (corners of the tietriangle). Each pair of adjacent corners of the tietriangle is connected by a binodal curve (as well as by a side of the tietriangle). Compositions between a side of the tietriangle and the adjacent binodal curve form two conjugate phases in equilibrium with each other; each such pair of phases is connected by a tieline. For two of the binodals the compositions of the conjugate phases can become closer and closer and their connecting tielines shorter and shorter, until the phase compositions and the

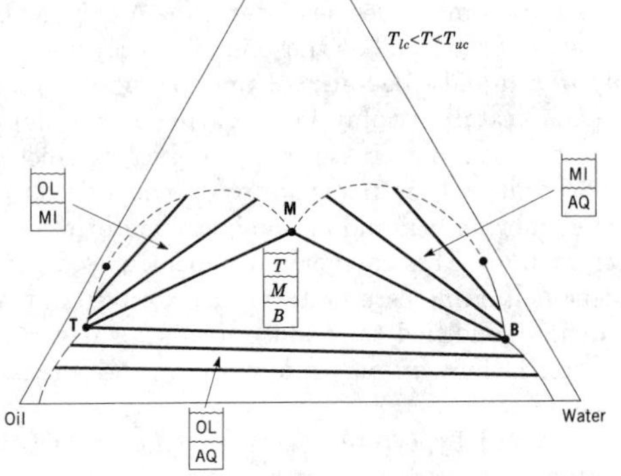

Figure 1. Phase diagram of an amphiphile-oil-water system that forms a middle-phase microemulsion, definition of microemulsion, and illustration of the pairs (and triad) of phases formed in the various multiphase regions of the diagram. Boundaries: —, aqueous (AQ); – – –, oleic (OL); – – – –, limiting microemulsion (MI).

end points of the tielines become identical at a plait point. Any composition outside of the tietriangle and the three binodal curves forms only a single liquid phase. On the amphiphile–water side of the phase diagram these single-phases contain only amphiphile and water (no oil); on the amphiphile–oil side of the phase diagram these single-phases contain only amphiphile and oil (no water).

By the most general definition of a microemulsion, every phase described by Figure 1 would be a microemulsion. When two (or even three) phases are simultaneously present, little (except confusion) is gained by giving the different phases the same name. Accordingly, along the binodal curves (which describe the compositions of conjugate phases) only compositions between the two plait points are termed microemulsions. Other conjugate phases are called oleic phases or aqueous phases, respectively, depending on whether their main component is oil or water.

In Figure 1, the pairs (or triad) of phases that form in the various multiphase regions of the diagram are illustrated by the corresponding test-tube samples. Except in rare cases, the densities of oleic phases are less than the densities of conjugate microemulsions and the densities of microemulsions are less than the densities of conjugate aqueous phases. Thus, for samples whose compositions lie within the oleic phase-microemulsion binodal, the upper phase (ie, layer) is an oleic phase and the lower layer is a microemulsion. For compositions within the aqueous phase-microemulsion binodal, the upper layer is a microemulsion and the lower layer is an aqueous phase. When a sample forms two layers, but the amphiphile concentration is too low for formation of a middle phase, neither layer is a microemulsion. Instead the upper layer is an oleic phase ("oil") and the lower layer is an aqueous phase ("water").

In three-phase systems the top phase, *T,* is an oleic phase, the middle phase, *M,* is a microemulsion, and the bottom phase, *B,* is an aqueous phase. Microemulsions that occur in equilibrium with one or two other phases are sometimes called "limiting microemulsions," because they occur at the limits of the single-phase region.

Temperature and Salinity Scans

The locations of the tietriangle and binodal curves in the phase diagram depend on the molecular structures of the amphiphile and oil, on the concentration of cosurfactant and/or electrolyte if either of these components is added, and on the temperature (and, especially for compressible oils, on the pressure).

Often the identities (aqueous, oleic, or microemulsion) of the layers can be deduced reliably by systematic changes of composition or temperature. Thus, without knowing the actual compositions for some amphiphile and oil of points *T, M,* and *B* in Figure 1, an experimentalist might prepare a series of samples of constant amphiphile concentration and different oil–water ratios, then find that these samples formed the series (*a*) 1 phase, (*b*) 2 phases, (*c*) 3 phases, (*d*) 2 phases, (*e*) 1 phase as the oil–water ratio increased. As illustrated by Figure 1, it is likely that this sequence of samples constituted (*a*) a "water-continuous" microemulsion (of normal micelles with solubilized oil), (*b*) an upper-phase microemulsion in equilibrium with an excess aqueous phase, (*c*) a middle-phase microemulsion with conjugate top and bottom phases, (*d*) a lower-phase microemulsion in equilibrium with excess oleic phase, and (*e*) an oil-continuous microemulsion (perhaps containing inverted micelles with water cores).

Physical Properties and Applications

The current or potential industrial applications of microemulsions include metal working, catalysis, advanced ceramics processing, production of nanostructured materials (see NANOTECHNOLOGY), dyeing, agrochemicals, cosmetics, foods, pharmaceuticals, and biotechnology. Environmental and human-safety aspects of surfactants are receiving considerable attention.

Microemulsions became well known from about 1975 to 1980 because of their use in "micellar-polymer" enhanced oil recovery (EOR). This technology exploits the ultralow interfacial tensions that exist among top, microemulsion, and bottom phases to remove large amounts of petroleum from porous rocks, that would be unrecoverable by conventional technologies. Since about 1990, interest in the use of this property of microemulsions has focused on the recovery of chlorinated compounds and other industrial solvents from shallow aquifers. The latter application is sometimes called surfactant-enhanced aquifer remediation (SEAR).

Microemulsions and Macroemulsions

Operationally, it is not always easy to determine whether a given sample is a microemulsion or macroemulsion. However, the formal differences between microemulsions and macroemulsions are well defined. A microemulsion is a single, thermodynamically stable, equilibrium phase; a macroemulsion is a dispersion of droplets or particles that contains two or more phases, which are liquids or liquid crystals.

From the definitions of microemulsions and macroemulsions and from Figure 1, it immediately follows that in many macroemulsions one of the two or three phases is a microemulsion. Until recently, it was thought that all nonmultiple emulsions were either oil-in-water (O/W) or water-in-oil (W/O). However, the phase diagram of Figure 1 makes clear that there are six nonmultiple, two-phase morphologies, of which four contain a microemulsion phase. These six two-phase morphologies are oleic-in-aqueous (OL/AQ, or O/W) and aqueous-in-oleic (AQ/OL, or W/O), but also, oleic-in-microemulsion (OL/MI), microemulsion-in-oleic (MI/OL), aqueous-in-microemulsion (AQ/MI), and microemulsion-in-aqueous (MI/AQ).

DUANE H. SMITH
Technical Solutions and West Virginia University

D. Robb, *Microemulsions*, Plenum Press, New York, 1982.

V. Degiorgio and M. Corti, *Physics of Amphiphiles: Micelles, Vesicles, and Microemulsions*, Elsevier Science Publishing Co., New York, 1985.

D. O. Shah, *Macro and Microemulsions: Theory and Applications*, American Chemical Society, Washington, D.C., 1985.

S. E. Friberg and P. Bothorel, *Microemulsions: Structure and Dynamics*, CRC Press, Boca Raton, Fla., 1987.

MICROENCAPSULATION

Microencapsulation is the coating of small solid particles, liquid droplets, or gas bubbles with a thin film of coating or shell material. Here, the term microcapsule is used to describe particles with diameters between 1 and 1000 μm. Particles smaller than 1 μm are called nanoparticles; particles greater than 1000 μm can be called microgranules or macrocapsules.

Many terms have been used to describe the contents of a microcapsule: active agent, actives, core material, fill, internal phase (IP), nucleus, and payload. Many terms have also been used to describe the material from which the capsule is formed: carrier, coating, membrane, shell, or wall. In this article the material being encapsulated is called the core material; the material from which the capsule is formed is called the shell material.

Table 1 lists representative examples of capsule shell materials used to produce commercial microcapsules along with preferred applications.

Microcapsules can have a wide range of geometries and structures. Figure 1 illustrates three possible capsule structures. Parameters used to characterize microcapsules include particle size, size distribution, geometry, actives content, storage stability, and core material release rate.

Table 1. Shell Materials Used to Produce Commercially Significant Microcapsules

Shell material	Regulatory status	Chemical class	Encapsulation process	Applications
gum arabic	edible	polysaccharide	spray drying	food flavors
gelatin	edible	protein	spray drying	vitamins
gelatin-gum arabic[a]	nonedible[b]	protein-polysaccharide complex	complex coacervation	carbonless paper
ethylcellulose	edible	cellulose ether	Wurster process or polymer—polymer incompatibility	oral pharmaceuticals
polyurea or polyamide	nonedible	cross-linked polymer	interfacial polymerization	agrochemicals and carbonless paper
aminoplasts	nonedible	cross-linked polymer	*in situ* polymerization	carbonless paper, fragrances, and adhesives
maltodextrins	edible	low molecular weight carbohydrate	spray drying and desolvation	food flavors
hydrogenated vegetable oils	edible	glycerides	fluidized bed	assorted food ingredients

[a] Treated with glutaraldehyde. [b] For intended application, ie, carbonless paper.

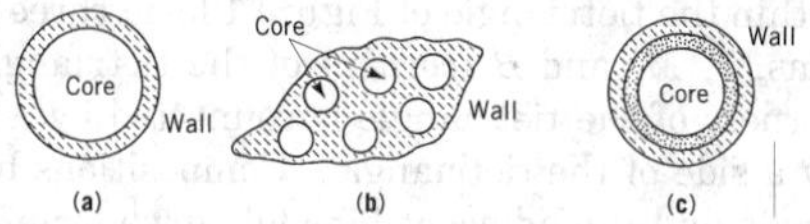

Figure 1. Schematic diagrams of several possible capsule structures: (**a**) continuous core/shell microcapsule in which a single continuous shell surrounds a continuous region of core material; (**b**) multinuclear microcapsule in which a number of small domains of core material are distributed uniformly throughout a matrix of shell material; and (**c**) continuous core capsule with two different shells.

Encapsulation Process

Classification of the many different encapsulation processes is useful. Previous schemes employing the categories chemical or physical are unsatisfactory because many so-called chemical processes involve exclusively physical phenomena, whereas so-called physical processes can utilize chemical phenomena. An alternative approach is to classify all encapsulation processes as either Type A or Type B processes. Type A processes are defined as those in which capsule formation occurs entirely in a liquid-filled stirred tank or tubular reactor. Emulsion and dispersion stability play a key role in determining the success of such processes. Type B processes are processes in which capsule formation occurs because a coating is sprayed or deposited in some manner onto the surface of a liquid or solid core material dispersed in a gas phase or vacuum. This category also includes processes in which liquid droplets containing core material are sprayed into a gas phase and subsequently solidified to produce microcapsules. Emulsion and dispersion stabilization can play a key role in the success of Type B processes also.

Many Type A and Type B processes are similar. For example, solvent evaporation is a key step in most spray dry encapsulation protocols (Type B) and protocols involving solvent evaporation from an emulsion (Type A). The difference in these protocols is that evaporation in the former case occurs directly from a liquid to a gas phase, whereas in the latter case evaporation involves transfer of a volatile liquid from a dispersed phase to a continuous liquid phase from which it is subsequently evaporated. Another example is encapsulation by gelation. In Type A gelation processes, the droplets that are gelled and become microcapsules are formed by dispersion in a liquid phase and are gelled in this phase. In Type B gelation processes, droplets formed by atomization or extrusion into a gas phase are subsequently gelled either in the gas phase or a liquid gelling bath.

Most Type A processes might be classified as chemical processes, whereas most Type B processes are classified as mechanical processes. Representative examples of both types of processes follow. Type B processes tend to be promoted by organizations that sell and service equipment for producing microcapsules. Most Type A processes are not promoted by equipment manufacturers, but are developed and used by organizations that produce microcapsules.

Type A processes	Type B processes
complex coacervation	spray drying
polymer—polymer incompatibility	fluidized bed
interfacial polymerization at liquid—liquid and solid—liquid interfaces	interfacial polymerization at solid—gas or liquid—gas interfaces
in situ polymerization	centrifugal extrusion
solvent evaporation or in-liquid drying	extrusion or spraying into a desolvation bath
submerged nozzle extrusion	rotational suspension separation (spinning disk)

Applications

Microcapsules are used in a number of pharmaceutical, graphic arts, food, agrochemical, cosmetic, and adhesive products. Other specialty products also exist, thus the concept of microencapsulation has been accepted by a wide range of industries. In order to illustrate how microcapsules are used commercially, it is appropriate to describe a number of commercial microcapsule-based products and the role that microcapsules play in these products.

Carbonless copy paper is by far the largest single commercial application of microcapsules. This product consumes thousands of tons of capsules annually. Figure 2, a schematic diagram of a three-part business form, illustrates the concept of carbonless copy paper.

Success of all carbonless paper products depends on the microcapsules, leuco dyes, and reactive coating. A number of leuco dyes are available.

The concept of microencapsulation has intrigued the pharmaceutical industry for many years, because it offers the possibility of providing a number of important new oral and parenteral dosage forms. Microcapsules in oral dosage forms could conceptually taste-mask bitter pharmaceuticals, provide extended release *in vivo,* provide enteric release, improve the stability of incompatible drug mixtures, provide resistance to oxidation, reduce volatility, and distribute a drug in many small carrier particles so that effects of the drug on the sensitive walls of the stomach are minimized. Microencapsulated parenteral formulations could provide prolonged delivery of drugs with

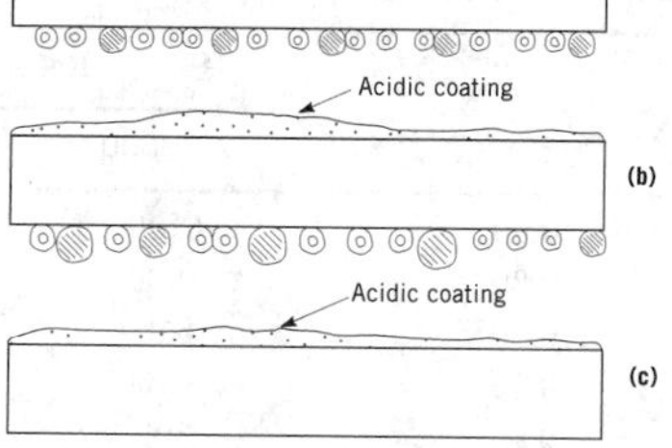

Figure 2. Cross section of a three-part business form prepared from carbonless copy paper where ◎ are microcapsules and ⊘ are starch: (**a**), CB sheet; (**b**), CFB sheet; and (**c**), CF sheet.

short half-lives *in vivo* and perhaps even achieve targeted drug delivery. For these reasons, microencapsulation has received much attention by pharmaceutical scientists. Several microcapsule-based oral pharmaceutical formulations which offer some of these features are available.

The use of microcapsules for a variety of biomedical and biological applications has been promoted for many years. Several biomedical microcapsule applications are in clinical use or have approached clinical use. One application is the use of air-filled human albumin microcapsules as ultrasound contrast agents. Another biomedical application of microcapsules is the encapsulation of live mammalian cells for transplantation into humans. The purpose of encapsulation is to protect the transplanted cells or organisms from rejection by the host.

A number of food ingredients or additives have been encapsulated and are available commercially. Solid ingredients encapsulated are typically water-soluble and are encapsulated with a hydrophobic or hydrophilic coating material usually applied by the Wurster process. Both types of coating materials are well-accepted food-grade products (see FOOD ADDITIVES).

The microencapsulation of pesticides (qv) and herbicides (qv) has been an active area of development that has produced several commercial products. The function of the microcapsules is to prolong activity while reducing mammalian toxicity, volatilization losses, phytotoxicity, environmental degradation, and movement in the soil. Ideally, encapsulation would also reduce the amount of agrochemical needed.

Advertising inserts that utilize encapsulated perfumes and flavors contain a coating of scent-filled capsules which break and release scent when the insert is torn open are widely used as a marketing tool, primarily for new perfumes. Children's crayons loaded with encapsulated scents are appearing on the market. The capsules break during the drawing process thereby releasing a scent characteristic of the drawn object.

Microcapsules are used in several film coatings other than carbonless paper. Encapsulated liquid crystal formulations coated on polyester film are used to produce a variety of display products including thermometers. Polyester film coated with capsules loaded with leuco dyes analogous to those used in carbonless copy paper is used as a means of measuring line and force pressures. Encapsulated deodorants that release their core contents as a function of moisture developed because of sweating represent another commercial application. Microcapsules are incorporated in several cosmetic creams, powders, and cleansing products.

A majority of the fasteners used in automobiles in the U.S. are coated with microcapsules loaded with an adhesive. Other uses include encapsulated ammonium polyphosphate incorporated in plastics that acts as a fire-retardant and microencapsulated oil-field chemicals for use by the oil industry.

CURT THIES
Washington University

A. Kondo, *Microcapsule Processing and Technology,* Marcel Dekker, Inc., New York, 1979.

P. B. Deasy, *Microencapsulation and Related Drug Processes,* Marcel Dekker, Inc., New York, 1984.

J. A. Bakan, in L. Lachman, H. A. Lieberman, and J. L. Kanig, eds., *The Theory and Practice of Industrial Pharmacy,* 3rd ed., Marcel Dekker, New York, 1986.

C. Thies, *How-to-Make Microcapsules: Lecture and Lab Manual,* Thies Technology, St. Louis, Mo., 1994.

MICROSCOPY

Microscopy involves the production and study of magnified images of objects too small to resolve with the unaided eye. There is no single microscope; rather, there are dozens of types, having in common only the ability to present an enlarged image. Microscopes vary in the mechanism by which enlarged images are formed, and in the information presented for each enlarged object. Some microscopes use visible light; others use infrared, ultraviolet, or x-ray radiation; electrons are also used. There are a variety of scanning microscopes, some of which scan the object with an electron beam, a laser beam, or ultraviolet light. Most recently, an entire family of microscopes has been developed that scan a surface, maintaining a constant, tiny distance from that surface and thus plotting a surface map of objects as small as single atoms.

Microscopes are also classified by the type of information they present: size, shape, transparency, crystallinity, color, anisotropy, refractive indices and dispersion, elemental analyses, and fluorescence, as well as infrared, visible, or ultraviolet absorption frequencies, etc. One or more of these microscopes are used in every area of the physical sciences, ie, biology, chemistry, and physics, and also in their subsciences, mineralogy, histology, cytology, pathology, metallography, etc.

Microscopy is an unusual scientific discipline, involving as it does a wide variety of microscopes and techniques. All have in common the ability to image and enlarge tiny objects to macroscopic size for study, comparison, evaluation, and identification. Few industries or research laboratories can afford to ignore microscopy, although each may use only a small fraction of the various types.

Microscopists in every technical field use the microscope to characterize, compare, and identify a wide variety of substances, eg, protozoa, bacteria, viruses, and plant and animal tissue, as well as minerals, building materials, ceramics, metals, abrasives, pigments, foods, drugs, explosives, fibers, hairs, and even single atoms. In addition, microscopists help to solve production and process problems, control quality, and handle trouble-shooting problems and customer complaints. Microscopists also do basic research in instrumentation, new techniques, specimen preparation, and applications of microscopy. The areas of application include forensic trace evidence, contamination analysis, art conservation and authentication, and asbestos control, among others.

The proliferation of microscopes and microscopy can be considered under the following three needs: (*1*) improved specimen contrast; (*2*) increased resolving power; and (*3*) obtaining more characterization data.

Improving Specimen Contrast

It was learned very early that the angular aperture of the substage condenser controls specimen contrast. Decreasing that aperture, usually with a continuously adjustable iris diaphram, greatly increases contrast. It was not, however, appreciated fully until Ernst Abbe's classic contributions in the period ca 1880–1889 that decreasing the aperture to increase contrast also decreases the resolving power of the microscope.

The choice of the mounting medium usually obviates the need for any new instrumental technique for enhancing contrast. A particle mounted in a liquid medium having the same refractive index (n_D) may disappear completely. The darkness of the borders (contrast), however, increases rapidly as the refractive index difference increases. Choosing the proper refractive index liquid is the first step in contrast enhancement. Unfortunately, biologists are often restricted by their specimens in their choice of mountant. Water (n_D = 1.33) is far superior to glycerol (n_D = 1.47) or canada balsam (n_D =) for most biological substances whose refractive indices lie in a narrow range near 1.54. More specialized contrast-enhancing microscopes have been developed during the twentieth century, especially after 1930. They are also useful to both biologists and materials scientists.

Techniques for improving specimen contrast are darkfield illumination, phase contrast, differential interference contrast, the use of an interference microscope, modulation contrast, video-enhanced contrast, and the use of confocal microscopy.

Improving Resolving Power

The resolution of the best of the early nineteenth century microscopes was not too different from the corresponding present-day microscope (about 225 nm), although few early microscopists were able to align their microscopes to achieve this resolution. Only late in the nineteenth century, with the invention of highly corrected (apochromatic) objectives by Ernst Abbe, and the sound reasoning of August Köhler, were microscopists generally able to optimize the illumination conditions to assure the best resolution possible. Ernst Abbe, in 1883, presented his diffraction theory of resolution for the light microscope. This made it possible for microscopists to understand and use the microscope as a high resolution instrument. His equation emphasizes that resolution (RP) improves with higher apertures (NA) and lower wavelengths,

$$RP = 0.61\lambda/NA \tag{1}$$

where RP = the resolving power, the distance between two points just resolved; 0.61 = a Rayleigh constant related to the detectability of resolution; λ = wavelength in nm; NA = the numerical aperture, in turn = n sin AA/2; AA = the average of the angular apertures of objective and condenser; and n = the refractive index of the medium between the objective and coverslip. During the 1800s, attempts had been made to increase resolution by using higher refractive index objective front lenses (even diamond) to increase NA; however, the difficulties outweigh the advantages. Table 1 shows the effects of NA on resolution for a common set of objectives at wavelengths of 500 and 225 nm.

Microscopes and techniques used are the ultraviolet light microscope, transmission electron microscope, scanning electron microscope, photon tunneling microscopy, x-ray microscopy, video-enhanced imaging, nuclear magnetic resonance, scanning tunneling microscopy, and other probe microscopes such as the near-field scanning optical microscope, field-emission electron microscope, and field ion microscope.

Obtaining Additional Characterization Data

The ability to enlarge tiny objects to macroscopic dimensions immediately suggests the need to make measurements and other observations helpful in documenting what is seen and thus enabling others to confirm that a specimen has been identified with certainty. Many physical and chemical properties of a microscopic substance can be measured, even on particles nearing atomic dimensions.

Polarized Light Microscopy. The most useful and characteristic properties presently employed by light microscopists are also among the earliest actual improvements, dating from the 1830s. These were made possible by adding polarizing elements and a rotating stage to a biological microscope to make a polarized light microscope (PLM), the most generally useful of all forms of microscope, whether light, electron, or probe.

Many things can be done with the polarized light microscope (PLM). In fact, no other analytical tool or technique yields the variety and extent of information about small objects. An emission spectrograph gives a spectrum identifying the chemical elements present; an absorption spectrograph gives only the functional groups present (eg, carbonyl, amino, hydroxyl, etc); a mass spectrometer

Table 1. Resolution as a Function of NA and Wavelength

		Resolution, nm	
Objective	NA	λ = 500	λ = 225
5×	0.05	6100	2740
10×	0.30	1000	460
20×	0.50	610	270
40×	0.65	470	210
40×[a]	0.85	390	160
100×	1.40	220	100

[a] The standard 40× objective is the 0.65-NA version, rather than the higher resolution but shorter working distance 0.85-NA objective.

gives only the nuclear masses of the constituent elements or their combinations; a scanning electron microscope shows an image only of the surface of a sample and, when properly modified, an elemental analysis; and a transmission electron microscope gives a silhouette and, with very thin substances, a transmission view and an electron diffraction pattern. It may also be fitted for elemental analysis (EDS) and electron energy loss spectra (EELS).

Thermal Microscopy. Otto Lehman, by 1880, had added a hotstage to his microscope and used it to study crystals. He measured melting points and polymorph transition temperatures, studied phase and composition diagrams, and published the first detailed studies of liquid crystals. Later, additional techniques and applications were developed. The behavior of crystals on heating is quite distinctive, and a rich variety of thermal and optical properties are quickly measured by thermal microscopy, sometimes termed fusion methods. It is also an excellent way to recrystallize a compound to yield well-formed crystals of different polymorphs or solvates for use in observing crystal morphology and optical properties. These include refractive indices, birefringence, extinction angles, crystal system, axes, and forms as well as optic sign, optical axial angle, and dispersion thereof. Although there are snowflakes that look alike, contrary to common belief, no two compounds will ever be identical in all of the parameters measured by PLM.

Fluorescence Microscope. A useful light microscope utilizes UV light to induce fluorescence in microscopic samples. Because fluorescence is often the result of trace components in a given sample rather than intrinsic fluorescence of the principal component, it is useful in the crime laboratory for the comparison of particles and fibers from suspect and crime scene. Particles of the same substance from different sources almost certainly show a different group of trace elements. It is also very useful in biology where fluorescent compounds can be absorbed on (and therefore locate and identify) components of a tissue section.

Dispersion Staining. Dispersion staining, introduced by Germain Crossmon during the 1940s, involved the observation of colored particle boundaries for small particles immersed in mounting media having suitable refractive indices and dispersion. Specifically, the dispersion of refractive index must be different for the particle and the mounting liquid, and the two curves relating refractive index to wavelength must cross within or close to the visible range (400–700 nm). Light rays of different wavelength are refracted differently at particle boundaries, and by suitable masking, images of the particle are observed and show edge colors corresponding to the matching wavelength.

Dispersion staining is useful for rapid determination of refractive index and dispersion. It is applied most often, however, for needle-in-a-haystack detection of any particular substance in a mixture such as chrysotile in insulation, cocaine in dust samples, quartz in mine samples, or any particular mineral, eg, tourmaline, in a forensic soil sample.

Schlieren Microscope. Other important developments were proposed during the 1960s and 1970s, including a unique schlieren microscope, a simple interference microscope, and a novel holographic microscope. These novel developments were, however, lost in the general eclipse of light microscopy that occurred as the transmission and scanning electron microscopes and their accessory techniques of elemental analysis by energy or wavelength dispersive analysis took precedence over PLM in most laboratories.

Interference Microscope. There are many variations in the design of interference microscopes, eg, Baker, Jamin-Lebedeff, and Mach-Zender. In general, they are relatively sophisticated, complex instruments.

Holographic Microscope. Holography (qv) can be done microscopically and produces three-dimensional images with excellent detail. The hologram stores both amplitude and phase information in a 2-D image which is then restored to a 3-D image by viewing that image with a reconstruction beam of light. A great advantage of holography is that the hologram can be studied later at various magnifications to make measurements of size, etc, as one focuses through or scans laterally any portion of the holographic image.

Laser Raman Microprobe. A more sophisticated microscope is the Laser Raman Microprobe, sometimes referred to as MOLE (the molecular orbital laser examiner). This instrument is designed around a light microscope to yield a Raman spectrum on selected areas or particles, often $<1\ \mu m^3$ in volume. The data are related, at least distantly, to infrared absorption, since the difference between the frequency of the exciting laser and the observed Raman frequency is the frequency of one of the IR absorption peaks. Both, however, result from rotational and vibrational states. Unfortunately, strong IR absorption bands are weak Raman scatterers and vice versa; hence there is no exact correspondence between the two.

Cathodoluminescence Microscope. There is also a cathodoluminescence microscope (CLM). An electron source built into the stage of a PLM bombards microscopic samples and yields a spectrum of the resulting characteristic luminescence for many substances.

Infrared Microscopy. Many substances, opaque or nearly so in the visible spectrum, are transparent in the near infrared (eg, silicon, iodine, and potassium permanganate). Silicon chips can be examined for internal defects or for strongly absorbing substances. Many normally opaque particles and thin films can be examined by infrared microscopy.

Acoustic Microscope. In acoustic microscopy, magnified images are formed using sound waves. This process replaces x-ray radiography as a way of seeing inside opaque objects such as metals, ceramics, polymers, etc. The acoustic microscope has the advantage of being able to detect and image voids, cracks, and phase and chemical regions in light-transmitting or opaque materials. It is nondestructive.

WALTER C. MCCRONE
McCrone Research Institute

G. L. Clark, ed. *The Encyclopedia of Microscopy,* Reinhold Publishing Corp., New York, 1961.

G. H. Needham, *The Practical Use of The Microscope,* Charles C. Thomas, Springfield, Ill., 1958.

C. W. Mason, *Handbook of Chemical Microscopy,* 4th ed., John Wiley & Sons, Inc., New York, 1983.

W. C. McCrone, L. B. McCrone, and J. G. Delly, *Polarized Light Microscopy,* McCrone Research Institute, Chicago, Ill., Ninth printing, 1995.

MICROWAVE TECHNOLOGY

The application of electrical or electromagnetic (EM) energy to materials as part of some chemical process is a broad subject that has a long history in chemical technology. Electrical energy can be delivered to materials through conductive, near-field coupling, or radiative techniques. The microwave portion of the electromagnetic spectrum in its role as a power source for chemical and materials processing is discussed herein. Microwave is the name applied to the central portion of the nonionizing radiation part of the electromagnetic spectrum. Nonionizing radiation has quantum energy too low to ionize an atom on a single event basis and conventionally is defined as ranging from d-c power to visible light. This portion of the spectrum can be divided into five regions in order of increasing frequency: static, quasistatic, microwave, quasioptical (nanowave), and optical.

Microwaves may be used to ionize gases when sufficient power is applied, but only through the intermediate process of classical acceleration of plasma electrons. The electrons must have energy values exceeding the ionization potential of molecules in the gas (see PLASMA TECHNOLOGY). Ionizing radiation exhibits more biological-effect potential whatever the power flux levels.

The microwave spectrum was defined as >1000 Mhz. Over the years, these definitions have been variously specified to include frequencies as low as 100 MHz to as high as 3000 GHz. The more scientific meaning of microwaves refers to the principles and techniques applying to electromagnetic systems where the principal dimensions are of the order of a wavelength, γ, or more broadly ca 0.1–10 γ.

For many materials, in particular biological tissue, maximum penetration of the electromagnetic energy irradiating objects of macroscopic size occurs in the microwave range. Interest in microwaves in chemical technology involves the fields of dielectric spectrometry, electron spin resonance (esr), or nuclear magnetic resonance (nmr) (see MAGNETIC SPIN RESONANCE).

Microwave power is an important factor in the commercial processing of materials. These processes almost exclusively utilize classical interactions such as dielectric heating in solids or plasma heating. Other interactions may also be possible. The field of microwave power applications as distinct from information processing is relatively new. Microwave power applications in a wide range of areas including food processing (qv), ceramics (qv) processing, biological tissue fixation, chemical analysis and processing, and plasma applications, as well as microwave power transmission, are found in the literature.

Microwave Power Applications

Frequency Allocations. Under ideal conditions, an optimum frequency or frequency band should be selected for each application of microwave power.

In 1979, the industrial, scientific, and medical (ISM) frequency allocations were revised as a result of the World Administrative Radio Conference (WARC). A considerable effort was made to increase the number and worldwide uniformity of ISM frequency allocations. Most of those proposals were rejected. The resulting allocations are listed in Table 1.

Principles in Processing Materials. In most practical applications of microwave power, the material to be processed is adequately specified in terms of its dielectric permittivity and conductivity. The permittivity is generally taken as complex to reflect loss mechanisms of the dielectric polarization process; the conductivity may be specified separately to designate free carriers. For simplicity, it is common to lump all loss or absorption processes under one constitutive parameter which can be alternatively labeled a conductivity, ς, or an imaginary part of the complex dielectric constant, ϵ_i, as expressed in the following equations for complex permittivity:

$$\epsilon = \epsilon_0(\epsilon_r + j\epsilon_i) = \epsilon_0(\epsilon_r + j\sigma/\omega\epsilon_0) \quad (1)$$

where ϵ is the complex dielectric permittivity in F/m, $\epsilon_0 = 8.86 \times 10^{-12}$ F/m, the permittivity of free space, ϵ_r, is the real part of the relative dielectric constant, and ς is the conductivity in S/m (mhos/m) which is equivalent to the following:

$$\epsilon_i = \sigma/\omega\epsilon_0$$

where ω is the assumed radian frequency of the fields. It is convenient to define auxiliary terms like the loss tangent, tan δ:

$$\tan\delta = \epsilon_i/\epsilon_r = \sigma/\omega\epsilon_r\epsilon_0 \quad (2)$$

From Maxwell's equation, the current density J in A/m^2 is related to the internal electric field, E_i, by equation 3:

$$J = (\sigma - j\omega\epsilon_r\epsilon_0)E_i \quad (3)$$

thus the rate of internal density of absorbed energy, or power, P, is given by equation 4, where rp = real part of:

$$P = \text{rp}(J)J\cdot E^* = \sigma|E_i|^2 \quad (4)$$

or simply,

$$P = \omega\epsilon_r\epsilon_0\tan\delta|E_i|^2 \quad (5)$$

Equation 5 is the practical equation for computing power dissipation in materials and objects of uniform composition adequately described by the simple dielectric parameters.

The internal field is that microwave field which is generally the object for solution when Maxwell's equations are applied to an object of arbitrary geometry and placed in a certain electromagnetic environment. The E_i is to be distinguished from the local field seen by a single molecule which is not necessarily the same.

Table 1. Frequency Allocations for ISM Applications[a]

Frequency, MHz	Region	Conditions
6.765–6.795	worldwide	special authorization with CCIR[b] limits; both in-band and out-of-band
13.553–13.567		
26.957–27.283	worldwide	free radiation bands
40.66–40.70		
433.05–434.79	selected countries in Region 1[c]	free radiation bands
433.05–434.79	rest of Region 1[c]	special authorization with CCIR[c] limits
902–928	Region[d]	free radiation band
2.40–2.50 × 10³	worldwide	free radiation band
5.725–5.875	worldwide	free radiation band
24.0–24.25	worldwide	free radiation band
61.0–61.5		special authorization with CCIR[b] limits; both in-band and out-of-band
122–123	worldwide	
244–246		

[a] ISM = industrial, scientific, and medical. [b] CCIR = International Radio Consultative Committee of the International Telecommunications Union (ITU). [c] Region 1 comprises Europe and parts of Asia; the selected countries are Germany, Austria, Lichtenstein, Portugal, Switzerland, and Yugoslavia. [d] Region 2 comprises the western hemisphere.

Instrumentation

Power Sources. The development of electron tubes, including those for the microwave range, is a mature field. It is feasible to generate almost any desired power for most microwave frequencies of practical interest, limited only by costs.

Power sources in the millimeter wave range are mostly in the category of extended interaction klystrons or narrow band backward wave oscillators. Power outputs of tens of watts or even more than 100 W are feasible but available tubes are quite expensive and suffer from low life and efficiency compared to what is available at low microwave frequencies. Because of the unavailability of inexpensive and efficient tubes for millimeter wave frequencies, ISM applications in that range are essentially nil.

Significant power is generated only below 300 GHz. Above this frequency, the expectation has always been that useful laser sources would be eventually developed. Thus the most difficult region for power generation appears to be that of submillimeter waves or the far infrared, ie, 300–3000 GHz (1000–100 μm).

The most dramatic evolution of a microwave power source is that of the cooker magnetron for microwave ovens. These magnetrons are air-cooled, weigh 1.2 kg, generate well over 700 W at 2.45 GHz into a matched load, and exhibit a tube efficiency on the order of 70%. Application is enhanced by the availability of comparatively inexpensive microwave power and microwave oven hardware. The cost of these tubes has consistently dropped since their introduction in the early 1970s.

The availability of a low cost source of microwave power has led to an explosion of work at 2.45 GHz on newer microwave power applications. For many applications at 2.45 GHz, it is feasible to utilize a number of low cost tubes to generate large total powers, eg, 25 or 50 kW. These tubes are designed to meet the requirements of government agencies on out-of-band spurious emissions. Hence, filter boxes are used around the high voltage terminals of the tubes.

Microwave tubes for other ISM bands are not commonly available as tubes designed specifically for ISM use. Use of traveling wave tube (TWT) amplifiers at power levels of hundreds of watts has been proposed for power applications, at least when the heating chamber is well shielded.

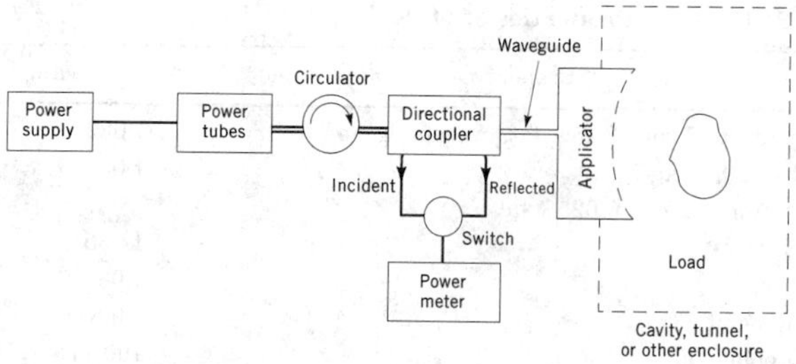

Figure 1. Basic elements of a microwave power system for processing of materials.

Applicators and Instruments. The basic elements of a microwave power system for materials processing are indicated schematically in Figure 1.

Health and Safety Factors

There are some unique safety considerations in microwave systems. Microwave voltage breakdown can occur in microwave systems and waveguides at power levels far below the theoretical values for ideal systems, ie, by a factor of at least 100 below theoretical breakdown. This is often the result of impurities or dirt particles that overheat and cause a breakdown or a spurious high quality factor resonance in the system which builds up high fields. In addition, the presence of sharp metal objects, accidental small gaps, and other situations often can induce localized arcing or corona which may or may not lead to a basic system breakdown. In this case, the plasma region of the breakdown travels down the feed waveguide toward the source and may cause failure of the tube through cracking of the output window. Therefore flammable materials should not be processed in microwave systems. Precautions can be taken, however.

The most serious hazard that can occur from leakage of microwave energy is interference with other systems. This could be caused by out-of-band radiation, ie, a violation of RFI regulations, or by high power effects where the offending radiation is out of the band of the affected system, but still effectively interferes because of its intense level. An example is the incidental interference with cardiac pacemakers. Concern has arisen over potential RFI from cellular phones disrupting medical electronics, both in the hospital and in portable units.

Radiated r-f energy is a hazard to systems containing flammable fuel or electroexplosive devices (EED) used for construction blasting or for military purposes. It is recommended that users of large amounts of microwave/r-f energy be aware of guidelines on safe distances of EEDs from sources of radiated power. The hazard of exposure of personnel to microwave energy has been thoroughly reviewed.

Uses

Food. The most successful application of microwave power is that of food processing (qv), cooking, and reheating. The consumer industry surpasses all other microwave power applications. Essentially all microwave ovens operate at 2450 MHz except for a few U.S. combination range models that operate at 915 MHz. The success of this appliance resulted from the development of low cost magnetrons producing over 700 W for oven powers of 500–800 W.

Advances in the technology of microwave ovens include techniques for achieving uniform heating, eg, mode stirrers and turntables; temperature and humidity monitoring; and the use of microprocessors in programming cooking time, defrost, and variable power levels. In addition, a wide variety of ceramic, glass, and plastic, as well as paperboard products have been developed for utensils, shelves, and food packaging (qv). A growing number of accessory products have been developed, eg, browning dishes or utensils, popcorn poppers, coffee-makers, etc. Combination ranges have been developed for microwave electric, microwave gas, and microwave convection, this last in counter-top arrangement.

A unique and successful innovation in food packaging for microwave heating is the microwave susceptor.

Only a few industrial food-processing applications are commercialized, although many have been investigated. Some, such as the drying of potato chips, were too costly; others, such as freeze drying, encountered technical problems like vacuum breakdown. There have been reports however of the successful resumption of a potato chip processing application. Microwave ovens have been used extensively for thawing. No effective solution has been found for the runaway heating caused by the great increase in dielectric loss at 0°C.

Nutritional quality of food cooked or heated by microwaves generally does not differ greatly from food cooked or heated by other means. Scorching is minimized in microwave heating. Underheating or undercooking can be avoided only in ovens having superior mode-stirrer techniques or through occasional manual rotation.

Biological, Medical, and Agricultural Applications. Diathermy at both 27.33 and 2450 MHz has been extensively used in physical therapy since the 1950s; its popularity has greatly declined in the 1990s. An extension of diathermy heating techniques has been investigated for application in hyperthermia as an adjunct to cancer therapy. The basis of preferential destruction of tumor cells was studied by hyperthermia alone, and in conjunction with ionizing radiation or chemotherapy. A variety of applicators have been designed, including multiple focused antennas, injected probe antennas, and contact applicators. Microwaves have successfully been used for rewarming of blood for medical applications. Reports of sterilization against bacteria by nonthermal effects have appeared, but it is generally believed that the effect is only that of heating (see STERILIZATION TECHNIQUES). Because microwave heating often is not uniform, studies in this area can be seriously flawed by simplistic assumptions of uniform sample temperature.

The use of microwaves has been investigated to affect plant growth by irradiating seeds or to achieve insect control. Most useful effects result from heating, however. Studies of agricultural applications of microwaves, eg, the drying of maize, exist. There has also been some investigation of the use of microwaves in the processing of pharmaceuticals (qv).

Chemical Applications. Chemical analysis can be conveniently accelerated by heating of samples in small pressure-tight plastic containers.

Catalysis can sometimes be improved through the use of microwaves, particularly pulsed microwaves. An important component of this process is believed to be an appropriate metallized combination catalyst-susceptor. Microwave catalysis is an active area of research.

Other chemical applications being studied include the use of microwaves in the petroleum industry, chemical synthesis, preparation of semiconductor materials, and the processing of polymers.

Other Applications. Microwave energy has been studied for the desulfurization of coal (qv) and treatment of wastes. Developments in microwave incinerators for medical and radioactive wastes have occurred.

Classical applications of microwave plasmas have been in the areas of torches, deposition of organic films (see THIN FILMS), fusion energy, and plasma chemistry, eg, synthesis or decomposition in nitrogen discharges. These developments continue, but emphasis has shifted to other areas, in particular the growing of diamond films and the development of efficient lamps for ultraviolet curing applications and visible light for illumination of large rooms, eg, those in museums, and outside areas. Another development is microwave excited discharges for lamps. One of the largest application areas for microwaves is in ceramics (qv) processing. Microwaves are used for sintering, drying, and enhancement of certain materials properties.

The use of lower frequency energy (rf) has been explored for *in situ* heating of oil shale (qv). Other power applications address the potential application to electrified roadways and microwave power transmission over large distances on land as well as from space to earth.

JOHN OSEPCHUK
Raytheon Company

J. M. Osepchuk, *IEEE Trans. Mic. Theory Tech.* **MTT-32,** 1200 (Sept. 1984).

J. Thuéry, *Microwaves: Industrial, Scientific and Medical Applications,* Artech House, Boston, Mass., 1992.

C. Buffler, *Microwave Cooking and Processing: Engineering Fundamentals for the Food Scientist,* Van Nostrand Reinhold, New York, 1992.

D. E. Clark and co-workers, eds., *Microwaves: Theory and Applications in Materials Processing* Vol. II, American Ceramics Society, Westerville, Ohio, 1993.

MILK AND MILK PRODUCTS

Milk has been a source for food for humans since the beginning of recorded history. Although the use of fresh milk has increased with economic development, the majority of consumption occurs after milk has been heated, processed, or made into butter. The milk industry became a commercial enterprise when methods for preservation of fluid milk were introduced. The successful evolution of the dairy industry from small to large units of production, ie, the farm to the dairy plant, depended on sanitation of animals, products, and equipment; cooling facilities; health standards for animals and workers; transportation systems; construction materials for process machinery and product containers; pasteurization and sterilization methods; containers for distribution; and refrigeration for products in stores and homes.

Composition and Properties

Milk consists of 85–89% water and 11–15% total solids (Table 1); the latter comprises solids-not-fat (SNF) and fat. Milk having a higher fat content also has higher SNF, with an increase of 0.4% SNF for each 1% fat increase. The principal components of SNF are protein, lactose, and minerals (ash). The fat content and other constituents of the milk vary with the animal species, and the composition of milk varies with feed, stage of lactation, health of the animal, location of withdrawal from the udder, and seasonal and environmental conditions. The nonfat solids, fat solids, and moisture relationships are well established and can be used as a basis for detecting adulteration with water (qv). Physical properties of milk are given in Table 2.

Nutritional Content. Vitamin D is added directly to processed milk to provide 400 *U.S. Pharmacopoeia* (USP) units/L. Vitamin A may be added to low fat skimmed milk to provide 1000 retinol equivalents (RE) per liter. Multivitamin, mineral fortified milk provides the recommended daily requirements.

Fat. Milk fat is a mixture of triglycerides and diglycerides (see FATS AND FATTY OILS). The triglycerides are short-chain, C_{24}–C_{46}; medium-chain, C_{34}–C_{54}; and long-chain, C_{40}–C_{60}. Milk fat contains more fatty acids than those in vegetables. In addition to being classified according to the number of carbon atoms, fatty acids in milk may be classified as saturated or unsaturated and soluble or insoluble. Fat carries numerous lipids (Table 3) and vitamins A, D, E, and K, which are fat soluble.

Table 1. Constituents of Milk from Various Mammals, Average, wt %

Species	Water	Fat	Protein	Lactose	Ash	Nonfat solids	Total solids
human	87.4	3.75	1.63	6.98	0.21	8.82	12.57
cows							
Holstein	88.1	3.44	3.11	4.61	0.71	8.43	11.87
Ayrshire	87.4	3.93	3.47	4.48	0.73	8.68	12.61
Brown Swiss	87.3	3.97	3.37	4.63	0.72	8.72	12.69
Guernsey	86.4	4.5	3.6	4.79	0.75	9.14	13.64
Jersey	85.6	5.15	3.7	4.75	0.74	9.19	14.34
goat	87.0	4.25	3.52	4.27	0.86	8.65	12.90
buffalo (India)	82.76	7.38	3.6	5.48	0.78	9.86	17.24
camel	87.61	5.38	2.98	3.26	0.70	6.94	12.32
mare	89.04	1.59	2.69	6.14	0.51	9.34	10.93
ass	89.03	2.53	2.01	6.07	0.41	8.49	11.02
reindeer	63.3	22.46	10.3	2.50	1.44	14.24	36.70

Table 2. Physical Properties of Milk

Property	Value
density at 20°C with 3–5% fat, average, g/cm^3	1.032
weight at 20°C, kg/L[a]	1.03
milk serum at 20°C, 0.025% fat	
density, g/cm^3	1.035
weight, kg/L[a]	1.03
freezing point, °C	−0.540
boiling point, °C	100.17
maximum density at °C	−5.2
electrical conductivity, S(= Ω^{-1})	(45−48) × 10^{-8}
specific heat at 15°C, kJ/(kg·K)[b]	
skim	3.94
whole	3.92
40% cream	3.22
fat	1.95
relative volumes	
4% milk at 20°C = 1, volume at 25°C	1.002
40% cream 20°C = 1.0010, volume at 25°C	1.0065
viscosity at 20°C, mPa·s(= cP)	
skim	1.5
whole	2.0
whey	1.2
surface tension of whole milk at 20°C, mN/m(= dyn/cm)	50
acidity, pH	6.3–6.9
titratable acid, %	0.12–0.15
refractive index at 20°C	1.3440–1.3485

[a] To convert kg/L to lb/gal, multiply by 8.34. [b] To convert kJ/(kg·K) to Btu/(lb·°F), divide by 4.183.

Processing

The processing operations for fluid or manufactured milk products include cooling, centrifugal sediment removal and cream (a mixture of fat and milk serum) separation, standardization, homogenization, pasteurization or sterilization, and packaging, handling, and storing.

Cooling. After removal from the cow by a mechanical milking machine, (at ~34°C), the milk is rapidly cooled to ≤4.4°C to maintain quality. At this low temperature, enzyme activity and microorganism growth are minimized. Commercial dairy production operations usually consist of a milking machine, a pipeline to convey the milk directly to the tank, and a refrigerated bulk milk tank in which the milk is

Table 3. Composition of Lipids in Cow Milk

Class of lipid	Range of occurrence
triglycerides of fatty acids, %	97.0–98.0
diglycerides, %	0.25–0.48
monoglycerides, %	0.016–0.038
keto acid glycerides, %	0.85–1.28
aldehydrogenic glycerides, %	0.011–0.015
glyceryl ethers, %	0.011–0.023
free fatty acids, %	0.10–0.44
phospholipids, %	0.2–1.0
cerebrosides, %	0.013–0.066
sterols, %	0.22–0.41
free neutral carbonyls, ppm	0.1–0.8
squalene, ppm	70
carotenoids, ppm	7–9
vitamin A, ppm	6–9
vitamin D, ppm	0.0085–0.021
vitamin E, ppm	24
vitamin K, ppm	1

cooled and stored for later pickup. Rancidity is avoided by preventing air from passing through the warm milk, via air leaks and long risers in the pipeline. The pipelines, made of glass or stainless steel, are usually cleaned by a cleaning-in-place (CIP) process. Bulk milk is pumped from the refrigerated bulk milk tank to a tanker and transported to a processing plant.

Centrifugation. Centrifugal devices include clarifiers for removal of sediment and extraneous particulates, and separators for removal of fat (cream) from milk (see SEPARATION, CENTRIFUGAL).

A standardizing clarifier removes fat to provide a certain fat content while removing sediment, a clarifixator partially homogenizes while separating the fat, and a high speed clarifier removes bacteria cells in a bactofugation process. Clarifiers have replaced filters in the dairy plant for removing sediment, although the milk may have been previously strained or filtered on the farm.

Bactofugation is not used for ordinary fluid milk, but for sterile milk or cheese. Although no longer used in the United States, bactofugation is a specialized process of clarification in which two high velocity centrifugal bactofuges operate at 20,000 rpm in series. The first device removes 90% of the bacteria, and the second removes 90% of the remaining bacteria, providing a 99% bacteria-free product.

Originally, cream separators were basic plant equipment, and dairy plants were known as creameries. The original gravity-fed units incorporated air to produce foam and separators developed 5,000–10,000 times the force of gravity to separate the fat (cream) from the milk. Skimmed milk was discarded or returned to the farm as animal feed, and the cream was used for butter and other fat-based dairy products. Separators in the 1990s are pressure- or forced-fed sealed airtight units. The separator removes all or a portion of the fat, and the skimmed milk or reduced fat milk is sold as a beverage or ingredient in other formulated foods.

Standardization. Standardization is the process of adjusting the ratio of butterfat and solids-not-fat (SNF) to meet legal or industry standards. Adding cream of high butterfat milk into serum of low butterfat milk might result in a product with low SNF, thus careful control must be exercised.

Homogenization. Homogenization is an integral part of continuous HTST pasteurization. It is the process by which a mixture of components is treated mechanically to give a uniform product that does not separate. In milk, the fat globules are broken up into small particles that form a more stable emulsion in the milk. In homogenized milk, the fat globules do not rise by gravity to form a creamline as with untreated whole milk. The fat globules in raw milk are 1–15 μm in diameter; they are reduced to 1–2 μm by homogenization. The U.S. Public Health Service defines homogenized milk as "milk that has been treated to insure the breakup of fat globule to such an extent that after 48 h of quiescent storage at 45°F (7°C) no visible cream separation occurs in the milk..." Most fluid milk is homogenized.

Pasteurization. Pasteurization is the process of heating milk to kill pathogenic bacteria, and most other bacteria, without greatly altering the flavor. It also inactivates certain enzymes, eg, phosphatase, thus the degree of pasteurization can be determined by measuring the phosphatase present. The principles were developed by and named after Pasteur and his work in 1860–1864. Since then, stringent codes have been developed to assure that pasteurization is done properly. The basic regulations are included in the U.S. Public Health Service Pasteurized Milk Ordinance which has been adopted by most local and state jurisdictions. The quality of milk depends on the care of the animals that produce it, the environment on the farm, and the care of the product throughout.

Pasteurization may be carried out by batch- or continuous-flow processes. In the batch process, each particle of milk must be heated to at least 63°C and held continuously at this temperature for at least 30 min. In the continuous process, milk is heated to at least 72°C for at least 15 s in what is known as high temperature—short time (HTST) pasteurization, the primary method used for fluid milk. For milk products having a fat content above that of milk or that contain added sweeteners, 66°C is required for the batch process and 75°C for the HTST process. For either method, following pasteurization the product should be cooled quickly to $\leq$7.2°C. Time—temperature relationships have been established for other products including ice cream mix, which is heated to 78°C for 15 s, and eggnog, which must be pasteurized at 69°C for 30 min or 80°C for 25 s. Another continuous pasteurization process, known as ultrahigh temperature (UHT), employs a shorter time (2 s) and a higher temperature (minimum 138°C). The UHT process approaches aseptic processing.

The principal continuous-flow process is the high temperature—short time (HTST) method. The product is heated to at least 72°C and held at that temperature for not less than 15 s.

The equipment needed includes a balance tank, regenerative heating unit, positive pump, plates for heating to pasteurization temperature, tube or plates for holding the product for the specified time, a flow-diversion valve (FDV), and a cooling unit (Fig. 1). Often the homogenizer and booster pump also are incorporated into the HTST circuit.

Cleaning Systems. Both manual and automatic methods are used for cleaning food processing (qv) equipment. In dairy plants, the equipment surfaces and pipelines are cleaned in place at least once every 24 hours. Cleaning-in-place (CIP) systems evolved from recirculating cleaning solutions in pipelines and equipment to a highly automatic system with valves, controls, and timers. The results of cleaning in place are influenced by equipment surfaces, time of exposure, and the temperature and concentration of the solution being circulated. Cleaning is a mechanical–chemical operation.

In the CIP procedure, a cold or tempered aqueous prerinse is followed by circulation of a cleaning solution for 10 minutes to one hour at 54–82°C. The temperature of the cleaning solution should be as low as possible, because hot water rinses may harden the food product on the surface being cleaned, but high enough to avoid excess cleaning chemicals. A wide variety of cleaning solutions may be used, depending on the food product, hardness of water, and equipment.

A CIP system includes pipelines, interconnected with valves to direct fluid to appropriate locations, and the control circuit, which consists of interlines to control the valves that direct the cleaning solutions and water through the lines, and air lines which control and move the valves. A programmer controls the timing and the air flow to the valves on a set schedule. The 3A Standards for CIP components, equipment, and installation have been developed.

Storage, Cooling, Shipping, and Packaging

Bulk Milk Tanks. Commercial dairy production enterprises generally employ tanks in which the milk is cooled and stored. In some operations, the warm milk is first cooled and then stored in a tank; 3A Standards have been established for their design and operation. Among other requirements, the milk must be cooled to 4.4°C within two hours after milking. The temperature must not be permitted to increase above 10°C when warm milk from the following milking is placed in the tank. Bulk milk tanks are classified according to method of refrigeration, ie, direct expansion (DX) or ice bank (IB); pressure in tank, ie, atmospheric or vacuum; regularity of pickup, ie, every day or every other day; capacity, in liters, when full or at amount which can be received per milking; shape, ie, cylindrical, half-cylindrical, or rectangular; position, ie, vertical or horizontal; and method of cooling refrigeration condenser, ie, by water, air, or both.

Cooling. A compression refrigeration system, driven by an electric motor, supplies cooling for either direct expansion or ice bank systems. Milk coming from cows may be rapidly cooled over a stainless steel surface cooler before entering a bulk tank. The cooler may either use compression refrigeration or have two sections, one using cold water followed by a section using compression refrigeration.

Shipping. Bulk milk is hauled to the processing plant in insulated tanks using truck tanks or trailer tankers. The milk is transferred from the bulk tank to the tanker with a positive or centrifugal-type pump. For routes of some distance, pick-up every other day reduces handling costs.

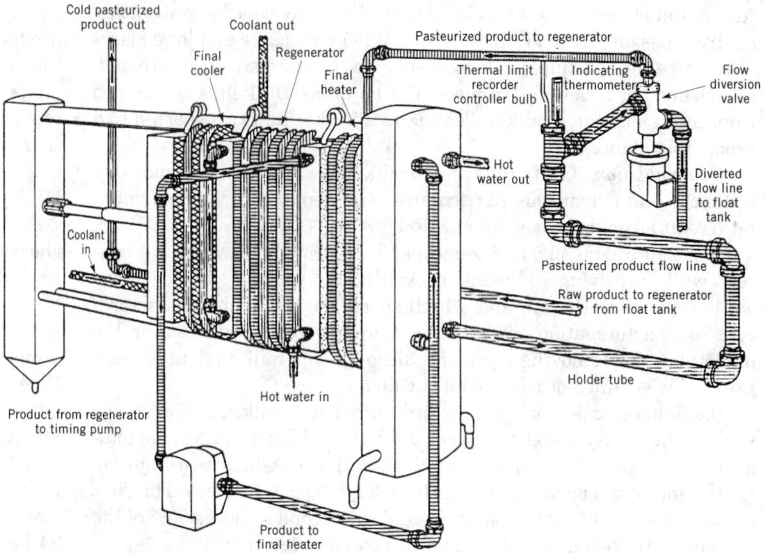

Figure 1. Flow through a typical HTST plate pasteurizer, where ▤ is raw milk, ▥ pasteurized, ▧ hot water, and ▨ coolant. Courtesy of St. Regis Crepaco (now APV Crepaco).

Bulk milk-receiving operations consist primarily of transferring milk from the tanker to a storage tank in the plant. Practically all Grade A milk is handled in bulk. The handling of milk in 38-L cans requires equipment and space for quality and quantity check of the product, washing of cans, and conveyors for moving and storing the cans.

Packaging. Aseptic packaging was developed in conjunction with high temperature processing and has contributed to make sterilized milk and milk products a commercial reality.

Health and Safety Factors

Milk may be a carrier of diseases from animals or from other sources to humans. To avoid contamination before pasteurization, healthy animals should be separated from sick animals or those with infected udders. The animals should be clean, kept in clean housing with clean air, and handled by workers and equipment under strictly sanitary conditions. Post-pasteurization contamination can occur as a result of improper handling, due to exposure to contaminated air, improperly sanitized equipment, or an infected worker.

Proper refrigeration prevents the growth of some microorganisms, such as Salmonella, and the production of toxins, such as *Staphylococcus aureus.* The growth of bacteria *Escherichia coli* and *Bacillus cereus* is substantially checked by proper cooling and handling of milk. Pasteurization is the best means of prevention.

Manufactured Products

In the United States, 62% of fluid milk production is used for manufactured products, mainly cheese, evaporated and sweetened condensed milk, nonfat dry milk, and ice cream. Evaporated and condensed milk and dry milk are made from milk only; other ingredients are added to make ice cream and sweetened condensed milk.

Evaporated and Condensed Milk. Evaporated milk is produced by removing moisture from milk, under a vacuum, followed by packaging and sterilizing in cans. The milk is condensed to half its volume in single- or multiple-effect evaporators. The final product has a fat to solids-not-fat ratio of 1:2.28, and is standardized before and after evaporation.

Large quantities of evaporated milk are used to manufacture ice cream, bakery products, and confectionery products (see BAKERY PROCESSES AND LEAVENING AGENTS). When used for manufacturing other foods, evaporated milk is not sterilized, but placed in bulk containers, refrigerated, and used fresh. This product is called condensed milk.

For sweetened condensed milk, unlike evaporated milk that is sterilized, sugar is added as a preservative and provides keeping quality. The equipment is similar to that used for evaporated milk, except that sugar is added in a hot well before condensing (evaporating) the liquid. Preheating pasteurizes the product and no sterilization is needed. According to standards, sweetened condensed milk must contain a minimum of 8.5% fat and 28% total milk solids, including fat (fat to solids-not-fat ratio = 1:2.3). The final product contains 43–45% sugar. Sweetened condensed skimmed milk has not less than 24% total milk solids, but up to 50% sugar may be added.

Dry Milk. Dry milk provides long-term storage capabilities, supplies a product that can be used for food manufacturing operations, and because of its reduced volume and weight, transportation and storage costs are reduced. Dry milk has been used for manufactured products, but is used to a much greater extent for beverage products.

Cream. Cream is a high fat product which is secured by gravity or mechanical separation through differential density of the fat and the serum. Fat content may range from 10 to 40%, depending on use and federal and state laws. The U.S. Public Health Service milk ordinance defines cream as a product that contains not less than 18% milk fat. Whipping cream has a fat content of 30–40%, and light cream has a fat content of 18–30%. Half-and-half, suggesting a mixture of cream and milk, has not less than 10.5% milk fat, and in some states up to 12%. Cream is standardized in the same manner as milk, following separation. The addition of whole milk rather than serum is preferred.

The sale of fresh cream as a table item for serving has decreased greatly over the last 30 years. A variety of cream and fat substitutes are available for spreads, toppings, whiteners, and cooking (see DAIRY SUBSTITUTES).

Anhydrous Milk Fat. One high milk-fat material is butter oil (99.7% fat), also called anhydrous milk fat or anhydrous butter oil if less than 0.2% moisture is present. Although the terms are used interchangeably, anhydrous butter oil is made from butter and anhydrous milk fat is made from whole milk. For milk and cream there is an emulsion of fat-in-serum, for butter oil and anhydrous milk fat there is an emulsion of serum-in-fat, such as with butter. It is easier to remove moisture in the final stages to make anhydrous milk fat with the serum-in-fat emulsion.

Butter. In the United States about 10 wt % of edible fats used are butter. Butter is defined as a product that contains 80% milk fat with not more than 16% moisture. It is made of cream with 25–40% milk fat. The process is primarily a mechanical one in which the cream, an emulsion of fat-in-serum, is changed to butter, an emulsion of serum-in-fat. The process is accomplished by churning or by a continuous operation with automatic controls.

Buttermilk. Buttermilk is drained from butter (churn) after butter granules are formed; as such, it is the fluid other than the fat which is removed by churning. Buttermilk may be used as a beverage or may be dried and used for baking. Buttermilk from churning is ~ 91% water and 9% total solids. Total solids include lactose, 4.5%; nitrogenous matter, 3.4%; ash, 0.7%; and fat, 0.4%.

Cultured buttermilk is that which is produced by the fermentation (qv) of skimmed milk, often with some cream added. The principal fermentation organisms used are *Lactococcus lactis* subsp. *cremoris, Lactococcus lactis* subsp. *lactis,* and *Leuconostic citrovorum.* The effect of the high processing temperature and the lactic acid provide an easily digestible product. Dried buttermilk is used primarily for baking, confectionery, and dairy products.

Cheese. The making of cheese is based on the coagulation of casein from milk, and to a minor extent the proteins of whey. The casein is precipitated by acidification which can be accomplished by natural souring of milk. The procedures for making cheese vary greatly and cheese products are countless. The composition and handling of the original milk, bacterial flora, and starter culture are the basis variables, which along with heat treatments, flavoring, salting, and forming, affect the final product.

Yogurt. Yogurt is a fermented milk product that is rapidly increasing in consumption in the U.S. Milk is fermented with *Lactobacillus bulgaricus* and *Streptococcus thermophilous* organisms that produce lactic acid. Usually some cream or nonfat dried milk is added to the milk in order to obtain a heavy-bodied product.

Frozen Desserts. Ice cream is the principal frozen dessert produced in the U.S. It is known as the American dessert and was first sold in New York City in 1777. Frozen yogurt is also gaining in acceptance as a dessert. The composition of various frozen desserts is given in Table 4. Although ice cream is by far the most important frozen dessert, other frozen desserts such as frozen yogurt, ice milk, sherbet, and mellorine-type products are also popular. The consumption of frozen yogurt has been increasing rapidly.

By-Products From Milk. Milk is a source for numerous by-products resulting from the separation or alteration of the components. These components may be used in other so-called nondairy manufactured foods, dietary foods, pharmaceuticals (qv), and as a feedstock for numerous industries, such as casein for glue.

Lactose (milk sugar), makes up about 5% of cow's milk and is used for foods and pharmaceutical products.

Milk contains proteins and essential amino acids lacking in many other foods. Casein is the principal protein in the skimmed milk (nonfat) portion of milk (3–4% of the weight). After it is removed from the liquid portion of milk, whey remains. Whey can be denatured by heat treatment of 85°C for 15 minutes.

Casein is used to fortify flour, bread, and cereals. Casein also is used for glues and microbiological media. Calcium caseinate is made from a pressed casein, by rinsing, treating with calcium hydroxide, heating, and mixing followed by spray drying. A product of 2–4% moisture is obtained. Casein hydrolyzates are produced from dried casein.

Many so-called nondairy products such as coffee cream, topping, and icings utilize caseinates (see DAIRY SUBSTITUTES). In addition to fulfilling a nutritional role, the caseinates impart creaminess, firmness, smoothness, and consistency of products. Imitation meats and soups use caseinates as an extender and to improve moistness and smoothness.

Nutritional Value of Milk Products. Milk is considered one of the principal sources of nutrition for humans. Some people are intolerant to one or more components of milk so must avoid the product or consume a treated product. One example is intolerance to lactose in milk. Fluid milk is available in which the lactose has been treated to make it more digestible. The consumption of milk fat, either in fluid milk or in products derived from milk, has decreased markedly. Complete information on the nutritive value of milk and milk products is provided on product labels.

The concern by consumers about cholesterol has stimulated the development of methods for its removal. Three principal approaches are in the pilot-plant stages: use of enzymes, supercritical fluid extraction, and steam distillation. Using known techniques, it is not possible to remove all cholesterol from milk. Therefore, FDA guidelines identify cholesterol-free foods as containing less than 2 mg cholesterol per serving, and low cholesterol foods as containing from 2 to 20 mg.

Table 4. Composition of Frozen Desserts,[a] %

Component	Ice cream: Premium[b]	Ice cream: Average	Ice milk	Sherbet	Ice	Soft-serve
milk fat[c]	16.0	10.5	3.0	1.5		6.0
milk solids, nonfat	9.0	11.0	12.0	3.5		12.0
sucrose	16.0	12.5	12.0	19.0	23.0	9.0
corn syrup solids		5.5	7.0	9.0	7.0	6.0
stabilizer[d]	0.1	0.3	0.3	0.5	0.3	0.3
emulsifier[d]		0.1	0.15			0.2
total solids, kg/L	41.1	39.9	34.45	33.5	30.3	33.5
	1.09	1.12	1.13	1.14	1.13	1.11
overrun, %	65–70	95–100	90–95	50	10	40
~kg/L, from freezer	0.64	0.55	0.57	0.74	1.01	0.77

[a] Frozen desserts containing vegetable fat (mellorine-type) are permitted in some states. A wide variation of composition exists depending on individual state standards. [b] To be classified as custard or French, product must contain ≥ 1.4% egg yolk solids. [c] Milk-fat content regulated by individual state. [d] Usage level as recommended by manufacturer.

Biotechnology

Biotechnology is being applied in the dairy industry. A significant and controversial development is the technique of producing transgenic animals, ie, animals in which hereditary deoxyribonucleic acid (DNA) has been augmented by DNA from another source, using recombinant DNA (rDNA) techniques.

CARL W. HALL
Engineering Information Services

Grade A Pasteurized Milk Ordinance, U.S. Department of Health and Human Services, Public Health Service Publication No. 229, Rev. ed., Washington, D.C., 1989.

R. T. Marshall, *Standard Methods for the Examination of Dairy Products,* American Public Health Association, Washington, D.C., 1993, 546 pp.

C. W. Hall, A. W. Farrall, and A. L. Rippen, eds., *Encyclopedia of Food Engineering,* 2nd ed., Avi Publishing Co., Westport, Conn., 1986, 882 pp.

D. R. Heldman and D. R. Lund, eds., *Handbook of Food Engineering,* Marcel Dekker, Inc., New York, 1992, 756 pp.

MINERAL NUTRIENTS

Minerals that are essential to life are the source of metals and other inorganic elements involved in the most fundamental processes. For example, oxygen, required by the cells of animals, is utilized with the aid of metal complexes. In humans both iron-containing hemoglobin and zinc-containing carbonic anhydrase play pivotal roles in binding oxygen and delivering it to the cells. Moreover, enzymes developed to protect cells from high levels of oxygen also contain metals. One such class of protective enzymes is known as the superoxide dismutases (SODs). These contain metals such as manganese, copper, zinc, and iron. Mutations in the copper- and zinc-containing superoxide dismutase gene have been linked to amyotrophic lateral sclerosis.

Table 1. Essential Mineral Nutrients

Element[a]	Body content, mg/kg body wt	Daily requirement, mg
	Principal elements	
calcium	14,000–20,000	800–1,200[b,c]
phosphorus	11,000–12,000	800–1,200[b,c]
sulfur	1,600–2,500	[d]
potassium	2,000–3,500	2,000
sodium	1,500–1,600	500
chlorine	1,200–1,500	750
magnesium	270–500	280[b,c,e];350[b,f]
	Trace and ultratrace elements	
iron	60–66	10[b,f]; 15[b,e]
fluorine	37	1.5–4.0[b]
zinc	33–50	12[b,c,e]; 15[b,f]
silicon	15–16	5–20
copper	1.0–2.5	1.5–3.0
boron	0.69	0.5–1.0
selenium	0.2–0.3	0.055[b,c,e]; 0.07[b,f]
iodine	0.2–0.4	0.15[b,c]
manganese	0.2–4.0	2.0–5.0[b]
molybdenum	0.1–0.5	0.075–0.25[b]
chromium	0.06–0.2	0.05–0.2[b]
cobalt	0.02	0.003[g]
tin	0.2	
vanadium	0.14	<0.01
nickel	0.07–0.14	<0.10

[a] Generally not ingested in elemental form. [b] Values are for adults. [c] Increased amounts are required during pregnancy and lactation. [d] Adequate intake with adequate intake of protein. [e] Value for females. [f] Value for males. [g] As vitamin B_{12}.

As for other biological substances, states of dynamic equilibrium exist for the various mineral nutrients as well as mechanisms whereby a system can adjust to varying amounts of these minerals in the diet. In forms usually found in foods, and under circumstances of normal human metabolism, most nutrient minerals are not toxic when ingested orally. Amounts considerably greater than the recommended dietary allowances (RDAs) can generally be eaten without concern for safety (Table 1).

Some elements found in body tissues have no apparent physiological role, but have not been shown to be toxic. Examples are rubidium, strontium, titanium, niobium, germanium, and lanthanum. Other elements are toxic when found in greater than trace amounts, and sometimes in trace amounts. These latter elements include arsenic, mercury, lead, cadmium, silver, zirconium, beryllium, and thallium. Numerous other elements are used in medicine in non-nutrient roles. These include lithium, bismuth, antimony, bromine, platinum, and gold. The interactions of mineral nutrients with carbohydrates, fats, and proteins, minerals with vitamins (qv), and mineral nutrients with toxic elements are areas of active investigation.

The amount of each element required in daily dietary intake varies with the individual bioavailability of the mineral nutrient.

The Principal Elements

Calcium. Calcium, the most abundant mineral element in mammals, comprises 1.5–2.0 wt % of the adult human body, over 99 wt % of which is present in bones and teeth. About 48% of serum calcium is ionic, ca 46% is bound to blood proteins, the rest is present as diffusible complexes, eg, of citrate. The calcium ion level must be maintained within definite limits.

Bones act as a reservoir of certain ions, in particular Ca^{2+} and PO_4^{3-}, which readily exchange between bones and blood. Bone structure comprises a strong organic matrix combined with an inorganic phase which is principally hydroxyapatite, $3\ Ca_3(PO_4)_2{\cdot}Ca(OH)_2$. Bones contain two forms of hydroxyapatite. The less soluble crystalline form contributes to the rigidity of the structure. The crystals are quite stable, but because of the small size present a very large surface area available for rapid exchange of ions and molecules with other tissues. There is also a more soluble intercrystalline fraction. Bone salts also contain small amounts of magnesium, sodium, carbonate, citrate, chloride, and fluoride. Osteoporosis is reported to result when bone resorption is relatively faster than bone formation. The calcium ion, necessary for blood-clot formation, stimulates release of bloodclotting factors from platelets (see BLOOD, COAGULANTS AND ANTICOAGULANTS).

In normal adults, the blood Ca_{2+} level is established by an equilibrium between blood Ca^{2+} and the more soluble intercrystalline calcium salts of the bone. Additionally, a subtle and intricate feedback mechanism responsive to the Ca^{2+} concentration of the blood that involves the less soluble crystalline hydroxyapatite comes into play. The thyroid and parathyroid glands, the liver, kidney, and intestine also participate in Ca^{2+} control.

In addition to hypocalcemia, tremors, osteoporosis, and muscle spasms (tetary), calcium deficiency can lead to rickets, osteomalacia, and possibly heart disease. These, as well as Paget's disease, can also result from faulty utilization of calcium. Calcium excess can lead to excess secretion of calcitonin, possible calcification of soft tissues, and kidney stones when combined with magnesium deficiency.

Phosphorus. Eighty-five percent of the phosphorus, the second most abundant element in the human body, is located in bones and teeth. Whereas there is constant exchange of calcium and phosphorus between bones and blood, there is very little turnover in teeth. The Ca:P ratio in bones is constant at about 2:1. Every tissue and cell contains phosphorus, generally as a salt or ester of mono-, di-, or tribasic phosphoric acid, as phospholipids, or as phosphorylated sugars. Phosphorus is involved in a large number and wide variety of metabolic functions. Examples are carbohydrate metabolism, adenosine triphosphate (ATP) from fatty acid metabolism, and oxidative phosphorylation.

The formation of phosphate esters is the essential initial process in carbohydrate metabolism (see CARBOHYDRATES). The glycolytic, ie, anaerobic or Embden-Meyerhof pathway comprises a series of nine such esters. The phosphogluconate pathway, starting with glucose, comprises a succession of 12 phosphate esters. Cyclic adenosine monophosphate (cAMP), produced from ATP, is involved in a large number of cellular reactions including glycogenolysis, lipolysis, active transport of amino acids, and synthesis of protein. Inorganic phosphate ions are involved in controlling the pH of blood. The principal anion of intercellular fluid is HPO_4^{2-}.

Phospholipids, components of every cell membrane, are active determinants of membrane permeability. They are sources of energy, components of certain enzyme systems, and involved in lipid transport in plasma. Because of their polar nature, phospholipids can act as emulsifying agents. The structure of most phospholipids resembles that of triglycerides except that one fatty acid radical has been replaced by a radical derived from phosphoric acid and a nitrogen base, eg, choline or serine.

Phosphorus is an essential component of nucleic acids, polymers consisting of chains of nucleosides, a sugar plus a nitrogenous base, and joined by phosphate groups. In ribonucleic acid (RNA), the sugar is D-ribose; in deoxyribonucleic acids (DNA), the sugar is 2-deoxy-D-ribose.

Phosphorus nutrient deficiency can lead to rickets, osteomalacia, and osteoporosis, whereas an excess can produce hypocalcemia. Faulty utilization of phosphorus results in rickets, osteomalacia, osteoporosis, and Paget's disease, and renal or vitamin D-resistant rickets.

Sulfur. Sulfur is present in every cell in the body, primarily in proteins containing the amino acids methionine, cystine, and cysteine. Inorganic sulfates and sulfides occur in small amounts relative to total body sulfur, but the compounds that contain them are important to metabolism. Sulfur intake is thought to be adequate if protein intake is adequate and sulfur deficiency has not been reported.

Although sulfur is in the same group of the Periodic Table, Group 16(VIA), as oxygen, sulfur functions much more like phosphorus,

Group 15(VA), in biological systems. In fat metabolism, sulfur plays a key role analogous to that of phosphorus in carbohydrate metabolism. Fatty acid synthesis and degradation begin and end with the same compound, acetyl-S coenzyme A (acetyl–SCoA).

Detoxification systems in the human body often involve reactions that utilize sulfur-containing compounds. For example, reactions in which sulfate esters of potentially toxic compounds are formed, rendering these less toxic or nontoxic, are common as are acetylation reactions involving acetyl–SCoA. Another important compound is *S*-adenosylmethionine (SAM), the active form of methionine. SAM acts as a methylating agent, eg, in detoxification reactions such as the methylation of pyridine derivatives, and in the formation of choline (qv), creatine, carnitine, and epinephrine. Sulfur nutrient deficiency results in retarded growth, and faulty utilization in homocystinuria.

Sodium and Potassium. Whereas sodium ion is the most abundant cation in the extracellular fluid, potassium ion is the most abundant in the intracellular fluid. Small amounts of K^+ are required in the extracellular fluid to maintain normal muscle activity. Some sodium ion is also present in intracellular fluid.

Sodium ion acts in concert with other electrolytes, in particular K^+, to regulate the osmotic pressure and to maintain the appropriate water and pH balance of the body. Homeostatic control of these functions is accomplished by the lungs and kidneys interacting by way of the blood. Sodium is essential for glucose absorption and transport of other substances across cell membranes. It is also involved, as is K^+, in transmitting nerve impulses and in muscle relaxation. Potassium ion acts as a catalyst in the intracellular fluid, in energy metabolism, and is required for carbohydrate and protein metabolism.

Maintenance of the appropriate concentrations of K^+ and Na^+ in the intra- and extracellular fluids involves active transport, ie, a process requiring energy. Sodium ion in the extracellular fluid (0.136–0.145 *M* Na^+) diffuses passively and continuously into the intracellular fluid (<0.01 *M* Na^+) and must be removed. This sodium ion is pumped from the intracellular to the extracellular fluid, while K^+ is pumped from the extracellular (ca 0.004 *M* K^+) to the intracellular fluid (ca 0.14 *M* K^+). The energy for these processes is provided by hydrolysis of adenosine triphosphate (ATP) and requires the enzyme Na^+-K^+ ATPase, a membrane-bound enzyme which is widely distributed in the body. In some cells, eg, brain and kidney, 60–70 wt % of the ATP is used to maintain the required Na^+-K^+ distribution.

Sodium and potassium ions are actively absorbed from the intestine. As a consequence of the electrical potential caused by transport of these ions, an equivalent quantity of Cl^- is absorbed. The resulting osmotic effect causes absorption of water.

Selective excretion and reabsorption of Na^+ and K^+ are accomplished by means of the kidney tubular cell membranes. The volume of extracellular fluid is directly related to the Na^+ concentration which is closely controlled by the kidneys. Homeostatic control of Na^+ concentration depends on the hormone aldosterone. The kidney secretes a proteolytic enzyme, rennin, which is essential in the first of a series of reactions leading to aldosterone. In response to a decrease in plasma volume and Na^+ concentration, the secretion of rennin stimulates the production of aldosterone resulting in increased sodium retention and increased volume of extracellular fluid.

Salt-free or low salt diets often are prescribed for hypertensive patients. However, sodium chloride increases the blood pressure in some individuals but not in others. Conversely, restriction of dietary NaCl lowers the blood pressure of some hypertensives, but not of others. Genetic factors and other nutrients, eg, Ca^{2+} and K^+, may be involved. The optimal intakes of Na^+ and K^+ remain to be established.

Potassium and/or sodium deficiency can lead to muscle weakness and sodium deficiency to nausea. Hyperkalemia resulting in cardiac arrest is possible from 18 g/d of potassium combined with inadequate kidney function. Faulty utilization of K^+ and/or Na^+ can lead to Addison's or Cushing's disease.

Chlorine. The chlorides are essential in the homeostatic processes maintaining fluid volume, osmotic pressure, and acid–base equilibria. Most chloride is present in body fluids; a little is in bone salts. Chloride is the principal anion accompanying Na^+ in the extracellular fluid. Less than 15 wt % of the Cl^- is associated with K^+ in the intracellular fluid. Chloride passively and freely diffuses between intra- and extracellular fluids through the cell membrane. If chloride diffuses freely, but most Cl^- remains in the extracellular fluid, it follows that there is some restriction on the diffusion of phosphate.

Some of the blood Cl^- is used for formation in the gastric glands of hydrochloric acid, HCl, required for digestion. Hydrochloric acid is secreted into the stomach where it acts with gastric enzymes in the digestive processes. The chloride is then reabsorbed with other nutrients into the blood stream. Chloride is actively transported in gastric and intestinal mucosa. In the kidney, chloride is passively reabsorbed in the thin ascending loop of Henle and actively reabsorbed in the thick segment of the ascending loop, ie, the distal tubule. In the chloride shift, Cl^- plays an important role in the transport of carbon dioxide (qv).

Numerous neurotransmitter receptors, eg, glutamate, γ-aminobutyric acid (GABA), and benzodiazepine (called the valium receptor), have been identified as chloride channel proteins. The genetic defect in cystic fibrosis involves defective-functioning chloride channel proteins with excessive Cl^- loss. Deficient Cl^- during development adversely affects language skills in humans, as well as impaired growth in infants and metabolic alkalosis.

Fruit and vegetable juices high in potassium have been recommended to correct hypokalemic alkalosis in patients on diuretic therapy. Apparently the efficacy of this treatment is questionable. A possible reason for ineffectiveness is the low Cl^- content of most of these juices. Because Cl^- is high only in juices in which Na^+ is high, these have to be excluded.

Magnesium. In the adult human, 50–70% of the magnesium is in the bones associated with calcium and phosphorus. The rest is widely distributed in the soft tissues and body fluids. Most of the nonbone Mg^{2+}, like K^+, is located in the intracellular fluid where it is the most abundant divalent cation. Magnesium ion is efficiently retained by the kidney when the plasma concentration of Mg^{2-} falls; in this respect it resembles Na^+. The functions of Na^+, K^+, Mg^{2+}, and Ca^{2+} are interrelated so that a deficiency of Mg^{2+} affects the metabolism of the other three ions.

Magnesium is essential in numerous metabolic processes. It is the activator of many enzymes, eg, adenyl cyclase, alkaline phosphatases, and the phosphokinases, pyrophosphatases, and thiokinases. Because the phosphokinases are required for the hydrolysis and transfer of phosphate groups, magnesium is essential in glycolysis and in oxidative phosphorylation. The thiokinases are required for the initiation of fatty acid degradation. Magnesium is also required in systems in which thiamine pyrophosphate is a coenzyme.

As an activator of the phosphokinases, magnesium is essential in energy-requiring biological processes, such as activation of amino acids, acetate, and succinate; synthesis of proteins, fats, coenzymes, and nucleic acids; generation and transmission of nerve impulses; and muscle contraction.

Regulation of serum Mg^{2+} appears to result from a balance among intestinal absorption, renal reabsorption, and excretion. The controlling factor is probably the renal threshold.

A severe magnesium deficiency in humans is seldom encountered except as a secondary effect resulting from numerous disease states, eg, chronic alcoholism with malnutrition, acute or chronic renal disease, long-term Mg^{2+}-free parenteral feeding, protein–calorie malnutrition, and hyperthyroidism. In these situations, it is difficult to attribute specific clinical manifestations to magnesium deficiency. The specific role of magnesium in cardiovascular disease, eg, arrythmia, spasms, or ischemia, remains a subject of conflicting research findings.

Neuromuscular irritability, convulsions, muscle tremors, mental changes such as confusion, disorientation, and hallucinations, heart disease, and kidney stones have all been attributed to magnesium deficiency. Excess Mg^{2+} can lead to intoxication exemplified by drowsiness, stupor, and eventually coma.

Trace Elements and Ultratrace Elements

Iron. The total body content of iron, ie, 3–5 g, is recycled more efficiently than other metals. There is no mechanism for excretion of iron and what little iron is lost daily, ie, ca 1 mg in the male and 1.5 mg in the menstruating female, is lost mainly through exfoliated mucosal, skin, or hair cells, and menstrual blood.

A large percentage of the iron in the human body is in hemoglobin: 85 wt % in the adult female, 60 wt % in the adult male. The remainder is present in other iron-containing compounds involved in basic metabolic functions, or in iron transport or storage compounds.

Absorption of iron from food to maintain homeostasis, tightly controlled, increases in instances of increased demands, such as during pregnancy and lactation, and iron-deficiency states which are the result of blood loss or iron-deficiency anemia resulting from inadequate iron intake. Iron absorption is greatly reduced in the normal individual when iron stores are adequate or excessive. Absorption is enhanced by acid conditions and reducing agents. Heme iron from animal sources is absorbed more readily than nonheme iron from cereals and vegetables.

A system of internal iron exchange exists which is dominated by the iron required for hemoglobin synthesis. For formation of red blood cells, iron stores can furnish 10–40 mg/d of iron, as compared to 1–3 mg from dietary sources. Only ca 10 wt % of ingested iron actually is absorbed. Transferrin is essential for movement of iron and without it, as in genetic absence of transferrin, iron overload occurs in tissues. This hereditary atransferrinemia is coupled with iron-deficiency anemia. The iron overload in hereditary or acquired hemochromatosis results in fully saturated transferrin and is treated by phlebotomy.

Iron deficiency is a significant worldwide nutritional problem and cause of anemia which can also lead to a decreased resistance to infection. Insufficient dietary iron intake; iron losses, eg, bleeding and parasite infestation; and malabsorption of iron are the principal causes. The groups at greatest risk for developing iron-deficiency anemia are menstruating females, pregnant or nursing females, and young children. Children can experience impaired psychomotor development and intellectual performance.

Iron toxicity resulting from excess absorbable iron ingestion is rare except in Africa where fermented beverages made in large iron pots have levels of iron approaching 80 mg/L in a brew where the pH is very low. This results in Bantu siderosis which can result in hemochromatosis, ie, damage to various organs from excessive storage of iron. This condition can cause numerous disease states, eg, hepatic fibrosis and diabetes in 80% of the cases of idiopathic hemochromatosis patients. Iron overload is frequently a complication of repeated blood transfusions in anemias, eg, thalassemia. The lethal dose of ferrous sulfate for a two-year old is 2 g; for an adult the lethal dose is from 200–300 g.

Fluorine. Fluoride is present in the bones and teeth in very small quantities. Human ingestion is from 0.7–3.4 mg/d from food and water.

Fluoridation of public water supplies, a common practice throughout much of the United States, may be an effective means of significantly reducing the incidence of dental caries (see DENTAL MATERIALS; FLUORINE COMPOUNDS, INORGANIC). Concern regarding the narrow range of safety between effective and toxic fluoride concentrations has been expressed and poisoning from excessive fluoride, fluorosis, added to public water has been reported. Assertions that fluoridation of water supplies increases the incidence of cancer have not been substantiated.

Excess fluoride ingestion damages developing teeth, causing mottling, chalky-white coloration, and pitting.

Zinc. The 2–3 g of zinc in the human body are widely distributed in every tissue and tissue fluid. About 90 wt % is in muscle and bone; unusually high concentrations are in the choroid of the eye and in the prostate gland. Almost all of the zinc in the blood is associated with carbonic anhydrase in the erythrocytes. Zinc is concentrated in nucleic acids, and found in the nuclear, mitochondrial, and supernatant fractions of all cells.

Zinc is essential for the function of many enzymes, either in the active site, ie, as a nondialyzable component, of numerous metalloenzymes or as a dialyzable activator in various other enzyme systems.

Zinc–hormone interactions include hormonal influence on absorption, distribution, transport, and excretion of zinc and zinc influence on synthesis, secretion, receptor binding, and function of numerous hormones (qv). Zinc enhances pituitary activity by increasing circulating levels of growth hormone, thyroid-stimulating hormone, luteinizing hormone, follicle-stimulating hormone, and adrenocorticotropin. The role of zinc in insulin action is recognized but not well understood. Zinc is required for maintenance of normal plasma concentrations of vitamin A and for normal mobilization of vitamin A from the liver.

Zinc was confirmed as essential for humans in 1956 and deficiency symptoms were reported in 1961. The size of the human fetus is correlated with zinc concentration in the amniotic fluid and habitual low zinc intake in the pregnant female is thought to be related to several congenital anomalies in humans. Low zinc intakes result in hypogonadism, dwarfism, mental retardation, low serum and red blood cell zinc in humans and animals, and retarded growth and teratogenic effects on the nervous system in rats.

In children suffering from marginal zinc deficiency, impaired taste acuity, poor appetite, and suboptimal growth can be reversed upon zinc supplementation. Accelerated wound healing occurs in humans upon zinc supplementation, suggesting that marginal zinc deficiency in humans may be more widespread than has been thought. Zinc supplementation has also been effective in alleviating symptoms of active rheumatoid arthritis in clinical trials. Acrodermatitis enteropathica, a hereditary disease that involves aberrant zinc metabolism, responds to oral zinc supplementation. Excessive zinc intake may interfere with copper metabolism.

Silicon. Silicon comes mainly from ingestion of silicates, primarily from vegetables. It is found in the serum as silicic acid, $Si(OH)_4$, and normal blood serum levels are ca 1 mg/100 mL regardless of intake because of efficient kidney excretion of excess. Silicon is necessary for calcification, growth, and as cross-linking material in mucopolysaccharide formation. Silicon is especially helpful in situations where the diet is low in calcium or high in aluminum, or thyroid function is inadequate. The human requirement may be 5–20 mg/d. Silicon deficiency may lead to altered metabolism of connective tissue and bone and/or aluminum accumulation in the brain.

Copper. All human tissues contain copper. The highest amounts are found in the liver, brain, heart, and kidney. In blood, plasma and erythrocytes contain almost equal amounts of copper, ie, ca 110 and 115 mg/100 mL, respectively.

In plasma, ca 90 wt % of copper is in the metalloprotein ceruloplasmin, also known as a_2-globulin, mol wt 151,000, which contains 8 atoms of copper per molecule. Ceruloplasmin has been identified as a ferroxidase(I) which catalyses the oxidation of aromatic amines and of Fe^{2+} to Fe^{3+}. The ferric ion is then incorporated into transferrin which is necessary for the transport of iron to tissues involved in the synthesis of iron-containing compounds, eg, hemoglobin. Lowered levels of ceruloplasmin interfere with hemoglobin synthesis.

Copper deficiency is characterized by poorly formed collagen which leads to bone fragility and spontaneous bone fractures in animals, and also results in cardiac hypertrophy. Abnormal electrocardiographs have been noted when low copper diets were fed to humans. Anemia, neutropenia, and bone disease have been reported in children having protein calorie malnutrition (PCM) and accompanying hypocupremia. At least two genetic diseases involving copper are known: Wilson's disease, an autosomal recessive disease, usually detected in adulthood, and Menke's kinky-hair syndrome.

Analytical data indicate that many diets contain less than the RDA for copper. Excessive copper has been reported to be fatal for oral dose levels of copper sulfate of 200 mg/kg body weight for a child and 50 mg/kg for adults.

Boron. The essentiality of boron, first accepted for higher plants in 1923, then for animals, was recognized in 1981 for human metabolism. Boron is reported to help maintain function or stability of cell membranes and is thought to be involved with hormone reception and transmembrane signaling. Enhanced need for boron may develop with nutritional or metabolic stress involving other nutrients, eg, magnesium deprivation or physiological changes in calcium metabolism. Boron depletion impairs cognitive function. Organs known to contain the highest levels of boron are bone, spleen, and thyroid. An excess of boron can, however, cause seizures in infants, riboflavinuria, and gastrointestinal upset.

Selenium. Selenium, thought to be widely distributed throughout body tissues, is present mostly as selenocysteine in selenoproteins or as selenomethionine. Animal experiments suggest that greater concentrations are in the kidney, liver, and pancreas and lesser amounts are in the lungs, heart, spleen, skin, brain, and carcass.

The most clearly documented role for selenium is as a necessary component of glutathione peroxidase. Selenium is also involved in the functions of additional enzymes, eg, type 1 iodothyronine deiodinase, leukocyte acid phosphatase, and glucuronidases. A role for selenium in electron transfer has been suggested as has involvement in nonheme iron proteins. Selenium and vitamin E appear to be necessary for proper functioning of lysosomal membranes. A role for selenium in metabolism of thyroid hormone has been confirmed.

Alkali disease and blind staggers of grazing livestock in the western U.S. were reported as the result of selenium poisoning. There are unusually high concentrations of selenium in certain plants, because of selenium-accumulating properties, or in ordinary plants growing on highly seleniferous soils. No treatment for this type of poisoning is known, thus excess selenium in the animal diet must be avoided. Prolonged ingestion of up to 600 mcg/d of selenium did not produce toxic effects in humans. Toxic effects in humans have been reported, however, from chronic ingestion of food in China supplying 5 mg/d of Se, and from supplements in the U.S. of 27–2387 mg/d of Se. Effects include hair loss, changes in the nails, gastrointestinal upset, and peripheral neuropathy.

Pure selenium deficiency, without concurrent vitamin E deficiency, is not generally seen except in animals on experimental diets. In China, selenium deficiency in humans has been associated with Keshan disease, a cardiomyopathy seen in children and in women of child-bearing ages, and Kashin-Beck disease, an endemic osteoarthritis in adolescents. Selenium may have anticarcinogenic effects possibly because of the antioxidant properties of selenium compounds.

Iodine. Of the 10–20 mg of iodine in the adult body, 70–80 wt % is in the thyroid gland (see THYROID AND ANTITHYROID PREPARATIONS). The essentiality of iodine, present in all tissues, depends solely on utilization by the thyroid gland to produce thyroxine and related compounds. Well-known consequences of faulty thyroid function are hypothyroidism, hyperthyroidism, and goiter. Dietary iodine is obtained from eating seafoods and kelp and from using iodized salt.

The functions of the thyroid hormones and thus of iodine are control of energy transductions. These hormones increase oxygen consumption and basal metabolic rate by accelerating reactions in nearly all cells of the body. A part of this effect is attributed to increase in activity of many enzymes. Additionally, protein synthesis is affected by the thyroid hormones.

In many parts of the world, simple goiter is endemic and usually results from dietary iodine deficiency or goiterogens in foods which bind iodine. Iodine deficiency disorders (IDD) include cretinism, myxedema, hypothyroidism, and goiter. In technologically advanced countries, the problem of iodine deficiency has been minimized by the use of iodized salt. Faulty utilization of iodine can lead to Grave's disease. A sodium iodide excess can also produce goiter and an excess of 500 mg/kg body weight can be fatal.

Manganese. The adult human body contains ca 10–20 mg of manganese widely distributed throughout the body. The largest Mn^{2+} concentration is in the mitochondria of the soft tissues, especially in the liver, pancreas, and kidneys. Manganese concentration in bone varies widely with dietary intake.

Manganese is essential for normal body structure, reproduction, normal functioning of the central nervous system, and activation of numerous enzymes. An excess of manganese can lead to neural damage and possible impaired insulin production.

In animals, manganese deficiency results in wide-ranging disorders, eg, impaired growth, abnormal skeletal structure, disturbances of reproduction, and defective lipid and carbohydrate metabolism. Although overt manganese deficiency has not been induced in humans, some forms of epilepsy in humans and animals and a decrease in glucose tolerance in animals have been linked to low levels of manganese in the tissue.

Molybdenum. Molybdenum is a component of the metalloenzymes xanthine oxidase, aldehyde oxidase, and sulfite oxidase in mammals. Two other molybdenum metalloenzymes present in nitrifying bacteria have been characterized: nitrogenase and nitrate reductase. The molybdenum in the oxidases, is involved in redox reactions. The heme iron in sulfite oxidase also is involved in electron transfer. Foods rich in molybdenum include legumes, dark green vegetables, liver, whole-grain cereals, and milk. Xanthine oxidase, mol wt ca 275,000, present in milk, liver, and intestinal mucosa, is required in the catabolism of nucleotides. Xanthine oxidase is also involved in iron metabolism.

A copper–molybdenum antagonism involving sulfate occurs in animals, ie, large amounts of molybdenum and sulfate can depress copper absorption. Cattle grazing on pasturage of high Mo content succumb to teart or peat scours, characterized by diarrhea and general wasting. Control involves increasing copper intake. The Cu—Mo antagonism has been observed in humans. Significant increases in urinary copper excretion have been observed with increasing Mo intake.

Molybdenum deficiency in humans results in deranged metabolism of sulfur and purines and symptoms of mental disturbances. Toxic levels produce elevated uric acid in blood, gout, anemia, and growth depression. Faulty utilization results in sulfite oxidase deficiency, a lethal inborn error.

Chromium. Chromium(III) potentiates the action of insulin and may be considered a cofactor for insulin. Chromium is thought to form a complex with insulin and insulin receptors.

Studies of elderly people and mildly diabetic patients showed significant improvement in the glucose tolerance test (GTT) when chromium supplementation of 150–200 μg/d was given. In other tests, these positive results were not obtained. It is possible that not all subjects are capable of utilizing inorganic chromium to the same extent. Some may require a preformed GTF (glucose tolerance factor). Chromium chloride supplementation has been effective in normalizing impaired glucose tolerance in malnourished children and in patients receiving total parenteral nutrition for a long time. The most available form of chromium is GTF obtained from brewer's yeast. Chromium deficiency may also lead to atherosclerosis and peripheral neuropathy.

Cobalt. Cobalt is nutritionally available only as vitamin B_{12}. Although Co^{2+} can function as a replacement *in vitro* for other divalent cations, in particular Zn^{2+}, no *in vivo* function for inorganic cobalt is known for humans. In ruminant animals, B_{12} is synthesized by bacteria in the rumen.

In pernicious anemia, the bone marrow fails to produce mature erythrocytes as a result of defective cell division, a consequence of impaired DNA synthesis which requires vitamin B_{12}. If the disease goes untreated, extensive neurological damage, eg, irreversible degeneration of the spinal cord by demyelinization, may occur because of faulty fatty acid metabolism.

Vitamin B_{12} deficiency commonly is caused by inadequate absorption resulting from a lack or insufficient intrinsic factor (IF). Intrinsic factor is a glycoprotein, mol wt ca 50,000, which binds vitamin B_{12} in a 1:1 molar ratio. The B_{12}–IF complex, formed in the stomach, is absorbed in the ileum. Absorption in this part of the intestine occurs because of the specific characteristics of the cells of the microvilli (brush border) of the ileum. The IF remains in the intestine attached to the epithelial cells. Transport of B_{12} into the blood stream requires Ca^{2+}. In the blood, B_{12} is bound to transcobalamin II (transport protein).

Whatever bound B_{12} is not utilized immediately is stored in the liver. With increasing quantities of dietary B_{12}, the fraction that is absorbed decreases. Generally, vitamin B_{12} is excreted in the urine, but with large intake, some is excreted in the bile. A nutritional excess of cobalt can lead to polycythemia.

Tin. The widespread use of canned foods results in a daily intake of tin that is ca 1–17 mg for an adult male. At this level it has not been shown to be toxic. Some grains also contain tin. Too much tin can adversely affect zinc balance and iron metabolism. Essentiality has not been confirmed for humans. It has been shown for the rat. An enhanced growth rate results from tin supplementation of low tin diets. Animals on deficient diets exhibit poor growth and decreased feed efficiency.

Vanadium. Vanadium is essential in rats and chicks. Estimated human intake is less than 4 mg/d. In animals, deficiency results in impaired growth, reproduction, and lipid metabolism, and altered thyroid peroxidase activities. Vanadium may play a role in the regulation of (NaK)–ATPase, phosphoryl transferases, adenylate cyclase, and protein kinases.

Nickel. There is considerable evidence for the essentiality of nickel in animals. Various pathological manifestations of nickel deficiencies have been observed in chicks, cows, goats, pigs, rats, and sheep. Average intake is reported to be about 60–260 μg/d, and a dietary requirement for humans of less than 100 μg/d has been suggested. *In vitro* studies have shown nickel to be an activator of several enzymes.

Arsenic. Arsenic is under consideration for inclusion as an essential element. No clear role has been established, but arsenic, long thought to be a poison, may be involved in methylation of macromolecules and as an effector of methionine metabolism.

Health and Safety Factors

Under unusual circumstances, toxicity may arise from ingestion of excess amounts of minerals. This is uncommon except in the cases of fluorine, molybdenum, selenium, copper, iron, vanadium, and arsenic. Toxicosis may also result from exposure to industrial compounds containing various chemical forms of some of the minerals. Aspects of toxicity of essential elements have been published.

Efficient homeostatic controls of mammalians generally prevent serious toxicity from ingestion of the mineral nutrients. Toxicity may occur under conditions far removed from those of nutritional significance or for individuals suffering from some pathological conditions. Because of very low concentrations in foods, the trace elements are not toxic under normal nutritional conditions. Exceptions are selenium and iron.

MARY G. ENIG
Enig Associates, Inc.

A. H. Ensminger, M. E., Ensminger, J. Konlande, and J. R. K. Robson, *Food and Nutrition Encyclopedia,* 2nd ed. CRC Press, Boca Raton, 1994.

M. C. Linder, ed., *Nutritional Biochemistry and Metabolism with Clinical Applications,* 2nd ed., Appleton & Lange, Norwalk, Conn., 1991.

W. Mertz, ed., *Trace Elements in Human and Animal Nutrition,* Vols. 1 and 2, Academic Press, San Diego, 1987.

M. E. Shils, J. A. Olson, and M. Shike, eds., *Modern Nutrition in Health and Disease,* 8th ed., Vols. 1 and 2, Lea & Febigher, Philadelphia, 1993.

MINERALS RECOVERY AND PROCESSING

Minerals, critically important to the development of civilization, are derived from the earth's crust, a relatively thin shell of siliceous material about 13 km deep. The art of minerals processing can be traced back to the beginnings of civilization as evidenced by distinct historic periods known as the Stone Age, Copper Age, Bronze Age, and Iron Age. Mining can be considered the second largest industry in the world, next to agriculture. In terms of importance, mining is considered equal to agriculture.

Table 1. Abundance of Metals in the Earth's Crust[a]

Element	Abundance, %	Amount in 3.5 km of crust, t	Element	Abundance, %	Amount in 3.5 km of crust, t
silicon	28.2		chromium	0.010	10^{14}–10^{15}
aluminum	8.2	10^{16}–10^{18}	nickel	0.0075	
iron	5.6		zinc	0.0070	
calcium	4.1		copper	0.0055	10^{13}–10^{14}
sodium	2.4		cobalt	0.0025	
magnesium	2.3	10^{16}–10^{18}	lead	0.0013	
potassium	2.1		uranium	0.00027	
titanium	0.57		tin	0.00020	
manganese	0.095	10^{15}–10^{16}	tungsten	0.00015	10^{11}–10^{13}
barium	0.043		mercury	8×10^{-6}	
strontium	0.038		silver	7×10^{-6}	
rare earths	0.023		gold	$<5 \times 10^{-6}$	
zirconium	0.017	10^{14}–10^{16}	platinum metals	$<5 \times 10^{-6}$	$<10^{11}$
vanadium	0.014				

[a] By comparison, oxygen has an abundance of 46.4%.

The occurrence of minerals or elements in the earth's crust is not uniform. Rather, minerals concentrate in particular areas (called deposits) as a result of geological conditions and activity. Such a deposit is referred to as an ore deposit, ore body, or when it is large enough, ie, the percentage of the useful component is high enough to make mining worthwhile for economic recovery of the mineral, simply as an ore. Most minerals or elements are present in the earth's crust in very small amounts. The concentrations of selected metals are given in Table 1. Eight elements, O, Si, Al, Fe, Ca, Na, Mg, and K, account for more than 99% of the crust. The abundance of a given element or a mineral often has no bearing on its industrial importance or its price (see also METALLURGY, EXTRACTIVE METALLURGY).

Ores which comprise a variety of minerals are, as a rule, heterogeneous. An ore body is usually named for the most important mineral(s) in the rock, referred to as value minerals, mineral values, or simply values. Some minerals contain metals, which are extracted by concentration and smelting. Other minerals, such as diamond, asbestos (qv), quartz (see SILICON COMPOUNDS), feldspars, micas (see MICA), gypsum, soda, mirabillite, clays (qv), etc, may be used either as found, with some or no pretreatment, or as stock materials for industrial compounds or building materials (qv).

Most ores as mined require some processing before they can be converted into usable, final mineral products. Processing of ores by physical or chemical methods is described as mineral processing. For convenience, it is often the practice to restrict minerals processing to physical methods, and to treat chemical methods under the realm of extractive metallurgy. This distinction is a fine one, however, and hydrometallurgy, which comprises chemical methods of minerals treatment in aqueous medium, has much in common with minerals processing. Therefore, hydrometallurgy can often be considered as minerals processing. Until the early part of the twentieth century, minerals processing was much more of an art than a science. Rapid developments in understanding the fundamentals of processing have given minerals processing a sound scientific basis.

The minerals processing industry has made contributions to all areas of technology, both in terms of products and processing. Technologies developed in the mineral industry are used extensively in the chemicals industry as well as in municipal and industrial waste treatment and recycling industry, eg, scrap recycling, processing of domestic refuse, automobiles, electronic scrap, battery scrap, and decontamination of soils.

Ores

Deposits. Ore deposits are classified as igneous, metamorphic, or sedimentary. A further classification is into sulfides, oxides, metallic, nonmetallics, etc, depending on the nature of occurrence of the value mineral and use of the mineral product. Considerable effort goes into estimating the size and geometry of ore deposits. These are important factors in economic assessment. The size of ore deposits is described in terms of proven or measured reserve which is well-outlined by extensive drilling; indicated reserve, implied by a limited amount of drilling; and potential reserve, based on geological data with no drilling information. Potential reserves could also include well-outlined deposits that are not economical to treat at present.

Ores are mined by open-pit, underground, alluvial, and solution mining methods.

Mineralogy. For the purposes of minerals processing, any ore can be considered to be made up of valuable minerals and gangue (waste) minerals, and these are associated with one another intimately in the deposit. The valuable minerals may contain more than one valuable metal. Ores may be designated simple or complex, depending on the ease and extent of mineral liberation and subsequent processing. The choice of a processing method is entirely dependent on ore mineralogy, ie, the specific minerals present, their relative proportions, composition, and the mode of physical occurrence of the various mineral constituents such as size and extent of liberation. A thorough knowledge of mineralogy is also important in monitoring the efficiency of day-to-day operation of the mill because mineralogical factors can change frequently.

Mineralogical information is routinely gathered by collecting representative samples of the ore, or of the various flow streams in an operating plant, and studying polished or thin sections of these samples using optical microscopy (qv) in either the reflected light mode or the transmitted light mode. The electron microprobe and x-ray diffraction are also used extensively in conjunction with the optical microscope (see X-RAY TECHNOLOGY). A more recent development, qem*sem, is a fully automated image analyzer combining a scanning electron microscope and electron microprobe. The type of information sought in mineralogical examination includes the grain size of the value minerals in the host matrix rock; associations or locking between the value minerals and gangue minerals and between value minerals; existence of trace elements in the lattice of the value minerals; presence of oxidation or alteration of the mineral surfaces; and the occurrence of minor amounts of potentially valuable metals or minerals.

Liberation. In most ores, the value minerals are intimately locked or associated with gangue minerals. One of the most important prerequisites to the success of any physical separation process is liberation of value minerals from the ore matrix. Liberation, achieved by size reduction (qv) operations, can occur either because of intergranular or transgranular fracture. The degree of liberation of a particular mineral species, ie, the percentage of that mineral occurring as free particles (in a macroscopic sense) in relation to the total quantity of the mineral, is dictated by the type, performance, and economics of the processing operation, ie, throughput rate, recovery, product grade, processing costs, and profits.

Processing

Flow Sheets. All minerals processing operations function on the basis of a flow sheet depicting the flow of solids and liquids in the entire plant. The complexity of a flow sheet depends on the nature of the ore treated and the specifications for the final product. The basic operations in a flow sheet are size reduction (qv) (comminution) and/or size separation (see SEPARATION, SIZE), minerals separation, solid–liquid separation, and materials handling. The overall flow sheet depends on whether the specification for the final mineral product is size, chemical composition, ie, grade, or both. Products from a quarry, for example, may have a size specification only, whereas metal concentrates have a grade specification.

In addition to encompassing all of the unit operations in the plant, the plant flow sheets may also include materials handling operations associated with the transport and storage of materials in and around the mill. Typically, flow sheets provide quantitative information regarding water and slurry flows, tonnages, and assays.

Metallurgical Performance. Most minerals separations, whether by size or composition, are not perfect. There are invariably misplaced values and gangue minerals. The performance of a separation unit is, therefore, described by both recovery and grade. These are similar to yield and purity used in the chemical industry. In addition, other parameters such as ratio of concentration, ie, ratio of weight of feed to weight of concentrate; enrichment ratio, ratio of grade of concentrate to grade of feed; economic efficiency; and separation efficiency, recovery of value minus recovery of gangue, are also used sometimes.

Material Balances. Material balances in the plant flow sheet provide an assessment of the efficiency of operation of individual unit operations and the plant on the whole. Information necessary for trouble-shooting is also provided. Basically, materials balance involves calculation of input, output, and accumulation around any given circuit of a flow sheet. Many unit operations are characterized by high accumulations (circulating loads), sometimes by design. Most units are described as separation or junction points (nodes). Materials balance equations are presented as two-, three-, or n-product formulas depending on the number of products a flow stream consists of or is split into. The bases for materials balance are usually mass, assay (grade), volume, mineral composition, or energy. Balancing a circuit is usually quite a tedious task. There is either inadequate information about flow streams (leading to errors) or too much data (leading to conflicting results). Computer programs are available to make the task easier and to handle a complex set of nodes that typically exist in a plant. These can also provide for reconciliation of excess data.

There are many sources of errors in the plant. The principal ones are related to sampling (qv), mass flow rates, assaying, and deviations from steady state. Collecting representative samples at every stage of the flow sheet constitutes a significant task. Numerous methods and equipment are available.

Environmental Aspects

Environmental issues, concerns, and management are a primary preoccupation of the mining and minerals industry. The goal is to minimize the impact of the industry on the world ecosystem.

Principal areas of environmental management in minerals processing are tailings and other waste treatment and disposal, water discharge and protection of ground and surface waters, acid mine drainage, land reclamation, restoration and abatement, mine subsidence, mine closure, dust control, air emissions, and the elements or minerals that cause environmental pollution in subsequent operations such as smelting or burning of coal. Such regulations as the Clean Air Act have high impact on the industry. Laws in the 1990s require that permits for new mine openings be covered by a huge performance bond. The release of the bond is contingent on the completion of all required reclamation, restoration, and abatement work on the permit area. A significant amount of research and development is being conducted to address the environmental issues in mineral processing. Many newer technologies have already become available.

Size Reduction

Size reduction (qv) or comminution is the first and very important step in the processing of most minerals. It also involves large expenditures for heavy equipment, energy, operation, and maintenance. Size reduction is necessary because the value minerals are intimately associated with gangue and need to be liberated, and/or because most minerals processing/separation methods require the ore mass to be of certain size and/or shape. Size reduction is also required in the case of quarry products to produce material of controlled particle size (see SIZE MEASUREMENT OF PARTICLES). In some instances, liberation of valuables or impurities from the ore matrix is achieved without any apparent size reduction. Scrubbers and attritors used in the industrial minerals plants, eg, phosphate, rutile, glass sands, or clay, are examples.

Size reduction is conducted in stages, the first occurring during mining of the ore body by using either explosives, eg, for hard rock mining or mechanical means, eg, for hydraulic mining of sedimentary

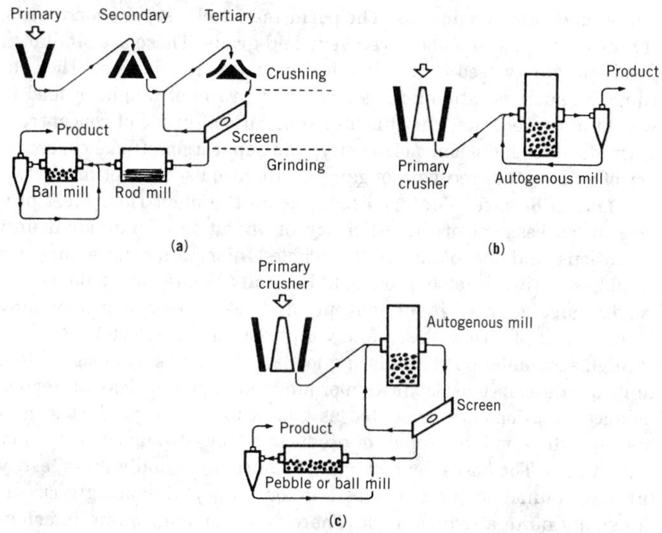

Figure 1. Three basic types of size reduction circuits: (**a**) conventional, (**b**) autogenous, and (**c**) autogenous plus separate fine grinding.

deposits such as clays (qv) and phosphates (see EXPLOSIVES AND PROPELLANTS). Subsequent size reduction stages involving crushing and grinding are grouped according to the particle size. General types of size reduction equipment include crushers, tumbling mills, impact mills, fixed-path mills, and fluid energy mills. A wide variety of crushers and grinding mills has been developed as necessitated by the ore type and rock hardness. The extent of size reduction achieved by any of these units is described by the reduction ratio, the ratio of the feed size to the product size. Particle size is defined as the size of the separator through which a certain percentage by weight, typically 80%, of the particles pass.

The three basic types of size reduction circuits used to produce a fine product are shown in Figure 1.

Size Separation

Sizing of the crushed and ground product is a necessary step prior to any mineral processing operation, and in the production of a product having a specific size. Controlling the size of material fed to other equipment is important. All equipment has an optimum size range of material that it can handle most efficiently. Size separation can be achieved either by screening (for coarser particles) or by classification (for fines) (see also SEPARATION, SIZE).

Minerals Concentration

Although the size separation/classification methods are adequate in some cases to produce a final saleable mineral product, in a vast majority of cases these produce little separation of valuable minerals from gangue. Minerals can be separated from one another based on both physical and chemical properties (Fig. 2). Physical properties utilized in concentration include specific gravity, magnetic susceptibility, electrical conductivity, color, surface reflectance, and radioactivity level. Among the chemical properties, those of particle surfaces have been exploited in physico-chemical concentration methods such as flotation and flocculation. The main objective of concentration is to separate the valuable minerals into a small, concentrated mass which can be treated further to produce final mineral products. In some cases, these methods also produce a saleable product, especially in the case of industrial minerals.

Selection of a particular concentration method depends entirely on the mineral and metal in question, the nature and mineralogy of the ore deposit, particle size at economic liberation, and the prevailing socio-economic factors. For many centuries, sorting by hand and gravity concentration were the only methods available. By the end of the nineteenth century magnetic and electrostatic methods had been introduced. Minerals processing underwent a revolution when the flotation method was developed in the early twentieth century because large tonnages of a wide variety of ores at various grades could be processed and complex mineral separations that were not possible by any other method could be performed.

Solid–Liquid Separation

Most minerals processing operations are conducted in large quantities of water. A typical copper ore flotation plant uses about 3800 L/t of ore treated. Water usage can be as high as 23,000 L/t in glass sands flotation. Dewatering (qv) of mineral slurries to varying degrees becomes necessary for a variety of reasons. For example, subsequent treatment by pyrometallurgical operations, such as pelletizing and smelting, transportation, disposal, water recovery for recycle, etc, require dewatering.

Dewatering is performed by using one or more of the following methods: sedimentation, also known as settling or thickening; filtration (qv); and thermal drying (qv). Frequently all three are used, in that order, on the same slurry to ensure that the final product has a low moisture content. Thickening is generally the most economical method for tailings dewatering before disposal in the tailings pond.

Materials Handling

The management of transportation, storage, feeding, washing, and packing of processing streams to final products in a mineral processing operation constitutes a significant effort and cost factor. Materi-

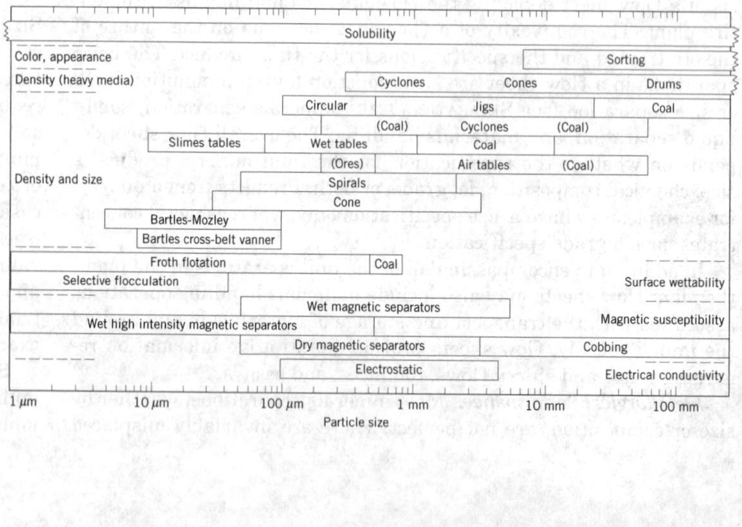

Figure 2. Particle size ranges for concentrating equipment based on mineral properties.

als handling comprises dry solids and slurry handling, and tailings disposal.

Dry solids, such as as-mined ore, crushed ore, and dried concentrates, are transported using trucks, rail cars, ore passes, conveyor belts (see CONVEYING), or slurry pipelines (qv) as dictated by the logistics, distances involved, and capacity. Within the mill, conveyor belts are more common, but for fine particles, tailings, and coal, slurry transportation is more typical.

Process Control

The most significant developments in minerals processing in the latter 1900s have been in the area of automation and computer control of minerals processing plants. Rapid advances in the electronics area are largely responsible for the introduction of on-stream analyzers and sensors (qv). Process control (qv) is an extremely difficult task because of the heterogeneous and complex nature of ores, the extreme variability in any given circuit, and the severity of conditions in the operation. Frequently data needed for process control are available only after the fact. Obtaining representative samples from various flow streams is a prerequisite to obtaining reliable data.

The primary goals in process control are to improve the efficiency and/or selectivity and to reduce the operating cost of each unit operation. Overall goals are also strongly influenced by economic factors prevailing in the market. Flexibility is therefore a key factor in process control. As for any process control system, the key elements are measurement or sensing, comparison to a target value, manipulation of the variable value, and feedback to the controller. Control strategies become more effective when predictions can be made for any unit operation at a high degree of confidence. The more modern control systems are based on multivariable control and model-based concept and digital instrumentation. Present trends are toward knowledge-based and artificial intelligence (see EXPERT SYSTEMS), controlled systems which optimize overall performance rather than performance of individual unit operations. These are rule-based systems that attempt to implement human expert knowledge or a rule-of-thumb approach and the uncertainty inherent in human decision making involving linguistic variables (fuzzy terms) and subjective interpretation.

A development in the 1960s was that of on-line elemental analysis of slurries using x-ray fluorescence. These have become the industry standard. Both in-stream probes and centralized analyzers are available. The latter is used in large-scale operations. Neutron activation analyzers are also available. These are especially suitable for light element analysis. On-stream analyzers are used extensively in base metal flotation plants as well as in coal plants for ash analysis. Although elemental analysis provides important data, it does not provide information on mineral composition which is most crucial for all separation processes. Devices that can give mineral composition are under development.

Dry ore tonnage is measured by belt scales mounted along the conveyor belt line. Slurry flow rates are measured using magnetic or ultrasonic flow meters. Slurry densities are routinely measured by batch operation by collecting a representative slurry sample in a liter vessel and weighing it on a density scale. Continuous pulp density measurement is made using a nuclear density meter or a gamma-gauge which measures the transmission of gamma rays from a radioactive source through the slurry using an ionization chamber-type detector. Transmittance is inversely proportional to the slurry density. Particle size is measured routinely, in a batch operation, by collecting a representative sample and using laboratory standard sieves. Numerous devices are also available for continuous particle size measurement. Various other components of plant control and automation include the elaborate alarm systems and shutdown mechanisms for crushers based on bearing pressure and temperature; crusher power and ore level; grinding and classifier controls comprising ore feed rate, water addition rate, classifier feed rate and pulp density, particle size distribution, mill power, and load; flotation controls comprising aeration rate, pulp and froth level; reagents addition rates; and pH and lime addition. Considerable effort is going into improving the performance of existing measurement devices and sensors, and developing new ones. Reliable redox control systems for sulfide flotation circuits and color (or vision) sensors for on-line analysis of flotation froths and slurries are also under development.

D. R. NAGARAJ
Cytec Industries

B. A. Wills, *Mineral Processing Technology,* 5th ed., Butterworth-Heinemann Ltd., Oxford, U.K., 1992.

E. G. Kelly and D. J. Spottiswood, *Introduction to Minerals Processing,* John Wiley & Sons, Inc., New York, 1982.

N. L. Weiss, ed., *SME Mineral Processing Handbook,* AIMME, New York, 1985.

R. Thomas, ed., *E/MJ Operating Handbook of Mineral Processing,* McGraw-Hill Book Publishing Co., Inc., New York, 1977.

MINERAL WOOL. See REFRACTORY FIBERS.

MISCHMETAL. See CERIUM AND CERIUM COMPOUNDS.

MIXING AND BLENDING

Fluid mixing is a unit operation carried out to homogenize fluids in terms of concentration of components, physical properties, and temperature, and create dispersions of mutually insoluble phases. It is frequently encountered in the process industry using various physical operations and mass-transfer/reaction systems (Table 1). These industries include petroleum (qv), chemical, food, pharmaceutical, paper (qv) and mining. The fundamental mechanism of this most common industrial operation involves physical movement of materials between various parts of the whole mass. This is achieved by transmitting mechanical energy to force fluid motion.

Mixing systems are broadly divided into single-phase systems involving miscible liquids, and multiphase systems such as solid–liquid, mutually insoluble liquids, and gas–liquid. This article discusses fundamental mixing concepts and design/scale-up issues for various mixing systems and equipment types (Fig. 1).

Table 1. Classes of Mixing Applications

Mixing class	Physical	Mass-transfer/reaction
miscible liquids	blending: lube oils, gasoline additives, for pH control, dilution	slow batch chemical reactions, fast reactions in in-line mixers
liquid–solid	preparing homogeneous slurries of light and heavy solids such as polymers, catalyst, etc	dissolving, crystallization, liquid-solid reactions, solvent extraction
immiscible liquids	washing liquids with immiscible solvents, cosmetics, salad dressing	hydrolysis/neutralization reactions, extraction, suspension polymerization
gas–liquid	gas scrubbing, steam heating of liquids	absorption, stripping, oxidizing liquids, hydrogenation, oxonation of olefins, chlorination, fermentation
viscous liquids Newtonian non-Newtonian	blending polymer solutions, paints and pigments, food products	solution polymerization, de-ashing of catalyst from polymers
heat transfer	heating and cooling through jackets and internal coils	

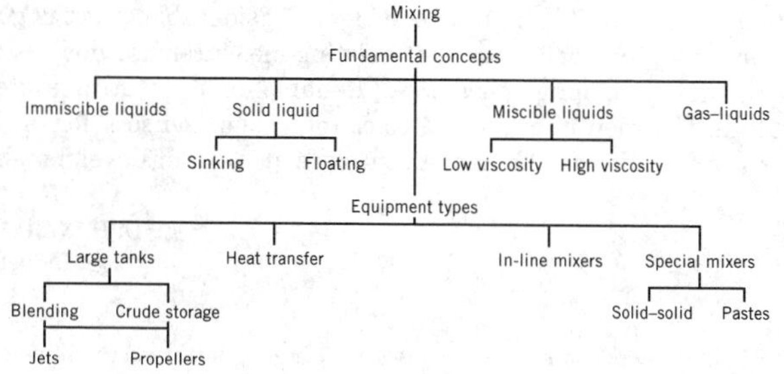

Figure 1. Design and scale-up issues for various mixing classes (see Table 1).

These mixing systems offer high flexibility because they can be operated in batch, semibatch, or continuous modes. Adequate mixing is a prerequisite for the success of chemical processes in terms of minimizing investment and operating costs. In addition, chemical reactions with mass-transfer limitation can be enhanced to provide high yields. Good mixing, therefore, plays a significant role in the profitability of the process industry.

The desired mixing in a commercial process is achieved with different types of equipment, eg, agitators, jets, static mixers, air lifts, etc. The design approach requires defining process mixing requirements; specifying mixer type and size, and other internals such as baffles; and designing mechanical components such as impeller blades, shaft, drive assembly, bearings and supports.

Fundamental Concepts

Pumping, Velocity Head, and Power. Mechanical mixers can be compared with pumps because they produce circulating capacity Q and velocity head H. Power input P to a pump is represented by

$$P = Q(P_1 - P_2)$$

The parameter Q in a mixer represents internal recirculation which is not confined and directed as in a pump. The pressure drop in a mixer is analogous to velocity head H which can also be considered degradation of kinetic energy. H is proportional to shear in mixing because head from kinetic energy generates shear through the jet or pulsating motion of the fluid. Q in a mixer is related to speed N and the impeller diameter D by

$$Q = N_q N D^3$$

The pumping number N_q is a function of impeller type, the impeller/tank diameter ratio (D/T), and mixing Reynolds number $Re = \rho N D^2/\mu$. The total flow in a mixing tank is the sum of the impeller flow and the flow entrained by the liquid jet. The entrainment depends on the mixer geometry and impeller diameter. For large-size impellers, enhancement of total flow by entrainment is lower.

Depending on the process needs, the combination of pumping and shear can be changed by merely changing the impeller diameter and the mixer speed. A larger diameter impeller at slow speed would provide high pumping action necessary for systems such as blending and solids suspension. On the other hand, a small diameter impeller and high mixer speed would be more suited for high shear systems where mass transfer is important.

Shear Rate/Shear Stress. Whenever there is relative motion of liquid layers, shearing forces exist that are related to flow velocities. These forces, represented by shear stress, carry out the mixing process and are responsible for producing fluid intermixing, dispersing gas bubbles, and stretching or breaking liquid drops. The shear stress is a complex function of shear rate defined by the velocity gradients $\Delta V/\Delta Y$. These velocity gradients can be caused by either entraining liquid with a propeller jet or by creating velocity profiles with rotating impellers.

There are different definitions of shear rate, depending on the regions in a mixing tank, eg, maximum near the blade tip, minimum near the liquid surface, average in the impeller region, and average in the entire tank volume.

Macromixing vs Micromixing. Mixing in an agitated tank is considered at two levels, macromixing and micromixing. Macromixing is established by the mean convective flow pattern and is sufficient for certain mixing duties such as blending liquids for concentration homogeneity. Micromixing, on the other hand, involves complete intermingling of molecules by turbulent diffusion. Micromixing is necessary for fast reacting systems or where enhancement of mass is desired.

Dimensionless Numbers. With impeller diameter D as length scale and mixer speed N as time scale, common dimensionless numbers encountered in mixing depend on several controlling phenomena (Table 2). These quantities are useful in characterizing hydrodynamics in mixing tanks and when scaling up mixing systems.

Flow Patterns. There are two main classes of turbine impellers based on the flow patterns they generate: axial flow and radial flow. Axial flow impellers produce a flow pattern involving full tank volume as a single stage. Radial flow impellers, on the other hand, produce two circulating loops, one below and one above the impeller. Mixing occurs between the two loops, but less intensely than within each loop. These differences in flow patterns cause variations in shear rate distributions in the tank. Therefore, the impeller flow patterns have a significant impact on the process result. The flow patterns within a given impeller are altered by parameters such as impeller diameter, liquid viscosity, and use of multiple impellers.

Wall Baffles. Vertical baffles located along the tank wall are necessary for providing top-to-bottom mixing without a swirl and for eliminating cavitation.

Scale-Up Principles. The main objective of scale-up is to achieve the same quality of mixing in a commercial size mixing tank as in a laboratory test tank. Several scale-up methods have been developed depending on the process type and mixing requirements; the most common one uses constant power per unit volume.

Impeller Types. There are literally hundreds of impeller types in commercial use. Only the most common and general types are shown in Figure 2.

Blending of Miscible Liquids

Mutually soluble liquids are blended to provide a desired degree of uniformity in an acceptable mixing time. Agitator effects are critical, particularly in commercial operations when both product quality and production rates are important. Although mixing time and pumping rate for a given impeller are related, this relationship can be different with different impeller types. For designing axial flow turbines for blending, there are two useful methods using number of tank turnovers and bulk fluid velocity, both based on impeller pumping rate.

Table 2. Dimensionless Numbers Used in Mixing

Number	Formula	Important for
flow, Fl	Q_g/ND^3	gas dispersion
Froude, Fr	N^2D/g	free surface or vortexing
Nusselt, Nu	hT/k	heat transfer
power, N_p	$P/\rho N^3D^5$	power consumption
Prandtl, Pr	$C_p\mu/k$	heat transfer
pumping, N_q	Q/ND^3	blending
Reynolds, Re	$\rho ND^2/\mu$	laminar or turbulent flow
Richardson, Ri	$g\Delta\rho l/\rho\mu^2$	blending
viscosity ratio, V_i	μ_1/μ_2	viscosity differences
Weber, We	$\rho N^2/D^3/\varsigma$	emulsification
Weissenberg, Wi	$\varsigma_1/N\mu$	viscoelastic effects

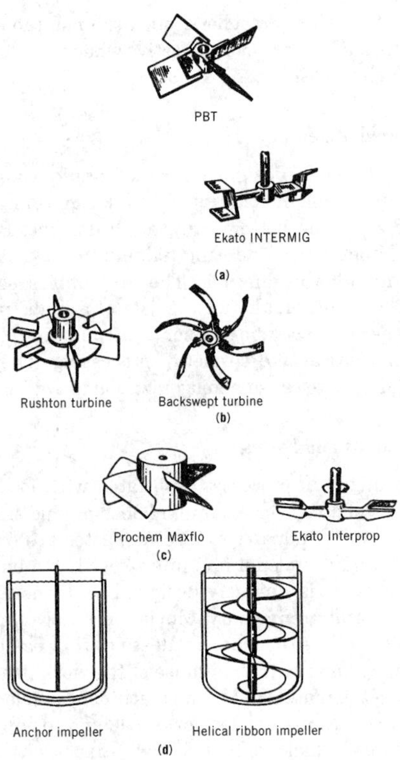

Figure 2. Common impeller types: (**a**) axial, (**b**) radial, (**c**) hydrofoil, and (**d**) close-clearance.

Feedpipe Backmixing. Backmixing of fluid into the feedpipe can result in lower yields of the desired product in a system with fast competitive/consecutive reactions. Therefore, it is important to design the feed injector nozzles to provide minimum jet velocity v_f to prevent backmixing into the feedpipe; v_f depends on impeller type, mixer speed, and the feedpipe location.

Viscous Liquid Blending. As liquid viscosity is increased, the discharge stream velocity from an impeller is more rapidly dissipated by liquid friction. As a result, the flow pattern flattens out and axial flow impellers generate radial flows. This change significantly affects blending quality.

The selection of mixer type and mixing system design method changes with ranges of viscosity. For low viscosity liquids, axial or radial flow patterns provide excellent mixing. Anchors are suitable for liquid viscosity between 5,000 and 50,000 mPa·s(= cP). Helical impellers should be used for liquid viscosity greater than 50,000 mPa·s.

Suspension of Solids

Solids are suspended in liquid-filled tanks for the purpose of dissolution, accelerating chemical reactions or preparing a homogeneous slurry. Typical industrial applications include catalyst suspension, rubber particle slurring, TiO_2 slurry for paper coatings (qv), activated carbon suspension in water treatment, or slurring in leaching of metals. The critical mixing parameters depend on whether the particles sink or float under unmixed conditions, and therefore can be discussed separately for the two types of solids.

Sinking Solids. Lifting and distribution of solids heavier than the liquid is accomplished by inducing necessary flow patterns by expending mechanical energy. This energy, supplied by a rotating impeller, is dissipated in turbulent eddies having a variety of sizes. When the velocity scale of eddies is higher than the free settling velocity of particles, entrainment occurs. The energy requirement for maintaining a suspension is much less than that required for suspending settled solids.

For most applications an off-bottom suspension is sufficient. Some processes require complete slurry uniformity which for practical purposes is achievable only up to 95–98% of liquid height. Axial flow impellers are more efficient for suspending solids than radial flow impellers. They can be designed by determining minimum mixer speed from impeller diameter and system physical properties.

Floating Solids There are several applications where solids are lighter than the carrying liquid and are also sticky at process conditions. These solids either have density less than that of the liquid or are highly porous. In order to keep these particles from rising to the surface and agglomerating, pulling down action must be provided by the mixer. Such floating solids are encountered in polymeric processes and in slurring of difficult-to-wet porous solids.

Mixing tanks equipped with modified baffles called partial baffles have been used successfully for entrainment of floating solids at reasonably low energy with few mechanical problems.

Immiscible Liquid–Liquid Mixing

Mixing of immiscible liquids is frequently encountered in several reacting and nonreacting systems such as extraction (qv), alkylation (qv), interfacial and suspension polymerization, emulsifications, and phase-transfer catalysis (see CATALYSIS, PHASE-TRANSFER). When mutually insoluble liquids are mixed in an agitated tank, a dispersion of one phase is produced, thereby increasing the interfacial area manyfold.

Turbines. Turbine agitators provide the desired mixing conditions for immiscible liquids. Even in high viscosity liquid emulsifications, turbines are more effective than agitators conventionally used for blending of viscous liquids. The design of mixing systems generally requires laboratory experimentation to determine effects of mixing parameters on the dispersed phase drop size and/or on the process result. The results are then scaled up on the basis of constant power per unit volume.

Phase Inversion. This is a phenomenon caused by changing mixing conditions so that the dispersed and continuous phases interchange. The phase to become dispersed depends on the volume concentration of the two liquids, their physical properties, and the dynamic characteristics of the mixing process. There is always a range of volume fractions throughout which either component remains stably dispersed, and this is called the range of ambivalence. The limits of this range are influenced by the size and shape of the vessel, mixer speed, and physical properties of the liquids.

Gas–Liquid Mixing

Mechanically agitated gas–liquid contactors are widely used in nonreacting processes such as absorption and stripping, and reacting systems including oxidation, hydrogenation, chlorination, etc. They are also used for carrying out biochemical processes such as aerobic fermentation (qv) manufacture of protein, and wastewater treatment. The fractional hold-up of gas ϕ in these contactors is a basic measure of their efficiency. This, in conjunction with Sauter mean bubble diameter, d_{32}, determines the interfacial area, $a = 6\phi/d_{32}$, and hence, the mass-transfer rate. Knowledge of ϕ also gives the residence time for each phase. One of the most commonly used impeller is the disk flat blade turbine (DFBT), also called the Rushton turbine, which can create large interfacial areas by means of the turbulent radial flow pattern. Concave blade radial flow impeller can be more effective for higher gas holding capacity. Axial flow impellers and hydrofoil impellers have also been used when high liquid recirculation is desired. Gas dispersion quality is characterized in terms of three hydrodynamic conditions: flooding, dispersion, and recirculation. Mixing systems must be designed for preventing flooding regime.

Blending in Large Tanks

Blending of miscible liquids in 6–90 m (20–300 ft) diameter tanks is carried out with jet mixers and side-entering propellers (SEP). Jet mixers are used in conjunction with a pump that serves as the source of the required mixing energy. They are attractive for use in liquefied natural gas (LNG), liquefied petroleum gas (LPG), gasoline, jet fuel, and distillate fuel storage and blending tanks. SEP mixers are commonly used for suspending sludge in crude oil storage tanks. The basic requirements for blending in large storage tanks are that the entire contents of the tank be mixed and that the mixing be completed in the desired time.

Jet Mixers. A jet mixer can be an inclined nozzle or an eductor installed in the side or center of a tank near the bottom or top. These mixers can be used for continuously blending miscible fluids as they enter a tank, or batchwise by recirculating a portion of the tank contents through the jet.

SEP Mixers. These mixers induce a spiral jet flow across the floor of the tank continually entraining liquid from other areas of the tank. This jet stream initially only agitates the denser liquid at the bottom, but gradually penetrates the upper layers of the tank with sufficient velocity to generate both full top-to-bottom flow and break the interface between various density strata to achieve a full homogeneous mix.

Crude Tank Sludge Control

Depending on the type and concentration of sludge in petroleum crude receipt, settling of sludge in crude storage can occur at a rate of 25–150 mm height per month. Without its suspension, this sludge can harden due to packing of top layers, and at high levels mounds form because of sludge shifting during crude receipts and pumpouts. Mixing energy, therefore, must be supplied on a continuous and frequent basis to maintain tank cleanliness. For achieving good on-stream sludge control, two competing technologies are available: SEP mixers and Butterworth P-43 rotating jet machines. Adequately designed and operated, SEP mixers can prevent sludge settling by establishing movement of crude throughout the tank. They cannot, however, be expected to resuspend sludge accumulated for an extended period. The P-43 submerged jet nozzle system can prevent sludge accumulation and resuspend large sludge volumes. Selection of the mixing system depends on the nature of existing facilities.

Inline Motionless Mixers

Inline motionless mixers derive the fluid motion of energy dissipation needed for mixing from the flowing fluid itself. These mixers include orifice mixing columns, mixing valves, and static mixers. A typical service of orifice mixing columns or mixing valve involves caustic washing of gas oil and water–oil mixing upstream of a desalter.

Static mixers are used in the chemical industries for plastics and synthetic fibers, eg, continuous polymerization, homogenization of melts, and blending of additives to extruders; food manufacture, eg, oils, juices, beverages, milk, sauces, emulsifications, and heat transfer; cosmetics, eg, shampoos, liquid soaps, cleaning liquids, and creams; petrochemicals, eg, fuels and greases; environmental control, eg, effluent aeration, flue gas/air mixing, and pH control; and paints, etc.

These mixers consist of repeated structures called mixing elements attached inside a pipe. The elements create shearing action in the fluid at the cost of pressure drop, which causes mixing of single- and multiphase systems. These mixers provide complete transverse uniformity and minimize longitudinal mixing, and therefore their performance approaches perfect plug flow conditions.

Static mixers are usually classified as operating either in laminar or turbulent flow regimes. There are many proprietary designs marketed.

Heat Transfer in Agitated Tanks

Agitators which force large amounts of fluid to circulate near the heat-transfer surfaces provide the most efficient heat transfer. In the case of close-clearance impellers used for high viscosity liquids, heat transfer is promoted by the thinning of the stagnant fluid layer near the wall. Rubber scrapers can be attached to these impellers to keep the wall surface free from deposits.

Commonly used heat-transfer surfaces are internal coils and external jackets. Coils are particularly suitable for low viscosity liquids in combination with turbine impellers, but are unsuitable with process liquids that foul. Jackets are more effective when using close-clearance impellers for high viscosity.

Specially Designed Mixers

There are many specially designed mixers for unique nonconventional systems. The air-lift mixer consists of a tank containing a draft tube. Air is injected from the bottom into the draft tube. Because of the buoyancy, air bubbles rise, inducing a liquid upflow. The liquid then flows down through the annulus. The air bubbles escape through the surface. The mixer emulsifier consists of a high speed rotor and stator, and generates shearing action and pumping. This device is used in pharmaceutical and cosmetics processing. The vortex mixer uses centrifugal flow to cause homogenization of two or more liquids.

Mixing of Dry Solids and Pastes

Dry solids are mixed in processes associated with food, pharmaceuticals, fertilizer, tobacco, cement (qv), rubber products, ceramics (qv), soap, and many other industries. Viscous pastes are frequently handled in polymer and petroleum (qv) processes. The mixing equipment used in such systems is uniquely designed to divide and recombine the materials to attain uniformity. Moving agitator components often scrape the walls because the material does not flow easily. The energy requirements may be very high because of the work involved in dividing and shearing the material. Mixing machinery is selected according to its capacity to shear material at low speed and to wipe, smear, fold, stretch, or knead the mass to be handled. Mixers with intermeshing blades are sometimes required to keep the material from clinging unmixed to the lee side of the blade. Wiping of heat-transfer surfaces promotes addition or removal of heat. Some mixing devices break down solids in pastes and thus have the character of mills.

Solids. For mixing of solids, the mixers can be categorized according to the mixing mechanism: tumbling, convection, ribbon type, spiral elevator, paddle type, planetary type, pan type, and fluidization.

Pastes. For blending of viscous pastes, mixers are classified as batch or continuous. Most convection-type mixers for dry solids are also used for thick pastes. Batch mixers include change-can mixers, stationary tank mixers, double-arm kneading mixers, intensive mixers, and roll mills. Continuous mixers include single-screw extruders and twin-screw mixers.

RAMESH R. HEMRAJANI
Exxon Research and Engineering Company

R. R. Hemrajani, D. L. Smith, R. M. Koros, and B. L. Tarmy, *6th European Conference on Mixing,* Pavia, Italy, May 24, 1988.

M. M. C. G. Warmoeskerken and J. M. Smith, *CES* **40**(11), 2063 (1985).

N. Harnby, M. F. Edwards, and A. W. Nienow, *Mixing in the Process Industries,* Butterworths, London, 1985.

J. Y. Oldshue, *Fluid Mixing Technology,* McGraw-Hill, Inc., New York, 1983.

MOLD RELEASE AGENTS. See RELEASE AGENTS.

MOLECULAR MODELING

Molecular modeling refers boldly to any study of molecules utilizing physical or theoretical models to explain an observed or predicted behavior. Whereas molecular modeling as a practice has its roots in the development of quantum theory at the turn of the twentieth century, it was the exponential growth in computing power between the mid-1970s and the mid-1990s that catalyzed the development and application of molecular modeling methods during that period. The spectrum of software systems available covers all aspects of modeling.

Molecular modeling can be defined as the application of computational techniques, grounded in theory, to predict or explain observable biological or physical chemical properties. Wherever molecular modeling is practiced using a computer, the technique then becomes computer-assisted (aided) molecular modeling, or CAMM. CAMM, is often used synonymously with CAMD, or computer-assisted molecular (materials) design/discovery. CADD refers to computer-assisted drug design/discovery. A computational technique as used herein is a mathematical model derived from principles of chemistry, physics, or statistics which facilitates molecular modeling. An entire branch of chemistry, ie, computational chemistry, is devoted to developing, benchmarking, and applying computational techniques in order that researchers may be able to better understand and predict properties. Some of the properties which may be calculated either exactly or approximately by computational methods are the following:

boiling points	dipole moments
melting points	quadrupole moments
crystallization energy	octupole moments
heat capacity	infrared spectra/intensities
heat of formation	nmr spectra/chemical shifts
heat of fusion	optical rotary dispersion
heat of sublimation	raman spectra
heat of vaporization	ultraviolet spectra
entropy	
molar refractivity	ionization potentials
molar volume	electron affinities
partition coefficients	protonation energies and pK_as
	ionic strength
radius of gyration	
elasticy	conformational energies
tensile strength	Boltzmann distributions

Molecular properties can be classified according to their end-point observables, such as chemical (reactivity, solubility, acid–base), physical (a function of physical state: gas, liquid, solid; thermodynamic), or biological (ligand or enzyme; agonist or antagonist). These properties reflect macroscopic, or bulk, properties, which exist only for the bulk material, eg, heat of crystallization, or microscopic properties, which exist for an ensemble of the molecule. As use of CAMM methods expands to address a broader horizon of applications beyond those in organic, medicinal, and biological chemistry, calculations on metals, semiconductors, and magnetic systems have become more common.

The specific process involved in a given molecular modeling study depends significantly on the nature of the primary objective of the task. However, for many small-molecule and even macromolecular studies, a number of authors have diagrammed the individual steps in the process, and the flowchart in Figure 1 summarizes their efforts. An initial step that is critical to any CAMM project is the generation or retrieval of the pertinent structures themselves. Structures may range from simple organic molecules or monomers of more complex polymers, to full proteins, enzymes, metal surfaces, or zeolites. The modeling process can be influenced by the initial structure and its geometry. Thus, the selection and development of the starting molecular geometry needs to be given particular attention. An important component of the quality control process, as well as in gaining an understanding of the molecules themselves, is the visual examination of structures involved in a modeling study.

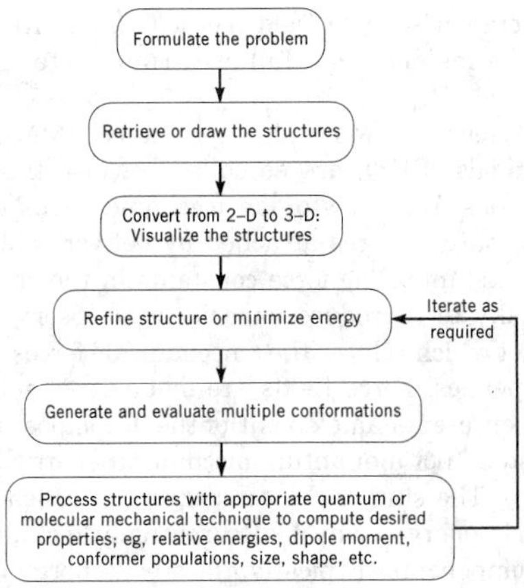

Figure 1. Flow chart for a typical small molecule modeling project.

Computer Graphics In Molecular Modeling

The goal of molecular modeling is to define clearly the relationship between chemical constitution, ie, the molecular formula or a topographic representation thereof, its geometric constitution or 3-D topology (the disposition of its atoms in Cartesian space), and its observed (or predicted) properties. The representation and facile manipulation of 3-D arrays of atoms comprise the domain of molecular or computer graphics. The growth of graphics tools has paralleled the evolution of computing hardware. Computing software and hardware systems have had a profound impact on the ability of modelers to compose a modeling study and address all aspects of the work, ranging from generating 2-D drawings of structures to statistical quality control of computed properties via visualization of multidimensional data.

One facet of chemistry which has benefited greatly from the use of computer graphics to enhance its own development is x-ray crystallography; computers are also used in x-ray structure refinement.

Among the available software systems which together help to exemplify the promise of computer graphics, Advanced Visualization Systems (AVS), stands out as one of the more extensible and practical of them to use. The premise behind development of the system was to provide modelers with a toolkit of modules having sophisticated intrinsics that would enable even casual programmers to link together multiple simple functionalities into a complex construct with which to accomplish exactly the types of visualization and manipulations that their work required.

Computational Methods for Molecular Modeling

The ultimate goal of quantum mechanical calculations as applied in molecular modeling is the *a priori* computation of properties of molecules with the highest possible accuracy (rivaling experiment), but utilizing the fewest approximations in the description of the wavefunction. *Ab initio,* or from first principles, calculations represent the current state of the art in this domain. *Ab initio* calculations utilize experimental data on atomic systems to facilitate the adjustment of parameters such as the exponents of the Gaussian functions used to describe orbitals within the formalism.

The performance of *ab initio* techniques distinguishes them significantly from their predecessors, semiempirical methods. Their consistent reproduction of data from structural, thermodynamic, and reaction sources to a range falling within the error limits of the experimental values provides scientists with an important tool with which to address various modeling problems. Whereas the absolute value of the relative performance of *ab initio* techniques varies for each structural or energetic feature examined, it is not unreasonable to suggest that if the quantity can be computed by both semiempirical and *ab initio* methods, the *ab initio* value will be closer to experiment or an ideal value than any other method.

Molecular Mechanics and Molecular Dynamics. In the realm of quantum mechanics, researchers deal explicitly with electrons, with their interactions, and with their attraction to nuclei, albeit in varying degrees depending on the rigor of the method chosen to solve the particular problem involved. These techniques were somewhat limited in their application prior to the age of large-scale computers and super workstations. Computers made possible the development of molecular mechanics, or empirical force field methods and, ultimately, the application of these methods to a full spectrum of studies in structure and energetics.

Molecular Mechanics. Molecular mechanics (MM), or empirical force field methods (EFF), are so called because they are a model based on equations from Newtonian mechanics. This model assumes that atoms are hard spheres attached by networks of springs, with discrete force constants. The force constants in the equations are adjusted empirically to reproduce experimental observations. The net result is a model which relates the "mechanical" forces within a structure to its properties. Force fields are made up of sets of equations each of which represents an element of the decomposition of the total energy of a system (not a quantum mechanical energy, but a classical mechanical one). The sum of the components is called the force field energy, or steric energy, which also routinely includes the electrostatic energy components. Typically, the steric energy is expressed as

$$E_{\mathrm{Total}} = E_{\mathrm{steric}} + E_{\mathrm{electrostatic}} = E_{\mathrm{bonds}} + E_{\mathrm{angles}} + E_{\mathrm{vdW}} + E_{\mathrm{torsion}} + E_{\mathrm{charge/dipole}}$$

The overall form of each of these equations is fairly simple, ie, energy = a constant times a displacement. In most cases the focus is on differences in energy, because these are the quantities which help discriminate reactivity among similar structures. The computational requirement for molecular mechanics calculations grows as n^2, where n is the number of atoms, not the number of electrons or basis functions. These calculations will be much faster than an equivalent quantum mechanical study. The size of the systems which can be studied can also substantially eclipse those studied by quantum mechanics.

In a force field calculation, a molecule in three dimensions is constructed using either Cartesian coordinates x, y, and z, or via an internal coordinate matrix consisting of bond distances, bond angles, and dihedral angles to specify the atoms' unique positions. Then the initial structure is evaluated to determine the extent to which each degree of freedom (bonds, angles, etc) deviates from the ideal (the zero-energy value) for the particular element and its hybridization. An energy minimization process follows wherein the energy associated with the distortions from ideal is minimized as the individual atomic positions or degrees of freedom are adjusted. Iteratively, this converges on a "minimum energy" or an "optimized" structure. This structure represents the best attempt of the minimization algorithm to render the smallest deviations in position of each of the atoms such that either the derivatives of the change in energy associated with the deviations are the smallest, or they satisfy either energetic convergence or coordinate change criteria from iteration to iteration. This process is analogous to the geometry optimization process within a quantum mechanical program, except that there the objective is to converge on a structure which yields the smallest energy derivatives and lowest total energy from solution of the SCF equations. Most simple molecular mechanics force fields include terms (Fig. 2) for bond stretching, bond angle distortion (bending), dihedral angles, van der Waals nonbonded interactions, Coulombic interactions (dipoles or charges).

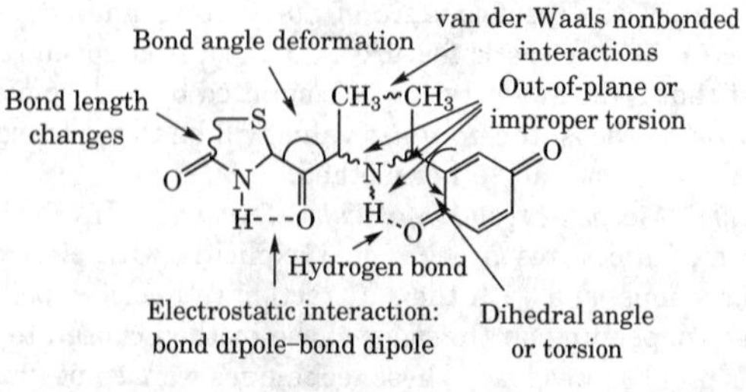

Figure 2. Structural representation of the energetic components of a typical molecular mechanics force field.

Force Fields, Molecular Dynamics, and Vibrational Spectroscopy. The link between molecular mechanics and molecular dynamics comes about through the force field itself. In molecular mechanics, the main interest is in computing the energy of molecules in the gas phase at room temperature in a single, discrete configuration and conformation; time is not a variable in the equations. From molecular dynamics, the objectives are properties which represent a time-averaged ensemble of states, including, for example, conformationally excited states, rotational states, interconversion rates, or inversion barriers for amines or amides. From this ensemble of states, characterizing the existence of the excited states and their contribution to the total energy of the system (from a Boltzmann distribution) is next. From an understanding of the vibrational spectroscopic roots of molecular mechanical force fields used in dynamics simulations, molecular dynamics may be seen as an extension bridge between theory and experiment, linking "static" molecular mechanical representations of properties with "dynamical" experimental properties. Applications of force field techniques to problems in environmental chemistry, materials science, and molecular biology demand that the methods go substantially beyond those for which reliable experimental data is available. Only since the time computers have made larger-scale *ab initio* quantum mechanical calculations practical for appropriate model systems have these techniques been reliable for such a broad spectrum of important applications.

Molecular Dynamics and Monte Carlo Simulations. At the heart of the method of molecular dynamics is a simulation model consisting of potential energy functions, or force fields. Molecular dynamics calculations represent a deterministic method, ie, one based on the assumption that atoms move according to laws of Newtonian mechanics. Molecular dynamics simulations can be performed for short time-periods, eg, 50–100 picoseconds, to examine localized very high frequency motions, such as bond length distortions, or, over much longer periods of time, eg, 500–2000 ps, in order to derive equilibrium properties. Those that can be evaluated by performing molecular simulations are conformational states and energetics, kinetic properties: rates of reaction and interconversion, reaction pathways, solubilities, diffusion rates, binding and complexation data, folding processes, transition temperatures, free energies for point mutations, and free energies of binding. The molecular simulations can also serve as an adjunct to x-ray and nmr for structure refinements.

As noted, force fields are a set of equations relating the total energy of the system to its individual interaction components. After selection of a force field simulation program which is appropriate to a given problem, the general procedure includes (*1*) initial structure equilibration, and (*2*) structure refinement.

Monte Carlo (MC) techniques for molecular simulations have been used to a great extent in studying the chemical physics of polymers. The majority of molecular modeling studies today do not involve the use of MC methods; however, the sampling capability provided by MC methods has gained some popularity among computational chemists.

Combined Quantum and Molecular Mechanical Simulations. In this technique, a molecular dynamics simulation includes the treatment of some part of the system with a quantum mechanical technique. This approach, QM/MM, is similar to programs that use quantum mechanical methods to treat the π-systems of the structures in question separately from the sigma framework. The results are combined at the end to render a structure which is optimized and energy-refined to satisfy both self-consistent field (SCF) and force field energy convergence.

SALVATORE PROFETA, JR.
Monsanto

W. Heisenberg, *Z. Phys.* **33,** 879 (1925).

E. Schrödinger, *Ann. Phys.* **79,** 361, 489; **80,** 437; **81,** 109 (1926).

J. D. Bolcer and R. B. Hermann, in K. B. Lipkowitz and D. B. Boyd, eds., *Reviews in Computational Chemistry,* Vol. 5, VCH Publishers, New York, 1994, pp. 1–63.

R. S. Mulliken, in D. A. Ramsey and J. Hinze, eds., *Selected Papers of Robert Mulliken,* University of Chicago Press, Chicago, 1975, pp. 39–42.

MOLECULAR RECOGNITION

Receptor–Substrate-/Host–Guest-Chemistry

Molecular recognition implies complementary lock-and-key type fit between molecules. The lock is the molecular receptor and the key is the substrate that is recognized and selected to give a defined receptor–substrate complex, a coordination compound or a supermolecule. Hence molecular recognition is one of the three main pillars, fixation, coordination, and recognition, that lay the foundation of what is known as supramolecular chemistry.

Supramolecular chemistry, the chemistry beyond the molecule, is a highly interdisciplinary field of science covering the chemical, physical, and biological features of chemical species of greater complexity than molecules themselves that are held together and organized by means of intermolecular (nonbinding) interactions. The chemistry of molecular recognition is also the core of host–guest chemistry, which is a subdiscipline or a particular aspect of supramolecular chemistry mostly involving inclusion and complex formation (see also INCLUSION COMPOUNDS).

Principles of Receptor Design

Information Storage and Read Out. Molecular recognition is defined by the energy and the information involved in the binding and selection of substrates by a given receptor molecule that may also involve a specific function. Mere binding is not recognition, although it is often taken as such. Instead, one may say that recognition is binding with a purpose, like receptors are ligands with a purpose. It implies a pattern recognition process through a structurally well-defined set of intermolecular interactions. Molecular recognition, thus, deals with the molecular storage and supramolecular readout of molecular information.

Information may be stored in the architecture of the receptor, in its binding sites, and in the ligand layer surrounding the bound substrate such as specified in Table 1. It is read out at the rate of formation and dissociation of the receptor-substrate complex. The success of this approach to molecular recognition lies in establishing a precise complementarily between the associating partners, ie, optimal information content of a receptor with respect to a given substrate.

Table 1. Structural Parameters for Storage of Information in a Chemical Receptor

Receptor	Parameter
architecture	size
	shape
	connectivity
	cyclic order
	conformation
	chirality
	dynamics
binding sites	electronic properties (charge, polarity, polarisability, van der Waals attraction and repulsion)
	size
	shape
	number
	arrangement
	reactivity (protoniziable, deprotonizable, reducible, oxidizable)
surrounding ligand layer	thickness
	overall polarity (lipophilic, hydrophilic)
	specific polarity(exo/endo-lipo/polarophilic)

Complementarity. To a first approximation, complementarity should take two forms. Firstly, the shape and size of the receptor cavity must complement the form of the substrate. Secondly, there must be a chemical complementarity between the binding groups lining the interior of the cavity and the external chemical features of the substrate.

The weak intermolecular forces that are principally involved in stabilizing receptor-substrate interactions and involved in molecular recognition processes are summarized in Table 2. Examples are shown in Figure 1.

Table 2. Types of Interactions in Molecular Recognition

hydrogen bonding between basic and acidic center
elastrostatic attraction between anionic and cationic centers
metal-ligand interaction
dipole-dipole interaction
π-stacking and charge-transfer interaction between aromatic residues in the receptor and delocalized regions of the substrate
van der Waals attraction between hydrophobic regions on the two components
covalent bonds, that can be reversibly formed and broken (eg, disulfides, borate esters).

Reorganization and Preorganization. On principle there are two different modes of receptor behavior. One of them is the so-called lock-and-key image involving complementary fit concept between rigid substrate and rigid receptor or rigid guest and rigid host relating to conformational flexibility of the molecular constituents forming the receptor–substrate (host–guest) complex. Receptors of this type are expected to present very efficient recognition between complementary partners, ie, both high stability and high selectivity of the receptor–substrate complex.

However, in most biological systems there is a degree of flexibility in the receptor. The approach of the substrate leads to conformational changes and an organization of the binding site around it. With this induced fit mechanism of binding, a higher entropy price is paid, but there are several advantages. A flexible receptor will permit a more wraparound interception or even complete encapsulation with the substrate involving many more potential binding interactions. This may lead to high selectivity of binding involving the amplification of molecular recognition interactions.

The balance between rigidity and flexibility is of particular importance for the binding and the dynamic properties of a receptor. It is, thus, a decisive structural design parameter of the receptor depending on the use. For instance, processes of exchange regulation, cooperativity and allostery connected with molecular recognition require a built-in flexibility so that the receptor may adapt and respond to changes, unlike rigid receptors.

Topology. This parameter may have reference to either the receptor as an individual molecular structure or to the receptor–substrate complex on a higher level of organization that is directly related to the mode and efficiency of molecular recognition.

A concave receptor is a favorable case. Under these circumstances the receptor cavity is lined with binding sites directed towards the bound species (see Fig. 1). This corresponds to Cram's definition

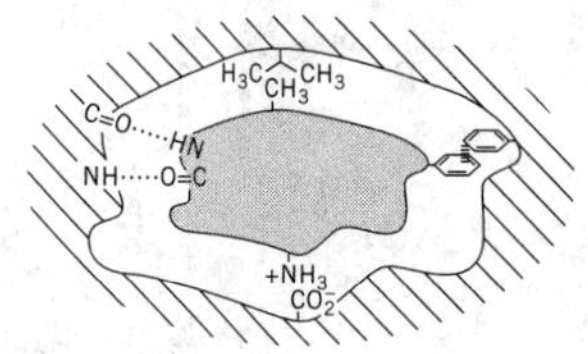

Figure 1. Schematic representation of a receptor–substrate (host–guest) complex involving cavity inclusion of the substrate and the formation of different types of weak supramolecular interactions between receptor (hatched) and substrate (dotted).

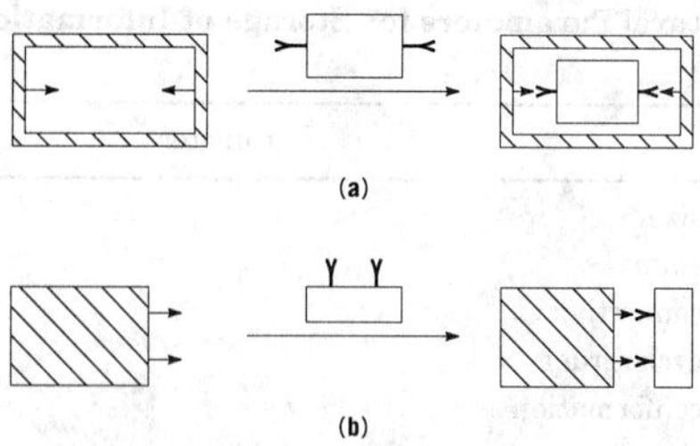

Figure 2. Diagram of (**a**) endo- and (**b**) exo-receptor recognition of substrates (dotted rectangles).

of a receptor (host) molecule providing binding sites that are convergent, as contrasted with the bound substrate (guest) featuring divergent complementary sites, ie, the substrate is more or less completely surrounded by the receptor, forming an inclusion complex. This widely used principle of convergence defines a convergent or endo-supramolecular chemistry (host–guest chemistry) with endo-receptors (endo-hosts) effecting endo-recognition (Fig. 2**a**).

The opposite procedure consists in making use of an external receptor surface rather than an internal cavity as substrate receiving site. This amounts to the passage from a convergent endo-supramolecular chemistry to a divergent or exo-supramolecular chemistry, and from endo- to exo-receptors (Fig. 2**b**). Here receptor–substrate binding occurs by surface-to-surface interaction which may be termed affixation as contrasted with inclusion. Exo-recognition with strong and selective binding, in particular, requires a large enough contact area and a sufficient number of complementary interactions along the interface. Such a mode of molecular recognition also finds biological analogies, for instance at the antibody–antigen interface of immunological importance. Metallo-exoreceptor aggregation, molecular recognition at organic and inorganic monolayers, films and solid surfaces bearing recognition groups, as well as the design of supramolecular solid architectures and materials, are other important instances of the exo-recognition principle.

Simple Modes of Molecular Recognition

Substrates involved in molecular recognition may feature a particular shape, size, state of charge, chemical affinity or optical specification. In general most of these parameters share. Nevertheless there may be dominating features of a certain substrate molecule to be used by a complementary receptor in the recognition process.

Size and Shape Dominated Substrate Recognition. Perhaps the simplest recognition process is that of a spherical substrate, in its most elementary form a ball-shaped metal ion of defined diameter. Three main classes of receptors provide the spherical recognition property. They are (*1*) macrocyclic polyethers, the well-known crown ethers and their derivatives; (*2*) the macropolycyclic cryptands; and (*3*) the acyclic analogues of crown compounds and cryptands, usually designated as podands. Prototypical compounds for each substance class are given by compounds (**1**)–(**3**) (Fig. 3). They all possess a spherical or quasi-spherical negatively polarized cavity prepared for the accommodation of alkali- and alkaline-earth metal ions that have complementary size, giving rise to a feature known as spherical recognition. Coronates,

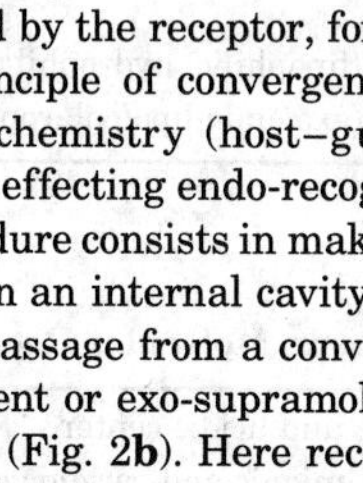

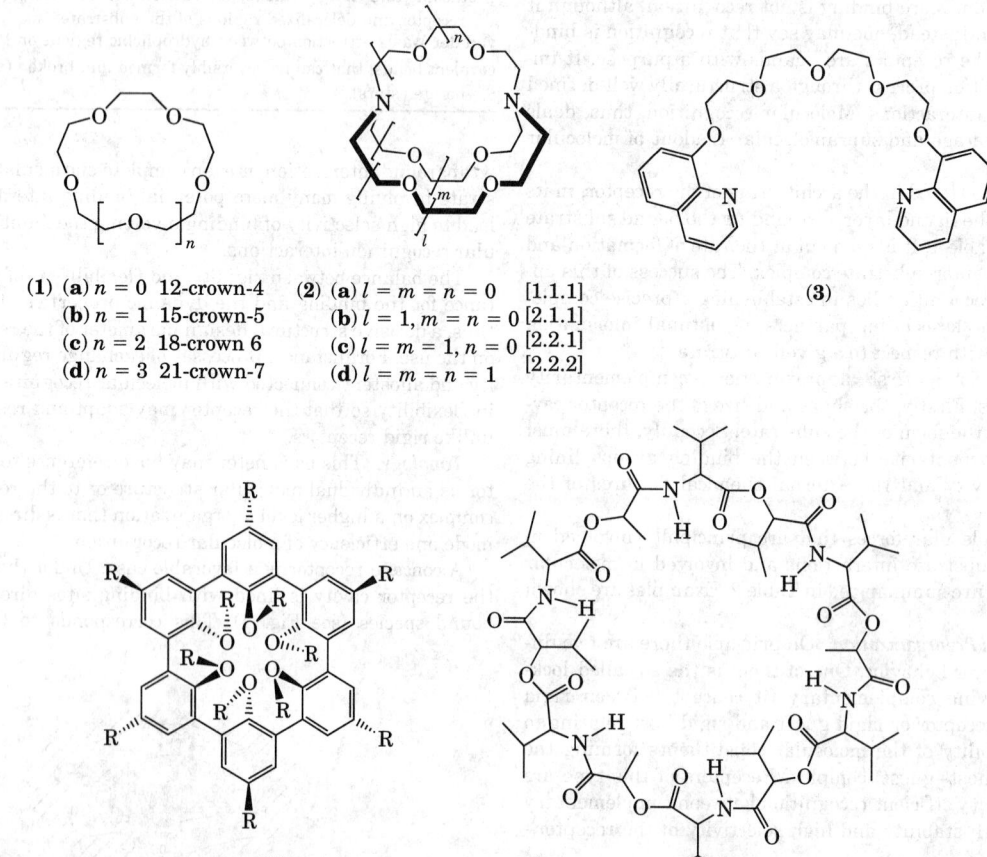

(**1**) (**a**) $n = 0$ 12-crown-4
(**b**) $n = 1$ 15-crown-5
(**c**) $n = 2$ 18-crown 6
(**d**) $n = 3$ 21-crown-7

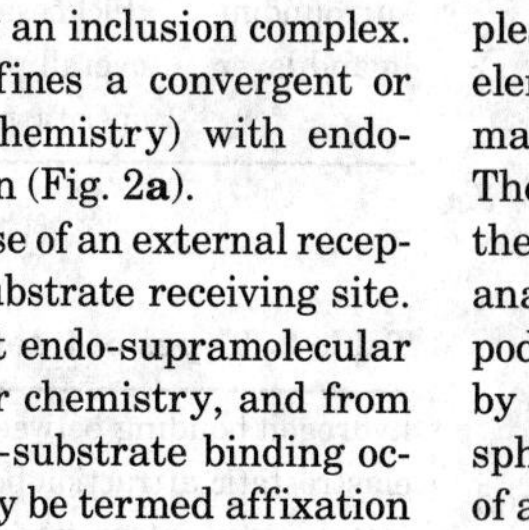

(**2**) (**a**) $l = m = n = 0$ [1.1.1]
(**b**) $l = 1, m = n = 0$ [2.1.1]
(**c**) $l = m = 1, n = 0$ [2.2.1]
(**d**) $l = m = n = 1$ [2.2.2]

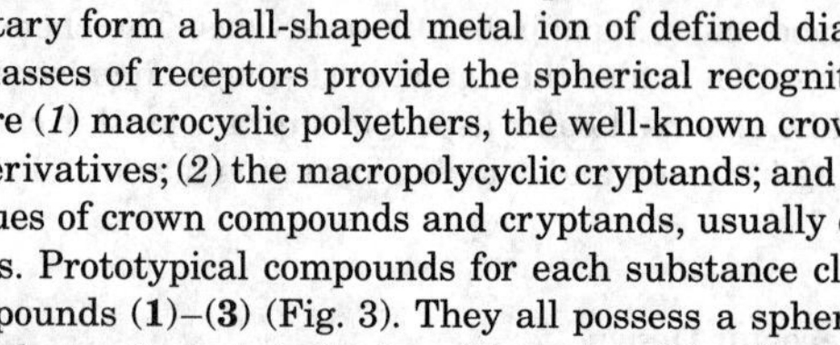

(**3**)

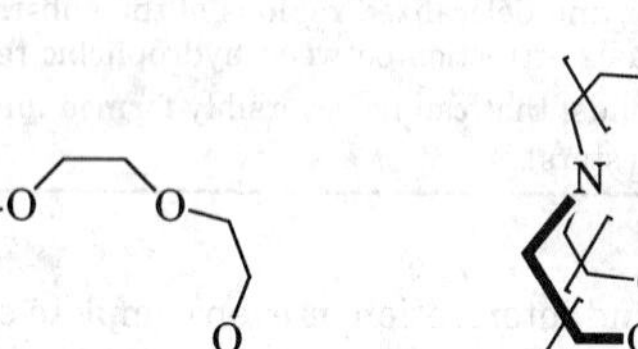

$R = CH_3$

(**4**)

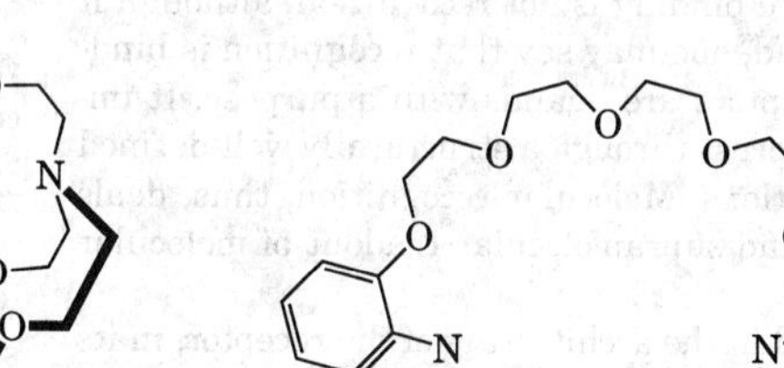

(**5**)

Figure 3. Crown type and analogous receptor molecules of different varieties; (**1**) crown ethers; (**2**) cryptands; (**3**) a podand; (**4**) a spherand; and (**5**) the natural depsipeptide valinomycin.

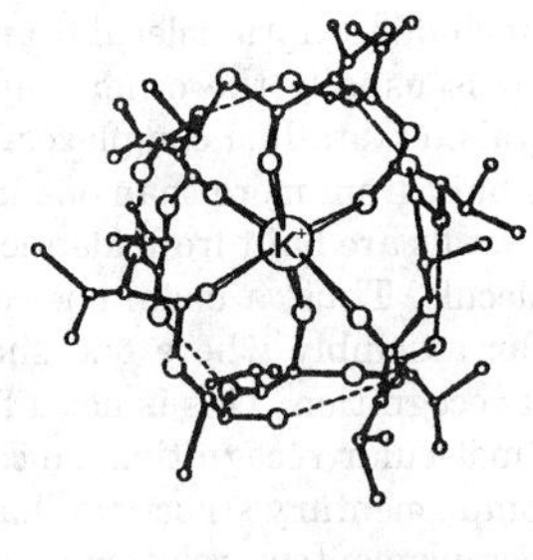

Figure 4. Spherical recognition of K^+ complex of valinomycin (**5**).

cryptates or podates are the names of the respective inclusion complexes (see INCLUSION COMPOUNDS).

Although acyclic podands do not provide a permanent cavity, they may create one by encircling a spherical cation with the length of the receptor molecular thread being the controlling parameter. Nevertheless, from what has already been said, low preorganization and topology of the podands handicap the substrate recognition which is increasingly higher in the circular crown and spheroidal cryptand case, but is most pronounced for the spherand type of receptor (eg, **4** in Fig. 3).

Natural macrocycles displaying antibiotic properties are also very efficient in the recognition of alkali metal ions. For instance, valinomycin (**5** in Fig. 3) gives a strong and selective complex in which a K^+ ion is included in the macrocyclic cavity in octahedral environment of six carbonyl oxygens (Fig. 4).

Recognition of a tetrahedral substrate geometry requires the construction of a receptor molecule with a tetrahedral recognition site. This may be realized by positioning four suitable binding sites at the corners of a tetrahedron and incorporating them into a bridged molecular framework.

Charge Attraction Dominated Recognition. Thus far, for recognition sizes and shapes of the substrate have been the focus. Nevertheless, charge attraction between the substrate and the receptor has also played a part, because cations such as metal ions or ammonium ions were complexed by negatively polarized cavities. But metal ions involved hard alkali and alkaline-earth cations rather than weaker transition metal ions. Replacing the oxygen sites of crown compounds and cryptands with nitrogen and sulfur atoms yields receptors that show marked preference for transition metal ions and may allow highly selective recognition of toxic heavy metal ions such as cadmium, lead or mercury according to the hard and soft acid and base (HSAB) principle. Others containing internally directed functionalized units form very strong and selective complexes with Fe^{3+} or actinide and lanthanide ions, while a similar receptor with hard endocarboxylic acid groups is efficient for hard Ca^{2+} and Mg^{2+} ions, showing again responsibility of a charge density effect in the receptor—substrate recognition. Thus, recognition of hard alkali and alkaline earth metal ions is determined by coulombic attraction, whereas the weak transition metal ions are mainly controlled by geometrical parameters of orbital overlap.

Hydrogen Bond Dominated Recognition. Recognition of bioactive compounds is largely determined by the use of hydrogen bonding between polar sites. Here substrate recognition results from the formation of specific hydrogen bonding pattern between complementary subunits, in a way reminiscent of base pairing in nucleic acids. Hence hydrogen bonding has also been determined an important parameter for the design of artificial receptors.

π-Stacking and Charge-Transfer Dominated Substrate Recognition. Nature's strategy for the recognition of substrates featuring a flat aromatic frame-work (planar recognition) affords another recognition element, namely $\pi-\pi$ stacking interactions between aromatic rings, ie, aromatic groups of receptor and substrate that meet a parallel face-to-face orientation, apart from hydrogen bonding being also typical of the nucleotide recognition.

In a way, $\pi-\pi$ stacking and charge transfer type of recognition have something in common. For example, a certain macrobicyclic intercaland and related receptors have been found to recognize flat shaped substrates through $\pi-\pi$ stacking and bind them to form a molecular cryptate, in particular if electron donating substrate species are involved to allow charge-transfer interaction, such as planar molecular anions or nucleic acids.

Lipophilic Interaction Dominated Substrate Recognition. Making recognition through lipophilic interaction possible requires receptors presenting large and more or less rigidly connected architectures of macrocyclic or cage-like nature.

The naturally occurring cyclodextrins having endo-lipophilic cone-shape are perhaps the most important and also the first receptor molecules whose selective inclusion properties towards lipophilic organic molecules were recognized. They comprise a family of cyclic oligosaccharides, composed of 6, 7, and 8 glucose units in its most familiar representatives (α, β, and γ-cyclodextrin, respectively) providing endo-lipophilic and exo-hydrophilic cone-shaped molecular cylinders of increasing size. Cyclodextrins form size and shape selective inclusion compounds with a wide variety of substrates including benzene derivatives, paraffins and noble gases.

Calixarenes (from the Latin *calix*) may be understood as artificial receptor analogues of the natural cyclodextrins. In its prototypical form they feature a macrocyclic metacyclophane framework bearing protonizable hydroxy groups made from condensation of *p*-substituted phenols with formaldehyde. Dependent on the ring size, benzene derivatives are the substrates most commonly included into the calix cavity, but other interesting substrates such as C_{60} have also been accommodated.

Multiple and Multisite, Coreceptor- and Coupled-System Substrate Recognition

Once recognition units for specific groups and individual features of a substrate have been identified, one may consider combining several of them within the same receptor. Thus far, though not carefully directed, the previous receptors in many cases already possess this property of nonindividual interaction modes. More carefully directed, this leads to multiple and multisite recognition depending on the design of binding subunits which may cooperate for the simultaneous complexation of several substrates or of a multiply bound polyfunctional species to yield polynuclear complexes (homo- or heteronuclear) and mononuclear polyhapto-type complexes, respectively.

Chiral Recognition

Enantiomers are perhaps the substrate type most difficult to distinguish. They are stereochemical species that have exactly the same structure except for their mirror image (chirality) relationship (see also CHIRAL SEPARATIONS). Chiral (enantiomer) recognition in complexation is one of the most important means by which receptor sites of biological systems such as in genes or enzymes act and regulate. From the principle point of view, recognition of a substrate enantiomer from racemic mixture (50:50 % mixture of enantiomers) requires an enantiomeric optically resolved receptor structure in order to make possible two diastereomeric receptor–substrate complexes allowing differentiation.

Following this line, a great variety of optically resolved (optically active) crown compounds were prepared for the resolution of racemic cationic substrates.

Artificial Receptors for Particular Substrate Recognition

Some particular substrates are biorelevant species or play central roles as drugs. Barbiturates are such an important family of drugs and are the target for molecular recognition. According to their structure, the barbiturate moiety essentially fuses two imide groups within a six-membered ring. Thus, two diaminopyridine units correctly positioned in a macrocyclic ring should bind to all six of the accessible hydrogen bonding sites in barbiturates. A crystal structure of a respective receptor–substrate complex has been performed that comes up to the expectations.

The structural and synthetic relationships shared between barbiturates and urea, which is another substrate of high physiological interest, suggest that the above receptor strategy could be modified for the selective complexation of urea. The designed modification for urea recognition involves replacement of the H-bond donating pyridine-6-amido groups in the previous barbiturate receptor by two H-bond accepting groups that differ by 120° in alignment to the substrate.

Receptors that are monomolecular species possessing a monomolecular cavity, pocket, cleft, groove or combination of it including the recognition sites to yield a molecular receptor–substrate complex can be assembled and preserved in solution. By way of contrast, molecular recognition demonstrated in the following comes from multimolecular assembly and organization of a nonsolution phase such as polymer materials and crystals.

Molecular Recognition in Polymers and Solids

If a polymer is prepared in the presence of molecules, the "print molecules" of which are extracted after polymerization, the remaining polymer may contain cavities, prints, or footprints that can recognize the print molecule. Actually, the cast relates to the matrix molecule like lock-and-key fit. (see Fig. 1).

A great many functionalized styrenes, including carboxylic acids, amino acids, Schiff bases, or specific compounds, eg, L-DOPA, have successfully been applied as print templates. Moreover, it has also been shown that silica gel can be imprinted with similar templates, and that the resulting gel has specific recognition sites determined by the print molecule.

Microporous inorganic materials dominated historically by the zeolites and alumosilicates, and the great variety of more recent nonoxide and coordination framework materials should also be mentioned here. This type of molecular recognition is usually known as molecular sieving.

Molecular Recognition at Interfaces and Surface Monolayers

There are three advantages to study molecular recognition on surfaces and interfaces (monolayers, films, membranes or solids): (*1*) rigid receptor sites can be designed; (*2*) the synthetic chemistry may be simplified; (*3*) the surface can be attached to transducers, which makes analysis easier and may transform the molecular recognition interface to a chemical sensor. This kind of molecular recognition involves outside directed interaction sites, ie, exo-receptor function (see Fig **2b**).

Molecular recognition of crystal interfaces makes possible the control of crystal growth processes in that suitably designed auxiliary molecules act as promoters or inhibitors of crystal nucleation inducing, for instance, the resolution of enantiomers or the crystallization of desired polymorphs and crystal habits.

Following another direction, it has previously been shown that alkanethiols spontaneously adsorb to Au from dilute solutions of ethanol and other nonaqueous solvents, and that the resulting self-assembling monolayers (SAMs) assume a close-packed overlayer structure on Au and other textured Au surfaces, being quite robust in aqueous solutions and vapor-phase ambients. This mode of self-assembly chemistry has been used to synthesize monolayer assemblies that function as molecular recognition interfaces based on the presence of recognizer end groups.

Self-Recognition

This mode of molecular recognition, on principle, is defined as the recognition of like from unlike or self from unself molecules, embodied in the spontaneous selection and preferential assembly of like components in a mixture.

So far this article has been concerned with interactions among chemically different species, which is true for most of the chemical recognition processes, including supramolecular and biomolecular processes. With crystals it is usually the other way around. Although some crystals, co-crystals, crystalline complexes, and crystalline inclusion compounds are built from more than one kind of molecule and are exceptions, most crystals are built from identical (or enantiomeric) copies of the same molecule. Thus, a usual one-component crystal is a macro-supramolecular assembly where one should more properly speak of molecular self-recognition. This is not a fact contradictory to the basic principles of molecular recognition, since the case might occur in which the two complementary structures happen to be identical in dealing with a self-complementary relationship. Even when all the molecules are identical (or enantiomeric), an acceptor part of one molecule can interact with a donor part of a second, and the acceptor part of the second can interact in exactly the same manner with the donor part of a third, and so on, giving rise to periodicity of the crystal and to the limited number of space groups used in molecular crystals. For instance, it is very uncommon for molecules in a crystal structure to be related by rotation axis or mirror planes, because identical parts of molecules avoid one another, except for molecular sites having a so-called self-complementary donor-acceptor group. Self-complementary groups form finite, one-dimensional tape, two-dimensional layer, or three-dimensional motifs of organic molecules mostly obtained from hydrogen bonding.

In solution, highly ordered structures created via self-recognition and self-assembly of a programmed H-bonding molecular component are also possible. With respect to inorganic self-recognition and self-assembly this would involve preferential binding of like metal ions by like ligands in a mixture of ligands and ions.

A particular point of interest included in these helical complexes concerns the chirality. The helicates obtained from the achiral strands are a racemic mixture of left- and right-handed double helices. This special mode of recognition where homochiral supramolecular entities, as a consequence of homochiral self-recognition, result from racemic components is known as optical self-resolution. It appears in certain cases from racemic solutions or melts (spontaneous resolution) and is often cited as one of the possible sources of optical resolution in the biological world.

EDWIN WEBER
Technische Universität Bergakademie
Freiberg Institut für Organische Chemie

M. I. Page, *The Chemistry of Enzyme Action,* Elsevier, Amsterdam, 1984.

E. C. Hulme, *Receptor Biochemistry,* Oxford University Press, New York, 1990.

S. M. Roberts, ed., *Molecular Recognition–Chemical and Biochemical Problems,* The Royal Society of Chemistry, Cambridge, 1989.

J.-P. Behr, ed., *The Lock and Key Principle, Perspectives in Supramolecular Chemistry,* Vol. 1, Wiley, Chichester, 1994.

MOLECULAR SIEVES

In this article, the term *molecular sieve* is restricted to inorganic materials that possess uniform pores with diameters in either the micro- (<2 nm) or meso- (2–20 nm) size range. The most technologically important molecular sieves are zeolites, ie, crystalline silicate or aluminosilicate framework structures with channels of diameters <1.2 nm. Several of these topologies, with boron, gallium, or iron replacing aluminum, or germanium replacing silicon, have also been prepared. The chemical composition of microporous framework structures has been expanded considerably with the substitution of phosphorus for silicon, and new families of aluminophosphate and silicoaluminophosphate structures have been synthesized in the laboratory. Some of these frameworks have zeolite analogues, whereas others are unique. The addition of elements such as Mg, Ti, Mn, Co, Fe, or Zn into these structures has made it possible to generate metalloaluminophosphates, metallosilicoaluminophosphates, etc. Microporous sulfide-based framework structures are also possible.

Considerable synthesis effort has been devoted to developing frameworks with pore diameters within the mesoporous range; the largest synthesized are the phosphate-based AlPO-8 (14-membered ring), VPI-5 (18-MR), and cloverite (20-MR), which have pore diameters within the 0.8–1.3 nm range. A new family of mesoporous molecular sieves designated M41S has been discovered. Although not framework structures like zeolites, silicate, and aluminosilicate, M41S materials possess very uniform mesopores.

The technological applications of molecular sieves are as varied as their chemical makeup. Heterogeneous catalysis and adsorption processes make extensive use of molecular sieves. The utility of the latter materials lies in their microstructures, which allow access to large internal surfaces, and cavities that enhance catalytic activity and adsorptive capacity.

Zeolites

Molecular-sieve zeolites of the most important aluminosilicate variety can be represented by the chemical formula $M_{2/n}O{\cdot}Al_2O_3{\cdot}ySiO_2{\cdot}wH_2O$, where y is 2 or greater, M is the charge balancing cation, such as sodium, potassium, magnesium, and calcium, n is the cation valence, and w represents the moles of water contained in the zeolitic voids. The zeolite framework is made up of SiO_4 tetrahedra linked together by sharing of oxygen ions. Substitution of Al for Si generates a charge imbalance, necessitating the inclusion of a cation. The structures contain channels or interconnected voids that are occupied by the cations and water molecules. The water may be removed reversibly, generally by the application of heat, which leaves intact the crystalline host structure permeated with micropores that may account for >50% of the microcrystal's volume. In some zeolites, dehydration may produce some perturbation of the structure, such as cation movement, and some degree of framework distortion.

Zeolite minerals are formed over much of the earth's surface, including the sea bottom. Most zeolites occurring in cavities of basaltic and volcanic rocks are exceedingly rare. However, several zeolite minerals generated by the natural alteration of volcanic ash in alkaline environments over long periods of time occur in recoverable deposits. Although such minerals are rarely useful for catalytic application, mainly because of iron impurities (exception: Chabazite of Bowie, Arizona), more abundant zeolites, eg, clinoptilolite, have found use as soil conditioners, additives to animal feed, as animal litter, in aquaculture to remove ammonia (phillipsite), and for ion exchange to remove heavy metals from industrial and mining effluents (chabazite).

Synthetic and mineral zeolites of primary importance are listed in Table 1.

Structure

Of the approximately 120 known framework aluminosilicates, ~ 50 occur naturally, the rest being synthetic. There are 56 structural types of zeolites known. Understanding the complexities of zeolite structures is made easier by recognizing three important structural keys: the basic arrangement of the individual structural units in space, which defines the framework topology; the location of the charge-balancing cations; and the channel-filling material, such as water or an organic template, which is incorporated as the zeolite is formed. After the channel-filling material is removed, the void space can be used for the adsorption of gases, liquids, salts, elements, metal complexes, etc. In turn, this void-filling property makes zeolites commercially useful in ion exchange, catalysis, etc.

There are two types of structures: one provides an internal pore system comprising interconnected cage-like voids; the second provides a system of uniform channels which, in some instances, are one-dimensional and in others intersect with similar channels to produce two- or three-dimensional channel systems. The preferred type has two- or three-dimensional channel systems to provide rapid intercrystalline diffusion in adsorption and catalytic applications.

In most zeolite structures, the primary structural units, tetrahedra, are assembled into secondary building units, which may be simple polyhedra such as cubes, hexagonal prisms, or truncated octahedra. The final framework structure consists of assemblages of the secondary units.

Table 1. Zeolite Compositions

Zeolite	Typical formula
Natural	
chabazite	$Ca_2[(AlO_2)_4(SiO_2)_8]{\cdot}13H_2O$
mordenite	$Na_8[(AlO_2)_8(SiO_2)_{40}]{\cdot}24H_2O$
erionite	$(Ca, Mg, Na_2, K_2)_{4.5}[(AlO_2)_9(SiO_2)_{27}]{\cdot}27H_2O$
faujasite	$(Ca, Mg, Na_2, K_2)_{29.5}[(AlO_2)_{59}(SiO_2)_{133}]{\cdot}235H_2O$
clinoptilolite	$Na_6[(AlO_2)_6(SiO_2)_{30}]{\cdot}24H_2O$
phillipsite	$(0.5Ca, Na, K)_3[(AlO_2)_3(SiO_2)_5]{\cdot}6H_2O$
Synthetic	
zeolite A	$Na_{12}[(AlO_2)_{12}(SiO_2)_{12}]{\cdot}27H_2O$
zeolite X	$Na_{86}[(AlO_2)_{86}(SiO_2)_{106}]{\cdot}264H_2O$
zeolite Y	$Na_{56}[(AlO_2)_{56}(SiO_2)_{136}]{\cdot}250H_2O$
zeolite L	$K_9[(AlO_2)_9(SiO_2)_{27}]{\cdot}22H_2O$
zeolite omega	$Na_{6.8}TMA_{1.6}[(AlO_2)_8(SiO_2)_{28}]{\cdot}21H_2O$[a]
ZSM-5	$(Na, TPA)_3[(AlO_2)_3(SiO_2)_{93}]{\cdot}16H_2O$[b]

[a] TMA = tetramethylammonium. [b] TPA = tetrapropylammonium.

Zeolite Minerals. Crystal structures of zeolite minerals are illustrated by the zeolite chabazite. The structure of chabazite is hexagonal and the framework consists of double six-membered rings of $(Si,Al)O_4$ tetrahedra arranged in parallel layers in an AABBCC sequence. These tetrahedra are cross-linked by four-membered rings, as shown in Figure 1.

Synthetic Zeolites. Many new crystalline zeolites have been synthesized and several fulfill important functions in the chemical and petroleum industries and in consumer products such as detergents. The structural formula of a zeolite is based on the crystal unit cell, the smallest unit of structure, represented by $M_{x/n}[(AlO_2)_x(SiO_2)_y]\cdot wH_2O$, where n is the valence of cation M, w is the number of water molecules per unit cell, x and y are, respectively, the number of AlO_4 and SiO_4 tetrahedra per unit cell, and y/x usually has values of 1–5. Examples of important synthetic zeolites are shown in Table 1.

The secondary structure unit in zeolites A, X, and Y is the truncated octahedron. These polyhedral units are linked in three-dimensional space through the four- or six-membered rings. The former linkage produces the zeolite A structure, and the latter the topology of zeolites X and Y and of the mineral faujasite.

The structure of the high silica zeolite ZSM-5 (y/x = 10to > 5000) contains a high concentration of five-membered rings and has two

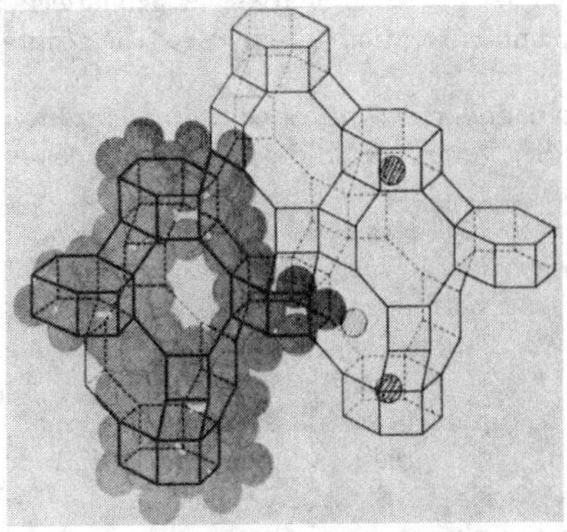

Figure 1. Structure of the mineral zeolite chabazite is depicted by packing model, left, and skeletal model, right. The silicon and aluminum atoms lie at the corners of the framework depicted by solid lines. In this figure the solid lines do not depict chemical bonds. Oxygen atoms lie near the midpoint of the lines connecting framework corners. Cation sites are shown in three different locations referred to as sites I, II, and III. Courtesy of *Scientific American*.

intersecting channels. It has become a very important catalyst for petrochemical reactions.

Structure Modification. Several types of structural defects or variants can occur which figure in adsorption and catalysis: *(1)* surface defects due to termination of the crystal surface and hydrolysis of surface cations; *(2)* structural defects due to imperfect stacking of the secondary units, which may result in blocked channels; *(3)* ionic species, eg, OH^-, AlO_2^-, Na^+, SiO_4^-, may be left stranded in the structure during synthesis; *(4)* the cation form, acting as the salt of a weak acid, hydrolyzes in aqueous suspension to produce free hydroxide and cations in solution; and *(5)* hydroxyl groups in place of metal cations may be introduced by ammonium ion exchange, followed by thermal deammoniation.

Properties

Adsorption. Although several types of microporous solids are useful as adsorbents for the separation of vapor or liquid mixtures, the distribution of pore diameters does not enable separations based on the molecular-sieve effect. The most important molecular-sieve effects are shown by crystalline zeolites. The sieve effect may be total or partial.

Activated diffusion of the adsorbate is of interest in many cases. As the size of the diffusing molecule approaches that of the zeolite channels, the interaction energy becomes increasingly important. If the aperture is small relative to the molecular size, then the repulsive interaction is dominant and the diffusing species needs a specific activation energy to pass through the aperture. Similar shape-selective effects are shown in both catalysis and ion-exchange, two important applications of these materials.

In order to utilize the absorption properties of the synthetic zeolite crystals in processes, the commercial materials are prepared as pelleted aggregates combining a high percentage of the crystalline zeolite with an inert binder. The formation of these aggregates introduces macropores in the pellet which may result in some capillary condensation at high adsorbate concentrations. In commercial materials, the macropores contribute diffusion paths. However, the main part of the adsorption capacity is contained in the voids within the crystals.

Zeolites are high capacity, selective adsorbents capable of separating molecules based on the size and shape of the structure. They adsorb molecules, in particular those with a permanent dipole moment which show other interaction effects, with a selectivity that is not found in other solid adsorbents. Separation may be based on the molecular-sieve effect or may involve the preferential or selective adsorption of one molecular species over another. These separations are governed by several factors. The basic framework structure, or topology, of the zeolite determines the pore size and the void volume. The exchange cations, in terms of their specific location in the structure, their population density, their charge and size, affects the molecular-sieve behavior and adsorption selectivity of the zeolite. By changing the cation types and number, the selectivity of the zeolite in a given separation can be tailored or modified, within certain limits.

The high silica version of ZSM-5, also known as silicalite, is a hydrophobic adsorbent capable of adsorbing, eg, ethanol from an aqueous solution.

Catalytic Properties. In zeolites, catalysis takes place preferentially within the intracrystalline voids. Catalytic reactions are affected by aperture size and type of channel system, through which reactants and products must diffuse. Modification techniques include ion-exchange, variation of Si/Al ratio, hydrothermal dealumination or stabilization, which produces Lewis acidity, introduction of acidic groups such as bridging Si(OH)Al, which impart Brønsted acidity, and introducing dispersed metal phases such as noble metals. In addition, the zeolite framework structure determines shape-selective effects. Several types have been demonstrated, including reactant selectivity, product selectivity, and restricted transition-state selectivity. Nonshape-selective surface activity is observed on very small crystals, and it may be desirable to poison these sites selectively, with bulky heterocyclic compounds unable to penetrate the channel apertures, or by surface silation.

Some current and possible future zeolite catalyst applications are as follows: alkylation, cracking, hydrocracking, dewaxing, isomerization, hydrogenation and dehydrogenation, hydrodealkylation, methanation, shape-selective reforming, dehydration, methanol to gasoline, methanol to olefins, organic catalysis, inorganic reactions, H_2S oxidation, NH_3 reduction of NO, $H_2O \rightarrow 1/2\ O_2 + H_2$, and CO oxidation.

Ion Exchange. The exchange behavior of nonframework cations in zeolites, eg, selectivity, and degree of exchange, depends on the nature of the cation, eg, the size and charge of the hydrated cation, on the temperature, the concentration, and, to some degree, on the anion species. Cation exchange may produce considerable change in various other properties, such as thermal stability, adsorption behavior, and catalytic activity.

Framework Modification

The zeolite framework can be stabilized by hydrothermal treatment, which removes aluminum from the framework and forms aluminum cations. During this steaming process, the tetrahedral vacancies left behind in the framework are gradually refilled with silicon, which appears to migrate as a form of silicic acid from other parts of the framework and contributes to stabilization by repairing the damaged framework. Simultaneously, such other parts of the framework disappear under formation of mesopores. Cationic aluminum can be extracted with an acid, and a subsequent steaming causes further dealumination of the framework and migration of silicon into the vacancies. Carefully controlled conditions can produce high silica forms of zeolites, eg, zeolite Y. Since the Si–O bond is shorter than the Al–O bond, hydrothermal dealumination causes the unit cell parameter to decrease. Mesopores can be avoided by replacing the aluminum directly with external silicon, eg, by treatment with silicon tetrachloride. In a reversal of the reaction with $SiCl_4$, aluminum can be introduced into the framework by reaction of the hydrogen or ammonium form with gaseous $AlCl_3$.

Manufacture

Zeolites are formed under hydrothermal conditions, defined here in a broad sense to include zeolite crystallization from aqueous systems containing various types of reactants. Most synthetic zeolites are produced under nonequilibrium conditions, and must be considered as metastable phases in a thermodynamic sense.

Many important types of zeolites have no natural mineral counterpart. Conversely, synthetic counterparts of many zeolite minerals are not yet known. The conditions generally used in synthesis are reactive starting materials such as freshly co-precipitated gels, or amorphous solids; relatively high pH introduced in the form of an alkali metal hydroxide or other strong base, including tetraalkylammonium hydroxides; low temperature hydrothermal conditions with concurrent low autogenous pressure at saturated water vapor pressure; and a high degree of supersaturation of the gel components, leading to nucleation of a large number of crystals.

A gel is defined as a hydrous metal aluminosilicate prepared from either aqueous solutions, reactive solids, colloidal sols, or reactive aluminosilicates such as the residue structure of metakaolin and glasses.

The gels are crystallized in a closed hydrothermal system at temperature varying from room temperature to about 200°C. The time required for crystallization varies from a few hours to several days. When prepared, the aluminosilicate gels differ in appearance, from stiff and translucent to opaque gelatinous precipitates and heterogenous mixtures of an amorphous solid dispersed in an aqueous solution. The alkali metals form soluble hydroxides, aluminates, and silicates. These materials are well suited for the preparation of homogeneous mixtures.

Gel preparation and crystallization is represented systematically using the $Na_2O—Al_2O_3—SiO_2—H_2O$ system as an example.

$$NaAl(OH)_4(aq) + Na_2SiO_3(aq) \xrightarrow{25°C} [Na_a(AlO_2)_b(SiO_2)_c \cdot NaOH \cdot H_2O]gel$$

$$\xrightarrow{25-175°C} Na_x[(AlO_2)_x(SiO_2)_y] \cdot mH_2O$$

Table 2. Some Synthetic Zeolites Prepared from Sodium Aluminosilicate Gels

Zeolite type	Typical composition, mol/mol Al_2O_3			Reactants	Reactant temp, °C	Zeolite product composition, mol/mol Al_2O_3		
	Na_2O	SiO_2	H_2O			Na_2O	SiO_2	H_2O
A	2	2	35	$NaAlO_2$ NaOH sodium silicate	20–175	1	2	4.5
X	3.6	3	144	$NaAlO_2$ NaOH sodium silicate	20–120	1	2.0–3.0	6
Y	8	20	320	$NaAlO_2$ colloidal SiO_2 NaOH	20–175	1	3.0–6.0	9
mordenite, Zeolon	6.3	27	61	$NaAlO_2$ diatomite sodium silicate	100	1	9–10	6.7
omega	5.60[a]	20	280	colloidal SiO_2 $Al(OH)_3$ TMAOH[b] NaOH	100	0.71 0.36 TMA	7.3	6.3
ZSM-5	10[c]	7.7	453	$NaAlO_2$ SiO_2 TPAOH[e]	150	0.89	31.1[d]	2.0

[a] Also 1.4 TMA_2O. [b] TMA = tetramethylammonium. [c] Also 8.6 TPA_2O. [d] After calcination at 1000°C. [e] TPA = tetrapropylammonium.

Table 3. Processes for Molecular Sieve Zeolites

Process	Reactants	Products
hydrogel	reactive oxides	high purity powders
	soluble silicates	gel preform
	soluble aluminates	zeolite in gel matrix
	caustic	
clay conversion	raw kaolin	
	meta-kaolin	
	calcined kaolin	low to high purity powder
	acid-treated clay	binderless, high purity preform
	soluble silicate	zeolite in clay-derived matrix
	caustic	
	sodium chloride	
other	natural SiO_2	
	amorphous minerals	low to high purity powder
	volcanic glass	zeolite on ceramic support
	caustic	binderless preforms

Typical gels are prepared from aqueous solutions of reactants such as sodium aluminate, NaOH, and sodium silicate; other reactants include alumina trihydrate ($Al_2O_3 \cdot 3H_2O$), colloidal silica, and silicic acid. Some synthetic zeolites prepared from sodium aluminosilicate gels are given in Table 2. The temperature strongly influences the crystallization time of even the most reactive gels.

Synthesis mechanisms of the typical low silica zeolites, such as A, X, and Y, are apparently different from the high silica zeolites such as ZSM-5. In the low silica zeolites, nuclei are formed consisting of alkali metal-ion complexes of the aluminosilicate species. Structural units consisting of four-membered rings, six-membered rings, and cages coordinated with cations are thought to be involved in the nucleation and crystallization. In the high silica zeolites, the mechanism appears to be a templating type where an alkylammonium cation complexes with silica by hydrogen bonding. These complexes cause the structures to replicate by hydrogen bonding of the organic cation with framework oxygen atoms.

Processes. Manufacturing processes for commercial molecular sieve products may be classified into three groups, as shown in Table 3.

Economic Aspects

The U.S. capacity for synthetic zeolites at the end of 1994 was 325×10^3 t for detergent use; $145 \times 10^3 hspsp = 0.167 > t$ for catalysts; and about 60×10^3 t for use as adsorbents and desiccants. As of mid-1995, the price of builder-grade zeolite A to the largest consumers was about \$0.66–0.68/kg on anhydrous basis (list price, \$0.77–0.79/kg), whereas zeolite 4A adsorbent sold for \$2.31–2.53/kg in 1995. The price of zeolite-containing FCC catalyst was in the range of \$1450–2000/t in 1991. Prices for catalysts tailored to individual customer requirements can exceed this range.

Health and Safety Factors and Toxicology

Zeolites have applications in food, drugs, cosmetic products, and detergents. Thus, extensive toxicological and environmental studies have been carried out. Feeding of 5.0 g/kg of body weight (powder form of type 4A, 5A, 13X, and Y) for seven days produced no ill effect in rats. There is no contraindication to the use of zeolite A (Sasil) in detergents. No negative effect on biological wastewater treatment was found, and zeolite A showed no evidence of acute toxicity to four species of freshwater fish.

Uses

In most cases, the water content of the commercial product is below 1.5–2.5 wt %; certain products, however, are sold as fully hydrated crystalline powders. Molecular sieve products are used for adsorption (eg, purification of water, carbon dioxide, and sulfur compounds); bulk separation of normal and isoparaffins, xylene, olefin, and oxygen from air; catalysis (eg, catalytic cracking, hydrocracking for fuels production, dewaxing of distillate fuel and lube basestocks, paraffin isomerization, catalysis of aromatic reactions (in selective toluene disproportionation, xylene isomerization, and ethylbenzene synthesis), synthesis of *p*-ethyltoluene, and the methanol-to-gasoline process; and ion exchange (eg, cesium and strontium radioisotopes, ammonium ion removal, and detergent builders).

New Trends

Aluminosilicate of faujasite topology, but higher Si/Al ratio (≤ 5), has been synthesized with the use of a crown ether, 15-crown-5, as the directing agent. The same Si/Al ratio was obtained when 18-crown-6 was applied, but the topology, although related to faujasite, had a different stacking order of the sodalite cages, so that the structure has hexagonal instead of cubic symmetry. This product, sometimes called hexagonal faujasite, has the designation EMT.

New directions in the preparation of framework structures of different chemical composition and of large-pore molecular sieves include the development of phosphate-containing molecular sieves and mesoporous molecular sieves.

GÜNTER H. KÜHL
CHARLES T. KRESGE
Mobil Research and Development Corporation

D. W. Breck, *Zeolite Molecular Sieves, Structure, Chemistry, and Use,* John Wiley & Sons, Inc., New York, 1974.

R. M. Barrer, *Zeolites and Clay Minerals as Sorbents and Molecular Sieves,* Academic Press, London, 1978.

W. M. Meier, D. H. Olson, and Ch. Baerlocher, *Atlas of Zeolite Structure Types,* Fourth Revised Edition, Elsevier, New York, 1996.

J. A. Rabo and G. J. Gajda, *Catal. Rev.-Sci. Eng.* **31,** 385 (1989–1990).

MOLLUSCILIDES. See PESTICIDES.

MOLYBDENUM AND MOLYBDENUM ALLOYS

Molybdenum was first identified as a discrete element in 1778. Along with the many metallurgical uses of the metal, molybdenum compounds (qv) are used in such chemical applications as catalysis (qv), corrosion protection (see CORROSION AND CORROSION CONTROL), and lubrication (see LUBRICATION AND LUBRICANTS).

Most of the world's supply of molybdenum comes as a by-product or co-product from copper (qv) mining. Only about one quarter of the supply comes from primary mines. A small but significant supply of molybdenum is also obtained from the processing of spent petroleum catalysts (see CATALYSTS REGENERATION; RECYCLING, METALS–NONFERROUS METALS). The most abundant mineral, and the only one of commercial significance, is molybdenite, MoS_2. The minerals powellite, $Ca(MoW)O_4$, and wulfenite, $PbMoO_4$, also are known but are not sources of the metal.

The largest share of molybdenum supply comes from North America, once the only significant source. Sizeable amounts also come from Latin America (mostly Chile), China, and countries of the former Soviet Union.

Molybdenite is concentrated by first crushing and grinding the ore, and passing the finely ground material through a series of flotation (qv) cells (see MINERALS RECOVERY AND PROCESSING). Operations which recover molybdenum as a by-product of copper mining produce a concentrate containing both metals. Molybdenite is separated from the copper minerals by differential flotation.

Molybdenite concentrate contains about 90% MoS_2. The remainder is primarily silica, with lesser amounts of Fe, Al, and Cu. The concentrate is roasted to convert the sulfide to technical molybdic oxide. Molybdenum is added to steel in the form of this oxide. In modern molybdenum conversion plants, the oxidized sulfur formed by roasting MoS_2 is converted to sulfuric acid.

Technical molybdic oxide can be reduced by reaction of ferrosilicon in a thermite-type reaction. The resulting product contains about 60% molybdenum and 40% iron. Foundries generally use ferromolybdenum for adding molybdenum to cast iron and steel.

Environmental and Safety Considerations

Because of its position in the Periodic Table, molybdenum has sometimes been linked to chromium (see CHROMIUM AND CHROMIUM ALLOYS) or to other heavy metals. However, unlike those elements, molybdenum and its compounds have relatively low toxicity. On the other hand, molybdenum has been identified as a micronutrient essential to plant life (see FERTILIZERS), and plays a principal biochemical role in animal health as a constituent of several important enzyme systems (see MINERALS NUTRIENTS). Information on the toxic effects of molybdenum in humans is scarce.

Physical Properties

Molybdenum has many unique properties, leading to its importance as a refractory metal (see REFRACTORIES). Molybdenum, atomic no. 42, is in Group 6 (VIB) of the Periodic Table between chromium and tungsten vertically and niobium and technetium horizontally. It has a silvery gray appearance. The most stable valence states are +6, +4, and 0; lower, less stable valence states are +5, +3, and +2.

Molybdenum, a typical transition element, has the maximum number, five, of unpaired 4*d* electrons, which account for its high melting point, strength, and high modulus of elasticity. There are many similarities between molybdenum and its horizontal and vertical neighbors in the periodic system. See Table 1 for selected properties of molybdenum.

Table 1. Selected Properties of Molybdenum

Property	Value
melting point, °C	2626 ± 9
heat of fusion,[a] kJ/mol[b]	28
boiling point, °C	5560
heat of vaporization, kJ/mol[b]	491
entropy of crystals, $S°_{298.16}$, J/(mol·K)[b]	28.6
vapor pressure, Pa[c]	
at 1725°C	3.95×10^{-9}
2225°C	1.72×10^{-4}
2725°C	8.71×10^{-3}
3725°C	
5225°C	5.57
diffusivity, cm^2/s	
at 200°C	0.43
540°C	0.40
870°C	0.38
specific heat at 100°C, kJ(kg·K)[b]	0.27
coefficient of linear expansion, %	
at 0–400°C	0.23
0–1200°C	0.72
thermal conductivity, W/(m·K)	
at 500°C	122
at 1500°C	82

[a] Estimated value. [b] To convert J to cal, divide by 4.186. [c] To convert Pa to mm Hg, multiply by 0.0075.

Chemical Properties

Molybdenum has good resistance to chemical attack by mineral acids, provided that oxidizing agents are not present. The metal also offers excellent resistance to attack by several liquid metals.

Molybdenum has high resistance to a number of alloys of these metals and also to copper, gold, and silver. Among the molten metals that severely attack molybdenum are tin (at 1000°C), aluminum, nickel, iron, and cobalt. Molybdenum has moderately good resistance to molten zinc, but a molybdenum–30% tungsten alloy is practically completely resistant to molten zinc at temperatures up to 800°C. Molybdenum metal is substantially resistant to many types of molten glass and to most nonferrous slags. It is also resistant to liquid sulfur up to 440°C.

Manufacture

Ammonium molybdate or molybdenum trioxide is reduced to molybdenum metal powder by hydrogen in a two-stage process. In the first stage, MoO_3 or ammonium molybdate is reduced to molybdenum dioxide, MoO_2, at temperatures around 600°C; in the second stage, the dioxide is reduced to metallic powder at temperatures near 1100°C. Both rotary and boat-and-tube types of furnaces are used for first-stage reduction. Boat-and-tube furnaces are used for the second stage.

Molybdenum wire is produced by a long-established powder metallurgy process (see METALLURGY–POWDER).

Two consolidation processes are used to produce molybdenum mill products such as forging billets, bars, rods, plate, sheet, and foil: powder metallurgy and arc-casting or vacuum-arc melting. In the powder metallurgy process, molybdenum powder is compacted isostatically in hydraulic pressure chambers to cylindrical bars or billets and to rectangular sheet bars. Pressed and sintered sheet bars are rolled directly to plate, sheet, and foil. As mechanical work proceeds, the density of powder metallurgy molybdenum improves to the full theoretical density. In the arc-casting process, a consumable electrode of compacted molybdenum powder is melted by an alternating current arc inside a water-cooled copper tube, or mold, to form an ingot. Ingots weighing 820 kg have been produced consistently in the pressing, sintering, and melting (PSM) machine.

Molybdenum metal can be mechanically worked by almost any process: forging, extrusion, rolling, bending, punching, stamping, deep drawing, spinning, conventional forming, and power roll forming. Except for fine wire and thin sheet, it is recommended that mill products be heated moderately for most shaping operations.

Uses

Molybdenum metal is the most widely used electrical resistance element in furnaces where temperatures beyond the limits of ordinary resistance alloys are required. Such furnaces are generally used for temperatures up to about 1650°C, but some are in successful operation at 2200°C. The elements, which may be wire, rod, ribbon, or expanded sheet, must be protected from oxidation by a reducing or inert atmosphere, or a vacuum. Hydrogen is commonly employed. Under these conditions, molybdenum has a long life and seldom is the limiting factor in the durability of the furnace. Molybdenum sheet for susceptors in high frequency units, radiation shields, baffles, structural supports, muffle liners, skids, hearths, boats, and firing trays is used in all types of high temperature vacuum and controlled atmosphere furnaces. Molybdenum metal is useful as support wires for tungsten filaments in incandescent light bulbs, as targets in x-ray tubes and in electrically heated glass furnaces.

Alloys

The strength of molybdenum depends on work-hardening. The greater the amount of cold-working, ie, percent reduction of area below the recrystallization temperature, the higher the yield and tensile strengths at all temperatures and the lower the creep rates at elevated temperatures. Exposure to temperatures high enough to cause recrystallization produces a drastic reduction in tensile properties, and a loss of ductility.

Additions of selected alloying elements raise the recrystallization temperature, extending to higher temperature regimes the tensile properties of the cold-worked molybdenum metal. The simultaneous additions of 0.5% titanium and 0.1% zirconium produce the TZM alloy, which has a corresponding recrystallization temperature of 1500°C and which cold-works to higher hardness and strength than unalloyed molybdenum. An alloy of molybdenum containing 1.2% hafnium with carbon at the level of 0.08–0.10% has a slight advantage over TZM.

Tungsten has little effect on recrystallization temperature or the high temperature properties of molybdenum. However, the Mo–30% W alloy is recognized as a standard commercial alloy for stirrers, pipes, and other equipment that is required to be in contact with molten zinc during processing of the metal and in galvanizing and die casting operations.

Molybdenum, imparts numerous beneficial properties to irons and steels and to some alloy systems based on cobalt, nickel, or titanium. Molybdenum is a potent contributor to hardenability, and has been shown to be even more effective in the presence of carefully selected amounts of other alloying elements.

Many steels used for gears and bearings are surface-hardened by carburizing, quenching, and tempering. Molybdenum is frequently used in carburized steels, and carburized Ni–Mo steels have been shown to provide optimum resistance to fatigue and impact effects.

Molybdenum is effective in reducing the susceptibility of Cr-Ni steels to embrittlement following tempering at >600°C or exposure to temperatures ca 500°C for extended times. Molybdenum-alloyed steels have been found to perform better than other steels in service in the oil industry (see PETROLEUM). Steels containing molybdenum in amounts as high as 0.80% can be heat-treated to high strength levels and resist sulfide stress cracking.

High strength, low alloy (HSLA) steels often contain 0.10–0.30% molybdenum and exhibit toughness at low temperatures and good weldability. They are used extensively for undersea pipelines (qv) transporting gas and oil from offshore wells to pumping stations on shore, and in remote Arctic environments.

Molybdenum improves the corrosion resistance of stainless steels that are alloyed with 17–29% chromium.

A simple, low cost steel for high temperature service in electric power generation (qv) is the C-0.5% Mo steel used by the power industry and oil refineries. Mo steels have improved resistance to graphitization and oxidation, as well as higher creep and rupture strength. Molybdenum is used also in cast irons and increases hardness, strength, and wear resistance.

D. V. DOANE
G. A. TIMMONS
Consultants
C. J. HALLADA
Climax Molybdenum Company

Metals Handbook, 9th ed., Vol. 2, American Society for Metals, Metals Park, Ohio, 1979,

A. Sutulov, *International MolybdenumEncyclopedia,* Intermet Publications, Santiago, Chile, 1980.

J. Shields, *Applications of Molybdenum Metal and Its Alloys,* CSM Industries, Cleveland, Ohio, 1996.

MOLYBDENUM COMPOUNDS

The chemistry of molybdenum, Mo, is among the most diverse of the transition elements. In its compounds, molybdenum exhibits coordination numbers from four to eight, oxidations numbers from −II to VI, and numerous states of aggregation (nuclearity). Molybdenum forms binary compounds with many nonmetallic elements, and a number of these, namely halides, oxides, sulfides, carbides, nitrides, and silicides, are of technological interest. In contrast to its congeners, chromium and tungsten, molybdenum is found naturally in the form of its sulfide molybdenite, MoS_2. Similarly, in the enzymes in which molybdenum is found, and in a number of its technological uses, the active form of Mo is bound by sulfur.

In biology, molybdenum is a component of fertilizer and nutrient formulations (see FERTILIZERS; MINERAL NUTRIENTS). In technology, various solids and soluble molybdenum compounds have found use in lubrication (see LUBRICATION AND LUBRICANTS); hydrodesulfurization, hydrogenation, and oxidation catalysis; anticorrosion and coatings (qv); flame and smoke retardancy (see FLAME RETARDANTS); and various forms of pigmentation.

The most important molybdenum oxidation states are VI, V, IV, III, II and 0. The high oxidation states are usually characterized by molybdenum binding to electronegative atoms, such as oxygen and the halogens. The lowest oxidation states are largely in the realm of organometallic chemistry, wherein the Mo is bound directly to the carbon atom of carbon monoxide (qv), to organic phosphines, and/or to a variety of unsaturated carbonaceous ligands.

Molybdenum (VI)

The chemistry of hexavalent molybdenum (Fig. 1) is very prominent in both biological and industrial systems. Oxygen coordination of molybdenum is most common in this oxidation state. Molybdenum trioxide, MoO_3, is a key intermediate in the technological utilization of molybdenum (Fig. 2). The structure of MoO_3 is a complex, layered arrangement in which each of the six-coordinate Mo(VI) atoms shares the face of an octahedron with another Mo(VI) atom. MoO_3 reacts with base to produce a variety of molybdate salts.

Molybdenum (V)

Molybdenum(V) compounds generally occur as mononuclear or dinuclear species. Molybdenum pentachloride, $MoCl_5$, formed by combination of the elements, serves as a useful and reactive starting material (Fig. 2). $MoCl_5$ has a dinuclear structure (Fig. 3) in the solid state but is mononuclear in the gas phase.

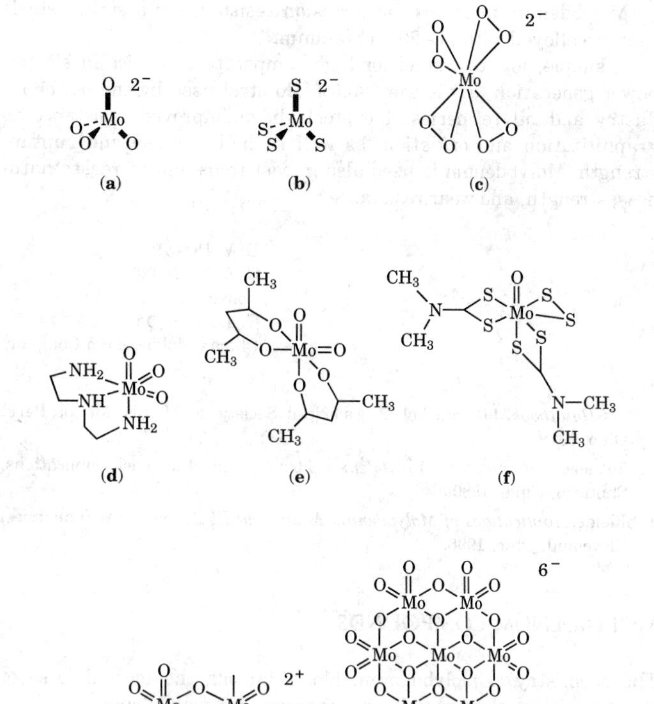

Figure 1. Representative structures for compounds of molybdenum(VI): (**a**) molybdate(VI), MoO_4^{2-}; (**b**) tetrathiomolybdate (VI), MoS_4^{2-}; (**c**) tetrakis(peroxo)molybdate(VI), MoO_{24}^{2-}; (**d**) *cis*-trioxodiethylenetriaminemolybdenum(VI), (MoO_3(dien)), $C_4H_{13}N_3MoO_3$; (**e**) *cis*-bis(acetylacetonato)dioxomolybdenum(VI), $MoO_2(C_5H_7O_2)_2$; (**f**) bis(dialkyldithiocarbamato)disulfidooxomolybdenum(VI), MoO(S2)$(S_2CNR_2)_2$ (R = alkyl); (**g**) the dinuclear core structure for $Mo_2O_5^{2+}$ complexes; (**h**) heptamolybdate(VI), $Mo_7O_{24}^{6-}$.

Molybdenum(IV)

Representative compounds for the +4 oxidation state are shown in Figure 4. The violet tetravalent molybdenum dioxide, MoO_2, is formed by the reduction of MoO_3 with H_2 at temperatures below which Mo metal is formed or MoO_3 is volatile (ca 450°C). $MoCl_4$ is formed upon treatment of MoO_2 at 250°C with CCl_4 (Fig. 2). The most important compound of Mo(IV) is molybdenum disulfide, MoS_2.

Molybdenum(III)

Molybdenum(III) complexes include the molybdenum trihalides. Molecular examples of trivalent molybdenum are shown in mononuclear, dinuclear, and tetranuclear complexes, as illustrated in Figure 5.

Molybdenum(II)

Divalent molybdenum compounds occur in mononuclear, dinuclear, and hexanuclear forms. Selected examples are shown in Figure 6.

Molybdenum(0)

Molybdenum hexacarbonyl (Fig 7**a**) is the starting material for the synthesis of most organometallic compounds of molybdenum.

Chemistry of Molybdenum Compounds

The degree of nuclearity exhibited as a function of the oxidation state of molybdenum is shown in Table 1.

The halides of molybdenum are solids that are quite reactive and useful starting points for further synthesis. The properties of the molybdenum halides have been described in some detail.

Molybdenum has well-characterized aqueous chemistry in the five oxidation states, VI, V, IV, III, and II. Except for the Mo(VI) species, all of the aqua ions are only soluble or stable in acidic media. The range of aqueous ions known for molybdenum is far broader than that of other elements.

Biological Aspects

Molybdenum, recognized as an essential trace element for plants, animals and most bacteria, is present in a variety of metallo enzymes. Indeed, the absence of Mo, and in particular its co-factor, in humans leads to severe debility or early death. Molybdenum in the diet has been implicated as having a role in lowering the incidence of dental caries and in the prevention of certain cancers. To aid the growth of plants, Mo has been used as a fertilizer and as a coating for legume seeds (see FERTILIZERS; MINERAL NUTRIENTS).

Environmentally, the presence of molybdenum has been of concern only in isolated instances. Reports of molybdenum toxicity have been rare. Molybdenum is involved in copper–molybdenum antagonism wherein excess molybdenum in the soil elicits a copper (qv) deficiency in animals (especially ruminants) that graze on the vegetation.

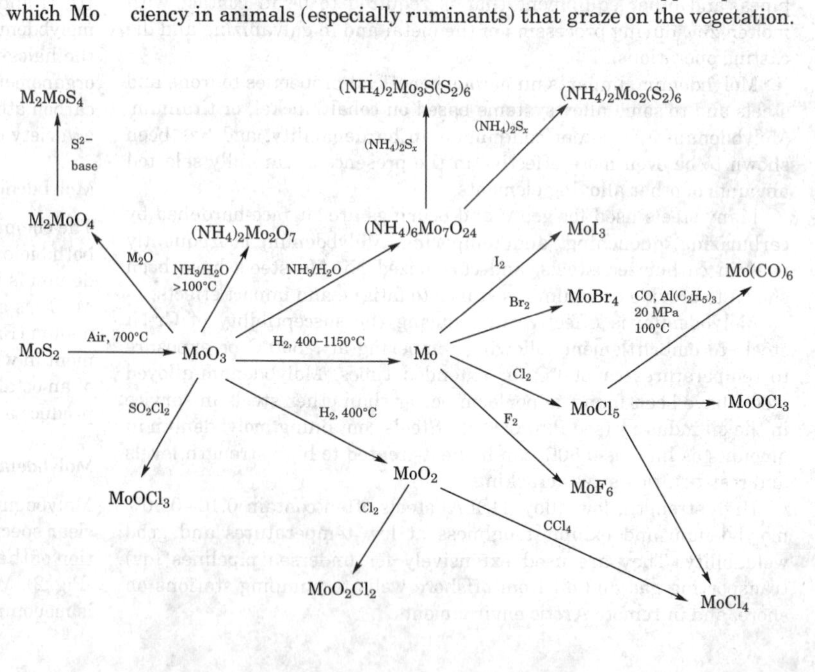

Figure 2. Scheme for the preparation of technologically important compounds of molybdenum, where M = Li, Na, K, Rb, Cs, and NH_4. To convert MPa to psi, multiply by 145.

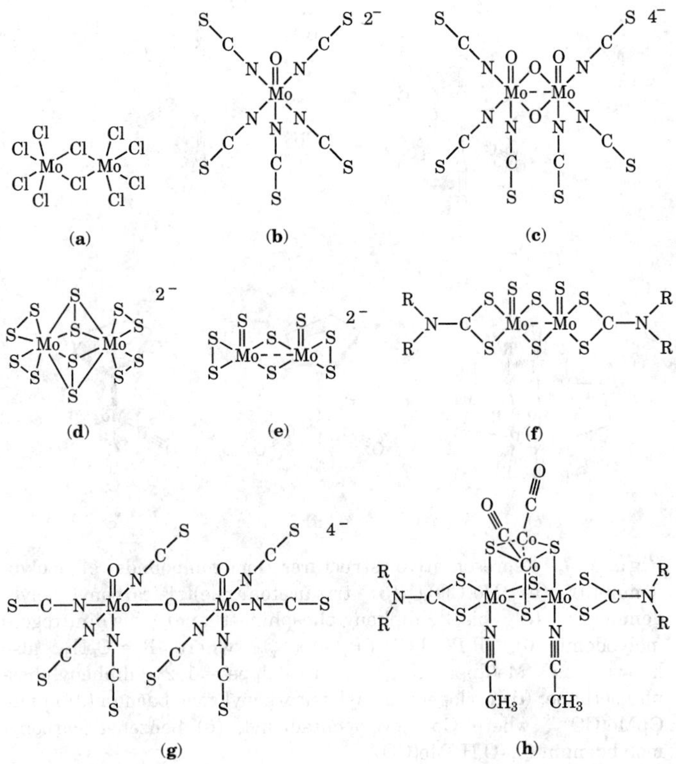

Figure 3. Representative structures for compounds of molybdenum(V): (**a**) dimolybdenum dodecachloride [26814-39-1], Mo_2Cl_{10}, the dimer of molybdenum pentachloride; (**b**) pentakis (thiocyanato)oxomolybdenum(V), $MoO(NCS)_5^{2-}$; (**c**) $Mo_2O_4(NCS)_6^{4-}$; (**d**) $Mo_2(S)_6^{2-}$2; (**e**) $Mo_2S_8^{2-}$; (**f**) $Mo_2S_4(S_2CNR_2)_2$; (**g**) $Mo_2O_3(NCS)_8^{4-}$; (**h**) $Mo_2Co_2S_4(S_2CNR_2)_2(CO)_2(CH_3CN)_2$.

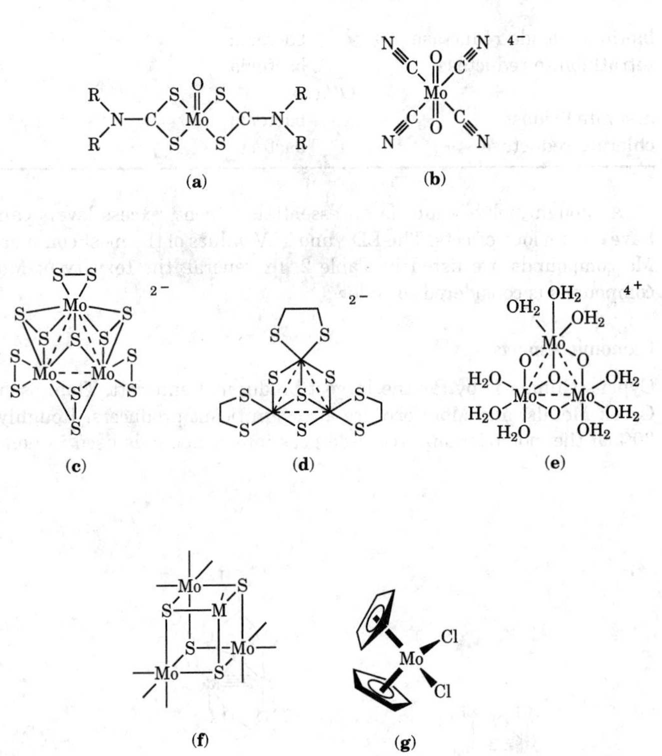

Figure 4. Representative structures for compounds of molybdenum(IV): (**a**) bis(dialkyl-dithiocarbamato)oxomolybdenum(IV), $MoO(S_2CNR_2)_2$, where R = alkyl; (**b**) *trans*-tetracyanodioxomolybdenum(IV), $MoO_2(CN)_4^{4-}$; (**c**) $Mo_3S_{13}^{2-}$; (**d**) $Mo_3S_4(SCH_2CH_2)_3^{2-}$; (**e**) $Mo_3O_4(H_2O)_9^{4+}$; (**f**) the $Mo_3M'S_4$ thiocubane core structure; (**g**) bis(cyclopentadienyl)dichloromolybdenum(IV), Cp_2MoCl_2, where Cp = cyclopentadienyl.

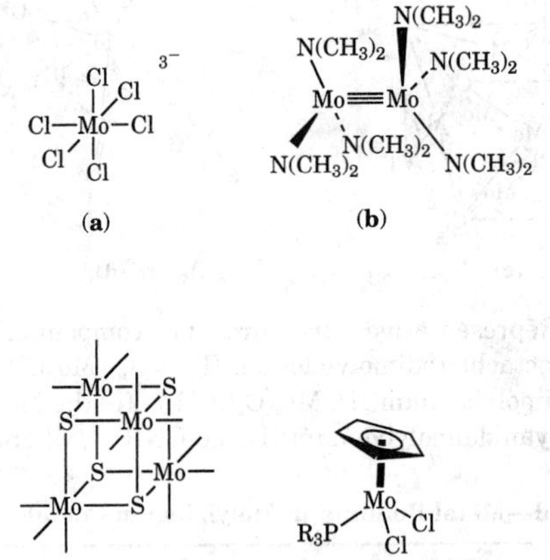

Figure 5. Representative structures for compounds of molybdenum(III): (**a**) hexacholoromolybdenum(III) ion, $MoCl_6^{3-}$; (**b**) hexakis (dimethylamido)dimolybdenum(III), $Mo_2(N(CH_3)_2)_6$; (**c**) the Mo_4S_4 thiocubane core structure; (**d**) dichlorocyclopentadienyl trialkylphosphinedichloromolybdenum(III), $CpMo(PR_3)Cl_2$, where Cp = cyclopentadienyl and R = alkyl.

Conversely, excess copper in the soil induces a molybdenum deficiency in ruminant animals that graze on the vegetation (see FEEDS AND FEED ADDITIVES, RUMINANT FEEDS).

Molybdate is also known as an inhibitor of the important enzyme ATP sulfurylase where ATP is adenosine triphosphate, which activates sulfate for participation in biosynthetic pathways. Molybdate is also a co-effector in the receptor for steroids (qv) in mammalian systems, a biochemical finding that may also have physiological implications. The clearest manifestation of molybdenum in biology is its presence in over 30 enzymes which participate in a wide variety of redox processes. Some of the Mo enzymes and their occurrence are as follows.

Enzyme	Occurrence
Nitrogen metabolism	
nitrogenase	bacteria (including symbionts)
nitrate reductase	plants, fungi, algae, bacteria
trimethylamine *N*-oxide reductase	bacteria
xanthine oxidase	cow's milk, mammalian liver, kidney
xanthine dehydrogenase	chicken liver, bacteria
quinoline oxidoreductase	bacteria
picolinic acid dehydrogenase	bacteria
Carbon metabolism	
aldehyde oxidase	mammalian liver
formate dehydrogenase	fungi, yeast, bacteria, plants
carbon monoxide oxidoreductase	bacteria
formylmethanofuran dehydrogenase	bacteria
Sulfur metabolism	
sulfite oxidase	mammalian liver, bacteria
dimethyl sulfoxide (DMSO) rectuctase	bacteria

biotin sulfoxide reductase	bacteria
tetrathionite reductase	bacteria
Others	
arsenite oxidase	bacteria
chlorate reductase	bacteria

Although molybdenum is an essential element, excess levels can have deleterious effects. The LD_{50} and TLV values of the most common Mo compounds are listed in Table 2. In general, the toxicity of Mo compounds is considered to be low.

Economic Aspects

Cyprus Climax is by far the largest producer. Kennecott, Thompson Creek Metals, and Molycorp are also significant producers. Roughly 30% of the molybdenum processed goes into compounds used in nonmetallurgical applications. The diversity of molybdenum compounds, coupled with potential environmental advantages and reduced costs of molybdenum relative to the noble metals, leads to projections for its increased use, especially in catalysis.

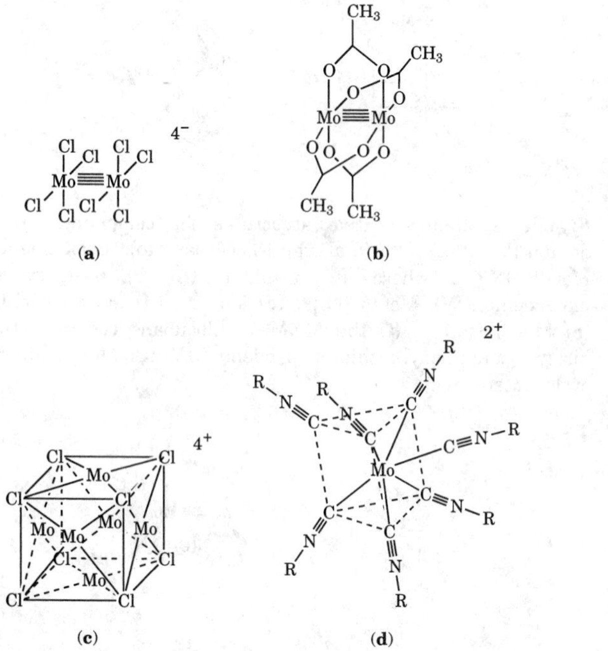

Figure 6. Representative structures for compounds of molybdenum(II):(**a**) octachlorodimolybdenum(II) ion, $Mo_2Cl_8^{4-}$; (**b**) tetrakis(acetato)dimolybdenum(II), $Mo_2(O_2CCH_3)_4$; (**c**) the $Mo_6Cl_8^{4+}$ core; (**d**) heptakis(isocyanide)molybdenum(II), $Mo(CNR)_7^{2+}$, where R = alkyl.

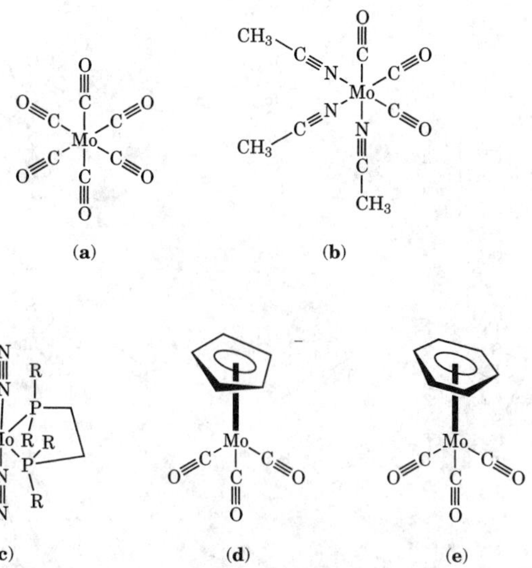

Figure 7. Representative structures for compounds of molybdenum(0): (**a**) $Mo(CO)_6$; (**b**) tris(acetonitrile)tris(carbonyl)molybdenum(0); (**c**) bis(1,2-diphenylphosphinoethane) bis(dinitrogen) molybdenum(0), $[R_2PCH_2CH_2PR_2]_2Mo(N_2)_2$, where R = C_6H_5, also known as $Mo(dppe)_2(N_2)_2$, where dppe = 1, 2 − diphenylphosphinoethane; (**d**) cyclopentadienyl tricarbonyl molybdenum(0) anion, $CpMo(CO)_3^-$, where Cp = cyclopentadienyl; (**e**) benzenetricarbonyl molybdenum(0), $(C_6H_6)Mo(CO)_3$.

Uses

In most of the nonmetallurgical uses of molybdenum compounds, the metal is coordinated by oxygen or sulfur ligands. Molybdenum nitrides, carbides, and silicides are, however, coming under increasing study for various applications. Roughly 75% of all molybdenum compounds are used as catalysts in the petroleum and chemicals industries. They are also used in: lubrication; advanced structural and heating materials; anticorrosion agents; coatings, paints, and pigments; flame and smoke retardants; pyrotechnics; battery electrodes; soil nutrients; and biomedical applications.

Table 1. Metal—Metal Bonding in Molybdenum Complexes and Clusters

	Oxidation state				
Metal	Mo(VI)	Mo(V)	Mo(IV)	Mo(III)	Mo(II)
d-electron configuration	d^0	d^1	d^2	d^3	d^4
number of metal—metal bonds	0	1	2	3	4
formulation mononuclear or dinuclear	Mo	Mo—Mo	Mo=Mo	Mo≡Mo	Mo≣Mo
polynuclear		Mo—	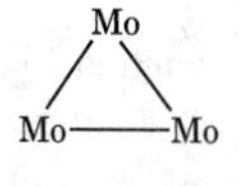	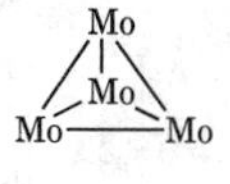	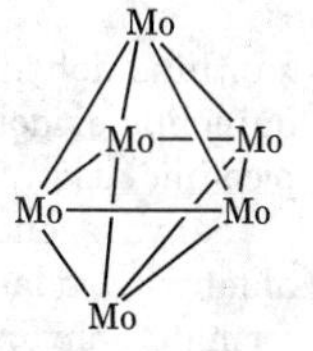
polynuclear geometry		linear	triangle	tetrahedron	octahedron

Table 2. Toxicity of Molybdenum Compounds

Compound	Molecular formula	LD_{50} (rat, oral), mg/kg	TLV(TWA),[a] mg/m^3
ammonium heptamolybdate	$(NH_4)_6Mo_7O_{24}4H_2O$	333	
molybdenum trioxide	MoO_3	2689	5
sodium molybdate	Na_2MoO_4	4000	5
molybdenum disulfide, molybdenite	MoS_2	nontoxic[b]	

[a] On the basis of weight of Mo. [b] Rats ingesting 500 mg/d for 44 d showed no toxic signs.

EDWARD I. STIEFEL
Exxon Research and Engineering Company

E. R. Braithwaite and J. Haber, eds., *Molybdenum: An Outline of Its Chemistry and Uses,* Elsevier Science, Inc., New York, 1994.

E. I. Stiefel, *Progr. Inorg. Chem.* **22,** 1 (1977).

E. I. Stiefel, D. Coucouvanis, and W. E. Newton, eds., *Molybdenum Enzymes, Cofactors, and Model Systems,* ACS Symposium Series, Vol. 535, Washington, D.C., 1993.

G. Wilkinson, R. D. Gillard, and J. A. McCleverty, eds., *Comprehensive Coordination Chemistry,* Vol. 3, Pergamon Press, Oxford, UK, 1987, pp. 1023–1444.

MONAZITE. See CERIUM AND CERIUM COMPOUNDS; THORIUM.

MONOCLONAL ANTIBODIES. See ENZYME APPLICATIONS; VACUUM TECHNOLOGY.

MONOSODIUM GLUTAMATE. See AMINO ACIDS (MSG).

MORPHOLINE. See AMINES.

MOUTHWASHES. See DENTIFRICES.

MUCILAGES. See GUMS.

MUNTZ METAL. See COPPER ALLOYS, WROUGHT COPPER AND ALLOYS.

MUSCLE RELAXANTS. See NEUROREGULATORS; PSYCHOPHARMACOLOGICAL AGENTS.

N

NANOTECHNOLOGY

Molecular nanotechnology is the production of functional materials and structures in the 0.1 to 100 nm range (the nanoscale) by any of a variety of physical and chemical methods. These methods include nanolithography, direct atomic and molecular manipulation with nanoscale probes, biotechnological selection and production of useful nanomaterials, and chemical synthesis and self-assembly of functional molecules and molecular aggregates. Nanoscale structures and devices may be constructed synthetically from their atomic or molecular constituents (the synthetic or "bottom-up" approach, also referred to as nanochemistry), or by fabrication techniques that use methods to form small structures from larger ones (the reductive or "top-down" approach). A natural analogy to nanotechnology can be found in the biosphere, wherein small and large molecules interact to form complex structures necessary for all the functions of living organisms.

Synthetic vs Reductive Technologies

Many of the devices that have thus far been envisioned as products of nanotechnology (eg, nanoscale environmental sensors, information processors, and actuators) cannot be produced by the large-scale microfabrication techniques currently in use. The further development of nanotechnology hinges on the understanding and manipulation of physical laws and processes at the nanometer level, such as electronic, interatomic, and intermolecular interactions that can be manipulated to allow efficient assembly of nanostructures.

Biological systems employ a variety of synthetic strategies and processes that make efficient use of these interactions to form highly ordered molecular and supermolecular units that perform a wide variety of functions. The distribution and functional success of biological nanostructures in living organisms are an evolutionally directed balance between the internal stability of the nanostructures (by atomic and molecular interactions) and their responses to the external environment. Thus, biological systems form important existence theorems and design criteria for the production of functional, nanoscale materials through nanotechnology employing the synthetic approach to nanostructures.

An important nature-mimicking methodology involves the use of covalent synthesis followed by molecular self-assembly of the synthesized molecules. These molecules are generally small mono- or oligomers that interact with each other and with other kinds of mono- or oligomers to form thermodynamically stable, nanoscale structures (see MOLECULAR RECOGNITION).

Existence Theorem: Nanobiology

Analogies: Structure and Function. Biology is replete with complex, functional nanoscale structures formed by directed synthesis and subsequent self-assembly of the component molecules. Initially, the small molecules are produced by covalent synthesis, with the larger, functional structures resulting from many weak, noncovalent interactions that energetically overcome interactions with the solvent and the entropic advantages of disintegration of the ordered aggregates. The final structure represents a thermodynamic minimum, and incorrect subunits are rejected in the dynamic, equilibrium assembly. In this process, complementarity in shape and polarity provides the foundation for the association (binding) between components (eg, phospholipids, polypeptide chains, proteins, nucleic acids). Shape-dependent association based on the nonspecific van der Waal's and hydrophobic interactions are made stronger and more specific by hydrogen bonds and electrostatic interactions, as well as, in some cases, covalent bonds (eg, disulfides). Often, positive cooperativity is displayed, ie, conformations of individual subunits change upon binding in such a manner that their affinity for other components of the final structure increases. Moreover, the amount of information required to execute the assembly of a particular structure (eg, a protein) is minimized by use of only a few types of molecules and a limited number of binding interactions.

For example, a polypeptide is synthesized as a linear polymer derived from the 20 natural amino acids by translation of a nucleotide sequence present in a messenger RNA (mRNA). The mature protein exists as a well-defined three-dimensional structure. The information necessary to specify the final (tertiary) structure of the protein is present in the molecule itself, in the form of the specific sequence of amino acids that form the protein. This information is used in the form of myriad noncovalent interactions (such as those in Table 1) that first form relatively simple local structural motifs (helix and sheet structures associated through networks of hydrogen bonds); these motifs then tend to aggregate in ways that associate hydrophobic regions with one another and out of reach of water, and to place hydrophilic regions close to each other and to water. Thus, proteins self-assemble by two types of processes after synthesis: formation of relatively simple local structures from an unfolded polypeptide chain; and more complex structure-specific association (ie, associations involving some form of molecular recognition) of these local structures.

In addition to self-assembly of protein structures, in living systems the complex maneuvers needed to achieve properly folded tertiary structures are facilitated by the function of a pre-existing protein machinery, of which, the molecular chaperones are an illustrative example. Chaperones are proteins that bind to and stabilize an otherwise unstable conformer of another protein, and by controlled binding and release, facilitate its correct fate *in vivo.* Molecular chaperones may be said to be the natural counterparts of assemblers and transporters envisaged as products of nanotechnology.

Analogies: Molecular Devices. Among the many examples in biology of complex nanoscale structures and devices, one of the most ubiquitous and versatile classes is that of the membrane proteins. Membrane proteins, ie, proteins that are associated with cellular and organellular membranes, serve myriad functions including recognition, adhesion, chemical triggering, ion and molecular transport, light harvesting, chemical and mechanical sensing, and actuation. Humans have attempted to mimic the function of some of these structures for a wide variety of uses, including chemical sensing, selective chemical transport (pumping), and tissue engineering.

Table 1. Types of Bonds and Interactions that are Potentially Useful in the Engineering of Functional Nanoscale Materials

Bond type	Examples: Natural	Examples: Synthetic
covalent bonds	peptide (amide), disulfide (RSSR), and phosphodiester (DNA) bonds	palytoxin, oligomers of peptides, nucleotides, and thiophenes
electrostatic interactions	salt bridges in proteins, zinc fingers	metal-coordinated crowns
hydrogen bonds	nucleotide base pairs, amide hydrogen bonds in proteins	peptide cylinders, DNA objects, replicating systems
hydrophobic interactions	hydrophobic cores of proteins; lipid—lipid and lipid—protein interactions	pockets within cyclodextrins and cyclophanes
van der Waals interactions	membranes, vesicles	Langmuir and self-assembled monolayers, lipid bilayers
aromatic π-stacking and charge transfer	nucleic acids	porphyrins

An interesting class of proteins which may prove to be an illustrative archetype for future nanotechnological devices, if not integral components of such, are a class of molecular devices collectively known as motor proteins. Motor proteins are enzymatic protein complexes whose catalytic function results in a distinct mechanical function; a variety of examples are known, including those that perform functions analogous to levers, rotary motors, pumps and springs. The structures and mechanisms of action of a variety of motor proteins are being investigated.

Reductive Approaches

Conventional methods of microfabrication of integrated circuits and devices constitute the reductive (top-down) approach to the construction of micrometer and submicrometer-scale structures. The smallest features in commercial integrated circuits have measured 0.35 μm across, and technologies for further reduction in size have succeeded in forming feature sizes as small as 0.18 μm. Further miniaturization, however, will require major technological breakthroughs in the processes underlying microfabrication, especially photolithography, the heart of microfabrication. The technological barriers faced by the lithographers may not be completely insurmountable.

Limitations. The number of transistors present on a chip has doubled approximately every 18 months since the integrated circuit was first developed. The main reason for this continuing decrease in the minimum feature sizes of transistors (and consequent increase in density of transistors on the chip) has been the development of photolithography. The most important limitation for further size reduction remains the development of new photo- and other lithographic techniques.

Current Advances. New light sources are currently being developed, such as the krypton–fluoride ultraviolet laser (wavelength 0.248 μm for features as small as 0.25 μm) and excimer lasers (wavelength 0.193 μm for features below 0.2 μm). But these technologies still need to overcome several obstacles before they can be implemented by the semiconductor industry. For example, for wavelengths below 0.2 μm, the current photoresists absorb so much light that throughput suffers. Also, the fused silica glass lenses used to demagnify the image absorb light and heat up, resulting in a degradation of the image. Problems associated with depth of focus become more acute, requiring further innovation in planarization technology. New photoresist materials, based on the deposition and patterning of monomolecular layers (eg, self-assembled monolayers and Langmuir-Blodgett films), are also being developed to address the issues of further reduction in the limits of photolithography.

Further reduction in feature size to achieve nanoscale structures by photolithography will necessitate the use of ever smaller wavelengths of light.

Focused particle-based lithography, such as ion-, neutral atom-, and electron-beam lithography, are capable of achieving very high resolutions. However, the method of writing each circuit feature separately is serial and inherently slow so that the technology cannot be used for simultaneous fabrication of many chips. To speed up the process of electron-beam lithography, methods are being explored to scan a broad electron beam across the entire chip by projecting the beam through an appropriate mask (in mimicry of photolithography).

The most recent approach to reductive nanofabrication that can indeed construct nanoscale structures and devices uses microscopic tools (local probes) that can build the structures atom by atom, or molecule by molecule. Optical methods using laser cooling (optical molasses) are also being developed to manipulate nanoscale structures.

Atomic and Molecular Manipulation

Scanning Probe Microscopy. The scanning tunneling microscope (STM) can image and manipulate matter on the atomic scale. In general, when a small conducting probe (the tip, consisting of one or a few atoms, or a metal) is placed close (less than 10 nm) to the surface of a conducting substrate, an electronic current results under a suitable bias due to the overlapping of the electronic wave functions of the probe and the surface. Because this tunneling current is exponentially dependent on the separation of the probe from the surface, imaging resolutions of fractions of an angstrom can be obtained. These images reflect both the topography and the electronic structure of the surface. The STM can also be used to modify surfaces locally. As a result, individual atoms and molecules can be manipulated with atomic-scale precision.

In parallel processes, eg, field-assisted diffusion and sliding, the bond between the surface and the adatom is never completely broken. Field-assisted diffusion of an adatom on the surface occurs due to the presence of the intense, inhomogeneous electric field between the probe tip and the surface, which gives rise to a potential gradient.

The second class of atomic manipulations, the perpendicular processes, involves transfer of an adsorbate atom or molecule from the STM tip to the surface or vice versa. The tip is moved toward the surface until the adsorption potential wells on the tip and the surface coalesce, with the result that the adsorbate, which was previously bound either to the tip or the surface, may now be considered to be bound to both. For successful transfer, one of the adsorbate bonds (either with the tip or with the surface, depending on the desired direction of transfer) must be broken. The fate of the adsorbate depends on the nature of its interaction with the tip and the surface, and the materials of the tip and surface. Directional adatom transfer is possible with the application of suitable junction biases.

Optical Manipulation. Laser beams provide another means of capturing individual atoms or molecules. When an atom is irradiated from both sides by laser light at a frequency slightly lower than the frequency at which the atom absorbs photons, then the atom loses some of its momentum. In particular, the laser beam propagating in a direction opposite that of the motion of the atom increases in frequency due to the Doppler effect, resulting in light absorption with subsequent isotropic emission (scattering). The light propagating in the same direction as the atom is not absorbed, so that the atom is pushed in a direction opposite its motion and slows down. By surrounding the atom with three sets of counterpropagating laser beams orthogonal to each other the atom can be cooled (ie, slowed) in all three dimensions. Because the light field acts as a viscous drag force, the combination of laser beams is known as optical molasses.

Synthetic Approaches

The synthetic (bottom-up) approach offers a level of control over the selection and placement of atoms and molecules that is ultimately much higher than that offered by other methods of large-scale microfabrication (eg, fabrication of integrated circuits). Such synthesis can employ a variety of chemical methods utilizing some or all of the forces and interactions listed in Table 1 to produce nanoscale molecules. Most chemical synthetic reactions that produce large molecules (eg, collections of atoms that can act as nanodevices) generate polydisperse materials. These materials are mixtures of oligo- or polymeric chains of varying molecular weights. For nanotechnological applications, it is important to have synthetic strategies that yield compounds of uniform length, size, and shape; inhomogeneities can be detrimental to the designated function of the molecule. Thus, the specific objectives of synthesis are to discover and develop rapid and efficient methods for the precise control of composition, molecular weight, stereochemistry, aggregation, and placement of functional molecules. Four strategies are currently in use (either separately or in combination with one another) for the fabrication of large molecules (substances with molecular weights of a few hundred to a few million): biotechnological synthesis, sequential covalent synthesis, covalent polymerization, and molecular self-assembly.

In nature, complex functional molecules are produced by using the chemical synthetic approach. The molecules are first formed by covalent synthesis. In particular, the polypeptides are formed by the directional joining of amino acids through amide bonds. The primary structure of the amino acids (ie, sequence) during synthesis is specified by the mRNA (a transcript of the original gene encoded in the

DNA). Once a polypeptide is produced, it then undergoes many conformational changes that reduce its size to a compact, native form that is the functional protein. These conformational changes are termed folding.

RNA is also capable of folding into specific shapes for ligand recognition and catalysis. Although there are only four naturally occurring bases available for the formation of tertiary structures through noncovalent interactions (while proteins have twenty different amino acids to choose from), myriad RNA shapes can be produced. Metal ions have been shown to confer extraordinary stability to RNA and RNA fragments.

Thus, protein and RNA folding studies form a fundamental paradigm for the design and synthesis of new, functional marcomolecules with both final structure and function built into the primary structure of the macromolecule. Such synthetic strategies could be applied to the rational design of functional nanostructures, including drugs, sensing elements, photonic and electronic components, catalysts, and even mechanical devices.

Biotechnological Synthesis. Biotechnological synthesis of new nanomaterials (eg, proteins) and biotechnological modification of living systems (eg, conferring specific therapeutic properties, new genetic traits) exploit the many ways in which this protein manufacturing machinery can be modified. In particular, the DNA of an organism can be altered in a specific manner, resulting in a modified mRNA, and consequently, a modified or new protein.

Recombinant DNA technology provides a powerful tool for analysis, synthesis, and alteration of genes and proteins. It is based on the ability to rapidly synthesize polynucleotides with any sequence using nucleic acid enzymology (eg, using DNA polymerases, restriction enzymes or endonucleases, DNA ligases). The unique base-pairing attributes of the constituents of the DNA and the ability to express the modified or synthetic DNA in microorganisms and eukaryotic cells result in a powerful tool for the production of synthetic molecules with specific biological functions.

Covalent Synthesis. The first strategy for chemical synthesis employs elaborate and sophisticated methods for assembling atoms into molecules based on the general strategy of sequential formation of covalent bonds. The atoms can also be assembled into subunits that are then reacted to form more complex, designed molecules (convergent synthesis).

Covalent synthesis of complex molecules involves the reactive assembly of many atoms into subunits with aid of reagents and established as well as innovative reaction pathways. These subunits are then subjected to various reactions that will assemble the target molecule. Very complex molecules can be synthesized in this manner.

Molecular Self-Assembly. Reductive techniques, such as those used in the microelectronics industry, can produce structural features smaller than about 200 nm. The use of proximal probes and other nanomanipulative techniques can be considered to be a hybrid of the reductive lithographic techniques and the synthetic strategies of assembling functional nanostructures atom by atom, or molecule by molecule. The organization of nanostructures and devices by the self-assembly of the component atoms and molecules, a ubiquitous phenomenon in biological systems, forms the noncovalent synthetic approach to nanotechnology.

In this approach, well-defined subunits (small molecules similar to, eg, nucleotides) are first formed through covalent synthesis. Second, these subunits aggregate with themselves or with other subunits through covalent or noncovalent (or both) interactions to form large, stable, structurally defined assemblies. For the final supramolecular structure to be stable and to have a well-defined shape, the noncovalent connections must be collectively stable. Therefore, molecules must be stabilized by many noncovalent interactions.

Conclusions and Outlook

Much of the current progress in nanotechnology is confined to understanding and harnessing pre-existing, exquisitely evolved nanomachinery, ie, natural living systems. Table 2 summarizes the various reductive and synthetic strategies that are employed.

Table 2. Some Approaches in Reductive (Top Down) and Synthetic (Bottom Up) Nanofabrication

Approach	Examples
reductive (also known as top-down)	conventional microfabrication
	photolithography
	x-ray lithography
	e-beam lithography
	imprint lithography
	local probe lithography
synthetic (also known as bottom-up)	nanochemical synthesis
	biotechnological expression
	molecular templating
	molecular self-assembly
hybrid (reductive/synthetic)	local probe-assisted synthesis
	microcontact printing
	optical-laser manipulation

Support for this work was provided by ONR Multidisciplinary University Research Initiative Grant N00014-95-1-1315, by ONR Grant N00014-95-1-0901, and by NSF Grant HRD-9450475.

RAJESH VAIDYA
GABRIEL LÓPEZ
University of New Mexico
JOSÉ A. LOPEZ
Baylor College of Medicine

P. Ball, *Designing the Molecular World: Chemistry at the Frontier,* Princeton University Press, Princeton, 1994.

W. M. Tolles, in G. M. Chow and K. E. Gonsalves, eds., *210th National Meeting of the ACS,* ACS, Chicago, Ill., 1996, pp. 1–15.

J. A. Stroscio and D. M. Eigler, *Science* **254,** 1319–1326 (1991).

G. M. Whitesides and co-workers, *Acc. Chem. Res.* **28,** 37–44 (1995).

NAPHTHALENE

The accepted configuration of naphthalene, ie, two fused benzene rings sharing two common carbon atoms in the ortho position, was established in 1869 and was based on its oxidation product, phthalic acid. Based on its fused-ring configuration, naphthalene is the first member in a class of aromatic compounds with condensed nuclei. Naphthalene is a resonance hybrid:

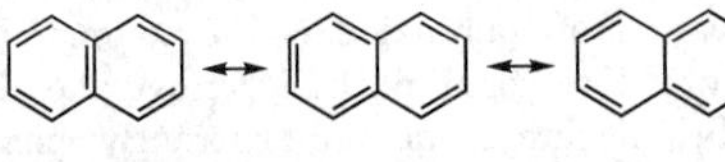

In chemical reactions, naphthalene usually acts as though the bonds were fixed in the positions, as shown in the first structure above at the left.

Some selected chemical and physical properties of naphthalene are given in Table 1. Naphthalene is very slightly soluble in water but is appreciably soluble in many organic solvents, eg, 1,2,3,4-tetrahydronaphthalene, phenols, ethers, carbon disulfide, chloroform, benzene, coal-tar naphtha, carbon tetrachloride, acetone, and decahydronaphthalene.

The ir, uv, mass, nmr and ^{13}C-nmr spectral data for naphthalene and other related hydrocarbons have been reported in the literature. Additionally, information regarding the properties of naphthalene has been published.

Table 1. Properties of Naphthalene

Property	Value
molecular wt	120.1732
mp, °C	80.290
normal bp at 101.3 kPa[a], °C	217.993
flash point (closed cup), °C	79
ignition temperature, °C	526
heat of vaporization, kJ/mol[b]	43.5
heat of fusion at triple point, kJ/mol[b]	18.979
heat of combustion, at 15.5 °C and 101.3 kPa[a], kJ/mol[b]	−5158.41
density at 25 °C, g/mL	1.175

[a] To convert kPa to atm, divide by 101.3. [b] To convert J to cal, divide by 4.184.

Reactions

Substitution. Substitution products are formed by the substitution of one or more hydrogen atoms with other functional groups. Substituted naphthalenes of commercial importance have been obtained by sulfonation and alkali fusion, alkylation, nitration and reduction, and chlorination.

Sulfonation. Sulfonation of naphthalene with sulfuric acid produces mono-, di-, tri-, and tetranaphthalenesulfonic acids (see NAPHTHALENE DERIVATIVES). Naphthalenesulfonic acids are important starting materials in the manufacture of organic dyes (see AZO DYES). They are also intermediates used in reactions.

Nitration. Naphthalene is easily nitrated with mixed acids, eg, nitric and sulfuric, at moderate temperatures to give mostly 1-nitronaphthalene and small quantities, 3–5%, of 2-nitronaphthalene.

Halogenation. Under mild catalytic conditions, halogen substitution occurs, and all of the hydrogen atoms of the naphthalene molecule can be replaced. The only commercially significant halogenated naphthalene products are the mixed chlorinated naphthalenes. Uses for the chlorinated naphthalenes include solvents, gauge and instrument fluids, capacitor impregnants, components in electric insulating compounds and electroplating stop-off compounds.

Alkylation. Naphthalene can be easily alkylated. Isopropylnaphthalenes produced by alkylation of naphthalene with propylene have gained commercial importance as chemical intermediates, eg, 2-isopropylnaphthalene, and as multipurpose solvents, eg, mixed isopropylnaphthalenes.

Chloromethylation. The reactive intermediate, 1-chloromethylnaphthalene, has been produced by the reaction of naphthalene in glacial acetic acid and phosphoric acid with formaldehyde and hydrochloric acid.

Addition. The most important addition products of naphthalene are the hydrogenated compounds used in solvents, paints and other products. Of less commercial significance are those made by the addition of chlorine.

Oxidation. The vapor-phase reaction of naphthalene over a catalyst based on vanadium pentoxide is the commercial route used throughout the world to form phthalic anhydride. In the United States, the one phthalic anhydride plant currently operating on naphthalene feedstock utilizes a fixed catalyst bed, the preferred route worldwide.

Manufacture

Two sources of naphthalene exist in the U.S.; coal tar and petroleum (qv). Coal tar was the traditional source until the late 1950s, when it was in short supply. In 1960, the first petroleum-naphthalene plant accounted for over 40% of total naphthalene production. The availability of large quantities of *o*-xylene at competitive prices during the 1970s affected the position of naphthalene as the prime raw material for phthalic anhydride. Production for 1992 was less than 50% of the levels in the early 1980s. The last dehydroalkylation plant for petroleum naphthalene was shut down late in 1991. Coal tar has stabilized at around 85 × 10³ t/yr, and petroleum-naphthalene production is around 6–8 × 10³ t/yr. The reduction of petroleum production has opened the door for imported naphthalene, mainly from Canada. The 1993 United States naphthalene capacities are given in Table 2.

Table 2. U.S. Naphthalene Capacities, 1993

Producer	Location	Coal tar or petroleum	Capacity, t
AlliedSignal, Inc.	Ironton, Ohio	coal tar	34,000
Crosscreek Industries	Baytown, Tex.	petroleum	10,000
Koppers Industries, Inc.	Follansbee, W. Va.	coal tar	80,000
Total			*124,000*

Coal-Tar Process. The largest quantities of naphthalene are obtained from the coal tar that is separated from the coke-oven gases. The coal tar first is processed through a tar-distillation step where ca the first 20 wt% of distillate, ie, chemical oil, is removed. The chemical oil contains practically all the naphthalene present in the tar. It is processed to remove the tar acids by contacting with dilute sodium hydroxide and, in a few cases, is next treated to remove tar bases by washing with sulfuric acid. Principal U.S. producers obtain their crude naphthalene product by fractional distillation of the tar acid-free chemical oil.

Economic Aspects

Total nameplace capacity for all U.S. naphthalene producers in 1993 was 124 × 10³ t, with 114 × 10³ t produced from coal tar and 10 × 10³ from petroleum.

The economics of naphthalene recovery from coal tar can vary significantly, depending on the particular processing operation used. A significant factor is the cost of the coal tar. As the price of fuel oil increases, the value of tar also increases.

The high price of the petroleum product results from its higher quality. The price of the crude coal-tar naphthalene is primarily associated with that of *o*-xylene, its chief competitor as phthalic anhydride feedstock.

The preferred route to higher purity naphthalene, either coal-tar or petroleum, is crystallization. This process has demonstrated significant energy cost savings and yield improvements. There are several commercial processes available: Sulzer-MWB, Brodie type, Betz, and Recochem.

Health and Safety Factors

Handling. Naphthalene is generally transported in molten form in tank trucks or tank cars that are equipped with steam coils. Storage tanks containing molten naphthalene have a combustible mixture in the vapor space and care must be taken to eliminate all sources of ignition. Naphthalene dust also can form explosive mixtures with air, which necessitates care in the design and operation of solid handling mixtures. Perhaps the greatest hazard to the worker is the potential for operating or maintenance personnel to be accidentally splashed with hot molten naphthalene while taking samples or disassembling process lines (ASTM D3438). Molten naphthalene tank vents must be adequately heated and insulated to prevent the accumulation of sublimed and solidified naphthalene.

Toxicology. The acute oral and dermal toxicity of naphthalene is low, with LD_{50} values for rats from 1780–2500 mg/kg orally and greater than 2000 mg/kg dermally. The inhalation of naphthalene vapors may cause headache, nausea, confusion, and profuse perspiration, and if exposure is severe, vomiting, optic neuritis, and hematuria may occur. Chronic exposure studies conducted by the NTP in mice for two years showed that naphthalene caused irritation to the nasal passages, but no overt toxicity was noted. Rare cases of such corneal epithelium damage in humans have been reported. Naphthalene can be irritating to the skin, and hypersensitivity does

Table 3. U.S. Naphthalene Consumption, 1992

Use	Consumption, 10^3t	% of Total
phthalic anhydride	69	63
surfactants	19	17
insecticides	13	12
moth repellants	6	6
miscellaneous	2	2
Total	*109*	*100*

occur. In other chemical carcinogen tests, little cancer risk was indicated. No incidents of chronic effects have been reported as a result of industrial exposure to naphthalene. Threshold limit value of 10 ppm (50 mg/m^3) has been set by the ACGIH.

Uses

The U.S. naphthalene consumption by markets for 1992 is listed in Table 3.

Alkylnaphthalenes

Methyl- and dimethylnaphthalenes are contained in coke-oven tar and in certain petroleum fractions in significant amounts. In the U.S., separation of individual isomers is seldom attempted; instead, a methylnaphthalene-rich fraction is produced for commercial purposes. Such mixtures are used for solvents for pesticides, sulfur, and various fluids. They also can be used as low freezing, stable heat-transfer fluids. Mixtures that are rich in monomethylnaphthalene content have been used as dye carriers (qv) for color intensification in the dyeing of synthetic fibers, eg, polyester. They also are used as the feedstock to make naphthalene in dealkylation processes. Phthalic anhydride also can be made from methylnaphthalene mixtures by an oxidation process that is similar to that used for naphthalene. A mixed monomethylnaphthalene-rich material can be produced by distillation and can be used as feedstock for further processing. Applications include use in solvents, drugs, polyesters, surfactants, and detergent products.

Acenaphthene. Acenaphthene is a hydrocarbon, $C_{12}H_{10}$, present in high temperature coal tar. Acenaphthene may be halogenated, sulfonated, and nitrated in a manner similar to naphthalene. Oxidation first yields acenaphthenequinone, followed by 1,8-naphthalenedicarboxylic acid anhydride and pesticides.

ROBERT T. MASON
Koppers Industries, Inc.

Naphthalene, American Petroleum Institute Monograph Series, Publication 707, API, Washington, D.C., Oct. 1978.

A. N. Sachanen, *Conversion of Petroleum,* 2nd ed., Reinhold Publishing Corp., New York, 1948, pp. 550–565.

N. Donaldson, *The Chemistry and Technology of Naphthalene Compounds,* Edward Arnold Publishers, London, 1958, pp. 455–473.

E. E. Sandmeyer, in G. D. Clayton and F. E. Clayton, eds., *Patty's Industrial Hygiene and Toxicology,* 3rd rev. ed., Vol. II, Wiley-Interscience, New York, 1981, Chapt. 46.

NAPHTHALENE DERIVATIVES

Naphthalenesulfonic Acids

Naphthalenesulfonic acids are important chemical precursors for dye intermediates, wetting agents and dispersants, naphthols, agricultural formulations, leather tanning agents, photographic materials, and air-entrainment agents for concrete. The production of many intermediates used for making azo, azoic, and triphenylmethane dyes (see TRIPHENYLMETHANE AND RELATED DYES). involves naphthalene sulfonation and one or more unit operations, eg, caustic fusion, nitration, reduction, or amination.

Generally, the sulfonation of naphthalene leads to a mixture of products. Naphthalene sulfonation at less than ca 100°C is kinetically controlled and produces predominantly 1-naphthalenesulfonic acid (**1**). Sulfonation of naphthalene at above ca 150°C provides thermodynamic control of the reaction and 2-naphthalenesulfonic acid as the main product. Reaction conditions for the sulfonation of naphthalene to yield desired products are given in Figure 1; alternative paths are possible. A list of naphthalenesulfonic acids and some of their properties are given in Table 1.

Table 1. Melting Points of Naphthalenesulfonic Acids

Compound	Mp, °C	Mp of corresponding sulfonyl chloride, °C
1-naphthalenesulfonic acid	139–140	68
1-naphthalenesulfonic acid dihydrate	90	
2-naphthalenesulfonic acid	139–140	76
2-naphthalenesulfonic acid hydrate	124–125	
2-naphthalenesulfonic acid trihydrate	83	
1,2-naphthalenedisulfonic acid		160
1,3-naphthalenedisulfonic acid		137.5
1,4-naphthalenedisulfonic acid	240–245 dec	162
1,5-naphthalenedisulfonic acid	125 dec	183
1,6-naphthalenedisulfonic acid		129
1,7-naphthalenedisulfonic acid		123
2,6-naphthalenedisulfonic acid	199 dec	228–229
2,7-naphthalenedisulfonic acid		159.5
1,3,5-naphthalenetrisulfonic acid		146
1,3,6-naphthalenetrisulfonic acid		194–197
1,3,7-naphthalenetrisulfonic acid		165–166
1,4,5-naphthalenetrisulfonic acid		156–157
1,3,5,7-naphthalenetetrasulfonic acid		261–262

Table 2. Melting Point of Nitronaphthalenes

Compound	Mp, °C
1-nitronaphthalene	52[a]; 57.8[b]
2-nitronaphthalene[c]	78.7[d]
1,2-dinitronaphthalene[c]	161–162
1,3-dinitronaphthalene[c]	148
1,4-dinitronaphthalene[c]	134
1,5-dinitronaphthalene	219
1,6-dinitronaphthalene[c]	166.5[e]
1,7-dinitronaphthalene[c]	156
1,8-dinitronaphthalene	172
2,3-dinitronaphthalene[c]	174.5–175
2,6-dinitronaphthalene[c]	279
2,7-dinitronaphthalene[c]	234
1,2,3-trinitronaphthalene[c]	190
1,2,4-trinitronaphthalene[c]	258
1,3,5-trinitronaphthalene	122
1,3,6-trinitronaphthalene[c]	186
1,3,8-trinitronaphthalene	218
1,4,5-trinitronaphthalene	149
1,3,5,7-tetranitronaphthalene[c]	260
1,3,5,8-tetranitronaphthalene	194–195
1,3,6,8-tetranitronaphthalene	203
1,4,5,8-tetranitronaphthalene	340–345 dec

[a] Metastable form. [b] Bp 304°C (169°C at 1.6 kPa (12 mm Hg)). [c] Made by indirect methods, not by the direct nitration of naphthalene or naphthalene-nitration products. [d] Bp 312.5°C at 97.8 kPa (733 mm Hg) and 165°C at 2.0 kPa (15 mm Hg). [e] Bp 370°C (235°C at 1.3 kPa (9.75 mm Hg)).

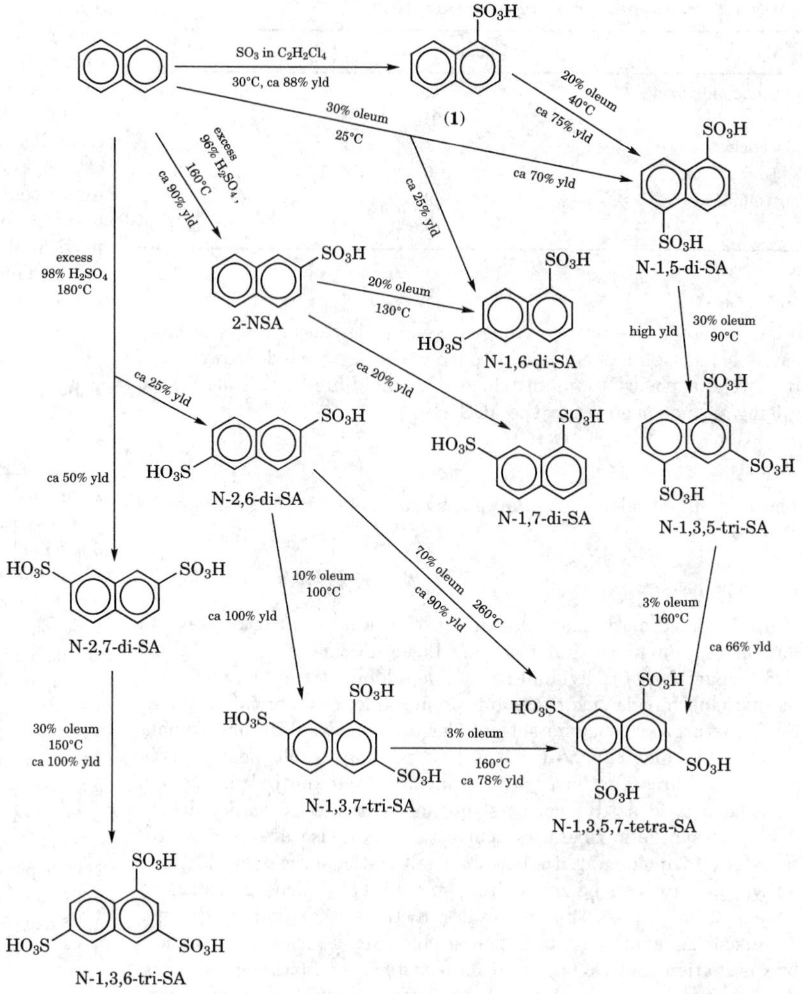

Figure 1. Selected paths to naphthalenesulfonic acids where N = naphthalene, SA = sulfonic acid, and yld = yield.

Table 3. Physical Properties of Naphthaleneamines and Naphthalenediamines

Compound	Mp, °C	Density	Other
1-naphthaleneamine	50	1.13_4^{14}	flash pt, 157°C; sol 0.496 g/L H_2O; vol with steam; bp 301°C (160°C at 1.6 kPa[a])
2-naphthaleneamine	111–113	1.061_4^{96}	sol hot water; vol with steam; bp 306°C (175.8°C at 2.7 kPa[a])
1,2-naphthalenediamine	96–98		sol hot water, alc, ether; bp at 0.01 kPa[a] 150–151°C
1,4-naphthalenediamine	120		sl sol hot water
1,5-naphthalenediamine	189.5		sol hot water, alc
1,6-naphthalenediamine	78	$1.147_4^{99.4}$	sol hot water, alc
1,7-naphthalenediamine	117.5		sol alc
1,8-naphthalenediamine	66.5	$1.127_4^{99.4}$	sol alc, ether; bp at 1.6 kPa[a] 205°C
2,3-naphthalenediamine	191		sol alc, ether
2,6-naphthalenediamine	216–218		sparingly sol alc, ether
2,7-naphthalenediamine	159		

[a] To convert kPa to mm Hg, multiply by 7.5.

Nitronaphthalenes and Nitronaphthalenesulfonic Acids

The nitro group does not undergo migration of the naphthalene ring during the usual nitration procedures. Therefore, mono- and polynitration of naphthalene is similar to low temperature sulfonation. The nitronaphthalenes and some of their physical properties are listed in Table 2. Many of these compounds are not accessible by direct nitration of naphthalene but are made by indirect methods, eg, nitrite displacement of diazonium halide groups in the presence of a copper catalysts, decarboxylation of nitronaphthalenecarboxylic acids, or deamination of nitronaphthalene amines. They are used in the manufacture of chemicals, dye intermediates, and colorants for plastics.

A by-product of some of these naphthalene derivatives, 2-naphthylamine, is carcinogenic. Respirators, protective clothing, proper engineering, controls and medical monitoring programs for workers involved with them should be used. The National Institute of Occupational Safety and Health (NIOSH) has published recommendations for working with this product.

Naphthaleneamines and Naphthalenediamines

Selected physical properties of naphthaleneamines and naphthalenediamines are listed in Table 3. They are used in rodenticides, rubber antioxidants, dye intermediates, insecticides, herbicides, pharmaceuticals, chemical manufacture, and colorants.

Aminonaphthalenesulfonic Acids

Many aminonaphthalenesulfonic acids are important in the manufacture of azo dyes (qv) or are used to make intermediates for azo acid dyes, direct and fiber-reactive dyes (see DYES, REACTIVE). Usually, the aminonaphthalenesulfonic acids are made by either the sulfonation of naphthaleneamines, the nitration–reduction of naphthalenesulfonic

Table 4. Manufacture, Production, and Application Data for Selected Aminonaphthalenesulfonic Acids

Acid	Trivial name
1-amino-2-naphthalenesulfonic	
4-amino-2-naphthalenesulfonic	
4-amino-1-naphthalenesulfonic	Piria's acid; naphthionic acid
5-amino-1-naphthalenesulfonic	Laurent's acid
5-amino-2-naphthalenesulfonic	1,6-Cleve's acid
8-amino-2-naphthalenesulfonic	1,7-Cleve's acid
5- and 8-amino-2-naphthalenesulfonic	Cleve's acid (mixed)
8-amino-1-naphthalenesulfonic	Peri acid
8-phenylamino-1-naphthalenesulfonic	Phenyl Peri acid
2-amino-1-naphthalenesulfonic	Tobias acid
6-amino-1-naphthalenesulfonic	Dahl's acid
6-amino-2-naphthalenesulfonic	Broenner's acid
7-amino-2-naphthalenesulfonic	F-acid
7-amino-1-naphthalenesulfonic	Badische acid
1-amino-2,7-naphthalenedisulfonic	Kalle's acid
4-amino-2,7-naphthalenesulfonic	1,3,6-Freund's acid
4-amino-2,6-naphthalenedisulfonic	1,3,7-Freund's acid
8-amino-1,6-naphthalenedisulfonic	amino-ϵ-acid
4-amino-1,7-naphthalenedisulfonic	Dahl's acid II
4-amino-1,6-naphthalenedisulfonic	Dahl's acid III
8-amino-1,5-naphthalenedisulfonic	
5-amino-1,3-naphthalenedisulfonic	
3-amino-2,7-naphthalenedisulfonic	amino-R-acid
3-amino-1,5-naphthalenedisulfonic	Cassella acid
6-amino-1,3-naphthalenedisulfonic	amino J-acid
7-amino-1,3-naphthalenedisulfonic	amino G-acid
4-amino-1,3,5-naphthalenetrisulfonic (as the sultam)	
8-amino-1,3,6-naphthalenetrisulfonic	Koch's acid
8-amino-1,3,5-naphthalenetrisulfonic	B-acid
6-amino-1,3,5-naphthalenetrisulfonic	
7-amino-1,3,6-naphthalenetrisulfonic	2R amino acid
6,8-di(phenylamino)-1-naphthalene-sulfonic	diphenyl-ϵ-acid

acids, the Bucherer-type amination of naphtholsulfonic acids, or the desulfonation of an aminonaphthalenedi- or trisulfonic acid. Most of these processes produce by-products or mixtures which often are separated in subsequent purification steps. A list of commercially important aminonaphthalenesulfonic acids is given in Table 4.

Naphthalenols and Naphthalenediols

Naphthalenols, naphthalenediols, and their sulfonated and amino derivatives are important intermediates for dyes, agricultural chemicals, drugs, perfumes, and surfactants. The methods of manufacture include caustic fusion of naphthalene-1-sulfonic acid, hydrolysis of 1-chloro- or bromonaphthalene, pressure hydrolysis of 1-naphthaleneamine, oxidation–aromatization of tetralin, and hydroperoxidation of 2-isopropylnaphthalene. As the toxic hazard of the 1-naphthaleneamine was recognized, its commercial use was minimized. The sulfonation–caustic fusion process is more difficult to operate than in the past because of increasing difficulties posed by product purity requirements, high investment and replacement cost, and by-product effluent handling problems. In the U.S., the naphthalenols are made by hydrocarbon oxidation routes.

The chemical properties of the naphthalenols are similar to those of phenol and resorcinol, with added reactivity and complexity of substitution because of the condensed ring system. Some of the naphthols and naphthalenediols are listed with some of their physical properties in Table 5.

Table 5. Properties of Naphthalenols and Naphthalenediols

Compound	Mp, °C	Density	Other
1-naphthalenol	95.8–96.0	1.224_4^4; 1.099_4^{99}	sublimes; sol 0.03 g/100 mL H_2O at 25°C; readily sol alc, ether, benzene; bp 280°C (158°C at 2.6 kPa[a])
2-naphthalenol	122	1.078_4^{130}; 1.22_4^{25}	sublimes; sol 0.075 g/100 mL H_2O at 25°C; readily sol alc, ether, benzene; flash pt 161°C; bp 295°C (161.8°C at 2.6 kPa[a])
1,2-naphthalenediol	103–104		
1,3-naphthalenediol	124		
1,4-naphthalenediol	195		heat of combusion 4.77 MJ[b]
1,5-naphthalenediol	258		sublimes; sparingly sol water; readily sol ether, acetone
1,6-naphthalenediol	137–138		
1,7-naphthalenediol	181		
1,8-naphthalenediol	144		
2,3-naphthalenediol	159		
2,6-naphthalenediol	222		
2,7-naphthalenediol	194	sol boiling water	

[a] To convert kPa to mm Hg, multiply by 7.5. [b] To convert MJ to kcal, divide by 4.184×10^{-3}.

Hydroxynaphthalenesulfonic Acids

Hydroxynaphthalenesulfonic acids are important as intermediates either for coupling components of azo dyes or azo components, as well as for synthetic tanning agents. Hydroxynaphthalenesulfonic acids can be manufactured either by sulfonation of naphthols or hydroxynaphthalenesulfonic acids, by acid hydrolysis of aminonaphthalenesulfonic acids, by fusion of sodium naphthalenepolysulfonates with sodium hydroxide, or by desulfonation or rearrangement of hydroxynaphthalenesulfonic acids (Table 6).

Aminonaphthols and Aminonaphtholsulfonic Acids

The aminonaphthols are of minor use but the aminohydroxynaphthalenesulfonic acids are intermediates for dyes, eg, fiber-reactive azo dyes and plain and metallized azo dyes (Table 7). A number of *N*-acyl-, *N*-alkyl-, and *N*-arylaminonaphthaleneosulfonic acids are used as couplers for azo dyes.

Naphthalenecarboxylic Acids

Physical properties for naphthalene mono-, di-, tri-, and tetracarboxylic acids are summarized in Table 8. Most of the naphthalene di- or polycarboxylic acids have been made by simple routes such as the oxidation of appropriate di- or polymethylnaphthalenes, or by complex routes, eg, the Sandmeyer reaction of the selected aminonaphthalenesulfonic acid, to give a cyanonaphthalenesulfonic acid followed by fusion of the latter with an alkali cyanide, with simultaneous or subsequent hydrolysis of the nitrile groups. These acids are used in the manufacture of dyes, photographic materials, pharmaceuticals, rodenticides, chemicals, polymers, and synthetic materials.

1- and 2-Naphthalenecarboxylic Acids. Naphthalenecarboxylic acids are useful intermediates for dyes and photographic materials. These acids are also used in the preparation of antitumor agents and also in the preparation of cholecystokinin-agonist tetrapeptide. The acids are prepared readily by the oxidation of 1- or 2-alkylnaphthalenes with dilute nitric acid, chromic acid, or permanganate. The oxygen or air

Table 6. Manufacture, Production, and Application Data for Selected Hydroxynaphthalenesulfonic Acids

Compound	Trivial name	Manufacturing method[abcdef]	Intermediate for
4-hydroxy-2-naphthalenedisulfonic acid	Armstrong & Wynne's acid; 1,3-oxy-acid	a,b	azo dyes, eg, CI Direct Blue 127
4-hydroxy-1-naphthalenesulfonic acid	Nevile-Winther acid; 1,4-oxy-acid	c,d	azo dyes, eg, CI Acid Red 14; tanning agents
5-hydroxy-1-naphthalenesulfonic acid	L-acid	d,e	azo dyes and pigments, eg, CI Pigment Red 54, toner; 1,5-naphthal enediol
8-hydroxy-1-naphthalenesulfonic acid		f	metallized *o,o'*-dihydroxyazo dyes, eg, CI Acid Blue 58
2-hydroxy-1-naphthalenesulfonic acid	oxy-Tobias acid	c	Tobias acid; J-acid
6-hydroxy-2-naphthalenesulfonic acid	Schaeffer's acid	c	azo dyes, eg, CI Acid Orange 12; synthetic tanning agents
7-hydroxy-2-naphthalenesulfonic acid	F-acid	e	azo dyes, eg, CI Direct Blue 128
7-hydroxy-1-naphthalenesulfonic acid	Crocein acid; Baeyer's acid	c	azo dyes, eg, CI Acid Red 70
4,5-dihydroxy-1-naphthalenesulfonic acid	dioxy S-acid	e	azo dyes, eg, CI Direct Blue 26
6,7-dihydroxy-2-naphthalenesulfonic acid	dioxy R-acid	e	2,3-dihydroxy-naphthalene
5-hydroxy-2,7-naphthalene-disulfonic acid	RG-acid; violet acid	e	azo dyes, eg, CI Acid Red 99
8-hydroxy-1,6-naphthalene-disulfonic acid	ϵ-acid; Andresen's acid	f	azo dyes, eg, CI Direct Blue 98
4-hydroxy-1,6-naphthalene-disulfonic acid	Dahl's acid; D-acid	a,d	nitro coloring matter, eg, CI Acid Yellow 1
4-hydroxy-1,5-naphthalene-disulfonic acid	Schoellkopf's acid; CS-acid; δ-acid	f	azo dyes, eg, CI Acid Blue 169
3-hydroxy-2,7-naphthalene-disulfonic acid	R-acid	c	azo dyes, eg, CI Acid Red 115, Acid Red 26
7-hydroxy-1,3-naphthalene-disulfonic acid	G-acid	c	azo dyes, eg, CI Acid Red 73; triphenyl-methane dyes
4,5-dihydroxy-2,7-naphthalenedisulfonic acid	chromotropic acid	a,e	azo dyes, eg, CI Acid Violet 3
8-hydroxy-1,3,6-naphthalene-trisulfonic acid	oxy-Koch's acid	a	azo dyes, eg, CI Direct Blue 27; chromotropic acid
7-hydroxy-1,3,6-naphthalene-trisulfonic acid		c	azo dyes, eg, CI Acid Red 41

[a] By hydrolysis of corresponding aminonaphthalenesulfonic acid. [b] By desulfonation of 8-hydroxy-1,6-naphthalenedisulfonic acid. [c] By sulfonation of appropriate (1- or 2-) naphthalenol. [d] By Bucherer reaction (with sulfite) of appropriate aminonaphthalenesulfonic acid. [e] By alkali fusion or alkaline hydrolysis under pressure of appropriate naphthalenedisulfonic or naphthalenetrisulfonic acid or hydroxynaphthalenedisulfonic acid. [f] By alkaline hydrolysis of sulfone formed on boiling aqueous solution of diazonium salt of 8-amino-1-naphthalenesulfonic acid or appropriate derivatives.

Table 7. Selected Aminonaphthalenols and Aminohydroxynaphthalenesulfonic Acids

Compound	Trivial name	Manufacturing method[abc]	Intermediate for
5-amino-1-naphthalenol	Purpurol	a	azo dyes, eg, CI Acid Blue 70; sulfur dyes
7-amino-2-naphthalenol	Cyanol	a	azo, dyes, eg, CI Mordant Brown 65
3-hydroxy-4-amino-1-naphthalenesulfonic acid	1,2,4-acid; Boeniger acid	b	azo dyes, eg, CI Acid Red 186, Mordant Red 7; chrome complex dyes
5-amino-6-hydroxy-2-naphthalene- sulfonic acid	Amino-Schaeffer acid	c	photographic developer; rarely used for dyes
4-hydroxy-8-amino-2-naphthalene- sulfonic acid	M-acid	a	azo dyes, eg, CI Direct Green 42
4-hydroxy-7-amino-2-naphthalene- sulfonic acid	J-acid	a	azo dyes, eg, CI Direct Blue 71, Direct Red 16; direct dyes using *N*-phenyl J-acid and J-acid imide
4-hydroxy-6-amino-2-naphthalene- sulfonic acid	γ-acid	a	azo dyes, eg, CI Direct Black 22
4-amino-5-hydroxy-2,7-naphthalene-disulfonic acid	H-acid	a	azo dyes, eg, CI Direct Black 19, Direct Blue 15
4-amino-5-hydroxy-1,3-naphthalene-disulfonic acid	Chicago acid; SS-acid; 2S-acid	a	azo dyes, eg, CI Acid Blue 42
4-amino-5-hydroxy-1,7-naphthalene-disulfonic acid	K-acid	a	azo dyes, eg, Sulfon Acid Blue G, CI 13400
3-amino-5-hydroxy-2,7-naphthalene-disulfonic acid	RR-acid; 2R-acid	a	azo dyes, eg, CI Direct Brown 31

[a] By alkali fusion or hydrolysis of appropriate aminonaphthalenesulfonic acid. [b] By nitrosation of 2-naphthalenol and reaction of nitroso compound with sodium bisulfite. [c] By nitrosation/reduction of 6-hydroxy-2-naphthalenesulfonic acid.

oxidation of alkylnaphthalenes in an alkanoic acid solvent in the presence of Ce-, Co-, or Mn-containing catalyst and a Br-containing catalyst gives good results. The direct carboxylation catalyst naphthalene with CO and oxygen in the presence of Pd-carboxylate catalysts has been patented. The photo carboxylation of naphthalene in the presence of carbon dioxide and an electron donor has been described. About 67% naphthoic acids were obtained by this method, upon visible light irradiation with phenazine as a sensitizer. Over 90% of the naphthoic acids was 1-naphthoic acid.

Table 8. Selected Properties of Naphthalenecarboxylic Acids

Compound	Mp, °C	Other
1-naphthalenecarboxylic acid	162	sol ethanol; sparingly sol water; $K_a = 2.04 \times 10^{-4}$ at 25 °C; bp at 6.7 kPa[a] 231°C
2-naphthalenecarboxylic acid	184–185	sol ethanol, ether, chloroform; $K_a = 6.78 \times 10^{-5}$ at 25°C; bp > 300°C
1,2-napthalenedicarboxylic acid	175 dec	sol ethanol, ether, acetic acid; mp anhydride 168–169 °C
1,3-naphthalenedicarboxylic acid	267–268	
1,4-naphthalenedicarboxylic acid	309	sol ethanol; insol boiling water
1,5-naphthalenedicarboxylic acid	315–320 dec	insol common solvents
1,6-naphthalenedicarboxylic acid	310	sol hot ethanol, acetic acid
1,7-naphthalenedicarboxylic acid	308	sol common organic solvents
1,8-naphthalenedicarboxylic acid	converts to anhydride (mp 274°C)	sol warm ethanol; bp anhydride at 440 Pa[b] 215°C
2,3-naphthalenedicarboxylic acid	239–241 dec	sol hot ethanol; mp anhydride 246°C
2,6-naphthalenedicarboxylic acid	310–313 dec	sol aq alc
2,7-naphthalenedicarboxylic acid	>300	sol ethanol
1,2,5-naphthalenetricarboxylic acid	270–272	sol methanol
1,3,8-naphthalenetricarboxylic acid		mp 1,8-anhydride 289–290°C
1,4,5-naphthalenetricarboxylic acid	forms anhydride (mp undefined)	mp 4,5-anhydride 274°C
1,2,4,5-naphthalenetetra-carboxylic acid	263	mp dianhydride 263°C
1,4,5,8-naphthalenetetra-carboxylic acid	forms anhydride	sol acetone; dianhydride sublimes >300°C

[a] To convert kPa to mm Hg, multiply by 7.5. [b] To convert Pa to mm Hg, divide by 133.3.

Table 9. Selected Properties of Hydroxynaphthalenecarboxylic Acids

Carboxylic acid	Mp, °C	Other
2-hydroxy-1-naphthalene-	157–159	sparing sol H_2O; sol alcohol, benzene
3-hydroxy-1-naphthalene-	248–249	
4-hydroxy-1-naphthalene-	188–188	
5-hydroxy-1-naphthalene-	236	
6-hydroxy-1-naphthalene-	213	
7-hydroxy-1-naphthalene-	256–257	
8-hydroxy-1-naphthalene-	1691	acetone, mp 108°C
1-hydroxy-2-naphthalene-	2000.55 wt %	sol in boiling water, alcohol, ether, benzene
3-hydroxy-2-naphthalene-	222–2230.1 wt %	sol in water at 25°C, ether, benzene chloroform
4-hydroxy-2-naphthalene	225–226	
5-hydroxy-2-naphthalene-	215–216	
6-hydroxy-2-naphthalene-	245–248	
7-hydroxy-2-naphthalene-	274–275	
8-hydroxy-2-naphthalene-	229	

Hydroxynaphthalenecarboxylic and Aminonaphthalenecarboxylic Acids

Some properties of selected hydroxynaphthalenecarboxylic acids are presented in Table 9. These acids are used in the manufacture of dyes (most importantly, naphthol AS dye stuffs), color film, and polyester.

MANNAN TALUKDER
CURTIS R. KATES
Advanced Aromatics, Inc.

N. Donaldson, *The Chemistry and Technology of Naphthalene Compounds*, E. Arnold Ltd., London, 1958.

H. Cerfontain, *Mechanistic Aspects in Aromatic Sulfonation and Desulfonation*, Wiley-Interscience, New York, 1968.

K. Venkataraman, ed., *The Chemistry of Synthetic Dyes*, 8 vols., Academic Press, Inc., New York, 1952–1978.

F. Radt, in F. Radt, ed; *Elsevier's Encyclopedia of Organic Chemistry*, Vol. 12B, Elsevier Publishing Co., New York, 1948, pp. 132–161.

NAPHTHENIC ACIDS

The term *naphthenic acid*, as commonly used in the petroleum industry, refers collectively to all of the carboxylic acids present in crude oil. Naphthenic acids are classified as monobasic carboxylic acids of the general formula RCOOH, where R represents the naphthene moiety consisting of cyclopentine and cyclohexane derivatives. Naphthenic acids are composed predominantly of alkyl-substituted cycloaliphatic carboxylic acids, with smaller amounts of acyclic aliphatic (paraffinic or fatty) acids. Aromatic, olefinic, hydroxy, and dibasic acids are considered to be minor components. Commercial naphthenic acids also contain varying amounts of unsaponifiable hydrocarbons, phenolic compounds, sulfur compounds, and water. The complex mixture of acids is derived from straight-run distillates of petroleum, mostly from kerosene and diesel fractions (see PETROLEUM).

Chemical Structure

Naphthenic acids are based on saturated single or multicyclic condensed ring structures. The low molecular weight naphthenic acids contain alkylated cyclopentane carboxylic acids, with smaller amounts of cyclohexane derivatives occurring. The carboxyl group is usually attached to a side chain rather than directly attached to the cycloalkane. The simplest naphthenic acid is cyclopentane acetic acid (**1**, $n = 1$).

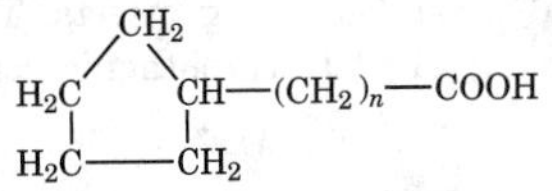

(**1**)

Naphthenic acids are represented by a general formula $C_nH_{2n-z}O_2$, where n indicates the carbon number and z specifies a homologous series. The z is equal to 0 for saturated, acyclic acids and increases to 2 in monocyclic naphthenic acids, to 4 in bicyclic naphthenic acids, to 6 in tricyclic acids, and to 8 in tetracyclic acids.

Physical and Chemical Properties

Naphthenic acids are viscous liquids, with phenolic and sulfur impurities present that are largely responsible for their characteristic odor. Their colors range from pale yellow to dark amber. Naphthenic acids have wide boiling point ranges at high temperatures (250–350°C). They are completely soluble in organic solvents and oils but are insoluble (50 mg/L) in water. Commercial naphthenic acids are available in various grades and are marketed by acid number, impurity level, and color. Chemically, naphthenic acids behave like typical carboxylic acids with similar acid strength as the higher fatty acids.

Naphthenic acid corrosion has been a problem in petroleum-refining operations since the early 1990s. Refineries processing highly naphthenic crudes must use steel alloys; 316 stainless steel is the material of choice. Conversely, naphthenic acid derivatives find use as corrosion inhibitors in oil-well and petroleum refinery applications.

Occurrence

Not all crudes contain sufficient quantities of usable acids to make recovery an economic process. Heavy crudes from geologically young formations have the highest acid content, and paraffinic crudes usually have low acid content. Typical concentrations of acids are shown in Table 1.

Manufacture

The commercial production of naphthenic acid from petroleum is based on the formation of sodium naphthenate. Naphthenic acids are recovered by caustic extraction of petroleum distillates rather than from crude petroleum. Crude naphthenic acid is obtained by acidulating the sodium naphthenate, and can be further refined to remove impurities.

Interest in synthetic naphthenic acid has grown as the supply of product has fluctuated. Oxidation of naphthene-based hydrocarbons, free-radical addition of carboxylic acids to olefins, and addition of unsaturated fatty acids to cycloparaffins have been studied but not commercialized.

Production

Nameplate capacities of naphthenic acid producers in North America are 9000 metric tons of crude and refined acid at Merichem (Tuscaloosa, Ala.), and 3600 t of crude acid at Hewchem (Gulfport, Miss.). However, actual production capacity may vary widely as a result of the mix of feedstocks being processed. Naphthenic acid products are shipped in tank cars, tank trucks, and drums under DOT 9137/UN 3082 identifications numbers.

Economic Aspects

Naphthenic acid availability exceeds demand, although some minor market disruptions occurred in North America during the early 1990s. Long-term yearly feedstock is expected to meet market growth projections with recent discoveries of high naphthenic-content oil.

Health and Safety Factors

Naphthenic acids are only slightly toxic to mammals but are toxic to fish, bacteria, and wood-destroying insects. The lethal oral dose for humans is approximately 1 L. Naphthenic acid is not listed as a carcinogen.

Commercial Uses

More than two-thirds of the naphthenic acids produced is used to make metal salts, with the largest volume being used for copper naphthenate consumed in the wood preservative industry (see WOOD). Oil field uses are primarily imidazolines for surfactant and corrosion inhibition (see PETROLEUM). Besides the lubrication market for metals salts, the miscellaneous market is comprised of free acids used in concrete additives, motor oil lubricants, and asphalt-paving applications (see ASPHALT; LUBRICATION AND LUBRICANTS).

Naphthenic acid is ideal for synthesizing metal carboxylates that require a ligand with some oxidative stability, solubility in hydrocarbons and oils, and insolubility in water.

Another market application for naphthenic acid is the tire industry, where cobalt naphthenate is used as an adhesion promoter (see ADHESIVES; TIRE CORDS). Naphthenic acid esters have been repeatedly cited as surfactants, lubricants, and replacements for phthalates as plasticizers for PVC resins.

Naphthenyl alcohols are formed by reduction of the acids or their simple esters. They are valuable as surfactants, solvents, and components of lubricants. The acid halides are of value mainly as chemical intermediates.

Table 1. Acid Content of Various Crudes

Crude oil source	Petroleum acids, wt %
Pennsylvania	0.03
West Texas	0.4
Gulf Coast	0.6
California	1.5
Russia, Balakhany light	1.0
Russia, Balakhany heavy	1.6
Romania, waxy	0.2
Romania, asphaltic	1.6
Venezuela, Lagunillas	1.2

JAMES A. BRIENT
PETER J. WESSNER
Merichem Company
MARY NOON DOYLE
Shepard Chemical Company

E. S. Lower, *Specialty Chem.* **7,** 76 (1987); **7,** 282 (1987); **8,** 174 (1988); **9,** 135 (1989); **9,** 267 (1989).

H. L. Lochte and E. R. Littmann, *The Petroleum Acids and Bases,* Chemical Publishing Co., Inc., New York, 1955.

G. Narmetova, B. Khamidov, N. Ryabova, and E. Aripov, *Purification, Identification, and Use of Naphthenic Acid,* Fan, Tashkent, former USSR, 1983.

W. Maass, E. Buchspiess-Paulentz, and F. Stinsky, *Naphthensâuren und Naphthenate,* Verlag für Chemische Industrie H. Ziolkowsky, Augsburg, Germany, 1961.

NEODYMIUM. See LANTHANIDES.

NEPTUNIUM. See ACTINIDES AND TRANSACTINIDES.

NEUROPEPTIDES. See MEMORY-ENHANCING DRUGS.

NEUROREGULATORS

Neuroregulators represent a diverse group of compounds that include both neurotransmitters and neuromodulators. Receptors for neuroregulators on both the cell surface and within the cell represent the molecular targets of the majority of drugs in clinical use. In order to classify an endogenous agent as a neurotransmitter, it must possess a number of general characteristics, which are described. Neurotransmitter receptors are grouped according to sequence homology and structures are given. In many cases human receptors have been cloned.

NICKEL AND NICKEL ALLOYS

Nickel occurs in the first transition row in Group 10 (VIIIB) of the Periodic Table. Some physical properties are given in Table 1. Nickel is a high melting point element having a ductile crystal structure. Its chemical properties allow it to be combined with other elements to form many alloys.

In the United States in 1992, 57% of the nickel consumed was used in stainless steels and alloy steels (see STEEL), 28% in nonferrous and high temperature alloys (qv), 9% in electroplating (qv) and the

Table 1. Physical Constants of Nickel

Property	Value
atomic weight	58.71
crystal structure	fcc
melting point, °C	1453
boiling point (by extrapolation), °C	2732
density at 20°C, g/cm^3	8.908
specific heat at 20°C, kJ/(kg·K)[a]	0.44
avg. coefficient of thermal expansion $\times 10^{-6}$, °C^{-1}	
at 20–100°C	13.3
20–500°C	15.2
thermal conductivity, W/(m·K)	
at 100°C	82.8
500°C	61.9
electrical resistivity at 20°C, $\mu\Omega$·cm	6.97
temperature coefficient of resistivity at 0–100°C, ($\mu\Omega$·cm)/°C	0.0071
Curie temperature, °C	353
saturation magnetization. T[b]	0.617
residual magnetization, T[b]	0.300
coercive force, A/m[c]	239
initial permeability, mH/m[d]	0.251
max permeability, mH/m[d]	2.51–3.77
modulus of elasticity $\times 10^3$, MPa[e]	
tension	206.0
shear	73.6
Poisson's ratio	0.30

[a] To convert J to cal, divide by 4.184. [b] To convert T to G, multiply by 1.0×10^4. [c] To convert, A/m to Oe, divide by 79.58. [d] To convert mH/m to G/Oe, multiply by 795.8. [e] To convert MPa to psi, multiply by 145.

remaining 6% consumed primarily as catalysts (see CATALYSIS) in ceramics (qv) in magnets (see MAGNETIC MATERIALS), and as nickel salts (see NICKEL COMPOUNDS). The U.S. markets for nickel alloys (wrought nickel, nickel alloys, and superalloys) in 1992 were ca 40% in the transportation and aircraft industries, 16% in the chemical–petrochemical industry, 18% in the electrical equipment industry, 11% in construction, machinery and fabricated metal products, and 15% in other uses. In the 1990's, these proportions remained quite constant, with transportation and aircraft usage decreasing slightly.

Reserves and Resources

Although nickel comprises ca 7% of the earth's core, it comprises only about 0.009% of the earth's crust, ranking 24th in order of abundance in the crust. Fortunately, ore forms amenable to economic mining exist.

Canada, Cuba, and Russia have the largest economic reserves, whereas the United States has less than 0.1% of the world's estimated reserves.

The world economic (proven) reserves are estimated at 47.0×10^6 t. If annual mine production increases at a rate that reflects a predicated increase in the world primary nickel consumption of 2% annually, these reserves would be depleted before 2030.

In addition to the reported economic reserves, there are substantial nickel resources which could be amenable to mining and refining once appropriate technology becomes available. The single largest such resource is seabed nodules which contain ca 1% nickel and which could represent 800×10^6 t of nickel (see OCEAN RAW MATERIALS).

Nickel Ores. The two types of nickel ore that can be mined economically are classified as sulfide and lateritic. As of 1992, the sulfide deposits accounted for just over 50% of the nickel produced worldwide. Important sulfide deposits are found in Canada, Russia, and Finland. Lateritic ores are distributed widely and constitute the largest nickel reserves.

Extraction and Refining

The treatments used to recover nickel from its sulfide and lateritic ores differ considerably because of the differing physical characteristics of the two ore types. The sulfide ores, in which the nickel, iron and copper occur in a physical mixture as distinct minerals, are amenable to initial concentration by mechanical methods, eg, flotation (qv) and magnetic separation (see SEPARATION, MAGNETIC). The lateritic ores are not susceptible to these physical processes and chemical means must be used to extract the nickel (MINERALS RECOVERY).

Sulfide Ores. The following procedures must be used to obtain pure nickel from these ores: pyrometallurgical processes, electrolytic refining, the carbonyl process, and hydrometallurgical processes.

Lateritic Ores. Nickel oxide ores are processes processed by pyrometallurgical or hydrometallurgical methods.

Commercial Forms of Nickel

The main commercial forms of nickel markets are those listed in Table 2.

Alloys

Properties. Selected properties of commercially available nickel alloys are given in Table 3. Nickel-base alloys provide excellent mechanical properties from cryogenic temperatures through temperatures in excess of 1000°C. Nickel alloys are strengthened by solid solution hardening, carbide strengthening, and precipitation hardening.

Nickel. Nickel metal is available in many wrought forms and usually is designated as Nickel 200 or Nickel 201 and according to the Unified Numbering System (UNS) as UNS NO2201, 205 (UNS NO2205), and 270 (UNS NO2270). Nickel 200 is the general-purpose nickel used in ambient-temperature applications in food processing (qv) equipment, chemical containers, caustic-handling equipment and plumbing, electromagnetic parts, and aerospace and missile components. Nickel 201 has a much lower trace carbon content than the 200 and is thus more suitable for elevated temperature applications. Nickel 205 is low in carbon but contains trace amounts of magnesium; Nickel 270 is one of the purest, ie, 99.98 wt%, commercial nickels.

Nickel has excellent corrosion-resistance properties. In general, nickel is very resistant to corrosion in marine and industrial atmosphere, in distilled and natural waters, and in flowing seawater.

Wrought and cast nickel anodes and sulfur-activated electrodeposited rounds are used widely for nickel electrodeposition onto many base metals. Nickel also can be plated by an electroless process (see

Table 2. Commercial Forms of Nickel

Type	Nickel content, wt %[a]	Uses
electrolytic (cathode)	>99.9	alloy production, electroplating
electrolytic rounds	>99.9	electroplating
carbonyl pellets	99.7	alloy production, electroplating
briquettes	99.9	alloy production
rondelles	99.3	alloy production
powder	99.74	sintered parts, battery electrodes
nickel oxide sinter	76.0	steel and ferrous alloy production
ferronickel[b]	20–50	steel and ferrous alloy production
nickel salts[c]		electroplating, catalysts
nickel chloride	24.70	
nickel nitrate	20.19	
nickel sulfate	20.90	

[a] Values are approximate.
[b] Different grades of ferronickel are produced, and the nickel content denoted includes 1–2 wt % Co.
[c] Nickel content is theoretical.

Table 3. Properties of Nickel Alloys

Alloy	UNS Number	Melting range, °C	Yield strength[a] MPa[b]				100-h rupture strength, MPa[b]		
			20°C	538°C	760°C	982°C	649°C	812°C	982°C
Nickel 200	N02200	1435–1446	103–931	139[c]					
MONEL alloy 400	N04400	1299–1349	172–1173	179[c]					
MONEL alloy K-500	N05500	1316–1349	241–1380	648[c]					
NIMONIC alloy 75	N06075	1340–1380	275	210	172	70	255	39	10
INCONEL alloy 600	N06600	1355–1415	285	220	180	41	160	55	19
INCONEL alloy 625	N06625	1290–1350	490	415	415	140	440	125	32
INCOLOY alloy 800	N08800	1355–1385	250	180	150		240	63	21
HASTELLOY alloy B-2[d]	N10665	1320–1350	412						
HASTELLOY alloy C-22	N06022	1357–1399	373–1391	214					
HASTELLOY alloy C-276	N10276	1323–1371	356	233					
HASTELLOY alloy G-3	N06985	1260–1343	311	186	165	94			
INCONEL alloy 718	N07718	1260–1335	1125	1020	800		725		
INCOLOY alloy 909	N19909	1395–1430	975	850	440		510		
B-1900[e]		1275–1300	825	870	808	415		505	170
MAR-M247		1221–1357	958	875	841	380		572	193
RENÉ 80								350[f]	165
WASPALOY	N07001	1330–1355	795	725	675	140	760	275	45
UDIMET 500	N07500	1300–1395	840	795	730	230	930	305	83
UDIMET 700		1205–1400	965	895	830	305	828	400	110
NIMONIC alloy 80A	N07080	1360–1390	620	530	505	62	595	195	14
NIMONIC alloy 115		1260–1315	865	795	800	240		400	110
INCONEL alloy MA754	N07754	1320–1390	662	504	262	166			131[g]

[a] Where two numbers appear, the first refers to the annealed or solution heat-treated condition, the second to the condition when maximum strength is achieved by cold-working or aging. Otherwise the number refers to the alloy heat-treated for optimum strength. [b] To convert MPA to psi, multiply by 145. [c] Value is at 316°C. [d] 3.18-mm sheet. [e] As cast. [f] Value is at 817°C. [g] Value is at 1093°C.

ELECTROLESS PLATING). Nickel plating provides resistance to corrosion for many commonly used articles, eg, pins, paper clips, scissors, keys, fasteners, etc, as well as for materials used in food processing (qv), the paper (qv), and pulp (qv) industries, and the chemical industry, each of which is often characterized by severely corrosive environments.

Porous nickel electrodes made from nickel powder are used in storage batteries (qv) and fuel cells (qv).

Nickel also is an important industrial catalyst. The most extensive use of nickel as a catalyst is in the food industry in connection with the hydrogenation or dehydrogenation of organic compounds to produce edible fats and oils (see FATS AND FATTY OILS).

Nickel Alloying. Nickel is alloyed into low alloy steels, ferritic alloy steels, and austenitic stainless steels through the conventional steelmaking processes, eg, open hearth, basic oxygen conversion, and the argon–oxygen decarbonization (AOD) processes.

The AOD process is used to produce a substantial quantity of the stainless steels in the world. EAF or AOD melting and air-induction melting (AIM) are used for some nickel-base alloys. Electroslag remelt (ESR) processing also is used to further refine these steels and nickel alloys.

Nickel alloys that are heavily alloyed with other elements including the nickel-base and iron-base superalloys, also are produced by vacuum-induction melting (VIM). In the VIM process, the melting, alloying, melt treatments, and ingot casting are carried out under vacuum.

Nickel–Copper. The nickel-rich, nickel–copper alloys are characterized by a good compromise of strength and ductility and are resistant to corrosion and stress corrosion in many environments, in particular water and seawater, nonoxidizing acids, neutral and alkaline salts, and alkalies.

Nickel–Chromium. Chromium is added to nickel to enhance strength, corrosion resistance, oxidation, hot corrosion resistance, and electrical resistivity.

Nickel–Iron. A large amount of nickel is used in alloy and stainless steels and in cast irons. Nickel is added to ferritic alloy steels to increase the hard-enability and to modify ferrite and cementite properties and morphologies, and thus to improve the strength, toughness, and ductility of the steel.

Many nickel–iron alloys have useful magnetic characteristics and are used in a wide range of devices in the electronics and telecommunication fields.

Demands for improved efficiency in aircraft gas turbines led to the use of a family of age hardenable nickel–iron controlled expansion superalloys for engine seals and casings.

Nickel–Molybdenum. Molybdenum in solid solution with nickel strengthens the latter metal and improves its corrosion resistance.

Nickel–Iron–Chromium. A large number of industrially important materials are derived from nickel–iron–chromium alloys. These alloys are within the broad austenitic, gamma-phase field of the ternary Ni–Fe–Cr phase diagram and are noted for good resistance to corrosion and oxidation and good elevated temperature strength (see HIGH TEMPERATURE ALLOYS).

Nickel-Base Superalloys. Superalloys, which are critical to gas-turbine engines because of their high temperature strength and superior creep and stress rupture-resistance, basically are nickel–chromium alloyed with a host of other elements. The alloying elements include the refractory metals tungsten, molybdenum, or niobium for additional solid-solution strengthening, especially at higher temperatures and aluminum in appropriate amounts for the precipitation of γ' for coherent particle strengthening (see REFRACTORIES). Titanium is added to provide stronger γ', and niobium reacts with nickel in the solid state to precipitate the γ''-phase; γ''.

JOHN H. TUNDERMANN
Inco Alloys International
JOHN K. TIEN
Columbia University
TIMOTHY E. HOWSON
Wyman-Gordon

BIBLIOGRAPHY

S. J. Rosenberg, *Natl. Bur. Stand. Monogr.* **106** (1968).

W. L. Mankins and S. Lamb, *Nickel and Nickel Alloys* in *ASM Handbook,* 10th ed., Vol. 2, ASM International, Materials Park, Ohio, 1990, pp. 428–445.

J. R. Boldt, Jr., and P. Queneau, *The Winning of Nickel,* D. Van Nostrand Co., Inc., New York, 1967.

High Temperature, High Strength Nickel Base Alloys, Nickel Development Institute, Toronto, Canada, 1987.

NICKEL COMPOUNDS

Nickel, Ni, has a [Ar] $3d^8 4s^2$ electronic configuration and forms compounds in which the nickel atom has oxidation states of −1 through +4. Whereas reagents yield an array of compounds in a variety of nickel oxidation states, Ni(II) represents the bulk of all known compounds. As of this writing, > 237,000 compounds of nickel have been reported. The primary uses for nickel compounds, aside from nickel refining and electroplating, are in steel (qv) making, catalysis (qv), storage batteries (qv), specialty chemicals, and specialty ceramics (qv).

Simple nickel salts form ammine and other coordination complexes (see COORDINATION COMPOUNDS). The octahedral configuration, in which nickel has a coordination number (CN) of 6, is the most common structural form.

Inorganic Compounds

Nickel Oxides. Nickel oxide, NiO, is a green cubic crystalline compound, mp 2090°C, density 7.45 g/cm^3, the properties of which are related to its method of preparation.

Several nickel oxides are manufactured commercially. A sintered form of green nickel oxide is made by smelting a purified nickel matte at 1000°C; a powder form is made by the desulfurization of nickel matte. Black nickel oxide is made by the calcination of nickel carbonate at 600°C.

Nickel oxides are also made by the calcination of nickel carbonate or nickel nitrate that were made from a pure form of nickel. A high purity, green nickel oxide is made by firing a mixture of nickel powder and water in air.

The sinter oxide form is used as charge nickel in the manufacture of alloy steels and stainless steels (see STEEL).

Green nickel oxide powder is used in the refining of nickel. Green and black nickel oxides are used in the ceramic industry for making frit, ferrites (qv), and inorganic colors (see CERAMICS; COLORANTS FOR CERAMICS). Black nickel oxide is used for the manufacture of nickel salts and speciadtyceramics.

Nickel oxide is used commercially to make nickel fibers in a process whereby a water slurry containing nickel oxide and a cellulose (qv) type binder is forced through tiny orifices to form green fibers. Subsequent steps include drying and reduction with hydrogen. The resulting nickel fibers are matted together and used for the filtration (qv) of gases.

Nickel Sulfate. Nickel sulfate hexahydrate, $NiSO_4 \cdot 6H_2O$, is a monoclinic emerald-green crystalline salt that dissolves easily in water and in ethanol.

The preferred method for making nickel sulfate is adding nickel powder to hot dilute sulfuric acid.

The principal use for nickel sulfate is as an electrolyte for the metal-finishing application of nickel electroplating.

Nickel Nitrate. Nickel nitrate hexahydrate, $Ni(NO_3)_2 \cdot 6H_2O$, is a green monoclinic deliquescent crystal, mp 56°C, density 2.05 g/cm^3, that is extremely soluble in water.

Nickel ammonium nitrate, $H_3N \cdot xHNO_3 \cdot xNi$, forms in the commercial methods that use nitric acid and metallic nickel. The relative concentrations of acid and metal control the ammonia formation. Nickel powder, added slowly to a stirred mixture of nitric acid and water, yields nickel nitrate containing the least ammonia.

Table 1. Properties of Anhydrous Nickel Halides

Compound	Mp, °C	Density, g/cm^3	Color	Solubility, 0°C, g/100 mL H_2O
nickel difluoride	1000[a]	4.63	light green	4
nickel dichloride	1001	3.56	yellow	64
nickel dibromide	963	5.10	orange	113
nickel diiodide	797	5.83	black	124

[a] Sublimes.

Nickel nitrate is an intermediate in the manufacture of nickel catalysts. Nickel nitrate also is an intermediate in loading active mass in nickel–alkaline batteries of the sintered plate type (see BATTERIES, SECONDARY CELLS).

Nickel Halides. Nickel forms anhydrous as well as hydrated halides. The properties of the anhydrous salts are given in Table 1.

Nickel chloride hexahydrate is an important material in nickel electroplating. Anhydrous nickel chloride is formed from a mixture of nickel powder and sodium chloride at the anode during recharging of sodium nickel chloride batteries, which have possible use as the power source in electric vehicles.

Nickel Carbonate. The addition of sodium carbonate to a solution of a nickel salt precipitates an impure basic nickel carbonate, $NiCO_3$. The commercial material is the basic salt $2NiCO_3 \cdot 3Ni(OH)_2 \cdot 4H_2O$. Nickel carbonate is prepared best by the oxidation of nickel powder in ammonia and CO_2.

Nickel carbonate is used in the manufacture of catalysts, in the preparation of colored glass (qv), in the manufacture of certain nickel pigments, and as a neutralizing compound in nickel electroplating solutions.

Nickel Hydroxides. Nickel hydroxide, $Ni(OH)_2$, is a light-green, microcrystalline powder, density 4.15 g/cm^3. A solution of nickel sulfate which is treated with sodium hydroxide yields the gelatinous nickel hydroxide which, when neutralized, forms a fine precipitate that can be filtered.

The principal use for nickel hydroxide is in the manufacture of nickel–cadmium batteries.

Nickel Fluoroborate. Fluoroboric acid and nickel carbonate form nickel fluoroborate, $Ni(BF_4)_2 \cdot 6H_2O$. Nickel fluoroborate is used as the electrolyte in specialty high speed nickel plating.

Nickel Cyanide. Nickel cyanide tetrahydrate, $Ni(CN)_2 \cdot 4H_2O$, forms apple-green plates which are, like other metal cyanides, highly poisonous. Nickel cyanide is made by the reaction of potassium cyanide and nickel sulfate. Nickel cyanide has been used in the Reppe process for the conversion of acetylene to butadiene (qv) and other products (see ACETYLENE-DERIVED CHEMICALS; CYANIDES).

Nickel Sulfamate. Nickel sulfamate, $Ni(SO_3NH_2)_2 \cdot 4H_2O$, commonly is used as an electrolyte in nickel electroforming systems, where low stress deposits are required. It is prepared by the reaction of fine nickel powder or black nickel oxide with sulfamic acid in hot water solution.

Nickel Sulfide. Nickel, like iron and cobalt, forms monosulfides which may show considerable deviation from stoichiometry without exhibiting heterogeneity. Nickel sulfide, NiS, occurs naturally as the mineral millerite, and has a trigonal crystalline form and a yellow metallic luster; density 5.65 g/cm^3, mp 797°C.

It can be prepared by the fusion of nickel powder with molten sulfur solution or by precipitation using hydrogen sulfide treatment of a buffered solution of a nickel(II) salt. Another naturally occurring sulfide is polydymite, Ni_3S_4.

Other Nickel Salts. Other nickel salts include nickel arsenate, nickel phosphate, and nickel double salts (eg, nickel ammonium chloride, $NiCl_2 \cdot NH_4Cl \cdot 6H_2O$; nickel ammonium sulfate, $NiSO_4 \cdot (NH_4)_2SO_4 \cdot 6H_2O$, and nickel potassium sulfate, $NiSO_4 \cdot K_2SO_4 \cdot 6H_2O$).

Table 2. Physical Properties of Nickel Carbonyl

Property	Value
melting point, °C	−17
crystallization point, °C	−25
density at 20.0°C, g/mL	1.3103
vapor pressure data, kPa[a] at 0°C	17.1
16.1°C	44.3
35.1°C	77.7
critical temperature, °C	200

[a] To convert kPa to psi, multiply by 0.145.

Nickel Amine Complexes. Nickel amine complexes include $Ni(NH_3)_6Cl_2$, $Ni(NH_3)_6Br_2$, $Ni(NH_3)_6I_2$, tetrakispyridienickel(II) dichloride, and the bidendate ethylenediamine (en) complex $[Ni(en)_3]Cl_2$.

Organic Compounds

Nickel plays a role in the Reppe polymerization of acetylene, the reduction of nickel halides by sodium cyclopentadienide, the synthesis of cyclododecatrienenickel and formation from elemental nickel powder and other reagents of nickel(0) complexes that serve as catalysts for oligomerization and other hydrocyanation reactions (see also ORGANOMETTALICS).

Nickel Carbonyl. The properties of nickel carbonyl are given in Table 2 (see also CARBONYLS). Thermodynamic properties, force constants, and infrared characteristics are documented.

Nickel carbonyl can be prepared by the direct combination of carbon monoxide and metallic nickel. Two commercial processes are used for large-scale production: an atmospheric method, whereby carbon monoxide is passed over nickel sulfide and freshly reduced nickel metal; and the second method which involves high pressure CO in the formation of iron and nickel carbonyls.

Substituted Nickel Carbonyl Complexes. The reaction of trimethyl phosphite and nickel carbonyl yields the monosubstituted colorless oil, $(CO)_3NiP(OCH_3)_3$, the disubstituted colorless oil, $(CO)_2Ni[P(OCH_3)_3]_2$, and the trisubstituted white crystalline solid, $(CO)Ni[P(OCH_3)_3]_3$ (mp 98°C). Liquid complexes result from the reaction of trifluorophosphine with nickel carbonyl yielding $(CO)_3Ni(PF_3)$, $(CO)_2Ni(PF_3)_2$, and $(CO)Ni(PF_3)_3$.

π-Cyclopentadienyl Nickel Complexes. Nickel bromide dimethoxyethane forms bis(cyclopentadienyl)nickel upon reaction with sodium cyclopentadienide.

Tetrakisligand Nickel(0) Complexes. Tetrakisligand nickel(0) complexes are made by several methods, eg, the substitution of CO in nickel carbonyl [yielding tetrakistrichlorophosphinenickel(0), $Ni(PCL_3)_4$, mp 120°C (dec)]; the substitution of more powerful donor ligands, such as triphenylphosphine, in other tetrakisligand nickel(0) complexes [yielding red solid tetrakistriphenylphosphinenickel(0), mp 125°C, and yellow solid tetrakistrimethylphosphinenickel(0), mp 185°C (dec)]; the direct reaction of nickel powder and ligands where halogens are on the donor atom [yielding $Ni(PF_3)_4$ and $Ni(CH_3PCl_2)_4$]; and the reaction of nickel halides and ligands in the presence of a reducing agent, eg, zinc metal powder.

Nickel Salts and Chelates. Nickel salts of simple inorganic acids can be prepared by reaction of the organic acid and nickel carbonate of nickel hydroxide; reaction of the acid and a water solution of a simple nickel salt; and, in some cases, reaction of the acid and fine nickel powder or black nickel oxide. Examples are nickel acetate tetrahydrate $[Ni(C_2H_3O_2)\cdot 4H_2O]$ and nickel formate dihydrate $[Ni(HCOO)_2\cdot 2H_2O]$.

Health and Safety Factors

Eye and Skin Contact. Some nickel salts and aqueous solutions of these salts, eg, the sulfate and chloride, may cause a primary irritant reaction of the eye and skin. The most common effect of dermal exposure to nickel is allergic contact dermatitis.

Protective equipment and clothing such as face shields and gloves should be worn and safety showers should be available wherever there is a possibility of being splashed or otherwise contacted by nickel containing solutions. If dermatitis should occur, the possibility that it is nickel related should be brought to the attention of a physician.

Inhalation. Nickel carbonyl is an extremely toxic gas. The permissible exposure limit (PEL) in the United States is 1 part per billion (ppb) in air. The American Conference of Governmental Industrial Hygienists (ACGIH) threshold limit value (TLV) for an 8-h, time-weighted average concentration is 50 ppb. Nickel carbonyl may form wherever carbon monoxide and finely divided nickel are brought together.

Nickel carbonyl should be used in totally enclosed systems or under good local exhaust. Plants and laboratories where nickel carbonyl is used should make use of air-monitoring devices, alarms should be present in case of accidental leakage, and appropriate personal respiratory protective devices should be readily available for emergency uses.

The potential chronic toxicity is of concern. Based on epidemiological and experimental results, the International Agency for Research on Cancer (IARC) has concluded that all nickel compounds are Category 1, ie, known human carcinogens, and has classified metallic nickel as a Category 2B carcinogen, ie, possibly carcinogenic to humans.

It is good practice to keep concentrations of airborne nickel in any chemical form as low as possible and certainly below the relevant standard. Local exhaust ventilation is the preferred method, particularly for powders, but personal respirator protection may be employed. In the United States, the Occupational Safety and Health Administration (OSHA) personal exposure limit (PEL) form all forms of nickel except nickel carbonyl is 1 mg/m^3.

Uses

Nickel compounds are used as catalysts and in electroplating, specialty ceramics, plastics additives, organic dyes and pigments, agricultural chemicals, and other specialty chemicals.

Nickel Chemical Waste Reduction

The obvious destination for nickel waste is in the manufacture of stainless steel, which consumes 65% of new refined nickel production. In 1996, 6×10^3 t of nickel from nickel-containing wastes were processed into 60×10^3 t of stainless steel remelt alloy (see RECYCLING, NONFERROUS METALS). This quantity is expected to increase dramatically as development of the technology of waste recycle collection improves.

D. H. ANTONSEN
International Nickel, Inc.

P. W. Jolly and G. Wilke, *The Organic Chemistry of Nickel,* Vol. II, Organic Synthesis, Academic Press, Inc., New York, 1975.

Metals Handbook, 9th ed. Vol. 7, *Powder Metallurgy: Production of Metal Powders,* American Society of Metals, Metals Park, Ohio, 1984.

G. P. Tyroler and C. A. Landolt, *Extractive Metallurgy of Nickel and Cobalt,* The Metallurgical Society, Inc., Warrendale, Pa., 1988.

H. Topsoe, B. S. Clausen, N. Topsoe, and J. Hyldtoft, *Symposium on the Mechanism of HDS/HDN Reactions,* Vol. 38, No. 3, Division of Petroleum Chemistry, Preprints, American Chemical Society, Chicago, Ill., July 1993.

NICOTINAMIDE. See VITAMINS.

NIELSBOHRIUM. See ACTINIDES AND TRANSACTINIDES.

NIOBIUM AND NIOBIUM COMPOUNDS

Niobium is most important as an alloy addition in steels (see STEEL). This use consumes over 90% of the niobium produced. Niobium is also vital as an alloying element in superalloys for aircraft turbine engines. Other uses, mainly in aerospace applications, take advantage of its heat resistance when alloyed singly or with groups of elements such as titanium, zirconium, hafnium, or tungsten. Niobium alloyed with titanium or with tin is also important in the superconductor industry (see HIGH TEMPERATURE ALLOYS; REFRACTORIES).

Properties. Elemental niobium, Nb, has a cosmic abundance of 0.9 relative to silicon = 106, an average value of 24 ppm in the earth's crust, and a comparable value on the lunar surface. Niobium-93 has a nuclear spin of 9/2 and a thermal neutron-capture cross section of $1.1 \pm 0.1 \times 10^{-28}$ m^2 (1.1 barns) which makes it of much interest to the nuclear industry (see NUCLEAR REACTORS).

Niobium, like vanadium, undergoes no phase transitions from room temperature to the melting point. It is a steel-grey, ductile, refractory metal having a higher melting point than molybdenum and a lower electron work function than tantalum, tungsten, or molybdenum.

The most common oxidation state of niobium is +5, although many anhydrous compounds have been made with lower oxidation states, notably +4 and +3. The aqueous chemistry primarily involves halo- and organic acid anionic complexes. Metal-metal bonding is common. Extensive polymeric anions form. Niobium resembles tantalum and titanium in its chemistry, and separation from these elements is difficult. Some properties of niobium are listed in Table 1.

Occurrence. Niobium and tantalum usually occur together. Niobium never occurs in the free state; most often it is combined with oxygen and another metal, forming a niobate or tantalate in which the niobium and tantalum isomorphously replace one another with little change in physical properties except density. Ore concentrations of niobium usually occur as carbonatites and are associated with tantalum in pegmatites and alluvial deposits. Principal niobium-bearing minerals can be divided into two groups, the titano- and tantalo-niobates.

Table 1. Properties of Niobium

Property	Value
atomic number	41
atomic weight	92.906
atomic volume, cm^3/mol	10.8
atomic radius, nm	0.147
electronic configuration	[Kr]$4d^45s^1$
ionization potential, eV	6.77
crystal structure	bcc
lattice constant at 0°C, pm	330.04
density at 20°C, g/cm^3	8.66
mp, °C	2468 ± 10
bp, °C	5127
latent heat of fusion, kJ/mol[a]	26.8
latent heat of vaporization, kJ/mol[a]	697
heat of combustion, kJ/mol[a]	949
heat capacity, J/(mol·K)[a]	
at 298 K	24.7
entropy, J/(mol·K)[a]	
at 298 K	36.5
vapor pressure at 2573 K, mPa[b]	22
evaporation rate at 2573 K, μg/(cm^2·s)	1.9
thermal conductivity at 298 K, W/(m·K)	52.3
coefficient of linear thermal expansion, 291–373 K, °C^{-1}	7.1×10^6
electrical resistivity, Ω·m	$13-16 \times 10^{-6}$
temperature coefficient of resistivity, °C^{-1}	3.95×10^{-3}
work function, eV	4.01
secondary emission (primary δ_{max} = 400 V), eV	1.18
positive ion emission, eV	5.52

[a] To convert J to cal, divide by 4.184.
[b] To convert mPa to μm Hg, divide by 133.3.

Extraction, Refining, and Metallurgy. The process of extracting and refining niobium consists of a series of consecutive operations. Frequently, several steps are combined; the upgrading of ores by preconcentration; an ore-opening procedure to disrupt the niobium-containing matrix; preparation of a pure niobium compound; reduction to metallic niobium; and refining, consolidation, and fabrication of the metal.

The most straightforward process is the direct conversion, ie, reduction, of the niobium concentrate to metallic niobium. The primary method is the aluminothermic reduction of a pyrochlore and iron–iron oxide mixture. Ferroniobium also is produced in an electric furnace procedure. Essentially, the same reactants are used as in the aluminothermic method.

In addition to the standard ferroniobium, there is a lesser but significant demand for high purity niobium alloys, mainly high purity ferroniobium and nickel–niobium. These high purity alloys are used in the fabrication of nickel- and cobalt-based superalloys.

Direct attack by hot 70–80 wt % hydrofluoric acid, sometimes with nitric acid, is effective for processing columbites and tantalocolumbites. Yields are >90 wt%. This method, used in the first commercial separation of tantalum and niobium, is used commercially as a lead-in to solvent extraction procedures.

Concentrated sulfuric acid (97 wt %) at 300–400°C has been used to solubilize niobium from columbite and pyrochlore. Fusion with caustic soda at 500–800°C in an iron crucible is an effective method for opening pyrochlores and columbites.

The reaction of chlorine gas with a mixture of ore and carbon at 500–1000°C yields volatile chlorides of niobium and other metals. These can be separated by fractional condensation.

Once the niobium ore has been opened, the niobium must be separated from the tantalum and/or impurities. The classical method of doing this is by means of repeated recrystallization. However, the recrystallization of complex fluoride salts has been replaced completely by solvent extraction techniques which are used extensively.

Once purification of the niobium has been effected, the niobium can be reduced to metallic form. Electrowinning has been used to obtain niobium from molten alkali halide electrolytes.

Niobium pentoxide also has been reduced to metal commercially by the aluminothermic process and then purified by melting. Are melting can be used, however, the most common method in commercial use is electron-beam melting.

Niobium metal is available as ingot, sheet, rod, and wire and can be fabricated and formed by most metallurgical and engineering techniques.

Economic Aspects. Information on the economic aspects of niobium and its compounds is available. Brazil has the most niobium-bearing ore reserves. The principal concentrate producers are the Araxa (Companhia Brasilira de Metallurgica e Mineracao) and Catala mines (Mineracao Catalao de Goias Ltds) in Brazil, plus Niobec (a Teck/Cambior 50/50 joint venture) in Canada.

Health and Safety Factors. Toxicity data on niobium and its compounds are sparse. The most common materials, eg, niobium concentrates, ferroniobium, niobium metal and niobium alloys, appear to be relatively inert biologically. Limited animal experiments show high toxicity for some salts which are related to disturbance of enzyme action. Niobium hydride has moderate fibrogenic and general toxic action. Recommended maximum allowable concentrations are 6 mg/m^3 Recommended maximum permissible concentration of Nb in reservoir water is 0.01 mg/L. The threshold for affecting clarity and biological oxygen demand (BOD) is 0.1 mg/L.

Unstable niobium isotopes that are produced in nuclear reactors or similar fission reactions have typical radiation hazards (see RADIOISOTOPES).

Fire fighting procedures for niobium and niobium hydride powder suggest letting the fire burn itself out. Small fires can be controlled by smothering with dry table salt or using Type D dry powder fire-extinguishing material. Under no circumstances should water be used, as a violent explosion may result.

Table 2. Properties of Niobium Compounds

Compounds	Molecular formula	Density, g/cm^3	Mp, °C	Bp, °C	Specific resistivity, $\mu\Omega\cdot cm$
niobium boride	NbB	7.5	2000		64.5[a]
niobium diboride	NbB_2	6.9[b]	3050		65[a,c]
diniobium carbide	Nb_2C	7.8	3090		
niobium carbide	NbC	7.788[d]	3600	4300	180 max[e]
niobium pentafluoride	NbF_5	3.54	79	234	
niobium fluorodioxide	NbO_2F				
niobium pentachloride	$NbCl_5$	2.74[f]	208.3	248.2	
niobium trichloromonoxide	$NbOCl_3$	3.72	vacuum sublimes at ca 200		
niobium pentabromide	$NbBr_5$	4.36	254	365	
niobium tribromomonoxide	$NbOBr_3$		vacuum sublimes at 180	ca 320 dec	
niobium pentaiodide	NbI_5		ca 200 dec		
niobium hydride	NbH	6–6.6			
diniobium nitride	Nb_2N	8.08	2050		
niobium nitride	NbN	8.4[a]		200[a], 450 (at mp)	
niobium oxide	NbO	7.30			
niobium dioxide	NbO_2	5.90			
α-niobium pentoxide	α-Nb_2O_5	4.55	1491 ± 2		

[a] At 25°C.
[b] Has a Mohs' hardness of 8+.
[c] Thermal conductivity value is 17 W/(m·k) at 23°C.
[d] Has a Mohs' hardness of 9+.
[e] Thermal conductivity value is 14 W/(m·k) at 23°C.
[f] Has a hardness of 208.3.

Niobium Compounds.

A summary of niobium compounds is given in Table 2.

JAMES H. SCHLEWITZ
Teledyne Wah Chang

R. J. H. Clark and D. Brown, *The Chemistry of Vanadium, Niobium and Tantalum,* Pergamon Press, Elmsford, N. Y., 1975.

T.I.C. Bulletin, Tantalum—Niobium International Study Center, Brussels, Belgium.

F. Fairbrother, *The Chemistry of Niobium and Tantalum,* Elsevier Publishing Co., Amsterdam, 1967.

S. Harry, ed. *Niobium, Proceedings of the International Symposium,* San Francisco, Calif., Nov. 8–11, 1981, The Metallurgical Society of AIME.

NITRATION

Nitration is defined in this article as the reaction between a nitration agent and an organic compound that results in one or more nitro ($-NO_2$) groups becoming chemically bonded to an atom in this compound. Nitric acid is used as the nitrating agent to represent *C*-, *O*-, and *N*-nitrations. *O*-Nitrations result in esters. *N*-nitrations are often used as a first step for production of nitramines.

For example, a nitro group is substituted for a hydrogen atom, and water is a by-product. Nitro groups may, however, be substituted for other atoms or groups of atoms. Nitro compounds can also be produced by addition reactions, eg, the reaction of nitric acid or nitrogen dioxide with unsaturated compounds such as olefins or acetylenes.

Nitrations are highly exothermic, ie, ca 126 kJ/mol (30 kcal/mol). However, the heat of reaction varies with the hydrocarbon that is nitrated. The mechanism of a nitration depends on the reactants and the operating conditions. The reactions usually are either ionic or free-radical. Ionic nitrations are commonly used for aromatics; many heterocyclics; hydroxyl compounds, eg, simple alcohols, glycols, glycerol, and cellulose; and amines. Nitration of paraffins, cycloparaffins, and olefins frequently involves a free-radical reaction.

Ionic Nitration Reactions

Acid mixtures containing nitric acid and a strong acid, eg, sulfuric acid, perchloric acid, selenic acid, hydrofluoric acid, boron trifluoride, or an ion-exchange resin containing sulfonic acid groups, can be used as the nitrating feedstock for ionic nitrations. These strong acids are catalysts that result in the formation of nitronium ions, NO_2^+. Sulfuric acid is almost always used industrially since it is both effective and relatively inexpensive.

Mechanism. The NO_2^+ mechanism has been accepted since about 1950 for the nitration of most aromatic hydrocarbons, glycerol, glycols, and numerous other hydrocarbons in which mixed acids or highly concentrated nitric acid are used. The mechanism has been discussed in detail and critically analyzed. NO_2^+ attacks an aromatic compound (ArH) as follows:

$$ArH + NO_2^+ \longrightarrow \left[Ar\begin{matrix} H \\ NO_2 \end{matrix} \right]^+ \longrightarrow ArNO_2 + H^+ \qquad (1)$$

Nitrosonium ions, NO^+, are, however, the ions employed to start the nitration sequence for easily nitratable aromatic compounds such as phenol.

Kinetics of Aromatic Nitrations. The kinetics of aromatic nitrations are functions of temperature, which affects the kinetic rate constant, and of the compositions of both the acid and hydrocarbon phase. In addition, a larger interfacial area between the two phases increases the rates of nitration since the main reactions occur at or near the interface. Larger interfacial areas are obtained by increased agitation and by the proper choice of the volumetric % acid in the liquid–liquid dispersion. The viscosities and densities of the two phases and the interfacial tension between the phases are important physical properties affecting the interfacial area.

Increased agitation of a given acid–hydrocarbon dispersion results in an increase in interfacial areas owing to a decrease in the average diameter of the dispersed droplets. As the droplets decrease in size, the ease of separation of the two phases, following completion of nitration, also decreases.

Industrial Applications. Significant process changes have occurred in many nitration plants. Continuous-flow units are now widely used in the 1990s, replacing batch nitrations. A well-designed continuous-flow plant often offers all of the following advantages, per unit weight of product, as compared to batch units; increased safety, decreased energy requirements, reduced amounts of undesired by-products, fewer environmental problems, reduced labor requirements, and lower operating expenses.

Many nitrated products are explosives, including DNT, TNT, and nitroglycerine (NG). To minimize the potential for run away reactions and explosions, the compositions of the feed acids and reaction conditions are currently better controlled than formerly. In some processes, 99% or more of the feed HNO_3 reacts. Dispersions (or mixtures) of such a waste acid and the nitration product are relatively safe to handle. Also, centrifugal separators are used in many modern processes to rapidly separate the hydrocarbon and used acid phases. Rapid separation greatly reduces the amounts of nitrated materials in the plant at any given time and reduces undesired reactions of the nitrated products.

Considerable effort has been made to minimize energy requirements in the nitrations plants too.

A significant concern in all nitration plants using mixed acid centers on the disposal method or use for the waste acids. They are sometimes employed for production of superphosphate fertilizers. Processes have also been developed to reconcentrate and recycle the acid.

Nitrations Using N_2O_5. Considerable worldwide interest has occurred in the late 1980s and the early and mid-1990s for nitrations using N_2O_5. Production of nitramines (or *N*-nitrations) is particularly promising, since these compounds are more stable in the presence of $N_2O_5–HNO_3$ solutions as compared to mixed acids containing H_2SO_4. Good results have been obtained for the production of the high explosives, cyclotetramethylenetetranitramine or HMX and DADN. Another high explosive, polynitrofluorene, has been produced via *C*-nitrations, for the first time. The overall exothermicities of the reactions are less when N_2O_5 is used, as compared to mixed acids.

Solutions of CH_2Cl_2 and N_2O_5 have only mild nitrating power. Yet some nitrations are rapid: the N_2O_5 reacts on an almost stoichiometric basis, and only minimal residual nitric acid is present upon completion of the nitration. Some nitrations having unique characteristics can be accomplished.

Free-Radical Nitrations of Paraffins

Both vapor-phase and liquid-phase processes are employed to nitrate paraffins, using either HNO_3 or NO_2. The nitrations occur by means of free-radical steps, and sufficiently high temperatures are required to produce free radicals to initiate the reactions steps.

Chemistry. Free-radical nitrations consist of rather complicated nitration and oxidation reactions. When nitric acid is used in vapor-phase nitrations, the main initiating reaction (eq. 2) produces either $\cdot NO_2$ or $\cdot ONO$. Temperatures of >ca 350°C are required to obtain a significant amount of initiation, and equation 2 is the rate-controlling step for the overall reaction. Reactions 3 and 4 are chain-propagating steps.

$$HNO_3 \rightarrow \cdot OH + NO_2 \qquad (2)$$

$$RH + \cdot OH \rightarrow R\cdot + H_2O \qquad (3)$$

$$R\cdot + HNO_3 \rightarrow RNO_2 + \cdot OH \qquad (4)$$

When nitrogen dioxide is used, the main reaction steps are as in equations 5 and 6.

$$RH + NO_2 \rightarrow R\cdot + HNO_2 \rightarrow R\cdot + \cdot OH + NO \qquad (5)$$

$$R\cdot + \cdot NO_2 \rightarrow RNO_2 \qquad (6)$$

An important side reaction in all free-radical nitrations is production, of unstable alkyl nitrites (eq. 7). They decompose to form nitric oxide and alkoxy radicals (eq. 8) which form oxygenated compounds and lower molecular weight alkyl radicals which can form lower molecular weight nitroparaffins by reactions 4 or 6. The oxygenated hydrocarbons often react further to produce carbon oxides and water.

$$R\cdot + \cdot ONO \rightarrow RONO \qquad (7)$$

$$RONO \rightarrow NO + RO\cdot \qquad (8)$$

Processes for Paraffin Nitrations. Propane is thought to be the only paraffin that is commercially nitrated by vapor-phase processes. Temperature control is a primary factor in designing the reactor, and several approaches have been investigated. A spray nitrator in which liquid nitric acid is sprayed into hot propane is used industrially. Relatively small-diameter tubular reactors, fluidized-bed reactors, and molten salt reactors have all been successfully used in laboratory units.

Health and Safety Factors

The danger of an explosion of a nitrated product generally increases as the degree of nitration increases. Nitroaromatics and some polynitrated paraffins are highly toxic when inhaled or when contacted with the skin. All nitrated compounds tend to be highly flammable.

LYLE F. ALBRIGHT
Purdue University

J. F. Fischer, in H. Feuer and A. T. Nielsen, eds., *Nitro Compounds: Recent Advances in Synthesis and Chemistry,* VCH Publishers, New York, 1990, Chapt. 3.

M. E. Hill and co-workers, in *ACS Symposium Series No. 22,* American Chemical Society, Washington, D.C., 1976, Chapt. 17, pp. 253–271.

T. Urbanski, *Chemistry and Technology of Explosives,* Vols. 1–3, The Macmillan Co., New York, 1964, 1965; Vol. 4, Permagon Press, Elmsford, N.Y., 1983.

E. Gilbert, in S. M. Kaye, ed., *Encyclopedia of Explosives and Related Items,* Vol. 9, U.S. Army Armament Research and Development Command, Dover, N.J., 1980, T235–286.

NITRIC ACID

Nitric acid, HNO_3, also known as *aqua fortis,* azotic acid, hydrogen nitrate, or nitryl hydroxide, is a chemical of major industrial importance. Because of its properties as a very strong acid and a powerful oxidizing agent, as well as its ability to nitrate organics, nitric acid is essential in the production of many chemicals (eg, pharmaceuticals, dyes, synthetic fibers, insecticides, and fungicides), but is used mostly in the production of ammonium nitrate for the fertilizer industry (see FERTILIZERS). Because of the increased popularity of urea as a fertilizer, production has leveled off in the 1990s. Most growth in demand has come from the production of polyurethanes, fibers, and ammonium nitrate-based explosives. Other uses for nitric acid are in the manufacture of explosives (trinitrotoluene, nitroglycerin, etc), metal nitrates, nitrocellulose, and nitrochlorobenzene, the treatment of metals (eg, the pickling of stainless steels and metal etching), as a rocket propellant, and for nuclear fuel processing.

Physical Properties

Crystals of pure nitric acid are colorless and quite stable. Above the melting point of −41.6 °C, nitric acid is a colorless liquid that fumes in moist air and has a tendency to decompose, forming oxides of nitrogen. The rate of decomposition is accelerated by exposure to light and increases in temperature. The normal boiling point of nitric acid is 83.4°C, but when heated the liquid gradually decomposes to form a maximum boiling azeotrope at 120°C and 69 wt % HNO_3. Nitric acid is completely miscible with water.

Chemical Properties

Nitric acid is a strong monobasic acid, a powerful oxidizing agent, and nitrates many organic compounds.

Acidic Properties. As a typical acid, it reacts readily with alkalies, basic oxides, and carbonates to form salts. However, because of its oxidizing nature, nitric acid does not always behave as a typical acid. Bases having metallic radicals in a reduced state (eg, ferrous and stannous hydroxide becoming ferric and stannic salts) are oxidized by nitric acid.

Oxidizing Properties. Nitric acid is a powerful oxidizing agent (electron acceptor) that reacts violently with many organic materials (eg, turpentine, charcoal, and charred sawdust). The concentrated acid may react explosively with ethanol (qv).

Concentrated nitric acid favors the formation of nitrogen peroxide, whereas low strength favors the generation of nitric oxide. As a general rule for metals, those below hydrogen in the electrochemical series yield nitrogen peroxide as nitric oxide. Those above hydrogen react to produce nitrogen, ammonia, hydroxyl amine, or nitric oxide when treated with nitric acid. Nitric acid reacts with all metals except gold, iridium, platinum, rhodium, tantalum, titanium, and certain alloys.

Organic Reactions. Nitric acid is used extensively in industry to nitrate aliphatic and aromatic compounds. In many instances, nitration requires the use of sulfuric acid as a dehydrating agent or catalyst; the extent of nitration achieved depends on the concentration of nitric and sulfuric acids used. This is of industrial importance in the manufacture of nitrobenzene and dinitrotoluene, which are intermediates in the manufacture of polyurethanes.

Manufacture and Processing

Almost all commercial quantities of nitric acid are manufactured by the oxidation of ammonia with air to form nitrogen oxides that are absorbed in water to form nitric acid. Because nitric acid has a maximum boiling azeotrope at 69 wt %, the processes are usually categorized as either weak (subazeotropic) or direct strong (superazeotropic). Typically, weak processes make 50–65 wt % acid and direct strong processes make up to 99 wt % acid. Direct strong nitric (DSN) processes differ from weak acid plants in the additional processing steps required to achieve superazeotropic strengths of nitric acid. DNS plants have found little application in the United States, but several have been built in Europe. Strong acid may also be made indirectly from the weak acid by using extractive distillation with a dehydrating agent. Nitric acid concentration processes use a dehydrating agent such as sulfuric acid or magnesium nitrate to enhance the volatility of HNO_3 so that distillation methods can surpass the azeotropic concentration of nitric acid.

NO_x Abatement. Source performance standards for nitric acid plants in the United States were introduced by the U.S. EPA in 1971. These imposed a discharge limit of 1.5 kg of NO_x as equivalent nitrogen dioxide per 1000 kg of contained nitric acid, which corresponds to about 200–230 ppmv of nitrogen oxides in vented tail gas, whereas concentrations after absorption may contain as much as 2000–3000 ppmv of nitrogen oxides. Regulations and a review of abatement methods used in the EC are available. Various methods have been used to reduce tail gas NO_x concentrations to an acceptable level for discharge to the atmosphere. The most commonly employed methods are extended absorption, selective catalytic reduction (SCR), and nonselective catalytic reduction (NSCR).

Other Processes

Most other routes for making nitric acid involve the formation of nitrogen oxide directly from the air using various energy sources, such as electric arc furnaces and shock waves. Alternative approaches to nitric oxide formation include irradiation of air in a nuclear reactor and the oxidation of ammonia to nitric oxide in a fuel-cell generating energy. Both methods indicate some potential for commercial application, but require further study and development.

Materials of Construction

Weak Acid. Stainless steels (SS) have excellent corrosion resistance to weak nitric acid and are the primary materials of construction for a weak acid process. Low carbon steels a preferred because of their resistance to corrosion at weld points. However, higher grade materials of construction are required for high temperature areas around the gauze (ca 900°C) and places in which contact with hot liquid nitric acid is likely to be experienced (the cooler condenser and tail gas preheater).

Typical materials of construction around the gauze are high strength alloys made of iron–nickel–chromium and nickel–chromium. Alloys made of nickel–chromium–tungsten–molybdenum are also finding application in such service.

Zirconium has seen increasing use with nitric acid (in cooler condensers and tail gas preheaters).

Duplex stainless steels (ca 4% nickel, 23% chrome) have been identified as having potential application to nitric acid service. The higher strength and corrosion resistance of duplex steel offer potential cost advantages as a material of construction for absorption columns (see CORROSION AND CORROSION CONTROL).

Strong Nitric Acid. Materials of construction commonly used in the production of strong acid (98–99 wt %) are aluminum, tantalum, borosilicate glass, glass-lined steel, high silica cast iron, and high silica stainless steels.

Process Performance and Economics

Unit consumption of the key raw material, ammonia (qv), is typically in the range of 0.28–0.29 t/t HNO_3. Nitric acid plant designs are adjusted to meet the economic requirements associated with specific sites. There are many different equipment and process design variation that result in different process performance. Therefore, it is not practical to report typical performance figures for each type of nitric acid process. Production economics are a combination of operating costs (determined by process performance and local material costs) and fixed costs (related to capital investment).

Economic comparisons for the production of strong nitric acid using the various direct strong processes and weak acid plants in combination with a concentration (NAC) process indicate that the newer DSN technologies such as SABAR are competitive with the older, more established processes such as HOKO and extractive distillation.

Process Licensors. Some of the well-known nitric acid technology licensors are Espindesa, Spain, Grande Paroisse, France, Hercules, USA, Humphreys & Glasgow, Ltd., Rhône Poulenc, France, Sumitomo Chem. Co., Japan, Uhde GmbH, Germany, Weatherly, USA, Zimmer AG, Germany, (A subsidiary of Lurgi AG, formerly Davy McKee AG (Bamag).)

Production. The rapid post-war growth of nitric acid production in the United States came to an end in 1970, by which time production had reached 7 million t/yr. Existing commercial production capacity exceeds demand by ca 1–2 million t/yr; an additional capacity of 1.5 million t/yr exists in U.S. Army facilities.

Prices. Most nitric acid produced in the United States is for captive consumption for which merchant market pricing does not apply.

Health and Safety Factors

Nitric acid and the oxides of nitrogen found in its fumes are highly toxic and capable of causing severe injury and death. It is corrosive and can destroy human tissue. Nitric acid is regulated by OSHA, which lists it as a Process Safety Hazardous Chemical and Air Contaminant. Under SARAH, the EPA lists it as an Extremely Hazardous Substance and Toxic Chemical. Per OSHA, the 1991 permissible exposure limits for nitric acid are 2 ppm (5 mg/m^3) for an 8-h time-weighted average and 4 ppm (10 mg/m^3) for a 15-min short-term exposure. Exposure limits may vary according to local and national regulations.

First-aid practices for the treatment of exposure to nitric acid should be obtained from a current version of the Material Safety Data Sheet or other appropriate safety literature.

Uses

The largest use of nitric acid (ca 74–78% of total U.S. production) is for the manufacture of ammonium nitrate. Partly because of the increased popularity of urea, the use of ammonium nitrate as a fertilizer has declined. This has been offset to some extent by the growth in the use of ammonium nitrate for explosives and other chemical uses. Overall, the production of ammonium nitrate in the United States is expected to remain flat for several years.

STEPHEN I. CLARKE
WILLIAM J. MAZZAFRO
Air Products and Chemicals Inc.

T. H. Chilton, *Strong Water*, The M.I.T. Press, Cambridge, Mass., 1968.

C. Keleti, ed., *Nitric Acid and Fertilizer Nitrates,* vol. 4, Marcel Dekker, Inc., New York, 1985.

NITRIDES

At elevated temperatures and pressures, nitrogen combines with most elements to form nitrogen compounds. In the presence of metals and semimetals, it forms nitrides where nitrogen has a nominal valence of −3. Atomic nitrogen, which reacts much more readily with the elements than does molecular nitrogen, forms nitrides with elements that do not react with molecular nitrogen even at very high pressures. The binary compounds of nitrogen may be classified, according to their chemical and physical properties, into four groups: saltlike, metallic, nonmetallic or diamondlike, and volatile nitrides.

Properties

Saltlike Nitrides. The nitrides of the electropositive metals of Group 1 (IA), 2 (IIA), AND 3 (IIIB) form saltlike nitrides having predominantly heteropolar (ionic) bonding and are regarded as derivatives of ammonia. The composition of these nitrides is determined by the valency of the metal. The thermodynamic stability of the saltlike nitrides increases with increasing group number. The saltlike nitrides are generally electrical insulators or ionic conductors. The nitrides of the Group 3 (IIIB) metals are metallic conductors or at least semiconductors, and thus, represent a transition to the metallic nitrides. The saltlike nitrides are characterized by sensitivity to hydrolysis.

Metallic Nitrides. The nitrides of the transition metals of Groups 6 and 7 (IVB–VIIB) are generally termed metallic nitrides because of metallic conductivity, luster, and general metallic behavior. These compounds, characterized by a wide range of homogeneity, high hardness, high melting points, and good corrosion resistance, are grouped with the carbides (qv), borides and silicides as refractory hard metals. Metallic nitrides can be alloyed with other nitrides and carbides of the transition metals to give solid solutions.

Although there are several hundred binary nitrides, only a relative few ternary bimetallic nitrides are known. A group of ternaries of the composition $M_xM'_yN_z$, where M is an alkali, alkaline-earth, or a rare-earth metal and M is a transition or post-transition metal, have been synthesized.

Metallic nitrides are wetted and dissolved by many liquid metals and can be precipitated from metal baths.

Nonmetallic (Diamondlike) Nitrides. The nitrides of some elements of Groups 13 (IIIA) and 14 (IVA) eg, BN, Si_3N_4, AIN, GaN, and InN, are characterized by predominantly covalent bonding. These are stable chemically, have high degrees of hardness (eg, cubic BN) and high melting points, and are nonconductive or semiconductive. The structural elements of diamondlike nitrides are tetrahedral, M_4N, which are structurally related to diamond.

Volatile Nitrides. The nitrogen compounds of the nonmetallic elements are generally not very stable. Exceptions are $(SN)_x$, which is polymeric, chemically stable, and has semimetallic properties; and $(PNCl_2)_x$, which has attracted some scientific interest as inorganic rubber. None of the volatile nitrides has obtained any substantial industrial application except ammonia (hydrogen nitride) and nitrogen oxide (oxygen nitride).

Preparation

Nitriding Metals or Metal Hydrides. Metals or metal hydrides may be nitrided using nitrogen or ammonia. Pure metal powders or pure metal hydride powders yield nitride products that are nearly as pure as the precursors.

Metal Oxides. A process based on the reaction of metal oxides rather than more expensive metal powders and nitrogen or ammonia in the presence of carbon is economical and has possibilities for large-scale production. However, the products, which contain oxygen and carbon, are not very pure.

Metal Compounds. Many nitrides, eg, BN, AIN, TiN, ZrN, HfN, CrN, Re_2N, Fe_2N, Fe_4N, and Cu_3N, may be prepared by the reaction of the corresponding metal halide and ammonia. Nitrides may also be obtained by the reaction of ammonia and oxygen-containing compounds, ammonium-oxo complexes, or oxides and ferrous metal oxides. These nitrides, however, are not very pure and may contain residual oxygen and halogen.

Precipitation from the Gas Phase. The van Arkel gas decomposition process gives especially pure nitrides and nitride films, which under certain conditions may precipitate as single crystals. The nitrides include TiN, ZrN, HfN, VN, NbN, BN, and AIN.

Other Methods of Preparation. The nitrides, Si_3N_4, Ge_3N_4, Zn_3N_2, Cd_3N_2, and Ni_3N, may also be produced by thermal decomposition of the corresponding metal amide of imide. Rb_3N and Cs_3N are obtained by azide decomposition. AIN and Si_3N_4 can be produced by the carbothermal reduction of intercalation compounds, magadiite- and montmorillonite-polyacrylonitrile. Nitrides low in nitrogen can be synthesized from nitrides having a higher nitrogen content by decomposition in a vacuum or by reduction with hydrogen.

The formation of nitrides from gaseous halides, ammonia, and nitrogen (atomic and molecular) in a plasma processing torch is possible by means of a specific type of plasma processing called cathodic arc plasma deposition (CAPD). Plasma nitriding offers several advantages: It is nonpolluting and energy efficiency, provides flexible deposition conditions without sacrifice of quality, minimizes distortion, and is easily applicable to compound film deposition. Ion implantation directly inserts nitrogen into metal surfaces.

Nitride-Containing Layers. The hardening, ie, increase in nitrogen content, achieved by nitriding special alloy steels is technologically significant in the heat treatment of high quality parts, such as gears. Hardness (qv) properties are imparted by the resulting coatings of needle-shaped precipitates of the nitrides and carbonitrides of iron, aluminum, chromium, molybdenum, etc. The hardness of these coatings exceeds that of the precipitation-hardened parts by ca 30%.

In nitriding or carbonitriding of condensed materials, molten cyanides are used at ca 570°C. This method produces fairly thick coatings of nitrides or carbonitrides after ca 1 hr without the risk of distortion during surface hardening.

Wear-resistant layers can be deposited on the surface of nearly every kind of material (eg, steel, cast iron, and cemented carbides) by a chemical vapor deposition (CVD) process (see THIN FILMS, FILM FORMATION TECHNIQUES).

Manufacture and Processing

Nitride Coatings. Carbide tips coated with titanium nitride or titanium carbonitride are usually manufactured by a CVD process using $TiCl_4$, H_4, and N_2 in a hot-wall reactor.

Silicon Nitride. Silicon nitride is manufactured either as a powder as a precursor for the production of hot-pressed parts or as self-bonded, reaction-sintered, silicon nitride parts.

Health and Safety Factors

Toxicology. As a chemical group, toxicity of nitrides generally stems from the possible reactions with water to form toxic fumes (especially ammonia) rather than from the nitride. There are, of course, exceptions: Fine powder or dust of the nitrides of the transition metals can be pyrophoric; nitrides of the actinide metals are carcinogenic.

The diamondlike nitrides, especially as dust, can irritate the lungs or cause scratching of the eyes owing to mechanical means. Nitrides of the 11(IB) and 12(IIB) metals and especially the volatile nitrides have to be handled with extreme care because of their instability and high degree of toxicity.

Uses Nitrides are used for their high strength and hardness, in nuclear applications, solid electrolytes, refractories, abrasives, coatings and lubrication, catalysis, and electronic and optoelectronic applications.

ERIC J. MARKEL
M. E. LEAPHART II
University of South Carolina

R. Freer, ed., NATO ASI Series: *The Physics and Chemistry of Carbides, Nitrides, and Borides,* Vol. 185, Kluwer Academic Publishers, Boston, Mass., 1990.

A. Rabenau, *Solid State Ionics* **6,** 277 (1982).

L. E. Toth, *Transition Metal Carbides and Nitrides,* Academic Press, Inc., New York, 1971.

G. V. Samsonov, *Nitridij,* Naukova Dumka, Kiev, USSR, 1969.

H. Goldschmidt, *Interstitial Alloys,* Butterworths, London, 1967.

NITRILES

Nitriles, or organic cyanides, are organic compounds which contain the cyano (ie, –CN) group. Nitriles are often considered derivatives of carboxylic acids and are named according to the carboxylic acid which is produced upon hydrolysis of the nitrile. For example, cyanomethane (methyl cyanide) is named acetonitrile, because hydrolysis of its cyano group yields acetic acid. Nitriles which contain additional functional groups are typically named as cyano-substituted compounds, (eg, cyanoacetic acid). Nitriles which contain a hydroxy (–OH) group on the carbon atom that is bonded to the cyano moiety are known as cyanohydrins (qv). Aliphatic nitriles are named as derivatives of the longest carbon chain and the carbon of the nitrile is included.

General Preparations and Chemical Properties

While nitriles may be prepared by several methods, the reaction of alkyl halides with sodium cyanide to produce nitriles (eq. 1) is a general reaction with wide applicability:

$$RX + NaCN \rightarrow RCN + NaX \quad (1)$$

where X = Cl, Br, or I. If dimethyl sulfoxide is used as solvent, high yields of nitriles can be obtained with both primary and secondary alkyl chlorides (see SULFOXIDES).

Ammoxidation, a vapor-phase reaction of hydrocarbon with ammonia and oxygen (air) (eq. 2), can be used to produce hydrogen cyanide (HCN), acrylonitrile, acetonitrile (as a by-product of acrylonitrile manufacture), methacrylonitrile, benzonitrile, and toluinitriles from methane, propylene, butylene, toluene, and xylenes, respectively (see ACRYLONITRILE; METHACRYLIC ACID AND DERIVATIVES).

$$RCH_3 + NH_3 + O_2 \xrightarrow{\text{catalyst}} RCN + H_2O \quad (2)$$

Addition of HCN to unsaturated compounds is often the easiest and most economical method of making organonitriles. However, the addition of HCN to unactivated olefins and the regioselective addition to dienes is best accomplished with a transition metal catalyst.

Chemistry and Uses of Nitriles

As a class of compounds, nitriles have broad commercial utility that includes their use as solvents, feedstocks, pharmaceuticals, catalysts, and pesticides. The versatile reactivity of organonitriles arises both from the reactivity of the C≡N bond, and from the ability of the cyano substituent to activate adjacent bonds, especially C–H bonds. Nitriles can be used to prepare amines, amides, amidines, carboxylic acids and esters, aldehydes, ketones, large-ring cyclic ketones, imines, heterocycles, orthoesters, and other compounds. Some of the more common transformations involve hydrolysis or alcoholysis to produce amides, acids and esters, and hydrogenation to produce amines, which are intermediates for the production of polyurethanes and polyamides.

Acrylonitrile is an important monomer both for plastics and synthetic fibers. Acetonitrile, a by-product of acrylonitrile manufacture, is commercially important for solvent extraction, reaction media, and as an intermediate in the preparation of pharmaceuticals (qv) and other organic chemicals (see EXTRACTION, LIQUID–LIQUID EXTRACTION). Propionitrile, a by-product of the electrodimerization of acrylonitrile to adiponitrile, is used as a chemical intermediate. Hydrogenation of organonitriles to amines provides important intermediates both for polyurethanes (by way of isocyanates) and polyamides (nylons); adiponitrile is used almost exclusively by the manufacturers in the production of 1,6-diaminohexane (hexamethylenediamine), an intermediate for nylon 6,6. Other nitriles that are produced in thousands of metric tons per year include acetone cyanohydrin, 2-amino-2-methylpropionitrile, and fatty acid nitriles. Acetone cyanohydrin is an intermediate for the preparation of methyl methacrylate and acrylic resins, (eg, lucite and plexiglas) and for 5,5-dimethylhydantoin, which is used to make commercial water treatment chemicals. 2-Amino-2-methylpropionitrile is an intermediate for the preparation of azobis(isobutyronitrile), which is a widely used polymerization initiator, (eg, Vazo 64) and in the production of some agrichemicals. Other aminonitriles are unisolated intermediates in the production of chelants such as ethylenediaminetetraacetate (EDTA) and nitrilotriacetate (NTA). The fatty acid nitriles are intermediates in the production of a large variety of commercial amines and amides.

General Health and Safety Factors

As a class of compounds, the two main toxicity concerns for nitriles are acute lethality and osteolathyrsm. Nitriles vary broadly in their ability to cause acute lethality and subtle differences in structure can greatly affect toxic potency. The biochemical basis of their acute toxicity is related to their metabolism in the body.

The propensity of nitriles to release cyanide subsequent to metabolism is the basis of their acute toxicity. Cyanohydrins are acutely toxic because they are unstable and release cyanide quickly. Persons handling nitriles should take precautions to prevent inhalation of fumes or skin contact.

Acetonitrile

Acetonitrile (ethanenitrile), CH_3CN, is a colorless liquid with a sweet, ethereal odor. It is completely miscible with water and its high dielectric strength and dipole moment make it an excellent solvent for both inorganic and organic compounds including polymers. Many gases also are highly soluble in acetonitrile. It forms low boiling azeotropes with many organics and high boiling azeotropes with BF_3, $SiCl_4$, and $(CH_3)_4Pb$.

Although acetonitrile is one of the more stable nitriles, it undergoes typical nitrile reactions and is used to produce many types of nitrogen-containing compounds.

Most, if not all, of the acetonitrile produced commercially in the United States recently was isolated as a by-product from the manufacture of acrylonitrile by propylene ammoxidation. The acetonitrile is recovered as the water azeotrope, dried, and purified by distillation.

Uses. Because of its good solvency and relatively low boiling point, acetonitrile is used widely as a recoverable reaction medium, particularly for the preparation of pharmaceuticals. Its largest use is for the separation of butadiene from C_4 hydrocarbons by extractive distillation (see DISTILLATION, AZEOTROPIC AND EXTRACTIVE).

Acetonitrile also is used as a catalyst and as an ingredient in transition-metal complex catalysts. There are many uses for it in the photographic industry and for the extraction and refining of copper. It also is used as a reagent for the preparation of a wide variety of compounds.

Adiponitrile

Adiponitrile (hexanedinitrile, dicyanobutane, ADN), $NC(CH_2)_4CN$, is manufactured mainly for use as an intermediate for hexamethylenediamine (1,6-diaminohexane), which is a principal ingredient for nylon-6,6. BASF has announced the development of a process to make caprolactam from adiponitrile. Caprolactam is used to produce nylon-6.

Pure adiponitrile is a colorless liquid and has no distinctive odor. It is soluble in methanol, ethanol, chloroalkanes, and aromatics but has low solubility in carbon disulfide, ethyl ether, and aliphatic hydrocarbons. At 20°C, the solubility of adiponitrile in water is ca 8 wt %; the solubility increases to 35 wt % at 100°C.

Adiponitrile undergoes the typical nitrile reactions, eg, hydrolysis to adipamide and adipic acid and alcoholysis to substituted amides and esters.

Adiponitrile is made commercially by several different processes utilizing different feedstocks. The reaction of adipic acid with ammonia in either liquid or vapor phase produces adipamide as an intermediate, which is subsequently dehydrated to adiponitrile. The most widely used catalysts are based on phosphorus-containing compounds. Vapor-phase processes involve the use of fixed catalyst beds; whereas, in liquid–gas processes, the catalyst is added to the feed. DuPont currently practices a butadiene-to-adiponitrile route based on direct addition of HCN to butadiene.

Uses. The principal use of adiponitrile is for hydrogenation to hexamethylene diamine leading to nylon-6,6. Adipoquanamine, prepared by the reaction of adiponitrile with dicyandiamide (cyanoguanidine), has typical liquid nitrile properties that suggest its use as an extractant for aromatic hydrocarbons.

α-Aminonitriles

α-Aminonitriles are compounds containing both cyano and amine substituents attached to the same carbon atom. They are versatile synthetic intermediates that are used to make amino acids, agrichemicals, chelants, radical initiators, and water-treatment chemicals. In some cases, aminonitriles produced as intermediates are not isolated, but immediately further reacted, for example by hydrolysis, as is the case in producing ethylenediaminetetraacetate (EDTA) or nitrilotriacetate (NTA). Isolated and commercially available aminonitriles include 2-amino-2-methylpropanenitrile (aminoisobutyronitrile, AN-64), 2-amino-2-methylbutanenitrile (AN-67), 2-amino-2,4-dimethylpentanenitrile (AN-52), and 1-aminocyclohexane carbonitrile (AN-88). The designations in parentheses arise from their identity as intermediates in the production of azo radical initiators.

In 1990, DuPont began practicing a one-step process in which a ketone is treated simultaneously with both HCN and ammonia at 40–60°C. This process (Fig. 1) is both faster and more selective than previous two-step processes.

Physical Properties. α-Aminonitriles are stable at modest temperatures (<70°C) in the absence of water; in the presence of water, they can degrade to their original constituents, ie, ketone (aldehyde), ammonia and hydrogen cyanide if insufficient ammonia is present. The aminonitriles based on ketones are clear colorless liquids, but sometimes appear yellow to brown depending on the synthetic procedure and the amount of decomposition. They are soluble in polar organic solvents and in aromatic solvents.

Uses. α-Aminonitriles may be hydrolyzed to amino acids, such as is done in producing ethylenediaminetetracetate (EDTA) or nitrilotriacetate (NTA). In these cases, formaldehyde is utilized in place of a ketone in the synthesis. The principal use of the ketone-based aminonitriles is in the production of azobisnitrile radical initiators.

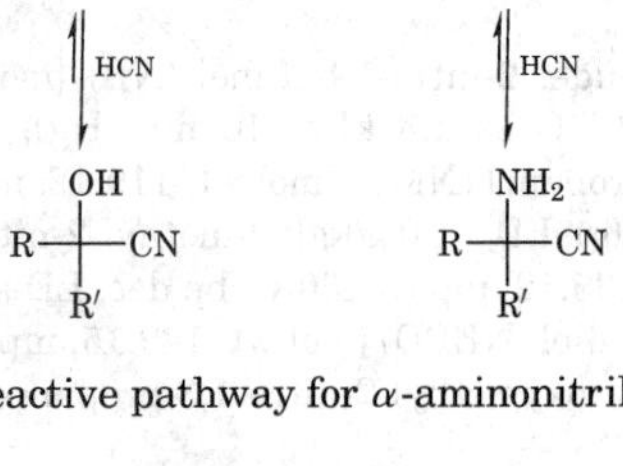

Figure 1. Reactive pathway for α-aminonitriles synthesis.

Azobisnitriles

Azobisnitriles are efficient sources of free radicals for vinyl polymerizations and chain reactions, eg, chlorinations (see INITIATORS). These compounds decompose in a variety of solvents at nearly first-order rates to give free radicals with no evidence of induced chain decomposition. They can be used in bulk, solution, and suspension polymerizations; and because no oxygenated residues are produced, they are suitable for use in pigmented or dyed systems that may be susceptible to oxidative degradation.

The structures of several members of this class of compounds are shown below. They are crystalline solids that are produced by hypochlorite oxidation of α-aminonitriles.

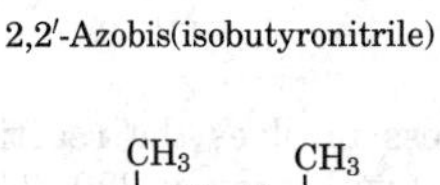

2,2′-Azobis(isobutyronitrile)

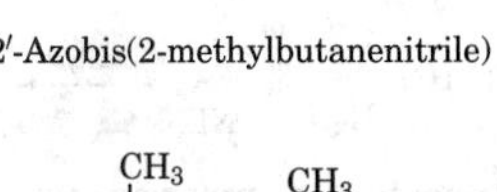

2,2′-Azobis(2-methylbutanenitrile)

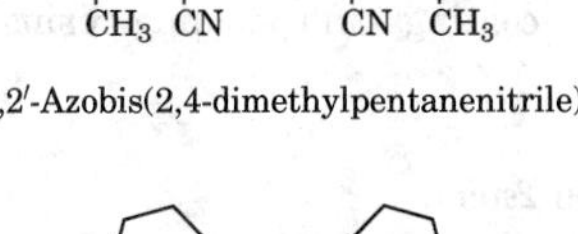

2,2′-Azobis(2,4-dimethylpentanenitrile)

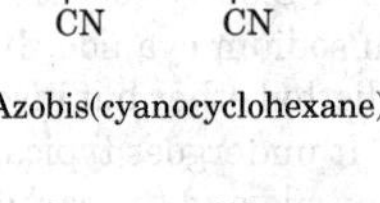

1,1′-Azobis(cyanocyclohexane)

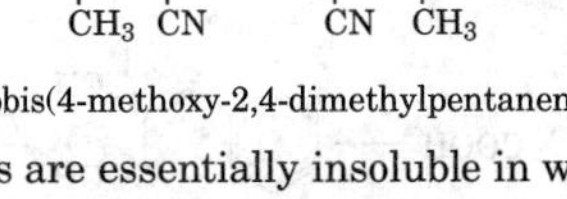

2,2′-Azobis(4-methoxy-2,4-dimethylpentanenitrile)

These compounds are essentially insoluble in water, sparingly soluble in aliphatic hydrocarbons, and soluble in functional compounds and aromatic hydrocarbons.

In solution, the azobisnitriles decompose on heating to form two free radicals with the liberation of nitrogen (eq. 1):

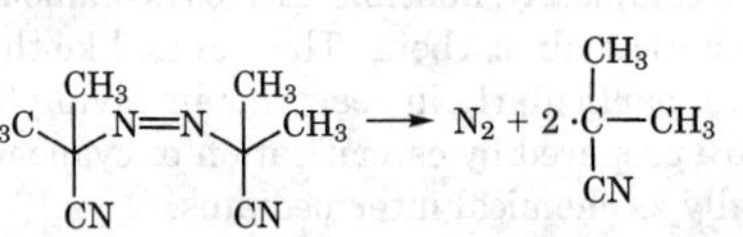 (1)

Uses. The azobisnitriles have been used for bulk, solution, emulsion, and suspension polymerization of all of the common vinyl monomers, including ethylene, styrene vinyl chloride, vinyl acetate, acylonitrile, and methyl methacrylate. The polymerizations of unsaturated polyesters and copolymerizations of vinyl compounds also have been initiated by these compounds.

Benzonitrile

Benzonitrile, C_6H_5CN, is a colorless liquid with a characteristic almondlike odor. It is miscible with acetone, benzene, chloroform, ethyl acetate, ethylene chloride, and other common organic solvents but is immiscible with water at ambient temperatures and soluble to ca 1 wt% at 100°C. It distills at atmospheric pressure without decomposition, but slowly discolors in the presence of light.

Like acetonitrile, benzonitrile is a powerful solvent for many inorganic and organic materials including some polymers. It can be converted to a large number and variety of derivatives by simple syntheses; eg, by hydrolysis, it can be converted to either benzoic acid or benzamide. The most important reaction is with dicyandiamide to produce 2,4-diamino-6-phenyl-1,3,5-triazine (benzoguanamine):

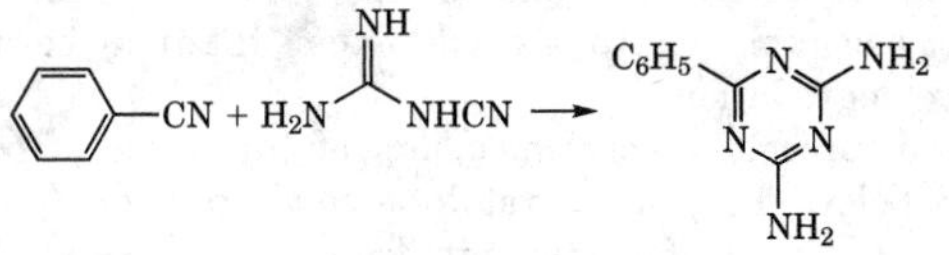

Benzonitrile can be produced in high yield by the vapor-phase catalytic ammoxidation of toluene:

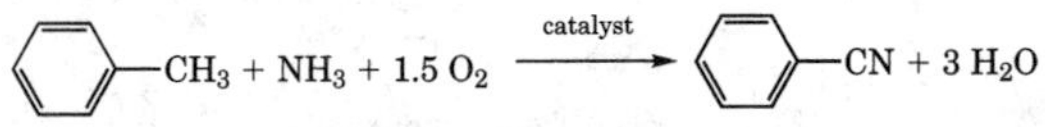

A more recent process involves the reaction of benzoic acid (or substituted benzoic acid) with urea at 220–240°C in the presence of a metallic catalyst.

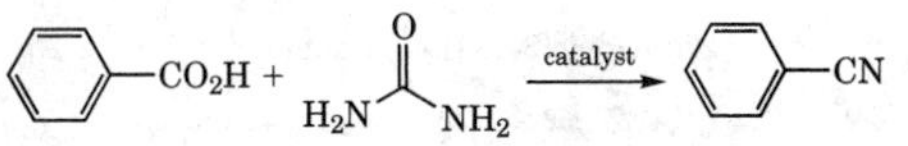

Uses. The most important commercial use for benzonitrile is the synthesis of benzoguanamine, which is a derivative of melamine and is used in protective coatings and molding resins (see AMINO RESINS; CYANAMIDES).

Cyanoacetic Acid and Esters

Cyanoacetic acid, $NCCH_2COOH$, is a strong organic acid with a dissociation constant at 25°C of 3.36×10^3. It is prepared by the reaction of chloroacetic acid with sodium cyanide. It is hygroscopic and highly soluble in alcohols and diethyl ether but insoluble in both aromatic and aliphatic hydrocarbons. It undergoes typical nitrile and acid reactions but the presence of the nitrile and the carboxylic acid on the same carbon cause the hydrogens on C-2 to be readily replaced. The resulting malonic acid derivative decarboxylates to a substituted acrylonitrile:

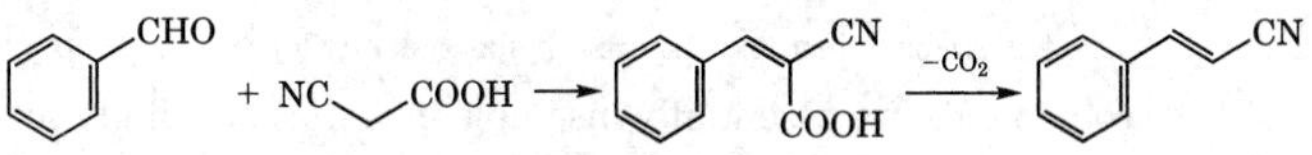

cinnamonitrile

The methyl and ethyl esters of cyanoacetic acid are slightly soluble in water but are completely miscible in most common organic solvents including aromatic hydrocarbons. The esters, like the parent acid, are highly reactive, particularly in reactions involving the central carbon atom. They are prepared by esterification of cyanoacetic acid and are used principally as chemical intermediates.

Uses. Although cyanoacetic acid can be used in applications requiring strong organic acids, its principal use is in the preparation of malonic esters and other reagents used in the manufacture of pharmaceuticals (see ALKALOIDS; HYPNOTICS; VITAMINS).

Isophthalonitrile

Isophthalonitrile (1,3-dicyanobenzene, IPN), is a white solid which melts at 161°C and sublimes at 265°C. It is slightly soluble in water but readily dissolves in dimethylformamide, *N*-methylpyrrolidinone and hot aromatic solvents. IPN undergoes the reactions expected of an aromatic nitrile. It is prepared by vapor-phase ammoxidation of *meta*-xylene. Its principal use is as an intermediate to amines. As a reagent, IPN can be used to convert aromatic acids to nitriles in near quantitative yields.

2-Methylglutaronitrile

Methylglutaronitrile (2,3-dicyanobutane) MGN, is a by-product of DuPont's adiponitrile process.

Uses. Methylglutaronitrile is readily hydrogenated to give 2-methyl-1,5-pentanediamine (DYTEK A, MPMD), used as a co-monomer in polyamide fibers and resins, as a curing agent for epoxy coatings, and as its isocyanate in specialty urethanes. A co-product of the DYTEK A process is 3-methylpiperidine, which can be used to produce vulcanization accelerators for rubber curing.

Pentenenitriles

Pentenenitriles are produced as intermediates and by-products in DuPont's adiponitrile process. 3-Pentenenitrile is the principal product isolated from the isomerization of 2-methyl-3- butenenitrile.

Uses. 3-Pentenenitrile (3PN) is used entirely by the manufacturers to make adiponitrile. *cis*-2-Pentenenitrile (2PN) can be cyclized catalytically at high temperature to produce pyridine, a solvent and agricultural chemical intermediate. 2PN is also used in the manufacture of pentachloropyridine, an intermediate in the insecticide Dursban, and 1,3-pentadiamine, which is used as a curing agent for epoxy coatings and as a chain modifier in polyurethanes.

Fatty Acid Nitriles

Fatty acid nitriles are produced as intermediates for a large variety of amines and amides (see also CARBOXYLIC ACIDS). Fatty acid nitriles are produced from the corresponding acids by a catalytic reaction with ammonia in the liquid phase. They have little use other than as intermediates.

DISCLAIMER

This article has been reviewed by the Office of Pollution Prevention and Toxics, U.S. Environmental Protection Agency, and approved for publication. Approval does not signify that the contents necessarily reflect the views and policies of the Agency, nor does mention of commercial products or synthesis constitute endorsement or recommendation for use.

RONALD J. MCKINNEY
E. I. du Pont de Nemours & Co., Inc.
STEPHEN C. DEVITO
U.S. Environmental Protection Agency

D. H. R. Barton and W. D. Ollis, *Comprehensive Organic Chemistry,* Vol. 2, Pregamon Press, Oxford, U.K., 1979, pp. 528–562.

D. T. Moury, *Chem. Rev.* **42,** 192 (1948).

U.S. Pat. 2,915,455 (Nov. 10, 1959), R. A. Smiley (to DuPont).

U.S. Pat. 2,481,826 (Sept. 13, 1949), J. N. Cosby (to Allied Chemical).

NITRO ALCOHOLS

A nitro alcohol is formed when an aliphatic nitro compound with a hydrogen atom on the nitro-bearing carbon atom reacts with an aldehyde in the presence of a base. In addition to the mononitro compounds, monohydric and dihydric dinitro alcohols have been prepared, but are not available commercially. The formation, properties, and reactions of nitro alcohols have been reviewed.

Physical Properties

Nitro alcohols include: 2-nitro-1-butanol (NB) [mol.wt. 119.12, mp -47 to -48°C, bp 105°C at 1.3 kPa (10 mm Hg), LD_{50} 1.2 g/kg]; 2-methyl-2-nitro-1-propanol (NMP) [mol.wt. 119.12, mp 90°C, bp 94°C at 1.95 kPa (15 mm Hg), LD_{50} 1.0 g/kg]; 2-methyl-2-nitro-1,3-propanediol (NMPD) [mol.wt. 135.12, mp ca 160°C, bp dec, LD_{50} 4.0 g/kg]; 2-ethyl-2-nitro-1,3-propanediol (NEPD) [mol.wt. 149.15, mp 56°C, bp dec, LD_{50} 2.8 g/kg]; and 2-hydroxymethyl-2-nitro-1,3-propanediol (TRIS NITRO) [mol.wt. 151.12, mp 175–176°C, bp dec, LD_{50} 1.9 g/kg]. Except for nitrobutanol, these nitro alcohols are white crystalline solids when pure. They are thermally unstable above 100°C.

The nitro alcohols are generally soluble in water and in oxygenated solvents, eg, alcohols. The monohydric nitro alcohols are soluble in aromatic hydrocarbons; the diols are only moderately stable even at 50°C; at 50°C, the trial is insoluble.

Chemical Properties

The nitro alcohols can be reduced to the corresponding alkanolamines (qv). Commercially, reduction is accomplished by hydrogenation of the nitro alcohol in methanol in the presence of Raney nickel. Production of alkanolamines constitutes the largest single use of nitro alcohols.

Nitro alcohols form salts upon mild treatment with alkalies. Acidification causes separation of the nitro group as N_2O from the parent compound, and results in the formation of carbonyl alcohols, ie, hydroxy aldehydes, from primary alcohols and ketols from secondary nitro alcohols.

Nitro alcohols react with amines to form nitro amines. The products of reactions between dihydric nitro alcohols and amines are nitrodiamines, many of which are good fungicides (qv).

Manufacture and Processing

The nitro alcohols available in commercial quantities are manufactured by the condensation of nitroparaffins with formaldehyde. These condensations are equilibrium reactions, and potential exists for the formation of polymeric materials.

The purification of liquid nitro alcohols by distillation should be avoided because violent decompositions and detonation have occurred when distillation was attempted. The only commercially produced liquid nitro alcohol, 2-nitro-1-butanol, is not distilled because of the danger of decomposition. Instead, it is isolated as a residue after the low boiling impurities have been removed by vacuum treatment at a relatively low temperature.

Economic Aspects

Nitro alcohols are manufactured in commercial quantities; however, three of the five of them are used only for the production of the corresponding amino alcohols. 2-Methyl-2-nitro-1-propanol (NMP) is available as the crystalline solid or as a mixture with silicon dioxide. 2-Hydroxymethyl-2-nitro-1,3-propanediol is available as the solid, a 50% solution in water, and a 25% solution in water.

Health, Safety and Environmental Factors

Because of their low volatility, the nitro alcohols present no vapor inhalation hazard. They are nonirritating to the skin and, except for 2-nitro-1-butanol, in the eye of a rabbit.

Because it is the nitro alcohol with greatest potential for human exposure, additional testing of 2-hydroxymethyl-2-nitro-1,3-propanediol has been conducted. It was not found to be mutagenic in *in vitro* tests and no teratogenic effects were noted in either rats or rabbits. It was also found to have low potential for harm in the environment.

Uses

The nitro alcohols are useful as intermediates for chemical synthesis. In particular, they are used to introduce a nitro functionality and, by reduction of the resultant intermediate, an amino functionality. They are used in the manufacture of polymers and stabilizers.

ALLEN F. BOLLMEIER, JR.
ANGUS Chemical Company

H. B. Hass and E. F. Riley, *Chem. Rev.* **32,** 373 (1943).

B. M. Vanderbilt and H. B. Hass, *Ind. Eng. Chem.* **32,** 34 (1940).

NP Series, Technical Data Sheet No. 15, ANGUS Chemical Co., Jan. 1989.

"Tris(hydroxymethyl)nitromethane," *Reregistration Eligibility Decision,* U.S. EPA, Washington, D.C., 1993.

NITROBENZENE AND NITROTOLUENES

Nitrobenzene

Nitrobenzene (oil of mirbane), $C_6H_5NO_2$, is a pale yellow liquid with an odor that resembles bitter almonds. Depending on the purity, its color varies from pale yellow to yellowish brown.

Physical Properties. Nitrobenzene is readily soluble in most organic solvents and is completely miscible with diethyl ether and benzene. Nitrobenzene is only slightly soluble in water with a solubility of 0.19 parts per 100 parts of water at 20°C. Nitrobenzene is a good organic solvent. Some of the physical properties of nitrobenzene are summarized in Table 1.

Chemical Properties. Nitrobenzene reactions involve substitution on the aromatic ring and reactions involving the nitro group. Nitrobenzene can undergo halogenation, sulfonation, and nitration. The reduction of the nitro group to yield aniline is the most commercially important reaction of nitrobenzene. Other reduction products include hydrazobenzene, azobenzene, *N*-phenylhydroxylamine, azoxybenzene, azobenzene, and sodium phenylsulfamate.

Manufacturing and Processing. Nitrobenzene is manufactured commercially by the direct nitration of benzene using a mixture of nitric and sulfuric acids, which commonly is referred to as mixed acid or nitrating acid. It can be produced by either a batch or continuous process.

Due to increasingly strict environmental regulations, effort has been put into reducing the amount of contaminants in the waste stream of the nitration process. For instance, residual nitrobenzene can be removed from the wastewater with a multistage extraction process in which the organic and aqueous phases are run countercurrent to each other. This method can extract up to 99.44% of nitrobenzene from the wastewater.

The need for neutralization of the organic phase with alkali can be reduced by extracting the acidic contaminants using molten salts and the acidic contaminants can also be removed by employing a system that utilizes extracting, precipitation, distillation, and other treatments for rendering the waste stream acceptable for current disposal standards.

Environmental aspects, as well as the requirement of efficient mixing in the mixed acid process, have led to the development of single-phase nitrations.

The use of vapor-phase nitrations has been one of the most active areas in aromatic nitration chemistry since the 1970s. Although several approaches have been reported, most of the patents issued have one technique in common: the use of a solid nitration catalyst through which the nitric acid-benzene mixture flows in a continuous process. Table 2 summarizes some of the reported results.

Economic Aspects. The two main areas affecting the economic aspects of nitrobenzene production are process related costs, including raw material costs, energy requirements, waste treatment, etc, and

Table 1. Physical Properties of Nitrobenzene

Property	Value
mp, °C	5.85
bp, °C, at 101 kPa[a]	210.9
density, g/cm^3, d_4^{25}	1.199
refractive index, $n^{20}{}_D$	1.55296
viscosity at 15°C, mPa·s(= cP)	2.17
dielectric constant at 25°C	34.82
autoignition temperature, °C	482

[a] To convert kPa to mm Hg, multiply by 7.5.

Table 2. Vapor-Phase Nitration of Benzene

Catalyst	Conversion, %
alumina—silica—metal oxide	99
acidic sheet clay, acidic composite oxides	91
solid supported sulfuric catalysts	96
silica–alumina zeolites	95
molecular sieves, $\geq 5 \times 10^{-4}$ μm	92
nitrate salts (KNO_3, $NaNO_3$, $LiNO_3$)	40

the U.S. and world demand for products made from nitrobenzene. Raw material costs typically make up at least 85% of the production costs for nitrobenzene.

Health and Safety Factors. Nitrobenzene is a very toxic substance; the maximum allowable concentration for nitrobenzene is 1 ppm or 5 mg/m^3. It is readily absorbed by contact with skin and by inhalation of vapor. The primary effect of nitrobenzene is the conversion of hemoglobin to methemoglobin; thus the conversion eliminates hemoglobin from the oxygen-transport cycle. Exposure to nitrobenzene may irritate the skin and eyes. Nitrobenzene also affects the central nervous system. In areas of high vapor concentrations (>1 ppm), full face masks with organic-vapor canisters or air-supplied respirators should be used. Clean work clothing should be worn daily, and showering after each shift should be mandatory. Nitrobenzene is classified as a moderate hazard when exposed to heat or flame and as a Class-B poisonous liquid by the ICC.

Uses. Approximately 95–98% of nitrobenzene is converted to aniline; its demand fluctuates with that for aniline. Nitrobenzene is also used to produce products such as *para*-aminophenol (PAP), an intermediate for acetaminophen, dyes, pigments, and solvents.

Derivatives

Mononitrochlorobenzenes. The physical properties of the ortho, meta, and para isomers of nitrochlorobenzene are summarized in Table 3.

The mononitrochlorobenzenes are moderate fire hazards when exposed to heat or flame. They are classified by the ICC as Class-B poisons. The same handling precautions should be used for these compounds as are used for nitrobenzene.

o-Nitrochlorobenzene is used in the synthesis of azo dye intermediates. It is also used in corrosion inhibitors, pigments, and agriculture chemicals. *p*-Nitrochlorobenzene is principally used in the production of intermediates for azo and sulfur dyes. Other nitrochlorobenzenes include 2,4-dinitrochlorobenzene, 3,4-dichloronitrobenzene, and 2,5-dichloronitrobenzene.

Nitrotoluenes

Mononitrotoluenes. The mononitration of toluene results in the formation of a mixture of the ortho, meta, and para isomers of nitrotoluene.

Table 3. Physical Properties of Mononitrochlorobenzenes

Property	*o*-Nitrochlorobenzene	*m*-Nitrochlorobenzene	*p*-Nitrochlorobenzene
melting point, °C	32.5	46 (stable) 24 (labile)	83
boiling point, $°C_{kPa}$[a]	246_{100} $119_{1.1}$	236_{101}	242_{101} $113_{1.1}$
density,[b] g/mL	1.368	1.534	1.520
flash point, closed cup, °C	123	103	110

[a] To convert kPa to mm Hg, multiply by 7.5.
[b] d^{22}_{4}.

Properties. *o*-Nitrotoluene is a clear yellow liquid. The solid is dimorphous and the melting points of the α- and β-forms are −9.55 and −3.85 °C, respectively. *o*-Nitrotoluene is infinitely soluble in benzene, diethyl ether, and ethanol. It is soluble in most organic solvents and only slightly soluble in water (0.065 g in 100 g of water at 30°C).

m-Nitrotoluene (mp 16.1°C, bp 231.9°C at 101 kPa) is a clear yellow liquid that freezes at 16.1°C. It is soluble in most organic solvents, such as ethanol, benzene, and diethyl ether, and is only sparingly soluble in water, 0.05 g/100 g of water at 30°C.

p-Nitrotoluene (mp 53.5°C, bp 238.5°C at 101 kPa) crystallizes in colorless rhombic crystals. It is only slightly soluble in water, 0.044 g/100 g of water at 30°C; moderately soluble in methanol and ethanol; and readily soluble in acetone, diethyl ether, and benzene.

Manufacture and Processing. Mononitrotoluenes are produced by the nitration of toluene in a manner similar to that described for nitrobenzene. They can be produced by either a batch of continuous process. The separation of the isomers is carried out by a combination of fractional distillation and crystallization.

Effort has focused on increasing the amount of the para isomer formed in the mononitration of toluene, because it is generally in the greatest demand of the three isomers.

If pure isomers are required, the ortho and meta compounds can be prepared by indirect methods. The mononitrotoluenes are manufactured by Du Pont and First Chemical Corp.

Health and Safety Factors. The toxic effects of the mononitrotoluenes are similar to but less pronounced than those described for nitrobenzene. The maximum allowable concentration for the mononitrotoluenes is 2 ppm (11 mg/m^3). The mononitrotoluenes represent moderate fire hazards when exposed to heat or flame. The same precautions used in handling nitrobenzene should be used for these compounds.

Uses. *o*-Nitrotoluene is used in the synthesis of intermediates for azo dyes, sulfur dyes, rubber chemicals, and agriculture chemicals.

Dinitrotoluenes. Dinitration of toluene results in the formation of a number of isomeric products, and with a typical sulfuric–nitric acid nitrating mixture the following mixture of isomers is obtained: 75 wt % 2,4-dinitrotoluene, 19 wt % 2,6,-dinitrotoluene, 2.5 wt % 3,4-dinitrotoluene, and 0.5 wt % 2,5-dinitrotoluene. The dinitrotoluenes are a moderate fire hazard and explosion hazard when exposed to heat or flame. The maximum allowable concentration in air is 1.5 mg/m^3 (0.2 ppm). Dinitrotoluenes are used as intermediates for the production of toluene diisocyanate and dyestuffs. They are also used as explosives.

RICK L. ADKINS
Bayer Corporation

N. R. Sax, *Dangerous Properties of Industrial Materials,* 8th ed., Van Nostrand Reinhold, New York, 1992.

L. F. Albright and C. Hanson, eds., *Industrial and Laboratory Nitrations,* American Chemical Society, Washington, D.C., 1976.

Y. A. Gawargious, *The Determination of Nitro and Related Functions,* Academic Press, Inc., New York, 1973.

NITROFURANS. See ANTIBACTERIAL AGENTS, SYNTHETIC.

NITROGEN

Nitrogen, atomic number 7, is a nonmetallic element situated between carbon and oxygen in the Periodic Table. It is the most abundant element accessible to human beings which exists uncombined with any other elements. Most of the nitrogen in the atmosphere occurs as a diatomic gas N_2, sometimes referred to as dinitrogen, and comprises 78.03% by volume and 75.45% by weight of the earth's atmosphere. Industry annually isolates millions of tons of nitrogen from air. Nitrogen-containing compounds are essential to all life.

In the 1990s, five of the 15 largest volume industrial chemicals produced in the United States contain nitrogen: ammonia, nitrogen, (gaseous and liquified), ammonium nitrate, nitric acid, and urea.

Physical Properties

Nitrogen is the lightest of the Group 15 elements and has an atomic weight of 14.008. There are two naturally occurring stable isotopes: ^{14}N (relative atomic natural abundance 0.99634%) and ^{15}N (relative atomic mass 15.000, natural abundance 0.366%). Both of these isotopes have a nuclear spin and are used in nmr experiments. Nitrogen can have oxidation states ranging from +5 to −3. Unlike the heavier elements of Group 15, nitrogen readily forms multiple bonds with itself and other atoms.

Molecular nitrogen, N_2, is a colorless, odorless, diamagnetic, noncombustible gas at standard pressure (101.3 kPa) and temperature (0°C). The gas condenses to a colorless liquid at −195.8°C at atmospheric pressure. Depending on the temperature, solid molecular nitrogen exists in one of two forms at atmospheric pressure, α and β, both of which are white. Other properties of molecular nitrogen is shown in Table 1. Density is represented by *Q*.

Chemical Properties

The chemistry of molecular nitrogen is marked by its relative inertness. The electronic configuration of the N_2 molecule in terms of molecular orbital theory is $1\varsigma^2{}_g 1\varsigma^2{}_u 2\varsigma^2{}_g 2\varsigma^2{}_u 1\pi^4{}_u 3\varsigma^2{}_g 1\pi^0{}_g$.

Despite the stability of molecular nitrogen, the chemistry of nitrogen is extremely important. There is essentially a limitless supply of nitrogen in the atmosphere. However, most biological species can only use nitrogen which has chemically combined with some other element of compound, so-called fixed nitrogen, and the abundance of this type of nitrogen is relatively limited. There is much research interest in elucidating inexpensive methods of converting molecular nitrogen to some form of fixed nitrogen (see NITROGEN FIXATION).

In most solvents, N_2 has a relatively low solubility. The solubility in water at 0°C and 101.3 kPa (1 atm) of N_2 is only 23.5 ppt (by volume). Nitrogen is slightly soluble in iron and steel through the reaction N_2 (gas) $\rightleftarrows$ 2 N (dissolved), where N (dissolved) represents atomic nitrogen dissolved in iron. Nitrogen has a significant alloying effect in steels. The presence of nitrogen can strengthen low carbon steels at the expense of increased strain aging properties.

Because of the inherent stability of the N_2 molecule, high temperatures are often used to coax its reactivity. Arguably, the most important industrial process which utilizes N_2 as a constituent is ammonia formation at high pressures from nitrogen and hydrogen in the presence of heat and a catalyst, known as the Haber-Bosch process.

Manufacture and Processing

Atmospheric air is the feedstock for all commercial nitrogen production processes. Nitrogen is separated from air commercially by cryogenic distillation, pressure swing adsorption, membrane permeation, or hydrocarbon combustion processes. Cryogenic distillation is the most cost-effective technology for production of large quantities of relatively pure nitrogen and is the most commonly used. Pressure swing adsorption and membrane permeation are the most economical processes for production of lower purity nitrogen in low to moderate volume ranges [25–500 m^3/h (1000–20,000 SCFH)]. All statements of volume (m^3) are at normal conditions (t = 25°C, pressure = 101.3 kPa). Both are rapidly growing technologies. Industry estimates indicate that noncryogenic separation will eventually account for greater than 30% of all commercial nitrogen production. Combustion-based processes are in decline in most applications due to displacement by noncryogenic processes but are still widely used in heat treatment where residual contaminants play an active process role. The choice of the most economic technology is principally driven by required nitrogen purity and flow rate. Liquid nitrogen is produced exclusively from cryogenic processes.

Table 1. Physical Properties of Molecular Nitrogen

Property	Value
molecular weight	28.0134
boiling point,[a] K	77.35
heat of vaporization, kJ/kg[b]	199
heat of fusion, kJ/kg[b]	25.8
Q,[c] g/L	1.2505
specific heat capacity,[d] J/(g·K)	1.039
dynamic viscosity, mPa·s(= cP)[d,c]	15.9 × 10^{-3}
thermal conductivity, mW/(m·K)	23.86

[a] At 101.3 kPa = 1 atm. [b] To convert kJ to kcal, divide by 4.184. [c] To convert MPa to psi, multiply by 145. [d] At 273.15 K and 101.3 kPa = 1 atm.

Shipment

Nitrogenic is shipped and stored in gaseous form in steel cylinders under high pressure and in liquid form in vacuum-insulated containers. Because of the weight of high pressure storage vessels and the volatile nature of liquid nitrogen, most nitrogen is produced and consumed locally within a radius of less than 250 km. An exception is the large-scale distribution of nitrogen at moderate pressures [4000 kPa (600 psig)] in extensive pipeline networks serving the petrochemical and petroleum refining industries along with U.S. Gulf Coast of Texas and Louisiana, and serving the steel and chemical industries in northern France, the Benelux countries, and in the Chicago area in the United States. Nitrogen with oxygen is produced by numerous cryogenic air separation plants located along the pipelines networks (see CRYOGENICS).

Economic Aspects

The principal international producers and distributors of nitrogen are l'Air Liquide S.A. (France), The BOC Group Plc (U.K.), Air Products and Chemicals, Inc (U.S.), and Praxair, Inc. (U.S.). There are many other smaller regional producers.

Nitrogen production in the United States has grown substantially because of growth in applications. Production gas increased an average of 7.1% per year since 1970 and the increase in noncryogenic production should spur continued production increases in the 5% per year range through the rest of the twentieth century.

Specifications, Standards and Quality Control

In the United States, the Compressed Gas Association lists nine grades of nitrogen, differentiated by oxygen content, dew point, total hydrocarbon content, and other contaminant levels. These grades, are more often specified in government than commercial contracts. Commercial cryogenically produced liquid nitrogen usually meets or exceeds Type II, Grade L.

Health and Safety Factors

Gaseous nitrogen is nontoxic and nonflammable, but does not support life. Nitrogen should be stored and used only in well-ventilated areas. Special care must be taken entering an enclosed area which may be enriched in nitrogen.

Liquid nitrogen and its vapor are extremely cold and can rapidly freeze human tissue. Liquid nitrogen spills should be flushed with water to accelerate evaporation. When exposed to liquid nitrogen, carbon steel, rubber, and plastic become embrittled and may fracture under stress. Copper, brass, bronze, Monel, aluminum, and 300 series austenitic stainless steels remain ductile and are acceptable for cryogenic service. Liquid nitrogen in poorly insulated containers can concentrate and condense atmospheric oxygen on the exterior surfaces which may cause a serious fire hazard. Storage vessels or handling equipment should be provided with multiple pressure relief devices to prevent the buildup of high pressure. A pressure relief valve for primary protection and a frangible disk for secondary protection are commonly provided for on commercial liquid nitrogen storage vessels.

Uses

Applications for nitrogen are widespread in both its gaseous and liquid phases. Gaseous nitrogen is usually used as an inert blanketing or carrier gas and as a foaming agent (eg, in metallurgy, electronics manufacturing, and oil, and natural gas production). Liquid nitrogen is used as an expendable nonreactive, nontoxic refrigerant. Very few applications, excepting the large-scale synthesis of ammonia from atmospheric nitrogen, use nitrogen as a reactant.

THOMAS L. HARDENBURGER
Air Liquide America Corporation
MATTHEW ENNIS
Stanford University

K. D. Timmerhaus and T. M. Flynn, *Cryogenic Process Engineering,* Plenum Press, New York, 1989.

R. G. Scurlock, ed., *History and Origins of Cryogenics,* Oxford University Press, New York, 1992.

Handbook of Compressed Gases, 3rd ed., Compressed Gas Association, Arlington, Va., 1990.

NITROGEN FIXATION

More than 99.9% of the nitrogen (qv) on earth is present as the dinitrogen molecule, N_2, of which somewhat more than 97% is trapped in primary and sedimentary rocks (2×10^{17} and 4×10^{14} metric tons, respectively) and about 2% (4×10^{15} t) is free in the atmosphere. Thus, only a very small proportion of the nitrogen present on earth is, at any time, in a usable or fixed form. Various transformations in the nitrogen cycle allow nitrogen to move between the atmospheric inert pool and the fixed, usable terrestrial pools.

Nitrogen fixation is involved with the atmosphere-to-terrestrial direction of the cycling. Nitrification and denitrification convert ammonia to nitrate and then, via nitrogen oxides, to dinitrogen which is lost to the atmosphere. Leaching and erosion of soils result in the movement of fixed nitrogen between land and sea. The biological world stays just ahead of a nitrogen deficiency because the fixation rate slightly exceeds the denitrification rate. Only about one-third of the available nitrate is assimilated by plants, one-third is leached away, and one-third is denitrified and lost to the atmosphere.

Dinitrogen is fixed either by natural processes or by industrial ammonia (qv) production. The estimates for the annual biological contribution range around $100-200 \times 10^6$ t. Industrial fixation contributes about 50×10^6 t/yr for fertilizer uses (see FERTILIZERS). Other processes, eg, lightning and combustion, are estimated to fix about 30×10^6 t/yr. Thus, the biological process represents the majority (ca 65%) of the total annual fixation rate, contributing about three times as much as the commercial production of fertilizer.

Plants depend on the availability of nitrogenous compounds produced from atmospheric N_2 either commercially or biologically. The availability of fertilizer nitrogen is almost always the limiting factor in crop productivity. In nature, however, only a relatively few species of bacteria have the capability of converting N_2 into ammonia, which can then be incorporated into amino acids (qv) and the precursors of nucleic acids (qv). These microbes usually do so for their own benefit. In certain crops, such as legumes, ie, peas, beans, alfalfa, etc, nature has provided a mechanism for biological interaction between the plant and nitrogen-fixing bacteria. The plant receives fixed nitrogen directly from the bacteria which are harbored in nodules on its roots. The most important food crops, however, such as cereal grains, ie, rice, wheat, and corn, and root and tuber crops, do not harbor symbiotic partners. Hence, for crop productivity to reach commercially acceptable levels, extensive augmentation by commercially fixed nitrogen is necessary.

Considerable progress in the understanding of biological nitrogen fixation has been made since the 1970s. Although a well-defined mechanism for biological nitrogen reduction is still lacking, a useful numerical model has been developed. Moreover, purely chemical processes have been devised that bind N_2 and in some cases activate the nitrogen sufficiently so that reduced nitrogen compounds, ie, ammonia, and/or hydrazine, are produced on protonation.

Industrial Processes

Until the early nineteenth century, the fixed usable nitrogen stockpiled over millions of years by various natural processes was enough to sustain the needs of the earth's population. Then, the dramatic growth of cities and populations led to the beginnings of the nitrogenous fertilizer industry. Guano, hardened by bird droppings, was imported into Europe from Peru, as was saltpeter (sodium nitrate) from Chile. These fertilizer forms were supplemented in the industrialized nations from the ammoniacal by-products from coal gas. Further increases in demand led to the invention of several nitrogen fixation processes. Some were exploited commercially, eg, the Frank-Caro cyanamide process and the Haber-Bosch process. The ammonia synthesis industry of the latter part of the twentieth century employs only the Haber-Bosch process (Fig. 1). As of this writing, the process suffers from the requirement for significant quantities of nonrenewable fossil fuels.

Biological Systems

In contrast to the large industrial facilities required to produce ammonia economically, some microorganisms are capable of diazotrophy, ie, the ability to use N_2 gas as the sole source of nitrogen for growth. Only prokaryotes, ie, those living things without an organized nucleus (Eubacteria, cyanobacteria, Archaea, and actinomycetes) can perform biological nitrogen fixation, the result of which is the reduction of N_2 to ammonia. Such bacteria can be either free-living, such as *Azotobacter* and *Clostridium,* or symbiotic, like the rhizobia. The latter group, in tight associations with higher leguminous plants, are much more important agriculturally. In exchange for the fixed nitrogen supplied by the bacterium, the legume supplies a protective environment in the form of the root nodule and energy in the form of carbohydrate generated by photosynthesis. Thus, renewable solar energy (qv) powers this fertilizer production system. As food demands increase and fossil fuel reserves deplete, the exploitation of biological nitrogen fixation becomes more and more attractive as an alternative to commercial fertilizer production. Research in this area ranges from employing molecular genetic techniques to engineer nonlegume cash crops such as corn and wheat (see WHEAT AND OTHER CEREAL GRAINS) to fix enough N_2 for its own requirements (see GENETIC ENGINEERING), through the increased use of associative symbioses to the development of catalysts based on nitrogenase for N_2-reducing processes.

Nitrogenase. The biological catalyst that reduces atmospheric N_2 to ammonia is the metalloenzyme nitrogenase which exists in three genetically distinct forms: the conventional Mo-based system, Mo-nitrogenase, and two more recently discovered alternative systems, V-nitrogenase and nitrogenase-3.

Regulation of Nitrogen Fixation. Both the synthesis and activity of nitrogenase are under tight genetic control in nitrogen-fixing cells,

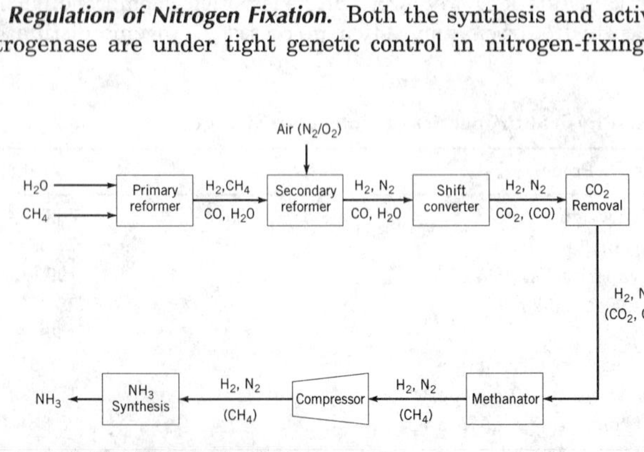

Figure 1. The Haber-Bosch process. Gases in parentheses are minor constituents of the mixture.

mainly because nitrogenase is a principal cellular component having a high energy demand. Regulation responds to three environmental factors: nitrogen status, O_2 tension, and metal-ion availability. The presence of fixed nitrogen, eg, nitrate, is a significant regulatory factor, and bacteria utilize the fixed nitrogen until it is totally depleted rather than synthesize nitrogenase to fix their own. This control mechanism is important for free-living N_2 fixers, such as *A. vinelandii,* but less so for symbiotic organisms, eg, *B. japonicum,* which are adapted to export fixed nitrogen to their host. Of primary importance is O_2 tension, because both nitrogenase proteins are extremely sensitive to oxidative damage. Finally, because nitrogenase consists of metal-containing proteins, the availability of certain metal ions becomes regulatory, particularly in those organisms that biosynthesize the alternative nitrogenases. Regulation of N_2 fixation can be exerted at several different levels in different organisms.

Chemical Approaches

Dinitrogen has a disassociation energy of 941 kJ/mol (225 kcal/mol) and an ionization potential of 15.6 eV. Both values indicate that it is difficult to either cleave or oxidize N_2. For reduction, electrons must be added to the lowest unoccupied molecular orbital of N_2 at −7 eV. This occurs only in the presence of highly electropositive metals such as lithium. However, lithium also reacts with water. Thus, such energetic interactions are unlikely to occur in the aqueous environment of the natural enzymic system. Even so, highly reducing systems have achieved some success in N_2 reduction even in aqueous solvents.

Dinitrogen-Reducing Systems. The binding of N_2 to a metal center is the first step in activating molecular nitrogen toward reduction. Since the first compounds of this type, $[Ru(NH_3)_5(N_2)]Br_2$ was synthesized most transition metals have been found to form similar compounds. Many dinitrogen compounds are so stable that they are unreactive toward reduction and so have little chance to form the basis of a catalytic system.

Nonaqueous Systems. The first nonbiological N_2-reducing system was reported in 1964 at about the same time that the first metal-N_2 complexes were prepared and when the first active N_2-fixing bacterial cell extracts were being produced. It used titanium tetrachloride, $TiCl_4$, or dichlorobis (η^5-cyclopentadienyl)titanium(IV), $(\eta^5\text{-}C_5H_5)_2TiCl_2$, with ethylmagnesium bromide or lithium naphthalenide as reductant in ethyl ether. Although the mechanism of reduction remains unclear, the conversion of N_2 into nitride is likely. These systems are not catalytic, however, because the solvolysis needed to liberate NH_3 also destroys the active species. A truly catalytic effect was demonstrated using a mixture of $TiCl_4$, metallic Al, and $AlBr_3$ at 50°C, when NH_3 was obtained at 200 mol/g atom Ti via the catalytic nitriding of aluminum.

The mono-N_2 Ti complexes apparently cannot produce hydrazine or ammonia directly, but in the presence of excess reductant do give NH_3. In contrast, the tri-N_2 complexes of both Zr and Ti react directly with HCl to liberate two N_2 molecules and produce hydrazine at a ratio of 0.9 mol/mol of complex from the third.

Early on, it appeared that no direct connection existed between the highly reducing systems that produce ammonia or hydrazine from N_2 and the well-defined metal-N_2 complexes. However, just as metal-N_2 compounds have been isolated from the reducing systems, so too have a number of metal-N_2 compounds been degraded to ammonia or hydrazine.

Aqueous systems that reduce N_2 to either hydrazine or ammonia are known. Aqueous or aqueous–alcoholic solutions of sodium molybdate(VI), Na_2MoO_4, or oxotrichloromolybdenum(V), $MoOCl_3$, with $TiCl_3$ as reductant and Mg^{2+} at pH 10–14, produce some N_2H_4 at 25°C and 0.1 MPa (1 atm) N_2. However, at 50–100°C and 5.1–15.2 MPa N_2, yields of hydrazine reach 100 mol/mol Mo. At the higher temperature, some NH_3 is also produced. Vanadium(II) or Cr(II) are equally effective as reductants.

A related homogeneous aqueous–alcoholic system, composed of V(II) complexes of catechol and its derivatives, reduces N_2 to ammonia and H_2. Only catecholates are active in this system, which is sensitive to pH. This system has been likened to nitrogenase by suggesting that both use a sequence of two four-electron reductions to evolve one H_2 for every N_2 reduced.

A third aqueous N_2-reducing system has been developed based on the knowledge that nitrogenase contains iron, molybdenum, sulfide, and thiol groups. It was reported that 3–5 μmol of NH_3 are produced from about 5 mmol Na_2MoO_4, 2.5 mmol thioglycerol, 0.1 mmol $FeSO_4 \cdot 5H_2O$, and 0.25 g $NaBH_4$ in 50 mL of borate buffer (pH 9.6) under 13.7 MPa (135 atm) N_2. In the absence of molybdate, no NH_3 is obtained. Yields up to ca 0.04 mol NH^3/mol Mo are obtained with a molybdenum–cysteine complex under 0.1 MPa (1 atm) N_2.

All the chemical N_2-reducing systems require further development to become important as N_2-reduction methods.

Outlook

The mature Haber-Bosch technology is unlikely to change substantially in the foreseeable future. The centers for commercial ammonia production may, however, relocate to sites where large quantities of natural gas are flared from crude oil production, eg, Saudi Arabia or Venezuela. Relocation would not offset the problems for agriculture of high transportation and storage costs for ammonia production and distribution. Whereas the development of improved lower temperature and pressure catalysts is feasible, none is on the horizon as of this writing.

The discovery of chemical N_2 fixation under ambient conditions is more compatible with a simple complementary, low temperature and low pressure system, possibly operated electrochemically and driven by a renewable energy resource (qv) such as solar, wind or water power, or other off-peak electrical power, located near or in irrigation streams. Such systems might produce and apply ammonia continuously, eg, directly in the rice paddy, or store it as an increasingly concentrated ammoniacal solution for later application. In fact, the Birkeland-Eyde process of N_2 oxidation in an electrical arc has been reconsidered in just such a context for areas where fertilizer production capacity of a few t/yr can make a significant impact on agriculture production. Thus simple, inexpensive, small-scale systems may have a place in areas where power is cheap and where the high capital investment of an ammonia factory cannot be justified.

Other important contributions could be made by improving the utilization of applied nitrogen fertilizer. Less than 50% of the nitrogen applied is actually assimilated by plants. To this end, slow-release fertilizer can make an impact as can development of nitrification and denitrification inhibitors (see CONTROLLED RELEASE TECHNOLOGY, AGRICULTURAL). All such strategies would prevent ammonia losses to the atmosphere and ground water. However, the effects of such inhibitors on the nitrogen cycle are unclear.

The exploitation of the benefits of biological fixation is expected to demand, in the shorter term, an increased use of legumes and other symbiotic systems in agriculture, taking care to match the most effective rhizobial strains with the appropriate cultivar. If these associations could be manipulated to start fixation earlier or to continue it later into the plant's growth, a substantial benefit could accrue. Another significant benefit would result if the ability to fix N_2 in the presence of fixed-nitrogen sources were to be conferred on bacteria.

Genetic manipulation of N_2 fixation appears to be the ultimate solution both for reducing fossil-fuel energy inputs to fertilizer production and for increasing food supplies.

WILLIAM W. NEWTON
Virginia Polytechnic Institute and State University

C. Elmerich, A. Kondorosi, and W. E. Newton, eds., *Biological Fixation for the 21st Century,* Kluwer Academic Press, Dordrecht, Boston, London, 1998.

G. Stacey, R. H. Burris, and M. J. Evans, eds., *Biological Nitrogen Fixation,* Chapman and Hall, New York, 1992.

"Opportunities for Biological Nitrogen Fixation in Rice and Other Non-Legumes," *Plant and Soil* **194**(1,2) (1997).

"Agriculture and Fertilizers" Agricultural Group Norsk Hydro a.s., Oslo, Norway.

NITROPARAFFINS

Nitroparaffins (or nitroalkanes) are derivatives of the alkanes in which one hydrogen or more is replaced by the electronegative nitro group, which is attached to carbon through nitrogen. The nitroparaffins are isomeric with alkyl nitrites, RONO, which are esters of nitrous acid. Nitroparaffins are classed as primary, RCH_2NO_2, secondary, R_2CHNO_2, and tertiary, R_3CNO_2, by the same convention used for alcohols.

The nitroparaffins are named as derivatives of the corresponding hydrocarbons by using the prefix "nitro" to designate the NO_2 group, eg, 1,1-dinitroethane, $CH_3CH(NO_2)_2$. The salts obtained from nitroparaffins and the so-called nitronic acids are identical and may be named as derivatives of either, eg, sodium salt of *aci*-nitromethane, or sodium methanenitronate.

Nitromethane, nitroethane, 1-nitropropane, and 2-nitropropane are produced by a vapor-phase process developed in the 1930s.

Physical Properties

The physical constants of the lower mononitroparaffins and of a number of polynitroparaffins are listed in Tables 1–3. Most polynitroparaffins are colorless crystalline or waxlike solids at or near room temperature. They are insoluble in water and alkanes but soluble in most other organic solvents. The lower nitroparaffins are colorless, dense liquids with mild odors. The boiling points of the mononitroparaffins are much higher than those of the isomeric nitrates.

Table 1. Physical Properties of the Lower Mononitroparaffins

Property	Nitromethane	Nitroethane	1-Nitropropane	2-Nitropropane
molecular weight	61.041	75.068	89.095	89.095
boiling point at 101.3 kPa,[a] °C	101.20	114.07	131.18	120.25
vapor pressure,[b] kPa[a]	3.64	2.11	1.01	1.73
freezing point, °C	−28.55	−89.52	−103.99	−91.32
density,[b] g/mL	1.138	1.051	1.001	0.988
refractive index, n_D^{20}	1.38188	1.39193	1.40160	1.39439
surface tension,[b] mN/m(= dyn/cm)	37.48	32.66	30.64	29.87
viscosity,[b] mPa·s(= cP)	0.647	0.677	0.844	0.770
heat of combustion (liq) at 25°C kJ/mol[c]	−708.4	−1362	−2016	−2000
specific heat at 25°C				
J/(mol·°C)[c]	106.0	138.5	175.6	175.2
J/(g·°C)[c]	1.74	1.85	1.97	1.97
dielectric constant at 30°C	35.87	28.06	23.24	25.52
solubility in water, wt % at 70°C	19.3	6.6	2.2	2.3
solubility[b] of water in nitroparaffin, wt % at 70°C	7.6	3.0	1.7	1.6
critical temperature, °C	315	388	402	344

[a] To convert kPa to mm Hg, multiply by 7.5.
[b] At 20°C.
[c] To convert J to cal, divide by 4.184.

Table 2. Physical Constants of C-4 and Higher Mononitroparaffins

Property	1-Nitrobutane	2-Nitrobutane	1-Nitro-2-methylpropane	2-Nitro-2-methylpropane	Nitrocyclohexane
freezing point, °C	−81.33	glass	−76.85	26.23	−34
boiling point, °C	152.77	139.50	141.72	127.16	205.5–206
vapor pressure at 20°C, kPa[a]	0.36	0.77	0.64	solid	
density at 25°C, g/mL	0.96848	0.96036	0.95848	solid	$1.0680^{19}{}_4$
refractive index, n_D^{20}	1.41019	1.40407	1.40642	1.39175^{30}	1.4608

[a] To convert kPa to mm Hg, multiply by 7.5.

Table 3. Physical Constants of Polynitro Compounds

Compound	Mp, °C	Boiling point, °C	Sp gr	Refractive index, n_D^t	Water solubility[a]
dinitromethane		39–40	1.524	1.4480^{20}	
trinitromethane	14.3 (dec)	45–47	$1.5967^{24}{}_4$	$1.445511^{24}{}_{He}$	sol
tetranitromethane	13.8	125.7	$1.6377^{21}{}_4$	1.43416^{21}	insol
1,1-dinitroethane		185–186	$1.3503^{23}{}_{23}$	1.4346^{20}	sl sol
1,2-dinitro-ethane	39–40	135	$1.4597^{20}{}_4$	1.4488^{20}	sl sol
1,1,1-trinitroethane	57	68	$1.4223^{77}{}_4$	$1.4171^{77}{}_\alpha$	insol
2,2-dinitropropane	54	185			insol
1,1-dinitrocyclohexane	36	142–143	$1.2452^{21}{}_4$	1.4732^{21}	insol

[a] All named compounds are soluble in ethanol and in ethyl ether.

This phenomenon may be attributed in large part to intermolecular hydrogen bonding.

Most organic compounds, including aromatic hydrocarbons, alcohols, esters, ketones, ethers, and carboxylic acids are miscible with nitroparaffins, whereas alkanes and cycloalkanes have limited solubility. The lower nitroparaffins are excellent solvents for coating materials, waxes, resins, gums and dyes.

Chemical Properties

The chemical reactions of the nitroparaffins have been discussed in depth and their utility for the synthesis of heterocyclic and other compounds has been noted.

Tautomerism. Primary and secondary mononitroparaffins are acidic substances which exist in tautomeric equilibria with their nitronic acids.

$$RCH_2NO_2 \rightleftharpoons RCH{=}NO_2H \text{ and } RR'CHNO_2RR'CHNO_2 \rightleftharpoons RR'C{=}NO_2H$$

Salts. Nitroparaffins dissociate to form ambidentate anions, which are capable of alkylation at either the carbon or oxygen atom. Reaction with nitrous acid can be used to differentiate primary, secondary, and tertiary mononitroparaffins. Primary nitroparaffins give nitrolic acids. Secondary nitroparaffins give alkali-insoluble nitroso derivatives known as pseudonitroles.

Acid Hydrolysis. With hot concentrated mineral acids, primary nitroparaffins yield a fatty acid and a hydroxylamine salt.

Halogenation. In the presence of alkali, chlorine replaces the hydrogen atoms on the carbon atom holding the nitro group. If more than one hydrogen atom is present, the hydrogen atoms can be replaced in stages.

Halonitroparaffins can be prepared in which the halogen and nitro groups are not on the same carbon atom. The direct chlorination of nitroparaffins to give nongeminal substitution is promoted by irradiation in anhydrous media.

Reaction with Carbonyl Compounds. Primary and secondary nitroparaffins undergo aldol-type reactions with a variety of aldehydes and ketones to give nitro alcohols. Those derived from the lower nitroparaffins and formaldehyde are available commercially (see NITRO ALCOHOLS). Nitro alcohols can be reduced to the corresponding amino alcohols (see ALKANOLAMINES).

Mannich-Type Reactions. Secondary nitroparaffins, formaldehyde (qv), and primary or secondary amines can react in one step to yield Mannich bases.

$$R_2CHNO_2 + CH_2O + R'_2NH \longrightarrow R_2C(NO_2)CH_2NR'_2 + H_2O$$

Reduction. The lower nitroparaffins are reduced readily to the corresponding primary amines with a number of reducing agents. Some of the products obtained are useful as pharmaceuticals.

Oxidation. Nitroparaffins are resistant to oxidation.

Addition to Multiple Bonds. Mono- or polynitroparaffins with a hydrogen on the carbon atom carrying the nitro group add to activated double bonds under the influence of basic catalysts.

Preparation and Manufacture

Synthetic methods suitable for preparation of a wide variety of nitroparaffins have been reviewed.

A general one-step method for preparation of primary and secondary nitroparaffins from amines by oxidation with *m*-chloroperbenzoic acid in 1,2-dichloroethane has been reported. This method is particularly useful for laboratory quantities of a wide variety of nitroparaffins.

Higher nitroalkanes are prepared from lower primary nitroalkanes by a one-pot synthesis. Successive condensations with aldehydes and acylating agents are followed by reduction with sodium borohydride. Overall conversions in the 75–80% range are reported.

The only method utilized commercially is vapor-phase nitration of propane, although methane, ethane, and butane can also be nitrated quite readily. Nitromethane, nitroethane, 1-nitropropane, and 2-nitropropane are made on a large scale at the Sterlington, Louisiana, plant of ANGUS Chemical Co.

Shipment and Storage

The four commercial nitroparaffins are available in drums; they are also available in bulk except for nitromethane, for which shipment in tank cars or trucks is prohibited.

Safety factors have been of prime consideration in the development of recommendations for the storage and safe handling of nitromethane during recovery operations and transfer in piping systems. Nitromethane preferably should be stored in the 208-L (55-gal) drums in which it is shipped. These containers are of lightweight construction and there is little possibility that they might develop sufficiently high internal pressure to either ignite the nitromethane or allow it to burn as a monopropellant.

Commercial-grade nitroparaffins are shipped and stored in ordinary carbon steel. However, wet nitroparaffins containing more than 0.1–0.2% water may become discolored when stored in steel for long periods, even though corrosion is not excessive. Aluminum and stainless steel are completely resistant to corrosion by wet nitroparaffins.

Because of their flash points, nitroparaffins are classified as flammable liquids under DOT regulations (hazard class 3, PG III). Nitromethane and nitroethane fires can be extinguished with water, CO_2, foam, or class ABC dry chemical extingishers. Nitroparaffins should not be exposed to dry caustic soda, lye or similar alkaline materials.

Table 4. Oral Toxicity[a] and Threshold Limit Values of Nitroparaffins

Nitroparaffin	LD_{50}, mg/kg	TLV, ppm	TLV, mg/m^3
nitromethane	1210 ± 322	20	50
nitroethane	1625 ± 193	100	307
1-nitropropane	455 ± 75	25	91
2-nitropropane	725 ± 160	10	36

[a] In rat.

Health and Safety Factors

Toxicology. The nitroparaffins have minimal effects by way of actual contact. Inhalation is the chief route of worker exposure. The 1997 threshold limit values as recommended by the American Conference of Governmental Industrial Hygienists (ACGIH) are given in Table 4.

The International Agency for Research on Cancer (IARC) found that there is sufficient evidence to conclude that 2-nitropropane causes cancer in rats but that epidemiologic data are inadequate to reinforce the conclusion in humans. The National Toxicology Program also concluded that it "may reasonably be anticipated to be a carcinogen".

Because of these findings with 2-nitropropane, lifetime inhalation studies have been conducted in rats at 200 ppm with nitromethane and nitroethane and at 100 ppm with 1-nitropropane. In no instance was cancer found in any exposed rats. Also, mutagenicity testing have provided clear evidence that the primary nitroparaffins are not genotoxic or mutagenic. 2-Nitropropane, in contrast, is mutagenic in the Ames test. An excellent review of the toxicity of 2-nitropropane has been published by the World Health Organization.

Safe Handling. Any work area where nitroparaffins are present should be adequately ventilated so as to maintain concentration levels below the accepted exposure limit. Fresh-air masks should be supplied to workers entering confined spaces, eg, storage tanks, containing a high concentration of nitroparaffin vapors. Nitroparaffins have high heats of adsorption on respirator canisters containing Hopcalite. The use of respirator masks containing these substances may lead to fire in the presence of high concentrations of the nitroparaffins.

The ignition temperature of the lower homologues are relatively high for organic solvents. Some dry-chemical fire extinguishers contain sodium or potassium bicarbonate; these should not be used on nitromethane or nitroethane fires.

Three conditions have been identified under which nitromethane can be detonated: (*1*) nitromethane can explode if subjected to a severe shock such as that of a high explosive with more power than a No. 8 blasting cap; (*2*) it can be initiated by a rapid compression under adiabatic conditions; and (*3*) nitromethane can be detonated by heating it under confinement to near the critical temperature (315°C). These conditions combining high pressure with high temperature are the same as those under which nitromethane burns as a monopropellant. Certain compounds, eg, amines or strong oxidizing agents, when present in admixture with nitromethane, can sensitize it to decomposition by strong shock.

The insensitivity of nitromethane to detonation by shock under normal conditions of handling has been demonstrated by a number of full-scale tests. Sensitivity to shock increases with temperature.

Environmental Concerns. Few data on the environmental effects of the nitroparaffins are available. Reviews of the available data on the environmental effects of nitromethane and 2-nitropropane have been published by the U.S. Environmental Protecting Agency.

When disposed of, all the nitroparaffins are considered to be hazardous waste.

Uses

The nitroparaffins have been utilized for many applications. Some of these uses have been discontinued because of economic and environmental considerations.

Nitromethane. The nitroparaffins are used widely as raw materials for synthesis. Nitromethane is used to produce the nitro alcohol (qv) 2-(hydroxymethyl)-2-nitro-1,3-propanediol.

Halogenation of nitromethane is utilized to produce two economically important pesticides, chloropicrin and bronopol, and in the synthesis of the antiulcer drug, ranitidine. Nitromethane finds used as fuel and in explosive applications.

Nitroethane. The principal use of nitroethane is as a raw material for synthesis of α-methyl dopa, a hypertensive drug and for the insecticide, *S*-methyl-*N*-[(methylcarbomoyl)-oxy]thioacetimidate.

1-Nitropropane. 1-Nitropropane is used in the production of the alkanolamines (qv), 2-amino-2-ethyl-1,3-propanediol and 2-amino-1-butanol. Though less important economically, solvent usage consumes a larger portion of the 1-nitropropane production than is consumed of the other nitroparaffins for this use.

2-Nitropropane. Derivatives such as 2-methyl-2-nitro-1-propanol (used in tire cord adhesive) and 2-amino-2-methyl-1-propanol (a pigment dispersant and buffer) have served as an outlet for 2-nitropropane production.

ALLEN F. BOLLMEIER, JR.
ANGUS Chemical Company

U.S. Pat. 1,967,667 (July 24, 1934), H. B. Hass, E. B. Hodge, and B. M. Vanderbilt (to Purdue Research Foundation); *Ind. Eng. Chem.* **28**, 339 (1936).

R. F. Purcell, in J. J. McKetta ed., *Encyclopedia of Chemical Processes and Design,* Vol. 31, Marcel Dekker, Inc., New York, 1990, pp. 267–281.

D. Seebach, E. W. Colvin, F. Lehr, and T. Weller, *Chimia* **33,** 1 (1979).

M. T. Shipchandler, *Synthesis,* 666 (1979).

N-NITROSAMINES

N-Nitrosodialkylamines (*N*-nitrosamines) were first characterized in the late nineteenth century. Although they have been used as synthetic intermediates and solvents, and possess interesting structural and spectroscopic properties, their potential uses have been eclipsed by their toxicity and especially their genotoxicity. Most of the *N*-nitrosamines are carcinogenic in animals. Nitrosamines have induced tumors in every animal species tested. The primary focus of research on these compounds in the 1990s is therefore on biochemistry, utility as experimental mutagens and carcinogens, and on potential human exposure to them.

Properties

Many of the chemical, physical and biological properties of more than 20 selected *N*-nitrosamines have been summarized. *N*-Nitrosamines encompass a wide range of structural types because the single feature common to them is the NNO functionality, and there are few restrictions on the groups than can be attached to the remaining two valences on the amine nitrogen, where R_1 or R_2 can be alkyl, aryl, or mixed.

Synthesis. The classic laboratory synthesis of *N*-nitrosamines is the reaction of a secondary amine with acidic nitrite at ca pH 3. Although nitrosations with primary and tertiary amines are generally slower and give lower yields than is the case with secondary amines, there are exceptions including some drugs such as aminopyrine.

Inhibition of nitrosation is generally accomplished by substances that compete effectively for the active nitrosating intermediate. *N*-Nitrosamine formation *in vitro* can be inhibited by ascorbic acid (vitamin C) and α-tocopherol (vitamin E), as well as by several other classes of compounds including pyrroles, phenols, and aziridines. Inhibition of intragastric nitrosation in humans by ascorbic acid and by foods such as fruit and vegetable juices or food extracts has been reported in several instances.

Reactions. Most of the reactions of the nitrosamines, with respect to their biological and environmental behavior, involve one of two main reactive centers, either the nitroso group itself or the C-H bonds adjacent (α) to the amine nitrogen. The nitroso group can be readily removed by a reaction which is essentially the reverse of the nitrosation reaction, or by oxidation or reduction.

Handling and Disposal

N-Nitrosamines are potentially hazardous and should be handled in designated hoods and with protective clothing. Nitrosamines can be destroyed by treatment with aluminum–nickel alloy under basic conditions.

Health and Safety Factors

Toxicity. Many *N*-nitrosamines are toxic to animals and cells in culture. Some of the toxicological properties of a selected group of nitrosamines are listed in Table 1.

Carcinogenicity. The number of nitrosamines that have been tested for carcinogenicity exceeds 200. Most are carcinogenic, although the potency varies dramatically with the series. Carcinogenicity has been observed both with single, relatively large doses and with long-term chronic exposure to lower doses. The *N*-nitrosamines are generally organ selective. Most nitrosamines are not direct-acting carcinogens, but require metabolic activation in order to exert their carcinogenic effect. The potency and the organ selectivity of the *N*-nitrosamines are therefore determined by complex interactions involving molecular structures and the spectrum of metabolizing enzymes in the test animal. There are consequently species- and sex-related differences in both potency and organ selectivity.

Mutagenicity. The *N*-nitrosamines, in general, induce mutations in standard bacterial-tester strains. As with carcinogenicity, enzymatic activation, typically with liver microsomal preparations, is required.

Human Exposure. *N*-Nitrosamines have been reported in pesticide preparations, corrosion inhibitors, lubricating fluids and cosmetics (qv), ie, *N*-nitrosodiethanolamine, sunscreens, rubber products including baby-bottle nipples and pacifiers, foods including cheese, processed meats, beer (qv), cooked bacon, and powdered milk, and tobacco products (see INSECT CONTROL TECHNOLOGY; CORROSION AND CORROSION CONTROL; LUBRICATION AND LUBRICANTS; FOOD TOXICANTS, NATURALLY OCCURRING). In addition to exposure by preformed *N*-nitroso compounds, there may be endogenous nitrosation of amines in the mouth, stomach, or other sites in the body. Formation of nitrosamines inside the body has been demonstrated unambiguously in humans and in experimental animals. These findings have stimulated a large number of studies concerning possible relationships between nitrosation in the body and elevated cancer risk, and whether this nitrosation can be blocked or inhibited, especially by dietary compounds.

Table 1. Toxicological Properties of Some Representative *N*-Nitrosamines in the BD Rat[a]

Compound	LD_{50}	$\log(1/D_{50})$[b]	Principal target organ
N-nitrosodimethylamine (NDMA)	40	2.3	liver
N-nitrosodiethylamine (NDEA)	280	3.2	liver, esophagus
N-nitrosodiethanolamine (NDELA)	7500	0.005	liver
N-nitrosodipropylamine	480	0.05	liver, esophagus
N-nitrosodiisopropylamine	850	2.1	liver
N-nitrosopyrrolidine (NPYR)	900	1.0	liver
N-nitrosomorpholine (NMOR)	320	1.9	liver
N-nitrosodicyclohexylamine		c	
N-nitrosoproline (NPRO)		c	
N nitrosomethyl(benzyl)amine	18	3.1	esophagus
N-nitrosopiperidine	200	1.9	liver, esophagus
N-nitrosonornicotine		d	

[a] BD rat represents a particular strain used to test carcinogenicity of some *N*-nitroso compounds.

[b] D_{50}: dose causing tumors in 50% of the test animals; increasing values for $\log(1/D_{50})$ represent higher carcinogenicity.

[c] Not carcinogenic to the BD rat.

[d] Not investigated in the BD rat. Suspected human carcinogen.

There is insufficient evidence to unequivocally link nitrosamine exposure to elevated risk for human cancer. There are, however, a number of specific cases, especially with respect to the tobacco-related nitrosamines, in which exposure to *N*-nitroso compounds is of concern. The strongest evidence in this context is probably that relating to elevated oral cancer rates among habitual users of smokeless tobacco (snuff).

JOHN S. WISHNOK
Massachusetts Institute of Technology

I. K. O'Neill, J. Chen, and H. Bartsch, eds., *Relevance to Human Cancer of N-Nitroso Compounds, Tobacco Smoke and Mycotoxins,* IARC Scientific Publication No. 105, International Agency for Research on Cancer, Lyon, France, 1991.

W. Lijinsky, *Chemistry and Biology of N-Nitroso Compounds,* Cambridge University Press, Cambridge, U.K., 1992.

D. Hoffmann, A. Rivenson, and S. S. Hecht, "The Biological Significance to Tobacco-Specific *N*-nitrosamines: Smoking and Adenocarcinoma of the Lung," *Critical Reviews in Toxicology* **26,** 199–211 (1996).

R. N. Loeppky and C. J. Michejda, eds., *Nitrosamines and Related N-Nitroso Compounds,* American Chemical Society, Washington, D.C., 1994.

NOBELIUM. See ACTINIDES AND TRANSACTINIDES.

NOMENCLATURE

Chemical nomenclature embraces several subcategories; names for chemical elements and compounds; names for classes of compounds and substances, such as mixtures and composites; names for particles, processes and transformations, properties, effects, units of measurements, techniques, instruments and apparatus, and even for theories and concepts. Only the first three are considered to be the heart of chemical nomenclature. The largest part of the subject is the nomenclature of organic compounds, simply because there are so many of them, and of such diverse nature. Concern with chemical nomenclature has grown on a broad international scale as the importance of consistent, uniform nomenclature is increasingly recognized. Various committees, both national and international, are working toward a consistent, systematic nomenclature. Among the areas in which nomenclature plays a key role are patent law, trade and customs regulations, identification of controlled substances, pharmaceutical and health information, and studies of the environment and pollution.

In the United States, the Committee on Nomenclature of the American Chemical Society is the clearinghouse for nomenclature recommendations and adoptions, aided by various divisional nomenclature committees of the Society. Close liaison is maintained with the various nomenclature bodies of the International Union of Pure and Applied Chemistry. Progress is being made not only in improved nomenclature, but also in the extension of nomenclature recommendations to newly developing areas of chemistry.

Important as names are, they cannot serve all purposes. There are other, complementary means of identifying chemical compounds, eg, structural formulas, notation systems, and registry numbers. None of these are nomenclature, however.

Although symbols are not a part of nomenclature, the two are closely related, and the former have played an extremely important role in chemistry. Because of the difficulty of establishing priority of discovery for most of the elements of atomic number above 100, and because of the need to refer to hypothetical elements with higher atomic numbers, IUPAC has developed interim systematic symbols and names for such elements.

Inorganic Nomenclature

Perhaps no subject in chemistry has undergone less change over the twentieth century than inorganic nomenclature. This longevity attests to the fundamental soundness of the original proposals of Guyton de Morveau that established it, but it also suggests why inconsistencies and confusions have remained as well, which have continued to disconcert chemists. The development of inorganic nomenclature has again accelerated, however, and the inconsistencies are being eliminated.

The System of Guyton de Morveau, Lavoisier, and Co-Workers. The first attempt toward a convenient nomenclature belongs to Guyton de Morveau. His pioneer work led to publication in 1787 of *Methode de Nomenclature Chimique*, written in collaboration with Lavoisier, Berthollet, and Fourcroy, which proved to be a landmark in the development of chemistry.

The fundamental principle of the new nomenclature was that the name of a compound should exhibit the elements involved and their relative proportions, if known. The combinations of oxygen with other elements played a dominant role. Thus, the product of the union of a simple nonmetallic substance with oxygen was called an acid, whereas that of the union of a metal with oxygen was called an oxide. The union of an acid and an oxide produced salt. The acids or oxides were given names in which the generic part was the word "acid" or "oxide" and the specific part was an adjective derived from the name of the other element. The same principle supplied names for sulfides and phosphides.

The names adopted for salts consisted of a generic part derived from the acid and a specific part from the metallic base: The names for salts of acids containing an element in different degrees of oxidation were given different terminations.

Berzelius divided the elements into metalloids (nonmetals) and metals according to their electrochemical character, and the compounds of oxygen with positive elements (metals) into suboxides, oxides and peroxides. His division of the acids according to degree of oxidation has been little altered. He introduced the terms anhydride and amphoteric and designated the chlorides in a manner similar to that used for the oxides.

Established Practice in the English Language. The nearly literal translation of the French terms in English, Russian and other languages resulted in the system whose use has become standard practice in English-speaking as well as other countries. The system has been molded by the fact that elemental composition and valence (or oxidation number) are the principal variables for most inorganic compounds other than the most complex, whereas connectivity and the possibility for isomers have been of little concern.

Modified Forms in Common Use. There are numerous situations in which the foregoing system does not meet all requirements. In the formation of binary compounds, several elements exhibit more than two states of oxidation. One method, recommended by the IUPAC, of handling these situations is the use of prefixes derived from Greek to indicate stoichiometric composition, eg, titanium dichloride, $TiCl_2$; and dinitrogen oxide (nitrous oxide) N_2O. Other accepted methods of indicating proportions of constituents are the Stock system (oxidation number) and the Ewens-Bassett (charge number) system.

Some elements form acids with more than four oxidation states, requiring other combinations of prefixes and suffixes: $H_4P_2O_6$, intermediate between H_3PO_3 and H_3PO_4, is known as hypophosphoric acid. Here again, the oxidation-number and charge-number systems offer advantages. Ortho-, meta-, and pyro- prefixes or numerical prefixes to denote stages of hydroxylation of acids also find use. In many instances, special names have been created to deal with unusual situations.

Systems of Compounds. The nomenclature system of Guyton de Morveau and co-workers was designed specifically for oxygen compounds. As early as 1826, it became evident that the halogens could play much the same role in many other compounds as oxygen does in the familiar oxygen salts. By 1840, Hare was writing of chloro acids and chloro bases, and recognized classes of salts: oxy-, sulfo- (now called thio-), seleni-, telluri-, chloro-, fluoro-, cyano-, etc. Remsen was a proponent of this system of nomenclature, but it received its fullest treatment from Franklin in connection with his concept of systems of compounds.

The analogies are shown by the following reactions:

$$K_2O + B_2O_3 \rightarrow K_2O{\cdot}B_2O_3 \text{ or } KBO_2$$

$$K_2S + B_2S_3 \rightarrow K_2S{\cdot}B_2S_3 \text{ or } 2\ KBS_2$$

$$KF + BF_3 \rightarrow KF{\cdot}BF_3 \text{ or } KBF_4$$

$$K_3N + BN \rightarrow K_3N{\cdot}BN \text{ or } K_3BN_2$$

The products resulting from such reactions should, therefore, have analogous names. If KBO_2 is a borate, then KBS_2 is a thioborate, and KBF_4 is a fluoroborate. Similarly, the replacement of an oxygen atom by a sulfur atom or two fluorine atoms is understandable. However, the relationship of K_3BN_2 is less obvious, until one considers the dehydration and deammoniation schemes:

$$B(OH)_3 \xrightarrow{-H_2O} OBOH$$

$$B(NH_2)_3 \xrightarrow{-NH_3} HNBNH_2$$

His scheme of nomenclature for nitrogen compounds, and the names thio-, chloro-, etc, did become widespread, especially for sulfur and halogen compounds.

Although the foregoing pattern of nomenclature is useful, it does lead to some difficulties. Many quaternary compounds contain oxygen and another electronegative element. In the series M_2CO_3, M_2CO_2S, M_2COS_2, and M_2CS_3, the names are carbonates, (mono)thiocarbonates, dithiocarbonates, and trithiocarbonates, respectively. However, in practice both the prefixes mon- and tri- are often omitted, and its is uncertain whether the omission signifies the mono- or the completely substituted compound. The situation is somewhat more complicated when oxygen and fluorine are present in the same compound, because one is bivalent and the other univalent, and the coordination number toward fluorine is different from that toward oxygen: H_3PO_4, H_2PO_3F, HPO_2F_2, and HPF_6. Furthermore, investigators have not always been consistent in choosing the same reference state for the names of the oxygen salts and the halogen salts.

Coordination Compounds. The approach of Werner to the problem of naming ternary and higher order compounds is based on an entirely different point of view. By considering all such substances as complex or coordination compounds, he succeeded in making a wide variety of them according to a single general pattern. To designate the oxidation state of the element serving as the center of coordination, Werner chose the characteristic endings suggested by Brauner, but these have been totally superseded by the oxidation-number and charge-number systems.

The Stock Oxidation-Number System. Stock sought to correct many nomenclature difficulties by introducing Roman numerals in parentheses to indicate the state(s) of oxidation.

The oxidation-number system is easily extended to include other coordination compounds. Even substances represented by the formulas $Na_4Ni(CN)_4$ and $K_4Pd(CN)_4$ create no nomenclature problem; they become sodium tetracyanonickelate(0) and potassium tetracyanopalladate(0), respectively.

The Charge-Number (Ewens-Bassett) System. The oxidation state of an atom as expressed by the oxidation number is a formal concept for partitioning the electric charge between atoms in a molecule or chemical structure. For many chemical structures, this formal procedure may lead to representations of charge distribution that are inconsistent with experiment. Therefore, Ewens and Bassett proposed to express only the total charge on an ion without representing valence and its associated arbitrariness of assigning electronic distribution within a given structure, eg, titanium(2+) chloride for $TiCl_2$; titanium(3+) for $TiCl_3$; potassium tetrachloroplatinate(2−) for K_2PtCl_4; and sodium tetracyanonickelate(4−) for $Na_4Ni(CN)_4$.

International Agreement. The first report of the Commission for the Reform of the Nomenclature of Inorganic Chemistry was written in 1926 by Délépine. Subsequent rules (1940, 1959) were expanded and improved in 1990 to provide the basis for naming inorganic compounds. They retain most of the well established names for binary and pseudobinary compounds and for the oxoacids of the nonmetals and derivatives.

The IUPAC Commission on Nomenclature of Inorganic Chemistry continues its work, which is effectively open-ended. Guidance in the use of the IUPAC rules as well as explanations of their formulation are available. A second volume on nomenclature of inorganic chemistry is in preparation; it will be devoted to specialized areas. Some of the contents have had preliminary publication in the journal, *Pure and Applied Chemistry*, eg, "Names and Symbols of Transfermium Elements" in 1944.

Organic Nomenclature

Modern organic nomenclature is such that it can be better understood by first tracing how it developed. Organic substances played a minor role in the *Methode de Nomenclature Chimique*. Eighteen organic acids were given their present names (succinic, malic, etc), and several other substances were mentioned, such as alcohol, ether (including esters as well as true ethers), starch, gluten, and camphor. Gaseous hydrocarbons, the only ones included, were lumped together as carbonated hydrogen gas. Thus a few common names were incorporated into the new method, but no systematic organic names were possible because of lack of knowledge. Little else could be done, for the basis for determining elemental composition, as in empirical formulas, did not yet exist.

The practice of assigning ad hoc names to organic compounds was neither avoidable, nor burdensome when only a small number of compounds were recognized. Such ad hoc names are termed "trivial" or "traditional," to indicate that they contain no encoded structural information. They are useful for common compounds, and many of them are retained to this day, but they are not helpful in understanding chemical relationships. As they proliferated, the number and variety of them became unmanageable. The development of systematic nomenclature was driven by this circumstance, and was made possible by advances in understanding and determining the structure of molecules.

Systematic nomenclature is in essence a scheme for encoding structural information in a name. For organic chemistry, it probably began in 1832, when Justus Liebig's journal, *Annalen der Chemie*, was born and when Liebig and Woehler published their memorable article on the radical of benzoic acid. This radical (C_6H_5CO in the modern formula) they termed "benzoyl," thus coining -yl (from the Greek *hyle*, meaning stuff or material), one of the most useful suffixes in chemistry. By radical or compound radical, they meant a group of atoms that remains unaltered in chemical transformations. The word group is used for almost any portion of a molecule considered as a unit for convenience in naming or otherwise. The name ethyl soon followed. These two names, one of an acid group (radical) and the other of a hydrocarbon group (radical), may be regarded as the progenitors of the host of group names used in the 1990s. From them it was an easy step to the combinations of benzoyl chloride, ethyl iodide, ethyl oxide, etc, many of which still survive. These binary names are analogous to the binary inorganic names introduced in 1787.

It was many years before organic nomenclature shook off the influence of electrochemical theory and its binary names. Gradually, as facts accumulated, it became clear that this theory must give way to a unitary conception of the molecule. At the same time, the phenomenon of substitution, or replacement of one atom or group of atoms by another, was recognized to be of central importance. Some binary names are still used, either as a true expression, as for salts, or for convenience, as in ethyl sulfide or acetyl chloride, but for the most part the principle of substitution is used without regard to whether such replacement can actually be effected experimentally. Usually the atom replaced is hydrogen, and the replacement may be indicated by either a prefix or a suffix. Thus, in naming CH_3Cl chloromethane rather than methyl chloride, the replacement of one atom of hydrogen in methane, CH_4, by chlorine is indicated. A third group of names is formed by combining a class name with a specific word, as in ethyl

Table 1. Prefixes and Suffixes for Some Principal Functional Groups[a]

Formula	Class name	Prefix	Suffix	Radicofunctional form
—COOH	carboxylic acids	carboxy	-carboxylic acid[b] or -oic acid	
$—SO_3H$	sulfonic acids	sulfo	-sulfonic acid	
—COOR	esters	alkoxy carbonyl	alkyl -oate or carboxylate[b]	
—COX	acid halides	halocarbonyl	-oyl halide or carbonyl halide[b]	
$—CONH_2$	amides	carbamoyl cyano	-amide or carboxamide[b]	
—CN	nitriles		-nitrile or -carbonitrile[b]	alkyl[c] cyanide
—CH═O	aldehydes	formyl	-al or -carbaldehyde[b]	
—C═O	ketones[d]	oxo	-one	dialkyl[c] ketone
—OH	alcohols	hydroxy	-ol	alkyl[c] alcohol
—SH	thiols (or mercaptans[e])	sulfanyl (mercapto[e])	-thiol	alkyl mercaptan
$—NH_2$	amines	amino	-amine	
═NH	imines	imino	-imine	
—OR	ethers	alkoxy or aryloxy		(di)alkyl ether
—Cl	chlorides	chloro		alkyl[c] chloride
$—NO_2$	nitro compounds	nitro		
$—SO_2$	sulfones	alkyl[c] sulfonyl		(di)alkyl[c] sulfone

[a] In order of precedence.
[b] The shorter form implies no additional carbon atoms, and is used when the group is part of a chain. The long form implies one more carbon atom than the parent structure, and is used when the group is attached to a ring, or for other reasons is not conveniently named as part of a carbon skeleton.
[c] Or aryl.
[d] Both bonds from the carbonyl group must be to a carbon atom.
[e] Has been widely used, but is no longer officially recommended.

alcohol or benzophenone oxime. Whatever the method or combination of methods used, there must be a name for a parent compound to form a basis for it.

By 1866, it was possible for Hofmann to arrange hydrocarbons in series by their empirical formulas, ie, methane, CH_4, and methene, CH_2; ethane, C_2H_6, ethene, C_2H_4, and ethine, C_2H_2; propane, C_3H_8, propene, C_3H_6, and propine, C_3H_4; and quartane, C_4H_{10}, quartene, C_4H_8, and quartine, C_4H_6. These are known as the Hofmann-Gerhardt names.

Hofmann's scheme has been modified by replacing quartane with butane and continuing the homologous series with the Greek forms pentane, hexane, etc, which are still used. For the C_nH_{2n} series, ie, the olefins, the names methylene, ethylene, propylene, etc, came into use instead of Hofmann's terms, but the names propene, butene, pentene, etc were revived in the Geneva system and are the preferred terms. For C_2H_2, ethine has never replaced the older term acetylene, but propine, butine, etc, reappeared in the Geneva system. The ending -yne is used, as in propyne, etc, to avoid confusion with the ending -ine of organic bases such as aniline.

The Hofmann-Gerhardt names did not distinguish between isomers. Different methods of distinguishing isomers arose: $CH_3CH_2CH_2CH_3$ became normal butane (abbreviated to *n*-butane), and $CH_3CH(CH_3)_2$ isobutane or trimethylmethane. Of olefins, $CH_2{=}CHCH_2CH_3$ became α-butylene or ethylethylene; $CH_3CH{=}CHCH_3$, β-butylene or symmetrical dimethylethylene; and $CH_2{=}C(CH_3)_2$, isobutylene or unsymmetrical dimethylethylene. It thus becomes evident that as the number of carbon atoms and therefore the number of isomers increases, the coining of such names meets with insuperable difficulties. The situation with regard to hydrocarbons had its parallel in the nomenclature of alcohols and other types of compounds.

The Geneva System. The Geneva Conference was strongly influenced by the need for names that would be suitable for systematic indexing of organic compounds. The groundwork was laid by a French subcommission. One of the chief principles of the system was the selection of the longest straight chain of carbon atoms in the molecule as a parent structure. Thus, the names butane, pentane, etc, would refer to the normal (unbranched) isomers only. The parent hydrocarbon could then be modified by attaching to its name one or more prefixes or suffixes to specify chemically characteristic features commonly termed functional groups). A representative selection is given in Lance 1 when two or more different positions of attachment of a prefixed or suffix exist, a position designator, called a locant, is necessary. These are arabic numerals, set off by hyphens, starting with 1 at an end of the chain. Accordingly, $CH_3CH{=}CHCH_3$ became 2-butene, and $CH_3CH_2CHOHCH_2CH_3$ became 3-pentanol. The position of the locants at the beginning or at the end was considered equally acceptable. In the 1990s however, the official IUPAC recommendation is to place the locant immediately before the feature that it locates, as in but-2-ene and pentan-3-ol for the foregoing examples.

The International Union and the Definitive Report. The next important step was the *Definitive Report of the Commission on the Reform of the Nomenclature of Organic Chemistry* in 1930 at a meeting in Liège. This report used the Geneva rules as a basis for modification, and many of the 68 Liège rules deal with topics not touched in the original Geneva report.

Table 2. Compounds Used as Parent Structures with Trivial Names

Chemical formula	Name	Systematic equivalents
C_6H_6	benzene	
$C_{10}H_6$	naphthalene	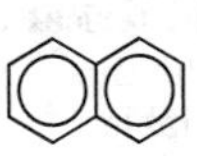
C_5H_5N	pyridine	
C_4H_4S	thiophene	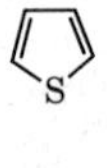
C_4H_4O	furan	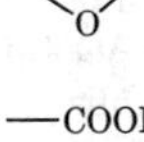
		—COOH
C_6H_5OH	phenol	benzenol
$C_6H_5NH_2$	aniline	benzenamine
$H_2N{=}NH_2$	hydrazine	diazane
NH_3	ammonia or amine	azane
HO═OH	hydrogen peroxide	dioxidane
CH_3COCH_3	acetone	propan-2-one
H_2NCONH_2	urea	carbamide

An important modification of the Geneva system is that the fundamental chain used as a basis in an aliphatic compound is not necessarily the longest chain in the molecule, but must be the longest chain of those containing the maximum number of occurrences of the principal functional group (Rule 18). This shifts the importance for naming from side chains such as methyl and ethyl to functional groups such as -COOH and -OH.

The concept of the principal function raises the question of how priority is determined when two or more different functional groups are present. No arbitrary rule can be entirely satisfactory, but an order has been codified in IUPAC recommendations, and an essentially similar order is used by Chemical Abstracts Service. In general, a higher state of oxidation takes precedence over a lower one (Table 1).

The use of prefixes and suffixes for distinguishing the various radicals, groups, and functions has caused some problems, because some groups, eg, HS—, have borne more than one name, and some names, eg, anisyl, have had more than one meaning. The *Definitive Report* included only a limited number of prefixes and suffixes. Chemical Abstracts Service publishes its own lists the most recent version can be found in the comprehensive *Guide to the Use of IUPAC Nomenclature of Organic Compounds*. An important departure from earlier recommendations is that the systematic names of acyl groups derived from carboxylic acids must end in -oyl, common traditional or trivial names, such as acetyl and oxalyl, excepted. The purpose of this rule is to distinguish unambiguously between hydrocarbyl groups and acyl groups. Thus anisyl can only mean methoxyphenyl, whereas anisoyl refers only to methoxybenzoyl.

The ending -yl (or -oyl) is standard for univalent groups (with certain traditional exceptions, such as succinoyl). It may be combined with a sign for unsaturation, as in propenyl, $CH_3CH{=}CH$ thynyl, $CH{\equiv}C$—. The ending -ylene is one device for denoting a bivalent group in which the two free valences are on different atoms, but, with the exception of methylene, —CH_2—, and ethylene, —CH_2CH_2—, the ending -diyl, with locants as appropriate, is preferred, as in propane-1,3-diyl, —$CH_2CH_2CH_2$—. When the two free valences of a bivalent group form a double bond, the ending is -ylidene, as in ethylidene, $CH_3CH{=}$. For a trivalent group forming a triple bond, the ending is -ylidyne, as in ethylidyne, $CH_3C{\equiv}$.

For indicating the number of groups of the same kind, the prefixes di-, tri-, tetra-, etc are used when the expressions are simple, and bis-, tris-, tetrakis-, etc when they are complex; for example, "dichloro," but "bis(dimethylamino)." The prefix bi- is used to denote the joining of two groups of the same kind together, as in biphenyl, C_6H_5—C_6H_5, or the doubling of a compound with loss of two hydrogen atoms, as in biarsine, H_2AsAsH_2.

A historical account of the development of organic nomenclature from the time preceding the Geneva Conference to fairly recent times is available.

The IUPAC Commission on Nomenclature of Organic Chemistry has continuing responsibility for revising and expanding the rules that appeared in the *Definitive Report*.

A considerable number of trivial or semitrivial (traditional) names have been retained by IUPAC for compelling practical reasons; the approved ones are available. A very brief selection is shown in Table 2. Other contributions to organic nomenclature are available.

Biochemical Nomenclature

The IUPAC Commission of Nomenclature of Biological Chemistry was established in 1921, along with the organic and inorganic commissions. It worked actively and closely with the organic commission. Early subjects of concern were carbohydrates, proteins, enzymes, and fats. More recently, this Commission shared its work with a corresponding Commission of the International Union of Biochemistry, which is not the International Union of Biochemistry and Molecular Biology (IUBMB); this led to the establishment of the Joint Commission on Biochemical Nomenclature (JCBN) in 1964.

The Joint IUPAC/IUBMB Commission has published many recommendations dealing with the nomenclature of natural products. The IUBMB Commission on Nomenclature has also issued a number of recommendations dealing with areas of a more biochemical nature and for naming enzymes.

The presence of many chiral centers in compounds of biochemical significance or natural-product interest has led to the use of stereoparents. These are parent structures having trivial names that imply (without explicitly expressing) a particular steric configuration. Common examples are the names of simple sugars, exemplified by glucose.

Although it is not strictly within the subject of biochemical nomenclature, it is appropriate to mention the existence of standardized generic names for pharmaceutical drugs. Such names are essentially coined, or trivial, names, but often include syllables from the systematic organic names, and endings that reflect a structural class, eg, -cillin (from penicillin), or an important area of medical application, eg, -vir (antiviral). Glossaries of these generic names are published periodically and a glossary of United States Approved Names (USAN) is published annually.

Macromolecular Nomenclature

In 1967, the Polymer Nomenclature Committee of the American Chemical Society published proposals for naming linear polymers on the basis of their chemical structure, which were then introduced into *Chemical Abstracts (CA) Indexes* and published in their final form in 1968.

A Macromolecular Division of IUPAC was created in 1967, and it created a permanent Commission on Macromolecular Nomenclature, parallel to the other nomenclature commissions. The Commission over the years has issued recommendations on basic definitions, stereochemical definitions and notations, structure-based nomenclature for regular single-strand organic polymers and regular single-strand and quasi-single-strand inorganic and coordination polymers, source-based nomenclature for polymers and abbreviations for polymers. All of these are collected in a compendium referred to as the IUPAC Purple Book.

Recommendations on additional aspects of macromolecular nomenclature such as that of regular double-strand (ladder and spiro) and irregular single-strand organic polymers continue to be published in *Pure and Applied Chemistry*. Recommendations on naming nonlinear polymers and polymer assemblies (networks, blends, complexes, etc) are expected to be issued in the near future.

Examples of the two macromolecular nomenclature systems are as follows. For source-based names for homopolymers and copolymers: polyacrylonitrile, poly(methyl methacrylate), poly(acrylamide-*co*-vinylpyrrolidinone), polybutadiene-*block*-polystyrene, and poly(propyl methacrylate)-*graft*-poly(1-vinylnaphthalene). Structure-based examples are as follows: poly(oxy-1,4-phenylene) (**1**), poly(oxyethyleneoxyterephthaloy) (**2**) and poly[imino(1-oxo-1,6-hexanediyl)] (**3**).

(1)

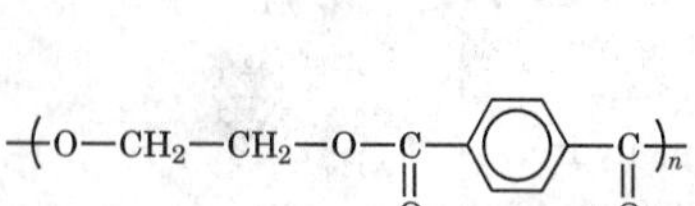

(2)

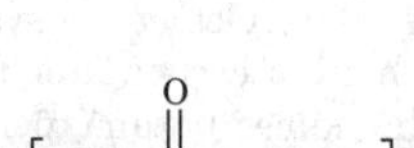

(3)

Nomenclature in Other Areas of Chemistry

A number of glossaries of terms and symbols used in the several branches of chemistry have been published. They include physical chemistry, physical-organic chemistry, and chemical terminology (other than nomenclature) treated in its entirety. IUPAC has also issued recommendations in the fields of analytical chemistry, colloid and surface chemistry, ion exchange, and spectroscopy, among others.

PETER A. S. SMITH
University of Michigan

J.-C. Richer, R. Panico, and W. H. Powell, *Guide to the Use of IUPAC Nomenclature of Organic Compounds*, Blackwell, Oxford and London, 1994.

International Union of Biochemistry and Molecular Biology, *Biochemical Nomenclature and Related Documents*, 2nd ed., Portland Press, London, 1992.

American Chemical Society, *Macromolecules* **1**, 193 (1968).

B. P. Block, W. H. Powell, and W. C. Fernelius, *Inorganic Chemical Nomenclature: Principles and Practices*, American Chemical Society, Washington, D.C., 1990.

NONDESTRUCTIVE EVALUATION

The technology of nondestructive evaluation (NDE) includes all nondamaging or nonintrusive methods for determining material identity; for evaluating material properties, composition, structure, or serviceability; and for detecting discontinuities and defects in materials. Nondestructive tests are commonly used for quality control process control (qv), and for reliability assurance of materials, protective coatings (qv), components, welds, assemblies, structures, and operating systems. The use of these tests helps prevent premature failure of materials during processing, manufacturing, or assembly, and during service under anticipated operating stresses and environments. Proper testing can lower manufacturing and operating costs, minimize insurance risks, and help prevent interruptions of service and potential disasters that might cause loss of life or usefulness of costly facilities.

Operating companies and regulatory agencies (qv) have a vital responsibility for specifying and managing the use of nondestructive testing during erection and service of costly facilities and systems such as chemical plants, petroleum refineries, off-shore drilling platforms, nuclear power plants, and transport systems where failures during services are potentially disastrous. Effective in-service nondestructive tests of materials and additional nondestructive tests made during maintenance (qv) shutdown periods can permit early detection and subsequent correction of hidden damage or deterioration (see also MATERIALS RELIABILITY; MATERIALS STANDARDS AND SPECIFICATION). Such test programs provide protection for the public, as well as for corporations and management from failure costs and penalties, which can include high litigation and insurance costs.

Nondestructive tests applied to consumer products ensure safety and proper operation for the purchaser. Tests might also be used periodically by the consumer during service to determine whether deterioration or conditions that might lead to premature failure have developed as a result of improper handling or storage, or misuse or abuse by operators. Inspection is specially important in cases where cracking or failure has already begun but has not progressed to the point of a disastrous sudden failure. Use of NDE can add to consumer satisfaction, eliminate the need for recalls and repairs, and help to protect both manufacturers and service industries from damage suits and litigation. The proper use of nondestructive tests can be a factor of economic survival for organizations manufacturing and servicing consumer products.

Program Requirements

Education and Certification. Nondestructive tests only provide indications which must be interpreted. An effective NDE program requires educated and experienced personnel at all levels. Management personnel who have technical degrees are essential. Decisions on inspection frequency, method, and the critical nature of the findings require engineering knowledge.

Test operators and engineers must evaluate indications from test results and make decisions concerning the suitability of the material for further processing or use. For the NDE team, this decision involves interpretation of the test indications based on prior knowledge of the nature, composition, and structure of the material and effects of prior process handling. If failure is thought to be possible, the dangerous material should be replaced or repaired, and further NDE made to confirm adequate repair or replacement.

An effective NDE program relies heavily on periodic certification of the competence of its personnel. Certification programs designate levels of competence for all levels of personnel.

Risk-Based Inspection. Inspection programs developed using risk analysis methods are becoming increasingly popular (see HAZARD ANALYSIS AND RISK ASSESSMENT). In this approach, the frequency and type of in-service inspection (ISI) is determined by the probabilistic risk assessment (PRA) of the inspection results. Implementation of a risk-based inspection systems should lead to an overall improvement in the inspection costs as well as in the safety in operation for a plant, component, or a system. Unless the database is well established, however, costs may fluctuate considerably.

Human Factors. Several nontechnical factors can significantly affect the results of a nondestructive inspection. Many of these are classified as human factors. Operator experience affects the probability of detection of most flaws. Operator fatigue, boredom, unfavorable environment and also negativity may affect performance.

Visual Inspection. Without scientific proof, it is estimated that 80% of defects are found by visual inspection. Human factor considerations are particularly important for the visual inspection process.

Automation and Control of Tests. Equipment must often operate under severe conditions of test material and instrumentation: contamination, vibration, temperature, corrosive exposure, and inaccessibility. In large installations, surveillance systems designed to detect human or equipment intrusions into critical locations may become a vital part of nondestructive tests systems designed to monitor plant or transportation equipment and prevent disastrous failure during its operating life.

Principles

Most nondestructive tests require use of suitable probing media to cause the test objects to emit signals that can be detected and interpreted in terms of material properties or defects. These resultant test signals may be converted to visible or audible indications, analogue electrical signals, meter or digital display readings, computer data, or images of many different types. Output signals, images, or data must then be evaluated by the human test operator or by automated test systems.

Each type of probe used in active nondestructive test systems has specific capabilities and limitations. No single type of inspection method is adequate to detect all of the material conditions and discontinuities that can influence serviceability of materials or systems. Thus, for reliable testing, two or more basically different types of test methods are usually necessary to detect and confirm conditions that may affect performance.

External Energy Tests. Material response to an external energy or probing medium input is the nondestructive test output signal. Many test systems involve measuring the reflection or scattering of the probing medium by the material of the test object.

Passive Tests. Passive nondestructive tests are those for which no specific probing medium, other than environmental or service loading conditions, need be applied to test objects to obtain test signals. Signals are emitted when changes in internal stress distributions, phase or structure, temperature, corrosion attack, external loading, or other operating conditions produce them.

Surface and Near-Surface Defect Detection

Inspecting external surfaces or walls for cracks and leaks, or gauging the dimensions and surface geometry of tests objects, requires techniques having special capabilities. Surface inspections typically involve techniques using gases, vapors, fluids, and particles. Motion may be imparted to such active, probing media by mechanical, hydraulic, or electromagnetic forces, or it may result from gas or vapor pressure, diffusion, permeation, osmosis solubility, evacuation, pumping, or other factors. In other cases, capillary action is all that is required. Additionally, devices such as calipers, mechanical gauges, and micrometers may be useful for surface inspections. Subsurface discontinuities and defects cannot, as a general rule, be detected by tests based on the active media techniques (see also SURFACE AND INTERFACE ANALYSIS). Surface evaluations include visual inspection and optical tests, liquid-penetrant inspection, filtered particle inspection, magnetic flux leakage tests, eddy-current and magnetic induction tests, radioactive krypton-85 gas penetrant inspection, electron surface-transit inspection with solid probe, electrified particle tests, and electron imaging tests.

Inspection for Internal Defects

Energy transmitted for use in nondestructive testing includes acoustic, static, electric, magnetic, electromagnetic, gravitational fields, dynamic electromagnetic fields, and photon beams. Such probing media may be capable of penetrating to great depths, as can high energy x- or y-rays. Among the advantages of the energy transmission methods is the possibility of visualizing tests object surfaces of interior volumes. This also makes possible the detection of objects at considerable distances, precise dimensional or displacement measurement, and the inspection of objects that are in motion, as in flash photography or radiography, or in motion picture or television imaging and recording. When imagining is feasible, there is a further psychological advantage. Visual information is often more convincing to management and workers. Digital imaging also allows test results to be transmitted to remote points, stored, and retrieved for use by others at later times. Further, image enhancement may be useful for special interpretation purposes. Energy transmission methods include ultrasonic tests (eg, immersion testing and angle beam testing), acoustic emission tests, penetrating eddy-current tests in nonmagnetic material, microwave tests, radiographic imaging tests (eg, x-ray and y-ray tests), neuron imaging tests, positron annihilation tests, infrared and thermal conduction tests, optical holography tests, and high-voltage probe and corona tests.

Materials Characterization and Identification

Nondestructive evaluation plays an important role in material property investigations. Among the many possible applications, the various techniques are able to determine the elastic constant and the texture as well as the chemical content necessary for alloy identification. Microwave tests are used most widely for measurements of moisture content of grains, foods, paper (qv), and ceramics, and in thickness tests of highway pavements. Thickness tests of metals employ radar transmitters and detectors on both sides of moving strips. Microwaves can respond to various surface coatings. Microwave phase and amplitude changes during transmission or reflection have been used for materials evaluation of ceramics, organic resin materials and composites, and rock samples from the earth. Microwaves are used to detect and evaluate faults in microwave plumbing or transmission systems.

Probing media that use chemical reactions to identify materials, evaluate surface coatings, detect leaks and surface connected discontinuities, or measure corrosion resistance are typically limited to surface and regions accessible to fluid chemical probing. Chemical spot tests and modifications involving electrochemical reactions are commonly used to identify materials or to sort accidentally mixed metals and alloys.

For steel and other ferromagnetic materials, property determination is more difficult. Other tests are made to measure the continuity of protective metallic coatings. Both ultrasonic and radiographic techniques have shown applications which are useful in determining residual stresses.

Inspections

Leak Detection and Measurement. Leak testing utilizes fluid probing media for detection and location of leaks and for measurements of leakage rates through individual complex systems. Significantly high leak testing sensitivity is attainable using gaseous tracers rather than liquid tracers. Leak detection and prevention is vital where systems contain poisonous or hazardous chemicals, fuels, oxidants, radioactive materials, high pressure, or vacuum.

High voltage probe and corona tests can be used to detect cracks and pinholes in dielectric coatings on conducting base metals by electric sparks from probe to base metal. Analogous tests are used to detect leaks into evacuated systems, where a high frequency spark coil may be used to create the electric field.

Flaw Sizing. Correct flaw sizing is one of the most crucial aspects of NDE. Using a confident assessment of the flaw size, the risks associated with continued operation may be determined. Whereas liquid penetrant and wet magnetic reveal surface defects, the length of tightly closed crack ends may be underestimated. Eddy-current techniques may be more sensitive to correctly locating crack ends. Wet magnetic or eddy-current are usually more sensitive than dry magnetic or penetrant. Depth of cracks is best determined using ultrasonics, even though this method gives rise to uncertainties. The electric current conduction method may also be used for depth determinations, but care must be exercised to avoid seriously faulty readings. For submerged, cracklike flaws, the technique through which the flaw was found is often best for size estimation. Complimentary methods are also useful. Length can often be confirmed using radiography, if that is possible.

Where ultrasonics is the best choice of technique, tip diffraction, where the probes are located are several positions, is helpful. Electric current conduction tests, in which current is introduced into test materials by direct electric contacts and potential drops are measured across specific zones of the test surface using additional contact electrodes, are used to measure effects of corrosion wall-thinning, as in petroleum product storage tank roofs, or to detect cracks in welds, metallic sheets and plate, and railroad rails. Difficulties in applying direct electric contact testing include the possibility of sparks or burns created by passing high current through poor electrode contact areas, and the thermoelectric and triboelectric effects that may produce false signal voltages owing to short-circuit paths that the current may find through touching crack surfaces. Alternatively, thermoelectric and triboelectric voltage signals can be used to identify and sort metals and alloys where surface contamination effects do not interfere.

Imaging Techniques. Image reconstruction techniques generally involve the placement of probe, source sensor, etc, at several locations around the part and the storage in a computer of the locations of the source and sensor and the observed response. Some surface techniques, such as penetrant and magnetic particles, are inherently imaging techniques. Little may be known about the shape below the surface however. Basic radiographic methods are, in some situations, able to produce good imaging. Depth resolution, again, may be lacking. Advancements in the imaging capabilities of radiography, ultrasound, and magnetic testing, have occurred throughout the 1990s (see also IMAGING TECHNOLOGY). Flaw imaging techniques using ultrasound, such as the computed tomography (CT) and the pitch-catch method are used industrially.

Thickness Measurement. Accurate thickness measurement is important in determining the remaining life in many piping and pressure vessel installations. Ultrasonics provide good thickness determination capability.

Radiographic imaging of corrosion wall thinning at tube bends and other large corrosion pitting regions may be applied to piping, storage tanks, and pressure vessels used in the chemical and petroleum industries.

Supplies of Equipment and Services. A successful NDE program depends on qualified suppliers of services, training, and equipment. Potential suppliers of these are listed in yearly update sources.

Economic Aspects

Nondestructive evaluation is a service industry, and NDE economic activity is directly related to the economy of the basic industries, ie, aerospace, utilities, petrochemical automotive, metals, or other. Overall, the NDE equipment market is small. Aerospace is at the top with 26% of the market; petrochemical is at the bottom with 10%. Ultrasonic equipment dominates worldwide equipment sales; x-ray film and equipment also show strong markets. Eddy-current, magnetic particle, dye penetrant, and other techniques represent small portions of total sales.

Growth areas in NDE appear to be in equipment improvement and the development of new engineering systems for more efficient and reliable NDE. NDE is seen to be moving to the role of a tool used by industry for economic benefit related to minimizing losses and maximizing income. Growth in the demand for NDE in engineering education is likely to continue.

DONALD E. BRAY
Texas A&M University

D. E. Bray and D. McBride, *Nondestructive Testing Techniques,* John Wiley & Sons, Inc., New York, 1992.

D. E. Bray and R. K. Stanley, *Nondestructive Evaluation,* Revised Edition, CRC Press, Boca Raton, FL, 1997.

ASM International, *Metals Handbook,* Vol. 17, 9th ed., ASM International, Materials Park, Ohio, 1989.

R. C. McMaster, ed., *Nondestructive Testing Handbook,* The Ronald Press Co., New York, 1959, 1963, The American Society for Nondestructive Testing, Columbus, Ohio, 1977, 1979.

NONLINEAR OPTICAL MATERIALS

Nonlinear optical (NLO) materials are the building blocks of the emerging technology of photonics, ie, the acquisition, transmission, processing, and storage of information using photons (light quanta). The interactions between photons are weak compared to interactions between electrons, thus long distance transmission of information over fiber optic lines (see FIBER OPTICS) with minimal signal degradation is facilitated. Moreover, compared to electronics, photonics offer advantages in terms of speed (bandwidth) of information processing.

Analogous to electronic materials, photonic materials can be divided into two classes: linear (passive) and nonlinear (active). Linear materials are employed in the construction of transmission lines (waveguides), passive directional couplers, gratings, etc; nonlinear materials are used in the construction of switches, modulators, frequency doublers and triplers, active beam steering elements, limiters, amplifiers, rectifiers, and transducers. Nonlinear optical phenomena arise when applied external fields, ie, either light or low frequency electrical fields, are sufficiently strong to compete with internal electrostatic interactions. Thus, nonlinear optical materials are typically those containing weakly bound (highly polarizable) electrons. Nonlinear optical phenomena include harmonic generation, sum- and difference-frequency generation, optical parametric oscillation, rectification, intensity-dependent refraction, phase conjugation, multiple wave mixing, Raman scattering, Brillouin scattering, induced opacity and reflectivity, and multiple photon absorption.

Further subclassification of nonlinear optical materials can be explained by the following two equations of microscopic, ie, atomic or molecular, polarization, p, and macroscopic polarization, P, as power series in the applied electric field, E (disregarding quadrupolar terms which are unimportant for device applications):

$$p = \alpha E + \beta EE + \gamma EEE + \cdots \quad (1)$$

where α is the linear polarizability, β and γ are the first and second (atomic or molecular) hyperpolarizabilities, respectively; and

$$P = \chi^{(1)}E + \chi^{(2)}EE + \chi^{(3)}EEE + \cdots \quad (2)$$

where $\chi^{(1)}$ is the linear optical susceptibility, $\chi^{(2)}$ the second-order NLO susceptibility, and $\chi^{(3)}$ the third-order NLO susceptibility. The second-order coefficients are zero for centrosymmetric symmetries.

Materials are also classified according to a particular phenomenon being considered. Applications exploiting off-resonance optical nonlinearities include electrooptic modulation, frequency generation, optical parametric oscillation, and optical self-focusing. Applications exploiting resonant optical nonlinearities include sensor protection and optical limiting, optical memory applications, etc. Because different applications have different transparency requirements, distinction between resonant and off-resonance phenomena are thus application specific and somewhat arbitrary.

Nonlinear optical materials can also be classified according to atomic composition into organic and inorganic materials. Inorganic materials range from crystalline lithium niobate, $LiNbO_3$, to amorphous semiconductor materials such as gallium arsenide, GaAs (see ELECTRONIC MATERIALS, SEMICONDUCTORS). Organic materials can be either crystalline such as polydiecetylene, or polymeric materials incorporating a variety of organic chromophores. For most organic materials, however, the optical nonlinearity can be associated with weakly bound π-electrons. As of this writing (ca mid-1995), no single material has been proven suitable for significant commercial applications, thus discussion herein focuses on the material requirements for various applications.

Second-Order Nonlinear Optical Materials

Many classes of second-order material applications can be envisioned by noting the sinusoidal nature of electromagnetic radiation and rewriting equation 2 as

$$P = \chi^{(1)}E_0 \cos(\omega t - kz) + (1/2)\chi^{(2)}[1 + \cos(2\omega t - 2kz)] + \cdots \quad (3)$$

where ω is the frequency of the electromagnetic radiation, t the time, and k and z are the wave vector and spatial coordinate (position), respectively.

Two applications can be identified for second-order NLO materials: frequency doubling, ie, the 2ω term; and electrooptic modulation, ie, the first term in brackets.

The most commonly encountered device configurations are shown in Figure 1.

The method of choice for the fabrication of organic EO materials appears to be the electric field poling of NLO chromophores in polymer matrices near the glass-transition temperature, T_g, of the polymer matrix. A deficiency of this approach is the relaxation of poling-induced order when the poling field is removed. To overcome this problem, a variety of lattice-hardening reactions have been developed. These have generally been successful in permitting the fabrication of NLO-active polymer lattices that can withstand temperatures on the order of 100°C for long (thousands of hours) periods of time and even higher temperatures for such short periods of time as associated with the deposition of metal electrodes. The final glass-transition temperature appears to be the critical parameter in defining the long-term stability of poling-induced optical nonlinearity both for chromophore–polymer composites and for chromophores covalently attached to the polymer host lattices.

Third-Order Nonlinear Optical Materials

Third-order processes include third harmonic generation (THG), self-focusing, self-defocusing, self-phase modulation, saturable absorption, and reverse saturable absorption. Because of the weak magnitude of third-order optical nonlinearities, practical applications are rare; the few applications that do exist typically exploit long interaction lengths.

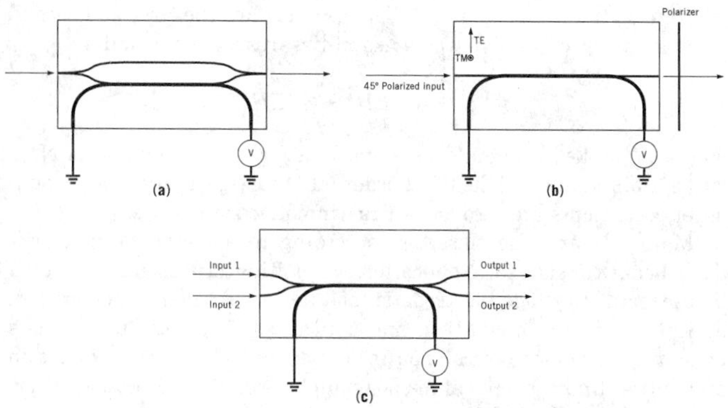

Figure 1. Representative device configurations exploiting electrooptic second-order nonlinear optical materials are shown. Schematic representations are given for (**a**) a Mach-Zehnder interferometer, (**b**) a birefringent modulator, and (**c**) a directional coupler. In (**b**) the optical input to the birefringent modulator is polarized at 45 degrees and excites both transverse electric (TE) and transverse magnetic (TM) modes. The applied voltage modulates the output polarization. Intensity modulation is achieved using polarizing components at the output.

A wide range of materials have been studied in the search for materials containing usable optical nonlinearities; such materials include inorganic crystals and glasses; doped glasses; simple gases and liquids (atomic and molecular as well as organic and inorganic); semiconductors; quantum dots, wells and wires; conductive particle composites; organic and polymer crystals; amorphous organic and inorganic polymers; and biological complexes. Mechanisms of optical nonlinearity can be found in the literature for many types of optical nonlinearity and various classes of materials contributing to these nonlinear phenomena.

Photorefractive Materials

Photorefractivity can be thought of as a four part process. Initially, pairs of spatial frequency modes of a single input beam interfere with the photorefractive material to produce a periodic intensity distribution. The input light causes impurities (inorganics) or electron donors/acceptors (organics) to release charges (electrons or holes) that migrate through the dark regions of the material and become trapped. This results in a periodic charge distribution in the material which, in turn, yields a periodic electrical field. If no external electric field is present, this periodic electrical field is phase shifted by $\pi/2$ from the original light interference pattern. Finally, the induced space-charge field alters the refractive index periodically through the linear electrooptic effect. The index of refraction grating created in this manner combines with the intensities of the spatial frequency modes and introduces a nonlinear phase for each mode. The overall effect is a Kerr-like phenomenon realized with a second-order nonlinear optical chromophore. Photorefractive materials range from crystalline inorganic materials such as lithium niobate to organic polymer composites and homopolymer materials.

Economic Aspects

As of the mid-1990s, large-scale commercial applications have yet to be found for nonlinear optical materials. This situation may change dramatically before the year 2000. Among the most promising classes of materials for extensive commercial application are second-order materials for use as signal transducers in cable television (CATV) and for real-time intercomputer communication applications; switches in local area optical networks; ultrafast digital-to-analogue converters; and alternatives to electronic switches for ultrafast switching operations, for radio frequency and microwave power distribution, and for remote voltage sensing. A small market also exists for frequency doublers based on inorganic crystalline materials. Similarly, a niche market application has been found for BBO and KTP materials for optical parametric oscillators and amplifiers. These devices are available through laser suppliers such as Coherent, Continuum, and Spectral Physics.

LARRY R. DALTON
University of Southern California

R. W. Boyd, *Nonlinear Optics,* Academic Press, Inc., New York, 1992.

D. L. Wise, G. E. Wnek, D. J. Trantolo, T. M. Cooper, and J. D. Gresser, *Electrical Optical Polymer Systems,* Marcel Dekker, New York, 1998.

H. S. Nalwa and S. Miyata, *Nonlinear Optics of Organic Molecules and Polymers* CRC Press, Boca Raton, Fla., 1997.

L. R. Dalton, *Chem. Ind.* **13,** 510(1997).

NONWOVEN FABRICS

STAPLE FIBERS

Nonwoven fabrics are similar to woven and knitted fabrics in that both are planar, inherently flexible, porous structures composed of polymer-based materials. They are technically sophisticated engineered structures that can be made to resemble in appearance, and exceed in properties, many woven or knitted fabrics. The main difference between nonwoven and woven fabrics is the manner in which the fabric is made.

Nonwoven Processes

A nonwoven fabric can be assembled by mechanically, chemically, or thermally interlocking layers or networks of fibers, filaments, or yarns. Fabrics made from textile fibers using textile technology have been classified as dry-laid nonwovens. A nonwoven fabric can also be made by suspending fibers in water or some other fluid (including air), controlling the way by which the fibers and suspending media are separated, and then mechanically, chemically, or thermally interlocking the fibers together. Fabrics made in this manner using papermaking technology have been classified as wet-laid nonwovens. A third way that nonwoven fabrics can be made is by extending the fiber extrusion process to include interlocking fibers or filaments concurrent with their extrusion, modifying the porosity of a film by perforating it, or modifying the film manufacturing process in order to form porous films concurrent with their extrusion. Fabrics made using extrusion technology are called polymer-laid nonwovens.

Nonwoven technologies that employ machinery and processing principles traditionally used to manufacture textile, paper, or extruded materials, when viewed collectively, form what may be termed the primary or basic nonwoven fabric manufacturing systems. These systems are or can be continuous processes. Common to each of these systems are four sequential phases: fiber selection and preparation, web formation, bonding, and finishing.

An outline of the three basic nonwoven manufacturing systems arranged according to parent technology and the four manufacturing phases common to each is given in Table 1.

Nonwoven hybrid technology includes (*1*) methods to combine two or more nonwoven fabrics made by any of the primary nonwoven manufacturing systems, (*2*) methods to provide a combination of fabric properties, and (*3*) methods to produce true composite nonwoven structures.

The various nonwoven processes and the fabrics made from each have a number of common characteristics. In general, textile technology-based processes provide maximum product versatility, because almost all textile fibers and bonding systems can be utilized and conventional textile fiber processing equipment readily adapted at minimal cost. Extrusion technology-based processes provide somewhat less versatility in product properties, but yield

Table 1. Basic Nonwoven Fabric Manufacturing Systems

	Textile			Paper		Extrusion		
System	Garnetting	Carding	Air-laid fiber	Air-laid pulp	Wet-laid	Spunbond	Meltblown	Film
fiber selection and preparation	Natural and manufactured textile fibers			Natural and manufactured fiber/pulp		Fiberforming polymer chips		
	mechanical opening and volumetric blending			mechanical opening				
				gravimetric feeding	wet slurry	mechanical, electrostatic, aerodynamic filament orientation	aerodynamic fiber orientation and shattering	perforate, cast; cast and aperture
web formation	Mechanical		Fluid					
	parallel fiber layers randomized batts cross-lapped layers		isotropic fiber layers	random fiber mattes	controlled fiber layers	pattern layering on conveyor screen	collection on conveyor screen or shape	heat, heat stretch, perforate, heat, stretch
web consolidation (bonding)	Mechanical				Mechanical	Mechanical	Cooling	
	stitchbonding, needle-punching, hydroentangling				hydroentangling	needle-punching		
	Chemical							
	sprayed latex or powder; saturated, printed, or frothed latex; solvent							
	Thermal							
	thermal calender, radiant or convection oven, vacuum drum or mold, laminating, sonic welding							
finishing	Slitting and winding							
	other application-dependent physical or chemical surface treatments							

fabric structures with exceptional strength-to-weight ratios, as is the case with spunbonds; high surface area-to-weight characteristics, a benefit of using meltblown technology; or high property uniformities per unit weight, as is the case with textured films, at modest cost. Paper technology-based nonwoven processes provide the least product versatility and require a high investment at the outset, but yield outstandingly uniform products at exceptional speeds. Hybrid processes provide combined technological advantages for specific applications.

Fibers for Nonwovens

The properties of nonwoven fabrics are highly influenced by the properties of their constituent fibers (see FIBERS–SURVEY). Discontinuous fibers, such as wool, cotton, and acrylic, used in nonwoven fabrics designed to serve as vibration felts, furniture batting, and blankets, respectively, are referred to as staple fibers. Staple fibers range in length from about 2 to 20 mm.

Virtually all fibers are composed of long-chain molecules or polymers arranged along the fiber axis. Essential requirements for fiber formation include long-chain molecules with no bulky side-groups, strong main-chain bonding, parallel arrangement of polymer chains, and chain-to-chain attraction or bonding. Basic phases in the fiber formation process are obtaining a suitable polymeric material, converting the material to liquid form, solidifying the material into fiber dimensions, and treating the fiber to bring about desired properties. These four phases are present in the formation of natural as well as manufactured synthetic fibers, the principal differences being the amount of time and energy required.

Web Formation

Web formation, the second phase in manufacturing nonwoven fabrics, transforms fibers or filaments from linear elements into planar arrays in the form of preferentially arranged layers of lofty and loosely held fiber networks termed webs, batts, matts, or sheets. Basic fabric parameters established at web formation, in addition to fiber orientation, are unfinished product weight and manufactured width. In all nonwoven manufacturing systems, the fiber material is deposited or laid on a forming or conveying surface.

Textile Carding. Textile fiber-processing machine design is based on fiber length and diameter. In the preparation of discontinuous fibers such as cotton and wool for conversion into yarns, the carding process transforms entangled fiber mats weighing about 500 g/m^2 into parallel strands or slivers weighing about 5 g/m. In carding, the fiber mats are decreased in mass per unit length, and individual fibers are provided a parallel orientation. The two primary actions involved are termed carding and stripping.

The carding action is the combining or working of fibers between fine surfaces or points oriented in opposing directions. Actual carding or parallelization of fibers occurs when one of the surfaces moves at a speed greater than the other. The stripping action occurs when the points are arranged in the same direction and the more quickly moving surface removes or transfers the fibers from the more slowly moving surface.

Nonwoven Cards. Modern, high speed cards designed to produce nonwoven webs show evidence of either a cotton or wool fiber-processing heritage and have processing rate capabilities comparable to those of garnetts. Contemporary nonwoven cards are available in widths up to 5 m and are configured with one or two main cylinders, roller or stationary tops, one or two doffers, or various combinations of these principal components.

Web Layering. Forming fibers into a web on carding or garnetting machinery is a mechanical transfer operation and takes place at the doffer. The web is formed as the doffer strips and accumulates fibers from the cylinders. The number of fibers accumulated and the mass of each fiber determine the weight of the web. For a given fiber orientation, web weight per unit area is limited by the ratio of the surface speed of the cylinder to the surface speed of the doffer.

Nonwoven fabrics are produced in weight ranging from less than 10 to several hundred grams per square meter, and fiber orientations ranging from parallel (a ratio of 6 to 10 times more fibers aligned in the machine direction as opposed to the cross-machine direction) to biaxial to isotropic or random. Web building or web-layering to achieve a desired weight can be accomplished by folding from one forming machine, collection from multiple forming machines, or cross-lapping.

Web Spreading and Web Drafting. Spreading layers of parallel fiber webs is a means of simultaneously increasing web width, decreasing web weight, and altering fiber orientation. Controlled stretching or drafting web layers is a means of simultaneously increasing web throughput, decreasing web weight, and altering fiber orientation.

Random Cards. Fiber orientation ratios as low as 3:1 can be achieved on cards by expanding the condensing action at doffing through the addition of scrambling or randomizing rolls operating at successively slower surface speeds.

Aerodynamic Web Formation. Air-laid nonwovens can be grouped into two categories: those formed from natural or synthetic textile fibers and those formed from natural or synthetic pulps. An aerodynamic web formation machine designed to process textile fibers is shown in Figure 1.

Short Fiber Systems. Traditional papers utilize a variety of woodpulps or other short (1–4-mm) cellulosic fibers which pack together to form relatively dense, nonporous, self-adhered sheets. The use of textile fibers, instead of cellulose-based materials, with papermaking machinery distinguishes wet-laid nonwoven manufacturing from traditional paper manufacturing. Both manufacturing methods, however, transport the fibers in a water slurry. The use of papermaking fibers on air-laid nonwoven machinery bridges a gap between textile and paper systems. In both technologies, the transport medium is a fluid; water in wet-laid nonwovens, and air in dry-laid pulps.

Web Consolidation

Nonwoven bonding processes interlock webs or layers of fibers, filaments, or yarns by mechanical, chemical or thermal means. The extent of bonding is a significant factor in determining fabric strength, flexibility, porosity, density, loft, and thickness. Bonding is normally a sequential operation performed in tandem with web formation, but it is also carried out as a separate and distinct operation. In some fabric constructions, more than one bonding process is used as a means to enhance the physical or chemical properties of the fabric.

Needle-Punching. In this method, sometimes called needle-felting, fiber webs are mechanically interlocked by physically repositioning some of the fibers from a horizontal to a vertical position. Fiber repositioning is achieved by intermittently passing a barbed needle into the web to move groups of fibers from one layer to another, and then withdrawing the needle without disturbing the newly oriented fibers.

Principal applications of needle-punched nonwovens during the 1990s include automotive, apparel components, blankets, carpeting, carpet padding, coating substrates, filtration, furniture, geotextiles,

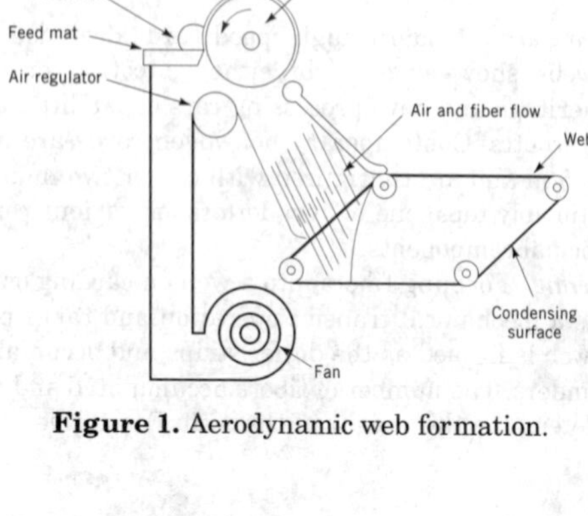

Figure 1. Aerodynamic web formation.

insulation, roofing substrates, and wall covering. In 1997, the production of needle-punched fabric was estimated by the author to be about 62,500 tons and 875 million square meters.

Stitchbonding. This is a mechanical bonding method that uses knitting elements, with or without yarn, to interlock fiber webs.

Stitchbonded fabrics are used in home furnishings, footwear, filtration, packaging, and coating. A variation of stitchbonding is used to make multiaxial-layered yarn and yarn-and-sheet structures for composite material reinforcement substrates.

Hydroentaglement. This is a generic term for a nonwoven process that can be used for either web consolidation, or fabric surface-texturing purposes, or both. The mechanism is one of fiber rearrangement within a preformed web by means of fluid forces.

Also termed spunlaced or jet-laced nonwovens, fabrics of this type have been sold commercially since the early 1970s and have been successfully used in applications such as interlinings, bedding, wound dressings, coating substrates, roofing, wipes, and surgical gowns.

Chemical Bonding. Sometimes called resin bonding, chemical bonding is a general term describing the technologies employed to interlock fibers by the application and curing of a chemical binder. The chemical binder most frequently used to bond nonwovens is a waterborne latex (see LATEX TECHNOLOGY). Most latex binders are made from vinyl materials. The versatility of a chemical binder system can be seen by considering a few of the factors involved in formulating such a system. The chemical composition of the monomer determines stiffness and softness properties, strength, water affinity (hydrophilic–hydrophobic balance), elasticity, and durability. The type and nature of functional side-groups determine solvent resistance, adhesive characteristics, and cross-linking nature. The type and quantity of surfactant used include the polymerization process and application method. The ability to incorporate additives such as colorants, water repellents, bacteriastats, flame retardants (qv), wetting agents, lubricants, and catalysts to enhance curing expands this inherent versatility even further.

Thermal Bonding. In thermal bonding, heat energy is used to activate an adhesive, which in turn flows to fiber intersections and interlocks the fibers upon cooling. The adhesive may be individual fibers, portions of individual fibers, or powders. In addition to fiber-to-fabric nonwovens, thermal bonding is used to consolidate spunbonds, meltblowns, dry-laid pulps, textured films, and combination nonwovens. Advantages of thermal bonding include low cost, the general availability of new binder materials and machinery, and process and product enhancement. Three basic methods of heating are used for thermal bonding: conduction, radiation and convection.

Finishing

Commercial nonwoven fabrics are shipped from manufacturing plant to customer in the form of rolls of varying dimensions to accommodate the fabric end use applications or subsequent conversion processes. Slitting and winding are finishing processes common to all nonwoven manufacturing methods. Roll width is determined at the slitting operation, and roll length is determined at the winding operation.

The fabric may also be given one or more of a number of other finishing treatments, either in tandem with web formation and bonding or off-line as a separate operation, as a means of enhancing fabric performance or aesthetic properties. Performance properties include functional characteristics such as moisture transport, absorbency, or repellency; flame retardancy; electrical conductivity or static propensity; abrasion resistance; and frictional behavior. Aesthetic properties include appearance, surface texture, and smell.

Generically, nonwoven finishing processes can be categorized as either chemical, mechanical, or thermomechanical. Chemical finishing involves the application of chemical coatings to fabric surfaces or the impregnation of fabrics with chemical additives or fillers. Mechanical finishing involves altering the texture of fabric surfaces by physically reorienting or shaping fibers on or near the fabric surface.

Thermomechanical finishing involves altering fabric dimensions or physical properties through the use of heat and pressure.

Production Trends

The nonwoven fabrics industry is relatively young, but has experienced rapid growth. In the 1990s, about half of worldwide fabric production takes place in North America, with one-third in Europe, and one-eighth in Japan. New nonwoven enterprises are in the works throughout Asia and South America; slightly more than half of all nonwovens are made directly from polymers.

Production of nonwoven fabrics according to manufacturing technology is as follows: spunbond, 43%; needle-punch, 7%; bonded pulp, 5%; card (thermal bond), 6%; hybrids, 9%; card (resin bond), 4%; spunlace, 7%; wet-laid, 5%; meltblown, 5%; stitchbond, 2%; and porous films, 2%.

Product Applications

Consumption of nonwoven roll goods is often reported in two broad areas, according to product application: disposables and durables. In general, disposable products account for ~80% of the volume and 65% of the value of nonwoven roll goods consumption. Items within each disposable or durable product category reflect the diversity of applications which utilize nonwovens.

Applications include as disposable products: absorbent covers; medical/surgical; wipes; filters; apparel (nonmedical); and laundry aids; a application as durable products include coating substrates; geotextiles and roofing; interlinings; bedding and home furnishings; carpet components; automotive trim; and electronic components.

E. A. VAUGHN
Clemson University

E. A. Vaughn, *Nonwovens World Factbook 1991,* Miller Freman, San Francisco, Calif., 1990.

E. A. Vaughn, *Nonwoven Fabrics Sampler and Technology Reference,* INDA, Cary, N.C., 1998.

E. A. Vaughn, *J. Nonwovens Res.* **4,** 1 (1992).

J. R. Starr and co-workers, *The Nonwoven Fabrics Handbook,* INDA, Cary, N.C., 1992.

FABRICS, SPUNBONDED

Spunbonded fabrics are distinguished from other nonwoven fabrics in their one-step manufacturing process, which provides either a complete chemical-to-fabric or polymer-to-fabric process. In either instance, the manufacturing process integrates the spinning, laydown, consolidation, and bonding of continuous filaments to form a fabric.

The large investment required for a spunbonded plant is offset by their high productivity. Most of the first plants constructed are still in operation, attesting to the usefulness of the method. New production plants continue to be built to supply the growing demand. Spunbonded production was originally limited to Western Europe, the United States, and Japan, but considerable activity has begun in other areas. Production lines, mainly nonproprietary, have been installed in China, Taiwan, Brazil, Argentina, Israel, and other countries that previously did not participate in the technology.

The area of largest growth for spunbonded fabrics has been disposable diaper component, which accounts for approximately 50% of the U.S. spunbonded market. Growth is forecast to generally exceed the growth of all nonwovens, which itself is expected to grow at 7% per annum. In addition to diaper coverstock and hygiene, growth is anticipated in geotextiles, roofing, carpet backing, medical wrap, and durable paper applications.

New plant construction will bring increased capacity to a level which will depend on real growth to keep sales abreast of production. It is anticipated that consolidation of ownership will continue and that the trend to specialized businesses supporting a plant facility will also continue. Pressures from environmental issues could change the cost of final products as well as mandate the use of post-consumer waste resin as feedstock for production. Any serious challenges to existing markets will likely come from film, foam, or advances in alternative technologies within a specific market segment.

General Characteristics

Fabric Structure. Spunbonded fabrics are filament sheets made through an integrated process of spinning, attenuation, deposition, bonding, and winding into roll goods. The fabrics are made up to 5.2 m wide and usually not less than 3.0 m in order to facilitate productivity. Fiber sizes range from 1.0 to 50 dtex, although a range of 2–20 dtex is most common. A combination of thickness, fiber fineness (denier), and number of fibers per unit area determines the fabric basis weight, which ranges from 10–800 g/m^2; is typical.

Most spunbonded processes yield a sheet having planar–isotropic properties owing to the random laydown of the fibers (Table 1). Unlike woven fabrics, spunbonded sheets are generally nondirectional and can be cut and used without concern for higher stretching in the bias direction or unraveling at the edges. It is possible to produce non-isotropic properties by controlling the orientation of the fibers in the web during laydown. Although it is not readily apparent, most sheets are layered or shingled structures with the number of layers increasing with higher basis weights for a given product. Fabric thickness varies from 0.1 to 4.0 mm; the range 0.2–1.5 mm represents the majority of fabrics in demand. The method of bonding greatly affects the thickness of the sheets, as well as other characteristics. Fiber webs bonded by thermal calendering are thinner than the same web that has been needle-punched, because calendering compresses the structure through pressure, whereas needle-punching moves fibers from the x–y plane of the fabric into the z (thickness) direction.

Fabric Composition. The method of fabric manufacture dictates many of the characteristics of the sheet, but intrinsic properties are firmly established by the base polymer selected. Properties such as fiber density, temperature resistance, chemical and light stability,

Table 1. Physical Properties of Spunbonded Products

Product	Basis weight, g/m^2	Tensile strength,[a] N[b]	Tear strength,[a] N[b]	Bonding method
Accord	69	144 MD	36 MD	point thermal
		175 XD	40 XD	
Bidim	150	495	280	needle-punch
Cerex	34	135 MD	40 MD	chemically induced area
		90 XD	32 XD	
Colback	100	300[c]	120	area thermal (sheath-core)
Corovin	75	130	15	point thermal
Lutradur	84	225 MD	85 MD	copolymer area thermal
		297 XD	90 XD	
Polyfelt	137	585	225	needle-punch
Reemay	68	225 MD	45 MD	copolymer area thermal
		180 XD	50 XD	
Terram	137	850	250	area thermal (sheath-core)
Trevira	155	630 MD	270 MD	needle-punch
		495 XD	248 XD	
Typar	103	540 MD	207 MD	undrawn segments
		495 XD	235 XD	area thermal
Tyvek	54	4.6[d] MD	4.5 MD	area and point
		5.1[d] MD	4.5 XD	thermal

[a] MD = Machine direction; XD = cross direction.
[b] Unless otherwise noted. To convert N to pound-force, divide by 4.448.
[c] 300 N/5 cm = 34.5 ppi.
[d] N/mm; to convert N/mm to ppi, divide by 0.175.

Table 2. Fibers for Spunbonded Nonwoven Fabrics

Fiber type	Breaking tenacity, N/tex[a]	Elongation, %	Specific gravity	Moisture regain,[b] %	Approximate melt point, °C
polyester	0.17–0.84	12–150	1.38	0.4	248–260
nylon-6,6	0.26–0.88	12–70	1.14	4.0	248–260
polypropylene	0.22–0.48	20–100	0.91	~0.0	162–171

[a] To convert N/tex to gf/den, multiply by 11.3
[b] At 21°C and 65% rh.

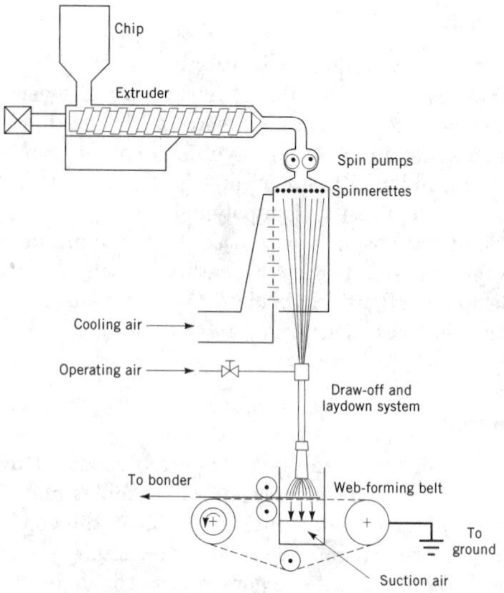

Figure 1. Melt spunbonded process.

ease of coloration, surface energies, and others are a function of the base polymer.

The majority of spunbonded fabrics are based on isotactic polypropylene and polyester. Small quantities are made from nylon-6,6, and a growing percentage of fibers are made from high density polyethylene.

Table 2 illustrates the basic characteristics of fibers made from different base polymers.

Polymer Combinations. Some fabrics are composed of combinations of polymers where a lower melting polymer functions as the binder element. The binder element may be a separate fiber interspersed with higher melting fibers, or the two polymers may be combined in one fiber type.

Fabric Properties. When taken together, properties such as tensile, tear and burst strength, toughness, elongation to break, basis weight, thickness, air porosity, dimensional stability, and resistance to heat and chemicals are often sufficient to uniquely describe one product. This is because these properties reflect both the fabric composition and its structure, the latter being defined by a manufacturing process unique to that fabric.

Diverse applications for the fabric sometimes demand specialized tests such as ones for moisture vapor, liquid transport, barrier to fluids, coefficient of friction, seam strength, resistance to sunlight, oxidation and burning, and/or comparative aesthetic properties. Most properties can be determined using standardized test procedures which have been published as nonwoven standards by INDA. A comparison of typical physical properties for selected spunbonded products is shown in Table 1.

Spinning and Web Formation

Spunbonded fabric production couples the fiber spinning operation with the formation of the web in order to maximize productivity. It is the coupling of these two processes that distinguishes the spunbonded process from traditional methods of fabric formation. If the bonding device is placed in line with spinning and web formation, the web is converted into bonded fabric in one step (Fig. 1). In some arrangements, the web is bonded off-line in a separate step which appears at first to be less efficient; however, this offers the advantage of being more flexible if more than one type of bonding is to be performed on the web being produced.

Spinnerette Process. The basic spinning process is similar to the production of continuous filament yarns and utilizes similar extruder conditions for a given polymer.

In traditional textile spinning, some orientation of fibers is achieved by winding up the filaments at a rate of approximately 3200 m/min to produce the so-called partially oriented yarns (POYs). In spunbonded production, filament bundles are partially oriented by being pneumatically accelerated at speeds of 6000 m/min or greater. Accelerating the filaments at such great speeds not only achieves a partial orientation but results in extremely high rates of web formation, particularly for lightweight structures (eg, 17 g/m^2). The formation of wide webs at high speeds results in a high efficiency of manufacture.

For many applications, this partial degree of orientation imparts a sufficient increase in strength and decrease in extendibility to make the final bonded fabric perfectly functional, eg, diaper coverstock. However, some applications, such as geotextiles (qv) and primary carpet backing, demand that the filaments achieve a very high tensile strength and low degree of extension. This requires subsequent additional attenuation, such as the mechanical drawing of filaments. Because drawing rolls cannot normally dispatch filaments as fast as pneumatic jets, the web-forming process is usually less rapid, although the resulting web has greater physical strength.

In order for the web to achieve maximum uniformity and cover, it is desirable for the individual filaments to be separated from each other prior to reaching the belt. Failure to sufficiently separate individual filaments results in the appearance of "ropes" in the web. One method used to effect this state of separation is to induce an electrostatic charge onto the bundle while still under tension and prior to pneumatic deposition. The charge may be induced either triboelectrically or by applying a high voltage charge, the former being a result of the rubbing of the filaments against a grounded but conductive surface. After deposition onto the moving belt, it is necessary to discharge the filaments; this is usually accomplished by bringing the filaments in contact with a conductive grounded surface. Other routes to reachieving filament separation have been described and rely on mechanical or aerodynamic forces to effect separation.

It is sometimes desirable to control the directionality of the splayed filaments on the laydown belt in order to achieve a particular characteristic in the final fabric. Directionality can be controlled by transversing the filament bundles either mechanically or aerodynamically as they travel downward toward the collecting belt.

Curtain Spin Process. The curtain spin process utilizes a single plate (or spinneret) the width of the desired web which has been drilled with holes for fiber formation. The advantage to this approach is that it results in a uniform distribution of filaments within the curtain of continuous fibers produced from the spinning plate. The use of the single uniform distribution of filaments within the curtain of continuous fibers is produced from the spinning plate. The use of the single spinning plate automatically places the fibers in a uniformly distributed array and thereby presents a curtain of high uniformity filaments to the fiber attenuation mechanism.

By comparison, the multiple spinnerette per bank process requires additional effort prior to laydown in order to compensate for the gaps between the individual spinnerettes.

In general, once the curtain of filaments has been produced, it is necessary to attenuate the filaments in order to provide strength and resistance to deformation. The most commonly practiced approach is to utilize a single slot, which is at least the width of the curtain, at a point below the spinning plate and above the laydown screen. One of the limitations of the curtain/slot draw process is that the amount of fiber attenuation is constrained due to the short distance generally allowed between the spinnerette and the venturi slot and the use of relatively low pressure air for drawing so as not to induce high turbulence in the area of the laydown. This has made the process difficult to adapt for the production of polyester fabrics which inherently require much higher fiber acceleration to attain the desired polyester fiber properties.

Bonding

Many methods can be used to bind the fibers in the spun web. Thermal and chemical/binder methods may bond the web by fusion or adhesion of fibers using either large or small regions, generally referred to as area bonding and point bonding, respectively. Point bonding results in the fusion of fibers at discrete points with fibers remaining relatively free in between the point bonds.

Of the three standard bonding methods used in spunbonded manufacturing, mechanical needling, also called needle-punching or needlebonding, is the simplest and least expensive. Although it is the oldest process, it continues to be widely used.

In the needle-punching process, a continuous filament web is subjected to barbed needles which are rapidly passed through the plane of the moving spun web (see NONWOVEN FABRICS–STAPLE FIBERS). The needles pass in and out of the web at frequencies up to 2200 strokes per minute which can result in as many as 500 penetrations per cm^2 depending on the needle density and the line speed, typically between 5 and 25 m/min. The effect of this operation is to interlace the fibers and thus bond the structure together, relying only on the mechanical entanglement and fiber-to-fiber friction. The fabric produced tends to be more conformable and bulky than fabrics bonded by thermal or chemical/binder methods. Because the fibers have freedom to move over each other, the fabric is easily deformed and exhibits a low initial modulus. The principal variables in needle-punching are the needle design, punch density, and depth of punch.

Both thermal and chemical/binder bonding depend on fiber-to-fiber attachment as the means of establishing fabric integrity. It is the degree and extent of attachment which determines many of the fabric qualities. Because point bonding can be accomplished using as little as 10% bonding area, ie, 90% unbonded area, such fabrics are considerably softer than area-bonded structures. Fiber mobility is retained, in part or in total, outside the areas of the point bonds. Thermal bonding is far more common than chemical/binder bonding and is generally more economical. Both area and point thermal bonding are rapid processes.

Area thermal bonding can be accomplished by passing the spun web through a source of heat, usually steam or hot air. Complete fusion of the fiber crossover points leads to a paper-like structure with low resistance to tearing.

The use of steam is generally limited to polypropylene and polyethylene fusion because impractical pressures are required to reach the temperature levels, eg, >200 °C, required for bonding polyesters.

Thermal point bonding utilizes both temperature and pressure to affect fiber-to-fiber fusion. Thus it is a simpler approach to bonding because it does not require the web to contain lower melting fibers or segments and is less demanding of the technology required to produce the web. Point bonding is usually accomplished by passing a preheated consolidated web through heated nip rolls, one of which contains a raised pattern on its surface. The degree of bonding between the points can be controlled by varying the ratio of heights of the raised points to the depth of the web. Typically only 10–25% of the surface available for bonding is converted to fused, compacted areas of bonding.

Chemical/binder bonding is used less frequently than thermal bonding in the production of spunbonded fabrics, whereas the opposite has been true for staple fiber nonwovens. Resin binders are occasionally used with spunbonded webs to achieve special characteristics which are unattainable thermally.

Resin binders may alternatively be applied in discreet points in a pattern so as to immobilize fewer fibers and produce a softer fabric; however, it is difficult to accurately control the diffusion of the resin, and the drying step requirements make it less attractive than thermal bonding.

Chemical bonding with hydrogen chloride gas has been used with spun webs of nylon-6,6 to commercially produce spunbonded nylon fabrics.

Bonding a web by any means allows for certain generalizations. If the web is highly bonded, most of the fibers are bonded to another fiber. The resulting structure is relatively stiff, paper-like, and has higher tensile and modulus but lower resistance to tear propagation. On the other hand, if the web is only slightly bonded fewer fiber-to-fiber bonds are present and the structure is more conformable with lower tensile and modulus but higher resistance to tear propagation. Webs which are only slightly bonded exhibit low surface abrasion resistance. Greater varieties of structures are achievable through point bonding because of the various bonding roll patterns available.

Meltblown Fabrics

Meltblown fabrics differ from the traditional spunbonded fabrics by having lower fiber denier (fineness) and by usually being composed of discontinuous filaments. The inherent fiber entanglement often makes additional bonding unnecessary. Fibers produced by meltblowing are extremely fine and largely unoriented, causing the webs to be quite weak and easily distorted. Most thermoplastic polymers can be meltblown, but the majority of commercial products are high melt flow polypropylene.

In the manufacture of meltblown fabrics, a special die is used in which heated, pressurized air attenuates the molten polymer filament as it exits the orifice of the die or nozzle. Air temperature ranges from 260–480°C with sonic velocity flow rates.

Sandwich structures have been created with the meltblown web in the middle between two conventional spunbonded webs. Mixtures of meltblown and crimped bulking fibers have been sold as thin thermal insulation for use in outdoor clothing and gear. Meltblown technology has also been adapted to produce nontraditional spunbonded fabrics, such as elastomeric webs.

The great quantity of very fine fibers in a meltblown web creates several unique properties, such as barriers and large surface areas for hospital gowns, sterile wrap, incontinence devices, oil spill absorbers, battery separators, and special requirement filters.

Flash Spun Fabrics

Flash spinning begins with a 10–15% polymer solution, prepared by dissolving a solid polymer, such as high density polyethylene, with a suitable solvent, such as trichlorofluoromethane, methylene chloride, or pentane, with the latter now preferred to CFCs. The solution is heated to approximately 200°C, pressurized to ~4.5 MPa (653 psi), and the pressurized vessel is connected to a spinnerette containing a single hole. When the pressurized solution is permitted to expand rapidly through the hole, the low boiling solvent is instantaneously flashed off, leaving behind a three-dimensional film–fibril network referred to as a plexifilament. The three-dimensionality results from the cross-linking interconnection of the fine fibers which produces a film thickness of 4 μm or less. Thus, many individual but interconnected fibers are created from a single-hole spinnerette.

When a multiplicity of single-hole spinnerettes are assembled across a width, the plexifilaments produced can form a wide web that can be thermally bonded to produce a flat sheet structure. The web-forming procedure is ameliorated by use of a baffle which deflects the stream of plexifilaments after they exit the spinnerette.

Unlike the fine fibers prepared by meltblowing, the plexifilaments from flash spinning are substantially oriented and possess relatively high tenacities (0.08 N/tex (>1 g/den)). The plexifilaments scatter light effectively as a result of high surface areas (ca 2 m^2/g) and thus form opaque webs. In addition, the fineness of the plexifilament fibrils also results in a web structure of exceptional softness. Webs are either area or point bonded to yield paper- or cloth-like aesthetics, respectively. The paper-like sheets are used as durable papers and may be printed using conventional inks (qv) and printing equipment, whereas the point-bonded structures are very soft and find use in disposable protective clothing.

Applications for Spunbonded Fabrics

Both the durable and disposable markets for spunbondeds have experienced dramatic growth (~6%/yr). Significant areas of durable growth have been in the building and construction industries, where spunbondeds are used in geotextiles and roofing membranes (see BUILDING MATERIALS). Growth has also been achieved in primary carpet backing in automotive carpets and carpet tiles, where moldability and high dimensional stability, respectively, were achieved through the use of spunbondeds. In the 1990s, new diaper designs using leg cuffs and backsheet have increased the demand for lightweight disposable spunbonded fabrics far beyond earlier predictions. With the possible exception of geotextiles and housewrap, however, there have been virtually no new markets established as a result of the special characteristics of spunbonded fabrics. Growth has come about in an evolutionary fashion where spunbonded fabrics were substituted for woven fabrics, other nonwoven fabrics (including knits), paper or film in previously existing applications, or where the cost–property relationship has permitted an extension of an existing application, such as the redesign of diapers.

Of the four basic polymer types available in spunbonded form, ie, polypropylene, polyethylene, polyester, and nylon, both polyester and nylon are more costly polymer forms than either of the olefins. It is possible for this cost advantage to be offset by other factors. For example, higher temperature resistance of polyester largely differentiates the opportunities for these spunbondeds versus olefinic counterparts.

RONALD SMORADA
BBA Nonwovens

R. L. Smorada, *INDA J. Nonwovens Res.* **3**(4) (Fall 1991).

Y. Ogawa, *Spunbonded Technology Today 2,* Miller-Freeman, San Francisco, Calif., 1992, p. 123.

U.S. Pat. 3,338,992 (Aug. 29, 1967), G. A. Kinney (to E. I. du Pont de Nemours & Co., Inc.).

U.S. Pat. 4,405,297 (Sept. 20, 1983), D. W. Appel and M. T. Morman (to Kimberly-Clark Corp.).

NUCLEAR MAGNETIC RESONANCE. See MAGNETIC SPIN RESONANCE.

NUCLEAR REACTORS

INTRODUCTION

The nuclear reactor is a device in which a controlled chain reaction takes place involving neutrons and a heavy element such as uranium. Neutrons are typically absorbed in uranium-235, ^{235}U, or plutonium-239, ^{239}Pu, nuclei. These nuclei split, releasing two fission fragment nuclei and several fast neutrons. Some of these neutrons cause fission in other uranium nuclei in a sequence of events called neutron multiplication. The fission fragments are stopped within the nuclear fuel, where their kinetic energy becomes thermal energy. The thermal energy is removed by a cooling agent and converted into electrical energy in a turbine-generator system. Many of the fission fragments are radioactive, releasing radiation and decay heat. Some of the radioactive materials have useful purposes; others form nuclear waste (see NUCLEAR REACTORS–WASTE MANAGEMENT).

Nuclear reactors as a source of heat energy and radiation were the outgrowth of World War II defense applications. Research and development were pursued on several fronts in the Manhattan Project.

A variety of nuclear reactor designs are possible, using different combinations of components and process features for different purposes (see NUCLEAR REACTORS–REACTOR TYPES). Two versions of the lightwater reactors were favored: the pressurized water reactor (PWR) and the boiling water reactor (BWR). Each requires enrichment of uranium in ^{235}U. To assure safety, careful control of coolant conditions is required (see NUCLEAR REACTORS–WATER CHEMISTRY OF LIGHTWATER REACTORS; NUCLEAR REACTORS–SAFETY IN NUCLEAR POWER FACILITIES).

Power Generation

The principal application of the nuclear reactor is as a heat source for electrical power generation. As of 1998, there were 105 nuclear power reactors in operation in the United States, generating almost 100 GW of electrical power. Outside of the United States, there were 328 reactors producing 250 GW. Table 1 lists power reactors by country. Table 2 gives the worldwide distribution by reactor type. The fraction of total electricity that is derived from nuclear reactors varies greatly among countries. Notable approximate figures are France, 75%; Japan 30%; and the United States, 21%. Some of the characteristics of the PWR and BWR, ie, the pressurized lightwater reactor and the boiling water reactor, which are the most widely used reactor types, are given in Table 3.

Safety. A large inventory of radioactive fission products is present in any reactor fuel where the reactor has been operated for lengths of time on the order of months. In steady state, radioactive decay heat amounts to about 5% of fission heat, and continues after a reactor is shut down. If cooling is not provided, decay heat can melt fuel rods, causing release of the contents. Protection against a loss-of-coolant accident (LOCA), eg, a primary coolant pipe break, is required. Power reactors have an emergency core cooling system (ECCS) that comes into play upon initiation of a LOCA.

Nuclear power has achieved an excellent safety record. Exceptions are the accidents at Three Mile Island in 1979 and at Chernobyl in 1986. In the United States, safety can be attributed in part to the strict regulation provided by the Nuclear Regulatory Commission, which reviews proposed reactor designs, processes applications for licenses to construct and operate plants, and provides surveillance of all safety-related activities of a utility. The utilities seek continued improvements in capability, use procedures extensively, and analyze any plant incidents for their root causes. Similar programs intended to ensure reactor safety are in place in other countries.

A technique called probabilistic safety assessment (PSA) has been developed to analyze complex systems and to aid in assuring safe nuclear power plant operation.

Reactors are designed to be inherently safe based on physical principles, supplemented by redundant equipment and special procedures.

Environmental Aspects

Emissions of radioactive materials during regular operations are within regulatory requirements based on medical knowledge. These emissions include radionuclides of the noble gases xenon and krypton, which readily disperse throughout the atmosphere. Small quantities of soluble radionuclides are released into lakes or streams that provide very large dilution factors. Plant and animal life are monitored regularly at such facilities. On the other hand, the potential, however small, of radioactive contamination of the environment in case of a reactor accident in which containment is breached does exist.

As the result of many years of nuclear reactor research and development and weapons production in U.S. defense programs, a large number of sites were contaminated by radioactive materials. A thorough cleanup of this residue of the Cold War is expected to extend well

Table 1. World Nuclear Power Plants[a]

Nation	Operative units	Power, net MWe	Total number of units	Power, net MWe
Argentina	2	935	3	1627
Armenia	1	376	1	376
Belgium	7	5,712	7	5,712
Brazil	1	626	3	3,100
Bulgaria	6	3,538	6	3,538
Canada	22	15,439	22	15,439
China	3	2,167	11	8,737
Cuba	0	0	2	834
Czech Republic	4	1,648	6	3,472
Finland	4	2,310	4	2,310
France	56	58,748	60	64,303
Germany	20	22,282	20	22,282
Hungary	4	1,731	4	1,731
India	10	1,695	16	3,443
Iran	0	0	1	950
Japan	53	43,414	58	48,295
Kazakhstan	1	70	1	70
Korea	12	9,770	20	16,770
Lithuania	2	2,370	2	2,370
Mexico	2	1,308	2	1,308
Netherlands	1	452	1	452
Pakistan	1	125	2	425
Romania	1	650	5	3,160
Russia	26	19,849	30	23,224
Slovakia	4	1,632	8	3,296
Slovenia	1	632	1	632
South Africa	2	1,842	2	1,842
Spain	9	7,207	9	7,207
Sweden	12	10,035	12	10,035
Switzerland	5	3,077	5	3,077
Taiwan	6	4,884	8	7,484
Ukraine	15	12,840	20	17,590
United Kingdom	35	12,930	35	12,930
United States	105	97,979	108	101,582
Total	*433*	*348,273*	*495*	*399,603*

[a] Courtesy of *Nuclear News,* March 1998.

Table 2. Worldwide Nuclear Power Units by Reactor Type[a]

Reactor type	Units of operation	Power, net MWe	Total number of units	Power, net Mwe
pressurized light-water reactors (PWR)	250	221,552	286	254,126
boiling light-water reactors (BWR)	93	79,803	99	87,004
gas-cooled reactors, all types	35	11,889	35	11,889
heavy-water reactors, all types	37	19,971	52	27,621
graphite-moderated light-water reactors (LGR)	15	14,195	16	15,120
liquid-metal-cooled fast breeder reactors (LMFBR)	3	863	7	3,843

[a] Courtesy of *Nuclear News,* March 1998.

into the twenty-first century and cost many billions of dollars. New technologies are needed to minimize the cost of the cleanup operation.

Wastes. Nuclear reactors produce unique wastes because these materials undergo radioactive decay and in doing so emit harmful radiation. Spent nuclear fuel has fission products, uranium, and transuranic elements. Plans calls for permanent disposal in underground repositories. Geological studies are in progress at the Yucca

Table 3. Characteristics of Reactors

Parameter	Reactor type: PWR	BWR
heat power, MWt	3425	3579
electrical power, MWe	1150	1120
coolant temperatures, °C	292 (326)[a]	216 (285)[a]
pressure, MPa[b]	15.5	7.0
reload fuel, wt % ^{235}U	4.0–5.0	3.5–3.8

[a] In (out).
[b] To convert MPa to psia, multiply by 145.

Mountain site in Nevada. Until a repository is completed, spent fuel must be stored in water pools or in dry storage casks at nuclear plant sites.

Nuclear electric power plants do not emit carbon dioxide, as do fossil-fueled plants. If concerns increase about global warming resulting from greenhouse gases, more nuclear plants may be built in the future as alternative power sources.

Economic Aspects

In the early years of reactor development, electricity from nuclear sources was expected to be much cheaper than that from other sources. Whereas nuclear fuel cost is low, the operating and maintenance costs of a nuclear facility are high. Thus on average, electric power from coal and nuclear costs about the same.

Inflation, high interest rates, long construction periods, and regulatory delays resulted in severe cost overruns on nuclear reactor development. Moreover, the reactor accidents of Three Mile Island and, later, Chernobyl produced an atmosphere of public concern. As a consequence, there is a general reluctance in the financial community to support the construction of new nuclear plants.

Resources

Uranium resources were originally expected to be rapidly depleted in a growing economy. There were, however, ample supplies of uranium as of 1998.

In the hope of stimulating interest in the building of nuclear power plants, the nuclear industry is designing advanced lightwater reactors (ALWR). These are of two types, known respectively as simplified and enhanced safety. The first has lower (ca 600-MW) power levels than the 1200-MW reactors of the 1970s and the 1980s. The second uses passive features such as natural convection and the force of gravity for enhanced safety. Both take advantage of operating experience of the current generation of reactors. The U.S. government is funding limited development of liquid-metal and gas-cooled advanced reactors.

RAYMOND L. MURRAY
Consultant

R. L. Murray, *Nuclear Energy: An Introduction to the Concepts, Systems, and Applications of Nuclear Processes,* Butterworth-Heinemann, Oxford, U.K., 1996.

A. E. Waltar, *America the Powerless: Facing Our Nuclear Energy Dilemma,* Cogito Books, Madison, Wis. 1995.

R. Rhodes, *Nuclear Renewal: Common Sense About Energy,* Penguin Books, New York, 1993.

Nuclear Power: Technical and Institutional Options for the Future, National Academy Press, Washington, D.C., 1992.

NUCLEAR FUEL RESERVES

Lightwater reactors, the primary type of nuclear power reactors operated throughout the world, are fueled with uranium dioxide, UO_2, enriched from the naturally occurring concentration of 0.71%

uranium-235, ^{235}U, to approximately 3% ^{235}U. Up to this writing (ca 1996), all civilian nuclear fuel has been produced by enriching natural uranium (see DIFFUSION SEPARATION METHODS). An additional source of enriched uranium for civilian nuclear reactor fuel is expected to become available from the dismantlement of nuclear weapons from the stockpiles of the United States and the Commonwealth of Independent States (former Soviet Union).

Uranium Mineral Resources

The Organization for Economic Cooperation and Development's Nuclear Energy Agency (OECD/NEA) and the International Atomic Energy Agency (IAEA) estimate uranium resources in four cost categories: \$40/kg U or less; \$80/kg U or less; \$130/kg U or less; and \$260/kg U or less. Previous NEA/IAEA evaluations employed only the last three cost categories.

The U.S. Department of Energy (DOE) and the NEA/IAEA employ similar terms to classify uranium resources, as reasonably assured, estimated additional (EA), or speculative. The NEA/IAEA divides the estimated additional resources into two types, EAR-I and EAR-II, describing known resources and undiscovered ones, respectively.

Geochemical Nature and Types of Deposits. The crust of the earth contains approximately 2–3 ppm uranium. Alkalic igneous rock tends to be more uraniferous than basic and ferromagnesium rocks. Elemental uranium oxidizes readily. The solubility and distribution of uranium in rocks and ore deposits depend primarily on valence state. The hexavalent uranium ion is highly soluble, and the tetravalent ion relatively insoluble. Uraninite, the most common mineral in uranium deposits, contains the tetravalent ion.

A classification system for the principal types of uranium ore deposits was revised by the IAEA in 1988–1989. This system assigns uranium resources to various categories on the basis of geological setting. There are 15 main categories or uranium ore deposits arranged according to approximate economic significance.

The first six geologic ore types, together with selected types from the seventh category, are considered conventional resources. These categories represented a majority of the uranium-producing geologic formations worldwide as of 1992. Very low grade resources, which were not economic as of the mid-1990s, or from which uranium is only recoverable as a minor by-product, are considered unconventional resources. The categories are unconformity-related deposits; sandstone deposits; quartz-pebble conglomerate deposits; vein deposits; breccia complex deposits; intrusive deposits; phosphorite deposits, collapse breccia pipe deposits; volcanic deposits; surficial deposits; metasomatite deposits; metamorphic deposits; lignite; black shale deposits; and other, unclassified deposits.

Uranium Reserves

Domestic. Estimates of U.S. uranium resources for reasonable assured resources, estimated additional resources, and speculative resources are given in Table 1.

Foreign. NEA/IAEA data for reasonable assured and estimated additional resources is available. These estimates incorporate data from former world outside centrally planned economies (WOCA) and non-WOCA nations. Other known uranium resources total about 1.4×10^6 t. However, these estimates are not strictly consistent with standard NEA/IAEA definitions.

Estimates of speculative resources (SR) at \$130/kg uranium and those having an unassigned cost range total about 11.28×10^6 t. Estimates of uranium resources from unconventional and by-product sources total about 7×10^6 t for phosphates, 0.013×10^6 t for nonferrous ores, 0.016×10^6 for carbonates, and 0.014×10^6 for lignites.

Resources

As of the beginning of 1993, reasonably assured resources (RAR) recoverable at costs of \$80/kg U or less were estimated at 1.53×10^6 t U. Estimated additional resources (EAR) in the same cost category were about 1.769×10^6 t U. Total RAR and EAR, recoverable at costs of \$130/kg U or less were estimated at 2.205×10^6 t U and 3.540×10^6 t U, respectively, as of the beginning of 1993. There remains good potential for the discovery of additional uranium resources of the conventional type, as reflected by estimates of speculative resources (SR). Based on reported estimates, this potential is about 13×10^6 t U.

Production

World reactor-related requirements are expected to increase from 57,182 t U in 1992 to about 75,673 t U by the year 2010. The total annual production capability is expected to reach a maximum of about 48,200 t U in the year 2000, when projected total world demand should be about 63,500 t U. The total annual production capability is then projected to fall steadily to about 30,500 t U in 2010.

Supply Projections. Additional supplies are expected to be necessary to meet the projected production shortfall. A significant contribution is likely to come from uranium production centers such as Eastern Europe and Asia, which are not included in the capability projections. The remaining shortfall between fresh production and reactor requirements is expected to be filled by excess inventory drawdown, utilization of low cost resources that could become available as a result of technical developments or policy changes, production from either low or higher cost resources not identified in production capability projections, recycled material such as spent fuel, and low enriched uranium converted from the high enriched uranium (HEU) found in warheads.

Demand. The demand for uranium in the commercial sector is primarily determined by the requirement of power reactors. World annual uranium requirements in 1993 are estimated at ca 58,382 t natural uranium equivalent. Reactor-related requirements are expected to rise about 1015 t/yr on the average, reaching 75,700 t U total requirements in the year 2010.

Alternative Sources

Alternative sources include low grade resources, eg, seawater. The world's oceans contains ca 4×10^9 t of uranium. Significant engineering development and associated environmental concerns have limited the development of an economic means of uranium extraction from seawater (see OCEAN RAW MATERIALS) and mill tailings and HEU deenrichment (Table 2).

Toxicology of Uranium

The two primary effects associated with the introduction of uranium species into the human body are the development of cancer, primarily

Table 1. U.S. Uranium Resources

	Resource category[a]			
Uranium cost category, \$/kg	RAR	EAR	SR	Total
80	114	850	502	1466
130	370	1314	889	2573
260	588	1893	1352	3833

[a] In 1000 metric tonnes U as of Dec. 31, 1992.

Table 2. Estimates of HEU, t of U

Country	HEU	Percent of total
China	15	1.15
France	15	1.15
United Kingdom	10	0.76
CIS	720	54.96
United States	550	41.98
Total	*1310*	*100.00*

from radiation-induced tissue damage in the lung, and renal damage accompanied by possible kidney failure owing to uranium ingestion.

The U.S. Nuclear Regulatory Commission (NRC) regulates the protection of the health and safety of the public by issuing standards. In addition, the U.S. Environmental Protection Agency (EPA) issues drinking water regulations that address radioactive contamination of drinking water supplies. The maximum allowable effluent concentrations for release of uranium to air, water, and sewer systems depends on the individual isotope of uranium.

Handling of soluble uranium compounds requires appropriate clothing to prevent skin contact and eye protection to prevent any possible eye contact. Respirators should always be worn to prevent inhalation of uranium dust, fumes, or gases.

DANIEL B. BULLEN
Iowa State University

M. Benedict, T. Pigford, and H. Levi, *Nuclear Chemical Engineering,* McGraw-Hill Book Co., Inc., New York, 1981, pp. 261–264.

Fuel and Heavy Water Availability, Report of Working Group 1, International Nuclear Fuel Cycle Evaluation, Vienna, Austria, International Atomic Energy Agency STI/PUB/534, UNIPUB, Inc., New York, 1980, pp. 174–175.

D. Albright, F. Berkhout, and W. Walker, *World Inventory of Plutonium and Highly, Enriched Uranium 1992,* Oxford University Press, Oxford, U.K., 1993, p. 198.

K. H. Wedepohl, ed., *Handbook of Geochemistry,* Vol. II-5, Springer-Verlag, New York, 1969, pp. 92-D-1, 92-D-4.

WATER CHEMISTRY OF LIGHTWATER REACTORS

As of 1994, there were 105 operating commercial nuclear power stations in the United States (see POWER GENERATION). All of these facilities were light, ie, hydrogen–water reactors. Seventy-one were pressurized water reactors (PWRs); the remainder were boiling water reactors (BWRs).

In a PWR, a closed circuit of high pressure, high temperature water transfers heat from the reactor core to once-through or recirculating U-tube steam generators (see HEAT-EXCHANGE TECHNOLOGY). The steam (qv), which is produced on the secondary side of the steam generator, is used to drive a turbine generator. In contrast to fossil-fired steam-generating equipment, steaming in a PWR occurs on the outside of the boiler tubes, ie, the secondary side, where a large number of tube-to-tube support plate and tube-to-tube sheet crevices exist. Extensive corrosion has been observed in such crevices and beneath sludge piles in the tube–tube support plates and tubesheet. Control of secondary chemistry is critical if high concentration of aggressive chemicals and accelerated local corrosion are to be avoided (see CORROSION AND CORROSION CONTROL; WATER–INDUSTRIAL WATER TREATMENT).

Boron, in the form of boric acid, is used in the PWR primary system water to compensate for fuel consumption and to control reactor power. The concentration is varied over the fuel cycle. Small amounts of the isotope lithium-7 are added in the form in lithium hydroxide to increase pH and to reduce corrosion rates of primary system materials. Primary-side corrosion problems are much less than those encountered on the secondary side of the steam generators.

In a BWR, steam is generated in the reactor core and used directly to drive a turbine generator. Radiation levels during operation near the turbine, condenser, and feedwater heaters are higher than in a PWR as a result of steam transport of short-lived activation products. Other than this consideration, the power-generating cycles of PWRs and BWRs are reasonably similar. No reactivity or pH control additives are used in BWRs. As a result, the corrosion behavior of the materials of construction is dependent primarily on coolant oxygen concentrations, which are governed by the radiolytic decomposition rates of water and steam-water equilibrium relations. In some cases, these parameters are controlled by the addition of dissolved hydrogen to the feedwater entering the reactor vessel.

Pressurized Water Reactors

Primary System. In a PWR, reactor coolant is circulated at ca 300°C and 15 MPa (150 bar). The primary water specifications for a PWR are O_2 <5 ppb; H_2 25–50 mL/kg H_2O; Cl <50 ppb; F <50 ppb; SO_4 <50 ppb; suspended solids <10 ppb; and SiO_2 <1000 ppb. Rigid controls are applied to the primary water makeup to minimize contaminant ingress into the system. In addition, a bypass stream of reactor coolant is processed continuously through a purification system to maintain primary coolant chemistry specifications. This system provides for removal of impurities plus fission and activated products from the primary coolant by a combination of filtration (qv) and ion exchange (qv). The bypass stream is also used both to reduce the primary coolant boron as fuel consumption progresses, and to control the ^7Li concentrations.

Oxygen is a prime factor in the corrosion of system materials and the release, activation, and redeposition of activated corrosion products. Dissolved hydrogen is maintained to promote rapid recombination of the oxygen whether radiolytically formed or introduced into the coolant from other sources, thereby minimizing corrosion rates.

Secondary System Cobalt-60 with a 5.3 yr half-life is the primary source of the feedwater out-of-core radiation levels in PWRs. The water quality specifications for the feedwater and blowdown water in a recirculating steam generator (RSG) and the feedwater for a once-through steam generator (OTSG) are given in Table 1.

Recirculating Steam Generator. The corrosion performance of many RSGs in commercial power stations in the United States has been marginal. Many tube bundles have had to be replaced. Many tubes have been plugged or sleeved with inserts as a result of excessive corrosion on the secondary side.

As of this writing (ca 1995), the treatment of choice is in all volatile treatment (AVT) employing ammonia (qv) for pH control of the feedwater (Table 1). Despite the use of AVT, inward deformation of the Al-

Table 1. PWR Steam Generator Water Specifications[a]

		Recirculating	
Parameter[b]	Once-through feedwater	Feedwater[c]	Blowdown water
components, ppb			
N_2H_4	20[d]	<100	
O_2	≤3	<5	
Na	≤3		<5
Cl	≤5		<10
SiO_2	≤10		
total Fe	≤5	<5	
total Cu	≤1	<1	
SO_4	≤3		<10
other	[e]	[f]	[f,g]
cation conductivity at 25°C, μS/cm corrected[h]	≤0.2	≤0.2	
ammonia AVT[i]			0.15
nonammonia AVT[i]			0.5

[a] Values correspond to normal power operation.
[b] ppb = parts per billion.
[c] The pH is plant specific, depending on additive used and secondary system materials. Feedwater generally should be equivalent to pH = 9.3 at 25°C for ammonia and carbon steel equipment.
[d] Value given is minimum.
[e] Organics related to use of amines for pH control. Value is plant specific.
[f] Boron related to boric acid treatment of ~5–10 ppm B in feedwater.
[g] For caustic crevice environment, a plant-specific chemical impurity molar ratio 0.5 is defined, eg, Na:Cl molar ratio 0.5.
[h] Corrected for the presence of organics such as acetates and formates.
[i] AVT = all-volatile treatment.

loy 600 tubing at the tube-to-tube support plate interface region was detected. This phenomenon, commonly referred to as denting, results from excessive corrosion of carbon steel in the crevice region between the tube and the tube support plate; as denting progresses, tube support plate cracking can occur. The tube stress corrosion cracking also continues.

Possible remedial and preventive actions are the following: (*1*) Use of elevated hydrazine, N_2H_4, treatment; (*2*) boric acid treatment; (*3*) alternative amine treatment (eg, morpholine or ethanolamine) in place of ammonia; (*4*) molar ratio management where the ratio of cations to anions in the steam generator water is controlled.

Before any remedial or preventive actions are implemented, an evaluation should be conducted as to applicability to the specific plant. The main action should be to take measures to reduce the ingress of contaminants into the steam generator by using more reliable materials, such as in the condenser tubes, to reduce leakages. Contaminant control equipment, such as full-flow condensate demineralizers, should also be employed.

Once-Through Steam Generator. The corrosion of OTSGs has not been as extensive or as serious a problem as that of RSGs. Design and operating differences between these systems may be responsible for the corrosion differences.

Feedwater quality control is the main method for controlling the chemistry environment in OTSGs (Table 2). Whereas the all-volatile treatment (AVT) utilizing ammonia was used first, a switch has been made to morpholine or other amines for better distribution in the steam plant. The amount of corrosion products (mainly iron) carried into the OSTGs has been greatly reduced.

OTSGs also experience deposition of material on the flow areas in the tube support plates, which causes an increase in pressure drop and eventual reductions in plant power production.

The possible remedial and preventive actions are hot soaks and drains during cooldown to help remove soluble deposited material, chemical cleaning to remove corrosion products and reduce the pressure drop (see METAL SURFACE TREATMENTS), and reduced corrosion product transport into OTSG using amines other than ammonia in feedwater.

Boiling Water Reactors

BWRs operate at ca 7 MPa (70 bar) and 288°C. A side stream is continuously purified using demineralizers and filters to control the water quality of the reactor water. Full-flow condensate demineralizers are also used to control the ingress of impurities into the reactor water from the steam plant.

The BWR water chemistry parameters are given in Table 2. The desired electrochemical potential (ECP) value to help control intergranular stress corrosion cracking (IGSCC) is not obtainable unless H_2 is injected into the feedwater.

Whereas addition of hydrogen to feedwater helps solve the O_2 or ECP problem, other complications develop. An increase in shutdown radiation levels results from the increased volatility of the short-lived radioactive product nitrogen-16, ^{16}N (7.1 s half-life), formed from the coolant passing through the core. Without H_2 addition, the ^{16}N in the fluid leaving the reactor core is in the form of nitric acid, HNO_3; with H_2 addition, the ^{16}N forms ammonia, NH_3, which is more volatile than HNO_3, and thus is carried over with the steam going to the turbine.

Although iron is the main corrosion product entering the reactor water, as for PWRs, ^{60}Co is the long-term source of shutdown radiation levels substantially lower than those of other plants having similar operating conditions and water chemistry. The common denominator for these low radiation levels was the use of tubes containing zinc in the condensers. It appears that whereas Zn inhibits the deposition of cobalt, there are also some side effects. Zinc addition produces ^{65}Zn hot spots within the plant. Zn-65 has a 244-d half-life.

The cobalt deposition rate on new, replacement, or decontaminated recirculation piping surface has been reduced by pretreating the piping using an atmosphere of oxygenated wet steam to form an oxide film. Studies have been conducted for both PWRS and BWRS to reduce the cobalt content of materials used in the nuclear parts of the plants. Some low cobalt materials have been developed; however, the use of the materials is limited to replacement parts or new plants.

Table 2. BWR Normal Water Chemistry Values[a]

Parameter[b]	Median value
Reactor water	
conductivity at 25°C, μS/cm	ca 0.11
Cl, ppb	ca 1
SO_4, ppb	ca 2
Zn, ppb	5–10[c]
electrochemical potential, V	[d]
O_2 in recirculation water, ppb	[e]
SiO_2, ppb	<100
Feedwater/condensate	
feedwater conductivity at 25°C, μS/cm	0.06
feedwater total Fe, ppb	ca 1–3[f]
feedwater total Cu, ppb	ca 0.1–0.2[f]
O_2, ppb	ca 30

[a] Values given correspond to normal power operation.
[b] ppb = parts per billion.
[c] Consistent with plant program for zinc injection.
[d] Corrective action should be initiated when value is >−0.23 V against the standard hydrogen electrode (SHE). Plant-specific values should be established for protection of stainless steels and nickel-based critical components.
[e] Plant specific. In range of 200 ppb without H_2 addition and in range of 10 ppb with H_2 addition.
[f] Values depend on plant-specific design and operating procedures for steam plant.

JACK H. HICKS
Consultant

P. Cohen, *Water Coolant Technology of Power Reactors,* American Nuclear Society, LaGrange Park, Ill., 1980.

"PWR Primary Water Chemistry Guidelines: Revision 3," Report IR-105714, Electric Power Research Institute, Palo Alto, Calif., Nov. 1996.

"PWR Secondary Water Chemistry Guidelines—Revision 4," Report TR-102134, Electric Power Research Institute, Palo Alto, Calif., 1996.

"BWR Water Chemistry Guidelines—1993 Revision, Normal and Hydrogen Water Chemistry," Report TR-103515, Electric Power Research Institute, Palo Alto, Calif., Feb. 1993.

ISOTOPE SEPARATION

The high cost of isotope separation has limited the use of separated isotopes in nuclear reactors to specific cases where substitutes that do not involve separated isotopes are not available.

Uranium-235

The separation of the isotope ^{235}U, which occurs in natural uranium to the extent of 0.72 atomic %, is discussed in detail elsewhere (see DIFFUSION SEPARATION METHODS). Uranium-235 is concentrated mostly by means of the gaseous diffusion process using uranium hexafluoride, UF_6, as the process gas.

The gaseous diffusion process is energy-intensive. As a consequence of this and other economic considerations, as of this writing (ca 1995), most new facilities for uranium enrichment utilize the gas centrifuge process. Several processes for uranium enrichment have been developed, including a chemical exchange process, an ion-exchange (qv) process, and several laser photochemical processes, including an atomic vapor laser isotope separation method (AVLIS). However, as of the mid-1990s the supply of enriched uranium exceeds demand. New facilities are not expected to be developed until the demand for enriched uranium increases. General discussions of uranium enrichment are available.

Deuterium

The use of deuterium in nuclear reactors is discussed in detail elsewhere (see DEUTERIUM AND TRITIUM). Methods of isotope separation include catalytic exchange between H_2 and H_2O, electrolysis, water distillation, hydrogen distillation, and various chemical-exchange systems.

Tritium

Ultimately, tritium could be used as fuel in thermonuclear reactors (see FUSION ENERGY). Tritium can be produced in a manner requiring very little isotopic separation by the reaction of neutrons with ^{6}Li.

$$^{6}_{3}\text{Li} + ^{1}_{0}n \rightarrow ^{3}_{1}\text{H} + ^{4}_{2}\text{He}$$

Boron

Boron-10 is used at 40–95 atomic % in safety devices and control rods of nuclear reactors. Its use is also intended for breeder-reactor control rods.

Examination of possible systems for boron isotope separations resulted in the selection of multistage exchange-distillation of boron trifluoride–dimethyl ether complex, $BF_3\ O(CH_3)_2$, as a method for ^{10}B production.

The search for a system with less decomposition and a higher separation factor has been summarized. The most promising system is the BF_3–anisole system, in which BF_3 (g) exchanges with the anisole (methyl phenyl ether) NF_3 complex (1).

Lithium-7

Lithium-7 has been produced in the United States by chemical exchange between lithium hydroxide and lithium amalgam.

EDWARD VON HALLE
Consultant

J. M. Lerat, in P. Louvet, P. Noe, and Soubbaramayer, eds., *Proceedings of the Second Workshop on Separation Phenomena in Liquids and Gases, Versailles, July, 1989,* Centre d'Études Nucléaires de Saclay and Cité Scientifique Parcs et Technopoles Ile de France Sud, Massy, France, 1989, pp. 555–594.

H. London, ed., *Separation of Isotopes,* George Newnes, Ltd., London, 1961.

M. Benedict, T. Pigford, and H. Levi, *Nuclear Chemical Engineering,* 2nd ed., McGraw-Hill Book Co., Inc., New York, 1981, Chapt. 14.

E. A. Evans, *Tritium and Its Compounds,* 2nd ed., John Wiley & Sons, Inc., New York, 1974.

CHEMICAL REPROCESSING

The process of separating the components of irradiated nuclear reactor fuel into several streams, usually uranium, plutonium and wastes, is called chemical reprocessing. Chemical separation processes of interest include precipitation, solvent extraction, ion exchange, zone melting/slagging, molten salt/pyrometallurgical, and fluoride volatility. These processes have attributes that influence selection for a specific application. Within the context of nuclear fuel cycle, synonyms for chemical reprocessing include reprocessing, fuel reprocessing, separations, and chemical separations. The term separations is also used to describe isotopic separation techniques, which fall outside the scope of this article. At times additional products have been recovered from the waste stream. Both flow sheet selection and the resulting economics depend on the context in which the reprocessing is performed. This context is commonly referred to as the fuel cycle. The fuel cycles of interest as of this writing (ca 1995) are illustrated in Figure 1.

The raw material for nuclear reactor fuel, uranium, exits the mining–milling sequence as uranium oxide. The uranium dioxide is converted to uranium hexafluoride and enriched in ^{235}U (see DIFFUSION SEPARATION METHODS; NUCLEAR REACTORS–ISOTOPE SEPARATION). The energy required by the enrichment process is measured in terms of separative work units (SWU). The enriched uranium hexaflouride is then reduced to the oxide and formed into pellets, which are fabricated into reactor-fuel elements. The cost of uranium-based reactor fuel is thus the sum of the costs for mining and milling uranium, uranium hexafluoride conversion, enrichment, and fuel fabrication. Other fuel-cycle costs are associated with the waste streams originating at each of the fuel-cycle facilities (see NUCLEAR REACTORS, WASTE MANAGEMENT).

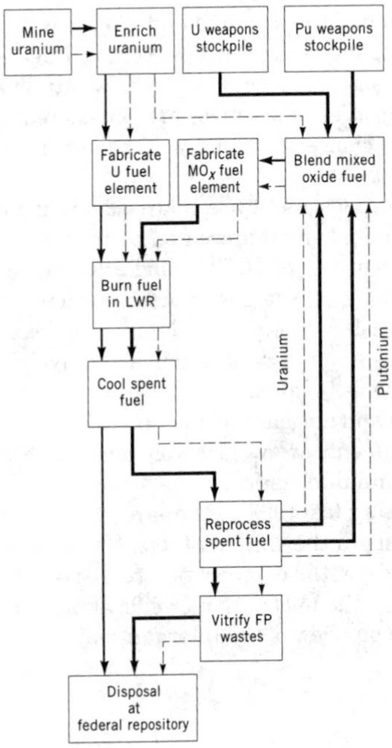

Figure 1. Alternative fuel cycles for nuclear fuel, where (– – –) corresponds to the classical fuel cycle, (—) the throwaway fuel cycle, and (—) the recycle weapons fuel cycle. FP = fuel processing; LWR = lightwater reactor.

The classical fuel cycle is based on the recovery and recycle of energy values contained in spent fuel, where energy values are defined in terms of the four components of reactor fuel cost; recovered fissile plutonium values reduce the demand for uranium mining/milling and enrichment.

As the recycled fuel composition approaches steady state after approximately four cycles, the heat and radiation associated with ^{236}U and ^{238}Pu build-up require more elaborate conversion and fuel fabrication facilities than are needed for virgin fuel. The storage, solidification, packaging, shipping, and disposal considerations associated with wastes that result from this approach are primarily concerned with the relatively short-lived fission products. The transuranic isotopes are recycled to and burned in the reactors. Commercial reactor fuel is being reprocessed and recycled in the U.K., France, Japan, India, and the former Soviet Union. Essentially all other nuclear nations, except the United States, have contracted to have these services performed.

A variation of the classical fuel cycle is the breeder cycle. Special breeder reactors are used to convert fertile isotopes into fissile isotopes, which creates more fuel than is burned (see NUCLEAR REACTORS–REACTOR TYPES). A breeder economy implies the existence of both breeder reactors that generate and nonbreeder reactors that consume the fissile material. The breeder reactor fuel cycle has been partially implemented in France and the U.K.

The throwaway fuel cycle (Fig. 1) does not recover the energy values present in the irradiated fuel. Instead, all of the long-lived actinides are routed to the final waste repository along with the fission products. Whether or not this is a desirable alternative is determined largely by the scope of the evaluation study. For instance, when only the value of the uranium and SWU equivalents are considered, the world market values for these commodities do not fully cover the cost of reprocessing. However, when costs attributable to the disposal of large quantities of actinides are considered, the classical fuel cycle has been the choice of virtually all countries except the United States.

The recycle weapons fuel cycle takes advantage of the SWU and virgin uranium equivalents represented by the fissile materials in decommissioned nuclear weapons. This variation impacts the prereactor portion of the fuel cycle. The post-reactor portion can be either classical or throwaway. Because the availability of weapons-grade fissile material for use as an energy source is a relatively recent phenomenon, it has not been fully implemented. Currently (ca 1998), U.S. reactor manufacturers and nuclear fuel manufacturers, in conjunction with the federal government, are working on core and fuel designs to use weapons-grade plutonium as fuel.

As of this writing (ca 1998), there are no nuclear fuel reprocessing plants operating in the United States. Other nuclear nations have constructed second- or third-generation reprocessing facilities. These nations have signed the nuclear nonproliferation treaty, and the facilities are under the purview of the International Atomic Energy Agency (IAEA).

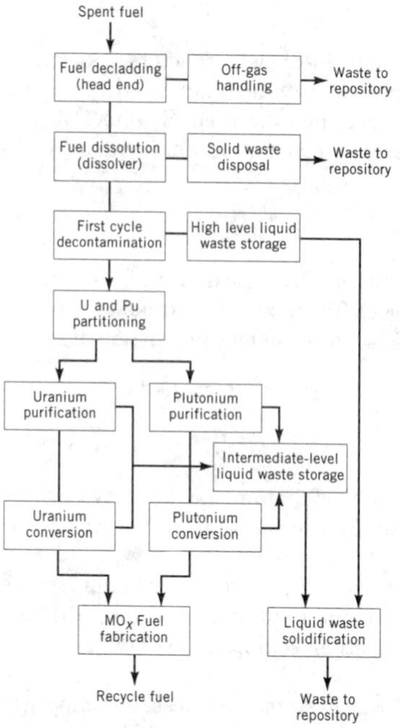

Figure 2. Processing steps for irradiated fuel.

Reprocessing Strategy

The radioactive isotopes associated with spent fuel emit substantial amounts of α-, β-, and γ-radiation as a result of radioactive decay. These radioisotopes (qv) have half-lives, $t_{1/2}$, up to thousands of years. The storage of spent fuel prior to reprocessing allows the shorter-lived isotopes to decay, thus reducing the costs associated with shielding and heat removal. These savings are offset by the high costs associated with maintaining a large spent fuel inventory.

A second consideration affecting reprocessing strategy is the preservation of separative work units (SWU), a measure of the relative value of enriched uranium. Fuel is campaigned through a reprocessing facility in batches having similar residual enrichments. However, a single utility rarely has enough fuel of the same enrichment and burn-up to avoid SWU degradation. Batches are therefore a mix of similar residual enrichments from a number of utilities, and the utility/reprocessor differences in fissile measurements become important. The amount of fissile material in spent fuel is calculated by the utility based on reactor physics models. The reprocessor determines fissile content based on analysis of the dissolved solutions and the dissolved solids. Reconciliation is an ongoing effort. The process steps involved in reprocessing irradiated fuel are illustrated in Figure 2.

Fuel Characteristics. Historically, chemical reprocessing of irradiated fuel was developed specifically to handle the U.S. government's defense-related fuels. These were of two types. There were the very low enrichments and burn-ups, eg, 0.9% ^{235}U and 2000 MW · d/t, as was used for plutonium production, and very high enrichments and burn-ups, eg, 97% ^{235}U and 10^5 MW·d/t, as is used in naval reactors. In both cases, metallic uranium is used.

By contrast, uranium fuels for lightwater reactors fall between these two extremes. A typical pressurized water reactor (PWR) fuel element begins life at an enrichment of about 3.2% ^{235}U and is discharged at a burn-up of about 30×10^3MW·d/t, at which time it contains about 0.8 wt% ^{235}U and about 1.0 wt % total plutonium. Boiling water reactor (BWR) fuel is lower in both initial enrichments and burn-up.

Head-End. A power reactor fuel assembly consists of uranium or mixed uranium/plutonium oxide pellets enclosed in a zirconium metal tube.

Modern facilities shear the spent fuel elements into segments several centimeters long to expose the oxide pellets to nitric acid for dissolution. This operation is often referred to as chop-leach. The design and operation of the shear is of primary importance because *(1)* the shear can be the production rate-limiting step, and *(2)* the shear is the point at which tritium and fission gases are released.

In the American and British plants, LWR fuel pieces typically fall directly from the shear into a dissolver basket, which fits inside the dissolver vessel. A soluble poison such as gadolinium is added to the nitric acid to prevent criticality. The massive end fittings are sometimes separated from the fuel pieces before the latter enter the dissolver.

Fast reactor fuel assemblies are shrouded with a relatively heavy metal envelope. This envelope is removed before shearing by either laser cutting or stress cutting.

The rate (kinetics) and the completeness (fraction dissolved) of oxide fuel dissolution is an inverse function of fuel burn-up. This phenomenon becomes a significant concern in the dissolution of high burn-up MO_x fuels. The insoluble solids are removed from the dissolver solution by either filtration or centrifugation prior to solvent extraction. Both financial considerations and the need for safeguards make accounting for the fissile content of the insoluble solids an important challenge for the commercial reprocessor.

Chemical Separation. A reprocessing facility typically utilizes multiple extraction/reextraction (stripping) cycles for the recovery and purification of uranium and plutonium. For example, a co-decontamination and partitioning cycle is followed by one or more cycles of uranium and plutonium purification.

Chemical separation is achieved by countercurrent liquid–liquid extraction and involves the mass transfer of solutes between an aqueous phase and an immiscible organic phase.

Product Conversion. *Uranium.* The uranium product from the PUREX process is in the form of uranyl nitrate which must be converted to some other chemical depending on anticipated use. One option to MO_x fuel is to mix uranium and plutonium nitrates and perform a co-precipitation step. The precipitate is heated to form a mixed U–Pu oxide. Another approach is to convert the uranium to an oxide, either directly or via an ammonium diuranate intermediate precipitation step, followed by a U–Pu oxide powder blending step. The resulting mixed oxide can then be reenriched by blending with more highly enriched uranium. The value of the uranium at this point is a function of the residual separative work (SWU) it contains and the cost of disposal. If the uranium has ^{235}U concentrations in excess of natural uranium, it can be recycled in lieu of virgin feedstock. This approach has both economic and environmental benefits, in that less uranium mining is required.

Plutonium. The plutonium nitrate product must be converted to MO_x fuel if it is to be recycled to lightwater reactors. The plutonium is typically precipitated as the oxalate and subsequently calcined to the oxide for return to the fuel cycle.

By-Products. The PUREX process is efficient at separating uranium and plutonium from everything else in the spent fuel. Within the high level waste stream are a number of components which have, from time to time, been sufficiently interesting to warrant their recovery. The decision to recover a particular isotope is usually based on a combination of market incentives and desired waste reduction.

Neptunium has been recovered during the reprocessing of defense-related fuels. The ^{237}Np is recycled back to a reactor where it is transmuted to ^{238}Pu. Plutonium-238 is most useful in space programs, but is also of interest as part of a proliferation-resistant fuel cycle.

The 30-yr half-lives of ^{137}Cs and ^{90}Sr make these attractive isotopic heat or radiation sources. However, ^{144}Ce and ^{134}Cs are shorter-lived alternatives. Rhodium, palladium, and technetium have also been recovered for potential catalytic or precious metal applications.

Waste Handling. *Off-Gas Treatment.* Newer shear designs contain the fission gases and provide the opportunity for more efficient treatment. The gaseous fission products krypton and xenon are chemically inert and are released into the off-gas system as soon as the fuel cladding is breached. Efficient recovery of these isotopes requires capture at the point of release, before dilution with large quantities of air. Two processes have been developed: cryogenic distillation and Freon absorption. Neither method has seen widespread use.

A large portion of the ^{14}C can be released from the reprocessing facility as CO_2. Thus, a recovery process has been developed using an alkaline absorbent column.

In the reprocessing environment, there are many ruthenium compounds, some of which are gaseous. Some reprocessing approaches, notably REDOX process, require a ruthenium removal step in the off-gas system. The PUREX process maintains ruthenium in one of its nonvolatile states.

In contrast, iodine recovery is both highly developed and universally used throughout the industry. Iodine can be absorbed on solid sorbents or in an off-gas scrub of caustic, mercuric nitrate solution, or hyperazeotropic nitric acid (qv). The last process, called Lodox, is effective, particularly for trace amounts of organic iodine. Solid absorbers, eg, silver nitrate deposited on Berl saddles or silver zeolite granules, are reliable and effective. The disposal of silver reactors has a negligible environmental impact. These are, however, costly, and have the potential for forming highly reactive compounds, eg, silver azide, if operated at excessively high temperatures.

Hulls Handling. After the fuel has been dissolved, the residual pieces of zirconium cladding, referred to as hulls, are rinsed and removed from the dissolver vessel.

The hulls are the first material removed from the process to merit concern in respect to safeguards. These represent a possible route for special nuclear material (SNM) to be clandestinely removed from the system. Special instrumentation is used to assay the hulls for residual SNM.

The volume of hulls generated is nominally 62 m^3/t of fuel, which is about 10 times the actual volume of metal. Whereas they are not yet in commercial use, both compaction and melting processes are being developed for volume reduction to improve waste handling economics.

High Level Waste Management. The U.S. experience with high level wastes (HLW) is limited to the interim storage of spent LWR fuel and liquid wastes from the weapons program (see NUCLEAR REACTORS–WASTE MANAGEMENT).

In the United States, liquid HLW from the reprocessing of defense program fuels was concentrated, neutralized with NaOH, and stored in underground, mild steel tanks pending solidification and geologic disposal (see TANKS AND PRESSURE VESSELS). These wastes are a complex and chemically active slurry. After the first few months, the decay heat in the HLW is dominated by ^{137}Cs and ^{90}Sr, which are found in the supernatant liquid and sludge, respectively. To ease the problems of heat removal and mitigate the consequences of spills, these isotopes can be removed and stored separately. The cesium capsules have been used as radiation sources for food preservation, and the strontium capsules are available for heat sources. Removal of these radioisotopes has simplified waste management and may position some of the waste for a less costly management option than geologic disposal.

By contrast, HLW from LWR fuel reprocessing is stored in cooled, well-agitated, stainless steel tanks as an acidic nitrate solution having relatively few solids. Modern PUREX flow sheets minimize the addition of extraneous salts, and as a result the HLW is essentially a fission-product nitrate solution. Dissolver solids are centrifuged from the feed stream and are stored separately. Thus, the HLW has a low risk of compromising tank integrity and has a favorable composition for solidification and disposal.

Reprocessing Equipment

Fuel Shear. Reprocessing facilities use equipment capable of shearing entire fuel assemblies. This accounts for a significant position of the reprocessing cost in terms of size of the equipment needed and the consequent size of the cell required to house it. Shear blade maintenance requirements are also important. There are active programs aimed at reducing these costs.

A number of alternatives to shearing have been investigated wherein the cladding is breached by lasers (qv), plasma torches (see PLASMA TECHNOLOGY), or inductive heating.

Liquid–Liquid Contactors. A variety of contactors have been developed for liquid–liquid extraction processes. These designs include equilibrium-stage contactors, such as mixer-settlers and centrifugal contactors, and differential contactors, such as packed and sieve-tray pulse columns. Each design has advantages and disadvantages.

Liquid Waste Storage Tanks. Reprocessing of high burn-up, short-cooled LWR fuel required improvements in the design of the liquid waste storage tanks. Modern tanks are fabricated from stainless steel, and are placed in underground, stainless steel-lined concrete vaults.

W. L. GODFREY
J. C. HALL
G. A. TOWNES
BE Incorporated

R. H. Rainey, W. D. Burch, M. J. Haire, and W. E. Unger, *Fuel Cycle for the 80's Conference,* CONF-800943, Gatlinburg, Tenn., 1980, pp. 155–158.

Safety Analysis Report, NFS Reprocessing Plant, Docket-50-201, Nuclear Fuel Services, Inc., Rockville, Md., 1973.

Safety Evaluation of the Midwest Fuel Recovery Plant, General Electric Co., Docket No. 50-268, United States Atomic Energy Commission, Washington, D.C., 1972.

Final Safety Analysis Report—Barnwell Nuclear Fuel Plant, Separations Facility, Docket-50-332, Allied-General Nuclear Services, Barnwell, S.C., 1973.

REACTOR TYPES

The minimum ingredients of a nuclear reactor, where the basic reactions are nuclear rather than chemical, are neutrons and a fuel such as uranium, the atoms of which can undergo fission. Products of the fission process, in order of importance, are (*1*) heat energy, originally in the form of kinetic energy of particles; (*2*) neutrons, originally of high energy, which can be slowed to lower energies; (*3*) radionuclides, originally as fission fragments and collectively called fission products; (*4*) beta and gamma radiation, released in the fission process and by decay of fission products, which contributes both to heat and hazard; and (*5*) neutrinos, which play no role, because of the ease with which these penetrate matter. The distribution of energy among these products of fission is as follows, for a total of 200 MeV: fission fragments (166 MeV); neutrons (5 MeV); prompt γ-rays (7 MeV); fission product γ-rays (7 MeV); beta particles (5 MeV); and neutrinos (10 MeV).

Reactor Components

Several components are required in the practical application of nuclear reactors. The first and most vital component of a nuclear reactor is the fuel, which is usually uranium slightly enriched with uranium-235, or plutonium produced by neutron absorption in uranium-238 (see NUCLEAR REACTORS–NUCLEAR FUEL RESERVES). The chemical form of the reactor fuel typically is uranium dioxide, UO_2, but uranium metal and other compounds have been used, including sulfates, silicides, nitrates, carbides, and molten salts.

The second important component is the cooling agent or reactor coolant which extracts the heat of fission for some useful purpose and prevents melting of the reactor materials. The most common coolant is ordinary water at high temperature and high pressure to limit the extent of boiling. Other coolants that have been used are liquid sodium, sodium–potassium alloy, helium, air, and carbon dioxide (qv).

The third component is the moderator, a substance containing light elements such as hydrogen, deuterium, or carbon. Because low (ca 0.025-eV) energy neutrons are much more effective in causing fission than high (ca 2-MeV) energy neutrons, such a medium is desirable to slow neutrons by causing multiple collisions.

The fourth component is the set of control rods, which serve to adjust the power level and then, when needed, to shut down the reactor. These are also viewed as safety rods. Control rods are composed of strong neutron absorbers such as boron, cadmium, silver, indium, or hafnium, or an alloy of two or more metals.

The fifth component is the structure, a material selected for weak absorption for neutrons, and having adequate strength and resistance to corrosion. In thermal reactors, uranium oxide pellets are held and supported by metal tubes, called the cladding. The cladding is composed of zirconium, in the form of an alloy called Zircaloy. Additional hardware is required to hold the bundles of fuel rods within a fuel assembly and to support the assemblies that are inserted and removed from the reactor core. Stainless steel is commonly used for such hardware. If the reactor is operated at high temperature and pressure, a thick-walled steel reactor vessel is needed.

The sixth component of the system is the shield, which protects materials and workers from radiation, especially neutrons and gamma rays. Concrete is commonly used, augmented by iron and lead for gamma rays and water for fast neutrons.

Classification and Uses

Nuclear reactors can be classified in a variety of ways: by purpose or use, key components, method of heat extraction, role in application, neutron energy, power level or neutron flux, arrangement of materials, stage of development, and manufacturer.

Herein, reactors are described in their most prominent application, that of electric power. Five distinctly different reactors, ie, pressurized water reactors, boiling water reactors, heavy water reactors, graphite reactors, and fast breeder reactors, are emphasized. A variety of other applications and types of reactors also exist.

Graphite Reactors

The first nuclear reactor made was composed of graphite, the only moderator available at that time for use with natural uranium. Reactors for the production of plutonium during World War II and for power in the United Kingdom also utilized carbon in the form of graphite. A modern helium-cooled graphite reactor has been tested. A distinct advantage of having carbon as the moderator is that it provides the ability to use natural uranium as fuel, avoiding the necessity of expensive and power-absorbing enrichment facilities.

Graphite reactors include: the Magnox and AGR reactors (Table 1), the sodium graphite reactor, the high temperature gas-cooled reactors, Chernobyl, and the Hanford N reactor.

Table 1. Operating Graphite Reactors of the U.K.[a]

Parameter	Reactor site: Oldbury	Reactor site: Heysham 2
reactor type	Magnox	AGR
startup date	Jan. 1968	July 1988
number of reactors	2	2
electrical output, MWe	435	1230
coolant	CO_2	CO_2
coolant temperatures, °C		
in	220	298
out	365	635
gas pressure, MPa[b]	2.52	4.57
pressure vessel, inside dimensions		
diameter, m	23.5	20.25
height, m	18.3	21.87
core dimensions		
diameter, m	14.2	9.46
height, m	9.8	8.31
fuel type	metal rods	UO_2 pellets
fuel can material	Magnox	stainless steel

[a] Courtesy of Nuclear Electric plc.
[b] To convert MPa to psi, multiply by 145.

Pressurized Water Reactors

Figure 1 shows a schematic diagram of the reactor vessel, heat exchanger, turbine-generator, and other equipment of a modern PWR.

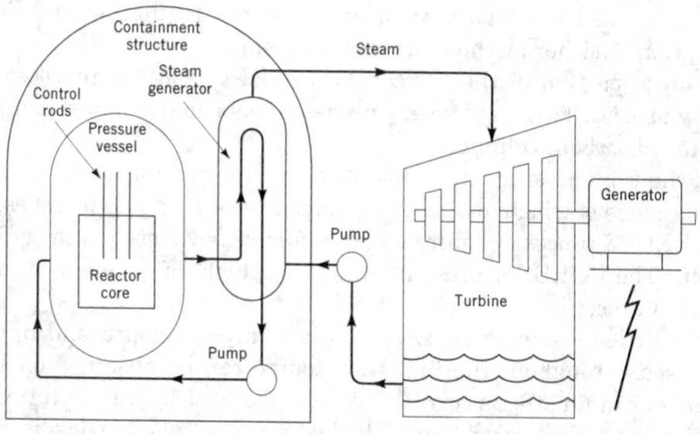

Figure 1. Schematic of a pressurized water reactor system. Fission heat is extracted by the lightwater coolant. The steam drives the turbine-generator. Courtesy of the Nuclear Energy Institute.

Also shown is a schematic of the containment, a large concrete and steel structure capable of withstanding a significant excess pressure from accidental release of hot steam from the reactor vessel.

The key feature of the pressurized water reactor is that the reactor vessel is maintained above the saturation pressure of water and thus the coolant-moderator does not boil. At a vessel pressure of 15.5 MPa (2250 psia), high water temperatures averaging above 300°C can be achieved, leading to acceptable thermal efficiencies of approximately 0.33.

About half of the world's nuclear power plants are from Westinghouse Electric Corp. or its licensees. One Westinghouse PWR design if the four-loop Model 412 (Table 2). The reactor uses several control mechanisms. The first is the control cluster which can be moved up and down, or released to shut down the reactor quickly. Other PWRs use an alloy of cadmium, indium, and silver, all strong neutron absorbers, as control material.

The second control mechanism is the soluble reactor poison boric acid, H_3BO_3. Natural boron contains 20% boron-10, ^{10}B, which has a thermal neutron cross section of ca 4.0×10^{-25} m^2 (4000 barns). As fuel is consumed and fission products build up during a year or so of operation, the concentration of boron is adjusted by dilution. Starting from an initial value of around 2000 ppm, the boron concentration goes to zero at the end of the cycle.

Table 2. Westinghouse Model 412 Pressurized Water Reactor

Parameter	Value
thermal power, MW	3425
electrical power, MWe	1150
reactor vessel ID, m	4.394
primary system pressure, MPa[a]	15.5
coolant flow rate, kg/s	17,438
coolant temperatures, °C	
inlet	291.9
outlet	325.8
rise	33.9
steam pressure, MPa[a]	6.9
fuel dimensions, mm	
fuel rod OD	9.14
Zircaloy-4 cladding thickness	0.572
diametral gap	0.157
UO_2 pellet diameter	7.844
lattice pitch	12.60
fuel assembly array	17 × 17
rods per assembly	264[b]
number of assemblies in core	193
rods per core	50,952
fuel total weight, kg	81,639
core dimensions, m	
effective diameter	3.38
fuel height	3.658

[a] To convert MPa to psia, multiply by 145.

[b] 25 spacase are taken by control rods, burnable poison rods, or neutron sources, or are plugged.

The third control is by use of a fixed burnable poison. This consists of rods containing a mixture of aluminum oxide and boron carbide. The burnable poison is consumed during operation, causing a reactivity increase that helps counteract the drop owing to fuel consumption. It also reduces the need for excessive initial soluble boron. Other reactors use gadolinium as burnable poison, sometimes mixed with the fuel.

In the startup of a reactor, it is sometimes necessary to have a source of neutrons other than those from fission. Otherwise, it might be possible for the critical condition to be reached without any visual or audible signal. Two types of sources are used to supply neutrons. The first, applicable when fuel is fresh, is californium-252, ^{252}Cf. The second, which is effective during operation, is a capsule of antimony and beryllium.

An engineered safety system is provided to protect against hazard from a loss-of-coolant accident (LOCA). If a coolant pipe should break, causing a drop in pressure in the vessel, an emergency core cooling system (ECCS) begins supplying auxiliary water from storage tanks to continue cooling the core.

A PWR can operate steadily for periods of a year or two without refueling. Uranium-235 is consumed through neutron irradiation: uranium-238 is converted into plutonium-239 and high mass isotopes. The usual measure of fuel burn-up is the specific thermal energy release. A typical figure for PWR fuel is 33,000 MW d/t. Spent fuel contains a variety of radionuclides: ^{235}U (0.81%); ^{236}U (0.51%); ^{238}U (94.3%); ^{239}Pu (0.52%); ^{240}Pu (0.21%); ^{241}Pu (0.10%); ^{242}Pu (0.05%) and fission products (3.5%).

The original fuel contained 3.3 wt % uranium-235 and 96.7 wt % uranium-238.

Boiling Water Reactors

Water is also used as moderator-coolant in the boiling water reactor (BWR). The principal distinguishing feature from the PWR is that in the BWR, steam is produced in the core and delivered directly to a steam turbine for the generation of electricity, eliminating the need for a heat exchanger. Benefits are that the system is simpler than the PWR and the capital equipment cost is lower.

Initial studies for the BWR were made at Argonne National Laboratory in the early 1950s. The first experiments used electrical heating for metal plates and tubes immersed in a water bath. Stable boiling, a high velocity of steam, and a very short time necessary for steam bubbles to form were observed. These results encouraged planning for a boiling water experiments involving fuel.

In the BWR, cooling water for fuel rods is provided by external recirculation pumps and internal jet pumps. As water rises in the reactor core, the steam volume fraction increases. Water droplets are removed from the steam by separators and a dryer. The reactor uses cross-shaped control rods inserted from the bottom of the reactor vessel. Their position can be controlled automatically or manually for both start-up and power adjustment during operation. For safety in case of accidental release of steam, a quenching water pool is provided.

Heavy Water Reactors

A heavy water reactor (HWR) uses deuterium oxide, D_2O, also called heavy water, as moderator. There was been relatively little experience using commercial heavy water moderated power reactors in the United States. Early experimental reactors were operated at Argonne National Laboratory in the 1950s. Additionally, a prototype power reactor was built in the 1960s, sponsored by the Carolinas–Virginia Power Associates. Several heavy water production reactors were operated at the Savannah River Plant. The most successful application of heavy water reactors has been in Canada.

The CANDU Reactors. The Canadian deuterium uranium (CANDU) reactors are unique among power reactors in several respects. Heavy water is used as moderator; natural uranium having ^{235}U isotopic content of 0.72 wt % is used as fuel, rather than the typical 2–4 wt % ^{235}U for lightwater reactors; the heavy water coolant flows through pressure tubes passing through the moderator tank; and continuous refueling is performed.

The 4-mm wall-thickness zirconium–4.5% niobium alloy pressure tubes are surrounded by heavy water moderator at much lower temperature and pressure. The reactor vessel, called a calandria, is a large cylinder 8.5 m in diameter and 6 m long, oriented horizontally. Because heavy water is expensive, costing around $50/kg, negligible leakage is mandatory. Refueling is done without shutting the reactor down, reducing outage times. Refueling machines are located at each end of the reactor vessel. A steam generator having light water in the secondary side supplies steam to the turbine.

CANDU has a unique negative pressure containment that functions if an accident such as a cooling-water pipe break should occur in the reactor building, resulting in a release of steam, hot water, and radioactive material. The increased pressure is relieved into a vacuum building, maintained at nearly zero pressure. The building is 50 m in diameter and in height, having one meter thick walls and roof. Inside is a large emergency storage tank that provides a water spray to quench the hot vapor and wash out radioactivity.

The Canadian nuclear power program was developed by Atomic Energy of Canada, Ltd. (AECL) of Ottawa. The CANDU reactors provide a large fraction of the electricity of that region. The price of electricity is much lower than in any other country because of low fuel cost, high capacity factor, and efficient operation. At full capacity the 22 heavy water reactors produced 15,442 MW of electrical power. Most of these are operated in the province of Ontario. The latest reactor was put in operation in 1993 and has a power of 881 MWe.

A variant of the HWR is the Fugen reactor developed by Japan. This reactor is heavy water-moderated but lightwater-cooled. It is fueled by mixed uranium–plutonium oxides.

Fast-Breeder Reactors

Breeding of nuclear fuel was recognized as having a potentially important impact on the availability of energy resources as soon as plutonium was discovered. The most likely nuclear reaction involved the absorption of a neutron in uranium-238 to form plutonium-239, a fissile nuclide with a half-life of approximately 24×10^3 yr. Fission in plutonium-239 by fast neutrons gives rise to about three fast neutrons per absorption. Thus in a reactor using plutonium as fuel, the chain reaction can be maintained and enough neutrons are left over to produce more fuel than is burned. At the same time, by consuming uranium-238 instead of merely burning uranium-235 as in converter reactors, the amount of uranium ore needed to produce a given energy is reduced by a factor as large as 50, thus extending the practical life of the uranium resource for thousands of years.

Full advantage of the neutron production by plutonium requires a fast reactor, in which neutrons remain at high energy. Cooling is provided by a liquid metal such as molten sodium or NaK, an alloy of sodium and potassium. The need for pressurization is avoided, but special care is required to prevent leaks that might result in a fire. A commonly used terminology is liquid-metal fast-breeder reactor (LMFBR).

Most fast reactors that use Na and NaK as coolant utilize an intermediate heat exchanger (IHX) that transfers heat from the radioactive core coolant to a nonradioactive liquid-metal loop, which has the reactor's steam generator. This helps minimize the spread of contamination in the event of a leak or fire.

One of the most advanced versions of a LMFBR was the French SuperPhénix, located at Creys-Malville but no longer operating.

The MONJU fast-breeder reactor is located on the northern coast of Japan. The only other fast-breeder reactors in operation in the world are the 233-MWe Phénix in France, the 135-MWe BN-350 in Kazakhstan, and the 560-MWe BN-600 Beloyarskiy in Russia.

In the twenty-first century, as fossil fuel resources become more scarce, fast-broader reactors may become important.

Other Reactors

Other reactors include a natural uranium and water fission reactor that existed long before humans appeared on Earth in the African state of Gabon, near Oklo; homogenous aqueous reactors; aircraft reactors; naval reactors; maritime reactors; package power reactors; space reactors (eg, electrical power and propulsion); research and training reactors; and advanced power reactors, the advanced light-water reactors (ALWR).

RAYMOND L. MURRAY
Consultant

A. V. Nero, Jr., *A Guidebook to Nuclear Reactors,* University of California Press, Berkeley, Calif. 1979.

R. A. Knief, *Nuclear Engineering: Theory and Technology of Commercial Nuclear Power,* Hemisphere Publishing Corp., Washington, D.C., 1992.

D. Bodansky, *Nuclear Energy: Principles, Practices, and Prospects,* AIP Press, Woodbury, N. Y., 1996.

The Westinghouse Pressurized Water Reactor Nuclear Power Plant, Westinghouse Electric Corp., Water Reactor Divisions, Pittsburgh, Pa., 1984.

WASTE MANAGEMENT

Radioactive wastes are generated in all parts of the fuel cycle supporting nuclear electric power, including mining and milling of uranium ore, chemical conversion, isotope separation, fuel fabrication, nuclear reactor operation, spent fuel storage, and waste disposal. Successful management of wastes from nuclear reactors is vital to continued use of nuclear power, and assurance of waste safety should aid in improving public acceptance.

Classification of wastes may be according to purpose, distinguishing between defense waste related to military applications and commercial waste related to civilian applications. Classification may be also by the type of waste, ie, mill tailings, high level radioactive waste (HLW), spent fuel, low level radioactive waste (LLW), or transuranic waste (TRU). Alternatively, the radionuclides and the degree of radioactivity can define the waste. Surveys of nuclear waste management and more technical information are available.

Sources

Radioactivity occurs naturally in earth minerals containing uranium and thorium. It also results from two principal processes arising from bombardment of atomic nuclei by particles such as neutrons, ie, activation and fission.

Radioactive waste is characterized by volume and activity, defined as the number of disintegrations per second, known as becquerels. Each radionuclide has a unique half-life, $t_{1/2}$, and corresponding decay constant, $\lambda = 0.693/t_{1/2}$. For a component radionuclide consisting of N atoms, the activity, A, is defined as $A = N\lambda$.

Radioactive wastes are generated in every step of the nuclear fuel cycle, including the extraction of uranium from the ore, its chemical conversion, isotope separation, fabrication of fuel assemblies, irradiation in a nuclear reactor, and treatment of residues. In the U.S. defense program, the production of plutonium as weapons material gave rise to transuranic wastes.

The principal legislative actions related to the management of radioactive waste are the 1980 Low Level Radioactive Waste Policy Act, which assigned responsibility for LLW disposal to the individual states, and the Nuclear Waste Policy Act of 1982, which set forth the procedure and schedule by which high level waste should be managed by the U.S. Department of Energy (DOE). Each of these acts has been amended.

Treatment

Several modes of waste management are available. The simplest is to dilute and disperse. This practice is adequate for the release of small amounts of radioactive material to the atmosphere or to a large body of water. Noble gases and slightly contaminated water from reactor operation are eligible for such treatment. A second technique is to hold the material for decay. This is applicable to radionuclides of short half-life such as the medical isotope technetium-99m ($t_{1/2} = 6$ h), the concentration of which becomes negligible in a week's holding period. The third and most common approach to waste management is to concentrate and contain. Various processes are applied to minimize volume and to prevent or delay access of water to the contents of waste containers.

Low-Level Waste Treatment. Methods of treatment for radioactive wastes produced in a nuclear power plant include *(1)* evaporation (qv) of cooling water to yield radioactive sludges, *(2)* filtration (qv) using ion-exchange (qv) resins, *(3)* incineration with the release of combustion gases through filters while retaining the radioactively contaminated ashes (see INCINERATORS), *(4)* compaction by presses, and *(5)* solidification in cement (qv) or asphalt (qv) within metal containers.

All processes in a nuclear power plant, in a treatment facility, or at a disposal site are governed by rules of the U.S. Nuclear Regulatory Commission (NRC). Radiation protection, ie, the limits on radiation dose to workers and the public, are specified. Exposure is maintained as low as reasonably achievable (ALARA).

Mixed wastes, ie, those having a hazardous material or feature plus a radioactive component, are subject to regulation by both the NRC and EPA. Standards differ significantly.

Radiation dose limits at a disposal site boundary are specified by the NRC as 25×10^{-5} Sv/yr (25 mrem/yr), a small fraction of the average radiation exposure of a person in the United States of 360×10^{-5} Sv/yr (360 mrem/yr). Protection against nuclear radiation has been fully described.

Nuclear utilities have sharply reduced the volume of low level radioactive waste over the years. In addition to treating wastes, utilities avoid contamination of bulk material by limiting the contact with radioactive materials. Decontamination of used equipment and materials is also carried out.

Recycling techniques for radioactive material are evolving, especially for scrap metals. Large quantities of scrap are expected from decommissioning operations. Advantage is being taken of successful German and Japanese experience. One important use of recycled steel from the nuclear industry is for waste containers and canisters. A slight residual radioactivity is immaterial.

Spent Fuel Treatment. Spent fuel assemblies from nuclear power reactors are highly radioactive because they contain fission products. Relatively few options are available for the treatment of spent fuel. The tubes and the fuel matrix provide protection against attack and release of nuclides. To minimize the volume of spent fuel that must be shipped or disposed of, the fuel rods in assemblies can be consolidated into compact bundles. Alternatively, intact assemblies can be encased in metal containers.

Reprocessing, as practiced outside the United States, involves chopping fuel rods into small pieces, leaching out the uranium oxide by nitric acid (qv), and applying suitable solvents to separate the uranium, plutonium, and fission products. A vitrification process is applied to fission product wastes.

Special chemical treatment can isolate the nuclides of intermediate half-life, ie, cesium-137 and strontium-90, ^{90}Sr, $t_{1/2}$ 29 yr. These provide most of the radioactivity, radiation, and heat during the early years of a disposal facility. Solid-phase extraction methods using macrocyles and membranes promise to yield waste of low volume and activity (see INCLUSION COMPOUNDS; MEMBRANE TECHNOLOGY). The separated intensely radioactive chemicals can be placed in separate storage or made available for industrial radiation use.

Storage and Transport

Storage. Storage of spent fuel assemblies in deep water pools at reactor sites serves several safety functions. Cooling by water prevents the fuel from melting from decay heat. The shielding effect of the water provides protection for workers in the vicinity against gamma-radiation. Moreover, the separation of assemblies by racks prevents a chain reaction from occurring. The reinforced concrete pools are designed to withstand earthquakes. Water passes through a heat exchanger to maintain a constant temperature, and the purity of the water is assured by use of a demineralizer.

The pools of most reactors were designed for a limited number of fuel assemblies. As a consequence, pools are filling up. The material awaits permanent disposal. Alternatives to pool storage are being sought. One relatively inexpensive alternative is dry storage in large sealed concrete containers.

The Nuclear Waste Policy Act of 1982 specified that DOE would begin accepting spent fuel from nuclear utilities in 1998. If DOE honors that requirement, it would have to store spent fuel at federal laboratories, awaiting construction of a monitored retrievable storage facility.

Waste by-products of the operation of plutonium-producing reactors beginning in World War II have been stored in underground tanks at Hanford (Washington state) and the Savannah River Plant (South Carolina). Some single-walled tanks (Hanford) have leaked. A remediation program for tank storage is in place and mechanical stirring pumps have removed the potentially explosive hydrogen from these Hanford tanks.

Plutonium itself, generated by production reactors during the Cold War for weapons purposes, may become a waste by national policy. The plutonium would then be disposed of along with other high level wastes. Alternatively, the plutonium could serve as fuel for reactors that generate electric power or could be bombarded by neutrons produced by high energy charged particles from an accelerator (see also PLUTONIUM AND PLUTONIUM COMPOUNDS).

Low level waste with its generally smaller radioactivity level can be stored in suitable containers in buildings. Protective shielding and handling equipment are required.

Transport. In the United States, waste transportation is regulated by the NRC and the Department of Transportation (DOT). Packaging and shipping must be in conformity with comprehensive rules. Shipping container classes are defined in accord with the amount of radioactivity involved.

A multipurpose canister (MPC) is planned for the transportation, storage, and disposal of spent fuel, minimizing the amount of handling required.

The safety record for transport of radioactive materials including spent fuel and wastes is excellent. Information about transportation of radioactive material including waste is managed by DOE.

Disposal

The disposal of radioactive waste is governed by rules of the NRC and the EPA. Regulations differ for low level waste and for high level waste, including spent fuel.

Isolation of radioactive wastes for long periods to allow adequate decay is sought by the use of multiple barriers. Barriers limit water access to the waste and minimize contamination of water supplies. The length of time that wastes must remain secure is dependent on the regulatory limit of the maximum radiation exposure of individuals in the vicinity of the disposal site.

Performance assessments are predictions of radioactivity releases, the rate of transfer of contaminants through various media, and the potential for hazard to the public. These are based on a combination of experimental data obtained in the process called site characterization and detailed computations about radionuclides and their effects.

High Level Waste. Many studies have been made of possible modes of disposal for high level waste including spent fuel. As of this writing (ca 1998), the preferred method is deep underground burial in the floor of a mined cavity.

Regulations on high level radioactive waste management have traditionally been provided by the NRC, and a number of specific requirements on the character of a disposal site are spelled out. The generic regulations planned by the EPA have been delayed by court actions and federal law requirements. Wastes are to be placed at least 300 meters below the earth's surface. The waste form must be free of liquid, noncorrosive, and noncombustible. The container is to remain intact for between 300 and 1000 years. The travel time of groundwater prior to waste emplacement is preferred to be greater than 10,000 years, but not less than 1000 years. The repository must be placed where there are no attractive resources and far from population centers. Wastes are to be retrievable for a period of 50 years. Finally, releases of radionuclides from the repository must be less than figures specified by EPA. Typical limits are 3.7×10^{12} Bq/10^3 t (100 Ci/10^3 t) of heavy metal. Regulations include guidelines on geologic conditions.

Tuff, a compressed volcanic material, is the primary constituent of Yucca Mountain, near Las Vegas, Nevada, the site selected by Congress in 1987 for assessment for spent fuel disposal.

Site characterization studies include a surface-based testing program, potential environmental impact, and societal aspects of the repository. Performance assessment considers both the engineered barriers and the geologic environment.

Low Level Waste. Regulations specify the nature of the protection required for waste containers. Class A wastes must meet minimum standards, including no use of cardboard; wastes must be solidified, have less than 1% liquid, and not be combustible, corrosive, or explosive. Class B wastes must meet the minimum standards but also have stability, ie, these must retain size and shape under soil weight, and not be influenced by moisture or radiation. Class C wastes must be isolated from a potential inadvertent intruder, ie, one who uses unrestricted land for a home or farm. Institutional control of a disposal facility for 100 years after closure is required.

The traditional method of disposal of low level radioactive waste has been shallow land burial, consisting of filling a deep trench and covering it with a layer of earth. Three of the original six commercial sites were closed owing to leaks of radioactivity. Two initiatives resulted: stricter regulations and the 1980 Low Level Radioactive Waste Policy Act, which called for each state to be responsible for wastes generated within its borders, but recommended the formation of compacts among states to build regional disposal facilities.

Shallow land burial is planned for disposal facilities in arid regions, notably in California and Texas. Public demand at locations in humid regions led to the adoption of greater confinement designs, which provide additional protection. Included are concrete vaults, overpacks, multiple-layered caps, and enhanced monitoring systems.

Methods to control infiltration of water into low level waste disposal facilities are being studied. Three techniques that may be employed separately, in sequence, or in conjunction are use of a resistive layer, eg, clay; use of a conductive layer, including wick action; and bioengineering, using a special plant cover.

Funding for developing commercial waste disposal facilities is to come from the waste generators.

Transuranic Waste. Transuranic wastes (TRU) contain significant amount (>3,700 Bq/g (100 nCi/g)) of plutonium. Experimental test of TRU disposal is planned for the Waste Isolation Pilot Plant (WIPP) site near Carlsbad, New Mexico. The geologic medium is rock salt, which has the ability to flow under pressure around waste containers, thus sealing them from water. Studies center on the stability of structures and effects of small amounts of water within the repository.

Other fuel besides that from U.S. commercial reactors may be disposed of in the ultimate repository. Possibilities are spent fuel from defense reactors and fuel from research reactors outside of the United States. To reduce the proliferation of nuclear weapons, the United States has urged that research reactors reduce fuel enrichment in uranium-235 from around 90% to 20%. The latter fuel could not be used in a weapon. The United States has agreed to accept spent fuel from these reactors.

Environmental Issues

There is an enormous amount of research literature about nuclear waste management, and nuclear scientists and engineers are generally convinced that wastes can be disposed of safely. Delays in implementing the objectives of the waste policy acts can be attributed to frequent changes in government policy and to public opposition based on concerns about radiation and economic effects.

RAYMOND L. MURRAY
Consultant

R. L. Murray, *Understanding Radioactive Waste,* 4th ed., Battelle Press, Columbus, Ohio, 1994.

Integrated Data Base for 1996: U.S. Spent Fuel and Radioactive Waste Inventories, Projections, and Characteristics, DOE/RW-0006, Rev. 13, Oak Ridge National Laboratory, Oak Ridge, Tenn., Dec. 1997.

A. A. Moghissi, H. W. Godbee, and S. A. Hobart, *Radioactive Waste Technology,* American Society of Mechanical Engineers, New York, 1986.

Nuclear Power: Technical and Institutional Options for the Future, National Academy Press, Washington, D.C., 1992.

SAFETY IN NUCLEAR POWER FACILITIES

Nuclear energy is a principal contributor to the production of the world's electricity. As shown in Table 1, many countries are strongly dependent on nuclear energy.

Table 1. Nuclear Power Units in Operation Worldwide

Nation	Number of units[a]	Net power, MWe[a]	Percent of total electricity generated[b]
Belgium	7	5,527	59
Bulgaria	6	3,420	34
Canada	22	15,439	16
China	3	2,100	
Czech Republic	4	1,632	
Finland	4	2,310	33
France	55	57,373	73
Germany	21	22,715	32
Hungary	4	1,729	46
India	9	1,620	2
Japan	49	38,859	24
South Korea	9	7,220	48
Lithuania	2	2,760	
Russia	25	19,799	
Slovakia	4	1,632	
South Africa	2	1,840	6
Spain	9	7,084	35
Sweden	12	10,075	52
Switzerland	5	3,025	40
Taiwan	6	4,884	38
Ukraine	14	12,095	
United Kingdom	34	11,540	19
United States	109	99,238	21
other[c]	9	3,602	
Totals	*424*	*337,518*	

[a] Status as of Dec. 31, 1994.
[b] Status as of Jan. 1, 1992.
[c] Nations having <1000 MWe capacity are combined under "other."

Safety has played a dominant role in the ability to generate electricity by nuclear means. As energy is generated, highly radioactive materials, which can be harmful to living organisms if not kept under strict control, are produced.

All safety provisions must meet strict limitations on allowable levels of radiation exposure, assuring that neither the public nor the plant workers are harmed as a result of operation of the nuclear power plant. Control of radioactive materials must be effected by (*1*) careful design and testing of the integrity and reliability of the components and systems which contain and control the radioactive materials in the nuclear power plant; (*2*) fabrication, installation, and construction of these components to meet high quality standards; and (*3*) thorough training of plant operators to assure that the systems and components function as designed and integrity is maintained.

The design of safety systems and components must also provide tolerance for human fallibility in achieving strict controls, ie, protect against design mistakes, equipment failure, and operational error. Redundancy and diversity are provided for key safety functions.

Whereas these design measures provide the primary assurance of protection from the harmful effects of radiation, additional protection is provided in the unlikely event that the integrity of the systems or components breaks down. The entire portion of the nuclear plant containing radioactive material is enclosed in a strong containment building. If a release of radioactivity from the plant were to occur, the radioactive material would be captured within the containment building. As a further precaution, these processes are subject to continual cross-checks in the form of design reviews, inspections, operations, and safety audits by people separate from those engaged in the design, fabrication, construction, and operational processes. The U.S. nuclear industry has set up the Institute of Nuclear Power Operations (INPO) to establish operations and training standards and to audit nuclear plant operations for compliance with those standards. In addition, a totally independent body responsible to the public, the U.S. Nuclear Regulatory Commission (NRC), establishes overriding safety regulation and monitors compliance.

Safety provisions have proven effective. The nuclear power industry in the Western world, ie, outside of the former Soviet Union, has made a significant contribution of electricity generation, while surpassing the safety record of any other principal industry. In addition, the environmental record has been outstanding.

Basic Safety Principles

The three fundamental safety objectives advocated by the International Atomic Energy Agency (IAEA) for all nuclear power plants worldwide are (*1*) to protect individuals, society, and the environment by establishing and maintaining in nuclear power plants an effective defense against radiological hazard; (*2*) to ensure in normal operation that radiation exposure within the plant as well as that resulting from any release of radioactive material from the plant is kept as low as reasonably achievable (ALARA) and below prescribed limits, and to ensure mitigation of the extent of radiation exposures owing to accidents; and (*3*) to prevent accidents in nuclear power plants with high confidence; to ensure that for all accidents taken into account in the design of the plant, even those of low probability, radiological consequences, if any, would be minor; and to ensure that the likelihood of severe accidents with serious radiological exposure is extremely small.

IAEA also defined the fundamental responsibilities for nuclear power plant safety as ultimately resting with the operating organization. Designers, suppliers, constructors, and regulators are also responsible for their separate activities. Responsibility is reinforced by the establishment of a safety culture, ie, "the personal dedication and accountability of all individuals engaged in any activity which has a bearing on the safety of nuclear power plants" (IAEA Safety Series #75). Safety design of nuclear power plants is founded on the defense-in-depth concept, which provides multiple levels of protection to both the public and the workers, in the form of physical barriers and levels of implementation of the associated defenses. Each of the multiple physical barriers prevents the release of radioactive materials, but all envelop a given number of the others so that if an inner barrier fails, the next outer barriers holds back the radioactive material. Figure 1 shows both the physical barriers and the multiple levels of protection in conceptual form.

The reliability of the physical barriers is assured by implementation of multiple levels of defense-in-depth, characterized by a sequence of concentric design features and operational defenses against the release of radiation from the plant. The first level is the design, fabrication, and construction of the plant to high quality standards together with its reliable operation and maintenance within the prescribed operational bands. The second level of defense is comprised of systems and operating procedures which control abnormal conditions, ie, transients beyond the prescribed bands, so that the basic integrity of the system is maintained. The third level of defense-in-depth is the provision of backup systems and emergency operating procedures that become operative in the event that there is a loss of integrity or a loss of a basic function of the normal nuclear systems, assuring that the radioactive material is not released from the nuclear systems. In each of these first three levels, a separate layer of multiple protection is provided through redundancy and diversity.

The fourth level of defense-in-depth is activated if all of the previous levels fail and radioactivity is released from the power-generating system. This level consists of containment systems and accident management processes that prevent the dissemination of radioactivity to the atmosphere even if it is released from the nuclear systems. The fifth level is the provision for emergency planning outside the plant boundary in the highly unlikely event that all of the first four defenses were to fail.

The defense-in-depth process requires that each physical barrier be designed conservatively, using substantial margins against failure, on-line monitoring instrumentation, off-line inspections to detect incipient failures, and highly trained operators and maintenance personnel guided by prudent procedures. In particular, the containment building is designed to withstand external assaults from earthquakes, hurricanes, tornadoes, floods, and flying objects such as crashing airplanes. The safety of the nuclear power plant and the integrity of its containment must also be maintained in the event of aggression from terrorists or saboteurs. Stringent security measures are provided at each plant to meet such a challenge. Strict personnel checks, emphasis on professional discipline, and the redundant, fail-safe design of the safety systems provide protection against internal sabotage. A prioritization process applies in the design of the barriers and the provisions for defense-in-depth that is based on the principle "prevention first." Thus, no compromise is permitted in the design integrity of the first three barriers and the first two levels of defense-in-depth which prevent an accidental release.

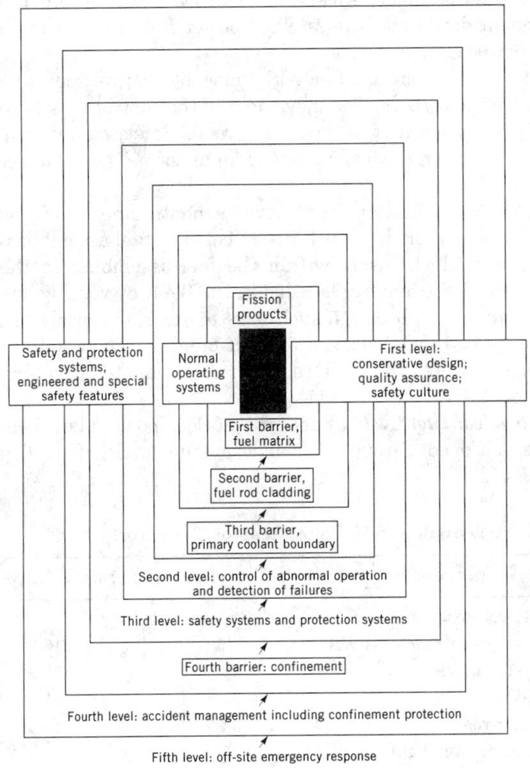

Figure 1. The relation between physical barriers and levels of protection in defense-in-depth design of a nuclear facility. Courtesy of IAEA.

Safety Design

Design Features. Design safety features are utilized at each of the concentric safety barriers. The most important of these safety features apply at the innermost barriers in what is called the reactor core, a cylindrical arrangement of bundles of nuclear fuel rods which are arranged to cause fission and spaced to permit the flow of cooling water through them. Sustained fission, also called critical mass or criticality, is possible because a chain reaction can be established.

First Barrier. The rate of fission must be kept under strict control so as to prevent a runaway power excursion, ie, an excessive increase in fission rate. Control is carried out in two ways: one, intrinsic to the chain reaction, involves a negative coefficient of reactivity; the other is external, through use of control rods. The fission rate is dependent upon the temperature of the fuel and the temperature and density of the coolant. Fuel composition and absorber materials, ratio of fuel to coolant, and geometrical arrangement of the fuel and the fuel rods can be designed so that the fission rate decreases as temperature, coolant density, or power increases. This intrinsic feature can be designed into the fuel system, ie, the core, to cause the fission rate to slow down when temperature, steam content, or power increases. This is called a negative temperature, void, or power coefficient of reactivity.

The external means of controlling the fissioning rate is through use of control rods. Metal rods composed of strong neutron absorbers can reduce or cut off the chain reaction. These can be inserted into the core to reduce or stop fission or alternatively pulled out of the core to start or increase fission. The control rods are moved by remote control by the operator for normal power control.

Cutting off the chain reaction removes concern that a runaway power excursion involving rapid melting of rods can occur, but does not eliminate the possibility of slow fuel melting.

Second Barrier. Safety design features at the second barrier involve the primary coolant circuit. These are derived from adherence to rigorous standards in the selection of materials and in conservative design of the coolant system and coolant pressure boundaries. The conservatism is provided by designing for forces, pressures, temperatures, fluid conditions, radiation levels, thermal transients, and fatigue cycles that are higher than are expected during power operation. This difference between the design levels and the actual levels is called margin.

Another design feature is the provision for on-line coolant leakage monitors that would signal incipient pressure boundary failure. Radiation monitors are installed in the containment building to detect airborne radiation, which would signal incipient loss of integrity of some part of the pressure boundary. Extensive inspection requirements are also stipulated.

Third Barrier. At the level of the third barrier, the key design feature is the provision of backup cooling systems, eg, gas-pressurized accumulated tanks and safety coolant injection systems, which continue to cool the core in the event of a significant loss of integrity which would disable the normal cooling functions of the primary circuit.

Fourth Barrier. The design feature of the fourth barrier is the containment building. It is designed to withstand the high temperatures, pressures, and radiation resulting from a severe accident entailing fuel meltdown. Supplementary features are utilized to redce the consequences of such severe conditions.

Assessment. It is important to verify that safety is actually being achieved by monitoring the operations of the nuclear plants. Both the U.S. NRC and the INPO perform key roles in this process. Each operating organization is given a performance rating, backed up by detailed critiques identifying operational strengths as well as weaknesses. The plant is then required to follow up on any corrective actions indicated by these safety audits. The NRC ratings are made public and have significant power in motivating corrective action when that is needed.

Audits by INPO and the U.S. NRC are a culmination of a high degree of self-auditing by the plant operators and the utilities themselves, often assisted by special third-party safety review boards set up to help carry out safety assessments.

Another element of safety monitoring is the requirement stipulated by the NRC that each utility report any operational event which is out of the ordinary or has safety implications. These licensing event reports (LERs) are placed in the public record. Both INPO and the NRC evaluate the LER and inform all the utilities of any event that has broad safety significance to the industry. If one of these field events is judged to be the precursor of a serious accident, all operating nuclear plants are made fully aware of its implications. Thus, steps can be taken to prevent the event from occurring, and if the event does occur, the plant should be prepared to stop the progress of the event before the stage of a severe accident is reached.

A further assessment is carried out through the definition and measurement of industry-average performance indexes related to safety. These indexes have been established by the utilities, working with INPO, the Electric Power Research Institute (EPRI), and the suppliers. Each index bears on some aspect of safety operation of the nuclear power plant. Five-year goals are established for average performance of all U.S. plants for each of these performance indexes. A substantial improvement has been made in all of these indexes since the early 1980s.

The World Association of Nuclear Operators (WANO) has been formed, consisting of nuclear plant operators over the entire world who have pledged to assist each other in the achievement of safe operations.

Comparative Risks. All efforts are directed to reducing to an extremely low level the chance of a severe nuclear accident that would harm the public. The question remains how low a level this should be. A safety goal stipulated by the U.S. NRC states that *(1)* the risk of prompt fatality to an average individual in the vicinity of a nuclear power plant that might result from reactor accidents should not exceed 0.1% of the sum of prompt fatality risks resulting from other accidents to which members of the U.S. population are generally exposed, and *(2)* the risk of cancer fatalities to the population in the area near a nuclear power plant in operation should not exceed 0.1% of the sum of cancer fatality risks resulting from all other causes.

All U.S. plants are assessed for the ability to meet this goal by carrying out what are called probabilistic safety assessments (PSA). These are sophisticated failure analyses which estimate the probability that a sequence of failures would result in incipient core melting, called the core degradation frequency (CDF). The detailed evaluation by this process of every nuclear power plant in the United States has led the U.S. NRC to judge that existing operating plants are adequately safe.

Analyses and experimental results used to assess the consequences of a severe potential accident have resulted in substantially reduced estimates of severe accident consequences.

The safety goal and the safety record of the nuclear power industry indicate much lower societal risks from commercial nuclear power than from a wide range of other common activities. If safety comparisons are focused on energy systems, nuclear power is also estimated to be safer than all electricity generation methods except for natural gas.

Safety Characteristics of the Nuclear Power Plant

The Reactor. The nuclear power plant reactor types used to produce electricity worldwide are listed in Table 2 (see NUCLEAR REACTORS–REACTOR TYPES). The lightwater reactor (LWR) is representative of commercial nuclear power plants.

Pressurized Water Reactor. The reactor core made of the nuclear fuel assemblies is installed within the reactor pressure vessel, into which coolant water is pumped. When the power plant is shut down for refueling or repair, a residual heat-removal system continues to cool the core, which would otherwise heat up from fission-product radioactive decay.

To assure continued cooling of the fuel in the event of loss of normal cooling, the emergency core cooling system is brought into play. These systems are activated automatically and are installed in multiple form to provide redundancy. Emergency electric power is provided from independent diesel generators when power from the reactor plant or off-site power is unavailable.

Control of the core is affected by movable control rods which contain neutron absorbers. The movement of the control rods is governed remotely by an operator in the control room. Safety circuitry automatically inserts the rods in the event of an abnormal power or reactivity transient.

Boiling Water Reactor. The movable neutron-absorbing elements are in the form of blades which insert between fuel assemblies, rather than in rods which insert within the fuel assemblies in the PWR. Containment heat removal is aided in the BWR by venting any steam issuing from a pipe break, if such were to occur, to a pool of water located at the bottom of the containment building. The resultant steam condensation would then reduce the pressure and temperature within containment.

The Nuclear Fuel Cycle. Fuel for a nuclear power plant is provided and dispositioned through the nuclear fuel cycle. The disposition

Table 2. Nuclear Power Units by Reactor Type Worldwide

Nuclear reactor	Number of units	Net power, MWe × 10^3
lightwater reactor		
pressurized water reactor (PWR)	245	215.7
boiling water (BWR)	92	75.9
gas-cooled reactor	35	11.7
heavy-water reactor	34	18.5
graphite-moderated lightwater reactor	15	14.8
liquid metal-cooled fast breeder reactor	3	0.9

process has two options, recycling, also called chemical reprocessing, or throwaway. In the former, the spent fuel is reprocessed in a chemical plant, where residual uranium and plutonium are separated from the fission products (see NUCLEAR REACTORS–CHEMICAL REPROCESSING). The uranium and plutonium are then recycled to fabricate mixed uranium–plutonium oxide pellets to be used in subsequent reactor refuelings. The fission product waste is then vitrified, ie, converted to a gaseous form, encapsulated in metal casks, and sent to a permanent repository (see NUCLEAR REACTORS–WASTE MANAGEMENT).

The recycling option is being utilized outside the United States. Whereas the technology of this option has been completely demonstrated in the United States, the economics have not been favorable. Moreover, concerns have been raised as to the diversion of the plutonium to weapons use. Thus, the throwaway option is the only one in use in the United States as of this writing (ca 1998).

The safety principles and criteria used in the design and construction of the facilities which implement the nuclear fuel cycle are analogous to those which govern the nuclear power plant. The principles of multiple barriers and defense-in-depth are applied with rigorous self-checking and regulatory overview. However, the operational and regulatory experience is more limited.

One feature of reprocessing plants which poses potential risks of a different nature from those in a power plant is the need to handle highly radioactive and fissionable material in liquid form. The liquid materials and the equipment with which it comes in contact need to be surrounded by 1.5–1.8-m thick high density concrete shielding and enclosures to protect the workers both from direct radiation exposure and from inhalation or airborne radioisotopes. Rigid controls must also be provided to assure that an inadvertent criticality does not occur.

Shielding protection entails design engineering and installation similar to that provided in a nuclear power plant. Additionally, to protect against exposure to airborne radioactivity, controlled air ventilation and air cleaning is provided.

The principal methods for preventing criticality are limitations on the mass of the fuel being handled, the equipment size, the concentration of nuclear material in solution, minimization of the presence of water or plastic that would reduce the margin to reaching a critical mass, and the addition of a neutron absorber, eg, cadmium or boron, either in solution or as a solid packing in vessels. At least two and sometimes more of these independent methods are usually employed at fuel-processing facilities to prevent criticality. In addition, control of other parameters individually or in combination permits the safe handling of quantities many times the critical mass.

In fuel fabrication facilities handling plutonium, Pu, protection must be afforded against both inadvertent criticality and inhalation of airborne particles. Air-flow controls are employed. Fabrication operations are carried out in glove boxes. Ventilation air for the boxes is cleaned through high efficiency particulate air filters. Inadvertent criticality is an even more sensitive issue in plutonium fuel fabrication. Strict control of Pu quantities, therefore, is enforced, limiting the amount of Pu handled in a single operation to less than that needed to start a chain reaction.

The sum total of risks of the nuclear fuel cycle, most of which are associated with conventional industrial safety, are greater than those associated with nuclear power plant operations. However, only 1% of the radiological risk is associated with the nuclear fuel cycle, so that nuclear power plant operations are the dominant risk. Public perception, however, is that the disposition of nuclear waste poses the dominant risk.

Spent Fuel and Radioactive Waste. The basic safety objective governing radioactive waste management is to protect the public and the workers from radiation, at a minimum meeting federal regulatory standards for maximum allowable radiation dosage. This protection is provided for both spent fuel and plant radioactive wastes during the transfer and treatment processes at the plants, temporary storage of wastes and the plants, and during transportation from the plants to storage sites and repositories. Plant radioactive wastes, called low level radwastes (LLW), are of lower radiation intensity than those arising from the nuclear fuel, called high level radwastes (HLW) (see NUCLEAR REACTORS–WASTE MANAGEMENT).

The safety objective for the final storage facilities for the low level wastes and permanent repository for the spent fuel or high level wastes is to package the waste in rugged containers and bury these containers in stable geologic formations far from ground water so that they do not come in contact with humans directly or indirectly. The specific means of meeting this objective can vary in regard to whether the spent fuel throwaway or recycling option is being utilized.

For the throwaway option in the United States, planning as of the mid-1990s was for storing the spent fuel at the reactor site initially in water pools where the relatively high decay heat is removed by natural circulation of the water in the pool. After about 10 years, the spent fuel is then moved to a dry storage facility where natural circulation of air provides sufficient cooling, all the while awaiting transfer of the fuel either to an interim centralized storage facility or to a permanent repository. Once it has been removed from the water pool, the spent fuel assemblies are placed in a stainless steel or titanium container, called a multipurpose canister, which provides an inner shell from which the fuel assembly need never be removed again. This shell is inserted into various other overpacks of concrete or steel, depending on whether the fuel is being stored on-site, is being transported, or is being placed in a permanent repository.

In keeping with the overall safety principles, the spent fuel repository is designed using concentric barriers. The first barriers are the solid, corrosion-resistant ceramic fuel pellets and the Zircaloy cladding which surrounds the pellets. The next barrier is the canister or inner shell, which becomes a permanent element and within which the fuel assemblies are placed. The third barrier is the overpack of concrete or steel. This set of barriers makes up the engineered package. The last barrier is the geologic surroundings within which the engineered package is buried.

The fundamental safety criterion for the permanent repository is that the engineered package retain complete integrity for at least 1000 yr. Rigorous engineering criteria and testing is needed to provide that assurance. The level of radioactivity is greatly reduced after 1000 yr and the need for complete integrity thus alleviated. The half-lives of the radioactive species (the time it takes for radioactive species to diminish by a factor of 2 through radioactive decay) are important characteristics of the evaluations. Half-lives determine the remaining content of the radioactive material in the repository over time. However, half-life in itself is not the dominant characteristic of concern. Rather, it is the toxic and chemical characteristics in combination with the radioactivity which determine the radioisotopes (qv) of dominant risk.

In all of the transportation and storage steps, sensitive radiation monitors are located at and around the spent fuel to detect incipient leakage. Whenever such leakage is detected, steps are taken to repair the defect. Even for the permanent repository, radiation monitoring should be kept up indefinitely and provision made for retrieving the spent fuel during a period of at least 50–100 years to effect repairs.

The primary issue is to prevent groundwater from becoming radioactively contaminated. Thus, the property of concern of the long-lived radioactive species is their solubility in water. The long-lived actinides such as plutonium are metallic and insoluble. Certain fission-product isotopes such as iodine-129 and technetium-99 are soluble, however, and therefore represent the principal although very low level hazard.

The high level waste of the recycling (qv) option is made up primarily of fission products having only residual amounts of plutonium and other actinides following the reprocessing. The fission-product wastes come from the chemical reprocessing plant in liquid form and have to be converted to a solid. Vitrification of the water is planned, so that the first barrier in radioactive containment design is a highly corrosion-resistant glass. The vitrified form is in pellets or logs stored in stainless steel or titanium canisters, which in turn are installed in an overpack to make up the engineered package. This package could then be buried in much the same manner as planned for the spent fuel.

Repository safety advantages exist in this option in that the bulk of the long-lived plutonium has been removed.

In addition, processes are under development to separate the other long-lived actinides from the fission products and recycle these materials into the reactor. This recycling is most effective in a liquid metal-cooled reactor because in its high energy neutron spectrum, neutrons are not absorbed appreciably by the actinides, and thus the efficiency of the chain reaction is maintained. By contrast, efficiency would be poor in a lightwater-cooled reactor, which has a low energy neutron spectrum, and actinides become strong neutron absorbers. Other improvements could be made to the waste by converting the soluble fission products into insoluble forms. A promising pyrometallurgical reprocessing method is under development for actinide separations.

If the economics of recycling were improved, that option would become preferable for spent fuel because the permanent repository issues of the residual fission products would be simpler.

Another safety issue to be considered which might be exacerbated in the reprocessing option is that the plutonium generated in power reactors, called reactor-grade plutonium because it is made up of a variety of plutonium isotopes, contains plutonium-241, which is subject to spontaneous fission. The mixture of isotopes makes it difficult to build an effective nuclear weapon. However, an explosive device could be built using this mixture if control of detonation is sacrificed.

When reactor-grade plutonium is left in spent fuel, the large size of the fuel assemblies and the lethal radiation fields make it extremely difficult to divert the material covertly. Once the reactor-grade plutonium is separated in the commercial reprocessing option, however, the radiation barrier is almost eliminated, and in certain steps of the process the plutonium is in powder or liquid form, which is much more easily diverted than large, bulky fuel assemblies. This issue is under study and strict standards of control of separated reactor-grade plutonium have been instituted.

Concern about the potential diversion of separated reactor-grade plutonium has led to a reduction in U.S. governmental support of development of both plutonium recycle and the liquid metal reactor. This latter ultimately depends on chemical reprocessing to achieve its long-range purpose of generating more nuclear fuel than it burns in generating electricity.

Radiation Exposure and Health Standards

In the United States, each person achieves an average of about 0.0036 Sv/yr of radiation exposure. Nuclear utility workers may be exposed to an additional occupational radiation exposure. To assure that this additional radiation exposure is not harmful, several measures are taken: (*1*) standards are set for the maximum allowable radiation dosage by national and international commissions of radiation health experts and incorporated into federal regulations; (*2*) dosimeters, ie, radiation monitors, are worn by all workers potentially exposed to occupational radiation and accurate records are kept of the accumulated exposure to each worker to assure that maximum allowable levels are not exceeded; and (*3*) the concept of keeping occupational radiation exposure to as low a level as reasonably achievable (ALARA) is practiced so that a relatively small number of workers come close to the maximum allowable levels.

High levels of radiation exposure received over a short period of time (minutes to hours) can cause both near- and long-term effects. Near-term effects include radiation sickness or death. Long-term effects predominantly involve the incidence of cancer.

For radiation doses <0.5 Sv, there is no clinically observable increase in the number of cancers above those that occur naturally. There are two risk hypotheses: the linear and the nonlinear. The former implies that as the radiation dose decreases, the risk of cancer goes down at roughly the same rate. The latter suggests that risk of cancer actually falls faster as radiation exposure declines. Because risk of cancer and other health effects are quite low at low radiation doses, the incidence of cancer cannot be clearly ascribed to occupational radiation exposure. Thus, the regulations have adopted the more conservative or restrictive approach, ie, the linear hypothesis. Whereas nuclear industry workers are allowed to receive up to 0.05 Sv/yr, the ALARA practices result in much lower actual radiation exposure.

In the decade between 1983 and 1993, the annual total radiation dosage receive by U.S. nuclear plant workers dropped by 54%, whereas the annual MW yr or electricity generated increased by 51%. Thus, the annual ratio of total occupational radiation exposure to total electricity generated dropped by almost a factor of 5. This achievement can be credited in part to improved management practices, but a series of technological innovations have also made a significant contribution.

The dominant sources of residual radiation in the primary circuit outside the reactor core in nuclear plants are cobalt isotopes: ^{60}Co and ^{58}Co form by neutron absorption in ^{59}Co and ^{58}Ni. These last two species are naturally occurring isotopes in commonly used plant construction materials. Technological approaches to reduce this residual radioactive cobalt are as follows. (*1*) Minimize the cobalt impurities in the structural materials, replacing the high cobalt hardfacing alloys where practicable. (*2*) Precondition out-of-core primary circuit surfaces to minimize the release of corrosion products and the resuspension of radioactive species. (*3*) Specify and control primary water chemistry to minimize corrosion and the transport of corrosion products into the core, the disposition and subsequent activation of these products, and resuspension in the coolant. (*4*) Remove the residual radiation in the out-of-core primary circuit by decontamination. Several decontamination processes, such as CITROX, CANDECON, and LOMI, have been developed. The last, LOMI, has been the most widely used.

Safety of Future Reactors

The development of computer capabilities in hardware and software, related instrumentation and control, and telecommunication technology represent an opportunity for improvement in safety (see COMPUTER TECHNOLOGY) in lightwater-cooled nuclear power plants. Plant operators can be provided with a variety of user-friendly diagnostic aids to assist in plant operations and incipient failure detection. Communications can be more rapid and dependable. The safety control systems can be made even more reliable and maintenance-free. Moreover, passive safety features to provide emergency cooling for both the reactor system and the containment building are being developed.

The Electric Power Research Institute (EPRI) initiated the advanced lightwater reactor (ALWR) program because lightwater reactors are expected to continue to be used. The ALWR program, supported by electric utilities in the United States, Europe and Asia, the U.S. suppliers, and the U.S. Department of Energy, is developing large-sized (1350-MWe) reactors and smaller sizes (600-MWe) of the passive safety feature type. Both PWR and BWR units are being developed.

The evolutionary reactors are based on the same design concept as is used in the lightwater reactors of the mid-1990s. Many significant improvements have been made, such as selection of alloys having more corrosion resistance, eg, Inconel 690, for steam generator tubes; a high pressure system for the removal of decay heat; and the reactor vessel materials and weldments chosen to reduce radiation embrittlement and shielded to reduce the fast neutron fluency.

Two smaller-sized passive reactors are under development in the United States: a PWR called AP600, designed by Westinghouse, and a BWR called SBWR, designed by GE. These designs combine the improvements of the larger reactors with passive emergency cooling features.

Containment integrity is ensured by cooling the containment shell through evaporation of water that is gravity-fed from a large tank located above the containment. The heat is ultimately removed to the atmosphere by a natural circulation air system. Only active automatic valve operations, ie, no operator action and no pump, diesel, or fan operations, are required to provide emergency core cooling and containment cooling after a significant energy release into containment from the maximum loss-of-coolant accident.

Safety objectives have been established to make both ALWR even safer than the plants of the early 1990s and safer than required by the safety goals established by the U.S. NRC. The ALWR safety objectives are that there would be only one chance in 10×10^4 reactor-years that a severe accident would be initiated, a factor of 10 better than the U.S. NRC goal. Mitigation of the accident through the containment systems would reduce the risk by another factor of 10, so that the chance that the radiation dose at the boundary of the plant would be as high as 0.25 Sv, the level below which there is no clinically observable effect, would be one in 1×10^6. An additional objective has been set to limit the level of occupation radiation exposure. No more than 1.00 Sv/yr occupational exposure should be received by all the workers in each plant, an average of about 0.001 Sv/yr. Improved performance objectives have also been set to provide an additional power margin. This places less burden on both the equipment and operators in running the plant, resulting in increased reliability and lower operating and maintenance costs.

Another overall objective of the ALWR Program is to achieve standardization of families of plants in design, construction, and operation. Two fundamental bases for that standardization are common owner–operator (utility) requirements and common regulatory requirements.

Two other more advanced systems offer some promise for nuclear energy. One such advanced system is the gas-cooled reactor, which can operate at significantly higher temperatures than existing designs and can therefore be used to provide the energy for many process heat applications in industry. The second system is cooled with liquid metal, ie, sodium, and can produce more nuclear fuel than it burns as it generates electricity. Reactors such as the liquid–metal type should be needed to generate abundant nuclear fuel for the production of electricity.

Nuclear power plants of the future are to be designed and operated with the objective of better fulfilling their role as a bulk power producer that, because of reduced vulnerability to severe accidents, should be more broadly accepted and implemented. Use of these plants could stem the tide of environmental damage caused by air pollution from fossil-fuel combustion products.

J. J. TAYLOR
Electric Power Research Institute

R. A. Knief, *Nuclear Engineering—Theory and Technology of Commercial Nuclear Power,* Hemisphere Publishing Corp, Washington, D.C., 1992.

Basic Safety Principles for Nuclear Power Plants, IAEA Safety Series #75, INSAG-3, IAEA, Vienna, Austria, 1988, pp. 6–8.

Reactor Safety Study: An Assessment of Accident Risks in U.S. Commercial Nuclear Power Plants, Report WASH-1400 (NUREG-75/014), U.S. Nuclear Regulatory Commission, Washington, D.C., Oct. 1975.

J. J. Taylor, *Science* **244,** 318325 (Apr. 1989).

NUCLEIC ACIDS

Nucleic acids are polymeric materials formed from nucleotides and essential to all organisms. Deoxyribonucleic acid (DNA), most often a double-helical biopolymer, encodes the genetic information contained in each cell. Ribonucleic acid (RNA) constitutes a more diverse class of biopolymers that are able to adopt both helical and other more complex tertiary structures. The structural diversity of RNAs enables these molecules to carry out a variety of intracellular functions, including transmitting the genetic message to the site of protein synthesis (see PROTEINS). Both DNA and RNA interact with a host of other molecules, eg, proteins, drugs, as well as other RNAs and DNAs. The specificity of these interactions is thought to be related to local sequence-dependent structural variation.

Development of techniques to synthesize oligonucleotides, ie, short, well-defined sequences of DNA or RNA, has provided the opportunity to study nucleic acid structure in detail. In addition, oligonucleotides have proved invaluable in analytical procedures used in genetic engineering (qv), protein engineering (qv), affinity chromatography, and forensics, as well as in medicine (see CHROMATOGRAPHY; FORENSIC CHEMISTRY). The unique ability of nucleic acids to bind to self-complementary sequences has been exploited in the design of oligonucleotide probes and in antisense drug strategies.

DNA Structure

The structure of DNA is characterized by its primary sequence, secondary helical structure, and higher order structure or topology. The primary sequence of DNA refers to the atomic connectivities required to construct the polynucleotide chain. The helical conformation of these polynucleotide chains constitutes the secondary structure of DNA. Sequence-dependent structural diversity and flexibility are important DNA characteristics and play a crucial role in biological processes. The organization of helical DNA in topologically distinct three-dimensional conformations represents the higher order structure. Higher order structural features, in particular the supercoiling of DNA, are thought to have a profound influence on the dynamic processes and biology of nucleic acids within living cells.

The DNA double helix was first identified by Watson and Crick in 1953. Not only was the Watson-Crick model consistent with the known physical and chemical properties of DNA, but it also suggested how genetic information could be organized and replicated, thus providing a foundation for modern molecular biology.

The primary structure of DNA is based on repeating nucleotide units, where each nucleotide is made up of the sugar, ie, 2′-deoxyribose, a phosphate, and a heterocyclic base, N. The most common DNA bases are the purines, adenine (A) and guanine (G), and the pyrimidines, thymine (T) and cytosine (C) (see Fig. 1). The base, N, is bound at the 1′-position of the ribose unit through a heterocyclic nitrogen.

The nucleotides are linked together via the phosphate groups, which connect the 5′-hydroxyl group of one nucleotide and the 3′-hydroxyl group of the next to form a polynucleotide chain (Fig. **1a**). DNA is not a rigid or static molecule; rather, it can adopt a variety of helical motifs.

In A- or B-form DNA, two self-complementary polynucleotide strands associate with one another to form a right-handed double helix. The two polynucleotide chains are antiparallel.

In addition to A- and B-form DNA, several other helical conformations have been identified. Among these, the most well-studied is Z-DNA, a left-handed helix first characterized by x-ray crystallographic analysis of the oligonucleotides d(CGCGCG) and d(CGCG). Other alternating purine–pyrimidine sequences, in particular alternating CG sequences, have been shown to adopt the Z-conformation at high ionic strength.

RNA Structure

RNA has a variety of functions within a cell; for each function, a specific type of RNA is required. Messenger RNA (mRNA) serves as intermediaries for carrying genetic messages from the DNA to the ribosomes where protein synthesis takes place. Ribosomal RNA (rRNA) serves both structural and functional roles in the ribosome; it is diverse, both in terms of its size and structure. Transfer RNAs (tRNAs) are small molecules that have a central role in protein synthesis. Other RNA molecules, called ribozymes, function as enzymes to catalyze chemical transformations. Although ribozymes most often catalyze cleavage of the RNA phosphodiester backbone, they have also been shown to participate in cleavage of DNA, replication of RNA, and reactions with phosphate monoesters. Other RNAs are associated with enzymes to form riboprotein complexes involved in many biological

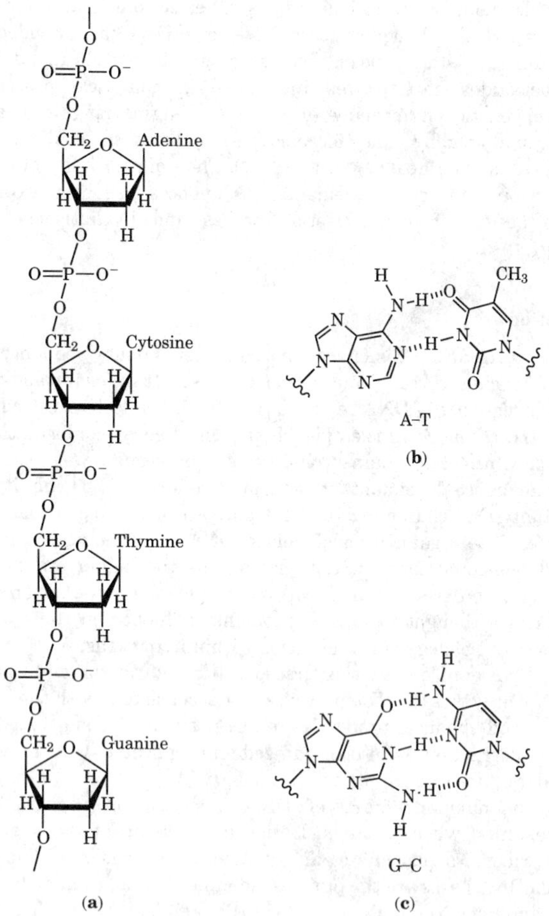

Figure 1. Elements of DNA structure: (**a**) a deoxypolynucleotide chain, which reads d(ACTG) from 3′ → 5′ or d(GTCA) from 3′ → 5′; and (**b**) and (**c**) the Watson-Crick purine–pyrimidine base pairs, A–T and G–C, respectively, where —⌇ represents attachment to the deoxyribose.

processes. The multifunctional character of RNA, particularly the involvement of RNA in enzymatic processes, has led to the hypothesis that life on earth evolved from RNA, and that RNA had both the genetic and catalytic functions commonly associated with DNA and proteins, respectively.

The primary structure of RNA is similar to that of DNA, but with a few notable exceptions. First, in RNA, instead of thymine, the pyrimidine base uracil (U) occurs, forming a complementary base pair with adenine in regions of double-stranded RNA. Also, a wide variety of ribonucleotides having modified or minor bases are found in naturally occurring RNA, one of the most common of which is pseudouridine. In human tRNAs, as many as 25% of the bases are nonstandard. Over 80 modified bases have been characterized in naturally occurring tRNA; although the role of base modification is not clear, it may be important for biological recognition.

The other important feature of the primary structure of RNA is the presence of the 2′-hydroxyl group in ribose. Although this hydroxyl group is never involved in phosphodiester linkages, it does impose restrictions on the helical conformations accessible to double-stranded RNA.

RNAs are single-stranded molecules that fold, allowing different regions of the ribonucleotide to form distinct secondary structural elements. When self-complementary regions of the RNA strand are aligned, duplex regions, which may have Watson-Crick base pairs, are formed. In contrast to DNA, double-stranded regions in RNA are much more likely to have unusual base-pairing between noncomplementary bases and to incorporate non-Watson-Crick base-pairing. Owing to the steric requirements of the 2′-hydroxyl group on the ribose sugar, these duplex regions are constrained to an A-form helix, ie, a 3′-endo sugar conformation. Although double-stranded RNA has the general features of an A-form helix, actual duplex characteristics, such as rise per base pair, groove dimensions, and base pair displacement from the helical axis, may vary.

The functional diversity of RNA is directly related to its structural diversity. In contrast to DNA, RNA molecules are synthesized as single-stranded polynucleotides that fold to give complex tertiary structures. These structures, which incorporate hairpins, loops, bulges, and junctions between single-stranded and double-stranded regions, exhibit long-range interactions within the folded tertiary structure. Long-range intramolecular interactions serve to stabilize the three well-characterized RNA structures.

Oligonucleotide Synthesis

Synthetic oligonucleotides are widely used in scientific investigations. Most synthetic oligonucleotides are produced for use as primers in the polymerase chain reaction (PCR), a widely used analytical technique having commercial applications in diagnostic medicine, genetic engineering (qv), and forensics (see Forensic chemistry). A large volume of oligonucleotides are also synthesized for use as primers in DNA-sequencing. Demands for sequencing primers have increased rapidly to support large-scale DNA mapping and sequencing efforts such as the human genome project. Although the quantities of oligonucleotides used are relatively small, these materials are essential in many areas of basic research.

In molecular biology, synthetic oligonucleotides are used as linkers in gene-cloning and to introduce site-directed mutations in genes. Synthetic oligonucleotides are required for structural, biochemical, and biophysical studies of DNA and RNA. Oligonucleotides also are important for examining the association of proteins and small molecules, eg, for intercalating drugs with nucleic acids on a molecular level.

The first procedures for oligonucleotide synthesis, typically carried out in solution, made use of *H*-phosphonate and phosphotriester chemistry. These approaches are useful in some large-scale syntheses and in syntheses of various oligonucleotide analogues. Most modern procedures, however, are based on solid-phase phosphoramidite chemistry. Automated oligonucleotide synthesizers are commercially available, as are the required reagents and phosphoramidites. Together these permit the rapid production of custom oligonucleotides and oligonucleotide analogues.

Modified Oligonucleotides. Much of the interest in modified oligonucleotides is related to use as antisense agents. Antisense agents are typically short (15–30 base pairs in length) oligonucleotides having sequences that are complementary to coding or regulatory regions within mRNA, although some antisense oligonucleotides have also been designed to target DNA. The antisense sequence recognizes and binds to a complementary sequence via the formation of a double-stranded duplex that have normal Watson-Crick base-pairing. Antisense oligonucleotides can inhibit gene expression at the translational level. The potential to design oligonucleotides having the ability to recognize and inhibit specific genes makes the antisense approach promising in the development of new therapeutic agents. In addition, antisense oligonucleotides can be used in research to elucidate gene function by providing a mechanism for regulating a gene artificially.

Although all natural antisense oligonucleotides are short RNA sequences, most of the synthetic antisense oligonucleotides are deoxyoligonucleotides. In the design of an effective antisense oligonucleotide, several factors must be considered. First, the oligonucleotide must be specific, binding with high affinity to a single sequence within the target RNA. A second consideration is stability within the cellular environment. Thus all unmodified oligonucleotides are degraded too rapidly to be used effectively as therapeutic agents. A significant research effort has been directed toward discovering chemical modifications that can increase the nuclease resistance of the oligonucleotide backbone.

An effective therapeutic agent must also have the ability to reach its target sequence *in vivo.* In order to enhance membrane transport, antisense oligonucleotides are frequently modified by covalent attachment of carrier molecules or lipophilic groups.

Antisense oligonucleotides are usually designed to inhibit gene expression by interfering with the translation of mRNA. One mechanism for this type of inhibition involves binding the oligonucleotide to the translation-initiation sequences of the mRNA, which prevents ribosome association and protein synthesis. Another potential mechanism involves hybrid formation at some other sequence within the mRNA, thus impeding translocation of the ribosome along the mRNA strand by steric blocking. These two mechanisms are based on blocking a sequence of RNA or DNA by double-stranded duplex formation using a specific antisense oligonucleotide.

A less specific mechanism based on the action of Rnase H, an enzyme catalyzing single-strand cleavage of RNA, may be predominate for unmodified, thioate and dithioate oligonucleotides. In the Rnase H mechanism, the duplex formed by the antisense oligonucleotide and the target RNA is a substrate for Rnase H. The enzyme cleaves the RNA at the complexes site rendering the RNA vulnerable to further degradation and inactivation by cellular exonucleases. The oligonucleotide, which is probably not a substrate for Rnase H, can target multiple copies of complementary RNA. Where applicable, antisense oligonucleotide action mediated by the Rnase H mechanism has been shown to be a potent inhibitor of gene expression.

Modified oligonucleotides can also be designed for binding to double-stranded DNA by forming a triple helix.

Oligonucleotides can also inhibit gene expression at the transcriptional level by binding to a single-stranded or open sequence of DNA. In this mechanism, the antigene oligonucleotide is designed to be complementary to a regulatory sequence preceding a gene. For example, an oligonucleotide complementary to the *lac* operator sequence (repressor protein binding site) has been found to inhibit specifically β-galactosidase synthesis in *E. coli.* Normally, expression of the lactose-metabolizing enzyme, β-galactosidase, is blocked at the transcriptional level by the repressor binding to the operator sequence. In the presence of a lactase metabolite, the repressor is converted to a nonbinding form and dissociates from the operator, which results in the transcription of the gene and β-galactosidase production. However, in the presence of the antisense oligonucleotide, the synthesis of β-galactosidase is inhibited. The antisense oligonucleotide can then act as a repressor by binding to an open or single-stranded region within the operator.

Although development of modified oligonucleotides as antisense and antigene agents is a principal focus of research in the 1990s, there are many other interesting applications that have greater immediate commercial significance. Included among these are applications using nucleic acid probes, which are oligonucleotides that have been modified by the attachment of a detectable chemical group. Probes can be designed to recognize RNA or DNA sequences characteristic of specific eukaryotic genes, viruses, or bacteria. Several analytical and diagnostic procedures have been developed based on the hybridization of the probe with its target sequence and the subsequent detection of the hybrid by the group attached to the oligonucleotide. Probes, particularly useful in automated sequencing protocols, may contain fluorescent groups, phosphors, radioactive tracers (qv), etc. In addition, probes can be designed to help elucidate the structure of biological molecules. DNA-binding molecules, including intercalators, alkylating agents, and photosensitive molecules, have also been linked to oligonucleotides as a way of directing a drug to a specific DNA sequence. These modifications often enhance binding as well.

Oligonucleotide Bioconjugates

Although many molecules, including proteins and small intercalating and groove-binding ligands, bind to RNA and DNA, only nucleic acids are able to bind with the high specificity required to recognize a single sequence within the 3×10^9 base-pairs of the human genome. The unique specificity of oligonucleotides can be exploited to direct a multitude of other chemical agents to a sequence of interest by attaching these agents to oligonucleotides through molecular linkers. Oligonucleotides labeled with fluorescent or other detectable groups provide nucleic acid probes that can be used to screen a large pool of DNA for a specific sequence. Cationic or lipophilic groups can be attached to improve the binding and bioavailability of antisense oligonucleotides. Bioconjugates have also been widely used in research because they enable scientists to learn more about the structure and function of nucleic acids and ligands that bind to them.

Several strategies have been devised to attach various chemical groups to oligonucleotides. Groups can be attached to the 3′- or 5′-terminus of the oligonucleotide, along the backbone through the phosphate or the 2′-hydroxy group of ribose, or to modified purines or pyrimidines.

JILL REHMANN
Fordham University

W. Saenger, *Principles of Nucleic Acid Structure,* Springer-Verlag, New York, 1984.

R. F. Gesteland and J. F. Atkins, eds., *The RNA World,* Cold Spring Harbor Laboratory Press, Cold Spring Harbor, New York, 1993.

M. H. Caruthers and co-workers, *Methods in Enzymology,* Vol. 154, Academic Press, Inc., New York, 1987, pp. 287–313; Vol. 211, 1992, pp. 3–20.

NUCLEOSIDE ANTIBIOTICS.

See ANTIBIOTICS, NUCLEOSIDES AND NUCLEOTIDES.

NUTS

Nuts, as generally defined, are hard-shelled seeds enclosing a single edible oily kernel. Most of the common nuts fall within this classification. Many species, however, differ greatly in size, structure, shape, composition, and flavor.

With the exception of peanuts, most of the important nuts from around the world are borne on trees, many of them from native seedlings. Among the latter group are the beechnut, Brazil nut, butternut, chestnut, filbert, hickory nut, pecan, pine nut, and black walnut. The pecan, English walnut, filbert, and almond are the four principal edible tree nuts produced in the United States, where the term English walnut is used synonymously with the Persian or Carpathian walnut.

Chemical Composition

Most nuts for commercial use are characterized by high oil and protein contents (see PROTEINS) as well as a low percentage of carbohydrates (qv). The proximate composition of a number of nuts and of some nut products are given in Table 1.

Chemical Changes during Development and Storage

Several investigations have been made on the chemical changes taking place during the development of various nuts; for instance, in macadamia, pecan, almond, English walnut, eastern black walnut, and tung. The levels of sugars and other carbohydrates decrease rapidly during oil synthesis, so that by the time the kernel is mature, most of the carbohydrates have disappeared.

Table 1. Composition of Nuts[a]

Name	Refuse, wt %	Water, wt %	Protein, g	Fat, g	Total carbohydrate, g	Fiber, g	Ash, g	Calcium, mg	Phosphorus, mg	Iron, mg	Sodium, mg	Potassium, mg	Magnesium, mg	Vitamin A, IU	Thiamine, mg	Riboflavin, mg	Niacin, mg	Ascorbic acid, mg	Fuel value, MJ[b]
acorns, raw		27.9		23.9	40.8	2.6	1.4	41.0	79		0	539	62		0.11	0.12	1.80	0.00	1.54
almond, dried, blanched	49	5.4	20.4	52.5	18.5	2.3	3.1	2.5	532	3.7	10	750	286		0.16	0.68	3.20	0.60	2.45
beechnut, dried	39	6.6	6.2	50.0	33.5	3.7	3.7	1.0											2.41
Brazil nut, dried	50	3.3	14.3	66.2	12.8	2.3	3.3	176.0	600		2	600	225		1.00	0.12	1.60	0.70	2.74
butternuts, dried		3.3	2.5	57.0	12.1	1.9	2.7	53.0	446	4.0	1	446	237						2.56
cashew nut, dry roasted		1.7	15.3	46.4	32.7	0.7	4.0	45.0	490	6.0	16	565		0	0.20	0.20	1.40	0.00	2.41
chestnut, European, dried, peeled	19	9.0	5.0	3.9	78.4	5.0	3.6	64.0	137	2.4	37	991			0.35	0.05			1.55
coconut																			
cream, expressed liquid		53.9	3.6	34.7	6.6		1.1	11.0	122	2.3	4	325		0	0.03	0.00	0.90	2.80	1.38
milk, expressed		67.6	229.0	23.8	5.5		0.7	16.0	100	1.6	15	263	37	0	0.03	0.00	0.80	2.80	0.96
hazelnut (filberts), dried, unblanched	53	5.4	13.0	62.6	15.3	3.8	3.6	188.0	312	3.3	3	445	285	67	0.50	0.11	1.10	1.00	2.64
hickory nut	80	2.7	12.7	64.4	18.3	3.2	2.0	61.0	336	2.1	1	436	173						2.75
macadamia nut, dried	69	1.7	7.3	76.5	12.9	1.7	1.7	45.0	200	1.8	7	329	117	9	0.22	0.11	2.00	0.00	3.01
peanut, kernels, dried		6.7	25.7	49.2	16.2	4.9	2.3	58.0	383	3.2	16	717	180	0	0.66	0.13	14.20	0.00	2.37
pecans, dried	56	4.8	7.8	67.6	18.2	1.6	1.6	36.0	291	4.8	1	392	128	128	0.85	0.13	0.89	2.00	2.79
pilinut		0.8	10.8	79.6	4.0	2.8	2.9	145.0	575	3.5	3	507		41	0.91	0.09	0.52		3.01
pinenut																			
pignolia		6.7	24.0	50.7	14.2	0.8	4.4	26.0	508	9.2	4	599			0.81	0.19	3.57		2.15
piñon		5.9	11.6	61.0	19.3	4.7	2.3	8.0	35	3.1	72	628	234	29	124.00	0.22	4.40	2.00	2.38
pistachio nut, dried	70	3.9	20.6	43.4	24.8	1.9	2.4	135.0	503	6.8	6	1093	158	233	0.82	0.17	108.00		2.42
walnuts, black	78	4.4	24.4	56.6	12.1	6.5	2.6	58.0	464	3.1	1	524	202	296	0.22	0.11	0.70		2.54
walnuts, English or Persian	55	3.7	14.3	61.9	18.3	4.6	1.9	94.0	317	2.4	10	502	169	124	0.39	0.15	1.05	3.20	2.68

[a] 100 portions are used for the calculations.
[b] To convert J to cal, divide by 4.184.

Considerable research has been done on the storage of nuts, especially of pecans, which have a high free fatty acid content. Most pecans are refrigerated at some time during storage and marketing. They may be held at 3°C for one year, at −4.5° C for three years, or at −17.5° C for more than five years. Most nuts cannot be held for more than three or four months at ordinary temperatures, especially during the summer, without developing rancidity; however, storage experiments have shown that nuts and nut products can be kept two to five years longer under refrigeration at 0–5°C. For instance, peanuts stored in the shell at 4.4°C have remained perfect for as many as six years with no detectable chemical changes. Nut kernels sealed in tins by a vacuum process also keep well (see VACUUM TECHNOLOGY). On the other hand, various tests with nuts stored in glass, plastic, or containers of other kinds, and sealed with nitrogen, carbon dioxide, or other gases have proved satisfactory only when the nuts are held under refrigeration. Slight ventilation must be provided in containers storing chestnuts because of the respiration of nuts.

An edible and nutritive coating of zein protects tree nuts and peanuts from developing rancidity, staleness, and sogginess during storage.

Nuts are susceptible to infestation by weevils in storage, especially in warm weather. Infestation is usually prevented or controlled by placing the nuts in cold storage or by fumigating them either in the open with gas, such as methyl bromide or hydrogen cyanide, or under vacuum with a mixture of carbon disulfide and carbon dioxide.

Although the storage of peanut butter is influenced by temperature, the extent of roasting, hydrogenation, and addition of salt have little effect on product stability if packages are sealed properly. This is because oxygen in the headspace is the main factor in reducing stability; consequently, peanut butter for the retail trade is packed in airtight containers or under vacuum.

Aflatoxins. Mycotoxins are toxic compounds elaborated by fungi. Interest in the effects of mycotoxins was renewed in recent years after the discovery that aflatoxins were the causative agents in turkey X disease, which killed thousands of turkeys in England in 1960. Dietary intake of aflatoxin-contaminated products (maize and peanuts) is associated with hepatitis, Indian childhood cirrhosis, and liver cancer.

Aflatoxins are fluorescent toxic factors elaborated by the common mold *Aspergillus flavus* during its growth on nuts or grains. Although other mycotoxins have been and are being discovered, aflatoxins retain a position of importance because of their high toxicity and common natural occurrence in such foods as cereal grains, oilseeds, and oilseed meals stored under adverse conditions.

Four main approaches are used to control the problem of aflatoxin in peanuts: (*1*) the development of cultivars that resist the invasion of aflatoxin-producing fungi; (*2*) cultural practices minimizing insect damage that facilitates fungal invasion; (*3*) detoxification of contaminated nuts and their products; and (*4*) separation of contaminated nuts.

Processing

Shelling, Cracking, and Bleaching. Very elaborate machinery is used for cleaning, grading, bleaching, cracking, and packing edible tree nuts such as pecan, English walnut, filbert, and almond. For other nuts, the processing remains at least in part a hand operation.

Blanching, Salting, and Roasting. Some of the nut kernels processed for market and for various nut products are blanched, ie, the skin or membrane covering the white meats is removed. Pecan meats, in contrast, are not blanched. Several plants in the United States and Europe are processing peanuts that are salted in the shell before roasting. Usually, the peanuts are impregnated with salt by using a saturated brine solution and vacuum. A great increase in peanut consumption in the 1990s has been due to the popularity of dry roasted peanuts. The shelled peanuts are heated for blanching; then blanched and coated with a mixture of proprietary ingredients consisting of gum, zein, acetylated monoglycerides, spices, an antioxidant, and a coloring agent; finally the coated peanuts are roasted for maximum flavor. These vacuum-packed products have excellent flavor and long shelf life.

Peanut Butter. By federal regulation, at least 90% of commercial peanut butter consists of shelled roasted peanuts that are ground and blended with salt, sweeteners, and emulsifiers. No artificial flavors and sweeteners, chemical preservatives, natural or artificial color, purified vitamins, or minerals are allowed. To meet consumer demand for low fat products, several U.S. manufacturers have created products in which the peanut content has been partially replaced by maltodextrins, corn syrup solids or similar starches, and soy protein.

Commercial manufacture of peanut butter varies considerably. Roasting time influences sensory attributes and chemical measurements of flavor components (see FLAVOR CHARACTERIZATION).

Uses

Nuts and Nut Products. Nuts are used mainly as edible products and marketed either with or without the shell, as the demand requires. The most popular nuts in the shell are English walnut, filbert, almond, Brazil nut, peanut, pistachio, and the improved, or papershell, pecan; the most popular salted and roasted nut kernels include these as well as the cashew, macadamia, and pignolia. Each year more nuts are shelled in centrally located plants and marketed as meats.

For convenience, meats are rated and sold by sizes and grades. Nut kernels are used extensively in confections and in the baking industry as ingredients in pies, cakes, cookies, and other products. Various nut products include salted and roasted nuts, nut butters such as peanut and almond butter, macaroon paste and powder from almonds, specialty oils, and various confections, notably pralines, peanut brittle, and nut bars.

Nuts have many uses, both industrial and domestic. For instance, the ivory nut, or tagua, is a source material for the manufacture of buttons and turnery articles. The kola nut supplies ingredients for popular cola beverages in the United States (see CARBONATED BEVERAGES). *Strychnos nux-vomica* provides the important medicine and poison, strychnine. The areca or betel nut is chewed by the Indian and Malayan people as a narcotic.

Peanuts. Peanuts are one of the three most important oil-bearing seed crops in the world, and the majority of their world production is crushed for the manufacture of peanut oil. Most of the countries with high levels of peanut production have very limited process and storage facilities for the manufacture and handling of other peanut products. However, more than half of the peanuts grown in the United States are used as peanut-food products. Only the surplus and mold-contaminated peanuts are crushed for oil.

Partially Defatted Nuts. There is considerable demand for nuts and nut products of reduced fat content. Almond meal and peanut meal are examples of products having low fat content achieved by pressing oil from the nuts and by grinding the cake. Much of the flavor is in the oil; defatted nuts are thus less tasty.

Defatted peanuts are high in protein, low in moisture, contain only 20% of the naturally occurring fat, and have better stability than whole peanuts. Monosodium glutamate (MSG) has been used as a flavor enhancer for defatted nuts, but the result has not been entirely satisfactory as the addition of MSG produces a meaty rather than nutty flavor. This meaty flavor is more compatible with salted butter and nuts than with candy.

Oil. Tung and oiticia are sources of quick-drying oils for the paint and varnish industry (see DRYING OILS). Coconut, babassu, and palm oils are used chiefly for the manufacture of margarine, soap, shaving cream, cosmetics (qv), and other domestic products. Walnut oil is used in the preparation of artists' colors (see PIGMENTS). Peanut oil is used as a lubricant and in shaving creams, shampoos, and cosmetics. It is also a good source of edible oil in the manufacture of shortening and margarine. Sweet almond oil is transparent, consisting chiefly of triolein. It has important uses as a laxative, in treating bronchitis and colds, and in fine soaps and cosmetics.

Formerly a waste product of the cashew kernel industry in southern India, cashew nutshell oil has become a valuable raw material in the manufacture of many industrial products.

Meal. The meal or press cake from oil extraction of pecan, walnut, almond, and other nuts is usually bitter because it contains skins and pieces of shell; when refined, however, it can be used in the baking industry and, more commonly, in animal feed. Almond, peanut, babassu, and other nut cakes and meals are extensively used in such feeds; in addition, ivory nut meal, a by-product of the button industry, also provides a valuable animal feed since most of its cell wall carbohydrate (92.5% mannose) can be converted into hexose sugar or its equivalent.

Shells and Hulls. Nutshell waste from shelling and processing plants is used extensively for a variety of purposes. Pecan and English walnut shells reduced to flour of various mesh sizes can be used as soft grit in blasting metals; as an ingredient for plastic fillers, battery cases, molding resin forms, and industrial tile; as an insecticide diluent; and for cleaning fur. Radio horns made from walnut-shell flour seem to filter vibrations more effectively than loudspeakers made from other materials.

Peanuts have been studied extensively for the development of products ranging from dyes and ink to artificial wool; peanut shells or hulls have been used in the manufacture of lacquers, linoleum, dynamite, guncotton, celluloid, artificial leather, photographic film, cellophane, and rayon. Peanut shells are also used as a stock feed by mixing ground hulls and molasses. Peanut shells, combined with an asphalt binder, can also produce insulation block. Finally, an artificial vegetable fiber that resembles wool is produced from peanut protein; it can be used for making clothing and other fabrics by weaving with it an equivalent amount of wool.

World Production

Although nuts have been a staple food in many countries for generations, their status in the United States as a chief food crop is relatively recent. The main supplier of English walnuts, filberts, and almonds had been Europe. However, pecans and black walnuts are indigenous to North America, and the United States is the principal producer of pecans. Other U.S. nuts, such as beech, butternut, white walnut, American chestnut, chinquapins, hickory, piñon, and northern California black walnut, are utilized mainly for local consumption. Chestnuts and chinquapins are susceptible to the chestnut blight fungus, *Endothia parasitica,* which has virtually destroyed the American chestnut.

Foreign production of almonds, filberts, and walnuts is concentrated in southern Europe, especially in countries with mild, dry climates bordering the Mediterranean Sea. In some of these countries, only part of the total production is on a commercial basis. In addition to orchards and groves, nut-bearing trees are found growing in scattered plantings amid field and other tree crops, along roads, and in yards. For this reason, and for the additional reason that crop-reporting methods have heretofore been inadequate, data on areas used for the growing of nuts are of limited usefulness.

Commercially important nuts in world trade include almond, Brazil nut, cashew, chestnut, coconut (copra), filbert, macadamia, palm nut, peanut, pecan, pignolia, pistachio, and English walnut. Coconut, palm nut, peanut, as well as babassu, oiticia, and tung, are important sources of oil for soap, paint, varnish, as well as many other domestic and industrial uses.

CLYDE T. YOUNG
North Carolina State University

J. A. Duke, *CRC Handbook of Nuts.* CRC Press, Inc., Boca Raton, Fla., 1989.

H. E. Pattee and C. T. Young, eds., *Peanut Science and Technology,* American Peanut Research and Education Society, Yoakum, Tex., 1982.

D. K. Salunke and co-workers, *World Oilseeds: Chemistry, Technology and Utilization,* Van Nostrand Reinhold Co., Inc., New York, 1992.

J. G. Woodroof, *Tree Nuts: Production, Processing, Products,* Avi Publishing, Westport, Conn., 1979.

NYLON. See POLYAMIDES, FIBER.

O

OCEAN RAW MATERIALS

The ocean is host to a variety and quantity of inorganic raw materials equal to or surpassing the resources of these materials available on land. Inorganic raw materials are defined here as any mineral deposit found in the marine environment. The mineral resources are classified generally as industrial minerals, mineral sands, phosphorites, metalliferous oxides, metalliferous sulfides, and dissolved minerals and include geothermal resources, precious corals, and some algae. The resources are mostly unconsolidated, consolidated, or fluid materials which are chemically enriched in certain elements and are found in or upon the seabeds of the continental shelves and ocean basins. These may be classified according to the environment and form in which they occur (Table 1) and with few exceptions are similar to traditional mineral deposits on land.

Unconsolidated Deposits

Deposits that can be recovered without having to use explosives or other primary energy sources to break up the material in place are called unconsolidated deposits. These may be found stratified or disseminated as surficial or subsurface deposits on the continental shelf (conshelf) or in deep ocean basins (see Table 1).

Consolidated Deposits

Consolidated deposits are those which occur as solid masses upon or within the structure of the seabeds. These may be removed only by fracturing, fluidizing, or dissolving the materials to be recovered.

Most consolidated mineral deposits found on the continental shelf are identical to those found on land and are only fortuitously submerged. Exceptions include those laid down in shallow marine seas or basins in earlier geochemical environments such as bedded ironstones, limestones, potash, and phosphorites.

Known consolidated mineral deposits in the deep ocean basins are limited to high cobalt metalliferous oxide crusts precipitated from seawater and hydrothermal deposits of sulfide minerals which are being formed in the vicinity of ocean plate boundaries.

Fluid Deposits

Fluid deposits are defined as those which can be recovered in fluid form by pumping, in solution, or as particles in a slurry. Petroleum products and Frasch process sulfur are special cases. At this time no valid distinction is made between resources on the continental shelf and in the deep oceans. However, deep seabed deposits of minerals which can be separated by differential solution are expected to be amenable to fluid mining methods in either environment.

Minerals Recovery

Technology. There are four basic methods of mining solid minerals: scraping the surface, excavating a pit or trench, removal through a borehole in the form of a slurry or fluid, and tunneling into the deposit. All deposits on land are mined by one or more adaptations of these methods. Marine mining is amenable to the same basic approaches whether on the continental shelf or in the deep seabeds. Each mining method has variations that may be tailored to a specific situation. Most of the deposit types can be mined by more than one method. Similarly, any one method can be applied to more than one deposit type.

Environmental Considerations. A significant advantage that appears to accrue from the recovery of mineral raw materials from the marine environment is the apparently benign effects of these activities when compared to the recovery of the same materials from land.

Table 1. Classification of Global Marine Mineral Resources

	Resource classification		
Location	Unconsolidated	Consolidated	Fluid
		Conshelf	
seabed	industrial materials	outcrops	seawater
	sand and gravel	exposures of veins, etc	magnesium
	shell sands		sodium
	argonite		uranium
	coral sands		bromide and salts of 26 other elements
	mineral sands		
	magnetite		
	ilmenite		
	rutile		
	chromite		
	monazite		
subseabed	mineral sands	vein, stratified, disseminated or massive deposits	freshwater springs
	gold		
	platinum		
	cassiterite	coal	
	gem stones	phosphates	
	bedded deposits	carbonates	
	phosphorites	potash	
		ironstone	
		limestone	
		metal sulfides	
		metal salts	
		Ocean basins	
seabed	muds or oozes	crusts	seawater
	metalliferous	phosphorite	magnesium
	carbonaceous	cobalt	sodium
	siliceous	manganese	uranium
	calcareous	mounds and stacks	bromine and salts of 26 other elements
	baritic	metal sulfides	
	nodules		
	manganese		
	cobalt		
	nickel		
	copper		
subseabed		vein, stockwork, stratbound or massive deposits	hydrothermal fluids
		metal sulfides	

Although a great deal of work remains to confirm this assumption, limited testing and monitoring on existing producing operations indicate an environmental advantage to ocean mining.

MICHAEL J. CRUICKSHANK
University of Hawaii at Manoa

M. J. Cruickshank, in *Mining Engineering Handbook,* Society of Mining Engineers, Littleton, Colo., 1992, Chapt. 28, pp. 1985–2028.

P. Hoagland and J. M. Broadus, *Seabed Commodity and Resource Summaries,* Technical Report WHOI 87–43, Woods Hole Oceanographic Institute, Woods Hole, Mass., 1987.

F. C. F. Earney, *Marine Mineral Resources,* Routledge, London, 1990.

M. J. Cruickshank and C. L. Morgan, *Synthesis and Analysis of Existing Information Regarding Environmental Effects of Marine Mining,* consulting report to Continental Shelf Associates for U.S. MMS, U.S. Dept. of the Interior, Washington, D.C., in MMS 93-0006, 392 pp.

OCTANE NUMBER. See GASOLINE AND OTHER MOTOR FUELS.

ODOR CONTROL. See AIR POLLUTION CONTROL METHODS; ODOR MODIFICATION.

ODOR MODIFICATION

Olfaction

Olfaction begins when an odorant stimulates the olfactory receptor cells, triggering the opening or closing of the ion channels, which in turn convert this stimulus to an electrical response to the olfactory bulb and ultimately to other parts of the brain. The olfactory neurons send messages to the olfactory bulbs, structures about the size and shape of peach pits, located on the underside of the large overhanging frontal lobes of the cerebrum. The fact that there is a gene responsible for odor receptor proteins was discovered in 1991.

Perception of odor is therefore a physical mechanism by which information is processed in the brain. Day by day new odors are appearing, all of which are immediately accepted and sorted within the seemingly unlimited categories of the olfactory brain. The brain not only recognizes this information, but evaluates it, sorts it, and associates it with experiences, events, likes, and dislikes.

Some substances are odorous, others are not. Humans can smell at a distance; if one smells the roses in a garden, it is not ordinarily considered that part of the rose is in contact with the nose. Substances of different chemical constitution may have similar odors.

The sense of smell is rapidly fatigued. Fatigue for one odor does not affect the perception of other dissimilar odors, but will interfere with the perception of similar odors. Two or more odorous substances may cancel each other out; this compensation means that two odorous substances smelled together may be inodorous.

Odor travels downwind. Many animals have a keener sense of olfaction than humans. Insects have such extraordinary keenness of smell that it may be a different modality of the chemical sense from that known to humans.

Odors

Odors have been classified according to Carolus Linnaeus, the eighteenth century Swedish botanist who proposed seven odoriferous qualities: aromatic, fragrant, musky, garlicky, goaty, repulsive, and nauseous. Later in the twentieth century, ethereal (fruity) and empyreumatic (burnt organic matter), together with subdivisions of Linnaeus' classification, were added. In the 1990s, researchers concentrate less on categorizing odors, and more on how people detect and interpret them. Although the average person can name only a handful of common odors, this limitation results from memory retrieval failure, rather than a failure to detect the differences.

Odors are measured by their intensity. The threshold value of one odor to another, however, can vary greatly. Detection threshold is the minimum physical intensity necessary for detection by a subject where the person is not required to identify the stimulus, but just detect the existence of the stimulus. Accordingly, threshold determinations are used to evaluate the effectiveness of different treatments and to establish the level of odor control necessary to make a product acceptable. Concentration can also produce different odors for the same material.

Evaluation Methodologies

Industry has standardized procedures for the quantitative sensory assessment of the perceived olfactory intensity of indoor malodors and their relationship to the deodorant efficacy of air freshener products. Synthetic malodors are used for these evaluation purposes. These malodors should be hedonically associated to the "real" malodor, and must be readily available and of consistent odor quality. These malodors should be tested in various concentrations and be representative of intensities experienced under normal domestic conditions. Panelists are trained to evaluate malodor intensity and the degree of modification.

Modification

Masking. Masking can be defined as the reduction of olfactory perception of a defined odor stimulus by means of presentation of another odorous substance without the physical removal or chemical alteration of the defined stimulus from the environment. Masking is therefore hyperadditive; it raises the total odor level, possibly creating an overpowering sensation, and may be defined as a reodorant, rather than a deodorant. Its end result can be explained by the simple equation of $1 + 1 => 2$.

Odor masking does little or nothing to control malodors; it merely covers them up. Many materials used in masking odors are aldehydes, which are very chemically reactive and usually comprise the top note of a fragrance. Odor masking is used in many areas of household, industrial, and institutional use via products that mask such malodors as pet smells, smoke, cooking, and numerous other odors. The forms by which masking is executed vary, and can be solid, liquid, and aerosol.

Counteraction. Counteraction, sometimes referred to as neutralization, occurs when two odorous substances are mixed in a given ratio and the resulting odor of the mixture is less intense than that of the separate components. The acceptable term to describe this occurrence is compensation. Materials that can accomplish this are basically organic odors which are highly polarized, have a strong affinity for each other, and may also have a low vapor pressure. Some of these molecules have the ability to compensate physiologically for certain malodor materials; others to react chemically with them. Counteraction occurs when the compensating substrate is able to form a coordinate bond with osmophoric sites unique to malodor molecules, such as amino- and thio- moieties. The result is overall reduction in odor; the malodor is transformed into an acceptable state, often with some residual freshening odor. This result lowers the total odor perception and can be exemplified by $1 + 1 =< 2$.

Commercial Aspects

Translating odor modifiers into consumer products results in forms, such as solids, liquids, and aerosols, for a market defined as products "for the nose." This includes products that cover up or eliminate odors, perfume the home, or cleanse the air. The categories of this market can be broken out as traditional air fresheners, cat litter products, aroma care, air purification, and disinfectant in both consumer and industrial applications.

Behavior Modification by Odor

Although odorous materials no doubt impact each other, much discussion centers around the ability of odorous materials to influence human behavior. In articles ranging from scientific journals to trade magazines, there is discussion on the potential of fragrances, ie, essential oils, to affect people's moods, their ability to focus and maintain attention, to relax and sleep, and even their sexual capability.

The words aromatherapy, aromachology, and aromakinetics are coinages of the 1990s. Aromatherapy, once based on a tradition of folklore and herbal medicine, is being investigated scientifically.

A technique known as contingent negative variation (CNV) measures brain-wave reaction to olfaction. These types of studies have shown the effect of materials such as lavender and nutmeg in reducing stress or anxiety, and the ability of oils such as peppermint to stimulate brainwave activity. CNV research was incorporated into the development of the fragrance for a consumer personal care product launched in the late 1980s.

The interrelationship between fragrance and psychology has been the subject of systematic investigation only recently. Consumer moods can be calculated in both positive and negative directions, and changes can be measured subjectively after exposure to fragrance.

Odors play a much greater role in human behavior than previously thought. The sense of smell provides a direct link with the function of the brain; therefore, the further study of olfaction can only advance the learning of causes and effects of stimuli to the brain.

The future in research will certainly lead to a better understanding of how odors are recognized, sorted, and classified. Studies promise, among other things, to determine whether perceptually similar, but structurally different, odors share the same class of receptor proteins, whether responses to odors can be modified, and possibly why olfactory neurons regenerate but other neurons do not.

YVETTE BERRY
Reckitt & Colman Inc.

C. P. McCord and W. N. Witheridge, *Odors Physiology and Control,* McGraw-Hill Book Co., Inc., New York, 1949.

S. Lord, *Vogue,* 171 (Dec. 1991).

H. T. Lawless, *Chem. Senses Flavor,* **14**(3), 349–360 (1989).

H. C. Zwaardemaker, *Arch. Anat. Physical,* 423–432 (1900).

OIL SANDS. See TAR SANDS.

OILS, ESSENTIAL

The volatile etherial fraction obtained from a plant or plant part by a physical separation method is called an essential oil. The physical method involves either distillation (including water, steam, water and steam, or dry) or expression (pressing). For the most part, essential oils represent the odorous part of the plant material, and therefore these oils have traditionally been associated with the fragrance and flavor industry (see PERFUMES). Since essential oils frequently occur as a very small percentage by weight of the original plant material, the processing of large quantities is often required to obtain usable amounts of oil. As a result, expression of an essential oil is only employed in those cases where both the form of the natural plant material, such as a citrus peel, and the quantity of oil present make the process feasible.

It has frequently been observed that the aroma of an essential oil is substantially different from that of the plant before processing. Because this phenomenon is largely the result of the treatment of the plant material with heat or hot water, various other methods have evolved over the years in an attempt to obtain a concentrate of the volatiles which more truly represents the aroma of the original. With the exception of the method of expression, almost all of these involve treatment of the plant material with one or more organic solvents (or mixtures thereof) followed by concentration of the extracted solute. Solvent extraction frequently yields, in addition to the volatile oil, various quantities of semi- or nonvolatile organic material such as waxes (qv), fats, fixed oils, high molecular weight acids, pigments (qv), and even alkaloidal material. However, because solvent extraction often results in a product with superior and more representative odor properties to that of a distilled oil, many natural products critically important to the flavor and fragrance industry are available as various extracts in addition to an essential oil.

Some of the commonly used botanical extracts include the following.

Absolute. This is concentrated extract obtained by treatment of a concrete or other hydrocarbon-type extract of a plant or plant part with ethanol.

Absolute Oil. This is the steam distillable portion of an absolute.

Aroma Distillate. Used by the flavor industry, aroma distillates are the product of continuous extraction of the plant material with alcohol at temperatures between ambient and 50°C followed by steam distillation, and, lastly, concentration of the combined hydro-alcoholic mixture.

Concrete. Hydrocarbon extracts of plant tissue, concretes are usually solid to semisolid waxy masses often containing higher fatty acids such as lauric, myristic, palmitic, and stearic as well as many of the nonvolatiles present in absolutes.

Infusion. Infusion botanical extracts are tinctures that have been concentrated by either total or partial removal of the alcohol by distillation.

Oleoresin. Natural oleoresins are exudates from plants, whereas prepared oleoresins are solvent extracts of botanicals, which contain oil (both volatile and, sometimes, fixed), and the resinous matter of the plant. Natural oleoresins are usually clear, viscous, and light-colored liquids, whereas prepared oleoresins are heterogeneous masses of dark color.

Pommade. These are botanical extracts prepared by the enfleurage method wherein flower petals are placed on a layer of fat which extracts the essential oil.

Resin and Resinoid. Natural resins are plant exudates formed by the oxidation of terpenes. Many are acids or acid anhydrides. Prepared resins are made from oleoresins from which the essential oil has been removed. A resinoid is prepared by hydrocarbon extraction of a natural resin.

Tincture. This is prepared by aqueous alcoholic extraction of the raw plant material. Since the extract is not further concentrated, the plant extract is not exposed to heat.

Essential oils are isolated from various plant parts, such as leaves (patchouli), fruit (mandarin), bark (cinnamon), root (ginger), grass (citronella), wood (amyris), heartwood (cedar), gum (myrrh oil), balsam (tolu balsam oil), berries (pimento), seeds (dill), flowers (rose), twigs and leaves (thuja oil), and buds (cloves).

Exceptions to the simple definition of an essential oil are, for example, garlic oil, onion oil, mustard oil, or sweet birch oils, each of which requires enzymatic release of the volatile components before steam distillation. In addition, the physical process of expression, applied mostly to citrus fruits such as orange, lemon, and lime, yields oils that contain from 2–15% nonvolatile material.

Economic Aspects

Essential oils are used as flavoring and fragrance agents in every possible application. Combinations have raised greatly the total sales volume; eg, mint and cinnamon are used in toothpaste, mouthwash, or lozenges. Combinations can be found in every fragranced product, such as fine fragrances, soaps, detergents, room fresheners, paper, printing ink, paint, candles, condiments, floor polishes, etc. Convenience foods and frozen foods are flavored best by essential oils or oleoresins. Although citronella oil was used as such as an insect repellant, synthetic repellants have, for the most part, taken their place. Flavor essential oils are encountered in baked goods, snack foods, soft drinks, liqueurs, tobacco, sauces, gravies, salad dressings, and other food products.

Composition

The volatile components of essential oils, for the most part, are made up of relatively low molecular weight ($\leq\sim 300-350$) organic molecules of carbon, hydrogen, and oxygen, and occasionally nitrogen and sulfur. By far the largest class of natural volatiles of plants is the terpenes, which consist of head-to-tail condensation products of unsaturated five-carbon isoprene units.

Other commonly occurring chemical groups in essential oils include aromatics such as β-phenethyl alcohol, eugenol, vanillin, benzaldehyde, cinnamaldehyde, etc; heterocyclics such as indole (qv), pyrazines, thiazoles, etc; hydrocarbons (linear, branched, saturated, or unsaturated); oxygenated compounds such as alcohols, acids, aldehydes, ketones, ethers; and macrocyclic compounds such as the macrocyclic musks, which can be both saturated and unsaturated.

An essential oil may contain >200 components, and often the trace substances (≤ppm) are essential to the odor and flavor of the oil. The absence or decreased presence of even one component may be cause for odor or flavor rejection of the oil. The same species of plant grown in different parts of the world usually contains the same chemical components, but the relative percentages may be different. Climatic and topographical conditions affect plant chemistry and can alter the essential oil content both qualitatively and quantitatively.

Commercial Essential Oils

Commercial essential oils include rose, jasmin, orange flower (neroli) oil, lavender and lavandin, geranium oil, citronella oil, bergamot oil, lime oil, orange oil, grapefruit oil, sandalwood oil (East Indian), patchouli oil, vetiver oil, galbanum oil, myrrh, oakmoss, tonquin musk, ambergris, tobacco, osmanthus, olibanium, amyris oil, anise oil, anise oil (star), sweet basil oil, bay oil, bitter orange oil, black pepper, bois de rose oil, cannaga oil, caraway oil, cardamom oil, cassia oil, cedarleaf oil, cedarwood oil, Roman chamomile oil, cinnamon bark oil, citronella oil, clove bud oil, coriander oil, cornmint oil, eucalyptus oil, ginger oil, juniper oil, labdanum oil, lemon oil, nutmeg oil, oregano oil (Spanish), orris, palmarosa oil, peppermint oil, petitgrain bigarade oil, pimento berry oil, pine oil, rosemary oil, sage oil (Dalmatian), sage (clary) oil, spearmint oil (native), tagetes oil, thyme oil, turpentine oil, wintergreen oil, and ylang ylang.

Safety and Regulatory Aspects

Essential oils possess a variety of biological properties which may result in varying responses by humans on exposure. An important factor in these effects is the dose to which one is exposed. Thus, essential oils may have both beneficial and toxic effects, depending on their dose. The potential for biological effects from essential oils is not surprising; many botanical species are known to contain substances that possess biological properties, and their identification has contributed significantly to knowledge of biochemistry and physiology as well as the development of therapeutic agents, eg, quinine and digitalis.

The toxicities of many essential oils have been reported in monographs. Most essential oils used by the flavor and fragrance industries are relatively nontoxic or slightly toxic on acute oral or dermal exposure, and are considered safe when used at levels present in consumer products. In general, the levels of fragrances and flavors in consumer products, and thus the levels of any essential oil ingredients, are relatively low. For example, a fragrance oil may typically be used in a soap at 0.5%. The oil may contain 5% of orange oil distilled. The final concentration of orange oil distilled in the soap therefore is 0.025%.

Because essential oils are used predominantly by the flavor and fragrance industries, these commercial oils must undergo the same scientific scrutiny as all other flavor and fragrance substances and must be in compliance with all applicable health, safety, and environmental regulations. Guidelines and regulations on the use of essential oils in fragrances differ from those applying to essential oils used in flavors.

Many essential oils have been designated by the FDA or by the expert panel of FEMA as Generally Recognized As Safe (GRAS) for their intended use in foods and flavors. The use and safety of these GRAS substances are continuously being reviewed and the list of GRAS substances updated. New essential oils intended to be used as a flavor ingredient must undergo extensive safety evaluations and scrutiny by one or more of these groups of experts before they may be used in flavors.

Many countries have adopted chemical substance inventories in order to monitor use and evaluate exposure potential and consequences. In the case of essential oils used in many fragrance applications, these oils must be on many of these lists. New essential oils used in fragrances are subject to premanufacturing or premarketing notification (PMN). PMN requirements vary by country and predicted volume of production. They require assessment of environmental and human health-related properties, and reporting results to designated governmental authorities.

Essential oils are also influenced by legislation that regulates specific products that may contain these oils, eg, the U.S. Food, Drug, and Cosmetic Act and the European Community Cosmetic Directive. Essential oils would not be anticipated to be of environmental concern, considering that they originate from botanical sources. Thus, natural processes exist to degrade essential oils and recycle their components effectively in the environment.

BRAJA D. MOOKHERJEE
RICHARD A. WILSON
International Flavors & Fragrances, Inc.

S. Arctander, *Perfume and Flavor Materials of Natural Origin,* 1960.

B. D. Mookherjee and R. A. Wilson in *On Essential Oils,* Synthite Industrial Chemicals Private Ltd., Synthite Valley, Kolenchery, India, 1986, pp. 281–329.

B. D. Mookherjee and C. J. Mussinan, eds., *Essential Oils,* Allured Publishing Corp., Wheaton, Ill., 1981.

S. R. Srinivas, *Atlas of Essential Oils,* the Bronx, N.Y., 1986.

OIL SHALE

Oil shale is a sedimentary mineral that contains kerogen, a mixture of complex, high molecular weight organic polymers. The solid kerogen is a three-dimensional polymer that is insoluble in conventional organic solvents. Upon heating, kerogen decomposes to form gas composed of hydrogen (qv), low molecular weight hydrocarbons (qv), and carbon monoxide (qv); liquids, composed of water and shale oil; and a solid char residue.

Oil shale deposits were formed in ancient lakes and seas by the slow deposition of organic and inorganic remains. The geology and composition of the inorganic minerals and organic kerogen components of oil shale vary with deposit locations throughout the world (see also FUEL RESOURCES; PETROLEUM).

Reserves

Estimates of oil shale deposits by continent are given in Table 1. Characteristics of many of the world's best known oil shales are summarized in Table 2. Oil shale deposits in the United States occur over a wide area.

General Properties

The thermal decomposition of oil shale, ie, pyrolysis or retorting, yields liquid, gaseous, and solid products. The amounts of oil, gas, and

Table 1. Shale Oil Resources, 10^9 m^3[a]

	Total resource[b,c]			Marginal or submarginal resources[c]		
Geographic area	21–42	42–104	104–417	21–42	42–104	104–417
Africa	71,500	12,700	636	small	small	14
Asia	93,800	17,500	874		2	11
Australia and New Zealand	15,900	3,200	159		small	small
Europe	22,260	4,100	223		1	6
North America	41,400	8,000	477	350	254	99
South America	33,400	6,400	318		119	small
Total	*278,260*	*51,900*	*2,687*	*350*	*376*	*130*

[a] To convert m^3 to bbl, divide by 0.159.
[b] Includes oil shale in known resources, in extensions of known resources, and in undiscovered but anticipated resources.
[c] Numbers represent shale oil yield range in L/t. To convert L/t to gal/short ton, multiply by 0.2397.

Table 2. Properties of World Wide Oil Shales

Property	Timahdit	Irati	Nagoorin	Kentucky	Maoming	Colorado	Condor	Alpha	New Brunswick	Israeli	Kunkersite
				Fischer assay, %							
oil weight	6–9	6–12	14.1	5.3	9.7	16.5	6.3	52.0	6–12	6.2	28.6
water, bound	2.1–2.7	0.2–2.1	6.9	1.9	3.8	1.0	1.9	4.0	0.9–1.4	2.8	2.5
spent shale	85–88	83–90	72.4	90.0	82.0	78.6	87.3	33.0	91.1–84.5	87.4	62.7
gas + loss	2.8–3.7	2–4	6.6	2.8	4.5	3.8	4.5	11.0	2.0–2.1	3.6	6.2
				Other properties							
moisture, wt %	6.7–9.8	0.2–6	23.2	2.8	11.3	0.7	7.7	2.8	5.4–6.7	8.1	5.8
specific gravity	1.88–1.99	1.9–2.1	1.47	2.22	1.73	1.94	2.05	1.16	2.32–1.97	1.57	1.60
gross heating value, J/g[a]	5.230–6.904	5.439–6.987	11.950	5.791	8.577	9.113	4.728	30.669	3.766–6.908	4.209	16.07
total carbon, wt %	14.78–19.46	12–17	25.67	12.82	18.74	23.45	10.50	70.54	10.75–16.58	15.5	36.8
total hydrogen, %	1.9–2.0	0.9–2.4	3.7	1.5	2.9	2.9	1.7	8.39	1.4–2.2	1.6	4.3
total sulfur, %	2.1–2.7	3.9–5.6	1.0	4.4	1.6	1.1	0.9	1.4	0.9–1.0	2.9	2.0
nitrogen, %	0.46–0.63	0.3–1.9		0.3	1.3	0.6	0.3	1.0			0.1
loss on ignition, at 950°C, %	31.4–38.9	20–24	41.9	21.3	32.1	38.0	18.6	91.7	21.1–28.5		56.2
				Ash composition, wt %							
SiO_2	31.6–37.5	5.0–5.6		64.8	57.2	45.2	73.2	53.4	54.8–55.7		33.2
Fe_2O_3	3.5–5.8	7.6–9.8		10.7	12.2	5.5	8.1	9.9	6.8–5.7		6.6
Al_2O_3	8.6–13.0	9.8–12.6		12.5	19.5	2.3	12.1	24.3	17.9–15.0		8.9
CaO	15.7–26.7	1.3–3.9		1.9	1.1	18.9	2.0	3.4	8.9–13.8		33.7
MgO	5.6–7.4	2.0–3.7		0.6	0.8	17.4	1.0	4.1	6.1–3.7		9.5
				Fischer assay oil							
specific gravity, 20°C	0.962	0.906	0.918	0.926	0.890	0.902	0.895	0.905	0.880	0.980	0.958
total carbon, wt%	78.73	84.60	83.40	84.95	84.81	84.21	84.72	84.32	85.6	80.8	83.4
hydrogen, %	9.69	12.50	11.37	11.85	11.65	11.29	12.54	11.89	12.3	10.4	10.7
sulfur, %	6.33	1.10	1.16	1.40	0.52	0.92	0.46	1.72	0.6	5.0	0.7
nitrogen, %	1.52	0.90	1.18	1.12	2.60	1.78	1.30	0.69	1.1	1.2	0.1
gross heating value, J/g[a]	40.074	42.547	43.070	41.773	42.447	42.723	42.677	42.539	43.932	39.748	39.790

[a] To convert J to cal, divide by 4.184.

coke which ultimately are formed depend on the heating rate of the oil shale and the temperature–time history of the liberated oil. There is little effect of shale richness on these relative product yields under fixed pyrolysis conditions.

Numerous kinetic mechanisms have been proposed for oil shale pyrolysis reactions. It has been generally accepted that the kinetics of the oil shale pyrolysis could be represented by a simple first-order reaction (kerogen → bitumen → oil), or

$$\text{sequential} A \rightarrow B \rightarrow C \qquad (1)$$

This sequential first-order reaction adequately describes the kinetics of pyrolysis of the Green River oil shale in western United States. Additional kinetic studies indicate that sequential reactions are inadequate to describe the kinetic reactions for the thermal decomposition of oil shales worldwide.

Most oil shale retorting processes are carried out at ca 480°C to maximize liquid product yield.

The carbonate content of Green River oil shale is high. In addition, the northern portion of the Pieceance Creek basin contains significant quantities of the carbonate minerals nahcolite and dawsonite. The decomposition of these minerals is endothermic and occurs at ca 600–750°C for dolomite, 600–900°C for calcite, 350–400°C for dawsonite, and 100–120°C for nahcolite. Carbon dioxide, a product of decomposition, dilutes the off-gases produced from retorting processes at the above decomposition temperatures.

Retorting

Oil shales are solid minerals, impervious to the flow of fluids, and are generally situated in deposits below the earth's surface. Therefore, several process steps must be undertaken to produce crude shale oil. In the case of the commonly used above-ground retorting (AGR), these steps involve mining, crushing, and heating (see MINERALS RECOVERY AND PROCESSING). The grade (volume of oil per weight of rock) of most oil shales is low, and large amounts of the oil shale rock must be processed to produce crude shale oil. Depending on the grade, 2 to 25 metric tons of oil shale must be processed to produce one cubic meter of crude shale oil (0.4–4.6 short tons per barrel of crude shale oil). In order to eliminate the costs of mining and material handling, direct underground retorting (*in situ* retorting) has been considered as an alternative to the conventional AGR.

Historically, direct combustion has been employed in which some of the organic matter of the kerogen is combusted to provide the heat necessary for retorting. Although these direct heat (DH) processes do not require a supplemental source of fuel, some of the kerogen is consumed and the gaseous products of the kerogen decomposition are diluted with the products of combustion. In order to obviate these shortcomings, indirect heat (IH) processes were developed in which the heat required for retorting was supplied by hot gases or solids that were heated externally. However, the IH processes do not utilize any of the solid residual carbon or char resulting from kerogen decomposition and they do require an external source of fuel.

There are numerous means of classifying the many processes that have been employed to retort oil shale. In addition to the types of retorting, the retorting process can be classified by the type of feed used and by the flows within the retort. A list of most of the oil shale retorting processes in use worldwide since the 1940s is provided in Table 3.

Retorting processes consist of several well-defined steps, or zones, within the retort. For DH systems the zones are the oil shale preheating or off-gas oil-mist cooling zone; the pyrolysis zone, where the solid organic kerogen is converted into gases, oil mists and vapors, and residual carbon; the combustion zone, where carbon is burned to provide heat; and the shale cooling zone, where the retorted shale is cooled and the incoming air is preheated.

Table 3. Retorting Technologies

Technology	Country	Heating process[a,b]	Feed	Flow[c]
		Above-ground retorting		
Chevron	United States	DH	fine	
FBC	Israel	DH	fine	CC
Fuschun	China	DH	coarse	CC
Galoter	Russia	IH	coarse	CO
Gas combustion	United States	DH	coarse	
Kiviter	Russia	IH and DH	coarse	CO and CC
LLNL/HRS	United States	IH (ash)	fine	CO
Lurgi	United States	IH (ash)	fine	CO
Paraho DH	United States	DH	coarse	CC
Paraho IH	United States	IH (gas)	coarse	CC
Petrosix	Brazil	IH (gas)	coarse	CC
Superior	United States	IH (gas)	coarse	
Taciuk	Australia	IH (gas)	fine	CC
TOSCO II	United States	IH (solids)	fine	CC
Unishale A	United States	DH	coarse	CC
Unishale B	United States	IH (gas)	coarse	CC
		In situ retorting		
Equity BX	United States	IH (steam)		
IGT	United States	IH (H_2/steam)		CO
LOFRECO	United States	DH		
MultiMineral	United States	DH		
RISE	United States	DH		
VMIS	United States	DH		

[a] DH = direct heat; IH = indirect heat.
[b] Heat-transfer medium is given in parentheses.
[c] CC = countercurrent; CO is concurrent.

Crude Shale Oil

Properties. The composition of shale oil has depended on the shale from which it was obtained as well as on the retorting method by which it was produced.

Shale oil contains large quantities of olefinic hydrocarbons, which cause gumming and constitute an increased hydrogen requirement for upgrading. High pour points prevent pipeline transportation of the crude shale oil (see PIPELINES). Arsenic and iron can cause catalyst poisoning.

The primary difference in shale oils produced by different processing methods is in boiling point distribution. Rate of heating, as well as temperature level and duration of product exposure to high temperature, affect product type and yield. Gas combustion processes tend to yield slightly heavier liquid products because of combustion of the lighter, ie, naphtha, fractions.

Upgrading Shale Oil. Crude shale oil has a high (~2 wt %) content of organic nitrogen which acts as a catalyst poison, contains a large (20–50 wt %) atmospheric residuum fraction, and has a high (<5°C) pour point. Prerefining crude shale oil to produce a synthetic crude that is compatible with typical refineries generally is necessary.

Shale oil has been refined to produce gasoline, kerosene, jet fuel, and diesel fuel. Different procedures have been tested to produce different product states, eg, hydrotreating followed by hydrocracking for jet fuel production, hydrotreating followed by fluid catalytic cracking for gasoline production, and coking followed by hydrotreating for diesel fuel production.

Alternative Uses

Oil shale is an energy resource that produces a liquid fuel that can be used to replace conventional crude oil or petroleum. However, the costs associated with processing oil shale into conventional refined products are significantly greater than that of processing conventional crude oil. In order to develop the oil shale resource, other uses have been considered. These functions include direct combustion to produce process heat for power generation (qv), direct gasification of the oil shale geological deposit, and special petrochemical production.

Environmental Issues

The plans to develop a commercial oil shale industry in the three-state region of Colorado, Utah, and Wyoming in the 1970s raised the possibility of significant adverse environmental, health, safety, and socioeconomic (EHSS) impacts. Processing oil shale to produce oil on a large-scale commercial basis requires a large amount of mining, crushing, material transport, and disposal operations.

Adverse EHHS impact could result from uncontrolled, or inadequately controlled, large-scale oil shale operations. Without controls, significant amounts of dust, ie, particulates, would be produced. Because the gas produced from kerogen breakdown contains significant amounts of hydrogen sulfide and ammonia, uncontrolled release, or direct combustion with no control technology, could pose adverse health impacts and air pollution. The liquids produced from retorting operations, ie, process water and crude shale oil, contain significant levels of toxic metals, suspected or known carcinogens, and other hazardous materials. Discharge of this water would thus require treatment. Combusting and/or refining the crude shale oil would also require adequate treatment and environmental controls. The large quantities of materials involved in oil shale development means that disposal of the retorted shale poses special problems. Proper controls are needed to avoid significant air pollution from dust emissions, and surface and groundwater contamination from leaching and runoff. The amount of water required for commercial oil shale operations poses water quality impact on the semiarid region of Colorado, Wyoming, and Utah. Engineering technology was thus developed for oil shale operations.

Commercial Operations

The number of commercial oil shale operations worldwide has decreased significantly since the decade 1975–1985 and are producing only a fraction of the world's liquid fuels' needs. Most commercial oil shale operations have been scaled back.

Petroleo Brasilerio (Petrobras) has a dedicated facility to produce crude shale oil from the Irati formation in southern Brazil. The facility is called the Oil Shale Industrialization Superintendency (SIX) and uses the PETROSIX retorting technology.

Oil shale, the only fossil fuel resource in Israel, is being used to generate electric power. The oil shale feed stock, typical of the low grade Israeli oil shale, is situated in a deposit overlying phosphate ore. The oil shale operations are being carried out because the oil shale has to be mined to obtain the phosphate ore.

United States. In 1980, Unocal began constructing the Parachute Creek Project, designed to produce 1600 m^3 (10,000 bbl) of upgraded shale oil per day. However, the Parachute Creek Project was shut down in mid-1991 for economic reasons.

The New Paraho Corp. has been conducting research on asphalt derived from shale oil, SOMAT, at its pilot plant (Rifle, Colorado). It is the only active oil shale operation in the United States as of 1995.

As of this writing, commercial production of shale oil is still being conducted in the People's Republic of China and Estonia. However, production rates continue to dwindle owing to the availability of conventional petroleum and other sources of energy as well as continued worldwide energy conservation.

Plans are underway to develop commercial shale oil operations in Australia. Southern Pacific Petroleum, N.L. is planning a commercial oil shale project utilizing the Stuart deposit (Brisbane, Australia). Favorable economics are attained by tax incentives to the Stuart project in the form of increased depreciation write-offs and exempting excise tax for gasoline produced from shale oil.

EDWIN M. PIPER
Piper Designs LLC
ROBERT N. HEISTAND
Consultant

G. L. Baughman, ed., *Synthetic Fuels Data Book,* 2nd ed., Vol. 4, Cameron Engineers (Division of The Pace Co.), Denver, Colo., 1978, pp. 67–104.

C. P. Reeg, A. C. Randle, and J. H. Duir, *Proceedings, 23rd Oil Shale Symposium,* Colorado School of Mines Press, Golden, Colo., 1990, pp. 68–95.

E. M. Piper and co-workers, *Proceedings of the 25th Oil Shale Symposium,* Colorado School of Mines Press, Golden, Colo., 1992, pp. 221–242.

J. E. Bunger and A. V. Deveni, *Proceedings, 25th Oil Shale Symposium,* Colorado School of Mines Press, Golden, Colo., 1992, pp. 281–294.

OLEFIN FIBERS. See FIBERS, OLEFIN.

OLEFIN POLYMERS

POLYETHYLENE

INTRODUCTION

Polyethylene (PE) is a generic name for a large family of semicrystalline polymers used mostly as commodity plastics. PE resins are linear polymers with ethylene molecules as the main building block; they are produced either in radical polymerization reactions at high pressures or in catalytic polymerization reactions. Most PE molecules contain branches in their chains. In very general terms, PE structure can be represented by the following formula:

$$(CH_2—CH_2)_x—branch_1—(CH_2—CH_2)_y—branch_2—(CH_2—CH_2)_z$$
$$—branch_3 \cdots$$

where the $—CH_2—CH_2—$ units come from ethylene, and *x, y,* and *z* values can vary from 4 or 5 to over 100. This allows the industry to produce a large variety of PE resins with different molecular weights and branching characteristics.

The total number of monomer units (which approximately equals $x + y + z + \ldots$ in the above formula) in PE chains is called the degree of polymerization. It can vary from small (about 10–20 in PE waxes) to very large (over 100,000 for PE of ultrahigh molecular weight (UHMW)).

In high pressure ethylene polymerization processes, the branches are formed spontaneously according to peculiarities of radical polymerization reactions (see OLEFIN POLYMERS, LOW DENSITY POLYETHYLENE). These branches are either linear or branched alkyl groups. Their lengths vary widely, sometimes even within a single polymer molecule, and can be both short (from methyl to isooctyl group, collectively known as short-chain branching) and long, up to several thousands of carbon atoms (long-chain branching). In catalytic polymerization processes, the branches are introduced deliberately by copolymerizing ethylene with α-olefins (see OLEFIN POLYMERS, LINEAR LOW DENSITY POLYETHYLENE). The structure of these branches is determined by the type of olefin used in the copolymerization reaction.

Some PE molecules, on the other hand, contain no branches at all. From a chemical standpoint, such resins can be regarded as polnymethylene, $H\!\!-\!\!(CH_2)\!\!-_n H$

The distributions of branches among different polymer molecules in PE resins can be quite different, depending on the method of PE production. Some PE resins have uniform branching distributions, which means that any given polymer molecule contains the same relative fraction of branches as all others. PE prepared in high pressure processes and most ethylene copolymers produced with metallocene catalysts belong to this group. In contrast, PE resins produced with heterogeneous titanium- and chromium-based catalysts are mixtures of polymer molecules with very different branching degrees.

Classification of PE resins

The classification of PE resins has developed in conjunction with the discovery of new catalysts for ethylene polymerization as well as new polymerization processes and applications. The classification (given in Table 1) is based on two parameters: the resin density and its melt index. This classification provides a simple means for a basic differentiation of PE resins, even though it cannot easily describe some important distinctions between the structures and properties of various resin brands.

Table 1. Commercial Classification of Polyethylenes

Designation	Acronym	Density, *d*, g/cm^3
high density polyethylene	HDPE	≥0.941
ultrahigh molecular weight polyethylene[a]	UHMWPE	0.935–0.930
medium density polyethylene	MDPE	0.926–0.940
linear low density polyethylene	LLDPE	0.915–0.925
low density polyethylene[b]	LDPE	0.910–0.940
very low density polyethylene	VLDPE	0.915–0.880

[a] Linear polymer with molecular weight of over 3×10^6.
[b] Produced in high pressure processes.

Synthesis Technologies

A variety of technological processes are used for polyethylene manufacture. They include polymerization in supercritical ethylene at a high ethylene pressure and temperature above the PE melting point (110–140°C), polymerization in solution at 120–150°C or in slurry, and polymerization in the gas phase.

Control of PE Properties

The tailoring of PE properties in commercial processes is achieved mostly by controlling the density, molecular weight, and molecular weight distribution, or by cross-linking. Successful control of all reaction parameters enables the manufacture of a large family of PE products with considerable differences in physical properties, such as the softening temperatures, stiffness, hardness, clarity, impact, and tear strength.

The Market

PE resins command a wide range of applications, both as commodity resins and as specialty polymers. Their uses include numerous film grades of LDPE, HDPE, and LLDPE for bags and packaging; coatings for paper, metal, wire, and glass; household and industrial containers such as bottles for various fluids, ie, water, food products, detergents, liquid fuels, etc; toys; and different types of pipe and tubing. Because of its versatility, PE has become the largest commercially manufactured polymer in the world.

YURY V. KISSIN
Mobil Chemical Company

LOW DENSITY POLYETHYLENE

The first high molecular weight crystalline polyolefin was produced in 1933 through the high pressure process. Initial production was targeted for use in specialized applications such as insulation for high voltage cable. Since that time, the number of applications of various grades of homopolymers and copolymers have expanded to cover many areas once dominated by paper (qv), glass (qv), steel (qv), and other polymers.

The molecular weight of low density polyethylene (LDPE) ranges from waxy products at about 500 mol wt to very tough products at about 60,000 mol wt. One unique feature of LDPE, as opposed to high density polyethylene (HDPE) or linear low density polyethylene (LLDPE), is the presence of both long- and short-chain branching along the polymer chain. Another important feature of LDPE is its ability to incorporate a wide range of comonomers that can be polar in nature along the polymer chain. Disadvantages of LDPE include the high capital investment for commercial plant construction, engineering problems related to high pressure operation, and high energy costs in production (see HIGH PRESSURE TECHNOLOGY).

LDPE, also known as high pressure polyethylene, is produced at pressures ranging from 82–276 MPa (800–2725 atm). Operating at 132–332°C, it may be produced by either a tubular or a stirred autoclave reactor. Reaction is sustained by continuously injecting free-radical initiators, such as peroxides, oxygen, or a combination of both, to the reactor feed.

Traditionally, LDPE has been defined as homopolymer products having a density between 0.915–0.940 g/cm^3 (products having a density above 0.940 g/cm^3 are considered HDPE). However, with the commercialization of LLDPE via the fluidized-bed or solution processes, this distinction is no longer valid.

Properties

The mechanical properties of LDPE fall somewhere between rigid polymers such as polystyrene and limp or soft polymers such as polyvinyls. LDPE exhibits good toughness and pliability over a moderately wide temperature range. It is a viscoelastic material that displays non-Newtonian flow behavior, and the polymer is ductile at temperatures well below 0°C. Table 1 lists typical properties.

Structure. The physical properties of LDPE depend on the molecular weight, the molecular weight distribution, as well as the frequency and distribution of long- and short-chain branching.

Molecular Weight Distribution. MWD offers a general picture of the range of long, medium, and short molecular chains in the polymer; the broad molecular weight distribution in LDPE, however, is attributed only to the presence of long branches on the polymer molecule. LDPE may have molecules that range in length from a few thousand carbons to a million or more carbons.

With increasing molecular weight, certain properties increase: melt viscosity, abrasion resistance, tensile strength, resistance to creep, flexural stiffness, resistance to brittleness at low temperature, shrinkage, warpage, and film impact strength. On the other hand, increasing mol wt results in reduced film transparence, freedom from haze, and gloss; drawdown rate; neck-in and beading; and adhesion.

Table 1. Properties of Low Density High Pressure Polyethylene

Property	ASTM Method	LDPE[a]
tensile yield stress, kPa[b]	D638	80–180
yield elongation, %	D638	10–40
tensile ultimate stress, kPa[b]	D638	100–170
ultimate elongation, %	D638	100–700
secant modulus of elasticity at 1% strain, kPa[b]	D638	900–5000
hardness, Rockwell	D785	D41–D60
dart drop 38-μm thickness, at g/25 μm	D1709	50–300
low temperature brittleness at F_{50}[c], °C	D746	<−76°C
dielectric constant at 60 Hz	D150	2.25–2.35
density, g/cm^3[d]	D1505	0.912–0.940
refractive index, n_D^{25}[d]	D542	1.51
thermal expansion, 10^{-5} cm/cm per °C[d]	D696	10–22
narrow angle scatter	D1746	4–80
haze, %	1003	40–50
gloss, %	2457	0–80
water absorption 24 h, %[d]	D570	<0.02

[a] LDPE homopolymers in the 0.2–150 melt index or 100,000–20,000 mPa·s(= cP) viscosity range. Specialty polymers such as greases and waxes or highly cross-linked polymers are not included.
[b] To convert kPa to psi, divide by 6.895.
[c] F_{50} = number of hours at which 50% fail.
[d] Taken at 25°C.

Melt Index or Melt Viscosity. Melt index describes the flow behavior of a polymer at a specific temperature under specific pressure. If the melt index is low, its melt viscosity or melt flow resistance is high; the latter is a term that denotes the resistance of molten polymer to flow when making film, pipe, or containers. ASTM D1238 is the designated method for this test.

Film Clarity. Slight haziness is a characteristic of all polyethylene resins. It may be caused either by a surface roughness which diffuses light passing through the film (surface roughness is a function of extrusion conditions and the fundamental structure of the polymer), or it may be caused by the partly crystalline structure of the polymer which has a larger index of refraction than the surrounding amorphous material.

Stress Crack Resistance. Failure caused by environmental stress cracking may be attributed to stored stresses acquired in the molding or extruding operation. These dormant stresses may release themselves by cracking under the combined influence of an adverse environment and polyaxial stretching during use. Polyethylene of narrow molecular weight distribution tends to crack less under environmental stress.

Yield Strength. Yield strength, tensile strength, and elongation are all functions of the basic molecular properties. In other words, a higher density LDPE homopolymer has higher yield strength, slightly lower tensile strength, and lower elongation; in contrast, a higher average molecular weight polymer has higher tensile strength and slightly higher yield strength.

Low Temperature Brittleness. Brittleness temperature is the temperature at which polyethylene becomes sufficiently brittle to break when subjected to a sudden blow. Because some polyethylene end products are used under particularly cold climates, they must be made of a polymer that has good impact resistance at low temperatures; namely, polymers with high viscosity, lower density, and narrow molecular weight distribution.

Film Appearance. It is important to minimize film imperfections or defects when LDPE is blown into film. The common defects are arrowheads, pinpoint gels, gels or "fish eyes", and oxidized gels or colored specks.

Film imperfections are one of the more serious quality problems for both the producer and the film converter. Film imperfections can arise from many sources. Most commonly, they are the result of contamination in the reaction system, in post-reactor handling, during shipping and unloading, or in the end users equipment.

Chemical Resistance. LDPE is highly resistant to penetration by most chemically neutral or reactive substances. This is a property of prime importance for all kinds of packaging applications. Because of its high impermeability, polyethylene containers can store and transport many kinds of chemicals without leak hazards. Likewise, easily spoiling foods such as vegetables or meats can be shelved and sold in polyethylene bags without the danger of water infiltration from the outside or irreplaceable moisture being lost from the inside; exchange of gases through the film can also be kept to a minimum. In addition, LDPE is resistant to penetration from most polar liquids, water, aqueous acids, and alkalies, as well as most metal plating solutions. However, it can be easily penetrated by nonpolar liquids such as hydrocarbons, and animal and vegetable oils.

Electrical Properties. LDPE's electrical properties make it extremely well suited for wire and cable insulation for electrical power supplies at high transmission, lower domestic voltage, and high frequency, very high frequency, or ultrahigh frequency applications in electronics (see INSULATION, ELECTRIC). As a dielectric, ie, electric insulator, LDPE for all practical purposes does not transmit electrical current.

Manufacture

LDPE is produced in either a stirred autoclave or a tubular reactor; total domestic production, divided between the two systems at 45% for tubular and 55% for autoclave, is estimated to be 3.4 million metric tons per year. Neither process has gained a clear advantage over the other, although all new or added capacity production in the 1990s has been through the autoclave.

Recycle and Polymer Collection. Due to the incomplete conversion of monomer to polymer, it is necessary to incorporate a system for the recovery and recycling of the unreacted monomer. Both tubular and autoclave reactors have similar recycle systems.

Additives. Compounds are often added to the polymer at the extruder or melt homogenizer. Common additives are antioxidants (qv), thermal stabilizers, slip agents, antiblock agents, and uv stabilizers.

Blending and Purging. Polymer pellets are air-conveyed to holding silos where they are purged with heated, filtered air and then blended to ensure uniformity. Purging is necessary for removing the entrapped monomer and comonomer.

Environmental Considerations. Good progress has been made in reducing emissions during production. Ethylene is by far the largest fugitive gas; purge from the recycle system is commonly recycled to the ethylene cracking unit for recovery. Owing to improved operation techniques and control equipment, runaway reactions or "decomps" that are vented to the atmosphere have also decreased substantially. In addition, many plants utilize particulate and noise reduction devices on the emergency venting systems.

Many units have waste heat recovery systems that generate low pressure steam from reaction heat. Such steam is often employed to drive adsorption refrigeration units to cool the reactor feed stream and to increase polymer conversion per pass, an energy-saving process that reduces the demand for electrical power.

Pellet degassing, however, still presents a problem. The volume of purge air required to degas the pellets makes recovery of ethylene from the purge stream using conventional methods uneconomical.

Applications

LDPE is used in a wide range of applications, the largest segment of which is taken up by end uses requiring processing into thin film (see FILM AND SHEETING MATERIALS).

Blown Film. Blown film has a number of advantages over flat film. For instance, in blown film extrusion, molecular orientation is achieved in both the machine and transverse direction, the relative degree depending primarily on the drawdown ratio (the die opening to film gauge) and the blow-up ratio. The result is a film with more uniform strength in both directions. Moreover, with proper extrusion conditions, the physical properties of the film are equal in both the machine and transverse direction; such even distribution provides maximum toughness. Another advantage of blown film or tubing is the absence of a lengthwise seam.

In blown film extrusion, the molten polymer enters a ring-shaped die either through the bottom or from the side. It is then forced around a mandrel inside the die, shaped into a sleeve, and extruded through the die opening in the form of a comparatively thick-walled tube.

Flat Film Extrusion. In flat film extrusion, the melt is extruded through a long slot in a "T" or coat hanger-type die, past the die lands. In this setup, the polymer melt is forced into the slot die at its center; it reaches the slot opening by way of a manifold and over the lands. The principal advantages of film casting are substantial improvements in the film's transparency, freedom from haze, improved gloss, and other optical properties.

Extrusion Coating. In extrusion coating, a thin film of molten polymer is pressed onto or into the substrate. Coating thickness may range from 6.5 μm or less to more than 100 μm. In polymer lamination, a related operation, two or more substrates, such as paper or aluminum foil, are combined by using the polymer film as adhesive and moisture barrier. In order to coat a substrate, the polymer must be extruded through a narrow slit in the extrusion coating die by an extruder screw.

Molding Applications. Molding is accomplished by three different methods: blow molding, injection molding, and rotational molding, although the use of LDPE in these applications has been declining since the introduction of LLDPE on the market.

Adhesives and Sealants. Dominated by copolymers, adhesives and sealants remain somewhat of a specialty market. These polymers usually contain high copolymer content and low viscosity, and often require blending with other compounds prior to final application. Their uses are numerous, eg, as seals for bottled drinks, as tie layers between incompatible polymers, or as automotive adhesives.

Health and Safety Factors

LDPE is nontoxic, and is commonly used in food packaging (qv) where Food and Drug Administration requirements must be met. It is also used in packaging pharmaceuticals and other medical applications such as iv bags and tubes.

Waste management of LDPE can be approached in several different ways, ie, via recycling, energy recovery, biodegradability, increased packaging efficiency, or uv degradability. Ultraviolet degradable grades for can carriers are available to processors; these products are mandated in several states.

Environmental Degradability. It has become increasingly clear that, for certain end uses at least, plastics with a limited lifetime are preferable because they help to solve problems of litter and waste disposal. These types of plastics are subject to biodegradation, which occurs when microorganisms such as fungi or bacteria secrete enzymes and chemically break down the polymer structure into small fractions that can be digested by other microorganisms.

Biodegradability has been approached in several areas. One attempt is to add fillers (qv) such as starches into the product.

Another approach is to replace petrochemical-based polymers with polymers made from carbohydrates. Unfortunately, approaches of this type have yet to produce economically competitive polymers.

LLOYD W. PEBSWORTH
Polyethylene Technology

R. A. V. Raff and K. W. Doak, *Crystalline Olefin Polymers,* John Wiley & Sons, Inc., New York, 1965, pp. 307,495,682.

Petrothene Polyolefins: A Processing Guide, 5th ed., Quantum Chemical Co., Cincinnati, Ohio, 1986.

F. Billmeyer, *Text Book of Polymer Science,* 3rd ed., Wiley-Interscience, New York, 1984.

H. Oosterwijk and H. Van Der Bend, *Akzo Chemie America Bulletin,* Initiations Seminar, 1980, New York, pp. 87–35.

HIGH DENSITY POLYETHYLENE

High density polyethylene (HDPE) is defined by ASTM D1248-84 as a product of ethylene polymerization with a density of 0.940 g/cm^3 or higher. This range includes both homopolymers of ethylene and its copolymers with small amounts of α-olefins. The first commercial processes for HDPE manufacture were developed in the early 1950s and utilized a variety of transition-metal polymerization catalysts based on molybdenum, chromium, and titanium. Commercial production of HDPE was started in 1956 in the United States by Phillips Petroleum Co. and in Europe by Hoechst. HDPE is one of the largest volume commodity plastics produced in the world. The term HDPE embraces a large variety of products differing predominantly in molecular weight, molecular weight distribution (MWD), and crystallinity.

Molecular Weight. The range of molecular weights of commercially produced HDPE is wide, from several hundreds for polyethylene (PE) waxes to several millions for ultrahigh molecular weight PE resins (UHMWPE). A parameter that is widely accepted, easily measured, and which provides information on molecular weight, is the rheological parameter called the melt index. Different HDPE resins have melt indexes ranging from over 500 (low molecular weight polymers) to less than 0.001.

Molecular Weight Distribution. The width of the molecular weight distribution (MWD) of PE resins is usually represented by the ratio of the weight-average and the number-average molecular weights, $\overline{M}_w/\overline{M}_n$, or by MFR value, which is the ratio of two melt indexes measured at two melt pressures that differ by a factor of 10. The range of MFR values for commercial HDPE resins is wide, from around 25 for injection molding resins with a narrow MWD, to over 150 for some HDPE film resins with a broad MWD.

Crystallinity and Density. Crystallinity and density of HDPE resins depend primarily on the extent of short-chain branching in polymer chains and, to a lesser degree, on molecular weight. The density range for HDPE resins is between 0.960 and 0.941 g/cm^3. UHMWPE is a completely nonbranched ethylene homopolymer, but, due to its very high molecular weight, it crystallizes poorly and has a density of 0.93 g/cm^3.

Molecular Structure and Chemical Properties

HDPE is a linear polymer with the chemical composition of polymethylene, $(CH_2)_n$. Depending on application, HDPE molecules either have no branches at all, as in certain injection molding and blow molding grades, or contain a small number of branches which are introduced by copolymerizing ethylene with α-olefins, eg, ethyl branches in the case of 1-butene and *n*-butyl branches in the case of 1-hexene. The number of branches in HDPE resins is low, at most 5 to 10 branches per 1000 carbon atoms in the chain.

HDPE is a saturated linear hydrocarbon and exhibits very low chemical reactivity. The most reactive parts of HDPE molecules are the double bonds at chain ends and tertiary CH bonds at branching points in polymer chains. Because its reactivity to most chemicals is reduced by high crystallinity and low permeability, HDPE does not react with organic acids, most inorganic acids, or with alkaline solutions.

At room temperature, HDPE is not soluble in any known solvent, but at a temperature above 80–100°C, most HDPE resins dissolve in some aromatic, aliphatic, and halogenated hydrocarbons.

HDPE is relatively stable under heat. Chemical reactions at high temperature in the absence of oxygen become noticeable only above 290–300°C. Thermocracking of HDPE is a free-radical C-C bond scission reaction.

At elevated temperatures, oxygen attacks HDPE molecules in a series of radical reactions. These reactions reduce the molecular weight of HDPE and introduce oxygen-containing groups, such as hydroxyl and carboxyl groups, into polymer chains. Other oxidation products are low molecular weight compounds such as water, aldehydes, ketones, and alcohols. Oxidative degradation in HDPE is initiated by impurities, which are mainly catalyst residues containing transition metals, eg, titanium and chromium. The protection from thermooxidative degradation is provided by antioxidants such as naphthylamines or phenylenediamines, hindered phenols, quinones, and alkyl phosphites, which are used in 0.1–1.0 wt % concentration.

Many commercial processes involving surface dyeing and printing (eg, on film and containers) employ thermooxidation as a pretreatment step. Dyes adhere poorly to HDPE surfaces but their adhesion can be improved by thermooxidation of the surface layer by treatment with an open flame or in a strong electric field.

Even though degradation of HDPE initiated by oxygen and light resembles thermooxidative degradation, it proceeds at a much lower temperature. Photooxidative degradation of HDPE causes aging, development of surface cracks, brittleness, change in color, and drastic deterioration of mechanical and dielectric properties. The reaction can be slowed down or prevented by utilizing light stabilizers that protect the resin and absorb uv radiation. Photooxidative degradation is the principal process responsible for gradual disintegration of discarded PE litter. The chemical industry manufactures a large number of antioxidants (qv) as well as uv stabilizers; their mixtures with other additives are used to facilitate resin processing.

Crystalline Structure and Physical Properties

HDPE is a semicrystalline plastic, whose crystallinity varies from 40 to 80%, depending on the degree of branching and molecular weight. Polymer chains in crystalline HDPE have a flat zigzag configuration. The principal crystalline form of HDPE is orthorhombic, with a density of 1.00 g/cm^3 and the cell parameters $a = 0.740$, $b = 0.493$, and $c = 0.2534$ nm. The polymer chains are aligned in the *c*-axis direction. HDPE crystallizes from the melt under typical conditions as densely packed morphological structures known as spherulites. Spherulites are small spherical objects (usually from 1 to 10 μm) visible only under high magnification. They are composed of even smaller structural subunits: rod-like fibrils that spread in all directions from the spherulite centers, filling the spherulite volume. These fibrils, in turn, are made up of the smallest morphological structures distinguishable, small planar crystallites called lamellae. Crystalline lamellae offer the spherulites rigidity and account for their high softening temperature, whereas the amorphous regions between lamellae provide flexibility and high impact strength to HDPE articles.

The extrapolated equilibrium melting point of orthorhombic HDPE crystals is 146–147°C. Actual measurements of slowly crystallized samples give the highest melting point, T_m, at 133–138°C.

Owing to the high crystallinity of HDPE, most articles made from HDPE resins are opaque. Thin HDPE film, in contrast, is translucent, but its transparency is significantly lower than that of LDPE or LLDPE film. The ultraviolet transmission limit of HDPE is around 230 nm.

Various properties of HDPE are listed in Table 1.

Catalysts for HDPE Production

HDPE resins are produced in industry with several classes of catalysts, ie, catalysts based on chromium oxides (Phillips), catalysts utilizing organochromium compounds, catalysts based on titanium or vanadium compounds (Ziegler), and metallocene catalysts.

Polymerization Processes and Processing

All technologies employed for catalytic polymerization processes in general are widely used for the manufacture of HDPE. The three most often used technologies are slurry polymerization, gas-phase polymerization and solution polymerization. Catalysts are usually fine-tuned for a particular process.

Most high density polyethylene processing technologies require the melting of HDPE. Typical HDPE melt viscosities are between 1,000 and 100,000 Pa·s(10,000 – 10^6 P); the melt viscosity of HDPE strongly depends on temperature and on the resin molecular weight. Some resins can have a viscosity 250 times greater than that of others. The effect of temperature on the HDPE melt viscosity is described by an exponential dependence similar to the Arrhenius equation with an activation energy of 25–29 kJ/mol (6–7 kcal/mol).

Because of its low melting point and high chemical stability, HDPE is easily processed by most conventional techniques (injection molding, blow molding, rotational molding, and extrusion). Blown HDPE film is manufactured on high stalk film lines; specialized techniques have also been described (see PLASTICS PROCESSING).

Because of high molecular weight and extremely high melt viscosity, UHMWPE cannot be readily processed by any technique involving melt extrusion or thermoforming. Instead, these resins are processed either by compression molding into sheet, block, and precision parts; by ram extrusion into board, rods, pipe, and profiles; or by forging into parts of complex configuration.

Recycling of HDPE. Polyolefins, including HDPE are the second most widely recycled thermoplastic materials after PET. A significant fraction of articles made from HDPE (mostly bottles, containers, and film) are collected from consumers, sorted, cleaned, and reprocessed. Processing of post-consumer HDPE includes the same operations as those used for virgin resins: blow molding, injection molding, and extrusion.

Specifications, Standards, and Quality Control

According to ASTM D1248, HDPE materials are divided into various classifications based on properties. Two of the most easily measured

Table 1. Physical, Thermal, Electrical, and Mechanical Properties of HDPE

Property	Highly linear	Low degree of branching[a]
	Physical	
density, g/cm^3	0.962–0.968	0.950–0.960
refractive index, n_D at 25°C	1.54	1.53
	Thermal	
melting point, °C	128–135	125–132
brittleness temperature, °C	−140 to −70	−140 to −70
heat resistance temperature, °C	~ 122	~ 120
specific heat capacity, kJ/(kg·K)[b]	1.67–1.88	1.88–2.09
thermal conductivity, W/(m·K)	0.46–0.52	0.42–0.44
temperature coefficient of linear expansion	$(1-1.5) \times 10^{-4}$	$(1-1.5) \times 10^{-4}$
of volume expansion	$(2-3) \times 10^{-4}$	$(2-3) \times 10^{-4}$
heat of combustion, kJ/g[b]	46	46
	Electrical	
dielectric constant at 1 MHz	2.3–2.4	2.2–2.4
dielectric loss angle, 1 kHz–1 MHz	$(2-4)\cdot 10^{-4}$	$(2-4)\cdot 10^{-4}$
volume resistivity, Ω· m	$10^{17}-10^{18}$	$10^{17}-10^{18}$
surface resistivity, Ω	10^{15}	10^{15}
dielectric strength, kV/mm	45–55	45–55
	Mechanical	
yield point, MPa[c]	28–40	25–35
tensile modulus, MPa[c]	900–1200	800–900
tensile strength, MPa[c]	25–45	20–40
notch impact strength, kJ/m^2[d]	~120	~150
flexural strength, MPa[c]	25–40	20–40
shear strength, MPa[c]	20–38	20–36
elongation, %		
at yield point	5–8	10–12
at break point	50–900	50–1200
hardness		
Brinell, MPa[c]	60–70	50–60
Rockwell	R55, D60–D70	

[a] 2–3 CH_3 per 1000 carbons.
[b] To convert J to cal, divide by 4.184.
[c] To convert MPa to psi, multiply by 145.
[d] To convert kJ/m^2 to ft·lbf/in.^{2}, divide by 2.10.

characteristics are density and melt index; the former determines the type of HDPE, the latter its category.

Health and Safety Factors

HDPE by itself is a safe plastic material on account of its chemical inertness and lack of toxicity. Film and containers made from HDPE are used on a large scale in food and drug packaging and HDPE has been used in prosthetic devices including hip and knee joint replacements. All these applications underscore polymer safety. If articles made of HDPE contain fillers, processing aids, and colorants, their toxic effects must be estimated separately.

HDPE can present health hazards when it burns. Heavy smoke, fumes, or potentially toxic decomposition products can result from incomplete combustion. Large-scale fire testing has shown that the products formed from HDPE present no greater hazard than those from cellulosic materials, wood, felt, or rubber.

A significant part of HDPE is collected from consumers for recycling; uncollected HDPE can be disposed of by landfill or incineration. In landfill, HDPE is completely inert, degrades very slowly, does not produce gas, and does not leach any pollutants into groundwater. When incinerated in commercial or municipal facilities, HDPE produces a large amount of heat (the same as heating fuel) and therefore should constitute less than 10% of the total trash.

Uses

Blow-molded products represent the biggest use of HDPE resins, at around 40%. Packaging applications account for by far the greatest share of this market. These include such products as bottles (especially for milk, juice, and soap), housewares, toys, pails, drums, and tanks.

Injection-molding products are the second largest application, with approximately 20% of the HDPE market. These products include housewares, toys, food containers, pails, crates, and cases.

Film is the third most important application for HDPE resins, accounting for nearly 15% of the total HDPE market. Its share of the market is increasing rapidly as it gradually replaces paper and glass and competes with LLDPE film for many uses.

HDPE pipes are used for transporting water, sewer wastes, and gas; they are also widely used in the chemical industry. Other significant applications include wire and cable coatings, foam, insulation for coaxial and communication cables, as well as those areas where high resistance to oil and chemicals is desirable.

Low molecular weight HDPE with molecular weight of several thousand (waxes) is widely used for paper coatings, spray coatings, emulsions, printing inks, crayons, and wax polishes. Waxes are also used as additives to butyl rubber and various higher molecular weight PE grades for improving melt flow characteristics, hardness, and resistance to abrasion and grease.

UHMWPE possesses a unique combination of mechanical and technological properties and enjoys a variety of special applications based on low friction (solid lubricant), wear resistance (protection of metal surfaces), excellent chemical stability, as well as radiation and neutron resistance. UHMWPE is used in chemical processing, food and beverage industries, foundries, the lumber industry; the electrical industry, as medical implants; and in mining and mineral processing, sewage treatment, and transportation.

YURY V. KISSIN
Mobil Chemical Company

H. V. Boenig, *Polyolefins: Structure and Properties,* Elsevier, Amsterdam, the Netherlands, 1966.

T. E. Nowlin, *Prog. Polym. Sci.* **11,** 29 (1985).

Y. V. Kissin, in N. P. Cheremisinoff, ed., *Encyclopedia of Engineering Materials,* Part A, Vol. I, Marcel Dekker, Inc., New York, 1988, p. 103.

LINEAR LOW DENSITY POLYETHYLENE

The chemical industry manufacturers a large variety of semicrystalline ethylene copolymers containing small amounts of α-olefins. These copolymers are produced in catalytic polymerization reactions and have densities lower than those of ethylene homopolymers known as high density polyethylene (HDPE). Ethylene copolymers produced in catalytic polymerization reactions are usually described as linear ethylene polymers, to distinguish them from ethylene polymers containing long branches which are produced in radical polymerization reactions at high pressures (see OLEFIN POLYMERS–LOW DENSITY POLYETHYLENE).

Densities and crystallinities of ethylene–α-olefin copolymers mostly depend on their composition. The classification in Table 1 is commonly used (ASTM D1248-48).

The large number of commodity and specialty resins collectively known as LLPDE are in fact made up of various resins, each different from the other in the type and content of α-olefin in the copolymer, compositional and branching uniformity, crystallinity and density, and molecular weight and molecular weight distribution (MWD).

Four olefins are used in industry to manufacture ethylene copolymers: 1-butene, 1-hexene, 4-methyl-1-pentene, and 1-octene. Copolymers containing 1-butene account for approximately 40% of all LLDPE resins manufactured worldwide, 1-hexene copolymers for 35%, 1-octene copolymers for about 20%, and 4-methyl-1-pentene copolymers for the rest. The type of α-olefin exerts a significant influence on the copolymer properties.

Table 1. Basic Classification of Copolymers

Resin	Designation	α-Olefin, mol %	Crystallinity, %	Density, g/cm^3
PE of medium density	MDPE	1–2	55–45	0.940–0.926
linear PE of low density	LLDPE	2.5–3.5	45–30	0.925–0.915
PE of very low density	VLDPE	>4	<25	<0.915

Two classes of LLDPE resins are on the market. One has a predominantly uniform compositional distribution (uniform branching distribution); that is, all copolymer molecules in these resins have approximately the same composition. Most commercially produced LLDPE resins, in contrast, have pronounced nonuniform branching distributions; there are significant differences in copolymer compositions among different macromolecules in a given resin.

Crystallinity and density of LLDPE, which are closely related, depend mostly on the amount of α-olefin in the copolymer. Both density and crystallinity of ethylene copolymers are also influenced by their compositional uniformity. As a rule, lower α-olefin content is needed in a uniformly branched ethylene copolymer to decrease its crystallinity and density to a given level.

The range of molecular weights of commercial LLDPE resins is relatively narrow, usually from 50,000 to 200,000. One accepted parameter that relates to the resin molecular weight is the melt index, a rheological parameter which, broadly defined, is inversely proportional to molecular weight. A typical melt index range for LLDPE resins is from 0.1 to 5.0, but can reach over 30 for some applications.

Molecular Structure and Properties

LLDPE resins are copolymers of ethylene and α-olefins with low α-olefin contents. Molecular chains of LLDPE contain units derived both from ethylene, —CH_2–CH_2—, and from the α-olefin, —CH_2–CHR—, where R is C_2H_5 for ethylene–1-butene copolymers, n-C_4H_9 for ethylene–1-hexene copolymers, —CH_2—$CH(CH_3)_2$ for ethylene–4-methyl-1-pentene copolymers, and n-C_6H_{13} for ethylene–1-octene copolymers. In a typical copolymer molecule containing 2–4 mol % of α-olefin, the majority of olefin units stand alone in the chain but most ethylene units form long sequences. As a rule, LLDPE resins do not contain long-chain branches. However, some copolymers produced with metallocene catalysts in solution processes can contain about 0.002 long-chain branches per 100 ethylene units.

LLDPE resins produced with different catalysts vary greatly in their compositional uniformity. The fastest way to evaluate the branching distribution of an LLDPE resin is to measure its melting point. Melting points of copolymers with uniform compositional distributions show a noticeable dependence on their composition and may vary widely: from ~120 °C for copolymers containing 1.5–2 mol % of α-olefin to ~110 °C for copolymers containing 3.5 mol % of α-olefin. A copolymer with a nonuniform compositional distribution, in contrast, is a mixture that contains copolymer molecules with a broad range of compositions, from almost linear macromolecules (usually of higher molecular weights) to short macromolecules with quite high α-olefin contents. Melting of such mixtures is dominated by their low branched fractions which are highly crystalline. They are not too sensitive to copolymer composition and usually fall in the temperature range of 125–128°C.

LLDPE is a saturated branched hydrocarbon. The most reactive parts of LLDPE molecules are the tertiary CH bonds in branches and the double bonds at chain ends. LLDPE is nonreactive with both inorganic and organic acids and is stable in alkaline and salt solutions. At room temperature, LLDPE resins are not soluble in any known solvent; at temperatures above 80–100°C, however, the resins can be dissolved in various aromatic, aliphatic, and halogenated hydrocarbons.

LLDPE is relatively stable to heat. Thermal degradation starts at temperatures above 250°C and results in a gradual decrease of molecular weight and the formation of double bonds in polymer chains.

Oxidation of LLDPE starts at temperatures above 150°C. To protect molten resins from oxygen attack, antioxidants must be used in concentrations of 0.1–0.5 wt %.

Photooxidative degradation of LLDPE at ambient temperature under sunlight is also a radical oxidation reaction. It causes change in color and drastic deterioration of mechanical and dielectric properties of LLDPE articles. Photooxidation can be prevented by using light stabilizers.

Physical Properties. LLDPE is a semicrystalline plastic whose chains contain long blocks of ethylene units that crystallize in the same fashion as paraffin waxes or HDPE. The degree of LLDPE crystallinity depends primarily on the α-olefin content in the copolymer and is usually below 40–45%. The principal crystalline form of LLDPE is orthorhombic (the same as in HDPE); the cell parameters of nonbranched PE are $a = 0.740$ nm, $b = 0.493$ nm, and c (the direction of polymer chains) $= 0.2534$ nm.

LLDPE rapidly crystallizes from the melt with the formation of spherulites, small spherical objects 1–5 μm in diameter visible only in a microscope. The elementary structural blocks in spherulites are lamellae, small flat crystallites formed by folded linear segments in LLDPE chains, which are interconnected by polymer chains that pass from one lamella to another (tie molecules). Crystalline lamellae within spherulites give LLDPE articles necessary rigidity, whereas the large amorphous regions between lamellae, constituting over 60% of the spherulite volume, provide flexibility.

The size of the lamellae for a copolymer of a given composition depends on the degree of branching uniformity. If an LLDPE resin is compositionally uniform, all its macromolecules crystallize poorly due to branching, forming very thin lamellae. Such materials have low rigidity (low modulus) and high flexibility. On the other hand, if an LLDPE resin is compositionally nonuniform, its least-branched components are able to form thicker lamellae; consequently, more branched fractions of the resin remain amorphous and fill the voids between the lamellae. Articles made from such resins are more rigid.

Optical properties of LLDPE resins also depend on the degree of branching uniformity. Resins with a uniform branching distribution make highly transparent film with haze as low as 3–4%. In contrast, film manufactured from compositionally nonuniform copolymers is much more opaque, with haze of over 10–15%; this is due to the presence of large crystalline lamellae consisting of nearly nonbranched PE chains.

Because it is a saturated aliphatic hydrocarbon, LLDPE does not conduct electricity, and so is widely used for wire and cable insulation. LLDPE is poorly permeable to water and inorganic gases and only slightly more so to organic compounds, whether liquid or gas.

Mechanical Properties. Mechanical characteristics of ethylene copolymers are functions of their structural characteristics, such as content and type of α-olefin, branching uniformity, molecular weight and width of molecular weight distribution (MWD), and orientation (see Table 2 for properties of films made from three grades of LLDPE).

Catalysts and LLDPE Polymerization Processes

LLDPE resins are produced in industry with three classes of catalysts titanium-based catalysts (Ziegler), metallocene-based catalysts (Kaminsky and Dow), and chromium oxide-based catalysts (Phillips).

Ziegler catalysts account for by far the greatest share of LLDPE resins manufactured. They consist of two components: the first contains as its active ingredient a derivative of a transition metal (usually titanium); the second is an organoaluminum compound such as triethylaluminum. The molar [Al]:[Ti] ratio in these catalyst systems is usually in the 50–500 range. Most commercially important catalysts are heterogeneous; a large number of them are supported. Both inorganic and organic supports are used, the most important are $MgCl_2$ and silica.

Table 2. Properties of Commercial LLDPE Film of Resins with Nonuniform Compositional Distribution[a].

Property	Copolymer of ethylene and:				
	1-butene		1-hexene		1-octene
density g/cm^3	0.918	0.918	0.918	0.918	0.919
melt index, g/10 min	2.0	1.0	1.0	0.5	1.0
dart impact strength, g[b]	110	150	250	300	350
puncture energy J/mm[c]	60	70	85	94	61
tensile strength,[d] MPa[e]					
MD	33	38	38	43	43
TD	25	31	32	43	34
elongation at break,[e] %					
MD	690	620	570		550
TD	740	760	790		660
modulus,[d] MPa[e]					
MD	210	230			
TD	250	260			

[a] Film thickness 37 μm

[b] Average weight of the dart sufficient to break the film in 50% of tests (ASTM D4272-90, Method A).

[c] To convert J/mm to ft·lbf/in., multiply by 18.73.

[d] MD = machine direction; TD = transverse direction.

[e] To convert MPa to psi, multiply by 145.

Typical heterogeneous Ziegler catalysts operate at temperatures of 70–100°C and pressures of 0.1–2 MPa (15–300 psi). The polymerization reactions are carried out in an inert liquid medium (eg, hexane, isobutane) or in the gas phase. Molecular weights of LLDPE resins are controlled by using hydrogen as a chain-transfer agent. Reactivities of α-olefins in copolymerization with ethylene depend on two factors: the size of the alkyl groups attached to their double bonds and the type of catalyst.

Three types of metallocene catalysts are presently used in industry: Kaminsky, ionic, and Dow constrained-geometry catalysts.

Chromium oxide-based catalysts, which were originally developed for the manufacture of HDPE resins, have been modified for ethylene–α-olefin copolymerization reactions. These catalysts use a mixed silica–titania support containing from 2 to 20 wt % of Ti.

Polymerization

The technologies suitable for LLDPE manufacture include gas-phase fluidized-bed polymerization, polymerization in solution, polymerization in a polymer melt under high ethylene pressure, and slurry polymerization. Most catalysts are fine-tuned for each particular process.

Processing

All LLDPE processing technologies involve resin melting; viscosities of typical LLDPE melts are between 5000 and 70,000 Pa·s (50,000–700,000 P). The main factor that affects melt viscosity is the resin molecular weight; the other factor is temperature. Its effect is described by the Arrhenius equation with an activation energy of 29–32 kJ/mol (7–7.5 kcal/mol).

LLDPE melts in the 140–250°C range (a typical processing range) are non-Newtonian liquids; their effective viscosity is significantly reduced when the melt flow speed is increased. This phenomenon is called shear-thinning; it plays an important role in resin processing. The resins with an expressed shear-thinning capability have decreased viscosities and hence a greatly reduced energy demand at high speed processing.

LLDPE is easily processed by most conventional techniques due to its low melting point and high chemical stability.

Most LLDPE produced worldwide is made into thin film, either melt blown or cast from the melt. Blown film is produced by extrusion of LLDPE melt through a circular die with a large diameter, up to 100–120 cm, and a narrow gap, usually less than 1 mm. The tube of molten polymer expands from internal air pressure and forms a bubble of a larger diameter. Typically, the blow-up ratio, the ratio of the final film tube diameter to the circular-die diameter, is between 1.5:1 and 4:1. LLDPE cast film is manufactured by depositing polymer melt on a rotating heated drum with a highly polished surface. Compositionally uniform LLDPE resins produced with metallocene catalysts can be easily processed into film by using standard equipment with minor modifications.

Injection molding is used for the manufacture of LLDPE articles of complex shapes. The duration of the molding cycle depends on the melt viscosity and the rate of polymer crystallization. Because LLDPE crystallizes rapidly, molding cycles are short, typically from 10 to 30 seconds. Molds usually accommodate up to 50 or more articles formed in a single shot. Bottles and simple containers are manufactured in large quantities by the blow molding technique. LLDPE resins with high molecular weights and high melt viscosities are used in this method. Large containers and some toys are manufactured from LLDPE powder with a specialized technique called rotational molding. Extrusion applications include resin pelletization after LLDPE synthesis, and manufacture of thick film, sheet, pipe, tubing, and insulated wire.

Health and Safety Factors

LLDPE by itself does not present any health-related hazard on account of its chemical inertness and low toxicity. Consequently, film, containers, and container lids made from LLDPE are used on a large scale in food and drug packaging. Some LLDPE grades produced with unsupported metallocene catalysts have an especially high purity due to high catalyst productivity and a low contamination level of resins with catalyst residue. FDA approved the use of film manufactured from these resins for food contact and for various medical applications. However, if LLDPE articles contain fillers, processing aids, or colorants, their health factors must then be judged separately.

LLDPE can present a certain health hazard when it burns, since smoke, fumes, and toxic decomposition products are sometimes formed in the process.

LLDPE can be disposed of by landfill or incineration. In landfill, the material is completely inert, degrades very slowly, does not produce gas, and does not leach any pollutants into ground water. When incinerated in commercial or municipal facilities, LLDPE produces a large amount of heat (the same as heating fuel) and should constitute less than 10% of the total trash.

Uses

By far the largest application for LLDPE resins (over 60% in the United States) is film. Because LLDPE film has high tensile strength and puncture resistance, it is able to compete with HDPE film for many uses. Bags manufactured from thin LLDPE film have excellent tensile strength, puncture resistance, and seal strength at thin gauges; they can be used either for packaging or as garment bags, laundry and dry-cleaning bags, and ice bags. Several LLDPE films are currently competing for a special film market: elastic stretch film for packaging. Both blown and die-cast film can be used for this application. The resins based on ethylene–1-hexene and ethylene–1-octene copolymers are particularly suited for these purposes. A significant volume of LLDPE film is used to manufacture large-size packaging material for food (eg, grocery sacks) and textiles, in addition to such applications as industrial sheeting and agricultural mulch film. LLDPE bags are thinner than LDPE bags; their thickness can be further reduced to around 25 μm, which makes them price-competitive with paper.

High oxygen-barrier properties of metallocene-derived resins make their film especially attractive for packaging poultry, frozen foods, and vegetables. Such properties, moreover, make these resins and their blends with HDPE suitable for producing film for blood bags, surgical disposable bags, and medical gowns.

An important property recommending the use of LLDPE in many packaging applications is their sealability. Compositionally uniform resins are especially attractive for such use because their melting and softening points are 15–20°C lower than those of commodity LLDPE resins.

Injection molding is the second largest market for LLDPE, accounting for over 10% of its total consumption. Over half of the LLDPE consumed in injection-molding applications is used for housewares.

LLDPE resins formulated for the blow molding applications have superior environmental stress-cracking resistance and low gas permeability. These features opened new bottle markets where such properties are important. A large variety of molded articles with a complex configuration are manufactured from LLDPE resins, including toys, large square-edged containers, as well as tanks for agriculture and water treatment.

The same qualities that make LLDPE attractive for blow molding applications also play a crucial role in its being adapted for pipe manufacture, an area that accounts for about 1% of the LLDPE market. LLDPE pipes provide flexibility, high burst strength, and high environmental stress-cracking resistance, and a high heat-distortion temperature. LLDPE tubing is used for drip piping, swimming pool tubing, household hoses, and in such specialty markets as medical tubing applications.

LLDPE is also widely used for wire and cable coating in electrical and telephone industry, which amounts to 2.5% of LLDPE production.

YURY V. KISSIN
Mobil Chemical Company

B. A. Krentsel, Y. V. Kissin, V. I. Kleiner, and L. L. Stotskaya, *Polymers and Copolymers of Higher α-Olefins,* Hanser Publishers, Germany, Munich, 1997, Chapt. 8.

T. E. Nowlin, *Prog. Polym. Sci.* **11,** 29 (1985).

Y. V. Kissin, in N. P. Cheremisinoff, ed., *Encyclopedia of Engineering Materials,* Part A, Vol. I, Marcel Dekker, Inc., New York, 1988, p. 103.

F. S. Dyachkovsky and A. D. Pomogailo, *J. Polym. Sci. Polym. Symp.* **68,** 97 (1980).

POLYPROPYLENE

Propylene polymerization processes have undergone a number of revolutionary changes since the first processes for the production of crystalline polypropylene (PP) were commercialized in 1957 by Montecatini in Italy and Hercules in the United States. These first processes were based on Natta's discovery in 1954 that a Ziegler catalyst could be used to produce highly isotactic polypropylene. The stereoregular, crystalline polymers produced by this technology had sufficiently attractive economic and property performance that they became significant commercial thermoplastics in a remarkably short period.

Properties

Structure and Crystallinity. The stereochemistry of propylene polymers was first studied by Natta, who defined three possible structures of polypropylene by the location of the pendent methyl groups relative to the polymer backbone. Isotactic polypropylene consists of molecules in which all methyl groups have the same stereochemistry as a result of all insertions of propylene monomer being identical. Syndiotactic polypropylene is produced by regular alternating stereochemistry of monomer insertion, resulting in alternating locations of the pendent methyl groups. Atactic polypropylene, which is noncrystalline, is the result of nonstereospecific monomer insertion and random location of the pendent methyl groups (Fig. 1).

Crystallinity of polypropylene is usually determined by x-ray diffraction. Isotactic polymer consists of helical molecules, with three monomer units per chain unit, resulting in a spacing between units of identical conformation of 0.65 nm.

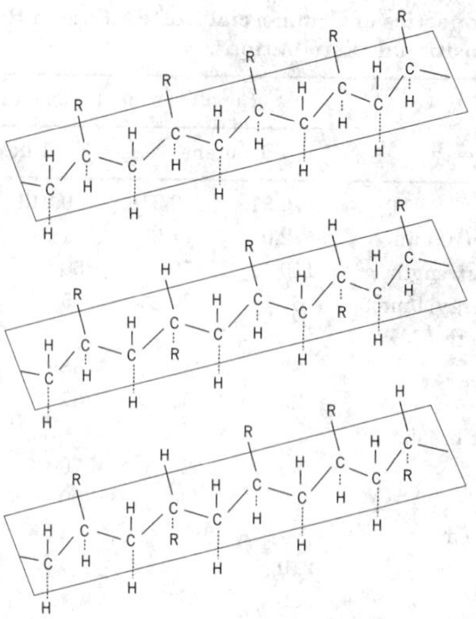

Figure 1. Polypropylene stereoisomers.

Syndiotactic polypropylene also forms helical molecules; however, each chain unit consists of four monomer units having a spacing of 0.74 nm. The unit cell is orthorhombic and contains 48 monomer units having a crystallographic density of 0.91 g/cm^3.

Molecular Weight. The molecular weight of polypropylene is typically determined by viscosity measurements. The melt viscosity, or melt flow rate, measured under standard conditions, can also be correlated to molecular weight. Because of the distribution of molecular weights in polypropylene, and the relationship of this distribution to important polymer properties, considerable effort has been exerted to measure the molecular weight distribution of these polymers.

Thermodynamic Properties. The thermodynamic melting point for pure crystalline isotactic polypropylene obtained by the extrapolation of melting data for isothermally crystallized polymer is 185°C. Under normal thermal analysis conditions, commercial homopolymers have melting points in the range of 160–165°C. The heat of fusion of isotactic polypropylene has been reported as 88 J/g (21 cal/g). The value of 165 ± 18 J/g has been reported for a 100% crystalline sample.

The value of the glass-transition temperature, T_g, is dependent on the stereoregularity of the polymer, its molecular weight, and the measurement techniques used. Transition temperatures from −13 to 0°C are reported for isotactic polypropylene, and −18 to 5°C for atactic.

Syndiotactic polypropylene has an ultimate melting point of 174°C, and extrapolated heat of fusion of 105 J/g (25.1 cal/g); both lower than those of isotactic polymer. The heat of fusion of the polymer produced using a metallocene catalyst is reported as 79 J/g (19 cal/g).

Physical Properties. Properties of various homopolymer grades are given in Table 1.

Polypropylene polymers are typically modified with ethylene to obtain desirable properties for specific applications. Specifically, ethylene–propylene rubbers are introduced as a discrete phase in heterophasic copolymers to improve toughness and low temperature impact resistance (see ELASTOMERS, SYNTHETIC-ETHYLENE-PROPYLENE-DIENE RUBBER).

Catalysts Used in Manufacture

$TiCl_3$-Based Catalysts. Isotactic polypropylene was first synthesized by Natta in 1954, using a catalyst system consisting of $TiCl_4$ and $Al(C_2H_5)_3$. This system, based on Ziegler's catalyst for polyethylene, produced a large fraction of polymer with poor structural uniformity and properties. These catalysts, activated with $Al(C_2H_5)_2Cl$ or $Al(C_2H_5)_3$, dramatically increased the percentage of isotactic polymer. Although all four crystal forms of $TiCl_3$, α, β, γ, and δ, are active as catalysts, the best results were obtained using δ-$TiCl_3$ activated by

Table 1. Properties of Homopolymers

Properties	ASTM test	Extrusion, sheet	General-purpose injection molding		Injection molding thin complex parts	
melt flow, g/10 min	D1238L	0.8	4	12	20	35
density, g/cm^3	D792A-2	0.903	0.903	0.903	0.903	0.902
tensile strength,[a] MPa[b]	D638	35	35	35	32	33
elongation,[a] %	D638	13	12	11	11	12
flexural modulus 1% secant, MPa[c]	D790B	1700	1700	1600	1500	1450
Rockwell hardness, R scale	D785A	95	99	100	100	98
deflection temperature at 455 kPa,[b] °C	D648	95	97	92	91	90
notched Izod impact at 23°C, J/m[d]	D256A	130	40	35	21	32

[a] At yield.
[b] To convert kPa to psi, multiply by 0.145.
[c] To convert MPa to psi, multiply by 145.
[d] To convert J/m to ft·lb/in., divide by 53.38.

$Al(C_2H_5)_2Cl$. These catalysts enabled rapid commercialization of the production of isotactic polypropylene.

$TiCl_3$ catalysts produced by the reduction of $TiCl_4$ with $Al(C_2H_5)_2Cl$, and subsequently treated first with an electron donor (diisoamyl ether), then with $TiCl_4$, are highly stereospecific and four to five times more active than δ-$TiCl_3$. These catalysts were a significant advance over the earlier $TiCl_3$ systems, because removal of atactic polymer was no longer required. They are often referred to as second-generation catalysts. The life of many older slurry process facilities has been extended by using these catalysts to produce "clean" polymers with very low catalyst residues.

$MgCl_2$-Supported Catalysts. Magnesium chloride, in active form as a support for $TiCl_4$, has been found to significantly increase catalyst activity, enabling the design of processes in which removal of catalyst residues from the polymer is not required. The use of electron donors as stereoregularity control agents reduces formation of the undesirable atactic polymer, giving stereoregularity sufficient for commercial production of polypropylene.

Active catalyst systems have also been produced using silica as a support for $MgCl_2$ and $TiCl_4$. A number of magnesium-containing compounds such as alkylmagnesium, magnesium hydrocarbyl carbonates, magnesium alkanoates, magnesium alkoxides and aryloxides, and alkyl magnesium chloride, can be used as the source of activated $MgCl_2$ in addition to the alcoholates of $MgCl_2$ or anhydrous $MgCl_2$. In all cases, these systems contain active centers consisting of $TiCl_4$ supported on activated $MgCl_2$.

Metallocene Catalysts. The use of the dicyclopentadienyl titanium dichloride—diethyl aluminum chloride system to catalyze the polymerization of ethylene homogeneously was first reported in 1957. Dramatic improvements in the performance of these metallocene catalysts were obtained by using methylalumoxane as cocatalyst and dicyclopentadienyl zirconium dichloride as catalyst. Modification of the zirconocene by the addition of substituents to the cyclopentadiene rings provided the capacity of achieving high molecular weight polyethylene at economical process temperatures. Modifications of the organic substituents have resulted in the development of metallocene catalysts capable of producing isotactic polypropylene with similar melting points and molecular weights to commercial Ziegler-Natta propylene.

Manufacturing Processes

Early Processes. The first commercial processes for the production of polypropylene were batch polymerization processes using $TiCl_3$ catalysts activated by $Al(C_2H_5)_2Cl$ in a hydrocarbon medium. As the demand for polypropylene increased, these batch polymerization processes were rapidly replaced by continuous ones.

Polymerization in liquid monomer was pioneered by Rexall Drug and Chemical and Phillips Petroleum (United States). Gas-phase polymerization of propylene was pioneered by BASF, who developed the Novolen process which uses stirred-bed reactors. Eastman Chemical has utilized a unique, high temperature solution process for propylene polymerization. In the 1970s, Solvay introduced an advanced $TiCl_3$ catalyst with high activity and stereoregularity.

Montedison and Mitsui Petrochemical introduced $MgCl_2$-supported high yield catalysts in 1975. These third-generation catalyst systems reduced the level of corrosive catalyst residues to the extent that neutralization or removal from the polymer was not required. Stereospecificity, however, was insufficient to eliminate the requirement for removal of the atactic polymer fraction. These catalysts are used in the Montedison high yield slurry process.

Current Processes. The development of superactive third-generation supported catalysts enabled the introduction of simplified processes, without sections for catalyst deactivation or removal of atactic polymer. By eliminating the waste streams associated with the neutralization of catalyst residues and purification of the recycled diluent and alcohol, these processes minimize any potential environmental impact. Investment costs are reduced by approximately one-third over slurry process plants. Energy consumption is minimized by elimination of the distillation of recycled diluent and alcohol. The total plant cost for the production of polymer is less than 130% of the monomer price, when a modern process is used, compared to 175% for a slurry process.

Processing

Polypropylene is produced in a variety of molecular weights, molecular weight distributions, and crystallinities; consequently, it can be used in most polymer processing technologies. The physical and mechanical properties of polypropylene in the end use product are a function of both the molecular structure and the processing conditions. The final crystalline morphology is a function of the melt temperatures, polymer orientation, and cooling temperatures and rates. The most commonly used processes are injection molding, blow molding, extrusion, and thermoforming (see PLASTICS PROCESSING).

Melt Spinning. This process is used to produce a broad range of polypropylene fibers ranging from fine, dtex (one denier) staple to coarse continuous filaments. Homopolymers are almost exclusively used to produce fibers, although copolymer blends are used in some special applications. Processing conditions and polymer melt flow vary with the desired fiber type.

Slit and Split Films. Thick industrial-grade yarns are often produced by slitting films, providing a less expensive alternative to melt spun fiber. Cast film is slit in the machine direction by parallel rotary knives. The resulting tape can then be cold drawn in an oven in a manner similar to melt spun fibers to produce the final fiber.

Melt Blowing. The melt blowing process uses very high melt flow polymers, sometimes in excess of 400 dg/min, and extrusion and die temperatures above 300°C to produce very fine fibers (<5 μm dia).

Spun Bonded Fabrics. Spun bonded fabrics are produced by depositing extruded, spun filaments onto a collecting belt in a uniform randomized manner. The fibers are separated during the web laying process by air jets and the collecting belt is usually perforated to prevent the air stream from deflecting and carrying the fibers in an uncontrolled manner. Spun bonded fabrics are used in a variety of applications requiring nonwoven fabrics (qv), competing with thermally bonded staple fiber.

Polypropylene (PP) films were first produced by extrusion casting. Polymer is extruded through a slit or tubular die and quenched by cooling on chill rolls or in a water bath. Cast films can be sealed over a wide range of temperatures and do not shrink in a steam autoclave. Polymers with melt flow rates below 5 dg/min are usually used to maintain the stability of the extrudate. Higher clarity films are produced using random copolymers.

Biaxially oriented polypropylene (BOPP) films have higher stiffness than cast films and consequently can be used in much thinner gauges. Homopolymers are used almost exclusively to provide maximum stiffness and water-vapor barrier. Oriented films are produced by the tenter frame and bubble processes.

Polypropylene is subject to attack by oxygen, radiation, and excessive heat causing a loss of molecular weight and physical properties. Stabilizers are added to the polymer to minimize these effects. Small quantities of hindered phenolic antioxidants (qv) are added in the polymerization plant, usually in the drying section, to protect the polymer against degradation during short-term storage. The bulk of the stabilizer is added during pelletization or fabrication to protect the polymer during processing or in the final application.

Uses

Polypropylene is extensively used in injection molding because of the wide range of physical properties and melt flow rates available. The principal markets served include transportation (primarily automotive), appliances, consumer products, rigid packaging, and medical products. Polypropylene use has increased in the automotive industry because of the wide availability of high melt flow rate impact copolymers for use in large thin parts, such as interior trim.

Polypropylene is frequently utilized in the design of container closures, lunchboxes, and similar articles. Polypropylene closures are used on a wide variety of containers, including child-proof caps and screw caps on plastic beverage bottles. Medical devices injection molded from polypropylene include syringes, pans, trays, and a variety of utensils. A large variety of toys, cups, dishes, and other household articles are molded from polypropylene. The development of high melt flow rate copolymer grades for thin-wall injection molding has increased the use of polypropylene in food containers.

Polypropylene blow-molded bottles are used to package a wide variety of products including foods, cleansers, shampoo, pharmaceuticals, and mouthwash.

Improved equipment and polymers have increased the capability to extrude and thermoform polypropylene. Drinking straws are commonly extruded from polypropylene, however most larger diameter tubes, such as pipes and conduits, are predominantly extruded from other thermoplastics. Extruded sheet is thermoformed into food containers and trays; polypropylene is used when microwavability is desired.

Polypropylene fibers are extensively used in carpeting.

Disposable polypropylene nonwoven fabrics are widely used as the cover-stock for disposable baby diapers, baby wipes, adult incontinence, and feminine hygiene products. Use of polypropylene nonwovens in disposable medical apparel, such as surgical gowns, has increased as a means of reducing the spread of infection.

Oriented polypropylene films are widely used in the packaging of snack foods, candy, and other products. The most common method of packaging using these films is the form and fill process in which the package is formed filled with product, and sealed in a continuous process.

Cast films provide a high clarity, heat sealable film and are primarily used as an overwrap for boxes and other packaging. These films have a lower density than cellophane and provide a longer product shelf life.

RICHARD B. LIEBERMAN
Montell Polyolefins

C. Maier and T. Calafut, *Polypropylene, The Definitive User's Guide and Databook,* Plastics Design Library, New York, 1998.

E. P. Moore, Jr., *Polypropylene Handbook,* Hanser Publishers, Munich, Germany, 1996.

S. Van der Ven, *Polypropylene and Other Polyolefins: Polymerization and Characterization,* Elsevier Science Publishers, B. V., Amsterdam, the Netherlands, 1990.

POLYMERS OF HIGHER OLEFINS

Crystalline polymers of α-olefins, ie, those with carbon numbers of four or higher, are stereoregular, ie, isotactic or syndiotactic polymers, such as polypropylene (PP). These polymers are produced with a number of different catalysts. Heterogeneous Ziegler-Natta catalysts and some soluble bridged metallocene catalysts produce isotactic polymers, other types of bridged metallocene catalysts produce crystalline syndiotactic polymers whereas nonbridged metallocene catalysts yield amorphous atactic polymers.

The synthesis of isotactic polymers of higher α-olefins was discovered in 1955, syndiotactic polymers of higher α-olefins were first prepared in 1990. The first commercial production of isotactic poly(1-butene) (PB) and poly(4-methyl-1-pentene) (PMP) started in 1965.

Higher α-olefins can also be polymerized with cationic initiators to liquid oligomeric materials with isomerized structures. These liquids are manufactured commercially and used as lubricating oils.

Cycloolefins are polymerized by means of two different mechanisms. In the first, catalysts based on tungsten and molybdenum compounds induce ring-opening polymerization (metathesis) of monocycloolefins with the formation of linear elastomers containing regularly spaced double bonds in polymer chains. If cyclodienes are used in this reaction, these catalysts then produce cross-linked resins. In the second mechanism, metallocene-based catalysts polymerize cycloolefins without ring opening into linear, stereoregular, highly crystalline polymers. Polymers of several cycloolefins, polydicyclopentadiene, polyoctenamers, and norbornene elastomers are all produced commercially.

Monomers

The monomers of the greatest interest are those produced by oligomerization of ethylene (qv) and propylene (qv). Some olefins are also available as by-products from refining of petroleum products or as the products of hydrocarbon (qv) thermal cracking.

Commercial production of 1-butene, as well as the manufacture of other linear α-olefins with even carbon atom numbers, is based on the ethylene oligomerization reaction.

4-Methyl-1-Pentene is produced commercially by dimerization of propylene in the presence of potassium-based catalysts at 150–160°C and ~10 MPa.

Linear α-olefins, such as 1-hexene and 1-octene, are produced by catalytic oligomerization of ethylene with triethylaluminum or with nickel-based catalysts (see OLEFINS, HIGHER). Olefins with branched alkyl groups are usually produced by catalytic dehydration of corresponding alcohols.

1-Butene is a colorless, flammable, noncorrosive gas. Because 1-butene has a very low flash point, it poses a strong fire and explosion hazard.

4-Methyl-1-pentene is a light, colorless, flammable liquid. It is an irritant and, in high concentrations, a narcotic. Like 1-butene, this chemical compound has a low flash point and represents a significant fire hazard when exposed to heat, flame, or oxidizing agents.

Higher α-olefins are exceedingly reactive in radical and ionic reactions. These olefins participate in numerous reactions, such as oxidations, hydrogenation, double-bond isomerization, complex formation with transition-metal derivatives, polymerization, and copolymerization with other olefins in the presence of Ziegler-Natta, metallocene, and cationic catalysts. All olefins readily form peroxides by exposure to air.

Polymer Properties

Chemical properties of most polyolefins resemble those of polypropylene. The resins resist most inorganic or organic acids and bases below 90°C as well as most salt solutions, solvents, soaps, and detergents. Properties of polyolefins rapidly deteriorate in contact with strong oxidizing agents. All polyolefins undergo peroxidation, halogenation, and halosulfonation reactions.

Commercially produced crystalline polyolefins, PB, and PMP exhibit high stability to inorganic substances, and excellent resistance in nonoxidative inorganic environments. PMP easily withstands prolonged boiling and autoclave treatment required for medical and pharmaceutical applications. Prolonged exposure of polyolefin specimens under stress to some hydrocarbon solvents and aqueous detergent solutions can cause cracks and eventual failure, a phenomenon usually referred to as environmental stress cracking.

Polymers of α-olefins are susceptible to thermal and thermooxidative degradation. Reactivity in degradation reactions is especially significant in the case of polyolefins with branched alkyl side groups.

Thermooxidative degradation of PMP is noticeable even at 140–150°C, and its photooxidative degradation, especially at wavelengths below 400 nm, also proceeds at a relatively high rate, thus limiting some of its outdoor applications.

Both thermooxidation and photooxidation of polyolefins can be prevented by using the same antioxidants as those employed for the stabilization of polypropylene, ie, alkylated phenols, polyphenols, thioesters, and organic phosphites in the amount of 0.2–0.5%.

Polybutene can be cross-linked by irradiation at ambient temperature with γ-rays or high energy electrons in the absence of air. PMP is relatively stable to β- and γ-radiation employed in the sterilization of medical supplies.

Highly crystalline isotactic polyolefins are not soluble in organic solvents at room temperature. However, most amorphous polyolefins and oligomers of α-olefins are easily soluble in saturated and aromatic hydrocarbons at ambient temperature. This difference in solubility can be used to separate amorphous atactic components of polyolefins from crystalline isotactic material in crude polyolefins mixtures.

Above 100°C, most polyolefins dissolve in various aliphatic and aromatic hydrocarbons and their halogenated derivatives.

Crystalline PMP is relatively highly permeable to various organic and inorganic gases. Permeabilities to oxygen, nitrogen, and light hydrocarbons are 20–30 times higher than those of HDPE.

Physical Properties. Table 1 lists physical properties of stereoregular polymers of several higher α-olefins.

Thermal Properties. Melting points of stereoregular PO resins depend on the size and shape of side groups in the polymer chains (Table 1). In the case of isotactic polymers of linear α-olefins, the melting points of the crystalline phase rapidly decrease with increasing side-chain length, isotactic poly(1-hexene) is amorphous, and isotactic polyolefins with longer linear side groups derive their crystallinity from the side chains rather than from the polymer backbone. Polymers of α-olefins with branched alkyl groups generally exhibit much higher melting points (Table 1); the melting points of isotactic poly(3-methyl-1-butene) and poly(vinylcyclohexane) are the highest, over 350°C.

Mechanical Properties. The side-group type in the polymer chains determines mechanical properties of stereoregular polyolefins. Resins with long linear side groups have low crystallinities and exhibit mechanical behavior typical for elastomers. On the other hand, PB and most polymers with branched side groups are highly crystalline and exhibit mechanical properties similar or superior to those of isotactic PP. Excellent mechanical and optical properties contribute to the industrial importance of PB and PMP. Although poly(3-methyl-1-butene) and poly(vinylcyclohexane) also exhibit good mechanical characteristics, their high melting points, poor oxidative stability, and brittleness still preclude them from finding industrial application.

All polyolefins have low dielectric constants and can be used as insulators; in particular, PMP has the lowest dielectric constant among all synthetic resins. As a result, PMP has excellent dielectric properties and a low dielectric loss factor, surpassing those of other polyolefin resins and polytetrafluoroethylene (Teflon). These properties remain nearly constant over a wide temperature range. The dielectric characteristics of poly(vinylcyclohexane) are especially attractive: its dielectric loss remains constant between −180 and 160°C, which makes it a prospective high frequency dielectric material of high thermal stability.

Table 1. Properties of Polyolefins

Polymer[a]	Melting point, °C	Crystal type	Helix type	Crystalline density, g/cm^3
Polymers of α-olefins				
iso-polybutene				
form I	138–142	hexagonal	3_1	0.951
form II	120–130	tetragonal	11_3	0.902
form III	101–109	orthorhombic		0.905
syndio-polybutene				
form I	~50		4_1	
form II	~50		10_3	
iso-poly(1-pentene)	105–115	monoclinic	3_1	0.92
	75–80	pseudo orthorhombic	4_1	0.90
iso-poly(3-methyl-1-butene)	350	monoclinic	4_1	0.93
iso-poly(1-hexene)	<20	monoclinic	7_2	0.83
iso-poly(3-methyl-1-pentene)	200		7_2	
iso-poly(4-methyl-1-pentene)	235–240	tetragonal	7_2	0.813
syndio-poly(4-methyl-1-pentene)	197		24_7	
iso-poly(1-heptene)	18		3_1	
iso-poly(4-methyl-1-hexene)	188–200	tetragonal	7_2	0.845
syndio-poly(4-methyl-1-hexene)	147			
iso-poly(5-methyl-1-hexene)	110–130	monoclinic	3_1	0.84
iso-poly(1-octene)	~20			
iso-poly(5-methyl-1-heptene)	130	tetragonal	3_1	
iso-poly(vinylcyclohexane)	376–385	tetragonal	4_1	0.95
syndio-poly(vinylcyclohexane)		amorphous		
iso-poly(1-decene)	22–27	side chain		
Polymers of cycloolefins				
diiso-polycyclobutene	485			
diiso-polycyclopentene	395			
polynorbornene	>600			

[a] *iso* designates isotactic; *syndio* designates syndiotactic.

Although most polyolefins are highly opaque, isotactic PMP possesses the outstanding feature: it has low haze (1.2–1.5%) and high optical transparency (~90–92%) comparable to that of polystyrene (88–92%) and acrylics (90–92%). Light transmittance of PMP in the near uv region is also excellent, higher than that of glass and inferior only to quartz. Optical clarity accounts for many applications of PMP.

Polymers of monocyclic olefins (cyclopentene, cyclooctene) produced by ring-opening metathesis are linear elastomers. Their properties are somewhat similar to those of poly(*cis*-1,4-butadiene). Polymers of dicyclopentadiene produced with the same catalysts are heavily cross-linked resins displaying high toughness and tensile strength as well as excellent impact strength at low temperatures (see CYCLOPENTADIENE AND DICYCLOPENTADIENE).

Polymerization of α-Olefins

Ziegler-Natta Catalysts. All isotactic polymers of higher α-olefins are produced with the same type of heterogeneous, titanium-based

Ziegler-Natta catalyst systems as that used for the manufacture of isotactic PP. The catalyst systems have two components, a solid catalyst containing a titanium compound and a co-catalyst containing an organoaluminum compound. Both 1-butene and 4-methyl-1-pentene are three to four times less reactive in the polymerization reactions with Ziegler-Natta catalysts than propylene. Modern, highly active catalysts are supported on $MgCl_2$. Some contain aromatic esters such as ethyl benzoate. These catalysts are employed with co-catalyst mixtures containing $Al(C_2H_5)_3$ or $Al(i\text{-}C_4H_9)_3$ and aromatic esters, ethyl benzoate or ethyl anisate. Another highly active type of the supported catalysts uses aromatic diesters (phthalates) and mixtures of $Al(C_2H_5)_3$ and phenylalkoxysilanes as co-catalysts.

Polymerization reactions with Ziegler-Natta catalysts are carried out at 40–80°C in pure monomers or in monomer mixtures with aliphatic solvents. Molecular weights of polymers are controlled by the addition of hydrogen, an effective chain-transfer agent. Some Ziegler-Natta catalysts polymerize linear α-olefins, such as 1-hexane or 1-decene, into linear polymers with ultrahigh molecular weights which are used as drag-reducing agents for hydrocarbon flow.

All higher α-olefins, in the presence of Ziegler-Natta catalysts, can easily copolymerize both with other α-olefins and with ethylene. In these reactions, higher α-olefins are all less reactive than ethylene and propylene.

Isotactic PB and PMP are produced commercially in slurry processes in liquid monomers or monomer mixtures (optionally diluted with light inert hydrocarbons) at 50–70°C.

Metallocene Catalysts. Higher α-olefins can be polymerized with catalyst systems containing metallocene complexes. The first catalysts of this type (Kaminsky catalysts) include metallocene complexes of zirconium such as biscyclopentadienylzirconium dichloride, activated by methylaluminoxane. These catalysts polymerize α-olefins with the formation of amorphous atactic polymers. Polymers with high molecular weights are produced at decreased temperatures and have rubber-like properties.

Zirconocene complexes containing two indenyl or tetrahydroindenyl groups bridged with short links such as —CH_2—CH_2— or —$Si(CH_3)$— produce isotactic polymers of higher α-olefins. To synthesize syndiotactic PO, bridged zirconocene complexes with rings of two different types are required, one example of which is isopropyl(cyclopentadienyl)(1-fluorenyl)zirconocene. These complexes are used for the synthesis of syndiotactic PB, PMP, and poly(4-methyl-1-hexene).

Cationic Polymerization Reactions. α-Olefins with linear and branched alkyl groups can be readily polymerized with cationic initiators. Olefins containing linear alkyl groups (1-pentene, 1-hexene, 1-octene, 1-decene, etc) and their mixtures are oligomerized by using BF_3, mixtures of BF_3 and alcohols, as well as $AlCl_3$ or $AlBr_3$-HBr systems at low temperatures with the formation of low molecular weight oils of an irregular structure. These oligomers are used as base stocks for synthetic lubricating oils.

Polymerization of Cycloolefins

Depending on the type of catalyst used, polymerization of cycloolefins proceeds through either ring opening or by opening of the double bond with the preservation of the ring.

Ring-Opening Polymerization. Ring-opening polymerization of cycloolefins in the presence of tungsten- or molybdenum-based catalysts proceeds by a metathesis mechanism.

Dicyclopentadiene is also polymerized with tungsten-based catalysts. Because the polymerization reaction produces heavily cross-linked resins, the polymers are manufactured in a reaction injection molding (RIM) process, in which all catalyst components and resin modifiers are slurried in two batches of the monomer. The first batch contains the catalyst (a mixture of WCl_6 and $WOCl_4$), additives, and fillers; the second batch contains the co-catalyst (a combination of an alkylaluminum compound and a Lewis base such as ether), antioxidants, and elastomeric fillers. Mixing two liquids in a mold results in a rapid polymerization reaction.

Metallocene Catalysts. Polymerization of cycloolefins with combinations of metallocenes and methylaluminoxane produces polymers with a completely different structure. The reactions proceeds via the double-bond opening in cycloolefins. If the metallocene complexes contain bridged and substituted cyclopentadienyl rings, such as ethylene(bisindenyl)zirconium dichloride, the polymers are stereoregular and have the *cis*-diisotactic structure.

Processing

Both PB and PMP melts exhibit strong non-Newtonian behavior: their apparent melt viscosity decreases with an increase in shear stress. Melt viscosities of both resins depend on temperature. Equipment used for PP processing is usually suitable for PB and PMP processing as well; however, adjustments in the processing conditions must be made to account for the differences in melt temperatures and rheology.

Extrusion. The main applications of this method include the production of film, sheet, pipe, and tubing. PB is usually extruded by using the same equipment (single- or twin-screw extruders) as that used for PP and HDPE, at melt and die temperatures of 170–190°C. PMP is processed on extruders with a high length-to-diameter ratio at temperatures of 240–300°C.

Injection molding of PB is carried out under conditions similar to those for PP at 145–190°C. Injection molding of PMP is carried out at melt temperatures of 260–330°C, mold temperatures of 30–80°C, injection pressures of around 30 MPa (300 atm), and at relatively low injection rates.

Film. The blown film process is most commonly used in the production of PB film from resins with melt indexes from 0.3 to 10 g/10 min at a melt temperature of 200–215°C using conventional equipment. Mechanical properties of blown PB film depend on the degree of orientation and other processing parameters. PB film can be sealed at 160–220°C. Another technique for the PB film production consists of film casting from the melt on polished chilled rolls and co-extrusion or lamination with other films.

Health and Safety Factors

Polymers and higher α-olefins are not toxic; their main potential health hazards are associated with residual monomer, antioxidants, and catalyst residues. In particular, PB and PMP are inert materials and usually present no health hazard. PMP is employed extensively for a number of medical and food packaging applications.

Uses

Polybutene. The largest share of commercially produced crystalline PB is used for manufacturing pipe and tubing. The advantages of PB in pipe applications include high flexibility, toughness, and high resistance to creep, environmental stress-cracking, wet abrasion, and various chemicals. Pipes manufactured from PB retain their properties at temperatures up to 85°C. PB pipe is used in residential and commercial hot- and cold-water plumbing (including chlorine-containing hot water), water wells, water manifolds, and fire sprinklers. Black-pigmented pipe grades are suitable for outdoor use.

Blown film manufactured from PB has a high tensile strength and exhibits good resistance to tear, impact, and puncture.

Poly(4-methyl-1-pentene). Most PMP applications capitalize on the resin's high optical transparency, excellent dielectric characteristics, high thermal stability, and good chemical resistance. The manufacture of medical equipment comprises about 40% of PMP production, including such articles as hypodermic syringes, needle hubs, blood collection and transfusion equipment, pacemaker parts, blood analysis cells, and respiration equipment. It is also used in chemical and biomedical laboratory equipment, eg, cells for spectroscopic and optical analysis, laboratory ware, and animal cages, and in a variety of injection-molded articles, such as caps for enclosures, ink cartridges for printers, light covers, tableware, and sight-glasses. PMP is suitable also for microwave oven cookware and service, and is used in food packaging (qv). In many applications, PMP replaces stainless steel trays. PMP is also utilized for wire and cable coating, as well as for film and paper coatings with good release properties.

Synthetic Lubricating Oils. Liquid oligomers of higher linear α-olefins such as 1-decene are produced with cationic initiators. They are the most versatile of all synthetic lubricants. They exhibit not only good lubricating properties over a wide temperature range (including excellent low temperature properties), but also high frictional and oxidative stability as well as low volatility, and are miscible with all mineral oils and most synthetic lubricants. They have found wide application as synthetic base oils in the formulation of various lubricants, including lubricating oils for cars, transformer oils, transmission and crankcase fluids, hydraulic fluids, and compressor oils (see LUBRICATION AND LUBRICANTS).

Polymers of Cycloolefins. Polyoctenamer elastomers are processed by extrusion, injection molding, and calendering into hoses, rubber coatings, and tire components. They are mostly used as components in rubber-, PVC-, and PS-based compositions.

Cured liquid-molding resins based on polydicyclopentadiene are uses in the manufacture of automotive parts for trucks, snowmobiles, wheel loaders, recreational vehicles, and also in other areas that require toughness and good all-weather impact resistance.

YURY V. KISSIN
Mobil Chemical Company

B. A. Krentsel, Y. V. Kissin, V. I. Kleiner, and L. L. Stotskaya, *Polymers and Copolymers of Higher α-Olefins,* Hanser Publishers, Munich, Germany, 1997.

I. D. Rubin, *Poly(1-Butene): Its Preparation and Properties,* Gordon & Breach, New York, 1968.

R. L. Shubkin, *Synthetic Lubricants and High Performance Functional Fluids,* Marcel Dekker, Inc., New York, 1993.

V. Dragutan, A. T. Balaban, and M. Dimonie, *Olefin Metathesis and Ring-Opening Polymerization of Cycloolefins,* John Wiley & Sons, Ltd., Chichester, U.K., 1985.

OLEFINS, HIGHER

Higher olefins are versatile chemical intermediates for a number of important industrial and consumer products, providing a better standard of living with low environmental impact (qv) in many commercial uses. These uses can be characterized by carbon number and by chemical structure.

The even-numbered carbon alpha olefins (α-olefins) from C_4 through C_{30} are especially useful. For example, the C_4, C_6, and C_8 olefins impart tear resistance and other desirable properties to linear low and high density polyethylene; the C_6, C_8, and C_{10} compounds offer special properties to plasticizers used in flexible poly(vinyl chloride). Linear C_{10} olefins and others provide premium value synthetic lubricants; linear C_{12}, C_{14}, and C_{16} olefins are used in household detergents and sanitizers. In addition, many carbon numbers from C_4 to C_{30+} are also utilized in specialty applications such as sizing agents to produce longer-lasting paper.

The C_6–C_{11} branched, odd and even, linear and internal olefins are used to produce improved flexible poly(vinyl chloride) plastics.

Physical Properties

For a listing of selected physical properties of linear alpha olefins, see Table 1.

Table 1. Properties of C_4 to C_{20} Linear 1-Olefins

Compound	Mol wt	Density, g/mL 20°C	Viscosity, mm²/s(= cSt) 20°C	Viscosity, mm²/s(= cSt) 100°C	Free energy of formation, kJ/mol[a]
1-butene	56.11	0.6012			72.09
1-pentene	70.13	0.6402	0.202		78.67
1-hexene	84.16	0.67317	0.39		87.61
1-heptene	98.19	0.69698	0.50		96.02
1-octene	112.2	0.71492	0.656	0.363	104.4
1-nonene	126.2	0.72922	0.851	0.427	112.8
1-decene	140.3	0.74081	1.09	0.502	121.3
1-undecene	154.3	0.75032	1.38	0.587	129.6
1-dodecene	168.3	0.75836	1.72	0.678	138.0
1-tridecene	182.3	0.7653	2.14	0.782	146.4
1-tetradecene	196.4	0.7713	2.61	0.894	154.8
1-pentadecene	210.4	0.7765	3.19	1.019	163.3
1-hexadecene	224.4	0.78112	3.83	1.152	171.7
1-heptadecene	238.4	0.7852	4.60	1.30	180.1
1-octadecene	252.5	0.7888	5.47	1.46	188.5
1-nonadecene	266.5	0.7920 f[b]		1.63	196.9
1-eicosene	280.5	0.7950 f[b]		1.82	205.3

[a] To convert kJ/mol to kcal/mol, divide by 4.184.
[b] f = frozen.

Chemical Properties

The general reactivity of higher α-olefins is similar to that observed for the lower olefins. However, heavier α-olefins have low solubility in polar solvents such as water; consequently, in reaction systems requiring the addition of polar reagents, apparent reactivity and degree of conversion may be adversely affected. Reactions of α-olefins typically involve the carbon-carbon double bond and can be grouped into two classes: (*1*) electrophilic or free-radical additions; and (*2*) substitution reactions.

Commercial Olefin Reactions. Some of the more common transformations involving α-olefins in industrial processes include the oxo reaction (hydroformylation), oligomerization and polymerization, alkylation reactions, hydrobromination, sulfation and sulfonation, and oxidation.

Commercial α-Olefin Manufacture

Most linear α-olefins are produced from ethylene.

Ethylene oligomerization can be accomplished in the following commercial processes: (*1*) stoichiometric chain growth on aluminum alkyls followed by displacement (Albemarle); (*2*) catalytic chain growth on aluminum alkyls (Chevron-Gulf); (*3*) catalytic chain growth using a nickel ligand catalyst (Shell); and (*4*) catalytic chain growth using a modified zirconium catalyst (Idemitsu). In the Albemarle (formerly Ethyl) process, stoichiometric quantities of aluminum alkyls are used with subsequent displacement of α-olefins from the aluminum, followed by separation of the α-olefins from the aluminum alkyls. In the Chevron-Gulf process, catalytic amounts of aluminum alkyl are used. The operating temperatures are higher than those in the stoichiometric process, thus favoring displacement reactions after a finite amount of chain growth. In the Shell process, a three-phase system is employed, which gives a high linearity at higher carbon numbers. A nickel ligand catalyst dissolved in a solvent forms one liquid phase, the produced olefins form a second liquid phase, and the ethylene forms a third. Once formed, the olefins usually do not engage in further reactions because most of them are not in contact with the catalyst. Shell practices isomerization and disproportionation to produce a narrow range of internal linear olefins for feed to their oxo-alcohol unit. In the Idemitsu process, a zirconium oligomerization catalyst is modified by adding an aluminum alkyl and a Lewis base or an alcohol in a solvent. Variations in the catalyst mix thus offer a variety of carbon-number distributions, some of which resemble those in the catalytic processes, others approaching those in the stoichiometric process. Although operating at lower pressures than the other ethylene-based oligomerizations, the Idemitsu process still produces high quality linear α-olefins.

Vista has offered for license a stoichiometric process, which has not yet been commercialized, although the related primary alcohol process has been described (see ALCOHOLS, HIGHER ALIPHATIC–SYNTHETIC PROCESSES).

One-Step Ziegler Process. Gulf Research and Development Corp. developed the one-step Ziegler process. This process is now owned by Chevron, which has two plants at Cedar Bayou, Texas. Plants based on licensing the technology are operated by Mitsubishi in Japan and at Neratovice in the Czech Republic. Chevron further improved the process around 1990 by reducing paraffin impurities.

Uses

The principal outlets for higher olefins are in the polymer, surfactant, and detergent industries (see also ALCOHOLS, HIGHER ALIPHATIC–SYNTHETIC PROCESSES). Generally, higher olefins are seldom incorporated directly into a product as an ingredient; rather, they are processed through at least one chemical reaction step before appearing in a finished product.

Polymers. The manufacture of alcohols from higher olefins via the oxo process for use in plasticizers is a significant outlet for both linear α-olefins and branched olefins such as heptenes, nonenes, and dodecenes. These olefins are converted into alcohols containing one more carbon number than the original olefin. The alcohols then react with dibasic anhydrides or acids to form PVC plasticizers. The plasticizers produced from the linear olefins have superior volatility and cold-weather flexibility characteristics, making them an ideal product to use in flexible PVC for automobile interiors.

Detergents. The detergent industry consumes a large quantity of α-olefins through a variety of processes. Higher olefins used to produce detergent actives typically contain 10–16 carbon atoms because they have the desired hydrophobic and hydrophilic properties.

Lubricants. Lubricants represent a significant and growing outlet for higher olefins. Both basestocks and lube additives are produced from higher olefins by a variety of processes (see LUBRICATION AND LUBRICANTS).

Other Uses. A small but growing outlet for C_{16} and higher linear olefins is the production of alkenylsuccinic anhydride (ASA) for the paper industry. ASA is an effective alkaline sizing agent and competes with alkylketene dimer (AKD) in this application.

Additional uses for higher olefins include the production of epoxides for subsequent conversion into surface-active agents, alkylation of benzene to produce drag-flow reducers, alkylation of phenol to produce antioxidants, oligomerization to produce synthetic waxes (qv), and the production of linear mercaptans for use in agricultural chemicals and polymer stabilizers. Aluminum alkyls can be produced from α-olefins either by direct hydroalumination or by transalkylation. In addition, a number of heavy olefin streams and olefin or paraffin streams have been sulfated or sulfonated and used in the leather (qv) industry.

Health and Safety Factors

Toxicological Information. The toxicity of the higher olefins is considered to be virtually the same as that of the homologous paraffin compounds. Based on this analogy, the suggested maximum allowable concentration in air is 500 ppm. Animal toxicity studies for hexene, octene, decene, and dodecene have shown little or no toxic effect except under severe inhalation conditions.

Handling. The main hazard associated with these olefins, especially the lighter homologues, is their low flash point. Although no special precautions are necessary with regard to fire extinguishing, these olefin products should be stored and shipped under an inert atmosphere to maintain product purity.

G. R. LAPPIN
L. H. NEMEC
J. D. SAUER
J. D. WAGNER
Albemarle Corp.

G. R. Lappin and J. D. Sauer, eds., *Alpha Olefins Applications Handbook,* Marcel Dekker, Inc., New York, 1989.

C. S. Read, R. Wilhalm, and Y. Yoshida, *SRI Chemical Economics Handbook: Linear Alpha Olefins,* SRI International, Menlo Park, Calif., Oct. 1993.

N. B. Godrej and co-workers, *Proceedings from* "Alpha Olefins from Oleochemical Raw Materials," *The Third World Detergent Conference,* Montreux, Switzerland, Sept. 1993.

Technical data, UOP Inc. Division of AlliedSignal, Des Plaines, Ill., 1966 to 1993.

OLIGOSACCHARIDES. See ANTIBIOTICS, OLIGOSACCHARIDES.

OPIATES. See ALKALOIDS; ANALGESICS, ANTIPYRETICS, AND ANTI-INFLAMMATORY AGENTS; HYPNOTICS, SEDATIVES, ANTICONVULSANTS, AND ANXIOLYTICS.

OPIOIDS, ENDOGENOUS

Decades of research in opioid analgesia culminated in the discovery of the endogenous opioid peptides (see ANALGESICS, ANTIPYRETICS, AND ANTIINFLAMMATORY AGENTS; NEUROREGULATORS). Early studies of the structure–activity relationships of opiate alkaloids (qv) had provided evidence of the stereospecificity and antagonist reversibility of opiate action, suggesting that these drugs acted through specific receptors.

The presence of specific opioid receptors in the vertebrate central nervous system suggested the existence of endogenous ligands for these receptors, a hypothesis which received considerable support from the finding that electrical stimulation of specific sites in the rat brain elicited profound analgesia. This stimulation-produced analgesia was naloxone reversible, and was subject to tolerance development and to cross-tolerance to morphine. These results were most readily explained by the electrically induced release of endogenous substances having morphinelike properties.

Evidence emerged that the endogenous opioids were peptides rather than simple morphine-like molecules.

The three principal classes of endogenous opioid peptides share one common characteristic: the pentapeptide structure of enkephalin, either the Met- or Leu-derivatives. Loss of any portion of that structure significantly reduces the affinity of β-endorphin, dynorphin, or enkephalin in binding to opioid receptors. (see Table 1).

Receptors for Opioid Peptides

Multiple Opioid Receptors. The concept of multiple opioid receptors was first postulated in 1976. Three distinct opioid receptors

Table 1. Structures of Endogenous Opioid Peptides

Compound	Structure
	Pro-opiomelanocortin-derived
β-endorphin	H-Tyr-Gly-Gly-Phe-Met-Thr-Ser-Glu-Lys-Ser-Gln-Thr-Pro-Leu-Val-Thr-Leu-Phe-Lys-Asn-Ala-Ile-Val-Lys-Asn-Ala-His-Lys-Lys-Gly-Gln-OH
	Pro-enkephalin-derived
Leu-enkephalin	H-Tyr-Gly-Gly-Phe-Leu-OH
Met-enkephalin	H-Tyr-Gly-Gly-Phe-Met-OH
octapeptide	H-Tyr-Gly-Gly-Phe-Met-Arg-Gly-Leu-OH
heptapeptide	H-Tyr-Gly-Gly-Phe-Met-Arg-Phe-OH
	Pro-dynorphin-derived
dynorphin A	H-Tyr-Gly-Gly-Phe-Leu-Arg-Arg-Ile-Arg-Pro-Lys-Leu-Lys-Trp-Asp-Asn-Gln-OH
dynorphin B	H-Tyr-Gly-Gly-Phe-Leu-Arg-Arg-Gln-Phe-Lys-Val-Val-Thr-OH
α-neoendorphin	H-Tyr-Gly-Gly-Phe-Leu-Arg-Lys-Tyr-Pro-Lys-OH

were postulated: mu (μ), kappa (κ), and sigma (ς). A fourth type of opioid receptor, the delta (δ) receptor, was postulated in 1977 after discovery of the endogenous opioid peptides. Originally, the prototype agonists for these receptors were morphine (μ), ketazocine (κ), *N*-allylnormetazocine (ς), and Met- and Leu-enkephalin (δ), although more selective compounds for each receptor type are available. The ς-receptor is no longer thought to be a receptor for the endogenous opioids. The classification of opioid receptor types is primarily based on the specific affinities displayed by various opioid drugs and peptides in radioligand-binding assays and on the potency of these compounds to inhibit smooth muscle contractions, or to block opioid inhibition. Confirmation of the original discoveries of multiple opioid receptor types is being obtained by molecular cloning studies.

Opioid Peptide Analogues Receptor Affinities. In an effort to develop nonaddictive and nontolerance-producing opioid analgesics numerous metabolically stable enkephalin analogues have been synthesized (see PSYCHOPHARMACOLOGICAL AGENTS). The most successful stability-enhancing techniques have included the replacement of naturally occurring L-amino acids with the D-isomer and amidation of the carboxyl terminal residue, to form compounds such as D-Ala^2,Met^5-enkephalinamide. These derivatives show little promise as nonaddictive analgesics, because they share the tolerance and dependence liabilities of the endogenous opioids. However, many enkephalin analogues show remarkable receptor selectivity compared to the naturally occurring peptides. The principal design strategies for analogues include (*1*) substitution, addition, or deletion of amino acid residues; (*2*) introduction of conformational restrictions; and (*3*) modification of peptide bonds. Compared to the native enkephalins, the modified peptide analogues can display increased receptor selectivity for one of three reasons: (*1*) decreased affinity for other sites along with unchanged affinity for the target site; (*2*) increased affinity for the target site with no change in affinity for other sites; or (*3*) a combination of the above.

Receptor Structure and Function. All of the known opioid receptor types belong to the superfamily of G protein-coupled receptors. These receptors reside on the plasma membrane and affect cell physiology by interacting with the signal-transducing guanosine triphosphate (GTP)-binding regulatory proteins (G proteins). In most cells, opioid receptors are coupled to G_i and G_o, a class of G proteins that are adenosine diphosphate (ADP)-ribosylated by pertussis toxin. It is the G protein, rather than the receptor itself, that determines which effector(s), an enzyme or ion channel, are affected by receptor activation. The effector activity can be stimulated or inhibited by the receptor, depending on the G protein involved.

Biological Activities

Soon after the identification of endogenous opioid peptides, studies were conducted to determine their contribution to physiological function. Morphine was a well-established analgesic drug with central actions mediated by an endogenous anatomical substrate. Further studies on analgesia demonstrated that β-endorphin was more potent than morphine in eliciting analgesia in a variety of species, including the human. Moreover, evidence accumulated to implicate endogenous opioid peptides in mediating stimulation-produced analgesia, especially when stimulation was applied to the periaqueductal gray (PAG), a region rich in opioid peptides and receptors.

In contrast to the potent, long-lasting analgesic effects of β-endorphin, the enkephalins are extremely weak analgesics in laboratory tests. This difference is likely a result of the relatively short (2–3 min) biological half-life of the enkephalins vs the long (2–3 h) half-life of β-endorphin.

The finding of analgesic activity for the endogenous opioids created a renewed but short-lived hope that these or related peptides might lead to an analgesic devoid of dependence liability. However, Met-enkephalin and β-endorphin produce symptoms of physical dependence and evidence of tolerance and morphine cross-tolerance in animals and in vitro. Furthermore, β-endorphin and the enkephalins are reinforcing stimuli in behavioral experiments. The effects of these peptides may be mediated in part by disinhibition of mesolimbic dopaminergic neurons, which have been implicated in mediating the reinforcing effects of morphine. Thus, the evidence indicates that the opioid peptides, including at least β-endorphin and the enkephalins, are similar to the opiate alkaloids in their reinforcing properties as well as in their ability to produce tolerance and dependence.

Although many studies have focused on the analgesic effects of opioids, the endogenous opioid peptides have been found to influence a wide range of physiological functions. Opioid peptides and receptors are found in brain areas that influence respiratory and cardiovascular function. One aspect of opioid function that has received a great deal of interest is the effect of endogenous opioid systems on immune function. Both β-endorphin and Met-enkephalin enhance the cytotoxicity of natural killer cells in a manner that is inhibited by naloxone. In contrast, the C-terminal fragment of β-endorphin reduces the activity of natural killer cells; however, this activity is not affected by naloxone. Endogenous opioid peptides may also influence reproductive behavior. POMC mRNA levels are also decreased by both estrogen and testosterone. In contrast, estradiol has been shown to increase proenkephalin mRNA levels in the hypothalamus in a manner that coincided with the display of lordosis. Another hypothalamic action of opioid peptides is thermoregulation. Hyperthermia occurs after the injection of a μ-agonist, whereas dynorphin decreases temperature by decreasing metabolic rate.

Metabolic Inactivation of Opioid Peptides

Several enzymes, none of which are completely specific for the enkephalins, are known to cleave Leu- and Met-enkephalin at various peptide bonds. The main enzymes that degrade enkephalin are zinc metallopeptidases. The first enkephalin-degrading enzyme to be identified, an aminopeptidase which cleaves the amino terminal Tyr-Gly bond, has been shown to be aminopeptidase-N (APN). It is a cytoplasmic enzyme which is uniformly distributed throughout the brain. The increased analgesic activity of synthetic enkephalins substituted by D-amino acids at position 2, eg, [D-Ala^2]-Met-enkephalin, is probably the result of increased stability toward this aminopeptidase. A second enkephalin-degrading enzyme, enkephalinase B, is a dipeptidylaminopeptidase (DAP) which cleaves the Gly-Gly bond of enkephalin. This membrane-bound enzyme has the least overall enkephalin-degrading activity in crude brain homogenates and is uniformly distributed throughout the brain.

Intensive research efforts have focused on the discovery of potent and specific inhibitors of the enkephalin-degrading enzymes for novel analgesic agents. A potent and specific NEP inhibitor (K_i of 4.7 nM), thiorphan, produced analgesia on its own and potentiated analgesia elicited by enkephalin analogues. Subsequently, a number of modifications were made in order to increase the selectivity and bioavailability of thiorphan. Other classes of thiol-based inhibitors, such as the *N*-mercaptoacetyldipeptides, also show high potencies as NEP inhibitors. Another important class of enkephalinase inhibitors is the N-protected amino acid hydroxamates.

APN inhibitors include substituted aminoethanols and phenyalanine-based compounds. In addition to their promise as analgesic agents, mixed peptidase inhibitors with specificity for NEP and ACE have been found to possess antihypertensive activity (see CARDIOVASCULAR AGENTS).

Endogenous Opiate Alkaloids

Although the opioid peptides have long been identified as the primary endogenous opioid ligands in brain, several groups have identified the opiate alkaloid morphine and related compounds in the tissues of several species. A nonpeptide opioid has been isolated from toad skin in sufficient quantity for purification and has the same profile as morphine in high performance liquid chromatography (hplc), gas chromatography/mass spectrometry, radioimmunoassay, opiate receptor binding assay and bioassay (see ANALYTICAL METHODS; CHROMATOGRAPY; IMMUNOASSAY; MASS SPECTROMETRY). A nonpeptide opioid was also identified in bovine brain and adrenal gland, as well as rabbit and rat skin, that corresponded to morphine in hplc analysis. However, the concentration of the compound in these tissues was too low for further purification. Morphine and codeine have been identified in bovine hypothalamus and adrenal gland, as well as rat brain, and the presence of 6-acetylmorphine has been demonstrated in the bovine brain. This latter compound is a metabolite of heroin that had not previously been identified in plants or animals. The potential biological importance of 6-acetylmorphine is that it readily enters the central nervous system, where it is then converted to morphine.

DANA E. SELLEY
LAURA J. SIM
STEVEN R. CHILDERS
Wake Forest University

A. Herz, ed., *Opioids I*, Springer-Verlag, New York, 1993.

J. M. Van Ree, A. H. Mulder, V. M. Weigant, and T. B. Van Wimersma Greidanus, eds., *New Leads in Opioid Research,* Excerpta Medica, Amsterdam, the Netherlands, 1990.

G. W. Pasternak, ed., *The Opiate Receptors,* Humana Press, Newark, N.J., 1988.

S. R. Childers, in A. Herz, ed., *Opioids I,* Springer-Verlag (Berlin), 1993, p. 189.

OPTICAL BLEACHES. See FLUORESCENT WHITENING AGENTS.

ORGANOLEPTIC TESTING. See FLAVOR CHARACTERIZATION; ODOR MODIFICATION; PERFUMES.

OSMIUM. See PLATINUM-GROUP METALS.

OSMOSIS AND OSMOTIC PRESSURE. See HOLLOW-FIBER MEMBRANES; MEMBRANE TECHNOLOGY; REVERSE OSMOSIS.

OXALIC ACID

Oxalic acid, HOOC–COOH, or ethanedioic acid, mol wt 90.04, is the simplest dicarboxylic acid. It is soluble in water, and acts as a strong acid. This acid does not exist in anhydrous form in nature and is available commercially as a solid dihydrate, $C_2H_2O_4{\cdot}2H_2O$, mol wt 126.07. The commercial product is packed in polyethylene-lined paper bags or flexible containers. Anhydrous oxalic acid can be efficiently prepared from the dihydrate by azeotropic distillation in a low boiling solvent that can form a water azeotrope, such as benzene and toluene.

Oxalic acid was synthesized for the first time in 1776 by Scheele through the oxidation of sugar with nitric acid. Then, Wöhler synthesized it by the hydrolysis of cyanogen in 1824.

The potassium or calcium salt form of oxalic acid is distributed widely in the plant kingdom. Oxalic acid is found in spinach, rhubarb, etc. Oxalic acid is a product of metabolism of fungi or bacteria and also occurs in human and animal urine; the calcium salt is a principal constituent of kidney stones.

Oxalic acid is used in various industrial areas, such as textile manufacture and processing, metal surface treatments (qv), leather tanning, cobalt production, and separation and recovery of rare-earth elements. Substantial quantities of oxalic acid are also consumed in the production of agrochemicals, pharmaceuticals, and other chemical derivatives.

Physical Properties

The physical and thermochemical constants of anhydrous oxalic acid and oxalic acid dihydrate are summarized in Table 1.

Table 1. Physical and Thermochemical Properties of Oxalic Acid and its Dihydrate

Property	Value
Oxalic acid, anhydrous, $C_2H_2O_4$	
melting point, °C	
α	189.5
β	182
density d_4^{17}, g/mL	
α	1.900
β	1.895
refractive index, β, n_4^{20}	1.540
vapor pressure (solid, 57–107°C), kPa[a]	$\log_{10} P = -(4726.95/T) + 11.3478$
specific heat (solid, −200 to 50°C), J/g	$C_p{}^b = 1.084 + 0.0318\ t$
heat of combustion, ΔE_c (at 25°C), kJ/mol[c]	−245.61
standard heat of formation, ΔH_f (at 25°C), kJ/mol[c]	−826.78
standard free energy of formation, ΔG_f (at 25°C), kJ/mol[c]	−697.91
heat of solution (in water), kJ/mol[c]	−9.58
heat of sublimation, kJ/mol[c]	90.58
heat of decomposition, kJ/mol[c]	826.78
specific entropy, S (at 25°C), J/(mol·K)[c]	120.08
logarithm of equilibrium constant, $\log_{10} K_f$	122.28
thermal conductivity (at 0°C), W/(m·K)[d]	0.9
ionization constant	
K_1	6.5×10^{-2}
K_2	6.0×10^{-5}
coefficient of expansion (at 25°C), nL/(g·K)	178.4
Oxalic acid dihydrate, $C_2H_2O_42H_2O$	
mp, °C	101.5
density d_4^{20}, g/mL	1.653
refractive index, n_4^{20}	1.475
standard heat of formation, ΔH_f (at 18°C), kJ/mol[c]	−1422
heat of solution (in water), kJ/mol[c]	−35.5
pH (0.1 M soln)	1.3

[a] To convert $\log_{10} P_{kPa}$ to $\log_{10} P_{mm\ Hg}$, add 0.875097 to the constant, T = K.
[b] To convert C_p, J/g, to C_p, cal/g, divide both terms of the equation by 4.184.
[c] To convert J to cal, divide by 4.184.
[d] To convert W/(m·K) to (Btu·in.)/(h·ft^2·°F), divide by 0.1441.

Reactions

The reactions of oxalic acid, including the formation of normal and acid salts and esters, are typical of the dicarboxylic acids class. Oxalic acid, however, does not form an anhydride.

On rapid heating, oxalic acid decomposes to formic acid, carbon monoxide, carbon dioxide, and water (qv). In aqueous solution, it is decomposed by uv, x-ray, or γ-radiation with the liberation of carbon dioxide. Photodecomposition also occurs in the presence of uranyl salts.

Oxalic acid is a mild reducing agent, and is oxidized by potassium permanganate in acid solution to give carbon dioxide and water. Oxalic acid is catalytically reduced by hydrogen in the presence of ruthenium catalyst to ethylene glycol, and electronically reduced to glyoxylic acid.

Oxalic acid reacts with various metals to form metal salts, which are quite important as the derivatives of oxalic acid. It also reacts easily with alcohols to give esters.

Manufacture

Many industrial processes have been employed for the manufacture of oxalic acid since it was first synthesized. The following processes are in use worldwide: oxidation of carbohydrates, the ethylene glycol process, the propylene process, the dialkyl oxalate process, and the sodium formate process. Sodium formate process is no longer economical in the leading industrial countries, except for China.

Nitric acid oxidation is used where carbohydrates, ethylene glycol, and propylene are the starting materials. The dialkyl oxalate process is the newest, where dialkyl oxalate is synthesized from carbon monoxide and alcohol, then hydrolyzed to oxalic acid. This process has been developed by UBE Industries in Japan.

Many attempts have been made to synthesize oxalic acid by electrochemical reduction of carbon dioxide in either aqueous or nonaqueous electrolytes.

Health and Safety Factors

Oxalic acid is caustic and corrosive to humans. The severity of symptoms associated with oxalic acid poisoning is related to the concentration and quantity ingested. Oxalic acid removes calcium in the blood, forming calcium oxalate, and severe damage to the kidney may occur because of the insoluble calcium oxalate.

Uses

Because rare-earth oxalates have low solubility in acidic solutions, oxalic acid is used for the separation and recovery of rare-earth elements. The oxalic acid process for anodizing aluminum was developed in Japan. In addition to oxalic acid, inorganic oxalate salts are also used in coloring anodic coatings (qv). Oxalic acid is a constituent of cleaners that are used for automotive radiators, boilers, and steel plates before phosphating. As a chelating agent, oxalic acid forms water-soluble complexes on metal surfaces during cleaning and rinsing.

In pulp bleaching, oxalic acid serves as a bleaching agent, but is often used together with other bleaching agents (qv) because of its relatively high cost. Oxalic acid is also used for the bleaching of cork, wood (particularly veneered wood), straw, cane, and natural waxes.

Oxalic acid has various uses in fabric cleaning, application of dyestuff, and modifying properties of cellulose fabrics. Oxalic acid is used as a pH modifier in leather tanning by tannin and basic chromium sulfate. It also functions as a bleaching agent for leather (qv). It is used for marble polishing especially in Italy. It not only removes iron veins by forming water-soluble iron oxalate, but also serves as a polishing auxiliary. Starch powder is heated together with oxalic acid and hydrolyzed to produce millet jelly. Oxalic acid functions as a hydrolysis catalyst, and is removed from the product as calcium oxalate. This application is carried out in Japan. Oxalic acid is also used for the production of cobalt, as a raw material of various agrochemicals, and pharmaceuticals, for the manufacture of electronic materials, for the extraction of tungsten from ore, for the production of metal catalysts, as a polymerization initiator, and for the manufacture of zirconium and beryllium oxide.

Derivatives

Oxalic acid forms neutral and acid salts, as well as complex salts.

Ammonium Oxalate. Anhydrous ammonium oxalate is obtained when the monohydrate is dehydrated at 65°C. It is used for textiles, leather tanning, and precipitation of rare-earth elements.

Ammonium Iron(III) Oxalate. This mixed salt is produced as an emerald-green crystalline trihydrate. The compound is not stable to light. It was once used extensively in the manufacture of blueprinting papers.

Potassium Hydrogen Oxalate. Potassium acid oxalate, exists as a monohydrate. It is of historical interest because it is the salt of sorrel found in vegetation and the first oxalate isolated.

Potassium Oxalate. The monohydrate is produced as a colorless crystalline material or a white powder. The anhydrous salt is obtained when the monohydrate is dehydrated at 160°C. The monohydrate is preferred as a reagent in analytical chemistry and in miscellaneous uses principally because of its high solubility as compared with other simple neutral oxalates; the saturated solution, at 0°C, contains about 20 wt %, and at 20°C, about 25 wt % $K_2C_2O_4$.

Sodium Oxalate. This salt is obtained in such high purity and is so stable that it is used as a titrimetric standard.

Calcium Oxalate. The monohydrate is of importance principally as an intermediate in oxalic acid manufacture and in analytical chemistry; it is the form in which calcium is frequently quantitatively isolated.

Nickel Oxalate. This salt is produced as a greenish white crystalline dihydrate. Nickel oxalate is used for the production of nickel catalysts and magnetic materials.

Yttrium Oxalate. This compound exists as a trihydrate, nonahydrate, or heptadecahydrate. The compound is used for the production of a red fluorescent material for color television.

Dialkyl Oxalates. Oxalic acid gives various esters. Dialkyl esters, ROOC–COOR, are industrially useful, but monoalkyl esters, ROOC–COOH, are not. The dialkyl esters are characterized by good solvent properties and serve as starting materials in the synthesis of many organic compounds, such as pharmaceuticals, agrochemicals, and fine chemicals (qv). Among the diesters, dimethyl, diethyl, and di-*n*-butyl oxalates are industrially important.

Oxamide. This diamide is sparingly soluble in water and insoluble in various organic solvents. It melts at about 350°C, with accompanying decomposition. Because of the low solubility in water, the compound is granulated and used as a slow-release nitrogen fertilizer. Conventional nitrogen fertilizers (qv), such as ammonium sulfate, urea, ammonium nitrate, and ammonium phosphate, are soluble in water, and thus are easily lost as run-off when it rains. On the contrary, oxamide stays in the soil longer. Therefore, it is gradually decomposed by microorganisms in the soil and utilized by plants for longer periods.

Oxalyl Chloride. This diacid chloride is produced by the reaction of anhydrous oxalic acid and phosphorus pentachloride. The compound vigorously reacts with water, alcohols, and amines, and is employed for the synthesis of agrochemicals, pharmaceuticals, and fine chemicals.

Reduction Products. Glyoxylic acid is produced as aqueous solution by the electrolytic reduction of oxalic acid. It is used for the manufacture of vanillin.

Glycolic acid can be obtained by the electrolytic reduction of oxalic acid or the catalytic reduction of oxalic acid with hydrogen in the presence of a ruthenium catalyst. Because of its acidity it is used as a cleaning agent for metal surface treatments and for boiler cleaning. It also serves as an ingredient in cosmetics (qv).

HIROYUKI SAWADA
TORU MURAKAMI
UBE Industries, Ltd.

U.S. Pat. 2,057,119 (1936), G. S. Simpson (to General Chemical (Allied Chemical)).

Jpn. Pat. 61-26977-B (1986), S. Tahara and co-workers (to UBE Industries).

L. A. Sarver and P. H. M. P. Briton, *J. Am. Chem. Soc.* **49,** 943 (1927).

S. Werneck and R. Pinner, *The Surface Treatment and Finishing of Aluminum and Its Alloys,* Robert Draper, Ltd., Teddington, U.K., 1972.

OXO PROCESS

The oxo process, also known as hydroformylation, is the reaction of carbon monoxide (qv) and hydrogen (qv) with an olefinic substrate to form isometric aldehydes (qv) as shown in equation 1. The ratio of isomeric aldehydes depends on the olefin, the catalyst, and the reaction conditions.

$$RCH{=}CH_2 + CO + H_2 \xrightarrow{catalyst} RCH_2CH_2CHO + R(CH_3)CHCHO \quad (1)$$

If a double-bond shift occurs, the number of aldehyde isomers is increased.

Synthesis gas, a mixture of CO and H_2, also known as syngas, is produced for the oxo process by partial oxidation (eq. 2) or steam reforming (eq. 3) of a carbonaceous feedstock, typically methane or naphtha. The ratio of CO to H_2 may be adjusted by cofeeding carbon dioxide (qv), CO_2, as illustrated in equation 4, the water gas shift reaction.

$$2\ CH_4 + O_2 \rightarrow 2\ CO + 4\ H_2 \tag{2}$$

$$CH_4 + H_2O \rightarrow CO + 3\ H_2 \tag{3}$$

$$CO_2 + H_2 \rightleftarrows CO + H_2O \tag{4}$$

$$2\ CH_4 + CO_2 + O_2 \rightarrow 3\ CO + 3\ H_2 + H_2O \tag{5}$$

The overall process for producing a 1:1 CO to H_2 ratio by partial methane oxidation and the water gas shift reaction is represented by equation 5.

The oxo reaction proceeds most frequently in the presence of a Group 8–10 (VIII) metal catalyst in the liquid phase, most particularly with members of Group 9, the Co-Rh-Ir triad. The earliest catalyst, hydrocobalt tetracarbonyl, $HCo(CO)_4$, was an outgrowth of Fischer-Tropsch investigations carried out prior to World War II on the effect of olefins on hydrocarbon synthesis. The hydroformylation reaction, as practiced in the early days using cobalt catalysis, presented formidable requirements of high pressure, containment of the hydrogen, containment of carbon monoxide, and handling of the toxic and unstable metal carbonyls.

The search for catalyst systems which could effect the oxo reaction under milder conditions and produce higher yields of the desired aldehyde resulted in processes utilizing rhodium. Oxo capacity built since the mid-1970s, both in the United States and elsewhere, has largely employed tertiary phosphine-modified rhodium catalysts.

Propylene (qv) is the predominant oxo process olefin feedstock. Ethylene (qv), as well as a wide variety of terminal, internal, and mixed olefin streams, are also hydroformylated commercially. Branched-chain olefins include octenes, nonenes, and dodecenes from fractionation of oligomers of C_3–C_4 olefins as well as octenes from dimerization and codimerization of isobutylene and 1- and 2-butenes (see BUTYLENES).

Linear terminal olefins are the most reactive in conventional cobalt hydroformylation.

Oxo aldehyde products range from C_3 to C_{15}, ie, detergent range, and are employed principally as intermediates to alcohols, acids, polyols, and esters formed by the appropriate reduction, oxidation, or condensation chemistry.

The classic challenges in oxo technology are simultaneously to achieve high reaction rate, high selectivity to the desired aldehyde, and to utilize a highly stable catalyst.

Catalysts

Unmodified Cobalt. Typical sources of the soluble cobalt catalyst include cobalt alkanoates, cobalt soaps, and cobalt hydroxide (see COBALT COMPOUNDS). These are converted *in situ* into the active catalyst, $HCo(CO)_4$, which is in equilibrium with dicobalt octacarbonyl.

Although largely supplanted by low pressure ligand-modified rhodium-catalyzed processes, the unmodified cobalt oxo process is still employed in some instances for propylene to give a low, eg, ~3.3–3.5 : 1 isomer ratio product mix, and for low reactivity mixed and/or branched-olefin feedstocks, eg, propylene trimers from the polygas reaction, to produce isodecanol plasticizer alcohol.

Ligand-Modified Cobalt. The ligand-modified cobalt process, commercialized in the early 1960s by Shell, may employ a trialkylphosphine-substituted cobalt carbonyl catalyst, $HCo(CO)_3P(n\text{-}C_4H_9)_3$, to give a significantly improved selectivity to straight-chain product. There has been large industrial usage of the Shell process since the 1960s, particularly for the preparation of detergent range alcohols (see ALCOHOLS, HIGHER ALIPHATIC). 2-Ethyl-1-hexanol can be produced in a single step from propylene by conducting the hydroformylation in the presence of caustic.

Ligand-Modified Rhodium. The triphenylphosphine-modified rhodium oxo process, termed the LP Oxo process, is the industry standard for the hydroformylation of ethylene and propylene as of this writing (ca 1995). It employs a triphenylphosphine (TPP) modified rhodium catalyst. The process operates at low (0.7–3 MPa (100–450 psi)) pressures and low (80–120°C) temperatures. Suitable sources of rhodium are the alkanoate, 2,4-pentanedionate, or nitrate. A low (60–80 kPa (8.7–11.6 psi)) CO partial pressure and high (10–12%) TPP concentration are critical to obtaining a high (eg, 10:1) normal-to-branched aldehyde ratio.

The first commercial LP Oxo process flow scheme (Fig. 1) used syngas and propylene feed.

Rhodium Modified with Ionic Phosphine Ligands. In 1984, a rhodium catalyst process employing a water-soluble ligand, triphenylphosphine-*m*-trisulfonic acid trisodium salt (TPPTS) was commercialized. Product recovery is achieved by decantation from the aqueous phase containing rhodium and ligand. An isomer ratio of 20:1 is obtained with the TPPTS-modified rhodium catalyst, but the catalyst activity is significantly lower, so higher temperatures, higher rhodium concentrations, and higher propylene pressures are employed.

Other Rhodium Processes. Unmodified rhodium catalysts, eg, $Rh_4(CO)_{12}$, have high hydroformylation activity but low selectivity to normal aldehydes.

Functional Olefin Hydroformylation. There are two commercially practiced oxo processes employing functionalized olefin feedstocks: allyl alcohol hydroformylation, and the production of 1,4-butanediol by successive hydroformylation of allyl alcohol aqueous extraction of the intermediate 2-hydroxytetrahydrofuran, and subsequent hydrogenation.

Hydroformylation Using Other Metals. Ruthenium, as a hydroformylation catalyst, has an activity significantly lower than that of rhodium and even cobalt.

Platinum catalysts that utilize both phosphine and tin(II) halide ligands give good rates and selectivities, in contrast to platinum alone, which has extremely low or nonexistent hydroformylation activity.

A further improvement in platinum catalysis is claimed from use of tin(II) halide and phosphine ligands which are rigid bidentates, eg, 1,2-bis(diphenylphosphinomethyl)cyclobutane.

Future Trends. In addition to the commercialization of newer extraction/decantation product/catalyst separations technology, there

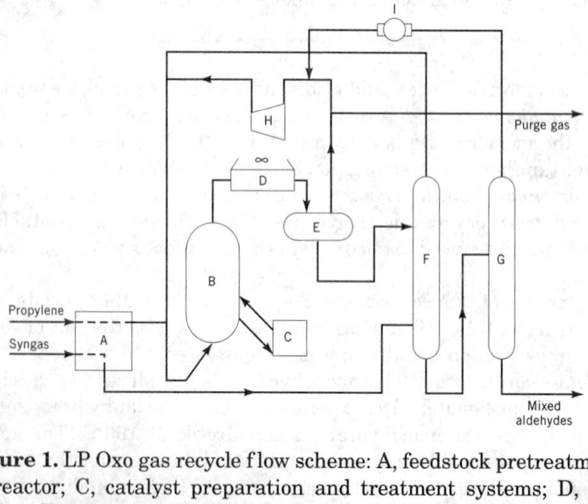

Figure 1. LP Oxo gas recycle flow scheme: A, feedstock pretreatment; B, reactor; C, catalyst preparation and treatment systems; D, condenser; E, separator; F, stripper; G, stabilizer; H, cycle compressor; and I, stabilizer overhead gas compressor.

have been advances in the development of high reactivity oxo catalysts for the conversion of low reactivity feedstocks such as internal and α-alkyl substituted α-olefins. These catalysts contain (as ligands) ortho-t-butyl or similarly substituted arylphosphites, which combine high reactivity, vastly improved hydrolytic stability, and resistance to degradation by product aldehyde, which were deficiencies of earlier, unsubstituted phosphites.

Uses

n-Propanol and n-propyl acetate account for about 70% of the U.S. propionaldehyde derivative market (see PROPYL ALCOHOLS). These compounds are used principally in flexographic and gravure inks (qv) which require volatile solvents to prevent smearing and ink accumulation on the printing presses (see PRINTING PROCESSES). Some propanol is also converted into n-propylamines which are important pesticide intermediates (see PESTICIDES). n-Propanol is also employed as a precursor for glycol ethers.

The highest volume oxo chemical in the United States, n-butyraldehyde, is converted mainly into n-butanol, employed chiefly to produce butyl acrylate and methacrylate (see ACRYLIC ACID AND DERIVATIVES). In contrast, the principal n-butyraldehyde derivative in Europe and Japan is 2-ethylhexanol, the precursor to the poly(vinyl chloride) (PVC) plasticizer, DOP.

1,4-Butanediol (BDO) goes primarily into tetrahydrofuran (THF) for production of polytetramethylene ether glycol (PTMEG), used in the manufacture of polyurethane fibers.

The principal C_5 valeraldehyde derivatives, n-amyl and 2-methylbutyl alcohols, are used predominantly to make zinc diamyldithiophosphate lube oil additives (see AMYL ALCOHOLS; LUBRICATION AND LUBRICANTS).

C_7–C_9 oxo-derived acids are the principal derivatives of the C_7–C_9 oxo aldehydes, and in analogy to C_5 oxo aldehyde market applications, are used chiefly to make neopolyol esters which are employed almost entirely in aeromotive applications.

Several alcohols in the C_6–C_{13} range are produced by oxo reactions and are used in both plasticizer and detergent applications. Linear C_{12}–C_{15} alcohols are employed primarily in detergent applications.

Safety, Health, and Environmental Concerns

Oxo plants employ mixtures of highly toxic, flammable gases under pressure at high temperatures and require strict adherence to established operating safety codes and emergency reporting procedures to local, state, and federal authorities. In the United States, carbon monoxide is classified as both an acute, fire, and sudden release hazard.

The carbon monoxide component of the oxo reactant gases presents the most immediate human health hazard.

ERNST BILLIG
DAVID R. BRYANT
Union Carbide Corporation

R. L. Pruett, *Adv. Organometal. Chem.* **17,** 1 (1979).

C. K. Brown and G. Wilkinson, *J. Chem. Soc. (A),* 1392 (1970).

Chemical Economics Handbook, Oxo Chemicals Report, SRI International, Menlo Park, Calif., Jan. 1991 and preliminary 1994 draft.

C. D. Frohning and C. W. Kohlpaintner, in *Applied Homogeneous Catalysis with Organometallic Compounds,* Vol. 1, B. Cornils and W. A. Herrmann, eds., VCH Publishers, Weinheim, Germany, 1996, pps. 29–90.

OXYGEN

Molecular oxygen, O_2, is a gaseous element constituting 20.946% of the earth's atmosphere. Oxygen is essential to respiration and life in animals and is formed as a waste product by most forms of vegetation. Oxygen supports the combustion of fuels that supply heat, light, and power and enters into oxidative combination with many materials. The speed of reaction and effectiveness of combination increase with oxygen concentrations greater than that of air. Industry has established a 99.5% purity for the majority of commercial product.

Oxygen in combination with hydrogen forms the waters of the earth's surface (89 wt % O_2). In combination with metals and nonmetals oxygen is contained in well over 98% of rocks, entering into a very large number of known minerals as well as a vast array of organic compounds. Together, free and combined oxygen constitute 46.6% of the mass of the earth's crust, making it the most abundant element.

In nature, oxygen occurs in three stable isotopic species: oxygen-16, ^{16}O, 99.76%; oxygen-17, 0.038%; and oxygen-18, 0.20%. Commercial fractional distillation of water produces concentrations of ^{18}O as high as 99.98%; ^{17}O concentrations up to 55% are also produced. The ^{18}O isotope has been used to trace mechanisms of organic reactions.

Physical Properties

Gaseous oxygen is colorless, odorless, and tasteless. Selected physical properties are listed in Table 1.

Table 1. Physical Properties of Oxygen

Property	Value
triple point	
temperature, K	54.359 ± 0.002
pressure, Pa[a]	146.4
density, g/L	
gas	0.0108
liquid	1306.5
solid	1300
boiling point, at 101.3 kPa, K	90.188
density, g/L	
gas	4.470
liquid	1141.1
melting point, K	54.22
critical point	
temperature, K	154.581
pressure, MPa[b]	5.043
density, g/L	436.1
gas, at 101.3 kPa	
density, g/L	
at 0°C	1.42908
at 21°C	1.327
heat capacity, J/(mol·K)[c]	
C_p, at 25°C	29.40
C_p/C_v, at 26°C	1.396
dielectric constant at 20°C	1.0004947
n°_D	1.0002639
viscosity at 25°C, μPa·s(= cP × 10^{-3})	20.639
thermal conductivity, at 0°C, mW/(m·K)	2.448
sound velocity, at 0°C, m/s	317.3
liquid	
heat capacity, sat liq, J/(mol·K)[c]	54.317
heat of vaporization, J/mol[c]	6820
viscosity, μPa·s(= cP × 10^{-3})	189.4
thermal conductivity, mW/(m·K)	149.87
sound velocity, at 87 K, m/s	904.6
surface tension, at 87 K, N/m	13.85×10^{-7}
volume ratio, gas at 21°C to liquid at bp	859.9
solid	
heat of sublimation, J/mol[c]	8204.1
heat capacity, J/(mol·K)[c]	46.40
heat of fusion, J/(mol)[c]	444.5

[a] To convert Pa to mm Hg, multiply by 0.0075.
[b] To convert MPa to psi, multiply by 145.
[c] To convert joule to cal, divide by 4.184.

Chemical Properties

Oxygen reacts with all other elements except the light, rare gases helium, neon, and argon. The reactants usually must be activated by heating before the reaction proceeds at appreciable rates, and if the final union releases more than enough energy to activate subsequent portions of both reactants, the overall process may be self-sustaining. The process is known as combustion when light and heat are evolved.

Oxygen usually exhibits a valence of −2 in combination with other chemical elements to form compounds such as oxides. Most elements combine with oxygen, which is highly electronegative, in more than one ratio because of the variety of valences exhibited by the other element, or because of the existence of complicated molecular structures.

Manufacture

Commercial oxygen, both gaseous and liquid, at about 99.5% purity is produced by cryogenic distillation in air separation plants. In these plants the air is cleaned, dried, compressed, and refrigerated until it partially liquefies at about 80 K (see CRYOGENICS). The air is then distilled into its components. Commercial gaseous oxygen at about 90–93% purity is produced from air by vacuum swing adsorption (VSA) processes (see ADSORPTION, GAS SEPARATION). The VSA method is the fastest growth portion of oxygen production.

Production, Pipelines, and Shipping

Oxygen production facilities for relatively large users generally fall into one of three categories: (*1*) a captive plant on the oxygen user's property owned and operated by the user; (*2*) a plant owned and operated by an industrial gas company that is on or adjacent to the oxygen user's property (on-site facility), where a long-term contract for the supply of oxygen usually exists between the industrial gas company and the oxygen user; and (*3*) a plant owned and operated by an industrial gas company that supplies oxygen to several users. In the first two cases, the gaseous oxygen is generally supplied to the user site via a pipe. In the last instance gaseous oxygen is carried via a pipeline having branches to the individual industrial users. The pipeline and the central production facility are typically owned by the same industrial gas company.

Gas by pipeline is the least expensive way to manufacture and supply oxygen. The energy of refrigeration is recovered by the heat exchangers at the point where ambient-temperature gas exits and ambient-temperature air enters the plant. There is no loss by evaporation, and the costs of truck delivery are eliminated. Between 80 and 90% of all oxygen is transported in gas pipelines.

Merchant oxygen gas and liquid is transported to the smaller oxygen users by tube (gas) or cryogenic tank (liquid) trailers and railroad car (liquid). Cryogenic liquid tanks or customer stations of appropriate sizes are permanently installed on the premises of large- and small-volume merchant users, including most hospitals. A liquid oxygen tank or customer station may be found outside nearly every hospital.

Health and Safety Factors

Hazards associated with the use of oxygen derive from the facts that the cryogenic liquid is very cold and increases in volume enormously when it vaporizes, that combustion rates accelerate with oxygen concentrations above 21%, and that the gas is transported and dispensed in high pressure cylinders. Liquid oxygen and other cryogenic liquids can inflict severe damage on human tissues. When oxygen from either gas or liquid sources is brought into the presence of oxidizable materials, mixtures ranging from combustible to explosive may result. Hazards are associated with atmospheres containing either excessive concentrations of oxygen or deficient concentrations of oxygen.

Uses

There are relatively few substitutes for oxygen in most of its uses. Oxygen cannot be reclaimed commercially or recycled except via the atmosphere. The spectacular increase of oxygen production since World War II has in a very large measure occurred because of availability and its usefulness in steelmaking (see STEEL) and in chemicals and petrochemicals production, high temperature production furnaces, water purification, and hazardous waste destruction. Other uses include in nonferrous metallurgy as an oxidizing agent, in the pulp and paper industry, in aquaculture, in treatment of municipal wastewater, in medical applications, and life support applications.

JAMES G. HANSEL
Air Products and Chemicals, Inc.

W. Braker and A. Mossman, eds., *Matheson Gas Data Book,* 6th ed., Matheson, Inc., Lyndhurst, N.J., 1980.

W. Stowasser, "Oxygen" in *Mineral Facts and Problems,* U.S. Bureau of Mines Bulletin 667, U.S. Government Printing Office, 1980.

Commodity Specification for Oxygen, Pamphlet G. 4.3, Compressed Gas Association, Inc., Arlington, Va., 1989.

A. Lapin, *Liquid and Gaseous Oxygen Safety Review,* NASA CR-120922 ASRDI, NASA Lewis Research Center, Cleveland, Ohio, 1972 (available from National Technical Service, U.S. Department of Commerce, Springfield, Va.).

OXYGEN-GENERATION SYSTEMS

Oxygen (qv) generation from oxygen-containing compounds is used for systems for respiratory support in submarines, aircraft, spacecraft, and bomb shelters, as well as in breathing apparatus. Convenience and reliability, rather than low cost, are stressed.

Chlorates and Perchlorates

The chlorates and perchlorates of lithium, sodium, and potassium evolve oxygen when heated. These salts may be compounded with a fuel to form a chlorate-based candle that produces oxygen by a continuous reaction. Components include the oxygen-producing material, a fuel, a material which fixes traces of chlorine, and usually an inert binder. Once the reaction begins, oxygen is released from the hot salt by thermal decomposition. A portion of the oxygen reacts with the fuel to produce more heat resulting in production of more oxygen, and so on.

Relevant properties of the chlorates are given in Table 1. Sodium chlorate is generally used.

Materials and Reactions. Candle systems vary in mechanical design and shape but contain the same generic components (Fig. 1). The candle mass contains a cone of material high in iron which initiates reaction of the solid chlorate composite. Reaction of the cone material is started by a flash powder train fired by a spring-actuated hammer

Table 1. Chlorates and Perchlorates as Sources for Oxygen

Substance	Molecular formula	Mp, °C	Decomp., °C[a]	Oxygen density g/g cmpd	Oxygen density g/cm³[b]
lithium chlorate	$LiClO_3$	129	270	0.53	1.39
sodium chlorate	$NaClO_3$	261	478	0.45	1.12
potassium chlorate	$KClO_3$	357	400	0.39	0.93
lithium perchlorate	$LiClO_4$	247	410	0.60	1.46
sodium perchlorate	$NaClO_4$	471	482	0.52	1.31
potassium perchlorate	$KClO_4$	585	400	0.46	1.17
oxygen	O_2				
liquid				1.0	1.14
at 52 MPa[c]				1.0	0.58
hydrogen peroxide	H_2O_2	-89		0.94	1.37
water	H_2O	0		0.89	0.80

[a] Without catalyst.
[b] Based on crystal densities.
[c] To convert MPa to psi, multiply by 145.

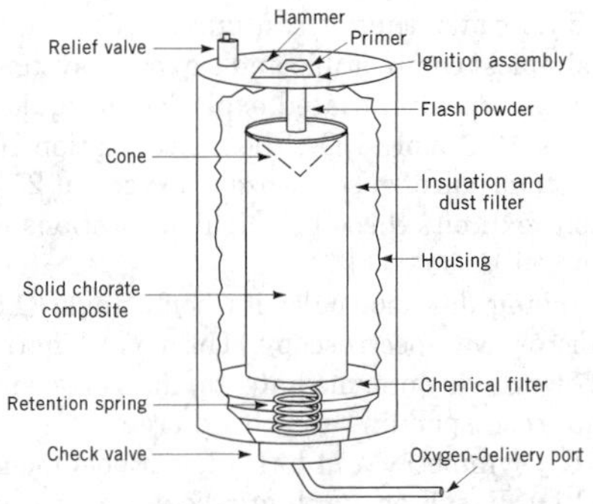

Figure 1. Cutaway view of generator housing.

against a primer. An electrically heated wire has also been used. The candle is wrapped in insulation and held in an outer housing that is equipped with a gas exit port and relief valve. Other elements of the assembly include gas-conditioning filters and chemicals and supports for vibration and shock resistance.

A fuel provides heat upon reaction with some of the generated oxygen. Whereas a variety of powdered elements, eg, Fe, B, Al, Co, etc, have been used with varying degrees of success, iron is the universal choice for commercial applications. Some of the oxygen from the chlorate decomposition combines with the iron to generate heat to effect decomposition of the chlorate, leaving a mixture of ironoxides.

Oxygen Purity. Impurities in the generated oxygen are chlorine-containing compounds, CO, CO_2, H_2O, and simple organics. All can be minimized by using high purity ingredients and control of moisture, or by gas conditioning. Aside from the chlorate and barium peroxide, materials are degreased by heating at 472°C before manufacture. Iron powders usually are reduced with hydrogen.

Candle Fabrication. All ingredients must be contaminant-free (especially grease) and the chemical materials must be dry. The oxygen-generating mass is made by mixing and then pressing or casting the ingredients. Care must be exercised to assure thorough mixing or reaction rates can vary throughout the candle. Shape can be varied as desired, especially if casting is used. With pressing, the shape is limited to some extent, although hydrostatic pressing provides freedom in candle form. All other factors being equal, the rate of oxygen evolution is directly proportional to the cross-sectional area of the unit. Some oxygen evolution rate control can be achieved using graded compositions along the length of the candle.

Operational Characteristics. Oxygen generation from chlorate candles is exothermic and management of the heat released is a function of design of the total unit into which the candle is incorporated. The oxygen release rate is directly proportional to the cross-sectional area of candle for a specific composition and also depends on the linear burn rate.

Uses. A primary early use was the incorporation of a small cast candle in a quick-start canister, which was filled with potassium superoxide and used in a portable breathing apparatus. The candle rapidly produced an initial supply of oxygen until the superoxide became fully activated, particularly at lower temperatures. Large candles, delivering 3–4 m^3 (120 ft^3) oxygen in 45 minutes, are used in long-duration submergence submarine operation. A furnace holds a stack of two candles; the upper one is ignited, which subsequently ignites the lower one. Together these furnish enough oxygen for 120 people for 1.5 hours.

A large-scale use of chlorate candles is oxygen supply in the event of decompression in passenger aircraft. Candles meet the requirements of no maintenance (qv), no oxygen leakage, high reliability, and long (15-yr min) storage life. Both percussion cap and electrical ignition are used.

In projecting uses for candles, the characteristics to be considered are volumetric oxygen density that is nearly equal to that of liquid oxygen, development of extremely high pressure, high oxygen purity, long storage life with no leakage, preprogrammed oxygen delivery, and heat release.

Peroxides and Superoxides

Devices made with peroxides and superoxides produce oxygen and absorb carbon dioxide. Potassium peroxide is difficult to prepare and lithium superoxide is very unstable. The ozonides, MO_3, of the alkali metals contain a very high percentage of oxygen, but are only stable below room temperature.

Peroxides. In the presence of lithium peroxide, both water and carbon dioxide react, resulting in evolution of oxygen. The reaction mechanism for sodium peroxide is the same as for lithium peroxide, ie, both carbon dioxide and moisture are required to generate oxygen. Sodium peroxide has been used extensively in breathing apparatus.

All the peroxides are colorless and diamagnetic when pure. Traces of the superoxide in technical-grade sodium peroxide impart a yellow color. Storage containers must be sealed to prevent reaction with atmospheric carbon dioxide and water vapor.

Superoxides. The superoxides are colored and paramagnetic: KO_2 is yellow, NaO_2 is orange-yellow, and $Ca(O_2)_2$ is red. In uses as oxygen suppliers and carbon dioxide scrubbers, these materials are demand chemicals, ie, react to the load imposed by generating more oxygen as more water is introduced.

Potassium superoxide, the most commonly used superoxide of the alkali metals, is produced by spraying the molten metal into dry air. Sodium superoxide is produced from an open-pore sodium peroxide which is produced by spraying liquid sodium into dry air. Calcium superoxide has been produced to 65% purity by careful dehydration of $CaO_2 \cdot 2H_2O_2$. Calcium superoxide is of interest for breathing apparatus because calcium hydroxide has a higher melting point than the hydroxides of potassium or sodium, and calcium superoxide is less sensitive to water.

Uses. The peroxides and the superoxides must be hermetically sealed for storage. The superoxides especially are strong oxidizing agents and should be kept away from grease, oil, and organic materials. In general, sodium peroxide is used more widely than lithium peroxide. A breathing apparatus based on peroxides often is supplied with bottled oxygen, because the peroxides are not very oxygen-weight efficient.

Superoxides are used in breathing applications requiring no auxiliary source of oxygen. Portable breathing apparatus are used by fire departments, damage-control teams, and workers in unbreathable atmospheres. The wearer uses a canister containing the chemical, a breathing bag, and a mask.

In air conditioning (qv) of closed spaces, a wider latitude in design features can be exercised. On-board oxygen-generation (OBOG) systems are used in military aviation. The OBOG systems obviate the use of stored liquid oxygen or high pressure gaseous oxygen. In OBOG systems, oxygen is separated from air using chelating agents (qv), ion exchange (qv), barium oxide–dioxide shift, or molecular sieves (qv). Commercial units supplying therapeutic oxygen are also available.

Other Chemical Systems

Regenerative systems that dissociate carbon dioxide to recover the oxygen are of interest to the U.S. space program and in long-duration habitat support. The Bosch process utilizes an iron catalyst for the single-step reaction of hydrogen, carbon dioxide, water, and carbon at 700°C. Oxygen is produced by water electrolysis, and the hydrogen is recycled to the Bosch reactor. Carbon is removed as a solid.

In the Sabatier reaction, methane and water are formed over a nickel–nickel oxide catalyst at 250°C. The methane is recovered and cracked to carbon and hydrogen, which is then recycled.

Hydrogen peroxide can be dissociated over a catalyst to produce oxygen, water, and heat. It is an energetic reaction, and contaminants can spontaneously decompose the hydrogen peroxide. Oxygen from water electrolysis is used for life support on submarines.

Health and Safety

Peroxides, superoxides, and chlorates are oxidizing compounds and should not contact organic materials, eg, oil, greases, etc. This is especially true while oxygen is being produced. Caustic residues that may remain after use of peroxides and superoxides require disposal appropriate to alkali metal hydroxides. Spent candles containing barium may require special disposal considerations.

Dusts associated with these oxidizing compounds produce caustic irritation of skin, eyes, and nasal membranes. Appropriate protection should be worn when handling.

J. WILSON MAUSTELLER
Consultant

W. J. O'Reilly and co-workers, *Development of Sodium Chlorate Candles for the Storage and Supply of Oxygen for Space Exploration Applications,* Rept. No. 69-4695, Air Research Corp., Los Angeles, Calif., July 1969.

Exploratory Study of Potassium and Sodium Superoxides for Oxygen Control in Manned Space Vehicles, Contract No. NASW-90, USA Research Corp., Evans City, Pa., March 1962.

A. J. Adduci, *Chemtech,* 575 (Sept. 1976).

OZONE

Ozone, O_3, is an allotropic form of oxygen first recognized as a unique substance in 1840. Its pungent odor is detectable at ~0.01 ppm. It is thermally unstable and explosive in the gas, liquid, and solid phases. In addition to being an excellent disinfectant, ozone is a powerful oxidant not only thermodynamically, but also kinetically, and has many useful synthetic applications in research and industry. Its strong oxidizing and disinfecting properties and its innocuous by-product, oxygen, make it ideal for the treatment of water. Indeed, the most important application of ozone is in the treatment of drinking water, which began in Europe. The treatment of swimming pool water was also developed in Europe (see WATER, TREATMENT OF SWIMMING POOLS, SPAS, AND HOT TUBS). Another important ozone application is for odor control in industrial processes and municipal wastewater-treatment plants. Ozone also is used on a large scale for the treatment of municipal secondary effluents (see WATER, MUNICIPAL WATER TREATMENT). Industrial high quality water supplies are also treated with ozone (see WATER, INDUSTRIAL WATER TREATMENT). In addition, ozone has applications in the treatment of cooling-tower water and in pulp bleaching. Advanced oxidation processes employing ozone in combination with uv, H_2O_2, and/or solid catalysts such as TiO_2 greatly improve the reactivity of ozone toward organic contaminants.

Ozone, which occurs in the stratosphere (15–50 km) in concentrations of 1–10 ppm, is formed by the action of solar radiation on molecular oxygen. It absorbs biologically damaging ultraviolet radiation (200–300 nm), prevents the radiation from reaching the surface of the earth, and contributes to thermal equilibrium on earth.

Properties

At ordinary temperatures, pure ozone is a pale blue gas (d = 2.1415 g/L at 0°C and 101.3 kPa (1 atm)) that can be condensed to an indigo blue liquid, which freezes to a deep blue-violet solid. The solubility of gaseous ozone at atmospheric pressure and 0°C is 1.1 g/L H_2O. Gaseous ozone can be adsorbed by porous solid substrates such as silica gel and is often used in this form in organic synthesis.

Ozone is endothermic, thus it can burn or detonate by itself and represents the simplest combustible and explosive system. The concentration threshold for spark-initiated explosion of liquid ozone in oxygen at −183 °C is 18.6 mol % O_3; the concentration limit for shock wave-initiated detonation of gaseous ozone-oxygen at 25°C is 9.2 mol % O_3. Gaseous ozone exhibits three principal absorptions in the infrared at 710, 1043, and 2105 cm^{-1}.

Ozone is a triangular molecule; its bond angle (116.8°) was established by microwave spectroscopy. The bond length of the ozone molecule (0.1278 nm) is intermediate to that of a single and double oxygen bond, corresponding to a bond order of 1.7. Ozone is diamagnetic with C_{2v} symmetry and has a low dipole moment of 1.77×10^{-30} C·m (0.53 D). Based on Pauling resonance concepts, the structure of ozone is a hybrid, principally of form (1), with a small contribution from (2).

Thermal Decomposition

Gas Phase. The decomposition of gaseous ozone is sensitive not only to homogeneous catalysis by light, trace organic matter, nitrogen oxides, mercury vapor, and peroxides, but also to heterogeneous catalysis by metals and metal oxides.

The calculated half-life of 1 mol % (1.5 wt %) of pure gaseous ozone diluted with oxygen at 25, 100, and 250°C is 19.3 yr, 5.2 h, and 0.1 s, respectively. Although pure ozone-oxygen mixtures are stable at ordinary temperatures in the absence of catalysts and light, ozone produced on an industrial scale by silent discharge is less stable due to the presence of impurities; however, ozone produced from oxygen is more stable than that from air. At 20°C, 1 mol % ozone produced from air is ~30% decomposed in 12 h.

Aqueous Phase. In pure water, the decomposition of ozone at 20°C involves a complex radical chain mechanism, initiated by OH^- and propagated by O_2^- radical ions and HO radicals.

Hydrogen peroxide greatly accelerates the decomposition of ozone in alkaline solutions because of formation of HO_2^-, which reacts rapidly with ozone to form the radical ion O_2^-.

Photochemical Decomposition

Gas Phase. Gaseous ozone is decomposed to oxygen atoms and molecules by absorbing radiation in the visible and uv spectrum: $O_3 + h\nu \rightarrow O_2 + O$.

Aqueous Phase. In contrast to photolysis of ozone in moist air, photolysis in the aqueous phase can produce hydrogen peroxide initially because the hydroxyl radicals do not escape the solvent cage in which they are formed. Hydrogen peroxide is photolyzed slowly to hydroxyl radicals, which decompose ozone.

Chemistry of Ozone

The inorganic chemistry of ozone is extensive, encompassing virtually every element except most noble metals, fluorine (qv), and the inert gases.

Ozone reacts rapidly with various free radicals and radical ions such as O, O_2^-, H, HO, N, NO, Cl, and Br. Some of these radicals (HO,NO, Cl, and Br) can initiate the catalytic decomposition of ozone.

The strong electrophilicity of ozone is manifested in its reaction with a wide variety of organic and organometallic functional groups, eg, olefins, acetylenes, aromatics (carbocyclic and heterocyclic), activated C—H bonds (acetals, alcohols, aldehydes, ethers, and glycosides), unactivated C—H bonds (alkanes, cycloalkanes, and alkyl aromatics), deactivated C—H bonds (carboxylic acids and ketones), C═N and N═N bonds, Si—H and Si—C bonds, organometallic bonds (eg, Grignard reagents), and nucleophiles (eg, ammonia, amines, amino acids, arsines, disulfides, hydroxylamines, nitriles, phosphites, selenides, sulfides, and thioethers). Ozone also acts as a nucleophile, eg, in its reaction with carbocations.

Atmospheric Ozone

Stratosphere. Ozone is formed rapidly in the stratosphere (15–50 km) by the action of short-wave ultraviolet solar radiation (<240 nm) on molecular oxygen, $O_2 + h\nu \rightarrow 2O$. At wavelengths above 175 nm, only ground-state (^3P) atoms are formed; whereas at wavelengths below 175 nm, one ground-state and one excited (^1D) atom are formed. Ground-state atoms also can be formed by the predissociation of electronically excited O_2. The oxygen atoms can react with molecular oxygen to yield ozone: $O + O_2 + M \rightarrow O_3 + M$. Ozone can be destroyed photochemically: $O_3 + h\nu \rightarrow O + O_2$; at 226 nm, however, this reaction also can produce vibrationally excited O_2 capable of forming ozone. In addition, ozone can be destroyed by reaction with oxygen atoms, as well as with excited O_2 molecules and other free radicals. Since the early 1960s, it has been recognized that radicals such as NO, OH, Cl, and Br affect the abundance and distribution of ozone in the stratosphere. Earlier studies simulating stratospheric chemistry concluded that ozone formation is significantly less than its destruction, hence the ozone deficit problem. However, studies indicate that this may not be the case.

Most ozone is formed near the equator, where solar radiation is greatest, and transported toward the poles by normal circulation patterns in the stratosphere. Consequently, the concentration is minimum at the equator and maximum for most of the year at the north pole and about 60°S latitude. The equilibrium ozone concentration also varies with altitude; the maximum occurs at about 25 km at the equator and 15–20 km at or near the poles. It also varies seasonally, daily, as well as interannually. Absorption of solar radiation (200–300 nm) by ozone and heat liberated in ozone formation and destruction together create a warm layer in the upper atmosphere at 40–50 km, which helps to maintain thermal equilibrium on earth.

Troposphere. Ozone and nitric oxide are transported from the stratosphere to the troposphere, the region of the atmosphere below 15 km. Though only about 10% of the atmospheric ozone is present in the troposphere, this small fraction plays a fundamental role in atmospheric chemistry because it leads to the formation of hydroxyl radicals. Hydroxyl radicals initiate the oxidation and prevent the buildup of many organic and inorganic pollutants in the atmosphere. The radical-dominated chemistry of the troposphere is complex, involving intertwining cycles of gas-phase, condensed-phase, and multiple-phase reactions. In unpolluted atmosphere, HO radicals react with naturally occurring CO and methane, resulting in a net increase in ozone concentration.

Although the naturally occurring concentration of ozone at the earth's surface is very low, this distribution has been altered by the emission of anthropogenic pollutants which increase the production of ozone via the above mechanism. Photochemical smog, an aerosol irritant gas mixture, occurs in urban industrialized areas where heavy motor vehicle traffic is common, especially those areas where temperature inversions are common. It forms at low altitudes by photolytic reactions involving nonmethane hydrocarbons, NO, and CO, resulting in low but potentially harmful concentrations of ozone and other irritating substances, such as aldehydes, ketones, acids, H_2O_2, organic peroxides, and peroxyacetyl nitrate.

Although the background concentration of ozone in surface air is ~0.01–0.03 ppm, during severe smog days in the Los Angeles area, for example, it has often reached 0.5 ppm, and a maximum of 1 ppm in 1957. In the early morning hours, NO is removed slowly by the oxygen atom chain, which is initiated by the photolysis of NO_2 and subsequently by the photolysis of ozone. Later in the day when the light intensity is higher, the hydroxyl chain causes the NO conversion to accelerate.

Ozone can react rapidly with NO to produce NO_2, which re-enters the ozone formation cycle: $O_3 + NO \rightarrow O_2 + NO_2$. This is the main ozone-depleting reaction in the absence of sunlight. Ozone also reacts with NO_2 (to form NO_3, which in turn reacts with NO_2 to form N_2O_5), C_2H_4, as well as HO and HO_2 radicals. Nitric acid formed by the reaction $HO + NO_2 \rightarrow HNO_3$ is removed from the atmosphere by rain-out.

Ozone Generation

Ozone can be generated by a variety of methods, the most common of which involves the dissociation of molecular oxygen electrically (silent discharge) or photochemically (uv). The short-lived oxygen atoms (lifetime ~10^{-5} s) react rapidly with oxygen molecules to form ozone. The widely employed technique of electric discharge produces much higher concentrations than the ultraviolet technique and is more practical and efficient for production of large quantities. A less common method of ozone formation is electrochemical generation.

Silent Electric Discharge. Commercial production and utilization of ozone by silent electric discharge consists of five basic unit operations: gas preparation, electrical power supply, ozone generation, contacting (ie, ozone dissolution in water), and destruction of ozone in contactor off-gases.

Ultraviolet Light. The mechanism of the practical photochemical production of ozone is similar to that in the stratosphere; that is, oxygen atoms, formed by the photodissociation of oxygen by short-wavelength uv radiation (≤240 nm), react with oxygen molecules to form ozone. In practice, ozone concentrations obtained by commercial uv devices are low. This is because the low intensity, low pressure mercury lamps employed produce not only the 185-nm radiation responsible for ozone formation, but also the 254-nm radiation that destroys ozone, resulting in a quantum yield of ~0.5 compared to the theoretical yield of 2.0. The low concentrations of ozone available from uv generators preclude their use for water treatment because the transfer efficiencies of ozone from air into water is low and large volumes of carrier gas must be handled.

Uses

Ozone is used in the treatment of drinking water and in industries where high purity water is required (e.g., breweries, pharmaceuticals, and electronics). Ozone is also used in industrial wastewater pollution control, wastewater disinfection, and odor control; in the treatment of process water, such as cooling tower water; in the treatment of swimming pools and spas; in pulp bleaching; and in organic synthesis, as a selective oxidant.

Among other uses, ozone therapy, employing O_3–O_2, is increasingly being employed and studied in dentistry, veterinary and sports medicine, and proctology. Ozone is used as an aquatic oxidant and disinfectant in zoos, large aquariums, as well as fish and shrimp hatcheries. Ozone also is used for food preservation, in cold storage rooms, brewery cellars, hotel and hospital air ducts, and air conditioning systems. Ozone has also been used in textile bleaching and in the bleaching of esters, oils, fats, waxes (qv), starch, flour, ivory, etc. Oxidation of Ag^+ by ozone is employed commercially to produce high purity AgO. The use of ozone as a chemical agent decontaminant has been patented.

Health and Safety

As a constituent of the atmosphere, ozone forms a protective screen by absorbing radiation of wavelengths between 200 and 300 nm, which can damage DNA and be harmful to life. Consequently, a decrease in the stratospheric ozone concentration results in an increase in the uv radiation reaching the earth's surfaces, thus adversely affecting the climate as well as plant and animal life. For example, the incidence of skin cancer is related to the amount of exposure to uv radiation. Ozone can be toxic to plants, animals, and fish.

The toxicity of ozone to humans is largely related to its powerful oxidizing properties. The odor threshold of ozone varies among individuals but most people can detect 0.01 ppm in air, which is well below the limit for general comfort. The symptoms experienced on exposure to 0.1–1 ppm ozone are headache, throat dryness, irritation of the respiratory passages, and burning of the eyes caused by the formation of aldehydes and peroxyacyl nitrates. Exposure to 1–100 ppm ozone can cause asthma-like symptoms such as tiredness and lack of appetite.

Short-term exposure to higher concentrations can cause throat irritations, hemorrhaging, and pulmonary edema.

Ozonation of drinking water produces various by-products such as aldehydes, ketones, carboxylic acids, organic peroxides, epoxides, nitrosamines, *N*-oxy compounds, quinones, hydroxylated aromatic compounds, brominated organics, and bromate ion. Although some of these compounds are potentially toxic or carcinogenic, most bioassay-screening studies have shown that ozonated water induces substantially less mutagenicity than chlorinated water.

J. A. WOJTOWICZ
Consultant

P. S. Bailey, *Ozonation in Organic Chemistry,* Vols. 1 and 2, Academic Press, Inc., New York, 1978–1982.

J. Hoigné and H. Bader, *Water Res.* **17,** 185 (1983).

P. O. Wennberg and co-workers, *Science* **266,** 398 (1994).

U. Kogelschatz, B. Eliasson, and M. Hirth, *Ozone Sci. Eng.* **10,** 367 (1988).

P

PACKAGING

CONTAINERS FOR INDUSTRIAL MATERIALS

In any operation involving the manufacturing, distribution, and use of chemical substances, it is essential that consideration be given to packaging at an early stage of the manufacturing process. Container systems for industrial chemicals must fulfill several important functions: they must contain the product in order to be able to move it safely from point of manufacture to use; protect both the product from contamination and the immediate surroundings (plant, people, equipment) from the potential harm caused by the product itself; provide features that aid users in the effective utilization of the product; and communicate valuable information such as product identity, potential hazards, and handling information, to shippers, carriers, and users.

The environmental impact of packages and packaging materials has come under increasingly vigorous scrutiny by all kinds of interests, ie, government and regulatory agencies, consumer groups, and environmentalists. It is becoming increasingly important that packaging for all kinds of products be developed in a rational manner, use less material, and have the ability to be recovered and reused whenever it is economically feasible and permitted by regulation.

Virtually any chemical can be stored and transported safely and effectively by using one of many package types. The choice of a container system, in general, is dictated by manufacturing, marketing, and economic considerations; for a chemical, however, the choice of packaging materials often is influenced primarily by safety and chemical compatibility factors. Both aspects can affect the cost of physical distribution, which often is comparable to the cost of the product being packaged.

Regulations

Regulations governing packaging and shipping of chemicals depend on the classification of the chemical as hazardous or nonhazardous. For nonhazardous chemicals, the packaging and shipping requirements are subject to the rules issued by the carrier. The most common of these rules are published in the *Uniform Freight Classification* for railroads and the *National Motor Freight Classification* for trucks. These rules are similar in that they both include sections listing the participating carriers, index to articles, article requirements, and packaging descriptions. The participating carriers have the right to collect a surcharge as well as to refuse handling or paying damage claims for articles not packaged according to the classification requirements.

There is also a procedure for trial shipments of new or improved packages.Regulations controlling the packaging and shipping of hazardous materials in the United States are prepared by the Research and Special Programs Administration (RSPA) of the U.S. Department of Transportation (DOT). The primary document is Hazardous Materials Regulations (HMR) 49 CFR, parts 171–179. The *Code of Federal Regulations* (CFR) has been extensively changed to bring it into agreement with the international rules recommended by the United Nations Committee of Experts.

In the words of RSPA, the changes to the regulations will (*1*) simplify and reduce the volume of the HMR, (*2*) enhance safety through better classification and packaging, (*3*) promote flexibility and technological innovation in packaging, (*4*) reduce the need for exemptions from the HMR, and (*5*) facilitate international commerce.

The primary change in the HMR is in replacing the specific container requirements for products with performance orientated packaging (POP). In general, this means any package can be used as long as it passes certain rigid test requirements.

Transportation and Storage. Three considerations apply to both transportation (qv) and storage: compliance with legal requirements; package compatibility with the product as well as manufacturing and physical distribution requirements, such as safety requirements; and selection of an optimal-cost packaging system consistent with the preceding considerations. The interrelated nature of these factors, the variety of products to be shipped, the numerous packaging methods, and the large costs which are often associated with such decisions are all details that should be evaluated by a packaging specialist.

Bulk Handling of Products

Liquids. Approximately 170,000 railroad tank cars are used in the United States. The interior surfaces of these cars are tailored to carry a wide variety of products and are constructed of steel which is either unlined or lined with materials to enhance the chemical compatibility with a specific product; these lining materials include synthetic rubber, phenolic or modified epoxy resins, or corrosion-resistant materials such as aluminum, nickel-bearing steel, or stainless steel.

For commodities that solidify at temperatures commonly encountered during shipping, tank cars are equipped with internal or external heating coils.

Tank cars have been constructed with capacities as great as 130,000 L (~34,000 gal) and weights as much as around 91 metric tons.

Solids. Increasing use of bulk cars, especially of covered hopper cars, has accompanied the expansion of the tank-car fleet. The principal drawback of bulk cars is the requirement for limited use, specialized cars, which necessitates a large investment. However, if such investment can be justified, the cost of transportation for dry bulk materials in hopper cars usually is less than those for goods in shipping containers.

Semibulk Containers. Use of semibulk containers falls between bulk handling, eg, accomplished by tank cars and hopper cars, and individual package handling, which is often performed manually. Semibulk containers are also known as intermediate bulk containers (IBCs), the provisions and requirements for the construction and testing of which can be found in the U.N. recommendations.

Industrial Packaging Materials

Industrial packaging materials include bulk containers (tank cars, bulk cars and semibulk containers), steel drums and pails, plastic drums, wooden barrels, fiber drums, bags (textile, laminated textile, multiwall paper, and plastic), carboys and bottles, and boxes and cartons.

David L. Olsson
A. Ray Chapman
Rochester Institute of Technology

DOT DOCKET HM-181, Performance Packaging, in *Fed. Reg.* **55**(246) (Dec. 21, 1990). *U.N. Recommendations on the Transport of Dangerous Goods,* United Nations, New York, 6th ed., 1990.

A. L. Brody and K. S. Marsh, eds., *The Wiley Encyclopedia of Packaging Technology,* 2nd ed., John Wiley & Sons, Inc., New York, 1997.

J. F. Hanlon, *Handbook of Package Engineering,* Technomic Publishing Co., Inc., Lancaster, Pa., 1992.

CONVERTING

Well over 90% of the consumer goods sold in the United States is shipped, stored and purchased in some form of packaging. Over 55% of the packaging is paper-based.

The selection of a specific package type by a product manufacturer is based on a number of factors, including the characteristics of the product being packaged; the nature of the shipping and storage environment to which it will be exposed; the physical strengths and properties of the package; regulatory requirements for packaging, eg, by the United Nations, U.S. Food and Drug Administration (FDA), U.S. Department

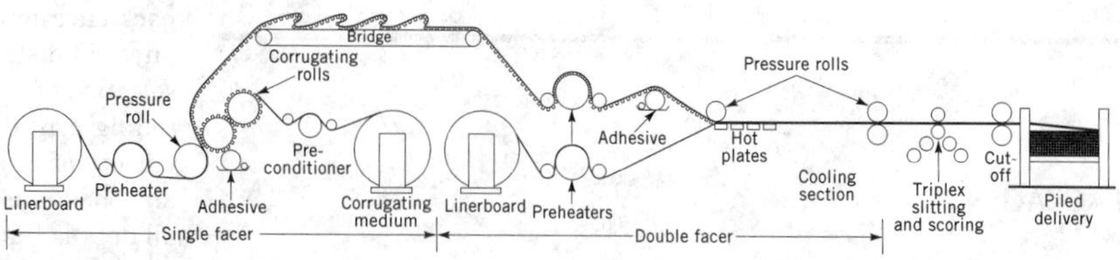

Figure 1. Schematic of a corrugator (simplified and condensed).

of Agriculture, U.S. Military Regulations, U.S. Department of Transportation, U.S. Uniform Classification Committee, U.S. National Motor Freight Classification Committee, and ISO; consumer preference; and the unit packaging cost. In many instances, different packaging types compete for the same packaging application. For example, glass jars, plastic jars, and metal cans may compete with each other, or corrugated boxes, solid fiber boxes, wooden crates, and paper or plastic bags may compete with each other. Assuming that there are no regulatory restrictions or consumer preference differences, the choice of a packaging type is dependent on the relative direct costs of the competing packaging products, and the relative indirect costs of the package types throughout the entire filling, shipping, storage, and usage cycle, for example, package-filling line jams, product damage, warehouse stacking life, retail shelf life, moisture resistance, and pilferage.

In order to understand the various packaging options available for a given packaging application, there must be an understanding of the converting process and materials used to produce the various packages and the range of characteristics that can be achieved with each package type.

Paper and Paperboard Materials

The paper and paperboard materials used to manufacture corrugated boxes, solid fiber boxes, folding cartons, paper bags, fiber drums, and fiber cans are generally made from natural cellulosic fibers, such as wood fiber, and are typically supplied to the package manufacturing plant in roll form. Although there are many specific grades of paper available commercially, they can be divided into specific categories: coated or noncoated, and regular or wet strength.

The strength properties of paper are dependent on the strength of the individual fibers and the bonding which occurs between the cellulosic fibers. Paper fibers readily absorb moisture when exposed to humid or moist conditions. This absorbed moisture reduces the strength of the paper. Typically, regular grades of paper lose 94% of their tensile strength when wet, and wet-strength grades of paper lose 61% of their tensile strength when wet. Wet-strength resins used in paper include urea–formaldehyde, melamine–formaldehyde, polyamide–polyamine–epichlorohydrin, polyethylenimine, dialdehyde starches, insolubilized protein, vegetable gums and extracts, and silicates and silicones.

Corrugated Paperboard Boxes

Corrugated paperboard is a sandwich structure formed by gluing a fluted corrugating medium ply to two linerboard facings. The first corrugated fiberboard shipping containers were used commercially in 1903. The corrugated paperboard sheets are produced in a single, continuous manufacturing process consisting of five operations (Fig. 1). The types of corrugated board are single-wall, double-wall, triple-wall, and laminated. The corrugator adhesive is a water slurry of approximately 18% uncooked raw pearl starch, 3% cooked and gelled pearl starch, 0.5% borax, and 0.5% caustic.

Surface coatings can be applied to the corrugated board to achieve water resistance, humidity resistance, oil and grease resistance, product abrasion resistance, corrosion resistance, adhesion release properties, flame-retardant properties, nonskid properties, and static electricity control properties. The coating materials used include silicone, antimony pentoxide, fluorocarbon, stearato chromic chloride, graphite, colloids of silica or alumina, and paraffin wax.

Poly(vinyl acetate) or hot melt adhesives are typically used to form the box from the folded corrugated sheet.

Solid Fiber Paperboard Boxes

A solid fiber paperboard package is a sandwich structure composed of inner plies called filler stock and outer plies called facings. Solid fiber differs from corrugated in that the inner plies are not fluted. Solid fiber packaging is typically used where toughness is important. Many solid fiber packages are reusable. The plies are bonded together on a nonheated Paster using a poly(vinyl alcohol)–clay–starch adhesive blend.

Paper Bags

Paper bag packaging includes multiwall bags, for consumer and commercial packaging applications, as well as grocery sacks and retail bags and sacks. Bags are manufactured in a continuous operation from rolls of paper. The paper web is folded and glued at a seam to form a tube, which is cut to the required length for the specific bag size. The bag bottom is then folded and glued closed. The adhesives used include unborated dextrin, borated dextrin, casein, latex–casein, latex, poly(vinyl acetate), vinyl acetate copolymers, and hot-melt materials.

Folding Cartons

Folding cartons are used extensively for point-of-purchase packaging and generally include high quality printing graphics. The paperboard material used typically has smooth surface properties and very often is clay coated. The paperboard roll stock is printed and/or coated, cut into sheets of an appropriate size, die cut into individual package blanks, which are then folded and glued to form the finished package. Common coating types include paraffin wax, polyethylene and vinyl acetate copolymers, and organic solvent-based coatings and lacquers. The adhesives used include latex, poly(vinyl acetate), vinyl acetate copolymer, and hot-melt.

Fiber Drums and Cans

Fiber drum packaging and fiber can packaging are included together because of the similarity of their production processes. Both types of packaging are produced by wrapping plies of paperboard into a tube form. Two basic winding processes, convolute and spiral, are in use. The adhesives used include silicates, poly(vinyl alcohol), and poly(vinyl acetate).

JOSEPH J. BATELKA
Consultant

Fiber Box Handbook, Fiber Box Association, Rolling Meadows, Ill., 1992.

J. J. Batelka, *Corrugating Medium—Its Influence on Box Plant Operations and Combined Board Properties and Package Performance,* Institute of Paper Science and Technology, Atlanta, Ga., 1993.

J. C. W. Evans, ed., *Trends in Paper and Paperboard Converting,* Lockwood Trade Journal Co., New York, 1965.

W. E. Scott, J. C. Abbott, *Properties of Paper: An Introduction,* 2nd ed., TAPPI Press, Atlanta, Ga., 1995.

COSMETICS AND PHARMACEUTICALS

Cosmetics (qv) and pharmaceuticals (qv) each have their own special packaging requirements. The primary purpose of packaging for both broad classifications is to provide a means to store and distribute the product until the contents of the package are used. Each product must be analyzed for stability in the package being considered for use by the manufacturer; changes in container material, resin formulation, color, and closure system can all affect product stability. Although the distribution function of the packaging is always important, each product has other objectives that packaging components must achieve.

Cosmetic packaging, also is used to enhance the image of the product. This can be accomplished by frosting the container, graphics, proprietary design of the package, or use of metallized closures. The display package, or other secondary packaging, is also used to promote the image of the product.

Although pharmaceutical packaging has the same basic objectives as cosmetic packaging, different parameters dictate product stability and safe packaging requirements. Both classes of products and their packaging are regulated by the U.S. Food and Drug Administration (FDA), but requirements for pharmaceutical packaging are more stringent because of product tampering prevention and child safety requirements of the FDA and the Consumer Product Safety Commission, respectively.

Child-Resistant Packaging

Under the Poison Prevention Packaging Act of 1970, any product that, if consumed by a child, could result in harm to the child must be packaged using components difficult for a child to open. This is referred to as child-resistant packaging.

The Consumer Product Safety Commission is responsible for administering the packaging rule under 16 CFR 1700, and the procedures for testing packages to assure compliance with the rule are included in the Code of Federal Regulations (CFR).

Product Tampering

In 1982, seven people died from consuming cyanide-laced Tylenol capsules. Since that time, the packaging industry has become visible to most consumers. This awareness has benefited the consumer by a reduction in loss of life due to consumption of adulterated products from tampering. Never before has an industry reacted so swiftly to resolve a problem.

Every developed nation has experienced product tampering incidents. The principal difference between domestic and foreign incidents is the motive of the tamperers. Most developed nations are either implementing or modifying their rules on the use of tamper-evident packaging. Some features as they are used in the United States would have to be modified or the use of a secondary feature required to meet the standards of various other countries.

The FDA Rule

The FDA has passed a rule (21 CFR 211.132) requiring the use of tamper-evident packaging on all over-the-counter (OTC) drugs and some cosmetics (qv), while ignoring other products they regulate. Table 1 offers examples of such packaging forms.

Table 1. FDA Examples of Tamper-Resistant Package Forms

Type	Description
film wrappers	transparent film[a] with distinctive design wrapped securely around a product or product container
blister or strip packs	dosage units individually sealed[b] in clear plastic or foil
bubble packs	product and container sealed in plastic[c] and mounted in or on display card
shrink seals and bands	bands or wrappers with distinctive design are shrunk by heat or drying to seal[a] union of cap and container
foil, paper, or plastic pouches	product enclosed in individual pouches[b]
bottle seals	paper or foil with distinctive design sealed[d] to mouth of container under cap
tape seals	paper or foil with distinctive design sealed[d] over all carton flaps or bottle cap
breakable caps	container sealed by plastic or metal cap[e] that either breaks away completely when removed from container or leaves part of cap attached to container
sealed tubes	mouth of tube is sealed and seal must be punctured to obtain product
sealed carton	all flaps of carton securely sealed and carton must be visibly damaged when opened to remove product
aerosol containers	inherently tamper resistant

[a] Must be cut or torn to open container and remove product.
[b] Must be torn or broken to obtain product.
[c] Must be torn or broken to remove product.
[d] Must be torn or broken to open container and remove product.
[e] Must be broken to open container and remove product.

Studies into consumer preferences for tamper-evident (TE) packaging have consistently revealed that consumers prefer products that are resistant to tampering and have shelf-visible features.

When compared to the potential expense for defending a single claim of tampering, the cost of effective tamper-evident packaging becomes insignificant.

Tamper-Evident Features

Selection of features to use should be done by objective testing during the package development stage. During the design stage, the package engineer should consider the function of the product and how the consumer intends to use it. Next, each tamper-evident feature that is usable on the package should be tested to determine which feature offers the greatest protection to the consumer. The test used should be objective, consistent, and replicable. Records of the test results should be retained indefinitely. If a feature selected for use achieves a lower value than others that were rejected, reasons for the selection should be recorded and retained with the test results. Cost should not be a factor in selecting which feature to use.

No single TE feature is best for all products. There are variations in effectiveness of similar features from different manufacturers, as well as variations in effectiveness where the product contributes to the effectiveness. The best feature for a product is the one that provides the greatest resistance to violation for the product in its current form and size. All features can be violated in some manner, but effective TE features provide greater difficulty in violating the product. In a particular instance a package was opened, the original product was replaced with a toxic substance, and no attempt was made to restore the package to its original appearance. The package worked as intended, ie, it showed it had been opened, but because there was no indication of violation to the actual product, the consumer still experienced injury.

Productiveness of Tamper-Evidence

Increased consumer awareness of packaging has led to an increase in the number of complaints of possible product tampering, although most are later dismissed as unfounded. Tamper-evident packaging prevents in-store tasting and violation, and if the feature is intact assures the consumer that the product is safe. Most experts agree that consumers should be more aware of what to look for in tamper-evident packaging. Educating the consumer may include pictures of the feature and product on the label and in media ads.

JACK L. ROSETTE
Forensic Packaging Concepts, Inc.

Food & Drug Packaging, published monthly.
Packaging, published monthly.
Packaging Digest, published monthly.
Packaging Technology & Engineering, published eight times per year.

ELECTRONIC MATERIALS

Electronic packaging refers to the placement and connection of many, sometimes thousands, electronic and electromechanical components in an enclosure that protects the system from the environment and provides easy access for routine maintenance. The packaging process starts with a chip, which is fabricated from a wafer. A chip is electrically connected to a chip carrier or substrate. This is the first level of interconnection, for which various interconnection methods are available, including wire-bonding, tape automated bonding (TAB), and flip-chip. The substrate-chip assembly is packaged in a case; the chip carriers are then mounted to a common base, normally a printed circuit board. Various chip carriers are electrically connected by metallized conductor paths, forming the second level of interconnection. Next, the various printed circuit boards are mounted on a backplane, forming the third level of interconnection. Several racks may be interconnected to form a cabinet. The fourth level of interconnection involves the interconnection between various cabinets.

Semiconductor Materials

Semiconductors (qv) are materials with resistivities between those of conductors and those of insulators (between 10^6 and 10^{-3} $\Omega\cdot$cm). The electrical properties of a semiconductor determine the functional performance of the device. Important electrical properties of semiconductors are resistivity and dielectric constant. The resistivity of a semiconductor can be varied by introducing small amounts of material impurities or dopants. Through proper material doping, electron movement can be precisely controlled, producing functions such as rectification, switching, detection, and modulation.

Silicon is most widely used in electronic applications as a material for dies, and also in the production of integrated circuits, rectifiers, diodes, transistors, and triacs.

Gallium arsenide is a dark gray material that is becoming increasingly popular as a semiconductor material. It is a compound semiconductor; the crystalline structure consists of gallium atoms alternating with arsenic atoms. Gallium arsenide is used in microwave devices, varactor diodes, schotty barrier diodes, light-emitting diodes, injection lasers, as well as Gunn-mode oscillators.

Substrate Materials

A substrate is a robust element that provides mechanical support for the die. It can be mounted with more than one die; such packages are called multichip modules. Because parasitic capacitance effects are directly proportional to the dielectric constant, substrate material should have a low dielectric constant. To minimize electrical losses, especially at high frequencies, a low dissipation factor is required. High volume resistance provides good insulation to prevent electrical current leakage between the conductor tracks. Since the substrate provides mechanical support to the die, its material should have good mechanical strength; it should also be thermally conductive to dissipate heat produced by the active devices. The coefficient of thermal expansion (CTE) of the substrate should match closely to that of the die to avoid thermomechanical stresses due to a CTE mismatch. The CTE of silicon is 2.3–4.7 ppm/°C and for GaAs, 5.4–5.72.

Materials used for substrates can be broadly classified into ceramics and metals. Commonly used ceramics, ie, alumina, aluminum nitride, and beryllia, can be easily incorporated into a hermetic package, ie, a package permanently sealed by fusion or soldering to prevent the transmission of moisture, air, and other gases.

In applications in which electrical conductivity is required, metals, copper, tungsten, molybdenum, and Kovar are the preferred chip-carrier materials. Metals have excellent thermal conductivities.

Attachment Materials

Die attach, the process by which the die is anchored to the substrate, can be accomplished by either (*1*) introducing an adhesive between the backside of the die and the substrate, or (*2*) using an electrical connection procedure, such as TAB or solder bumping. One of the most important properties of the attachment material is its bonding strength, which ensures that the die and the substrate stay in place when subjected to the stresses imposed during manufacture, storage, and operation. The other important properties for attachment materials are tensile strength, shear strength, and fatigue endurance. If the attachment material must transmit heat from the die to the substrate, its thermal conductivity is also a critical material property. The electrical properties of the attachment materials assume importance when the attachment material serves as an ohmic contact.

Case Materials

A case provides mechanical support and protection for the devices, interconnects, and substrate mounted in it; it also helps to dissipate heat during component operation and offers protection to the contents of the package from environmental stresses, contaminants, and, in the case of hermetic packages, moisture.

Cases can be classified as either hermetic or nonhermetic, based on their permeability to moisture. Ceramics and metals are usually used for hermetic cases, whereas plastic materials are used for nonhermetic applications. Cases should have good electrical insulation properties. The coefficient of thermal expansion of a particular case should closely match those of the substrate, die, and sealing materials to avoid excessive residual stresses and fatigue damage under thermal cycling loads. Moreover, since cases must provide a path for heat dissipation, high thermal conductivity is also desirable.

Among ceramics, the most commonly used material is alumina, which has good electrical resistivity (10^{12} $\Omega\cdot$cm). Kovar, one of the most commonly used metals for hermetic cases, is composed of 54% iron, 29% nickel, and 17% cobalt.

A PEM encapsulant is generally an electrically insulating plastic material formulation that protects an electronic device and die-leadframe assembly from the adverse effects of handling, storage, and operation (see EMBEDDING). Various molding compounds are used as encapsulants. The molding compound is a proprietary multicomponent mixture of an encapsulating resin with various types of additives. The principal active and passive (inert) components in a molding compound include curing agents or hardeners, accelerators, inert fillers, coupling agents, flame retardants, and stress-relief additives.

Lead Materials

Leads serve as the input–output interconnections between the component package and the mounting platform. Sometimes leads also aid in the dissipation of heat generated in the package. In the case of plastic packages, leads are formed from the leadframe, which also acts as a heat-dissipation path and a mechanical support for the die.

Because they serve as a path for electrical signals, leads should have good electrical conductivity. They should also have good corrosion resistance, since corrosion products can change the electrical properties of the lead, and because leads are soldered to the board, they should be wettable by solder. The coefficient of thermal expansion should closely match that of the die material for leadframes, substrate, and sealing glass. Other important properties of leads are yield strength and fatigue properties. Good thermal conductivity is also desirable to enhance the path for heat dissipation.

Alloy 42 (42% Ni–58% Fe) and Kovar are lead materials commonly used in ceramic chip carriers.

Solder Materials

Solders are alloys that have melting temperatures below 300°C, formed from elements such as tin, lead, antimony, bismuth, and cadmium. Tin–lead solders are commonly used for electronic applications, showing traces of other elements that can tailor the solder properties for specific applications.

Alloy selection depends on several factors, including electrical properties, alloy melting range, wetting characteristics, resistance to oxidation, mechanical and thermomechanical properties, formation of intermetallics, and ionic migration characteristics. These properties determine whether a particular solder joint can meet the mechanical, thermal, chemical, and electrical demands placed on it.

Printed Circuit Board Materials

A printed circuit board (PCB) typically consists of a copper circuit pattern created on a copper-clad composite laminate by lithography and electroplating technologies. A typical printed circuit board serves the electronic system in three ways: mechanically, by providing support to the components and the conductors; thermally, by offering a thermal conduction path to dissipate heat generated by the devices mounted on it; and electrically, by serving as an insulator for the conductors. Chemical properties also need to be considered, since moisture absorption and chemical attack can both degrade the material properties.

Most of the laminates used for rigid printed circuit boards have been classified, by the National Electrical Manufacturers Association (NEMA), according to the combination of properties that determine the suitability of a laminate for a particular use. Fiber reinforcements make laminate-effective properties orthotropic.

E-glass–epoxy laminates, by far the most widely used circuit board materials, have been designated FR-4 (fire retardant epoxy-glass cloth) by NEMA.

Ceramic laminates withstand a harsher environment than organic laminates. They also have good thermal conductivities, higher flexural strength, and a better match of coefficients of thermal expansion with the components mounted on them, ensuring good board interconnection reliability under cyclic thermal loading.

Conformal Coating Materials

Conformal coatings are protective coatings applied to circuit board assemblies. They protect the interconnect conductors, solder joints, components, and the board itself; they reduce permeability to moisture, hostile chemical vapors, and solvents in the coating. Use of conformal coatings eliminates dendritic growth between conductors, conductor bridging from moisture condensation, and reduction in insulation resistance by water absorption.

The critical property for conformal coatings is resistance to chemicals, moisture, and abrasion. Other properties, such as the coefficient of thermal expansion, thermal conductivity, flexibility, and modulus of elasticity, are significant only in particular applications. The dielectric constant and loss tangent of the conformal coating are important for high speed applications.

Materials for coating must last as long as the product itself, be easy to apply and rework, and be cost-effective. Various materials are used for conformal coatings. Polyurethanes are the most widely used and offer good resistance to moisture, fungus, abrasion, solvent and chemical. In addition, they have good adhesion, low shrinkage, flexibility, and elasticity, and are particularly suited to applications requiring good humidity resistance. Also used are acrylic resins, epoxies, and silicone–based coatings.

Connector Materials

Connectors are third-level interconnections between daughter boards, between subassemblies, between systems, and to peripherals. The connector assembly normally consists of insert, pin, and contact materials. Platings and/or lubricants are often used to reduce wear and improve contact resistance between the pin and the contact. The insert material must provide support and insulation for the pin and contacts. The coefficient of thermal expansion of the material and the mold shrinkage should be low to ensure pin-to-pin dimensional stability in high density connectors. At higher pin densities, the volumetric and surface resistivity should be high for proper electrical performance. Flexibility is essential for a snap fit. Properties required of metals used for pin and contacts are low electrical resistivity, high strength, wear resistance, corrosion resistance, and a high modulus of elasticity. A low dielectric constant is required for high speed connectors.

Polymers are used as inserts for pins and contacts. Commonly used materials for pins and contacts are brass, beryllium copper, phosphor–bronze, and copper–nickel.

Cable and Flex Circuit Materials

A cable or flex circuit is often employed as an interconnection between electronic circuits that are not easily connected by other means. The interconnection density obtained with a cable or flex circuit is lower than that achieved with a backpanel.

Properties desired in cable insulation and flexible circuit substrate materials include mechanical flexibility, fatigue endurance, and resistance to chemicals, water absorption, and abrasion. Both thermoplasts and thermosets are used as cable-insulating materials. Thermoplastic materials possess excellent electrical characteristics and are available at relatively low cost.

Copper is by far the most widely used conductor material. Commonly used materials for cable insulation are poly(vinyl chloride) (PVC) compounds, polyamides, polyethylenes, polypropylenes, polyurethanes, and fluoropolymers.

MICHAEL PECHT
University of Maryland
at College Park
ASOHK PRABHU
Nitto Denko America Inc.

M. Pecht, *Integrated Circuit, Hybrid, and Multichip Module Package Design Guidelines: A Focus on Reliability,* John Wiley & Sons, Inc., New York, 1994.

C. A. Harper, *Electronic Packaging and Interconnection Handbook,* McGraw-Hill Book Co., Inc., New York, 1991.

A. H. Kumar and R. R. Tummala, *Int. J. Hybrid Microelectronics,* **14**(4), 137–150 (Dec. 1991).

M. G. Pecht, L. Nyugen, and E. B. Hakim, *Plastic Encapsulated Microelectronics: Materials, Processes Quality, Reliability and Applications,* John Wiley & Sons, Inc., New York, 1995.

PACKING MATERIAL. See PUMPS.

PAINT

ARCHITECTURAL

Paint is defined as "any liquid, liquefiable, or mastic composition designed for application to a substrate in a thin layer which is converted to an opaque solid film after application." Paint is "used for protection, decoration, or identification, or to serve some functional purpose such as the filling or concealing of surface irregularities, the modification of light and heat, etc" (S. LeSota, *Paint/Coatings Dictionary*). Paints discussed herein are commonly referred to as architectural coatings, house paints, and trade sales paints.

Paint provides protection for materials that would have a much shorter life span without a protective coating. Paints are most often described according to the type of binder or solvent employed, eg, latex-, alkyd-, or oil-based paint describes the type of binder or resin system used. Paints are also described as water- or solvent-based, referring to the type of solvent used in the formulation. In the paint industry, the word solvent usually refers to organic hydrocarbon solvents and does not include water. The conditions for which a paint is developed is also an important part of its description, ie, a paint recommended for interior or exterior application. Properties such as color and appearance are also terms for describing paints. Flat, satin, semigloss, and gloss paints refer to how dull or shiny the dried paint film can be expected to appear. The terminology for paint color is so varied that color swatches or chips often accompany a paint purchase so the user can visualize how the color will appear.

Chemical and Physical Properties

Most paints are made up of four basic groups of chemical raw materials: binders or resins, pigments, solvents, and additives. When a paint is applied to a surface, the solvents begin to evaporate while the binder, pigments, and additives remain on the surface to form a hard, dry solid film. The paint formulator selects the proper type and concentration of raw material from each of these groups that will provide paint with the desired end use properties.

Examples of binders and resins used in paints include latex emulsions based on acrylic and vinyl acetate, as well as styrene polymer and copolymer systems. Alkyds, linseed oil, and oil-modified epoxy and polyurethane resins are common binders in solvent-based paints. Water-reducible alkyd and oil systems are also available. The types of pigments used include organic and inorganic colored pigments as well as inorganic extenders and filler pigments. Solvent choice is limited mainly to a solvent that is compatible with the binder and has the desired evaporation rate and toxicity profile. Additives include thickeners, biocides, driers, pigment dispersants, surfactants (qv), defoamers, and other specialty ingredients used at relatively low levels in a paint formulation.

Paint Formulations. Tables 1 and 2 provide examples of generic water-based latex and solvent-based alkyd oil paints. These formulations exhibit typical proportions of paint ingredients.

Application and Appearance Properties

A well-formulated paint that is properly manufactured and packaged should have good product uniformity in the container without either pigment settling, phase-separation of the liquids, or color float. The paint should also have an acceptable viscosity and rheological profile. The expectation for most paints are that they will have a degree of structure and viscosity in the can, and load heavily but easily on a brush or roller with minimum dripping, making it easy to apply heavy coats quickly and uniformly.

Table 1. Exterior Acrylic Latex Flat Water-Based House Paint

Raw material ingredients	Weight, kg	Volume, L
Grind portion[a]		
water	144.1	144.1
propylene glycol	72.3	69.2
in-can preservative	2.0	2.0
cellulosic thickener, 100%	3.6	2.5
dispersant, 25%	14.7	13.3
surfactant	2.4	2.2
defoamer	2.4	2.6
titanium dioxide	210.8	52.5
zinc oxide	30.1	5.3
extenders	192.8	80.7
Let-down portion		
latex emulsion, 53.3%	391.0	365.8
polymeric opacifier	79.5	76.7
texanol	11.9	12.4
defoamer	2.4	2.6
mildewcide	2.4	2.3
polyurethane thickener, 25%	12.0	12.5
aqueous ammonia, 28%	2.7	2.9
water	150.4	150.4
Total	*1327.5*	*1000.0*
Properties		
pigment volume concentration, %	49	
volume solids, %	37.5	
weight solids, %	51.5	
density, g/L	1.3	
VOC, g/L	200	

[a] Pigment, dispersion, or millbase.

Table 2. Exterior Alkyd-Linseed Oil Flat Solvent-Based House Paint

Raw material ingredients	Weight, kg	Volume, L
Grind portion[a]		
mineral spirits	78.3	100.0
clay thickener	13.5	9.2
mildewcide	13.5	8.4
alkyd resin, 60%	186.7	200.0
surfactant	12.8	12.5
titanium dioxide	321.0	80.0
extenders	319.3	120.0
Let-down portion		
alkyd resin, 60%	166.3	177.0
linseed oil, 80%	119.9	130.0
mineral spirits	121.7	155.4
organometallic driers	6.0	5.5
antiskinning agent	1.8	2.0
Total	*1360.8*	*1000.0*
Properties		
pigment volume concentration, %	40	
volume solids, %	54	
weight solids, %	72	
density, g/L	1.4	
VOC, g/L	360	

[a] Pigment, dispersion, or millbase.

For the most part, additives control the application or rheological properties of a paint. The volume solids of a paint is an equally important physical property affecting the application and rheological properties. Without adequate volume solids, the desired application and rheological properties may be impossible to achieve, no matter how much or many additives are incorporated into the paint.

The color and appearance of an object are among the first things that cause people to evaluate and judge it as acceptable or not. Thus, color and appearance of a paint are critical properties for analyzing product quality.

Light reflection and absorption are the properties that give paints their particular color. The overall appearance of a paint is a combination of its surface qualities and its color. The surface qualities of a paint are described in terms of gloss and texture. Gloss is a function of light reflection off a surface and can be defined as the degree of approach to that of a mirror surface. Smooth paint surfaces have a high gloss that are mirror-like in appearance; dull or rough paint surfaces have a nonglossy or flat appearance. Paints of the same color but different surface qualities, eg, one gloss and one flat, appear different to an observer and are often mistakenly described as being slightly different in color.

Manufacture and Processing

Along with the choice of a binder for the paint formula, the selection of pigments and their volume relationship to the binder govern the performance properties of the resultant paint. When evaluating the properties of paint formulations, volume relationships of the pigments to the binder are used in preference to weight relationships because of the wide density range of various binders and pigments used in paints. The volume relationships between pigments and binder in a paint formula provide more precise correlation to paint performance than do weight relationships.

Pigment Volume Concentration. The volume relationship of pigments to binders is known as the pigment volume concentration (*PVC*). An important aspect in the *PVC* determination of a paint formula is the knowledge of the critical *PVC* (*CPVC*). The *CPVC* of a paint is the pigment volume concentration at which there is just a sufficient amount of binder to fill all the voids between the pigment particles in a dry paint film.

Knowledge of the *CPVC* of a paint formula is important because dramatic changes in the behavior and properties of the paint occur when the *PVC* is in the *CPVC* region. In a paint formula where the *PVC* is below the *CPVC,* the paint is characterized by having an airless film, higher gloss, lower porosity, good weather resistance, flexibility, and abrasion resistance. When the *PVC* of a paint is greater than its *CPVC,* the resultant properties are air voids in the dry film, greater porosity, poor weatherability, low gloss, and poor flexibility and abrasion resistance.

Pigment Dispersion and Paint Volume Solids. Another important aspect in pigment–binder relationships is the degree of dispersion of pigment particles in the binder. Pigment particles tend to agglomerate both in their dry state before manufacture and later when they are part of the paint formula. As a result, energy in the form of paint-mixing equipment is required to disperse the pigment agglomerates in the paint vehicle (binder and solvent). Along with proper mixing, chemical additives such as dispersants, surfactants, and thickeners are necessary ingredients of the paint formulation for controlling and inhibiting pigment agglomeration during and after the mixing process.

A third criterion for the proper formulation of house paints is volume solids level. Paint with low volume solids are characterized by poor adhesion and poor exterior durability.

Paint Processing. The manufacture of house paints involves mixing together the raw material ingredients in such a way that the finished paint is a homogeneous mixture of evenly distributed ingredients. The particular raw material ingredients used in house paints are chosen not only for their appearance and performance attributes, but also for their relative compatibility with each other and their ability to be mixed together to produce a relatively stable and homogenous paint product without significant chemical reactions during processing. Most paint manufacturing processes result in a 100% yield of final product, mostly due to the fact that no chemical reactions take place that can result in an unwanted chemical by-product. Therefore, the manufacture of house paints is concerned mostly with proper blending and mixing of the raw material ingredients in order to ensure that the ingredients are evenly dispersed in the finished paint. Most house paints manufactured in the 1990s rely on high speed mixers or dispersators.

Both water-based latex paint and solvent-based alkyd and oil paint have most manufacturing processes in common. The paint formulations in Tables 1 and 2 list the paint raw material ingredients according to the ingredients in both the grind portion of the paint and the let-down portion of the paint. The order of adding these ingredients into the paint mixing vessel essentially follows the order of ingredient listing. The grind portion of a paint formulation is typically made in the vessel by using a high speed disperser, whereas the let-down portion of the paint is typically made in another vessel that may have a smaller motor with a paddle-type mixing blade to keep the liquids stirring but not necessarily under shear. Liquids in the form of water, solvents, and/or resins are added to the grind mixing vessel first. Additives that assist pigment dispersion and wetting are added next. Pigments are then added to the liquid; the motor speed and the height of the blade in the mixing vessel are adjusted as more pigments are added. For efficient mixing it is important to have the mixing blade submerged approximately halfway in the grind volume of the liquid–pigment mixture. Once all the materials in the grind portion are added to the mixing vessel the high speed mixing should continue for 15–20 minutes to ensure complete dispersion of the pigment particles.

Paint Types and End Uses

Because a paint's appearance is mostly a function of *PVC,* paints can be described in reference to their *PVC* as in Figure 1.

Exterior House Paints. Quality exterior flat paints, which provide protection for the exterior surfaces of the house, are characterized by proper choice of binder, lower *PVC*s formulated below *CPVC,* and higher volume solids. Exterior paints are more likely to be applied at

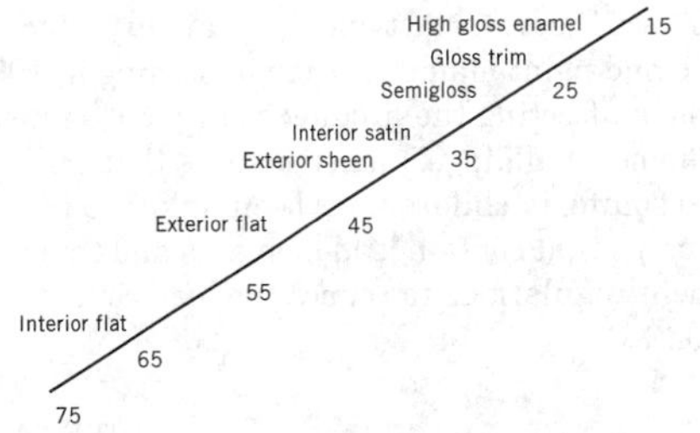

Figure 1. Formulation types where each number represents the percent of pigment volume concentration for the paint types shown.

marginal temperature, subjected to rainfall soon after the paint is applied, or applied to deteriorating surfaces. In order to protect the substrate, the paint film must adhere to it well and have the flexibility to expand and contract as the substrate does.

Interior House Paints. Interior flat paints range in *PVC* from below to well above *CPVC,* depending on the quality and intended service. Ceiling paints, for example, are not usually scrubbed, and can be formulated well above *CPVC* to obtain good hiding at low cost. Trim paints such as semigloss and gloss paints are lower in *PVC* and usually receive the most wear and abuse on doors, windows, and surrounding trim. They must resist permanent staining, make removal of stains relatively easy, and not be marred by the washing or scrubbing necessary to remove dirt or stains. This is a principal reason that gloss and semigloss paints are used for trim. Increased consumer awareness of chemical hazards and the development of new paint technology for solvent-free water-based latex paints is likely to accelerate this trend, which is already underway in Europe.

Specialty House Paints and Finishes. Several other related types of house paint finishes not already mentioned are formulated for specific end uses or appearance characteristics. These products include primers, sealers, and opaque solid stains, and are available from most paint manufacturers in either water-based latex or solvent-based alkyd and oil.

Application Methods and Surface Preparation. For good durability and performance, proper surface preparation and correct application of house paints are as important as the formulation of high quality paint. Proper surface preparation prior to painting involves several considerations. For new construction, proper installation and protection of the substrate material are necessary. For previously painted surfaces, preparation involves mostly cleaning and removing any existing paint that is unstable. Once surface preparation is complete, the application process can begin.

The most common application tools for house painting are brushes, spray equipment, rollers, and pad applicators.

Environmental, Health, and Safety Factors

The most significant environmental and health issues affecting the paint and coatings industry in the 1990s are regulations to lower the VOC content for virtually all types of paints and to restrict the use of certain solvents known as hazardous air pollutants (HAPs) under the federal Clean Air Act. Except for the water in a latex paint or in other water-based coatings, solvents used in house paints are mostly all VOCs.

Health and safety issues affect both the professional painter and paint manufacturer who as part of the occupation can be exposed to high concentrations of organic solvent for extended periods of time. Environmental issues focus on the contribution of organic solvents to air pollution and other issues such as hazardous waste disposal.

Restriction on the use of certain types of solvents, listed as HAPs under the Clean Air Act, are forcing paint manufacturers not only to lower the limits on the amount of organic solvents in a paint, but also to eliminate certain types of solvents. Thus paint manufacturers are challenged to comply simultaneously with both VOC and HAP regulations. These Clean Air Act mandates are expected to affect most types of paints and paint manufacturers beginning in 1996.

Another issue affecting the architectural paint industry is the remediation of homes, buildings, and structures that contain lead-based paint. Lead poisoning in children has been linked to ingestion of paint dust or paint chips that contain lead pigments and this has resulted in U.S. government regulations to reduce the lead content in paint to no more than 0.06%.

ARTHUR A. LEMAN
Rohm and Haas Company

S. LeSota, *Paint/Coatings Dictionary,* Federation of Societies for Coatings Technology, Blue Bell, Pa., 1978.

Federation Series on Coatings Technology, Federal of Societies for Coatings Technology, Blue Bell, Pa.

Journal on Coatings Technology, Federation of Societies for Coatings Technology, Blue Bell, Pa.

Annual Book of ASTM Standards, Section 6: Paints and Related Coatings, American Society for Testing and Materials, Philadelphia, Pa.

PAINT AND FINISH REMOVERS

The term finish denotes the final process of manufacturing. Finishing operations include such processes as clear coating (varnishes and lacquers), painting, plating, anodizing, phosphatizing, galvanizing, and blueing, all of which take place at the terminal point of manufacturing. Finishing is defined as the process of coating or treating a surface for the purpose of protecting and/or decorating the product. The useful life of most usable objects is greater than the finish. This results in a periodic need to remove and replace the finish.

The physical properties of finish removers vary considerably due to the diverse uses and requirements of the removers. Finish removers can be grouped by the principal ingredient of the formula, method of application, method of removal, chemical base, viscosity, or hazardous classification. Except for method of application, a paint remover formulation usually has one aspect of each group, by which it can be used for one or more applications.

Finish removers are applied by brushing, spraying, troweling, flowing, or soaking. Removal is by water rinse, wipe and let dry, or solvent rinse. Removers may be neutral, basic, or acidic. The viscosity can vary from water thin, to a thick spray-on, to a paste trowel-on remover. The hazard classification, such as flammable or corrosive, is assigned by the U.S. Department of Transportation (DOT) for the hazardous materials contained in the remover.

Organic Finish Removers

Methylene Chloride Finish Removers. The major ingredient of a formula is the chemical of greatest volume. Methylene chloride formulas are the most common organic chemical removers. The low molar volume of methylene chloride allows it to rapidly penetrate the finish by entering the microvoids of the finish. When the solvent reaches the substrate, the remover releases the adhesive bond between the finish and the substrate and causes the finish to swell. The result is a blistering effect and an efficient rapid lifting action. Larger molecule solvents generally cannot cause this lifting action and must dissolve the finish. When methylene chloride is used in amounts of 78% or more, even with flammable cosolvents, the mixture is nonflammable. A typical methylene chloride base remover includes cosolvents, activators, evaporation retarders, corrosion inhibitors, thickeners, and wetting agents.

Typical cosolvents include methanol, ethanol, isopropyl alcohol, or toluene. The selection of cosolvents depends on the requirement of the formula and their interaction with other ingredients. Methanol is a common cosolvent in methylene chloride formulas since it has good solvency and is needed to swell cellulose-type thickening agents. A typical methylene chloride formula used to strip wood is as follows: methylene chloride, 81.1%; toluene, 2.1%; paraffin wax (ASTM 50–53°C mp), 1.6%; methycellulose, 1.2%; methanol, 7.8%; and mineral spirits, 6.2%.

Health and Safety. Remover formulas that are nonflammable may be used in any area that provides adequate ventilation. The vapor of methylene chloride produces hydrogen chloride and phosgene gas when burned. Methylene chloride-type removers should not be used in the presence of an open flame or other heat sources such as kerosene heaters.

Persons exposed to methylene chloride removers should wear protective clothing and eye protection.

Environmental Impact. Methylene chloride is nonphotochemically reactive and is not listed as an ozone (qv) depleter. Methylene chloride removers can easily be recovered from paint chips and other residue sedimentation, thus allowing recovery of remover and its continued use.

Petroleum and Oxygenate Finish Removers. Many older finishes can be removed with single solvents or blends of petroleum solvents and oxygenates. Varnish can be removed with mineral spirits, shellac can be stripped with alcohols, and lacquers can be removed with blends of acetates and alcohols (lacquer thinners). The removal mechanism is one of dissolving the coating, then washing the surface or wiping away the finish. This method is often used to reamalgamate or liquefy old finishes on antique items of furniture.

In petroleum and oxygenate finish removers, the major ingredient is normally acetone, methyl ethyl ketone, or toluene.

Health and Safety. Petroleum and oxygenate formulas are either flammable or combustible. Flammables must be used in facilities that meet requirements for hazardous locations.

Adequate ventilation that meets the exposure level for the major ingredient must be attained. Extreme caution must be taken to prevent the possibility of fire when using flammable removers.

Environmental Impact. Most petroleum and oxygenate removers are photochemically reactive and classed as volatile organic compounds (VOCs). Disposal of this type of remover is difficult because the dissolved finish cannot be separated from the spent remover and the whole mixture must be disposed as a liquid hazardous waste. Distillation to recover the solvents is dangerous because the nitrocellulose from lacquer finish may cause autoignition in the still. Several states restrict the use of these products.

Other Organic Finish Removers. Concerns over the reported toxicity and carcinogenicity of methylene chloride have stimulated research for alternative solvents in remover formulas. *N*-Methylpyrrolidinone and dibasic esters (dimethyl glutarate or dimethyl adipate) have been used in removers. They remove single-component finishes but work much more slowly than methylene chloride, petroleum, and oxygenate group removers. They have little success on epoxy and catalyzed finishes.

Health and Safety. Both *N*-methylpyrrolidinone and dibasic esters have very low vapor pressure which limits worker exposure to vapors. Manufacturers recommend that the same safety precautions be taken as with other organic solvents. Hazardous location requirements must be considered if the formula is flammable. Ventilation that reduces vapors to manufacturer's recommended exposure levels should be used. Protective clothing must be worn during use.

Environmental Impact. The volume of waste remover from these products is remarkably increased when compared to methylene chloride, petroleum, and oxygenate removers, since both *N*-methylpyrrolidinone and dibasic esters have low vapor pressures. Recovery of the remover after use is difficult because the finish is resolubilized by the remover.

Inorganic Finish Removers

Liquid Alkaline Removers. This group consists of alkaline materials that are dissolved in water then heated to an appropriate temperature to remove finishes. In a typical application, a hot water bath large enough to submerge an item is used. Various alkaline materials may be used to provide the desired alkalinity. Of these, sodium compounds are preferred, such as sodium hydroxide, sodium carbonate, sodium silicates, mono-, di-, and trisodium phosphates, tetrasodium pyrophosphate, and sodium tripolyphosphate. Compounds of other metals, such as potassium or lithium, may be used.

This aqueous alkaline remover is used for stripping the finish from wood or ferrous metals at a mix ratio of 30–600 g/L (0.25–5 lbs/gal).

Paste-Type Alkaline Removers. Sodium hydroxide, potassium hydroxide, or other caustic compounds are blended to make these types of removers. Polymer-type thickeners are added to increase the viscosity that allows the remover to be applied with a brush, trowel, or spray. Some of these products use a paper or fabric covering to allow the remover finish mixture to be peeled away. The most common application for this group of removers is the removal of architectural finishes from the interior and exterior of buildings. The long dwell time allows for many layers of finish to be removed with one thick application of remover.

Sodium hydroxide and salt can be heated to a fused state in baths to allow the removal of finishes from ferrous metals. The most common use of this method is the removal of heavy concentrations of paint on conveyer parts and hangers used in production spray systems.

Health and Safety. Protective clothing that is compatible with the remover formula must be worn. Caustic soda baths should be ventilated to remove vapors from the work area. Most caustic removers are corrosive and cause severe burns with minimal contact to the skin.

The liquid from spent caustic soda baths must be disposed of or treated as a hazardous waste. The finish residue may contain heavy metals as well as caustic thus requiring treatment as a hazardous waste.

Manufacturing and Processing

Finish removers are manufactured in open or closed kettles. Closed kettles are preferred because they prevent solvent loss and exposure to personnel. To reduce air emissions from the solvents, condensers are employed on vent stacks.

Standard 0.25 or 0.50 lb (227 g) tin coated cans are used for packaging liquid with neutral and mildly alkaline base formulas; polypropylene is used for acid–base removers. Steel and polypropylene drums are used for industrial removers. Viscous removers are packaged in removable top containers. Dry caustic removers are packaged in bag-lined boxes or fiber drums.

The DOT has established standards for the packaging and labeling of hazardous materials offered for shipment by public transportation. The Consumer Product Safety Commission (CPSC) has set standards for retail labeling and packaging. OSHA and the U.S. EPA have labeling requirements.

DAVID L. WHITE
Kwick Kleen Industrial Solvents, Inc.
JAY A. BARDOLE
Vincennes University

Industrial Users of Paint and Finish Removers, Paint Remover Manufacturer's Association, 1992.

Solvents Used in Paint Removers, Paint Remover Manufacturer's Association, Sept. 1991.

Chemical Protective Clothing for Furniture Stripping, Department of Health and Human Services, Washington, D.C., Mar. 11, 1991.

Methylene Chloride Consumption By Paint and Coating Removal Groups, Paint Remover Manufacturer's Association, 1992.

PAPER

Paper consists of sheet materials that are comprised of bonded small discrete fibers. The fibers usually are cellulosic in nature and are held together by hydrogen bonds (see CELLULOSE). The fibers are formed into a sheet on a fine screen from a dilute water suspension. The word paper is derived from papyrus, a sheet made in ancient times by pressing together very thin strips of an Egyptian reed (*Cyperus papyrus*).

Paper is made in a wide variety of types and grades to serve many functions. Writing and printing papers constitute ca 30% of the total production. The balance, except for tissue and toweling, is used primarily for packaging (qv). Paperboard differs from paper in that it generally is thicker, heavier, and less flexible than conventional paper.

More than 95% of the base material used in paper and board manufacture is fibrous. Whereas a large (90%) percentage originates from wood (qv), the filler contents of some grades of paper approach 30%. Many tree species encompassing both hardwood and softwood are used to produce pulp. In addition to the large number of wood types, there are many different manufacturing processes involved in the conversion of wood to pulp. These range from mechanical processes, by which only mechanical energy is used to separate the fiber from the wood matrix, to chemical processes, by which the bonding material, ie, lignin (qv), is removed chemically (eg, the kraft process). Many combinations of mechanical and chemical methods also are employed (see PULP). Pulp properties are determined by the raw material and manufacturing process, and must be matched to the needs of the final paper product.

Reclaimed fiber accounts for ca 40% of the total fiber used in the United States. A variety of sequences are used to disperse and clean the waste fiber. Emphasis is on mechanical screening and cleaning, washing, and flotation. The properties of these pulps depend largely on the input raw material and generally are lower in strength and brightness than a comparable virgin pulp. Typical yields for de-inking processes range from 60 to 80%.

Nonwood fibers are used in relatively small volumes. Examples of nonwood pulps and products include cotton linters for writing paper and filters, bagasse for corrugated media, esparto for filter paper, or Manila hemp for tea bags. Synthetic pulps which are based on such materials as glass (qv) and polyolefins also are used (see OLEFIN POLYMERS). These pulps are relatively expensive and usually are used in blends with wood pulps where they contribute a property such as tear resistance, stiffness, or wet strength which is needed to meet a specific product requirement.

Physical Properties

Most properties of paper depend on direction, ie, the machine, cross-machine, and thickness directions. For example, strength is greater if measured in the machine direction, ie, the direction of manufacture, than in the cross-machine direction.

Because the fibers generally are anisotropic, they tend to be deposited on the wire in layers under shear. The layered structure results in the different properties measured in the thickness direction as compared to those measured in the in-plane direction. The orthotropic behavior of paper is observed in most paper properties and especially in the electrical and mechanical properties.

The basis weight, W, is the mass in g/m^2 (TAPPI T410). It can also be expressed as pounds of a ream of 500 sheets of a given size, but the sheet sizes are not the same for all kinds of paper.

The caliper is the thickness in μm of a single sheet measured under specified conditions (TAPPI T411). Calipers for a number of common paper and board grades are capacitor tissue, 7.6 μm; facial tissue, 65 μm; newsprint, 85 μm; off-set bond, 100 μm; linerboard, 230–640 μm; and book cover, 770–7600 μm.

The tensile strength is the force per unit width parallel to the plane of the sheet that is required to produce failure in a specimen of specified width and length under specified conditions of loading (TAPPI T404). The strength of paper also is expressed in terms of a breaking length, ie, the length of paper that can be supported by one end without breaking. Breaking lengths for typical papers are from ca 2 km for newsprint to 12 km for linerboards.

Other physical properties of paper include stretch, bursting strength, tearing strength, stiffness, folding endurance, moisture content, and water-vapor permeability.

The common optical properties of paper are brightness, color, opacity, transparency, and gloss.

Chemical Properties

The chemical composition of paper is determined by the types of fibers used and by any nonfibrous substances incorporated in or applied to the paper during the papermaking or subsequent converting operations. Paper usually is made from cellulose fibers obtained from the pulping of wood. Occasionally, synthetic fibers and cellulose fibers from other plant sources are used. Paper properties affected directly by the fibers' chemical composition include color, opacity, strength, permanence, light fastness, and electrical properties. Development of interfiber bonding during papermaking also is strongly influenced by the composition of the fibers.

Mechanical permanence depends principally on the pH of paper. Whereas lignin-containing pulps tend to yellow with aging, experimental evidence demonstrates that lignin does not cause strength loss with aging. Papers that are manufactured under neutral or mildly alkaline conditions can maintain their physical properties for hundreds of years. In order to prevent the acidificiation of papers by the absorption of atmospheric pollutants such as sulfur dioxide and nitrous oxide, it is also important that the paper contain an alkaline reserve, usually in the form of calcium carbonate used as a filler. Fine paper is made predominantly under alkaline conditions in order to take advantage of low cost bright minerals, such as ground and precipitated calcium carbonates, as fillers (qv). In the United States, the transition to alkaline papermaking has been dramatic in the 1990s.

Hemicelluloses in chemical pulps contribute to bonding; therefore, pulps containing hemicelluloses are used for wrapping papers and other grades which require bonding for strength, and in glassine which requires bonding for transparency. In most papers, the chemical composition largely reflects those nonfibrous materials that were added to the paper to achieve the desired physical, optical, or electrical properties (see HEMICELLULOSE; PAPERMAKING ADDITIVES). Examples of chemicals and resultant properties are dyes and optical brighteners to enhance appearance, resins to impart wet strength, rosin or starch size to reduce penetration of aqueous liquids, pigment coatings to provide a smooth surface for printing, mineral fillers to increase opacity, polymers applied by saturation or extrusion to impart mechanical or barrier properties, and cationic polyelectrolytes and resistive polymers used in the interior and on the surface, respectively, of papers for dielectric recording. The performance of a limited number of papers depends on chemical reactions of noncellulosic additives, eg, photographic, thermal, and carbonless copy papers (see MICROENCAPSULATION; PHOTOGRAPHY), and strips of paper saturated with color-forming reagents for urinalyses (see AUTOMATED INSTRUMENTATION, CLINICAL CHEMISTRY).

Manufacture and Processing

Stock Preparation. During the stock preparation steps, papermaking pulps are most conveniently handled as aqueous slurries. In the case of adjacent pulping and papermaking operations, pulps usually are delivered to the paper mill in slush form directly from the pulping operation. Purchased pulps and waste paper are received as dry sheets (laps) and must be slushed before use.

Beating and Refining. Virtually all pulps are subjected to certain mechanical actions before being formed into a paper sheet. Such treatments are used to improve the strength and other physical properties of the finished sheet and influence the behavior of the system during the sheet-forming and drying steps. During refining, the cellulose fibers are swollen, cut, macerated, and fibrillated.

Batch systems have been replaced largely with continuous, pump-through equipment.

Sheet Forming, Pressing, and Drying

Continuous sheet forming and drying came into use in ca 1800. The equipment was of two types: the cylinder machine and the Fourdrinier machine. In the former, a wire-covered cylinder is mounted in a vat containing the fiber slurry. As the cylinder revolves, water drains inward through the screen and the paper web is formed on the outside. The wet web is removed at the top of the cylinder, passes through press rolls for water removal, and then passes into steam-heated, cylindrical drying drums. The Fourdrinier is more complex and basically consists of a long continuous wire screen which is supported by various devices that improve drainage. The fiber slurry, which is introduced at one end through a headbox and slice, loses water as it progresses down the wire, thereby forming the sheet. It then passes to presses and dryers as in the cylinder machine.

Continuous paper machines have undergone extensive mechanical developments since the 1950s, although the principles employed have changed little.

Sheet Pressing. The sheet leaving the wet end contains on the order of five parts of water per part of fiber; however, it is possible to remove additional water mechanically without adversely affecting sheet properties. This is achieved in rotary presses, of which there may be one or more on a given paper machine. The press rolls may be solid or perforated and often suction is applied through the interior. The sheet is passed through the presses on continuous felts (one for each press), which act as conveyors and porous receptors of water. They are essential to the efficiency of the papermaking process since it is much more expensive to remove water from paper by drying with heat than by dewatering with pressure. The water content of the sheet usually can be reduced by pressing to 1.9–1.2 parts of water per part of fiber without deleteriously affecting product quality.

Sheet Drying. At a water content of ca 1.2–1.9 parts of water per part of fiber, additional water removal by mechanical means is not feasible and evaporative drying must be employed. This is at best an efficient but costly process and often is the production bottleneck of papermaking. The dryer section most commonly consists of a series of steam-heated cylinders. Alternate sides of the wet paper are exposed to the hot surface as the sheet passes from cylinder to cylinder. The water vapor is removed by way of elaborate air systems. Most dryer sections are covered with hoods for collection and handling of the air, and heat recovery is practiced in cold climates. The final moisture content of the dry sheet usually is 4–10 wt %.

Other types of dryers may be employed for special products or situations. For example, the Yankee dryer, a steam-heated cylinder, 3.7–6.1 m dia, dries the sheet from one side only. It is used extensively for tissues, particularly where creping is accomplished as the sheet leaves the dryer, and to produce machine-glazed papers where intimate contact with the polished dryer surface produces a high gloss finish on the contact side.

High velocity air drying, in which jets of hot air are directed against the sheet in a normal direction, is used in Yankee dryers and in combination with a percolation through-drying process. The latter technology is commonly used for special, high quality tissue products. Infrared and other radiant drying techniques also are utilized in special cases.

There are significant differences in physical properties of machine-made paper in the manufacture direction (MD) and in the cross direction (CD) that result from restraint during drying. For example, modulus of elasticity and tensile strength are typically more than twice as high in the MD. Hygrostability of paper also depends strongly on the degree of restraint during drying, therefore it is of interest to design drying sections which hold the paper under CD restraint as far as possible.

Converting. Almost all paper is converted by undergoing further treatment after manufacture. Among the many converting operations are embossing, impregnating, saturating, laminating, and the forming of special shapes and sizes, eg, bags and boxes (see PACKAGING, CONVERTING).

Pigment Coatings. Pigment coatings are compositions of pigments and binders with small amounts of additives and are applied to one or both sides of a paper sheet. These generally are designed to mask or change the appearance of the base stock, improve opacity, impart a smooth and receptive surface for printing, or provide special properties for particular purposes.

Application. Pigment coatings normally are applied to the base paper in the form of water suspensions and are referred to as coating colors. Paper may be coated either on equipment that is an integral part of the paper machine, ie, on-machine coating, or on separate converting equipment. Many plants include both types of coating equipment and utilize each to its maximum advantage for paper and paperboard.

Pigments. Pigments comprise 70–90% of the dry solids in paper coatings. In nearly all cases, the individual particles of the pigments are less than 5 μm in equivalent spherical diameter and average less than 1 μm. These particles can fill the spaces between fibers on the sheet surface and form a nearly uniform surface mat. Pigments also control opacity, gloss, and the color of the raw stock. Refractive index, particle size, crystal structure, light scattering and absorption, and adhesive demand are important characteristics. Minimum amounts of binders are used to bind the pigments, because the binders are more expensive than the pigments and excessive amounts can adversely affect the opacity and brightness of the coatings by filling in the air–pigment interfaces that scatter light.

Clays are the most common and most widely used pigments employed in paper coating.

In many coating formulations, a combination of binders is used to maximize the properties of various types of binder systems. The amount of binder in a coating formulation varies from about 5 to 25%, based on the pigment. The actual amounts depend on properties desired in the coating, specific use of the sheet, and types of binders to be used (see ADHESIVES). Binders include acid-hydrolyzed starches that have been hydroxy ethylated or acetylated, hypochlorite-oxidized starches, poly(vinyl alcohol) (PVA), various rubber latexes and other emulsions, acrylic-based emulsions, and poly(vinyl acetate).

Additives. Additives control coating behavior during application or they can be used to alter the properties of the finished product. They include dispersing agents, foam-control agents, lubricants, plasticizers (qv), and flow modifiers.

Barrier Coatings. In packaging applications, a barrier may be needed against water, water vapor, oxygen, carbon dioxide, hydrogen sulfide, greases, fats and oils, odors, or miscellaneous chemicals. Polyethylene and other barrier polymers such as ethylene–vinyl alcohol are applied by extrusion or coextrusion with intervening tie layers.

Economic Aspects

The United States and Western Europe, which represent ca 13% of the world population, consumed about 60% of production.

The domestic production of paper and board plus imports and minus exports maintains a remarkably constant ratio with real gross domestic product in the United States. One sector of the paper industry that has grown at a higher rate than GDP is recycled papers and boards which is projected to grow at 6.8% annually. The industry as a whole is expected to achieve 50% recovery rate for paper and board products by the end of the twentieth century (see RECYCLING, PAPER).

Printing and writing papers continue to gain share at the expense of packaging and industrial papers. This trend is likely to continue as digital printing creates markets.

Environmental Issues and Plant Efficiency

Modern practice is to recycle as much water as is compatible with efficient machine operation. It is common to design a fiber-recovery system into the white water cycle. The three general types of save-all fiber recovery are based on filtration (qv), flotation (qv), and sedimentation (qv). If these are operated efficiently, the net fiber loss can be less than 1%.

In order to conform to environmental quality guidelines, mills have installed a number of primary and secondary treatment systems to control effluents. The primary treatment is composed of settling basins and/or tanks, ie, clarifiers. These remove ca 85–100 wt % of solids, eg, fibers and clay. The secondary treatment generally consists of a biological treatment followed by secondary clarifying.

As more process water is recycled to reduce overall water consumption and wastewater discharge volumes, more nonpathogenic microbial growth, ie, slime, occurs in the mill system. The efficiency of any form of slime control is greatly increased by ordinary good mill cleaning procedures. The application of chlorine or chloramine with or without frequent cleaning is effective in many cases. Antiseptics and disinfectants reduce or inhibit slime formation (see DISINFECTANTS AND ANTISEPTICS).

Sludge Handling and Disposal. Two kinds of sludges are generated by pulp and paper mills: primary sludges contain fibers, clay filler materials, and other chemical additives; secondary sludges are largely biological in nature and harder to handle and dewater.

Generally, sludge handling processes include thickening, stabilization, conditioning, dewatering, incineration, and disposal. Most sludges are disposed in landfills.

Water Quality Assessment. Pressure is being exerted to go back to the receiving water system as the ultimate test for new and more stringent discharge-control measures. The increasing knowledge of the interrelationships between the various biological, chemical, and physical components of aquatic systems has provided significant restructuring of field assessment programs which are designed to analyze effluent impact.

M. BRUCE LYNE
Paper Science, International Paper

D. Hunter, *Papermaking,* 2nd ed., Alfred A. Knopf, Inc., New York, 1947.

J. A. Bristow and P. Kolseth, *Paper Structure and Properties,* Marcel Dekker, Inc., New York, 1986.

H. F. Rance, *Handbook of Paper Science,* Elsevier, Amsterdam, the Netherlands, 1980.

Tappi Press Books, Technical Association of the Pulp and Paper Industry, Atlanta, Ga.

PAPERMAKING ADDITIVES

In papermaking, chemicals can be added either to the pulp slurry prior to sheet formation, ie, internal or wet-end addition, or to the resulting sheet after complete or partial drying, ie, surface or dry-end addition.

Papermaking additives can be categorized either as process additives that improve the operation of the paper machine, or as functional additives that enhance or alter specific properties of the paper product.

Environmental constraints on the paper industry have resulted in drastic processing changes, primarily because very large amounts of water are used to produce paper. Numerous regional, state, local, and foreign national regulations exist concerning emissions to air, and discharges to water and sludge. OSHA workplace regulations may have also altered the additive process and the choice of additives.

In addition, many grades of paper and paperboard are used in direct or indirect contact with foods. Thus, many mills only use paper chemicals that have been cleared for use by the U.S. Food and Drug Administration (FDA), so that it is not necessary to segregate machine broke (off-grade paper and edge clippings that are reclaimed for their fiber value) and white water. Most of the chemicals discussed in this article are approved by the FDA for use in paper and paperboard that are intended for applications in food processing and packaging. However, there are various restrictions on both the specific functional uses and amounts of paper chemical additives which can be used, so the FDA status should be confirmed by the supplier before use.

It is also important to study the interactions of papermaking additives in the paper machine water system. Optimization of the addition points and usage rates of the entire additive system is necessary in order to maximize performance of the chemical additives and the paper sheet properties, and to minimize cost and negative interactions both on the paper machine and in the white-water system.

Lists of the manufacturers of each type of product used by the paper industry are available. Surveys of chemical suppliers for the U.S. and Canadian pulp and paper industries have been published.

Process Aids

Retention and drainage additives are vital to the use of recycled fibers. Papermakers consider recycled fibers to behave like virgin fines, while recycled fines behave like filler. Drainage on the paper machine can be impeded and first-pass retention reduced by the use of recycled fiber. Additionally, the negative impact of contaminants found in recycled fibers can be minimized by the appropriate use of dispersants and other pitch-control additives.

Retention Aids. In an aqueous suspension, most fillers, like most paper-pulp fibers and fines, develop a negative surface charge which prevents coflocculation of fillers with fines. Cationic retention aids encourage coflocculation by two mechanisms: they neutralize the negative charges on fillers, fibers, or fines so that van der Waals forces can hold them together, and they form molecular bridges between two particles to which they are adsorbed.

Molecular weights of the charge-biasing polymeric retention aids are ca 10^3–10^5. These aids usually contain amine or quaternary ammonium groups, and include condensation polymers, but the simplest charge-biasing agent is alum, which is hydrated aluminum sulfate. It is used less as a sole retention aid than as an adjunct to other retention aids, especially the bridging types.

Molecular weights of polymers that function as bridging agents between particles are ca 10^6–10^7. Ionic copolymers of acrylamide are the most significant commercially (see ACRYLAMIDE POLYMERS).

Synergistic improvements in filler retention have been achieved through the use of combinations of additives, in which addition of a low molecular weight cationic polymer, often referred to as a coagulant, is followed by that of a high molecular weight anionic polymer, or flocculant. These probably function through a combination of charge biasing and bridging.

Another type of dual-retention system has emerged; in these systems, a synergistic combination of a small, inorganic particle and a polymer enhances retention. These microparticulate retention systems are composed of a high molecular weight cationic flocculant and an anionic particle. The most common of these systems consist of combinations of cationic starch with anionic aluminum hydroxide or colloidal silica, or high molecular weight cationic polyacrylamides with bentonite or colloidal silica.

Salts, eg, alum or calcium chloride, and cationic polyacrylamides are effective retention aids in bleached and unbleached kraft pulp.

Formation Aids. The repulsion of negatively charged fibers in water is not always sufficient to prevent all flocculation, which can result in uneven fiber density in paper. Dispersants can prevent flocculation prior to immobilization of the wet fiber mat on the wire. Polyacrylamide, poly(ethylene oxide), and natural gums, eg, guar gum and locust bean gum, promote even fiber distribution. High molecular weight anionic polymers, such as polyacrylates, lignin sulfonates, and naphthalene sulfonates, can also act as dispersants and enhance sheet formation, but they may impede drainage in some systems.

Drainage Aids. For machines where dewatering on the Fourdrinier wire is the slow step, dewatering or drainage aids can effect faster machine operation and, therefore, can increase production. An ideal drainage aid alters the surface properties of the fibers so that they hold less water, and limits the amount of fiber flocculation. This balance of properties can be achieved by combining a drainage aid with sufficient agitation to redisperse the fiber flocs immediately prior to sheet formation.

Defoamers. Foam, a common problem in papermaking systems, is caused by surface-active agents which are present in the pulp slurry or in the chemical additives.

Liquid defoamers have become the preferred form of defoamer. They include preemulsified versions of paste defoamers, fatty alcohols, and ultrapure mineral oil. In addition, the trend in paper machine defoamers is toward water-based defoamers.

The defoamer formulations mentioned so far consist of fairly inexpensive raw materials, but several more costly defoaming materials have come into use in paper mills, eg, hydrophobicized silica particles and silicone solutions and emulsions.

Wet-Web Strength Additives. A number of water-soluble chemical additives have been shown to enhance wet-web strength, including synthetic, eg, anionic polyacrylamides, and natural, eg, locust bean and guar gum, polymers. Several modified natural polymers have also shown promise, including chitosan, blocked reactive group starches with acetal protected aldehydes, and cationic aldehyde starches produced from them.

Pitch Control Agents. Successful at controlling both natural pitch (from wood pulp) and so-called "white pitch" from recycled fibers, the most widely used pitch control agents are pitch dispersants, which can be either organic, ie, typically anionic polymers such as naphthalene sulfonates, ligninsulfonates, and polyacrylates, or inorganic, ie, typically clay or talc. Low molecular cationic polymers or alum can also be used.

Effluent Treatment. The paper industry must minimize the amount of suspended solids in mill effluents. Flocculants are used in save-alls to help recover finely divided solids, both organic and inorganic, from the white water. Alum is used widely as a coagulant for effluent treatment in paper mills. Most of the low and high molecular weight polymers, which serve as retention aids, also can be used as coagulants and flocculants (see FLOCCULATING AGENTS).

Slimicides and Biocides. Slimicides and biocides are frequently added at various points in the papermaking process. This is of crucial importance when the paper is produced for food contact or medical applications. Two categories of biocides are in use in paper mill systems: oxidizing biocides, including chlorine, hypochlorite, hypobromous acid, and chlorine dioxide, and nonoxidizing biocides, ie, methylene bisthiocyanate, carbamates, and quaternary ammonium compounds.

Creping Aids. Principal creping adhesives include animal glues, starch, neutral-cure wet-strength resins, specialized polyamines, and high molecular weight retention aids. Among the agents that can be added to facilitate release from the roll are emulsified paraffin oil, silicone oils, or poly(ethylene glycol)s. Residual hemicellulose in the pulp affects adhesion significantly.

Functional Internal Additives

Sizing Agents. Paper sizing provides paper and paperboard with controlled resistance to wetting by liquids.

The contact angle formed between water and the paper surface is the primary factor determining the extent of wetting. When properly oriented on the fiber surfaces, purely hydrophobic materials such as wax, and amphipathic, ie, polar–nonpolar, materials such as rosin and so-called synthetic materials such as alkenyl succinic anhydride (ASA) or alkyl ketene dimer (AKD), provide a low surface energy coating which gives the high contact angle necessary for sizing.

Wax emulsions have been widely used to impart special resistance to functional penetrants, such as oil, grease, and blood. *Fluorochemicals* have been used in the manufacture of oil-resistant paper and paperboard, and other specialty grades.

The general mechanism of effective sizing involves the following sequential steps. (*1*) Efficient retention of the sizing agent in the sheet, preferably as tiny particles. (*2*) Uniform distribution of sizing agent over fiber surfaces. (*3*) Firm anchoring to fiber surfaces so that impinging liquids do not overturn the molecules which present a hydrophobic surface to the exterior.

Dry-Strength Additives. An increasing amount of dry-strength additives is being used. This expanded use is driven largely by the need to utilize higher levels of weaker fiber sources, ie, recycled paper and paperboard. Modified starches are the most widely used, with the cationic and amphoteric starches dominant for wet-end application. Polyacrylamides, both anionic and cationic, are the most important of the synthetic dry-strength additives, and are used in a variety of grades. Vegetable gums, both natural and modified, and sodium carboxymethylcellulose (CMC) are also widely used.

Utility. Increased sheet strength through increased fiber refining generally results in an increase in sheet density; thus sheet strength is accompanied by a reduction in density-related properties, such as opacity, porosity and bulk. Dry-strength additives increase sheet strength with little or no change in sheet density, for faster drainage during paper formation, allowing for separate optimization of sheet strength and density-related properties. These include reduced refining energy, higher filler levels, increased use of weaker, less expensive fibers, and reductions in basis weight.

Wet-Strength Additives. In the presence of water, the cellulose hydrogen bonds that are augmented by dry strength additives are disrupted, and the paper does not have an appreciable amount of wet strength, ie, strength in the presence of water. With few exceptions, wet-strength additives are capable of covalently bonding in order to preserve paper strength in the presence of water. Tissue and toweling, linerboard, medium, carrierboard, coffee filter, and bleached carton are some of the principal grades that require an amount of wet strength to be functional.

Wet-strength additives include urea–formaldehyde and melamine–formaldehyde resins, aminopolyamide–epichlorohydrin resins, polymeric amine–epichlorohydrin resins, and aldehyde-modified resins.

There are two predominant theories that attempt to explain the mechanism of wet-strength development in paper. The protection theory proposes that the wet-strength resin forms a restraining network, by cross-linking either with itself or bonding with the cellulose. Thus, the network protects a fraction of the hydrogen bonding in the dry sheet by limiting the swelling of the cellulose and hemicelluloses in the presence of water. Alternately, the reinforcement theory proposes that the new covalent bonds formed by the wet-strength resin are the ones that remain unbroken by water.

Concern about formaldehyde in the workplace has caused a permanent shift from UF and MF resins to polyamide–epichlorohydrin resins.

Fillers. Opacity must be high enough in paper with print or writing on both sides of the sheet to prevent images from showing through from the back side. Opacity is increased by incorporating particulate materials or fillers into the sheet in order to increase the scattering of light which passes through the sheet (see FILLERS).

Though functionally and chemically similar, fillers and pigments are distinguished from one another in that fillers are added at the wet end of the paper machine, and serve to fill the sheet; pigments are added at the size press and serve to alter the surface of the sheet. The most common fillers are mineral pigments, eg, clay, titanium dioxide, calcium carbonate, silica, hydrated alumina, and talc. Overall, the use of these fillers in paper has been increasing because they are less expensive than pulp fiber. Also, as the trend toward lower basis weight continues, the use of filler increases to compensate for opacity losses.

Mineral fillers function as opacifiers primarily by increasing the amount of surface area in the paper sheet and thus increasing the scattering of light.

Functional Surface Treatments

Although many functional chemicals can be added to the wet end of the paper machine, some grades of paper require special properties that cannot be provided by the low levels of wet-end additives that are retained in the interior of the sheet of paper. To achieve the properties required for these grades of paper, it is necessary to apply special chemicals to the surface of the preformed paper web.

Processes. The most common method for the application of chemicals to the surface of a paper web is by a size press. In the size press, dry paper, which usually is sized to prevent excess water and chemical penetration, is passed through a flooded nip or pond, and a solution or dispersion of the functional chemical contacts both sides of the paper. Excess liquid is squeezed out in a press and the paper is redried.

Sizing. The most commonly used materials for surface sizing are starches and modified starches, including oxidized, enzyme-converted, hydroxyethylated, and cationic starches. They are used not only for sizing, but also to improve strength, especially surface strength, and to impart smoothness.

Synthetic polymeric sizing agents contain hydrophobic elements and water-soluble functionalities. The two most popular classes of synthetic polymeric sizing agents are styrene–maleic anhydride copolymers and polyurethane dispersions.

Other sizing agents include alkylketene–dimer emulsion sizes and fluorochemical emulsion sizing agents.

Application of Dry-Strength Additives. Starches and modified starches, especially cationic starches, are used in large quantities to improve the surface strength of paper. The natural gums, eg, guar, locust bean, and tamarind, also can be applied to the surface of the paper to enhance strength.

Application of Wet-Strength Resins. One of the most commonly used resins for creping is an aminopolyamide–epichlorohydrin resin.

Judicious application of water to the opposite side of the dry sheet followed by redrying may correct curling.

Pigmented Coatings. A large volume of paper is coated in order to improve printability. The main function of the coating is to provide a smooth surface for printing. Other properties that are important to the coating are receptivity to inks and sufficient surface strength to withstand the forces of the printing process on the coated paper.

Paper coatings are applied as coating colors, which are aqueous slurries containing 35–65 wt % solids. There are three main components of the solids; pigments, binders, and minor additives. The pigment is the primary component of a paper coating and consists of small, white, particulate material. Pigments usually are minerals, eg, clay, calcium carbonate, or titanium dioxide. The packed pigment particles fill pitted areas of the rough paper surface, thereby providing a suitable surface for printing. Binders are the resins or polymers that function as the glue that binds the pigment particles to each other and to the paper substrate. The level of binder is low in a paper coating, typically 5–30 parts by weight per 100 parts of pigment. This low level of binder distinguishes paper coatings from paints, which are pigment-filled polymer films. Minor additives are used to modify the properties of the coating color, primarily before and during the coating operation.

Synthetic Fibers

A variety of wet-laid felts and nonwoven fabrics are produced on Fourdrinier-type paper machines (see NONWOVEN FABRICS). Noncellulosic materials may be included as part or all of the fiber furnish; latexes, water-soluble polymers, or other adhesives are used as bonding agents. Synthetic fibers can make paper highly resistant to wetting; chemical attack; mechanical wear, eg, folding; weathering; and biological degradation. Synthetic fiber-containing papers are used as backings for carpets and vinyl floor coverings, industrial filters, disposable bed linens and hospital garments, heavy-duty wiping materials, tea bags, tissues, labels, and embossable wall papers.

Heavy-duty wiping materials and some disposable garments are made from nonwoven cellulose mats which are reinforced with a synthetic fiber.

Owing to concern about its long-term hazard to health, asbestos fiber, traditionally used with a latex binder in backings for roll-vinyl floor coverings, has been replaced by glass wool, rock wool, polyolefin fibers, and cellulose.

MARGARET A. DULANY
UCB Chemicals Corporation
GEORGE L. BATTEN, JR.
MICHAEL C. PECK
CHARLES E. FARLEY
Georgia-Pacific Resins, Inc.

R. M. Husband, ed., *Survey of Paper Additives*, 2nd ed., H&H Consulting Group, Trunball, Conn., 1989.

J. Marton and D. D. Jarrell, *Sizing Short Course 1987*, Tappi Press, Atlanta, Ga., pp. 53–62.

F. M. K. Werdouschegg, in W. F. Reynolds, ed., *Dry Strength Additives*, Tappi Press, Atlanta, Ga., 1980, pp. 67–93.

Tappi Wet and Dry Strength Short Course 1988, Tappi Press, Atlanta, Ga., 1988.

PATENTS AND TRADE SECRETS

Practice and Management

The significance of aggressive patent and trade secret protection to the economic well-being of a business or organization cannot be underestimated. Without patents and trade secrets, the marketplace is reduced to competition on the basis of price as profit margins diminish with the inability to protect market share.

Patents and trade secrets are protected by securing rights to ideas and the application of ideas that have commercial worth. The grant of rights in patents and trade secrets is based on an appreciation of development, advancement, and invention that will stimulate innovation by advancing technology. Patents and trade secrets are two distinct mechanisms for protecting invention vis-á-vis the application of ideas. Both are supported by the policies and laws of the United States.

Compiling a portfolio of patents provides an organization with an offensive weapon with which to protect and ensure profitability. If the leading technology is protected by patents, the owner of this technology has an excellent tool with which to prevent others from making, using, or selling this technology in the marketplace. A patentee may in turn realize a return on the time and energy invested in obtaining this protection by securing a principal interest in the marketplace, royalties from competitors, or even damages from those who choose to ignore the rights flowing from the patent grant (see LICENSING).

A patent also serves a defensive function. Publication of the issued patent may be used to preclude others from obtaining patents on the same or similar technology.

Trade secret rights are based on the complete absence of disclosure of the invention to anyone other than the owner. Oftentimes ideas, developments, and advances that are the subject of trade secret protection are those which may not be patentable, for any of a number of reasons.

Some factors to consider when evaluating patent and trade secret protection include (*1*) the form and content of the technological advance, idea, development, or application; (*2*) the desired term of protection; (*3*) the potential for the technological advance, idea, development, or application to be the subject of a commercial product; (*4*) work done previously; (*5*) events which have publicized or publicly disclosed the technological advance, idea, development, or application; and (*6*) factors that may be critical to keeping the technological advance, idea, development, or application confidential, and what events may necessitate disclosure.

Harmonization

Two principal conventions, the General Agreement on Tariffs and Trade (GATT) as well as the North American Free Trade Agreement (NAFTA) have effected a change in the term of patents issued from the United States Patent and Trademark Office (USPTO); a change in type of patents that may be filed in the USPTO; and prospective changes that will internationalize U.S. patent law.

By the turn of the century the USPTO may be operating under a system that includes (*1*) publication of patent applications; (*2*) opposition of allowed applications for purposes of testing validity; and (*3*) the dawn of first-to-file priority examination. Legislation implementing many of these changes is pending before the U.S. Congress.

The Origin of Patent Rights

A patent is a bundle of affirmative rights granted by the U.S. Federal Government. The published written document, referred to as a patent, provides a full and complete description of the invention. The affirmative rights which stem from the issuance of a patent allow the owner of that patent to prevent other parties from making, using, or selling what the patent covers in the United States. The coverage of a patent is the actual property of the patent owner and is defined by the patent claims, which are like the legal description of real estate in a real property deed. Ultimately, the printed published document which represents the patent rights granted by the Federal Government can be a complex literary work. Interpretation of the patent claims involves answering complex legal questions and is dependent on, among other things, the written description in the patent.

A patent is intended to further the development of science and technology by providing a published record of technological developments for all to read, consider, and discuss. At the same time, a patent provides a delineation or definition of the rights which the patent owner considers its own through the claims appended to the patent. The publication of a description of the invention in conjunction with the claimed limits of the invention provides the public with notice of the patent owner's affirmative rights to the invention.

The process of invention generally starts with the transcribing of ideas that may, or may not, result in an advance or a solution to a recognized problem. Once an inventor is satisfied that the development has attained the desired level of usefulness, a summary of the inventive concept may be prepared. From this summary, including any appropriate data or laboratory work, an application for a U.S. patent may be written. The patent application is generally written by, or at least under the supervision of, a patent attorney or patent agent with one or more inventors.

Once the patent application is complete and the inventor has made a formal declaration of inventorship, the application is filed with the USPTO. In the USPTO, the application is the subject of a thorough, formal and substantive examination by a patent examiner. Once the patent examiner is convinced that the patent application satisfies the statutory requirements provided for under the laws of the United States, the patent application will be issued as a U.S. patent. Issuance takes the form of a publication provided by the U.S. Government. The publication of patents occurs only on Tuesdays that are not federal holidays. At the time of issuance, the patent is assigned a number and made public in a form which allows all interested parties to obtain access to it.

The term of a patent depends on the date on which the application for patent is filed with the USPTO. Patents filed and issued before June 8, 1995 had a term which is the longer of 20 years from the filing date or 17 years from issuance. The "filing date" in this context is the earliest date which a patent application is filed, (U.S. or otherwise) and relied on by the applicant. Patent applications filed before June 8, 1995 that issue after that date also have a term which is the longer of 20 years from filing or 17 years from issuance. Any original or follow-on patent application, ie, continuation, divisional, or continuation-in-part applications, filed after June 8, 1995 have a term of 20 years from filing, once it is issued, as a U.S. patent.

The Nature of Invention

Invention may result from many different types of scientific or engineering efforts and advances. However, invention can also arise through the simple application of an idea that improves, refines, or otherwise modifies something that had been done previously. The simplest and most common area in which invention arises is in the development of products.

The nature of product development is such that it consists more of a process than a single discrete event. The process of developing a product may result in one large breakthrough that could be considered a broad invention. This breakthrough may result in a new product that is useful and has many of the benefits that the inventor desired at the outset of the developmental project. However, the product still may or may not be suitable for commercial introduction or various other intended applications. As a result, further efforts may need to be expended toward refinement of the product so that it may take on its ultimate commercial form. Each of these potential refinements may also represent one or more patentable inventions that, while narrower in their intended usefulness than the original product, are still commercially valuable in their own right. In the search for improvements, refinements, and further solutions, invention thus may result either from a developed research effort or through the simple discovery of a solution to the original problem which is arrived at completely outside of the research context.

The resulting discoveries may provide a broad range of solutions or products. Alternatively, invention may result in products or applications, which add value to basic commercial products that are already in existence. Inventions may also be used to assist an individual or company in commercial efforts toward developing a defensive posture in any given marketplace. When patented, applications may also provide an extended opportunity to license or market the patent without the actual production of a product by the inventor.

Reading a Patent. Reviewing patent documents requires the skill of understanding the significance of what is being disclosed. Legal counsel should always assist in interpreting the legal effect of any patent on commercial activity. However, a patent attorney or patent agent often must seek the assistance of technical personnel to gain a full understanding of the technology disclosed and claimed in a given patent. Further, an understanding of the form, content, and function of the various sections of a U.S. patent assist the nonlawyer in understanding the commercial importance of any issued patent.

The Technical Subject Matter of Patents. A fundamental requirement for obtaining a patent is defining an advance, development, or invention which is within those classes of "subject matter" which the law of the United States regards as patentable. Two classes of patentable subject matter, ie, computer software and biotechnology, are the subject of relatively new and evolving law. However, other types of subject matter rest on fairly certain ground as to patentability. Examples of patents of subject matter are described in the following.

Composition of Matter. This is the subject matter category into which many chemical and biochemical (and biotechnology) inventions often fall.

Article of Manufacture. An article of manufacture is an invention such as a two-headed tooth brush, an intravenous fluid bag, or an optical fiber manufactured or "made" by a machine.

Machines. A machine is a device which is capable of manufacturing a product or completing a task such as removing hydrocarbon contaminants from silica and dirt.

Processes. Methods or processes represent patentable subject matter regardless of whether the invention represents a method of using or a method of manufacturing an article, composition, or device.

Design. Ornamental designs are also a legally recognized class of patentable subject matter. The design must be embodied in an article of manufacture, such as a concrete masonry block or a sun screen for a car window.

Plants. Asexually reproducing plants also represent a legally recognized class of patentable subject matter under U.S. patent laws.

The Origin of Invention. Invention results from the application of an idea or concept. The idea itself is generally not patentable. An application of the idea may be patentable if it falls within one of the categories of subject matter previously discussed.

Although it is not always necessary, a practical application of a concept may move through a series of steps or stages. Indeed, the recognized pathway to invention involves at least two factors, ie, conception of the invention and reducing the invention to practice.

The initial research effort may prove to be a broad spectrum of applications or solutions to the original problem that in turn provide any number of inventions. When efforts move toward reducing the invention to practice and refining the invention so that it proves to be commercially marketable, certain applications may prove to be unfeasible or commercially impractical. As a result, only one application may ultimately prove commercially marketable. However, all the solutions which are developed and considered over the research and development process may comprise inventions that are worthy of disclosure and claiming in a patent. An application which is not commercially viable today may become viable within the lifetime of a patent.

Unlike other countries, where award of patent rights is based on the date on which a patent application is filed, in the United States the patent grant is based on the first date of invention; although patent term is based upon patent application filing date.

Inventorship. Those who may deserve to be considered inventors include all those who have contributed to conception of the invention. Further, those who have provided contributions which would be considered something above and beyond textbook knowledge in the reduction of the invention to practice may also deserve to be listed as inventors.

The legal guidelines which direct inventorship determinations are some of the most stringent and complex in modern patent law. Many factual and legal analyses may justify the listing of an individual as an inventor, and certain levels of contribution do merit this designation. However, one is not an inventor based merely on employment status such as by being a laboratory supervisor or a supervised laboratory assistant or technician.

Developing the Record of Invention

Developing the record of invention is an important, if not a fundamental, point in the process of securing protection over the invention. Generally, there are two stages in the development of the record of invention. The first stage is the laboratory or experimental work that is done in conceiving the invention and applying it within the intended area of use. This may take place over a period of weeks, if not months, with continual or intermittent work toward the ultimate production of an advance, development, or application which solves various problems. The second stage of developing a record of invention is the actual process of defining the invention along with noting any events or facts that may limit, expand, or define the invention.

The Laboratory Notebook Page. Most engineers, scientists, and technicians make a record of their work. A common form of record keeping is the use of a laboratory notebook. The laboratory notebook should reflect elements, parameters, conditions, and thoughts that were material factors in the completion of a given experiment.

Parameters that were thought to be completely irrelevant may become relevant once the scientist has reflected upon notations made over the course of repeated experiments. Testing of quality or efficacy should also be recorded in detail. In any case, the notebook should memorialize the experimentation in a manner which enables any future reader to reproduce the work and exploit its benefits. Once the experiment is completed, the scientist, engineer, or technician who has undertaken the work should confirm completion of the work by signing and dating the record. The record should also be witnessed by at least one other person who reads and understands it, and did not take part in the experimentation.

The Record of Invention. The second phase of developing a record of the invention is to condense the record into a summary form which serves several purposes. Specifically, the record of invention establishes a date of invention through attached copies of notebook records, spectra, and the like which all prove that the invention has in fact been conceived and reduced to practice in some form having practical utility.

Along with other elements of the invention, it is good practice to include within the record of invention any first written descriptions or drawings of the invention. It is also prudent to attach photocopied notebook pages that evidence this development. Another helpful part of the record of the invention is to list the names of individuals who worked on the invention and enumerate their respective contributions.

One further component of the record of invention is a list of any uncovered publications or patents which are relevant to the invention. Such a listing should also include any disclosures made by any of those who worked on the invention to other parties inside or outside the organization. All inventors should sign the record of invention. At least two witnesses who are not inventors should also read and understand the record of invention so that they can sign and date this document.

The importance of an accurate and complete record of invention cannot be underestimated. The record of invention should serve as the basic document for establishing the date of conception and reduction to practice of the invention. The USPTO issues patents to those who are first to invent.

Determining the Scope of the Invention

Once the record of invention has been written, an evaluation of the invention should be undertaken. A careful evaluation of the record of invention is usually best completed by a committee of individuals from technical, commercial, and legal disciplines. It is important to include the viewpoint of those scientists working in the field, those commercial or sales people who will be responsible for selling any products which stem from the invention, and those individuals who may be able to offer a legal opinion given the insights of commercial and scientific personnel.

First the committee should consider the technical merit of the invention. Specifically, is it reasonable from a scientific or engineering standpoint? Further, is there a clear advance in technology that has not been previously undertaken or achieved by another party?

It is also important to ascertain the commercial significance of the invention. Even if there is a meritorious technical development, can this development be reasonably and practically put forth in a commercial product?

The legally trained member of the interdisciplinary committee should provide insight as to the significance of the technological advance and as to whether any commercial product ultimately derived from the invention could be protected by an issued patent. Another important function of this person is to determine the scope of the invention based on preceding events, publications, or activities which may have otherwise limited the breadth of the invention. To this end, U.S. law requires that an invention satisfy a number of prerequisites or requirements before issuing a patent: novelty, nonobviousness, utility, and disclosure.

Novelty. A fundamental statutory prerequisite to patentability is novelty. A lack of novelty occurs when each and every element of the invention, as recited in the claims, is found in a single disclosure which occurs before the date of invention. Such a disclosure may occur in any of a number of forms. To be an adequate disclosure, it should be catalogued or inventoried as a book might be in a reference library and open to public dissemination. The novelty requirement presents the inventor with an extensive list of "cans" and "cannots." Unfortunately, the natural course of research and development often leads to activities which are much more readily categorized as "cannots" than "cans." Ultimately these activities may even proscribe the issuance of a patent if an application is not filed in a timely fashion.

Questions that should be considered when determining whether an invention is novel include the following. Has the invention been publicly known or publicly used by others or the subject of a patent or publication anywhere in the world prior to the applicant's actual invention date? Has someone other than the inventor published a journal article, received a patent, or used the invention publicly? If so, the inventor will not be able to receive a patent on the invention.

Events which may destroy novelty are often also referred to as "prior art," given their nature as an earlier event which is relevant to the technical art.

Some examples of prior events or prior art which may destroy novelty are as follows: Graduate school dissertations. Abstracts of meetings of technical organizations. Approved or published grant proposals. Published articles in the popular press. Prior products manufactured by the client. Prior company literature related to the invention. Prior publications of the inventor in the area. Third party research and commercial activity in the area of invention. Any patents in the area.

Economic reality dictates that the invention must eventually be commercially exploited. Experimental trials are a natural follow-on to laboratory work and are often necessary to further refine or otherwise reduce the invention to practice. The application of further refinements to the invention, the facts surrounding the trial, and the ultimate timing of the filing of the patent application may all be determinative of whether or not the novelty of the invention survives.

In any event, the highest importance should be accorded to coordinating events which may affect the novelty of the invention. Careful consideration should be given to the importance and timing of promotional events. It is often the case that patent applications can be filed and drafted well before announcements occurring at technical conferences. Further, technical publications often have an extended lead time before they are actually published. In any instance, the filing date of a patent application retains extremely great importance, being a determining factor in the timing of any disclosure.

Nonobviousness. The grant of a patent is also dependent on whether the advance, application, development, or invention is obvious. If an invention is obvious, it is not patentable. The legal qualification of obviousness is a very difficult concept to understand.

An initial determination on the degree to which an invention may be "obvious" can be obtained by answering the following questions: What do prior patents, publications, and public activity disclose relative to the invention? What are the differences between all of this prior activity and the new invention? Would the skilled technician, engineer, or scientist consider the newer invention unexpected or surprising in view of this previous work?

However, even if there is some disclosure of the invention in the prior activity, the law of patents in the United States requires a high level of detail concerning the invention. Some factors to consider in establishing that an invention is not obvious is as follows: The results achieved by the invention are new, unexpected, or superior; up to now, the techniques used in the invention were unworkable; Up to now, problems solved by the invention were not solvable; the invention has attained commercial success; the problem solved by the invention was never recognized before; and that the invention has been copied by an infringer are only among the few reasons.

Utility. Aside from designs and plants, inventions are required to exhibit usefulness or utility to be patentable. In fact, issued patents for processes, machines, compositions, and articles are often commonly referred to as "utility" patents. Depending on the nature of the technology, a single assertion of utility may suffice. Although utility may be supported by an assertion of use, application, or benefit, the assertion must be accurate and credible to ensure the enforceability of any patent relied upon to cover the invention.

An inventor may establish utility by providing several working examples which disclose preparation, application, and even some or all of the benefits of the invention.

Disclosure. An additional statutory requirement is that of disclosure. A patent must provide the public with a disclosure which is enabling, definite, and shows the best mode for practicing the claimed invention.

Enablement. The patent has to enable any person reading the disclosure who has skill in the relevant technical area to make and use the invention. The enablement requirement mandates that the applicant provide a description of the process of manufacturing given invention. Also, the patent provides an adequate description of the process of using the invention.

The disclosure requirement provides that the patent is a teaching document, and that it enhances the breadth of knowledge held by the public.

Problems with enablement arise when the patent fails to provide an adequate disclosure of parameters or materials for use in producing or performing the invention. The enablement requirement may, however, be satisfied by relying on and referencing a particular level of experience or knowledge in the given field of technology and incorporating that reference directly into the patent application.

Definiteness. Adequate description or definiteness requires that the patent claims provide an outline of those elements which are integral to the application's invention. The applicant cannot make a claim of right to the invention where essential elements of the invention are not disclosed in the patent.

The definiteness requirement serves notice to potential infringers as to the exact boundaries of the patentee owner's rights. Thus, a patent provides a record of what the inventor has brought to the technological field, and also provides other parties with notice as to what conduct is permissible in view of the patent claims.

Best Mode. The patent applicant must disclose the best mode of practicing the invention known to the inventor at the time the application is filed. Concerns over best mode often arise when a patent applicant seeks patent protection for an invention but, at the same time, desires to keep as a trade secret one aspect of the invention necessary to the production of a commercial product. This action denies the public access to this information and undermines the policies of the patent system.

Drafting the Patent Application

Once the record of invention has been assembled and evaluated, a decision should be made as to whether to move forward and draft a patent application.

In drafting the patent application, the inventor may work alone gathering the elements of the disclosure which the inventor deems relevant and material to the invention. It is more advisable for the inventor to retain a patent agent or attorney. In order for a patent agent or attorney to represent inventors before the USPTO, these individuals must have a degree in one of the sciences or in the field of engineering. Further, a patent agent or attorney must have demonstrated a proven competence in understanding the procedures and rules of the USPTO by obtaining admission to practice before this office.

In drafting a patent application, proceeding methodically through the several steps necessary to produce the type of disclosure legally and technically sufficient to satisfy the requirements of the laws of the United States is absolutely essential to the successful grant of a patent. A first step is to outline those elements of the invention which are absolutely essential to its practice.

Once this process has been completed for each of the essential elements, patent claims may be drafted which cover the invention. These claims will cover, in the broadest sense, only those elements of the invention which are essential.

Claims should also be drafted to cover alternative forms of the invention. Alternative forms of the invention may not necessarily be considered to be preferred commercially, but they may present an area where a competitor could attempt to "engineer around" the invention.

Once the claims have been written, a fuller disclosure of the invention may be drafted. This description of the invention will generally follow the outlines of the essential and optional elements. Such an outline will include a functional description of elements including relevant broad and preferred parameters for each of the elements. The description of the invention also should explain the intended interrelationship of the elements that is needed to produce the invention.

Other known embodiments of the invention should also be disclosed to the extent practical. These embodiments can prevent future patenting by third parties if they are published in the applicant's issued patent.

The patent application should also provide a thorough description of the benefits and advantages of the invention and the manner in which it advances to technology.

The Provisional Application

The provisional application provides an applicant the opportunity to gain an early U.S. filing date for a relatively low filing fee without commencing the patent term. Design applications are not included in the provisional application system. The provisional application is not examined by the USPTO except for compliance with formalities, and it has a nonextendable life of one year from the filing date. Drawings must be included if they are necessary for the understanding of the disclosed subject matter.

To maintain the benefit of the provisional application filing date, a regular utility application must be filed during the pendency of the provisional application, ie, within one year of its filing date, and must include at least one inventor in common with the provisional application. Provisional applications also start the time period (one year in length) during which counterpart foreign applications must be filed to perfect a claim to the initial filing date.

Filing and Examination of the Application

Once the application has been finalized, it should be reviewed by all inventors to make sure that it is a complete teaching of the invention and that the level of disclosure satisfies the legal requirements of U.S. law. The inventors then execute an oath or declaration to this effect. The inventors may in addition have an obligation to assign the rights for the invention as embodied in the patent application to their employer. In such cases, it is usually appropriate to secure the execution of an assignment by the inventors.

Once the patent application has been reviewed and all formal documents executed, all paperwork including the application is filed with the USPTO. Legal regulations govern how a patent application should be filed which is not a simple matter.

Once it has been filed, the patent application enters the domain of the USPTO, which is organized by technical discipline into various groups, eg, polymer chemistry, biotechnology, inorganic solid chemistry, as well as organic chemistry. Within each group are specific art units handling areas of technology which are even further focused on specific advances and developments within their respective technical fields.

Figure 1 depicts a generic step-by-step process of examination as it generally occurs within the USPTO.

Correction of Errors in Issued Patents. A patentee should review the issued patent to ensure that the patent grant is free of errors and contains the intended claims. Errors may arise in a patent application or issued patent during the writing of the patent application, examination of the patent application, or the printing of the issued patent document. The errors may be inconsequential, stemming from misspellings, misprintings, as well as insertions or deletions of text. These errors may have occurred through sections taken by the applicant or the USPTO in transforming the patent application to a printed patent document. The issued patent should also be reviewed for compliance with the formal and substantive requirements of United States law and the regulations of the USPTO.

The patentee may ask the USPTO to correct this error. There are four administrative vehicles for correcting errors in issued patents. The application of each of these mechanisms is dependent on the nature and severity of the error, as well as the source of its creation.

The Notice of Errors. The first mechanism for correction of errors is called a "Notice of Errors." This document may be filed by the patentee after issuance of the patent with the USPTO.

The Notice of Errors should resolve those problems which are evident on the face of the patent but which also may be, by their nature, obvious and correctable problems to someone reading only the patent. The Notice of Errors does not result in a further publication by the USPTO.

The Certificate of Correction. Another mechanism for correcting the patent is the "Certificate of Correction," which is essentially a petition filed by the patentee to correct minor errors in the patent produced either by the USPTO or inadvertently by the applicants. Unlike the Notice of Errors, a Certificate of Correction does result in an additional publication from the USPTO.

Patent Reissue and Reexamination. Reissue and reexaminations proceedings may require the resubmission of the issued patent to the

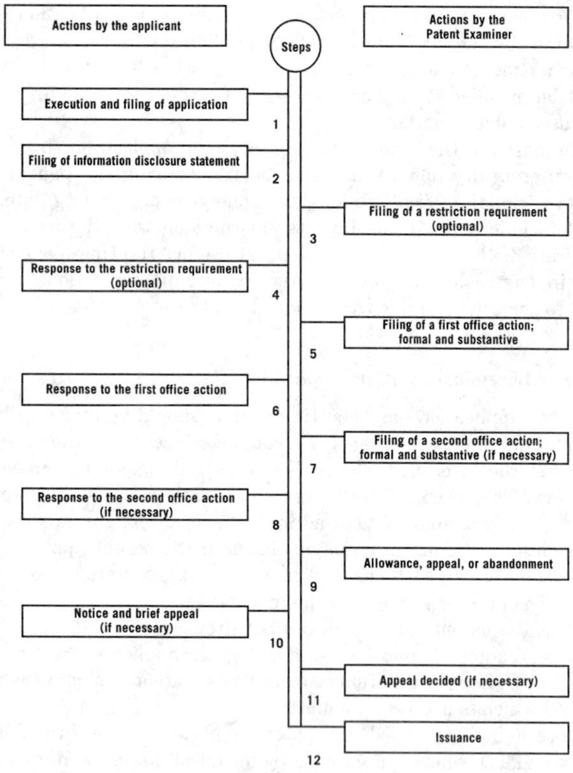

Figure 1. Timeline for examination in the United States Patent Office.

USPTO. Each requires the additional expenditure of substantial funds and a loss of time in the active life of the patent.

A reissue may be ordered to correct any minor or major mistake which occurred during prosecution of a patent, but the mistake must be one that makes the patent partially or wholly inoperable. Inoperable essentially means that the patent cannot be enforced.

A request for reexamination may be made by the patentee, a third party, or the Commissioner of the Patent and Trademark Office whenever a new question of patentability arises. This new question of patentability has to be raised in the form of a publication such as a journal article or a patent which was not considered during the prior prosecution. Reexamination is a more expedited and economical means of receiving a judgment on whether or not a patent is valid in advance of litigation.

By requirement, the patent generally must be resubmitted to the USPTO for reissuance. Reexamination and reissue proceedings allow for varying levels of participation by third parties. As a result, before undertaking any such proceedings a patentee should consult legal counsel to define a legal strategy and choose an appropriate forum for correction of the patent in question.

Patent Maintenance Fees. On the date a patent issues, it has a 20-yr life measured from the earliest filing date relied on for priority. Under current U.S. law, the patentee is required to pay maintenance fees, a policy stemming from an interest in the public in practicing the technology covered or claimed in the patent. After a patent issues, the claims are generally very important commercially and provide the patentee with relatively easily exercised rights to prevent others from making, using, or selling that which is found in the claims. However, as the patent grows older, the public interest in practicing the technology grows stronger. Often the claims become less important commercially and the commercial value of the claims then needs to be assessed in view of the expense of maintaining the patent.

Payment of maintenance fees is required at the fourth-, eighth-, and twelfth-year anniversaries of the date of issuance of the patent. The costs of these maintenance fees vary from year to year depending on the regulations of the USPTO.

The patentee should develop an implement a policy for auditing its patent portfolio in the process of paying maintenance fees to the USPTO. This practice should also be used to justify the further payment of annuities to foreign national patent offices.

Patent Interference. An interference is a contested action in the USPTO to determine inventorship between two or more patent applicants or between at least one patentee and one or more patent applicants. The principal contest in an interference concerns the right to claim the invention. In the simplest situation, an interference occurs when a pending application discloses and claims the same invention which is claimed in at least one other copending application or issued patent.

The interference proceeding is declared b the patent examiner and occurs in the USPTO. A patentee or patent applicant may then win an interference proceeding by proving the right to the invention as the first inventor. Alternatively, a patentee or patent applicant may win an interference proceeding by default. If the invention was known to the public prior to the first date of invention, none of the parties to the interference have a right to claim the invention.

Legal Actions Based on Patents. The issuance of a patent initiates a term during which the patentee may enforce its rights, ie, the patentee may prevent others from making, using, or selling that which the patent claims. To literally infringe upon the patentee's rights, another person, business, or organization must make, use, or sell something which has each and every element found within the claims of the patent in question. Patent infringement may also occur if the action in question contributes to, or otherwise induces the making, using, or selling of something which contains each and every element of that which is found in the claims of the patent.

An action for patent infringement may be based on one or more claims of the patent. If infringement of patent rights is found by court of law, the patentee may receive remedies that include monetary damages, attorney's fees, and injunctive relief.

One accused of infringing on patent rights may defend against the action by showing that they have not made, used, or sold something which includes each and every element found in a claim of an issued patent. Further, one may also defend against a legal action for infringement of patent rights by clearly showing that the patent in question is invalid, ie, that it lacks novelty, is obvious, has not complied with the formal disclosure requirements of the USPTO, or does not designate the proper inventor.

Legal actions based on patents almost always have tremendous commercial significance to the parties involved. The factual and legal issues surrounding and relating to these legal actions can be complex, burdensome, and not easily resolved. One considering, or threatened by, a legal action involving a patent should retain competent legal counsel.

Foreign Patent Prosecution

The foreign filing of a patent application is an immensely complex task, requiring retention of foreign patent lawyers or patent agents, complying with highly specific rules of foreign practice, and usually requiring a significant expenditure of capital. However foreign patents can provide significant commercial opportunities in valuable international markets. Further, various systems for obtaining patent protection, put in place by multinational treaties, are allowing most organizations to operate on a commercial level which is not national or regional, but global.

For many years, the method of obtaining foreign patent protection corresponding to a U.S. patent application was to file separate, individual patent applications in selected foreign countries. Each of the applications had to be written to conform with the national requirements of the country in which it was filed.

The national laws of most countries are unique to the particular country. However, most industrialized countries are parties to one or more International Conventions which provide for the filing of foreign patent applications.

Trade Secret Rights

An alternative to patent protection for advances, developments, ideas, and applications is to treat such information as a trade secret. The protection of trade secrets relies on the development of ideas, applications, and advances which are not found in the public domain. Trade secrets, by definition, are kept in confidence by their owner, disseminated only to those who accept an absolute obligation of confidentiality, and then only for purposes of which the trade secret owner knows and approves.

If a trade secret is believed to have been violated, a judge must initially decide whether or not it actually existed.

The life of a trade secret may extend indefinitely if the owner of the secret has taken the proper steps to safeguard the invention, in contrast to a 17-yr patent term, after which time the invention is in the public domain. Traditionally, trade secrets have been protected by confidentiality or nondisclosure agreements, and employment agreements.

The Creation of a Trade Secret

Because there is no "federal law of trade secrets," protection of trade secrets is often left to the variability of the criminal and civil laws of the 50 states. To the extent that a trade secret is property, violation, theft, or misappropriation of the trade secret may be the subject of criminal penalty. To the extent that a trade secret is bound to rights, violation or misappropriation of the trade secret may be the subject of civil penalty. Significant effort, however, has been made in developing a uniform body of law to apply to ideas and innovations which may be the subject of this form of protection.

To summarize, in order to be considered a trade secret, the information (*1*) must not be generally known or readily ascertainable; (*2*) must provide a competitive advantage; (*3*) must have been developed, maintained, or acquired at the trade secret owner's expense; and (*4*) must be the subject of the trade secret owner's intent and efforts to keep it confidential.

The protection of a trade secret is a complex task dependent upon any number of factors. The mere formation of an intention to maintain information as secret is not enough; actual safeguards must be put into place. The owner of a trade secret must identify the information as a trade secret and protect the information from disclosure. Means used to prevent disclosure might include the following: Guarding entrances to the facility in which the information is kept. Using employee and visitor identification badges. Limiting access to trade secrets to those having an obligation to maintain it as confidential. Destroying trade secret information by means that will prevent its disclosure, eg, incineration or shredding. Monitoring or clearing employee activities which involve removing any business-related information or objects from the facility. Using photographic, electronic, or keyed access and monitoring equipment. Consistently protecting all trade secret information to the same level, including consistently investigating any concerns over the theft or breach of trade secret protection.

Exploitation of Trade Secrets

Trade secrets become unprotectable when they are found in the public domain, are independently developed, or are disclosed out of confidence. Events of the latter type may occur in any number of controlled or uncontrolled situations which may lead to a disclosure of trade secrets. Idle correspondence, conversations, or communications with sales associates, suppliers, or distributors may also result in disclosure of trade secret information.

A trade secret owner may also beneficially exploit the trade secret through licensing, sales, or various other business ventures based on the confidential information.

In business transactions the parties should have a clear understanding of exactly what constitutes trade secret information and consider how the information will be used and who will retain ownership rights.

When licensing or otherwise undertaking a joint venture based on trade secret rights occurs, other considerations arise. For example, research efforts invariably give rise to additional information which may be the subject of trade secret or even patent protection. If this additional information is derived from the licensed or shared body of initial information, consideration should be given to ownership, further protection, eg, who files and pays for patent protection, and at the end of the agreement how, or even if, this information should be divided. Commercial partners of the trade secret owner should not be provided this information except under the strictest obligations of confidentiality. License and joint venture agreements regularly contain confidentiality provisions with substantial penalties for any violations that may occur.

In the case of contract and noncontract employees, a rigid program devised for the identification and protection of trade secret information should be implemented.

Confidentiality agreements should be signed by all employees at the time of hire. The employee should be given tools for maintaining information as a trade secret; for example, the simple use of bound notebooks for maintaining laboratory experiments is almost a universally accepted standard practice. The use of a resource person for questions on identification and protection of established and newly developed trade secret information is also a good practice.

Employees should be regularly briefed on the organization's trade secret program.

Trade secret information should be disseminated only when commercially necessary, only under obligations of strict confidentiality, and only with definite penalty provisions for improper use or further dissemination.

In instances where publication of a trade secret is necessary for commercial exploitation, the filing of a patent application may be an adequate substitute for the complete dedication of rights to the public. If the information would satisfy the requirements of U.S. patent law, then, despite perceived difficulties in enforcement of any patent rights obtained, the best defense against theft or unauthorized use may be obtaining patent rights covering this information.

Violation of Trade Secrets

Trade secret rights are generally violated through an unauthorized use by someone other than the owner. This use may take the form of theft or misappropriation for later use in a commercial product. The unauthorized use can also take the form of an unauthorized disclosure to a third party who is not bound to keep the information confidential.

Concern for trade secret owners is that like other legal actions, there is a definite limitation to the time period for bringing an action for misappropriation of trade secrets. As a general rule, such a legal action must be brought within three years after the misappropriation is discovered. Remedies for trade secret misappropriation can include injunctive relief and money damages, as well as attorneys' fees for bringing the action.

JOHN J. GRESENS
Merchant & Gould

J. L. White, *Chemical Patent Practice,* Patent Resources Group, Inc., Apr. 1993.

H. L. Hanson, *Creativity, Innovation and Intellectual Asset Management,* Honeywell, Inc., 1984.

General Information Concerning Patents, U.S. Department of Commerce, Washington, D.C., Apr. 1989.

J. W. Baxter, *World Patent Law and Practice,* Matthew Bender and Co., 1991.

PATENTS, LITERATURE

The second half of the twentieth century has witnessed a sharp increase of activity in research and development, as well as an increased internationalization of technology-based industries. As a result, there have been significant changes in the patent literature, the chief literature of technology. The number of countries publishing patent documents has increased as former communist and Third World countries have enacted patent laws. The number of patent-issuing authorities is also growing rapidly as a result not only of the emergence of new nations formerly embedded within the Soviet Union and other communist countries of central and eastern Europe, but also of the enactment of new patent laws by other countries in response to the intellectual property provisions of the General Agreement on Tariffs and Trade (GATT), which established the World Trade Organization (WTO), and to the North American Free Trade Agreement (NAFTA). The ideal of full harmonization of patent laws among countries has often been discussed, but seems far from being realized; nevertheless, significant changes have been made in the patent laws and procedures of individual countries.

Table 1 shows some of the milestones in the development of the primary patent literature during the last third of the twentieth century.

As the number of patent applications filed during the middle of the twentieth century grew, the time required to notify the public that an invention had been claimed in a pending patent application was seen as a serious inconvenience. Laws were introduced in some countries to inform the public about potential patents during their pendency.

Once granted, patents are in force for a term prescribed by law. Patent terms are not renewable. Most countries have established a term of 20 years, measured from their national filing date, but patent laws enacted before the latter third of the twentieth century vary considerably in the length of the patent term.

The expiration date of a patent is not normally printed on its face and must be calculated on the basis of the applicable national laws. Exceptions can occur in the United States when a term is foreshortened because of that patent's close relationship to a previously issued patent or, under the new law, when a term is extended because of delays in the course of patent prosecution. In addition, patents issued by most countries are kept in force by payment of periodic maintenance fees. Because some products cannot be marketed without the approval of governmental regulatory agencies, the owners of patents on drugs, medical devices, and agricultural chemicals have long complained that the effective term of their patents is less than the term of unregulated products. Some countries have provisions for the extension of the patent term for products approved for marketing under regulatory laws. A compilation of national and international laws regarding patent expiration has been published by Derwent Information Ltd.

Table 1. Recent Milestones in the Development of Primary Patent Literature

Year	Country or authority	Significant development
1964	the Netherlands	first principal examining office to switch to universal publication and deferred examination
1968	FRG	switch to universal publication; huge backlog of pending cases published, often at rate of over 1000/wk, which strained documentation services
1971	Japan	switch to universal publication, output rose quickly above 100,000/yr (~350,000/yr in 1992–1994); language and numbers make quality documentation a substantial problem
1979	European Patent Office (EPO)	single patent covering multiple countries; tended to supplant national patent offices; increased share of English-language publications
1979	World Intellectual Property Organization (WIPO)	single application submitted to multiple countries and regional offices; further increased share of English-language documents; has become very significant in 1990s
1980	United States	periodic maintenance payments required for granted patents having file dates later than Dec. 12, 1980
1995	United States	switch to 20-year term from file date; may begin publishing unexamined applications

Patenting Procedures. Procedural pathways followed by patent applications filed in various countries and resulting in the publication of patent documents are shown in Figure 1.

Patent Documents. The internal structure of patent documents has been standardized and the amount of bibliographic detail recorded in a patent document has increased. Early patent documents included rudimentary information about the patent's filing details. Most patent documents published during the 1990s begin with an informative cover page. The front page of a modern patent provides key information about the patent that aids the reader greatly in determining the patent's potential relevance. The cover page provides a title, gives the name of the patent owner, inventors, and other individuals involved in the issuance of the patent, and offers serial numbers and dates that identify the document and relate it to other patent documents covering the same invention. An abstract is provided by the patentee. Where appropriate, the abstract may include structural diagrams for chemical species important to the invention, and a representative drawing. National and International Patent Classification appropriate to the patent are shown, as is the list of classes searched by the examiner in determining patentability. Patent and other publications deemed by the examiner to be related to the invention are listed, and are referred to as examiner's citations. These bibliographic data have been standardized according to Internationally agreed Numbers for Identification of Data (INID) codes established by WIPO. The INID codes provide a means whereby the various data appearing on the first page of a patent and other similar documents can be identified without knowledge of the language used and the laws applied. They are used by most patent offices and have been applied to U.S. patents since August 4, 1970.

Because each country has its own patent laws, the precise meaning of the bibliographic data and the legal significance of the published patent document vary from country to country. The Patent Cooperation Treaty (PCT) provides a recommended code to distinguish the various types of documents and to simplify storage and retrieval of patent data, but the code is implemented differently by different countries. For example, in the United States an A-document in 1995 was a patent; in the Netherlands, an A-document was a published unexamined application. It is essential to understand each country's system to interpret the status of its patent documents.

The invention covered by the patent is defined in the patent claims, which appear at the end of patents by most countries and at the beginning of patents published by a few others. The majority of patents are known as utility patents. The claims of these patents may relate to new products, including new chemical compounds and compositions, to processes for making or using new or previously known products, and to machines for making or using such processes.

In addition to utility patents, some countries publish patent documents under different or less stringent standards for patentability and with shorter patent terms.

The bulk of the patent specification is the disclosure, the text and illustrations that describe the claimed invention in detail and explain how the claimed invention differs from the prior art. Modern patent disclosures contain a summary of the claimed invention, a description of the background of the invention, a general description of the way in which the invention is made and used, specific examples, and, where applicable, drawings of the invention in general or specific

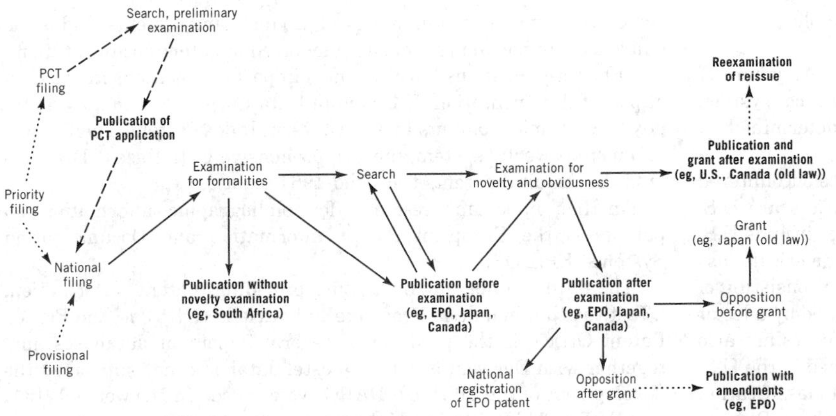

Figure 1. Procedures for publication of patent documents. Publications are shown in boldface. Dotted lines indicate events that may take place before a national application has been filed or after a patent is granted. Dashed lines indicate events that take place only when PCT filing is chosen.

embodiments. The technical information provided in a patent specification may be used without infringing the patent; only practicing the invention defined in the claims within the term and territory of the patent grant is forbidden. Because much of the information in patent specifications is never published in refereed journals or other non-patent media, patent disclosures are an invaluable part of the technical literature.

Patent documents differ from journal literature in several ways. First of all, they are legal documents whose disclosures support one or more claims that define an area of property rights. The language in patent documents can therefore be quite convoluted "patentese" as the applicant strives to achieve the broadest possible scope of coverage. Examples provided in patents may never have happened. Paper examples are generally written in the present tense. They lack hard data, and can provide grounds for attacking the patent should they prove to be inoperable. Finally, chemical patent disclosures and claims can be written in terms of generic structures, or the so-called Markush structures, in which one or more portions of a chemical entity can vary, including functional groups, numbers of substituents, and points of attachment. Markush structures are used as one method of obtaining the broadest possible claims in a patent. Markush structures can be simple, describing just a handful of chemical compounds, or highly complex, encompassing thousands, millions, even infinite numbers of compounds. The effective indexing and searching of Markush structures provides a significant challenge to those concerned with chemical patents.

Patent Families. Patent specifications are published as individual documents in the language of the originating country, but many inventions are claimed in patents issued by more than one country. These patents form a family of equivalent patent documents, which usually disclose the same information but may differ somewhat in the scope of their claims. When filing in more than one country, an applicant establishing priority under the Paris Convention is generally required to submit a copy of the original application to each national or regional patent office selected, sometimes with a translation. A simple patent family is based on a single priority application, in which each family member discloses the same information and cites the same priority application number. When the technical content of the patent rather than its legal scope is of interest, any member of the family can be substituted for another, thus often obviating the need for translation.

Patent Searches. Because valid patent claims can only be issued on an invention that is novel and innovative in light of prior art, it is necessary to search the prior art for previous references either to the composition of matter, process, or machine defined in the claims of a patent application, or to any similar composition, process, or apparatus that would render the claimed invention obvious to a person skilled in the field of the invention. Inventions that have been described in a publication or embodied in a product are said to have been anticipated in the prior art and are not patentable. Patentability searches are performed by examiners employed by the national and regional patent offices and are an important step in the examination of patent applications. Patentability searches should also be performed by the representatives of inventors prior to the filing of a patent application so that the claims will not overlap with any publication in the prior art. These searches may encompass the full scope of the published literature, including patents, technical journals, gray literature, and even catalogs. Individuals or organizations who are making plans to introduce a new product or process must conduct infringement searches to ensure that they will not infringe patents that belong to others. Infringement searches need only consider patents in force and pending applications that may result in patents in countries where manufacturing or marketing are contemplated. After a patent application has been published and/or a patent has been granted, organizations that wish to practice the invention may also conduct validity searches to be used as ammunition for opposition proceedings or invalidity lawsuits. Validity searches, like patentability searches, should include all forms of published literature, but are limited to publications with effective dates earlier than the filing date of the patent application being challenged.

Searches of scientific and technical literature are performed using any of the information retrieval tools suitable for searches done for other purposes (see INFORMATION RETRIEVAL). Patent offices have devised special classification systems to facilitate searches among the individual patent documents in their collections. Patents are assigned classification codes by the examining office and the relevant primary classification and any cross-reference classifications are printed on the first page of the patent, eg, INID codes. Although patent classifications originated as tools for manual searches, they can be searched through printed or electronic indexes as well.

National patent offices created patent classification systems for internal use without correlating their guidelines for subdividing technologies or the symbols used to identify classifications with those of other countries. The assignment of national classification codes to patents of the issuing country facilitates manual searching for inventions claimed in national patents, but is not helpful for prior art searches that must include patents issued by patent offices that use different classification systems. The internationalization of commerce has led to the internationalization of patent classifications. The International Patent Classification (IPC) has been adopted by most of the patent-issuing countries of the world. Even countries such as the United States, which continues to use a national classification system to organize their patent search files, print a corresponding IPC classification on the patent documents. Although the IPC is used by most countries, these countries do not all follow the same guidelines for applying the codes, nor do they all use the finest divisions of the classification system. Differences in the scope of the patent claims, on which the IPC classification is based, as well as differences in the classifying examiner's interpretation of the novel features of the invention, also contribute to differences in patent classification among countries. It is not unusual for patents having identical claims to be classified differently in each country where the patent application was filed.

IPC codes, which have the format *ANNA NNN/NN,* where *A* stands for a letter and *N* a numeral, represent a hierarchical system. The hierarchical relationships within the groups are determined by the relationships published in the IPC manual.

The most common national patent classification codes encountered outside the public search rooms of national patent offices are U.S. classes, which are indexed in many patent databases that include U.S. patents. They are formatted as a one- to three-digit numerical class code, followed by a slash or hyphen and a subclass code consisting of from one to three numbers, which are occasionally followed by a letter or by a decimal point and additional numerals. These codes are also arranged hierarchically according to the scheme published in the U.S. *Manual of Classification.* Unlike IPC codes, U.S. patent classification codes do not contain clues to their technological relationships. Patents are given a single original classification and usually one or more cross-reference classifications. The U.S. system designation corresponding to the IPC C07c 45/50 would be 568-451.

Secondary Sources of Patent Information

Patent systems were conceived as a means for promoting technical progress by encouraging the dissemination of information on technological developments. Information dissemination is therefore essential for the patenting process. Patent offices have traditionally announced the issuance of new patents in bulletins and gazettes. Other organizations, notably scientific and technical societies and for-profit publishers, have produced value-added patent information services. These secondary sources of patent information serve multiple purposes, among which are current awareness alerting, document delivery, and retrospective searching. Increasingly they have found second use in electronic form in on-line databases, and in the 1990s there has been rapid growth of optical storage of information, especially as Compact Disk-Read Only Memory (CD-ROM) products. Patent documentation is a field in considerable ferment, with rapid introduction of new products and capabilities.

Printed Patent Office Gazettes. The issuance of patents is announced by patent offices in publications typically known as gazettes and bulletins, which are published most commonly at the time of the patent's publication, but there are exceptions. Advance information is published in a patent gazette by some countries prior to the publication of patent documents, typically as a notification of filing details. However, some patent gazettes do not appear until well after the effective publication date of the patents they announce. The amount of information included in patent gazettes varies.

In addition to announcements of new patents and applications, the various gazettes typically include listings of patents that have been rejected, challenged, or disclaimed, patents that have been allowed to lapse, and in some instances even listings of new applications that have been made but that will not be published for some time, if ever. Gazettes often include indexes to the information they contain; the amount of indexing available varies from country to country.

Information from Other Sources. Some of the abstracting and indexing services produced by scientific and technical societies have traditionally included patent information, especially in the field of chemistry. For example, *Chemical Abstracts* (CA) has always covered patents, as did the discontinued *Chemisches Zentralblatt* and *British Chemical Abstracts.* On the other hand, some notable information services have not included patent coverage. *Science Abstracts* has not covered patents since 1976. *RAPRA Abstracts* resumed the coverage in 1994. However, even where patents are covered, the focus may not be ideal for those concerned with the legal aspects of patents. Thus, CA in its patent coverage documents the new chemistry involved, but shies away from the legal aspects of patents. For these and other reasons, others have stepped in to develop a variety of patent information services, eg, Derwent Information Ltd. of London.

The Derwent organization has broadened its country coverage and improved its capabilities for information retrieval in many ways. It continues to work on new and improved products and systems, and is the single most important organization involved in patent documentation.

Other organizations have assumed important positions in the field of patent documentation. IFI/Plenum Data Corp. began in 1955 to index U.S. chemical patents by the Uniterm Index system. Their newer and more powerful system, the Comprehensive Data Base (CDB), covers U.S. chemical patents from mid-1964 to date.

Another important resource for bibliographic information on patents is the European Patent Information and Documentation Systems (EPIDOS).

There are other organizations providing patent information. L'Institut National de la Propriété Industrielle (INPI), ie, the French Patent Office, is the producer of several important databases and, together with Derwent and the Questel databank, has supported the development of the Markush DARC system used in Derwent's WPIM and INPI's PHARMSEARCH databases. The American Petroleum Institute's (API) Central Abstracting and Information Service has since 1964 produced APIPAT, a database covering patents on petroleum refining, petrochemicals, and related technology. API also produces APILIT and APIBIZ, complementary databases covering published technical literature and business information.

Other specialized patent information products, as well as general information products that include patent and other information, are produced by a variety of organizations, most notably in the area of pharmaceuticals. Such products include *Approved New Drugs* (U.S. FDA), *Drugs Under Patent* (FOI Services, Inc.).

Advances in Patent Documentation

The last half of the twentieth century has seen a strengthening of patent coverage by some traditional abstracting and indexing services whose patent coverage extends back for many years, as well as the establishment of an increasing number of specialty services for the documentation and manipulation of patent information. Advances have involved traditional printed products as well as various electronic forms of information. Computerized databases have become increasingly important to users of patent information, and new and modified information tools continue to appear and develop.

Countries covered by patent databases vary from one database to another; some databases provide complete information about patents published by a single patent-issuing authority, others attempt to catalog the world's entire patent output. Multinational patent databases have historically provided good coverage of heavily industrialized countries and lesser coverage of less industrialized countries. Users' manuals from such database producers as Derwent and EPIDOS include detailed lists of country coverage ranges.

Derwent Information Ltd. Derwent Information Ltd., previously known as Derwent Publications Ltd., provides a wide spectrum of information products and services, many of them relating to patents. Derwent also produces important databases (qv) of nonpatent information from the pharmaceutical and agricultural chemical literature. These products and services encompass alerting tools for current awareness, systems for retrospective search and retrieval, and means for document delivery and archiving.

Derwent has gradually moved into covering the full range of chemistry (in depth that differs from one subfield to another) and to nonchemical patents. The basic framework of the *Chemical Patents Index* (CPI) was established in 1970, and the overall *World Patents Index* (WPI), encompassing CPI as well as nonchemical patents, was established in 1974. The CPI is divided into 12 sections by technology, as shown in Table 2.

The Derwent patent database is based on records covering a family of equivalent patents. New patent publications are checked against the existing database to see if they are equivalent to previously published references. This is done by comparing priority application details or, if priority is not claimed, by comparing inventor or patentee and technical content with known references. Those publications determined to be new to the system are considered to be basic patents,

Table 2. Scope of Derwent CPI and WPI

Section	Subject content	Number of basic patent references, ×10³ 1973	1983	1993
A	PLASDOC: polymers	28.3	39.7	63.1
B	FARMDOC: pharmaceuticals	6.8	10.1	18.9
C	AGDOC: agricultural chemicals	4.0	5.1	6.2
D	food, biotechnology, detergents, cosmetics, etc	8.3	15.5	29.4
E	CHEMDOC: general chemicals	15.8	19.6	27.8
F	textiles, paper, cellulose	11.6	10.1	15.1
G	printing, coating, photographic chemistry	5.4	10.0	21.2
H	petroleum	4.3	8.1	9.0
J	chemical engineering	6.6	14.4	19.6
K	nuclear, explosives, protection	2.2	3.4	4.3
L	glass, refractories, ceramics, electrochemistry	9.0	26.6	47.8
M	metallurgy	13.1	25.2	32.3
Total CPI		*81.0*	*128.0*	*183.5*
Total WPI		*81.0*	*302.4*	*413.5*

and are assigned to one or more sections of the Derwent system according to their technical content. Chemical patents may appear in as many sections of the CPI as needed, although a limit of four sections per patent was applied in the past.

In addition to alerting abstracts, documentation abstracts, formerly called basic abstracts, are produced for all basic chemical patents, except for those from the handful of title-only countries. These documentation abstracts frequently provide a substantial amount of technical details beyond those included in the alerting abstract. Documentation abstracts are produced in a format that highlights key features, including claimed matter, uses, and detailed examples. These abstracts include coding by Derwent's manual code system, which uses a vocabulary of several thousands of keyword-like codes to identify the key aspects of the patent. Documentation abstracts are published in documentation journals covering each CPI section, as well as in profile booklets covering selected segments of polymer and other technology. Microfilm and CD-ROM collections of documentation abstracts are produced for archival purposes.

Besides the worldwide WPI database, Derwent provides on the ORBIT system the USPatents database, a bibliographic file of patent front page and claim information for U.S. patents since 1971. Derwent also produces a biotechnology database, GENESEQ, that indexes sequence structures of proteins or nucleic acids disclosed specifically or generically in patents. This database is searchable with special sequence software on the IntelliGenetics system, and is a new addition to STN's database catalog.

Abstracts can serve as a pointer to patents of potential interest, and in some cases may provide sufficient information to judge the relevance of a patent, but there is no true substitute for the examination of complete patent specifications, particularly when legal decisions must be made.

Derwent has developed, either on its own or through contractors, a number of computer aids to information processing. For example, the TOPFRAG series of programs, available as MARKUSH TOPFRAG, aids users in searching chemical structure information. A new initiative introduced by Derwent during 1995 is the Patents Citation Index, an on-line database of patent citations that includes both examiners' citations and patentees' citations to prior art from patent specifications.

Chemical Abstracts Service. The Chemical Abstracts Service (CAS), a division of the American Chemical Society, has produced *Chemical Abstracts* (CA) since 1907. CA has been the preeminent medium for documenting new publications in the field of chemistry and chemical engineering. CA documents chemical publications of all types. It is not a patent database per se, but its patent component is larger than most databases devoted entirely to patents. Thus, for example, the number of patent references in CA for the years 1991–1993 ranged from 95,500–99,400 per year.

Derwent products have always been targeted at those in and near the legal profession. CA has the mission of documenting chemistry for chemists and chemical engineers. Therefore, CA abstracts of patents have emphasized what was actually done in examples that provide hard data, and have avoided discussing the purpose or scope of a patent or prophetic paper examples.

There is an important difference, however, between the CAS and the Derwent treatment of members of a patent family. Derwent abstracts the first member of a family that it sees, then adds bibliographic data for all equivalent patents to the record, so that the record for a given invention in the WPI database can be accessed by bibliographic information on any member of the family. Although CAS also abstracts the first member it sees of a family, it enters subsequent equivalents into only the printed CA patent index, not the on-line CA or CASearch databases. Thus a searcher looking for a given patent number in CA will find it only if it was the member of a family that happened to be covered first. Priority application details common to a basic patent and its equivalents are thus more reliable than patent numbers as an access point to the CA database, although indirect access based on patent number can also be achieved via the cross-file techniques of various on-line systems.

EPIDOS (Formerly INPADOC). The International Patent Documentation Center (INPADOC) was created as a result of agreements reached in 1972 between WIPO and the government of Austria. The INPADOC operation has now become part of the division of the European Patent Office known as the European Patent Information and Documentation Systems (EPIDOS). Information is obtained directly from national and international patent offices, which in the 1990s number around 60 and include more countries than any other patent information service.

EPIDOS issues printed and microfiche compilations of its data; in addition, its database can be searched on its own computer or on several on-line host systems. In general, EPIDOS provides the most complete patent family information of any service, although Derwent tends to include more information on intellectual (nonconvention) families, whereas the French Patent Office's EDOC file on the Questel system includes information on granted Japanese patents that is not readily available elsewhere. Also available from the EPO is the EPIDOS Register file, which is unique in providing information on the course of prosecution of European patent applications.

An important new service from EPIDOS is the series of ESPACE CD-ROM products, providing document delivery of full patent specifications from the EPO, PCT, and a lengthening list of individual countries. Approximately 1000 full specifications can be contained on an individual CD-ROM for printed documents such as European or U.S. patents.

IFI/Plenum Data Corp. IFI/Plenum's predecessor company, Information for Industry (IFI), began indexing U.S. patents by its Uniterm system in 1955. The Uniterm system (in 1972) was complemented by a more powerful retrieval system called the Comprehensive Data Base and is available only to subscriber organizations. With the advent of on-line databases, these chemical indexing systems were augmented by bibliographic information, including bibliographic data for nonchemical patents going back to 1963.

Besides its on-line databases, IFI produces magnetic tape versions of the databases which some users choose to run in-house. Other IFI patent products include the *IFI Assignee Index* and the *Patent Intelligence and Technology Report.*

L'Institut National de la Propriété Industriele (INPI). The French Patent Office (INPI), is a principal provider of patent and trademark databases, all of them accessible on Questel. FPAT, EPAT, and PCT-PAT have full bibliographic data, abstracts, and claim text for French, European, and PCT patent documents, including information about changes in the status of the applications after their original publication.

American Petroleum Institute (API). In 1972, the API reached an agreement with Derwent to use repackaged Derwent alerting abstracts for its printed patent bulletins and to do its patent indexing from Derwent documentation abstracts. This enabled API to discontinue patent abstracting for the most part, and the documentation abstracts provided richer material for indexing than did the relatively brief API abstracts. Cross-referencing from the APIPAT database to WPI enabled on-line searchers to move from hits in the APIPAT database to the corresponding WPI references, including their complete patent families. Ultimately, the APIPAT and WPI databases were merged on the ORBIT system, which enables a searcher to combine API and Derwent retrieval parameters in a search. APIPAT and WPI remain separate databases on the DIALOG and STN systems.

Types of Patent Information Searches

There are many different reasons to search for information about or related to patents. The methods, sources, and techniques vary widely, depending on the purpose and the complexity of the individual situation. An ill-conceived computer search strategy can produce mountains of expensive output that can require a huge outlay to analyze. An inadequate search strategy can lead to even greater costs if the result is patent infringement. Anyone performing patent searches must have a sound understanding of the costs and benefits that may be involved.

Novelty Searching. At the heart of the patenting process is the novelty or patentability search. A novelty search should be carried out by an inventor or representative before a patent application is drafted in order to help ascertain whether the invention is indeed patentable and, if so, what its limits might be. A novelty search is carried out by a patent office examiner to make a decision on the patentability of an invention. A novelty search is normally focused sharply on the specific details of the invention, but broader searches that provide a context for the invention in relation to the state of the art can be justified when an invention gives promise of having wide application and high value.

A valid patent covering any claimed invention can be obtained only if the wording of the claims defines an invention that has never been used or described before the filing of the patent application, and that is not an obvious variation of something that has been described in the prior art. However, the standard for prior art references that may be brought to bear against a patent differs from country to country.

Traditionally, novelty searches have been performed by leafing through stacks of patents in those divisions of a classification that seem best to categorize the invention. It excludes from consideration those patents that might be relevant to the search but which, for hierarchical or other reasons, are classified elsewhere. Searching based solely on classification risks the omission of useful references.

Searching of one or more on-line databases is a technique increasingly used in novelty studies. The use of such databases enables the searcher to combine indexing parameters, including national and international classifications; natural language words in the full text of patents, in their claims, or in abstracts supplied by inventor and by professional documentation services; and indexing systems of various sorts. The use of multiple databases is thus prudent, and is facilitated by multifile and cross-file techniques provided by the various on-line hosts.

Novelty searches are not necessarily limited to patent information. The anticipation of a purportedly novel idea can occur in journals, books, magazines, etc. Thus, the potential scope for a novelty search is essentially infinite, and one of the challenges to the searcher is to devise an effective strategy whose cost is commensurate with the potential value of the invention.

Infringement Searching. An individual or organization found to be infringing the patent rights of others is subject to penalties that can be extremely costly. It is essential for anyone contemplating a commercial venture that is technology-dependent to find out first whether or not the proposed venture falls within the area covered by adversely held patents.

Whereas the potential field of search for the novelty search is essentially limitless, there are certain limits that can be placed on an infringement search. An infringement search can be limited to the content of the claims of patents, and only to the country or countries in which manufacturing, sale, or use of the invention is contemplated. Only patents that are in force or that are potentially in force need to be considered. Patents that have expired, that have been invalidated, or that have lapsed because of failure to pay maintenance fees can be excluded from consideration. A searcher must be alert, however, to patents that are potentially in force.

Since the exact language of claims is vital to matters of infringement, the search of full patent specifications remains the most reliable method of infringement searching. The full claims text of all U.S. patents is available in several on-line databases covering a time span longer than the life of a U.S. patent. On-line databases are sometimes used for infringement searches by carrying out careful searches of parameters other than the claim language. However, reliance on computerized databases lacking full claims text for infringement searches involves compromises; at the very least the searcher must obtain the full claims text, including any associated drawings or chemical structure diagrams, of all patents of potential interest that are disclosed by the computer search.

Validity and Opposition Searches. Given the identification of a patent that presents a potential infringement risk, an individual or organization may choose to obtain a validity study in the hope that references can be located which show that an invention was either anticipated or obvious, and that the patent should not have been granted. As was the case with novelty searches, the potential scope for a validity search is broad. It can include both patent and nonpatent literature. In particular, a disclosure in a patent specification not closely tied with that patent's claims can often be useful in invalidating a patent. Since such disclosures are typically not reflected in the classification of the patent, which is tied to the claims, classified sets of patents are not necessarily effective for validity searching. Deep-indexed databases on the other hand can be useful, and full-text patent databases also promise great utility in validity searching. The stakes involved in gaining freedom from blocking patents can be substantial; the cost and effort expended in a validity search can be correspondingly large.

Closely akin to validity searching is searching for the purpose of opposition. Validity and opposition searches have the same requirements as novelty searches, ie, any reference that would render an invention unpatentable under national laws is relevant in an opposition search. Opposition searches are performed long enough after the filing of the patent application so that all of the prior art published before the date of filing is made available for searching.

State-of-the-Art Searches. State-of-the-art searches are typically carried out when research in a newer area is to begin in order to identify what has previously been done, what is known, and where fruitful opportunities might be found. Typically, a state-of-the-art search is broad and general, although tighter and more focused follow-up searches are often carried out once the areas of potential interest are identified.

Alerting Searches. Various means are available for keeping up with the latest in patents, and it can be effective to use a combination of these methods. Thus, computer profiles created to represent individual, group, or organization interests can be run against databases as they are updated, and specific searches can also be run against these databases. There are databases that are updated promptly.

Traditional paging through patent office gazettes and printed abstract bulletins still serves a useful purpose in patent alerting. It can be difficult to frame a query for a computer search on all the subject matter that might be of interest to an organization.

Most new patent cases of interest are published by at least one of the U.S., European, or Japanese patent offices, and WIPO (PCT). Japan presents problems for those not able to read Japanese, but the U.S. *Official Gazette* (with representative claims) and *PCT Gazette* (with English-language abstracts) can be in one's hands within a week of patent publication. Similar timing is available for the *European Patent Office Bulletin* which contains trilingual titles and the on-line EPAT file and various CD-ROM products. A highly effective alerting program can be developed from a combination of these methods.

Family and Equivalent Searches. A wide range of inquiries fall under the category of patent family searches. It may be desirable to find an equivalent to a known patent in a given language, typically but not necessarily English. It may be necessary to find whether an invention is protected in a given country. It may be desirable to estimate a patentee's interest in an invention on the basis of how broadly it has been filed, or to know in detail all the countries in which an invention has been patented, including the legal status of each. Or it may be necessary to trace the entirety of a complex extended family. All of these tasks have become relatively simple because of the efforts of Derwent, EPIDOS, and INPI. Derwent's WPI database covers 40 patenting authorities in 1995, and provides information on multiple stages of publication in many of them. It identifies many intellectual patent families, and provides data links that make tracing the web of extended families possible. Family information goes back to 1970 for chemistry and several years earlier than that for pharmaceuticals, agriculturals, and polymers. The ORBIT version merged with APIPAT also includes family information from the 1960s relating to petroleum and petrochemicals.

Citation Searching. In the scholarly literature, authors cite earlier publications that relate to the work being reported, thus a subject relationship exists between the citing and cited literature. This relationship has formed the basis for the *Science Citation Index* and related products, developed by the Institute for Scientific Information. Known as Scisearch in its on-line version, the *Science Citation Index* has become an important information retrieval tool in the second half of the twentieth century. It has been used for straightforward subject searching, in which mode it complements traditional indexed databases and indexes. It has also become a popular tool for bibliometric studies of various sorts, such as attempts to measure the relative impact of research carried out by different individuals or organizations, or the relative impact of publications in different journals.

Citations appear in patents as well as in the journal literature, and it has been proposed that they should become searchable to provide different types of useful information, such as research trends and estimates of the effectiveness of research organizations.

Citation searching of patents offers a perspective different from either traditional class searching or traditional subject index searching. A citation search on known fundamental patents can lead directly to improvement patents, even when those patents are so new that they have not yet been indexed. This technique can be especially effective when working in an unfamiliar area, or one which is difficult to index.

The availability of citation searching tools is on the increase. In 1995, Derwent introduced the Patents Citation Index covering inventors' and examiners' citations from 16 countries, and enabling searches to be carried out for citing patents as well as for cited patent and nonpatent references.

On-Line Database Searching Methods

Coordinate Indexing and Boolean Logic. Three methods of indexing have been prominent in the chemical literature in recent times. (*1*) Articulated indexing, has been used in printed *Chemical Abstracts* subject indexes from their earliest days until well into the 1990s. A number of important concepts are identified as permissible index entries, including specific compounds, material types, reactions, and processes. A basic index entry with modifying statement is **Hydrocarbons**, pyrolysis of, in plasmas. (*2*) The keyword-in-context (KWIC) index, arose during the early days of computer processing. The same entry that would appear in a KWIC index is PYROLYSIS OF **HYDROCARBONS** IN PLASMAS. (*3*) The coordinate index, in which all of the individually indexable concepts of a document are posted to the record for that document, ie, HYDROCARBONS, PYROLYSIS, and PLASMAS.

A number of methods have been developed to introduce context to on-line databases, enabling searches to be refined to minimized false retrieval. One of the earliest techniques is proximity searching, in which two words are required to be adjacent, or within a limited distance from each other in text. The assignment of roles to chemical substances is a method of precoordinating concepts. A substance can be identified as a reactant, as a product, and in some systems in a number of additional roles.

Another source of context comes from links between index concepts. In a database that describes chemical compounds in terms of their fragments, it is important that those fragments are tied together, and that the fragments of compound A are tied separately from the fragments of compound B.

Subject-Based Retrieval Parameters. There are numerous means by which the subject content of a patent can be expressed, and which a searcher can use in developing a search strategy. Different databases offer differing subsets of these means. Effective strategies should in general not be limited to a single type of retrieval parameter; rather, they should be built from different parameters and modified as needed to provide the strategy best fitted to the subject at hand.

Patent titles are usually short, and sometimes extremely uninformative. Because a well-written abstract highlights the most important concepts in a document, words in abstracts can be highly valuable retrieval terms; however, abstracts can vary greatly in format and quality.

The text of patent claims is especially important for infringement searching. Several databases make available the complete claims of U.S. patents covering a period exceeding the life span of U.S. patents. Representative claims for European, French, and German patents, as well as PCT applications, are available on-line.

Controlled indexing can help overcome the vagaries of free text. The structures of indexing languages can differ sharply, and thus have a substantial effect on retrieval techniques.

A different situation pertains to databases such as CA or the indexed CLAIMS files. In CA, there are generic and specific terms, but a broad generic term cannot normally be searched with confidence that all of the specifics that fall into the class will be retrieved. It is necessary to build up groups of homologues and synonyms in a search strategy, although the introduction of polymer class terms in the CAS Registry is a great help in carrying out broad searches of polymer information.

Patent classification codes are another subject-search parameter available in most patent databases. IPC codes are usually present and U.S. codes exist in a number of files; in the case of Japan Patent Information Organization (JAPIO), Japanese codes too are available.

Structure Searching. Fragmentation systems have been the traditional means for indexing and searching generic and Markush chemical structures in patents. Derwent's FARMDOC–AGDOC–CHEMDOC code is such a system, as are the systems used in CLAIMS-Uniterm and -CDB and in APIPAT, but there are important differences among the systems.

Fragmentation systems, useful as they are, describe molecular structures incompletely. Topological indexing systems, typified by the CAS Registry, are used to identify unambiguously each of the more than 13 million substances covered in CA, and can be searched for specific complete structures as well as for substructures. With the advent of the MARPAT system in 1988 the Registry began to handle generic and Markush structures as well. Recent databases enable the compounds in the CAS Registry to be searched by a combination of parameters such as name fragments, molecular formulas, and ring system identifiers. Dictionary searching may lack some of the power of the full Registry database but it is a highly useful technique in its own right, and can be used in combination with topological searching in the STN version of the Registry.

Cross-File and Multifile Techniques

Databases differ in their strengths and weaknesses, as well as in their focus. As a result, duplicate searches carried out on different databases generally produce different results. This has been demonstrated in comparative studies of retrieval results for a group of patent databases. Searchers are counseled to use multiple databases whenever possible.

Cross-file techniques permit searchers to combine the approaches and capabilities of different databases and achieve a synergistic result.

Multifile searching differs from cross-file searching in that it permits a single strategy to be brought to bear on more than one file at the same time, but the individual files are searched independently without interaction. An advantage of multifile searching is that it is possible to create and use in one step a single strategy; a disadvantage is that the single strategy may not be optimum for all of the files used. ORBIT's version of multifile searching, called PowerSearch, can introduce an element of cross-file searching to multifile situations. Different databases can be searched separately, each by whatever strategy desired, and the outputs can be merged.

Term Extraction and Analysis Software. Closely allied to the software used in cross-file searching is the software that extracts terms and provides statistical analysis of their occurrence within the set being analyzed.

Patent Databases

The following are the important patent databases available: Derwent World Patents Index (WPI) and WPI Markush, Chemical Abstracts and CAS Registry, INPADOC and EPIDOS Register, EDOC, and CLAIMS databases.

Full-text patent databases such as LEXPAT and DIALOG's PATFULL are useful in locating minute disclosure.

Archiving and Document Delivery

Many organizations have traditionally maintained in-house collections of patent specifications in areas of interest. Microforms and, more recently, CD-ROMs have taken the place of paper copy as the volume of the patent literature grows. CD-ROMs can each hold the images for approximately 1000 U.S. or European patent documents, and high quality copies can be produced with laser printers. CD-ROMs are available soon after patent publication, and represent a significant advance over earlier patent copy delivery media, but they too may be supplanted in the future.

Most users of patent information, occasionally even those that have extensive patent copy collections, must order copies from the outside, and national patent offices such as the U.S. PTO have been important suppliers of copies, typically for their own country and other countries as well. In the United States, there are more than 70 patent depository libraries that serve as regional patent information resources. In addition, a number of other organizations supply patent copies, including Derwent, CAS, DIALOG, Rapid Patent, The Library Connection, the British Library, EPIDOS, and others. Electronic means of document delivery are becoming more common; telefacsimile (FAX) are used by many suppliers, and satellite transmission has been discussed by some potential suppliers. The Internet has been used for the transmission of some patent information.

Other Technological Initiatives

Strategies developed using Boolean logic have been central to the searching of on-line databases, but during the early 1990s it has been suggested that effective searches could be carried out with natural language input by using systems based on artificial intelligence. Two of these in particular are related to patent searching. DIALOG's TARGET software is aimed not particularly at patents, but at any database containing substantial amounts of text, especially full-text files. Full text is desirable because the system is based in part on term-frequency counts. A second system is Patent Analyzer, offered jointly by Derwent and Electronic Data Systems. Both are so new, but it would appear that such systems might be less appropriate for searching the chemical arts than for those functions in which the recognition of chemical structures is not required.

CASLINK, a software feature from STN, carries out searches of several structure files, including Registry, MARPAT, and MARPAT Previews, collects the results, runs these against CA bibliographic databases, and identifies and eliminates duplicate records. The ORBIT PowerSearch software is probably the most effective at recognizing patent family relationships, but competition among the on-line hosts and database producers will lead to continued improvement of these capabilities.

EDLYN S. SIMMONS
Hoechst Marion Roussel, Inc.
STUART M. KABACK
Exxon Research and Engineering Co.

J. Maynard and H. Peters, *Understanding Chemical Patents,* 2nd ed., American Chemical Society, Washington, D.C., 1991.

E. S. Simmons in C. J. Armstrong and J. A. Large, eds., *Manual of Online Search Strategies,* 3rd ed., Ashgate, Aldershot, U.K. (in press).

S. Van Dulken, ed., *Introduction to Patents Information,* 3rd ed., British Library Science Reference and Information Service, London, 1998.

W. Warr and C. Suhr, *Chemical Information Management,* VCH Publishers, New York, 1992.

PCBS. See CHLOROCARBONS AND CHLOROHYDROCARBONS, TOXIC AROMATICS.

PECTIC SUBSTANCES. See GUMS.

PELLETING AND BRIQUETTING. See SIZE ENLARGEMENT.

PENICILLINS. See ANTIBIOTICS, β-LACTAMS–PENICILLINS AND OTHERS.

PENTANES. See HYDROCARBONS.

PEPTIDE ANTIBIOTICS. See ANTIBIOTICS, PEPTIDES.

PERCHLORIC ACID AND PERCHLORATES

When in a +7 valence state and combined with oxygen, chlorine forms a family of compounds known as the perchlorates. The perchlorate anion, ClO_4^-, as progenitor is derived from perchloric acid, $HClO_4$, one of the strongest of the mineral acids. The perchlorates are more stable than the other chlorine oxyanions (see CHLORINE OXYGEN ACIDS AND SALTS). Essentially, all of the commercial perchlorate compounds are prepared either directly or indirectly by electrochemical oxidation of chlorine compounds (see ALKALI AND CHLORINE PRODUCTS; ELECTROCHEMICAL PROCESSING). The perchlorates of practically all the electropositive metals are known, except for a few cations having low charges.

The most outstanding property of the perchlorates is their oxidizing ability. On heating, these compounds decompose into chlorine, chlorides, and oxygen gas. Aqueous perchlorate solutions exhibit little or no oxidizing power when dilute or cold. However, hot concentrated perchloric acid is a powerful oxidizer and whenever it contacts oxidizable matter extreme caution is required. The acidified concentrated solutions of perchlorate salts must also be handled with caution.

Actual perchlorate production is difficult to determine in any given year, because AP is classified as a strategic material. Future production is expected to depend mostly on space programs.

Properties

Chlorine Heptoxide. The anhydride of perchloric acid is chlorine heptoxide, Cl_2O_7, also known as dichlorine heptoxide. It is obtained as a colorless oily liquid by dehydration of perchloric acid using a strong dehydrating agent such as phosphorus pentoxide, P_2O_5:

$$2\ HClO_4 + P_2O_5 \rightarrow Cl_2O_7 + 2\ HPO_3 \quad (1)$$

Chlorine heptoxide is more stable than either chlorine monoxide or chlorine dioxide; however, the Cl_2O_7 detonates when heated or subjected to shock. It melts at −91.5°C, boils at 80°C and has a molecular weight of 182.914. It is soluble in benzene. It explodes on contact with a flame or by percussion. Reaction with olefins yields the impact-sensitive alkyl perchlorates.

Perchloric Acid. Pure anhydrous perchloric acid, $HClO_4$, is quite unstable. In aqueous solution, however, $HClO_4$ is a familiar and useful reagent. Perchloric acid is commonly obtained as an aqueous solution, although the pure anhydrous compound can be prepared by vacuum distillation as a colorless liquid, which freezes at −112°C and boils at 16°C at 2.4 kPa (18 mm Hg) without decomposition.

A number of hydrates of perchloric acid, $HClO_4 \cdot nH_2O$, where $n = 1, 2, 2.5, 3$, and 3.5, are known.

The combination of oxidizing effect, acidic strength, and high solubility of salts makes perchloric acid a valuable analytical reagent. It is often employed in studies where the absence of complex ions must be ensured.

Ammonium Perchlorate. Ammonium perchlorate is a colorless, crystalline compound having a density of 1.95 g/mL and a molecular weight of 117.5. It is prepared by a double displacement reaction between sodium perchlorate and ammonium chloride, and is crystallized from water as the anhydrous salt. The perchlorates, especially those of the light metals and ammonium ion, are favored as solid oxidizers for rocket propellants.

A newer approach developed for producing commercial quantities of high purity AP involves the electrolytic conversion of chloric acid to perchloric acid, which is neutralized by using ammonia gas.

Alkali Metal Perchlorates. The anhydrous salts of the Group 1 (IA) or alkali metal perchlorates are isomorphous with one another as well as with ammonium perchlorate.

The alkali metal perchlorates are either white or colorless, and have increasing solubility in water in the order of Na > Li > NH_4 > K > Rb > Cs. The high solubility of sodium perchlorate, $NaClO_4$, makes this material useful.

Group 11 (IB) Perchlorates. Copper and silver perchlorates have been studied quite extensively. Gold forms organic perchlorate complexes as well as complexes with silver.

Alkaline-Earth Perchlorates. Anhydrous alkaline-earth metal perchlorates can be prepared by heating ammonium perchlorate in the presence of the corresponding oxides or carbonates.

Group 12 (IIB) Perchlorates. The zinc perchlorate, cadmium perchlorate, mercury(I) perchlorate, and mercury(II) perchlorate all exist.

Group 13 (IIIA) Perchlorates. Boron perchlorates occur as double salts with alkali metal perchlorates. Aluminum perchlorate, $Al(ClO_4)_3$, forms a series of hydrates.

Group 3 (IIIB) and Inner Transition-Metal Perchlorates. The rare-earth metal perchlorates of yttrium and lanthanum have been reported, as have tetravalent cerium perchlorate, $Ce(ClO_4)_4$, and uranium perchlorate.

Group 14 (IVA) Perchlorates. Perchlorates containing organic carbon have been reported, as have diazonium perchlorates, oxonium perchlorates, and the perchlorate esters. Extreme caution must be used in working with organic perchlorates; many decompose violently when heated, contacted with other reagents, or subjected to mechanical shock.

Group 4 (IVB) Perchlorates. Titanium tetraperchlorate is known.

Group 15 (VA) Perchlorates. Nitrogen perchlorates have been used as oxidizers in rocket propellants. Hydrazine perchlorate, $NH_2NH_3ClO_4$, and hydrazine diperchlorate, $ClO_4NH_3ClO_4$, have been investigated as oxidizers for propellant systems.

Other Group 15 perchlorates include nitronium perchlorate, NO_2ClO_4; nitrosyl perchlorate, $NOClO_4$; and phosphonium perchlorate, $P(OH)_4ClO_4$.

Group 5 (VB) Perchlorates. Vanadyl perchlorate, $VO(ClO_4)_3$, has been prepared.

Group 16 (VIA) Perchlorates. A perchlorate compound perchloryl sulfate, $SO_4(ClO_4)_2$, is a strong oxidizer.

Group 6 (VIB) Perchlorates. Both divalent and trivalent chromium perchlorate compounds have been reported. Chromyl perchlorate has been suggested for a gas-generating system operating at −45°C.

Group 17 (VIIA) Perchlorates. Fluorine perchlorate, $FClO_4$, is normally a gas. It melts at −167.5°C and boils at −15.9°C. It is extremely reactive and explosive in all states.

The perchloryl fluoride, $FClO_3$, the acyl fluoride of perchloric acid, is a stable compound. Normally a gas having a melting point of −147.7°C and a boiling point of −46.7°C, it can be prepared by electrolysis of a saturated solution of sodium perchlorate in anhydrous hydrofluoric acid. Some of its uses are as an effective fluorinating agent, as an oxidant in rocket fuels, and as a gaseous dielectric for transformers.

Other Transition Element Perchlorates. Both divalent and trivalent manganese perchlorate compounds are known. Perchlorates of Fe, Co, Ni, Rh, and Pd have been produced as colored crystals.

Manufacture

Perchloric Acid. Several techniques have been employed in the manufacture of perchloric acid, including thermal decomposition of chloric acid, anodic oxidation of chloric acid, irradiation of chlorine dioxide solutions, electrolysis of hydrochloric acid, oxidation of hypochlorites by ozone (qv), ion exchange (qv), and electrodialysis of perchlorate salts.

Perchlorates. Historically, perchlorates have been produced by a three-step process: (*1*) electrochemical production of sodium chlorate; (*2*) electrochemical oxidation of sodium chlorate to sodium perchlorate; and (*3*) metathesis of sodium perchlorate to other metal perchlorates. The advent of commercially produced pure perchloric acid directly from hypochlorous acid means that several metal perchlorates can be prepared by the reaction of perchloric acid and a corresponding metal oxide, hydroxide, or carbonate.

Shipping and Handling

Perchloric acid and perchlorates are classified as strong oxidizers and emit toxic fumes when decomposed; contact with combustible, flammable, or reducing materials must be avoided. Perchloric acid and perchlorates must be shipped in accordance with the U.S. Department of Transportation hazardous material regulations. Handling these compounds requires the procedures and safety precautions specified by the product supplier. Perchlorates contain a self-sustaining source of oxygen, thus fires involving perchlorates must be extinguished with water. A class of more hazardous compounds is formed by mixing inorganic perchlorates with finely divided metals, sulfur, or organic compounds and must be handled with the same precautions as explosives.

Economic Aspects

Anhydrous perchloric acid is not sold commercially. Aqueous solutions of perchloric acid are sold at low concentrations for analytical standard applications and at concentrations up to 70%. The price for 70% perchloric acid starts at $2.70/kg, depending on the quantity and level of impurities. The U.S. domestic capacity of ammonium perchlorate is roughly estimated at 31,250 t/yr. The actual production varies, based on the requirements for solid propellants. Environmental effects of the decomposition products, which result from using solid rocket motors based on ammonium perchlorate-containing propellants, are expected to keep increasing public pressure until consumption is reduced and alternatives are developed. Approximately 450 t/yr of NH_4ClO_4-equivalent cell liquor is sold to produce magnesium and lithium perchlorate for use in the production of batteries. Total U.S. domestic sales and exports for sodium perchlorate are about 900 t/yr. In 1995, a solution containing 64% $NaClO_4$ was priced at ca $1.00/kg; dry product was also available at $1.21/kg.

Uses

Perchloric acid is used in analytical chemistry for the determination of trace metal constituents in oxidizable substances as well as in the production of high purity metal perchlorates; it has also been introduced as a stable reaction media in the thermocatalytic production of chlorine dioxide. Perchlorates are primarily used in ammonium perchlorate as an oxidizer in the formulations of propellant for solid rocket motors. Perchlorates are used in the production of explosives, pyrotechnics, and in solid, slurried, and gelled blasting formulations. Both magnesium and lithium perchlorates are used in dry batteries. Other perchlorates have found application in oxygen-generation systems (qv), adhesive bonding of steel plates, and the recovery of potassium from brines such as $KClO_4$ (see CHEMICALS FROM BRINE).

SUDHIR K. MENDIRATTA
RONALD L. DOTSON
ROBERT T. BROOKER
Olin Corporation

R. C. Rhees, ed., *McGraw-Hill Encyclopedia of Science and Technology,* 2nd ed., McGraw-Hill Book Co., Inc., New York, 1966, p. 9.

J. C. Schumacher, *Perchlorates, Their Manufacture & Uses,* ACS Monograph 146, Reinhold Publishing Corp., New York, 1960.

A. A. Shilt, *Perchloric Acid and Perchlorates,* G. F. Smith Chemical Co., Columbus, Ohio, 1979.

G. F. Smith, *Perchloric Acid,* 2nd ed., G. F. Smith Chemical Co., Columbus, Ohio, 1951.

PERCHLORO COMPOUNDS. See CHLOROCARBONS AND CHLOROHYDROCARBONS.

PERFLUORO COMPOUNDS. See FLUORINE COMPOUNDS, ORGANIC.

PERFUMES

Perfumes are mixtures created for use in a wide variety of applications, ranging from expensive couturier perfumes to cosmetics, personal grooming products, laundry products, household cleaning products, and many others. They are created from a palette of several thousand materials, most of which are manufactured by chemical processing methods. Until late in the nineteenth century, fragrances were derived from natural sources, which put a limitation on where and how they could be used.

Creation of Perfumes

Perfumes are usually considered in two broad categories, as either fine or functional (household product) fragrances. Fine fragrances include perfumes, colognes, men's colognes, aftershaves, and fragrances for cosmetic products. For the purposes of this article, functional products include all personal and household cleaning products that are perfumed. The investment in each fragrance that reaches the marketplace is quite substantial. For these reasons, fragrance formulas are held as trade secrets.

Often a fragrance creation begins with a concept inspired by an existing perfume, a newly available ingredient, or a newly discovered odor facet of an existing material. The perfumer creates an accord based on the inspiring note. Computer-assisted design methods are utilized by perfumers to help with formulation changes and to keep track of the large material base available to them.

Table 1. Perfumery Descriptions

Floral		Citrus	Woody	Green	Fruity
carnation	lilac	bergamot	cedar	basil	apple
chrysanthemum	lily	grapefruit	fir	cucumber	apricot
gardenia	marigold	lemon	hickory	grass	banana
honeysuckle	muguet	lime	patchouli	parsley	black currant
hyacinth	narcissus	mandarin	pine	rhubarb	cherry
iris	orange flower	orange	sandal	string bean	fig
jasmine	rose	tangerine		violet	grape
jonquil	violet	verbena		watercress	melon
lavender	mimosa				peach
					pineapple
					prune
					raspberry
					strawberry

The performance of a fragrance over time in its intended application is an important consideration. A perfume can be viewed as a blend with a top note continuing into a middle and on to an end note. As might be expected, this is a function of relative volatility and odor strengths of the materials used.

There is a generally agreed-upon odor vocabulary that is used to characterize individual ingredients and finished fragrances. Table 1 shows some commonly used odor descriptors grouped into five general classifications.

Fine Fragrances

Fine fragrances must work on the skin and blend with body odor. They must be pleasant, diffusive, and substantive (long-lasting), and have the quality of genuine beauty, as well as signatures that distinguish them from each other. For most fine fragrances, the perfumes are themselves the products. They are sold to the consumer at various concentrations in alcoholic or aqueous–alcoholic solutions, depending on the type of application intended. For example, women's perfumes are typically 20–35% fragrance oil in 95% ethanol. Women's colognes are offered in the range of 15–22% fragrance oil, whereas men's colognes and aftershaves are usually in the range of 2–12%. Fine fragrances often set trends that eventually find their way into other kinds of products. Perfumes can be grouped into broad odor categories in an attempt to show their relationships to each other and sometimes indicate the progress of creative evolution as new fragrances are built on the foundations laid by older ones. Following are a number of fine fragrances grouped by a widely used classification scheme.

Women's Fragrances. *Straight Floral Family.* The straight floral family contains a large and popular group of flowery odors. This group includes carnation [Bellodgia (Caron 1927)], jasmine [Honeysuckle (Avon 1963)], rose [Tea Rose (Workshop 1972)], muguet [diorissimo (Dior 1956)], and tuberose [Fracas (Piguet 1945)].

Floral Bouquet Family. In the floral bouquet family, fantasy accords are blended into the floral. Examples include White Shoulders (tuberose), Fidji (floral, green), Joy (rose, jasmin, muguet), and L'Air du Temps (spicy carnation).

Aldehydic Floral Family. This is an important family of fragrances, the typical odor of which is the class odor of the aldehydes, which are added in small, but effective amounts. Examples are Chanel No. 5 (floral, aldehydic) and Madame Rochas (woody, mossy, peach).

Oriental Family. In these perfumes, a mossy, woody, and spicy accord combines with the sweetness of vanilla or balsam and is accented with animal notes. The most important floral accords used are rose and jasmine. This family includes oriental [Youth Dew (Lauder 1953)], sweet vanilla [Emeraude (Coty 1921)], and orange flower spice [Après L'Ondée (Guerlain 1906)].

Chypre Family. The fragrances of this large and important group are warm, mossy, and long-lasting, having rose, jasmine, and animal notes. Examples are chypre [Chypre (Coty 1917)], oriental [Pavilion (Lauder 1978)], bandit (woody amber) [Bandit (Piguet 1944)], Miss Dior (chypre patchouli aldehyde green) [Miss Dior (Dior 1947)], Halston (amber, woody) [Halston (Halston 1975)], Mitsouko (chypre peach) [Mitsouko (Guerlain 1919)], and Crêpe de Chine (chypre aldehyde) [Crêpe de Chine (Millot 1928)].

Woody Family. These combine woody odors such as sandal, cedar, or patchouli in harmony with sweet notes, florals, and animal accords. Included in this family are orris [Chamade (Guerlain 1970)], and patchouli [Shocking (Schiaperelli 1935)].

Green Family. This group includes Vent Vert (Balmain 1945) Aliage (Lauder 1972) and Cristalle (Chanel 1974).

Citrus Family. In the 1980s, this type experienced a revival in which citrus was blended with florals and sweet notes. It includes Jean Marie Farina (Roger & Gallet 1806), and Ô De Lancôme (Lancôme 1975).

Musk Family. Important examples are Musk Oil (Caswell-Massey 1950) and Musk Oil (Jovan 1972).

Leather Family. This group includes Tabac Blond (Caron 1919) and Cuir de Russe (Chanel 1924).

Men's Fragrances. Earlier in the twentieth century, men's fragrances were expected to have a masculine direction, such as tobacco or leather. However; since the 1970s, men's perfumes have allowed much more creative use of rich woody, ambery, and green notes.

Green Family. The green family is increasingly popular, but is still regarded as rather exclusive. It includes green [Old Spice Herbal (Shulton 1974)] and herbal [Grey Flannel (Beene 1975)].

Citrus Family. Lemon, lime, orange, and bergamot are important ingredients. These oils combine well with lavender and amber accords. Examples are lavender [English Lavender (Yardley 1770)], pine [Pino Silvestre (Vidal 1948)], and Eau Sauvage [Eau Sauvage (Dior 1966)].

Fougere Family. This family has a typically accepted masculine note reminiscent of fern, tonka, and moss. It includes Fougere Royal (Houbigant 1822) and Jicky (Guerlain 1889).

Canoe Family. This is one of the most popular fragrance families dating back to the 1940s. Also liked by women, it has a typical unisex note, eg, Canoe (Dana 1935).

Spice Family. This is an easily recognizable fragrance that has a strong, spicy character, eg, Old Spice (Shulton 1937).

Woody Family. This group of fragrances owes its character to woody naturals, such as vetivert and patchouli, but has become more complex over the years. It includes amber [Halston Z-14 (Halston 1976)], sandalwood [Arden for Men (Arden 1955)], and patchouli [Aramis 900 (Lauder 1970)].

Musk Family. Included in this family are Musk for Men (Yardley 1971) and Musk for Men (Jovan 1973).

Leather Family. Examples are Knize Ten (Caswell-Massey 1927) and Ted (Lapidus 1978).

Oriental Family. Sweet, balsamic notes are typical for this group, which includes Habit Rouge (Guerlain 1964) and Pierre Cardin (Cardin 1972).

Chypre Family. This is an extremely popular group that was well received at the end of the 1940s. It is not exclusively masculine since many women's perfumes have been used as models for fragrances in this family. This group represents creativity at its best. It includes Zizanie (Fragonard 1932) and Royal Copenhagen (Swank 1971).

Functional Fragrances

Functional fragrances are incorporated into a variety of media in relatively small amounts. Perfuming of cleaning products probably began with efforts to cover the undesirable odors that accompanied the rendering of tallow to make soap. Although consumers have come to enjoy pleasant-smelling personal and household cleaning products, the covering malodors in the product bases is still a significant challenge to the perfumer (see ODOR MODIFICATIONS). Product bases may also contain ingredients that react with certain perfume ingredients to alter or destroy their odors, or to cause discoloration problems. Thus the task of the functional products perfumer is dominated from the outset by the nature and economics of the product to be fragranced. Aroma chemicals have important advantages over essential oils and other naturals in functional applications because the former are much better characterized in terms of chemical type and reactivity. Product bases are frequently changed and their complete compositions are generally not revealed. Therefore, a certain amount of empirical testing in product bases is necessary. The following are brief descriptions of the fragrance requirements for a number of household products and the perfumery approaches used for them.

Detergent Fragrances. The incorporation of bleaching agents into laundry products and the advent of highly concentrated detergents present new and increasingly difficult challenges to the perfumer. Several factors play important roles. Most critical are chemical stability of the fragrance material in detergent and the rate of evaporation from the sales package. Also important is performance in the product's end use, ie, in the wash water and on the laundered cloth. To some degree, these factors can be predicted from chemical principles and physical properties, but testing of individual materials under use conditions is often necessary. The country of destination is important in determining stability test conditions. Clearly the customer requirements for laundry detergents in Canada are different from those in Brazil.

Detergent fragrances must be particularly powerful and effective because they are incorporated into the final product at rather low levels. Typical detergent powders can contain as little as 0.3% fragrance, although this may be higher in concentrated products. Substantive ingredients such as Galaxolide, Lyral, Lilial, and Ambroxan are used to obtain residual fragrance on the cloth.

Soap Fragrances. The function of soap is to clean; however, the fragrance, at a dosage of 1–2%, plays a large role in the perceived quality of the soap bar. Besides the aesthetic quality of the soap fragrance, there are a number of technical complications that the perfumer has to deal with: limitations on cost, odor quality and other characteristics of the soap base, the presence of additives, and the high pH of most soaps (often between 9.5 and 11.0) which may lead to hydrolysis or discoloration problems.

Liquid Fabric Softeners. In these laundry additives, the fragrance must reinforce the sense of softness that is the desired result of their use. Most fabric softeners have a pH of about 3.5, which limits the materials that can be used in the fragrances. A special requirement of perfumes for fabric softeners is the ability to leave a residual of odor on fabric after line- or machine-drying.

Tumble-Dryer Softeners. In these products, which are designed for machine drying, the carrier contains the active softener ingredients and the fragrance. The fragrance partly evaporates with the hot drying air and is partly absorbed into the fabrics. Aesthetically, the fragrance should support a sense of softness and caring for fine laundry.

Bleach Products. Hypochlorite bleaches impose severe limitations on the fragrance materials that can be used due to their oxidizing power and high pH of 12.5. These problems have been largely overcome through knowledge of the chemical properties of available aroma chemicals and the selective development of new ones.

Shampoo Perfumes. The stability of perfume in shampoo is usually not a problem because shampoo pH is near neutral; it is only when special additives are used that stability and performance tests may be required. However, in some cases the addition of perfume can affect properties of the shampoo such as viscosity. Testing is therefore often necessary to identify interactions that can negatively affect the product. Fragrance dosages are generally 0.5–1.0% for normal shampoos, but can be up to 1.5%.

Deodorants and Antiperspirants. Fragrances for these products have stability requirements similar to soap fragrances, and may withstand have to withstand the relatively high temperatures (~ 60°C) encountered during manufacture of these products in stick form. Antiperspirants usually contain aluminum or zirconium salts that can reduce the pH to about 2.3, which makes acid-sensitive fragrance materials unsuitable.

Deodorant fragrances must be long-lasting in order to help maintain a pleasant body odor for as long as possible. In the 1990s when naturalness is the trend, a certain amount of fresh sweat odor is acceptable and the body odor is used as an animal accord that blends well with the rest of the fragrance.

Talcs and Powders. In the perfuming of talcum powders and face powders, stability is the most important factor (see also COSMETICS). Even the finest talc contains alkaline impurities that can cause decomposition and discoloration of fragrance ingredients. Systematic screening programs are often necessary to obtain a list of stable fragrance raw materials for these applications.

Perfume Ingredients

The classical materials of perfumery are natural products. These are mostly of vegetative origin, with some obtained from animal secretions, e.g., tincture of tonquin musk, civet gum, and beaver castoreum. Such materials have been or are being replaced by synthetic substitutes for environmental, political, or economic reasons.

Natural Products. Various methods are employed to obtain useful materials from various parts of plants. Essences from plants are obtained by distillation (often with steam), direct expression (pressing), collection of exudates, enfleurage (extraction with fats or oils), and solvent extraction. Solvents used include typical chemical solvents such as alcohols and hydrocarbons. Liquid (supercritical) carbon dioxide has come into commercial use in the 1990s as an extractant to produce perfume materials.

Concretes. Concretes are produced by extraction of flowers, leaves, or roots, usually with hydrocarbon solvents.

Absolutes. Absolutes are prepared from concretes by further processing to remove materials that can cause solubility problems in perfumes. This is done by dissolution in alcohol, filtering, and removal of the solvent, usually at reduced pressures. The resulting products are viscous, oily materials which may be diluted with low odored substances such as diethyl phthalate.

Essential Oils. Essential oils are produced by distillation of flowers, leaves, stems, wood, herbs, roots, etc. Distillations can be done directly or with steam.

Naturally Derived Materials. The following are descriptions of some of the most important naturally derived materials in use.

Bergamot. Bergamot oil is produced by cold expression from peels of fruits from the small citrus tree, *Citrus bergamia.* Bergamot is grown mainly in southern Italy and northern and western Africa. Its largest chemical constituent, to the extent of 35–40%, is linalyl acetate, with a much smaller amount of citral.

Bois de Rose. Bois de rose oil is obtained by steam distillation of wood chips from South American rosewood trees, *Aniba rosaeodora.* It is used as such in perfumes or as a source of linalool, which is used directly in perfumery and for conversion to esters, eg, the acetate.

Cedarwood. Many varieties of cedarwood oil are obtained from different parts of the world. They are produced mainly by steam distillation of chipped heartwood, but some are also produced by solvent extraction. The principal constituents of these oils are cedrene, thujopsene, and cedrol.

Citronella. Citronella oil is produced in Ceylon, China, Java, and Brazil by steam distillation of similar, but not identical, grasses. The main constituents of the oil are citronella, geraniol, and citronellol.

Clove Leaf Oil. Clove leaf oil is produced mainly in Madagascar and Indonesia. It is obtained by distillation of leaves and twigs of *Eugenia caryophyllata.* Main constituents are eugenol, caryophyllene, and humulene.

Galbanum. Galbanum gum is an exudate collected from large umbelliferous plants of the Ferula species, which grow wild in the Middle East.

Geranium. The most important geranium product by far is geranium bourbon, an oil produced by steam distillation of *Pelargoneum graveolens,* leaves and branches. Most current production is from China and Egypt. This material is comprised mainly of *l*-citronellol.

Jasmine. Jasmine is one of the most precious florals used in perfumery. The concrete of jasmine is produced by hydrocarbon extraction of flowers from *Jasminum officinale* (var. *Grandiflorum*). The concrete is then converted to absolute by alcoholic extraction. It is produced in many countries, the most important of which is India, followed by Egypt. Four of the principal odor contributors to jasmine are *cis*-jasmone, methyl jasmonate, benzyl acetate, and indole.

Lavandin. Lavandin is cultivated mainly in southern France. The flowering tops of the *Lavandula hybrida* shrub are used to produce a concrete, an absolute, and a steam-distilled oil; the last is by far the most used. Chemically it is comprised of 30–32% linalool and linalyl acetate, along with numerous other substances, mostly terpenic.

Oakmoss or Mousse de Chene. Oakmoss, *Evernia prunastri,* is a lichen that grows mostly on oak and spruce trees. It is collected mostly in the Czech Republic, Croatia, and Morocco. Oakmoss is worked into a variety of products, including a concrete, resinoid, and absolute, the last of which is the most used. The main odor constituent of oakmoss is methyl 2,4-hydroxy-3,6-dimethylbenzoate.

Orris. Orris is produced from rhizomes of *Iris pallida* and *Iris germanica.* The plants are found and cultivated mostly in Italy, but also in Morocco and China. It is used in perfumery as an absolute, a steam-distilled essential oil, and a concrete. Its most important odor contributors are the irones.

Orange Flower. Extraction of freshly picked flowers of the bitter orange tree, *Citrus aurantium* (subspecies *amara*), for the production of concrete is carried out mainly in Morocco and Tunisia. Methyl anthranilate, linalool, methyl jasmonate, and indole are important odor contributors.

Patchouli. Patchouli oil is produced by steam distillation of the dry leaves of *Pogostemon cablin.* Most of the production in the 1990s is from Indonesia, although some is also produced in China. Its main odor-donating constituents are polycyclic sesquiterpenic alcohols, including patchouli alcohol, norpatchoulenol, and nortetrapatchoulol.

Petitgrain. Petitgrain oils are produced by steam distillation of leaves and twigs of the bitter orange tree, *Citrus aurantium.* Petitgrain Paraguay, by far the most used material of this type, is produced from the bitter-sour variety in South America. Important odor constituents are linalyl acetate, linalool, methyl anthranilate, geraniol, and nerol, 3,7-dimethyl-2,6-octaden-1-ol.

Rose. The most valuable derivatives are produced from *Rosa damascena,* which is grown principally in Bulgaria, but also an Russia, Turkey, Syria, India, and Morocco. The concrete, absolute, and steam-distilled essential oil (rose otto) are particularly valuable perfume ingredients. They are complex mixtures of which citronellol, geraniol, phenethyl alcohol, and β-damascenone (trace component) are important odor constituents.

Sandalwood. Sandalwood is one of the oldest materials in fragrance use. Its oil is produced by steam distillation of coarsely ground wood and roots of *Santalum album.* World production is concentrated in India and Indonesia, the latter being of less preferred quality. In order to obtain a good oil and high yield, only trees that are over 30 years in age are used. This limits the supply of the oil and makes it expensive. Much of the sandalwood odor of perfumes produced in the 1980s and 1990s is a result of the excellent synthetic materials. The main odor contributors to sandalwood oil are alpha-santalol and beta-santalol.

Vetivert. Vetivert oil is steam-distilled from cleaned, dried, and chopped rootlets of *Vetiveria zizanoides*, a tall perennial grass grown in Haiti, Indonesia, Reunion, and, of a poorer quality, in China. β-Vetivone is probably the main odor contributor to this essential oil.

Violet Leaf. Violet Leaf absolute is produced by the usual extraction methods from *Viola odorata* (var. *Victoria*) in the south of France and Egypt. The principal odorant in violet leaf absolute is 2-*trans*-6-*cis*-nonadienal.

Ylang-Ylang. Flowers from the cultivated *Cananga odorata* tree, grown mostly in the Comoro islands and Madagascar, are the starting material for the concrete, absolute, and steam-distilled oil of ylang-ylang. Benzyl acetate is the largest component (~30%) of the oil but is responsible for only a small part of its odor profile, which is the result of numerous minor constituents.

Aroma Chemicals. The use of aroma chemicals in perfumery has been growing since they were first introduced. A number of practical advantages account for this trend. Probably foremost among them is that the growing use of fragrance in the world outstripped the ability to produce enough natural materials, particularly the aesthetically important concretes and absolutes. Another main reason why aroma chemicals have grown to be such a large part of the fragrance industry is their availability. This has been possible because synthetic fragrance materials are produced from a wide variety of starting materials, from both petrochemical and renewable sources. The most important renewable source is turpentine, followed at some distance by cedarwood oil.

Aroma chemicals are not limited to any particular functional group, some being more common than others.

Table 2 presents a number of widely used aroma chemicals by chemical type.

They are used in perfumes over a wide range of concentrations. The importance of a material to the overall creation should not be judged by the amount used. It is characteristic of chemical manufacturing in the fragrance industry that many materials are produced over a wide range of production volumes.

Table 2. Typical Aroma Chemicals by Functional Group

Name	Odor type
Hydrocarbons	
caryophyllene	woody, spicy (cloves)
β-farnesene	mild, sweet, warm
limonene	orange, citrus
α-pinene	piney, woody
β-pinene	
Alcohols	
Bacdanol	sandalwood
citronellol	rosy, citrus
linalool	floral, citrus
phenethyl alcohol	floral, rosy
α-terpineol (R = H)	floral, lilac
Aldehydes	
2-methyl undecanal	fresh, citrus (orange)
citral	citrus, lemon
hexyl cinnamic aldehyde	floral, jasmine
Isocyclocitral	floral, carnation
Lilial	floral, muguet
10-undecenal	citrus, waxy
Ketones	
Cashmeran	musky, sweet
α-ionone	floral, violet
Isocyclemone E	amber, woody
Koavone	woody, ambery, floral
muscone	musk
Tonalide	musk
Esters	
benzyl acetate	floral, jasmine
4-t-butylcyclohexyl acetate	
cis	woody, floral
trans	
cedryl acetate	woody, cedar
Cyclacet	green, woody
isobornyl acetate	pine needles
α-terpinyl acetate	herbaceous, piney
(R = acetyl)	
Lactones	
coumarin	sweet, hay
jasmine lactone	floral, jasmine
muskalactone	musk
peach aldehyde	fruity, peach
Ethers	
Ambroxan	amber, woody
Anther	floral, hyacinth
Galaxolide	musk
Nitriles	
cinnamonitrile	cinnamic, balsamic
geranonitrile	citrus, lemon
Polyfunctionals	
amyl salicylate	floral, jasmine
isoeugenol	warm, spicy, floral
Hedione	jasmin, lemon
heliotropine	sweet, floral
Lyral	floral, muguet
vanillin	sweet, vanilla

Manufacture and Quality Control

Perfumes are manufactured by blending ingredients as called for by the perfumers' formulas. During this operation, some protection from air oxidation is advisable for safety reasons and to assure the quality of the finished product. Batch sizes range from several kilograms to several tons.

Quality Control. Reproducible production of perfumes requires careful quality control of all materials used as well as the compounding process itself.

The use of analytical tools has increased over the years with their availability, but there can be no substitute for organoleptic evaluation. The human nose is far more sensitive than any analytical instrument for certain materials, yet it is also limited as a quantitative tool and is subject to fatigue. The fragrance industry therefore relies on both odor evaluation and analytical methods for control of ingredient and product quality.

Research

Analytical Chemistry. Chemists have long analyzed essential oils and other fragrant materials derived from nature in order to determine their compositions and in particular to identify the odoriferous principles. Analysis of naturals has also been used to allow the preparation of synthetic reconstitutions or duplications, which can have great commercial importance if a material is in short supply or becomes very expensive. Duplications also allow important natural odor notes to be used in functional perfumery where supply or discoloration problems can arise.

There have been moves in the 1990s to replace natural materials of animal origin for humanitarian or economic reasons. One of the first and best known perfume ingredients to be eliminated from use was tincture of ambergris. The starting material for this was ambergris, a principal by-product of the whaling industry. Its use was therefore eliminated as part of international efforts to preserve whale populations. Tincture of ambergris has been replaced by formulas that include the most important contributors to ambergris odor, namely, α-ambrinol and dihydro-γ-ionone. These are believed to form via oxidation and cyclization from the principal component, ambrein.

In order to obtain chemical structural information on the hundreds of materials that may be present in a complex mixture, gas chromatography is coupled with mass spectrometry (gc/ms). For those components that are not readily identified from their mass spectra alone, gas chromatography coupled with infrared spectroscopy (gc/ir) has been applied.

Head space analysis techniques have been used to investigate fragrant materials and have produced fascinating technical and commercial results. It was found that these techniques could be applied directly to flowers and other parts of plants. This avoids the heating and other processing involved in producing natural extracts, so that duplications based on these results differ significantly from the older standard products. It also allows the duplication or imitation of flowers, herbs, fruits, etc, which do not produce satisfactory extracts.

It has been found that flowers and other plant parts can be analyzed by using head space techniques without removing them from the living plant. It was immediately observed that there are remarkable differences in the volatile compositions observed from live and picked flowers. Reconstitutions produced from this information have provided perfumers with novel and fresh notes for use in their creations.

Synthesis. Exploratory research has produced a wide variety of odorants based on natural structures, chemicals analogous to naturals, and synthetic materials derived from available raw materials and economical processing. In the search for new aroma chemicals, many new materials are prepared for screening each year. Initial evaluations of chemicals produced for screening are performed by smelling them from paper blotters. However, more information is necessary given the time and expense required to commercialize a new chemical. Aroma chemicals must be stable in use if their desirable odor properties are to reach the consumer. Therefore, testing in functional product applications is an important part of the evaluation process. Other properties that can be important for new aroma chemicals are substantivity on skin and cloth, and the ability to mask certain malodors.

Structure–Odor Correlations and Olfactory Receptors. The issue of structure–odor correlation is one that continues to fascinate and frustrate fragrance chemists. An important new theory suggests that there may be more than one combination of receptors that can produce a given type of odor perception, or more than one way to stimulate a given combination of receptors.

Process Research and Development. The fragrance industry has invested much effort in process research and development. In addition to looking for ways to improve yields and throughputs, significant changes have taken place in the scale of manufacture, the equipment used, and the manufacturing operations themselves. As these changes occur, maintaining product quality has been a critical concern.

Physiological and Psychological Effects of Fragrance. Considerable research effort has been and continues to be made to evaluate the role of fragrance in human behavior. Faced with the limits of physiological measurements in determining the effects of fragrance on humans, researchers have turned to psychological measurements. The results indicate that odors have small priming effects on mood and can affect behavior in predictable ways. Pleasant odors tend to improve mood. Subjects exposed to pleasant odors have been found to be more cooperative and less prone to confrontational approaches. By studying and quantifying the effects of fragrance and fragrance ingredients on mood, it may be possible to add a new dimension to the performance of perfumes.

Economic and Market Information

The fragrance industry has enjoyed fairly rapid growth rates, but figures from year to year can be highly distorted by relative currency fluctuations. For purposes of this discussion, only the value to the fragrance supplier is considered. Retail sales of fine perfumes would give substantially higher numbers. For 1996, worldwide sales of compounded fragrances have been estimated at $6000 million.

Safety, Regulatory, and Environmental Aspects of the Industry

The fragrance industry has a long record of safety, largely on account of the nature and sources of its ingredients, and how its products are used. By examining individual ingredients and setting appropriate limits on their use, it is possible to ensure the safety of fragrances as they are created. This approach is accepted by relevant governmental agencies around the world, such as the U.S. Food and Drug Administration.

The industry supports two key organizations that strengthen scientific criteria and develop guidelines for safe and environmentally sound use of fragrances: the Research Institute for Fragrance Materials (RIFM) and the International Fragrance Association (IFRA).

Fragrances must comply with all applicable regulations and legislation that address occupational and consumer health, safety, and environmental concerns.

Perfumes are also impacted by legislation that regulates specific products which may contain fragrances; such legislation includes the U.S. Food, Drug and Cosmetic Act and the European Community Cosmetic Directive. Thus, the manufacture and use of perfumes must comply with a growing body of environmental and human health-related regulations worldwide.

WILLIAM L. SCHREIBER
International Flavors & Fragrances, Inc.

E. T. Theimer, ed., *Fragrance Chemistry: The Science of the Sense of Smell,* Academic Press, Inc., San Diego, Calif., 1982.

S. Van Toller and G. H. Dodd, eds., *Fragrance: The Psychology and Biology of Perfume,* Elsevier Science Publishing Co., Inc., New York, 1992.

P. Z. Bedoukian, *Perfumery and Flavoring Synthetics,* Allured Publishing Corp., Wheaton, Ill., 1986.

R. R. Calkin and J. S. Jellinek, *Perfumery Practice and Principles,* John Wiley & Sons, Inc., New York, 1994.

PEROXIDES AND PEROXIDE COMPOUNDS

INORGANIC PEROXIDES

A peroxide or peroxo compound contains at least one pair of oxygen atoms, bound by a single covalent bond, in which each oxygen atom has an oxidation number of −1. The peroxide group can be attached to a metal, M, through one (**1**) or two (**2**) oxygen atoms, or it can bridge two metals (**3**):

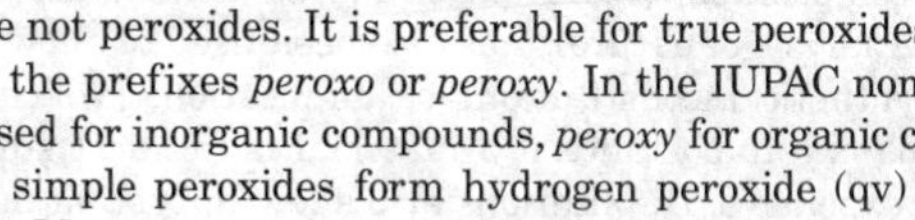

(1) (2) (3)

Peroxides should be distinguished from several other types of compounds having similar names. The higher oxides of lead, manganese, and other elements, although sometimes called peroxides, are not peroxides as defined herein because these contain no oxygen–oxygen bond. Similarly, compounds such as the perchlorates and permanganates are not peroxides. It is preferable for true peroxides to be designated by the prefixes *peroxo* or *peroxy*. In the IUPAC nomenclature, *peroxo* is used for inorganic compounds, *peroxy* for organic compounds.

All the simple peroxides form hydrogen peroxide (qv) on contact with water. Many inorganic peroxides tend, as does H_2O_2, to decompose evolving oxygen.

Group 1 (1A) Peroxides

Peroxides of all the alkali metals having the formula M_2O_2 are known. There are several general methods of preparation: reaction of the metal and oxygen, reaction of the metal monoxide and oxygen, thermal decomposition of the superoxide, and reaction of alkaline solutions of the metal and hydrogen peroxide.

Alkali metal peroxides are stable under ambient conditions in the absence of water. They dissolve vigorously in water, forming hydrogen peroxide and the metal hydroxide. They are strong oxidizing agents and can react violently with organic substances. Only lithium peroxide and sodium peroxide have been commercialized.

Lithium Peroxide. Lithium peroxide, Li_2O_2, is used in space technology because it absorbs carbon dioxide and liberates oxygen (see

OXYGEN-GENERATION SYSTEMS). This peroxide, also used for hardening certain plastics, is a white or pale yellow solid, stable at ambient temperature, and not hygroscopic.

Lithium peroxide is a strong oxidizer and can promote combustion when in contact with combustible materials. It is a powerful irritant to skin, eyes, and mucous membranes. Five grams of many lithium compounds can be fatal.

Sodium Peroxide. Sodium peroxide, Na_2O_2, is a pale yellow solid, stable at ambient temperature, and hygroscopic. Its melting point is 460°C.

The commercial product is a powder containing a minimum of 96% Na_2O_2 and approximately 20% active oxygen. It is made commercially by oxidizing the molten metal with either oxygen or air enriched in oxygen. As of the mid-1990s, sodium peroxide has only a few special applications, including chemical analysis and the extraction of platinum from its ores by the Leidie process.

Although neither inflammable nor self-igniting, sodium peroxide is highly inflammable when mixed with oxidizable substances. Such mixtures burn violently, even in the absence of air.

Sodium peroxide is a powerful irritant to skin, eyes, and mucous membranes; protective clothing should be worn when handling it.

Group 2 (IIA) Peroxides

All the elements of Group 2 form peroxides, with the exceptions of beryllium and radium. There are two general methods of preparation: reaction of the metal or monoxide with oxygen, and reaction of the hydroxide with aqueous hydrogen peroxide. These peroxides are more stable in the presence of water than the Group 1 peroxides, primarily because of insolubility in water. Calcium peroxide is used on a large scale; magnesium, strontium, barium, and zinc peroxides have small-scale uses; whereas cadmium and mercury peroxides have no commercial uses at all. A general account of these peroxides is available.

The materials are generally made by triturating the oxides, or hydroxides, with aqueous hydrogen peroxide and drying the solid products. The commercial products are typically mixtures of the peroxides with varying amounts of hydroxides, oxides, carbonates, hydrates, and peroxohydrates.

Magnesium Peroxide. Magnesium peroxide and MgO_2, used in medicine as a stomach antacid and as an antiseptic (see DISINFECTANTS AND ANTISEPTICS), has not been prepared in the pure state. The product is a white powder containing about 25% MgO_2 and 7% active oxygen. This material is sparingly soluble in water but reacts with water slowly, forming hydrogen peroxide and liberating oxygen gas.

There are minor uses for magnesium peroxide in household products, veterinary medicine, and metallurgy (qv). Magnesium peroxide is a strong oxidizer and can cause fire when in contact with combustible materials. It is a powerful irritant to skin, eyes, and mucous membranes.

Calcium Peroxide. Commercial material contains either 60 or 75% CaO_2; the remainder is a poorly defined mixture of calcium oxide, hydroxide, and carbonate.

An important application of calcium peroxide is for curing the polysulfide sealants (qv) used in double-glazing window units. Calcium peroxide is also used at several gold mines in Australia to increase the recovery of gold and reduce the consumption of cyanide. Solid calcium peroxide can also be used in the heap-leaching of lean gold ores. A proprietary form of calcium peroxide for this purpose is sold by FMC (United States) under the trademark PermeOx.

PermeOx is also used to improve the bioremediation of soils contaminated with creosote or kerosene (see BIOREMEDIATION), to deodorize sewage sludges and wastewater (see ODOR MODIFICATION), and to dechlorinate wastewater and effluents.

Calcium peroxide has several horticultural and agricultural applications, particularly in Japan. Usually used in the form of granules, it acts by providing extra oxygen for germinating plants and other organisms.

Calcium peroxide has been used for many years as a dough conditioner in the United States, but not in Europe, where this use is not permitted. Another industrial application of calcium peroxide is as an oxidizing agent in the production of certain titanium–aluminum alloys.

Calcium peroxide is among the safest of the inorganic peroxides, presenting no significant hazard with regard to skin contact or absorption, inhalation, and ingestion; but it may be irritating to the skin under humid conditions. Airborne dust is irritating to the eyes, nose, throat, and lungs, but poses no significant long-term inhalation hazard.

Strontium Peroxide. Commercial strontium peroxide contains about 85% SrO_2 and 10% active oxygen. The only substantial application for this compound is in pyrotechnics (qv). Strontium peroxide produces a red color in flames.

Strontium peroxide is a strong oxidizer and can cause fire when in contact with combustible materials. It is a powerful irritant to skin, eyes, and mucous membranes.

Barium Peroxide. The commercial product is a dull yellow powder containing about 90% BaO_2 and about 8.5% active oxygen. The principal use is in pyrotechnics, but there are also small uses in the curing of polysulfide rubbers and in the production of certain titanium–aluminum alloys.

Barium peroxide is a strong oxidizer and can cause fire when in contact with combustible materials. It is a powerful irritant to skin, eyes, and mucous membranes. Consequently, it is also toxic via the subcutaneous route; protective clothing should be worn during handling. The LD_{50} value (mouse, oral) is 50 mg/kg.

Group 12 (IIB) Peroxides

Zinc Peroxide. The commercial product is a pale yellow powder containing about 55% ZnO_2 and 9% active oxygen. It is stable in dry air but loses its oxygen in moist air and on heating. It is insoluble in water but dissolves in dilute acid, liberating hydrogen peroxide.

Zinc peroxide is used as an accelerator in rubber-compounding, as a curing agent for synthetic elastomers, and as a deodorant for wounds and skin diseases. Zinc peroxide is a powerful irritant to skin, eyes, and mucous membranes. The systemic toxicity is similar to that of zinc oxide, for which the LD_{50} (rat, oral) is 7950 mg/kg.

Zinc peroxide is a strong oxidizer and can cause fire when in contact with combustible materials.

Group 13 (IIIB) Peroxides

Boron Compounds. *Nomenclature.* The naming of sodium perborate, one of the most important commercial boron compounds, has long been confused. The tetrahydrate has more recently come to be called the hexahydrate. The crystallographically derived names are used to avoid confusion. The commercial or common names are also given.

Sodium Peroxoborate Hexahydrate. The compound sodium peroxoborate hexahydrate (sodium perborate tetrahydrate), $Na_2[B_2(O_2)_2(OH)_4]\cdot 6\,H_2O$, was formerly written as $NaBO_3\cdot 4\,H_2O$. This material has been an important commercial bleaching agent for many years (see BLEACHING AGENTS). The commercial product is a white, crystalline powder having an active oxygen content of at least 10%. It melts at about 60°C; however, if water vapor is free to escape during heating, the crystals do not melt but are converted to the anhydrous peroxoborate.

Sodium peroxoborate hexahydrate is an important ingredient of many household detergents, working best at temperatures above 60°C. It is also used in dishwasher detergents, denture cleaners, as well as foot and bath salts. Organic chemists have been using sodium peroxoborates as oxidants since the 1980s.

The toxicity of sodium peroxoborate hexahydrate in solution is equivalent to those of sodium borate and hydrogen peroxide. The LD_{50} (mouse, oral) is 1060 mg/kg. Local use of high concentrations in the mouth can cause chemical burns and other problems.

The product is considered nonhazardous for international transport purposes. However, it is an oxidizing agent sensitive to decomposition by water, direct sources of heat, catalysts, etc.

Sodium Peroxoborate Tetrahydrate. The compound sodium peroxoborate tetrahydrate (sodium perborate trihydrate), $Na_2B_2(O_2)_2[(OH)_4]\cdot 4H_2O$, was formerly written as $NaBO_3\cdot 3H_2O$.

Sodium peroxoborate tetrahydrate is the most stable of the three peroxoborate hydrates under ambient conditions. It has, however, never been commercialized because it is slow to dissolve in water.

Sodium Peroxoborate. Sodium peroxoborate (sodium perborate monohydrate), $Na_2[B_2(O_2)_2(OH)_4]$, formerly written as $NaBO_3\cdot H_2O$, is known only as a microcrystalline powder, made by dehydrating the hexahydrate.

The commercial product has an active oxygen content of at least 15%. This product has replaced the hexahydrate in some household detergents and other domestic products because it dissolves faster and has a greater content of active oxygen per unit volume of granular product.

The toxicity of sodium peroxoborate is similar to that of the hexahydrate. Sodium peroxoborate is a severe eye irritant, but not a skin irritant. Absorption through large areas of abraded or damaged skin can give systemic boron poisoning. The maximum eight-hour time-weighted average exposure is 5 mg/m^3.

The product is considered nonhazardous for international transport purposes. However, it is an oxidizing agent sensitive to decomposition by water, direct sources of heat, catalysts, etc. Decomposition in the presence of organic material is rapid and highly exothermic.

Anhydrous Sodium Perborate Anhydrous sodium perborate, $NaBO_3$, is an ill-defined, powdery material. It should perhaps be regarded more as an amorphous assemblage of radicals than as a defined compound.

Anhydrous sodium perborate effervesces in water. It is used mainly as an ingredient in denture-cleaning formulations. No toxicological data have been reported on this product, except that in humans, swallowing large amounts can cause nausea, vomiting, and diarrhea. Anhydrous sodium perborate is irritating to eyes, skin, and mucous membranes. It is also mutagenic to *E. coli*.

Group 14 (IVB) Peroxides

Peroxocarbonates. Peroxocarbonates contain the C—O—O— group and should be distinguished from the carbonate peroxohydrates.

There are international transport regulations controlling the transport of sodium percarbonate, which assigned it to Class 5.1, oxidizing substances, however, no such compound has ever been commercialized, and sodium carbonate peroxohydrate is treated as nonhazardous. The origin of this item is not known.

Peroxosilicates. No solid peroxosilicates are known.

Peroxotin Compounds. Older literature records some tin peroxides or peroxohydrates, but these claims have not been substantiated. In contrast, organometallic peroxotin compounds are well established.

Group 15 (VB) Peroxides

Peroxonitrous Acid and Its Salts. Peroxonitrous acid HOONO, is an isomer of nitric acid, HNO_3, to which it rapidly converts. The half-life of peroxonitrous acid at 0°C is 10 s; at 27°C, 0.23 s. It has been known since 1904 that the yellow solution made by mixing nitrous acid and hydrogen peroxide at low temperature contains a stronger oxidant than either ingredient alone, but the chemistry involved was not put on a sound basis until 1994. Additional preparatory methods are also available.

Peroxonitrous acid can decompose by two pathways: isomerization to nitric acid, and dissociation into the hydroxyl radical and nitrogen dioxide.

Peroxonitrite is believed to be present in the crystals of nitric acid trihydrate that form in the stratosphere and in Martian soil (see EXTRATERRESTRIAL MATERIALS). Peroxonitrous acid may be present in mammalian blood and other biochemical systems.

Peroxophosphoric Acids and Their Salts. In its usual impure form (H_3PO_4 is the main contaminant), peroxomonophosphoric acid is a viscous, coloress liquid. It is not produced or used commercially and the salts that have been prepared are unstable and impure.

Pure peroxodiphosphoric acid, $H_4P_2O_8$, has not been obtained, but its properties in aqueous solution are understood.

Tetrapotassium peroxodiphosphate, $K_4P_2O_8$, is a colorless, crystalline solid, soluble in water to 42.2 wt % at 0°C and 51.2 wt % at 40°C.

Tetrapotassium peroxodiphosphate is being investigated as an ingredient in toothpaste as an anticalculus agent and bactericide. However, the peroxodiphosphates are not useful commercial products at present.

Arsenic Peroxides. Arsenic peroxides have not been isolated; however, elemental arsenic, and a great variety of arsenic compounds, have been found to be effective catalysts in the epoxidation of olefins by aqueous hydrogen peroxide. Transient peroxoarsenic compounds are believed to be involved in these systems.

Group 16 (VIB) Peroxides

Peroxosulfuric Acids and Their Salts. Two kinds of peroxosulfuric acid are known: peroxomonosulfuric and peroxodisulfuric acids. Neither is available commercially in the pure state.

Peroxomonosulfuric acid, H_2SO_5, when pure, forms colorless crystals that melt with decomposition at 45°C. Peroxomonosulfuric acid is a strong oxidizing agent. It hydrolyzes rapidly at pH <2 to hydrogen peroxide and sulfuric acid. It is usually made and used in the form of Caro's acid.

Caro's Acid. Caro's acid is the equilibrium mixture that results from mixing hydrogen peroxide and sulfuric acid. These liquids mix instantly, generating a considerable amount of heat.

Because the product is decomposed by heat, it is essential either to remove the heat of reaction quickly or to use the product quickly. The first option is known as the isothermal process; the second option, perfected and commercialized in the early 1990s, is known as the adiabatic process.

Caro's acid is finding increasing application in hydrometallurgy, pulp bleaching, effluent treatment, and electronics.

Peroxomonosulfates. When oleum is mixed with hydrogen peroxide and the mixture is partially neutralized by potassium hydroxide, a triple salt crystallizes out. The commercial product is a white, finely crystalline powder containing a minimum of 4.7% active oxygen. It is used because it is stable and safe, despite being a powerful oxidant. Its main use is in denture cleaners. It is also used in dishwashing detergents and toilet bowl cleaners, in the metal-fabricating industry as a mild etchant and pickling agent (see METAL SURFACE TREATMENTS), in the electroplating (qv) industry for detoxifying cyanide solutions, and in the textile industry for rendering wool (qv) shrink-resistant and nonfelting.

In general, peroxomonosulfates have fewer uses in organic chemistry than peroxodisulfates. However, the triple salt is used for oxidizing ketones (qv) to dioxiranes, which in turn are useful oxidants in organic chemistry.

Potassium hydrogen monoperoxosulfate monohydrate $KHSO_5\cdot H_2O$, related to the triple salt, is not made commercially. This compound is reported as toxic and irritating to eyes, skin, and mucous membranes. Although undoubtedly correct, this description probably better relates to the triple salt.

Peroxodisulfuric Acid. Also called persulfuric acid, and Marshall's acid, peroxodisulfuric acid, $H_4S_2O_8$, when pure, forms colorless crystals that melt with decomposition at 65°C. Peroxodisulfuric acid is a strong acid but not stable. It is seldom isolated but is synthesized and used in solution.

Peroxodisulfates. The salts of peroxodisulfuric acid are commonly called persulfates, three of which are made on a commercial scale: ammonium peroxodisulfate, $(NH_4)_2S_2O_8$; potassium peroxodisulfate, $K_2S_2O_8$; and sodium peroxodisulfate, $Na_2S_2O_8$. The peroxodisulfates are all colorless, crystalline solids, stable under dry conditions at ambient temperature but unstable above 60°C. All the peroxodisulfates are made commercially by electrolytic processes.

The peroxodisulfate ion in aqueous solution is one of the strongest oxidizing agents known. The principal use of the peroxodisulfate salts is as initiators (qv) for olefin polymerization in aqueous systems, particularly for the manufacture of polyacrylonitrile and its copolymers (see ACRYLONITRILE POLYMERS). Etching of printed circuit boards and removal of photoresists are also important applications (see ELECTRONIC MATERIALS; INTEGRATED CIRCUITS). Bleaching of textiles and natural fibers and finishing of furs are both long-established applications. Other established applications include curing grouts for soil stabilization (qv), initiating polymerization of graphite filament coatings, cleaning metal surfaces prior to plating or adhesive bonding, and regenerating active carbon.

An expanding development is the use of peroxodisulfates as oxidants in organic chemistry.

The three peroxodisulfates are all toxic and irritating to skin, eyes, and mucous membranes. The LD_{LO} value for sodium peroxodisulfate using iv administration in rabbits is 178 mg/kg.

Other Metal Peroxides

Transition-Metal Peroxides. Transition-metal peroxides, as isolated species, have no place in chemical technology because they are too dangerously explosive.

Transition metals can be divided into two groups according to the characteristics of their peroxides. The first group comprises those metals that, in their highest oxidation states, have no *d* electrons, eg, Ti^{4+} and W^{6+}. These metals form peroxides from hydrogen peroxide. The peroxo species act as electrophiles.

The other group of transition metals comprises those metals that retain *d* electrons in their normal valence states, eg, Co^{3+} and Pt^{2+}. These metals form peroxides from dioxygen or from hydrogen peroxide.

Actinide Peroxides. Many peroxo compounds of thorium, protactinium, uranium, neptunium, plutonium, and americium are known. Uranium peroxide has found several applications in the nuclear energy industry.

Peroxohydrates

Peroxohydrates are crystalline adducts containing molecular hydrogen peroxide. Peroxohydrates are usually made by simple crystallization from solutions of salts or other compounds in aqueous hydrogen peroxide. They are fairly stable under ambient conditions, but traces of transition metals catalyze the liberation of oxygen from the hydrogen peroxide.

Sodium Carbonate Peroxohydrate. Known commercially as sodium percarbonate, sodium carbonate peroxohydrate does not contain the C–O–O–C group and is not a peroxocarbonate. The stoichiometry is $2Na_2CO_3{\cdot}3H_2O_2$. The material is made commercially by three processes: batch crystallization, continuous crystallization, and fluid-bed reaction.

The commercial product is a white powder containing a minimum of 13% of active oxygen and up to 15% of anhydrous sodium carbonate. The solubility in water at 20°C is about 150 g/L.

The principal use of sodium carbonate peroxohydrate is as a bleaching agent in domestic and laundry detergents. It is used also for industrial textile-bleaching, tripe-bleaching, and in denture cleaners. It can also be used as a convenient oxidant in organic chemistry.

The LD_{50} (rat, oral) of sodium carbonate peroxohydrate is 1034 mg/kg. The occupational exposure limit is 10 mg/m^3 per 40-hour week. The compound is a skin and eye irritant; inhalation of dust can cause irritation to the mucous membranes and the respiratory system. Decomposition in the presence of organic material can be rapid and highly exothermic.

Other Peroxohydrates. Other peroxohydrates include those of potassium, rubidium, and cesium carbonates, $M_2CO_3{\cdot}3H_2O$; ammonium carbonate peroxohydrate, $(NH_4)_2CO_3{\cdot}H_2O_2$; and urea peroxohydrate, $CO(NH_2)_2{\cdot}H_2O_2$.

Urea peroxohydrate is an irritant to skin, eyes, and mucous membranes. The U.S. Food and Drug Administration approves it as an over-the-counter drug.

Peroxopolyoxometallates

Polyoxometallates, derived from both isopoly acids and heteropoly acids, are important homogeneous oxidation catalysts. The metals involved are vanadium, niobium, tantalum, molybdenum, and tungsten. The reactions involved are the oxidation of a wide range of organic compounds by hydrogen peroxide or organic hydroperoxide.

Superoxides

The superoxides are ionic solids containing the superoxide, O_2^-. Superoxides of all of the alkali metals have been prepared. Alkaline-earth metals, cadmium, and zinc all form superoxides, but these have been observed only in mixtures with the corresponding peroxides. The tendency to form superoxides in the alkali metal series increases with increasing size of the metal ion.

Metal superoxides are yellow-to-orange solids. Strong oxidizing agents, they react vigorously with most organic materials and reducing agents, and oxidize many metals to their highest oxidation states.

Sodium superoxide, NaO_2, is a yellow solid. No applications are known.

Potassium superoxide, KO_2, is a canary yellow solid that melts at 450–500°C when pure. Potassium superoxide, a strong oxidizing agent, is similar to the Group 1 metal peroxides. Potassium superoxide is produced commercially by spraying molten potassium into an air stream, which may be enriched with oxygen.

Mine Safety Appliances Co. (MSA) manufactures potassium superoxide in the United States for use in self-contained breathing equipment (see OXYGEN GENERATION SYSTEMS). There are several published uses for potassium superoxide in organic chemistry, eg, for oxidizing aromatic compounds and for initiating anionic polymerization.

On contact with skin and mucous membranes, potassium superoxide is converted to potassium hydroxide, which is corrosive and irritating. The reaction with moisture is exothermic and may induce further decomposition with the production of oxygen.

Other superoxides include rubidium superoxide, RbO_2; cesium superoxide, CsO_2; calcium superoxide, $Ca(O_2)_2$; strontium superoxide, $Sr(O_2)_2$; and barium superoxide, $Ba(O_2)_2$. These superoxides are not produced commercially.

Ozonides

The ozonides are characterized by the presence of the ozonide ion, O_3^-. They are generally produced by the reaction of the inorganic oxide and ozone (qv). Sodium ozonide, NaO_3; potassium ozonide, KO_3; rubidium ozonide, RbO_3; and cesium ozonide, CsO_3, have all been reported. Ammonium ozonide, NH_4O_3, and tetramethylammonium ozonide, $(CH_3)_4{\cdot}NO_3$, have been prepared at low temperatures. Whereas the inorganic ozonides are of potential importance as solid-oxygen carriers in breathing apparatus, they are not produced commercially.

Economic Aspects

All of the large-tonnage peroxo compounds, eg, sodium peroxoborate hexahydrate, sodium peroxoborate, and sodium carbonate peroxohydrate, are made by hydrogen peroxide producers using captive hydrogen peroxide. The world demand for active oxygen provided by these products is fairly stable, rising with the gross national product.

ALAN E. COMYNS
Solvay Interox

W. Gerhartz, ed., *Ullman's Encyclopedia of Industrial Chemistry*, 5th ed., Vol. A19, VCH, Weinheim, Germany, 1991.

C. A. Morgan, *Mellor's Comprehensive Treatise on Inorganic and Theoretical Chemistry*, Suppl., Vol. 5. Longman, London, 1980.

A. McKillop and W. R. Sanderson, *Tetrahedron* **51**(22), 6145 (1995).

B. Bertsch-Franck and co-workers, in R. Thompson, ed., *Industrial Inorganic Chemicals: Production and Use*, Royal Society of Chemistry, Cambridge, U.K., 1995, pp. 188–198.

ORGANIC PEROXIDES

Organic peroxides are compounds possessing one or more oxygen–oxygen bonds. They are derivatives of hydrogen peroxide, HOOH, in which one or both hydrogens are replaced by a group containing carbon (R, R′), ie, ROOH or ROOR′. The ultimate source of the oxygen–oxygen linkage in organic peroxides is oxygen; either from direct air oxidation or from reactions of organic compounds with peroxidic materials derived from oxygen, eg, hydrogen peroxide, alkali metal peroxides, ozone (qv), or other organic peroxides. Organic peroxides are intermediates or products in air oxidation of many synthetic and natural organic compounds. They are involved in many biological processes including development of rancidity in fats, loss of activity of vitamin products, and firefly bioluminescence. Some biological products contain a peroxide group. Organic peroxides are also involved in gum formation in lubricating oils, prepolymerization of some vinyl monomers, and degradation of olefin polymers.

Almost all organic peroxides are thermally and photolytically sensitive owing to the facile cleavage of the weak oxygen–oxygen bond. This cleavage is a unimolecular (first-order) reaction. The thermal decomposition rates are affected by the structure of the organic peroxide and the decomposition conditions.

Thermal decomposition of peroxides initially forms oxygen-centered free radicals from the oxygen–oxygen bond homolysis. These radicals are reactive intermediates generally having very short lifetimes, ie, half-life times less than 10^{-3} s. Because they form useful free radicals, they are used commercially as initiators for free-radical reactions.

Approximately 100 different organic peroxides in well over 300 formulations are commercially produced throughout the world as free-radical initiators for polymerizing vinyl monomers, grafting of monomers onto polymers, curing agents for unsaturated resins, rubber, and elastomers, cross-linking of thermoplastics (eg, polyethylene), modification/degradation of polypropylene, halogenations, anti-Markovnikov additions to terminal olefins (eg, formation of primary mercaptans), and telomerizations. Some are used as bleaching agents (qv) (ie, for grain flours and fabrics), olefin epoxidizing agents, and active species in a variety of other applications, eg, the use of BPO as the active antibacterial component in acne medications.

Organic peroxides can be classified according to peroxide structure. There are seven principal classes: hydroperoxides; dialkyl peroxides; α-oxygen substituted alkyl hydroperoxides and dialkyl peroxides; primary and secondary ozonides; peroxyacids; diacyl peroxides (acyl and organosulfonyl peroxides); and alkyl peroxyesters (peroxycarboxylates, peroxysulfonates, and peroxyphosphates).

Hydroperoxides

There are two main subclasses of hydroperoxides: organic (alkyl) hydroperoxides, ie, ROOH, and organomineral hydroperoxides, ie, $R_mQ(OOH)_n$, where Q is silicon, germanium, tin, or antimony. The alkyl group in ROOH can be primary, secondary, or tertiary. Except for ethylbenzene hydroperoxide, only *tert*-alkyl hydroperoxides are commercially important.

Physical Properties. Some physical properties of alkyl hydroperoxides (in order of increasing carbon content) are listed in Table 1.

Alkyl hydroperoxides can be liquids or solids. Those having low molecular weight are soluble in water and are explosive in the pure state. Alkyl hydroperoxides are stronger acids than the corresponding alcohols and have acidities similar to those of phenols. *tert*-Alkyl hydroperoxides can be purified through their alkali metal salts.

Hydroperoxides exist as hydrogen-bonded dimers in nonpolar solvents and readily form hydrogen-bonded associations with ethers, alcohols, amines, ketones, sulfoxides, and carboxylic acids. Other physical properties of hydroperoxides have been reported in the literature.

Chemical Properties. Hydroperoxides can react with or without cleavage of the oxygen–oxygen bond. Reactions resulting in scission of the oxygen–oxygen bond involve heterolytic, homolytic, or metal-promoted oxidation–reduction reactions.

Alkyl hydroperoxides are reduced readily to the corresponding alcohols; many such reductions are quantitative and useful for analytical methods. Alkyl hydroperoxides have been used as oxidizing or hydroxylating reagents in organic syntheses.

Bases, such as potassium or sodium hydroxide, piperidine, and pyridine, react with primary and secondary hydroperoxides to form aldehydes or ketones. *tert*-Alkyl hydroperoxides form stable alkali metal salts with caustic; however, when equimolar amounts of the hydroperoxide and its sodium salt are present in aqueous solution, rapid decomposition to *tert*-alcohol and oxygen occurs.

Acids react with alkyl hydroperoxides in two different ways, depending on the hydroperoxide structure and the acid strength.

$$R_3COOH \xrightarrow[+]{H} R_3C\text{—}OOH \rightarrow H_2O_2 + R_3\overset{+}{C} \rightarrow \textit{tert}\text{-alcohol or olefin}$$

$$R_3COOH \xrightarrow{H^+} R_3CO\text{—}OH \rightarrow H_2O$$

$$+ R_2\overset{+}{C}OR \rightarrow \text{alcohol or phenol and a carbonyl compound}$$

Hydroperoxides are photo- and thermally sensitive and undergo initial oxygen–oxygen bond homolysis, and they are readily attacked by free radicals undergoing induced decompositions.

Hydroperoxides are decomposed readily by multivalent metal ions, ie, Cu, Co, Fe, V, Mn, Sn, Pb, etc, by an oxidation-reduction or electron-transfer process. Depending on the metal and its valence state, metallic cations either donate or accept electrons when reacting with hydroperoxides. Either one or two electrons may be transferred depending on the metal. With most transition metals, eg, Cu, Co, and Mn, both valence states react with hydroperoxides via one electron transfer. Thus, a small amount of transition-metal ion can decompose a large amount of hydroperoxide and, consequently, inadvertent contamination of hydroperoxides with traces of transition-metal impurities should be avoided.

The reactions of *tert*-alkyl hydroperoxides with ferrous ion generate alkoxy radicals. These free-radical initiator systems are used industrially for the emulsion polymerization and copolymerization of vinyl monomers, eg, butadiene–styrene. Alkyl hydroperoxides are among the most thermally stable organic peroxides. However, hydroperoxides are sensitive to chain decomposition reactions initiated by radicals and/or transition-metal ions. Such decompositions, if not controlled, can be autoaccelerating and sometimes can lead to violent decompositions when neat hydroperoxides or concentrated solutions of hydroperoxides are involved.

Organomineral hydroperoxides undergo thermal and photolytic homolyses:

$$R_3QOOH \xrightarrow[h\nu]{\Delta \text{ or}} R_3QO + OH$$

Synthesis. Hydroperoxides have been prepared from several types of peroxygen compounds including hydrogen peroxide or sodium peroxide, ozone, oxygen, and other organic peroxides. Hydrogen peroxide (H_2O_2) and its anions are powerful nucleophiles and react with reagents RX to form ROOH and HX, where X can be sulfate, acid sulfate, alkane- and arenesulfonate, chloride, bromide, hydroxyl, alkoxide, perchlorate, etc. RX can also be an alkyl orthoformate or *tert*-alkyl carboxylate.

Electron-rich olefins react with hydrogen peroxide under acidic conditions to form hydroperoxides, presumably by means of a carbonium ion intermediate, eg, *tert*-butyl hydroperoxide from isobutylene.

Organomineral hydroperoxides have been prepared from hydrogen peroxide and organomineral halides, hydroxides, oxides, peroxides, and amines. If HX is an acid, ammonia is used to prevent acidic decomposition.

$$R_mQX_n + nH_2O_2 \rightarrow R_mQ(OOH)_n + nHX$$

Table 1. Properties of Some Alkyl Hydroperoxides

Hydroperoxide	Structure	Bp, °C (kPa)[a]	Mp, °C	n_D^{20}
methyl	CH_3–OOH	45.5–46.5 (24.53)		1.3654[b]
ethyl	C_2H_5–OOH	43–44 (6.67)		
isopropyl	$(CH_3)_2CH$–OOH	38–38.5 (2.67)		
n-butyl	*n*-C_4H_9–OOH	40–42 (1.07)	1.4057	
sec-butyl	*sec*-C_4H_9–OOH	41–42 (1.47)	1.4050	
tert-butyl	*t*-C_4H_9–OOH	33–42 (2.27)	4.0–4.5	1.3983[c]
2-methoxy-2-propyl	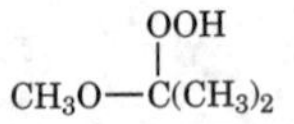	61–63 (2.40)		
tert-amyl	*t*-C_5H_{11}–OOH	34–35 (0.93)	1.4120[c]	
1,1-di-methylpropynyl	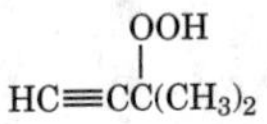	42 (2.27)		
3-hydroxy-1,1-dimethylbutyl	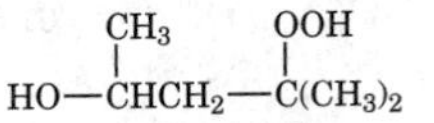			1.4418[d]
cyclohexyl	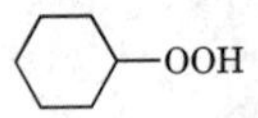	57 (0.16)	−20	1.4622
n-heptyl	*n*-C_7H_{15}–OOH	42–43 (0.008)	35.5	1.4269
3-ethyl-3-pentyl	$(C_2H_5)_3C$–OOH	71–73 (2.27)	2–3	1.4379
1-methylcyclohexyl	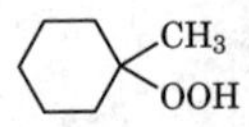	38 (0.004)		1.4652
1-methoxycyclohexyl	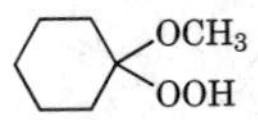	54.5–55 (0.027)		
ethylbenzene	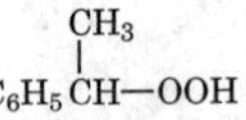	48.2 (0.027)		1.5265
1,1,3,3-tetramethylbutyl	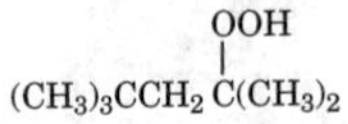	44–45 (0.12)		
2,5-dimethyl-2,5-dihydroperoxyhexane	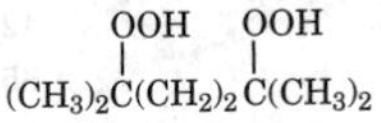		105	
2,5-dimethyl-2,5-dihydroperoxy-3-hexyne	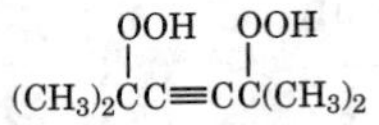		107–109	
α-cumyl	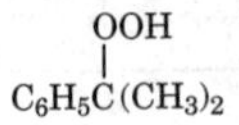	60 (0.027)		1.5242
1,2,3,4-tetrahydronaphthalene	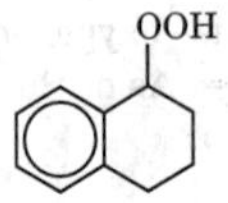	120–125 (0.027)	56	
p-menthane	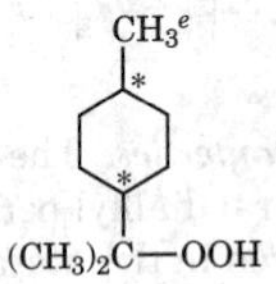			1.4558[f]
pinane	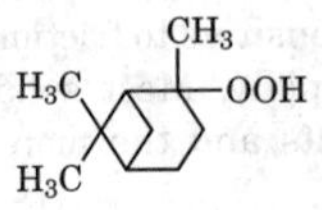	57 (0.013)		
p-diisopropylbenzene mono-hydroperoxide	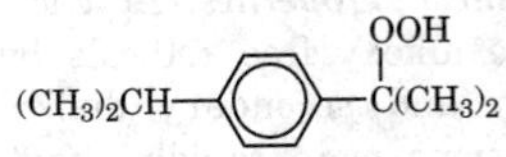		33–34	1.5134

Table 1. Properties of Some Alkyl Hydroperoxides *(continued)*[e]

Hydroperoxide	Structure	Bp, °C (kPa)[a]	Mp, °C	n_D^{20}
p-diisopropylbenzene dihydroperoxide	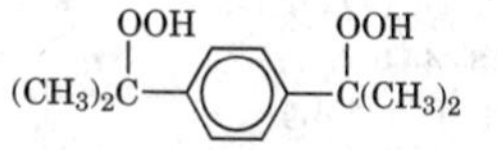		140–141.5	
stearyl	n-$C_{18}H_{37}$–OOH		49–50	

[a] To convert kPa to mm Hg, multiply by 7.5. [b] At 21°C. [c] At 25°C. [d] 94% assay material. Courtesy of Elf Atochem North America, Inc. [e] OOH group may alternatively be at positions marked by asterisk. [f] At 25°C for 54% *p*-menthane hydroperoxide in *p*-menthane.

Table 2. Properties of Some Dialkyl Peroxides

Dialkyl peroxide	Structure	Bp, °C (kPa)[a]	Mp, °C
dimethyl peroxide	CH_3–OO–CH_3	13.5 (98.66)	
perfluoro dimethyl peroxide	CF_3–OO–CF_3	−37 (101.32)	
diethyl peroxide	C_2H_5–OO–C_2H_5	62–63 (101.32)	
1,2-dioxane	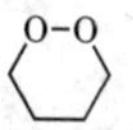	61.5 (14.67)	
tert-butyl methyl peroxide	t-C_4H_9–OO–CH_3	23 (2.53)	
tert-butyl 2-hydroxyethyl peroxide	t-C_4H_9–OO–CH_2CH_2OH	37–38 (0.27)	
diisopropyl peroxide	i-C_3H_7–OO–i-C_3H_7		
3,3,5,5-tetramethyl-1,2-dioxolane	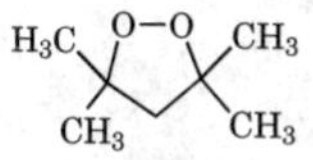	55–58 (29.73), 46 (3.33)	14
di-*tert*-butyl peroxide	t-C_4H_9–OO–t-C_4H_9	109 (101.32)	−18
perfluoro-di-*tert*-butyl peroxide	$(CF_3)_3C$–OO–$C(CF_3)_3$	99 (101.32)	
3,3,6,6-tetramethyl-1,2-dioxane	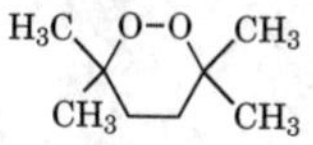	44–45 (1.5)	−26
di-*tert*-amyl peroxide	t-C_5H_{11}–OO–t-C_5H_{11}	44 (1.33)	
tert-butyl *tert*-cumyl peroxide	t-C_4H_9–OO–$C(CH_3)_2C_6H_5$	40 (0.027)	13
9,10-dihydro-9,10-epidi-oxyanthracene	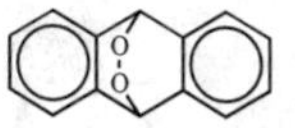		120[b]
2,5-dimethyl-2,5-di(*tert*-butyl-peroxy) hexane	$[t$-C_4H_9–OO–$C(CH_3)_2CH_2\!\!+_2$	42 (0.008)	8
2,5-dimethyl-2,5-di(*tert*-butylperoxy)-3-hexyne	t-$C_4H_9OOC(CH_3)_2$–C≡C–$C(CH_3)_2OO$–t-C_4H_9	65–67 (0.27)	
dicumyl peroxide	$C_6H_5(CH_3)_2C$–OO–$C(CH_3)_2C_6H_5$		40–41
1,4–di(2-*tert*-butylperoxyisopropyl)benzene	1,4-$[t$-C_4H_9–OO–$C(CH_3)_2\!\!+_2C_6H_4$		79

[a] To convert kPa to mm Hg, multiply by 7.5. [b] Explodes at 120°C.

Many hydroperoxides have been prepared by autoxidation of suitable substrates with molecular oxygen. These reactions can be free-radical chain or nonchain processes, depending on whether triplet or singlet oxygen is involved.

Many organic peroxides of metals have been hydrolyzed to alkyl hydroperoxides. Saponification of *tert*-alkyl peroxyesters yields alkyl hydroperoxides and carboxylic acids or their alkali metal salts.

Dialkyl Peroxides

Dialkyl peroxides have the structural formula R–OO–R′, where R and R′ are the same or different primary, secondary, or tertiary alkyl, cycloalkyl, and aralkyl hydrocarbon or hetero-substituted hydrocarbon radicals. Organomineral peroxides have the formulas $R_mQ(OOR)_n$ and R_mQOOQR_m, where at least one of the peroxygens is bonded directly to the organo-substituted metal or metalloid, Q. Dialkyl peroxides include cyclic and bicyclic peroxides where the R and R′ groups are linked, eg, endoperoxides and derivatives of 1,2-dioxane. Also included are polymeric peroxides, which usually are called poly(alkylene peroxides) or alkylene–oxygen copolymers, and poly(organomineral peroxides), where Q = As or Sb.

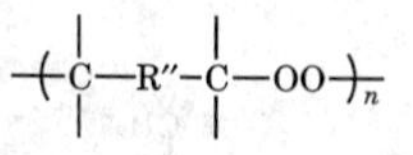

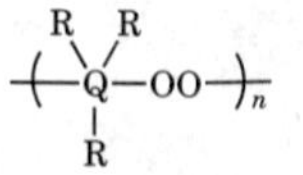

Physical Properties. The structures and the boiling and melting points of several dialkyl peroxides are listed in Table 2; a comprehensive list is given in the literature.

Metalloid peroxides behave as covalent organic compounds and most are insensitive to friction and impact but can decompose violently if heated rapidly. Most solid metalloid peroxides have well-defined melting points and the more stable liquid members can be distilled (Table 3).

Chemical Properties. Acyclic di-*tert*-alkyl peroxides efficiently generate alkoxy free radicals by thermal or photolytic homolysis. Primary and secondary dialkyl peroxides undergo thermal decompositions more rapidly than expected owing to radical-induced

Table 3. Properties of Some Organomineral Peroxides

Organomineral peroxide	Structure	Bp, °C (kPa)[a]	Mp, °C
diethoxyaluminum *tert*-cumyl peroxide	$(C_2H_5O)_2Al{-}OO{-}C(CH_3)_2{-}C_6H_5$		113 dec
tri (*tert*-butylperoxy)borane	$(t\text{-}C_4H_9{-}OO)_3B$	60–70 (0.0013)	18
tert-butyl triethylgermanium peroxide	$(C_2H_5)_3Ge{-}OO{-}t\text{-}C_4H_9$	78 (1.87–2.0)	
dioxybis[triethylgermane]	$(C_2H_5)_3Ge{-}OO{-}Ge(C_2H_5)_3$	56–57 (0.0067)	
(*tert*-butyl-dioxy)triethylplumbane	$(C_2H_5)_3Pb{-}OO{-}t\text{-}C_4H_9$		34–36
tetra(*tert*-butylperoxy)silane	$(t\text{-}C_4H_9OO)_4Si$	78 (0.067)	35–40
dioxybis[trimethylsilane]	$(CH_3)_3Si{-}OO{-}Si(CH_3)_3$	36–38 (4.0)	
tert-butylperoxytrimethylsilane	$(CH_3)_3Si{-}OO{-}t\text{-}C_4H_9$	78 (28.66)	
dioxybis[triethylstannane]	$(C_2H_5)_3Sn{-}OO{-}Sn(C_2H_5)_3$		60[b]
tert-butylperoxytrimethylstannane	$(CH_3)_3Sn{-}OO{-}t\text{-}C_4H_9$	56 (1.60)	

[a] To convert kPa to mm Hg, multiply by 7.5. [b] Explodes at 60°C.

decompositions. Such radical-induced peroxide decompositions result in inefficient generation of free radicals.

The low molecular weight primary dialkyl peroxides are shock-sensitive and explosive, with sensitivity decreasing with increasing molecular weight. Decomposition products from primary and secondary dialkyl peroxides include aldehydes, ketones, alcohols, hydrogen, hydrocarbons, carbon monoxide, and carbon dioxide.

Because di-*tert*-alkyl peroxides are less susceptible to radical-induced decompositions, they are safer and more efficient radical generators than primary or secondary dialkyl peroxides. They are the preferred dialkyl peroxides for generating free radicals for commercial applications.

The susceptibility of dialkyl peroxides to acids and bases depends on peroxide structure and the type and strength of the acid or base. In acidic environments, unsymmetrical acyclic alkyl aralkyl peroxides undergo carbon–oxygen fission, forming acyclic alkyl hydroperoxides and aralkyl carbonium ions. The latter react with nucleophiles, X^-.

Substitution reactions on dialkyl peroxides without concurrent peroxide cleavage are known eg, the nitration of dicumyl peroxide and the chlorination of di-*tert*-butyl peroxide.

The polymeric peroxides, $\mathord{+}OOCH_2CXH\mathord{+}_n$, where X = H, C_6H_5, $CH{=}CH_2$, etc, are viscous liquids or amorphous solids having as many as 10 repeating units. These compounds usually explode when heated. The products obtained from the thermal or photodecomposition show that cleavage of both oxygen–oxygen and carbon–carbon bonds occurs. The type and amounts of products formed depend on the decomposition conditions and the structure of the peroxide.

Unsaturated aliphatic endoperoxides form bis(epoxides) and/or epoxy aldehydes upon thermolysis. The endoperoxides of polynuclear aromatic compounds are crystalline solids that extrude singlet oxygen when heated, thus forming the parent aromatic hydrocarbon. Endoperoxides undergo carbon–oxygen cleavage in acids and oxygen–oxygen bond cleavage in bases, and they are more easily reduced than dialkyl peroxides.

1,2-Dioxetanes have very low activation enthalpies (ca 109 kJ/mol), therefore, they are unstable at low temperatures and generally cleave thermally or photochemically at the oxygen–oxygen and carbon–carbon bonds. Upon fragmentation, chemiluminescence occurs and two carbonyl compounds are produced in the absence of trapping agents. 1,2-Dioxetanes are reduced to diols, epoxides, or allylic alcohols; the dioxetane structure and the reducing system determine which product forms or predominates.

Dioxiranes are three-membered cyclic ring peroxides that are expected to be very unstable owing to ring strain. They are effective oxygenating agents for epoxidations of olefins, allenes, polycyclic aromatic hydrocarbons, enols, and α,β-unsaturated ketones; for insertions of oxygen into X–H bonds of alkanes, primary and secondary alcohols, aldehydes, and silanes; and for oxidations of sulfides (to sulfoxides and sulfones), imines (to nitrones), and primary amines (to nitro compounds). In these reactions, the dioxirane transfers oxygen to the substrate and generates the ketone from which the dioxirane was derived.

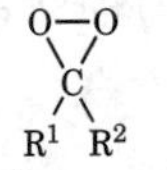

Most organomineral peroxides are hydrolytically unstable and readily hydrolyze to alkyl hydroperoxides or hydrogen peroxide:

$$R_mQ{-}OO{-}QR_m \xrightarrow{H_2O} 2R_mQ{-}OH + H_2O_2$$

Consequently, most organomineral peroxides must be prepared and stored under anhydrous conditions.

Basic hydrolysis of secondary alkyl-substituted silicon and germanium peroxides results in oxygen–oxygen bond cleavage.

The reduction of alkyl-substituted silicon and tin peroxides with sodium sulfite and triphenylphosphine has been reported. Alkyl-substituted aluminum, boron, cadmium, germanium, silicon, and tin peroxides undergo oxygen-to-metal rearrangements, as in the following equations:

$$R_3Si{-}OO{-}SiR_3 \rightarrow R_2Si(OR)OSiR_3$$

$$R_2B{-}OO{-}R \rightarrow RB(OR)_2$$

Organomineral peroxides also undergo thermal and photo-induced homolysis, yielding free radicals that are effective for initiating polymerization of vinyl monomers.

Synthesis. Dialkyl peroxides are prepared by the reaction of various substrates with hydrogen peroxide, hydroperoxides, or oxygen. They also have been obtained from reactions with other organic peroxides.

α-Oxygen-Substituted Hydroperoxides and Dialkyl Peroxides

Dialkyl peroxides and hydroperoxides which have either a hydroxy, hydroperoxy, alkoxy, or alkylperoxy group on the carbon adjacent to the parent peroxide group are considered separately from the parent compounds due to their unique reactions and properties, but mainly because of their unique syntheses. Their primary preparation from aldehydes and ketones via reaction with hydrogen peroxide, alkyl hydroperoxides and peroxyacids is unique and makes it almost impossible to discuss them without referring to the parent carbonyl compound(s).

The α-oxygen-substituted hydroperoxides and dialkyl peroxides comprise a great variety as shown in Figure 1. When discussing peroxides derived from ketones and hydrogen peroxide, (1) is often referred to as a ketone peroxide monomer and (2) as a ketone peroxide dimer.

Syntheses, Physical and Chemical Properties

An example of the complex equilibrium that exists for mixtures of carbonyl compounds and hydrogen peroxide is that from aldehydes and hydrogen peroxide. Hydroxyalkyl hydroperoxides (**1**, X = OH, R^3 =

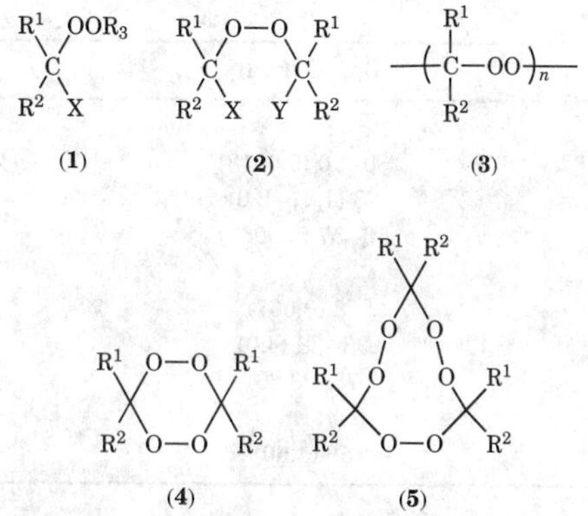

(1) (2) (3) (4) (5)

Figure 1. Varieties of α-oxygen-substituted hydroperoxides and dialkyl peroxides. R^1, R^2, R^3 = H or alkyl; X, Y = OH, OOH, OR^4, $OSiR_3$, or OOR^5; R^4, R^5 = alkyl; and R^3 and R^5 may also be acyl, $C(=O)R^6$.

H) and di(hydroxyalkyl) peroxides (**2,** X = Y = OH) are formed; cyclic diperoxides (**4**) are formed in some cases, eg, from benzaldehyde with concentrated sulfuric acid. Hydroxyalkyl hydroperoxides are the principal products when equimolar amounts of aldehyde and hydrogen peroxide are used at low temperatures. Di(hydroxyalkyl) peroxides are obtained by using excess aldehyde or higher temperatures. These reactions occur without catalysts but occur at much faster rates in the presence of acids. The peroxides (**1**) and (**2**) from most straight-chain aldehydes, ie, C_1–C_{11}, have been characterized, and a few of these and some from other aldehydes are listed in Table 4.

Starting with ketones and hydrogen peroxide in the presence of a catalytic amount of acid, mixtures of up to eight components have been identified, ie, (**1,** X = OH, R^3 = H), (**1,** X = OOH, R^3 = H), (**2,** X = Y = OH), (**2,** X = Y = OOH), (**2,** Y = OH, Y = OOH), (**3**), (**4**), and (**5**). The ketone structure and reaction conditions, ie, acid strength, reactant molar ratios, temperature, and time, determine which compounds form and predominate. Mixtures of several peroxide structures usually are present. Individual peroxides have been isolated from several ketones under different conditions (Table 5). The pure peroxides should be handled with extreme caution since most, especially those derived from the low molecular weight ketones, are shock- and friction-sensitive and can explode violently. Methyl ethyl ketone peroxide (MEKP) mixtures are produced commercially only as solutions containing <40 wt% MEKPs in solvents, commonly dialkyl phthalates.

Hydroxyalkyl Hydroperoxides. These compounds, represented by (**1,** X = OH, R^3 = H), may be isolated as discreet compounds only with certain structural restrictions, eg, that one or both of R^1 and R^2 are hydrogen, ie, they are derived from aldehydes, or that R^1 or R^2 contain electron-withdrawing substituents, ie, they are derived from ketones bearing α-halogen substituents. Other hydroxyalkyl hydroperoxides may exist in equilibrium mixtures of ketone and hydrogen peroxide.

Alkoxyalkyl Hydroperoxides. These compounds (**1,** X = OR^4, R^3 = H) have been prepared by the ozonization of certain unsaturated compounds in alcohol solvents. Alkoxyalkyl hydroperoxides are more commonly called ether hydroperoxides. They form readily by the autoxidation of most ethers containing α-hydrogens, eg, dioxane, tetrahydrofuran, diethyl ether, diisopropyl ether, di-*n*-butyl ether,

Table 4. Melting Points of Some Peroxy Compounds from Aldehydes and Hydrogen Peroxide

Peroxy compound	R^{1a}	Mp, °C
hydroxymethyl hydroperoxide	H	oil
1-hydroxyethyl hydroperoxide	CH_3	oil
2,2,2-trichloro-1-hydroxyethyl hydroperoxide	CCl_3	122
1-hydroxypentyl hydroperoxide	n-C_4H_9	oil
1-hydroxyoctyl hydroperoxide	n-C_7H_{15}	46
1-hydroxynonyl hydroperoxide	n-C_8H_{17}	50–54
di(hydroxymethyl) peroxide	H	63–64
di(1-hydroxyethyl) peroxide	CH_3	
di(2,2,2-trichloro-1-hydroxyethyl) peroxide	CCl_3	
di(1-hydroxypentyl) peroxide	n-C_4H_9	
di(1-hydroxyoctyl) peroxide	n-C_7H_{15}	72
di(1-hydroxynonyl) peroxide	n-C_8H_{17}	74

[a] See Fig. 1; R^2 = R^3 = H and X = Y = OH.

Table 5. Peroxy Compounds from Ketones and Hydrogen Peroxide

Peroxy compound	Structure	Mp, °C
2-chloro-1-hydroperoxycyclohexanol[a]	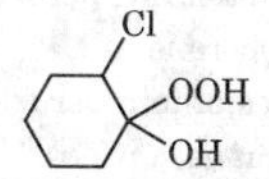	76
1,1-dihydroperoxycyclododecane[b]	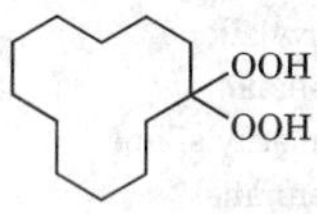	140
3,5-dihydroxy-3,5-dimethyl-1,2-dioxolane[c]	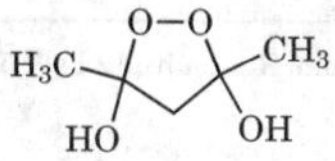	90–91
di(1-hydroxycyclohexyl) peroxide[d]	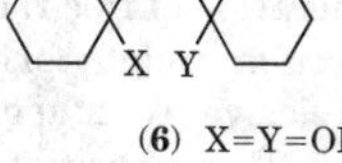 (**6**) X=Y=OH	69–71
1-hydroxycyclohexyl 1-hydroperoxycyclohexyl peroxide	(**6**) X = OH Y = OOH	76–77
di(1-hydroperoxycyclohexyl) peroxide	(**6**) X = Y = OOH	82–83
di(2-hydroperoxy-2-butyl) peroxide	$R^1 = CH_3$ (**2**) $R^2 = C_2H_5$ X = Y = OOH	39–42
3,3,6,6-tetramethyl-1,2,4,5-tetroxane	(**4**) $R^1 = R^2 = CH_3$	131–133
3,6-diethyl-3,6-dimethyl-1,2,4,5-tetroxane	(**4**) $R^1 = CH_3$ $R^2 = C_2H_5$	[e]
7,8,15,16-tetraoxadispiro-[5.2.5.2]-hexadecane[f]	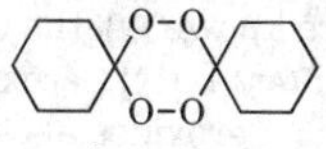	127–128
3,3,6,6,9,9-hexamethyl-1,2,4,5,7,8-hexoxononane	(**5**) $R^1 = R^2 = CH_3$	96–97
3,6,9-triethyl-3,6,9-trimethyl-1,2,4,5,7,8-hexoxononane	(**5**) $R^1 = CH_3$ $R^2 = C_2H_5$	30–32
7,8,15,16,23,24-hexaoxatrispiro-[5.2.5.2.5.2] tetracosane[g]	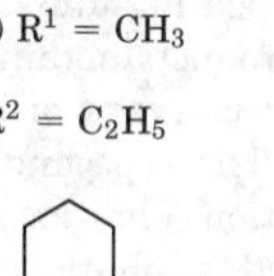	93

[a] Type (**1**) R^1 and R^2 are the ring; R^3 = H; X = OH. [b] Type (**1**) R^1 and R^2 are the ring; R^3 = H; X = OOH. [c] Type (**2**) $R^1 = CH_3$; R^2 is $-CH_2-$; X = Y = OH. [d] Structure (**6**) is type (**2**) wherein R^1 and R^2 are the ring and X and Y are specified. [e] Cis compound has an mp of 12–14°C; trans compound has mp = 23–25°C. [f] Type (**4**) R^1 and R^2 are the ring. [g] Type (**5**) R^1 and R^2 are the ring.

and diisoamyl ether. From certain ethers, eg, diethyl ether, the initially formed ether hydroperoxide can yield alcohol on standing, or with acid treatment form dangerously shock-sensitive and explosive polymeric peroxides.

Hydroxyalkyl Alkyl Peroxides and Hydroxyalkyl Peroxyesters. Hydroxyalkyl alkyl peroxides (**1,** X = OH, R^3 = alkyl) are reasonably stable and usually can be distilled under a vacuum; the boiling points and structures of representative compounds are listed in Table 6.

Alkoxyalkyl Alkyl Peroxides. *tert*-Butyl tetrahydropyran-2-yl peroxide (**1**), where R^3 = *tert* − butyl, X = OR^4, R^1 = H, R^2 and R^4 = 1, 4 − butanediyl, has been isolated. This is one of many examples of alkoxyalkyl alkyl peroxides which may be prepared by reaction of hydroperoxides with vinyl ethers.

1,2,4-Trioxacycloalkanes. 1,2,4-Trioxanes (**1**), X = OR^4; R^3 and R^4 = alkylene) are generally prepared by the interaction of aldehydes with zwitterionic intermediates made from reaction of singlet oxygen with olefins. They can also be prepared by catalyzed reaction of ketones or aldehydes with 1,2-dioxetanes or endoperoxides, and they can be prepared directly from certain hydroperoxides.

Geminal Dihydroperoxides. These dihydroperoxides as described previously (**1,** X = OOH, R^3 = H) can be made from many different carbonyl compounds. These peroxides can also be synthesized by perhydrolysis of ketals. Low molecular weight dihydroperoxides are soluble in water and are explosive when pure. They have been reduced to the corresponding ketones with hydriodic acid or zinc and acetic acid. Hydrolysis also gives the corresponding ketones. In the presence of catalytic amounts of acids or on prolonged storage, solutions of dihydroperoxides form equilibrium amounts of hydrogen peroxide and di(hydroperoxyalkyl) peroxides and ultimately equilibrium amounts of cyclic triperoxides.

Diperoxyketals and Diperoxyacetals. Aromatic aldehydes react with alkyl hydroperoxides in the presence of strong acid catalysts such as sulfuric acid to form diperoxyacetals (**1,** X = OOR^5; R^1 = H, R^2 = Ar, R^3 = R^5 = alkyl). Diperoxyketals (**1,** X = OOR^5; R^1, R^2, R^3, R^5 = alkyl) are generally prepared by acid-catalyzed reaction of a ketone with two equivalents of an alkyl hydroperoxide.

Diperoxyketals are solids or colorless liquids and are soluble in common organic solvents and insoluble in water. The physical properties and structures of some diperoxyketals are listed in Table 7. In the pure state, the low molecular weight compounds can decompose violently when heated, and addition of concentrated sulfuric acid can result in flaming decompositions. There are many commercial diperoxyketals, and they are usually diluted with solvents for improved safety.

Tertiary diperoxyketals (**1,** X = OOR^5, R^1, R^2 = alkyl, R^3, R^5 = tertiary alkyl) are excellent free-radical initiators. Such diperoxyketals are stable, especially those with R^3 = R^5 = *tert* − butyl. Less thermally stable diperoxyketals are those derived from cyclic ketones and those with bulkier *tert*-alkyl groups, eg, *tert*-amyl, *tert*-octyl, *tert*-cumyl. Commercial members of this group all have R^3 = R^5, and thermally decompose to free radicals by cleavage of only one oxygen–oxygen bond initially, usually followed by β-scission of the resulting alkoxy radicals. For acyclic diperoxyketals, β-scission produces an alkyl radical and a peroxyester. Owing to similarity of thermal stability, the peroxyester decomposes almost simultaneously.

Table 6. Boiling Points of Some Hydroxyalkyl Alkyl Peroxides[a]

Hydroxyalkyl alkyl peroxide	R^3	R^1	Bp, °C (kPa)[b]
hydroxymethyl methyl peroxide	CH_3	H	45 (2.27)
tert-butyl hydroxymethyl peroxide	t-C_4H_9	H	52–53 (1.07)
1-hydroxyethyl methyl peroxide	CH_3	CH_3	25–27 (2.27)
1-hydroxyethyl ethyl peroxide	C_2H_5	CH_3	48–50 (8.67)
tert-butyl 1-hydroxyethyl peroxide	t-C_4H_9	CH_3	30–31.5 (0.13)
tert-butyl 1-hydroxybutyl peroxide	t-C_4H_9	n-C_3H_7	34–37 (0.13)

[a] Structure (**1**), R^2 = H; X = OH; R^1 and R^3 are specified.
[b] To convert kPa to mm Hg, multiply by 7.5.

Diperoxyketals, and many other organic peroxides, are acid-sensitive, therefore removal of all traces of the acid catalysts must be accomplished before attempting distillations or kinetic decomposition studies. The low molecular weight diperoxyketals can decompose with explosive force and commercial formulations are available only as mineral spirits or phthalate ester solutions.

Di(hydroxyalkyl) Peroxides. The lowest molecular weight member of this group (**2,** X = Y = OH), di(hydroxymethyl) peroxide (R^1 = R^2 = H) is a dangerously explosive solid. With increasing molecular weight, di(hydroxyalkyl) peroxides become liquids and eventually solids of decreasing explosive nature and water solubility. In solution, these dialkyl peroxides exist in equilibrium with other α-oxygen-substituted peroxides, carbonyl compounds, and hydrogen peroxide.

Formaldehyde reacts with di(hydroxymethyl) peroxide and phosphorus pentoxide to form di(hydroxymethoxymethyl) peroxide (**2**), where X = Y = OCH_2OH, R^1 = R^2 = H.

Reaction of 1,3- and 1,4-diketones (n = 1 or 2) with hydrogen peroxide yields cyclic di(hydroxyalkyl) (X = OH) or di(hydroperoxyalkyl) (X = OOH) peroxides (**7**).

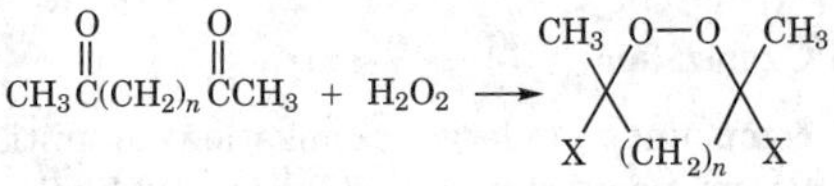

(**7**)

The di(hydroxyalkyl) peroxide (**2**) from cyclohexanone is a solid which is produced commercially. The di(hydroxyalkyl) peroxide (**2**) from 2,4-pentanedione (**7,** n = 1; X = OH) is a water-soluble solid which is also produced commercially (see Table 5). Both these peroxides are used for curing cobalt-promoted unsaturated polyester resins.

Hydroxyalkyl Hydroperoxyalkyl Peroxides. There is evidence that hydroxyalkyl hydroperoxyalkyl peroxides (**2,** X = OH, Y = OOH) exist in equilibrium with their corresponding carbonyl compounds and other α-oxygen-substituted peroxides. Thermal decomposition of hydroxyalkyl hydroperoxyalkyl peroxides produces mixtures of starting carbonyl compounds, mono and dicarboxylic acids, cyclic diperoxides, carbon dioxide, and water.

Di(hydroperoxyalkyl) Peroxides. Low molecular weight di(hydroperoxyalkyl) peroxides (**2,** X = Y = OOH) are dangerously prone to explosive decomposition when they are pure. Some have been characterized by acylation to the corresponding diperoxyesters.

Cyclic Peroxides. Cyclic diperoxides (**4**) and triperoxides (**5**) are solids and the low molecular weight compounds are shock-sensitive and explosive. The melting points of some characteristic compounds of this type are given in Table 5.

Polymeric α-Oxygen-Substituted Peroxides. Polymeric peroxides (**3**) are formed from the following reactions: ketone and aldehydes with hydrogen peroxide, ozonization of unsaturated compounds, and dehydration of α-hydroxyalkyl hydroperoxides; consequently, a variety of polymeric peroxides of this type exist. Polymeric peroxides are generally viscous liquids or amorphous solids, are difficult to characterize, and are prone to explosive decomposition.

Miscellaneous α-Substituted Peroxides. 3-Aryl-3-(*tert*-alkylperoxy)phthalides (**8**) are prepared from the corresponding 3-chlorophthalides and *tert*-alkyl hydroperoxide. 2-Methyl-2-(*tert*-alkylperoxy)-1,3-benzodioxan-4-ones (**9**) are obtained from *o*-acetylsalicyloyl chloride and *tert*-alkyl hydroperoxides. Trisubstituted 2-(*tert*-alkylperoxy)-1,3-dioxolan-4-ones (**10**) are synthesized from sterically favored α-acyloxy acid chlorides and *tert*-alkyl hydroperoxides.

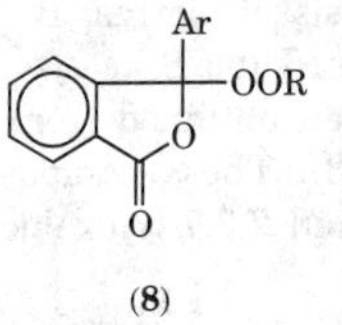

(**8**)

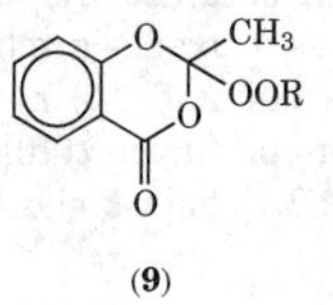

(**9**)

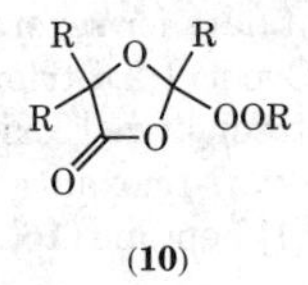

(**10**)

Table 7. Boiling Points of Some Diperoxyketals

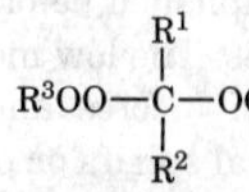

Diperoxyketal	R^3, R^5	R^1	R^2	Bp, °C (kPa)[a]
2,2-di(*t*-butylperoxy)propane	*t*-C_4H_9	CH_3	CH_3	69–70 (2.0)
2,2-di(*t*-amylperoxy)propane	*t*-C_5H_{11}	CH_3	CH_3	68 (0.17)
2,2-di(*t*-butylperoxy)butane	*t*-C_4H_9	CH_3	C_2H_5	50 (0.27)
1,1-di(*t*-butylperoxy)cyclohexane	*t*-C_4H_9	$(CH_2)_5$		52–54 (0.02)
2,2-bis[4,4-di(*t*-butylperoxy)cyclohexyl]propane	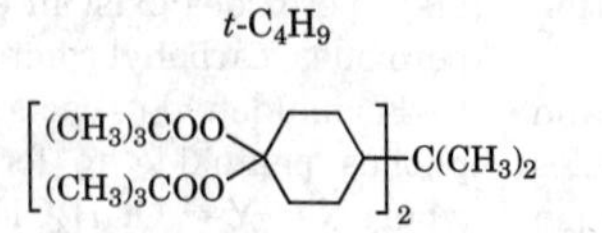			117–120[b]

[a] To convert kPa to mm Hg, multiply by 7.5.
[b] Mp, °C.

Ozonides and Ozonization

Unsaturated compounds undergo ozonization to initially produce highly unstable primary ozonides (**11**), ie, 1,2,3-trioxolanes, also known as molozonides, which rapidly split into carbonyl compounds (aldehydes and ketones) and 1,3-zwitterion (**12**) intermediates. The carbonyl compound-zwitterion pair then recombines to produce a thermally stable secondary ozonide (**13**), also known as a 1,2,4-trioxolane.

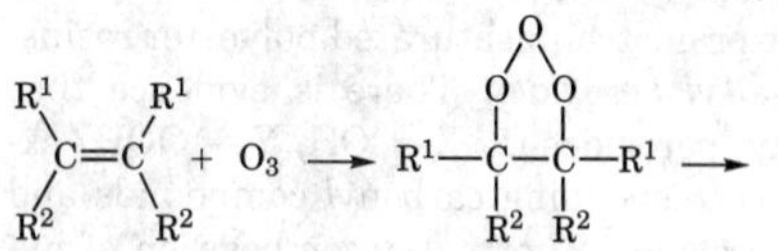

(**11**)

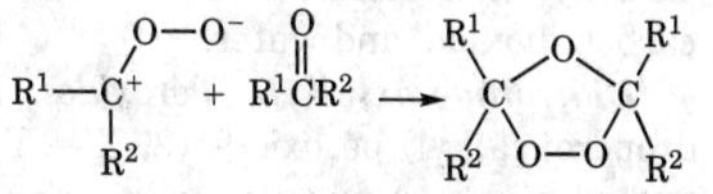

(**12**) (**13**)

Most ozonolysis reaction products are postulated to form by the reaction of the 1,3-zwitterion with the extruded carbonyl compound in a 1,3-dipolar cycloaddition reaction to produce stable 1,2,4-trioxanes (ozonides) (**13**) as shown; with itself (dimerization) to form cyclic diperoxides (**4**); or with protic solvents, such as alcohols, carboxylic acids, etc, to form α-substituted alkyl hydroperoxides. The latter can form other peroxidic products, depending on reactants, reaction conditions, and solvent.

In the presence of alcohols, the ozonization products are alkoxyalkyl hydroperoxides (**1**, X = OR^4, R = H3):

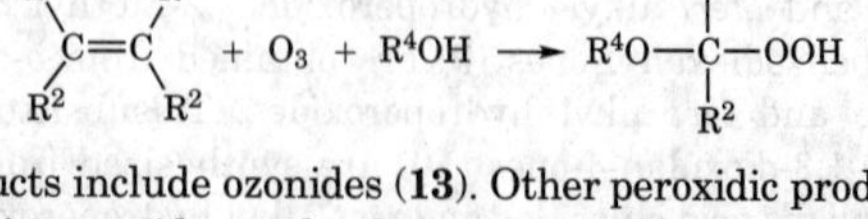

By-products include ozonides (**13**). Other peroxidic products including polymeric peroxides and polymeric ozonides can form, depending on reaction conditions, solvent, and olefin used. A variety of cyclic diperoxides (**4**) have been obtained by ozonolysis of olefins. Boiling point data for several 1,2,4-trioxanes are listed in Table 8.

Cyclic 1,2,4-trioxanes (**14** and **15**) have been obtained from the photosensitized oxidation of furans (10,44,163). These compounds are 2,3,7-trioxabicyclo [2.2.1] hept-5-ene (**14**) and 2,3,7-trioxabicyclo [2.2.1] heptane (**15**).

Table 8. Boiling Points of Some 1,2,4-Trioxolanes[a]

1,2,4-Trioxolanes	Structure	Bp, °C (kPa)[b]
1,2,4-trioxolane	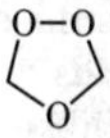	18 (2.13)
3,5-dimethyl-1,2,4-trioxolane	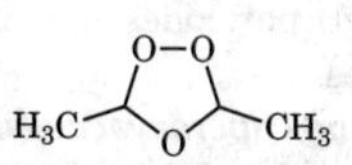	15 (2.67)
3,3-dimethyl-1,2,4-trioxolane	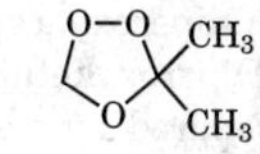	42–42.5 (18.67)
1,5-dimethyl-6,7,8-trioxabicyclo [3.2.1] octane	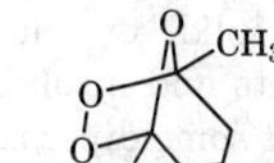	58.8 (2.0)
3,5-diphenyl-1,2,4-trioxolane	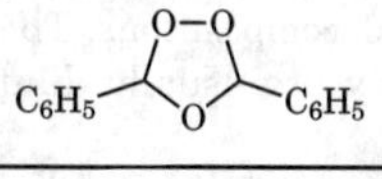	

[a] Secondary ozonides. [b] To convert kPa to mm Hg, multiply by 7.5.

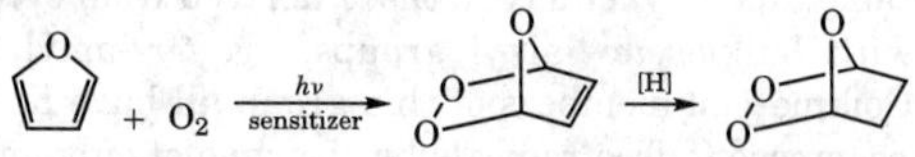

(**14**) (**15**)

Peroxyacids

There are two broad classes of organic peroxyacids: peroxycarboxylic acids, $R[C(O)OOH]_n$, where R is an alkyl, aralkyl, cycloalkyl, aryl, or heterocyclic group and n = 1 or 2, and organoperoxysulfonic acids, RSO_2–OOH.

Three peroxyacids are produced commercially for the merchant market: peroxyacetic acid as a 40 wt % solution in acetic acid, *m*-chloroperoxybenzoic acid, and magnesium monoperoxyphthalate hexahydrate. Other peroxyacids are produced for captive use, eg, peroxyformic acid generated *in situ*, as an epoxidizing agent.

Physical Properties. Physical properties of peroxyacids have been extensively reviewed in the literature. The melting points of some peroxycarboxylic acids are listed in order of increasing number of carbon atoms in Table 9. Aliphatic peroxyacids are characterized by sharp un-

Table 9. Properties of Some Organic Peroxyacids

Peroxyacid	Structure	Mp, °C
peroxyformic acid	HCO_3H	$-18^{a,b}$
peroxyacetic acid	CH_3CO_3H	0^{d}
peroxypropionic acid	$C_2H_5CO_3H$	-13^{d}
peroxybutyric acid	$n\text{-}C_3H_7CO_3H$	-10^{e}
monoperoxysuccinic acid	$HO_2C(CH_2)_2CO_3H$	107, dec
peroxyhexanoic acid	$n\text{-}C_5H_{11}CO_3H$	15^{f}
peroxybenzoic acid	$C_6H_5CO_3H$	41–42
m-chloroperoxybenzoic acid	$m\text{-}Cl\text{-}C_6H_4CO_3H$	88
diperoxyhexanedioic acid	$HO_3C(CH_2)_4CO_3H$	116–117 dec
peroxyoctanoic acid	$n\text{-}C_7H_{15}CO_3H$	31
4-methylperoxybenzoic acid	$4\text{-}CH_3C_6H_4CO_3H$	95–96
monoperoxyphthalic acid	$2\text{-}HO_2CC_6H_4CO_3H$	110^{b}
peroxynonanoic acid	$n\text{-}C_8H_{17}CO_3H$	35
peroxycinnamic acid	CHCH*nf* > H56	67–68 dec
diperoxynonanedioic acid	$HO_3C(CH_2)_7CO_3H$	90
peroxydecanoic acid	$n\text{-}C_9H_{19}CO_3H$	41
diperoxydecanedioic acid	$HO_3C(CH_2)_8CO_3H$	98
peroxydodecanoic acid	$n\text{-}C_{11}H_{23}CO_3H$	50
peroxytetradecanoic acid	$n\text{-}C_{13}H_{27}CO_3H$	56
peroxyhexadecanoic acid	$n\text{-}C_{15}H_{31}CO_3H$	61
peroxyoctadecanoic acid	$n\text{-}C_{17}H_{35}CO_3H$	65
magnesium monoperoxyphthalate hexahydrate	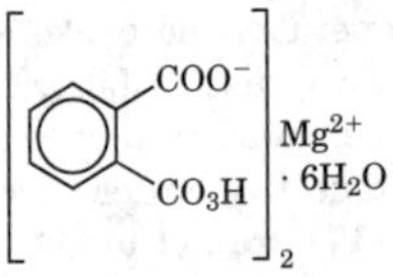	93

[a] 90 wt % melts at the given temperature. [b] Bp = 50°C at 13.33 kPa[c]. [c] To convert kPa to mm Hg, multiply by 7.5. [d] Bp = 25°C at 1.6 kPa[c] for peroxyacetic acid, and 2.67 kPa[c] for peroxypropionic acid. [e] Bp = 26−29°C at 1.6 kPa[c]. [f] Bp = 41−43°C at 0.067 kPa[c].

pleasant odors, the intensity of which decreases with increasing chain length. They also are irritating to the skin and mucous membranes.

Chemical Properties. Organic peroxyacids are not noted for their stability and many lose active oxygen during storage at room temperature. Those that are water soluble hydrolyze slowly to the parent acid and hydrogen peroxide; however, peroxyformic acid hydrolyzes more rapidly. The longer-chain aliphatic members decompose rapidly in methanol. Stabilizers are commonly used for peroxycarboxylic acid solutions, eg, dipicolinic acid, phytic acid, and pyro- and metaphosphates. Stability of peroxycarboxylic acids increases with increasing molecular weight. The stabilities of peroxybenzoic acids are enhanced when ring substituents are present.

Peroxycarboxylic acids and precursors to peroxycarboxylic acids are used as bleaches for removal of stains and soils from textiles. Precursors to peroxycarboxylic acids are nonperoxidic compounds possessing a reactive acyl group and a good leaving group, L. These precursors react under basic conditions with hydrogen peroxide, inorganic perborates, and inorganic percarbonates to generate peroxycarboxylic acids or salts:

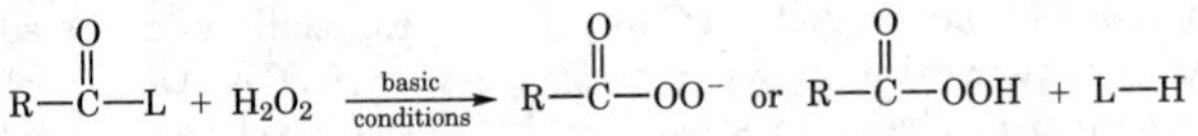

Thermal decompositions of peroxycarboxylic acids and their salts can proceed by free-radical and nonradical paths. Often the decomposition products and the rate are affected by the nature of the solvent. Peroxycarboxylic acids undergo photodecomposition and radical-induced decomposition. They also are decomposed by a variety of metals, metal ions, and complexes.

Peroxycarboxylic acids are among the most powerful organic peroxide oxidizing agents. The main industrial uses of these acids are in the manufacture of epoxides, synthetic glycerol (qv), and epoxy resins (qv). They also have been used as disinfectants, fungicides, and bleaching agents and for shrink-proofing wool.

Synthesis. Many different methods for the preparation of peroxyacids have been described. The most widely used method is the direct, acid-catalyzed equilibrium reaction of 30–98 wt % hydrogen peroxide with carboxylic acids: The equilibrium also can be shifted to the right by removing water azeotropically and/or under a vacuum. Chelating agents may be added during processing to reduce metal-catalyzed decompositions. Sulfuric acid, methanesulfonic acid, and sulfonic acid ion-exchange resins are the most commonly used acid catalysts.

Other methods for preparing peroxycarboxylic acids include: (*1*) autoxidation of aldehydes, (*2*) reaction of acid chlorides, anhydrides, or boric-carboxylic anhydrides with hydrogen or sodium peroxide, and (*3*) basic hydrolysis or perhydrolysis of diacyl peroxides.

Organoperoxysulfonic acids and their salts have been prepared by the reaction of arenesulfonyl chlorides with calcium, silver, or sodium peroxide; treatment of metal salts of organosulfonic acids with hydrogen peroxide; hydrolysis of di(organosulfonyl) peroxides, $R^2S(O)—OO—S(O)R^2$, with hydrogen peroxide; and sulfoxidation of saturated, nonaromatic hydrocarbons, eg, cyclohexane.

Other Peroxyacids. Benzeneperoxyseleninic acid has been prepared *in situ* from benzeneseleninic acid and hydrogen peroxide and is used to epoxidize terpenic olefins and Baeyer-Villiger oxidation of cyclic ketones.

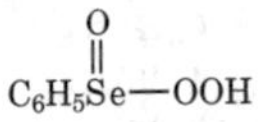

Acyl Peroxides

The acyl peroxide class is characterized by the following structures:

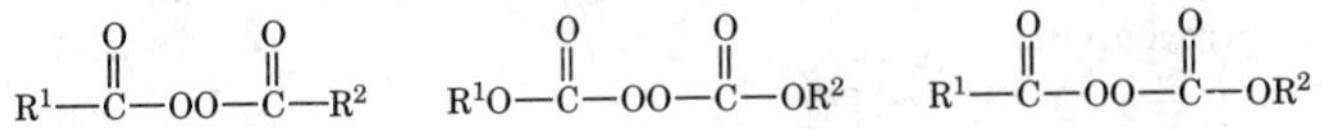

(16) **(17)** **(18)**

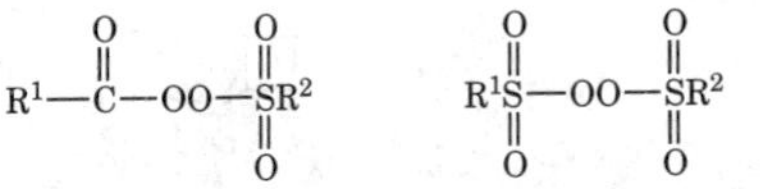

(19) **(20)**

Acyl peroxides of structure (**16**) are known as diacyl peroxides. In this structure R^1 and R^2 are the same or different and can be alkyl, aryl, heterocyclic, imino, amino, or fluoro. Acyl peroxides of structures (**17**), (**18**), (**19**), and (**20**) are known as dialkyl peroxydicarbonates, *OO*-acyl *O*-alkyl monoperoxycarbonates, acyl organosulfonyl peroxides, and di(organosulfonyl) peroxides, respectively. R^1 and R^2 in these structures are the same or different and generally are alkyl and aryl. Many diacyl peroxides (**16**) and dialkyl peroxydicarbonates (**17**) are produced commercially and used in large volumes.

Physical Properties. Almost all liquid diacyl peroxides (**16**) and concentrated solutions of the solid compounds are unstable to normal ambient temperature storage; many must be stored well below 0°C. Most of the solid compounds are stable at ca 20°C but many are shock-sensitive. Other physical constants and properties have been reviewed. The melting points and refractive indexes of some acyl peroxides are listed in Tables 10–12.

Chemical Properties. Diacyl peroxides (**16**) decompose when heated or photolyzed (<300 mm). Although photolytic decompositions generally produce free radicals, thermal decompositions can produce nonradical and radical intermediates, depending on diacyl peroxide structure. Symmetrical aliphatic diacyl peroxides of certain structures, ie, diacyl peroxides (**16**, $R^1 = R^2 =$ alkyl) without α-branches or with a mono-α-methyl substituent, and diaroyl peroxides (**16**, $R^1 = R^2 =$ aryl) thermally decompose almost exclusively by homolysis.

Table 10. Properties of Some Diacyl Peroxides

$$R^1{-}\overset{\overset{\displaystyle O}{\|}}{C}{-}OO{-}\overset{\overset{\displaystyle O}{\|}}{C}{-}R^2$$

Diacyl peroxide	R group	Mp, °C[a]
	Symmetrical, $R^1 = R^2$	
diacetyl peroxide	CH_3	30
di(chloroacetyl) peroxide	$ClCH_2$	85
dipropionyl peroxide	C_5H_2	oil
diisobutyryl peroxide	i-C_7H_3	80–80.5
di(3-carboxypropionyl) peroxide	$H_2OC(CH)_2$	132–133
dipentanoyl peroxide	n-C_9H_4	oil
di(4-carboxybutyryl) peroxide	$H_2OC(CH)_3$	104
di(2-furanylcarbonyl) peroxide		86–87
di(2-thienylcarbonyl) peroxide	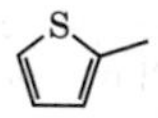	92–93
dinicotinoyl peroxide	3-cyclo-C_4H_5N	88–89
diheptanoyl peroxide	n-$C_{13}H_6$	[b]
di(cyclohexylcarbonyl) peroxide	cyclo-$C_{11}H_6$	oil
dibenzoyl peroxide	C_5H_6	106–107
di(4-chlorobenzoyl) peroxide	4-ClC_4H_6	137–138
di(4-nitrobenzoyl) peroxide	4-$NO_4C_6H_2$	157–158
di(2-methylbenzoyl) peroxide	2-$CH_4C_6H_3$	54
di(2-carboxybenzoyl) peroxide	2-$HOC_4C_6H_2$	156
dioctanoyl peroxide	n-CH	29
di(phenylacetyl) peroxide	$CH_2C_5H_6$	41
dinonanoyl peroxide	n-C_7H_8	13.0–13.5
di(3,5,5-trimethylhexanoyl) peroxide		[c]
dicinnamoyl peroxide	$CHCH{=}C_5H_6$	133–134
di(benzocyclobutene-4-carbonyl) peroxide	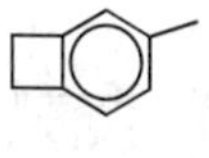	130–132
didecanoyl peroxide	n-$C_{19}H_9$	44–45
di(2-naphthaleny carbonyl) peroxide	2-C_7H_{10}	138–140
didodecanoyl peroxide	n-$C_{23}H_{11}$	54.7–55
dihexadecanoyl peroxide	n-$C_{31}H_{15}$	71.4–71.9
dioctadecanoyl peroxide	n-$C_{35}H_{17}$	76.5–76.9
	Unsymmetrical, $R^1 = CH_3$; R^2 as given	
acetyl propionyl peroxide	C_5H_2	[d]
acetyl cyclohexyl carbonyl peroxide	cyclo-$C_{11}H_6$	42
acetyl benzoyl peroxide	C_5H_6	37–39
adipoyl bis(acetyl peroxide)	$CH_3\overset{O}{\overset{\|}{C}}O\overset{O}{\overset{\|}{C}}(CH_2)_4{-}$	61–62

[a] Most of these peroxides decompose on melting, some violently. [b] $n_D^{25} = 1.4340$. [c] $n_D^{20} = 1.4382$. [d] $n_D^{20} = 1.4069$.

Diaroyl peroxides and diacyl peroxides without α-branches are significantly more thermally stable than those with mono- or di-α-substituents. The primary use of most commercial diacyl peroxides (**16**, $R^1 = R^2$ = alkyl or aryl) is initiation of free-radical reactions.

Diacyl peroxides (**16**, $R^1 = R^2$ = alkyl or aryl) also undergo radical induced decomposition either by direct radical displacement on the oxygen–oxygen bond, or by radical additaon to, or abstraction from, the hydrocarbyl group adjacent to the peroxide.

Diacyl peroxide decompositions also are catalyzed by the metal ions of copper, iron, cobalt, and manganese:

$$R^1{-}\overset{O}{\overset{\|}{C}}{-}OO{-}\overset{O}{\overset{\|}{C}}{-}R^2 + Cu^+ \longrightarrow R^1{-}\overset{O}{\overset{\|}{C}}O\cdot + R^2{-}\overset{O}{\overset{\|}{C}}O^- + Cu^{2+}$$

This radical-generating reaction has been used in synthetic applications, eg, aroyloxylation of olefins and aromatics, oxidation of alcohols to aldehydes, etc.

Hydrolysis and perhydrolysis of diacyl peroxides yields peroxycarboxylic acids. Carbanions react by displacement on oxygen.

Amines also react with diacyl peroxides by nucleophilic displacement on the oxygen–oxygen bond forming an ion pair intermediate.

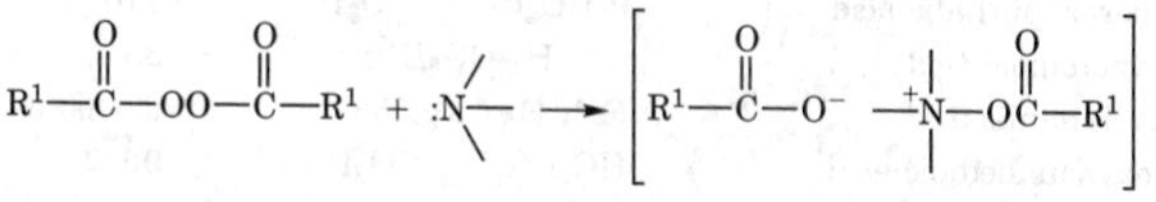

Dialkyl peroxydicarbonates (**17**) undergo thermolysis to form two alkoxycarbonyloxy radicals that subsequently undergo β-scission to form CO_2 and alkoxy radicals:

$$R^1O{-}\overset{O}{\overset{\|}{C}}{-}OO{-}\overset{O}{\overset{\|}{C}}{-}OR^1 \longrightarrow 2\,R^1O{-}\overset{O}{\overset{\|}{C}}{-}O\cdot \longrightarrow 2\,R^1O\cdot + 2\,CO_2$$

These low temperature peroxides are susceptible to radical-induced decompositions. This susceptibility largely accounts for the hazards associated with their production and storage. In contrast to diacyl peroxides (**16**); the true first-order decomposition rates for dialkyl peroxydicarbonates (**17**) are not affected by the nature of the R group. In free-radical scavenging solvents, eg, trichloroethylene, the decomposition rates of di-n-propyl, diisopropyl, di-*sec*-butyl, dicyclohexyl, di-2-ethylhexyl, and deheaxadecyl peroxydicarbonates are all essentially the same.

Dialkyl peroxydicarbonates are used primarily as free-radical initiators for vinyl monomer polymerizations. Dialkyl peroxydicarbonate decompositions are accelerated by certain metals, concentrated sulfuric acid, and amines. Violent decompositions can occur with neat or highly concentrated peroxides.

Acyl organosulfonyl peroxides (**19**) such as acetyl cyclohexanesulfonyl peroxide are efficient radical initiators for vinyl chloride polymerization.

Di(arenesulfonyl) peroxides (**20**, $R^2 = R^1$ = aryl) react with aromatic solvents to form aryl arenesulfonates. These peroxides also form 1:1 adducts with styrene and form hydrobenzoin diarenesulfonates with stilbenes. Di(benzenesulfonyl) peroxide decomposes in water to phenol and sulfuric acid.

Synthesis. Symmetrical diacyl peroxides (**16**, $R^2 = R^1$ = alkyl or aryl) are prepared by the reaction of an acyl chloride or anhydride with sodium peroxide or hydrogen peroxide and a base. Unsymmetrical diacyl peroxides (**16**, $R^2 \neq R^1$ = alkyl or aryl) are prepared by the reaction of acid chlorides or anhydrides with peroxycarboxylic acids in the presence of a base. Polymeric diacyl peroxides can be prepared from the reaction of dibasic acid chlorides, eg, succinoyl, fumaryl, sebacoyl, and terephthaloyl chlorides, with sodium or hydrogen peroxide. Cyclic diacyl peroxides can be generated from suitable dibasic acid chlorides and sodium or hydrogen peroxide, especially in dilute solutions. Symmetrical or unsymmetrical diacyl peroxides (**16**, R^1, R^2 = alkyl or aryl) can be synthesized directly from carboxylic acids and hydrogen peroxide or from peroxycarboxylic acids with dicyclohexylcarbodiimide or *N,N*-dicarbonyldiimidazole as condensing agents.

Diacyl peroxides (**16**, $R^2 = R^1$ = alkyl or aryl) have been obtained from the oxidation of carboxylic acid potassium salts by Kolbe electrolysis or by elemental fluorine.

Dialkyl peroxydicarbonates (**17**) are produced by reaction of alkyl chloroformates with sodium peroxide. *OO*-Acyl *O*-alkyl monoperoxycarbonates (**18**) are obtained from the reaction of alkyl chloroformates with peroxycarboxylic acids in the presence of a base. Symmetrical

Table 11. Properties of Some Dialkyl Peroxydicarbonates

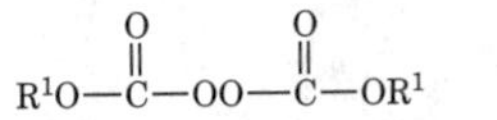

Peroxydicarbonate	R^1	n_D^{20}	Mp, °C[a]
diethyl	C_5H_2—	1.4065	
di-*n*-propyl	*n*-C_7H_3—	1.4091	
diisopropyl	*i*-C_7H_3—	1.4034	8–10
dibutyl	*n*-C_9H_4—	1.4129	
di-*sec*-butyl	*sec*-C_9H_4—	1.4112	
dicyclohexyl	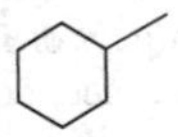		46
dibenzyl	$CH_2C_5C_6$—		101–102
di(2-ethylhexyl)	$CH_3(CH_2)_3CH(C_2H_5)CH_2$—	1.4366	
di(2-phenoxyethyl)	$CH_2OCH_5CH_6$—		97–100
di(*cis*-3,3,5-trimethylcyclohexyl)	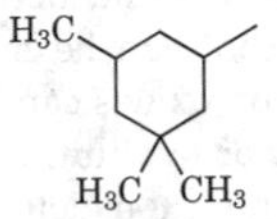		78–79
di(4-*tert*-butylcyclohexyl)	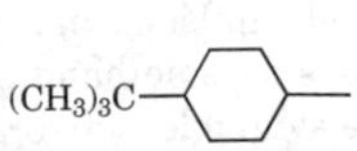		91–92
di(isobornyl)	isobornyl—		92–93
didodecyl	*n*-$C_{25}H_{12}$—		28–30
ditetradecyl	*n*-$C_{29}H_{14}$—		40–42
dihexadecyl	*n*-$C_{33}H_{16}$—		50–53

[a] All listed peroxides are unstable in liquid state >20°C; some decompose violently.

Table 12. Melting Points of Some Organosulfonyl Peroxides

Organosulfonyl peroxide	Structure	Mp, °C[a]
di(methanesulfonyl) peroxide	$[CH_3—S(O)_2—O—]_2$—	77
acetyl *tert*-butanesulfonyl peroxide	*t*-C_4H_9–$S(O)_2$–OO–C(O)CH_3	35–37
acetyl cyclohexanesulfonyl peroxide	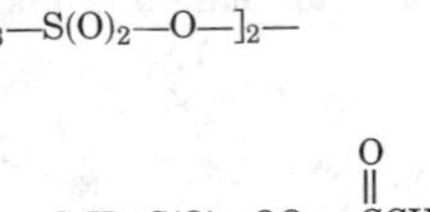	35–36
acetyl *sec*-heptanesulfonyl peroxide	*sec*-$C_7H_{15}S(O)_2$–OO–C(O)CH_3	liquid
acetyl (1-methycyclohexane)-sulfonyl peroxide	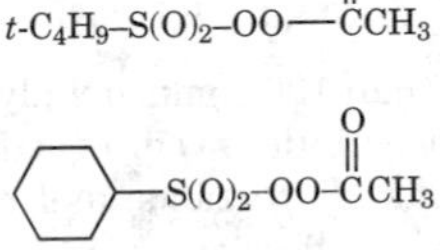	liquid
di(benzenesulfonyl) peroxide	$[C_5H_6—S(O)_2—O—]_2$—	53–54
di(*p*-toluenesulfonyl) peroxide	$[p\text{-}C_4H_6—CH_3—S(O)_2—O—]_2$—	50

[a] Most listed peroxides decompose at mp; some decompose violently.

di(organosulfonyl) peroxides (**20**, R = R21) have been prepared by the reaction of organosulfonyl chlorides with sodium peroxide or hydrogen peroxide in the presence of a base. Acyl organosulfonyl peroxides (**19**) are prepared from the organosulfonyl chlorides and a metal salt of a peroxycarboxylic acid. Acetyl cyclohexanesulfonyl peroxide has been produced commercially by the sulfoxidation of cyclohexane, C_6H_{12}, in the presence of acetic anhydride.

Potassium salts of the peroxides (**21–23**) are prepared from the reaction of Caro's acid, H_2SO_5, with acyl chlorides, chloroformates, or organosulfonyl chlorides in the presence of potassium hydroxide.

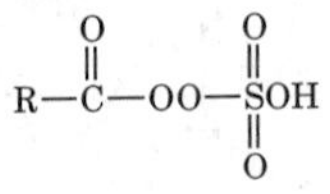

(**21**)

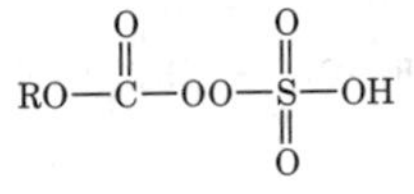

(**22**)

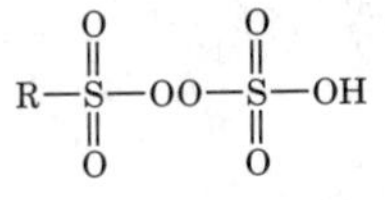

(**23**)

Alkyl Peroxyesters

Peroxyesters include the alkyl esters of peroxycarboxylic acids; monoperoxydicarboxylic acids; diperoxycarboxylic acids; monoperoxy- (**24**) and diperoxycarbonic (**25**) acids; monoperoxy- (**26**) and diperoxy-oxalic (**27**) acids; peroxycarbamic acids (**28**); peroxysulfonic acids (**29**); and peroxyphosphoric acids.

Synthesis. Peroxyesters are prepared by the reaction of alkyl hydroperoxides R′OOH, with acylating agents, eg, acid chlorides, anhydrides, ketenes, organosulfonyl chlorides, phosgene, alkyl chloroformates, oxalyl chloride, alkyl chlorooxalates, isocyanates, carbamoyl chlorides, carboxylic acids, and esters, under appropriate reaction conditions, according to Figure 2. Reactions with acylating agents that generate hydrogen chloride are carried out in the presence of a base, eg, pyridine or sodium hydroxide, or by using the sodium or potassium salt of the hydroperoxide.

Physical Properties. Properties of some *tert*-alkyl peroxyesters are listed in Table 13 and the properties of some *tert*-alkyl areneperoxysulfonates are given in Table 14.

Chemical Properties. Alkyl peroxyesters are hydrolyzed more readily than the analogous nonperoxidic esters and yield the original acids and hydroperoxides from which they were prepared rather than alcohols and peroxyacids. The *tert*-alkyl peroxyesters undergo homolysis, thermally and photochemically, to generate free radicals.

Primary and secondary alkyl peroxyesters thermally decompose by a nonradical process, giving almost quantitative yields of carboxylic acids and carbonyl compounds. *tert*-Alkyl peroxyesters are much less sensitive to radical-induced decompositions than diacyl peroxides. Induced decomposition is only significant in peroxyesters containing nonhindered α-hydrogens or α,β-unsaturation.

Peroxyesters decompose by an electron-transfer process catalyzed by transition metals. This reaction has been used synthetically to bond an acyloxy group to appropriate coreactive substrates.

Criegee rearrangement competes with homolysis in *tert*-alkyl peroxyesters, RC(O)–OOC$R^1R^2R^3$, in which R is strongly electron-withdrawing and the *tert*-alkyl group, ie, C$R^1R^2R^3$, contains a group with high migratory aptitude and ability to stabilize adjacent carbonium ions. The rearrangement converts the peroxyester to a nonperoxidic ester.

The main industrial use of *tert*-alkyl peroxyesters is in the initiation of free-radical chain reactions, primarily for vinyl monomer polymerizations.

Manufacture and Processing

Owing to the inherent hazards of organic peroxides, they are almost never distilled or confined during manufacture. Generally, open reactors are employed that can be easily vented and deluged with water if an unanticipated exotherm occurs. The preferred materials of reactor construction are 316 stainless steel, plastic, and glass. Significant cooling capacity is required to handle reaction exotherms and to maintain temperature. Because over 100 different organic peroxides are produced commercially, organic peroxide producers manufacture many organic peroxides in the same equipment. Batch processing

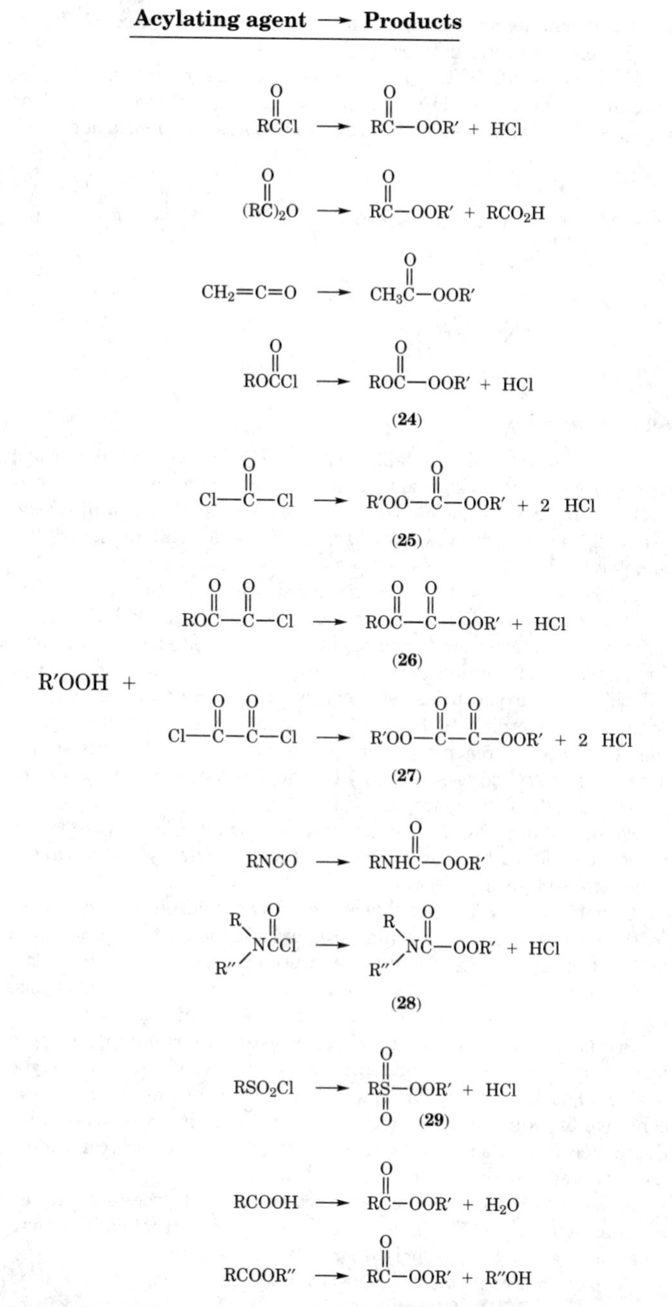

Acylating agent ⟶ Products

$$\underset{RCCl}{\overset{O}{\parallel}} \longrightarrow \underset{RC—OOR'}{\overset{O}{\parallel}} + HCl$$

$$(RC)_2O \longrightarrow \underset{RC—OOR'}{\overset{O}{\parallel}} + RCO_2H$$

$$CH_2=C=O \longrightarrow \underset{CH_3C—OOR'}{\overset{O}{\parallel}}$$

$$\underset{ROCCl}{\overset{O}{\parallel}} \longrightarrow \underset{ROC—OOR'}{\overset{O}{\parallel}} + HCl \qquad (24)$$

$$Cl—\underset{\overset{\parallel}{O}}{C}—Cl \longrightarrow R'OO—\underset{\overset{\parallel}{O}}{C}—OOR' + 2\ HCl \qquad (25)$$

$$RO\underset{\overset{\parallel}{O}}{C}—\underset{\overset{\parallel}{O}}{C}—Cl \longrightarrow ROC—\underset{\overset{\parallel}{O}}{C}—OOR' + HCl \qquad (26)$$

$$Cl—\underset{\overset{\parallel}{O}}{C}—\underset{\overset{\parallel}{O}}{C}—Cl \longrightarrow R'OO—\underset{\overset{\parallel}{O}}{C}—\underset{\overset{\parallel}{O}}{C}—OOR' + 2\ HCl \qquad (27)$$

R'OOH +

$$RNCO \longrightarrow \underset{RNHC—OOR'}{\overset{O}{\parallel}}$$

$$\underset{R''}{\overset{R}{>}}N\underset{\overset{\parallel}{O}}{C}Cl \longrightarrow \underset{R''}{\overset{R}{>}}N\underset{\overset{\parallel}{O}}{C}—OOR' + HCl \qquad (28)$$

$$RSO_2Cl \longrightarrow \underset{\overset{\parallel}{O}}{\overset{\overset{\displaystyle O}{\parallel}}{RS}}—OOR' + HCl \qquad (29)$$

$$RCOOH \longrightarrow \underset{RC—OOR'}{\overset{O}{\parallel}} + H_2O$$

$$RCOOR'' \longrightarrow \underset{RC—OOR'}{\overset{O}{\parallel}} + R''OH$$

Figure 2. Synthetic routes to alkyl peroxyesters. The acylating agent reacts with R'OOH in each case.

is generally employed when relatively small production volumes are required, whereas semicontinuous and continuous processing are employed when larger production volumes are required and when safety is a primary issue. Continuous processes are significantly safer to operate than batch processes as smaller amounts of organic peroxides are continuously in process. Besides safer continuous processing, another trend has been the use of reactants of higher purity. These process improvements have resulted in reduced environmental impact as unplanned process decompositions have been decreased and waste streams have been reduced.

Economic Aspects

Prices of commercial organic peroxides range from ca $2.50 to >$35/kg, depending on peroxide type, production volume, assay, nature of formulation, cost of raw materials, and degree of special processing and handling requirements.

Health and Safety Factors

Toxicology. In general, organic peroxides are characterized by a low order of acute toxicity. Most organic peroxides have some oxidizing properties and are irritants. Most of the available toxicity data on commercial organic peroxides are summarized in the literature. There is limited evidence to suggest that organic peroxides are carcinogenic.

Decomposition Hazards. The main causes of unintended decompositions of organic peroxides are heat energy from heating sources and mechanical shock, eg, impact or friction. In addition, certain contaminants, ie, metal salts, amines, acids, and bases, initiate or accelerate organic peroxide decompositions at temperatures at which the peroxide is normally stable. These reactions also liberate heat, thus further accelerating the decomposition. Commercial products often contain diluents that desensitize neat peroxides to these hazards. Commercial organic peroxide decompositions are low order deflagrations rather than detonations.

The organic peroxides and peroxide compositions produced commercially are those that can be manufactured, shipped, stored, and used safely. Organic peroxides can be thermally and mechanically desensitized by wetting or by dilution with suitable solvents, inert solid fillers, or insoluble liquids (suspension of solid peroxides in liquid plasticizers or water, and emulsions of liquid peroxides in water).

Recommendations for safe handling and storage of commercial organic peroxides are available from organic peroxide manufactures.

In 1984 the United Nations (UN) Committee on the Transportation of Dangerous Goods, made up of experts from Prins Maurits Laboratory (TNO), Bundesanstalt für Materialprüfung (BAM), and the organic peroxide producers, developed a test procedure for the classification of organic peroxide compositions for transport purposes. The test procedure was accepted by most of the industrial countries of the world. The Department of Transportation (DOT) mandated that the United States peroxide industry would comply with the UN classification system by October 1993. Material Safety Data Sheets (MSDS) and the organic peroxides producers' recommendations should be followed carefully for handling and storage of organic peroxide compositions.

Uses

There are more than 100 commercially available organic peroxides in well over 300 formulations, eg, neat liquids and solids, and pastes, powders, solutions, dispersions, and emulsions, that have utility in many commercial applications.

Excluding the peroxyacids, which are used primarily as epoxidizing and bleaching agents, approximately 90% of the commercial organic peroxides are consumed by the polymer industry.

They are used in the polymer industry as thermal sources of free radicals. They are used primarily to initiate the polymerization and copolymerization of vinyl and diene monomers.

Organic peroxides also are used as flame-retardant synergists for polystyrene, for preparing block and graft copolymers, for reactive processing, for reducing the molecular weight of polypropylene (ie, controlled rheology or vis-breaking), for curing adhesives, for drying alkyd resin films, and for initiating cationic polymerization with cyclic ethers and maleic anhydride.

BPO is the preferred bleaching agent for flour and has been used to bleach gums, waxes, fats, and oils. It is the active ingredient in many acne medications. Diacyl peroxides have been used as burnout agents for acetate yarn, drying agents for Chinawood oils, and as free-radical sources in many organic syntheses. Di-*tert*-butyl peroxide is used as an ignition accelerator for diesel fuels and has been used in many organic syntheses either as a source of *tert*-butoxy (photo) or methyl (thermal) radicals.

Table 13. Properties of Some *tert*-Alkyl Peroxyesters

Name	R	Mp, °C	Bp, °C (kPa)[a]
	R′ = tert-butyl$(CH_3)_3C$—		
tert-butyl peroxyacetate	CH_3—		22 (0.133)
tert-butyl *N,N*-dimethyl peroxycarbamate	$(CH)_2N_3$—		43–45 (0.013–0.027)
OO-tert-butyl *O*-isopropyl monoperoxycarbonate	*i*-C_7H_3O—		52–55 (0.133)
OO-tert-butyl *O*-hydrogen monoperoxymaleate	HOCCH=CH_2—	114–116	
tert-butyl peroxypivalate	*t*-C_9H_4—	oil	
di-*tert*-butyl diperoxycarbonate	*t*-C_9H_4OO—		54–55 (0.067)
di-*tert*-butyl diperoxyoxalate	*t*-C_9H_4OO-C(O)—	50.5–51.5	
tert-butyl peroxybenzoate	C_5H_6—	8	75–77 (0.267)
tert-butyl 2-ethyl-peroxy-haxonoate	$CH(CH)_3C_5H_2(C_2H_3)$—	<−30	
tert-butyl 2-carboxy-peroxy-benzoate	2-$HOC_4C_6H_2$—	104–104.5	
tert-butyl peroxydecanoate	*n*-$C_{19}H_9$—	−6.5	
di-*tert*-butyl diperoxyadipate	*t*-C_4H_9—OO—C(=O)$(CH_2)_4$—	42–43	
di-*tert*-butyl diperoxyphthalate	2-(*t*-C_4H_9—OO—C(=O))C_6H_4—	57.0–57.5	
tert-butyl peroxystearate	*n*-$C_{35}H_{17}$—	38.9–39.3	
	R′ = tert-cumyl$(C_6H_5C(CH_3)_2$—		
tert-cumyl peroxyacetate	CH_3—		67–68 (0.0067)
tert-cumyl peroxypivalate	*t*-C_9H_4—	−18	
tert-cumyl peroxybenzoate	C_5H_6—	45	
	R′ = other		
tert-amyl[b] peroxyacetate	CH_3—		65–66 (2.0)
3-hydroxy-1,1-dimethyl-butyl peroxyneodecanoate[c]	*t*-$C_{19}H_9$—	oil	
2,5-dimethyl-2,5-di(benzoyl-peroxy)-hexane[d]	C_5H_6—	118	

[a] To convert kPa to mm Hg, multiply by 7.5.
[b] R = *tert*-amyl/ (t-C_5H_{11}).
[c] R′ = 3-hydroxy-1, 1-dimethylbutyl($CH_3CHOHCH_2C(CH_3)_2$−).
[d] R′ = —C$(CH_3)_2$—CH_2CH_2—C$(CH_3)_2$— in the diester R—C(=O)—OO—C$(CH_3)_2$—$(CH_2)_2$—C$(CH_3)_2$—OO—C(=O)—R.

Table 14. *tert*-Butyl Areneperoxysulfonates

Benzeneperoxy sulfonate	Structure	Mp, °C
tert-butyl	$C_6H_5S(=O)_2$-OO-*t*-C_4H_9	[a]
tert-butyl *p*-chloro-	*p*-Cl-$C_6H_4S(=O)_2$-OO-*t*-C_4H_9	30–35[b]
tert-butyl *p*-methyl-	*p*-CH_3-$C_6H_4S(=O)_2$-OO-*t*-C_4H_9	36.5–37
tert-butyl *p*-methoxy-	*p*-CH_3O-$C_6H_4S(=O)_2$-OO-*t*-C_4H_9	47[b]

[a] n_D^{25} = 1.4629. [b] Explodes at mp.

JOSE SANCHEZ
TERRY N. MYERS
Elf Atochem North America, Inc.

D. Swern, ed., *Organic Peroxides,* Vols. I–III, Wiley-Interscience, New York, 1970–1972.

S. Patai, ed., *The Chemistry of Peroxides,* John Wiley & Sons, Inc., New York, 1983; S. Patai, ed., *Supplement E2: The Chemistry of Hydroxyl, Ether and Peroxide Groups,* John Wiley & Sons, New York, 1993.

W. Ando, ed., *Organic Peroxides,* John Wiley & Sons, Inc., New York, 1992.

PERSULFATES. See PEROXIDES AND PEROXIDE COMPOUNDS, INORGANIC.

PERVAPORATION. See HOLLOW-FIBER MEMBRANES.

PESTICIDES

The term pesticide refers to any agents used to control pests, and includes insecticides, herbicides, fungicides, defoliants, disinfectants, fumigants, plant and insect growth regulators, repellents, wood preservatives, and others.

The 1994 Pesticide Manual has 725 entries for chemicals in use as pesticides, plus 560 additional chemicals believed to be no longer manufactured or marketed for crop protection use. The entries include many different classes of chemicals used as the active ingredients (AIs) in at least 35,000 formulated pesticide products worldwide. Many of these chemicals have more than one type of action from among those listed as regulated under pesticide laws. Each pesticide product used in the United States must be evaluated and registered by the EPA. Pesticides are used for many diverse purposes, including the following: crop protection; disinfection; domestic animals; forest restoration; fuel preservation; fumigation; indoor pests; outdoor pests; pests in aquatic sites; post-harvest treatment; seed treatment; transport equipment; tree preservation; turf protection; vector control and plagues; vegetation control; water purification; and wood preservation.

Production

The EPA has approved a number of biological agents to replace some of the more toxic chemicals. Such biological agents, often called natural insecticides, are derived from living organisms. For example the bacterium *Bacillus thurengiensis* (B.t) produces a toxin that injures the gut of insect larvae. Different strains of B.t. can be selected to control various insects in home gardens, agriculture, and forestry. Many urban areas rely on spraying with B.t. *israelensis* to control mosquitoes.

Figure 1 compares pesticide markets in the United States and worldwide. A primary factor in raising prices in the United States has been the need for manufacturers to generate extensive additional data to support reregistration of all active ingredients (AIs) contained in products registered in the United States prior to November 1984. These new data must meet the standards for registration of new pesticides. Almost 80% of pesticide sales are in developed countries. Some governments such as Denmark have introduced legislation to reduce pesticide use by 50% or more by the year 2000, primarily because of concern about potential contamination of drinking water.

In developing countries, 90% of pesticide use is on agricultural crops, particularly on cotton and rice. The remaining 10% is used in vector control of human diseases such as malaria in the tropics and river blindness in Africa.

Only very large companies can afford the high costs of developing pesticides, estimated to be more than U.S.$70 $\times 10^6$ for a successful new active ingredient in 1994 and is much higher now (Table 1 lists principal agrochemical companies by sales).

Discovery. Optimization in selection of discovery pathways has been enhanced through use of tools such as quantitative structure activity relationship (QSAR) and computer-aided molecular design (CAMD).

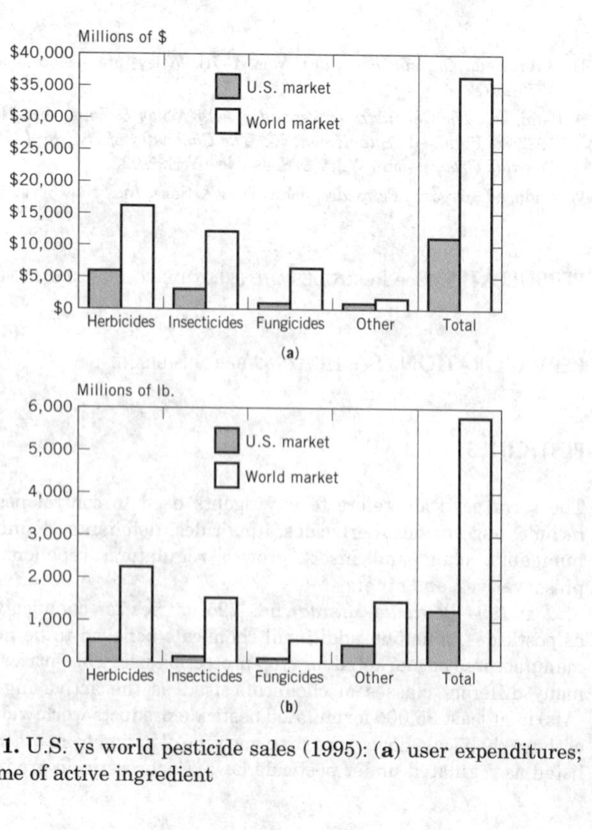

Figure 1. U.S. vs world pesticide sales (1995): **(a)** user expenditures; **(b)** volume of active ingredient

Table 1. Ranking of Principal Agrochemical Companies by Sales in 1997

Rank	Company	U.S.$ $\times 10^6$
1.	Monsanto/Cyanamid	5245
2.	Novartis	5127
3.	Zeneca	2674
4.	Du Pont	2518
5.	AgroEvo	2352
6.	Bayer	2254
7.	Dow	2200
8.	Rhône-Poulenc	2202
9.	BASF	1855
10.	Makhteshim/Agan	649

Table 2. Examples of Simple Early Pesticides

Compound	Pesticide class
formaldehyde	bactericide
bromomethane	fumigant
dimethylarsinic acid	herbicide
dalapon	herbicide
trichloroacetic acid	herbicide
acrolein	herbicide
2-phenylphenol	fungicide
biphenyl	fungicide
diphenylamine	fungicide
mercuric oxide	fungicide
mercurous chloride	fungicide
sodium fluoride	insect bait

Manufacturing. Early pesticides (Table 2) were simple compounds, often easy to make. Some of these have a high mammalian toxicity and present unacceptable hazards to farmers and other agricultural workers. In contrast, the manufacture of new chemical classes of pesticides having complex structures generally requires multistep synthesis processes, any of which can lead to side products or impurities. Also, concern about the fate of bioactive chemicals introduced into the environment has led to strict regulations on release of vapors into air or of manufacturing waste into effluent water, as well as for proper disposal of containers and wastes from pesticide use.

Formulation, Packaging, and Distribution. Advanced technology is needed to formulate pesticides to meet the needs of farmers and commercial applicators who must operate under regulatory constraints for protection of the environment. Each AI and pesticide product formulated in the United States must be registered by the EPA before it can be distributed and sold. It must have approved labeling for each recommended use, and must also be registered by each state where it is sold and used. Newer innovations include premeasured amounts of concentrates in water-soluble pouches packaged in recyclable paper that applicators can handle without contacting the product. Some companies have also developed proprietary encapsulated formulations and slow-release products that allow pesticides to become activated in the soil by trigger mechanisms such as moisture or temperature.

Inert ingredients include solvents, emulsifiers, surfactants, dispersants, stabilizers, preservatives, sequestrants, and other substances. These but aid in ensuring consistent action of one or more AIs in a formulated product. The products must remain stable during storage and distribution under a variety of environmental conditions, ranging from extreme heat in southern regions to subzero temperatures in northern areas. Containers must not rust or leak, must withstand rough handling, and must not collapse when stacked in warehouses.

The openings must be childproof and liquid contents should not gurgle or splash when being poured.

In 1987, EPA issued a policy statement regarding inert ingredients in formulated pesticide products, and in 1992, published a list of inerts in four categories according to the degree of toxicological concern.

As more effective pesticides were discovered, application rates dropped from pounds per acre (lb/A) of formulated products per acre to grams per hectare (g/ha). Highly active, low volume products are easier to ship to dealers, take up less storage space until sold to customers, and are easy to deliver to farmers.

Application of Pesticides. Older, less active pesticides were often applied using backpack tanks and hand-held wands to direct the spray onto target weeds and brush. For somewhat larger projects, the sprays were applied from tractor-drawn rigs. Formulations that provide large droplets are necessary for herbicides applied by aircraft. Fogging sprays can be used to control insect pests such as mosquitoes in urban and recreational areas, and destructive insects like gypsy moths in forests. Very dilute ready-to-use products are packaged in aerosol spray cans and are registered for use around homes and institutions.

More efficient application systems have been developed, including longer spray booms to cover a greater area per pass and installation of movable or permanent tanks to dispense the prepared sprays. Some products have also been approved by the EPA for drip application in irrigation water, or from central pivot sprayers above artesian wells on the circular fields of the arid Great Plains area in the United States.

Regulation

Pesticides are more closely regulated than other chemicals because pesticides are intentionally applied in the environment, often repeatedly at relatively high rates. In the United States, pesticides are regulated under the Federal Insecticide, Fungicide and Rodenticide Act (FIFRA), and residues from uses of pesticides in food or feed crops are regulated under Sections 408 and 409 of the Federal Food, Drug and Cosmetics Act (FFDCA). Both were amended in 1996 with the passage of the Food Quality Protection Act (FQPA) which is more restrictive and could well eliminate many crop protection agents.

Requirements for Pesticide Registration in the United States. Pesticide registration decisions are based primarily on EPA evaluations of test data provided by applicants to ensure that, when used according to label directions, the pesticide does not cause unreasonable adverse effects to human health or the environment. Testing is needed to show whether the pesticide has the potential to cause adverse effects to humans, wildlife, fish, and plants, including endangered species. Potential human risks, which are identified using laboratory tests in animals, include acute toxic reactions such as poisoning and skin and eye irritation, as well as possible long-term effects such as cancer, birth defects, and reproductive disorders. Data on the fate of pesticides in the environment are also required so that scientists in the EPA's office of Pesticide Programs can determine, among other things, whether a pesticide poses a threat to groundwater or surface water (lakes, rivers, and streams). Extensive analytical studies are also required to establish maximum residue levels anticipated from recommended uses in food or feed crops.

Cost Estimates for U.S. Registration of Pesticides. Some environmental fate and some chronic toxicology studies cost more than a million US dollars per study, and require as many as four years to complete, representing a large investment of both money and time for pesticide manufacturers. There is no assurance that registration can be obtained soon enough to recoup expenses amounting to up to $\$70 \times 10^6$ before the patent runs out. Furthermore, many new requirements and costs have been added and are still being added for reviewing studies in support of registration applications.

Large fees are also assessed for review of studies submitted in support of a petition for residue tolerances in foods or feeds. Petition fees increased to more than $70,000 on May 27, 1998.

Individual registrations are needed for each formulated product containing one or more approved AIs, and each product must bear a label approved for each use that is recommended on that label. The EPA charges an ever-increasing annual maintenance fee for each product label registration and each product must be registered in each state where it is sold and used, with appropriate fees charged by each state. States can also require a registrant to submit studies to be evaluated independently in that state, such as in California which sets its own standards for registration.

The EPA also requires the registration of any establishment that manufactures or formulates pesticides in the United States. Special labeling is required for export products that are not registered for use in the United States. These same products can be produced in other countries and exported more expediently. Thus, U.S. manufacturers are at a disadvantage in competing for international trade.

Reregistration of Existing Pesticides in the United States. The EPA was to revaluate the scientific database underlying some 600 cases representing about 1300 AIs contained in 45,000 pesticide products registered prior to November 1, 1984. A deadline of 1997 was given to complete the process. The concern was that some older agrochemicals might not have been tested adequately, and might be hazardous to human health or the environment. Any studies that did not meet these stringent standards and requirements for registration of a new chemical would have to be repeated, submitted for review by scientists in the EPA's Office of Pesticide Programs (OPP), and accepted as meeting all requirements for new registration. This has now been extended to the year 2000 to allow for new requirements imposed by FQPA passed unanimously by the U.S. congress in 1996.

In 1993, the U.S. General Accounting Office (GAO) estimated that EPA is not likely to complete the reregistration program until the year 2006. It will, however, take much longer because the EPA is required to re-assess all tolerances under much more stringent FQPA criteria.

Role of International Organizations in Pesticide Regulation. Most developed countries have established laws and regulations that outline policies for the production, registration, and use of pesticides. Much as in the United States, these determine the risks and benefits associated with pesticides, and promote safe and effective use. The harmonization of regulatory standards has become of greater importance with the expansion of world agricultural trade and the movement of agricultural commodities among nations, particularly in compliance with the General Agreement on Tariffs and Trade (GATT) adopted in 1994.

Benefit and Risk Issues

As world population increases, urban expansion encroaches more and more on productive land areas used for growing food and fiber to feed and clothe all these people. Whereas crop yields may continue to increase with advances in agricultural technology, crop protection agents are still needed to avoid losses owing to weeds, insects, and fungi. Pest infestations can affect the economies of entire regions. In the early to mid-1990s devastating insect-related cotton crop failures occurred in China, and in Pakistan and India. In the United States, cotton is making a comeback, because the boll weevil has been largely eradicated through the development of improved pesticides. Instead of spraying twice a week, each application costing $7 to $8 an acre, farmers might spray only twice a year.

Pesticides are subjected to extensive testing (93 tests) for residues in food, toxicology in laboratory animals, and fate in the environment before being registered for use. There are 18 listed for residue chemistry, 25 for environmental chemistry, 28 for toxicology, and 22 for wildlife and aquatic organisms. There have been requirements added for testing for endocrine disruptors, but have no guidelines. Moreover, uses are closely regulated by governmental agencies worldwide.

MARGUERITE L. LENG
Leng Associates

Proceedings of the 9th International Congress of Pesticide Chemistry, (IUPAC), London, 1999, American Chemical Society, Washington, D.C.

M. L. Leng, E. M. K. Leovey, and P. L. Zubkoff, eds., *Agrochemical Environmental Fate: State of the Art,* Lewis Publishers, CRC Press, Boca Raton, Fla., 1995, 410 pp.

C. Tomlin, ed., *The Pesticide Manual,* 10th ed, incorporating the *Agrochemicals Handbook,* Royal Society of Chemistry, Cambridge, U.K., 1994, 1341 pp.

G. Ekstrom, ed., *World Directory of Pesticide Control Organisations,* 2nd ed., The British Crop Protection council and the Royal Society of Chemistry, Crop Protection Publications, Cambridge, U.K., 1994.

PET AND LIVESTOCKS FEEDS. See FEEDS AND FEED ADDITIVES.

PETROLATUM WAXES. See WAXES.

PETROLEUM

NOMENCLATURE IN THE PETROLEUM INDUSTRY

Crude oils, complex mixtures of naturally occurring organic liquids, are difficult to characterize in detail. Thus, many of the definitions used by the exploration, production, and refining sectors of the petroleum industry to describe petroleum and its products often lack precision. Even the term petroleum itself is poorly defined, and although it is often used synonymously with crude oil, it is also frequently used to include natural gas (see GAS, NATURAL) and even solid hydrocarbons. Traditionally the unit of crude oil production has been the barrel (bbl), equal to 42 U.S. gallons, 5.61 ft^3, 158.8 L, or 0.159 m^3. Increasingly, petroleum reserves are given in metric tons, but because one unit is a volume and the other a weight, there can be no unique conversion factor for a material with a range of densities. Oil density may be reported in any appropriate units, and although metric units are used it is more common to report densities as degrees API, °API, or API gravity (where API stands for American Petroleum Institute). The relationship between density and API gravity is an inverse one, defined by the relationship:

$$°API = [141.5/\text{specific gravity at} 60°F] - 131.5$$

Water corresponds to an API gravity of 10, and crude oils fall between 10 and 60 °API, with the commonest values being in the 35–40° range.

Other terms relating to physical properties include viscosity; refractive index; pour point, ie, the lowest temperature at which the oil will flow; flash point, ie, the temperature at which the oil will ignite; and aniline point, ie, the minimum temperature at which equal volumes of oil and aniline are completely miscible. They are determined under defined conditions established by ASTM.

Natural gas production is generally given in cubic feet or cubic meters (1000 ft^3 = 1 Mcf = 28.3 m^3. Natural gas is called dry when methane is the dominant hydrocarbon, and wet if it contains more than 4 L/100 m^3 of natural gas liquids (>0.3 gallons per 1000 ft^3). When gas (or oil) has a bad odor due to high concentrations of hydrogen sulfide and volatile sulfur compounds it is called sour. Sweet gas has no noticeable odor.

Crude oils contain a wide range of hydrocarbons including straight and branched chains, ring compounds, and aromatics, as well as more complex compounds that incorporate nitrogen, sulfur, and oxygen (often called the NSOs), and some nickel and vanadium. Petroleum chemists still use the obsolete word naphthenes for the compounds that organic chemists call alicyclics. Aromatic hydrocarbons form a minor but important group of compounds in crude oils and range from single-ring to multiring compounds.

Most crude oil is refined to provide useful products and the dominant process is distillation (qv) (Table 1). Petroleum products produced by simple distillation without the use of pressure, cracking, or catalysts are called straight run. A number of other words that have traditionally been used in the petroleum industry are difficult to define precisely. They refer partly to specific boiling ranges, but also to certain intended uses.

Table 1. Generalized Distillation Ranges for Products Obtained During Crude Oil Refining

Product	Temperature range, °C	Carbon number range
gasoline	30–210	5–12
naphtha	100–200	8–12
kerosene and jet fuel	150–250	11–13
diesel and fuel oils	160–400	13–17
atmospheric gas oil	220–345	
heavy fuel oils	315–540	20–45
atmospheric residue	≥450	30+
vacuum residue	≥615	60+

Gas oil is a product boiling slightly higher (235–425°C, or sometimes wider) than kerosene. The main feedstock to the catalytic cracking units (see FEEDSTOCKS), it received its name from use as an enriching agent in the production of city or manufactured gas. It is often used as diesel fuel.

Cylinder oil is a viscous oil used for lubricating the cylinders and valves of steam engines (see LUBRICATION AND LUBRICANTS). It is prepared from cylinder stock. The product from cylinder stock, when filtered and processed, is bright stock.

Cycle stock (recycle stock) denotes any product that is recycled, that is, taken back to an earlier stage in the process. But the term cycle stock is also used for the gas oil-like product of catalytic cracking.

The term distillate is sometimes used to denote distillate fuel oil as opposed to residual fuel oil.

In the petroleum industry the International Union of Pure and Applied Chemistry (IUPAC) system is in widespread use for naming organic compounds. However, the system does not always result in convenient terms for groups of compounds. Because hydrocarbons with the same number of carbon atoms are apt to have boiling points within a small range, it would be convenient to have words that would refer to C_4, C_5, C_6, ... saturates, and C_4, C_5, C_6, ... monounsaturates, etc. The IUPAC system goes by the number of carbon atoms in the longest straight chain. The situation is different when naming the ethylenic hydrocarbons, because the IUPAC has provided names such as propene, butene, and pentene, which are different from the former names ending in -ylene.

In names such as isobutane, isopentane, isobutyl alcohol, and isoamyl alcohol, the prefix iso has a precise meaning, ie, one methyl group attached to the next-to-terminal carbon atom, and no other branch. This notation is frequently used by petroleum chemists with a much wider meaning, denoting nothing more than branched-chain.

COLIN BARKER
University of Tulsa

N. J. Hyne, *Dictionary of Petroleum Exploration, Drilling, and Production,* PennWell Books, Tulsa, Okla., 1991, 625 pp.

J. G. Speight, *The Chemistry and Technology of Petroleum,* 2nd ed., Marcel Dekker, Inc., New York, 1991, 760 pp.

ORIGIN OF PETROLEUM

Petroleum is a naturally occurring complex mixture made up predominantly of carbon and hydrogen compounds, but also frequently containing significant amounts of nitrogen, sulfur, and oxygen, together with smaller amounts of nickel, vanadium, and other elements. It may occur in solid, liquid, or gaseous form as asphalt (qv), crude oil, or natural gas (see GAS, NATURAL), respectively. Commercial petroleum accumulations are the result of generation in a source rock, migration to a reservoir, and compositional modifications in the reservoir.

The evidence supporting a biological source for the material that generates petroleum is extensive. Organisms produce a wide range of organic compounds including significant amounts of biopolymers like proteins (qv), carbohydrates (qv), and lignins (see LIGNIN), together with a wide variety of lower molecular weight lipids. Survival of these organic materials depends on many factors, but preservation is strongly favored in anoxic sediments. However, the formation of a petroleum accumulation requires more than just a concentration of the hydrocarbons that are initially incorporated in sediments. Although C_2–C_{10} hydrocarbons are present in extremely low (parts per billion (ppb) level) concentrations in organisms and sediments, these can account for up to 50% or more of the volume of some crude oils.

Compounds that are not synthesized by organisms are also reported in crude oils. As the organic matter in sediments is buried in a reducing environment and subjected to gradually increasing temperature, petroleum is generated as an intermediate in a transformation process that ultimately leads to methane and graphite (see CARBON, NATURAL GRAPHITE; HYDROCARBONS). The nature of the of the complex, insoluble organic material (kerogen) in source rocks controls whether oil or gas is generated, and also controls the composition of these products. The generation of petroleum is nonbiological, induced by temperature, and influenced by available time. It follows first-order kinetics with an increase of 10°C roughly doubling the reaction rate at low temperatures. The petroleum generation process can be duplicated by laboratory pyrolysis, but higher temperatures are needed to produce reactions in a few hours rather than the millions of years that it takes in nature. The petroleum generation process can be treated quantitatively using models based on parallel first-order kinetic reactions.

An important exception to thermal generation is the bacterial formation of methane. The bacteria are anaerobic and are effective in sulfate-free, anoxic conditions and have long been recognized for their role in forming marsh gas. Bacterially produced methane is isotopically light for both carbon and hydrogen.

Most petroleum is found in reservoir rocks that have high permeabilities and porosities, but low contents of organic materials. The amount of organic matter is insufficient to generate commercially significant quantities of petroleum (see OIL SHALE), and it is now believed that petroleum generation occurs in organic-rich source rocks, and that part of the bitumen then migrates to accumulate in reservoir rocks. This is the source rock concept. Migration has a critical role in linking the organic-rich source rocks to the reservoir. Buoyancy is the main driving force through the carrier beds, and oils continue to move upward (toward shallower areas) until stopped at a slope reversal in a structural trap, or where permeability decreases as in a stratigraphic trap. Migration distances can be in excess of 100 km. Oil may be remobilized after its initial accumulation in the reservoir. Although in the simplest case this may involve only a simple relocation, it can lead to major compositional changes if both gas and oil are involved.

The composition of petroleum changes and evolves in the reservoir in response to changing conditions. Thermal maturation of crude oil is brought about by the increasing temperature that accompanies increasing depth of burial and produces lighter oils of higher quality and eventually gas. Large changes in petroleum composition can be produced by contact with flowing water. As the water moves past the oil in the reservoir it removes the most soluble components. These include the light ends, particularly the small aromatics, and their loss leaves a tar layer at the oil–water interface. If the water brings bacteria and oxygen into contact with the oils at temperatures below approximately 70°C, substantial changes in crude oil composition can result, producing heavier oils enriched in sulfur and nitrogen compounds.

Oils contain chemical fossils, now usually called biomarkers, which are compounds with characteristic molecular structures that can be related to those in living systems. The compounds include isoprenoids, porphyrins, steranes, hopanes, and many others. Although biomarkers form a small percentage of bitumen and crude oils, their relative distributions and complex structures are modified by the various processes involved during petroleum generation and accumulation. They are widely used for correlation studies, and for recognition and documentation of the progress of generation and maturation.

The overall petroleum system that leads to the accumulation of oil and gas in natural reservoirs can be summarized as follows: organic matter is incorporated into sediments as they are deposited; possible shallow generation of biogenic methane is effected; organic matter is converted to petroleum-like materials (bitumen) by the influence of increasing temperature, lower temperatures being partially offset by longer times; part of the lower molecular weight material that is generated subsequently migrates from the source rock through permeable carrier beds to the reservoir; and after the oil reaches the reservoir, significant compositional changes may be produced by increasing temperature, water washing, and bacterial degradation.

COLIN BARKER
University of Tulsa

C. Barker, *Thermal Modeling of Petroleum Generation: Theory and Applications,* Elsevier, Amsterdam, The Netherlands, 1996, 512 pp.

J. M. Hunt, *Petroleum Geochemistry and Geology,* 2nd ed., W. H. Freeman, San Francisco, Calif., 1995, 743 pp.

M. H. Engel and S. A. Macko, eds., *Organic Geochemistry, Principles and Applications,* Plenum Press, New York, 1993, 861 pp.

L. B. Magoon and W. G. Dow, *The Petroleum System—From Source to Trap,* AAPG Memoir 60, American Association of Petroleum Geologists, Tulsa, Okla., 1994, 644 pp.

COMPOSITION

Petroleum, literally rock oil, describes a myriad of hydrocarbon-rich fluids present in source rocks and accumulated in subterranean reservoirs (see HYDROCARBONS). Petroleum can include three phases: gaseous (natural gas), liquid (crude oil), and solid or semisolid (bitumens, asphalt (qv), tars, and pitches) (see COAL; GAS, NATURAL; TAR AND PITCH). The molecular composition of the liquid portion of petroleum contributing to the crude oil properties and behavior is discussed herein. Crude oils vary dramatically in color, odor, and flow properties. These properties often reflect the origin of the crude. Historically, physical properties such as boiling point, density (gravity), odor, and viscosity have been used to classify oils. Crude oils may be called light or heavy in reference to relative density (or specific gravity). Light crude oils are rich in low boiling and paraffinic hydrocarbons; heavy crude oils contain greater amounts of high boiling and asphalt-like molecules. The heavy oils tend to be more viscous, higher boiling, more aromatic, and contain larger amounts of heteroatoms. Likewise, odor is used to distinguish between sweet or low sulfur, and sour or high sulfur, crude oils.

Petroleum is thought to be derived from a variety of living organisms buried with sediments in previous geological eras.

The distribution of biomarker isomers, molecules that retain the basic carbon skeletons of biological compounds from living organisms, serves not only as a set of fingerprints for oil–oil and oil–source correlation (to relate the source and reservoir for exploration), but also to give geochemical information on organic source input (marine, lacustrine, or terrigenous source), age, maturity, depositional environment (clay or carbonate, oxygen levels, salinity, etc), and alteration (water washing, biodegradation, etc).

Knowledge of the composition of petroleum allows the refiner to optimize conversion of raw petroleum into high value products.

A knowledge of the molecular composition of a petroleum also allows environmentalists to consider the biological impact of environmental exposure.

Crude oils contain an extremely wide range of organic functionality and molecular size. The variety is so great that a complete compound-by-compound description for even a single crude oil is not likely. The composite molecular composition of petroleum can, however, be described in terms of three classes of compounds: saturates, aromatics, and compounds bearing the heteroatoms sulfur, oxygen, or nitrogen. Within each of these classes there are several families of related compounds. Some of the compounds typically found in petroleum crude oils are *n*-octane, 2-methyloctane, propylcyclohexane, *n*-butylbenzene, 1-methylnaphthalene, 9-methylphenanthrene, propyl mercaptan, methyl propyl sulfide, dibenzothiophene, phenol, 2-phenanthrene carboxylic acid, cyclohexyl carboxylic acid, quinoline, carbazole, and 2(1*H*)-quinolinone.

Elemental Composition

On an atomic basis, H/C ratios range from 1.5–2.0. The range of elemental composition (wt %) of crude oil may be given as follows: carbon (84–87%); hydrogen (11–14%); sulfur ($<$0.1–8%); oxygen ($<$ 0.1$-$1.8%); nitrogen ($<$0.1$-$1.6%); nickel (0–0.1%); and vanadium (0–0.5%).

Molecular Classes

The molecules in crude oil include several basic structural types. Because they may contain from 1 to 100+ carbon atoms and may occur in combination, the statistical potential for isomeric structures is staggering. For example, whereas there are just 75 possible paraffinic structures for C_{10}, there are $> 10^5$ isomers for C_{20}. A few structures tend to dominate the distributions of each isomer group, however.

Molecular characterization of a whole oil is beyond the capability of most analytical techniques. Distillation (qv), however, can separate petroleum into molecular weight fractions that simplify the task. Chromatography (qv) allows oils to be fractionated by polarity as a second dimension. Individual compounds have been isolated and quantified from increasingly higher boiling fractions. Techniques have been developed that use combinations of classical open-column adsorption chromatography, gel permeation chromatography, and ion-exchange (qv) separations to isolate fractions in which compounds could be identified by mass spectrometry (qv).

Whereas neither distillation nor chromatography achieves perfect separations among groups, the fractions generated are amenable to molecular characterization.

About the same amounts are distilled into the middle distillate and vacuum gas oil from conventional crude oils. More naphtha is distilled from light crude oils and more vacuum residuum is obtained from heavy crude oils. The typical distribution of classes of petroleum compounds shows a significant shift with boiling point (Fig. 1).

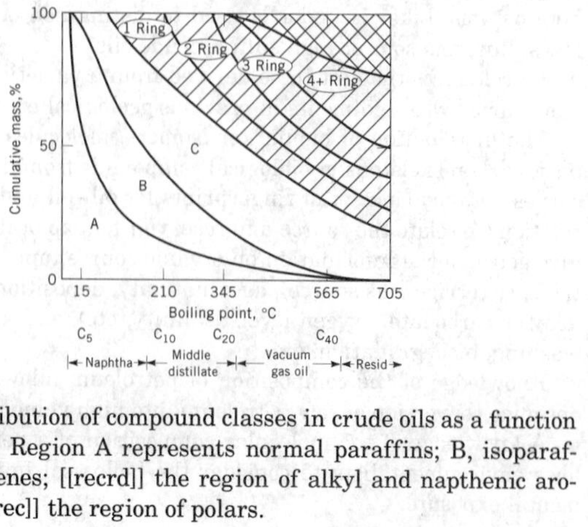

Figure 1. Distribution of compound classes in crude oils as a function of boiling point. Region A represents normal paraffins; B, isoparaffins; C, naphthenes; [[recrd]] the region of alkyl and napthenic aromatics; and [[ccrec]] the region of polars.

Petroleum Gases and Naphtha. Methane is the main hydrocarbon component of petroleum gases. Lesser amounts of ethane, propane, butane, isobutane, and some C_4+ light hydrocarbons also exist. Other gases such as hydrogen, carbon dioxide, hydrogen sulfide, and carbonyl sulfide are also present.

The naphtha fraction is dominated by saturates but does contain some aromatics, including benzene (Fig. 1). Most raw naphthas are too low in octane to be used directly as gasoline components.

Mid-Distillates. As is indicated in Figure 1, saturates remain the primary component in the mid-distillate fraction of petroleum, but aromatics, which include simple compounds having up to three aromatic rings, and heterocyclics are present and represent a larger portion of the total. Some raw middle distillates are used directly as kerosenes, jet fuels, and diesel fuels; others are cracked and hydroprocessed before use.

Vacuum Gas Oils

As is indicated in Figure 1, saturates contribute less to the vacuum gas oil (VGO) than the aromatics, but more than the polars (NSO/heterocyclics). VGO itself is occasionally used as a heating oil but most commonly it is processed by catalytic cracking to produce naphtha or by extraction to yield lubricant oils.

Vacuum Residua

The vacuum residua or vacuum bottoms is the most complex fraction. Vacuum residua are used as asphalt and coker feed. Both the metals (Ni/V) and the NSO/heterocyclics concentrate into these bottoms. Molecular weights range from 400 to $>$ 2000; this is so high that characterization of individual species is virtually impossible. Separations by group type become blurred by sheer mass of substitution around a core structure and by the presence of multiple functionalities in a single molecule. Furthermore, this fraction lacks volatility and is characterized only by nontraditional gc or ms techniques.

WINSTON K. ROBBINS
CHANG SAMUEL HSU
Exxon Research and Engineering Company

B. P. Tissot and D. H. Welte, *Petroleum Formation and Occurrence,* Springer-Verlag, New York, 1978.

J. G. Speight, *Fuel Science Technology Handbook,* Marcel Dekker, New York, 1990.

K. E. Peters and J. M. Moldowan, *The Biomarker Guide: Interpreting Molecular Fossils in Petroleum and Ancient Sediments,* Prentice-Hall, Englewood Cliffs, N.J., 1993.

K. H. Algelt and M. M. Boduszynski, *Composition and Analyses of Heavy Petroleum Fractions,* Marcel Dekker, New York, 1994.

DRILLING FLUIDS

The drilling fluid performs a variety of functions that influence the drilling rate, and the cost, efficiency, and safety of the drilling operation. The drilling fluid or mud, as it is commonly called, is pumped down a hollow drill string through nozzles in the bit at the bottom of the well, and back up the annulus formed by the hole or casing and the drill string to the surface. The bit is turned by rotating the entire drill string from the surface or by using a downhole motor to only rotate the bit. After reaching the surface, the drilling fluid is passed through a series of vibrating screens, settling tanks or pits, hydrocyclones, and centrifuges to remove formation material brought to the surface. It is then treated with additives to obtain a set of desired physical and chemical properties. Once treated, the fluid is pumped back into the well and the cycle repeated.

Drilling fluids generally are composed of liquids, eg, water, petroleum oils, and other organic liquids; dissolved inorganic and organic additives; and suspended, finely divided solids of various types. Drilling fluids were commercialized about 1926.

Since 1980 over 1000 patents have been issued for drilling fluid systems and materials in the United States alone. A 1994 listing of products from 117 suppliers offers ca 3000 trade names.

Classification

Drilling fluids are classified as to the nature of the continuous phase: gas, water, oil, or synthetic. Within each classification are divisions based on composition or chemistry of the fluid or the dispersed phase.

Gas-Based Muds. Gas-based drilling fluids are used mostly for hard-rock drilling. These fluids range from compressed dry air or natural gas (see GAS, NATURAL) to water-based mist or stable foams (qv). Foam is considered gas-based.

Chemical additives for gas-based drilling fluids are limited to surfactants (qv), certain polymers, and occasionally salts such as sodium or potassium chloride. No additives are used in dry air or gas drilling operations.

Water-Based Muds. About 85% of all drilling fluids are water-based systems. The types depend on the composition of the water phase (pH, ionic content, etc), viscosity builders (clays or polymers), and rheological control agents (deflocculants or dispersants (qv)).

Freshwater fluids can range from clear water having no additives to high density muds containing clays, barite, and various organic additives. Onshore wells typically use freshwater muds, as do some offshore wells where highly weighted muds are needed.

Many offshore wells are drilled using a seawater system because of ready availability. Seawater muds generally are formulated and maintained in the same way that a freshwater mud is used. However, because of the presence of dissolved salts in seawater, more additives are needed to achieve the desired flow and filtration (qv) properties.

In many drilling areas both onshore and offshore, salt beds or salt domes are penetrated. Mud saturated with the salt present in the formation is used to reduce the hole enlargement that would result from salt dissolution by contact with an undersaturated liquid.

The high salinity of salt water muds may require different clays and organic additives than those used in fresh- or seawater muds.

Fresh- or seawater muds may be treated with gypsum or lime to alleviate drilling problems that may arise from drilling water-sensitive shale or clay-bearing formations.

Potassium treated systems combine one or more polymers and a potassium ion source, primarily potassium chloride, in order to prevent problems associated with drilling certain water-sensitive shales.

Fresh water, clay, and polymers for viscosity enhancement and filtration control make up low solid/nondispersed muds. Low solids muds are maintained using minimal amounts of clay and require removal of all but modest quantities of drill solids. These are called nondispersed systems because no additives are used to further disperse or deflocculate the viscosity building clays. Most water-based muds are considered dispersed because deflocculating additives are used to control the flow properties.

Oil-Based Muds. Oil-based drilling fluids have diesel or mineral oil as a continuous phase with both internal water and solid phases. Fluids having no or very low water content are usually called oil-base muds or all oil muds; fluids having higher water contents are called invert oil-emulsion muds, or simply inverts. Most oil muds maintain a fixed oil-water ratio depending on the desired properties. Oil muds are employed for high angle wells where good lubricity is required, for high temperature wells where water-based systems may be thermally unstable, for drilling water-sensitive shale formations, or where corrosive gases such as hydrogen sulfide and carbon dioxide may be encountered. Environmental restrictions and cost often limit use, although higher drilling rates achievable using oil muds and polycrystalline diamond compact (PDC) bits can often offset the high fluid and disposal costs.

Synthetic-Based Muds. A new class of drilling muds, the synthetic-based muds, has been introduced to counteract the high costs associated with disposal of drill cuttings generated when diesel or mineral oil-based muds are used. These newer fluids, similar in formulation and performance to oil-based muds, have a continuous phase that consists of a synthetic organic liquid. Because of the similarity, synthetic-based muds are often called pseudo-oil muds outside the United States. A variety of fluids have been used as the continuous synthetic phase including an ester of a vegetable oil fatty acid, a poly(alpha-olefin), an acetal, linear alkyl benzenes, a low viscosity nonpetroleum hydrocarbon, etc. As of this writing (ca 1995) synthetic-base fluids are being developed and field tested at a rapid rate.

Properties

Density. The density of the drilling fluid is adjusted using powdered high density solids or dissolved salts to provide a hydrostatic pressure against exposed formations in excess of the pressure of the formation fluids. In addition, the hydrostatic pressure of the mud column prevents collapse of weak formations into the borehole. Fluid densities may range from that of air to >2500 kg/m^3 (20.8 lb/gal). Most drilling fluids have densities >1000 kg/m^3 (8.33 lb/gal), the density of water. The hydrostatic pressure imposed by a column of drilling fluid is expressed as follows: $P = 0.098\ L\rho_m (= 0.052\ L\rho_m)$, where P = the hydrostatic pressure in kPa (psi); ρ_m = the drilling fluid density, kg/m^3 (lb/gal); and L = the column length or well depth, m (ft).

Flow Properties. The fluid viscosity and annular flow velocity must be high enough to remove cuttings generated by the drill bit and other formation material that may fall into the wellbore. These solids are carried up the annulus to the surface where they are separated with varying degrees of efficiency. In order to accomplish this, low viscosity drilling fluids are circulated at high flow rates or high viscosity fluids at low flow rates. In addition, for maximum drilling rate, a low effective viscosity is desired at shear rates generated through the bit nozzles (10,000–100,000/s).

The varying demands on the flow properties are best met by fluids exhibiting non-Newtonian rheological characteristics.

Filtration Properties. Drilling fluids have a natural tendency to flow into permeable formations because the borehole pressure is generally higher than that in the formation. To prevent excessive leak-off, a thin, low permeability filter cake is formed using additives. Filtration (qv) occurs under both dynamic (during circulation) and static (no circulation) conditions. Additives may affect each of these filtration conditions differently. The filtration rate is adjusted using colloidal solids and organic polymers to reduce loss of filtrate to the formation and prevent buildup of a thick filter cake which would restrict the wellbore. Excessive decrease of the filtration rate can be costly and may result in a viscous fluid that may affect the drilling rate adversely.

Water Chemistry. Water is present in all but purely gaseous or oil drilling fluids, both of which comprise only a small percentage of drilling fluid applications. Water added to drilling fluids may be fresh water, seawater, or saturated salt solutions. The salt in the last is normally sodium chloride but may be another halide or alkali or alkaline-earth salt. The nature of the dissolved salts affects colloidal clays and other additives and thus must be monitored together with properties such as salinity, total hardness (calcium plus magnesium), pH, and alkalinity. Drilling fluids are nearly always basic. The pH ranges from 6 to 13 depending on the type of system. Concentrations of soluble carbonates, sulfide, sulfite, etc, may also be determined. These ions may be added intentionally or incorporated during drilling.

Drilling Fluid Materials

Density Control. The pressure exerted by the column of drilling fluid in the well balances formation pressures to prevent uncontrolled influx of formation fluids which may result in a blowout. The mud density must be controlled accurately by suitable weighting materials that do not adversely affect the other properties. Most important is the

Table 1. API Specifications for Barite and Hematite

Assay	Barite	Hematite
specific gravity[a]	4.20	5.05
wet-screen analysis, % residue[b]		
>75 μm	3.0	1.5
>45 μm		15
particles <6 μm, %[b]	30	15
soluble alkaline-earth metals as calcium, mg/kg[b]	250	100

[a] Value given is minimum.
[b] Values given are maximum.

specific gravity of the weighting agent as well as its water insolubility and chemical inertness. The weighting material should be ground to the preferred particle-size distribution and be relatively nonabrasive. As of 1995, little weighting material other than barite was used in the United States.

Barite, predominately $BaSO_4$, is virtually insoluble in water and does not react with other mud constituents. Most operators prefer barite that meets API specifications (Table 1). The barite content in mud depends on the desired density but can be as high as 2000 kg/km^3 (700 lb/bbl).

The only other high specific gravity material that is used to any degree as of 1995 was hematite. Ilmenite, used for a short time, is no longer available commercially.

The specifications for drilling fluid hematite have also been set by the API and are listed in Table 1. Hematite is used most frequently in high density oil-based muds to minimize the total volume percent solids. The abrasivity of hematite limits its utility in water-based muds.

Calcite and siderite are used occasionally because of their solubility in hydrochloric acid which offers a method of removing mud filter cake deposited on productive formations.

Weighted fluids without solids are provided by solutions of various water-soluble salts. Aside from the density required by a specific application, cost and corrosion have to be considered. Brine sources include seawater, natural brine, and manufactured salts.

Viscosity Buildup. The drilling fluid removes cuttings from the wellbore as drilling progresses. This process is governed by the angle of the hole and the velocity at which fluid travels up the annulus, as well as by the fluid viscosity or flow properties, and fluid density. The cuttings removal efficiency usually increases with increasing viscosity and density, although at high wellbore angles a less viscous fluid may be desirable provided high flow rates can be achieved. Viscosity depends on the concentration, quality, and state of dispersion of suspended colloidal solids.

Although numerous mud additives aid in obtaining the desired drilling fluid properties, water-based muds have three basic components: water, reactive solids, and inert solids. The water forming the continuous phase may be fresh water, seawater, or salt water. The reactive solids are composed of commercial clays, incorporated hydratable clays and shales from drilled formations, and polymeric materials, which may be suspended or dissolved in the water phase. Solids, such as barite and hematite, are chemically inactive in most mud systems. Oil and synthetic muds contain, in addition, an organic liquid as the continuous phase plus water as the discontinuous phase.

The most important commercial clays used for increasing the viscosity of drilling fluids are bentonite, attapulgite, and sepiolite. For oil-base and synthetic-base muds, organophilic clays are used.

The most commonly used polymeric viscosity builders are the cellulosics, xanthan gum, and polyacrylamides.

Viscosity Reduction. Proper control of viscosity and gel strengths is essential for efficient cleaning of the borehole, suspension of weight material and cuttings when circulation is interrupted, and to minimize circulating pressure losses and swab/surge pressures owing to axial movement of the drill string. Viscosity may be increased as previously indicated, but there is often the necessity of reducing the viscosity. A reduced viscosity can be achieved by thinning or deflocculating clay-water suspensions. Thinning is measured as a reduction of plastic viscosity, yield point, or gel strength, or a combination of these properties. Typical mud-thinning chemicals are polyanionic materials that are adsorbed on positive edge sites of the clay particles, thereby reducing the attractive forces between the particles without affecting clay hydration.

Thinners or deflocculants for clay–water muds include polyphosphates, tannins, lignites, lignosulfonates, and low molecular weight polyacrylates and their derivatives.

Filtration Control. Filtration control is particularly important in permeable formations where the mud hydrostatic pressure exceeds the formation pressure. Proper filtration control reduces drill-string sticking and drag, and rotary torque, as well as minimizing damage to protective formations; in some formations it improves borehole stability. Several types of materials are available for water-based muds and application varies according to the type and the chemical environment of the mud. These include clays, organic polymers, and lignite derivatives. The bentonite present in the system often acts as the primary filtration control agent. It not only develops viscosity, but also lowers the filtration rate, particularly in freshwater muds. The ability of bentonite clay to control filtration is attributed to the flat, plate-like particle shape, the capacity to disperse and hydrate, the ability to form a compressible filter cake, and the colloidal to near-colloidal particle size.

Although a combination of bentonite clay and an organic thinner provides filtration control in many water-based muds, additional control generally is needed. Filtration additives for both fresh- and salt water muds are usually organic polymers and lignites.

Starches, used first in the late 1930s for filtration control, are still in use in the 1990s. Corn starch is most commonly used in the United States.

Numerous modifications and derivatives of starch have been made for application in drilling and workover fluids.

Carboxymethyl cellulose (CMC) and polyanionic cellulose (PAC) are available in several viscosity grades and high and low purity grades. All grades can be effective filtration control agents depending on the well conditions.

Acrylate and acrylamide polymers have several uses in drilling fluids, one of which is for filtration control.

Lignite products, mined, ground, and possibly treated with sodium or potassium hydroxide, are economical filtration control additives for some water-based muds, in addition to improving flow properties.

A number of synthetic polymers having the ability to control filtration rates at high temperature and in the presence of calcium and magnesium have also been developed.

Cellulosic fibers, powdered limestone, gilsonite, and asphalt are frequently added to both water and oil muds at levels of 10 to 25 kg/m^3 (4–10 lb/bbl) when high differential pressures are encountered to control seepage losses to the formation. This treatment also is used to improve the quality of the mud filter cake to reduce the chance of differential pressure sticking.

Alkalinity Control. Water-base drilling fluids are generally maintained at an alkaline pH. Most mud additives require a basic environment to function properly and corrosion is reduced at elevated pH. The primary additive for pH control is sodium hydroxide in concentrations from 3 to 14 kg/m^3 (1–5 lb/bbl).

The second most common alkalinity control agent is lime, normally in the form of calcium hydroxide, used in both water and oil muds. Potassium hydroxide is occasionally used for alkalinity control. A fourth alkalinity control additive is magnesium oxide, which is used in clay-free polymer-base fluids.

Removal of Contaminants. A drilling fluid contaminant is any material or condition encountered during drilling operations that adversely affects the performance of the fluid. Elevated temperatures and drill solids are encountered in every drilling operation. In most wells these are handled easily, but in some wells one or both can seriously reduce drilling efficiency. Temperature problems normally are treated using viscosity or filtration control additives, material having better thermal stability, or possibly by replacement of the mud system with an oil or synthetic mud. Drill solids are removed mechanically by various combinations of screens, hydrocyclones, and centrifuges, or chemically by flocculants. Dilution or replacement of part or all of the mud system may reduce drill solids to tolerable levels.

Various inorganic chemicals remove soluble contaminants encountered during drilling. Salt, NaCl, is a common contaminant that can be removed only by dilution. The adverse effects of salt, primarily clay flocculation, can be overcome by a deflocculant such as a lignosulfonate or sulfomethylated tannin.

Stabilization of Water-Sensitive Formations. Many subsurface formations encountered during drilling are water-sensitive shales containing various amounts of clay minerals. The clay mineral components may include a highly swelling smectite or less water-sensitive illite, mixed layer smectite–illite, kaolinite, or chlorite. All shales appear to swell to some extent when contacted by fresh water. Those containing smectite or mixed layer clays are much more sensitive than illitic shales. The uptake of water by shales has two effects: a volume change owing to swelling, and a strength reduction as the water content increases. This may result in flow of plastic shale into the wellbore, softening and erosion of the exposed shale, or spalling of hard shale, all of which can cause expensive operational problems. In the latter case, large hard pieces of formation fall around the drill string, the removal of which may become difficult.

A variety of methods have been devised to stabilize shales. The most successful method uses an oil or synthetic mud that avoids direct contact between the shale and the emulsified water.

Sodium chloride has long been used as a shale stabilizer because of low cost, wide availability, and its presence in many subsurface formations. The inhibitive nature of salt muds increases as the salt content increases from seawater to saturated sodium chloride.

A variety of shale-protective muds are available which contain high levels of potassium ions. The reaction of potassium ions with clay, well known to soil scientists, results in potassium fixation and formation of a less water-sensitive clay.

Ammonium chloride, ammonium sulfate, and di-ammonium phosphate have also been used for shale stabilization.

In Europe where magnesium-bearing salt formations are encountered, magnesium chloride is used, but in the United States it is used only on a small scale.

A number of nonionic and anionic polymers are employed in water-based muds to stabilize shales. These may be added to a freshwater mud or to a system containing one of the salts mentioned.

The method of action of the polymers is thought to be encapsulation of drill cuttings and exposed shales on the borehole wall by the nonionic materials, and selective adsorption of anionic polymers on positively charged sites of exposed clays which limits the extent of possible swelling. The latter method appears to be true particularly for certain anionic polymers because of the low concentrations that can be used to achieve shale protection.

A number of cationic muds have been developed and used. These are formulated around quaternary amines or positively charged polymers.

A number of glycol and glycerol-base additives are being used to formulate shale protective muds usually in conjunction with a salt and/or a polymer. The glycols (qv) consist of a number of combinations of ethylene oxide (qv) and propylene oxide depending on the purpose of the additive, not all of which are strictly for shale stabilization, and the desired cloud point.

A more recent addition to the list of shale protective water-base muds is a system developed around concentrated solutions of methyl glucoside. At concentration of 25% by weight and above, methyl glucoside appears to stabilize water-sensitive shales on par with a typical oil- or synthetic-base mud.

Solid materials, such as gilsonite and asphalt, and partially soluble sulfonated asphalt may also be added to plug small fractures in exposed shale surfaces and thereby limit water entry into the formation.

Surfactants. Surfactants (qv) perform a variety of functions in a drilling fluid. Depending on the type of fluid, a surfactant may be added to emulsify oil in water (o/w) or water in a nonaqueous liquid (w/o), to water-wet mud solids or to maintain the solids in a nonwater-wet state, to defoam muds, or to act as a foaming agent.

Foaming agents maintain stable drilling foams in areas where minimal bottomhole pressures are required. A large number of chemicals generate drilling foams, including anionics, such as alkyl and alkylaryl sulfates and alkylarylsulfonates; nonionics, such as ethoxylated fatty alcohols and alkylaryl alcohols; and cationics, such as imidazolines and tertiary amines. The foaming agent must be chosen to handle a variety of possible contaminants (salt, crude oils, solids) and downhole temperatures.

Lost Circulation Control. To function properly, a drilling fluid must be circulated through the well and back to the surface. Occasionally, highly permeable or cavernous formations and fractured zones, both natural and induced by the mud pressure, are encountered and circulation is partially or completely lost. Loss of drilling fluid, owing to openings in the formation, can result in loss of hydrostatic pressure at the bottom of the hole and allow influx of formation fluids and possibly loss of well control. It is essential that circulation be regained for drilling to continue. A wide variety of materials can be added to the drilling fluid to seal off the lost circulation zones. Lost circulation materials are flake, fiber, or granular-shaped particles. Each type is sold individually, often in two or more size grades, or two or more materials of different shapes may be sold as a blend. Materials of different shapes and sizes are often blended into the mud at the well site.

Some common flake-shaped LCMs consist of shredded cellophane and paper, mica (qv), rice hulls, cottonseed hulls, or laminated plastic. Fibrous additives include a variety of cellulose fibers, sawdust, sugarcane bagasse, paper, straw, leather (qv), and many others of similar size, shape, and availability.

Granular LCMs generally are much stronger than the other types and include ground rubber, nylon, plastics, limestone, gilsonite, asphalt, and ground nut shells, eg, walnut and pecan (see NUTS).

Concentrations of LCMs are 14–143 kg/m^3 (5–50 lb/bbl). The higher concentrations are used in the form of a 10–20 m^3 (60–120 bbl) pill that is placed across the loss zone until a seal is established that allows circulation to be regained.

Removal of Solids. Solids incorporated in the mud during drilling generally are separated mechanically, reduced by dilution, or removed chemically by flocculation. It is desirable to maintain a low concentration of drill solids (4–8 vol %) and in some cases total removal is required.

Lubricants and Spotting Fluids. The frictional resistance generated by the rotating drill string against the formation or casing may require extra torque if the hole is crooked or being drilled directionally. Considerable frictional resistance to raising and lowering the drill string may also occur; this is referred to as drag.

Drilling fluids are treated with a variety of mud lubricants. They are mostly general-purpose, low toxicity, nonfluorescent types that are blends of several anionic or nonionic surfactants and products such as glycols and glycerols, fatty acid esters, synthetic hydrocarbons, and vegetable oil derivatives.

Corrosion Control. Drill string and casing corrosion can present serious problems in some drilling operations (see CORROSION AND CORROSION CONTROL). Corrosion is generally caused by oxygen dissolved or entrained in the mud as it is circulated through the well. Acid gases, such as carbon dioxide and hydrogen sulfide, contribute to corrosion, particularly hydrogen sulfide which can produce catastrophic drill string failures through stress cracking. Preventing entry of these corrosive gases into the well is the most effective method of control, but is not always feasible. Additives are employed to counteract corrosive attack.

Maintaining a high pH is the most common means of corrosion control. Other methods include oxygen removal by scavengers such as sodium sulfite or ammonium bisulfite, or protecting the pipe from attack by coating with amines.

Drilling Rate Enhancers. Drilling rates can be severely reduced when drilling formations that tend to stick to the surface of the drill bit, a situation known as bit balling. If the bit cannot be kept clean and free of sticky material, drilling performance suffers and the operational cost can increase substantially. The new generation synthetic polycrystalline diamond compact (PDC) bits are capable of achieving high penetration rates if the cutters are kept clean and the proper hydraulics and bit rotary speeds can be maintained (see TOOL MATERIALS). Additives being used to enhance the drilling performance of PDC bits in water-base muds include water-soluble glycols and various terpenes. The use of a cationic mud has been reported to improve drilling rates as well.

Environmental Aspects

The disposal of waste drilling fluids and drill cuttings in the United States has long been regulated either by local authorities, the individual states, or by the federal government. These regulations continue to change. The offshore disposal of both diesel and mineral oil drilling fluids and associated cuttings has always been prohibited in U.S. waters. However discharge of mineral oil mud cuttings has been permitted in the North Sea and elsewhere as long as the oil content of the cuttings was below some regulatory limit. The regulatory oil-on-cuttings limit in some sectors of the North Sea is, as of this writing, being lowered significantly. There is a definite move toward alternative fluid systems, many of which are used in U.S. offshore areas.

Other regulations apply in different offshore drilling areas in the United States and around the world. All have had a profound effect on drilling fluid technology. Very few instances of water-base muds failing the regulatory tests exist in the 1990s. Operators and service companies have eliminated use of the more toxic additives, reformulated old mud systems, and developed new ones to ensure acceptable environmental performance based on pertinent regulations.

R. K. CLARK
Shell E&P Technology Company

G. R. Gray and H. C. H. Darley, *Composition and Properties of Oil Well Drilling Fluids,* 5th ed., Gulf Publishing Co., Houston, Tex., 1988.

World Oil, 51 (June 1994).

Specification for Drilling Fluid Materials, API Spec. 13A, 15th ed., Sect. 2, American Petroleum Institute, Washington, D.C., May 1, 1993.

ENHANCED OIL RECOVERY

Enhanced oil recovery (EOR) requires the successful application of chemical, chemical engineering, and petroleum engineering technologies. As the petroleum industry becomes more dependent on increasing production from existing fields, the use of EOR is expected to grow.

The Nature of Oil Reservoirs

Oil reservoirs are layers of porous sandstone or carbonate rock, usually sedimentary. Impermeable rock layers, usually shales, and faults trap the oil in the reservoir. The oil exists in microscopic pores in rock. Various gases and water also occupy rock pores and are often in contact with the oil. These pores are interconnected with a complicated network of microscopic flow channels. The weight of overlaying rock layers places these fluids under pressure. When a well penetrates the rock formation, this pressure drives the fluids into the wellbore. The flow channel size, wettability of flow channel rock surfaces, oil viscosity, and other properties of the crude oil determine the rate of this primary oil production.

As reservoir pressure is reduced by oil production, additional recovery mechanisms may operate. One such mechanism is natural water drive.

Primary production typically recovers 10–25% of the oil originally in the reservoir. Efficiency of primary production is related to oil properties, reservoir properties, geometric placement of oil wells, and the drilling and completion technology used to drill the wells and prepare them for production. Pumping the well can maintain production at economic levels for years.

Waterflooding. Injection wells are used when the natural pressures driving fluids to production wells are depleted and pumping is no longer economical. Fluid injection repressurizes the reservoir, restoring a driving force and promoting oil production. Water injection or waterflooding is usually termed secondary oil recovery. It accounts for about 40% of the total U.S. oil production. Additional oil recovery by waterflooding is typically 15–25% of the oil originally in the reservoir.

Oil Recovery Mechanisms

There are two principal mechanisms of enhanced oil recovery: increasing volumetric sweep efficiency of the injected fluid and increasing oil displacement efficiency by the injected fluid. In both, chemicals are used to modify the properties of an injected fluid whether water, steam, a miscible gas such as CO_2 or natural gas, or an immiscible gas, usually nitrogen. Poor reservoir volumetric sweep efficiency is the greatest obstacle to increasing oil recovery.

Wettability is defined as the tendency of one fluid to spread on or adhere to a solid surface (rock) in the presence of other immiscible fluids. As many as 50% of all sandstone reservoirs and 80% of all carbonate reservoirs are oil-wet. Strongly water-wet reservoirs are quite rare. Rock wettability can affect fluid injection rates, flow patterns of fluids within the reservoir, and oil displacement efficiency. Rock wettability can strongly affect its relative permeability to water and oil. When rock is water-wet, water occupies most of the small flow channels and is in contact with most of the rock surfaces as a film. Crude oil does the same in oil-wet rock. Alteration of rock wettability by adsorption of polar materials, such as surfactants and corrosion inhibitors, or by the deposition of polar crude oil components, can strongly alter the behavior of the rock.

When water is injected into a water-wet reservoir, oil is displaced ahead of the injected fluid. Injection water preferentially invades the small-and medium-sized flow channels or pores. As the water front passes, unrecovered oil is left in the form of spherical, unconnected droplets in the center of pores or globules of oil extending through interconnected rock pores. In both cases, the oil is completely surrounded by water and is immobile. There is little oil production after injection water breakthrough at the production well.

In an oil-wet rock, water resides in the larger pores, oil exists in the smaller pores or as a film on flow channel surfaces. Injected water preferentially flows through the larger pores and only slowly invades the smaller flow channels resulting in a higher produced water:oil ratio and a lower oil production rate than in the water-wet case.

Injection Well Considerations. Fluid injection rate can have a significant effect on oil recovery economics. Flow is radial from the wellbore into the reservoir. Thus the region near the injection wellbore acts as a choke for the entire reservoir.

Injection Fluids. Whereas water is the most commonly used injection fluid, other fluids can provide higher oil recovery efficiency. Injecting gases miscible with reservoir crude oil can result in low interfacial tension promoting a high oil displacement efficiency. The WAG process of miscible gas flooding uses alternate injection of water and carbon dioxide (qv). Other suitable gases include natural gas and flue gas. Injection of hydrocarbon miscible gas is used in the Alaskan North Slope and is under study for use in the North Sea.

Nonmiscible gases such as nitrogen have been used as EOR injection fluids. Oil recovery mechanisms include volatilization of low molecular weight components of the crude oil and displacement of oil from the top of the reservoir.

Gas injection into a gas cap overlaying an oil reservoir is considered an EOR method. The resulting repressurization of the reservoir promotes additional oil production.

High temperature steam (qv) is also used for recovery of viscous crude oils.

The injection of large volumes of steam, steam flooding, is used to mobilize oil which is produced at offset production wells. Smaller volumes of steam are injected in the cyclic steam stimulation or huff'n'puff process.

Improving Volumetric Sweep Efficiency. Volumetric sweep efficiency is determined by the permeability and wettability distribution in the reservoir and by the properties of injected fluids. High permeability rock streaks or layers (thief zones) and natural or induced rock fractures can channel the injected fluid through a small portion of the reservoir, resulting in a low rock volumetric sweep efficiency. Low viscosity injection fluids exhibit poor volumetric sweep efficiency and lead to low oil production. Thus, proper diagnosis of the cause of poor volumetric sweep efficiency is critical in designing a successful well treatment.

Both sodium silicate gelation and *in situ* cross-linking of organic polymers can reduce the permeability of fractures and high permeability streaks. Polymers are usually injected at concentrations of 1000–5000 ppm. The most commonly used polymers are partially hydrolyzed polyacrylamides. Chromium(III) Cr(III), compounds have largely replaced Al(III) compounds as cross-linkers. Sequential injection of partially hydrolyzed polyacrylamide and cross-linker is giving way to treatments in which the two components are injected together.

Polymer Flooding. Even in the absence of fractures and thief zones, the volumetric sweep efficiency of injected fluids can be quite low. The poor volumetric sweep efficiency exhibited in waterfloods is related to the mobility ratio, M, the mobility of the injected water in the highly flooded (low oil saturation) rock, m_w, divided by the mobility of the oil in oil-bearing portions of the reservoir, m_o. The mobility ratio is related to the rock permeability to oil, k_{ro}, and injected water, k_{rw}, and to the viscosity of these fluids by the following equation: $M = m_w/m_o = (k_{rw}/\eta_w)/(k_{ro}/\eta_o)$. The terms η_w and η_o represent the viscosity of the aqueous and oil phases, respectively.

The polymer flooding method requires a preflush to condition the reservoir, the injection of a polymer solution for mobility control to minimize channeling, and a driving fluid (water) to move the polymer solution and resulting oil bank to production wells.

Each EOR polymer type has important advantages and significant disadvantages.

Polymer Interactions. Various methods of utilizing polymer interactions to modify solution viscosity are under study. Polymer association complexes, which substantially increase water viscosity at quite low polymer concentrations, offer potentially improved cost effectiveness compared to the flood polymers. However, use has not progressed beyond the laboratory testing stage.

Surfactants for Mobility Control. Water, which can have a mobility up to 10 times that of oil, has been used to decrease the mobility of gases and supercritical CO_2 (mobility on the order of 50 times that of oil) used in miscible flooding. Gas:oil mobility ratios, M, can be calculated by the formula $M = [(k_g/\mu_s) + (k_w/\mu_w)]/[k(o/\mu_o) + (k_w/\mu_w)]$ where k refers to permeability, μ to viscosity, and the subscripts g, s, o, and w to gas, miscible solvent, oil, and water, respectively. The water may be injected simultaneously with the gas or in alternate slugs with the gas (WAG process). X-ray computerized tomography of core floods has demonstrated the increased volumetric sweep efficiency attained in the WAG process compared to injection of CO_2 alone. The design parameters most affecting WAG CO_2 flood oil recovery are CO_2 and water slug sizes, produced gas:oil ratio as a function of time, and total volume of injected CO_2.

The WAG process has been used extensively in the field, particularly in supercritical CO_2 injection, with considerable success. One means of increasing the viscosity of CO_2 is through the use of supercritical CO_2-soluble polymers and other additives. The use of surfactants to form low mobility foams or supercritical CO_2 dispersions within the formation has received more attention. Foam has also been used to reduce mobility of hydrocarbon gases and nitrogen. Among the classes of surfactants studied for this application are alcohol ethoxylates and their sulfate and sulfonate and carboxylate derivatives, alkylphenol ethoxylates, alpha-olefin sulfonates, and alkylated diphenylether disulfonates. In addition to the mobility control characteristics of surfactants, critical issues in gas mobility control processes are surfactant salinity tolerance, hydrolytic stability under reservoir conditions, surfactant propagation through the reservoir, and foam stability in the presence of crude oil saturations. Lignosulfonate has been reported to increase foam stability and function as a sacrificial adsorption agent. Addition of sodium carbonate or sodium bicarbonate to the surfactant solution reduces surfactant adsorption by increasing the aqueous-phase pH.

Additives can improve surfactant propagation. Both anionic surfactant partitioning and precipitation increase with increasing calcium ion concentration so minimizing divalent metal ion concentration in the surfactant solution is desirable. Injection of a surfactant preslug containing NaCl converts clays from the calcium to the sodium form and reduces later ion-exchange processes that add Ca^{2+} ions in the surfactant solution. The use of a hydrotrope such as sodium xylene sulfonate has been reported to increase oil recovery in laboratory steam-foam flood tests. Hydrotropes are additives that increase surfactant solubility. They also may function as sacrificial adsorption agents or act as foam stabilization agents.

Thermal stability of the foaming agent in the presence of high temperature steam is essential. Alkylaromatic sulfonates possess superior chemical stability at elevated temperatures.

Water-soluble polymers (qv) can increase the viscosity of the foam external phase. This improves foam stability and reduces mobility. Gelation of the foam external phase can reduce chemical requirements to plug thief zones and fractures.

Improving Oil Displacement Efficiency. The use of relatively large (ca 2–5 wt%) concentrations of surfactants to increase oil displacement efficiency has been studied extensively. This method, called the micellar flooding or surfactant–polymer flooding, usually involves the injection of a brine preflush to adjust reservoir salinity. The preflush is followed by injection of a micellar slug comprised of the surfactant, a cosurfactant (usually a C_{4-6} alcohol), and a hydrocarbon. A polymer solution is then injected to reduce viscous fingering of the drive fluid into and through the micellar slug. Viscous fingering causes dilution of the surfactant, reduced contact of the micellar slug with the crude oil, and trapping of some of the micellar slug in the reservoir. These effects reduce oil recovery. A freshwater buffer to protect the polymer follows, prior to addition of the driving fluid, ie, water, to move the chemicals and the resulting oil bank to the well.

Alkaline Flooding. Alkaline or caustic flooding involves injection of high pH agents such as sodium hydroxide, sodium carbonate, or sodium silicate solutions. At equivalent Na_2O levels, the three alkaline agents give equivalent recovery of each of nine different crude oils in laboratory core floods. However, the use of buffered sodium carbonate rather than strong alkali can result in reduced interaction with mineral surfaces. The lower reagent consumption can reduce the amount of sodium carbonate required.

Caustic flooding chemicals are relatively inexpensive. Including a surfactant in the caustic formulation (surfactant-enhanced alkaline flooding) can increase optimal salinity of a saline alkaline formulation. This can reduce interfacial tension and increase oil recovery. Encouraging field test results have been reported. Both nonionic and anionic surfactants have been evaluated in this application.

Surfactants evaluated in surfactant-enhanced alkaline flooding include internal olefin sulfonates, linear alkylxylene sulfonates, petroleum sulfonates, alcohol ethoxysulfates, and alcohol ethoxylates/anionic surfactants. Water-thickening polymers, either xanthan or polyacrylamide, can reduce injected fluid mobility in alkaline flooding and surfactant-enhanced alkaline flooding. The combined use of alkali, surfactant, and water-thickening polymer has been termed the alkali–surfactant–polymer (ASP) process. Cross-linked polymers have been used to increase volumetric sweep efficiency of surfactant–polymer–alkaline agent formulations.

Steam flooding can greatly increase the recovery of high viscosity crude oils by heat thinning. This increases oil mobility in the reservoir. The addition of urea and iron sulfate or nickel compounds is said to further lower the viscosity of the crude oil. Surfactant foaming agents can be used to reduce the mobility of the high temperature steam. Because some heavy crude oils have relatively high acid numbers, it is not surprising that addition of alkaline agents to high temperature steam can increase recovery of these oils.

Other Technologies

Microbial-enhanced oil recovery involves injection of carefully chosen microbes. Subsequent injection of a nutrient is sometimes employed to promote bacterial growth. Molasses is the nutrient of choice owing to its low (ca $100/t) cost. The main nutrient source for the microbes is often the crude oil in the reservoir. A rapidly growing microbe population can reduce the permeability of thief zones improving volumetric sweep efficiency. Microbes, particularly species of Clostridium and Bacillus, have also been used to produce surfactants, alcohols, solvents, and gases *in situ*. These chemicals improve waterflood oil displacement efficiency (see also BIOREMEDIATION).

The *in situ* combustion method of enhanced oil recovery through air injection is a chemically complex process. There are three types of *in situ* combustion: dry, reverse, and wet. In the first, air injection results in ignition of crude oil and continued air injection moves the combustion front toward production wells. Temperatures can reach 300–650°C. Ahead of the combustion front is a 90–180°C steam zone, the temperature of which depends on pressure in the oil reservoir. Zones of hot water, hydrocarbon gases, and finally oil propagate ahead of the steam zone to the production well.

The oil zone is fairly cool, and in a viscous oil reservoir this can result in little oil movement (liquid blocking). Reverse combustion, in which oil ignition occurs near the production well, can avoid this problem. The combustion zone moves countercurrent to the flow of air from the injection well. Oil flows through heated rock and remains mobile. Reverse combustion requires more air and consumes more oil than forward combustion.

In wet combustion, water is injected concurrently and alternately with air, extending the steam zone and aiding heat transfer to the crude oil reducing oil viscosity. This can decrease injected air:produced oil ratio and improve project economics.

JOHN K. BORCHARDT
Shell Chemical Company

G. Moritis, *Oil Gas J.*, 51 (Sept. 26, 1994).

D. H. Smith, *Surfactant-Based Mobility Control—Progress in Miscible-Flood Enhanced Oil Recovery,* ACS Symposium Series No. 373, American Chemical Society, Washington, D.C., 1988.

K. S. Sorbie, *Polymer-Improved Oil Recovery,* Blackie and Son, Ltd., London, 1991.

L. L. Schramm, ed., *Foams: Fundamentals and Applications in the Oil Industry,* American Chemical Society, Washington, D.C., 1994.

REFINERY PROCESSES, SURVEY

Petroleum refining, also called petroleum processing, is the recovery and/or generation of usable or salable fractions and products from crude oil, either by distillation or by chemical reaction of the crude oil constituents under the effects of heat and pressure. Crude petroleum is a mixture of compounds boiling at different temperatures that can be separated into a variety of different generic but often overlapping fractions (Table 1). The amounts of these fractions produced by distillation depend on the origin and properties of crude petroleum.

When petroleum occurs in a reservoir that allows the crude material to be recovered by pumping operations as a free-flowing dark-to-light colored liquid, it is often referred to as conventional petroleum.

Table 1. Distillation Fractions of Petroleum

Fraction	Boiling, °C
light naphtha	−1 to 150
gasoline	−1 to 180
heavy naphtha	150–205
kerosene	205–260
stove oil	205–290
light gas oil	260–315
heavy gas oil	315–425
lubricating oil	>400
vacuum gas oil	425–600
residuum	>600

Heavy oil differs from conventional petroleum in that its flow properties are reduced and it is much more difficult to recover from the subsurface reservoir. These materials have a much higher viscosity and lower API (American Petroleum Institute) gravity than conventional petroleum.

Heavy oil generally has an API gravity of less than 20 degrees and usually, but not always, a sulfur content of >2% by weight. Extra heavy oil occurs in the near-solid state and is virtually incapable of free flow under ambient conditions. Bitumen, often referred to as native asphalt, is a subclass of extra heavy oil and is frequently found as the organic filling in pores and crevices of sandstones, limestones, or argillaceous sediments.

A residuum, often shortened to resid, is the residue obtained from petroleum after nondestructive distillation has removed all the volatile materials. The temperature of the distillation is usually below 345°C because the rate of thermal decomposition of petroleum constituents is substantial above 350°C. Temperatures as high as 425°C can be employed in vacuum distillation. When such temperatures are employed and thermal decomposition occurs, the residuum is usually referred to as pitch.

Asphalt, prepared from petroleum, often resembles bitumen. When asphalt is produced by distillation, the product is called residual, or straight-run, asphalt. However, if the asphalt is prepared by solvent extraction of residua or by light hydrocarbon (propane) precipitation, or if it is blown or otherwise treated, the name should be modified accordingly to qualify the product, eg, propane asphalt.

Sour and sweet are terms referring to a crude oil's approximate sulfur content, which relates to odor. A crude oil that has a high sulfur content usually contains hydrogen sulfide, H_2S, and/or mercaptans, RSH; it is called sour. Without this disagreeable odor, the crude oil is judged sweet.

General refinery steps are given in Figure 1.

Desalting and Dewatering

Crude oil is recovered from the reservoir mixed with a variety of substances: gases, water, and dirt (minerals). Refining actually commences with the production of fluids from the well or reservoir and is followed by pretreatment operations that are applied to the crude oil either at the refinery or prior to transportation.

Field separation, which occurs at a field site near the recovery operation, is the first attempt to remove the gases, water, and dirt that accompany crude oil coming from the ground.

Desalting is a water-washing operation performed at the production field and at the refinery site for additional crude oil cleanup.

The usual practice is to blend crude oils of similar characteristics, although fluctuations in the properties of the individual crude oils may cause significant variations in the properties of the blend over a period of time. Blending several crude oils prior to refining can eliminate the frequent need to change the processing conditions that may be required to process each of the crude oils individually.

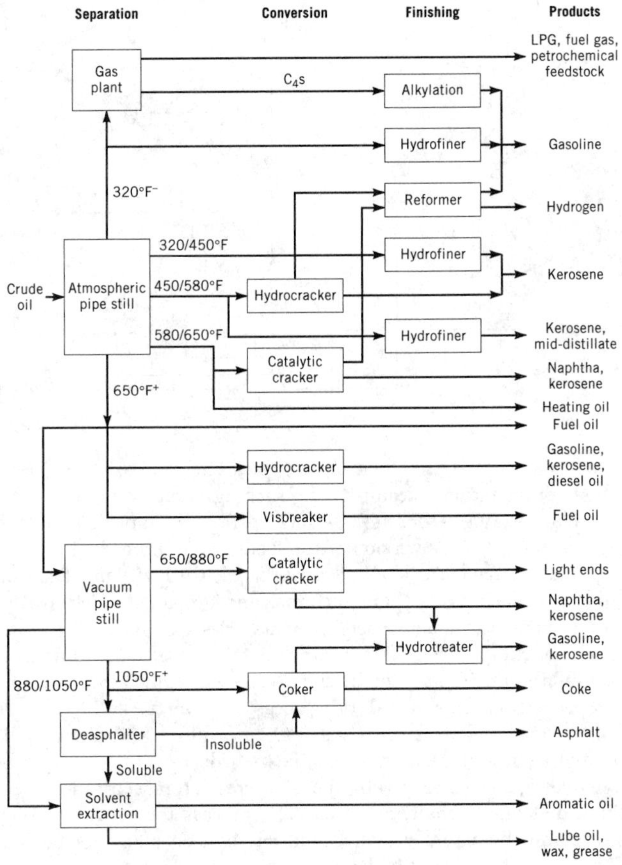

Figure 1. General refinery operations.

Feedstock Evaluation

Three frequently specified properties are density–specific gravity–API gravity, characterization factor, and sulfur content.

The API (American Petroleum Institute) gravity is a measure of density or specific gravity (sp gr):

$$°\text{API} = \frac{141.5}{\text{sp gr}} - 131.5$$

For a wide-boiling-range material such as crude oil, the boiling point is taken as an average of the five temperatures at which 10, 30, 50, 70, and 90% of the material is vaporized. A highly paraffinic crude oil can have a characterization factor as high as 13, whereas a highly naphthenic crude oil can be as low as 10.5, and the breakpoint between the two types of crude oil is approximately 12.

Refining

A refinery is a group of manufacturing plants that vary in number according to the variety of products produced. Refinery processes must be selected to convert crude oil into products according to demand. A refinery must also be flexible and be able to change operations as needed, especially if heavier oils are the primary feedstocks. This is accomplished through two basic process concepts: carbon rejection, eg, coking processes, and hydrogen addition, eg, hydroprocesses. However, certain downstream processes, such as catalytic reforming, applied to the product streams do not fit into either of these categories (see CATALYSIS).

In general, when the product is a fraction from crude oil that includes a large number of individual hydrocarbons, the fraction is classified as a refined product. Examples of refined products are gasoline, diesel fuel, heating oils, lubricants, waxes, asphalt, and coke. In contrast, when the product is limited to, perhaps, one or two specific hydrocarbons of high purity, the fraction is classified as a petrochemical product. Examples of petrochemicals are ethylene (qv), propylene (qv), benzene (qv), toluene, and xylene (see BTX PROCESSING).

The application designed for a product requires detailed specifications for various properties; such specifications are set by organizations varying from country to country.

Distillation. This is the point at which refining begins and was the first method by which petroleum was refined.

Atmospheric Distillation. The petroleum distillation unit in the 1990s brings about a fairly efficient degree of fractionation (separation). The feed to a distillation tower is heated by flow through pipes arranged within a large furnace. The heating unit is known as an atmospheric pipe still heater or pipe still furnace, and the heating unit and fractional distillation tower make up the essential parts of a distillation unit or pipe still. Vapors pass up the tower to be fractionated into gas oil, kerosene, and naphtha.

The primary fractions from a distillation unit are equilibrium mixtures and contain some proportion of the lighter constituents characteristic of a lower boiling fraction. The primary fractions are stripped of these constituents (stabilized) before storage or further processing.

Vacuum Distillation. Vacuum distillation evolved as the need arose to separate the less volatile products, such as lubricating oils, from petroleum without subjecting these higher boiling materials to cracking conditions. The boiling point of the heaviest cut obtainable at atmospheric pressure (101.3 kPa = 760 mm Hg) is limited by the temperature (ca 350°C) at which the residue starts to decompose or crack. It is at this point that distillation in a vacuum pipe still is initiated.

Azeotropic and Extractive Distillations. Effective as they are for producing various liquid fractions, distillation units generally do not produce specific fractions. In order to accommodate the demand for such products, refineries have incorporated azeotropic distillation and extractive distillation in their operations (see DISTILLATION, AZEOTROPIC AND EXTRACTIVE).

Thermal Cracking. *Visbreaking.* Viscosity breaking (reduction) is a mild cracking operation used to reduce the viscosity of residual fuel oils and residua. The process, evolved from the older and now obsolete thermal cracking processes, is classed as mild because the thermal reactions are not allowed to proceed to completion.

Residua are sometimes blended with lighter heating oils to produce fuel oils of acceptable viscosity. By reducing the viscosity of the nonvolatile fraction, visbreaking reduces the amount of the more valuable light heating oil that is required for blending to meet the fuel oil specifications. The process is also used to reduce the pour point of a waxy residue. Visbreaking conditions range from 455–510°C and 345–2070 kPa (50–300 psi) at the heating coil outlet. Liquid-phase cracking takes place under these low severity conditions. In addition to the primary product, fuel oil, material in the gas oil and gasoline boiling range is produced. Gas oil can be used as additional feed for catalytic cracking units or as heating oil.

Coking Processes. Coking is a generic term for a series of thermal processes used for the conversion of nonvolatile heavy feedstocks into lighter, distillable products. The feedstock is typically a residuum and the products are gas, naphtha, fuel oil, gas oil, and coke. Gas oil can be the primary product of a coking operation and serves primarily as a feedstock for catalytic cracking units. The coke obtained is usually used as fuel, but specialty uses, such as electrode manufacture and the production of chemicals and metallurgical coke, are also possible, thus increasing the value of the coke.

Delayed coking is a semicontinuous process in which the heated charge is transferred to large soaking, or coking, drums, which provide the residence time needed for the cracking reactions to proceed to completion. The feed to these units is normally a vacuum residuum, although residua from other thermal processes are also used.

Fluid coking is a continuous process that uses the fluidized solids technique to convert atmospheric and vacuum residua to more

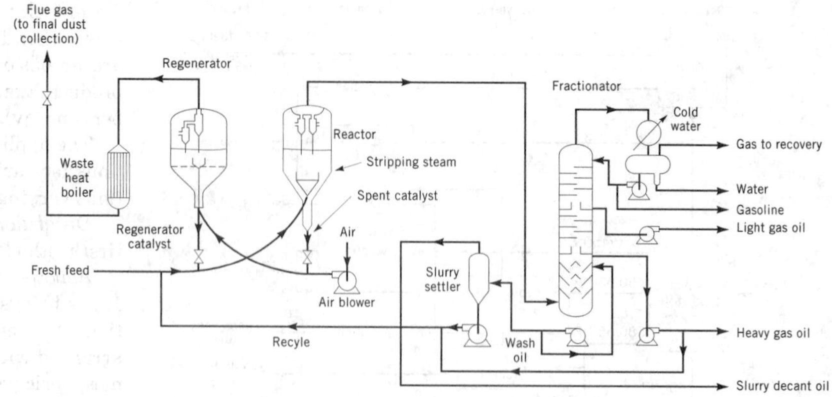

Figure 2. Fluid-bed catalytic cracking with product separation.

valuable products. The residuum is converted to coke and overhead products by being sprayed into a fluidized bed of hot, fine coke particles, which permits the coking reactions to be conducted at higher temperatures and shorter contact times than they can be in delayed coking. Moreover, these conditions result in decreased yields of coke; greater quantities of more valuable liquid product are recovered in the fluid coking process.

Catalytic Cracking. Catalytic cracking (Fig. 2), which has progressively supplanted thermal cracking, is the thermal decomposition of petroleum constituents in the presence of a catalyst. The acid catalysts first used in catalytic cracking were designated low alumina catalysts; amorphous solids composed of approximately 87% silica, SiO_2, and 13% alumina, Al_2O_3. Later, high alumina catalysts containing 25% alumina and 75% silica were used. However, this type of catalyst has largely been replaced by catalysts containing crystalline aluminosilicates (zeolites) or molecular sieves (qv).

The catalyst is employed in bead, pellet, or microspherical form and can be used as a fixed bed, moving bed, or fluid bed. The fixed-bed process was the first process used commercially and employs a static bed of catalyst in several reactors, which allows a continuous flow of feedstock to be maintained. The cycle of operations consists of (*1*) the flow of feedstock through the catalyst bed; (*2*) the discontinuance of feedstock flow and removal of coke from the catalyst by burning; and (*3*) the insertion of the reactor back on-stream. The moving-bed process uses a reaction vessel, in which cracking takes place, and a kiln, in which the spent catalyst is regenerated and catalyst movement between the vessels is provided by various means.

Hydroprocessing. In this group of refining processes, which includes hydrotreating and hydrocracking, the feedstock is heated with hydrogen at high temperature and under pressure. The outcome is the conversion of a variety of feedstocks to a range of products. The purpose of hydroprocessing is (*1*) to improve existing petroleum products or develop new products or uses; (*2*) to convert inferior or low grade materials into valuable products; and (*3*) to transform near-solid residua to liquid fuels. Products are as follows: from naphtha, reformed feedstock and liquefied petroleum gas (LPG); from atmospheric gas oil, diesel and jet fuel, petrochemical feedstock, and naphtha; from vacuum gas oil, catalytic cracker feedstock, kerosene, diesel and jet fuel, naphtha, LPG, and lubricating oil; and from residuum, catalytic cracker and coker feedstock, diesel fuel, and others.

Hydroprocesses for the conversion of petroleum and petroleum products can be classified as destructive or nondestructive. The former (hydrogenolysis and hydrocracking) is characterized by the rupture of carbon–carbon bonds and is accompanied by hydrogen saturation of the fragments to produce lower boiling products. Such treatment requires rather high temperatures and high hydrogen pressures, the latter to minimize coke formation.

Hydrogenolysis is analogous to hydrolysis and ammonolysis, which involve the cleavage of a bond induced by the action of water and ammonia, respectively.

Hydrocracking. Hydrocracking is a catalytic process (> 350°C) in which hydrogenation accompanies cracking. Relatively high pressures, 6,900–21,000 kPa (1000–3000 psi), are employed and the overall result is the conversion of the feedstock to lower boiling products. An attractive feature of hydrocracking is the low yield of gaseous components, such as methane, ethane, and propane, which are less desirable than the gasoline components. Essentially all the initial reactions of catalytic cracking occur, but some of the secondary reactions are inhibited or stopped by the presence of hydrogen.

Hydrotreating. This catalytic process converts sulfur- and/or nitrogen-containing hydrocarbons into low sulfur low nitrogen liquids, hydrogen sulfide, and ammonia. A wide variety of metals are active hydrogenation catalysts; those of most interest are nickel, palladium, platinum, cobalt, and iron. The process temperature affects the rate and the extent of hydrogenation as it does any chemical reaction. Practically every hydrogenation reaction can be reversed by increasing temperature.

Reforming. Thermal reforming, less effective and less economical than catalytic processes, has been largely supplanted. Like thermal reforming, catalytic reforming converts low octane gasolines into high octane gasolines, ie, reformate. Whereas thermal reforming produces reformate having research octane numbers in the 65–80 range, depending on the yield, catalytic reforming produces reformate having octane numbers on the order of 90 –105. Catalytic reforming is conducted in the presence of hydrogen over hydrogenation – dehydrogenation catalysts, eg, in the platforming process. Catalytic reformer feeds are saturated, ie, not olefinic, materials. Catalytic cracker naphtha and hydrocracker naphtha that contains substantial quantities of naphthenes are also suitable reformer feedstocks.

Isomerization. Isomerization is used with the objective of providing additional feedstock for alkylation units (isobutane) or high octane fractions for gasoline blending (pentane and hexane). The latter application is useful in the production of reformulated gasoline by increasing the octane number while converting or removing benzene. Isobutane is also used for the synthesis of methyl *t*-butyl ether (MTBE), an additive that maintains the octane ratings of gasoline in the absence of added lead.

Alkylation. The combination of olefins with paraffins to form higher isoparaffins is termed alkylation (qv). Alkylate is a desirable blendstock because it has a relatively high octane number and serves to dilute the total aromatics content. Reduction of the olefins in gasoline blendstocks by alkylation also reduces tail pipe emissions. In refinery practice, butylenes are routinely alkylated by reaction with isobutane to produce isobutane–octane. In some plants, propylene and/or pentylenes (amylenes) are also alkylated.

Polymerization. In the petroleum industry, polymerization is the process by which olefin gases are converted to higher molecular weight liquid products which may be suitable for gasoline (polymer gasoline) or other liquid fuels.

The feedstock, usually consisting of propylene and butylenes (various isomers of C_4H_8) from cracking processes, may even consist of selective olefins for dimer, trimer, or tetramer production. The molecular size of the product is limited insofar as the reaction is terminated at the dimer or trimer stage. Thus the process is more properly termed oligomerization. The four- to twelve-carbon compounds required as the constituents of liquid fuels are the prime products.

Thermal polymerization is not as effective as catalytic polymerization but has the advantage that it can be used to polymerize saturated materials that cannot be induced to react by catalysts.

Treating

Since the original crude oils contain some sulfur compounds, the resulting products and gasolines also contain sulfur compounds, including hydrogen sulfide, mercaptans, sulfides, disulfides, and thiophenes. The processes used to sweeten, ie, desulfurize, the products depend on the type and amount of the sulfur compounds present and the specifications of the finished gasoline or other stocks.

Hydrotreating is the most widely practiced treating process for all types of petroleum products. However, there are other treating processes suitable for the removal of mercaptans and hydrogen sulfide; such processes are necessary and are performed as part of the product improvement and finishing procedures. For example, mercaptan, removal is achieved by using regenerative solution processes, in which the treatment solutions are regenerated rather than discarded. Hydrogen sulfide, H_2S, is removed by a variety of processes, of which one is a regenerative solution process using aqueous solutions of sodium hydroxide, NaOH, calcium hydroxide, $Ca(OH)_2$, sodium phosphate, Na_3PO_4, and sodium carbonate, Na_2CO_3.

Alkali Treatment. Caustic washing is the treatment of materials, usually products from petroleum refining, with solutions of caustic soda. The process consists of mixing a water solution of lye (sodium hydroxide or caustic soda) with a petroleum fraction. The treatment is carried out as soon as possible after the petroleum fraction is distilled, since contact with air forms free sulfur, which is corrosive and difficult to remove. The lye reacts either with any hydrogen sulfide present to form sodium sulfide, which is soluble in water, or with mercaptans, followed by oxidation, to form the less nocuous disulfides.

Nonregenerative caustic treatment is generally economically applied when the contaminating materials are low in concentration and waste disposal is not a problem. However, the use of nonregenerative systems is on the decline because of the frequently occurring waste disposal problems that arise from environmental considerations and because of the availability of numerous other processes that can effect more complete removal of contaminating materials.

Steam-regenerative caustic treatment is directed toward the removal of mercaptans from such products as gasoline and low boiling solvents (naphtha).

Acid Treatment. The treatment of petroleum products with acids has been in use for a considerable time in the petroleum industry. Various acids such as hydrofluoric acid, hydrochloric acid, nitric acid, and phosphoric acid have been used in addition to the most commonly used sulfuric acid, but in most instances there is little advantage in using any acid other than sulfuric.

Sulfuric acid also has been employed for refining kerosene distillates and lubricating oil stocks. It is used for desulfurizing high boiling fractions of cracked gasoline distillates, for refining paraffinic kerosene, for manufacturing low cost lubricating oils, and for making specialty products such as insecticides, pharmaceutical oils, and insulating oils.

Clay Treatment. The original method of clay treating was to percolate a petroleum fraction through a tower containing coarse clay pellets. The use of clay treating has been superseded by other processes, in particular, by the use of inhibitors. Nevertheless, clay treating is still used as a finishing step in the manufacture of lubricating oils and waxes. The clay removes traces of asphaltic materials and other compounds that give oils and waxes unwanted odors and colors.

Solvent Treatment. Solvent processes can be divided into two main categories, solvent extraction and solvent dewaxing. The solvent used in the extraction processes include propane and cresylic acid, 2,2′-dichlorodiethyl ether, phenol (qv), furfural, sulfur dioxide, benzene, and nitrobenzene. In the dewaxing process, the principal solvents are benzene, methyl ethyl ketone, methyl isobutyl ketone, propane, petroleum naphtha, ethylene dichloride, methylene chloride, sulfur dioxide, and *N*-methylpyrrolidinone.

Gas Processing

The gas streams produced during petroleum refining usually contain many noxious constituents that have an adverse effect on the use of the gas for other purposes, eg, as a fuel or as a petrochemical feedstock, and some degree of cleaning is required.

Gas purification processes fall into three categories: the removal of gaseous impurities, the removal of particulate impurities, and ultrafine cleaning. The extra expense of the last process is only justified by the nature of the subsequent operations or the need to produce a pure gas stream. Several factors must be considered: (*1*) the types and concentrations of contaminants in the gas; (*2*) the degree of contaminant removal desired; (*3*) the selectivity of acid gas removal required; (*4*) the temperature, pressure, volume, and composition of the gas to be processed; (*5*) the carbon dioxide-to-hydrogen sulfide ratio in the gas; and (*6*) the desirability of sulfur recovery on account of process economics or environmental issues.

Process selectivity indicates the preference with which the process removes one acid gas component relative to or in preference to another. One of the principal aspects of refinery gas cleanup is the removal of acid gas constituents, ie, carbon dioxide, CO_2, and hydrogen sulfide, H_2S. Treatment of natural gas to remove the acid gas constituents is most often accomplished by contacting the natural gas with an alkaline solution. The most commonly used treating solutions are aqueous solutions of the ethanolamines or alkali carbonates.

Products

Liquefied Petroleum Gas (LPG). Certain specific hydrocarbons, such as propane, butane, pentane, and their mixtures, exist in the gaseous state under atmospheric ambient conditions but can be converted to the liquid state under conditions of moderate pressure at ambient temperature. This is termed liquefied petroleum gas (LPG). Liquefied petroleum gas (qv) is a refinery product and the individual constituents, or light ends, (Table 2), are produced during a variety of refining operations.

Gasoline. The naphtha fraction from crude oil distillation is ultimately used to make gasoline. The two streams are isolated early in the refining scheme so that each can be refined separately for optimum blending in order to achieve the required specifications (see GASOLINE AND OTHER MOTOR FUELS).

Gasoline is a complex mixture of hydrocarbons that boils below 200°C. The hydrocarbon constituents in this boiling range are those that have four to twelve carbon atoms in their molecular structure. Gasolines can vary widely in composition, even those having the same octane number can be quite different. Because of the differences in composition of the various gasolines, gasoline blending is necessary. The physical process of blending the components is carried out by simultaneously pumping all the components of a gasoline blend into a pipeline that leads to the gasoline storage, and the pumps must be set to deliver automatically the proper proportion of each component. Sophisticated instrumentation is employed to achieve the desired blends.

Aviation gasolines, usually used in light aircraft and older civil aircraft, have narrower boiling ranges (38–170°C) than automobile gasolines (0–200°C).

A condition to keep gasoline engines running smoothly is to allow the fuel–air mixture to start burning at a precise time in the combustion cycle. An electrical spark starts the ignition. The remainder of the fuel–air mixture should be consumed by a flame front moving out from the initial spark. Sometimes a portion of the fuel–air mixture ignites spontaneously instead of waiting for the flame front from the carefully times spark.

Table 2. Constituents of Light Ends

Hydrocarbon	Carbon atoms	Molecular weight	Boiling point, °C	Uses
methane	1	16	−182	fuel gas
ethane	2	30	−89	fuel gas
ethylene	2	28	−104	fuel gas, petrochemicals
propane	3	44	−42	fuel gas, LPG
propylene	3	42	−48	fuel gas, petrochemicals, polymer gasoline
isobutane	4	58	−12	alkylate, motor gasoline
n-butane	4	58	−1	motor gasoline
isobutylene	4	56	−7	synthetic rubber and chemicals, polymer gasoline, alkylate, motor gasoline
butylene-1[a]	4	56	−6	synthetic rubber and chemicals,
butylene-2[a]	4	56	1	alkylate, polymer gasoline, motor gasoline
isopentane	5	72	28	motor and aviation gasolines
n-pentane	5	72	36	motor and aviation gasolines
pentylenes	5	70	30	motor gasolines
isohexane	6	86	61	motor and aviation gasolines
n-hexane	6	86	69	motor and aviation gasolines

[a] Numbers refer to the positions of the double bond; for example, butylene-1 (or butene-1 or but-1-ene) is $CH_3CH_2CH{=}CH_2$ and butylene-2 (or butene-2 or but-2-ene) is $CH_3CH{=}CHCH_3$.

Octane number is a measure of a fuel's ability to avoid knocking.

By defining isooctane as having an octane number of 100 and *n*-heptane as having an octane number of 0, the volumetric percentage of isooctane in heptane that matches the knock from the unknown fuel can be calculated as the octane number of the fuel. For example, 90 vol % isooctane and 10 vol % normal heptane produce a 90-octane-number reference fuel.

Reformulated gasoline is believed to be the answer to many environmental issues that arise from the use of automobiles and there has been a serious effort to produce reformulated gasoline components from a variety of processes. However, reformulation may increase gasoline consumption, when in fact the converse is preferable. It has also been claimed that methyl *t*-butyl ether (MTBE), an additive that maintains the octane ratings of gasoline in the absence of added lead, can reduce the emissions of unburned hydrocarbons during gasoline use through more efficient combustion of the hydrocarbons. However, the ether, MTBE, is believed to have an adverse effect insofar as it appears that aldehyde emissions are increased.

Solvents. Petroleum naphtha is a generic term applied to refined, partly refined, or unrefined petroleum products. Naphthas are prepared by any of several methods, including fractionation of distillates or even crude petroleum, solvent extraction, hydrocracking of distillates, polymerization of unsaturated (olefinic) compounds, and alkylation processes.

The main uses of petroleum naphtha fall into the general areas of solvents (diluents) for paints, etc, dry-cleaning solvents, solvents for cutback asphalt, solvents in rubber industry, and solvents for industrial extraction processes.

Kerosene. Kerosene (kerosine) originated as a straight-run (distilled) petroleum fraction that boiled over the temperature range of 205–260°C. Kerosene is believed to be composed chiefly of hydrocarbons containing twelve to fifteen carbon atoms per molecule. Low proportions of aromatic and unsaturated hydrocarbons are desirable to maintain the lowest possible level of smoke during burning. Although some aromatics may occur within the boiling range assigned to kerosene, excessive amounts can be removed by extraction.

The significance of the total sulfur content of kerosene varies greatly with the type of oil and the use to which it is put. Sulfur content is of great importance when the kerosene to be burned produces sulfur oxides, which are of environmental concern.

Diesel fuel, jet fuel, kerosene (range oil), no. 1 fuel oil, no. 2 fuel oil, and diesel fuel are all popular distillate products coming from the kerosene fraction of petroleum.

The cetane number of a diesel fuel is a number that indicates the ability of a diesel engine fuel to ignite quickly, and burn smoothly, after being injected into the cylinder.

Cetane has a short delay period during ignition and is assigned a cetane number of 100; heptamethylnonane has a long delay period and has a cetane number of 15. Just as the octane number is meaningful for automobile fuels, the cetane number is a means of determining the ignition quality of diesel fuels and is equivalent to the percentage by volume of cetane, in the blend with heptamethylnonane, that matches the ignition quality of the test fuel. The cetane number of diesel fuel usually falls into the 30–60 range; a high cetane number is an indication of the potential for easy starting and smooth operation of the engine.

Fuel Oil. Fuel oil is classified in several ways, but generally into two main types: distillate fuel oil and residual fuel oil. Distillate fuel oil is vaporized and condensed during a distillation process; it has a definite boiling range and does not contain high boiling oils or asphaltic components. A fuel oil that contains any amount of the residue from crude distillation hydrocracking is a residual fuel oil. However, the terms distillate fuel oil and residual fuel oil are losing their significance because fuel oils are made for specific uses and can be either distillates, residuals, or mixtures of the two. The terms domestic fuel oil, diesel fuel oil, and heavy fuel oil are more indicative of the uses of fuel oil.

Domestic fuel oils are those used primarily in the home and include kerosene, stove oil, and furnace fuel oil. Diesel fuel oils are also distillate fuel oils, but residual oils have been successfully used to power marine diesel engines, and mixtures of distillates and residuals have been used on locomotive diesels. Heavy fuel oils include a variety of oils, ranging from distillates to residual oils, that must be heated to 260°C or higher before they can be used. In general, heavy fuel oil consists of residual oil blended with distillate to suit specific needs. Heavy fuel oil includes various industrial oils and, when used to fuel ships, is called bunker oil.

Stove oil is a straight-run (distilled) fraction from crude oil whereas other fuel oils are usually blends of two or more fractions. The straight-run fractions available for blending into fuel oils are heavy naphtha, light and heavy gas oils, and residua. Cracked fractions such as light and heavy gas oils from catalytic cracking, cracking coal tar, and fractionator bottoms from catalytic cracking may also be used as blends to meet the specifications of different fuel oils.

Heavy fuel oil usually contains residuum that is mixed (cut back) to a specified viscosity with gas oils and fractionator bottoms. For some industrial purposes in which flames or flue gases contact the product (eg, ceramics, glass, heat treating, and open hearth furnaces), fuel oils must be blended to low sulfur specifications; low sulfur residues are preferable for these fuels.

Lubricating Oil. Lubricating oils are distinguished from other fractions of crude oil by their usually high (>400°C) boiling point as well as their high viscosity. Lubricating oil may be divided into many categories according to the types of service; however, there are two main groups: oils used in intermittent service, such as motor and aviation oils, and oils designed for continuous service, such as turbine oils.

Wax. Petroleum waxes are of two general types: paraffin wax in distillates and microcrystalline was in residua. The melting point of wax is not directly related to its boiling point because waxes contain hydrocarbons of different chemical structure. Nevertheless, waxes (qv) are graded according to their melting point and oil content. Paraffin wax is a solid crystalline mixture of straight-chain (normal) hydrocarbons ranging from mostly C_{20} to C_{30} and higher. Wax constituents are solid at ordinary temperatures (25°C) whereas petrolatum (petroleum jelly) contains both solid and liquid hydrocarbons.

Asphalt. This is a distillation residuum that can also be produced by propane deasphalting and thereafter modified to meet specifications. For example, asphalt (qv) can be made softer by blending hard asphalt with the extract obtained in the solvent treatment of lubricating oils. On the other hand, soft asphalts can be converted into harder asphalts by oxidation (air blowing).

Road oils are liquid asphalt materials intended for easy application to earth roads. They provide a strong base or a hard surface and maintain a satisfactory passage for light traffic.

Cutback asphalts are mixtures in which hard asphalt has been diluted with a lighter oil to permit application as a liquid without drastic heating. Asphalt can be emulsified with water to permit application without heating. Such emulsions are normally of the oil-in-water type.

Coke. This is the residue left by the destructive distillation (coking) of residua. Petroleum coke is employed for a number of purposes; its principal use is in the manufacture of carbon electrodes for aluminum refining, which requires a high purity carbon that is low in ash and free of sulfur. In addition, coke is employed in the manufacture of carbon brushes, silicon carbide abrasives, structural carbon (eg, pipes and Rashig rings), as well as calcium carbide manufacture from which acetylene is produced. Coke produced from low quality crude oil is mixed with coal and burned as a fuel. Flue gas scrubbing is required. Coke is used in fluidized-bed combustors or gasifiers for power generation.

Petrochemicals

Petrochemicals are those chemicals produced from petroleum or natural gas and can be generally divided into three groups: (*1*) aliphatics, such as butane and butene; (*2*) cycloaliphatics, such as cyclohexane, cyclohexane derivatives, and aromatics (eg, benzene, toluene, xylene, and naphthalene); and (*3*) inorganics, such as sulfur, ammonia, ammonium sulfate, ammonium nitrate, and nitric acid.

Aliphatics. Methane, obtained from crude oil or natural gas, or as a product from various conversion (cracking) processes, is an important source of raw materials for aliphatic petrochemicals (see HYDROCARBONS). Ethane, also available from natural gas and cracking processes, is an important source of ethylene, which, in turn, provides more valuable routes to petrochemical products.

Ethylene (qv), an important olefin, is usually made by cracking gases such as ethane, propane, butane, or a mixture of these as might exist in a refinery's off-gases. Propane is usually converted to propylene by thermal cracking, although some propylene is also available from refinery gas streams. The various butylenes are more commonly obtained from refinery gas streams. Butane dehydrogenation to butylene is known, but is more complex than ethane or propane cracking, and its product distributions are not always favorable. The production of gasoline and other liquid fuels consumes large amounts of butane.

The gaseous constituents produced in a refinery give rise to a host of chemical intermediates that can be used for the manufacture of a wide variety of products. Synthesis gas (carbon monoxide, CO, and hydrogen, H_2) mixtures are also used to produce valuable industrial chemicals.

Cycloaliphatics and Aromatics. Cyclic compounds (cyclohexane and benzene) are also important sources of petrochemical products. Aromatics are in high concentration in the product streams from a catalytic reformer. When aromatics are needed for petrochemical manufacture, they are extracted from the reformer's product using solvents such as glycols (eg, the Udex process) and sulfolane.

The mixed monocyclic aromatics are called BTX as an abbreviation for benzene, toluene, and xylene (see BTX PROCESSING). The benzene and toluene are isolated by distillation, and the isomers of the xylene are separated by superfractionation, fractional crystallization, or adsorption (see XYLENES AND ETHYLBENZENE). Benzene is the starting material for styrene (qv), phenol (qv), and a number of fibers and plastics. Toluene (qv) is used to make a number of chemicals, but most of it is blended into gasoline. Xylene use depends on the isomer: *p*-xylene goes into polyester and *o*-xylene into phthalic anhydride. Both are involved in a wide variety of consumer products.

Benzene, toluene, and xylene are made mostly from catalytic reforming of naphthas. As a gross mixture, these aromatics are the backbone of gasoline blending for high octane numbers. However, there are many chemicals derived from these same aromatics; thus many aromatic petrochemicals have their beginning by selective extraction from naphtha or gas–oil reformate. Benzene and cyclohexane are responsible for products such as nylon and polyester fibers, polystyrene, epoxy resins (qv), phenolic resins (qv), and polyurethanes (see FIBERS; STYRENE PLASTICS; URETHANE POLYMERS).

Inorganics. Of the inorganic chemicals, ammonia is by far the most common. Ammonia is produced by the direct reaction of hydrogen with nitrogen; air is the source of nitrogen: $N_2 + 3\ H_2 \rightarrow 2\ NH_3$. Refinery gases, steam reforming of natural gas (methane) and naphtha streams, and partial oxidation of hydrocarbons or higher molecular weight refinery residual materials (residua, asphalts) are the sources of hydrogen. Ammonia (qv) is used predominantly for the production of ammonium nitrate, NH_4NO_3, as well as other ammonium salts and urea (qv), H_2NCONH_2, which are primary constituents of fertilizers.

Carbon black, also classed as an inorganic petrochemical, is made predominantly by the partial combustion of carbonaceous (organic) material in a limited supply of air. Carbonaceous sources vary from methane to aromatic petroleum oils to coal tar by-products. Carbon black is used primarily for the production of synthetic rubber (see CARBON, CARBON BLACK).

Sulfur, another inorganic petrochemical, is obtained by the oxidation of hydrogen sulfide. Hydrogen sulfide is a constituent of natural gas and also of the majority of refinery gas streams, especially those off-gases from hydrodesulfurization processes. A majority of the sulfur is converted to sulfuric acid for the manufacture of fertilizers and other chemicals. Other uses for sulfur include the production of carbon disulfide, refined sulfur, and pulp and paper industry chemicals.

JAMES SPEIGHT
Consultant

M. R. Gray, *Upgrading Petroleum Residues and Heavy Oils,* Marcel Dekker, Inc., New York, 1994.

J. G. Speight, *The Chemistry and Technology of Petroleum,* 3rd ed., Marcel Dekker, Inc., New York, 1998.

G. W. Mushrush and J. G. Speight, *Petroleum Products; Instability and Incompatibility,* Taylor & Francis, Washington, D.C., 1995.

PETROLEUM RESOURCES

Petroleum resources are distributed widely in the earth's crust as gases, liquids, and solids. The products derived from these naturally occurring resources are used principally as energy sources, although substantial volumes serve as feedstocks in the chemical, plastics, and other industries (see FEEDSTOCKS). Petroleum resources are found as natural gas, as a variety of liquids that are usually classified as normal or heavy crude oils, and as semisolid and solid substances such as asphalt (qv), tar, pitch, gilsonite, and many others. The petroleum resources considered here are those liquid crude oils that can be produced through a conventional wellbore by current primary, secondary, or tertiary (enhanced recovery) production techniques and those unconventional crude oils that may be captured and converted into conventional sources of crude petroleum by advancing production technologies.

No method has been devised to estimate with complete accuracy the amount of crude petroleum that ultimately will be produced from the world's conventional oil and gas fields. Degrees of uncertainty, therefore, should be attached to all such estimates.

Resources represent the total amount (including reserves) of petroleum that exists in a form and amount such that economic extraction is currently or potentially feasible.

Reserves constitute the petroleum that has been discovered and can be produced at the prices and with the technology that exist when the estimate is made.

Proved reserves are estimates of petroleum reserves contained primarily in the drilled portion of fields.

Indicated reserves constitute known petroleum that is currently producible but cannot be estimated accurately enough to qualify as proved.

Inferred reserves are producible, but the assumption of their presence is based on limited physical evidence and considerable geologic extrapolation. This places them on the borderline of being considered undiscovered, and the accuracy of the estimate is very poor.

Subeconomic resources constitute the petroleum in the ground that cannot be produced at present prices and technology but may become producible at some future date at higher prices or by improved technology.

Undiscovered resources are estimated totally by geological reasoning; no evidence through drilling is available.

World Reserves

Most of the large volume of crude petroleum consumed in the world is extracted from only a small fraction of the total number of oil fields discovered. The concentration of crude petroleum in a few large fields is a consequence of the interaction of the geologic processes that create and trap petroleum. Even though commercial quantities of petroleum have been discovered in many localities around the world, there are enormous volume differences in fields present in a single region and in the total volume of petroleum present in different regions.

By far the largest known concentrations of conventional petroleum reserves are in the Middle East, particularly in Saudi Arabia, the United Arab Emirates, and Kuwait. The largest concentration of reserves is in the Burgan field (10.2×10^9 m^3(64.2×10^9 bbl)) in Kuwait, which contains about 68% of that country's reserves. The second largest concentration of reserves is in the Ghawar field (7.4×10^9 m^3(46.5×10^9 bbl)) in Saudi Arabia, which is about 18% of that country's reserves.

The world's reserves of conventional petroleum have increased from 91.7×10^9 m^3(577×10^9 bbl) in 1978 to 160.1×10^9 m^3(1006.8×10^9 bbl) in 1991. These increases are the result of recording of additional reserves in known fields as well as some new field discoveries, principally in the Middle East.

The U.S. proved reserves of crude petroleum and natural gas liquids (NGL) together are 5.1×10^9 m^3(32.1×10^9 bbl) and constituted 3.2% of the world's proved reserves in 1991. The U.S. position in proved reserves has fallen since 1978, when it reported 5.4×10^9 m^3(34×10^9 bbl) and constituted 6% of the world's proved reserves. Canada's proved reserves declined slightly between 1978 and 1991, whereas Mexico reported a large increase in crude petroleum reserves development over the same period, from 4.5×10^9 m^3(28.3×10^9 bbl) to 8.2×10^9 m^3(51.6×10^9 bbl), thereby surpassing the United States and becoming the country with the largest proved reserves in North America.

In South America, Venezuela continues to dominate in the proved reserve and the production categories. Since 1978, reserves of crude petroleum in Venezuela have increased from 2.9×10^9 m^3(18.2×10^9 bbl) to 10.0×10^9 m^3(62.6×10^9 bbl), nearly doubling its share of the world's proved reserves from 3.2 to 6.2%.

The 1991 petroleum resources of Western Europe were almost identical to what they were in 1978 (2.5×10^9 m^3(16.1×10^9 bbl) vs 2.6×10^9 m^3(15.8×10^9 bbl)).

At the end of 1991, the reserves of crude petroleum in Africa were 9.8×10^9 m^3(61.9×10^9 bbl), or only slightly higher than those in 1978, when they were 8.9×10^9 m^3(56.3×10^9 bbl). Algeria, Libya, and Nigeria account for over 80% of these reserves and over 65% of the production from Africa.

The reserves of crude petroleum in Asia stood at 16.5×10^9 m^3(103.8×10^9 bbl) in 1991. This is an increase of 10% since 1978; most of this increase was accounted for by China, India, and Brunei/Malaysia. Levels of proved reserves fell during this period in Australia/New Zealand, Indonesia, and several other Asian countries. In the countries that formerly composed the Soviet Union, reserves decreased slightly (2.5%) between 1978 and 1991, whereas annual production decreased 19%. For many years, the Soviet Union had been the leading producer of crude petroleum in the world, a position it still held in 1991, when it produced 527×10^6 m^3(3.3×10^9 bbl). This level is only slightly higher than production levels in the United States (514×10^6 m^3(3.2×10^9 bbl)) and Saudi Arabia (507×10^6 m^3(3.2×10^9 bbl)).

The proved reserves and levels of production for Japan, Myanmar (formerly Burma), Pakistan, Taiwan, and Thailand are insignificant by world standards. In 1979, the Philippines established the first commercial production in the small offshore South Nido field. This success came after more than 75 years of wildcat drilling in the Philippines. After several additional discoveries, production rose to 0.3×10^6 m^3(1.7×10^6 bbl) in 1991.

U.S. Reserves

Between 1978 and 1991, U.S. proved reserves of crude petroleum decreased by 21.3% from 5.0×10^9 m^3(31.4×10^9 bbl) to 3.9×10^9 m^3(24.7×10^9 bbl). During this same period, NGL reserves increased by 33% from 0.9×10^9 m^3(5.9×10^9 bbl) to 1.2×10^9 m^3(7.5×10^9 bbl). Despite small net additions in several U.S. states, eg, Colorado and New Mexico, the conventional crude petroleum reserves of the United States were depleted rapidly between 1978 and 1991. Even with this decline in proved reserves, the United States was the second largest producer of crude petroleum in the world in 1992 after the former Soviet Union. Although much crude petroleum in the United States in recent years has been credited to the proved inventory through the extension and revision development processes, many of the newer discoveries of conventional hydrocarbon have been natural gas (see GAS, NATURAL).

Ultimate Petroleum Resources of the World

Since the late 1960s, the ultimate amount of crude petroleum in the world that is producible through conventional production techniques has been estimated to be about 350×10^9 m^3(2.2×10^{12} bbl). By the end of 1991, cumulative world production was 103.8×10^9 m^3(652.9×10^9 bbl), and world proved reserves were estimated to be 160.1×10^9 m^3(1006.8×10^9 bbl). Thus, by the end of 1991, 263.9×10^9 m^3(1659.7×10^9 bbl) of crude petroleum had been discovered, which is more than 75% of the estimated 350×10^9 m^3(2200.0×10^9 bbl) of conventional crude petroleum estimated to be ultimately recoverable.

World Petroleum Supply and Consumption

The 1992 world consumption of petroleum was nearly 10.4×10^6 m^3/d(65.4×10^6 bbl/d), which is slightly higher, at 3.6%, than in 1978. In most regions, consumption and production levels are not in balance.

Outlook

Petroleum displaced coal (qv) as the principal source of energy in the United States by 1948 and in the world by 1965. In 1992, petroleum satisfied over 40% of the world's energy needs, while coal filled only 28% of needs, barely ahead of natural gas at 23%. The spectacular growth in consumption of crude petroleum in the world during the middle and late twentieth century is directly attributable to the ease with which petroleum can be discovered, produced, transported, processed, and utilized (see PETROLEUM, ENHANCED OIL RECOVERY). This growth has been so rapid that as much crude petroleum (55.5×10^9 m^3(349.4×10^9 bbl)) was taken from the ground between 1976 and 1992 as was produced during the entire previous 119-yr period (1857–1975). This rapid rate of expansion in production and consumption, coupled with the finiteness of the conventional petroleum resource base, has from time to time led some analysts to conclude that world petroleum production will peak in the near future. Other analysts who examine such data forecast impending global crisis as crude petroleum consumption declines and coal reclaims its former position as the principal source of fossil energy.

Perhaps the biggest contribution that technological advancement in petroleum production will make is bringing large volumes of unconventional petroleum resources, eg, heavy oil and tar sands, into a viable economic realm by lowering the unit cost of production. Compared to the inventory of conventional petroleum reserves and undiscovered resources, the physical inventories of such unconventional petroleum resources are extremely large.

Large unconventional resources of petroleum also occur as extra heavy crude oils in the Orinoco belt, Venezuela, and in oil shale in the western United States. Petroleum resources in the unconventional category, such as tar sands, heavy crude oils, and oil shales, are located mostly in the Western Hemisphere, as opposed to the conventional resources, which are located mostly in the Middle East. Also, the in-place resources of these unconventional resources are about twice as large as the in-place resources of conventional crude petroleum. Although the recovery rates from these resources are low, improving technology may capture increasing volumes of these unconventional petroleum resources, thereby converting them into conventional petroleum resources.

The world will never "run out" of petroleum, simply because there is so much of it in the ground in so many different forms. However, the resources of conventional crude petroleum are finite. These are the petroleum resources that are very inexpensive to produce because they flow to the wellbore either directly or by pumping after the application of standard well completion methods. There is a more or less general agreement among analysts that the size of the inventory of these resources is about 350×10^9 m^3(2200×10^9 bbl); the world is consuming these resources at about 1%/yr. Analysis of the pattern of world energy consumption shows that the world consumption of crude petroleum may gradually increase even with increased efficiency in the use of energy, simply as a result of population growth. However, these developments could be dramatically altered by an increase in the price of energy.

LAWRENCE J. DREW
U.S. Geological Survey

U.S. Geological Survey—U.S. Minerals Management Service, *Estimates of Undiscovered Conventional Oil and Gas Resources in the United States—A Part of the Nation's Energy Endowment,* unnumbered report, U.S. Geological Survey and the U.S. Minerals Management Service, Washington, D.C., 1989, 44 pp.

Energy Information Administration, *U.S. Crude Oil, Natural Gas, and Natural Gas Liquids Reserves, 1991,* DOE/EIA-0216(91), U.S. Department of Energy, Washington, D.C., 1991, 129 pp.

D. Gautier and co-workers, *Estimates of Undiscovered Conventional Oil and Gas Resources in the United States—A Part of the Nation's Energy Endowment,* USGS Research on Energy Resources (V. E. McKelvey Forum), Washington, D.C., Feb. 1995.

M. Grenon, *World Oil Resources—Assessment and Potential for the 21st century,* Preprint for U.S. Geol. Int. Resources Symposium, Reston, Va., Oct. 1979.

PHARMACEUTICALS

Pharmaceuticals are best viewed as drug-containing products in dosage forms. These forms are designed and manufactured to deliver safe and effective therapeutic responses each time administered within appropriate regimens and even after storage under well-documented conditions in scientifically designed packaging for designated time periods (see PACKAGING, COSMETICS AND PHARMACEUTICALS). Thus, pharmaceuticals are actual drug delivery systems (qv).

Various technologies are required to produce drug products. Both federal and state laws and regulations exist in the United States to control the manufacture and distribution of pharmaceuticals. The U.S. drug distribution system is multifaceted including drug usage within the community and hospitals, under long- or short-term home health care or pharmacy practice. Individual pharmaceuticals are covered elsewhere.

In the United States, there is no national qualifying or licensing body for pharmacists. Licensure requirements are promulgated by State boards of pharmacy that administer examinations, issue internship requirements, and oversee the practice of pharmacy. The National Association of Boards of Pharmacy serves the collective needs of the state boards. This organization has no licensure authority. However, it has developed a standardized licensure examination (NABPLEX), which as of this writing (ca 1995) is used by 48 states (see LICENSING).

Several national organizations serve the professional needs of U.S. pharmacists. The American Pharmaceutical Association (APhA), founded in 1852, is composed of the Academy of Pharmaceutical Research and Science, Academy of Pharmaceutical Practice and Management, and the Academy of Students of Pharmacy. Other organizations include the American Society of Health-Systems Pharmacists (ASHP), National Association of Chain Drug Stores (NACDS), and National Association of Retail Druggists (NARD).

The American College of Apothecaries represents pharmacists whose practices can best be described as emphasizing prescription and related products.

The pharmaceutical industry is represented by several organizations. Examples are the Pharmaceutical Research and Manufacturers of America, the Non-Prescription Drug Manufacturers Association, and the National Pharmaceutical Council. The schools and colleges of pharmacy are organized as the American Association of Colleges of Pharmacy, representing both schools and colleges, and faculty members.

Each state has a professional pharmacy organization, some of which are affiliated with the American Pharmaceutical Association. Similarly, state organizations of hospital pharmacists exist in affiliation with the ASHP. Likewise, local or county associations exist in most instances. Each national association publishes a journal as do most state organizations. The *Federal Register* reports proposed and enacted federal regulatory occurrences several times a week. Each state has a similar publication to report its legislation and regulatory developments, eg, *The Pennsylvania Bulletin.*

Drugs and Drug Products

The U.S. Food and Drug Administration (FDA) approved 22 new drugs and one biotech medicine during 1994. These new drug entities had an adjusted average review time of 19.7 months, from filing of the New Drug Application (NDA) at the FDA to time of approval. This was down from the 25.6 months for the 26 new entities approval in 1993. In the total drug development and approval process it takes approximately 12 years for an experimental drug to go from the lab to the medicine chest. Only about 5 in 5000 new chemical entities that enter preclinical (lab and animal studies) testing reach human clinical testing (Phase I, II, and III) and only one of the five tested clinically is approved. On average a pharmaceutical manufacturer invests ~$360 million to get one new drug to the consumer or patient.

In the United States, through the NDA review process, pharmaceutical companies that seek FDA approval for new drug products are assessed user fees by FDA to gain faster approval, by virtue of the U.S. Prescription Drug User Fee Act of 1992. In 1962, amendments to the U.S. Federal Food, Drug and Cosmetic Act promulgated regulations concerning the requirements for premarketing approval by the FDA. This legislation established requirements of proof of both safety and therapeutic efficacy and strict control of human clinical testing, for example, which have extended the time and cost to market a new drug.

The increase in time to bring a new drug to the point of FDA approval, that the 1962 amendments generated, reduced the length of the effectiveness of the patent period. During the same period, the 1960s, availability of generic drug products began to increase significantly. The FDA at that time utilized the Abbreviated New Drug Approval (ANDA) process developed in the 1962 amendments for review and approval of generic drug products that were to be marketed after FDA approval and patent expiration of the originator new drug entity. Some legal questions, however, arose as to the use of the ANDA procedure for generic approval for both pre-1992 and post-1962 new drug approvals (see PATENTS AND TRADE SECRETS).

The world trade agreement, the General Agreement on Tariffs and Trade (GATT), resulted in a U.S. federal law, the Uruguay Round Agreement Act (URAA), that became effective in June 1995. Under this Act, numerous drugs are projected to gain months or even years of additional patent protection, depending on current patent expiration dates. The GATT provides new prescription drugs with 20 years of patent protection from the patent application date.

The 1962 Amendments also mandated a review of safety and therapeutic efficacy for U.S. nonprescription, ie, over-the-counter or proprietary, products. There are an estimated 125,000–300,000 U.S. OTC products covering a variety of sizes, dosage form types, and dosage form strengths. The FDA has increased its approval rate for the switch of prescription drugs to nonprescription status in the 1990s. This procedure has gained impetus as more than 450 OTC products in 1994–1995 used ingredients and dosages only available by prescription in 1974–1975.

The principal OTC pharmaceutical products include cold remedies, vitamins and mineral preparations, antacids, analgesics, topical antibiotics, antifungals and antiseptics, and laxatives. Others include suntan products, ophthalmic solutions, hemorrhoidal products, sleep aids, and dermatological products for treatment of acne, dandruff, insect parasites, burns, dry skin, warts, and foot care products. More recent prescription-to-OTC switches have included hydrocortisone, antihistamine and decongestant products, antifungal agents, and, as of 1995, several histamine H_2-receptor antagonists.

Personnel. A large number of personnel trained in a wide range of special skills are needed for the development of a new drug. Skills include organic synthesis, medical and analytical chemistry, microbiology and immunology, biochemistry, physiology, pharmacology, toxicology, and pathology. Likewise, in the development of safe, stable, and therapeutically effective drug products various physical chemistry principles apply and specialists trained in this phase of development, pharmaceutics, assume such responsibility. These people become involved in the preformulation studies that investigate the properties of the new drug for inclusion in dosage forms, in the scale-up procedures that are needed to transfer dosage form preparation from laboratory batch sizes to manufacture batch sizes, and in the actual manufacture of the product. These specialists work closely with chemical engineers, especially during the scale-up phase.

Concepts and Processes. Contemporary dosage forms are drug delivery systems, designed and manufactured to achieve safe and effective therapeutic responses each time the forms are used as part of an appropriate regimen. Each drug product involves several interrelated concepts that must be considered in its design and manufacture. Examples include the following:

Component/concept	Requirement
drug (active ingredient)	purity, stability, accuracy in measurement
nontherapeutic ingredients (excipients)	needed for safe and effective delivery of the active ingredient
unit process/manufacturing technology	procedures needed to ensure batch-to-batch, dose-to-dose reliability of safe and effective response
packaging/labeling	designed for patient compliance and product stability
quality assurance procedures	to protect the drug product throughout its projected shelf-life
storage	to ensure stability and safety/efficacy

Attention to various physiochemical parameters of the drug moiety, such as particle size, crystalline form, and solubility, is vital to the design of a dosage form, as are its purity and accurate measurement. Nontherapeutic or excipient ingredients are selected to ensure stability (buffers, chelating agents (qv), antioxidants (qv), antimicrobial preservatives), and accuracy and precision of dosage (diluents, vehicles). Various types of excipients are used for specific types of dosage forms in order to permit their manufacture and desired therapeutic performances. Other excipients function as processing aids. Lubricating agents are solids used in tablet compression to lubricate the die-walls and punch faces to prevent sticking, capping, and/or excessive die-wall wear. Polymers find wide excipient use in dosage form design as viscosity-building agents in suspensions and emulsions and in the control of drug release in products prepared to achieve longer (8–12 h) than usual therapeutic periods. Various excipients are used to provide drug palatability for patients, eg, colorants (see COLORANTS FOR FOOD, DRUGS, COSMETICS, AND MEDICAL DEVICES) and flavoring agents (see FLAVORS AND SPICES).

The selection of excipient ingredients is important. These must be both chemically and physically compatible with the drug moiety and cannot negatively affect product stability or therapeutic performance, ie, bioavailability.

The various preparation processes and technologies used in drug product manufacture also can effect product safety, stability, and performance, eg, compression during tablet manufacture. The principal processes used in dosage form manufacture are as follows.

Dosage form types	Processes
liquid solutions	dissolution and filtration
parenterals	sterilization, lyophilization
liquid dispersion (suspensions, emulsions)	dispersion/wetting of solids, homogenization
semisolid dispersions (ointments, creams)	levigation, melting
liquid/semisolid capsules	soft gelatin encapsulation
suppositories	molding
solids (granules, capsules, tablets)	comminution, blending, granulation, compression, coating
aerosols	specialized packaging under pressure
general	heating, cooling, mixing

The therapeutically active drug can be extracted from plant or animal tissue, or be a product of fermentation (qv), as in the case of antibiotics.

Biological characterization includes toxicological studies, dose relationships, routes of administration, identification of side effects, and absorption, distribution, metabolism, and excretion patterns. If the results are still acceptable, product formulation and dosage form are developed.

Application for discovery and product patents must be made early in the process. Appropriate labels are designed and the product is submitted to the FDA for approval to begin human testing in the form of an Investigational New Drug Application (INDA). When such approval is granted, a clinical evaluation is developed which includes general testing for human pharmacology in healthy volunteers; clinical studies for therapeutic safety and efficacy in volunteer patients who are suffering from the disease for which the drug has therapeutic promise; and drug samples are made available to select clinicians for use on large numbers of patients.

Manufacturing, analytical, and quality control procedures are established. Specifications for raw and in-process materials, as well as for final products per USP/NF and in-house standards are also determined. Process and formula validation assures that each technological procedure in manufacture accomplishes its purpose most efficiently, eg, blending times for powdered mixtures in tableting, and that each formula ingredient is present in optimal concentrations. Thus, it serves to ensure process control (qv), reproducibility, and content uniformity.

Stability studies are developed to assure a desirable shelf-life period. These also establish limits of acceptability for impurities and degradation compounds, when present, and determine acceptable storage conditions for raw materials and the manufactured products. Stability studies are thus important to the determination of expiration dates for drug products.

Finally, all data, including the results of the clinical investigation, are collected in a New Drug Application (NDA) and sent to the FDA. Once approved, the new drug goes into production. After manufacturing begins, the new drug products must be monitored in clinical use in the marketplace for reports of untoward reactions. This amounts to post-approval surveillance known as Phase IV. All such reports must be submitted to the FDA in a timely manner.

Bioavailability, Bioequivalence, and Pharmacokinetics. Bioavailability can be defined as the amount and rate of absorption of a drug into the body from an administered drug product. It is affected by the excipient ingredients in the product, the manufacturing technologies employed, and physical and chemical properties of the drug itself, eg, particle size and polymorphic form. Two drug products of the same type, eg, compressed tablets, that contain the same amount of the same drug are pharmaceutical equivalents, but may have different degrees of bioavailability. These are chemical equivalents but are not necessarily bioequivalents. For two pharmaceutically equivalent drug products to be bioequivalent, they must achieve the same plasma concentration in the same amount of time, ie, have equivalent bioavailabilities.

Bioavailability, important to the design and preparation of drug products, can be affected adversely by the selection of excipients and/or the manufacturing processes used. Excessive pressure used in the compression of tablets, for example, could cause a tablet to pass through the gastrointestinal tract with no therapeutic effect.

Pharmacokinetics is the study of how the body affects an administered drug. It measures the kinetic relationships between the absorption, distribution, metabolism, and excretion of a drug. To be a safe and effective drug product, the drug must reach the desired site of therapeutic activity and exist there for the desired time period in the concentration needed to achieve the desired effect. Too little of the drug at such sites yields no positive effect (<MEC); too much (>MTC) leads to toxicity. For intravenous administration there is no absorption factor. Total body elimination includes both metabolic processing and excretion.

In cases of all but intravenous administration, dosage forms must make the active moiety available for absorption, ie, for drug release. This influences the bioavailability and the drug's pharmacokinetic profile. Ideally the drug is made available to the blood for distribution and elimination at a rate equal to those processes. Through technological developments drug product design can achieve release, absorption, and elimination rates resulting in durations of activity of 8–12 hours, ie, prolonged action/controlled release drug products.

Manufacturing

Compressed Tablets. This popular type of dosage form offers convenience, stability, accuracy and precision, and good bioavailability of active ingredients. After the best formulation has been established, compressed tablets can be manufactured at high rates of speed on advanced equipment. Tablets can be made to achieve rapid drug release or to produce delayed, repeat, or prolonged therapeutic action (CONTROLLED RELEASE TECHNOLOGY, PHARMACEUTICAL). Tablets are produced directly by compression of powder blends or granulations, which include a small percentage of fine, particle-sized powders.

Granulation. Granulation methods can be wet or dry. Wet granulation cannot be used for drugs that are sensitive to moisture and heat. The powered drug and diluent are blended with a dispersion of the binder excipient, eg, gelatin, to a consistency that can be screened to 840–1800-μm granules (10–20 mesh). These granules are dried on trays in hot-air ovens or fluid-bed dryers. Dry granulation is used when the drug is not stable under the conditions of wet granulation and when the combined powders of a formulation cannot be compressed directly.

Direct Compression. This process is relatively simple and time saving. All the ingredients are blended and then compressed into the final tablet. This is an excellent method, but encumbered by a number of problems. Not all substances can be compressed directly, necessitating a granulation step. Likewise, the flow properties of many blends of fine, particle-sized powders are not such as to ensure even filling of the die cavities of tablet presses. In addition, air entrapment can occur.

The availability of spray-dried lactose, microcrystalline cellulose, and other excipients allows for the use of granular rather than powdered phases. This eliminates some of the problems of particle segregation according to size (demixing) and even flow to the die. Direct compression eventually may be the preferred method of tablet preparation.

Tablet Press. The main components of a tablet compression machine (press) are the dies, which hold a measured volume of material to be compressed (granulation), the upper punches which exert pressure on the down stroke, and the lower punches which move upward after compaction to eject the tablets from the dies. Mechanical components deliver the necessary pressure. The granulation is fed from a hopper with a feed-frame on rotary-type presses and a feeding shoe on single-punch presses. A smooth and even flow ensures good weight and compression uniformity. Using the proper formulation, demixing in the hopper is minimized.

Compressed tablets that are composed of several layers require specially adapted presses designed with several fed hoppers. For a two-layer tablet, one granulation is first fed to a die and partially compressed into a soft tablet. The second granulation is added, and the total die components then are compressed fully. Such procedures are used when the tablet ingredients may be incompatible, which requires separate granulations. If needed, a layer of inert ingredient, eg, lactose, is inserted between the two.

Layered tablets are also used for a prolonged or sustained therapeutic effect. In this case, one layer disintegrates and dissolves rapidly to provide the initial dosing, whereas the other is designed for controlled release.

Formulation. Compressed tablet formulations contain several types of inert, adjuvant ingredients necessary for proper preparation and therapeutic performance. Tablets designed to be swallowed need diluent, disintegrating, binding (adhesive), and lubricating inert ingredients, whereas troches or lozenges intended to be dissolved slowly in the mouth should not disintegrate quickly, need more binder, and no disintegrant. Lactose or dicalcium phosphate are common diluents, whereas starch and cellulose derivatives are used as disintegrating agents.

Glidants are needed to facilitate the flow of granulation from the hopper. Lubricants ensure the release of the compressed mass from the punch surfaces and the release/ejection of the tablet from the die. Combinations of silicas, corn starch, talc (qv), magnesium stearate, and high molecular weight poly(ethylene glycols) are used. Most lubricants are hydrophobic and may slow down disintegration and drug dissolution.

Colors and flavors increase the elegance and acceptability of the product. Sometimes colors are used for identification.

Effervescent tablets disintegrate by virtue of the chemical reaction occurring in water between component ingredients, such as sodium bicarbonate and citric or tartaric acid, to achieve release of carbon dioxide.

Coating. Sugar or film coatings offer protection from moisture, oxygen, or light and mask unpleasant taste or appearance. Enteric coatings delay the release of active ingredients in the stomach and may prolong the onset of therapeutic activity. The latter are used for drugs that are unstable to gastric pH or enzymes, cause nausea and vomiting or irritation to the stomach, or should be present in high concentration in the intestines, eg, preoperative sterilization of the gut or as anthelmintics. Effectiveness depends on the varying pH patterns of the gastrointestinal tract and the enzymes present for dissolution and aqueous solubility.

Enteric coating is also used for repeat-action tablets, which contain an enteric-coated core tablet and a sugar or film-coated second dose, permitting the administration of two doses simultaneously. The core dose is released several hours after the initial, outer dose.

Some tablets that provide a sustained period (up to 8–12 h) of therapy may be coated during processing. A portion is released first to bring the drug to the desired blood concentration (onset of activity), whereas a sustained-release portion maintains an effective level for a prolonged period of time (duration of activity), eg, by coating erosion or diffusion of drug through it.

A more recent development in tablet coating involves the use of gelation as the coating material to produce geltabs. If a tablet is compressed as a capsule-shaped unit prior to gelatin coating it is called a gelcap.

Capsules. Capsules are made in two types. In hard-gelatin capsules, powders or granules are enclosed in rigid gelatin shells. Soft-gelatin capsules contain glycerol as well as gelatin and maintain plastically even when dried. Hard-gelatin capsules are made in two sections, cap and body, which are then filled, whereas soft-gelatin capsules are formed and filled in succession in one manufacturing procedure. Soft-gelatin capsules are generally filled using nonaqueous solutions, although powders can also be used. Most drug companies buy the hard-gelatin shells from external sources. These are made by dipping precisely tooled pins into controlled solutions of gelatin. A film of gelatin adheres to the pins. Upon drying, the units are trimmed to specified length, removed from the pins, and the cap and body portions are joined. Various colors can be incorporated (see GELATIN).

The formulations of filled, hard-gelatin capsules are generally less complex than those of compressed tablets, and require no binders or disintegrators. Upon swallowing, the capsule shell dissolves quickly and the powder ingredients are available for dissolution. Because no initial disintegration step is needed, bioavailability of drugs in capsule formulations is generally better than that of compressed tablets. The capsules are filled by various high speed machines. Occasionally the pharmacist has to perform this procedure manually.

Prolonged Action/Controlled Release, Orally Administered Solid Dosage Forms. The therapeutic purpose of prolonged action and controlled release solid, oral drug products is to maintain safe and effective concentrations of the drug in the blood for 2–4 times longer than those times achieved using regular compressed tablets or capsules. This is accomplished by releasing one portion of the drug quickly, whereas the remaining portion is released at a rate that approaches the elimination rate. Ideally, the second portion should be released at a zero-order rate to achieve this profile. The technologies used for such controlled release only approach such a rate, but do accomplish the increased therapeutic period. These oral products mainly use diffusion-controlled or dissolution-controlled release profiles. The more recognized technologies used to achieve these methods include ion-exchange (qv) resins, coated micropellets, barrier coatings (see BARRIER POLYMERS), drug embedment in either slowly eroding or plastic matrices, swelling hydrogels of various polymer resins, drug complexation, and osmotic pressure controlled tablets. Other technologies that have been attempted or tested include altered density micropellets, prodrugs, and bioadhesives.

The best drug candidates for incorporation into prolonged action systems are uniformly absorbed throughout the gastrointestinal (GI) tract, have medium (2–8 h) biological half-lives, and are prescribed for chronic maintenance use. Drugs in large doses are difficult to formulate into such products (see CONTROLLED RELEASE TECHNOLOGY, PHARMACEUTICAL).

Liquid Dosage Forms. Simple aqueous solutions, syrups, elixirs, and tinctures are prepared by dissolution of solutes in the appropriate solvent systems. Adjunct formulation ingredients include certified dyes, flavors, sweeteners, and antimicrobial preservatives. These solutions are filtered under pressure, often using selected filtering aid materials. The products are stored in large tanks, ready for filling into containers. Quality control analysis is then performed.

Dosage forms of naturally occurring materials having therapeutic activity are prepared by extractive processes, especially percolation and maceration. Examples of such dosage forms have included certain tinctures, syrups, fluid extracts, and powdered extracts.

Solutions for external or oral use do not require sterilization but generally contain antimicrobial preservatives. Ophthalmic solutions and parenteral solutions require sterilization (see STERILIZATION TECHNIQUES).

For the preparation of suspensions and emulsions, colloid mills and homogenizers, respectively, are used. Ultrasonic mills that utilize vibrating reeds in restricted chambers to reduce the particle size of the dispersed ingredients can also be employed (see COLLOIDS).

Semisolid Dosage Forms. The ingredients that constitute the base of ointments, eg, petrolatum and waxes, are melted together, powdered drug components are added, and the mass stirred with cooling. Generally, the product then is passed through a roller mill to achieve the particle-size range desired for the dispersed solid. Pastes are ointments having relatively large, dispersed solid content, and are prepared similarly.

Creams are semisolid emulsions either water-in-oil (w/o) or oil-in-water (o/w).

Suppositories are semi-rigid, plastic dosage forms are designed to deliver a unit dose of medication to body cavities, ie, rectum, vagina, or urethra. Depending on the base, suppositories either melt (cocoa butter) at body temperature or dissolve (poly(ethylene glycol)s, glycerogelatin) in the fluids of the cavity. They can be used for systemic therapy (rectal suppositories) or for localized treatment. Rectal suppositories are a route of administration in comatose conditions or after gastrointestinal surgery, and for pediatric patients. On a large scale, suppositories are produced by molding.

Parenteral Dosage Forms. The most commonly used forms for drug products designed and manufactured for injection through the skin include those meant for subcutaneous, intramuscular, and intravenous administration.

Intravenous aqueous injections provide an excellent means of achieving a rapid therapeutic response. Parenteral product design, eg, vehicle and other excipient selection, as well as choice of route of administration, can prolong therapeutic activity and increase onset times. Thus, oily solutions, suspensions, or emulsions can be administered by subcutaneous or intramuscular routes to create prolonged effect, ie, depot injection.

Several factors of design and manufacture are of great importance: sterility, absence of pyrogens and foreign particulate matter, and tonicity. The last, when adjusted to the osmotic pressure of body fluids in the case of aqueous solutions, reduces the risk of tissue irritation and pain.

Lyophilization. Lyophilization is essentially a drying technology. Some drugs and biologicals are thermolabile and/or unstable in aqueous solution. Utilization of freeze drying permits the production of granules or powders that can be reconstituted by the addition of water, buffered solution, or mixed hydrophilic solvents just prior to use, eg, certain antibiotic suspensions.

Ophthalmic Dosage Forms. Ophthalmic preparations can be solutions, eg, eye drops, eyewashes, ointments, or aqueous suspensions. They must be sterile and any suspended drug particles must be of a very fine particle size. Solutions must be particle free and isotonic with tears. Thus, the osmotic pressure must equal that of normal saline (0.9% sodium chloride) solution. Hypotonic solutions are adjusted to be isotonic by addition of calculated amounts of tonicity adjusters, eg, sodium chloride, boric acid, or sodium nitrate.

Radiopharmaceuticals. Radioactive isotopes for human use in the diagnosis and treatment of disease states are called radiopharmaceuticals (qv). Whereas the dosage form types used, eg, solutions or injections, are traditional, special handling of these products during compounding, transport, and use is vital. Most are administered intravenously and shortly after preparation. Specialized pharmacies prepare these products overnight and transport them to hospitals for early administration by members of nuclear medicine departments.

Aerosols. Pressurized containers to deliver aerosolized drug products through appropriate systems of valves and actuators have been available since the 1950s (see AEROSOLS). Such dosage forms are used as external applications of lotions and creams, for oral inhalation, or for treatment of the vaginal cavity, eg, contraceptive foams. Aerosols contain two- or three-phase systems, wherein a volatile liquid or admixture of liquids is sealed in a container in equilibrium with a vapor phase (propellant). Upon actuation and delivery of the product, the propellant evaporates quickly, and fine dispersion of the drug settles on the area of application. For aerosol products that need accurate dosing, metered valves are used with the valve chamber being recharged between each actuation or dose.

The popularity of aerosols has been declining. A widely used group of propellants, the fluorinated hydrocarbons, have been restricted in use since it was found that they can harm the environment by reducing the ozone layer of the upper atmosphere (see AIR POLLUTION; ATMOSPHERIC MODELING; OZONE).

Biotechnology and Dosage Forms. In drug development, biotechnology (qv) generally is recognized as a term that identifies those technologies that utilize living organisms in the production and/or alteration of chemical entities that have potential therapeutic activity. Besides the production of pharmacologically or biochemically active moieties, these technologies also have been used to produce food ingredients, vaccines, diagnostic testing reagents, and agricultural products (see FERMENTATION; MEDICAL DIAGNOSTIC REAGENTS; VACCINE TECHNOLOGY).

Packaging. The packaging components of pharmaceutical products are vital to their safe and effective use. Besides serving the patient as a convenient unit of use, the composite package (unit container, labeling, and shipping components) must provide appropriate identification and necessary information for proper use (including warnings and cautions) and preservation of the product's chemical and physical integrity (see PACKAGING, COSMETICS AND PHARMACEUTICALS).

Labeling. Labeling, controlled by FDA regulations, includes not only the affixed labels, but also the package inserts that provide more detailed information. Trade, generic, or common name, dose, number of dose units present, and name and address of manufacturer and distributor are required. For nonprescription products, adequate directions for use are required. Prescription products must bear the phrase, "Caution: Federal law prohibits use without a prescription" on their labels.

All drug labels must include batch or lot numbers. The nature of the drug product may require special cautionary phrases, eg, "store in cool place or refrigerator," "protect from light," and "shake well before using." In the 1990s, labels also carry the expiration date, ie, shelf-life. This information is expected to become mandatory.

Labeling information also includes warnings as to possible side effects, eg, drowsiness, and potential harm if used with other drugs or certain foods (drug–drug or drug–food interactions). Inserts are generally intended for use by physicians or pharmacists and give name and description of the product, mode of administration, dosage regimen, therapeutic indications and contraindications, precautions and side effects, units of supply, and literature citations. All labeling must be approved by the FDA as part of the New Drug Application.

Containers. The USPXXIII–NFXVIII lists container requirements such as well-closed, tight, or light-resistant. Most containers are light-resistant (amber) glass or plastic. The latter is break-resistant and lightweight, which reduces shipping costs and increases safety.

In hospitals and long-term care units, unit-dose packages are used more and more. This system allows better control of the dispensed drugs in institutional settings and precludes the dispensing of larger numbers of doses than needed.

Quality Control and Quality Assurance

Quality control (QC) involves the regular, daily assessment and/or analysis, according to established protocols and standards, of all ingredients, processes, and finished products. Official USP/NF monographs, for example, provide various chemical, physical, and biological tests and specifications for assurance of purity, potency, and stability of component ingredients used to prepare and package drug products. The FDA requires process validation procedures as QC constituents. The FDA also monitors QC standards through the requirements of the Current Good Manufacturing Procedures regulations.

PAUL ZANOWIAK
Temple University

Trends in U.S. Pharmaceutical Sales and R & D: 1990–93 PMA Annual Survey Report, Pharmaceutical Manufacturers Association, Washington, D.C., 1993.

H. C. Ansel and N. G. Popovich, *Pharmaceutical Dosage Forms and Drug Delivery Systems,* 5th ed., Lea & Febiger, Philadelphia, Pa., 1990, pp. 92–133.

G. S. Banker, in G. S. Banker and C. T. Rhodes, eds., *Modern Pharmaceutics,* 2nd ed., Marcel Dekker, New York, 1990, pp. 15–20.

P. Zanowiak, in *Ullmann's Encyclopedia of Industrial Chemistry,* VA19, VCH Verlagsgesellschaft, MbH, Weinheim, Germany, 1991, pp. 241–271.

PHARMACEUTICALS, CHIRAL

Stereoisomers are compounds which have the same molecular formula but differ in the arrangement of their atoms in space. Chiral compounds are compounds which have nonsuperimposable mirror images. Enantiomers are pairs of stereoisomers which are nonsuperimposable mirror images; they possess identical physical and chemical properties within an achiral environment. Stereoisomers other than enantiomers, ie, diastereomers, are identified by distinct physical and chemical properties including melting points, spectral characteristics, and rates of reaction with both chiral and achiral reactants. Enantiomers, however, are only distinguished when in the presence of a homochiral environment such as polarized light, chiral solvents, chiral reagents, or chiral molecules such as biomolecules, eg, nucleic acids (qv), proteins (qv), and carbohydrates (qv). The two molecules in a pair of enantiomers rotate a plane of polarized light with equal intensities, but in opposite directions. The dextrorotatory isomer (+ or *d*) rotates the plane of polarized light clockwise; the levorotatory isomer (− or *l*) rotates the plane of polarized light counterclockwise. An equal mixture of (+) and (−)-enantiomers is a racemic mixture or racemic compound and does not rotate a plane of polarized light. Optical rotation, an intrinsic property of the substance, has no bearing on drug-macromolecule interactions. It is the absolute configuration of the homochiral compound that is important for its interaction with biomolecules.

Absorption, metabolism, and biological activities of organic compounds are influenced by molecular interactions with asymmetric biomolecules. These interactions, which involve hydrophobic, electrostatic, inductive, dipole–dipole, hydrogen bonding, van der Waals forces, steric hindrance, and inclusion complex formation give rise to enantioselective differentiation. Within a series of similar structures, substantial differences in biological effects, molecular mechanism of action, distribution, or metabolic events may be observed. For example, (*R*)-carvone (**1**) has the odor of spearmint whereas (*S*)-carvone (**2**) has the odor of caraway.

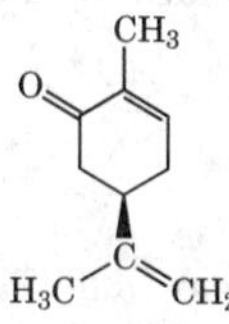

(1)

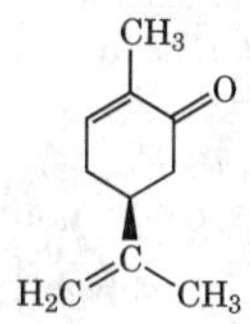

(2)

The amino acids L-leucine, L-phenylalanine, L-tyrosine, and L-tryptophan all taste bitter, whereas their D-enantiomers taste sweet (see AMINO ACIDS).

The importance of optical isomers with regard to biological effect has a long history beginning with the observations of Pasteur. The U.S. FDA requires that both enantiomers of a drug be individually tested when associated toxicities occur near the effective dose of the racemic substance. It has been suggested that the use of racemic drugs in human subjects cannot be justified until both enantiomers are tested thoroughly, both individually and in composite mixtures. Often, side effects of therapeutics are not discovered until after large-scale marketing. The distomer (therapeutically inactive enantiomer) may be at best a nontoxic impurity, but is often associated with dangerous side effects as exemplified by the thalidomide problem.

The thrust toward homochiral drugs by leading researchers and organizations such as the FDA, the rapidly expanding technology of asymmetric syntheses and chiral separations, the decreased side effects found with homochiral drugs, and the potential financial benefits are expected to ensure that the majority of chiral synthetic drugs will, in the future, be available in enantiomerically pure form (see PHARMACEUTICALS).

Background

Nomenclature. Compounds which have tetrahedral atoms having four different substituents are often chiral. These tetrahedral atoms are referred to as stereocenters or stereogenic atoms; the terms asymmetric atom or asymmetric center are considered misnomers. The letters D and L are used to denote the absolute configurations of amino acids and sugars according to Fischer-Rosanoff nomenclature (qv) (see SUGAR). In this system, dextrorotatory glyceraldehyde (**3**) is arbitrarily assigned an absolute configuration of D. In a Fischer projection, the most highly oxidized carbon is placed on top and the last stereocenter determines the absolute configuration L or D. Examples of Fischer projections are shown for D- and L-glyceraldehyde (**4**) and D- (**5**) and L-glucose (**6**) where the arrow denotes the determining stereocenter.

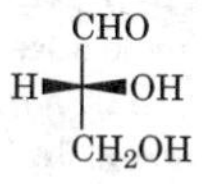

(3)

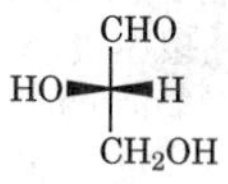

(4)

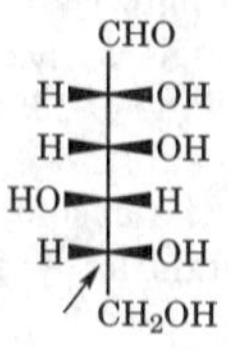

D-glucose [492-62-6]

(5)

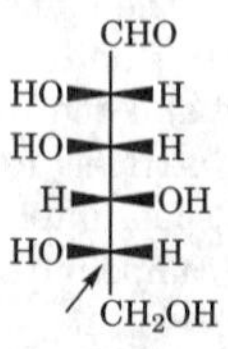

L-glucose [921-60-8]

(6)

If the heteroatom attached to the last stereocenter projects to the right, the compound is of the D-configuration; if the heteroatom points to the left, the compound is of the L-configuration. There is no simple relationship between sign of rotation (d (+) or l (−)) and the absolute configuration D or L. However, optical activity may be related empirically to absolute configuration by observing changes in optical rotation with varying wavelength, ie, optical rotatory dispersion (ord) and circular dichroism (cd).

Because of ambiguities involved in this nomenclature, the Cahn-Ingold-Prelog rules were introduced and are widely used to designate the absolute configuration of stereocenters. Groups attached to the stereogenic atom are assigned priorities according to the atomic number of the atom attached to the stereogenic center. Highest priority is given to the atom with the highest atomic number. The molecule is drawn such that the function of lowest priority (d) is directed away from the viewer:

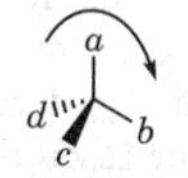

(R)-enantiomer

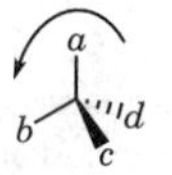

(S)-enantiomer

If the observed order of priority of the remaining three functions ($a > b > c$) is in a clockwise direction, the absolute configuration is designated R (rectus or right); if counterclockwise, the configuration is S (sinister or left). Diastereomers which differ at a single stereocenter are called epimers.

Enantiomeric purity, measured as the enantiomeric excess (ee) of an isomer, is determined by the formula (% major isomer) −(% minor isomer). Thus, if a chiral drug is said to be of 50% ee, the composite mixture contains 75% of one enantiomer and 25% of the other. Enantioselectivity refers to the greater activity of one enantiomer over its mirror image. Enantiospecificity is rarely observed and implies that one enantiomer possesses 100% of the observed activity; in most cases it is more accurate to use the term highly enantioselective. The pharmacologically more active enantiomer is termed the eutomer and the less active enantiomer is referred to as the distomer.

The therapeutic efficacy of a drug is generally measured in terms of ED_{50} or ID_{50} which represent the concentration of drug which produces 50% of the maximum effect or 50% of maximum inhibition.

Role of Homochiral Molecular Building Blocks. Generally L-amino acids and D-sugars are found in biological systems.

The formation of D-amino acids in polypeptides and in monomeric form during processing of proteinaceous foods has raised considerable concern about associated nutritional and toxicity effects. Racemization of amino acids occurs under strongly basic or acidic conditions, which are conditions used in food processing (qv). The presence of aldehydic contaminants enhances the rate of amino acid racemization through formation of stereogenically labile α-imino acids. D-Amino acids may be utilized in a nutritional manner if they are converted to L-amino acids. For the most part, D-amino acids are generally no more toxic than their L-enantiomers. Thus, foods containing proteins with high concentrations of D-amino acid residues may be useful for weight management.

Modeling of Drug-Receptor Interactions. The identification of molecular interactions between drugs and their receptor or enzyme targets and the three-dimensional spatial requirements of the macromolecular binding pocket are important for the rational design of new, more selective, and potent pharmaceuticals. Methods used to explore such interactions include nmr spectroscopy of receptor-ligand complexes, molecular modeling, point mutation analysis, and binding assays of conformationally constrained, stereochemically defined small molecules (see MAGNETIC SPIN RESONANCE; MOLECULAR MODELING). X-ray crystal structure analysis of macromolecule-ligand complexes provides information concerning important molecular interactions which give rise to the observed affinity between the macromolecule and ligand. Computers are used to graphically display calculated crystal structures. Numerous computer programs have been developed and refined which are capable of determining the energy minimized structures of such complexes.

Molecular modeling techniques, although aesthetically pleasing, are far from reliable owing to the large number of associated variables.

Methods for the Preparation of Homochiral Drugs

Resolution Methods. Chiral pharmaceuticals of high enantiomeric purity may be produced by resolution methodologies, asymmetric synthesis, or the use of commercially available optically pure starting materials. Resolution refers to the separation of a racemic mixture. Classical resolutions involve the construction of a diastereomer by reaction of the racemic substrate with an enantiomerically pure compound. The two diastereomers formed possess different physical properties and may be separated by crystallization (qv), chromatography (qv), or distillation (qv). A disadvantage of the use of resolutions is that the best yield obtainable is 50%, which is rarely approached. However, the yield may be improved by repeated racemization of the undesired enantiomer and subsequent resolution of the racemate. Resolutions are commonly used in industrial preparations of homochiral compounds.

Three general methods exist for the resolution of enantiomers by liquid chromatography (qv). Conversion of the enantiomers to diastereomers and subsequent column chromatography on an achiral stationary phase with an achiral eluant represents a classical method of resolution. Diastereomeric derivatization is problematic in that conversion back to the desired enantiomers can result in partial racemization. Direct resolution of the enantiomers without derivatization is performed by use of an achiral stationary phase with a chiral mobile phase or, more commonly, by use of a chiral stationary phase and an achiral mobile phase. Ligand-exchange chromatography, using Cu^{2+} and proline, and chiral ion pair chromatography, which involves use of a chiral counterion in the mobile phase, exemplify the former method. Chromatographic resolution of enantiomers using chiral stationary phases is advantageous over the previously described methods in that no derivatization is required, there is no need to employ expensive mobile phases, and the method does not require complicated product analysis. Stationary phases include cyclodextrins, protein bonded supports, chiral polymers, and the Pirkle type (see POLYMERS). The three-point rule for chiral recognition is used to rationalize the separation of enantiomers by use of a homochiral stationary phase and an achiral eluent. The (*S*)-enantiomer possesses three favorable interactions with the stationary phase and is therefore expected to traverse the column at a slower rate than its (*R*)-enantiomer, which maintains only two favorable interactions.

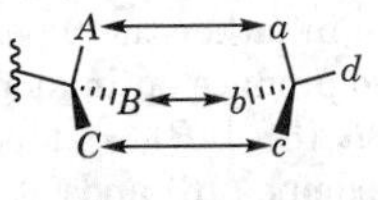

(*S*)-enantiomer

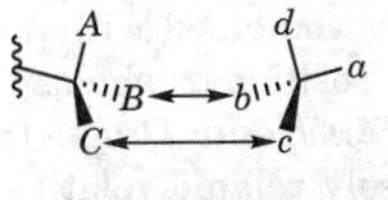

(*R*)-enantiomer

Methods Employing Enantiomerically Pure Starting Materials. A large number of optically pure natural products are commercially available and relatively inexpensive. Amino acids, carbohydrates, and terpenes are some of the homochiral building blocks used routinely in enantiomeric syntheses. The stereocenter(s) in such building blocks are used as either chiral synthons (chirons) or as chiral auxiliaries. Chirons are an inexpensive source of chirality and their use in organic synthesis generally produces enantiomerically pure compounds of known absolute configuration. The commercially available polyether antibiotic monensin, which contains 17 stereocenters, is synthesized by the preparation and coupling of three fragments, each prepared from commercially available optically active starting materials.

Many notable examples of the synthesis of complex natural products from optically pure starting materials have been reported. One synthesis of considerable interest is that of taxol, a potent antitumor agent used clinically. The starting material used in the first total synthesis of taxol is produced in enantiomerically pure form from inexpensive and readily available *l*-camphor.

Asymmetric Induction Methodologies. Asymmetric synthesis is defined as the construction of new chiral centers within a prochiral molecule, with the condition that one optical isomer is formed to a greater extent than the other. The most common type of asymmetric induction involves the conversion of a trigonal carbon atom to a tetrahedral carbon atom by use of a reagent that is biased toward preferential attack from one side or face of the prochiral molecule. Generally, asymmetric induction involves diastereotopic transition states wherein one transition state is favored due to steric and electronic effects which govern the selective formation of one enantiomer. The primary advantage of asymmetric synthesis resides in the stereoselective production, in general, of either enantiomer of a compound; synthesis of both enantiomers is not always possible using chiral synthons or auxiliaries. Asymmetric syntheses avoid the use of inefficient resolutions, and often the reagent or catalyst is recyclable making the synthetic process both material and cost efficient. Asymmetric reduction of ketones, epoxidation of allylic alcohols, hydroboration, hydrogenation, and dihydroxylation reactions, as well as asymmetric cycloaddition reactions, allyl borations, and aldol condensations represent several classes of the numerous enantioselective reactions developed since the 1960s.

Analysis of Synthetic Homochiral Drugs

Determination of Absolute Configuration. X-irradiation of a crystal produces a diffraction pattern from which the relative spatial orientation of the atoms that make up the molecule may be determined. If the crystal is made of homochiral molecules, the absolute configuration of the compound may be deduced.

Chemical conversion of compounds to intermediates of known absolute configuration is a method routinely used to determine absolute configuration. This is necessary because x-ray analysis is not always possible; suitable crystals are required and determination of the absolute configuration of many crystalline molecules cannot be done because of poor resolution. Such poor resolution is usually a function of either molecular instability or the complex nature of the molecule.

ORD and CD also provide a basis by which the absolute configuration of a compound may be correlated with that of a known compound of similar structure by observing changes in degree of rotation with wavelength.

Determination of Enantiomeric Purity. In order to analyze the biological properties of a single enantiomer, the optical purity of the compound should be enantiomerically pure, ie, 100% ee. Contrasting reports on the differences in pharmacological activity of single enantiomers, as well as the misinterpretation of data, are often a result of unknowingly testing enantiomerically impure material. The oldest and perhaps easiest method for determining optical purity is by measuring optical rotation and comparing the value with that reported for the enantiopure compound. There are several drawbacks to this method. The assumption must be made that the reported literature value is without error, and truly represents the optically pure compound. Numerous examples exist in which unambiguous methods, ie, chiral gc, hplc, nmr, for determination of optical purity reveal that the previously reported values for optically pure compounds were in error. Variables such as temperature, solvent, concentration, purity of the compound, type of cell, and even differences between polarimeters employed in the measurement influence the observed degree of rotation. Therefore, polarimetry measurements for determination of optical purity deviate by at least $\pm 4\%$.

^{1}H-nmr is commonly used to determine enantiomeric purity and is reliable to above 98% ee.

Chiral Pharmaceuticals

Enantiomeric Pairs. Enantioselective differences in absorption, metabolism, clearance, drug–macromolecule binding affinity, and other factors, which culminate in the observed enantioselective efficacy of chiral drugs, are considered.

Antihypertensive Agents. Hypertension (high blood pressure) is a significant risk factor for cardiovascular diseases such as angina, heart attacks, and strokes. β-Adrenoceptor (adrenergic nervous system receptors of the β-type) antagonists (β-blockers), calcium channel blockers, angiotensin-converting enzyme (ACE) inhibitors, and potassium channel activators (KCAs) are among the numerous classes of drugs developed to control hypertension (see CARDIOVASCULAR AGENTS; ENZYME INHIBITORS; NEUROREGULATORS). β-Adrenoceptor antagonists exemplified by the phenoxypropanolamine derivatives propranolol and alprenolol and or the phenethanolamine drugs such as sotalol and require for activity both the ethanolamine portion and an aromatic ring. Furthermore, the correct spatial arrangement of the phenyl, ethylamine, and hydroxyl moieties is critical for β-blockade.

It has been demonstrated that the β_1-selectivity is due to the para-substituents of β-adrenoceptors. In contrast, (−)-erythro-isoetharine, a bronchodilator, is 80 times more selective for β_2-adrenergic receptors than for β_1-receptors. Isoetharine contains an α-alkyl substituent, thus producing four isomeric compounds. The (−)-erythro isomer is 100-fold more active than the (−)-threo isomer and has more than 500 times the activity of either of the (+)-isomers and in blocking electrically stimulated spasms. In general, introduction of α-alkyl substituents on both β-blockers and agonists provides diastereomers with increased β_2-selectivity, but often with compromised potency.

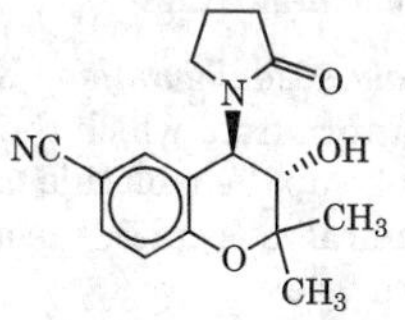

(7)

Cromakalim (**7**) is a potassium channel activator commonly used as an antihypertensive agent. The rationale for the design of cromakalim is based on β-blockers such as propranolol and atenolol. Conformational restriction of the propanolamine side chain as observed in the cromakalim chroman nucleus provides compounds with desired antihypertensive activity free of the side effects commonly associated with β-blockers. Enantiomerically pure cromakalim is produced by resolution of the diastereomeric (*S*)-α-methylbenzylcarbamate derivatives. X-ray crystallographic analysis of this diastereomer provides the absolute stereochemistry of cromakalim. Biological activity resides primarily in the (−)-(3*S*,4*R*)-enantiomer. In spontaneously hypertensive rats, the (−)-(3*S*,4*R*)-enantiomer, at dosages of 0.3 mg/kg, lowers the systolic pressure 47%, whereas the (+)-(3*R*,4*S*)-enantiomer only decreases the systolic pressure by 14% at a dose of 3.0 mg/kg.

Nonsteroidal Antiinflammatory Drugs. Nonsteroidal antiinflammatory drugs (NSAIDs) include, among the numerous agents of this class, aspirin (acetylsalicylic acid), the arylacetic acids indomethacin and sulindac, and the arylpropionic acids, (*S*)-(**8**) and (*R*)-(**9**) ibuprofen, (*S*)-(**10**) and (*R*)-(**11**), flurbiprofen naproxen, and fenoprofen (see ANALGESICS, ANTIPYRETICS, AND ANTIINFLAMMATORY AGENTS; SALICYLIC ACID AND RELATED COMPOUNDS).

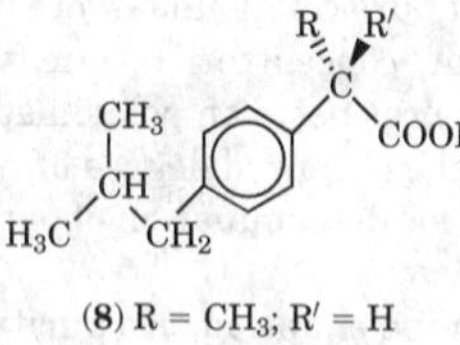

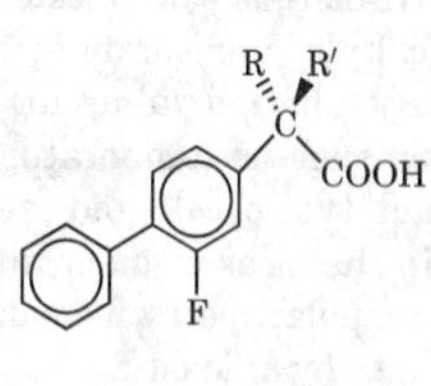

(**8**) R = CH_3; R′ = H
(**9**) R = H; R′ = CH_3

(**10**) R = CH_3; R′ = H
(**11**) R = H; R′ = CH_3

Although the arylpropionic acids contain a stereogenic center they are generally marketed as racemic mixtures. The only exception is naproxen, which is marketed as its (*S*)-enantiomer. NSAIDs produce their antiinflammatory effects by inhibiting cyclooxygenase (COX), the enzyme which catalyzes the first transformation in the biosynthetic conversion of arachidonic acid to the 20 carbon prostaglandins.

CNS Depressant Drugs. Central nervous system (CNS) depressant drugs including antianxiety agents (benzodiazepines), sedative-hypnotics, general anesthetics, and certain spasticity agents all demonstrate high degrees of enantioselective activity (see HYPNOTICS, SEDATIVES, ANTICONVULSANTS, AND ANXIOLYTICS; PSYCHOPHARMACOLOGICAL AGENTS). Barbiturates are commonly prescribed for their sedative-hypnotic activities. In general, the (*S*)-(−)-enantiomers possess CNS depressant activities, whereas the (*R*)-(+)-isomers *R*)-(+)-isomers often produce an excitatory effect. In humans, (*R*)-(+)-pentobarbital is found bound to human plasma proteins to a lesser extent than the (*S*)-(−)-isomer (36.6% free vs 26.5% free) and is subsequently cleared 14% faster. This increased rate of clearance is not sufficient to account for the two- to threefold greater duration of action of (*S*)-(−)-pentobarbital, and suggests that the difference in activity between the enantiomers is due to the pharmacodynamics of the more potent (*S*)-isomer. (*S*)-(+)-Hexobarbital, the eutomer, is eliminated about 2.5 times more slowly than the inactive (*R*)-(−)-isomer, a result of differences in hepatic metabolism. Diazepam (**12**), an achiral benzodiazepine, undergoes stereoselective metabolism to (*S*)-(+)-oxazepam (**13**) in the liver. (*S*)-(+)-Oxazepam produces antianxiety effects to a greater degree than the mirror image isomer (**14**).

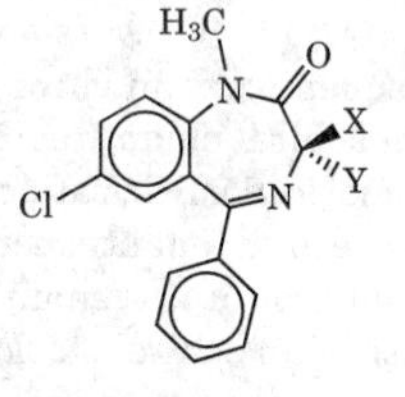

(**12**) X, Y = H
(**13**) X = H; Y = OH
(**14**) X = OH; Y = H

Antibiotic and Antimicrobial Drugs. The antimicrobial agents flumequine and methylflumequine (S-25930) effectively eliminate a number of microbial pathogens via inhibition of the topoisomerase II enzyme of c-DNA containing bacteria (see ANTIBACTERIAL AGENTS, SYNTHETIC). The (*S*)-enantiomers of both drugs are much more potent than the (*R*)-enantiomers. The potent analogue (*S*)-(−)-ofloxacin is 8–125 times more potent than its enantiomer although it is sold only as the racemate. In humans the disposition of (*R*)- and (*S*)-enantiomers of ofloxacin is stereoselective due to differences in renal clearance rates. This difference, however, does not fully explain the large enantioselective difference in antibacterial potency. β-Lactam antibiotics (see ANTIBIOTICS, B-LACTAMS), such as the penicillins and cephalosporins, require the (3*S*,5*R*,6*R*)-configuration of the β-lactam functionality combined with a D-amine in either the 6-position (penicillins) or 7-position (cephalosporins) to produce optimal activity.

Opioid Analgesic Drugs. (5*R*,6*S*,9*R*,13*S*,14*R*)-(−)-Morphine (**15**) and its closely related relatives (−)-codeine (**16**) and (−)-heroin (**17**) are potent analgesics, while their (+)-isomers possess no analgesic effects. α-Dextropropoxyphene (DARVON) (**18**) is a marketed analgesic whereas its enantiomer, α-levopropoxyphene (NOVRAD) (**19**) is sold as an antitussive devoid of analgesic activity (see EXPECTORANTS, ANTITUSSIVES, AND RELATED AGENTS). These analgesics produce their biological effects via stimulation of the opioid receptor subclasses mu-, delta-, kappa-, and sigma.

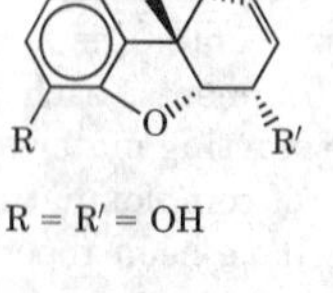

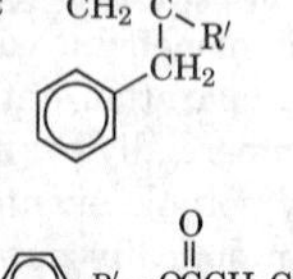

(**15**) R = R′ = OH

(**16**) R = —OCH_3; R′ = OH

(**17**) R, R′ = —$\overset{O}{\overset{\|}{C}}$O—$CH_3$ (printed as —OC(=O)—CH_3)

(**18**) R = —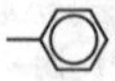; R′ = $O\overset{O}{\overset{\|}{C}}CH_2CH_3$;

(**19**) R = $O\overset{O}{\overset{\|}{C}}CH_2CH_3$; R′ = —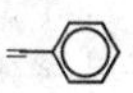

Anticoagulant Drugs. Warfarin, a potent anticoagulant, was first isolated from spoiled clover hay and identified as the agent responsible for the hemorrhagic symptoms associated with the death of livestock in the 1930s. This functionalized coumarin derivative exerts its effect via competitive inhibition of vitamin K-dependent carboxylation of blood clotting factors. Warfarin is generally administered as the racemate, even though (*S*)-warfarin is fivefold more active than (*R*)-warfarin in both rats and humans. The (*S*)-enantiomer is eliminated at a higher rate than its antipode in humans, but in rats, (*R*)-warfarin is more rapidly eliminated.

Neurotransmitters. Histamine receptors are found in at least two subtypes designated H_1 and H_2 (see HISTAMINE AND HISTAMINE ANTAGONISTS). H_1-receptor antagonists produce vasoconstriction, while H_2-receptor antagonists inhibit gastric secretion. Neobenodine (**20**) and (**21**), the chiral *p*-methylphenyl analogue of benadryl (**22**), is an antihistamine marketed as the racemate. The (*R*)-(+)-isomer (**20**) is 65 times more potent than its (−)-enantiomer (**21**) when tested in guinea pig ileum. Chlorpheniramine (**23**) and (**24**) is also an enantioselective H_1-antagonist, wherein the (*S*)-enantiomer (**23**) is most potent. It has been demonstrated that the more potent enantiomer of diphenhydramine and pheniramine drugs is the one in which the aryl moiety, alkylamine group, and the *p*-substituted aryl functionality occur in a clockwise orientation.

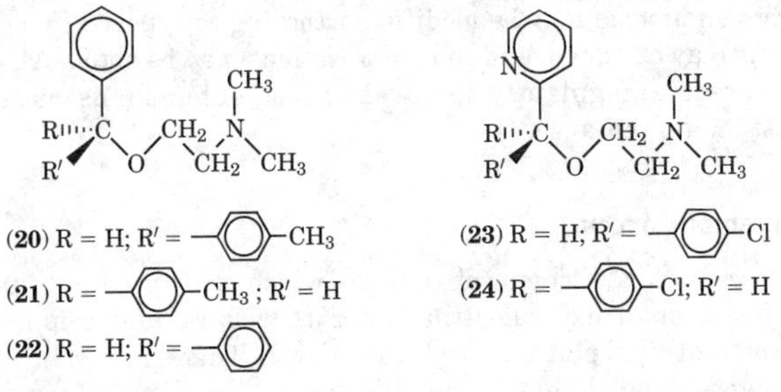

Antineoplastic Drugs. Cyclophosphamide (**25**) produces antineoplastic effects (see CHEMOTHERAPEUTICS, ANTICANCER) via biochemical conversion to a highly reactive phosphoramide mustard (**26**); it is chiral owing to the tetrahedral phosphorus atom. The therapeutic index of the (*S*)-(−)-cyclophosphamide (**25**) is twice that of the (+)-enantiomer due to increased antitumor activity; the enantiomers are equally toxic. The effectiveness of the DNA intercalator drugs adriamycin (**27**) and daunomycin (**28**) is affected by changes in stereochemistry within the aglycon portions of these compounds. Inversion of the carbohydrate C-1 stereocenter provides compounds without activity. The carbohydrate C-4 epimer of adriamycin, epirubicin, is as potent as its parent molecule, but is significantly less toxic. (*R*)-3-Ethyl-3(4-pyridyl)piperidine-2,6-dione (**29**), useful in the treatment of certain breast cancers, is a 20-fold more potent aromatase inhibitor (IC_{50} = 10 μM) than is its (*S*)-enantiomer.

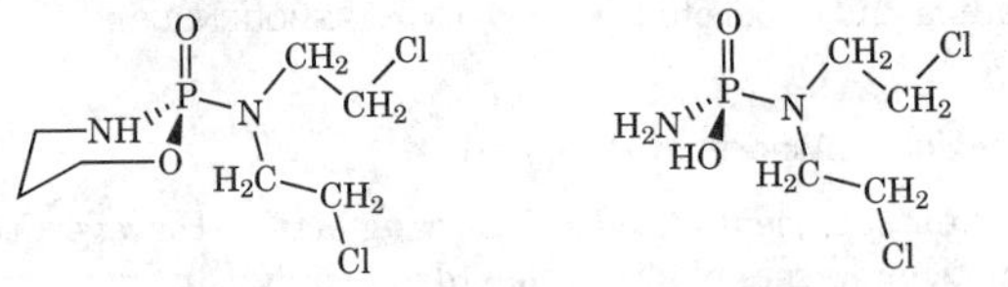

(25) **(26)**

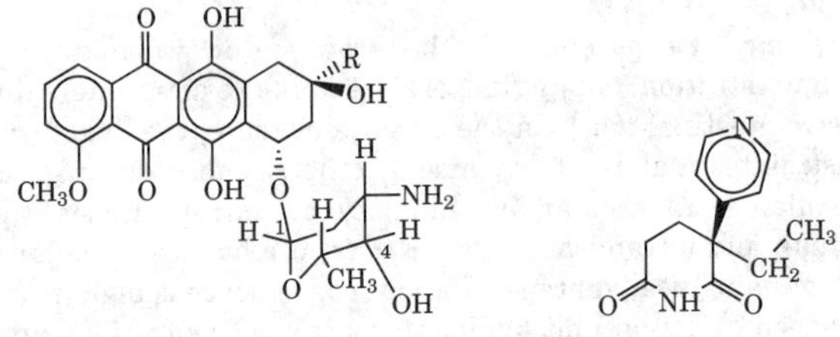

(27) R = —COCH3
(28) R = —COCH2OH
(29)

Peptidomimetics. Many drugs mimic natural small peptides. For example, morphine **(15)** is believed to be a natural peptidomimetic for the enkephalins. Similarly, FK-506 mimics the binding of peptidal FK-506 to the intracellular receptor, FKBP12. Numerous small endogenous peptides have been characterized which possess potent cellular signaling and homeostatic regulating activities. The regulation of glycolysis, growth, mitosis, and apoptosis, as well as the maintenance of blood pressure and the natural relief of pain, exemplify a few of the regulatory actions of peptide hormones (qv). Exogenous control, through the use of synthetic compounds, of the activities of these regulatory elements is highly desirable as demonstrated by the use of drugs such as the ACE inhibitor captopril. The design of peptidomimetics is complicated owing to the flexibility and stereochemical complexity of such hormones. Numerous small peptides have been synthesized which possess tremendous enzyme inhibitory and receptor binding activities *in vitro*; HIV protease inhibitors are one example. Unfortunately, the use of such peptides *in vivo* generally is not successful, as peptidase enzymes rapidly degrade synthetic peptides.

Several methods are being studied to enhance the stability of peptide mimics and improve their stereochemical similarity to the endogenous peptides.

Economic Aspects

Drugs classified as either natural or semisynthetic in origin accounted for ~22% of the market share in 1991. Nearly 94% of the agents are chiral compounds and are sold as single enantiomers. Chiral synthetic drugs make up 38% of the market share and 43% of these are sold as single enantiomers, a two- to threefold increase since 1981. Achiral or symmetrical synthetic drugs make up 40% of the drug market. The vast number of marketed racemic drugs are being reinvestigated and newer pharmacological data as well as production technology are being patented.

A steady increase in the number of homochiral drugs on world markets has created an increased demand for enantiomerically pure intermediates as well as for enantioselective technologies. Many pharmaceutical companies are pursuing new financial opportunities and gaining improved bargaining positions by producing patent protected and more expensive enantiomerically pure drugs from unprotected racemic pharmaceuticals.

DONALD T. WITIAK
ALLEN T. HOPPER
University of Wisconsin-Madison

R. J. Ott and K. M. Giacomini, in I. W. Wainer, ed., *Drug Stereochemistry Analytical Methods and Pharmacology Second Edition, Revised and Expanded,* Marcel Dekker, Inc., New York, 1993, pp. 281–314.

S. Levin and S. Abu-Lafi, in P. R. Brown and E. Grushka, eds., *Advances in Chromatography,* Vol. 33, Marcel Dekker, Inc., New York, 1993, pp. 233–266.

M. Hyneck, J. Dent, and J. B. Hook, in C. Brown, ed., *Chirality in Drug Design and Synthesis,* Academic Press, Inc., San Diego, Calif., 1990, pp. 1–28.

R. A. Sheldon, *Chirotechnology: Industrial Synthesis of Optically Active Compounds,* Marcel Dekker, Inc., New York, 1993, pp. 271–341.

PHARMACEUTICALS, CONTROLLED RELEASE. See CONTROLLED RELEASE TECHNOLOGY; DRUG DELIVERY SYSTEMS.

PHARMACODYNAMICS

Pharmacodynamics is the study of drug action primarily in terms of drug structure, site of action, and the biochemical and physiological consequences of the drug action. The availability of a drug at its site of action is determined by several processes, including absorption, metabolism, distribution, and excretion. These processes constitute the pharmacokinetic aspects of drug action. The onset, intensity, and duration of drug action are determined by these factors as well as by the availability of the drug at its receptor site(s) and the events initiated by receptor activation (see DRUG DELIVERY SYSTEMS).

Both pharmacokinetic and pharmacodynamic processes are involved in mediating nonconstant expressions of drug action. Thus, resistance to the actions of a drug, eg, in the development of antibiotic-resistant bacteria or of barbiturate tolerance, can arise from changes in drug metabolism and/or alterations in the receptor target site. Factors controlling drug resistance may be whole-body, cellular, or individual events. Decreased absorption, increased metabolism, or increased elimination reduce circulating drug levels and affect the whole body. Increased drug metabolism, increased concentration of an agent that antagonizes drug action, decreased affinity or concentration of a drug receptor, and depletion of an agent that mediates drug action are examples of cellular events; and genetic factors controlling metabolism, receptor alterations, and disease states are examples of individual events. Individual variation in the susceptibility to a particular drug or class of drugs also may arise from genetically based pharmacokinetic factors as well as from specific receptor-linked changes.

For a large number of drugs, including neurotransmitters, peptide and protein hormones (qv), and their analogues and antagonists, the cell membrane is the principal locus of action. Concepts of cell membrane structure are derived from the original Davson-Danielli lipid bilayer hypothesis. More specifically, the membrane is viewed as a dynamic fluid mosaic or a matrix of fluid bilayer in which there are asymmetrically inserted proteins (qv) and glycoproteins. Phospholipids and proteins diffuse laterally and the resultant protein-protein communication is of considerable importance to the understanding of membrane-receptor function. Despite the dynamic nature of the membrane and the absence of global organization, local organization is possible through the local assembly of individual protein components and the attachment of membrane proteins to the subcellular structure of contractile proteins. However, the cell membrane is not the site of action of all drugs. A number of drugs, including steroid and thyroid hormones, exert their effects intracellularly at the level of the genetic material as well as at the plasma membrane (see STEROIDS; THYROID AND ANTITHYROID PREPARATIONS). Other agents, including polypeptide growth factors, exert their effects not only at the plasma membrane through tyrosine kinase receptors, but also on cell growth and differentiation at the genetic level.

Drug Discovery and Regulation

In the United States and elsewhere, the introduction of a new drug is subject to a sequence of well-defined stages of development and approval. Each stage involves either scientific testing or submission and preparation of data and analysis review (Fig. 1).

An investigational new drug (IND) application usually initiates the process for drug approval. The IND derives from the concept that

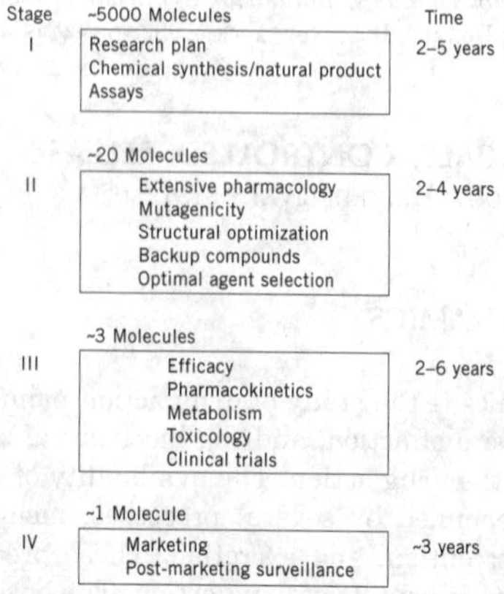

Figure 1. Pathway for drug development.

a specific molecule or molecules may have a particular therapeutic benefit. Preclinical data are analyzed to determine the implications of such molecules for human pharmacology, chemical composition, manufacturing processes, and the protocols for subsequent clinical work. Clinical trials are usually carried out in at least three phases. Phase one involves a small number of individuals and is designed to find information about basic safety and response issues. In phase two studies, the drug is employed on a larger number of individuals (100–200) who suffer from the condition that the drug is designed to treat. Phase three studies involve a much larger group of patients and are designed to assess safety, efficacy, and dosage regimens in a broad range of patients across lines of age, race, and gender. Phase three studies may involve several thousand patients and be carried out at several sites.

New drug application (NDA) is the process through which the U.S. Food and Drug Administration (FDA) authorizes the marketing of a new drug. In the NDA, the data are intended to demonstrate the safety and efficacy of the drug in its intended application. After approval, the drug becomes available to the public. Subsequently, dosage amounts and forms may be modified according to experience, new indications may be added, and contraindications may be noted. All of the changes require regulatory approval. A drug in human use is subject to constant surveillance.

The Receptor Concept

Drug receptors are chemical entities which are typically, but not exclusively, small molecules that interact with cellular components, frequently at the plasma membrane level. There are many types of receptors; heat, light, immune, hormone, ion channel, toxin, and virus are but a few that can excite a cell. The receptor concept can be applied generally to signal recognition processes where a chemical or physical signal is recognized. This recognition is translated into response and the process can be seen as a flow of information.

Elucidation of the structural requirements for drug interaction at the recognition site is by the study of structure-activity relationships (SAR), in which, according to a specific biologic response, the effects of systematic molecular modification of a parent drug structure are determined. Such studies have permitted the classification of discrete classes of pharmacological receptors.

The demonstration of the existence of strictly defined SARs, which is perhaps the most important criterion of drug action at a specific receptor site, has made possible the most important pharmacologic discoveries. For example, the analgesic actions of morphine and related agents, which are indicative of specific receptors, led to the discovery of endogenous opiate peptides, ie, the leucine and methionine enkephalins and endorphins (see OPIOIDS, ENDOGENOUS).

Pharmacokinetic Aspects of Drug Action

The receptor represents the locus of drug action. However, the pharmacokinetic processes of absorption (drug entry), distribution, metabolism, and excretion play principal roles in determining *in vivo* time courses and concentrations of drugs and thus modify actions initiated at receptors.

Drug Entry. Drugs enter the body by one of two routes. In enteral administration (sublingual, oral, rectal), the drug enters directly the gastrointestinal tract. In the parenteral route, the drug bypasses the gastrointestinal tract by, among others, subcutaneous (sc), intramuscular, intravascular (iv), inhalational, intraperitoneal (ip), intravaginal, and intranasal routes. Each route has a particular set of advantages and disadvantages. Patient convenience is high in the oral route; speed of action and ability to control concentrations are high in the iv route; and nonoral routes are best for unstable or insoluble drugs.

In light of the recognized importance of achieving stable, reproducible plasma concentrations of drugs, particular attention is given to pathways and devices, including sustained-release formulations, pumps, and transdermal entry processes that ensure such properties (see CONTROLLED RELEASE TECHNOLOGY, PHARMACEUTICAL).

Drug Distribution. After administration, a drug may be distributed either generally or selectively in the body. The distribution pattern depends on many factors, including the pattern and time-course of blood flow, diffusion of drugs into tissues, binding of drugs to plasma proteins and cellular compartments, and elimination kinetics and mechanisms.

Drug Metabolism. Generally, metabolism (biotransformation) of drugs increases their water solubility as well as the rate and ease of elimination, but reduces their volume of distribution. Many drug-metabolizing pathways have arisen during evolution to deal with foreign compounds present in food materials. Although metabolism generally leads to more polar and less active compounds, there are exceptions. Metabolic pathways have also been exploited to design prodrugs, materials that are converted to active species through biotransformation.

Biotransformation reactions can be classified as phase I and phase II. In phase I reactions, drugs are converted to product by processes of functionalization, including oxidation, reduction, dealkylation, and hydrolysis. Phase II or synthetic reactions involve coupling the drug or its polar metabolite to endogenous substrates and include methylation, acetylation, and glucuronidation (Table 1).

The biotransforming pathways are subject to manipulation and modification in a variety of ways. Drug metabolism also depends on age and sex. Drug metabolism may also produce toxic materials.

Drug Elimination. Drugs are removed from their sites of action through metabolism, storage, and excretion. These processes are not necessarily independent and drugs are frequently metabolized prior to excretion. Indeed, for lipophilic drugs this is virtually a necessity. Drugs are excreted via the kidneys, biliary systems, intestines, and lungs.

The process of reabsorption depends on the lipophilic-hydrophilic balance of the molecule. Charged and ionized molecules are reabsorbed slowly or not at all. Reabsorption of acidic and basic metabolites is pH-dependent, an important property in detoxification processes in drug poisoning. Both passive and active carrier-mediated mechanisms contribute to tubular drug reabsorption.

Clinical Pharmacokinetics. Clinical pharmacokinetics attempts to define the relationship between drug concentration and therapeutic response. The underlying assumption is that response is proportional to drug concentration at the site of action. This concentration is dependent on many factors that are frequently pharmacokinetic determinants. The most important factors are defined as clearance, bioavailability, and volume of distribution.

Clearance, *CL*, is defined by $DL = CL{\cdot}C_{SS}$, where *DR* represents dosing rate and C_{SS} the steady-state concentration of the drug.

Once the steady-state concentration is known, the rate of drug clearance determines how frequently the drug must be administered. Because most drug elimination systems do not achieve saturation under therapeutic dosing regimens, clearance is independent of plasma concentration of the drug.

The half-life, $t_{1/2}$, for a drug in plasma, ie, the time it takes for the concentration of a drug to be reduced by 50%, is determined by both volume of distribution, *V*, and clearance: $t_{1/2} = 0.693{\cdot}V/CL$. The bioavailability of a drug can be defined as the fraction of a dose, *F*, that reaches the systemic circulation. When $F < 1$, $F{\cdot}DR = CL{\cdot}C_{SS}$.

Table 1. Biotransformation Reactions

Pathways	Reactions types	Examples
	Phase I reactions	
oxidative	aliphatic and aromatic oxidation	phenobarbital, phenytoin
	N- and *O*-dealkylation	desipramine, phenacetin
	N-oxidation	guanethidine
	oxidative deamination	amphetamine
	desulfuration	thiobarbitol
	dehalogenation	chloroform
hydrolytic	esters and amides	procaine, lidocaine
reductive	azo reduction	prontosil
	nitro reduction	chloramphenicol
nonmicrosomal oxidative	alcohol and aldehyde oxidation	ethanol
	purine oxidation	6-mercaptopurine
	oxidative deamination (monoamine oxidase)	serotonin
	Phase II reactions	
coupling	glucuronidation	acetaminophen
	acetylation	isoniazid
	glycine conjugation	salicylic acid
	sulfate conjugation	steroids, phenols
	methylation	norepinephrine

Pharmacodynamic Aspects of Drug Action

Although the same general principles of chemical specificity apply to all ligand-macromolecular interactions, the term receptor is generally applied to those cellular macromolecules and macromolecular complexes with which ligands, physiological or synthetic, interact both to complex and to initiate a physiological response. Receptors are conveniently viewed as existing in several principal classes, ie, G-protein-coupled receptors, ligand-gated ion channels, voltage-gated ion channels, tyrosine kinase receptors, guanylyl cyclase receptors, and steroid hormone receptors. All of these receptors form homologous classes according to structure and mechanisms of action. G-protein-coupled receptors form a homologous class of membrane proteins characterized by seven transmembrane domains and the ability to couple to guanine (G) nucleotide-binding proteins.

Ligand-gated ion channels represent a significant family of ion channels that feature as an integral component of their multimeric subunit organization a receptor site for either acetylcholine (nicotine acetylcholine receptor (AChR)), amino acids including glycine and γ-aminobutyric acid (GABA) (inhibitory transmitters), or glutamic acid (excitatory transmitter). The interaction of the ligand with the endogenous receptor site causes channel opening or closing.

An important characteristic of both classes of ion channel is that they possess multiple drug binding sites. Many of the channel-active drugs have achieved particular therapeutic importance, including, for example, the Ca^{2+} antagonists, widely used for a number of cardiovascular disorders, such as hypertension.

Structure–Activity Relationships. Until the mid-1980s, the attempted correlation of chemical structure and biological activity was the only available approach to the definition of receptor site structures. The basic assumption in the analysis of structure-activity relationships (SAR) is the existence of a definable mutual complementarity between the structure of the drug and its corresponding binding site. This application is limited when applied in empirical fashion. Many drug molecules are flexible structures and, although conformations in the solution and solid states can be determined by spectroscopic and crystallographic methods, these bear no necessary relationship to those adopted at the receptor site. The possibility of mutual conformational adaptation of both the drug and the receptor site during the binding process adds a further complication. Furthermore, there may exist multiple drug-binding modes at the receptor such that transitions in binding modes occur at some point in a structurally related series. An additional problem in the quantitative interpretation of SAR is that of the relationship between biological response and drug-receptor interaction. Despite these limitations, SARs have been of great value in providing qualitative concepts of binding site geometry, classifying receptors, furnishing evidence for the existence of new classes of receptor-specific drugs, and generating new and therapeutically effective compounds.

The simplest SARs occur in homologous series of compounds. Thus a linear relationship exists between carbon chain length and biological activity in 1-alkanol-mediated anesthesia (see ANESTHETICS). The activity can be related to the water:cell partition coefficient. For other homologous series, however, such linear relationships may not be observed; for example, in the antagonistic activity of α,ω-bistrimethylammonium alkanes at acetylcholine receptors where binding to sites of defined anionic site geometry probably is involved.

Relatively unambiguous monotonic SARs also occur where activity depends on the ionization of a particular functional group. A classic example is that of the antibacterial sulfonamides where activity is exerted by competitive inhibition of the incorporation of *p*-aminobenzoic acid into folic acid. The bell-shaped relationship is consistent with the sulfonamide acting as the anion but permeating into the cell as the neutral species.

The SAR is also determined at the level of stereochemistry of interaction. In principle, three limiting situations can apply to the stereochemistry of drug-receptor interactions: the enantiomers may not differ in activity; the species may differ quantitatively; or they may differ qualitatively.

The issue of drug stereoselectivity has become one of both developmental and regulatory significance. In principle, a racemic drug possesses only 50% of the active ingredient, and the rest may have other or interacting pharmacologic activities, which may contribute a metabolic burden or be inert. Over 50% of clinically available drugs have chiral centers and only about 10% of synthetic chiral drugs are marketed in homochiral (enantiomerically pure) form. In contrast, drugs that are naturally occurring substances, obtained from or related to naturally occurring molecules, are frequently homochiral.

There is increasing pressure to develop homochiral drugs.

Often pharmacologic agonist activity decreases and is lost with progressive structural change.

Increasing attention has been paid to the generation of quantitative structure—activity relationships in which the effects of molecular substitution on pharmacologic activity can be interpreted in terms of the physicochemical properties of the substituents. These approaches are based on the extrathermodynamic analysis of substituent effects.

Advancing technology permits increasing attention to the definition of the three-dimensional structure of the ligand in its bioactive conformation as it binds to the receptor or active site. This bioactive conformation is not necessarily the solution or the crystal structure of the ligand, which is often the most experimentally accessible structure. It is of critical importance to define the three-dimensional structure of the ligand complexed with its target. This resolution permits not only the understanding of a particular ligand–macromolecule, but also the *in vivo* design of ligand homologues that may have tighter or more selective affinities for the site.

Considerable effort must be applied to obtaining adequate quantities of the protein target and its structural solution, together with the structural solution of the complexed ligand, either by x-ray or solution nmr techniques. Alternatively, homology modeling may be possible when the structure of a homologue protein is already available. Although many examples of ligand–protein structure determinations are available, some of the most interesting targets, eg, membrane-bound receptors, defy structural solution at the necessary resolution. The examination of the real structure of ligand–receptor complexes should be an increasingly important and integral part of the drug discovery process.

Quantitative Aspects of Drug–Receptor Interactions. As a general rule, pharmacological responses are graded and a defined relationship exists between the concentration of a drug and the receptor response. This usually is expressed as a concentration–response (A–R) relationship in linear or semilogarithmic coordinates and usually is referred to as a dose–response curve. The shape of these curves offers a clear analogy to processes of physical adsorption but, because of the complexity of the sequence of events between drug–receptor interaction and the response, the interpretation of dose–response curves is not simple. A quantitative understanding of drug–receptor interactions is crucial both to the nontrivial interpretation of structure–activity relationships and to the determination of the mechanisms by which drug–receptor complexes initiate pharmacological response.

Nonreceptor-Mediated Drug Action. At least one important class of drugs, the general anesthetics (qv), has been assumed not to owe its therapeutic activities to a specific receptor process. Anesthetic potency shows an excellent linear correlation with partition coefficient and this has been extrapolated to a definition of action at a lipid site. The phospholipids of cell membranes, particularly nerve cells, have been considered as principal targets for general anesthetic action. It has been hypothesized that anesthetics may disrupt phospholipid structure by fluidizing or expanding the cell membrane or by altering the phase relationships of the phospholipids. However, it is possible that anesthetics bind to hydrophobic sites on proteins and thus affect directly excitable cell behavior. This latter proposal is consistent both with the activity of the gaseous general anesthetics and with the activity of structurally more complex agents, eg, 3α-hydroxy-5α-pregnane-11,20-dione, 3α-hydroxy-5-pregn-16-ene-11,20-dione, and 1,5-desmethyl-5-cyclohexenylbarbituric acid.

Although most anesthetics are achiral or are administered as racemic mixture, the anesthetic actions are stereoselective. This property can define a specific, rather than a nonspecific, site of action. Stereoselectivity is observed for such barbiturates as thiopental, pentobarbital, and secobarbital. The (*S*)-enantiomer is modestly more potent. Additionally, the volatile anesthetic isoflurane also shows stereoselectivity. The (*S*)-enantiomer is the more active. Further evidence that proteins might serve as appropriate targets for general anesthetics come from observations that anesthetics inhibit the activity of the enzyme luciferase. The potencies parallel the anesthetic activities closely.

It is likely that a principal target of the general anesthetics is neuronal ion channels of both voltage-gated and ligand-gated classes. Interactions at GABA-mediated inhibitory channels is a significant, but not exclusive, target. Thus, a general anesthetic may have specific but multiple, rather than nonspecific, sites of action.

Receptor–Effector Coupling. The informational signal initiated by drug–receptor interaction must be translated to biological response. This is activated by a variety of effector-coupling processes that lead to ionic or biochemical changes, including ion channel opening and closing; the formation of second messengers such as cyclic adenosine-3′-5′-monophosphate (cAMP) and inositol-1,4,5-triphosphate (IP_3); and protein phosphorylation through protein kinase A (cAMP-dependent) and protein kinase C (CA^{2+}-dependent), or through autophosphorylation (tyrosine kinase receptors). In these systems, it is increasingly clear that the individual components of a receptor system may be linked in multiple ways. The virtue of this organization lies in the multiple coupling processes permitted beyond a set of components.

These cascades serve as operational amplifiers of the initial ligand–receptor interaction. In each step of the process, amplification by several powers of 10 may occur so that an original signal may be multiplied several millionfold.

G-Protein Coupling. The heterotrimeric guanosine triphosphate (GTP) binding proteins, known as G-proteins, are a principal family of proteins serving to couple membrane receptors of the G-protein family to ionic and biochemical processes. The G-proteins are heterotrimers made of three families of subunits, α, β, and γ, which can interact specifically with discrete regions on G-protein-coupled receptors. This includes most receptors for neurotransmitters and polypeptide hormones (see NEUROREGULATORS). G-protein-coupled receptors also embrace the odorant receptor family and the rhodopsin-linked visual cascade.

The underlying coupling mechanisms are defined by the enzymatic activity of the G-protein, that of hydrolyzing GTP, ie, GTPase activity. In the inactive state, the heterotrimeric G-protein is liganded to the diphosphate GDP. Receptor activation reduces the affinity of the α-subunit for GDP and increases the affinity for GTP. The GTP-liganded complex then dissociates to the GTP-bound activated α-subunit and the β- and γ-subunits. These dissociated subunits then interact with the corresponding effectors. The effectors include adenylyl cyclase, phospholipase C, cGMP phosphodiesterase, some ion channels (K^+, Ca^{2+}), and receptor kinases. These signals may be excitatory or inhibitory according to the class of G-protein, some of which are listed for G-protein-linked adenylyl cyclase:

Stimulation (Gs)	Inhibition (Gi, Go)
β-adrenergic	opiate
H_2-histamine	muscarinic
dopamine	α_1-adrenergic
polypeptide hormones (glucagon, ACTH, etc)	adenosine (fat cells) A_1, prostaglandins (fat cells)
adenosine (platelets, lymphocytes) A_2	
prostaglandins (platelets)	polypeptide hormones
serotonin (5-$HT_{1\alpha}$)	somatostatin, neuropeptide Y, atriopeptin

A critical component of the G-protein effector cascade is the hydrolysis of GTP by the activated α-subunit (GTPase). This provides not only a component of the amplification process of the G-protein cascade but also serves to provide further measures of drug efficacy. The coupling process also depends on the stoichiometry of receptors and G-proteins. A reduction in receptor number should diminish the efficacy of coupling and thus reduce drug efficacy.

The ability of receptors to couple to G-proteins and initiate GTPase activity may also be independent of ligand.

The principal intracellular messengers derived from activation of G-protein-coupled receptors are cAMP and IP_3. cAMP may be degraded by phosphodiesterase (PDE) or it may activate cAMP-dependent protein kinase (PKA). The activation of this enzyme involves dissociation of the inactive form. (R_2C_2) into the active form which subsequently phosphorylates specific proteins. In contrast, IP_3, one of the products of receptor-mediated phospholipase C breakdown of phosphatidylinositol (PI), acts on specific receptors in the endoplasmic reticulum to release Ca^{2+} from intracellular sources. The other product of PI turnover is a 1,2-diacylglycerol that activates protein kinase C (PKC). This is also the receptor for the tumor-promoting phorbol esters. These diacylglycerols can be cleaved by monoacyl- or diacylglycerol kinases to yield arachidonic acid, a precursor to the prostaglandins (qv) and thromboxanes.

Ion Channels. The excitable cell maintains an asymmetric distribution across both the plasma membrane, defining the extracellular and intracellular environments, as well as the intracellular membranes which define the cellular organelles. This maintained asymmetric distribution of ions serves two principal objectives. It contributes to the generation and maintenance of a potential gradient and the subsequent generation of electrical currents following appropriate stimulation. Moreover, it permits the ions themselves to serve as cellular messengers to link membrane excitation and cellular response. In some instances, the current itself may be the response, as, for example, in the electric organ of electric fishes. In most instances, however, the current serves to initiate or modulate another cellular response, including propagation of impulses in nerve fibers, and alteration of the sensitivity of membranes to other stimuli or coupling to cellular responses such as contraction and secretion. In the latter examples, a role for calcium is particularly prominent because Ca^{2+} can serve as both a current-carrying and a messenger species.

Regulation of ion channels by drugs may have excitatory or inhibitory effects according to the channels affected.

Channels may be regulated exclusively by electrical or chemical signals corresponding to purely voltage-gated or ligand-gated channels, respectively. Regardless of regulatory mechanism, ion channels may be regarded as allosteric enzymes. The function is to accelerate the transit of ions across an essentially impermeable barrier and to be responsive to a variety of heterotropic signals.

Ion channels may be regarded as pharmacological receptors frequently possessing a multiplicity of drug binding sites. These sites may be for endogenous physiological regulators or for endogenous or synthetic agents.

Tyrosine Kinase Receptors. The polypeptide growth factors control cell proliferation, differentiation, and survival. Several distinct subfamilies of receptor tyrosine kinases exist and at least nine have been characterized. These include families for epidermal growth factor, insulin and insulin-related factors, fibroblast growth factors, and neurotrophin receptors such as nerve growth factor and brain-derived neurotrophic factor. All of these receptors have kinetics that share certain fundamental signaling properties. Ligand binding to the extracellular domain activates a tyrosine kinase of the cytoplasmic domain. Subsequently, a variety of downstream signaling molecules are activated. These include phospholipase C, GTPase activating factor (GAP), *Ras,* and MAP kinases.

Guanylyl Cyclase Receptors. Cyclic GMP concentrations (cGMP) rise in response to a number of cell signals. Membrane-associated guanylyl cyclase catalyzes the conversion of guanosine triphosphate (GTP) to cGMP. This enzyme resembles in organization the tyrosine kinases having an intracellular protein kinase-like domain and a cyclase catalytic domain. The enzymes are activated by several distinct species that include atrial natriuretic peptide (ANF) and peptides related to the heart-stable enterotoxins.

In contrast, the soluble guanylyl cyclases are regulated by nitric oxide and NO-forming drugs through the Ca^{2+} calmodulin-dependent nitric oxide synthase.

Receptor Regulation and Defects. Specific recognition and the initiation of response are the accepted attributes of the drug–receptor interaction. However, target cells can alter on both short- and long-term time scales their sensitivity to drugs. Such regulation, achieved by altering the number and/or affinity of receptors, is well established for all receptor systems and can be viewed as an integral component of the drug–receptor interaction. In this view, subsequent to the formation of the drug–receptor complex with agonist, the continued existence of the drug–receptor complex may lead to one or more phases of desensitization, according to which there may occur initially transient and subsequently prolonged phases of reduced or lost sensitivity. Occupancy by antagonist, in contrast, leads to an increased number of receptors and increased drug sensitivity. This phenomenon may contribute to clinical rebound during abrupt withdrawal from drugs, including β-blockers. Additional to this homologous regulation, receptor sensitivity may be controlled through heterologous influences, whereby hormones, including thyroid and corticosteroids, regulate other receptors. These regulatory events are made possible because pharmacologic receptors, in common with other cellular components, are in dynamic balance between synthesis and degradation. This balance is sensitive to a number of influences that include agonist and antagonist presence.

There are probably several processes that contribute to the total desensitization process and these may be directed homologously (to own receptor) or heterologously (to other receptor). Additionally, the influences may be directed at the receptor itself and affect only that receptor, ie, specific desensitization, or may affect other receptor processes as well, ie, nonspecific desensitization.

An increasing number of diseases are known to be linked to defects in receptor structure, function, or coupling. The defects may lie at several locations: in the structure of the receptor, which may alter its ability either to bind drugs, to be inserted into the membrane, or to couple to effectors (including G-proteins); in the coupling protein; or in the presence of autoantibodies, which can proceed to activate, block, or lyse the receptors and its components.

Components of Drug Action and Responses to Drugs. The response to a drug can vary among race, gender, and age groups. It may vary according to disease state and age, and it may vary according to the time of administration. These factors may have several origins, including (*1*) compliance, the ability or desire of the subject to take a drug according to a specific regimen; (*2*) pharmacokinetic, disease-, age-, race-, and gender-based factors that contribute to variable absorption, distribution, metabolism, and excretion of a drug; and (3) pharmacodynamic, disease-, age-, race-, and gender-based factors that contribute to variable drug–receptor interactions.

DAVID J. TRIGGLE
State University of New York at Buffalo

W. B. Pratt and P. Taylor, eds., *Principles of Drug Action: The Basis of Pharmacology,* 3rd ed., Churchill Livingstone, New York, 1990.

A. G. Gilman and co-workers, eds., *The Pharmacological Basis of Therapeutics,* 8th ed., Pergamon Press, New York, 1990.

C. G. Wermuth, ed., *The Practice of Medicinal Chemistry,* Academic Press, San Diego, London, and New York, 1996.

M. E. Wolff, ed., *Burger's Medicinal Chemistry and Drug Discovery,* Vol. I., Principles and Practice, Wiley-Interscience, New York, 1995.

PHARMACOKINETICS. See PHARMACODYNAMICS.

PHASE EQUILIBRIUM. See EXTRACTION; HIGH PRESSURE TECHNOLOGY; SEPARATIONS PROCESS SYNTHESIS.

PHENAZINE ANTIBIOTICS. See ANTIBIOTICS.

β-PHENETHYL ALCOHOL. See BENZYL ALCOHOL AND β-PHENYL ALCOHOLS; PERFUMES.

PHENETIDINES. See ANALGESICS, ANTIPYRETICS, AND ANTIINFLAMMATORY AGENTS; XANTHENE DYES.

PHENOL

Phenol is the common name of hydroxybenzene, C_6H_5OH, and belongs to the class of compounds, commonly referred to as phenols, containing one or more hydroxyl groups attached to an aromatic ring. Phenol has also been called carbolic acid, phenic acid, phenylic acid, phenyl hydroxide, or oxybenzene. More than 99% of phenol produced worldwide in the 1990s is from synthetic processes.

In 1993, worldwide phenol production was more than 5.2 million metric tons. The predominant uses of phenol are in phenolic resins (qv), bisphenol A, caprolactam (qv), aniline (see AMINES–AMINES, AROMATIC–ANILINE AND ITS DERIVATIVES), and alkylphenols (qv).

Physical Properties

At room temperature phenol is a white, crystalline mass. It has a distinctive sweet, tarry odor, and burning taste. Phenol has limited solubility in water between 0 and 65°C. Above 65.3°C phenol and water are miscible in all proportions. It is very soluble in alcohol, benzene, chloroform, ether, and partially disassociated organics in general. The important physical properties of phenol are listed in Table 1.

Chemical Properties

Phenol's chemical properties are characterized by the influences of the hydroxyl group and the aromatic ring upon each other. Although the structure of phenol is similar to cyclohexanol, phenol is a much stronger acid.

Beside being acidic, a significant industrial chemical property of phenol is the extremely high reactivity of its ring toward electrophilic substitution.

Table 1. Physical Properties of Phenol

Property	Value
molecular weight	94.11
boiling point at 101.3 kPa[a]	181.75
freezing point, °C	40.91
vapor pressure at 25°C, MPa[b]	46.84
flash point (closed cup), °C	79
density at 20°C (solid), g/cm^3	1.0722

[a] To convert kPa to mm Hg, multiply by 7.5.
[b] To convert MPa to atm, divide by 0.1013.

The most important commercial chemical reactions of phenol are condensation reactions. Phenolic resins and bisphenol A account for more than two-thirds of U.S. phenol consumption.

Manufacture

The cumene oxidation route is the leading commercial process of synthetic phenol production, accounting for more than 95% of phenol produced in the world. The remainder of synthetic phenol is produced by the toluene oxidation route via benzoic acid. Other processes including benzene via cyclohexane, benzene sulfonation, benzene chlorination, and benzene oxychlorination have also been used in the manufacture of phenol.

Economic Aspects

Because the cumene process accounts for more than 95% of the world's phenol supply, the economics of phenol production are closely tied to this production method. In the cumene process 615 kg of acetone are coproduced with each ton of phenol produced. Thus, the economics of phenol production are influenced by acetone (qv).

There is a disparity in the growth rates of phenol and acetone, with phenol demand projected at 3.0%/yr and acetone demand at 2.0%/yr. If this continues, the coproduct supply of acetone will exceed the total acetone demand and on-purpose production of acetone will be forced to shut down; the price of acetone is expected to fall below the floor price set by the on-purpose cost production. Projections indicate that such a situation might occur in the world market by 2010. To forestall such a situation, companies such as Mitsui Petrochemical and Shinnippon (Nippon Steel) have built plants without the coproduction of acetone.

Specifications and Standards

DOT's Hazardous Materials Regulations classifies phenol as a Class B poison.

The *U.S. Pharmacopeia* (USP) specification for phenol includes (*1*) purity is to be no less than 98 wt %, (*2*) clear solubility of 1 part of phenol in 15 parts of water, (*3*) a congealing temperature to be not lower than 39°C, and (*4*) a content of nonvolatiles of no more than 0.05 wt %. Commercially, phenol specifications far exceed the USP requirement.

Storage. Phenol is shipped in drums, tank trucks, and tank cars. It is loaded and shipped at elevated temperatures as a bulk liquid. In storage, phenol may acquire a yellow, pink, or brown discoloration which makes it unusable for some purposes. When stored as a solid in the original drum or in nickel, glass-lined, or tanks lined with baked phenolic resin, phenol remains colorless for a number of weeks.

Storage tanks should be equipped with heating coils that pass upward through the entire vessel. Both horizontal and vertical tanks are suitable for phenol storage.

Health and Safety Factors

Phenol fumes are irritating to the eyes, nose, and skin. According to the National Institute for Occupational Safety and Health (NIOSH), exposure to phenol should be controlled so that no employees are exposed to phenol concentrations > 20 mg/m^3, which is a time-weighted average concentration for up to a 10-h work day, 40-h work week. Phenol is very toxic to fish and has a nearly unique property of tainting the taste of fish if present in marine environments at 0.1–1.0 ppm. Phenol presents no unusual fire hazard when handled at ambient temperatures, but burns if ignited.

Phenol is a general protoplasmic poison that is corrosive to any living tissue it contacts.

Personnel who handle phenol should wear protective clothing, safety goggles, and rubber gloves, depending on the working conditions and amount of phenol handled.

JIM WALLACE
M. W. Kellogg Company

S. E. Howland and coworkers, "Phenol—A World Outlook," presented at the *1994 DeWitt Petrochemical Review*, Houston, Tex., 1994.

The Merck Index, 11th ed., Merck and Co., N.J., 1989.

CRC Handbook of Chemistry and Physics, 57th ed., CRC Press, Boca Raton, Fla., 1976–1977.

Material Safety Data Sheets Collection: Sheet No. 355, Revision C, Genium Publishing Corp., Schenectady, N.Y., Nov. 1990.

PHENOLIC FIBERS. See PHENOLIC RESINS.

PHENOLIC RESINS

Phenolic resins are a large family of polymers and oligomers, composed of a wide variety of structures based on the reaction products of phenols with formaldehyde. Phenolic resins are employed in a wide range of applications, from commodity construction materials to high technology applications in electronics and aerospace. Generally, but not exclusively, thermosetting in nature, phenolic resins provide numerous challenges in the areas of synthesis, characterization, production, product development, and quality control.

Early phenolic resins consisted of self-curing, resole-type products made with excess formaldehyde, and novolaks, which are thermoplastic in nature and require a hardener. The early products produced by General Bakelite were used in molded parts, insulating varnishes, laminated sheets, and industrial coatings. These areas still remain important applications, but have been joined by numerous others such as wood bonding, fiber bonding, and plywood adhesives. The number of producers in the 1990s is approximately 20 in the United States and over 60 worldwide.

Monomers

Phenol. This is the monomer or raw material used in the largest quantity to make phenolic resins (Table 1).

The most widely used process for the production of phenol is the cumene process developed and licensed in the United States by AlliedSignal (formerly Allied Chemical Corp.). Other commercial processes for making phenol include the Raschig process, using chlorobenzene as the starting material, and the toluene process, via a benzoic acid intermediate. In the United States, ~ 35–40% of the phenol produced is used for phenolic resins.

Table 1. Properties of Phenol

Property	Value
mol wt	94.1
mp, °C	40.9
bp, °C	181.8
flash point, °C	79.0
autoignition temperature, °C	605.0
explosive limits, vol %	2–10
vapor pressure at 20°C, Pa[a]	20

[a] To convert Pa to mm Hg, multiply by 7.5×10^{-3}.

Table 2. Substituted Phenols Used for Phenolic Resins

Substituted phenol	Resin application
cresol (*o*-, *m*-, *p*-)	coatings, epoxy hardeners
p-*t*-butylphenol	coatings, adhesives
p-octylphenol	carbonless paper, coatings
p-nonylphenol	carbonless paper, coatings
p-phenylphenol	carbonless paper
bisphenol A	low color molding compounds, coatings
resorcinol	adhesives
cashew nutshell liquid	friction particles

Substituted Phenols. Phenol itself is used in the largest volume, but substituted phenols are used for specialty resins (Table 2). Substituted phenols are typically alkylated phenols made from phenol and a corresponding α-olefin with acid catalysts.

Formaldehyde. In one form or another, formaldehyde is used almost exclusively in the production of phenolic resins, regardless of the type of phenol. It is frequently produced near the site of the resin plant by either of two common processes using methanol (qv) as the raw material: the silver catalyst process and the more common metal oxide process.

Other Aldehydes. The higher aldehydes react with phenol in much the same manner as formaldehyde, although at much lower rates. Examples include acetaldehyde, CH_3CHO; paraldehyde, $(CH_3CHO)_3$; glyoxal, OCH-CHO; and furfural.

Hexamethylenetetramine. When used either as a catalyst or a curative, hexa contributes formaldehyde-residue-type units as well as benzylamines.

Other Reactants. Other reactants are used in smaller amounts to provide phenolic resins that have specific properties, especially coatings applications. Other materials include rosin (abietic acid), dicyclopentadiene, unsaturated oils such as tung oil and linseed oil, and polyvalent cations for cross-linking.

Polymerization

Phenolic resins are prepared with strong acid or alkaline catalysts. Occasionally, weak or Lewis acids, such as zinc acetate, are used for specialty resins.

Strong-Acid Catalysts, Novolak Resins. Phenolic novolaks are thermoplastic resins having a molecular weight of 500–5000 and a glass-transition temperature, T_g, of 45–70°C. The phenol-formaldehyde reactions are carried to their energetic completion, allowing isolation of the resin. The properties of an acid-catalyzed phenolic resin are shown in Table 3.

The typical acid catalysts used for novolak resins are sulfuric acid, sulfonic acid, oxalic acid, or occasionally phosphoric acid. Hydrochloric acid, although once widely used, has been abandoned because of the possible formation of toxic chloromethyl ether by-products. The type of acid catalyst used and reaction conditions affect resin structure and properties.

Neutral Catalysts, High Ortho Novolaks. In the range of pH 4–7, formaldehyde substitution of the phenolic ring is possible, using divalent metal catalysts containing Zn, Mg, Mn, Cd, Co, Pb, Cu, and Ni; certain aluminum salts are also effective. Organic carboxylates are required as anions in order to obtain sufficient solubility of the catalyst in the reaction medium, as well as to provide a weak base. Acetates are most convenient and economical. Zinc and calcium salts are probably the most widely used catalysts.

High ortho novolaks have faster cure rates with hexa. Typical properties of a zinc acetate-catalyzed high ortho novolak are also shown in

Table 3. Novolak Resin Properties

Property	Catalyst Acid	Zn acetate[a]
formaldehyde-phenol molar ratio	0.75	0.60
nmr analysis, %		
2,2′	6	45
2,4′	73	45
4,4′	21	10
gpc analysis		
phenol, %	4	7
M_n	900	550
M_w	7300	1800
water, %	1.1	1.9
T_g, °C	65	48
gel time, s	75	25

[a] High ortho.

Table 4. Properties of Resole Resins

Property	Catalyst NaOH	Hexa
concentration, pph	3	10
formaldehyde-phenol ratio	2.0	1.5
water solubility, %	100	swells
gpc analysis phenol, %	6	8
M_n	280	900
M_w	500	3000
T_g, °C	35	47
gel time, s	65	110

Table 4. The gel time with hexa is one-third of that with a strong acid-catalyzed novolak.

Alkaline Catalysts, Resoles. Resole-type phenolic resins are produced with a molar ratio of formaldehyde to phenol of 1.2:1 to 3.0:1. For substituted phenols, the ratio is usually 1.2:1 to 1.8:1. Common alkaline catalysts are NaOH, $Ca(OH)_2$, and $Ba(OH)_2$. Whereas novolak resins and strong acid catalysis result in a limited number of structures and properties, resoles cover a much wider spectrum. Resoles may be solids or liquids, water-soluble or -insoluble, alkaline or neutral, slowly curing or highly reactive.

Although monomeric methylolated phenols are used in certain applications, such as in fiber bonding, higher molecular weight resins are usually desirable. Molecular weight is increased by further condensation of the methylol groups, sometimes after the initial pH has been reduced. Dibenzyl ether and diphenylmethylene formation are shown in the following. The formation of diphenylmethylene bridges is favored above 150°C and under strongly alkaline conditions; dibenzyl ether formation is favored at lower temperatures and near neutral pH.

Special resoles are obtained with amine catalysts, which affect chemical and physical properties because amine is incorporated into the resin. In practice, ammonia is most frequently used.

The physical properties of a resole resin prepared with hexa catalyst are shown in Table 4.

Manufacture

The final state of a phenolic resin varies dramatically from thermoplastic to thermoset and from solid to liquid, and includes solutions and dispersions. With a bulk process, resole resins, in neat or concentrated form, must be produced in small batches (ca 2–9.5 m^3) in order to maintain control of the reaction and obtain a uniform product. On the other hand, if the product contains a large amount of water, such as liquid plywood adhesives, large reactors (19 m^3) can be used. Melt-stable products such as novolaks can be prepared in large batches (19–38 m^3) if the exotherms can be controlled.

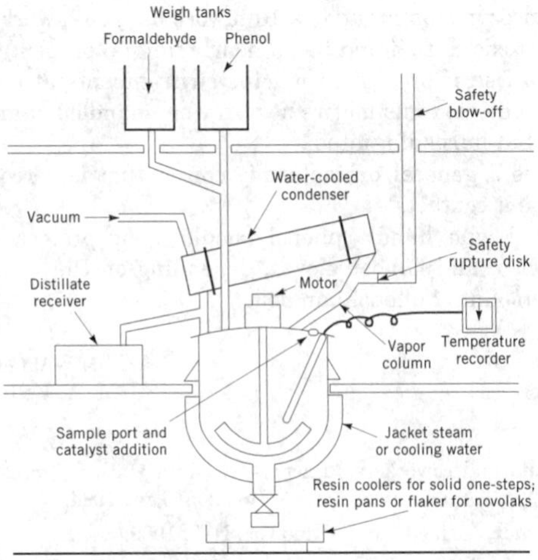

Figure 1. Typical phenolic resin production unit.

Batch processes for most phenolic resins employ the equipment shown in Figure 1. Liquid reactants are metered into the stirred reaction vessel through weigh tanks, whereas solid reactants such as bisphenol A and $Ba(OH)_2$ present handling problems. Facilities are provided to carry out the reaction under a vacuum or an inert gas.

Materials of Construction. Compatibility of the materials of construction and the process chemicals is extremely important. The reactors are usually made of stainless steel alloys. Copper is avoided because of the possible presence of amines. Glass-lined reactors are occasionally used for nonalkaline resins. Because the use of HCl has been largely discontinued, material requirements are less stringent.

Novolak Resins. In a conventional novolak process, molten phenol is placed into the reactor, followed by a precise amount of acid catalyst. The formaldehyde solution is added at a temperature near 90°C and a formaldehyde-to-phenol molar ratio of 0.75:1 to 0.85:1. The heat of reaction is removed by refluxing the water combined with the formaldehyde or by using a small amount of a volatile solvent such as toluene. Toluene and xylene are used for azeotropic distillation. Following decantation, the toluene or xylene is returned to the reactor.

The reaction is completed after 6–8 h at 95°C; volatiles, water, and some free phenol are removed by vacuum stripping up to 140–170°C. For resins requiring phenol in only trace amounts, such as epoxy hardeners, steam distillation or steam stripping may be used. Oxalic acid (1–2 parts per 100 parts phenol) does not require neutralization because it decomposes to CO, CO_2, and water; furthermore, it produces milder reactions and low color. Sulfuric and sulfonic acids are strong catalysts and require neutralization with lime; 0.1 parts of sulfuric acid per 100 parts of phenol are used. A continuous process for novolak resin production has been described. An alternative process for making novolaks without acid catalysis has also been reported, which uses a peroxidase enzyme to polymerize phenols in an aqueous solution. The enzyme can be derived from soybeans or horseradish.

High Ortho Novolaks. The process for high ortho novolaks is similar to the one used for those catalyzed by strong acid. Overall, the process is more expensive because of higher raw material costs, lower yields, and longer cycle times.

Another process employs a pH maintained at 4–7 and a catalyst that combines a divalent metal cation and an acid.

Resoles. Like the novolak processes, a typical resole process consists of reaction, dehydration, and finishing.

Phenolic Dispersions. These systems are predominantly resin-in-water systems in which the resin exists as discrete particles.

In the post-dispersion process, the solid phenolic resin is added to a mixture of water, cosolvent, and dispersant at high shear mixing, possibly with heating. The cosolvent, frequently an alcohol or glycol ether, and heat soften the resin and permit small particles to form. On cooling, the resin particles, stabilized by dispersant and perhaps thickener, harden and resist settling and agglomeration. Both resole and novolak resins have been made by this process.

The *in situ* process is simpler because it requires less material handling; however, this process has been used only for resole resins.

Resole dispersions intended for isolation as discrete particles can be used as flatting agents in coatings. Particles larger than 1000 μm are used in friction-element compositions. A-stage, thermosetting phenolic particles have been isolated from dispersion. These A-stage products (gel time at 150°C, 50–100 s) are suitable in applications where pulverized phenolic resins are being used, as well as in applications that take advantage of their spherical nature.

Spray-Dried Resins. Spray drying produces resins in particulate form. Spray-drying a resole solution containing a blowing agent produces phenolic microballoons. Spray drying also produces A-stage resins. The principal application for this type of product is believed to be wood binding, especially for waferboard applications.

Cure

A typical resin has an initial molecular weight of 150 to perhaps 1500. For systems of unsubstituted phenols, the final cross-link density is 150–300 atomic mass units (amu) per cross-link. In other words, 25–75% of the ring-joining reactions occur during the cure phase.

Resoles. The advancement and cure of resole resins follow reaction steps similar to those used for resin preparation; the pH is 9 or higher and reaction temperature should not exceed 180°C. Methylol groups condense with other methylols to give dibenzyl ethers and react at the ortho and para positions on the phenol to give diphenylmethylenes. In addition, dibenzyl ethers eliminate formaldehyde to give diphenylmethanes.

In some resole applications, such as foam and foundry binders, a rapid cure of a liquid resin is obtained at RT with strong acid.

At pH 4–6, the cure is slower than it is at pH 8 and higher. Reactions at pH 4–6 resemble those on the more alkaline side, but with a substantial increase in side-products.

Novolaks. Novolak resins are typically cured with 5–15% hexa as the cross-linking agent. As much as 75% of nitrogen is chemically bound. The cure begins with the formation of benzoxazine, progresses through a benzyl amine intermediate, and finally forms (hydroxy)diphenylmethanes (DPM).

Decomposition of Cured Resoles and Novolaks. Above 250°C, cured phenolic resins begin to decompose. Substantial decomposition of phenolic resins begins above 300°C.

Analysis and Characterization

The principal techniques for determining the microstructure of phenolic resins include mass spectroscopy, proton, and ^{13}C-nmr spectroscopy, as well as gc, lc, and gpc. The softening and curing processes of phenolic resins are effectively studied by using thermal and mechanical techniques, such as tga, dsc, and dynamic mechanical analysis (dma). Infrared (ir) and electron spectroscopy are also employed.

Control Tests. Numerous chemical and physical tests are used in the manufacture of phenolic resins to ensure correct properties of the finished resins, including the following: refractive index is used to estimate the dehydration during manufacture and is proportional to the solids content; viscosity is used to determine molecular weight and solids content; nonvolatiles content is roughly proportional to polymer content; miscibility with water depends on the extent of reaction in resoles; specific gravity is measured for liquid resins and varnishes; melting point of novolaks and solid resoles affects application performance; gel times determine the reactivity of the resins; resin flow is a measure of melt viscosity and molecular weight; particle size affects performance and efficiency; and flash point and autoignition temperature provide flammability-characteristic measurements required by government agencies regulating safety and shipping.

Health, Safety, and Environmental Factors

The factors contributing to the health and safety of phenolic resin manufacturing and use are those primarily related to phenol (qv) and formaldehyde (qv). The toxicity of the resins is significantly lower than that of the phenol and formaldehyde starting materials. No detrimental toxicological effects have been reported for cured phenolic resins, which can be used in direct contact with food as in can coatings.

Uncured resins are skin sensitizers and contact should be avoided, as well as breathing the vapor, mist, or dust. Novolak-based pulverized products generally contain hexamethylenetetramine, which may cause rashes and dermatitis. Phenolic molding compounds and pulverized phenolic adhesives must be controlled as potentially explosive dusts. In addition, they contain irritating or toxic additives.

Phenol. Phenol monomer is highly toxic and absorption by the skin can cause severe blistering. Large quantities can cause paralysis of the central nervous system and death. The threshold limit value (TLV) for phenol is 5 ppm. The health and environmental risks of phenol and alkylated phenols, such as cresols and butylphenols, have been reviewed.

Formaldehyde. Formaldehyde is classified as a probable human carcinogen by the International Agency for Research on Cancer (IARC) and as a suspected human carcinogen by the American Conference of Governmental and Industrial Hygienists (ACGIH); the latter has lowered its TLV to 0.3 ppm. The Occupational Safety and Health Administration (OSHA) has set its time-weighted average for eight hours at 1.0 ppm and its short-term exposure level at 2.0 ppm.

Gaseous formaldehyde in concentrations above 1 ppm is extremely irritating to the mucous membranes. Aqueous formaldehyde is a protoplasmic poison and causes irritation of skin, eyes, nose, and throat. As a preservative it attacks bacteria and reacts with the amino groups of proteins.

Wastewater. Phenol is a toxic pollutant to the waterways and has an acute toxicity (~5 mg/L) to fish. Chlorination of water gives chlorophenols, which impart objectionable odor and taste at 0.01 mg/L. Biochemical degradation is most frequently used to treat wastewater containing phenol.

Flammability. Phenolics have inherently low flammability and relatively low smoke generation. For this reason they are widely used in mass transit, tunnel-building, and mining. Fiber glass-reinforced phenolic composites are capable of attaining the 1990 U.S. Federal Aviation Administration (FAA) regulations for total heat release and peak heat release for aircraft interior facings.

Recycling. Thermosets are inherently more difficult to recycle than thermoplastics and thermosetting phenolics are no exception. However, research in this area has been reported, and molded parts have been pulverized and incorporated at 10–15% in new molding powders. Both German and Japanese groups had instituted this type of practice in 1992 (see RECYCLING).

Economic Aspects

In 1993, worldwide consumption of phenolic resins exceeded 3×10^6 t; slightly less than half of the total volume was produced in the United States. The largest-volume application is in plywood adhesives, an area that accounts for ca 49% of U.S. consumption.

As a mature industry, U.S. production and application of phenolic resins have paralleled the growth in the GNP. Only the consumption of phenolic resins for coatings and molding powders has decreased in recent years.

U.S. phenolic resin manufacturers include AlliedSignal Inc./Bendix; Ashland Chemical, Inc.; Borden, Inc.; Dexter Corp.; Dyno Polymers; Georgia-Pacific Corp.; Neste Resins Corp.; Occidental Chemical Corp.; Owens-Corning Corp.; Plastics Engineering Co.; PMC, Inc.; Resinoid Engineering Co.; Spurlock Adhesives, Inc.; Stuart-Ironsides, Inc.; and Valite Division of Valentine Sugars, Inc.

Prices of phenolic resins vary substantially depending on the application.

Applications

Phenolic resins are used in coatings, dispersions, adhesives, carbonless copy paper, molding compounds, abrasives, friction materials, foundry resins, laminates, air and oil filters, wood bonding, fiber bonding, composites (eg, carbon-fiber composites and carbon–carbon composites), liquid-injection molding, foam, spheres, and fibers.

PETER W. KOPF
Arthur D. Little, Inc.

R. W. Martin, *The Chemistry of Phenolic Resins*, John Wiley & Sons, Inc., New York, 1956.

N. J. L. Megson, *Phenolic Resin Chemistry*, Butterworth & Co. Ltd., Kent, U.K., 1958.

E. Dradi and G. Casiraghi, *Macromolecules* **11**, 1295 (1978).

T. Liu and S. Rhee, *Wear* **76**, 213 (1978).

PHENOLSULFONIC ACIDS. See SULFONIC ACIDS

PHENYLACETIC ACID. See BENZYL ALCOHOL AND β-PHENETHYL ALCOHOL; PERFUMES.

PHENYLENEDIAMINES AND TOLUENEDIAMINES. See AMINES, AROMATIC–PHENYLENEDIAMINES; AMINES, AROMATIC–DIAMINOTOLUENES.

PHEROMONES. See INSECT CONTROL TECHNOLOGY; HORMONES, SURVEY.

PHOSGENE

Phosgene (carbonyl chloride, carbon oxychloride, chloroformyl chloride), Cl_2CO, is a colorless, low boiling liquid. Phosgene may be formed at elevated temperatures by oxidation of chlorinated solvents. Phosgene has been used in the preparation of a great variety of chemical intermediates. It is widely used in the preparation of isocyanates which are used in the preparation of polyurethanes (see URETHANE POLYMERS), in the manufacture of polycarbonate, and in the synthesis of chloroformates and carbonates which are used as intermediates in the synthesis of pharmaceuticals and pesticides (see CARBONIC AND CARBONOCHLORIDIC ESTERS). Because of its toxicity, a high level of safety technology has been developed to help ensure the safe handling of phosgene.

Properties

Some physical properties of phosgene are listed in Table 1. At room temperature and normal pressure, it is a colorless gas. Impurities may cause discoloration of the product to pale yellow to green. Phosgene has a characteristic odor; the odor of the gas can be detected only briefly at the time of initial exposure. In general, phosgene is soluble in aromatic and aliphatic hydrocarbons, chlorinated hydrocarbons, and organic acids and esters.

Reactions. Phosgene interacts with many classes of inorganic and organic reagents. The reactions have been described extensively. Phosgene is an excellent chlorinating agent. Ammonia reacts vigorously with phosgene. The products are urea, biuret, ammelide (a polymer of urea), cyanuric acid, and sometimes cyamelide (a polymer of cyanic acid).

Phosgene reacts with a multitude of nitrogen, oxygen, sulfur, and carbon centers. Reaction with primary alkyl and aryl amines yield carbamoyl chlorides which are readily dehydrohalogenated to isocyanates. This reaction is the basis for the manufacture of isocyanates.

Table 1. Some Physical Properties of Phosgene

Properties and characteristics	Value
molecular weight	98.92
melting point, °C	−127.84
boiling point, °C[a]	7.48
density at 20°C, g/cm^3	1.387
vapor pressure at 20°C, kPa[b]	161.68
critical temperature, °C	182
critical pressure, MPa[c]	5.68
surface tension, mN/m(= dyn/cm) at 0.0 °C	34.6
34.5°C	17.6

[a] At 101.3 kPa[b] = 1 atm.
[b] To convert kPa to psi, multiply by 0.145.
[c] To convert MPa to psi, multiply by 145.

The reaction of phosgene with alcohols yields chloroformates, and with a basic catalyst present, carbonates are formed. This reaction is commercially important because it serves as a basis for the manufacture of polycarbonate.

Carboxylic acids react with phosgene to give acid chlorides (see CARBOXYLIC ACIDS). Amides react with phosgene to yield nitriles (qv). Phosgene also can initiate ring opening.

Although $POCl_3$ is the traditional reagent in the Vilsmeier aldehyde synthesis, phosgene may be employed.

Manufacture

Phosgene is manufactured by reaction of carbon monoxide with chlorine over activated carbon. Depending on the quantity needed and availability of the raw materials, numerous variations of the basic synthetic process are being practiced. Continuous processing and a high degree of automation are required for phosgene purification, condensation, and storage. Because of its toxicity, careful and extensive safety procedures and safety equipment are incorporated in plant design and operation. The manufacture of phosgene consists of preparation and purification of carbon monoxide, preparation and purification of chlorine, metering and mixing of reactants, reaction of mixed gases over activated carbon catalyst, purification and condensation of phosgene, and recovery of traces of phosgene to assure worker and environmental safety.

Waste Gas Streams. Several methods of decomposing phosgene in waste gas streams are used. The outlet gas from the phosgene decomposition equipment is continuously monitored for residual phosgene content to ensure complete decomposition. Methods include decomposition by caustic scrubbing, decomposition with moist activated carbon, and combustion.

Storage and Handling

All phosgene containers require a Class A, poison gas label as well as a corrosive label: Phosgene is transported in steel cylinders which conform to rigid safety design specifications.

Because phosgene reacts with water, great care must be taken to prevent contamination with traces of water since this could lead to the development of pressure by hydrogen chloride and carbon dioxide. Wet phosgene is very corrosive; therefore phosgene should never be stored with any quantity of water.

Health and Safety Factors

The odor threshold for phosgene is ca 0.5–1 ppm, but it varies with individuals and is higher after prolonged exposure. Phosgene may irritate eyes, nose, and throat. The permissible exposure TLV by volume in air is 0.1 ppm. Long-term exposure to phosgene has been reviewed, and potential hazards may exist at concentrations slightly higher than the TLV. Medical problems and adverse health effects associated with phosgene exposure have been reviewed, and therapy for phosgene poisoning has also been reviewed.

Breathing phosgene causes pulmonary edema which may be characterized by a delayed onset. Exposed persons must be removed immediately from the contaminated area. Rescue workers should wear self-contained breathing apparatus. Injured persons should not be allowed any physical activity, and a physician should be consulted immediately.

In handling phosgene, extensive safety precautions and procedures are required to prevent exposure to phosgene. The first point is to design the phosgene system to prevent phosgene emissions from the closed equipment detect leaks and contain or decompose escaped phosgene.

In case of extensive leaks or spills, immediate evacuation upwind of the phosgene source is necessary. Water should not be used on the source of a phosgene leak because the resulting corrosion enlarges the leak. Suitable personal protective equipment includes respiratory equipment and eye protection. In case of fire, it is essential to cool all phosgene-containing vessels. Leaks of liquid phosgene or phosgene solutions can be effectively combated by covering the phosgene-containing liquid with an absorbent and decomposition agents.

Waste Disposal. If recycle of phosgene is not feasible, phosgene waste can be handled by one of the decomposition methods mentioned above, ie, caustic scrubbing, moist activated carbon towers, or combustion.

KENNETH L. DUNLAP
Bayer Corporation

Chemical Safety, Data Sheet SD-95, Manufacturing Chemists Association, Washington, D.C., rev. 1978.

Phosgene Criteria Document, NIOSH 76-137, Washington, D.C.

PHOSPHINE AND ITS DERIVATIVES

Manufacture of Phosphine

Two processes have been used to manufacture gaseous phosphine on a large scale. These are commonly known as the alkaline and acid processes.

The acid process has three advantages over the alkaline process, ie, (*1*) higher yield of phosphine (60 vs 25%); (*2*) more pure gas for use in subsequent reactions (95 vs 40%); and (*3*) by-product phosphoric acid is relatively valuable and can be sold into a number of markets, eg, in the manufacture of fertilizers and flame retardants. There is no ready outlet for the mixture of phosphites produced via the alkaline route and additional processing by oxidative spray drying is needed to produce phosphates for sale.

The principal disadvantage of the acid process is the higher capital cost involved; mainly because of more processing steps and the corrosivity of hot, concentrated phosphoric acid which requires a reactor built from dense graphite.

Typical properties of phosphine are given in Table 1.

Table 1. Properties of Phosphine

Property	Value
appearance and odor	colorless gas with garlic odor
freezing point, °C	−133.8
boiling point, °C	−87.8
critical temperature, °C	51
critical pressure, MPa[a]	6.485
heat of fusion, kJ/mol[b]	1.13
heat of vaporization, kJ/mol[b]	14.6
heat of formation, kJ/mol[b]	9.59
solubility[c] in water, mL/100 mL H_2O	26
heat capacity (liquid) at 0°C J/g	2.43
viscosity (gas) at 0°C, mPa(= cP)	0.01
viscosity (liquid) at 25°C, mPa(= cP)	0.05
surface tension (liquid) at 0°C, mW/m (= dyn/cm)	70

[a] To convert MPa to psi, multiply by 145.
[b] To convert kJ to kcal, divide by 4.184.
[c] At 101.3 kPa (= 1 atm).

Health and Safety Factors

Toxicity. Phosphine may be fatal if inhaled, swallowed, or absorbed through skin. All phosphine-related effects seen at sublethal inhalation exposure concentrations are relatively small and completely reversible. The symptoms of sublethal phosphine inhalation exposure include headache, weakness, fatigue, dizziness, and tightness of the chest. Convulsions may be observed prior to death in response to high levels of phosphine inhalation.

Safety. The pyrophoric and toxic nature of phosphine requires the adoption of special precautions to ensure safety during manufacture on a commercial scale. Of particular note are the provisions of flame retardant, protective clothing for operating personnel, and strategically located breathing-air stations equipped with in-line respirators.

Uses

Apart from the manufacture of derivatives, there are only two known uses for phosphine itself, ie, in the preparation of semiconductors and as a fumigant.

Phosphine Derivatives

Commercial phosphine derivatives are produced either by the acid-catalyzed addition of phosphine to an aldehyde or by free-radical addition to olefins, particularly α-olefins. The reactions usually take place in an autoclave under moderate pressures ($\leq$4 MPa (580 psi)) and at temperatures between 60 and 100°C.

Textile Flame Retardants. The first known commercial application for phosphine derivatives was as a durable textile flame retardant for cotton and cotton–polyester blends. The compounds are tetrakis(hydroxymethyl)phosphonium salts which are prepared by the acid-catalyzed addition of phosphine to formaldehyde.

After application to the fabric, the compounds are polymerized by reaction with gaseous ammonia, then oxidized to phosphine oxides by reaction with hydrogen peroxide.

This provides a durable finish which, unlike many other flame retardants, can withstand repeated launderings without a loss of efficiency. An added advantage is that the feel of the cloth (hand) is little effected. Principal markets are in the treatment of industrial protective clothing, military uniforms, and, in Europe, for furnishings. These products are available from Albright & Wilson Ltd. and Cytec Industries Inc.

Flotation Reagents. Only one sulfide mineral flotation collector is manufactured from phosphine, ie, the sodium salt of bis(2-methylpropyl)-phosphinodithioic acid. It is available commercially from Cytec Industries Inc. as a 50% aqueous solution and is sold as AEROPHINE 3418A promoter. The compound is synthesized by reaction of 2-methyl-1-propene with phosphine to form an intermediate dialkylphosphine which is subsequently treated with elemental sulfur and sodium hydroxide to form the final product.

AEROPHINE 3418A promoter is widely used in North and South America, Australia, Europe, and Asia for the recovery of copper, lead, and zinc sulfide minerals (see FLOTATION).

Phase-Transfer Catalysts. The use of phase-transfer catalysts to improve kinetics and yields in heterogeneous reactions has been growing rapidly since the 1960s. The principal areas of application are in the preparation of polymers, accounting for 50% of catalyst consumption, followed by pharmaceuticals (20%) and agricultural chemicals (10%). Details of the chemistry and applications have been given elsewhere (see CATALYSIS, PHASE-TRANSFER). The most common phase-transfer catalysts are quaternary ammonium salts containing either alkyl or mixed alkaryl groups. However, these compounds are being displaced in some applications by the corresponding phosphonium salts mainly because of the enhanced thermal stability of the phosphorus compounds. Additionally, the phosphonium salts tend to be more efficient than the nitrogen-based analogues and can promote more rapid reaction kinetics.

Phosphonium salts are readily prepared by the reaction of tertiary phosphines with alkyl or benzylic halides; eg, the reaction of tributylphosphine with 1-chlorobutane to produce tetrabutylphosphonium chloride at 60°C, $(C_4H_9)_3P + C_4H_9Cl \rightarrow (C_4H_9)_4P^+Cl^-$.

Biocides. Two phosphine derivatives are in commercial use as biocides. These are tetrakis(hydroxymethyl)phosphonium sulfate and tributyl(tetradecyl)phosphonium chloride. These compounds are sold by Albright and Wilson Ltd. and FMC, respectively. The preparation of the hydroxymethylphosphonium salt has been discussed (see FLAME RETARDANTS). Synthesis of the tetraalkylphosphonium chloride follows the reaction described above except that 1-chlorotetradecane is employed in place of 1-chlorobutane.

Ultraviolet Photoinitiators. Ciba-Geigy has introduced a type of phosphine-based photoinitiator. In general, the compound can be described as a bis(acyl)phosphine oxide and is prepared by the reaction of a monoalkylphosphine with a substituted benzoyl chloride. The composition of the first commercial product is proprietary.

Solvent Extraction Reagents. The large commercial uses of phosphine derivatives in this area involve the separation of cobalt from nickel and the recovery of acetic acid and uranium.

WILLIAM A. RICKELTON
Cytec Canada Inc.

Fr. 1,352,605 (Feb. 14, 1964), (to Albright & Wilson Ltd. and Hooker Chemical Corp.).

Phase Transfer Catalysis in Industry, PTC Interface, Inc., Marietta, Ga.

K. Dietliker and co-workers, "Novel High Performance Bisacylphosphine Oxide (BAPO) Photoinitiators," paper presented at *RadTech'94, Orlando, Florida*, May 1–5, 1994.

W. A. Rickelton, D. S. Flett, and D. W. West, *Solv. Extr. Ion Exch.* **2**(6), 815–838 (1984).

PHOSPHORIC ACIDS AND PHOSPHATES

Phosphoric acids and the phosphates may be defined as derivatives of phosphorus oxides where the phosphorus atom is in the +5 oxidation state. These are compounds formed in the M_2O-P_2O_5 system, where M represents one cation equivalent, eg, H^+, Na^+, 0.5 Ca^{2+}, etc.

Orthophosphoric acid, H_3PO_4, can be considered the building block from which other phosphoric acids and the phosphate salts are derived through the basic reactions of polymerization and/or neutralization. Polymerization occurs via dehydration, hence the polymers are known generally as condensed phosphates.

Traditionally, phosphates have been represented as stoichiometric combinations of oxides.

Phosphoric Acids

Orthophosphoric Acid. Phosphoric acid is a tribasic acid, in which the first hydrogen ion is strongly ionizing, the second moderately weak, and the third very weak.

$$H_3PO_4 \overset{-H^+}{\rightleftharpoons} H_2PO_4^- \overset{-H^+}{\rightleftharpoons} HPO_4^{2-} \overset{-H^+}{\rightleftharpoons} PO_4^{3-}$$

Phosphoric acid, aside from its acidic behavior, is relatively unreactive at room temperature. At higher temperatures, the acid reacts with most metals and their oxides. Phosphoric acid is stronger than acetic, oxalic, silicic, and boric acids, but weaker than sulfuric, nitric, hydrochloric, and chromic acids.

Table 1. Physical Properties of Aqueous Solutions of Phosphoric Acid

Concentration, wt %					Viscosity, mPa·s(= cP)		
H_3PO_4	P_2O_5	Density at 25°C, g/cm³	Boiling point, °C	Freezing point, °C	20°C	60°C	100°C
0	0	0.997	100.0	0	1.0	0.48	0.30
5	3.62	1.025	100.1	−0.8	1.1	0.54	0.33
10	7.24	1.053	100.2	−2.1	1.2	0.61	0.38
20	14.49	1.113	100.8	−6.0	1.6	0.78	0.48
30	21.73	1.182	101.8	−11.8	2.2	1.0	0.62
50	36.22	1.333	108	−44.0	4.3	1.8	1.1
75	54.32	1.573	135	−17.5	15	4.8	2.4
85	61.57	1.685	158	21.1	28	8.1	3.8
100	72.43	1.864	261	42.35	140	25	9.2
105	76.10	1.925	>300	16.0	600	70	1.9
115	83.29	2.044	>500			1500	250

Physical properties of phosphoric acid solutions of various concentrations are listed in Table 1.

Manufacture. Phosphoric acid, H_3PO_4, is the second largest volume mineral acid produced; sulfuric acid is the first. The greatest consumption of phosphoric acid is in the manufacture of phosphate salts, as opposed to direct use as acid. Markets are differentiated according to the purity of the acid.

Phosphoric acid is produced commercially by either the wet process or the thermal (furnace) process. Thermal acid, manufactured from elemental phosphorus, is more expensive and considerably purer than wet-process acid. Wet-process acid is used primarily in the production of fertilizers and animal feed supplements. Wet-process acid may be purified for the manufacture of technical- and food-grade phosphate salts, usually employing a solvent extraction process. Both thermal and purified wet-process phosphoric acid (WPA) are used almost exclusively in various technical and food applications where fertilizer-grade wet acid is not suitable.

Thermal Process. In the manufacture of phosphoric acid from elemental phosphorus, white (yellow) phosphorus is burned in excess air, the resulting phosphorus pentoxide is hydrated, heats of combustion and hydration are removed, and the phosphoric acid mist collected. The principal process types (Fig. 1) include the wetted-wall, water-cooled, or air-cooled combustion chamber, depending on the method used to protect the combustion chamber wall.

Wet Process. Over 90% of the phosphoric acid produced, both in the United States and worldwide, is wet-process phosphoric acid.

Wet-process acid is manufactured by the digestion of phosphate rock (calcium phosphate) with sulfuric acid. Phosphoric acid is separated from the resultant calcium sulfate slurry by filtration. To generate a filterable slurry and to enhance the P_2O_5 content of the acid, much of the acid filtrate is recycled to the reactor. Two main categories of the wet process exist, depending on whether the calcium sulfate is precipitated as the dihydrate or the hemihydrate. For more detailed discussion of the wet-process acid, see FERTILIZERS.

Purification. Process development for the purification of wet-process acid has taken place primarily outside North America where the cost differential between the sulfur used in manufacture of wet-process acid and the electricity needed for thermal acid has been large. Decline of the market for technical-grade phosphates in detergents, along with the escalating cost of electric power for elemental phosphorus production, has resulted in the closing of less efficient elemental phosphorus facilities and the introduction of wet-acid purification into the United States.

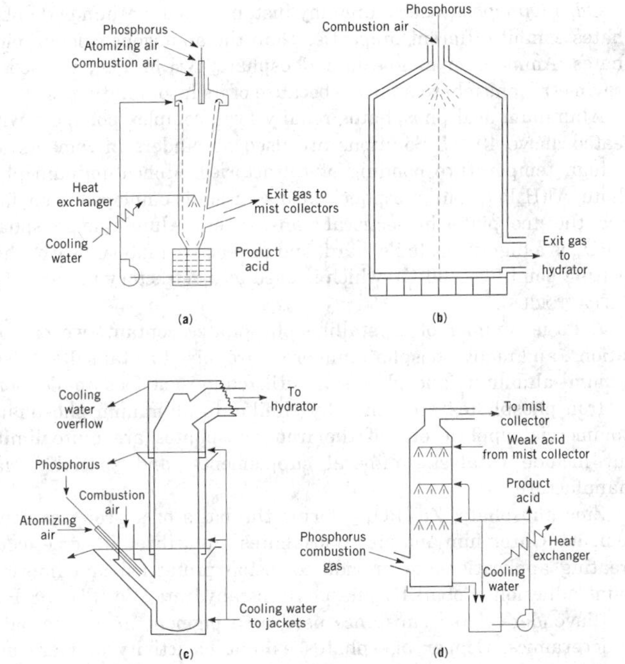

Figure 1. Thermal phosphoric acid processes: (**a**) wetted-wall combustion chamber; (**b**) air-cooled combustion chamber; (**c**) water-cooled combustion chamber; and (**d**) hydrator-absorber.

Chemical precipitation and solvent extraction are the main methods of purifying wet-process acid, although other techniques such as crystallization and ion exchange (qv) have also been used.

Condensed Phosphoric Acids. Commercial condensed phosphoric acids are mixtures of linear polyphosphoric acids made by the thermal process either directly or as a by-product of heat recovery. Wet-process acid may also be concentrated to ~70% P_2O_5 by evaporation. Linear phosphoric acids are strongly hygroscopic and undergo viscosity changes and hydrolysis to less complex forms when exposed to moist air. Upon dissolution in excess water, hydrolytic degradation to phosphoric acid occurs; the hydrolysis rate is highly temperature-dependent.

Pyrophosphoric (diphosphoric) acid, $H_4P_2O_7$, is the only condensed phosphoric acid definitely obtainable in crystalline form. The term metaphosphoric acid should technically be reserved for ring-structured acids. Two cyclic acids are reasonably well-defined, tri- and tetrametaphosphoric acids, $H_3P_3O_9$ and $H_4P_4O_{12}$, respectively. Tetrametaphosphoric acid is the main product resulting from the hydrolysis of phosphorus pentoxide and an excess of water in organic medium. Trimetaphosphoric acid can be prepared by ion exchange of its sodium salt, a commercially available material. The cyclic metaphosphoric acids are strong acids having a single, strong inflection in their pH titration curves.

Uses. Owing to extensive use in fertilizers, wet-process phosphoric acid is the largest source of phosphoric acid in the United States, accounting for more than 90% of total acid demand. The remaining phosphoric acid is technical- and food-grade acid supplied by the furnace process or purified wet-process acid. Most of this relatively pure material is marketed in the various forms of phosphate salts. Technical- and food-grade phosphoric acid is used in a variety of applications, including metal treatment, refractories (qv), catalysts, foods, and beverages (see FOOD ADDITIVES).

Phosphates

Orthophosphates. Orthophosphate salts are generally prepared by the partial or total neutralization of orthophosphoric acid. The commercial phosphates include alkali metal, alkaline-earth, heavy metal, mixed metal, and ammonium salts of phosphoric acid. Sodium phosphates are the most important, followed by calcium, ammonium, and potassium salts.

Sodium Phosphates. Elementary chemical considerations might predict three simple sodium phosphates resulting from successive neutralization of the acidic protons of phosphoric acid; ie, monosodium dihydrogen phosphate (MSP) NaH_2PO_4; disodium monohydrogen phosphate (DSP) Na_2HPO_4; and trisodium phosphate (TSP), Na_3PO_4. There are double salts as well as several hydrate forms.

Both mono- and disodium phosphates are prepared commercially by neutralization of phosphoric acid using sodium carbonate or hydroxide. Crystals of a specific hydrate can then be obtained by evaporation of the resultant solution within the temperature range over which the hydrate is stable.

The trisodium phosphate system is the most complex and the commercial product is generally of variable composition and often contains excess sodium hydroxide. Trisodium phosphate readily forms a variety of double salts with other sodium compounds. The double salt of trisodium phosphate and sodium hypochlorite is a source of both alkalinity and active chlorine in disinfectant cleaners and automatic dishwashing formulations.

Uses. The principal use of monosodium phosphate is as a water-soluble solid acid and pH buffer, primarily in acid-type cleaners. Mixtures of mono- and disodium phosphates are used in textile processing, food manufacture, and other industries to control pH at 4–9. Monosodium phosphate is also used in boiler-water treatment, as a precipitant for polyvalent metal ions, and as an animal-feed supplement.

The single largest use for disodium phosphate is as an emulsifying aid for pasteurized process cheese. Other food-related uses are in ham curing, starch processing, and as an ingredient in instant cereals and evaporated milk (see MILK AND MILK PRODUCTS). Disodium phosphate is also used in the preparation of certain ceramic glazes and enamels, in leather (qv) tanning, textile dyeing, pigment manufacture, water (qv) treatment, and detergents (see ENAMELS, PORCELAIN OR VITREOUS; PIGMENTS; TEXTILES).

Trisodium phosphate is strongly alkaline; many of its applications depend on this property. Traditionally, trisodium phosphate has been used in water softening to remove polyvalent metal ions by precipitation as insoluble phosphates. Because the hypochlorite complex of trisodium phosphate provides solutions that are strongly alkaline and contain active chlorine, it is used in disinfectant cleaners, scouring powders, and automatic dishwashing formulations.

Potassium Phosphates. The K_2O-P_2O_5-H_2O system parallels the sodium system in many respects. In addition to the three simple phosphate salts obtained by successive replacement of the protons of phosphoric acid by potassium ions, the system contains a number of crystalline hydrates and double salts.

Although the cost of the potassium phosphates is higher than the corresponding sodium salts, the former have applications utilizing their higher solubility and nutrient value. Monopotassium phosphate (MKP) is also used in various buffering systems and in paper (qv) processing. The piezoelectric effect of MKP has led to its use in sonar systems and other electronic applications.

Dipotassium phosphate (DKP) and tripotassium phosphate (TKP) are marketed both as solids and in 50% active solution. Most of the commercial output is used in conjunction with borates, nitrites, nitrates, and/or silicates as the corrosion inhibitor system in ethylene glycol antifreeze formulations (see ANTIFREEZES AND DEICING FLUIDS). The second largest use for DKP is as a buffer in coffee creamers to prevent casein protein coagulation and precipitation by coffee acids. Tripotassium phosphate is utilized in the polymerization of styrene-butadiene rubber to control the polymerization rate and latex stability.

Ammonium Phosphates. Ammonium phosphates are comparatively unstable. Monoammonium and diammonium phosphates are produced on a large scale as fertilizers.

Owing to the thermally unstable nature of ammonium phosphates, other applications are related to flame retarding and fire extinguishing.

Evolution of ammonia from a boiling dilute solution of diammonium phosphate gradually reduces the pH. This process is used commercially to control the precipitation of alkali-soluble—acid-insoluble colloidal dyes on wool.

Calcium Phosphates. Calcium phosphates include the most abundant natural form of phosphorus, ie, apatites, $Ca_{10}(PO_4)_6X_2$, where X = OH, F, Cl, etc. Apatite ores are the predominant basic raw material for the production of phosphorus and its derivatives. Calcium phosphates are the main component of bones and teeth. After sodium phosphates, the calcium salts are the next largest volume technical- and food-grade phosphates. Many commercial applications of the calcium phosphates depend on their low solubilities.

Many orthophosphate salts, in particular those of polyvalent cations, exhibit incongruent solubility where disporportionation occurs in solution to yield a more basic orthophosphate salt and phosphoric acid. Hydrolytic disproportionation is probably one of the mechanisms related to the formation of bone and naturally occurring apatites.

For many years, the hygroscopic nature of anhydrous monocalcium phosphate limited its commercial applications. However, the addition of small amounts of K^+, Na^+, and Al^{3+} ions to the crystallization mother liquor followed by heating to >200°C results in a mixed metal-polyphosphate coating on the calcium phosphate. This glassy coating protects the calcium phosphate from moisture and greatly improves its handling properties and performance in several applications, most notably as a leavening agent.

Crystalline $CaHPO_4 \cdot 2H_2O$ loses both water molecules in a single step at moderately elevated temperature or upon storage to yield the anhydrous salt. Addition of a few percent of tetrasodium pyrophosphate or trimagnesium phosphate, $Mg_3(PO_4)_2$, stabilizes the dihydrate. These materials are used widely to stabilize , particularly for toothpaste applications.

Tricalcium phosphate, $Ca_3(PO_4)_2$, is formed under high temperatures and is unstable toward reaction with moisture below 100°C. Commercial tricalcium phosphate prepared by the reaction of phosphoric acid and a hydrated lime slurry consists of amorphous or poorly crystalline basic calcium phosphates close to the hydroxyapatite composition and has a Ca/P ratio of approximately 3:2.

Hydroxyapatite, $Ca_{10}(PO_4)_6(OH)_2$, may be regarded as the parent member of a whole series of structurally related calcium phosphates that can be represented by the formula $M_{10}(ZO_4)_6X_2$, where M is a metal or H_3O^+; Z is P, As, Si, Ga, S, or Cr; and X is OH, F, Cl, Br, 1/2 CO_3, etc. The apatite compounds all exhibit the same type of hexagonal crystal structure. Included are a series of naturally occurring minerals, synthetic salts, and precipitated hydroxyapatites. Highly substituted apatites such as Francolite, $Ca_{10}(PO_4)_{6-x}(CO_3)_x(F, OH)_{2+x}$, are the principal component of phosphate rock used for the production of both wet-process and furnace-process phosphoric acid.

Uses. Commercial monocalcium phosphate is available as both the anhydrous and the monohydrate salts. Most uses are based on acidic properties. Monocalcium phosphate is used to control acidity in powdered drink mixes, as an ingredient in effervescent tablets, as a plastics stabilizer, and in ceramics. Its single largest application is as a leavening agent in bread, cake mixes, and self-rising flour.

The main use for dicalcium phosphate on a tonnage basis is as an animal feed supplement (see FEEDS AND FEED ADDITIVES), for which it is produced from defluorinated wet-process phosphoric acid. Food-grade dicalcium phosphate is used as a dental polishing agent in toothpastes. Dicalcium phosphate also is used as a leavening agent, plastics stabilizer, and in the manufacture of glass, medicines, and phosphors.

Commercial tricalcium phosphate is an effective flow conditioner for food products such as sugar and salt. The product is also used as a whitening agent in the manufacture of ceramics, as a mordant in dyeing, and as a polishing agent. Considerable research on apatites has been sparked in the 1980s and 1990s by the desire for biocompatible bone and tooth enamel replacements (see PROSTHETIC AND BIOMEDICAL DEVICES).

Other Orthophosphates. In many instances, magnesium orthophosphates exhibit different properties than the analogous calcium phosphates. Ammonium magnesium phosphate, NH_4MgPO_4, is used for gravimetric phosphate analysis because of its insolubility in water.

Aluminum acid phosphates readily form complex polymers when heated above 400°C. Solutions are used as binders in cements and in high temperature bonding of refractories. Monoaluminum phosphate, $Al(H_2PO_4)_3$, in phosphoric acid solution is employed to surface-treat the steel plates in electrical transformers. Aluminum phosphate, $AlPO_4$, is a highly insoluble, hard, and unreactive material with a high melting point ((>1800°C)) which is used as a refractory material (see REFRACTORIES).

A large number of crystalline phosphates contain two or more cations, and many phosphate minerals are mixed metal salts. Mixed-sodium–aluminum phosphates are utilized in some food applications.

Iron phosphates are generally similar to aluminum phosphates. Commercial applications of the iron phosphates are quite limited but include catalysts, mineral supplements, and specialty glass manufacture.

Zinc phosphate, $Zn_3(PO_4)_2$, forms the basis of a group of dental cements. Chromium and zinc phosphates are utilized in some metal-treating applications to provide corrosion protection and improved paint adhesion. Cobalt(II) phosphate octahydrate, $Co_3(PO_4)_2 \cdot 8\,H_2O$, is a lavender-colored substance used as a pigment in certain paints and ceramics. Copper phosphates exhibit bioactivity and are used as insecticides and fungicides. Zinc, lead, and silver phosphates are utilized in the production of specialty glasses. The phosphate salts of heavy metals such as Pb, Cr, and Cu, are extremely water insoluble.

The tertiary metal phosphates are of the general formula MPO_4, where M is B, Al, Ga, Fe, Mn, etc. Boron phosphate has limited use as a catalyst support, in ceramics, and in refractories.

Many phosphates exhibit two- or three-dimensional structures. The titanium and zirconium phosphates, $M(HPO_4)_2 \cdot nH_2O$, form inert, high temperature-stable ion-exchange agents possessing a layered structure. Both α-$Zr(HPO_4)_2 \cdot H_2O$ and γ-$Zr(HPO_4)_2 \cdot 2\,H_2O$ are of particular interest. Proposed uses include high temperature processing of nuclear waste and kidney dialysis.

Condensed Phosphates. Condensed phosphates are derived by dehydration of acid orthophosphates. The resulting polymeric structures are based on a backbone of P—O—P linkages where PO_4 tetrahedra are joined by shared oxygen atoms.

Pyrophosphates. The simplest linear condensed phosphates are pyrophosphates, which can be considered as the dehydration product of two orthophosphate groups. Many pyrophosphates can be prepared by thermal treatment of the acid orthophosphates.

A large number of pyrophosphate salts have been prepared. In addition to individual metal salts, ammonium pyrophosphates and many mixed-metal pyrophosphates are known. Pyrophosphates of notable commercial importance include sodium, potassium, and calcium salts.

Commercially important sodium pyrophosphates include tetrasodium pyrophosphate (TSPP), $Na_4P_2O_7$, and disodium pyrophosphate, $Na_2H_2P_2O_7$, commonly referred to as sodium acid pyrophosphate (SAPP). These are prepared industrially by thermal dehydration of disodium and monosodium orthophosphate, respectively.

Tetrasodium pyrophosphate is a builder in detergent and cleaning formulations. Food applications include consistency control in buttermilk (thinning), chocolate milk (thickening), and instant puddings. Tetrasodium pyrophosphate is widely used as an effective deflocculant, eg, in kaolin clays, drilling muds, dyes, and inks. TSPP stabilizes hydrogen peroxide through chelation of heavy-metal ion impurities that catalyze peroxide decomposition. TSPP is used as an anticalculus agent in toothpastes and mouthwashes (see DENTIFRICES). Commercial $Na_2H_2P_2O_7$ typically contains small amounts of added potassium, calcium, and aluminum to reduce the rate of reaction with sodium bicarbonate in leavening, the largest application of SAPP. SAPP is also used to eliminate darkening of cut potatoes to prevent the formation of highly colored iron–tannin complexes. Other applications are in acid-cleaning formulations and in electroplating.

Tetrapotassium pyrophosphate has been used as a highly soluble detergent builder. The calcium pyrophosphates are utilized primarily as dental abrasives in fluoride-containing toothpastes.

Tripolyphosphates. The most commercially important tripolyphosphate salt is sodium tripolyphosphate (STP), $Na_5P_3O_{10}$. Sodium tripolyphosphate is produced by calcination of an intimate mixture of orthophosphate salts containing the correct overall Na/P mole ratio of 1.67. The solubility and hydration behavior of sodium tripolyphosphate are of particular importance in many of its industrial applications.

Uses. As a builder in cleaning formulations, sodium tripolyphosphate is used in household laundry products, automatic dishwashing formulations, car washes, and numerous industrial cleaners.

Food-grade sodium tripolyphosphate is used for the curing of hams and bacon. Treatment with STP improves the quality of poultry and seafood products. Uses for technical-grade material include clay processing, water softening, textile processing, paper pulping, rubber and paint manufacture, drilling muds, and ore flotation.

Long-Chain Polyphosphates and Metaphosphates. The composition of long-chain polyphosphates approaches that of metaphosphate, $(MPO_3)_n$, and long-chain polyphosphates may commonly be referred to as metaphosphates, although this term should be reserved for cyclic anions of the exact (PO_{3n}^-) composition. Most polyphosphates are amorphous glasses, but several high molecular weight polyphosphates occur as crystalline substances. Both types are used commercially.

Thermal dehydration of monosodium phosphate gives rise to numerous condensed polyphosphates. Structures are diverse and can be controlled by manipulating the conditions of dehydration, ie, temperature, water vapor, and tempering.

Potassium Kurrol's salt, potassium polymetaphosphate, $(KPO_3)_n$, is easily obtained by thermal dehydration of KH_2PO_4.The potassium salt has limited commercial usage in sausage processing in Europe.

Ammonium polyphosphate, $(NH_4PO_3)_n$, is most easily obtained by heating a mixture of $NH_4H_2PO_4$ and urea in an atmosphere of NH_3. There are at least five crystalline forms under the generic term ammonium polyphosphate. Form I, NH_4PO_3-I, is used as a water-insoluble fire retardant in intumescent paints and mastics. Form II, having a higher temperature stability than Form I, is used as a fire retardant in thermoplastics. Ammonium polyphosphate liquid fertilizers are made from wet-process acid.

Properties of Condensed Phosphates. Condensed phosphates all exhibit hydrolytic instability of the P—O—P linkages and, under the appropriate conditions, can all be cleaved, ultimately affording the monomeric orthophosphate ion. Like the orthophosphates, condensed phosphates may also exhibit hydrolytic disproportionation, ie, incongruent solubility.

The hydrolysis rates of polyphosphates are mainly affected by temperature, pH, and the location of the phosphate group in a condensed phosphate. Hydrogen ion is a good catalyst for hydrolytic degradation of the P—O—P linkage. Many cations have a catalytic effect on hydrolysis, although generally less than that exhibited by hydrogen ions.

Long-chain polyphosphate hydrolysis is more complex than that of the shorter chains on account of additional mechanistic pathways and the accompanying formation of cyclic metaphosphates. Three mechanisms for hydrolysis of polyphosphate chains in solution are generally recognized: clipping of a monomeric unit from the end of the chain, concurrent loss of three units from the end of the chain by splitting off a trimetaphosphate ion from the end or interior of the chain, and random cleavage from within the interior of the chain to afford shorter chains.

Phosphates form water-soluble complex ions with metallic cations, a phenomenon commonly called sequestration. The amount of metal ion that can be sequestered by polyphosphates generally increases with increasing chain length. Sequestration forms the basis for detergent and water-treatment applications of polyphosphates.

Polyphosphates are strongly sorbed onto a variety of surfaces where they alter the charge. As a result, the properties of colloidal systems can be dramatically changed by the addition of small amounts of polyphosphate. A striking example is the deflocculation of clays.

Calcium carbonate (calcite) scale formation in hard water can be prevented by the addition of a small amount of soluble polyphosphate in a process known as threshold treatment.

Manufacture of Phosphate Salts

Condensed phosphates are prepared from the appropriate orthophosphate or mixture of orthophosphates. Phosphoric acid is neutralized to form a solution or slurry with a carefully adjusted acid/base ratio according to the desired orthophosphate product. The orthophosphate may be recovered either by crystallization from solution, or the entire solution or slurry may be evaporated to dryness. Acid orthophosphate salts may be converted to condensed phosphates by thermal dehydration (calcination).

Orthophosphates. *Alkali Metal Phosphates.* Alkali metal orthophosphates generally exhibit congruent solubility and are therefore usually manufactured by either crystallization from solution or drying of the entire reaction mass. Alkaline-earth and other phosphate salts of polyvalent cations typically exhibit incongruent solubility and are prepared either by precipitation from solution having a metal oxide/P_2O_5 ratio considerably lower than that of the product, or by drying a solution or slurry with the proper metal oxide/P_2O_5 ratio.

Alkali metal phosphates include monosodium phosphate (NaH_2PO_4), disodium phosphate (Na_2HPO_4), and trisodium-phosphate (TSP).

Potassium Phosphates. Potassium phosphate salts are analogous to the sodium salts and share many of the same functional properties.

Ammonium Phosphates. In the manufacture of ammonium phosphates, an atmosphere of ammonia may need to be maintained because the partial pressure of ammonia rises rapidly as either the temperature or the NH_3/P_2O_5 mole ratio of the reaction mass increases.

Monoammonium phosphate (MAP), $NH_4H_2PO_4$, is produced by reaction of anhydrous ammonia and phosphoric acid in batch or continuous reactors and crystallized in conventional crystallizers because the partial pressure of ammonia over this acidic solution is relatively low (see CRYSTALLIZATION). Diammonium phosphate (DAP), $(NH_4)_2HPO_4$, solutions, on the other hand, have a high partial pressure of ammonia and the reaction is carried out in a two-stage reactor system in which the feed acid passes counter-currentwise to the flow of ammonia gas.

Calcium Phosphates. Because calcium phosphates and calcium bases as raw materials have low solubilities, the manufacture of calcium phosphates must therefore deal with nonequilibrium chemistry in a heterogenous system. Calcium phosphates also exhibit incongruent solubility, which means that the mother liquor necessarily has a different CaO/P_2O_5 ratio than the product. As a result, most commercial calcium phosphates are mixtures of several salts having an average composition approximating that of the pure material.

For fertilizer and animal nutrition uses, the primary concern is the CaO and P_2O_5 analysis of the product. For industrial, dentifrice, food, and pharmaceutical uses, important functional properties relating to composition or solids characteristics may significantly differ in two products having nearly identical CaO and P_2O_5 content. Manufacture is something of an art and details are often proprietary.

Monocalcium phosphate (MCP) is generally made as a composition equivalent to the monohydrate, $Ca(H_2PO_4)_2 \cdot H_2O$. The monohydrate is manufactured by several methods. Phosphoric acid and hydrated lime may be mixed in a pan or other heavy-duty mixer that allows the rapid escape of steam. The product is a paste that is dried and sized. The reaction may also be carried out in a more dilute system in conventional mixing equipment to produce a pumpable slurry that is then spray dried to give a lighter and more rapidly soluble product. Commercial MCP monohydrate usually contains several percent of dicalcium phosphate, which is acceptable in baking powder as an anticaking agent. MCP may also be manufactured by crystallization from solution.

Anhydrous monocalcium phosphate, $Ca(H_2PO_4)_2$, can be made in a pan mixer from concentrated phosphoric acid and lime. A small amount of aluminum phosphate or a mixture of sodium and potassium phosphates is added in the form of proprietary stabilizers for coating the particles.

Formation of minor amounts of more basic calcium phosphates results in rendering soluble fluoride in the dentifrice formulation as an inactive and insoluble fluorapatite. A much more dilute slurry reaction is used to obtain good mixing. Conventional mixing equipment is used for the reaction followed by centrifugation, drying, and milling to the desired particle size. Stability of the DCP dihydrate (DCPD) against reaction with fluoride in toothpaste formulations is of tantamount importance. The dihydrate is often stabilized by sodium pyrophosphate and/or trimagnesium phosphate.

Commercial tricalcium phosphate (TCP) is actually a basic calcium phosphate close to hydroxyapatite in composition. TCP is separated by drum-, spray-, or flash-drying the TCP slurry, with or without intermediate sedimentation or filtration steps. It is used as an industrial-grade flow conditioner and parting agent.

Condensed Phosphates. Condensed phosphates are prepared by calcining an orthophosphate composition having the proper metal oxide/P_2O_5 mole ratio. Rotary calciners, either gas- or oil-fired, are preferred. Following calcination, the product is water-cooled in screw or tube coolers and then sized to granular or powder specifications by screening, air separation, and milling techniques.

Three crystalline polyphosphates can be produced by thermal dehydration of monosodium phosphate at successively higher temperatures: sodium acid pyrophosphate (SAPP), insoluble metaphosphate (IMP), and sodium trimetaphosphate (STMP).

The poorly soluble, long-chain crystalline ammonium polyphosphate $(NH_4PO_3)_n$, is used in intumescent fire-retardant coatings and paints where resistance to leaching by water is required. It is manufactured by heating a mixture of urea (qv) and ammonium phosphate or polyphosphoric acid under a controlled atmosphere of NH_3 and water at ~300°C.

Calcium pyrophosphate, $Ca_2P_2O_7$, is manufactured by high temperature calcination of DCP in a rotary calciner.

Economic Aspects

Over 90% of phosphoric acid production is wet-process (agricultural-grade) acid; the remainder is industrial-grades (technical, food, pharmaceutical, etc) made by the thermal route or by the purification of wet-process acid.

U.S. consumption of industrial-grade phosphoric acid and phosphates in 1993 according to product categories was phosphoric acid, at 29%; sodium phosphate, 52%; calcium phosphate, 7%; potassium phosphate, 3%; ammonium phosphate, 5%; and others, 4%. They are used in builders and water treatment, and food and metal production.

The production of industrial-grade acid and phosphates has been influenced since the 1970s by environmental concerns and related legislation. Industry consolidation and rationalization have resulted. A shift in production toward wet-acid purification routes has occurred and only the most economically viable elemental phosphorus and thermal acid producers remain in business. There has also been an increased focus on higher value phosphate products.

Safety and Environmental Considerations

Inorganic phosphates present little hazard to humans and are mineral nutrients essential to life processes. Attention must be given to the acidity of phosphoric acid, the alkalinity of the bases with which it reacts and the heat released upon neutralization. Appropriate protective gear should be worn when in close contact. Some phosphate salts are reasonably acidic or basic.

Larger environmental issues are associated with the manufacture of wet-process acid and elemental phosphorus, than with the manufacture of technical- or food-grade acids and salts from these raw materials.

Because of the nutritive value, phosphates have been implicated in promoting the growth of algae in lakes. Considerable controversy has centered on the contribution of phosphate-built detergents to excessive algae growth and subsequent eutrophication of natural receiving water. Legislation against the use of phosphates in detergents has resulted in a patchwork of restrictions worldwide. Societal pressure has resulted in the voluntary reduction or elimination of phosphates in many cleaning products by the manufacturers. A more logical but also more costly approach is phosphorus removal during sewage treatment. Excellent reviews of this area are available.

DAVID R. GARD
Solutia Inc.

J. R. Van Wazer, ed., *Phosphorus and Its Compounds*, Vol. 1 (1958) and Vol. 2 (1961), Interscience Publishers, Inc., New York.

D. E. C. Corbridge, *Phosphorus: An Outline of Its Chemistry, Biochemistry, and Technology*, 4th ed., Elsevier, Amsterdam, the Netherlands, 1990.

K. Schrödter and co-workers, in B. Elvers, S. Hawkins, and G. Schulz, eds., *Ullmann's Encyclopedia of Industrial Chemistry*, VCH, Weinheim, Germany, 1991, Vol. A19, 465–503.

A. D. F. Toy, *Phosphorus Chemistry in Everyday Living*, American Chemical Society, Washington, D.C., 1987.

PHOSPHORS. See See LUMINESCENT MATERIALS; PHOTODETECTORS.

PHOSPHORUS

Phosphorus is a nonmetallic element having widespread occurrence in nature as phosphate compounds (see PHOSPHORIC ACIDS AND PHOSPHATES). Fluorapatite, $Ca_5F(PO_4)_3$, is the primary mineral in phosphate rock ores from which useful phosphorus compounds (qv) are produced. The recovery from the ore into commercial chemicals is accomplished by two routes: the electric furnace process, which yields elemental phosphorus; and the wet acid process, which generates phosphoric acid. Less than 10% of the phosphate rock mined in the world is processed in electric furnaces. Over 90% is processed by the wet process, used primarily to make fertilizers (qv).

Most of the phosphorus produced as the element is later converted to high purity phosphoric acid and phosphate compounds; the remainder is used in direct chemical synthesis to produce high purity products. Recently, a small portion of wet acid is purified in a second process and then also used in high purity acid and phosphate compound applications.

Elemental phosphorus is produced and marketed in the α-form of white or yellow phosphorus, the tetrahedral P_4 allotrope. A small amount of red amorphous phosphorus, P, is produced by conversion from white phosphorus. White phosphorus as the element is characterized by its combustion in air to form phosphorus pentoxide. Consequently, white phosphorus is generally stored and handled under water. Elemental white phosphorus is also highly toxic, and suitable precautions are required by those who manufacture or handle it. North American production of elemental phosphorus reached its zenith in the early 1970s. Production has declined since then because of reformulations of detergent phosphates.

Physical Properties

White phosphorus is a soft waxy solid often compared to paraffin wax. Phosphorous has low solubility in most common solvents, but is quite soluble in carbon disulfide and some other special solvents.

Some of the more common physical properties of α-white phosphorus are given in Table 1.

Upon heating, α-white phosphorus first metals, then either vaporizes or converts to amorphous red phosphorus. The conversion to red P proceeds slowly in one to two days at temperatures slightly below the 280°C boiling point of liquid P_4.

Several allotropes of black phosphorus have also been reported.

Table 1. Physical Properties of α-White Phosphorus, P_4

Parameter	Value
melting point, °C	44.1
boiling point, °C	280.5
density, g/cm^3	
solid	1.83
liquid at 50°C	1.74
liquid viscosity, at 50°C, mPa·s(= cP)	1.69

Chemical Properties

Phosphorus shows a range of oxidation states from −3 to +5 by virtue of its electronic configuration. Elemental P_4 is oxidized easily by nonmetals such as oxygen, sulfur, and halides to form compounds such as P_2O_5, P_2S_5, and PCl_3. It is also reduced upon reaction with metals to generate phosphides. Certainly the more useful and common reactions of phosphorus involve oxidation with air, sulfur, or chlorine. The largest volume products are phosphoric acid and phosphate derivatives of phosphoric acid (see PHOSPHORUS COMPOUNDS).

Elemental phosphorus reacts with oxidizing acids such as nitric or strong sulfuric. It also reacts with alkali, forming a combination of phosphine, hypophosphite, and phosphite at increasing rates as the pH increases.

Manufacture of White Phosphorus

As of the mid-1990s all commercial phosphorus is manufactured at a few sites around the world. Significant production occurs in Idaho in the United States, in the Netherlands, and in China; smaller production occurs in Russia, and India.

Elemental phosphorus is produced from a phosphorus-rich ore mostly recovered by strip mining. This ore usually contains fluorapatite, plus some silica and silicates. When a carbon source, usually coke, is added to the ore at temperatures greater than 1100°C, the following overall reaction occurs:

$$2\ Ca_5F(PO_4)_3 + 9\ SiO_2 + 15\ C \rightarrow 9\ CaSiO_3 + CaF_2 + 15\ CO(g) + 3\ P_2(g)$$

As the gas cools, dimers combine as:

$$2\ P_2 \rightarrow P_4$$

Current phosphorus production uses a submerged arc furnace. The submerged arc furnace performs three functions: chemical reactor, heat-exchanger, and gas–solid filter, respectively, each of which requires a significant amount of preparation for the solid furnace feed materials.

The two ore preparation technologies in common use are moving grate calciners and rotary kilns. The primary purposes of this preparation step are to produce strong feed agglomerates, often called nodules; to provide a consistently sized material; and to remove energy-consuming impurities that disrupt furnace operation. Some common preparation steps for the silica include crushing, screening, washing, and drying. The coke may be metallurgical, petroleum, or formed coke, but must be size-controlled and is often dried before use.

Once each of these materials has been prepared for furnace use, a furnace charge (burden) is produced by mixing and proportioning the three components. The burden is then transported to the furnace charge bins. As the furnace is operated, feed falls continuously by gravity from the charge bins through a system of vertical chutes into the furnace. Two molten furnace by-products, slag and ferrophos, are tapped (removed) from the furnace using either interval or continuous tapping. The furnace off-gas containing primarily phosphorus and carbon monoxide leaves the furnace by flowing up through the porous burden while exchanging heat and dust. Next, the off-gas is usually cleaned of dust particles, using an electrostatic precipitator. Lastly, P_4 is separated from the CO using a water-spray condenser, and the molten P_4 is pumped to storage tanks.

Mining. With the closure of all North American white phosphorus production facilities outside of the western United States, the only remaining mines utilized for U.S. P_4 production are located along the southern Idaho–Wyoming border. The two remaining white phosphorus producers as of 1998, Monsanto (Soda Springs, Idaho) and FMC (Pocatello, Idaho), operate mines near Soda Springs, Idaho. However, some ore mined in Florida is still shipped overseas for use in elemental phosphorus plants.

Electric Furnace. Present day phosphorus furnaces use a three-electrode, symmetrical triangular configuration with a three-phase Y or delta electrical connection operating in the 45–65 MW range at potentials of 200–650 V. These furnaces are circular, or of a rounded triangular cross-section. The furnaces are basically run continuously, except during repairs, process upsets, power curtailments, or electrode building. Typical operating characteristics for a large (60-MW), phosphorus furnace are shown in Table 2.

Product Recovery. At standard conditions (25°C, 101.3 kPa (1 atm)) typical furnace off-gas compositions are about 86% CO, 7.5% P_4, 5% H_2, 1% N_2, and traces of PH_3, CO_2, F, and S; large furnaces generate off-gas at a rate of about 120–180 m^3/min. In most installations the off-gas is passed through a series of Cottrell electrostatic precipitators which remove 80–95% of the dust particles. The phosphorus is typically condensed in closed spray towers. The condensed product along with the accompanying spray water is processed in sumps where the water is separated and recycled to the spray condenser, and the phosphorus and impurities are settled for subsequent purification.

Although most of the particulate in the off-gas from the furnace can be captured by the electrostatic precipitators before condensing the phosphorus, some carryover into the product P_4 is inevitable. This particulate is partly separated into the condenser water. The remainder reports to the phosphorus to yield either dirty product or a stable emulsion called phosphorus mud or sludge. Over many years a variety of approaches have been used to minimize the formation of sludge and to recover phosphorus product from the sludge.

Product Quality and Specifications. Most of the elemental phosphorus produced is converted to derivatives by the manufacturer. Some white phosphorus is sold on the open market.

Shipping and Handling

Phosphorus is stored and handled under a protective layer of water. Production quantities are transferred as a liquid by either water displacement or pumps, with water recycle to maintain the water balance and cover. As of the 1990s storage is limited to tanks located inside

Table 2. Operating Characteristics of a 60-MW Phosphorus Furnace

Parameter	Value
average potential between electrodes, V	250–350
power factor	0.96–0.98
raw materials consumed[a]	
power, kWh	13–15
fluorapatite ore, kg	10–13
coke, kg	1.2–1.7
silica, kg	1–2
products formed[a]	
slag	8–9
ferrophos	0.1–0.2
furnace off-gas	2.6–2.9
recovery, based on P_4 charged to the furnace, %	84–90
temperature of off-gas, °C	300–450
temperature of slag at tapping, °C	1400–1500

[a] Per kilogram of elemental P_4 produced.

diked areas that are accessible on the outside for safety and leakage control.

For off-site transportation, the phosphorus is loaded into railcars for transfer to the sites where it is used directly as a raw material or burned and hydrated to phosphoric acid.

Smaller amounts of phosphorus, or elemental phosphorus-containing materials, are also shipped in 115-L (30-gal) drums that are DOT regulated (U.S. DOT 1A1 or 1A2 classification) and have thick shells and special gaskets and fittings for protection. All air transportation of elemental P_4, both U.S. and international, was prohibited beginning in 1992.

A DOT regulation covers both domestic and international shipping. For transportation safety, the DOT has information for first responders to incidents involving elemental phosphorus. In addition, the Chemtrec phone number 1-800-424-9300 accesses DOT emergency information and assistance in the United States. Also, the phosphorus producers in the United States have established a Phosphorus Emergency Response Team (PERT) to assist in handling P_4 emergencies.

Health and Safety

At ambient temperatures white phosphorus spontaneously ignites when exposed to air. It has an autoignition temperature of 30°C. As a result, any human exposure to white phosphorus can cause severe thermal burns to the skin and eyes.

Phosphorus production plants and users should ensure that processing of the material is contained and that potential high exposure areas are well ventilated. Workers wear aluminized fiber glass or Kevlar flame-retardant full protective clothing, face shield with hard hat, rubber boots, and heavy rubber gloves when handling or transferring the product. Workers who have had dental surgery and pregnant women should be kept away from phosphorus exposure areas completely. Medical assistance should be obtained as soon as possible after any instance of phosphorus exposure.

By-Products

The electric furnace process generates four streams that can be considered by-products: slag, ferrophos, precipitator dust, and carbon monoxide off-gas. The approximate composition of the slag and precipitator dust are given in Table 3. These vary somewhat among different phosphorus manufacturers.

Environmental Control

Pollution control has become a primary concern and expense of elemental phosphorus producers since the 1970s. Problem areas include mining dusts, feed preparation and feed-handling dusts, slag dusts and emissions, precipitator dust, off-gas flare emissions, and phossy water and sludge treatment. The specific contaminants include elemental phosphorus, dust particulates, leachable hazardous metals, and low level radioactivity. The final discharges from the phosphorus plants are being confined to slag, ferrophos, and precipitator dusts, which are all nonhazardous wastes. Commercial application of the slag and precipitator solids is a point of contention owing to low level radioactivity. Ferrophos is often sold for its iron, vanadium, and chromium content.

Table 3. Composition of Phosphorus Manufacture By-Products, Wt%

By-product	P_2O_5	CaO	SiO_2	Al_2O_3	F	Fe_2O_3	K_2O	ZnO	C
slag	1.0–2.5	40–50	38–44	3–7	2–3	0.1–0.5			
precipitator dust	20–30	5–15	15–30	2–4	3–5		5–20	5–15	0–10

Economic Aspects

Beginning in 1969, a movement to restrict and then legally ban the use of phosphates in detergents led to the closing of significant amounts of plant capacity.

A second pressure on elemental P_4 production was the development of processes which remove impurities from phosphoric acid made by the wet process, to generate acid of equivalent purity to that obtained by the electric furnace route.

Outside of North America, the total number of plants, not including an undetermined number of smaller plants in China, has declined to less than a dozen.

Through March 1998 the list price of white phosphorus has remained stable for several years at $2.00/kg, freight equalized, in tank cars. After rationalization stabilizes the industry, the expectation is that the demand for white phosphorus should have grown 2% annually.

Uses

About 85% of the elemental P_4 is burned to P_2O_5 and hydrated to phosphoric acid. Part of the acid (ca 21%) is used directly, but the biggest part is converted to phosphate compounds. Final applications include home laundry and automatic dishwasher detergents, industrial and institutional cleaners, food and beverages, metal cleaning and treatment, potable water and wastewater treatment, antifreeze, and electronics. The purified wet acid serves the same markets.

The remaining 15% of the elemental P_4 is used in P_4-dependent applications which require the element as a direct reactant. Final applications include flame retardants (qv), lubricant additives, insecticides, herbicides, water treatment, cleaning compounds, plasticizers, and semiconductors.

Manufacture of Red Phosphorus

Red phosphorus is manufactured from white phosphorus for applications such as striking surfaces for matches, fireworks (see PYROTECHNICS), flame retardants in polymers, semiconductors, and PH_3 used to manufacture semiconductors. Manufacturers include Hoechst (Germany), United Phosphorus (India), Nippon (Japan), Rinkagaku Kogya (Japan), and Italmatch Srl (Italy).

J. R. BRUMMER
J. A. KEELY
T. F. MUNDAY
FMC Corporation

J. R. Van Wazer, *Phosphorus and Its Compounds,* Vol. II, Wiley-Interscience, New York, 1961.

H. Diskowski and T. Hofmann, in *Ullmann's Encyclopedia of Industrial Chemistry,* 5th ed., Vol. A 19, VCH Verlagsgesellschaft, Weinheim, Germany, 1991, pp. 505–525.

D. C. DeWitt, in *Encyclopedia of Chemical Processing and Design,* Vol. 36, Marcel Dekker, Inc., New York, 1991, pp. 1–33.

J. R. Van Wazer, *Phosphorus and Its Compounds,* Vol. I, Interscience Publishers, Inc., New York, 1958.

PHOSPHORUS COMPOUNDS

Phosphorus compounds exhibit an enormous variety of chemical and physical properties as a result of the wide range in the oxidation states and coordination numbers for the phosphorus atom. The most commonly encountered phosphorus compounds are the oxide, halide, sulfide, hydride, nitrogen, metal, and organic derivatives, all of which are of industrial importance. The halide, hydride, and metal derivatives, and to a lesser extent the oxides and sulfides, are reactive intermediates for forming phosphorus bonds with other elements.

The largest-volume phosphorus compounds are the phosphoric acids and phosphates (qv), ie, the oxide derivatives of phosphorus in the +5 oxidation state. With the exception of the phosphoric acid anhydride, P_4O_{10}, and the phosphate esters, these materials are discussed elsewhere (see PHOSPHORIC ACIDS AND PHOSPHATES). An overview of phosphorus compounds other than the phosphoric acids and phosphates is given herein. These phosphorus compounds are manufactured only from elemental phosphorus (qv) obtained by reduction of naturally occurring phosphate rock (calcium phosphate).

Because of the high stability of the P—O and P═O bonds, the largest group of phosphorus compounds in existence is the oxides. The oxyacids form the basis for the most systematic nomenclature. Table 1 lists the well-characterized lower molecular weight oxyacids and the corresponding structures. The basicity of the acid is related to the P—OH moiety providing the acid function. Acids and salts containing more than one phosphorus atom of the same or different oxidation state are also known, such as diphosphoric(III,V) acid, $H_4P_2O_6$. These lower oxidation state acids or salts containing more than one phosphorus atom are encountered infrequently and are often formed, eg, as somewhat metastable intermediates in the hydrolysis of phosphorus halides. They may possess either P—P or P—O—P linkages. Salts and esters are described by naming the cations or organic group and changing the phosphorus suffix from *-ic* to *-ate* and *-ous* to *-ite*. Compounds are named according to the acids from which they are derived, eg, $(C_2H_5O)_2P({=}O)H$ is diethyl phosphonate, and $(C_6H_5O)_3P$ is triphenyl phosphite.

Other phosphorus compounds may also be considered as derivatives of the oxyacids. The P—O or P═O moieties may be replaced with isoelectronic groups to yield halo, P—X; amide, P(O)—NR_2; thio, P═S; imide, P═NH; etc. However, many phosphorus compounds are named as salts with phosphorus as the metallic or electropositive element, eg, phosphorus trichloride, PCl_3.

Some compounds are named as derivatives of the simple phosphorus hydrides (phosphines), eg, dimethylphosphine, $(CH_3)_3PH$.

Chemical Properties

Oxidation States, Coordination Numbers, and Geometries. Phosphorus has electronic structure $1s^22s^22p^63s^23p^3$. There are thus five valence electrons, three of which are unpaired. Phosphorus bonding is primarily covalent because of its intermediate electronegativity, $X = 2.1$. Oxidation states range from −3 to +5, but phosphorus exists in nature almost exclusively as phosphates or phosphate derivatives. The +5 oxidation state is the most stable for the oxide derivatives in both acidic and basic media. The oxidation potentials indicate the tendency for intermediate oxidation states to disproportionate.

The most common coordination numbers for the phosphorus atom are three, four, or five, although covalent linkages can range anywhere from one to six. The quadruply connected compounds appear to exhibit the highest stability, followed by the coordination numbers three and then five. Singly or doubly connected phosphorus compounds are typically unstable, and fewer examples of the five- and six-coordinate compounds are known.

The geometries of the phosphorus atom are related to the hybridization and the coordination number. Some of the more commonly encountered hybridizations and their corresponding spatial arrangements include the following.

Hybridization	Coordination number	Geometry
p^3	three	orthogonal trigonal
sp^3	four	tetrahedron
sp^3d	five	trigonal bipyramid
sp^3d^2	six	octahedron

Hybridization can be of mixed character and the bond angles in compounds can vary from ideal orientations of the pure hybrid.

Table 1. Oxyacids of Phosphorus

Name and CAS Registry Number	Oxidation state	Molecular formula	Structure	Basicity and salts
(ortho)phosphoric acid	+5	H_3PO_4	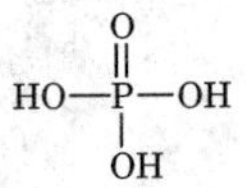	tribasic; salts are called phosphates polymeric forms, eg, $H_4P_2O_7$, $H_5P_3O_{10}$, $(HPO_3)_n$, etc
hypophosphoric acid	+4, +4	$H_4P_2O_6$	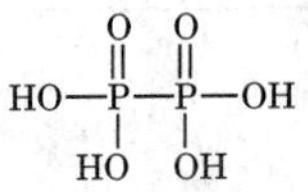	tetrabasic; salts are called hypophosphates
isohypophosphoric acid	+3, +5	$H_4P_2O_6$	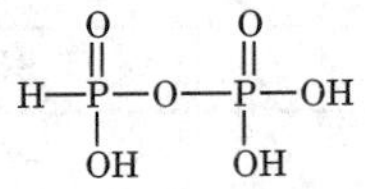	tribasic; salts are called isohypophosphates
phosphonic acid	+3	H_3PO_3	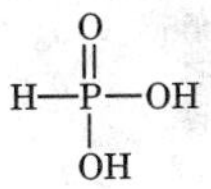	dibasic; salts are called phosphites
diphosphonic acid (pyrophosphorous acid)	+3, +3	$H_4P_2O_5$	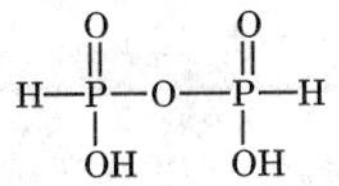	dibasic; salts are called pyrophosphites
hypophosphonic acid	+2, +2	$H_4P_2O_4$	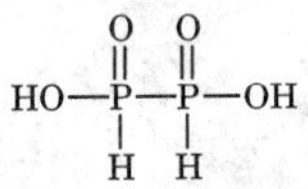	monobasic; salts are called hypodiphosphites
phosphinic acid (hypophosphorous acid)	+1	H_3PO_2	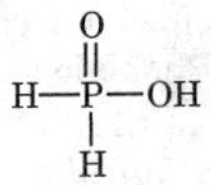	monobasic; salts are called hypophosphites

Bond Properties. Bond strengths, bond lengths, and atom electronegativity differences of various phosphorus—atom linkages are given in Table 2.

Phosphorus Sulfides

Properties and Reactions. Phosphorus combines with sulfur to form the binary tetraphosphorus trisulfide (phosphorus sesquisulfide), P_4S_3; tetraphosphorus pentasulfide, P_4S_5; tetraphosphorus heptasulfide, P_4S_7; and phosphorus(V) sulfide (tetraphosphorus decasulfide), P_4S_{10}. Further, tetraphosphorus enneasulfide, P_4S_9, has also been reported. In addition, a stable oxysulfide, $P_4O_6S_4$, exists as a colorless, deliquescent crystalline solid which has a melting point of 102°C. Some physical constants and thermodynamic data for these compounds are presented in Table 3.

The hydrolysis of phosphorus sulfides has been studied quantitatively. A number of products are formed. Whereas phosphorus(V) sulfide reacts slowly with cold water, the reaction is more rapid upon heating, producing mainly hydrogen sulfide and orthophosphoric acid, H_3PO_4. At high pH, P_4S_{10} hydrolyzes to a mixture of products containing thiophosphates and sulfides.

Phosphorus(V) sulfide reacts with olefins, amines, Grignard reagents, and terpenes. Dialkyl and diaryl dithiophosphoric acids are the bases of many high pressure lubricants, oil additives (see

Table 2. Properties and Electronegativity Differences of Phosphorus—Atom Bonds

Bond, P—X	Bond energy, kJ/mol[a]	Bond length, nm	X_x[b]
P—O	360	0.15–0.17	1.4
P=O	544	0.141–0.151	1.4
P—H	322	0.140–0.146	0.0
P—C	272	0.183–0.194	0.4
P—F	527	0.150–0.160	1.9
P—Cl	331	0.204–0.205	0.9
P—Br	264	0.215–0.220	0.7
P—I	184	0.248–0.252	0.4
P—N	230	0.17–0.18	0.9
P—P	209	0.217–0.227	0.0
P—S	230	0.200–0.215	0.4
P=S		0.187–0.1970.4	0.4

[a] To convert J to cal, divide by 4.184.
[b] Pauling's electronegativity difference, where X_x represents the electronegativity of atom X.

Table 3. Properties of Phosphorus Sulfides

Property	P_4S_3	P_4S_5	P_4S_7	P_4S_9	P_4S_{10}
mp, °C	173	170–220	307	240–270	285
bp, °C	407	dec[a]	523		515
density, g/mL	2.03	2.17	2.19	2.08	2.09
color of solid	yellow	light yellow	almost white		yellow
solubility in CS_2, g/100 g					
0°C	27		0.005		0.18
17°C	100[b]	10	0.029		0.22

[a] Decomposes to P_4S_3 and P_4S_7.
[b] The solubility in C_6H_6 at 17°C is 2.5 g/100 g; at 80°C, 17 g/100 g.

LUBRICATION AND LUBRICANTS), and ore flotation chemicals (see MINERALS RECOVERY AND PROCESSING). Organophosphorus insecticides such as Parathion also are produced (see INSECT CONTROL TECHNOLOGY).

Manufacture. Phosphorus sulfides are manufactured commercially by direct reaction of the elements.

Phosphorus(V) sulfide, an important commodity in the United States since about 1920, is the dominant commercial material.

Shipping and Storage. Phosphorus(V) sulfide is stored and shipped in 208-L (55-gal) drums, special tote-bins, and railcars. P_4S_{10} is classified as a flammable solid having the international shipping code of UN No. 1340.

Health and Safety Factors, Toxicology. One source of danger in the handling of phosphorus(V) sulfide is hydrogen sulfide. The OSHA exposure limit to hydrogen sulfide gas, which has a rotten egg odor at 0.1 ppm, is 15 ppm for 15 minutes or 10 ppm for 8 hours. Respiratory problems occur for exposure above 50 ppm; death on exposure to concentration above 1000 ppm. Phosphorus(V) sulfide combines with atmospheric moisture to release hydrogen-sulfide via gradual hydrolysis. Care should be taken (*1*) to store P_2S_5 in well-sealed containers; (*2*) to convey P_2S_5 in well-sealed materials-handling systems containing dry, inert atmospheres; and (*3*) to handle phosphorus(V) sulfide in well-ventilated areas. Another source of danger in handling phosphorus(V) sulfide is fire and explosion. Ignition sources such as spark, static electricity, and heat should also be eliminated.

Phosphorus(V) sulfide is a mild skin irritant and may cause dermatitis in sensitive individuals. The oral LD_{50} of P_4S_{10} in rats is 389 mg/kg; the OSHA standard time-weighted average (TWA) is 1 mg/m^3.

Uses. Phosphorus(V) sulfide is used in the manufacture of lubricating oil additives, insecticides, ore flotation agents, and specialty chemicals. Phosphorus sesquisulfide, P_4S_3, has been used extensively in the manufacture of strike-anywhere matches (qv). In addition, small quantities are used in fireworks (see PYROTECHNICS).

Phosphorus Halides

Phosphorus forms well-defined halogen compounds of the types PX_3, PX_5, POX_3, and PSX_3, all of which except the pentaiodide and the oxy- and sulfoiodides are known. The commercially important phosphorus halides are phosphorus trichloride, phosphorus oxychloride, phosphorus pentachloride, and phosphorus sulfochloride. A few other phosphorus halides, eg, PI_3, PBr_3, PBr_5, PF_3, and PF_5, are marketed as reagent chemicals.

The trihalides of phosphorus usually are obtained by direct halogenation under controlled conditions. Physical properties some phosphorus halides are listed in Table 4.

Phosphorus Trichloride. *Properties and Reactions.* Phosphorus trichloride can be prepared either by direct chlorination of elemental phosphorus, by reduction of $POCl_3$ by CO, or by chlorination of ferrophosphide.

Phosphorus trichloride, PCl_3, is a clear, volatile liquid having a pungent, irritating odor. The compound PCl_3 is an excellent chlorination reagent for various hydrocarbons.

Although PCl_3 is nearly insoluble in water, it hydrolyzes rapidly.

Manufacture. Phosphorus trichloride is made by direct union of the elements.

Storage, Shipping, and Handling. Phosphorus trichloride is classified by the ICC as a corrosive liquid and poison inhalation hazard. U.S. Department of Transportation (DOT) white acid label and red poison label are required by law on individual containers: DOT UN No. 1809. Alloy or glass-lined vessels are used for shipping storage, and reactors.

Health and Safety Factors, Toxicology. Phosphorus trichloride severely burns skin, eyes, and mucous membranes. Delayed, massive, or acute pulmonary edema and death can develop as consequences of inhalation exposure.

Phosphorus trichloride is highly toxic by ingestion and slightly toxic by single dermal applications. It reacts violently with water and can generate gases sufficient to cause rupture of closed or inadequately vented containers. The OSHA standard TWA is 0.5 ppm.

Table 4. Physical Properties of Phosphorus Halides

Compound	Melting point, °C	Boiling point, °C	Specific gravity[a]	Physical state at STP[b]
PCl_3	−93.6	76.1	1.575	liquid
PCl_5	167[c]	159 subl	2.114	solid[d]
$POCl_3$	+1.2	106.5	1.68	liquid
$PSCl_3$	−36	125	1.668	liquid
PBr_3	−41.5	173.3	2.880	liquid
PBr_5	83.8 (subl)	>106 dec		solid[e]
PI_3	61.2	120 dec		solid[f]
P_2I_4	125.5	dec		solid[g]
PF_3	−151.5	−101	3.907	gas[h]
PF_5	−93.7	−84.5	5.84	gas
POF_3	−39.4	−39.7	4.65	gas
PSF_3	−148.8	−52.3		gas
PF_2Cl	−164.8	−47.3		gas
PF_3Cl_2	−124	2.5		gas
PF_4Cl	−132	−43.4		
PF_2Br	−133.8	−16.1		gas
PF_2Br_3	−20	~106		liquid
$PFCl_2$	−144.1	13.9		gas
$PFBr_2$	−145	78.4		liquid
PF_2I	−93.8	267		

[a] At 298.15 K and 101.3 kPa (1 atm). [b] Colorless unless otherwise noted. [c] At 122 kPa (919 mm Hg). [d] White to pale yellow. [e] Red-brown. [f] Yellow. [g] Dark red. [h] Light orange.

Uses. The largest usage of PCl_3 is to produce phosphonic acid, H_3PO_3. Phosphorus trichloride is also a convenient chlorinating reagent for producing various acyl and alkyl chlorides.

Phosphorus Oxychloride. *Properties and Reactions.* Several methods of preparation are available for $POCl_3$, including partial hydrolysis of PCl_5 by heating in the presence of oxalic or boric acid, chlorination and hydrolysis of PCl_3 in the presence of H_3PO_4, heating a mixture of P_2O_5 and PCl_5, controlled oxidation and chlorination of elemental phosphorus, oxidation of PCl_3 using ozone (qv) or oxygen (qv), and heating calcium phosphate in a mixture of chlorine and carbon monoxide.

Phosphorus oxychloride (phosphoryl chloride), $POCl_3$, is a colorless fuming liquid having a pungent, disagreeable odor and is reactive with water. The reactions of $POCl_3$ resemble those of PCl_3. Physical Properties of Phosphorus Oxyhalides include the following for POF_3, $POCl_3$, $POBr_3$, POF_2Cl, $POFCl_2$, POF_2Br, and $POFBr_2$, respectively. Boiling point, (°C): −40, 107, 193, 3.1, 52.9, 30.5, and 110.1; melting point (°C): −68, 1.25, 56, −96.4, −80.1, −84.4, and −117.2; apical angle, °, 107, 103.5, 108, 106, and 106; bond length (pm) for P═O: 155, 145, 141, 155, and 154; for P—F: 151, 151 (POF_2Cl), and 151 ($POFCl_2$); for P—Cl: 202 ($POCl_3$), 202 (POF_2Cl), and 200 ($POFCl_2$); and for P—Br: 206 ($POBr_3$).

Manufacture. Phosphorus oxychloride has been manufactured by oxidizing phosphorus trichloride. A manufacturing method consists of the chlorination reaction of the trichloride with the pentoxide.

Storage, Shipping, and Handling. Phosphorus oxychloride is classified by the ICC as a corrosive liquid and a poisonous inhalation hazard. Shipment of $POCl_3$ must be in conformance with ICC regulations, and individual containers must be affixed with the DOT white acid label and red poison label: DOT UN No. 1810. Glass and glass-lined steel equipment frequently is used for shipping and storage, as well as for reaction vessels.

Health and Safety Factors, Toxicology. Phosphorus oxychloride vapors are extremely irritating to the eyes, skin, and mucous membranes. Direct contact with the liquid can produce severe burns. Inhalation of $POCl_3$ vapors can cause pulmonary edema and temporary eyesight problems.

The liquid reacts violently with water, releasing HCl and other gases in sufficient amounts to cause sudden rupture of closed or inadequately vented containers. The acid reaction products can react with metals to generate hydrogen, which is flammable and explosive. The oral LD_{50} in rats is 380 mg/kg.

Uses. Phosphorus oxychloride is used extensively to manufacture alkyl and aryl orthophosphate triesters.

Phosphorus Sulfochloride. *Properties and Reactions.* Phosphorus sulfochloride (thiophosphoryl chloride), $PSCl_3$, is a colorless fuming liquid and is made by the reaction of phosphorus trichloride with sulfur and by the reaction of PCl_5 with P_2S_5. Some physical properties of the phosphorus sulfohalides follow for PSF_3, $PSCl_3$, $PSBr_3$, PSF_2Cl, $PSFCl_2$, PSF_2Br, and $PSFBr_2$, respectively. Boiling point (°C): −52.9, 125, 175 dec, 6.3, 64.7, 35.5, and 125.3; melting point (°C): −148.8, −36.2, 39, −155.2, −96.0, −136.9, and −75.2; apical angles (°): 100, 101, 106, 106 (PSF_2Br), and 100; bond length (pm) for P═S: 185, 194, 189, 187 (PSF_2Br), and 187; for P—F: 153, 145 (PSF_2Br), and 150; for P—Cl: 202 ($PSCl_3$); and P—Br: 213 ($PSBr_3$), 214 (PSF_2Br), and 223.

Manufacture. Phosphorus sulfochloride is manufactured by the direct addition of sulfur to phosphorus trichloride. Phosphorus sulfochloride is used primarily in the manufacture of insecticides, such as Parathion.

Phosphorus Pentachloride. *Properties and Reactions.* Phosphorus pentachloride, PCl_5, is a pale, greenish yellow solid having a pungent odor (see Table 4). It is made from PCl_3 and chlorine.

Manufacture. Phosphorus pentachloride is manufactured by either batch or continuous processing.

Storage, Shipping, and Handling. Phosphorus pentachloride is in the EPA extreme hazardous substance list. It is treated as a flammable solid, and containers in which it is stored or shipped must be affixed with a yellow acid label: DOT UN No. 1806. In general, the pentachloride should be handled with the same precautions that are used with the trichloride.

Health and Safety Factors, Toxicology. Because of its fuming and deliquescent properties, PCl_5 is irritating and corrosive to skin, eyes, and mucous membranes. Inhalation symptoms range from coughing, delayed sneezing, to pulmonary edema. The pentachloride is toxic; its TLV is 1 mg/m^3 of air. The OSHA standard in air TWA is 1 mg/m^3.

Uses. Phosphorus pentachloride is used in the manufacture of chlorophosphazenes, and serves as a catalyst and a chlorinating agent in organic syntheses.

Phosphorus Oxides

There are five well-defined oxides of phosphorus: phosphorus(III) oxide, P_4O_6; phosphorus(V) oxide (phosphorus pentoxide), P_4O_{10}, phosphorus tetroxide, P_2O_4; tetraphosphorus heptoxide, P_4O_7; and tetraphosphorus nonaoxide, P_4O_9. A rare and higher oxide is the highly oxidizing deep violet solid P_2O_6, which decolorizes and loses oxygen rapidly when heated to 130°C. The structures of P_4O_6 and P_4O_{10} are related to that of the phosphorus molecule, P_4.

All phosphorus oxides are obtained by direct oxidation of phosphorus, but only phosphorus(V) oxide is produced commercially. Besides the oxides mentioned above, there are commonly termed lower oxides of phosphorus (LOOPs) which are mixtures of usually water-insoluble, yellow-to-orange, and poorly characterized polymers, often formed as a disproportionation by-product in a number of reactions.

Phosphorus(V) Oxide. *Properties and Structure.* Phosphorus(V) oxide, the extremely hygroscopic acid anhydride of the phosphoric acids, exists in several forms but is often referred to by its empirical formula, P_2O_5. Some properties of the various forms of phosphoric oxide are listed in Table 5.

The best characterized form of phosphorus pentoxide is the volatile, metastable, and crystalline *H* (hexagonal) or α-modification.

Manufacture. Phosphorus(V) oxide is made by burning elemental phosphorus in a controlled excess of dry air in a stainless steel, externally cooled combustion chamber similar to that for producing thermal phosphoric acid. The combustion gases are then cooled in a large chamber or barn in which the P_4O_{10} gas is condensed.

Health and Safety; Storage and Handling. Phosphorus(V) oxide is extremely hygroscopic. It reacts with explosive violence when in contact with water or aqueous solutions. Phosphorus(V) oxide is a local corrosive and irritates skin, eyes, and mucous membranes. Phosphorus(V) oxide is sold in small glass bottles contained in boxes or hermetically sealed metal cans. Larger quantities are shipped in metal barrels or drums. In addition to the yellow label, the Chemical Manufacturers Association (CMA) suggests that a special label warning against burns be affixed to the container.

Table 5. Properties of Allotropic Forms of Phosphorus(V) Oxide

Allotrope	Mp, °C	Triple point, °C/ kPa[a]	Specific gravity	ΔH_{vap}, kJ/ mol[b] P_4O_{10}	Structure
H (hexagonal)	420	420/480	2.30	95[c]	P_4O_{10} molecules
O (orthorhombic)	550–570	562/58.3	2.72	152.4	sheets of interlocking rings
O′ (orthorhombic)		580/74.0	2.89	141.9	sheets of interlocking rings
fused hexagonal (metastable)				67.8	
liquid (stable above 580°C)				78.3	

[a] To convert kPa to psi, multiply by 0.145.
[b] To convert J to cal, divide by 4.184.
[c] Standard heat of formation is −3009.9 kJ/mol[b] P_4O_{10}.

Uses. The most important chemical property of phosphorus pentoxide is its avidity for water. It is used as a drying agent for liquids and gases with which it does not react, especially for removing traces of water from vacuum systems (see VACUUM TECHNOLOGY). A drawback in its use as a drying agent is its tendency to coat with a layer of polyphosphoric acid, which prevents further moisture absorption.

Phosphorus(V) oxide is used in the manufacture of phosphorus oxychloride, as a catalyst in air-blowing of asphalt (qv), and as an intermediate for phosphate esters.

Phosphorus compounds are effective flame retardants for oxygenated synthetic polymers such as polyurethanes and polyesters.

Mixed mono- and dialkyl phosphates are used as catalysts for resin curing and as intermediates for fire retardants, oil additives, antistatic agents (qv), and extraction solvents.

Phosphorus(III) Oxide. Phosphorus(III) oxide, the anhydride of phosphonic acid, is formed along with by-products such as phosphorus pentoxide and red phosphorus when phosphorus is burned with less than stoichiometric amounts of oxygen. Phosphorus(III) oxide is a poisonous, white, wax-like, crystalline material, which has a melting point of 23.8°C and a boiling point of 175.3°C.

Phosphonic Acid and P(III) Derivatives. Phosphonic or phosphorous acid is a white deliquescent crystalline compound having a melting point of 73.6°C. Phosphonic acid is prepared by the dissolution of phosphorus(III) oxide or by the hydrolysis of phosphorus trichloride. This reaction can be violent partly because of the heat liberated in the solvation of the hydrogen chloride. Additional phosphonic acid is derived from by-product streams.

Phosphonic acid and hydrogen phosphonates are used as strong but slow-acting reducing agents.

The dimer of phosphonic acid, diphosphonic acid (pyrophosphorus acid), $H_4P_2O_5$, is formed by the reaction of phosphorus trichloride and phosphonic acid in the ratio of 1:5. Phosphonic acid is an intermediate in the production of alkylphosphonates that are used as herbicides and as water treatment chemicals for sequestration, scale inhibition, deflocculation, and ion-control agents in oil wells, cooling tower waters, and boiler feed waters.

1-Hydroxyethane-1,1-diphosphonic acid (HEDP) is produced by hydrolysis of the reaction product of phosphonic acid and acetic anhydride.

Glyphosate, ie, *N*-carboxymethylaminomethanephosphonic acid (*N*-phosphonomethyl glycine) is a large-volume, biodegradable, total herbicide sold as the isopropylammonium salt by Monsanto under the trade name of Roundup. 1,1-Bisphosphonates are becoming an important class of pharmaceuticals (qv) for inhibiting bone resorption (calcium regulator). A commercial example is disodium clodronate.

Monoesters of the phosphonic acids are little used in industry. The diesters, $O{=}PR\,(OR)_2$, of phosphonic acid are commonly prepared in industry from trialkyl phosphites in a Michaelis-Arbusov reaction.

Trialkyl esters of phosphonic acid exist in two structurally isomeric forms. The trialkylphosphites, $P(OR)_3$, or the more stable phosphonates, $O{=}PR\,(OR)_2$. The dialkyl alkylphosphonates are used as flame retardants, plasticizers, and intermediates.

Phosphinic Acid. Phosphinic acid (hypophosphorus acid) is a deliquescent crystalline solid that melts at 26.5°C.

The acid can be prepared by the oxidation of phosphine by iodine and water.

Commercially, phosphinic acid and its salts are manufactured by treatment of white phosphorus with a boiling slurry of lime.

Phosphinic acid and its salts are strong reducing agents, especially in alkaline solution. A principal commercial application of the hypophosphites is in the electroless plating (qv) process.

Phosphazenes and Other Phosphorus–Nitrogen Compounds

Phosphazenes. Phosphazenes, $(NPX_2)_n$, constitute a class of linear and cyclic compounds having an unsaturated skeleton of alternating phosphorus and nitrogen atoms. Both pure and mixed halophosphazenes and many of their substituted derivatives have been characterized. Most of the species are toxic and irritating. The lower linear chlorophosphazenes are oily substances, whereas most of the cyclic members are white crystalline solids.

Hexachlorocyclotriphosphazene and octachlorocyclotetraphosphazene are made by the slow addition of phosphorus pentachloride to an excess of ammonium chloride in a solvent such as chlorobenzene or tetrachloroethane.

The halophosphazenes are hydrolyzed by water. The lower chlorophosphazenes are characterized by high vapor pressures at room temperature and small concentrations cause prolonged irritation of eye membranes. Temporary throat and lung irritation following inhalation of these compounds also occurs. Protective clothing is recommended when handling large quantities of chlorophosphazenes.

Poly(dichlorophosphazene), $Cl\,{+}PCl_2{=}N{+}_n PCl_3^+X^-$, where X is Cl or PCl_6, is formed by heating purified hexachlorocyclotriphosphazene in the molten state at 210–250°C while protecting the reaction from moisture.

Although polyphosphazenes exhibit interesting properties as elastomers, commercial importance of the unsubstituted chloro polymers has been limited by hydrolytic instability (see ELASTOMERS, SYNTHETIC–PHOSPHAZENES). Polymers substituted with organic side groups, however, are generally stable to water. Properties of the polymers such as crystallinity, hydrophilicity, and electrical conductivity can be controlled over a wide range by selection of the substituted side groups. Properties of the substituted polymers include low temperature flexibility and elasticity; thermal stability in excess of 200°C; high stability to various kinds of radiation; resistance to water, solvents, and oils; nonflammability; and flame retardancy. However, in spite of their variety of controllable properties, phosphazene polymers have not yet achieved widespread commercial application.

Other Phosphorus–Nitrogen Compounds. Other compounds include triphosphorus pentanitride, P_3N_5; trichlorophosphineimide, $Cl_3P{=}NH$; and phospham, $(PN_2H)_n$.

Inorganic Phosphines and Phosphides

Properties and Reactions. Phosphine is prepared commercially from the acid- or base-catalyzed reaction of elemental phosphorus with water. Phosphine is also made as a by-product of the commercial calcium hypophosphite. Calcium phosphite is also produced.

Another approach for the production of phosphine is an aqueous electrolytic process, whereby nascent hydrogen reacts with elemental phosphorus. Phosphine is produced at the cathode.

The standard heat of formation of PH_3 (g) is 22.89 kJ/mol (5.47 kcal/mol).

Phosphides. A large number of binary phosphides as well as many ternary mixed-metal phosphides, metal phosphide nitrides, etc, are known.

The phosphides are usually made by direct combination of the elements at elevated temperature. The reactive phosphorus is typically red phosphorus, white phosphorus, or phosphorus vapor. Lithium phosphide, Li_3P; sodium phosphide, Na_3P; potassium phosphide, KP_{15}; iron(III) phosphide, FeP; and diiron phosphide Fe_2P, are all made in this manner.

Phosphides have varying degrees of metallic, covalent, and ionic characters to the bonding.

The usual alkali metal phosphides are reddish brown to black and are stable up to 650°C but react instantly with moisture to form phosphine. Magnesium phosphide, Mg_3P_2, and aluminum phosphide, AlP, or the mixed compounds, are stable in dry air but decompose on contact with water or humid air. These compounds, prepared by direct union of the elements, are used with an igniting agent, eg, 1 wt % nitric acid or nitric oxide, in sea flares. Aluminum phosphide has been used since the 1930s for the generation of phosphine in grain fumigation (see WHEAT AND OTHER CEREAL GRAINS). Magnesium phosphide, calcium phosphide, Ca_3P_2, and zinc phosphide, Zn_3P_2, are also used as fumigants and/or rodent poisons. Calcium phosphide is prepared commercially by heating quick lime in phosphorus vapor.

Phosphides of the less electropositive metals and the metalloids may be considered more as metal–phosphorus alloys. These are thermally stable and typically resistant to attack by water, even at 100°C. Transition-metal-rich phosphides are comparable in structure and properties to transition-metal borides and silicides.

Phosphides are sometimes classified according to stoichiometry into metal-rich phosphides, ie, $M/P > 1$; monophosphides, $M/P = 1$; and phosphorus-rich phosphides, $M/P < 1$. The metal-rich phosphides are typically hard and brittle, having a metallic luster and processing high electrical and thermal conductivities. The monophosphides are also typically hard, chemically inert, and possess a luster. Many of the monophosphides and the phosphorus-rich phosphides are semiconductors (qv), eg, gallium phosphide, GaP, and indium phosphide, InP.

Ferrophosphorus is produced as a by-product in the electrothermal manufacture of elemental phosphorus, in which iron is present as an impurity in the phosphate rock raw material. The commercial product contains ca 23–29% P and is composed primarily of Fe_2P and Fe_3P, along with impurities such as Cr and V. Ferrophosphorus is used in metallurgical processes for the addition of phosphorus content.

Copper and tin phosphides are used as deoxidants in the production of the respective metals, to increase the tensile strength and corrosion resistance in phosphor bronze, and as components of brazing solders (see SOLDERS AND BRAZING FILLER METALS).

Organophosphines and Derivatives

Preparation and Properties of Organophosphines. Aliphatic phosphines can be gases, volatile liquids, or oils. Aromatic phosphines frequently are crystalline, although many are oils. Some physical properties are listed in Table 6. The most characteristic chemical properties of phosphines include their susceptibility to oxidation and their nucleophilicity. The most common derivatives of the phosphines include halophosphines, phosphine oxides, metal complexes of phosphines, and phosphonium salts. Phosphines are also raw materials in the preparation of P^I derivatives, ie, derivatives of the isomers phosphinic acid, $HP(OH)_2$, and phosphonous acid, $H_2P(=O)OH$.

There are a few economical routes that can be employed for production of the largest-volume phosphines as specialty chemicals. The preparation of alkyl phosphines, where $R \geq C_2H_5$, employs the addition of lower phosphines across an olefinic double bond. The reaction may be either acid-, base-, or radical-catalyzed.

Table 6. Physical Properties of Phosphines

Compound	Bp, °C	Density at 20°C, g/cm^3	Dipole moment, 10^{-30} Cm[a]	Index of refraction, n_D^{20}
PH_3	−88	1.529[b]		
CH_3PH_2	−14		3.7	
$(CH_3)_2PH$	21		4.10	
$(CH_3)_3P$	38–41			1.192
$C_2H_5PH_2$	25		3.87	
$(CH_3CH_2)_2PH$	85–86	0.7862	4.7	1.447
$(CH_3CH_2)_3P$	129–130	0.7999	4.7–9.7	1.456
n-$C_4H_9PH_2$	86–88	0.7693	4.54	1.4372
$(n$-$C_4H_9)_2PH$	178–186	0.8083		1.456
$(n$-$C_4H_9)_3P$	240–242	0.817–0.82	5.0–7.3	1.4635
iso-$C_4H_9PH_2$	60–80			
tert-$C_4H_9PH_2$	66–67			
$(n$-$C_8H_{17})_3P$	291[c]			1.4683
CF_3PH_2	−25.5		6.41	
$HOCH_2CH_2PH_2$	139–140	1.004		1.4950
$HO_2CCH_2PH_2$	85–86[d]			
$H_2NCH_2CH_2PH_2$	110			
$C_6H_5PH_2$	157–160	1.001[e]	3.70	1.5796
$(C_6H_5)_2PH$	280			
$(C_6H_5)_3P$	384[f]		4.74–5.14	

[a] To convert Cm to debye, divide by 3.336×10^{-30}.
[b] Value is in g/L at 0°C and 101.3 kPa (1 atm).
[c] At 6.7 kPa (50 mm Hg); melting point = 48°C.
[d] At 1.3 kPa (10 mm Hg).
[e] At 15°C.
[f] Melting point = 79–81°C.

Reactions of Phosphines. Phosphines are generally subject to air oxidation at ambient temperatures via a free-radical mechanism. Phosphines also react with elemental sulfur and other common oxidizing agents such as H_2O_2 or HNO_3.

Because relatively mild oxidizing agents react with phosphines, the latter are convenient deoxidizers or desulfurizers. Functional groups within the substituents in a phosphine usually behave similarly to a hydrocarbon, provided that they do not react with the phosphine group.

Primary and secondary phosphines can be treated with halogenating agents to produce halophosphines. Phosgene, $COCl_2$, is a useful chlorinating reagent, but PCl_5, $SOCl_2$, SO_2Cl_2, $ClSO_3H$, and $SiCl_4$ also are used.

Tertiary phosphines and primary and secondary phosphines can be oxidized by elemental halogen to halophosphine halides.

Dihalophosphines or halophosphites, prepared from phosphorus trichloride, are used in the synthesis of organophosphinates. Organophosphinates may also be prepared by the oxidation of secondary phosphines or halophosphines with hydrogen peroxide or sulfur:

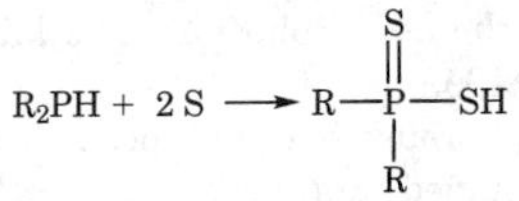

A useful application of phosphines for replacing a carbonyl function with a carbon–carbon double bond is the Wittig reaction.

Health and Safety Factors, Toxicology. Because low molecular weight phosphines generally are spontaneously flammable, they must be stored and handled in an inert atmosphere. The higher and less volatile homologues are more slowly oxidized by air and present less of a problem.

Phosphine is a central nervous system and liver toxin. Phosphine is one of the most toxic of the simple gases; it is lethal to adults in a 0.5–1-h exposure at 0.05 mg/L. The corresponding value for H_2S is 9.6 mg/L; for HCN, 0.12 mg/L. Federal specifications regarding exposure are 400 mg/m^3 TWA and the TLV is 0.3 ppm. The lowest reported LC_{50}s for phosphine are 2500 ppm/20 min for rabbits, and 1000 ppm/5 min for humans. For dibutylphenylphosphine, the lowest reported LC_{50} is 1100 mg/m^3 in 10 min. Phosphine is unacceptable for transport on either passenger or cargo aircraft.

Phosphine Oxides. Controlled oxidation of secondary or tertiary phosphines using H_2O_2 yields the corresponding phosphine oxides. Trioctylphosphine oxide (TOPO) may be manufactured by the radical-catalyzed addition of 1-octene to trioctyl phosphine, followed by peroxide oxidation. Phosphine oxides having higher alkyl substituents are also prepared industrially using Grignard reagents. Phosphine oxides may be prepared by the acid-catalyzed reaction of phosphine with carbonyl compounds such as ketones. Because of their relative instability, primary phosphine oxides cannot be isolated and must be converted directly to derivatives.

Phosphonium Salts. The most common route to phosphonium salts is the reaction of tertiary phosphines with alkyl or aryl halides in polar solvents. Phosphonium salts may also be prepared by the addition of tertiary phosphines to carbonyl compounds or olefins. Tetrakis(hydroxymethyl)phosphonium hydroxide, used for flameproofing cellulosic fabrics, is manufactured in a two-step process.

$$PH_3 + CH_2O + HCl \xrightarrow{H_2O} (HOCH_2)_4P^+Cl^-$$

$$(HOCH_2)_4P^+Cl^- + NaOH \rightarrow (HOCH_2)_4P^+OH^- + NaCl$$

Curing the treated fibers with ammonia chemically attaches the compound to the cloth. The corresponding sulfate has replaced much of the hydroxide because under certain conditions of manufacture or use the carcinogen bis(chloromethyl) ether may form.

Phosphonium salts are typically stable crystalline solids that have high water solubility. Uses include biocides, flame retardants, the phase-transfer catalysts.

Economic Aspects

Phosphorus compounds are manufactured for a variety of uses, either directly or as intermediates in the production of other compounds. Phosphorus trichloride and phosphorus pentasulfide are the compounds in highest demand. Up to 36% of PCl_3 is used for pesticide products.

Twenty-five percent of PCl_3 is used for the manufacture of surfactants and sequestrants. In Europe, phosphonate surfactants have been used in some household detergent formulations.

Organophosphorus compounds, primarily phosphonic acids, are used as sequestrants, scale inhibitors, deflocculants, or ion-control agents in oil wells, cooling-tower waters, and boiler-feed waters. Organophosphates are also used as plasticizers and flame retardants in plastics and elastomers, which accounted for 22% of PCl_3 consumed. Phosphites, in conjunction with liquid mixed metals, such as calcium–zinc and barium–cadmium heat stabilizers, function as antioxidants and stabilizer adjutants. Because PVC production is expected to increase, the use of phosphorus additives should increase 3% annually through 1999.

Phosphonic (phosphorous) acid, produced by hydrolysis of PCl_3, is for the most part consumed captively.

Phosphorus trichloride is also used in the manufacture of antifoam agents, catalysts, dyes and pigments, as well as pharmaceutical and quaternary compounds, and is commonly used as a chlorinating agent. Phosphorus trichloride is also used to make phosphorus oxychloride, which is used in the manufacture of adsorbents.

Liquid phosphate esters, eg, tricresyl phosphate, are one of two types of fire-resistant hydraulic fluids (qv).

Approximately 35% or 220,000 t of the total U.S. 1992 pesticide production were phosphorus-containing products. Organophosphate insecticides offer broad spectrum pest control, and for this reason are under environmental pressure. Further, organophosphate insecticides face strong competition from pest control agents having higher specificity and much lower application rates, such as synthetic pyrethroids (see INSECT CONTROL TECHNOLOGY; PESTICIDES). As a result, the use of P_2S_5 for insecticides is expected to be flat or declining in the United States.

The lubricating oil additive market accounted for about 36,000 t (57%) of the U.S. P_2S_5 production in 1993, an increase from ca 32,000 t in 1980. The use of P_2S_5 for oil additives is expected to grow with the economy.

Phosphorus pentoxide, which is used to make asphalt-blowing agents and in water treatment, amounted to 1.0×10^3 t in 1992.

Approximately 4500 tons of sodium hypophosphite, NaH_2PO_2, was produced in 1990. This material is used principally in electroless nickel plating of plastic objects. Of the secondary products made from primary phosphorus compounds, phosphorus oxychloride is manufactured in the largest volume. Phosphorus pentachloride and phosphorus sulfochloride are made from phosphorus trichloride.

DARRELL C. FEE
DAVID R. GARD
CHEN-HSYONG YANG
Solutia Inc.

G. Bettermann and co-workers, "Phosphorus Compounds, Inorganic," and J. Svara, N. Weferling, and T. Hofmann, "Phosphorus Compounds, Organic," in B. Elvers, S. Hawkins, and G. Schulz, eds., *Ullmann's Encyclopedia of Industrial Chemistry,* Vol. A19, VCH Verlagsgesellschaft mbH, Weinheim, Germany, 1991.

D. E. C. Corbridge, *Phosphorus: An Outline of Its Chemistry, Biochemistry, and Technology,* 4th ed., Elsevier, Science, Inc., New York, 1990.

A. A. Eldridge, G. M. Dyson, A. J. E. Welch, and D. A. Pantony, eds., *Mellor's Comprehensive Treatise on Inorganic and Theoretical Chemistry,* Vol. VIII, Suppl. III, Longman, London, U.K., 1971.

J. R. Van Wazer, *Phosphorus and Its Compounds,* Vols. 1 and 2, Interscience Publishers, Inc., New York, 1958 and 1961.

PHOTOCHEMICAL TECHNOLOGY

SURVEY

Intense research in photochemistry has provided a substantial base of photochemical technologies for industrial application. The oldest recorded continuously practiced-application of photochemical technology is the processing of natural dyes of the indigo class.

Photoimaging is another use of photochemical technology (see IMAGING TECHNOLOGY). Photochemical innovation gave rise to modern photopolymer technology which is responsible for the environmentally significant solventless coatings (qv) industry.

Photochemical technology has been developed so as to increasingly exploit inorganic and organometallic photochemistries, recognizing the importance of photoinduced electron transfer as the phenomenological basis of a majority of commercially successful photochemical technologies. Use of coherent light sources in industrial applications has led to the field of photodynamic therapy as a photochemically based medical technology. The application of photochemistry to information storage and communication processes is expected (see INFORMATION STORAGE MATERIALS).

Light Sources

The exploitation of coherent sources has been a much awaited advance in photochemical technology. Application of laser sources has proved revolutionary in photoimaging, and enabled innovations in photodynamic therapy and photochemical memory technologies. Lasers (qv) have not yet proved to be of significant importance for large-scale industrial synthetic applications, but have been applied to effect various photochemistries, eg, photodecomposition of polymers, surface treatments, high precision machining of synthetic polymeric structures, and medical surgery.

An excimer lamp is a gas-phase fluorescent lamp powered by an electrical or microwave discharge. It produces monochromatic, incoherent radiation in a quasi-continuous mode, ie, pulses of light at a relatively high (ca 100 Hz) repetition rate. Wavelength of emission is the same as for the corresponding excimer laser, eg, 248 nm for KrF, 193 nm for ArF, and 157 nm for F_2. Efficiencies of these lamps are theoretically as high as 40% and under practical operating conditions efficiencies of ca 10% based on microwave power are obtained for the more efficient, microwave discharge powered lamps. One application for excimer lamp photochemistry is the hardening and/or drying on-press of uv-curable inks used in the printing industry, eg, for newspaper printing (see RADIATION CURING). Excimer lamps are also used to cure epoxy adhesives (qv) incorporating, eg, diaryliodonium or analogous traryl-sulfonium salts as photo-acid sources, in the process of making laminated products such as industrial wipers, work wear, surgical drapes, etc. Principal application of excimer lamp technology, however, has been in the photochemical deposition of insulating layers in microelectronic fabrication.

Photophysics

Photochemical Laws and Sensitization. All photochemical technologies are practiced in accord with the fundamental laws of photophysics, relating to the nature of light and its interaction with matter. Of interest herein are the following: (*1*) the Bunsen-Roscoe law, ie, only light which is absorbed by the reactive system is useful; (*2*) Einstein's law, ie, light absorption is quantized and each quantum absorbed activates one molecule; and (*3*) the lowest energy excited state of a given spin multiplicity is the starting point for practical photochemical processes. A corollary of the first principle is that not only must the desired component of a photochemical reaction mixture absorb light from the source to be exploited but also other components of the mixture must not absorb in the same spectral regime.

The reactive species does not absorb light from available sources. In these cases a sensitizer is used that is capable of absorbing available radiation and subsequently transferring either the excitation energy or, in many cases, an electron to the reactive system. In principle the sensitizer does not participate in the reaction and should not affect its course. In practice this is often not the case. Diaryliodonium salts are commonly used photoinitiators for either free-radical or cationic polymerizations (see INITIATORS).

Previous expositions of photochemical laws have distinguished prominently between states of singlet and triplet multiplicity. This distinction continues to be important with respect to photophysics of small organic molecules, but among inorganic and organometallic compounds, states of other multiplicities, eg, doublet and quartet states, play an important role. Spin conservation characterizes electronic molecular excitations and localized exciton formation in solids. Thus initially formed excited states exhibit the same multiplicity as the ground state. In the presence of heavier atoms, eg, organometallic compounds, increased spin-orbit coupling enhances facility of interconversion of spin states. Spectroscopic studies suggest that triplet manifold states of organic compounds with their characteristic photochemistries can also be accessed efficiently, with respect to internal conversion within the same spin manifold, as a consequence of multiphoton laser excitation.

Multiphoton Effects. Intense laser sources are capable of producing multiphotonic excitation, allowing different photochemistries to be obtained from higher excited states than from the lowest excited state usually accessible with conventional light sources. Generation of radicals from higher excited states of diketones and photoactivation of primary radicals from monophotonic decomposition reactions of organic precursors, eg, benzophenone, benzil, and the aryl-substituted cyclohexanones, represent newer routes to laser specific photoinitiation of polymerization in radiation curing applications, particularly for rapid curing of thick polymer specimens, as well as high contrast, submicrometer resolution microlithography.

Multiphoton processes are also undoubtedly involved in the photodegradation of polymers in intense laser fields, eg, using excimer lasers. Multiphoton excitation may be viewed as an exception to the Bunsen-Roscoe law.

Excited-State Relaxation. A further photophysical topic of intense interest is pathways for thermal relaxation of excited states in condensed phases. According to the Franck-Condon principle, photoexcitation occurs with no concurrent relaxation of atomic positions in space, either of the photoexcited chromophore or of the solvating medium. Subsequent to excitation, but typically on the picosecond time scale, atomic positions change to a new equilibrium position, sometimes termed the *thexi*-state. Relaxation of the solvating medium is often more dramatic than that of the chromophore itself. Photochemical reactions generally occur from the *thexi*-state. This relaxation can have practical consequences, eg, with respect to choice of the optimum medium in which to effect a particular photochemical reaction.

Electron-Transfer Dynamics. To a large extent, commercial photochemical technology, including radiation curing, photomedical applications, photochemical information storage, and silver halide photography, involve photoinduced electron transfer. Electron transfer is also the primary process occurring in organic photoconductors as used in office copying applications (see ELECTROPHOTOGRAPHY); the basis for other photoinitiated, heterogeneous-phase polymerization processes; and a crucial player in photocatalysis. An understanding of the theory and phenomenology of photoinduced electron transfer is de rigeur for development of photochemical technology in the contemporary environment. Dynamics of light-induced electron-transfer processes in both homogeneous and heterogenous phases have been reviewed.

Applications

Photohalogenation. Photochemical chlorination of aliphatic hydrocarbons has been the textbook example of industrial photochemistry for decades. It is still commercially important. In most examples of historical importance, ultraviolet radiation was used to dissociate Cl_2 to yield atomic chlorine which, in turn, abstracts a hydrogen atom from the hydrocarbon substrate yielding an organic free radical and HCl. Recombination of this radical with atomic chlorine yields a monochlorocarbon which is more susceptible to the hydrogen abstraction reaction than the parent hydrocarbon. A mixture of chlorohydrocarbons usually results by this method (see CHLOROCARBONS AND CHLOROHYDROCARBONS). Production of trichloroethane occurs in this manner. Photohalogenation is also used for production of the insecticide lindane (γ-hexachlorocyclohexane) (see INSECT CONTROL TECHNOLOGY).

A more energy-efficient variation of photohalogenation, which has been used since the 1940s to produce chlorinated solvents, is the Kharasch process. Ultraviolet radiation is used to photocleave benzoyl peroxide (see PEROXIDES AND PEROXIDE COMPOUNDS). The radical products react with sulfuryl chloride (from SO_2 and Cl_2) to liberate atomic chlorine and initiate a radical chain process in which hydrocarbons become halogenated. Thus, for Ar = aryl,

$$(ArCOO)_2 + h\nu \rightarrow 2\ ArCOO\cdot$$

$$ArCOO \rightarrow Ar\cdot + CO_2$$

$$Ar\cdot + SO_2Cl_2 \rightarrow ArCl + SO_2 + Cl\cdot$$

$$Cl\cdot + RH \longrightarrow HCl + R\cdot$$

$$R\cdot + SO_2Cl_2 \longrightarrow RCl + SO_2Cl\cdot \longrightarrow SO_2 + Cl\cdot$$

The most innovative photohalogenation technology developed in the latter twentieth century is that for purposes of photochlorination of poly(vinyl chloride) (PVC). More highly chlorinated products of improved thermal stability, fire resistance, and rigidity are obtained. In production, the stepwise chlorination may be effected in liquid chlorine which serves both as solvent for the polymer and reagent. A solid-state process has also been devised in which a bed of microparticulate PVC is fluidized with Cl_2 gas and simultaneously irradiated. In both cases the reaction proceeds, counterintuitively, to introduce Cl exclusively at unchlorinated carbon atoms on the polymer backbone.

Laser Photochemical Vapor Deposition. Laser pyrolytic and laser photochemical vapor deposition (LPCVD) technologies based on pyrolysis and photolysis of organometallic precursors are important to the deposition of thin inorganic films and surface patterning for the microelectronics industry. Owing to use of organometallics this technology is sometimes called metal organic chemical vapor deposition (MOCVD).

Vitamin D. Vitamin D is synthesized by photochemical means (see VITAMINS). The term vitamin D is actually applied to several isomers wherein the side chain in the parent structure varies. Vitamin D_2 and D_3 are the most important, commercially, owing to use as animal feed additives. These compounds are synthesized from the steroids ergosterol and 7-dehydrocholesterol, respectively. The synthesis involves a photochemical ring opening (photo-Cope rearrangement) to yield the corresponding pre-vitamin D.

Polymer-Based Technologies. One example of materials science is the use of photoinitiated polymerization in three-dimensional engineering prototyping. In contemporary computer-aided design (CAD) a three-dimensional object is created on the computer. In this process the designer is guided by various digitally generated projections of the object on the monitor. In order to have a solid prototype for visual inspection of the final design, a photopolymerization technique, sometimes called stereolithography, is employed.

Photochemical Therapies. A particularly successful therapeutic application of photochemistry has been the treatment of hyperbilirubinemia, otherwise known as newborn jaundice. The basis for photochemical treatment of newborn jaundice stems from the discovery in the 1930s that bilirubin is efficiently decomposed when exposed to light. Devices for effecting whole-body irradiation of newborns are commercially available. These utilize cool white or blue fluorescent lamps, or appropriately filtered tungsten–halogen lamps.

New impetus was given to photomedicine by development of lasers that are compatible with the clinical environment. These include HeNe, Ar ion, ruby, and tunable dye lasers operating in the continuous wave (cw) mode. Prior to the advent of lasers in medicine, only the treatment of newborn jaundice, and the application of long wavelength uv irradiation in conjunction with administration (or topical application) of psoralen class sensitizers to treatment of skin diseases, principally psoriasis, were clinically important phototherapies.

A principal application for photomedicine is the photodynamic treatment of cancer. Direct irradiation of tumors coupled with administration of a sensitizer is used to effect necrosis of the malignancy.

As penetration of tissue by light increases more or less monotonically beyond 450 nm, it is desirable for sensitizer absorption bands and actinic laser lines to be located as far into the red as possible.

M. R. V. SAHYUN
3M Center

A. Gilbert and J. E. Baggott, *Essentials of Molecular Photochemistry,* CRC Press, Boca Raton, Fla., 1991.

J. Michl, *Electronic Aspects of Organic Photochemistry,* John Wiley and Sons, Inc., New York, 1990.

J. D. Scaiano, *Handbook of Organic Photochemistry,* CRC Press, Boca Raton, Fla., 1989.

N. J. Turro, *Modern Molecular Photochemistry,* University Science Books, 1991.

PHOTOCATALYSIS

Catalysis (qv) refers to a process by which a substance (the catalyst) accelerates an otherwise thermodynamically favored but kinetically slow reaction and the catalyst is fully regenerated at the end of each catalytic cycle. When photons are also implicated in the process, photocatalysis is defined without the implication of some special or specific mechanism as the acceleration of the rate of a photoreaction by the presence of a catalyst. The catalyst may accelerate the photoreaction by interaction with a substrate either in its ground state or in its excited state and/or with the primary photoproduct, depending on the mechanism of the photoreaction. Therefore, the nondescriptive term photocatalysis is a general label to indicate that light and some substance, the catalyst or the initiator, are necessary entities to influence a reaction.

Photochemical processes provide an alternative to the traditional U.S. EPA recognized routes to water (qv) purification: air stripping (removal of volatile components) and carbon adsorption (removal of both volatile and nonvolatile contaminants). A weakness of these traditional processes is that pollutants are not destroyed; rather, the pollutants are moved from one phase to another. Ultraviolet (uv) light in combination with oxidation processes can remove bacterial substances and dissolved organics from solution. Advanced oxidation processes (AOPs), such as uv/O_3, uv/H_2O_2, and heterogeneous photocatalytic methods, result in relatively rapid and complete destruction of numerous organics, including halogenated hydrocarbons (see CHLOROCARBONS AND CHLOROHYDROCARBONS). Heterogeneous photocatalysis uses air or oxygen, rather than O_3 or H_2O_2, and a semiconductor photocatalyst at low (ambient) temperature (see SEMICONDUCTORS). This process leads to total mineralization of organic pollutants to CO_2 without significant formation of photocyclized intermediate products. Also, the photocatalyst employed, eg, TiO_2 or ZnO, is inexpensive and can be supported on suitable materials (see CATALYSTS, SUPPORTED).

Advanced Oxidation Processes

Homogeneous Photocatalysis. Remediation of wastewaters from organic contaminants presupposes some oxidative process to burn off the dissolved carbonaceous species utilizing such oxidizers as ozone (qv) or hydrogen peroxide (qv). It is important, however, that the oxidative process lead to total destruction of the contaminants. Used alone, neither oxidizing agent is efficient. The simultaneous coupling of light and an oxidant, such as O_3, H_2O_2, TiO_2, or others, however, has led to mineralization of organic carbon to CO_2.

The common thread in AOP is the presence and action of the hydroxyl radical, ·OH. The steps implicated in two AOPs, O_3/uv and H_2O_2/uv, are summarized in Figure 1. Two other systems, O_3/H_2O_2/dark and O_3/H_2O_2/uv light are also shown.

Heterogeneous Photocatalysis. Heterogeneous photocatalysis is a technology based on the irradiation of a semiconductor (SC) photocatalyst, for example, titanium dioxide, TiO_2, zinc oxide, ZnO, or cadmium sulfide, CdS. Semiconductor materials have electrical conductivity properties between those of metals and insulators, and have narrow energy gaps (band gap) between the filled valence band and the conduction band (see ELECTRONIC MATERIALS; SEMICONDUCTORS).

Titanium dioxide has been used extensively to photocatalyze the mineralization of a large number of organic pollutants in aqueous dispersions. This photocatalyzed mineralization typically proceeds via formation of a series of intermediates of progressively higher oxygen-to-carbon ratios which eventually are oxidized quantitatively to CO_2 and H_2O. In the case of phenol, a model compound often used in heterogeneous photocatalysis studies, mineralization proceeds via formation of several hydroxylated intermediates that include predominantly catechol and hydroquinone. Organic compounds containing phosphorus, sulfur, and nitrogen atoms are oxidized quantitatively to PO_4^{3-}, SO_4^{2-}, and NO_3^-, respectively, in addition to CO_2 (29); halocarbons yield X^- ions. In some cases, N atoms are converted reductively to NH_3 and oxidatively to NO_3^- ions.

Photooxidations. Some of the compounds which have been successfully photocatalytically degraded are pentachlorophenol, *m*-fluorophenol, 2,4,5-trichloro-phenoxyacetic acid, 4,4′-dichlorodiphenyl-trichloroethane (4,4′-DDT), 3,3′-dichlorobiphenyl (3,3′-DCB), 2,7-dichlorodibenzo-*p*-dioxin, and 6-chloro-*N*-ethyl-*N*′-(1-methylethyl)-1,3,5-triazine-2,4-diamine (atrazine).

Studies carried out on anionic, cationic, and nonionic surfactants have shown that the aromatic and hydrophilic portions of molecules are easily oxidized, whereas the long hydrocarbon chains are converted at slower rates. Surfactant activity does, however, disappear upon loss of the aromatic portion, thereby reducing the nuisance of the

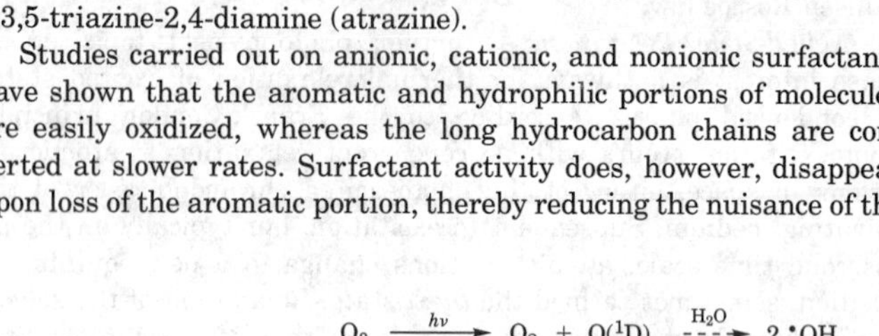

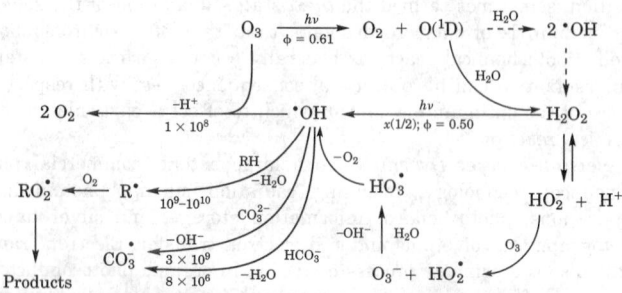

Figure 1. Steps in advanced oxidation process (AOPs) involving ozone, hydrogen peroxide, and uv light of 254 nm. (^{1}D) represents the doublet state; ϕ represents quantum yield, and the other numbers associated with the reaction arrows are rate constants in units of $(M\cdot s)^{-1}$. Dashed arrows indicate processes less likely to occur.

reactants. Total mineralization to CO_2 has been demonstrated for nonionic polyethoxylated 4-nonylphenols having average numbers of 2, 5, and 12 ethoxy units.

Water Treatment. Several components must be treated simultaneously in a multicomponent mixture as available in wastewaters to prove the technology of heterogeneous photocatalysis. The formation and subsequent elimination of intermediates in the photooxidative process must be monitored, identifying all intermediates and final products. Actual water treatment challenges are multicomponent.

Mechanism of Heterogeneous Photocatalysis

A debate centers on the mechanistic details of heterogeneous photocatalysis. The goal is to improve the photocatalytic activity of TiO_2, and understand the role and importance of mineralization by (*1*) free versus surface bound oxidizing radicals, ·OH, and (*2*) by surface ·OH radicals versus direct hole oxidation.

Two principal pathways have been established in mineralization of organic substrates and oxidation of inorganic materials, eg, CN^-. One considers surface OH^- groups or H_2O on TiO_2 as the primary target(s) for the reaction of photogenerated holes, a reaction which yields ·OH radicals.

$$TiO_2 + h\nu \rightarrow TiO_2\{e^- \cdots h^+\} \rightarrow e^-_{CB} + h^+_{VB}$$

$$\{Ti^{4+} - O^{2-} - Ti^{4+}\} - OH^- + h^+_{VB} \rightarrow \{Ti^{4+} - O^{2-} - Ti^{4+}\}\text{-}\dot{}$$

$$\{Ti^{4+} - O^{2-} - Ti^{4+}\} - OH_2 + h^+_{VB} \rightarrow \{Ti^{4+} - O^{2-} - Ti^{4+}\}\dot{\text{-}}OH + H^+$$

The prevailing view favors these radicals as primary oxidizing species. The alternative route implicates direct hole oxidation of the organic substrate.

$$h^+_{VB} + D_{ads} \rightarrow D^+_{ads}$$

$$e^-_{CB} + A_{ads} \rightarrow A^-_{ads}$$

Another problem is whether the primary oxidation event of ·OH radicals occurs on the surface of the photocatalyst or whether ·OH desorb and react with oxidizable substrates in solution.

Complete destruction of organics by an oxidative path over light-activated aqueous TiO_2 suspensions (TiO^*_2) does not occur if either H_2O and/or molecular O_2 are absent. Partial oxidations, and not mineralization, are the rule when photooxidations are carried out in redox-inert solvents such as acetonitrile and dichloromethane. In the absence of water, mineralization of the organic substrate does not occur, only partially oxidized products, often involving photooxygenation, have been isolated. In these cases the primary oxidizing species may be the photohole, h^+, but the intervention of ·OH species is by no means precluded.

NICK SERPONE
Concordia University

N. Serpone and E. Pelizzetti, eds., *Photocatalysis—Fundamentals and Applications,* Wiley-Interscience, New York, 1989.

G. R. Helz, R. G. Zepp, and D. G. Crosby, eds., *Aquatic and Surface Photochemistry,* Lewis Publishers, Boca Raton, Fla., 1994.

D. F. Ollis and H. Al-Ekabi, eds., *Photocatalytic Purification and Treatment of Water and Air,* Elsevier, New York, 1993.

D. F. Ollis, E. Pelizzetti, and N. Serpone, *Environ. Sci. Technol.* **25,** 1522 (1991).

PHOTOCONDUCTIVE POLYMERS

Polymers are, in general, insulators. Some conjugated polymers, such as polyacetylene, can be made into conductors by chemical doping (see ELECTRICALLY CONDUCTIVE POLYMERS). Strong oxidizing agents (eg, AsF_5) and reducing agents (eg, Na) have been used as dopants. The function of doping can also be performed by photons if photoactive molecules are present. Certain polymers are insulators in the dark but become conductive when irradiated by light. Poly(*N*-vinylcarbazole) (PVK) was the first known photoconductive polymer.

Photoconductive polymers are widely used in the imaging industry as either photosensitive receptors or carrier (electron or hole) transporting materials in copy machines and laser printers. This is still the only area in which the photoelectronic properties of polymers are exploited on a large-scale industrial basis. It is also one electronic application where polymers are superior to inorganic semiconductors.

A good photoconductive polymer should exhibit the following properties. First, it has to be a good insulator in the dark and be capable of sustaining a high electric field. The superior dielectric strength of polymers, along with their good film-forming properties, are important reasons for their success in electrophotography. Secondly, when irradiated by light, the material has to generate carriers with high quantum efficiency. The charge generation efficiencies of most polymers are low and usually have to be enhanced by doping with electron donors or acceptors. Finally, the generated carriers have to move through the polymer film without being significantly trapped. Almost all known photoconductive polymers transport holes only.

The sensitivity of electrophotography is not as high as that of silver halide but is better than many other techniques, such as photopolymerization.

Classification of Materials

Photoconductive polymers can be conveniently classified into five categories based on their structures and modes of photoconduction.

Polymers With Pendent Groups. Poly(*N*-vinylcarbazole) (PVK) belongs in this category. In this class of material, the polymer backbone does not participate in carrier transport directly. Instead, the carriers move by hopping along the electroactive pendent groups such as carbazole. These pendent groups are covalently attached to the polymers. With electron donors as the pendent group, the polymer conducts holes; with electron acceptors as the pendent group, the polymer conducts electrons. In general, only hole-conducting polymers have been successfully made with good enough properties for practical applications. The hole mobilities of this class of polymers are low, around 10^{-6}–10^{-7} $cm^2/V{\cdot}s$. Their intrinsic charge generation efficiency and spectral sensitivity range are also limited. Sensitizers such as dyes, 2,4,7-trinitro-9-fluorenone, and fullerenes enhance the charge generation efficiency and extend the spectral range. Most of the sensitizers are electron acceptors, which form charge-transfer complexes with the donor groups of the polymers. The excitation of the charge-transfer complexes then leads to the generation of electrons and holes.

Molecularly Doped Polymers. Many small molecules such as aromatic amines, eg, triphenylamine (TPA), are excellent hole transport materials. They can be used alone as the hole transporting layer in a device if they can be deposited as amorphous thin films. In general, however, it is more advantageous to mix them with polymers with high mechanical strength and good film-forming properties. A high concentration of hole transport molecules must be used so that the latter form an interconnecting conductive network. In this case, polymers merely act as the binders. They do not participate in carrier transport directly, but can affect the carrier mobility by modifying the trap depth and the distance between traps. The carrier mobility of this class of polymers is sensitive to the volume concentration of hole transport molecules present. Usually, the higher the concentration, the larger the hole mobility.

This class of polymeric photoconductors is distinguished from the pendent group-containing polymers, eg, PVK, by the fact that the active hole transport groups are not covalently bonded to the polymer backbone, but are merely dissolved in the polymer. This provides great flexibility for sample preparation. Different polymer matrices with different hole-transporting molecules can be combined without the need for difficult chemical synthesis. Two commonly used polymers are polycarbonate and polystyrene.

Backbone Conjugated Polymers. Polysilanes, $(RR'Si)_n$, are a unique class of polymers with the backbone consisting entirely of tetrahedrally coordinated silicon atoms. Extensive delocalization of σ-electrons takes place along the silicon chain, giving rise to many interesting electronic properties. Because of this σ-conjugation, carrier transport along the silicon backbone is very efficient. The hole mobility of polysilanes, ca 10^{-4} $cm^2/V{\cdot}s$ is among the highest observed for polymers. Because the hole transport is through the σ-conjugated Si backbone, the hole mobility is insensitive to the substituent on the backbone. The hole mobilities of (phenylmethyl)polysilane (PMPS), poly(n-dodecylmethylsilane), poly(n-propylmethylsilane), and poly(methylcyclo-hexylsilane) are essentially the same. The charge generation efficiency and the spectral sensitivity range of polysilanes are, however, limited. Both can be enhanced by doping with sensitizers such as fullerenes.

Other polymers in this category include σ-conjugated polygermylenes and π-conjugated polyacetylene, polythiophene, and poly(p-phenylenevinylene). The photoconductivity of many π-conjugated polymers can be enhanced by doping with fullerenes.

Liquid Crystalline Systems. Conventional photoconductive polymers are amphorous or systems with low order. In the case of PVK, the hole moves by hopping between the pendent carbazole groups. The hole mobilities are usually low, $\sim 10^{-6}$ $cm^2/V{\cdot}s$, due to a trap-dominated hopping transport. One approach to enhancing the hole mobility is to use conjugated polymers such as (phenylmethyl)polysilane (PMPS). Another approach is the use of liquid crystalline systems where, in principle, transport can occur between ordered mesogenic groups. For example, the ordered columnar arrangement of the hexapentyloxyttriphenylene molecules provides good overlap of the p-electrons of the triphenylene moieties along the director axis. This results in efficient hole transport in the mesophase. The hole photocurrent shows nondispersive transport with a high mobility up to 1×10^{-3} $cm^2/V{\cdot}s$.

Nanoclusters/Polymer Composites. The foundation for this new class of material is based on the ability to synthesize small semiconductor particles, typically in the nanometer-size regime. The structures of these semiconductor nanoclusters are usually the same as those of the bulk crystals, yet their properties are remarkably different. The electronic properties of these clusters depend on the cluster size, a phenomenon commonly referred to as the quantum size effect. It is manifested as a blue-shift in the exciton energy and enhancement in the volume-normalized oscillator strength as the cluster size decreases. With the proper surface-capping agents, clusters of varying sizes can be isolated as powders and redissolved into various organic solvents in the same manner as molecules. By co-dissolving these clusters with the polymer, a thin film of nanocluster-doped polymer can be easily made by spin-coating. Alternatively, semiconductor nanoclusters can be directly synthesized in the polymer film.

So far polymers such as PVK, polysilane, and amine-doped polycarbonate have been used as the charge-transporting matrices. A wide variety of semiconductor nanoclusters have been synthesized within these polymers. Many narrow gap and ir-sensitive semiconductors such as InAs normally cannot be made into high field, room temperature photoconductors for electrophotography purposes. Other than the typical difficulty of growing good quality large area thin film, the main problem is the dark decay owing to thermal excitation of carriers. By dispersing nanometer-sized InAs in charge-transporting polymers, the charge-generation efficiency of InAs is retained, but the dark decay problem is removed. An additional benefit is the ease of thin-film preparation with polymers.

Charge Transport

For imaging applications such as electrophotography, the speed with which the carriers (electrons or holes) move through the photoconductor is less critical than for those applications involving serial processing such as photodetectors, where nanosecond or picosecond time resolution is often required. For electrophotography, typically the carriers need to move through a ca 10-μm film in milliseconds or tens of milliseconds. This requires the polymer to have a mobility of $\geq 10^{-6}$ $cm^2/V{\cdot}s$, which is easily achievable. More importantly, carriers have to move through the film without being trapped. It is quite remarkable that modern photoconductive polymers have been developed to the extent that polymer thin film without permanent traps can be easily fabricated. This is not the case for crystalline inorganic semiconductors, where carrier transport is extremely sensitive to the presence of impurities and fabrication of good quality thin film requires elaborate procedures.

Carrier transport in polymers is characterized by a succession of hops from site to site. The distances between various neighboring sites and the energetics of each site are different from one another. These distributions (dispersions) in energy and distance cause different hopping rates between different sites. This is called dispersive transport. One of the central issues in photoconductive polymer research has been to develop a theoretical framework for understanding hopping transport in polymers. Studies using the Monte-Carlo simulation technique have shown great success in describing the charge transport properties of polymers. In this model, charge transport occurs by hopping through a manifold of localized states with both energy and positional disorder.

This equation has been used to analyze many experimental mobility data successfully. The model predicts that the high field mobility follows the following equation where $\hat{\sigma} = \sigma/kT$ (σ is the width of the Gaussian distribution density of states), Σ is a parameter that characterizes the degree of positional disorder, E is the electric field, μ_0 is a prefactor mobility, and C is an empirical constant given as $2.9 \times 10^{-4}(\text{cm/V})^{1/2}$.

$$\mu(\hat{\sigma}, \Sigma, E) = \mu_0 \exp\left[-\left(\frac{2\hat{\sigma}}{3}\right)^2\right] \exp[C(\hat{\sigma}^2 - \Sigma^2)E^{1/2}]$$

This equation has been used to analyze many experimental mobility data successfully.

Charge Generation

Another important property of a photoconductor is the efficiency with which it converts photons into electrons and holes. Most of the photoconductive polymers have low charge-generation efficiencies. This can be a result either of an intrinsically low charge-generation efficiency or of a low absorption coefficient in the interesting spectral region. To enhance the charge-generation efficiency, sensitizers have to be added. Photoexcitation of the sensitizer generates an excited state, which can be either a singlet, triplet, or charge-transfer state. The excited state can relax radiatively or nonradiatively, or undergo electron-transfer reaction. If it accepts an electron from the surrounding polymer, a hole is then generated in the polymer. This initially generated electron-hole pair may recombine to the neutral ground state, or separate under the electric field into free carriers for conduction.

The theory developed by Onsager has been the standard model to use for analyzing the electric field dependence of the charge-generation efficiency. The model solves the diffusion equation of the relative motion of an electron-hole pair, bounded by their Coulomb interaction, under an electric field. The origin of the electron-hole pair and the pathway by which it is generated are not considered in this model.

The inadequacy of the Onsager model has been recognized for many years. Recently a new model incorporating both the Onsager and Marcus electron transfer theory has been developed to successfully account for the field-dependent charge generation efficiency of fullerene-doped PVK photo conductor.

Applications and Related Technologies

The most important industrial application of photoconductive polymers is electrophotography (qv). This is a billion dollar industry and one of the few electronic areas where polymeric material excels.

The availability of photoconductive polymers opens up many areas for research, in addition to electrophotography.

Electroluminescence. Photoconductivity is based on the conversion of light to electricity. The reverse phenomenon, electroluminescence, is based on the conversion of electricity to light. Electroluminescence is useful for flat-panel display and II-VI semiconductors such as ZnS are employed for this purpose. The current trend is toward the development of polymeric electroluminescent material for their processing flexibility. Hole-transporting polymers such as poly(*p*-phenylenevinylene) and PMPS have been used in such devices. Semiconductor nanocluster-doped polymers represent another interesting class of materials for the exploration of electroluminescent phenomenon. It has already been demonstrated that properly doped semiconductor nanoclusters such as $Zn_xMn_{1-x}S$ emits light efficiently. With the demonstration of photoconductivity these nanocluster-doped polymers can become possible candidates for electroluminescent materials.

Photorefractive Effect. If a material possesses second-order optical nonlinearity and is photoconductive, it may be photorefractive. Photorefractivity is a third-order nonlinear optical, $x^{(3)}$, phenomenon. Photorefractive materials provide a medium in which holographic gratings can be reversibly written and are useful as optical interconnects. A good photorefractive material requires large second-order nonlinearity and high charge-generation efficiency.

Recently photorefractivity in photoconductive polymers has been demonstrated.

Data Storage. An interesting extension of photoconductor technology to optical data storage has been reported. The device consists of a solid thin film of the photoconductor, zinc-octakis(β-decoxyethyl)porphyrin, sandwiched between two transparent electrodes. Irradiation by light (550 nm) under an applied electric field generates electron-hole pairs which are separated within the photoconductive layer. When the irradiation is interrupted, these electron-hole pairs become trapped within the film because of the low dark conductivity. This corresponds to the data-writing step. The written information can be read by irradiation of the device with a read beam (at 550 nm) under short circuit conditions. The basic principle behind this data storage scheme is similar to that of xerography, where the readout method is by toning with carbon particles to form an image. The reported method represents a digital way of reading the electrostatic image (see INFORMATION STORAGE MATERIALS, OPTICAL).

YING WANG
DuPont Company

W. D. Gill, in J. Mort and D. M. Pai, eds., *Photoconductivity and Related Phenomena,* Elsevier, Amsterdam, the Netherlands, 1976, p. 303.

H. Bässler, *Phys. Stat. Sol. (b)* **175**, 15 (1993).

J. Mort and G. Pfister, in J. Mort and G. Pfister, eds., *Electronic Properties of Polymers*, John Wiley & Sons, Inc., New York, 1982, p. 215.

Y. Wang and A. Suna, *J. Phys. Chem.* **101**, 5627 (1997).

PHOTODEGRADABLE POLYMERS.
See POLYMERS, ENVIRONMENTALLY DEGRADABLE.

PHOTODETECTORS

Photodetector devices convert electromagnetic radiation or photons to electric signals which can be processed to obtain the spectral, spatial, and temporal information inherent in the radiation. Photodetectors may be operated in many modes. The more popular ones are photoconductors, photodiodes, charge-transfer devices, and the pyroelectrics. The detectors may be used as single elements such as in street light controls, film camera exposure control, or motion detectors for security, or in the form of linear arrays used in analytical spectrometers, night-vision equipment, or configured as large matrix arrays found in video cameras. Photodetectors are expected to appear in small, low cost spectrometers for the control of building ventilation and environmental pollution monitoring. Detectors using artificially structured materials such as semiconductor superlattices and high temperature superconductors are in the early development stage.

Principles

The basic detection process is the generation of free electrons, holes, or both by the absorption of photon energy. The absorption may be intrinsic, ie, creating a free electron–hole pair, or extrinsic, creating a free hole or electron. The process can be indirect whereby the absorbed photon energy raises the lattice temperature (thermal detection) and a phonon generates the free charged particle. The change in free charge is sensed in an amplifier circuit as a signal voltage or current. Random generation of free charge is the source of detector noise. The detector geometry, spectral response, electrical bias, and temperature are adjusted to optimize the signal-to-noise ratio. The absorption efficiency (photon to electron conversion) is a critical issue. The detector surface is typically coated to minimize reflections at a particular wavelength and the internal efficiency is given by $1 - \exp(-\alpha x)$. Because the absorption coefficient α is typically 3000 cm^{-1} for intrinsic detection and 3 cm^{-1} for the extrinsic case, detector thickness, x, has a large range of values.

Important figures of merit that describe the performance of a photodetector are responsivity, noise, noise equivalent power, detectivity, and response time. However, there are several related parameters of measurement, eg, temperature of operation, bias power, spectral response, background photon flux, noise spectra, impedance, and linearity. Operational concerns include detector-element size, uniformity of response, array density, reliability, cooling time, radiation tolerance, vibration and shock resistance, shelf life, availability of arrays, and cost.

Photodetector Modes of Operation

The need for detectors with high performance and low cost has resulted in fewer than two dozen types of detectors which are available as commercial products. These are made from only 10 basic semiconductor elements or compounds. The development of the silicon charge-coupled device (CCD) detector has been an important consequence of the microprocessor industry. Silicon photovoltaic and charge-coupled detectors operate near room temperature. Cadmium sulfide and germanium detectors also require no cooling. Wide acceptance of the ternary compound semiconductor mercury–cadmium–telluride (MCT), began in 1972 as a photoconductor. Requirements for large focal planes has since led to significant developments of the MCT photodiode focal plane. The controllable (by composition) band gap of MCT in each case results in a cutoff wavelength that is tailored for a specific application. Lead sulfide, indium arsenide, and MCT (40% at. wt CdTe) detectors can be operated at 300 K but perform much better at lower temperatures.

Four principal modes of semiconductor-based photodetectors are the charge-coupled device (CCD), photoconductor, photodiode, and bolometer. The Schottky barrier mode using internal photoemission has been well developed using platinum silicide, but low quantum efficiency restricts the range of uses. The quantum well infrared photodetector (QWIP) is included herein because artificially structured materials such as this superlattice detector are representative of a new metallurgy. The semiconductor materials for most applications are Si, Ge, GaAsP, InSb, PbS, CdS, and HgCdTe. These materials grown as single crystals and thin polycrystalline or amorphous films are doped with various elements such as boron, phosphorus, indium, gold, etc, to control the polarity of the conductivity and for selective optical absorption by the introduction of impurity states in the forbidden energy gap. Photodetection using these materials extends from the ultraviolet (0.3 μm) to long wavelength infrared (200 μm). Higher sensitivity, especially in the infrared, can be achieved by cooling the detector and its immediate surroundings to low temperature, typically 77 K. Interfacing with signal processing electronics can be difficult when the photon signal is weak and imaging systems require sophisticated signal processing to handle the hundreds of thousands of pixels on a focal plane.

Charge Mode Detector. Charge mode devices are a group of detectors which utilize the metal–insulator–semiconductor (MIS) capacitor as their basic constituent. An MIS capacitor is a device consisting of a nondegenerately doped semiconductor substrate, a thin dielectric layer, and a metal gate (see THIN FILMS).

Photoconductors. The photoconductor is a semiconductor resistor that lowers in resistance when photons generate excess carriers. The two general classes of photoconductor are extrinsic, which is characterized by impurity states that are emptied by photons, and intrinsic, where the photons excite electrons directly from the valence band to the conduction band. The photoconductor is connected in series with a constant load resistor and current source. The ambient light or background radiation sets up a steady-state resistance in the detector. Changes in illumination produce changes in resistance and thereby changes in the voltage across the load resistor. When the load resistance is much larger than the detector resistance, signal voltage is given by

$$V_S = \frac{\eta \phi_s \tau V}{nt}$$

where the signal flux, ϕ_s, is caused by the changing background. At low bias the signal is linear with the bias voltage, V, but at high bias the minority carrier sweep time becomes less than the lifetime, τ, and the signal saturates. The majority carrier density, n, is typically 1×10^{15} to $1 \times 10^{16}/\text{cm}^3$. The thickness, t, is typically one to three absorption lengths (8–20 μm).

Photodiodes. Photovoltaic detectors generate a voltage (open circuit) or a current (closed or integrating circuit) when receiving radiation and therefore do not require an external bias for operation. The charge integrated on a capacitor of the integrating circuit is proportional to the intensity of radiation. The current generated by the photovoltaic cell, I_s, is given by $I_s = \eta \phi_s q A$.

The quantum efficiency can be determined using the measured current and the aid of a calibrated photon source, eg, a blackbody.

Bolometers. The bolometer has made a comeback as a popular detector thanks to advances in micromachining technology. When applied to silicon it is feasible to fabricate large arrays of bolometers and detecting elements having very small thermal mass. The imaged infrared radiant power slightly heats the bolometer film a few millidegrees kelvin causing a lowering of the electrical resistance. The resulting change of the bias current is the signal. The noise components are l/f, Johnson-Nyquist, thermal conductance, and photon. The photon noise is actually the radiative part of the thermal conductance noise.

The peak to peak signal voltage, V_s, is proportional to the bias power and can be expressed as follows: $V_s = \alpha (PR_b)^{1/2} \Delta T_b$, where α is the temperature coefficient of resistance and is 2.8%/K for α-Si, P is the bias power for the bolometer element (typically 0.3 μW), and R_b is the bolometer resistance (typically 3×10^7 Ω for α-Si and 1×10^4 Ω for VO).

The noise components may be expressed as voltage.

The noise components may be combined with the responsivity to give the noise equivalent power (NEP) equation:

$$\text{NEP} = \frac{(V_{\text{JN}}^2 + V_{\text{EX}}^2 + V_{\text{TC}}^2)^{1/2}}{R_V}$$

Detector Fabrication and Performance

Charge-Coupled Devices and Imaging Arrays. The single most popular type of visible photon detector is the charge-coupled device (CCD). The CCD has been a resounding commercial success become the detector of choice in the video camera market, and demonstrating great potential as the optimum detector for future electronic still photographyPhotography (qv) systems. Furthermore, since its inception the CCD has also been an invaluable tool in a wide range of scientific applications involving the detection of near-infrared, visible, ultraviolet, and x-ray photons, as well as in the detection of ionizing radiation and charged particles.

CCDs are sensitive photon detectors. The physical mechanism governing photon detection in silicon depends on the energy of the incident photons.

Utilization. CCDs are often grouped into two broad categories: commercial-grade and scientific-grade. For both categories the format and required performance of the CCD are driven by the application for which the array is designed. In general, commercial-grade CCDs are designed to optimize image resolution and color fidelity. On the other hand, scientific-grade CCDs are designed to maximize photon quantum efficiency and minimize device noise. The vast majority of available CCDs have been designed for commercial applications.

Most CCDs are specifically designed for video camera applications, which detect photons in the visible portion of the electromagnetic spectra. Video camera CCDs have specific camera formats compatible with standard video display systems.

Color applications require more complex image detection schemes owing to the need for spectral differentiation. Signal samples from at least three distinct spectral bands are required to accurately reproduce a color image. Typically a monolithic color filter array consisting of alternating windows of appropriately colored filter media is attached to the CCD to accomplish color differentiation.

Video camera applications obligate the CCD imager to have an internal, electronic, image shuttering capability in order to clearly differentiate the data in one image frame from that of the next. The two most common forms of internal shuttering are frame transfer and interline transfer.

CCDs are being actively developed for use in the field of electronic still photography to provide a means of electronically gathering high quality images. The ideal CCD-based still camera would gather an image of comparable quality to that of a 35-mm film camera. However, unlike film-based cameras, the image would be stored on electronic media such as a floppy or optical disk (see INFORMATION STORAGE MATERIALS). The image could then be read into a computer-based system, edited or enhanced as desired, and stored again onto the digital media. Hard copies would be obtained on photographic film by a separate image transfer system.

Whereas commercial video CCDs can sometimes be adapted for scientific endeavors, these devices have format and operability constraints which negatively impact their performance in many applications. Scientific CCDs are somewhat more difficult to categorize because of the wide range of capabilities available in such devices. The performance features of a specific scientific CCD are determined by the device design. A few of the more significant performance features which may be found in a given scientific CCD include the following: (*1*) spectral response extending over a wide range of photon and charged particle energies: (*2*) noise levels as low as one electron per pixel, which result in very high signal-to-noise ratios; (*3*) very high charge-transfer efficiencies to achieve no measurable degradation in spatial or spectral information upon image readout; (*4*) dark current generation rates permitting several minute integration times at room temperature and multiple hour integration times at cryogenic temperatures; (*5*) high photon detection efficiency; and (*6*) high frame rate operability. In addition, scientific CCDs are typically operated in the full-frame readout mode, in which an external mechanical shutter shields the imager during scene readout.

Scientific CCDs have been used in a wide range of applications. In the visible and uv portions of the spectrum, scientific CCDs have proven to be effective tools for low photon flux applications. Specifically, CCDs have been utilized as image detectors in ground-based telescopes as well as in space-borne systems such as the Wide-Field Planetary camera of the earth-orbiting space telescope, the Giotto mission to Halley's comet, and the Galileo mission to Jupiter.

The same features that make scientific CCDs excellent devices for astronomy, that is, high photon collection efficient and low readout noise, also make CCDs excellent tools for chemical analysis. CCDs can be utilized in many forms of spectroscopy (see SPECTROSCOPTY, OPTICAL), including absorption, fluorescence, luminescence, emission and Raman, over a spectral range of 0.1 to 1100 nm.

In the x-ray portion of the spectrum, scientific CCDs have been utilized as imaging spectrometers for astronomical mapping of the sun, galactic diffuse x-ray background, and other x-ray sources. Additionally, scientific CCDs designed for x-ray detection are also used in the fields of x-ray diffraction, materials analysis, medicine, and dentistry. CCD focal planes designed for infrared photon detection have also been demonstrated in InSb and HgCdTe but are not available commercially.

Fabrication. Although CCDs have been fabricated in many semiconducting materials such as Ge, InP, and HgCdTe, by far the most readily available devices are those which utilize Si as the semiconductor. There are several common types of silicon CCDs. All share certain processing steps, and most utilize p-type silicon as the semiconducting material. Silicon single crystals are grown in conventional Czochralski vertical pullers using a single-crystal silicon seed dipped and rotated in a silicon melt. The large boules of silicon are sawed into wafers.

Device Type vs Application. The application for which the CCD is designed dictates the variants to the process that are utilized to provide the desired performance enhancements. A useful CCD variant for such applications is the back-side illuminated CCD. Back-side illuminated CCDs undergo additional processing to remove the underlying $p+$ substrate from the p-type epitaxial layer. As the name implies, photons impinge upon the device from the back side, thereby avoiding the absorption layers of gate electrodes present in front-side illuminated devices. In this manner the photon collection efficiency of the device can be improved in the blue, uv, and low energy x-ray photon regimes.

Other CCDs have special processing steps that lower the rate at which surface dark current is generated during the interval in which signal charge is collected. One such device is known as the virtual phase CCD (VPCCD).

A second device having additional specialized ion-implanted layers is the multipinned phase (MPP) CCD. The MPP device is a compromise between a multiphase and a virtual phase CCD, and is becoming the CCD of choice for scientific applications. As in the generic multiphase CCD, the entire charge storage region is covered by polysilicon gates. Although front-side illuminated devices are available, the MPP CCD is typically back-side illuminated in order to achieve state-of-the-art photon collection performance. Even with this stipulation the MPP CCD is still a popular detector due to its availability in various array sizes and formats specifically designed for scientific applications.

Silicon Photodiodes. The popularity of silicon photodiodes is directly related to the ability to detect photons over a spectral range spanning the near-infrared to low energy x-ray regimes. The fast response time of less than 1 μs is attractive when compared to the response times of photoconductive or bolometer-based devices. The silicon photodiode has a responsivity ca 0.4 A/W at the peak response wavelength. The photodiode has proven to be a useful tool for photon-counting and imaging applications over the entire range of spectral sensitivity and has been utilized in the visible portion of the spectrum for power generation.

Silicon photodiodes are available in both discrete and array formats. The simplicity of the discrete photodiode makes this device one of the least expensive photon detectors available. Discrete photodiodes can be used in a myriad of applications including high speed optical switching, intensity determination for automatic exposure control circuitry in film cameras, and photon counting for spectroscopic analyses.

Photodiode arrays are more complex than their discrete counterparts due to the difficulty of directing the signal information from each diode to off-chip electronics. The more common linear arrays contain internal multiplexing circuitry located on the periphery of the imaging area. This circuitry amplifies and buffers the signal from each diode, presenting the information from each pixel through a single output amplifier in a controlled, time-sequenced fashion. The performance characteristics of linear photodiode arrays typically rival those obtained from discrete diodes. Linear photodiode arrays are commonly found in high resolution image scanning applications such as photocopiers and facsimile (FAX) machines.

As for linear photodiode arrays, two-dimensional photodiode arrays require internally integrated circuitry to mediate the signal information from each pixel. However, with the exception of the outermost rows and columns, the pixels in two-dimensional arrays are surrounded on all sides by other pixels. Thus the required circuitry cannot reside solely on the periphery of the array but must be integrated into the actual pixel site. Recently CCDs and metal oxide semiconductor (MOS) arrays, two-dimensional imaging arrays which utilize p–n diodes as photosites, and complimentary metal oxide semiconductor (CMOS)-based components for read-out circuitry, have emerged as strong competitors in this arena. Nevertheless some two-dimensional standard video format photodiode arrays are still manufactured. These devices are most useful in situations unsuited for CCD and CMOS-based imagers, such as ionizing radiation environments.

In addition to being a popular image detector, the silicon homojunction photodiode and avalanche photodiode can also be used for power generation (qv) by operating the device in the photovoltaic mode (see PHOTOVOLTAIC CELLS).

Fabrication. Photodiodes are made by a process similar to that used to manufacture bipolar integrated circuits (qv). For a p- on n-diode formation process, the starting wafer is an n-type silicon substrate of roughly 500-μm thickness. The silicon has been doped with either P, As, or Sb during the wafer formation process to a nominal resistivity of 10 $\Omega\cdot$cm.

The method of diode formation as well as the density and profile of the impurity ions determines the specific optical and electrical performance parameters of the photodiode.

Cadmium Sulfide Photoconductor. CdS photoconductive films are prepared by both evaporation of bulk CdS and settling of fine CdS powder from aqueous or organic suspension followed by sintering. The evaporated CdS is deposited to a thickness from 100 to 600 nm on ceramic substates. The evaporated films are polycrystalline and are heated to 250°C in oxygen at low pressure to increase photosensitivity. Copper or silver may be diffused into the films to lower the resistivity and reduce contact rectification and noise. The copper acceptor energy level is within 0.1 eV of the valence band edge. Sulfide vacancies produce donor levels and cadmium vacancies produce deep acceptor levels.

The films have an area resistance of 100,000–300,000 Ω/square. Most applications, such as switching on outdoor lights at twilight, require a detector resistance of near 1000 ohms to operate the switching circuit without the need of impedance matching electronics. This is accomplished by depositing the contacts in an interdigitated geometry. A protective film is deposited over the detector and contacts to provide for long-term stability or the detector structure is mounted in a hermetic package as shown.

GaAsP and InGaAs Photodiodes. Gallium–arsenic and gallium–arsenic–phosphorus diodes are fabricated as photodiodes as well as light emitters (see LIGHT GENERATION, LIGHT-EMITTING DIODES). Fabrication is typically with mesa etch technology of the films of GaAsP or InGaAs grown by the vapor-phase epitaxial process using metal organic chemical vapor deposition (MOCVD) (see THIN FILMS, FILM FORMATION TECHNIQUES). This growth technique results in impurity densities less than 1×10^{14} atoms/cm^3. The spectral cutoff range extends from 500 to 900 nm depending on the phosphorus (qv) content. These detectors can be used for color discrimination and do not require expensive interference filters. Emitters–diode pairs are utilized for very high impedance signal coupling in high speed integrated circuits.

PbS and PbSe Photoconductors. The lead chalcogenides, PbS, PbSe, and PbTe, were among the first infrared-detector materials to have been investigated. Although photovoltaic effects are observed with p–n junctions in single-crystal material the response is quite poor and not reproducible. However, very sensitive photoconductors are prepared as polycrystalline thin films, ca 1-μm thick, which are deposited on glass (qv) or quartz substrates between gold or graphite electrodes.

Detector elements are prepared either by sublimation in the presence of a small partial pressure of O_2 or by chemical deposition from alkaline solution containing a lead salt and thiourea or selenourea. Lead sulfide and lead selenide deposit from solutions as mirror-like coatings made up of cubic crystallites 0.2–1 μm on a side. The reaction may nominally be represented by the following: $Pb^{2+} + SC(NH_2)_2 + 2\ OH^- \rightarrow PbS + C(=NH)_2 + 2\ H_2O$. The actual reaction probably is more complex.

Platinum Silicide Schottky Barrier Arrays. The Pt:Si detector essentially is a metal semiconductor barrier whereby the platinum silicide is a quasi-metal that generates a small energy barrier to electrons. The effective photons are absorbed in a very thin region of the silicide next to the barrier and generate free electrons that flow over the barrier and tunnel through it into the n-type silicon. The efficiency of this process is only a few percent even at high energies (short wavelengths) because of the low electron diffusion coefficient in the silicide.

Techniques of platinum deposition vary but sputtering and annealing of a very thin layer of platinum produces a uniform platinum silicide film with less than 0.3% variation in responsivity in an area of 2×2 cm. Details of deposition and annealing processes are considered trade secrets by the manufacturers.

InSb Photodiode Detectors and Arrays. Sensitive photodiodes have been fabricated from single-crystal InSb using cadmium or zinc to form a p-type region in bulk n-type material. High quality InSb crystals can be grown by the infinite-melt process where an InSb film is grown epitaxially (from the liquid phase) on a slice of InSb which was prepared in a conventional Czochralski vertical puller. The diode formation process typically is a closed-tube diffusion.

Mercury Cadmium Telluride. HgCdTe has proven to be an excellent infrared detector material where the CdTe content can be readily adjusted to obtain cutoff wavelengths from 2 to 20 μm. The benefit is high spectral sensitivity of the photon detector, low defect density, and high cooling efficiency. The dependence of energy gap on mole fraction is linear. For $Hg_{1-x}Cd_xTe$, the x values of most interest lie between 0.17 and 0.50. The need for large focal planes up to 2×2 cm has dramatically changed the direction of single-crystal HgCdTe growth technology.

Crystal Growth. Large-area CdZnTe substrates form the basis for liquid-phase epitaxy (LPE) growth of mid- and long wavelength ir HgCdTe detector material. Substrates are obtained from 3.5 kg CdZnTe ingots grown in graphite boats in sealed quartz ampuls.

Liquid-phase epitaxial films are grown in production prototype dipping reactors using the CdZnTe substrates. Film growth is both from tellurium and mercury solutions.

Films grown in Hg are usually structured to make heterojunction photodiode arrays. The first or base layer is narrower band gap HgCdTe, grown on CdZnTe substrates, doped with indium for excess electrons (n-type) in the $3 \times 10^{16}/cm^3$ range and is 10-μm thick. The second or cap layer is wider band gap HgCdTe, doped with arsenic for excess holes (p-type) in the $5 \times 10^{15}/cm^3$ range and is 4-μm thick. The composite HgCdTe film is photolithographically etched part way into the base layer to form an array of mesas, each one being a photodiode detector element. The p–n junction is close to or coincident with the metallurgical heterojunction. For infrared detection in the 8–12 μm atmospheric spectral window the base layer CdTe content is ca 20% and the cap layer CdTe content is ca 20% and the cap layer CdTe content is ca 30%.

Detectors Arrays. Greater than 50,000 HgCdTe linear arrays have been produced in the United States since 1972 for the Department of Defense infrared systems ranging from night vision for M1 tanks to targeting sights for laser guided bombs. These common module detector arrays consist of 180 elements photoetched on a 50-μm pitch. Virtually all of the material used for these arrays was prepared by the solid-state recrystallization process.

Detectors, Arrays, and Focal Planes. The two popular types of photodiode arrays in HgCdTe are based on homojunction and heterojunction technologies. Homojunction diode arrays are fabricated with p-type epitaxial HgCdTe.

Heterojunction diode arrays utilize the grown p–n junction and mesa etch technology. HgCdTe photodiode performance for the most part depends on high quantum efficiency and low dark current density.

Doped Germanium and Silicon Photoconductors. The extrinsic photoconductors are typically single-crystal germanium doped with zinc, cadmium, mercury, boron, and gold and silicon doped with indium, gallium, or arsenic. The doping density ranges from 1×10^{15} to 1×10^{17} impurity atoms/cm^3 leading respectively from low (1 cm^{-1}) to higher absorption coefficient (50 cm^{-1}).

GaAs-AlGaAs Quantum Well Arrays. The quantum well infrared detector is a newer technology based on the artificial structure called a superlattice. The quantum well detector array is an attempt to achieve a monolithic architecture and thereby very large size with high reliability. The idea is to build the long wavelength detector array directly on the readout integrated circuit. Infrared detection out to 12-μm wavelength can be achieved using engineered material having a controlled energy gap. The technique is to fabricate multilayers of semiconductors having alternating band gaps such that a series of potential wells exists in the direction normal to the layers. Some success has been achieved using very thin layers of GaAs and AlGaAs grown by molecular beam epitaxy (MBE).

Semiconductor Bolometer Arrays. The use of bolometers to sense infrared radiation is not new. What is different is that, rather than having a single cell of large dimension, a matrix of miniature cells or microbolometers is created. Rather than a single amplifier connected externally, a custom integrated circuit is built under each cell to form a totally integrated focal plane array. Key elements of the structure are the thin (100 nm) amorphous silicon (α:Si) or vanadium oxide (VO) thermally sensitive membrane, the thermally insulating support arms, and the integrated circuit underlying this structure.

Two technological advances have occurred that make such a structure feasible to build. The first is the development of microetching techniques that can be used to form microscopic structures in silicon and its coatings. This makes formation of the membrane and support arms possible with thickness control in the 10-nm range. Cell dimensions of less than 50 μm have been demonstrated. The second critical factor is the increased circuit density in silicon integrated circuits, making it possible to build a small circuit for each cell of the array. By doing this, the noise bandwidth of each cell can be minimized, thereby maximizing performance. This device is a thermal detector in the infrared and is therefore independent of wavelength.

Health and Safety Factors

The completed photodetector usually is packaged hermetically in inert glasses or plastics or is enclosed in an evacuated metal or glass container. Although most detector materials are toxic, the means taken to passivate and isolate these materials are often adequate to protect the user. However there are exceptions. The preparation of detector materials and detector fabrication can present considerable hazards. Ampul explosions do occur. The electrical circuitry required to operate photodetectors almost always couples to detector devices at low voltages, eg, >10 V; therefore, electrical hazards are minimal.

SEBASTIAN R. BORRELLO
MARK V. WADSWORTH
Texas Instruments, Inc.

R. A. Smith, F. E. Jones, and R. P. Chasmar, *The Detection and Measurement of Infrared Radiation,* Oxford University Press, London, 1968.

W. L. Wolfe and G. J. Zissis, eds., *The Infrared Handbook, Rev. Ed.,* Environmental Research Institute of Michigan, Ann Arbor, 1985.

P. W. Kruse, L. D. McGlauchlin, and R. B. McQuistan, *Elements of Infrared Technology,* John Wiley and Sons, Inc., New York, 1962.

P. W. Kruse, "The Photon Detection Process," in R. J. Keyes, ed., *Topics in Applied Physics, Optical and Infrared Detectors,* Springer-Verlag, New York, 1980.

PHOTOGRAPHY

The unique light-sensing properties of silver halide crystals have been recognized since the 1500s. In spite of many technical advances in nonsilver halide (eg, electronic) technologies, chemically based silver halide systems continue to dominate in the ability to record images of superb image quality and archival characteristics. Photochemical reduction in which the silver ion, Ag^+, in the ionic silver halide crystal is reduced to elemental silver, Ag^0 was first observed by the alchemist Fabricius in 1556. As photochemical reduction continues, elemental silver atoms aggregate and grow into clusters of a colloidal size sufficient to scatter light and produce hue shifts. The science of photography uses this photochemical property of silver halide to form images and record scenes.

The daguerreotype process and the calotype process were among the first photographic techniques to produce continuous-tone images as reproductions of scenes. The steps of a typical daguerreotype process include polishing and cleaning a silver-plated copper plate; treating the silver side of the plate with iodine vapors to convert silver into light-sensitive silver iodide, AgI; exposing the plate through the optics of a camera that projects and focuses a scene on the plate, ie, where light strikes these plates, silver ions are photochemically reduced to silver metal; and treating the exposed plate with mercury. The mercury reacts with silver metal to produce a silver amalgam. White silver amalgam appears in areas of the plate exposed by light; the unexposed areas remain dark, thereby producing a positive image.

Daguerrotype images were produced primarily by the action of light. For the production of satisfactory images, long exposures to light, on the order of minutes, were required. The calotype process reduced exposure times to seconds and produce visible images without dependence on the action of light. In the calotype process, the exposure of silver halide produced an invisible latent image that was composed of only trace amounts of reduced silver. This acted as a catalyst for subsequent chemical reduction, ie, a nonphotochemical continuation of the light-initiated reduction process that eventually produced a visible silver image. The use of chemical amplification after low level image exposure is a conceptual approach applied in modern photography.

Unexposed silver halide in these early recorded images photolytically darkened upon repeated exposure to light, thus photographic images were not permanent. In 1839 Herschel discovered that unexposed silver halide could be dissolved with sodium thiosulfate and washed away, whereas metallic silver was relatively unaltered by sodium thiosulfate treatments. This process, called fixation, along with chemical and spectral sensitization, were necessary to the foundation of modern photography. Sensitization improvements led to the development of high sensitivity available-light photographic systems. Before such sensitivity improvements, high energy illumination was required to give enough exposure intensity.

Without special sensitizing treatments, silver chloride, AgCl, microcrystals, which do not significantly absorb light having wavelengths greater than 400 nm, are virtually insensitive to visible light. Similarly, silver bromide, AgBr, and Ag(Br,I) and Ag(Br,Cl,I) microcrystals popularly used in modern photography are effectively insensitive to electromagnetic radiation of wavelengths longer than 500 nm. Photographically these crystals are said to be blue sensitive, but green and red insensitive. Blue sensitivity refers to the intrinsic sensitivity of the silver halide crystals. To reduce the number of blue photons required to produce a developable latent image, ie, a catalytic center, the silver halide crystals are treated with materials called chemical sensitizers, that absorb to the crystal surfaces and may or may not react with them. Chemical sensitizers do not significantly alter the light-absorption properties of the silver halide crystals, but they do alter the efficiency with which the latent image is formed. Sulfur- and gold-containing compounds are among the most popular chemical sensitizers. In 1925 it was demonstrated that in certain samples of gelatin (qv), sulfur could increase the intrinsic sensitivity of silver halide microcrystals. Gelatin has been used since 1847 as a protective colloid to prevent the silver halide microcrystals from aggregating or coalescing before and after coating on paper, film, or glass supports. The use of gold salts as chemical sensitizers was discovered in 1936 at the Agfa film plant in Germany. However the mechanisms by which these salts enhance photographic sensitivity continue to be investigated.

To achieve photographic sensitivity in the green (500–600 nm) and red (600–700 nm) regions of the visible spectrum, silver halide crystals are spectrally sensitized with dyes (see DYES, SENSITIZING). In spectral sensitization, dye molecules are absorbed to the silver halide surfaces. The transition from glass plates to film bases for supporting light-sensitive emulsions represents the final step that made photography a popular and commercial success.

The first satisfactory photographic film was produced in 1888 when gelatin-dispersed microcrystals of silver halide were coated on celluloid sheets. Within a year George Eastman prepared and marketed roll films on a base produced by dissolving nitrocellulose with camphor and amyl acetate in methanol (qv).

A broad range of photographic materials exists in the 1990s including x-ray films (see X-RAY TECHNOLOGY), graphic arts films, microfilms, and complex multilayer coatings for color films that provide low granularity, available-light sensitivity, and rapid access. Many modern color films have more than 15 separate layers. For such films, coatings having as much silver as 10 g/m^2 and as many as one hundred different chemical compounds may be necessary to provide the desired image quality (granularity and sharpness), color reproduction, image permanence, and light sensitivity (see COLOR PHOTOGRAPHY).

The modern preparation of photographic films, papers, and plates begins with the growth of silver halide microcrystals (Fig. 1). Commercially, reaction vessels having capacities as large as 2000 L are used to produce microcrystals (grains) ranging in size from tens of nanometers for microfilms up to micrometers for the highly sensitive crystals required in available-light photography. Once the grains have been precipitated and chemically and spectrally sensitized, the emulsions are ready for coating on a support. The choice of support depends on usage requirements. Paper, glass (qv), and polymeric films are most popular because of the dimensional stability, chemical inertness, archival properties, flexibility, and convenience they exhibit. Before coating, spreading agents (surfactants) and high molecular weight polymers are added to the emulsions to facilitate coating operations, antifoggants are added to improve the signal-to-noise performance of the photographic film, hardeners are added to produce thermally stable gelatin matrices, emulsion stabilizers are added to extend shelf-life properties, and in the case of color photography (qv), organic compounds called couplers or dye-release materials are added to allow for the production of colored images.

Once coated, the photographic material is in an appropriate form to be exposed. The optics of the camera focus an image of a given scene onto the emulsion grains. Development is an amplification stage following exposure in which the relatively small number of silver atoms comprising the latent-image center is magnified by factors as large as 10^{10}. Development can occur within seconds after exposure, as in instant photography (see COLOR PHOTOGRAPHY, INSTANT), or it can be months or years after exposure, as in conventional photography. When amplification occurs in the exposed regions of a negative film, a signal composed of elemental silver is produced; however, when amplification occurs in the unexposed regions of such a film, the relative signal intensity is diminished. Such unwanted amplification in the unexposed regions (or in the exposed regions of a reversal film) is called fog or D-min (minimum density). In a conventional black-and-white film, the developed silver produces visual darkening by scattering and absorbing light. After development the coatings are transported to a bath containing thiosulfate, where the remaining undeveloped silver halide is removed by fixation. During fixation, water-soluble complexes are formed between silver and thiosulfate ions, and are later washed from the coatings. The silver may be recovered for future use. After the undeveloped silver ions are removed, a stable silver record of the original scene remains. In instant color films there is no fixation.

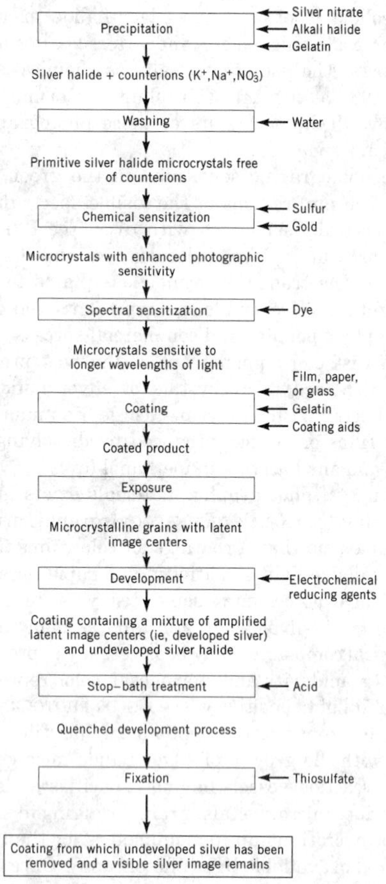

Figure 1. Flow chart of the photographic process.

The extent of silver-ion reduction during development depends on the development time, the composition and temperature of the developer solution, and the level of the original exposure to light.

The Photographic Crystal

The preparation of light-sensitive photographic materials begins with the precipitation of silver halide grains. Various process control parameters are rigorously regulated during crystal growth to achieve the desired grain morphologies, size-frequency distributions, solid-state properties, light sensitivity, and catalytic activity (developability). In addition to the use of mechanical process control parameters, eg, flow and mixing rates, various chemicals are added during precipitation to control silver halide growth rates, ripening characteristics, stability, and even light sensitivity. Common to all precipitations is the controlled mixing of solutions of a halide, such as an alkali halide salt, and a silver salt, usually silver nitrate, in the presence of a peptizing agent. Gelatin is one of the best peptizing agents known; however, others have been used. During mixing, a chemical reaction occurs and a suspended solid phase, in the form of microscopic silver halide crystals, separates from the liquid phase. As the reactants are added to the reaction vessel, the concentrations of the soluble counterions (alkali cations and nitrate anions) increase. If the ionic strength of the solution in the reaction vessel is sufficiently high, the double-layer repulsion force between grains can be reduced to less than the van der Waals attraction force, a condition that encourages the coagulation of the silver halide particles. The presence of a peptizing agent that adsorbs to the grain surfaces but does not inhibit continued growth is essential to prevent coagulation and to maintain a uniform dispersion of microcrystalline grains.

For some applications, inorganic impurities are intentionally added to emulsions during precipitation to achieve certain desired photographic responses. These impurities or dopants are usually incorporated into the silver halide grains at very low (ie, in the ppm or ppb range) concentrations. In spite of these low concentrations, dopants can have significant effects on the solid-state properties of the grains as well as on light sensitivity, contrast, and developability.

After crystal growth is completed, the resulting photographic emulsion is a dispersion of silver halide microcrystalline grains in an aqueous gelatin phase. Precipitation by-products such as counterions, ripeners, and others are also present in solution. If these by-products are not removed, adverse crystallization may occur when the emulsion is coated on a support and dried, and certain by-products may interfere with subsequent chemical and spectral sensitization. When photographic materials are coated on a paper support, some of the by-products are adsorbed on the fiber matrix of the paper, which can minimize or eliminate the adverse effects of reaction by-products. However, for many photographic products, such as those coated on film, glass, and water-impermeable resin-treated paper, the adsorbent properties of the substrate are not sufficient, and the amount of water present after crystal growth is often inconveniently high for coating operations.

Response Enhancement

Chemical Sensitization. After the photographic microcrystals are precipitated but before they are coated on a support, the crystals are treated to enhance their sensitivity to light. Chemical sensitization is a process which improves that ability of the emulsion grains to use the absorbed photons, independent of the wavelength. Various methods of post-precipitation chemical sensitization have been developed to reduce the number of photons required to produce a developable latent-image center. Typical chemical sensitizers include sulfur-containing compounds such as thiourea and sodium thiosulfate, gold-containing complexes such as gold thiocyanate and potassium tetrachloroaurate, and chemical reducing agents such as hydrogen gas and *tert*-butylamine borane. These sensitizers can be used either alone or in combination to increase sensitivity, and produce optimum photographic properties when used in trace amounts, ie, μmole sensitizer per mole Ag. At these low concentrations the sensitizers have advantageous effects on the solid-state phenomena during exposure without adversely affecting subsequent amplification processes during development.

Silver sulfide is generally considered to be the active chemical species resulting from sulfur sensitization treatments with either sodium thiosulfate or thiourea.

The reaction of thiosulfate with silver halide crystals to form adsorbed sulfide on the grain surfaces is activated thermally. If the reaction is allowed to continue too long before quenching or if excessive concentrations of sodium thiosulfate are used, the emulsion grains become spontaneously developable, ie, no exposure is required to induce catalytic activity, and image discrimination is lost. The sulfiding reaction is a two-step process that involves physical adsorption followed by a thermally activated chemical reaction.

$$(AgBr)_n + S_2O_3^{2-} \rightarrow (AgBr)_{n-1}[Ag(S_2O_3)]^-_{ads} + Br^-$$

$$(AgBr)_{n-1}[Ag(S_2O_3)]^-_{ads} + Ag^+ + H_2O \rightarrow (AgBr)_{n-1}Ag_2S + SO_4^{2-} + 2\,H^+$$

Gold is often used in combination with sulfur for an additional increase in sensitivity, particularly for high intensity exposures. Gold enhances catalytic activity of the latent image center with which it is associated. In this capacity, gold reduces the total number of photochemically reduced atoms (Ag and/or Au) required for developability and therefore increases the light sensitivity of the grain. Treatment of emulsion grains with chemical reducing agents also enhances sensitivity. Reduction sensitization can be achieved by high pH treatments, low pAg treatments, or treatment with chemical reducing species such as H_2, stannous ions, or hydrazine (see HYDRAZINE AND ITS DERIVATIVES). Reduction sensitization contributes to enhanced sensitivity by either hole-trapping processes, electron-trapping processes, or some combination thereof, depending on the particular technique used.

Spectral Sensitization. The intrinsic absorption, and therefore the intrinsic photographic sensitivity, of silver bromide and silver iodobromide microcrystals falls off rapidly for wavelengths greater than 500 nm. In fact, silver chloride crystals have almost no sensitivity in the visible regions of the spectrum. The need to extend silver halide sensitivity into the green (ca 500–600 nm) and red (ca 600–700 nm) regions of the visible spectrum is obvious for the production of color photographic products. Furthermore, even in black-and-white materials, extension of the photographic response beyond 500 nm is necessary for optimum effective light sensitivity. The process of expanding the wavelength sensitivity beyond the intrinsic region is called spectral sensitization. It is usually done after precipitation but before coating, and usually is achieved by adsorbing certain organic dyes to the silver halide surfaces. Once the dye molecule is adsorbed to the crystal surface, the effects of electromagnetic radiation absorbed by the dye can be transferred to the crystal. As a result of this transfer, mobile electrons are produced in the conduction band of the silver halide grain. Once in the conduction band, the electrons are available to initiate latent-image formation.

Many spectral-sensitizing dyes can be classified according to molecular structures. The structural part of a dye molecule that enables the molecule to absorb visible or infrared radiation is called a chromophore. The resonance structure for three common chromophores is shown.

Amidinium ion system

$$\overset{+}{N}=CH-(CH=CH)_n-N \rightleftharpoons N-CH=(CH-CH)_n=\overset{+}{N}$$

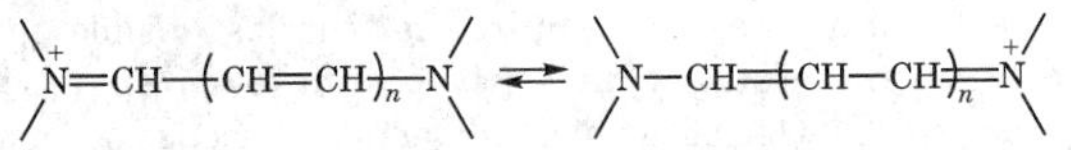

Carboxyl ion system

$$O=CH-(CH=CH)_n-O^- \rightleftharpoons {}^-O-CH=(CH-CH)_n=O$$

Dipolar amidic system

$$\overset{+}{N}=CH-(CH=CH)_n-O^- \rightleftharpoons N-CH=(CH-CH)_n=O$$

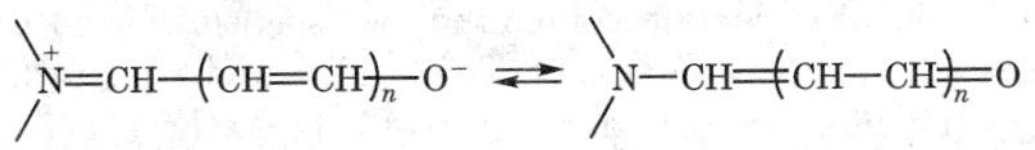

Other Emulsion Additives

In addition to chemical and spectral sensitizers, several other classes of chemical compounds are added to emulsions before coating. Additives are used to facilitate coating operations, eg, surfactants (qv) and viscosity enhancers; to reduce spontaneous development in unexposed regions, eg, tetraazaindenes and mercaptotetrazoles; and to reduce abrasion and permit high temperature processing, eg, aldehydes (qv).

For certain component compositions the viscosity and surface tension of the melted emulsion may not allow adequate emulsion spreading on the support during the coating procedures. For these situations, various surfactants that act as spreading agents are available to control the surface tension. Stabilizers include halide ions, acid, benzimidazoles, benzotriazoles, benzothiazolium salts, and mercaptotetrazoles. Many of these compounds adsorb to silver and complex with silver ions. Specifically as a result of these interactions, phenylmercaptotetrazole restrains development and enhances sensitivity even in freshly coated samples. Gelatin cross-linking agents (hardeners) represent another class of materials that may be added before coating. These compounds render the coated emulsion layers more resistant to abrasion during handling and improve the thermal stability of the gelatin. Gelatin must be hardened before development. The enhanced thermal stability and improved mechanical durability produced by hardeners result from the formation of three-dimensional bridging of various sites within the gelatin molecules. Both inorganic, eg, chromium salts, and organic, eg, aldehydes, compounds have been used as hardeners.

In most color photographic products, organic compounds such as couplers or redox dye releasers are added to the melted emulsions before coating. These compounds are essential to the development reactions that produce the dye molecules composing color images.

Coating the Emulsion

The Support. For most practical applications, the sensitized emulsions must be coated on a base or support to permit convenient handling. There are three basic classes of supports: glass, plastic, and paper. Supports are chosen on the basis of dimensional stability, low water permeability, flexibility, freedom from surface irregularities, compactness, cost, and safety. Clear plastic film supports are the most commonly used bases in modern photography. These materials are designed and selected based on safety, environmental concerns, and how they are to be used. Cellulose nitrate supports have been replaced by solvent-cast materials, eg, cellulose triacetate, and extruded materials, eg, poly(ethylene terephthalate) (PET). These materials are not only safe but also strong and dimensionally and chemically stable. Paper supports are commonly used in products that are viewed in the reflection mode, such as color or black-and-white print materials.

Coating Techniques. There are rigid constraints on the technology of coating photographic materials. The emulsion coatings must be uniform in thickness and composition and free from streaks. Compounding the difficulty is the need to produce a multilayered coating often composed of more than ten separate chemical-containing layers. Furthermore, light-sensitive materials must be coated in near or total darkness. Most of the coating application techniques for photographic materials employ a flexible support which is transported on rollers past a coating station where the emulsion is delivered. Once the emulsion has been spread over the support, the support is conveyed to a cooling chamber where the emulsion gels, then into a drying chamber where much of the water is removed from the coated gelatin. During drying, the emulsion thicknesses are reduced to ca 10% of the originally coated thicknesses.

Coating Structures. Light-sensitive photographic elements can be produced by coating an emulsion layer directly on the support; however such simple structures rarely have practical application. Generally, the support is electrostatically or chemically treated to improve the adhesion of the hydrophilic gelatin layer to the more hydrophobic support. In some products intermediate layers are coated between the emulsion and base to facilitate spreading. These intermediate layers often contain light-absorbing materials to prevent stray light from reflecting back into the emulsion layer during exposure. The stray light is reflected at the film–air interface on the back of the support because of the refractive index change at that surface. In the early days of photography the reflected light often produced photographs showing halos around small bright images, hence the use of light-absorbing materials is referred to as antihalation and the light-absorbing layers, antihalation layers. Several different light-absorbing materials have been used in antihalation layers, including finely divided carbon particles, dyes, and colloidal silver. The antihalation materials must be removed during processing as they can interfere with subsequent viewing.

Most films also are overcoated with a gelatin layer that protects sensitive emulsion layers from the image-degrading effects of pressure and abrasion.

Exposure and Latent-Image Formation

For photographic materials in which the image is produced with uv, visible, or ir radiation, optical lens systems are required. The lens system focuses the image of a scene on the emulsion layers of the photosensitive coating. The degree of magnification is, therefore, a function of the effective focal length of the optical lens system. The quality of the image as recorded within the coating depends not only on the optical and chemical properties of the coated light-sensitive material, but also on the lens optics. Accordingly, lens flare, astigmatism, and chromatic and spherical aberration all must be considered in designing a photographic system. Certain photographic systems do not use lenses during imaging, eg, x-ray and laser-scanning systems. In medical radiography, x-ray images are recorded by placing the photosensitive coating next to an x-ray absorbing screen. The phosphors of the screen are electronically excited upon x-ray absorption and emit visible light during subsequent de-excitation. The original x-ray pattern is recorded as the photographic grains absorb the photons emitted from the screen.

Positive imaging can be accomplished by several different techniques. In positive imaging, the density produced by developed silver or dye molecules decreases with increasing exposure. Negative-working silver halide grains are also used to produce the positive images provided by instant photography products. After exposure in instant photography systems, the film is passed between a pair of rollers which rupture a reagent-containing pod. The reagents include development initiating chemicals that are released from the pod and uniformly spread within the film structure (see COLOR PHOTOGRAPHY, INSTANT).

Other reversal imaging materials make use of specially designed reversal emulsion grains. Reversal grains are conveniently divided into two classes: photobleach reversal grains and internal image grains. Photobleach grains are chemically treated to have surface catalytic activity before exposure.

The developed image produced in a given photographic film is not uniquely determined by the exposure. Exposure is a measure of the total incident light energy and is therefore equal to the mathematical product of the light irradiance, I, and the exposure time, t.

Special Exposure Effects

Solarization, the photobleach effect, the Herschel effect, and the Clayden effect are all exposure-related phenomena that produce positive photographic images without the need for special development solutions or processing sequences. For negative-working emulsions that are conventionally processed, the developed density increases and then becomes constant with increasing exposure.

Development

Composition of Developer Solutions. In most practical photographic materials, exposure of the silver halide grains does not produce visible images. The growth of the small clusters of silver atoms into silver centers visible to the unaided eye is achieved with developer solutions. At least three significant classes of development components are present in most practical developer solutions: reducing agents, restrainers, and preservatives. The most important component is the chemical reducing agent.

For negative films, the electrochemical reduction properties of the reducing agents must be properly positioned to provide rapid amplification of exposed. The ability to discriminate between exposed and unexposed grains is a well-known property of chemical reducing agents that possess the Kendall structure, represented by $A{+}CH{=}CH{+}_nB$ where n may have either zero or integral values and where A and B may be hydroxyl, amine, or substituted amino groups. Most of the useful chemical reducing agents are benzene derivatives having Kendall structures. Examples of effective compounds that have Kendall structures include hydroquinones (1,4-dihydroxybenzene), catechols (1,2-dihydroxybenzene), *p*-aminophenols, *p*-phenylenediamines, and ascorbic acid. Phenidone developing agents and certain thiadiazoles discriminately reduce silver ions also.

Hydroquinone is in a class of commercially important black-and-white chemical reducing agents (see HYDROQUINONE, RESORCINOL, AND CATECHOL).

Stop-Bath Treatment and Fixation

Once development has progressed to some desired extent, amplification usually is quenched by a rapid decrease in pH. In conventional photographic films, the pH is decreased by mechanically or manually transferring the coatings from the alkaline developer solution to an acidic stop bath. Quenching usually can be done by a brief treatment with a stop bath composed of a dilute solution of a weak acid, eg, 0.5% acetic acid. For both the silver diffusion-transfer films and the dye-transfer films used in instant photography, development is initiated by spreading a viscous alkaline reagent between an emulsion layer and an image-receiving layer. To remove undeveloped silver halide, development and quenching are followed by fixation. Most fixing baths are composed of thiosulfate ions formed by dissolving the corresponding sodium or ammonium salt in water. The thiosulfate ions convert the remaining silver halide to water-soluble complexes such as argentodithiosulfate and argentotrithiosulfate. Once bound in these complexes, the silver can be washed readily from the gelatin coating. For the fixation of silver chloride, silver bromide, and silver bromochloride, thiosulfate is effective. However, for the relatively insoluble silver iodide with a solubility product of ca 10^{-16}, dissolution rates with thiosulfate solutions are significantly diminished.

In black-and-white photography, fixation generally is conducted under acidic conditions. The hardening provided by alum helps prevent physical damage, eg, scratches on the coating surfaces, during handling of the wet gelatin materials in the course of various post-development processes. The acidity can be maintained if, in addition to the salt of a strong base (eg, the sodium ion from sodium thiosulfate and sodium sulfite), a weak acid (eg, acetic acid) is present to provide a buffered solution. In alkaline thiosulfate fixing baths, alum forms an undesirable white sludge of aluminum compounds.

The primary component of a fixation bath, thiosulfate, tends to decompose in acidic environments: $S_2O_3^{2-} + H^+ \rightarrow HSO_3^- + S$ The decomposition can be retarded significantly by additions of bisulfite.

The rate of fixation is monitored in terms of the clearing time, which is defined as the time required for the last visible opacity to disappear. At various stages during the post-development process, the coatings are rinsed. One of the most important reasons for rinsing is to reduce chemical carryover from one treatment bath to another, which increases solution lifetime. The most important washing occurs at the end of the processing, just before drying. The purpose of this washing is to eliminate all soluble compounds from the coated gelatin layers. Efficient removal of certain compounds is essential for good keeping properties and image permanence. Thiosulfate and argentothiosulfate complexes have particularly damaging effects on the keeping properties of the final print or film.

Stabilization

Stabilization is an alternative to fixation for the production of a permanent although probably not archival image. In stabilization, the undeveloped silver halide is not removed but rather is converted to a compound that is relatively insensitive to light and stable to heat, humidity, and atmospheric gases. As in fixation, stabilization is achieved by the use of complexing agents to transform silver halide into silver complexes. Because the silver complexes and the unreacted complex agents are retained in the coated gelatin layer, the stability of these compounds is critical to the utility of the stabilization process. Stabilization is used where rapid access is needed, eg, news film and oscillograph tracing. Thus the complexing agents should react rapidly and completely with the silver halide contained in the gelatin layers. Furthermore, the agents cannot be toxic and should not soften the gelatin, bleach the image silver, or form colored complexes with silver.

Stabilizing agents can be classified into two categories: those agents that form water-soluble silver complexes and those that form insoluble complexes. The former category includes thiocyanate, thiosulfate, and thioureas.

Environmental Aspects of Processing

The quality of industrial effluents discharged into public sewage systems is specified by regulatory agencies. For certain quality parameters, photographic processing effluents fall within the required range without special treatment. Lubrication oils and greases are not present in photographic wastes, and the concentrations of suspended solids are too low (generally <20 mg/L) to be significant. Occasionally, specific processing baths must be maintained in a range of 38–51.5°C. However, when the total effluent is mixed, the temperatures are usually below 32°C, and thermal effects are generally of little consequence. Despite these qualities of photographic waste, processing effluents that ultimately are discharged into streams almost invariably require special treatment to meet stream standards. These treatments normally include silver removal, settling, biochemical degradation, aeration, and finally chlorination. Settling removes the solid wastes, biochemical degradation and aeration reduce the biochemical oxygen demand (BOD) of the waste, and chlorination destroys any pathogenic organisms remaining after treatment.

The concentrations of chemicals found in photographic processing effluents generally are not toxic to the bacteria and other treatment plant microorganisms necessary for biological degradation. However, hexavalent chromium used in dichromate bleaches as an oxidizing agent to remove developed silver from color coatings can be toxic to bacteria. Some processing components that may have undesirable environmental effects can be recycled during photographic processing and are not discharged into waste-disposal systems. Wash water, developer, and fixing and bleach-fix solutions are often regenerated, recycled, and reused. Silver can be electrochemically removed from thiosulfate fixing solutions which can then be chemically readjusted and reused. Photofinishing businesses occasionally install aerated lagoons capable of reducing the oxygen demand of their waste before discharge into community sewer systems. Such lagoons must be carefully engineered to guarantee that subsurface soil and streams do not become chemically contaminated.

Trace quantities of certain metal dopants occasionally are used to impart desired solid-state and photographic properties to emulsion grains. Because of its toxicity to aquatic life and microorganisms, the metal of primary ecological concern in the photographic industry is silver. Silver that is not salvaged during recovery operations may be present in the waste from photographic processing. The discharged silver usually is bound in thiosulfate complexes, which are not detrimental to the essential microorganisms in sewage treatment plants. The silver thiosulfate complexes can be converted to insoluble silver sulfide and removed as a solid sludge. Optimization of silver recovery is not only an important conservation measure but is also an economic benefit to photographic processing businesses. Various techniques for silver recovery are in common use including electrolysis, metallic replacement, ion exchange (qv), and silver precipitation.

The Silver Image

Stability. Image permanence and archival quality of recorded information is of increasing interest. The effects of residual thiosulfate and silver thiosulfate, the humidity, temperature, and chemistry of the environment can also adversely affect image permanence. The elemental silver composing the image is not completely stable and may be oxidized by the atmosphere to silver sulfide or silver oxide. Because such oxidation is initiated on the surface of the developed silver image, image degradation by air oxidation is of particular concern in photographic materials containing grains with high surface-to-volume ratios, especially if keeping properties are desired. However, various degrees of image fading can still occur even in large grains with gelatin protection. If the silver image is exposed to oxidizing fumes from such sources as automobile exhaust, freshly painted rooms, nitric oxides, and peroxides, then the possibility of image fading is enhanced. To improve silver image resistance to oxidation, photographic coatings can be treated with solutions such as gold chloride-plus-thiourea solutions or with iodide either during or after the process. Adsorbed iodide appears to stabilize the silver image by reducing the surface energy of the silver filaments. Selenium toning of microfilm products has been shown to enhance image stability (see SELENIUM AND SELENIUM COMPOUNDS).

Image Tone. Once the silver is developed, changes in image tone are sometimes desired. Treatment with any of a wide variety of toning baths can alter image tone. Two types of toning baths are used: one type involves the conversion of the image silver to silver sulfide, and the other involves the substitution of silver in the image with yet another metal. Metal toning makes possible a wide range of image colors.

Image Intensification and Reduction. In addition to tone changes after processing, it may also be necessary to alter the optical density or contrast of the original negative. Such alterations are rarely applied to the print. The process of increasing the image density is called intensification, whereas the process of lowering image density is called reduction. An image is intensified by adding a metal, eg, mercury, chromium, silver, or copper, to the image silver. If a given negative had sufficiently detailed information of an original scene but was underdeveloped during processing, then intensification may be useful. However, in underexposed films, the density of the original negative may not only be low but also the information may not have been recorded satisfactorily. In the latter case, intensification increases the density, but cannot provide missing details from shadow areas, ie, the toe region of the original negative. In contrast, reduction selectively removes silver from the developed image. Reduction can be used to correct overdevelopment (overexposure) or to alter the contrast of a negative. Lowering the density of a negative allows printing exposure intensities or printing exposure times to be reduced. Exposure time reduction becomes of practical importance when multiple printed copies of an original negative are required.

Image Evaluation. The subjective quality of a developed silver image depends on the color tone of the developed silver, brightness reproduction of the original scene, and perceived graininess and sharpness. Certain objective measurements and analyses correlate with these subjective qualities of a developed image.

Throughout the history of silver halide-based photography, much research has been directed toward improving the efficiency and therefore the speed without increasing the size of the silver halide grains. These research efforts have produced several improvements. Through advances in emulsion-making and sensitizing techniques, emulsion quantum efficiency has been increased. Because these advances have rendered grains of a given size more sensitive to light, the speed–grain relationship has been improved correspondingly. Furthermore, alterations in coating formats and modifications of development chemistry also have contributed significantly to improved speed–grain positions.

The visual sharpness of a recorded image is yet another subjective measure of image quality. The impression of sharpness or crispness is achieved when the boundaries and edges of the objects composing the image are clear and well defined. When high resolution of fine detail is required in the image, then, in addition to contrast and granularity, sharpness becomes a particularly important image quality parameter. Sharpness is evaluated by a number of methods. It is often measured as the ability of a recorder to produce an image of very narrow and closely spaced lines. The resolving power of a photographic material is determined by granularity and contrast as well as by effects of image spread. The modulation transfer function (MTF) is a more objective and quantitatively interpretable measure of the quality of sharpness. Analysis of MTF data can also be used to deduce the image density distribution produced by knife-edge exposures; it is in general more reliable than direct analyses of such edge exposures.

DAVID J. LOCKER
Kodak Manufacturing Research and Engineering

M. A. Kriss, in T. H. James, ed., *The Theory of the Photographic Process,* 4th ed., Macmillan Publishing Co., Inc., New York, 1977, pp. 592–635.

C. Kittel, *Introduction to Solid State Physics,* John Wiley & Sons, Inc., New York, 1971, Chapt. 9.

N. B. Hannay, *Solid-State Chemistry,* Prentice-Hall, Inc., Englewood Cliffs, N.J., 1967, pp. 80–98.

E. J. Birr, *Stabilization of Photographic Silver Halide Emulsions,* Focal Press, Inc., New York, 1974.

PHOTOMULTIPLIER TUBES. See PHOTODETECTORS.

PHOTOVOLTAIC CELLS

A photovoltaic (PV) solar power system is a complete electrical source that uses solar cells to directly convert light energy into electricity. The system can be self-contained and completely autonomous or it can work in tandem with other conventional fuel-based sources of power to offer robust power availability.

A solar cell is a semiconductor device that can convert light instantaneously into direct-current (d-c) electricity. A number of cells are typically connected together in series in a weather-resistant package such that enough voltage is generated to recharge a 12-volt lead–acid storage battery, the most common storage device used in conjunction with solar power (see BATTERIES, LEAD-ACID). Such a package of cells is designated a PV module, which is often constructed of an external sheet of strengthened glass and polymeric encapsulation. The most common size module is 0.5–1 m^2 in area and delivers between 25 and 150 watts of power.

The advantages of photovoltaic cells as a source of electric power over alternative power sources may be characterized as follows: solar cells capture sunlight, an essentially inexhaustible and nonpolluting energy source which is freely distributed, and directly convert that light into electricity; photovoltaic generation of electricity requires no machinery with moving parts and produces no noise, waste, or polluting by-products; photovoltaic systems are modular and therefore can be adapted for a variety of applications. Solar power systems are particularly useful in areas where power lines cannot be readily or inexpensively routed.

Solar cells have been used extensively and successfully to power satellites in space since the late 1950s. On earth, where electrical systems typically provide large amounts of power at reasonable costs, three principal technical limitations have thus far impeded the widespread use of photovoltaic products: solar cells are expensive, sunlight has a relatively low power density, and commercially available solar cells convert sunlight to electricity with limited efficiency.

The power density of sunlight is about 1350 W/m^2 at elevations just above the earth's atmosphere. Less than 1000 W/m^2 is typically incident on earth after filtering through the atmosphere. Due to the low power density of sunlight and limited conversion efficiencies, the most efficient solar modules can generate about 250 W/m^2 in peak sunlight conditions. The maximum power output of a solar cell or module is defined in peak watts (W_{peak}), a rating based on a standard measurement method established by international consensus. A solar panel of one square meter area nominally produces one kilowatt hour of electricity per day. For most large-scale, power-producing applications, solar modules have conversion efficiencies above 10% in order to minimize the total cost of a generating system.

Chemistry

Crystalline silicon *p–n* junction solar cells are the principal commercially available type and are used here to illustrate the operation of a solar cell. When sunlight falls on a solar cell, a voltage is induced and an electric current flows in an external circuit that is connected to the cell. Each atom in the silicon crystal lattice is surrounded by and bound to four equidistant neighboring atoms. The outermost shell of electrons of each silicon atom contains four valence electrons, and each of the four valence electrons in the crystal lattice is shared in a bonding orbital with an electron from one of its four nearest neighbors. This electron pair or covalent bond firmly binds the crystal. The energy required to break a covalent bond is the bond energy or energy gap, E_g. In silicon, E_g is ca 1.1 eV.

The absence of an electron from a covalent bond leaves a hole and the neighboring valence electron can vacate its covalent bond to fill the hole, thereby creating a hole in a new location. The new hole can, in turn, be filled by a valence electron from another covalent bond, and so on. Hence, a mechanism is established for electrical conduction that involves the motion of valence electrons but not free electrons. Because holes and electrons move in opposite directions under the influence of an electric field, a hole has the same magnitude of charge as an electron but is opposite in sign. The energy in light also can break the bonds of silicon valence electrons.

If the hole and electron are not kept apart, they recombine to produce a small amount of thermal energy within the crystal and no net current flow. When the holes and electrons are kept apart, collected, and made to flow in a circuit outside the crystal, they produce electric current in that circuit. Solar cells are equipped with a barrier or a junction which provides an internal electric field that segregates photogenerated electrons and holes. Thus, although unmodified silicon has an equal number of holes and electrons, a *p–n* junction silicon solar cell consists of two charge-dissimilar regions which are separated by a junction: one region is rich in holes (positive), ie, *p*-type silicon, and the other is rich in electrons (negative), ie, *n*-type silicon. Such regions do not occur naturally; they are fabricated by doping, ie, replacing some silicon atoms in the lattice with atoms having a valence other than four. Replacement of a few silicon atoms, ie, ca one in several million, causes large increases in the electrical conductivity of the resultant doped crystal.

Junctions

Four different types of junctions can be used to separate the charge carriers in solar cells: (*1*) a homojunction joins semiconductor materials of the same substance, eg, the homojunction of a *p–n* silicon solar cell separates two oppositely doped layers of silicon; (*2*) a heterojunction is formed between two dissimilar semiconductor substances, eg, copper sulfide, Cu_xS, and cadmium sulfide, CdS, in Cu_xS–CdS solar cells; (*3*) a Schottky junction is formed when a metal and semiconductor material are joined; and (*4*) in a metal–insulator–semiconductor junction (MIS), a thin insulator layer, generally less than 0.003-μm thick, is sandwiched between a metal and semiconductor material.

Fabrication methods that are generally used to make these junctions are diffusion, ion implantation, chemical vapor deposition (CVD), vacuum deposition, and liquid-phase deposition for homojunctions; CVD, vacuum deposition, and liquid-phase deposition for heterojunctions; and vacuum deposition for Schottky and MIS junctions.

Efficiency

The most efficient silicon cells produced are based on *p–n* homojunctions and convert 23.1% of the energy in incident light set to simulate the global air mass (AM) 1.5 spectrum, an artificial reference spectrum used to standardize measurement of PV power, with an intensity of 1000 W/m^2 at 25°C. This is the definition of peak sunlight test conditions. In theory, silicon *p–n* junction solar cells can convert a maximum approaching 26% of the energy in AM 1.5 sunlight to electricity. Approximately 75% of the energy in sunlight is lost to factors intrinsic to the silicon material.

In comparison, *p–n* homojunction cells made of more costly semiconductor materials, eg, indium phosphide, InP, and gallium arsenide, GaAs, which have energy gaps of 1.2–1.4 eV, and maximum theoretical conversion efficiencies of ca 28–30%, depending on the device construction and layering of junctions.

Commercial Silicon Solar Cells

Silicon cells are hundreds of micrometers (μm) thick in order to facilitate handling with minimal breakage, although most solar radiation is absorbed in the first 20–30 μm. The junction in a silicon cell usually is ca 0.2–0.5 μm from the surface of the cell. The crystal surface has many broken bonds that act as recombination centers. In conventional silicon cells, a comb or narrow metal grid lattice is connected to a current-carrying bus to collect charge carriers from the side of the cell facing the sun. The fingers are small enough in total area so that minimal cell area is in their shadow.

Antireflection coatings are used over the silicon surface which, without the coating, reflects ca 35% of incident sunlight. Materials such as titanium dioxide, TiO_2, tantalum pentoxide, Ta_2O_5, or silicon nitride, Si_3N_4, ca 0.08-μm thick are common.

Types of Solar Cells. There are three basic technology options for making solar cells with dozens of variations on each. These approaches are conveniently grouped as follows: thick (~300 μm) crystalline materials, concentrator cells, and thin (~1 μm) semiconductor films.

Thick Crystalline Materials. Crystalline silicon technology is the worldwide industry standard. The total cost of solar cells made from ingots reflects the costs of the silicon raw material used in forming an ingot, cutting and etching thin silicon wafers from the ingot, fabricating and encapsulating the cells, and assembling them into modules. An attractive cost-reducing approach is to grow good quality crystalline sheets directly from molten silicon. Smoothly grown sheets ca 100-μm thick require little or no cutting and polishing and incur little waste.

Gallium arsenide is a promising material for gaining the advantages of high efficiency. It is superior to silicon in several respects. The E_g of GaAs, ca 1.4 eV, is higher than that of silicon and is in the range that provides the highest calculated conversion efficiency for a single-junction cell. Because of this high efficiency and the fact that it does not decline as rapidly as that of silicon cells with increasing temperature, GaAs single-crystal cells are attractive for use as concentrator cells.

Gallium arsenide solar cells advanced in the 1980s for space use because they weighed much less than silicon cells of similar output, since GaAs absorbs sunlight much more strongly than silicon.

Concentrator Cells and Systems. Concentrators circumvent the problem of high semiconductor material cost by using mirrors or lenses to concentrate sunlight on small surface areas of more expensive solar cells. Concentration allows more power to be produced from a given amount of photosensitive material.

Concentrator optics vary from low ratio designs, eg, concentration of sunlight of an order of magnitude by Winston collectors, which do not require elaborate tracking of the sun, to much higher ratio systems based on parabolic mirrors or Fresnel lenses and which require precise, two-axis tracking. Three types of concentrator systems are being developed which operate at low level (<30 times), mid-level (100–400 times), and high level (>400 times) sunlight concentrations. The cell specifications and engineering requirements for each of these types of systems are quite different. Specially designed silicon has shown potential for use in concentrator systems.

Thin Film. In the thin-film approach, raw material usage is generally more than two orders of magnitude less and patterning is more direct.

Good solar cell results have been obtained from cells of materials, including polycrystalline silicon, amorphous silicon–hydrogen (α-Si:H) alloys, Cu_xS-CdS, $CuInSe_2$–CdS, and CdTe.

Electrochemical Photovoltaic Cells. The application of photoelectrochemistry in solar energy conversion technologies includes biomass conversion, photoelectrolysis, photogalvanic cells, electrochemical photovoltaic cells, etc. In electrochemical photovoltaic cells, electric energy is converted directly from sunlight by absorption of light in a semiconductor electrode. In many respects, these cells closely resemble conventional solid-state cells, except that the charge-separating barrier layer is formed at the interface between a semiconductor surface with a liquid electrolyte. When sunlight is incident on the semiconductor electrode, free holes and electrons are created. The relevant minority carriers must migrate to the interface and be separated; these carriers then react with the electrolyte either through oxidation or reduction. The counterelectrode reverses the reaction, thereby maintaining the electrolyte balance. The semiconductor electrode material may be either polycrystalline or amorphous material because in some cases the poorer material properties cause relatively little degradation of conversion efficiencies. In addition, incorporation of a third electrode may make possible *in situ* storage. The main disadvantage of these cells is the instability of the semiconductor electrode, especially under sunlight, for extended periods of operation. Electrochemical cells could be inexpensive, since the electrode–electrolyte barriers usually are easy to form, but appropriate deployment strategies have not yet been identified. The stability problems encountered to date have been extensive.

Balance of Systems. A solar photovoltaic system contains, in addition to solar cells and module(s), an array structure to support the modules, power-conditioning circuitry for control and modification of the output, and a means of storing energy if required. All elements beyond the module are referred to as balance-of-system (BOS) components. The cost of BOS items is nominally about equal to the cost of the PV module. However, the BOS fractional cost contribution can vary from one- to two-thirds of the total installed cost of a system, depending on application.

Material Availability and Environmental Impact

Photovoltaic systems must satisfy four principal requirements before solar photovoltaic conversion can provide a significant portion of general energy needs. The system costs must be low enough to be competitive with other means of energy generation, the amount of energy generated during the life cycle of a photovoltaic system must be substantially greater than the energy required to fabricate the system to meet the criteria of a sustainable technology, the materials used in the cells must be available to generate a substantial portion, ie, at least a few percent of world energy needs, and the fabrication and utilization of the conversion systems should not cause more environmental problems than other competing energy systems.

Silicon is the second most abundant element in the world and is not toxic. Inherent in the use of materials other than silicon for solar cells are challenges of material availability and environmental safety. In terms of production of CdS-based cells, sulfur is abundant, but the world's resources of cadmium, tellurium, selenium, and indium are much less than those of silicon. However, these resources are several orders of magnitude greater than the amount needed to provide photovoltaic power production of 50,000 MW/yr. Similarly, although arsenic is plentiful, the supply of gallium for GaAs cells is limited. However, studies have concluded that the gallium supply also is sufficient for substantial manufacturing scale.

Although photovoltaic conversion is nonpolluting, environmental, health, and safety aspects must be considered, especially with regard to harmful emission and waste products resulting from the production of the solar cell modules. It has been shown that, with proper encapsulation and a proactive recycling program, it should be possible to minimize environmental concerns.

Photovoltaic Markets

In the mid-1990s, utility applications have once again begun receiving a great deal of attention due to a profound paradigm shift that appears to be taking place in the utility industry. Rather than replacing or adding large central fossil-fueled or nuclear generation facilities, small PV systems deployed at the outer extremities of the grid can be cost-effectively used to manage demand profiles, defer transmission hardware upgrades, and support electrical service quality (voltage, power factor, etc) during periods of peak demand in locations where the utility grid transmission is unidirectional.

PV Market Segment Categories. Solar modules are used to provide power to a broad range of industrial, commercial, and consumer systems and products. Most participants in the PV industry use the following categories to describe the various market segments, which group applications by functional product requirement, system type, sales channel, and client base. These include the following: specialties, eg, spacecraft circuits, calculator chips, automobile sunroofs, and building facades; industrial power, ie, telecommunications, warning/signal lights, and remote data gathering; rural and off-grid electrification, eg, lighting, water pumping and purification, refrigeration, and recreational travel and boating; consumer convenience, eg, garden and security lighting and small battery charging; and grid-connected power, ie, distributed grid support and peaking power augmentation.

CHARLES F. GAY
National Renewable Energy Laboratory
CHRIS EBERSPACHER
UNISUN

Maintenance and Operations of Stand-Alone Photovoltaic Systems, Naval Facilities Engineering Command, Southern Division, rev. 1991.

K. Smith, *Survey of U.S. Line-Connected Photovoltaic Systems,* EPRI GS-6306, Palo Alto, Calif., 1989.

R. H. Annan, W. L. Wallace, T. Surek, E. Boes, and L. O. Herwig, *Department of Energy Review of the U.S. Photovoltaic Industry,* Report ST-211-3488, Solar Energy Research Institute, Golden, Colo., 1989.

G. D. Cody and T. Tiedje, in B. Abeles, A. Jacobson, and P. Sheng, eds., *Energy and the Environment,* World Scientific, Teaneck, N.J., 1994.

PHTHALIC ACIDS AND OTHER BENZENEPOLYCARBOXYLIC ACIDS

Physical and Chemical Properties

The physical properties of the benzenepolycarboxylic acids and the most important anhydrides are summarized in Tables 1 and 2.

The chemistry of benzenecarboxylic acids generally is the same as that of other carboxylic acids, which can be converted into esters, salts, acid chlorides, and anhydrides. Each carboxyl group can react separately, so that compounds in which carboxyl groups are converted into different derivatives can be prepared. Because there are aromatic hydrogens available in most of these acids, they also undergo reactions characteristic of the benzene nucleus.

Phthalic Acid and Phthalic Anhydride

The first of the benzene polycarboxylic acids to become a commercial product was phthalic acid, mostly in the form of the anhydride. The anhydride is obtained by the catalytic vapor-phase air oxidation of *o*-xylene or naphthalene. The IUPAC name of phthalic anhydride is 1,3-isobenzofurandione.

Manufacture and Processing. Until World War II, phthalic acid and, later, phthalic anhydride, were manufactured primarily by liquid-phase oxidation of suitable feedstocks.

Naphthalene (qv) from coal tar was the feedstock of choice in both the United States and Germany until the late 1950s, when a shortage of naphthalene coupled with the availability of xylenes from a burgeoning petrochemical industry forced many companies to use *o*-xylene.

Fixed-Bed Vapor-Phase Oxidation of o-Xylene. Well in excess of 90% of the phthalic anhydride produced is obtained by oxidizing *o*-xylene in the vapor phase over a fixed bed of catalyst. In the 1960s, there were two types of fixed-bed processes, low temperature/low space velocity, and high temperature/high space velocity. Catalyst development resulted in higher allowable space velocities for the low temperature case while high yields were maintained. Consequently, use of the low temperature process which runs at $< 400°C$ has predominated. A commercially viable plant must operate at high selectivity of at least 75 mol % with a feed *o*-xylene concentration of 60 g/m^3. This concentration is above the lower explosion limit of 43 g/m^3. The catalyst should last at least three years.

Some reactors are designed specifically to withstand an explosion. The multitube fixed-bed reactors typically have ca 2.5-cm inside-diameter tubes, and heat from the highly exothermic oxidation reaction is removed by a circulating molten salt.

The catalyst combines two essential ingredients found in earlier catalysts, vanadium oxide and titanium dioxide, which are coated on an inert, nonporous carrier in a layer 0.02- to 2.0-mm thick.

There are thermal gradients along the catalyst tube, and a temperature maximum or hot spot develops which tends to place limits in the feed rate owing to possible catalyst damage and to decrease the yield. Various methods are used to reduce this hot spot, including two or more catalysts in series to limit the reaction initially. This use of catalysts could be higher rubidium or potassium concentrations in the first bed, with higher phosphorus levels in the second. Different phosphorus levels may also be used. Placing varying concentrations of inert packing along the tube length is also possible. Another method is to have two temperature zones, the first zone being cooler than the second.

Other processes include fixed-bed vapor-phase oxidation of naphthalene, fluidized-bed vapor-phase oxidation, and liquid-phase oxidation of *o*-xylene.

Phthalic Anhydride Recovery and Purification. The accepted method of recovering phthalic anhydride from vapor-phase oxidation processes is first to recover some of the heat of oxidation, eg, by using the reactor effluent to generate steam. Then the vapors are passed through automatically cycled switch condensers where up to 99.5% of the phthalic anhydride solidifies on cooled, finned tubes. The off-line condenser is then charged with hot oil to melt the anhydride into a storage tank. Vent gases from the switch condensers contain maleic anhydride, citraconic acid, and benzoic acid as well as phthalic anhydride. The gas can be scrubbed with water or incinerated, either thermally or catalytically, before being released into the atmosphere. The water used to scrub the gases contains these mixed acids, and maleic anhydride can be recovered.

The crude phthalic anhydride is subjected to a thermal pretreatment or heat soak at atmospheric pressure to complete dehydration of traces of phthalic acid and to convert color bodies to higher boiling compounds that can be removed by distillation. Use of potassium hydroxide and sodium nitrate, carbonate, bicarbonate, sulfate, or borate has been patented. Purification is by continuous vacuum distillation. The most troublesome impurity is phthalide (1(3)-isobenzofuranone), which is structurally similar to phthalic anhydride. Reactor and recovery conditions must be carefully chosen to minimize phthalide contamination. Phthalide is also reduced by adding potassium hydroxide during the heat soak.

Storage and Shipment. Storage and shipment are preferably molten. Insulated and heated tanks are used, as are insulated rail tank cars and tank trucks. The molten form can be handled and pumped in bulk form, and as a result is priced lower than the solid. Solid phthalic anhydride is available as flake in 1-t and 0.5-t super sacks, and 22.7-kg multiwall bags.

Health and Safety Factors. Phthalic anhydride is a severe irritant to the eyes, respiratory tract, and skin, especially to moist tissue. There are explosion hazards with phthalic anhydride, both as a dust or vapor in air and as a reactant.

Uses. Phthalic anhydride is used mainly in plasticizers, unsaturated polyesters, and alkyd resins (qv). The plasticizers (qv) are used mainly with poly(vinyl chloride) to produce flexible sheet such as wallpaper and upholstery fabric from normally rigid polymers. The plasticizers are of two types: diesters of the same monohydric alcohol such as dibutyl phthalate, or mixed esters of two monohydric alcohols.

The second largest use at 21% is for unsaturated polyester resins, which are the products of polycondensation reactions between molar equivalents of certain dicarboxylic acids or their anhydrides and glycols.

The manufacture of alkyd resins (qv), which are obtained by the reactions of polybasic acids or anhydrides, polyhydric alcohols, and fatty oils and acids, consumes about 17% of the phthalic anhydride demand.

The remaining uses are better considered as derivatives of phthalic anhydride and consume less than 10% of the demand, but they provide a diverse group of products. Examples of dyes derived from phthalic anhydride are phthalocyanine blues, quinoline yellow, and anthracene brown (see DYES AND DYE INTERMEDIATES). Another use for phthalic anhydride is in the production of isatoic anhydride, a raw material used in the production of saccharin. Tetrachloro- and tetrabromophthalic anhydrides are manufactured by the reaction of phthalic anhydride with chlorine and bromine, respectively, at high temperatures. The halogenated forms impart fire resistance

Table 1. Physical Properties of Benzenepolycarboxylic Acids

			Dissociation constants in aqueous solution, 25°C					Solubility, g/100 g water	
Common name	Formula weight	Melting point, °C	pK_1	pK_2	pK_3	pK_4	ΔH_f° at 25°C kJ/mol[a]	at 25°C	at 100°C
phthalic	166.14	211[b]	2.95	5.41			−782	0.7	19.0
isophthalic	166.14	384	3.62	4.60			−803	0.012	0.32
terephthalic	166.14	402[c]	3.54	4.46			−816	0.0017	0.033
hemimellitic[d]	210.15	197[b]	2.80	4.20	4.87		−1160	v sol	v sol
trimellitic	210.15	238[b]	2.52	3.84	5.20		−1179	2.1	60
trimesic	210.15	380	2.12	3.89	4.70		−1190	0.24	6.4
mellophanic	254.16	241[b]	2.06	3.25	4.73	6.21	−1562	sol	v sol
prehnitic	254.16	238	2.38	3.51	4.44	5.81	−1549		
pyromellitic	254.16	282	1.92	2.87	4.49	5.63	−1571	1.5	>30
benzenepentacarboxylic[e]	298.17	228	1.80	2.73	3.97	5.25	−1930	sol	v sol
mellitic[f]	342.18	288[b]	1.40	2.19	3.31	4.78	−2299	sol	v sol

[a] To convert J to cal, divide by 4.184. [b] Decomposes at mp. [c] Sublimes. [d] Hemimellitic acid usually is handled as the dihydrate; formula wt = 246.18, mp = 191°C, decomposes. [e] $pK_5 = 6.46$. [f] $pK_5 = 5.89$; $pK_6 = 6.96$.

Table 2. Physical Properties of Anhydrides of the Benzenepolycarboxylic Acids

Common name	Formula weight	Mp, °C	Bp, °C
phthalic anhydride	148.12	131	284.5
hemimellitic anhydride	192.14	196	
trimellitic anhydride	192.14	168	390
mellophanic dianhydride	218.13	198	
pyromellitic dianhydride	218.13	285	390
benzene-1,2,4,5-tetracarboxylic dianhydride-3-carboxylic acid	262.14		
mellitic trianhydride	288.14	320 dec	

to polyester resins, polyurethane foams, and surface coatings. Phenolphthalein is the condensation product of phthalic anhydride and phenol in the presence of a dehydrating agent and is a pH indicator. Pesticides and anthranilic acid can be made from phthalimide which is in turn produced from phthalic anhydride.

Terephthalic Acid and Dimethyl Terephthalate

Purified terephthalic acid and dimethyl terephthalate are used as raw materials for the production of saturated polyesters. Terephthalic acid is also produced in technical or crude grades which are not pure enough for manufacture of poly(ethylene terephthalate). In almost all cases, the technical-grade material is immediately converted to purified terephthalic acid or dimethyl terephthalate, which together are the articles of commerce.

Manufacture and Processing. Terephthalic acid and dimethyl terephthalate did not become large-volume industrial chemicals until after World War II. In 1953 commercialized fibers were made from poly(ethylene terephthalate). Dimethyl terephthalate and ethylene glycol were the comonomers used to make these fibers (see FIBERS, POLYESTER).

Initial production of the dimethyl terephthalate started with the oxidation of *p*-xylene to terephthalic acid using nitric acid.

p-Xylene is the only feedstock used for either product. However, purified terephthalic acid has replaced dimethyl terephthalate as the leading terephthalate source for poly(ethylene terephthalate). Specifically, polyester processes that are based on dimethyl terephthalate must have equipment for recovery of methanol, which is the by-product of the transesterification with ethylene glycol. Use of the pure acid produces water as a by-product of direct esterification. In addition, terephthalic acid provides a higher yield of polyester per kilogram of starting feedstock. The need for transesterification catalysts with dimethyl terephthalate introduces metals into the polyester which can cause undesirable side reactions.

Technical-Grade Terephthalic Acid. All technical-grade terephthalic acid is produced by catalytic, liquid-phase air oxidation of *p*-xylene. Several processes have been developed, but they all use acetic acid as a solvent and a multivalent heavy metal or metals as catalysts. Cobalt is always used. In the most popular process, cobalt and manganese are the multivalent heavy-metal catalysts and bromine is the renewable source for free radicals. This catalyst system is used in about 70% of the *p*-xylene oxidations, and the percentage is increasing as new plants almost invariably employ it. Process conditions are highly corrosive owing to the acetic acid and bromine, and titanium must be used in contact with some parts of the process.

On a worldwide basis, the Hercules Inc./Dynamit Nobel AG process is the dominant technology for the production of dimethyl terephthalate. Modifications in commercial practice have occurred over the years, with several variations being practiced commercially. The reaction to dimethyl terephthalate involves four steps, which alternate between liquid-phase oxidation and liquid-phase esterification. Two reactors are used. First, *p*-xylene is oxidized with air to *p*-toluic acid in the oxidation reactor, and the contents are then sent to the second reactor for esterification with methanol to methyl *p*-toluate. The toluate is isolated by distillation and returned to the first reactor where it is further oxidized to monomethyl terephthalate, which is then esterified in the second reactor to dimethyl terephthalate.

Esterification of terephthalic acid is also used to produce dimethyl terephthalate commercially, although the amount made by this process has declined.

The Amoco process is used to purify terephthalic acid produced by the bromine-promoted air oxidation of *p*-xylene. The main impurity in the oxidation product is 4-formylbenzoic acid and the Amoco process removes this to less than 25 ppm. Metals and colored organic impurities are also almost completely removed by the purification.

Since the mid-1970s, starting in Japan, several companies have developed oxidation processes to yield relatively pure forms of terephthalic acid without a separate purification. These products, normally called medium purity terephthalic acids, contain 200–300 ppm 4-formylbenzoic acid and trace amounts of acetic acid and thus do not meet normal specifications for the highest purity grades available.

Hoechst Celanese and Formosa Chemical Fibers Corp. produce a polymer-grade terephthalic acid by hydrolysis of high purity dimethyl terephthalate.

Modern terephthalic acid plants usually produce at least 250×10^3 t annually. Storage of terephthalic acid is in silos, and the preferred method for dimethyl terephthalate storage is molten in insulated and heated tanks. The huge-volume use of these chemicals means that bulk shipment is preferred. Rail hopper cars or hopper trucks for terephthalic acid or insulated rail tank cars or tank trucks for dimethyl terephthalate are used where possible.

Health and Safety Factors. Terephthalic acid has a low order of toxicity. High doses of terephthalic acid lead to formation of calcium terephthalate at levels exceeding its solubility in urine. This insoluble material leads to the calculi and provides a threshold below which cancer is not observed. Normal precautions used in handling industrial chemicals should be observed with terephthalic acid. If ventilation is inadequate, a toxic-dust respirator should be used to avoid prolonged exposure.

Dimethyl terephthalate also shows low toxicity. As in the case of terephthalic acid, a toxic-dust respirator should be worn when ventilation is inadequate.

Molten dimethyl terephthalate burns if ignited. Dimethyl terephthalate vapor and dust, and terephthalic acid dust form explosive mixtures with air.

Uses. Essentially all polymer-grade terephthalic acid and dimethyl terephthalate are used to make saturated polyesters, the great majority being poly(ethylene terephthalate). Poly(ethylene terephthalate) is employed to make fiber and is the largest-volume synthetic fiber in the world. Fiber use makes up 73% of worldwide poly(ethylene terephthalate) production. It is used for woven and knitted fabrics for clothing, draperies, upholstery, and carpeting. In clothing applications, it is usually blended with other fibers, primarily cotton (qv). Applications for high strength polyester continuous fibers are reinforcing cord for tires, V-belts, conveyor belts, and hoses (see FIBERS, POLYESTER).

Polyester film consumes 7% of production.

The fastest growing application for poly(ethylene terephthalate) is in packaging, especially bottles. It currently accounts for 15% of production. Polyester bottles have almost displaced glass carbonated beverage bottles due to their clarity, light weight, and shatter-resistant properties. Applications in food packaging are expanding rapidly for the same reasons (see POLYESTERS, THERMOPLASTIC).

Small amounts of polymer-grade terephthalic acid and dimethyl terephthalate are used as polymer raw materials for a variety of applications.

Derivatives. In general, the esters of terephthalic acid derived from saturated alcohols undergo the same reactions as dimethyl terephthalate. The di-*n*-butyl and di-2-ethylhexyl esters find use as plasticizers (qv). Terephthaloyl chloride, which is prepared by reaction of terephthalic acid and thionyl chloride, is used to prepare derivatives of terephthalic acid.

Isophthalic Acid

Like terephthalic acid, isophthalic acid is used as a raw material in the production of polyesters. Much of the isophthalic acid is used for unsaturated polyesters, whereas terephthalic acid is used almost exclusively in saturated (thermoplastic) polyesters. However, a considerable amount of isophthalic acid is used as a minor comonomer in saturated polyesters, where the principal diacid is terephthalic acid. The production volume of isophthalic acid is less than 2% that of terephthalic. Isophthalic acid was formerly produced in technical or crude grades and only a small amount was purified. Now, however, it is all purified to a standard similar to that of terephthalic acid.

Manufacture and Processing. Isophthalic acid is synthesized commercially by the liquid-phase oxidation of *m*-xylene. The chemistry of the oxidation is almost identical to that of *p*-xylene oxidation to terephthalic acid, and production facilities can be used interchangeably for these two dicarboxylic acids. Because isophthalic acid is more soluble than terephthalic acid in reaction solvents, crystallization equipment is more important in isophthalic acid facilities.

Vapor phase oxidation of *m*-xylene is not satisfactory.

Production, Storage, and Shipment. There are currently (ca 1998) several isophthalic acid plants under construction, all but one of which has an annual capacity in excess of 70,000 t. Storage of isophthalic acid is in silos. Shipment is in 22.7- and 25-kg bags, 0.5-t and 1-t bags, or hopper trucks. The far lower production quantity of isophthalic acid and its more varied applications vs terephthalic acid mean that high volume rail hopper cars are not used to any extent, and hopper truck shipment is also small.

Health and Safety Factors. Isophthalic acid has a low order of toxicity. As with terephthalic acid, isophthalic acid was found to form urinary tract calculi in rats in 90 d when it constituted 3% of their diet. This led to some cancer owing to the presence of the calculi. Some mild eye irritation is possible, so eye protection should be worn. Otherwise, normal precautions used in handling industrial chemicals should be observed with isophthalic acid.

Isophthalic acid dust forms explosive mixtures with air at certain concentrations. Fires can be extinguished with dry chemical, carbon dioxide, water or water fog, or foam.

Uses. About 35% of the isophthalic acid is used to prepare unsaturated polyester resins. These are condensation products of isophthalic acid, an unsaturated dibasic acid, most likely maleic anhydride, and a glycol such as propylene glycol. The polymer is dissolved in an inhibited vinyl monomer, usually styrene with a quinone inhibitor. When this viscous liquid is treated with a catalyst, heat or free-radical initiation causes cross-linking and solidification. A range of properties is possible depending on the reactants used and their ratios. The second application of isophthalic acid, with about 30% of the output, is for alkyd coatings.

The third, and fastest growing, area of isophthalic acid use is in other types of polymers, primarily as a minor comonomer with terephthalic acid in saturated polyesters. Over 20% of the isophthalic acid is sold in this application. One rapidly expanding use is in polyester beverage bottles where addition of up to 3% isophthalic acid to the terephthalic acid allows faster production of more complex shapes.

Isophthalic acid is also used in formulations for adhesives, inks (qv), wire enamels, and dental materials (qv). Copper isophthalate is an ingredient in algicides and fungicides.

Derivatives. Commercially significant derivatives of isophthalic acid are its diesters with saturated alcohols and isophthaloyl chloride. This derivation is similar to that of terephthalic acid. Plasticizers form the main use of the diesters, these being the dimethyl, dioctyl, and di(2-ethylhexyl) isophthalates. Diallyl isophthalate is a cross-linking agent for high temperature resistant polybenzimidazoles. Isophthaloyl chloride is used in the manufacture of high temperature-resistant polyamide fibers, films, dyes, and protective coatings.

Trimellitic Acid and Trimellitic Anhydride

Of the three benzenetricarboxylic acids, only trimellitic acid as the anhydride is commercially produced in large volume, by liquid-phase air oxidation of either pseudocumene or dimethyl benzaldehyde. The acid is available as a laboratory chemical. The IUPAC name of trimellitic anhydride is 5-isobenzofurancarboxylic acid (1,3-dihydro-1,3-dioxo).

Owing to the dual functional groups of acid and anhydride, trimellitic anhydride imparts performance enhancements to its end-use applications, many of which are similar to those of phthalic anhydride. In many cases, products made from trimellitic anhydride exhibit properties superior to those of phthalic anhydride.

Physical and Chemical Properties. Trimellitic acid and trimellitic anhydride are odorless white crystalline solids in their pure form. The acid is reasonably stable up to the melting point, where dehydration to the anhydride occurs. The anhydride reacts with atmospheric moisture, even at room temperature, to revert to the acid.

Manufacture and Processing. Amoco Chemical Co. is the sole U.S. producer of trimellitic anhydride. The Amoco process is the liquid-phase air oxidation of pseudocumene using acetic acid as a solvent and a cobalt—manganese—bromine catalyst system. Trimellitic acid is recovered from the oxidation reactor effluent by a suitable means, and the acetic acid solvent is sent to the solvent tower to remove the water of reaction and for recycling. Trimellitic acid is thermally dehydrated to form trimellitic anhydride and further purified by fractional distillation to obtain high purity trimellitic anhydride.

Health and Safety Factors. Trimellitic anhydride may cause respiratory irritation and, in some cases, individuals exposed over long periods may become sensitized and experience mild to severe reactions upon subsequent exposure.

Hazardous Material Identification System/National Fire Protection Association (HMIS/NFPA) codes for trimellitic anhydride are health, 3; flammability, 1; and reactivity, 1. Flaked or molten trimellitic anhydride will burn if ignited. High dust concentrations in the air from either trimellitic anhydride or acid have a potential for combustion or explosion. High voltage static electricity buildup is possible when handling trimellitic anhydride; therefore, adequate precautions, including bonding and grounding of equipment, as well as use of inert gas purge, should be observed. The vapor from molten trimellitic anhydride forms explosive mixtures with air. Other health- and safety-related properties are flash point, 227°C (ASTM D1310); and minimum explosive dust concentration in air, 35 g/m^3.

Uses. The largest end-use application of trimellitic anhydride is as high performance poly(vinyl chloride) plasticizers in the form of triesters of aliphatic alcohols. The second most important use is in coatings applications; conventional solvent-borne, water-borne, and powder coatings. The third largest market is the use in wire enamels for high temperature performance. These three end uses make up about 95% of total trimellitic anhydride usage. Other minor applications are as epoxy curing agent, textile sizing agent, rubber curing accelerator, electrostatic toner binder, and a vinyl cross-linking agent as triallyl trimellitate.

Derivatives. The dual functionality of trimellitic anhydride makes it possible to react either the anhydride group, the acid group, or both. Derivatives of trimellitic anhydride include ester, acid esters, acid chloride, amides, and amide–imides.

4-Chlorocarbonyltrimellitic acid 1,2-anhydride is used in the preparation of esters and amide–imide polymers. Triallyl trimellitate is used as a cross-linking or co-curing agent for ethylene-derived rubbers and plastics.

Trimesic Acid

Trimesic acid is also referred to as 5-carboxyisophthalic acid trimesinic acid, or trimesitinic acid. It is a small-volume, synthetic chemical and is sold commercially.

Manufacture. The only current U.S. manufacturer of trimesic acid is Amoco Chemical Co. It is produced by oxidation of mesitylene (1,3,5-trimethylbenzene) via the liquid-phase oxidation in acetic acid using the cobalt–manganese–bromine catalyst system.

Health and Safety Factors. Trimesic acid is an irritant to the skin, eyes, and respiratory system. It is mildly toxic when ingested. Trimesic acid is flammable.

Uses. The only significant commercial use for trimesic acid is as a cross-linking agent in solid rocket fuels. Other reported uses include cross-linking agents for polymers, as the base acid in making triesters as plasticizers, as a component of electrostatic toners, and as a stationary phase for gas chromatography. Trimesic acid is sold as a research chemical.

Pyromellitic Acid and Pyromellitic Dianhydride

Pyromellitic acid is a commercial product, and it forms a dianhydride which has specialized commercial applications, primarily as an ingredient in the preparation of high temperature polymers.

Manufacture and Processing. Pyromellitic acid and its dianhydride can be synthesized by oxidizing durene (1,2,4,5-tetramethylbenzene). Liquid-phase oxidation using strong oxidants such as nitric acid, chromic acid, or potassium permanganate produces the acid which can be dehydrated to the dianhydride in a separate step.

Health and Safety Factors. Both pyromellitic acid and its dianhydride irritate skin, eyes, and mucous membranes, and they cause skin sensitization. Precautions against fire and dust explosions should be taken.

Uses. Pyromellitic dianhydride imparts heat stability in applications where it is used. Its relatively high price limits its use to these applications. The principal commercial use is as a raw material for polyimide resins (see POLYIMIDES). These polypyromellitimides are condensation polymers of the dianhydride and aromatic diamines such as 4,4′-oxydianiline.

Polyimides are used for film, semiconductor coatings, molding applications, and wire enamel, respectively. They have excellent thermal, electrical, and physical properties. Pyromellitic dianhydride is used to cross-link epoxy resins for elevated temperature service. Such epoxies may be used as an insulating layer in printed circuit boards to improve heat resistance. Other uses include inhibition of corrosion, hot melt traffic paints, azo pigments, adhesives, and photoresist compounds. Other derivatives, none manufactured commercially, are hemimellitic acid; mellophanic, prehnitic, and benzenepentacarboxylic acid; and mellitic acid.

CHANG-MAN PARK
RICHARD J. SHEEHAN
Amoco Chemical Company

V. Nikolov, A. Anastanov, and K. Kussurski, *Chim. Ind. (Milan)* **73**(2), 111 (1991).

W. Partenheimer, *Catalysis Today*, Vol. **23**, Elsevier Science B. V., Amsterdam, the Netherlands, 1995, pp. 69–158.

PHTHALOCYANINE COMPOUNDS

Phthalocyanine, $C_{32}H_{18}N_8$, compounds have found widespread acceptance in a variety of applications. The discovery of iron phthalocyanine and the elucidation of its structure led to the commercial application of copper phthalocyanine.

Copper phthalocyanine (**1**) was developed in the 1930s and is the most commonly used blue organic pigment in the coatings (qv), paint (qv), and printing inks (qv) industry. Phthalocyanine forms complexes with numerous metals. Various complexes with 66 chemical elements are known. Phthalocyanines are structurally related to naturally occurring dyes such as hemoglobin and chlorophyll A.

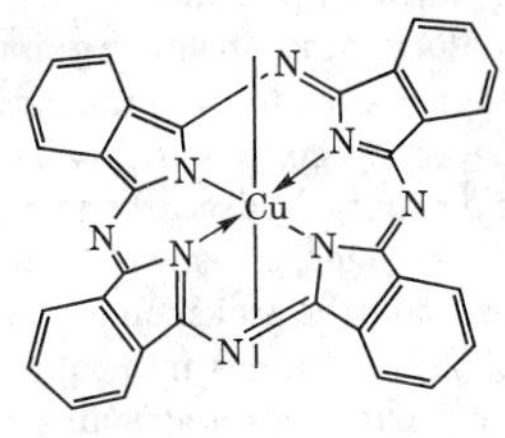

(**1**)

Physical Properties

The density of β-phthalocyanine, H_2Pc, is 1.43 g/cm^3; β-copper phthalocyanine, CuPc, 1.61 g/cm^3; and polychloro-copper phthalocyanine, 2.14 g/cm^3. The color of most phthalocyanines ranges from blue-black to a metallic bronze, depending on the manufacturing process and the chemical and crystalline form of the material. The colors of the finely divided pigment forms vary from dark blue to green, as phthalocyanines absorb in the visible region at 600–700 μm. Most compounds do not melt but sublime above 200°C. CuPc can be sublimed without decomposition at 500–580°C under an inert gas and normal pressure and at 900°C under vacuum. It decomposes vigorously, however, at 405–420°C in air and in nitrogen between 460–630°C. The thermodynamic stability of the five crystalline forms of CuPc increases in the sequence $\alpha = \gamma < \delta < \epsilon < \beta$. The solubility of most phthalocyanines in water and organic solvents is very low. The α-form, however, is slightly soluble in polar solvents and converts rapidly to the β-form.

Chemical Properties

The chemical properties of phthalocyanines depend mostly on the nature of the central atom. Phthalocyanines are stable to atmospheric oxygen up to approximately 100°C. Mild oxidation may lead to the formation of oxidation intermediates that can be reduced to the original products. In aqueous solutions of strong oxidants, the phthalocyanine ring is completely destroyed and oxidized to phthalimide. Oxidation in the presence of ceric sulfate can be used to determine the amount of copper phthalocyanine quantitatively.

Phthalocyanine compounds exhibit favorable catalytic properties which makes them interesting for applications in dehydrogenation, oxidation, electrocatalysis, gas-phase reactions, and fuel cells (qv).

Manufacturing and Processing

Phthalocyanine compounds have been synthesized with various metals. The most important metal phthalocyanines are derived from phthalodinitrile, phthalic anhydride, Pc derivatives, or alkali metal Pc salts.

The route from *o*-phthalodinitrile can be represented 4 $C_8H_4N_2$ + M → MPc, where M is a bivalent metal, metal halide, metal alcoholate, or an equivalent amount of metal of valence other than two in a 4:1 molar ratio. If a solvent, eg, trichlorobenzene, benzophenol, pyridine, nitrobenzene, or quinoline, is used, the reaction takes place at approximately 180°C. Without a solvent the dry mixture must be heated to ca 300°C to initiate the exothermic reaction.

The synthesis from phthalimide derivatives, eg, diimidophthalamide (or phthalimide) is usually carried out in a solvent such as formamide. Metal phthalocyanines may also be prepared using alkali metal salts or from metal-free phthalocyanine by boiling the latter in quinoline with metal salt.

Industrial production of copper phthalocyanine usually favors either the phthalic anhydride–urea process or the *o*-phthalodinitrile process. Both can be carried out continuously or batchwise in a solvent or bake process of the solid reactants.

Crude copper phthalocyanine must be treated to obtain a satisfactory pigment in regard to the crystal modification and optimal particle size (see PIGMENTS). The particle size of crude phthalocyanine can be reduced by chemical or mechanical methods.

The second process to finish phthalocyanine, which is more important for β-copper phthalocyanine, involves grinding the dry or aqueous form in a ball mill or a kneader. Agents such as sodium chloride, which have to be removed by boiling with water after the grinding, are used. Solvents like aromatic hydrocarbons, xylene, nitrobenzene or chlorobenzene, alcohols, ketones, or esters can be used.

Incorporation of less than a stoichiometric amount of alkyl sulfonamides of copper phthalocyanines into copper phthalocyanine improves the pigment's properties in rotogravure inks.

Performance in ink and coatings can be improved by addition of surfactants (qv), dispersants, resins, or copper phthalocyanine derivatives with long aliphatic chains, $CuPc(CH_2{-}NHR)_3$, to stabilize the pigment in the binder system. Another possibility is wet-milling of aqueous pigment dispersions incorporating an organic medium, eg, glycols, polyethers, or surfactants.

Some references cover direct preparation of the different crystal modifications of phthalocyanines in pigment form from both the nitrile–urea and phthalic anhydride–urea process. Metal-free phthalocyanine can be manufactured by reaction of *o*-phthalodinitrile with sodium amylate and alcoholysis of the resulting disodium phthalocyanine. The phthalic anhydride–urea process can also be used. Other sodium compounds or an electrochemical process have been described. Production of the different crystal modifications has also been discussed.

Perchloro- and perchlorobromo copper phthalocyanine are important organic green pigments. They are accessible through direct chlorination of copper phthalocyanine in a eutectic melt of aluminum and sodium chloride or in a chlorosulfonic acid medium. Bromine can be used instead of chlorine in the $AlCl_3$–NaCl melt to obtain polybromochloro copper phthalocyanine.

Phthalocyanine sulfonic acids, which can be used as direct cotton dyes, are obtained by heating the metal phthalocyanines in oleum.

Polymeric phthalocyanines, which possess a higher stability compared to the monomers, can be obtained by combining a phthalocyanine with a polymer. The linking of the polymeric chain can occur at the central metal atom, the phenyl rings, through bridging or attachment to a polymeric chain.

Uses

Approximately 90% of the phthalocyanines (predominantly copper phthalocyanine) are used as pigments (qv). In addition, they have found acceptance in many types of dyestuffs, eg, direct and reactive dyes, water-soluble and solvent-soluble dyes with physical and chemical binding, azo-reactive dyes, azo nonreactive dyes, sulfur dyes, and vat dyes (see DYES AND DYE INTERMEDIATES; DYES, REACTIVE).

Available Forms. Phthalocyanines are available as powders, in paste, or liquid forms. They can be dispersed in various media suitable for aqueous, nonaqueous, or multipurpose systems, eg, polyethylene, polyamide, or nitrocellulose. Inert materials like clay, barium sulfate, calcium carbonates, or aluminum hydrate are the most common solid extenders. Predispersed concentrates of the pigments, like flushes, are interesting for manufacturers of paints and inks, who do not own grinding or dispersing equipment. Pigment–water pastes, ie, presscakes, containing 50–75% weight of water, are also available.

Colorants. The pigmentary forms of copper phthalocyanine are by far the most important commercial products of that class. They provide excellent color properties, excellent resistance to heat and light, acid and alkali, and are extremely insoluble in most solvents. They are less expensive than other organic pigments and color practically every type of printing ink, paint, plastic, and textile. Other uses include the coloring of roofing granules, cements and plasters, fine art paint, soaps, detergents, and other cleaning products. The two principal classes of copper phthalocyanine pigments are the blues and the greens. The blues may be further classified as the α- and β-crystal types, and the greens as the chlorinated and brominated derivatives.

Phthalocyanines have interesting properties as catalysts, lasers (qv), semiconductors, lubricants, or as photographic components.

Health and Safety Factors

Phthalocyanines do not pose any significant risk to human health in the environment or the workplace. In several studies, no carcinogenic risk or toxicity to humans was revealed. The FDA approved the use of CuPc in general and ophthalmolic surgery, for contact lenses, and food packaging (qv). Phthalocyanine Blue may be used as a colorant for coatings that are used in manufacturing, packing, processing, preparing, treatment, packaging, transporting, or holding food. The TLV value for CuPc is 10 mg/m^3.

Polychlorinated biphenyls (PCBs) have been detected in pigments manufactured in trichlorobenzene, but not in those made with nonchlorinated solvents. High boiling hydrocarbons or esters are suitable replacements.

GERD LOEBBERT
BASF Corporation

G. Booth, in K. Venkataraman, ed., *The Chemistry of Synthetic Dyes,* Vol. V, Academic Press, Inc., New York, 1971, p. 241.

C. C. Leznoff and A. B. P. Lever, eds., *Phthalocyanines: Properties and Applications,* VCH Verlagsgesellschaft, Weinheim, Germany, Vol. 1, 1989; Vols. 2 and 3, 1993.

F. H. Moser and A. L. Thomas, *Phthalocyanine Compounds,* Reinhold Publishing Co., New York, 1963.

A. B. P. Lever, in H. J. Emeleus and A. G. Sharpe, eds., *Advances in Inorganic Chemistry and Radiochemistry,* Vol. 7, Academic Press, Inc., New York, 1965, pp. 27–113.

PIGMENT DISPERSIONS

A pigment dispersion in a concentrated form is a uniform distribution of very fine color pigment particles in a suitable medium or carrier. Such a dispersion is normally used for applying color to the surface of a substrate, such as an ink film on paper or a paint film on a steel surface. It is also used for mass coloring, as in the case of plastics. Considering the high cost and specialized equipment in its preparation, a dispersion is manufactured in relatively small batches in highest concentration of pigment. The concentrate made in such a manner is usually diluted, reduced, or extended to produce the finished product.

Dispersion

Organic and inorganic pigment powders are finely divided crystalline solids that are essentially insoluble in application media such as ink or paint. The carrier used for dispersion of a pigment is usually a liquid or solid, such as a polymer, that is deformable at the processing conditions of high temperature and/or shear. The color strength of the dispersed pigment increases markedly with decrease in particle size. Optimum color strength from a given pigment in practice requires a mean particle size of the order of 0.1 μm or less, which is half the wavelength of the light involved. Therefore, the dispersion process involves size reduction of the pigment particle to the smallest particle size, reasonably complete wetting of its solid surfaces by the carrier, and stabilization of the resulting dispersion.

Because the intensity and color strength of pigments are largely dependent on the exposed surface, it is desirable to reduce the particles to primary particle size. This is the size of the solid pigment crystals as they are precipitated in their synthesis. The maximum aggregate size permissible in a given dispersion system depends on the thickness of the film or the coating. Any dispersion system, however, is expected to contain a very small number of these largest aggregates. Generally, it is important to reduce most aggregates to the smaller size to achieve color strength, gloss, film integrity, and durability.

In a dispersed pigment system, a primary pigment particle refers to an individual crystal and a loosely formed association of the pigment crystals from the manufacturing process. Size reduction beyond primary particle size requires excessive energy, but it also has an adverse effect on the visual properties of the pigment. Generally, the particle size of most organic pigments is much smaller initially by precipitation than optimum primary particles, but the particles tend to grow to a much larger size when their formation is complete.

Organic pigments, such as the azo red and yellow pigments, in the process of striking the color undergo definite crystal growth following their precipitation from the aqueous media (see AZO DYES). The individual crystals are joined together due to forces on the crystal surfaces to form the aggregate. These are held together as static systems by van der Waals forces. Subsequent processing to recover the pigment product results in the formation of agglomerates, which are large associations of pigment crystals and aggregates. These formations are loosely held together and are usually easy to break down by application of shear. Various surface treatments are used to suppress the formation of large aggregates and, thereby, ease the dispersion process. These treatments range from the classical approach of rosination to additions of a variety of surface-active agents at the synthesis step. However, occasionally large agglomerates, several millimeters in diameter, form during the initial stages of dispersion in a highly viscous system. The commercial processes used in dispersion manufacturing may not fully eliminate the aggregates.

Wetting of the pigment surface constitutes a critical step in achieving a stable and uniform pigment dispersion. Wetting refers to displacement of adsorbed gases (usually air) on the surface of pigment particles, followed by attachment of a vehicle system to the pigment surface. The system of wetted fine primary pigment particles must be stabilized to prevent reversal of the dispersion process. It is usually done by surrounding the particles with a protective colloid or buffer which blocks the reagglomeration action of particles. In some cases, the stabilization is attained by addition of ions to establish similar charges on all particles.

Flushing

Flushing processes are used extensively in preparing organic pigment dispersion concentrates for color printing ink applications. The process can be described as a direct transfer of pigment from an aqueous phase to an oil or nonaqueous phase without drying. When the pigment presscake is mixed with an oil-based vehicle or a carrier, water is separated from the pigment surface and replaced by the vehicle. Most organic pigments demonstrate an affinity for hydrocarbon oils and lend themselves to easy dispersion in oil by the process of flushing (see PIGMENTS, ORGANIC). Inorganic pigments, on the other hand, have to be treated with cationic surfactants to make their surface lipophilic. The majority of inorganic pigments are usually dried and dispersed as dry powders in the carrier, as opposed to being flushed. Techniques used for dispersion of these pigments are different and should be treated as special cases (see PIGMENTS, INORGANIC).

The process of flushing typically consists of the following sequence: phase transfer; separation of aqueous phase; vacuum dehydration of water trapped in the dispersed phase; dispersion of the pigment in the oil phase by continued application of shear; thinning the heavy mass by addition of one or more vehicles to reduce the viscosity of dispersion; and standardization of the finished dispersion to adjust the color and rheological properties to match the quality to the previously established standard.

Flushing is frequently used for the manufacture of large quantities of a dispersion having a specific pigment in a compatible vehicle system. The flushed products, typically containing 28–40% pigment, offer sufficient flexibility to the formulator to produce the finished offset ink. The flushed products exhibit superior gloss, transparency, and strengths, compared to those produced by dispersing the dried pigment. Flushing is particularly important to dispersions of organic pigments, such as Diarylide Yellow (CI Pigment Yellow 12, CI 21090) and Alkali Blue (Pigment Blue 61, CI 42765) because the drying process is detrimental to the product quality of these pigments.

Equipment

Various types of equipment are used commercially to manufacture dispersed pigment concentrates or finished dispersion products used by printing ink, and the coatings and plastics industry. These include kneaders or internal mixers; close tolerance mills; high speed fluid energy mills; ball and pebble mills; sand, bead, and shot mills.

Uses

The formulation of dispersed pigment concentrates is influenced by the manufacturing process, as well as the performance parameters desired in the final application. The finished product in many cases is significantly different in formulation than the concentrate to achieve desired properties. One of the principal factors to be considered is the concentration of pigment in the dispersion concentrate. Compatibility of the carrier (solvent additives, etc) used in the preparation of concentrated dispersion and that used in the finished color product also plays an important role. In some cases this can be difficult because the carriers having the best performance, from the standpoint of processing, could be poor in the application systems. However, in the majority of the applications, particularly in coatings and colored plastics, the concentration of the pigment in the finished product is quite low, and the incompatibility problem is easily overcome.

Generally, the pigment dispersion concentrates are formulated for specific end use. They can be supplied as flushed pigments, dispersions or pastes for offset inks, chip dispersions for solvent and aqueous inks, and color concentrates for coloring large quantities of plastics. Although it is feasible for the end user to prepare the pigment dispersion concentrates, it is usually more cost effective and technologically advantageous to manufacture these dispersions by the pigment manufacturers of specialty dispersion houses. Three significant areas of application for concentrated dispersions are printing inks, coatings, and plastics.

GERD LOEBBERT
ANAND S. G. SHARANGPANI
BASF Corporation

T. Patton, ed., *Pigment Handbook,* Vols. I, II, and III, John Wiley & Sons, Inc., New York.

R. B. McKay and F. M. Smith, *Dispersion of Powders in Liquid,* 3rd ed., G. D. Parfitt, ed., Applied Science Publication, London, 1981.

P. A. Lewis, ed., *Pigment Handbook,* Vol. 1, 2nd ed., John Wiley & Sons, Inc., New York, 1988.

T. C. Patton, *Paint Flow and Pigment Dispersion,* 2nd ed., John Wiley & Sons, Inc., New York, 1979.

PIGMENTS

INORGANIC

Inorganic pigments, black, white, or colored inorganic substances produced and marketed as fine powders, are an integral part of many decorative and protective coatings (qv) and are used for the mass coloration of plastics, fibers (qv), paper (qv), rubber, glass (qv), cement (qv), glazes, and porcelain enamels (see ENAMELS, PORCELAIN OR VITREOUS). These materials are colorants in printing inks (qv), cosmetics (qv), and markers, eg, crayons. In all these applications the pigments are dispersed, ie, they do not dissolve, in the media forming a heterogeneous mixture (see PIGMENT DISPERSIONS). In nature, inorganic pigments contribute to the color of some rocks and minerals (see COLORANTS FOR CERAMICS; COLORANTS FOR FOOD, DRUGS, COSMETICS, AND MEDICAL DEVICES; COLORANTS FOR PLASTICS; PAINT–ARCHITECTURAL).

Originally, only fine powders used for coloring various media were defined as pigments. This definition has been expanded to include many powdery materials, eg, metallic powders and powders having magnetic or anticorrosive properties, which are intentionally dispersed (not dissolved) into media to increase value and/or impart some special properties.

Chemically, inorganic pigments are quite simple materials and include elements, their oxides, mixed oxides, sulfides, chromates, silicates, phosphates, and carbonates. The application usefulness of inorganic pigments is determined by physical as well as chemical properties.

Properties

The value of pigments results from their physical–optical properties. These are primarily determined by the pigments' physical characteristics (crystal structure, particle size and distribution, particle shape, agglomeration, etc) and chemical properties (chemical composition, purity, stability, etc). The two most important physical–optical assets of pigments are the ability to color the environment in which they are dispersed and to make it opaque.

The opacity of a pigment lies in its ability to prevent a transmission of light through the medium. White pigments disperse the whole visible light spectrum more effectively than they absorb it; black pigments do the opposite. Color results when pigment particles absorb only certain portions of the visible light spectrum while dispersing the rest of it.

The opacity of pigments is a function of the pigment particle size and the difference between the pigment's refractive index and that of the medium in which pigment particles are dispersed. The multiple light dispersion in the pigment–medium interface results in the appearance that the light is transmitted through a much thicker layer than it actually is. A pigment having a particle size between 0.16–0.28 μm gives the maximum dispersion of the visible light.

The most common measurements of pigment properties comprise elemental analysis, impurity content, crystal structure, particle size and shape, particle size distribution, density, and surface area. These parameters are measured so that pigments' producers can better control production, and set up meaningful physical and chemical pigments' specifications. Measurements of these properties are not specific only to pigments. The techniques applied are commonly used to characterize powders and solid materials, and the measuring methods have been standardized in various industries.

Coloristic properties of pigments are best evaluated by dispersing them into the media they were developed to color, eg, plastics, glass enamels, glazes, etc. The measured characteristics include color, color strength, opacity, lightfastness, weathering, heat stability, chemical stability, and rheological properties. The dispersing media and the processing conditions can strongly influence the results. Because pigments, as any fine powders, have a tendency to segregate by size during transportation and handling, the use of proper sampling (qv) methods is critical for getting meaningful physical and chemical data (see POWDERS, HANDLING).

Chemical Properties. Elemental profile, impurity content, and stoichiometry are determined by chemical or instrumental analysis. Instrumental analytical methods (qv) are usually faster, can be automated, and can be used to determine very small concentrations of elements (see TRACE AND RESIDUE ANALYSIS). Atomic absorption spectroscopy and x-ray fluorescence methods are the most useful instrumental techniques in determining chemical compositions of inorganic pigments. Chemical analysis of principal components is carried out to determine pigment stoichiometry.

Crystal Structure. Crystal structure, the information about compounds such as impurities or unreacted materials present in the pigment, the presence of various crystal phases, and the degree of the crystallinity, can be resolved using an x-ray diffractometer. In most cases, analyzed pigments are not completely unknown, and the powder diffraction pattern is sufficient in finding present phases or unreacted starting materials in the pigment. The x-ray analysis has become an indispensable tool of the inorganic pigments' development and production (see X-RAY TECHNOLOGY).

Physical Properties. Particle size and distribution are the most fundamental measured properties of powders (see SIZE MEASUREMENT OF PARTICLES). These properties impact a number of pigment characteristics. Those affected the most are the color, color strength, hiding power, and rheological properties. Actual powders consist of a population of particles of many different shapes. To permit a good description of powder population, a representative sample of the powder must be collected, measured, and the results interpreted using statistical methods. For inorganic pigments to be useful in most applications, they must have an average particle size between 0.1 and 10 μm. A character, ie, color or pattern, of a substrate becomes obscured when coated with a pigment containing film such as a paint or a ceramic glaze. The degree of the obscuration (opacity) depends on the amount and type of pigment used and the thickness of the applied film.

The ability of a coating to hide the substrate is called its hiding power. Hiding power of a uniform coating is expressed as the area of substrate that can be hidden by a unit volume of the coating (ft^2/gal or m^2/L).

The ability of a pigment to change the color of an opaque film is known as its tinting strength. Both the hiding power and tinting strength are the fundamental pigment properties. Hiding power and tinting strength can be determined visually or instrumentally.

Color matching is a process in which a technician prepares a formulation, ie, a mixture of pigments in a desired medium, that has the desired color effects. A good color match in one medium, eg, plastic, is not always a good match in another medium, eg, ceramic glaze.

Experienced color matchers can achieve a good color match by trial and error without using any instrumentation. In some cases, however, this technique can be a lengthy process. To get the most cost-effective match in the shortest possible time, the use of a computer color matching system is preferable.

Many pigments, when exposed to high intensity light such as direct sunlight or uv lamp, can get darker, change their shade or lose the color saturation. The color and its saturation change mainly for organic pigments. Inorganic pigments, particularly those containing ions that can exist in several oxidation states, usually get darker. Some color changes can be reversible; others are permanent.

Lightfastness is measured by exposing pigmented film to an artificial or natural light for a predetermined time. It is a relative term where the color of a sample exposed to a known light source is compared to its original color values.

Weathering is the ability of the colored system, ie, the coating, paint, etc, not the pigment alone, to resist light and environmental conditions. Changes in color and gloss are two main factors that are evaluated in weathering tests.

Heat stability is measured as a change in the hue of the colored system and a degree of yellowing of the white system after exposure to a desired high temperature for a certain time. This property can also be expressed as the maximum temperature at which the color of the system does not change.

In determining the chemical resistance, color changes of pigmented binder surfaces are measured after their exposure to various chemicals, such as water–sulfur dioxide or water–sodium chloride systems. These systems imitate the environment to which the colored articles could become exposed.

The surfaces of pigment particles can have different properties and composition than the particle centers. This disparity can be caused by the absorption of ions during wet milling, eg, the —OH groups, on the surface.

Most inorganic pigments are hydrophilic and therefore can be readily wetted only by polar solvents, eg, water. The wettability and dispersion of inorganic pigments in an organic matrix (polymer, solvent) can be improved by the physical or chemical absorption of surface-active compounds containing polar groups, such as $—NH_2$, —OH, or longer aliphatic chains on pigment particles. The absorption of these compounds makes the pigment surface hydrophobic. Compounds that help to form a bridge between inorganic particles and an organic polymeric matrix are called coupling agents, the most common being tetrafunctional organometallic compounds.

Specifications, Standards, and Quality Control

Production and product quality of most pigment producers are controlled through pigments' standards. Whenever a pigment is developed or significantly improved, a new standard is set that represents the average production results, not necessarily the best ones. If all production processes are under control, the properties of the produced pigment's lots are evenly distributed around those of the standard.

White Pigments

The most common white pigments are titanium dioxide, zinc oxide, leaded zinc oxide, zinc sulfide, and lithopone, a mixture of zinc sulfide and barium sulfate. The use of lead whites and antimony oxides has been decreasing steadily for environmental reasons.

Titanium Dioxide. Chemically, titanium white is titanium dioxide either in an anatase or rutile form.

Properties. Crystals of titanium dioxide, TiO_2, can exist in one of the three crystal forms: rutile, anatase, and brookite. Only anatase and rutile forms have good pigmentary properties, and rutile is more thermally stable. Compared to other white pigments, titanium dioxide has the highest refractive index, giving white paints formulated with these pigments the highest coverage, ca 38 m^2/g.

Titanium whites resist various atmospheric contaminants such as sulfur dioxide, carbon dioxide, and hydrogen sulfide. Owing to its chemical inertness, titanium dioxide is a nontoxic, environmentally preferred white pigment.

Titanium is the seventh most common metallic element in the earth's crust. Titanium minerals are plentiful in nature. The most common mineral/raw materials used for the production of titanium dioxide pigments are shown in Table 1.

Table 1. Mineral/Raw Materials for TiO_2 Production

Mineral/raw material	Main composition	TiO_2, %
ilmenite	$FeO–TiO_2$	35–65
leucoxene	$Fe_2O_3–TiO_2(+TiO_2)$	60–90
rutile	TiO_2	90–98
rutile, synthetic	TiO_2	85–96
anatase	TiO_2	80–90
titanium slag	TiO_2(Fe)	70–85

Use. Titanium dioxide is used mainly in the production of paints and lacquers, plastics, and paper. Other applications include the pigmentation of printing inks, rubber, textiles (qv), leather, synthetic fibers, ceramics, white cement, and cosmetics.

Other White Pigments.

Zinc Oxide. By volume, zinc oxide is the second most significant white pigment. Its pigmentary properties are good, providing good coverage. It has a good lightfastness and is well miscible with other pigments. With the increasing popularity of titanium dioxide white pigment in the twentieth century, the pigmentary use of zinc oxide has been declining (see ZINC COMPOUNDS).

Whereas zinc oxide was originally used as a pigment, its most important modern application is to aid in vulcanizing synthetic and natural rubber. Paint and coating industries use zinc white mainly as an additive to improve anticorrosion properties, mildew resistance, and durability of external coatings.

Zinc Sulfide. Whereas zinc sulfide is important mainly as a component of the composite white pigment lithopone, it also has a limited use as a single pigment. After titanium white, it has the second highest refractive index of all the white pigments. However, its chemical and thermal resistances are inferior to those of TiO_2.

Zinc sulfide is used in applications where white color shade and low abrasivity are required. In printing inks and paints it also contributes to stability and good rheological and printing properties.

Lithopone. Lithopone is a mixture of ZnS and $BaSO_4$. The pigmentary properties of the mixture are determined by zinc sulfide, and therefore lithopone pigments are characterized by the amount of ZnS present in the mixture. The amount of ZnS in commercial lithopones varies from 15 to 60%.

Lithopones are used in water-based paints because of their excellent alkali resistance, in paper manufacturing as a filler and opacifying pigment, and in rubber and plastics as a whitener and reinforcing agent.

Lead Whites. Basic lead carbonate, sulfate, silicosulfate, and dibasic lead phosphite are commonly referred to as lead whites. Usage is limited because of environmental restrictions placed on the use of lead-containing compounds.

Colored Pigments

Iron Oxide Pigments. In general, all iron pigments are characterized by low chroma and excellent lightfastness. They are nontoxic, nonbleeding, and inexpensive. They do not react with weak acids and alkalies, and if they are not contaminated with manganese, do not react with organic solvents. However, properties vary from one oxide to another.

Natural Iron Oxides. The earth's crust contains about 7 wt % iron oxides, but only a few deposits are rich enough in iron to be suitable for mining pigmentary-quality iron oxides. Deposits that are a suitable source of natural iron oxide pigments are usually hydrated aluminum silicates that contain various amounts and forms of iron oxide.

Iron oxides are supplied to the market as red, ocher, sienna, and umber natural pigments. The hue of the natural iron oxide pigments is determined by raw material composition and processing.

About 60% of the natural iron oxide pigments is used to color cement and other building materials (qv). About 30% is consumed in the production of paints. For coloring plastics and rubber, synthetic iron oxide pigments are preferred. The main advantage of the natural iron oxide pigments, as compared to the synthetic ones, is cost. However, the quality is inferior, and in most cases, they are consumed in close proximity to the mines.

Synthetic Iron Oxides. Advantages of synthetic iron oxides over their natural counterparts include chemical purity, more uniform particle size and size distribution, and in the case of precipitated oxides the ability to prepare the pigment in predispersed vehicle systems by flushing techniques.

Iron Oxide Reds. From a chemical point of view, red iron oxides are based on the structure of hematite, α-Fe_2O_3, and can be prepared in various shades, from orange through pure red to violet. Different shades are controlled primarily by the oxide's particle size, shape, and surface properties.

Synthetic red iron oxides are prepared in a variety of grades from light to dark. These are sold under a variety of names, eg, Indian red, Turkey red, and Venetian red.

Iron Oxide Yellows. From a chemical point of view, synthetic iron oxide yellows, also known as iron gelbs, are based on the iron(III) oxide–hydroxide, α-FeO(OH), known as goethite. Color varies from light yellows to dark buffs and is primarily determined by particle size, which is usually between 0.1 and 0.8 μm. Because of their resistance to alkalies, these are used by the building industry to color cement.

Iron Blacks. Chemically, iron blacks are based on the binary iron oxide, $FeO{\cdot}Fe_2O_3$. Most of the black iron oxide pigments contain iron(III) oxide impurities, giving a higher ratio of iron(III) than would be expected from the theoretical formula.

Iron Browns. Iron browns are often prepared by blending red, yellow, and black synthetic iron oxides to the desired shade. The most effective mixing can be achieved by blending iron oxide pastes, rather than dry powders.

Complex Inorganic Color Pigments. Based on the crystal structure, the Color Pigments Manufacturers' Association (CPMA) has classified 53 key inorganic pigments into 14 categories; these inorganic colorants are known as complex inorganic color pigments. The original name, mixed-metal oxide are pigments (MMO), did not accurately describe the chemical nature of all the classified pigments.

Mixed-Metal Oxide Pigments. Mixed-metal oxide pigments can be considered a subcategory of complex inorganic color pigments. In reality these pigments are not mixtures but rather solid solutions or compounds consisting of two or more metal oxides. Structurally, mixed-metal oxide pigments belong to one of 14 structure types. The most common ones are rutile and spinel. The commercial significance is in their thermal, chemical, and light stability, combined with low toxicity. When these are employed for coloring glass enamels and ceramics, they are sometimes referred to as colors or stains; when used to color paints and plastics, they are known as pigments.

The color of mixed-metal oxide pigments results from the incorporation of chromophores, into the structure of stable host oxides.

Pigments having a spinel structure are widely used by the ceramic and plastic industries (see COLORANTS FOR CERAMICS). They cover a wide range of colors and many are thermally stable up to 1400°C and are resistant to molten glass. Another advantage is their intermiscibility, allowing the user a choice of creating many intermediate colors.

Spinel compounds have a common chemical formula, AB_2X_4. Structurally they have a cubic symmetry and are derived from magnesium aluminate, $MgAl_2O_4$, a naturally occurring mineral.

Structurally, all rutile pigments are derived from the most stable titanium dioxide structure, ie, rutile. The crystal structure of rutile is very common for AX_2-type compounds such as the oxides of four-valent metals, as well as halides of divalent elements.

Zircon pigments are derived from the tetragonal zirconium silicate, $ZrSiO_4$. Because of the high temperature (up to 1600°C) and chemical stability of zirconium silicate, zircon pigments can be used in the formulations of high temperature (1300–1400°C) glazes. Zirconium silicate is also used as an opacifier in porcelain and vitreous enamels.

In pigments, zirconium silicate serves as the host lattice for various chromophores, such as vanadium, praseodymium, iron, etc.

Bismuth Vanadate. The use of bismuth vanadate, $BiVO_4$, as a nontoxic, yellow pigment with good hiding strength and lightfastness was patented by DuPont in 1978. At least two pigment producers, Ciba and BASF, are marketing this pigment primarily for plastic and paint applications. Some users have already replaced the toxic pigment lead chromate in their paint formulations with a combination of organic pigments and bismuth vanadate (see BISMUTH COMPOUNDS).

Chromium(III) Pigments. There are two green pigments based on chromium in the +3 oxidation state. The first one is chromium oxide, Cr_2O_3; the second is hydrated chromium oxide, $Cr_2O_3{\cdot}xH_2O$.

Chromium(III) Green Pigment. Chromium oxide green is characterized by outstanding lightfastness and has excellent resistance to acids, alkalies, and high temperatures. Because it weathers extremely well, chromium oxide green is applied as a colorant for roofing granules, cement, concrete, and outdoor industrial coatings. It is also used in ceramic applications.

Hydrated Chromium(III) Green Pigment. Hydrated chromium oxide has a brilliant green color and is referred to as Gingnet's green. It exhibits a limited hue range, is semitransparent, and has a low opacity, but provides excellent lightfastness and alkali resistance.

Ultramarine Pigments. Ultramarines are derived from lazurite (Lapis Lazuli), a semiprecious stone; they can be prepared in many shades.

Chemically, ultramarines are complex sodium aluminates having a zeolite structure. Composition varies within certain wt % ranges, ie, Na_2O, 19–23; Al_2O_3, 23–29; SiO_2, 37–50; and S, 8–14.

Ultramarine pigments are used in printing inks, textiles, rubber, artists' colors, cosmetics, and laundry bluing. Because of their thermal stability they are also used to color roofing granules.

Cyanide Iron Blues. Cyanide iron blue, also known as Prussian blue, is one of the oldest industrially produced, inorganic pigments. Chemically, cyanide iron blues are based on the $\{Fe^{2+}[Fe^{3+}(CN)_6]\}^-$ anion. The charge is balanced by sodium, potassium, or ammonium cations.

Iron blues are mainly used by the printing industry for coloring printing inks. In Europe, cyanide blues are used for coloring fungicides (see IRON COMPOUNDS).

Cadmium Pigments. Historically, cadmium pigments have been very important, providing a range of clean, bright shades of yellow, orange, red, and maroon colors. This importance, however, has decreased because of environmental issues. Only a few pigment producers are willing to continue cadmium pigment production (see CADMIUM COMPOUNDS).

Lead Chromate Pigments. Lead chromate, $PbCrO_4$, occurs in nature as the orange-red mineral crocoite. Synthetically prepared lead chromate and its solid-state solutions with lead sulfate, $PbSO_4$, or lead molybdate, $PbMoO_4$, are known to have excellent pigmentary properties. The usage of these pigments has been steadily decreasing because of environmental regulations restricting the production and the use of lead-containing products.

Black Pigments

Black pigments can be divided into two basic groups. The first group is represented by carbon blacks. Many other inorganic black pigments, called noncarbon blacks, also are available. These belong chemically to the colored pigment category. Examples are spinel and rutile blacks, iron blacks, and some inclusion zircon pigments.

Carbon Blacks. Carbon black is one of the oldest pigments known. More than 90% of the production of this pigment is consumed by the rubber industries, in particular, by the tire industry as a reinforcing agent. The rest is used for coloring plastics, printing inks, and paints. Particle size of carbon blacks varies from 5 to 500 μm and can be controlled by the process conditions and feedstock (see CARBON–CARBON BLACK).

Environmentally, carbon blacks are relatively stable and unreactive. There is no evidence that these materials are toxic to humans or animals.

Extenders and Opacifiers

Extender pigments are low-cost, generally colorless or white pigments with a refractive index less than 1.7. Sometimes these pigments are also referred to as fillers (qv). Many extenders are derived from natural sources and display many diverse properties. They are added to various formulations to improve technical and application properties and to reduce costs. Like pigments, extenders are dispersed in media in which they do not dissolve, but compared to pigments they do not have any significant coloristic properties.

In coating applications, extender pigments control gloss, viscosity, texture, suspension, and durability (see COATINGS). Extender pigments also enhance the opacity of white hiding pigments, eg, TiO_2. In plastics applications, extenders influence numerous properties of the resin including melt viscosity, thermal conductivity, and electrical properties, tensile strength, and moisture resistance.

Opacifiers are fine inorganic powders, usually white, that are used to reduce the transparency of ceramic glazes and porcelain enamels. The coating becomes opaque because the particles of the opacifier scatter and reflect the incident light. When inorganic pigments are combined with white opacifiers, pastel colors are obtained.

Commercially, the most important opacifiers for glazes are ZrO_2, $ZrSiO_4$, and SnO_2.

Miscellaneous Pigments

Luminescent Pigments. Luminescence is the ability of matter to emit light after it absorbs energy (see LUMINESCENT MATERIALS). Materials that have luminescent properties are known as phosphors, or luminescent pigments. If the light emission ceases shortly after the excitation source is removed ($> 10^{-8}$ s), the process is fluorescence. The process with longer decay times is referred to as phosphorescence.

Semiconducting sulfides that can be represented by the formula nZnS(1$-n$)CdS:A, where A stands for an activator, and $n = 0.15-1$, are typical of fluorescence pigments. Phosphorescence pigments can be expressed by the general formula nZnS(1$-n$)CdS:Cu, where $n = 0.78-1.0$ and the amount of the Cu^+ activator is only a few hundredths of a percent.

Phosphorescent pigments are used in military applications, plastics, and paints. Zinc sulfide doped with Ag^+ (blue) cations, or with Cu^+ (green) cations are important pigments for the production of color television screens.

Metal Effect Pigments. Some metals, when prepared as small flakes, impart a special metallic appearance to the coatings and plastics in which they are dispersed. Metals most often used in these applications are aluminum (aluminum bronzes), copper and copper-zinc alloys (gold bronzes), and in smaller amounts zinc, tin, nickel, gold, silver, and stainless steel.

Nacreous Pigments. Nacreous, ie, pearlescent pigments are used for creating special decorative effects typical of natural pearls. Nacreous pigments are fine, thin, plate-like transparent particles having a high refractive index. Because of these physical characteristics, when dispersed in a transparent film, they produce a silky appearance.

Manufacture of the most popular nacreous pigments involves coating mica (qv) with 50–300-nm films of TiO_2, Fe_2O_3, or Cr_2O_3. The mica, which alone does not have a high enough refractive index for creating nacreous luster, provides the required transparent platelet base. The oxide coating provides the necessary high refractive index.

Transparent Pigments. Pigments having chemical composition corresponding to colored or white opaque pigments can, under certain circumstances, appear transparent in a medium. This happens when the particle size of these pigments becomes very small (2–15 nm), and if the particle refractive index is comparable to the refractive index of the media in which the particles are dispersed. Because of the very small particle size, the preparation of these pigments is much more complicated than the preparation of their nontransparent analogues. Their large surface area makes their dispersion difficult and they have a strong tendency to agglomerate.

Environmental Aspects

Some inorganic pigments contain heavy metals. Thus production, use, and disposal are becoming more and more regulated. In the United States there are several federal regulations that control the use and disposal of heavy metals.

The Resource Conservation and Recovery Act (RCRA) controls the disposal of hazardous waste. SARA Title III governs the toxic inventory and emission reporting; the Clean Water Act (CWA) sets the limits for metals that can be present in water discharge; and the Clear Air Act (CAA) Amendments of 1990 control the abatements of all materials in the air.

The Occupational Safety and Health Administration (OSHA) regulates the exposure to chemicals in the workplace. From the point of view of the inorganic pigments industry, the limits established for lead and cadmium exposure are particularly important. A comprehensive lead standard adopted by OSHA in 1978 has been successful in reducing the potential for lead contamination in the workplace.

Table 2 lists those metals regulated by federal law that are or might be present in inorganic pigments.

Packaging. Products packaging, which constitutes about one-third of the municipal solid waste, is usually decorated with various colors (see PACKAGING). Some of these colors contain heavy metals. As landfill construction and placement became more complicated, the cost of garbage disposal escalated. The Coalition of Northeast Governors (CONEG) developed a model state legislation to regulate cadmium, hexavalent chromium, lead, and mercury in packaging to a gradually decreasing content, not leachability, of 100 ppm of all four metals combined in the four years after adoption. This legislation had a significant effect on the inorganic pigments industry that has relied heavily on the bright red and orange colors of cadmium- and lead-containing pigments.

Responsible Care is a voluntary incentive sponsored by the Chemical Manufacturers' Association (CMA). Any CMA company must embrace the philosophy of continuous improvements of health, safety, and environmental efforts accompanied by an open communication to the public about products and their production. Thus the total impact of any product on the environment, from the extraction of raw materials, their beneficiation, transportation, production of final product, and disposal of the product at the end of its useful life, must be taken into consideration.

Table 2. Elements Potentially Present in Inorganic Pigments

	Federal regulation[a]				
Element	RCRA	SARA	OSHA	CWA	CAA
Al				+	
Ag	+	+			
As	+	+		+	+
Ba	+	+		+	
Be		+		+	
Cd	+	+	+	+	+
Co		+		+	+
Cr	+	+		+	+
Cu		+		+	
Hg	+	+		+	+
Mn		+			+
Mo				+	
Ni		+		+	
Pb	+	+	+	+	+
Sb	+	+		+	
Se	+	+		+	
Ti				+	
Tl		+			
Zn				+	

[a] A + indicates that the element is regulated by the particular act.

To assure the future of inorganic pigments, research efforts are directed toward the development of environmentally acceptable pigments, pigments that when produced under well-controlled conditions do not release any toxic materials into the environment whether during production, use, or disposal.

MIREK NOVOTNY
Cerdec Corporation
Z. SOLC
M. TROJAN
University of Pardubice

P. A. Lewis, ed., *Pigment Handbook*, Vol. 1, John Wiley & Sons, Inc, New York, 1987.

C. Patton, ed., *Pigment Handbook*, Vol. III, John Wiley & Sons, Inc., New York, 1973.

ORGANIC

Pigments are colored, colorless, or fluorescent particulate organic or inorganic finely divided solids which are usually insoluble in, and essentially physically and chemically unaffected by, the vehicle or medium in which they are incorporated. They alter appearance either by selective absorption and/or scattering of light. They are usually incorporated by dispersion in a variety of systems and retain their crystal or particulate nature throughout the pigmentation process. The large number of systems vary widely from paints to plastics to inks and fibers.

Dyes, on the other hand, are colored substances which are soluble or go into solution during the application process and impart color by selective absorption of light. In contrast to dyes, whose coloristic properties are almost exclusively defined by their chemical structure, the properties of pigments also depend on the physical characteristics of its particles.

The description of colored organic pigments excludes consideration of inorganic pigments, as well as black pigments which consist of specially treated forms of carbon and white pigments which are entirely of inorganic origin.

Pigments are categorized according to their generic name and chemical constitution in the *Color Index* (CI), published by the Society of Dyers and Colourists, and the American Association of Textile Chemists and Colorists.

Significant pigment attributes are tinctorial strength, durability (photochemical stability), hiding power, transparency, and heat and solvent resistance. Other properties include brightness (saturation), gloss, rheology, crystal stability, bleed resistance, flocculation resistance, and other properties associated with specialized applications.

The development of modern organic pigments started with the synthesis of dyestuffs for the textile industry. The period up to 1900 was characterized by the discovery and development of many dyes derived from coal-tar intermediates. Rapid advances in color chemistry were initiated after the discovery of diazo compounds and azo derivatives (shown to be largely hydrazone derivatives). The wide color potential of this class of pigments and their relative ease of preparation led to the development of azo colors, which represent the largest fraction of manufactured organic pigments (see AZO DYES).

The most important advance in pigment technology after World War I was the discovery of the relatively complex structure but easily synthesized copper phthalocyanines, which were characterized by excellent brightness, strength, bleed resistance, and lightfastness (see PHTHALOCYANINE COMPOUNDS).

After World War II the most important discovery was the family of red-violet quinacridone pigments, followed by the mostly yellow-orange benzimidazolone, the isoindolinone pigments, and the red diaryl pyrrolopyrroles.

Color and Constitution

The term chromophore is used to designate π-electron-containing moieties (conjugated double bonds) which contribute to the selective absorption of visible light. Generally, organic compounds absorb light in the ultraviolet (210–400 nm) and visible (400–750 nm) region of the spectrum at characteristic wavelengths. The intensities of these absorptions vary due to the excitation of the more loosely held electrons in the molecule. All unsaturated groups have remarkably similar $\pi-\pi^*$ transitions regardless of the atoms contained in the common chromophores.

The presence of $\pi-\pi$ or $n-\pi$ conjugated systems does not assure absorption of visible light or generation of color. However, all colored organic compounds, including pigments, possess extended conjugated resonance systems. Thus, whereas 1,4-diphenylbutadiene is colorless, 1,6-diphenylhexatriene is colored.

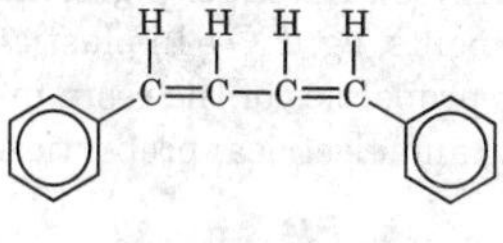

colorless

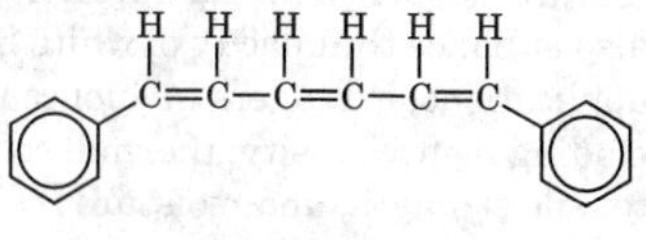

colored

All absorbed light is complementary to reflected light which produces observed color. All colors are a function of the wavelength of absorbed light as shown in Table 1 (see COLOR). Thus, if a pigment absorbs only blue light it imparts an orange color, whereas when it absorbs orange light the observed color is blue.

Properties of Pigments

The physical and chemical characteristics that control and define the performance of a commercial pigment in a vehicle system include its chemical composition, chemical and physical stability, solubility, particle size, shape and particle size distribution, degree of dispersion, crystal morphology including polymorphic forms, refractive index, specific gravity, electronic spectra with particular emphasis on extinction coefficients in the visible spectrum, surface area, and the presence of impurities, extenders, and surface modifying agents. Invariably a pigment is used in a vehicle system, therefore its ultimate performance in use derives from both physical and chemical pigment-vehicle interaction. Performance of most pigments is system dependent.

Unlike inorganic pigments, organic pigments are relatively strong and bright (saturated), but their fastness properties, though adequate for the purposes for which they are used, vary widely from poor to outstanding.

Strength. The inherent strength of a pigment depends on its light-absorbing characteristics, which are related to its molecular and crystalline structure. In addition, strength is a function of particle size or surface area. The ability of a pigment to absorb light increases with decreasing particle size or increasing surface area, until the particles become entirely translucent or transparent to incident light. Being finely divided pigment particles have a great tendency to aggregate and agglomerate into crystal assemblies. To obtain the inherent strength of a pigment the aggregates must be completely broken down to individual crystals by application of work and their reagglomeration or flocculation prevented. Total breakdown to single crystals in practical systems seldom happens.

Pigment strength in a vehicle also depends on the typical character of other components in a pigmented system insofar as they absorb or

Table 1. Colors of Absorbed Light and the Corresponding Complementary Colors as a Function of Wavelength

Wavelength, nm	Color of absorbed light	Complementary color
400–420	violet	yellow-green
420–450	indigo blue	yellow
450–490	blue	orange
490–510	blue-green	red
510–530	green	purple
530–545	yellow-green	violet
545–580	yellow	indigo blue
580–630	orange	blue
630–720	red	blue-green

scatter light. Strength comparisons are usually made with a series of samples featuring varying amounts of a pigment incorporated in a vehicle with a corresponding series in the same vehicle containing a reference pigment. Instrumental comparisons are commonly practiced.

A good absolute theoretical comparison of strength is represented by the area under the absorption bands in the visible spectrum, or less accurately by the molecular extinction coefficients at the maximum wavelength of absorption.

Brightness or Saturation. The saturation of a colored pigment is a measure of its brightness or cleanliness as opposed to dullness of hue. Generally, if a pigment absorbs light over a wide range of wavelengths, ie, shows broad absorption bands, or contains more than one chromophore, the pigment is likely to be duller than a pigment with sharp absorption bands due to a single chromophore. Because pigments are frequently used in combinations or blends, the brightness is determined by the selective absorption of the individual pigments and this significantly affects the brightness of the reflected color.

Fastness. Fastness describes the characteristics of a pigment in terms of its color stability in a pigmented system upon exposure to light, weather, heat, solvents, or various chemical agents. Ideally, a pigment should be insoluble and chemically and photochemically inert. Only a few organic pigments approach such perfection.

The development of resins, plastics, fibers, elastomers, etc that are processed at progressively higher operating and curing temperatures has created a need for pigments that stand up for relatively long periods of time to a hostile environment. They must remain essentially unaltered when incorporated into plastics such as polypropylene, ABS, or nylon at relatively high temperatures.

Once a pigment is incorporated into a system, it is expected to be durable and withstand the combined chemical and physical stresses of weather, solar radiation, heat, water, and industrial pollutants. Because a pigment is totally enveloped by the medium which is itself not inert, various pigments perform differently in different systems. Thus, a pigment may be lightfast or weatherfast in one system and fail in another.

Dispersibility. The dispersibility of a pigment is measured by the effort required to develop the full tinctorial potential of a pigment in a vehicle system. Dispersibility differs from system to system depending on pigment–medium interaction and compatibility.

Small particle size pigments, especially the very small crystals, seldom exist as individual entities, but as strongly coherent aggregates or less firmly bound agglomerates. A wide variety of additives are being used to reduce aggregation or agglomeration and flocculation to improve dispersibility and color strength of organic pigments. These include resins, especially those related to abietic acid, aliphatic amines, amides, substituted derivatives of pigments themselves, and various combinations thereof. The additives are most effective if they are present when the pigment crystals are being generated.

Hiding Power and Transparency. Hiding power of a pigment is a function of its strength, that is, its absorption coefficients, its particle size, or light-scattering coefficient, and relative refractive indexes of pigment and vehicle. Light scattering has a powerful influence on opacity and goes through a maximum as a function of particle size. The maximum occurs at a particle size which is approximately half the wavelength of absorbed visible light.

Similarly, a pigment which absorbs much light increases hiding even when light scattering is insufficient, and the higher the refractive index the greater the hiding power of a pigment.

Conversely, to increase transparency light scattering must be minimized by particle size reduction. The smaller the particle size and the better the dispersion the greater the transparency.

Other Working Properties. Other properties which facilitate pigment incorporation and use include compatibility with a system, oil absorption, rheological characteristics, gloss, distinctness of image, wettability, migration fastness, plate-out, polymer distortion, etc. Most of these properties are controlled during pigment manufacture or formulation, others require special treatments to overcome undesirable effects.

Types of Pigments

All organic pigments have to be synthesized and nearly all have to be conditioned or finished. The physical conditioning is as important as its chemical constitution, and it has become an important separate process step in the manufacture of organic pigments.

Azo Pigments. Azo pigments provide good examples of materials in which conditioning is an integral part of the synthesis process. The coupling process for azo components is simple. An aromatic amine is diazotized by treatment with nitrous acid under conditions which vary from dilute mineral acid to concentrated sulfuric acid, depending on the basicity of the amine (see AZO DYES). The simplest method, ie, direct coupling, involves running the diazo solution directly into the solution or suspension of the coupling component. In inverse coupling, the coupling component is run into the diazo solution. The most elegant technique, simultaneous coupling, entails running both the diazo and coupling component simultaneously into water or a dilute buffer.

The monoazo and disazo pigments contain one or more chromophoric groups usually referred to as the azo $-N{=}N-$ group. However, it has been shown by x-ray diffraction analysis and nuclear magnetic resonance (nmr) techniques that azo pigments exist in the hydrazone rather than the azo tautomeric form. The hydrazone form, which has three intramolecular hydrogen bonds, renders the molecule planar (with the exception of the aniline moiety) which is a stabilizing influence.

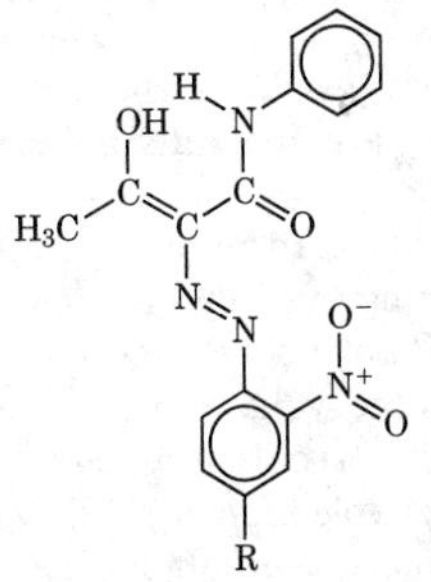

azo form

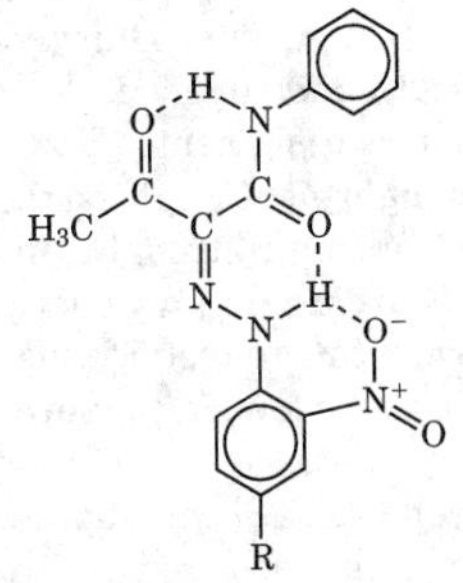

hydrazone form

Azo pigments, one of the oldest and most diverse group of pigments, comprise two types. One type consists of pigments that are insoluble in the aqueous reaction medium in which they are synthesized. Most simple azo pigments show poor bleed characteristics, but relatively good acid and alkali resistance. They show acceptable lightfastness in deep shades but poor tint lightfastness. The second type are laked or precipitated azo pigments derived from components substituted with sulfonic and/or carboxylic acid groups. These pigments are characterized by good to excellent bleed resistance, poor acid and alkali resistance, fair to good lightfastness in deep shades, and poor tint lightfastness. Also available are special azo pigments which show very good overall properties and therefore find applications in fairly demanding systems.

Monoazo Pigments. Monoazo yellow pigments are represented by the following general formula:

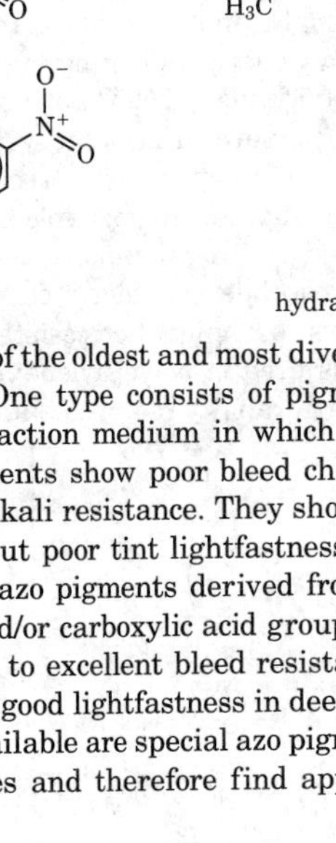

Many of the pigments carry a nitro group in the diazonium component, usually in the ortho position ($R = NO_2$). Among the acetoacetarylide components the *o*-methoxy derivative ($R_2 = OCH_3, R_3 = H$)

is one of the most important in the production of azo pigments. The colors of these pigments range from red to green-shade yellows.

These pigments are sensitive to heat and bleed in most paint solvents. They are, however, resistant to acids and bases. They are used extensively in emulsion paints, paper coating compositions, inks (qv), and, depending on particle size, can in some cases be used outdoors because of excellent lightfastness in full shades.

Benzimidazolones. This class of pigments derives its name from 5-aminobenzimidazolone, which upon reaction with diketene or 2-hydroxy-3-naphthoyl chloride leads to compounds that can be coupled with a variety of diazotized amines.

The acetoacetarylides yield yellow to orange pigments and the naphthoic acid amides yield red and brown pigments.

Diarylide Yellows. Diarylide or disazo yellow pigments are represented by the following general structural formula:

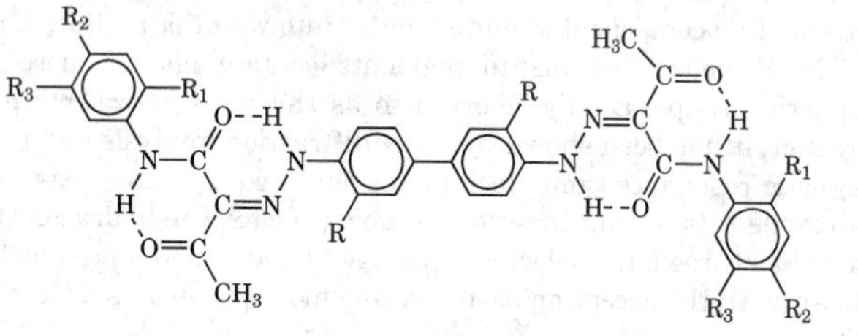

The chemistry and process of manufacture are very similar to the monazo pigments. Diarylides show significantly greater tinctorial strength and superior bleed and heat resistance than the conventional monoazo pigments. However, they are generally inferior to the monoazo pigments in lightfastness.

Based on high strength and versatile transparency most of the diarylides are used in a variety of printing inks, and in some plastics where temperature restrictions of 200°C have been imposed.

Monoazo Yellow Salts. Several monoazo yellow salts have gained popularity since the 200°C temperature restriction on the use of diarylide yellows has been imposed. An example Pigment Yellow 168, the calcium salt of diazotized 3-nitro-4-aminobenzenesulfonic acid coupled with acetoaceto-2-chloroanilide, provides a clean, somewhat greenish yellow color which shows good migration resistance but relatively poor tinctorial strength. It is used in polyethylene and inexpensive industrial finishes where the durability requirements are not high.

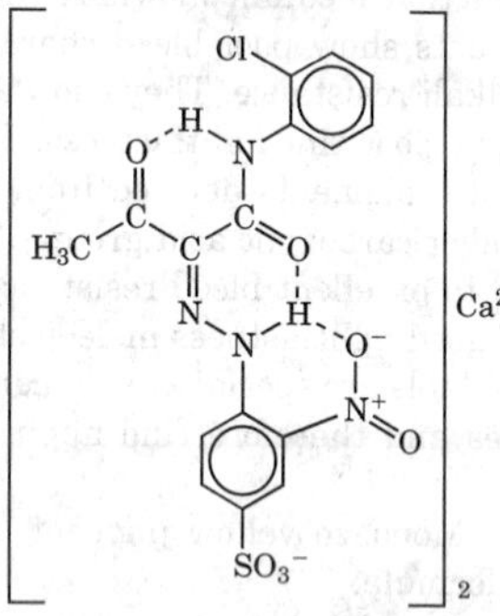

PY 168

Dinitraniline Orange. Dinitraniline Orange or Pigment Orange 5 is a strong and bright orange pigment with relatively low hiding power and good lightfastness in full shades, but poor tint lightfastness. It shows poor bleed but acceptable base resistance and finds principal application in air-drying systems, including a variety of printing inks.

Pyrazolone Orange. Pyrazolone Orange or Pigment Orange 13 is a disazo pigment of high strength and bright color with good lightfastness in full shades and poor tint lightfastness. It is characterized by fair base and chemical resistance and is used primarily in inks, with limited application in paints and plastics.

Azo Reds and Maroons. Pigment Red 3 is one of the most popular organic red pigments used in industrial finishes. Its hue varies considerably with particle size and therefore several shades are commercially available. Its principal application is in air drying paints, and to a limited extent in printing inks.

Para Red or Pigment Red 1 is an intense, reasonably opaque red which shows poor lightfastness, particularly in tints. The related pigment Parachlor Red (Pigment Red 4) is an intense yellowish red. Both pigments show poor bleed and bake resistance and a tendency to bloom in enamels. The pigments are used in some inks and in low cost articles such as detergents, floor polishes, colored pencils, etc. The use of these pigments has declined markedly as a result of greater quality demands by the coatings industry.

Lithol Red or Pigment Red 49:1 is one of the most important of the precipitated salt pigments. A family of sodium (PR 49), barium (PR 49:1), calcium (PR 49:2), and strontium (PR 49:3) salts of diazotized Tobias acid or 2-naphthylamine-1-sulfonic acid coupled with 2-naphthol. These reds are used where brightness, bleed resistance, and low cost are of primary importance.

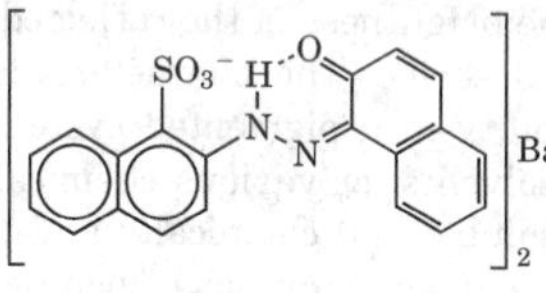

PR 49:1

The BON or BONA Reds and Maroons derive their name from β-hydroxynaphthoic acid, also known as 3-hydroxy-2-naphthoic acid. BON is used as a general coupling component for the entire group with various diazotized amines containing salt-forming groups.

Lithol Rubine (Pigment Red 57) is the calcium salt of diazotized 2-amino-5-methylbenzenesulfonic acid coupled with 3-hydroxy-2-naphthoic acid. It ranks high among organic pigments in production volume and use.

Lithol Rubine is characterized by high tinctorial strength, good bleed, and bake resistance but poor alkali, soap, and acid resistance. Its lightfastness is considered fair and varies within a wide range of shades obtained by inclusion of auxiliary agents.

Red 2B defines the important barium (PR 48:1), strontium (PR 48:3), calcium (PR 48:2), and manganese (PR 48:4) salts of diazotized 2-amino-4-chloro-5-methylbenzenesulfonic acid coupled with 3-hydroxy-2-naphthoic acid, bright red pigments. They exhibit high strength, good bleed, and bake resistance, but poor resistance to alkali, soap, and acids, and fair lightfastness. The main fields of application are printing inks, plastics, and inexpensive industrial paints.

The manganese salt, ie, manganese 2B (PR 48:4) is a bluish red, characterized by superior masstone lightfastness and outdoor durability. It finds use in some automotive and other high quality industrial finishes, as well as in some plastics and a variety of printing inks.

The pyrazolone reds are disazo pigments which provide high color strength and reasonable lightfastness in full shades but poor tint lightfastness, good bake, bleed, and chemical resistance. Some find application in plastics such as poly(vinyl chloride) where they show good dielectric properties, making them useful for cable insulations, in rubber, and specialized printing inks.

Naphthol Reds and Maroons. Naphthol Reds and Maroons are monoazo pigments which provide a wide range of colors from yellowish and medium red to bordeaux, maroon, and violet, and are characterized by high strength but marginal migration resistance. Depending on the substitution pattern some are strongly migrating and others are more or less resistant to migration. Lightfastness is generally marginal to good. Pigment Red 112 is a brilliant medium red pigment, approaching the shade of Toluidine Red, which is used in a variety of printing inks, air drying, and emulsion paints.

Another important pigment in this class is Pigment Red 170 which provides medium shades of red, and when particle-grown produces an opaque modification which shows improved migration resistance and lightfastness. It is used in high grade industrial paints and, in combination with high performance pigments, in automotive finishes. The transparent type which is tinctorially strong finds applications in a variety of printing inks.

Azo Condensation Pigments. A further improvement in heat stability of azo pigments was achieved by the condensation disazo pigments due to an enlarged molecular framework and higher molecular weight. Formally they are composed of two monoazo or more accurately two monohydrazone units, which are attached to each other by an aromatic dicarbonamide bridge.

The pigments are used primarily in plastics, including polypropylene fibers, because of very good bleed resistance, heat stability, and lightfastness. The reds also find use in printing inks, primarily for high quality products.

Lakes. Lakes are either dry toner pigments that are extended with a solid diluent, or an organic pigment obtained by precipitation of a water-soluble dye, frequently a sulfonic acid, by an inorganic cation or an inorganic substrate such as aluminum hydrate.

Basic dyes are characterized by bright shades and high strength but poor lightfastness. However, when laked by precipitation with soluble salts of organic acids such as tannic acid, or inorganic heteropolyacids like phosphotungstic (PTA; M = W) and phosphomolybdic (PMA; M = Mo), and the combined phosphotungstomolybdic acid (PTMA), the resulting pigments retain the dyes' tinctorial attributes, but become insoluble and show improved lightfastness.

Copper Phthalocyanines. Copper phthalocyanine (CPC) approximates an ideal pigment (Pigment Blue 15). This class of pigments offers extreme brightness, tinctorial strength, bleed and chemical resistance, stability to heat, and migration. The pigments show excellent weatherfastness but are restricted to the blue and green regions of the spectrum. Phthalocyanine blue and green are among the most important organic pigments on the worldwide market.

Copper Phthalocyanine Blue. CPC blue exists in several polymorphic modifications, two of which, the red-shade blue alpha and green-shade blue beta form, are of great commercial significance. Beta is the thermodynamically more stable phase and is the product resulting from manufacture by the two basic processes using either phthalonitrile or phthalic anhydride as starting materials. The alpha form is usually obtained by conversion from the beta form and has to be stabilized to prevent phase reconversion.

Copper Phthalocyanine Green. CPC green is obtained by electrophilic substitution of CPC blue with chlorine. The typical polychloro-CPCs are blue-shade green pigments. To provide yellower shades of green, bromine is substituted for chlorine. Like CPC blue the green pigments show outstanding pigmentary properties, but are lower in tinting strength with progressive halogen substitution, particularly with bromine (see PHTHALOCYANINE COMPOUNDS).

Quinacridones. Quinacridone pigments offer generally outstanding fastness properties across the visible spectrum from red-shade yellows to scarlet, maroon, red, magenta, and violet color ranges. The pigments are practically insoluble in most common solvents and therefore show excellent migration resistance in most application media. The low solubility is attributed to effective intermolecular hydrogen bonding which is also responsible for the photochemical stability. The various available colors are the result of polymorphism and various substitution patterns. The parent compound Pigment Violet 19 exists in three polymorphic modifications. The red gamma and violet beta forms are commercial pigments, whereas the red alpha form is metastable.

Due to the excellent pigmentary properties, quinacridones are used in many industries but particularly in automotive finishes, emulsion paints, plastics, and fibers.

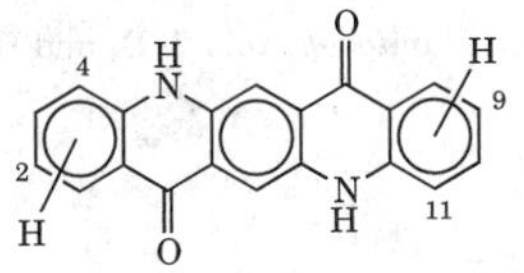

Pigment Violet 19

Diaryl Pyrrolopyrroles. The 1,4-diketo-3,6-diarylpyrrolo(3,4-c)pyrroles are the most recently discovered class of pigments ranging in color from orange to bluish reds. These pigments are synthesized by base-catalyzed condensation of higher diakyl esters of succinic acid with aromatic nitriles. One important member of this class is Pigment Red 254, which is a very opaque yellowish red pigment of outstanding durability, brightness, and chemical resistance. Another is the parent compound Pigment Red 255, which is a high performance orange pigment. Both are used in automotive finishes, and a higher strength variation of PR 254 is used in plastics applications.

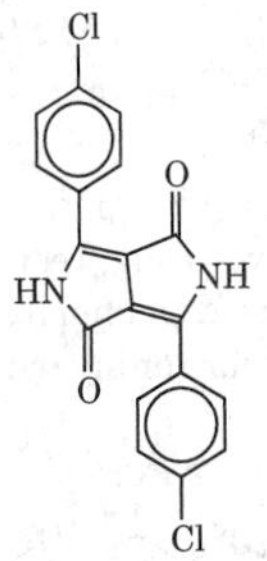

PR 254

Vat Dye Pigments. Vat dyes have been used for a long time for coloring textile fibers. As pigment technology evolved, new methods of particle size reduction have been successfully applied to largely insoluble dyes. Only a few of the very large number of vat dyes have found application in the pigment field.

Perylenes. Perylene pigments are either the 3,4,9,10-tetracarboxylic dianhydride or more often N,N′-substituted diimides.

The pigments are manufactured either by reaction of the dianhydride with an amine or N,N′-dialkylation of the diimide. They are characterized by high tinctorial strength, excellent solvent stability, very good weatherfastness, moderate brightness, and range in color from red to violet.

Most applications are in high grade industrial paints, especially automotive finishes. Some types are used primarily in plastics and fibers.

Perinones. The most important pigment in this family is the orange perinone, Pigment Orange 43; is fairly weatherfast and heat stable, and is used primarily in plastics and fiber applications.

Thioindigo. The most important thioindigo pigment is the red-violet Pigment Red 88. Although still used in some paint and plastic systems, it is being replaced by pigments of higher quality.

Aminoanthraquinone Pigments. Pigment Red 177 has the chemical structure of 4,4′-diamino-1,1′-dianthraquinonyl. It is the only known pigment with unsubstituted amino groups which are involved in both intra- and intermolecular hydrogen bonding. The bluish red pigment is used in plastics, industrial and automotive paints, and specialized inks (see DYES, ANTHRAQUINONE).

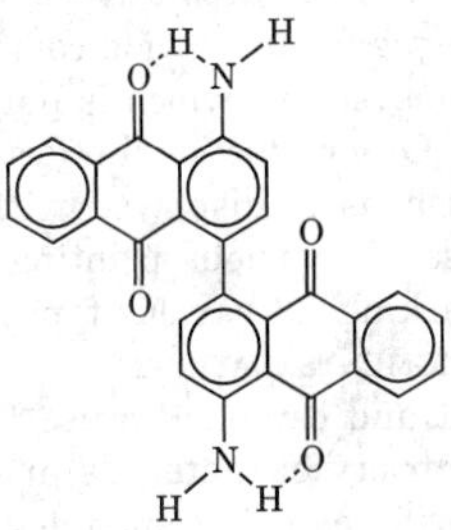

PR 177

Pigment Yellow 147 is a reddish shade yellow pigment used primarily in certain plastics and in polyester and polypropylene fibers.

Indanthrone. Pigment Blue 60 is a very red-shade blue pigment that shows outstanding weatherfastness in full shade as well as in light white reductions. It is used primarily in metallized automotive finishes where it is sometimes more weatherfast than some copper phthalocyanine blues.

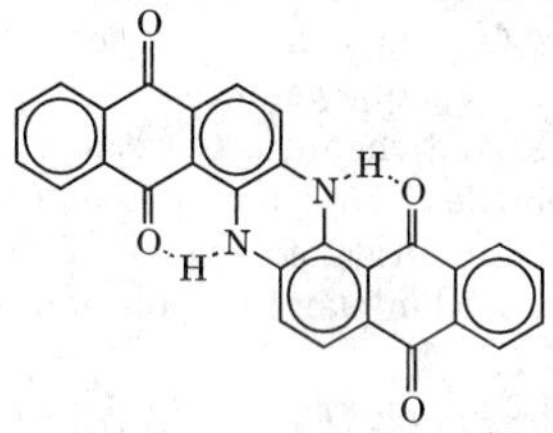

PB 60

Dioxazine. Carbazole Violet (Pigment Violet 23) is a bluish violet pigment that is uncommonly strong, resistant to solvents, and shows fair weatherfastness. It is used primarily as a shading pigment with copper phthalocyanines and for toning whites in a variety of systems.

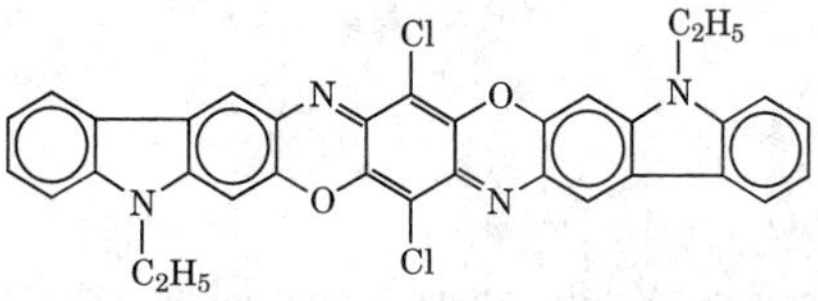

Isoindolinones and Isoindolines. Tetrachloroisoindolinone pigments are characterized by very good lightfastness, heat stability, migration resistance, and chemical inertness. Although Pigment Yellow 110, a red-shade yellow, is relatively weak, it finds extensive use in automotive and other high grade finishes and in a variety of plastics and ink applications.

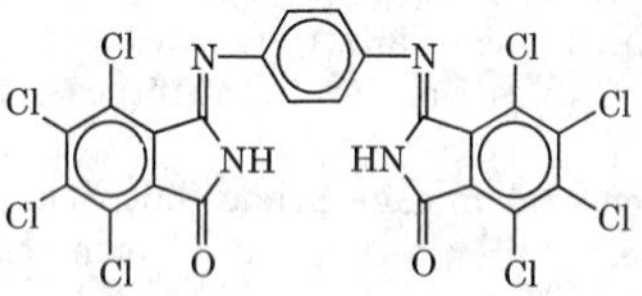

PY 110

Among the isoindoline pigments is Pigment Yellow 139, a reddish yellow pigment which differs in color as a function of particle size. The opaque version is the reddest. Although marginal in chemical resistance the pigment is used in the paint and plastics industries.

Quinophthalones. The quinophthalone pigments are prepared by condensation of quinaldines with a variety of aromatic anhydrides. One pigment in this series, Pigment Yellow 138, is a reasonably weatherfast greenish yellow pigment of good heat stability. The main field of application is paints and plastics.

Uses

Organic pigments are used for decorative and/or functional effects. In paints, for example, pigments provide color and contribute to exposure durability of the systems, which is particularly true for high performance pigments. Other functional effects include hiding power and high visibility, such as is displayed with daylight fluorescent pigments. They are used in various printing processes for textiles, plastics, and safety markings of various types (see LUMINESCENT MATERIALS–FLUORESCENT PIGMENTS (DAYLIGHT)).

The most important and established use for pigments is the imparting of color to a variety of materials and compositions. Examples are surface coatings for exteriors and interiors of automobiles and houses with oil- or water-based paints; wood stains, leather and artificial leather finishes, printing inks and many other applications.

Testing and Standardization

Pigments are subjected to a number of tests before they are released to customers. Testing is complicated because of the great diversity of pigment types and uses. A given pigment may be dispersible in one system but poorly dispersible in another, and can exhibit different durability depending on the system; performance is system dependent. Standardization is carried out against a standard sample for coloristics and a variety of working properties. Among the tests, depending on the pigment type, may be thermal stability, hiding power, rheology, migration, chemical stability, gloss, distinctness of image, durability, etc.

In the process of testing, color deviations are expressed in the CIELAB system or the equivalent polar LHC system. In either case tested samples must fall within acceptable ranges or limits established versus a standard by the pigment manufacturer and accepted by the pigment user.

In dispersing a pigment by an established method, acceptable pigment strength vs a standard must be achieved, even though an ideal dispersion normally is rarely realized.

Health and Safety Factors

Since pigments are generally insoluble, unlike most dyes, they are usually not bioavailable and consequently are generally not absorbed or metabolized. Nevertheless many health-related studies have been carried out and reported in the literature.

Acute toxicity of organic pigments has been studied extensively. The most common measure of toxicity is LD_{50} expressed in mg/kg of body weight which has a lethal effect on 50% of test animals after a single (oral, dermal, etc) administration. These tests assess toxicity vs other known compounds. A large LD_{50} value represents a low degree of toxicity. Pigments in general have very low levels of acute toxicity.

Chronic toxicity defines a specific dose or exposure level that will produce measurable, long-term toxic effects, including carcinogenicity.

One area which requires special comment is a study which showed that certain diarylide pigments processed in polymers above 200°C and particularly above 240°C decompose to give off 3,3-dichlorobenzidine, an animal carcinogen. As a consequence diarylide pigments (not, however, condensation disazo pigments) are not recommended for use in any applications where they might be exposed to temperatures exceeding 200°C.

The hazards associated with handling pigments are specified by an OSHA Hazards Communication Standard, which also requires labeling and employee information and training.

Ecological Effects. The starting materials for manufacture of organic pigments are as diverse as the pigments themselves. However, most starting materials are derived from petroleum or natural gas sources. Although many pigments are synthesized in water, a variety of organic solvents are also employed by the industry. The effective utilization of all starting materials and solvents, and reduction of undesirable by-products, is a primary objective of the organic pigment industry.

E. E. JAFFE
Ciba-Geigy Corporation

Color Index, 3rd ed., Vol. 4, The Society of Dyers and Colourists, Bradford, Yorkshire, England; American Association of Textile Chemists and Colorists, Research Triangle Park, N.C.

W. Herbst and K. Hunger, *Industrial Organic Pigments*, VCH Publishers, Inc., New York, 1993.

E. A. Clark and P. Anliker, in O. Hutzinger, ed., *Organic Dyes and Pigments, The Handbook of Environmental Chemistry*, Vol. 3, Springer-Verlag, Berlin, 1980.

T. C. Patton, ed., *Pigment Handbook*, Vols. I, II, and III, John Wiley and Sons, Inc., New York, 1973.

PILOT PLANTS

The design or substantial modification of a new plant or process, its subsequent construction, and start-up represent a tremendous investment of time and money. The rewards are great if a significant improvement is realized; the risks are also great if a costly commercial plant fails to produce as expected. To reduce the degree of risk, lengthy and expensive research programs are often undertaken.

A pilot plant is a collection of equipment designed and constructed to investigate some critical aspect(s) of a process operation or perform basic research. A pilot plant can range in size from a laboratory bench-top unit to a facility only marginally smaller than a commercial unit. The purposes for its construction and operation can vary widely: confirming feasibility of a proposed process; providing design data; determining the economic feasibility of a new process; determining optimum materials of construction; testing operability of a control scheme; determining the extent of plant maintenance (qv); producing sufficient quantities of product for market evaluation; obtaining kinetic data; screening catalysts; proving areas of advanced technology; providing data for solutions to scale-up problems; providing technical support to an existing process or product; assessing process hazards; determining operating costs; optimizing an existing process; evaluating alternative processes, feedstocks, or operating conditions; and performing basic process research.

Categorizing Pilot Plants

Pilot plants can be categorized by a number of different methods. Size is the most common classification as it is the most uniformly proportional to construction and operating costs; some typical characteristics are shown in Table 1. These characteristics are generalizations and may not be applicable in any specific case.

The pilot plant's degree of automation is the next most common classification, because the instrumentation is usually a large fraction of the initial construction and annual operating costs. The high cost of operating labor and the difficulties in manually controlling a process while taking accurate data have forced virtually all but the simplest new pilot plants to increase their level of automation, so this classification has become less useful.

Scale-Up

Scale-up is the process of developing a plant design from experimental data obtained from a unit many orders of magnitude smaller. This activity is considered successful if the commercial plant produces the product at planned rates, for planned costs, and of desired quality.

Scale-up problems exist for many reasons including basic equations unsolvable by known mathematical techniques, interconnected physical and chemical aspects of the process resulting in coupled basic equations, solutions used on the pilot plant not suitable on the commercial scale, and unknown equipment performance at sizes never used before. Although successful pilot-plant operation does not guarantee successful commercial plant operation it considerably increases the success factor.

Planned Pilot-Plant Experimentation

Planning the experimental program is an important part of the decision process concerning the type of pilot plant required. This step should be performed as soon as the research program objectives are formulated because the type of experimental program affects the total program cost, the type of pilot plant required, and the way the pilot plant needs to operate. Planned experimentation also helps reduce the number of tests required, a key factor in controlling costs.

Pilot-plant design specifications should be established only after careful consideration of the experimental program because decisions on the accuracy of instruments, analyzers, and other equipment should be based on the requirements of the experiments planned for the unit.

Statistical designs for experiments maximize information and reduce research time and costs. These techniques increase the overall confidence level in the experimental results. (see DESIGN OF EXPERIMENTS).

Table 1. Typical Classes of Pilot Plants by Size

Class	Characteristics
microunit or bench-top unit	small size (<2 m^2) typically found in hood or on bench-top; small tubing used throughout (3–6 mm dia); limited automation
integrated pilot plant	larger size (~10–25 m^2) found in open bay, walk-in hood, or containment cell; tubing and small (13–25 mm dia) pipe used; fully automated feed and product systems with limited capacity usually included
demonstration or prototype unit	very large size (>250 m^2) usually in dedicated building or area; pipe used extensively; fully automated usually with dedicated computer system; feed and product systems, including significant storage capacity

Pilot-Plant Design

The first step in pilot-plant design is to determine whether to design a pilot plant for process modeling or problem investigation. Investigating the problem involves designing a pilot plant to look at a specific area of interest. In this case, the pilot plant may not resemble the commercial operation.

Two approaches are commonly used for pilot-plant design. The first uses conventional design techniques that mimic commercial process design. This approach is not always feasible for a number of reasons; for example, copying the commercial design may carry some inherent limitations that adversely impact operations at ranges or conditions which are of interest to the pilot plant but not to the commercial plant. The alternative approach is to use a design methodology oriented to the pilot plant using many conventional design techniques but trying to maximize the advantages of the pilot-plant operations with regard to scale, technique, and operation. The quality of the final design is very dependent on the skill level and experience of the design engineer.

Methods of Estimating Pilot-Plant Costs

There are three basic methods for estimating the costs to design and construct a pilot plant: similarity, cost ratios, and detailed labor and materials.

Similarity involves estimating the cost of the pilot plant based on the costs to design and construct a similar unit. This is the fastest method, but it is the least accurate, because few similar units are ever identical. Cost ratios develop the cost estimate by relating the overall cost of the pilot plant or a part of the pilot plant to a known factor. The cost estimate is built up by using the ratios to develop the cost of the entire unit, individual equipment, or separate subsystems depending on how detailed a cost ratio estimate is made. Unfortunately, cost ratio information is rarely available for pilot-plant-scale equipment and the lack of a large data pool renders much of what is available suspect regarding its overall accuracy.

Detailed labor and material estimating involves breaking the pilot-plant construction down into a detailed series of small tasks and estimating the labor and materials required for each separate task. The difference between a general and detailed cost ratio cost estimate is the amount of detail involved.

The costs to operate a pilot plant are a summary of the costs of the feedstock, product disposal, utilities, operating labor, spare parts, maintenance, and support services.

Pilot-Plant Space

In general, pilot-plant space can be divided into five basic types: separate buildings, containment cells or barricades, open bays, walk-in hoods, and laboratory areas.

The space required for a pilot plant varies tremendously with its size and type. A small unit may require only part of a laboratory, whereas an average pilot plant may require a large room or building, excluding extended feed or product storage.

Scheduling. A significant concern in all pilot-plant work is minimizing the time involved between project inception and meaningful data generation. Typically, an average pilot plant requires from six to eighteen months to progress through this process: two to eight months for design, materials procurement, and scheduling; two to six months for construction; and one to four months for start-up (up to and including first actual run). Design time can be reduced if standardized designs have been developed in advance for common subsystems. Construction time can be reduced by various methods.

Pilot-Plant Control Systems

Defining the requirements for a pilot-plant control system is often difficult because process plant experience for comparison and evaluation is commonly lacking and the design is frequently performed by personnel inexperienced in either instrumentation systems or pilot-plant operations.

Automatic instrumentation systems for pilot plants was originally based on analogue devices. Computer-based control systems are available in sizes and costs suitable for most pilot-plant operations. Computer-based systems are growing smaller and less expensive; analogue equipment is being redesigned to use microprocessors. This is blurring the distinction between the two systems (see PROCESS CONTROL).

Three types of computer control systems are commonly used for pilot-plant instrumentation. The first is a centralized system, usually based on a minicomputer or occasionally a mainframe, that typically controls all the pilot plants in an area or facility.

The second is a stand-alone computer system, usually based on a personal computer (PC) or programmable logic controller (PLC), that provides a separate computer system for each pilot plant.

The third type comprises distributed control systems (DCS), a hybrid of the previous systems, having stand-alone computers for control and data gathering at each pilot plant together with one or more higher level computer(s) for data storage and work-up.

Instrumentation. Pilot plants are usually heavily instrumented compared to commercial plants. It is not uncommon for a pilot plant to have an order of magnitude more control loops and analytical instruments than a commercial plant because of the need for additional information no longer required at the commercial stage.

Analytical instrumentation for pilot plants is divided into two classes: off-line and on-line. Off-line analysis is a batch operation requiring sample taking, handling, and storage followed by analysis at a later date, frequently in another location.

On-line analysis is often more expensive and difficult to set up initially but can be more accurate and reliable if performed properly.

Feed and Product Handling

One of the most vexing aspects of pilot-plant work can be feed and product handling as a pilot plant is neither designed nor operated as a closed-loop system like a commercial plant.

The toxicity of both feeds and products must be carefully considered during the preliminary design stages, especially if the feeds or products contain known or suspected carcinogens.

Provision must be made for sufficient storage of both feeds and products, and the storage system should be designed to minimize operator handling time.

Pilot-Plant Start-Up

Pilot-plant start-up is different from principal process plant start because of the smaller scale of the unit, smaller resources committed, lack of advance start-up planning, and limited experience with the pilot-plant process and operation.

The key to a successful pilot-plant start-up is advance planning, at least six to eight weeks before project completion, to allow identification of problem areas and concerns in time for successful resolution. A detailed start-up sequence should be developed listing each task to be performed in chronological order. The start-up sequence then allows the development of a list of required resources and a tentative start-up schedule. A successful start-up requires this advance planning as well as adequately trained and experienced personnel with a variety of skills. Many companies have found that a specialized start-up group is a primary asset if pilot-plant work is regularly done. Safety is also a significant concern during start-up because interlock systems are not fully functional and equipment and subsystems are being energized for the first time.

Pilot-plant start-up costs vary widely. The costs booked to the start-up as well as the extent of the commissioning activities labeled as "start-up" vary widely among organizations. Typical costs range from 5 to 50% of construction cost; 10 to 20% is typical. Start-up durations also vary, but a range of one to three months is common depending on the personnel training and expertise.

Safety in Pilot-Plant Design and Operations

Pilot plants are often more hazardous than process plants, even though they are smaller in size, for many reasons. These include a tendency to relax standard safety review procedures based on the small scale, exceptionally qualified personnel involved, and the experimental nature of the research operations; the lack of established operational practice and experience; lack of information regarding new materials or processes; and lack of effective automatic interlocks due to the frequently changing nature of pilot-plant operations, the desire for wide latitude in operating conditions, and the lack of full-time maintenance personnel.

To minimize these concerns, most organizations require a formal series of safety reviews and hazard analyses for each new or modified pilot plant. At a minimum, this involves analyzing the proposed pilot plant before construction to identify and eliminate any potential hazards, including toxicity or flammability concerns, feed or product handling, disposal problems, relevant government regulations, and potentially harmful reactions.

The pilot plant must also be carefully designed so that its control and safety systems are "fail-safe" and any unexpected equipment or utility failure brings the unit into a safe and de-energized condition.

RICHARD P. PALLUZI
Exxon Research and Engineering Company

R. P. Palluzi, *Pilot Plants: Design, Construction and Operation*, McGraw Hill Book Co., Inc., New York, 1992.

PIPELINE HEATING. See PIPELINES.

PIPELINES

Pipelines or pipe lines are continuous large-diameter piping systems, usually buried underground where feasible, through which gases, liquids, or solids suspended in fluids are transported over considerable distances. They are used to move water, wastes, minerals, chemicals, and industrial gases, but primarily crude oil, petroleum products, and natural gas. In the oil and gas business, a pipeline system consists of a trunkline, ie, the large-diameter, high pressure, long-distance portion of the piping system through which crude oil is shipped to refineries, or natural gas and oil products, respectively, are transported to distribution points, and smaller low pressure gathering lines that transport oil or gas from wells to the trunkline. Smaller lines used by natural gas distributors are not considered part of a gas pipeline system (see GAS, NATURAL)

Pipeline transport involves the application of force to the material being moved, either through the use of pumps to transport liquids, compressors to move gases, or flowing water to move solids. In some applications, vacuum may create the pressure differential.

Pipeline Transport of Gases

Essentially all substances that are gases at standard conditions of temperature and pressure are transported commercially by pipeline, this includes ammonia, carbon dioxide, carbon monoxide, chlorine, ethane, ethylene, helium, hydrogen, methane (natural gas), nitrogen, oxygen, and others. Gases with moderate boiling points can be pipelined in either gaseous or liquid form; liquefied petroleum gases (LPG), carbon dioxide, ammonia, and chlorine are usually shipped as liquids because of the smaller pipeline volume for liquids.

Methane (Natural Gas). Natural gas is conveyed in strong, thin-walled, long-distance pipelines in virtually all principal countries of the world.

Polyethylene (PE), poly(vinyl chloride) (PVC), or polypropylene plastic pipe is being used in increasingly greater amounts throughout the world in both gas gathering systems and gas distribution systems.

Transport of natural gas starts with small-diameter gathering lines that convey the raw gas from individual wells to a collection point and from there to a gas treating plant, where heavier hydrocarbons (qv), particulates, and water are separated from the gas. The hydrocarbons known as natural gas liquids (NGL) are transported as raw mix in liquid pipelines to chemical plants, where they are fractionated into lighter and heavier fractions that are transported to market areas in products pipelines. The methane-rich gas is compressed to the appropriate pressure for transmission by large-diameter pipeline. Pressure is maintained by large compressors at compressor stations along the length of the pipeline.

Ammonia Pipelines. Ammonia, a commodity produced from natural gas, is transported to the market areas by tank car, barge, or pipeline. The pipelines are made of high strength pipeline steel, and operate with pressure sufficient to maintain the ammonia in the liquid state.

Hydrogen Pipelines. The energy carrying capacity of a given size pipeline is approximately the same when it is carrying either hydrogen or natural gas, provided that it is operating in turbulent flow with the same pressure drop along its length and at the same operating pressure. Hydrogen compressors must handle 3.8 times more gas than natural gas compressors for the same energy throughput, thus indicating higher pipeline transmission costs for hydrogen than for natural gas.

Industrial Gases. Industrial gas (oxygen, nitrogen, etc) pipelines are short compared with long-distance pipelines that transport crude oil, natural gas, or petroleum products; however, more than 80% of the oxygen and more than 60% of the nitrogen produced in the United States by air separation is transported by pipeline. Carbon dioxide and hydrogen are also moved by pipeline in the Houston area. Multiple users of oxygen (qv), nitrogen (qv), hydrogen (qv), and carbon monoxide (qv) near Rotterdam, the Netherlands, are supplied by pipeline from a nearby industrial gas complex.

The application that has led to increased interest in carbon dioxide pipeline transport is enhanced oil recovery (see PETROLEUM).

Helium, extracted from natural gas in the southwestern United States, is moved by a pipeline to storage in a partially depleted gas field.

Cryogenic Gases. Some of the most sophisticated pipeline technology deals with the transport of liquefied gases with very low boiling points, ie, cryogenic gases such as oxygen −182.96°C), argon (−185.7°C), nitrogen (−195.8°C), hydrogen (−252.8°C), and helium (−268.9°C). These gases are liquefied by modern cryogenic methods (see CRYOGENICS). Development of a system for piping them as liquids has been brought about by the needs of the space program, superconducting magnets, and other high technology areas.

Pipeline Transport of Liquids

The main technical difference between liquid and gas pipeline transport is the compressibility of the fluid being moved and the use of pumps, rather than compressors, to supply the pressure needed for transport. The primary use for liquids pipelines is the transport of crude oil and petroleum products.

Crude Oil and Products Pipelines. The crude oil delivery system starts with relatively small-diameter gathering lines from individual producing wells to a main-line pump station, from where it is pumped through a larger transmission trunkline to a refinery or other destination. The refined products are transported by products pipelines to markets, storage, shipping terminals, etc. In modern lines, all inputs and outputs are metered, monitored, and remotely controlled by supervisory control and data acquisition (SCADA) computer systems.

Several different refined products are shipped in the same pipeline (batching) by using control methods to minimize intermingling at different interfaces. This is achieved by maintaining high turbulent flow in the pipe. At the terminal, the comingled fraction is separated from other products and either blended or returned to the refinery.

Sulfur and Chlorine Pipelines. Underground sulfur is melted by superheated water and then piped as liquid to the surface with compressed air. At the surface, molten sulfur is transported by heated pipeline to a storage or shipping terminal.

Chlorine is shipped by pipeline in either gaseous or liquid form; however, great care must be taken to ensure that liquid pipelines are operated only in the liquid phase and gaseous pipelines operated only in the gaseous phase.

Pipeline Transport of Solids

Pipelines to transport solids are called freight pipelines, of which three different types exist: pneumatic pipelines, the use of which is known as pneumotransport or pneumatic conveying; slurry pipelines, which may also be called hydrotransport or hydraulic conveying; and capsule pipelines. When air or inert gas is used to move the solids in the pipeline, the system is called a pneumatic pipeline and often involves a wheeled vehicle inside the pipeline, propelled by air moving through the pipe. Slurry pipelines involve the transport of solid particles suspended in water or another inert liquid. Hydraulic capsule pipelines transport solid material within cylindrical containers, using water flow through the pipeline for propulsion.

Pneumatic Pipelines. Pneumatic pipe systems are used to move blood samples, medicine, and supplies between buildings in hospital complexes; cash and receipts in drive-up banks; parts and materials in factories; refuse from apartment complexes; and grain, cement, and many other materials. Most of these are small diameter and usually short.

Slurry Pipelines. Finely divided solids can be transported in pipelines as slurries, using water or another stable liquid as the suspending medium. Flow characteristics of slurries in pipelines depend on the state of subdivision of the solids and their distribution within the fluid system. Although slurry flow and conventional liquid flow are both divided into laminar, transitional, and turbulent flow, the contribution of the solids-to-slurry flow results in a further characterization as either homogeneous or heterogeneous flow. Homogeneous flow of slurries occurs when very finely divided solids are distributed uniformly throughout the suspending medium, eg, bentonite slurries as drilling muds. Heterogeneous flow occurs when larger, irregular solid particles are slurried and distribution is not uniform throughout the suspension.

Slurry pipeline design is similar to the design of conventional liquid pipelines, except for the slurry preparation stage and, if necessary, an additional step for separating the suspended solids from the suspending liquid at the point of use.

Capsule Pipelines. Capsule pipelines involve the transport of material inside a closed cylindrical container propelled by water flowing through the pipeline. This is called hydrotransport and the diameter of the cylindrical container is approximately 90–95% the diameter of the pipeline.

Pipeline Technology

Pipeline technology involves design, construction, maintenance (qv), and operation. Although certain aspects of the technology differ under different climatic conditions, whether above or below ground or under water, etc, the basic steps are the same for liquids pipelines as for gas pipelines.

Design. Pipeline design begins with a preliminary mapping of the proposed route, noting areas to be avoided and such obstacles as rivers, railroads, and highways. For cross-country pipelines, aerial and ground-control surveys are made to establish the final alignment of the pipeline. Permission to cross all land parcels must be obtained from landowners, and permits obtained from the proper authorities to cross rivers, highways, and railroads; where permission is not obtained, appeal is made to the courts under the right of eminent domain. In the United States, approval for gas pipeline construction must be obtained from the Federal Energy Regulatory Commission (FERC) and/or state public utility commissions, and an environmental impact statement must be filed. Liquid pipelines do not require FERC approval for construction; common carrier pipelines are generally accorded the right of eminent domain and are regulated by FERC or state authority.

While detailed routing is being established, mechanical design begins by establishing pipe size, based on the volume and type of material being transported, design pressures, pipeline length, and spacing between compressor or pumping The strongest steel possible is used and pipe wall thickness is determined by a code-specified design formula that recognizes pipe as an unfired pressure vessel; the design formula for natural gas pipelines in the United States is given by the industry code ASME B31.8 and also by federal regulation 49 CFR 12.

Specifications for valves, fittings, etc, must be written as part of the design phase. Shut-off valves are required at specific intervals to isolate any pipeline damage; powered valve-closing devices are commonly installed to shut off valves automatically in the event of a line break and isolate the damaged portion of the line. Specifications are made for exterior coatings and cathodic protection to minimize external corrosion, and for interior coatings or additives to minimize interior corrosion, as well as decrease flow resistance and improve transmission of odorants.

Construction. Pipeline construction is a continuous activity required by supply changes, obsolete pipelines are decommissioned, additions are made to existing pipelines, or new pipelines are designed and laid to take advantage of new supply sources. Construction of a cross-country pipeline may be divided into segments (spreads) and proceeds by the following steps: right-of-way clearing and grading, trenching, pipe stringing along the right-of-way, pipe bending to fit the trench bottom, welding, cleaning and priming the exterior surface for coating, coating and wrapping for cathodic protection unless it is mill-coated and only the weld areas need coating, lowering pipeline into trench, and backfill of earth over the pipeline.

Alternative construction methods are used for crossing obstacles. For river crossings, floats may be used to carry the pipe across the river with concrete coatings or weights, flooding the floats to sink the pipe to the river bottom where the weights hold it in place. Land under highways is bored with an earth auger followed closely by the pipe or casing, if required, advanced by the same machine.

When construction is complete, the pipeline must be tested for leaks and strength before being put into service; industry code specifies the test procedures.

Construction of underwater (submarine) pipelines does not take place under water. Pipelines are welded onshore and dragged into position by powerful winches on ships floating on the water surface (for short lines), welded on a specially constructed lay barge, and lowered to the ocean floor by a stinger from one end of the barge or welded onshore, floated on pontoons, and towed to the offshore area where they are lowered into position. Submarine pipelines are being used regularly to transport oil, natural gas, and other commodities to shore from offshore locations, such as the Gulf of Mexico, the North Sea, and the Arabian Gulf.

Maintenance. After construction, a pipeline must be tested, inspected, cleaned, mapped for bends, dents, or ovalties, maintained during operation, and monitored for leaks or corrosion to ensure safety of operation. These operations involve pigging, moving devices called pigs through the pipeline that can carry out these missions. These may be sophisticated electronic devices or something as simple as a rubber ball pumped through the pipeline to displace fluids.

Anticorrosion measures have become standard in pipeline design, construction, and maintenance in the oil and gas industries; the principal measures are application of corrosion-preventive coatings and cathodic protection for exterior protection and chemical additives for interior protection. Pipe for pipelines is available with a variety of coatings, either pre-applied or coated and wrapped on the job with special machines as the pipe is lowered into the trench.

Cathodic protection is provided by inducing an electric current on the pipeline to ensure that the electric potential of the metallic pipe is less than the earth surrounding the pipeline after it is laid, by using a sacrificial anode bed with a more electropositive metal than iron, or by thermoelectric protection, using heat to generate a direct current to lower the potential of the pipe (see CORROSION AND CORROSION CONTROL).

Operation. Operations are controlled from a central computerized office that maintains communication with compressor stations and flow stations along the route; reports are also received from weekly or biweekly aircraft flyovers to inspect for potentially threatening conditions, such as soil erosion, floods, approaching excavation, or any external evidence of leaks. Flow computers along the pipeline route monitor and quantify flow from producers entering the pipeline or flow leaving the line to customers. Operations and maintenance of pipelines in the United States are covered in ASME B31.8 and federal regulation 49 CFR 192.

Cross-country gas pipelines generally must odorize the normally odorless, colorless, and tasteless gas in urban and suburban areas, as is required of gas distribution companies. Organosulfur compounds, such as mercaptans, are usually used for this purpose.

Economic Aspects

Pipelines usually represent the least costly way to move fluid products, including solid slurries, wherever they compete with truck or rail transportation.

The cost of building a pipeline is usually divided into four categories: materials (line pipe, fittings, coatings, cathodic protection, etc), right-of-way and damages, labor, and miscellaneous (surveying, engineering, supervision, administration and overhead, interest, contingencies, afudc, and FERC filing fees). The relative contribution of each category depends on the size of pipe, the difficulty of terrain, and delays and added costs due to environmental factors, but generally the costs for materials and labor account for about 75% of the entire cost of the pipeline, with each accounting for about half this amount. The remainder is engineering, overhead, fees, interest, etc.

When a natural gas pipeline has been commissioned and is operating, its transmission costs are the operating expenses plus maintenance expenses.

Safety, Environmental, and Ecological Aspects

Data compiled by the U.S. Department of Transportation (DOT) indicate that pipeline transport is the safest materials transport mode, particularly over long distances.

To ensure the safety of gas pipelines in the United States, the Natural Gas Safety Act of 1968 was created to mandate federal regulation of gas storage facilities and pipeline transport of natural gas, with the DOT given exclusive authority to regulate safe operation of natural gas pipeline systems that fall within the jurisdiction of the Federal Power Commission under the Natural Gas Act.

Most ecological issues involve the pipeline's construction phase and the effect that it will have on vegetation, wildlife, topography, population density, and land use. All pipeline design and construction programs must include an environmental impact statement which must be submitted and evaluated before permission to proceed with construction is granted.

THOMAS P. WHALEY
Consultant
GEORGE M. LONG
Institute of Gas Technology

J. Watts, ed., *Pipeline Gas J.*, 12–36 (Sept. 1992).

"Transportation of Natural and Other Gas by Pipeline," *Code of Federal Regulations*, Hazardous Materials Regulations Board, U.S. Dept. of Transportation, Minimum Federal Safety Standards, U.S. Printing Office, Washington, D.C., Title 49, Chapt. 1, Part 192.

PIPING SYSTEMS

A piping system provides a conduit for the safe and economical transfer of fluid or fluid-like materials from one location to another, often with provisions for controlling the rate of flow. Piping system design requires specialized knowledge of the selection and application of materials of construction, fluid mechanics (qv), mechanical and structural design, facility safety requirements, and an understanding of the potential hazards and environmental consequences of accidental release of the material being transferred.

Piping must be suitable for the temperature, pressure, and corrosivity of the flowing material; the size must be adequate for the flow rate desired; and valving must be provided to enable the flow to be controlled.

Materials

Pipe materials are metallic, nonmetallic, or lined metallic. The most common material, carbon steel, is generally the least expensive. In many cases, however, it cannot be used because of the corrosivity or the temperature of the flowing medium. When temperature, rather than corrosivity, is the limiting factor, the lowest grade of steel (qv) that provides the required service life at the design temperature is used. Corrosion rate is usually temperature-dependent; perceptive process design and plant layout (qv) can result in significant cost savings.

For moderate temperatures and pressures and if corrosivity eliminates carbon steel, nonmetallic piping offers an alternative. Plastic and reinforced plastic piping and fittings are available in various forms. Generally, nonmetallic piping is limited to situations where the risk of fire is low. Glass piping can be used to temperatures of about 230°C and to about 700 kPa (100 psig). When product purity is of prime importance or the fluids are highly corrosive, glass piping has no equal.

For pressures above the capabilities of nonmetallic piping and where corrosion or temperature is controlling, internally lined metal piping may be used.

Design parameters as a function of temperature and design temperature limits are set forth in the ANSI/ASME B31 Piping Codes for a very broad range of materials. These codes, and the additional information available from manufacturers, vendors, and technical societies provide ample data for the selection of materials and safety considerations for piping systems.

Pipe Size

With respect to pipe size, piping systems may be divided into three classes. The first class comprises piping connecting equipment where pressure is set by process requirements and where the pressure loss in the piping system does not determine the pressure-rise requirements of pumps (qv) or compressors or the elevation of towers, drums, or other pieces of equipment. Pipe sizes are generally chosen as the minimum size able to carry the maximum flow rate required at a pressure loss equal to the pressure available.

The second class comprises piping systems where pressure loss determines all or part of the pressure rise developed by pumps or compressors. In these systems the choice of pipe size is strongly influenced by the overall pump (or compressor) and piping economics. Increasing the pipe size reduces the pressure loss, thereby reducing the investment and operating costs of the pump (or compressor), but increasing the piping cost. Reducing the pipe size obviously has the opposite effect.

The third class comprises piping systems where pressure loss contributes to setting the elevations of towers and drums; for example, piping systems connecting tower-bottom draw-offs to pump suctions and piping systems between towers and reboilers. Pumps require a specific minimum net positive suction head (NPSH), and for liquids at their boiling or bubble-point temperature, the NPSH available consists of the static head of the liquid above the pump suction less the velocity and the friction head loss in the piping. Increasing the tower elevation increases the costs for the tower support, whereas increasing the available NPSH enable a less expensive pump to be used. Here again, an economical balance must be found that minimizes the total system cost.

The driving force causing circulation in thermosiphon reboilers is derived from the difference in density between the liquid in the piping from the tower to the reboiler and that of the liquid–vapor or vapor in the piping from the reboiler to the tower. Sufficient pressure must be developed to overcome the pressure loss in the reboiler and in the piping.

In order to select the pipe size, the pressure loss is calculated and velocity limitations are established. Analytic equations (such as Darcy) are based on the pipe length and diameter, flow velocity, and a friction factor. For laminar flow (Reynold's number, $Re < 2000$), generally found only in circuits handling heavy oils or other viscous fluids f the Fanning friction factor, $f = 16/Re$. For turbulent flow, the friction factor is dependent on the relative roughness of the pipe and on the Reynolds number.

There are also many empirical formulas used for calculating the friction head loss in piping systems. These must be used carefully because many are based on the properties of specific fluids and are not applicable over a broad range of fluids, temperatures, and pressures. For example, the Hazen and Williams formula is widely used for water flow.

Valves

Valves in piping systems are employed for on-off service (gate, plug, and ball valves), for controlling the fluid-flow rate (globe, needle, angle, butterfly, and diaphragm valves), and for ensuring unidirectional flow (check valves).

For any given application of any type of valve, temperature, pressure, and corrosivity must be considered in the same manner as for the piping system itself. Valve vendors specify the temperature, pressure, and general service limitation for their valves and these are indicated in the manufacturers' catalogues.

The operation of system valves (also starting and shutdown of pumps) has a significant effect on the transient fluid pressures in the piping system because of the acceleration and deceleration of the fluid as it changes its velocity. The basic wave velocity α is a function of the bulk modulus and density of the fluid, where K = bulk modulus, Pa and ρ = fluid density, (kg/m^3): $\alpha = (K/\rho)^{1/2}$, and the maximum head rise caused by the instantaneous closing of a valve is given by $h_{\max} = \alpha v/g_c$, where $h_{\max}$ = maximum head rise of fluid, m, and α = pressure wave velocity in the fluid, m/s.

Mechanical and Structural Standards

The design code to which the piping system must usually conform is the American National Standards Institute/American Society of Mechanical Engineers (ANSI/ASME) B31 Code for Pressure Piping. The section of the code that must be followed is determined by the general service for which the piping is intended. These sections establish definite rules as to the design formula that must be used. They specify allowable stresses for the piping material, joint efficiencies, and minimum allowances for threading, mechanical strength, and corrosion or minimum thickness. The codes also specify how the combined effects of different loading conditions, eg, pressure, thermal expansion, weight, wind, and earthquake, should be evaluated.

Pipe-Wall Thickness. Once the design pressure and temperature have been established and the pipe material and size selected, the wall thickness is calculated using the appropriate section of the code. All codes prescribe essentially the same design formula for metallic hollow circular cylinders under internal pressure.

Supports and Restraints

In addition to wide ranges of movements, piping systems may be subject to operating conditions involving cyclic or local temperature variations or may be influenced by unusual startup conditions or emergency shutdowns. Therefore, the support design must be reasonably adequate and effective for any set of circumstances.

Supports. The spacing of supports is governed by the hot allowable stress of the piping materials; stability, in the case of large-diameter thin-wall pipe; deflection to avoid sagging or pocketing; and the natural frequency of the unsupported length to avoid susceptibility to undesirable vibration. A particular type of support assembly is selected according to the amount of restraint tolerable by the piping system and the movement to be allowed at each location.

Restraints. A restraint is a device preventing, resisting, or limiting the free thermal movement of a piping system. Because the application of a restraint reduces the inherent flexibility of the piping, its effect on the system is established through calculation.

Anchors. Anchors provide full restraint against the three deflections and three rotations relative to the principal axis. They usually are subject to large loadings and are often required to develop the full strength of the attached pipe. Their design must, therefore, be sufficiently rigid.

Attachments. Connections or devices attached directly to the surface of the pipe transmit thermal as well as dynamic or weight loads from the pipe shell to the restraining, bracing, or supporting fixtures. Pipe attachment devices are either integral or nonintegral with the shell of the pipe. Nonintegral pipe attachments are used widely for support. Integral pipe attachments are welded directly to the pipe and are usually of special design. They are used in anchors and in high temperature services, with moderate or severe loads, and in conjunction with supports, braces, and restraints where rigidity with the pipe shell is desired.

Piping supports, guides, and anchors increase local stresses on the pipe wall at the point of attachment. Code-allowable stresses are conservative with respect to structural failure that occurs when the limit load is reached.

System Flexibility

The need to ensure that the stresses in piping systems meet the appropriate code requirements and the concern that cyclic stresses resulting from events such as periodic heating and cooling of the piping may lead to fatigue failures, make accurate evaluation of the stresses and strains in piping systems a necessity.

Rigid Systems. The theoretical analysis essentially involves the analysis of a three-dimensional statically indeterminate structure. The piping system is simply a structure composed of numerous straight and curved sections of pipe. Although, for straight pipe, elementary beam theory is sufficient for the solution of the problem, it is not adequate for curved pipe. By the introduction of a flexibility factor, k, to account for increased flexibility of curved pipe over straight pipe, and a stress intensification factor, i, to account for the increase in stress in a curved pipe or any other piping component over that predicted by beam theory, the elementary beam analysis can be used.

A piping system can be evaluated for its displacement and stress either by manual methods (charts, tables, hand calculation) or computerized solution. The latter has become the standard approach; manual methods are used only for rough estimates on very simple systems.

Computerized Solutions. The systematic analysis of a piping system by the flexibility matrix method using elbows and straight pipe sections as building blocks was developed long before the advent of computer technology (qv). In the flexibility method, a basic anchor is assigned and the entire system is allowed to move relative to this basic anchor. The forces and moments that must be applied to the other anchors and restrained points to force those points back to their installed locations are then determined.

In contrast to the flexibility method, the stiffness method considers the displacements as unknown quantities in constructing the overall stiffness matrix.

This method has a simple straightforward logic for even complex systems. Multinested loops are handled like ordinary branched systems, and it can be extended easily to handle dynamic analysis.

The most recent developments in computational structural analysis are almost all based on the direct stiffness matrix method, for example, piping stress computer programs such as SIMPLEX, ADLPIPE, NUPIPE, PIPESD, and CAESAR.

Flexibility and Stress-Intensification Factors. The flexibility factor k is defined as the ratio between the rotation per unit length of the part in question produced by a given moment to the rotation of a straight pipe (of the same size and schedule) produced by the same moment.

The Stress-Range Concept. The solution of the problem of the rigid system is based on the linear relationship between stress and strain. If this relationship is nonlinear, an elementary problem, such as a single-plane two-member system, can be solved, but only with considerable difficulty. Using linear analysis in an apparently nonlinear problem is justified by the stress-range concept as it is applied to piping systems. Because of the cyclic nature of the piping stress problem, stress levels should be based on fatigue considerations. The ANSI B31.3 code for piping assigns stress levels for the allowable stress range based on the total number of cycles.

Semirigid and Nonrigid Systems. The calculated results of flexibility analysis with which the designer is concerned are the stress levels and movements at significant points within the piping system, and the magnitude and direction of forces and moments at terminal equipment. When stresses or reactions exceed their allowable limits, modification of the piping arrangement or of the physical cross section of the pipe is necessary.

The simplest method of reducing stresses and reactions is to provide additional pipe in the system in the form of loops or offset-bends.

If a rigid system is impractical, the piping configuration may be made more flexible with hinge, rotation, or translatory joints. Systems using such devices are commonly classified as semirigid or nonrigid depending on the degree of flexibility.

The movement-absorbing devices used in semirigid and nonrigid piping systems are called expansion joints. They are either of the packed type or the packless or bellows type. Selection depends not only on the required movement but also on the severity of service.

The selection and application of an expansion joint is not as simple as selecting a pipe fitting or a valve and requires a sound understanding of the joint's capabilities and limitations. Improper application of any type of joint can result in serious or damaging effects. However, when properly selected and integrated into the piping system, satisfactory service and safe operation can be expected. Selection and application of bellows expansion joints require special attention to design and installation.

Bellows can vibrate, both from internal fluid flow and externally imposed mechanical vibrations. Internal flow liner sleeves prevent flow-induced resonance, which can produce bellows fatigue failure in minutes at high flow velocities. Bellows should be selected with a natural from any expected forcing frequencies, if known. Bellows installations should always be inspected to ensure that temporary shipping bars are removed and that flow liners are installed in the right direction. Piping systems containing bellows are always pressure tested (hydraulic or pneumatic) without restraining the bellows in order to test main anchors and guides as well.

Environmental Aspects

The ANSI/ASME B31.3 Code for Pressure Piping recognizes that special provisions are needed for piping conveying toxic materials.

The Clean Air Act requires control of fugitive emissions from the piping system. Estimates of fugitive emissions, and controlled and uncontrolled process emissions, are generally required by the U.S. EPA in permitting new facilities.

STANLEY E. HANDMAN
Consultant

Flow of Fluids, Technical Paper 410, 1991: *Crane Companion*, ABZ, Inc. Chantilly, Va.

M. L. Nayyar and co-workers, *Piping Handbook*, 6th ed., McGraw-Hill Book Co., Inc., New York, 1992, pp. A227–A277.

The M. W. Kellogg Co., *Design of Piping Systems*, rev. 2nd ed., John Wiley and Sons, Inc., New York, 1964.

A. Ghali and A. M. Neville, *Structural Analysis—A Unified Classical and Matrix Approach*, 3rd ed., Chapman Hall, New York, 1990.

PLANT LAYOUT

Plant layout can be the single most important part of the overall design. A good layout can result in significant savings in erected plant cost. A layout can make the difference between a facility that is easy to operate and maintain, and a plant that is a nightmare. Many concepts are involved in designing a good layout. There are several different modeling concepts, including plastic scale models, three-dimensional computer-aided design (CAD) models, and isometric and orthographic presentations.

Importance of a Good Layout

Equipment layout and the design of the associated piperack (the elevated supporting structure used to convey piping between equipment) are critical elements of the plant design and require interfacing among many disciplines. A good layout engineer or designer can help provide realistic spacing and select equipment locations that do not need to be changed throughout the design. A cost-effective layout that provides for easy operability and quick turnarounds is often designed by someone with a background in the piping discipline (see PIPING SYSTEMS).

Development of a preliminary layout requires the following information: (*1*) process flow diagrams; (*2*) plot limits; (*3*) process unit rough area requirements; (*4*) storage and tankage requirements; (*5*) expected expansion requirements; (*6*) waste treating area, usually located at the low point of the plot; (*7*) rerun and product storage location (preferably close to process unit); (*8*) locations of roads, rail spurs, and pipeline tie-in points; (*9*) product and blending area (best if located close to sales loading area); (*10*) product tankage and loading (should be located near blending); (*11*) plant roads and access ways for considerations of maintenance access and plant constructability; (*12*) utility and steam generations (should be adjacent to process plot) if a significant requirement exists; and (*13*) cooling towers and electrical distribution substations (must be at plot periphery).

Plot Plans

A plot plan is the scaled drawing of the processing facility (Fig. 1). The preliminary plot plan is the first step in the layout process. The plot plan shows the location of the following items which are known or approximated: main equipment, main pipeway, major structures, housed electrical switch, control room, and tie-in connection locations. By using a computer-aided design (CAD) system, unit plot plans are easily developed from the plot plan file.

Initial Sketch. The process engineer develops a sketch on which is indicated which pieces of equipment will be located in elevated structures and shows which equipment should be located close by other equipment. Primary instrumentation is also shown. This information provides the designer with data to start the first cut of the layout in a rough sketch form.

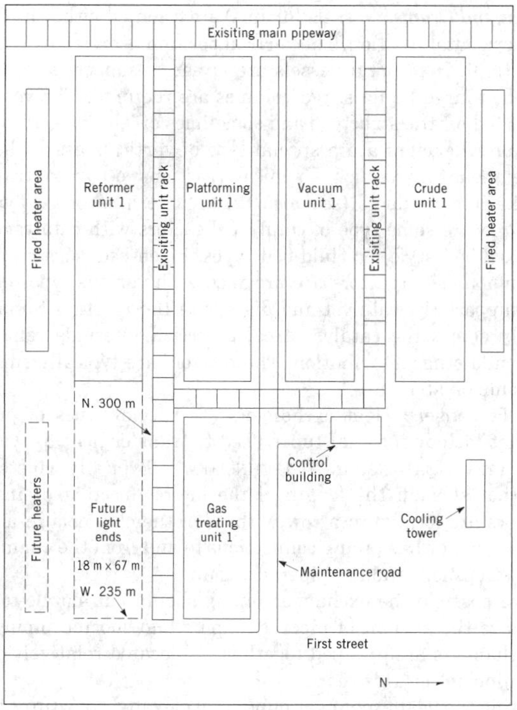

Figure 1. Overall plot plan.

The first-cut sketch is usually not drawn to scale but is roughly spaced out. The utilization of CAD on the first-cut layout can be beneficial when optimization of the plot is necessary.

Scaling the Sketch Example. Once the initial sketch is completed, it can be refined to include space requirements and shown to the approximate scale. The designer uses knowledge of spacing and clearances for safety and maintenance considerations. There are several rules of thumb on how much space to leave between each type of equipment. There is also a way of approximating the amount of total space needed in a plot area to provide for all the equipment.

Cut-and-Paste Method. Some experienced layout people go directly to a scaled preliminary plot plan by using a technique known as cut-and-paste. In this method, the background of the plot area is sketched on a blank drawing to scale. This background shows plot limits as well as existing roads and other objects within the facility of the proposed plot. All the equipment items are then cut out to a scale that matches the background plot scale.

Once the equipment has been located on the plot, it can be photographed for reference. Because many cases can be evaluated, references to these need to be documented. Client approval is also needed. Once the plot is finalized, a first pass can then be made at transposing the principal piping lines to determine if any significant piperack problem exists and whether equipment relocation is required.

Equipment Considerations

Every type of equipment requires certain layout considerations. These requirements address both operational and maintenance issues. Practical rules have been provided to guide the designer on the spacing requirements needed to address these concerns. Most of these are rule-of-thumb information based on reviews of existing designs from a maintenance and operation standpoint as well as practical engineering design rules and safety-in-design practices.

Fractionating Towers. Most petrochemical facilities multiple fractionating towers that perform separation of the products by boiling range. These towers are usually vertical vessels that contain trays or packing to perform the separation; they are supported by skirts or other structural means to provide the required vertical height.

Vessels and Drums. Vessels (drums) are oriented either in the vertical or horizontal position. The vertical position is preferred if the plot space is tight. Horizontal vessels are easier to support and are preferred when large liquid surge volumes are required. The vessel can be supported off the structure and sometimes off the rack.

Reactors. Reactors are a special type of vertical vessel; they provide the means by which chemical reactions occur to transform feedstocks into products (see also REACTOR TECHNOLOGY). Typically, reactors require some type of catalyst. Reactors with catalyst can be of the fixed-bed style for fluid-bed types. Fixed-bed reactors are the most common. The reactors are provided with various types of internals to support the catalyst and distribute the reaction components uniformly across the catalyst area; collection internals remove the products and other distribution. The reactors are typically supported from a common structure.

Heat Exchangers. Heat exchangers are special types of pressure vessel that include internal tubes used to transfer heat between two streams (see HEAT-EXCHANGE TECHNOLOGY). Typically clustered in groups that shorten the length of the interconnecting piping, heat exchangers are located in a row with all of the channels in a line. A typical exchanger has piping connections to and from the channel, the piperack, the shell, and the piperack again.

Where possible the exchanger piping should run nozzle-to-nozzle to minimize the amount of piperack piping and shorten piping runs. Heat exchangers are grouped together and located relatively close to other equipment for this reason.

Locating the heat exchanger bundles in elevated structures is often done where plot space is tight. These heat exchanger systems are often complex and the piping between the heat exchanger bundles is tightly compacted.

Heat exchangers are usually supported by two pedestals attached to the exchanger saddles. The saddle closest to the channel is the fixed support, which has the saddle bolted tight to the foundation pedestal. About two-thirds of the exchanger weight usually rests on this pedestal because it carries the weight of both the channel and part of the tube bundle as well as half of the shell. The rear end support saddle carries the rest of the weight. The saddle on the end away from the channel end is usually bolted loosely to the pedestal to allow the exchanger shell to grow thermally and expand as the unit heats up.

The location of exchangers is the key to maintenance. Access equipment must be able to get in and remove the shell cover and flange head.

Fired Heaters. In many European refineries, the overall plant layout is designed so that all heater exhaust gas is ducted to a centrally located 100-m high stack; in this manner release to the atmosphere is dispersed more broadly. Usually, fired heaters are located a minimum of 15 m from the hydrocarbon-containing equipment.

Heaters basically come in two types. The vertical cylindrical heater, which is usually bottom-fired and often has a self supporting stack, is popular from a space-saving standpoint. Larger-sized heaters are usually horizontal box heaters. The radiant coils can be located either on the side walls so that the units are fired from underneath, or in a center row of tubes in which the heater is fired from both sides. Environmental considerations may require a structure to provide a sampling access platform on the stack and the necessary analyzer housings.

Piping for snuffing steam injection into a heater firebox is required to help put out a fire if a tube rupture occurs. The snuffing steam isolation valve needs to be located at an accessible spot remote from the heater.

Compressors. There are two basic types of compressors: centrifugal and reciprocating. The centrifugal compressor is usually used for higher volume low head applications. The other type of compressor is the reciprocating machine; it is generally used for lower volumetric flow rates or when higher differential pressures are required. Compressors are expensive equipment items; they should be protected from the elements, and they are usually separated from the rest of the process facilities.

Pumps. Each pump is located as close to the suction source as possible. Both the main pump and the spare pump are frequently located on the same support foundation. Most in-house pumps require a pedestal for support. Space has to be provided around the pump for the piping and isolation valves as well as the electric supply.

The layout specialist should be aware of any special space requirements for a pump. Otherwise, pumps are usually fitted into a small area normally considered adequate for a general pump service.

Storage Tankage. Most tankages are located away from the main process area. On plot process, tankage is usually used only for nonflammable substances and the storage volumes are kept small to minimize the plot space taken up. Each site location must comply with local regulations in addition to general guidelines.

Tank storage of liquids at atmospheric conditions is generally classified into three classes: flammable liquids having a flash point below 100°F (38°C), flammable liquids having a flash point above 100°F but below 140°F (60°C), and combustible liquid having a flash point above 140°F and below 200°F (93°C). The National Fire Protection Association (NFPA) has developed criteria based on these classifications and the type of tankage used for storage (fixed roof, floating roof, and no roof). These guidelines must be followed in the early layout. It can be costly to move a tank after it has been constructed.

Cooling Towers. The cooling tower location relative to the prevailing wind direction should be such that the wind hits the short side or the side perpendicular to the inlet louvers. This helps balance the air flow to the two inlet sides.

The direction of the outlet plume relative to local roads should be considered. In foggy weather, the plume can cause the roadway to become a visual problem and a driving safety hazard.

Air to the cooling tower should be as cool as possible. Equipment that gives off heat should not be located upwind of the tower. Vented risers should be provided on most cooling towers to release only light hydrocarbon leakage from the cooling water before the spray header. No ignition or source of spark should be within 30 m of the vented riser.

Control Room. The control room location can be critical to the efficient operation of a facility. One prime concern is to locate it the maximum distance from the most hazardous units.

A central location where instrument leads are short is preferred. In modern facilities with distributed control systems, all units are controlled from a central control room with few operators.

Piperack Considerations

Piperack is considered the arterial system of the process plant (see PIPING SYSTEMS). All of the nonnozzle-to-nozzle pipe runs are made by running pipe to and from the piperack. Utility lines, electric lines, and instrumentation lines are often run on the piperack as well.

Pumps are usually located under the piperack. An open slot in the piperack has to be provided at the location of the pump suction and pump discharge lines so that the piping can make a straight run down to the pump suction and then the discharge line can be run directly back up to the piperack. The location of nozzles on the pump can affect the piping configuration.

Whenever a change in piping direction occurs, the elevation of the pipe run should also change. The piperack can also provide the support for air coolers and other equipment such as elevated drums.

CAD Models

Computer-aided design (CAD) (qv) technology has steadily improved. CAD models show only two-dimensional representations of three-dimensional models. However, the integration of all equipment, piping, and structural steel makes it easier to detect and interface between two components occupying the same space through interference detection checks.

Three-dimensional CAD model development requires some two-dimensional preliminary work such as plot plans. Individual pieces of equipment need to be dimensionally defined by creating an equipment model file with the special and geometric configuration generation. To these models the locations of externals are added, depending on the desired complexity of the model. This information is entered into the computer in various layers of data which permits the CAD designer to display any or all of these layers. Geometry of the structural steel, piperacks, and foundations are entered. Ultimately, piping instrumentation and electrical information is entered to the level of detail desired. The 3-D models can be viewed from almost any perspective by using the zoom-in display and walking through the model.

Scale Models. Replicas of a plant are often prepared as part of the design and can be in several forms: equipment only, equipment and major piping, or complete equipment and piping. Scale models are quickly being replaced by 3-D CAD models.

Noise Abatement

One of the prime OSHA requirements is control of the noise level, which can usually be met by using proper specifications and noise attention devices. Noise can cause hearing impairment and be hazardous to health. It also interferes with work efficiency.

Plant layout and noise suppression material are two general noise abatement methods. Plant layout does not affect noise levels at any given point; however, noise can be abated by screening off a section of the plant. The plot plan should be evaluated for noise generation and for finding the means of alleviating or moving noise to a less sensitive area.

ROLAND E. MEISSNER III
The Ralph M. Parsons Company

R. Weaver, *Process Piping Design*, Vol. 1, Gulf Publishing Co., Houston, Tex., 1973, p. 57 ff.

R. Kern, "Chemical Engineering Refresher: Plant Layout, Part 4: How to Find the Optimum Layout for Heat Exchangers," *Chem. Eng.* **84**(19), 169–178 (Sept. 12, 1977).

R. E. Meissner III and D. C. S. Shelton, *Chem. Eng.* **99**(4), 81–85 (Apr. 1992).

PLANT LOCATION

Selecting a plant site is critical to the financial success of a plant. Several factors must be considered in selecting a general plant site location.

Siting Factors

The primary siting factors which influence the selection of a plant location are as follows: environmental considerations, labor availability and productivity, raw material availability, proximity to market, property cost, accessibility to transportation, tax incentives, electric power availability and cost, and living conditions.

In the selection of a plant site, it is a good idea to get broad-based input, including information from sales, production, plant engineering, and from the general manager. The first objective is to narrow the range of possible choices. This involves focusing on the most important criteria, which differ widely for each type of facility. Table 1 presents some of the factors to consider for plant siting.

Environmental Considerations

No matter how advantageous a site location may be, if a permit to build cannot be obtained or the uncertainties in getting the necessary permits jeopardize the timing of a project, then it may be necessary to choose another site. Thus, environmental considerations may be, overall, the most important siting factor.

Many companies are opting to increase the efficiency of existing facilities and add more capacity to selected plants rather than to try to site new facilities, because of the difficulties of the environmental permitting process.

Labor Availability and Productivity

The availability of employees can constitute the overriding consideration in certain businesses, in relation to siting. Labor-intensive businesses have to either move to a location where labor is available or move their employees to the new plant site.

It is preferable to have access to a work force with developed skills that can be readily converted to suit the special requirements of a new business. Usually the factors that make a site desirable also include incentives which have already attracted a skilled work force that can be hired away from other, similar employers.

Raw Material Availability

For many industries, such as the petroleum (qv) and petrochemical industries, the accessibility of raw materials is the overwhelming factor in selecting a plant site.

Other industries that are traditionally located close to the source of raw materials include the steel (qv) industry, located close to iron ore; the flour industry, close to wheat fields; the meat-packing industry, close to grazing land for cattle; pulp and paper, close to forests; and the mining industry, close to mines. The inorganic industry, including salts, ash, borax, and gypsum, has always been located near the source of the needed raw material.

Accessibility to Transportation

The plant needs to be located close to the market. The cost of shipping by tanker is lowest, pipeline is next lowest (see PIPELINES), and truck and railcar shipment is the highest. These last two means are sometimes the only options for some products.

Property Cost

Land cost in certain highly desirable petrochemical manufacturing areas can be high. Where a lot of land is needed for feed and product storage, the cost of land can be significant. Plants tend to locate away from areas where encroaching residential homes drive up the property cost.

Tax

When considering taxes, all types need to be considered: initial fees, capital value, corporate rate, personal income tax, sales tax, property tax, unemployment insurance, workmen's compensation, and nuisance tax. During the construction phase, several types of taxes may be levied. These include building permits, special fees, assessments, and sewer connection fees.

Electric Power Availability and Cost

Several industries are highly dependent on cheap electric power. These include the aluminum industry, the Portland cement industry, electrochemical industries such as plating and chlorine production, the glass industry, and the pulp and paper industry. Other industries such as the petrochemical industry, which is highly competitive, depend on low priced power. About two-thirds of the cost of producing ammonia is electrical cost.

These industries try to locate near a source of hydropower or near a source of excess nuclear power. They generally work out arrangements to get power at a reduced cost based on being the first one cut off when electric load shedding is required.

Site Purchase

Once the site is selected, the land purchase must be made. Usually a third party such as a real estate agent is employed to do this work.

Table 1. Factors to Consider for Plant Siting

Plant requirements	Labor	Transportation	Raw materials	Markets	Power
acreage[a]	supply	railroads	haul	schedule/rates	fuel supply
present/future expansion	proximity available	rates, sidings[b], shifting, transfers, belt line	quantity	shipping	electricity
building[a]	character	waterways	permanency	storage	water supply
drainage, foundation, subsoil, maintenance[b]	wages	boat lines, canals	quality	requirements	quantity/quality, boiler purposes, condenser water
natural features	h/d, overtime, bonus	highways	annual	(inventory)	water power
topography, climate, local supplies	disposition	motor traffic	consumption	competition	(hydroelectric)
public utilities[b]	union, content, strikes	passenger buses, freight lines	storage	shipping costs, custom charges	
streets, phones, light, heat, gas, power, telegraph, newspaper, fire protection	housing	street/railways	requirements		
water supply[a]	health, sanitation, markets, shopping, schools, churches, recreation	present/future schedules, fares, freight, express			
quantity/quality					
waste disposal[b]					
sanitary/industrial					
advertising value					
government					
character, legal permits, police protection, taxes, insurance[b]					
legal phases					
laws, status, ordinances, board of health requirements					

[a] Represents an investment cost.
[b] Represents an operating cost.

Contact between the plant owner and utilities, railroads, pipelines, and the local community need to be finalized in writing before a final purchase.

Environmental Permitting. An environmental baseline is usually required to establish how the plant would affect the surroundings. Air quality issues can be the deciding factor in siting, as air permits often take the most time to obtain. Water pollution also needs to be addressed. Federal, state, and regional regulations are often in conflict with each other. What the environmental discharges are and how best to mitigate their effects by providing a plant that is designed to minimize the impact are key issues.

Local Site Condition Evaluation. In addition to visiting the site, drawing up a contour map and geology reports, acquiring soil-bearing information, and a knowledge of boundaries, setbacks, local requirements, utility tie-in locations, sewer connections, access to roadways, pipelines, railroads, etc, may be needed to make a full assessment.

Specific Plant Site Considerations

Once a general plant location has been established, a specific plant site location is determined by developing a list of requirements for the plant. One approach is to create an ideal list of specifications that define what would be required to put the plant in any open area. This approach should be carried out by devising a preliminary plant layout (qv). Some of the considerations that should be listed as requirements are as follows: number of employees; size of process units; utility requirements, ie, for cooling water, power, water, fuel, and steam; shipping requirements, ie, road access, and access to railway lines, waterways, and pipelines; disposal requirements for solids waste, chemical sewer loads, and sanitary sewer loads; and permitting factors such as air emission, water emission, and those that are applicable to the particular type of product.

ROLAND E. MEISSNER III
The Ralph M. Parsons Company

W. B. Speir, "Choosing and Planning Industrial Sites" *Chem. Eng.* (Nov. 30, 1970).

D. V. Bierwert and F. A. Krone, "How to Find Best Site for New Plant" *Chem. Eng.* (Dec. 1995).

PLANT SAFETY

Safety assessments of entire processes began with quantification of over-pressure potential and flammability hazards, by measurements of vapor pressure and of flash points and flammability limits, respectively. Process designers make use of data pertaining to reaction rates and energies for exothermic reactions and unstable chemicals: temperature limits beyond which explosive decompositions or other undesirable behavior can occur; rates of gas or vapor generation for proper design of emergency pressure-relief devices; recommended limits for exposures to toxic materials, radiation, noise, and heat; and strengths and corrosion rates of materials of construction (see CORROSION AND CORROSION CONTROL). The application of fault-tree analysis to chemical processes provides a means for quantitatively combining characteristics of process hazards with component and human failure rates to obtain a safety assessment of a process (see HAZARD ANALYSIS AND RISK ASSESSMENT).

Pollution Prevention

The responsibility of chemical process managers for preventing air, water, and soil pollution has indirectly influenced plant safety by requiring better control of plant processes to prevent releases of hazardous materials. Regulatory legislation was introduced by the Health, Education, and Welfare Department (Health and Human Services) and the U.S. Environmental Protection Agency (EPA) to require (*1*) improvements in air quality; (*2*) better waste disposal practices (see WASTES, INDUSTRIAL; WASTE TREATMENT, HAZARDOUS

WASTE); (3) reduced noise levels; (4) improved control of the manufacture and use of toxic materials; and (5) assignment of responsibility to manufacturers for product safety.

Process Safety Management

In the United States, the Occupational Safety and Health Act (OSHA) established standards for several types of occupational hazards, including toxicity, noise, equipment guarding and protection against falling, and electrical shock. It also promulgated other consensus standards for exit facilities and fire and explosion control. Personnel from OSHA and the National Institute for Occupational Safety and Health (NIOSH) are available for consultation to identify, evaluate, and correct workplace hazards.

Regulations have been promulgated to codify process hazards management, and the Clean Air Act has been amended to provided further protection of the public.

Process and Production Hazards

Much of the legislation aimed at minimizing the consequences of occurrence of hazards has resulted in the adoption of a hierarchy of controls essentially falling into two categories, ie, engineering and administrative.

Engineering controls include those providing inherent safety and those involving process equipment and conditions.

The administrative controls are concerned with operating procedures, maintenance programs, process hazards analysis, limiting personnel exposure, and emergency procedures for escape and evacuation.

Chemical Hazards. Chemical manufacturers and employees contend with various hazards inherent in production of even commonplace materials.

OSHA regulations require that material safety data sheets (MSDS) be developed for all process materials, so that the hazard data can be communicated to employees. Characteristics of toxicity, flammability, chemical instability, reactivity and reaction energy, operating conditions, and corrosive properties of construction materials must all be considered in analyzing hazard potentials of chemicals and chemical operations.

Threshold Limit Value. The American National Standards Institute (ANSI) has published standards regarding the maximum acceptable concentration for certain gases and vapors in the air at work locations. A list of threshold limit values (TLVs), published annually by the American Conference of Governmental Industrial Hygienists (ACGIH), provides the concentrations of dust, mist, or vapor believed to be harmless to most workers when exposed for 5 8-h days per week.

Control of Exposure Potential. Exposure to toxic materials can be controlled by a number of methods, eg, substitution, removal, enclosure, and personal protection. The best method of protecting workers is by substituting a less toxic material for a more toxic substance having equal effectiveness.

Flammability. Engineering and operational controls are usually effective in preventing fires involving flammable materials. The basic method of fire prevention is to avoid situations in which flammable materials, air, and ignition sources are in the same place at the same time.

Hazard identification of the contents of in-plant bulk storage tanks, warehouses, etc, may be achieved by a system developed by the National Fire Protection Association. The system makes use of three diamond-shaped areas, which are marked with numbers 0, 1, 2, 3, or 4 indicating increasing hazards of toxicity, flammability, and reactivity, respectively.

Design of Facilities

Plant Site and Layout. The choice of a location for a chemical plant depends on a number of factors, including effects on plant personnel and the surrounding community (see PLANT LAYOUT; PLANT LOCATION). The assessment of hazards, based on the flammability of materials, reaction energy, and presence of highly toxic materials, is important. Consideration is given to possible effects on plant personnel and the community from the worst possible incident (see HAZARD ANALYSIS AND RISK ASSESSMENT). An adequate water supply for process cooling and fire fighting is a vital necessity. Prevailing winds should also be considered.

Open areas around the operating units of a plant act as buffers to the surrounding community. Sufficient clearance should be allowed so that, if tall structures collapse, other on-site buildings or equipment, or off-site properties are not affected. Adequate roadways providing entry to the plant are extremely important, and multiple entries and exits are advisable. An overcrowded plant can lead to damage or shutdown of adjacent units and may impede the movement of vehicles and materials in case of emergency. Another consideration is community fire-fighting assistance, first aid, and medical facilities.

Hazard potentials should always be segregated from nonhazardous operations such as offices, laboratories, and warehouses.

Utilities. Principal electric power lines that are run underground reduce the probability of damage from exterior causes.

Generally, it is more economical to prevent explosive atmospheres in rooms than to try to provide explosion-proof electrical equipment. Personnel should never be allowed to work in a hazardous atmosphere.

Water mains should be connected to plant fire mains at two or more points, so that a sufficient water supply can be delivered in case of emergency.

For large plants located at natural water sources, special water mains can be used to supply untreated water for emergency use as well as for cooling, process water, or general plant uses other than human consumption and emergency washing facilities.

For indoor operations it may be necessary to remove toxic or flammable gases or vapors or process-generated atmospheric heat. Ventilation and heat-stress standards are intended to avoid hazards associated with high body temperature, heat exhaustion, heat stroke, or discomfort in processes generating high ambient temperatures.

Pressure Vessels and Piping. Some of the most critical components of a chemical plant involve pressure vessels. A thorough knowledge of the American Society of Mechanical Engineers (ASME) Pressure Vessel Code is essential for design and maintenance of chemical plants. Some states have their own codes, which usually conform closely to the ASME version (see HIGH PRESSURE TECHNOLOGY; TANKS AND PRESSURE VESSELS).

Materials Handling.

Liquids. Liquids usually are moved through pipelines (qv) by pumps. Special alloys, plastic pipe and liners, glass, and ceramics are widely employed in the chemical industry for transport of corrosive liquids.

Piping design requires consideration of the maximum pressures, temperatures, and flows that might be attained, the corrosive and erosive nature and the viscosities of the materials passing through the piping, the distances between the inlet and discharge points, and the external force and vibrational stresses to which the piping might be subjected.

Valves used to isolate sections of piping and to control the flow of materials through piping must be compatible with the materials being handled.

Drainage valves should be provided at the low points of the plant system, and all piping should slope downward toward them.

Solids. Equipment for transporting and feeding solid materials include belt, flight, and screw conveyors (see CONVEYING), and pneumatic systems. Common hazards associated with solids handling are dust explosions and the escape and dispersion of noxious or combustible dusts (see POWDERS, HANDLING). Two items are of critical importance in designing pneumatic conveying systems. First, if the material is combustible, inert gas should be used as the conveying fluid, or blowout panels or vents should be provided to avoid explosion damage. Second, where poisonous or noxious materials are being transported, special attention must be given to recovery of fines from the exit air.

Electric or fuel-powered means of transporting solid materials, such as fork-lift trucks, should be employed only when full consideration has been given to any hazardous atmospheres in which these might be used. Such transport must be properly maintained to preserve the integrity of built-in safety devices.

Operation of Facilities

Normal Operation. The designer of a chemical plant must provide an adequate interface between the process and the operating employees. This is usually accomplished by providing instruments to sense pressures, temperatures, flows, etc, and automatic or remote-operated valves to control the process and utility streams (see FLOW MEASUREMENT; TEMPERATURE MEASUREMENT).

Maintenance. Maintenance (qv) in the chemical industry differs from that in other industries because of the nature of the materials, processes and types of equipment used. Because much chemical work involves the movement of fluids, gases, and powdered solids from one piece of equipment to another, many pipelines (qv), conveyors, forklift trucks, and other material-handling devices are used. Prior to maintenance of inside equipment, all lines and equipment containing hazardous materials should be effectively separated, disconnected, blanked, or purged in order to minimize the possibility of release of harmful materials.

Product Handling

Labeling. In the United States the Federal Hazardous Substance Labeling Act requires that all containers sold to consumers be labeled with appropriate precautionary wording to protect the user and employees from injury resulting from contact with the chemical.

Sampling. The first consideration in sampling (qv) is protection of the person performing the sampling. Line sampling usually is carried out at a suitable valve. The sampler should be aware of the possibility of a sudden increase in flow or of the flow of high pressure or high temperature material when the valve is opened. Operating personnel should always be told that a sample is being taken.

Storage. Liquid products may be stored in tanks at isolated tank farms, or in drums or cans in warehouses (see TANKS AND PRESSURE VESSELS). A source of danger in bulk storage occurs when workers enter tanks or silos for maintenance or other duties. Gases or vapors above the stored materials can cause intoxication, asphyxiation, or explosions, and collapse of the stored materials could cause engulfment and suffocation.

In any warehousing operation, it is essential that incompatible substances be isolated to avoid a reaction in case of a spill or fire.

In the design of warehouses in which flammable or combustible materials are to be stored, consideration should be given to the installation of fire walls, fire doors, and duct shut-off dampers. Automatic sprinklers are standard equipment in such locations. Fire-detection devices such as flame-sensing or ionization-interference types operate much more rapidly than sprinkler heads and are used extensively both as alarms and to activate fixed fire-extinguishing systems.

Disposal. Disposal of hazardous waste must be carried out in accordance with precautions against fire and explosion hazards, severe corrosion, severe reactivity with water, toxic effects, and groundwater pollution. The chemical industry strives to minimize the generation of waste materials.

Human Relations

Personnel Selection and Training. The quality of operating personnel is of paramount importance to the safe operation of a chemical plant. Operators must be intelligent and emotionally stable. One successful method of safety training involves the preparation of a manual of standard operating procedures for each unit or plant. Another method involves job-safety analysis.

Medical Programs. Large chemical plants have at least one full-time physician who is at the plant five days a week and on call at all other times. Smaller plants either have part-time physicians or take injured employees to a nearby hospital or clinic by arrangement with the company compensation-insurance carrier.

Clinical tests can and should be made prior to employment or work assignment and at frequent intervals thereafter, when employees are exposed to hazardous operations.

First Aid and Rescue. Thorough knowledge of first aid, as taught in courses by the American Red Cross or the U.S. Bureau of Mines, should be a primary part of chemical plant training programs. Rescue techniques also should be taught and practiced.

Fire and Explosion Prevention and Protection

Fire and Explosion Prevention. Prevention of fire and explosion is an important consideration in the design of chemical plants. Such prevention involves the study of material characteristics and processing conditions to determine appropriate hazard avoidance methods.

Fire and Explosion Protection. Extinguishment or control of fire is essential. Exposure of personnel to thermal-radiation hazards must be minimized and property protected. Water is the preferred fire-control medium. Designs for automatic sprinkler protection against specific hazards and general area coverage have been well developed and tested. Foams (qv) generally are formed by adding natural proteins or similar synthetic materials and aerating at nozzles to make a blanket, which floats on flammable materials. Because the foam excludes air and reduces volatilization, it can be used to cover spills and, thus reduce the potential for fire, as well as to extinguish existing fire.

Protection against explosions is typically provided by explosion-venting, using panels or membranes which vent an incipient explosion before it can develop dangerous pressures. Protection from explosions can also be provided by isolation, either by distance or barricades.

Disaster Planning. Plant managers should recognize the possibility of natural and industrial emergencies and should oversee formulation of a plan of action in case of disaster. The plan should be well documented and be made known to all personnel critical to its implementation. Practice fire and explosion drills should be carried out to make sure that all personnel, ie, employees, visitors, construction workers, contractors, vendors, etc, are accounted for, and that the participants know what to do in a major emergency.

Control of Process Hazards

Process Hazards Analysis. Analysis of processes for unrecognized or inadequately controlled hazards (see HAZARD ANALYSIS AND RISK ASSESSMENT) is required by OSHA. The principal methods of analysis, in an approximate ascending order of intensity, are what-if; checklist; failure modes and effects; hazard and operability (HAZOP); and fault-tree analysis.

Accident Investigation. A study of all accidents and injuries with the objective of determining the cause or causes can lead to correction of unsafe practices or conditions and prevent recurrence of the accident.

Center for Chemical Process Safety. The Center for Chemical Process Safety (CCPS) of the American Institute of Chemical Engineers has published a substantial library of "Guideline" books, covering many chemical-plant safety subjects.

RICHARD W. PRUGH
Process Safety Engineering, Inc.

Annual Loss Prevention Symposium, American Institute of Chemical Engineers, New York, 1st, 1967; 32nd, 1998.

Loss Prevention, Vols. 1–14, American Institute of Chemical Engineers, New York, 1967–1981.

Plant/Operations Progress, Vols. 1–11, American Institute of Chemical Engineers, 1982–1992; *Process Safety Progress*, Vol. 12, 1993—present.

International Symposium on Loss Prevention and Safety Promotion in the Process Industries, 1st, European Federation of Chemical Engineering, The Hague/Delft, the Netherlands, 1974; 2nd, Heidelberg, Germany, 1977; 3rd, Basel, Switzerland, 1980; 4th, Harrogate, U.K., 1983; 5th, Cannes, France,

1986; 6th, Oslo, Norway 1989; 7th, Taormina, Italy, 1992; 8th, Antwerp, Belgium, 1995.

PLASMA TECHNOLOGY

Plasma can be broadly defined as a state of matter in which a significant number of the atoms and/or molecules are electrically charged or ionized. The generally accepted definition is limited to situations wherein the numbers of negative and positive charges are equal, and thus the overall charge of the plasma is neutral. This limitation on charge leaves a fairly extensive subject area. The vast majority of matter in the universe exists in the plasma state. Interstellar space, interplanetary space, and even the stars themselves are plasmas.

Plasma technology involves the use of natural or artificially produced plasmas. Many different types of plasmas exist or can be naturally or artificially created by various energy sources. Whereas gas plasmas are the most common and the most commonly used, plasmas also exist and find application within solids as well as liquids.

Definitions. When positive charges are fixed in a solid, but the electrons are free to move about, the system is called a solid-state plasma. In a liquid-state plasma, both the positive and negative charges are fully mobile. These solid-state and liquid system are examples of condensed matter plasmas as opposed to gaseous plasmas.

Gaseous plasmas sustained by electric fields at reduced pressure, either under direct or alternating current, are sometimes referred to as glow discharges, because they emit light. There is, in fact, a slight difference between the terms plasma and discharge. However, in general, the two terms are used interchangeably throughout the plasma technology literature.

An important characteristic of plasma is that the free charges move in response to an electric field or charge, so as to neutralize or decrease its effect. Reduced to its smallest components, the plasma electrons shield positive ionic charges from the rest of the plasma. The Debye length, λ_D, given by $\lambda_D = (kT/4\pi ne^2)^{1/2}$ where n is the electron density and e is the electron charge, is a measure of the extent to which plasma electrons collect in the vicinity of a charge and create a shielded potential. Ordinary, nonplasma gases have an average interparticle spacing of $d = n^{-1/3}$ and a mean free path between collisions of $\lambda = (\pi n D^2)^{-1}$ where D is the atomic or molecular particle size. The number of electrons within a Debye sphere, N_D, around an ion is then $N_D = 4\pi\lambda_D^3 n/3$. The volume of a plasma must be significantly larger than N_D or else the shielding would not be complete, a charge imbalance would exist, and the definition of a plasma having overall charge neutrality would be violated.

Among the special frequencies associated with plasmas, the most notable is the plasma frequency $\omega p = (4\pi ne^2/m)^{1/2}$, where m represents the mass of the particles. This frequency is a measure of the vibration rate of the electrons relative to the ions which are considered stationary. For true plasma behavior, plasma frequency, ωP, must exceed the particle-collision rate, f. This plays a central role in the interactions of electromagnetic waves with plasmas. The frequencies of electron plasma waves depend on the plasma frequency and the thermal electron velocity. They propagate in plasmas because the presence of the plasma oscillation at any one point is communicated to nearby regions by the thermal motion. The frequencies of ion plasma waves depend on the electron and ion temperatures as well as on the ion mass. Both electron and ion waves, ie, electrostatic waves, are longitudinal in nature; that is, they consist of compressions and rarefactions (areas of lower density, eg, the area between two compression waves) along the direction of motion.

Gaseous Plasmas. Gaseous plasmas are sometimes classified as equilibrium or nonequilibrium referring to the electron temperature as compared to the gas temperature. Low pressure glow discharges are generally classified as nonequilibrium plasmas because the electron temperature is significantly greater than the gas temperature. Commonly, glow discharges have electron temperatures in the 10^4–10^5 K (1–10 eV) range, whereas the gas temperature in those discharges is generally less than 5×10^2 K or near ambient. Low and high pressure arc discharges, also known as plasma jets, have no such large difference and the electron and gas temperatures are both in the 5×10^3–10^4 K range, thus being called equilibrium plasmas. The areas of most interest to plasma chemistry are the glow discharges and arcs.

Gaseous plasmas are often far from equilibrium and therefore can exhibit microscopic or particle instabilities, and macroscopic or hydromagnetic instabilities. Microscopic instabilities are caused by departures from the equilibrium Maxwellian distributions for the electrons or ions. Examples of such situations include a plasma expanding while cooling, anisotropies in the velocity distribution caused by applied magnetic fields, or the motion or streaming of a particle beam through a plasma. Macroscopic instabilities produce the motion of the plasma as a whole. Causes include pressure or density gradients or magnetic field curvature. All instabilities represent the tendency of plasmas to reach equilibrium more quickly than is possible by ordinary collisions alone. Instabilities can reduce plasma confinement times by many orders of magnitude, a significant problem in fusion research.

Production

Sources of matter and energy are necessary for the production of gaseous plasmas, and such plasmas serve as sources of matter and energy in their applications; ie, gaseous laboratory plasmas can be viewed as transducers of matter and energy. The initial and final forms of the material that enter a plasma and the requisite energy vary widely, depending on the particular plasma source and its utilization.

The molecules that are dissociated and the atoms that are ionized during plasma production can be in any state at the start. Steady-state plasmas are formed most often from gases, although liquids, such as volatile organics, and solids are also used. Gases and solids routinely serve as sources of material in pulsed plasma work.

The energy for plasma formation may be supplied in a variety of ways. The source may be internal, eg, the release of chemical energy in flames. Another energy source involves electrical excitation. Electromagnetic fields, eg, in the radio frequency (rf) or microwave range, can be used to form plasmas that interact with or feed back into the source. A fourth kind of energy source includes externally produced beams of photons, eg, laser beams or other energetic particles, that create a plasma by impact and absorption independent of the source. Plasmas may be produced in a fifth manner by strong shock waves.

Modification and Dissipation. Changes in the composition or energy of a plasma after it is formed are often desired. For example, materials can be introduced into a plasma and excited, thereby producing information for spectrochemical analysis. Plasma heating is a more common modification. Plasma heating, ie, raising the energies of all the particles comprising the plasma, is of primary importance or interest in fusion science and technology. Many energy sources and coupling mechanisms can be employed to heat plasmas. Laser beam and particle beam sources often are used to heat plasmas produced by electric discharges.

Restraining a gaseous plasma from expanding and compressing is also a form of plasma modification. Two reasons for plasma confinement are maintenance of the plasma and exclusion of contaminants. Plasmas may be confined by surrounding material. A second approach to confinement involves the use of magnetic fields. The third class of confinement schemes depends on the inertial tendency of ions and associated electrons to restrain a plasma explosion for a brief but useful length of time.

Low density plasmas are confined magnetically by a variety of field configurations that are designed to prevent particle losses and overall fluid instabilities. Magnetron sputtering sources include a variety of magnetic field configurations designed to restrain plasma particles near the cathode from which atoms are sputtered by impact of ions.

High density plasmas can be confined and compressed magnetically by fields produced by strong electric currents flowing in and heating the plasmas, as well as by externally applied fields.

Diagnostics

Plasma diagnostics, the determination of conditions within plasmas, also refers to the broad collection of experimental techniques and associated calibration and analytical methods used to assess the characteristics of plasmas, such as the identities, concentrations, and energy distributions of the various particle species and their velocity distributions as functions of space and time. Most diagnostic methods have limited resolution spatially, temporally, and spectrally. Therefore, plasma characteristics that are derived from measurements generally are averaged. Measured quantities in plasma diagnostics usually are integrated along a line of sight through the plasma to the instrumentation, yielding spatially integrated results. A variety of diagnostic tools is needed to characterize a plasma empirically and to compare its empirical and theoretical characteristics.

Diagnostic methods can be categorized most broadly according to those that involve external probes of plasmas or those that rely only on plasma self-emission. Probes of plasmas include solid instruments inserted into gaseous plasmas and beams of photons or charged particles, which are shot through plasmas. Electric voltage and current measurements for discharge-heated plasmas provide useful information, thus the energy source also is a plasma probe. Self-emission can include radiated fields, photons, electrons, ions, and neutrons, ie, essentially any plasma effect or constituent. Plasma diagnostics is, by necessity, a key field of plasma technology.

Types of Plasmas

Natural Gaseous Plasmas. Lightning is the most common atmospheric, plasma related phenomenon. Spectroscopic data have yielded considerable information on plasma conditions within a lightning discharge.

Meteors produce atmospheric plasmas as their kinetic energy is converted to thermal energy.

Auroras are observed primarily at polar latitudes near the geomagnetic poles and result from impact on the atmosphere of energetic particles that are guided by the earth's electromagnetic field. Ionization densities that produce auroras generally reach plasma levels following intense solar flare activity.

Absorption of solar uv radiation high in the atmosphere produces a tenuous but important plasma, the ionosphere. The ionosphere is part of the larger magnetosphere, a cavity in the stream of particles from the sun. The cavity is produced by the earth's magnetic field.

Magnetospheric plasmas are produced and heavily influenced by solar emissions and activity and by magnetic fields of the planets. Interplanetary plasmas result from solar emission processes alone.

Plasmas in Condensed Matter. In contrast to gaseous plasmas, which can be described by kinetic and fluid theories, plasmas in condensed matter are intimately related to the theory of solids. The formation of a diatomic molecule is a good example of the freedom and high velocities that electrons experience in the condensed state. Electrons that initially are confined to either atom, once the atoms bond, are free to range over the larger molecular volume. In the buildup of larger aggregates of condensed matter, addition of more atoms similarly expands the accessible volume. Bonding electrons increase their velocities when atoms form into molecules and condensed matter.

Because electrons are fermions, no two can occupy the same spatial and energy states simultaneously, according to the Pauli exclusion principle. Energy bands develop as more atoms agglomerate, forming a solid. These may be continuous, as for metals, or have an energy gap. Such gaps are narrow for semiconductors and wide for insulators.

In perfect semiconductors, there are no mobile charges at low temperatures. Temperatures or photon energies high enough to excite electrons across the band gap, leaving mobile holes in the Fermi distribution, produce plasmas in semiconductors. Thermal or photoexcitation produces equal numbers of electrons and holes, similar to the neutral charge situation in gaseous plasmas where the charges of electrons and ions usually balance.

Impurity-produced plasmas in semiconductors do not have to be compensated by charges of the opposite sign. These plasmas can be produced by introduction of either electron donors or electron scavengers, ie, hole producers, into semiconductor lattices. Plasmas in semiconductors generally are dilute, so that the Pauli principle cannot be used to determine their energy distribution. These plasmas are Maxwellian and have the same temperature as the lattice in equilibrium.

Plasmas in gaseous and condensed states are related by more than the principles that govern them. For gaseous matter, there is a continuum of behavior from a low density Maxwellian plasma to a high density fermion plasma.

Fundamental differences between gaseous and condensed plasmas include their states of excitation and characteristic lengths. Gaseous plasmas are produced by classical collisional effects. In contrast, solid-state metallic plasmas are produced by quantum effects, can be in or near the ground state, and can exist in the absence of photons. Instabilities are common in gaseous plasmas but can be avoided or excited as desired in condensed-state plasmas.

Uses

Radiation Sources. Most ion sources involve plasmas.

Pulsed plasmas containing hydrogen isotopes can produce bursts of alpha particles and neutrons as a consequence of nuclear reactions. The neutrons are useful for radiation-effects testing and for other materials research.

Plasmas frequently are used as sources of incoherent and coherent electro-magnetic radiation. Visible light sources involving steady-state plasmas, in which the electrons are hot compared to the ions, are common. Multimillon-degree plasmas provide uniquely short pulses of incoherent uv and soft x-radiation for spectroscopy, materials analysis, and other applications.

Chemistry. The material and energy available in plasmas can be used to excite materials and drive chemical reactions. The unique characteristics of plasmas, especially their abundance of energetic species, have been exploited in plasma chemical applications.

The analysis of existing materials and the production of new chemicals and materials involve the gamut of gaseous plasma sources. Nonequilibrium or cold plasmas, in which the ion or gas temperature is much less than the electron temperature, are widely used. Equilibrium or hot plasmas, in which the electrons and ions are characterized by approximately the same temperature, also are used in plasma chemistry, especially for spectrochemical analysis and the processing of refractory materials (see REFRACTORIES).

Chemical Analysis. Plasma oxidation and other reactions often are used to prepare samples for analysis by either wet or dry methods. Plasma excitation is commonly used with atomic emission or absorption spectroscopy for qualitative and quantitative spectrochemical analysis.

Use of glow-discharge and the related, but geometrically distinct, hollow-cathode sources involves plasma-induced sputtering and excitation. Such sources are commonly employed as sources of resonance-line emission in atomic absorption spectroscopy.

In plasma chromatography, molecular ions of the heavy organic material to be analyzed are produced in an ionizer and pass by means of a shutter electrode into a drift region. The velocity of drift through an inert gas at approximately 101 kPa (1 atm) under the influence of an applied electric field depends on the molecular weight of the sample.

Plasma Processing. Plasma processing is an extremely broad and growing field. Radio-frequency gas plasma processing, is an environmentally conscious surface modification and material production technology.

Plasmas can accelerate reactions that are otherwise slow to the point of impracticality. Moreover, plasmas are often used to accomplish processes not possible by other means, eg, providing atomically clean surfaces of materials that would be damaged by high temperature or wet chemical cleaning. In the plasma processing generally used in industry, no hazardous wastes are generated and, in fact, plasma usage has proven effective at decomposing hazardous waste materials.

Plasma processing also offers several operational and cost advantages, eg, replacement of batch with flow processes. Plasma equipment often is smaller than other process hardware, resulting in associated savings in capital expenditure and floor space.

Surface Modification. Plasma surface modification can include surface cleaning, surface activation, heat treatments, and plasma polymerization. Surface cleaning and surface activation are usually performed for enhanced joining of materials (see METAL SURFACE TREATMENTS). Plasma polymerization crosses the boundaries between surface modification and materials production by producing materials often not available by any other method. In many cases these new materials can be applied directly to a substrate, thus modifying the substrate in a novel way.

Treating a surface with activated gas plasma is a dry process requiring no solvents or rinses of any kind. Using the correct choice of gases and process parameters, plasma cleaning can render a surface atomically clean of organic contaminants and/or activate a surface for enhanced bonding without damage to that surface.

Plasma Polymerization. Plasma polymerization has become a commercially viable method for modifying low cost organic substrates. Organic and other complex molecules that are exposed to low temperature nonequilibrium plasmas can be affected in terms of polymerization, rearrangements (isomerizations), surface activation, elimination of constituent parts, and total destruction of the original molecules accompanied by the generation of atoms and ions. The production of polymers and other heavier molecules from gaseous monomers is an attractive application of plasmas in organic chemistry. A wide variety of chemically inert, adherent films such as plasma-polymerized fluorocarbon films can be produced using simple, commercially available apparatus.

Materials Production. Substances not producible by conventional means can be made using plasmas. Plasma materials production and modification embraces processes such as production of thin coatings (qv), heat treatment, and the joining of materials.

Inorganic small molecules are produced in glow discharges and rf-induced plasmas by chemical vapor deposition (CVD), generally termed plasma-assisted CVD (PACVD) or plasma-enhanced CVD (PECVD). The molecules are introduced in the gaseous state and the products usually deposit on a chosen substrate.

Thin surface films can be produced by plasma-sputtering deposition, without chemical reaction (see THIN FILMS). Both single- and multiple-layer materials are produced. Especially important is the use of sputtering to produce multilayered microstructures consisting of dozens of layers of two elements or compounds.

The manufacture of semiconductor chips, wafers, and devices makes extensive use of plasma processing. Another common and important use of plasma is spray coating of materials with plasma-melted substances. In addition to coating existing structures by plasma spraying, it also is possible to build composite materials by spraying fibers or whiskers with a binding substance (see COMPOSITE MATERIALS).

Guns and Missiles. The rapid burning of powder in a gun barrel produces relatively cold plasmas which eject the projectile on a ballistic trajectory. Missiles carry a propellant which burns during flight, generating motion by high velocity ejection of mass. Modern missiles contain liquid or solid propellants having high energy densities. The chemical reactions that occur during their burning produces plasmas in the reaction chamber and exhaust nozzles.

Most modern projectiles and virtually all missiles contain explosives. The plasmas that result from explosives are intrinsic to operation of warheads, bombs, mines, and related devices. Nuclear weapons and plasmas are intimately related. Plasmas are an inevitable result of the detonation of fission and fusion devices and are fundamental to the operation of fusion devices.

MARK D. SMITH
AlliedSignal Aerospace Company

H. V. Boenig, *Plasma Science and Technology*, Cornell University Press, Ithaca, N.Y., 1982.

J. L. Cecchi, in S. M. Rossnagel, J. J. Cuomo, and W. D. Westwood, eds., *Handbook of Plasma Processing Technology*, Noyes Publications, Park Ridge, N.J., 1990.

F. F. Chen, *Introduction to Plasma Physics*, Plenum Publishing Corp., New York, 1974.

H. J. Oskam, ed., *Plasma Processing of Materials*, Noyes Data Corp., Park Ridge, N.J., 1985.

PLASTIC BUILDING PRODUCTS. See BUILDING MATERIALS, PLASTIC.

PLASTICIZERS

A plasticizer is a material which, when added to a polymer, causes an increase in the flexibility and workability, brought about by a decrease in the glass-transition temperature, T_g, of the polymer. The most widely plasticized polymer is poly(vinyl chloride) (PVC) due to its excellent plasticizer compatibility characteristics, and the development of plasticizers closely follows the development of this commodity polymer. However, plasticizers have also been used and remain in use with other polymer types.

Types of Plasticizers

Two principal methods exist for softening a polymer achieve plasticization. A rigid polymer may be internally plasticized by chemically modifying the polymer or monomer so that the flexibility of the polymer is increased. Alternatively, a rigid polymer can be externally plasticized by the addition of a suitable plasticizing agent, ie, by preparing a product consisting of a resin and a plasticizer. The external plasticizing route is the more common principally because of lower overall costs and also the fact that the use of external plasticizers allows the fabricator of the final article a certain degree of freedom in devising formulations for a range of products.

Internal Plasticizers. There has been much research involving the possibility of internally plasticized PVC. However, in achieving this by copolymerization significant problems exist. Thus, since standard external plasticizers are relatively cheap they are normally preferred.

External Plasticizers. There are two distinct groups of external plasticizers. A primary plasticizer, when added to a polymer, increases the properties of elongation and softness. A secondary plasticizer, when added to the polymer alone, does not bring about these changes and may have limited compatibility with the polymer; but when added to the polymer in combination with a primary plasticizer, it enhances the plasticizing performance of the primary plasticizer.

Commodity Phthalate Esters. The family of phthalate esters is by far the most abundantly produced worldwide. Both orthophthalic and terephthalic acid and anhydrides are manufactured. The plasticizer esters are produced from these materials by reaction with an appropriate alcohol. Phthalate esters are manufactured from methanol (C_1) up to C_{17} alcohols, although phthalate use as PVC plasticizers is generally in the range C_4 to C_{13}.

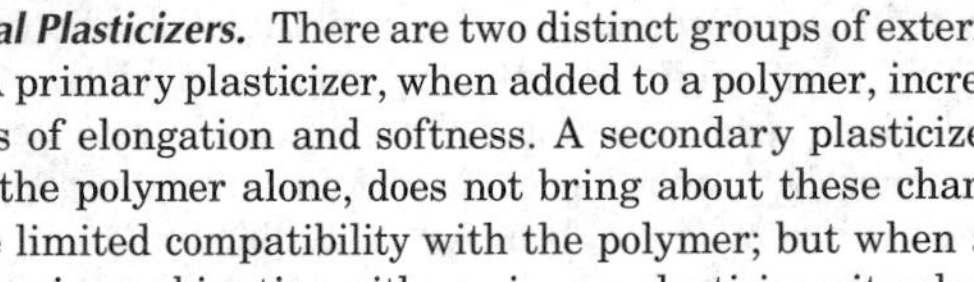

Di-2-Ethylhexyl Phthalate. Di-2-ethylhexyl phthalate (DEHP), also known as dioctyl phthalate (DOP), is in the mid-range of plasticizer properties. DEHP (or DOP) is the phthalate ester of 2-ethylhexanol, which is normally manufactured by the dimerization of butyraldehyde, the butyraldehyde itself being synthesized from propylene. It possesses reasonable plasticizing efficiency, fusion rate, and viscosity. Some concerns have been raised periodically as to the possible toxicity of this material, but it is believed that these concerns are often related to the vast and widespread study of the toxicity of DEHP.

Diisononyl Phthalate and Diisodecyl Phthalate. These primary plasticizers are produced by esterification of oxo alcohols of carbon chain length nine and ten. These alcohols are termed iso-alcohols and the subsequent phthalates iso-phthalates, an unfortunate designation in view of possible confusion with esters of isophthalic acid.

The C_9 and C_{10} iso-phthalates (DINP and DIDP) generally compete with DEHP as commodity general-purpose plasticizers. Other iso-phthalates are available at opposite ends of the carbon number range, but these serve more speciality markets. The C_8 iso-phthalate, diisooctyl phthalate (DIOP), has also been sold in the commodity plasticizer markets, where it is considered as equivalent to DEHP.

The Specialty Plasticizers. For the purpose of this article, the term specialty plasticizer refers to any plasticizer other than DEHP (DOP), DIOP, DINP, or DIDP.

Specialty Phthalates. These comprise the fast-fusing, low carbon number phthalates dibutyl phthalate (DBP), diisobutyl phthalate (DIBP), benzylbutyl phthalate (BBP), and diisoheptyl phthalate (DIHP); the low volatility isophthalates diisoundecyl phthalate (DIUP) and diisotridecyl phthalate (DTDP); and also the linear and semilinear phthalates for low viscosity applications. In each case these materials are the phthalate esters of alcohols of varying chain length.

Adipate Esters. Alcohols of chain length similar to those used in phthalate manufacture can be esterified with adipic acid rather than phthalic anhydride to produce the family of adipate plasticizers. The family of adipic acid esters in PVC applications possesses the significant properties of improved low temperature performance relative to phthalates and lower plastisol viscosities in plastisol applications, due to the lower inherent viscosities of the plasticizers themselves. Adipates used are typically in the C_8 to C_{10} range.

Trimellitate Esters. These materials are produced by the esterification of a range of alcohols with trimellitic anhydride (TMA), which is similar in structure to phthalic anhydride with the exception of the third functionality (COOH) on the aromatic ring.

The principal feature of these esters, when processed with PVC, is their low volatility; and consequently large volumes of trimellitate esters are used in high specification electrical cable insulation and sheathing.

Phosphate Esters. The principal advantage of phosphate esters is the improved fire retardancy relative to phthalates. The fire performance of PVC itself, relative to other polymeric materials, is very good due to its high halogen content, but the addition of plasticizers reduces this. Consequently there is a need, in certain demanding applications, to improve the fire-retardant behavior of flexible PVC.

Sebacate and Azelate Esters. Esters produced from 2-ethylhexanol and higher alcohols with linear aliphatic acids are used in some PVC applications where superior low temperature performance is required. Their usage is generally limited to extremely demanding low temperature flexibility specifications, eg, underground cable sheathing in arctic environments.

Polyester Plasticizers. These materials have found widespread use due to their exceedingly low volatility and high resistance to chemical extraction. Polyester plasticizers are based on condensation products of propanediols or butanediols with adipic acid or less commonly phthalic anhydride.

Sulfonate Esters. These are marketed as efficient and easily processible plasticizers with good resistance to extraction. They are typically aryl esters of a C_{13} to C_{15} alkanesulfonic acid.

Secondary Plasticizers. Also known as extenders, secondary plasticizers play a significant role in flexible PVC formulations. They do not impart flexibility to the PVC resin alone, but when combined with a primary plasticizer they add flexibility to the final product. The majority of secondary plasticizers in use are chlorinated paraffins, which are hydrocarbons chlorinated to a level of 30–70%. For a given hydrocarbon chain, viscosity increases with chlorine content, as does the fire retardancy imparted to the formulation.

Other materials that are often referred to as secondary plasticizers include materials such as epoxidized soybean oil (ESBO) and epoxidized linseed oil (ELO) and similar materials. These can act not only as lubricants but also as secondary stabilizers to PVC due to their epoxy content, which can remove HCl from the degrading polymer.

The Mechanism of Plasticizer Action

This discussion refers to external plasticization only. Several theories, varying in detail and complexity, have been proposed in order to explain plasticizer action. Although each theory is not exhaustive, an understanding of the plasticization process can be gained by combining ideas from each theory.

The steps involved in the incorporation of a plasticizer into a PVC product can be divided into five distinct stages: **1.** Plasticizer is mixed with PVC resin. **2.** Plasticizer penetrates and swells the resin particles. **3.** Polar groups in the PVC resin are freed from each other. **4.** Plasticizer polar groups interact with the polar groups on the resin. **5.** The structure of the resin is re-established, with full retention of plasticizer.

Steps 1 and 2 can be described as physical plasticization. The rate at which step 2 occurs depends on the physical properties of plasticizer visocity, resin porosity, and particle size.

Steps 3 and 4 can be described as chemical plasticization since the rate at which these processes occur depends on the chemical properties of molecular polarity, molecular volume, and molecular weight.

The importance of step 5 cannot be stressed too strongly, since no matter how rapidly and easily steps 1–4 occur, if plasticizer is not retained in the final product the product will be rendered useless.

The Lubricity Theory. This is based on the assumption that the rigidity of the resin arises from intermolecular friction binding the chains together in a rigid network. On heating, these frictional forces are weakened, allowing the plasticizer molecules between the chains. Once incorporated into the polymer bulk the plasticizer molecules shield the chains from each other, thus preventing the reformation of the rigid network. Although attractive in its simplicity, the theory does not explain the success of some plasticizers and the failure of others.

The Gel Theory. This extends the lubricity theory in that it deals with the idea of the plasticizer acting by breaking the resin–resin attachments and interactions and by masking these centers of attachment from each other, preventing their reformation. Such a process by itself is insufficient to explain a completely plasticized system, because although a certain concentration of plasticizer molecules provides plasticization by this process, the remainder act more in accordance with the lubricity theory, with unattached plasticizer molecules swelling the gel and facilitating the movement of plasticizer molecules, thus imparting flexibility. If plasticization took place solely by this method it would not be possible to explain the ability of PVC resins to accept their own weight in plasticizer without exudation, ie, large amounts of additional space (free volume) are created which other plasticizer molecules can occupy.

The Free Volume Theory. This extends the lubricity and gel theories and also allows a quantitative assessment of the plasticization process.

The free volume, V_f, of a polymer is described by the equation $V_f = V_t - V^0$ where V_t = specific volume at a temperature t and V^0 = specific volume of an arbitrary reference point, usually taken as zero degrees Kelvin. Free volume is a measure of the internal space available in a polymer for the movement of the polymer chain, which imparts flexibility to the resin. A rigid resin, eg, unplasticized PVC, is seen to possess very little free volume whereas resins which are flexible in their own right are seen as having relatively large amounts of free volume. Plasticizers increase the free volume of the resin and also ensure that free volume is maintained as the resin–plasticizer mixture is cooled from the melt. Combining these ideas with the gel and lubricity theories, it can be seen that plasticizer molecules not interacting with the polymer chain must simply fill free volume created by those molecules that do. These molecules may also be envisaged as providing a screening effect, preventing interactions between neighboring polymer chains thus preventing the rigid polymer network from reforming on cooling.

For the plasticized resin, free volume can arise from motion of the chain ends, side chains, or the main chain. These motions can be increased in a variety of ways.

The introduction of a plasticizer, which is a molecule of lower molecular weight than the resin, has the ability to impart a greater free volume per volume of material because there is an increase in the proportion of end groups and the plasticizer has a glass-transition temperature, T_g, lower than that of the resin itself.

Plasticizer molecules, or at least some of them, are not bound permanently to the polymer as in an internally plasticized resin; rather an exchange–equilibrium mechanism is present. This implies that there is no stoichiometric relationship between polymer and plasticizer levels, although some quasi-stoichiometric relationships appear to exist.

Generalized Structure Theories and Antiplasticization. These theories are based on the concept that if a small amount of plasticizer is incorporated into the polymer mass it imparts slightly more free volume and gives more opportunity for the movement of macromolecules. Many resins tend to become more ordered and compact as existing crystallites grow or new crystallites form at the expense of the more fluid parts of the amorphous material. For small additions of plasticizer, the plasticizer molecules may be totally immobilized by attachment to the resin by various forces. These tend to restrict the freedom of small portions of the polymer molecule necessary for the absorption of mechanical energy, resulting in a more rigid resin with a higher tensile strength and base modulus than the base polymer itself. This phenomenon is known as antiplasticization.

Interaction Parameters. Early attempts to describe PVC–plasticizer compatibility were based on the same principles used to describe solvation, ie, like dissolves like. To obtain a quantitative measure of PVC—plasticizer compatibility a number of different parameters have been used; However, in all cases it is not possible adequately to predict the behavior of polymeric plasticizers in all cases.

Specific Interactions. Some mechanism of attraction and interaction between PVC and plasticizer must exist for the plasticizer to be retained in the polymer after processing.

The role of specific interactions in the plasticization of PVC has been proposed from work on specific interactions of esters in solvents (eg, hydrogenated chlorocarbons), work on blends of polyesters with PVC, and work on plasticized PVC itself. This research uses newer analytical techniques, in particular molecular modeling and solid-state nuclear magnetic resonance (nmr) spectroscopy.

Molecular Modeling. The computer modeling of molecules has become an important branch of chemistry. High resolution graphics and fast computers allow the operator to build molecules in minimum energy configurations and view them in real time. Such models can be constructed from crystallographic coordinates available from databases (qv) or by simple intervention from the operator. Molecular mechanics or quantum mechanics programs are then used to arrive at a likely structure.

Solid-State Nuclear Magnetic Resonance Spectroscopy. Advances in technology have made the study of solids by nmr techniques considerably easier than in previous years. For the accumulation of solid-state ^{13}C-nmr spectra, cross-polarization magic angle spinning (CP-MAS) can be utilized to significantly reduce signal broadening effects present in solid state but not in the liquid state. The technique has been used to study the molecular effects of plasticization by comparing spectral shifts of PVC and plasticizer under various degrees of processing.

Plasticized Polymers: The Dominance of PVC

Well over 90% of plasticizer sales by volume are into the PVC industry. The reason for such a concentration of sales is that the benefits imparted by the plasticization of PVC are far greater than those imparted to other polymers. PVC stands alone among polymers in its ability both to accept and retain large concentrations of plasticizer. This is due in part to a morphological form comprising highly amorphous, semicrystalline, and highly crystalline regions. The development of PVC as a commodity polymer is fundamentally linked to the development of its additives.

Different types of PVC exist on the market. The two principal types are suspension and paste-forming PVC; the latter includes the majority of emulsion PVC polymers. The plasticizer applications technologies associated with these two forms are distinctly different.

Applications Technology

Suspension PVC. These polymers are produced by suspending vinyl chloride in water and polymerizing this monomer using a monomer-soluble initiator. PVC polymers produced via a suspension polymerization route have a relatively large particle size (typically 100–150 μm). Additionally, suspension polymers produced for the flexible sector have particles that are highly porous and are therefore able to absorb large amounts of liquid plasticizer during a formulation mixing cycle. A typical flexible PVC formulation (Table 1) using a suspension polymer is typically processed by a dry-blend cycle, during which all formulation ingredients are heated (typically 70–110°C) and intimately mixed to form a dry powder (the PVC dry blend or powder blend). This dry blend can be either stored or processed immediately. Processing of suspension resin formulations is performed by a variety of techniques such as extrusion, injection molding, and calendering to totally fuse the formulation ingredients and therefore produce the desired product.

Paste-Forming Polymers. Paste- or plastisol-forming PVC polymers differ from their suspension analogues in that after mixing with plasticizer they produce a paste or plastisol, similar in appearance to paint, rather than a dry blend. In this respect these polymers are used for flexible applications only. The plastisol can then be spread, coated, rotationally cast, or sprayed for processing. Plastisol-forming polymers are produced by microsuspension polymerization or emulsion polymerization. Microsuspension produces very fine particles of monomer to ensure that small particle sizes of polymer are produced. In emulsion polymerization the vinyl chloride is polymerized using a water-soluble initiator; the vinyl chloride particles are small and stabilized using surfactants.

Much lower particle size resins are produced by these routes, relative to suspension resins, but they also differ in that some residual surfactant from the polymerization process is retained on the polymer. The low particle size imparts a lack of porosity to the resin and thus the mixing of formulation ingredients using a dry-blending cycle is not possible. The demands on plasticizer behavior in a plastisol tends to be more complex than those in suspension technology since choice of plasticizer is made with consideration given to the required viscosity of the plastisol and also the required rheology of the plastisol. Each paste-forming polymer shows individual characteristics with respect to particle size and particle size distribution, and therefore plastisol viscosity and rheology, and no two polymers behave the same way, thus making the choice of plasticizer for this sector somewhat complex.

Table 1. Typical Flexible PVC

Ingredient	Parts by weight	Parts per hundred resin (phr)	% By weight
PVC	75	100	50
plasticizer	45	60	30
filler	26.25	35	17.5
stabilizer	3	4	2
lubricant	0.75	1	0.5
Total	*150*	*200*	*100*

Effect of Plasticizer Choice on the Properties of Flexible PVC

A change in plasticizer affects the properties of a flexible PVC article. Certain properties are more important for some applications than others and hence some plasticizers find more extensive use in some application areas than others.

Plasticizer Efficiency. This is a measure of the concentration of plasticizer required to impart a specified softness to PVC. Such a softness of material may be measured as a British Standard Softness (BSS) or a Shore hardness. For a given acid constituent of plasticizer ester, such as phthalate or adipate, plasticizer efficiency decreases as the carbon number of the alcohol chain increases. In addition to the value of the carbon number of the alcohol chain, the amount of branching is also significant; the more linear isomers are of greater efficiency. Choice of the acid constituent can also be significant. For equivalent alcohol constituents, phthalate and adipate esters are approximately equivalent but both are considerably more efficient than the trimellitate equivalent.

High Temperature Performance. High temperature performance in flexible PVC and its production are related to plasticizer volatilization and plasticizer degradation. Plasticizer volatilization, both from the finished article during use at elevated temperatures, eg, in electrical cable insulation, and also during processing, ie, release of plasticizer fume is directly related to the volatility of the plasticizer in use. Hence the higher molecular weight plasticizers give superior performance in this area. Higher molecular weight esters such as trimellitates are very stable thermally and find extensive use in the demanding cable specifications.

Polyester plasticizers give the best performance in this area, with performance increasing with molecular weight.

For the generation of fume in the workplace, the same structure relationships apply. Not only does excessive plasticizer volatilization have environmental consequences, but inaccuracies on formulation can be incurred since not all the plasticizer in use is entering the PVC resin, resulting in a material harder than calculated. As a result of environmental protection legislation, more end users are looking to means of recovering and re-using plasticizer fume and breakdown products, either through re-use in the processing operation or as an alternative fuel.

Low Temperature Performance. The ability of plasticized PVC to remain flexible at low temperatures is of great importance in certain applications, eg, external tarpaulins or underground cables. For this property the choice of the acid constituent of the plasticizer ester is also important. The linear aliphatic adipic, sebacic, and azeleic acids give excellent low temperature flexibility compared to the corresponding phthalates and trimellitates.

There is also a significant contribution to low temperature performance from the alcohol portion of the ester, the greater the linearity of the plasticizer the greater the low temperature flexibility.

Gelation Properties. The gelation characteristics are a measure of the ability of a plasticizer to fuse with the polymer so as to yield a product of maximum elongation and softness, ie, maximum plasticization properties. Gelation properties are often measured either as a processing temperature, the temperature to which the plasticizer and polymer must be heated in order to obtain these properties, or as a solution temperature, the temperature at which one grain of polymer dissolves in excess plasticizer, giving a measure of the solvating power of the plasticizer. Ease of gelation is related to plasticizer polarity and molecular size. The greater the polarity of a plasticizer molecule the greater the attraction it has for the PVC polymer chain and the less additional energy, in the form of heat, is required to cause maximum plasticizer–PVC interactions. The most active plasticizers are able to bring about these effects soon after the T_g of the polymer (70–80°C) is reached, whereas the less active plasticizers require temperatures on the order of 180°C in order for the maximum elongation properties to be obtained. The polarity of the plasticizer is determined by both acid type and alcohol chain length. Aromatic acids, being of greater polarity, tend to show greater ease of gelation than aliphatic acid-based esters. Molecular size also has a key contribution. The smaller the plasticizer molecule the easier it is for it to enter the PVC matrix; larger molecules require more thermal energy to establish the desired interaction with the polymer.

Migration and Extraction. When plasticized PVC comes into contact with other materials, plasticizer may migrate from the plasticized PVC into the other material. The rate of migration depends not only on the plasticizer employed but also on the nature of the contact material.

Plasticizer can also be extracted from PVC by a range of solvents including water. The aggressiveness of a particular solvent depends on its molecular size and its compatibility with both the plasticizer and PVC. Water extracts plasticizer very slowly, oils are slightly more aggressive, and low molecular weight solvents are the most aggressive.

The key characteristic for migration and extraction resistance is molecular size. The larger the plasticizer molecule the less it tends to migrate or be extracted. The greater the linearity of the ester the greater its migration and extraction rate in comparison to the more branched isomers.

Plastisol Viscosity and Viscosity Stability. After the primary contribution of the resin type in terms of its particle size and particle size distribution, for a given PVC resin, plastisol viscosity has a secondary dependence on plasticizer viscosity. The lower molecular weight and more linear esters have the lowest viscosity and hence show the lowest plastisol viscosity. In spite of these viscosity differences, however, if plastisols are being formulated to equal softness, ie, taking into account the efficiency of the plasticizers involved, more of the less efficient plasticizer has to be employed in the plastisol so as to impart the same softness to the product being manufactured. The addition of this extra liquid to the plastisol may produce a viscosity equivalent to that of the plastisol with the less viscous plasticizer. Esters based on aliphatic acids, being of lower viscosity than the corresponding aromatic acids, show lower plastisol viscosities.

Automotive Windshield Fogging. The phenomenon of car windscreen fogging has been known for some time. The term fogging relates to the condensation of volatile material on the car windshield causing a decrease in visibility. Although this volatile material may arise from a variety of sources, eg, exhaust fume being sucked in through the ventilation system, material from inside the car, eg, crash pads or rear shelves, may also contribute to windscreen fogging on account of the high temperatures that can be encountered inside a car when standing in sunlight. In the case of flexible PVC such a contribution may arise from emulsifiers in the polymer, stabilizers, and plasticizers.

In each case manufacturers have studied their products in detail and recommend low fogging polymers, stabilizers, and plasticizers. Tests have been designed to assess the fogging performance of both the PVC sheet and the raw materials used in its production. The higher molecular weight and more linear plasticizers give superior performance.

The Plasticization of Polymers Other Than PVC

The plasticization of PVC accounts for the vast majority of plasticizer sales. However, significant amounts of plasticizers are used in non-PVC polymers. Although PVC stands alone in its ability to accept and retain large quantities of commercial plasticizer, effective plasticization of other resins using slightly modified plasticizers may be possible if certain conditions specific to the polymer of interest are met.

The first factor to be considered when looking at the plasticization of a polymer is the need. Other factors are short- and long-term compatibility, ie, the ability of a polymer to accept and retain the plasticizer.

In order for a plasticizer to enter a polymer structure the polymer should be highly amorphous. Crystalline nylon retains only a small quantity of plasticizer if it retains its crystallinity. Once it has penetrated the polymer the plasticizer fills free volume and provides polymer chain lubrication, increasing rotation and movement.

Acrylic Polymers. The plasticization of acrylic resins is complicated by the fact that acrylic resins constitute a large family of polymers rather than a single polymeric species. An infinite variation in physical properties may be obtained through copolymerization of two or more acrylic monomers selected from the available esters of acrylic and methacrylic acid.

Plasticizers are used in the acrylics industry to produce tough, flexible coatings.

Plasticizers for acrylics include all common phthalates and adipates. There has been interest in the development of acrylic plastisols similar to those encountered with PVC. The same aspects of both plastisol viscosity and viscosity stability are important.

Nylon. The high degree of crystallinity in nylon means that plasticization can occur only at very low levels. Plasticizers are used in nylon but are usually sulfonamide based since these are generally more compatible than phthalates.

Poly(ethylene terephthalate). PET is a crystalline material and hence difficult to plasticize. Additionally, since PET is used as a high strength film and textile fiber, plasticization is not usually required although esters showing plasticizing properties with PVC may be used in small amounts as processing aids and external lubricants. Plasticizers have also been used to aid the injection molding of PET, but only at low concentrations.

Due to its lack of hydrogen bonds PET is relatively difficult to dye. Plasticizers used in this process can increase the speed and intensity of the dyeing process.

Polyolefins. In the plasticization of polyolefins, plasticizer use generally results in a reduction of physical properties; compatibility can be achieved only up to 2 wt %. Most polyolefins give adequate physical properties without plasticization, but plasticizers have been used with polypropylene to improve its elongation at break.

Polystyrene. Polystyrene shows compatibility with common plasticizers but modification of properties produced is of little value. Small amounts of plasticizer (eg, DBP) are used as a processing aid.

Fluoroplastics. Conventional plasticizers are used as processing aids for fluoroplastics up to a level of 25% plasticizer.

Rubbers. Plasticizers have been used in rubber processing and formulations for many years, although phthalic and adipic esters have found little use since cheaper alternatives, eg, heavy petroleum oils, coal tars, and other predominantly hydrocarbon products, are available for many types of rubber. It has been noted that the more polar elastomers such as nitrile rubber and chloroprene are insufficiently compatible with hydrocarbons and require a more specialized type of plasticizer, eg, a phthalate or adipate ester.

Health and Safety Aspects

Numerous toxicological studies have been conducted on a variety of plasticizers.

Acute Toxicity. Plasticizers possess an extremely low order of acute toxicity. In addition to their low acute toxicity, many years of practical use coupled with animal tests show that plasticizers do not irritate the skin or mucous membranes and do not cause sensitization.

Chronic Toxicity. The effects of repeated oral exposure to phthalates for periods ranging from a few days to two years have been studied in a number of animal species.

A large number of investigations on a variety of plasticizers and different animal species have revealed the following. *(1)* Plasticizers are not genotoxic. *(2)* Oral administration of plasticizers, fats, and other chemicals to rodents causes a proliferation of microbodies in the liver (peroxisomes) which may be considered to be linked to the formation of liver tumors. *(3)* Administration of plasticizers, fats, and hypolipidemic drugs to nonrodent species such as marmosets and monkeys does not lead to peroxisome proliferation and liver damage. *(4)* These species differences have also been observed in *in vitro* studies. Phthalates, their metabolites, and a variety of other peroxisome proliferators caused peroxisome proliferation in rat and mouse liver cells but not in those of humans, marmosets, or guinea pigs.

On the basis of these differences in species response it was concluded that phthalates do not pose a significant health hazard to humans.

Phthalates have been shown to cause reproductive effects in rats and mice but primates are resistant to these effects. This may be due in part to pronounced differences in the way in which phthalates are metabolized by rodents and primates, including humans.

The reproductive toxicity of some phthalate esters has been reviewed by the Commission of the European Communities. This review concludes that testicular atrophy is the most sensitive indicator of reproductive impairment and that the rat is the most sensitive species.

Comparing estimates of the average human daily lifetime exposure to DEHP (0.3–6 μg/kg body weight per day) to the level at which no effects are observed in rats indicates that the margin of safety for the general public is more than 10,000.

The Effect of Plasticizers on the Environment

Several hundred plasticizers are in commercial use in the world, but only relatively few (ie, phthalates) are used in amounts that make them significant in tonnage terms, and hence in their likely environmental input and impact.

Estimated Emissions to the Environment. Phthalates may be emitted to the environment during their incorporation into PVC and from the finished PVC article during its use or after its final disposal. However, because their purpose is to make PVC flexible and for it to remain so over long periods of time, plasticizers are of very low volatility relative to many other commonly used products, for example solvents.

Emissions During Plasticizer Production and Distribution. Phthalate plasticizers are produced by esterification of phthalic anhydride in closed systems; hence losses to atmosphere are minimal.

The transport of phthalates by road tankers and ships within Europe is carried out by international companies with sophisticated tank cleaning facilities. Wash waters from these modern facilities are passed through a series of separators to remove any residual plasticizer which is then incinerated.

Emissions During Processing. During the production of flexible PVC products, plasticizers are exposed for up to several minutes to temperatures of ~180°C. The exact conditions depend on the processing technique employed, but it is evident that the loss of plasticizer by evaporation and degradation can be significant.

Of the various processing techniques used, injection molding and extrusion involve little or no exposure of hot product to the surrounding air, hence they give rise to no significant emission of plasticizer to the atmosphere. This is not the case in the production of sheet and film by calendering or spread coating.

Knowledge of the quantity of plasticizer used in each application together with the level of exhaust air treatment allows estimation of the level of plasticizer lost to atmosphere during these processes.

Emissions During Interior End Use. The majority of flexible PVC is used indoors in applications such as flooring, wall covering, upholstery, wire and cable, etc.

Some products, particularly flooring, may lose plasticizer not only by evaporation but also through extraction by soapy water during cleaning. Wastewater associated with the cleaning process typically goes to the municipal sewage system. Thus, the phthalates are biodegraded and do not end up in the environment.

Emissions During Exterior End Use. When flexible PVC is used in exterior applications plasticizer loss may occur due to a number of processes which include evaporation, microbial attack, hydrolysis, degradation, exudation, and extraction. It is not possible, due to this wide variety of contribution processes, to assess theoretically the rate of plasticizer loss by exposure outdoors. Little suitable data has been published with the exception of some studies on roofing sheet, which has been used to estimate the plasticizer losses from all outdoor applications. This estimate may well be too high because of the extrapolation involved.

Emissions During Disposal and Incineration. The increasing use of modern incinerators to dispose of domestic waste results in complete combustion of plasticizers to carbon dioxide and water. The preponderance of plasticizer going into landfills is as plasticized PVC. Once a landfill has been capped anaerobic conditions prevail and it is biologically relatively inactive. Under these conditions the main route by which organic components are removed from the landfill contents is by ingress of water, extraction, and subsequent loss of water from the site to the environment.

Occurrence of Plasticizers in the Environment. The contamination of laboratory chemicals and equipment causes problems in the analysis of very low concentrations of phthalates in environmental samples. While efforts have been made to overcome these difficulties in recent studies, the results of earlier investigations must be treated with caution.

Phthalates in Air. Atmospheric levels of phthalates in general are very low.

Phthalates in Water. Reported levels of phthalates in natural waters are, in general, low. Concentrations found in fresh waters range from nondetectable up to 10 μg/L.

Phthalates in Sediments. Phthalates are lipophilic and hence partition onto organic-rich particulate matter in water. This particulate matter on settling gives rise to sediments which contain higher levels of phthalate than the overlying water.

By taking sections from a sediment core sample of 120 cm in depth it was found that the phthalate level in the mid-1990s was only 15% of what it had been in 1972–1978, despite the fact that the total usage of plasticizers has continued to increase annually.

Wastewater Treatment Plants. Numerous studies have shown that phthalates in wastewater systems are removed to a significant extent by treatment plants.

Environmental Modeling. The estimated plasticizer emissions outlined earlier have been entered into the HAZCHEM model developed by the European Centre for Ecotoxicology and Toxicology of Chemicals (ECETOC). This enables the levels of plasticizer in the various environmental compartments to be estimated. The advantage of using environmental models in conjunction with emission estimates is that they give an overview of the concentrations present in any chosen region. This is helpful in conducting realistic environmental risk assessments.

Environmental Effects of Plasticizers. Measurement of the effect of phthalates on environmental species is difficult because standard test methods are not designed to deal with poorly water-soluble substances.

Atmospheric Toxicity. The only known atmospheric toxicity effect of phthalates is the phytotoxicity arising from the use of DBP plasticized glazing bars in greenhouses. General atmospheric concentrations of phthalates are extremely low and it is concluded that they pose no risk to plants or animals.

Aquatic Toxicity. The standard tests to measure the effect of substances on the aquatic environment are designed to deal with those that are reasonably soluble in water.

The majority of studies on the acute and chronic toxicity of phthalates to aquatic organisms show no toxic effects at concentrations 200–1000 times the water solubility.

ECPI has commissioned further aquatic toxicity studies in order to clarify the situation. Data from biodegradation studies have confirmed that the phthalates commonly used in the plasticization of PVC do not require classification "Dangerous for the Environment."

Sediment Toxicity. Because of their low solubility in water and lipophilic nature, phthalates tend to be found in sediments. Unfortunately little work has previously been carried out on the toxicity of phthalates to sediment dwelling organisms. For this reason the European Council for Plasticizers and Intermediates (ECPI) has commissioned some sediment toxicity studies designed to measure the effect of DEHP and DIDP in a natural river sediment on the emergence of the larvae of the midge, *Chironomus riparius*, an organism that is distributed throughout North America and Europe in a wide variety of freshwater habitats.

Storage and Handling

Plasticizer esters are relatively inert, thermally stable liquids with high flash points and low volatility. Consequently they can be stored safely in mild steel storage tanks or drums for extended periods of time. Exposure to high temperatures for extended periods, as encountered in drums in hot climates, is not recommended since it may lead to a deterioration in product quality with respect to color, odor, and electrical resistance.

DAVID F. CADOGAN
European Council for Plasticizers and Intermediates
CHRISTOPHER J. HOWICK
European Vinyls Corporation

J. K. Sears and J. R. Darby, *The Technology of Plasticizers*, John Wiley & Sons, Inc., New York, 1982.

M. J. Bunten, M. W. Newman, P. V. Smallwood, and R. C. Stephenson, in J. I. Kroschwitz, ed., *Encyclopedia of Polymer Science and Engineering*. Vol. 17 and Supplement, John Wiley & Sons, Inc., New York, 1989, pp. 241–392.

J. P. Sibilia, *A Guide to Materials Characterisation and Chemical Analysis*, VCH, Weinheim, Germany, 1988.

S. S. Kurtz, J. S. Sweely, and W. J. Stout in P. F. Bruins, ed., *Plasticizer Technology*, Reinhold Publishing Corp., New York, 1965.

PLASTICS PROCESSING

Plastics are classified as thermoplastic or thermosetting resins, depending on the effect of heat. Thermoplastic resins, when heated during processing, soften and flow as viscous liquids; when cooled, they solidify. The heating/cooling cycle can be repeated many times with little loss in properties. Thermosetting resins liquefy when heated and solidify with continued heating; the polymer undergoes permanent cross-linking and retains its shape during subsequent cooling/heating cycles. Thus, a thermoset cannot be reheated and molded again. Thermoplastics can be melt-reprocessed, and hence readily recycled (see RECYCLING, PLASTICS).

Thermoplastic Resins

Almost 85% of the resins produced are thermoplastics. They can be divided into two broad classes: amorphous and crystalline. Amorphous thermoplastics are characterized by their glass-transition temperature, T_g, a temperature above which the modulus decreases rapidly and the polymer exhibits liquid-like properties; amorphous thermoplastics are normally processed at temperatures well above their T_g. Semicrystalline resins can have different degrees of crystallinity ranging from 50 to 95%; they are normally processed above the melting point, T_m, of the crystalline phase.

Over 70% of the total volume of thermoplastics is accounted for by the commodity resins: polyethylene, polypropylene, polystyrene, and poly(vinyl chloride) (PVC) (see OLEFIN POLYMERS; STYRENE PLASTICS; VINYL POLYMERS). Because of their low cost they are the first choice for a variety of applications. Next in performance and in cost are acrylics, cellulosics, and acrylonitrile-butadiene-styrene (ABS) terpolymers (see ACRYLIC-ESTER POLYMERS; ACRYLONITRILE POLYMERS; CELLULOSE ESTERS). Engineering plastics (qv) such as acetal resins (qv), polyamides (qv), polycarbonates (qv), polyesters (see POLYESTERS, THERMOPLASTIC), and poly(phenylene sulfide), and advanced materials such as liquid crystal polymers, polysulfone, and polyetheretherketone are used in high performance applications (see POLYMERS CONTAINING SULFUR).

Thermoplastics are usually marketed in the form of pellets. They are shipped in containers of various sizes, from 25-kg bags to railroad hopper cars.

The packaging (qv) requirements for shipping and storage of thermoplastic resins depend on the moisture that can be absorbed by the resin and its effect when the material is heated to processing temperatures.

A variety of equipment and methods are available to fabricate the desired thermoplastic product. Extrusion is the most popular; it is used to produce profiles, pipe and tubing, film, sheet, wire, and cable.

Injection molding follows as a preferred processing method. Other common methods include blow molding, rotomolding, thermoforming, calendering, and, to some extent, compression molding. Computer-aided design software for molds and extruder screws is commercially available (see COMPUTER-AIDED DESIGN AND MANUFACTURING (CAD/CAM); COMPUTER-AIDED ENGINEERING (CAE)).

Extrusion. Extrusion is defined as continuously forcing a molten material through a shaping device. Because the viscosity of most plastic melts is high, extrusion requires the development of pressure in order to force the melt through a die.

To provide a homogeneous product, incorporation of any additives, such as antioxidants (qv), colorants, and fillers, requires mixing them into the plastic when it is in a molten state. This is done primarily in an extruder (Fig. 1). In the act of melting, and in subsequent sections along the barrel, the required amount of mixing is usually achieved. The final portion of the extruder (L) is used to develop the pressure for pumping the homogenized melt through a filtering screen (optional) and then through a shaping die attached to the end of the extruder.

Extruders are defined by their screw diameter and length, with the length expressed in terms of the length-to-diameter ratio (*L/D*). Single-screw extruders range from small laboratory size (6-mm *D*) to large commercial units (450-mm *D*) capable of processing up to 20 t/h. Reactive extrusion is the term used to describe the use of an extruder as a continuous reactor for polymerization or polymer modification by chemical reaction.

An extruder employs drag flow to perform a conveying action that depends on the relative motion between the screw and the barrel. With higher friction on the barrel than on the screw, the solids are conveyed almost as solid plug in the deep-feed channel section (J in Fig. 1). As the channel depth becomes shallower, the compressive action causes more frictional heat, which, combined with the conduction supplied by the barrel heaters, causes the plastic to melt. The molten plastic then enters a constant shallow depth section of the screw called the metering section, where the pumping pressure necessary for extrusion through the final shaping die is developed.

In addition to the conveying and melting steps, extruders perform the vital task of homogenization of additives into the base resin.

The blown film process uses a tubular die from which the extrudate expands in diameter while traveling upward to a film tower. The top of the tower has a collapsing frame followed by guide and pull rolls to transport the collapsed film to subsequent slitting and windup rolls. The tubular bubble from the die is inflated to the desired diameter by air passing through the center of the die. Polyethylene is the primary plastic used in most films, especially for packaging and trash bags.

The cast film process provides a film with gloss and sparkle and can be used with various resins. Figure 2 is an illustration of the essential features of the extrusion equipment.

The process used to make an extruded plastic sheet produces sheeting with thicknesses of 0.25–5 mm and widths as great as 3 m.

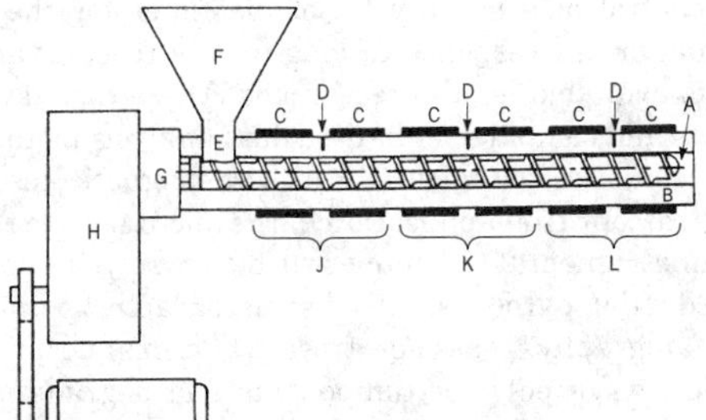

Figure 1. Parts of an extruder: A, screw; B, barrel; C, heater; D, thermocouple; E, feed throat; F, hopper; G, thrust bearing; H, gear reducer; I, motor; J, deep channel feed section; K, tapered channel transition section; and L, shallow channel metering section.

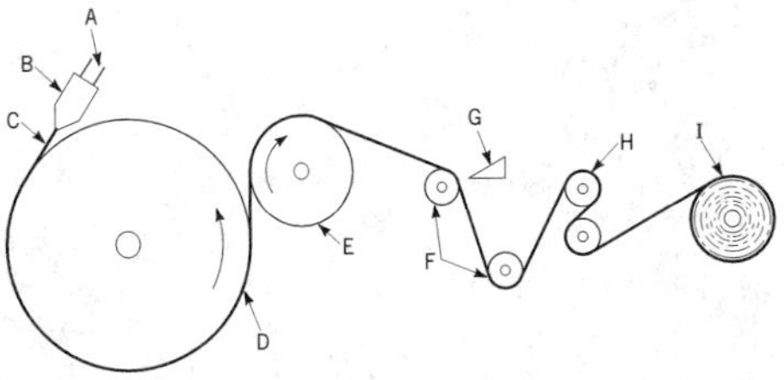

Figure 2. Extrusion of roll-cast film: A, die inlet; B, cast-film die; C, air gap with molten web; D, casting roll; E, stripping roll; F, idler roll; G, edge-trim slitter; H, pull rolls; and I, windup roll.

There usually is no need for a high melt temperature to obtain flow through a sheeting die, because die openings are large. Most sheeting is used for thermoforming.

A coating of an appropriate thermoplastic, such as polyethylene, may be applied to a substrate of paper, thin cardboard, or foil to provide a surface property which enables heat sealing or better barrier performance (see BARRIER POLYMERS). In the extrusion-coating process, a molten web of resin is extruded downward, and the web and substrate make contact at the nip between a pressure roll and a chill roll. Pressure and high melt temperatures are needed for adhesion of resin and substrate.

In contrast to most extrusion processes, extrusion coating involves a hot melt, ca 340°C. The thin web cools rapidly between the die and nip even at high linear rates. Both mechanical and chemical bonding to substrates are involved.

Protective and insulating coatings can be applied continuously to wire as it is drawn through a cross-head die. A typical wire coating line consists of a wire payoff, wire preheater, extruder, die, cooling trough, capstan, and wire takeup.

Foamed thermoplastics provide excellent insulating properties because of their very low thermal conductivity, good shape retention, and good resistance to moisture pickup. The blowing agent is dissolved and held in solution by the pressure developed in the extruder. As the molten thermoplastic exits the extruder die, the pressure release causes instantaneous foaming.

Molding. *Injection Molding.* In injection molding a molten thermoplastic is injected under high pressure into a steel mold. After the plastic solidifies, the mold is opened and a part in the shape of the mold cavity is removed.

The molds are custom-machined from steel. Cavities must be polished to a very high gloss, since the plastic reproduces the surface in every detail.

Mold designs must take into account that cooled moldings are always smaller than the cavity, owing to shrinkage. Amorphous plastics shrink less than crystalline plastics.

Structural Foam Molding. This is a modified injection molding process for large articles having a cellular core and an integral solid skin with an overall 20–50% reduction in density, compared to their solid counterparts. Structural foam molding is most frequently used with polyethylene, high impact polystyrene, polypropylene, and several engineering resins. Some modifications to the resin, machine, and mold are required, and a blowing agent must be added to the resin. A chemical blowing agent, which releases gas when heated, is commonly used.

Blow Molding. Blow molding is the most common process for making hollow thermoplastic components. In extrusion blow molding a molten tube of resin called a parison is extruded from a die into an open mold (Fig. 3**a**). The mold is closed around the parison (Fig. 3**b**), and the bottom of the parison is pinched together by the mold. Air under pressure is fed through the die into the parison, which expands to fill the mold. The part is cooled as it is held under internal air pressure. Figure 3**c** shows the open mold with the part falling free.

Rotational Molding. Hollow articles and large, complex shapes are made by rotational molding, usually from polyethylene powder of relatively low viscosity. A measured quantity is placed inside an alu-

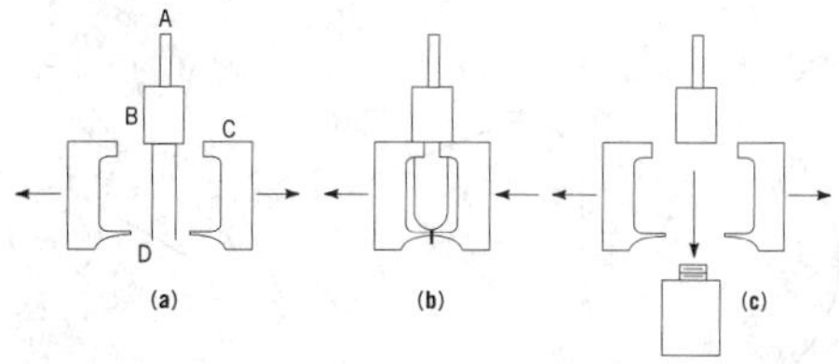

Figure 3. Three stages of blow molding (**a–c**): A, air line; B, die; C, mold; and D, parison.

minum mold and the mold is heated in an oven and rotated at low speed. The resin sinters and fuses, coating the inside of the mold. The mold is then cooled by water spray and the part solidifies, duplicating the inside of the mold.

Expandable Polystyrene Molding. Molding expandable polystyrene gives foamed products such as insulation board shapes for packaging and disposable food and cup containers.

Thermoforming. Thermoforming is a process for converting a preform, usually an extruded plastic sheet, into an article such as a thin-wall container or a tray for packaging meat. Under vacuum the process is called vacuum forming. The sheet is clamped in a frame and exposed to radiant heaters. The sheet softens to a formable condition, is moved over a mold and sucked against the mold by vacuum.

Amorphous resins such as styrenics, acrylics, PVC, and some modified crystalline resins are used for thermoforming.

Calendering. Calendering is a process uniquely applied to rubbery polymers, mainly semirigid and flexible PVC, for making sheeting of uniform thickness from 0.75–0.05 mm after stretching. A calender has four heavy, large steel rolls, which are usually assembled in an inverted "L" configuration.

Casting. Casting refers to the formation of an object in a batch process by pouring a fluid monomer–polymer solution into a mold where it solidifies or by continuously pouring the liquid onto a moving belt. The casting process is most frequently used with acrylics. Cast acrylic sheeting is made by polymerization of methyl methacrylate in a cell assembled from two glass plates and a flexible gasket.

Thermosetting Resins

Common thermosetting resins are unsaturated polyesters, phenolic resins (qv), amino resins (see AMINO RESINS AND PLASTICS), polyurethanes, epoxy resins (qv), and silicones (qv) (see POLYESTERS UNSATURATED; URETHANE POLYMERS). Thermosetting resins are usually low viscosity liquids or low mol wt solids that are formulated with suitable additives known as cross-linking agents to induce curing; curing involves permanent chemical changes resulting in infusible, insoluble products with excellent thermal and dimensional stability. Thermosetting resins are commonly used in combination with fillers or fibrous reinforcements; as a result, processing methods are often quite different from those employed for thermoplastics.

Compression, Injection, and Transfer Molding. Compression molding is common, although many thermosetting materials are also injection molded. The equipment consists of a vertical hydraulic press with platens for mold attachment. Mold cavities are hardened and highly polished, similar to those of an injection mold. A measured quantity of thermosetting resin compound in granular, sheet, or other form is placed in the hot mold. The mold is closed and the liquified resin through pressure fills the cavity. Continued heating cures the resin within a few minutes, and the part is removed from the mold. Compression molding is commonly employed for phenol–formaldehyde, urea–formaldehyde, melamine–formaldehyde, and polyester molding compounds.

Open-mold Processing. Common open-mold processes are hand or mechanized methods such as lay-up and spray-up, that use a single cavity mold and produce one finished surface. These techniques are used to produce fiber-reinforced structures containing the reinforcement in the form of cloth, chopped strands, mat, continuous roving, woven roving, etc. Thermosetting resins of choice are liquid unsaturated polyester or epoxy resins.

Pultrusion. In pultrusion, reinforcing fibers in a combination of styles are wetted by a resin containing a high temperature cross-linking agent and then pulled through a forming system that positions the fibers into the desired packed structural arrangement; the material then moves through a heated die where curing takes place.

Reaction Injection Molding. Reaction injection molding (RIM) is used for the production of solid or partially foamed polyurethane moldings by rapid injection of metered streams of polyol and isocyanate into a mold.

Resin Transfer Molding. In resin transfer molding (RTM) a piston-type positive displacement pump injects a premixed resin–catalyst stream into a closed mold thoroughly impregnating a preplaced reinforcement pack. RTM differs from structural RIM in that it uses slower-reacting formulations and injects under lower pressures components that are premixed or homogenized by passing through static mixers.

Polyurethane Foam Processing. Flexible foams are resilient open-cell structures with densities varying from 25–650 kg/m^3, depending on the choice of the raw materials. Most flexible foams are produced in the form of a slab or bun in a continuous process. A liquid foamable mixture is pumped onto a conveyor, which moves through a tunnel where reaction and foaming occur.

These are closed-cell foams with excellent thermal insulation characteristics. Rigid foams can be made as continuous slabs that are cut into panels. Other production methods include coating suitable substrates to produce laminated products and pouring or pumping the foamable mixture into place.

M. XANTHOS
D. B. TODD
Polymer Processing Institute

Z. Tadmor and C. G. Gogos, *Principles of Polymer Processing*, John Wiley & Sons, Inc., New York, 1979.

C. Rauwendaal, *Polymer Extrusion*, Hanser Publishers, Munich, Germany, 1990.

M. L. Berins, ed., *Plastics Engineering Handbook of the Society of Plastics Industry*, 5th ed., Van Nostrand Reinhold Co., Inc., New York, 1991, Chapt. 4.

I. I. Rubin, *Injection Molding—Theory and Practice*, John Wiley & Sons, Inc., New York, 1972.

PLASTICS TESTING

Plastics testing encompasses the entire range of polymeric material characterizations, from chemical structure to material response to environmental effects. Whether the analysis or property testing is for quality control of a specific lot of plastic or for the determination of the material's response to long-term stress, a variety of test techniques is available for the researcher. These include instrumental techniques and computer-enhanced data analysis methods. Fourier transform techniques for infrared, nuclear magnetic resonance, and Raman spectroscopy find application in the micro sample region. Combinations of instrumental techniques such as pyrolysis–Fourier transform infrared (ftir), pyrolysis–gas chromatography (pyrolysis–gc), or liquid chromatography–mass spectroscopy (lc–ms) permit separation and identification of polymer components and degradation products from complex systems. The use of microprocessors to capture transient signals from physical test methods, ie, high speed impact, allows the graphing of the complete impact event as a stress–strain curve.

Composition and Structure

Infrared techniques, particularly Fourier transform infrared, are the most commonly used analysis methods for determining the composition of polymers. The ftir technique, which corresponds to the vibrational energies of atoms or specific groups of atoms within a molecule as well as rotational energies, identifies components by comparing the spectrum of a sample to reference spectra.

Combination techniques such as microscopy–ftir and pyrolysis–ir have helped solve some particularly difficult separations and complex identifications.

Raman spectroscopy is an emission phenomenon as opposed to infrared absorption, and results from vibrations caused by changes in polarizability. It can be complementary to infrared and may be more useful when the sample for analysis is a filled polymer or composite containing an inorganic filler.

Nuclear magnetic resonance (nmr) requires an atomic nuclei that can absorb a radio-frequency signal impinging it in a strong magnetic field to give a spectrum. The field strength at which the nucleus absorbs is a function of both the nucleus and its immediate electronic environment. The atoms most commonly used in polymer analyses are ^{1}H and ^{13}C. Fourier transform has also been applied to nmr, allowing increased sensitivity on microsamples.

Other techniques that have been shown effective in the determination of particular polymer compositions employ uv methods, pyrolysis–gc, or diffraction and scattering studies.

Structure determinations in polymers have involved the use of small-angle neutron scattering (sans) techniques to evaluate not only the radius of polymer molecule gyration but also phase dimensions in multicomponent polymers.

Molecular Weight

The molecular weight and the distribution of multiple molecular weights normally found within a commercial polymer influence both the processibility of the material and its mechanical properties. For a few well-defined homopolymers, an analysis of composition and molecular weight is sufficient to define the likely mechanical properties of the polymer.

Low angle laser light scattering (turbidity) has been found to be the most accurate and reproducible technique for measuring polymer molecular weights (average molecular weight) as a primary method. Most turbidimetric techniques use polymer solubility as a function of temperature for defining molecular weights. Dynamic light scattering has been used to obtain the distribution of hydrodynamic radius, which can be converted to molecular weight distribution. Gel-permeation chromatography (gpc), although a secondary technique requiring careful calibration and interpretation of data, is the most frequently used commercial technique because of its ease of use, low cost, short time for analysis, and generation of excellent comparative data.

Thermal Properties

Thermal analysis involves techniques in which a physical property of a material is measured against temperature at the same time the material is exposed to a controlled temperature program. A wide range of thermal analysis techniques is available. Of these the best known and most often used for polymers are thermogravimetry (tg), differential thermal analysis (dta), differential scanning calorimetry (dsc), and dynamic mechanical analysis (dma).

Processing Properties

The flow properties of polymers, whether the viscosity of liquid thermosets prior to gelation or of molten thermoplastics, are important parameters for the proper processing of materials. Several test methods have been developed to predict or correlate to actual processing conditions or for quality control testing of polymers. For thermoplastics, the commonly used quality control method is melt flow rate; this technique can give values at melt index conditions that are inversely proportional to the molecular weight of homopolymers. Melt flow rate values, although useful for quality control, are not indicative of the actual response of the material during processing, primarily due to the viscoelastic nature of polymers.

Because processing conditions cover a wide range of shear rates, tests that can simulate both temperature and shear rate conditions are more useful in predicting flow properties. Tests have been developed based on correlating common processing techniques that use torque rheometers, capillary rheometers, and oscillating disk rheometers for both thermoplastics and thermosets.

Standard tests for determining pressure–volume–temperature (PVT) properties of materials using rheometers have been established to determine the correlation for the fundamental properties of plastics. These in combination with no-flow temperature, ejection temperature, and capillary rheology data are useful in predicting the behavior of molten plastic as it enters and fills a mold.

Mechanical Properties

Mechanical properties of plastics can be determined by short, single-point quality control tests and longer, generally multipoint or multiple condition procedures that relate to fundamental polymer properties. Single-point tests include tensile, compressive, flexural, shear, and impact properties of plastics; creep, heat aging, creep rupture, and environmental stress-cracking tests usually result in multipoint curves or tables for comparison of the original response to post-exposure response.

Tensile properties are those of a plastic being pulled in an uniaxial direction until sufficient stress is applied to yield or break the material. For many materials, Hooke's law is valid for a portion of the stress–strain curve. Tensile curves can be used as an indication of polymer strength and toughness. Figure 1 shows the relationship normally seen for the high stress necessary for yield or break with strength, whereas high elongation beyond yield shows ductility (toughness). Similar curves can be generated for tests in comparison, flex, shear, and some forms of impact.

Mechanical properties are determined on solid polymers in arbitrary forms defined precisely by standard test methods in ISO, ASTM, or other national standards organizations. Several types of impact testing have been developed to measure a plastic's response to a high rate of strain. Drop-weight impacts, either manual or instrumented, give practical information because these are multiaxial test procedures and closer to normal impact seen in part applications. Instrumented impact allows the recording of the full impact event as a stress–strain curve, showing similar characteristics to tensile or other modes of stress–strain tests. Instrumented impact is preferred for monitoring ductile–brittle transitions and determining the effects of polymer composition on material toughness.

Creep, creep rupture, and stress relaxation tests are multiple-point tests requiring long periods of time (1000 h min) to generate useful

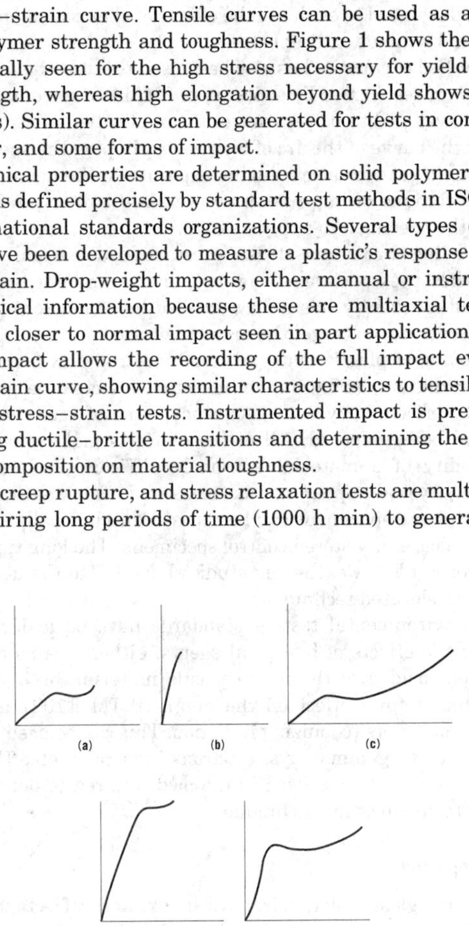

Figure 1. Types of stress-strain curves: (**a**) soft and weak; (**b**) hard and brittle; (**c**) soft and tough; (**d**) hard and strong; and (**e**) hard and tough.

data; these are standard tests for determining more fundamental polymer properties.

Fatigue testing of polymers may consist of static fatigue, ie, creep rupture, or dynamic fatigue.

As materials deform in mechanical testing, they emit stress waves having frequencies from 10 kHz to 10 MHz called acoustic emission (AE) waves. These are detectable by piezoelectric transducers as electric signals and are used to determine the initiation and growth of cracks in materials.

Environmental Effects Tests

In practical applications of plastic materials, their mechanical properties can be significantly influenced by environmental factors of chemical exposure, temperature, radiation (photon and gamma ray), biological agents, and/or combinations of several of these factors at once. Few standard tests have been developed in these areas. Mechanical properties can be determined by using ovens or cold-temperature chambers under the same procedures described in ASTM or ISO short-term tests to evaluate the effect of heat or cold. ASTM has published D5870-95 for comparative indexing of heat-aged materials to original properties, but it does not indicate a minimum exposure time. Thermal expansion testing also addresses the physical dimensional changes in plastic parts exposed to temperature changes.

Chemical exposure of plastics may exhibit a wide range of effects with one chemical and one type of polymer, depending on the concentration of chemical agent, temperature of exposure, molded-in stresses within the specimen, applied stress level to the part during exposure, and formulation of the specific polymer grade. Standardized tests have been developed to evaluate the effects of chemical agents on plastics with and without external stress.

The method of caustics has also been used to study the formation of cracks and crazes formed by exposure of PMMA to solvents. ISO 4599 has been developed to better control the application of stress using a jig having the curve of the arc of a circle for shaping the specimen and maintaining a set curvature during exposure to the agent. ISO 4600 uses the technique of impressing an oversized ball or pin into a hole drilled in the specimen to apply a strain.

A test using a constant stress (constant load) normally by direct tension takes the specimen to failure, or a minimum time without failure, and frequently has a flaw (drilled hole or notch) to act as a stress concentrator to target the area of failure.

Weathering of plastic materials combines complex factors of temperature, radiation, oxidation, and moisture effects on a plastic part. Weathering effects vary with geography, time of year, and position of material being exposed. All of these effects make predicting the weatherability of a material extremely difficult. Frequently, only changes in physical appearance are measured after weathering; however, appropriate specimens can be exposed, mechanically tested, and compared to carefully stored control specimens. The long time periods required for outdoor weathering studies have led to the development of several accelerated techniques.

Other environmental testing standards have been developed for evaluating the effects of biological agents, either strains of fungi or bacteria, on mold growth on the plastic material, or if the plastic material has a toxic effect on the fungi. ASTM E1027 is used for exposure of polymers to ionizing radiation. This can be used for testing the resistance to gamma rays, electrons, neutrons, etc. These tests are important for plastics used in the medical area to determine the resistance to sterilization techniques.

Optical Properties

Transparency, gloss, color, refractive index, and reflectance are the properties normally associated with aesthetics of plastic materials. In some areas, changes in optical properties, increases in haze after abrasion testing, color differences after weathering, and birefringence analysis of residual stress within a transparent part are all used to measure the effects of applied stresses.

One of the most widely accepted measurements of color in plastics is based on standards developed by the Commission Internationale de l'Eclairage (CIE) for illuminants and observers to establish tristimulus values. A common color scale is used to describe color in numeric terms of lightness and hue. Spectrophotometers are used to measure the full reflectance of a colored material over the visible range and convert to the tristimulus values via microprocessors.

Electrical Properties

Excellent insulating properties, along with the ability to be structural components, make plastics the ideal candidate materials for electrical applications. Although generally used as insulators, carbon black or carbon fiber can be added to make plastic materials electrically conductive, thereby expanding their usefulness in the electronics area.

Standard testing of electrical properties of plastics includes dielectric strength, permittivity, dissipation factor, surface and volume resistivity, and arc resistance.

Electromagnetic interference (EMI) testing has become more prevalent for materials that either emit or are affected by EMI. Shielding efficiency (SE) of materials is determined by measuring electric field strength between a transmitter and receiver with or without the presence of the material under test.

Underwriters' Laboratories (UL) is an independent, nonprofit organization that develops standards for safety in electrical products. UL 746 A, B, and C describe tests and limits for materials used in electrical equipment; UL 746 D lists test requirements for the fabricated plastic parts.

Flammability Properties

Plastics have become an important material in the construction industry and are used in areas of insulation, wire coating, flooring materials, and piping; they are also considered for structural components as foam panels and plastic wood. As use in the industry has grown, so has concern about the fire properties of these materials. Tests to determine ignition resistance, flame spread, heat release, and smoke or toxic gas release have been developed. However, the results of these tests are not applicable to the performance of the materials under actual fire conditions. These small-scale tests are intended for quality control and potential hazard ranking. Several organizations that go beyond testing standard bodies such as ASTM or the International Electrotechnical Commission (IEC) have developed tests and/or regulations for the use of plastics in electrical appliances or buildings.

Rate of heat release data is highly important in determining fire behavior, but few tests have been established. Smoke generation tests have used changes in light absorption of a photoelectric cell within a confined chamber as the primary measurement. Results are reported as percent obscuration or smoke density.

Nondestructive Testing

Nondestructive testing can include any test that does not damage the plastic piece beyond its intended use, such as visual and, in some cases, mechanical tests (see NONDESTRUCTIVE EVALUATION). The term is normally used to describe x-ray, nuclear source, ultrasonics, atomic emission, as well as some optical and infrared techniques for polymers. Nondestructive testing is used to determine cracks, voids, inclusions, delamination, contamination, lack of cure, anisotropy, residual stresses, and defective bonds or welds in materials.

Thermal imaging is sensitive to infrared radiation that detects temperature changes over the surface of a part when heat has been applied. Thermal diffusion in a solid is affected by variation in composition or by the presence of cracks, voids, delaminations, etc; the effects are detected by surface temperature changes.

Other Tests

There are tests for physical properties such as density and hardness (qv) of plastics. Microscopy (qv) is important in fracture analysis as well as in analysis of the morphology of polymer systems for an understanding of polymer blend performance.

BARBARA J. FURCHES
The Dow Chemical Company

M. P. Sepe, in N. P. Cheremisinoff, ed., *Elastomer Technology Handbook*, CRC Press, Inc., Boca Raton, 1993, p. 139.

A. R. McGhie, in R. Linford, ed., *Electrochemical Science and Technology of Polymers-2*, Elsevier Applied Science Publishers, Ltd., London, 1990, p. 214.

R. P. Brown, ed., *Handbook of Plastics Test Methods*, 3rd ed., John Wiley & Sons, Inc., New York, 1988, p. 84.

F. Settle, ed., *Handbook of Instrumental Techniques for Analytical Chemistry*, Prentice Hall PTR, Upper Saddle River, N.J., 1997.

PLASTISOLS. See VINYL POLYMERS, VINYL CHLORIDE AND POLY(VINYL CHLORIDE).

PLATINUM-GROUP METALS

The platinum-group metals (PGMs), which consist of six elements in Groups 8–10 (VIII) of the Periodic Table, are often found collectively in nature. They are ruthenium, Ru; rhodium, Rh; and palladium, Pd, atomic numbers 44 to 46, and osmium, Os; iridium, Ir; and platinum, Pt, atomic numbers 76 to 78. Corresponding members of each triad have similar properties, eg, palladium and platinum are both ductile metals and form active catalysts. Rhodium and iridium are both characterized by resistance to oxidation and chemical attack (see PLATINUM-GROUP METALS, COMPOUNDS).

The PGMs are of significant technological importance. They are also extremely rare, owing in part to low natural abundance and in part to the complex processes required for extraction and refining. The PGMs are used primarily in industrial applications.

The PGMs have become widely established in chemical, electrical, and electronic engineering. The most widely used unit of mass for the PGMs is the troy ounce (1 troy oz = 0.0311 kg). However, herein masses are given in metric tons.

Properties

Physical and Mechanical Properties. Whereas there are some similarities in the physical and chemical properties between corresponding members of the PGM triads, eg, platinum and palladium, the PGMs taken as a unit exhibit a wide range of properties. Some of the most important are summarized in Table 1.

Ruthenium and osmium have properties similar to the refractory metals, ie, they are hard, brittle, and have relatively poor oxidation resistance (see REFRACTORIES). Platinum and palladium, with properties akin to gold, are soft, ductile, and have excellent resistance to oxidation and high temperature corrosion.

Many of the properties of rhodium and iridium, Group 9 metals, are intermediate between those of Group 8 and Group 10. The mechanical and many other properties of the PGMs depend on the physical form, history, and purity of a particular metal sample.

Chemical Properties and Corrosion Resistance. Among the outstanding characteristics of the PGMs are exceptional resistance to corrosive attack by a wide range of liquid and gaseous substances, and stability at high temperatures under conditions where base and refractory metals are easily oxidized. This is owing to thermodynamic stability over a wide range of conditions and the formation of thin protective oxide films in aqueous media under oxidizing or anodic conditions. The PGMs are often used as sheaths, linings, electrodeposits, or other thin coatings on strong supporting structures. In many cases the strength, rigidity, hardness, and resistance to corrosion of the PGMs can be further improved by alloying, particularly with a second metal of the same group.

The PGMs are extremely resistant to corrosion by aqueous solutions of alkalies and salts and by dilute acids, and are generally quite resistant to more concentrated acids and halogens. In concentrated acids at high redox potentials, ie, under oxidizing conditions, there is a zone of corrosion.

High Temperature Properties. There are marked differences in the ability of PGMs to resist high temperature oxidation. Many technological applications, particularly in the form of platinum-based alloys, arise from the resistance of platinum, rhodium, and iridium to oxidation at high temperatures. Osmium and ruthenium are not used in oxidation-resistant applications owing to the formation of volatile oxides.

Sources and Production

Total known world resources of platinum-group metals have been variously estimated as between 68,000 and 96,000 metric tons.

Almost all known deposits of platiniferous ores are related to basic igneous rocks. In the deposits found in South Africa, Canada, and Russia, the PGM-containing ores occur in association with nickel, copper (qv), and iron sulfides. There are over 90 known minerals of the PGMs, of which the most numerous are the minerals of palladium, followed by platinum.

South Africa is by far the largest producer of primary (newly mined) PGMs. In South Africa, PGM deposits are found within a geological area known as the Bushveld Igneous Complex, an area covering approximately 650,000 km^2 in the Central Transvaal. Canada and Russia are also major producers of PGMs. The most significant PGM deposit in the United States is at Stillwater, Montana, where PGMs are mined as the primary product.

Recovery and Refining Techniques

In order to separate the PGMs from each other and from other metals at high yield, high percentage recovery, and high purity, a multistage refining process has been developed. The actual series of processing steps that the ore undergoes varies according to the composition and grade of that ore, and the source from which it was obtained. The refining processes are tailored to each particular material. Every refinery has developed its own specific reagents and separation technology.

PGM Concentration. The aim of a concentration process is to separate a crude metal concentrate from the ore. The majority of other metals, such as nickel and copper, are separated out at this stage for further refining.

An example of a typical concentration process is that used for the ore of South Africa's Merensky Reef. The crude ore is crushed and pulverized, using a series of jaw crushers followed by primary and secondary ball mills. The metal sulfide particles are then separated from the gangue by froth flotation. Conventional gravity concentration is used to separate out free, large, and dense PGM particles from the flotation concentrate, and these particles can go straight to the refining process. The PGM sulfide pulp is smelted in an electric furnace to produce a matte containing copper, nickel, cobalt, and PGMs. Nickel and copper are separated magnetically. The nonmagnetic component is pressure-leached to yield the final concentrate ready for refining.

Conventional Refining Process. The conventional refining process is based on complex selective dissolution and precipitation techniques. In a typical scheme, the PGM concentrate is attacked with aqua regia to dissolve gold, platinum, and palladium. The more insoluble metals, iridium, rhodium, ruthenium, and osmium remain as a residue. Gold is recovered from the aqua regia solution either by reduction to the metallic form with ferrous salts or by solvent-extraction methods. The solution is then treated with ammonium chloride to produce a precipitate of an ammonium platinum salt. Calcination of the precipitate gives Pt sponge, which can undergo further purification. The remaining aqua regia solution is treated with ammonium hydroxide, followed by hydrochloric acid, to precipitate a palladium salt. Calcination of this gives impure palladium sponge.

Table 1. Properties of Platinum-Group Metals

Parameter	Ruthenium	Rhodium	Palladium	Osmium	Iridium	Platinum
atomic number	44	45	46	76	77	78
atomic weight	101.07	102.91	106.40	190.20	192.20	195.09
number of stable isotopes	7	1	6	7	2	6
elemental abundance, ppm	1×10^{-3}	1×10^{-3}	5×10^{-3}	1×10^{-3}	1×10^{-3}	1×10^{-2}
usual valency	3,4,6,8	3	2,4	4,6,8	3,4	2,4
ionic radius, pm	67	68	65	69	68	65
crystal structure[a]	hcp	fcc	fcc	hcp	fcc	fcc
lattice constant, a, pm	270.56	380.3	389	273.41	384	392.31
color	white	silvery	steel white	bluish	yellowish white	silvery
reflectance, %	63	79	54	64	55	
melting point, °C	2310	1960	1552	3050	2443	1769
vapor pressure at mp, Pa[b]	1.31	0.133	3.47	1.8	0.467	0.0187
density, g/cm^3	12.45	12.41	12.02	22.61	22.65	21.45
heat capacity at 25°C, J/(°C·mol)[c]	24.06	24.98	25.98	24.70	24.50	25.85
thermal expansion, °C$^{-1} \times 10^6$	9.1	8.3	11.11	6.10	6.8	9.1
magnetic susceptibility, cm^3/g	4.27×10^{-7}	9.9×10^{-7}	5.23×10^{-6}	5.2×10^{-8}	1.33×10^{-7}	9.71×10^{-7}
work function, eV		4.8	4.99			5.27
Young's modulus, kN/m^2	4.85×10^8	3.86×10^8	1.24×10^8	5.56×10^8	5.28×10^8	1.71×10^8
ultimate tensile strength, MPa[d]	500–600	400–560	180–200		400–500	120–160
Poisson's ratio	0.31	0.36	0.39	0.28	0.28	0.36
Vickers' hardness (VPN)	200–250	100–120	40	300–670	220	40
elongation, %	10	6.5	20–35			40
specific strength per unit cost, MPa·cm^3/(g·\$)[d]	6.66	0.17	0.26		0.72	0.06
temperature coefficient of resistance (TCR), K^{-1}	4.2×10^{-3}	4.6×10^{-3}	3.8×10^{-3}	4.2×10^{-3}	4.3×10^{-3}	3.9×10^{-3}
electrical resistivity at 0°C, $\mu\Omega$·cm	6.8	4.33	9.93	8.12	4.71	9.85
thermal conductivity, W/(m·K)	119	153	75	88	147	73

[a] hcp = hexagonal close-packed; fcc = face-centered cubic.
[b] To convert Pa to mm Hg, multiply by 0.075.
[c] To convert J to cal, divide by 4.184.
[d] To convert MPa to psi, multiply by 145.

The Ir, Ru, Rh, and Os-containing residue from the original aqua regia is subjected to further refining steps that eventually yield the pure metals.

The disadvantage of the conventional refining process is that any single dissolution–reprecipitation step does not give a complete separation of the metals; hence the separation efficiency is low. Precious metals are thus tied up for long periods of time.

Solvent Extraction Technology. The use of solvent extraction technology to replace traditional processes has been the subject of considerable research and development effort since the 1970s. This technique is being used commercially in several of the principal refineries.

Solvent extraction is a relatively high cost process, owing to the specialty organic extractants required and the expenses of recovery and storage of organic solvents. However, in a precious-metal recovery operation, these costs are easily outweighed by the increased efficiency and PGM recovery as well as shortened metal-in-process time.

The actual solvent extraction processes used, including the specific extractants and the order in which the components are separated, vary from refinery to refinery.

Analysis

Like the refining of the PGMs, the analysis is complicated by the chemical similarity of the metals. The techniques used depend on the elements present and their concentration in the sample.

Colorimetric and fluorimetric techniques are sensitive and accurate but have generally been superseded by other methods in specialist analytical laboratories. Gravimetric techniques are still widely used for all the PGMs. The examination of solid materials is carried out by using either x-ray fluorescence spectrometry or dc-arc spectrography and spectrometry. The most widely used method for determining low concentrations of the PGMs in solution is atomic absorption spectrometry. Plasma emission spectrometry is the method of choice for rapid multielement analysis.

Health and Safety Factors

In bulk metallic form, the PGMs are not hazardous to health. In common with many other metals, however, the PGMs in finely divided form can be hazardous to handle. For example, powdered iridium can ignite in air, palladium dust is combustible, and powdered platinum is a powerful catalyst and liable to ignite combustible materials. Fumes of ruthenium can harm the eyes and lungs, and can produce nasal ulcers. Some PGM compounds are toxic. In particular, platinum halogeno salts are highly allergenic.

Uses

The principal applications of the PGMs are summarized in Table 2.

Catalytic Applications. The PGMs are widely used as catalysts for a variety of chemical reactions, such as hydrogenation, oxidation, dehalogenation, dehydrogenation, isomerization, and cyclization. The main areas of commercial application are automotive emission control catalysts (autocatalysts), oil refining, ammonia oxidation, liquid-phase catalysis, and fuel cells.

High Temperature Applications. Exceptional mechanical as well as corrosion- and oxidation-resistant properties of PGMs are particularly pronounced at elevated temperatures. This has led to a substantial number of applications for the PGMs and their alloys in high temperature process industries. For example, platinum and platinum–rhodium alloys are used for glass (qv) industry equipment, eg, crucibles, stirrers, tubes, furnace linings, and bushings.

Table 2. Applications of the Platinum-Group Metals

Application	Pt	Pd	Ru	Rh	Ir
automotive catalysts	+	+		+	
industrial emission catalysts	+				
fuel cells	+				
gas sensors	+	+			
jewelry	+	+			
investment	+				
biomedical devices	+				+
chemotherapeutics, anticancer	+				+
dental materials		+			
electronics	+	+	+		+
electrochemical	+		+		+
chemical[a]	+	+	+	+	+
petroleum refining	+	+			
glass	+			+	
crucibles	+				+
coatings	+				
spark plugs	+				+

[a] Osmium may be used as a catalyst for the chemical and pharmaceutical industries.

Coatings of PGM on a refractory ceramic substrate allow the mechanical properties of the substrate to be combined with the corrosion resistance of the PGM and help overcome the inherent brittleness and poor thermal shock characteristics of the ceramic.

Electronic Applications. The PGMs find important and diverse applications in the electronics industry. Palladium or palladium–silver thick-film pastes are used in multilayer ceramic capacitors and conductor inks for hybrid integrated circuits (qv). Palladium salts are increasingly used to plate edge connectors and lead frames of semiconductors (qv).

Increasing data storage requirements for computer applications have created demands for advanced magnetic and magnetooptic storage media. Platinum cobalt multilayers demonstrate good perpendicular magnetic anisotropy, making them excellent candidates for information storage materials (qv).

Temperature Measurement. PGM thermocouples are widely used for high temperature measurement (qv) in the glass, steel, and semiconductor industries owing to their linear, high thermoelectric voltages. In the semiconductor and glass industries, permanently installed profiling thermocouples are used to control furnace temperatures (see FURNACES, ELECTRIC).

Jewelry and Investment. Platinum is widely used in jewelry. As pure platinum is too soft for jewelry applications, it is alloyed with other PGMs such as iridium or ruthenium to increase its hardness and wear resistance. Pure platinum (99.95%) in the form of coins and ingots is used as an investment metal.

Medical and Dental Applications. The most important medical use of platinum is in the anticancer drugs, carboplatin and cisplatin (see CHEMOTHERAPEUTICS, ANTICANCER; PLATINUM-GROUP METALS, COMPOUNDS). However, the PGMs are also used in metallic form in a range of biomedical devices (see PROSTHETICS AND BIOMEDICAL DEVICES). The PGMs are biocompatible as well as highly corrosion resistant and workable.

In dentistry, palladium alloys are widely used as alternatives to base metal alloys in the manufacture of crowns and bridges as well as the replacement of lost or damaged teeth (see DENTAL MATERIALS).

RICHARD J. SEYMOUR
JULIA I. O'FARRELLY
LOUISE C. POTTER
Johnson Matthey Public Limited Company

Y. M. Savitskii and A. Prince, *Handbook of Precious Metals*, Hemisphere Publishing, New York, 1989.

L. S. Benner and co-workers, eds., *Precious Metals Science and Technology*, IPMI, Allentown, Pa., 1991, pp. 375–399.

Platinum Metals Review, published quarterly since 1957 by Johnson Matthey, London, U.K.

E. Savitsky and co-workers, *Physical Metallurgy of Platinum Metals*, MIR Publishers, Moscow, CIS, 1978.

PLATINUM-GROUP METALS, COMPOUNDS

The platinum-group metals (qv) (ruthenium, osmium, rhodium, iridium, palladium and platinum) are kinetically inert relative to other transition-metal ions. All of the platinum-group metals (PGMs) have several readily accessible stable oxidation states. The extreme case is that of Ru and Os, which exhibit oxidation states ranging from −2 and 0 to +8. The relative ease of conversion between oxidation states gives rise to a rich catalytic chemistry. The lower oxidation states of the PGMs tend to enable them to form complexes with soft ligands such as sulfur and phosphorus whereas higher oxidation states tend toward hard ligands such as oxygen and fluorine (see COORDINATION COMPOUNDS). PGMs form binary compounds, coordination compounds, and organometallic compounds.

Ruthenium Compounds

The most common oxidation states and the corresponding electronic configuration of ruthenium are +2 (d^6) and +3 (d^5). Compounds are usually octahedral. Compounds in oxidations states from −2 (d^{10}), and 0 (d^8) to +8 (d^0) have various coordination geometries. Important applications of ruthenium compounds include oxidation of organic compounds and use in dimensionally stable anodes (DSA).

Synthesis. The most important starting material for the synthesis of ruthenium compounds is the commercial trichloride trihydrate.

Osmium Compounds

The most common oxidation states and the corresponding electronic configurations of osmium are +2 (d^6) and +3 (d^5), which are usually octahedral. Stable oxidation states that have various coordination geometries include −2 (d^{10}), and 0 (d^8), to +8 (d^0). The single most important application is OsO_4 oxidation of olefins to diols. Enantioselective oxidations have also been demonstrated.

Synthesis. The most important starting material for synthesis of osmium complexes is OsO_4.

Rhodium Compounds

The most common oxidation states and corresponding electronic configurations of rhodium are +1 (d^8), which is usually square planar although some five coordinate complexes are known, and +3 (d^6) which is usually octahedral. Dimeric rhodium carboxylates are +2 (d^7) complexes. Compounds in oxidation states −1 (d^{10}) to +6 (d^3) exist. Significant industrial applications include rhodium-catalyzed carbonylation of methanol to acetic acid and acetic anhydride, and hydroformylation of propene to *n*-butyraldehyde. Enantioselective catalytic reduction has also been demonstrated.

Synthesis. The most important starting material for rhodium compounds is rhodium(III) chloride hydrate, $RhCl_3 \cdot nH_2O$. Other commercially available starting materials useful for laboratory-scale synthesis include $[Rh_2(OOCCH_3)_4]$, $[Rh(NH_3)_5Cl]Cl_2$, $[Rh_2Cl_2(CO)_4]$, and $[Rh_2Cl_2(cod)_2]$.

Iridium Compounds

The most common oxidation states, corresponding electronic configurations, and coordination geometries of iridium are +1 (d^8) usually square plane although some five-coordinate complexes are known, and +3 (d^6) and +4 (d^5), both octahedral. Compounds in every oxidation state between −1 (d^{10}) and +6 (d^3) are known. Iridium compounds are used primarily to model more active rhodium catalysts.

Synthesis. The principal starting material for synthesis of iridium compounds is iridium trichloride hydrate, $IrCl_3 \cdot xH_2O$. Another useful material for laboratory-scale reactions is $[Ir_2Cl_2(cod)_2]$.

Palladium Compounds

The most common oxidation state of palladium is +2 which corresponds to a d^8 electronic configuration. Compounds have square planar geometry. Other important oxidation states and electronic configurations include 0 (d^{10}), which can have coordination numbers ranging from two to four and is important in catalytic chemistry; and +4 (d^6), which is octahedral and much more strongly oxidizing than platinum (IV). The chemistry of palladium is similar to that of platinum, but palladium is between 10^3 to 5×10^5 more labile. A primary industrial application is palladium-catalyzed oxidation of ethylene (see OLEFIN POLYMERS) to acetaldehyde (qv). Palladium-catalyzed carbon–carbon bond formation is an important organic reaction.

Synthesis. The most common starting materials for palladium complexes are $PdCl_2$ and $[PdCl_4]^{2-}$. Commercially available materials useful for laboratory-scale synthesis include $[Pd_3(OOCCH_3)_6]$, $[PdCl_2(NCC_6H_5)]$, $[Pd(acac)_2]$, $[PdCl_2(cod)]$, and $[Pd(P(C_6H_5)_3)_4]$.

Platinum Compounds

The most common oxidation states and corresponding electronic configurations of platinum are +2 (d^8), which is square planar, and +4 (d^6), which is octahedral. Compounds in oxidation states between 0 (d^{10}) and +6 (d^4) exist. Platinum hydrosilation catalysts are used in the manufacture of silicone polymers. Several platinum coordination compounds are important chemotherapeutic agents used for the treatment of cancer.

Synthesis. The most important starting materials for platinum compounds are potassium tetrachloroplatinate(II), $K_2[PtCl_4]$, and $H_2[PtCl_6]$. Other useful starting materials are $PtCl_2(CCN_6H_5)_2$, $[PtCl_2(cod)$; cod = 1, 5-cyclooctane], and $[PtCl_2(P(C_6H_5)_3)_2]$.

Health and Safety

Compounds of most PGMs are only slightly to moderately toxic by oral ingestion. The most serious acute hazard arises from exposure to volatile RuO_4 (bp 40°C) and OsO_4 (bp 131°C), which deposit black RuO_2 or OsO_2 upon contact with tissue. These substances are especially hazardous to the eyes and respiratory system. Other volatile Os compounds, eg, OsF_6 (bp 45.9°C) and $OsCl_4$ (bp 130°C), pose a similar hazard. Finely divided osmium metal can react in air to form OsO_4. The acceptable 8-h exposure for OsO_4 is 2 $\mu g/m^3$.

Work using PGM compounds should be carried out in a properly functioning hood. Special care should be taken to avoid inhalation or contact of fumes with the eyes. Exposure to some anionic salts of platinum can lead to serious allergic reaction and sensitization. Symptoms include rhinitis, conjuctivitus, asthma, urticara, and contact dermatitis. Whereas the allergic reaction can be life-threatening, there is no evidence of long-term effects if a sensitized individual is removed from exposure to platinum salts. Appropriate protective clothing should be worn and precautions should be taken to avoid inhalation or contact with the skin when working with these compounds. Compounds such as *cis*-$[PtCl_2(NH_3)_2]$, $[Pt(NH_3)_4]Cl_2$, and $K_2[Pt(NO_2)_4]$ are not allergenic. Platinum antitumor compounds are toxic. There have been no reports of adverse toxic effects of platinum compounds owing to environmental exposure arising from the use of automobile catalysts.

CHRISTEN M. GIANDOMENICO
Johnson Matthey

E. A. Seddon and K. R. Seddon, *The Chemistry of Ruthenium*, Elsevier Science Publishing Co., Inc., New York, 1984.

R. F. Heck, *Palladium Reagents in Organic Synthesis* Academic Press, New York, 1985.

G. Wilkinson, R. D. Gillard, and J. A. McCleverty, eds., *Comprehensive Coordination Chemistry* Pergamon Press, Oxford, U.K., 1987.

G. Wilkinson, F. Gordon, A. Stone, and E. W. Abel, eds., *Comprehensive Organometallic Chemistry* Pergamon Press, Oxford, U.K., 1982.

PLUTONIUM AND PLUTONIUM COMPOUNDS

Plutonium, Pu, atomic number 94, is the fifth member of the actinide series and is metallic (see ACTINIDES AND TRANSACTINIDES). Isotopes of mass number 232 through 246 have been identified. All are radioactive. The most important isotope is plutonium-239, ^{239}Pu; also of importance are ^{238}Pu, ^{242}Pu, and ^{244}Pu.

The large energy release that accompanies the nuclear fission reaction is the most significant property of plutonium. Upon fission, one gram of plutonium-239 releases energy equivalent to that produced by combustion of three metric tons of coal (qv). Much of the world's separated plutonium has been used for nuclear weapons.

In plutonium-fueled breeder power reactors, more plutonium is produced than is consumed (see NUCLEAR REACTORS, REACTOR TYPES). Thus the utilization of plutonium as a nuclear energy or weapon source is especially attractive to countries that do not have uranium-enrichment facilities.

The isotope plutonium-238, ^{238}Pu, is of technical importance because of the high heat that accompanies its radioactive decay. This isotope is used as fuel in small terrestrial and space nuclear-powered sources.

Plutonium was the first element to be synthesized in weighable amounts. Its metallurgy and chemistry are complex. Metallic plutonium exhibits seven allotropic modifications. Five different oxidation states are known to exist in compounds and in solution.

Sources

Plutonium occurs in natural ores in such small amounts that separation is impractical. The atomic ratio of plutonium to uranium in uranium ores is less than $1:10^{11}$. Traces of primordial plutonium-244 have been isolated from the mineral bastnasite. The content of plutonium-239, ^{239}Pu, in uranium minerals is given in Table 1.

With the exception of ^{244}Pu, the plutonium isotopes are synthetic, and all are radioactive (see RADIOISOTOPES).

The technologically most important isotope, ^{239}Pu, has been produced in large quantities since 1944 from natural or partially enriched uranium in production reactors. This isotope is characterized by a high fission reaction cross section and is useful for fission weapons, as trigger for thermonuclear weapons, and as fuel for breeder reactors.

Commercial electric power generating reactors generally produce plutonium by irradiating uranium fuels to a total neutron exposure of more than 5000 MW · d/t. The recoverable plutonium contains a larger fraction of heavier isotopes.

Plutonium Metal

Preparation. Reductions of PuF_4 or PuF_3 on the gram or multigram scale have been carried out in centrifugal bombs and stationary bombs

Table 1. Plutonium in Natural Sources

Ore	Uranium content, wt %	Mass ratio, ^{239}Pu/ore
pitchblende		
Canadian	13.5	9.1×10^{-12}
Zaire	38	4.8×10^{-12}
Colorado	50	3.8×10^{-12}
Zaire concentrate	45.3	7.0×10^{-12}
monazite		
Brazilian	0.24	2.1×10^{-14}
North Carolina	1.64	5.9×10^{-14}
fergusonite	0.25	$<1.0 \times 10^{-14}$
carnotite	10	$<4.0 \times 10^{-14}$
bastnasite		1×10^{-18}[a]

[a] Value corresponds to a mass ratio of ^{244}Pu/ore.

in order to produce coherent, pure Pu in high yield. The direct oxide reduction (DOR) process reduces PuO_2 using calcium metal in a molten salt bath of $CaCl_2$ or $CaCl_2-CaF_2$.

Physical Properties. The element exists in six allotropic modifications under ordinary pressure, and there is a high pressure allotrope.

The expansion coefficient of α-plutonium is exceptionally high for a metal, whereas those of δ- and δ'-plutonium are negative. The net linear increase in heating a polycrystalline rod of plutonium from room temperature to just below the melting point is 5.5%.

The electrical resistivity of plutonium is high in all modifications as a result of the band structure in metallic plutonium.

Because of α-emission radioactivity, all plutonium isotopes generate heat. The self-heating of ^{239}Pu metal is $(1.923 \pm 0.019) \times 10^{-3}$ W/g$(1.824 \pm 0.18$ Btu/(s·g)). Large ^{239}Pu samples have a temperature higher than that of the human body.

Storage and Handling. Plutonium can be stored safely in dry air. Because of self-heating, storage accompanied by heat removal is advisable.

Chemical Properties of Plutonium Metal. Plutonium is a reactive metal. It forms compounds with all the nonmetallic elements except the rare gases. The halogens and halogen acids form Pu halides, other chalcogens form chalcogenides, and CO forms a carbide. Nitrides (qv) are formed with NH_3 and N_2 and hydrides (qv) with H_2.

The metal dissolves readily in concentrated HCl, H_3PO_4, HI, or $HClO_4$. α-Plutonium oxidizes very slowly in dry air. The rate is accelerated by water vapor. Plutonium is similar to uranium with respect to corrosion characteristics.

Properties of Atomic Plutonium and Plutonium Ions

Spectroscopic. The electronic configuration of Pu vapor is [Xe] $4f^{14}5d^{10}5f^6 6s^2 6p^6 6d^0 7s^2$. The singly charged ion Pu^+ and the doubly charged ion Pu^{2+} are produced in emission spectroscopy. Summaries and comprehensive listings and assignments of spectra have been published; a comprehensive and interpretive review has also been published.

Aqueous Solution Chemistry. The aqueous solution chemistry of plutonium is complex. Plutonium can exist in acidic aqueous solutions in oxidation states III, IV, V, and VI. Additionally plutonium can be oxidized to Pu(VII) in alkaline solutions. Aqueous solutions of each oxidation state can be prepared by chemical oxidants or reductants, but the best methods appear to be electrochemical.

The ability of the various oxidation states of Pu to form complex ions with simple hard ligands, such as oxygen, is, in order of decreasing stability, $Pu^{4+} > PuO_2^{2+} > Pu^{3+} > PuO_2^{+}$.

Complexes formed by Pu ions with OH^- represent hydrolysis reactions. There is extensive interaction between Pu^{x+} and water.

Because carbonate and bicarbonate are commonly found under environmental conditions in water, and because carbonate complexes Pu readily in most oxidation states, Pu carbonato complexes have been studied extensively. The reduction potentials vs the standard hydrogen electrode of Pu(VI)/(V) shifts from 0.916 to 0.33 V and the Pu(IV)/(III) potential shifts from 1.48 to −0.50 V in 1 *M* carbonate. These shifts indicate strong carbonate complexation.

Analytical Chemistry

It is possible to analyze a plutonium-containing sample gravimetrically by precipitating it as oxalate, calcining to PuO_2, and weighing. Laser photoacoustic spectroscopy has been developed for both elemental analysis and speciation (oxidation state) at concentrations of $10^{-4}-10^{-5} M$.

X-ray fluorescence, mass spectroscopy, emission spectrography, and ion-conductive plasma–atomic emission spectroscopy (icp–aes) are used in specialized laboratories equipped for handling radioisotopes with these instruments.

A large number of radiometric techniques have been developed for Pu analysis on tracer, biochemical, and environmental samples.

Separation Chemistry and Extractive Metallurgy. The principal problem in the purification of metallic plutonium is the separation of small amounts of plutonium (ca 200–900 ppm) from large amounts of uranium, which contain intensely radioactive fission products. The plutonium yield or recovery must be high and the plutonium relatively pure with respect to fission products and light elements.

Production methods include liquid–liquid extraction (qv), ion exchange (qv), pyrometallurgical processing, and nuclear waste reprocessing.

Storage, Usage, and Disposal of Excess Weapons Plutonium

The production of plutonium for nuclear weapons in the United States and Russia appears to have ended. A vast amount, approximately 50 metric tons in each of the two countries, is being removed as metal from nuclear weapons. Criteria for storage facilities have been published. A number of options for long-term disposition that are feasible from an engineering viewpoint and environmentally appropriate are under study.

Plutonium Compounds

Plutonium forms compounds with many of the metallic elements and all of the nonmetallic elements, except the helium-group gases (qv).

The most important oxide is PuO_2. The high melting point (2390 ± 20°C), chemical stability, radiation stability, and similarity to UO_2 characterize plutonium dioxide, PuO_2, as an attractive reactor fuel. Plutonium dioxide also is an important intermediate in processing operations such as the preparation of PuF_4.

Pu(IV) peroxide is an important intermediate in the conversion of plutonium nitrate solution to metal.

The following binary plutonium halides have been characterized: PuF_6, PuF_4, PuF_3, $PuCl_3$, $PuBr_3$, and PuI_3.

Only Pu(III) oxyhalides (PuOF, PuOCl, PuOBr, and PuOI) and Pu(VI), oxyhalides (PuO_2F_2, $PuOF_4$, and $PuO_2Cl_2 \cdot 6\ H_2O$); are known. Of these the most important are PuOCl and PuO_2F_2.

Plutonium hydrides are important because they display unique involvement of the 5*f* electrons in bonding, thereby giving these compounds interesting magnetic and transport properties.

Plutonium carbides and silicides have received consideration as advanced reactor fuels. They can be prepared from the elements.

Plutonium nitride, PuN, has been studied as a possible fast-reactor fuel. The pnictides are also interesting for their solid-state magnetic and electrical properties.

All of the chalcogenides can be prepared from the elements. The chalcogenides are also of interest because of their solid-state magnetic and electrical behavior.

Plutonium in the Environment

It has been estimated that 1.3×10^{16} Bq of 239 + 240 Pu has been released to the environment from atmospheric detonation of nuclear weapons; that 7.9×10^{14} Bq of ^{238} Pu has been released, mostly from burn-up of the nuclear powered satellite SNAP-9a; and that 3.7×10^{13} Bq of 239 + 240 Pu was released by the Chernobyl accident.

Health and Safety Factors

The principal hazards of plutonium are those posed by its radioactivity, nuclear critical potential, and chemical reactivity in the metallic state. ^{239}Pu is primarily an α-emitter. Thus, protection of a worker from its radiation is simple and usually no shielding is required unless very large (kilogram) quantities are handled or unless other isotopes are present.

Protection Against Nuclear and Chemical Hazards. In a laboratory where plutonium is processed and workers are present, a ventilation rate of 8–10 air changes per hour has been recommended. The design and operation of hoods, glove boxes, and laboratories for work with plutonium are specialties that have been highly developed as a result of years of experience.

Plutonium metal and air-sensitive plutonium compounds must be isolated for chemical as well as nuclear safety. They are handled in airtight glove boxes containing an inert atmosphere.

Although massive pieces of plutonium and its alloys are safe to handle in air, a few Pu compounds and finely powdered Pu metal are pyrophoric. The preferred method of extinguishing a Pu fire is by excluding oxygen. Controlled burning of Pu metal also has been recommended as a way to convert it to an inert oxide.

The presence of a critical mass of Pu in a container can result in a fission chain reaction. The quantity of ^{239}Pu required for a critical mass depends on several factors: the form and concentration of the Pu, the geometry of the system, the presence of moderators (water, hydrogen-rich compounds such as polyethylene, cadmium, etc), the proximity of neutron reflectors, the presence of nuclear poisons, and the potential interaction with neighboring fissile systems.

LESTER R. MORSS
Argonne National Laboratory

J. M. Cleveland, *The Chemistry of Plutonium*, 2nd ed., American Nuclear Society, La-Grange Park, Ill., 1979.

M. Taube, *Plutonium—A General Survey*, Verlag Chemie, Weiheim, Germany, 1974.

F. Weigel, J. J. Katz, and G. T. Seaborg, in J. J. Katz, G. T. Seaborg, and L. R. Morss, eds., *The Chemistry of the Actinide Elements*, 2nd ed., Chapman & Hall, London, 1986, Chapt. 7.

O. J. Wick, ed., *Plutonium Handbook*: A Guide to the Technology, Vols. I and II, Gordon and Breach, New York, 1967.

POLISHES

Polishes are used to maintain a glossy finish on surfaces as well as to prolong the useful lives of these surfaces. Appearance enhancement provided by polishes generally results from the presence of components that leave a glossy coating, and/or materials that smooth and clean surfaces. Furniture, shoe, and most floor polishes rely on the deposition of a film. In addition to providing glossy protective films, car polishes contain abrasives (qv) to remove weathered paint (qv) and soils. Metal polishes are based on either abrasive smoothing and cleaning or tarnish-removing chemicals, and sometimes deposit materials that retard future tarnishing (see COATINGS; METAL SURFACE TREATMENTS).

Furniture

Furniture polishes contain one or two classes of film-forming ingredients, solvents, and various stabilizers. Natural and synthetic waxes are the main film formers in many formulations. Typically, these waxes are long-chain molecules that do not evaporate easily. They contain few polyunsaturated compounds that might be susceptible to atmospheric degradation, and the waxes may be immune to enzymatic breakdown by microbes.

The different types of furniture polishes include liquid or paste solvent waxes, clear oil polishes, emulsion oil polishes, emulsion wax polishes, and aerosol or spray polishes. Paste waxes contain ca 25 wt % wax, the remainder being solvent. Clear oil polishes contain 10–15 wt % oil and a small amount of wax, the rest being solvent. Aerosol or spray products may contain 2–5 wt % of a silicone polymer, 1–3 wt % wax, 0–30 wt % hydrocarbon solvent, and ca 1 wt % emulsifier. The remainder is water.

Floor

Floor polishes are subject to more mechanical abuse than other polish films and, therefore, should be resistant to abrasion, soiling, water, and detergents. Polymers, which generally are of higher molecular weight than natural waxes, can be formulated to provide films having these properties. In addition, the ability to tailor the properties of synthetic polymers has produced self-polishing aqueous formulations. These formulations result in the deposition of films having high gloss, low color, durability, little tendency to powder, and built-in mechanisms for easy removal.

Aqueous, self-polishing, polymeric formulations generally contain two or three polymeric film formers, coalescing agents, leveling aids, plasticizers, zinc complexes, ammonia, and wetting and emulsifying agents.

Waxes are added to formulations in the form of emulsions. Waxes provide buffability to polish films and increase abrasion resistance.

Water-clear floor finishes contain little or no wax, whereas buffable products contain relatively large amounts of wax. Sealers contain little wax and relatively large amounts of emulsion polymers.

Floor polishes typically are evaluated for gloss, application and leveling properties, discoloration, slip resistance, scratch resistance, heel-mark resistance, scuff resistance, damp-mopping and detergent resistance, repairability, lack of sediment, and removability.

Automobiles

Automobile polishing is designed to remove road film and oxidized paint, and to lay down a continuous, glossy film, which resists removal by water and car-wash detergents. Much of the market is represented by one-step products which generally contain four functional ingredients. Abrasives are the principal cleaning ingredients but must not be so aggressive as to scratch the paint film. Representative types are fine grades of aluminum silicate, diatomaceous earth, and silicas (see ABRASIVES; DIATOMITE; SILICA; SILICON COMPOUNDS). Modern acrylic automobile paints, in contrast to older alkyd or nitrocellulose types, do not oxidize as rapidly and so reduce the need for highly abrasive polishes.

Car polishes can be solid, semisolid, or liquid. They can be solvent-based or emulsions. A representative liquid emulsion product may contain 10–15 wt % abrasive, 0–30 wt % solvent, 2–12 wt % silicone, and 0–4 wt % wax; an emulsion paste product may contain 3–15 wt % wax with other ingredients at similar levels.

Metal

In industrial metal finishing, polishing is an abrading operation involving the use of coarse abrasives, which remove significant amounts of metal from a surface and leave visible line patterns; buffing is the smoothing of the resultant surface and involves the use of fine abrasives to reduce the dimensions of the polishing patterns. Nonindustrial polishing does not remove large amounts of metal, but cleans and buffs to remove tarnish, oxides, and stubborn soils. The exposed metal is buffed to a high luster.

Formulated metal polishes consist of fine abrasives similar to those involved in industrial buffing operations, ie, pumice, tripoli, kaolin, rouge and crocus iron oxides, and lime. Other ingredients include surfactants (qv), chelating agents (qv), and solvents.

A problem associated with the use of abrasive metal polishes is that the fresh metal, which has been exposed by the cleaning, rapidly oxidizes or tarnishes. Thus, many modern polishes contain inhibitors.

Metal polishes may contain emulsifiers and thickeners for controlling the consistency and stabilization of abrasive suspensions, and the product form can be solid, paste, or liquid.

Shoe

Use of a shoe polish imparts high gloss, maintains the supple hand of the leather (qv), and increases the weather resistance of the leather. Three general types of polishes are produced: solvent pastes, self-polishing liquids, and emulsion creams. Solvent pastes are similar to the paste furniture and floor polishes, except that the former contain higher wax-to-solvent ratios and high levels of dye. The resulting film dries to a dull finish and must be buffed to a high shine. Liquid self-polishing products contain a soft polymer and coloring agents. Shoe creams can be made in any consistency. They are emulsions of waxes, solvents, and water.

Health, Safety, and Environmental Factors

Liquid polishes and waxes containing 10 wt % or more petroleum distillates must be contained in childproof packaging. General experience indicates that natural waxes and polyethylene waxes are nontoxic. Although nonsolvent floor polishes are relatively nontoxic, concern for floor waxes continues to be slip-resistance.

FRANCIS J. RANDALL
SEÁN G. DWYER
S. C. Johnson & Son, Inc.

E. W. Flick, *Household and Automotive Chemical Specialties: Recent Formulations*, Noyes Data Corp., Park Ridge, N.J., 1979.

1994 Waxes, Polishes, and Floor Finishes Survey by the Chemical Specialties Manufacturers Association, CSMA, Washington, D.C.

W. J. Hackett, *Maintenance Chemical Specialties*, Chemical Publishing Co., New York, 1972.

POLLUTION. See AIR POLLUTION; ENVIRONMENTAL IMPACT.

POLYACETALS. See ACETAL RESINS.

POLYACRYLAMIDES. See ACRYLAMIDE POLYMERS.

POLYACRYLATES. See ACRYLIC ESTER POLYMERS, SURVEY.

POLACRYLONITRILE. See ACRYLONITRILE POLYMERS.

POLYAMIDES

GENERAL

Polyamides, often also referred to as nylons, are high polymers which contain the amide repeat linkage in the polymer backbone. They are generally characterized as tough, translucent, semicrystalline polymers that are moderately low cost and easily manipulated commercially by melt processing. However, significant exceptions to all these attributes occur. The regularity of the amide linkages along the polymer chain defines two classes of polyamides: AB and AABB.

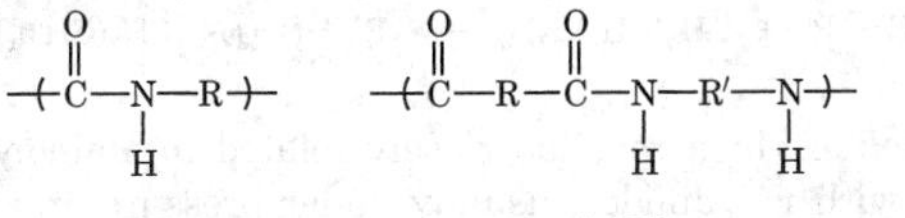

AB AABB

Type AB, which has all the amide linkages with the same orientation along the backbone, can be viewed as being formed in a polycondensation reaction from ω-amino acids to give a polymer with the repeat unit AB. Type AABB, where the amide linkages alternate in orientation along the backbone, can be viewed as being formed from diacids and diamines in a polycondensation reaction to form a polymer with the repeat unit AABB. The R and R′ groups in these structures are hydrocarbon radicals and can be aliphatic, aromatic, or mixed. Because the chemical and physical properties of polyamides differ drastically between those which contain greater than ~15% aliphatic character, ie, nylons, and those which are predominately aromatic, ie, aramids, the aromatic polyamides are treated separately in this article.

Nomenclature

The nomenclature (qv) of polyamides is fraught with a variety of systematic, semisystematic, and common naming systems used variously by different sources. In North America the common practice is to call type AB or type AABB polyamides nylon-*x* or nylon-*x,x*, respectively, where *x* refers to the number of carbon atoms between the amide nitrogens. For type AABB polyamides, the number of carbon atoms in the diamine is indicated first, followed by the number of carbon atoms in the diacid. For example, the polyamide formed from 6-aminohexanoic acid is named nylon-6; that formed from 1,6-hexanediamine or hexamethylenediamine and dodecanedioic acid is called nylon-6,12. In Europe, the common practice is to use the designation "polyamide," often abbreviated PA, instead of "nylon" in the name. Thus, the two examples above become PA-6 and PA-6,12, respectively. PA is the International Union of Pure and Applied Chemistry (IUPAC) accepted abbreviation for polyamides.

Economic Aspects

Since its introduction in the 1940s as the polymer base for the first synthetic fiber, nylon has been a significant commercial polymeric material.

Table 1 shows the principal end uses for polyamides and the approximate relative percent of the total consumption, where the "other" category includes monofilaments and nonwovens. Synthetic fiber, comprising almost 75% of the total, still represents the principal end use for nylon polymer.

Physical Properties

Crystallinity. Linear polyamide homopolymers consist of crystalline and amorphous phases and are termed semicrystalline. Crystallinity enhances yield strength, hardness (qv), abrasion resistance, tensile strength, elastic and shear modulus, and probably resistance to thermooxidation, but it decreases moisture absorption and impact strength. Most commercial samples of nylon-6,6 and nylon-6 are 40–50% crystalline by weight, as determined by density measurements.

Solubility. In general, the homopolymer aliphatic polyamides are insoluble in common organic solvents at room temperature. However, they are soluble in formic acid, phenols, chloral hydrate, mineral acids, and fluorinated alcohols. At higher temperatures, lithium or calcium chloride–methanol mixtures are effective solvents, as are benzyl alcohol, unsaturated alcohols, alcohol–halogenated hydrocarbons, and nitro alcohols. Copolymers of aliphatic polyamides and polyamides with substitution on the amide nitrogen, both of which significantly reduce the degree of crystallinity, are more soluble, and methanol–chloroform mixtures can often be used. Predominately or wholly aromatic polyamides require powerful solvents, such as trifluoroacetic acid (TFA) or concentrated sulfuric acid. Lower molecular weight aromatic polyamides are often soluble in basic solvents such as *N,N*-dimethylacetamide, usually with the addition of lithium chloride or calcium chloride.

Piezoelectric Effect. The electrical properties of piezoelectricity, ie, the ability to generate an electrical signal in response to a mechanical stress; pyroelectricity, ie, the ability to generate an electrical signal in response to a temperature change; and ferroelectricity, ie, the ability to respond repeatedly to reversing external electric fields, have been

Table 1. Percentage of World Consumption of Polyamides by End Use

End use or type	Percentage
fibers	
carpet	34
textile	27
industrial	13
engineering resins	15
aramid	4
films/coatings	4
adhesives	1
other	2

recognized in polymers for many years. Materials with these properties find application in microphones, tone generators, hydrophones, ir detectors, electromechanical transducers, and in numerous other devices. The odd-numbered nylons possess a strong piezoelectric effect. The phenomenon has been observed in nylon-11, nylon-9, nylon-7, and nylon-5.

The odd-numbered nylons exhibit an additional useful property in that their piezoelectric constants increase with increasing temperature almost to the melting point of the polymer.

Pyroelectricity has been observed in nylon-11 and nylon-5,7.

Chemical Properties

Direct Amidation. The direct reaction of amino acids to form Type AB polyamides (eq. 1) and diacids and diamines to form type AABB polyamides (eq. 2) are two of the most commonly used methods to produce polyamides.

$$n\,H_2N{-}R{-}\overset{\overset{O}{\|}}{C}{-}OH \rightleftharpoons H_2N{-}R{+}\overset{\overset{O}{\|}}{C}{-}NH{-}R{+}_{(n-1)}\overset{\overset{O}{\|}}{C}{-}OH + (n-1)H_2O \qquad (1)$$

$$n\,H_2N{-}R{-}NH_2 + n\,HO{-}\overset{\overset{O}{\|}}{C}{-}R'{-}\overset{\overset{O}{\|}}{C}{-}OH \rightleftharpoons H_2N{-}R{-}NH{+}\overset{\overset{O}{\|}}{C}{-}R'{-}\overset{\overset{O}{\|}}{C}{-}NH{-}R{-}NH{+}_{(n-1)}\overset{\overset{O}{\|}}{C}{-}R'{-}\overset{\overset{O}{\|}}{C}{-}OH + (2n-1)H_2O \qquad (2)$$

The integer n is called the degree of polymerization (DP). The average DP is approximately 200 for a typical nylon-6, or about 100 for nylon-6,6; thus the number-average molecular weight is approximately equal for both, since the monomer, hexamethyleneadipamide, for nylon-6,6 has twice the unit weight as the monomer, ϵ-aminocaproamide, has in nylon-6. Ideally for the amino acids, only one homologous series of linear polymers is formed, each member of which possesses one amino and one carboxyl end group, as shown in equation **2**. However, for the type AABB polymers, two additional homologous series of linear polymers are possibly one with two amino end groups and one with two carboxyl end groups:

$$H_2NRNH{+}\overset{\overset{O}{\|}}{C}R'\overset{\overset{O}{\|}}{C}NHRNH{+}_{(n-1)}H \qquad HO{+}\overset{\overset{O}{\|}}{C}R'\overset{\overset{O}{\|}}{C}NHRNH{+}_{(n-1)}\overset{\overset{O}{\|}}{C}{-}R'{-}\overset{\overset{O}{\|}}{C}{-}OH$$

Acid Chloride Reaction. In situations where the reactants are sensitive to high temperature or the polymer degrades before the melt point is reached, the acid chloride route is often used to produce the polyamide.

Ring-Opening Polymerization. Ring-opening polymerization is the method used to convert lactams to polyamides. There are several variations of the method, but the most commonly practiced method in industry is hydrolytic polymerization, in which lactams containing six or more carbons in the ring are heated in the presence of water above the melting point of the polyamide.

Other Preparative Reactions. Many of the methods used to prepare simple amides are applicable to polyamides. Polyamides of aromatic diamines and aliphatic diacids can also be made by the reaction of the corresponding aromatic diisocyanate and diacids.

Reactions of Polyamides

Acidolysis, Aminolysis, and Alcoholysis. When heated, polyamides react with monofunctional acids, amines, or alcohols, especially above the melt temperature, to undergo rapid loss of molecular weight, eg, as in acidolysis (eq. **3**) with acetic acid or aminolysis (eq. **4**) with an aliphatic amine:

$$R{-}\overset{\overset{O}{\|}}{C}{-}NH{-}R' + CH_3\overset{\overset{O}{\|}}{C}{-}OH \longrightarrow R{-}\overset{\overset{O}{\|}}{C}{-}OH + CH_3\overset{\overset{O}{\|}}{C}{-}NH{-}R'$$

$$R{-}\overset{\overset{O}{\|}}{C}{-}NH{-}R' + CH_3(CH_2)_xNH_2 \longrightarrow R'{-}NH_2 + CH_3(CH_2)_xNH{-}\overset{\overset{O}{\|}}{C}{-}R$$

Ammonolysis. In a reaction closely related to aminolysis, ammonia reacts with polyamides, usually under pressure and at elevated temperatures. The ammonolysis of polyamides obtained from post-consumer waste has been used to cleave the polymer chain as the first step in a recycle process in which mixtures of nylon-6,6 and nylon-6 can be reconverted to diamine.

Transamidation and Transesteramidation. Transamidation is the mutual exchange of chain fragments in a polyamide, shown as follows where R, R″ and R′, R‴ represent polymer chain fragments of any length.

$$R{-}\overset{\overset{O}{\|}}{C}{-}NH{-}R' + R''{-}\overset{\overset{O}{\|}}{C}{-}NH{-}R''' \longrightarrow R{-}\overset{\overset{O}{\|}}{C}{-}NH{-}R''' + R''{-}\overset{\overset{O}{\|}}{C}{-}NH{-}R'$$

Transamidation is an important process in the melt phase for polyamides because it is usually the process by which an equilibrium molecular weight distribution is reestablished and, in the case of the melt blending of two or more polyamides to form a copolymer, it is the process by which randomization of the individual monomers along the chain is effected.

Transesteramidation is a process similar to transamidation, except that a polyamide is mixed with a polyester rather than another polyamide. This is often a convenient route to produce polyesteramides.

Grafting. Grafting is the process of chemically bonding additional polymeric units, usually not polyamides, to the nylon polymer chain. This is most often initiated by reaction of the grafting substrate directly with the polyamide backbone, but grafting can also be achieved by introducing non-amide reactive sites into the polymer chain, through the incorporation of reactive comonomers during polymerization.

Grafting can be used to change the surface properties of the final polyamide article, especially film or fiber; its hydrophilic, antistatic, frictional, or other characteristics are altered. Also, grafting can be made to occur in the bulk polymer; this alters, for example, the mechanical (toughening) or optical (delustering) properties of the polyamide.

Degradation of Polyamides

Hydrolysis. Hydrolysis is the reverse of the amidation reaction (Eqs. **1–2**). The hydrolysis reaction is generally slow at room temperature, but it is accelerated at higher temperatures and when catalyzed by acids or bases. An example is the rapid loss of strength (degradation) in tire cord that occurs as water comes in contact with nylon fiber in a tire carcass.

Thermal Degradation. The degradation that occurs in the absence of oxygen affects all polyamides at a sufficiently high temperature and is usually significant above 300°C. Thermooxidation reactions often occur simultaneously owing to the presence of small amounts of air, which can lead to a confusion of the two processes. The general thermal decomposition reaction in polyamides, which is the cleavage of the amide bond to eventually form an olefin and a nitrile, results in chain cleavage and thus a loss in molecular weight.

Thermooxidation. This is an autoxidation process that occurs in all polyamides. It is significantly accelerated at elevated temperatures and can lead to carbonization of the polymer, but it also occurs during ambient temperature storage unless the polymer is protected with an antioxidant or the storage temperature is reduced. The principal effects of thermooxidation are a loss in molecular weight, increase in acid ends, decrease in amine ends, and the generation of color.

Photodegradative Processes. Polymers can undergo two types of photodegradative processes; one in the presence of oxygen, photooxidation, and one in its absence, photodegradation.

The mechanism for photodegradation at short wavelengths is generally believed to be initiated by the photolytic cleavage of the amide bond. The product radicals are then consumed by recombination or by the reaction with oxygen when the sample is exposed to air.

The initial steps in the mechanism for photooxidation are generally accepted as being the same as for thermooxidation. This is supported by the facts that similar degradation products have been detected and the effectiveness of similar stabilizers. Dyes and pigments can act as either prodegradants or as stabilizers, presumably as a result of their propensity to produce free radicals or to act as excited-state quenchers, respectively.

The generation of color during photooxidation, known as photoyellowing, has long been recognized as a source of color in nylon-6,6 and nylon-6. This effect has been shown to occur in all aliphatic polyamides at wavelengths between 320 and 350 nm.

Bio-, Environmental, and Mechanical Degradation. Pressure on industry to reduce or remove plastic materials from waste streams is increasing. One approach to meeting this expectation is to manufacture plastic materials that degrade in the environment (see POLYMERS, ENVIRONMENTALLY DEGRADABLE). Unfortunately, polyamides, like almost all synthetic polymers, are not directly biodegradable. However, if the polymeric material is reduced to low molecular weight oligomers, then many can be metabolized by microorganisms. Polyamide materials can be fragmented and then reduced in molecular weight by a process of mechanical destruction, photooxidation, and hydrolysis. This process can occur in a managed waste treatment facility, but it is expensive. An alternative approach has been to incorporate naturally occurring amino acids into polyamide polymer chains to provide sites for enzymatic attack.

Polyamides, like other macromolecules, degrade as a result of mechanical stress either in the melt phase, in solution, or in the solid state.

Principal Commercial Nylons: Nylon-6,6 and Nylon-6

Nylon-6,6. Nylon-6,6 is a tough, translucent white, semicrystalline, high melting (T_m = 265°C) material. The common physical properties are shown in Table 2 for nylon-6,6 and nylon-6.

Ingredients. Nylon-6,6 is made from the reaction of adipic acid and hexamethylenediamine. The manufacture of intermediates for

Table 2. Physical Constants of Nylon-6,6 and Nylon-6

Property	Nylon-6,6	Nylon-6
melting point, °C	255	220
at equilibrium		231
crystalline	270	260
specific gravity	1.14	1.13
density, g/cm^3		
crystalline[a]		
α	1.22–1.24 T	1.21–1.24 M
β	1.248 T	
γ		1.13 H
γ		1.17 M
amorphous		
γ		1.09
α		1.11
heat of fusion, kJ/kg[b]		
crystalline form ΔH_m	196	190
α-crystalline form ΔH_m		240–260
amorphous, annealed 8 h at 50°C		45
entropy of fusion, J/(mol·K)[b]		
crystalline	83–86	44–47.5
heat capacity, J/(mol·K)[b]		
crystalline, 20°C	374	204
specific heat, J/(g·K)[b]	1.67	1.67
heat of crystallization, kJ/kg[b]	−54	−46.5
heat of combustion, kJ/kg[b]	−31.4	−31.4
coefficient of thermal expansion		
linear at 20°C, m/m/K × 10^5	7–10	7–10
volume at 20°C m^3/m^3/K × 10^4	2.8	2.7
thermal conductivity, W/(m·K)		
crystalline (wet) at 30°C	0.43	0.43
amorphous (wet) at 30°C	0.36	0.36
melt at ~250°C	0.15	0.21
moldings	0.23	0.23
refractive index (n_D)		
single crystals		
α (calc)	1.475	
β (calc)	1.565	
γ (obs)	1.58	
moldings	1.53	1.53

[a] T = triclinic; M = monoclinic; and H = hexagonal.
[b] To convert J to cal, divide by 4.184.

polyamides is extremely important; not only is the quality of the polymer, such as color, degree of polymerization, and linearity, strongly dependent on the ingredient quality, but also the economic success of the producer is often determined by the yields and cost of manufacture of the ingredients.

Adipic acid (qv) has a wide variety of commercial uses besides the manufacture of nylon-6,6, and thus is a common industrial chemical.

Polymer Production. Three processes are used to produce nylon-6,6. Two of these start with nylon-6,6 salt, a combination of adipic acid and hexamethylenediamine in water; they are the batch or autoclave process and the continuous polymerization process. The third, the solid-phase polymerization process, starts with low molecular weight pellets usually made via the autoclave process, and continues to build the molecular weight of the polymer in a heated inert gas, the temperature of which never reaches the melting point of the polymer.

Nylon-6. Nylon-6, like nylon-6,6, is a tough, white translucent, semicrystalline solid, but melts at a lower temperature (T_m = 230°C). The physical properties of nylon-6 are listed in Table 2.

Ingredients. Nylon-6 is produced commercially from caprolactam, which is the most important lactam industrially. All industrial production processes for caprolactam are multistep and produce ammonium sulfate or other by-products. Approximately 95% of the world's caprolactam is produced from cyclohexanone oxime.

Polymer Production. Commercially the ring-opening polymerization of caprolactam to nylon-6 is accomplished by both the hydrolytic and anionic mechanisms. However, the hydrolytic process is by far the most predominantly used method because it is easier to control and better adapted for large-scale production. Like nylon-6,6, the polymerization process for nylon-6 via the hydrolytic mechanism can be batch or continuous.

Batch processing of nylon-6 is generally used only for the production of specialty polymers. A typical modern batch process caprolactam is mixed in a holding tank with the desired additives and then charged to an autoclave with a small amount (2–4%) of water. During the two-stage polymerization cycle, the temperature is raised from 80 to 260°C. In the first stage, water is held in the reactor, the pressure rises, and the hydrolysis and addition steps occur. After a predetermined time the pressure is released and the final condensation reaction step occurs.

In the continuous polymerization process for nylon-6 the three steps of polymerization can be made to take place in a series of connected vessels or in a single long, vertical, tubular reactor.

COMPARISON OF NYLON-6,6 AND NYLON-6 The chemical and physical properties of nylon-6,6 and nylon-6 are almost identical, as the similarity of their chemical structure might suggest: the amide functions are oriented in the same direction along the polymer chain for nylon-6, but are alternating in direction for nylon-6,6.

Nylon-6

Nylon-6,6

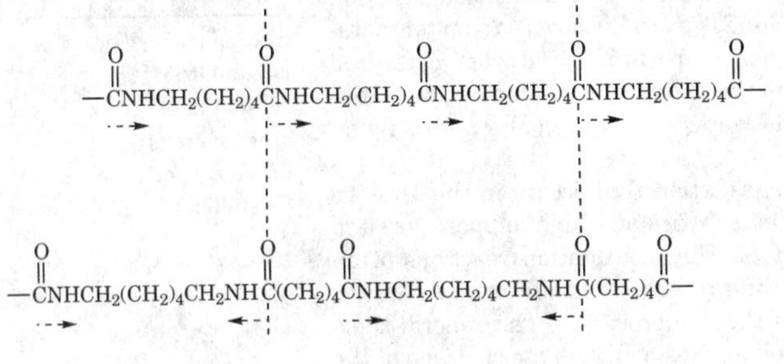

Table 2 shows the similarity of properties; however, a few differences between the two polyamides do exist.

Aromatic Polyamides

Polyamides that contain 85% or greater of the amide bonds attached to aromatic rings are classified as aramids.

The aramids were first introduced into the marketplace by DuPont in 1961 with the commercialization of Nomex fiber, which is used in flame-resistant fabrics and as an industrial fiber, such as in powerhouse filtration systems. The *p*-aramids were also first introduced commercially by DuPont as Kevlar fiber in 1972. They possess outstanding tensile strength and modulus and are stronger than fiber glass or steel on a performance-per-weight basis. Their end uses include tire cord, industrial fibers, and plastic reinforcement; they are used as pulp for vehicle brake linings and in woven and nonwoven blends. *m*-Aramids are used as fire-blocking materials for aircraft seating.

The diacid components for the manufacture of poly(*m*-phenyleneisophthalamide) and poly(*p*-phenyleneterephthalamide) are produced by one of two processes. In the first, the diacid chlorides are produced by the oxidation of *m*-xylene or *p*-xylene followed by the reaction of the diacids with phosgene. In the second, process *m*- or *p*-xylene reacts with chlorine initiated by ultraviolet light to form the *m*- or *p*-hexachloroxylene. This then reacts with the respective aromatic dicarboxylic acid to form the diacid chloride.

Health and Environmental Aspects

Health. As is the case for almost all commercial polymers, there appears to be no significant recognized health hazard for additive-free polyamides in their normal fiber, film, or bulk plastic end uses, and the same also appears to apply to their higher oligomers. However, the manufacturer's or supplier's Material Safety Data Sheet (MSDS) should always be consulted for the following reasons: *(1)* many commercial products contain additives which could significantly alter the health risks associated with the use of polyamides; *(2)* polyamides comprise an active area of research and changes in recognized levels of safety can occur; and *(3)* some potential end uses exceed the range of the manufacturer's intended applications for a given product. During incomplete combustion, polyamides can emit toxic products such as carbon monoxide, hydrogen cyanide, and NO_x, as well as other less hazardous products.

Environmental Aspects. In general, the polymerization processes for nylon-6,6 and nylon-6 generate little waste. However, because of economic advantages and governmental regulations, there has been a substantial increase in recycling of the starting materials for polyamides and of the energy, most often as steam.

Polymer recycle has been practiced as part of the manufacturing process for nylon-6,6 and nylon-6 almost from the beginning of the industry.

JOSEPH N. WEBER
DuPont Company, Inc.

J. Brandrup and E. H. Immergut, eds., *Polymer Handbook*, 3rd ed., John Wiley & Sons, Inc., New York, 1989.

H. Sekiguchi and B. Coutin, in H. R. Kricheldorf, ed., *Handbook of Polymer Synthesis*, Part A, Marcel Dekker, New York, 1992, pp. 807–939.

D. B. Jacobs and J. Zimmerman, in C. E. Schildknecht and I. Skeit, eds., *Polymerization Processes*, Wiley-Interscience, New York, 1977, pp. 424–467.

W. Sweeny and J. Zimmerman, "Polyamides," in N. Bikales, ed., *Encyclopedia of Polymer Science and Technology*, Vol. 10, John Wiley & Sons, Inc., New York, 1969, pp. 483–597.

FIBERS

Polyamide fibers are spun from linear thermoplastic polymers having recurring amide groups made from diamines and dicarboxylic acids $(CONH—R—NHCO—R)_n$ or lactams $(RCONH)_n$. Polyamides are generally referred to as nylons when R and R′ are essentially aliphatic, alicyclic, and less than 85% of the amide linkages are attached directly to two aromatic rings. When these linkages are equal or greater than 85% aromatic, the fibers are referred to as aramids. Polyamides from the condensation of diamines and dicarboxylic acids are termed AABB type; from lactams, AB type. Aliphatic polyamides are identified by numerals that indicate the number of carbon atoms in the monomer. One number is used for the AB type and two numbers, the first designating the diamine, for the AABB type. For example, the polyamide made from caprolactam (six carbons) is nylon-6; from hexamethylenediamine and adipic acid, nylon-6,6.

Properties of Nylon-6 and Nylon-6,6

The properties of textile fibers can be divided into three categories: geometric, physical, and chemical.

All synthetic fibers that are useful for textile applications are linear, semicrystalline, oriented polymers, whose properties are defined by molecular structure and molecular organization. The chemical and macromolecular structures of polyamide fibers can be determined by chemical and instrumental analysis. The supermolecular organization of nylon-6 and nylon-6,6 have been defined in terms of crystallinity, crystalline and amorphous orientation, and chain folds by x-ray diffraction, fiber density, (dsc), and nmr methods.

Tensile Properties. Tensile properties of nylon-6 and nylon-6,6 yarns (Table 1) are a function of polymer molecular weight, fiber spinning speed, quenching rate, and draw ratio. The degree of crystallinity and crystal and amorphous orientation obtained by modifying elements of the melt-spinning process have been related to the tenacity of nylon fiber.

Tensile properties of synthetic and natural fibers (or yarn) are measured from stress-strain curves. These measurements are important not only in determining the suitability of a nylon fiber for a specific end use, but also in comparing it to other fiber types.

Table 1. Tensile Properties of Nylon-6 and Nylon-6,6[a] Continuous-Filament Yarns

Property	Nylon-6 Regular tenacity	Nylon-6 High tenacity	Nylon-6,6 Regular tenacity	Nylon-6,6 High tenacity
breaking tenacity, N/tex[b]				
standard	0.35–0.64	0.57–0.79	0.20–0.53	0.52–0.86
wet	0.33–0.55	0.51–0.72	0.18–0.48	0.45–0.71
loop	0.34–0.49	0.45–0.89	0.18–0.45	0.44–0.67
knot	0.34–0.48	0.42–0.59	0.18–0.45	0.44–0.67
tensile strength, MPa[c]	503–690	703–862	275–731	593–924
breaking elongation, %				
standard	17–45	16–20	25–65	15–28
wet	20–47	19–33	30–70	18–32
average modulus (stiffness), N/tex[b]	1.6–2.0	2.6–4.2	0.44–2.1	1.9–5.1
average toughness, N/tex[b]	0.06–0.08	0.06–0.08	0.07–0.11	0.07–0.11
elastic recovery, %	98–100 at	99–100 at	88 at 3%	89 at 3%
moisture regain at 21°C, %	1–10%	2–8%		
65% rh	2.8–5.0	2.8–5.0	4.0–4.5	4.0–4.5
95% rh	3.5–8.5	3.5–8.5	6.1–8.0	6.1–8.0

[a] Conditioned at 65% rh and 21°C.
[b] To convert N/tex to g/den (gpd), multiply by 11.33.
[c] To convert MPa to psi, multiply by 145.

Definitions of the commonly measured tensile properties include *linear density* (tex), the weight in grams of 1000 m of yarn; *tenacity*, the tensile stress at break and is expressed in force-per-unit linear density of unstrained specimen, N/tex; *breaking strength*, the maximum load required to rupture a fiber, fiber expressed in grams; *tensile strength*, the maximum stress or load per unit area in units of Pa (1 MPa = 145 psi) *elongation* at break, the increase in sample length during a tensile test, expressed as a percentage of original length; *tensile modulus* (Young's modulus), the load required to stretch a specimen of unit cross-sectional area by a unit amount, expressed as the ratio of the tensile stress divided by the tensile strain, *creep*, the change in shape of a material while subject to a stress, which is also time-dependent; *elasticity*, the ability of a fiber to return to its original dimensions upon release of a deformation stress; *toughness*, the ability of a fiber to absorb work before it ruptures, in units of gram-per-unit linear density; and *creep and recovery*, the plasticity or plastic flow of a fiber in relation to load and time. Elastic recovery is the ability of a fiber to regain its original form after being stretched at a specified percentage of its length.

Temperature and Moisture Properties. Thermal treatment and moisture affinity significantly influence the physical properties of nylon fibers and fabrics. The absorption of water causes the fiber to swell, which alters its dimensions and in turn changes the size, shape, stiffness, and permeability of yarns and fabrics. It also alters the frictional and static behavior of yarns in mill processing and the hand and performance of fabrics in use.

Nylon-6 melts at 215–220°C and softens at 170°C. Nylon-6,6 melts at 255–260°C and softens at 234°C. Nylons also have thermal behavior properties between the melting and glass-transition temperatures that are important in fiber melt spinning and drawing and in fiber-to-fabric processing.

Electrical Properties. Nylon has low electrical conductivity (high electrical resistivity) and behaves like an insulator.

Because of its insulating nature, nylon can accumulate positive or negative electrical charges on its surface when rubbed or in contact with other substances followed by separation. These charges can cause problems in fiber processing.

Optical Properties. When light falls on a nylon fiber, it can be partially transmitted, absorbed, reflected, or scattered, depending on the cross-section shape and the nature of any second substance added during polymerization or melt spinning. Fibers that need to be opaque are melt-spun from polymer containing TiO_2 delusterant.

Chemical Properties. The chemical reactivity of nylon is a function of the amide groups and the amine and carboxyl ends. The aliphatic segment of the chain is relatively stable.

Generally, nylon is insoluble in organic solvents, but soluble in formic acid, some phenolic solvents, and fluorinated alcohols. Nylon is inert to alkalies. Dilute solutions of strong mineral acids, such as sulfuric and hydrochloric, weaken nylon fibers and hydrolyze them at high concentration and elevated temperature. Strong oxidizing agents and mineral acids such as potassium permanganate solution and nitric acid degrade nylon. Nylon, however, can be bleached in most bleaching solutions.

Thermal degradation of polyamides involves complex chemical reaction paths that are a function of the chemical structure, temperature, length of time exposed, and levels of moisture and oxygen.

The preferred and also the most effective thermal stabilizers for nylon are salts and organic derivatives of copper.

The extent of photodegradation of nylon-6 and nylon-6,6 depends on the intensity and spectral distribution of the light, the length of time exposed, humidity, air quality, and the presence of photosensitizers on the surface or in the fiber. Photosensitizers include the degradation products created during melt spinning and high temperature fiber processing, TiO_2 (the standard fiber delusterant), iron salt contamination, as well as certain dyes and dye bath chemicals.

Staple Properties. In contrast to continuous filament, spun nylon staple yarns have a softer, warmer tactile hand and a more natural appearance. Nylon staple is used as the pile material in plush carpets and woven velour automotive upholstery fabrics because of its high strength, abrasion resistance, and ease of recovery from pile crush and distortion.

Manufacture of Nylon-6 and Nylon-6,6 Fibers and Yarns

Polyamide fibers are manufactured by melt spinning (or extrusion) followed by drawing (or stretching).

In addition to continuous-filament yarn, nylon is also offered in staple, tow, and flock forms. Staple is made by cutting crimped continuous-filament yarn into 3–20-cm lengths. In manufacturing, tow is made by combining many yarn ends, either flat or crimped, to give a total tow size of 6–111 ktex. Flock is made by precision-cutting tow into 0.5–3-mm fiber lengths.

Texturing Processes. Texturing is the conversion of flat to crimped continuous-filament yarns to simulate properties inherent in natural and synthetic spun staple yarns, such as thermal insulation, fullness, cover (bulk), softness, and moisture transport. In texturing, the geometry and, to a degree, the surface of the filaments are mechanically deformed by bending, twisting, or compression to introduce permanent waviness (crimp) loops and coils.

Staple Fiber. Staple manufacturing consists of spinning, drawing, crimping, cutting, and baling.

Staple is used directly in the manufacturing of nonwoven fabrics (qv) and spun into yarn through the cotton, worsted, and woolen systems in 100% form or in blends with other synthetic or natural fibers.

Flock. Flocking is the mechanical and/or electrostatic application of finer fiber particles to adhesive-coated fabrics, paper, yarns, plastic, or metal objects. Flocking offers a soft velvety surface for decorative and visual appeal and has a variety of functions: sound dampening, thermal insulation, friction reduction of sliding surfaces, increased surface exposure for evaporation and filtering, buffing and polishing, liquid retention or dispersal, and cushioning of heavy objects.

Nylon is the preferred fiber for flocking because of its good chemical bonding to a wide range of adhesives, its toughness, and its ease of dyeability and printability.

Finishes. Fiber finishes are designed to provide fiber cohesion, lubricity, and static-free operability at low and high traverse speeds over a variety of metallic and ceramic surfaces encountered in fiber plant and mill operations. Finishes, consisting primarily of lubricants, emulsifiers, and antistatic agents, are generally applied as aqueous emulsions at concentrations giving a finish-on-yarn level of 0.3–1% after the evaporation of water. Lubricants can consist of mineral, vegetable, and animal oils or waxes (triglycerides), or of such synthetic types as esters, polyethers, silicones, and ethoxylated esters.

In formulating any new finish, environmental issues such as biodegradability, water and air pollution must be considered.

Modified Cross Sections. Nylon filaments are spun in a variety of cross-section shapes that include the conventional round to irregular solid and hollow shapes. The cross-section shape is an important variant in designing the functionality and luster of fibers.

In spinning the different fiber cross sections, spinnerette orifice shape and dimensions must be made to exacting tolerances, and changes in polymer melt viscosity, block temperature, and air quenching conditions must be anticipated to assure the desired shape, spinning continuity, and quality yarn. Various specific orifice shapes have been developed and fiber cross sections have been defined using mathematical relationships.

Additives. Additives can be introduced in salt preparation, polymerization, or the molten polymer stream prior to extrusion of the filaments. Additives are used to alter the basic characteristics and performance of the fiber and can be classified by function.

Delusterants reduce the transparency, increase the whiteness, and alter the fiber's reflectance of light.

Colorants can be introduced into the fiber by adding dyes and pigments in salt preparation, during polymerization, or into the molten polymer just before spinning (see COLORANTS FOR PLASTICS; DYES, APPLICATION AND EVALUATION).

Antioxidants (qv) are used to prevent thermal and oxidative degradation of nylon.

Antistats are added to nylon to reduce static charge and improve moisture transport and soil release in fabrics.

Antimicrobial agents are used where there is a need to inhibit bacterial and fungal growth (see INDUSTRIAL ANTIMICROBIAL AGENTS).

Flame retardants designated for nylon include halogenated organic compounds, phosphorous derivatives, and melamine cyanurate.

Dyeability

Because of its physical and chemical structure, nylon has an affinity for every dye class: disperse, direct, vat, fiber reactive, chrome, acid, premetallized, and cationic dyes with special nylons (see DYES, APPLICATION AND EVALUATION). Commonly used dyes are the disperse, acid, and premetallized.

Modified Nylon-6 and Nylon-6,6 Fibers

Bicomponent and Biconstituent Fibers. Bicomponent fibers consist of two polymers of the same generic class, eg, nylon-6 and nylon-6,10; biconstituent fibers consist of two dissimilar generic polymers, eg, nylon-6,6 and polyester. Both fiber types are made by melt-spinning separately the two different polymers through a common, specially designed spinneret.

Microdenier fibers, ie, very fine fibers, can be made by spinning biconstituent conjugate fibers.

Microdenier nylon and polyester, have the finer-than-silk fibers, added a new dimension to fabric aesthetics, comfort, and performance. Microdenier nylons are used in weaving, warp knits, and weft knits.

Copolymers. There are two forms of copolymers, block and random. Block copolymerization is a way of introducing a new variant into a base polymer without grossly affecting the spinning performance and physical properties of the yarn. Random copolymers are made by combining three or more monomers in the polymerization process. The general properties of random copolyamides are high dyeability, lower melting and softening points, reduced dry and wet strength properties, high creep failure, and high shrinkage.

Other Nylons

Other nylons include Stanyl, a yarn highly suited for industrial yarn applications, and miscellaneous polyamides. Nylons have also been tested extensively over the years for apparel and carpets.

Applications

Most of the end uses for nylon-6 and nylon-6,6 filament, tow, staple, and flock have been categorized into four major application areas, in descending order of market size: carpets, industrial, apparel, and home furnishings.

In the category of industrial applications, nylon is the predominant fiber used in the carcass of bias truck, racing car, and airplane tires because of its excellent strength, adhesion to rubber, and fatigue resistance. Automobile air bag fabric is woven with high tenacity nylon-6 or nylon-6,6 industrial yarns.

Recycling

The commitment to recycle and protect the environment in compliance to regulatory legislation will have a profound effect on how fiber producers design their offerings for the future. Recycling can occur in many forms. A fabric or article of commerce can be reconditioned for reuse, mechanically converted to a new use (eg, shredded for filler material), burned for fuel value, or converted to its original raw materials (see RECYCLING).

ANTHONY ANTON
E. I. du Pont de Nemours & Company, Inc.

M. Lewin and E. M. Pearce, *Handbook of Fiber Science and Technology*, Vol. IV, Marcel Dekker, Inc., New York, 1985 pp. 75–169.

W. E. Morton, J. W. S. Hearle, *Physical Properties of Textile Fibers*, The Textile Institute, Manchester, U.K., 1993.

H. F. Mark, S. M. Atlas, and E. Cernia, eds., *Man-Made Fibers Science and Technology*, 3 vols., Wiley-Interscience, New York, 1967–1968.

H. L. Needles, *Textile Fibers, Dyes, Finishes and Processes, A Concise Guide*, Noyes Publications, New Jersey, 1986.

PLASTICS

Polyamides are considered to have been the first engineering plastics. They possess an excellent combination of mechanical and thermal properties. Introduced in the 1930s, these materials have retained their vitality; new applications and new types of nylon continue to be developed.

Amide groups along the backbone of long-chain molecules link the monomeric units in polyamides. Polyamide is generally referred to as nylon and was first introduced in 1938 by DuPont. Nylons comprise a range of materials, depending on the monomers employed, and can be prepared by either a condensation reaction between a diacid and a diamine or by a ring-opening addition of a lactam.

Nylon-6,6 and nylon-6 are the most popular types, accounting for approximately 90% of nylon use. There are a number of different nylons commercially available; Table 1 gives a summary of the properties of the more common types.

Properties

Originally, the main attraction of nylon-6,6 was as a fiber-forming material whose strength, elasticity, and high dye uptake its were considered most important properties, along with the ability to withstand ironing temperatures. However, the properties of nylon-6,6 hold many advantages for use as a plastic. In particular, the relatively high tensile strength and stiffness, together with good toughness, high melting point (and therefore temperature stability), and good chemical resistance, all have combined to allow a wide range of applications. The material is now considered an engineering plastic that can be used for metal replacement in structural or semistructural end uses. These properties are present to a greater or lesser extent in the entire semicrystalline polyamide family, together with the "Achilles heel" of nylon, ie, the hygroscopic nature that leads to moisture uptake, change of properties, and the potential for hydrolysis.

Appropriate choice of monomer can provide a balance of properties to meet particular types of applications.

Crystallinity. The presence of the polar amide groups allows hydrogen bonding between the carbonyl and NH groups in adjacent sections of the polyamide chains. For common nylons such as nylon-6,6 and nylon-6, the regular spatial alignment of amide groups allows a high degree of hydrogen bonding to be developed when chains are aligned together, giving rise to a crystalline structure in that region. These nylons are semicrystalline materials that can be thought of as a combination of ordered crystalline regions and more random amorphous areas having a much lower concentration of hydrogen bonding. This semicrystalline structure gives rise to the good balance of properties. The crystalline regions contribute to the stiffness, strength, chemical resistance, creep resistance, temperature stability, and electrical properties; the amorphous areas contribute to the impact resistance and high elongation.

Thermal Properties. The high melting point of polyamides such as nylon-6,6 is a function of the strong hydrogen bonding between the chains and the crystal structure. This also allows the materials to retain significant stiffness above the glass-transition temperature (T_g) and almost up to the melting point. The effect is further increased when reinforcements such as glass fiber are added, giving a high deflection temperature under load even at high loading.

Moisture Absorption. A characteristic property of nylon is the ability to absorb significant amounts of water. This is related to the polar amide groups around which water molecules can become coordinated. Water absorption is generally concentrated in the amorphous regions of the polymer where it has the effect of plasticizing the material by interrupting the polymer hydrogen bonding, making it more flexible (with lower tensile strength) and increasing the impact strength. The T_g is also reduced.

Electrical Properties. Nylons are frequently used in electrical applications mainly for their combination of mechanical, thermal, chemical, and electrical properties. They are reasonably good insulators at low temperatures and humidities and are generally suitable for low frequency, moderate voltage applications.

Flammability. Most nylons are classified V-2 by the Underwriters' Laboratory UL-94 test, which means that these nylons are self-extinguishing within a certain time-scale under the conditions of the test. They achieve this performance by means of giving off burning drips. Inclusion of reinforcement such as glass fiber converts this behavior to one where the sample continues to burn as a result of the reinforcement holding it together.

Mechanical Properties. The semicrystalline structure of most commercial nylons imparts a high strength (tensile, flexural, compressive, and shear) as a result of the crystallinity and good toughness (impact strength) due mainly to the amorphous region. The properties of nylon are affected by the type of nylon (including copolymerization), molecular weight, moisture content, temperature, and the presence of additives. Strength and modulus (stiffness) are increased by increasing density of amide groups and crystallinity in aliphatic nylons; impact strength and elongation, however, are decreased. Nylon-6 having a lower crystallinity than nylon-6,6 has a higher impact strength and slightly lower tensile strength. Nylons containing aromatic monomers tend to have increased stiffness and strength by virtue of the greater rigidity of the chains.

Generally, nylon is notch-sensitive and the unnotched impact strength is dramatically reduced when a notch or flaw is introduced into the material. This needs to be considered when designing parts so that sharp angles are avoided where possible.

Properties such as stiffness and strength can be considerably increased by adding a reinforcing agent to the polymer, particularly glass or carbon fiber.

As with most plastics the properties of nylons are time-dependent. The strain in a molding constantly under load increases with time (creep); equally, the load or stress required to maintain a constant deformation decays with time (stress relaxation).

Two more properties for which nylon shows particular advantages are abrasion resistance and coefficient of friction. These properties make the material suitable for use in, for example, unlubricated bearings and intermeshing gears.

Hydrolysis and Polycondensation. The polymerization of nylon is a reversible process and the material can either hydrolyze or polymerize further, depending on the conditions.

In the melt the material is in a dynamic situation and only at a certain (equilibrium) moisture content does the rate of hydrolysis equal the rate of polymerization. This equilibrium moisture content (in a sealed system) depends on the polymer, the temperature, the molecular weight, and the end group balance of the polymer. Nylons that absorb lower amounts of moisture have improved hydrolysis resistance.

Thermal Degradation. Although nylons have good thermal stability, they tend to degrade in the melt when held for long periods of time or at high temperatures. This is particularly the case for nylons containing adipic acid such as nylon-6,6. As well as reduction of molecular weight, cross-linking also occurs, and the material eventually sets into an intractable gel.

Oxidation. All polyamides are susceptible to oxidation. Similarly, nylon parts exposed to high temperature in air lose their properties with time as a result of oxidation.

Table 1. Properties of the More Common Nylons, Dry as Molded

Property	Nylon-6,6	Nylon-6	Nylon-11	Nylon-12	Nylon-6,9	Nylon-6,12
specific gravity	1.14	1.13	1.04	1.02	1.09	1.07
water absorption, wt %						
24 h	1.2	1.6	0.3	0.25	0.5	0.25
equilibrium at 50% rh	2.5	2.7	0.8	0.7	1.8	1.4
saturation	8.5	9.5	1.9	1.5	4.5	3.0
melting point, °C	255	215	194	179	205	212
tensile yield strength, MPa[a]	83	81	55	55	55	61
elongation at break, %	60–90	50–150	200	200	125	150
flexural modulus, MPa[a]	2800	2800	1200	1100	2000	2000
Izod impact strength, J/m[b]	53–64	55–65	40–68	95	58	53
Rockwell hardness, R scale	121	119	108	107	111	114
deflection temperature under load, °C						
at 0.5 MPa[a]	235	185	150	150	150	180
1.8 MPa[a]	90	75	55	55	55	90
dielectric strength, kV/mm						
short time	24	17	16.7	18	24	16
step by step	11	15		16	20	
dielectric constant at 60 Hz	4.0	3.8	3.7	4.2	3.7	4.0
10^3 Hz	3.9	3.7	3.7	3.8	3.6	4.0
10^6 Hz	3.6	3.4	3.1	3.1	3.3	3.5
starting acid[c] or lactam	adipic acid[c]	caprolactam	11-aminoundercanoic acid	dodecano-lactam	azaleic acid[c]	dodecanedioic acid[c]

[a] To convert MPa to psi, multiply by 145.
[b] To convert J/m to ft·lbf/in., divide by 53.38.
[c] The starting amine is hexamethylenediamine for nylon-6,6, nylon-6,9, and nylon-6,12.

Ultraviolet Aging. Nylon parts exposed to sunlight and uv rays undergo a free-radical aging process. This can be reduced with appropriately stabilized materials.

Effect of Chemicals and Solvents. Nylons have excellent resistance to many chemicals, although the effect varies depending on the nature of the nylon. Generally, polyamides tend to be particularly resistant to nonpolar materials such as hydrocarbons. Resistance is least to strong acids and phenols.

Manufacture and Processing

Injection Molding. This is the largest single processing route for nylon, taking more than 60% of the material produced. This technique is generally carried out using a screw preplasticizing injection-molding machine (see PLASTICS PROCESSING). As melt accumulates at the screw front, the screw is forced back against a residual oil pressure (screw back pressure). When enough melt has accumulated, the screw stops rotating and moves forward, injecting material into the mold.

Developments in injection include gas injection and fusible-core molding. Gas injection enables a saving in part weight and avoidance of sink marks in thick sections. Fusible-core technology involves overmolding a low melting metal alloy core that is subsequently melted out in an oil bath, a technique that allows much more complex moldings to be produced.

Extrusion. Extrusion accounts for about 30% of nylon produced. Nylons can be extruded on conventional equipment with an extruder drive that is capable of continuous variation over a range of screw speeds. Extruded nylons are used to produce films, tubing and pipe, monofilaments, and to coat wire and cable.

Additives and Modifications. For plastics uses, nylon is only rarely employed as the pure polymer, and is almost always modified to some extent. Additives used to modify nylons, alone or in combination, include lubricants, nucleants, stabilizers, impact modifiers, flame retardants, plasticizers, and reinforcements. It is not unusual to find formulations that contain less than 50% nylon and half a dozen or more additives.

Polymer Blends. Commercial blends of nylon with other polymers have also been produced in order to obtain a balance of the properties of the two materials or to reduce moisture uptake.

Uses

More than 60% of nylon is used in injection-molding applications. About 55% of this use is in the transportation industries, and most of this use is concerned with automobile production. Nylons are used throughout automobiles in both underhood and exterior applications, and the amount used is continually increasing as a result of the drive to reduce weight.

About 16% of injection-molded nylon has been used in the electronic and electrical industries. The remaining applications include domestic appliances, power tools, consumer products, and various industrial parts. About 30% of nylon is used in applications involving extruded monfilaments and film.

ROBERT J. PALMER
DuPont de Nemours Int. SA

M. I. Kohan, *Nylon Plastics*, John Wiley & Sons, Inc., New York, 1973.

D. B. Jacobs and J. Zimmerman, in C. E. Schildknecht and I. Skeist, eds., *Polymerisation Processes, High Polymers*, Vol. XXIX. Wiley-Interscience, New York, 1977, pp. 424, 467.

R. S. Williams and T. Daniels, *Rapra Rev. Rep.* **3**(3), 33/1–33/116 (1990).

M. I. Kohan, *Nylon Plastics Handbook*, Carl Hanser Verlag, Munich, Germany, 1995.

POLYBLENDS. See POLYMER BLENDS.

POLY(BUTYLENE TEREPHTHALATE). See POLYESTERS, THERMOPLASTIC.

POLYCARBONATES

Polycarbonates are an unusual and extremely useful class of polymers. The vast majority of polycarbonates are based on bisphenol A (BPA) and sold under various trade names. BPA polycarbonates, having glass-transition temperatures, in the range of 145–155°C, are widely regarded for optical clarity and exceptional impact resistance and ductility at room temperature and below. Other properties, such as modulus, dielectric strength, or tensile strength are comparable to other amorphous thermoplastics at similar temperatures below their respective glass-transition temperatures. Whereas below their T_gs most amorphous polymers are stiff and brittle, polycarbonates retain their ductility.

Important products are based on polycarbonate in blends with other materials, copolymers, branched resins, flame-retardant compositions, and foams (see FLAME RETARDANTS).

Properties and Characterization

Solubility and Solvent Resistance. The majority of polycarbonates are prepared in methylene chloride solution. Chloroform, *cis*-1,2-dichloroethylene, *sym*-tetrachloroethane, and methylene chloride are the preferred solvents for polycarbonates. Hydrocarbons (qv) and aliphatic alcohols, esters (see ESTERS, ORGANIC), or ketones (qv) do not dissolve polycarbonates. Acetone (qv) promotes rapid crystallization of the normally amorphous polymer, and causes catastrophic failure of stressed polycarbonate parts.

In general, polycarbonate resins have fair chemical resistance to aqueous solutions of acids or bases, as well as to fats and oils. Chemical attack by amines or ammonium hydroxide occurs, however, and aliphatic and aromatic hydrocarbons promote crazing of stressed molded samples. BPA polycarbonate has excellent resistance to hydrolysis.

Certain blends and copolymers of polycarbonate demonstrate dramatically improved solvent resistance. The blend of polycarbonate and poly(butylene terephthalate) combines the toughness of polycarbonate with the solvent resistance of the semicrystalline polyester.

Copolycarbonates of BPA and hydroquinone (HQ) can be prepared via the intermediacy of cyclic oligomeric cocyclics. Although hydroquinone linear oligomers having degrees of polymerization greater than two are insoluble in CH_2Cl_2, the cyclic analogues remain soluble when randomly cyclized with BPA.

Molecular Weight and Viscosity. BPA polycarbonates are commercially available in a wide range of molecular weights. As the molecular weight increases, melt and solution viscosities increase proportionally. Molecular weights may be determined or inferred by several means, including gel-permeation chromatography, light-scattering chromatography, measurement of intrinsic or inherent viscosity, and measurement of melt viscosity and flow. Correlation of intrinsic viscosity (IV), $[\eta]$, with weight-average mol wt (M_w) has been carried out on carefully characterized polycarbonate samples. The following relationship exists when $[\eta]$ is in mL/g. $[\eta] = 41.2 \times 10^{-3} \cdot M_w^{0.69}$.

The mechanical properties of polycarbonate, eg, tensile strength, impact resistance, flexural strength, elongation, etc, improve dramatically with increasing polymer intrinsic viscosity up to a value of about 0.45 dL/g. After that point, slight increases in mechanical properties are seen with increasing molecular weight, but melt viscosity continues to climb. At IV values greater than 0.6 dL/g, the melt viscosity becomes so high that processing is very difficult.

Spectroscopy and Analysis. Polycarbonates have a strong C═O stretching-band at 1770 cm^{-1}, and strong C—O stretching bands at 1220 and 1235 cm^{-1}, distinguishing them from polyesters.

Differential scanning calorimetry reveals a T_g at around 154°C, shifting slightly with molecular weight or level of branching. End group and impurity analysis is best revealed by hydrolysis of the polycarbonate using KOH-methanol in tetrahydrofuran under nitrogen, followed by reversed-phase hplc analysis or by spectroscopic techniques. Trace levels of impurities, are determined by standard analytical techniques.

Structure and Crystallinity. The mechanical–optical properties of polycarbonates are those common to amorphous polymers. The polymer may be crystallized to some degree by prolonged heating at elevated temperature (8 d at 180°C), or by immersion in acetone (qv). Powdered amorphous powder appears to dissolve partially in acetone, initially becoming sticky, then hardening and becoming much less soluble as it crystallizes. Enhanced crystallization of polycarbonate can also be caused by the presence of sodium phenoxide end groups.

Film or fibers derived from low molecular weight polymer tend to embrittle on immersion in acetone; those based on higher molecular weight polymer (>0.60 dL/g) become opaque, dilated, and elastomeric.

Glass-Transition Temperature and Melt-Behavior. The T_g of BPA polycarbonate is around 150°C, which is unusually high compared to other thermoplastics. The high glass-transition temperature can be attributed to the bulky structure of the polymer, which restricts conformational changes, and to the fact that the monomer has a higher molecular weight than the monomer of most polymers. The high T_g is important for the utility of polycarbonate in many applications, because, as the point which marks the onset of molecular mobility, it determines many of the polymer's properties such as dimensional stability, resistance to creep, and ultimate use temperature. Polycarbonates of different structures may have significantly higher or lower glass-transition temperatures.

BPA polycarbonate becomes plastic at temperatures around 220°C. The viscosity decreases as the temperature increases, exhibiting Newtonian behavior, with the melt viscosity essentially independent of the shear rate. At the normal injection molding temperature of 270–315°C, the melt viscosity drops from 1,100 to 360 Pa · s (11,000 to 3,600 poise). Because the viscosity of polycarbonate can only be reduced by increasing the temperature, the ultimate limit on molecular weight is controlled by the processing conditions and the thermal stability of the polymer.

Thermal, Flame-Retardant, and Hydrolytic Behavior. BPA polycarbonate exhibits excellent thermal stability, especially in the absence of oxygen and water. At temperatures above 400°C, rapid decomposition and cracking occur. BPA has an oxygen index of 26; this indicates that under test conditions, an atmosphere of 26% oxygen is required for combustion. Owing to thermal–oxidative stability, polycarbonate has some inherent flame resistant properties and can be classified as V-2 according to UL94 of the Underwriters Laboratory. Several polycarbonate grades have additives to increase the flame-retardant properties, and to decrease smoke.

Because of the low solubility of water in the resin, BPA polycarbonates are inherently resistant to aqueous acid and base, although strong nucleophilic bases can cause hydrolysis.

Optical Properties. Polycarbonate is a transparent colorless polymer, making it attractive for glass replacement. Visible light transmission is about 90%, and haze is minimal (1–2%). Absorption in the ultraviolet region is essentially complete. Polycarbonate's high (1.584) refractive index and light weight relative to glass make it attractive for eyewear.

Special polycarbonate grades have been developed for the optical information storage market eg, compact disks (see INFORMATION STORAGE MATERIALS).

Mechanical Properties. The room temperature modulus and tensile strength are similar to those of other amorphous thermoplastics, but the impact strength and ductility are unusually high. Whereas most amorphous polymers are glass-like and brittle below their glass-transition temperatures, polycarbonate remains ductile to about −10°C. The stress–strain curve for polycarbonate typical of ductile materials, places it in an ideal position for use as a metal replacement. Weight savings as a metal replacement are substantial, because polycarbonate is only 44% as dense as aluminum and one-sixth as dense as steel.

Impact strength can be measured by a variety of methods, including notched Izod, tensile impact, and falling dart impact. Polycarbonates are among the highest rated engineering polymers for impact resistance, and are the toughest transparent materials known (see ENGINEERING PLASTICS).

Glass-reinforced polycarbonates are sold as high modulus materials having properties approaching those of metals, while retaining the basic plastic attributes of low cost processing, dielectric character, resistance to corrosion, and inherent color.

Preparation

Interfacial Polymerization. Most BPA polycarbonate is produced by an interfacial polymerization process utilizing phosgene. The interfacial process for polycarbonate preparation involves stirring a slurry or solution of BPA and 1–3% of a chain stopper, such as phenol, *p-t*-butylphenol, or *p*-cumylphenol, in a mixture of methylene chloride and water, while adding phosgene (qv) in the presence of a tertiary amine catalyst.

Transesterification. The transesterification process is an environmentally friendly process that utilizes no solvent during polymerization, producing neat polymer directly.

Processing. Polycarbonates may be fabricated by all conventional thermoplastic processing operations, of which injection molding is the most common.

Injection blow molding of polycarbonates produces an assortment of containers from 20-L water bottles and 0.25-L milk bottles to outdoor lighting protective globes.

Conventional thermoforming of sheet and film is applicable to the production of skylights, radomes, signs, curved windshields, prototype production of body parts for automobiles, skimobiles, boats, etc. Because BPA polycarbonate is malleable, it can be cold-formed like metal, and may be cold-rolled, stamped, or forged.

Health and Safety Factors

Polycarbonate is considered a slight or nonexistent fire hazard. Odor and volatiles are negligible. Processing fumes, which include water, carbon dioxide, diphenyl carbonate, methylene chloride, and phenol, are not formed in levels considered to be hazardous. Polycarbonate has very low acute oral and dermal toxicity, is not a primary skin irritant, and does not cause systemic or local sensitization. Polycarbonate does not degrade during storage, and no heating or cooling requirements are necessary.

Uses

Extreme toughness, transparency, low color, resistance to burning, and maintenance of engineering properties over a wide thermal range are the outstanding properties of polycarbonate that make it useful for a variety of applications. Glazing and sheet are the largest markets for polycarbonate resins. Other uses for polycarbonates include automotive components; packaging; electrical, electronic, and technical applications; and medical and health-care related applications.

Polycarbonate is popular because of its clarity, impact strength, and low level of extractable impurities.

Other Polycarbonates, Blends, and Copolymers

Blends of polycarbonate with ABS and with poly(butylene terephthalate) (PBT) have shown significant growth since the mid-1980s.

Copolymers. The copolymer of tetrabromoBPA and BPA was one of the first commercially successful copolymers. Low levels of brominated comonomer lead to increased flame resistance.

Polyester carbonates can be prepared by the copolymerization of BPA with diacyl chlorides, leading to high heat materials with $T_g \sim$ 190°C.

Some of these block copolymers have improved low temperature impact strength and higher stress–crack resistance than neat BPA polycarbonate.

Blends. The concept of blending two or more commercially available materials to create a new material having properties different from either starting material has generated a great deal of interest. Polycarbonate blends are used to tailor performance and price to specific markets.

Fundamental studies of blends of polycarbonate with acrylonitrile–butadiene–styrene (ABS) indicate that the presence of ABS greatly decreases the melt viscosity in the blend, enhancing processibility. A synergistic improvement of the notched impact strength at low temperature is also seen for polycarbonate–ABS blends.

Polycarbonate–polyester blends are used on exterior parts for the automotive industry. Such blends combine the toughness and impact strength of polycarbonate with the crystallinity and inherent solvent resistance of PBT, PET, and other polyesters.

The most significant blends are with polyurethanes, polyetherimides, acrylate–styrene–acrylonitrile (ASA), acrylonitrile–ethylene–styrene (AES), and styrene–maleic anhydride (SMA).

DANIEL J. BRUNELLE
General Electric

H. Schnell, *Ang. Chem.* **68**, 633 (1956).

U.S. Pat. 3,028,365 (1962), H. Schnell, L. Bottenbruch, and G. Grimm (to Bayer AG).

U.S. Pat. 3,153,008 (1964), D. W. Fox (to General Electric).

C. S. Read, *CEH Marketing Research Report*, SRI International, 1993. Latest production figures available are for 1992.

POLYCHLOROTRIFLUOROETHYLENE. See FLUORINE COMPOUNDS, ORGANIC.

POLYELECTROLYTES. See IONOMERS.

POLYENE ANTIBIOTICS. See ANTIPARASITIC AGENTS, ANTIMYCOTICS.

POLYESTER FIBERS. See FIBERS, POLYESTER.

POLYESTERS, THERMOPLASTIC

Taken collectively, thermoplastic polyesters constitute a significant item of commerce, entering into almost every imaginable end use: fibers, textiles, industrial yarns, tire cord (qv), ropes, molded items, consumer goods, medical accessories, automotive and electronic items, photographic film, magnetic tape filmbase for audio and video recording, packaging materials, bottles, containers, etc. The broad span of their applications illustrates the wide utility of such materials. This article considers thermoplastic polyesters as materials for injection molding and similar applications and is confined to semicrystalline thermoplastic polyesters. Thermoset, unsaturated polyesters, extruded or melt-spun polyester films and fibers, and thermoformed sheets are covered elsewhere (see FIBERS, POLYESTER; FILM AND SHEETING MATERIALS; POLYESTERS, UNSATURATED). Blow-molded polyester containers are considered.

The principal thermoplastic polymers are poly(butylene terephthalate) (PBT) and poly(ethylene terephthalate) (PET). Increasingly commercially important are the high performance liquid crystalline all-aromatic polyesters.

Manufacture of Raw Materials and Monomers

Terephthalic Acid and Dimethyl Terephthalate. PET and PBT can both be made from terephthalic acid or its dimethyl ester (see PHTHALIC ACIDS AND OTHER BENZENEPOLYCARBOXYLIC ACIDS).

Liquid Crystal Polyesters. These high performance, high added-value products are derived from all-aromatic precursors.

Manufacture of Polyesters and Polymerization Processes

Thermoplastic polyesters are step-growth polymers that need to be made to high molecular weight (12,000–50,000) to be useful. The first stage is an esterification (qv) or ester-exchange stage where the diacid or its dimethyl ester reacts with the appropriate diol to give the bis(hydroxyalkyl)ester and some linear oligomers. Water or methanol is evolved at this stage and is removed by fractional distillation, often under reduced pressure at the conclusion of the cycle.

The final polymerization stage is usually done in an autoclave fitted with a powerful mechanical stirrer to handle the viscous melt under high vacuum at a temperature above the melting point of the final polymer.

As the polymer molecular weight increases, so does the melt viscosity, and the power to the stirrer drive is monitored so that an end point can be determined for each batch. When the desired melt viscosity is reached, the molten polymer is discharged through a bottom valve, often under positive pressure of the blanketing gas, and extruded as a ribbon or as thick strands which are water-quenched and chopped continuously by a set of mechanical knives. Large amounts of PET are also made by continuous polymerization processes. PBT is made both by batch and continuous polymerization processes. The polymer is then dried thoroughly and stored for subsequent processing.

Liquid crystal polyesters are made by a different route. The usual method is the so-called reverse ester exchange or acidolysis reaction where the phenolic hydroxyl groups are acylated with a lower aliphatic acid anhydride, and the acetate or propionate ester is heated with an aromatic dicarboxylic acid. The phenolic polyester forms readily. Many liquid crystal polymers are derived formally from hydroxyacids and their acetates readily undergo self-condensation in the melt, stoichiometric balance being automatically obtained.

PET Blow-Molded Bottles

One of the largest uses of PET resin is the stretch blow-molded PET soda bottle. Blow-molded thermoplastic soda bottles are lightweight, shatterproof, and potentially reusable.

The principal technical problem facing any thermoplastic bottle manufacturer is the permeability of the bottle wall to oxygen and carbon dioxide, which affects the shelf-life of the contents.

In the blow-molded PET bottle process, the polymer bottle wall is subjected to a rapid biaxial drawing stage which greatly increases its molecular orientation. This increases the mechanical strength of the bottle and reduces the permeability of the walls to carbon dioxide diffusing out.

The Blow-Molding Process. Blow-molding thermoplastic hollow articles is a highly specialized process. Of the various processes in use the most popular is the two-stage (reheat) blow-molding process. A bottle preform is molded from PET by a conventional injection molding process. The preform looks like a large test tube with thick walls and the screw-cap threads and neck molded in place. In the second stage, the preforms are heated in a mold cavity to a carefully controlled temperature above the glass–rubber transition temperature. The inside of the mold cavity is the size and shape of the finished bottle and the preform is subjected to a biaxial stretching process by a combination of mechanical deformation and air pressure inflation. A hollow metal mandrel passes into the preform and partially elongates it in the axial direction; simultaneously dry air is applied to blow the walls of the softened preform outward to fill the mold, thus giving radial stretching. The mold opens to allow the bottle to cool. This combined process results in both radial and axial drawing of the bottle walls causing stress-induced crystallization and giving a container with superior strength, clarity, and freedom from environmental stress cracking.

Properties of Bottle Resin. PET bottle resins are usually made to high molecular weight intrinsic viscosity (IV) 0.75–0.90 dL/g) so that the preforms can be blow molded without problems. The base polymer is made in a continuous melt polymerization plant using either direct esterification of ethylene glycol with terephthalic acid or by ester interchange using purified dimethyl terephthalate (DMT), often recovered from recycled PET bottles.

The number-average molecular weight for typical bottle resin is between 24,000 and 31,000 daltons/mol.

PET Molding Resins

Both PET and PBT engineering resins have good resistance to chemicals, and because they are crystalline do not suffer from the solvent stress cracking problems that plague amorphous materials. Polyesters in general are attacked only by severe chemicals such as powerful acidic or phenolic solvents; hot, strong aqueous alkali; and certain bases, such as hydrazine.

In the unfilled state, PET is not a good molding resin and all commercial grades are filled with either chopped glass strand, 3–4-mm long (1/8 in.); mineral fillers, usually mica (qv); or a mixture of the two. Some grades also have longer glass fibers.

Properties of PET Molding Resins. The full crystal structure of poly(ethylene terephthalate) has been established by x-ray diffraction. It forms triclinic crystals with one polymer chain per unit cell.

Thermochemical data depend on the degree of crystallinity in the polymer and a very highly annealed polymer sample can have T_m = 280°C, much higher than the usual value of 260–265°C. The heat of fusion is about 140 J/g, (33.5 cal/g). The glass–rubber-transition temperature (T_g) depends both on the method of measurement and the state of the polymer.

As a step-growth polymer made under equilibrium conditions, PET has a molecular weight distribution very close to the theoretical value of 2.0.

PET does not crystallize well in the unoriented state even in a hot mold unless nucleating agents and/or plasticizers are added. Commercial PET molding-grade polymers are nearly always filled. Typical compounded polymer properties are shown in Table 1.

Safety and Environmental Factors. PET polymer is safe and poses no threat to animals or humans. PET fibers have been in use since the 1950s and PET has U.S. Food and Drug Administration (FDA) approval for use as a food packaging (qv) material. PET fibers have

Table 1. Typical Properties of PET Molding Resins

	Glass			
Property	30%	45%	35% min	
specific gravity	1.58	1.70	1.60	1.60
tensile strength, MPa[a]	166	197	97	103
elongation at break, %	2.0	2.0	2.2	2.1
flexural strength at 5%, MPa[a]	245	310	148	152
flexural modulus, GPa[b]	9.66	14.5	9.66	9.66
notched Izod, J/m[c]	80.1	107	58.7	58.7
heat deflection temperature at 1.82 MPa[a], °C	224	229	202	216
flammability[d]	HB	HB	HB	HB
dielectric strength, V/25 μm				
5.2 mm	565	540	500	450
1.6 mm	904	631	550	575
0.8 mm	975	951	810	860
volume resistivity at 23°C, 50% rh, Ω·cm	3.0×10^{15}		1.0	1.0
dielectric constant, ε				
10^3 Hz	3.2	3.5	3.8	3.8
10^5 Hz	3.1	3.4	3.6	3.7

[a] To convert MPa to psi, multiply by 145.
[b] To convert GPa to psi, multiply by 145,000.
[c] To convert J/m to ft·lbf/in., divide by 53.38.
[d] HB = Brinell hardness.

been used in internal arterial prostheses. The only significant hazard in handling PET resins is dust associated with mineral or glass fillers during chip grinding or compounding operations.

PCT Molding Resins

The thermoplastic polyester, poly(cyclohexyldimethylene terephthalate) (PCT), was first produced as a polyester fiber. PCT is a copolymer of terephthalic acid and the cis and transforms of cyclohexanedimethanol. It was introduced as a molding resin in 1987 in glass-filled and flame-retarded grades with specific end uses. One specific advantage of PCT is that it has similar flow characteristics (although at higher temperatures) during molding to both PET and PBT which means that extensive mold redesign is not necessary. PCT has low moisture uptake and is not affected by changes in humidity as is nylon.

PBT Molding Resins

Poly(butylene terephthalate) is the oldest of the crystalline thermoplastic polyester molding resins. PBT has a unique ability to crystallize rapidly, even in a cold mold, to give tough, distortion-free moldings without special additives or nucleants. Although the unmodified polymer has very good flow properties and is used in electrical connectors and fiber optical cable buffer tubes, it performs even better if reinforced with inorganic fillers, notably chopped glass fiber.

The filled grades of PBT are tougher, stiffer, and stronger materials. Even when unfilled, the plastic has good strength, rigidity, and toughness, low creep and minimal moisture absorbance, and does not undergo dimensional changes with fluctuations in humidity. It is characterized by excellent electrical and dielectric properties and high surface finish.

Physical Properties of PBT. Unlike PET, the polymer PBT exists in two polymorphs called the α- and β-forms, which have distinctly different crystal structures. The two forms are interconvertible under mechanical stress.

The melting point of PBT is 222–224°C depending on the degree of crystallization and annealing conditions. The heat of fusion is about 140 J/g; and the T_g, usually quoted at about 45°C, depends on the physical nature of the sample.

Chemical Properties. PBT is highly crystalline and, as in the case of PET, does not suffer from solvent stress corrosion cracking as do amorphous materials. It is resistant at room temperature to most common chemicals and solvents, lubricants, greases, and automotive fluids. Ketones attack PBT at temperatures above ambient.

Mechanical Properties. Properties of typical grades of PBT, either as unfilled neat resin, glass-fiber filled, and FR-grades, are set out in Table 2.

Table 2. Mechanical Properties of PBT

Property	Unfilled, grade low mol wt	30% Glass: General purpose	30% Glass: Flame retardant	30% Glass: High impact
specific gravity	1.31	1.54	1.66	1.53
tensile strength, MPa[a]	57	135	135	97
tensile modulus, GPa[b]	2.5	9.7	11.7	8.3
elongation, %	5	2	1.5	3.1
flexural strength, MPa[a]	85.5	193	193	152
flexural modulus, GPa[b]	2.5	8.3	10.3	6.9
notched Izod, J/m[c]	37.4	90.7	69.4	160
unnotched Izod, J/m[c]	1228	240	214	641
HDT at 1.82 MPa[b], °C	51	206	208	191
volume resistivity, $\Omega \cdot$ cm	10^{15}	10^{16}	5×10^{15}	4×10^{14}
dielectric strength, V/25 μm	420	560	490	500
dielectric constant, ϵ, 100 Hz	3.2	3.7	3.9	4.3
flammability UL94, at 0.8 mm	HB	HB	V0	HB

[a] To convert MPa to psi, multiply by 145.
[b] To convert GPa to psi, multiply by 145,000.
[c] To convert J/m to ft·lbf/in., divide by 53.38.

Processing. PBT is one of the easiest thermoplastics to injection mold provided the polymer is thoroughly dried before melting. Owing to its good flow properties and extremely high rate of crystallization in the mold, cycle times are short (5–45 s).

Health and Safety Aspects. PBT resins are not harmful or hazardous when handled at room temperature under normal conditions. No problem regarding contact with the pellets has been encountered under normal conditions. Glass fines can, however, cause skin irritation.

Thermoplastic Copolyester Elastomers

Thermoplastic copolyester elastomers are generally block copolymers produced from short-chain aliphatic diols, aromatic diacids, and polyalkylene ether-diols. They are often called polyesterether or polyester elastomers. The most significant commercial product is the copolymer from butane-1,4-diol, dimethyl terephthalate, and polytetramethylene ether glycol, which produces a segmented block copolyesterether with the following structure.

The polymer with this structure has "hard" crystallizable segments of poly(tetramethylene terephthalate) (PBT or 4GT). The "soft" segment phase is poly(tetramethylene ether glycol terephthalate), called PTMEG-T. The ratio of soft to hard segments determines the elastomeric nature of the copolymer.

Copolyesterether elastomers are considered to be high performance elastomers. In general they are used in applications which involve some type of repeated mechanical movement such as bending, flexing, pushing, rotating, pulsing, impacting, or recoiling.

Physical Properties and Morphology. Elastomers require a system of cross-bonding which links the flexible molecular chains to each other to give a network structure. Without this network, the elastomer would not have the properties to make it commercially useful. Thermoplastic elastomers do not undergo a vulcanization step and the network is produced by reversible physical bonds between the polymer chains. Copolyesterether elastomers are cross-bonded through a crystallization process. These block copolymers contain a crystalline phase which physically locks the flexible soft segments into an elastic network below the crystalline melting point of this phase.

Mechanical Properties. Commercial grades of copolyesterether elastomers are mechanically durable with high tensile strength and high load-bearing capabilities for an elastomer. They are very resilient, have low hysteresis, and excellent creep resistance. In addition they have high tear strength, good abrasion resistance, and a long flex life.

Polyester elastomers are resistant to a variety of common solvents including aqueous acids or bases.

Manufacture. Polyesterether elastomers are made by a polycondensation reaction, either batchwise or by continuous polymerization processes. The reaction proceeds in two steps, the first stage being an ester interchange followed by a polycondensation step, exactly analogous to the steps used in the manufacture of PET or PBT.

Processing. Injection molding is probably the most important process for shaping polyesterether elastomer parts. However, extrusion and blow molding are also important.

Processing conditions for blow molding and extrusion are similar to those for injection molding, but cooler temperatures are required to achieve higher viscosities for extrusion and blow molding. Other methods of fabrication are rotational molding, film extrusion, and melt casting.

Health and Safety Issues. Polyesterether elastomers derived from dimethyl terephthalate, butanediol, and PTMEG are not hazardous according to the published Materials Safety Data Sheets (MSDS) for this elastomer. Polymers of a similar structure containing isophthalic acid are also not considered hazardous. One environmental advantage of thermoplastic elastomers of this type is that they are melt-reprocessible and thus scrap and off-specification material and even obselete parts can be easily recycled (see RECYCLING, PLASTICS).

Uses. Specific applications for polyesterether elastomers are too numerous to mention in detail; they include wire and cable, automotive, footware, hose and tubing, and other industrial components.

Liquid Crystal Polymers

Some molecules, both small molecules and polymers, can exist in an ordered liquid phase or mesophase. Such substances are called liquid crystals (see LIQUID CRYSTALLINE MATERIALS). The mesophase state is intermediate between a structured solid and an isotropic or disordered liquid. There are many types of mesophase, but they all fall into two main categories: those which possess some measure of three-dimensional order and those which have only rotational order and are therefore fluid. It is this dual property of behaving like a liquid and a solid that gives rise to many interesting properties.

Most polymeric liquid crystals are based on stiff rod-like molecular units which are called calamitic mesogens.

Mesophases-which exist only in solution are called lyotropic mesophases. Mesophases that can exist by altering the temperature are called thermotropic mesophases. Lyotropic mesophases typically form from solutions of stiff rod-like polymers in solvents above a certain critical concentration.

Polymers have been described exhibiting all types of mesophase characters, but the only ones to have practical commercial utility as structural materials are rod-like polymers. These polymers are aromatic polyesters (or polyesteramides). The abbreviation LCP refers to thermotropic liquid crystalline aromatic polyesters.

Molecular Structure of LCPs. The persistence length of LCPs is greater by a factor of 2–3 than that of random coil polymers. Most LCPs have a large polarizability along the rigid chain axis compared to that in a transverse direction.

During thermoplastic processing (eg, extrusion or injection molding) the polymer flows in the liquid crystal (nematic) phase. The relatively rigid rod-like polymer chains align themselves in the flow direction and there is little or no entanglement, giving a very fluid melt. As a result, such polymers are ideally suited to molding intricate and finely detailed components.

LCPs as Molding Resins. Some of the principal advantages of LCPs as molding resins include *(1)* extremely low shrinkage and warpage and exceptional dimensional repeatability, *(2)* high melt strength for versatility of fabrication, *(3)* low melt viscosity for high flow rates in thin sections and intricate molds, *(4)* low heat of fusion for very fast cooling, short cycle times, *(5)* low flash due to low injection pressures and shear-sensitive viscosity, and *(6)* high stiffness at high temperatures which allows parts to be ejected while hot.

Rheology of LCPs. LCPs show evidence of shear thinning at much lower shear rates than those of most conventional thermoplastics. Injection-molded LCP parts have a distinct fibrous morphology that can be obtained only in an extensional flow process. Key processing and property advantages of LCPs in injection molding are the ability to fill long flow paths, thin walls, multiple cavities, and multiple inserts; fast mold cycles; reduced injection and clamp pressures; low shrinkage and warpage; and low thermal expansion coefficients in the machine direction due to the orientation of the molecules.

Chemical Resistance of LCPs. Certain liquid crystal polymers have extremely high chemical resistance to a variety of aggressive chemicals and solvents.

Applications of LCPs. The most advanced application for LCPs is as injection-molding compounds for electrical interconnect devices. In the electrical and electronics area, surface mount components, connectors, chip carriers, ceramic replacements, bobbins, electric motor insulation, fiber optic components, closures, and fuse-holders are some of the applications. Other applications are bearings, bushings, gears, cams, microwave components, under-the-hood automotive components, aerospace applications, and aircraft components.

ANTHONY J. EAST
MICHAEL GOLDEN
Hoechst Celanese Corporation
SUBHASH MAKHIJA
Consultant

I. Goodman, in J. I. Kroschwitz, ed., *Encyclopedia of Polymer Science and Engineering*, 2nd ed., Vol. 12, John Wiley & Sons, Inc., New York, 1988, pp. 1–75; J. Y. Jadhav and S. W. Kantor, in Vol. 12, pp. 217–256.

B. M. Walker and C. P. Rader, eds., *Handbook of Thermoplastic Elastomers*, 2nd ed., Van Nostrand Rheinhold Co., New York, 1988 pp. 181–223.

L. M. Sherman, *Plast. Technol.*, 62 (Mar. 1996). *Chem. Eng. News*, 10 (Apr. 1996).

POLYESTERS, UNSATURATED

Low molecular weight polyester polymers derived from unsaturated dibasic acids (or anhydrides) dissolved in unsaturated vinyl monomers, comprise a versatile family of thermosetting materials known generally as unsaturated polyester resins.

The dominant applications for these resins are in conjunction with glass fiber reinforcement to form laminar composites known generically as fiber glass-reinforced plastic (FRP) in the United States and glass fiber-reinforced plastic (GRP) in Europe and elsewhere. Resins have also evolved for use in casting processes, which usually contain high loadings of fillers or mineral aggregate and are defined as one form of polymer concrete.

Raw Materials

The properties of polymers formed by the step growth esterification of glycols and dibasic acids can be manipulated widely by the choice of coreactant raw materials (Table 1). The reactivity fundamental to the majority of commercial resins is derived from maleic anhydride (MAN) as the unsaturated component in the polymer, and styrene as the coreactant monomer. Propylene glycol (PG) is the principal

Table 1. Ingredients of Polyester Resin Formulations in Descending Order of Commercial Significance

Glycol	Dibasic acid or anhydride	Unsaturated acid or anhydride	Unsaturated monomer
propylene glycol	phthalic anhydride	maleic anhydride	styrene
diethylene glycol	dicyclopentadiene–maleic anhydride[a]	furamic acid	vinyltoluene
ethylene glycol		methacrylic acid	methyl methacrylate
neopentyl glycol	isophthalic acid	acrylic acid	diallyl phthalate
dipropylene glycol	adipic acid	itaconic acid	α-methylstyrene
dibromoneopentyl glycol	chlorendic anhydride[b]		triallyl cyanurate
			divinylbenzene
bisphenol A diglycidyl ether	tetrabromophthalic anhydride		
bisphenol A dipropoxy ether	tetrabromophthalic anhydride		
tetrabromobisphenol diethoxy ether	terephthalic acid		
	tetrachlorophthalic anhydride		
propylene oxide			
1,4-butanediol			

[a] Acid addition product, formed *in situ*.
[b] 1,4,5,6,7,7-Hexachlorobicyclo(2,2,1)-5-heptene-2,3-dicarboxylic anhydride.

glycol used in most compositions, and (*ortho*)-phthalic anhydride (PA) is the principal dibasic acid incorporated to moderate the reactivity and performance of the final resins (see PHTHALIC ACID AND OTHER BENZENEPOLYCARBOXYLIC ACIDS).

Other glycols (qv) can be used to impart selective properties to these simple compositions.

Process Equipment

The polyesterification reaction is normally carried out in stainless steel vessels ranging from 8,000–20,000 liters, heated and cooled through internal coils. Blade agitators revolving at 70–200 rpm are effective in stirring the low viscosity mobile reactants, which are maintained under an inert atmosphere during the reaction at temperatures up to 240°C.

Polyesterification

The reaction of glycols with dibasic acid anhydrides proceeds at above 100°C, ending with the exothermic formation of the acid half-ester produced by the opening of the anhydride ring. The reaction temperature rises to over 150°C, at which point the half-esters condense into polymers with the evolution of by-product water.

Maleic Isomerization. Polyester polymers are formulated using maleic anhydride as the common unsaturated moiety. During the course of the polyesterification reaction at 200°C, the *cis*-maleate ester isomerizes to the *trans*-fumarate. The fundamental reactivity of the final polyester with styrene is directly proportional to the degree of isomerization and the level of fumarate polymers formed during the course of esterification.

Molecular Weight. Unsaturated polyester resins are relatively low in molecular weight, and are formulated to achieve low working viscosities when dissolved in styrene. The number-average molecular weight (M_n) normally falls between 1800–2500, although dicyclopentadiene and orthophthalic resins can be useful below this range.

Formulation

ortho-Phthalic Resins. Resins based on *ortho*-phthalic anhydride comprise the largest group of polyester resins and are used in a variety of commercially significant applications.

Isophthalic Resins. Isophthalic acid (IPA) can be substituted for phthalic anhydride to enhance mechanical and thermal performance and improve resistance to corrosive environments. Although phthalic resins find wide application in ambient fabrication processes, isophthalic resins are more widely used in products employing high temperature forming processes such as pultruded profile and electrical-grade laminate.

Dicylopentadiene Resins. Dicyclopentadiene (DCPD) can be used as a reactive component in polyester resins. The addition reaction of maleic anhydride in the presence of an equivalent of water produces a dicyclopentadiene acid maleate that can condense with ethylene or diethylene glycol to form low molecular weight, highly reactive resins. These resins have largely displaced *ortho*-phthalic resins in marine applications because of beneficial shrinkage properties that reduce surface profile. The inherent low viscosity of these polymers also allows for the use of high levels of fillers to extend the resin-enhancing, flame-retardant properties.

Flame-Retardant Resins. Flame-retardant resins are formulated to conform to fire safety specifications developed for construction as well as marine and electrical applications.

Methyl methacrylate is often used in combination with styrene to improve light transmission and uv stability in flame-retardant glazing applications.

Bisphenol Resins. Derivatives of bisphenol A form the basis for two distinct resin resins groups that demonstrate superior thermal and corrosion resistance. The addition product of propylene oxide and bisphenol A, reacted with fumaric acid and dissolved in styrene monomer is used in applications involving extreme corrosive environments. The resins are known generically as bisphenol fumarates.

Bisphenol A diglycidyl ether reacts readily with methacrylic acid in the presence of basic catalysts to form the corresponding methacrylate hydroxy esters. Unlike polyesters that rely on fumarate unsaturation distributed within the polymer chain to form the cross-linked network, the bisphenol A epoxy dimethacrylates (vinly esters) have two reactive sites on the extremities of the polymer, and the low poly-dispersity defines a reactive polymer having almost uniform molecular weight. The resulting products have high flexural properties and inherent resiliency characterized by high tensile elongation.

Stabilizers. Hydroquinone is widely used in commercial resins to provide stability during the dissolution of the hot polyester resin in styrene during the manufacturing process. At levels of 150 ppm, a shelf life of over six months can be expected at ambient temperatures.

Performance Characteristics

Polyester resins undergo a rapid transformation from a viscous liquid to a solid plastic state that comprises a three-dimensional cross-linked polymer structure.

The cross-linked polymers form a thermoset plastic which cannot be changed or returned to its original condition by heating as it can with thermoplastics. This thermoset characteristic is beneficial in providing high temperature properties, good solvent and chemical resistance, and high flexural modulus. Reinforcing with glass fiber produces a composite plastic, which has exceptional strength characteristics suitable for replacing conventional fabricating materials such as wood, steel, and concrete.

Mechanical Properties. The performance of various polyester resin compositions can be distinguished by comparing the mechanical properties of thin castings (3 mm) of the neat resin defined in ASTM testing procedures. This technique is used widely to characterize subtle changes in flexural, tensile, and compressive properties that are generally overshadowed in highly filled or reinforced laminates.

Resins of higher molecular weight demonstrate higher tensile strength, whereas high fumarate resins have higher flexural modulus. The strength of all polyester resins is enhanced significantly by glass and other fibrous reinforcements.

Thermomechanical Properties. The highly cross-linked structure of cured unsaturated polyester resins produces thermoset characteristics in which the resistance to softening and deformation is greatly enhanced at elevated temperatures. Resins containing high adipic acid levels display rubbery or elastomeric properties at below ambient temperatures. Glass-reinforced products using these resins have exceptional impact properties and demonstrate high tolerance to low temperature cryogenic applications. Deformation at higher temperatures is moderated by fibrous reinforcements.

Dielectric Properties. Polyester resins are nonconductors, have relatively low dipolar characteristics, and provide high dielectric strength and surface resistivity.

Chemical Properties. The three-dimensional cross-linked network resists penetration and attack by most corrosive chemicals and nonpolar solvents. Water has wide ranging effects on different resin compositions as it penetrates into the plastic network. Cross-linking density and the presence of steric constituents local to the ester groups can enhance water resistance.

Flammability. Polyester resin products ignite and burn by emitting sooty smoke. Flammability can be reduced significantly through halogen-modified components. Additives such as phosphate esters are frequently used to enhance flame retardance, whereas antimony trioxide at levels of 5% on resin provides optimum retardance in combination with halogenated intermediates.

Weathering. Polyester resins in the form of laminates, coatings (gel coats), and castings perform well in outdoor exposures; marine craft, tanks, pipes, and architectural facia produced in the 1960s were still in service in the late 1990s. Most fiber-reinforced plastic (FRP) products are designed with protective and decorative gel coats which, in combination with uv stabilizers, provides improved weather resistance.

Application Processes

Polyester resins are fabricated easily in open molds at room temperature. Such processes account for over 80% of production volume, the remaining being fabricated using matched metal dies in high temperature semiautomated processes.

Closed-mold systems contain two mating dies. In resin-transfer molding (RTM), glass reinforcement is placed in the open mold; once the molds are in place, precatalyzed resin is injected into the cavity under pressure. The process is adaptable to large components and can be used to encapsulate foam, aluminum, and wood components into the structure.

Matched die molding is the most efficient process to produce high volumes of relatively large parts.

High strength composites with linear symmetry can be produced by the pultrusion process using continuous glass fiber reinforcements in the form of rovings.

JEFFREY SELLEY
Consultant

H. Boenig, *Unsaturated Polyesters Structures and Properties*, Elsevier Science, Inc., New York, 1964.

The American Standards and Testing Methods Manual, The American Society for Testing and Materials, Philadelphia, Pa., 1992.

POLYETHER ANTIBIOTICS. See ANTIBIOTICS, POLYETHERS.

POLYETHER ELASTOMERS. See ELASTOMERS, SYNTHETIC-POLYETHERS.

POLYETHERS

AROMATIC

Aromatic polyethers are best characterized by their thermal and chemical stabilities and mechanical properties. The aromatic portion of the polyether contributes to the thermal stability and mechanical properties, and the ether functionality facilitates processing but still possesses both oxidative and thermal stability. With these characteristic properties as well as the ability to be processed as molding materials, many of the aromatic polyethers can be classified as engineering thermoplastics (see ENGINEERING PLASTICS).

One class of aromatic polyethers consists of polymers with only aromatic rings and ether linkages in the backbone, for example, the poly(phenylene oxide)s. A second type contains a wide variety of other functional groups in the backbone, in addition to the aromatic units and ether linkages.

Poly(phenylene oxide)s

The poly(phenylene oxide)s are also known as polyoxyphenylenes and poly(phenylene ether)s. Variations in the configuration of the ether group and in the extent and type of substitution on the aromatic backbone give rise to a large number of possible homo- and copolymers. Poly(2,6-dimethyl-1,4-phenylene oxide) (DMPPO) is marketed as PPO resin. Blends of DMPPO with polystyrene and additives are marketed under the trade name of Noryl thermoplastic resin.

Chemical Properties. The phenolic end groups in poly(phenylene oxide)s react with oxidizing agents in a variety of ways; the type of product depends in part on other reagents that may be present. Thus, in the presence of other phenols, a catalytic amount of oxidizing agent generates aroxy radicals and the ensuing coredistribution produces low molecular weight products.

Synthesis. Many poly(1,4-phenylene oxide)s have been prepared in a one-step polymerization of 2,6-disubstituted phenols by oxidative coupling.

Poly(phenylene oxide)s can also be prepared from 4-halo-2,6-disubstituted phenols by displacement of the halogen to form the ether linkage.

An unusual aspect of the oxidative coupling of substituted phenols is the formation of DMPPO from 2,4,6-trimethylphenol.

Copolymers of poly(phenylene oxide)s can be prepared in several ways. Oxidative coupling of mixtures of phenols usually provides random copolymers. With a pair of phenols that have different oxidation potentials or that coredistribute at different rates, block copolymers can form. Another route is the oxidation of mixed dimers which forms random copolymers. Copolymers can also be produced by allowing only some of the rings to undergo reaction in a substitution reaction. Block copolymers have been prepared by condensing the phenolic end groups of DMPPO with other polymers bearing reactive leaving groups on their end groups.

Polyether Blends

DMPPO and polystyrene form compatible blends. The two components are miscible in all proportions. Tensile strength and modulus of blends of DMPPO and crystal polystyrene reach a maximum with a composition containing about 80 wt % DMPPO, but most properties of blends are close to the weighted average for the two polymers. Blends with rubber-modified polystyrene also have intermediate property values, but the ductile PPO matrix is toughened more effectively by rubber than is the brittle polystyrene.

Blends with good mechanical properties can be made from DMPPO and polymers with which DMPPO is incompatible if an appropriate additive, compatibilizing agent, or treatment is used to increase the dispersion of the two phases.

Blends have also been prepared by dissolving DMPPO in a monomer and then polymerizing the monomer. The solutions can be applied to glass cloth before curing to produce prepregs for composites in applications such as printed circuit boards.

Noryl. Noryl engineering thermoplastics are polymer blends formed by melt-blending DMPPO and high impact polystyrene (HIPS) or other polymers such as nylon with proprietary stabilizers, flame retardants, impact modifiers, and other additives. Because the rubber characteristics that are required for optimum performance in DMPPO–polystyrene blends are not the same as for polystyrene alone, most of the HIPS that is used in DMPPO blends is designed specifically for this use. Noryl is produced as sheet and for vacuum forming, but by far the greatest use is in pellets for injection molding.

Principal application areas for Noryl are in water distribution, electrical–electronic applications, business machines, and automobiles. Noryl is used in many small appliances and personal care products, especially those involving exposure to heat and moisture.

Health and Safety Factors. Animal-feeding studies of DMPPO itself have shown it to be nontoxic on ingestion. The solvents, catalyst, and monomers that are used to prepare the polymers, however, should be handled with caution.

Polyethersulfones

The aromatic sulfone polymers are a group of high performance plastics, many of which have relatively closely related structures and similar properties (see POLYMERS CONTAINING SULFUR, POLYSULFONES). Chemically, all are polyethersulfones, ie, they have both aryl ether (ArOAr) and aryl sulfone ($ArSO_2Ar$) linkages in the polymer backbone. The simplest polyethersulfone consists of aromatic rings linked alternately by ether and sulfone groups.

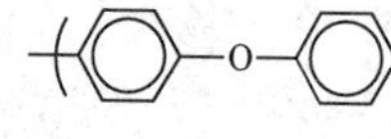

Synthesis. Aromatic polyethersulfones can be prepared by two different routes. In polyetherification, the sulfone group is present in one of the monomers and the ether linkage is formed in the polymerization step. In polysulfonylation, the alternative approach is used and the aryl ethers are coupled through a reaction that forms the sulfone linkage.

Polyetherketones

The polyetherification route to polyethersulfones can be adapted to the synthesis of polyethers containing strongly electron-withdrawing groups other than sulfone groups.

A polyetheretherketone (PEEK) introduced in 1978 is crystalline with a melting point of 334°C and a glass-transition temperature of ca 145°C. It can be molded and is used in applications such as liquid chromatography fittings. It is also used for coatings, in electrical insulation for high temperature service, and in composites.

Polyetherimides

An all aromatic polyetherimide (Kapton) is made by DuPont from reaction of pyromellitic dianhydride and 4,4′-oxydianiline. It possesses excellent thermal stability, mechanical characteristics, and electrical properties.

Certain aromatic polyetherimides are characterized by a combination of properties that makes them potential engineering thermoplastics. One of these polymers, with an isopropylidene unit in the backbone to enhance the solubility, is a molding material produced by General Electric, as Ultem resin. Attractive features include high temperature stability, flame resistance without added halogen or phosphorus, high strength, solvent resistance, hydrolytic stability, and injection moldability.

Syntheses. The presence of the ether and imide functionalities provides two general approaches for synthesis. Polyetherimides can be prepared by a nucleophilic displacement polymerization similar to the halide displacement in polysulfone synthesis or by a condensation of dianhydrides and diamines that is similar to normal polyimide synthesis (see POLYIMIDES).

Properties. In flammability testing, the oxygen index of many polyetherimides is high and they are self-extinguishing (V-0), nondripping, and generate little smoke.

Ultem polyetherimides have applications in areas where high strength, dimensional stability, creep resistance, and chemical stability at elevated temperatures are important.

DWAIN M. WHITE
General Electric Company

C. E. Schildknecht and I. S. Skeist, eds., *Polymerization Processes, High Polymers*, Vol. 29, John Wiley & Sons, Inc., New York, 1977.

S. E. McGrath, L. M. Robeson, and M. Matzner, in L. M. Sperling, ed., *Recent Advances in Polymer Blends, Grafts and Blocks*, Plenum Press, New York, 1974.

ETHYLENE OXIDE POLYMERS

Poly(ethylene oxide) (PEO) is a water-soluble, thermoplastic polymer produced by the heterogeneous polymerization of ethylene oxide. The white, free-flowing resins are characterized by the following structural formula: $-(CH_2CH_2O)_n-$.

Although most commonly known as poly(ethylene oxide) resins, they are occasionally referred to as poly(ethylene glycol) or poly(oxyethylene) resins.

Physical Properties

Crystallinity. At molecular weights of 105–107, poly(ethylene oxide) forms a highly ordered structure. This has been confirmed by nmr and x-ray diffraction patterns and by the sharpness of the crystalline melting point (62–67°C). The molecular conformation of poly(ethylene oxide), as determined by the use of x-ray diffraction, ir, and Raman spectroscopic methods, is shown in Figure 1.

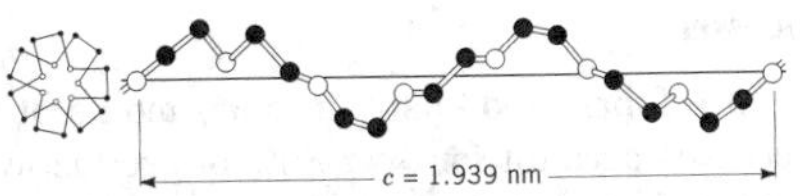

Figure 1. Molecular conformation of poly(ethylene oxide).

Density. The density is 1.15–1.26 g/cm^3.

Glass-Transition Temperature. The glass-transition temperature, T_g, of poly(ethylene oxide) has been measured over the molecular weight range of 10^2–10^7. These data indicate a rapid rise in the transition temperature to a maximum of −17C for a molecular weight of 6000.

Solubility. Poly(ethylene oxide) is completely soluble in water at room temperature. However, at elevated temperatures (>98C) the solubility decreases. It is also soluble in several organic solvents, particularly chlorinated hydrocarbons (see WATER-SOLUBLE POLYMERS). Aromatic hydrocarbons are better solvents for poly(ethylene oxide) at elevated temperatures.

The viscosity of aqueous solutions of poly(ethylene oxide) depends on the concentration of the polymer solute, the molecular weight, the solution temperature, concentration of dissolved inorganic salts, and the shear rate. Viscosity increases with concentration, and this dependence becomes more pronounced with increasing molecular weight.

Near the boiling point of water, the solubility–temperature relationship undergoes an abrupt inversion. Over a narrow temperature range, solutions become cloudy and the polymer precipitates; the polymer cannot dissolve in water above this precipitation temperature.

The viscosity of the aqueous solution is also significantly affected by temperature. In polymers of molecular weights ($1-50 \times 10^5$), the solution viscosity may decrease by one order of magnitude as the temperature of measurement is increased from 10 to 90°C.

The presence of inorganic salts in aqueous solutions of poly(ethylene oxide) reduces the upper temperature limit of solubility and viscosity.

Concentrated aqueous solutions of poly(ethylene oxide) are pseudoplastic. The degree of pseudoplasticity increases as the molecular weight increases.

Thermoplasticity. High molecular weight poly(ethylene oxide) can be molded, extruded, or calendered by means of conventional thermoplastic processing equipment. Films of poly(ethylene oxide) can be produced by the blown-film extrusion process; they tend to orient under stress, resulting in high strength in the draw direction. Melt viscosities are relatively unaffected by temperature changes but are directly proportional to the molecular weight of the polymer.

Chemical Properties

Association Complexes. The unshared electron pairs of the ether oxygens, which give the polymer strong hydrogen bonding affinity, can also take part in association reactions with a variety of monomeric and polymeric electron acceptors.

Oxidation. Because of the presence of weak C–O bonds in the backbone, high molecular weight polymers of ethylene oxide are susceptible to oxidative degradation in bulk, during thermoplastic processing, or in solution. During thermoplastic processing at elevated temperature, oxidative degradation is manifested by a rapid decrease in melt viscosity with time.

Stabilizers are useful in minimizing oxidative degradation during thermoplastic processing or in the bulk solid.

Manufacture and Processing

Heterogeneous Catalytic Polymerization. The polymerization of ethylene oxide to produce high molecular weight polymer involves heterogeneous reaction with propagation at the catalyst surface. The polymerization can involve anionic or cationic reactions of ethylene oxide that generally produce lower molecular weight products.

Catalysts capable of polymerizing ethylene oxide to high molecular weight polymers include many metal compounds.

Polymer Suspensions. Poly(ethylene oxide) resins are commercially available as fine granular solids. However, the polymer can be dispersed in a nonsolvent to provide better metering into various systems.

Thermoplastic Processing. Poly(ethylene oxide) resins can be thermoplastically formed into solid products. Through the use of plasticizers (qv), poly(ethylene oxide) can be extruded, molded, and calendered on conventional thermoplastic processing equipment.

Irradiation and Cross-Linking. Exposure of poly(ethylene oxide) to ionizable radiation can result in molecular weight breakdown or cross-linking, depending on the environmental conditions. If oxygen is present, hydroperoxides are formed and chain scission leads to an overall decrease in molecular weight. However, in the absence of oxygen, cross-linking becomes the preferred reaction. The resulting polymer network exhibits hydrogel properties of high water capacity.

Analytical and Test Methods

Molecular Weight. Measurement of intrinsic viscosity in water is the most commonly used method to determine the molecular weight of poly(ethylene oxide) resins.

A number of techniques, including static and dynamic light scattering, viscometry, and gel-permeation chromatography (gpc) with low angle laser light-scattering detection, have been used to study the behavior of poly(ethylene oxide) in solution.

Analysis for Poly(Ethylene Oxide). A special analytical method takes advantage of the fact that poly(ethylene oxide) forms a water-insoluble association compound with poly(acrylic acid). This reaction can be used in the analysis of the concentration of poly(ethylene oxide) in a dilute aqueous solution.

Health and Safety Factors, Toxicology

Poly(ethylene oxide) resins are safely used in numerous pharmaceutical and personal-care applications. Poly(ethylene oxide) resins show a low order toxicity in animal studies by all routes of exposure. Because of their high molecular weight, they are poorly adsorbed from the gastrointestinal tract and completely and rapidly eliminated. The resins are not skin irritants or sensitizers, nor do they cause eye irritation.

Uses

Significant use properties of poly(ethylene oxide) are complete water solubility, low toxicity, unique solution rheology, complexation with organic acids, low ash content, and thermoplasticity.

Pharmaceutical and Biomedical Applications. Because of its low toxicity and unique properties, poly(ethylene oxide) is utilized in a variety of pharmaceutical and biomedical applications. These include denture adhesives, adhesives for mucosal surfaces, ophthalmic solutions, wound dressings, oral drug release, biomaterials with low thrombogenicity, and lubricious coatings for biomaterials (see DENTAL MATERIALS; CONTACT LENSES).

Industrial Applications. Poly(ethylene oxide)s also have numerous industrial uses. These include flocculation; drag reduction; binders in ceramics, powder metallurgy, and water-based coatings of fluorescent lamps; detergents and lotions; adhesives; acid cleaners; jet cutting and control; construction; batteries; and various other applications (see BATTERIES; FLOCCULATING AGENTS).

DARLENE M. BACK
ELKE M. CLARK
RAMESH RAMACHANDRAN
Union Carbide Corporation

P. J. Flory, *Principles of Polymer Chemistry*, Cornell University Press, Ithaca, N.Y., 1953, pp. 266–313.

F. W. Stone and J. J. Stratta, in N. Bikales, ed., *Encyclopedia of Polymer Science and Technology*, Vol. 6, Wiley-Interscience, New York, 1967, pp. 103–145.

W. R. Sorenson and T. W. Cambell, *Preparative Methods of Polymer Chemistry*, Interscience Publishers, New York, 1961.

B. Scrosati, ed., *Second International Symposium on Polymer Electrolytes*, Elsevier Science Publishing Co. Inc., New York, 1990.

PROPYLENE OXIDE POLYMERS

Propylene oxide and other epoxides undergo homopolymerization to form polyethers. In industry the polymerization is started with multifunctional compounds to give a polyether structure having hydroxyl end groups. The hydroxyl end groups are utilized in a polyurethane forming reaction. This article discusses mainly propylene oxide (PO) and its various homopolymers that are used in the urethane industry.

Poly(propylene oxide) is usually abbreviated PPO and copolymers of PO and ethylene oxide (EO) are referred to as EOPO. Diol poly(propylene oxide) is commonly referred to by the common name poly(propylene glycol) (PPG). Propylene oxide and poly(propylene oxide) and its copolymers, with ethylene oxide, have by far the largest volume and importance in the polyurethane (PUR) and surfactant industry compared to all other polyepoxides.

Uses of Poly(propylene oxide). The vast majority of uses of PPO and EOPO copolymers are in polyurethanes and surfactants. Therefore, there are many applications in both the industrial and medical areas.

Propylene Oxide Monomer

There are two principal processes for producing PO, the chlorohydrin process and indirect oxidation (see PROPYLENE OXIDE).

Propylene Oxide Polymers

Propylene oxide and other epoxides polymerize by ring opening to form polyether structures. Either the methine, CH–O, or the methylene, CH_2–O, bonds are broken in this reaction. If the epoxide is unsymmetrical (as is PO) then three regioisomers are possible: head-to-tail (H–T), head-to-head (H–H), and tail-to-tail (T–T) dyads, ie, two monomer units shown as a sequence. The anionic and coordination polymerization of PO results in nearly all (95–98%) H–T sequences.

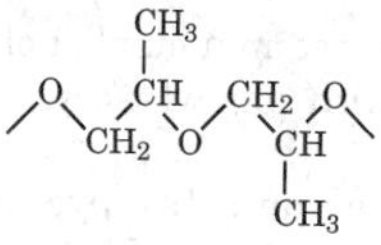

H–T

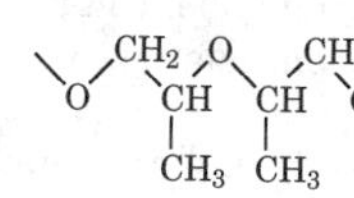

H–H

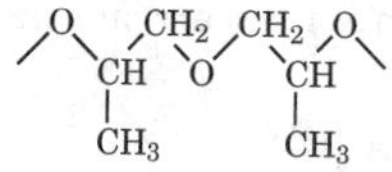

T–T

Base-Catalyzed Polymerization of Propylene Oxide. Most polyether polyols used commercially for urethanes and surfactants are produced by anionic polymerization. The bases of choice are potassium hydroxide or sodium hydroxide.

Tetrabutylammonium benzoate has been used as a catalyst for the polymerization of PO over the temperature range 40–108°C and the yield of polymer was typically low (2–78%); a large amount of unsaturation was present due to chain transfer. When synthetic hydrotalcite, $Mg_6Al_2(OH)_{16}CO_3 \cdot 4\ H_2O$, is used to polymerize PO and is activated by calcining at 450°C, a quantitative yield of PPO is obtained at 50°C in two hours. At Olin, POLY-L polyols have been produced with reduced unsaturation, but the catalyst used to produce them has not been disclosed. The use of zinc hexacyanocolbaltate to prepare low unsaturation polyols has been reported.

This yields a 3000 number-average molecular weight triol. In order to make this polyol in a reasonable amount of time, high temperature and consequently high pressure are required; therefore a stainless steel autoclave reactor is employed instead of a glass apparatus. The reactor is nominally 3.78 L (1 gal) in size, and has the following features: an oxide addition tube which extends to the bottom of the vessel and is pointed toward a high speed stirrer; a means of adding the oxide at a constant rate such as a pump or a flow controller; an inlet for vacuum or inert gas; a means to monitor the temperature and pressure (the oxide feed rate should also be monitored to give reproducible results); a charge port to add starter and catalyst; a water and steam inlet and outlet for cooling and heat; a high speed stirrer; a water jacket to help control the temperature; and a discharge port.

The charges for this polyol are shown in Table 1. Glycerol and 90% KOH are charged to the autoclave which is then purged with nitrogen. The charge of glycerol is only 3% of the total charge and may not be enough material for efficient stirring. A 4 or 5 mol PO adduct of glycerol can be made and used as the starter. The reactor is pressurized with nitrogen to 450 kPa (50 psig), where it is held for 15 minutes to check for leaks. The pressure is relieved and the reactor is heated to 105°C. Then the reactor is evacuated to 8 kPa (60 mm Hg), and the required amount of water is removed by stripping. The oxide is then added at a constant rate (600–900 g/oxide per mole initiator) in five hours at 105°C. The pressure is not allowed to exceed 722 kPa (90 psig) during the addition.

The mixture is kept for three hours at 105°C after the oxide addition is complete. By this time, the pressure should become constant. The mixture is then cooled to 50°C and discharged into a nitrogen-filled bottle. The catalyst is removed by absorbent (magnesium silicate) treatment followed by filtration or solvent extraction with hexane. In the laboratory, solvent extraction is convenient and effective, since polyethers with a molecular weight above about 700 are insoluble in water. Equal volumes of polyether, water, and hexane are combined and shaken in a separatory funnel. The top layer (polyether and hexane) is stripped free of hexane and residual water. The hydroxyl number, water, unsaturation value, and residual catalyst are determined by standard titration methods.

Hydroxyl Number. The molecular weight of polyether polyols for urethanes is usually expressed as its hydroxyl number or percent hydroxyl.

The hydroxyl number can be determined in a number of ways such as acetylation, phthalation, reaction with phenyl isocyanate, and ir and nmr methods.

Starters. Nearly any compound having an active hydrogen can be used as starter (initiator) for the polymerization of PO.

Unsaturation Value. The reaction temperature, catalyst concentration, and type of counterion of the alkoxide affect the degree of unsaturation. The tendency for rearrangement of PO to allyl alcohol is greatest with lithium hydroxide and decreases in the following order (100): $Li^+ > Na^+ > K^+ > Cs^+$.

Acid Catalysis. The ring-opening polymerization of PO using acid catalysts yields products that range from isomerization of PO to low molecular weight oligomers.

Coordination Polymerization of PO. Ring-opening polymerization catalysts, called coordination catalysts, include organoaluminum and organozinc compounds that have been modified with alcohols, ketones, phenols, and others. These polymerizations are characterized by controlled molecular weight with narrow molecular weight distribution and result in some amount of stereoregular polymer. The process is described as living polymerization, defined as consisting only of initiation and propagation reactions with no termination or chain-transfer reactions; that is, polymerization can be stopped by cutting off the flow of monomer and can be restarted by adding a new monomer. In what is known as immortal polymerization, the mixture continues to initiate polymerization until the reaction is specifically quenched.

Autoxidation of PPO. The oxidation of PPO is initiated by the formation of a radical on the carbon (usually secondary) α to the ether oxygen. The radical is then trapped by oxygen to form an α-alkoxy hydroperoxide. The hydroperoxide decomposes unimolecularly to give an oxy radical and a hydroxyl radical.

The autoxidation of PPO is characterized by having an induction period which becomes longer upon addition of increasing amounts of an antioxidant (see ANTIOXIDANTS).

Table 1. Charges for a 3000 Molecular Weight Glycerol-Initiated PPO Triol

Charges	Wt, g	Moles of hydroxyl	Hydroxyl equivalents	Hydroxyl number contribution
glycerol	178.0	1.93	5.80	54.23
potassium hydroxide, 90%	36.0	0.58	0.58	
water from KOH	−14.0			
propylene oxide (PO)	5822			
water from PO	0.6	0.03	0.06	0.56
unsaturation[a]		0.14	0.14	1.35
−K + H[b]	−22.0	−0.58	−0.58	
Total[c]	*6000*	*2.10*	*6.00*	*56.14*

[a] Unsaturation is normally expressed in meq/g but it is convenient to convert it to hydroxyl units for charge calculation.
[b] The replacement of K by H in the equation $ROK + H_2O \rightarrow ROH(polyol) + KOH$.
[c] The functionality can be calculated from the hydroxyl equivalents and hydroxyl moles: $f = 6.00/2.10 = 2.86$.

Characterization and Properties of Polyethers

Viscosity. In the molecular weight range of 200–6000, PPO polyols are liquids. The viscosity depends on the functionality. Polyols with higher functionality have higher viscosity at a given equivalent weight. At low equivalent weight the viscosity depends strongly on the initiator.

In measurements of the viscosity–temperature–mol wt relationship for PPO diols and triols the viscosity of PPG diols was found to be independent of shear rate, that is they are Newtonian fluids.

Nmr Studies. ^1H- and ^{13}C-nmr has been valuable in elucidating the structure of PPO and copolymers of EO and PO, especially since high field nmr has become widely available.

^1H-nmr (300 and 500 MHz) has been used to determine the number-average molar masses and molar ratio of the double-bond content of anionically polymerized PO over a range of conversions. ^{13}C-nmr has been used to differentiate between random and block copolymers, and to study persistence ratio (a measure of the deviation from fully random statistics), mean sequence length of EOPO sequences, triad probabilities, and starter and end groups.

Refractive Index. The effect of mol wt (1400–4000) on the refractive index (RI) increment of PPG in benzene has been measured. Generally, the RI decreases with temperature, with the rate of change increasing as the concentration increases.

Infrared Spectroscopy. The following bands are seen in the ir spectrum of PPG: 2970, 2940, 2880 cm^{-1} (C–H stretch, m); 1460, 1375 cm^{-1} (C–H bend, m); 1100, 1015 cm^{-1} (C–O stretch, m) of which the 2940 and 1015 band are specific.

Chromatography. One gpc study of low molecular weight polyethers used two systems: THF solvent and PLgel columns and water with TSK gel column sets. In THF the elution volume depends predominantly on chain length, whereas in water the composition as well as chain length influences the elution volume. Gpc calibration is typically done with poly(ethylene glycol) (PEG) or polystyrene standards, but the latter tend to overestimate the mol wt of PPO.

The composition of PPG–PEG blends has been determined using gpc with coupled density and RI detectors. An hplc system has been used to gather information about the functionality of PPO.

Reversed-phase hplc has been used to separate PPG into its components using evaporative light scattering and uv detection of their 3,5-dinitrobenzoyl derivatives. Polymer glycols in PUR elastomers have been identified by pyrolysis-gc.

Solubility. PPO polyols with a molecular weight below 700 are water soluble. Polyethers prepared from propylene oxide are soluble in most organic solvents. Water solubility of PPO has been determined using turbidimetric titration.

Mass Spectrometry. Field desorption mass spectrometry has been used to analyze PPO. Laser desorption Fourier transform mass spectrometry was used to measure PPG ion.

Density. At low equivalent weight, the specific gravity (density) of polyethers depends on the initiator and at high equivalent weight it depends on the alkylene oxide.

Other Properties. The glass-transition temperature for PPO is $T_g \sim 190$ K and varies little with molecular weight. The thermal conductivity of PPO is approximately 0.16 W/(m·K) for a 3000 mol wt polyol and 0.15 W/(m·K) for a 5000 mol wt polyol. The specific heat of PPO varies with temperature but not with the molecular weight.

Health and Safety

Propylene oxide is highly reactive. It reacts exothermically with any substance that has labile hydrogen such as water, alcohols, amines, and organic acids; acids, alkalies, and some salts act as catalysts.

Propylene oxide is a primary irritant, a mild protoplasmic poison, and a mild depressant of the central nervous system. Skin contact, even in dilute solution (1%), may cause irritation to the eyes, respiratory tract, and lungs. Propylene oxide is a suspected carcinogen in animals.

PPO and EOPO copolymers are low hazard–low vapor pressure liquids.

STEVEN D. GAGNON
BASF Corporation

S. D. Gagnon, pp. 273–307; N. Clinton, and P. Matlock, pp. 225–273; and R. W. Body and V. L. Kyllingstad, pp. 307–322; in J. I. Kroschwitz, ed., *Encyclopedia of Polymer Science and Engineering*, 2nd ed., Vol. 6, John Wiley and Sons, Inc., New York, 1986.

F. E. Bailey, in B. Elvers, S. Hawkins, and G. Schulz, eds., *Ullmann's Encyclopedia of Industrial Chemistry*, 5th ed., Vol. A21; VCH Publishers, Inc., New York, 1992, pp. 579–589.

L. C. Pizzini, and J. T. Patton, Jr., pp. 145–167; F. W. Stone, and J. J. Stratta, pp. 103–145; J. Furukawa, and T. Saegusa, pp. 175–195; and L. C. Pizzini, J. T. Patton, Jr., pp. 168–175, in H. F. Mark, N. G. Gaylord, and N. M. Bikales, eds., *Encyclopedia of Polymer Science and Technology*, Vol. 6, John Wiley and Sons, Inc., New York, 1967.

TETRAHYDROFURAN AND OXETANE POLYMERS

The polymerizations of tetrahydrofuran (THF) and of oxetane (OX) are classic examples of cationic ring-opening polymerizations. Under ideal conditions, the polymerization of the five-membered tetrahydrofuran ring is a reversible equilibrium polymerization, whereas the polymerization of the strained four-membered oxetane ring is irreversible.

Tetrahydrofuran Polymers

Physical Properties. Tetrahydrofuran polymers crystallize readily near ambient temperature. Moderately high molecular weight polymers turn into waxy solids, whereas high molecular weight polymers display thermoplastic behavior.

The only THF polymers of commercial importance in the late 1990s were diprimary low molecular weight poly(tetramethylene ether) (PTME) glycols or their derivatives. These materials generally are waxy solids when crystallized and colorless viscous fluids when melted.

Chemical Properties. The most important tetrahydrofuran polymers are the hydroxy-terminated polymers, that is, the α,ω-poly(tetramethylene ether) glycols used commercially to manufacture polyurethanes and polyesters (see URETHANE POLYMERS; POLYESTERS, THERMOPLASTIC).

End-Group Reactions. PTME glycols are normally the primary THF polymerization products, but THF polymers can be prepared with other end groups, either by direct polymerization of THF, or by chemical transformation of the hydroxy groups of the preformed polymer.

Commercially, the most important reaction of PTME glycols is the reaction with diisocyanates. Reaction with an excess of diisocyanate yields a prepolymer having isocyanate end groups, which can further react with short-chain diols or amines to give high molecular weight polyurethanes or polyurethane ureas.

Main-Chain Reactions. The backbone of PTME, polytetrahydrofuran (PTHF), consists of a series of linear aliphatic ether sequences. Like monomeric ethers, it is subject to oxidation to hydroperoxides and subsequent thermal degradation. The addition of common antioxidants, such as amines or hindered phenols, inhibits these reactions and thereby imparts adequate stability to the polymer. PTHF is quite stable to attack by bases but can be degraded by strong acids.

The PTHF chain is subject to attack and reaction during the polymerization process. These are reactions of the active end group of the growing polymer chain in the polymerizing mixture and an oxygen atom in the main chain.

Polymerization. The THF ring contains an oxygen atom with two unshared pairs of electrons. Therefore THF is a nucleophilic monomer having little steric interference toward potential electron acceptors. The polymerization is characterized by modest rates and heats of polymerization. Simply substituted tetrahydrofurans, in general, do not polymerize, although oligomers have been reported for monomethyl derivatives. Copolymerization of these substituted THFs is possible.

Cationic ring-opening polymerization is the only polymerization mechanism available to tetrahydrofuran. The propagating species is a tertiary oxonium ion associated with a negatively charged counterion:

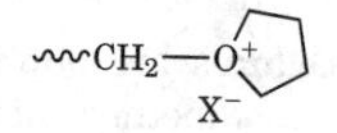

For continuing polymerization to occur, the ion pair must display reasonable stability.

The basic requirement for polymerization is that a THF tertiary oxonium ion must be formed by some mechanism. If a suitable counterion is present, polymerization follows. The requisite tertiary oxonium ion can be formed in any of several ways.

Direct alkylation or acylation of the oxygen of THF by exchange or addition occurs with the use of trialkyloxonium salts, carboxonium salts, super-acid esters or anhydrides, acylium salts, and sometimes carbenium salts.

The tertiary THF oxonium ion undergoes propagation by an S_N2 mechanism as a result of a bimolecular collision with THF monomer. Only collisions at the ring α-carbon atoms of the oxonium ion result in chain growth.

THF can be polymerized in the virtual absence of termination and transfer reactions. Under these conditions a living polymerization results and the number-average molecular weight of the polymer produced can be calculated from the number of active sites introduced and the amount of polymer produced at equilibrium.

A number of materials act as true transfer agents in THF polymerization; notable examples are dialkyl ethers and orthoformates. In low concentrations, water behaves as a transfer agent, and hydroxyl end groups are produced. The oxygen of dialkyl ethers are rather poor nucleophiles compared to THF and are therefore not very effective as transfer agents. On the other hand, orthoformates are effective transfer agents.

Copolymerization. THF copolymerizations are of interest for several reasons. Random copolymerization provides a way of reducing the melting temperature of THF polymers to room temperature and below. The crystallization tendency of THF segments in products is thereby reduced and mechanical properties of the products are often improved.

A host of copolymers of these types have been prepared. They include block copolymers from ϵ-caprolactam and PTMEG as well as block copolymers from PTHF and other cationically polymerizable heterocycles. Block copolymers from polystyrene and PTHF have been prepared. One-, two-, three-, and four-arm stars have been prepared with PTHF arms. Graft copolymers with PTHF branches have been made from a variety of hydrocarbon backbones, and graft copolymers with poly(vinyl chloride) branches have been prepared from PTHF backbones.

Manufacture and Processing. THF can be polymerized by many strongly acidic catalysts, but not all of them produce the required bifunctional polyether glycol with a minimum of by-products. Several large-scale commercial polymerization processes are based on fluorosulfonic acid, $HFSO_3$, catalysis, which meets all these requirements.

The primary polymerization product in a number of proprietary processes has a relatively wide molecular weight distribution, and a separate step is often used to narrow the polydispersity.

Storage and Handling. PTMEG is normally available in 20-kg steel pails, 200-kg steel drums, and in stainless steel tank cars or tank trucks, which are insulated and equipped with heating coils. Shipping temperature for tank trucks is normally 80°C (175°F) to maintain the product in liquid state. If subjected to low temperatures, it has to be reheated and melted.

PTMEG is not regulated as a hazardous material by the U.S. Department of Transportation. Liquid spills may be absorbed with a material such as vermiculite and handled as nontoxic waste. Larger spills, if fluid, may be pumped into drums or, if solid, shoveled into drums for later recovery or disposal by burning under controlled conditions.

Specifications, Standards, and Analysis. The standard commercial molecular weight grades for polytetramethylene ether glycol are 650, 1000, 1800, and 2000, but other molecular weight grades, such as 1400 and 2900, are available for special applications. Commercial poly(tetramethylene ether) glycols are waxy, white solids that melt over a temperature range near room temperature to clear, colorless, viscous liquids.

General analytical procedures are applicable in most cases, although a number of specific test methods have been developed for the analysis of polyether glycols. One of the most important tests is the determination of the hydroxyl number, ie, the number of milligrams of KOH (formula weight = 56.1) equivalent to the hydroxyl content of 1 g of the polymer diol sample.

Other important tests are for acid and alkalinity number and for water content, because water content and alkalinity of the polyether glycol can influence the reaction with isocyanates.

The most important general test methods are issued as ASTM Test Methods and are periodically updated.

Health and Safety Factors. Poly(tetramethylene ether) glycols were found to have low oral toxicity in animal tests. No adverse effects on inhalation have been observed. The polymer glycols are mild skin and eye irritants, and contact with skin, eyes, and clothing should be avoided.

Tests with bacterial or mammalian cell cultures demonstrated no mutagenic activity. PTMEG is not listed as a carcinogen. Additional data on safety of PTMEG may be found in the material safety and data sheets provided by the manufacturers.

Uses. The most important use area for poly(tetramethylene ether) glycols is polyurethane technology. Polyurethanes based on PTMEG have some outstanding properties that set them apart from polyurethanes based on other soft segments. They have excellent hydrolytic stability, high abrasion resistance, and excellent elastomeric properties.

The largest polyurethane end use areas are in spandex fibers for apparel (see FIBERS, ELASTOMERIC).

Oxetane Polymers

Physical Properties. In contrast to THF, substituted oxetanes polymerize readily as a result of the added ring strain associated with the smaller ring size. A large number of polyoxetanes have been prepared and partially characterized. The properties of the polymers vary greatly with the symmetry, bulk, and polarity of the substituents on the chain. The polymers range from totally amorphous liquids to highly crystalline, high melting solids. The unsubstituted oxetane polymer (POX) has a melting temperature, T_m, of 35°C, not far above ambient temperature.

Chemical Properties. The chemistry of polymerization of the oxetanes is much the same as for THF polymerization. The ring-opening polymerization of oxetanes is primarily accomplished by cationic polymerization methods, but because of the added ring strain, other polymerization techniques.

Manufacture and Processing. The only commercially important products in this area are the low molecular weight polyethers with hydroxyl end groups resulting from THF polymerization.

Most of the analytical and test methods described for THF and PTHF are applicable to OX and POX with only minor modifications.

Health and Safety Factors. Toxicity and hazards of handling oxetanes and their polymers are influenced markedly by their substituents. For many of the monomers, these factors are described in detail in material safety data sheets.

Uses. A large number of uses have been explored worldwide for high performance, but none of these uses were adequate in the marketplace to support its high price.

GERFRIED PRUCKMAYR
E. I. du Pont de Nemours & Co., Inc.
P. DREYFUSS
M. P. DREYFUSS
Consultants

S. Penczek, P. Kubisa, and K. Matyjaszewski, *Adv. Polym. Sci.* **68,69** (1985).

P. Dreyfuss, *Poly(tetrahydrofuran)*, Gordon & Breach, New York, 1982.

P. Dreyfuss, M. P. Dreyfuss, and G. Pruckmayr, in J. I. Kroschwitz, ed., *Encyclopedia of Polymer Science and Engineering*, 2nd ed., Vol. 16, Wiley-Interscience, New York, 1989, pp. 649–681.

Test Methods for Polyurethane Raw Materials, 2nd ed., The Society of the Plastics Industry, New York, 1992.

POLYETHYLENE. See OLEFIN POLYMERS.

POLY(ETHYLENE OXIDE). See POLYETHERS, ETHYLENE OXIDE POLYMERS.

POLYFLUOROSILICONES. See FLUORINE COMPOUNDS, ORGANIC.

(POLYHYDROXY)BENZENES

Polyhydric phenols with more than two hydroxy groups (ie, the three positional isomers of benzenetriol, the three isomeric benzenetetrols, benzenepentol, and benzenehexol) are discussed in this article. The benzenediols are catechol, resorcinol, and hydroquinone (see HYDROQUINONE, RESORCINOL, AND CATECHOL).

The following names of the benzenetriols have been used.

Common (trivial) name	Chemical Abstracts	Other usage
pyrogallol (pyrogallic acid)	1,2,3-benzenetriol	1,2,3-trihydroxybenzene
hydroxyhydroquinone	1,2,4-benzenetriol	1,2,4-trihydroxybenzene
phloroglucinol	1,3,5-benzenetriol	1,3,5-trihydroxybenzene

Derivatives of these compounds or their corresponding quinones are of widespread occurrence in nature. They are abundant in plants and fruits as glucosides, chromones, coumarin derivatives, flavonoids, essential oils, lignins, tannins, and alkaloids (see ALKALOIDS; COUMARIN; LIGNIN; OILS, ESSENTIAL). They also occur in microorganisms and animals. Many of these compounds have distinct properties and uses, eg, antibiotics (qv), plant-growth factors, insecticides, astringents, antioxidants (qv), toxins, sweeteners (qv), pigments (qv) and dyes, drugs, and many others (see DYES AND DYE INTERMEDIATES; INSECT CONTROL TECHNOLOGY; PHARMACEUTICALS). The most recent applications of these compounds are as components of photosensitive compounds in high resolution heat-resistant photoresist compositions.

The biochemical activity of the benzenepolyols is at least in part based on their oxidation–reduction potential.

Pyrogallol

Pyrogallol (**1**) is of widespread occurrence in nature; it is incorporated in tannins, anthocyanins, flavones, and alkaloids.

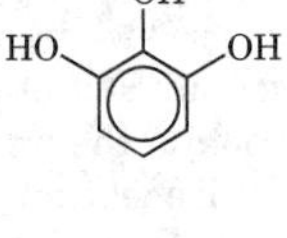

(1)

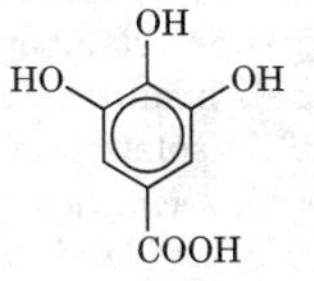

(2)

Properties. Pyrogallol (mp 133–134°C) forms colorless needles or leaflets which gray on contact with air or light. Its boiling point at atmospheric pressure with partial decomposition is 309°C; at 13.3 kPa (100 mm Hg), 232°C; and at 1.3 kPa (10 mm Hg), 168°C. When heated slowly, pyrogallol sublimes without decomposition; sp gr at 4°C, 1.453; heat of combustion, 2.673 MJ/mol (638.9 kcal/mol); solubility in parts per 100 parts solvent: 40 in water at 13°C, 62.5 in water at 25°C, 100 in alcohol at 25°C, 83.3 in ether at 25°C, slightly soluble in benzene, chloroform, and carbon disulfide. Pyrogallol is the strongest reducing agent among the benzenepolyols; it is oxidized rapidly in air. Its aqueous alkaline solution absorbs oxygen from the air and darkens rapidly. Sodium sulfite retards such oxidation.

Pyrogallol oxidized is a brownish black to black lustrous powder and is almost insoluble in water, alcohol, or ether but is soluble in alkalies.

Manufacture and Synthesis. The commercial manufacturing process starts with crude gallic acid, which is extracted from nutgalls or tara powder. It proceeds according to the following equation: $C_6H_2(OH)_3COOH \rightarrow C_6H_3(OH)_3 + CO_2$.

Because of the continuing uncertainties of supply of plant materials for gallic acid–pyrogallol manufacture, and because of valuable uses for pyrogallol, there is much interest in the development of synthetic processes.

Health and Safety Factors. Pyrogallol is extremely toxic. Extensive exposure of the skin may cause discoloration, local irritation, eczema, or death if it is absorbed. The principal symptom of poisoning attributable to pyrogallol is its effect on the red blood corpuscles which break down and lose their hemoglobin. Severe pyrogallol poisoning also leads to degeneration of the liver and kidneys. A yeast test has proved useful in checking acute toxicity of a number of chemicals including pyrogallol.

Uses. The main commercial applications of pyrogallol are in pharmaceuticals (qv) and pesticides (qv). Pyrogallol is the oldest and one of the more versatile of the photographic developing agents in use (see PHOTOGRAPHY). Pyrogallol is used to demonstrate chemiluminescence and traces of chromium (III) are determined with a pyrogallol chemiluminescence system (see LUMINESCENT MATERIALS, CHEMILUMINESCENCE).

Derivatives. Gallic acid (**2**) is the most important derivative of pyrogallol. Other derivatives include propyl gallate, gallein (pyrogallolphthalein), gallacetophenone (4-acetylpyrogallol), mescaline, and colchicine.

Hydroxyhydroquinone

Hydroxyhydroquinone (**3**) occurs in many plants and trees in the form of ethers, quinonoid pigments, coumarin derivatives, and complex compounds. It has strong reducing properties.

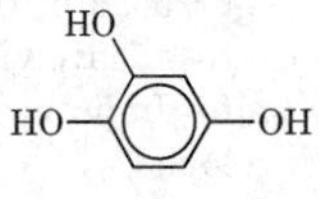

(3)

Properties. Hydroxyhydroquinone forms platelets or prisms (mp 140.5°C). The compound is easily soluble in water, ethanol, diethyl ether, and ethyl acetate and is very sparingly soluble in chloroform, carbon disulfide, benzene, and ligroin.

Hydroxyhydroquinone reacts as a typical oxidizable polyhydric phenol, but also undergoes certain keto-group reactions. In aqueous alkaline solution, it absorbs oxygen as effectively as pyrogallol.

Synthesis. The most convenient preparation of hydroxyhydroquinone is the reaction of *p*-benzoquinone with acetic anhydride in the presence of sulfuric acid or phosphoric acid. The resultant triacetate can be hydrolyzed to hydroxyhydroquinone.

Analysis. Thin-layer chromatography and liquid chromatography are well suited for the qualitative and quantitative estimation of hydroxyhydroquinone.

Health and Safety Factors. The LD_{50} of 1,2,4-trihydroxybenzene in mice after intracutaneous injection is 371 $\mu g/g$. Contact with hydroxyhydroquinone may blacken skin and fingernails.

Uses. Hydroxyhydroquinone has been used in hair and mordant dyes, for healing plant wounds, and in corrosion inhibitors and adhesives.

Derivatives. Derivatives include Scopoletin (6-methoxyumbelliferone), primin (2-methoxy-6-pentyl-1,4-benzoquinone), versicolin (1,2,4-trihydroxy-3-methylbenzene), an epoxygeranyl ether of 3,4-methylenedioxyphenol, rotenone, precocene-2 (2,2-dimethyl-6,7-dimethoxy-2*H*-chromene, and Maesanin.

Phloroglucinol

Phloroglucinol (**4**), a colorless and odorless solid only sparingly soluble in cold water, occurs in many natural products in the form of derivatives such as flavones, catechins, coumarin derivatives, anthocyanidins, xanthins, and glucosides.

Phloroglucinol is of low toxicity, but complex natural products containing a phloroglucinol moiety range in biological properties from antibiotic and antimitotic to potently carcinogenic.

Properties. Phloroglucinol forms odorless, colorless, sweet-tasting, rhombic crystals which tend to discolor on exposure to air or light. The dihydrate loses its water of crystallization at about 110°C (mp 113–116°C on quick heating); the anhydrous material melts at 217–219°C when heated rapidly. Phloroglucinol sublimes at higher temperatures with partial decomposition.

Although most of the physical and chemical properties of phloroglucinol characterize it as a polyhydric phenol, in many cases it reacts in a tautomeric keto form or as the β-triketone, 1,3,5-cyclohexanetrione.

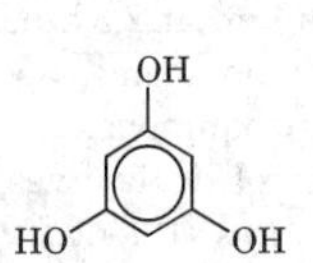

(4)

Friedel-Crafts acylation with acid chlorides and aluminum chloride in carbon disulfide gives the nuclear monoacylated phloroglucinols in good yield (see FRIEDEL-CRAFTS REACTIONS).

Manufacture and Synthesis. The only commercial process in use in the United States through the 1970s involved the oxidation of

2,4,6-trinitrotoluene (TNT) with dichromate in sulfuric acid to 2,4,6-trinitrobenzoic acid, followed by reduction and simultaneous decarboxylation with iron and hydrochloric acid to give 1,3,5-trianimobenzene. Acid hydrolysis at ca 108°C gave phloroglucinol in ca 75% yield. The process involved some explosion hazard in the initial stages, and it is no longer used in the United States because of the problem with waste disposal involving acid liquors and iron, chromium, and ammonium salts. Improved versions of the acid hydrolysis process have been developed.

Analysis. The instrumental methods of analysis are applicable, especially gas chromatography, with possible derivatization, and liquid chromatography.

Health and Safety Factors. Phloroglucinol has low toxicity by ingestion. Prolonged severe overexposure may disrupt the thyroid function. High dust concentration may cause respiratory irritation; the product is irritating to eyes and skin. Toxicity data include LD_{50} oral (rat) = 5800 mg/kg; Ames test = negative.

Uses. Two of the principal commercial applications of phloroglucinol, ie, in the diazotype copying process and textile dyeing processes, are based on the ability of each mole of phloroglucinol to couple rapidly with 3 mol of diazo compound. The azo dyes (qv), which are produced, give fast superior black shades. Phloroglucinol also is used in resins and adhesives, as a plastics component or additive, as an intermediate for hydraulic fluids, as a rubber additive, as a photographic chemical, and as a starting material for priming compositions.

Phloroglucinol is particularly valuable in the dyeing of acetate fiber but also has been used as a coupler for azoic colors in viscose, Orlon, cotton (qv), rayon, or nylon fibers, or in union fabrics containing these fibers.

Phloroglucinol and its derivatives are used as developers for light-sensitive planographic plates and for other photographic purposes.

Bischromonyl compounds derived from phloroglucinol are valuable in the treatment of asthma (see ANTIASTHMATIC AGENTS).

Derivatives. Derivatives of phloroglucinols include cotoin, griseofulvin, uvaretin, aflatoxins, bioflavanoids, and hesperidin.

Benzenetetrols

1,2,3,4-Benzenetetrol. 1,2,3,4-Tetrahydroxybenzene (apionol) forms needles from benzene (mp 161°C). It is easily soluble in water, diethyl ether, ethanol, and glacial acetic acid and is sparingly soluble in benzene.

1,2,3,4-Benzenetetrol is best prepared by the hydrolysis of 4-aminopyrogallol hydrochloride.

Derivatives. The most important derivatives of 1,2,3,4-benzenetetrol are the ubiquinones, eg, coenzyme Q, which are dimethoxytoluquinones with polyisoprenoid side chains. They occur in plants and animals.

Derivatives of ubiquinones are antioxidants for foodstuffs and vitamins (qv).

1,2,3,5-Benzenetetrol. 1,2,3,5-Tetrahydroxybenzene forms needles (mp 165°C) from water. The compound is easily soluble in water, alcohol, and ethyl acetate and is insoluble in chloroform and benzene.

1,2,3,5-Benzenetetrol has been prepared by the hydrolysis of 2,4,6-triaminophenol with dilute hydrochloric acid and by heating aqueous solutions of <0.2 *M* 2,4,6-triaminophenol at >130°C. The acid hydrolysis is improved by copper.

Derivatives. Derivatives of 1,2,3,5-benzenetetrol include 3,6-dihydroxy-2,4-dimethoxyacetophenone, 3,4,5-trimethoxyphenol (antiarol), and 2,6-dimethoxybenzoquinone.

Many 1,2,3,5-benzenetetrol derivatives are used medicinally (see ANALGESICS, ANTIPYRETICS, AND ANTIINFLAMMATORY AGENTS).

1,2,4,5-Benzenetetrol. 1,2,4,5-Tetrahydroxybenzene forms leaflets from glacial acetic acid (mp 215–220°C). It is easily soluble in water, ethanol, and diethyl ether but is not quite as soluble in concentrated hydrochloric acid and glacial acetic acid.

1,2,4,5-Benzenetetrol is obtained by the reduction of 2,5-dihydroxyl-1,4-benzoquinone, which is readily made by oxidation of hydroquinone dissolved in strong aqueous sodium hydroxide with hydrogen peroxide, with stannous chloride and hydrochloric acid or by catalytic hydrogenation.

Phosphorus derivatives of 1,2,4,5-benzenetetrol are effective antiwear and antioxidant additives for lubricating oils and also have flame-retardant properties (see FLAME RETARDANTS; LUBRICATION AND LUBRICANTS).

Benzenepentol

Benzenepentol (pentahydroxybenzene) has been prepared by boiling 2,4,6-triaminoresorcinol diethyl ether with water, followed by ether cleavage with HI. The product is very soluble in water but sparingly soluble in organic solvents.

Benzenehexol

Properties. Benzenehexol (hexahydroxybenzene) forms snow-white crystals when freshly prepared and collected in an inert atmosphere. Benzenehexol of good purity does not melt up to at least 310°C. It is sparingly soluble in water, ethanol, diethyl ether, and benzene.

Synthesis. The simplest laboratory preparation involves the aeration of the glyoxal–bisulfite addition product in sodium carbonate solution at 40–80°C, isolation of the sodium salt of tetrahydroxybenzoquinone, followed by acidification to obtain the free tetrahydroxy-*p*-benzoquinone; the latter is reduced to benzenehexol.

Derivatives. A considerable number of compounds that contain the benzenehexol structure possess therapeutic activity. Derivatives of benzenehexol include the hexaesters, which are antiatherogenics; tetroquinone, used for oral treatment of keloids; the dipotassium salt of rhodizonic acid and related derivatives; and the inositols.

GERD LESTON
Consultant

B. Z. Shakhashiri, *Chemical Demonstrations*, Vol. 1, The University of Wisconsin Press, Madison, 1983, p. 175.

POLYIMIDES

Polyimides (PI) are polycondensation products (**1**) prepared from derivatives of tetracarboxylic acids and primary diamines.

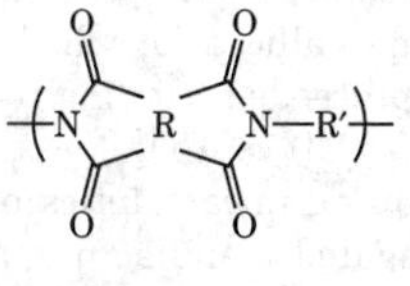

(**1**)

The main chain of these polymers contains, as the principal component, five- or six-membered heteroaromatic rings, ie, imides, which are usually present as condensed aromatic systems, such as with benzene (phthalimides, **2**) and naphthalene (naphthalimides, **3**) rings.

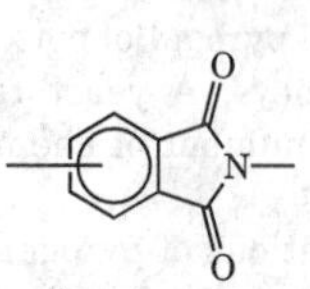

(**2**)

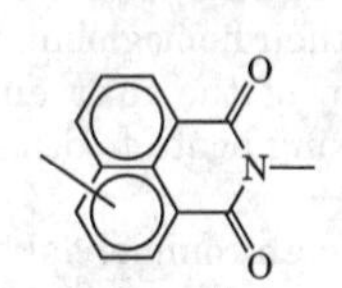

(**3**)

Among imide-containing polymers, polyimides derived from aromatic tetracarboxylic acids and aromatic diamines are of primary importance and represent typical high performance specialty polymers that are commercially employed in various applications. In structures (**1–3**), the oxidatively unstable amino group has been converted to an imino group that is stabilized by two carbonyl groups directly attached to it. The powerful electron-withdrawing effect of the carbonyl groups results in the formation of heteroaromatic systems that are low in electron density. Because oxidation is an electron-abstracting process, such heteroaromatic systems are generally resistant to oxidation. Aromatic polyimides exhibit outstanding mechanical and electrical properties as well as high thermoxidative and chemical resistance. Polyimides are widely used in critical components in aerospace, automotive, electrical, electronic, film, and coating applications.

Synthetic Methods

Since successful commercialization of Kapton by Du Pont Company in the 1960s, numerous compositions of polyimide and various methods of syntheses have been reported. A successful result for each method depends on the nature of the chemical components involved in the system, including monomers, intermediates, solvents, and the polyimide products, as well as on physical conditions during the synthesis. Properties such as monomer reactivity and solubility, and the glass-transition temperature, T_g, crystallinity, T_m, and melt viscosity of the polyimide products ultimately determine the effectiveness of each process. Accordingly, proper selection of synthetic method is often critical for preparation of polyimides of a given chemical composition.

Two-Step Poly(Amic Acid) Method. The two-step poly(amic acid) process is the most commonly practiced procedure. In this process, a dianhydride and a diamine react at ambient temperature in a dipolar aprotic solvent such as *N,N*-dimethylacetamide (DMAc) or *N*-methylpyrrolidinone (NMP) to form a poly(amic acid), which is then cyclized into the polyimide product.

One-Step Method. A single-stage homogeneous solution polymerization technique can be employed for polyimides that are soluble in organic solvents at the polymerization temperature. In this process, a stoichiometric mixture of monomers is heated in a high boiling solvent or a mixture of solvents at a temperature of 140–250°C, where the imidization reaction proceeds rapidly.

When use of one-step solution polymerization is an available option, generally it is the superior method to synthesize structurally pure polyimides because complete imidization and quantitative capping of end groups can be readily achieved in solution.

Tetracarboxylic Acids with Diamines. Early methods of polyimide synthesis used tetracarboxylic acids and their ester derivatives. The carboxylic acids were combined with diamines to form salts that were thermally imidized to form polyimides. For intractable polyimides such as those based on pyromellitic acid and aromatic diamines, high molecular weight poly(amic acid) intermediates have to be made which require the use of dianhydrides as monomer. For more tractable polyimides, however, one-step solution or melt polymerization can be employed as long as the reaction system can be maintained in solution or above the glass-transition temperature in the melt. Although the majority of literature reports that dianhydrides and diamines are used for such step-growth high temperature processes, the tetracarboxylic acids can be used in place of dianhydrides without significant differences in the overall results.

Other Synthetic Methods. Other synthetic methods are based on diesters of tetracarboxylic acids with diamines, polyisoimides as precursors, ester derivatives of poly(amic acid)s, polymerization via nucleophilic substitution reaction, polymerization of dianhydrides and diisocyanates, polymerization by C–C coupling, and polymerization by cycloaddition.

Thermoset Polyimides

Various thermoset polyimide resins are used as matrix resins for advanced composites (see COMPOSITE MATERIALS, POLYMER-MATRIX). In general, low molecular weight oligomers with functional end groups (cross-linkable groups) are used in impregnating the fiber fabrics. Such resins possess low viscosities which facilitate flow of resins and good wetting of the fibers. The resulting preimpregnated fabrics (prepregs) are consolidated in the mold and thermally cross-linked (cured). Thermoset polyimides produce creep-resistant composites of high rigidity. One of the more widely used thermoset polyimides is bismaleimide resin (BMI). A commonly used BMI component is a bismaleimide, 4,4′-bis(maleimido)diphenylmethane, derived from the lowest cost aromatic diamine, 4,4′diaminodiphenylmethane or 4,4′-methylenedianiline (MDA).

A great many modifications have been applied to BMIs in order to improve toughness and strength as well as processibility. Acetylene-terminated oligoimides have been reported as useful thermoset matrix resins and adhesives.

Properties and Applications of Polyimides

Because a wide range of properties are realized with various compositions, polyimides are used in diverse areas of application.

Aromatic high temperature polyimides based on dianhydrides are commercially produced, mostly in the form of films and therefore the films are produced via the two-stage poly(amic acid) process. Kapton films are available in various thicknesses and also as surface-modified forms and others containing various inorganic fillers. Because of their outstanding thermal, mechanical, and electrical properties as well as radiation resistance, these films are widely used as insulation materials in aerospace, electric, and electronic components.

In assembling high temperature composites and composites with other materials such as ceramics and metals, high temperature polyimide adhesives have become important. Another interesting application of polyimides is gas-separation membranes, because of their high strength and high temperature capability.

Health and Safety Factors

Handling of monomers, solvents, and solutions required in the preparation of polyimides should be practiced under standard safe environmental conditions for chemical processes. Some of the aromatic diamines are suspected carcinogens.

TOHRU TAKEKOSHI
General Electric Company

M. K. Ghosh and K. L. Mittal, *Polyimides, Fundamentals and Applications*, Marcel Dekker, New York, 1996.

D. Wilson, H. D. Stenzenberger, and P. M. Hergenrother, eds., *Polyimides*, Chapman and Hall, New York, 1990.

C. E. Sroog, *Prog. Polym. Sci.* **16**, 561–694 (1991).

M. I. Bessonov, M. M. Koton, V. V. Kurdryavtsev, and L. A. Laius, *Polyimides, Thermally Stable Polymers*, Plenum Press, New York, 1987.

POLYISOBUTYLENE. See ELASTOMERS, SYNTHETIC–BUTYL RUBBER.

POLYISOPRENE. See ELASTOMERS, SYNTHETIC–POLYISOPRENE.

POLYMER BLENDS

Mixing of two or more polymers of different chemical composition offers a powerful way of tailoring performance and economic relationships using existing materials. The area of polymer blends or alloys is important for both scientific investigation and commercial product development. Fundamental issues that affect the properties of blends include equilibrium phase and interfacial behavior, physical and chemical interactions between the components, phase morphology, and rheology, all of which relate to issues of compatibility. One of the most important examples of polymer blends is the judicious incorporation of an elastomeric phase in a rigid matrix to enhance mechanical toughness. Discussion of polymer blends is typically limited to those containing only two different components.

Equilibrium-Phase Behavior

A mixture of two amorphous polymers may form a single phase of intimately mixed segments of the two macromolecular components or separate into two distinct phases consisting primarily of the individual components; which occurs at equilibrium is dictated by the principles of solution thermodynamics. The older polymer literature often refers to the molecularly mixed blends as compatible, but the modern literature uses the more scientifically precise term miscible to avoid confusion with other uses of the former term. It is important to note that amorphous polymers form glasses on sufficient cooling and that a homogeneous (or miscible) polymer blend exhibits a single, composition-dependent glass-transition temperature, T_g, whereas an immiscible blend has separate glass transitions associated with each phase. If one of the polymers in a blend can crystallize, a separate crystalline phase of that component can form even when the two polymers are miscible in the melt.

Experimental Techniques for Characterizing Phase Behavior

A variety of experimental techniques have been used to prepare and characterize polymer blends; some of the more important ones for establishing the equilibrium-phase behavior and the energetic interactions between chain segments are described here.

Blend Preparation. The most common techniques for preparing blends are melt mixing and solution casting. For immiscible pairs, the details of the mixing process determine the morphology of the resulting composite. To determine whether the components are miscible or not, considerable care must be exercised in the preparation stage, to assure that a physical equilibrium has in fact been achieved.

Transition Behavior. Immiscibility of a blend is usually readily apparent, because phase separation causes light scattering or limited transparency. However, simple visual inspection may not be reliable.

A simple and usually reliable approach for determining whether a blend system is miscible or not is to examine its glass-transition behavior using thermal, mechanical, or dielectric techniques. Miscible blends show a single, composition-dependent glass transition reflecting the mixed environment of the blend, whereas two-phase blends show two T_gs characteristic of each phase.

Sorption. Measurement of the equilibrium sorption of vapors or gases in miscible blends can, in principle, give information about the interaction energy parameter. The concept of inverse gas chromatography has been one of the most popular methods of obtaining measurements.

Analogue Calorimetry. Basically, the interactions between polymer segments are the same as those between lower molecular weight compounds of similar molecular structures, and the heats of mixing for these liquid analogues can be measured by direct calorimetry.

Spectroscopy. A variety of spectroscopic techniques, including nmr, eximer fluorescence, and nonradiative energy transfer, have been used to obtain information about blend phase structure and the underlying molecular interactions. Fourier transform infrared (ftir) has been most extensively used to learn about mechanisms of specific interactions involved in blend miscibility.

Scattering. The classical techniques of small-angle scattering of light and x-rays have been used to study polymer blends. Scattering techniques provide a sensitive method for determining whether blends are homogeneous or heterogeneous and to follow the kinetics of phase-separation processes.

Critical Molecular Weight. Two polymers that form immiscible blends at molecular weight levels typical of commercial materials may show miscibility if the molecular weights of one or both components can be reduced enough. Quantitative determination of these critical molecular weights can then be used to compute the interaction energy for this pair using an appropriate mixing theory.

Effect of Molecular Structure on Polymer–Polymer Interactions

The literature contains extensive reports on investigations of the equilibrium-phase behavior for an enormous number of polymer-polymer pairs. The number of blends known to be miscible has grown so rapidly since the mid-1980s that it is more instructive to attempt to understand these observations in terms of the molecular structures of the components rather than to catalog them. This involves studies of dispersive interactions, hydrogen bonding, copolymer models, mean field approximations, and copolymer models.

Property Relationships

The relationships between the physical properties of a blend and those of its components can depend on the thermodynamic interaction between the components and many other factors. Some generalizations are possible, but exceptions are common and fundamental understanding for some properties remains incomplete in spite of the central importance of this issue in blend technology.

Fully miscible blends generally represent the simplest case. In the absence of crystallinity, most properties follow some additive relationship; miscible blends are similar to random copolymers in this regard. The glass-transition temperature, and hence the softening point, is generally a monotonic function of composition. As a rule, most mechanical properties, permeation (logarithmic scale) to small molecules, etc, follow nearly linear relations with composition in such systems. If the blend converts from rubbery to glassy with composition, the usual changes in properties at the glass transition become superimposed on this relationship. Crystallization of one of the components changes properties in similar ways.

For blends where the components form separate phases, properties depend on the arrangement of these phases in space and the nature of the interface between the phases. Immiscible blends behave like composite materials (qv) in many respects. Properties like softening temperature, modulus, permeation, etc, are dominated by the properties of the component that forms the continuous phase. Failure properties, especially those related to ductility, eg, elongation at break and impact strength, often depend on the dimensions of the phases and the degree of interfacial adhesion between the components.

Compatibilization

Incompatible blends often have little commercial value because of the deficiencies in ductility-related properties such as impact resistance, elongation at break, and strength at break. Two general methods are used to remedy these problems: copolymer addition and reactive compatibilization.

Copolymer Addition. Addition of block or graft copolymers to improve the mechanical properties of immiscible polymer blends has been used since the mid-1970s with varying degrees of success. Interfacial adhesion and mechanical compatibility can be improved by the addition of appropriate block and graft copolymers.

Reactive Compatibilization. Polymers functionalized with anhydride, carboxylic acid, amine, hydroxyl, epoxide, etc, groups have been utilized in reactive compatibilization. These groups can react by means of condensation chemistry to give block or graft copolymers whose constituent chains are joined by ester, amide, and imide linkages during the melt-blending process.

Grafting reactions are often employed to chemically modify polymer melts so as to achieve a particular desired functionality.

Blend Morphology

In many instances, phase-separated blends are the preferred means of achieving useful results. For example, such polymer–polymer composites yield materials whose stiffness can be adjusted, in principle, to any value between those of the component polymers. However, tailoring blends to achieve this or any other characteristic requires, among other things, control over the spatial arrangement or morphology of the phases, and some degree of stability once they are formed. These arrangements may consist of one phase dispersed as simple spheres in a matrix of the other polymer, as shown in Figure **1a**. On the other hand, the dispersed phase may take the form of platelets or fibrils

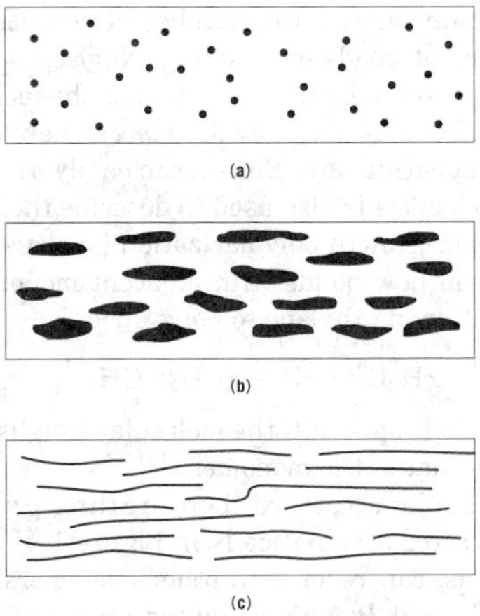

Figure 1. Types of dispersion of a polymer (dark regions) in the matrix of an immiscible polymer. The spherical droplets (**a**) are progressively extended into (**b**) platelets (biaxial) or (**c**) fibrils (uniaxial) by deformation.

with varying aspect ratios (Figure **1b–c**). Another distinct morphology consists of both phases simultaneously having a continuous character or an interpenetrating network of phases. The typical dimensions of the phases are important in all these morphologies.

Morphology Generation and Control. Homogeneous mixtures of two polymers may phase-separate upon a change in the temperature or removal of the solvent. In most commercially significant blends, however, the component polymers are immiscible, and the morphology of the dispersed phase is often generated by either added block and graft copolymer compatibilizers or chemical reactivity of the two principal components. Such blends are typically prepared by extrusion-melt blending.

Characterization. Electron microscopy has become one of the most widely used techniques for characterizing blend morphology. Scanning electron microscopy (sem) offers the simplest procedure. Because it reveals only surface features, the internal structure of blends is investigated by viewing fracture surfaces created at ambient or cryogenic temperatures.

Transmission electron microscopy (tem) requires viewing thin sections of material, and staining methods to enhance contrast are often used.

Toughened Polymers

Improvement of toughness is frequently a reason for blending. Usually, this is accomplished by addition of a small amount of an elastomer as a discrete particulate second phase. When it is properly done, this can result in a significant improvement of impact strength, with only a small reduction of modulus and tensile strength.

The matrix polymers can be divided into brittle or ductile categories, each having specific requirements for achieving toughness. Numerous variations are possible.

Many polymers may be toughened by the addition of a dispersed elastomer phase, and the growth of the rubber-toughened versions of these polymers has been large. The polymers that are most frequently rubber-modified include PS, PPO, PVC, and ABS.

High Impact Polystyrene. The toughening of brittle, glassy PS was introduced in the late 1940s. High impact polystyrene (HIPS) exhibits improved toughness and satisfactory rigidity, and in the 1990s it accounted for about 50% of all commercial polystyrene production.

The toughness as well as the other mechanical and rheological properties of HIPS are strongly affected by the nature of the rubber phase. Some of the variables that control HIPS performance include rubber composition and concentration, rubber-phase volume, particle-size distribution, degree of grafting and cross-linking, and molecular weight and molecular weight distribution of the matrix PS. Antioxidants (qv), plasticizers (qv), flame retardants (qv), and other additives also influence performance.

HIPS may be modified by mechanical blending with elastomeric polymers such as styrene–butadiene–styrene block copolymers, for the improvement of toughness and stress-crack resistance.

Acrylonitrile–Butadiene–Styrene. ABS is an important commercial polymer, with numerous applications. In the late 1950s, ABS was produced by emulsion grafting of styrene–acrylonitrile copolymers onto polybutadiene latex particles. This method continues to be the basis for a considerable volume of ABS manufacture. More recently, ABS has also been produced by continuous mass and mass-suspension processes. The various products may be mechanically blended for optimizing properties and cost. Flame retardancy of ABS is improved by chlorinated PE and other flame-retarding additives.

Poly(vinyl chloride). PVC is one of the most important and versatile commodity polymers. It is inherently flame retardant and chemically resistant and has found numerous and varied applications, principally because of its low price and capacity for being modified. Without modification, processibility, heat stability, impact strength, and appearance all are poor. Thermal stabilizers, lubricants, plasticizers, impact modifiers, and other additives transform PVC into a very versatile polymer.

Poly(2,6-dimethyl-1,4-phenylene oxide). PS and PPO are miscible in all proportions, and the rubber particles from HIPS are distributed uniformly throughout the new mixed matrix.

Because of the miscibility and the significantly different properties of component polymers, production of a large number of products with well-balanced properties is possible. Blends containing large amounts of PPO are particularly useful because of their high heat-deformation temperature; those containing less PPO are less expensive. They have an excellent combination of strength and toughness and are easily processed.

Polypropylene. PP is a versatile polymer that possesses excellent performance characteristics. New PP-blend formulations exhibit improved toughness, particularly at low temperatures. PP is blended mechanically with various elastomers to reduce low temperature brittleness.

Nylon. Many commercial nylon resins are modified by additives in order to improve toughness, heat fabrication, stability, flame retardancy, and other properties.

In rubber modification, an elastomeric second phase is incorporated, usually by mechanical blending with a variety of chemically modified elastomers to effect interaction between the nylon matrix and the elastomer phase.

Polycarbonates. Bisphenol A polycarbonate (PC) is used in a wide variety of applications because of its excellent balance of properties, including optical clarity, high heat-deformation temperature, toughness, and electrical properties. However, PC has some characteristics that deter its use in some areas. A number of approaches have been used to enhance its end-use properties. For example, polycarbonate blends with acrylonitrile-butadiene-styrene materials are commercially important plastics that have been found to be useful in many molding applications, particularly in the automotive industry and as engineering plastics. These blends economically combine some of the best properties of the components, eg, excellent impact strength (including improved notch sensitivity and thick section toughness), high heat distortion temperature, and relatively low melt viscosity for improved processibility.

Polyesters. Many polyester blends are commercially available, and most of these are rubber-toughened, typically by the incorporation of core-shell impact modifiers. For example, in a blend of a core-shell poly(methyl methacrylate) (PMMA) modifier with PBT, the emulsion-made impact modifier particles are not well-dispersed in the PBT matrix. The addition of a few percent of PC to the blend improves the rubber phase dispersion significantly and renders the system super-tough.

Epoxy Resins. Low molecular weight liquid carboxy-terminated butadiene-acrylonitrile rubbers are widely used for modification of epoxy resins (qv). Toughened epoxy resins are typically prepared *in situ* by quiescent bulk polymerization of epoxy in the presence of dissolved rubber.

Commercial Blends and Applications

The commercial development of polymer blends is strongly influenced by a set of more favorable economics than those affecting the more conventional chemical routes to new material systems. Blend systems, comprising pre-existing materials, can be developed much more quickly than newer polymers. Because the properties of an existing blend system are functions of the blend composition, an existing blend can also be easily and quickly modified to meet different performance objectives relative to cost required by new or changing markets.

It is anticipated that advances in understanding of the influence of molecular structure on polymer–polymer interactions, design of copolymers and terpolymers to achieve specific miscibility, interfacial, or compatibility effects in commercial products will become more common. Concern about the recycling or disposal of products after their useful life will play an increased role in the selection or design of blends that may be used in commercial applications.

H. Keskkula
D. R. Paul
J. W. Barlow
University of Texas at Austin

D. R. Paul and S. Newman, eds., Polymer Blends, Vols. I and II, Academic Press, Inc., New York, 1978.

D. J. Walsh, J. S. Higgins, and A. Maconnachie, eds., *Polymer Blends and Mixtures*, NATO ASI Series, Series E, Applied Sciences, No. 89, Martinus Nijhoff Publishers, Dordrecht, the Netherlands, 1985.

L. A. Utracki, *Polymer Alloys and Blends*, Hanser, Munich, Germany, 1989.

A. A. Collyer, ed., *Rubber-Toughened Engineering Plastics*, Chapman and Hall, London, 1994.

POLYMERS

Polymers are very large molecules made by covalently binding many smaller molecules. The word polymer is derived from the Greek *poly* (many) and *meros* (part). The size of polymer molecules imparts many interesting and useful properties not shared by low molecular weight materials. Polymers are the fundamental materials of plastics, rubbers and most fibers, and surface coatings and adhesives, and as such are essential to modern society. Also, many important constituents of living organisms, eg, proteins (qv) and cellulose (qv), are biopolymers (qv).

Classification and Nomenclature

Polymers were initially classified according to their response to temperature. Those that are softened (plasticized) reversibly by heat are known as thermoplastics. Others, though they might initially be liquid or soften once upon heating, undergo a curing (setting) reaction that solidifies them, and further heating leads only to degradation. These are known as thermosets. The ability of polymers to soften and flow at least once is one of their most valuable assets, as it allows them to be formed into complex shapes easily and inexpensively.

In general, polymers are formed by two types of reactions: condensation and addition. The formation of a polyester by polycondensation may be illustrated as follows.

$$\underset{\text{diol}}{x\,\mathrm{HOROH}} + \underset{\text{diacid}}{x\,\mathrm{HOOCR'COOH}} \longrightarrow \underset{\text{polyester}}{\mathrm{H{+}ORO{-}\overset{O}{\overset{\|}{C}}R'\overset{O}{\overset{\|}{C}}{+}_{x}OH}} + (2x-1)\,\mathrm{H_2O}$$

In the polyester formula shown, parentheses enclose the repeating unit. The quantity x is the degree of polymerization, sometimes also called the chain length, the number of repeating units strung together like identical beads on a string. Neglecting the ends of the molecule, which is usually justified for large x, the molecular weight M of the polymer molecule is given by $M = mx$, where m is the molecular weight of the repeating unit. Since x can easily be in the thousands, the term macromolecules is also used to describe these materials.

Addition or chain-growth polymerization involves the opening of a double bond to form new bonds with adjacent monomers, as typified by the polymerization of ethylene to polyethylene:

$$x\mathrm{H_2C{=}CH_2} \longrightarrow {+}\mathrm{CH_2{-}CH_2}{+}_x$$

Because no molecule is split out, the molecular weight of the repeating unit is identical to that of the monomer.

In terms of molecular structure, there are three principal categories of polymers, illustrated schematically in Figure 1. If each monomer is difunctional, that is, can react with other monomers at two points, a linear polymer is formed. Polymers that contain two different repeating units, say A and B, are known as copolymers (qv). A linear polymer with a random (AABBABAAABABB) arrangement of the repeating units is a random or statistical copolymer, or just copolymer. It is termed poly(A-*co*-B), with the primary constituent listed first. A molecule in which the two repeating units are arranged in long, contiguous blocks ($[A]_x[B]_y$) is a block (*b*) copolymer, poly(A–*b*–B).

A few points of tri- or higher functionality introduced along the polymer chains, either intentionally or through side reactions, give a branched polymer. Branches may grow from a linear backbone. A branched structure with the backbone consisting of one repeating unit (A) and the branches of another (B), is a graft (*g*) copolymer, poly(A–*g*–B).

As the length and frequency of branches increase, they may ultimately reach from chain to chain. If all the chains are connected together, a cross-linked or network polymer is formed. Cross-links may be built in during the polymerization reaction or may be created chemically or by radiation between previously formed linear or branched molecules (curing or vulcanization).

Structure and Properties

Various levels of structure ultimately determine the properties of a polymer.

Molecular Weights. With the exception of some naturally occurring polymers, all linear and branched polymers consist of molecules with a distribution of molecular weights. Two average molecular weights are commonly defined; the number-average, $\overline{M}_n$, and the weight-average, $\overline{M}_v$.

It may be shown that $\overline{M}_w \geq \overline{M}_n$. The two are equal only for a monodisperse material, in which all molecules are the same size. The ratio $\overline{M}_w/\overline{M}_n$ is known as the polydispersity index and is a measure of the breadth of the molecular weight distribution.

Most molecular weight characterization now is done by size-exclusion chromatography (sec), also known as gel-permeation chromatography (gpc).

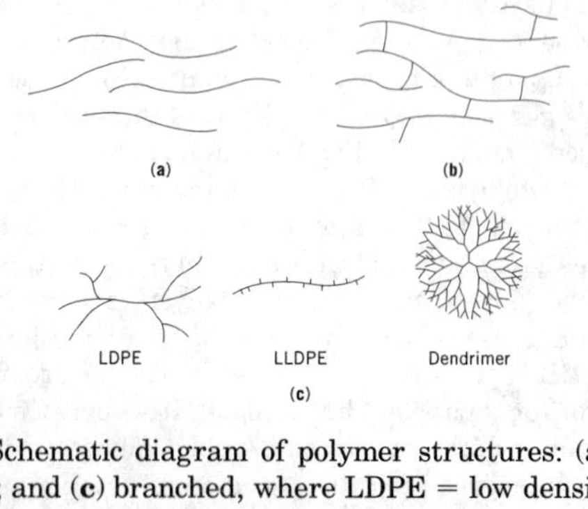

Figure 1. Schematic diagram of polymer structures: **(a)** linear; **(b)** cross-linked; and **(c)** branched, where LDPE = low density polyethylene and LLDPE = linear low density polyethylene.

Size-exclusion chromatography easily and rapidly gives the complete molecular weight distribution and any desired average.

Secondary Bonding. The atoms in a polymer molecule are held together by primary covalent bonds. Linear and branched chains are held together by secondary bonds: hydrogen bonds, dipole interactions, and dispersion or van der Waal's forces. By copolymerization with minor amounts of acrylic ($CH_2{=}CHCOOH$) or methacrylic acid followed by neutralization, ionic bonding can also be introduced between chains. Such polymers are known as ionomers (qv).

$$\sim COOH + M(OH)_2 + HOOC \sim \rightarrow \sim COO]^{-+}[M]^{+-}[OOC \sim +2\ H_2O$$

Secondary bonds are considerably weaker than the primary covalent bonds. When a linear or branched polymer is heated, the dissociation energies of the secondary bonds are exceeded before the primary covalent bonds are broken, freeing up the individual chains to flow under stress. When the material is cooled, the secondary bonds reform. Thus, linear and branched polymers are generally thermoplastic. On the other hand, cross-links contain primary covalent bonds like those that bond the atoms in the main chains. When a cross-linked polymer is heated sufficiently, these primary covalent bonds fail randomly, and the material degrades. Therefore, cross-linked polymers are thermosets.

Stereoisomerism. Vinyl monomers, $CH_2{=}CHR$, generally polymerize in a head-to-tail fashion, placing the R group on every other carbon atom in the chain backbone. If a chain is conceptually stretched out, the carbon atoms in the backbone will lie in a plane. The arrangement in which the R groups are all on one side of that plane is the isotactic stereoisomer. Regular alternation of the R groups from side to side is the syndiotactic form. Random placement of the R groups is the atactic (without order) polymer. Stereoisomers are formed during polymerization, and cannot be altered subsequently by rotation about the bonds.

Crystallinity. Crystals are an ordered, regular arrangement of units in a repeating, three-dimensional lattice structure. Small molecules, which in the liquid state have three-dimensional mobility, crystallize readily when cooled. It is not so easy for polymers, because a repeating unit cannot move independently of its neighbors in the chain. Nevertheless, some polymers can and do crystallize, though never completely.

Liquid-crystal polymers exhibit considerable order in the liquid state, either in solution (lyotropic) or melt (thermotropic). When crystallized from solution or melt, they have a high degree of extended-chain crystallinity, and thus have superior mechanical properties.

The Amorphous Phase and T_g. Not all polymers crystallize, and even those that do are not completely crystalline. Noncrystalline polymer is termed amorphous. Four types of molecular motion have been identified in amorphous polymers. Listed in order of decreasing activation energy, they are *(1)* translational motion of entire molecules, *(2)* coiling and uncoiling of 40–50 C-atom segments of chains, *(3)* motion of a few (five to six) atoms along the main chain or on side groups, and *(4)* vibrations of individual atoms. Type 1 motions are responsible for flow. Type 2 motions give rise to rubber elasticity. The temperature below which type 1 and 2 motions are frozen out is known as the glass-transition temperature, T_g. Below its T_g, an amorphous polymer is a glass; hard, rigid, and often brittle. Above T_g, it becomes rubbery, and at still higher temperatures, if it is not cross-linked, it flows easily.

Effects of Crystallinity on Properties. In polymers that can crystallize, the ratio of crystalline to amorphous material has a profound effect on properties. Because the chains are packed more tightly and efficiently in the crystalline areas than in the amorphous, the crystalline phase has a higher density and greater mechanical strength. In fact, density is a common measure of degree of crystallinity.

The stiffest polymers are both crystalline and have a glassy amorphous phase. They are often useful as engineering (structural) plastics (qv).

Solubility. Cross-linking eliminates polymer solubility. Crystallinity sometimes acts like cross-linking because it ties individual chains together. Thus, there are no solvents for linear polyethylene at room temperature, but as it is heated toward its crystalline melting point, T_m (135°C), it dissolves in a variety of aliphatic, aromatic, and chlorinated hydrocarbons. A rough guide to solubility is that like dissolves like, ie, polar solvents tend to dissolve polar polymers and nonpolar solvent dissolve nonpolar polymers.

Polymer Synthesis

Step-Growth Polymerization. Step-growth polymerization is characterized by the fact that chains always maintain their terminal reactivity and continue to react together to form longer chains as the reaction proceeds, ie, x − mer + y − mer → $(x + y)$-mer. Because there are reactions that follow this mechanism but do not produce a molecule of condensation, the terms step-growth and polycondensation are not exactly synonymous.

Chain-Growth Polymerization. Chain-growth polymerizations are characterized by chains that propagate by adding one monomer molecule at a time, ie, x − mer + monomer → $(x + 1)$-mer. There are, however, several mechanisms by which this occurs.

Free-Radical Addition. In free-radical addition polymerization, the propagating species is a free radical. The free radicals are most commonly generated by the thermal decomposition of a peroxide or azo initiator (see INITIATORS, FREE-RADICAL INITIATORS).

Unlike step-growth polymerization, free-radical chains do not continue to grow as the reaction proceeds. The average lifetime of a growing chain, from initiation to termination, is typically less than a second. Thus, high molecular weight polymer is produced right from the beginning.

Polymerization Processes. Free-radical polymerization is carried out in a variety of ways.

Bulk polymerization involves only monomer and initiator. It gives the greatest polymer yield per unit of reactor volume and a very pure polymer.

In solution polymerization an inert solvent is added to the reaction mass. The solvent adds its heat capacity and reduces the viscosity, facilitating convective heat transfer.

In suspension polymerization, the organic reaction mass is dispersed in the form of droplets 0.01–1 mm in diameter in a continuous aqueous phase. Each droplet is a tiny bulk reactor. Heat is readily transferred from the droplets to the water, which has a large heat capacity and a low viscosity, facilitating heat removal through a cooling jacket.

In emulsion polymerization the organic monomer is emulsified with soap in an aqueous continuous phase.

Ionic Polymerization. Addition polymerization may also be initiated and propagated by anions. Ionic polymerizations are almost exclusively solution processes.

There are some important differences between anionic and free-radical addition. First, unlike free-radical initiators, which decompose and start chains randomly throughout the course of the reaction, anionic initiators ionize readily in fairly polar organic solvents or at low concentrations in hydrocarbons, and chains are started immediately, one for each molecule of initiator. Second, in the absence of impurities, there is no termination.

When the initial monomer supply is exhausted, the anionic chain ends retain their activity. Thus, these anionic chains have been termed living polymers. If more monomer is added, they resume propagation. If it is a second monomer, the result is a block copolymer.

Cationic polymerization has been used commercially to polymerize isobutylene and alkyl vinyl ethers, which do not respond to free-radical or anionic addition (see ELASTOMERS, SYNTHETIC-BUTYL RUBBER).

Stereospecific Polymerization. In the early 1950s, Ziegler observed that certain heterogeneous catalysts based on transition metals polymerized ethylene to a linear, high density material at modest pressures and temperatures. Natta showed that these catalysts also could produce highly stereospecific poly-α-olefins, notably isotactic polypropylene, and polydienes. They shared the 1963 Nobel Prize in chemistry for their work. More recently, metallocene catalysts that provide even greater control of molecular structure have been introduced.

STEPHEN L. ROSEN
University of Missouri–Rolla

J. Brandrup and E. H. Immergut, eds., *Polymer Handbook*, 3rd ed., Wiley-Interscience, New York, 1989.

S. L. Rosen, *Fundamental Principles of Polymeric Materials*, 2nd ed., Wiley-Interscience, New York, 1993.

G. Odian, *Principles of Polymerization*, 3rd ed., Wiley-Interscience, New York, 1991.

P. J. Flory, *Principles of Polymer Chemistry*, Cornell UP, Ithaca, N.Y., 1953.

POLYMERS, CONDUCTIVE.

See ELECTRICALLY CONDUCTIVE POLYMERS; PHOTOCONDUCTIVE POLYMERS.

POLYMERS CONTAINING SULFUR

POLY(PHENYLENE SULFIDE)

Poly(*p*-phenylene sulfide) (PPS), (**1**), also known as poly(thiophenylene) or poly(thio-1,4-phenylene), has a structure that consists of alternating para-disubstituted aromatic rings (*p*-phenylene moieties) and divalent sulfur atoms (sulfide linkages). PPS is a semicrystalline polymer possessing a desirable combination of characteristics that include good mechanical properties, excellent electrical and thermal properties, as well as inherent flame resistance. Combined with the ease of molding, PPS plays an important role in the class of materials known as engineering thermoplastics. End uses include coatings, injection molding, film, fiber, pipe, and advanced composite materials (qv).

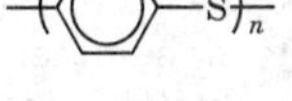

(**1**)

Polymerization Processes

The neat resin preparation for PPS is quite complicated. However, many commercial PPS polymerization processes have been developed.

Properties of PPS Neat Resins

Highly desirable properties of PPS include excellent chemical resistance, high temperature thermal stability, inherent flame resistance, good inherent electrical insulating properties, and good mechanical properties.

Thermal Properties. Thermodynamic stability of the chemical bonds comprising the PPS backbone is quite high. The large expenditure of energy required to dissociate these bonds (and therefore initiate thermal degradation) results in PPS having excellent thermal stability.

At temperatures between the glass transition temperature and the average equilibrium melting point, PPS crystallizes readily. However, because the rate of crystallization is slow, rapid cooling from the melt can result in a molded part that is not fully crystallized, but nearly amorphous.

Melt and Solution Properties. Melt rheology of PPS is complicated by the high temperatures needed and the requirement that air be excluded to avoid adventitious oxidative curing.

PPS exhibits exceptional chemical resistance. There are no known solvents for PPS below 200°C.

Properties of PPS Injection-Molding Compounds

PPS injection-molding compounds are distinguished among other thermoplastic compounds for their high performance properties, including excellent high temperature resistance, dimensional stability, and chemical resistance; inherent flame resistance; high electrical resistance properties; and a good balance of mechanical properties.

Applications

Inherent properties of PPS that determine the utility of PPS resins and compounds include excellent electrical insulation, excellent long- and short-term thermal stability, inherent flame resistance, easy moldability, outstanding chemical resistance, good mechanical properties, and dimensional stability of molded parts. PPS finds acceptance in a wide variety of applications, including coatings, injection-molding compounds, fiber, film, composites, blends, and alloys.

Coatings. Corrosion-resistant, pinhole-free, thermally stable coatings of PPS having good release characteristics can be applied to steel, aluminum, and other metals from aqueous slurries of PPS, by electrostatic powder coating, fluidized-bed coating, and powder spraying and flocking. Low molecular weight linear PPS is conventionally used in coatings.

Injection-Molding Compounds. PPS injection-molding compounds typically contain reinforcing fibers such as glass or carbon fiber, often with coupling agents, and/or mineral fillers to enhance specific properties. Filled compounds also have good processibility and can be used for molding of intricate parts.

Fiber. High molecular weight linear PPS is well-suited for fiber applications. The inherent properties of PPS (flame resistance, chemical resistance, and thermal stability) make PPS fiber highly desirable in textile applications (see HIGH PERFORMANCE FIBERS).

PPS fiber has excellent chemical resistance, and it is an excellent electrical insulator.

Composites. High molecular weight PPS can be combined with long fiber to produce advanced composite materials. Such materials having PPS as the polymer matrix have been developed by using a variety of reinforcements, including glass, carbon, and Kevlar fibers as mat, fabric, and unidirectional reinforcements (see COMPOSITE MATERIALS–POLYMER-MATRIX).

Film. High molecular weight PPS is suitable for film-making applications. PPS film is amenable to biaxial orientation. Biaxially oriented film is manufactured by extruding sheet that must be amorphous to accommodate the drawing operation. After biaxial stretching, the drawn film is heat-set to allow polymer crystallization.

Biaxially oriented PPS film is transparent and nearly colorless. It has low permeability to water vapor, carbon dioxide, and oxygen. PPS film has a low coefficient of hygroscopic expansion and a low dissipation factor, making it a candidate material for information storage devices and for thin-film capacitors.

Health and Safety Considerations

For personal protection when using PPS, adequate ventilation should be employed to control airborne powder concentration and off-gases from molding and extruding processes. No respiratory protection is generally required unless needed to control respiratory irritation from dust or off-gases. To control off-gases during molding, an appropriate air purifying respirator equipped with an organic vapor cartridge and face mask should be used. For eye protection, use safety glasses with side shields and provide eyewash stations in the work area. No special garments are required for skin protection. Heat-resistant gloves should be used when handling hot or molten material.

The principal decomposition products released during molding of PPS are carbon dioxide and carbon monoxide.

JON GEIBEL
JOHN LELAND
Phillips Petroleum Company

J. P. Blackwell, D. G. Brady, and H. W. Hill, *J. Coatings Tech.* **50**, 62–66 (1978).

J. F. Geibel and R. W. Campbell, in G. C. Eastmond, and co-workers, eds., *Comprehensive Polymer Science*, Vol. 5, Pergamon Press, Oxford, U.K. 1989, pp. 543–560.

J. F. Geibel and R. W. Campbell, in J. J. McKetta and W. A. Cunningham, eds., *Encyclopedia of Chemical Processing and Design*, Marcel Dekker, Inc., New York, 1992, pp. 94–125.

J. W. Cleary, in B. M. Culbertson and J. E. McGrath, eds., *Advances in Polymers Synthesis*, Plenum Press, New York, 1985, pp. 159–172.

POLYSULFIDES

Polysulfide polymers have the following general structure: $HS{+}R{-}S_x{+}_nSH$ where x is referred to as the rank and represents the average number of sulfur atoms in the polysulfide unit. This article is limited to polymers of this type where R is an aliphatic group and $x > 1$. The rank, x, usually ranges from slightly less than two to about four.

Polysulfides have unusually good resistance to solvents and to the environment and possess good low temperature properties. This makes them particularly useful in a variety of sealant applications. For example, the outstanding resistance of polysulfides to petroleum (qv) products has made them the standard sealant for virtually all aircraft integral fuel tanks and bodies. Another important application is in insulating glass window sealants (qv).

Polysulfides became the first high performance elastomeric sealants to be used in building construction (see BUILDING MATERIALS).

Physical Properties

The commercial polysulfides are made from the reaction of bischloroethylformal formal $ClCH_2CH_2OCH_2OCH_2CH_2Cl$, with sodium polysulfide, (NaS_x).

The solid polysulfide products are light brown millable rubbers. They have excellent resistance to a wide range of chemicals. They also have low permeability to gases, water, and organic liquids, excellent low temperature flexibility, and superior resistance to the effects of sunlight, ozone (qv), aging, and weathering.

Chemical Properties

Oxidative Curing. The rich chemistry of the thiol end group provides versatility in modifying and curing polysulfide polymers. The most common means of curing polysulfides is by chain extension with oxidizing agents, eg, as in the following reaction, where R–SH represents liquid polysulfide; O, oxidizing agent; and the product is a disulfide. $2\ R{-}SH + O \rightarrow R{-}S{-}S{-}R + H_2O$. Because thiols are easily oxidized, a host of organic and inorganic oxidants may be used.

The inorganic peroxide curing agents for liquid polysulfides are activated by water. By formulating and packaging polysulfides under anhydrous conditions, one-part sealants are prepared. These cure when exposed to atmospheric moisture and are used in construction sealant applications.

Polysulfides may also be cured by reaction with epoxy resins (qv). Amines or other catalysts are used and often primary or secondary amine resins are cured together with the polysulfide.

Diisocyanates or Polyisocyanates. The thiol end groups of the liquid polysulfides are quite reactive with isocyanates. Typical diisocyanates such as 1,3-toluene diisocyanate (m-TDI) or diphenylmethane-4,4′-diisocynate (MDI) are effective in curing liquid polysulfides. Using liquid polysulfides instead of the common hydroxy-terminated polymers brings the advantages of the polysulfide to the cured product. Thus, good chemical and solvent resistance, weatherability, adhesion, etc, can be attained.

Phenolic Resins. At elevated temperatures, phenolic resins are cured with polysulfide resins through a condensation reaction. The product may be considered a block copolymer of the rigid phenolic resin and the flexible polysulfide.

Miscellaneous Curing Reactions. Other functional groups can react with the thiol terminal groups of the polysulfides to cross-link the polymer chains and build molecular weight.

Reactions of the Disulfide Group. Besides the thiol end groups, the disulfide bonds also have a marked influence on both the chemical and physical properties of the polysulfide polymers. One of the key reactions of disulfides is nucleophilic attack on sulfur. Various thiophiles are capable of splitting the disulfide bond and thus reducing the molecular weight of the polymers.

Disulfide interchange also affects the physical properties of the cured polysulfide polymers. Polysulfide polymers undergo stress relaxation in a manner markedly different from conventional rubbers. Stress applied to stretch a sample of polysulfide rubber rapidly falls to zero. There is no change in the chemical and physical properties of the polymer recovered after the tests. The polysulfide polymer can be repeatedly recycled through the relaxation process.

Manufacture and Processing

In the commercial production of polysulfide polymers, bis-chloroethylformal is the monomer of choice because of its favorable economics, minimal competition with ring formation, and the desirable physical properties of the resulting polymer.

Formulation. Polysulfide-based sealants are formulated with appropriate ingredients to obtain the desired properties for a particular application. A typical formulation contains liquid polysulfide polymer, curing agent, cure accelerators (bases) or retarders (acids), fillers, plasticizers, thixotropes, and adhesion promoters.

For a two-part (A and B) sealant, Part A consists of liquid polysulfide, filler, plasticizer, thixotrope, and adhesion promoter. Part B contains the curing agent, plasticizer, a small amount of filler, and the accelerator or retarder. The one-part sealant bases require drying the ingredients.

Specifications and Testing

Specifications for the sealants vary widely depending on the specific application and the needs of the applicators. The tests should simulate the environment the sealants will be exposed to and their performance should be measured under these conditions.

Health and Safety

The polysulfides have a characteristic odor that is somewhat objectionable. The lower molecular weight polymers exhibit the strongest odor. Toxicity tests conducted on a representative liquid polysulfide used in Morton sealant applications indicate that the polymers are not eye irritants and have a low order of oral toxicity. Tests on the lower molecular weight liquid polysulfide products show similar findings. When used in accordance with prescribed procedures, liquid polysulfide products do not pose a health hazard.

Uses

Insulating Glass Sealants. One of the largest scale applications of polysulfide polymers is as a sealant for insulating glass windows. The window consists of two panes of glass separated by a hollow spacer that is filled with desiccant to remove moisture or volatiles from the air space. This prevents condensation and fogging of the window at low temperature (see SEALANTS).

Aircraft Sealants. Polysulfides have been used for sealing fuel tanks and aircraft structural components since the 1940s. There are stringent requirements for these sealants. They must have outstanding resistance to fuels, excellent adhesion to many different materials, and must perform in extremely variable weather conditions.

Construction Sealants. Curtain wall construction in high rise buildings requires sealing of the joints against wind and weather. Polysulfide-based sealants were developed for this application and have served in many large-scale projects.

Below-Ground Sealants. Polysulfide-based sealants have proved themselves useful in water purification plants and wastewater treatment plants, which have special demands for physical, chemical, and microbiological properties.

The chemical and fuel resistance of polysulfides makes them useful as sealants and coatings for secondary containment areas and for bridges, air fields, and road construction.

Epoxy Flexibilizers. Polysulfides are useful as flexibilizers in epoxy resin formulations. Compounders can target the properties desired for a particular application through the selection and balance of the epoxy, the liquid polysulfide-epoxy ratio, curing agent, and filler.

Water Dispersions. Polysulfide products are offered as aqueous dispersions. These are useful for applying protective coatings to line fuel tanks, and for concrete, wood, and in some cases fabrics, felt, leather (qv), and paper (qv).

DAVID VIETTI
MICHAEL SCHERRER
Morton International, Inc.

M. B. Berenbaum, in N. G. Gaylord, ed., *Polyethers*, Part III, Vol. 13, Interscience, New York, 1962, p. 43.

E. R. Bertozzi, *Rubber Chem. Technol.* **41**, 114 (1968).

H. Lucke, *ALIPS Aliphatische Polysulfide*, Hüthig & Wepf, Basel, Switzerland, 1992; English transl. in press.

D. E. Vietti, *Comprehensive Polymer Science*, Vol. 5, Pergamon, London, 1989, p. 533.

POLYSULFONES

A polysulfone is characterized by the presence of the sulfone group as part of its repeating unit. Whereas polysulfones may be aliphatic (**1**) or aromatic (**2**), the term polysulfones is used almost exclusively to denote aromatic polysulfones.

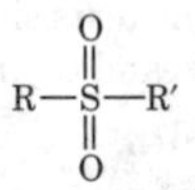

(**1**)

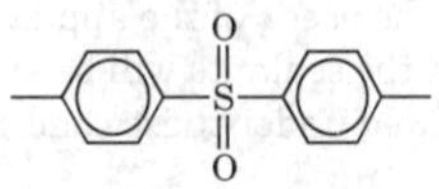

(**2**)

Polysulfones are a class of amorphous thermoplastic polymers characterized by high glass-transition temperatures, good mechanical strength and stiffness, and outstanding thermal and oxidative resistance. These polymers are characterized by the presence of the para-linked diphenylsulfone group (**2**) as part of their backbone repeat units. By virtue of their mechanical, thermal, and other desirable characteristics, these polymers enjoy an increasingly wide and diversified range of commercial applications. The basic repeat unit of any polysulfone always contains sulfone, aryl, and ether units as part of the main backbone structure and are thus often referred to in the polymer literature as poly(arylethersulfone)s.

In addition to sulfone, phenyl units, and ether moieties, the main backbone of polysulfones can contain a number of other connecting units. The most notable such connecting group is the isopropylidene linkage which is part of the repeat unit of the well-known bisphenol A-based polysulfone. In order to clearly describe the chemical makeup of polysulfones it is necessary to refer to the chemistry used to synthesize them. There are several routes for the synthesis of polysulfones, but the one which has proved to be most practical and versatile over the years is by aromatic nucleophilic substitution. This polycondensation route is based on reaction of essentially equimolar quantities of 4,4′-dihalodiphenylsulfone (usually dichlorodiphenylsulfone (DCDPS)) with a bisphenol in the presence of base thereby forming the aromatic ether bonds and eliminating an alkali salt as a by-product. This route is employed almost exclusively for the manufacture of polysulfones on a commercial scale.

Table 1. Chemical Structures of PSF, PES, and PPSF[a]

Polymer	Repeat unit structure
polysulfone (PSF)	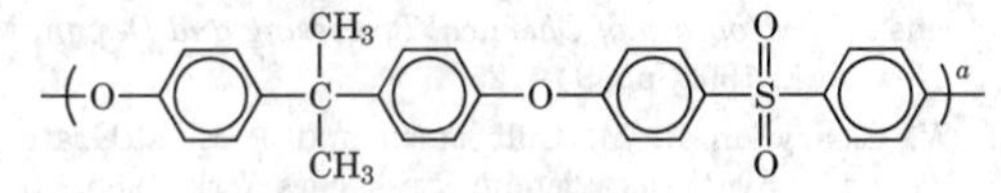
polyethersulfone (PES)[b,c]	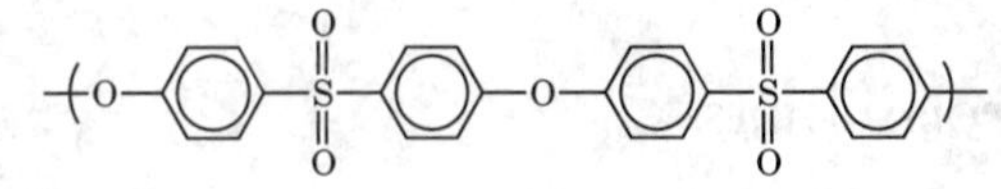
polyphenylsulfone (PPSF)[d]	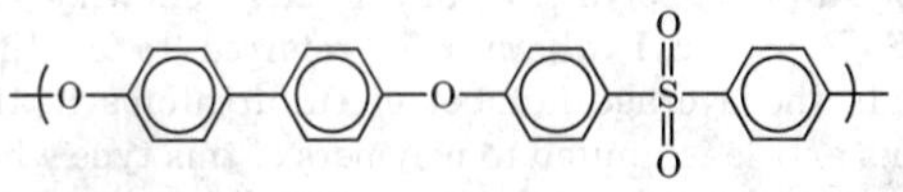

[a] Bisphenol A polysulfone.
[b] PES repeat unit structure can alternatively be drawn as

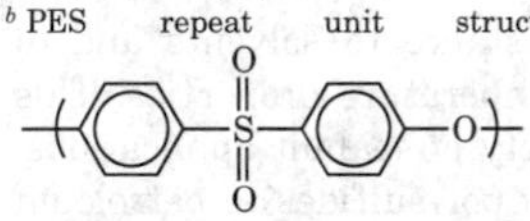

[c] Victrex polyethersulfone.
[d] RADEL R polyphenylsulfone.

There are three commercially important polysulfones referred to generically by the common names polysulfone (PSF), polyethersulfone (PES), and polyphenylsulfone (PPSF). The repeat units of these polymers are shown in Table 1.

Polymerization

Nucleophilic Substitution Route. Commercial synthesis of poly(arylethersulfone)s is accomplished almost exclusively via the nucleophilic substitution polycondensation route. This synthesis route involves reaction of the bisphenol of choice with 4,4′-dichlorodiphenylsulfone in a dipolar aprotic solvent in the presence of an alkali base.

The rate of polymerization in this type of reaction depends on both the basicity of the bisphenol salt and the electron-withdrawing capacity of the activating group (in this case sulfone) in the dihalide monomer.

Properties

Structure–Property Relationships. The characteristic feature of each of the polymers in Table 1 is the highly resonant diaryl sulfone grouping. As a consequence of the sulfur atom being in its highest state of oxidation and the enhanced resonance of the sulfone group being in the para position, these resins offer outstanding thermal stability and resistance to thermal oxidation. The thermal stability is further augmented by the high bond dissociation energies inherent in the aromatic backbone structure. As a result, these polymers can be melt fabricated at temperatures of up to 400°C with no adverse consequences.

Mechanical properties of aromatic polysulfones are intimately tied to backbone structure. For the achievement of good strength and toughness together with favorable melt processing characteristics, the first and foremost requirement is a linear (unbranched) and para-linked structure for the aryl groups in the backbone.

Physical, Chemical, and Optical Properties. Aromatic polysulfones possess several common key attributes including high glass-transition temperatures (generally >170°C) and a high degree of thermal oxidative stability (Table 2). Because PSF, PES, and PPSF are fully amorphous, these resins exhibit optical transparency. The glass-transition temperature of polysulfones produced via nucleophilic polycondensation can be tailored by the choice of the bisphenol. By virtue of the chemically nonlabile aromatic ether backbone, these polymers exhibit superb resistance to hydrolysis in hot water and steam environments.

Table 2. Physical and Thermal Properties of PSF, PES, and PPSF

Property	PSF	PES	PPSF
color	light yellow	light	light
haze[a], %	<7	<7 amber	<7 amber
light transmittance[b], %	80	70	70
refractive index	1.63	1.65	1.67
density, g/cm^3	1.24	1.37	1.29
glass-transition temperature[c], °C	185	220	220
heat deflection temperature[d], °C	174	204	207
continuous service temperature[e], °C	160	180	180
coefficient of linear thermal expansion	5.1×10^{-5}	5.5×10^{-5}	5.5×10^{-5}
specific heat at 23°C, J/(g·K)[f]	1.00	1.12	1.17
thermal conductivity, W/(m·K)[g]	0.26	0.18	0.35
water absorption, % in 24 h	0.22	0.61	0.37
at equilibrium	0.62	2.1	1.1
mold shrinkage, cm/cm	0.005	0.006	0.006
temperature at 10% weight loss (tga)[h] in nitrogen	512	547	550
in air	507	515	541

[a] As measured on 3.1-mm thick specimens.
[b] Typical values; varies with color. All three resins are transparent.
[c] Onset value as measured by differential scanning calorimetry.
[d] As measured on 3.1-mm thick ASTM specimens under a load of 1.82 MPa (264 psi).
[e] Practical maximum long-term use temperatures for PSF and PES based on UL 746 thermal rating data; value for PPSF is estimated.
[f] To convert J(g·K) to Btu/(lb·°F), divide by 4.184.
[g] To convert W/(m·K) to Btu/(h·ft·°F), multiply by 1.874.
[h] Thermogravimetric analysis (tga) run at heating rate of 10°C/min and 20 mL/min gas (nitrogen or air) flow rate.

Table 3. Room Temperature Mechanical Properties of PSF, PES, and PPSF

Property	PSF	PES	PPSF
tensile[a] (yield) strength, MPa[b]	70.3	83.0	70.0
tensile modulus, GPa[c]	2.48	2.60	2.30
elongation at yield, %	5.7	6.5	7.2
elongation at break, %	75	40	90
flexural strength, MPa[b]	106	111	91
flexural modulus, GPa[c]	2.69	2.90	2.40
compressive strength, MPa[b]	96	100	99
compressive modulus, GPa[c]	2.58	2.68	1.73
shear (yield) strength, MPa[b]	41.4	50	62
notched Izod impact, J/m[d],[e]	69	85	694
tensile impact, kJ/m^2[f]	420	340	400
Poisson ratio, at 0.5% strain	0.37	0.39	0.42
Rockwell hardness	M69	M88	M86
abrasion resistance,[g] mg/1000 cycles	20	19	20

[a] Tensile, flexural, and impact properties based on 3.1-mm thick ASTM specimens.
[b] To convert MPa to psi, multiply by 145.
[c] To convert GPa to psi, multiply by 145,000.
[d] To convert J/m to ft·lbf/in., divide by 53.38.
[e] No break for unnotched samples.
[f] To convert kJ/m^2 to ft·lbf/in^2, divide by 2.10.
[g] Taber abrasion test using CS-17 wheel and 1000-g load for 1000 cycles.

Furthermore, they can withstand acidic and alkali media over a wide range of concentrations and temperatures.

Mechanical Properties. Polysulfones are rigid and tough, with practical engineering strength and stiffness properties even without reinforcement. Their strength and stiffness at room temperature are high compared to traditional aliphatic backbone amorphous plastics. The polymers exhibit ductile yielding over a wide range of temperatures and deformation rates.

The room temperature mechanical properties of bisphenol A, bisphenol S, and biphenol-based polysulfones are given in Table 3.

The tensile and flexural properties as well as resistance to cracking in chemical environments can be substantially enhanced by the addition of fibrous reinforcements such as chopped glass fiber.

Flammability. Polysulfones exhibit excellent inherent burning resistance characteristics compared to many engineering thermoplastics. The wholly aromatic polysulfones such as PES and PPSF possess particularly outstanding flame retardancy and very low smoke release characteristics. (see FLAME RETARDANTS).

Electrical Properties. Polysulfones offer excellent electrical insulative capabilities. The resins exhibit low dielectric constants and dissipation factors even in the GHz (microwave) frequency range. This performance is retained over a wide temperature range and has permitted many applications.

Resistance to Chemical Environments and Solubility. Amorphous plastics are susceptible, to various degrees, to cracking by certain chemical environments when the plastic material is placed under stress, a phenomenon known as environmental stress cracking (ESC), the resistance of the polymer to failure by this mode is known as environmental stress cracking resistance (ESCR). Polysulfones, being completely amorphous, exhibit susceptibility to stress cracking by some organic environments. For example, PSF resists aliphatic hydrocarbons and most alcohols but readily undergoes stress cracking by ketones.

The exact mechanism of environmental stress cracking of a polymer is still not completely understood. A number of crazes are generated at stressed polymer surfaces as a result, and if the number of crazes is large the stress level is moderated and cracking is prevented, or at least delayed. If only a few crazes form, on the other hand, they tend to grow readily and propagate as cracks resulting in rupture.

In terms of ESCR performance, the polymers PSF, PES, and PPSF are highly resistant to hydrolysis by hot aqueous media, including boiling water, high pressure steam, mineral acids and alkalies, and salt solutions. This resistance is usually a key reason behind the selection of polysulfones over the other engineering plastics.

Radiation Resistance. Polysulfones exhibit resistance to many electromagnetic frequencies of practical significance, including microwave, visible, and infrared. Especially notable is the excellent resistance to microwave radiation, which has contributed to the use of polysulfones in cookware applications. Polysulfone also shows good resistance to x-rays, electron beam, and gamma radiation.

Polysulfones exhibit poor resistance to ultraviolet light. They absorb in the uv region, with attendant discoloration and losses in mechanical properties due to polymer degradation at and directly beneath the exposed surface.

Fabrication

Polysulfones are fully thermoplastic materials and readily flow at temperatures ≥ 150°C above their respective glass-transition temperatures. The backbone structure is extremely thermally stable during melt processing, remaining unchanged even when subjected to several melt fabrication cycles. Injection molding is the most common fabrication technique.

Polysulfones are easily processible by other thermoplastic fabrication techniques, including extrusion, thermoforming, and blow molding. A common fabrication technique is sheet extrusion followed by thermoforming.

Once formed, parts made of polysulfones (particularly those produced by injection molding) can be annealed to reduce molded-in stress.

Blends and Alloys

The blending of two or more polymers is used to tailor existing commercial polymers to specific end use requirements. The blending of polysulfones with polymers presents opportunities, but at the same time poses some significant technical challenges. Miscibility of PSF or

PES with any nonsulfone-based polymer is extremely rare. As a result, blends of PSF or PES with other nonsulfone-based polymers generally rely on interfacial adhesion and good shear mixing during compounding to produce an intimately mixed blend with good mechanical compatibility. None of the binary blends comprising PSF, PES, and PPSF are miscible, although their blends form mechanically compatible mixtures with relatively stable phase morphologies.

Health and Safety

Polysulfones are chemically inert polymers for the most part, and as of 1996 no known negative health effects were reported. These polymers have been used for many years in applications where safety is of the utmost importance. Numerous grades comply with U.S. and international governmental regulations for direct food contact.

The thermally and oxidatively stable backbones of polysulfones preclude development of any significant amount of toxic volatile degradation by-products when the resins are heated during melt processing. As with other plastic materials, adequate ventilation of the molding area is recommended when injection-molding polysulfones.

Uses

Polysulfones are used in a wide variety of applications that take advantage of hydrolytic and acid/caustic stability, clarity, and high heat deflection temperatures. These application areas include consumer items, electrical and electronic packaging and substrates, automotive, aerospace, and a host of industrial and plumbing uses. The resistance of polysulfones to chemical attack has resulted in their use in chemical processing equipment. Glass-reinforced grades can be used in very severe chemical environments for enhanced resistance and long service life.

Polysulfone, polyethersulfone, and polyphenylsulfone may be used interchangeably in many applications. However, PES represents an incremental improvement in performance over PSF, and PPSF is generally regarded as the highest performing member of this class of polymers.

M. JAMAL EL-HIBRI
Amoco Polymers, Inc.

U.S. Pat. 3,332,909 (July 25, 1967), A. G. Farnham and R. N. Johnson (to Union Carbide Corp.).

R. N. Johnson, A. G. Farnham, R. A. Clendinning, W. F. Hale, and C. N. Merriam, *J. Polym. Sci., Part A-1* **5**, 2375 (1967).

R. N. Johnson and J. E. Harris, in J. I. Kroschwitz, ed., *Encyclopedia of Polymer Science and Engineering*, 2nd ed., Vol. 13, John Wiley & Sons, Inc., New York, 1988, pp. 196–211.

M. J. El-Hibri, J. Nazabal, J. I. Eguiazabal, and A. Arzak, "Poly(aryl ether sulfone)s" in O. Olabisi, ed., *Handbook of Thermoplastics* Marcel Dekker, New York, 1997, Chapt. 36.

POLYMERS, ENVIRONMENTALLY DEGRADABLE

Interest in environmentally degradable plastics began in the early 1960s with the recognition that the common packaging plastics were accumulating in the environment, contributing to landfill depletion and litter problems resulting from careless disposal after use.

More recently, it has become increasingly apparent that in addition to the primary synthetic plastics, water-soluble and other specialty polymers and plastics, and even some modified natural polymers also potentially contribute to environmental problems and thus become targets for environmentally degradable substitutes. These polymers are widely used as coatings additives, pigment dispersants, temporary coatings, and detergents, in many applications. This article discusses polymers in a general sense rather than focusing on commodity plastics. The term polymer will refer to both water-soluble polymers and plastics, unless there is a need to differentiate the two principal polymer types that are the subject of most of the attention given to environmentally degradable polymers. The discussion covers all polymeric materials, natural, synthetic, and modified natural, designed to be degradable in the environment by any of the accepted degradation pathways: photodegradation, biodegradation, and chemical degradation which is hydrolytic or oxidative degradation. Of these pathways, biodegradation is recognized as by far the most important, insofar as it is the only one that can lead to complete removal from the environment, and accordingly it has received the greatest attention. The other degradation pathways are more appropriately described as biodeterioration or biodisintegration because their products are left in the environment unless they are biodegradable.

Development of environmentally degradable polymers is no longer considered a total solution to the polymer and plastic waste management issue. Alternative technologies such as recycling of plastic materials, including recycling of plastics; recycling of plastics to monomers and subsequent repolymerization to the same or new polymers; recycling to olefinic feedstocks by pyrolysis; continued burial in landfill sites; and incineration are recognized as viable options, along with environmental degradation (see RECYCLING, PLASTICS).

Definitions

There have been numerous definitions proposed for environmentally degradable plastics and polymers. Those developed by the American Society for Testing and Materials (ASTM) for degradable, biodegradable, hydrolytically degradable, and oxidatively degradable plastics and given here are probably the most widely accepted, either as written or in some slightly modified form. They are equally applicable to polymers, in general.

Degradable plastic is a plastic designed to undergo a significant change in its chemical structure under specific environmental conditions, resulting in a loss of some properties that may vary as measured by standard test methods appropriate to the plastic and the application in a particular period of time that determines its classification.

Biodegradable plastic is a degradable plastic in which the degradation results from the action of naturally occurring microorganisms such as bacteria, fungi, and algae.

Hydrolytically degradable plastic is a degradable plastic in which the degradation results from hydrolysis.

Oxidatively degradable plastic is a degradable plastic in which the degradation results from oxidation.

Photodegradable plastic is a degradable plastic in which the degradation results from the action of natural daylight.

The definitions indicate only the mechanism that is operating to promote degradation. Although this is acceptable in a scientific sense in order to define the chemical process, it does not really go far enough to satisfy the requirements for environmentally acceptable polymers, a key issue in the minds of legislators and lay people. If environmentally degradable polymers and plastics are to be acceptable as a waste management option, definitions must be more practical and descriptive in conveying the assurance that no harmful residues are left in the environment after degradation has occurred. The ASTM definitions, therefore, require elaboration. The environmental degradation processes are interrelated, as shown schematically in Figure 1.

All four degradation pathways, ie, biodegradation, oxidation, hydrolysis, and photodegradation, initially give intermediate products or fragments that may biodegrade further to some other residue, biodegrade completely and be removed from the environment entirely (being converted into biomass and carbon dioxide and/or methane, depending on whether it is an aerobic or anaerobic system) and ultimately mineralized, or remain unchanged in the environment. The term mineralization of a polymer is used here loosely to indicate complete or total removal from the environment to carbon dioxide or methane, water, and biomass. In the cases where residues remain in the environment, they must be established as harmless by suitably rigorous fate and effect evaluations. Only biodegradation has the potential to remove plastic and polymers completely from the environment. Thus, when developing and designing polymers and plastics for

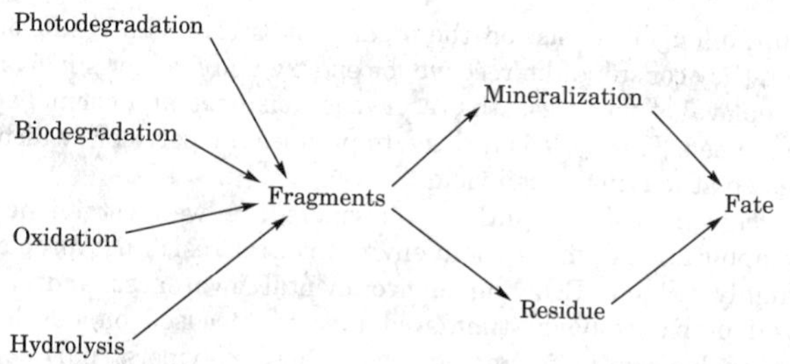

Figure 1. Interrelationships of processes for environmentally degradable polymers.

degradation in the environment by other pathways, the final stage preferably should be complete biodegradation and removal from the environment, with ultimate mineralization. In this way, the polymers are essentially recycled through nature into microbial cells, plants, and higher animals, and thence back into chemical feedstocks. Therefore, an environmentally acceptable degradable plastic or polymer may be defined as one which degrades by any of the above-defined mechanisms.

Opportunities for Environmentally Degradable Plastics and Polymers

The conditions providing a significant stimulus for the development of environmentally degradable polymers and plastics are the same ones that govern waste management programs and the restriction of the uses for nondegradable polymers and plastics in the environment.

Water-soluble polymers (qv) after use are usually very dilute solutions and are preferably disposed of through wastewater treatment facilities or sometimes directly into the aquatic environment. There is a preference for biodegradable water-soluble polymers, as it is unlikely that the other degradation paths would be applicable in dilute aqueous solutions.

The advantage of complete biodegradation is that it can be established with a high degree of certainty with the appropriate test methods, whereas the assessment of environmental fate and effect of any residue is always a risk assessment.

Plastic materials offer more disposal options than water-soluble polymers, because they are usually solid, handleable materials and are recoverable in most cases after use for several disposal options, including land-filling, recycling, incineration, and composting. Where recovery of current plastics is not economically feasible, viable, controllable, or attractive, the plastics remain as litter.

For water-soluble polymers, there is a well-established disposal infrastructure, with the widely available wastewater treatment plants, whereas plastics being developed for composting require large-scale implementation of a composting infrastructure.

Test Methods

Photodegradation. Test methods for measuring photodegradation are usually a combination of an exposure to some form of radiation and subsequent measurement of property loss in another test. Several standard test practices have been developed for plastic exposure (Table 1). Subsequent property testing by the standard ASTM test methods may be done at various time intervals throughout the exposure to assess the rate of degradation.

Biodegradation. Recent test methodology for biodegradation stresses the proper selection of environment to reflect probable disposal sites for a given polymer or plastic and the need for quantitative testing as the most important aspect of assessing acceptability for environmental disposal. Qualitative tests are recognized as important for indicating the rate of disintegration of plastics, which has a bearing on such disposal methods as composting for compaction and volume reduction of the compost.

Some of the earlier tests were developed for assessing the resistance of plastics to fungal and bacterial growth and included Growth

Table 1. ASTM Standard Practices for Photodegradation

Standard test practice
degradation end points using a tensile test
operation of a xenon arc ARC-type exposure apparatus
operation of a fluorescent ultraviolet (uv) and condensation apparatus
outdoor exposure testing of photodegradable plastics

Ratings for growth on suitable plastics. After a given time period, growth is assessed by a subjective numerical rating in which higher numbers are considered to correlate with the susceptibility of the plastic to biodegradation:

Rating	Growth
0	no visible growth
1	<10% of surface with growth
2	10–30% surface with growth
3	30–60% surface with growth
4	60–100% surface with growth

The advantage of such a test is that it is quick and easy to do, and gives an indication of biodegradation potential; however, the test is not definitive. Other simple tests include the soil burial test used to demonstrate the biodegradability of polycaprolactone, following its disappearance as a function of time; and the clear zone method, which indicates biodegradation by the formation of a clear zone in an agar medium of the test polymer or plastic as it is consumed. The burial test is still used as a confirmatory test method in the real-world environment after quantitative laboratory methods indicate biodegradation.

Biodegradation in aerobic and anaerobic environments may be described by chemical equations [eqs. 1 (aerobic) and 2 (anaerobic)]. Most of the testing reported in the literature has been with aerobic biodegradation. However, anaerobic degradation, which is particularly pertinent to water-soluble polymers that may enter anaerobic digestors in sewage treatment facilities, is the object of growing interest. Unless otherwise specified, this article considers biodegradation as it occurs in an aerobic environment.

$$\text{polymer} + O_2 \rightarrow CO_2 + H_2O + \text{biomass} + \text{residue} \qquad (1)$$

$$\text{polymer} \rightarrow CO_2 + CH_4 + H_2O + \text{biomass} + \text{residue} \qquad (2)$$

To assess the degree of biodegradation quantitatively, analytical techniques are needed for any or all of the reactants and products; the polymer; oxygen uptake, known as biochemical oxygen demand (BOD); and the residue. The more rigorous the analysis, the more reliable the measurement of the extent of biodegradation and the greater the acceptability of the data and the conclusions drawn. For total biodegradation, there should be no residue remaining in the environment.

Standard testing protocols for biodegradation of plastics are shown in Table 2.

These test protocols are only screening tests for readily biodegradable polymers and plastics. Failure in these tests does not exclude the possibility of biodegradation; it merely indicates that under the environmental conditions evaluated there is no biodegradation. Repeated tests, particularly in other environments, are recommended before acceptance of nonbiodegradability. Biodegradation has no value in itself; it is a part of fate and effects to be utilized in the important Environmental Safety Assessment.

Degradation Mechanisms

Photodegradation. Photodegradable polymers degrade in the environment by chain scission promoted by natural daylight and usually

Table 2. Biodegradation Test Protocols

Environment	Measurement
aerobic sewage sludge	CO_2
anaerobic sewage sludge	CO_2 and CH_4
aerobic specific microorganisms	molecular weight
aerobic activated sewage sludge	O_2 and CO_2
aerobic controlled composting	CO_2
marine floating conditions	physical properties
mixed microbial	O_2

oxygen to yield low molecular weight fragments that are more susceptible to biodegradation than the original high molecular weight polymer. The polymers are generally structurally similar to currently used environmentally stable polymers but have been modified during synthesis or post-treatment to insert photochemically active groups. The ultimate fate of the fragments produced is not yet fully established; in most cases the argument is put forward that if the molecular weight of the degradation products is low enough, then they will biodegrade.

Biodegradation. Biodegradable polymers and plastics are readily divided into three broad classifications: *(1)* natural, *(2)* synthetic, and *(3)* modified natural.

Natural Polymers. Natural polymers (biopolymers) are produced in nature by all living organisms. As a class they represent truly renewable resources since they are biodegradable, even if slowly in some cases. Because they are produced in nature there is no concern about this slow rate of biodegradation, contrary to concerns about synthetic polymers. Biopolymers are considered environmentally acceptable degradable polymers. The most widespread natural polymers are the polysaccharides, such as cellulose and starch. Other important classes include polyesters such as polyhydroxyalkanoates, proteins like silk and poly(γ-glutamic acid), and hydrocarbons such as natural rubber. With few exceptions, natural polymers have not been considered as suitable for practical applications either because they lack the property requirements or because they are too expensive for other than specialty markets such as biomedical applications.

Polysaccharides are largely limited to starch and cellulose derivatives for practical applications either in plastics or as water-soluble polymers. Both these polymers are composed of D-glycopyranoside repeating units to very high molecular weight, thousands of units. They differ in that starch is poly (1,4-α-D-glucopyranoside) and cellulose is poly(1,4-β-D-glucopyranoside) and cellulose is poly(1,4,-β-glucopyranoside). This difference in structure controls biodegradation rates and properties of the polymers.

Complex carbohydrates (qv), eg, microbially produced xanthan, are accepted as biodegradable and are finding uses where cost is not an impediment (see GUMS).

Proteins (qv) have not found widespread use as plastic materials because they are not soluble or fusible without decomposition; hence, they must be used as found in nature. They are widely used as fibers.

Polyesters are known to be produced by many bacteria as intracellular reserve materials for use as a food source during periods of environmental stress. They have received a great deal of attention since the 1970s because they are biodegradable, can be processed as plastic materials, are produced from renewable resources, and can be produced by many bacteria in a range of compositions. The thermoplastic polymers have properties that vary from soft elastomers to rigid brittle plastics in accordance with the structure of the pendent side-chain of the polyester.

Polyesters derived from bacteria are produced in some stressed conditions in which the bacteria are deprived of some essential component for their normal metabolic processes. Under normal conditions of balanced growth the bacteria utilize any substrate for energy and growth, whereas under stressed conditions they utilize any suitable substrate to produce polyesters as reserve material. When the bacteria can no longer subsist on the organic substrate as a result of depletion, they consume the reserve for energy and food for survival; or upon removal of the stress, the reserve is consumed and normal activities resumed. This cycle is utilized to produce the polymers which are harvested at maximum cell yield.

Synthetic Polymers. Synthetic polymers are well established in many applications where their environmental resistance properties are highly valued. Evolving environmental awareness and waste-disposal problems have stimulated research focused on developing biodegradable synthetic analogues of these polymers, particularly water-soluble polymers and plastics. The search for synthetic polymeric structures that can be biodegraded has progressed from minor modification of the nondegradables in use to structures that mimic nature.

Some guidelines based on polymer structure, polymer physical properties, and environmental conditions at the exposure site have emerged for predicting the biodegradability of synthetic polymers. In considering polymer structure, the following generalizations can be made: A higher hydrophilic/hydrophobic ratio is better for biodegradation. Carbon chain polymers are unlikely to biodegrade. Chain branching is deleterious to biodegradation. Condensation polymers are more likely to biodegrade. Lower molecular weight polymers are more susceptible to biodegradation. Crystallinity slows biodegradation.

Favorable polymer physical properties include water solubility and sample purity. Environmental conditions to consider in evaluating biodegradability are temperature, pH, moisture, oxygen, nutrients, suitable microbial population (fungal, algae, bacterial), concentration, and test duration.

Modified Natural Polymers. Modifying natural polymers offers a way of capitalizing on their well-accepted biodegradability in the development of polymers that might be environmentally acceptable. The modification must not interfere with the biodegradation process and the product must meet guidelines for environmental acceptability, ie, they must be either demonstrated to be totally biodegraded and removed from the environment, or be biodegradable to the extent that no environmentally harmful residues remain. The approaches that have received the most attention include blends with other natural and synthetic polymers, grafting of another polymeric composition, and chemical modification to introduce some desirable functional group by oxidation or some other simple chemical reaction, such as esterification or etherification.

Related patents to water-soluble carboxylated derivatives of starch are Hoechst's on the oxidation of ethoxylated starch and another on the oxidation of sucrose to a tricarboxylic acid. All the oxidations are specific to primary hydroxyls and are with a platinum catalyst at pH near neutrality in the presence of oxygen. Polysaccharides as raw materials in the detergent industry have been reviewed.

Production of Environmentally Degradable Polymers. There are signs that the use of environmentally degradable polymers and plastics is expanding. As the market begins to become aware of the availability of these new materials, it is expected that they will move into niche opportunities. When this occurs, production will increase, and costs, the biggest barrier to acceptance, should begin to come down.

GRAHAM SWIFT
Rohm and Haas Company

G. S. Kumar and co-workers, *J. Macromol. Sci. Rev., Macromol. Chem. Phys.* **C22**(2), 225 (1981–1983).

G. Swift, *Accounts Chem. Res.* **26**, 105–110 (1993).

A. L. Andrady, *J. Mater. Sci. Rev. Macromol. Chem. Phys.* **C34**(1), 25–76 (1994).

Y. Doi, *Microbial Polyesters*, VCH Publishers, New York, 1990.

POLYMERS OF HIGHER OLEFINS. See OLEFIN POLYMERS.

POLYMETHACRYLATES. See METHACRYLIC POLYMERS.

POLYMETHINE DYES

Polymethine dyes (PMD) represent a large class of organic colored compounds that contain a chain of methine groups (—CH=) as the basic constitutive elements. According to S. Daehne's triad theory, polymethines, together with polyenes and aromatics, are the three main types of conjugated systems. The term polymethine dyes was introduced by W. Koenig in 1922. The formula of PMDs having the stable, closed electron shell, in its general form, can be written as two resonance structures, where n is the number of vinylene groups in the polymethine chain,

$$[G_1^{\pm}\!\!-\!\!(CH{=}CH)_n\!-\!CH{=}G_2]X^{\mp} \leftrightarrow [G_1{=}CH\!\!-\!\!(CH{=}CH)_n\!-\!G_2^{\pm}]X^{\mp}$$

(1)

and G_1 and G_2 are terminal or end groups of various chemical structure, ie, acyclic, carbo-, or heterocyclic residues. Polymethine chains of symmetrical dyes ($G_1 = G_2$) consist of an odd number of carbon atoms, and these systems carry charges, ie, they exist as either cations or anions. In formula (**1**), end groups differ by a number of π-electrons, and thus one is written in electron donor, and the other in electron acceptor, form.

The simplest PMDs include the polymethines (**2**), streptocyanines (**3**), oxonols (**4**), and merocyanines (**5**).

$$H_2C{=}CH\!\!-\!\!(CH{=}CH)_n\!-\!CH_2^{\pm} \qquad R_2N^+{=}CH\!\!-\!\!(CH{=}CH)_n\!-\!NR_2$$

(2) **(3)**

$$O{=}CH\!\!-\!\!(CH{=}CH)_n\!-\!O^- \qquad R_2N\!\!-\!\!(CH{=}CH)_n\!-\!CH{=}O$$

(4) **(5)**

A great number of different heterocyclic residues have been used as the terminal groups of PMDs. PMDs containing residues with quaternary nitrogen atoms are traditionally called cyanine dyes (qv).

Polymethines with branched polymethine chains also exist. Among these, PMDs with symmetrically branched chains are the best known; they are referred to as trinuclear polymethine dyes (TPMD) (**6**).

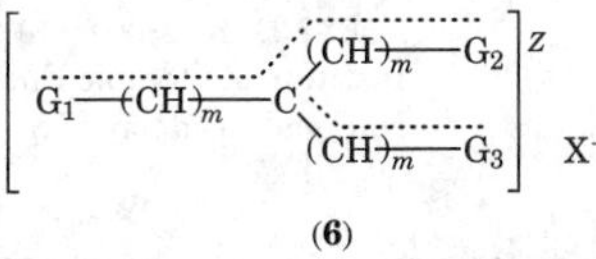

(6)

PMDs demonstrate pronounced absorption and contain fluorescence bands that are relatively narrow and highly intense, which arise from electron transitions occurring within the polymethine chromophore $\!-\!\!(\!-\!(CH{-}CH)_n\!-\!CH\!-\!)\!-\!$. These spectral properties give rise to a wide range of applications of PMDs such as silver halide sensitizers, laser media components, polymerization initiators, etc.

According to one classification, symmetrical dinuclear PMDs can be divided into two classes, A and B, with respect to the symmetry of the frontier molecular orbital (MO). Thus, the lowest unoccupied MO (LUMO) of class-A dyes is antisymmetrical and the highest occupied MO (HOMO) is symmetrical, and the π-system contains an odd number of π-electron pairs. On the other hand, the frontier MO symmetry of class-B dyes is the opposite, and the molecule has an even number of π-electron pairs.

For convenience, unsymmetrical PMDs should be considered as the derivatives of the corresponding symmetrical polymethines, commonly called parent dyes. In contrast to symmetrical compounds, unsymmetrical PMDs can contain an even number of methine groups in the polymethine chains, for example, styryls (**7**), where X = S, O, NCH_3, $C(CH_3)_2$, or CH=CH; and (**8**), where Y = O, S, Se or NCH_3.

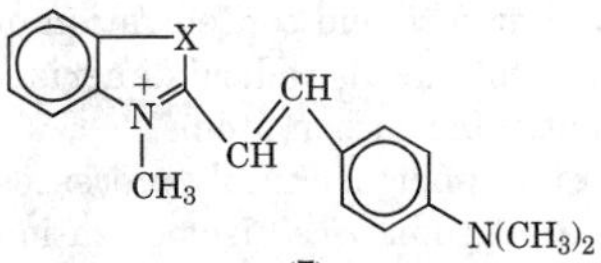

(7)

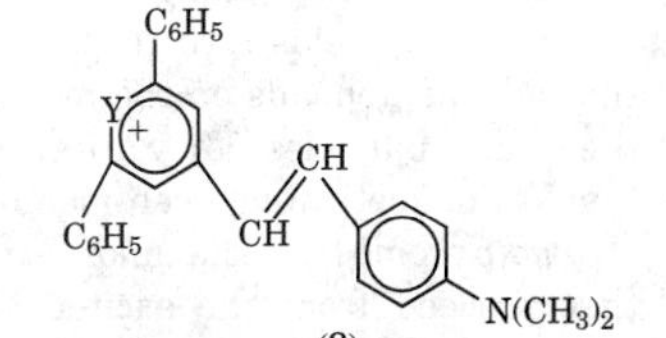

(8)

Unsubstituted PMDs (**2**) or dyes containing odd alternate hydrocarbon residues as end groups can exist in two relatively stable forms distinguished by a π-electron pair, eg, α,ω-diphenylpolymethines (**9**).

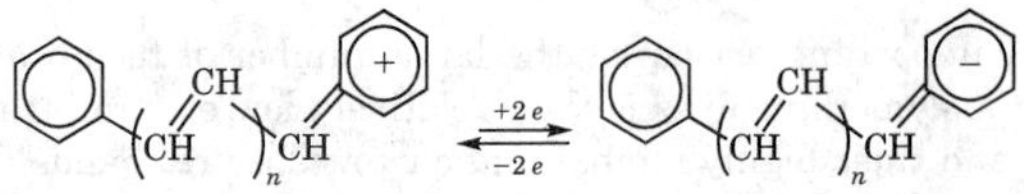

(9a) **(9b)**

Electron Structure

A considerable number of experiments have shown that symmetrical PMDs in the ground state have an all-trans configuration and are nearly planar with practically equalized carbon–carbon bonds and slightly alternating valence angles within the polymethine chain.

Electron Transitions

PMD color or the nature of the electron transitions produces the widest application for PMDs. Depending on the polymethine chain length, the end-group topology, and the electron shell occupation, polymethines can absorb light in uv, visible, and near-ir spectral regions.

Chemical Reactivity

As conjugated systems with alternating π-charges, the polymethine dyes are comparatively highly reactive compounds. Substitution rather than addition occurs to the equalized π-bond. If the nucleophilic and electrophilic reactions are charge-controlled, reactants can attack regiospecifically.

Protonation. As expected from π-electron distribution, the proton attacks the negatively charged odd positions in the polymethine chain, eg,

$$G^+{-}CH{=}CH{-}CH{=}CH{-}CH{=}G \xrightarrow{H^+} G^+{-}CH_2{-}CH{=}CH{-}CH{=}CH{-}G^+$$

The destruction of the total π-system causes the color to vanish; the protonated molecule absorbs light in the uv region. Protonation has been proved to be reversible.

Other Electrophilic Reactants. Reversibility of the electrophilic reactions enables substituted dye derivatives to be obtained. Halogen atoms are mobile in the polymethine chain, and the derivatives themselves can function as halogenation reagents.

The dye can be formulated by means of a phosgene (qv). Chloromethylation leads to the alkylated polymethines; nitration of cyanines results in mononitro-substituted compounds.

Nucleophilic Reagents. In contrast to electrophilic reactions, nucleophiles attack positively charged, even carbons in the chain. The reactions lead to the exchanging of substituents or terminal residues. Thus, SR and OR groups, or halogen atoms can be exchanged by other suitable nucleophiles.

If the dye contains no mobile substituents in the chain, nucleophiles attack primarily the end carbon atoms (changing of terminal residues). Nucleophilic reactions with the methylene bases of the corresponding heterocycles result in polymethines containing new end groups.

The asymmetrical polymethines appear to be ambivalent systems, and the number of possible reaction paths increases considerably as a result.

Reactions with Parting of Radicals. The one-electron oxidation of cationic dyes yields a corresponding radical dication. The stability of the radicals depends on the molecular structure and concentration of the radical particles. They are susceptible to radical–radical dimerization at unsubstituted, even-membered methine carbon atoms.

Photochemistry. The most important photochemical processes that proceed from the excited state are geometrical isomerization and photochromic reactions. Photoisomerization of polymethines is a reversible trans-cis transfer. The cis-isomer absorbs at longer wavelength with a smaller intensity than the trans-isomer.

Applications of Polymethines

The most important reason for the large number of technical applications of polymethine dyes is their relatively low electron-transition energies and their highly intense and narrow spectral bands. Indeed, polymethines display strong light absorption and emission, between 300 and 1600 nm. These dyes have been used as photographic sensitizers and desensitizers, as laser dyes, as probes of membrane potentials, and in other applications where the theoretical aspects of polymethines are useful.

Spectral Sensitization. Photographic silver halide emulsions are active with light only up to about 500 nm. However, their sensitivity can be extended within the whole visible and near-ir spectral region up to about 1200–1300 nm. This is reached by the addition of deeply colored dyes that transfer excited electrons.

According to the electron-transfer mechanism of spectral sensitization, the transfer of an electron from the excited sensitizer molecule to the silver halide and the injection of photoelectrons into the conduction band are the primary processes. Thus, the lowest vacant level of the sensitizer dye is situated higher than the bottom of the conduction band. The regeneration of the sensitizer is possible by reactions of the positive hole to form radical dications. If the highest filled level of the dye is situated below the top of the valence band, desensitization occurs because of hole production.

Based on correlations between energy level positions and electrochemical redox potentials, it has been established that polymethine dyes with reduction potentials less than −1.0 V (vs SCE) can provide good spectral sensitization. On the other hand, dyes with oxidation potentials lower than +0.2 V are strong desensitizers.

Improvement of spectral sensitization can be accomplished by dye combinations. The effect has been found to often be greater than the predicted additive sensitivity increase. This phenomenon is called supersensitization, which is applied most effectively to polymethine aggregates. The opposite phenomenon, a decrease of sensitivity, is known as desensitization. The main reasons for desensitization are the results of relative electron level positions as well as the secondary processes of the photoelectrons.

Quantum Electronics and Laser Dyes. In quantum electronics, PMDs are usually applied as mode-locking compounds in passive mode-locked lasers as well as active laser media. The required characteristics of dyes used as passive mode-locking agents and as active laser media differ in essential ways. For passive mode-locking dyes, short excited-state relaxation times are needed; dyes of this kind are characterized by low fluorescence quantum efficiencies caused by the highly probable nonradiant processes. On the other hand, the polymethines to be applied as active laser media are supposed to have much higher quantum efficiencies, approximating a value of one.

Photopolymerization. In many cases polymerization is initiated by irradiation of a sensitizer with ultraviolet or visible light. The excited state of the sensitizer may dissociate directly to form active free radicals, or it may first undergo a bimolecular electron-transfer reaction, the products of which initiate polymerization.

Synthesis

By varying the molecular structure, it is possible to synthesize dye initiators with the required characteristics. Polymethine dyes with different chain length, end groups, and substituents, or with other variations of the chromophore, have been synthesized (see also CYANINE DYES).

Polymethine dyes consist of three main structural elements: two identical or different end groups and a conjugated chain containing an odd number of methine groups. There are many possibilities for changing the chromophore constitution: using new heterocyclic systems for the end groups, introducing specific substituents in either the chain or in the residues, branching of the polymethine chromophore, replacement of the methine groups by heteroatoms, and cyclization of the chain by conjugated or unconjugated bridges.

General Aspects. As a rule, the end-group synthones have the following reactive centers: an activated methyl or methylene group with high CH acidity, a functional group (OR, SR, X(halide), NR_2) leaving as an anion in the reaction, and a carbonyl or heteroanalogous group as a leaving group. Complementary reactive centers are needed in the chain synthones in the α- and ω-positions. In particular, derivatives of formic acid are used to prepare monomethine dyes; for dyes with longer chromophores, the application of vinylogous aminals or ω-methylpolyenals are preferred.

Molecular Design

Extension of the applications of polymethine dyes has required special spectral and other characteristics. As a rule, the search for and synthesis of promising new compounds having desired properties imply the preliminary estimation of their most important parameters on the basis of elaborated theoretical conceptions. Thus, an effective way of governing electron properties consists of the variation of molecular topology of polymethines.

The first encouraging results for engineered dyes having desired ground- and excited-state properties were reported in 1992. To design effective spectral sensitizers, it is necessary to engineer dyes having special positions of their frontier levels, desired wavelengths of the absorption band, and special thermodynamic stability after light excitation.

Also, using dyes as laser media or passive mode-locked compounds requires numerous special parameters, the most important of which are the band position and bandwidth of absorption and fluorescence, the luminescence quantum efficiency, the Stokes shift, the possibility of photoisomerization, chemical stability, and photostability. Applications of PMDs in other technical or scientific areas have additional special requirements.

ALEXY D. KACHKOVSKI
Institute of Organic Chemistry
National Academy of Sciences of the Ukraine

J. Fabian and S. Daehne, *J. Mol. Structure (Theochem.)* **92**, 217 (1983).

N. Tyutyulkov and co-workers, *Polymethine Dyes: Structure and Properties*, St. Kliment Ohridski University Press, Sofia, Bulgaria, 1991.

F. M. Hamer, *The Cyanine Dyes and Related Compounds*, Interscience Publishers, New York, 1964.

S. F. Mason, in K. Venkataraman, ed., *The Chemistry of Synthetic Dyes*, Vol. 3, Academic Press, Inc., New York, 1970, p. 169.

POLYMETHYLBENZENES

Polymethylbenzenes (PMBs) are aromatic compounds that contain a benzene ring and three to six methyl group substituents (for the lower homologues see BENZENE; TOLUENE; XYLENES AND ETHYLBENZENE). Included are the trimethylbenzenes, C_9H_{12} (mesitylene (**1**), pseudocumene (**2**), and hemimellitene (**3**)), the tetramethylbenzenes, $C_{10}H_{14}$ (durene (**4**), isodurene (**5**), and prehnitene (**6**)), pentamethylbenzene, $C_{11}H_{16}$ (**7**), and hexamethylbenzene, $C_{12}H_{18}$ (**8**). The PMBs are primarily basic building blocks for more complex chemical intermediates.

Physical Properties

The structures of the eight PMBs are shown here and their physical and thermodynamic properties are given in Table 1.

Table 1. Physical and Thermodynamic Properties of Polymethylbenzenes

Property	Systematic (benzene) name							
	1,3,5-Trimethylbenzene	1,2,4-Trimethylbenzene	1,2,3-Trimethylbenzene	1,2,4,5-Tetramethylbenzene	1,2,3,5-Tetramethylbenzene	1,2,3,4-Tetramethylbenzene	Pentamethylbenzene	Hexamethylbenzene
mol wt	120.194	120.194	120.194	134.221	134.221	134.221	148.248	162.275
bp, °C	164.74	169.38	176.12	196.80	198.00	205.04	231.9	263.8
flash point, °C	43.0	46.0	51.0	67.0	68.0	73.0		
density, g/cm^3								
at 20°C	0.8651	0.8758	0.8944	0.8875[a]	0.8903	0.9052	0.917[b]	solid
25°C	0.8611	0.8718	0.8905	0.8837[a]	0.8865	0.9015	0.913[b]	solid
freezing point, °C in air at 101.3 kPa[b]	−44.694	−43.881	−25.344	79.240	−23.689	−6.229	54.35	165.7
refractive index, n_D at 25°C	1.49684	1.50237	1.51150	1.5093[a]	1.5107	1.5181	1.525[a]	solid
surface tension, mN/m (= dyn/cm), at 20°C	28.84	29.72	31.28	solid	33.51	35.81	solid	solid
critical temperature, °C	364.20	376.02	391.32	401.85	405.85	416.55		
critical pressure, kPa[b]	3127	3232	3454	2940	2860	2860		
critical volume, cm^3/mol	427	427	427	482	482	482		
heat of vaporization at bp, kJ/mol[c]	39.0	39.2	40.0	45.52	43.81	45.02	45.1	48.2
heat of formation at 25°C, liquid, kJ/mol[c]	−63.4	−61.8	−58.5	−119.87[d]	−96.35	−90.20	−135.1[d]	−171.5[b]
heat of combustion, kJ/mol[c] at 25°C	5193.1	5194.8	5198.0	5816.0[a]	5839.6	5845.7	6490.8[a]	
dielectric constant, at 20°C		2.383	2.636					
specific heat, C_p, liquid, at 25°C, J/(mol·K)[c]	200.5	214.9	216.4		240.7	238.3		

[a] Supercooled liquid.
[b] To convert kPa to atm, divide by 101.3.
[c] To convert J to cal, divide by 4.184.
[d] Crystal.

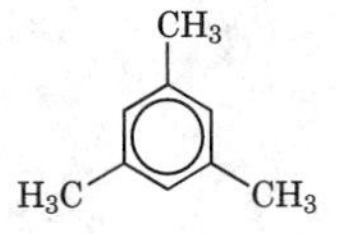

(1)

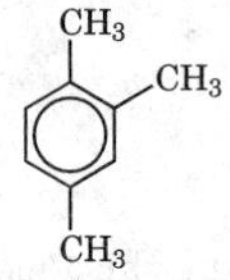

(2)

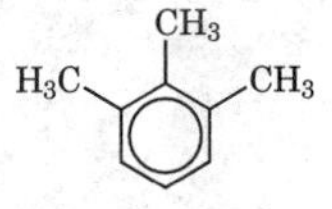

(3)

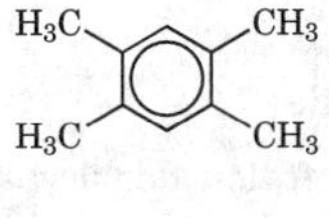

(4)

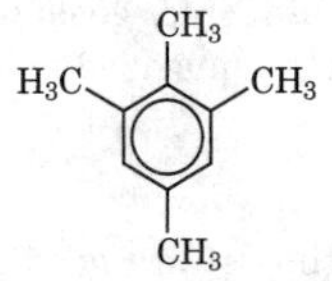

(5)

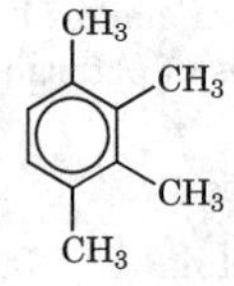

(6)

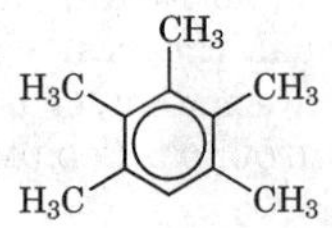

(7)

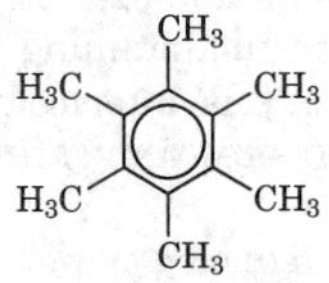

(8)

Manufacture

High purity mesitylene, hemimellitene, and durene are often produced synthetically, whereas pseudocumene is obtained from extracted C_9 reformate by superfractionation.

Koch Chemical Company is the only U.S. supplier of all PMBs (except hexamethylbenzene).

Health and Safety Factors

The PMBs, as higher homologues of toluene and xylenes, are handled in a similar manner, even though their flash points are higher (see Table 1). Containers are tightly closed and use areas should be ventilated. Breathing vapors and contact with the skin should be avoided.

Uses

Pseudocumene is used as a component in liquid scintillation cocktails for clinical analyses. Pseudocumene and durene are oxidized to trimellitic anhydride and pyromellitic dianhydride, respectively. Mesitylene is a key building block for important antioxidants and agricultural chemicals. Prehnitene, isodurene, pentamethylbenzene, and hexamethylbenzene have no significant commercial uses. The higher polymethylbenzenes show potential as highly regiospecific methylation agents for methylation of 4-alkylbiphenyls to form 4,4′-alkylmethylbiphenyls which can be oxidized to the monomer 4,4′-biphenyldicarboxylic acid (see LIQUID CRYSTALLINE MATERIALS).

H. W. EARHART
Consultant
ANDREW P. KOMIN
Koch Chemical Company

H. W. Earhart, *The Polymethylbenzenes*, Noyes Development Corp., Park Ridge, N.J., 1969.

U.S. Pat. 3,542,890 (Nov. 24, 1970), H. W. Earhart and G. Sugerman (Sun to Koch Industries Inc.).

POLYOXETANE. See POLYETHERS, TETRAHYDROFURAN AND OXETANE POLYMERS.

POLYPEPTIDE ANTIBIOTICS. See ANTIBIOTICS, PEPTIDES.

POLYPEPTIDES. See PROTEINS.

POLY(PHENYLENE OXIDE). See POLYETHERS, AROMATIC.

POLY(PHENYLENE SULFIDE). See HIGH PERFORMANCE FIBERS; POLYMERS CONTAINING SULFUR, POLY(PHENYLENE SULFIDE.

POLYPHOSPHAZENES. See ELASTOMERS, SYNTHETIC-PHOSPHAZENES; INORGANIC HIGH POLYMERS.

POLYPROPYLENE. See OLEFIN POLYMERS.

POLYPROPYLENE FIBERS. See FIBERS, OLEFIN.

POLY(PROPYLENE OXIDE). See POLYETHERS, PROPYLENE OXIDE POLYMERS.

POLYSACCHARIDES. See CARBOHYDRATES; MICROBIAL POLYSACCHARIDES; STARCH.

POLYSTYRENE. See STYRENE PLASTICS.

POLYSULFIDES. See POLYMERS CONTAINING SULFUR, POLYSULFIDES.

POLYSULFONES. See POLYMERS CONTAINING SULFUR, POLYSULFONES.

POLYTETRAFLUOROETHYLENE. See FLUORINE COMPOUNDS, ORGANIC-POLYTETRAFLUOROETHYLENE.

POLYTETRAHYDROFURAN. See POLYETHERS, TETRAHYDROFURAN AND OXETANE POLYMERS.

POLYURETHANES. See URETHANE POLYMERS.

POLY(VINYL ACETALS). See VINYL POLYMERS, POLY(VINYL ACETALS).

POLY(VINYL ACETATE). See VINYL POLYMERS, POLY(VINYL ACETATE).

POLY(VINYL ALCOHOL). See VINYL POLYMERS, POLY(VINYL ALCOHOL).

POLY(VINYL CHLORIDE). See VINYL POLYMERS, VINYL CHLORIDE AND POLY(VINYL CHLORIDE).

POLY(VINYL ETHERS). See VINYL POLYMERS, VINYL ETHER MONOMERS AND POLYMERS.

POLY(VINYL FLUORIDE). See FLUORINE COMPOUNDS, ORGANIC-POLY(VINYL) FLUORIDE).

POLY(VINYLIDENE CHLORIDE). See VINYLIDENE CHLORIDE AND POLY(VINYLIDENE CHLORIDE).

POLY(VINYLIDENE FLUORIDE). See FLUORINE COMPOUNDS, ORGANIC-POLY(VINYLIDENE FLUORIDE).

POLY(*N*-VINYLPYRROLIDINONE). See VINYL POLYMERS, N-VINYL MONOMERS AND POLYMERS.

POLY(XYLENE). See XYLYENE POLYMERS.

PORCELAIN. See CERAMICS; DENTAL MATERIALS; ENAMELS, PORCELAIN OR VITREOUS.

POROMERIC MATERIALS. See LEATHER-LIKE MATERIALS.

POTASSIUM

Potassium, K, is the third element in the alkali metal series. Potassium and sodium share the position of the seventh most abundant element on earth. Common minerals such as alums, feldspars, and micas are rich in potassium. Potassium metal, a powerful reducing agent, does not exist in nature.

Physical Properties

Potassium, a soft, low density, silver-colored metal, has high thermal and electrical conductivities, and very low ionization energy.

Potassium has three naturally occurring isotopes: ^{39}K (93.08%), ^{40}K (0.01%), and ^{41}K (6.91%). The radioactive decay of ^{40}K to argon (^{40}Ar), half-life of 10^9 years, makes it a useful tool for geological dating. Some physical properties of potassium are summarized in Table 1.

Chemical Properties

All the alkali metals (Li, Na, K, Rb, Cs) are good reducing agents. Potassium is the most electropositive reducing agent used in industry.

The alkali metals share many common features, yet differences in size, atomic number, ionization potential, and solvation energy leads to each element maintaining individual chemical characteristics. Sodium chemistry is intermediate between that of potassium and lithium (see LITHIUM AND LITHIUM COMPOUNDS; SODIUM AND SODIUM ALLOYS).

The superb reducing power of potassium metal is clearly demonstrated by its facile displacement of protons in the weakly acidic hydrocarbons (qv), amines, and alcohols. Reactions with inorganics and gaseous elements are summarized in Table 2.

Table 1. Physical Properties of Potassium

Property	Value
atomic weight	39.09
atomic radius, pm	227
ionic radius, pm	138
Pauling electronegativity	0.8
crystal lattice	body-centered cubic
analytical spectral line, nm	766.4
melting point, °C	63.2
boiling point, °C	765.2
density, at 20°C, g/cm^3	0.856
specific heat, J/(g·K)[a]	0.741
heat of fusion, kJ mol^{-1}[a]	2.39
heat of vaporization, kJ mol^{-1}[a]	79
electrical conductance, at 20°C, μS	0.23
surface tension, at 100°C, mN/m(= dyn/cm)	86
thermal conductivity, at 200°C, W/(m·K)	44.77
$E°$,/V for m$^+$	−2.925
ΔH dissoc, kJ mol^{-1} (m$_2$)[a]	49.9

[a] To convert J to cal, divide by 4.184.

Table 2. Chemical Reactions of Potassium

Reactant	Reaction	Product
H_2	begins slowly at ca 200°C; rapid above 300°C	KH
O_2	begins slowly with solid; fairly rapid with liquid	K_2O, K_2O_2, KO_2
H_2O	extremely vigorous and frequently results in hydrogen–air explosions	KOH, H_2
$C_{(graphite)}$	150–400°C	KC_4, KC_8, KC_{24}
CO	forms unstable carbonyls	(KCO)
NH_3	dissolves as K; iron, nickel, and other metals catalyze in gas and liquid phase	KNH_2
S	molten state in liquid ammonia	K_2S, K_2S_2, K_2S_4
F_2, Cl_2,	violent to explosive	KF, KCl, KBr
I_2 Br_2	ignition	KI
CO_2	occurs readily, but is sometimes explosive	CO, C, K_2CO_3

Preparation and Manufacture

On the laboratory scale, potassium can be prepared by the following reactions, however, these reactions are not easily adaptable to a commercial scale.

$$K_2CO_3 + 2C \xrightarrow{\text{heat}} 3CO + 2K^0 \quad (1)$$

$$2\ KCl + CaC_2 \xrightarrow{\text{heat}} CaCl_2 + 2\ C + 2\ K^0 \quad (2)$$

$$2\ KN_3 \xrightarrow{\text{heat}} 3\ N_2 \uparrow + 2\ K^0 \quad (3)$$

In industry, chemical reduction is preferred over electrolytic processes for potassium production. Potassium–sodium alloy is easily prepared by the reaction of sodium with molten KCl, KOH, or solid K_2CO_3 powder (see SODIUM AND SODIUM ALLOYS).

Health and Safety Factors

Reactions of potassium with water and oxygen are hazardous and safe handling is a concern. Potassium oxidizes slowly in air at room temperature, and it usually ignites if it sprays hot into the air. The peroxide and superoxide products may explode in contact with free potassium metal or organic materials including hydrocarbons. Packaging (qv) under oils is less desirable than packaging under an inert cover gas or in a vacuum. Potassium encrusted with a peroxide and superoxide layer should be destroyed immediately by careful, controlled disposal.

Airborne potassium dusts or potassium combustion products attack mucous membranes and skin causing burns and skin cauterization. Inhalation and skin contact must be avoided. Safety goggles, full face shields, respirators, leather gloves, fire-resistant clothing, and a leather apron are considered minimum safety equipment.

Fire Fighting. Potassium metal reacts violently with water releasing flammable, explosive hydrogen gas. Dry soda ash, dry sodium chloride, or Ansul's Met-L-X must be used for potassium or potassium alloy fires.

Uses

Historically, potassium metal was used by the Mine Safety Appliances Company (parent company of Callery Chemical Company) to develop potassium superoxide, KO_2, for use as an oxygen source in self-contained breathing equipment (see OXYGEN-GENERATION SYSTEMS). Numerous additional and important industrial applications have been developed. Potassium, potassium–sodium alloys, and potassium derivatives such as alkoxides, amides, and the hydride are extensively used, both industrially and academically, to synthesize organic and inorganic materials.

JOSEPH A. GORELLA
Callery Chemical Company

J. W. Mellor, *Supplement III to Mellor's Comprehensive Treatise on Inorganic and Theoretical Chemistry*, Vol. II, John Wiley & Sons, Inc., New York, 1963.

O. J. Foust, ed., *Sodium and Sodium–Potassium Engineering Handbook*, Vol. I, Gordon and Breach, New York, 1972, Chapt. 2.

J. W. Mausteller, F. Tepper, and S. J. Rodgers, *Alkali Metal Handling and Systems Operating Techniques*, Gordon and Breach Scientific Publishers, Inc., New York, 1967.

A. A. Morton, *Solid Organoalkali Metal Reagents*, Gordon & Breach, New York, 1964.

POTASSIUM COMPOUNDS

Occurrence

Commercial production of potassium compounds is generally limited to the extraction of ores from underground deposits containing significant concentrations of soluble potassium salts. Exceptions include commercial potassium chemical operations on the Dead Sea and the Great Salt Lake (see CHEMICALS FROM BRINE). These natural brine refining operations exploit the highly favorable climatic conditions at those locations for the utilization of solar energy (qv).

Canada leads the world in both reserves and production, operating at about 57% of estimated production capacity.

Four minerals are the principal commercial sources of potash: sylvite, KCl; carnallite, $KCl{\cdot}MgCl_2{\cdot}6H_2O$; kainite, $KCl{\cdot}MgSO_4{\cdot}2.75H_2O$; and langbeinite, $K_2SO_4{\cdot}2MgSO_4$. In all ores, sodium chloride is the principal soluble contaminant.

Approximately 98% of the potassium recovered in primary ore and natural brine refining operations is recovered as potassium chloride. The basic raw material for all of the potassium compounds discussed in this article, except potassium tartrate, is potassium chloride. Physical properties of selected potassium compounds are listed in Table 1.

Potassium Chloride

Potassium chloride, or muriate of potash (MOP) as it is known in the fertilizer industry (at about 97% purity), is the world's most commonly used potash (see FERTILIZERS).

Mining. Potassium chloride is produced mostly from solid ores occurring in underground deposits 300–1700-m deep. Conventional mining methods adapted from coal (qv) and hard-rock mining operations and machine mining are employed to mine ores to a depth of ca 1100 m. Solution mining is employed to recover potassium chloride

Table 1. Physical Constants of Potassium Compounds

Potassium compound	Formula	Mol wt	Form	Specific gravity	Melting point, °C
acetate	$KC_2H_3O_2$	98.14	white powder	1.57	292
bromide	KBr	119.01	cubic	2.75	734
carbonate	K_2CO_3	138.20	monoclinic	2.428	891
bicarbonate	$KHCO_3$	100.11	monoclinic	2.17	100–200 dec[a]
chlorate	$KClO_3$	122.55	monoclinic	2.32	356
chloride	KCl	74.55	cubic	1.984	770
formate	$KHCO_2$	84.11	rhombic	1.91	167.5
hydroxide	KOH	56.10	rhombic	2.044	360.4 ±0.7
iodide	KI	166.02	cubic	3.13	681
nitrate	KNO_3	101.10	rhombic, trigonal	2.109	334
nitrite	KNO_2	85.10	colorless, prism	1.915	440
orthophosphates normal phosphate	K_3PO_4	212.27	rhombic	2.564	1340
monohydrogen phosphate	K_2HPO_4	174.18	amorphous		dec[a]
dihydrogen phosphate	KH_2PO_4	136.09	tetragonal	2.338	252.6
sulfate	K_2SO_4	174.26	rhombic or hexagonal	2.662	1069
bisulfate	$KHSO_4$	136.17	monoclinic	2.24–2.61	210
			rhombic	2.322	214
sulfite	K_2SO_3	194.29	monoclinic		dec[a]
acid tartrate	$KHC_4H_4O_6$	188.1	rhombic	1.984	

[a] Decomposes.

from deposits exceeding ca 1100 m or from deposits that cannot be mined conventionally because of geological anomalies (see MINERALS RECOVERY AND PROCESSING).

Refining. Process selection for the separation of potassium chloride as a relatively pure product from other constituents is based on the physical and chemical characteristics of a given ore. Ores amenable to treatment by the physical separation methods commonly used in other nonmetallic minerals processing industries generally are chosen to recover the potassium chloride. These methods include heavy-media and froth-flotation separations. Physical separation processes are less energy-intensive than fractional crystallization (qv), which is the traditional method of producing potassium chloride.

Potassium Sulfate

Compared to potassium chloride, potassium sulfate, K_2SO_4, and its complexes with magnesium sulfate play a minor role as sources of potassium in agriculture. However, potassium sulfate in simple form or combined with $MgSO_4$ as a double salt is an essential source of potassium for crops that are chloride-sensitive. In arid parts of the world and places where saline water is used for irrigation, potassium sulfate must be used to provide potassium in order to avoid chloride toxicity.

Potassium sulfate is a source of soluble sulfur, an essential element for plant growth.

Other Potassium Compounds

Potassium Acetate. Potassium acetate, $KC_2H_3O_2$, is usually made from carbonate and acetic acid. It is also made industrially by the simple reaction of acetic acid and potassium hydroxide. Potassium acetate is very soluble and is used in the manufacture of glass (qv), as a buffer (see HYDROGEN-ION ACTIVITY) or a dehydrating agent, and in medicine as a diuretic (see DIURETIC). It is deliquescent and is used as a softening agent for papers and textiles.

Potassium Bromide. Potassium bromide, KBr, can be prepared by a variation of the process by which bromine is absorbed from ocean water. Potassium carbonate is used instead of sodium carbonate: $3\ K_2CO_3 + 3\ Br_2 \rightarrow KBrO_3 + 3\ CO_2 + 5\ KBr$. Potassium bromate, K_2BrO_3, is much less soluble than the bromide and can mostly be removed by filtration.

Potassium bromide is extensively used in photography (qv) and engraving. It is the usual source of bromine in organic synthesis. In medicine, it is a classic sedative.

Potassium Carbonate. Except for small amounts produced by obsolete processes, potassium carbonate, K_2CO_3, is produced by the carbonation, ie, via reaction with carbon dioxide, of potassium hydroxide. Impurities are small amounts of sodium and chloride plus trace amounts (<2 ppm) of heavy metals such as lead. Heavy metals are a concern because potassium carbonate is used in the production of chocolate intended for human consumption.

In many heavy-chemical manufacturing operations requiring an intermediate alkaline metal carbonate reactant, potassium carbonate and sodium carbonate can be used with equal effectiveness. Potassium carbonate possesses properties for some applications that preclude the substitution of sodium carbonate, eg, in television glass. Uses of potassium carbonate other than glass include applications in ceramics (qv), engraving processes, finishing leather (qv), chemical dyes and pigments, foods, cleaners, and gas purification, eg, CO_2 and H_2S.

Potassium Bicarbonate. Potassium bicarbonate, $KHCO_3$, is made by absorption of CO_2 in a carbonate solution, ie, potassium hydroxide is carbonated to K_2CO_3, which in turn is carbonated to $KHCO_3$. Potassium bicarbonate is used in foods and medicine.

Potassium Formate. Potassium formate, HCOOK, is made by the reaction of carbon monoxide and potassium hydroxide. Potassium formate melts at 167°C and decomposes almost entirely to oxalate at ca 360°C.

Potassium Hydroxide. Potassium hydroxide, KOH, is produced industrially by the electrolysis of potassium chloride (see ELECTROCHEMICAL PROCESSING–INORGANIC).

Principal uses of KOH include the production of potassium carbonate and potassium permanganate, pesticides (qv), fertilizers (qv), and other agricultural products; soaps and detergents; scrubbing and cleaning operations, eg, industrial gases; dyes and colorants; and rubber chemicals (qv).

Potassium Iodide. Some potassium iodide, KI, is made by the iron and carbonate process similar to that used for the bromide. However, most U.S. production involves absorption of iodine in KOH.

$$3\ I_2 + 6\ KOH \rightarrow 5\ KI + KIO_3 + 3\ H_2O$$

Approximately half of the iodine consumed is used to make potassium iodide (see IODINE AND IODINE COMPOUNDS). The main uses of KI are in animal and human food, particularly in iodized salt, pharmaceuticals (qv), and photography (qv).

Potassium Nitrate. Potassium nitrate, KNO_3, is produced commercially in the United States based on the reaction of potassium chloride and nitric acid (qv). Nitrosyl chloride, a by-product of the basic reaction, has no commercial value and is converted to salable chlorine and to nitric acid for recycling.

The main uses of potassium nitrate include fertilizer, fireworks, steel (qv) making, the food industry, and in a eutectic mixture with $NaNO_2$, service as a heat-transfer agent (see HEAT-EXCHANGE TECHNOLOGY).

Potassium Phosphates. Phosphoric acid is the source of phosphate for the production of potassium phosphates (see PHOSPHORIC ACID AND PHOSPHATES).

Condensed potassium phosphates have been used as builders in liquid detergents. Potassium tripolyphosphate is also used in detergents.

Potassium Acid Tartrate. The monopotassium salt of tartaric acid, $KHC_4H_4O_6$, also called potassium acid tartrate or cream of tartar, is a common ingredient in baking powders and leavening systems (see BAKERY PROCESSES AND LEAVENING AGENTS). Cream of tartar, tartaric acid, and its Rochelle salt are manufactured from the by-products of the wine (qv) industry.

Biological Role of Potassium

Potassium ions are essential to both plants and animals (see MINERAL NUTRIENTS). Within cells, potassium serves the critical role as counterion for various carboxylates, phosphates, and sulfates, and stabilizes macromolecular structures.

In plants, potassium is involved in water and nutrient transport and has a role in photosynthesis. Although sodium and potassium are similar in their inorganic chemical behavior, these ions are different in their physiological activities. In fact, their functions are often mutually antagonistic. For example, K^+ increases both the respiration rate in muscle tissue and the rate of protein synthesis, whereas Na^+ inhibits both processes.

Analytical Methods

Potassium is analyzed in chemicals that are used in the fertilizer industry and in finished fertilizers by flame photometric methods or volumetric sodium tetraphenylboron methods. Other methods used for control purposes and special analyses include atomic absorption spectrophotometry, inductively coupled plasma (icp) emission spectrophotometry, and a radiometric method.

Health and Safety Factors

Potash mining and refining operations in the United States are strictly regulated by appropriate federal and state agencies. Field studies conducted by the National Institute for Occupational Safety and Health (NIOSH) failed to disclose any evidence of predisposition of undergound miners to any of the diseases evaluated, including lung cancer.

Potassium compounds commonly used in fertilizers, eg, KCl and K_2SO_4, are not considered to be hazardous substances.

MARK B. FREILICH
RICHARD L. PETERSEN
The University of Memphis

V. A. Zandon, E. A. Schoeld, and J. McManus, *Minerals Processing Handbook*, 3rd ed., Society of Mining Engineers of AIME, Denver, Colo., 1982, Section 22.

G. T. Austin, ed., *Shreve's Chemical Process Industries*, 5th ed., McGraw-Hill Book Co., Inc., New York, 1984, pp. 288–302.

O. Braitsch, *Salt Deposits, Their Origin and Composition*, Springer-Verlag, New York, 1971.

J. J. R. Frausto da Silva and R. J. P. Williams, *The Biological Chemistry of the Elements: the Inorganic Chemistry of Life*, Oxford University Press, Oxford, U.K., 1991, pp. 223–242.

POWDER COATINGS. See COATING PROCESSES, POWDER TECHNOLOGY.

POWDER METALLURGY. See METALLURGY, POWDER.

POWDERS, HANDLING

DISPERSION OF POWDERS IN LIQUIDS

Suspensions (dispersions or slurries) of powders in liquids (vehicles) are involved in many commercial processes and products, eg, inks, paints, asphalt, and pharmaceuticals (qv). Surface forces control the flow and sedimentation characteristics of particles whose diameters are 0.02–200 μm. The preparation of a dispersion requires special equipment and chemicals to facilitate the steps of dispersion, ie, wetting the external surface of the particles, breaking up clumps, and preventing reagglomeration. A surfactant (surface-active agent) is a chemical that concentrates in the interfacial region between the solid and the liquid. A surfactant that creates an interparticle repulsion large enough to overcome normal interparticle attractions may be called a dispersant for that particular solid–liquid pair.

Improved methods for polymer synthesis have led to a proliferation of highly effective dispersants; larger and faster computers have facilitated modeling details of interactions between particles, dispersants, and the liquid; and highly sophisticated instruments are available to measure particle–particle and particle–dispersant interactions. There are three stages of dispersing a powder in a liquid: (*1*) wetting the powder into the liquid, (*2*) deagglomerating the wetted clumps, and (*3*) preventing reagglomeration (flocculation) (see COLLOIDS; DISPERSANTS; EMULSIONS; SURFACTANTS).

Wetting Powders Into Liquids

The atoms and molecules at the interface between a liquid (or solid) and a vacuum are attracted more strongly toward the interior than toward the vacuum. The material parameter used to characterize this imbalance is the interfacial energy density γ, usually called surface tension. It is highest for metals ($\geq$1 J/m^2) (1 J/m^2 = N/m), moderate for metal oxides ($\leq$0.1 J/m^2), and lowest for hydrocarbons and fluorocarbons (0.02 J/m^2 minimum). The International Standards Organization (ISO) describes well-established methods for determining surface tension.

Fundamentals. *Contact Angle.* When a drop of liquid is placed on a flat, horizontal, solid surface it spreads out until it attains an equilibrium shape with a fixed angle from the solid surface to the gas surface (through the liquid). This is the contact angle. Figure 1 shows how the contact angle, θ_{SLG} is related to the balance of surface tension forces at the drop perimeter where the three phases (solid, S; liquid; L; gas, G) meet the shoreline. In cases where neither gas nor solid dissolves in the liquid and there are no surface-active solutes present, the contact angle is related to the interfacial energy densities γ at the solid–gas (SG), solid–liquid (SL), and liquid–gas (LG) interfaces.

A liquid is called wetting if the contact angle from solid to liquid through gas is $< 90°$ and nonwetting if it is $> 90°$

Wetting Single Particles. When a particle is submerged in a liquid, the work of wetting a surface is the change in interfacial energy $\gamma_{SG} - \gamma_{LG}$ density times the particle surface area.

A buoyant spherical particle floating in a liquid which has no tendency to wet, ie, pull the shoreline up, or to reject, ie, push the shoreline down, the solid submerges to a certain depth at which the downward force of gravity on the sphere equals the upward buoyancy force of the displaced liquid.

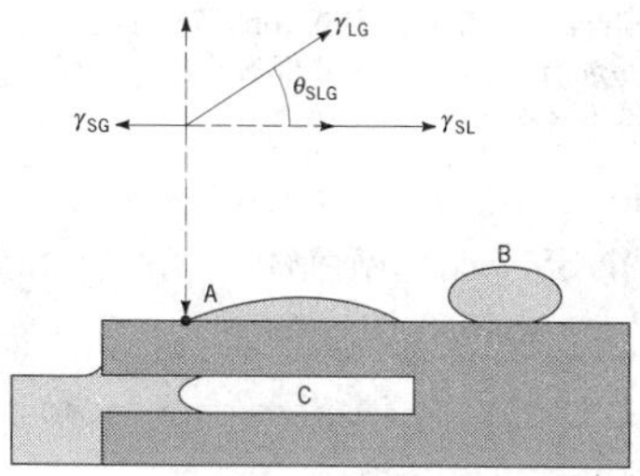

Figure 1. Contact angles. The shapes of drops that are A, wetting and B, nonwetting with respect to the solid (■), and C, penetration of a wetting liquid into a pore to compress the gas (□) inside. The vector diagram shows the balance of forces at the perimeter of the liquid drop on the solid plate.

If the liquid wets the solid, surface tension pulls the solid down farther into the liquid. Centrifugal force does not affect the depth of submergence because it increases the buoyancy force as much as it increases the settling force. A nonbuoyant particle submerges fully unless the liquid is sufficiently nonwetting.

Wetting Clumps. The density of a clump is related to the fraction of void space within the perimeter of the clump (porosity). If the liquid is nonwetting it does not enter the pores to displace the gas unless the hydrostatic pressure at the bottom of the submerged clump exceeds the pressure required to drive it into the pores. Gravity or centrifugation rarely provide enough hydrostatic pressure to force liquid into nonwetting pores. If the liquid wets the solid, the clump density increases as gas is displaced from the interior. It is best if submersion does not occur until the liquid has completely displaced gas from the pores.

Flushing a Dispersion Into a Second Liquid. If a dispersion in one liquid is mixed with a second liquid which is immiscible with the first but which preferentially wets the powder, the powder transfers into and becomes dispersed in the second liquid. This process, called flushing, is sometimes used to prepare an oil-based suspension directly from a water-based dispersion.

Practical Matters and Process Equipment. The rate of wet-in may be enhanced by adding a wetting agent to the liquid or by exposing the powder, treated or not, to water or some other vapor. Some powders sold commercially have already been given surface treatments so they readily wet into the liquid used in a particular market application.

Gas that is drawn beneath the liquid surface and enters a high shear region is dispersed into small bubbles which may be stabilized by a surfactant and rise to the surface to form a foam which is hard to dissipate. Foam can be avoided even in the presence of a foaming agent by preventing air from being drawn beneath the liquid surface during the wet-in process (see FOAMS).

Commercial equipment designed to wet powders into liquids exploits the factors that lead to rapid wet-in: low gas pressure, high centrifugal force, distribution of powder in a thin and deagglomerated layer on the liquid, and liquid surface refreshing to expose a large area for powder addition.

Deagglomerating Wetted Clumps

Fundamentals. Individual particles in a dry powder are held together in clumps by polarizability attraction of adjoining particles, surface tension at wetted contacts, or moderately strong sintering or precipitation (particle or soluble salt) bonds. These bonds are formed at the contact points during drying or because of exposure to humidity or temperature cycles. Breaking these bonds to fully disperse the submerged and wetted clumps into individual particles may require high shear stress or mechanical impact. In contrast to milling, deagglomeration generates only a small amount of new surface area and thus uses only moderate energy per unit mass of material. The surface area newly created by deagglomeration may have higher adsorptivity and reactivity than the conditioned surface of the clump.

High shear can be achieved in high viscosity systems using slowly turning mixing paddles. For intermediate viscosity systems, the suspension may be passed through the gap between two mill rolls rotating with different surface velocities. For low viscosity systems, suspensions may be circulated through the small gaps between a slotted stator and a close-fitting slotted rotor moving at a high rpm or forced at high pressure through a very small clearance in a spring-loaded valve.

Impact can be achieved using grinding media beads set in motion by stirring with bars or disks, by vibration, or by cascading within a cylindrical container rotating on a horizontal axle.

Ultrasonic probes, which create alternating pulses of vacuum (that pull the liquid apart to form evacuated cavities) and pressure (that cause the cavities to collapse) at up to millions of cycles per second, are commonly used to prepare laboratory dispersions. Although large ultrasonic generators are available, industrial applications are limited.

Practical Matters and Process Equipment. The strength of bonds holding clumps together may increase with storage time because of surface migration of material into the high energy contact region. Migration results from solubilization and mobilization caused by adsorbed vapors or surface-treatment chemicals. Some powders sold commercially have been coated with a solid or liquid to decrease the likelihood of such changes and to keep the clumps soft or easy to deagglomerate.

Typical commercial devices for deagglomerating clumps in low viscosity suspensions use slotted rotors, multiple teeth, and clearances that decrease as the suspension is pumped by centrifugal force through an annular gap.

Surface Modification. Reaction or adsorption at the solid surface can alter its properties and lead to a surface charge or steric stabilization.

The surfaces of metal oxides and hydroxides can take up or release H^+ or OH^- ions and become charged. Potentials as high as ±100 mV may be sustained in aqueous solutions. For aqueous solutions this is a function of the pH.

For ionic salts, one of the ions making up the lattice may be complexed or hydrated more readily than the other, resulting in an imbalance in particle charge. For example, in the case of silver iodide, dispersion in water leads to a net positive particle charge as the iodide ion is hydrated to a greater extent and thus goes into solution more readily than the silver ion.

Polymer chains which are soluble in the suspending liquid may be grafted to the particle surface to provide steric stabilization. The most common technique is the reaction of an organic silyl chloride or an organic titanate with surface hydroxyl groups in a nonaqueous solvent (see DISPERSANTS).

A dense surface coating (encapsulation) that contains no occluded solvent decreases interparticle attraction provided that the coating has a Hamaker constant intermediate between the particle and the liquid. This is called semisteric stabilization (ST).

Additives Used To Stabilize Dispersions. A dispersant for a particular solid-liquid pair consists of molecules having one region that is soluble in the liquid and a second region that is not very soluble. The insolubility of the second region causes it to adsorb on the particles and, at concentrations above the critical micelle concentration (CMC), to force the dispersant molecules into multimolecular clusters with the soluble regions on the outside. The adsorbed dispersant may prevent agglomeration by keeping the particles from coming close enough to be strongly attracted (steric stabilization) or by creating a large enough charge to overwhelm interparticle attraction (electrostatic stabilization).

Because of electrostatic attraction, positive ions are attracted to negatively charged surfaces and have a higher concentration near the surface than in the bulk. Negative ions are repelled from the negative surface and have a lower concentration near that surface.

Adsorption of an organic dispersant can reduce polarizability attraction between particles, ie, provide semisteric stabilization.

Ionizable organic dispersants can lose or gain protons (H^+) in aqueous solution if the pH is in the right region for that particular dispersant. The ionic species may adsorb on an oppositely charged (typically inorganic) surface and provide semisteric stabilization or may adsorb on an uncharged (typically organic) surface to create a surface potential that can stabilize the suspension.

Some polymeric units can be hydrated; thus they are water soluble or adsorb strongly to highly polar materials. Other polymeric units are oil soluble and adsorb strongly on nonpolar materials. Homopolymers consist of a series of identical segments. If the segments adsorb strongly on the particle, the polymer layer will have a compressed conformation (little occluded solvent) and provide only semisteric stabilization. Less strongly adsorbed homopolymers provide a coating whose off-surface sequences may be anchored to the surface at both ends (loops) or at only one end (tails) to provide a steric layer.

Random copolymers consist of random length sequences of soluble segments interspersed with less soluble segments. The less soluble segments adsorb on the particle and the more soluble sequences form loops or tails extending into the liquid. The provision of separate anchor segments and soluble segments provides better control over adsorbed layer depth and desorption than can be achieved using a homopolymer (see POLYMERS).

Associative Thickeners. Although low viscosity is convenient for mixing and pumping operations during manufacture, high viscosity is desirable to deter the formation of a dense sediment bed and possible agglomeration during transportation and storage. Associative thickeners are polymers which have insoluble (with respect to the suspending liquid) end groups and a long soluble backbone. The insoluble end groups form weakly bonded clusters which are kept apart by the backbones. The three-dimensional network of linked micelles forms quickly in the absence of shear to prevent particles from settling when the suspension is quiescent, but breaks up readily when subjected to low shear forces.

Independent Floc Network. A space-filling structure similar to that created by an associative thickener can be provided by a loose network of weakly bonded particles. The addition of clay to an aqueous suspension of particles that are not stable to reagglomeration can prevent the particles from coming into contact with each other and forming large flocs.

Adsorption of Microparticles. Microparticles, having diameters less than one-tenth that of the particles to be dispersed, can be attached to the larger particles by sintering prior to wet-in. The attached microparticles prevent the surfaces of the larger particles from coming close together, thus reducing the maximum attractive energy in essentially the same way that encapsulation does.

Evaluating Dispersions

A variety of methods are available for evaluating dispersions and the chemical character of the particle surface. Numerous standards are available that provide the details of well-established methods for evaluating dispersions.

RALPH D. NELSON, JR.
E. I. du Pont de Nemours & Co., Inc.

R. J. Hunter, *Introduction to Modern Colloid Science*, Oxford University Press, Oxford, U.K., 1993.

H. Lyklema, *Fundamentals of Interface and Colloid Science*, Academic Press, London, Vol. 1, 1991; Vol. 2, 1995; further volumes in press.

R. D. Nelson, *Dispersing Powders in Liquids*, 2nd ed., Elsevier, Amsterdam, the Netherlands, 1995.

P. C. Hiemenz and R. Rajagopalan, *Principles of Colloid and Surface Chemistry*, 3rd ed., Marcel Dekker, New York, 1997.

BULK POWDERS

During the late 1950s and early 1960s A. W. Jenike employed a soil mechanics continuum approach to powder handling, developing a logical, theoretical basis to bulk solids flow. Testing equipment and methods were developed along with design techniques. The basic concepts of bulk solids flow have since been expanded to allow design of bins, hoppers, and feeders.

Definitions

The following definitions are specific to bulk powders handling.

Bin or silo	container for bulk solids having one or more outlets for withdrawal of solids either by gravity alone or by flow-promoting devices which assist gravity
Bulk solid	material consisting of discrete solid particles, which are submicrometer to several centimeters in size, handled in bulk form, as opposed to unit handling
Cylinder	vertical part of a bin (constant cross-sectional area)
Discharger	device used to enhance material flow from a bin but which is not capable of controlling the rate of withdrawal
Feeder	device for controlling the rate of withdrawal of bulk solid from a bin
Flow channel	space in a bin through which a bulk solid is actually flowing during withdrawal
Hopper	converging part of a bin (changing cross-sectional area)

Flow Problems

Bulk solids do not always discharge reliably. Five common flow problems can occur when handling bulk solids.

No-flow. The problem of no flow may occur when an attempt is made to initiate flow, such as by opening a gate or starting a feeder. One of two things may happen. In the first, an arch (bridge, dome) forms over the outlet (Fig. **1a**). Sometimes the only way to initiate flow is to use force greater than that of gravity to overcome the arch and force material flow. The second form of a no-flow problem is commonly referred to as a stable rathole (pipe, core) (Fig. **1b**). In this case, some material discharges as the feeder or gate is operated; but as the flow channel empties out, the resulting hole becomes stable, and material stops flowing.

Erratic flow. A combination of the two no-flow conditions can lead to erratic flow. If flow has been initiated but a stable rathole develops, then when the rathole collapses using vibration, the material may

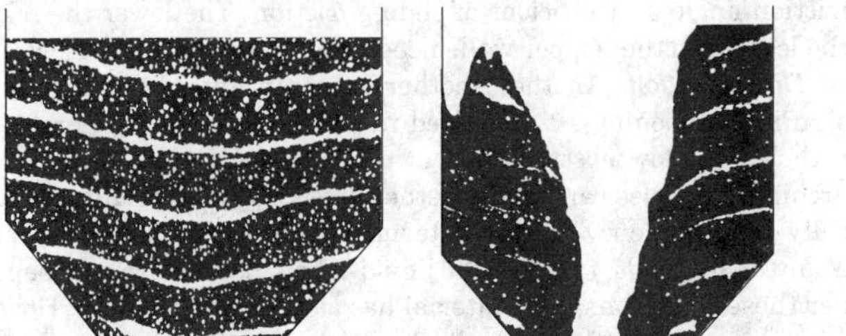

Figure 1. Examples of no-flow situations where the darkened areas represent material within the bin: (**a**) cohesive arch at the outlet of a bin, and (**b**) stable rathole formed within bin.

arch as it impacts the outlet. Flow may be restarted by vibration and maintained for a short time until the rathole forms again.

Flooding. When a stable rathole forms in a bin and fresh material is added, a flood can occur if the bulk solid is a fine powder. As the powder falls into the channel, it becomes entrained in the air in the channel and becomes fluidized (aerated). When this fluidized material reaches the outlet, it is likely to flood from the bin, because most feeders are designed to handle solids, not fluids (see FLUIDIZATION).

Limited discharge rate. Bulk solids, especially fine powders, sometimes flow at a rate lower than required for a process. This flow rate limitation is often a function of the material's air or gas permeability.

Segregation. The problem of segregation occurs when a bulk solid composed of different particle sizes or densities separates.

Flow Patterns

Funnel Flow. A funnel flow pattern occurs when some of the material in a bin or hopper moves toward the outlet while the rest remains stationary. This happens when the walls of the hopper section at the bottom of the bin are not sufficiently steep or smooth to cause the material to flow along them. As a result, the material flows only in a narrow channel, usually directly over the outlet.

Funnel flow bins are suitable only for bulk solids that are coarse, free flowing, and do not degrade, and for use when segregation is not important.

Mass Flow. A material is in mass flow when all of it is in motion whenever any is withdrawn. This means that material flows along the walls. The walls of the hopper section are thus required to be steep and smooth. As long as the outlet is large enough to prevent arching, all the material starts to move as discharge begins keeping the contents of the bin fully live. Stable ratholes cannot form.

Fine powders are easily handled in a mass flow bin where the flow channel is stable and predictable. However, the maximum flow rate of a fine powder through the outlet of a mass flow bin is low compared to that of a coarse, granular solid.

Expanded Flow. Expanded flow uses the best aspects of funnel flow and mass flow by attaching a mass flow hopper section below one that exhibits funnel flow. The flow pattern expands sufficiently at the top of the mass flow hopper to prevent a stable rathole from forming in the funnel flow hopper above it.

Measurement of Flow Properties

In order to develop the proper flow pattern, knowledge of a material's flow properties is essential. Standard test equipment and procedures for evaluating solids flow properties are available. Direct shear tests, run to measure a material's friction and cohesive properties, allow determination of hopper wall angles for mass flow and the opening size required to prevent arching. Other devices available to evaluate solids flowability include biaxial and rotary shear testers.

Wall Friction Angles. Wall friction values, important when characterizing the flow properties of a bulk solid, are expressed as the wall friction angle or coefficient of sliding friction. The lower the friction, the less steep the hopper walls need to be to achieve mass flow.

Flow Function. Another Another important consideration in bin design is the opening size required to prevent formation of arches and ratholes. A flow obstruction can occur in a bin caused by cohesive arching. Particles can bond together physically, chemically, or electrically. This tendency to bond is termed a material's cohesive strength. Many bulk solids flow like a liquid when poured from a bag. Under these conditions, the material has no cohesive strength. However, when squeezed in the palm of a hand, the material may gain enough cohesive strength to retain the shape of the palm once the hand is opened.

In order to characterize this bonding tendency, the flow function of a material must be determined. Data on flow function can be generated in a testing laboratory by measuring the cohesive strength of the bulk solid as a function of consolidation pressure applied to it.

Compressibility. The bulk density of a solid is an essential value used in the analysis of its flow properties, such as when calculating mass flow hopper angles, opening sizes, bin loads, etc. Loose and/or packed density values are not sufficient. Bulk solids exhibit a range of densities that vary as a function of consolidating pressure. This range of densities, called the compressibility of the solid, can often be expressed on a log–log plot as a line or relationship.

Permeability. Two-phase (gas–solid) interactions can be analyzed by considering how gas flows through a bed of powder in the presence of a pressure differential across the bed. When flow through the bed is laminar, Darcy's law, which can be used to relate gas velocities to gas pressure gradients within or across the bed, can be written as in equation 1, where K = permeability factor of the bulk solid; u = superficial relative gas velocity through the bed of solids; γ = bulk density of the solid in the bed; and dp/dx = gas pressure gradient acting at the point in the bed of solids where the velocity is being calculated.

$$u = -K\left(\frac{dp/dx}{\gamma}\right) \qquad (1)$$

The permeability factor, K, has units of velocity and is inversely proportional to the viscosity of the gas. A permeability test is run by passing air through a representative column of solids.

Equipment Design

Hopper Angles for Mass Flow. The wall friction angle for a given bulk material/wall surface combination can be calculated from the results of a wall friction test. This angle, the tangent of which is the coefficient of sliding friction, often varies with the pressure acting normal to the wall surface. From this angle the hopper angles compatible with mass flow can be determined. This is most easily done for hoppers which are either conical or wedge-shaped. There is no magic angle, because mass flow is dependent on both the smoothness and steepness of the hopper wall and the properties of the bulk material involved.

Outlet Size Determination. The second consideration for proper design of a mass flow bin is the size of the outlet required to prevent arching and to achieve the required discharge rate.

There are two mechanisms by which arching can occur: particle interlocking and cohesive strength. The minimum outlet size required to prevent mechanical interlocking of particles is directly related to the size of the particles.

If most of the particles are less than ca 0.6 cm in size, flow obstructions can occur by physical, chemical, or electrical bonds between particles. This cohesiveness is characterized by the bulk material's flow function. The forces acting to overcome a cohesive arch and cause flow are described by a hopper's flow factor, which can be obtained from the design charts.

Sizing the outlet of a funnel flow bin involves consideration of both arching and ratholing. Minimum dimensions to overcome both can be calculated from the material's flow function.

Structural Considerations. Silos, bins, and hoppers fail, in one way or another, each year. Such failures can range from a complete and dramatic structural collapse, to cracking in a concrete wall, or denting of a steel shell. This last is often a danger signal indicating that corrective measures are required.

Silo design requires knowledge of the material's flow properties, flow channel geometry, flow and static pressure development, and dynamic effects. Problems like ratholing and vibration have to be prevented, while assuring reliable discharge at the required rate.

The designer considers load combinations, load paths, primary and secondary effects on structural elements, and the relative flexibility of the elements.

In the construction phase there are two ways problems can arise. The more common of these is poor workmanship. The other is the introduction of badly chosen, or even unauthorized, changes during construction in order to expedite the work.

If a bulk material other than the one for which the silo was designed is placed in the silo, the flow pattern and loads may be completely different. The load distribution can be radically changed if alterations to the outlet geometry are made, if a side outlet is put in a center discharge silo, or if a flow controlling insert or constriction is added.

There are two types of maintenance work required. The first is regular preventative work, such as periodic inspection and repair. The second area of maintenance involves looking for signs of distress. If evidence of a problem appears, expert help should be summoned immediately.

Feeders

Most flow problems can be overcome by using a mass flow design if the mass flow pattern developed by the bin is not disturbed. Thus a properly designed feeder or discharger must be employed. A feeder is used whenever there is a requirement to transfer solids at a controlled rate from the bin to a process or a truck. A discharger is used when there is a need to discharge solids, not control the rate of discharge.

To be consistent with a mass flow pattern in the bin above it, a feeder must be designed to maintain uniform flow across the entire cross-sectional area of the hopper outlet. In addition, the loads applied to a feeder by the bulk solid must be minimized. Accuracy and control over discharge rate are critical as well. Knowledge of the bulk solid's flow properties is essential.

Volumetric Feeders. Examples of volumetric feeders are screws, belts, rotary valves, louvered, and vibratory.

Screws are used primarily when feed over a slotted outlet is required. Screws are a good choice when an enclosed feeder is required, when space is restricted, when handling dusty or toxic materials, or when attrition (particle breakage) is not a problem.

Belts are used to feed over long slotted openings. Typically, belt feeders are used to handle friable, coarse, fibrous, elastic, sticky, or very cohesive solids.

Rotary valve feeders are commonly used for circular or square configured outlets. These are particularly useful when discharging materials to a pneumatic conveying system where a seal is required to prevent air flow through the hopper outlet. The discharge rate is set by the speed of rotation of the vanes or pockets of the valve.

Louvered feeders, designed to withdraw material uniformly across the entire outlet cross section, can control discharge from a circular, square, or rectangular cross section using the material's natural angle of repose. When the drive is energized, the material's angle of repose is overcome and material discharges usually very evenly and accurately. When the drive is stopped, the material stops flowing.

Vibratory pan devices act much in the same way as louvered feeders in that they can feed material gently and accurately. They are limited primarily to applications involving round or square outlets.

Gravimetric Feeders. Examples of gravimetric feeders are weigh belts, loss-in-weight systems, and gain-in-weight systems. Gravimetric feeders rely on weighing the material to achieve the required discharge rate. A gravimetric feeder would be used when accuracies of less than 5% are required, particularly over short time periods; when the material's bulk density varies; or when the weight of material used for a particular process needs to be recorded. There are basically two systems: continuous and batching. A continuous gravimetric system controls the weight/unit time. A batch system simply controls the weight of material discharged to a mixer.

Gain in weight feeders are used only for batching applications.

Special Considerations

Particle Segregation Mechanisms. Segregation is the process by which an assembly of solid particles separates as it is being handled.

Segregation problems occur in a wide range of industries handling materials as diverse as coal (qv) and pharmaceuticals (qv). The cost implications can be great, even when handling small quantities of material.

The primary mechanisms responsible for most particle segregation problems are sifting, particle velocity, air entrainment, particle entrainment, and dynamic effects.

The movement of smaller particles through a mixture of larger ones (sifting) is the most common mechanism by which particles segregate.

Sifting may occur while a pile is being formed. Fine particles are concentrated in the center under the fill point. However, as the pile is formed, the slope stability is such that layers of finite thickness intermittently move from the central fill point, carrying some of the finer particles with them.

Smaller particles, those that are more irregular in shape and/or those that have a higher surface roughness, typically have a higher frictional drag on a hopper or chute surface.

On a chute, higher drag results in lower particle velocity, which can be accentuated by stratification on the chute surface because of the sifting mechanism.

Fine particles generally have a lower permeability than coarse particles, and therefore tend to retain air longer in void spaces. When a mixture of particles is charged into a bin, it is not uncommon to find a vertical segregation pattern, where the coarser, heavier particles concentrate at the bottom of the bed and the finer, lighter particles concentrate near the top.

The lighter the particle and the finer its size, the longer it may remain suspended in an airstream such as upon filling of a bin. Secondary air currents can carry airborne particles away from a fill point into outer areas of a bin, scattering them in a way that bears no resemblance to the calculated trajectory.

Particles often differ in their resilience, inertia, and other dynamic characteristics which can cause them to segregate, particularly when they are forming a pile such as when charged into a bin or discharged from a chute.

Correcting Particle Segregation. The main techniques to consider when segregation problems are present are to change the material, change the process, or change the design of the equipment.

A common characteristic of most highly segregating materials is that they are free flowing, and therefore the particles easily separate from each other. One way to decrease segregation tendencies of a material is to increase its cohesiveness by, for example, adding water or oil. Another technique is to change the particle size distribution.

If a mixture is handled that consists of several ingredients, which are more or less uniform in themselves but vary distinctly from one to another, each ingredient should be handled separately up to the final processing step, then proportioned and mixed just before this step.

Increasing the height-to-diameter ratio of the cylinder section of a mass flow bin above 1.0 usually results in a uniform velocity pattern across the top surface.

An alternative to traditional mass flow bin design is to use a patented BINSERT, which consists of a hopper-within-a-hopper below which is a single-hopper section. The velocity pattern in such a unit is controlled by the position of the bottom hopper.

Fine Powder Flow Phenomena. A fine powder is a material in which the flow behavior is significantly influenced by the effects of entrained gas in its void spaces. This is in contrast to a coarse material, in which the voidage is so great that gas effects can be neglected. For a fine powder such effects are extremely important, causing, for example, the maximum flow rate of such a material through a mass flow bin outlet to be several orders of magnitude lower than predicted by theory.

When handling a fine powder in a bin, funnel flow, mass flow, and fluidized handling can be considered.

Sometimes it is more practical to handle fine powders in a fluidized state rather than to deaerate the particles and handle by gravity alone. Through the use of air pads, air slides, and/or air nozzles, some or all of the contents of a bin can be fluidized. This overcomes the common no-flow problems of arching and ratholing, and allows discharge rates through bin outlets several orders of magnitude faster than if the powder is deaerated.

Vessels designed for processing solids are often adaptations of conventional storage bins modified to achieve the desired process activity. A wide variety of solids including chemicals, plastics, and sugar (qv) are processed in this way.

Purge and conditioning vessels often present problems caused by nonuniform purge of conditioning.

The key to solving these problems is to design the vessel for a mass flow pattern. This involves consideration of both the hopper angle and surface finish, the effect of inserts used to introduce gas and control the solids flow pattern, and sizing the outlet valve to avoid arching and discharge rate limitations.

Mixing and Blending. Quality control is important, and blending is a useful technique by which quality is improved. Static in-bin blenders are sometimes used to dampen upsets either going into or coming from a process, while batch and tumble blenders are sometimes used for close, well-defined quality control (see MIXING AND BLENDING).

An in-bin blender consists of basically a storage bin or silo which doubles as a blender. Requirements for an effective blender include no stagnant regions, large velocity gradients throughout the blender, minimum need for recirculation, and blending uniformity.

These blenders are widely used to blend materials that are free-flowing, uniform in size, and have low angles of internal friction.

The basic components of a typical batch blending system are a blender, one or more portable or stationary containers, and a chute to a process. A typical operation consists of mixing the ingredients in the blender, discharging the blend into a container, and gradually discharging the blend into the process.

Solid–solid blending can be accomplished by a number of techniques. Some of the most common include mechanical agitation and fluidization.

Tumble blending has a number of advantages over other common blending techniques. First, it eliminates having to discharge the blender into a portable or stationary container. Second, tumble blending eliminates the costly step of blender cleaning between batches. All the material is confined within the portable container, so when cleaning is required only the container needs to be cleaned, not the stationary blender. Third, tumble blending eliminates downtime required to fill and empty a blender, because the tumble station is ready to accept a new container as soon as the previous one has finished tumbling.

Flow Aids. Flow aids are devices used to assist in discharging materials from a bin or other storage container. The best use of such a device is when gravity alone is insufficient or impractical to provide reliable discharge.

Flow aids can generally be divided into three categories: mechanical, eg, those that rely on vibration or agitation of the material; introduction of air, eg, air cannons, air slides, or air nozzles; and chemical, eg, fumed silica. Some flow aids rely on a combination of these types.

These mechanical devices, sometimes called bin activators, consist of an outer shell to which one or more motors with eccentric weight are mounted. By hanging the unit from the bin and using a rubber gasket to prevent spillage, the unit vibrates in a horizontal motion to assist in discharging material. The vibration is transmitted to material inside the unit through the use of a central baffle.

Air- and electrically operated mechanical vibrators are sometimes placed on the exterior of hoppers and chutes. In general, such devices are better suited to cleaning off chutes than for use on hoppers to be filled with bulk solids.

Air cannons, or air blasters, consist of a cylinder of compressed air that has been pressurized, typically to plant air conditions. When the unit is fired, a quick-acting valve is opened allowing the air to quickly exit the blaster and enter the bin. Such devices can be effective in collapsing a stable arch, but are usually far less effective in breaking up a stable rathole.

Devices consisting of a horizontal or vertical shaft with arms may be used to break up material and thereby cause it to flow. Other types of mechanical agitation consist of vibrating screens or expanded metal panels.

JOSEPH MARINELLI
Peabody SolidsFlow
JOHN W. CARSON
Jenike & Johanson, Inc.

A. W. Jenike, *Storage and Flow of Solids*, Bulletin No. 123, University of Utah Engineering Experiment Station, Salt Lake City, Nov. 1964.

R. Kulwiec, ed., *Materials Handling Handbook*, John Wiley and Sons, Inc., New York, 1985.

C. R. Woodcock and J. S. Mason, *Bulk Solids Handling*, Chapman and Hall, London, 1987.

M. E. Fayed and L. Otten, *Handbook of Powder Science and Technology*, Van Nostrand Reinhold Co., New York, 1984.

POWER GENERATION

Power, P, defined as the rate at which work is performed, is expressed in terms of energy divided by time and is most commonly given in units of horsepower, as for the power supplied by mechanical devices such as diesel engines, or in the SI units of watts, especially when measuring electrical power. One horsepower is equivalent to the amount of power needed to lift 33,000 pounds (14,982 kg) one foot (30.5 cm) in one minute. One watt is equivalent to the power required to perform one joule of work per second. In a simple direct-current circuit where potential is represented by E: $P = EI = I^2R$. One watt of power is required to force one ampere of electric current, I, through a length of conductor having a resistance, R, to flow of one ohm. The calculation of power in alternating-current circuits is more complex because of the characteristics of these circuits and their loads. For a single-phase circuit having an effective line voltage, E, and an effective line current, I: $P = EI \times PF$, where PF is the power factor corresponding to the lag between the voltage and current wave forms. For multiphase systems, a factor is introduced on the right-hand side of the equation. The factor is two for a two-phase system; 1.73 for a three-phase system.

Combustion Fundamentals

The combustion of fossil fuels, typically coal, oil (see PETROLEUM), or natural gas, is central to the energy conversion process by which most electric power is generated worldwide, particularly in the United States. A fundamental goal of power plant designers and operators is to ensure that the maximum economical amount of energy is extracted for practical use from a given amount of fuel burned (see COMBUSTION TECHNOLOGY).

Combustion is basically an exothermic chemical reaction by which the potential chemical energy stored in a fuel is released during its reaction with oxygen in the presence of heat. Thus, the basic constituents of combustion are fuel, oxygen, and heat. At normal temperatures, the oxygen, usually from air, and fuel can typically mingle without any combustion effect. However, upon the addition of sufficient heat, the fuel and air are excited to a point at which high velocity molecular collisions occur. If the velocity of these collisions is high enough, the bonds holding the various molecules together break and release heat.

The point at which enough heat has been added to start combustion is known as the ignition point. Once initiated, external heating sources are typically not required to maintain the combustion process, because most fuels release sufficient heat during the combustion process.

The main fossil fuel constituents enabling combustion are carbon, hydrogen, and various hydrocarbons. The main products resulting from combustion are water and carbon dioxide. However, a variety of trace compounds and unburned matter may also be released during the combustion of power plant fuels. These materials vary, depending on the specific fuel being burned and whether complete combustion occurs. For example, coal and fuel oil may contain small amounts of sulfur. This is an undesirable constituent, despite its nominal heat content, because sulfur reacts during combustion to form acidic sulfur dioxide.

Steam Cycles

Conventional fossil fuel-fired power plants, nuclear power facilities, cogeneration systems, and combined-cycle facilities all have one key feature in common: some type of steam generator is employed to produce steam. Except for simple-cycle cogeneration facilities, the steam is used to drive one or more rotating turbines coupled to rotating electric generators for electricity production. The thermodynamic cycle by which water is boiled in a steam generator (heat addition), forced through a turbine (expansion), and then condensed (heat rejection) before being pumped back to the steam generator is known as the Rankine cycle.

Because of the simplicity and reliability of the Rankine cycle, facilities employing this method dominated the power industry in the twentieth century and typically play an important role in most modern combined-cycle facilities. Water is the working fluid of choice in nearly all Rankine cycle power plants because water is nontoxic, abundant, and low cost.

Under standard pressure and temperature conditions of 101.3 kPa and 15°C (14.7 psia and 60°F), the amount of heat (or thermal energy in transition) that must be added to one pound of water to raise the temperature of the pound of water 1°F is known as one British thermal unit (Btu). One Btu, a common unit of measure in the energy field, is equivalent to 1.05435 kJ.

The generation of steam (qv) is essentially a two-stage process. Through the addition of heat, the temperature of water is first raised to its boiling point, ie, 100°C at atmospheric pressure. After the boiling point is reached, the temperature at the contact point between the liquid and vapor remains constant as long as the pressure is not permitted to build, such as in a partially closed vessel, until vaporization is completed. When water is boiled in a closed vessel, such as in the steam drum of a utility boiler, vapor generation increases the pressure within the vessel. As the pressure increases, the boiling point (or saturation temperature) increases. Thus, in a boiler operating at 8.27 MPa (1200 psia), water must be heated to nearly 299°C to achieve boiling.

In most utility boilers, steam pressure regulation is achieved by the throttling of turbine control values where steam generated by the boiler is admitted into the steam turbine. Some modern steam generators have been designed to operate at pressures above the critical point where the phase change between liquid and vapor does not occur.

Rankine Cycle Power Plants

The most common heat sources employed by Rankine cycle power plants are either fossil fuel-fired or nuclear steam generators. The former are the most widely used.

Fossil Fuel-Fired Plants.

In modern, fossil fuel-fired power plants, the Rankine cycle typically operates as a closed loop. In describing the steam–water cycle of a modern Rankine cycle plant, it is easiest to start with the condensate system. Condensate is the water that remains after the steam employed by the plant's steam turbines exhausts into the plant's condenser, where it is collected for reuse in the cycle. Many modern power plants employ a series of heat exchangers to boost efficiency. As a first step, the condensate is heated in a series of heat exchangers by steam extracted from strategic locations on the plant's steam turbines (see HEAT-EXCHANGE TECHNOLOGY).

Another stage of heating occurs in a direct contact, deaerating heater before the water is directed to the boiler feed pumps. The deaerating heater serves a dual role: it not only adds heat to the water, but also improves the water quality by removing dissolved oxygen and other noncondensable gases in the condensate that can cause corrosion within the plant's steam and water piping, steam generator, and steam turbines.

The deaerator is typically considered the dividing point between the condensate and feedwater systems. Water exiting the deaerator is fed directly to a boiler's high speed centrifugal feedwater pump, which boosts the water's pressure to a level high enough to enable its ultimate introduction into the steam drum where boiling occurs.

After exiting the feedwater pump and before entering the boiler's steam drum, feedwater usually goes through additional high pressure feedwater heaters (heated by steam turbine extraction steam) as well as another stage of heating in the plant's economizer. The economizer consists of a bank of heat exchanger tubes located in the boiler's backpass section, where heat is transferred from the hot exhaust gas exiting the boiler to the incoming feedwater flowing through the economizer tubes. Most modern fossil fuel-fired power plants share these design features. However, many different types of feedwater heater and boiler arrangements are available, based on the specific steam needs of an application.

Water Treatments. The goal of water treatment is to ensure that the water used for steam generation is pure, such that any contaminants that could lead to corrosion, erosion, or scaling of boiler components are absent. The deaerating feedwater heater is one means by which noncondensable gases are removed from the system. Oxygen scavenging chemicals are also dosed into the feedwater system to eliminate any oxygen not removed by the deaerator or air ejectors.

Demineralizers are often used to treat raw makeup water or condensate where high purity is required. Demineralizers employ a combination of cation and anion exchange to remove additional material, including sodium and ammonium cations.

Fuel System. The main goals of any firing system are the same, ie, to ensure that fuel and air are delivered in such a way as to promote safe, efficient, smooth combustion, and to minimize pollutant emission formation and maximize heat transfer to the waterwall tubes. Properly controlling the air-to-fuel ratio is central to these goals. Whether a boiler fires liquid fuel, coal, or natural gas, there are typically multiple burners which serve to direct the fuel into the furnace for combustion and which ensure that proper fuel–air mixing occurs so that combustion takes place at the desired rate and in the desired location.

Emission Control. Power industry emissions include the air pollutants sulfur dioxide (SO_2), nitrogen oxides (NO_x) and carbon dioxide (CO_2). In oil-, gas-, and coal-fired facilities, low NO_x burners have been successfully applied to reduce NO_x emissions by limiting the peak burner flame temperatures either by rapid premixing of fuel and air, or by staging the introduction of air or fuel to achieve a longer, cooler flame.

Emissions control systems play an important role at most coal-fired power plants. For example, pulverized coal-fired plants sited in the United States require some type of sulfur dioxide control system to meet the regulations set forth in the Clean Air Act Amendments of 1990, unless the boiler burns low sulfur coal or benefits from offsets from other highly controlled boilers within a given utility system. Flue-gas desulfurization (FGD) is most commonly accomplished by the application of either dry- or wet-limestone systems (see SULFUR REMOVAL AND RECOVERY).

Similar to oil-fired plants, either low NO_x burners, selective catalytic reduction or selective noncatalytic reduction can be applied for NO_x control at PC-fired plants. Likewise, fabric filter baghouses or electrostatic precipitators can be used to capture flyash (see AIR POLLUTION CONTROL METHODS).

Nuclear Fuel-Fired Plants. There are some basic design differences of the nuclear plants steam cycles (see NUCLEAR REACTORS) as compared to conventional fossil fuel plants. In nuclear power plants, thermal energy is released during the fissioning of a nuclear fuel (or fissile), such as uranium-235. Fission, the splitting of a heavy nucleus, is typically initiated when the material is struck by a neutron. When the nucleus of a fissile material's atom is split, the energy associated with the atom is released. This energy, approximately 100 billion times that released during the combustion of one carbon atom in fossil fuels, is significant. In addition, extra neutrons are released that impact other nuclei, creating chain reaction.

Extremely safe means of controlling the nuclear reaction process have been devised by introducing materials that can moderate the production and absorption of neutrons released during fission.

Cogeneration

In the power industry, cogeneration refers to the simultaneous generation of heat and power. One example of a cogeneration plant is the central utility boiler that provides steam for both electric power generation and supply to a local district heating system where it can be used for facilities or process heating, or cooling, via absorption coolers or steam-turbine-driven refrigeration and/or air conditioning systems. Such a plant can use a noncondensing steam turbine to supply some or all of the steam required by the district heating system. A well-designed cogeneration facility can convert 80% or more of the fuel energy input for useful purposes, ie, power generation and process heating. In a conventional Rankine cycle power plant, the thermal energy remaining in the steam as it enters the condenser can range up to 50% or more of the fuel energy input, boiler associated losses are ca 15%, and other losses about 2%. Thus, a conventional Rankine cycle plant may convert only 35% of the fuel energy for power generation. In a modern coal-fired cogeneration system, heat losses can be cut to 16%, 15% of which are boiler-associated. Such a system, where the waste heat is recovered from the main power generation cycle for reuse, is often referred to as a topping cycle (see ENERGY MANAGEMENT; PROCESS ENERGY CONSERVATION).

Gas Turbines and Combined-Cycle Power Plants

Gas turbine engines have become increasingly popular for power generation and cogeneration.

Gas turbines are based on the Brayton thermodynamic cycle (Fig. 1). Most modern units operate in the following manner. Combustion and cooling air is first drawn in and compressed in a multistage axial compressor located on the cold end of the gas turbine's rotor (point A of Fig. **1b**). The compressed air is injected into a combustor section where it is combined with fuel for combustion, generating hot, expanding gaseous exhaust. The expanding combustion gases are then directed axially along the rotor into a power turbine or turbines. There the gases impart torque on the tangential turbine blades in a manner similar to a steam turbine, before exhausting to the atmosphere or a heat recovery steam generator.

Natural Gas for Power Plant Use

Most modern gas-fired, heavy-duty gas turbines operate at gas pressures between 1.2 and 1.7 MPa (180–250 psig). However, aeroderivative gas turbines and newer heavy-duty units can have such high air-inlet compression ratios as to require booster compressors to raise gas inlet pressures, in some cases as high as 5.2 MPa (750 psig).

All gas-fired power plants require oxygen analyzers to ensure that air has not been drawn into the piping system. Oxygen intake can lead to the presence of an explosive mixture in the pipeline before the fuel reaches the burner or combustor zone. When gas-fired units are located in an enclosed area, multiple ultraviolet flame detectors

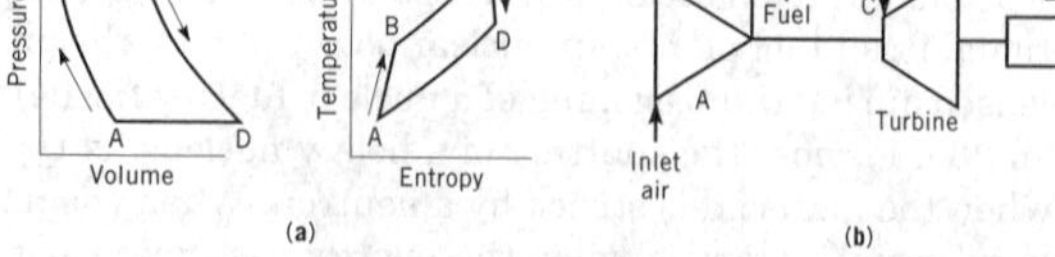

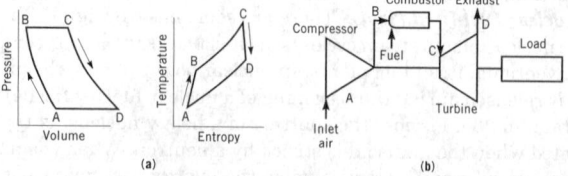

Figure 1. Brayton cycle, where A = compressor inlet, B = combustor inlet, C = power turbine inlet, and D = exhaust: (**a**) thermodynamic relationships and (**b**) schematic of a simple-cycle, single-shaft gas turbine. Courtesy of General Electric.

are used to shut down equipment and flood the area with CO_2 or a chemical fire suppressant whenever a spark or flame is detected.

Processed gas is compressed for transmission through interstate and intrastate pipelines (qv).

Modern pipeline companies control gas flow through their systems in response to and anticipation of demand swings via computerized operations centers. These facilities typically use satellite or modem-based telemetry systems to monitor gas flow and automatically regulate valves and compressor stations located along the pipeline and at gas storage facilities.

The demand for gas is highly seasonal. Thus pipeline companies economize by sizing production facilities to accommodate less than the system's maximum wintertime demand. Underground storage facilities are used to meet seasonal and daily demand peaks. In North America, gas is stored in three main types of underground formations: depleted oil or gas fields, aquifers that originally contained water, and caverns formed by salt domes or mines.

Other Generation Options

Reciprocating Engines. Reciprocating engines, particularly medium- and slow-speed diesels similar to those used for shipboard propulsion and power generation, have been used for land-based power generation since the 1940s. Because of their relatively high efficiency (up to 45%), reliability, and quick start-up capability, these units have been popular for peaking, emergency, and base-load power generation.

Nuclear Reactors. Nuclear power facilities account for about 20% of the power generated in the United States. Although no new plants are planned in the United States, many other countries, particularly those that would otherwise rely heavily on imported fuel, continue to increase their nuclear plant generation capacity.

Fuel Cells. Fuel cells (qv) are essentially batteries (qv) that run on fuel and therefore do not run down. Advanced fuel cells feature efficiencies above 40% and consist of a series of porous, conducting electrode layers (a layer corresponding to one anode and one cathode), separated by an electrolyte (an ionic charge carrier). The multiple cells are configured in series to achieve the voltages required for industrial and commercial use.

Fuel cells generate significant levels of waste heat that can be captured and utilized to improve overall plant efficiency (see PROCESS ENERGY CONSERVATION).

STEVEN COLLINS
Empire State Electric Energy Research Corporation

T. Elliott, *Standard Handbook of Powerplant Engineering*, McGraw-Hill Book Co., Inc., New York, 1989.

Steam: Its Generation and Use, 40th ed., The Babcock and Wilcox Co., Barberton, Ohio, 1992.

B. H. Bunch and A. Hellemans, *The Timetables of Technology*, Simon and Schuster, New York, 1993.

PRASEODYMIUM. See LANTHANIDES.

PRESERVATIVES. See ANTIOXIDANTS; ANTIOZONANTS; COATINGS; FOOD ADDITIVES; PAINT; WOOD.

PRESSURE MEASUREMENT

Pressure measurement is important in the chemical process industries (CPI) and in laboratories for a number of reasons: differential pressure is the driving force in fluid dynamics; product quality frequently depends on certain pressures (or vacuums) being reached and accurately maintained for specific lengths of time during a process; and pressure is a crucial safety consideration in the operation of process equipment particularly where boiler or reactor pressures must not exceed certain limits (see FLUID MECHANICS; HIGH PRESSURE TECHNOLOGY; VACUUM TECHNOLOGY).

Units of Measurement

Pressure is defined as force per unit of area. The International System of Units (SI) pressure unit is the pascal (Pa), defined as 1.0 N/m^2 (see UNITS AND CONVERSION FACTORS; front matter).

Definition of Terms

Absolute pressure is pressure measured relative to a perfect vacuum, an absolute zero of pressure. Like the absolute zero of temperature, perfect vacuum is never realized in a real world system but provides a convenient reference for pressure measurement. The acceptance of strain gauge technology in the fabrication of pressure sensors is resulting in the increased use of absolute pressure measurement in the CPI (see SENSORS). The pressure reference for most of the modern pressure gauges used in the CPI has been atmospheric or local barometric pressure. Barometric pressure varies with elevation and weather. These variables have been eliminated by establishing a standard atmospheric pressure of 101,325 Pa (14.696 psi) as a basis for correcting gauge indication for variations in barometric pressure. One standard atmosphere is equal to the pressure exerted by a column of mercury 760 mm high at a temperature of 0°C where the acceleration owing to gravity is 9.80665 m/s^2. The Pa is an absolute pressure unit. SI conventions make no provisions for differentiating between absolute and gauge pressure. Absolute and gauge pressure are often differentiated in psi by the notation psia and psig, respectively.

Gauge pressure is equal to absolute pressure minus barometric pressure. Absolute pressure is gauge pressure plus barometric pressure. Gauge pressure can be either positive or negative. When the term pressure gauge is used, the reference is almost always to a gauge that is used to measure positive pressures, ie, pressures that exceed local barometric pressure. A vacuum gauge is used to measure negative pressures. A compound gauge is designed to measure both positive and negative pressures and indicates gauge pressure and vacuum on the same scale. A negative gauge pressure indicates that the system is operating under a vacuum, ie, the absolute pressure is less than barometric. For systems that operate under negative pressures, ie, under vacuum, absolute pressure is equal to the barometric pressure minus the vacuum.

Mechanical Gauges

Where pressure is monitored as opposed to being controlled, the pressure is generally measured by directly actuated mechanical elements. Mechanical gauges dominate process applications in older plants. Moreover, the sensing elements for many of the more modern and sophisticated electronic transmitters are simple mechanical elements such as Bourdon tubes or diaphragms. Mechanical gauges may be divided into two groups. The first, which includes liquid manometers, bell gauges, and slack diaphragm gauges, measure pressure by balancing an unknown force against a known force. The second, Bourdon gauges, diaphragm gauges, and bellows elements, rely on elastic deformation of a sensing element for pressure measurement.

Liquid Manometers. The typical liquid manometer consists of a cyclindrical glass U-tube partially filled with liquid. One end is connected to the process; the other can be either open or closed.

An open manometer is normally used to measure pressure relative to local barometric pressure or to measure differential pressure, ie, the difference between two pressures. The principal use of liquid column manometers has been as a primary standard for calibrating other gauges.

Inverted Bell-Type Pressure Element. An inverted bell manometer consists of two inverted bells immersed in oil. The oil provides a liquid seal. The bells are suspended from opposite ends of a balance beam and are arranged so that pressure, P, can be introduced under each bell. One of the lines is usually open to atmospheric pressure, the other to the pressure to be measured. The bell subjected to the higher pressure rises in the oil, tilting the beam which moves a pointer on a scale. This instrument responds to a pressure difference, ΔP, as small as 0.1 Pa (0.0004 in. H_2O). Inverted bell manometers are used for measuring very low positive pressures, such as those found in furnace kiln drafts and conveyor dryers (see DRYING; FURNACES, FUEL-FIRED).

Bourdon Tube. A Bourdon tube is made from a flattened or elliptical tube, where one end is sealed, the other open to the process. All Bourdon tubes are based on the simple principle that a closed-end, flattened or elliptical coiled tube tends to straighten out when a gas or a liquid under pressure is allowed to enter the tube. The Bourdon tube responds to the pressure difference between the inside and the outside of the tube. Bourdon tubes are used extensively in pressure gauges, in vacuum gauges, and in compound gauges.

Diaphragm Gauges. The sensing element for a diaphragm gauge is a flexible disk, either flat or having concentric corrugations, made of sheet metal. Some gauges use the diaphragm itself as the pressure sensor; others use it as the basic component for a capsule, manufactured by fusion-welding two diaphragms together at their peripheries.

Diaphragm elements are sensitive to small pressure changes and are therefore particularly useful in the measurement of low pressures.

Slack diaphragm gauges use diaphragms made from an elastomer (see ELASTOMERS, SYNTHETIC). A slack diaphragm gauge does not rely on elastic deformation of the diaphragm for pressure measurement. The pressure-sensing element is a calibrated spring. The gauge measures pressure by balancing an unknown force against a known force.

Bellows Elements. In a spring-and-bellows pressure element, the bellows is enclosed in a metal housing connected by piping to the process and restrained at the top by a form-fitted nut. A rod resting on the bottom of the bellows transmits any vertical motion of the bellows through a suitable linkage to a pointer or pen. As the pressure inside the bellows increases, the bellows compresses the spring. The stiffness of the bellows is small compared to the stiffness of the spring, and therefore the pressure range is primarily a function of the stiffness of the spring.

Meters. Diaphragms, diaphragm capsules, and bellows elements are used extensively in meter bodies designed to measure differential pressure. These units can be used to measure the differences in pressure from 25 Pa (0.1 in. H_2O) to 0–4826 kPa (0–700 psi) and are designed to operate at pressures as great as 69 MPa (10,000 psi).

Electronic Sensors

Electronic sensors and electronic control systems have displaced many of the mechanical sensors and pneumatic control systems in the CPI. This change, occurring in the 1980s and 1990s, resulted from the superiority of electronic sensor technology, as well as the superiority of electronic control systems.

It is especially difficult in discussing electronic instruments to distinguish between a pressure sensor, a pressure transducer, and a pressure transmitter. Sensor and transducer are synonymous. The distinction between transducer and transmitter is, however, fundamental. All transmitters are transducers; not all transducers are transmitters. An electronic transducer converts movement of a mechanical element, or the output from an electronic sensing element, to an electrical quantity such as resistance, capacitance, or voltage. In its simplest form, a pressure transmitter is a combination of a pressure transducer and a signal-conditioning circuit that outputs a current proportional to the process pressure. An electronic transmitter converts the movement of a mechanical element or the output from an electronic sensing element to a 4–20-mA signal, or to digital information, for transmission to an indicator, a recorder, a controller, or to a distributive control system (DCS).

Piezoelectric Elements. Designs of piezoelectric pressure elements are based on the principle that certain crystalline insulators, such as quartz (see SILICA, SYNTHETIC QUARTZ SILICA), when properly cut and oriented with respect to their crystallographic axes, generate a small electric charge when stressed. In practice, a stack of quartz plates is mounted in a housing, which has a thin diaphragm at one end. The housing is usually designed to be mounted in the wall of a pressure vessel. The diaphragm is exposed to the pressure and deflects, thereby applying a compressive force to the stack, which in turn generates a charge that is directly proportional to the force. Piezoelectric sensors cannot measure static or absolute pressures for more than a few seconds, but this automatic elimination of static signals allows drift-free operation. Because piezoelectric sensors are very rugged are used extensively in demanding applications. Such devices are available in pressure ranges between 0–345 kPa (0–50 psi) and 0–69 MPa (0–10,000 psi).

Linear-Variable-Differential-Transformer and Reluctive Pressure Transducers. In a linear-variable-differential-transformer (LVDT) pressure transducer, the pressure to be measured is fed to a Bourdon tube or diaphragm. The motion of this element is transferred to the magnetic core of a transformer, and a varying voltage is produced in the secondary coil as the core is moved. When the core is moved from center, a differential voltage is produced. That is proportional to the movement of the Bourdon tube or diaphragm, which is proportional to the pressure.

Measurement by a reluctive transducer is based on the ratio of the reluctance of the magnetic flux path of two coils. The LVDT and reluctive pressure transducers are particularly well suited for low pressure measurement.

Strain Gauges. The simplest version of a strain gauge uses the change in the electrical resistance of a metal wire under strain to measure pressure. A bonded strain gauge is constructed by using an adhesive to bond a metal wire (or a metal foil) to an elastic sensing element, usually a Bourdon tube, diaphragm, or a bellows element. The gauge converts movement of the sensor to an electrical signal. When the length of the wire is changed by tension or compression, the result is a change in the diameter of the wire and, hence, a change in electrical resistance. The change in resistance is a measure of the pressure.

Piezoresistive Sensors. The distinction between strain-gauge sensors and piezoresistive (integrated-circuit) sensors is minor. Both function by measuring the strain on an elastic element as it is subjected to pressure. A piezoresistive transducer is a variation of the strain gauge that uses bonded single-crystal semiconductor wafers.

Capacitive Pressure Transducers. In all capacitive pressure detectors, the basic operating principle is that a change in capacitance occurs owing to the movement of an elastic element. In a conventional capacitance-type pressure transducer, the sensing element is a diaphragm. The process comes in contact with an isolating diaphragm, a metal diaphragm used to isolate the sensing element from the process. Process pressure is hydraulically coupled to the sensing element. Diaphragm movement changes the capacitance between the plates and the diaphragm; one side increases while the other decreases. Signal conditioning converts the capacitance change to a stable direct current or voltage signal.

Vacuum Measurement

Vacuum measurement spans the range from 10^5 Pa (atmospheric pressure) to pressures that are less than 10^{-10} Pa (7.5×10^{-13} torr), ie, more than 16 orders of magnitude.

The pressure sensors discussed herein are reasonably accurate for measurement into the low vacuum region. Upon some modification, these sensors can be extended into the lower end of the range, to approximately 1.3×10^{-1} Pa(10^{-3} torr). Vacuum gauges are required for accurate measurement of lower pressures.

Vacuum gauges may be broadly classified as either direct or indirect. Direct gauges measure pressure as force per unit area. Indirect gauges measure a physical property, such as thermal conductivity or ionization potential, known to change in a predictable manner with the molecular density of the gas.

Capacitance Manometers. The development of capacitance manometers designed specifically for the harsh environments that characterize chemical processes is a fairly recent development. The near-phenomenal accuracy of capacitance manometers, typically 0.10% of a reading ≥13.3 Pa (≥0.1 torr) to 3% of a reading at 10^{-2} Pa (10^{-4} torr), excellent linearity over a wide range of pressures, and the development of sensors that are rugged and reliable, increasingly makes this the technology of choice for vacuum measurements in the range $10^{-3}-10^{3}$ Pa ($10^{-5}-10$ torr).

The capacitance manometer is in essence an electronic diaphragm gauge. It differs from the mechanical diaphragm gauge in the way diaphragm deflections are translated into pressure readings. The mechanical gauge employs a mechanical linkage. The capacitance manometer uses capacitance changes to measure diaphragm deflections.

Thermal Conductivity Gauges. Thermal conductivity gauges measure thermal conductivity and are thus indirect gauges. Pressure is measured indirectly. The thermal conductivity of a gas is essentially independent of pressure for pressures greater than approximately 1.33 kPa (10 torr).

The sensor for a thermal conductivity gauge is a thin metal wire or heat-sensitive element enclosed in a metal cylinder. The element is electrically heated. If a constant current is maintained across the element, the temperature of the element increases as the pressure in the metal cylinder is reduced below 1.33 kPa (10 torr). The rate at which heat is transferred from the element to the cylinder wall decreases, because the thermal conductivity of the gases surrounding the element decreases with pressure. The thermal conductivity of the gases surrounding the element, and indirectly the pressure, can therefore be measured by measuring the temperature of the element.

Hot-Cathode Ionization Gauges. The operation of the ion gauge is based on ionization of gas molecules as a result of collisions with electrons. These ions are then subsequently collected by an ion collector. Ionization gauges, used almost exclusively for pressure measurement in high, very high, ultrahigh, and extreme ultrahigh vacuums, measure molecular density or particle flux, not pressure itself.

Cold-Cathode Ionization Gauges. The cold-cathode gauge is more rugged, but less accurate, than the hot-filament ionization gauge. A potential of 2–10 kV is applied between the cathodes and the anode. Free electrons are formed by the collision of positive ions with the cathodes. Electrons ejected from the cathodes collide with gas molecules to produce additional positive ions, setting up a Townsend discharge between the cathode and the anode.

Electrons emitted by the cathode pass through the anode and continue until repelled by the other cathode. The electrons oscillate between the cathodes and eventually are captured by the anode. Sensor output, the sum of the positive ion current to the cathode and the electron current leaving the cathode, is proportional to pressure over the range $1-10^{-4}$ Pa ($10^{-2}-10^{-6}$ torr), the nominal pressure range for Penning cold-cathode ionization gauges.

Smart Pressure Transmitters

Microprocessors mated to pressure transmitters produced the first smart transmitters in the mid-1980s. Conventional electronic pressure transmitters are strictly analogue in nature, converting the motion or changes in the electrical resistance of a sensor to standard 20–100-kPa (3–15-psig) pneumatic signals or to 4–20-mA d-c electrical signals. Smart transmitters convert the response of the sensor to a high resolution digital signal.

Intelligent or smart transmitters feature digital electronics, remote communication and configuration, and high turndown. A smart transmitter performs continuous diagnostics of its sensing element, ie, the meter body, and electronics and of the loop power supply and wiring. Smart transmitters have the advantage, as compared to conventional analogue electronic transmitters, of greater accuracy and stability, and much greater range, ie, one model can be installed anywhere, and respanned (without recalibration) at any time.

The advent of the smart transmitter revolutionized the CPI's approach to process instrumentation (see PROCESS CONTROL).

J. L. RYANS
Eastman Chemical Company

B. G. Liptak, *Instrument Engineers' Handbook—Process Measurement and Analysis*, 3rd ed., Chilton Book Co., Radnor, Pa., 1995, pp. 523–601.

B. G. Liptak, *Instrument Engineers' Handbook—Process Control*, 3rd ed., Chilton Book Co., Radnor, Pa., 1995, pp. 252–284.

S. Dushman, in J. M. Lafferty, ed., *Scientific Foundations of Vacuum Techniques*, 2nd ed., John Wiley & Sons, Inc., New York, 1962, pp. 258–370.

J. F. O'Hanlon, *A Users's Guide to Vacuum Technology*, 2nd ed., John Wiley & Sons, Inc., New York, 1989, pp. 75–100.

PRESSURE VESSELS. See HIGH PRESSURE TECHNOLOGY; TANKS AND PRESSURE VESSELS.

PRINTING INK. See INKS.

PRINTING PROCESSES

In the latter part of the twentieth century, revolutionary changes took place in printing processes that involved a shift in the printing and publishing industry to electronic prepress, electronic printing, and alternative media such as compact disk–read only memory (CD-ROM) and multimedia. These developments could conceivably rival the cultural and commercial impact of the introduction of printing to the Western world. Just as the fast, economical printing production processes to which Johannes Gutenberg contributed around 1450 served to spread knowledge and advance civilization in the Renaissance, the tools of electronic printing have helped to enhance global education and change the ways in which information is communicated. Printing and mass distribution are being replaced by electronic distribution of information and local printing for specialized, sometimes even individual, interests.

The role of chemical technology in printing has also changed. Whereas a need exists for hard copy, whether for visual, legal, or historical reasons, the hard copy must meet revised standards of performance. Many traditional printing processes have become unacceptable in the workplace, and are being replaced by processes that are water-based, dry, desktop, or in some other ways more convenient.

There are four main printing processes: planography or lithography, intaglio or gravure, porous or screen, and relief (flexography or letterpress). Use of letterpress, which once excelled in the reproduction of text and pictures, has rapidly waned as first photolithography, and then electronic prepress systems, have increased in ability to provide text material without the setting of metal type.

In general, the process of printing involves generating two physically different areas: the printing or image area, and the nonprinting or nonimage area. In relief printing, whether flexographic or letterpress, the image or printing area is raised above the nonprinting area. Ink is applied to the raised surface, which is brought into direct contact with the paper (qv) or other surface upon which the print is to appear. The relief printing process is used to print on a variety of paper and plastic packaging (qv) materials as well as for some magazines and newspapers, labels, and business forms. Water-based or solvent inks are used. Letterset describes the use of relatively thin relief plates for printing by the offset principle (see INKS).

In the intaglio process, the nonprinting area is at a common surface level but the printing area is recessed, consisting of wells etched or engraved, usually to different depths. The most typical method of intaglio printing is the gravure process. Solvent inks with the consistency of light cream are transferred to the whole surface and a metal doctor blade is used to remove excess ink from the nonprinting surface. Ink is transferred directly to the substrate, usually with an electrostatic assist.

In the planographic or lithographic process, the image and nonimage areas are on the same plane and the difference between image and nonimage areas is maintained by the physicochemical principle that oil and water do not mix. The image area is oil-receptive and water-repellent; the nonimage area is water-receptive and oil-repellent. Therefore, the ink adheres only to the image areas, from which it is transferred to the surface to be printed, usually by the offset method.

In the stencil or screen printing process, a stencil representing the non-printing areas is applied to a silk, nylon, or stainless-steel fine-mesh screen to which ink having the consistency of paint is applied and transferred to the surface to be printed by scraping with a rubber squeegee.

Direct printing is the transfer of the image directly from the image carrier to the paper. Most letterpress and gravure and all screen printing are done by this method. In indirect or offset printing, the image is transferred from the image carrier to an intermediate rubber-covered blanket cylinder, from which it is transferred to the paper. Because most lithography is printed in this way, lithography is usually referred to as offset printing. Letterpress and gravure can also be printed by the offset method.

Images are defined for these printing processes in a number of different ways. Letterpress uses cast-metal type for printing. The other processes produce images on a support by manual, chemical, mechanical, or increasingly by electronic imaging means. At the end of the 20th century, the greatest number of plates and images are made by photomechanical methods. These systems are characterized by photographic images and light-sensitive coatings that, by using chemical etching or other treatments, lead to the formation of a printing surface. Increasingly, this printing surface is produced directly by electronic imaging without the traditional photographic intermediates. In some cases the final creation of a printing surface is electronic. These processes are termed computer-to-plate or direct-to-press.

A typical workflow involves image creation, capture, assembly, storage, approval, duplication, output, delivery, and distribution.

Image Creation

Photography. Most press-printed illustrations are reproductions of photographic originals. Press-printed photographic images fall into three main photography (qv) categories based on intended use: commercial, editorial, and fine art.

Commercial photography typically illustrates a product or other salable item for a package, advertisement, catalog, or brochure. Commercial originals must be technically excellent and are often retouched to improve or enhance the image, or to remove blemishes. Editorial photography usually illustrates a magazine or newspaper story, thus pictorial content is more important than technical excellence. Fine-art photography is regarded as attractive, collectible, or salable as works of art, and may be press-printed in the form of posters, books, or prints.

Both the camera and the media play a role. The photographer chooses a film and camera, adjusting the camera variables to achieve particular effects and meet the requirements of the project.

Black-and-white photographs are usually reproduced from photographic prints.

Most color photographic images reproduced on press are made from positive color transparencies known as slides.

Choice of camera and camera settings affects the final appearance of the image. Variables include the lens, lens aperture, depth of the image field; and shutter speed all combine to produce the desired artistic effect.

A rapidly emerging type of photography is the direct acquisition of monochrome and red–green–blue (RGB) images electronically using digital cameras.

Creative Processes. Use of computer technology has propelled the printing industry into the digital era. Work processes have been radically transformed, giving the creators more prepress control of the printing process.

Creative processes have four distinct workflow steps: conceptualization, electronic design and initial layout, preparation of graphic and photographic elements, and finalization of electronic mechanical.

During conceptualization, one or more creative solutions are generated. Text, headlines, and graphic elements can be prepared either conventionally or electronically. The creator then begins to develop a tight comprehensive of the design. The first layout is output in either black and white or representative color. In the third step, supporting graphic elements used within the design are created using drawing software applications. In the last step, all text and graphic elements are finalized and placed in position, and an electronic file is prepared for output.

Synthetic Image Creation. Two types of synthetic image creation programs are available: bit map-based and vector-based (object oriented). Ultimately, all digital images are converted into bit-mapped (raster) images for display or output. The distinction between bit-mapped and vector is the form of the image in the creating application or program.

A bit map is a grid pattern composed of tiny cells or picture elements (pixels). Each pixel has two attributes: a location and a value or set of values. Location is defined as the address of the cell in a Cartesian, ie, x and y coordinate, system. Value is defined as the color of the pixel in a specified color system. Users create bit-mapped images by accessing each cell in the bit map and assigning a color value to it. Software programs provide users with the tools to accomplish this. Conceptually, a vector image is not an image at all, but a set of equations defined by a user that directs the way a computer creates or alters a raster image. By storing only the equations and not the pixel information, vector images take up much less space than bit-mapped images. Software tools are used to modify images in just about any way imaginable. The user can make the image lighter or darker, or alter the red, blue, and green channels to change the color in a desired way. Images can also be resized, cropped, or inserted into other images.

Natural images typically originate as photographs and are invariably represented in some type of bit-mapped form.

Natural images and synthetic bit-mapped images can be merged directly to form a single image. A common software tool, for example, allows cutting of images from one file and pasting of the same image into another file.

Creative Tools. Software applications, designed to perform specific functions, have changed the methods of producing a layout from conventional cut-and-paste techniques to electronic ones. Image creators are likely to have available powerful desktop publishing capabilities, an array of sophisticated software applications, and digital monochrome or color printers for digital proofing and review.

Image Capture

Color and Color Separation. Printers use colored materials, eg, inks, that absorb or subtract regions of the visible spectrum from white light. Subtractive color is usually represented by the three printer's primaries: cyan, magenta, and yellow (CMY). Cyan absorbs red light, magenta absorbs green, and yellow absorbs blue light (see COLOR).

Each color of the primary set must be printed sequentially, one on top of the other, building up a color image one color at a time. The basic task of color prepress work is converting a color image into color separations that represent each of the CMY color components of an image. In addition, the color planes must be adjusted for the characteristics of the printing system. Inks used in printing are made from real pigments and must satisfy many requirements, not the least of which is cost. The colors show overlap in their visible spectra, referred to as color contamination. Thus areas that are printed by using equal amounts of CMY, instead of appearing neutral, ie, grey or black, are actually colored. Also, the maximum amount of ink that can be printed plays a role. Ink quantity does not allow for a true deep black.

Some of the color deficiencies can be overcome by using a fourth ink, black, which allows printing neutral tones and dark blacks and colors. Black, usually referred to as K to distinguish it from blue, makes up the fourth member of the printer's primaries, CMYK.

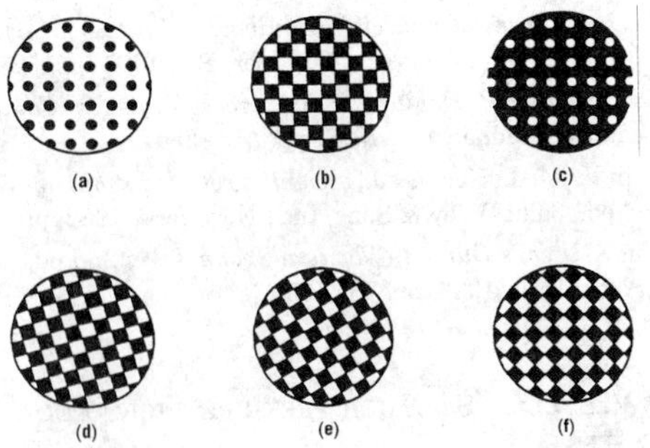

Figure 1. Halftone dots magnified: (**a**), (**b**), and (**c**) are 20, 50, and 80% dots at 0°, respectively; (**d**), (**e**), and (**f**) are 50% dots at 15, 30, and 45°, respectively.

The color-separation process involves taking an image, usually analyzed in the red, green, blue (RGB) primary color space, and converting that image to the CMYK primary color space, while accommodating the deficiencies of the printing process.

Color Scanning. Historically, the process of color separation was done by using a process camera. The color-separation process has been automated through the use of electronic color scanners. These devices convert color images to electronic data, which are used either to image directly monochrome photographic film that represents the cyan, magenta, yellow, and black content, or to provide CMYK or RGB image data files to computer-based pagination and retouching systems.

Process Camera. Although replaced by the digital scanner and image-setter to a large extent, a process camera is still used in many shops to convert originals to film. The camera consists of a movable glass-covered copyboard to hold the original, a movable lens board, and a stationary vacuum back to hold the film. The degree of enlargement or reduction is governed by the lens and the distance from copyboard to lens and lens to film plane.

Image Assembly

Page Creation. *Traditional Methods.* In the preparatory stages of printing, the traditional process starts with one or more originals in the form of written matter, line art (diagrams, drawings), and photographs. These originals are photographed to make films that, in a final step, are used to expose a printing plate. Because most printing processes cannot print continuous-tone photographs directly, these photographs are converted to halftones in which the intermediate tones of the original are represented by solid dots on the film, which are spaced equally but vary in area.

Halftone Screens. Halftone screens are used in cameras to convert a continuous-tone photograph to halftone dots (Fig. 1). Most screens, sometimes called contact screens, are made from film and consist of evenly spaced rows of vignetted dots. The screen ruling, governs the level of detail of the printed halftone. Finer rulings give greater detail.

To create the halftone image, the screen is contacted against the film in the camera. Exposure of the original through this screen creates a pattern of dots on the film that correspond in size to the different tonal areas in the original.

Typesetting. Typesetting or typography is the process of arranging, composing, and placing type onto the printed page. Typesetting has evolved through three important stages: hot metal, phototypesetting, and electronic. The roots of typesetting began in hot metal with Gutenberg, who cast the earliest metal type used in Western civilization.

The basic hot-metal process remained in use from Gutenberg's time to the late 1960s, but over the centuries became automated with the advent of equipment that allowed an operator to enter characters through a typewriter-like keyboard, which caused type to be cast one line at a time. These lines, or slugs, were then placed together to form a page.

Early phototypesetters used an optical process, whereby a disk of characters, in different sizes and typefaces, was spun under computer control. Each character was projected in turn onto photosensitive film or paper. This was followed by systems where characters drawn on a cathode ray tube (CRT) exposed the photosensitive material. In each case, the operator interacted with the system at a video screen that showed only the characters of the text (the information content) and codes that indicated how the characters were to look on paper. An experienced operator was required to obtain high quality results.

Conventional typesetting has been largely superseded by electronic typesetting using the PostScript page description language (Adobe Systems, Inc.). Fonts and typefaces in PostScript, as well as in a similar system, TrueType (Microsoft Corporation), actually exist as mathematical descriptions of the characters. This allows each character to be rendered at arbitrary sizes and resolutions for different output devices. Electronic typographic programs are usually implemented in such a way that the operator sees a close approximation of the final output on a computer screen.

The page layout program, which is used to prepare the text, generates a representation of the printed page, including text in the PostScript language. A PostScript raster image processor (RIP) interprets the PostScript commands and renders the image into a bit map, which can then be output on a printer or imagesetter.

Digital Page Creation. Historically, designers created the page, a layout artist recreated the page to exact specification as a mechanical, and then the prepress house, referred to as a trade shop, created the films for producing the printing plates. Individual page elements, type, and photographic images were transferred to color-separated film by the use of cameras. Strippers in the trade shop gathered the pieces of film for a given page and recomposed them for plate creation, one for each color printing plate that needed to be generated. In the mid-1990s, many shops still used this technique, but usually the film has been electronically generated.

Using color scanners, photographs are digitized directly onto a computer system. These digital images are then positioned in an electronic page layout document that contains type and other page elements, created by the designer or production artist. Finally, the page is imaged to color-separation film or directly to the printing plate.

Image Storage and Movement

Electronic Data Storage. Modern color electronic prepress systems rely on high capacity magnetic hard disks for quick data storage and retrieval (see INFORMATION STORAGE MATERIALS, MAGNETIC).

Compact disk–read only memory (CD-ROM) technology has become a commonly used medium for the distribution of large image files. Magnetic tape is still used for inexpensive long-term bulk storage and backup of files stored on other media (see MAGNETIC MATERIALS).

Networks. Image files are typically transferred among many different stages in the prepress process. This is most often done over a network that connects the various computer work stations used in the process.

The same concepts of a network as the local area network (LAN) in an office can be extended over a broader area, eg, wide area network (WAN), effectively to extend the network around the world.

The physical medium that links the LAN at each end of the network is typically supplied by a telecommunications provider. The user is able to move images among locations as easily as moving them on a personal computer.

Image Compression. Data compression encodes the information contained so that the resulting file is smaller. Two general types of compression are used.

Lossless compression preserves all of the information in a file, so the original file can be reconstructed bit for bit. Lossy methods, on the other hand, do not preserve all the information. The original file cannot be reconstructed bit for bit.

Image Files. The format in which the data is actually stored or transmitted is another part of the process. There are a number of file formats, many of which are proprietary, developed to suit the needs of a particular computer platform or software application.

Image Manipulation

After creating the set of halftones, line art, and type for a given page, the craftsperson may have to manipulate the images to produce the page as designed.

Once films corresponding to all of the separate components of a page, eg, text, line art, and halftone images, have been made in their final forms, they are assembled onto a carrier sheet to create a flat. This stage of image assembly is commonly known as stripping, because this process at one time involved stripping film emulsions from a glass plate. The flat is used to expose the printing plate to form a printable image.

Electronic Film Output. Imagesetters are precision electromechanical devices that image monochromatic color separations onto photographic film or paper. Each separation is used to expose a printing plate for a single ink color. All modern imagesetters use a laser light source to expose the media (see LASERS).

Imagesetter output requires that the data be converted from full-color pictorial information to its individual color components. Additionally, the data must undergo a significant format conversion.

The PostScript page description language was developed Inc. in the 1980s to provide a common format for describing page contents. It has been adopted almost universally by desktop publishing vendors, as well as many of the high end proprietary page makeup systems.

PostScript, an intermediate language, is converted into the appropriate imagesetter data stream by a device called the raster image processor (RIP).

The primary objective of screening is to use area coverage algorithms to represent different shades of color, despite the fact that the imagesetter laser is typically a binary (on/off) device. The biggest difficulty in screening is the prevention of moiré patterns, which are interference patterns that occur when two or more color separations overlap.

Stochastic screening is a way to eliminate problems with moiré. Instead of using a fixed angle for each screen and varying the size of each dot, a random distribution of small, fixed-size dots is used. Different amounts of color are represented by more or fewer dots in a particular region.

Image Approval

For both analogue and digital prepress processes it is frequently necessary to check the appearance of an image and then to gain customer approval to proceed with the expensive step of image duplication by printing. For this purpose, proofs are made at several steps of the workflow.

Analogue Proofing. Analogue proofing can be described as the process of making an image, either monochrome or color, by photomechanical methods. The proof is a representation of the appearance of the final image off-press.

Digital Proofing. In a modern electronic prepress environment, much of the page makeup is done using computer systems. Images separated on a scanner may be stored electronically; images created using an electronic drawing system may exist only as a digital file.

An ideal digital proofer would take digital files and output a hard copy that looks exactly like the printed page, including accurate color to judge the adequacy of the color-separation process. Several digital proofing technologies involve technologies that are also used as printing technologies. They include thermal printing, electrophotographic printing, and ink-jet printing.

Monitors. Cathode ray tube (CRT) monitors may also be used to judge general adequacy of color in a process called soft proofing. The displayed color is also operator-dependent unless a calibration scheme is used to measure the output of the CRT. Soft proofing using a calibrated monitor is accurate for position and color when used by a trained operator.

Image Duplication and Output

The principal processes for making many printed copies from a prepress image include lithography, gravure, flexography, letterpress, and screen processes, as well as the newer technologies of thermal printing, electrophotography, and ink-jet printing (see IMAGING TECHNOLOGY).

Lithography. While lithography is by far the most widely used of the principal printing processes, the mechanism of the process is not well understood. This reflects the complexity of the various interactions of ink, plate, and water that come into play whenever a lithographic plate runs on press.

Lithography is a planographic process. Image, or printing areas, and non-image, or nonprinting areas, reside in the same plane and are differentiated by the extent to which these areas accept printing ink. Nonprinting areas are hydrophilic, accepting water and repelling ink; printing areas are oleophilic, repelling water and accepting ink. During printing, water is applied to the surface of the plate as an aqueous solution of surface-active agents, generally referred to as a fountain solution, and the ink is applied to the plate surface through a roller train.

Offset plates usually comprise a support, most frequently from metal, and one or more layers of a radiation-sensitive composition. The image must be strongly bonded to the support in order for it to resist the powerful abrasion forces which come into play during printing.

Gravure. The gravure printing process, sometimes called intaglio or rotogravure, utilizes a recessed image plate cylinder to transfer the image to the substrate. The plate cylinder can be either chemically or mechanically etched or engraved to generate the image cells. The volume of these cells determines the darkness or lightness of the image. If an area is darker, the cells are larger; if the area is lighter, the cells are smaller.

In gravure, all elements within the image are screened. This is in contrast to flexographic and lithographic plates, which can contain true solids as well as halftones.

The gravure printing process is based around an inking system that is extremely simple, giving the process a high degree of consistency, particularly with regard to color printing. This consistency is difficult to match using other printing techniques. The system utilizes a liquid ink that has traditionally been solvent-based, although environmental pressures have resulted also in the development of aqueous-based inks.

Flexography. Flexography is a variation of letterpress printing used mainly for packaging applications. It is characterized by the use of an elastomeric printing plate, fast drying inks, and an anilox roll ink-metering system.

The heart of the flexographic printing system is the anilox roll, a steel cylinder optionally coated with ceramic and engraved with a pattern of pits or cells. The function of the anilox roll is to meter a uniform film of ink from the ink fountain to the printing plate without the need for continuous adjustment.

There are three primary types of flexographic printing plates: molded rubber, solid-sheet photopolymer, and liquid photopolymer.

Conventional flexography uses low viscosity fast-drying inks.

Letterpress. Letterpress is the oldest and, until the mid-1960s, the dominant and most versatile printing process. By the late 1990s, it was used mainly in high quality design work, fine books, and quality stationery. Unlike lithography, which prints best on coated papers, letterpress can print on any paper.

Letterpress is printed directly by the relief method from cast metal or plates on which the image or printing areas are raised above the nonprinting areas. Ink rollers apply ink to the surface of the raised areas, which transfer it directly to paper.

Screen Printing and Stencil Processes. There are two stencil processes in general use: screen printing and stencil duplicating. Screen printing used for art reproduction is called seriography.

Screen-printing image carriers can be produced manually or by photomechanical means. The screens can consist of silk cloth nylon, or metal. The screen material is attached to a rigid frame and stretched tightly so that it is level and smooth. The stencil is applied to the bottom side of the screen, ie, the side in contact with the surface to be printed. Ink having a consistency similar to thick paint is used in the screen. Ink is transferred by rubbing on the screen surface using a rubber squeegee.

Manual stencils are made by knife-cutting special film stencil materials. These consist of two plastic layers. Manual stencils can also be produced by drawing directly on the screens using special materials.

Photomechanical stencils are of two types: direct coatings and transfer films.

There are four transfer-film methods for making screens: carbon tissue, unsensitized film, presensitized film, and photographic transfer film.

Thermal Printing. Thermal printing is a generic name for methods that mark paper or other media with text and pictures by imagewise heating of special-purpose consumable media. Common technologies are direct thermal; thermal, ie, wax, transfer; and dye-sublimation, ie, diffusion, transfer. Properties and preferred applications are diverse, but apparatus and processes are similar.

Electrophotography. The term electrophotography (qv) describes both the product, ie, an optical copy or photograph of an original, and the key to its process, ie, electrostatics.

The electrophotographic system involves two key physicochemical elements: a photoreceptor and a toner. The minimum steps the process are (*1*) to charge a photoconductive photoreceptor uniformly; (*2*) to illuminate selectively the photoreceptor to form a latent electrostatic image; (*3*) to develop the image by applying charged toner; and (*4*) transfer the toner to paper or other receptor.

Some photoreceptors are used with a positive surface charge; others are used with a negative charge. By electrically biasing the toner development module with respect to the photoreceptor and using toner of the proper charge, toner can be attracted to either charged or discharged portions of the latent image.

Ink-Jet Printing. Ink jet is a digital printing process in which small drops of ink are propelled from a nozzle to a receiving surface without contact between the ink source and the surface. There are two types of ink jets: continuous and impulse. The former generate a continuous stream of ink drops from each nozzle. The size and frequency of drops in a continuous ink-jet printer are determined by pumping pressure, ink viscosity, and nozzle size. Drops not needed for printing are electrostatically charged and deflected into a sump. Impulse printers generate drops only in response to a computer signal.

There are two types of impulse printers. A piezoelectric ink jet propels a drop by flexing one or more walls of the firing chamber to decrease rapidly the volume of the firing chamber. Thermal impulse ink jets also propel one drop at a time, but these use rapid bubble formation to force part of the ink in a firing chamber out the orifice.

Inks for continuous ink-jet printers typically comprise dyes dissolved in water or solvent having salts added to make the ink conductive for electrostatic charging. Whenever waterproof printing is required, low boiling solvent inks are used. For printers that are used in office environments, water is used as the ink solvent.

Finishing. Most printing must be converted from printed press sheets to a finished piece through various finishing and bindery processes. Finishing is a general term that includes a number of different and often specialized operations that transform printed press sheets into their final form. Some of the most common finishing operations include cutting, folding, stitching, collating, binding, scoring, perforating, round cornering, drilling or punching, die-cutting, embossing, laminating, and padding.

Binding. The work required to convert press sheets into finished books, magazines, and catalogs is called binding. Binding includes processes such as saddle and side stitching, perfect binding, mechanical or coil binding, and book binding. All binding methods usually begin by folding a single press sheet printed on both sides down to a signature. Signatures vary from 4–64 pages on a single sheet to maximize press sheet usage, but are usually 16 or 32 pages.

The simplest and least expensive method of binding is the saddle stitch method, in which signatures are collated together and placed on a saddle beneath a mechanical stitching head. Staples are forced through the spine of the book.

Perfect binding or adhesive binding has become a popular alternative to stitched binding. Mechanical or coil bindings are used for notebooks and calendars.

Edition binding or case binding is the most common method used for school textbooks and other hardbound books.

Environmental Aspects

Printing processes generate waste. With the common goal of protecting the environment, printers must balance and carefully analyze all waste minimization, treatment, and recycling programs (see RECYCLING; WASTES, INDUSTRIAL).

Solid and Hazardous Wastes. Among the most stringent environmental regulations impacting the printing industry in the United States are the hazardous waste control standards issued under the Resource Conservation and Recovery Act (RCRA). Printers must segregate hazardous materials on-site and ship them via an approved hazardous waste hauler to a permitted facility. Many solid wastes generated from the printing process are not hazardous and often may be recycled locally, eg, corrugated paperboard, aluminum printing plates, and paper from scrap product.

Employee Safety and Health Issues. Printers, like other industrial employers, must be concerned with employee safety and occupational health regulations and reduction of hazardous chemicals, especially carcinogens, in the workplace.

ARTHUR J. TAGGI
PETER WALKER
Du Pont Printing and Publishing

J. Sturge, V. Walworth, and A. Shepp, *Imaging Processes and Materials*, 8th ed., Van Nostrand Reinhold Co., Inc., New York, 1989.

T. M. Destree, ed., *The Lithographers Manual*, 9th ed., Graphic Arts Technical Foundation, Inc., Pittsburgh, Pa., 1994.

J. M. Adams, D. D. Faux, and L. J. Rieber, *Printing Technology*, 4th ed., Delmar Publishers Inc., Albany, N.Y., 1996.

Pocket Pal: A Graphic Arts Production HandBook, 15th ed., International Paper Co., Memphis, Tenn., 1992.

PROCESS CONTROL

Basic Elements and Equipment

In order to operate a process facility in a safe and efficient manner, it is essential to be able to control the process at a desired state or sequence of states. This goal is usually achieved by implementing control strategies on a broad array of hardware and software. The state of a process is characterized by specific values for a relevant set of variables, eg, temperatures, flows, pressures, compositions, etc. Both external and internal conditions, classified as uncontrollable or controllable, affect the state. Controllable conditions may be further classified as controlled, manipulated, or not controlled.

Process Systems. Because of the large number of variables required to characterize the state, a process is often conceptually broken down into a number of subsystems which may or may not be based on the physical boundaries of equipment. Generally, the definition of a system requires both definition of the system's boundaries, ie, what is part of the system and what is part of the system's surroundings; and knowledge of the interactions between the system and its environment, including other systems and subsystems. The system's state is governed by a set of applicable laws supplemented by empirical relationships. These laws and relationships characterize how the system's state is affected by external and internal conditions. Because conditions vary with time, the control of a process system involves the consideration of the system's transient behavior.

Process systems are broadly categorized as self-regulatory and nonself-regulatory. The former is one in which a change in an external condition can cause the system to move from an initial steady state to another steady state without additional external intervention. The latter, a nonself-regulatory process system, does not achieve another steady state without additional control action once the first external change occurs.

Controlled Conditions, Correcting Conditions, and Control Algorithm. The basic elements of process control are the conceptual definition of the process system; the selection of the controlled conditions, the correcting conditions, and the disturbance sources to be addressed; and the selection of the control algorithm. The goal of process control is achieved by adjusting the values of an appropriate subset of process variables, ie, the correcting conditions or manipulated variables, so as to change the values of other process variables, ie, the controlled conditions or variables, to compensate for variations and disturbances in the process system. The controlled variables are selected so that their values characterize the state of the process system and the process and operating objectives. The manipulated variables are selected so that these can easily be manipulated to affect the controlled variables. The control algorithm defines how the manipulated variables are to be adjusted to bring the controlled variables to their desired values, ie, to bring the process system to its desired state.

Generic Control Strategies. The two generic strategies for process control are feedback and feedforward control. Most process control strategies are based on one or a combination of these strategies.

The conceptual structure of the feedback control strategy is shown in Figure **1a**. A feedback control strategy measures the controlled variable after it has been affected and, therefore, after a deviation between the controlled variable measurement and its desired setpoint value may have occurred. The deviation is compensated for by adjusting the manipulated variable(s). In this strategy, the flow of information is from the output of the process, namely, the measured controlled variable, to the process's manipulated variable inputs. By continuing to measure the output response as well as the manipulated variable compensating changes, the feedback control strategy in essence continues to seek, by trial and error, the values of the manipulated variables to bring the process into balance.

The feedforward cgntrol strategy (Fig. **1b**) measures the disturbance before it affects the output of the process. A model of the process determines the adjustment in the manipulated variable(s) to compensate for the disturbance. The information flow is therefore forward from the disturbances, before the process is affected, to the manipulated variable inputs.

Control Equipment. Process control computers (PCC) to perform direct digital control (DDC) and supervisory process control (SPC) were

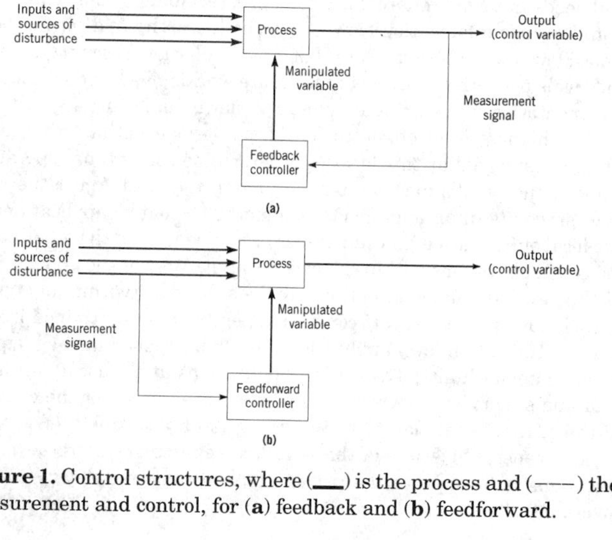

Figure 1. Control structures, where (—) is the process and (----) the measurement and control, for (**a**) feedback and (**b**) feedforward.

introduced during the late 1960s and early 1970s. Regulatory control was provided by analogue controllers, which did not require backup, but the operator's attention was split between the control panel and the computer screens. The terminal displays provided the operator interface when DDC/SPC was being used, but the control panels were still located in the control house for the times when the analogue backup was necessary. Within the PCC environment, the use of advanced control techniques such as feedforward control, model-based predictive control, interaction decoupling control, multilevel cascades, and dead-time compensation was broadened.

Since the early 1980s, there has been a general increase in the use of digital communications technology within process control. Some advanced control strategies, previously only implemented on a PCC, are implemented within the DCS. Most local control units (LCU) perform their own analogue-to-digital (A/D) and digital-to-analogue (D/A) conversion. The LCUs can be located in equipment rooms closer to the process. Digital communications via a coaxial or fiber optic cable send information back to the control room. With this trend toward increased use of digital communications technology, smart transmitters and smart actuators have also gained in popularity (see SMART MATERIALS). These devices, equipped with their own microprocessor, perform tasks such as autoranging, autocalibration, characterization, signal conditioning, and self-diagnosis at the device location.

Initially, programmable logic controllers (PLCs) were dedicated, stand-alone, microprocessor-based devices executing straightforward binary logic for sequencing and interlocks. However, PLCs have become increasingly more powerful in terms of calculational capabilities and integration with other digital hardware and software.

Safety interlocking, shutdown, and sequencing logic is used to protect personnel and equipment from potentially dangerous situations.

There is a trend in process control technology toward increased use of digital technology. Digital communication occurs over a field bus, ie, a coaxial or fiber optic cable, to which intelligent devices are directly connected and transmitted to and from the control room or remote equipment rooms as a digital signal.

Facility Control Hierarchy. The hierarchy of control levels in a process-related plant is outlined in Table 1. Advanced higher level controls are a means to achieve and maintain consistent and improved quality. Advances in the capabilities of DCSs make the development of advanced higher level controls within the DCS feasible. Modern process facilities are often designed with a relatively high degree of process integration in order to minimize the theoretical cost of producing the product. From an operation's standpoint, however, this integration gives rise to relatively complex interactions between the operating variables. Thus, it can be relatively complex to determine how to adjust the plant to optimize the operation.

Each of the four conceptual control levels has its own requirements and needs in terms of hardware, software, techniques, and customization. Because information flows up in the hierarchy and control decisions flow down, effective control at a particular level occurs only if all the levels beneath the level of concern are working well. Thus, good instrumentation is the foundation of good hierarchial control.

The highest level of control is the plant-wide optimization level. The primary goal at this level is to determine the optimal operating point of the plant's mass and energy balance and to adjust the relevant setpoints in an appropriate manner. The control applications at the local optimization and supervisory control level, on the other hand, focus on subsystems within the overall plant. Most of the applications at this level are aimed at optimizing the subsystem within an operating window defined by soft constraints, eg, values determined by the plant optimization level applications, and hard constraints, eg, equipment material limits. The control applications at the local optimization and supervisory level typically provide setpoints for the controls at the advanced regulatory and basic regulatory control levels.

The general objective of the advanced regulatory control level applications is to improve the performance of basic regulatory control level controllers.

Instrumentation

Components of a Control Loop. Instrumentation, which provides the direct interface between the process and the control hierarchy, serves as the fundamental source of information about the process state and the ultimate means by which corrective actions are transmitted to the process. The function of the process measurement device is to sense the value, or changes in value, of process variables. The actual sensing device may generate a physical movement, pressure signal, millivolt signal, etc. The transducer transforms the measurement signal from one physical or chemical quantity to another. The transduced signal is then transmitted to a remote location through the transmission line. Most modern control equipment requires a digital signal for displays and control algorithms, thus the analogue-to-digital converter transforms the transmitter analogue signal to a digital format.

Process Measurements. The most commonly measured process variables are pressures, flows, levels, and temperatures (see FLOW MEASUREMENT; LIQUID LEVEL MEASUREMENT; PRESSURE MEASUREMENT; TEMPERATURE MEASUREMENT). When appropriate, other physical properties, chemical properties, and chemical compositions are also measured. The most common physical–chemical property analyzers include density, viscosity, moisture, heating value, and pH. Spectrographic or chromatographic analyzers are used for compositional analysis. The selection of the proper instrumentation for a particular application is dependent on factors such as the type and nature of the fluid or solid involved; relevant process conditions; rangeability, accuracy, and repeatability required; response time; installed cost; and maintainability and reliability.

Signal Transmission and Conditioning. A wide variety of physical and chemical phenomena is used to measure the many process variables required to characterize the state of a process. Because most processes are operated from a control house, these values must be available there. Hence, the measurements are usually transduced to an electronic form, and then transmitted to the control house or to a remote terminal unit and then to the control house.

Control Valves and Other Final Control Elements. Good control at any hierarchial level requires good performance by the final control elements in the next lower level. Ultimately, the control command will affect the process through the final control elements at the regulatory control level, eg, control valves.

A control valve consists of two principal assemblies, a valve body and an actuator.

The valve body, the portion that contains the process fluid, consists of the internal valve trim, packing, and bonnet. The internal trim determines the relationship between the flow area and the stem position, which is usually proportional to the air signal.

The actuator provides the force to move the stem or rotary shaft in response to changes in the controller output signal.

Devices other than control valves are also used as final control elements, eg, dampers, louvers, feeders, and pumps (qv) with variable speed drives. In these cases, the pump speed adjustment is achieved by adjusting the speed of the prime mover, eg, electric motor and steam turbine, or through the transmission linkage.

Generic Control Techniques and Strategies

Classical Feedback Control. The majority of controllers in a continuous process plant are of the linear feedback controller type. These controllers utilize one or more of three basic modes of control: proportional (P), integral (I), and derivative (D) action. In the days of pneumatic or electrical analogue controllers, these modes were implemented in the controller by hardware devices. These controllers implemented all or parts of a control algorithm. With the advent of digital control devices, the basic control algorithm was implemented as a digital approximation. The proportional, the integral, and the derivative modes which make up the PID algorithm are almost always used in combinations in actual process control situations. The proportional mode computes the magnitude of the output, as proportional to the deviation of the

Table 1. Conceptual Levels in a Process Facility Control Hierarchy

Conceptual control level[a]	Class of plant functions	Operation	Communication time	Data level	Class of disturbances[b]	Frequency: Action	Frequency: Scan
plant-wide optimization		directions for local optimization and supervisory level	hours		low or intermittent frequency[c]	low or event-driven	low
	stewardship			CIP[d]			
local optimization and supervisory		optimum settings on local basis, eg, unit operation	minutes to hours		low to medium frequency	low	medium
	analysis and coordination			PCC[e] simulation and database management			
advanced regulatory		assist in achieving better regulatory control of operations	seconds to minutes		medium to high frequency	medium	medium
	operations			DCS[f]/PCC and other transaction or event-driven system			
basic regulatory		stabilize operations at F, P, L, T[g] setpoints	0.1 s to seconds		high to medium frequency	high	high

[a] Control increases from top to bottom; information from bottom to top.
[b] Amplitudes vary unless otherwise noted.
[c] Amplitude is usually large.
[d] CIP = computer-integrated processing.
[e] PCC = process control computer.
[f] DCS = distributed control system.
[g] F = flow, P = pressure, L = level, and T = temperature.

controlled variable from the setpoint. It acts as soon as a deviation is detected.

The derivative term of the algorithm anticipates future deviations by considering the rate of the deviation change. The greater the rate of deviation change, the stronger the derivative action. The PID controller is a flexible, effective, and reliable controller for the process industries. A considerable range of controller actions is possible by selecting tuning parameters to provide different weights to the present (proportional), the past (integral), and the projected future (derivative).

Cascade Control. Cascade control is a control strategy where the control output of a primary (master) controller is used to set the setpoint of the secondary (slave) controller. Generally, the controlled variable of the primary controller is the primary controlled variable of interest. The secondary controlled variable is selected to recognize the effect of particular disturbances to the process faster than the primary controller, and to compensate before these have had a chance to disturb significantly the primary controlled variable.

Feedforward Control. A feedforward control strategy is one in which a variable characterizing a particular disturbance (the disturbance variable) is measured or computed. This variable is then used in a model to determine the adjustment appropriate for the manipulated variable to mitigate the impact of the disturbance on the controlled variable. In other words, the measurement of a source of disturbance is used to anticipate the need for a corrective control action. Most feedforward control applications are combined with an appropriate feedback controller. The feedback controller (trim) compensates for deviations in the controlled variable.

Ratio and Multiplicative Feedforward Control. In many physical and chemical processes and portions thereof, it is important to maintain a desired ratio between certain input (independent) variables in order to control certain output (dependent) variables.

There are two fundamental ways to implement ratio controls. In one approach the disturbance (or load) variable and the variable to be manipulated to maintain the ratio are measured, the ratio is calculated and compared with the desired value, and, based on the deviation, the manipulated variable is adjusted. In the other approach, the disturbance (or load) variable is measured and multiplied by the value of the desired ratio, and the result is used as a setpoint for the controller for the variable being manipulated.

Selective Control. Selective control strategies can be used to implement overrides, prioritized control actions, and constraint control applications. In selective control, signal select functions, or their equivalent in custom code, pick a signal or group of signals from a larger group of signals, eg, the highest, the lowest, or the median.

Multivariable Control. A situation where the adjustment of one input variable affects the values of a number of output variables is common in processes and process equipment. Using the proper pairing of controlled and manipulated variables, a great majority of the controlled variables can be adequately controlled by using single input-single output (SISO) controllers. However, there are situations where the process and control objectives cannot be met by using SISO controllers alone. Multivariable control strategies can maintain good control performance in these instances.

Multivariable control strategies utilize multiple input–multiple output (MIMO) controllers that group the interacting manipulated and controlled variables as an entity.

Model-Based Control. A number of the control strategies utilize models as part of the strategy, eg, feedforward control and adaptive control. The fundamental concept underlying these model-based predictive control strategies is a feedback control strategy with a feedback signal of some measure of the deviation between the actual controlled variables and the predicted values over some time horizon. Algorithms utilizing upon discrete convolution models are being increasingly used in industry (eg, dynamic matrix control).

Adaptive Control. An adaptive control strategy is one in which the controller characteristics, ie, the algorithm or the control parameters within it, are automatically adjusted for changes in the dynamic characteristics of the process itself. The incentives for an adaptive control strategy generally arise from two factors common in many process plants: (*1*) the process and portions thereof are really nonlinear; and (*2*) the process state, environment, and equipment's performance all vary over time.

Two general classes of adaptive control strategies are programmed adaptation and self-adaptation. If the adaptation occurs primarily as a result of the measurement of a factor that has caused the need for adaptation, the strategy is referred to as programmed adaptation. If, on the other hand, the adaptation that occurs is based on a measure of the controller's performance, and which, therefore, requires a measurement of the controlled variable, then the strategy is referred to as self-adaptive.

Programmed adaptation can be straightforward to implement. At times, only relatively simple logic is required. Self-adaptation, on the other hand, often requires more sophisticated logic and algorithms.

Constraint and Optimizing Control. A process plant is a dynamic, integrated environment where external and internal conditions can cause the optimal operating point for each operating objective to vary from time to time. Therefore, the two higher levels of control, ie, the local and the plant-wide optimization levels (see Table 1), can generate significant credits from following the optimal operating point.

Often, in operating facilities, the optimal operating points at the local optimization level lie along a constraint. When such is the case, a constraint control strategy is a straightforward way to achieve and maintain operation close to the optimal operating point. A constraint control strategy monitors the proximity to relevant constraints, identifies which are active, and incrementally adjusts the manipulated variables to control and maintain operation close to a constrained condition. Constraint control strategies may be generally classified as steady-state or dynamic.

Discrete Logic and Control. Discrete logic and control functions are used to implement overrides, interlocks, sequencing, signal selection, etc, based on inputs that have a discrete number of states.

Batch and Sequence Control. In batch processes the product is made in discrete batches by sequentially performing a number of processing steps on the raw materials and intermediate products. In a batch process, the path followed in state space is often important. Compared to a continuous process, batch process control requires a greater percentage of discrete logic and sequential control than regulatory control loops.

In order to describe what must be done, structural models are commonly used to represent the required batch processing actions, the batch equipment, and the combination of components. The Instrument Society of America has been developing a standard, referred to as ISA-S88, for standarizing batch control concepts, terminology, data structures, configuration, and programming.

JOHN PAUL SAN GIOVANNI
Jockey Hollow Technologies

D. E. Seborg, T. F. Edgar, and D. A. Mellichamp, *Process Dynamics and Control,* John Wiley & Sons, Inc., New York, 1989.

B. G. Liptak, ed., *Instrument Engineer's Handbook,* 3rd ed., Vols. 1 and 2, Chilton Publishing Co., Philadelphia, Pa., 1995.

T. G. Fisher, *Batch Control Systems,* Instrument Society of America, Research Triangle Park, N.C., 1990.

A. B. Corripio, *Tuning of Industrial Control Systems,* Instrument Society of America, Research Triangle Park, N.C., 1990.

PROCESS ENERGY CONSERVATION

The main driving force for increased energy conservation, which continues in times of both rising and falling energy prices, is broadscale technological process. Advances in technology are responsible for the historical rise in energy efficiency of 1–3% per year achieved by process industries. A wide range of big and little steps have contributed to these advances, such as improved gas turbine efficiency, structured packing in distillation, computer control (see PROCESS CONTROL), variable speed drives, computer design tools, and improved catalysts (see CATALYSIS) and synthetic processes for a variety of materials, eg, low density polyethylene (see OLEFIN POLYMERS), acrylonitrile (qv), ammonia (qv), and acetic acid (see ACETIC ACID AND DERIVATIVES).

The second force that has driven increased energy conservation is the trade of capital for energy. This trade is optimized within an existing technology and nets large increases when energy prices rise rapidly compared to capital price.

Energy Balance

The energy balance for the overall process has become a document almost as important as the material balance. The overall energy balance serves as an evergreen framework during design to highlight the areas having greatest potential for improvement. Moreover, this document serves as a tool for plant-operating personnel after start-up, to aid optimization of energy use.

The energy balance should analyze the energy flows by type and amount, ie, present summaries of electricity, fuel gas, steam level, heat rejected to cooling water, etc. It should include realistic loss values for turbine inefficiencies and heat losses through insulation.

Exergy, Lost Work, and Second-Law Analysis. When energy is critically important to process economics, the simple energy balance is sometimes carried into an analysis of lost work. This compares the actual design against the theoretical ideal at each step and defines where the true energy use, or lost work, is occurring. In the discussions herein of reaction, separation, heat exchange, compression, refrigeration, and steam systems, the importance of this concept is illustrated. A few terms are defined below.

Exergy, *E,* is the potential to do work. It is also sometimes called availability or work potential. Thermodynamically, this is the maximum work a stream can deliver by coming into equilibrium with its surroundings: $E = (H - H_0) - T_0(S - S_0)$, where E = the maximum theoretical work potential; H and S = enthalpy and entropy of the stream at its original conditions; H_0 and S_0 = enthalpy and entropy of the same stream at equilibrium with the surroundings; and T_0 = temperature of the surroundings (sink).

Free energy, *G,* is a related thermodynamic property. It is most commonly used to define the condition for equilibrium in a processing step. It is identical to ΔE if the processing step occurs at T_0. $\Delta G - \Delta H - T\Delta S$. Lost work, *LW,* is the irreversible loss in energy that occurs because a process operates with driving forces or mixes material at different temperatures or compositions: $LW = E_{in} - E_{out}$. Second-law analysis looks at the individual components of an overall process to define the causes of lost work. Sometimes it focuses on the efficiency of a step and ratios the theoretical work needed to accomplish a change, eg, a separation, to that actually used.

Reactor Design for Energy Conservation

Maximizing Yield. Often the greatest single contribution to reduced energy cost is increased yield. High yield reduces the amount of material to be pumped, heated, and cooled while also simplifying downstream separation. This says nothing about the indirect energy reduction achieved through reduced raw material use. On average, the chemical industry uses almost as much energy in its raw materials as it does in direct purchases of fuel.

Minimizing Diluent. The case concerning diluent is less clear. A careful balance must be made of the benefits a diluent gives in higher yield against the costs needed for mass handling and separation.

Heat Recovery and Feed Preheating. The objective is to bring the reactants to and from reaction temperature at the least utility cost, and to recover maximum waste heat at maximum temperature. The impact of feed preheating always merits a careful look. In an exothermic reaction, preheated feed permits the reactor to act as a heat pump, ie, to buy low and sell high. The most common example is combustion-air preheating for a furnace.

Batch vs Continuous Reactors. Usually, continuous reactors yield much lower energy use because of increased opportunities for heat interchange. Sometimes the savings are even greater in downstream separation units than in the reaction step itself. Especially for batch reactors, any use of refrigeration to remove heat should be critically reviewed. Batch processes often evolve little from the laboratory-scale glassware setups where refrigeration is a convenience.

Separation

About one-third of the chemical industry's energy is used for separation. A correlation exists between selling price and feed concentration as well as between selling price and product purity.

Relative work is an important consideration when comparing separation techniques. Some leave much of the work undone, as, for example, in crystallization involving an unseparated eutectic mixture.

Distillation. Distillation is by far the most common separation technique because of its inherent advantages. Its phase separation is clean, its equilibrium is closely approached in each stage, and its multistage countercurrent device is relatively easy to build.

Distillation is generally the preferred separation method for feed concentrations of 10–90%, but is probably a poor choice for feed concentrations of less than 1%. Techniques such as adsorption, chemical reaction, and ion exchange (qv) are chiefly used to remove impurity concentrations of $< 1\%$ (see DISTILLATION).

The losses for ΔT are typically far greater than those for reflux beyond the minimum. The economic optimum for temperature differential is usually under 15°C. This is probably the best opportunity for improvement in the practice of distillation. A specific example is the replacement of direct-fired reboilers with steam heat.

Different ways to increase the options include using multieffect distillation, using waste heat for reboil, and recovering energy from the condenser.

Often it is possible to heat the feed with a utility considerably less costly than that used for bottom reboiling. Sometimes the preheating can be directly integrated into the column-heat balance by exchange against the condensing overhead or against the net bottoms from the column.

The penalty for column pressure drop is an increase in temperature differential:

$$\Delta T = \left(\frac{dT}{dP}\right)\Delta P$$

$$\frac{dT}{dP} = \frac{R}{\Delta H}\frac{T^2}{P}$$

penalty associated with this ΔT is approximately defined by the following ratio.

$$\frac{\Delta T_{\text{pressuredrop}}}{T_{\text{reboiler}} - T_{\text{condenser}}} = \text{fraction of work potential for } \Delta P$$

A powerful technique for cutting ΔP is the use of packing. Conventional packings can achieve a factor of four reductions over trays, and structured packing can achieve a factor of 10 reductions. However, structured packing is more vulnerable to mistakes in detailed engineering and much less tolerant of fouling than trays.

Relative volatility increases as pressure drops. For some systems, a 1% drop in absolute pressure cuts the required reflux by 0.5%. Again, if operating at reduced pressure looks promising, the process can be evaluated by simulation.

Other Separation Techniques. Under some circumstances, distillation is not the best method of separation. A variety of other techniques may be more applicable.

Reaction Purification by reaction is relatively common when concentrations are low (ppm) and a high energy but low value molecule is present. Some examples are the hydrogenation of acetylene and the oxidation of waste hydrocarbons:

$$C_2H_2 + H_2 \rightarrow C_2H_4$$

$$\text{waste hydrocarbon} + O_2 \rightarrow H_2O + CO_2$$

Absorption. As a separation technique, absorption (qv), also called extractive distillation shares most of the advantages of distillation. Additionally, because it separates by molecular type, it can be tailored to obtain a high relative volatility (see DISTILLATION, AZEOTROPIC AND EXTRACTIVE).

Extraction. The advantage of extraction is that a liquid is purified rather than a vapor, allowing operation at lower temperatures and the removal of a series of similar molecules at the same time, even though these molecules differ widely in boiling point (see EXTRACTION, LIQUID–LIQUID).

The disadvantage of extraction relative to extractive distillation is the greater difficulty of getting high efficiency countercurrent processing.

Adsorption. Adsorbents can achieve even more finely tuned selectivity than extraction. The most common application is the fixed bed with thermal regeneration. An example is gas drying (see ADSORPTION).

Another approach is the simulated moving-bed system, which has large-volume applications in normal-paraffin separation and *para*-xylene separation. (see ADSORPTION, LIQUID SEPARATION).

Melt Crystallization. Crystallization (qv) from a melt is inherently more attractive than distillation because the heat of fusion is much lower than that of evaporation. However, it is rarely used because of various disadvantages, particularly the difficulty of physical separation and the requirement of a second separating process for eutectic mixture.

Membranes. Liquid separation via membranes, ie, reverse osmosis, is used in production of pure water from seawater.

Membranes are also used to separate gases (see MEMBRANE TECHNOLOGY).

Heat Exchange

Heat exchangers use energy two ways: as frictional pressure drop, and as the loss in ability to do work when heat is degraded.

$$\text{lost work} = QT_{\text{sink}}\left(\frac{1}{T_{\text{cold}}} - \frac{1}{T_{\text{hot}}}\right) + \text{frictional work for } \Delta P$$

In an optimized system, the lifetime value of the lost work associated with ΔT typically exceeds the cost of the heat exchanger. The lifetime value of the ΔP lost work in an optimized system is typically one-third as great as the heat exchanger capital. This means that when the costs for pumping power to overcome the heat exchanger ΔP (for the lifetime of the heat exchanger) are discounted to the time of heat exchanger purchase, their sum approximates one-third the heat exchanger cost. (This assumes a large heat exchanger designed for optimum pressure drop.)

The selection of design numbers for ΔP and ΔT is frequently the most important decision the process designer makes.

Heat-Exchange Networks. A basic theme of energy conservation is to look at a process broadly, ie, to look at how best to combine process elements. The heat-exchange network analysis can be a useful part of this optimization (see HEAT-EXCHANGE TECHNOLOGY).

Network analysis, or pinch technology, has become an increasingly powerful approach to process design that includes most of the virtues of second-law analysis.

ΔT and ΔP Optimization. Ideally, ΔT and ΔP are optimized by trying several values, making preliminary designs, and finding the point where savings in utility costs just balance the incremental surface costs.

There are several cases of high importance of which the waste-heat boiler, in which only one fluid involves sensible heat transfer, is the simplest.

Waste-Heat Boiler. In a waste-heat boiler, the optimum occurs when

$$\Delta T_{\text{approach}} = \frac{K_l}{K_v}\frac{1.33}{U}$$

where K_l = annualized cost per unit of surface, \$/(m^2·yr); K_v = annualized cost per unit of utility saved, \$/(W·yr); and U = heat-transfer coefficient, W/(m^2·K).

Optimum Pressure Drop. For most heat exchangers there is an optimum pressure drop. This results from the balance of capital costs against the pumping (or compression) costs. The total cost curve is fairly flat within ±50% of the optimum, but the incremental costs of power are roughly one third of those for capital on an annualized basis. This simple relationship can be extremely useful in quick design checks.

However, the best approach is to have a computer program check a series of pressure drops and see how energy requirements decrease as surface increases.

Fired Heaters. The fired heater is first a reactor and second a heat exchanger. Often, in reality, it is a network of heat exchangers.

Use of unpreheated air in the combustion step is probably the biggest waste of thermodynamic potential in industry (Table 1).

Advantages of air preheating include a direct cut in fuel consumed and an increase in the heat-input capability of the firebox.

The most common type of air preheater on new units is the rotating wheel. On retrofits, heat pipes or hot-water loops are often more cost-effective because of ductwork costs or space limits.

Limitations in the material of construction make it difficult to use the high temperature potential of fuel fully. This restriction has led to the insertion of gas turbines into power generation steam cycles and even to the use of gas turbines in preheating air for ethylene-cracking furnaces.

Improved efficiency in fired heaters has tended to focus on heat lost with the stack gases, but other losses can be much bigger when viewed from a lost-work perspective. For example, a reformer lost-work analysis is shown in Table 1.

Losses for ΔT in the convection section are almost twice those for the hot exit flue gas. These losses can be cut by adding surface to the convection section and shifting load from the radiant section, as well as by looking at the overall process (including steam generation) for streams to match the cooling curve of the flue gases.

Dryers. A drying operation can be viewed as both a separation and a heat-exchange step. When it is seen as a separation, the obvious perspective is to cut down the required work, accomplished by mechanically squeezing out the water.

When the dryer is seen as a heat exchanger, the obvious perspective is to cut down on the enthalpy of the air purged with the evaporated water, achieved by using the minimum amount of air and cooling as low as possible.

Optimum Design of Pumping, Compression, and Vacuum Systems

Pumping. In an optimized system, the annualized cost for pumping power should be one-seventh the annualized cost of piping. (see PIPING SYSTEMS).

Compression. The work of compression is typically compared against the isentropic–adiabatic case, where η is the efficiency.

$$\eta_{\text{comp}} = \frac{W_{\min}}{E_{\text{out}} - E_{\text{in}}}$$

Efficiencies should always exceed 0.6, and 1.00 is approachable in reciprocating devices.

Table 1. Lost-Work Analysis for a Fired Heater

Parameter	Lost-work potential, %
combustion step	54
radiant section ΔT	7
convection section ΔT	24
stack losses (exit temp 225°C)	13
wall losses	2

Thermocompressors. A thermocompressor is a single-stage jet using a high pressure gas stream to supply the work of compression. One application is in boosting waste-heat-generated steam to a useful level. Thermocompressors can also be used to boost a waste combustible gas into a fuel system by using high pressure natural gas. The net efficiency for such a device can be respectable (25–30%).

Vacuum Systems. The most common vacuum system is the vacuum jet. Because of the high ratio of motive pressure to suction pressure, the efficiency of vacuum systems is generally only 10–20% (see VACUUM TECHNOLOGY).

Because of the low efficiency of steam-ejector vacuum systems, there is a range of vacuum above 13 kPa (100 mm Hg) where mechanical vacuum pumps are usually more economical. As pressure falls, the capital cost of the vacuum pump rises more swiftly than the energy cost of the steam ejector, which increases as $(1/P)^{0.3}$. Usually below 1.3 kPa (10 mm Hg), the steam ejector is more cost-effective.

Refrigeration

Refrigeration is a high value utility (see REFRIGERATION AND REFRIGERANTS). The value of heat in a hot stream is the amount of work it can surrender and the value of refrigeration is the work required to heat-pump it to the sink temperature.

Because of its value, refrigeration justifies thicker insulation and lower ΔT values in heat exchange.

Steam and Condensate Systems

Many process plants employ accounting systems where all steam carries the same price regardless of temperature or pressure. This may be appropriate in a polymer or textile unit where there is no special use for the high temperature; it is wrong in a petrochemical plant.

The best system is one that relates the value of steam to that at the generation pressure by its work potential (exergy content).

$$\frac{\text{value at pressure}}{\text{value at highest generation pressure}} = \frac{\text{exergy at pressure}}{\text{exergy at highest generation pressure}}$$

Cooling-Water Systems

Cooling water is a costly utility, it can cost one-fifth as much as the primary fuel. Heat exchangers should be designed to use the available pressure drop. A heat exchanger that is designed for 10 kPa (1.45 psi) when 250 kPa (36 psi) is available will have five times the design flow.

Other energy considerations for cooling towers include the use of two-speed or variable-speed drives on cooling-tower fans, and proper cooling-water chemistry to prevent fouling in users (see WATER, INDUSTRIAL WATER TREATMENT).

Special Systems

Heat Pumps. A heat pump is a refrigeration system that raises heat to a useful level. Its application hinges primarily on low cost power relative to the alternative heating media.

Energy Management Systems. The reduction in computing costs has made it possible to do a wide range of routine monitoring and controlling. For example, a distillation system can be monitored continuously, the energy use can be compared against an optimum, and the cost-per-hour deviation from the optimum setpoint can be displayed.

Existing Plants

Good design ideas for new plants are also good for existing plants, but there are three basic differences. (*1*) Because a plant already exists, the capital–operating cost curve differs. (*2*) The real economic justification for change is more likely to be obscured by the plant accounting system and other nontechnical inputs. (*3*) The real process needs are measurable and better defined.

Energy Audit

The energy audit has seven components: as-it-is balance, field survey, equipment tests, checking against optimum design, idea-generation meeting, evaluation, and follow-up.

The as-it-is balance a mandatory first step; it permits the targeting of major potential savings.

The field survey is often done by a team of two: one who knows what to look for and one who knows the process.

Equipment tests should be rigorous, but can be done with relatively simple measurements. For fired heaters, stack temperature and excess O_2 in stack should be measured; for turbines, pressures (in and out) and temperatures (in and out).

Checking against optimum design attempts to answer the question whether a balance needs to be as it is. The first thing to compare against is the best current practice, and the second step is to look for obvious violations of good practice.

The evaluation of each idea should include a technical description as well as its economic impact and technical risk. Solutions to an energy waste problem must fit into the over plant-energy balance.

DAN STEINMEYER
Monsanto Company

D. E. Steinmeyer, *Chemtech,* **12**(3), 188 (1982).

B. Linnhoff, *Chem. Eng. Prog.* **90**(8), 33 (1994).

R. Smith, *Chemical Process Design,* McGraw-Hill Book Co., Inc., New York, 1995.

PROCESS RESEARCH AND DEVELOPMENT.

See RESEARCH/TECHNOLOGY MANAGEMENT.

PRODUCT LIABILITY

In the early 1960s, the law of American products liability underwent significant change. Until that time, the ability of claimants to recover for personal injury resulting from defective and dangerous products was severely restricted by doctrines that either barred or limited the prosecution of successful tort actions. Since the early 1960s until the mid-1990s, the most dynamic and explosive area of tort law has been the field of products liability. However, even the most dynamic areas of the law tend, after a period of expansion, to settle down doctrinally. Thus, though products liability cases continue to be brought across a broad range of products, it has become increasingly clear since the early 1980s that the period of doctrinal expansion that eased recovery for claimants has slowed down considerably. Courts throughout the United States have entered a period of consolidation.

Historical Overview

Prior to 1960, an injured plaintiff seeking recovery in a products liability action could bring a case under either of two theories. The plaintiff could allege that the product seller was negligent, or that the seller breached a warranty that attended every product sale provided for under the Uniform Sales Act and, later, under Section 2-314 of the Uniform Commercial Code (UCC), which stipulated that for a product to be merchantable it must be "reasonably fit for the ordinary purposes for which such goods are used." Each of these theories came with a distinct disadvantage.

Negligence. Early in the development of the law of negligence-based liability for defective products, the courts almost universally held that the negligent supplier of a defective product could be held liable only to an injured person with whom the supplier had directly contracted. This rule, which limited a product supplier's tort liability, is generally referred to as the privity rule.

Although the privity rule generally came to be recognized by the American courts by the end of the nineteenth century, a number of exceptions were developed judicially.

In MacPherson vs Buick Motor Company, Justice Cardozo effectively eliminated privity as an obstacle to recovery against negligent manufacturers.

The MacPherson ruling abolishing privity in negligence actions was widely accepted, and in the 1990s was the law in all jurisdictions. Though privity was abolished and a plaintiff was free to sue a manufacturer directly for negligence, a substantial obstacle to recovery remained. A plaintiff was still required to prove that the defendant was negligent in bringing about the defective product. In order to assist plaintiffs in establishing their cases, the court increasingly relied on the doctrine of *res ipsa loquitur.* The evidentiary doctrine of *res ipsa loquitur* (the thing speaks for itself) allows an inference of negligence to be drawn from the occurrence of an accident involving an instrumentality (in a products liability case, the product itself) within the defendant's control, under circumstances where such an accident would not ordinarily occur in the absence of negligence. Thus, from the fact that a defective product failed in use and caused an accident, courts allowed triers of fact to infer that the manufacturer of the product negligently caused the defect to occur.

The abolition of privity opened manufacturers to liability for negligence. Plaintiffs, however, could not establish claims merely by proving that they were harmed by defective products from a manufacturer. The requirement that classic fault be established often stood as a formal barrier to a successful tort action.

Implied Warranty of Merchantability. An alternative method of recovery that required the plaintiff to establish only defect (no fault) was available. The plaintiff could bring an action for breach of the implied warranty of merchantability. Under the Uniform Sales Act and later under the UCC Section 2-314, a warranty accompanied every sale of goods (unless disclaimed) stating that the product is "reasonably fit for the ordinary purposes for which such goods are used." Defective products fail to meet the statutory definition and provide a predicate for a cause of action. The implied warranty of merchantability can be characterized as strict liability in contract. Disadvantages of the Code warranty were quite substantial, however.

Combining Tort and Contract Advantages. Two methods were available to allow plaintiffs an easier road to recovery. Courts either stripped the tort action of the necessity for establishing fault, or interpreted the UCC in such a way that privity was not necessary and the other Code defenses were not applicable to cases involving personal injury or property damage. Either way a manufacturer would be open to direct suit without the need to prove fault.

The Second Restatement

All discussion of the post-1965 era must begin with the Restatement (Second) of Torts, Section 402A. The person responsible for this section was Dean William Prosser, who became the reporter for American Law Institute's Restatement (Second) of Torts at a time when change was on the horizon in the field of products liability. His initiative in drafting Section 402A provided the courts with a ready-made formulation for the adoption of strict tort liability.

Prima Facie Case and Affirmative Defenses. The imposition of strict tort liability did not mean that a plaintiff was entitled to automatic recovery when injured by a product.

A plaintiff must establish that the product which injured him was defective. A product can be defective in three ways: it may be defectively manufactured, defectively designed, or sold with inadequate warnings.

Manufacturing defects are those that arise during the production process of the product. Quality control should eliminate most manufacturing defects from reaching the market (see QUALITY ASSURANCE).

Unlike manufacturing defects that are idiosyncratic and random, design defects arise on the drawing board and are generic. If the product is defectively designed, every unit that is in the marketplace suffers from the same defect.

A product may also be defective because it was sold with inadequate warnings.

It is not sufficient that a product is found to be defective. In order to be successful in a products liability action, a plaintiff must establish that the product defect was causally related to the harm the plaintiff suffered. If the selfsame harm would have occurred even if the product had not been defective, then plaintiff's harm cannot be linked to the product defect and recovery is barred.

Even when there is a causal connection between the defect and the harm, it is still necessary for a plaintiff to establish that the nature of the plaintiff's harm, together with the circumstances of its occurrence, were reasonably foreseeable. The law, in general, does not protect against remote unforeseeable risks.

Until the early 1970s, the majority rule in the United States was that if a plaintiff's fault was a contributing cause to the injury, recovery was barred. Since that time a revolution of sorts has taken place. Almost all states follow a rule that reduces a plaintiff's recovery by the percentage of fault attributed to the plaintiff. Thus juries are asked to compare the conduct of the plaintiff and the defendant and then to assign a percentage of fault to each. Plaintiff's recovery is then reduced by a given percentage. Many jurisdictions take the position that if the fault assigned to the plaintiff is over 50% then recovery is barred.

The Third Restatement

New Definitions of Defect. The concept espoused by Section 402A of the Restatement (Second) of Torts, namely, that a manufacturer was strictly liable for selling a defective product and that proof of fault was unnecessary, worked well with regard to products that contained manufacturing defects.

If the plaintiff can demonstrate that a particular product unit came off the assembly line with a manufacturing defect which made the product significantly more dangerous than other similar units, then the conclusion can be drawn that the defect rendered the product legally unacceptable.

In examining whether a design was defective, most courts found it necessary to revert back to a test analogous to negligence. The question that had to be confronted was whether the manufacturer had acted reasonably in designing the product.

In June 1992, the American Law Institute undertook the task of drafting the Restatement (Third) of Torts: "Products Liability." Tentative Draft No. 2 takes the position that different liability rules must apply for manufacturing defects and defects based on inadequate design or failure to warn.

A large number of U.S. states are in agreement with this new formulation. In cases where the claim is based on design defect, the need for plaintiff to establish the availability of a reasonable alternative design gives content to the liability rule. In cases based on inadequate warning, the rule requiring the plaintiff to establish that the harm could have been avoided by the provision of a reasonable instruction or warning is not an easy one to administer. Careful balance is necessary to make certain that claims of inadequate warnings are reality-based and will actually render products safer.

The Role of Foreseeability. In cases based on design defect and failure to warn, defect is established by showing the presence of foreseeable risks of harm which warranted either a better design or a warning. The requirement that foreseeability be established is at odds with the concept of strict liability. However, a overwhelming majority of courts has refused to impose liability for risks that were unforeseeable at the time of product sale.

Defendant Identification. It is normally incumbent on a plaintiff to establish which defendant's product caused harm. In a small class of toxic tort cases, courts have relaxed this requirement. A problem developed in the 1980s in connection with which plaintiffs had great difficulties identifying the defendant that caused the relevant harm. The problem centered on the distribution and sale of the drug diethylstilbesterol (DES) from 1941 to 1971. DES was prescribed to help prevent miscarriage. Daughters of mothers who consumed DES during pregnancy claimed that as a result they suffered various forms of cancerous and precancerous conditions and required surgery to prevent the spread of the disease.

DES was a generic drug without any clearly identifiable shape, color, or markings, and was produced with a single chemical formula. Over three hundred companies at one time or another produced or marketed DES. Because of the long latency period between exposure and injury, it was often impossible to locate records with which to identify which manufacturers' drugs were dispensed to the mothers of particular plaintiffs. Several leading courts recognized the exceptions to the normal rule requiring the plaintiff to bear the burden of establishing the defendant's identity. Instead, they imposed liability in proportion to the defendant's share of the relevant DES market.

Other courts addressing this problem have refused to adopt a rule of proportional recovery. Some have asserted that such a fundamental change of a basic tort principle is more appropriately a legislative function.

After a period of turbulence, the field of products liability began to settle down in the 1990s. The most explosive aspects of the field, ie, litigation based on design defect and failure to warn, are recognized to be closely analogous to traditional rules of negligence. Liability will attach only if there was a reasonable alternative design or warning that would have prevented the plaintiff's harm. Over the years numerous bills have been introduced in Congress to federalize aspects of products liability law. However, the law of products liability is part of the great common law tradition administered by state courts throughout the United States. It is likely to remain so for the foreseeable future.

AARON TWERSKI
Brooklyn Law School

A. W. Weinstein and co-workers, *Products Liability and the Reasonably Safe Product,* John Wiley & Sons, Inc., New York, 1978.

PROLINE. See AMINO ACIDS.

PROMETHIUM. See RARE-EARTH ELEMENTS.

PROPELLANTS. See EXPLOSIVES AND PROPELLANTS.

PROPYL ALCOHOLS

ISOPROPYL ALCOHOLS

Isopropyl alcohol, $CH_3CHOHCH_3$, also known as isopropanol, 2-propanol, dimethyl-carbinol, and *sec*-propyl alcohol, is a colorless, volatile, and flammable liquid. It is the lowest member of the class of secondary alcohols, and is generally known as the first petrochemical. Isopropyl alcohol is used for the manufacture of agricultural chemicals, pharmaceuticals, process catalysts, and solvents.

Physical Properties

Physical properties of isopropyl alcohol are characteristic of polar compounds because of the presence of the polar hydroxyl, –OH, group. Isopropyl alcohol is completely miscible in water and readily soluble in a number of common organic solvents such as acids, esters, and ketones. It has solubility properties similar to those of ethyl alcohol (qv). Isopropyl alcohol has a slight, pleasant odor resembling a mixture of ethyl alcohol and acetone; but unlike ethyl alcohol, isopropyl alcohol has a bitter, unpotable taste.

Physical and chemical properties of isopropyl alcohol reflect its secondary hydroxyl functionality. For example, its boiling and flash

Table 1. Physical Properties of Isopropyl Alcohol

Property	Anhydrous	91 Vol %
molecular weight	60.10	
boiling point, at 101.3 kPa[a], °C	82.3	80.4
freezing point, °C	−88.5	−50.0
specific gravity, 20/20	0.7864	0.8183
density, at 20°C, g/cm^3	0.7854	0.8173
surface tension, at 20°C, mN/m(= dyn/cm)	21.32	21.40[b]
specific heat, liquid at 20°C, J/(kg·K)[c]	2510.4	
refractive index, n_D^{20}	1.3772	1.3769
heat of combustion, at 25°C, kJ/mol[c]	2005.8	
latent heat of vaporization, at 101.3 kPa[a], kJ/mol[c]	39.8	
vapor pressure, at 20°C, kPa[a]	4.4	4.5
critical temperature, °C	235.2	
critical pressure, at 20°C, kPa[a]	4764	
viscosity, mPa·s(= cP) at 0°C	4.6	
20°C	2.4	2.1[b]
40°C	1.4	
flammability limit in air, vol %		
lower	2.5	
upper	12	
flash point, °C		
Tag open cup	17.2	21.7
closed cup	11.7	18.3

[a] To convert kPa to mm Hg, multiply by 7.50. [b] At 25°C. [c] To convert J to cal, divide by 4.184.

points are lower than n-propyl alcohol, whereas its vapor pressure and freezing point are significantly higher. Isopropyl alcohol boils only 4°C higher than ethyl alcohol.

Anhydrous and 91 vol% alcohol, the two main grades of isopropyl alcohol marketed in the United States, differ mainly in water content. The latter represents the azeotrope with water and is usually referred to as constant boiling mixture (CBM) isopropyl alcohol. Some important physical constants of anhydrous and CBM isopropyl alcohol are given in Table 1. Because of its tendency to associate in solution, isopropyl alcohol forms azeotropes with compounds from a variety of classes, including hydrocarbons, esters, halocarbons, amines, ketones, and aromatics.

Chemical Properties

Except for the production of acetone, most isopropyl alcohol chemistry involves the introduction of the isopropyl or isopropoxy group into other organic molecules by the breaking of the C—OH or the O—H bond in the isopropyl alcohol molecule.

Isopropyl alcohol undergoes reactions typical of an active secondary alcohol. It can be dehydrogenated, oxidized, esterified, etherified, aminated, halogenated, or otherwise modified at the OH moiety more readily than primary alcohols such as n-propyl or ethyl alcohol. Manufacture of the commercially important aluminum isopropoxide and isopropyl halides illustrates this reactivity. The aluminum isopropoxide reaction involves the aluminum replacement of the hydroxyl hydrogen atom and concomitant hydrogen evolution; the isopropyl halides reaction involves hydroxyl group displacement (see ALKOXIDES, METAL).

Dehydrogenation. Before the large-scale availability of acetone as a co-product of phenol in some processes, dehydrogenation of isopropyl alcohol to acetone was the most widely practiced production method.

Oxidation. Isopropyl alcohol can be catalytically oxidized using air or oxygen at high temperatures to give acetone and water.

Isopropyl alcohol can be partially oxidized by a noncatalytic, liquid-phase process at low temperatures and pressure to produce hydrogen peroxide and acetone. This process is normally employed when hydrogen peroxide is desired, in which case acetone and unreacted isopropyl alcohol are recycled.

Isopropyl alcohol can be oxidized by reaction of an α, β-unsaturated aldehyde or ketone at high temperature over metal oxide catalysts.

Esterification. Isopropyl alcohol is esterified readily by treatment of carboxylic acids in the presence of an acidic catalyst, eg, p-toluenesulfonic acid. An equilibrium is established in the reaction typically carried out at 100–160°C and atmospheric pressure, using an excess of alcohol and in presence of acid: $RCOOH + (CH_3)_2CHOH \rightleftarrows RCOOCH(CH_3)_2 + H_2O$.

Energy is supplied to remove the water as an azeotrope, thus forcing the reaction in the desired direction. Excess alcohol is distilled and recycled, and yields of ester are nearly quantitative. Esterification of isopropyl alcohol with myristic acid forms isopropyl myristate, an emollient and lubricant in various cosmetic products and topical medicinals (see COSMETICS).

Xanthate esters are prepared by reaction of isopropyl alcohol and carbon disulfide.

Phosphite esters are formed readily by the reaction of phosphorus halides and isopropyl alcohol.

Similarly, another important esterification reaction of isopropyl alcohol involves the production of tetraisopropyl titanate, a commercial polymerization catalyst, from titanium tetrachloride and isopropyl alcohol.

Isopropyl nitrate can be prepared by the reaction of isopropyl alcohol with nitric acid: $(CH_3)_2CHOH + HNO_3 \rightarrow (CH_3)CHONO_2 + H_2O$. Isopropyl nitrate is a valuable engine-starter fuel and can be used in explosives (see EXPLOSIVES AND PROPELLANTS). The nitrite ester, isopropyl nitrite, is used as a jet engine propellant.

Etherification. Isopropyl alcohol can be dehydrated in either the liquid phase over acidic catalysts, eg, sulfuric acid, or in the vapor phase over acidic aluminas to give diisopropyl ether (DIPE) and propylene. Either product can be favored over the other by proper selection of catalyst and reaction conditions. However, the principal source of DIPE is as a by-product from isopropyl alcohol production.

The 1990 Clean Air Act mandates for blended oxygenates in gasoline created a potentially large new use for DIPE as a fuel oxygenate. Isopropyl alcohol can react with propylene over acidic ion-exchange (qv) catalysts at low temperatures, which favor high equilibrium conversions per pass to produce DIPE.

Glycol ethers can be prepared from isopropyl alcohol by reaction of olefin oxides, eg, ethylene oxide or propylene oxide. Reactions such as that to produce 2-isoproxyethanol (isopropyl Cello-solve) are generally catalyzed by an alkali hydroxide.

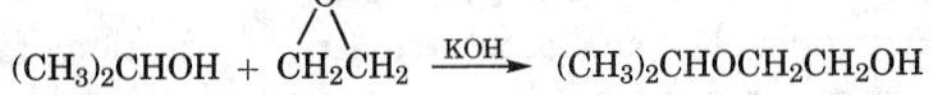

Higher alkoxylated products, ie, oligomers, are formed by secondary reaction of oxide and the hydroxy group of the previous product.

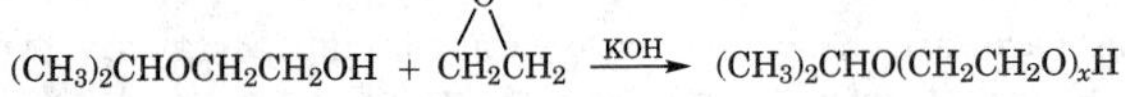

Glycol ethers are used as solvents in lacquers, enamels, and water-borne coatings to improve gloss and flow.

Amination. Isopropyl alcohol can be aminated by either ammonolysis in the presence of dehydration catalysts or reductive ammonolysis using hydrogenation catalysts. Either method produces two amines: isopropylamine and diisopropylamine.

Isopropylamine is the most widely used of the propylamines. Most of it is consumed in herbicide manufacture, primarily in production of 2-chloro-4-ethyl-6-isopropylamino-sym-triazine. A smaller quantity is used for pesticide manufacture. Diisopropylamine is used chiefly in pesticides (qv) and as a corrosion inhibitor, eg, diisopropylammonium nitrate (see CORROSION AND CORROSION CONTROL).

Halogenation. Normally, 2-halopropane derivatives are prepared from isopropyl alcohol most economically by reaction with the corresponding acid halide. However, under appropriate conditions, other reagents, eg, phosphorus halides and elemental halogen, also react by replacement of the hydroxyl group to give the halide: $3\ (CH_3)_2CHOH + PBr_3 \rightarrow 3\ (CH_3)_2CHBr + H_3PO_3$. Halogenated 2-propanol derivatives, eg, 1,3-dichloro-2-propanol, are generally prepared from glycerol. These materials are used in the preparation of halogen-containing phosphates to plasticize and lower the flammability of plastics, eg, polyurethanes and cellulosics.

Manufacture

The indirect hydration, also called the sulfuric acid process, practiced mainly by three U.S. domestic producers, was the only process used worldwide until ICI started up the first commercial direct hydration process in 1951. Both processes use propylene and water as raw materials. In the indirect hydration process, C_3-feedstock streams from refinery off-gases containing only 40–60 wt % propylene are often used in the United States.

Indirect Hydration. Indirect hydration is based on a two-step reaction of propylene and sulfuric acid. In the first step, mixed sulfate esters, primarily isopropyl hydrogen sulfate, but also diisopropyl sulfate, form. These are then hydrolyzed, forming the alcohol and sulfuric acid.

Diisopropyl ether is the principal by-product formed by reaction of the intermediate sulfate esters with isopropyl alcohol.

Other by-products include acetone, carbonaceous material, and polymers of propylene. Minor contaminants arise from impurities in the feed.

In addition to generating malodorous sulfur dioxide, the acetone formed can undergo further condensation in the acidic medium to generate mesityl oxide $(CH_3)_2C{=}CHCOCH_3$, and higher products.

In a typical indirect hydration process propylene reacts with sulfuric acid (> 60 wt%) in agitated reactors or absorbers at moderate (0.7–2.8 MPa (100–400 psig)) pressure. The isopropyl sulfate esters form are maintained in the liquid state at 20–80°C. The reaction is exothermic and internal cooling coils or external heat exchangers are used to control the temperature.

There are two general operational modes for conducting the reaction. In the two-step strong acid process, separate reactors are used for the propylene absorption and sulfate ester hydrolysis stages. The weak acid process is conducted in a single stage at low acid (60–80 wt %) concentration and at higher (2.5 MPa (350 psig)) pressure and (60–65°C) temperature. Chemical selectivity to isopropyl alcohol and diisopropyl ether are above 98% for each process.

Direct Hydration. The acid-catalyzed direct hydration of propylene is exothermic and resembles the preparation of ethyl alcohol from ethylene.

$$CH_3CH{=}CH_2 + H_2O \overset{\text{catalyst}}{\rightleftarrows} (CH_3)CHOH\ \Delta_H = -50\ \text{kJ/mol}(-12\ \text{kcal/mol})$$

There are three basic processes in commercial operation: (*1*) vapor-phase hydration over a fixed-bed catalyst of supported phosphoric acid (Veba-Chemie) or silica-supported tungsten oxide with zinc oxide promoter (ICI); (*2*) mixed vapor—liquid-phase hydration at low (150°C) temperature and high (10.13 MPa (100 atm)) pressure using a strongly acidic cation-exchange resin catalyst (Deutsche Texaco AG) and (*3*) liquid-phase hydration at high (270°C) temperature and high (20.3 MPa (200 atm)) pressure in the presence of a soluble tungsten catalyst (Tokuyama Soda).

The principal difference between the direct and indirect processes is the much higher pressures needed to react propylene directly with water. Products and by-products are also similar, and refining systems are essentially the same.

Storage and Shipping

The U.S. domestic shipping name of isopropyl alcohol is UN No. 1219 Isopropanol. Anhydrous as well as water solutions to 91 vol % alcohol are considered flammable liquid materials by the DOT.

Analytical and Test Methods

Purity of commercial aqueous isopropyl alcohol mixtures is most simply determined by specific gravity measurement. When impurities are not present in significant quantities. Gas chromatography is an excellent technique for determining isopropyl alcohol in the presence of other organic substances, eg, ethyl alcohol, methyl alcohol, and acetones.

Health and Safety Factors

Isopropyl alcohol is about twice as toxic as ethyl alcohol, but less toxic than methyl alcohol. There is no known systematic investigation of the effects of inhalation, eg, from aerosols, of isopropyl alcohol in humans. The known human toxicity is based on numerous cases of accidental ingestion or topical application. Toxic doses of ingested isopropyl alcohol, usually as rubbing alcohol, may produce narcosis, anesthesia, coma, and death. The single lethal dose for humans is about 250 mL, although as little as 100 mL can be fatal.

Use of isopropyl alcohol in industrial applications does not present a health hazard.

Uses

Uses of isopropyl alcohol are chemical, solvent, and medical.

Isopropyl alcohol is used as a feedstock for the production of acetone; it is also consumed in the production of other chemicals.

Because of its balance between alcohol, water, and hydrocarbon-like characteristics, isopropyl alcohol is an excellent, low cost solvent free from the government regulations and taxes that apply to ethyl alcohol.

Many aerosol products contain isopropyl alcohol solvent in their formulations (see AEROSOLS).

Isopropyl alcohol is also used as an antiseptic and disinfectant for home, hospital, and industry (see DISINFECTANTS AND ANTISEPTICS). Rubbing alcohol, a popular 70 vol % isopropyl alcohol-in-water mixture, exemplifies the medicinal use of isopropyl alcohol. Isopropyl alcohol solutions are used in a variety of pharmaceuticals.

JOHN E. LOGSDON
RICHARD A. LOKE
Union Carbide Corporation

L. F. Hatch and W. R. Fenwick, *Isopropyl Alcohol,* Enjay Chemical Co., New York, 1966.

S. Zakhori and co-workers, in L. Golberg, ed., *Isopropanol and Ketones in the Environment,* CRC Press, Inc., Cleveland, Ohio, 1977.

J. C. Fielding, in E. C. Hancock, ed., *Propylene and Its Industrial Derivatives,* John Wiley & Sons, Inc., New York, 1973.

n-PROPYL ALCOHOL

n-Propyl alcohol 1-propanol, $CH_3CH_2CH_2OH$, mol wt 60.09, is a clear, colorless liquid having a typical alcohol odor; it is miscible in water, ethyl ether, and alcohols. 1-Propanol occurs in nature in fusel oils and forms from fermentation and spoilage of vegetable matter.

Properties

A number of physical and chemical properties of 1-propanol are listed in Table 1. The chemistry of 1-propanol is typical of low molecular weight primary alcohols (see ALCOHOLS, HIGHER ALIPHATIC). Biologically, 1-propanol is easily degraded by activated sludge and is the easiest alcohol to degrade.

Table 1. Physical and Chemical Properties of 1-Propanol

Property	Value
freezing point, °C	−126.2
boiling point, °C	97.20
vapor pressure, kPa[a]	
at 20°C	1.987
40°C	6.986
60°C	20.292
80°C	50.756
vapor density (air = 1)	2.07
density, at 20°C, g/cm^3	0.80375
refractive index, n_D^{20}	1.38556
viscosity, at 20°C, mPa·s(= cP)	2.256
surface tension, at 20°C, mN/m(= dyn/cm)	23.75
critical temperature, °C	263.65
critical pressure, kPa[a]	5169.60
critical density, g/cm^3	0.275
heat capacity, liquid at 25°C, J/(mol·K)[b]	141
heat of vaporization, kJ/mol[b] at 25°C	47.53
heat of combustion, liquid at 25°C, kJ/mol[b]	2033
flash point, Tag open cup, °C	28.9
autoignition temperature, °C	371.1
explosive limit, in air, vol %	
lower	2.2
upper	14.0

[a] To convert kPa to mm Hg, multiply by 7.5.
[b] To convert J to cal, divide by 4.184.

Manufacture

1-Propanol has been manufactured by hydroformylation of ethylene (see OXO PROCESS) followed by hydrogenation of propionaldehyde or propanal and as a by-product of vapor-phase oxidation of propane (see HYDROCARBON OXIDATION). Hydroformylation or oxo technology has been the principal process for commercial manufacture of 1-propanol in the United States and Europe. Sasol in South Africa makes 1-propanol by Fischer-Tropsch chemistry.

Hydroformylation and Hydrogenation. The production of 1-propanol by hydroformylation or oxo technology is a two-step process in which ethylene is first hydroformylated to produce propanal. The resulting propanal is hydrogenated to 1-propanol (eqs. 1 and 2).

$$CH_2{=}CH_2 + CO + H_2 \xrightarrow[\Delta,\ \text{pressure}]{\text{catalyst}} CH_3CH_2CHO \quad (1)$$

$$CH_3CH_2CHO + H_2 \xrightarrow[\Delta,\ \text{pressure}]{\text{catalyst}} CH_3CH_2CH_2OH \quad (2)$$

The rhodium-triphenylphosphine catalyst system is generally used instead of cobalt in oxo processes because the former gives higher reaction rates, greater stability, lower operation pressure, and lower by-product production.

Analytical and Test Methods, and Purity Specifications

The separation and analysis of 1-propanol are straightforward. Gas chromatography is the principal method employed. Other instrumental techniques, eg, nmr, ir, and classical organic qualitative analysis, are useful. Molecular sieves have been used to separate 1-propanol from ethanol and methanol. Commercial purification is accomplished by distillation.

Health and Safety Factors

1-Propanol is defined as a flammable liquid. This alcohol is only slightly toxic to animals. 1-Propanol gives negative results in the Ames test and in the Mouse Lymphoma Forward Mutatation Assay. The National Toxicology Program (NTP) and International Agency for Research on Cancer (IARC) do not list 1-propanol as a carcinogen.

Eye contact can cause irritation or burns. Repeated skin contact can result in dermatitis. Exposure to excessive vapor concentrations irritates the eyes and respiratory tract. Very high concentrations have a narcotic effect (43).

Uses

1-Propanol is used mainly as a solvent and as a chemical intermediate. The largest uses are as a specialty solvent in flexographic printing inks, particularly for printing on polyolefin and polyamide film.

J. D. UNRUH
D. PEARSON
Celanese, Ltd.

D. Evans, G. Yagupsky, and G. Wilkinson, *J. Chem. Soc. A,* 2660 (1968).

D. Evans, J. A. Osborn, and G. Wilkinson, *J. Chem. Soc. A,* 3133 (1968).

J. A. Monick, *Alcohols, Their Chemistry, Properties and Manufacture,* Van Nostrand Reinhold, New York, 1968, pp. 117–119.

PROPYLENE

Propylene, $CH_3CH{=}CH_2$, is perhaps the oldest petrochemical feedstock and is one of the principal light olefins (see FEEDSTOCKS). It is used widely as an alkylation (qv) or polymer–gasoline feedstock for octane improvement (see GASOLINE AND OTHER MOTOR FUELS). In addition, large quantities of propylene are used in plastics such as polypropylene, and in chemicals, eg, acrylonitrile, propylene oxide, 2-propanol, and cumene (see OLEFIN POLYMERS, POLYPROPYLENE; PROPYL ALCOHOLS). Propylene is produced primarily as a by-product of petroleum (qv) refining and of ethylene (qv) production by steam pyrolysis.

Physical Properties

Some of the physical properties of propylene are listed in Table 1 (1).

Chemistry

The chemistry of propylene is characterized both by the double bond and by the allylic hydrogen atoms. Propylene is the smallest stable unsaturated hydrocarbon molecule that exhibits low order symmetry, ie, reflection only along the main plane. Carbon atoms 1 and 2 have trigonal planar geometry identical

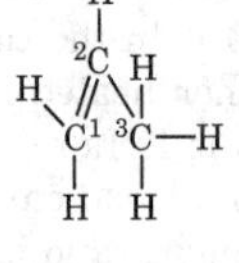

to that of ethylene. Generally, these carbons are not free to rotate, because of the double bond. Carbon atom 3 is tetrahedral, like methane, and is free to rotate. The hydrogen atoms attached to this carbon are allylic.

The propylene double bond consists of a σ-bond formed by two overlapping sp^2 orbitals, and a π-bond formed above and below the plane by the side overlap of two p orbitals. The π-bond is responsible for many of the reactions that are characteristic of alkenes. It serves as a source of electrons for electrophilic reactions such as addition reactions. Simple examples are the addition of hydrogen (eq. 1) or a halogen, eg, chlorine (eq. 2).

$$CH_3CH{=}CH_2 + H_2 \xrightarrow{\text{catalyst}} CH_3CH_2CH_3 \quad (1)$$

$$CH_3CH{=}CH_2 + Cl_2 \xrightarrow{\text{catalyst}} CH_3CHClCH_2Cl \quad (2)$$

Table 1. Selected Physical Properties of Propylene

Property	Value
mol wt	42.081
freezing temperature, K	87.9
bp, K	225.4
critical temperature, K	365.0
critical pressure, MPa[a]	4.6
liquid density, at 223 K, g/cm^3	0.612
heat of vaporization at bp, kJ/mol[b]	18.41
solubility, at 20°C, 101.3 kPa (1 atm), mL gas/100 mL solvent	
in water	44.6
in ethanol	1250
in acetic acid	524.5
refractive index, n_D	1.3567

[a] To convert MPa to atm, divide by 0.1013.
[b] To convert J to cal, divide by 4.184.

The presence of allylic hydrogens in propylene often serves to distinguish its chemistry from that of ethylene. For example, these hydrogens cause cross-linked, gummy materials to form when propylene polymerizes in the presence of peroxide initiators. The effect of the allyl hydrogens on propylene reactions can be explained by the stability of allyl radicals and allyl carbocations. When an allylic hydrogen is abstracted from propylene, the sp^3 hybridized carbon of the methyl group changes to sp^2. The p -orbital of this carbon can then overlap with the p-orbital involved in the π-bond, forming a new π-bond, where the bonding overlap is over all three carbon atoms. The electrons from the alkene π-bond and the free-radical electron are delocalized over the entire molecule. In molecular-orbital terms, the allyl radical can be represented as stabilized by overlap between the three adjacent atomic p-orbitals, forming a molecular orbital that extends over three atoms.

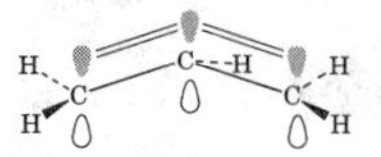

Polymerization Reactions. Polymerization addition reactions are commercially the most important class of reactions for the propylene molecule. Many types of gas- or liquid-phase catalysts are used for this purpose. Most recently, metallocene catalysts have been employed commercially.

Manufacture

Steam Cracking. In steam cracking, a mixture of hydrocarbons and steam is preheated to ca 870 K in the convective section of a pyrolysis furnace. Then it is further heated in the radiant section to as much as 1170 K (see ETHYLENE; PETROLEUM, REFINERY PROCESSES, SURVEY). Steam reduces the hydrocarbon partial pressure in the reactor. The steam-to-hydrocarbon weight ratio is generally a function of the feedstock and ranges from ca 0.2 for ethane to ≥2.0 for gas oils. The amount of steam used is probably a compromise between yield structure (olefin selectivity), energy consumption, and furnace run length, which is limited by coking. The residence time in the radiant section varies from ca 1 s in older plants to as low as 0.1 s in some newer furnaces. The residence time influences olefin selectivity. Generally, selectivity to ethylene improves as residence time decreases.

In the radiant section, the hydrocarbon mixture undergoes reactions involving free radicals. These mechanisms have been generalized to include the molecular reactions shown below:

$$\textit{Chain initiation reactions}: \ R{-}R' \rightarrow R\cdot + R'\cdot$$

$$\textit{Hydrogen-abstraction reactions}: \ R\cdot + R'H \rightarrow RH + R'\cdot$$

$$\textit{Radical-decomposition reactions}: \ R\cdot \rightarrow RH + R'\cdot$$

$$\textit{Radical-addition reactions to unsaturated molecules}: \ RH + R'\cdot \rightarrow R''\cdot$$

$$\textit{Chain-termination reactions}: \ R\cdot + R'\cdot \rightarrow R{-}R'$$

$$\textit{Molecular reactions}: \ RH + R'H \rightarrow R''H + R'''H$$

$$\textit{Radical-isomerization reactions}: \ R'\cdot \rightarrow R''\cdot$$

The total number of reactions depends on the number of constituents present in the hydrocarbon feedstock. As many as 2000 reactions can occur simultaneously.

The yield of propylene produced in a pyrolysis furnace is a function of the feedstock and the operating severity of the pyrolysis. In an olefins plant separation train, propylene is obtained by distillation of a mixed C_3 stream, ie, propane, propylene, and minor components, in a C_3-splitter tower. Propylene is produced as the overhead distillation product, and the bottoms are a propane-rich stream.

Refinery Production. Refinery propylene is formed as a by-product of fluid catalytic cracking of gas oils and, to a far lesser extent, of thermal processes, eg, coking. The total amount of propylene produced depends on the mix of these processes and the specific refinery product slate.

In fluid catalytic cracking, a partially vaporized gas oil is contacted with zeolite catalyst.

Converted feedstock forms gasoline-boiling-range hydrocarbons, C_4 and lighter gas, and coke. Propylene yield varies.

Two thermal-cracking processes, ie, delayed coking and Exxon's proprietary process, Flexicoking or fluid coking, are used to convert residuum into more valuable products. In delayed coking, residuum and steam are heated in a furnace and then fed into an insulated drum where the free-radical decomposition of the feedstock takes place. In fluid coking, a residuum feed is injected into a reactor, where it cracks thermally.

Refinery propylene is recovered at the vapor-recovery unit. A chemical- or polymer-grade propylene can be made by further distillation in a propylene concentration unit.

Storage and Handling

Precautions must be taken to avoid health and fire hazards wherever propylene is handled. Equipment capable of causing ignition should be shut down while connecting, disconnecting, loading, and unloading equipment. No part of any cylinder containing propylene should be subjected to temperatures above 325 K.

Health and Safety Factors

Propylene is a colorless gas under normal conditions, has anesthetic properties at high concentrations, and can cause asphyxiation. It does not irritate the eyes and its odor is characteristic of olefins. Propylene is a flammable gas under normal atmospheric conditions. Vapor-cloud formation from liquid or vapor leaks is the main hazard that can lead to explosion. The autoignition temperature is 731 K in air and 696 K in oxygen. Evaporation of liquid propylene can cause skin burns. Propylene also reacts vigorously with oxidizing materials.

Uses

Propylene has many commercial uses. The actual utilization of a particular propylene supply depends not only on the relative economics of the petrochemicals and the value of propylene in various uses, but also on the location of the supply and the form in which the propylene is available.

The uses of propylene may be loosely categorized as refinery or chemical purpose. In the refinery, propylene occurs in varying concentrations in fuel-gas streams. As a refinery feedstock, propylene is alkylated by isobutane or dimerized to produce polymer gasoline for gasoline blending. Commercial chemical derivatives include polypropylene, acrylonitrile, propylene oxide, isopropyl alcohol, and others.

Polypropylene. One of the most important applications of propylene is as a monomer for the production of polypropylene. Propylene is polymerized by Ziegler-Natta coordination catalysts, carried out either in the liquid phase where the polymer forms a slurry of particles, or in the gas phase where the polymer forms dry solid particles.

The focus of commercial research is on catalysts that give desired and tailored polymer properties for improved processing, such as the metallocene catalyst systems.

NARASIMHAN CALAMUR
MARTIN CARRERA
Amoco Corporation

R. J. Hengstebeck, *Petroleum Processing, Principles and Applications*, McGraw-Hill Book Co., Inc., New York, 1959.

D. J. Hadley, R. E. Saunders, and P. T. Mapp, in E. G. Hancock, ed., *Propylene and its Industrial Derivatives*, Earnest Benn Ltd., London, 1973, pp. 416 ff.

R. C. Reid, T. M. Prausnitz, and T. K. Sherwood, *The Properties of Gases and Liquids*, 3rd ed., McGraw-Hill Book Co., Inc., New York, 1977.

PROPYLENE OXIDE

Propylene oxide (methyloxirane, 1,2-epoxypropane) is a significant organic chemical used primarily as a reaction intermediate for production of polyether polyols, propylene glycol, alkanolamines (qv), glycol ethers, and many other useful products (see GLYCOLS).

Physical Properties

Propylene oxide is a colorless, low boiling (34.2°C) liquid. Table 1 lists some general physical properties.

Chemical Properties

Propylene oxide (**1**) is highly reactive owing to the strained three-membered oxirane ring. Although some reactions, such as those with hydrogen halides or ammonia, proceed at adequate rates without a catalyst, most reactions of industrial importance employ the use of either acidic or basic catalysts.

Table 1. Selected Physical Properties of Propylene Oxide

Property	Value
molecular weight	58.08
boiling point at 101.3 kPa[a], °C	34.2
freezing point, °C	−111.93
critical pressure, MPa[a]	4.92
critical temperature, °C	209.1
explosive limits in air, vol %	
upper	36
lower	2.3
flash point, Tag closed cup, °C	−37
heat of fusion, kJ/mol[b]	6.531
heat of vaporization, 101.3 kPa[a], kJ/mol[b]	27.8947
heat of combustion, kJ/mol[b]	1915.6
specific heat, at 20°C, J/(mol·K)[b]	122.19
autoignition temperature, at 101.3 kPa[a], °C	465

[a] To convert kPa to psi, multiply by 0.145.
[b] To convert J to cal, divide by 4.184.

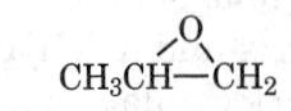

(**1**)

Ring Opening. The epoxide ring of propylene oxide may open at either of the C—O bonds. In anionic (basic) catalysis, the bond preferentially opens at the least sterically hindered position, resulting in mostly (95%) secondary alcohol products. Cationic (acidic) catalysts provide a mixture of secondary and primary alcohol products.

The ring-opening reactions of epoxides take place by nucleophilic substitution, ie, a S_N^2 mechanism, on one of the epoxide carbon atoms with displacement of the epoxide oxygen atom. The orientation of ring opening in propylene oxide is determined primarily by the steric hindrance of the substituent methyl group and secondarily by the electron-releasing effect of the methyl group. Thus, acid catalysis increases substitution on the secondary carbon by increasing the positive charge on this carbon.

Polymerization to Polyether Polyols. The addition polymerization of propylene oxide to form polyether polyols is very important commercially. Polyols are made by addition of epoxides to initiators, ie, compounds that contain an active hydrogen, such as alcohols or amines. The polymerization occurs with either anionic (base) or cationic (acidic) catalysis. The base catalysis is preferred commercially.

Some of the simplest polyols are produced from reaction of propylene oxide and propylene glycol and glycerol initiators. Polyether diols and polyether triols are produced, respectively (see GLYCOLS).

Reactions. Propylene oxide reacts with water to produce propylene glycol, dipropylene glycol, tripropylene glycol, and higher molecular weight polyglycols. This commercial process is typically run using an excess of water (12–20 mol water/mol propylene oxide) to maximize the production of the monopropylene glycol.

Isopropanolamine is the product of propylene oxide and ammonia in the presence of water (see ALKANOLAMINES). Propylene oxide reacts with isopropanolamine or other primary or secondary amines to produce *N*- and *N,N*-disubstituted isopropanolamines.

Propylene oxide and carbon dioxide react in the presence of tertiary amine, quaternary ammonium halides, or calcium or magnesium halide catalysts to produce propylene carbonate. Use of catalysts derived from diethylzinc results in polycarbonates.

Similarly, carbon disulfide and propylene oxide reactions are catalyzed by magnesium oxide to yield episulfides, and by derivatives of diethylzinc to yield low molecular weight copolymers.

Propylene oxide reacts with the hydroxyl group of alcohols and phenols to produce monoethers of propylene glycol. Use of basic catalysts results in mostly secondary alcohol products; use of acidic catalysts gives a mixture of primary and secondary alcohols.

Propylene oxide and carboxylic acids in equimolar ratios produce monoesters of propylene glycol. Higher ratios of oxide to acid produce polypropylene glycol monoesters.

Many natural products, eg, sugars, starches, and cellulose, contain hydroxyl groups that react with propylene oxide. Base-catalyzed reactions yield propylene glycol monoethers and poly(propylene glycol) ethers. Reaction with fatty acids results in a mixture of mono- and diesters.

2-Phenylpropanol results from the catalytic ($AlCl_3$, $FeCl_3$, or $TiCl_4$) reaction of benzene and propylene oxide at low temperature and under anhydrous conditions (see FRIEDEL-CRAFTS REACTIONS).

Grignard reagents, RMgX, produce a mixture of secondary alcohols, $RCH_2CHOHCH_3$, and propylene halohydrin, $CH_3CHOHCH_2X$, upon reaction with propylene oxide (see GRIGNARD REACTIONS).

Isomerization of propylene oxide to propionaldehyde and acetone occurs over a variety of catalysts. Hydrogenolysis of propylene oxide yields primary and secondary alcohols as well as the isomerization products of acetone and propionaldehyde. Reduction of propylene oxide to propylene is accomplished by use of metallocenes and sodium amalgam.

Cyclic ketals and acetals (dioxolanes) are produced from reaction of propylene oxide with ketones and aldehydes, respectively. Suitable catalysts include stannic chloride, quaternary ammonium salts, glycol sulfites, and molybdenum acetyl acetonate or naphthenate. Lactones come from Ph_4SbI-catalyzed reaction with ketenes.

Propylene oxide reacts with hydrogen halides to give the corresponding isomeric halohydrins; with sodium bisulfite to give the sodium salt of 2-hydroxypropanesulfonic acid; with nitric acid to produce isomeric nitrate esters; with hydrogen cyanide to give 1-cyano-2-propanol; and with boric acid, boron trichloride, or diborane to give a variety of substituted boranes and borates.

2-Dioxolanimines, 2-oxathiolanimines, and 2-oxazolidinimines result from the reaction of propylene oxide with isocyanates, isothiocyanates, and carbodiimides, respectively.

Manufacture

Propylene oxide is produced by one of two commercial processes: the chlorohydrin process or the hydroperoxide process.

Chlorohydrin Process. The chlorohydrin process is fairly simple, requiring only two reaction steps, chlorohydrination and epoxidation, followed by product purification and effluent wastewater treatment. Propylene gas and chlorine gas in about equimolar amounts are mixed with an excess of water to generate propylene chlorohydrin and a small amount of chlorinated organic coproducts, chiefly 1,2-dichloropropane. Epoxidation, also called saponification or dehydrochlorination, is accomplished by treatment of the chlorohydrin solution with caustic soda or milk of lime (aqueous calcium hydroxide).

Hydroperoxide Process. The hydroperoxide process to propylene oxide involves the basic steps of oxidation of an organic to its hydroperoxide, epoxidation of propylene with the hydroperoxide, purification of the propylene oxide, and conversion of the coproduct alcohol to a useful product for sale. Incorporated into the process are various purification, concentration, and recycle methods to maximize product yields and minimize operating expenses.

Transportation, Specifications, and Analysis

Propylene oxide is classified as a flammable liquid and hazardous substance in the U.S. Department of Transportation (DOT) Hazardous Materials Table.

Specifications and Analysis

Propylene oxide is a high purity product. Thus only the impurities are analyzed and reported. Detection of propylene oxide has practical applications in the manufacturing process, in quality control of reaction products, and in environmental monitoring. Propylene oxide content in manufacturing streams is determined by chemical methods.

Health and Safety Factors

Propylene oxide has a variety of toxic effects on humans. Thus, exposure to propylene oxide during manufacture, storage and handling, and use should be minimized. Potential for high exposure to propylene oxide can occur during such routine activities as sampling, analysis, and maintenance, and in disconnecting hoses used for product transfer. Exposure is first minimized through proper design of sampling devices and the handling equipment and use of job procedures for both routine work and maintenance activity. A last resort is the use of personal protective equipment such as respirators, breathing air, gloves, and chemical suits.

Propylene oxide has been studied extensively for its effects on humans and animals. Accordingly, it is regulated under several U.S. Federal statutes and agencies. Exposure to vapors above the permissible exposure limit can be irritating to the eyes and respiratory tract. Low concentrations can cause nausea; high concentrations can cause pulmonary edema. Propylene oxide is a possible human carcinogen and it is classified as such by NIOSH, IARC, and NTP.

Uses

Propylene oxide is a useful chemical intermediate. Additionally, it has found use for etherification of wood to provide dimensional stability for purification of mixtures of organosilicon compounds, for disinfection of crude oil and petroleum products, for sterilization of medical equipment and disinfection of foods, and for stabilization of halogenated organics.

Propylene oxide has found use in the preparation of polyether polyols from recycled poly(ethylene terephthalate), halide removal from amine salts via halohydrin formation, preparation of flame retardants, alkoxylation of amines, modification of catalysts, and preparation of cellulose ethers.

Derivatives. Polyether polyols produced by polymerization of propylene oxide on polyhydric alcohols account for the largest use of propylene oxide.

Propylene glycol, the second largest use of propylene oxide, is produced by hydrolysis of the oxide with water. Propylene glycol has very low toxicity and is, therefore, used directly in foods, pharmaceuticals (qv), and cosmetics, and indirectly in packaging (qv). Propylene glycol also finds use as an intermediate for numerous chemicals, in hydraulic fluids (qv), in heat-transfer fluids (antifreeze), and in many other applications.

DAVID L. TRENT
The Dow Chemical Company

Safe Handling and Storage of DOW Propylene Oxide, Product Bulletin, Form 109-609-788 SMG, The Dow Chemical Co., Midland, Mich., 1988.

Propylene Oxide, Product Safety Bulletin, ARCO Chemical Co., Newton Square, Pa., Mar. 24, 1992.

PROSTAGLANDINS

Prostaglandins (PGs), typified by prostaglandin E_2 (**1**), comprise a family of naturally occurring substances found in animals and humans. They play important and diverse roles in both health and disease. Prostaglandins are biosynthesized from 20 carbon polyunsaturated fatty acids. In humans, the predominant precursor of prostaglandins is arachidonic acid (**2**) which is available either from the diet or by anabolic conversion of linoleic acid, an essential fatty acid.

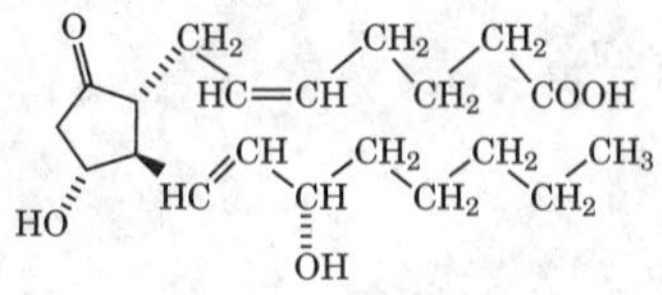

(**1**)

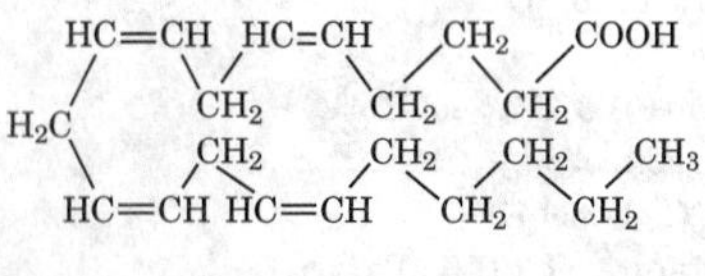

(**2**)

The enzyme system responsible for the biosynthesis of PGs is widely distributed in mammalian tissues. It is referred to as prostaglandin H synthase (PGHS) and exhibits both cyclooxygenase and peroxidase activity. In addition to the classical PGs two other prostanoid products, thromboxane A_2 (TxA_2) (**3**) and prostacyclin (PGI_2) (**4**) are also derived from the action of the enzyme system on arachidonic acid.

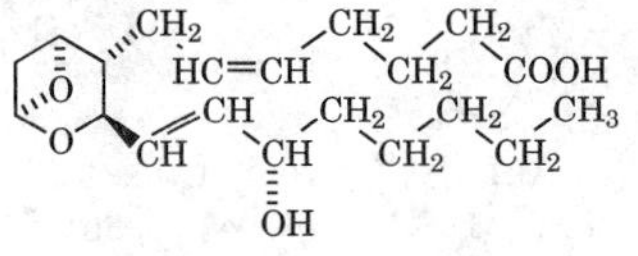

(**3**)

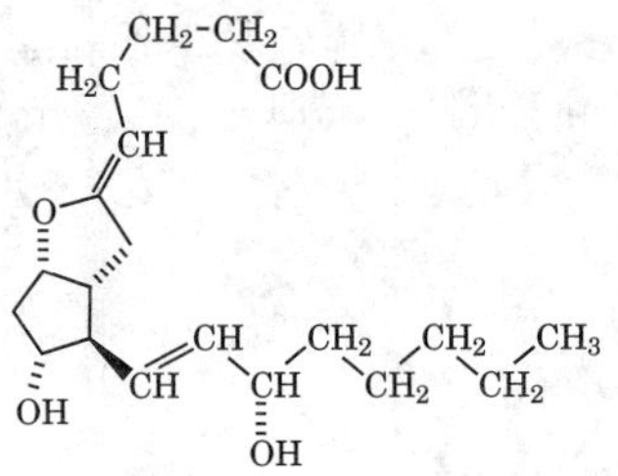

(**4**)

Prostaglandins were discovered in the 1930s when it was noted that fresh human semen caused strips of human uterine tissue to either relax or contract depending on whether or not the tissue donor had borne children. Later work led to the isolation and structural elucidation of prostaglandin E_1 (PGE_1) (**5**) and prostaglandin $F_{1\alpha}$ ($PGF_{1\alpha}$) (**6**).

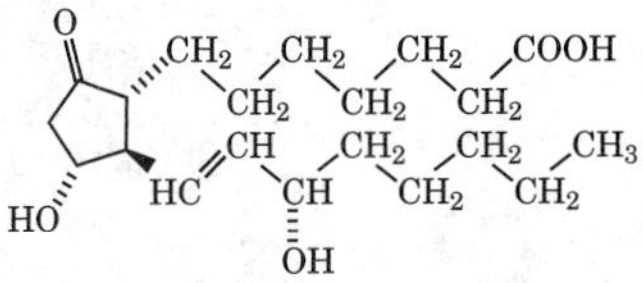

(**5**)

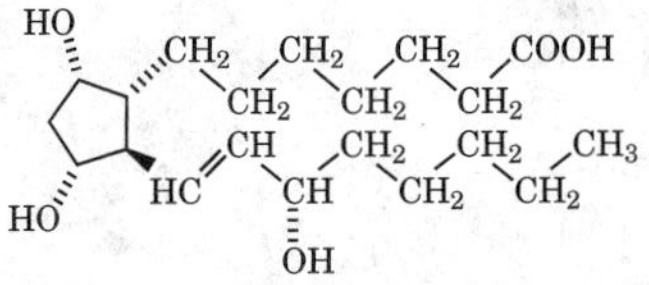

(**6**)

The absolute stereochemical configuration of PGs is based on the configuration of L-2-hydroxyheptanoic acid, obtained by oxidative ozonolysis of acetylated PGE_1 methyl ester. In 1983 Bergström and Samuelsson shared (with John Vane) the Nobel Prize for their contributions to this field.

The PGs TxA_2 and PGI_2, commonly referred to as prostanoids, may be thought of as hormone-like substances which are produced on demand rather than stored and which modulate cellular functions at or near their site of generation (see HORMONES). Unlike typical circulating hormones, which are released from one principal tissue site, prostanoids are synthesized and released by virtually all tissues. They are extremely potent and exert regulatory influences on the endocrine, reproductive, nervous, gastrointestinal, cardiovascular, renal, and immunological systems. They are also short lived because of rapid metabolism to inactive species. Their effects are generally considered to be beneficial and homeostatic in nature. In some cases, however, their biological effects can be detrimental and disease producing; for example, PGs can exert inflammatory actions and produce fever, and TxA_2 is a pro-aggregatory factor for platelets.

The nomenclature of prostaglandins and prostacyclins is based on the basic prostane skeleton (**7**), whereas thromboxane (**8**) is the parent for the thromboxanes.

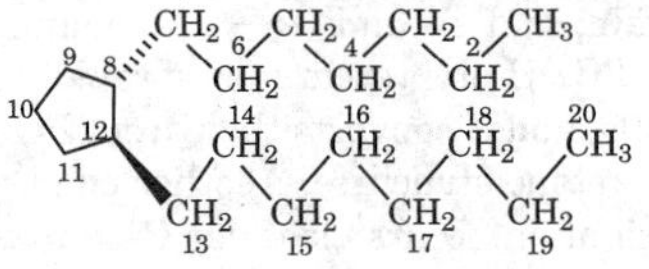

(**7**)

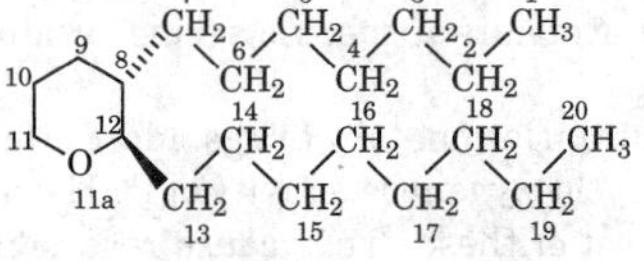

(**8**)

Implicit in the base names are the absolute configurations at carbons 8 and 12 and the indicated numbering systems. Derivatives of these parent structures are named according to terpene and steroid nomenclature rules (see STEROIDS; TERPENOIDS). The lengthy and awkward nature of the chemical abstract systematic nomenclature for these compounds has resulted in the development and use of simplified nomenclature based on common names.

The PGs are grouped into several basic families which differ from each other in the nature of the five-membered ring functionalities (Fig. 1). The principal families are designated as PGA through PGJ. The letters E and F originated from the finding that PGE and PGF compounds partition differently. The compound that was more soluble in ethyl ether was called prostaglandin E and the one that was more soluble in phosphate buffer (*fosfate* in Swedish) was termed prostaglandin F (**6**). The letters A and B refer to the formation of these derivatives from PGE compounds by treatment with acid and base, respectively. As indicated in the general structures of Figure 1, the carboxylic acid side chain is referred to as the alpha chain R_α, and the hydroxy-bearing chain as the omega chain, R_ω. The number of double bonds in the molecule are denoted by subscript numerals appearing after the name; for example, PGE_1 contains one double bond at C_{13}; PGE_2 has an additional double bond at C_5. The stereochemistry of substituents on the cyclopentane ring is designated α- or β- depending on whether the substituent is above or below the plane of the paper.

Biosynthesis and Metabolism

Under normal circumstances arachidonic acid (AA) is the most abundant C-20 fatty acid *in vivo*. The biosynthesis reactions of the two-series prostaglandins from AA make up a portion of what is known as the arachidonic acid cascade. Other lipid products of the cascade include the leukotrienes, lipoxins, and the hydroxyeicosatetraenoic acids (HETEs). Collectively, these substances are termed eicosanoids.

The primary prostaglandins are believed to be mediators of inflammation, and their synthesis can be inhibited by both

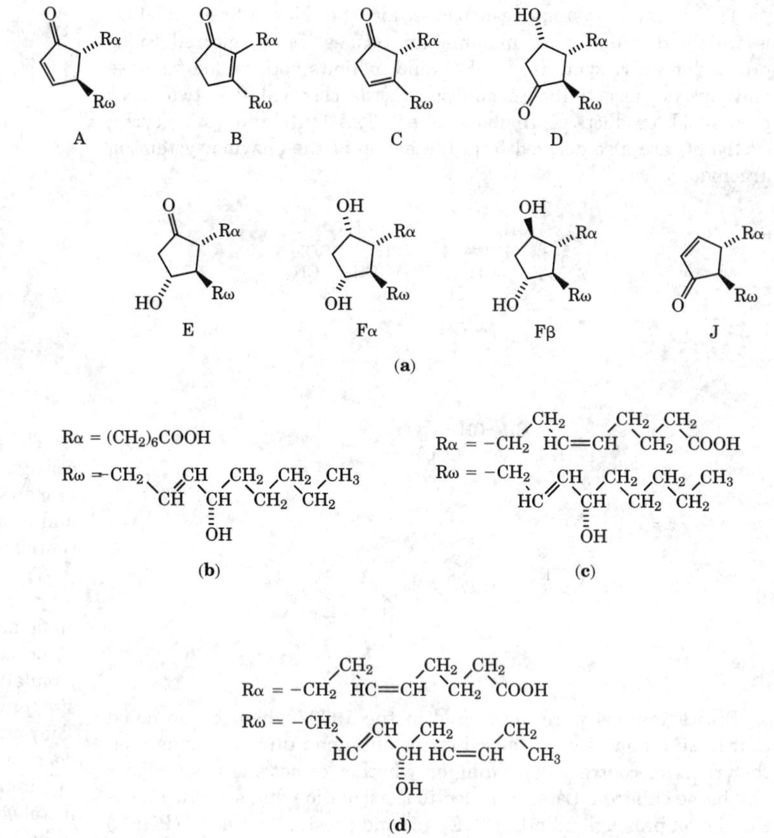

Figure 1. (**a**) The basis for prostaglandin nomenclature, where the letters A–F and J define principal families; (**b**) defines the side chains for PG_1 derived from dihomo-γ-linolenic acid; (**c**) PG_2 derived from arachidonic acid; and (**d**), PG_3 derived from eicosapentaenoic acid.

nonsteroidal antiinflammatory drugs (NSAIDS) and antiinflammatory steroids. The best known of the NSAIDS is aspirin (see ANALGESICS, ANTIPYRETICS, AND ANTIINFLAMMATORY AGENTS; SALICYLIC ACID AND RELATED COMPOUNDS). Aspirin competes with arachidonic acid (AA) for binding to the cyclooxygenase active site, but the binding of AA is about 10,000 times more efficient than that of aspirin. However, once bound, aspirin can acetylate a specific serine residue of PGH synthase: Ser^{530}.

There are many nonsteroidal antiinflammatory drugs. Most other NSAIDS also act by inhibiting the cyclooxygenase activity of PGH synthase. However, unlike aspirin, most of these drugs cause reversible enzyme inhibition by competing with AA for binding. Well-known examples of reversible NSAIDS are ibuprofen (Advil, Whitehall), indomethacin (Indocin, Merck) and naproxen (Naprosyn, Syntex).

There are two isozymes of PGH synthase. PGHS-1 (or COX-1) is constitutively expressed in most tissues, and is responsible for the production of PGs involved in cellular housekeeping functions such as coordinating the actions of circulating hormones and regulating vascular, gastric, intestinal, and renal homeostasis. PGHS-2 (COX-2), which shares about 59% amino acid homology with PGHS-1, is expressed only following cell activation. Its expression is stimulated by inflammatory mediators and inhibited by antiinflammatory steroids. All of the available nonsteroidal antiinflammatory drugs inhibit both forms of the enzyme. As a result they produce side effects in tissues which are dependent on homeostatic levels of PGs. The gastrointestinal tract and kidney are particularly sensitive to PG synthesis inhibition. NSAIDS can cause damage and ulceration in the stomach and intestinal tract and can compromise kidney function especially in people with pre-existing renal impairment.

Acetylation of PGH synthase by aspirin has important pharmacological consequences. Besides the analgesic, antipyretic, and antiinflammatory actions of aspirin, low dose aspirin treatment is a useful antiplatelet cardiovascular therapy (see CARDIOVASCULAR AGENTS).

Once prostanoids are formed, they exit the cell. They act very near their sites of synthesis, then are rapidly inactivated by metabolic enzymes and excreted. The prostanoids are subject to four principal metabolic transformations, as illustrated in Figure 2 for PGE_2: oxidation catalyzed by C-15 prostaglandin dehydrogenase (15-PGDH); reduction by 13,14-reductase; β-oxidation of the carboxylic acid side chain; and ω- and (ω − 1)-oxidation of the aliphatic side chain. The 15-PGDH step occurs most rapidly in humans and most other mammals, and it converts the parent PG molecule into its corresponding C-15 ketone structure. Another point of attack is the ω-chain terminus. Oxidation occurs either at C-20 to give the alcohol and subsequently the acid (**9**) or at C-19 to produce the 19-hydroxy metabolite. The primary urinary metabolite of natural PGs is (**10**) which is the final keto product of all of these processes plus the reduction of the C-9 carbonyl group.

Physical and Chemical Properties

The melting points, optical rotations, and uv spectral data for selected prostanoids are provided in Table 1.

Both thromboxane A_2 (TxA_2) and prostacyclin (PGI_2) are extremely unstable compounds. The instability of TxA_2 (**3**) is due to the strained bicyclic acetal system. Hydrolysis of the acetal to give TxB_2 (**11**) releases the strain.

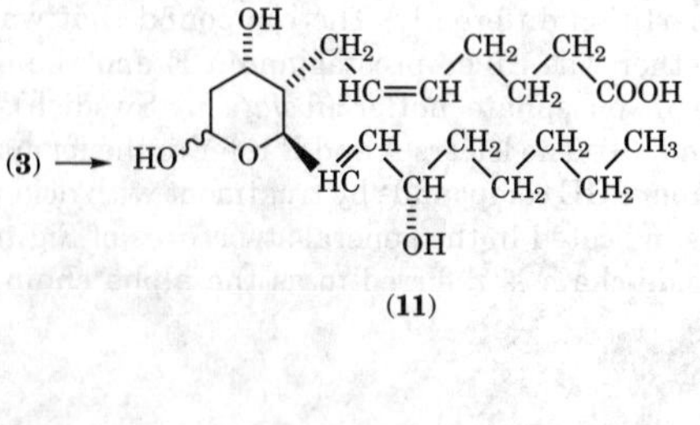

(**11**)

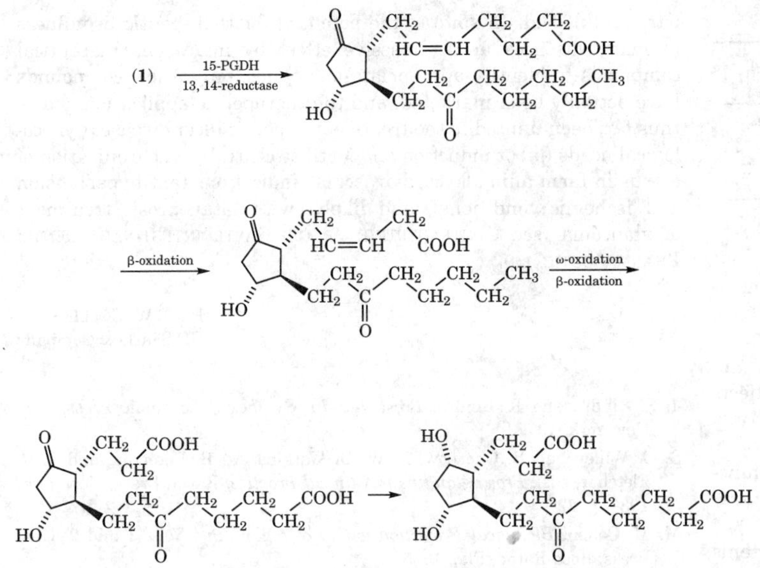

(9) (10)

Figure 2. Metabolism of natural prostaglandins.

TxB_2 (**11**) is a stable but biologically inactive compound, and its isolation and characterization were essential to the discovery of TxA_2. TxA_2 has never been isolated and characterized directly. Its structural assignment was based on TxB_2 and its more stable synthetic analogues.

PGI_2 (**4**) contains an acid-labile enol ether which is readily hydrolyzed to generate 6-keto-$PGF_{1\alpha}$ (**12**).

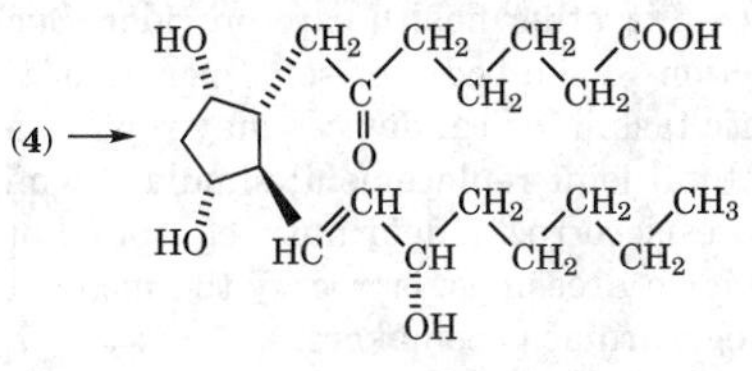

(**12**)

As with TxA_2, the reactivity of PGI_2 ($t_{1/2}$ = 3 min at pH 7.6 and 37°C) made isolation of the natural substance difficult, and a pure chemical sample was obtained only through chemical synthesis. PGI_2 is stable under more alkaline conditions and can be isolated and stored as a salt.

Table 1. Physical Properties of Selected Prostanoids Derived from Arachidonic Acid

Compound	Mp, °C	$[\alpha]_D$ (Solvent), deg
PGA_2[a]	pale yellow oil	145 (chloroform)
PGB_2[b]	30–34	16–18
PGC_2[c]	oil	
PGC_2, methyl ester[d]	oil	
PGD_2	62.8–63.3	13 (chloroform)
PGE_2	65–66	−61 (tetrahydrofuran)
	62–64	−52 (tetrahydrofuran)
	65.0–67.5	
$PGF_{2\alpha}$	30–35	26 (ethanol)
		23.8 (tetrahydrofuran)
PGI_2 sodium salt	166–168	88 (chloroform)
	116–124	97 (ethanol)
methyl ester	30–33	78 (chloroform)
6-*oxo*-$PGF_{1\alpha}$	75–78	
TXB_2	91–93	57.4 (ethyl acetate)
	92–94	
	89–90	

[a] In C_2H_5OH, λ_{max} is 217 nm and $\epsilon = 10,300(M \cdot cm)^{-1}$.
[b] In C_2H_5OH, λ_{max} is 278 nm and $\epsilon = 26,000(M{\cdot}cm)^{-1}$.
[c] In CH_3OH, λ_{max} is 234 nm and $\epsilon = 17,000(M{\cdot}cm)^{-1}$.
[d] In CH_3OH, λ_{max} is 229 (sh) and 234 nm.

Biological Properties

The PGs, PGI_2 and TxA_2 collectively exhibit a wide variety of biochemical and pharmacological activities and are involved in both physiological and pathophysiological processes.

Cardiovascular System. In most species and vascular beds PGEs and PGAs are potent vasodilators, whereas responses to $PGF_{2\alpha}$ vary. Prostacyclin, PGI_2, is a potent vasodilator with five times the potency of PGE_2 and causes prominent hypotension in animals and humans following intravenous administration. The opposing effects of PGI_2 and TxA_2 on vascular tone and platelet aggregatory state are believed to be important in vascular homeostasis, and disruption of this balance can lead to cardiovascular disease (see CARDIOVASCULAR AGENTS). Aspirin is used prophylactically to prevent heart attacks because it can selectively reduce TxA_2 levels relative to PGI_2 levels.

Pulmonary System. In general PGFs contract and PGEs relax bronchial and tracheal muscle. Asthmatics are particularly sensitive to PGs and $PGF_{2\alpha}$ can cause severe bronchospasm in these patients. In contrast, PGE_1 and -E_2 are potent bronchodilators when given by aerosol to asthmatic patients (see ANTIASTHMATIC AGENTS).

Reproductive System. The primary PGs are intimately involved in reproductive physiology. PGE_2 and $PGF_{2\alpha}$ are potent contractors of the pregnant uterus and intravenous infusion of either of these compounds to pregnant humans produces a dose-dependent increase in frequency and force of uterine contraction. PGI_2 and TxA_2 have mild relaxant and stimulatory effects, respectively, on uterine tissue. The primary PGs also play a role in parturition, ovulation, luteolysis, and lactation and have been implicated in male infertility.

Gastrointestinal System. PGEs, PGAs, and PGI_2 inhibit gastric acid secretion stimulated by feeding, histamine, or gastrin. The volume of secretion, acidity, and content of pepsin are all reduced, probably by an action exerted directly on the secretory cells. In addition, these PGs are vasodilators in the gastric mucosa. PGs cause substantial movement of water and electrolytes into the intestinal lumen. Such effects contribute to the diarrhea noted in animals and humans following the oral or parenteral administration of PGs.

Table 2. Prostaglandin Receptor Subtypes

Receptor subtype	Most potent natural PG agonist	Usual response on smooth muscle
EP_1	PGE_2	contraction
EP_2	PGE_2	relaxation
EP_3	PGE_2	inhibition

Nervous System. PGEs produce sedation and catatonia when injected into the cerebral ventricles of cats (see HYPNOTICS, SEDATIVES, ANTICONVULSANTS, AND ANXIOLYTICS; PSYCHOPHARMACOLOGICAL AGENTS). In humans, PGEs cause pain when injected intradermally or applied to facial skin. They also potentiate the pain-producing effects of bradykinin and histamine (see HISTAMINE AND HISTAMINE ANTAGONISTS).

Kidney Function. Prostanoids influence a variety of kidney functions including renal blood flow, secretion of renin, glomerular filtration rate, and salt and water excretion.

Metabolic and Endocrine Effects. The role of PGs in these systems is complex and generally modulatory in nature. PGE_2 is synthesized by adipocytes and is a potent inhibitor of lipolysis. It is also a potent inducer of bone resorption and of calcium release from bone. PGE_2 increases circulating levels of adrenocorticotropic hormone (ACTH), growth hormone, prolactin, and the gonadotropin hormones (see GROWTH REGULATORS; HORMONES, HUMAN GROWTH HORMONE). The effect of PGEs on insulin and glucose levels is also complex and regulatory in nature.

Inflammatory and Immune Responses. PGE_2 and PGI_2 are present in inflamed tissues in sufficient concentrations to account for the erythema and increased sensitivity characteristic of acute inflammation. PGEs are vasodilatory and hyperalgesic, ie, increase sensitivity to pain, and in concert with other mediators such as bradykinin and histamine, PGEs increase vascular permeability.

Cytoprotection. One of the most intriguing properties of prostanoids is their ability to protect cells and tissues from various damaging agents. Protective effects of PGs have also been documented in the colon, liver, kidney, and pancreas, and for injury caused by radiation and chemotherapeutic agents (see CHEMOTHERAPEUTICS, ANTICANCER; RADIOPROTECTIVE AGENTS). The mechanisms underlying the protective properties of prostanoids remain unclean.

Health and Safety. The prostanoids are extremely potent substances with a wide variety of biological effects. Therefore utmost caution should be used in their handling to avoid adverse effects. The LD_{50} (the dose lethal to 50% of the treated population) for PGE_1 in rats is 228 mg/kg of body weight by oral administration and 19.2 mg/kg intravenously. Prostanoids that are marketed as drugs, are regulated by the U.S. FDA and corresponding agencies in other countries.

Prostanoid Receptors

Characterization of prostanoid receptors and quantification of the action of ligands at these receptors have been hampered by lack of selectivity of the natural prostanoids and most synthetic agonists and antagonists for specific receptors, the ability of the prostanoids to induce opposing actions in the same tissue, and the multiplicity of prostanoid receptor subtypes in most tissues. The development of a classification system of prostanoid receptors (Table 2) provides a working framework toward understanding these interactions. PGE receptors are pharmacologically divided into three subtypes, EP_1, EP_2, and EP_3, which differ in their mode of signal transduction; binding at these receptors is believed to lead to elevation of intracellular calcium levels, stimulation of adenylate cyclase, and inhibition of adenylate cyclase, respectively.

Therapeutic Role of Prostanoids

The early promise and anticipation of a large and varied therapeutic role for the prostanoids and their analogues has gone, for the most part, unfulfilled. In spite of the potential for therapeutic usefulness in many diseases and a massive effort by many pharmaceutical companies to identify appropriate analogues, only a few compounds have actually been marketed, and the therapeutic applications have, thus far, been limited to the treatment of peptic ulcer disease, gynecological needs (labor induction and fertility control), synchronization of estrus in farm animals, cardiovascular indications (antihypertension and ischemic conditions), and likely, with latanoprost, treatment of glaucoma (see CONTRACEPTIVE DRUGS; GASTROINTESTINAL AGENTS; PHARMACEUTICALS).

PAUL W. COLLINS
GD Searle & Company

J. S. Bindra and R. Bindra, *Prostaglandin Synthesis,* Academic Press, Inc., New York, 1977.

S. C. Willey and B. Chernow, in W. D. Watkins, M. B. Peterson, and J. R. Fletcher, eds., *Prostaglandins in Clinical Practice,* Raven Press, New York, 1989, p. 227.

M. M. Cohen, *Biological Protection with Prostaglandins,* Vols. 1 and 2, CRC Press, Boca Raton, Fla., 1985.

R. A. Coleman, I. Kennedy, P. P. A. Humphrey, K. Bunce, and P. Lumley, in C. Hansch, P. G. Sammes, J. B. Taylor, and J. C. Emmeth, eds., *Comprehensive Medicinal Chemistry,* Vol. 3, Pergamon Press, Oxford, U.K., 1989.

PROSTHETIC AND BIOMEDICAL DEVICES

Prosthetics or biomedical devices are objects which serve as body replacement parts for humans and other animals or as tools for implantation of such parts. An implanted prosthetic or biomedical device is fabricated from a biomaterial and surgically inserted into the living body by a physician or other health care provider. Such implants are intended to function in the body for some period of time in order to perform a specific task. Medical devices may replace a damaged part of anatomy, eg, total joint replacement; simulate a missing part, eg, mammary prosthesis; correct a deformity, eg, spinal plates; aid in tissue healing, eg, burn dressings; or rectify the mode of operation of a diseased organ, eg, cardiac pacemakers.

Prosthetics and biomedical devices are composed of biocompatible materials, often called biomaterials. Polymers, metals, and ceramics originally designed for commercial applications have been adapted for prostheses, opening the way for implantable pacemakers, vascular grafts, diagnostic/therapeutic catheters, and a variety of other orthopedic devices. The term prosthesis encompasses both external and internal devices. This article concentrates mainly on implantable prostheses.

Biomaterials

A biomaterial is defined as a systemic, pharmacologically inert substance designed for implantation or incorporation within the human body. A biomaterial must be mechanically adaptable for its designated function and have the required shear, stress, strain, Young's modulus, compliance, tensile strength, and temperature-related properties for the application. Moreover, biomaterials ideally should be nontoxic, ie, neither teratogenic, carcinogenic, or mutagenic; nonimmunogenic; biocompatible; biodurable, unless designed as bioresorbable; sterilizable; readily available; and possess characteristics allowing easy fabrication. The traditional areas for biomaterials are plastic and reconstructive surgery, dentistry, and bone and tissue repair. Artificial organs play an important role in preventive medicine, especially in the early prevention of organ failure.

To be biocompatible is to interact with all tissues and organs of the body in a nontoxic manner, not destroying the cellular constituents of the body fluids with which the material interfaces. In some applications, interaction of an implant with the body is both desirable and necessary, as, for example, when a fibrous capsule forms and prevents implant movement.

Polymers (qv), an important class of biomaterials, vary greatly in structure and properties. The fundamental structure may be one of a carbon chain, eg, in polyethylene or Teflon, or one having ester, ether, sulfide, or amide bond linkages. Polysilicones, having a –Si–O–Si– backbone, may contain no carbon.

Plastics are found in implants and components for reconstructive surgery, as components in medical instruments, equipment, packaging materials, and in a wide array of medical disposables.

Metals are used when mechanical strength or electrical conductivity is required of a device. Titanium and titanium alloys are well tolerated in the body. This is partly the result of the strongly adhering oxide layer that forms over the metal surface, making the interface between the body and biomaterial effectively a ceramic rather than a metal.

Stainless steel alloys are also useful in orthopedic applications. Stainless steel alloys are used in the manufacture of staples, screws, pins, etc. These alloys are used primarily in applications requiring great tensile strength.

Ceramics include a large number of inorganic nonmetallic solids that feature high compressive strength and relative chemical inertness. Aluminum oxide, Al_2O_3, forms the basis of dental implants (see DENTAL MATERIALS).

Bioglasses are surface-active ceramics that can induce a direct chemical bond between an implant and the surrounding tissue.

Cardiovascular Devices

Cardiovascular Disease. Cardiovascular disease is a progressive condition which can eventually block the flow of blood through the coronary arteries to the heart muscle, thereby causing heart attacks and other life-threatening situations. The same plaque deposits occur in the peripheral arteries, leading to gangrene, amputations, aneurysms, and strokes.

Cardiovascular disease is the leading cause of premature death in the United States. The leading immediate cause of death, referred to as sudden death, in most cases is a massive electrical failure of the heart within the first hour following onset of symptoms. Most patients afflicted with peripheral vascular disease (PVD) already have heart disease, owing to the same underlying causes, namely coronary artery disease (CAD).

Cardiovascular devices are identified with the heart, the lungs, and the circulatory road map of the vascular system. This last is comprised of the 96,000 km of arteries, veins, and capillaries which transport blood continuously throughout the body (Fig. 1). Blood serves as the vehicle to carry oxygen, which is exchanged for carbon dioxide in the capillaries, bringing other vital substances to the body's cells, as well as removing waste products.

In addition to its internal blood flow operation, the heart has its own system of blood vessels to keep the muscle wall of the heart, the myocardium, supplied with oxygenated blood (Fig. 2**a**). The coronary arteries, which branch from the aorta to the right and left sides of the heart, are vital to maintaining that supply.

An intricate wiring system of muscle fibers conducts electrical impulses which cause the heart to contract and pump blood (Fig. 2**b**). The number and frequency of the impulses determine the heart rate which is automatically adjusted to speed up during exertion or, conversely, to slow down during sleep.

Cardiovascular Problems. Despite its durability and resilience, different aspects of the cardiovascular system can malfunction. Some problems are congenital; many are inherited. Diseases can also be caused by infection such as damaged heart valves owing to rheumatic fever. Cardiomyopathy, a diseased heart muscle which may become enlarged, can result from infection or an unknown cause. Other problems may be a function of age. In cardiovascular problems, the main culprit, regardless of its origin, is atherosclerosis. Atherosclerosis is a disease of the arteries resulting from the deposit of fatty plaque on the inner walls.

The first solution to the problem of atherosclerosis was the coronary artery bypass graft (CABG) procedure, in which a graft is taken

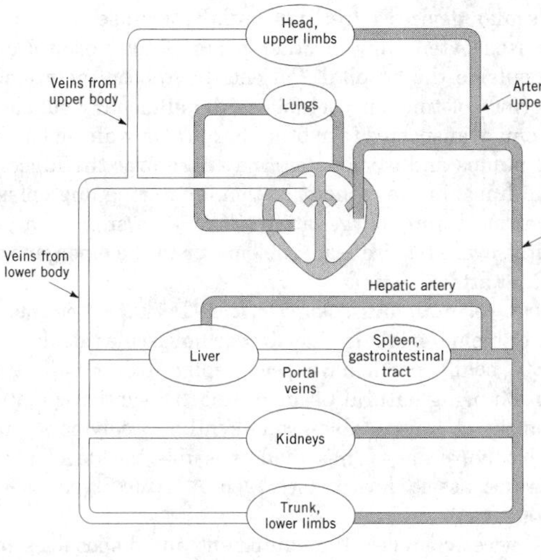

Figure 1. An overall view of the cardiovascular system.

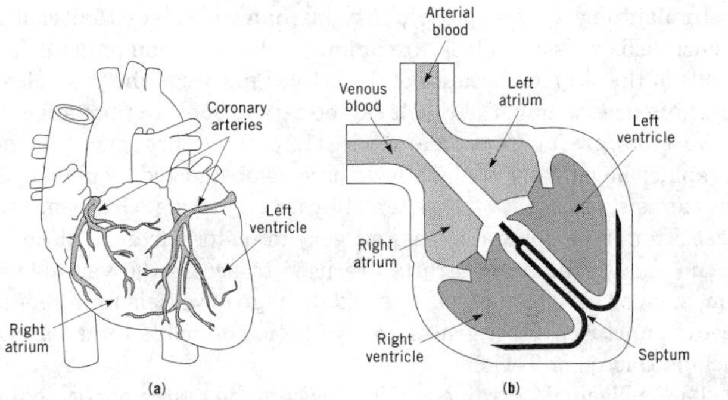

Figure 2. (**a**) Coronary arteries which form the heart's own blood supply; (**b**) electrical conduction system which powers the human heart.

from the patient's own saphenous vein. The graft is attached to the aorta where the coronary arteries originate and the opposite end is connected to the artery below the blocked segment. Blood can then bypass the obstructed area and reach the surrounding tissue below.

The second step toward solving cardiovascular disease from atherosclerosis, was angioplasty. Percutaneous transluminal coronary angioplasty (PTCA), a nonsurgical procedure, opens up blocked arteries.

The first implantable pacemaker, introduced in 1960, provided a permanent solution to a chronic bradyarrhythmia condition. This invention had a profound impact on the future of medical devices. The pacemaker was the first implantable device which became intrinsic to the body, enabling the patient to lead a normal life.

The primary solution to valve problems has been implantable replacement valves. The introduction of these devices necessitates open-heart surgery. There are two types of valves available: tissue (porcine and bovine) and mechanical. The disadvantage of tissue valves is that these have a limited life of about seven years before they calcify, stiffen, and have to be replaced. The mechanical valves can last a lifetime, but require anticoagulant therapy.

Interventional Procedures. The emergence of angioplasty created a specialty called interventional cardiology. Interventional cardiologists not only implant pacemakers and clear arteries using balloon catheters, but they also use balloons to stretch valves (valvuloplasty). In addition, they work with various approaches and technologies to attack plaque, including laser energy, mechanical cutters and shavers, stents to shore up arterial walls and deliver drugs, and ultrasound to break up plaque or to visualize the inside of the artery (see MEDICAL IMAGING TECHNOLOGY).

Cardiovascular devices developed initially for use in open-heart surgery are used extensively in other parts of the hospital and, in many cases, outside the hospital. Patients have been maintained for prolonged periods of time on portable cardiopulmonary support systems while being transported to another hospital or waiting for a donor heart. Blood pumps and oxygenators may take over the functions of the heart and lungs in the catheterization lab during angioplasty, in extracorporeal membrane oxygenation (ECMO) to support a premature baby with severe respiratory problems, or in the emergency room to assist a heart attack victim.

Biomaterials for Cardiovascular Devices. Perhaps the most advanced field of biomaterials is that for cardiovascular devices. Since the mid 1960s bodily parts have been replaced or repaired by direct substitution using natural tissue or selected synthetic materials. The development of implantable-grade synthetic polymers, such as silicones and polyurethanes, has made possible the development of advanced cardiac assist devices (see SILICON COMPOUNDS, SILICONES; URETHANE POLYMERS).

Surgical devices comprise the equipment and disposables to support surgery and to position implantable valves and a variety of vascular grafts. Central to open-heart surgery is the heart–lung machine and a supporting cast of disposable products.

Implantable valves, particularly mechanical valves that continue to encroach on tissue valves, are unique. Valve selection remains in the hands of the surgeon because of the critical nature of the procedure. If anything goes wrong, the result can be catastrophic to the patient.

Vascular grafts are tubular devices implanted throughout the body to replace blood vessels which have become obstructed by plaque, atherosclerosis, or otherwise weakened by an aneurysm. Grafts are used most often in peripheral bypass surgery to restore arterial blood flow in the legs. In addition, grafts are used to access the vascular system, such as in hemodialysis to avoid damage of vessels from repeated needle punctures. Most grafts are synthetic and made from materials such as Dacron or Teflon.

The principal cardiac-assist device, the intra-aortic balloon pump (IABP), is used primarily to support patients before or after open-heart surgery, or patients who go into cardiogenic shock. Other devices, which can completely take over the heart's pumping function, are the ventricular assist devices (VADs), supporting one or both ventricles.

Specialized applications of cardiac arrest devices include extracorporeal membrane oxygenation (ECMO) which occurs when the lungs of a premature infant cannot function properly.

Cardiac transplantation has become the treatment of choice for medically intractable congestive heart failure (CHF). Although the results of heart transplantation are impressive, the number of patients who might benefit far exceeds the number of potential donors.

In contrast, the total artificial heart (TAH) is designed to overtake the function of the diseased natural heart. While the patient is on heart–lung bypass, the natural ventricles are surgically removed. Polyurethane cuffs are then sutured to the remaining atria and to two other blood vessels that connect with the heart.

One successful total artificial heart is ABIOMED's electric TAH. This artificial heart consists of two seamless blood pumps which assume the roles of the natural heart's two ventricles. The pumps and valves are fabricated from a polyurethane, Angioflex. Small enough to fit the majority of the adult population, the heart's principal components are implanted in the cavity left by the removal of the diseased natural heart. A modest sized battery pack carried by the patient supplies power to the drive system. Miniaturized electronics control the artificial heart which runs as smoothly and quietly as the natural heart. Once implanted, the total artificial heart performs the critical function of pumping blood to the entire body.

The most commonly used heart valves in the mid-1990s included mechanical prostheses and tissue valves.

Surgical centers have a device, called the Cell Saver (Haemonetics), that allows blood lost during surgery to be reused within a matter of minutes, instead of being discarded. This device collects blood from the wound, runs it through a filter that catches pieces of tissue and bone, then mixes the blood with a salt solution and an anticoagulant. The device then cleanses the blood of harmful bacteria. Subsequently the blood is reinfused back to the same patient through catheters inserted in a vein in the arm or neck, eliminating the worry of cross-contamination from the HIV or hepatitis viruses (see BLOOD, COAGULANTS AND ANTICOAGULANTS; FRACTIONATION, BLOOD).

Polyurethanes as Biomaterials. Much of the progress in cardiovascular devices can be attributed to advances in preparing biostable polyurethanes, which offer significant advantages for important long-term products, such as implantable ports, hemodialysis, and peripheral catheters; pacemaker interfaces and leads; and vascular grafts.

The safest method of accessing the vascular system is by means of a vascular access device (VAD) or port. Ports allow drugs and fluids to be delivered directly into the bloodstream without repeated insertion of needles into a vein.

Vascular access ports typically consist of a self-sealing silicone septum within a rigid housing which is attached to a radiopaque catheter.

Technical advances in programmable pacemakers that assist both the tachycardia and bradycardia have led to the requirement of implanting a two-lead system. Owing to the ridigity and size of silicones, the only material that fulfills this possibility without significantly impeding blood flow to the heart is polyurethane.

Although the use of vascular grafts in cardiovascular bypass surgery is widely accepted and routine, numerous problems exist in these surgeries for the materials available. The primary needs are materials that can match compliance to native vessels, having a lesser diameter for small-bore grafts which would serve as a replacement for the saphenous vein in coronary bypass, thinner walls, biostability, controlled porosity, and greater hemocompatibility for reduced thrombosis. It is anticipated that newer polyurethane materials will be developed to address such problems.

Orthopedic Devices

Bone, or osseous tissue, is composed of osteocytes and osteoclasts embedded in a calcified matrix. Hard tissue consists of about 50% water and 50% solids. The solids are composed of cartilaginous material hardened with inorganic salts, such as calcium carbonate and phosphate of lime.

A living bone consists of three layers: the periosteum, the hard cortical bone, and the bone marrow or cancellous bone. The periosteum is a thin collagenous layer, filled with nerves and blood vessels, that supplies nutrients and removes cell wastes. Because of the extensive nerve supply, normal periosteum is very sensitive. When a bone is broken, the injured nerves send electrochemical neural messages relaying pain to the brain.

Joints are structurally unique. They permit bodily movement and are bound together by fibrous tissues known as ligaments.

Ligaments are composed of bands of strong collagenous fibrous connective tissue. Ligaments function to tie two bones together at a joint, maintain joints in position preventing dislocations, and restrain the joint's movements. Ligaments may be reattached to bone by the use of an orthopedic anchor.

Bones function as levers; joints function as fulcrums; muscle tissues, attached to the bones via tendons, exerts force by converting electrochemical energy (nerve impulses) into tension and contraction, thereby facilitating motion.

Bone Fractures. Bone fractures are classified into two categories: simple fractures and compound, complex, or open fractures. In the latter the skin is pierced and the flesh and bone are exposed to infection. A bone fracture begins to heal nearly as soon as it occurs. Therefore, it is important for a bone fracture to be set accurately as soon as possible.

Fracture Treatment. The movement of a broken bone must be controlled because moving a broken or dislocated bone causes additional damage to the bone, nearby blood vessels, and nerves or other tissues surrounding the bone. Emergency treatment requires splinting or bracing a fracture injury before further medical treatment is given.

Typically, x-rays determine whether there is a fracture, and if so, of what type (see MEDICAL IMAGING TECHNOLOGY).

Treatments used for various types of fractures are cast immobilization, traction, and internal fixation. A plaster or fiber glass cast is the most commonly used device for fracture treatment. This type of cast or brace is known as an orthosis. It allows limited or controlled movement of nearby joints.

Traction is typically used to align a bone by a gentle, constant pulling action. The pulling force may be transmitted to the bone through skin tapes or a metal pin through a bone.

In internal fixation, an orthopedist performs surgery on the bone. During this procedure, the bone fragments are repositioned (reduced) into their normal alignment and then held together with special screws or by attaching metal plates to the outer surface of the bone. The fragments may also be held together by inserting rods (intramedullary rods) down through the marrow space into the center of the bone.

Joint Replacement. The most frequent reason for performing a total joint replacement is to relieve the pain and disability caused by severe arthritis. The surface of the joint may be damaged by osteoarthritis, ie, a wearing away of the cartilage in a joint. The joint may also be damaged by rheumatoid arthritis, an autoimmune disease, in which the synovium produces chemical substances that attack the joint surface and destroy the cartilage. When arthritis has caused severe damage to a joint, a total joint replacement may allow the person to return to normal everyday activities.

A total joint replacement is a radical surgical procedure performed under general anesthesia, in which the surgeon replaces the damaged parts of the joint with artificial materials. For example, in the knee joint the damaged ends of the bone that meet at the knee are replaced, along with the underside of the kneecap. In the hip joint, the damaged femoral head is replaced by a metal ball having a stem that fits down into the femur. A new plastic socket is implanted into the pelvis to replace the old damaged socket. This is shown schematically in Figure 3. Whereas hips and knees are the joints most frequently replaced, because the scientific understanding of these is best, total joint replacement can be performed on other joints as well, including the ankle, shoulder, fingers, and elbow.

The materials used in a total joint replacement are designed to enable the joint to function normally. The artificial components are generally composed of a metal piece that fits closely into bone tissue. The metals are varied and include stainless steel or alloys of cobalt, chrome, and titanium. The plastic material used in implants is a polyethylene that is extremely durable and wear-resistant. Also, a bone cement, a methacrylate, is often used to anchor the artificial joint materials into the bone. Cementless joint replacements have more recently been developed. In these replacements, the prosthesis and the bone are made to fit together without the need for bone cement. The implants are press-fit into the bone.

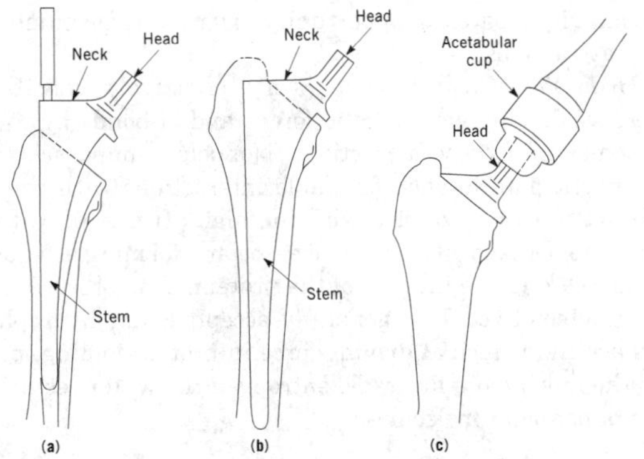

Figure 3. Three views of a hip implant for joint replacement: (**a**) insertion of the implant into the femur; (**b**) implant in place; and (**c**) femur and implant connected to plastic socket fitted into the pelvis.

Prosthesis Design. The challenge in prosthesis design is to create an implant that mimics the material characteristics and the exact anatomical functions of the joint. Stress and loading forces on the hip joint and femur are extraordinary. Stresses on the hip joint exceed 8.3 MPa (1200 psi). Standing on one leg produces a loading force on the hip joint of 250% of the total body weight. Running increases these forces to five times the total body weight. The hip joint is surrounded by the most powerful muscle structure in the body enabling movement while supporting sufficient structural force and loads. Proper surgical technique is as critical to the success of an implant procedure as is the design of the device itself.

A significant aspect of hip joint biomechanics is that the structural components are not normally subjected to constant loads. Rather, this joint is subject to unique compressive, torsion, tensile, and shear stress, sometimes simultaneously. Maximum loading occurs when the heel strikes down and the toe pushes off in walking. When an implant is in place its ability to withstand this repetitive loading is called its fatigue strength. If an implant is placed properly, its load is shared in an anatomically correct fashion with the bone.

Resorption of bone tissue occurs in total hip joint replacement patients if sufficient stresses are not adequately transmitted to the remaining bone in exactly the same way that the bone transmitted those stresses originally. Therefore, the design and proper placement of the neck collar and hip stem must be effective in recreating anatomical structure.

Bone remodeling is the ability of bone to change its size, shape, and structure by adapting to mechanical demands that are placed on it. Bone grows where it is needed, and resorbs where it is not needed.

Biomaterials. Pure metals are usually too soft to be used in prosthesis. Alloys which exhibit improved characteristics of fatigue strength, tensile strength, ductility, modulus of elasticity, hardness, resistance to corrosion, and biocompatibility are used.

Titanium alloy, composed of titanium, aluminum, and vanadium, is preferred by some orthopedic surgeons primarily for its low modulus of elasticity, which allows for transfer of more stress to the proximal femur. This alloy also exhibits good mechanical strength and biocompatibility. The stem flexibility optimizes the transfer of stress directly to the bone, and offers adequate calcar loading to minimize femoral resorption.

Composites used in orthopedics include carbon–carbon, carbon–epoxy, hydroxyapatite, ceramics, etc.

Tools and Procedures. Arthroscopy is a surgical procedure used to visualize, diagnose, and treat injuries within joints. The term arthroscopy literally means to look inside the joint. During this procedure the orthopedic surgeon makes an incision into the patients skin and inserts a pencil-shaped arthroscope. An arthroscope is a miniature lens and lighting systems that magnifies and illuminates the structures inside the joint. A television screen which is attached to the arthroscope displays the image of the joint on screen.

This is a minimally invasive procedure (MIP) resulting in a shorter hospital stay, faster recovery, and less evident scar in comparison to other types of surgery. Arthroscopic surgery gives the surgeon a precise, direct view of the affected bones and soft tissues. This procedure allows the surgeon to see areas of the joint that are difficult to see on x-rays and more of the joint than is possible even after making a large incision during open surgery.

Bioresorbable Polymers

Research by biomaterials scientists has focused on synthesis of polymeric structures which exhibit biocompatibility and long-term biostability. Devices made from these polymers are intended to be implanted in the body for years, and in some cases decades.

The concept of using biodegradable materials for implants which serve a temporary function is a relatively new one. This concept has gained acceptance as it has been realized that an implanted material does not have to be inert, but can be degraded and/or metabolized *in vivo* once its function has been accomplished. Resorbable polymers have been utilized successfully in the manufacture of sutures, small bone fixation devices, and drug delivery systems (qv).

MICHAEL SZYCHER
PolyMedica Industries, Inc.

Szycher's Dictionary of Biomaterials and Medical Devices, Technomic Publishing Co., Inc., Lancaster, Pa., 1992.

Technical data, Lifequest Medical, Inc., San Antonio, Tex., 1994.

U.S. Pat. 5,254,662 (1993), M. Szycher (to PolyMedica Industries, Inc.).

U.S. Pat. 4,209,607 (June 24, 1980), W. Shalaby (to Ethicon, Inc.).

PROTACTINIUM. See ACTINIDES AND TRANSACTINIDES.

PROTEIN ENGINEERING

Protein engineering encompasses a wide variety of techniques, ranging from the rational modification of existing proteins to the de novo design of novel proteins. Protein engineering is most commonly carried out by manipulation of the protein at the genetic level; that is, by mutagenesis of the gene (deoxyribonucleic acid (DNA)) which codes for the protein (see GENETIC ENGINEERING; NUCLEIC ACIDS). Other methods, such as chemical modification and chemical synthesis of proteins (qv), are also included within the scope of protein engineering.

One of the principal goals of protein engineering is to provide insight into the complex relationship between protein structure and function. Many protein engineering studies have involved changing specific amino acid residues within proteins of known crystal structures in order to test predictions as to how such changes may affect the structure or function of the protein.

Protein Structure

Much of protein engineering concerns attempts to explore the relationship between protein structure and function. Proteins are polymers of amino acids (qv), which have general structure $^{+}H_3N$—CHR—COO^-, where R, the amino acid side chain, determines the unique identity and hence the structure and reactivity of the amino acid. Formation of a polypeptide or protein from the constituent amino acids involves the condensation of the amino-nitrogen of one residue to the carboxylate-carbon of another residue to form an amide. The linear order in which amino acids are linked in the protein is called the primary structure of the protein or, more commonly, the amino acid sequence. Only 20 amino acid structures are used commonly in the cellular biosynthesis of proteins (qv).

Discrete segments of the polypeptide chain can fold into regular, repeating structural motifs, known as secondary structures. It is the identity and sequence of amino acids within a polypeptide chain that dictate its secondary structure.

The way in which the elements of secondary structure fold upon each other to form compact globular structures is referred to as tertiary structure. Noncovalent forces such as hydrogen bonds, electrostatic interactions, and hydrophobic interactions play a dominant role in determining protein tertiary structure. Often the active form of a protein is a complex of polypeptides held together by noncovalent or covalent (disulfide) interactions. The arrangement of these polypeptide subunits relative to one another defines the quaternary structure of the protein.

Noncovalent Forces Stabilizing Protein Structure. Much of protein engineering concerns attempts to alter the structure or function of a protein in a predefined way.

Through combined effects of noncovalent forces, proteins fold into secondary structures, and hence a tertiary structure that defines the native state or conformation of a protein. The native state is then that three-dimensional arrangement of the polypeptide chain and amino acid side chains that best facilitates the biological activity of a protein. Through protein engineering subtle adjustments in the structure of the protein can be made that can dramatically alter its function or stability.

Table 1. Names and Codes of the Natural Amino Acids

Amino acid	Three-letter code	One-letter code
alanine	Ala	A
arginine	Arg	R
asparagine	Asn	N
aspartate	Asp	D
cysteine	Cys	C
glutamate	Glu	E
glutamine	Gln	Q
glycine	Gly	G
histidine	His	H
isoleucine	Ile	I
leucine	Leu	L
lysine	Lys	K
methionine	Met	M
phenylalanine	Phe	F
proline	Pro	P
serine	Ser	S
threonine	Thr	T
tryptophan	Trp	W
tyrosine	Tyr	Y
valine	Val	V

Electrostatics. Electrostatic interactions, such as salt bridges, result from the attraction that occurs between oppositely charged molecules. In proteins these interactions usually involve the amino acids Lys, Arg, Glu, and Asp. This attractive force is inversely proportional to the distance between the charges and the dielectric constant of the solvent, as described by Coulomb's law.

Hydrogen Bonds. The hydrogen-bond (H-bond), consists of a hydrogen donor to which the H-atom is covalently bound and an acceptor containing either a partial or full negative charge that attracts the electropositive H-atom. Eleven of the 20 amino acids can participate in H-bonds, Arg, Asp, Asn, Cys, Glu, Gln, His, Lys, Ser, Thr, and Tyr. Stronger than van der Waals bonds but much weaker than covalent bonds, hydrogen bonds constitute a primary determinant of biological specificity. These bonds are sufficiently strong to direct molecular interactions such as the attraction between an enzyme and its substrate, but sufficiently weak to be reversibly made and broken, as in enzyme catalysis.

Van der Waals Interactions. Van der Waals interactions result from the asymmetric distribution of electronic charge surrounding an atom, which induces a complementary dipole in a neighboring atom, resulting in an attractive force. In general, the attractive force of van der Waals interactions is very weak but may become significant if steric complementarity creates an opportunity to form a large number of van der Waals attractions.

The Hydrophobic Effect. Water very effectively solvates polar molecules, weakening electrostatic forces and H-bonds by competing for the elements of those attractions. Nonpolar amino acids cannot, however, participate in such favorable interactions with water. This absence of interaction, coupled with the high affinity of water for itself, forces the nonpolar molecules to associate, forming a sequestered, hydrophobic core in the interior of the protein. This process is termed the hydrophobic effect. It is generally accepted that hydrophobic attractions are the principal driving force in protein folding, primarily owing to the much more favorable entropy of the water resulting from exclusion of nonpolar molecules.

The Protein Engineering Process

Identification and Characterization of the Target Protein. Prior to commencement of the process of protein engineering, a target pro-

Table 2. The Genetic Code

First position (5′-end)	Second position U	C	A	G	Third position (3′-end)
U	Phe	Ser	Tyr	Cys	U
	Phe	Ser	Tyr	Cys	C
	Leu	Ser	STOP	STOP	A
	Leu	Ser	STOP	Trp	G
C	Leu	Pro	His	Arg	U
	Leu	Pro	His	Arg	C
	Leu	Pro	Gln	Arg	A
	Leu	Pro	Gln	Arg	G
A	Ile	Thr	Asn	Ser	U
	Ile	Thr	Asn	Ser	C
	Ile	Thr	Lys	Arg	A
	Met (START)	Thr	Lys	Arg	G
G	Val	Ala	Asp	Gly	U
	Val	Ala	Asp	Gly	C
	Val	Ala	Glu	Gly	A
	Val	Ala	Glu	Gly	G

tein must be selected and some structural information on the natural form of that protein determined. At a minimum, some amino acid sequence information must be known in order to design the oligonucleotide probes that are used to identify the gene for the target protein. Additionally, information on other properties of the target protein, such as molecular weight and isoelectric point are useful to confirm the identity of the cloned protein at a later date.

Whereas most proteins are soluble in the cell cytosol, many proteins are associated with the various membranes found in cells. Membrane proteins present additional challenges and one must therefore know in advance whether or not the target protein is membrane-associated. Finally, it is important to know whether the active form of the target protein is a single, monomeric polypeptide, or if it consists of multiple subunits.

Cellular Protein Biosynthesis. The process of cellular protein biosynthesis is virtually the same in all organisms. The information which defines the amino acid sequence of a protein is encoded by its corresponding sequence of DNA (the gene). The DNA is composed of two strands of polynucleotides, each comprising some arrangement (sequence) of the four nucleotide building blocks of the nucleic acids: adenine (A), thymine (T), guanine (G), and cytosine (C). In ribonucleic acid (RNA), T is replaced by uracil (U). The two strands of DNA are held together by hydrogen-bonding patterns between specific nucleotide bases, ie, base pairing. The structures of the bases are such that A always pairs with T, G with C. Hence, the two strands of DNA are complementary; that is, the nucleotide sequence of one strand defines exactly the sequence of the other strand. This complementarity provides a convenient means for living organisms to replicate their genetic information and forms the technical basis for protein engineering.

The relationship between the nucleotide sequence of the DNA and the amino acid sequence of the protein is known as the genetic code (Table 2). In this code, a sequence of three nucleotides (a codon) specifies a particular amino acid. The genetic code consists of 64 codons, of which 61 code for amino acids and three are translation stop signals (stop codons). Because the genetic code is nearly the same in all organisms, it is possible to insert genetic information from one organism into a different host organism. The latter can then use its biosynthetic machinery to produce the foreign protein. This process is referred to as heterologous protein expression.

Cellular protein biosynthesis involves the following steps. One strand of double-stranded DNA serves as a template strand for the synthesis of a complementary single-stranded messenger ribonucleic acid (mRNA) in a process called transcription. This mRNA in turn serves as a template to direct the synthesis of the protein in a process called translation. The codons of the mRNA are read sequentially by transfer RNA (tRNA) molecules, which bind specifically to the mRNA via triplets of nucleotides that are complementary to the particular codon, called an anticodon. Protein synthesis occurs on a ribosome, a complex consisting of more than 50 different proteins and several structural RNA molecules, which moves along the mRNA and mediates the binding of the tRNA molecules and the formation of the nascent peptide chain. The tRNA molecule carries an activated form of the specific amino acid to the ribosome where it is added to the end of the growing peptide chain. There is at least one tRNA for each amino acid.

Cloning of the Target Protein. Cloning procedures have been made possible by the availability of several key types of enzymes, including restriction endonucleases, enzymes that cleave double-stranded DNA at specific sites; DNA polymerases (especially thermostable forms), which replicate DNA templates; and DNA ligases, which covalently link (ligate) fragments of DNA. In general, fragments of DNA are ligated into vectors that can autonomously replicate their DNA in the host organism. The two most common vectors used in bacteria are plasmids and phage. Plasmids are naturally occurring circular DNA molecules that can act as accessory chromosomes. Phage, such as lambda phage, are viruses which can both stably integrate into the chromosome of the host (lysogenic pathway) or use the host machinery to produce more viral particles, which in turn lyse the host cell and destroy it (lytic pathway).

Making a cDNA Library. Within the genes of eukaryotes, extraneous stretches of nucleotide sequences that do not code for amino acid residues in the protein (introns) are often interspersed among the coding sequences (exons). When DNA is transcribed, these introns are excised from the mRNA to form a contiguous sequence of exons that translate into the contiguous amino acid sequence of the protein. Eliminating the introns from the DNA directly would provide a convenient tool for cloning purposes. Therefore, complementary DNA (cDNA) libraries are constructed from cellular mRNA. This form of RNA has already been processed by the cell and therefore the DNA corresponds to the actual nucleotide sequence that codes for the protein without the interrupting introns.

Screening of a cDNA Library. The bacteria containing individual cDNA recombinant plasmids are grown on an agar plate where individual colonies can be visualized, and these colonies are screened for the presence of the desired DNA, using a probe specific to the DNA of interest. The probe is generally an oligonucleotide where the sequence complements a segment of the nucleotide sequence of the cDNA of interest. The probe is usually radiolabeled to allow easy identification of its location. The bacterial cells on an agar plate are replica-plated onto nitrocellulose filters, where the cells are immobilized. The filter thus becomes an exact replica of the arrangement of colonies on the agar plate. The cells are lysed and the DNA inside is denatured into single-stranded forms. The probe, which is present in excess, can then hybridize (base-pair) to complementary DNA on the nitrocellulose. The location of the hybridizing probe is then identified by autoradiography, and the colony corresponding to the positive signal is isolated and further tested. After a plasmid containing the cDNA is identified by hybridization techniques, the DNA sequence is determined.

Manipulating the Recombinant DNA Clone. Cloned DNA can be expressed or the gene can be altered by deletions, insertions, or substitutions in the cloned DNA. Deletion and insertion generally involve the use of restriction endonucleases to cut the DNA at a specific site or sites and then remove the desired segment or ligate an additional segment to the gene. Single site-specific substitution is a more common practice for the study of protein structure and function. The general procedure for substituting one amino acid residue for another involves a technique termed oligonucleotide-directed mutagenesis. In this procedure, a mutagenic oligonucleotide primer of 20–30 nucleotides, containing usually one to three nucleotide mismatches to change a codon that specify the desired amino acid substitution, is incorporated into the DNA and specifically selected for further replication. Two particularly popular procedures for this replication are use of single-stranded circular phage DNA and use of the polymerase chain reaction (PCR) using double-stranded DNA.

Gene Expression. Once the cDNA for the natural protein (wild type) or a mutant thereof is cloned, the cDNA is inserted into an appropriate vector containing a ribosome binding site and a promoter that can direct synthesis of the desired protein. The choice of vector depends on the choice of host organism in which to express the protein. The most common expression systems include bacteria, such as *E. coli;* mammalian cells, such as Chinese hamster ovary (CHO) and monkey kidney (COS) cells; baculovirus-infected insect cells; and yeast (see CELL CULTURE TECHNOLOGY; YEASTS).

Purification of Expressed Proteins. Once an appropriate host has been selected and the protein has been expressed, isolation of that protein in pure form is needed. Methods for general protein purification typically involve a combination of chromatographic separations based on physicochemical properties of the protein, such as molecular weight (size exclusion chromatography), electrostatic charge (ion exchange and chromatofocusing), or specificity of interaction with another protein or small molecule (affinity-based chromatography). Such methods are generally applicable to all proteins, whether expressed recombinantly or naturally occurring. For engineered proteins, additional purification strategies are available that take advantage of the ability to manipulate the cDNA of the target protein.

Applications

Studies of Protein Stability. Protein engineering has provided a means of assessing the contributions of various noncovalent interactions to protein stability. In general, one amino acid is replaced by another differing from the original by only one functional aspect such as size, charge, or H-bonding ability. By making a series of systematic mutations and determining the effect on protein stability, the importance of the deleted functional group can be assessed. Whereas results vary from protein to protein and depend on the details of the local environment experienced by the amino acid residue, similar results from studies of several different proteins have provided some encouraging generalizations regarding the contributions of noncovalent interactions to protein stability.

Protein stability is usually measured quantitatively as the difference in free energy between the folded and unfolded states of the protein. These states are most commonly measured using spectroscopic techniques, such as circular dichroic spectroscopy, fluorescence (generally tryptophan fluorescence) spectroscopy, nmr spectroscopy, and absorbance spectroscopy.

The importance of the hydrophobic core to protein stability is probed by substituting a nonpolar residue within the core with a smaller nonpolar residue. The resultant difference in free energy of unfolding can then be measured.

The importance of H-bonds to the stability of folded protein had been thought of as minor. However, mutational analysis has revealed that the sum of individual intramolecular H-bond energies can contribute a significant stabilization energy, approaching that of the hydrophobic effect.

The introduction of disulfide bonds can have various stabilizing or destabilizing effects on proteins. Stabilization is thought to result from reduction of the conformational entropy of the unfolded state, whereas in most cases the cause of destabilization is the introduction of dihedral angle stress.

The nature of the amino acid side chain can affect protein stability, not only by effecting changes in cavity sizes and hydrophobic or electrostatic interactions, but also because some side chains are not tolerable in certain secondary structure forms.

De Novo Designed Proteins. To further elucidate the forces which direct the protein folding pathway and stabilize the final native state, several laboratories are studying de novo (from first principles) designed proteins, that is, design of a particular protein structural motif using the level of understanding of forces believed to promote and stabilize a desired structural element. The method is then tested by evaluating the extent to which the designed protein matches the expected structure. De novo protein design has been successfully used to create proteins having all α-helix, all β-sheet, and mixed α/β secondary structures.

Studies of Protein Function. For enzymes to catalyze reactions, and receptors to transport molecules and transmit signals across cellular membranes, both classes of proteins require a ligand-binding pocket, the structure of which is highly complementary to that of the ligand. It is this complementarity of structure that provides the high degree of ligand specificity characteristic of these two classes of proteins. The binding-pocket structure depends on the overall tertiary structure of the protein as well as the specific amino acid side chains that line the binding pocket. In the case of enzymes, changes in the residues within the binding pocket can greatly affect the rate or nature of the reaction catalyzed. For both enzymes and receptors, changes of amino acid residues within the binding pocket can also dramatically alter the ligand specificity. Understanding these structure–function relationships allows the rational design of a binding pocket to alter the ligand specificity of these proteins for specific purposes, such as altered catalysis by an enzyme, or transport of unnatural molecules into cells by an engineered receptor.

Expanded Genetic Code. Protein engineering techniques have employed site-directed mutagenesis to elucidate the role of a particular residue in protein structure or function. A technique that uses natural protein synthesis machinery site-specifically to incorporate unnatural amino acids into the protein has also been adapted. Thus, instead of being limited to the 20 naturally occurring amino acids specified by the genetic code, an amino acid incorporating novel steric, electronic, or spectroscopic properties can be employed permitting a more systematic variation in the properties of the mutated residue. Size, shape, acidity, and nucleophilicity, hydrogen-bonding strength, and hydrophobicity can all be probed. Additionally, biophysical probes such as spin and isotopic labels can be incorporated into a protein at the site of interest. This technique has been applied to the study of protein stability, enzymatic catalysis, and signal transduction.

The procedure for incorporating a site-specific unnatural amino acid involves the following steps. The codon for the amino acid of interest is replaced with the stop codon UAG, using standard oligonucleotide-directed mutagenesis. A suppressor tRNA that recognizes this codon is then chemically acylated using the desired unnatural amino acid. Transcription and translation of the mutagenized mutated gene is performed *in vitro* using an *E. coli* or rabbit reticulocyte extract to synthesize the mutant protein containing the unnatural amino acid at the specified position.

JUNE P. DAVIS
ROBERT A. COPELAND
The Du Pont Merck Research Laboratories

A. R. Rees, M. J. E. Sternberg, and R. Wetzel, *Protein Engineering: A Practical Approach,* IRL Press at Oxford University Press, New York, 1992.

V. W. Cornish and P. G. Schultz, *Curr. Opin. Struct. Biol.* **4,** 601–607 (1994).

J. Sambrook, E. F. Fritsch, and T. Maniatis, *Molecular Cloning,* 2nd ed., Cold Spring Harbor Press, Cold Spring Harbor, N.Y., 1989.

Methods in Enzymology **185** (1990). The entire volume covers gene expression.

PROTEINS

Proteins, ubiquitous to all living systems, are biopolymers (qv) built up of various combinations of 20 different naturally occurring amino acids (qv). Proteins are encoded by the deoxyribonucleic acid (DNA) that is present in all living cells.

The large number of biochemical reactions within the living cell are catalyzed by enzyme proteins (see ENZYME APPLICATIONS; HORMONES; PHARMACEUTICALS). Traditional food processes such as baking (see BAKERY PROCESSES AND LEAVENING AGENTS), brewing (see

BEER), and cheese-making (see MILK AND MILK PRODUCTS) involve the action of enzymes found in different microorganisms. Other proteins are involved in the transport of electrons, ions, and small molecules. Proteins are also key components of the immune system (see IMMUNOTHERAPEUTIC AGENTS) and control the genetic expression of other proteins (see GENETIC ENGINEERING).

The study of proteins has a long history dating from the late 1700s. Great advances have taken place in protein research since the 1950s through biochemistry and molecular biology (see PROTEIN ENGINEERING).

Properties

The unique structure (conformation) and hence, function, of a given protein molecule is determined by its amino acid sequence, under normal physiological conditions. Although the size and chemical properties of the amino acid side chains vary greatly, these building blocks can be broadly classified into three categories: hydrophobic, charged, and uncharged polar.

Purification and Sequencing. The properties of a protein such as size, net electric charge, and solubility can be exploited in protein purification. Centrifugation is used to separate proteins based on density and shape. Electrophoresis exploits the net charge on a protein as well as its mass and shape (see ELECTROSEPARATIONS). Electrophoretic separation is usually carried out by passing the mixture through a gel medium in an electric field. In chromatography (qv), separation of proteins is carried out by passing them through a column packed with some porous material. The mixture is separated according to size in size-exclusion chromatography, ionic charge in ion-exchange chromatography, or selective adsorption in affinity-labeled chromatography.

Protein sequences are inferred from their corresponding nucleotide sequences since they are easier to obtain. Large databases exist for the sequences of proteins from different organisms.

Biosynthesis

The whole of molecular biology is based on the central dogma that DNA → ribonucleic acid(RNA) → protein. Each amino acid residue is encoded by a triplet of nucleotides which form the genetic material of chromosomes. The DNA and, in some cases the RNA, is transcribed onto a strand of messenger-RNA (mRNA) and its translation to the protein occurs on ribosomes, specialized organelles that are present in the cytoplasm of the cell. This is a complex process, the main events of which occur in the following series of steps: (*1*) the ribosome binds to one end of the mRNA strand; (*2*) the first triplet of the gene coding for a polypeptide, which always corresponds to the amino acid methionine (Met), binds to its specific transfer-RNA (tRNA) molecule, activating methionine; (*3*) the chain is elongated when the previous amino acid comes off its tRNA and combines with the next amino acid to yield a peptide bond; and (*4*) elongation continues until a specific termination triplet is encountered, when the polypeptide chain dissociates from the last tRNA and the ribosome.

Post-Translational Modifications. Proteins are often synthesized having 15–26 residue long signal peptides, usually at the amino terminus. The signal peptide serves to translocate the protein across a membrane when it is required in a location other than the cytoplasm. Once the protein has arrived, the signal peptide is cleaved. Certain enzymes and hormones undergo further cleavage to attain a specific functional form. Many viruses synthesize polyproteins which are then cleaved into individual polypeptide chains.

Proteins also undergo other covalent modifications after synthesis. These include (*1*) acetylation of the amino-terminal and amidation of the carboxy-terminal residue; (*2*) addition of fatty acid groups (myristoyl) or lipids, eg, farnesyl, for anchoring proteins to membranes; (*3*) glycosylation, one of the most common modifications, which serves many purposes such as cell–cell recognition through interaction with other molecules on the surface of cells; (*4*) phosphorylation, resulting in the addition of a negative charge, which serves as a switch to control the affinity for activators and inhibitors particularly in signal transduction; and (*5*) disulfide bond formation, in which cysteine (Cys) residues separated in sequence are brought spatially close together through covalent linkage.

Principles of Protein Structure

To understand the function of a protein at the molecular level, it is important to know its three-dimensional structure. The diversity in protein structure, as in many other macromolecules, results from the flexibility of rotation about single bonds between atoms. Each peptide unit is planar, ie, $\omega = 180°$, and has two rotational degrees of freedom, specified by the torsion angles ϕ and ψ, along the polypeptide backbone. The number of torsion angles associated with the side chains, R, varies from residue to residue. The allowed conformations of a protein are those that avoid atomic collisions between nonbonded atoms.

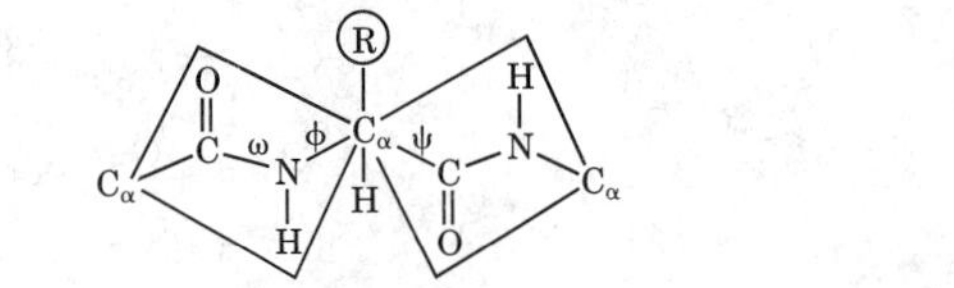

For any given protein, the number of possible conformations that it could adopt is astronomical. Yet each protein folds into a unique structure which is at a free energy minimum.

One of the principal driving forces determining the folded structure of a protein is the maintenance of peptide backbone hydrogen bonds on the removal of these bonds from the solvent to the protein interior. This is a force complementary to the hydrophobic effect that forces the nonpolar amino acid side chains away from the solvent into the interior. It is therefore natural to ask what types of structures allow for the maintenance of these hydrogen bonds. In the structure shown it is clear that on opposite edges of the peptide plane there is a hydrogen bond donor (N–H) and a hydrogen bond acceptor (O). Therefore if any set of peptide chains were stretched out side by side, a set of parallel hydrogen bonds between the chains could form. Another type of regular structure that can form is when the peptide chain twists into a long helix such that as one goes up the helix, one edge of the peptide plane always points upward (carboxy terminus) and the other points in the downward direction (amino terminus). The two basic structures of this kind, called the secondary structure, are the β-sheet and α-helix, respectively, shown in Figure 1.

Building on the two secondary structures and using the fact that amino acids can, to a first approximation, be classified as either hydrophilic (polar) or hydrophobic (nonpolar), allows for a simple understanding of the basic architecture of most proteins. Protein folding seeks to maximize the number of backbone hydrogen bonds while minimizing the number of exposed nonpolar amino acid side chains. Making the assumption that by starting with only helices and sheets the first part of the score has been optimized, only the number of solvent-exposed nonpolar residues has to be minimized.

There are basically three kinds of helices: hydrophobic, hydrophilic, and amphipathic. In the latter, the two faces have opposite charge distributions. Sheets can be either hydrophobic or amphipathic. These structural units are strung together by a set of small connecting loops. The number of ways that these units can pack against each other, subject to the constraints of charge complementarity and exposure to solvent, gives rise to the various topologies of protein structures, referred to as its tertiary structure.

Depending on the predominance of the type of secondary structure, proteins can fall into any of three classes; all-α-proteins, all-β-proteins, and α/β-proteins.

All-α-Proteins. The globins are well-known examples of all-α-proteins; they consist of eight helices packed together around a heme group to form a box-like structure. The helices are so arranged that the ridges formed by the side chains on one helix fit into the grooves between the side chains of another helix. This packing arrangement has been proved to be true in the packing of helices in proteins.

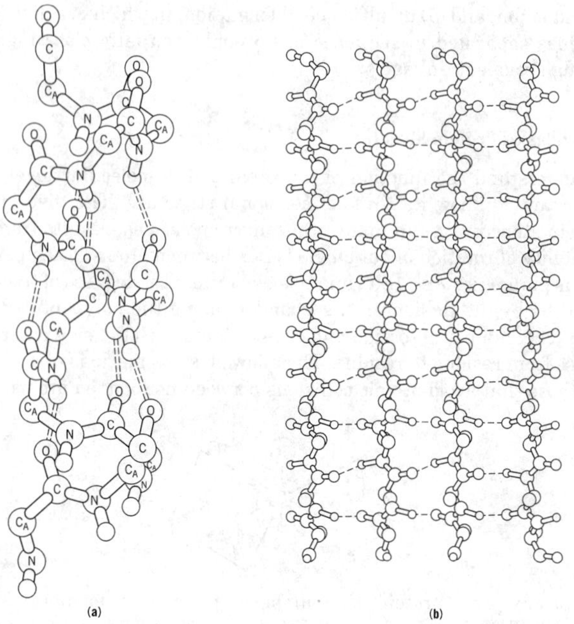

Figure 1. The two principal elements of secondary structure in proteins. (**a**) The α-helix stabilized by hydrogen bonds between the backbone of residue i and $i + 4$. There are 3.6 residues per turn of helix and an axial translation of 150 pm per residue. C_A represents the carbon connected to the amino acid side chain, R. (**b**) The β-sheet showing the hydrogen bonding pattern between neighboring extended β-strands. Successive residues along the chain point alternately up and down the plane of the paper.

Other examples are hemerythrin, α-keratin, and most of the membrane-spanning regions of integral membrane proteins.

All-β-Proteins. The general topology of all-β-proteins as a structural class of proteins consists of two or more β-sheets packed against each other, the mode of packing depending on the nature of the faces of the sheets. The immunoglobulin fold falls into this class.

α/β-Proteins. The most frequent domain structures observed in proteins are those having a mixture of helices and sheets. A large number of proteins have the α/β-sandwich structures in which a β-sheet is packed against either a single row of helices or is sandwiched between two such rows.

Larger proteins usually have two or more structural units termed domains, each domain having structures similar to single-domain proteins.

In some proteins, such as hemoglobin, separate polypeptide chains must associate for the chains to be functional. This forms a quaternary structure.

Cofactors. Frequently proteins exist in their native state in association with other nonprotein molecules or cofactors, which are crucial to their function. These may be simple metal ions, such as Fe^{n+} in hemerythrin or Ca^{2+} in calmodulin; a heme group, as for the globins; nucleotides, as for dehydrogenases, etc.

Native Conformation. Proteins assume their unique native conformation either during or soon after biosynthesis. This process, known as protein folding, can usually be reproduced *in vitro* under suitable conditions such as pH and temperature. It is believed that in arriving at the native state, proteins pass through a transient intermediate state, the so-called molten globule state. The precise structural features of this state are not yet known.

Many proteins frequently require the assistance of other protein molecules called molecular chaperonins, for assuming the final tertiary structure *in vivo*.

Protein Function

Proteins can be broadly classified into fibrous and globular. Many fibrous proteins such as collagen and α-keratin, serve a structural role. Others, like muscle fibers, have a motile function.

Globular proteins have biological function which they carry out by direct interaction with ligands that may be small atoms or large macromolecules. The protein without the bound ligand is known as the apo form, whereas the bound state is called the holo form. Many proteins are capable of binding more than one ligand, but usually do so on separate domains. Sometimes the binding of a ligand at a particular site of a protein can have an effect on the affinity of binding of another ligand at a second site. This is termed allostery.

The allosteric effect is seen in hemoglobin, which can exist in two quaternary structural states: oxygenated (R) or deoxygenated (T).

Binding sites comprise only a small fraction of the structure of most proteins. Most of the rest of the protein is responsible for providing the framework of the protein and in the process offering it stability. Consequently, functionally related proteins have the same overall topology, yet differ in the residues that form their binding sites, providing them their substrate specificity. During the folding process, functional residues distant in sequence are brought spatially close together.

Enzymes. Enzymes increase the rate of equilibrium of a chemical reaction by decreasing the barrier between the free energies of the products and reactants. An enzyme can function at very low concentrations by catalyzing the same reaction numerous times. The course of an enzymatic reaction can be studied using steady-state kinetics, where the dependence of the velocity of a reaction on substrate concentration is followed (see KINETIC MEASUREMENTS). An alternative method is relaxation spectrometry, in which the reaction is allowed to come to equilibrium, after which a rapid shift is made to one of the thermodynamic variables. The time course of the relaxation to the new equilibrium state can be followed using a variety of techniques.

Molecules of the Immune System. Vertebrates possess the capacity to recognize and bind foreign molecules called antigens for removal from the system. This is carried out by antibodies or immunoglobulin molecules, remarkable proteins in that having a limited gene pool, ie, about 1000 segments in humans, the body is able to generate millions of such molecules.

The major histocompatibility complex (MHC) molecules are responsible for binding foreign antigens and presenting these to T-cell receptor (TCR) molecules, which trigger a series of events leading to eventual destruction of the foreign invader.

DNA-Binding Proteins. The process of differentiation, by which a single cell multiplies to form different types of cells at different generations, is accomplished by the turning on and off of genes at various stages. The expression and regulation of the genetic information are carried out by proteins which bind to segments of DNA that surround the gene coding for the protein to be synthesized. These can either act as repressors or activators, depending on whether they prevent or help the RNA polymerase from binding to the DNA and initiate its transcription.

Another class of DNA-binding proteins are the polymerases. These have a nonspecific interaction with DNA because the same protein acts on all DNA sequences. DNA polymerase performs the dual function of DNA replication, in which nucleotides are added to a growing strand of DNA, and acts as a nuclease to remove mismatched nucleotides.

Other common DNA-binding motifs are the zinc finger found in many gene-regulating proteins and the Leucine zipper found in many transcription factors.

Signal Transduction Proteins. Biological systems have the ability to respond to changes in external concentration of hormones (qv), growth factors, or other molecules or stimuli. These ligands bind to specific transmembrane receptors initiating a cascade of biochemical processes that produce an intracellular signal.

Growth factors function by activating their receptors to phosphorylate other proteins. This results in the initiation of genetic transcription, thereby causing the cell to grow.

Homology

Proteins that carry out the same or similar functions in different organisms generally have very similar structures. Such proteins also are often encoded by similar sequences of amino acids. These similarities are a result of the shared evolutionary history of the organisms and of their genes. Such similar proteins are termed homologous, and are believed to have a common ancestor. There is no direct way of determining whether two proteins are truly homologous. However, when the sequences of any group of proteins are recognizably similar it seems much more likely that they are homologues, having a common ancestor, than that they independently evolved to look the same. There are also cases where the last common ancestor appears to have been so long ago that the sequences are no longer recognizably similar, yet they are believed to be homologous.

Structure Determination

The most common methods used to study the structure of proteins are as follows: electron microscopy (low resolution 3-D structure); x-ray and neutron diffraction (high resolution 3-D structure); electron diffraction (medium resolution 3-D structure); nmr (high resolution 3-D structure); cd/ord (secondary structure); infrared spectroscopy (secondary structure).

The techniques giving the most detailed 3-D structural information are x-ray and neutron diffraction, electron diffraction and microscopy, and nuclear magnetic resonance spectroscopy (nmr) (see ANALYTICAL METHODS; MAGNETIC SPIN RESONANCE; X-RAY TECHNOLOGY).

Structure Prediction

Although analytical techniques have resulted in the determination of many protein structures, the number is only a small fraction of the available protein sequences. Theoretical methods aimed at predicting the 3-D structure of a protein from its sequence form a very active area of research. This is important both to understanding proteins and to the practical applications in biotechnology and the pharmaceutical industries.

The most obvious approach for predicting the folded structure of a protein would be to search for its lowest energy conformation. In principle, knowledge of quantum chemistry should allow the necessary calculations to be carried out. However, the sheer size of the problem involving hundreds of thousands of interatomic interactions makes this extremely difficult. Simpler approaches, which fall into two basic categories, are used. The first is based on simplifying the energy calculations and conformational search, which usually involves reducing all the atomic level forces to simple classical mechanical forces. The second approach is not to attempt directly to predict the structure, but instead to use existing knowledge of protein structures to propose models most compatible with a given sequence.

Comparative Modeling. Given the limited success of prediction schemes, much effort has been turned to comparative modeling, ie, for a given sequence, identifying an approximately correct fold from among the different possible folds observed in nature. This technique is comparatively easy when homologous proteins having a known structure can be identified.

Nonhomologous Extension Modeling. In the case of protein sequences that have no clear sequence homologues with known structure, the problem becomes more challenging. One approach is to identify only the tertiary structural class to which a protein sequence belongs rather than its detailed structure. In some cases this is done from an analysis of the difference in properties of amino acids in the various classes. More recently a method has been developed which, instead of estimating the likelihood of a single or short segment of amino acids adopting a particular folded structure, estimates it for the entire sequence. This holds promise because any relevant structural information about the amino acids can be included in the calculations to increase the accuracy of prediction.

A related approach seeks to determine the compatibility of a given sequence by threading it through a structure.

Practical Applications

Study of the structure and function of proteins is possible because of the revolution in molecular biology, which has enabled the cloning of genes and the expression of their products in large quantities. Great strides have also been made in the area of peptide synthesis, improvements in experimental techniques for structure determination, and computational speed. Areas of molecular and structural biology are rapidly expanding (see PROTEIN ENGINEERING).

Development of Drugs. Most available drugs were derived from large-scale screenings. In many cases little is known about the target protein or other macromolecule, or the mode of action. The type of molecule for use as a drug is inferred from the biochemical mechanism of a disease. This information is then used to screen databanks containing thousands of compounds. Modifications to the leading candidates are then made and tested further for efficacy. It can take from 6 to 12 years for a drug to come to the market.

Rational drug design seeks to decrease this time lag and the resulting cost of development by rejecting unsuitable compounds. The first step is to obtain the quantitative structure-activity relationships (QSAR) between the biological activity of the protein and its chemical properties, particularly in its active site. The leading candidates are then examined further. The increasing computing power and the resulting advance in the development of molecular modeling techniques have made *de novo* design of novel molecules a distinct possibility.

RAMAN NAMBUDRIPAD
Beth Israel Hospital
TEMPLE F. SMITH
Boston University

C. Branden and J. Tooze, *Introduction to Protein Structure*, Garland Publishers, Inc., New York, 1991.

R. Scopes, *Protein Purification: Principles and Practice*, 2nd ed., Springer-Verlag, Berlin, 1987.

G. D. Fasman, ed., *Prediction of Protein Structure and the Principles of Protein Conformation*, Plenum Press, New York, 1989.

C. R. Cantor and P. R. Schimmel, *Biophysical Chemistry*, Part II, *Techniques for the Study of Biological Structure and Function*, W. H. Freeman, New York, 1980.

PROTOZOAL INFECTIONS, CHEMOTHERAPY. See ANTIPARASITIC AGENTS, ANTIPROTOZOALS.

PSYCHOPHARMACOLOGICAL AGENTS

Until the early 1950s only rudimentary pharmacotherapy was available for the treatment of significant psychiatric illnesses. By the middle 1990s, agents effective in treating prevalent psychiatric disorders, such as mood disorders, schizophrenia, anxiety disorders, insomnia, and substance use disorders, as well as dementias of diverse etiologies, had become available. The therapy of cognitive disorders, particularly age-related dementia which is only one area receiving increased medical attention (see ANTIAGING AGENTS; MEMORY-ENHANCING DRUGS).

Despite recognized successes of psychopharmacology, significant challenges remain. The causative factors underlying most neuropsychiatric disorders are only poorly understood. Missing is the discovery of psychopharmacological agents which directly impact on the etiological or pathophysiological processes underlying psychiatric illnesses.

Anxiolytics, Sedatives, and Hypnotics

Anxiety disorders are serious medical problems affecting not only quality of life, but additionally may indirectly result in considerable morbidity owing to association with depression, cardiovascular disease, suicidal behavior, and substance-related disorders. Insomnia is a related psychiatric illness having potentially serious consequences.

Pharmacological Profiles. Historically, chemotherapy of anxiety and sleep disorders relied on a wide variety of natural products such as opiates, alcohol, cannabis, and kawa pyrones. Upon the discovery of barbiturates, numerous synthetic compounds rapidly became available for the treatment of anxiety and insomnia. At present, barbiturates are in use primarily as injectable general anesthetics (qv) and as antiepileptics.

Beginning in the 1960s, benzodiazepine anxiolytics and hypnotics rapidly became the standard prescription drug treatment. In the 1980s, buspirone (**1**), which acts as a partial agonist at the serotonin (5-hydroxytryptamine, 5-HT) type 1A receptor, was approved as treatment for generalized anxiety. More recently, selective serotonin reuptake inhibitors (SSRIs) have been approved for therapy of panic disorder and obsessive—compulsive behavior.

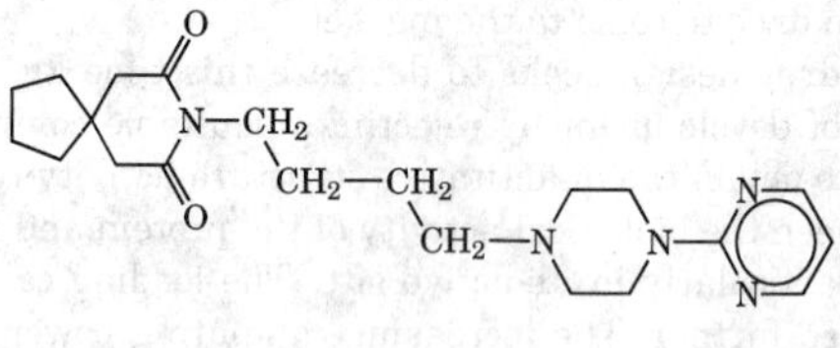

(**1**)

β-Adrenergic blockers, particularly propranolol, have sometimes been used off-label to treat anxiety, especially in those patients in which cardiovascular symptoms predominate (see CARDIOVASCULAR AGENTS). Clinical results also suggest the possible value of the monoamine–oxidase inhibitor phenelzine in ameliorating panic disorder with agoraphobia. In the middle 1990s, benzodiazepines continued to dominate the therapy of certain anxiety disorders, eg, generalized anxiety disorder and panic disorder, as well as insomnia.

Barbiturates and benzodiazepines have been demonstrated to act by modulating neurotransmission within the γ-aminobutyric acid (GABA) system, which is the primary inhibitory neurotransmitter system within the central nervous system.

Barbiturates. Barbiturates, which by common practice include both oxybarbiturates and thiobarbiturates, are effective as sedative—hypnotics and as anticonvulsants. The clinical effects are the result of allosteric modulation of sites on the $GABA_A$ receptor complex. Barbiturates are still occasionally used as alternatives to benzodiazepines in the therapy of anxiety and sleep disorders. Barbiturates interact with ethyl alcohol and other central nervous system depressants.

The ultrashort-acting barbiturates such as sodium salts of methohexital, thiopental, and thiamylal, are typically used as intravenous anesthetics, often in combination with inhalation agents. The short- to intermediate-acting barbiturates such as secobarbital, pentobarbital, amobarbital, allobarbital, and aprobarbital are prescribed as hypnotics. The long-acting barbiturates phenobarbital, metharbital, and methylphenobarbital are most commonly used as antiepileptics. Barbiturates are often components in combination preparations together with paracetamol (acetaminophen), acetylsalicylic acid (aspirin), caffeine, and/or codeine.

Benzodiazepines. The marketed benzodiazepine anxiolytics and sedative—hypnotics act via agonism at benzodiazepine receptors (BZRs) within the central nervous system to yield a wide spectrum of therapeutic actions including anxiolytic, hypnotic, muscle relaxant, and anticonvulsant effects. The BZR is a distinct binding site on the $GABA_A$-receptor complex. The BZR can exert either positive or negative modulation on the $GABA_A$ receptor depending on the ligand. Ligands having positive allosteric modulatory activity are classified as BZR agonists. These increase the affinity of GABA to its binding site. Compounds having structures different from those of benzodiazepines can also act via the BZR, for example, the full agonists zopiclone (**2**) and zolpidem (**3**).

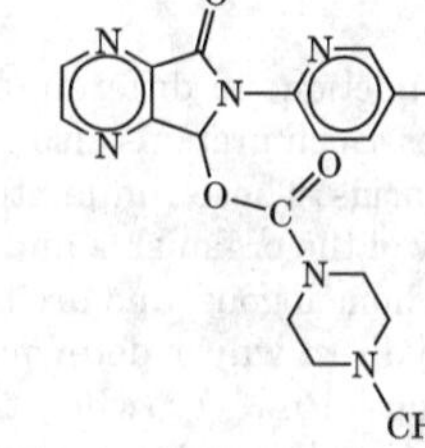

(**2**)

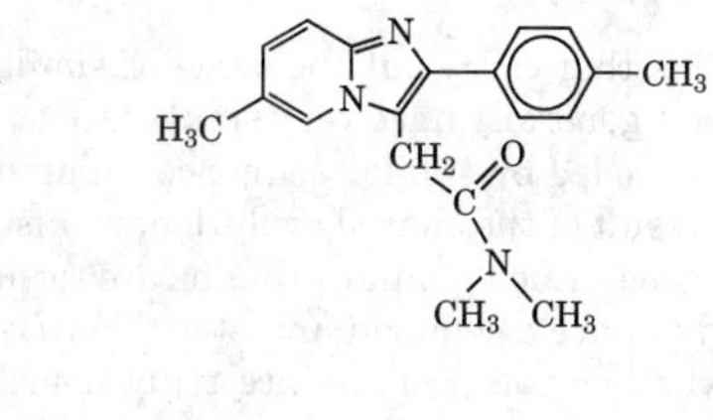

(**3**)

Other BZR ligands bind to the allosteric site without inducing any appreciable alteration of the $GABA_A$ receptor gating function. However, these compounds selectively and competitively block the action of both BZR agonists and inverse agonists. Accordingly, they are named BZR antagonists. The imidazobenzodiazepine flumazenil (**4**) is the only representative in clinical use. Flumazenil, a specific BZR antagonist having exceptionally good tolerance and effectiveness as an antidote of mono-overdose with BZR agonists, is also used for shortening post-operational unconsciousness following anesthesia involving a BZR agonist.

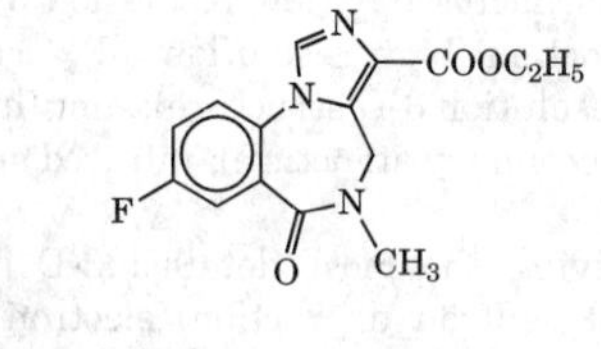

(**4**)

Benzodiazepines, ie, the full BZR agonists, are prescribed for anxiety, insomnia, sedation, myorelaxation, and as anticonvulsants. Those benzodiazepines most commonly prescribed for the treatment of anxiety disorders are lorazepam (**5**), alprazolam, diazepam, bromazepam, chlorazepate, and oxazepam. These drugs together represent about 70% of total worldwide tranquilizer unit sales. Those benzodiazepines used most frequently in treating insomnia are triazolam, temazepam, flunitrazepam, lormetazepam, estazolam, and flurazepam. Zopiclone and zolpidem are approved as hypnotics.

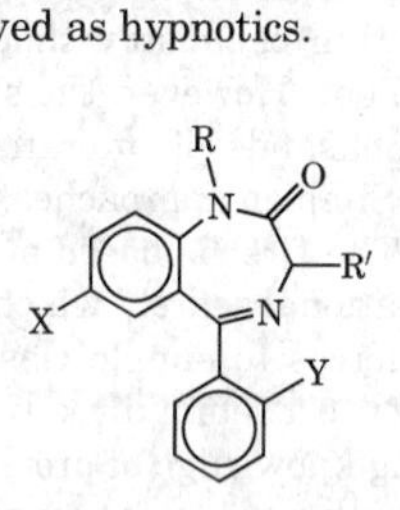

(**5**) R, R′ = H; X = Cl; Y = Cl

Buspirone. Buspirone (**1**) hydrochloride has been approved for the symptomatic management of generalized anxiety disorder. This drug is of special interest because it does not exert its therapeutic actions via modulation of the $GABA_A$ receptor complex. This compound is structurally unrelated to the benzodiazepines, barbiturates, or the other anxiolytics and sedative–hypnotics discussed. The anxiolytic effect of buspirone may result from partial agonism at the $5\text{-}HT_{1A}$ receptor, although its pharmacological profile exhibits some similarity to that of dopaminergic receptor antagonists. Antipsychotics are sometimes used to treat anxiety disorders.

Selective Serotonin Reuptake Inhibitors. In view of the mechanism of action of selective serotonin reuptake inhibitors (SSRIs), it appears that the resulting increased availability of the neurotransmitter serotonin within the synaptic cleft is responsible for the pharmacological effects of this drug class. Although originally developed and predominantly used as antidepressants, SSRIs have been increasingly used in treating panic disorder, eg, fluoxetine (Prozac) (**6**), or obsessive–compulsive disorder, eg, fluvoxamine (**7**).

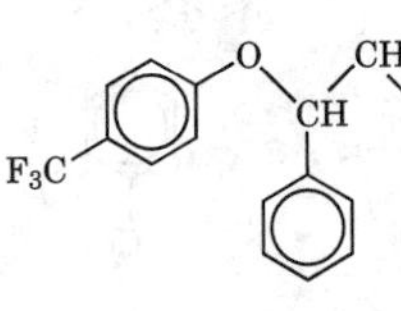

(**6**)

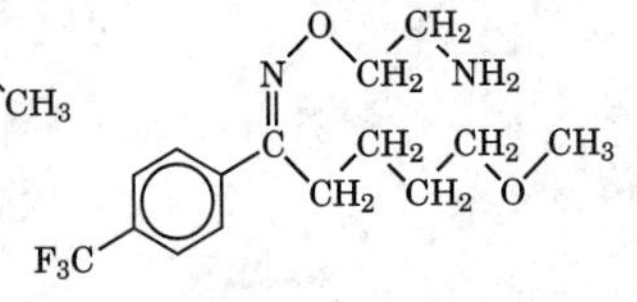

(**7**)

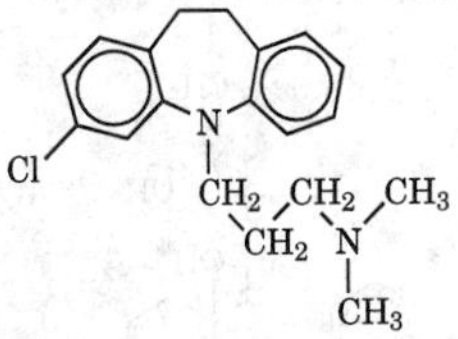

(**8**)

SSRIs are well tolerated. In addition, the tricyclic antidepressant clomipramine (Anafranil), (**8**), which is a potent nonselective serotonin reuptake inhibitor, is approved for treatment of obsessive–compulsive disorder.

Use of Sedative–Hypnotics in General Anesthesia. In addition to use in the therapy of anxiety and insomnia, sedative–hypnotics are widely used, when given intravenously in appropriate dosages, to induce general anesthesia. Local or regional anesthesia is induced differently by agents of the cocaine class. The optimal anesthetic agent would combine unconsciousness, amnesia, analgesia, loss of both sensory and autonomic reflexes, and muscle relaxation having minimal cardiovascular and respiratory disturbance. No single anesthetic agent combines all these desired features, thus anesthesiologists often employ a cocktail of drugs.

Pharmacological Profiles of Sedatives Used in Anesthesiology. Anesthetic management of a surgical patient includes premedication, induction, maintenance, emergence, and recovery. Light premedication provides sedation, decreased anxiety, amnesia, and decreased parasympathetic outflow. Oral or parenteral benzodiazepines are sedative–hypnotics used preoperatively for premedication when strong analgesia is not required. The rapid onset of action of intravenous anesthetics makes them ideally suited for induction of anesthesia. Agents having a short plasma elimination half-life such as midazolam (**9**) and propofol are often used as maintenance anesthetics.

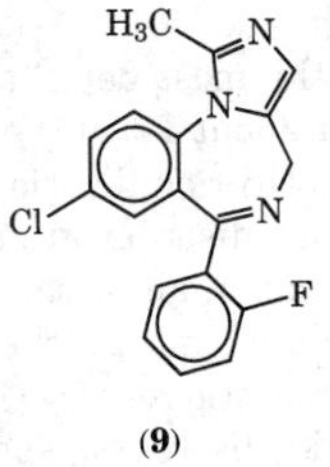

(**9**)

Combinations of barbiturates and benzodiazepine tranquilizers or even antihistaminergics having sedative properties are sometimes used. The antagonist flumazenil (**4**) is available to reverse the effects of anesthetics of the benzodiazepine class.

Barbiturates. The ultrashort-acting barbiturates methohexital, thiopental, and thiamylal are used for induction and maintenance of anesthesia.

Benzodiazepines. Benzodiazepine derivatives have gained popularity as intravenous anesthetics because of their limited effects on the cardiovascular and respiratory systems. Intravenous benzodiazepines can be used alone for minor procedures causing discomfort but not pain (eg, endoscopy), or in combination with local anesthetics. Midazolam (**9**) is the most widely utilized benzodiazepine in intravenous anesthesia. Diazepam, lorazepam (**5**), and flunitrazepam are less widely used than midazolam intravenously.

Antidepressants

Depression is a common psychiatric disorder. The most common mood disorders are major or unipolar depression and manic–depressive illness or bipolar depression. There are other affective disorders, such as dysthymia, which are generally treated with available antidepressants.

Pharmacological Profiles. Depression is believed to result from a decreased neurotransmission, ie, a lack of sufficient concentration of noradrenaline, serotonin, and/or dopamine at critical synapses in the brain, particularly within the limbic system. The various classes of antidepressant treatments, by blocking transmitter uptake by the neuron, or by slowing down their degradation, can increase neurotransmitter concentration to a normal level and consequently ameliorate depression.

Treatment of Major Depression. Drugs commonly used for the treatment of depressive disorders can be classified heuristically into two main categories: first-generation antidepressants with the tricyclic antidepressants (TCAs) and the irreversible, nonselective monoamine–oxidase (MAO) inhibitors, and second-generation antidepressants with the atypical antidepressants, the reversible inhibitors of monoamine–oxidase A (RIMAs), and the selective serotonin reuptake inhibitors (SRRIs).

First-Generation Antidepressants. Imipramine (**10**) was introduced in the late 1950s as one of the first pharmacotherapies for depression. Over the years, other congeners, such as desipramine (**11**), amitriptyline (**12**), and dothiepin (**13**), were synthesized and shown to be clinically efficacious antidepressant drugs. These substances, known under the general rubric of tricyclic antidepressants, share a basic chemical structure comprising

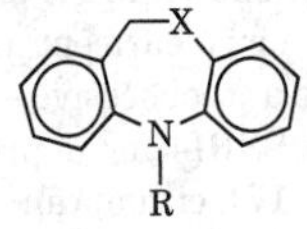

a three-ring core. TCAs remain the mainstay of treatment for depression despite some drawbacks.

In the mid-1950s, tuberculosis patients with depression being treated with iproniazid (**14**) were occasionally reported to become euphoric. This observation led to the discovery of irreversible monoamine–oxidase (MAO) inhibiting properties. Hydrazine and nonhydrazine-related MAO inhibitors were subsequently shown to be antidepressants. Three other clinically effective irreversible MAO inhibitors have been approved for treatment of major depression: phenelzine, isocarboxazid, and tranylcypromine.

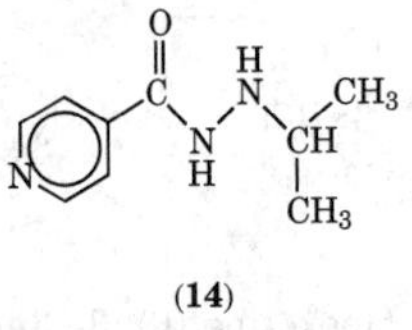

(**14**)

Interactions with tyramine contained in food and other drugs have severely limited use of irreversible MAO inhibitors.

Second-Generation Antidepressants. The frequency of adverse effects associated with first-generation antidepressants and the lack of patient compliance arising from such adverse effects led to the development of a number of second-generation antidepressants.

Structurally diverse drugs such as the tetracyclic mianserin and various bicyclic and tricyclic compounds such as trazodone, venlafaxine (Effexor) (**15**), nefazodone, and amfebutamone are atypical antidepressants. The exact mechanism of action is unclear but probably

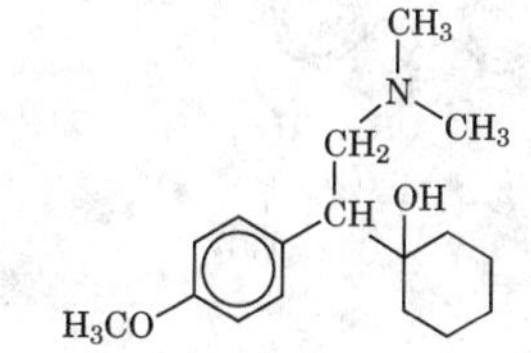

(**15**)

involves actions at serotonin, noradrenaline, and/or dopamine synapses. Such drugs exhibit reduced toxicity, improved patient compliance, and lower toxicity in overdose than first-generation antidepressants.

Nefazodone is a phenylpiperazine derivative exhibiting a pharmacological profile that is distinct from that of first-generation agents, as well as the more selectively acting second-generation agents, ie, serotonin or noradrenaline reuptake inhibitors. Nefazodone acts both as a serotonin receptor type 2 antagonist and as a serotonin reuptake inhibitor.

Venlafaxine (**15**) is a structurally novel phenylethylamine derivative that strongly inhibits both noradrenaline and serotonin reuptake. It lacks anticholinergic, antihistaminergic, and antiadrenergic side effects.

Serotonin plays a pivotal role in the physiological regulation of mood. The selective serotonin reuptake inhibitors (SSRIs) resulted from an effort by the pharmaceutical companies to develop drugs exhibiting a higher degree of selectivity for the central serotonin transporter. Owing to serotonergic uptake blocking properties, a net enhancement of serotonergic function results from the administration of SSRIs. The drug fluoxetine (**6**), the first SSRI approved by the U.S. Food and Drug Administration for the treatment of major depression, was rapidly followed by several other SSRIs. The synthesis of fluoxetine (Prozac), marketed as a racemate and the commercially most important antidepressant, is shown in Figure 1.

SSRIs are widely used for treatment of depression, as well as, for example, panic disorders and obsessive—compulsive disorder. The most commonly prescribed SSRIs for depression are fluoxetine (**6**), fluvoxamine (**7**), sertraline (**17**), citalopram, and paroxetine.

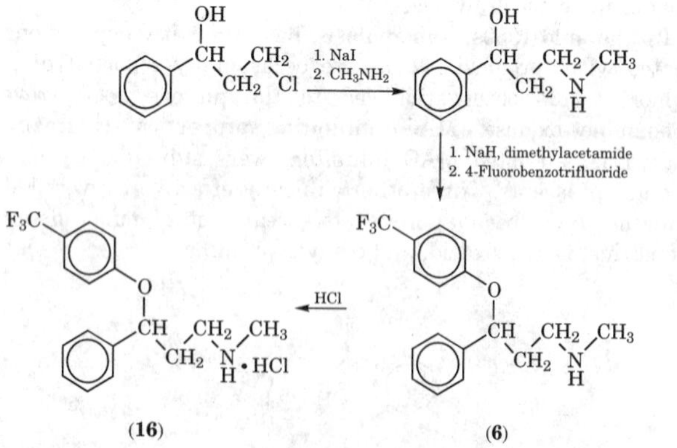

(**16**) (**6**)

Figure 1. Synthesis of fluoxetine (**6**). 3-Chloro-1-phenyl-1-propanol reacts with sodium iodide to afford the corresponding iodo derivative, followed by reaction with methylamine, to form 3-(methylamino)-1-phenyl-1-propanol. To the alkoxide of this product, generated using sodium hydride, 4-fluorobenzotrifluoride is added to yield after workup the free base of the racemic fluoxetine (**6**), thence transformed to the hydrochloride (**16**).

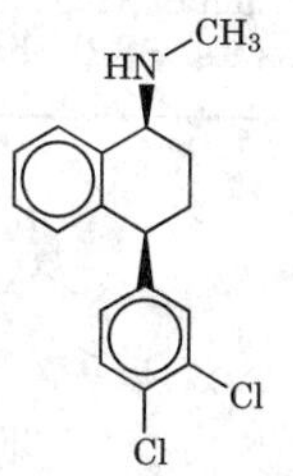

(**17**)

Selective MAO-A inhibitors, which are reversible (so-called RIMAs), have also been developed, therefore substantially reducing the potential for food and drug interactions. The RIMAs represent effective and safer alternatives to the older MAO inhibitors. The only marketed RIMAs are toloxatone and moclobemide.

Treatment of Manic–Depressive Illness. Since the 1960s, lithium carbonate and other lithium salts have represented the standard treatment of mild-to-moderate manic–depressive disorders. It is effective in about 60–80% of all acute manic episodes within one to three weeks of administration. Lithium ions can reduce the frequency of manic or depressive episodes in bipolar patients providing a mood-stabilizing effect.

The antiepileptic valproic acid and its salts have also been reported to be therapeutically useful in treating mania, possibly via enhancement of GABA metabolism in the brain. Valproate semisodium (divalproex sodium) has been approved in the United States for treatment of manic episodes associated with bipolar depression.

Other agents are also used for the treatment of manic–depressive disorders. The antiepileptic carbamazepine has been reported in some clinical studies to be therapeutically beneficial in mild-to-moderate manic depression. Severe manic depression is often treated with antipsychotics or benzodiazepine anxiolytics.

Future Outlook. Third-generation antidepressants are expected to combine superior efficacy and improved safety, but are unlikely to reduce the onset of therapeutic action in depressed patients.

Antipsychotics

Schizophrenia is perhaps the most debilitating psychiatric illness in modern medicine, affecting about 1% of the general population. Many of those affected require institutionalization. Unfortunately, the compounds available to treat this disorder are not fully effective in treating the spectrum of symptoms in all patients. Adverse effects are also a problem. In addition, available antipsychotic (neuroleptic) drugs can at most only provide symptomatic relief.

As for most psychiatric disorders, speculations concerning the pathology of the disease come from the limited understanding of the pharmacology of drugs effective in symptomatically treating the disorder.

Pharmacological Profiles. Most compounds used for antipsychotic therapy can be assigned to one of four structurally distinct groups. These are phenothiazines, eg, chlorpromazine (**18**); thioxanthenes, eg, chlorprothixene (**19**); diphenylbutylpiperidines, eg, pimozide (**20**); and butyrophenones, eg, haloperidol (**21**). These compounds represent more than 97% of total units sold worldwide, but only about half of the antipsychotics clinically available.

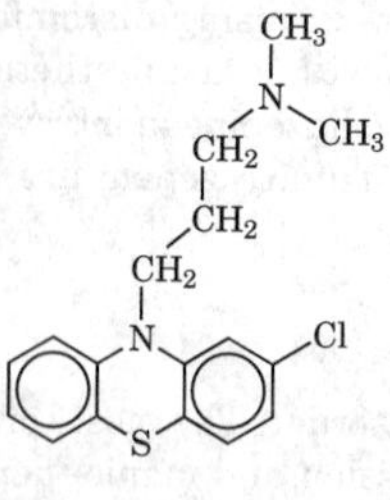

(**18**)

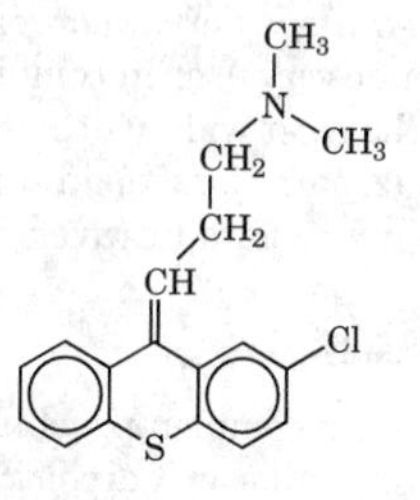

(**19**)

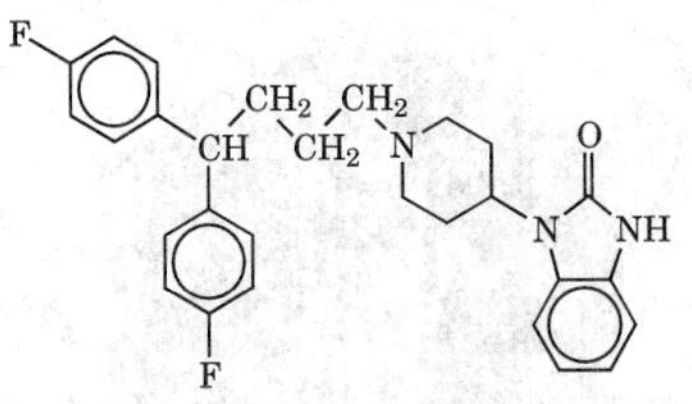

(20)

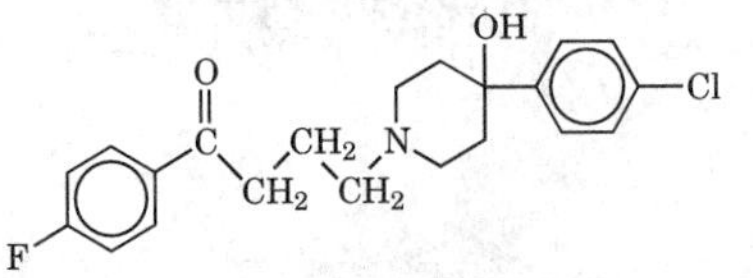

(21)

There is a good correlation between the affinity of antipsychotics for the dopamine D_2 receptor and their clinically effective dose used in the treatment of schizophrenia, suggesting the special importance of this neurotransmitter system with respect to schizophrenia. There are two principal dopaminergic pathways in the brain which project forward from cell bodies in the midbrain: the nigrostriatal dopaminergic pathway and the mesocorticolimbic dopaminergic pathway. It has been hypothesized that antipsychotic activity results from the blockade of dopaminergic transmission in the mesolimbic pathway, whereas blockade of dopaminergic transmission in the nigrostriatal pathway gives rise to extrapyramidal side effects such as tardive dyskinesia, akathisia, and Parkinsonian-like symptoms often associated with antipsychotic treatment. Therefore, antidopaminergic properties may account for both therapeutic actions and adverse effects of these drugs. The target of preclinical research is an antipsychotic drug that lacks extrapyramidal side effects. One such atypical antipsychotic is clozapine (**22**) which is effective in treating most of the symptoms of schizophrenia and has minimal or no concomitant extrapyramidal side effects.

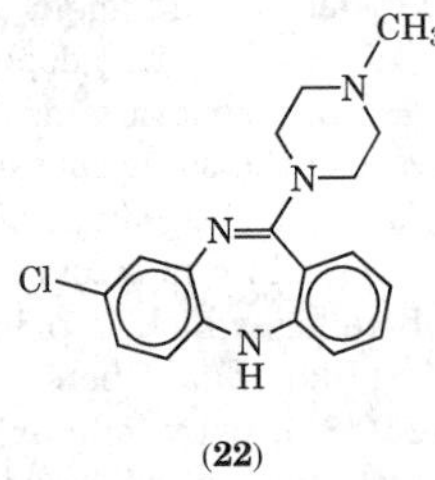

(22)

Future Outlook. Preclinical work is focused on the development of atypical antipsychotics having a favorable clinical profile similar to that of clozapine (**22**).

Drugs for Treating Substance Use Disorders

The abuse and dependence on ethyl alcohol, nicotine, cocaine, and heroin (diacetylmorphine), in particular, have produced devastating socioeconomic consequences. The similarity of psychosocial factors leading to abuse of these drugs, despite their diverse pharmacological mechanisms, suggests the predominant role played by the former in both the development and maintenance of substance abuse disorders, and consequently has resulted in an emphasis on social and behavioral treatment approaches. Aside from the clear therapeutic value of pharmacological treatment for both severe intoxication and acute withdrawal phenomena, protracted substance abuse and dependence may also prove amenable to therapy with pharmacologic agents.

Pharmacological Profiles. The subjective effects of drugs leading to their abuse are the result of modulation of brain neurotransmitter systems. Alcohol, as well as a number of other central nervous system (CNS) depressants such as benzodiazepines and barbiturates, enhances central GABAergic activity which normally serves as the main inhibitory neurotransmitter system in the brain. Nicotine exerts its effects on brain function via nicotinic acetylcholine receptors. Cocaine and other CNS stimulants facilitate the presynaptic release of catecholamines and block their reuptake by neurons.

One approach taken to the treatment of chronic alcoholism relies on a form of aversion therapy involving prophylactic administration of inhibitors of aldehyde dehydrogenase, such as disulfiram (**23**) and calcium carbimide, $CaCN_2$. Any consumption of alcohol then results in an accumulation of acetaldehyde yielding flushing hypotension, tachycardia, nausea, and vomiting. More recently, the μ-opiate antagonist naltrexone (**24**) was approved in the U.S. for the treatment of alcohol abuse.

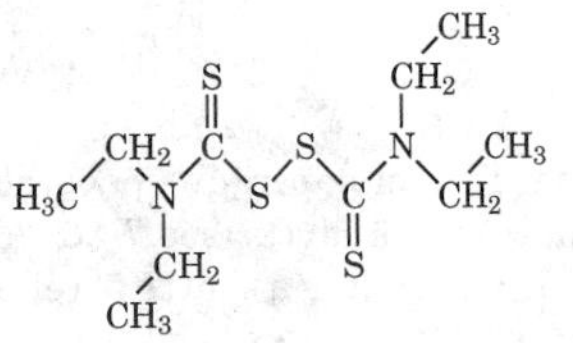

(23)

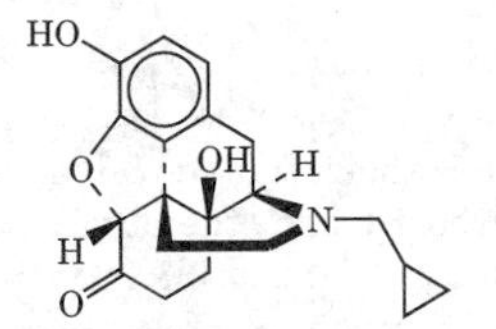

(24)

An approved pharmacologic approach to stopping tobacco smoking, ie, nicotine dependence, involves the substitution of nicotine-containing chewing gum or a dermal patch with subsequent gradual reduction of the nicotine dose until dependence has been eliminated.

A common strategy for treating chronic opiate addiction involves the substitution of methadone which can either be provided as maintenance therapy or tapered until abstinence is achieved. Naltrexone and buprenorphine have also been used in this manner.

Cognition Enhancers

Cognitive impairment can arise through diverse pathophysiological mechanisms. In view of the devastating socioeconomic consequences of senile dementia of Alzheimer's type (SDAT), this has received considerable attention, although cognitive disorders can occur throughout the life span owing to diverse etiologies. SDAT is the most common dementing disease, affecting one of every six individuals over 65 years of age. A common feature is the dysfunction or death of selective populations of neurons.

The acetylcholinesterase inhibitor tacrine was approved for the treatment of mild-to-moderate SDAT in the United States in 1993 followed by several other countries. The acetylcholinesterase inhibitor galanthamine, which has long been in clinical use in Austria for the treatment of indications such as facial neuralgia and residual poliomyelitis paralysis, has also been approved for use in Alzheimer's disease in that country. There are a variety of other drugs marketed in one or more countries for this or related indications. Codergocrine mesylate, a mixture of three dihydrogenated ergot alkaloids, has been marketed worldwide since the 1950s for treating cognitive impairment or dementia. Other drugs such as meclophenoxate, bifemelane, vincamine, vinpocetine, cyclandelate, naftidrofuryl, pyritinol, idebenone, and indeloxazine are used in some countries. Piracetam and a number of other structurally related pyrrolidinones, constitute the nootropic class of cognition enhancers used in treating dementia (see MEMORY-ENHANCING DRUGS).

Pharmacological Profiles. The common denominator of most of the cognition enhancers marketed or in clinical development is the ability to reverse learning and memory deficits in animal models. However, the putative neurobiological mechanisms of action of these compounds cover a wide range of mechanisms. A variety of drug development approaches focusing on cholinomimetic mechanisms has been clinically investigated.

J. R. MARTIN
T. GODEL
W. HUNKELER
F. JENCK
J.-L. MOREAU
A. J. SLEIGHT
U. WIDMER
F. Hoffmann-La Roche, Ltd.

W. Haefely, R. Schaffner, P. Polc, and L. Pieri, in F. Hoffmeister and G. Stille, eds., *Psychotropic Agents, Part II: Anxiolytics, Gerontopsychopharmacological Agents, and Psychomotor Stimulants*, Springer-Verlag, Berlin, 1981.

Diagnostic and Statistical Manual of Mental Disorders, 4th ed., American Psychiatric Association, Washington, D.C., 1994.

International Statistical Classification of Diseases and Related Health Problems (ICD-10), 10th Rev., World Health Organization, Geneva, Switzerland, 1992.

PULP

Pulp is the raw material for the production of paper (qv), paperboard, fiberboard, and other similar manufactured products (see WOOD-BASED COMPOSITES AND LAMINATES). In purified form, it is a source of cellulose (qv) for rayon (see FIBERS, REGENERATED CELLULOSICS), cellulose esters (qv), and other cellulose-derived products. Pulp is obtained from plant fiber and is therefore a renewable resource (see CHEMURGY). The principal wood-pulping processes are stone groundwood, soda, SO_2 or acid sulfite, and the sulfate or kraft processes. Since their development, the basic processes have been modified numerous times and the technology has been highly refined.

Although virtually any wood can be pulped by some process, there are certain species commonly used for pulp because of desirability of fiber, ease of pulping, availability, or less competition with other wood products.

Wood

In terms of abundance and suitability for pulping, there are two chief botanical classifications of trees: the softwoods or evergreens, which are gymnosperms, and the hardwoods or broad-leaved deciduous trees, which are dicotyledon angiosperms. The chemistry and anatomy of wood vary somewhat with the species of tree, but there are gross similarities between the two classifications. The softwoods, which are preferred for most pulp products because of their longer fibers, generally contain a higher percentage of lignin (qv) and a lower percentage of hemicellulose (qv) than the hardwoods.

Anatomy and Morphology. A cross section of pine is shown in Figure **1a** as a representation of the anatomy of softwoods. The main cell type is the axially aligned tracheid (TR). Although in botanical terminology tracheids are not considered true fibers, these are the papermaking fibers from softwoods and are referred to herein as fibers, according to the common practice of the industry. Other cell types in softwoods are the ray cells, ie, the fusiform wood ray (FWR) and wood ray (WR) cells, and the longitudinal and epithelial parenchyma, ie, the cells surrounding the horizontal and vertical resin ducts (HRD and VRD, respectively).

Hardwoods have a more varied and complex arrangement of cells than softwoods. Figure **1b** is a cross section of the structure of yellow poplar, a typical hardwood. The main structural element of hardwoods is the wood fiber (F), which is significantly shorter than the softwood tracheid (1–2 mm vs 3–6 mm) and generally thinner, ca 20 μm in diameter. The true fibers are uniform throughout the annual ring.

Other distinct classes of wood in a tree include the portion formed in the first 10–12 years of a tree's growth, ie, juvenile wood, and the reaction wood formed when a tree's growth is distorted by external forces. In certain tree stands, such as short-rotation coppice or high elevation regions, the juvenile wood or the reaction wood becomes a significant portion of the total, and allowance must be made for the different pulping and fiber characteristics. In some cases, specific

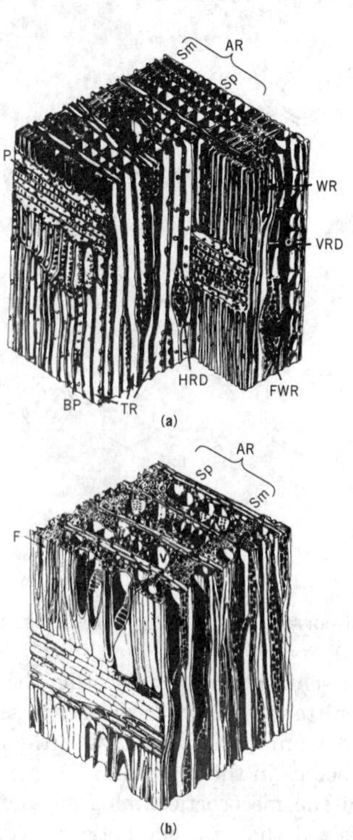

Figure 1. Schematic section, where AR = annual ring, BP = bordered pits, F = wood fiber, FWR = fusiform wood ray, HRD = horizontal resin ducts, P = primary wall, SC = scalariform plate, Sm = summerwood, Sp = springwood, TR = tracheid, VRD = vertical resin ducts, and WR = wood ray, of (**a**) pine, a softwood, and (**b**) yellow poplar, a hardwood.

parts of the tree are deliberately used to provide for desired pulp properties in the final product.

Chemical Composition. The chemical components found in wood fall into two categories: low molecular weight substances and macromolecular material. The exact composition depends on the species and age of the tree but considerable variation also exists within a single tree or a single stand of trees of the same species. The basic structural element of the cell wall is cellulose. Lignin and hemicelluloses are distributed throughout the cell wall in an incompletely understood manner. The intercellular substance, which is primarily lignin, must be softened or dissolved to free individual fibers.

Wood also contains 3–10% of extracellular, low molecular weight constituents, many of which can be extracted from the wood using neutral solvents and therefore are commonly called extractives.

Many of these chemicals are recovered as by-products of the pulping operation, eg, tall oil (qv) and turpentine (see TERPENOIDS). All trees contain inorganic minerals as essential nutrients. Generally, minerals amount to less than 0.5% of the weight of the wood. A larger amount occurs in the bark. The principal constituents are calcium, magnesium, and phosphorus. Heavy metals and many other elements are present in smaller amounts. They may also be associated with the carboxylic acids of lignin and carbohydrates in the cell wall (see MINERAL NUTRIENTS).

Other Fiber Sources. Wood is the primary source of fiber for pulp, but other resources are also used in areas where wood supply is sparse. Plant sources of fiber other than wood include grasses, canes, and straws. Pulp is produced from such nonwood sources as straws and grasses; canes and reeds; several varieties of bamboo; bast fibers; leaf fibers; and seed fibers, eg, cotton (qv) and cotton linters. Cotton and cotton linters, the very first fibers used for commercial papermaking, are still used in the manufacture of high quality paper. The introduction of synthetic fibers such as rayon and nylon and the use of certain other noncellulosic fibers such as polyethylene and polypropylene have further widened the range of raw materials and characteristics of the final paper products.

Secondary Fiber. Increasing costs of raw fiber, legislative mandates for recycling (qv), and availability of inexpensive waste papers have contributed to the increased use of recycled fibers. Recycled fibers are sometimes used in special writing papers, but the principal use is for the manufacture of linerboard, newsprint, tissue, cereal boxes, towels, and molded paper products such as paper plates and egg cartons (see PACKAGING, CONTAINERS FOR INDUSTRIAL MATERIALS).

Pulp Fibers

Pulp fibers are classified according to the method of manufacture as mechanical, chemical, chemimechanical, and semichemical. Mechanical pulps are produced by mechanical difibrillization of wood and are characterized by low fiber strength and small particle size. Virtually all of the wood's chemical components remain in the pulp.

Chemical pulps are produced in a digester where the wood is cooked in pressurized vessels using heat and chemicals to break the intercellular structure of the wood and extractives. The objective is to remove the lignin from the fibers without degrading the carbohydrate content of the wood.

Even though most of the cell wall lignin has been removed from the tracheids, the physical structure remains virtually unaffected. The long, narrow shape and tapered ends are characteristic of the structure in wood with the bordered pits visible on the cell wall. When made into paper, the fiber is collapsed into a ribbon-like structure. The cellular elements are bonded together by hydrogen bonding at the crossovers or intersection of the fibers.

The fibers of a hardwood kraft pulp are finer and shorter than the softwood tracheid. Noticeable in a porous hardwood pulp are the short and wide vessels. These vessel segments are not long enough to bond across numerous fibers (few crossovers) and, therefore, are more easily lost from a paper sheet. Papers made from hardwood fibers are weaker than those made from softwoods, but the former provide better formation or uniformity, smoothness, bulk, and opacity. Pulps of different types are frequently blended to produce specific properties for a given product.

Wood Procurement and Preparation

Wood procurement and preparation for use in the manufacture of pulp begins with the purchase of roundwood or wood chips from large woodlots or tree plantations. Wood is processed to produce pulp under specified conditions such as tree species and size, cleanliness, and needs of the product. Attempts are made to utilize as much of the tree as possible. Different species of trees are often mixed to give optimum properties in the pulp.

Operations of cleaning, debarking, and production of wood chips of uniform dimension all depend on the pulping process to be used. Grinding wood against a stone requires clean roundwood bolts of the same length as the width of the stone. Chemical and semichemical pulping call for a chip about 25 mm (1 in.) in length and about 3–7 mm (0.125–0.25 in.) in thickness, as does refiner pulping in which chips enter between ribbed rotating disks to be defibered. Wood subdivision and dimensioning operations include merchandizing, slashing, debarking, chipping, and screening.

Harvesting. Many factors influence the procedures and the choice of equipment used to harvest wood. These include physical considerations such as area of land to be harvested, tree size, terrain, climate, and environmental constraints, as well as commercial considerations such as demand for fresh wood, the need to supply a continuous flow to the mill, and maintaining inventory and a healthy corps of suppliers.

Sustainable Forest Management. The forest products industry has placed a high priority on developing sustainable forest practices that optimize yield and fiber properties while preserving soil, wildlife habitat, and species diversity. The industry is also focusing research on genetic engineering (qv), land management to protect biodiversity of species and gene pools, and conservation of areas having unique ecosystems.

Slashing. Wood is slashed or cut to provide a fit for subsequent conveying, debarking, chipping, and other processing equipment. Slashing is done in the woods or the woodyard and typically uses either circular or chainsaws to cut single large logs or several smaller stems bundled together.

Debarking. The bark of most pulpwood species is dark, containing extractives and little fiber, often contaminated with dirt or grit. Bark accompanying the wood into the mill causes wear in the equipment and lowers the strength and appearance of the pulp.

Specifications on bark content for debarked wood typically vary between 0.5–2.0% based on the dry weight of wood and bark. To achieve these standards, the efficiency of bark removal must be high for most processes.

Chipping. Wood is chipped to reduce it to a size that maximizes penetration of processing chemicals without excessive cutting or damage to the fibers. In the case of mechanical pulp produced from refiners, the refiner disks only accept a piece of wood small enough to enter between them.

Chippers may be classified as disk or drum chippers, depending on mechanical design.

Chip Screening. Because of considerable variation in dimension, chips must be screened before proceeding to the digester where they are cooked under conditions set for average-sized chips. Screening removes oversized and undersized chips, narrowing the size distribution and removing undesirable foreign matter, eg, knots, bark particles, and dirt.

In conventional systems, chips are fed to the screen and separated into oversize (slivers), acceptable chips, and dust (pins). Disk screens have been developed to give chips of uniform thickness.

Wood Residuals. Chips, sawdust, and other residuals such as planar shavings are used as a primary source of fiber for some pulp mills. Chips are screened; sawdust and other residuals obtained from wood processing plants must be cooked separately and require special digesters and handling equipment.

Mechanical Pulp

Mechanical pulp is a generic term covering a variety of pulps. Mechanical pulps are categorized by size reduction equipment used in manufacture. Either grinding on a pulpstone or in a rotating-disk attrition mill called a refiner is employed. Chemical pretreatment of wood causes a drop in yield, the magnitude of which depends on the severity of treatment and also on the manufacturing process. Chemithermomechanical pulp (CTMP) is obtained by a comparatively mild chemical treatment followed by pressurized refining. If a stronger chemical treatment is applied at an elevated temperature, the yield may drop to as low as 85%. The resultant chemimechanical pulp (CMP) is obtained when the treated chips are refined under either atmospheric or pressurized conditions.

Different manufacturing processes yield pulps with different properties. A common feature in mechanical pulps is a woody character caused by the presence of lignin in the fibers. Mechanical treatment disrupts the wood structure into fragments of variable dimensions. A rough classification of the pulp according to particle size gives three main types of material: fines, fibers, and shives. The final properties of mechanical pulp depend on size distribution and the formation of a fine fraction having high specific surface area which increases rapidly with decreasing particle size.

Mechanical pulp is composed mainly of fiber bundles and fragments, some whole individual fibers, and high quantities of fines, which are particles less than ca 76 μm (−200 mesh). Mechanical pulps exhibit special properties because of the mixed nature of the particle sizes present and because of the presence of lignin. Desirable properties include a small average particle length and a relatively stiff fiber that prevents close packing. Undesirable properties of mechanical pulps are relatively low tensile strength, a harsh feel, and lack of permanence.

Methods of Production. In the grinding process, debarked logs are forced against a revolving abrasive stone that pulls the fibers off the log, separating them from the wood matrix. In a refiner, process wood chips are fed between two metal disks, of which at least one disk is rotating, and the wood fibers are separated by the action of grooves and bars located on the surface of the two disks. In both types of processes, the crude pulp must be upgraded by removing undesirable materials before the accepted fiber can be used.

Bleaching Mechanical Pulps. Mechanical pulps are bleached to a higher brightness using oxidative or reductive bleaching agents. The bleaching is done in a lignin-conserving manner, called brightening, in which the chromophores, ie, molecules having conjugated double bonds, are modified and little bond cleavage or solubilization of the wood lignin occurs. Oxidative bleaching may use hydrogen peroxide (*P*) and sodium hypochlorite (*H*) solutions. In reductive bleaching, sodium hydrosulfite (dithionite, *Y*) or formamidine sulfinic acid (FAS) may be used. (see BLEACHING AGENTS, PULP AND PAPER).

Pulp Characteristics. Properties of mechanical pulp vary significantly. Pulps that depend on grinding have a higher content of fine material. Refiner pulps, notably refiner mechanical pulps (RMP) and thermomechanical pulps (TMP), generally have a smaller content of fine material, and the long fibers tend to be more ribbon-like. Chemithermomechanical (CTMP) and chemimechanical (CMP) pulps generally have higher quantities of longer fiber and less fines than TMP and RMP. The characteristics of a mechanical pulp also depend on the wood species used, the quality of the wood, the processing conditions, and the amount of mechanical energy applied.

Stone Groundwood Pulping. In a typical stone groundwood process, the peeled logs go through a drum washer to remove any remaining bark or dirt. The clean, wet wood is pressed against a wetted, rotating grindstone, and the axis of the wood is parallel to the axis of the stone. After grinding, the pulp is passed on first to coarse (bull) screens, which remove large splinters, and then to finer screens that remove bundles of fibers or shives. The accepted material is cleaned in centrifugal cleaners to remove short, stubby material called chop, as well as inorganic debris (grit) that comes from the grindstone. The final accepted pulp, which must be thickened for storage, may be bleached or used directly on the paper machine.

Refiner Mechanical Pulping. The conventional groundwood process requires bolts of roundwood as raw material. The refiner mechanical pulping (RMP) process produces a stronger pulp and utilizes various supplies of wood chips. However, the energy requirement of RMP is higher, and the pulp does not have the opacity of stone groundwood fibers. The process is similar to that for the stone groundwood pulping process.

The refiner plates have a construction of the type shown in Figure 4. The plates are paired face-to-face with a small interval or gap between them.

One disk rotates against a stationary disk, or both disks move in a counterrotating manner. Chips are fed through channels near the shaft in one of the disks and move toward the periphery of the plates while undergoing attrition. The chips are broken down first into matchstick-like fragments by the action of the breaker bars, then into progressively smaller bundles as the chips move through the intermediate and fine bar sections. They emerge from the periphery as single fibers or fiber fragments. This process is termed fibrillation. These thin, flexible elements considerably improve the bonding properties of the mechanical pulps.

Thermomechanical Pulping. A modification of RMP is thermomechanical pulping (TMP). If chips are presteamed to 110–150°C, they become malleable and do not fracture readily under the impact of the refiner bars. Thermoplasticization of the wood occurs when the wood is heated to a temperature approximating the glass-transition point of wet lignin. When these chips are fiberized in a refiner at high (30%) consistency, whole individual fibers are released. The increased proportion of long fibers improves the strength properties of TMP pulps, but the fibers in this fraction are stiff and contribute little to bonding.

Energy usage is high in the TMP process. This makes energy recovery a significant factor in the pulping processes.

Chemithermomechanical and Chemimechanical Pulping. The strength properties of thermomechanical pulps (TMP) can be increased by a mild pretreatment with sodium sulfite at pH 9–10 or by using other chemicals.

A range of properties can be obtained by adjusting process variables that affect pulp yield and particle size. In general, a CTMP has a greater long-fiber fraction and lower fines fraction than a comparable TMP. The intact fibers are more flexible than TMP fibers, giving better sheet-forming and bonding properties. Chemithermomechanical and chemimechanical pulpings are particularly suitable for pulping high density hardwoods.

Chemical Pulps

Processing chemical pulps normally involves wood and liquor preparation steps, delignification (pulping), pulp washing, knot removal, screening, cleaning, thickening, and, in the case of bleachable grades, residual lignin removal and brightening to eliminate chromophoric or light absorbing groups. Often, more than 50% of the mass of the wood is dissolved in chemical pulping. This allows recovery of these solids to generate steam and electrical energy for use in the mill. Spent chemicals are recovered for reuse in the process.

In chemical pulps, sufficient lignin is dissolved from the middle lamella to allow the fibers to separate, using little, if any, mechanical action. However, a portion of the cell wall lignin is retained in the fiber, and an attempt to remove this during digestion can result in excessive degradation of the pulp. For this reason, a small amount of lignin is normally left the pulps. The lignin is subsequently removed by bleaching in separate processing if completely delignified pulps are to be produced.

Kraft Process. The dominant chemical wood pulping process is the kraft or sulfate process. The alkaline pulping liquor or digesting solution contains about a three-to-one ratio of sodium hydroxide, NaOH, and sodium sulfide, Na_2S. The name kraft, which means strength in German, characterizes the stronger pulp produced when sodium sulfide is included in the pulping liquor, compared to the pulp obtained if sodium hydroxide alone is used, as in the original soda process. The alternative term, ie, the sulfate process, is derived from the use of sodium sulfate, Na_2SO_4, as a makeup chemical in the recovery process. Sodium sulfate is reduced to sodium sulfide in the recovery furnace by reaction with carbon in black liquor.

Chemical penetration and topochemical processes are important in all chemical and semichemical pulping operations because of the solid nature of the wood components and the liquid or gaseous state of the pulping reagents. The initial liquid transport through wood chips is through the bordered pits of softwoods and the vessels of hardwoods. In general, penetration into softwoods is faster than into hardwoods, especially in the cross-fiber direction. For the strongly alkaline cooking liquors of the kraft process, diffusion takes place at nearly equal rates in all directions.

Pulping. Solutions of sodium sulfide and sodium hydroxide are in equilibrium:

$$H_2O + Na_2S \leftrightarrows NaHS + NaOH$$

or

$$H_2O + S^{2-} \rightarrow HS^- + OH^-$$

Aqueous sodium sulfide is therefore itself a source of hydroxide ions. This must be considered in adjusting the chemical charge. The actual concentration of pulping chemicals relative to lignin is determined by the liquor-to-wood ratio (L/W) as the total liquid weight, which includes the moisture originally present in the chips, divided by the dry wood weight.

Chemical charge, liquor composition, time of heatup, and time at temperature of reaction are all functions of the wood species or species mix being digested and the intended use of the pulp.

Both batch and continuous digesting systems are in operation. The batch digestion method has advantages that include flexibility to process and product changes, lower maintenance costs, and higher turpentine yields. The advantages of continuous digesters include uninterrupted process flow, higher pulp yields and heat recovery, relative ease of automation, and in-line processing.

Batch Cooking. The primary disadvantage of conventional batch cooking has been relatively higher steam consumption, high labor, and air pollution (qv) effect. During the late 1980s and early 1990s, there was a resurgence in batch cooking because of advances in materials of construction, control technology, and energy conservation through liquor displacement.

Continuous Cooking. In the widely used continuous kraft pulping system, a single-vessel hydraulic digester has two stages of diffusion washing and high density brown-stock pulp storage. The chips are continuously steamed first in a chip bin at atmospheric pressure and then at low pressure in a steaming vessel where turpentine and gases are vented to the condenser. The chips are then brought to a digester pressure of ca 1000 kPa (150 psi) via a high pressure feeder and picked up in a stream of recycled liquor to which white liquor, ie, fresh pulping solution, makeup has been added. This stream carries through to the top of the digester, where the recycling feed liquor is extracted from the chips. The chips, having the balance of the liquor, flow continuously down through the digester, first passing through an impregnation zone where liquor is impregnated into the chips for a period of 0.5–1.0 h. The temperature is then raised to ca 170°C and subsequently, the chips are held at about 170°C for 1.5–2.5 h while passing through a cooking zone. At this point, the digestion is essentially complete. The chips next pass continuously through a high heat, countercurrent washing zone.

Modified Continuous Cooking. Theoretical studies have demonstrated that improved kraft pulping can be obtained by several process modifications: (*1*) low initial alkali concentrations; (*2*) high sulfidity at the outset of the bulk phase of delignification; (*3*) lowered temperature in the cooking zone; and (*4*) reduced concentration of dissolved lignin and sodium salts in the later part of the cook. Modifications to the continuous digester were developed to produce pulp under these specified theoretical conditions.

Pulp Processing. Following pulping, the acceptable pulp fibers must be separated from a variety of contaminants. Uncooked knots and fiber bundles (shives) are removed by screening, pulping liquor is removed by washing, and denser particles separated by centrifugal cleaning. Different grades of pulp, eg, higher yield, medium yield, and bleachable grades, require different processing.

From unbleached storage, the pulp is sent to the paper mill for brown grades or to the bleach plant for bleached grades. A variety of equipment is used for thickening, depending on the level of consistency required.

Recovery System. Kraft pulping depends on its associated recovery process for producing the digestion liquor. There are four important functions in kraft recovery: (*1*) the organic substances, ie, lignin and carbohydrate fragment molecules, dissolved during pulping are destroyed, thus eliminating a major potential source of stream pollution; (*2*) large quantities of steam are generated by burning the dissolved organic solids, thus supplying a large portion of the steam used in the pulp mill; (*3*) sodium hydroxide is regenerated for reuse in the pulping process; and (*4*) sulfur compounds present in the black liquor are converted to sodium sulfide, which is reused in the pulping process.

After washing, the black liquor is subjected to multiple-effect evaporation. Under conditions of direct-contact evaporation, residual sodium hydrosulfide partially hydrolyzes to hydrogen sulfide and sodium hydroxide, whereas sodium methanethiolate, partially hydrolyzes to methyl mercaptan and sodium hydroxide. In order to prevent the escape of the volatile hydrogen sulfide and methyl mercaptan with the flue gas, the black liquor is oxidized using air or oxygen, giving sodium thiosulfate and dimethyl disulfide.

Evaporation of the black liquor by direct contact with the hot recovery-furnace flue gases has proved to be a principal source of air pollution, and the use of direct-contact evaporators for concentrating black liquor has generally been eliminated in favor of the use of a concentrator (see AIR POLLUTION CONTROL METHODS). Concentrators, sometimes referred to as finishers, are steam-heated evaporators designed to produce liquor suitable for firing directly.

Uses. The kraft process provides wide adaptability to the pulping of virtually any wood species and yields pulps that are suitable for a broad range of products such as packaging products, bags and multiwalled sacks, and in newsprint. The brown color of kraft pulps is undesirable for applications such as writing and printing papers, but kraft pulps can be bleached to remove the residual lignin and other chromophores.

By-Products. There are three stages within the pulping operation at which wood-derived chemicals can be recovered as by-products. Turpentine is obtained from the relief of gases after an initial steaming of chips in the digester.

In the initial black liquor concentration, saponified fatty and resin acid salts separate as tall oil soaps. Tall oil is fractionated primarily into fatty acids (see CARBOXYLIC ACIDS).

The final source of by-products is the spent liquor. The organic constituents are mainly the various fragments and degradation products of lignin and of the carbohydrates. The dissolved alkali lignin, which is still polymeric, can be separated from black liquor by acidification and can be further purified or chemically modified to give useful materials.

Semichemical Pulping

A semichemical process is essentially a chemical delignification process in which the chemical reactions are stopped at a point where mechanical treatment is necessary to separate fibers from the partially cooked chips. Any chemical pulping process can be used to produce semichemical pulp. Reagents most commonly used are sodium sulfite, sodium hydroxide, sodium carbonate, and kraft green and white liquors. The usual procedure in semichemical pulping is to treat the wood chips using a pulping reagent in batch or continuous digesters. Pulping time, pulping temperature, and/or chemical-to-wood ratio are adjusted to obtain the desired pulp yield. Semichemical pulps, although less flexible, resemble chemical pulps more than mechanical pulps because they are not as dependent on rupture of the fiber wall for bonding.

The traditional method for producing semichemical pulps has been the neutral sulfite semichemical (NSSC) process. However, many other processes exist for producing the same products. The main difference is the composition of the pulping liquor. A few of the more important technologies are as follows. (*1*) In the acid sulfite process, the cooking liquor consists of a solution of bisulfite plus free sulfur dioxide. Delignification is the predominate reaction. (*2*) In the bisulfite process, bisulfite pulping liquor contains no free sulfur dioxide. The mechanism of the delignifying reactions is similar to that in acid sulfite pulping. (*3*) In the alkaline sulfite process, the cooking liquor contains sodium sulfite and any or all of the following: sodium carbonate, sodium hydroxide, and sodium sulfide. (*4*. Finally, in the kraft—semichemical process, semichemical pulps are readily produced by the kraft process.

Bleaching

The objective of bleaching is to whiten the pulp without reducing the intrinsic strength of the fibers comprising the pulp, or as a means to change other pulp properties. Pulps vary considerably in their color after pulping, depending on the wood species, method of processing, and extraneous components (see BLEACHING AGENTS, PAPER AND PULP).

Brightness Scales. There are numerous brightness scales used in commercial operations. The three most common are General Electric (GE), electronic reflectance photometer (Elrepho), and the International Standards Organization (ISO). In the United States, the TAPPI or GE scale is widely used on paper grades, whereas ISO is commonly used on pulp. The GE scale is gradually being replaced by the ISO scale.

Bleaching Methods. There are basically two types of bleaching operations or methods: those that chemically modify the chromophoric groups by oxidation or reduction but remove very little lignin or other substances from the fiber, and those that complete the delignification and remove pitch and some carbohydrate material. In the special case of dissolving-pulp production, bleaching is a final purification of cellulose, and most of the residual hemicellulose is removed (see also BLEACHING AGENTS). Both techniques result in the absorption coefficient for the pulp, being reduced and the reflectance factor raised.

Mechanical pulps are bleached in a lignin-conserving manner, known as brightening. The chromophores are treated with oxidizing agents such as hydrogen peroxide (*P*) and hypochlorite (*H*) solutions, and reducing agents such as sodium hydrosulfite (dithionite, *Y*) and formamidine sulfinic acid (FAS).

Removal of lignin is the goal of bleaching chemical pulp. It may be viewed as an extension of the pulping process. Pulping is discontinued before lignin removal is complete because the selectivity of pulping, ie, the rate of lignin removal relative to the rate of carbohydrate degradation, decreases at low lignin contents. Bleaching is inherently more selective than pulping and can be safely used to remove the last traces of lignin from the pulp.

Environmental Management and Pollution Prevention

Environmental protection in the pulp industry manages the amount and type of discharges to soil, water, and air. Management begins in the field when the fiber source is planted to provide high yield trees that can be harvested with minimum impact to soil resources and surrounding areas. Management of clear cutting and protection of critical habitat are considerations and techniques of silviculture. Biogenetic engineering researchers are working to develop fiber resources that grow quickly and produce high quality fibers. In addition, the rapid development of processes to use secondary fiber effectively has placed the industry at the forefront of global efforts to recycle waste paper, conserve forest resources, and protect watersheds.

Discharges to receiving water bodies must meet water quality standards to protect drinking water and aquatic life.

Treatment of Liquid Effluents. Before pulp mill effluents can be released to the environment they must be treated. Primary treatment involves the use of settling ponds or tanks in which suspended solids settle out of the liquid effluent. Solids can be composted and spread on land, converted to other useful products, or incinerated. Secondary effluent treatment includes oxidation and aeration in shallow basins having wide areas or in smaller areas using mechanical agitators and spargers to oxygenate fluids before release. Biological filter systems can be used to remove organic compounds and heavy metals, and often the process can be accelerated by adding nutrients and by using oxygen rather than air.

Air Pollutants. Air pollution regulations are concerned primarily with minimizing particulates, odor, ground-level ozone precursors, acid rain, and smog. Priority pollutants include chemicals such as acetone, methanol, sulfuric acid, sulfur dioxide, hydrogen sulfide and other reduced sulfur compounds, and nitrogen dioxide emissions, which are monitored closely. Pollution control equipment includes catalytic and thermal systems to destroy pollutants, electrostatic precipitators to remove particulate matter, cyclonic separators, venturi scrubbers, and fabric filters. Pollutants in molecular form are removed primarily by adsorption (qv), ie, dry scrubbing, and packed bed and venturi-type scrubbers, ie, wet scrubbing, from the process stream for reuse or for destruction.

Energy. The pulp and paper industry is a significant industrial consumer of energy. Much of the energy used, however, is self-generated from combustion of residues such as spent pulping liquors, bark, hogged wood, and other nonprocessed wood.

Significant changes are likely to occur in the energy self-sufficiency of the industry, resulting in an increased dependency on purchased power. Whereas new installations and retrofits of existing technologies can result in greater energy efficiency, demands from process changes, oxygen/ozone generation requirements, and increased environmental regulations may create even greater demand. Industry efforts in the area of energy use are expected to focus on increased energy conservation, greater utilization of biomass and other renewable energy sources, and process improvements such as better drying efficiency.

JOSEPH M. GENCO
University of Maine

D. Fengel and G. Wegener, *Wood-Chemistry, Ultrastructure, Reactions*, Walter De Gruyter Inc., Berlin, 1984.

R. A. Leask and M. J. Kocurek, *Pulp and Paper Manufacture*, Vol. 2, *Mechanical Pulping*, 3rd ed., The Joint Committee of the Paper Industry, TAPPI/CPPA, Atlanta, Ga., 1987.

T. M. Grace, E. W. Malcolm, and M. J. Kocurek, *Pulp and Paper Manufacture*, Vol. 5, *Alkaline Pulping*, 3rd ed., TAPPI/CPPA, Atlanta, Ga., 1989, p. 121.

C. W. Dence and D. W. Reeve, *Pulp Bleaching*, TAPPI Press, Atlanta, Ga., 1996.

PUMPS

Pumps are used in a wide range of industrial and residential applications. Pumping equipment is extremely diverse, varying in type, size, and materials of construction. The passage of the Clean Air Act of 1990 by the U.S. Congress, a heightened attention to a safe workplace environment, and users' demand for better equipment reliability have all led to improved mean time between failures (MTBF) and scheduled maintenance (MTBSM).

Classification

One general source of pump terminology, definitions, rules, and standards is the Hydraulic Institute (HI) Standards, approved by the American National Standards Institute (ANSI) as national standards.

Pumps are divided into two fundamental types based on the manner in which pumps transmit energy to the pumped media: kinetic or positive displacement. In the first type, a centrifugal force of the rotating element, called an impeller, impels kinetic energy to the fluid, moving the fluid from pump suction to the discharge. The second type uses the reciprocating action of one or several pistons, or a squeezing action of meshing gears, lobes, or other moving body, to displace the media from one area into another, ie, moving the material from suction to discharge. The pumped medium is usually liquid. However, many designs can handle solids in suspension, entrained or dissolved gas, paper pulp, mud, slurries, tars, and other exotic substances, which, at least by appearance, do not resemble a liquid. Nevertheless, an overall liquid behavior must be exhibited by the medium in order to be pumped.

The Hydraulic Institute (HI) classifies pumps by type, not by application. The user, however, must ultimately deal with specific applications.

Operating Conditions

Before a pump selection can be made, the duty conditions must be specified. These include type of fluid, density or specific gravity, temperature, viscosity, flow, inlet and outlet pressures, and presence of solids or corrosive/erosive material in the liquid. For a typical process installation, an accurate estimate of the pumping system is required, starting with the source of the liquid (tank, vessel, pipeline, basin, etc) through the planned system layout to the terminal point (see PIPELINES; PIPING SYSTEMS). Pressure drops through the entire piping system must be estimated, including all valves, fittings, process equipment such as heat exchangers, heaters, and boilers, and any losses through orifices or control valves.

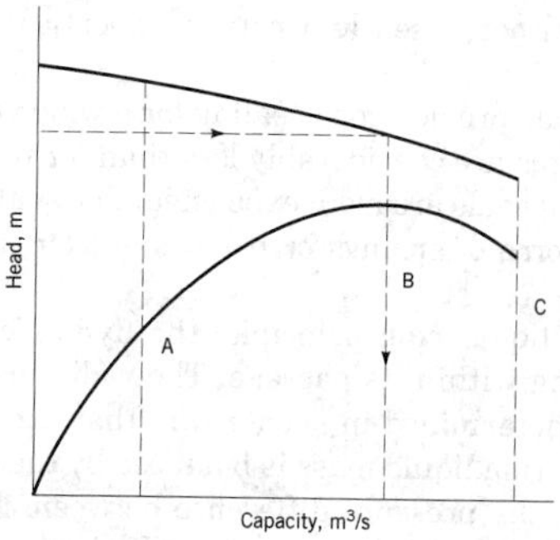

Figure 1. Head–capacity curve, where A represents the minimum allowable capacity; B, the best efficiency point (BEP); and C, the maximum allowable capacity.

Most process plant engineers utilize some form of preprinted pump calculation worksheet or a computer to aid in the collection of information.

Capacity. Pumps deliver a certain capacity, sometimes referred to as flow, which can be measured directly (see FLOW MEASUREMENT). The indirect way to determine capacity is often used. Whereas this method is less accurate than applying a flow meter, it often is the only method available in the field. The total head is measured and the capacity found from the pump head–capacity curve (Fig. 1).

Head. The true meaning of the total developed pump head is the amount of energy received by the unit of mass per unit of time. This concept is traceable to compressors and fans, where engineers operate with enthalpy, a close relation to the concept of total energy. However, because of the almost incompressible nature of liquids, a simplification is possible to reduce enthalpy to a simpler form known as a Bernoulli equation.

Power. There are two main ways to measure the power delivered by the driver to the pump. The first method is to install a torque meter between the pump and the driver. A torque meter is a rotating bar having a strain gauge to measure shear deformation of a torqued shaft. The benefit of this method is direct and accurate measurements. The power delivered to the pump from the driver is calculated from torque and speed (rpm) in units of brake horsepower.

This method is used only at a manufacturer's test facilities or research laboratories. It is not used in the field.

In the second method, the pump and the motor are coupled directly, and either power (in kilowatts) or the current and voltage are measured at the motor terminals. To determine the power actually transmitted into a pump, the motor power factor and efficiency must be known. These values are usually taken from the motor manufacturer's calibration curves.

Types of Pumps

Kinetic. Kinetic pumps, which act by impelling a fluid from one location to another, can be centrifugal, regenerative turbine, or special effect.

Overhung Impeller. A closed-coupled pump has an impeller mounted directly on the shaft of the driver, thus eliminating the need for pump-bearing housing. The driver bearings take all pump loads. These pumps are used for relatively light-duty services. They are often applied for sanitary and corrosive pumping requirements because of a clean-in-place (CIP) capability, ie, the pump can be flushed (cleaned) without much disassembly.

Close-coupled pump designs have been largely replaced by separately coupled pumps. The latter eliminate hydraulic loads on the motor bearings and are also less prone to hazard if pumpage splashes on the motor, potentially causing fire in the event of a seal failure. These centrifugal pumps find applications in petrochemical and refinery plants.

For medium and severe applications, when nozzle loads and thermal transients tend to impose high stresses and internal deflections inside the pump, frame-mounted designs are used. Typically, these designs use conventional impellers, but recessed impeller designs are also available.

In 1961, the American National Standards Institute (ANSI) introduced a chemical pump standard that defined common pump envelope dimensions, connections for the auxiliary piping and gauges, seal chamber dimensions, parts runout limits, and baseplate dimensions. This definition was to ensure the user of the availability of interchangeable pumps produced by different manufacturers, as well as to provide plant designers with standard equipment.

For refinery applications, temperatures, pipe loads, and product flammability are the prime factors in selecting a pump. The pump-mounting feet are located close to the centerline. As a result, the pump thermal growth is uniform on both sides of the centerline, leading to minimal distortions.

For sump pumping applications, three basic approaches are taken. In the first, called a wet-pit design, the pump is lowered to the pumping depth, but the motor is mounted above the highest anticipated liquid level and coupled to the pump impeller via a long shaft or a series of connected shaft sections.

For greater depths or as an alternative to the wet-pit design, a submersible pump having a special water-tight motor can be used.

Another option is the dry-pit design, a pump installed in a dry pit and connected to a well via a pipe.

The axial-flow propeller-type pumps are used to handle large volumes of pumped liquid at low head.

Impeller Between Bearings. These pumps are grouped into single- or multistage designs, and each type is available in either axially or radially split configurations.

Axially split pumps are designed for convenient maintenance, inspection, replacement of the impeller or wear parts, and easy access to pump internals. Owing to the nature of the axially split configuration and the large area of the gasket required to seal the two halves, these pumps are limited to relatively moderate developed heads and temperatures under 175°C (350°F).

There are traditional differences in the design philosophy between the axially and the radially split multistage pumps.

An axially split pump typically has a volute, and offers the advantage of easy maintenance because of quick upper-half casing and rotor removal that is similar to a single-stage version.

Radially split construction is typically a diffusor design, offering the advantage of reduced radial thrust, which is better from the standpoint of lowering shaft deflections, resulting in longer life of seals and bearings.

Verticals- or Turbine-Type. The deep-well pump is submerged into the well, and liquid comes into contact with the pump body and the well walls. Alternatively, a pump pit may be of dry design, and a can, ie, an enclosure around the pump, having connections to the source of liquid is provided.

Specialty Pumps. There is a multitude of other pump designs in the family of kinetic pumps. In this group are vacuum pumps, self-priming pumps, disk pumps, and jet pumps. In this group are vacuum pumps, self-priming pumps, disk pumps, and jet pumps.

Positive Displacement Pumps. Positive displacement pumps follow HI convention. As a rule, these pumps work against significantly higher pressures and lower flows than do kinetic, particularly centrifugal, pumps. Positive displacement pumps also operate at lower rotational speeds. There are many types of positive displacement pumps, for which designs are constantly being developed. Some of these are discussed herein.

Reciprocating Pumps. A classic example of a positive displacement pump is that of the reciprocating pump. These pumps are the most familiar to a nonspecialist. The energy of the reciprocating piston of this pump generates pressure, causing liquid to move from the chamber in which the fluid is acted upon by the piston, to the discharge piping. On the return stroke, the space vacated by the retrieving piston is filled by the liquid, which is moved by the suction pressure.

Rotary Vane Pumps. The rotary vane pump features an eccentric rotor having multiple plates (vanes). The rotation of the rotor translates into a reciprocating movement of pistons, resulting in pumping action.

Gear Pumps. Liquid enters the inlet port, fills the space between gear teeth, and is trapped in these spaces by the gear teeth and pump housing. It is carried around to where the gear teeth mesh, which forces the liquid out through the discharge opening. The rotor speeds are greater than those for reciprocating pumps, leading to a smaller pump size for the same throughput. Pressure pulsations are also significantly reduced.

Rotary Lobe Pumps. Rotary lobe pumps, similar to gear pumps in principle, have an added advantage of noncontacting metal parts by use of external gears, which reduces the wear, but adds complexity.

Screw Pumps. Constructed by using external timing gears and bearings, two-screw pumps are suited for capacities between 0.1–35 m^3/min (20–9000 gal/min), high suction lift capabilities, and differential pressures of up to 10 MPa (1500 psig).

Other positive displacement pump designs include progressive cavity pumps, flexible tube pumps, flexible impeller pumps, diaphragm pumps, and condensed recovery pumps.

Application Examples

Selection of pump for a given application is not a trivial task. Often more than one pump type can accomplish the required job. Thus a final choice on a pump type is often a result of personal experience and usage history. As a rule of thumb, the choice of a kinetic, such as centrifugal, or a positive displacement pump is made on the basis of the specific speed. Whereas specific speed is applicable primarily for centrifugal but not positive displacement pumps, the specific speed value can be used as a guide.

Reciprocating pumps operate at high pressures and low flow rates. Conversely, centrifugal pumps are applied at lower pressures and higher flow rates. Many rotary pumps are selected for viscous liquids having pressures equal to or less than, and capacities lower than, centrifugal pumps. However, these limits are relative and a gray area exists; some pump types cross boundaries into the domain of other types.

There are available special designs, including those for corrosive liquids, leakage prevention, corrosive or toxic applications, slurries, trash, metering, and for nonclogging applications.

Couplings and Seals

Various coupling designs are available to transmit torque from the driver, eg, electric motor, to a pump. In order to contain the pumped fluid inside the pump and prevent the pumpage from leaking, several types of sealing methods are used.

Couplings. Gear-type couplings are usually used for heavy-duty applications having high torques.

Diaphragms are used to transmit torque between the inside and outside diameters of the flexible element in disk coupling. Whereas no oil lubrication is required as compared to gear couplings, disks are limited to moderate torques.

The grid type couplings are applied for moderate loads.

Many other coupling types are available. These vary in degree of complexity, maintenance time, torque capabilities, and price. An elastomer-type coupling is one of the most popular types used, particularly in the chemical industry.

A common requirement for all coupling types, although to a different degree, is the need for good alignment. Alignment minimizes coupling strain and prevents shaft bending.

Sealing. The oldest method of sealing is through the use of packings. A packing material is inserted into a stuffing box. To prevent the shaft from rubbing wear by packing material, hardened sleeves are often used over the shaft. Some packing designs include graphite or Teflon particles for better self-lubricating properties to minimize wear damage.

Mechanical seals provide good sealing for a wide variety of applications. Leakage rates are considerably less than for packing materials, but the mechanical seals are more expensive. The sealed faces are kept together by the force of springs or bellows, or, for unbalanced seals, hydraulically.

Based on the liquid ring principle, the dynamic seal or expeller keeps a liquid ring within its passage. The radial height of this ring is automatically determined in such a way that the centrifugal force caused by the rotating liquid mass is balanced by the differential pressure across it, ie, the pressure difference between the expeller outer diameter, which is equal to the pump impeller back hub pressure, and the sealing chamber pressure. If the suction pressure increases, the amount of liquid in the expeller correspondingly increases.

Application Guidelines

Pumps are designed to give trouble-free operation for a long period of time. The ANSI B73.1M pumps are designed for a bearing life of no less than two years, and American Petroleum Institute (API) 610 pumps for a minimum of five years. However, in real applications, a typical mean time between failures (MTBF) is often found to be significantly less, and sometimes it is as short as a few weeks. The reason for such wide variations is often not poor pump design but equipment misapplication.

LEV NELIK
Roper Pumps Company

Hydraulic Institute Standards for Centrifugal, Rotary & Reciprocating Pumps, Hydraulic Institute, Parsippany, N. J., 1994.

A. J. Stepanoff, *Centrifugal and Axial Flow Pumps*, 2nd ed., John Wiley & Sons, Inc., New York, 1948.

H. H. Anderson, *Centrifugal Pumps*, 3rd ed., Trade & Technical Press Ltd., London, 1980.

L. Nelik, *Proceedings of 2nd International Conference on Reliability*, Houston, Tex., 1993.

PURGATIVES. See GASTROINTESTINAL AGENTS.

PURINES. See ANTIBIOTICS, NUCLEOSIDES AND NUCLEOTIDES; CHEMOTHERAPEUTICS, ANTICANCER; NUCLEIC ACIDS.

PYRAZOLES, PYRAZOLINES, AND PYRAZOLONES

The compounds of this article, ie, five-membered heterocycles containing two adjacent nitrogen atoms, can best be discussed according to the number of double bonds present. Pyrazoles contain two double bonds within the nucleus, imparting an aromatic character to these molecules. They are stable compounds and can display the isomeric forms, (**1**) and (**2**), when properly substituted. Pyrazoles are scarce in nature when compared to the imidazoles (**3**), which are widespread and have a central role in many biological processes.

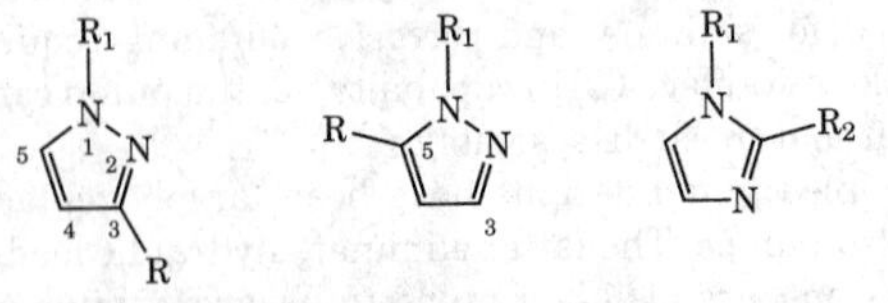

(**1**) (**2**) (**3**)

Pyrazolines have only one double bond within the nucleus and, depending on the position of the double bond, can exist in three separate forms: 1-pyrazoline (**4**), 2-pyrazoline (**5**), and 3-pyrazoline (**6**).

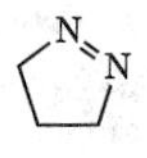
(4)
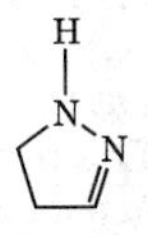
(5)
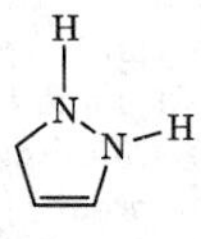
(6)

Pyrazolones, contain two double bonds, and are predominantly in the keto form (**7**), although they can also exist in the enol form (**8**).

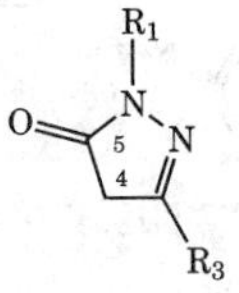
(7)
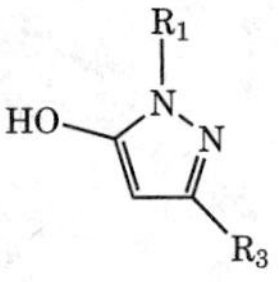
(8)

Neither pyrazolidines (**9**), which have no double bonds, nor pyrazoline diones (**10**), with two double bonds, and pyrazolidine triones (**11**), which have three double bonds, are covered in this article.

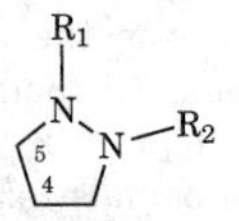
(9)
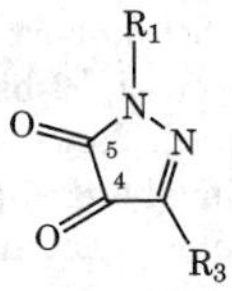
(10)
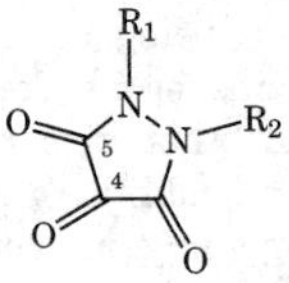
(11)

Despite their scarcity in nature, the title compounds have found use in many applications, including pharmaceuticals, agricultural chemicals, and dyes.

Theoretical Methods

A number of theoretical studies on the reactivity of pyrazoles have been published. However, due to the difficulties involving these calculations, the studies often only approximate the actual reactions occurring in the laboratory.

Structural Elucidation

Among the modern procedures utilized to establish the chemical structure of a molecule, nuclear magnetic resonance (nmr) is the most widely used technique. Mass spectrometry is distinguished by its ability to determine molecular formulas on minute amounts, but provides no information on stereochemistry. The third most important technique is x-ray diffraction crystallography, used to establish the relative and absolute configuration of any molecule that forms suitable crystals. Other physical techniques, although useful, provide less information on structural problems.

Nuclear Magnetic Resonance Spectroscopy. The main application of nmr in the field of pyrazolines is to determine the stereochemistry of the substituents and the conformation of the ring. For pyrazolones, nmr is useful in establishing the structure of the various tautomeric forms.

X-Ray Diffraction.

Because of the rapid advancement of computer technology, this technique has become almost routine and the structures of moderately complex molecules can be established sometimes in as little as 24 hours.

Miscellaneous Techniques. The use of ultraviolet (uv) and infrared (ir) spectroscopy has diminished drastically as newer and more powerful procedures have been introduced. However, uv is still useful in studying the tautomeric structures and ionization constants of pyrazoles.

Physical Properties

Pyrazoles in general are stable compounds, as demonstrated by pyrazole itself, which distills at 186°C at atmospheric pressure. The boiling point (bp) increases with an increase in the number of alkyl substituents on carbon. *N*-Methylation decreases both the bp and the melting point (mp) as a result of the elimination of hydrogen bonding. Pyrazoles with substituents at C_3 (C_5) are tautomeric mixtures and form azeotropes. The solubility of pyrazole in H_2O is about 1 g/mL, but it is much less soluble in organic solvents. Pyrazole is a weak base ($pK_a = 2.5$) and can be protonated by strong acids; strong bases yield metal salts. The pyrazolines resemble the pyrazoles in their physical properties. They are liquids with a high bp or low mp. Pyrazolines are basic and the ease of protonation is dependent on the position of the double bond. Most pyrazolones are solids and the mp usually decreases in the presence of substituents at N_1. Simple low molecular weight pyrazolones are soluble in hot water and the higher mol wt materials are soluble in most organic solvents. Hydrogen bonding has strong influence on the predominant tautomeric form. 3-Pyrazolones are more basic than the isomeric 5-pyrazolones.

Chemical Reactivity

Pyrazoles. The chemical reactivity of the pyrazole molecule can be explained by the effect of individual atoms. The N-atom at position 2 with two electrons is basic and therefore reacts with electrophiles. The N-atom at position 1 is unreactive, but loses its proton in the presence of base. The combined two N-atoms reduce the charge density at C_3 and C_5, making C_4 available for electrophilic attack. Deprotonation at C_3 can occur in the presence of strong base, leading to ring opening. Protonation of pyrazoles leads to pyrazolium cations that are less likely to undergo electrophilic attack at C_4, but attack at C_3 is facilitated. The pyrazole anion is much less reactive toward nucleophiles, but the reactivity to electrophiles is increased. Chlorination of pyrazole yields 4-chloropyrazole (**12**) and bromination can produce mono-, di-, or tribromo pyrazoles (**13**). 3-Methylpyrazole on treatment with chlorine in acetic acid yields the pentachloropyrazole derivative (**14**).

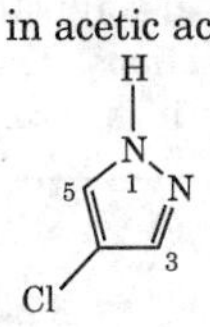
(12)
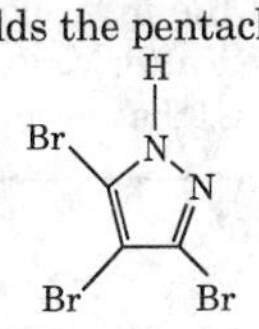
(13)
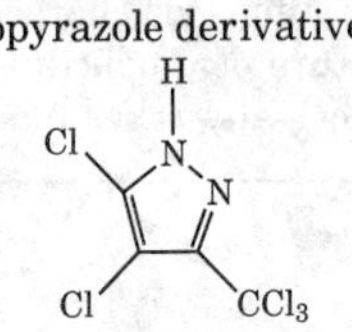
(14)

The pyrazole ring is resistant to oxidation and reduction. Only ozonolysis, electrolytic oxidations, or strong base can cause ring fission. On photolysis, pyrazoles undergo an unusual rearrangement to yield imidazoles via cleavage of the N_1–N_2 bond, followed by cyclization of the radical intermediate to azirine.

Oxidation of N_1-substituted pyrazoles to 2-substituted pyrazole-1-oxides using various peracids facilitates the introduction of halogen at C_3, followed by selective nitration at C_4. The halogen atom at C_3 or C_5 is easily removed by sodium sulfite and acts as a protecting group. Formaldehyde was used to direct the selective introduction of electrophiles at C_5 in a simple one-pot procedure.

Pyrazolines. The chemical properties of pyrazolines are governed by their relative instability. They readily undergo ring cleavage, and are easily reduced and oxidized. Loss of nitrogen occurs in pyrazolines lacking a substituent at N_1 to give a mixture of olefins and cyclopropanes, the latter being predominant. This elimination occurs near the mp and can be catalyzed by uv light, aluminum oxide, and many other substances. Mild reduction of pyrazolines leads to pyrazolidines. Sodium–alcohol, tin–HCl, or Raney nickel cause ring cleavage, yielding diamines or aminonitrile derivatives. Pyrazolines are easily oxidized to pyrazoles by many reagents, such as bromine, permanganate, and lead tetraacetate. Besides pyrazole formation, rearrangements or side-chain oxidations may also occur. Oxidation with peracids produce N-oxides. Pyrazolines lacking a substituent at N_1 undergo reactions typical of secondary amines, such as acylation, benzoylation, nitrosation, carbamate, and urea formation.

Pyrazolones. The oxo derivatives of pyrazolines, known as pyrazolones, are best classified as follows: 5-pyrazolone, also called 2-pyrazolin-5-one (**15**); 4-pyrazolone, also called 2-pyrazolin-4-one (**16**); and 3-pyrazolone, also called 3-pyrazolin-5-one (**17**). Within each class of pyrazolones many tautomeric forms are possible; for simplicity only one form is shown.

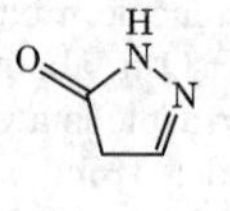
(15)

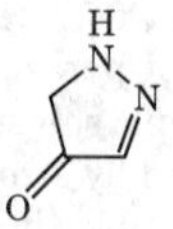
(16)

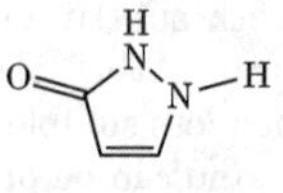
(17)

Substitution at N_1 decreases the possible number of tautomers: for 3-pyrazolones, two tautomeric forms are possible, (**18**) and (**19**), which in nonpolar solvents are both present in about the same ratio. 5-Pyrazolones exhibit similar behavior.

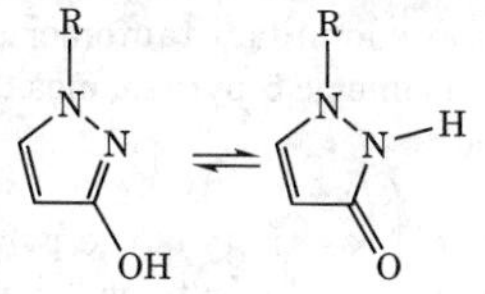
(18) **(19)**

In 4-pyrazolones, the enol form predominates, although the keto form has also been observed.

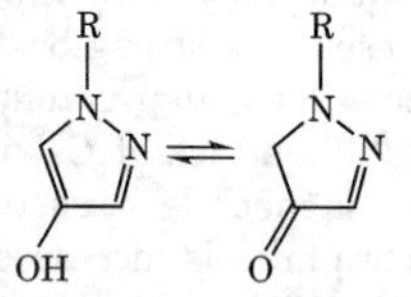
enol keto

The tautomeric character of the pyrazolones is also illustrated by the mixture of products isolated after certain reactions. Thus alkylation normally takes place at C_4, but on occasion it is accompanied by alkylation on O and N. Similar problems can arise during acylation and carbamoylation reactions, which also favor C_4. Pyrazolones react with aldehydes and ketones at C_4 to form a carbon–carbon double bond, eg (**20**). Coupling takes place when pyrazolones react with diazonium salts to produce azo compounds, eg (**21**).

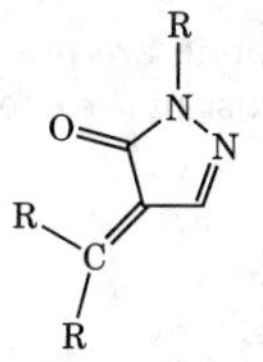
(20)

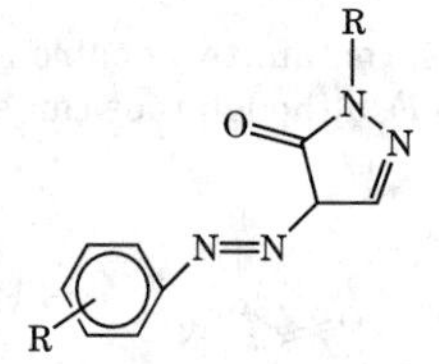
(21)

Compounds of type (**21**) are widely used in the dye industry (see AZO DYES).

Synthesis

In general, the synthesis of pyrazoles and related compounds can be classified into one of four principal categories, with the first two classes being by far the most important: (*1*) from the reaction of hydrazine or its derivatives with β-bifunctional compounds, or compounds that give rise to such functionality (eq. **1**) (*2*) by 1,3-dipolar cycloaddition, usually involving diazo compounds (eq. **2**) (*3*) by ring-opening of more complex systems already containing the pyrazole nucleus; and (*4*) by chemical, thermal, or photochemical rearrangement of other monocyclic heterocycles. Examples from each class follow.

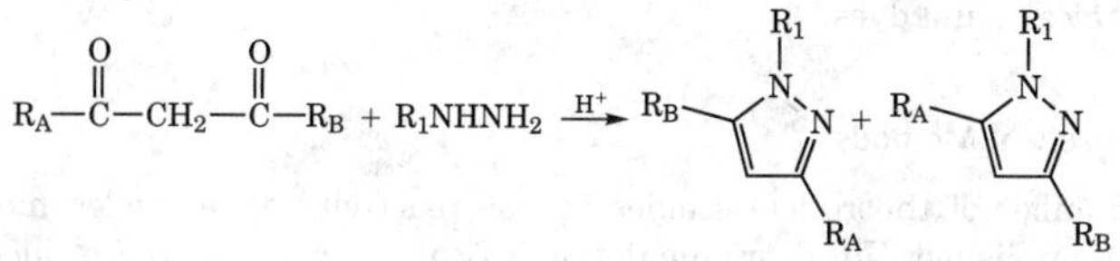

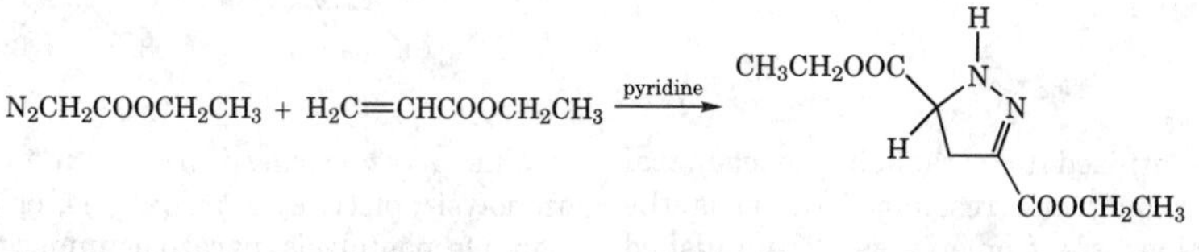

Health Factors

Pyrazole is considered a toxic material because in rats it causes hepatomegaly, anemia, and atrophy of the testis. It also inhibits the enzyme alcohol dehydrogenase, leading to severe hepatotoxic effects and liver necrosis when administered in combination with alcohol. Pyrazolones with a free NH group are easily nitrosated and give rise to nitrosamines, which cause tumors in the liver of test animals. The analgesics antipyrine (**22**) and aminopyrine (two pyrazolones), (**23**), if admixed with nitrites, are mutagenic when tested *in vitro*; however, when tested in the absence of nitrites, negative results are obtained.

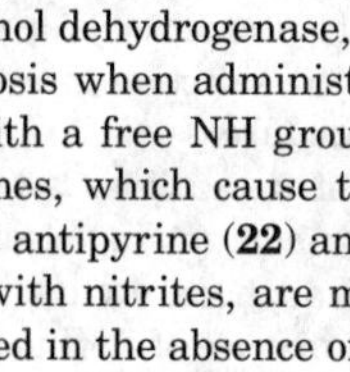
(22)

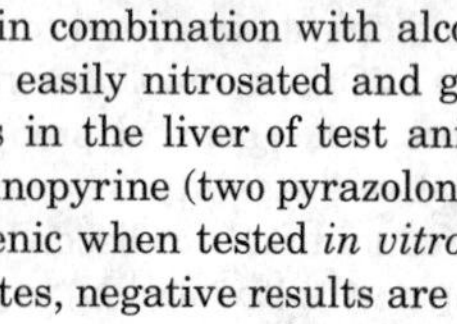
(23)

Pyrazolone-type drugs, such as phenylbutazone and sulfinpyrazone, are metabolized in the liver by microsomal enzymes, forming glucuronide metabolites that are easily excreted because of enhanced water solubility.

Applications

Pyrazoles, pyrazolines, and pyrazolones have all found wide use in many fields. Their greatest utility resides in pharmaceuticals, agrochemicals, dyes (textile and photography), and to a lesser extent in plastics. The main uses of the pharmaceuticals that incorporate the pyrazole nucleus are as antipyretic, antiinflammatory, and analgesic agents. To a lesser extent, they have shown efficacy as antibacterial/antimicrobial, antipsychotic, antiemetic, and diuretic agents.

Compounds containing the pyrazole nucleus have also found utility in agriculture. The organophosphate and carbamoyl functionalities, which impart insecticidal activity through linkage to many organic molecules. These compounds act by interfering with acetylcholinesterase in the cholinergic synapses.

Pyrazole derivatives have also considerable herbicidal activity.

GABE I. KORNIS
Pharmacia & Upjohn Inc.

T. L Jacobs, in R. C. Elderfield, ed., *Heterocyclic Compounds*, Vol. 5, John Wiley & Sons, Inc., New York, 1957.

L. C. Behr, R. Fusco, and C. H. Jarboe, in R. H. Wiley, ed., "Pyrazoles, Pyrazolines, Pyrazolidines, Indazoles and Condensed Rings," Vol. 22 of A. Weissberger, ed., *The Chemistry of Heterocyclic Compounds*, Wiley-Interscience, New York, 1967.

J. Elguero, in A. R. Katritzky and C. W. Rees, eds., *Comprehensive Heterocyclic Chemistry*, Vol. 4, Pergamon Press, Oxford, U.K., 1984.

A. R. Katritzky, *Handbook of Heterocyclic Chemistry*, Pergamon Press, Oxford, U.K., 1985.

PYRIDINE AND PYRIDINE DERIVATIVES

Pyridine compounds are defined by the presence of a six-membered heterocyclic ring consisting of five carbon atoms and one nitrogen atom. The carbon valencies not taken up in forming the ring are satisfied by hydrogen atoms. The arrangement of atoms is similar to benzene except that one of the carbon–hydrogen ring sets has been replaced by a nitrogen atom. The parent compound is pyridine itself (**1**). Substituents are indicated either by the numbering shown, 1 through 6, or by the Greek letters, α, β, or γ. The Greek symbols refer to the position of the substituent relative to the ring nitrogen atom, and are usually used for naming monosubstituted pyridines. The ortho, meta, and para nomenclature commonly used for disubstituted benzenes is not used in naming pyridine compounds.

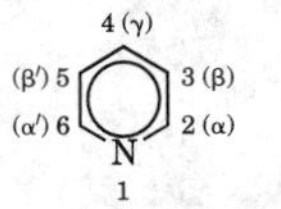

(**1**)

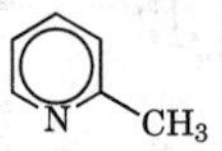

(**2**)

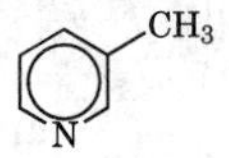

(**3**)

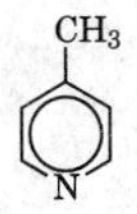

(**4**)

Important commercial alkylpyridine compounds are α-picoline (**2**), β-picoline (**3**), γ-picoline (**4**), 2,6-lutidine (**5**), 3,5-lutidine (**6**), 5-ethyl-2-methyl-pyridine (**7**), and 2,4,6-collidine (**8**). In general, the alkylpyridines serve as precursors of many other substituted pyridines used in commerce. These further substituted pyridine compounds derived from alkylpyridines are in turn often used as intermediates in the manufacture of commercially useful final products.

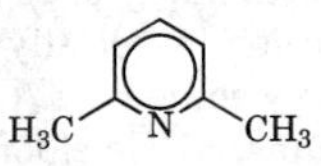

(**5**)

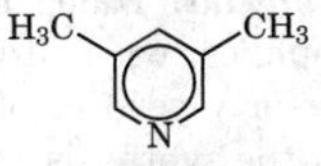

(**6**)

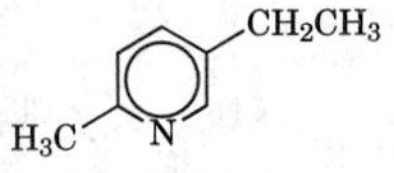

(**7**)

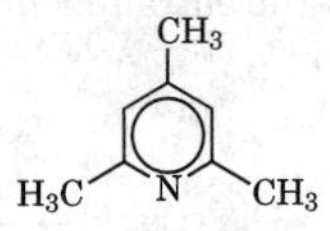

(**8**)

Pyridine was first synthesized in 1876 from acetylene and hydrogen cyanide. However, α-picoline (**2**) was the first pyridine compound reported to be isolated in pure form. There are few selective commercial processes for pyridine (**1**) and its derivatives, and almost all manufacturing processes produce (**1**) along with a series of alkylated pyridines in admixture. The chemistry of pyridines is significantly different from that of benzenoids.

Physical Properties

The physical properties of pyridines are the consequence of a stable, cyclic, 6-π-electron, π-deficient, aromatic structure containing a ring nitrogen atom. The ring nitrogen is more electronegative than the ring carbons, making the two-, four-, and six-ring carbons more electropositive than otherwise would be expected from a knowledge of benzenoid chemistries. The aromatic π-electron system does not require the participation of the lone pair of electrons on the nitrogen atom; hence the terms weakly basic and π-deficient used to describe pyridine compounds. The ring nitrogen of most pyridines undergoes reactions typical of weak, tertiary organic amines such as protonation, alkylation, and acylation.

Liquid pyridine and alkylpyridines are considered to be dipolar, aprotic solvents. Most pyridines form a significant azeotrope with water, allowing separation of mixtures of pyridines by steam distillation that could not be separated by simple distillation alone.

Pyridine. Many physical properties of pyridine are unlike those of benzene, its homocyclic counterpart. For instance, pyridine has a boiling point 35.2°C higher than benzene (115.3 vs 80.1°C), and unlike benzene, it is miscible with water in all proportions at ambient temperatures. Both benzene and pyridine are miscible with most other organic solvents. Pyridine is a weak organic base ($pK_a = 5.22$), being both an electron-pair donor and a proton acceptor. Table 1 lists some physical properties of pyridine.

Other Pyridine Bases. The nucleophilicity and basicity of pyridines can be reduced by large, sterically bulky groups around the nitrogen atom, such as *tert*-butyl in the 2- and 6-positions. Sterically undemanding groups like methyl tend to increase basicity relative to parent pyridine. Electron-withdrawing substituents can also reduce pyridine basicity and nucleophilicity.

Chemical Properties

Chemical reactivity of pyridines is a function of ring aromaticity, presence of a basic ring nitrogen atom, π-deficient character of the ring, large permanent dipole moment, easy polarizability of the π-electrons, activation of functional groups attached to the ring, and presence of electron-deficient carbon atom centers at the α- and γ-positions. Depending on the conditions of the chemical transformation, one or more of these (factors can give rise to the observed chemistry. The chemistry of pyridines can be divided into two categories: reactions at the ring-atomic centers, and reactions at substituents attached to the ring-atomic centers.

Reactions at Ring Atoms. Ring-atomic centers can undergo attack by electrophiles, easily at the ring nitrogen and less easily at ring carbons. Nucleophilic attack is also possible at ring carbons or hydrogens.

Electrophilic Attack at Nitrogen. The lone pair on pyridine (**1**) reacts with electrophiles under mild conditions, with protonic acids to give simple salts, with Lewis acids to form coordination compounds such as (**9**), and with transition metals to form complex ions, such as (**10**). The complex ion pyridinium chlorochromate (**11**) is a mild oxidizing agent suitable for the conversion of alcohols to carbonyl compounds.

Table 1. Physical Properties of Pyridine

Physical property	Value
enthalpy of fusion at −41.6°C, kJ/mol[a]	8.2785
enthalpy of vaporization, kJ/mol[a]	
at 25°C	40.2
115.26°C	35.11
critical temperature, °C	346.8
critical pressure, MPa[b]	5.63
enthalpy of formation, gas at 25°C, kJ/mol[a]	140.37
Gibbs free energy of formation, gas at 25°C, kJ/mol[a]	190.48
heat capacity, gas at 25°C, J/(K·mol)	78.23
ignition temperature, °C	550
explosion limit, %	1.7–10.6
surface tension, liquid at 25°C, mN/m(= dyn/cm)	36.6
viscosity, liquid at 25°C, mPa(= cP)	0.878
dielectric constant, liquid at 25°C, ϵ	13.5
thermal conductivity, liquid at 25°C, W/(K·m)	0.165

[a] To convert kJ to cal, divide by 4.184.
[b] To convert MPa to atm, multiply by 9.87.

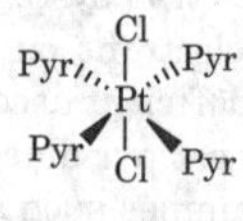
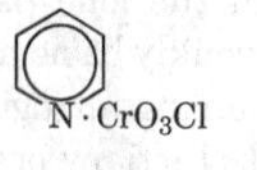

(9) (10) (11)

Reactive halogen compounds, alkyl halides, and activated alkenes give quaternary pyridinium salts, such as (**12**). Oxidation with peracids gives pyridine *N*-oxides, such as pyridine *N*-oxide itself (**13**), which are useful for further synthetic transformations.

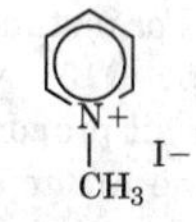

(12) (13)

Electrophilic Attack at Carbon. Electrophilic attack at a C-atom in pyridines is particularly difficult unless one or more strong electron-donating substituents are attached to the ring. Knowledge of this fact has resulted in the widespread use of pyridine as a solvent for reactions involving electrophilic species. Pyridine undergoes nitration in low yield (15%) to give 3-nitropyridine whereas pyridine *N*-oxide is nitrated at position 4 to give 4-nitropyridine *N*-oxide in high yield, and 2- and 4-aminopyridines may be dinitrated. The chlorination of pyridine and picolines (**2**), (**3**), and (**4**), on the other hand, is of great commercial importance.

Nucleophilic Attack at Carbon or Hydrogen. Only the strongest of nucleophiles (eg, $1NH_2$) can replace a hydrogen in pyridine. However, N-oxides and quaternary salts rapidly undergo addition, followed by subsequent transformations.

Free-Radical Attack at Carbon. Homolytic substitution of pyridines has not been as thoroughly studied as heterolytic processes have been, owing to low conversions and poor selectivity. However, some pyridinium salts have been found to undergo potentially useful regiospecific reactions. Protonated pyridines readily undergo acylation by acyl radicals, generated by abstracting a hydrogen atom from an aldehyde, eg, 2-acetyl-4-cyanopyridine from protonated 4-cyanopyridine.

Reactions of Substituted Pyridines. *Carbon Substituents.* Alkyl groups at positions 2 and 4 of a pyridine ring are more reactive than either those at the 3-position of a pyridine ring or those attached to a benzene ring. Carbanions can be formed readily at alkyl carbons attached at the 2- and 4-positions.

Cyanopyridines are usually manufactured from the corresponding picoline by catalytic, vapor-phase ammoxidation (eq. **1**) in a fixed-or fluid-bed reactor. 3-Cyanopyridine (**14**) is the most important nitrile, as it undergoes partial or complete hydrolysis under basic conditions to give niacinamide (**15**) or niacin (nicotinic acid) (**16**), respectively.

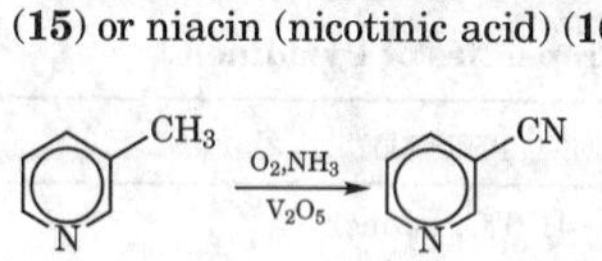

(3) (14)

Nitrogen Substituents. 4-Aminopyridine (**17**) and 2-aminopyridine (**18**) react with cold nitric acid to give the corresponding nitramines, which are insoluble in the media. On heating, these nitramines rearrange intermolecularly to nitroaminopyridines having the nitro group mainly adjacent to the amino group. From (**17**) the products are the intermediate 4-nitraminopyridine and 4-amino-3-nitropyridine (eq. **2**).

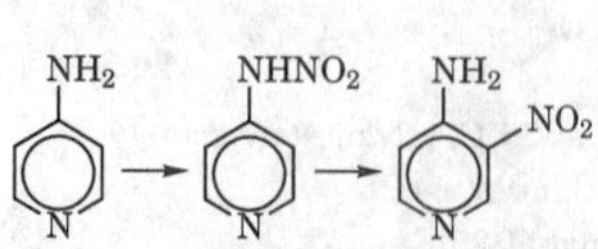

(17) (18)

Reaction of 2-aminopyridine (**18**) with *N*-acetylsulfanilyl chloride, followed by hydrolysis, gives sulfapyridine (**19**), an antibacterial.

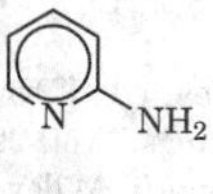
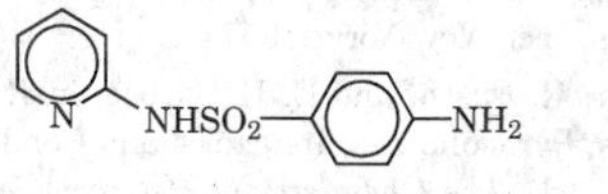

(18) (19)

Oxygen Substituents. The presence of oxygen or sulfur attached to the ring can affect the chemistry of those compounds through tautomerism. An example of 2-pyridone–2-pyridinol tautomerism is shown in equation **3**, compound (**20**).

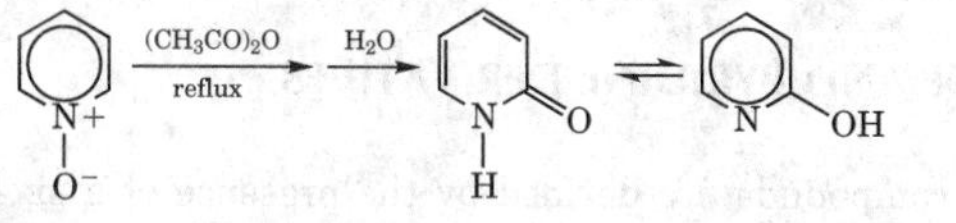

(13) (20)

The compounds 2- (**20**) and 4-pyridone (**21**) undergo chlorination with phosphorus oxychloride; however, 3-pyridinol (**22**) is not chlorinated similarly. The product from (**21**) is 4-chloropyridine. The 2- (**20**) and 4-oxo (**21**) isomers behave like the keto form of the keto–enol tautomers, whereas the 3-oxo (**22**) isomer is largely phenolic-like, and fails to be chlorinated.

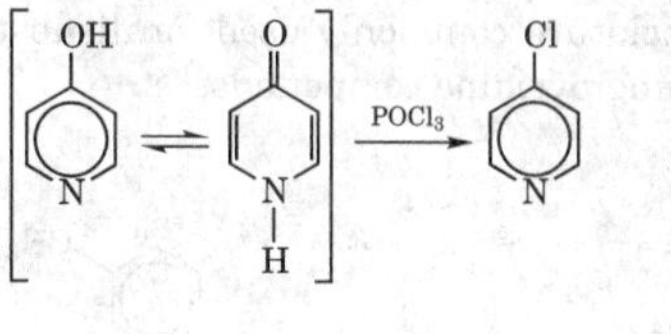

(21)

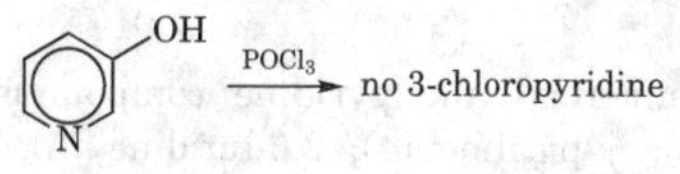

(22)

Sulfur Substituents. Acetylation and alkylation of pyridinethiones usually take place on sulfur. An exception to this is 4-pyridinethione, which is acetylated on nitrogen. Displacement of thioethers can be achieved with hydroxide or amines.

Halogen Substituents. Halogen functional groups are readily replaced by nucleophiles, eg, hydroxide ion, especially when they are attached at the α- or γ-position of the pyridine ring. This reaction has been exploited in the synthesis of the insecticide chlorpyrifos and the insecticide triclopyr.

Organometallics. Pentachloropyridine (**23**) forms Grignard reagent (**24**) by the entrainment method (eq. **4**).

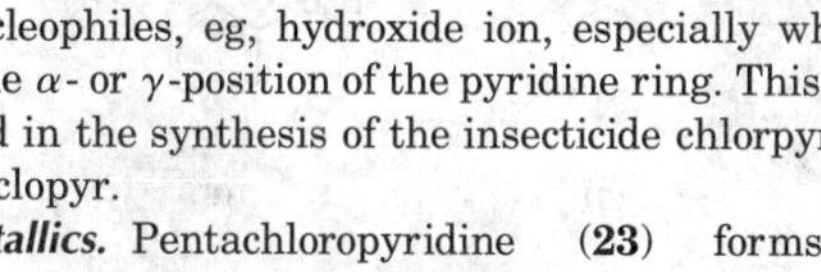

(23) (24)

Synthesis

Pyridine ring syntheses can be classified into essentially two categories: ring synthesis from nonheterocyclic compounds, and synthesis from other ring systems.

Ring Synthesis From Nonheterocyclic Compounds. These methods may be further classified based on the number of bonds formed during the pyridine ring formation. Synthesis of α-picoline (**2**) from 5-oxohexanenitrile is a one-bond formation reaction (eq. **5**). The nitrile is obtained by reaction between acetone and acrylonitrile. If both reaction steps are considered together, the synthesis must be considered a two-bond forming one, ie, formation of (**2**) from acetone and acryloni-

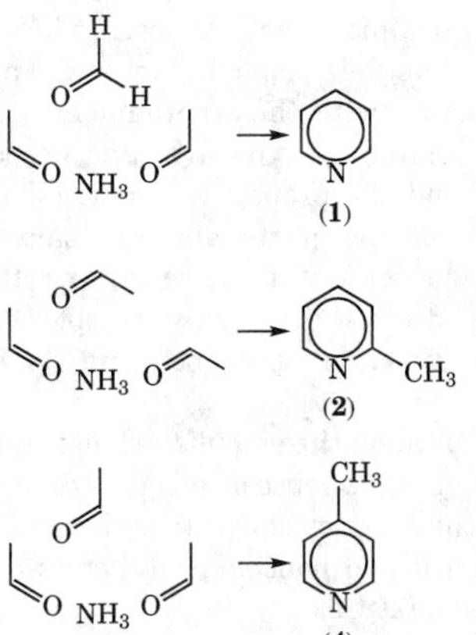

Figure 1. Four-bond reactions: formaldehyde, acetaldehyde, and ammonia mainly give pyridine (**1**), and acetaldehyde and ammonia give α- (**2**) and γ-picoline (**4**).

trile in a single step comes under the category of two-bond formation reaction.

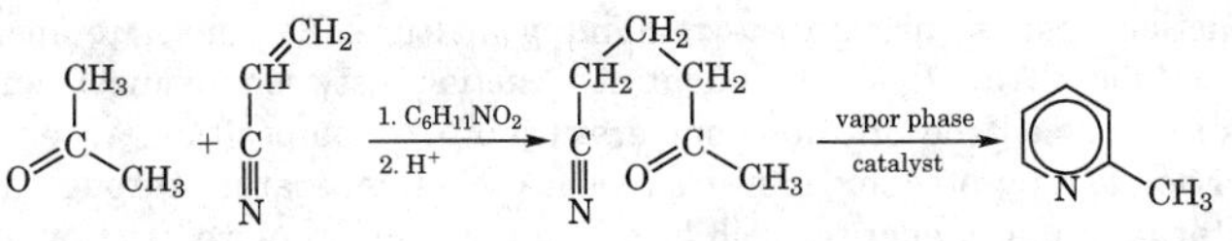

(**2**)

The vapor-phase synthesis of pyridines and picolines from formaldehyde, acetaldehyde, and ammonia falls in the category of four-bond formation reactions (Fig. 1).

Synthesis From Other Ring Systems. These syntheses are further classified based on the number of atoms in the starting ring. Ring expansion of dichlorocyclopropane carbaldimine (**25**), where R = H and R′ = aryl, on pyrolysis gives 2-arylpyridines. Thermal rearrangement to substituted pyridines occurs in the presence of tungsten(VI) oxide. In most instances the nonchlorinated product is the primary product obtained.

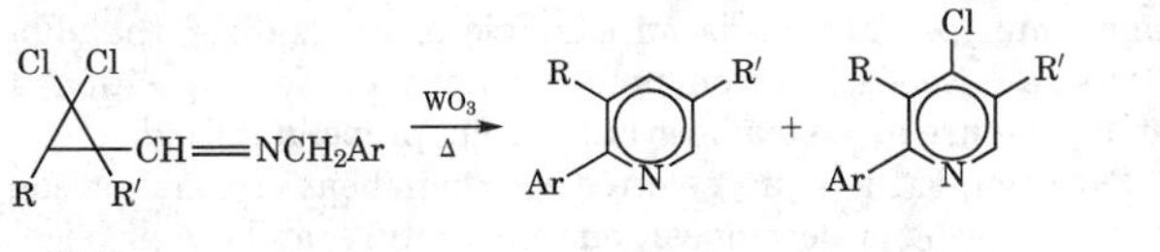

(**25**)

Furfurylamine reacts with hydrogen peroxide and acid to give 3-hydroxypyridine (**26**).

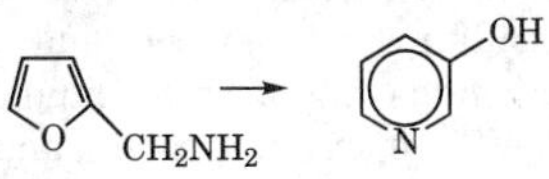

(**26**)

2-Alkyl-3-pyridinols (**27**) are reported to be formed from acyl furans and ammonia under pressure.

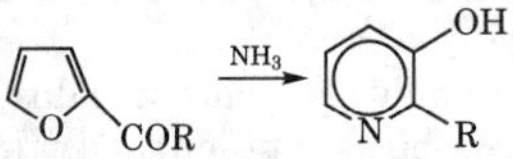

(**27**)

Manufacture and Processing

There are no natural sources of pyridine compounds that are either a single pyridine isomer or just one compound. Few commercial synthetic methods produce a single pyridine compound; most produce a mixture of alkylpyridines, usually with some pyridine (**1**). Those that produce mono- or disubstituted pyridines as principal components also usually make a mixture of isomeric compounds along with the desired material.

Most modern processes make pyridines by condensation of ammonia with aldehydes or ketones either in the vapor phase or in the liquid phase.

Commercial Manufacture of Pyridine. There are two vapor-phase processes used in the industry for the synthesis of pyridines. The first process utilizes formaldehyde and acetaldehyde as a co-feed with ammonia, and the principal products are pyridine (**1**) and 3-picoline (**3**). The second process produces only alkylated pyridines as products.

Commercial Manufacture of Specific Pyridine Bases. Condensation of paraldehyde with ammonia at 230°C and autogenous pressure is used to manufacture 5-ethyl-2-methylpyridine (**7**). This is one of the few liquid-phase processes used in the industry to make relatively simple alkylpyridines, and one of the few processes known to make a single alkylpyridine product selectively.

By-Products. Almost all commercial manufacture of pyridine compounds involves the concomitant manufacture of various side products.

Raw Material and Energy Aspects to Pyridine Manufacture. The majority of pyridine and pyridine derivatives are based on raw materials like aldehydes or ketones. These are petroleum-derived starting materials, and their manufacture entails cracking and distillation of alkanes and alkenes, and oxidation of alkanes, alkenes, or alcohols.

Health and Safety Factors

Pyridine Acute Toxicology. Pyridine causes gastrointestinal upset and central nervous system (CNS) depression at high levels of exposure.

Pyridine Chronic Toxicology. All mutagenicity tests have been negative and (**1**) is not considered a carcinogen or potential carcinogen. There have been no reports of adverse health effects on long-term exposure to (**1**) at low concentrations.

Safety Aspects in Handling and Exposure. Pyridine compounds are ubiquitous in the natural environment, and are often found in foods as minor flavor and fragrance components. Some synthetic pyridines are used as food additives. A high proportion of pyridine compounds show some type of bioactivity, albeit mostly minor, such as herbicidal, insecticidal, or medicinal activity. Therefore, all the normal precautions should be exercised when handling pyridines that would be used when handling other organic products that are potentially bioactive.

Pyridine and alkylpyridines are excellent solvents for many materials, a property that must be taken into account when selecting O-rings, gaskets, and other sealants that are in contact with liquids.

Uses

Many pyridines of commercial interest find application in market areas where bioactivity is important, as in medicinal drugs and in agricultural products such as herbicides, insecticides, fungicides, and plant growth regulators. However, pyridines also have significant market applications outside the realm of bioactive ingredients. For instance, polymers made from pyridine-containing monomers are generally sold on the basis of their unique physical properties and function, rather than for any bioactivity. Pyridines can be classified as specialty chemicals because of a relatively lower sales volume than commodity chemicals. They are most often sold in the marketplace as chemical intermediates used to manufacture final consumer products.

Pyridine and Picolines. These have been widely used as solvents in organic chemistry and, with increasing frequency, in industrial practice. Pyridine itself is a good solvent that is rather unreactive. The basic nature of pyridine and the picolines makes them ideal acid scavengers. Typically, pyridine is the solvent of choice for acylations.

The primary use of α-picoline (**2**) is as a precursor of 2-vinylpyridine. It is also used in a variety of agrochemicals and pharmaceuticals, such as nitrapyrin to prevent loss of ammonia from fertilizers; picloram a herbicide; and amprolium, a coccidiostat.

The predominant use of β-picoline (**3**) is as a starting material for agro-chemicals and pharmaceuticals.

The main use of γ-picoline (**4**) is in the production of the anti-tuberculosis agent, isoniazid. Compound (**4**) is also used to make 4-vinylpyridine, and subsequently polymers.

5-Ethyl-2-methylpyridine (**7**) is used as starting material for niacin.

ERIC F. V. SCRIVEN
JOSEPH E. TOOMEY, JR.
RAMIAH MURUGAN
Reilly Industries, Inc.

A. R. Katritzky and C. W. Rees, eds., *Comprehensive Heterocyclic Chemistry*, Vol. 2, Pergamon Press, Oxford, U. K., 1984.

A. R. Katritzky, C. W. Rees, and E. F. V. Scriven, eds., *Comprehensive Heterocyclic Chemistry II*, Vol. 5, Pergamon, Exeter, U. K., 1996.

R. A. Abramovitch, ed., *Pyridine and Its Derivatives*, Suppl., John Wiley & Sons, Inc., New York, 1974.

M. H. Palmer in S. Coffey, ed., *Rodd's Chemistry of Carbon Compounds*, 2nd ed., Vol. IV, Part F, Elsevier Scientific Publishing Co., Amsterdam, The Netherlands, 1976.

PYRIDOXINE, PYRIDOXAL, AND PYRIDOXAMINE. See VITAMINS.

PYRITE. See IRON; PIGMENTS, INORGANIC; SULFUR; SULFURIC ACID AND SULFUR TRIOXIDE.

PYROCATECHOL. See HYDROQUINONE, RESORCINOL, AND CATECHOL.

PYROGALLOL. See (POLYHYDROXY)BENZENES.

PYROMETALLURGY. See METALLURGY, EXTRACTIVE.

PYROMETRIC CONES. See CERAMICS.

PYROMETRY. See TEMPERATURE MEASUREMENT.

PYROTECHNICS

Pyrotechnics is the field of technology that combines science and art to chemically generate heat, and from that heat create light, color, audible effects, and gas pressure for entertainment, emergency signaling, and military applications. The civilian side of pyrotechnics includes fireworks, highway flares (fusees), air bag inflators, and special effects devices for the entertainment industry.

Military and aerospace pyrotechnics include a wide range of devices for illumination, signaling, obscuration, and gas generation. These devices are characterized by rugged construction and greater resistance to adverse environmental conditions with associated higher cost, reliability, and safety than most civilian pyrotechnic devices (see EXPLOSIVES AND PROPELLANTS).

Principles

The usefulness of pyrotechnic reactions derives from their being exothermic (heat-releasing), self-sustaining, and self-contained. Many chemical reactions require a sustained input of energy or they occur by interaction with other substances provided from external sources. Most pyrotechnic reactions occur independently of any external oxidizer, although in some instances the pyrotechnic effect is enhanced by interaction with the environment. Such characteristics of pyrotechnics are shared by explosives and propellants, which also involve exothermic and self-propagating reactions. Most pyrotechnic devices contain no moving parts and are small and lightweight. Compared with their mechanical analogues, pyrotechnic devices tend to be inexpensive and often are highly reliable. On the other hand, pyrotechnic devices function only once and do not normally lend themselves to reuse.

Pyrotechnics is based on the established principles of thermochemistry and the more general science of thermodynamics. There has been little work done on the kinetics of pyrotechnic reactions, largely due to the numerous chemical and nonchemical factors that affect the burn rate of a pyrotechnic mixture.

A pyrotechnic composition contains one or more oxidizers in combination with one or more fuels. Oxidizers used in pyrotechnics, such as potassium nitrate, KNO_3, are solids at room temperature and release oxygen when heated to elevated temperatures. The oxygen then combines with the fuel, and heat is generated by the resulting chemical reaction. Chemicals that release fluorine or chlorine on heating, such as polytetrafluoroethylene (Teflon) are also capable of serving as oxidizers, particularly with a metallic fuel. If the released heat is efficiently captured by adjacent pyrotechnic composition, further reaction occurs between oxidizer and fuel, and a self-propagating reaction ensues through the remaining mixture. The onset temperature for a rapid, self-propagating reaction between oxidizer and fuel is termed the ignition temperature of the composition. When a portion of a pyrotechnic mixture reaches this temperature through an external input of energy, such as a burning fuse, hot wire, friction, or impact, the composition undergoes a rapid chemical reaction that is self-sustaining through the remaining composition.

The oxidizers used in pyrotechnics are normally ionic solids such as nitrate or perchlorate salts; oxides, chlorates, and chromates are also used. The selection of oxidizer is determined by the desired heat output, reaction rate, and the physical state of the anticipated reaction products. A slower reaction occurs if the oxidizer releases its oxygen only at high temperatures, and if the oxidizer requires a net heat input (is endothermic) in its decomposition. Potassium chlorate ignites and readily reacts with a wide range of fuels, whereas iron(III) oxide can sustain only a self-propagating reaction with the most energetic fuels. Ammonium perchlorate is an example of an oxidizer that liberates extensive gaseous reaction products upon its decomposition. It has found wide use as an oxidizer in the solid propellant field.

Pyrotechnic fuels are selected for their heat of combination with oxygen, melting or decomposition temperature, and the physical state of their reaction products. Common pyrotechnic fuels include (*1*) metal powders, eg, aluminum, magnesium, titantium, magnesium–aluminum alloy (magnalium), and zirconium; (*2*) elemental fuels, such as carbon (charcoal), sulfur, boron, silicon, and phosphorus; and (*3*) carbon–hydrogen compounds (organic compounds), ie, starch, plastics (poly(vinyl chloride) (PVC)), epoxy, polyesters, and tree gums.

Extensive use has been made of natural products such as starches and gums, and the use of these materials continues to be substantial in the fireworks industry. Military pyrotechnics have moved away from the use of natural products due to the inherent variability in these materials.

Pyrotechnic mixtures may also contain additional components that are added to modify the burn rate, enhance the pyrotechnic effect, or serve as a binder to maintain t1he homogeneity of the blended mixture and provide mechanical strength when the composition is pressed or consolidated into a tube or other container. These additional components may also function as oxidizers or fuels in the composition.

In general, substances that attract water from the atmosphere are avoided in pyrotechnic formulations.

The process of designing a pyrotechnic mixture begins with the selection of oxidizer and fuel, and proceeds to incorporate additional components to achieve the exact pyrotechnic effect and burn rate desired in the end item. It is at this point that pyrotechnics takes on the dual nature of an art and science, and experience is often the only thing that can be relied upon for the solution of a difficult problem.

The classic example of a pyrotechnic composition is black powder, a blend of potassium nitrate, sulfur, and charcoal in a 75:10:15 ratio by weight. This composition has been used as a propellant for cannons and muskets as well as in fireworks for centuries, and is still used in the 1990s in significant amounts as a propellant, fuse powder, and bursting charge by the fireworks industry.

Color and Sound Production

Emission of a specific color can be achieved by the presence of a specific atomic or molecular species in the vapor state in the pyrotechnic flame. The atom or molecule undergoes electronic excitation due to the elevated flame temperature, and the atom or molecule subsequently returns to its ground electronic state with the emission of a photon of light of specific wavelength. If this specific emission happens to fall in the visible region of the electromagnetic spectrum, color is perceived by onlookers. The production of color by pyrotechnic means requires the pyrotechnic composition to generate the color-emitting species as a reaction product.

Another unique pyrotechnic effect, a shrill whistle, can be achieved if certain pyrotechnic mixtures are pressed into narrow-diameter tubes and ignited. The escaping gas pulsing out of the tube creates the whistling phenomenon.

Civilian Pyrotechnics

Black powder remains a key component of fireworks. It serves as the composition for fireworks fuses; as the propelling charge in sky rockets, Roman candles, and aerial shells; and as the bursting charge to explode aerial shells high in the air. Modern public displays are largely based on the use of mortar-fired aerial shells which explode into beautiful patterns of color high in the sky.

Types of Fireworks. Fireworks are conveniently classified as consumer or display.

Consumer Fireworks. An assortment of small fireworks devices are permitted for use by private citizens in many areas in the United States and elsewhere throughout the world. These devices consist of items such as wire sparklers, fountains, Roman candles, sky rockets, mines, and small aerial shells.

Wire sparklers are wires coated with pyrotechnic composition which are hand-held and produce a gentle spray of gold sparks from iron filings. Fountains are cardboard tubes filled with chemical mixtures that produce a spray of color and sparks extending 2–5 m into the air. Roman candles are cylindrical tubes which repeatedly fire colored stars distances of 5–20 m into the air.

Sky rockets are tubes with a stick attached for guidance and stability and which contain a pressed black powder propellant. They rise high into the air when ignited and a burst of color or a report, an audible bang, is normally produced in the air. Mines are aerial items that use a black powder propelling charge to fire a burst of colored stars, whistles, or firecrackers into the air from a cardboard tube. A barrage of color and noise results. Small aerial shells are plastic or cardboard spheres that are launched from a mortar tube by a black powder propelling charge.

Display Fireworks. Larger versions of the devices sold as consumer fireworks are used for public fireworks displays. The principal item used in fireworks displays is the aerial shell, a sphere or cylinder typically 8–20 cm (3–8 in.) in diameter that is launched several hundred meters into the air (~100–300 m), from a mortar tube, by a propelling charge of black powder. A time fuse burns as the device climbs into the air, and a bursting charge explodes the device high in the air, lighting a shower of stars to produce a spectacular visual effect. Lances are small, cigarette-sized tubes that burn with specific flame colors. Pyrotechnicians attach hundreds of lances to wooden frames to create fire pictures, called set pieces.

Theatrical Pyrotechnics or Special Effects. Many spectacular visual and audible effects are produced for stage presentations of both music and drama, and many motion pictures and television shows incorporate pyrotechnic and explosive special effects to liven up the presentation. These spectacular effects are a combination of pyrotechnics, explosives, combustion, and electronics.

Model Rockets and Missiles. Model rockets are another type of pyrotechnic device. These items are most often sold as kits with modular, preassembled, solid propellant engines. The engines have traditionally been constructed of nonmetallic casings, clay nozzles, and a compressed black powder charge, which is ignited by an electric match.

Other Civilian Devices. Pyrotechnic devices also serve an assortment of civilian uses, primarily for signaling. Highway flares, often known as fusees, are made chiefly of strontium nitrate mixed with sawdust, wax, sulfur, and potassium perchlorate and contained in a waterproof cardboard tube. Other hand-held and aerial devices are used for marine distress signals.

The air bag industry has become one of the principal users of pyrotechnic compositions. Most of the current air bag systems are based on the thermal decomposition of sodium azide, NaN_3, to rapidly generate a large volume of nitrogen gas, N_2.

Regulations. The manufacture, transportation, storage, and use of pyrotechnics are all very highly regulated fields, and the volume and scope of the regulations covering the industry, both military and civilian, has increased enormously since the 1980s. In the United States, the transportation of all commercial pyrotechnic articles, for both consumer and military use, is under the control of the U.S. Department of Transportation (DOT).

Military Pyrotechnics

Pyrotechnic devices are used for light generation in flares, tracers, and flash cartridges; for smoke generation for signaling and obscuration; for heat production for time delay components, incendiary applications, and ignition; and for gas-generation applications. The propellants used in many military devices are also essentially pyrotechnic compositions in nature, containing an oxidizer and fuel designed to burn at a high rate with the generation of a large quantity of hot gas to produce significant thrust. Military pyrotechnics are required to meet rigorous standards of performance, reliability, and storage lifetime, and the materials of construction are generally substantially more sturdy than those used for civilian pyrotechnics.

Many aspects of the performance of military pyrotechnics are measured and analyzed by modern instrumental techniques such as spectrophotometers for light intensity studies, replacing the qualitative, visual evaluations that were formerly used for acceptance of these devices. Military pyrotechnics are also experiencing a shift away from a concentration on the visible region of the electromagnetic spectrum, and are moving into the generation and obscuration of the infrared and microwave/millimeter regions, as modern techniques such as heat-seeking missiles, thermal imaging systems, and night vision equipment have dramatically altered the battlefield scenario.

Military flare technology has concentrated in recent years on the development and production of decoy flares to protect aircraft against heat-seeking missiles. These flares, which emit a considerable amount of infrared radiation, use a composition based on magnesium metal and polytetrafluoroethylene PTFE (trade names Teflon). Here, Teflon serves as the oxidizer, releasing fluorine atoms that energetically combine with magnesium to form magnesium fluoride.

The intensity of a flare is largely determined by its flame temperature, which depends on the stability of the reaction products. A flare temperature greater than 3000 K is required to generate gray-body radiation, which is optimum for the spectral sensitivity of the human eye. Metallic fuels such as magnesium, aluminum, zirconium, and titanium are necessary to achieve these high temperatures.

Military signal flares are designed to burn with clearly distinguishable colors, and the chemistry of these compositions is quite similar to that used for color effects in civilian pyrotechnics. Photoflash bombs are explosive devices that are used in night photography and that, for time intervals as short as 0.1 s, provide intense light output. Flash charges are loosely packed powdered mixtures of aluminum powder with potassium perchlorate and barium nitrate, and initiation is by an explosive charge. With advances in the technology of low light photography, the need for these devices is diminishing.

Smoke-Generating Devices. Smoke generators are used by the military for daytime obscuration and signaling. For field use where portable stable systems are required, pyrotechnic devices are often employed. The primary composition since the 1940s has been HC smoke, which generates a cloud of zinc chloride, $ZnCl_2$, smoke by a series of reactions between hexachloroethane, C_2Cl_6(HC), zinc oxide, and aluminum.

This mixture produces a dense, gray-white cloud that is largely zinc chloride. Concerns about the toxicity of the smoke from HC mixtures has led to efforts to find a replacement.

Signaling smokes may be white or colored. White signal smoke has traditionally been derived from red phosphorus or an HC mixture. New smokes based on the sublimation of terephthalic acid (TA) have been developed by the U.S. military to replace HC smoke for training purposes.

Grenades and smoke marker devices that produce highly visible plumes of brightly colored smoke are used by the military for signaling and marking purposes. The production of a distinct color is critical in these devices, and the color is produced by the sublimation of an organic dye. Some of the organic dyes formerly used in these devices have been replaced due to carcinogenicity concerns (see DYES, NATURAL).

Tracer Munitions. Tracer bullets guide the direction of the fire, aid in range estimation, mark target impact, and act as incendiaries. Tracers can, through preselected tracer colors, also serve for nighttime identification of the combatants. Red strontium-containing tracers are more visible under adverse atmospheric conditions, therefore these are preferred although green tracers based.

Incendiary Devices. Incendiary devices are used to initiate destructive fires in a variety of targets. Small-arms incendiaries are used primarily for starting fires in aircraft fuels. Whereas they are highly effective against subsonic aircraft, such as helicopters, the problem of defeating supersonic aircraft by incendiary action alone is more difficult owing to the high flash point of jet fuels.

Ignition

Pyrotechnic devices are initiated by some type of external energy input. This can range from a match lighting a piece of black powder-containing fuse to a battery sending a surge of current through a circuit, creating a hot spot on a narrow, high resistance section of wire that is coated with a thermally sensitive material (see BATTERIES). Alternatively, impact can be used to ignite a primer, as in a shotgun shell, or the friction generated by rubbing two surfaces together can create a hot spot; a highway flare uses this technique. The choice of the ignition method depends on the permissible ignition delay. Igniters containing primary explosives function in microseconds, whereas purely pyrotechnic igniters require milliseconds.

Within a given pyrotechnic device there may be several components such as a propelling charge, a delay column, and the main pyrotechnic effect (perhaps a flare).

Delay Elements. Many delay compositions contain lead, barium, and chromium compounds. There is a need for new delay compositions that are more environmentally compatible yet have the excellent reliability and performance properties of the current compositions. In some pyrotechnic devices, a precise time delay is required between the operating stages of the device. An example is the several-second delay designed into hand grenades to provide a delay between the release of the safety pin and the functioning of the grenade. Delay elements are self-contained pyrotechnic devices consisting of an initiator, a pressed column of pyrotechnic composition, and an output charge.

Safety Concerns

Safety concerns permeate all aspects of pyrotechnics. Great caution is needed in the manufacture of pyrotechnic mixtures to avoid possible ignition sources.

Wherever possible, operations involving the mixing and processing of pyrotechnic compositions should be performed remotely and monitored by video camera. Where this is not possible, the protection of personnel should include shielding for the eyes, face, and hands as well as the use of antistatic clothing.

Because pyrotechnic compositions contain their own oxygen, suffocation methods of fire fighting are not effective. Instead, the temperature of the burning material must be brought down below the ignition point of the material. A water deluge is perhaps the best way to accomplish this, as evaporating water is a very effective heat remover.

Many pyrotechnic mixtures potentially can be ignited by electrostatic discharge during the manufacturing process. The risk of electrostatic discharge can be minimized by grounding and bonding materials, containers, and personnel.

Separation of plant areas and the minimization of in-process material in each area are two key points of plant safety (qv). Separating the stored materials from work-in-progress helps to minimize the consequences of any event that might occur in manufacturing.

JOHN A. CONKLING
American Pyrotechnic Association

A. A. Shidlovski, *Principles of Pyrotechnics*, Publ. No. AD AOO1 859, Mashinostroyeniye Press, Moscow, Russia, 1973; trans. 1974.

H. Ellern, *Military and Civilian Pyrotechnics*, The Chemical Publishing Co., New York, 1968.

T. Shimizu, *Fireworks, the Art, Science, and Technique*, Maruzen Co., Ltd., Tokyo, Japan, 1981.

J. Conkling, *The Chemistry of Pyrotechnics*, Marcel Dekker, Inc., New York, 1985.

PYRROLE AND PYRROLE DERIVATIVES

Pyrrole, a five-membered, heterocyclic system, is a fundamental structural subunit of many of the most important biological molecules, eg, heme, the chlorophylls, the bile pigments, some naturally occurring antibiotics, many alkaloids, and some enzymes. Ring positions in pyrrole (**1**) are designated by number or Greek letter.

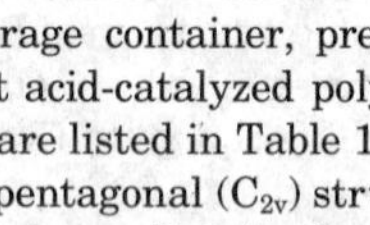

(**1**)

Physical Properties

Pyrrole is a colorless, slightly hygroscopic liquid which, if fresh, emits an odor like that of chloroform. On exposure to air, it darkens and eventually produces a dark brown resin. It can be preserved by excluding air from the storage container, preferably by displacement with ammonia to prevent acid-catalyzed polymerization. Some physical properties of pyrrole are listed in Table 1.

Pyrrole has a planar, pentagonal (C_{2v}) structure and is aromatic in that it has a sextet of electrons. It is isoelectronic with the cyclopentadienyl anion. The π-electrons are delocalized throughout the ring system, thus pyrrole is best characterized as a resonance hybrid, with contributing structures (**1–5**). These structures explain its lack of basicity (which is less than that of pyridine), its

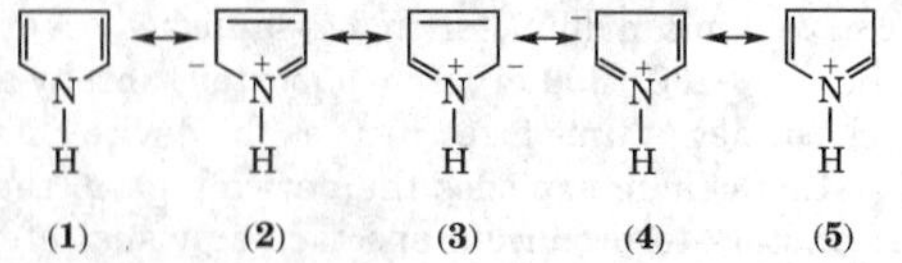

(**1**) (**2**) (**3**) (**4**) (**5**)

unexpectedly high acidity, and its pronounced aromatic character. The resonance energy which has been estimated at about 100 kJ/mol

Table 1. Physical Properties of Pyrrole

Property	Value
melting point, °C	−23.4
boiling point, °C	129.8
critical temperature, °C	366
density, d_4^{20}, g/mL	0.970
refractive index, n_D^{20}	1.5085
dielectric constant at 20°C, ϵ	8.00
flash point, closed cup, °C	39

(23.9 kcal/mol) is intermediate between that of furan and thiophene, or about two-thirds that of benzene.

Many of the physical characteristics of pyrrole indicate at least partial association. In particular, the boiling point is 98°C higher than that of furan.

Pyrrole is soluble in alcohol, benzene, and diethyl ether, but is only sparingly soluble in water and in aqueous alkalies. It dissolves with decomposition in dilute acids. Pyrroles with substituents in the β-position are usually less soluble in polar solvents than the corresponding α-substituted pyrroles. Pyrroles that have no substituent on nitrogen readily lose a proton to form the resonance-stabilized pyrrolyl anion, and alkali metals react with it in liquid ammonia to form salts.

Syntheses of Pyrroles

Knorr Synthesis. The Knorr reaction, condensation of an α-aminoketone with a carbonyl compound, and its modifications are among the most important and widely used methods for the synthesis of pyrroles.

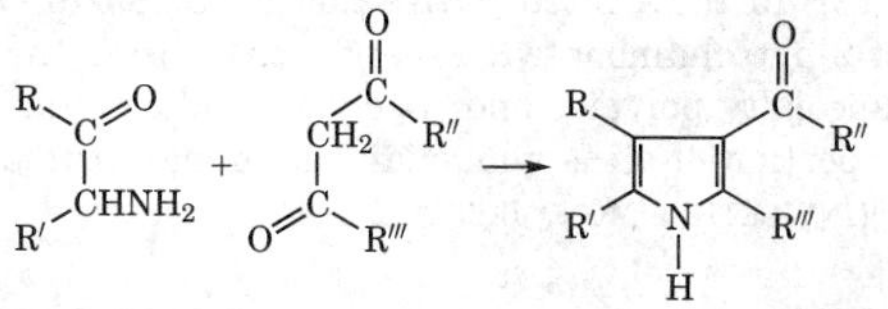

Because the α-aminoketone is subject to self-condensation, the condensation with a β-dicarbonyl derivative (**6**) is usually carried out by generating the α-aminoketone *in situ* through reduction of an oximino derivative (**7**).

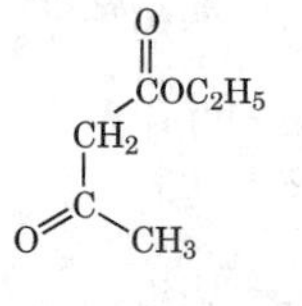
(**6**)

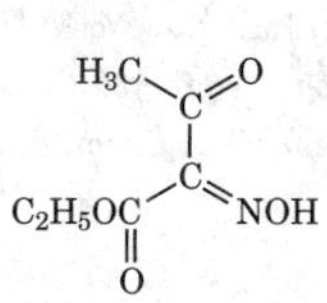
(**7**)

The Knorr synthesis is not particularly sensitive to the nature of R and R‴, ie, they may be alkyl, acyl, aryl, or carbalkoxy without significantly affecting the yield.

Hantzsch and Feist Syntheses. The Hantzsch synthesis of pyrroles involves condensation of an α-haloketone with a β-keto ester in the presence of ammonia or an amine.

The Feist synthesis is similar to the Hantzsch method and involves condensation of acyloins with aminocrotonic esters in the presence of zinc chloride.

Paal-Knorr Synthesis. The condensation of a 1,4-diketone, for example, with ammonia or a primary amine generally gives good yields of pyrroles; many syntheses have been reported.

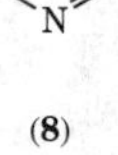
(**8**)

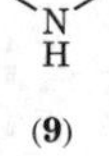
(**9**)

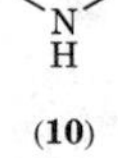
(**10**)

Other Methods. Newer methods for forming pyrrole and related heterocyclic rings include the formation of substituted pyrrole 2-carboxylate esters by condensation of β-dicarbonyl compounds with glycinate esters.

Acetylenic compounds have often been used as precursors to certain pyrroles. Thus, 2-butyne-1,4-diol reacts with aniline in the presence of alumina to produce *N*-phenylpyrrole.

Pyrrolines and Pyrrolidines

The pyrrolines or dihydropyrroles can exist in three isomeric forms: 1-pyrroline (**8**) is an unstable material that resinifies upon exposure to air; 2-pyrroline is even (**9**) is even more unstable; only 3-pyrroline (**10**) is reasonably stable. 3-Pyrroline boils at 91°C and has a density of 0.9097 g/cm^3 and a refractive index of 1.4664.

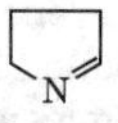
(**8**)

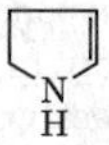
(**9**)

(**10**)

Pyrrolidine (tetrahydropyrrole) (**11**) is a water-soluble strong base with the usual properties of a secondary amine. An important synthesis of pyrrolidines is the reaction of reduced furans with excess amine or ammonia over an alumina catalyst in the vapor phase at 400°C. However, if labile substituents are present on the tetrahydrofurans, pyrroles may form.

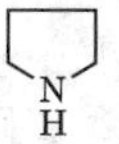
(**11**)

Pyrroles can also be catalytically hydrogenated to pyrrolidines.

Reactions of Pyrroles

In keeping with its aromatic character, pyrrole is relatively difficult to hydrogenate, it does not ordinarily serve as a diene for Diels-Alder reactions, and does not undergo typical olefin reactions. Electrophilic substitutions are the most characteristic reactions, and pyrrole has often been compared to phenol or aniline in its reactivity. Acids strong enough to form salts with pyrrole destroy the aromaticity and cause polymerization.

N-Alkylpyrroles may be obtained by the Knorr synthesis or by the reaction of the pyrrolyl metallates, ie, Na, K, and Tl, with alkyl halides. Alkylation of pyrroles at the other

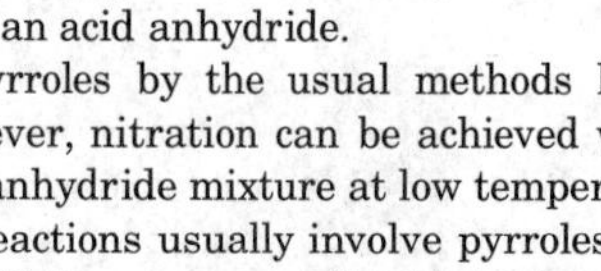

ring positions can be carried out under mild conditions with allylic or benzylic halides or under more stringent conditions (100–150°C) with CH_3I.

N-Acylation is readily carried out by reaction of the alkali metal salts with the appropriate acid chloride. C-Acylation of pyrroles carrying negative substituents occurs in the presence of Friedel-Crafts catalysts. Pyrrole and alkylpyrroles can be acylated noncatalytically with an acid chloride or an acid anhydride.

Nitration of pyrroles by the usual methods leads to extensive degradation. However, nitration can be achieved with an equimolar nitric acid–acetic anhydride mixture at low temperatures.

Halogenation reactions usually involve pyrroles with electronegative substituents. Mixtures are usually obtained and polysubstitution products, ie, tetrahalopyrroles, predominate. Pyrrole oxidizes in air to red or black pigments of uncertain composition.

Analytical and Test Methods. In addition to the modern spectroscopic methods of detection and identification of pyrroles, there are several chemical tests. The classical Runge test with HCl yields pyrrole red, an amorphous polymer mixture. In addition, all pyrroles with a free α- or β-position or with groups, eg, ester, that can be converted to such pyrroles under acid conditions undergo the Ehrlich reaction with *p*-(dimethylamino)benzaldehyde to give purple products.

Both pyrrole and indole react with selenium dioxide in the presence of nitric acid to give a deep violet solution. Very small quantities (ca 4×10^{-5} g) of pyrrole can be detected by this method.

Functional Derivatives

Hydroxypyrroles. Pyrroles with nitrogen-substituted side chains containing hydroxyl groups are best prepared by the Paal-Knorr cyclization. Pyrroles with hydroxyl groups on carbon side chains can be made by reduction of the appropriate carbonyl compound with hydrides, by Grignard synthesis, or by insertion of ethylene oxide or formaldehyde.

Aldehydes and Ketones. Pyrrole aldehydes and ketones are somewhat less reactive than the corresponding benzenoid derivatives. The aldehydes condense with a variety of compounds that contain active methylene groups. They also react with pyrroles under acidic conditions to form dipyrrylmethenes.

Pyrrole Carboxylic Acids and Esters. The acids are considerably less stable than benzoic acid and often decarboxylate readily on heating. The pyrrole esters are important synthetically because they stabilize the ring and may also act as protecting groups.

Vinyl Pyrroles. Synthetic routes based on a one-pot reaction between ketoximes and acetylene in an alkali metal hydroxide–dimethyl sulfoxide (DMSO) system have made vinyl pyrroles accessible.

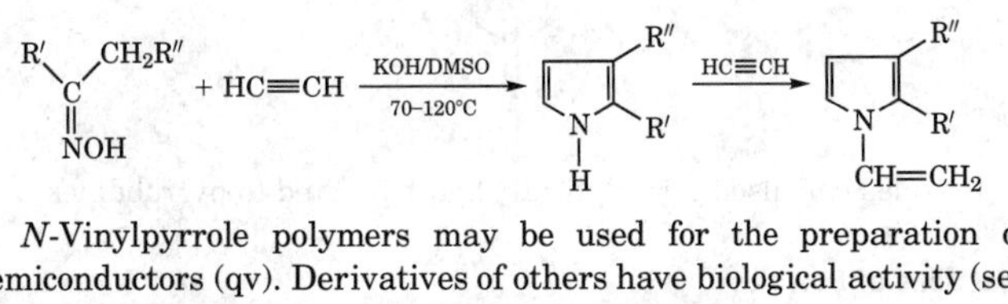

N-Vinylpyrrole polymers may be used for the preparation of semiconductors (qv). Derivatives of others have biological activity (see VINYL POLYMERS, *N*-VINYL AMIDES).

Condensed Pyrroles. Pyrroles can be condensed to compounds containing two, three, or four pyrrole nuclei. These are important in synthetic routes to the tetrapyrrolic porphyrins, corroles, and bile pigments and to the tripyrrolic prodigiosins. The pyrrole nuclei are joined by either a one-carbon fragment or direct pyrrole–pyrrole bond.

Polypyrroles. Highly stable, flexible films of polypyrrole are obtained by electrolytic oxidation of the appropriate pyrrole monomers. The films are not affected by air and can be heated to 250°C with little effect. Copolymerization of pyrrole with *N*-methylpyrrole yields compositions of varying electrical conductivity, depending on the monomer ratio.

Because of its physical properties, polypyrrole has been cited as a unique building block for intelligent polymeric materials, ie, it has characteristics which make it capable of sensing, information processing, and response actuation.

Pyrrolidinones and Derivatives

The simplest of the pyrrolidinones, 2-pyrrolidinone, is prepared by condensation of butyrolactone with ammonia at high temperatures. It melts at 25.6°C and has a normal boiling point of 245°C. It is miscible with water, lower alcohols, lower ketones, ether, ethyl acetate, chloroform, and benzene. It is soluble to ca 1 wt % in aliphatic hydrocarbons. It undergoes the reactions of a typical lactam, and can be polymerized to polypyrrolidinone (Nylon-4). Strong acids or bases catalyze the hydrolysis of 2-pyrrolidinone.

2-Pyrrolidinone forms alkali metal salts by reaction with alkali metal hydroxides under conditions in which the water of reaction is removed. The potassium salt prepared *in situ* serves as the catalyst for the vinylation of 2-pyrrolidinone in the commercial production of *N*-vinylpyrrolidinone.

1-Methyl-2-pyrrolidinone (NMP) is prepared commercially by the condensation of butyrolactone with methylamine and is usually considered a stable and unreactive solvent. It has a freezing point of −24.4°C and a normal boiling point of 202°C. 1-Methyl-2-pyrrolidinone is less toxic than many other dipolar aprotic solvents. Large amounts of NMP are consumed in the polymer industry as a medium for polymerization.

Higher 1-alkyl-2-pyrrolidinones, which combine the hydrophilicity of the pyrrolidinone ring system with the hydrophobicity of a longer chain alkyl group, are compounds with surfactant properties which enhance the solubility of hydrophobic materials. Two of these, 1-octyl-2-pyrrolidinone and 1-dodecyl-2-pyrrolidinone, have been made commercially. These two surface-active pyrrolidones (Surfadones) are finding applications as wetting agents and are also used in hard-surface cleaners, in fountain solutions, and for pigment dispersions.

1-Vinyl-2-pyrrolidone (VP) (1-ethenyl-2-pyrrolidinone), is manufactured by ISP in the United States and by BASF in Germany by vinylation of 2-pyrrolidinone with acetylene. It forms the basis for a significant specialty polymer and copolymer industry and consumes the primary portion of all 2-pyhrrolidinone manufactured (see VINYL POLYMERS, *N*-VINYL ETHER MONOMERS AND POLYMERS).

LOWELL RAY ANDERSON
KOU-CHANG LIU
ISP Corporation

R. A. Jones, ed., *Pyrroles in Heterocyclic Compounds*, Vol. 48, Pts. I and II, John Wiley and Sons, Inc., New York, 1990.

R. J. Sundberg, *Prog. Heterocycl. Chem.* **3**, 90–108 (1991).

A. H. Jackson, *Compr. Org. Chem.* **4**, 275 (1979).

Q

QUALITY ASSURANCE

The objective of chemical manufacturing is to provide products that perform to expectation. A manufacturing unit is responsible for producing a product. The quality assurance (QA) and quality control (QC) units are designated to assure the product not only meets its stated specification but also performs up to customer expectation. The activities typically performed by the QC laboratory ensure through testing that a product conforms to specification. The QA unit operates in support of the lab activities to assure the correctness of the results and the consistency of the product. The relationship between QA and QC is shown in Figure 1.

Quality Techniques.

Statistical Process Control. A properly running production process is characterized by the random variation of the process parameters for a series of lots or measurements. The statistical process control (SPC) approach is a statistical technique used to monitor variation in a process. If the variation is not random, action is taken to locate and eliminate the cause of the lack of randomness, returning the process or measurement to a state of statistical control, ie, of exhibiting only random variation.

Just-in-Time. Just-in-time (JIT) should be used to focus management attention on quality problems.

Prior to JIT, defective parts could be discarded and a replacement taken from inventory. Under JIT there is insufficient inventory to replace defective parts, often leading to a shutdown of the assembly process. The Japanese use this impact to heighten the awareness of nonconforming material and to ensure that the causes of such defective parts are eliminated. U.S. industry, however, often sees the disadvantageous tradeoff between interrupting production to improve quality vs the economic savings from reduced inventory.

Zero Defects. Whereas zero defects (ZD) was often interpreted to be a quality goal, its full meaning is to encourage continuous quality improvement. When ZD was treated only as a slogan, it failed to have a lasting impact.

As of 1996 the term ZD was no longer heard. One U.S. manufacturing company, however, successfully changed its culture using this approach.

Quality Function Deployment. Sometimes referred to as the House of Quality, quality function deployment (QFD) is a technique for translating the voice of the customer into design requirements. This is a systematic approach identifying customer expectations and relating the expectations to product properties. In the chemical industry, QFD results in chemical specifications optimized to assure the material is suitable for its intended use and performs up to customer expectations.

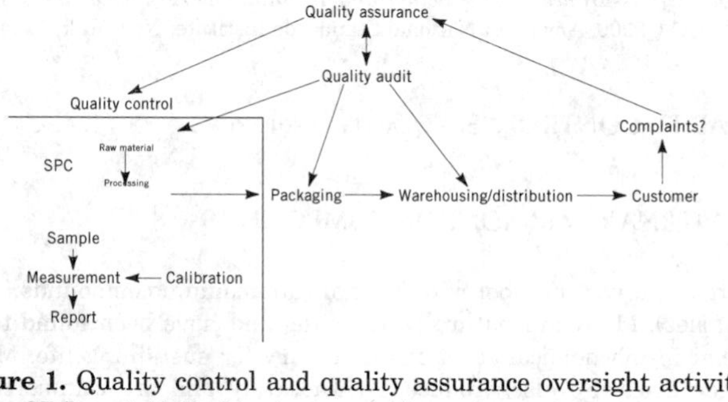

Figure 1. Quality control and quality assurance oversight activities where SPC = statistical process control.

Total Quality Management. Total quality management (TQM) is a term that encompasses all of the continuous improvement activities with the goal of world class quality. This corporate culture sets up the conditions for a climate favorable to companywide improvement.

Companies following a TQM philosophy place heavy emphasis on employee training for quality improvement. In the chemical industry, TQM continues to evolve from a corporate-run program to a decentralized one. Responsibility for implementation has been turned over to local sites, so that often a small corporate staff can provide resources and guidance.

In the chemical industry, it is often difficult to provide product specifications comprehensive enough to ensure product performance in all applications. Therefore, the manufacture of product having a minimum of lot-to-lot variability allows the customer to use the product without modifying their formulation or process to accommodate such variation.

Quality Control

Within the chemical industry, quality control (QC) is the systematic monitoring of product conformance to specification through testing. The reliability of laboratory testing is essential to effective operation. Therefore proper sampling (qv) and accurate and precise measurement are important. This relatively narrow focus differs from quality assurance, where the role involves monitoring all activities that impact product and service quality, including quality control (see Fig. 1).

The purpose of the QC laboratory is to monitor product quality through sampling and analysis.

Sampling Plan. The first step to assuring accuracy, ie, the conformity of the measured value to the true or expected value, is to obtain a representative sample. A written sampling plan approved by QC should be followed.

A thorough sampling plan should describe when the sample is to be taken and how many samples are required. It also specifies from what location within the equipment, to take the sample. The plan should also indicate what sampling equipment and sample container should be used, as well as the type of tests to be performed and the acceptance criteria.

Samples can be analyzed individually or may be combined into a homogeneous composite sample and then analyzed. In either case, only a portion of the sample is customarily used for a given test; this material must be representative of the entire lot. (see MIXING AND BLENDING; POWDERS, HANDLING; SIZE MEASUREMENT OF PARTICLES).

Calibration. Calibration of lab instruments is important to the accuracy of test results. Calibration, the use of an accepted standard to adjust an instrument or measurement standard so as to improve the accuracy of the instrument or measurement, is an essential requirement of both the U.S. Food and Drug Administration (FDA) Good Manufacturing Practice (GMP) (24) and the ISO 9000 standards.

Replicate Analyses. Confidence in the test result is improved by reducing the measurement variability. This variability in repeat analyses is known as precision. One method to improve the precision of the measurement is to perform complete replicate analyses of the same sample beginning with the sample preparation. This is appropriate when the sample is known to be representative of the material sampled. When this is not the case, multiple samples should be taken for analysis.

Statistical Control. Statistical quality control (SQC) is the application of statistical techniques to analytical data. Statistical process control (SPC) is the real-time application of statistics to process or equipment performance. Applied to QC lab instrumentation or methods, SPC can demonstrate the stability and precision of the measurement technique. The SQC of lot data can be used to show the stability of the production process.

Statistical control of an analysis or instrument is best demonstrated by SQC of a standard sample analysis. The preferred approach to demonstrate statistical control is to use a reference sample of the subject material that has been carefully analyzed or, alternatively, to use a purchased reference standard.

Statistical quality control charts of variables are plots of measurement data, preferably the average result of replicate analyses, vs time.

There are several rules applied to control charts to spot a lack of randomness. The most obvious is a point outside the control limits. A trend such as a run, where at least seven consecutive data points are either above or below the average line, or a trend of seven consecutive points either increasing or decreasing in value, also indicates an out of control situation.

The value of control charts is to provide early warning to lab personnel of changes affecting the test results.

Laboratory Information Management System. The QC lab must analyze raw material, in-process, and finished product samples; adhere to calibration schedules; record data; and perform statistical analyses. These activities lend themselves to the application of software packages such as a laboratory information management system (LIMS).

The LIMS software is essentially a database for tracking, reporting, and archiving lab data as well as scheduling and guiding lab activities.

Method Transfer. Method transfer involves the implementation of a method developed at another laboratory. Typically the method is prepared in an analytical R&D department and then transferred to quality control at the plant.

The Tools of Quality. Quality assurance plays an important role in problem solving and process improvement. To do so, QA personnel must be knowledgeable in the many so-called tools of quality (TOQ) and their application.

For example, a flow chart is used to help understand the organizational flow of a procedure or process. Its principal benefit is to enable teams, such as problem-solving or productivity improvement teams, to reach a common vision of the work flow. Figure 2 contains an example for manufacture of a polymeric material.

Other TOQs are the cause-and-effect diagram, the control chart, the histogram, the check sheet for recording data, the pareto chart (special type of histogram), and the scatter (correlation) diagram.

Proper application of one or more of the tools of quality should lead to the elimination of the causes effecting off-standard results and thus to improvement of the process under investigation.

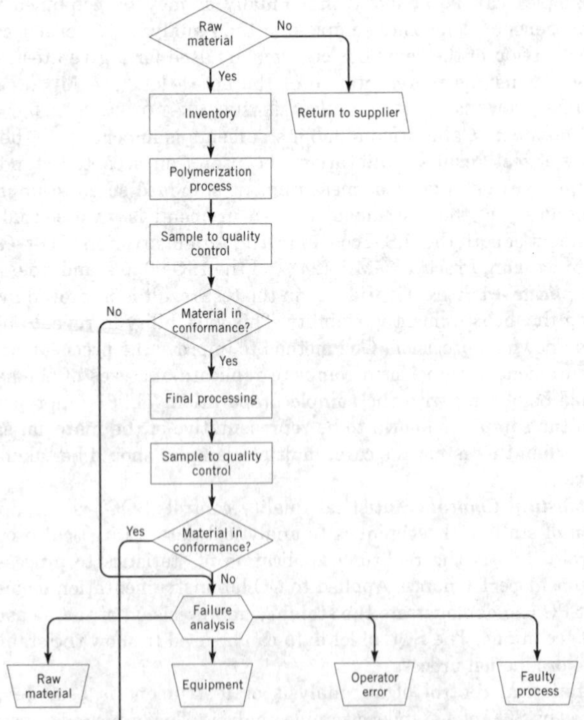

Figure 2. Flow chart of polymer quality control.

Quality Audit. Another important responsibility of quality assurance is the audit function. Using the quality audit as a tool, QA can monitor the operation of the manufacturing facility; a toll, ie, contract, manufacturer; or raw material supplier to assure that written procedures are in place and that there is documentation to indicate the procedures are being followed.

Customer Complaints. A failure in a company's quality system often shows up in the form of a customer complaint. Quality assurance tracks the progress and coordinates the complaint investigation.

Quality Systems

Besides internal quality audits, there are audits conducted by external authorities for conformance to established quality systems. The two chief standards affecting the chemical industry are the U.S. Food and Drug Administration Current Good Manufacturing Practice (GMP) regulation and the International Organization for Standardization ISO 9000 series. A quality system performance-related standard is the Malcolm Baldrige National Quality Award (MBNQA).

Good Manufacturing Practice. The GMPs were issued by the U.S. FDA in 1978 to provide minimum quality standards in the production of pharmaceuticals (qv) for the finished dosage form as well as their ingredients. The standard has been updated periodically.

ISO 9000. The ISO 9000 standard describes the selection criteria for four standards. Conformance to ISO 9000 by U.S. companies was led by the chemical industry as a result of the importance of international trade to chemical companies. Certification is considered an important supplier selection criterion by U.S. chemical companies.

Malcolm Baldrige National Quality Award. The most stringent and comprehensive quality system criterion is described in the Malcolm Baldrige National Quality Award. The award was created by an act of U.S. Congress in 1987 and is given annually. The award program is managed by the National Institute of Science and Technology (NIST) and administered by the American Society for Quality Control (ASQC).

Economic Aspects

Quality professionals use the term quality cost when discussing waste in a company. Quality cost includes any form of waste associated either with an activity that is unnecessary or an effort that must be corrected. Chemical companies have come to recognize the costs associated with wasteful efforts and are attempting to identify and eliminate waste.

IRWIN SILVERSTEIN
International Specialty Products

J. M. Juran, *A History of Managing for Quality: The Evolution, Trends, and Future Directions of Managing for Quality*, ASQC Quality Press, Milwaukee, Wis., 1995, p. 597.

J. M. Juran and F. M. Gryna, *Juran's Quality Control Handbook*, 4th ed., McGraw-Hill Book Co., Inc., New York, 1988, pp. 28.4–28.38.

A. D. Stratton, *Quality Progress*, ASQC, Milwaukee, Wis., Apr. 1990.

International Organization for Standardization, *Quality Systems—Model for Quality Assurance in Production and Installation, ISO 9000:1994*, ANSI/ASQC Q9000, American National Standards Institute, New York, 1994.

QUALITY CONTROL. See QUALITY ASSURANCE.

QUATERNARY AMMONIUM COMPOUNDS

There are a vast number of quaternary ammonium compounds (quaternaries). Many are naturally occurring and have been found to be crucial in biochemical reactions necessary for sustaining life. Many quaternaries are also produced synthetically and are commercially available.

Most quaternary ammonium compounds have the general formula $R_4N^+X^-$ and are a type of cationic organic nitrogen compound. The nitrogen atom, covalently bonded to four organic groups, bears a positive charge that is balanced by a negative counterion. Heterocyclics, in which the nitrogen is bonded to two carbon atoms by single bonds and to one carbon by a double bond, are also considered quaternary ammonium compounds. The R group may either be equivalent or correspond to two to four distinctly different moieties. These groups may be any type of hydrocarbon: saturated, unsaturated, aromatic, aliphatic, branched chain, or normal chain. They may also contain additional functionality and heteroatoms. Examples include methylpyridinium iodide (**1**); benzyldimethyloctadecylammonium chloride (**2**); and di(hydrogenated tallow)alkyldimethylammonium chloride (**3**), where R = $C_{14} - C_{18}$.

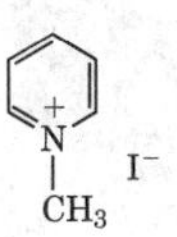

(**1**)

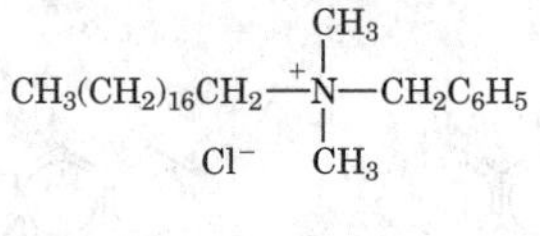

(**2**)

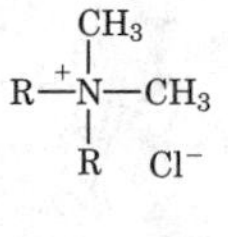

(**3**)

Nomenclature

Quaternary ammonium compounds are usually named as the substituted ammonium salt. The anion is listed last. Substituent names can be either common (stearyl) or IUPAC (octadecyl). If the long chain in the compound is from a natural mixture, the chain is named after that mixture, eg, tallowalkyl. Prefixes such as di- and tri- are used if an alkyl group is repeated. Complex compounds usually have the substituents listed in alphabetical order.

Properties

Physical Properties. Most quaternary compounds are solid materials that have indefinite melting points and decompose on heating. Physical properties are determined by the chemical structure of the quaternary ammonium compound as well as any additives such as solvents. The simplest quaternary ammonium compound, tetramethylammonium chloride, is very soluble in water and insoluble in nonpolar solvents. As the molecular weight of the quaternary compound increases, solubility in polar solvents decreases and solubility in nonpolar solvents increases.

The ability to form aqueous dispersions is a property that gives many quaternary compounds useful applications.

Higher order aliphatic quaternary compounds, where one of the alkyl groups contains ~ 10 carbon atoms, exhibit surface-active properties. These compounds compose a subclass of a more general class of compounds known as cationic surfactants.

Chemical Properties. Reactions of quaternaries can be categorized into three types: Hoffman eliminations, displacements, and rearrangements. Thermal decomposition of a quaternary ammonium hydroxide to an alkene, tertiary amine, and water is known as the Hoffman elimination (eq. 1a).

$$CH_3CH_2CH_2\overset{+}{N}(CH_3)_3 + {}^-OH \xrightarrow{\text{elimination}} CH_3CH_2{=}CH_2 + N(CH_3)_3 + H_2O \quad (1a)$$

$$CH_3CH_2CH_2\overset{+}{N}(CH_3)_3 + {}^-OH \xrightarrow{\text{displacement}} CH_3CH_2CH_2N(CH_3)_2 + CH_3OH \quad (1b)$$

Displacement of a tertiary amine from a quaternary (eq. 1b) involves the attack of a nucleophile on the α-carbon of a quaternary and usually competes with the Hoffman elimination.

The Stevens rearrangement (eq. 2) is a base-promoted 1,2-migration of an alkyl group from a quaternary nitrogen to carbon. The Sommelet-Hauser rearrangement (eq. 3) is a base-promoted 1,2-migration of a benzyl group to the *ortho*-position of that benzyl group.

$$C_6H_5COCH_2\overset{+}{N}(CH_3)_2(CH_2C_6H_5) \xrightarrow{^-OH} C_6H_5CO\underset{CH_2C_6H_5}{CH}N(CH_3)_2 + H_2O \quad (2)$$

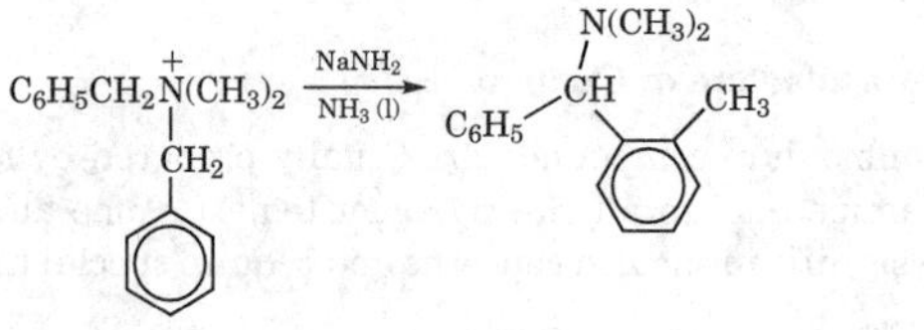

(3)

Naturally Occurring Quaternaries

Many types of aliphatic, heterocyclic, and aromatic derived quaternary ammonium compounds are produced both in plants and invertebrates. Examples include thiamine (vitamin B_1) (**4**) (see VITAMINS); choline (**5**); and acetylcholine (**6**). These have numerous biochemical functions.

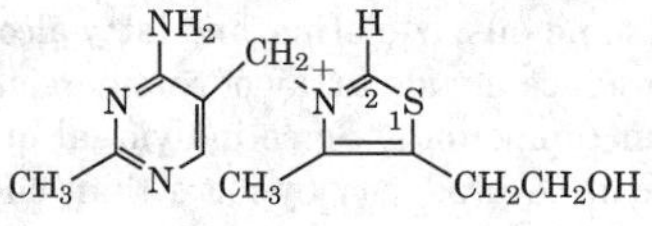

(**4**)

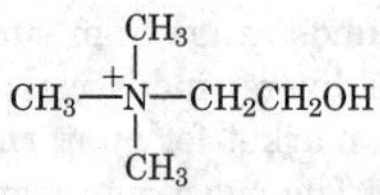

(**5**)

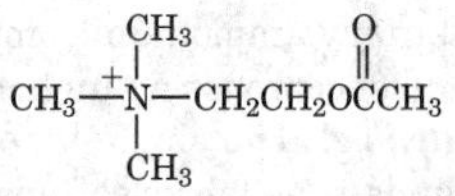

(**6**)

Biochemically, most quaternary ammonium compounds function as receptor-specific mediators; they also function biochemically as messengers.

Analytical Test Methods

There are no universally accepted wet analytical methods for the characterization of quaternary ammonium compounds. The American Oil Chemists' Society (AOCS) has established, however, a number of applicable tests.

The chain length composition of quaternaries can be determined by gas chromatography.

Mass spectral analysis of quaternary ammonium compounds can be achieved by fast-atom bombardment (fab) ms.

Liquid chromatography has been widely applied for analysis of quaternaries. Modified reverse-phase columns can provide chain length information, whereas normal-phase chromatography results in groupings of alkyl distributions.

Nuclear magnetic resonance (nmr) spectroscopy is useful for determining quaternary structure.

Toxicology and Environmental Fate

Some quaternary ammonium compounds are potent germicides, toxic in small (mg/L range) quantities to a wide range of microorganisms. Bactericidal, algicidal, and fungicidal properties are exhibited. Many quaternaries are considered to be moderately to severely irritating to the skin and eyes.

Most uses of quaternary ammonium compounds can be expected to lead to these compounds' eventual release into wastewater treatment systems except for those used in drilling muds. Useful properties of the quaternaries as germicides can make these compounds potentially toxic to sewer treatment systems. It appears, however, that quaternary ammonium compounds are rapidly degraded in the environment and strongly sorbed by a wide variety of materials.

The threat of accidental misuse of quaternary ammonium compounds coupled with potential harmful effects to sensitive species of fish and invertebrates has prompted some concern. Industry has responded with an effort to replace the questionable compounds with those of a more environmentally friendly nature. A newer class of compounds containing an ester linkage has been developed which are more readily biodegraded.

Synthesis and Manufacture of Quaternaries

Quaternary ammonium compounds are usually prepared by reaction of a tertiary amine and an alkylating agent (eq. **4**). Some alkylating reagents pose significant health concerns and require special handling techniques.

$$R\!-\!\underset{}{\overset{R'}{\overset{|}{N}}}\!-\!R'' + R'''X \longrightarrow R\!-\!\underset{\underset{R'''}{|}}{\overset{R'}{\overset{|}{N^{+}}}}\!-\!R'' \quad X^{-}$$

Synthesis and Manufacture of Amines. The chemical and business segments of amines and quaternaries are closely linked. The majority of commercially produced amines originate from three amine raw materials: natural fats and oils, α-olefins, and fatty alcohols. The amines are then used to produce a wide array of commercially available quaternary ammonium compounds. Some individual quaternary ammonium compounds can be produced by more than one synthetic route (see also AMINES).

Uses

Uses of quaternary ammonium compounds range from surfactants to germicides and encompass a number of diverse industries.

Fabric Softening. The single largest market for quaternary ammonium compounds is as fabric softeners. The use of quaternary surfactants as fabric softeners and static control agents can be broken down into three main household product types: rinse cycle softeners; tumble dryer sheets; and detergents containing softeners, also known as softergents.

Hair Care. Quaternary ammonium compounds are the active ingredients in hair conditioners. Quaternaries are highly substantive to human hair because the hair fiber has anionic binding sites at normal pH ranges.

Other Uses. An important market for quaternaries is sanitation. Quaternaries find use as disinfectants and sanitizers in hospitals, building maintenance, and food processing (qv); in secondary oil recovery for drilling fluids; and in cooling water applications (see DISINFECTANTS AND ANTISEPTICS; PETROLEUM). Additional applications of quaternaries include the manufacture of organo-modified clays and use as phase-transfer catalysts.

Other important classes of quaternaries are the polyamine-based (polyquats) and the perfluorinated quaternaries, both of which have a number of applications.

MAURICE DERY
Akzo Nobel Chemicals Inc.

E. Jungermann, ed., *Cationic Surfactants*, Marcel Dekker, Inc., New York, 1969.

Specialty Surfactants Worldwide in *Specialty Chemicals*, SRI International, Menlo Park, Calif., 1989, pp. 81–94.

J. Salamone and W. Rice, in J. I. Kroschwitz, ed., *Encyclopedia of Polymer Science and Engineering*, 2nd ed., John Wiley & Sons, Inc., New York, 1988.

QUINOLINE DYES. See QUINOLINES AND ISOQUINOLINES.

QUINOLINES AND ISOQUINOLINES

The isomeric heterocycles quinoline (**1**) and isoquinoline (**2**) possess structures that occur frequently in alkaloids and pharmaceuticals for example, quinine and morphine (see ALKALOIDS).

Table 1. Physical Properties of Quinoline and Isoquinoline

Property	Value	
	Quinoline	Isoquinoline
mp, °C	−15.6	26.5
bp, °C	238	243
ΔH_{vap}, kJ/mol[a]	46.4	49.0
n_D^{20}	1.6268	1.6148
d^{20}, g/cm^3	1.0929	1.0986
K_a	1.25×10^{-5}	3.80×10^{-6}
viscosity at 30°C, mPa·s(= cP)	2.997	3.2528
T_c	509	530

[a] To convert J to cal, divide by 4.186.

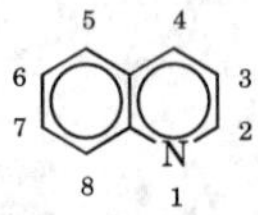
(**1**)

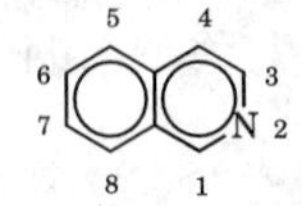
(**2**)

Quinoline and isoquinoline are aromatic, but less intensely than benzene.

Comparative Properties

Physical Properties. Both (**1**) and (**2**) are weak bases, showing pK_a 4.94 and 5.40, respectively. Selected physical data for quinoline and isoquinoline are given in Table 1.

Chemical Properties. The presence of both a carbocyclic and a heterocyclic ring facilitates a broad range of chemical reactions for (**1**) and (**2**). Quaternary alkylation on nitrogen takes place readily, but unlike pyridine both quinoline and isoquinoline show addition by subsequent reaction with nucleophiles. Nucleophilic substitution is promoted by the heterocyclic nitrogen. Electrophilic substitution takes place much more easily than in pyridine, and the substituents are generally located in the carbocyclic ring. Their facile formation of crystalline salts with either inorganic or organic acids and complexes with Lewis acids is in each case of considerable interest.

Quinoline

Reactions. Quinoline exhibits the reactivity of benzene and pyridine rings, as well as its own unique reactions.

As an aromatic system, (**1**), shows important synthetic and mechanistic nitro group chemistry (see NITRATION). The experimental conditions employed usually determine the product structure.

The main sulfonation product of quinoline at 220°C is 8-quinolinesulfonic acid; at 300°C it rearranges to 6-quinolinesulfonic acid.

Unlike pyridine, quinoline undergoes facile addition to the nitrogen-containing ring. Allylmagnesium chloride reacts with quinoline in deoxygenated tetrahydrofuran to produce 80% 2-allyl-1,2-dihydroquinoline. Similar results are observed with vinyl Grignard reagents and with alkyllithium reagents.

Treatment of quinoline with cyanogen bromide, the von Braun reaction, in methanol with sodium bicarbonate produces a high yield of 1-cyano-2-methoxy-1,2-dihydroquinoline (**3**).

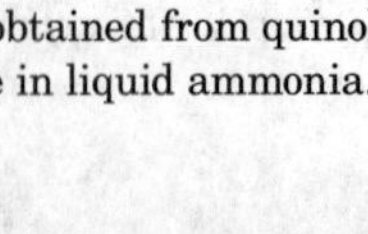
(**3**)

2-Aminoquinoline is obtained from quinoline in 80% yield by treatment with barium amide in liquid ammonia.

As with nitration, halogenation under acidic conditions favors reaction in the benzenoid ring, whereas reaction at the 3-position takes place in the neutral molecule.

The synthesis of quinolinic acid and its subsequent decarboxylation to nicotinic acid has been accomplished directly in 79% yield using a nitric–sulfuric acid mixture above 220°C. A wide variety of oxidants have been used in the preparation of quinoline *N*-oxide (see AMINE OXIDES).

The ring nitrogen of quinoline reacts with a wide variety of alkylating and acylating agents to produce useful intermediates, for example, *N*-benzoylquinolinium chloride (**4**).

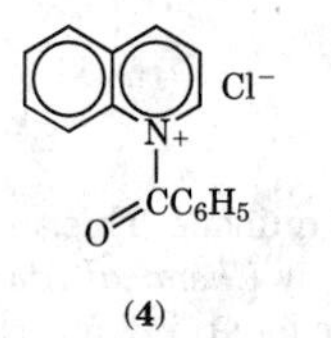

(**4**)

The direct introduction of carbon–carbon bonds in quinoline rings takes place in low yield and with little selectivity.

Quinoline may be reduced rather selectively, depending on the reaction conditions. Catalytic reduction with platinum oxide in strongly acidic solution at ambient temperature and moderate pressure gives a 70% yield of 5,6,7,8-tetrahydroquinoline. Further reduction of this material with sodium–ethanol produces 90% of *trans*-decahydroquinoline.

Manufacture From Coal Tar. Commercially, quinoline is isolated from coaltar distillates. Tar acids are removed by caustic extraction, and the oil is distilled to produce the methylnaphthalene fraction (230–280°C).

Syntheses of Quinolines. *Skraup Synthesis.* This general method, used for many quinolines, consists of heating a primary aniline with glycerol, concentrated sulfuric acid, and an oxidizing agent. Often the nitrobenzene corresponding to the aniline employed is used as the oxidant, and iron(II) sulfate is added to moderate the often violently exothermic process. The use of compounds related to acrolein, such as crotonaldehyde and methyl vinyl ketone, allow substituents to be placed in the heterocyclic ring. With ortho- and para-substituted anilines, a single product is usually found; meta-derivatives produce mixtures.

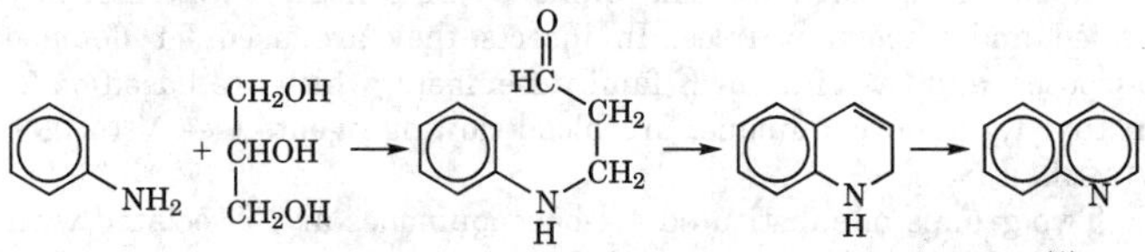

(**1**)

Döbner-von Miller Synthesis. A much less violent synthetic pathway, the Döbner-von Miller, uses hydrochloric acid or zinc chloride as the catalyst. α,β-Unsaturated aldehydes and ketones make the dehydration of glycerol unnecessary, and allow a wider variety of substitution patterns. No added oxidant is required.

Combes Synthesis. When aniline reacts with a 1,3-diketone under acidic conditions a 2,4-disubstituted quinoline results, eg, 2,4-dimethylquinoline from 2,4-pentadione.

Conrad-Limpach-Knorr Synthesis. When a β-keto ester is the carbonyl component of these pathways, two products are possible. Aniline reacts with ethyl acetoacetate below 100°C to form 3-anilinocrotonate, which is converted to 4-hydroxy-2-methylquinoline by placing it in a preheated environment at 250°C. If the initial reaction takes place at 160°C, acetoacetanilide forms and can be cyclized with concentrated sulfuric acid to 2-hydroxy-4-methylquinoline. This example of kinetic vs thermodynamic control has been employed in the synthesis of many quinoline derivatives. They are useful as intermediates for the synthesis of chemotherapeutic agents (see CHEMOTHERAPEUTICS, ANTICANCER).

Pfitzinger Reaction. Quinoline-4-carboxylic acids are easily prepared by the condensation of isatin with carbonyl compounds. The products may be decarboxylated to the corresponding quinolines.

Friëlander Synthesis. The methods cited thus far all suffer from the mixtures which usually result with meta-substituted anilines. The use of an ortho-disubstituted benzene for the subsequent construction of the quinoline avoids the problem. In the Friëlander synthesis a starting material like 2-aminobenzaldehyde reacts with an α-methyleneketone in the presence of base. The difficulty of preparing the required anilines is a limitation in this approach.

New Synthetic Approaches. There have been a number of efforts to prepare quinolines by routes quite different from the traditional methods. In one, the cyclization of 3-amino-3-phenyl-2-alkenimines using alkali metals leads to modest yields of various 4-arylaminoquinolines. Because this structure is found in many natural products and few syntheses of it exist, the method merits further investigation.

The importance of quinolinium salts to dye chemistry accounts for the long, productive history of their synthesis. The reaction of *N*-methylformanilide with ketones, aldehydes, ketone enamines, or enol acetates in phosphoryl chloride leads to high yields of *N*-methylquinolinium salts.

Toxicology. Quinoline is a poison when it enters the body by any of the normal routes, ie, ingestion, or subcutaneous or intraperitoneal injection. Even contact with the skin produces a moderate toxic reaction, and can result in severe irritation. There is evidence that quinoline is mutagenic, and long exposure can produce lung problems.

Uses.

Antioxidants. The 1,2-dihydroquinolines have been used in a variety of ways as antioxidants. These compounds react with aldehydes, and the products are useful as food antioxidants.

Corrosion Inhibitors. Steel-reinforcing wire and rods embedded in concrete containing quinoline or quinoline chromate are less susceptible to corrosion. Treating the surface of metals with 8-hydroxyquinoline makes them resistant to tarnishing and corrosion. Ethylene glycol-type antifreeze may contain quinoline or its derivative to prevent corrosion.

Agricultural Chemicals. A herbicide possessing activity comparable to 2,4-D is found in compounds like quinolyl esters of N-substituted dithiocarbamic acids. A wide variety of compounds containing the quinoline system are herbicides. Derivatives and salts of 8-quinolinecarboxylic acid as well as quinolyl carbamates are each useful insecticides. The copper salt of 8-hydroxyquinoline is an effective fungicide (see FUNGICIDES, AGRICULTURAL).

Polymers. Quinoline and its derivatives may be added to or incorporated in polymers to introduce ion-exchange properties.

Platinum-group metals form complexes with chelating polymers with various 8-mercaptoquinoline derivatives (see CHELATING AGENTS).

Metallurgy. The extraction and separation of metals and plating baths have involved quinoline and certain derivatives (see ELECTROPLATING; METAL SURFACE TREATMENTS; EXTRACTION). The extraction of metal ions depends on the chelating ability of 8-hydroxyquinoline. Dilute solutions of heavy metals such as mercury, cadmium, copper, lead, and zinc can be purified using quinoline-8-carboxylic acid adsorbed on various substrates.

Catalysts. Acrolein and methacrolein 1,4-addition polymerization is catalyzed by lithium complexes of quinoline. The peracetic acid epoxidation of a wide range of alkenes is catalyzed by 8-hydroxyquinoline.

Medicine. In addition to the naturally occurring compounds, a large number of synthetic quinolines have been prepared and studied for use in medicine.

Quinoline Dyes. The reaction of 2-methylquinoline with phthalic anhydride produces a 2:1 mixture of 2-(2-quinolinyl)-1,3-indandione ((**5**), R = H) and 2-(6-methyl-2-quinolinyl)-1,3-indandione ((**5**), R = CH_3).

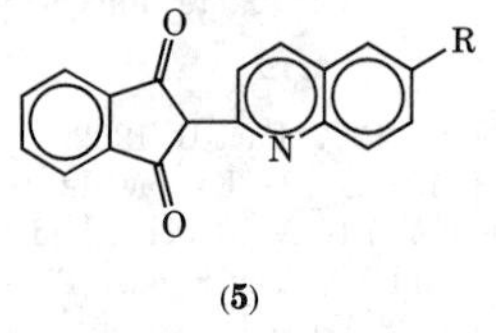

(**5**)

This mixture is known as Quinoline Yellow A and is most widely used with polyester fibers. Several other quinoline dyes are commercially available and find applications as pigments, biological stains, and analytical reagents.

Isoquinoline

The widespread occurrence of the isoquinoline (**2**) structure in such important alkaloids as those found in cactus, opium, and curare has created a long-standing interest in its synthesis and properties.

Reactions. In general, isoquinoline undergoes electrophilic substitution reactions at the 5-position and nucleophilic reactions at the 1-position. Nitration with mixed acids produces a 9:1 mixture of 5-nitroisoquinoline and 8-nitroisoquinoline. Sulfonation of isoquinoline gives a mixture with 5-isoquinolinesulfonic acid as the principal product.

Amination of isoquinoline with sodamide in neutral solvents gives 1-aminoisoquinoline.

Direct bromination of isoquinoline hydrochloride in a solvent like nitrobenzene gives an 81% yield of 4-bromoisoquinoline.

The oxidation of isoquinoline has also been examined using ruthenium tetroxide. In this instance, the observation that phthalic acid is the only significant product (58%) was made; this fact is both important and difficult to explain. Isoquinoline is also oxidized to its N-oxide by peracids. The N-oxides of isoquinolines have proved to be excellent intermediates for the preparation of many compounds.

Isoquinoline can be reduced quantitatively over platinum in acidic media to a mixture of *cis*-decahydroisoquinoline and *trans*-decahydroisoquinoline.

Synthesis of Isoquinoline and Isoquinoline Derivatives. *Bischler–Napieralski Reaction.* This synthetic method involves the cyclodehydration of N-acyl derivatives of β-phenethylamines to 3,4-dihydroisoquinolines, such as 1-methyl-3,4-dihydroisoquinoline.

Pictet-Spengler Synthesis. An acidic catalyst results in the condensation of β-phenethylamines with carbonyl compounds to give 1,2,3,4-tetrahydroisoquinolines.

Pomeranz-Fritsch Synthesis. Isoquinolines are available from the cyclization of benzalaminoacetals under acidic conditions.

Miscellaneous Synthetic Reactions. A number of *o*-disubstituted benzenes have been used to prepare isoquinolines. For example, the Radziszewski method and subsequent dehydration converts *o*-cyanomethylbenzoic acid to homophthalimide in 90% yield.

Toxicology. Isoquinoline is a poison when ingested or injected intraperitoneally. Even in cases of skin contact it is moderately toxic. Its vapors are irritating to the eyes, nose, and throat.

Uses. Isoquinoline and isoquinoline derivatives are useful as corrosion inhibitors, antioxidants, pesticides, and catalysts. They are used in plating baths and miscellaneous applications, such as in photography, polymers, and azo dyes (qv). Numerous derivatives have been prepared and evaluated as pharmaceuticals. Isoquinoline is a main component in quinoline still residue bases, which are sold as corrosion inhibitors and acid inhibitors for pickling iron and steel.

4-Aminoisoquinoline is a component of an ethylene glycol-based corrosion inhibiting antifreeze agent (see ANTIFREEZES AND DEICING FLUIDS).

A great many alkaloids and synthetic medicinal compounds are isoquinoline derivatives.

K. THOMAS FINLEY
State University of New York, Brockport

G. Jones, "Quinolines," Part I, 1977; Part II, 1982; Part III, 1990; G. Grenthe, "Isoquinolines," Part I, 1981; F. G. Kathawala, G. M. Coppola, and H. F. Schuster, Part II, 1990; and G. M. Coppola and H. F. Schuster, Part III, 1995, in A. Weissberger and E. C. Taylor, eds., *The Chemistry of Heterocyclic Compounds*, Vols. 32 and 38, Wiley-Interscience, New York.

QUINONES

These useful compounds have played a central role in both theoretical and practical organic chemistry since the 1840s. The compound 1,4-benzoquinone (**1**)

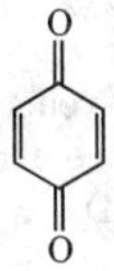

(**1**)

provides the generic name quinone. This simple, descriptive nomenclature has been abandoned by *Chemical Abstracts*, but remains widely used. The systematic name for (**1**) is 2,5-cyclohexadiene-1,4-dione. Several examples of quinone synonyms are given in Table 1. Common names are used in this article.

Simple quinones have two notable physical properties: odor and color. The 1,4-benzo- and 1,4-naphthoquinones and many of their derivatives have high vapor pressures and pungent, irritating odors. The single-ring compounds are often found as constituents of insects' chemical defense against predators. In general, the 1,2-quinones are vibrant in color, ranging from orange to red, whereas the 1,4-quinones are usually lighter, ie, yellow to orange.

The quinones have excellent redox properties and are thus important oxidants in laboratory and biological synthons. The presence of an extensive array of conjugated systems, especially the α,β-unsaturated ketone arrangement, allows the quinones to participate in a variety of reactions. Characteristics of quinone reactions include nucleophilic substitution; electrophilic, radical, and cycloaddition reactions; photochemistry; and normal and unusual carbonyl chemistry.

Physical Properties

Selected physical data for various quinones are given in Table 2.

Chemical Properties

Biochemical Reactions. The quinones in biological systems play varied and important roles. In insects they are used for defense purposes, and the vitamin K family members, which are based on 2-methyl-1,4-naphthoquinone, are blood-clotting agents (see VITAMINS, VITAMIN K).

Two groups of substituted 1,4-benzoquinones are associated with photosynthetic and respiratory pathways; the plastoquinone, eg, plastoquinone (**2**), and the ubiquinones, eg, ubiquinone (**3**), are involved in these processes. Although they are found in all living tissue and are central to life itself, a vast amount remains to be learned about their biological roles.

Table 1. Quinone Nomenclature

Common name	Synonym
1,4-benzoquinone	*p*-benzoquinone
1,2-benzoquinone	*o*-benzoquinone
1,4-napthoquinone	α-naphthoquinone
4,4′-diphenoquinone[a]	diphenoquinone
o-chloranil[b]	tetrachloro-*o*-quinone
p-chloranil[c]	tetrachloro-*p*-quinone
2,3-dichloro-5,6-dicyano-1,4-benzoquinone[d]	DDQ

[a] Bis-2,5-cyclohexadien-1-ylidene,4,4′-dione; 4-(4-oxo-2,5-cyclohexadiene-1-ylidene)-2,5-cyclohexadien-1-one is also used.
[b] 3,4,5,6-Tetrachloro-3,5-cyclohexadiene-1,2-dione.
[c] 2,3,5,6-Tetrachloro-2,5-cyclohexadiene-1,4-dione.
[d] 4,5-Dichloro-3,6-dioxo-1,4-cyclohexadiene-1,2-dicarbonitrile.

Table 2. Physical Properties of Selected Quinones

Name	Color	Mp, °C	Crystalline form	Solubility	
				Soluble	Insoluble
1,4-benzoquinone	yellow	113, 116	monoclinic prisms	alcohol, ether	water, pentane
1,2-benzoquinone	red	60–70 dec	plates or prisms	ether, benzene	pentane
1,2-naphthoquinone	yellow-red orange	145–147	needles	water, alcohol	ligroin
1,4-naphthoquinone	bright yellow	125, 128.5	needles	alcohol, benzene	water, ligroin
3,4,5,6-tetrachloro-1,2-benzoquinone	orange-red	133, 122–127			
2,3,5,6-tetrachloro-1,4-benzoquinone	yellow	290, 294	monoclinic prisms	ether	water, ligroin
2,3-dichloro-5,6-dicyano-1,4-benzoquinone	bright yellow	201–202 dec	plates		
2-chloro-1,4-benzoquinone	yellow-red	57	rhombic-hexagonal	water, alcohol	
2,5-dichloro-1,4-benzoquinone	pale yellow	161–162	monoclinic prisms	ether, chloroform	water, alcohol
2,5-dimethyl-1,4-benzoquinone	yellow	125		ether, alcohol	water, alcohol
2-methyl-1,4-benzoquinone	yellow	69	plates or needles	ether, alcohol	water
3-chloro-1,2-naphthoquinone	red	172 dec	needles	alcohol, benzene	water
2,3-dichloro-1,4-naphthoquinone	yellow	193, 195	needles	benzene, choloroform	water, alcohol
2-methyl-1,4-naphthoquinone	yellow	105–107	needles	ether, benzene	water, alcohol

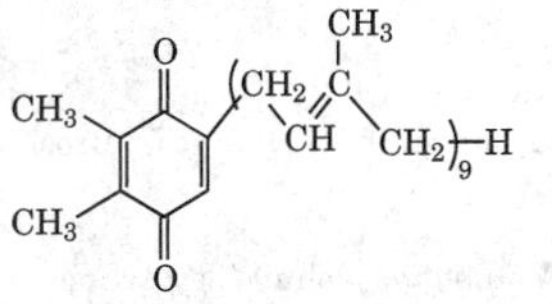

(2)

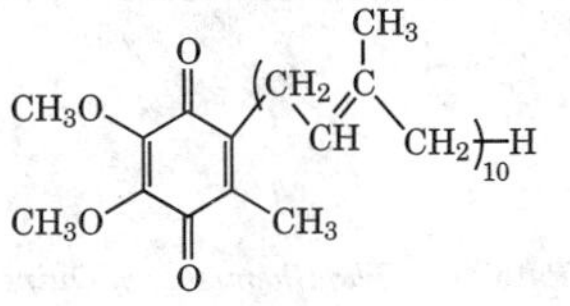

(3)

Quinones of various degrees of complexity have antibiotic, antimicrobial, and anticancer activities.

Dehydrogenation. The oldest and still important synthetic use of quinones is in the removal of hydrogen, especially for aromatization. This method has often been applied to the preparation of polycyclic aromatic compounds. Quinones are used extensively in the dehydrogenation of steroidal ketones. Such reactions are marked by high yield and selectivity. Generally, the results when using nonsteroidal ketones are disappointing.

Oxidation. The use of 1,4-benzoquinone (**1**) in combination with palladium(II) chloride converts terminal alkenes such as 1-hexene to alkyl methyl ketones in high yield (81%). The quinone appears to reoxidize the palladium.

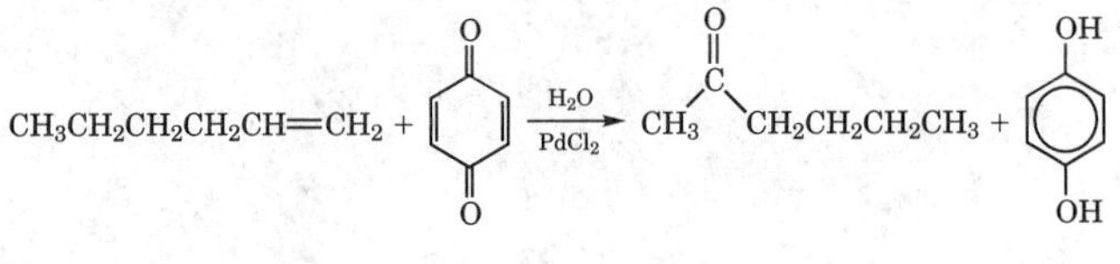

(**1**)

Photochemical Reactions. Increased knowledge of the centrality of quinone chemistry in photosynthesis has stimulated renewed interest in their photochemical behavior. Synthetically interesting work has centered on the 1,4-quinones and the two reaction types most frequently observed, ie [2 + 2] cycloaddition and hydrogen abstraction.

Addition Reactions. The addition of nucleophiles to quinones is often an acid-catalyzed, Michael-type reductive process. The addition of benzenethiol to 1,4-benzoquinone (**1**) was studied by A. Michael for a better understanding of valence in organic chemistry. The presence of the reduced product thiophenylhydroquinone, the cross-oxidation product 2-thiophenyl-1,4-benzoquinone, and multiple-addition products such as 2,5-(bis(thiophenyl)-1,4-benzoquinone and 2,6-bis(thiophenyl)-1,4-benzoquinone, is typical of many such transformations.

Nucleophilic Substitution Reactions. Many of the transformations realized through Michael additions to quinones can also be achieved using nucleophilic substitution chemistry. In some instances the stereoselectivity can be markedly improved in this fashion, eg, in the reaction of benzenethiol with esters (R^3 = $CH_3C{=}O$) and ethers (R^3 = CH_3) of 1,4-naphthoquinones. 2-Bromo-5-acetyloxy-1,4-naphthoquinone, R^1 = Br, yields 75% of 2-thiophenyl-5-acetyloxy-1,4-naphthoquinone, R^1 = SC_6H_5. 3-Bromo-5-methoxy-1,4-naphthoquinone, R^2 = Br, yields 82% of 3-thiophenyl-5-methoxy-1,4-naphthoquinone R^2 = SC_6H_5.

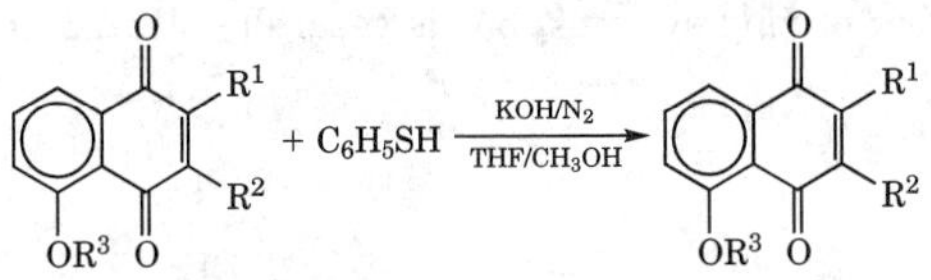

Syntheses

Syntheses of quinones often involve oxidation, because this is the only completely general method. Thus, in several instances, quinones are the reagents of choice for the preparation of other quinones. Oxidation has been especially useful with catechols and hydroquinones as starting materials. The preparative utility of these reactions depends largely on the relative oxidation potentials of the quinones.

For the preparation of ≤10 g of a quinone, the oxidation of a phenol with Fremy's salt in the Teuber reaction is the method of choice. A wide range of phenols has been used, including some having 4-substituents.

In small-scale syntheses, a wide variety of oxidants has been employed in the preparation of quinones from phenols. Of these reagents, chromic acid, ferric ion, and silver oxide show outstanding usefulness in the oxidation of hydroquinones. Thallium(III) trifluoroacetate converts 4-halo- or 4-*tert*-butylphenols to 1,4-benzoquinones in high yield. For example, 2-bromo-3-methyl-5-*t*-butyl-1,4-benzoquinone (**4**) has been made by this route.

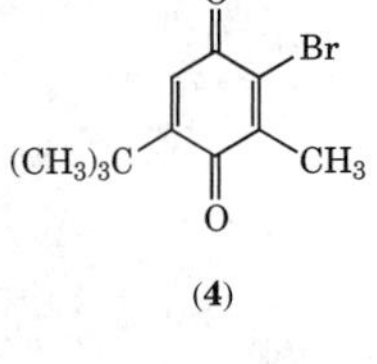

(**4**)

Thallium trinitrate oxidizes naphthols and hydroquinone monoethers, respectively, to quinones and 4,4-dialkoxycyclohexa-2,5-dienones, eg, 4,4-dimethoxy-2-methyl-2,5-cyclohexadienone (**5**). The yield of (**5**) is 89%.

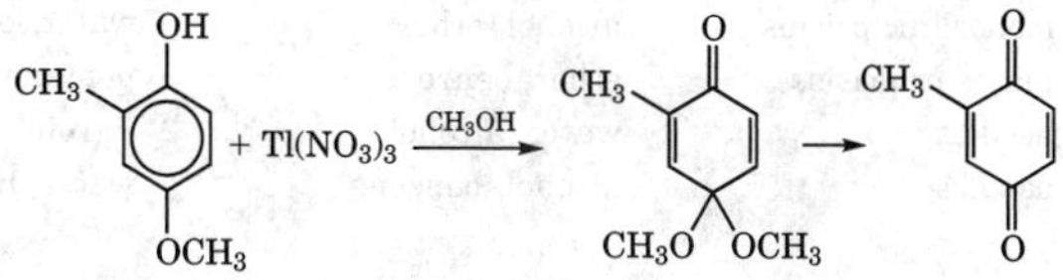

(**5**)

The oxidation of 4-bromophenols to quinones can also be accomplished using periodic acid.

The anodic oxidation of hydroquinone ethers to quinone ketals yields synthetically useful intermediates that can be hydrolyzed to quinones at the desired stage of a sequence.

Manufacture

With the exceptions of 1,4-benzoquinone and 9,10-anthraquinone, quinones are not produced on a large scale, but a few of these are commercially available (see ANTHRAQUINONE). The few large-scale preparations involve oxidation of aniline, phenol, or aminonaphthols, eg, (**6**), from which 1,2-napthoquinone (**7**) is obtained in 93% yield.

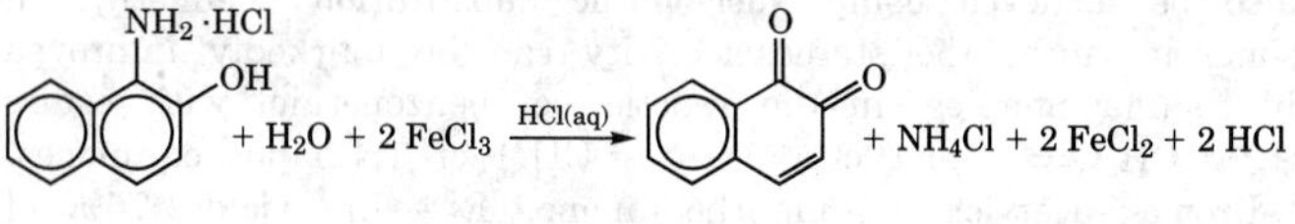

(**6**) (**7**)

In the case of 1,4-benzoquinone (**1**), the product is steam-distilled, chilled, and obtained in high yield and purity. Direct oxidation of the appropriate unoxygenated hydrocarbon has been described for a large number of ring systems, but is generally utilized only for the polynuclear quinones without side chains. A representative sample of quinone uses is given in Table 3.

Table 3. Uses of Selected Quinones

Quinone	Use
1,4-benzoquinone	oxidant, amino acid determination
2-chloro-, 2,5-dichloro-, and 2,6-dichloro-1,4-benzoquinone	bactericides
2,3-dichloro-5,6-dicyano-1,4-benzoquinone	oxidation and dehydrogenation agent
2-methyl- and 2,3-dimethyl-1,4-naphthoquinone	vitamin K substitutes, antihemorrhagic agents

Health and Safety Factors

Because of the high vapor pressure of the simple quinones and their penetrating odor, adequate ventilation must be provided in areas where these quinones are handled or stored. Quinone vapor can harm the eyes. In either solid or solution form quinone can cause severe local damage to the skin and mucous membranes. Swallowing benzoquinones may be fatal. The higher quinones are less of a problem because of their decreased volatility.

K. THOMAS FINLEY
State University of New York, Brockport

S. Patai, ed., *The Chemistry of Quinonoid Compounds*, John Wiley & Sons, Inc., New York, 1974.

S. Patai and Z. Rappoport, eds., *The Chemistry of Quinonoid Compounds*, Vol. 2, John Wiley & Sons, Inc., New York, 1988.

R. H. Thomson, *Naturally Occurring Quinones*, 3rd ed., Chapman and Hall, London, 1987.

R

R-ACID. See AZO DYES; NAPHTHALENE DERIVATIVES.

RADIATION CURING

The interaction of electromagnetic radiation with organic substrates, that is, the use of electromagnetic radiation to alter the physical and chemical nature of a material is sometimes known as radiation curing technology. In radiation curing, electromagnetic radiation interacts with organic substrates to develop cross-linked or solvent-insoluble network structures. For example, a preformed thermoplastic polymer that interacts directly with certain types of ionizing (high energy) radiation from a given source of energy can develop into cross-linked or network structures having higher melting points, improved heat resistance, and improved chemical resistance than the original thermoplastic polymer starting materials (Fig. 1).

Radiation and Electromagnetic Radiation Sources

Radiation curing, as applied to the cross-linking of polymers or coating materials, involves the full spectrum of electromagnetic radiation energies to effect chemical reactions. These forms of radiation energy include ionizing radiation, ie, α-, β-particles and γ-rays from radioactive nuclei; x-rays; high energy electrons; and nonionizing radiation such as are associated with uv, visible, ir, microwave, and radiofrequency wavelengths of energy (Table 1).

Mechanisms of Radiation Energy—Organic Substrate Interaction

High energy interaction with organic substrates produces excited states which undergo secondary reactions, eg, electron capture, charge neutralization, intermolecular and intramolecular energy-transfer processes, ion formation, and molecular dissociation to produce free-radical intermediate species. The resulting chemical reactions are caused by the excited species and the formation of reactive intermediates.

In the case of photochemical reactions, light energy must be absorbed by the system so that excited states of the molecule can

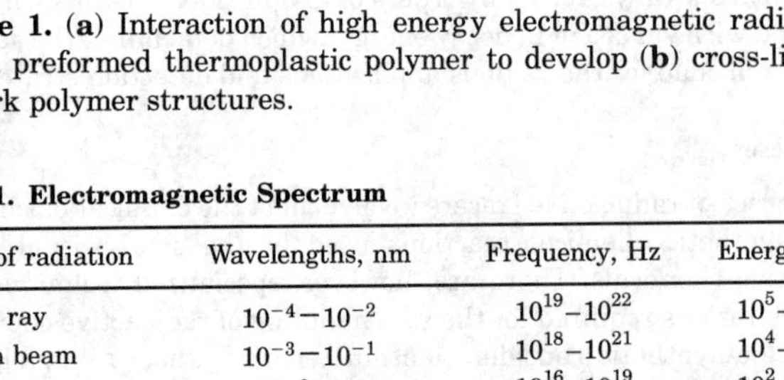

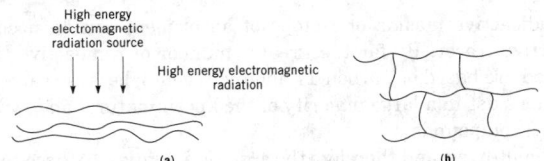

Figure 1. (**a**) Interaction of high energy electromagnetic radiation with a preformed thermoplastic polymer to develop (**b**) cross-linked network polymer structures.

Table 1. Electromagnetic Spectrum

Types of radiation	Wavelengths, nm	Frequency, Hz	Energy, eV
gamma ray	10^{-4}–10^{-2}	10^{19}–10^{22}	10^5–10^8
electron beam	10^{-3}–10^{-1}	10^{18}–10^{21}	10^4–10^7
x-ray	10^{-2}–10	10^{16}–10^{19}	10^2–10^5
ultraviolet	10–400	10^{15}–10^{16}	5–10^2
visible	400–750	10^{15}	1–5
infrared	750–10^5	10^{12}–10^{14}	10^{-2}–1
microwave	$>10^6$	10^{11}–10^{12}	$<10^{-2}$
radio frequency	$>10^6$	$<10^{11}$	$<10^{-2}$

form and subsequently produce free-radical intermediates (see PHOTOCHEMICAL TECHNOLOGY).

Curing Polymers with γ-Rays, X-Rays, and High Energy Electrons

Radiation curing of preformed polymers using ionizing radiation processing equipment can result in two types of chemical change that are associated with cross-linking and degradation reaction mechanisms. Cross-linking reaction mechanisms on preformed polymer substrates usually involve removal of hydrogen atoms to form a macroradical intermediate. These macroradical intermediates can then couple to form a single molecule, resulting in an increase in the original average molecular weight of the starting polymer.

If irradiation continues, the original polymer substrate is transformed into one gigantic molecule of infinite molecular weight having lower solvent solubility, higher melting point, and improved physical properties over the original material.

Radiation-induced degradation reactions are in direct opposition to cross-linking or curing processes, in that the average molecular weight of the preformed polymer decreases because of chain scission and without any subsequent recombination of its broken ends. In order for efficient radiation curing of a polymer to take place, these degradation processes must be minimized in favor of the desired cross-linking reaction.

Curing of Coatings with Electron Beams, γ-Ray, X-Ray, and Planar Cathodes

In conventional gas oven and other heat energy sources associated with the thermal curing of coatings, a mixture of polymers, cross-linking oligomers, catalysts, additives, pigments, and fillers are dissolved or dispersed in organic or water-based solvents to form a coating system. The coating is applied to a substrate, and the solvents are removed thermally. The coating cross-links into a three-dimensional network by an energy-rich chemistry, which requires a high degree of thermal energy to convert the polymers into those having useful commercial properties. Much of the energy is absorbed by the substrate before heat reaches the polymers to initiate the curing chemistry.

High energy electron-curable coatings generally consist of multifunctional acrylate or methacrylate unsaturated polymers. They differ from conventional coatings in that the solvents for the polymers are high boiling, usually nonvolatile, and 100% coreactive with themselves and with other organic components in the film. The curing process for these coatings is a free-radical chain reaction. Ionizing radiation from the processing equipment is absorbed directly in the coating and generates the free radicals that initiate the curing process. Because electron energies of only 100 eV or less are required to break chemical bonds and to ionize or excite components of the coating system, the shower of scattered electrons produced in the coating leads to a intense population of free radicals throughout the coating. These initiate the polymerization reaction and this polymerization process results in a dry, three-dimensional cross-linked coating. In this process, most of the energy is absorbed into the coating and is not lost to the substrate, as is the case in thermal curing reactions. Neither cobalt-60 nor x-ray energy sources are used in radiation curable coating systems; X-rays are used in photolithographic processes (see LITHOGRAPHIC RESISTS).

Electron Beam. An electron beam processing unit consists mainly of a power supply and an electron beam acceleration tube. The beam is produced when high voltage energizes a tungsten filament, thereby causing electrons to be produced at high rates. These fast electrons are concentrated to form a high energy beam and are accelerated to full velocity inside the electron gun.

Electrocurtain. The electrocurtain processor (Energy Sciences) is a high (150 kV) voltage electron tube that provides a continuous strip of energetic electrons from a linear filament or cathode, which is on the axis of symmetry of the system. The cylindrical electron gun shapes and processes the electron stream, in a grid-controlled structure. The energetic electrons from the processor are injected directly in the coating, where they create the initiating free radicals. These energetic electrons can penetrate many different types of pigmented coatings and are capable of producing through-cure to the substrate–polymer coating interface.

Multiple-Planar Cathode Processors. The design criterion for this electron accelerator system is a planar array of concentrated cathode control grid elements. The modular cathode construction allows for broad-beam (250-cm wide) processing of materials using powers of 30 kGy (3 Mrad) at 300 m/min.

Coatings Ingredients. Ingredients of liquid high energy electron radiation-curable coatings are analogues of components contained in conventional solvent-based thermal-curing coating systems. In conventional solvent-based coatings systems, a preformed polymer (usually 3,000–25,000 mol wt) is dissolved in an organic solvent (30–80% solids), and a cross-linking oligomer and various flow agents, catalysts, pigments, etc, are added. The coating is applied to a substrate by conventional methods, eg, spray, roll coating, and flow coating, and subsequently is cured in gas or ir thermal ovens.

Curing with Ultraviolet, Visible, and Infrared Processing Equipment

Polymers. Upon direct absorption of uv or visible wavelengths of light, polymer substrates undergo chain scission and cross-linking. Cross-linking or curing of preformed polymeric materials, ie, thermoplastics, can be markedly enhanced through the use of special photosensitive molecules that are mixed into the polymer matrix or that chemically attach to the backbone of the polymer chains. These special photosensitive molecules absorb uv or visible light energies much more efficiently than the polymer; they rapidly form excited states that undergo photochemical reactions which in turn form reactive free-radical intermediates that effect polymer dimerization or cross-linking. When compounded into the preformed polymer matrix, these special photosensitive molecules, eg, benzophenone, can undergo light-induced radical abstraction or insertion reactions that result in coupling of the polymer chains and in network formation.

In order to cure, ie, form three-dimensional network structures through chemical changes on polymer systems with ir radiation, it is necessary to design a reactive functionality within the polymer structure so that coupling reactions can take place between the polymer chains.

Certain polymeric structures can also be blended with other coreactive polymers or multifunctional reactive oligomers that affect curing reactions when exposed to ir radiation.

Light Source. The light source normally used in commercial photocuring reactions is the medium pressure mercury arc lamp having a quartz or Vicor envelope. These lamps may contain electrodes for electrical-to-light energy conversion or may be electrodeless, in which case a radio-frequency wave causes mercury atom excitation and subsequent light emission.

Photoinitiators. There are two general classes of photoinitiators: those that undergo direct photofragmentation upon exposure to uv or visible light irradiation and produce active free-radical intermediates, and those that undergo electron transfer followed by proton transfer to form a free-radical species. The absorption bands of the photoinitiators should overlap the emission spectra of the various commercial light sources.

Photoactive Catalysts for Acid or Cation Generators

The photoactive catalyst systems commonly used to cure epoxy resins and multifunctional vinyl ether materials include aryldiazonium salts ($ArN{\equiv}N^+X^-$), and triarylsulfonium and diaryliodonium salts ($Ar_3S^+X^-$ and $Ar_2I^+X^-$). Other cation-intermediate-generating catalyst systems are cyclopentadienyl iron(II)–arene hexafluorophosphate complexes, phenylphosphonium benzophenone salts, and pentafluoro phenyl borate anions associated with aryl sulfonium cations (see EPOXY RESINS).

Formulation Design for Free-Radical Cured Systems

Light-induced (free-radical intermediates), radiation-curable coating systems are similar to those used in high energy, electron-radiation-cured coating materials. The reactive coating ingredients in both the light and high energy electron curing processes utilize combinations of single vinyl unsaturated monomers, multifunctional vinyl-substituted cross-linking oligomers, and a variety of unsaturated polymer structures. The only significant difference between high energy cured coatings and light energy cured coatings is the use of a photoinitiator that absorbs the light energy and initiates the start of the curing process.

VINCENT D. MCGINNISS
Battelle Columbus Laboratory

Proceedings of RadTech 92 North American Conference, Vols. 1 and 2, Boston, Mass., published by RadTech International North America, Northbrook, Ill., 1992.

Proceedings of RadTech 94 North American Conference, Vols. 1 and 2, Orlando, Fla., published by RadTech International North American, Northbrook, Ill., 1994.

Suppliers of radiation-curable materials: CIBA-GEIGY (photoinitiators), Aceto Corp. (photoinitiators), BASF (photoinitiators), Henkel (photoinitiators), Sartomer (monomers, cross-linking agents/oligomers, polymers), Henkel (cross-linking agents/oligomers, polymers), Cargill (polymers).

RADIOACTIVE TRACERS

Radiochemical tracers, compounds labeled with radioisotopes (qv), have become one of the most powerful tools for detection and analysis in research, and to a limited extent in clinical diagnosis (see MEDICAL IMAGING TECHNOLOGY). A molecule or chemical is labeled using a radioisotope either by substituting a radioactive atom for a corresponding stable atom in the compound, such as substituting ^3H for ^1H, or by adding a radioactive atom to a molecule, such as iodinating a protein or peptide using ^{125}I. In some cases radioactive labeling results in substituting an atom using a noncorresponding, but chemically similar, radioactive isotope, such as replacing ^{16}O with ^{35}S.

Radiometric detection technology offers high sensitivity and specificity for many applications in scientific research. The radioactive emission of the labeled compound is easily detected and does not suffer from interference from endogenous radioactivity in the sample. Because of this unique property, labeled compounds can be used as tracers to study the localization, movement, or transformation of molecules in complex experimental systems.

Properties

Any radioactive nuclide or isotope of an element can be used as a radioactive tracer. By far the greater number of radioactive tracers produced are based on carbon-14 and hydrogen-3, because carbon and hydrogen exist in a large majority of the known natural and synthetic chemical compounds.

The half-lives, and therefore the specific activities, of these radioactive nuclides vary over many orders of magnitude. It is this variation, coupled with variation in decay energy, which determines the suitability of a nuclide for the various applications and detection strategies.

Syntheses

Syntheses of radioactive tracers involve all of the classical biochemical and synthetic chemical reactions used in the synthesis of nonradioactive chemicals. There are, however, specialized techniques and considerations required for the safe handling of radioactive chemicals, strategic synthetic considerations in terms of their relatively high cost, and synthesis scale constraints governed by specific activity requirements.

The radioactive isotopes available for use as precursors for radioactive tracer manufacturing include barium [^{14}C]-carbonate, tritium gas, [^{32}P]-phosphoric acid or [^{33}P]-phosphoric acid, [^{35}S]-sulfuric acid, and sodium [^{125}I]-iodide. It is from these chemical forms that the corresponding radioactive tracer chemicals are synthesized.

A multistep synthesis is strategically designed such that the labeled species is introduced as close to the last synthetic step as possible in order to minimize yield losses and cost.

Biosynthetic techniques utilizing enzymes isolated from plants, animals, or microorganisms have made possible the synthesis of many labeled compounds of biological interest. Enzymes are uniquely suited for such syntheses because the enzymes are biological catalysts that act with high specificity and at low substrate concentrations.

Microbiological procedures that exploit the ability of bacteria and photosynthetic algae to incorporate exogenous labeled precursors such as $^{14}CO_2$, $^{35}SO_4^{2-}$, and $^{32}PO_4^{3-}$ can be used to label complex molecules in cells such as proteins (qv) and nucleic acids (qv), which are then processed to give labeled constituents such as uniformly labeled ^{14}C-amino acids, ^{14}C-nucleotides, ^{14}C-lipids, ^{35}S-amino acids, etc.

Even higher organisms can be used for the production of labeled compounds. Plants, when grown in an exclusive atmosphere of radioactive carbon dioxide, [$^{14}CO_2$], utilize the labeled precursor as the sole source of carbon for photosynthesis. After a suitable period of growth, almost every carbon atom in the plant is radioactive. Thus, plants can serve as an available source of ^{14}C-labeled carbohydrates.

Purification

The small synthetic scale used for production of many labeled compounds creates special challenges for product purification. Numerous separation techniques are used in purifying labeled compounds.

Decomposition

Decay products of the principal radionuclides used in tracer technology are not themselves radioactive. Therefore, the primary decomposition events of isotopes in molecules labeled with only one radionuclide/molecule result in unlabeled impurities at a rate proportional to the half-life of the isotope. For ^{14}C and ^{3}H, impurities arising from the decay process are in relatively small amounts. For the shorter half-life isotopes the relative amounts of these impurities caused by primary decomposition are larger, but usually not problematic because they are not radioactive and do not interfere with the application of the tracer compounds.

More problematic is the decomposition of the labeled compounds caused by radiolysis during storage. This phenomenon is the result of the dissipation of the energy released in the surrounding media by decay of a radionuclide forming reactive species such as free radicals, which can cause chemical bond cleavage in other labeled molecules in the surrounding microenvironment. This mechanism of decomposition is usually referred to as secondary decomposition.

Detection and Quantitation

The methods for detection and quantitation of radiolabeled tracers are determined by the type of emission, the energy of the emission, and the efficiency of the system by which it is measured. Detection of radioactivity can be achieved in all cases using the Geiger counter. However, in the case of the radionuclides that emit low energy beta radiation such as ^{3}H, large amounts of isotopes are required for detection and accurate quantitation of a signal. This is in most cases undesirable and impractical. Thus, more sensitive and reproducible methods of detection and quantitation have been developed.

Liquid scintillation counting is by far the most common method of detection and quantitation of β-emission. This technique involves the conversion of the emitted β-radiation into light by a solution of a mixture of fluorescent materials. The sensitive detection of this light is affected by a pair of matched photomultiplier tubes (see PHOTODETECTORS) in the dark chamber. This signal is amplified, measured, and recorded by the liquid scintillation counter.

The nonquantitative detection of radioactive emission often is required for special experimental conditions. Autoradiography, which is the exposure of photographic film to radioactive emissions, is a commonly used technique for locating radiotracers on thin-layer chromatographs, electrophoresis gels, tissue mounted on slides, whole-body animal slices, and specialized membranes.

Gas-flow counting is a method for detecting and quantitating radioisotopes on paper chromatography strips and thin-layer plates.

Other methods of sensitive detection of radiotracers have been developed. For example, Fourier transform nmr can be used to detect ^{3}H (nuclear spin 1/2). Field-desorption mass spectrometry (fdms) and other mass spectral techniques can be applied to detection of nanogram quantities of radiolabeled tracers.

Health and Safety Factors

All isotopes present toxicity problems if taken into the body. Personal safety precautions are related to the relative quantities of radioactive materials handled. Basic laboratory procedures to be followed protect the user from oral ingestion, skin contact, self-injection, and inhalation.

Protection of the environment from uncontrolled radioactive release is also a consideration in the use of radiotracers. In the United States, the quantity and concentration of radionuclides that may be discharged into sewer systems is limited by regulations of the Nuclear Regulatory Commission (NRC). Similarly, airborne emission limits have been established by the NRC for nonrestricted areas.

Uses

The detectability of minute quantities of radiolabeled tracers makes possible the determination of microquantities of substances. The most effective use of radiotracers has been in biomedical research.

Labeled drugs and ligands using ^{3}H, ^{14}C, and ^{125}I, are also widely used for screening for potential new drugs. In this procedure the labeled drug is incubated with a receptor preparation *in vitro* and the radioactivity bound to the receptor is quantitated using liquid scintillation counting. By including an unlabeled drug candidate compound in the mixture and measuring the extent of displacement or inhibition of the binding of the radioactive tracer, the potential pharmacologic activity of the candidate compound can be assessed.

Radiotracers have also been used extensively for the quantitative microdetermination of blood serum levels of hormones, proteins, neurotransmitters, and other physiologically important compounds. Radioimmunoassay, which involves the competition of a known quantity of radiolabeled tracer, usually ^{125}I or ^{3}H, with the unknown quantity of serum component for binding to a specific antibody that has been raised against the component to be determined, is used in the microdetermination of physiologically active materials in biological samples (see IMMUNOASSAY).

Radioactive tracers also are used in agriculture. A test field containing a food crop is sprayed with either an organic fertilizer, pesticide, or fungicide that is laced with the appropriate radioactive tracer. Run off, leaching, or contamination of the water table can then be determined by measuring radioactivity in local ponds or rivers. (see FERTILIZERS; FUNGICIDES, AGRICULTURAL; PESTICIDES; SOIL CHEMISTRY OF PESTICIDES).

In the petroleum industry, the size of an underground oil deposit is determined by the injection of radiolabeled substances into a well head.

LASZLO BERES
DuPont—NEN Products

C. Filer, S. Hurt, and Y. P. Wan, in M. Williams, R. A. Glennon, and P. B. M. W. M. Timmermans, eds., *Receptor Pharmacology and Function*, Marcel Dekker, Inc., New York, 1989, pp. 105–135.

Y. Kobayashi and D. V. Maudsley, *Biological Applications of Liquid Scintillation Counting*, Academic Press, Inc., New York, 1974.

J. Shapiro, *Radiation Protection*, 3rd ed., Harvard University Press, Cambridge, Mass., 1990.

G. H. Keller and M. M. Manak, *DNA Probes*, 2nd ed., Stockton Press, New York, 1993.

RADIOCHEMICAL TECHNOLOGY. See RADIATION CURING.

RADIOGRAPHY. See X-RAY TECHNOLOGY.

RADIOIMMUNOASSAY. See IMMUNOASSAY; MEDICAL DIAGNOSTIC REAGENTS; RADIOACTIVE TRACERS.

RADIOISOTOPES

A radioisotope is an atom the nucleus of which is not stable and decays to a more stable state by the emission of various radiations. Radioactive isotopes, also called nuclides or radionuclides, are important to many areas of scientific research, as well as in medical and industrial applications (see RADIOACTIVE TRACERS; RADIOPHARMACEUTICALS).

Knowledge about the radiations from each isotope is important; as the uses of the radioisotopes have increased, it has become necessary to develop sensitive and accurate detection methods designed to determine both the presence of these materials and the amount present.

Although early researchers had reason to believe that only elements above Pb in the Periodic Table were radioactive, the development of far more sensitive measurement methods has shown that natural radioactivity exists throughout the Periodic Table.

Nuclear Physics Properties

The Atom. To include a matrix or wave mechanics theory that has developed into quantum mechanics, in which all of these properties are included. In this theory the state of the electron is described by a wave function from which the electron's properties can be deduced.

The theory of quantum mechanics requires that nuclear states have discrete energies. This is in contrast to classical mechanical systems, which can have any of a continuous range of energies. This difference is a critical fact in the applications of radioactivity measurements, where the specific energies of radiations are generally used to identify the origin of the radiation. Quantum mechanics also shows that other quantities have only specific discrete values, and the whole understanding of atomic and nuclear systems depends on these discrete quantities.

Half-Lives and Decay Constants. Each nuclear state, whether an unstable ground state or an excited level, has a characteristic probability of decay per unit time, λ, which is known as the decay constant.

Different isotopes have different λ values.

Each nucleus of a given radioactive species has the same probability for decay per unit time.

Experimental measurements have shown that the following description of the decay is correct. If at any time, t, there are a large number of nuclei, $N(t)$, in the same state which has the decay constant, λ, then the change in the number of nuclei in this state in a short time, dt, is

$$dN = \lambda N(t)dt$$

If N_0 is the number of nuclei in this state at time $t = 0$, the number of nuclei in this state at any later time, t, is

$$N(t) = N_0 e^{-\lambda t}$$

and the associated activity, or the decay rate, is

$$A(t) = |dN/dt| = \lambda N(t) = \lambda N_0 e^{\lambda t}$$

In a description of nuclear properties, the half-life, $t_{1/2}$, is quoted rather than the decay constant. This quantity is the time it takes for one-half of the original nuclei to decay. That is,

$$N(t_{1/2}) = N_0/2 = N_0 e^{-\lambda t_{1/2}}$$

In almost all cases λ is unaffected by any changes in the physical and chemical conditions of the radionuclide, so in all applied uses the decay constants and half-lives can be considered to be independent of the physical and chemical environment.

Atomic Levels and Their Decay. There are many commonalities between the properties of atomic and nuclear levels and between their respective decays. Each level has a quantum mechanical wave function which describes its properties.

The electrons in each atomic level have specific spins or angular momenta and parities. These properties define the characteristics of the level and the x-ray spectrum produced.

Modes of Nuclear Decay and Radiation

The decay of radioisotopes involves both the decay modes of the nucleus and the associated radiations that are emitted from the nucleus. In addition, the resulting excitation of the atomic electrons, the deexcitation of the atom, and the radiations associated with these processes all play a role.

There are four modes of radioactive decay that are common and that are exhibited by the decay of naturally occurring radionuclides. These four are α-decay, β^--decay, electron capture and β^+-decay, and isomeric or γ-decay. In the first three of these, the atom is changed from one chemical element to another; in the fourth, the atom is unchanged. In addition, there are three modes of decay that occur almost exclusively in synthetic radionuclides. These are spontaneous fission, delayed-proton emission, and delayed-neutron emission. Lastly, there are two exotic, and very long-lived, decay modes. These are cluster emission and double β-decay. In all of these processes, the energy, spin and parity, nucleon number, and lepton number are conserved.

α-Decay. In α-decay the parent atom of atomic number Z and mass A emits an α-particle, a ^{4}He nucleus having $Z = 2$ and $A = 4$, and becomes an atom having atomic number $Z - 2$ and mass $A - 4$.

In this decay process, only one particle is emitted and, because energy is conserved, for each level in the daughter nucleus there is a unique α-particle energy.

β^--Decay. In this decay, a β^--particle is emitted from the nucleus and the parent atom with Z and A is transmuted into the daughter atom of $Z + 1$ and A. The emitted β^--particle is an ordinary electron having the properties listed in Table 1. This decay mode is equivalent to converting a neutron into a proton. The complete process therefore is $n \rightarrow p + e^- + \bar{v}$.

The decay energy is shared between the electron and the antineutrino and each particle has an energy distribution that extends from 0 to the total energy available.

Electron Capture and β^+-Decay. These processes are essentially the inverse of the β^--decay in that the parent atom of Z and A transmutes into one of $Z - 1$ and A. This mode of decay can occur by the capture of an atomic electron by the nucleus, thereby converting a proton into a neutron. The loss of one lepton (the electron) requires the creation of another lepton (a neutrino) that carries off the excess energy, namely $Q - E_1 - E(e^-)$, where the last term is the energy by which the elec-

Table 1. Properties of Stable Particles Associated With Radioactive Decay

Name	Symbol	Energy, MeV	Charge	Spin
proton	p	938	+1	1/2
neutron[a]	n	940	0	1/2
electron	e^- or β^-	0.511	−1	1/2
positron	e^+ or β^+	0.511	+1	1/2
electron neutrino	v	ca 0.0	0	1/2
electron antineutrino	$\bar{v}$	ca 0.0	0	1/2
photon	γ or x	0.0	0	1

[a] A free neutron has a half-life of 10.4 minutes, but a neutron is stable when bound in a nucleus.

tron was bound to the atom before it was captured. So the process is equivalent to $p + e^- \rightarrow n + v$. The experimental signature for this process is the emission of x-rays from the atom as the resulting hole in an atomic shell is filled.

In this process only one particle is emitted, so the energy spectrum of the neutrinos consists of discrete lines.

γ-Decay. In the γ-decay mode, a nucleus in an excited state decays to a lower energy state via the emission of a γ-ray, and the Z and the A are unchanged. Although to first order the γ-ray carries off all of the available energy, a small amount is transferred to the atom, which recoils. An isomeric transition or isomeric decay is a γ-ray transition from a nuclear state that has a significantly long half-life, for example longer than 1 μs.

Spontaneous Fission. The spontaneous splitting of a heavy nucleus into two large fragments, usually accompanied by release of a few neutrons, is termed spontaneous fission. This process occurs only for very heavy nuclei where the mass of the original nucleus is significantly larger than that of all the products. This excess mass appears as kinetic energy and in the excitation energy of the products. The large fragments emit γ-rays until the ground states are reached. The resulting fission-product nuclei are generally radioactive also. This decay mode is observed in several U (Z = 92) and Pu (Z = 94) isotopes and many of the more neutron-rich isotopes of the elements having higher atomic number. Except for U, all of these elements are synthetic.

Delayed Proton and Neutron Decays. By means of a variety of nuclear reactions, as well as the spontaneous fission of synthetic nuclides, large numbers of isotopes of some elements have been produced.

Although protons and neutrons are not emitted from the ground states of these isotopes, there are many cases where particles are emitted from excited states.

Exotic Decays. In addition to the common modes of nuclear decay, two exotic modes have been observed. These decay modes are of theoretical interest because their long half-lives place strict constraints on the details of any theory used to calculate them.

Cluster emission is an exotic decay that has some commonalities with α-decay. In α-decay, two protons and two neutrons that are moving in separate orbits within the nucleus come together and leak out of the nucleus as a single particle. Cluster emission occurs when other groups of nucleons form a single particle and leak out; the emitted clusters include ^{14}C, ^{20}O, ^{24}Ne, ^{28}Mg, and ^{32}Si.

Double β-decay is the other class of exotic decay. The members of this class that are of interest are those in which the parent nuclide is stable against the single β-decay because the corresponding decay energy is negative, or because the single β-decay is very highly hindered by spin-selection rules. This decay occurs by the emission of two beta particles.

Combined Nuclear and Atomic Processes

There are two processes where nuclear and atomic contributions are interrelated. These are the emission of electrons from the atomic shells as an alternative to the emission of a photon and the emission of bremsstrahlung photons in the β-decay process.

Internal Conversion. As an alternative to the emission of a γ-ray, the available energy of the excited nuclear state can be transferred to an atomic electron and this electron can then be ejected from the atom. The kinetic energy of this electron is $E_\gamma - E_b$ where E_b is the energy by which the electron was bound to the nucleus. Because the atomic electrons exist in a series of discrete levels for each γ-ray, there is a series of discrete internal-conversion lines that can be observed in an appropriate electron spectrometer.

Internal Bremsstrahlung. Another type of radiation that originates in the atomic electron cloud as a direct consequence of a nuclear process is a continuous photon spectrum known as bremsstrahlung. This radiation is caused by the sudden change in the electric field on each electron around the atom resulting from the change in the nuclear charge associated with α, β^-, or electron-capture decay. This sudden change produces a continuous bremsstrahlung spectrum of photons that range in energy from 0 up to the decay energy.

Atomic Decays and Radiations

Of the modes of nuclear decay discussed, two produce excitation of the atomic electron system. In electron capture an electron goes into the nucleus, leaving a hole in the electron shell. In internal conversion, energy is transferred to an electron, which is thereby ejected from the atom. The electron system promptly returns to its ground state by a series of processes that radiate away the excess energy. This energy can be evidenced in a series of electromagnetic photons, called x-rays, or by transfer of energy to additional electrons, which are then ejected.

X-Rays. If an x-ray is emitted, it has an energy equal to the difference in the binding energies of the two atomic shells. Because the hole can be filled by an electron from any of the several outer shells, x-ray spectra contain a large number of discrete lines.

Auger Electrons. The fraction of the holes in an atomic shell that do not result in the emission of an x-ray produce Auger electrons. In this process a hole in an inner shell is filled by an electron from the next shell, and the available energy is transferred to the outer shell electron, which in turn is ejected from the atom with a kinetic energy that represents the differences in the binding energies of these shells.

Because the energies of the atomic electrons are discrete, the Auger electrons appear as discrete lines, although there are many lines which may be difficult to distinguish.

Secondary Radiations

The previously discussed radiations have their origin in the atom in which the original decay took place. If the radiation reaching a detector is measured, there are other radiations that are observed.

External Bremsstrahlung. When a charged particle is decelerated or accelerated, it produces a continuous photon spectrum called external bremsstrahlung. When α-, β^-, or β^+-particles are emitted, they are scattered in the surrounding material, and in the process they produce bremsstrahlung radiation. This spectrum is quite similar to the internal bremsstrahlung spectrum from β- or electron-capture decay. It is continuous, ranging from 0 to the decay energy.

Compton Scattered Photons. When a γ-ray interacts in material, there are several processes that may occur. One of these is Compton scattering, in which part of the γ-ray energy is transferred to an atomic electron, and the part remaining is carried off by a secondary photon. Compton-scattered photons are generated in the surrounding material and observed in any γ-ray detector.

X-Rays and Annihilation Radiation. The interaction of γ-rays with matter produces the x-rays that are characteristic of the atoms in the material in which the interactions take place. Such x-rays appear in measured spectra.

The β^+-particles that are emitted in the β^+-decay mode are slowed down in the material around the source. When these reach very low velocities they interact with an ordinary electron and the pair is annihilated. The corresponding energy, is normally released in the form of two photons, emitted in opposite directions.

Background Radiation. If the radiation from a radioactive source is measured, the spectrum also includes contributions from the radiations from the surrounding environment. There is also cosmic radiation that comes from space and interacts with the earth and atmosphere to produce radiations that may enter the detector, and thus is observed.

Decay Data

Most areas of research and applications involving the use of radioisotopes require a knowledge of what radiations come from each isotope. The particular application determines what type of information is needed. If the quantity of a radionuclide in a particular sample or at a particular location is to be determined and this value is to be determined from the γ-ray spectrum, the half-life of the nuclide and the energies and intensities or emission probabilities of the γ-rays of interest must be known. Usually it is preferable to use the γ-rays for an assay measurement because the α- and β-rays are much more readily

Table 2. Radionuclides Useful for Determining the Age of Materials

Parent	Half-life, yr	Stable daughter	Natural abundance of daughter, %	Intermediate nuclides
^{14}C	5.73×10^3	^{14}N	0.0	no
^{235}U	7.1×10^8	^{207}Pb	20.8	yes
^{238}U	4.5×10^9	^{206}Pb	26.3	yes
^{40}K	1.274×10^{10}	^{40}Ar	99.6	no
^{232}Th	1.39×10^{10}	^{208}Pb	51.6	yes
^{87}Rb	5.0×10^{10}	^{87}Sr	7.0	no

absorbed by the source material, and may not reach the sample surface having their original energies. Once these energies are altered they cannot be used to identify the parent radionuclide.

Applications

Dating. There are two methods of using radioactive decays to determine the age of an object. A method applicable to formerly living organisms is based on the assumption that the specific activity, eg, disintegrations per gram of sample of a particular isotope, is known for the time the organism was alive. The time since death is then determined by comparing this to the measured specific activity. The other method, applicable to nonliving objects, determines age from the amounts of both the radioactive parent and the usually stable daughter isotope present in a sample. In this latter case all of the daughter atoms must be assumed to come from the parent, or an independent method of determining what fraction of the daughter atoms are from the decay of the parent must be available. Table 2 lists half-lives of several radioisotopes that have been used for dating.

The radioisotope ^{14}C is used to determine the lapse of time since a living plant or animal died. ^{14}C, produced in the atmosphere by the $^{14}N + n \rightarrow {}^{14}C + p$ reaction, becomes distributed in the air and ground along with the stable isotopes of carbon, ^{12}C (98.9%) and ^{13}C (1.1%). The ^{14}C is then taken up by living bodies in food along with the stable isotopes of carbon. When the organism dies, however, it ceases taking up ^{14}C, so the amount present decreases with the characteristic half-life. In the dating of old trees, it can be assumed that the specific activity of ^{14}C is the same in trees, or at least in the specific types of trees, in the 1990s as it was a few thousand years ago.

The other radioisotopes in Table 2 have much longer half-lives and can be used to determine the ages of materials that are on the order of a million years old. These are especially useful in dating rocks.

In these cases what is usually measured is not the time of the original formation of the rocks, but the time at which the parent and daughter elements were last separated. That is, if the rocks were remelted at some point in their history in a manner that removed the daughter elements, this would be the age measured.

Medical Uses. There are many radioisotopes that are used for medical diagnosis (see RADIOACTIVE TRACERS; RADIOPHARMACEUTICALS). An example is the monitoring of blood flow to various regions of the heart using ^{201}Tl, which has a half-life of 3.0 d.

The radioisotopes ^{99m}Tc and ^{131}I are often used for medical purposes. ^{99m}Tc has a half-life of only 6 h, which would normally make it difficult to transport from a production facility to the medical facility. However, one can supply the longer-lived 2.7-d ^{99}Mo in a chemical form that allows one to separate out, generate or milk, the daughter ^{99m}Tc when the latter is needed.

Another medical use of radioisotopes, such ^{60}Co, is to irradiate certain tissues within the body. An intense source of ^{60}Co in a heavily shielded facility provides a highly collimated beam of γ-rays that impinge on a tumor in order to kill its cells.

RICHARD G. HELMER
Idaho National Engineering Laboratory

E. Browne and R. B. Firestone, in V. S. Shirley, ed., *Table of Radioactive Isotopes*, John Wiley & Sons, Inc., New York, 1986.

RADIOPAQUES

Medical examination of soft tissues or organs by nonsurgical means often requires the introduction of a special agent which makes the detection system responsive to detail in the tissue of interest. Diagnostic imaging agents include those used in magnetic resonance, ultrasound, radionuclide imaging, and x-ray technology (qv) (see MEDICAL IMAGING TECHNOLOGY). Radiopaques for x-ray imaging, more commonly called x-ray contrast media or radiographic contrast agents, are examples of such diagnostic agents. These chemicals absorb x-rays strongly. When they accumulate in the target area, they create a contrast in the x-ray image thereby permitting visual examination of the target organ.

Absorption of x-radiation is an atomic phenomenon related to the atomic number of the absorbing atom; thus the heavier elements are, in general, more efficient at absorbing x-rays. Except for barium sulfate, all other radiopaque agents presently in use are organic derivatives of iodine. The iodine atoms function as the x-ray absorbers, and the organic moiety can be manipulated to provide desirable characteristics of a contrast agent and to decrease toxicity and physiological side-effects. Table 1 contains a summary of the more important radiographic procedures and contrast agents for x-ray visualization of various tissues and organs.

Table 1. Radiographic Procedures and Corresponding Radiopaques

Procedure	Organ/region	Radiopaque agents
angiography	blood vessels	
arteriography	arteries	sodium or meglumine salt of diatrizoic acid, iothalamic acid, metrizoic acid, and ioxaglic acid; iopamidol, iohexol, ioversol, iopromide, iomeprol, iopentol, ioxilan, iobitridol
aortography	aorta	
ventriculography	ventricles of the heart	
venography (phlebography)	veins	
urography	urinary tract	
computed tomography	body, head	
myelography	subarachnoid space of the spinal cord	meglumine salt of iothalamic acid and iocarmic acid, metrizamide, iohexol, iopamidol, iotrol, iodixanol
cholecystography	gallbladder	iopanoic acid, iocetamic acid, sodium or calcium iopodate, sodium tyropanoate, meglumine iodipamide, ioglycamide, iodoxamate
cholangiography	bile ducts	
gastrointestinal radiography	alimentary tract	barium sulfate, sodium or meglumine diatrizoate, iohexol
arthrography	joints	meglumine diatrizoate, meglumine iothalamate, sodium and meglumine salt of ioxaglic acid, iohexol
hysterosalpingography	uterus and fallopian tubes	ethiodol, meglumine diatrizoate-meglumine iodipamide mixture, sodium and meglumine salts of iothalamic, diatrizoic and ioxaglic acids, iohexol

Angiography and Urography

Angiographic contrast media (CM) are administered intravascularly for the radiographic visualization of blood vessels to evaluate vascular abnormalities. Because of the very high concentrations of contrast media required for angiography, such materials must have high water solubility. More dilute formulations of these same agents are used for urography. In urographic procedures, the CM are injected intravenously and their excretion via the kidneys is visualized radiographically as an evaluation of renal function (excretory urography). Alternatively, the CM are instilled via catheters directly into the lower urinary tract (retrograde pyelography). Following intravascular administration, the angiographic–urographic CM are distributed in the extracellular space and subsequently excreted unchanged, principally in the urine.

High Osmolality Contrast Media. An important advance in radiopaques came with the synthesis of aminotriiodobenzoic acid and its acetylated derivative, acetrizoic acid (**1**). Aqueous solutions of sodium acetrizoate possessed the thermal stability so that they could be autoclaved with minimal decomposition. The higher iodine content, ie, 3 atoms/molecule, increased the contrast efficiency, and the clinical safety of acetrizoate was improved over that of the earlier urographic agents.

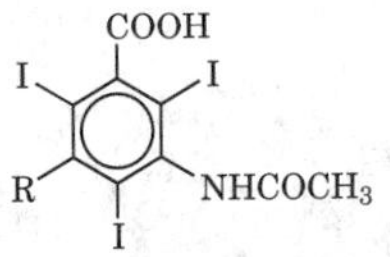

(**1**) R = H
(**2**) R = $NHCOCH_3$
(**3**) R = $CONHCH_3$
(**4**) R = $N(CH_3)COCH_3$

Further improvements in the late 1950s and early 1960s led to the development of three derivatives of acetrizoate that comprise a group of important angiographic and urographic ionic contrast media: diatrizoic acid (**2**), iothalamic acid (**3**), and metrizoic acid (**4**). These compounds, in which the hydrogen on the ring is replaced by more hydrophilic moieties, were found to be less toxic than acetrizoate.

Osmolality, a measure of the number of particles in a solution, is approximately proportional to the sum of the concentrations of all molecular and ionic particles present. These ionic compounds contribute to the overall osmolality of the diagnostic solution in a ratio of three iodine atoms delivered as two particles, the cation and the anion. In the range of concentrations required for good x-ray visualization, the high osmolality of these ionic agents relative to plasma and surrounding tissues causes leaching of water across semipermeable membranes, resulting in undesirable physiological effects. This class of agents is known as high osmolality contrast media (HOCM).

Low Osmolality Contrast Media. An ideal intravascular CM possesses several properties: high opacity to x-rays, high water solubility, chemical stability, low viscosity, low osmolality, and high biological safety. The newer nonionic and low osmolar agents represent an advanced class of compounds in the development of x-ray contrast media.

Development of nonionic compounds to eliminate ionicity, reduce osmolality, and hence minimize adverse effects, such as painful reactions associated with the injection of ionic agents, was proposed in 1968. These triiodobenzene derivatives contain no ionizable carboxyl moiety, and their high water solubility and biological safety are achieved by employment of highly polar, hydrophylic groups. Because these compounds do not dissociate in solution, approximately half the osmolality of the ionic agents results at equivalent concentrations. These nonionic agents are referred to as low osmolality contrast media (LOCM) or ratio-3 CM, ie, three iodine atoms per particle. The first clinically successful nonionic agent metrizamide (**5**) was introduced in 1975. Intensive research in the following decades produced safer and more stable agents, eg, iopamidol (**6**) iohexol (**7**), and ioversol (**8**). These three nonionic LOCM have become the primary products utilized for angiographic–urographic procedures.

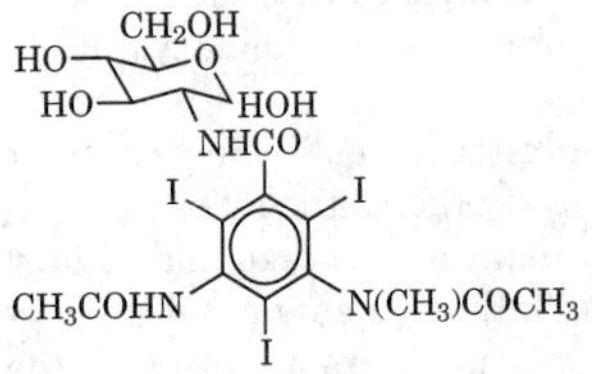

(**5**)

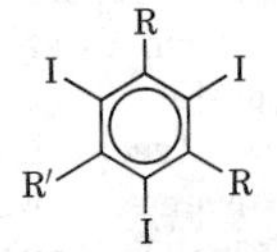

(**6**) R = $CONHCH(CH_2OH)_2$
R′ = $NHCOCH(OH)CH_3$ (L-form)
(**7**) R = $CONHCH_2CH(OH)CH_2OH$
R′ = $N(COCH_3)CH_2CH(OH)CH_2OH$
(**8**) R = $CONHCH_2CH(OH)CH_2OH$
R′ = $N(COCH_2OH)CH_2CH_2OH$

Contrast materials of low osmolality can be classified into three chemical types: (*1*) nonionic monomers, (iopamidol, iohexol, ioversol, iopromide (**9**), iomeprol (**10**) iopentol (**11**), ioxilan (**12**) and iobitridol) (**13**) (*2*) monoionic dimers, (ioxaglic acid) (**14**) and (*3*) nonionic dimers. (iotrolan and iodixanol).

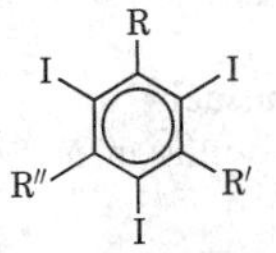

(**9**) R = $CON(CH_3)CH_2CH(OH)CH_2OH$
R′ = $CONHCH_2CH(OH)CH_2OH$
R″ = $NHCOCH_2OCH_3$

(**10**) R = R′ = $CONHCH_2CH(OH)CH_2OH$
R″ = $N(CH_3)COCH_2OH$

(**11**) R = R′ = $CONHCH_2CH(OH)CH_2OH$
R″ = $N(COCH_3)CH_2CH(OH)CH_2OCH_3$

(**12**) R = $CONHCH_2CH(OH)CH_2OH$
R′ = $CONHCH_2CH_2OH$
R″ = $N(COCH_3)CH_2CH(OH)CH_2OH$

(**13**) R = R′ = $CON(CH_3)CH_2CH(OH)CH_2OH$
R″ = $NHCOCH(CH_2OH)_2$

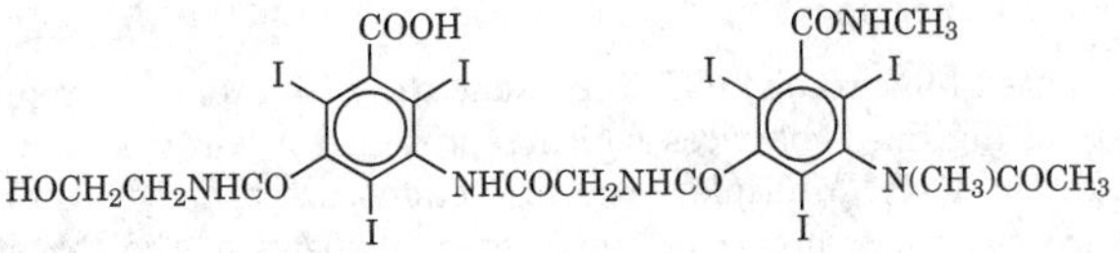

(**14**)

Myelography

The administration of a contrast agent into the subarachnoid space permits delineation of the spinal cord and is used for diagnosis of diseases of the nervous system and spinal canal. CM agents developed over the years were accompanied by significant adverse effects. The second-generation water-soluble nonionic CM, iopamidol (**6**) and iohexol (**7**) the approved and most widely used myelographic CM at the present time.

Cholecystography and Cholangiography

Radiographic studies of the gallbladder and bile duct with radiopaques are called cholecystography and cholangiography, respectively. Cholecystographic agents are administered orally for the evaluation of gallbladder abnormalities, such as gallstones. Cholangiographic agents are administered intravenously to produce opacification of the cystic and common bile ducts. Because of different biochemical transformations, oral cholecystographic agents and intravenous cholangiographic agents have distinct chemical requirements and pharmacokinetic properties.

Oral Agents. The orally administered media are absorbed through the gastrointestinal tract and, after entering the portal venous circulation, are taken up by the liver and excreted through bile into the gallbladder. Suitable oral cholecystographic CM are monomers of amino-triiodobenzene derivatives having alkanoic acid substituents.

These agents each contain an aliphatic carboxylic acid group capable of ionizing to form a water-soluble salt for increased solubility and intestinal absorption.

Intravenous Agents. Intravenous administration of biliary contrast agents circumvents the relatively slow absorption of the intestinal system and allows for rapid and efficient heptocyte uptake and biliary excretion. Structurally, the intravenous biliary agents are dimers of triiodobenzene derivatives and differ only in the composition of the methylenic linkage.

Although still valuable for selected studies, cholecystography and cholangiography have been largely replaced by other diagnostic modalities. Methologies include ultrasound, computed tomography, magnetic resonance, and radionuclide techniques (see MAGNETIC SPIN RESONANCE; MEDICAL IMAGING TECHNOLOGY; RADIOACTIVE TRACERS).

Gastrointestinal Radiography

In the early development of radiopaques, barium sulfate was introduced for use in imaging the gastrointestinal (GI) tract. This compound has remained the agent of choice for gastrointestinal radiography. Barium sulfate forms a colloidal suspension and is administered orally when the regions of interest reside in the upper GI tract, and rectally when the lower GI tract is the focus. Being chemically inert and practically insoluble in water, barium sulfate demonstrates negligible absorption from the digestive system and produces few physiological side effects. It is excreted in the feces unchanged.

Two types of imaging techniques are routinely used. Single-contrast imaging is performed using a large volume of low density barium sulfate preparation to fill the entire lumen of the GI segment, to produce full-column opacification. Double-contrast imaging utilizes a smaller amount of a high density, low viscosity barium preparation to coat the mucosal surface. Air or carbon dioxide is then administered to distend the region and provide a negative contrast. In this way, surface detail of the GI tract is finely delineated.

Computed Tomography

In computed tomography (CT) the usual x-ray film image is replaced by sets of digitized matrices which represent the x-ray attenuation through the body. Multiple x-ray projections are utilized. After the data are computer-analyzed, cross-sectional views of the target organ(s) can be generated. The advantage of CT over the more conventional x-ray imaging technique is the greater contrast sensitivity to attenuation changes. However, because film is a continuous medium whereas the CT images are derived from digital picture elements (pixels), resolution of very small structures generated from a finite number of pixels can be limited using CT, as compared to conventional film-screen radiography.

The CT procedure can be performed with or without the use of intravenous contrast media. Contrast-enhanced CT involves the administration of a radiopaque to increase the degree of contrast between anatomical structures and to improve the differentiation between pathological and physiological phenomena. In general, because of the increased sensitivity of CT compared to film methods, lower concentrations of CM are indicated. The same water-soluble CM used in angiography and urography are successfully utilized to enhance contrast in CT (see Table 1).

Arthrography

The radiological visualization of joint cavities using contrast media is termed arthrography. Single-contrast arthrographic techniques utilize direct injection of a water-soluble contrast agent that readily mixes with the synovial fluid, producing opacification of the joint surfaces and cavity. The CM is then rapidly absorbed and excreted in the urine. Double-contrast arthrography involves the removal of the joint fluid and injection of a water-soluble positive contrast agent to coat the surfaces of the joint, followed by the introduction of air or carbon dioxide as a negative contrast medium to fill the cavity. Double-contrast techniques can result in a finer delineation of surface contours than the single-contrast method, especially when combined with computed tomography.

Hysterosalpingography

Hysterosalpingography describes the radiological examination of the uterus and fallopian tubes for the purpose of detecting structural abnormalities and for the evaluation of fallopian tube patency. The CM for intrauterine administration include the oily agent Ethiodol, which consists of a mixture of ethyl esters of the iodinated fatty acids of poppy seed oil. The two main iodinated components of Ethiodol are diiodoethylstearate and monoiodoethylstearate.

YOULIN LIN
Mallinckrodt Medical, Inc.

1. Ackerman, *Diagnostic Agents*, in A. Burger, ed., *Medicinal Chemistry*, 3rd ed., Part II, John Wiley & Sons, Inc., New York, 1970, Chapt. 67, pp. 1686–1699.
2. Z. Parvez, ed., *Contrast Media: Biologic Effects and Clinical Applications*, 3 vols., CRC Press, Boca Raton, Fla., 1987.
3. M. Sovak, ed., *Radiocontrast Agents*, Springer-Verlag, New York, 1984.
4. D. P. Swanson, H. M. Chilton, and J. H. Thrall, eds., *Pharmaceuticals in Medical Imaging*, Macmillan Co., New York, 1990.

RADIOPHARMACEUTICALS

Radiopharmaceuticals form the chemical basis of the medical specialty of nuclear medicine, a group of techniques used primarily for diagnosis, but also to a lesser degree in the treatment of disease. *In vivo* diagnostic information is obtained by intravenous injection of compounds tagged with rapidly decaying radioactive isotopes. The biological distribution of these compounds is then determined using a gamma-camera. This distribution usually takes a form that is organ- and lesion-specific, although this is not always the case. From the distribution of radioactivity and its behavior over time, it is possible to obtain information about the presence, progression, and state of disease. Localized changes in the shape or concentration of radioactivity in a given organ or structure reflect alterations in either organ anatomy or local function (see RADIOACTIVE TRACERS; RADIOISOTOPES).

Unlike other medical imaging modalities such as x-rays and x-ray computed tomography (see X-RAY TECHNOLOGY), ultrasound, or magnetic resonance imaging (mri) (see MAGNETIC SPIN RESONANCE), nuclear medicine studies provide information about the functional state of tissues rather than primarily anatomy or shape of regions of differing fixed properties (see MEDICAL IMAGING TECHNOLOGY).

One of the great advantages of techniques using radiopharmaceuticals is extreme sensitivity. Compared to other *in vivo* diagnostic methodologies, nuclear medicine techniques can provide critical information upon introduction of many orders of magnitude less compound into the body.

Efficacy

In common with other pharmaceuticals, the two primary requirements for success are safety and efficacy. Safety is determined both by chemical toxicity and radiation dose delivered to the patient.

Apart from the physical factors related to the radioisotope used, the only general characteristic that is important in defining the efficacy of these materials is the macroscopic distribution in the body, or biodistribution. This time-dependent distribution at the organ level is a function of many parameters which may be divided into four categories: factors related to delivery of the radiopharmaceutical to a particular tissue; factors related to the extraction of the compound from circulation; factors related to retention of the compound by that tissue; and factors determined by clearance. The factors in the last set are rarely independent of the others.

Nuclear Medicine Studies

Nuclear medicine studies may reveal information that is primarily anatomic in nature, or indicate the function of an organ on a regional basis (Table 1). These studies may be intended to identify new disease, confirm or deny suspected disease, or follow the progress of treatment or the course of disease.

Table 1. Medicine Diagnostic Studies and Associated Radiopharmaceuticals

Study	Disease target(s)	Organ	Radiopharmaceutical
myocardial perfusion	coronary artery disease, myocardial infarction	heart	^{201}Tl–Cl
			^{82}Rb–Cl
			^{99m}Tc-sestamibi
			^{99m}Tc-teboroxime
			^{99m}Tc-tetrafosmin
bone	bone metastases, osteomyelitis, stress fractures	skeleton	^{99m}Tc-medronate
			^{99m}Tc-oxidronate
gallium study	primary tumors and metastases, infection, inflammation	any	^{67}Ga-citrate
white cell study	infection, inflammation	any	^{111}In-WBC[a]
			^{99m}Tc-WBC[b]
renal study	impaired renal function	kidney	^{111}In-DTPA
			^{131}I-iodohippurate
			^{99m}Tc-DTPA
			^{99m}Tc-gluceptate
			^{99m}Tc-mertiatide
brain perfusion	stroke, other perfusion abnormalities	brain	^{123}I-iofetamine
			^{99m}Tc-exametazime
			^{99m}Tc-bicisate
hepatobiliary study	acute cholecystitis	gall bladder	^{99m}Tc-disofenin
			^{99m}Tc-mebrofenin
lung perfusion–ventilation (V/Q)	pulmonary embolism	lung	^{99m}Tc-albumin aggregated (perfusion) and
			^{133}Xe (ventilation)
tumor localization	certain neuroendocrine tumors	any	^{111}In-pentetreotide
tumor localization	nonliver metastases of colorectal or ovarian cancer	any	^{111}In-satumomab pendetide
liver–spleen	metastatic disease of the liver	liver, spleen	^{99m}Tc-albumin colloid
tumor localization	pheochromocytoma neuroblastoma	any	^{131}I-iobenguane sulfate
thyroid study	thyroid carcinoma, hyperthyroidism	thyroid	^{131}I-NaI, ^{123}I-NaI
cardiac ejection fraction/wall motion	coronary artery disease, myocardial infarction, cardiomyopathy, other diseases affecting muscle function	heart	^{99m}Tc-red blood cells[c]
			^{99m}Tc-sestamibi

[a] ^{111}In-Oxine.
[b] ^{99m}Tc-Exametazime.
[c] Using 99mTc$-$TcO$_4^-$ and stannous pyrophosphate.

Isotopes for Nuclear Medicine

Radioactive isotopes are characterized by a number of parameters in addition to those attributable to chemistry. These are radioactive half-life, mode of decay, and type and quantity of radioactive emissions. The half-life, defined as the time required for one-half of a given quantity of radioactivity to decay, can range from milliseconds to billions of years. Except for the most extreme conditions under very unusual circumstances, half-life is independent of temperature, pressure, and chemical environment.

Diagnostic Radioisotopes. In order to be useful as an *in vivo* diagnostic agent, a radioisotope needs to possess a number of characteristics. (*1*) It must be practical to produce in sufficient quantities so that the cost of a patient dose is within acceptable limits. (*2*) It must be available at a concentration, ie, specific activity, sufficient to allow labeling reactions to proceed and to permit practical volumes of injectate for the required amount of radioactivity. (*3*) Its half-life must be sufficiently short so that the patient receives a radiation dose that is within acceptable limits. The half-life must be long enough, however, so that there is not unacceptable loss of isotope from decay during transportation between the site of production and the patient. It must also be long enough that, for a given chemical form and biological system of interest, there is time for the radioactivity to reach the required distribution in the organs of interest. (*4*) It must yield an acceptable radiation dose in patients for the injected chemical form and amount needed for a particular patient procedure. *In vivo* studies always require consideration of radiation dose. It must be possible to manufacture the radioisotope in a form that is compatible with labeling methodologies and be of sufficient radiophysical, ie, isotopic, and radiochemical purity. (*6*) It must emit radiation that can be detected readily in a way that the required information can be extracted.

Isotopes used in nuclear medicine may be characterized by the source used to produce the radioactive isotope, by whether the isotopes are produced at a central location and shipped or at the clinic, or by the type of emission and thus the equipment used to detect them. Some isotopes may be produced by more than one method.

Isotopes arising from reactors are produced by the addition of neutrons; relatively few diagnostically useful isotopes are produced by reactor methods.

Accelerator-produced radionuclides are formed by the addition of protons. These tend to lie above the line of stability, resulting in a preference for decay by positron, β^+, emission and electron capture. The remainder are produced by generators. The parent isotopes of those produced by generator are often reactor products.

Therapeutic Radioisotopes. Isotopes used for therapy must possess many of the same characteristics as those used for diagnosis (Table 2). Because the purpose is to deliver radiation dose to a tissue, radiation dosimetry considerations are different from those of the diagnostic isotope. However, the criteria given for diagnostic radioisotopes remain true with the exception of number six and with a redefinition of acceptable dose. Acceptable dose for a therapeutic radioisotope means a radiation dose sufficient to cause the required level of tissue damage in the tissue of interest and as little damage as possible in all others.

Labeling Chemistry of Radiopharmaceuticals

There are three general types of radiopharmaceuticals: elemental radionuclides or simple compounds, radionuclide complexes, and radiolabeled biologically active molecules. Among the first type are radionuclides in their elemental form such as ^{81m}Kr and ^{127}Xe or ^{133}Xe, and simple aqueous radionuclide solutions such as ^{125}I or ^{131}I-iodide, ^{201}Tl-thallous chloride, ^{82}Rb-rubidium(I) chloride, ^{89}Sr-strontium(II) chloride, and ^{99m}Tc-pertechnetate. These radiopharmaceuticals are either used as obtained from the manufacturer in a unit dose, ie, one dose for one patient, or dispensed at the hospital from a stock solution that is obtained as needed from a chromatographic generator provided by the manufacturer.

The second type of radiopharmaceuticals are radionuclide complexes. These are almost exclusively metal radionuclides complexed to

Table 2. Isotopes Used for Internal Therapy

Isotope	Therapeutic application
^{131}I	thyroid carcinoma, hyperthyroidism, antibody therapies
^{32}P	polycythemia vera
^{89}Sr	pain palliation in metastatic carcinoma
^{186}Re	pain palliation in metastatic carcinoma
^{153}Sm	pain palliation in metastatic carcinoma
^{165}Dy	synovectomy

a ligand or ligand system. The use of a ligand or ligand system allows the physical and chemical properties of the radionuclide to be modified. These properties include the stability, charge, oxidation state, and lipophilicity.

Biologically active molecules may be labeled either directly or indirectly. For direct labeling, that is, incorporating the radiolabel directly into the compounds, the labeling may be isotopic or nonisotopic. For isotopic labeling, one group already present in the molecule is substituted with or exchanged for the radioisotope. For nonisotopic labeling, that is, incorporating the radiolabel into the compounds through a chelator or other functional group which has been incorporated into the compounds, the radioisotope is added to the molecules without substituting with or exchanging for an already existing group.

Generally, labeled compounds are prepared by procedures which introduce the radionuclide at a late stage of the synthesis. This allows for maximum radiochemical yields, and reduces the handling time of radioactive material.

An important consideration for all radiopharmaceuticals and especially radiolabeled biologically active molecules is specific activity. There are two types of specific activity: radionuclidic and biological. Radionuclidic specific activity refers to the ratio of the number of atoms of a particular radioisotope to the total number of atoms of the element. Because all isotopes of an element are chemically identical, a low specific activity may lead to a low yield in the synthesis of a radiopharmaceutical if a significant proportion of the reagents is consumed by the undesired isotopes.

Biological specific activity refers to the ratio of the number of radiolabeled biologically active molecules to the total number of biologically active molecules. If the biological activity is dependent on the interaction of the molecule with a receptor or binding site, the excess unlabeled molecules may compete with the labeled molecules, limiting the number of detectable molecules that can be localized at the receptor or binding site. Both a high radionuclidic and a biological specific activity is preferred.

Instrumentation for Imaging and Counting

Detection of the emissions from radiopharmaceuticals is dependent on the type of emission. In practice, only those isotopes emitting either γ-ray or x-ray photons, either directly from the decay reaction or indirectly from annihilation of an emitted positron, are useful for *in vivo* imaging, although β-emissions may be of use for *in vitro* purposes. These high energy photons are ionizing radiation and interact with matter via only two mechanisms. For those photons emitted by commonly available radionuclides, the predominant mechanisms are the photo-electric effect and Compton scattering.

Measurement of Injected Patient Dose. The patient's injected dose, as distinguished from the radiation dose, must be within 10% of the prescribed dose for a given procedure by law in the United States. This is measured using a large ionization chamber of cylindrical geometry.

The modern ionization chamber, called a dose calibrator in this application, is capable of linear measurements of radioactivity having a precision in the range of several percent coefficient of variation over a range of 370 kBq (10 μCi) to at least 370 GBq (10 Ci). This extraordinary range is the chief advantage of this instrument. It may only be used when the sample is known to have only a single isotope.

Counting Systems. Counting of small samples of γ-emitting radioisotopes is commonly accomplished in biology and medicine using a scintillation detector. This is a block of crystalline sodium iodide with a small proportion of thallium hermetically sealed in an aluminum enclosure having one glass wall on a cylinder end. The glass panel of the crystal enclosure is coupled to a photomultiplier tube that records the bursts of light produced each time a γ-ray interacts to produce ionization in the crystal.

Samples in tubes are placed in the well of the crystal and the number of light bursts observed over a given time recorded. This value is proportional to the number of disintegrations that occurred during that time and therefore to the amount of radioactivity in the sample. This is made possible by the proportionality between the intensity of each light burst with the energy deposited by the photon that created it. If the photon was completely absorbed in the detector, either in a single event or several successive ones, then the recorded light burst yields information about the γ-ray energy and thereby the isotope that produced it.

Single-Photon Imaging Systems. There are many materials and schemes for detecting photons, but for nuclear medicine one instrument, the γ-camera, is used almost exclusively. This device consists of a disk or rectangle of crystalline sodium iodide with a small proportion of thallium, hermetically sealed in an aluminum enclosure with one glass wall on a large surface. γ-Rays absorbed in the crystal result in the emission of a burst of light photons over a period of several hundred nanoseconds. These are detected by an array of photomultiplier tubes and the centroid of the light distribution, an estimate of the γ-ray interaction point, may be determined by hardware interpolation among this array. For each γ-ray detected, a point on an image plane is incremented in intensity.

In order to produce an image, photons must escape the patient without interaction. Because scattering yields a nearly isotropic distribution of scattered photon directions, any photon undergoing scatter within the patient no longer carries information about its point of origin and must be rejected by an imaging apparatus. The system is able to reject γ-rays having energy lower than expected, an indication of scatter prior to detection.

The collimator, a tall lead honeycomb between the crystal and the patient but very close to the crystal functions as a lens. Rather than refracting the light to form an image on a focal plane, as does a light camera, the collimator simply discards all rays having a direction not perpendicular to (close to) the crystal surface. This results in a photon flux at each point on the crystal that is proportional to the integral of the radioactivity concentration along a line perpendicular to the crystal surface, with appropriate modification for attenuation in the body.

Although performance varies with the isotopes for which they are intended, and with the balance in the design between resolution and efficiency, the overall sensitivity of a γ-camera collimator is on the order of 5000 counts/(MBq·min) (several hundred counts/(μCi·min)). About two photons out of 10,000 emitted arrives at the crystal. This necessitates exposure times that range from several minutes to the better part of an hour.

Positron Imaging. Creating images of distributions of positron emitters requires a somewhat different type of apparatus. Positron cameras use many of the same technologies as do cameras for other isotopes, but there is a broader array of methods and physical arrangements. All of these systems take advantage of the physical characteristics of positrons.

Positrons are emitted having energies that can range from several tens of keV to several MeV. As for all charged particles, they travel a distance that is proportional to their starting energy, leaving a trail of ionization as they slow down. As the positron energy decreases, its probability of undergoing an annihilation reaction with an electron increases. This reaction results in the emission of two 511 keV γ-rays traveling essentially in opposite directions, yielding conservation of both momentum and energy.

Detectors are arranged on opposite sides of the patient and circuitry detects coincidence events in which photons are detected simultaneously within the time resolution of the system. These events mark annihilations that have occurred somewhere along the line connecting the detectors. If there are many detectors at different angles, this data may then be used to reconstruct the three-dimensional distribution of the radioisotope. Because virtually all positron imaging is tomographic, that is, a three-dimensional reconstruction or literally "slice writing," positron imaging is normally called positron emission tomography (PET).

Other Instrumentation. Minimization of radiological hazards to personnel and patients in a nuclear medicine environment requires appropriate instrumentation. The general armamentarium of radiation monitoring includes a survey meter or a small scintillation detector if additional sensitivity is required. Also needed is a system for assay of wipe tests; a defined area of the bench, floor, and countertop is wiped, and the wipes are assayed using a survey meter or specialized system for the detection of removable radioactive contamination. Studies involving absolute thyroid uptake can be performed without imaging using small amounts of ^{131}I or ^{123}I and a simple scintillation probe.

Safety

For virtually all radiopharmaceuticals, the primary safety consideration is that of radiation dosimetry. Chemical toxicity, although it must be considered, generally is a function of the nonradioactive components of the injectate.

Radiation Dosimetry. Radioactive materials cause damage to tissue by the deposition of energy via their radioactive emissions. Thus, when they are internally deposited, all emissions are important. When external, only those emissions that are capable of penetrating the outer layer of skin pose an exposure threat. The biological effects of radiation exposure and dose are generally credited to the formation of free radicals in tissue as a result of the ionization produced.

By definition, radiation dose represents the energy deposited per mass of tissue and has the units of Gray, where $1\text{Gy} = 10\ \mu\text{J/g}$ ($1\text{rad} = 0.01\text{Gy}$). Exposure is a material-independent measure of the radiation incident on a volume as defined by the electrical charge it produces in air. Exposure is measured in charge per mass, ie, C/kg (one Röentgen ($1\text{R} = \text{esu/g}$ dry air at STP) $= 2.58 \times 10^{-4}$ C/kg). Effective dose is a measure of radiation effect that includes biological factors, weighted on an organ-by-organ basis, and factors related to the microscopic distribution of the energy deposition and the effects of dose rate. Its unit is the Sievert (same dimensions as the Gray) or the rem (same dimensions as the rad).

Radiation dose for radiopharmaceuticals in a given patient can only be estimated. There are significant variations in dose from internally deposited radionuclides arising from the shape and location of organs and of the circulation within them. Thus, over the years, computer modeling methods have been developed to obtain estimates of dose, primarily under the auspices of the Medical Internal Radiation Dose (MIRD) Committee of the Society of Nuclear Medicine.

Chemical Toxicity. Radiopharmaceuticals are subject to the same requirements for safety as are other pharmaceuticals, and are tested for chemical toxicity in much the same manner. It is generally understood, however, that patients are likely to receive relatively few doses of any given radiopharmaceutical so that the effects of long-term chronic exposure to the compound rarely need be assessed.

Formulation and Packaging

The diversity of radionuclide half-life and chemical nature of commonly used radiopharmaceuticals demands a variety of formulation matrices, packaging containers, and storage conditions. The containers, ingredients, and used in these products must meet the stringent requirements for processes, parental pharmaceuticals, as well as provide safe conditions for storage, handling, and disposal of the radioactive material.

In the United States, the labeling on the containers and packages for use and storage are regulated by the Nuclear Regulatory Commission. There are additional regulations at state and local levels, as well as those by the Federal Drug Administration and the U.S. Public Health Service.

Direct. Some radionuclides are packaged in solution for direct sampling via a septum and injection into the patient.

Kits. Kits for the preparation of radiopharmaceuticals are a convenient solution to synthesis of products containing short-lived radionuclides (eg, ^{111}In, ^{123}I, ^{99m}Tc) bound to a nonradioactive moiety. The labeling step is performed either at a commercial radiopharmacy, or within the institutional nuclear medicine laboratory. The kits are usually stored as a frozen solution or lyophilized product. The material of interest is then metered out into kit dosages. The kit vials are thawed or reconstituted and mixed with the appropriate radionuclide.

On-Site. Positron-emitting radionuclides are short-lived and are produced either by a generator or within a cyclotron at a given institution for immediate formulation and use. Because carbon, nitrogen, and oxygen can all be produced as positron emitters, the technique is amenable to following biochemical processes directly.

JOEL L. LAZEWATSKY
PAUL D. CRANE
D. SCOTT EDWARDS
Du Pont Merck Pharmaceutical Company

P. J. Early and D. B. Sodee, *Principles and Practice of Nuclear Medicine*, 2nd ed., Mosby, St. Louis, Mo., 1995.

N. Alazraki and F. S. Mishkin, *Fundamentals of Nuclear Medicine*, The Society of Nuclear Medicine, New York, 1984.

M. P. Iturralde, *Dictionary and Handbook of Nuclear Medicine and Clinical Imaging*, CRC Press, Boca Raton, Fla., 1990.

G. Saha, *Fundamentals of Nuclear Pharmacy*, 2nd ed., Springer-Verlag, New York, 1984.

RADIOPROTECTIVE AGENTS

There has been substantial progress in the area of radioprotective agents since the early 1980s, especially in terms of the biological and mechanistic evaluation of the large number of compounds isolated or synthesized prior to that time. This information has also begun to direct the development of a new generation of derivative compounds. Advances have been made in understanding the fundamental mechanisms of chemical radioprotector action, an area long dominated by thiol compounds (see THIOLS). Nonthiol protectors, including protease inhibitors, vitamins (qv), metalloelements, and calcium antagonists have begun to plat a larger role. There has been an explosion of interest in biological, as opposed to chemical, modifiers of radiation injury. Some of these biologics act best when given prior to irradiation; many can modulate radiation injury when given after irradiation, presumably by affecting the recovery and repopulation of critical tissue elements. Biologics thus afford an opportunity for therapeutic intervention following accidental radiation exposure as well as in radiation therapy (XRT).

Thiols

Some important radioprotective thiols are listed in Table 1. The most effective compounds have a sulfhydryl, −SH, group at one terminus and a strong basic function, usually an amino group, at the other. The general structure of these aminothiols is $H_2N(CH_2)_xNH(CH_2)_ySH$, where x is optimally 3 and y is optimally 2 or 3. Because of lower toxicity, clinical interest has focused on prodrugs in which the −SH group has been derivatized. Initially developed for military applications, the emphasis for the thiols has shifted to their potential use in XRT to protect against damage to normal tissues, based on reports that concentrations of WR-2721 that are relatively well tolerated by mice can protect normal tissues more than tumors.

General Mechanisms of Radioprotection by Thiols. Thiols protect mammalian cells from radiation effects primarily by reducing the

Table 1. Radioprotective Thiols and Phosphorothioates

Compound[a]	Structure
Thiols	
dithiothreitol (DTT)	$HSCH_2CH(OH)CH(OH)CH_2SH$
2-mercaptoethanol (WR-15504)	$CH_2(SH)CH_2OH$
cysteamine (MEA, WR-347)	$H_2N(CH_2)_2SH$
2-((aminopropyl)amino)ethanethiol (WR-1065)	$H_2N(CH_2)_3NH(CH_2)_2SH$
WR-255591	$CH_3NH(CH_2)_3NH(CH_2)_2SH$
WR-151326	$CH_3NH(CH_2)_3NH(CH_2)_3SH$
Phosphorothioates	
WR-638	$H_2N(CH_2)_2SPO_3H_2$
WR-2721	$H_2N(CH_2)_3NH(CH_2)_2SPO_3H_2$
WR-3689	$CH_3NH(CH_2)_3NH(CH_2)_2SPO_3H_2$
WR-151327	$CH_3NH(CH_2)_3NH(CH_2)_3SPO_3H_2$

[a] WR = Walter Reed Army Institute of Research

severity of the initial damage inflicted to genomic DNA. The hydroxyl radical, $OH^{\bullet}$;, is a primary cause of damage to cellular DNA. Thiols, RSH; thiolate anions, RS^-; and disulfides, RSSR; react rapidly with $OH^{\bullet}$; and may thus protect cells by scavenging $OH^{\bullet}$; before they can react with DNA. Scavenging of secondary radicals may also contribute to protection. Thiols may also enhance cell survival by chemically repairing DNA radicals caused by both $OH^{\bullet}$; and the direct ionization of DNA. Either an H-atom or an electron can be transferred from the thiol to a DNA radical (see eq. 1). These reactions are embodied in the fixation-repair model, the cornerstone of which is the hypothesis that, in the absence of oxygen, DNA radicals, which are presumed to be responsible for radiation lethality, are either inherently irreparable (eq. 2); can be chemically repaired, usually by reaction with a thiol (eq. 1); or can be fixed by a competing reaction with oxygen (eq. 3). Direct evidence for such a competition between oxygen and thiols for DNA radicals in mammalian cells has come from fast-kinetic studies.

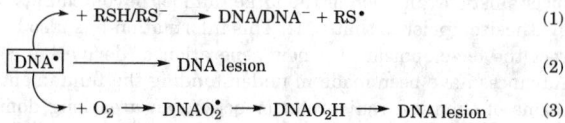

$$DNA^{\bullet} + RSH/RS^- \longrightarrow DNA/DNA^- + RS^{\bullet} \quad (1)$$

$$DNA^{\bullet} \longrightarrow \text{DNA lesion} \quad (2)$$

$$DNA^{\bullet} + O_2 \longrightarrow DNAO_2^{\bullet} \longrightarrow DNAO_2H \longrightarrow \text{DNA lesion} \quad (3)$$

Modulation of the Killing of Mammalian Cells by Thiols. Important aspects of the effects of exogenous thiols on clonogenic cell survival following exposure to low linear energy transfer (LET) radiations include the following.

(*1*) Thiols must be added before or within a very short time after irradiation to protect against cell killing.

(*2*) Subcellular distribution may be an important aspect of the biological activity of thiols of different net charge.

(*3*) Protection by thiols depends on the concentration of oxygen and of the thiol. Thiols generally protect cells at intermediate oxygen concentrations to a greater extent than fully oxic or anoxic cells.

(*4*) Alteration in the levels of endogenous thiols is not required for protection by exogenous thiols. A number of thiols elevate intracellular glutathione (GSH), although this is not always the case.

(*5*) Different types of cells can be protected by thiols to different degrees; for example, WR-1065 protects normal human fibroblasts, but not fibrosarcoma tumor cells, against the DNA-damaging and lethal effects of x irradiation *in vitro*.

(*6*) Thiols can protect against the lethal effects of high LET radiations. Pretreating cells with thiols such as WR-255591 protects against injury by high LET radiations such as neutrons, but invariably the magnitude of protection is lower than with x- or γ-rays.

(*7*) Thiols are effective modulators of apoptosis and oxidative injury to the cell membrane.

Modulation of Cellular Recovery and Stress Responses. Thiols have frequently been suggested to alter cellular radiosensitivity on a longer time scale than can be explained by free-radical events. This may be because thiols can also produce profound changes in cell metabolism, including effects on cell progression, DNA synthesis, and protein synthesis. These combined effects are often referred to as biochemical shock and, in some cases, they may contribute to protection by allowing additional time for the repair of DNA damage. However, the lack of protection against cell killing by thiols added immediately after irradiation is inconsistent with such a scenario. Biochemical shock may therefore be more important for antimutagenic or anticarcinogenic activity which, in contrast to effects on survival, are manifested when the thiol is added after irradiation.

Modulation of Mutation. Because mutations in cells that survive the radiation exposure may be expressed as undesirable late effects in XRT, the potential use of thiols to protect against mutation and transformation has generated considerable interest. Studies have shown several important aspects of such effects. (*1*) Timing. Adding aminothiols to V79 cells up to 3 h after irradiation has no effect on their survival following either γ-rays or neutrons but markedly reduces the induction of hprt mutations for both radiation types. (*2*) Concentration. WR-1065 protects against transformation and hprt mutations at much lower concentrations than are necessary to protect against cell killing. (*3*) LET dependency. Thiols are equally effective against neutron and γ-ray-induced hprt mutations *in vitro*.

Preclinical and Clinical Studies with Phosphorothioates. Because radiation does not discriminate between normal and tumor cells, XRT may cause adverse normal tissue reactions that can limit the intensity of treatment and also be life-threatening. An improved therapeutic index would be possible if the differential response between the tumor and irradiated normal tissues could be increased. An enormous literature has accumulated relating to the effects of phosphorothioates on animals and humans. On the basis of these studies, several generalizations can be made.

(*1*) Phosphorothioates generally protect normal tissues more than tumors.

(*2*) Thiols protect different normal tissues to different degrees.

(*3*) Fractionated and low dose-rate XRT schedules may be less amenable to modification.

(*4*) Clinical studies of WR-2721 with XRT for some types of cancer show promise, but toxicity remains a concern.

(*5*) Although WR-2721 is generally superior to other phosphorothioates in protecting normal tissue in mice, other drugs show some potential.

(*6*) Alternative methods of delivering radioprotective agents may provide an improved therapeutic index. Topical or regional application can circumvent problems related to systemic toxicity and tumor protection, eg, during irradiation of the chest wall after breast tumor surgery.

(*7*) Phosphorothioates attenuate the effects of neutrons *in vivo*, but are generally less protective than for x-rays.

Other Sulfur-Containing Compounds. Endogenous thiols such as GSH and cysteine are present in all mammalian cells and perform a variety of metabolic functions, including protecting cellular components from the damaging effects of radiation and other oxidative stresses. Radioprotection *in vitro* can generally not be achieved by adding cysteine or GSH *per se*, as these agents can be quite cytotoxic and often have irregular uptake into cells.

An alternative method for elevating intracellular GSH levels is to use agents that promote GSH synthesis, such as 2-oxothiazolidine-4-carboxylate (OTZ), a cysteine delivery system, or GSH ethyl ester or methyl ester, which is readily taken up by a variety of cell types and converted to GSH. Another alternative is *N*-acetylcysteine (NAC) which enters cells and is rapidly hydrolyzed to cysteine, which can then enter the GSH-synthetic pathway.

A variety of additional sulfur-containing organic compounds have been synthesized and screened for radioprotective activity. Compounds with a high ability to protect mice against lethal doses of total-body γ-radiation were identified among a series of quinolinium and pyridinium bis (methylthio) and methylthio amino derivatives. Several 1,2-dithiol-3-thione and dithioester compounds are also highly effective protectors of EMT6 cells *in vitro*, and in some cases are differentially cytotoxic toward hypoxic cells.

Other Classes of Protectors

Tempol. Nitroxides are low molecular-weight stable free radicals that protect against various biological manifestations of oxidative stress, including mutagenic effects. Tempol, a water-soluble piperidine nitroxide derivative with nonspecific radical-scavenging and superoxide dismutase (SOD) activity, protects cultured aerobic, but not hypoxic, cells against radiation-induced killing. Protection does not depend on intracellular thiols and does not involve O_2-depletion. Tempol reacts with peroxyl radicals and may also oxidize DNA-bound metal ions, thereby interfering with $OH^{\bullet}$; generation.

Captopril. The antihypertensive agent captopril is an inhibitor of angiotensin-converting enzyme (ACE) (see CARDIOVASCULAR AGENTS; ENZYME INHIBITORS). ACE converts angiotensin I to angiotensin II, a potent vasoconstrictor. Captopril is well tolerated and approved by the U.S. FDA for chronic use in humans. ACE inhibitors protect against radiation-induced injury to lung and skin and against early hemodynamic changes after local renal irradiation.

Metallothioneins. The metallothioneins, a group of low molecular-weight proteins (<10 kD) containing ~30% cysteine residues, are efficient $OH^{\bullet}$;/superoxide scavengers. Transcription of metallothionein genes can be induced by a variety of agents, including heavy metals such as cadmium, Cd; glucocorticoids (GC); interferon (IFN); Ha-*ras*; and mediators of the inflammatory stress response such as tumor necrosis factor (TNF) and interleukin 1 (IL-1).

Protease Inhibitors. Protease inhibitors such as antipain, a tripeptide analogue derived from actinomycetes, and the soybean-derived Bowman-Birk inhibitor family exhibit antitransforming and anticarcinogenic activity *in vivo* and *in vitro* at nontoxic doses but generally do not modifying cell killing.

Although the mechanism by which protease inhibitors block carcinogenesis is unknown, they appear to be able to reverse the initiating event, presumably by stopping an ongoing process begun by the carcinogen exposure.

Antioxidant Vitamins. The natural antioxidant vitamins A, C, and E (retinoic acid, ascorbic acid and α-tocopherol, respectively), as well as the vitamin A dietary precursor β-carotene, exhibit a range of radioprotective effects (see VITAMINS). These protect against lethality, mutation, and transformation, in a variety of cell types and laboratory animals. Effects have been reported for both pre- and post-irradiation treatments. Although vitamins protect much less efficiently than WR-2721, the vitamins have low toxicity and may thus be useful in some situations.

Metalloelements. Radioprotection by the essential metals Cu, Fe, Mn, and Zn and their intracellular complexes has been studied. Metalloelement-dependent enzymes include SOD, catalase, metallothioneins, lipoxygenases and cyclo-oxygenases involved in arachadonic acid metabolism, alkaline phosphatase, DNA polymerase and gyrase involved in DNA synthesis, and Zn-finger proteins involved in the regulation of transcription. These may modulate recovery from radiation-induced injury, and the depletion of their activity may partially account for the biological effects of ionizing radiation, as well as explaining the radioprotective activity of exogenous metal compounds. Cytokine-mediated redistribution of metalloelements may also be an important factor in radiation response.

Calcium Antagonists. The potential use of Ca^{2+} antagonists as radioprotective agents has been suggested based on the importance of maintaining Ca^{2+} homeostasis for cell viability following a variety of cytotoxic insults. Alterations of cytosolic Ca^{2+} levels can result in changes in the activity of Ca^{2+}-dependent degradative enzymes, which may contribute to cell death, and of the protein kinase C (PKC)-mediated signal transduction pathway. Although the role of perturbations in Ca^{2+} homeostasis in cellular radiation response remains poorly defined, the realization that modulation of intracellular Ca^{2+} can prevent apoptosis in some cell types has stimulated even greater interest in Ca^{2+} antagonists.

Adenosine Analogues. Exogenous adenine nucleotides are moderately radioprotective when given to animals shortly before irradiation; protection appears to be mediated by extracellular adenosine receptors which are coupled to the inhibition/activation of adenylate cyclase, which in turn regulates intracellular cyclic adenosine monophosphate (cAMP), which is itself radioprotective *in vitro*.

Priming with Other Antitumor Agents. Priming animals with low doses of some chemotherapeutic agents enhances their resistance to subsequent exposures to radiation (see CHEMOTHERAPEUTIC AGENTS, ANTICANCER). Priming mice with vincristine increases the radioresistance of the GI epithelium and of the bone marrow (BM), primarily by accelerating post-irradiation recovery of the hemopoietic stem- and progenitor cells, rather than altering their intrinsic radiosensitivity. Giving vincristine 24 h prior to irradiation provides optimal radioprotection of 12-day spleen colony-forming units (CFU-S), possibly by initiating a recruitment of stem cells into the cycle prior to irradiation. At the time of irradiation, cells may be in a more radioresistant phase (S) of the cell cycle, or have a decreased tendency to apoptose or increased radical-scavenging capacity. Many biologics that prime against radiation are also potent immunostimulants that activate macrophages and cells of the immune system to release cytokines, such as IL-1, that can confer radioprotection and enhance hemopoietic stem cell recovery.

Methylxanthines. Pentoxifylline (Trental), a synthetic dimethylxanthine derivative, is a hemorrheologic agent that can prevent or ameliorate late radiation injury to soft tissues and lung in animals and humans, and may be useful in the management of fibrotic sequelae, particularly if administered prophylactically or in the inflammatory early stages of fibrosis. In humans, pentoxifylline is used as an interventional therapy for persistent soft tissue ulceration or necrosis. The drug treatment is initiated ~ 30 weeks after irradiation. Pentoxifylline may influence late radiation injury via several mechanisms, including protection of endothelial cells soon after radiation exposure by enhancing blood flow in injured microvasculature and down-regulating TNF.

Pentoxifylline is structurally related to other methylxanthine derivatives such as caffeine (1,3,7-trimethylxanthine), theobromine (3,7-dimethylxanthine), and theophylline (3,7-dihydro-1,3-dimethyl-1*H*-purine-2,6-dione or 1,3-dimethylxanthine), which also show radioprotective activity in some instances, suggesting that methylxanthines as a drug class may radioprotect through a common mechanism (see ALKALOIDS).

Superoxide Dismutase. Superoxide dismutase (SOD) exhibits radioprotective activity in a variety of systems, including protecting against 30-day lethality in mice and some late radiation effects in humans when given after irradiation. The mechanistic basis for these effects is controversial. The pre-irradiation activity of SOD, and its activity in cultured cells, has generally been attributed to radical-scavenging effects, whereas its activity when given to animals or patients after irradiation is probably related to its antiinflammatory and/or immunostimulatory properties.

Chinese Herbal Medicines. Many traditional Chinese medicines have been screened for radioprotective activity in experimental animals. In one study of more than a thousand Chinese herbs, a number of agents increased the survival rate of dogs exposed to a lethal dose of γ-rays by 30–40%, and some symptoms of radiation injury were ameliorated. These effects are potentially related to stimulation of the hemopoietic and immune systems.

Antibiotics. Although not strictly speaking radioprotective agents, antibiotics (qv) are an important component of the treatment of radiation injuries. By preventing or delaying the onset of systemic infection owing to endogenous and exogenous organisms, they may allow greater recovery of tissues such as BM and intestine. Antibiotics such as penicillin and synthetic antibacterials such as quinolone (see ANTIBACTERIAL AGENTS, SYNTHETIC), alone and in combination, can reduce bacterial translocation from the intestine, treat the subsequent sepsis, and reduce mortality.

Other Radioprotective Chemicals. Glutamine has been widely examined as a potential agent for enhancing intestinal repair following radiation injury, although its value in this regard remains to be clearly established.

Several antiulcer drugs have been shown to protect against acute skin reactions, such as erythema and moist desquamation, which are major problems during XRT of superficially-located tumors.

Cytokines and Related Factors as Radioprotective Agents

Although thiols dominated the field of radioprotection from the 1950s through the 1980s, the radioprotective, or rather radiomodulating, activity of biologics, and especially cytokines, received considerable attention in the 1990s. The term cytokines herein means proteins that modify cellular responses through ligand-receptor interactions, including growth factors and interleukins, and also immunomodulating agents, which function mainly through the generation of cytokines. Eicosanoids are included because they can be generated during cytokine responses and mediate such diverse processes as vasoregulation and inflammation. These are potent protectors of jejunal and hemopoietic stem cells when given prior to irradiation, and are therefore implicated as potential mediators of radioprotection by cytokines, although this does not seem to be the case for IL-1. Some cytokines are most effective as radioprotectors when given between one and several days prior to irradiation; others work best when administered after irradiation. Biologics generally protect tissues by 1.3-fold or less, whereas WR-2721 can protect by as much as 2.7-fold. However, clinically, protection as low as 1.1-fold could be beneficial. The ability to increase the tumor dose by 10% while maintaining the same level of complications would translate into a significant increase in tumor control rates. Certain cytokines, such as granulocyte colony stimulating factor (G-CSF), stem-cell factor (SCF), and granulocyte-macrophage colony stimulating factor (GM-CSF), are well-tolerated at effective doses. Others, such as TNF-α and IL-1, are more toxic, especially if administered systemically, as is consistent with their roles in inflammation and septic shock.

G-CSF and GM-CSF. Because of availability in recombinant form and their enormous clinical potential in both XRT and chemotherapy, hemopoietic growth factors have moved rapidly from the laboratory to the clinic, in spite of their expense. Clinical use of cytokines is largely limited to accelerating lympho/hemopoietic regeneration using exogenous colony stimulating factors, in particular G-CSF and GM-CSF. Such agents have the potential to selectively protect normal hemopoietic and other tissues, allowing higher doses of radiation or chemotherapy to be delivered to the tumor. They also have roles in the management of disease.

Clinically, GM-CSF or G-CSF have been used to accelerate recovery after chemotherapy and total-body or extended-field irradiation, situations that cause neutropenia and decreased platelets and possibly lead to fatal septic infection and diffuse hemorrhage, respectively. G-CSF and GM-CSF reproducibly decrease the period of granulocytopenia, the number of infectious episodes, and the length of hospitalization in such patients, although it is not yet clear that dose escalation of the cytotoxic agent and increased cure rate can be reliably achieved. One aspect of the effects of G-CSF and GM-CSF is that these agents can activate mature cells to function more efficiently. This may, however, also lead to production of cytokines, such as TNF-α, that have some toxic side effects. In general, both cytokines are reasonably well-tolerated. The side effect profile of G-CSF is more favorable than that of GM-CSF. Medullary bone pain is the only common toxicity.

Interleukin-1 α and β. IL-1 has radioprotective activity toward BM and other tissues. IL-1 is produced in response to endotoxin, other cytokines, and microbial and viral agents, primarily by monocytes and macrophages. IL-1 appears to play an important role in the regulation of normal hemopoiesis directly by stimulating the most primitive stem cells and indirectly by stimulating other hemopoietic factors, including G-CSF, GM-CSF, macrophage colony stimulating factor (M-CSF), and IL-6.

Although IL-1 protects a number of normal tissues against radiation injury, its effects on BM are the best characterized. IL-1 provides varying degrees of protection, depending on the timing; ~20 − h prior to irradiation is optimal. Synergy was found with other cytokines, most notably TNF-α, IL-6, and SCF. Multiple daily injections of IL-1 α preceding irradiation are more effective than single doses in promoting both BM progenitor cell survival and granulocyte recovery. Protection may involve a number of mechanisms. One is the stimulation of BM progenitor cells such that more of these are in the radioresistant S-phase of the cell cycle. Another is the induction by IL-1 of a number of radioprotective substances such as prostaglandins (PGs), metallothionein, scavenging acute-phase proteins, GSH, and SOD, as well as other hemopoietic growth factors. The protective effects of pre-irradiation IL-1 in murine BM cells and human cell lines correlate closely with the induction of MnSOD, and growth factors induced from accessory cells that constitute the hemopoietic microenvironment can enhance repopulation of the immune and hemopoietic systems after irradiation.

Stem-cell factor. Stem-cell factor (SCF), also known as c-kit ligand, is the ligand for a tyrosine kinase-associated receptor encoded by the c-kit protooncogene, and stimulates primitive multipotential hemopoietic stem cells. It does not act as a colony-stimulating factor for these cells but rather primes cells to respond to other cytokines, such as IL-1. Mutations in white spotting and steel mice that affect hemopoiesis have been influential in characterizing SCF. These strains display an increased sensitivity to lethal doses of irradiation, which is thought to be the result of a lack of the apoptosis-suppressing effects of SCF, in addition to a lack of proliferative effects. SCF is radioprotective in mice on its own and in concert with other cytokines.

Mice treated with SCF show improved long-term survival, more rapid hemopoietic recovery after irradiation, and a much reduced incidence of septicemia.

Other Lympho/Hemopoietic Cytokines. IL-3 shares a common signaling receptor chain with GM-CSF, but stimulates the proliferation, differentiation, and function, of a less mature, multipotential, myeloid progenitor cell population. The broader activity of IL-3 suggests that it may be of greater benefit than the more lineage-restricted G-CSF and GM-CSF in accelerating BM recovery after injury. Administration of IL-3 to sublethally-irradiated mice induces cell recovery in the thymus, and in primates, IL-3 effectively decreases the period of thrombocytopenia after drug or radiation-induced BM aplasia, a feature that G-CSF or GM-CSF do not possess, although IL-1 and IL-6 do.

Other Cytokines. Basic fibroblast growth factor (b-FGF or FGF-2) belongs to a class of heparin-binding growth factors associated with growth and differentiation of a number of cell types, including endothelial cells and fibroblasts, and a number of *in vivo* processes including angiogenesis and wound healing. Its presence is required for endothelial cell culture *in vitro* and protects against the ability of radiation to induce apoptosis. Basic FGF has no effect on the repair of radiation-induced DNA strand breaks but is required for the shoulder on the survival curve, in essence allowing repair of potentially-lethal damage. Protection appears to be mediated by PKC activation. Radiation induces bFGF production by these cells, allowing a survival-associated positive autocrine loop to develop.

Other Agents. Many agents that nonspecifically enhance immunological and hemopoietic responses can also function as radioprotectants. Immunomodulatory substances such as endotoxin, glucan, muramyl dipeptide, BCG, and OK-432 (a lyophilized preparation of an attenuated strain of Streptococcus haemolyticus available commercially as Picinabil) given to mice prior to, and in many cases after, irradiation result in an increased survival beyond 30 days. Such substances are believed to protect by activating macrophages and inducing the production of endogenous cytokines such as IL-1, TNF-α, G-CSF, GM-CSF, and IFN in irradiated animals that subsequently act synergistically to stimulate the proliferation of the target cells.

Arachidonic Acid Metabolites. *Prostaglandins.* Various bioactive lipids, especially the prostaglandins (qv) (PGs) and leukotrienes (LTs), the principal eicosanoid products of the arachidonic acid cascade, are radioprotective. Indeed, protection of hemopoietic tissue by eicosanoids can approach that achieved with WR-2721. The mechanisms by which eicosanoids protect against radiation damage are obscured by their wide range of both pathological and normal physiological effects on tissues and animals. Protection by PGs is seen only when administered prior to irradiation.

Leukotrienes. Leukotrienes, products of the lipoxygenase pathway, are generally less radioprotective than the PGs with the exception of LTC_4, which is among the most potent of the naturally-occurring eicosanoids.

Other Bioactive Lipids. Linoleic acid (*cis*-9, *cis*-12-octadecadienoic acid), an essential fatty acid, can act as a radioprotective agent of BM while being toxic to certain tumor cells. Both irradiation (0.5 Gy (50 rad)) and exogenous linoleate generate increased levels of the oxygenated product of linoleate, 13-hydroxyoctadecadienoate (13-HODE), by macrophages, most likely through a 15-lipoxygenase-mediated pathway. This pathway seems to serve not only as a means of dealing with free radicals that are generated in cells but also as a mediator of oxidative stress responses.

Eicosanoid Blocking Agents. A number of studies have documented the radioprotective effects of eicosanoid blocking agents such as the nonsteroidal antiinflammatory agents. Prophylactic administration of indomethacin, an inhibitor of PG synthesis, delays or reduces radiation mucositis owing to head and neck and thoracic XRT and experimental radiation esophagitis. Whereas indomethacin radiosensitizes murine solid tumors that have a high level of eicosanoid production, apparently by stimulating antitumor immune reactions, it has little effect on tumors with low eicosanoid production.

Steroids and Glucocorticoids. Glucocorticoid (GC) steroids, potent antiinflammatory and immunosuppressive agents, inhibit the hydrolysis of membrane phospholipids by phospholipase A2, which is the initial step in the generation of both lipoxygenase and cyclo-oxygenase products of arachidonic acid. GCs are used clinically to treat radiation pneumonitis, although the response rate is variable.

Miscellaneous Radioprotective Agents. Steroid hormones have been extensively studied for their ability to protect against infertility in males receiving XRT (see HORMONES). A variety of results have been reported. Pretreatment using a gonadotrophin-releasing hormone antagonist protects spermatogonial stem-cell function from single doses of x-rays. Pretreatment using testosterone also protects spermatogonial stem cells from four daily x-ray fractions, whereas sub-q medroxyprogesterone and testosterone pretreatments protect against a single 3-Gy (300-rad) dose of x-rays.

Combinations of Biologics and Thiols or Other Agents

Significant clinical promise lies in the concept of combining low (nontoxic) doses of radioprotective drugs having different, but complementary, mechanisms of action to achieve better protection than using either agent alone. In particular, combining an agent such as IL-1 that has the potential to stimulate stem-cell proliferation before or after irradiation with an agent such as WR-2721 that protects the individual stem cells is appealing. An important general observation is that a combined effect will be seen only following radiation doses that do not eradicate most of the stem cells. Several biologics have now been shown to exhibit additive or synergistic effects in combination with aminothiols, and may thus be potentially useful for reducing the risks associated with myelosuppression induced by XRT.

DAVID MURRAY
University of Alberta,
Cross Cancer Institute
WILLIAM H. MCBRIDE
UCLA Medical Center

D. Murray, "Aminothiols," in E. A. Bump and K. Malaker, eds., *Radioprotectors: Chemical Biological and Clinical Perspective*, CRC Press, Boca Raton, Fla., 1998, pp. 53–109.

J. F. Weiss, K. S. Kumar, T. L. Walden, R. Neta, M. R. Landauer, and E. P. Clark, "Advances in Radioprotection through the Use of Combined Agent Regimens," *Int. J. Radiat. Biol.* **57**, 709 (1990).

R. Neta and P. Okunieff, "Cytokine-Induced Radiation Protection and Sensitization," *Semin. Radiat. Oncol.* **6**, 306–320, (1996).

W. R. Hanson, "Eicosanoid-Induced Radioprotection and Chemoprotection of Normal Tissue during Cancer Treatment, in J. E Harris, D. P. Braun, and K. M. Anderson, eds., *Prostaglandin Inhibitors in Tumor Immunology and Immunotherapy*, CRC Press, Boca Raton, Fla., 1994, pp. 171–186.

RADIUM. See RADIOACTIVITY, NATURAL.

RADON. See HELIUM GROUP, GASES.

RARE-EARTH ELEMENTS. See LANTHANIDES.

RAYON. See FIBERS, REGENERATED CELLULOSICS.

REACTOR TECHNOLOGY

Reactor technology comprises the underlying principles of chemical reaction engineering (CRE) and the practices used in their application. The focuses of reactor technology are reactor configurations, operating conditions, external operating environments, developmental history, industrial application, and evolutionary change.

Besides stoichiometry and kinetics, reactor technology includes requirements for introducing and removing reactants and products, efficiently supplying and withdrawing heat, accommodating phase changes and material transfers, assuring efficient contacting of reactants, and providing for catalyst replenishment or regeneration. Consideration must be given to physical properties of feed and products (vapor, liquid, solid, or combinations), characteristics of chemical reactions (reactant concentrations, paths and rates, operating conditions, and heat addition or removal), the nature of any catalyst used (activity, life, and physical form), and requirements for contacting reactants and removing products (flow characteristics, transport phenomena, mixing requirements, and separating mechanisms).

All the factors are interdependent and must be considered together.

Reactor Types and Characteristics

Specific reactor characteristics depend on the particular use of the reactor as a laboratory, pilot plant, or industrial unit. All reactors have in common selected characteristics of four basic reactor types: the well-stirred batch reactor, the semibatch reactor, the continuous-flow stirred-tank reactor, and the tubular reactor. A reactor may be represented by or modeled after one or a combination of these.

Batch Reactor. In a batch reactor, is a feed material is treated as a whole for a fixed period of time. Batch reactors may be preferred for small-scale production of high priced products, when multiple, low volume products are produced in the same equipment, or when continuous flow is difficult.

Almost all batch reactors are well stirred; thus, ideally, compositions are uniform throughout and residence times of all contained reactants are constant.

Semibatch Reactor. The semibatch reactor is similar to the batch reactor but has the additional feature of continuous addition or removal of one or more components. Use of a semibatch reactor permits more stable and safer operation than in a batch operation.

Continuous-Flow Stirred-Tank Reactor. Industrial practice generally favors processing continuously rather than in single batches because overall investment and operating costs usually are less. In a continuous-flow stirred-tank reactor (CSTR), reactants and products are continuously added and withdrawn. In practice, mechanical or hydraulic agitation is required to achieve uniform composition and temperature, a choice strongly influenced by process considerations, ie, multiple specialty product requirements and mechanical seal pressure limitations. The CSTR is the idealized opposite of the well-stirred batch and tubular plug-flow reactors. Analysis of selected combinations of these reactor types can be useful in quantitatively evaluating more complex gas-, liquid-, and solid-flow behaviors.

Tubular Reactor.

The tubular reactor is a vessel through which flow is continuous, usually at steady state, and configured so that conversion and other dependent variables are functions of position within the reactor rather than of time. In the ideal tubular reactor, the fluids flow as if they were solid plugs or pistons, and reaction time is the same for all flowing material at any given tube cross section; hence, position is analogous to time in the well-stirred batch reactor. Tubular reactors resemble batch reactors in providing initially high driving forces, which diminish as the reactions progress down the tubes.

Multiphase Reactors. The overwhelming majority of industrial reactors are multiphase reactors. Some important reactor configurations are illustrated in Figures 1 and 2. The presence of more than one phase, whether or not it is flowing, confounds analyses of reactors and increases the multiplicity of reactor configurations. Gases, liquids, and solids each flow in characteristic fashions, either dispersed in other phases or separately. Flow patterns in these reactors are complex.

A fixed-bed reactor is packed with catalyst. If a single phase is flowing, the reactor can be analyzed as a tubular plug-flow reactor or modified to account for axial diffusion. If both liquid and gas or vapor are injected downward through the catalyst bed, or if substantial amounts of vapor are generated internally, the reactors are mixed-phase, downflow, and fixed-bed reactors. If the liquid and gas rates are so low that the liquid flows as a continuous film over the catalyst, the reactors are called trickle beds.

At higher total flow rates, particularly when the liquid is prone to foaming, the reactor is a pulsed column. This designation arises from the observation that the pressure drop within the catalyst bed cycles at a constant frequency as a result of liquid temporarily blocking gas or vapor pathways.

Moving beds are fixed-bed reactors in which spent catalyst or reactive solids are slowly removed from the bottom and fresh material is added at the top.

In bubble column reactors, gas bubbles flow upward through a slower moving liquid. The bubbles, which rise in essentially plug flow, draw liquid in their wakes and thereby induce back-mixing in the liquid with which they have come in contact. Analogously, in spray columns, liquid as droplets descend through a fluid, usually a gas. Both bubble and spray columns are used for reactions where high interfacial areas between phases are desirable.

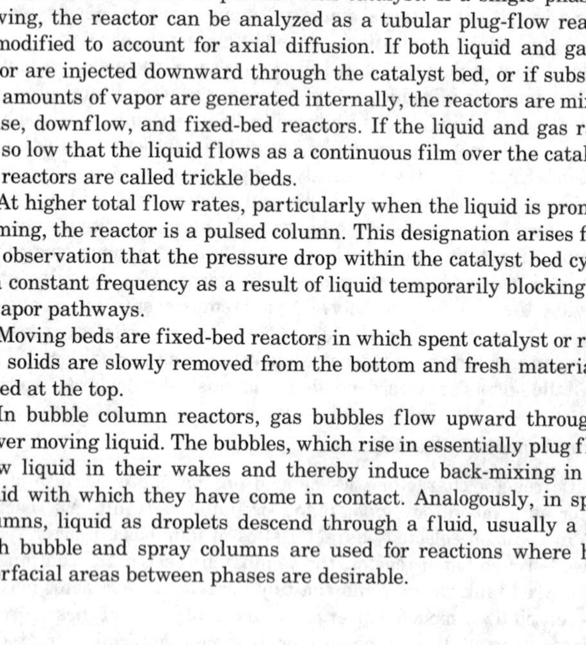

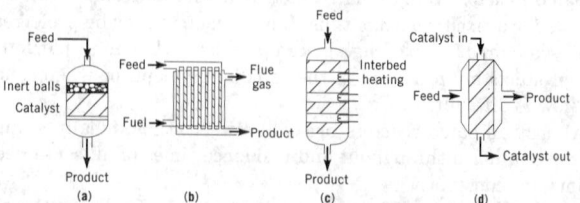

Figure 1. Multiple fixed-bed configurations: (**a**) adiabatic fixed-bed reactor, (**b**) tubular fixed beds, (**c**) staged adiabatic reactor with interbed heating (cooling), (**d**) moving radial fixed-bed reactor.

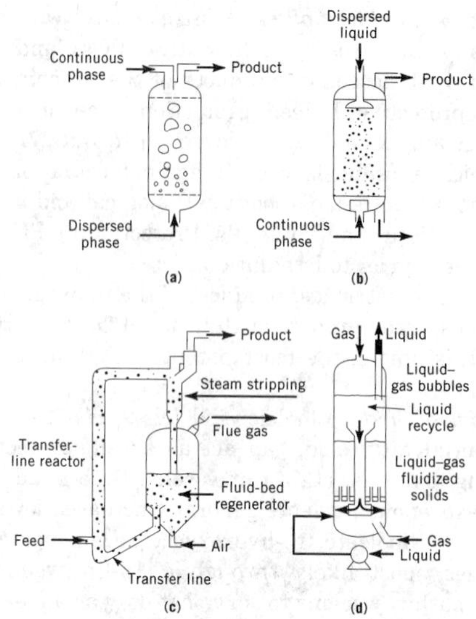

Figure 2. Multiphase fluid and fluid—solids reactors: (**a**) bubble column, (**b**) spray column, (**c**) fluidization unit, (**d**) gas—liquid—solid fluidized reactor.

Reactors are termed fluidized or fluid beds if upward gas or liquid flows, alone or in concert, are sufficiently high to suspend the solids and make them appear to behave as a liquid. This process is usually referred to as fluidization. The most common fluid bed is the gas-fluidized bed. With gas feeds, the excess gas over the minimum required for fluidization rises as discrete bubbles, through which the surrounding solids circulate.

A reactor is termed a radial or panel-bed reactor when gas or vapor flow perpendicular to a catalyst-filled annulus or panel. Similar cross-flow configurations also are used for processing solids moving downward under gravity while a gas passes horizontally through them. Rotary kilns, belt dryers, and traveling grates are examples. Cross-flow reactors are not restricted to solids-containing systems. Venturis, in which atomized liquids are injected across the gas stream, are effective for fast reactions and similarly for generating small gas bubbles in downward-flowing liquids where mass transport across the gas—liquid interface is limiting.

Reactor Selection

Selection of a reactor, whether for a new application or a changing situation, is often determined by economics, reliability, or availability of a proven system that is amenable to extension in a new service. Introduction of new catalysts, improvements in process and equipment design and operations, transient reactor behavior, heat generation and removal, reduced emissions and waste minimization, and safety-related hazards are other issues that influence the continued and extended use of a reactor configuration.

Coupling Reaction Kinetics with Transport

The interacting effects of reaction kinetics with mass, momentum, and heat transport and the mixing and interchange of components are essential elements in establishing reactors as process systems. Reactor response to various design and operating variables largely depends on the relative kinetic and transport rates at the physical scales of the phenomena being addressed. Three levels of scale are normally addressed: micro-scale or molecular level, meso-scale or catalyst particle or eddy size, macro-scale or reactor scale.

Back-Mixing, Staging, and Recycle. The extent of back-mixing and degree of staging or recycle directly affects overall reactor size and catalyst requirements for reactions with greater than zero-order dependencies on limiting reactants. Back-mixing does not necessarily imply reduced conversions. Complete back-mixing can decrease catalyst deactivation by diluting catalyst poisons, if these are limiting constituents. Because total conversion is a function of the kinetic behavior of the entire reactor system, the overall effects of reactor back-mixing must be considered.

Reaction and Transport Interactions. The importance of the various design and operating variables largely depends on relative rates of reaction and transport of reactants to the reaction sites. If transport rates to and from reaction sites are substantially greater than the specific reaction rate at meso-scale reactant concentrations, the overall reaction rate is uncoupled from the transport rates and increasing reactor size has no effect on the apparent reaction rate, the macro-scale reaction rate. When these rates are comparable, they are coupled, that is they affect each other. In these situations, increasing reactor size alters mass- and heat-transport rates and changes the apparent reaction rate.

For well-defined reaction zones and irreversible, first-order reactions, the relative reaction and transport rates are expressed as the Hatta number, Ha, which equals $(k_1 D_A/k_L)^{-1}$ where k_1 = reaction rate constant, D_A = diffusivity of reactant A, and k_L = mass-transfer coefficient. Reaction dominates with $Ha < 0.3$ and transport is limiting with $Ha > 2$.

Mass- and Heat-Transport. Phenomenological correlations describing the interrelated effects of fluid dynamic and geometric parameters on transport rate are commonly used to estimate mass- and heat-transport rates. Such generalized correlations have been developed from results obtained in a variety of systems. These are based on gradients linear, across the resistance paths (film theory), or nonlinear because of insufficient time for reaching steady state (penetration theory). Mathematical methods for analyzing mass- and heat-transport phenomena in reacting systems are available.

Complex Flow Behavior

The concepts of well-mixed and plug-flow become inadequate when flow patterns deviate significantly from ideal behavior. Adsorption and subsequent desorption from within catalyst pellets in a trickle bed, solids recirculating downward to compensate for those entrained into rising bubbles in a fluid bed, and liquid held in the wake of a solid slurry exemplify back-mixing or bypassing in mixed-phase tubular reactors. Practical considerations can result in substantial deviations from idealized flow behavior.

Residence Time Distributions. A range of residence time distributions results from departures from ideal behavior. Residence time distributions can be coupled with reaction kinetics and used in assessing reactor performance.

Cold-Flow Models. Residence time distributions are often obtained using tracers in cold flow models, replicas of portions of reactors, but operating at ambient conditions for detailed observations of flow patterns.

Computer Simulation

Computer simulation of the reactor kinetic hydrodynamic and transport characteristics reduces dependence on phenomenological representations and idealized models and provides visual representations of reactor performance. Modern quantitative representations of laminar and turbulent flows are combined with finite difference algorithms and other advanced mathematical methods to solve coupled nonlinear differential equations.

Computational Fluid Dynamics. Detailed quantitative predictions of free, wall, and recirculating flows, both turbulent and laminar, are possible in representations of fluctuations and other transient behavior. Three-dimensional flows have been modeled according to a finite-difference procedure, in which local turbulent energy and its dissipation are characterized by the $k - \epsilon$ model, where k is kinetic energy per unit mass of mean velocity field and ϵ is the dissipation rate.

Validation and Application. Validated examples of computational fluid dynamics have emerged. Realism depends on the adequacy of the physical and chemical representations, the scale of resolution for the application, numerical accuracy of the solution algorithms, and skills applied in execution.

A three-dimensional $k - \epsilon$ model has been formulated for fully baffled mixing tanks in which periodic flows produced by moving blades are time averaged. The model simulates both radial flow and axial recalculation patterns produced in these tanks.

Nonintrusive Instrumentation. Essential to quantitatively enlarging fundamental descriptions of flow patterns and flow regimes are localized nonintrusive measurements. Laser-Doppler and hot film anemometers, conductivity probes, and optical fibers are used to capture time-averaged turbulent fluctuations.

Hydrodynamics and Chemical Kinetics. Capabilities for specifying arrays of reactions and using these representations in analyzing reactor behavior have emerged from the union of high speed computing with modern analytical chemistry methods. Kinetic models generally contain minimal descriptions of hydrodynamics, usually as idealized reactors; similarly hydrodynamic models usually contain over-simplified kinetic descriptions. Whereas many reactants and products, particularly inorganic and specialty organic compounds, are readily processed as individual species, hydrocarbon mixtures, such as those derived from petroleum, must be divided into lumped groups and treated mathematically as invariant species.

Laboratory Reactors

Fundamental reaction kinetics and chemical and physical properties, where necessary, are best obtained with experimental reactors that are different from those used during process development or as part of commercial operations. It is important that the reactor be similar to one of the basic types, have known flow patterns, and in the case of multiphase reactors, have known flow regimes, operate isothermally, and provide data over wide ranges of variables. The range of the conditions to be studied must be sufficiently large for adequately defining the effects of variables, but the rate equations need not accurately reflect true reaction mechanisms. Extracting the intrinsic chemical kinetics leads to a better understanding of the process and more assured extrapolations of results to conditions that have not been studied.

Scale-Up

The objective of scale-up is to select designs and operating conditions so that similar responses, comparable yields, and product distributions are obtained in the different sized units. Scaling from laboratory units or pilot plants to industrial-sized reactors utilizes methodologies similar to those used in extrapolating to new situations for existing reactor operations, but with greater uncertainty and complexity. An evolving trend is the use of scale-independent unified models which incorporate reaction kinetics, thermodynamics, hydrodynamics, and hardware representations despite the source. The maximum advantages in assessing reaction paths and complex flow behavior and transport, and using computer simulation in projecting reactor performance. Effective scale-up requires disciplined assessment of micro-, meso-, and macro-responses of reactor configurations relative to process requirements.

Minimum Pilot-Plant Size

Economy of time and resources dictate using the smallest sized facility possible to assure that projected larger scale performance is within tolerable levels of risk and uncertainty. Minimum sizes of such laboratory and pilot units often are set by operability factors not directly involving internal reactor features. These include feed and product transfer line diameters, inventory control in feed and product separation systems, and preheat and temperature maintenance requirements. Most of these extraneous factors favor large units. Large industrial plants can be operated with high service factors for years, whereas it is not unusual for pilot units to operate at sustained conditions for only days or even hours.

Small Unit Characteristics. Small-diameter lines are prone to plugging and pressure drop buildup. Extraneous reactions occurring in feed and product inventories confound any data analysis. In smaller reactors, greater portions of the total system inventory tend to be heated or cooled and so are more likely to have thermal and catalytic reactions outside the reactor confines.

Highly Integrated Processing. Integrated multiple process and multiple process functions within individual reactors are growing considerations and where employed must be taken into account when designing a pilot-plant program. These close couplings are driven by economic and environmental requirements.

Industrial Reactor Development and Application

Most reactors have evolved from concentrated efforts focused on one type of reactor. Some processes have emerged from parallel developments using markedly different reactor types. In most cases, the reactor selected for laboratory study has become the reactor type used industrially, because further development usually favors extending this technology. Following are illustrative examples of reactor usage, classified according to reactor type.

Batch Reactors. The batch reactor is frequently encountered in petrochemical, pharmaceutical, food, and mining processes. The processes often require achieving uniform dispersions of micrometer-sized drops and providing adequate exothermic heat removal.

Batch reactors are used in manufacturing plastic resins, eg, polyesters, phenolics, alkyds, urea—formaldehydes, acrylics, and furans. Such reactors generally are 6–40 m^3 (ca 200–1400 ft^3) baffled tanks, in which there are blades or impellers connected from above by long shafts, and heat is transferred either through jacketed walls or by internal coils.

Semibatch Reactors. Semibatch reactors are the most versatile of reactor types. Thermoplastic injection molds are semibatch reactors in which shaped plastic articles are produced from melts. In molding thermoplastics, large clamping forces of up to 5000 metric tons are needed to keep molds together, while highly viscous polymers are forced into their cavities.

Reaction injection molding (RIM) provides the technology for fabricating large articles, such as polyurethane automobile fenders and bumpers.

Continuous-Flow Stirred-Tank Reactors. The switch from the conventional cobalt complex catalyst to a new rhodium based catalyst represents a technical advance for producing oxo aldehydes using water-soluble $HRh(CO)[P(m\text{-sulfophenyl-Na})_3]_3$ as the catalyst to produce butyraldehydes from propylene. Having both organic and water phases results in higher selectivity and activity achievable with an aqueous media, and complete catalyst–product separation, rendering easier waste product disposal. The system's flexibility permitted a 10,000-fold scale-up and the later accommodation of a 100-fold increase in catalyst activity obtained using superior rhodium-based ligands.

Thermal Tubular Reactors. Tubular reactors have been widely used for low temperature, liquid-phase noncatalytic oxidation. Generally, conversion and selectivity to any given product are low for these oxidations. Because runaway branch-chain reactions are possible, heat dissipation must be assured and oxygen concentrations controlled. These considerations often favor the use of a series of tubular reactors in plug flow with some back-mixing in each reactor to maintain sufficient radical concentrations to propagate the reactions.

Bubble Columns. Bubble columns are finding increasing industrial application such as in ethylene dimerization, polymer manufacture, and liquid-phase oxidation. Bubble column processing offers the advantages of simplicity, favorable operating costs, and potentially superior product quality.

Airlift Reactors. Airlift reactors are hydrodynamic variants of the bubble column in which the liquid or slurry circulates between two physically separated zones as a result of sparged gas in one zone inducing a density difference between the zones. The reactors have well-established uses in producing industrial and pharmaceutical chemicals, and in treating industrial and municipal wastes.

Spray Columns. Spray columns have diverse specialized uses in biotechnology processing, catalyst manufacture, and minimization of waste products. Spray columns are used in the production of milk powder, cheese, and other fermentation products for direct heat-induced conversion of proteins, microorganisms, and enzymes, thus affecting color, flavor, nutritive value, and biological safety.

Tubular Fixed-Bed Reactors. Bundles of downflow reactor tubes filled with catalyst and surrounded by heat-transfer media are tubular fixed-bed reactors. Such reactors are used most notably in steam reforming and phthalic anhydride manufacture.

Fixed-Bed Reactors (Single-Phase). Fixed-bed reactors supplied with single-phase reactants are used extensively in the petrochemical industry for ammonia synthesis, catalytic reforming, other hydroprocesses, eg, hydrocracking and hydrodesulfurization, and oxidative dehydrogenation. The feeds in these processes are gases or vapors. The reactors generally are of large diameter, operate adiabatically, and often house multiple beds in individual pressure vessels.

Fixed-Bed Reactors (Multi-Phase Flow). Flow regimes and contacting mechanisms in fixed-bed reactors that operate with mixtures of liquids and gases are totally different from those with single-phase feeds. Nevertheless, mixed-phase, downflow fixed-bed reactor designs are extensions of single-phase, fixed-bed hydroprocessing technology and outwardly resemble such reactors. The most generally used mixed-phase reactor is the trickle bed.

Fluid-Bed Reactors. The range of fluid-bed applications is large and diverse. One of these is fluid catalytic cracking (FCC). The reactors are dense-phase fluid beds, dilute-phase transfer lines, or combinations of the two fluidization regimes. The advent of high activity catalysts containing zeolites widened the range of design configurations, feed properties, and operating conditions. These units are large, and most feature short residence time, transfer line reactor sections.

Gas–Liquid–Solid Fluidization Reactors. The logical extensions of gas-fluidization technology are the ebullating bed gas–liquid–solid and slurry reactors, where both liquid and gas fluidize the solids. Such reactors can be used for catalytically hydrodesulfurizing and hydrodemetalizing residual fuels, upgrading heavy gas oils and residue for further processing, and converting nonpumpable heavy crudes and bitumens into transportable lighter crudes suitable for conventional processing.

BARRY L. TARMY
TBD Technology

G. F. Froment and K. B. Bischoff, *Chemical Reactor Analysis and Design*, John Wiley & Sons, Inc., New York, 1979.

K. R. Westerterp, W. P. M. van Swaaij, and A. A. C. M. Beenackers, *Chemical Reactor Design and Operation*, John Wiley & Sons, Inc., New York, 1984.

O. Levenspiel, *Chemical Reaction Engineering*, 2nd ed., John Wiley & Sons, Inc., New York, 1972.

H. S. Fogler, *Elements of Chemical Reaction Engineering*, 2nd ed., P.T.R. Prentice Hall, Englewood Cliffs, N.J., 1992.

RECORDING DISKS. See INFORMATION STORAGE MATERIALS.

RECREATIONAL SURFACES

Recreational surfaces are synthetic, durable areas of consistent properties designed for various activities, including the high performance requirements of many sports. The category also includes indoor—outdoor carpets and similar materials designed for low maintenance in home or light recreational service. The characteristics of the artificial playing surface may be selected to match natural surfaces under ideal conditions or may have special features for specific sports purposes. The intent is to provide appropriate functionality for the activity combined with good durability of the product. In most cases, the artificial surface permits greatly increased utilization compared with natural grass.

In terms of surface area covered, artificial turf is the dominant commercial product in the category of artificial surfaces described in this article. Light-duty surfaces are the largest representative, followed by tennis court and multipurpose recreational surfaces.

Types of Surfaces

Light-Duty Recreational Surfaces. Artificial surfaces intended for incidental recreational use, eg, swimming pool decks, patios, and landscaping, are designed primarily to provide a practical, durable, and attractive surface. Most surfaces in this category utilize polypropylene ribbon and a tufted fabric construction (see OLEFIN POLYMERS, POLYPROPYLENE).

Single-Use Athletic Surfaces. Included here are installations designed for a particular sport or recreational use. Specific performance criteria are important and differ depending on the application.

Multipurpose Recreational Surfaces. The performance demands control the design for artificial surfaces in this category. The shock absorbency of the system affects player safety and long-term performance under very heavy, usually multipurpose use. The grass-like fabrics used for these applications are made from various pile materials, including polypropylene, nylon-6,6, nylon-6, and polyester (see FIBERS, OLEFIN; FIBERS, POLYESTER; POLYAMIDES). The fabric may be woven, knitted, or tufted. The underpad is derived from various materials, representing a compromise of properties.

Performance Characteristics

User-Related Properties. The most important element in the player's contact with the surface is traction. Shoe traction for light-duty consumer purposes need address only provision of reasonable footing. The frictional characteristics are of much greater importance in surfaces designed for athletic use. With grass-like surfaces, traction is significantly affected by pile density and height, and other aspects of fabric construction.

The coefficient of static friction between the playing surface and the shoe determines traction.

Static friction coefficients demonstrate that the absolute traction values for synthetic surfaces are satisfactory in comparison with natural turf, provided that shoes with the appropriate surfaces are employed. Synthetic surfaces by virtue of their construction are to a degree directional, a characteristic which, when substantial, can significantly affect both player performance and ball roll.

Abrasiveness of an artificial turf surface upon contact with bare skin is a performance criterion to be considered. Artificial surfaces are more abrasive than natural grass in good condition, although the latter typically contains much higher levels of bacteria.

In general, user-related properties should encompass a good balance of traction, comfort, and safety.

Game-Related Properties. For some activities, such as running and wrestling, the only consideration is the direct impact by the player. For others, eg, tennis, baseball, or soccer, the system must also provide acceptable ball-to-surface contact properties. Important ball-response properties on the artificial surface are coefficients of restitution and friction, because these directly determine the angle, speed, and spin of the ball.

The coefficient of restitution is defined as the ratio of the vertical components of the impact and rebound velocities resulting when a ball is dropped or thrown onto a playing surface.

The coefficients of static friction between a ball and the playing surface are the ratios of the horizontal forces necessary to initiate a sliding or rolling motion across the surface to the normal forces (wt) perpendicular to the surface. The sliding and rolling coefficients of dynamic friction are similarly defined in terms of the forces necessary to sustain uniform motion across the playing surface. These friction coefficients determine slip or retention of inertial effects present upon impact.

As a general rule, artificial turf surfaces tend to be somewhat livelier in ball response, velocity, and distance of roll, with coefficients of friction lower than those for natural grass.

Impact Properties. Artificial playing surfaces for moderate to heavy use must provide shock absorbency for player comfort and safety. This is achieved by incorporation of a resilient layer, usually a shock-absorbing underpad.

An ideal shock-absorbing medium, eg, for football in the United States, would combine a reasonable softness in normal shoe contact with a high capacity for dissipation or distribution of kinetic energy involved in the impact of a player's fall. Various foamed elastomers are suitable for this purpose (see ELASTOMERS, SYNTHETIC).

Durability. Grass-like surfaces intended for heavy-duty athletic use should have a service life of at least eight years, a common warranty period provided by suppliers. Lifetime is more or less proportional to the ultraviolet (uv) exposure (sunlight) and to the amount of face ribbon available for wear, but pile density and height also have an effect. Color is a factor; generally uv absorption is highest with red fabrics and least with blue. In addition, different materials respond differently to abrasive wear. These effects cannot be measured except in simulated field use and controlled laboratory experiments, which do not necessarily reflect field conditions.

Materials and Components

A grass-like recreational surface system includes the top material directly available for use and observation, backing materials that serve to hold together or reinforce the system, fabric-backing finish, a shock-absorbing underpad system if any, and adhesives (qv) or other joining materials. The system is installed over a subbase, usually of asphalt or concrete.

Surface Materials. Pile materials used in grass-like surfaces may be selected from fiber-forming synthetic polymers, such as polyolefins, polyamides, vinyl polymers, and many others. These polymers exhibit good mechanical strength in the necessary direction.

Backing Materials. Any fiber-forming polymer with reasonable tenacity may be used in backing materials. The backing provides strength and offers a medium to which the pile fibers can be attached. It provides dimensional stability and prolongs service life.

Backing Finish. The backing material must be consolidated with the pile ribbon. Backing materials are usually applied as a coating which is subsequently heat-cured.

Underpads. Shock-absorbing underpad material is usually made of foamed elastomer, which provides good energy absorption at reasonable cost (see FOAMED PLASTICS).

Coating, Adhesives, and Joining Materials. Grass-like surfaces are employed over substantial areas, and lengths of rolls must be joined, glued, or sewn together. A variety of adhesives, ranging from low cost poly(vinyl acetate) materials to cross-linked epoxy cements, is utilized. Sewing threads may be selected from the group of drawn, high tenacity yarns such as nylon-6,6 and polyester.

The common asphalt tennis courts have been improved significantly by synthetic all-weather coatings with superior appearance and characteristics. Typical coatings are vinyl or acrylic compositions in various colors. Poured-in-place and preformed systems of polyurethane, vinyl, and rubber, although often used, are more expensive than coated asphalt or concrete. More recently, open molded mats that are easily installed as interlocking tiles have been used to construct new or to repair tennis courts.

Fabrication of Grass-Like Surfaces

Tufting. The tufting process is frequently employed in the construction of grass-like surfaces. The manufacturing techniques are essentially those developed for the carpet industry with characteristics of high speed and economy. Pile yarn is inserted into the back side of a woven or nonwoven fabric constituting the primary backing by a series of needles, each creating a loop or tuft as the yarn penetrates the backing, and forms the desired pattern on the other side. For artificial surfaces, the looped tufts that form in this process are cut to provide the desired individual blades in the playing surface.

Knitting. The knitting process, as applied to manufacture of artificial turf and related products, provides a high strength, interlocked assembly of pile fibers and backing yarns. Pile yarn, stitch yarn, and stuffer yarn are assembled in one operation. The pile and stitch yarns run in the machine or warp direction. The stuffer yarns interlock the rows (wales) formed by the pile and stitch yarns, knotting the system together in the width direction. Knitted fabrics typically possess high strength and high tuft bind.

Weaving. The weaving process consists of a two- or three-dimensional meshing of warp, pile, and fill yarns that may be of different types. The pile yarns are cut by a series of wires that are continuously assembled into and withdrawn through the fabric loops. The weaving technique is little used for recreational surfaces.

Finishing. In each of the processes discussed above, the artificial turf fabric is subjected to a finishing operation in which an adhesive is applied to the back side, with or without the optional reinforcing secondary backing, bonding the components and stabilizing the material. The finish may be applied with a knife or roll, in paste or foam form, followed by a heating and drying stage.

Underlayment. An installed artificial turf system may or may not include components between the fabric and the subbase. Such components are not required for light-duty applications, but are essential in attaining the shock-absorbing properties required by heavy-duty surfaces. The foam underpads for shock-absorbing systems are made by incorporating a chemical blowing agent into the foam latex or plastisol.

Installation

Grass-like surfaces for heavy-duty athletic use usually are glued to or laid over a subbase of asphalt or other permanent foundation material. The shock-absorbing underpad component is in contact with the subbase layer, and the turf component is placed on top of the underpad. Turf panels are fastened together by sewing or gluing, and the entire perimeter of the grass-like surface is securely anchored to the subbase.

Newer variations of the normal asphaltic subbase include a permeable version which allows rainwater to trickle through holes punched into the underpad and to drain away through the asphaltic subbase to a pipe grid system under the base (Fig. 1).

Bonding of the underpad to the subbase can vary from full glue-down or strip gluing, to loose-laying. The latter technique, without glue-bonding to asphalt, is employed in a float-drain construction in which rainwater drains laterally from the field between the pad and the asphalt. Glue is also omitted at the subbase interface in constructions where the pad is permanently bonded to fabric in the factory, and the composite is installed in the field. Loose-laying again is employed on convertible fields, such as indoor stadiums, where the turf can be removed to permit other sports uses of the subbase.

For other recreational surfaces, such as running tracks, the installation techniques are quite different. Most are poured-in-place. An interlocking tile technique may be employed for tennis courts. In all cases, adequate provision for weathering and water drainage is essential. In general, the resilient surfaces are installed over a hard base (see Fig. 1) that contains the necessary curbs to provide the finished level. Outdoors, asphalt is the most common base, and indoors, concrete. A poured-in-place polyurethane surface is mixed on-site and cast from at least two components, an isocyanate and a filled polyol of the polyether or polyester type.

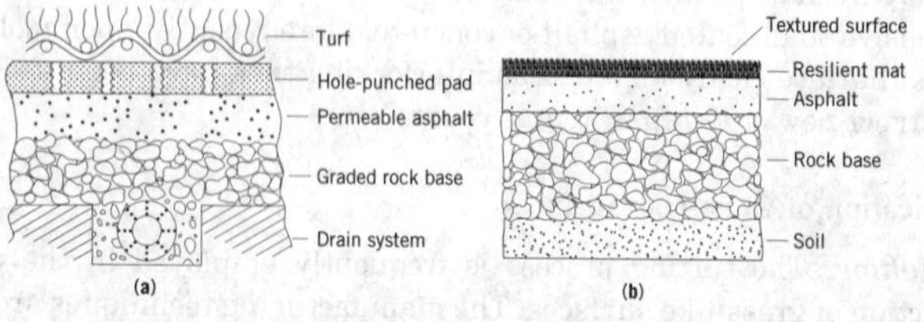

Figure 1. Cross sections of (**a**) typical artificial turf and (**b**) resilient track surfaces.

The mixed liquid is pumped into the area, where it cures and forms a slab.

T. A. OROFINO
H. G. SWEENIE
AstroTurf Industries

T. A. Orofino and J. W. Leffingwell, *ASTM STP 1073—1990*, 166–175 (1990).
T. A. Orofino, *Polym. News* **10**, 294 (1985).
E. M. Milner and J. R. Gilliam, technical data, Monsanto Co., Dalton, Ga., 1977.
R. D. Breland, *ASTM STP 1073*, 176–182 (1990).

RECYCLING

INTRODUCTION

Recycling is the process by which materials are separated from waste destined for disposal and remanufactured into usable or marketable materials. While the amount of public attention given to recycling has increased, recycling itself is an age-old process. However, widespread public interest in recycling is largely a modern phenomenon.

The demand for post-consumer materials has not kept pace with the increases in curbside collection programs. In many regions of the United States and elsewhere, the supply of recyclable materials is so great that cities have been forced to either store the materials or curtail the number of items collected.

As a result of this oversupply, scrap values for many recyclable materials fell noticeably.

Industrial Materials

Industrial recycling is the recovery for reuse or sale of materials from what otherwise would be wastes destined for disposal. Typically, the reclaimable materials employed in industrial recycling may consist of obsolete products, spent materials, industrial by-products or residues, or pollution control products.

The actual processing of industrial discards varies in complexity by material type. Typically, the hulk of the automobile is shredded and the pieces separated into ferrous and nonferrous metals (RECYCLING, FERROUS METALS; NONFERROUS METALS). The separated materials are then sent to be resmelted or are exported. For example, processing industrial by-products and pollution control products can be very complicated. Because these materials often consist of complex mixtures of metals or chemicals, recycling must take place in several stages. The industrial recycling of these materials, many of which are considered hazardous, has both environmental and economic benefits. Not only does recycling separate valuable constituents, it also removes hazardous materials. Thus, industrial recycling removes the threat that these materials pose to the environment and public health.

Municipal Solid Waste

Municipal solid waste (MSW) is most often defined as post-consumer solid waste generated by households, commercial establishments, and institutions. Normally, MSW is classified as either material waste, ie, items such as paper, yard waste, metals, and glass, or product waste, which encompasses both durable and nondurable goods as well as packaging (qv). Beyond these simple classifications, defining MSW has been problematic because of disagreements regarding specific materials and the proper classification of pre- and post-consumer waste.

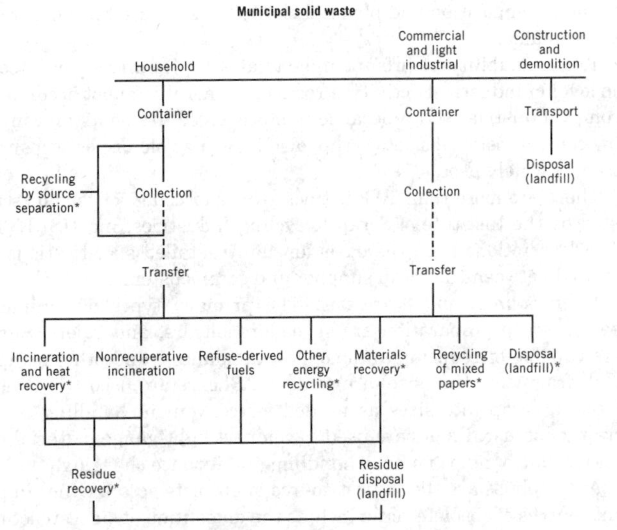

Figure 1. Municipal solid waste management system where (*) indicates recycling options and (— — —), optional transfer.

Figure 1 schematically depicts the system that has developed to manage solid waste. Both materials recycling and energy recovery are viable options to either landfilling or nonrecuperative incineration. Composting, which is not present in Figure 1, is not widely used in the United States as a method of handling MSW. This is primarily because composted material contains relatively high concentrations of heavy metals and supplies very few plant nutrients.

Quantity and Composition. The actual quantity and composition of waste are highly dependent on local use habits, income, as well as the degree of urbanization.

Estimates of per capita waste generation can be misleading. Although often reported in the popular press, the magnitude of these figures depends on the size of the community, how the statistics are gathered, and the percentage of waste in a given residential region. The use of national averages in designing facilities to handle MSW can result in large economic mistakes due to inaccurate estimates regarding the amount and type of waste to be received.

By far the largest contributors to MSW are paper and paperboard products (37.5% by weight) (see RECYCLING, PAPER). Yard waste, including leaves, grass clippings, weeds, and prunings, represents the second largest category of waste. The yard waste proportion of total discards has declined steadily, however, and this decline will likely accelerate because many individual states have banned yard wastes from municipal trash. The percentages of glass, metals, and food waste in MSW have likewise declined somewhat since the 1970s due in large measure to lightweighting and substitution by other materials.

On the other hand, the fraction of plastics has grown, an increase that corresponds to the substitution of other materials with plastics and the greater reliance on plastics as a source of packaging material (see RECYCLING, PLASTICS).

As with generation rates, the chemical composition of MSW varies significantly with local socioeconomic and demographic conditions.

Processing Recyclable Materials. The rate at which MSW is recovered for recycling varies by region. The best estimate for the average recovery of MSW in the United States is approximately 15%. This figure underestimates total recovery because it excludes both domestic reuse, such as washing and reusing plastic containers, as well as the reuse of industrial scrap.

Projected national recovery rates much beyond 20% are questionable. Although close to 85% (by weight) of MSW is composed of potentially recyclable material (paper and paperboard, metals, glass, plastics, wood, and yard waste), far less is economically recyclable.

Preparation of Collected Materials. The actual amount of recovered MSW that can be recycled to meet buyers' quality specifications is highly dependent upon how the material is collected and processed. There are primarily three methods available to collect MSW for recycling: mixed waste, waste with commingled recyclables, or waste with separated recyclables. Which method of collection is chosen, in turn, determines the amount of preparation that is needed prior to reclamation and reuse.

The preparation of mixed MSW for recycling is essentially a four-step process. In the first stage, commonly referred to as previewing, the mixed waste is dumped on a tipping floor where oversized materials, potential explosives and readily flammable materials, and any other items that could damage processing equipment are removed. The waste is then crushed or shredded. After the segregation process, valuable materials are normally baled or otherwise prepared for transportation to market. Residue waste is either incinerated or landfilled. In many systems, the leftover material is shipped to a waste-to-energy facility where it is converted into refuse-derived fuel.

The principal advantage of handling mixed MSW is that it requires no change in the existing waste collection system.

The technology required to process commingled recyclables is dependent upon the types of materials collected. As a general rule, however, commingled materials are first dumped into a receiving pit. After initial inspection, they are loaded onto conveyors and separated. As with mixed waste processors, handlers of commingled materials normally employ a combination of automated and manual systems.

Even when initial separation of recyclables takes place at the household level, the separated materials still require some preparation before being sent to market. By visually inspecting materials, facility processors are able to remove any remaining contaminants. This ensures that recyclables will be of sufficient quality to meet buyer specifications. In addition, materials are often shredded, crushed, or baled to facilitate cost-effective shipping.

Refuse-Derived Fuel. Many processing facilities divert a portion of the material that is not recovered for recycling to waste-to-energy plants, also referred to as resource recovery facilities, where the material is employed as fuel. The processes involved in the production of refuse-derived fuel (RDF) are varied. Nine different RDFs have been defined, as listed in Table 1. There are several ways to prepare RDF-3, which is perhaps the most popular form and is the feed used in the preparation of densified refuse-derived fuel (d-RDF). All forms of RDF are part of the broader set of waste-derived fuels (WDF), which includes various waste biomass (see FUELS FROM BIOMASS; FUELS FROM WASTE).

RDF-3 is intended for use as a supplement with coal for semisuspension or suspension firing or for use by itself in similar boilers. d-RDF is intended as a supplement with stoker coal or for use by itself in stoker biolers.

Mechanical and Chemical Recycling. The vast majority of recovered materials which are not burned for energy are simply remanufactured into second-generation products. Such mechanical recycling works primarily by applying heat (or in the case of paper, various chemicals) to the sorted and cleaned waste and then refashioning the liquid material into new products. For many materials (especially plastics), however, mechanical recycling has significant drawbacks. As a result, a number of projects are seeking to develop chemical technologies that

Table 1. Definitions of Refuse-Derived Fuels

RDF-1	wastes used as fuel in discarded form
RDF-2	wastes processed to coarse particle size with or without removal of magnetic metals
RDF-3	as MSW-derived shredded fuel which has been processed for the removal of metal, glass, and other entrained inorganic material; generally, this material has a particle size such that 95 wt % passes through a 5-cm mesh screen
RDF-4	combustible waste processed into powder form; 95 wt % passes through a 2.0-mm (10-mesh) screen
RDF-5	combustible waste compressed into pellets, slugettes, cubettes, or briquettes
RDF-6	combustible waste processed into gaseous fuel

can convert recovered wastes back into the higher value raw materials from which they were made.

Economic Aspects of Recycling

Production. Several key components of MSW enjoy relatively high rates of recycling, for example, aluminum. The primary reason for the success of aluminum recycling is that collecting and reprocessing post-consumer aluminum is more cost effective than mining and processing bauxite. Similarly, paper and paperboard is recycled at approximately a 29% rate.

Other MSW discards such as glass, ferrous metals, and plastics may begin to exhibit marked improvement in their recycling rates due to the explosion in curbside recycling programs and in construction of reprocessing facilities.

Economic Analysis. The economic success of recycling programs is subject to the following inequality where X = the cost to recover recyclable materials, Y = the cost of disposal, and Z = the value of the resource recovered: $X - Y \leq Z$.

Basic economic theory suggests that, in the earliest periods, society will rely primarily on virgin material because it is cheaper than collecting and recycling post-consumer goods. As the stock of virgin material is consumed over time, however, a point is reached when the costs of extraction and the price of this material will begin to rise. With the rise in virgin prices, consumer demand for alternative materials including recyclables will slowly increase as will Z.

Concurrently, increased investment in technology will likely lower the cost (X) to recover and recycle post-consumer materials. Eventually, the inequality above is satisfied for some materials and recycling becomes economically viable.

CHRISTOPHER BOERNER
KENNETH CHILTON
Washington University

J. H. Alexander, *In Defense of Garbage*, Praeger Publishing, Westport, Conn., 1993.

C. Boerner and K. Chilton, *Environment* **36**(1), 7–15, 32–33 (Jan./Feb. 1994).

W. Rathje and C. Murphy, *Rubbish: The Archaeology of Garbage*, Harpers Collins, New York, 1992.

METALS

FERROUS METALS

Recycling of ferrous scrap is a principal worldwide activity. Numerous grades of scrap and many technical and economic complexities are involved. The term ferrous metals refers to iron (qv) and steel (qv) scrap materials of three origins: home scrap, generated within the steel mill or foundry, and being a known material of high quality; prompt industrial scrap, produced from trimmings and discards during product manufacture; and obsolete scrap, ie, old scrap from discarded or rejected items.

Primary consumers for ferrous scrap are the iron and steel mills and foundries. Minor consumers include ferroalloy producers, copper producers for use in copper precipitation (see RECYCLING, NONFERROUS METALS), and the chemical industry. The steel industry consumes about three-fourths of the total.

Sources and Types of Ferrous Scrap

Home scrap, because it is generated within the plant during the production of steel or cast iron, has a known composition and is always recycled. Manufacturing facilities and steel service centers produce prompt industrial scrap during preparation to fabricator's specifications and during the fabrication of various industrial, commercial, or consumer products. Prompt industrial scrap is recycled because its chemical composition and physical characteristics are known or readily obtainable.

The availability of prompt industrial scrap is directly related to the level of industrial activity. Producers generally do not accumulate prompt industrial scrap because of storage requirements and inventory control costs. Obsolete scrap, also known as old or post-consumer scrap, is widely used.

There are more than 100 distinct grades or codes for ferrous scrap listed by the Institute of Scrap Recycling Industries, Inc. (ISRI). The ISRI descriptions form the basis for many detailed specifications negotiated between the scrap supplier and scrap customer.

Scrap Sources and Processing. The primary types of equipment used in scrap preparation are shredders, shears, and balers. Shredders capable of reducing automobiles, appliances, and other scrap to small fragments (fist-size) are common. Shears are used to cut large pieces of scrap into sizes as needed for convenient handling by the scrap consumer. Balers are used to compact light-gauge material into rectangular bundles for easy handling and furnace charging.

Automobiles are the largest source of obsolete scrap. Other important sources of obsolete scrap include the demolition of steel structures and railroad companies. The latter provide a steady flow of scrap from their fabricating shops and from the recovery of worn out or abandoned track and railroad cars. All iron and steel products are recyclable if economically retrieved when scrapped.

Appliances (white goods) are being recycled in increasing quantities.

Municipal solid waste (MSW) provides some forms of recyclable ferrous scrap, but the quality and markets are limited.

Automobile and industrial oil filters are another source of ferrous scrap. Processing equipment and melting procedures are being developed for routine melting in steel furnaces.

Construction and demolition (C&D) debris is a potentially large source of recyclables. Ferrous materials in C&D debris are typically reinforcing bars, wire mesh, and structural steel.

Primary Uses of Ferrous Scrap

All wrought steel is produced by basic oxygen furnace (BOF) and electric-arc furnace (EAF) processes. The steel industry consists of two principal types of steelmaking facilities: integrated plants, which use blast furnaces, BOFs, and EAFs, and specialty plants and minimills, which use EAFs exclusively. The integrated steel plants begin the steelmaking process with iron ore, which is reduced in blast furnaces and converted to molten pig iron, also known as hot metal. Hot metal is transferred to a BOF where scrap is added and refining takes place. Final refining and alloy additions to produce the desired grade of steel follow. The proportion of scrap added to the BOF is typically 10–30% of the total metal charge. Some integrated plants also have electric-arc furnaces for converting scrap to certain specialized grades of steel.

The EAF-based minimill steel industry has assumed a principal role in production of a variety of carbon-steel grades, once the exclusive domain of integrated steel producers. EAFs are used as rapid scrap melting devices producing steel from nearly 100% scrap charges. Refining is limited and obtaining the desired grade of steel product depends on the quality of scrap and other charge materials. Addition of pig iron or direct reduced iron (DRI) (see IRON BY DIRECT REDUCTION) during EAF steelmaking may be required to obtain the desired chemical compositions and limit residual elements.

Ferrous foundries consist of two types: steel foundries in which electric furnaces (EAF and induction) are used, and iron foundries in which hot-blast cupolas and/or electric furnaces are used. Electric furnaces use virtually 100% scrap charges. Cupolas are shaft furnaces which use preheated air, coke, fluxes, and metallic charge. Scrap is over 90% of the metallic charge. Iron foundry products have a high carbon content and the scrap charge usually contains a high percentage of cast iron or is used in combination with pig iron.

Residual Elements. Steelmakers need to control several elements associated with scrap in order to meet the quality requirements of their final products. Copper and tin are of particular concern, because these are not volatile, oxidizable, or otherwise refinable by normal furnace practice. Dilution using virgin materials or prime quality scrap may be required to obtain the desired chemical composition. Copper and tin in relatively small concentrations, degrade the hot workability and deep drawing quality of many steels. Other elements such as chromium, nickel, and molybdenum affect various mechanical properties. Cast irons are produced in several classes, such as gray iron, malleable iron, and ductile iron, each having specific microstructural characteristics that are critical for meeting product specifications. Low concentrations of lead, tin, or arsenic, for example, inhibit desirable graphite structure in ductile iron. Whereas the effects of individual elements may be known, interactions of two or more elements are difficult to determine. Relatively large samples of scrap (eg, ≥20 t lots have been melted to obtain homogenized melt analysis for the presence and concentrations of residual elements. Analyses of scrap have been published by various investigators over many years; these can be misleading, however, if used to indicate the expected analysis of a given grade of scrap from all suppliers and at different periods of time.

The steel mills and foundries consume many different grades of scrap. Such factors as metal yield during melting and uniformity of quality regarding desired physical form and chemical analysis are important. The primary users' products must meet strict chemical and physical quality requirements and, although scrap generally offers the most economical form of iron units, proper selection and blending is essential to assure acceptable final product quality. Obsolete scrap is typically very heterogeneous.

Factors Influencing Ferrous Scrap Recycling

Supply, Demand, and Prices. The economics of ferrous scrap recycling involve a complex variety of factors related to the demand of consuming industries and cost of supplying scrap. Cost of scrap is affected by such factors as collection, regional availability, processing and upgrading to acceptable quality, transportation, environmental controls, and export demand.

Environmental and Regulatory Aspects. Ferrous scrap recycling provides many well-documented environmental benefits including reduced roadside litter, landfill requirements, and pollution and energy consumption compared to use of virgin materials. Use of scrap for steelmaking results in large reductions in air pollution, water use and pollution, mining wastes, and energy consumption while also conserving iron ore, coal, and limestone. The savings in landfill space is also considerable. Recycling operations do, however, generate certain emissions and waste streams that are being subjected to increasingly stricter environmental controls, thus increasing the cost of recycling in many cases. Regulations affecting metal recycling are numerous.

Standards

The most comprehensive set of descriptions of ferrous scrap are published by ISRI. Individual steel mill and foundry consumers usually follow the ISRI specifications, although many also incorporate specific requirements tailored to the needs of the consuming facility.

Iron and steel products must meet increasingly strict quality standards; thus iron and steel producers continually seek scrap of uniformly consistent quality.

HARRY V. MAKAR
Consultant

R. E. Brown, *Metal Prices in the United States Through 1991*, U.S. Bureau of Mines Report, Washington, D.C., 1993, pp. 73–80.

A. J. Stone and P. H. Meyst, *Ferrous Scrap Materials Manual*, ICRI Report No. 517, Iron Casting Research Institute (ICRI), Columbus, Ohio, Mar. 31, 1988.

R. D. Burlingame, *Ferrous Scrap Explained*, Luria Brothers & Co., Inc., Cleveland, Ohio, 1981.

R. R. Jordan and G. L. Crawford, *The McGraw-Hill Recycling Handbook*, McGraw-Hill Book Co., Inc., New York, 1993, pp. 15.1–15.27.

NONFERROUS METALS

A nonferrous metal is any metal other than iron or one of the many iron alloys. For the nonferrous metals, recycling of aluminum, copper, lead, and zinc is of primary economic importance. The principal factors that determine recyclability include availability of both primary and secondary sources, purity, cost of energy, cost of waste generated, and regulatory environment. As natural ore bodies are depleted, the relative availability of waste materials is expected to increase. Two key advantages of recycling are the conservation of resources and protection of the environment (see ALUMINUM AND ALUMINUM ALLOYS; ZINC AND ZINC ALLOYS).

Nonferrous metals are primarily used as the metal or in an alloyed form, facilitating the recycle of the metal or alloy because much of the recycling process requires only that the scrap metal be sorted, melted, and cast (pyrometallurgical processing). The scrap is classified into two general categories, new and old. New scrap derives from material that never reaches the market, such as scrap generated during the manufacture of products. Old scrap is obtained from worn out or discarded products. Often, the recycling process consumes only 5–50% of the energy needed to produce from an ore and in most cases generates only a small portion of the waste.

The term recycling has no regulatory significance. Some recycling is regulated, some is not. Regeneration, a regulatory subset of reclamation, which itself is a subset of recycling, is regulated (see REGULATORY AGENCIES, CHEMICAL PROCESS INDUSTRY).

Metal finishing and plating industries produce wastewater treatment sludges. Although the metal content of these wastes has not been specified, it is expected that reclamation of metals from such sources could account for significant additional resource recovery (see WASTES, HAZARDOUS WASTE TREATMENT).

Metal and metallic salt wastes are generated primarily by the electronic, electroplating, metal finishing, and machining industries, and as consumer waste. Significant quantities of metal-containing spent catalysts are also produced by the organic chemical industry. Much of the metal-bearing waste is classified as hazardous by the U.S. EPA under the Resource Conservation and Recovery Act (RCRA). Often, wastes such as salts, basic salts, and hydroxides are more amenable to hydrometallurgical methods of reclamation.

Classification of Recycled Materials

The initial step in the recycle of metals is the physical segregation of the metals from other materials. This classification and segregation of scrap is of importance to the producers of the metals from secondary materials. Historically, much of the classification has relied on hand sorting which can be reliable, but it is labor intensive. The recycling of automotive scrap is illustrative of the techniques that can be used to separate nonferrous metals into broad categories.

Automobile scrap generates almost 20% of old scrap copper yet only contains 10–35 kg per unit. The average mass of an automobile, which has a useful life expectancy of 10 years, is 1100 kg. The unit operations normally used in the recycling of automotive scrap are illustrated in Figure 1.

Pyrometallurgical Methods of Recycling

Much of the technology used in the reclamation of metals from metal-bearing wastes was developed by the mining industries. The primary means of recycling metal from metal and alloyed scrap is via pyrometallurgy (see METALLURGY).

Aluminum. The recycling of aluminum scrap offers significantly reduced energy requirements when compared to refining from ore.

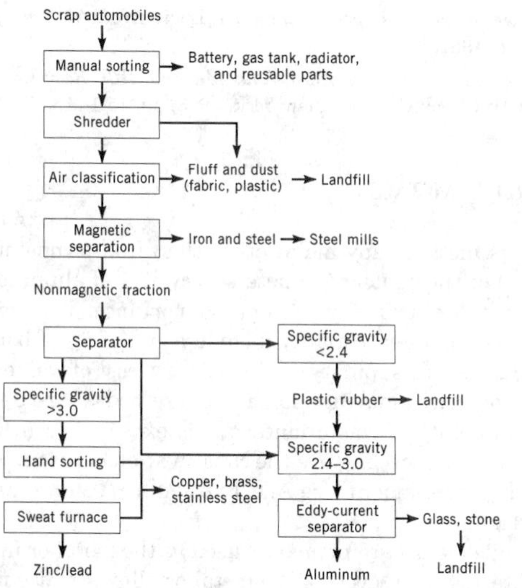

Figure 1. Unit operations and industrial usage for the recycling of automotive scrap.

(see ALUMINUM AND ALUMINUM ALLOYS). There are several melting techniques used, each having advantages and limitations. Aluminum is very reactive and has a strong tendency to form a dross, ie, an oxide containing entrained metallics, at the surface of the metal. In order to recover the aluminum from the dross, salts, usually chlorides, must be added. This treatment produces a secondary dross. Disposal of this waste, undergoing increasing restrictions, is becoming a less attractive alternative. Therefore, new technologies are being pursued such as centrifugation, flotation, plasma, and arc melting.

Rotary-barrel or vertical cement mixer-type rotary gas- and oil-fired furnaces offer several advantages in the processing of scrap. These can be operated in batch or continuous mode.

The processing of used beverage containers represents a significant portion of the recycled aluminum. The material is separated by the consumer and compacted into bales for shipping to the secondary recycler. The bales are crushed and shredded followed by removal of magnetic materials and gravel (see SEPARATIONS, MAGNETIC SEPARATIONS).

Copper. Metals that cannot be separated from copper (qv) by traditional physical methods, eg, alloys of zinc, tin, and lead, can be eliminated by melting and oxidation. The following general conclusions can be made: zinc is removed by reduction to zinc metal followed by fume oxidation and collection; aluminum and iron are removed by air oxidation to slag; lead partitions between the alloy, slag, and vapor, as does tin (vapors are a mixture of lead(II) oxide, sulfide, and chloride, and tin(II) oxide and sulfide and tin chloride); nickel partitions between the slag and metal (outlet is the electrolyte tankhouse bleed); and precious metals are removed during electrorefining.

Low grade scrap (15–70% Cu) requires complete smelting, converting, and refining. The smelting operation consists of a blast furnace operated under strongly reducing conditions. Alternatively, the converting furnace (Pierce-Smith, Hoboken, New Jersey) or top-blown rotary is distinctly more oxidizing, but produces a high copper-containing slag (see COPPER). This slag is usually recycled to the blast furnace.

Zinc. Secondary zinc metallics are usually melted and selectively drossed or vaporized. Lead is removed from the bottom of the melt as an insoluble and iron forms the intermetallic $FeZn_3$ at the zinc–lead interface. Aluminum forms a top dross intermetallic with iron, $FeAl_3$. Further purification is obtained through distillation. The zinc is separated from the intermetallic by the effects of the agitation. Air is introduced to oxidize the surface of the intermetallic, further separating the zinc from the dross. The dross contains about 10% each of aluminum and iron. The zinc is advanced to a reverbatory furnace for alloying with aluminum.

Hydrometallurgical Methods of Recycling

Changes in environmental regulations, notably RCRA, have brought about increased implementation of hydrometallurgical means for metals recycling. Because many of the wastes are classified as hazardous, there are economic incentives to recycle rather than discard such materials. Significant improvements have been realized in the efficiency of hydrometallurgical or aqueous solution-based processing.

In the mining industry, there has been a shift in emphasis from pyrometallurgy to hydrometallurgy. This shift in emphasis has led to the development and use of a variety of improved techniques, in particular the commercial availability of several metal specific extractants. These techniques are particularly useful in the separations and recycling of metals from metal sludges and metal salt solutions.

Precipitation. Precipitation is one of the oldest techniques used for metal–metal and metal–solution separations. Precipitation can be illustrated by the following reactions:

$$Cu^{2+} + 2\ Fe^{3+} + 6\ OH_- \xrightarrow{H^+ \text{ to pH } 3.5} Fe_2O_3(s) + 3\ H_2O + Cu^{2+} \quad (1)$$

$$Cu^{2+} + Fe^0 \rightarrow Cu^0(s) + Fe^{2+} \quad (2)$$

The first equation is an example of hydrolysis and is commonly referred to as chemical precipitation. The second equation is known as reductive precipitation and is an example of an electrochemical reaction. The use of more electropositive metals to effect reductive precipitation is known as cementation. Precipitation is used to separate impurities from a metal in solution such as iron from copper (eq. 1), or it can be used to remove the primary metal, copper, from solution (eq. 2). Precipitation is commonly practiced for the separation of small quantities of metals from large volumes of water, such as from industrial waste processes.

Electrowinning is the deposition of metals from a leach solution by electrolysis where the metal is produced by cathodic reduction. Cementation relies on galvanic potential differences to effect metal reduction. Electrowinning can be considered a form of reductive precipitation.

Solvent Extraction. The selective partitioning of metals by liquid–liquid solvent extraction is one of the most powerful methods of hydrometallurgical separation. Solvent extraction can also be used to concentrate dilute metal streams to facilitate further processing. The metal extraction process can be made highly selective or exclusive either by design of the extractant or by the chemical conditions imposed during extraction.

In practice the extractant is carried by an aqueous immiscible solvent. The aqueous layer containing the metal ions is contacted (mixed) with the solvent. Depending on conditions, typically pH, certain of the metal ions are extracted into the organic layer. The mixed organic aqueous system is allowed to separate and the metal-laden organic is then contacted by an aqueous solution that strips the metal ion from the organic. Several stages of solvent–aqueous contact are usually required in order to achieve the degree of separations and efficiencies desired.

Waste Management

Total elimination of metal-bearing wastes is not a likely occurrence owing to process and product requirements and consumer disposal tendencies. On-site waste minimization efforts have been extensive in the industrial sector, however, owing to economic and regulatory driving forces. Local and state programs have had a significant impact on the residential recycling of metal, primarily aluminum, and on glass, plastics, and paper products.

Although it is generally not practical to eliminate production of the wastes, it is often possible to produce a waste that is more attractive to a reclaimer. Some factors that can be considered are segregation of waste streams, maximization of recoverable metal content and quantity, and minimization of secondary metals, leach insolubles, organics, and chelating agents (see WASTES, INDUSTRIAL).

H. Wayne Richardson
Phibro-Tech, Inc.

H. W. Richardson, "Hydrometallurgical Reclamation of Copper from Metal-Bearing Waste," in H. Wayne Richardson, ed., *Handbook of Copper Compounds and Applications*, Marcel Dekker, New York, 1997, pp. 337–393.

P. B. Queneau, B. J. Hansen, and D. E. Spiller, in "Recycling Lead and Zinc in the United States," *Proceedings of the Milton E. Wadsworth 4th International Symposium on Hydrometallurgy*, AIME SME/TMS, Salt Lake City, Utah, Aug. 1–5, 1993.

B. Kos, "Waste Management and Recycling," *International Symposium*, CIM, Vancouver, B.C., Canada, Aug. 1995.

C. S. Brooks, *Metal Recovery from Industrial Waste*, Lewis Publishing, Chelsea, Mich., 1991.

GLASS

As some glass products reach the end of their useful life and are discarded, there is often the opportunity to have the glass recycled into other useful products. In many respects, this alternative is preferred over the glass entering a municipal waste stream for landfill disposal.

Glass chemistry is typically categorized by its oxide components (see Glass). Table 1 compares oxide components in the principal categories of glass products, which can potentially be recycled. In most instances, color separation of recycled glass is necessary to avoid color quality concerns upon remelting. Chemical content differences between glass product categories limit the opportunity to recycle between glass categories.

Levels of Recycling

Used and recovered, ie, post-consumer, commercial glass, as well as off-specification glass, suitable for remelting is referred to as cullet. Recycled glass (cullet) is not only made into new bottles and jars, but also used for secondary markets such as fiber glass and glasphalt, ie, paving asphalt utilizing crushed cullet as a grog constituent, replacing stone aggregate.

Commercial glass can be recycled when sufficient quantities can economically justify the development of a processing infrastructure. Typically, only container and flat post-consumer glass have been recycled commercially. A significant proportion of past cullet recycling has been in the glass container industry.

Cullet is one of the four principal ingredients in container glass. The other three are sand, limestone, and soda ash, which are all relatively plentiful and inexpensive. Cullet, by melting at a lower temperature than glass-forming raw materials, allows for a reduction of energy input to melting furnaces. This prolongs furnace life. Substituting cullet for raw materials can also reduce emissions into the atmosphere. From a nonmanufacturing perspective, using cullet conserves landfill space for disposal of nonrecyclable materials.

Cullet for container production is a mixture of in-house rejects and post-consumer glass that has been beneficiated to remove contaminants, such as content residues, bottle tops, and labels that can cause imperfections in the finished containers. Cullet colors must also be segregated to meet manufacturer's specifications. The quality of the cullet is an important factor in determining both the quantity to be used and the quality of the finished containers.

Table 1. Oxide Components in Glass, %

			Fiber glass		
Oxide	Glass containers	Flat for window and automotive	Textile	Wool	Lighting
SiO_2	72.5	70.7	55.0	57.0	76.4
B_2O_3			7.0	5.2	14.6
Al_2O_3	1.5	0.7	14.8	8.0	2.0
CaO	10.0	9.5	20.5	8.1	
MgO	0.5	3.8	0.5	4.2	
Na_2O	14.5	13.3	1.0	14.5	5.4
K_2O	0.5	0.9	1.0	2.1	1.6

Optimal recycling is properly defined as the remanufacturing or reprocessing of discarded materials into the same product. The definition of primary recycling was originally adopted by the American Society of Testing Materials in 1981. In primary recycling, ie, closed-loop recycling, a material is fabricated back into the same material product. Making glass containers into new glass containers or newspapers into newspapers are examples of this.

Secondary recycling, or cascading, reprocesses discarded materials into other materials or products. Examples of secondary uses for glass include road aggregate, reflector beads, and fiber glass. Closed-loop glass recycling is preferred by the glass container industry. However, when this is not economically or technically feasible, secondary recycling is critical to avoid landfilling.

Economics

Recycling of glass containers saves energy, but not as significant a quantity as compared to reuse. The primary energy saved includes energy required for the entire product life cycle, starting with raw materials in the ground and ending with either final waste disposition in a landfill or recycled material collection, processing, and return to the primary manufacturing process. Actual savings depend on many factors, including population density; locations of landfills, recovery facilities, and glass plants; and process efficiencies at the specific facilities available.

Quality and Specifications

Cullet quality specifications are manufacturing company-specific. Typically, the specification includes a representative sampling technique, particle size gradation, color mix ratio, allowable organic and moisture content, and the specific absence of nonglass contaminants.

Glass manufacturers, as well as other industries such as insulating fiber glass, asphalt grog, structural clay and brick additives, and foam glass, all use cullet as a raw material. The recovery and recycled content requirements for each product demands that cullet meet specific quality requirements because it becomes a larger percentage of the raw material blend.

Many collection systems have not adequately prevented nonglass contaminants, such as ceramic tableware, metals, foils, and other foreign articles, from becoming incorporated in the glass stream destined for melting into new containers. A number of processing schemes have been explored to separate various nonglass contaminants from crushed cullet. Magnetic metals can be easily removed. Methods for nonmetal separation have been found expensive, complicated, marginally effective, and not readily adaptable to high tonnage requirements.

Compared to ore minerals, glass can be crushed using less required force and has few equipment abrasion wear concerns.

The process of pulverized cullet reduction yields a product having near-batch equivalent sizing (−12 mesh (≤1.7 mm)) and in a furnace-ready condition. Foil-backed paper, lead and other metals, and some tableware ceramics can be removed in an oversized scalping operation after the first pass through the system. Other contaminants are reduced to a fine particle size that can be assimilated into the glass composition during melting.

Sources of Recoverable Glass

Cullet recyclers implement the following procedures: (*1*) purchase of scrap glass from a variety of sources; (*2*) supply of transportation to and from the beneficiation facility; (*3*) removal of foreign objects, ie, nonglass materials, including plastic, metal, and ceramic; (*4*) crushing of glass to a specified size; and (*5*) selling of cullet as raw material to a glass manufacturer. Recovery of post-consumer cullet is typically obtained from two categories: municipal solid waste (MSW) and direct resource recovery glass.

Generally, recyclables are either collected at curbside or deposited by consumers at various types of drop-off locations. Curbside collections of recyclables can be accomplished either in conjunction with the pickup of all MSW or as a separate activity. Co-collection systems range from complete commingling of all waste for later separation at a mixed waste processing facility to transporting essentially source-separated recyclables in the same truck as MSW.

C. PHILIP ROSS
Creative Opportunities, Inc.

C. P. Ross, *Glass Researcher*, **3**(2), 10–11 (1994).

Glass Containers: Current Industrial Reports, M32G(91)-13, U.S. Department of Commerce, Bureau of the Census, Washington, D.C., May 1992.

Solid Waste Management Policy, Glass Packaging Institute, Feb. 24, 1994.

L. L. Gaines and M. M. Mintz, *Energy Implications of Glass-Container Recycling*, Argonne National Laboratory, Argonne, Ill., Mar. 1994.

OIL

The term *oil* includes a variety of liquid or easily liquefiable, unctuous, combustible substances that are soluble in ether but not in water and that leave a greasy stain on paper and cloth. These substances can include animal, vegetable, and synthetic oils, but usually the word oil refers to a mineral oil produced from petroleum (qv). An oil that has been used or contaminated, or both, but not consumed, can often be recycled to regain a useful material, regardless of its origin. For the purposes of this article, only the recycling of used petroleum oils is considered. The following definitions are useful to the discussion of oil recycling.

Used oil: oil whose characteristics have changed since original manufacture and which is suitable for recycling. This term includes used lubricating oils of all types as well as other oils that can be economically recycled.

Waste oil: oil having characteristics that make it unsuitable for further use or economic recycling, but may be usable as a fuel.

Oil recycling: acquisition and processing of oil that has become unsuitable for its intended use in order to regain useful material.

Oil reclaiming or laundering: use of cleaning methods during recycling to render the oil suitable for further use.

Oil re-refining: use of refining processes during recycling to produce high quality lubricating base stock for lubricants or other petroleum products.

Oil reprocessing: preparation of used oil for application as a fuel.

Characteristics of Used Oils

Used petroleum oils to be recycled can be obtained from a variety of sources. The main types of used petroleum oils that are recycled are internal combustion engine lubricants and hydraulic and industrial oils. The additives and contaminants typical in these oils can cause both performance-related and environmental problems. Used oils may often be commingled with each other and with water, solvents, and other chemicals before being collected for recycling.

Technology

In the recycling of used oils to regain useful products, a number of stages are possible, depending on the original source of the used oil, the level of contamination, and the sophistication of the recycling technology utilized. Three levels of recycling are reprocessing, reclaiming, and re-refining.

Reprocessing. Combustion of used oil as burner fuel has often been condemned because it destroys a valuable resource and can cause substantial environmental pollution through widely dispersed distribution of metal oxides and stable organic contaminants. However, under certain conditions and using suitable precautions, the recovery of the energy as heat is a valid used-oil disposal option.

Processing techniques for the recycling of used oil into fuel include pretreatment of the used oil to remove all or most of the contaminants that cause environmental or operational concerns. An alternative approach is to subject the used oil to minimal reprocessing followed by burning in specialized facilities using acceptable environmental control.

Reclaiming. Used oils can be reclaimed within the user facility or sent outside the facility to a commercial reclaimer. A general description of reclaiming processes may include any or all of the following: (*1*) removal of solid particles by settling, centrifuging, or filtering; (*2*) neutralization of acidic components with clay or alkalies, and removal of resulting soaps by washing; (*3*) heating/distillation to remove volatile solvents, gasoline, and water; (*4*) clay contacting to remove oxygenated components and spent additives or for decolorization; (*5*) aerations and use of biocides to reduce bacterial levels; and (*6*) replenishment of additives.

The product of a reclaiming procedure is a reclaimed oil that often meets original specifications for such uses such as hydraulic fluids, gear lubricants, etc. The most important requirement for effective reclaiming is segregation of the used oils according to type.

Re-Refining. Petroleum refining processes are employed for re-refining used lubricating oil to produce clean, high quality lubricating base oil. These processes often include a pretreatment to reduce the impurity content by one or more of the following methods: application of heat, filtration, and treatment using acid, caustic, solvents, and/or other chemicals. In developed countries, pretreatment is usually followed by multistage vacuum distillation using clay or hydrogen finishing.

Regulations and Specifications

A significant source of concern for potential users of recycled petroleum products has been the lack of specifications or certifications related to the quality of the material and the consistency in producing oil of high quality. This perception of possible inferiority has been exacerbated by the reluctance of some equipment manufacturers to state whether they would honor warranties if recycled lubricants were used.

Based on the evidence that acceptable recycled petroleum products can be produced, there is a considerable legislative record encouraging the recycling of used oil. Starting with the Resource Conservation and Recovery Act in 1976, used oil was held apart from the normal hazardous waste system because the oil was viewed as a valuable commodity.

The U.S. EPA issued regulations implementing the legislation in stages, starting in 1988 with a Guideline for Federal Procurement of Lubricating Oils Containing Re-refined Oil and concluding with Management Standards in 1992. As part of the latter, the EPA decided not to list used oil as a hazardous waste in order to encourage its collection and recycling.

The Federal Trade Commission announced a rule, effective November 30, 1995, that set test procedures and labeling standards for recycled oil used as engine lubricating oil. The rule states in effect that if recycled oils meet the requirements of the API Certification System, such oils are substantially equivalent to new oil for use as engine oil. This federal rule preempts certain state recycled oil rules.

DONALD A. BECKER
National Institute for Standards and Technology
DENNIS W. BRINKMAN
Safety-Kleen Corporation

Used Oil Recycling Act of 1980, Public Law 94-463, U.S. Congress, Washington, D.C.

Guideline for Federal Procurement of Lubricating Oils Containing Re-Refined Oil, 42USC252, U.S. Congress, Washington, D.C., June 30, 1988.

Fed. Reg. **16**(311), 55414–55422 (Oct. 31, 1995).

Engine Oil Licensing and Certification System, 13th ed., API Publication 1509, American Petroleum Institute, Washington, D.C., 1995.

PAPER

Paper recycling mills encompass a range of unit operations. The choice and sequence of these operations is determined by the types of recovered paper being processed, the types of paper products produced, availability of process water, economic considerations, and environmental considerations. The type of recovered paper being processed determines the contaminants that must be removed. The degree to which these contaminants are removed is determined by the types of paper products being produced. Contaminants include ink, adhesives and glue, rosin, wax, starches and gums, coatings, paper fillers, styrofoam, and plastic from bags and tape. Adhesives, glue, and waxes are grouped together and called stickies.

Unit Operations

Pulping. The first of the unit operations is pulping. Pulping disintegrates paper into individual fibers dispersed in water. It may be a batch or continuous process. Both mechanical and chemical action promote ink detachment from cellulose fibers during pulping. Mechanical action includes interfiber abrasion and fiber flexing and bending. Chemical action includes fiber swelling at high pH and surfactant-promoted ink particle emulsification and solubilization.

Mild pulping conditions preserve stickies as relatively large particles, permitting their later removal by screens and mechanical cleaners. Another technique to reduce the problems caused by stickies is to use additives to reduce the tackiness of these particles.

High Density Cleaning and Screening. Usually as the first step after pulping, cleaners remove high and medium density large particles, eg, rocks and dirt, nuts, bolts, nails, paper clips, and other objects often found in wastepaper. Screening, which usually follows high density cleaning, removes relatively large ink and other contaminant particles from the pulp slurry. Coarse screens are fitted with holes which permit the passage of cellulose fibers and liquids while holding back large particles. Fine screens fitted with small slits separate smaller ($\geq 250\ \mu m$) contaminant particles and toner ink particles from the pulp.

Washing. Washing removes dispersed ink particles from the pulp slurry. This process is comparable to home laundering in many ways. The surfactant, such as alcohol ethoxylate or alkylphenol ethoxylate, is added to the pulper. Surfactants promote ink particle dispersion by increasing the hydrophilic nature of the ink particles on which they absorb. Washing is most effective in removing small, dispersed ink particles; it is less effective on large, poorly dispersed ink particles.

Water removed from the pulp slurry during washing passes through a mat of paper fibers. As more water is removed from the pulp, the pulp consistency at the washer discharge increases. The effluent from washers is heavily laden with ink, mineral coating and filler particles, and small cellulose fibers. As a result, it can be difficult to clarify.

Flotation. Flotation (qv) is used alone or in combination with washing and cleaning to deink office paper and mixtures of old newsprint and old magazines. An effective flotation process must fulfill four functions: (*1*) the process must efficiently entrain air; (*2*) ink must attach to air bubbles; (*3*) there must be minimal trapping of cellulose fibers in the froth layer; and (*4*) the froth layer must be separated from the pulp slurry before too many air bubbles collapse and return ink particles to the pulp slurry.

Mechanical Cleaning. A cleaner is a hydrocyclone device utilizing fluid pressure to create rotational fluid motion. Pulp is introduced tangentially near the top of the cleaner. Contaminants denser than water such as chemically treated toner inks and sand migrate toward the outer wall of the cleaner and exit in a separate (reject) stream.

Reverse cleaners operate on the same principles as forward cleaners. Contaminants less dense than water migrate toward the center of the cleaner and exit as a separate (reject) stream from the pulp slurry. Reverse cleaners are used to remove adhesive and plastic particles as well as paper filler particles and lightweight particles formed from paper coatings.

Dispersion and Kneading. These are mechanical processes designed to reduce dispersed ink particle size through fiber–ink abrasion processes. Fiber–fiber abrasion can detach additional ink from fibers. Dispersion is often performed at elevated temperatures above the softening point of toner inks and adhesive particles. Particle softening can aid in reducing particle size. Kneading may be performed at ambient temperature. Both processes are performed at high consistency and reduce the number of visible ink particles.

Dispersion and kneading are often used for pulp containing toner inks because these inks tend to form large particles during pulping. Dispersion and kneading reduce toner ink particle size below the visible range. With pulp made from old newsprint and magazines, dispersion is sometimes used after flotation and washing to improve optical homogeneity of paper made from the deinked pulp.

Bleaching. Bleaching was first developed for cellulose fibers prepared directly from wood (see BLEACHING AGENTS, PULP AND PAPER). In paper recycling, bleaching agents are used both during pulping and in separate bleaching operations performed after removal of inks and other contaminants from the pulp.

Bleaching as a separate operation whitens fibers and removes the coloring effect of dyes. As practiced in paper mills, bleaching is a multistage operation using different bleaching agents in each stage.

There are two types of bleaching: oxidative and reductive. Oxidative bleaching cleaves carbon–carbon double bonds thus destroying chromophores. Reductive bleaching reduces carbon–carbon double bonds to single bonds. The loss of extended conjugation in the chromophore leads to loss of color.

Refining and Fractionation. These processes are used to alter and select cellulose properties so the final sheet has the desired properties. Properties of recycled fibers differ from those of fibers prepared directly from wood. For example, recovered chemical fibers have lower freeness, an apparent viscosity leading to different water drainage characteristics on paper machines. Recovered fibers also have increased apparent density, lower sheet strength, increased sheet opacity, inferior fiber–fiber bonding properties, lower fiber swelling, lower fiber flexibility, lower water retention, reduced fiber fibrillation, and much lower internal fiber delamination.

Refining is used to develop the desired pulp drainage properties and control sheet properties, such as bulk and density, strength, surface smoothness, porosity, and printing characteristics.

Fractionation separates fines and short, weak cellulose fibers from longer, stronger cellulose fibers. Its primary application is in processing old corrugated containers into new packaging products.

Water Clarification. Process water that needs to be clarified comes from several different sources in the recycling mill: rejects from screens and mechanical cleaners; rejects from washers, thickeners, and flotation cells; water that drains from the pulp as it is converted into paper on the paper machine (white water); and water from felt washers. These waters contain different dissolved chemicals and suspended solids and are usually processed separately.

Mills have three principal options in handling their process water: (*1*) discharge the water to a municipal treatment facility, (*2*) use sedimentation tanks to allow solids to settle out of the liquid, and (*3*) dissolved air flotation. Dissolved air flotation is faster than settling, and the water loses less heat before being returned to mill operations. The sludge removed from the dissolved air flotation unit is higher in consistency than sludge from settling tanks.

Water from screens, cleaners, washers, thickeners, and flotation cells contain relatively high levels of ink. These waters also contain valuable chemicals, ie, sodium hydroxide and surfactants. Recycle of this water can save up to 10% in chemical costs.

Rejects and Sludge Handling. Efficient dewatering minimizes sludge volumes sent to landfills thus reducing hauling costs and tipping fees. Efficient sludge dewatering also increases the efficiency of heat utilization during incineration. Inclined screw thickeners are often used to thicken the rejects from dissolved air flotation clarifiers.

Economic Aspects

The growth of paper recycling will substantially increase demand for surfactants, bleaching agents, complexing agents, and other chemicals. Recycled paper contains more fines, short fibers, and anionic trash. This will increase demand for process chemicals such as drainage aids and both wet- and dry-strength additives.

JOHN K. BORCHARDT
Shell Chemical Company

R. W. J. McKinney, *Technology,* Chapman & Hall, New York, 1995.

M. R. Doshi and J. M. Dyer, eds., *Paper Recycling Challenge,* Vol. 1, *Stickies,* Doshi & Associates, Appleton, Wis., 1996.

M. R. Doshi and J. M. Dyer, eds., *Paper Recycling Challenge,* Vol. 2, *Deinking and Bleaching,* Doshi & Associates, Appleton, Wis., 1997.

R. J. Spangdenberg, ed., *Secondary Fiber Recycling,* TAPPI Press, Altanta, Ga., 1993, p. 101.

PLASTICS

Energy costs associated with plastics recycling are almost always less than in manufacture of products from virgin materials. Plastics recycling takes only 10–15% of the energy needed to refine petroleum and manufacture virgin resins. Incineration of plastics is a less efficient means of saving energy. Both economic and environmental factors have led to government regulations designed to promote recycling. In some areas, the number of landfill sites is becoming limited. Although the number of landfills in the United States is declining, the remaining sites are large, modern facilities. However, the effect of plastic wastes on the environment is a growing concern.

Separation of Commingled Materials

Random mixing of plastics leads to a significant adverse effect on properties. Hence different types of plastics must be separated from each other. Solid wastes, particularly from residential curbside collection programs, arrive at material recovery facilities (MRF) as a complex mixture. Unit operations are summarized in Figure 1. The wastes are dumped on a tipping floor. There paper products are separated from metals and plastics, mostly containers, which are pushed onto a conveyer belt. Two types of magnetic separators remove steel and aluminum from plastics and glass. Density differences or manual sorting are used to separate glass from plastics. The glass containers are hand sorted by color. The plastics are separated into individual polymer types. Plastic bottles are classified into clear poly(ethylene terephthalate) (PET) soft drink bottles; green PET soft drink bottles; translucent high density polyethylene (HDPE) milk, water, and juice bottles; pigmented high density polyethylene detergent bottles, poly(vinyl chloride) (PVC) water bottles, and food containers such as polypropylene ketchup bottles.

When processing municipal solid wastes, an eddy current separation unit is often used to separate aluminum and other nonferrous metals from the waste stream. This is done after removal of the ferrous metals. The eddy current separator produces an electromagnetic field through which the waste passes. The nonferrous metals produce currents having a magnetic moment that is phased to repel the moment of the applied magnetic field. This repulsion causes the nonferrous metals to be thrown out of the process stream away from nonmetallic objects.

Separation of Impurities. After separation of the plastics, a number of impurities may still be present. Washing technology has been used to remove inks, labels, and encrusted dirt from plastics, particularly bottles. Various technologies have been used to separate other materials from plastics.

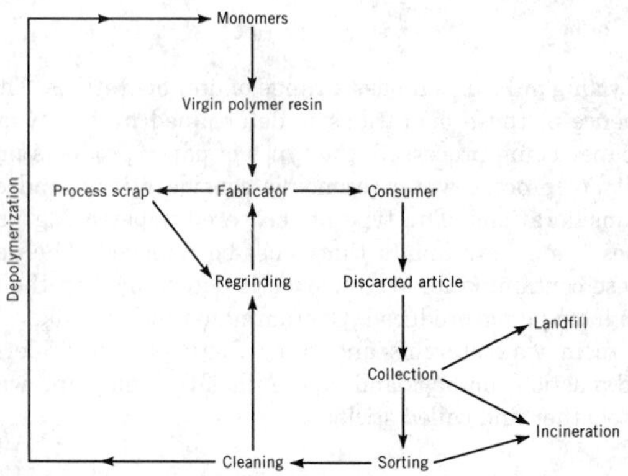

Figure 1. Process steps in plastics recycling.

Plastics

In May 1992, the U.S. Food and Drug Administration established the following guidelines to help assure the consumer safety of plastics recycling processes. Primary recycling is the recycling of plastics that are plant scrap and have not been sold for consumer use. Secondary recycling is the physical cleaning and processing of post-consumer plastic products. Tertiary recycling is the chemical treatment of polymers. This treatment is usually depolymerization to produce monomers which are purified and then polymerized to produce new polymer. Using tertiary recycling, materials such as fillers and fibers can be physically removed from the monomer. The monomers can also be purified by distillation and other processes prior to polymerization.

Sorted plastic packaging materials are shipped, usually in bales, to processing plants to be converted to polymer resins. The bales are broken and the bottles sorted to ensure that only one type of polymer is further processed. Processing consists of chopping and grinding the bottles into flakes, which are washed. Processing steps such as flotation are used to remove polymeric contaminants from the flakes. The flakes are melted and converted into pellets.

For high value food packaging applications, minimal migration of contaminants into food products is critical. Currently the FDA requirement is a maximum 0.5 parts per billion (ppb) of noncarcinogenic compounds by dietary exposure.

Poly(ethylene terephthalate). About 1.6 billion pounds of PET are used in food packaging applications annually in the U.S. About 42% of produced PET is recycled, mainly soft drink bottles. Cleaning of the recovered plastic comprises washing, rinsing, and drying. PVC is a common impurity in PET. Froth flotation has been shown to be an effective means of separating these two polymers. For food applications, improved cleaning of PET produced by secondary recycling is needed.

Polyethylene. About 24% of produced HDPE is recycled in the U.S., mainly milk and water jugs and liquid laundry detergent bottles. Cleaning of the recovered plastic comprises washing, rinsing, and drying.

During the rinse cycle, polyethylene particles float to the surface of the water bath. The higher density PET and PVC particles sink to the bottom of the bath and can be separated from the polyethylene.

HDPE has been molded into products such as plastic wood and the boards used in outdoor furniture. Other current uses include reuse container lids, truck bed liners, and pallets.

Polypropylene. Polypropylene (PP) is used in packaging applications as films and in rigid containers. Steps in polypropylene recycling include size reduction of grinding, washing, rinsing, and drying to remove contaminants and produce PP flakes. After extrusion, molten polymer is filtered through screen packs. The polymer may be separated into different melt flow ranges to produce more uniform product grades.

Polystyrene. Polystyrene (PS) is widely used in many packaging applications. Polystyrene items separated from the solid waste stream are subjected to one or more of the following unit operations: densification (for foam), granulation to reduce particle size, washing, drying, extrusion, and pelletizing. The finished pellets have properties similar to the virgin resin. High density baling is used to increase the bulk density of polystyrene, often by a factor of two. Contaminants are more easily removed before this densification step than after. Polystyrene has a high heating value, 46,000 kJ/kg compared to heating oil, 44,000 kJ/kg. Thus, incineration for its energy value is another possible application for recovered polystyrene.

Other Plastics. A relatively small amount of poly(vinyl chloride) goes into packaging applications and appears in municipal solid waste. The greatest concern with PVC is as a contaminant in other polymers being recycled, particularly PET. Applications for recycled PVC include use as an inner layer sandwiched between two virgin PVC layers in pipe and sheet for blister packaging and other packaging applications. Polyurethane is pulverized to increase its bulk density, mixed with 30–80% of a thermoplastic molding material, gelled, and then granulated to give coated urethane foam particles 0.1 to 0.15 mm in size. This material may be injection molded or extrusion molded into various products.

Commingled Plastic Wastes. Owing to the property deteriorations that usually occur on polymer mixing, commingled plastics are useful and economic only for low value applications in which mechanical properties are not demanding. Such applications include park benches and parking barriers. The supply of commingled plastics is much greater than the demand. Therefore, a critical issue in recycling commingled plastic wastes is the identification and separation of the plastics that are present. A spectrograph with an InGaAs-array detector has been developed to record spectra from post-consumer packaging materials located on conveyer belts. Atomic absorption spectroscopy can be used when one of the polymers has a different atomic composition than other polymers in a mixture.

One alternative to identifying and separating different types of plastics is using commingled plastics directly. Since the composition and physical properties of commingled plastics can vary from day to day, applications are limited. One such product is a building material containing Portland cement as the binder, a filler (sand, gravel, or stone), and a plastic with a maximum particle size of 5–10 mm. Concrete made using this material performs as well as standard concrete not containing plastic.

Fiber-Reinforced Plastics and Composites. It is usually too expensive to separate fillers and fibers from recovered polymers. Hence, the recycled use of these polymers must tolerate the presence of fillers or fibers. Thermoset matrix composites are ground and used as filler for polymers. Remolding is usually by injection or compression molding.

JOHN K. BORCHARDT
Shell Chemical Company

Sadler, and R. F. Stockel, eds., *Plastics, Rubber and Paper Recycling: A Pragmatic Approach,* American Chemical Society, Washington, D.C., 1995.

R. E. Landreth and P. E. Rebers, eds., *Municipal Solid Wastes* Lewis, Boca Raton, Fla., 1997.

G. D. Andrews and P. M. Subramian, eds., *Emerging Technologies in Plastics Recycling,* American Chemical Society, Washington, D.C., ACS Symposium Series, 1992, Vol. 513.

J. Brandrup, ed., *Recycling and Recovery of Plastics* Hanser Publications, Munich, 1996.

RUBBER

With fees collected by recyclers, tire recycling has become more profitable. Tires discarded in landfills tend to float on top of the ground; mosquito infestation and illegal tire disposal cause problems which can be alleviated by recycling.

Scrap Rubber as Fuel Source

The use of scrap rubber for fuel offers the best alternative for reusing rubber. There are four main markets for scrap rubber fuel, including fuel for cement kilns, electric utilities, pulp and paper mills, and dedicated tire-to-energy plants. Scrap tires used as supplemental fuel by these industries reduce solid waste and air pollution.

Tire-derived fuel (TDF), either as whole tires or processed into chips, is the largest single current and potential market for scrap tires. Tires have a higher energy value than most coals and are cleaner burning. The materials used to construct a tire have positive energy value. With lower emissions than coal, use of TDF has an added bonus of helping reduce air pollution (qv). The final advantage for most fuel users is that TDF can be obtained at a much lower cost per Btu than conventional fuels. From an environmental standpoint, the use of TDF also helps solve a significant solid waste management problem. Large industrial facilities, particularly those using cyclone boilers or fluidized-bed boilers, are potential markets. In addition, several vendors of small- and medium-sized industrial energy and steam facilities are marketing units capable of using TDF. As the availability of TDF expands with new producers entering the market, it is hoped that the industrial use of TDF will also expand.

Pyrolysis

Scrap tire pyrolysis has been the subject of several studies by rubber, oil, and carbon black industries. In the Tosco II process pyrolysis, chopped tires are fed into a rotary drum containing hot ceramic balls at 480°C in a reducing atmosphere. The rubber pyrolyzes and forms a solid residue, oil vapor, and off-gases; the condensed oil separates in a fractionator and the gas is used to heat the ceramic balls. A trommel screen separates the fine solid residue from the ceramic balls, and this residue is pelletized after steel, fiber glass, and other contaminants have been removed. The off-gas is a combination of ethylene, propylene, and butylene; the oil contains about 1% sulfur and can be substituted directly for fuel oil. Higher temperatures produce more gas and less liquid. Foster-Wheeler has two pyrolysis plants operating in Germany; other European countries and Japan also pyrolyze scrap rubber.

Rubber, Asphalt Modification

Ground recycled rubber has been developed as a modifier for asphalt paving materials since the 1950s in both the United States and Europe. The original developmental work was undertaken in an effort to improve the performance of asphalt paving materials. Two basic processes, wet and dry, have been developed, differing in the point at which the ground rubber is introduced.

In the wet process, ground rubber is prereacted with the hot liquid asphalt cement for a specified time, depending on the size of the rubber particles. During this process, the rubber particles react with and change the asphalt cement. The rubber-modified asphalt can then be used in one of several ways: as a crack sealant in pavement repair or with an aggregate chip as a cape sealant to effect temporary repair; or, in the rehabilitation of cracked pavements to prevent or reduce subsequent reflexive cracking. The dry process involves using slightly larger rubber particles mixed with the dry aggregate materials. The aggregates, which must be gap-graded to allow room for the rubber particles, are mixed with normal asphalt cement to form a modified asphalt concrete.

Cryogenic Pulverizing and Mechanical Tire Shredding

Tires must be pulverized or shredded before they can be reclaimed by devulcanization or used in asphalt and other recycling processes. The tires are mechanically ground, sometimes using cryogenic or solvent-swelling techniques to enhance grinding efficiency. Ground tire crumb rubber is commonly referred to as rubber reclaim, even though the rubber has not been devulcanized.

Cryogenics (qv) in conjunction with mechanical action has been used to make crumb rubber. Liquid nitrogen or other cryogenic fluids cool the rubber below the glass-transition temperature, and the brittle rubber is pulverized in a grinding mill. The morphology of ground rubber particles plays an important role in the ability to incorporate them into new compounds. Ambiently ground particles tend to have a more heterogeneous surface than cryogenically ground particles, the surfaces of which are more regular and cube-shaped. In practice, the adhesion of ground rubber in new compounds is a mechanical process. As a result, ambient ground materials appear to incorporate more easily into new compounds. In order to increase the reincorporation of cryogenically produced materials, work has been done to modify the surface of these particles to make them more reactive when incorporated into new compounds.

Reclaiming

Originally only solid rubber scrap was reclaimed, but with the advent of pneumatic tires and fiber-reinforced rubber, methods for removing the fiber had to be developed. Although the reclaiming processes were of limited use in the 1990s recycling market, modified reclaiming processes may become more prevalent as recycling increases.

The three rubber-reclaiming processes are digester, heater or pan, and reclaimator processes. Tires are most commonly reclaimed by digesting. Grinding devices reduce whole tires to uniform particles, and fiber is mechanically separated from the rubber with hammer mills, blown into collectors, and baled. Oils and processing aids are blended with the crumb rubber in ribbon blenders or similar mixers and are transferred to a digester, a steam-pressurized tank equipped with horizontal mixing paddles. The blend is mixed continuously at steam pressures of 1.01–1.70 MPa (10–17 atm) for four to six hours. The pressurized digester batch is forced into a blowdown tank, washed, and dried. High friction refining mills provide the smoothness and physical properties needed in the final product. The reclaim is baled, extruded into pellets, or formed into slabs for shipment.

Inner tubes of butyl and natural rubber and other fiber-free scrap rubbers are reclaimed by the heater or pan process. Brass tube fittings and other metals are removed from the scrap which is mechanically ground, mixed with necessary processing aids, loaded into pans or devulcanizing boats, and autoclaved at steam pressures of 1.01–1.42 MPa (10–14 atm) for three to eight hours. The product is refined by milling, extruded, and milled again in much the same way as in the digester process.

In the Reclaimator, a high pressure extruder, fiber-free rubber is heated to 175–205°C with oils and other ingredients. High pressure and shear between the rubber mixture and the extruder barrel walls effectively devulcanize the mixture in one to three minutes.

Tires, natural rubber tubes, and butyl tubes are the main sources of scrap and reclaim (see ELASTOMERS, SYNTHETIC–POLYISOPRENE). Specialty reclaim materials are made from scrap silicone, chloroprene (CR), nitrile–butadiene (NBR), and ethylene–propylene–diene–terpolymer (EPDM) rubber scraps (see ELASTOMERS, SYNTHETIC–POLYCHLOROPRENE; ELASTOMERS, SYNTHETIC–ETHYLENE–PROPYLENE–DIENE RUBBER).

Civil Engineering Market

The civil engineering market for scrap tires encompasses several distinct uses. Whole tires have been used to construct retaining walls and crash barriers. Whole tires have been used in erosion control, and to construct breakwaters and artificial reefs. A broader use of scrap tires as an engineering material is in the formation of chipped or shredded scrap tires as a fill material, especially useful where light weight is needed.

Small tire chips have also been utilized as a soil amendment to improve athletic playing fields (see RECREATIONAL SURFACES). A patented process marketed under the trade name Rebound (Jai Tire) combines crumb rubber from scrap tires with composted organic material to reduce soil compaction, resulting in better athletic playing surfaces.

Scrap whole tires have been used to form artificial fishing reefs, oyster beds, and as a floating breakwater. The tire splitting industry cuts tires into pieces for gaskets, shims, dock bumpers, shock absorbers, blasting mats, and other articles.

JOHN P. PAUL
Carter & Burgess, Inc.

F. T. Ryan, *Scrap Tires: Alternatives and Markets in the United States,* Goodyear Tire and Rubber Co., Washington, D.C., 1994.

J. R. Serumgurd, "A New Future for Old Tires—Recycle," presented at the *Chemical Marketing Research Association,* Oct. 1994.

M. B. Sikora, *Tire Recovery and Disposal, A National Problem With New Solutions,* Resource Recovery Report, Washington, D.C., June 1986.

J. R. Sarumgard, *Ground Rubber and Civil Engineering Markets for Scrap Tires,* Aug. 1994.

RECYCLING, WATER. See WATER, REUSE.

REFRACTION. See ANALYTICAL METHODS; SPECTROSCOPY, OPTICAL.

REFRACTORIES

Refractories are materials that resist the action of hot environments by containing heat energy and hot or molten materials. There is no well-established line of demarcation between those materials that are and those that are not refractory. The ability to withstand temperatures above 1100°C without softening has, however, been cited as a practical requirement of industrial refractory materials (see CERAMICS). The type of refractories used in any particular application depends on the critical requirements of the process.

Physical Forms

Refractories may be preformed (shaped) or formed and installed on-site. Castables, gunning mixes, and plastic and ramming mixes are used either for repair or for complete new construction of what is known as monolithic linings. The use of monolithics instead of constructions using shaped products has increased; monolithic installations have become as common as conventional shaped product construction.

Brick. The standard dimensions of a refractory brick are 229 mm (9 in.) length, 114 mm (4.5 in.) width, and 64 mm (2.5 in.) thickness. This is known as a standard straight. Quantities of bricks are given in brick equivalents, that is, the number of standard straight bricks that have a volume equal to that of the particular installation.

Other common refractory forms include setter tile and kiln furniture, fusion-cast shapes, cast and hand-molded refractories, insulating refractories, castables and gunning mixes, plastic refractories and ramming mixes, mortars, composite refractories, and refractory coatings (see REFRACTORY COATINGS).

Raw Materials

In the past, refractory raw materials were used essentially as mined minerals. Selective mining yielded materials of the desired properties and only in cases of expensive raw materials, such as magnesite, was a beneficiation process required. However, demand for high purity natural raw materials and synthetically prepared refractory grain made from combinations of high purity and beneficiated raw materials has increased (Table 1). The material produced upon firing raw as-mined minerals or synthetic blends is called grain, clinker, co-clinker, or grog. Recycled materials produced by the manufacturers and recovered from

Table 1. Physical Properties of Refractory Raw Materials

Material	Main crystalline phases	Specific gravity, g/cm^3	
		Bulk	True
	Silica		
ganister	quartz		2.66
gravel	quartz		2.61
	Clays		
flint clays	kaolinite, quartz, illite	2.55	
plastic clays	kaolinite, quartz, illite		
kaolin	kaolinite		
fireclay	kaolinite		
plastic	kaolinite, illite		
	High alumina		
natural siliceous bauxite,[a] ca 70% Al_2O_3	mullite	2.85–2.95	3.1–3.2
ca 60% Al_2O_3	mullite	2.75–2.85[b]	2.95–3.05
ca 50% Al_2O_3		2.65	
South American bauxite[a]	corundum, mullite	3.1	3.6–3.7
Chinese bauxite[a]		3.20	
kyanite[c]	kyanite		3.5–3.7
sillimanite[c]	sillimanite		3.23
synthetic fused alumina, black	α-alumina	3.87	4.01
gray	α-alumina	3.95	3.98
sintered alumina	α-alumina	3.45–3.6	3.65–3.80
sintered mullite		2.85	
sintered magnesium aluminate	spinel, periclase	3.33	
fused mullite	mullite	3.1	3.45
calcium aluminate cement, low purity	calcium monoaluminate		
high purity	α-alumina, calcium monoaluminate		
	Zirconium		
zircon	zircon		4.2–4.6
baddeleyite, ZrO_2	baddeleyite		5.5–6.5
	Basic raw materials		
calcined magnesias			
natural magnesite	periclase	3.2	
	periclase	3.4	
	magnesite, dolomite, calcite	3.40	
	[d]	3.39	
seawater	periclase	3.44	
brine	periclase	3.41	
dolomite[e]	periclase + CaO		
chrome ore	chromite spinel	4.2	

[a] Calcined. [b] Another type of ca 60% Al_2O_3 natural siliceous bauxite has a bulk density of 2.70 g/cm^3. [c] Raw. [d] Unclassified. [e] Both regular and low flux dolomite.

users are also used to reduce waste. Use of recycled materials is limited in the United States. It is more common in Europe and is expected to increase (see RECYCLING).

Silica. The most common refractory raw materials are ganister, which is a dense quartzite, and silica gravels (see SILICA; SILICON-COMPOUNDS).

Fireclay. Fireclays consist mainly of the mineral kaolinite, $Al_2O_3 \cdot 2SiO_2 \cdot 2H_2O$, with small amounts of other clay minerals, quartzite, iron oxide, titania, and alkali impurities.

Alumina. The naturally occurring raw materials are bauxites, sillimanite group minerals, and diaspore clays (see ALUMINUM COMPOUNDS).

Mullite. Although mullite is found in nature, for example, as inclusions in lava deposits on the island of Mull, Scotland, no commercial natural deposits are known. It is made by burning pure sillimanite minerals or sillimanite–alumina mixtures.

Zirconia. Zircon (zirconium silicate), the most widely occurring zirconium-bearing mineral, is dispersed in various igneous rocks and in zircon sands. Zircon can be used as such in zircon refractories or as a raw material to produce zirconia, ZrO_2 (see ZIRCONIUM AND ZIRCONIUM COMPOUNDS).

Basic Raw Materials. Basic raw materials include magnesite, dolomite, forsterite, chrome ore, silicon carbide, beryllia, thoria, and carbon (graphite).

General Properties

Oxides. Beryllium and magnesium oxides are stable to very high temperatures in oxidizing environments. Beryllia has good electrical insulating properties and high thermal conductivities; however, its high toxicity restricts its use. Stabilized or partially stabilized cubic ZrO_2 is the most useful simple oxide for operations above 1900°C.

Carbon, Carbides, and Nitrides. Carbon (graphite) is a good thermal and electrical conductor. It is not easily wetted by chemical action, which is an important consideration for corrosion resistance. As an important structural material at high temperature, pyrolytic graphite has shown a strength of 280 MPa (40,600 psi). It tends to oxidize at high temperatures, but can be used up to 2760°C for short periods in neutral or reducing conditions. When heated under oxidizing conditions, silicon carbide and silicon nitride, Si_3N_4, form protective layers of SiO_2 and can be used up to ca 1700°C (see NITRIDES). Silicon carbide has very high thermal conductivity and can withstand thermal shock cycling without damage. It also is an electrical conductor and is used for electrical heating elements.

Borides and Silicides. These materials do not show good resistance to oxidation.

Metals. The highest melting refractory metals are tungsten (3400°C), tantalum (2995°C), and molybdenum (2620°C). All show poor resistance to oxidation at high temperatures.

Phase Equilibria. Phase diagrams represent the chemical equilibria that exist among one, two, or three components of a system under the influence of temperature and pressure. Reference to a phase diagram permits the determination of the amount and composition of solid and liquid phases that coexist under certain specified conditions of temperature and pressure for a particular system. Using such information, the occurrence of physical and chemical changes within a system or between systems at high temperatures can be predicted.

Physical Properties. Brick bulk density depends on the specific gravity of the constituents and the porosity. Usually the highest density possible is desired. Upon firing, the grains and matrix form glassy, direct, or solid-state ceramic bonds. Sintering is generally accompanied by shrinkage. Particle size distribution, forming method, and firing process contribute to texture, whereas permeability is related to porosity, which in turn is dependent upon texture.

Mechanical Properties. The transverse strength (modulus of rupture) at room temperature is related to the degree of bonding. Fine-grained refractories generally are stronger than coarse-grained types and those having a low porosity are stronger than those of high porosity.

Generally, high temperature strength is lower than room temperature strength. The development of solid-state or direct-bonded basic brick requires high firing temperatures, and is impeded by glassy phases. By referring to phase diagrams, refractory compositions may be designed that avoid the development of such phases.

The modulus of elasticity (MOE) is related to the strength and can be used as a nondestructive quality control test on high cost special refractory shapes.

Thermal Properties. Refractories, like most other solids, expand upon heating, but much less than most metals. The degree of expansion depends on the chemical composition.

Reheat Change. Most refractory bricks are not chemically in equilibrium before use. During the prolonged heating in service, additional reactions occur that may cause the brick to shrink or expand. Considerable expansion may be caused by gas formation during heating. In general, basic brick exhibits good volume stability at high temperatures.

Thermal Conductivity. The refractory thermal conductivity depends on the chemical and mineral composition of the material and increases with decreasing porosity.

Thermal Spalling. The susceptibility to thermal cracking and spalling depends on certain characteristics of the raw material and the macrostructure of the particular refractory. Spalling resistance may be increased by either preventing cracks from forming or preventing cracks from growing. The approach used to effect spall resistance determines which properties of the refractory are optimized.

Refractoriness. Refractoriness is the resistance to physical deformation under the influence of temperature. It is determined by the pyrometric cone equivalent (PCE) test for aluminosilicates and resistance to creep or shear at high temperature (see ANALYTICAL METHODS).

Manufacture

Processing. Some materials can be used without further processing, although many must be subjected to heat treatment. In the case of synthetic grain, the selected and beneficiated raw materials are blended in the desired proportions and formed into suitable shapes for calcination by briquetting, pelletizing, or extrusion. The term *Calcination* is used to indicate heat treatment to sinter or burn (dead burn) the refractory grain to a stable dense material as well as to decompose minerals. Calcination may be carried out to rotary kilns, shaft kilns, multiple-hearth furnaces, or fluidized-bed reactors. Both raw and processed materials can be fused or melted in electric-arc furnaces.

Crushing and Grinding. Some raw materials, such as hard clay and quartzite, must first be crushed to grains small enough for the grinding equipment. (see SIZE REDUCTION). Almost all raw materials require grinding after primary crushing.

Screening. To obtain a high density product, the mix is made from materials that have been sized into classes by means of standard screens. In a continuous screening operation, the ground raw materials are generally fed to vibrating high capacity screens that may be heated. Material that does not pass the screen is returned to the grinding system for further size reduction. Coarse and medium fine-grain sizing is accomplished by the aforementioned methods, whereas fine-sized materials generated in rod mills, ball mills, ring-roll mills, etc, are classified by air separators.

Mixing. As in other ceramic processes, more than one type of raw material is often required for a refractory product. The purpose of mixing is to uniformly distribute the various ingredients (see MIXING AND BLENDING). Although the specific steps and equipment involved in the mixing of batches for fireclay, high alumina, and basic refractories are somewhat different, the general principles are similar. Mixes that are to be dry-pressed contain 2–6% binding liquid, depending on the plasticity of the raw material bond system and the fineness of the mix. The ingredients may be blended in a pug mill, dry pan, or other type mixer and tempered with the bonding ingredients. Tempering, in the sense used here, denotes the kneading action produced on the mix, usually in a muller mixer.

Forming. Most refractory shapes are formed by mechanical equipment, but some very large or intricate shapes require hand molding in wooden, steel-lined molds with loose liners to permit easy removal of plaster of Paris molds.

Refractory shapes are generally produced on a mechanical toggle press, screw press, or hydraulic press. Some special shapes are produced by air-ramming which is similar to hand molding, except that reinforced steel molds are required. Special shapes can also be formed by slip casting and hot pressing.

Drying. The drying (qv) step for large shapes is critical. Extremely large fireclay and silica shapes are sometimes allowed to dry on a temperature-controlled floor heated by steam or air ducts embedded in the concrete. Smaller shapes are generally dried in a tunnel dryer. The ware is placed on cars that enter the cold end and exit at the hot end.

Curing. Some chemically bonded bricks require some elevated heat treatment that is typically higher than the tempering process, but at a lower temperature than that required to form ceramic bonds.

Bricks are fired or burned in kilns to develop a ceramic bond within the refractory and attain certain desired properties. This step does not apply to chemically or organically bonded products. Burned brick may be impregnated with tar or pitch to improve corrosion resistance.

Specialty Refractories. Bulk refractory products include gunning, ramming, or plastic mixes, granular materials, and hydraulic setting castables and mortars. These products are generally made from the same raw materials as their brick counterparts.

Economic Aspects

The principal consumers of refractories are the iron (qv) and steel (qv) industries. There has been a decrease in refractories consumption that coincided with technological changes in the manufacture of steel. Steady improvements in basic oxygen furnace (BOF) practice and improvements in refractory composition and design led to improved refractory performance. More sophisticated ladle metallurgical practice has been employed, leading to improved steel quality and improved refractory performance.

ASTM Classifications and Specifications

Classifications. In addition to testing methods, the American Society for Testing and Materials (ASTM) publishes a list of classifications

Table 2. ASTM Test Methods for Refractories

Material	Test identification	Properties
burned brick	C20, C830	apparent porosity, water adsorption, bulk density
brick, various shapes	C133, C607, C93	crushing strength, modulus of rupture
basic brick	C456	hydration resistance
brick and tile	C154	warpage
granules	C357, C493	bulk density
periclase grains	C544	hydration
mortar	C198	cold-bonding strength
air-setting plastics	C491	modulus of rupture
castables	C298	modulus of rupture
granular dead-burned dolomite	C492	hydration
fireclay plastics	C181	workability index
castables	C417	thermal conductivity
plastics	C438	thermal conductivity
general refractories	C288	disintegration in CO atmosphere
	C135	true specific gravity
	C201	thermal conductivity
	C92	sieve analysis and water content

covering a wide variety of refractory types. The various brands from numerous producers and producing districts are grouped into classes using a nomenclature indicative of chemical composition, heat resistance, and service properties. The classifications may be inadequate to encompass all refractories encountered, but these represent the only universally accepted standards specifying composition or properties.

Specifications. Among the many specifications covering refractory products, the best known are those published by ASTM. In addition, specifications are issued by the U.S. Government and the armed forces. The International Organization for Standardization (ISO) has issued the ISO 9000 series standards for quality management and quality assurance requirements and guidance. These standards are being rapidly adopted as national standards (see MATERIALS STANDARDS AND SPECIFICATIONS; QUALITY ASSURANCE).

Analytical and Test Methods

The test methods applicable to refractories are summarized in Table 2.

Health and Safety Factors

Industrial refractories are by their very nature stable materials and usually do not constitute a physiological hazard. This is not so, however, for unusual refractories that might contain heavy metals or radioactive oxides, such as thoria and urania, or to binders or additives that may be toxic. Inhalation of certain fine dusts may constitute a health hazard. For example, exposure to silica, asbestos, and beryllium oxide dusts over a period of time results in the potential risk of lung disease.

Selection and Uses

Any manufacturing process requiring refractories depends on proper selection and installation. When selecting refractories, environmental conditions are evaluated first, then the functions to be served, and finally the expected length of service. All factors pertaining to the operation, service design, and construction of equipment must be related to the physical and chemical properties of the various classes of refractories.

By far the most common industrial refractories are those composed of single or mixed oxides of Al, Ca, Cr, Mg, Si, and Zr. These oxides exhibit relatively high degrees of stability under both reducing and oxidizing conditions. Carbon, graphite, and silicon carbide have been used both alone and in combination with the oxides. Refractories made from these materials are used in ton-lot quantities, whereas silicides are used in relatively small quantities for specialty application in the nuclear, electronic, and aerospace industries.

The common industrial refractories are classified into acid, SiO_2 and ZrO_2; basic, CaO and MgO; and neutral, Al_2O_3 and Cr_2O_3. Oxides within each group are generally compatible with each other, whereas mixtures of acid and basic oxides often give low melting products. Neutral oxides are generally compatible with both acidic and basic oxides.

Reactions Between Refractories and Liquids. The response of a refractory to a chemical environment generally depends on its slag resistance which, in turn, depends on the compositions and properties of slag and refractory. Other factors include temperature, severity of thermal cycling or shock of the process, velocity and agitation of the slag in contact with the refractory, and the abrasion to which the refractory is subjected. Thus similar refractories placed in similar furnaces can wear at vastly different rates under different operation practices.

Reactions Between Refractories and Gases. Reactions of refractories and gases can be quite destructive. The gases generally penetrate the pores of the refractory destroying its structure. An example is the disintegration of aluminosilicates in blast furnaces caused by the deposition of carbon from carbon monoxide. The growth of the carbon deposit causes the brick to rupture. Therefore, a brick of low iron and alkali content having a dense, low permeability is preferred.

Reactions Between Refractories. Dissimilar refractories can react vigorously with each other at high temperatures.

H. DAVID LEIGH III
Clemson University

F. Singer and S. S. Singer, *Industrial Ceramics,* Chemical Publishing Co., Inc., New York, 1964.

Manual of ASTM Standards on Refractory Materials, 8th ed., American Society for Testing and Materials, Philadelphia, Pa., 1957.

S. C. Carniglia and G. L. Barna, *Handbook of Industrial Refractories Technology—Principles, Types, Properties and Applications,* Noyes Publications, Park Ridge, N.J., 1992.

P. A. Janeway, ed., *Bull. Am. Ceram. Soc.* **73**(10), 46–55 (1994).

REFRACTORY COATINGS

Refractory coatings are metallic, refractory compounds, ie, oxides, carbides (qv), and nitrides (qv), and metal-ceramic coatings associated with high temperature service as contrasted to coatings (qv) used for decorative or corrosion-resistant applications. Coatings of high melting materials that are used in other than high temperature applications are also considered to be refractory. A coating may be defined as a near-surface region having properties that differ significantly from the bulk of the substrate (see CERAMICS; METALLIC COATINGS; METAL SURFACE TREATMENTS).

The highest melting refractory metals are tungsten, tantalum, molybdenum, and niobium, although titanium, hafnium, zirconium, chromium, vanadium, platinum, rhodium, ruthenium, iridium, osmium, and rhenium may be included (see REFRACTORIES). Many of these metals do not resist air oxidation. Hence, very few, if any, are used in elemental form for high temperature protection. Some modern high temperature oxidation- and corrosion-resistant coatings have compositions similar to high temperature bulk alloys (see HIGH TEMPERATURE ALLOYS) and are applied by thermal spraying, evaporation, or sputtering. The protection mechanism for these high temperature alloy coatings is based on adherent impervious surface films of Al_2O_3, SiO_2, CrO_2, or a spinel-type material that grow upon high temperature exposure to air.

Refractory coatings also include materials having high melting points, eg, silicides, borides, carbides, nitrides, or oxides, and combinations such as oxycarbides, etc. In addition, mixtures of metals and refractory compounds, sometimes called metallides, of various microstructural configurations, ie, laminates, dispersed phases, etc, can also be classified as refractory coatings.

All coating methods consist of three basic steps: synthesis or generation of the coating species or precursor at the source, transport from the source to the substrate, and nucleation, growth, or buildup of the coating on the substrate. These steps can be completely independent of each other or may be superimposed on each other, depending on the coating process (see COATING PROCESSES).

Numerous schemes can be devised to classify deposition processes. A scheme based on the dimensions of the depositing species uses the classes atomic deposition, particulate deposition, bulk coating, and surface modification.

Atomic Deposition Processes

Electrodeposition. Of the numerous metals used in electrodeposition, only platinum, rhodium, iridium, and rhenium are refractory (see ELECTROPLATING). Cermets, ie, materials containing both ceramic and metal, eg, TiC–Ni and Al_2O_3–Cr, can be deposited from plating baths if the particulate matter is suspended by air agitation or stirring (see CERAMICS; COMPOSITE MATERIALS, CERAMIC-MATRIX).

Electroless Deposition. Electroless plating (qv) is defined as a controlled, autocatalytic chemical reduction process for depositing metals. It resembles electroplating because it can be run continuously to build up a thick coating. It does not involve a chemical reaction with the substrate metal.

Physical Vapor Deposition Processes. The three physical vapor deposition (PVD) processes are evaporation, ion plating, and sputtering (see THIN FILMS, FILM FORMATION TECHNIQUES).

The materials deposited by PVD techniques include metals, semiconductors (qv), alloys, intermetallic compounds, refractory compounds, ie, oxides, carbides, nitrides, borides, etc, and mixtures thereof. The source material must be pure and free of gases and inclusions, otherwise spitting may occur.

Chemical Vapor Deposition and Plasma-Assisted Chemical Vapor Deposition. In chemical vapor deposition (CVD), thin films or bulk coatings up to 2.5 cm in thickness are deposited by means of a chemical reaction between gaseous reactants passing over a substrate. The optimum temperature for a given reaction often lies within a very narrow range, and the process needs to be tailored to the substrate and the intended application. Coatings have application in a wide array of corrosion- and wear-resistant uses, but also in decorative layers, semiconductors, and magnetic and optical films.

Plasma-assisted CVD processes use deposition temperatures lower than CVD processes. The desired deposition reaction is aided by the energy present in the plasma. The plasma greatly extends the utility of CVD processes.

Particulate Deposition Processes

Thermal Spraying. In thermal spraying, particles are heated and melted in combustion flame or electric arc heats particles of the refractory-coating material to a temperature sufficient to achieve sintering or cohesive solidification when the particles impinge on a substrate. Penetration of the substrate rarely accompanies the coalescence and bonding. In thermal spraying, small-diameter refractory particles are heated and melted in a combustion flame or electric are heated gas and projected onto the substrate at high velocity. Particles flatten and bond to the substrate and each other by mostly mechanical means.

Electrophoretic Processes. Electrophoretic coatings are obtained through the migration of charged particles when a potential is applied to electrodes immersed in a suitable suspension of the particles. This process is particularly suited for applying uniform layers on complex bodies. Particles, not ions, are deposited. Electrophoresis can be used to apply metals and alloys, ceramics, and cermets. The three steps are preparation of the dispersion, deposition and conversion, and bonding. (see ELECTROSEPARATIONS, ELECTROPHORESIS).

Bulk Coating

In bulk coating processes, bulk materials are joined to the substrate either by a surface melt process or by attachment of the solid material.

In cladding, one metal is coated with another by rolling or extruding the two metals in close contact to each other. Coherence of the two metals is induced by soldering, welding (qv), or casting one in contact with the other prior to the rolling operation. Other bulk coating methods include immersion (hot dipping), welding, and enameling (see ENAMELS, PORCELAIN OR VITREOUS).

In the microelectronics industry, powdered metals and insulating materials that consist of nonnoble metals and oxides are deposited by screen printing in order to form coatings with high resistivities and low temperature coefficients of resistance.

Surface-Modification Processes

Cementation or Diffusion Coatings. Cementation is defined as the introduction of one or more elements into the outer portion of a metal object by means of diffusion at elevated temperatures. The coating produced by cementation is formed by an alloying or chemical combination of the diffusing elements and the substrate level. The most advanced cementation coatings are the intermetallic coatings, specifically silicides and aluminides, that protect refractory metals.

Metalliding. Metalliding, a General Electric Company process, is a high temperature electrolytic technique in which an anode and a cathode are suspended in a molten fluoride salt bath. As a direct current is passed from the anode to the cathode, the anode material diffuses into the surface of the cathode, which produces a uniform, pore-free alloy rather than the typical plate usually associated with electrolytic processes. The process is called metalliding because it encompasses the interaction, mostly in the solid state, of many metals and metalloids ranging from beryllium to uranium.

Microstructure of Coatings

The microstructure of bulk coatings resembles the normal microstructure of metals and alloys produced by melt solidification. The microstructure of particulate-deposited materials resembles a cross between rapidly solidified bulk materials having severe deformation and powder compacts produced by pressing and sintering. A special feature of particulate coatings is a significant degree (ca 2–20 vol %) of porosity that strongly affects the properties of the deposit.

The microstructure and imperfection content of coatings produced by atomistic deposition processes can be varied over a very wide range to produce structures and properties similar to or totally different from bulk processed materials.

PVD Condensates. Physical vapor deposition condensates can deposit as single-crystal films on certain crystal planes of single-crystal substrates, ie, by epitaxial growth or, in the more general case, the deposits are polycrystalline. In the case of films deposited by evaporation techniques, the main variables are the nature of the substrates; the temperature of the substrate during deposition; the rate of deposition; and the deposit thickness.

CVD Coatings. As in PVD, the structure of the deposited material depends on the temperature and supersaturation. In the case of chemical vapor deposition (CVD), however, the effective supersaturation, ie, the local effective concentration in the gas phase of the materials to be deposited, relative to its equilibrium concentration, depends not only on concentration, but on temperature. The reaction is thermally activated. Growth of columnar grains is characteristic of many materials in certain ranges of conditions. This structure results from uninterrupted growth toward the source of supply.

Electrodeposits. Columnar structures are characteristic of deposits from solutions, especially acid solutions, containing no additives, high

Table 1. Characteristics of Deposition Processes

Characteristic	Evaporation	Ion plating	Sputtering	CVD	Electrodeposition	Thermal spraying
mechanism of production of depositing species	thermal energy	thermal energy	momentum transfer	chemical reaction	deposition from sol	from flames, arcs, plasmas
deposition rate, nm/min	≤75,000	≤25,000	low except for pure metals[a]	eg, 20–2500	low to high	very high
depositing species	atoms and ions	atoms and ions	atoms and ions	atoms	ions	droplets
throwing power[b]		good[c]	good[c]	good	good	none
into small blind holes	poor	poor	poor	limited	limited	very limited
deposition of metal	positive	positive	positive	positive	positive but limited	positive
alloy	positive	positive	positive	limited	limited	positive
refractory compound	positive	positive	positive	positive	limited	positive
energy, depositing species, eV	ca 0.1–0.5	1–100	1–100	high for PACVD	can be high	can be high
bombardment of substrate and deposit by inert gas ions	normally not	yes	depends on geometry	possible	none	positive
growth interface perturbation	normally not	yes	yes	yes, by rubbing	none	none
substrate heating by external means	normally yes	yes	generally not	no	none	normally not

[a] For copper, 1000 nm/min. [b] Except by gas scattering. [c] Thickness distributions may be nonuniform.

metal-ion concentration solutions with high deposition rates, or from low metal-ion concentration solutions at low deposition rates. These usually exhibit lower tensile strength, elongation, and hardness than other structures, but are generally more ductile. Such deposits are usually of highest purity (high density) and low electrical resistivity.

Selection Criteria

The selection of a particular deposition process depends on the material to be deposited and its availability; rate of deposition; limitations imposed by the substrate, eg, maximum deposition temperature; adhesion of deposit to substrate; throwing power; apparatus required; cost; and ecological considerations. Criteria for CVD, electrodeposition, and thermal spraying are given in Table 1.

Applications

Coatings can be classified into six categories: chemically functional, mechanically functional, optically functional, electrically functional, biomedical, and decorative. In addition, there are some unique applications in the aerospace program, such as the ablative coatings of pyrolytic carbon and graphite- and silica-based materials for protection of nose cones and the space shuttle during reentry (see ABLATIVE MATERIALS). Another unique energy-related application is the coating of low atomic number (low-Z) elements such as TiC for the first wall of thermonuclear reactors to minimize contamination of the plasma.

MERLE THORPE
Thorpe Thermal Technologies

B. Chapman and J. C. Anderson, eds., *Science and Technology of Surface Coatings,* Academic Press, Inc., New York, 1974.

R. Bakish, ed., "Electron and Ion Beam Science and Technology," *International Conferences.*

R. S. Holmes and R. G. Loasby, *Handbook of Thick Film Technology,* Electrochemical Publications, 1976.

M. L. Thorpe, *Adv. Mater. Proc.,* 50–61 (May).

REFRACTORY FIBERS

Refractory fibers are generally used in industrial applications at temperatures between 1000°C and 2800°C. These fibers may be oxides or nonoxides, vitreous or polycrystalline, and may be produced as whiskers, continuous filaments, or loose wool products (see FIBERS, SURVEY).

Refractory fibers generally have diameters ranging from submicrometer to 10 μm and lengths, as manufactured, may range from millimeters to continuous filaments.

Oxide fibers are manufactured by thermal or chemical processes into a loose wool mat, which can then be fabricated into a flexible blanket; combined with binders and formed into boards, felts, and rigid shapes; or fabricated into ropes, textiles and papers. The excellent thermal properties of these products make them invaluable for high temperature industrial applications.

Nonoxide fibers, such as carbides, nitrides, and carbons, are produced by high temperature chemical processes that often result in fiber lengths shorter than those of oxide fibers. Mechanical properties such as high elastic modulus and tensile strength of these materials make them excellent as reinforcements for plastics, glass, metals, and ceramics. Because these products oxidize at high temperatures, they are primarily suited for use in vacuum or inert atmospheres, but may also be used for relatively short exposures in oxidizing atmospheres above 1000°C.

Properties

Refractory fibers are most often used in applications above 1000°C. For short exposures, however, some fibers can be used with little degradation at temperatures within 100°C of their melting points.

The most important properties of refractory fibers are thermal conductivity, resistance to thermal and physical degradation at high temperatures, tensile strength, and elastic modulus. Thermal conductivity is affected by the material's bulk density, its fiber diameter, the amount of unfiberized material in the product, and the mean temperature of the insulation.

For reinforcement, room temperature tensile strength and Young's modulus (stress–strain ratio) are both important.

Heat treatment of vitreous refractory fibers often results in crystallization of the fiber. The rate of crystallization (qv), sometimes called devitrification, depends on temperature and time. Shrinkage and degradation of mechanical strength are attributable to this crystallization. Sintering, the bonding together of refractory fibers at temperatures below their softening points, is also responsible for shrinkage. At fiber contact points, solid-state diffusion of molecules joins the fibers together and transforms the previously flexible structure into a rigid mass.

Nonoxide Refractory Fibers

The most important nonoxide refractory fiber is silicon carbide. Silicon nitride fibers are a by-product in the production of silicon nitride powder from the reaction of silicon metal in a nitrogen atmosphere at high temperatures. Boron nitride fibers are produced by nitriding boron filaments obtained during the chemical vapor deposition of boron on a heated tungsten wire.

Nonoxide fibers in continuous form, eg, carbon, graphite, and boron, are often used in filament winding and in the manufacture of high strength, high modulus fabrics (see CARBON AND GRAPHITE FIBERS).

Shorter, nonoxide filaments are primarily used as strength-enhancing reinforcements in resins, ceramics, and metals. Silicon carbide whiskers (single-crystal fibers) are of special interest because they offer not only high strength and stiffness but also temperature resistance to 1800°C. Silicon carbide and silicon nitride fibers can be dispersed in a number of organic resins and then cast into shapes. Boron nitride fibers are of particular value for the reinforcement of cast aluminum parts because the material can be wetted by molten aluminum. Silicon carbide and boron nitride are also used to reinforce gold and silver castings.

Oxide Fibers

High Purity Silica Fibers. Leached glass fibers having extremely fine diameters are currently produced using flame attenuation processing (see GLASS). A glass composed of 75 wt % SiO_2 and 25 wt % Na_2O is melted in a typical glass furnace at 1100°C. Filaments having diameters of 0.3 mm are drawn from orifices in the bottom of the furnace. These filaments are passed through a gas flame and attenuated to a diameter of approximately 1.5 μm. The loose fibers are then subjected to an acid leaching process to remove the Na_2O, thoroughly rinsed, and dried. The resulting fibers have a purity of >99% silica.

High thermal efficiency insulating felts having very low bulk densities produced by rod-drawing are essentially free of unfiberized material and can be used not only for jet engine insulation but also for space vehicles (see ABLATIVE MATERIALS). High purity silica fibers (Q Fiber) are used for the tiles of the reusable thermal protection system on the space shuttle.

In addition to aerospace uses, silica fibers can be twisted into sewing threads and yarns for weaving into fabrics. These fabrics are used extensively for heat-resistant clothing, flame curtains for furnace openings, thermocouple protection, and electrical insulation.

Chemically Produced Oxide Fibers. Refractory fibers from oxides of alumina or zirconia are difficult to manufacture by conventional melt technologies because of high melting points and low viscosities (see CERAMICS). Several companies have developed technologies for chemical production of high temperature refractory fibers, including a sol-gel process (see SOL-GEL TECHNOLOGY).

Although more expensive than melt fiberization, the sol processes offer advantages in fiber chemistry selection. In melt fiberization, viscosity and surface tensions are greatly influenced by additions of small quantities of metallic oxides. In the sol process, where viscosity can be controlled independently, any number of metal salts may be added without adverse effects. These salts can serve as grain growth inhibitors, sintering aids, phase stabilizers, or catalysts.

Aluminosilicate Fibers. Vitreous aluminosilicate fibers, more commonly known as refractory ceramic fibers (RCF), belong to a class of materials known as synthetic vitreous fibers. Fiber glass and mineral wool are also classified as synthetic vitreous fibers.

RCF is produced by melting a combination of alumina and silica in a approximately equal proportions or by melting kaolin clay in an electric resistance furnace. The molten mixture is formed into fiber either by blowing an air stream onto the molten material flowing from an orifice in the bottom of the furnace, or by directing the molten material onto a series of spinning wheels.

RCF is primarily used in industrial applications where high temperature resistance, light weight, low thermal conductivity, and low heat storage are required. Oxide and nonoxide refractory fibers have become essential materials for use in modern high temperature industrial processes and advanced commercial applications.

Research is ongoing, but epidemiological evidence has not demonstrated a causal relationship between exposure to refractory ceramic fibers and the development of respiratory ailments, including cancer, fibrosis, and parenchymal disease. State-of-the-art inhalation toxicology research indicated that at extremely high doses (200+ times maximum recommended exposure levels) RCF is an animal carcinogen, but that a critical dose level may exist below which neither fibrosis nor tumors occur. Manufacturers' Material Safety Data Sheets (MSDS) should be consulted prior to handling RCF materials.

R. A. WAUGH
Thermal Ceramics

L. Olds, W. Miller, and J. Pallo, *Am. Ceram. Soc. Bull.* **59**(7), 739 (1980).

M. S. Reisch, *Chem. Eng. News,* **72**(48), 23 (Nov. 28,1994).

The Epidemiology and Toxicology of Exposure to Refractory Ceramic Fibers, RCF Coalition, Washington, D.C., Dec. 1993.

REFRIGERATION

Refrigeration, as defined in this article, is the process of cooling materials using mechanical means and spans the temperature range of −157° to +4°C. Human comfort cooling, which may be considered high temperature refrigeration in the temperature range of 4–13°C, is commonly referred to as air conditioning (qv). Ultralow temperature cooling from absolute zero, −273°C to −157°C, is commonly referred to as cryogenics (qv). This article focuses on refrigeration as defined above, specifically those applications using vapor-compression technology.

Types of refrigeration may be categorized within five broad areas of application: (*1*) domestic appliances, ie, household refrigerator/freezers; (*2*) commercial systems, ie, supermarket display cases, restaurants, and cafeterias; (*3*) cold storage and food processing, ie, refrigerated warehouses; (*4*) industrial, ie, liquefaction of gases, chemical process cooling, and crystallization; and (*5*) transportation, ie, refrigerated truck or trailer and marine containers.

Basic Principles

All refrigeration systems are composed of five fundamental components: compressor, evaporator, condenser, expansion device, and refrigerant. A typical arrangement of these components is shown in Figure 1. The compressor raises the pressure of the refrigerant vapor so that its saturation temperature is above the temperature of the available cooling medium, ie, air or water. This difference in temperature allows transfer of heat from the vapor to the cooling medium so that the vapor can condense. The liquid flows through the expansion device, and in doing so a portion of the liquid flashes into a vapor, cooling the remaining liquid refrigerant below the temperature of the product to be cooled. This resulting difference in temperature allows heat to be transferred from the product to the refrigerant, causing the refrigerant to evaporate. The vapor formed must be removed by the compressor at a rate sufficient to maintain the low pressure in the evaporator and keep the refrigerant flowing. The resulting continuous flow process is referred to as the refrigeration cycle, shown schematically in Figure 2. This cycle is also called the reverse Rankine cycle and is governed by the principles of thermodynamics. For a review of these principles and the reverse Rankine cycle see the articles on HEAT-EXCHANGE TECHNOLOGY and THERMODYNAMICS.

The standard unit of measure for refrigeration capacity is known as the refrigeration ton. It represents the amount of heat that must

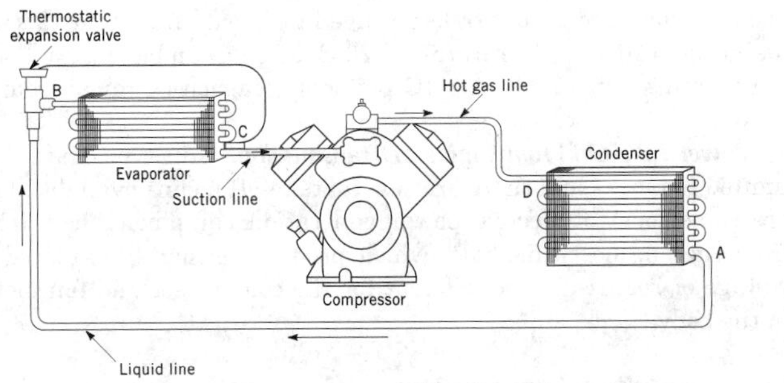

Figure 1. Components of a refrigeration system: A, condenser outlet; B, evaporator inlet; C, evaporator outlet; D, compressor discharge.

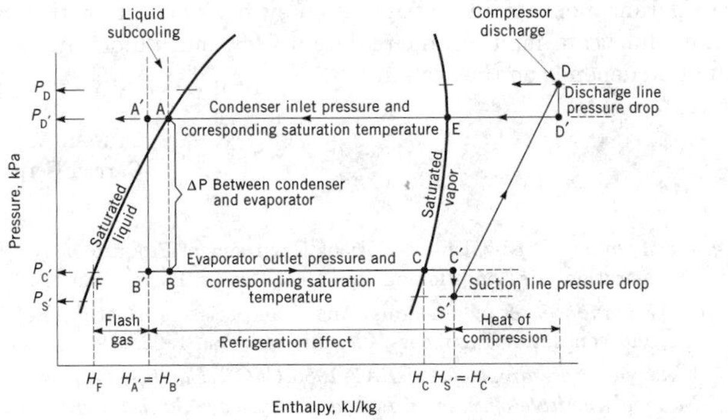

Figure 2. Basic refrigeration cycle. A or A′, condenser outlet; B or B′, evaporator inlet; C or C′, evaporator outlet; S′, compressor inlet; D, compressor discharge; D′, condenser inlet. To convert kPa to atm, divide by 101.3. To convert kJ/kg to Btu/lb, multiply by 0.4302.

be removed from a short ton (909 kg) of water to form ice in 24 h. Its value is 3.51 kWt (12,000 Btu/h(= 12.7 MJ/h)). It is conventional to designate a kilowatt of refrigeration as a thermal kilowatt (kWt) to distinguish it from the amount of electricity (kWe) required to produce the refrigeration.

Refrigerant Nomenclature

The specific fluid used to produce refrigeration is a very important consideration. The refrigeration industry uses a wide variety of compounds that fall into three general categories: halocarbons, hydrocarbons, and inorganic fluids. For saturated and unsaturated aliphatic hydrocarbons and aliphatic halogenated hydrocarbons designated as refrigerants, a numerical coding system has been developed that describes the refrigerant molecular structure by the use of a general designation having the form ABCD, in which A is the number of double bonds, B the number of carbon atoms less one, C the number of hydrogen atoms plus one, and D the number of fluorine atoms.

Inorganic refrigerants are described somewhat differently. They are assigned three-digit numbers the first of which is 7, with the following two numbers comprising the molecular weight.

Refrigerant mixtures are divided into two categories, azeotropes and zeotropes. In zeotropes, the equilibrium vapor and liquid compositions are different, and composition shifting can occur in a refrigeration cycle. Refrigerants are placed into one of six safety categories, shown in Figure 3.

Environmental Factors

Within the family of halocarbon refrigerants are compounds that contain chlorine, fluorine, and carbon, the chlorofluorocarbons (CFCs). Hydrochlorofluorocarbons (HCFCs) contain chlorine, fluorine, carbon, and hydrogen. Halocarbons that contain only carbon, fluorine, and hydrogen are called hydrofluorocarbons (HFCs) (see FLUORINE COMPOUNDS, ORGANIC-FLUORINATED ALIPHATIC COMPOUNDS). In the mid-1980s

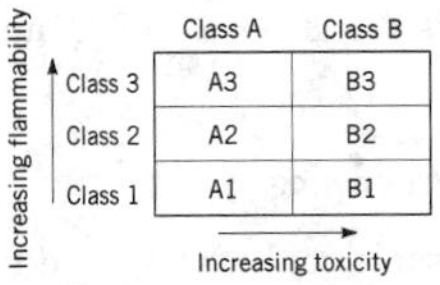

Figure 3. Refrigerant safety group classifications established by the American Society of Heating, Refrigerating, and Ventilating Engineers (ASHRAE).

scientific investigations confirmed that chlorine from compounds like CFCs and HCFCs were contributing to ozone depletion in the stratosphere, and a production phaseout schedule for these compounds was put in place in 1987 through the initialing of an international agreement known as the Montreal Protocol. Although originally scheduled for 50% production phaseout by the year 2000 in developed countries, the worsening ozone depletion forced acceleration of the CFC phaseout. Alternatives to CFCs and HCFCs must be sought and thus there has been increased interest in hydrocarbon and inorganic refrigerants such as propane and ammonia for applications where they have not traditionally been considered.

The selection of a refrigerant is based on many factors related to the specific application, including Ozone Depletion Potential (ODP), Global Warming Potential (GWP), toxicity, flammability, availability, cost, pressure, density, and theoretical cycle efficiency. The term ODP refers to the relative ozone depletion potential of a refrigerant compared to that of CFC-11, which is arbitrarily assigned a value of 1.0. It is the numerical quantity describing the extent of ozone depletion calculated to arise from the release to the atmosphere of 1 kg of a compound relative to the ozone depletion calculated to arise from a similar release of CRC-11. The term GWP is an index providing a simplified means of describing the relative ability of each greenhouse gas emission to affect radiative forcing and, thereby, the global climate. When written as GWP, the value is relative to carbon dioxide, CO_2, which is arbitrarily assigned a value of 1.0.

Household Refrigerator/Freezers

The vast majority of domestic refrigeration units have used CFC-12 as their working fluid, and in general these systems usually operate satisfactorily for 20 years or more. Over this period of time, an average compressor might operate $60-90 \times 10^3$ h without failure.

There are several common layouts for refrigerator/freezers, eg, single-door refrigerators, side-by-side combinations, top-mount combinations, chest freezers, etc. In a basic refrigerator circuit the typical design objective for performance might be to achieve average air temperatures of −19°C in the freezer section and −1°C in the general refrigerated space with continuous operation of the system in a 43°C ambient room. A system of ducts delivers proportional amounts of refrigerated air to the two compartments. The refrigerated air is derived from air blown through a fin-and-tube heat exchanger where the CFC-12 saturation temperature might be at −28 C and the entering air to the coil at −18°C.

Refrigerator/freezers require considerable wall insulation to minimize heat loss, and CFCs have been used as the blowing agent for foam insulation used in these systems. The insulation is typically a rigid polyurethane foam blown with CFC-11. The three basic components of polyurethane foam are an isocyanate, a polyol, and a blowing agent. The isocyanate and polyol are mixed, react exothermically, and the resulting heat boils the blowing agent. The resulting small bubbles froth the polyol–isocyanate mixture, which polymerizes, trapping the CFC-11 in the closed-cell structure (see FOAMED PLASTICS; URETHANE POLYMERS).

Commercial Refrigeration

The design of commercial refrigeration systems and the type of refrigerant used depend on the desired operating temperatures. Fresh fruit and vegetables require air temperatures ranging from 0–13°C, fresh meat and dairy products require −2 to 2°C, and ice cream and frozen food −18 to −32°C.

Cold Storage and Food Processing

Refrigeration for large-scale cold storage and food processing involves very large refrigeration capacities (1000 kW) and a large refrigerant charge. Such systems are typically designed for a 20–30-yr service life.

Ammonia is the predominant refrigerant in this application because of several excellent thermodynamic properties. Liquid ammonia has a high specific heat, reasonable density and viscosity, plus high thermal conductivity. These factors make it a good heat-transfer fluid. Combining these factors with its high latent heat effect per unit weight and a high vapor density that provides a high volumetric capacity (kJ/m^3) makes it an efficient refrigerant, and less ammonia is required for a given system than other refrigerants. Therefore, system charge requirements are lower, which helps to reduce both initial and long-term operating costs.

As CFC refrigerants are phased out of production, the use of ammonia, which is considered an environmentally benign refrigerant, will continue to play an ever-increasing role.

Industrial Refrigeration

Of all the application areas of refrigeration, the industrial sector uses the widest range of refrigerants and has the largest variety of cycles and systems, but uses the lowest total quantity of refrigerant compared to other areas of refrigeration.

Refrigeration can be accomplished in either closed-cycle or open-cycle configurations. In the closed-cycle, the refrigerant fluid is confined within the system and recirculates through the process. In the open-cycle, the fluid used as the refrigerant passes through the system once on its way to be used as a product or feedstock outside the refrigeration process, eg, the cooling of natural gas to separate and condense heavier components.

In addition to the distinction between open- and closed-cycle systems, refrigeration processes are also described as simple cycles, compound cycles, or cascade cycles. Simple cycles employ one set of components and a single refrigeration cycle. Compound and cascade cycles employ multiple sets of components and two or more refrigeration cycles. The cycles interact to accomplish cooling at several temperatures or allow a greater span between the lowest and highest temperatures in the system than can be achieved with the simple cycle.

Indirect Refrigeration (Brine). The process fluid is cooled by an intermediate liquid, water or brine, that is itself cooled by evaporating the refrigerant. The brine is cooled in the refrigeration evaporator and then pumped to the process load. The brine system may include a tank, either open or closed but maintained at atmospheric pressure through a small vent pipe at the top, or may be a closed system pressurized by an inert, dry gas.

Transport Refrigeration

Refrigeration is used in trucks, truck/trailers, intermodal containers, railcars, and ships. The majority of refrigerated vehicles fall into one of three refrigeration classifications: (*1*) 0 to 4°C, perishable produce; (*2*) −2 to 0°C, fresh meats; and (*3*) −43 to −17°C, frozen foods. Although the basic principles of the Rankine cycle apply to transport refrigeration, it has unique methods for powering the compression process.

Independent Engine- or Electric Motor-Driven. Many styles of independent engine- or electric motor-driven refrigeration units are available. The one-piece, plug-type design is a self-contained unit that mounts in an opening provided in the front wall of the vehicle box. The condensing section is on the outside, and the evaporator on the inside. The two sections are separated by an insulating plug attached to the vehicle wall that supports the various parts of the refrigeration unit. Self-contained, independent engine-driven plug-type units used in truck and truck/trailers range in capacity from 6 to 12 kW to maintain trailer temperatures of 20°C in 38°C ambients. At trailer temperatures of −18°C in 38°C ambients, capacities range from 2–6 kW.

Power Take-Off From Engine or Transmission. This type of system is limited to trucks and there are several take-off means available. Most are some form of electric power generation equipment, belt-driven from the engine crankshaft, which produces either a regulated a-c voltage or rectified direct current for the compressor and fan motors in the body.

Lubrication System Considerations

Compressors move refrigerant around the system and require a lubricant to cool and reduce friction of moving parts. Although desirable, it is impractical in many instances to attempt complete isolation of the refrigerant; hence many refrigeration systems operate with a refrigerant–lubricant mixture in circulation. How much lubricant is in circulation depends on the system.

HOWARD W. SIBLEY
Carrier Corporation

Technical Options Report, United Nations Environment Programme (UNEP) on Refrigeration, Air Conditioning, and Heat Pumps, Paris, France, 1991.

ASHRAE Handbook, Fundamentals, American Society of Heating, Refrigerating, and Ventilating Engineers, Inc., Atlanta, Ga., 1993.

M. J. Kurylo, *Proceedings of ASHRAE 1989 CFC Technology Conference at the National Institute of Standards and Technology, Gaithersburg, Md., Sept. 27–28, 1989,* ASHRAE Publications Dept., Atlanta, Ga., 1990.

Montreal Protocol, United Nations Environmental Program, Pts. I, II, III, United Nations Ozone Secretariat, Nairobi, Kenya, 1994.

REGENERATED CELLULOSIC FIBERS. See FIBERS, REGENERATED CELLULOSICS.

REGULATORY AGENCIES

SURVEY

In the United States, the two main federal agencies involved in the protection of human health and the environment are the Environmental Protection Agency (EPA) and the Occupational Safety and Health Administration (OSHA). EPA's principal concern is the protection of the environment, in most cases, the area outside of an industrial facility. The principal function of OSHA is the protection of people, eg, employees, visitors, and temporary help, in the workplace.

There are a number of other federal agencies involved in related work. Pertinent agencies and their areas of concern are listed in Table 1.

In addition to the federal agencies, there are many state agencies, as well as county and municipal agencies, that regulate the environmental and health areas. International laws and regulations must also be taken into account.

The difference between laws and regulations is important. For the former, the U.S. Congress first passes a bill, which is then signed by the President, and thereby made a law or act. The act describes what Congress wants regulated, the general method to be used, and the ultimate results expected. It is then the responsibility of the designated agency to write and administer regulations to meet these requirements.

The Federal Register, a daily document published by the General Services Administration, contains, in addition to proposed and final regulations, notices for all of the federal agencies. Twice a year, a regulatory agenda is published in *The Federal Register,* listing all regulatory activities, from pre-proposed activities through proposed and final rulemaking.

Table 1. Federal Agencies and Their Functions

Agency	Alias	Function
Agriculture Dept.	USDA	agriculture, animal and plant health inspection, forest service, food safety, Rural Electrification Administration, soil conservation service
Commerce Dept.	Commerce	Census Bureau, economics, trade, National Oceanic and Atmospheric Administration, Patent and Trademark Office
Dept. of Defense	DOD	Engineers Corps, dredge and fill operations, waterways, etc
Energy Dept.	DOE	all aspects of energy use, conservation, costs, etc
Health and Human Services Dept.	HHS	Food and Drug Administration, National Institute for Occupational Safety and Health (NIOSH)
Interior Dept.	Interior	land management, fish and wildlife, Geological Survey, mines, surface mining and reclamation
Justice Dept.	Justice	Antitrust Division, enforcement activities
Labor Dept.	Labor	employment standards and statistics, OSHA, Mine Safety and Health Administration (MSHA)
Transportation Dept.	DOT	Coast Guard; Federal Aviation Administration (FAA); highway, railroad, and maritime administration; hazardous material shipping
Treasury Dept.	Treasury	Alcohol, Tobacco, and Firearms Bureau (ATF), Customs Service, Internal Revenue Service
Council on Environmental Quality	CEQ	provides policy advice on environmental matters; implements National Environmental Policy Act (NEPA); coordinates environmental concerns among other agencies
Nuclear Regulatory Commission	NRC	ensures that radioactive materials are used and nuclear facilities are operated with regard for environment and public health, safety, and security
Office of Management and Budget	OMB	reviews regulations to ensure that they are cost-effective; approves paperwork or other information-gathering requirements

An important aspect of environmental, health, and safety laws and regulations is enforcement. Federal, state, and local regulatory authorities usually have large enforcement sections.

Regulations change continuously with updates and reauthorizations, and the specifications of these regulations quickly become outdated. Current copies of the laws and regulations must be consulted in order to deal with specific situations.

NANCY R. PASSOW
Lonza Inc.

Standard Industrial Classification Manual, Executive Office of the President, Office of Management and Budget, 1987; for sale by National Technical Information Service, 5285 Port Royal Road, Springfield, Va., 22161, Order No. PB87-100012.

CHEMICAL PROCESS INDUSTRY

The chemical process industry is highly regulated in terms of environmental, health, and safety. In addition to the regulatory requirements of government agencies, the chemical industry is developing standards of its own to ensure proper protection of the environment, employees, and the community. These include Responsible Care (registered by the Chemical Manufacturers Association (CMA)); the International Standards Organization (ISO); Environmental Management Systems; the sustainable development program of International Chamber of Commerce (ICC); and others.

Environmental Protection

Water. The first step in water quality standards is stream use classification. The four categories, as defined by the EPA, are Class A, primary water contact recreation; Class B, propagation of desirable aquatic life; Class C, public water supplies prior to treatment; and Class D, agricultural and industrial uses. The second step is to develop water-quality criteria. This is the specific concentration of a pollutant that is allowable for the designated use.

The Clean Water Act (1977) requires specific levels of control for dischargers. Outlined in the Effluent Guidelines and Standards for various industrial categories, these standards limit the discharge of pollutants, usually in terms of a unit weight of pollutant per unit of either product or raw material, rather than a concentration in the discharge stream, in order to eliminate the use of dilution to meet limits. The effluent standards require two levels of treatment: best practicable control technology (BPCT) and best available control technology (BACT).

Air. Various laws have been passed in the United States to control air pollution. The first law that had any real effect was the Clean Air Act of 1970 (CAA), which was followed by the Clean Air Act Amendments of 1977. Most recently, the Clean Air Act Amendments (CAAA) of 1990 further changed and updated the requirements.

Under the Clean Air Act, six criterion pollutants, ie, pollutants of special concern, have been established by the EPA: sulfur oxides (SO_x), particulates, carbon monoxide (CO), nitrogen oxides (NO_x), ozone (photochemical oxidants), and lead. National Ambient Air Quality Standards (NAAQS) were developed by EPA based on threshold levels of air pollution below which no adverse effects could be experienced on human health or the environment.

In order to have a nationwide basis for air pollution emission controls and to set a minimum emission limit, the EPA developed New Source Performance Standards (NSPS). The NSPS set specific pollutant emission limits or describe the best available control technology (BACT) that should be applied at that source.

The National Emission Standards for Hazardous Air Pollutants (NESHAP) have been expanded to cover a list of 189 pollutants, many of which are associated with chemical operations. Facilities are required to install maximum achievable control technology (MACT), standards issued by EPA.

Solid and Hazardous Waste. The Resource Conservation and Recovery Act (RCRA) of 1976 (6) authorizes substantial controls of pollution resulting from solid waste disposal water pollution.

The main objectives of RCRA are to protect public health and the environment and to conserve natural resources. The act requires EPA to develop and administer the following programs: solid waste disposal practices providing acceptable protection levels for public health and the environment; transportation, storage, treatment, and disposal of hazardous wastes practices that eliminate or minimize hazards to human health and the environment; the use of resource conservation and recovery whenever technically and economically feasible; and federal, state, and local programs to achieve these objectives.

The section of the RCRA of most concern to the chemical industry is Subtitle C, the hazardous waste management regulations. The purpose of this section is to regulate hazardous wastes from their generation to their disposal. Facilities that generate, treat, store, or dispose of hazardous wastes are covered by these regulations.

Emergency Planning and Community Right-to-Know Act. EPCRA, promulgated as part of the Superfund Reauthorization Act and originally known as SARA, Title III, requires facility emergency plans, spill and release reporting, annual inventory reporting, and annual toxic chemical emissions and pollution prevention reporting. The annual toxic emissions reports are called Toxic Release Inventory (TRI) reports; they must include information on any on-site or off-site releases, disposal, treatment, or recycling of the listed toxic chemicals.

Product Safety

Toxic Substances Control Act. EPA regulates the manufacture, use, and exposure to hazardous or toxic chemicals under a number of laws. For the chemical industry, the law of prime concern is the Toxic Substance Control Act (TSCA). The two main goals of TSCA are acquisition of sufficient information to identify and evaluate potential hazards from chemical substances, and regulation of the production, use, distribution, and disposal of these substances.

As part of TSCA, EPA can require the testing of any chemical if there is the possibility of an unreasonable risk to health or environment or if there is significant human or environmental exposure. If the substance poses an unreasonable risk, EPA can prohibit the manufacture, processing, or distribution of the substance; limit the amount of the substance that can be manufactured, processed, or distributed; prohibit a particular use for the substance; limit the concentration of the substance during manufacture, processing, or distribution; regulate disposal methods for the substance; and require manufacturers to maintain records of process and to conduct tests to assure compliance with EPA rules.

Federal Insecticide, Fungicide, and Rodenticide Act. FIFRA covers the use and manufacture of pesticides, rodenticides, etc, and includes such things as biocides and preservatives. More details on these regulatory requirements are given in the articles FUNGICIDES, PESTICIDES, INSECT CONTROL TECHNOLOGY, and INDUSTRIAL ANTIMICROBIAL AGENTS.

Transportation. The U.S. Department of Transportation regulates the transport, packaging, and handling of hazardous substances, including chemicals. In addition, there are international laws and regulations.

Employee Safety

Occupational Safety and Health Act. OSHA has broad responsibilities for protecting the workplace. The Occupational Safety and Health Act is administered by the Occupational Safety and Health Administration under the U.S. Department of Labor. The act covers all health and safety aspects of a worker's environment, including allowable employee exposure to many different chemical substances. These are given as ambient air concentrations over a certain time period, which usually is an 8-h time-weighted average. Sometimes a ceiling concentration is given as well. The monitoring of employee exposure is called industrial hygiene (qv).

The National Institute of Occupational Safety and Health (NIOSH), under the Department of Health and Human Services, works with OSHA. It is NIOSH's responsibility to determine safe exposure limits for chemical substances and to recommend to OSHA that these limits be adopted as standards.

Conservation, Land Use, and Other Areas Subject to Regulation

A new facility should be located in an area that is economically suitable (see PLANT LOCATION). At the same time, the social, environmental, and aesthetic effects must be considered. These decisions are part of land-use policies. While there is no national land-use program in the United States, there are certain laws that cover portions of this decision-making process.

National Environmental Policy Act. The principal goal of the NEPA is to establish a national policy in the United States that ensures continued growth and technological advancement while maintaining the quality of the environment. Any federal agency sponsoring action that significantly affects the environment, eg, granting a construction permit or introducing new legislation, must issue a detailed report describing the environmental impact of the proposed action and alternatives. This report is called an environmental impact statement (EIS).

NANCY R. PASSOW
Lonza Inc.

The Rivers and Harbor Act of 1899, 33 USC §§401–413, and Executive Order 11574.

Clean Water Act of 1977, 33 USC §§1251–1387: 33 CFR Parts 320–330, 335–338; 40 CFR Parts 104–140, 230–233, 401–471.

Safe Drinking Water Act, 42 USC §§300f–300j–26; 40 CFR Parts 141–149.

Environmental Management Source Book, 1994–1995, Environment Today, Enterprise Communications Inc., Marietta, Ga.; annual issue as part of monthly magazine subscription.

PHARMACEUTICALS AND COSMETICS

The U.S. Food and Drug Administration (FDA) is the primary federal scientific and regulatory government agency that monitors drug, biologic, medical devices, food, veterinary, and cosmetic products.

FDA's mission has traditionally been consumer protection. However, the manner in which the agency accomplishes its regulatory mission has been changing with the explosion of advances within analytical chemistry and toxicology. Because new and better compounds are continuously being developed, increased attention is being devoted to the dilemma of speeding their entry into the market while ensuring their safety and effectiveness. Because no compound is completely safe, FDA must make an assessment of the intended use and determine whether the level of risk is acceptable.

FDA Organization and Roles

The FDA is headed by the Commissioner of Food and Drugs. This position is not a Cabinet-level office but falls within the Public Health Service (PHS), a division within the U.S. Department of Health and Human Services (HHS).

The FDA's approval and enforcement programs are administered by five centers organized along product lines. Although all five centers must follow the general provisions of the Act, each center is governed by its own unique and distinctive set of laws and regulations. The five centers are as follows.

(*1*) Center for Drug Evaluation and Research (CDER). (*2*) Center for Biologics Evaluation and Research (CBER). (*3*) Center for Devices and Radiological Health (CDRH). (*4*) Center for Food Safety and Applied Nutrition (CFSAN). (*5*) Center for Veterinary Medicine (CVM).

Independent of these centers, the FDA also has overlapping jurisdiction with several other federal agencies, including the U.S. Department of Agriculture; the Bureau of Alcohol, Tobacco, and Firearms; the Federal Trade Commission; and the Environmental Protection Agency.

Regulating Drug Products. The FDA has the authority to regulate new drugs from early laboratory research through clinical testing and market approval. In order to be approved for marketing, a new drug must be shown to be both safe and effective for its intended use.

The FDA new drug approval process, which is the way most products enter the market, begins with clinical investigations. However, before clinical investigations can commence, FDA requires significant preclinical investigations involving tests on laboratory animals in order to determine the nature of the chemical and to establish evidence concerning the toxicity of the substance. If, through animal studies, the drug is determined to be safe enough for human experimentation, an Investigational New Drug (IND) Application must be submitted to the Center for Drug Evaluation and Research before beginning human clinical trials.

Clinical investigations are conducted under the oversight of an Institutional Review Board (IRB) to ensure that appropriate safeguards exist to protect the rights and welfare of the research subjects.

After all clinical trials are completed, the sponsor of the drug submits a New Drug Application (NDA) to FDA. The NDA consists of the IND data, manufacturing information, and data on drug stability. In order to be approved, an NDA must include data which demonstrate that the drug is both safe and effective. Several classifications exist within the broad category of pharmaceuticals, each of which has its own definition and form of regulation. A distinct class of drugs comprises those requiring a prescription or a written order from a physician or health professional.

Drugs which are available without a prescription are readily available to consumers over-the-counter (OTC). An OTC drug is low in toxicity, has low potential for harm, can be labeled for safe use without a doctor's supervision, is not habit-forming, and can be taken under easily understood conditions.

A subcategory of drugs comprises those that are reviewed under the Abbreviated New Drug Application (ANDA) process. These drugs are usually called generic drugs. A generic drug is one that is equivalent to a pioneer or brand-name drug but is not marketed until the brand-name drug's patent and exclusivity periods have expired.

Although generic drugs must meet the same standards as new drug products for identity, strength, purity, stability, adequate labeling, and bioequivalence, they need not go through the extensive clinical trials of a NDA. Instead, these generic drugs must show bioequivalence to the pioneer drug and fall into acceptable parameters set for bioavailability, which is the extent and rate at which the body absorbs the drug.

Finally, a different set of rules is applied to antibiotics and insulin-containing drugs. These categories are regulated under a monograph system mandated by statute. Thus, when a drug in these categories has been demonstrated to be safe and effective to the satisfaction of FDA, the agency promulgates a regulation of general applicability, describing in detail the required specifications of the drug.

Regulating Biological Products. The process for gaining FDA approval for a biological product is similar to that for a drug product. The FDA regulations require that the person or entity, eg, manufacturer, sponsoring or conducting a clinical study for the purpose of investigating a potential biological drug product's safety and effectiveness submit an IND to the Center for Biological Evaluation and Research. Clinical trials are subject to IRB review, just as are drug studies. After completing the IND studies, the manufacturer submits the safety and effectiveness data generated by the studies to FDA in the form of a product license application (PLA). If both product and facility meet all standards and regulations, FDA will approve a PLA for the product and an establishment license application (ELA) for the facility.

Regulating Medical Devices. A person or company engaged in the manufacture, preparation, compounding, assembly, or processing of a device intended for human use must follow the regulations enforced by FDA's Center for Devices and Radiological Health. The level of FDA regulation or control is governed by the class in which the device is placed by the agency, ie, Class I, II, or III. Class I devices are those requiring the lowest level of regulation and are subject to general control requirements. Class II devices are subject to special controls as well as the general control requirements. Class III devices are subject to general controls and cannot be marketed until they have an approved Premarket Approval Application (PMA) or, as a result of premarket notification (510(k)) submission, until they have been found by FDA to be substantially equivalent to preamendment devices.

Regulating Food Products. The mandate of the Center for Food Safety and Applied Nutrition (CFSAN) includes U.S. food processors, dietary supplement manufacturers, food warehouses, and cosmetic products.

FDA regulates not only the finished food product, but also the ingredients that are added to food. These ingredients may be either intentionally added to food or the unintended result of materials leaching to food from product packaging. Ingredients that are intentionally added directly to food fall into two separate categories: (*1*) pre-1958 substances and substances generally recognized as safe (GRAS) by scientific experts, and (*2*) food additives. There are two types of food additives, these that are added directly to food and those that are not intentionally added directly to food but come into contact with food. The latter are considered indirect food additives.

The Delaney Clause was included in the Food Additives Amendment of 1958; it states that no food additive or color additive can be deemed safe if it has been found to induce cancer when ingested by humans or animals.

Regulating Veterinary Products. Animal drug controls are similar to those for human drugs. The sponsor of a new animal drug must demonstrate both safety and effectiveness of the drug for a particular intended use before a New Animal Drug Application (NADA) is approved. Manufacturers of generic animal drugs may submit Abbreviated New Animal Drug Applications (ANADAs), which are comparable to abbreviated new drug applications submitted by manufacturers of human generic drugs.

Regulating Cosmetics. Cosmetics are among the least regulated of all FDA product categories. The FDA has no statutory preapproval authority over cosmetics. FDA's enforcement mechanism against cosmetics stems from the adulteration and misbranding sections of the Federal Food, Drug, and Cosmetic Act.

GARY L. YINGLING
SUZAN ONEL
McKenna & Cuneo, LLP

W. Schultz, testimony before the Subcommittee on Health and the Environment of the House Committee on Commerce, 104th Congress, 1st session, Washington, D.C., Feb. 2, 1995.

Federal Food, Drug, and Cosmetic Act of 1938 (as amended), 21 USC §§321–395 (1995).

D. O. Beers, *Generic and Innovator Drugs: A Guide to FDA Approval Requirements,* 4th ed., Aspen Law and Business/Aspen, Engelwood Cliffs, N.J., 1995.

Food and Drug Administration, Center for Devices and Radiological Health, *An Introduction to Medical Device Regulations,* HHS Publication No. 92-4222, U.S. Department of Health and Human Services, Public Health Service, Rockville, Md., 1992.

POWER GENERATION

Power, P, defined as the rate at which work is performed, is expressed in terms of energy divided by time and is most commonly given in units of horsepower, as for the power supplied by mechanical devices such as diesel engines, or in the SI units of watts, especially when measuring electrical power. One horsepower is equivalent to the amount of power needed to lift 33,000 pounds (14,982 kg) one foot (30.5 cm) in one minute. One watt is equivalent to the power required to perform one joule of work per second. In a simple direct-current circuit where potential is represented by E: $P = EI = I^2R$. One watt of power is required to force one ampere of electric current, I, through a length of conductor having a resistance, R, to flow of one ohm. The calculation of power in alternating-current circuits is more complex because of the characteristics of these circuits and their loads. For a single-phase circuit having an effective line voltage, E, and an effective line current, I: $P = EI \times PF$, where PF is the power factor corresponding to the lag between the voltage and current wave forms. For multiphase systems, a factor is introduced on the right-hand side of the equation. The factor is two for a two-phase system; 1.73 for a three-phase system.

Combustion Fundamentals

The combustion of fossil fuels, typically coal, oil (see PETROLEUM), or natural gas, is central to the energy conversion process by which most electric power is generated worldwide, particularly in the United States. A fundamental goal of power plant designers and operators is to ensure that the maximum economical amount of energy is extracted for practical use from a given amount of fuel burned (see COMBUSTION TECHNOLOGY).

Combustion is basically an exothermic chemical reaction by which the potential chemical energy stored in a fuel is released during its reaction with oxygen in the presence of heat. Thus, the basic constituents of combustion are fuel, oxygen, and heat. At normal temperatures, the oxygen, usually from air, and fuel can typically mingle without any combustion effect. However, upon the addition of sufficient heat, the fuel and air are excited to a point at which high velocity molecular collisions occur. If the velocity of these collisions is high enough, the bonds holding the various molecules together break and release heat.

The point at which enough heat has been added to start combustion is known as the ignition point. Once initiated, external heating sources are typically not required to maintain the combustion process, because most fuels release sufficient heat during the combustion process.

The main fossil fuel constituents enabling combustion are carbon, hydrogen, and various hydrocarbons. The main products resulting from combustion are water and carbon dioxide. However, a variety of trace compounds and unburned matter may also be released during the combustion of power plant fuels. These materials vary, depending on the specific fuel being burned and whether complete combustion occurs. For example, coal and fuel oil may contain small amounts of sulfur. This is an undesirable constituent, despite its nominal heat content, because sulfur reacts during combustion to form acidic sulfur dioxide.

Steam Cycles

Conventional fossil fuel-fired power plants, nuclear power facilities, cogeneration systems, and combined-cycle facilities all have one key feature in common: some type of steam generator is employed to produce steam. Except for simple-cycle cogeneration facilities, the steam is used to drive one or more rotating turbines coupled to rotating electric generators for electricity production. The thermodynamic cycle by which water is boiled in a steam generator (heat addition), forced through a turbine (expansion), and then condensed (heat rejection) before being pumped back to the steam generator is known as the Rankine cycle.

Because of the simplicity and reliability of the Rankine cycle, facilities employing this method dominated the power industry in the twentieth century and typically play an important role in most modern combined-cycle facilities. Water is the working fluid of choice in nearly all Rankine cycle power plants because water is nontoxic, abundant, and low cost.

Under standard pressure and temperature conditions of 101.3 kPa and 15°C (14.7 psia and 60°F), the amount of heat (or thermal energy in transition) that must be added to one pound of water to raise the temperature of the pound of water 1°F is known as one British thermal unit (Btu). One Btu, a common unit of measure in the energy field, is equivalent to 1.05435 kJ.

The generation of steam (qv) is essentially a two-stage process. Through the addition of heat, the temperature of water is first raised to its boiling point, ie, 100°C at atmospheric pressure. After the boiling point is reached, the temperature at the contact point between the liquid and vapor remains constant as long as the pressure is not permitted to build, such as in a partially closed vessel, until vaporization is completed. When water is boiled in a closed vessel, such as in the steam drum of a utility boiler, vapor generation increases the pressure within the vessel. As the pressure increases, the boiling point (or saturation temperature) increases. Thus, in a boiler operating at 8.27 MPa (1200 psia), water must be heated to nearly 299°C to achieve boiling.

In most utility boilers, steam pressure regulation is achieved by the throttling of turbine control values where steam generated by the boiler is admitted into the steam turbine. Some modern steam generators have been designed to operate at pressures above the critical point where the phase change between liquid and vapor does not occur.

Rankine Cycle Power Plants

The most common heat sources employed by Rankine cycle power plants are either fossil fuel-fired or nuclear steam generators. The former are the most widely used.

Fossil Fuel-Fired Plants.

In modern, fossil fuel-fired power plants, the Rankine cycle typically operates as a closed loop. In describing the steam–water cycle of a modern Rankine cycle plant, it is easiest to start with the condensate system. Condensate is the water that remains after the steam employed by the plant's steam turbines exhausts into the plant's condenser, where it is collected for reuse in the cycle. Many modern power plants employ a series of heat exchangers to boost efficiency. As a first step, the condensate is heated in a series of heat exchangers by steam extracted from strategic locations on the plant's steam turbines (see HEAT-EXCHANGE TECHNOLOGY).

Another stage of heating occurs in a direct contact, deaerating heater before the water is directed to the boiler feed pumps. The deaerating heater serves a dual role: it not only adds heat to the water, but also improves the water quality by removing dissolved oxygen and other noncondensable gases in the condensate that can cause corrosion within the plant's steam and water piping, steam generator, and steam turbines.

The deaerator is typically considered the dividing point between the condensate and feedwater systems. Water exiting the deaerator is fed directly to a boiler's high speed centrifugal feedwater pump, which boosts the water's pressure to a level high enough to enable its ultimate introduction into the steam drum where boiling occurs.

After exiting the feedwater pump and before entering the boiler's steam drum, feedwater usually goes through additional high pressure feedwater heaters (heated by steam turbine extraction steam) as well as another stage of heating in the plant's economizer. The economizer consists of a bank of heat exchanger tubes located in the boiler's backpass section, where heat is transferred from the hot exhaust gas exiting the boiler to the incoming feedwater flowing through the economizer tubes. Most modern fossil fuel-fired power plants share these design features. However, many different types of feedwater heater and boiler arrangements are available, based on the specific steam needs of an application.

Water Treatments. The goal of water treatment is to ensure that the water used for steam generation is pure, such that any contaminants that could lead to corrosion, erosion, or scaling of boiler components are absent. The deaerating feedwater heater is one means by which noncondensable gases are removed from the system. Oxygen scavenging chemicals are also dosed into the feedwater system to eliminate any oxygen not removed by the deaerator or air ejectors.

Demineralizers are often used to treat raw makeup water or condensate where high purity is required. Demineralizers employ a combination of cation and anion exchange to remove additional material, including sodium and ammonium cations.

Fuel System. The main goals of any firing system are the same, ie, to ensure that fuel and air are delivered in such a way as to promote safe, efficient, smooth combustion, and to minimize pollutant emission formation and maximize heat transfer to the waterwall tubes. Properly controlling the air-to-fuel ratio is central to these goals. Whether a boiler fires liquid fuel, coal, or natural gas, there are typically multiple burners which serve to direct the fuel into the furnace for combustion and which ensure that proper fuel–air mixing occurs so that combustion takes place at the desired rate and in the desired location.

Emission Control. Power industry emissions include the air pollutants sulfur dioxide (SO_2), nitrogen oxides (NO_x) and carbon dioxide (CO_2). In oil-, gas-, and coal-fired facilities, low NO_x burners have been successfully applied to reduce NO_x emissions by limiting the peak burner flame temperatures either by rapid premixing of fuel and air, or by staging the introduction of air or fuel to achieve a longer, cooler flame.

Emissions control systems play an important role at most coal-fired power plants. For example, pulverized coal-fired plants sited in the United States require some type of sulfur dioxide control system to meet the regulations set forth in the Clean Air Act Amendments of 1990, unless the boiler burns low sulfur coal or benefits from offsets from other highly controlled boilers within a given utility system. Flue-gas desulfurization (FGD) is most commonly accomplished by the application of either dry- or wet-limestone systems (see SULFUR REMOVAL AND RECOVERY).

Similar to oil-fired plants, either low NO_x burners, selective catalytic reduction or selective noncatalytic reduction can be applied for NO_x control at PC-fired plants. Likewise, fabric filter baghouses or electrostatic precipitators can be used to capture flyash (see AIR POLLUTION CONTROL METHODS).

Nuclear Fuel-Fired Plants. There are some basic design differences of the nuclear plants steam cycles (see NUCLEAR REACTORS) as compared to conventional fossil fuel plants. In nuclear power plants, thermal energy is released during the fissioning of a nuclear fuel (or fissile), such as uranium-235. Fission, the splitting of a heavy nucleus, is typically initiated when the material is struck by a neutron. When the nucleus of a fissile material's atom is split, the energy associated with the atom is released. This energy, approximately 100 billion times that released during the combustion of one carbon atom in fossil fuels, is significant. In addition, extra neutrons are released that impact other nuclei, creating chain reaction.

Extremely safe means of controlling the nuclear reaction process have been devised by introducing materials that can moderate the production and absorption of neutrons released during fission.

Cogeneration

In the power industry, cogeneration refers to the simultaneous generation of heat and power. One example of a cogeneration plant is the central utility boiler that provides steam for both electric power generation and supply to a local district heating system where it can be used for facilities or process heating, or cooling, via absorption coolers or steam-turbine-driven refrigeration and/or air conditioning systems. Such a plant can use a noncondensing steam turbine to supply some or all of the steam required by the district heating system. A well-designed cogeneration facility can convert 80% or more of the fuel energy input for useful purposes, ie, power generation and process heating. In a conventional Rankine cycle power plant, the thermal energy remaining in the steam as it enters the condenser can range up to 50% or more of the fuel energy input, boiler associated losses are ca 15%, and other losses about 2%. Thus, a conventional Rankine cycle plant may convert only 35% of the fuel energy for power generation. In a modern coal-fired cogeneration system, heat losses can be cut to 16%, 15% of which are boiler-associated. Such a system, where the waste heat is recovered from the main power generation cycle for reuse, is often referred to as a topping cycle (see ENERGY MANAGEMENT; PROCESS ENERGY CONSERVATION).

Gas Turbines and Combined-Cycle Power Plants

Gas turbine engines have become increasingly popular for power generation and cogeneration.

Gas turbines are based on the Brayton thermodynamic cycle (Fig. 1). Most modern units operate in the following manner. Combustion and cooling air is first drawn in and compressed in a multistage axial compressor located on the cold end of the gas turbine's rotor (point A of Fig. 1**b**). The compressed air is injected into a combustor section where it is combined with fuel for combustion, generating hot, expanding gaseous exhaust. The expanding combustion gases are then directed axially along the rotor into a power turbine or turbines. There the gases impart torque on the tangential turbine blades in a manner similar to a steam turbine, before exhausting to the atmosphere or a heat recovery steam generator.

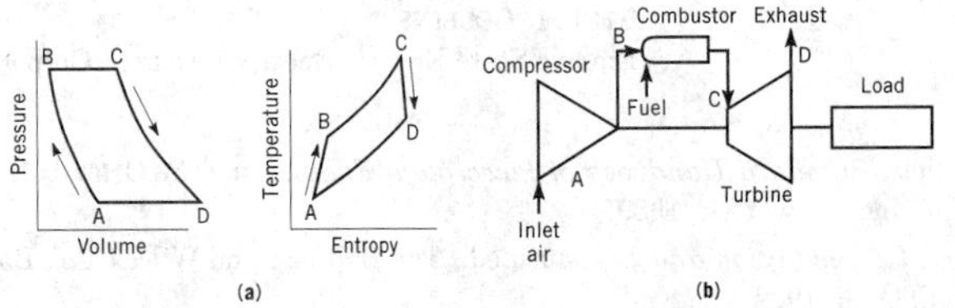

Figure 1. Brayton cycle, where A = compressor inlet, B = combustor inlet, C = power turbine inlet, and D = exhaust: (**a**) thermodynamic relationships and (**b**) schematic of a simple-cycle, single-shaft gas turbine. Courtesy of General Electric.

Natural Gas for Power Plant Use

Most modern gas-fired, heavy-duty gas turbines operate at gas pressures between 1.2 and 1.7 MPa (180–250 psig). However, aeroderivative gas turbines and newer heavy-duty units can have such high air-inlet compression ratios as to require booster compressors to raise gas inlet pressures, in some cases as high as 5.2 MPa (750 psig).

All gas-fired power plants require oxygen analyzers to ensure that air has not been drawn into the piping system. Oxygen intake can lead to the presence of an explosive mixture in the pipeline before the fuel reaches the burner or combustor zone. When gas-fired units are located in an enclosed area, multiple ultraviolet flame detectors are used to shut down equipment and flood the area with CO_2 or a chemical fire suppressant whenever a spark or flame is detected.

Processed gas is compressed for transmission through interstate and intrastate pipelines (qv).

Modern pipeline companies control gas flow through their systems in response to and anticipation of demand swings via computerized operations centers. These facilities typically use satellite or modem-based telemetry systems to monitor gas flow and automatically regulate valves and compressor stations located along the pipeline and at gas storage facilities.

The demand for gas is highly seasonal. Thus pipeline companies economize by sizing production facilities to accommodate less than the system's maximum wintertime demand. Underground storage facilities are used to meet seasonal and daily demand peaks. In North America, gas is stored in three main types of underground formations: depleted oil or gas fields, aquifers that originally contained water, and caverns formed by salt domes or mines.

Other Generation Options

Reciprocating Engines. Reciprocating engines, particularly medium- and slow-speed diesels similar to those used for shipboard propulsion and power generation, have been used for land-based power generation since the 1940s. Because of their relatively high efficiency (up to 45%), reliability, and quick start-up capability, these units have been popular for peaking, emergency, and base-load power generation.

Nuclear Reactors. Nuclear power facilities account for about 20% of the power generated in the United States. Although no new plants are planned in the United States, many other countries, particularly those that would otherwise rely heavily on imported fuel, continue to increase their nuclear plant generation capacity.

Fuel Cells. Fuel cells (qv) are essentially batteries (qv) that run on fuel and therefore do not run down. Advanced fuel cells feature efficiencies above 40% and consist of a series of porous, conducting electrode layers (a layer corresponding to one anode and one cathode), separated by an electrolyte (an ionic charge carrier). The multiple cells are configured in series to achieve the voltages required for industrial and commercial use.

Fuel cells generate significant levels of waste heat that can be captured and utilized to improve overall plant efficiency (see PROCESS ENERGY CONSERVATION).

STEVEN COLLINS
Empire State Electric Energy Research Corporation

T. Elliott, *Standard Handbook of Powerplant Engineering*, McGraw-Hill Book Co., Inc., New York, 1989.

Steam: Its Generation and Use, 40th ed., The Babcock and Wilcox Co., Barberton, Ohio, 1992.

B. H. Bunch and A. Hellemans, *The Timetables of Technology*, Simon and Schuster, New York, 1993.

PRASEODYMIUM. See LANTHANIDES.

PRESERVATIVES. See ANTIOXIDANTS; ANTIOZONANTS; COATINGS; FOOD ADDITIVES; PAINT; WOOD.

REINFORCED PLASTICS

Reinforced plastics are commonly referred to as composites or, more specifically, polymer composites. Not all composites are reinforced plastics; ceramic/metal-matrix composites and concrete are good examples of nonpolymeric composites (see COMPOSITE MATERIALS). Reinforced plastics are also referred to as RP, FRP (fiber glass-reinforced plastic), and GRP (glass-reinforced plastic) interchangeably.

Common to all reinforced plastics are two ingredients, resin and reinforcement. Resin is an organic material, usually of high molecular weight, that can be molded and set into a final shape. Resins are of two basic types. Thermoplastic resins soften upon heating, are shaped in a mold, and retain that shape when cooled. Thermosetting resins are placed in a mold and cured by the use of a catalyst, heat, or both, until they harden in the shape of the mold.

The second main ingredient in reinforced plastic is the reinforcement, eg, fibers of glass, carbon, boron, mineral, cellulose, or polymers. Reinforcements can be configured in many ways, such as continuous or chopped strands, milled fibers, rovings, tows, mats, braids, and woven fabrics.

Reinforced plastics may also include fillers (qv), which are inexpensive materials such as calcium carbonate used to displace resin and reduce cost; curing agents (catalysts), promoters, inhibitors, and accelerators, which affect thermosetting resin cure; colorants; release agents (qv) to facilitate removal from the mold; and other additives which can impart a wide variety of properties to the finished part, such as fire resistance, electrical conductivity, static dissipation, and ultraviolet resistance.

It is important to note that reinforced plastics remain a combination of materials differing in form or composition on a macro scale. The main constituents (resin, reinforcement, and filler) retain their identities and do not dissolve or merge into each other; rather, they act in concert. These components can be physically identified and exhibit an interface between each other.

Manufacturing Method Selection Factors

There are many different manufacturing methods available for reinforced plastics. The selection process for a given application requires simultaneous consideration of the product design; the constituent materials; required physical, mechanical, and chemical properties; and production requirements. Selection factors that help to define the appropriate processes to consider for a given application may be categorized as product design factors, material factors, mechanical property factors, and production factors.

Product design factors
- overall part size
- shape complexity
- critical dimensions
- weight limitations
- potential for parts integration
- assembly requirements
- secondary operations

Material factors
- polymer type, thermoplastic or thermosetting
- reinforcement type, amount, and orientation
- surface requirements
- chemical exposure
- thermal properties
- electrical properties
- weather resistance
- fire performance

Mechanical property factors
- reinforcement type, amount, and orientation
- loads loads to be applied
- static
- dynamic
- impact
- assembly
- shipping
- deflection limitations
- operating temperature range

Production factors
- total design life
- volatility of design, design modifications
- production rate buildup over time
- production rate (highest monthly rate)
- shipping range

Manufacturing Methods

Hand Lay-Up and Spray-Up. In hand lay-up, fiber reinforcements in mat or woven form are placed on the mold surface and then saturated with a liquid polymer, typically a polyester resin, that has been chemically activated to polymerize (cure) without the addition of heat.

In the spray-up process a reinforcement, usually glass fiber, is substituted for the mat and a special spray gun simultaneously chops the glass fiber and applies it with catalyzed resin to the mold surface. Hand rolling techniques then consolidate the fiber and resin to conform to the mold surface contours. Both processes rely heavily on the operators' skills for product quality.

Gel coats are typically used to provide a part with a finished surface directly from the mold. Final edge trimming is accomplished with a variety of tools such as routers, water jet cutters, or abrasive grinders with fixtures to define required dimensions.

Vacuum Bag, Pressure Bag, and Autoclave Molding. These thermoset processes are variations of the hand lay-up process where pressure is applied to the reinforcement and resin during the cure step. With vacuum bag molding, a plastic film is placed over the resin-saturated reinforcement and carefully sealed around the perimeter of the part to form the bag. Air is then drawn from within the bag, allowing atmospheric pressure to be applied to the bag and the enclosed saturated reinforcement.

Pressure bag molding substitutes air or liquid pressure for the atmospheric pressure, allowing higher pressures and higher fiber contents to be achieved. Autoclave molding uses a pressurized heating chamber to apply both heat and pressure to cure the product after it has been prepared using the vacuum bag method.

A special pre-impregnated (prepreg) reinforcement system usually used with these processes combines the reinforcement, typically in a cloth or tape form, with a polymer that has been partially polymerized (B-staged). The prepreg is cut into the required shape and positioned onto the mold surface prior to application of pressure and heat.

Resin-Transfer Molding and Cold Press Molding. Resin-transfer molding (RTM) and cold press molding are thermoset processes utilizing a clamping frame or press and matching male and female molds. The molds fully contain the reinforcement and resin during the consolidation and curing process which is accomplished at a relatively low pressure (>345 kPa (50 psi)). In both processes the reinforcement, either as a precut mat or preformed shape (preform), is placed in the open mold. In RTM the mold is closed and a catalyzed resin is then pumped into the mold cavity to saturate the reinforcement. The perimeter of the mold has a sealing gasket to contain the resin until it cures. In cold press molding the reinforcement and catalyzed resin are placed into the open mold and the resin is distributed through the reinforcement as the mold is closed and pressurized.

The preform is a form of reinforcement also utilized in the other liquid molding processes, ie, structural reaction injection molding (S-RIM) and wet system compression molding. An innovation to the preform process is robot-controlled fiber placement that results in rapid, uniform preforms for the liquid molding processes.

A special attribute of these processes is the ability to pre-position reinforcement, inserts, and core materials for stiffening ribs. Gel coatings can be applied to the mold surface to eliminate post-mold finishing.

Reaction Injection Molding. Two variations of conventional reaction injection molding (RIM) are used for reinforced plastic parts: reinforced RIM (R-RIM) and structural RIM (S-RIM). With R-RIM, very short reinforcing fibers are added to one of the resin components prior to being injected under high pressure through an impingement mixer into the closed mold. The mold is held closed during injection by a special clamping device that can be articulated to orient the mold for proper injection and to open the mold for part removal.

S-RIM utilizes the same polymers, molds, and process equipment. However, the reinforcement is in the form of a mat or specially shaped preform. The reinforcement is positioned in the mold prior to mold closure and resin injection.

Compression Molding. Compression molding processes dominate the higher production volume reinforced plastics applications. These processes use a hydraulically operated press to form the part in matching metal molds and to hold the densified molding material in the desired shape until the resin system has cured (if a thermoset) or cooled sufficiently (if a thermoplastic) to permit part removal. This process uses different input material forms such as bulk molding compound, sheet molding compound, wet system mat molding, and reinforced plastic sheet. Each form provides different composite properties and product shape capabilities.

Injection Molding. Injection molding is a thermoplastic or thermoset process that provides very high volume production capability. The basic process for injection molding reinforced plastics is similar to unreinforced plastics (see PLASTICS PROCESSING). The reasons for reinforcement differ depending on the specific polymer, but generally reinforcement increases certain mechanical properties, improves dimensional stability, and increases heat resistance, allowing the reinforced product to operate at higher temperature.

Pultrusion. Pultrusion is the reinforced plastic process that produces continuous profiles. In this process, reinforcements are pulled through a liquid thermoset resin bath into a heated mold where the cure reaction occurs. The pulling device has specially shaped gripping surfaces that continuously pull the cured profile out of the mold, thereby activating and controlling the speed of the entire process. A cut-off device trims the pultruded product to the required lengths. The pultrusion process is highly automated and once the process is started, very little direct labor is required.

Filament Winding. This is a process for products that are surfaces of revolution. Reinforcing fibers are drawn through a liquid resin bath and applied to a rotating mold surface or mandrel. For many applications, the reinforcement is precombined with the liquid resin to form a prepreg product.

Other Processes. There are a variety of other processes that have been developed to suit specific product needs.

Continuous laminating is a process in which glass fibers are chopped onto a moving film and then saturated with resin. Another film is applied to the surface before shaping and curing in an oven.

Centrifugal casting saturates a reinforcement with thermosetting resin within a mold that is then rotated at high speed to consolidate the laminate before curing.

Rotational molding involves placement of a resin and chopped glass in the metal mold that is then rotated in an oven where the thermoplastic resin melts and deposits the fiber on the metal surface. When cooled, the mold is opened and the part is removed.

The *rigidized thermoplastic sheet* process combines thermoplastic sheet vacuum forming with a spray-up or cold press molding process to add a thermoset composite structural backing to a decorative thermoplastic skin. There are also hybrid processes that have evolved to meet specific product needs.

Design Benefits. Design benefits include ways of making a finished part that eliminate fabrication steps needed for traditional materials. Parts with complex surface shapes are often good candidates for composite materials because composites can be made to virtually any predetermined shape from a mold. Composites are superior materials for applications that require large part size.

Strength can be engineered to the magnitude and in the direction needed for the application. Strength can be tensile, such as oil-field sucker rods; flexural, such as sporting goods; compressive, such as large pipes; or impact, such as in automotive wheels. Stiffness can also be engineered as needed.

Some design factors, however, work against composites. For example, glass fiber-reinforced plastics generally have lower modulus (stiffness) than metals. Composites cannot generally perform as well as metals or ceramics in very high temperature applications, but they can be made fire resistant to meet most construction and transportation codes.

Performance Benefits. Fiber-reinforced plastic (FRP) components are lightweight and deliver more strength per unit of weight, or less weight for a given strength requirement than unreinforced plastics and most metals. In transportation applications light weight means better fuel economy, and in marine and aerospace usage, light weight is critical for flotation and aerodynamic performance, respectively.

Reinforced plastics provide long-term corrosion resistance to many chemical and temperature environments, making them especially suited for chemical and power plants, and oil-field collection and distribution. Composites are usually excellent electrical insulators and do not absorb moisture. Composites can have low thermal conductivity. Under mechanical and environmental stresses, composites are dimensionally stable.

Economic Benefits. There are no simple rules of thumb in defining the cost of reinforced plastic components. Their successful use has resulted from proper design, utilizing the benefits these materials offer, process selection, tooling cost advantages that fit the production needs, and consideration of life cycle economics. Each existing application illustrates the cost-performance advantage of reinforced plastic over the traditional material that is displaced.

ANDREW L. BASTONE
STEVEN G. KATZ
ISORCA, Inc.

K. H. G. Ashbee, *Fundamental Principles of Fiber Reinforced Composites,* Technomic Publishing Co., Inc., Lancaster, Pa., 1989.

M. L. Berins, ed., *Plastics Engineering Handbook,* 5th ed., Society of the Plastics Industry, Inc., Van Nostrand Reinhold, New York, 1991.

Introduction to Composites, 3rd ed., Society of the Plastics Industry, Inc., Washington, D.C., 1995.

G. Lubin, ed. *Handbook of Composites,* Van Nostrand Reinhold, New York, 1982.

RELEASE AGENTS

Release agents are substances that control or eliminate the adhesion between two surfaces. They are used to expedite many industrial handling and processing operations, particularly of polymers (see PLASTICS PROCESSING). They are known by a variety of terms descriptive of their effect, including abherents, abhesives, antiblocking agents, antistick agents, external or surface lubricants, mold-release agents, parting agents, and slip agents. They find considerable use in the adhesive, food, furniture, glass, metal, plastics, and rubber industries.

Release agents function by either lessening intermolecular interactions between the two surfaces in contact or preventing such close contact. Thus, they can be low surface-tension materials based on aliphatic hydrocarbon, fluorocarbon groups, or particulate solids. The principal categories of material used are waxes, fatty acid metal soaps, other long-chain alkyl derivatives, polymers, and fluorinated compounds.

Some of these processing aids are incorporated into the bulk of the material rather than being directly applied to the surface and are known as internal release agents. In the process of migration to the surface they have an inevitable internal lubrication effect. Most internal lubricants also have an external lubricant effect, particularly at higher concentrations because usually some material finds its way to the exterior.

Product Types and Requirements

Release agents are available in a wide variety of forms and are formulated for numerous modes of application. Product types include neat liquids, solutions, powders, flakes, pastes, emulsions, dispersions, sprays, and films. Some are general-purpose inert products intended for a broad range of applications, including home consumer uses, whereas others are highly specific, designed to react *in situ* with particular substrates. There is a trend away from products containing volatile organic compounds (VOCs) and ozone-depleting substances, and toward water-based and high solids formulations. Users are also switching from the heavier metal soaps. Many products serve more than one processing function and contain other additives; such products are usually proprietary.

The choice of a release agent depends on the process conditions involved and the nature of the contacting substrates. Apart from the obvious ease of release, other important requirements are minimal buildup of residues on mold substrate, minimal effect on the molded article, adequate film-forming ability, compatibility with secondary operations and other processing parameters, health and safety requirements, and cost.

Classification of Release Agents

The diversity of release products and the wide range of release problems make classification difficult. One approach is by product form, with subdivisions such as emulsions, films, powders, reactive or inert sprays, reactive coatings, and so on. Another approach is by application, eg, metal casting, rubber processing, thermoplastic injection molding, and food preparation and packaging.

A classification by chemical type is given in Table 1. It does not attempt to be either rigorous or complete. The broad classes of release materials available are given in the chemical class column, the principal types in the chemical subdivision column, and one or two important selections in the specific examples column.

Table 1. Chemical Classification of Release Agents

Chemical class	Chemical subdivisions	Specific examples
waxes	petroleum waxes	paraffin wax, microcrystalline wax
	vegetable waxes	carnauba wax
	animal waxes	lanolin
	synthetic waxes	polyethylene wax, Fischer-Tropsch wax
fatty acid metal soaps	metal stearates	magnesium stearate, zinc stearate
	other	calcium ricinoleate
other long-chain alkyl derivatives	fatty esters	diethylene glycol monostearate, hydrogenated castor oil
	fatty amides and amines	ethylenebis(stearamide), oleyl palmitamide
	fatty acids and alcohols	palmitic acid, oleic acid
polymers	polyolefins	polypropylene
	silicones	polydimethylsiloxane, polymethyl(nonafluorohexyl)-siloxane
	fluoropolymers	polytetrafluoroethylene, poly(fluoroethers)
	natural polymers	cellophane
	other	polyoxalkylenes, poly(vinyl alcohol)
fluorinated compounds	fluorinated fatty acids	perfluorolauric acid
inorganic materials	silicates	talc
	clays	kaolin, mica
	other	silica, graphite

Mechanism of Release

Release agents are used to reduce the adhesion between materials. The main mechanisms of adhesion are interdiffusion, electrostatic attraction, surface energetics and wettability, and mechanical interlocking. The important surface and interfacial properties include surface topography (composition, structure, and roughness), surface tension or energy and its effect on substrate and adhesive wettability, the thermodynamic work of adhesion, and chemical reactivity at the interface. Nonsurface properties of the thin-film interfacial phase include its ability to set to a cohesive solid after wetting the substrate and its viscoelastic response to deformation controlled by factors such as degree of crystallinity, molecular weight and distribution, number of cross-links, and presence of fillers. Surface treatments that enhance adhesion do so by removing weak boundary layers, changing surface topography, or changing the chemical nature of the surface and by introducing strongly attracting functional groups.

An inversion of these arguments indicates that release agents should exhibit several of the following features: (*1*) act as a barrier to mechanical interlocking; (*2*) prevent interdiffusion; (*3*) exhibit poor adsorption and lack of reaction with at least one material at the interface; (*4*) have low surface tension, resulting in poor wettability, ie, negative spreading coefficient, of the release substrate by the adhesive; (*5*) low thermodynamic work of adhesion; (*6*) low intermolecular forces across the interface, eg, an absence of electrostatic or polar attractions; (*7*) display nonsetting or low cohesive interactions within the release phase; and (*8*) provide a weak boundary layer.

Many of these features are interrelated. Finely divided solids such as talc are excellent barriers to mechanical interlocking and interdiffusion. They also reduce the area of contact over which short-range intermolecular forces can interact. Because compatibility of different polymers is the exception rather than the rule, preformed sheets of a different polymer usually prevent interdiffusion and are an effective way of controlling adhesion, provided no new strong interfacial interactions are thereby introduced. Surface tension and thermodynamic work of adhesion are interrelated, as shown in equations 1, 2, and 3, and are a direct consequence of the intermolecular forces that also control adsorption and chemical reactivity.

The work of adhesion, W_A, is the change in energy per unit surface area when two interfaces come into contact, as given in equation 1 where σ_1 and σ_2 are the surface energy of each phase and σ_{12} the interfacial energy between them.

$$W_A = \sigma_1 + \sigma_2 - \sigma_{12} \quad (1)$$

For liquid systems these surface energies expressed in mJ/m^2 are numerically equivalent to the surface tensions in mN/m(= dyn/cm). If the adhesive is phase 1 and the release coating is phase 2, then the spreading coefficient, S, of 1 on 2 is as given in equation 2.

$$S = \sigma_2 - \sigma_1 - \sigma_{12} \quad (2)$$

Because the work of cohesion, W_C, of the adhesive is 2 σ_1, equation 3 follows.

$$W_A - W_C = S \quad (3)$$

This simple yet fundamental relationship states that when the forces of adhesion are less than those of cohesion, ie, when W_A is less than W_C, the spreading coefficient is negative and failure will be at the interface between the adhesive and release coating, the desired situation in a release coating application. Conversely, when the spreading coefficient is positive, undesirable cohesive failure in the adhesive or release coating will be the case. These equations clearly demonstrate the importance of the release coating surface tension being lower than that of the adhesive.

The intermolecular forces of adhesion and cohesion can be loosely classified into three categories: quantum mechanical forces, pure electrostatic forces, and polarization forces. Quantum mechanical forces give rise both to covalent bonding and to the exchange interactions that balance the attractive forces when matter is compressed to the point where outer electron orbits interpenetrate. Pure electrostatic interactions include Coulomb forces between charged ions, permanent dipoles, and quadrupoles. Polarization forces arise from the dipole moments induced in atoms and molecules by the electric fields of nearby charges and other permanent and induced dipoles.

The forces involved in the interaction at a good release interface must be as weak as possible. They cannot be the strong primary bonds associated with ionic, covalent, and metallic bonding; neither are they the stronger of the electrostatic and polarization forces that contribute to secondary van der Waals interactions. Rather, they are the weakest of these types of forces, the so-called London or dispersion forces that arise from interactions of temporary dipoles caused by fluctuations in electron density. They are common to all matter. The surfaces that are solid at room temperature and have the lowest dispersion-force interactions are those comprised of aliphatic hydrocarbons and fluorocarbons.

Among the hydrocarbons, the lowest surface tension values are found for surfaces comprising closely packed methyl groups. This is a characteristic of waxes and oriented long-chain alkyl derivatives.

Because the forces of attraction prevail when molecules are brought into sufficiently close proximity under normal conditions, release is best effected if both the strength of the interaction and the degree of contact are minimized. Aliphatic hydrocarbons and fluorocarbons achieve the former effect, finely divided solids the latter. Materials such as microcrystalline wax and hydrophobic silica combine both effects.

Industrial Applications

Metal and Glass. Mineral oils are widely used in molding and shaping glass, despite the considerable quantity of smoke and buildup of residues entailed. Heat-resistant, graphite-containing resins are also used. Phenolic resins are used in this application and also in mold fabrication for metal castings.

Rubber and Plastics. Release agents are widely used in the rubber and plastic industry to achieve release of polymers and release from polymers. They are useful in many polymer-processing applications, to prevent the polymer from sticking and building up on process equipment and thereby eliminating the accumulation of rejects. A significant release from polymer application is antideposition coatings to reduce accumulation of pollutants and other undesirable matter.

Molding operations such as the manufacture of automobile seat foams and reactive injection molding (RIM) consume considerable quantities of release agents. Many polymers are molded, but polyurethanes present one of the most significant challenges in this area, owing to the likelihood of adhesive bonding of isocyanates to the mold. Polymer films are often used to allow castings to separate from the mold.

Furniture. The furniture industry uses a lot of auxiliary release agents in the general gluing and veneering of wood and wood-based materials, as well as in a variety of coating processes. Mineral oils, paraffin waxes, fatty esters, and specialized paintable silicones are all used for this purpose.

Food. Familiarity with household food uses of release agents, such as in polyolefin refrigerator ice trays, gelatin dessert molds, wax paper, release-coated saucepans and bakeware, and nonstick cooking sprays, is nearly universal. Natural abherent products include wheat flour, confectioners' sugar, rice flour, and rice paper. The use of abherent surfaces in the food industry not only decreases product loss, but also increases the ease with which the equipment is cleaned.

Health and Safety Factors

Most general-purpose release agents, because of their low toxicity and chemical inertness, do not usually present health and safety problems. Some of the solvent dispersions require appropriate care in handling volatile solvents, and many suppliers are offering water-based alternatives. Some of the solids, particularly finely divided hydrophobic solids, can also present inhalation problems. Some of the metallic soaps are toxic, although there is a trend away from the heavier, more toxic metals such as lead. The reactive type of release coating with monomers, prepolymers, and catalysts often presents specific handling difficulties. The potential user with health and safety questions is advised to consult the manufacturer directly.

MICHAEL J. OWEN
Dow Corning Corporation

H. Lammerting, "Release Agents," in W. Gerhartz, ed., *Ullmann's Encyclopedia of Industrial Chemistry,* Vol. A23, VCH Publishers Inc., New York, 1993.

D. Satas, *Coatings Technology Handbook,* Marcel Dekker, Inc., New York, 1991.

J. Stepek and H. Daoust, *Additives for Plastics,* Springer-Verlag, New York, 1983.

M. J. Owen and J. D. Jones, "Silicone Release Coatings," in J. C. Salamone, ed., *Polymeric Materials Encyclopedia,* Vol. 10, CRC Press, Boca Raton, Fla., 1996.

RENEWABLE ENERGY RESOURCES

The term *renewable energy resources* implies technologies based on harnessing the sun, wind, falling water, plant matter, and heat from the earth. Renewable energy resources can complement fossil fuels and, eventually, may emerge as a significant energy source.

The characteristics of each technology are described in separate discussions of photovoltaics, solar-thermal power, and wind, biomass, waste-to-energy, geothermal, hydropower, and wave energy.

Photovoltaic Cells

Taken as a group, photovoltaic (PV) cells comprise solid-state devices in which photons of light collide with atoms and transfer their energy to electrons. These electrons flow into wires that are connected to the cells, thereby providing current to electrical loads (see PHOTOVOLTAIC CELLS).

PV systems consist of arrays of cells that are interconnected in panels or modules to increase total power output. Often the systems include sun-tracking equipment, as well as power-conditioning equipment to convert dc to ac. The systems can range in size from a simple one-panel, fixed-orientation unit to a vast field of modules that accurately track the movement of the sun.

PV cells may be fabricated using one of four production methods. In the traditional method, which still accounts for a large portion of annual production, PV cells are made from slices of single-crystal silicon. In this process, crystalline silicon is grown in pure ingots made from molten silicon, and is then cut into wafers. The wafers are polished and processed into PV cells, which are mounted in modules.

A second method grows silicon ribbons or sheets directly, bypassing the ingot stage. The sheets are cut into cell-size pieces that are processed to make cells.

Still another method used to produce PV cells is provided by thin-film technologies. Thin films are made by depositing semiconductor materials on a solid substrate such as glass or metal sheet.

A fourth category of PV technologies, concentrator photovoltaics, uses small but very efficient cells illuminated with concentrated sunlight. PV concentrators use lenses or reflective devices and track the sun through daily cycles. The tracking maintains the concentrated light at intensities up to several hundred times normal sunlight.

Solar-Thermal Power

Solar-electric technology, under favorable circumstances, can be cost-effective, as evidenced by the fact that solar-thermal gas-hybrid plants produce over 350 MW of commercial power in southern California. This power is used during peak demand to supplement that available from conventional generation.

Solar-thermal technology uses tracking mirrors to concentrate sunlight onto a receiver. In turn, the receiver absorbs solar energy as heat, warming a fluid that then drives a turbine generator. Most solar-thermal plants require cooling water.

Solar-thermal receivers may be centralized or distributed. Central receiver systems use fields of tracking mirrors, or heliostats, to focus sunlight onto a tower-mounted receiver. Distributed receivers use point-focusing parabolic dishes or line-focusing parabolic troughs to concentrate sunlight. Solar-thermal power plants may be entirely solar or they may be hybrids that use fossil fuels to boost power output or extend operating hours.

Wind Power

Wind turbines have had a varied history. Once widely used for electric power generation in remote areas in the United States, for example, they eventually fell into disuse by the 1940s as a result of rural electrification. In the 1970s, interest in wind revived in the face of the energy shortages.

In the early 1980s, favorable tax credits and energy rates for independent power producers encouraged the development of wind farms in California based on 50–100-kW turbines. Simpler and relatively easy to design, build, install, and repair compared to the earlier, larger-size turbines, this new generation of wind machines was also relatively reliable and provided lower cost electricity than the previous generation.

Typical wind turbines consist of rotor blades mounted atop a tower and connected by gears to a drive shaft that spins a generator. Another common design is the vertical axis turbine, which has an eggbeater-shaped rotor attached directly to a vertical shaft. The rotor blade length and wind speed determine the amount of electric power that can be delivered. In general, wind speeds of at least 6.7 m/s are sought for power generation.

Most of the best locations for wind projects lie outside California. Several northern Rocky Mountain states and Northern Plains states possess substantial resources. Also, the Northeast and Texas have considerable wind resources.

Traditionally, wind turbines have operated at constant rpm to produce 60 Hz a-c power. Because the extra torque generated by wind gusts must be absorbed by the drive trains of constant speed wind turbines, they require heavier designs than comparable variable speed models. By contrast, variable speed turbines employ a power electronic converter between the generator and the utility power line. The converter allows the rotor and generator to speed up with gusts or stronger winds, without increasing drive-train torque. The increased rotational energy is converted into additional electricity. Energy capture increases by 10% or more, and stresses on the turbine are reduced. In Europe, government policies are calling for a steadily increasing commitment to wind power.

Biomass Fuel

Concern over possible global warming trends has been linked with steady increases in greenhouse gases, such as carbon dioxide, CO_2. This gas is emitted whenever fossil fuels or biomass materials, such as wood and agricultural wastes, are burned. When used as a renewable fuel, however, the CO_2 released by biomass during combustion ideally equals that consumed when the fuel was grown. Thus, biomass should not contribute to CO_2-related global climate change.

As global demand for electricity increases substantially, it is anticipated that only a massive expansion in the use of biomass and other nonfossil fuel sources can slow the annual increase in global CO_2 emissions.

Also, wood fuel is low in sulfur, ash, and trace toxic metals. Wood-fired power plants emit about 45% less nitrogen oxides, NO_x, than coal-fired units. Legislation intended to reduce sulfur oxides, SO_x, and NO_x emissions may therefore result in the encouragement of wood-burning or cofiring wood with coal.

Much of the energy needs of many nations are met by biofuels, including wood and wood waste, spent pulping liquors, bagasse, and municipal waste. Some use is also made of dried corn cobs, rice hulls, and a wide variety of agricultural wastes used in niche applications.

Extensive efforts are underway to use wood and agricultural wastes as fuel, including an increasing emphasis on capturing these materials. Also, traditional direct combustion technologies used for electricity generation have been joined by developments in the use of circulating fluidized-bed technology for biofuels, and cofiring as a technology for using biofuels in pulverized coal and cyclone boilers. Existing and emerging gasification technologies are also under study for use in supporting conventional combustion turbine technology or for integrated gasification–combined cycle (IGCC) systems. Experiments continue with direct-fixed gas turbines.

Waste-to-Energy

Large quantities of municipal solid waste (MSW), comprised mostly of residential and light commercial waste, are generated in the United States at the rate of approximately 180×10^6 t/yr. Although the composition of MSW varies according to the source and time of year in which it is generated and collected, MSW generally comprises paper, cardboard, wood, plastic, garbage, food and beverage containers, metal, glass, organic material from yards, appliances, and miscellaneous materials, such as rugs, blankets, shoes, mattresses, and telephone wire.

The conventional means of disposing of MSW is by landfilling. However, because landfills are becoming a less acceptable solution, alternative means of disposing MSW have been advanced.

In the United States, MSW represents an available energy source of about 1.5×10^{15} kJ/yr (1.4×10^{15} Btu/yr) that could provide about 5% of the nation's annual electricity consumption. However, compared to coal, MSW is a poor fuel. Although MSW has a low sulfur content, it may be high in chlorine, aluminum, and some trace metals, including lead and cadmium. The nonhomogeneity of MSW and its high and variable moisture and ash content cause difficulties in maintaining good combustion on a continuous basis.

Nevertheless, plants are disposing of MSW through combustion and recovering energy in the form of steam for electric power generation. Two commercial technologies are (*1*) mass burning of MSW as received, using a facility designed for nonuniform, high moisture slow burning materials and (*2*) processing MSW into refuse-derived fuel (RDF), which is then burned at a cofired or dedicated facility. In cofired plants, RDF and coal are fired simultaneously, whereas at dedicated waste-fired plants, RDF is burned and coal is used only as a backup. Both methods begin by removing unacceptably large items, as well as others, such as discarded gas cans, that might explode during waste processing.

The manufacture of RDF entails moving the waste through a series of separators that successively reduce the feed by size and weight; a magnetic separator removes ferrous metals for recycling. Much of the heavier material comprises metal and glass, which can also be recycled. The lighter fraction contains most of the combustible material in the form of a light uniform fluff, RDF, which is conveyed to the power plant for burning. The small residue of unburned waste (ash) can be further treated to remove aggregates and remaining metals before placement in a landfill.

By substantially reducing the volume of landfill space needed for MSW disposal, converting waste to energy (WTE) can extend the life of a landfill. In fact, most of the revenues that go to WTE plant operators come from the payments for waste disposal services called tipping fees, and not from electric sales (see RECYCLING).

Tires. In the United States, scrap tires are generated at the rate of one tire per person per year, and only 20% are reused or recycled in some fashion (see RECYCLING, RUBBER). Scrap tires make good fuel, either whole or as shredded chips, commonly called tire-derived fuel (TDF). Also, tires are moderate in both sulfur and ash compared with bituminous coal and do not adversely affect emissions quality.

As an interim measure, burning scrap tires and recovering the energy assists in the solution to the problem of growing mountains of tires, as it reduces the use of nonrenewable fossil fuels. However, the utility industry has long been reluctant to burn materials other than conventional fuels. In general, the potential savings in fuel costs failed to provide sufficient incentives to offset possible environmental concerns, operational difficulties, and costly equipment modifications. That situation is changing as utilities and other industries are required to adapt to a changing social and economic climate, and several utilities are burning or have successfully test-burned TDF.

Landfill Gas Recovery. This process has emerged from the need to better manage landfill operations. Landfill gas is produced naturally: anaerobic bacteria convert the disposed organic matter into methane, carbon monoxide, and other gases. The quantity of methane gas is substantial and could be utilized as fuel, but generally is not. Most of the methane simply leaks into the surrounding atmosphere. Not all of the gas is wasted. About 300 MW of electricity is generated from landfills. A variety of electric generation systems have been employed by a small number of developers.

Geothermal Power Because the geologic systems may contain steam or both steam and liquid water, the means to recover geothermal energy varies accordingly. If only steam is present, the least common geothermal resource, the steam is fed from the well directly to a steam turbine, which drives an electric generator. If water is also present in a compressed state, a flash-steam cycle may be used: the liquid water flashes into a mixture of hot water and steam by the time the water reaches the lower pressure surface. The steam is then fed to a turbine, and the heat in the hot water may be used elsewhere. Alternatively a newer binary technique uses moderate-temperature geothermal fluids to vaporize low boiling-point organic fluids, which then drive turbogenerators.

The most widespread geothermal resource is hot dry rock (HDR), consisting of rocks rich in thermal energy but lacking entrapped water or steam. Harnessing HDR energy would entail drilling a deep well into the hot rock, which is then fractured by the injection of water under high pressure. The fractured zone becomes an engineered reservoir into which surface water is continuously injected and then recovered as hot water or steam. HDR systems are technically feasible, but not yet economical.

Two other localized regions of concentrated heat that are potentially extractable are geopressured geothermal systems and magma. The geopressured geothermal systems comprise hot, high pressure brines containing dissolved methane. Most known geopressured systems are not economical at current natural gas prices.

Geothermal Heat Pumps. Also called ground-source heat pumps, geothermal heat pumps (GHPs) use the earth as a heat source for heating or as a heat sink for cooling. This arrangement reduces electricity consumption by about 30% compared to air-source heat pumps.

Hydropower

Falling water has been used to generate electricity for over 100 years. Hydro plants have been built at natural waterfalls, as well as artificial ones created by dams.

In the United States, hydropower provides an essential contribution to the national power grid: its capability to respond in seconds to large and rapidly varying loads, which other baseload plants with steam systems powered by combustion or nuclear processes cannot accommodate. Also, ownership is spread over a broad base.

The amount of electricity that can be generated at a hydro plant is determined by two factors: head and flow. Head is the distance in elevation between the highest level of the damned water to the point where it goes through the power-producing turbine. Conventional hydro plants must have a head of water that is at least 3 m high to provide sufficient water pressure to operate the turbine. Flow is the rate of water moving through the system. In general, a high head plant needs less water flow than a low head plant to produce the same amount of electricity.

Some hydro plants use pumped storage systems, among the most reliable energy storage systems available. Pumped storage systems use recycled water instead of tapping free-flowing water. After flowing through the turbine, the water resource is pumped, usually through a reversible turbine, from a lower reservoir back to an upper reservoir. Whereas pumped storage facilities are net energy consumers, ie, more energy in total is required for pumping than is generated by the plant, they are valuable to a utility because they operate in a peak-power production mode, when electricity is most costly to produce. The pumping to replenish the upper reservoir is performed during off-peak hours using the utility's least costly resources.

Wave Energy

Ocean waves are formed by the wind driving water toward shore. The wave energy depends strongly on wind speed; the energy is a fifth-power function of speed. Most methods to convert this irregular and oscillating low frequency energy source to grid power employ pneumatic, hydraulic, or hydropower technology.

Pneumatic systems use the wave motion to pressurize air in an oscillating water column (OWC). The pressurized air is then passed through an air turbine to generate electricity. In hydraulic systems, wave motion is used to pressurize water or other fluids, which are subsequently passed through a turbine or motor that drives a generator. Hydropower systems concentrate wave peaks and store the water delivered in the waves in an elevated basin. The potential energy supplied runs a low head hydro plant with seawater.

Tidal Power. Tidal power is caused by the gravitational pull of the sun and especially the moon, as they pull at the earth. Reacting to this pull, the ocean's waters rise, causing a high tide where the moon is closest. The difference between low and high tide can range from a few cm to several meters. Harnessing tidal power for electricity production by the use of dams requires a tidal difference of at least 4.5 m, a requirement met at few locations in the United States. Thus, the principal demonstration sites of tidal power are in Canada, China, and France.

Information for this article came from numerous sources, including F. E. Porretto, D. M. DiMeo, and S. Collins (Empire State Electric Energy Research Corp.), E. A. DeMeo and E. E. Hughes (Electric Power Research Institute); J. M. Cohen and D. Entingh (Princeton Economic Research, Inc.); and G. Sommers and J. Renner (Idaho National Engineering Laboratory).

EDWARD A. TORRERO
Empire State Electric Energy Research Corporation

C. J. Winter, R. L. Sizman, L. I. Vant-Hull, eds., *Solar Power Plants: Fundamentals, Technology, Systems, Economics,* Springer-Verlag, New York, 1991.

R. Golob and E. Brus, *The Almanac of Renewable Energy,* World Information Systems, Henry Holt and Co., New York, 1993.

J. Jayadev, *IEEE Spectrum,* 78–83 (Nov. 1995).

E. Easwaran, D. Entingh, and D. Diachok, *United States Geothermal Technology: Equipment and Services for Worldwide Applications,* DOE/EE-0044, U.S. Dept. of Energy, Washington, D.C., 1995.

REPELLENTS

Repellents are materials that affect insects and other organisms by disrupting their natural behavior of bloodseeking through biting of humans and animals, and are the first line of defense that can be readily used for this purpose. The best overall standard repellent is *N,N*-diethyl-*m*-toluamide (DEET), systematically named *N,N*-diethyl-3-methylbenzamide (**1**); however, many other compounds are presently in use. Repellent-toxicant or biting depressant systems are available which are reasonably comfortable for the user and can completely protect against a number of pests for an extended period of time.

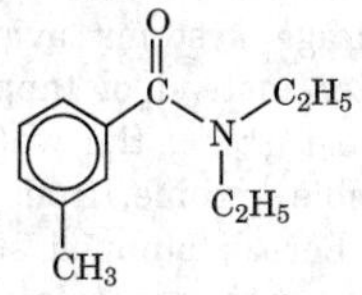

(**1**)

Newer, cosmetically appealing formulations of chemical repellents have become popular in the United States since the mid-1980s, and interest in personal protection against biting arthropods has been renewed. Children and adults may be exposed daily to risks of Lyme disease, and those exposed should be particularly interested in reducing exposure to nymphs of the deer tick *Ixodes scapularis,* which carries the bacterium *Borellia burgdorferi.* This disease difficult to diagnose and treat, and treatment of Lyme disease using antibiotics has no effect on the next infective bite.

The use of insecticides has led to the rise of widespread resistance in areas where residual insecticides were applied for malaria eradication (see also INSECT-CONTROL TECHNOLOGY), and in some locations that have widespread irrigation. Mosquito resistance to insecticides is prevalent in Southeast Asia, India, and East Africa. Use of personal protection and repellents is recommended for travelers to malarious areas for avoiding serious health risks. Insect species that can be vectors of diseases are listed in Table 1.

Table 1. Insect Vectors of Disease

Insect	Scientific name	Disease
biting midges	*Ceratopogonidae*	bluetongue virus
blackflies	*Simulium*	river blindness
sandflies	*Phlebotamus, Leptoconops*	leishmaniasis
tsetse flies	*Glossina*	African trypanosomiasis or sleeping sickness
tabanids or horseflies	*Tabanus*	
clegs	*Haemotopota*	
deerflies	*Chrysops*	
snipe flies and stableflies	*Stomoxys*	
cattle ticks	*Rhipicephalus*	heartwater fever and blackwater fever
soft ticks	*Dermacentor, Amblyomma*	Rocky Mountain and spotted fever
mites or chigger mites	*Trombidiidae*	scrub typhus
bedbugs	*Cimex*	
lice	*Pediculus*	typhus
kissing bugs	*Pentatomid*	Chagas disease
fleas	*Xenopsylla, Pulex*	plague
mosquitoes	*Aedes, Anopheles, Culex*	malaria, encephalitis, yellow fever, dengue

Evaluation of Repellents

Repellents on Skin. The candidate chemical is dissolved in ethanol and spread over one forearm of the human subject, as DEET (**1**) is similarly applied to the other forearm. Each arm is then exposed to 1500 avid *A. aegypti* female mosquitoes for 3 min at 30-min intervals. Effectiveness is based on complete protection.

Repellents on Cloth. Each candidate repellent is applied to a knit cotton stocking or cloth patch at 3.3 g/m^2 cloth, usually as a 1% solution of active ingredient (AI) in acetone. Two hours later, the stock or cloth patch is placed over an untreated nylon stocking on the arm of a subject, the hand covered, and the arm exposed to 1500 female mosquitoes for one minute. If fewer than five bites are counted, the test is repeated at 24 h, then weekly until failure, which is, by definition, five bites per minute.

Space-Borne Repellents. Air is drawn over a human arm through a 9.5-cm disk of cotton netting having 0.64-cm holes and treated with a solution of the candidate repellent; air is then drawn into an olfactometer cage containing 125 avid female *A. aegypti.* The number of days the repellent prevents >10% of the mosquitoes from passing through the netting constitutes effectiveness.

Skin Patch-Tested Repellents. Small areas of human forearms are marked and treated with small amounts of repellent on a unit area basis to ensure that the treatment rate is always the same between subjects. The patches are tested at 0 and 4 hours against small numbers (ca 15) of mosquitoes.

Repellents Not Using Human Bait (No Attractant). A treated strip of fabric and a control strip are lowered into a container of crawling arthropods such as ticks, fleas, and mites. After a predetermined time, the strips are lifted, the animals remaining are counted, and the percentage repellency is determined.

Repellents Tested with an Inanimate Attractant. Machines have been constructed by several groups to measure the intrinsic (initial) repellency of a compound when it is added to a warm, moist airstream to overcome the attractiveness of the airstream to mosquitoes.

Repellents Tested with Animal Attractants. Numerous methods have involved the use of animals as attractants, followed by evaluation of repellents as skin treatments or attached cloth treatments, often against crawling arthropods such as fleas, ticks, and mites.

Arthropod Repellents

Mosquito Repellents. Clothing impregnants may be applied to apparel by dipping into emulsions, manual application of liquid, or surface spraying with aerosols (qv). Repellents may be applied to cotton cloth using technical material dissolved in acetone. Repellent jackets made of 0.64-cm cotton mesh are impregnated with a repellent such as DEET (**1**), or other repellents more effective than DEET, against biting flies or other mosquitoes. An effective skin repellent such as DEET is lost much more slowly from the mesh jacket than from skin.

Several studies of cloth treatment for toxicity and repellency have been conducted using impregnation with permethrin emulsifiable concentrate (EC), together with residue studies. Permethrin-treated military uniforms, treated by aerosol spray, pressurized spray, or impregnation during the washing process, have been found effective in tests.

Standard Mosquito Repellents. Since its initial report as a promising repellent in 1954, DEET has been considered the best all-around repellent having generally acceptable characteristics, despite a continuing search for a superior chemical. Improvements include many commercial products with added cosmetic agents that use slow release technology. However, the use of 100% DEET on skin is usually not justified because of the possibility of skin irritation, and not for reasons of better or worse repellent activity.

Dimethyl phthalate (DMP), a clear oil insoluble in water and soluble in organic solvents, is an effective repellent, used as a standard against *A. aegypti,* and is effective for 11–22 d on cloth. The repellent Rutgers 6-12, 2-ethyl-1,3-hexanediol, is a clear oil that is insoluble in water and miscible with organic solvents such as ethanol. Screened during World War II, this repellent is exceptionally effective against *A. aegypti.* Tests have been run against the newer pests *A. albopictus* and *A. aegypti,* including five repellents containing DEET (test standard), a controlled release formulation containing DEET, two dosages of DEET in ethanol, and Avon Skin-So-Soft, a scented mineral oil cosmetic. On the skin, the repellent chemicals provide significant protection from biting; however, *A. albopictus* is more sensitive to repellents than *A. aegypti.*

Biting Midge Repellents. The genus *Culicoides* is found in fresh water, salt water, and tide water environments in the southeastern United States and Caribbean, where they may be properly called biting midges, or more commonly sandflies, sand fleas, sand gnats, punkies, or flying teeth. Because of their small size, they can pass easily through ordinary mosquito screens. Commonly used repellents are of varying effectiveness, depending on the species involved.

Among the 117 compounds that have been synthesized, structures that have been found especially active include the 1-piperidyl- and 1-[hexahydro-1*H*-azeipinyl] derivatives where the cyclohexenyl ring can either be monounsaturated or have methyl branches. The use of Avon Skin-So-Soft has been claimed to repel biting midges at Parris Island, South Carolina, and it is reportedly widely used for mosquitoes as well. When applied liberally, it apparently traps the midges or fouls their mouthparts and thus inhibits biting. Tests of several commercial mineral oil preparations, both scented and unscented, show interference with biting behavior and therefore a repellent effect.

Phlebotomine Sandfly Repellents. Phlebotomine sandflies are found primarily in relatively underdeveloped tropical and subtropical regions of the world. They are vectors of the various forms of leishmaniasis, bartonellosis, and numerous arboviruses, including the medically important sandfly fever group. However, personal protection against phlebotomine sandflies has had little attention as compared to repellents for mosquitoes. Only a few tests of repellents for Old World phlebotomine flies have been documented, none of them as recently as the 1990s.

Skin applications of five selected repellents including DEET provided, a mean coefficient of protection (CP) of 99.2% against the attack of at least three species of *Lutzomyia.* All of these repellents tested at the highest dosage give good protection from the bites of at least three species of phlebotomine sandflies, two of which are important vectors of leishmaniasis. However, the azepine was found to give complete protection and warrants further study.

Horseflies, Greenheadflies, Deerflies, Stableflies. Field trials of permethrin-treated clothing against highly susceptible tsetse flies have shown good effectiveness. Bath oils are commonly used by lifeguards on the Florida Gulf Coast for protection against biting stableflies, and have been used effectively on horses; however, repellents fail at high fly populations. It is likely that use of repellent-treated mesh jackets and hats for mosquitoes and sandflies would be most effective against these large biting flies.

Tick and Chigger Repellents. Prevention of tick attachment is possible by mechanically preventing access of the ticks to bare skin bordering or beneath the clothing, such as zippered long pants with bloused cuffs tucked into boot or sock tops and a long-sleeved shirt tucked into the pants. Repellents are best impregnated into clothing, on wrist skin under the sleeve, on and above the socks, and around the neck on the exposed skin and under the collar. DEET products are available as a 50% liquid and may be mixed with isopropyl alcohol (59 mL DEET and 1 L 2-propanol) to produce sufficient material to impregnate clothing with a 5% solution and, depending on tick density, give 80–98% protection.

A topical permethrin treatment of clothing has shown good effectiveness against crawling insects when applied as a water-based formulation of 0.5 wt % permethrin. It gives extremely effective protection against ticks.

Permethrin as a clothing treatment acts more as a toxicant than as a repellent, and though ticks may crawl on the clothing, the visit is only temporary and usually fatal within a few minutes. Permethrin materials are likewise effective repellents against chigger mites (*Trombiculidae*), also called chiggers or redbugs.

Body and Head Lice. Most populations of body lice are resistant to the organochlorine insecticides, including lindane. Fabric patches treated with permethrin have been evaluated against natural and laboratory strains of human body lice in Peru. Permethrin-treated fabric is toxic to lice on contact and quickly affects feeding behavior, even when washed up to 20 times. Thus permethrin-treated clothing interrupts disease transmission, and offers a passive louse control not previously feasible.

Cockroach Repellents. Transport of goods and materials provides rapid transport of cockroaches. Repellents may be helpful in preventing transport of cockroaches into uninfested areas. Some logical uses of repellents are on cardboard cartons for food and soft drinks, on beer crates, and in coin-operated vending machines, all of which provide excellent shelter and food for cockroaches. A good repellent can be used either alone or in conjunction with an insecticide as a residual treatment in business establishments or homes. Many repellents are found among amides, sulfonamides, cyanoacetic acids, and carboxamides, but two good ones are *N,N*-diethylcyclohexaneacetamide (**2**, $n = 1$) and *N,N*-diethylcyclohexanepropaneamide (**2**, $n = 2$).

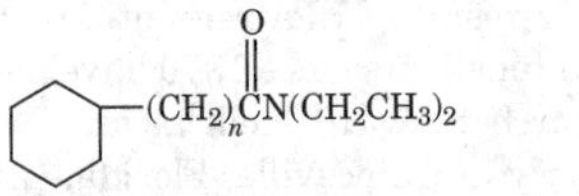

(2)

A problem lies in the overlap of repellent–toxicant definition, in that many toxicants are known to have repellent effects. Pyrethrins are often used on ships to flush cockroaches from harborages during a treatment with another, less activating toxicant. Tests show that pyrethrins which have been considered repellents for some years, MGK 264 and the emulsifier Triton X100, are noticeably repellent to both German and American cockroaches.

Other Insects. Bark beetle management in European forests has been successful using combinations of sex pheromones and tree volatiles. Repellents that were tested in Louisiana to deter attacks of the southern pine beetle afforded protection of high value loblolly pines by using the host tree compound 4-allylanisole.

Bird Repellents

Blackbirds, starlings, and sparrows are North American birds that cause serious damage to growing crops. Nonchemical techniques using repelling devices such as propane cannons, shiny Mylar ribbons, scarecrows, metallic pinwheels, and recorded distress calls give only temporary results.

Some bird repellents are composed of viscous, sticky materials that birds dislike having on their feet. These are often based on incompletely polymerized.

Intoxicating chemicals are those that are not necessarily lethal (see PESTICIDES) but operate as primary repellents or secondary repellents, eg, emetics causing sickness or distress. Primary bird repellents are those whose mode of action is having a bad taste; immediate rejection of food is the desired result. However, they are effective only if other foods are available; they are not effective in times of food shortages, because large flocks of migrating birds would be forced to feed or starve.

Avitrol (4-aminopyridine), has repellent–toxicant properties for birds and is classed as a severe poison and irritant. This secondary bird repellent can be used as a broadcast bait, causing uncoordinated flight and distress calls and escape responses in nearby birds.

Methyl anthranilate, the grape flavoring used in food products, has been shown effective as a waterfowl repellent. Anthrahydroquinones have been patented in Japan as bird repellents, and anthraquinone (qv) is used widely in Europe as a spray to protect growing crops and as a wood dressing.

Mammalian Pests

Feeding inhibitors and modern lethal treatments are considered practical solutions to control wildlife depredation. A food repellent has been defined as "a compound or combination of compounds that, when added to a food source, acts to produce a marked decrease in use. A useful repellent is meant to stop a hungry animal from feeding on a readily accessible, abundant, and palatable food, forcing the pest animal to leave the area or make a change in food habits, both unlikely choices.

Although no consistently effective chemical repellent has been developed for vertebrate pests, some promising materials have been tested as repellents that are based on predator avoidance, specifically compounds from the secretions of predators. Although few controlled tests have been run on predator scent materials in the past, more recent investigations of such odors have shown promise.

Health and Safety Factors

Toxicology. Toxicological testing has been carried out on many of the older, widely used materials, all of which require re-registration with the EPA. As a result of EPA regulations, many of the materials submitted as cloth repellents since 1970 have been tested at the USDA Agriculture Research Service, Medical and Veterinary Entomology Research Laboratory in Gainesville, Florida. Effective compounds, after further testing, are then submitted to the U.S. Army Environmental Health Agency for extensive toxicological testing. Compounds are tested as repellents on human skin only after passing the four standard toxicological tests: rabbit eye irritation, rabbit skin dermal, rat inhalation, and rat acute ingestion. All of these, plus EPA regulations in the United States classifying repellents as pesticides, have drastically reduced the number of candidate chemicals submitted to the USDA laboratory in Gainesville for general screening since about 1975, and virtually eliminated chemicals submitted as candidate repellents.

Hazard Assessment of Chemical Repellents. Labels for repellent products sold in the United States are recommended for purposes of efficacy and safety of use. Newer products containing DEET may contain less active ingredient but feature a cosmetic that makes the compound less objectionable on the skin and more acceptable to use.

D. A. CARLSON
U.S. Department of Agriculture

C. E. Schreck, in P. S. Auerbach, ed., *Wilderness Medicine: Management of Wilderness and Environmental Emergencies,* Mosby Co., St. Louis, Mo., 1995.

C. E. Schreck, in J. Adams, ed., *Insect Potpourri: Adventure in Entomology,* Sandhill Press, Inc., Gainesville, Fla., 1992.

C. F. Curtis and co-workers, in C. F. Curtis, ed., *Appropriate Technology in Vector Control,* CRC Press, Boca Raton, Fla., 1989.

D. A. Carlson, "Insect Repellents" in *Pest Management in the Subtropics: Integrated Pest Management—A Florida Perspective,* Intercept Ltd., Andover, U.K., 1996, Chapt. 20, 1996.

REPROGRAPHY. See ELECTROPHOTOGRAPHY; THERMOGRAPHY.

RESEARCH MANAGEMENT. See RESEARCH/TECHNOLOGY MANAGEMENT.

RESEARCH/TECHNOLOGY MANAGEMENT

Organizations that create and preserve value in their marketplaces prosper and grow. Those that do not run the risk of disappearing. Creating and preserving value requires continuous technology renewal and leadership capable of responding to new challenges in today's environment.

In many organizations today, a Chief Technology Officer (CTO) provides this leadership, acting as a technical businessperson deeply involved in shaping and implementing corporate strategy.

The Challenges Faced

These technology leaders face special challenges.

Fewer Resources. Investment in U.S. industrial R&D has grown steadily, but the effort in constant dollars peaked, leveled off, and started to decline in the early 1990s. This decline is a source of some concern as the cost of doing research and development is also rising.

Integration with the Business. Within U.S. firms, R&D is becoming more integrated with the business and is being managed as a system within the larger context of the business (Fig. 1). Within the larger system, research and especially development processes have become business processes, with teams from marketing, manufacturing, and finance participating from the start.

Complexity. Technology management has become more complex with global competition in ever more fragmented markets and technologies. A greater array of inputs from many different disciplines, sources, and locations must be brought together in order to develop new and improved products, processes, services and applications.

Changing Work Relationships. R&D involves a greater variety of working relationships, collaborative efforts involving colleagues from around the world in multifunctional teams.

Government Policies and Social Concerns. The creation of economic value through technology development is primarily the responsibility of the private sector. Government has a significant role, however, in creating an environment that fosters such technological innovation. Decreasing scientific literacy among the U.S. general public, in the national media, and in state and federal government also raises concerns over the society's ability to assess and balance technological risks and rewards objectively, and to put them in perspective with other needs.

Leadership Roles

Technology leaders have assumed several important roles. They must provide strategic focus, build on strengths, promote innovation, maximize human capital, be a spokesperson for technology, and understand the historical context.

Strategic Focus. As an integral part of the business, the technology leader must see that there is a consistent strategic focus and that technology development is linked to that strategy.

Building On Strengths. Technology development must be built on organizational and technological strengths. Technology leaders must understand the basis for these strengths, take responsibility for maintaining them, and ensure that the business recognizes their strategic implications. Leaders must also develop a critical perspective on competition, markets, and customers and place it in a technical context for the business.

Promoting Innovation. The innovation process involves the search for changes or the causes of changes, and then the analysis and exploitation of the opportunities that are made possible by these changes (Fig. 2).

Technology leaders must pioneer discontinuous/radical changes along with continuous/incremental innovations. Radical innovation establishes and periodically renews the competitive advantage; incremental innovations sustain it. This dual responsibility can be enabled

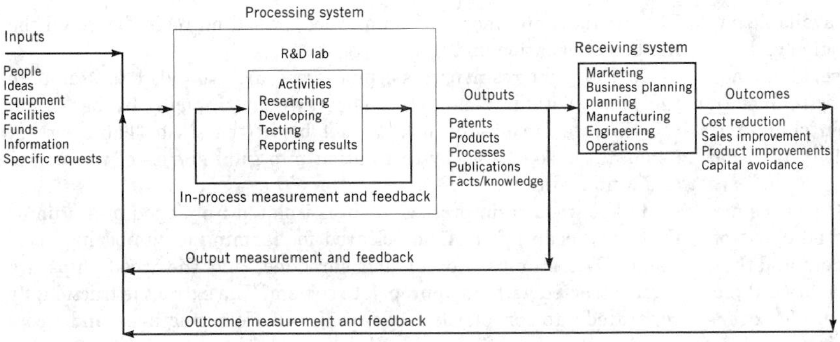

Figure 1. R&D as a system.

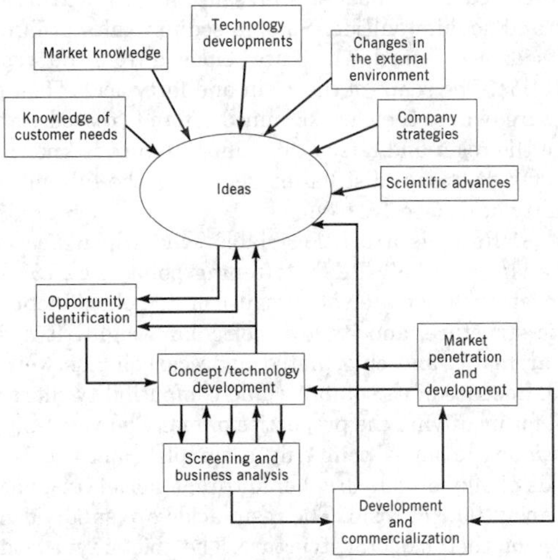

Figure 2. The innovation process model seen as a series of feedback loops involving technology transfer and the merging of cooperative efforts on a global scale.

by the existence of a laboratory that operates with a certain degree of independence within the corporation to (*1*) keep the science base renewed and vibrant, maintaining the base of science experts and state-of-the-art equipment; (*2*) leverage competencies, expertise, and resources across the firm in an affordable way, ensuring knowledge flow in shared areas of technical competencies: (*3*) explore and develop new technologies, competencies, and business options beyond the constraints of individual businesses; (*4*) maintain a longer-term focus, separate from the day-to-day concerns of the business; and (*5*) provide access to the world's best hires, technology leaders, consultants, and collaborators.

These scientists and engineers represent a special challenge to leadership in that the values and motivations may at times be at odds with corporate cultures that emphasize seniority, authority based on hierarchical influence, allegiance to corporate direction, a strict proprietary view of the results of science and technology, and expectations of instantaneous organizational response to changes in direction.

Speaking for Technology. The technology leader must be concerned with all aspects of the system for invention through commercialization, as well as with factors that enable and foster scientific and technological literacy. A spokesperson for the value and contributions of technology, a lobbyist for governmental policies that support technology development, and an educator to the public in promoting the need for increased scientific literacy.

Understanding the Historical Context. Technology leaders must also understand their corporate and business history and use the understanding as the foundation on which to build the mission and strategic plans. Technology management involves five functions: (*1*) setting the context, ie, positioning R&D as part of the business system; (*2*) planning; (*3*) organizing; (*4*) motivating; and then (*5*) assessing and improving. Each of these functions must be directed toward effective performance at each of three levels, i.e., the entire organization, various work processes, and individual efforts. A number of critical questions must be answered (Fig. 3).

Setting the Context. A key responsibility for technology management is positioning the R&D/technology function in the context of business, i.e., clarifying the mission and role of technology and establishing the necessary understanding in the entire business.

Planning. Through planning, technology managers establish objectives, develop the rationale for these objectives, and estimate the resources required over time to succeed at the strategic business, R&D organizational, and project levels.

Organizing. Organization of the R&D effort involves seeking out, bringing together, integrating, and structuring the company and its resources, the work processes, and the effort of individuals to accomplish the set goals and objectives.

Today's technology manager faces the special challenge of orchestrating a vast array of resources, including many that can be located outside the manager's organizational unit.

Technology managers are devoting more attention to developing and improving their organizational work processes and the particular practices used in these processes, which range from recruiting, to budgeting, to all the interconnected innovation processes. Furthermore, technology managers are harnessing technologies and entirely new tools to increase the productivity of R&D work. Computational chemistry and modeling eliminate some laboratory work altogether, new analytical tools arrive at answers faster, experimental equipment ranging from miniature computerized semiworks to fully automated laborato-

			Performance Needs:		
Levels of Performance	Context	Goals	Structure	Motivation	Improvement
R&D organization	Is R&D's role in the business understood?	Are technology plans developed and broadly understood?	Are R&D resources organized in the best way to carry out its mission?	Is the organizational climate assessed?	Does the organization have measures for success? Does the organization learn? Audits in place?
R&D work processes	Are R&D practices linked to business practices?	Are work processes those needed to implement the plans?	Are work processes defined, disciplined, and structured?	Is the human side of work practices understood?	Are there metrics related to the goals/results of the work processes?
Scientist/ engineer effort	Are global technology networks and collaborations fostered?	Are individual objectives supporting the plan?	Is individual work organized? Are teams adequately chartered?	Are individual motivation factors understood and acted upon?	Are researchers becoming active learners?

Figure 3. Managing technology: some key questions. Performance needs are examined for each of the five functions of technology management, at three different levels within the business.

ries feeds information directly to databases, and readily available online information sources prepare the researcher more effectively.

Networking within the technical community represents an additional general process which needs to be fostered by a technology manager. Technical networks are powerful forces for nourishing and applying key technologies. They are important technology transfer vehicles.

Motivating. Motivation within technical organizations involves not only energizing individuals but also lifting the morale and spirit of the organization as a whole. Technology leaders must understand that many values and expectations of the technical professional are unique and differ from those typical of others in the workplace. Moreover, technical organizations have unique cultures developed in the scientific tradition and characterized by collegiality, rigorous exploration, forthright challenges to one another, continuous development, and a high degree of skepticism.

Assessing and Improving. As the costs of R&D rise and as R&D becomes more integrated into the business, a number of constituencies or stakeholders are questioning the value, the effectiveness, and the relative productivity of R&D. Technology leaders find that they must define measures for R&D not only to measure effectiveness, but also to justify R&D efforts to others. Technology leaders have found that greatest impact comes from improving nine R&D processes and practices: (*1*) setting integrated technology strategies, plans, and objectives; (*2*) identifying customer needs; (*3*) identifying technical possibilities; (*4*) selecting programs; (*5*) assuring technical excellence; (*6*) planning staffing and terminating projects; (*7*) measuring effectiveness; (*8*) understanding core competencies; and (*9*) maintaining external awareness.

Looking Forward. The basic reason for industrial R&D will continue to be the necessity to understand customer needs and develop solutions linked to business strategies. Creativity will be at the heart of research and development processes. The continued development of critical science-based technologies will generate new ways for meeting existing needs and create new opportunities for emerging needs, and technology leaders who are articulate, courageous, and who can deal successfully with change will continue to be needed as an integral part of U.S. businesses.

PARRY M. NORLING
J. A. MILLER
J. W. COLLETTE
DuPont Central Research and Development

M. G. Brown and R. A. Svenson, *Res. Technol. Mgmt.*, 67–71 (July–Aug. 1988).

P. M. Norling, *Res. Technol. Mgmt.*, 42–48 (Jan.–Feb. 1996).

P. M. Norling, *CHEMTECH 1997*, **27**(10), 12–16.

C. W. Prather and L. K. Gundry, *Blueprints For Innovation: How Creative Processes Can Make You And Your Company More Competitive*, American Management Association, New York, 1995.

RESINS, NATURAL

Natural resins are generally described as solid or semisolid amorphous, fusible, organic substances that are formed in plant secretions. They are usually transparent or translucent yellow-to-brown colored, and are soluble in organic solvents but not in water. The term *natural resins* includes tree and plant exudates, fossil resins, mined resins, and shellac. For some applications, the resins have been chemically modified to increase their industrial utility.

Natural resins, except for shellac, are mixed condensation products of naturally occurring terpenoids and flavonoids contained in trees. Shellac is a product of insect secretion.

Rosin and Modified Rosins

Production. Rosin is isolated from pine trees, principally from longleaf *Pinus palustris*, slash *Pinus ellioti*, and loblolly pine *Pinus taeda*. The products are known as gum, wood, or tall oil rosin, based on the method of isolation and the source.

In the gum rosin process, pine trees are wounded to stimulate the flow of gum. V-shaped slashes are cut through the bark, and the exudate is collected in a bucket below the slash. The oleoresin (exudate) is separated by distillation into gum spirits of turpentine and gum rosin.

In the wood rosin process, rosin is isolated from aged pine stumps that have been left in fields cleared for farming or lumbering operations. The stumps are cut and shredded, and the wood chips are then extracted with an appropriate solvent. The extract is fractionally separated into nonvolatile crude rosin, volatile extractibles, and recovered solvent.

Tall oil rosin is a by-product of paper manufacturing. Raw wood chips are digested under heat and pressure with a mixture of sodium hydroxide and sodium sulfide. Soluble sodium salts of lignin, rosin, and fatty acids are formed, which are removed from the wood pulp as a dark solution. The soaps of the rosin and fatty acids float to the top of the mixture, where they are skimmed off and treated with sulfuric acid to free the rosin and fatty acids. This mixture is known as crude tall oil (CTO); fractional distillation separates the tall oil rosin acids from the fatty acids (see TALL OIL).

Properties. Rosin is a brittle, friable solid which has a T_g of ca 30°C and a ring and ball (R&B) softening point of ca 75°C. Rosin is compatible with many materials because of its polar functionality, cycloaliphatic structure, and its low molecular weight. It is soluble in aliphatic, aromatic, and chlorinated hydrocarbons, as well as esters and ethers. Because of its solubility and compatibility characteristics, it is useful for modifying the properties of many polymers.

Composition. Rosin is primarily a complex mixture of monocarboxylic acids of alkylated hydrophenanthrene nuclei (resin acids,) that represent about 90% of rosin. The resin acids are subdivided into two types, based on their skeletal structure. The abietic-type acids contain an isopropyl group pendent from the hydrophenanthrene nucleus. The pimaric-type acids have a methyl and vinyl group replacing the isopropyl group on the same carbon atom. The remaining 10% of commercial rosin consists of neutral materials that are either hydrocarbons or saponifiable esters.

Modified Rosins. Natural rosins are often modified to improve their stability to long-term aging and oxidative degradation. Pale color and color stability are also important. Stabilization methods involve the double bonds in the resin acid molecule that can form a conjugated system highly susceptible to degradation. The double bonds can be stabilized by hydrogenation, disproportionation, or polymerization.

Uses. The largest commercial use of rosin is in sizing paper to improve its water resistance, but it is being displaced by synthetic sizes. Paper size consists of rosin or modified rosins that have been partially saponified with sodium carbonate or sodium hydroxide. Sizes prepared from rosin may be modified with maleic anhydride to increase efficacy.

Modified rosins, metal resinates, and resin acid esters find application in printing ink formulations. Rosin-derived resins dispersed in linseed oil are used as vehicles for letterpress inks. Resins with high acid numbers are used as vehicles for flexographic inks, which are designed for printing on paper, metal, and plastic films. Metal salts of rosins, modified rosins, and polymerized rosins are used in inexpensive gravure inks. Rosin ester resins are used extensively in pressure-sensitive adhesives as tackifiers. Rosin, modified rosins, and derivatives are used in hot-melt adhesives. Rosin ester resins are used as modifiers in the formulation of chewing gum.

Traditional Natural Resins

Natural resins have been collected by hand throughout recorded history and used with minimal processing. Following is a description of commercial natural resins that are available in the United States.

Manila Copal. The Manilas, collected in Indonesia and the Philippines, are soluble in alcohols and ketones, and insoluble in hydrocarbons and esters. The resins soften between 81–90°C and have acid numbers of 110–141.

Dammar Resins. These resins are tapped from trees in Indonesia and Malaysia. They have low acid numbers, typically 17–35, and softening points of between 67–75°C. They are soluble in aliphatic and aromatic hydrocarbons and in terpenes. Dammar resins are used primarily in protective-coating formulations.

Gum Elemi. This resin, tapped from trees in the Philippines, contains a higher concentration of essential oils than other natural resins. Gum elemi is a film-forming plasticizing resin used in lacquers.

Sandarac. This resin, which originates in Morocco, is a polar, acidic, hard resin with a softening point of 100–130°C, an acid number of 117–155, and a saponification number of 145–157. Sandarac is soluble in alcohols and insoluble in aryl and aliphatic hydrocarbons. It is used in varnishes and lacquers for coating paper, wood, and metal.

Mastic. Mastic is a soft resin with a softening point of 55°C. It has an acid number of 50–70 and a saponification number of 62–90. It is soluble in alcohols and aryl hydrocarbons. Mastic is used in wood coatings, lacquers, adhesives, and printing inks.

Natural Resins in Medicines, Flavors, and Fragrances

With advances in pharmacology, the medicinal uses of natural resins have mostly disappeared. The only remaining significant application is the use of guaiac-impregnated paper to detect hemoglobin in stool. This widely used test is helpful in the early diagnosis of colorectal cancer. Use of the resins in cosmetics (qv) has come into question because of reports of contact dermatitis and skin sensitivity. Natural resins used as flavors or fragrances include Balm of Gilead, Balsam of Peru, and Balsam of Tolu.

Asphaltites

Mining operations are required to bring asphaltites to the market, principally for use in coatings and inks. They were formed in the same manner as oil and coal, below the surface of the earth from vegetation that had been subjected to high temperatures and pressures on a geological time scale. The most widely known asphaltite is gilsonite, which is found in the Uinta Basin in eastern Utah. Gilsonite has a softening point of 110–121°C. It has been used in protective coatings, battery cases, asphalt floor tiles, brake linings, and inks. Glance pitch, or manjak, and grahamite are related to gilsonite, but have higher specific gravities.

Shellac

Unlike other natural resins, shellac is derived from the hardened secretion of the lac insect (species *Laccifer* (Tachardia) *lacca Kerr* (family Coccidae), also known as *Kerris lacca* (Kerr)). Shellac is a refined grade of the crude lac secretion and is the most widely known lac product. Therefore, shellac has been accepted as the common generic term. Over 50% of the world's supply is produced in the Indian provinces of Bihar and Orissa, with the remainder originating in adjacent areas of southeast Asia such as Sri Lanka, China, Thailand, and Myanmar.

Raw lac is first treated to remove water-soluble carbohydrates and the dye that gives lac its red color. Also removed are woody materials, insect bodies, and trash. It is further refined by either hot filtration or a solvent process.

Composition. Shellac is primarily a mixture of aliphatic polyhydroxy acids in the form of lactones and esters. It has an acid number of ca 70, a saponification number of ca 230, a hydroxyl number of ca 260, and an iodine number of ca 15. Its average molecular weight is ca 1000. Shellac is a complex mixture of which only some of its constituents have been identified.

Uses. Synthetic resins have taken over a large share of the market for shellac. Unpigmented shellac is used on floors, woodwork, and paneling. White pigmented shellac is used as a primer–sealer for interior applications. Shellac is used as a protective coating for pharmaceuticals to maintain the potency of medication.

Candy is coated with shellac to seal in moisture and keep the product fresh. The coating provides a high gloss to the confection, which improves its appearance. Citrus fruits and some apples are often coated with shellac. Shellac is also used as a stiffener for felt hat bodies, primarily for recreational hats.

Health and Safety Information

Rosin has a low order of toxicity following ingestion or skin contact. Rosin and its numerous derivatives have a number of permitted food packaging and other direct and indirect food contact uses throughout the world. Natural resins such as dammar and Manila copal have been described as nontoxic and nonallergenic. Shellac has been affirmed by the U.S. FDA as GRAS for food applications.

JAY B. CLASS
Hercules Incorporated

N. Heaton, *Outlines of Paint Technology,* Charles Griffin & Co., London, 1947.

C. L. Mantell, C. W. Kopf, J. L. Curtis, and E. M. Rogers, *The Technology of Natural Resins,* John Wiley & Sons, Inc., New York, 1942.

H. Abraham, *Asphalts and Allied Substances,* 5th ed., D. Van Nostrand Co., New York, 1945.

RESINS, WATER-SOLUBLE. See WATER-SOLUBLE POLYMERS.

RETROVIRUSES. See ANTIVIRAL AGENTS; IMMUNOTHERAPEUTIC AGENTS.

RESORCINOL. See HYDROQUINONE, RESORCINOL, AND CATECHOL.

REVERSE OSMOSIS

Reverse osmosis (RO) is a fairly mature technology as used in the area of seawater and brackish water desalination (see WATER, SUPPLY AND DESALINATION). Use of reverse osmosis as a separation tool is, however, a relatively young technology.

Since the early 1960s, the industrial development of thin-film composite membranes (see MEMBRANE TECHNOLOGY) and improvements in polymer materials have widened applications of RO. The driving force for the development and use of RO membranes is the advantage these systems have over traditional separation processes such as distillation (qv), extraction, ion exchange (qv), and adsorption (qv). Because reverse osmosis is a pressure-driven process, no energy-intensive phase changes or potentially expensive solvents or adsorbents are needed for RO separations. Reverse osmosis is a process that is inherently simple to design and operate compared to many traditional separation processes. Also, simultaneous separation and concentration of both inorganic and organic compounds are possible using the RO process. In addition, using nanofiltration (loose RO), membrane-selective solute separations based on charge and molecular weight and size differences are possible. Finally, reverse osmosis technology can also be combined with ultrafiltration (qv), pervaporation, distillation, and other separation techniques to produce hybrid processes that result in highly efficient and selective separations.

Membrane Materials and Modules

Reverse osmosis membrane separations are governed by the properties of the membrane used in the process. These properties depend on the chemical nature of the membrane material, which is almost always a polymer, as well as its physical structure. Properties for the ideal RO membrane include low cost, resistance to chemical and microbial attack, mechanical and structural stability over long operating periods and wide temperature ranges, and the desired separation characteristics for each particular system. Few membranes satisfy all these criteria, and compromises must be made to select the best RO membrane available for each application.

Most commercially available RO membranes fall into one of two categories: asymmetric membranes containing one polymer, or thin-film composite membranes consisting of two or more polymer layers. Asymmetric RO membranes have a thin (~100 nm) permselective skin layer supported on a more porous sublayer of the same polymer. The dense skin layer determines the fluxes and selectivities of these membranes whereas the porous sublayer serves only as a mechanical support for the skin layer and has little effect on the membrane separation properties.

Thin-film, composite membranes consist of a thin polymer barrier layer formed on one or more porous support layers, which is almost always a different polymer from the surface layer. The surface layer determines the flux and separation characteristics of the membrane. The porous backing serves only as a support for the barrier layer, it has almost no effect on membrane transport properties. The barrier layer is extremely thin, thus allowing high water fluxes. The most important thin-film composite membranes are made by interfacial polymerization, a process in which a highly porous membrane, usually polysulfone, is coated with an aqueous solution of a polymer or monomer and then reacts with a cross-linking agent in a water-immiscible solvent.

Theoretical Aspects

A reverse osmosis membrane acts as the semipermeable barrier to flow in the RO process, allowing selective passage of a particular species, usually water, while partially or completely retaining other species, ie, solutes such as salts. Chemical potential gradients across the membrane provide the driving forces for solute and solvent transport across the membrane. The solute chemical potential gradient, $-\Delta\mu_s$, is usually expressed in terms of concentration; the water (solvent) chemical potential gradient, $-\Delta\mu_w$, is usually expressed in terms of pressure difference across the membrane.

Measurable Process Parameters. The RO process is relatively simple in design. It consists of a feed water source, feed pretreatment, high pressure pump, RO membrane modules, and in some cases, post-treatment steps. A schematic of the RO process is shown in Figure **1a.**

The three streams and associated variables of the RO membrane process are shown in Figure **1b:** the feed; the product stream, called

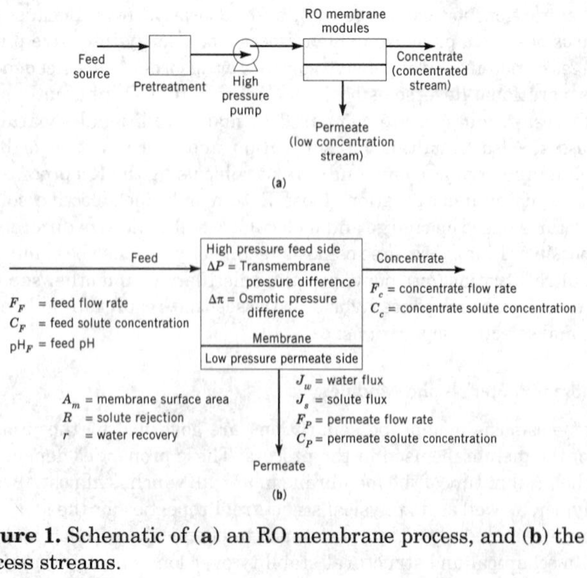

Figure 1. Schematic of **(a)** an RO membrane process, and **(b)** the RO process streams.

the permeate; and the concentrated reject stream, called the concentrate or retentate. The water flow through the membrane is reported in terms of water flux, J_w.

$$J_w = \frac{\text{volumetric or mass permeation rate}}{\text{membrane area}}$$

Solute passage is defined in terms of solute flux, J_s.

$$J_s = \frac{\text{mass permeation rate}}{\text{membrane area}}$$

Transport Models. Many mechanistic and mathematical models have been proposed to describe reverse osmosis membranes. Some of these descriptions rely on relatively simple concepts; others are far more complex and require sophisticated solution techniques. Models that adequately describe the performance of RO membranes are important to the design of RO processes. Models that predict separation characteristics also minimize the number of experiments that must be performed to describe a particular system.

Study of RO Variables and Typical Experimental Setup

Factors affecting RO membrane separations and water flux include feed variables such as solute concentration, temperature, pH, and pretreatment requirements; membrane variables such as polymer type, module geometry, and module arrangement; and process variables such as feed flow rate, operating time and pressure, and water recovery.

Fouling. One of the primary concerns in reverse osmosis processes is the irreversible fouling of the membrane surface. Fouling is a broad term that refers to the deposition or association of feed stream constituents on the membrane surface or within the membrane pores, consequently leading to a decrease in the membrane performance. The cause of fouling can generally be traced to constituents in the membrane feed stream.

The success of a reverse osmosis process hinges directly on the pretreatment of the feed stream. If typical process streams, without pretreatment to remove partially some of the constituents listed, were contacted with membranes, membrane life and performance would be unacceptable. There is no single pretreatment for all types of foulants. Pretreatment methods range from pH control, adsorption (qv), to filtration (qv), depending on the chemistry of the particular foulant.

Applications

Developments and advances in both membrane materials and reverse osmosis modules have increased the range of applications to which RO can be applied. Whereas the RO industry has developed around water desalination, RO has become a significant cornerstone in other industries. These include wastewater cleanup for electroplating, radioactive processing landfill leachate and municipal wastewater; ultrapure water production for electronics-grade, laboratory-grade, and pharmaceutical-grade materials; and food processing (qv).

Nanofiltration

Nanofiltration, also called loose RO, membranes are a more recent development in the field of RO membrane separations. These membranes typically have much higher water fluxes at low pressures when compared to traditional RO membranes. Nanofiltration (NF) membranes are usually charged, ie, have carboxylic groups, sulfonic groups, etc, and as a result, ion repulsion (Donnan exclusion) is the primary factor in determining salt rejection. More highly charged ions such as SO_4^{2-} are rejected more fully than monovalent ions such as Cl^- by a negatively charged nanofiltration membrane. Another characteristic of charged membranes is the tendency to foul in the presence of components having opposite charge to that of the membrane.

Nanofiltration membranes usually have good rejections of organic compounds having molecular weights above 200–500. NF provides the possibility of selective separation of certain organics from concentrated monovalent salt solutions such as NaCl. The most important nanofiltration membranes are composite membranes made by interfacial polymerization. Polyamides made from piperazine and aromatic acyl chlorides are examples of widely used nanofiltration membrane. Nanofiltration has been used in several commercial applications, among which are demineralization, organic removal, heavy-metal removal, and color removal.

Design Considerations

Design of the membrane module system involves selection of the membrane material; the module geometry, eg, spiral-wound or hollow-fiber; product flow rate and concentration; solvent recovery; operating pressure; and the minimum tolerable flux. The effects of these variables can be obtained from laboratory or pilot experiments using different membranes and modules. The membrane module as well as the solvent recovery can be chosen to minimize fouling. Spiral-wound modules are widely used because these offer both high surface area as well as a lower fouling potential.

D. BHATTACHARYYA
W. C. MANGUM
University of Kentucky
M. E. WILLIAMS
EET Corporation

C. Reid and E. Breton, *J. Appl. Polym. Sci.* **1**, 133 (1959).
S. Talati, *Desalination,* **97,** 353–361 (1994).
S. Ebrahim and M. Abdel-Jawad, *Desalination,* **99,** 39–55 (1994).

REYNOLDS NUMBER. See FLUIDIZATION; FLUID MECHANICS; RHEOLOGICAL MEASUREMENTS; SEDIMENTATION.

RHENIUM AND RHENIUM COMPOUNDS

Rhenium

Rhenium is the 75th element in the Periodic Table and the heaviest element in Group 7 (VIIB). Its congeners in Group 7 are manganese and technetium, elements 25 and 43, respectively.

Rhenium, atomic wt 186.2, occurs in nature as two nuclides: ^{185}Re atomic mass 184.9530, in 37.500% abundance; and ^{187}Re, atomic mass 186.9560, in 62.500% abundance. The latter isotope is radioactive, emitting very low energy radiation and having a half-life estimated at 4.3(±0.5) × 10^{10} yr. Radioactivity associated with ^{187}Re can be detected only by using sophisticated laboratory equipment because of the low energy of the emitted β-particles. This radioactivity poses no health or safety hazards. The radioactive decay of this isotope has been used to date accurately the time of the earth's formation.

$$187\ \text{RE} \xrightarrow{187} \text{Os} + 0.002\ \text{MeV}\ \beta$$

In addition to ^{185}Re and ^{187}Re, 16 other radioactive rhenium isotopes are known; none has an appreciable half-life.

Occurrence and Recovery. Rhenium is one of the least abundant of the naturally occurring elements. Various estimates of its abundance in the earth's crust have been made. The most widely quoted figure is 0.027 atoms per 10^6 atoms of silicon (0.05 ppm by wt). However, this number, based on analyses for the most common rocks, ie, granites and basalts, has a high uncertainty.

Rhenium and molybdenum occur together in nature, a consequence of the similarities of these elements. Both have a high affinity for sulfide ion and the radii of Re^{4+} and Mo^{4+} are almost identical; ReS_2 and MoS_2 have similar crystal structures with almost identical dimensions (see MOLYBDENUM COMPOUNDS).

Rhenium is obtained from molybdenite concentrates obtained from porphyry copper (qv) ores. Rhenium, present in small amounts in these ores, is concentrated in the molybdenite by-products to values as high as 2000 ppm, making recovery realistic.

Rhenium is obtained by roasting the molybdenite concentrate, which converts most of the rhenium to Re_2O_7. This oxide is volatile and exits the system along with some MoO_3 and sulfur oxides. These water-soluble compounds are concentrated in a wet-scrubbing system. Anion exchange or solvent extraction permits separation of ReO_4^- from other materials.

Table 1. Selected Physical Properties of Rhenium

Property	Value
atomic number	75
atomic weight	186.2
mp, °C	3180
bp, °C	5926[a]
density, g/cm^3	21.02
crystal structure	hexagonal close-packed
metal radius, 12 coordinate, nm	0.1372
ΔH_{subl}, kJ/mol[b]	791
specific heat, 20°C, J/(mol·K)[b]	25.1
ionization potentials, kJ[b]	
$Re \rightarrow Re^+ + e^-$	757
$Re^+ \rightarrow Re^{2+} + e^-$	1597
$Re^{2+} \rightarrow Re^{3+} + e^-$	2502
electrical resistivity, 20°C, $\mu\Omega$·cm	19.3
thermal coefficient of electrical resistivity, °C^{-1}	3.95 × 10^{-3}
thermal conductivity, 0–100°C, W/(cm·K)	0.400
module of elasticity, Pa[c]	0.46

[a] Estimated value. [b] To convert J to cal, divide by 4.184. [c] To convert Pa to mm Hg, multiply by 0.0075.

Physical Properties

Selected physical properties of rhenium are summarized in Table 1.

Chemical Properties

Rhenium does not react with atmospheric oxygen at ambient temperatures, but at higher temperatures it burns, giving Re_2O_7. This volatile yellow crystalline compound dissolves in water to give perrhenic acid, $HReO_4$. Reactions of the metal with halides produce Re_2Cl_{10}, Re_2Br_{10}, ReF_6, and ReF_7. Rhenium is not affected by water and hydrohalic acids but reacts quickly with HNO_3 to produce $HReO_4$.

Alloys

A significant property of rhenium is its ability to alloy with molybdenum and tungsten. Molybdenum alloys containing up to 50 wt % Re are body-centered cubic (bcc) solid solutions. An alloy having 50 wt % Re can be fabricated by either warm or cold working, and the alloy can be welded.

Metal and Alloy Processing. Rhenium and alloys of rhenium with molybdenum and tungsten can be consolidated by powder metallurgy (see METALLURGY, POWDER METALLURGY).

Uses of Rhenium-Containing Catalysts. Rhenium finds use as filaments in electron tubes, light bulbs, and in photoflash bulbs. It is particularly useful for filaments used in mass spectrometers. The largest use of rhenium by far has been in bimetallic petroleum (qv) reforming catalysts used in the production of unleaded and low lead gasoline (see GASOLINE AND OTHER MOTOR FUELS). These catalysts contain approximately 0.3 wt % platinum and 0.3 wt % rhenium. In newer catalyst systems a somewhat higher proportion of rhenium is used.

Rhenium Compounds

The properties of rhenium differ considerably from those of manganese. Many more rhenium compounds than manganese compounds

Table 2. Examples of Rhenium Compounds

Oxidation state	Electronic configuration	Coordination number	Metal atom geometry	Example[a]
−I	d^8	5	trigonal bipyramid	$Na(Re(CO)_5)$
O	d^7	6	octahedral	$Re_2(CO)_{10}$
I	d^6		octahedral	$ReCl(CO)_5$
				$K_5[Re(CN)_6]$
				$(Re(CNC_6H_5)_6)I$
				$Re(CO)_3(\eta\text{-}C_5H_5)$
				$ReCl(dppe)_2(N_2)$
II	d^5	6	octahedral	$ReCl_2(dppe)_2$
III	d^4	6	octahedral	$ReCl_3(P(CH_3)_2C_6H_5)_3$
			trigonal prism	$Re(S_2C_2(C_6H_5)_2)_3$
		7		$K_4(Re(CN)_7)\cdot 2\ H_2O$
			square pyramid	$K_2(Re_2Cl_8)$
		7	capped octahedral	$ReH_3(dppe)_2$
IV	d^3	6	octahedral	K_2ReCl_6
				$ReCl_4$
				$K_4(Re_2OCl_{10})$
V	d^2	6	octahedral	Re_2Cl_{10}
				$ReOCl_3(P(C_6H_5)_3)_2$
				$K_2(ReOCl_5)$
		8		$ReH_5(P(C_6H_5)(C_2H_5)_2)_3$
VI	d^1	5	square pyramid	$ReOCl_4$
		6	octahedral	ReF_6
				$Re(CH_3)_6$
		8	square antiprism	K_2ReF_8
VII	d^0	4	tetrahedral	$KReO_4$
				ReO_3Cl
		4, 6	tetrahedral and octahedral	$(Re_2O_7)_x$
		7	pentagonal bipyramid	ReF_7
		9	tricapped trigonal prism	K_2ReH_9

[a] dppe = 1,2-bis(diphenylphosphino)ethane.

having oxidation states of +4, +5, +6, and +7 are known. Although rhenium(II) compounds are known to exist as solids and in nonaqueous solution, no aqueous chemistry of the Re^{2+} ion is known. The +2 oxidation state is common for manganese.

A particularly significant part of rhenium chemistry involves cluster compounds in which there is metal–metal bonding. This chemistry centers largely around the +3 oxidation state. A survey of known types of rhenium compounds is presented in Table 2. The selected examples include a number of commonly encountered compounds.

PAUL M. TREICHEL, JR.
University of Wisconsin, Madison

F. A. Cotton and G. Wilkinson, *Advanced Inorganic Chemistry,* 5th ed., Wiley-Interscience, New York, 1988.

D. F. C. Morris and E. L. Short, *Handbook of Geochemistry,* Springer-Verlag, Heidelberg, Germany, 1965.

R. D. Peacock, *The Chemistry of Technetium and Rhenium,* Elsevier, Amsterdam, the Netherlands, 1966.

K. A. Conner and R. A. Walton, *Rhenium,* in G. Wilkinson, R. D. Gillard, and J. A. McCleverty, eds., *Comprehensive Coordination Chemistry,* Vol. 4, Pergamon Press, London, 1987.

RHEOLOGICAL MEASUREMENTS

Rheology is the science of the deformation and flow of matter. It is concerned with the response of materials to applied stress. That response may be irreversible viscous flow, reversible elastic deformation, or a combination of the two. Control of rheology is essential for the manufacture and handling of numerous materials and products, eg, foods, cosmetics, rubber, plastics, paints, inks, and drilling muds. Before control can be achieved, there must be an understanding of rheology and an ability to measure rheological properties.

Deformation is the relative displacement of points of a body. It can be divided into two types: flow and elasticity. Flow is irreversible deformation; when the stress is removed, the material does not revert to its original form. This means that work is converted to heat. Elasticity is reversible deformation; the deformed body recovers its original shape, and the applied work is largely recoverable. Viscoelastic materials show both flow and elasticity; they provide special challenges in terms of modeling behavior and devising measurement techniques.

The flow properties of a liquid are defined by its resistance to flow, ie, viscosity, and may be measured by determining the rate of flow through a capillary, the resistance to flow when the fluid is sheared between two surfaces, or the rate of motion of a bubble or ball moving through the fluid. The mechanical properties of an elastic solid may be studied by applying a stress and measuring the deformation or strain. For viscoelastic materials additional techniques beyond those indicated for solids and liquids are needed for complete characterization.

Viscosity

A liquid is a material that continues to deform as long as it is subjected to a tensile and/or shear stress. The latter is a force applied tangentially to the material.

For a liquid under shear, the rate of deformation (shear rate) is a function of the shearing stress. The original exposition of this relationship is Newton's law, which states that the ratio of the stress to the shear rate is a constant, ie, the viscosity. Under Newton's law, viscosity is independent of shear rate. This is true for ideal or Newtonian liquids, but the viscosities of many liquids, particularly a number of those of interest to industry, are not independent of shear rate. These non-Newtonian liquids may be classified according to their viscosity behavior as a function of shear rate. Many exhibit shear thinning, whereas others give shear thickening. Some liquids at rest appear to behave like solids until the shear stress exceeds a certain value, called the yield stress, after which they flow readily.

Some commonly observed types of flow behavior are shown in Figure 1, in which the shear stress is plotted against shear rate. These plots are called flow curves and are frequently used to express the rheological behavior of liquids. Newtonian flow is shown by a straight line, and shear thinning and thickening by curves. Yield stresses, τ_0, are shown by intercepts on the stress (y) axis.

Viscosity is equal to the slope of the flow curve, $\eta = d\tau/d\dot{\gamma}$. The quantity $\tau/\dot{\gamma}$ is the viscosity η for a Newtonian liquid and the apparent viscosity η_a for a non-Newtonian liquid. The kinematic viscosity is the viscosity coefficient divided by the density, $\nu = \eta/\rho$. The fluidity is the reciprocal of the viscosity, $\phi = 1/\eta$. The common units for viscosity, dyne seconds per square centimeter ((dyn·s)/cm^2) or grams per centimeter second ((g/(cm·s)), called poise, which is usually expressed as centipoise (cP), have been replaced by the SI units of pascal seconds, ie, Pa·s and mPa·s, where 1 mPa·s = 1 cP. In the same manner

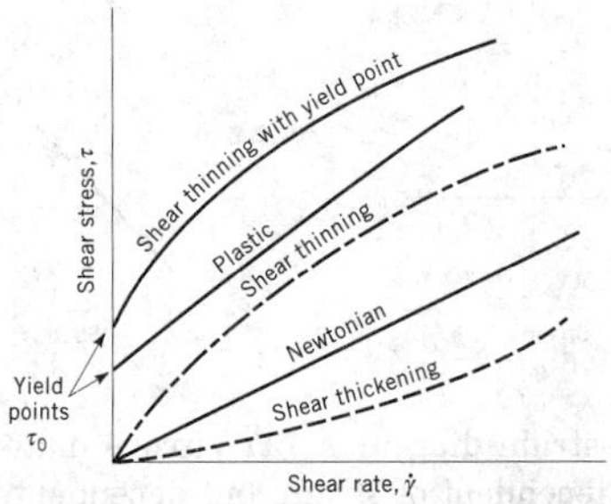

Figure 1. Flow curves (shear stress vs shear rate) for different types of flow behavior.

the shear stress units of dynes per square centimeter, dyn/cm^2, have been replaced by Pascals, where 10 dyn/cm^2 = 1 Pa, and newtons per square meter, where 1 N/m^2 = 1 Pa. Shear rate is $\Delta V/\Delta X$, or length/time/length, so that values are given as per second (s^{-1}) in both systems. The SI units for kinematic viscosity are square centimeters per second, cm^2/s, ie, Stokes (St), and square millimeters per second, mm^2/s, ie, centistokes (cSt).

Flow Models. Many flow models have been proposed; they are useful for the treatment of experimental data or for describing flow behavior (Table 1). No given model fits the rheological behavior of a material over an extended shear rate range; however, these models are useful for summarizing rheological data and are frequently encountered in the literature.

Thixotropy and Other Time Effects. Many fluids exhibit time-dependent effects. Some fluids increase in viscosity (rheopexy) or decrease in viscosity (thixotropy) with time when sheared at a constant shear rate. These effects can occur in fluids with or without yield values. Rheopexy is a rare phenomenon. Thixotropic fluids are common; examples are starch pastes and drilling muds. The thixotropic effect is shown in Figure 2, where the curves are for a specimen exposed first to increasing and then to decreasing shear rates. Because of the decrease in viscosity with time as well as shear rate, the up-and-down flow curves do not superimpose. Instead, they form a hysteresis loop, often called a thixotropic loop.

Table 1. Flow Equations for Flow Models

Flow model	Flow equation
Newtonian	$\tau = \eta\dot{\gamma}$
plastic (Bingham) body	$\tau - \tau_0 = \eta\dot{\gamma}$
power law	$\tau = k\lvert\dot{\gamma}\rvert^n$
power law with yield value	$\tau - \tau_0 = k\lvert\dot{\gamma}\rvert^n$
Casson fluid	$\tau^{1/2} - \tau_0^{1/2} = \eta_\infty^{1/2}\dot{\gamma}^{1/2}$
Williamson	$\eta = \eta_\infty + \frac{(\eta_0 - \eta_\infty)}{1 + \frac{\lvert\tau\rvert}{\tau_m}}$
cross	$\eta = \eta_\infty + \frac{(\eta_0 - \eta_\infty)}{1 + \alpha\dot{\gamma}^n}$

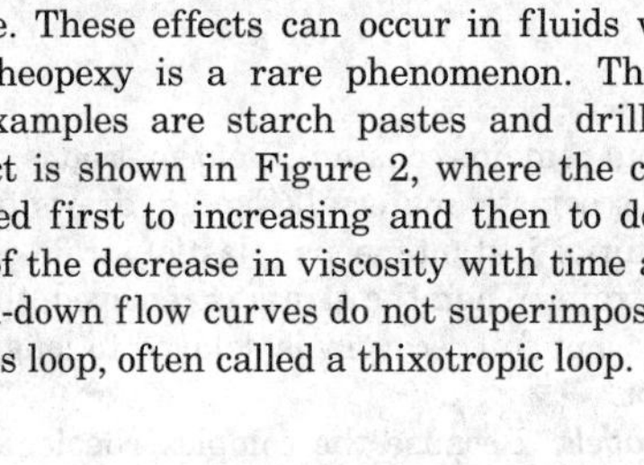

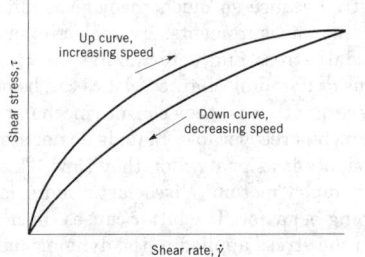

Figure 2. Flow curves (up and down) for a thixotropic material: hysteresis loop.

Experimentally, it is sometimes difficult to detect differences between a shear-thinning liquid in which the viscosity decreases with increasing shear, and a thixotropic material in which the viscosity decreases with time, because of the combined shear and time effects that occur during a series of measurements. This is especially true if only a few data points are collected. In addition, most materials that are thixotropic are also shear thinning. In fact, one definition of a thixotropic fluid limits it to materials whose viscosity is a function of both shear rate and time.

Viscosity–time measurements during or after shearing can be used to show time-dependent effects. Time-dependent effects are measured by determining the decay of shear stress as a function of time at one or more constant shear rates. A rotational viscometer connected to a recorder is used. After the sample is loaded and allowed to come to mechanical and thermal equilibrium, the viscometer is turned on and the rotational speed is increased in steps, starting from the lowest speed. The resultant shear stress is recorded with time. On each speed change the shear stress reaches a maximum value and then decreases exponentially toward an equilibrium level. The peak shear stress, which is obtained by extrapolating the curve to zero time, and the equilibrium shear stress are indicative of the viscosity–shear behavior of unsheared and sheared material, respectively. The stress–decay curves are indicative of the time-dependent behavior.

Results from measurements of time-dependent effects depend on the sample history and experimental conditions and should be considered approximate. However, measurements of time-dependent behavior can be useful in evaluating and comparing a number of industrial products and in solving flow problems.

Effect of Temperature. In addition to being often dependent on parameters such as shear stress, shear rate, and time, viscosity is highly sensitive to changes in temperature. Most materials decrease in viscosity as temperature increases. The dependence is logarithmic and can be substantial, up to 10% change/°C. This has important implications for processing and handling of materials and for viscosity measurement.

Dilute Polymer Solutions. The measurement of dilute-solution viscosities of polymers is widely used for polymer characterization. Very low concentrations reduce intermolecular interactions and allow measurement of polymer–solvent interactions. These measurements are usually made in capillary viscometers, some of which have provisions for direct dilution of the polymer solution. The key viscosity parameter for polymer characterization is the limiting viscosity number or intrinsic viscosity, $[\eta]$. It is calculated by extrapolation of the viscosity number (reduced viscosity) or the logarithmic viscosity number (inherent viscosity) to zero concentration.

Concentrated Polymer Solutions. Knowledge of the viscosity behavior of concentrated solutions is important to the manufacture and application of caulks, adhesives, inks, paints, and varnishes. It is also useful for designing and controlling polymer manufacturing processes, fiber spinning, and film casting. Viscosity behavior can be investigated by a variety of methods, including the use of simple capillary viscometers, extrusion rheometers, and rotational viscometers. Unlike dilute solutions, concentrated polymer solutions show a vast amount of interaction between the macromolecules. The degree of interaction is governed by the concentration, the characteristics of the chains, and the nature of the solvent.

Melt Viscosity. The study of the viscosity of polymer melts is important for the manufacturer who must supply suitable materials and for the fabrication engineer who must select polymers and fabrication methods. Thus melt viscosity as a function of temperature, pressure, rate of flow, and polymer molecular weight and structure is of considerable practical importance. Polymer melts exhibit elastic as well as viscous properties. This is evident in the swell of the polymer melt upon emergence from an extrusion die, a behavior that results from the recovery of stored elastic energy plus normal stress effects.

A number of experimental methods have been applied to measure the melt viscosity of polymers, but capillary extrusion techniques probably are generally preferred. Rotational methods are also used, and some permit the measurement of normal stress effects resulting from elasticity as well as of viscosity. Slit rheometers can also be used to measure normal stress. Oscillatory shear measurements are useful for measuring the elasticity of polymer melts. Controlled stress methods have also been applied.

Extensional Viscosity. In addition to the shear viscosity η, two other rheological constants can be defined for fluids: the bulk viscosity, K, and the extensional or elongational viscosity, η_e. The bulk viscosity relates the hydrostatic pressure to the rate of deformation of volume, whereas the extensional viscosity relates the tensile stress to the rate of extensional deformation of the fluid. Extensional viscosity is important in a number of industrial processes and problems. Shear properties alone are insufficient for the characterization of many fluids, particularly polymer melts.

Unlike shear viscosity, extensional viscosity has no meaning unless the type of deformation is specified. The three types of extensional viscosity identified and measured are uniaxial or simple, biaxial, and pure shear. Uniaxial viscosity is the only one used to characterize fluids. It has been employed mainly in the study of polymer melts, but also for other fluids. The two other extensional viscosities are used to study elastomers in the form of films or sheets. Uniaxial and biaxial extensions are important in industry, the former for the spinning of textile fibers and roller spattering of paints, and the latter for blow molding, vacuum forming, film blowing, and foam processes.

Electrorheological Behavior. Electrorheological (ER) fluids are colloidal suspensions whose properties change strongly and reversibly upon application of an electric field. There is substantial interest in ER fluids because they appear to offer a means for rapid processes or operations that take advantage of the calculation and control speeds of modern computers. Manufacturers hope to use these so-called smart materials in applications such as fast acting switches and valves, robot operations, safety catches and other mechanical release devices.

When an electric field is applied to an ER fluid, it responds by forming fibrous or chain structures parallel to the applied field. These structures greatly increase the viscosity of the fluid, by a factor of 10^5 in some cases. At low shear stress the material behaves like a solid. The material has a yield stress, above which it will flow, but with a high viscosity. The force necessary to shear the fluid is proportional to the square of the electric field.

Elasticity and Viscoelasticity

Elastic deformation is a function of stress and is expressed in terms of relative displacement or strain. Strain may be expressed in terms of relative change in volume, length, or other measurement, depending on the nature of the stress. An ideal elastic body is a material for which the strain is proportional to the stress (Hooke's law) with immediate recovery to the original volume and shape when the stress is released. The relationship between the stress ς and strain ϵ may be written as $\sigma = K\epsilon$, where K is a proportionality constant called the modulus of elasticity.

Materials such as metals are nearly elastic and show almost no flow or viscous component. Polymers and many of their solutions are both viscous and elastic, and both types of deformation must be taken into account to explain their behavior.

Mechanical Behavior of Materials. Different kinds of materials respond differently when they undergo basic mechanical tests. This is illustrated in Figure 3, which shows stress–strain diagrams for purely viscous and purely elastic materials. With the former, the stress is relieved by viscous flow and is independent of strain. With the latter, there is a direct dependence of stress on strain and the ratio of the two is the modulus E (or G).

The response of different materials to a stress applied at time $t = t_0$ followed by removal of that stress at $t = t_1$, ie, creep and recovery, is shown in Figure 4. For an elastic material (Fig. 4**a**), the resulting strain is instantaneous and constant until the stress is removed, at which time the material recovers and the strain immediately drops back to zero. In the case of the viscous fluid (Fig. 4**b**), the strain increases linearly with time. When the load is removed, the strain does not recover but remains constant. Deformation is permanent. The response of the viscoelastic material (Fig. 4**c**) draws from both kinds of behavior. An initial instantaneous (elastic) strain is followed by a time-dependent strain. When the stress is removed, the initial strain recovery is elastic, but full recovery is delayed to longer times by the viscous component.

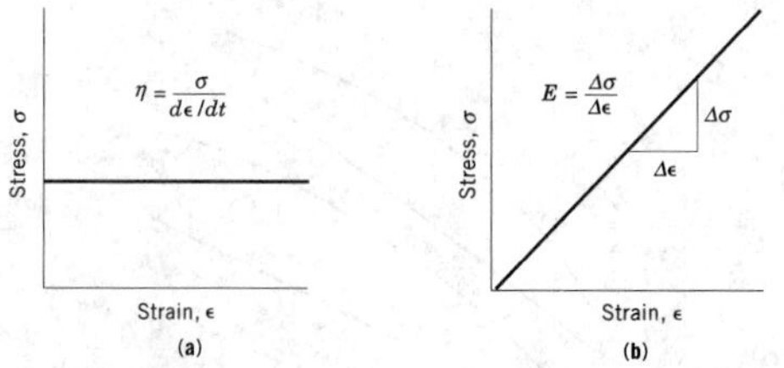

Figure 3. Stress–strain diagrams. (**a**) Viscous material of viscosity η; the stress is independent of strain, but dependent on the speed of testing, (**b**) Elastic material of modulus E; the slope is the modulus which is independent of the speed of testing.

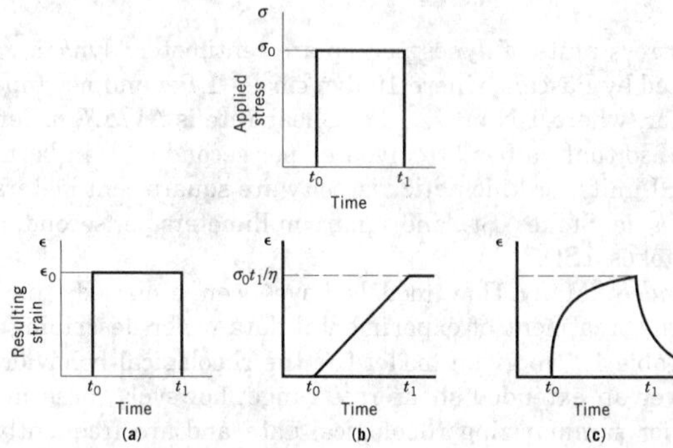

Figure 4. Response (strain) of different idealized materials to an instantaneous application of a stress at time $t = t_0$: (**a**) elastic, (**b**) viscous, and (**c**) viscoelastic.

Mechanical Models. Because the complex rheological behavior of viscoelastic bodies is difficult to visualize, mechanical models are often used. In these models the viscous response to applied stress is assumed to be that of a Newtonian fluid and is represented by a dashpot, ie, a piston operating in a cylinder of Newtonian fluid. The elastic response is idealized as an ideal elastic (Hookean) solid and is represented by a spring. The dashpot represents the dissipation of energy in the form of heat, whereas the spring represents a system that stores energy. These models are largely being replaced by mathematical (integral and differential) models.

Dynamic Behavior. Knowledge of the response of materials to stress–strain, creep, and stress–relaxation measurements are useful to define material properties. The dynamic response of viscoelastic materials to cyclic stresses or strains is also important, partly because cyclic motion occurs in many processing operations and applications, and partly because so much rheological information can be gained from dynamic measurements. By subjecting a specimen to oscillatory (sinusoidal) stress and determining the response, both the elastic and viscous or damping characteristics can be obtained. Elastic materials store energy; that is, they convert mechanical work into potential energy, which is recoverable. Liquids do not store energy when stressed, but dissipate it as heat when they flow. This dissipation results in highly damped motion. Viscoelastic materials exhibit both elastic and damping behavior. The latter causes the deformation to be out of phase with the stress applied in the dynamic measurement.

Normal Stress (Weissenberg Effect). Many viscoelastic fluids flow in a direction normal (perpendicular) to the direction of shear stress in steady-state shear. Examples of the effect include flour dough climbing up a beater, polymer solutions climbing up the inner cylinder in a concentric cylinder viscometer, and paints forcing apart the cone and plate of a cone–plate viscometer. The normal stress effect has been put to practical use in certain screwless extruders designed in a cone–plate or plate–plate configuration, where the polymer enters at the periphery and exits at the axis.

Viscometers

To solve a flow problem or characterize a given fluid, an instrument must be carefully selected. Many commercial viscometers are available with a variety of geometries for wide viscosity ranges and shear rates. In choosing a commercial viscometer a number of criteria must be considered. Of great importance is the nature of the material to be tested, its viscosity, its elasticity, the temperature dependence of its viscosity, and other variables. The degree of accuracy and precision required, and whether the measurements are for quality control or research, must be considered. The viscometer must be matched to the materials and processes of interest.

Viscometers may be separated into three main types: capillary, rotational, and moving body. There are other kinds, usually designed for special applications. The choice depends on the particular requirements of the investigator and the price range.

Capillary Viscometers. In capillary flow measurement, a liquid drains or is forced through a fine-bore tube, and the viscosity is determined from the measured flow, applied pressure, and tube dimensions. Capillary viscometers are useful for measuring precise viscosities of a large number of fluids, ranging from dilute polymer solutions to polymer melts.

Rotational Viscometers. Rotational viscometers consist of two basic parts separated by the fluid being tested. The parts may be concentric cylinders (cup and bob), plates, a low angle cone and a plate, or a disk, paddle, or rotor in a cylinder. Rotation of one part against the other produces a shearing action on the fluid. The torque required to produce a given angular velocity or the angular velocity resulting from a given torque is a measure of the viscosity. Rotational viscometers are more versatile than capillary viscometers. They can be used with a wide range of materials because opacity, settling, and non-Newtonian behavior do not cause difficulties. Viscosities over a range of shear rates and as a function of time can be measured. Therefore, they are useful for characterizing shear thinning and time-dependent behavior.

Rotational viscometers have been developed that are either integral or closely interfaced with computers for operation and control of the instrument as well as for data collection, reduction, and storage.

Moving Body Viscometers. In moving body viscometers, the motion of a ball, bubble, plate, needle, or rod through a material is monitored. Falling ball viscometers are based on Stokes' law, which relates the viscosity of a Newtonian fluid to the velocity of the falling sphere.

Other Viscometers. A number of other viscometers are built for specific research or product applications. In one design, vibrational techniques are used to measure viscosity. Because the rate of shear is not easily determined or changed, these instruments are best used for controlling or studying processes in which viscosity changes with time or temperature. In such cases they can be useful, because wide ranges of viscosity can be measured without changing sensors.

Measurement Techniques

Extensional Viscosity. All three types of extensional viscosity can be measured: uniaxial, biaxial, and pure shear. Only a few commercial instruments are available, however, and most measurements are made with improvised equipment.

Viscoelastic Measurement. A number of methods measure the various quantities that describe viscoelastic behavior. Some require expensive commercial rheometers, others depend on custom-made research instruments, and a few require only simple devices. Even qualitative observations can be useful in the case of polymer melts, paints, and resins, where elasticity may indicate an inferior batch or unusable formulation. For example, the extrusion swell of a material from a syringe can be observed with a microscope. The Weissenberg effect is seen in the separation of a cone and plate during viscosity measurements or the climbing of a resin up the stirrer shaft during polymerization or mixing.

Creep experiments are among the simplest for describing viscoelastic behavior. They involve the measurement of deformation as a function of time after a given load has been applied. Such measurements may be made on specimens in tension, compression, or shear.

In a stress–relaxation test the deformation is held constant, and the resulting stress is measured as a function of time. Deformation produces an initial stress that decays with time in the case of viscoelastic materials.

Penetration and indentation tests have long been used to characterize viscoelastic materials such as asphalt, rubber, plastics, and coatings. The basic test consists of pressing an indentor of prescribed geometry against the test surface. Most instruments have an indenting tip, eg, cone, needle, or hemisphere, attached to a short rod that is held vertically. The load is controlled at some constant value, and the time of indentation is specified; the size or depth of the indentation is measured.

Indentation is not usually thought of as a method for measuring viscoelastic properties, but the technique can be useful particularly for measurements of coatings on rigid substrates. Modern instrumentation allows for precise, computer-controlled loading and unloading of force. There are several types of instruments that have modes by which periodic and very short indentation creep measurements can be made while the specimen is being heated, thereby allowing estimates of the changes in modulus with temperature.

The most widely used instrument for measuring the viscoelastic properties of solids is the tensile tester or stress–strain instrument, which extends a sample at constant rate and records the stress. Creep and stress–relaxation can also be measured. Numerous commercial instruments of various sizes and capacities are available. They vary greatly in terms of automation, from manually operated to completely computer controlled. Some have temperature chambers, which allow measurements over a range of temperatures.

Measurement of the propagation of ultrasonic acoustic waves has been found useful for determining the viscoelastic properties of thin films of adhesives. In this method, the specimen is clamped between transmitting and receiving transducers. The change in pulse shape between successive reverberation of the pulse is dependent on the viscoelastic properties of the transmitting material. Modulus values can be calculated.

Dynamic Measurements. Dynamic methods are required for investigating the response of a material to rapid processes, studying fluids, or examining a solid as it passes through a transition region. Such techniques impart cyclic motion to a specimen and measure the resultant response. Dynamic techniques are used to determine storage and loss moduli, G' and G'', respectively, and the loss tangent, tan δ. Some instruments are sensitive enough for the study of liquids and can be used to measure the dynamic viscosity η'. Measurements are made as a function of temperature, time, or frequency, and results can be used to determine transitions and chemical reactions as well as the properties noted above. Dynamic mechanical techniques for solids can be grouped into three main areas: free vibration, resonance-forced vibrations, and nonresonance-forced vibrations.

Fluids. While the previous methods were designed for solid specimens, some can be used for fluids if a solid support or carrier is used. The fluid must be highly viscoelastic for data to register, and absolute modulus values are difficult to determine because of the presence of the support.

Viscoelastic measurements of dilute solutions are useful because they allow studies of the conformational dynamics of long-chain macromolecules in solution without interference from intermolecular interactions. Such measurements require special instrumentation and techniques. Viscoelastic fluids that are more concentrated are characterized with devices that are similar to the rotational viscometers described previously. However, instead of constant rotational motion in one direction, a sinusoidal oscillatory motion is provided. Some instruments allow both viscosity and viscoelastic measurements.

Viscoelasticity can also be determined by a controlled stress rheometer. The shape of a creep curve can show that a fluid is viscoelastic, and the amount of recovery after the stress is removed gives a measure of elasticity.

Time–Temperature Superposition and Master Curves. Because the modulus of a viscoelastic material varies with time and temperature, measurements must be made over wide ranges of these variables for full characterization. Particularly for very long and very short times, the amount of experimental work involved would be prohibitive. Techniques have been developed to determine modulus values and curves at times or temperatures not attainable experimentally. A series of stress–relaxation, indentation, or modulus curves measured at different temperatures can be shifted on a log time axis to give a single modulus time master curve that covers a wide time range. A creep–compliance–time master curve can be constructed in the same fashion, as can a similar curve by using the families of modulus curves at different frequencies, generated by dynamic mechanical instruments with multiplexing capabilities. These curves can be shifted to a single modulus–frequency master curve, because time and temperature have equivalent effects on modulus and other viscoelastic quantities. Master curves can be used to predict creep resistance, embrittlement, and other property changes over time at a given temperature, or the time it takes for the modulus or some other parameter to reach a critical value.

Practical Rheology

A good working rule in the correlation of viscosity measurements with processing conditions is to select and compare viscosities of materials at the shear rate that corresponds to the process in question. However, problems are encountered with high shear processes, because high shear rates can give heating effects, slip, and erroneous viscosity values. These difficulties can be solved by careful extrapolation from lower shear measurements.

Few processes involve a single shear rate or set of mechanical conditions. Typical processes involve low, intermediate, and high shear regions or sections or stages. Thus a complete rheological evaluation of a material to determine its processing characteristics requires consideration of the viscosity of the material over an extended shear rate range.

CLIFFORD K. SCHOFF
PETER KAMARCHIK, JR.
PPG Industries, Inc.

J. R. Van Wazer and co-workers, *Viscosity and Flow Measurement, A Laboratory Handbook of Rheology,* Wiley-Interscience, New York, 1963.

H. A. Barnes, J. F. Hutton, and K. Walters, *An Introduction to Rheology,* Elsevier Applied Science, New York, 1989.

C. W. Macosko, ed., *Rheology: Principles, Measurements, and Applications,* VCH Publishers, Inc., New York, 1994.

M. Rosen, *Polym. Plast. Technol. Eng.* **12**(1), (1979).

RIBOFLAVIN. See VITAMINS.

RISK ASSESSMENT. See HAZARD ANALYSIS AND RISK ASSESSMENT.

RODENTICIDES. See PESTICIDES.

ROOFING MATERIALS

Roofs have evolved from a simple covering to roofing systems designed to perform a number of functions to separate indoor and outdoor environments. A modern roof design normally includes a structure to carry loads, insulation to control heat flow, a barrier to control air and vapor flow, and a roofing element to prevent water penetration.

For low slope commercial roofing, bituminous-based roof coverings are the most common systems in the United States. Asphalt-based materials predominate over coal-tar based materials in these systems. For residential roofing materials, various types of roofing products, including asphalt, wood, and tile, are used for both new construction and reroofing.

Continuous or Jointed Systems

Built-Up Roofing. Built-up roofing (BUR) is a continuous-membrane covering manufactured on-site from alternate layers of bitumen, bitumen-saturated or coated felts, or asphalt-impregnated glass mats and surfacings. These membranes are generally applied with hot bitumens or cold applied bituminous adhesives (qv).

The deck may be nailable, eg, wood or light weight concrete, or not, eg, steel or structural concrete. The felts or mats may be organic (cellulose), or fiber glass. The roof slope ranges from dead level (0–2.1 cm/m), to flat (2.1–12.5 cm/m), to steep (12.5–25 cm/m).

Common BUR systems are installed in different ways: membrane adhered/attached to deck without insulation; insulation adhered/attached to deck with membrane applied to insulation; base sheet adhered/attached to deck, insulation to base sheet, and top membrane to insulation; and membrane adhered/attached to deck and insulation applied over the membrane, the so-called protected-membrane roof. The components must be anchored as protection against wind uplift, slippage, and membrane movement.

Protected-Membrane Roofs. In construction of protected-membrane roofs (PMRs), thermal insulation that is unaffected by water or that can be kept dry in some manner is required. Extruded polystyrene (XEPS) foam insulation boards are commonly employed (see INSULATION, THERMAL). They are placed on top of the waterproofing roof membrane, which is next to the deck. The insulation should not be adhered to the membrane. Ballast at the rate of $\geq$48.8 kg/m^2 (1000 lb/100 ft^2) holds the insulation in place and offers protection from the sun. The insulation joints are open and drainage must be provided. Various other materials, eg, patio blocks and concrete slabs, are also used as surfacings and ballast. The extra weight imposes more exacting requirements on construction.

Asphalt Roofing Components. Asphalt (qv) is a unique building material which occurs both naturally and as a by-product of crude-oil refining. The chemical composition of crude oils differs from source to source; the physical properties of asphalts derived from various crudes also differ. However, these properties can be tailored by further processing to fit the application for which the asphalt will be used. Softening point, ductility, flash point, and viscosity–temperature relationship are only a few of the asphalt properties that are important in the fabrication of roofing products.

Asphalt intended for roofing can be tailored to perform two separate functions. The first is to saturate the organic fiber-based material. This requires that the asphalt be very fluid at processing temperatures so that it can impregnate the base material completely. The second is to coat the saturated roofing to serve as the medium for adhering mineral granule surfacing to the roofing. In the manufacture of roofing on a fiber glass base material, the saturation step is eliminated.

Single-Ply Roofing Membranes. The single-ply roofing membrane category derives its name from the installation technique where a single-ply sheet membrane is applied over the roofing deck/substrate, giving continuous water-tight coverage. These flexible membranes offer light weight, excellent chemical and weather resistance, ease of application and repair, and multiple colors. They can be installed in a number of ways, including the traditional fully adhered, the mechanically fastened, and the ballasted roofing systems. They have been designed to comply with various wind and fire code requirements of the construction industry. Because of the light weight of the adhered and mechanically fastened systems, they sometimes can be installed directly over an existing roof system, thus eliminating the waste that would normally be generated from a roof tear-off. Proper precautions should be taken to remove saturated areas of the old roof before overlaying the new system.

Modified Bitumen Membrane Systems. The third main type of roofing system available is a hybrid between the conventional built-up roofing products and the elastomeric single-membrane materials. Modified bituminous membranes are made by adding a polymer to an asphalt, which then raises the softening point of the mixture, along with fillers and other processing ingredients, to form a waterproofing sheet. These products are primarily reinforced with polyester or fiber glass nonwoven mats. Some products can have both fiber glass and polyester mats to form a more specialized modified bitumen product. Two basic polymers are used to manufacture modified bitumen products: atactic polypropylene (APP) and styrene–butadiene–styrene (SBS) block copolymers.

Steep Roofing Products

Asphalt roofing shingles and related products are classified under three broad groups: shingles, roll roofing, and underlayment. Shingles and roll roofing are outer roof coverings, ie, they are exposed to the weather and are designed to withstand the elements. Underlayments are inner roof coverings that provide the necessary protection beneath the exposed roofing materials. Asphalt shingles and roll roofing contain two basic components: (*1*) a base material made of an organic felt or fiber glass mat which serves as the matrix that supports the other components and gives the product the strength to withstand manufacturing, handling, installation, and service conditions; and (*2*) a specially formulated asphalt coating that provides the long-term weatherability and stability under service temperature extremes.

Shingles. Asphalt shingles are manufactured as strip shingles, laminated (multithickness) shingles, interlocking shingles, and large individual shingles in a variety of weights and colors.

Strip shingles are typically rectangular and may have as many as five cut-outs along the long dimension. Cut-outs separate the shingle tabs, which are exposed to the weather, and give the roof the appearance of being comprised of a larger number of individual units.

Most of the shingles are available with strips or spots of a factory-applied, self-sealing adhesive, which is a thermoplastic material activated by heat from the sun after the shingle is on the roof. Exposure to the sun's heat bonds each shingle securely to the one below for greater wind resistance. This self-sealing action varies, depending on the geographic location, roof slope, season of the year, and solar orientation.

Weather-resistant mineral granules applied to the top surface of strip shingles during the manufacturing process make possible a wide range of colors.

Roll Roofing. Roll roofing is manufactured, packaged, and shipped in rolls. It comes in a wide range of weights and measures. Roll roofing products are produced by applying an asphalt coating to the reinforcing core, and finished with either a smooth or mineral granule surface. Some varieties of mineral-surface roll roofing are manufactured with a granule-free selvage edge that indicates the amount each succeeding course should overlap the preceding course. The amount of overlap determines how much of the material is exposed to the weather and the extent of coverage to the roof surface; ie, whether most of the surface has a single or a double layer of roll roofing. In addition to its use as a roof covering, roll roofing is also important as a flashing material.

Saturated Felts and Underlayments. These products consist of a dry felt which may be impregnated or coated with an asphalt saturant. They are used primarily as underlayment for asphalt shingles, roll roofing, and other types of roofing materials. These types of felts are also useful as sheathing paper. Saturated felts are manufactured in a variety of weights for use as underlayment or heavy-duty underlayment.

Specialty Eaves Flashing Membranes. Specialty eaves flashing membranes are polymer-modified bituminous sheet materials that are used as underlayment to resist leakage of shingle roofs as a result of ice dams. These sheet materials contain an adhesive layer which is exposed by removing its protective sheet. Special eaves flashing membranes can also be useful in warmer climates where a similar backup of water can occur from an accumulation of pine needles and leaves.

Fire Resistance. Asphalt roofing shingles and rolls are tested by independent laboratories in accordance with established fire-resistance standards. The most widely accepted standard for fire resistance in building materials is ASTM E108, Standard Test Methods for Fire Tests of Roof Coverings. If the material meets the standard, the product may carry the testing laboratory's label indicating its class of fire resistance in accordance with the named standard.

Wind Resistance. Asphalt shingles are certified to wind performance test standards on a continuous basis through independent third-party testing laboratories.

Other Shingle Products. Other shingles for steep roofs are wood shingles and shakes, concrete and clay tiles, natural and mineral fiber slates, and various styles of metal products.

Roofing System Performance

Performance (effectiveness) of a roof covering is determined not only by environmental conditions, but also by the individual components that make up the roofing system. Each component should be durable to contribute to the overall roof performance. Roof coverings are subjected to external abuse, eg, foot or other traffic, in addition to weather exposure. The majority of roof leaks are attributed to the defects at penetrations or flashings, so care must be taken to design and install flashings properly.

Weathering. Bitumen hardens progressively owing to ultraviolet attack, oxidation reactions, and, to some extent, loss of plasticizing oils, especially with coal tars. Temperature changes cause expansion and contraction of the roof covering. Other weather factors include moisture, ice, and hail. Wind affects shingles as well as BUR membrane performance. In an accelerated weather study of 15 coating-grade asphalts, a sixfold variation in durability was reported. Performance is correlated to properties and composition. Low asphaltene and high resin contents are desirable. As weathering progresses, the brittle-point temperature, taken by the Fraass procedure, increases. Blending of fractions or base stocks from different crudes improves performance more than antioxidants.

Thermal Effects. Temperature influences the oxidation rate, whereas temperature changes impart a mechanical stress to roof coverings. Daily temperature changes, sometimes quite abrupt, can result in a fatigue mechanism causing tensile failure. Repeated cycles of straining, especially at a high level, diminish the strength of roofing felts to the rupture point. Strain reduction by partial fixing of the membrane to the deck offers the biggest improvement in performance.

Water Effects. Water in its different forms, ie, liquid, vapor, hail, and ice, profoundly influences the performance of roof coverings. Moisture migrations through roof insulation, eg, vapor that accumulates and later liquefies or water that leaks through the roof covering, reduces insulating efficiency and leads to physical deterioration of the roofing material. Roof coverings should be able to withstand 1 1/2 in. diameter hailstones without damage. Icing is a function of the roof slope, roof covering composition, and amount of sun exposure.

Fire and Wind Hazards. Weather resistance of roof coverings is not necessarily correlated to fire and wind resistance. Underwriters' Laboratory (UL) and the Factory Mutual System test and rate fire and wind hazard resistance, and some durability tests. Organic felt or fiber glass mat base shingles are commonly manufactured to meet minimum UL requirements, which, in addition to minimum mass, require wind and fire resistance properties.

Fire Ratings. Above-deck fire hazards are rated by following ASTM E108 for propagation of the flame along the surface of the roof covering and on the penetration of the fire into the deck or structure.

Surface-burning characteristics are measured by either the spread-of-flame test, for noncombustible decks, or the burning brand and intermittant fire tests for combustible decks (see FLAME RETARDANTS).

Wind Testing. UL employs two wind resistance tests, one for shingles and another for BUR assemblies. ASTM D3161 is a standard procedure for measuring the wind resistance of asphalt shingles. In ASTM D3161 and the UL shingle test, the conditioned test deck is placed at a specified slope and exposed to a wind velocity of 97 km/h (60 mph) for two hours or until either failure by tab or shingle lifting is known.

Health, Safety, and Environmental Factors

The materials used by the roofing industry are constantly being scrutinized for health, safety, and environmental requirements. Because the regulations governing the use of these materials change rapidly, it is difficult to indicate where all of the materials stand. Any study of this industry should always involve a review of the latest regulations regarding the use of any products.

RUBEN G. GARCIA
RAY L. CORBIN
Schuller International, Inc.
RICHARD J. GILLENWATER
Carlisle Syntec Systems
JIM COMPTON
Globe Building Materials Company
CLIFF PATENAUDE
Bird, Inc.
BRIAN ANTHONY
The Brewer Company
JIM HUNTER
Gardner Asphalt Corporation

H. O. Laaly, *The Science and Technology of Traditional and Modern Roofing Systems*, Los Angeles, Calif., 1992.

Residential Asphalt Roofing Manual, Asphalt Roofing Manufacturers Association, Rockville, Md., 1993.

R. H. Herbert III, *Roofing: Design, Criteria, Options Selection*, R. S. Means Co., Inc., Kingston, Mass., 1989.

Factory Mutual Approval Guide, Factory Mutual System, Norwood, Mass., published annually.

ROSANILINE. See TRIPHENYLMETHANE AND RELATED DYES.

ROSIN AND ROSIN DERIVATIVES. See RESINS, NATURAL; TERPENOIDS.

RUBBER CHEMICALS

Rubber chemicals are materials that are added in minor amounts to rubber formulations in order to improve their properties and make them commercially useful. Raw rubber polymer has very limited practical applications because of tackiness, flow, and other undesirable features. Rubber chemicals are added to assist processing, promote cross-linking, and provide longevity to the part in service. Vulcanizing adjacent polymer chains together by cross-links prevents flow, increases strength, and provides recovery from deformation.

Accelerators of Vulcanization

Mercaptobenzothiazoles. 2-Mercaptobenzothiazole (MBT) (**1**) is prepared by heating aniline, carbon disulfide, and sulfur in an autoclave at elevated temperature and pressure. MBT can be oxidized to the disulfide (MBTS) (**2**) or it can react with zinc oxide to form the zinc salt (ZMBT) (**3**).

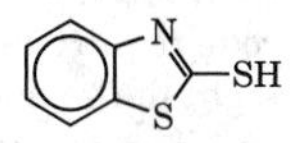

(1)

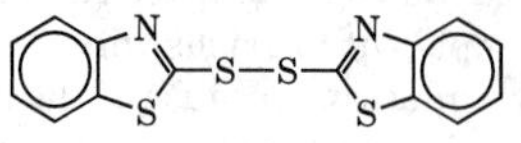

(2)

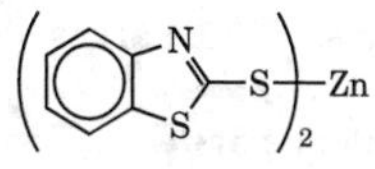

(3)

Sulfenamides. Sulfenamides (**4**) are often produced by oxidizing an equimolar mixture of MBT and an aliphatic amine. One sulfenamide, OTOS (**5**), uses a thiocarbamyl functionality in place of the benzothiazole group.

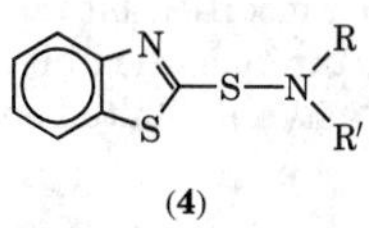

(4)

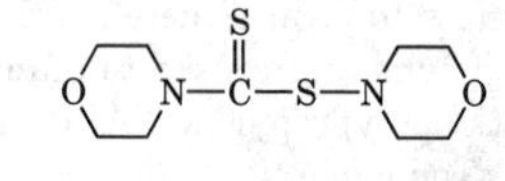

(5)

Sulfenamides do not become active accelerators until the sulfur–nitrogen bond thermally dissociates, providing delayed action before cross-linking occurs. Delayed action is useful in applications which require lengthy processing operations such as joining many layers together to make a tire.

Dithiocarbamates. These compounds, (**6**) and (**7**), are prepared by the reaction of dialkyl amines with carbon disulfide in the presence of sodium hydroxide. For use as rubber accelerators, the sodium salts initially prepared are converted to water-insoluble salts of zinc or other metals. Dithiocarbamates provide the most rapid cross-linking and, together with the related thiurams, are known as ultra-accelerators.

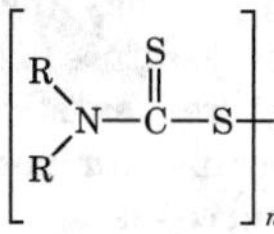

(6)

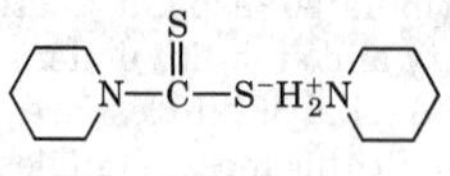

(7)

Thiuram Sulfides. An important class of accelerator, thiurams, (**8**) and (**9**) are produced by the oxidation of sodium dithiocarbamates. The use of these compounds at relatively high levels with little or no elemental sulfur provides articles with improved heat resistance. Ultra-accelerators have been introduced that are based on longer-chain and branched-chain amines; they are less volatile and less toxic. This development also minimizes airborne nitrosamines.

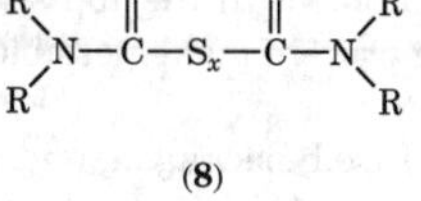

(8)

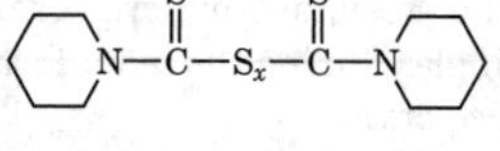

(9)

Guanidines. Guanidines (**10**) are formed by reaction of two moles of an aromatic amine with one mole of cyanogen chloride. When used alone, they show too little activity to be extensively used as primary accelerators.

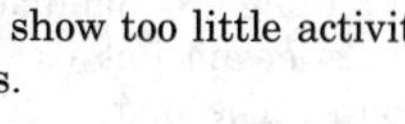

(10)

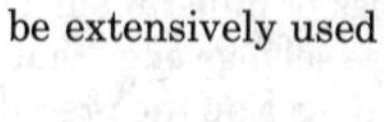

(11)

Thioureas. Thioureas (**11**), typically made from primary amines and carbon disulfide, are often used as accelerators for polychloroprene (Neoprene) or other rubbers that contain chlorine as reactive cross-linking sites.

Xanthates. These compounds (**12**) are relatively fast accelerators which are used at low temperature. Xanthates (qv) are produced by reaction of equimolar amounts of alcohol and carbon disulfide in the presence of caustic.

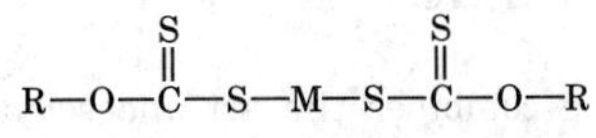

(**12**)

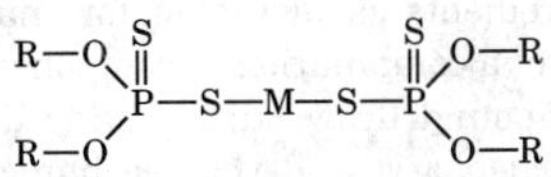

(**13**)

Dithiophosphates. These compounds (**13**) are made by reaction of an alcohol with phosphorus pentasulfide, then neutralization of the dithiophosphoric acid with a metal oxide. Like xanthates, dithiophosphates contain no nitrogen and do not generate nitrosamines during vulcanization.

Aldehyde–Amines. At one time, aldehyde–amines were used in making hard rubber (ebonite), but this use of rubber has largely been replaced by plastics. Aldehyde–amines find use as accelerators for thiadiazole cures of halogenated polymers.

Other Accelerators. Amine isophthalate and thiazolidine thione, which are used as alternatives to thioureas for cross-linking polychloroprene (Neoprene) and other chlorine-containing polymers, are also used as accelerators. Phosphonium salts are used as accelerators for the bisphenol cure of fluorocarbon rubbers.

Within a family of accelerators, materials of low molecular weight usually provide more cross-links at a given dosage. In general, rubber chemicals with low density are preferred, because rubber chemicals are purchased by the kilogram but rubber articles are sold by the item (volume).

Health and Safety Factors. To minimize exposure to potentially toxic materials, powders are generally offered with an antidusting agent applied. Many chemicals, including accelerators, can cause skin irritation in sensitive individuals. Pellets, prills, and polymer-encapsulated master batches to minimize worker exposure to chemicals are offered by several suppliers. Nitrosamine levels in rubber articles for use by infants are regulated in several countries. Manufacturers of these articles have developed nitrosamine-free compounds. However, great care must be taken in manufacturing and packaging to prevent recontamination of the articles with nitrosamines which are ubiquitous in the environment.

Cross-Linking Agents

Sulfur. The most widely used and economical cross-linking agent is sulfur. It provides rubber articles with high strength and excellent resistance to failure when flexed. Improved resistance to heat can be obtained by reducing the amount of sulfur and increasing the amount of accelerator. This change provides a greater proportion of monosulfide cross-links, which have higher thermal stability than di- and polysulfide cross-links. For optimum dispersion of small amounts of sulfur, some suppliers offer sulfur preblended with magnesium carbonate or in polymeric master batch forms.

In natural rubber compounds, insoluble sulfur is used for adhesion to brass-coated wire, a necessary component in steel-belted radial tires. The adhesion of rubber to the brass-plated steel cord during vulcanization improves with high sulfur levels (~3.5%).

Sulfur donors can donate one atom of sulfur from their molecular structure for cross-linking purposes. Monosulfide cross-links provide better thermal stability than the sulfur–sulfur bonds in di- and polysulfide cross-links, which predominate when elemental sulfur is used.

Peroxy Compounds. Peroxides give carbon–carbon cross-links, which provide rubber articles with the maximum resistance to heat, oxygen, and compression set. Resistance to compression set is important in applications such as gaskets and O-rings, which must maintain a seal by recovering their initial dimensions after deformation. Peroxides are beneficial in polymer blends or fully saturated polymers that cannot be cross-linked by other methods. Dialkyl peroxides are the principal type of peroxide used in the rubber industry. Peroxides (qv) are oxidizing materials and should be stored away from other rubber chemicals and kept away from heat sources.

Multifunctional Hydroxy, Mercapto, and Amino Compounds. These are used to cross-link halogenated polymers. Depending on the lability of the halogen, the cross-linking agents can be capped to reduce reactivity or used in combination with accelerators to increase the rate of reaction. Benzoyl capping is common with hydroxy and mercapto compounds; forming the carbamate by reaction with one equivalent of carbon dioxide is used with diamines.

Activators and Retarders

Zinc oxide is a common activator in rubber formulations. It reacts during vulcanization with most accelerators to form the highly active zinc salt. Translucent articles such as crepe soles can use a zinc carboxylate or employ zinc carbonate as a transparent zinc oxide.

Magnesium oxide is a typical acid scavenger for chlorinated rubbers. Retarders were originally arenecarboxylic acids. These acidic materials not only delay the onset of cross-linking but also slow the cross-linking reaction itself. The acidic retarders do not function well in black-filled compounds because of the high pH of furnace blacks. Modern retarders are designed to scavenge the most reactive accelerator precursors and are known as prevulcanization inhibitors. This technique delays the onset of cure without affecting the speed of the main cross-linking reaction.

Overview of Vulcanization Chemistry

Sulfur vulcanization leads to a variety of cross-link structures as shown in Figure 1. All the sulfur does not result in cross-links; some of it remains as pendent accelerator polysulfide groups and internal cyclic polysulfides. These alternative structures do not contribute to load bearing or strength properties and are more prevalent in unaccelerated or weakly accelerated vulcanization systems. Additional heating can also reduce the polysulfide rank of the cross-links. In some elastomers, this leads to a larger number of cross-links. However, in natural rubber or its synthetic polyisoprene equivalent, the overall result is a loss of cross-links, especially at temperatures over 160°C.

Although it has been studied since the 1950s, the exact mechanism of accelerated sulfur vulcanization remains unresolved, including whether it proceeds by a radical or ionic process. Sulfur vulcanization employs a combination of zinc oxide, fatty acid, sulfur, and at least one accelerator. These materials react to form a complex in which the eight-member sulfur ring has been opened. The complex then abstracts a hydrogen on an alpha-carbon position to the double bond in the polymer to form a rubber-bound pendent accelerator polysulfide, a cross-link precursor. A zinc complex and an adjacent polymer chain then react with the pendent polysulfide to form the polysulfide cross-link while regenerating the zinc–accelerator complex. The cycle is continued until all the sulfur is consumed.

Processing Agents

Natural rubber must be reduced in viscosity in order to obtain workable compounds. Many different chemical peptizers have been em-

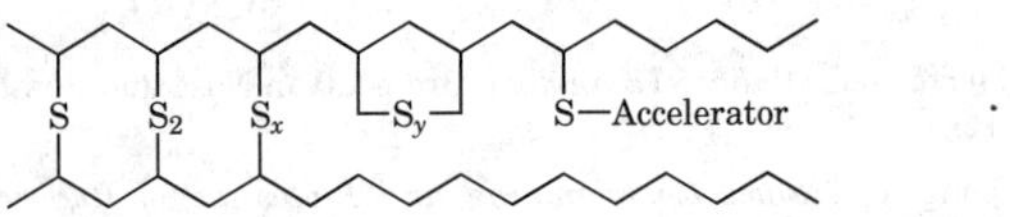

Figure 1. Structures formed during sulfur vulcanization of elastomers.

ployed over the years for this purpose. Dithiobisbenzanilide with an activator and clay diluent is the preferred peptizing agent for natural and synthetic rubbers.

Processing aids assist flow by physical rather than chemical means. Processing aids may be separated into two general categories: external and internal. External processing aids have a polar functionality at one end of the molecule and a long-chain, oleophilic tail. These materials tend to migrate to surfaces where the polar head becomes adsorbed and the oleophilic tail allows the hydrocarbon polymer to slide along the coated surface with less effort. Internal processing aids such as oils or waxes increase melt flow. They reduce the viscosity of the rubber matrix by allowing polymer chains to slide past one another with greater ease.

Homogenizing agents are higher molecular weight versions of external processing aids. They are useful in polymer blends to reduce the domain size of the dispersed polymer phase.

Vulcanized vegetable oils are processing aids that flow under shear but not under heat. They are made by heating unsaturated soybean or rapeseed oil with either sulfur monochloride, S_2Cl_2, or elemental sulfur. Other processing aids are tackifiers or antitack agents. Tackifiers are resinous products which help a multilayered article stick together before it is vulcanized. A variety of additives are used to keep unvulcanized sheets of rubber from sticking together.

Other Rubber Chemicals

Blowing agents are used to create cellular products. Closed-cell sponge is used for minimal water absorption in applications such as wet suits or automotive weather-strip. For closed-cell sponge, three nitrogen-releasing blowing agents are used: azodicarbonamide, *p,p'*-oxybis (benzenesulfonyl hydrazide), and dinitrosopentamethylenetetramine. Open-cell sponge is used for cushioning when the product must deform without much effort. The basic blowing agent for open-cell sponge is sodium bicarbonate.

Adhesion promoters are used to bond rubber to brass-plated steel cord in radial tires or conveyor belting. A gradual decrease in stiffness from steel to rubber is desired. Rubber latex is a liquid, oil-in-water emulsion which is used to make foam or thin-walled rubber articles. The same accelerators and antidegradants used in dry rubber are used in latex, with longer-chain versions preferred for greater oil solubility.

Dispersing agents are surface-active agents which reduce the surface tension and are used when a large quantity of surface area must be treated with surfactant without the mixture entrapping bubbles. Wetting agents more strongly reduce surface tension and are used to adjust the amount of fabric penetration or the flow of the latex mixture. Thickeners are used to prevent the settling of additives and can be organic polyelectrolytes or inorganic smectites. Various antifoams prevent air entrapment and preservatives help avoid growth of microbes that could create slime or odors (see INDUSTRIAL ANTIMICROBIAL AGENTS).

Organofunctional silanes are used to promote polymer-to-filler bonding with clay or silica fillers. Polysulfide silanes are used to reduce rolling resistance of tires. Consumer articles often use colorants (qv), reodorants, or finishing agents. Reodorants can be used to provide a more pleasing aroma for consumer articles. Finishing agents can be applied for a more attractive surface appearance.

ROBERT F. OHM
R. T. Vanderbilt Company, Inc.

M. Morton, ed., *Rubber Technology,* 3rd ed., Van Nostrand Reinhold, New York, 1987.

H. Long, ed., *Basic Compounding and Processing of Rubber,* ACS Rubber Division, Washington, D.C., 1985.

R. F. Ohm, ed., *The Vanderbilt Rubber Handbook,* 13th ed., R. T. Vanderbilt Co., Inc., Norwalk, Conn., 1990.

R. F. Mausser, ed., *The Vanderbilt Latex Handbook,* 3rd ed., R. T. Vanderbilt Co., Inc., Norwalk, Conn., 1987.

RUBBER COMPOUNDING

Rubber compounding combines the art and science of selecting the best combination of elastomer and other ingredients that meet the processing, cost, and performance quality, and cost requirements for rubber goods made and used in commerce. A wide variety of elastomers and ingredients is available for manufacturing rubber products. The choice of a combination of components that will produce a usable rubber compound draws upon many disciplines, including chemistry, physics, rheology, and the mechanics of polymers. Also important is an understanding of how the components will interact. Environmental concerns are creating an interest in utilizing used rubber products in virgin rubber compounds (see RECYCLING, RUBBER).

Compounding Hierarchy. Ingredients for use in a rubber compound, are generally selected in the following order: (*1*) polymer (natural or synthetic rubber), (*2*) vulcanization system (curing agent, accelerator(s), or coagent), (*3*) reinforcing agent and fillers, (*4*) plasticizers or oil, (*5*) antioxidant and antiozonant, (*6*) bonding agent or adhesive, and (*7*) tackifer if needed.

Elastomers Used in Rubber Compounding

The principal component of a rubber compound is the elastomer or blend of elastomers chosen for a specific component application. There are 25–30 different chemical classifications of elastomers; six of these classes represent over 90% of all elastomers used (see ELASTOMERS, SYNTHETIC).

Natural rubber generally comes from southeast Asia. Synthetic rubbers are produced from monomers obtained from the cracking and refining of petroleum (qv). The most common monomers are styrene, butadiene, isobutylene, isoprene, ethylene, propylene, and acrylonitrile. Monomers used for specialty elastomers include acrylics, chlorosulfonated polyethylene, chlorinated polyethylene, epichlorohydrin, ethylene–acrylic, ethylene octene rubber, ethylene–propylene rubber, fluoroelastomers, polynorbornene, polysulfides, silicone, thermoplastic elastomers, urethanes, and ethylene–vinyl acetate.

Elastomers are classified by their heat resistance and resistance to swelling. Properties for evaluating elastomers include hardness, tensile strength, change in tensile strength, elongation and hardness, change in volume, tensile strength, and hardness after exposure to IRM 903 oil. Table 1 summarizes some physical properties of elastomers.

Natural Rubber. To obtain natural rubber (NR), the *Hevea brasiliensis* tree is tapped for its sap. The off-white sap is collected and coagulated. This process produces a high molecular weight substance which is natural rubber (see RUBBER, NATURAL).

There are several systems that define the quality and uniformity of natural rubber. One system of grading natural rubber is based on form and visual observation of color and cleanliness. This is known as the International Natural Rubber Specification. The principal types and grades are as follows. There are five other types of rubber classified by this system and many other grades not listed here.

Type	Grade	Example
1	ribbed smoked sheet	RSS 1 or 3
2	pale and white crepe	1, 2, 3
3	estate brown crepe	1X, 2X

In addition to the solid form of natural rubber it is available as a solid suspended in water, known as latex. Synthetic rubbers are also available in latex form.

Table 1. Physical Properties of Elastomers

Elastomers[a]	Specific gravity	Hardness, Shore A	Tensile strength,[b] MPa[c]
NR	0.93	30–100	27.6
isoprene	0.92	30–100	24.1
SBR	0.94	35–100	20.7
butyl	0.92	30–90	17.2
butadiene	0.91	45–80	17.2
EPDM	0.86	30–90	20.7
chloroprene	1.23	35–95	20.7
nitrile	1.00	30–100	20.7
thiokol	1.25–1.35	20–80	10.3
urethane	1.02–1.25	55–100	55.2
silicone	0.98–1.60	25–90	10.3
CSM	1.12–1.28	40–95	20.7
acrylic	1.09	40–90	13.8
fluorocarbon	1.85	55–95	20.7
epichlorohydrin	1.27–1.36	40–90	17.2
chlorinated PE	1.16–1.25	45–95	24.1
cross-linked PE	0.92	90+	20.7

[a] NR = natural rubber; SBR = styrene–butadiene rubber; EPDM = ethylene–propylene–diene monomer; CSM = chlorosulfonated polyethylene; and PE = polyethylene. [b] Maximum value at room temperatuare. [c] To convert MPa to psi, multiply by 145.

Styrene–Butadiene Rubber. Styrene–butadiene rubber (SBR) is made either as an emulsion or solution product. Solution forms are finding widespread use in original equipment passenger tires. The principal use of SBR is in the tire industry as a major component of all passenger tires and a significant portion of all tire products, except aircraft and large off-the-road tires.

Polybutadiene. The many forms that can result from the polymerization of butadiene, depending on the catalysts used, include high cis, medium cis, low cis, and high vinyl polybutadiene (PBD). The property of polybutadiene of most interest to the rubber compounder is excellent abrasion resistance coupled with excellent resilience.

Ethylene–Propylene Rubber. Ethylene and propylene copolymerize to produce a wide range of elastomeric and thermoplastic products. Often a third monomer such as dicyclopentadiene, hexadiene, or ethylene norbornene is incorporated at 2–12% into the polymer backbone and leads to the designation ethylene–propylene–diene monomer (EPDM) rubber. The third monomer introduces sites of unsaturation that allow vulcanization by conventional sulfur cures.

Butyl and Halobutyl Rubber. Butyl rubber is made by the polymerization of isobutylene; a small amount of isoprene is added to provide sites for curing. It is designated IIR because of these monomers. Halogenation of butyl rubber with bromine or chlorine (halobutyl; HIIR) increases the reaction rate for vulcanization. Butyl polymers are about 8–10 times more resistant to air permeability compared to natural rubber and have excellent resistance to heat and steam or water. Halobutyl rubbers can be blended with natural rubber, polychloroprene, and EPDM to greatly enhance their permeability resistance.

Nitrile Rubber. Nitrile rubbers are made by the emulsion copolymerization of acrylonitrile (9–50%) and butadiene and designated NBR. The ratio of acrylonitrile (ACN) to butadiene has a direct effect on the properties on the nature of the polymers. As the ACN content increases, the oil resistance of the polymer increases. As the butadiene content increases, the low temperature properties of the polymer are improved. Nitrile rubber compounds have good abrasion and water resistance. They can have compression set properties as low as 25% with the selection of a proper cure system.

Specialty Elastomers. Since their invention in the 1930s, the total number of classes of synthetic rubbers has grown to almost 30. The following lists standard acronyms by the International Institute of Synthetic Rubber Producers (IISRP) and the American Society for Testing and Materials (ASTM) for several specialty elastomers.

Elastomer	Designation
acrylic elastomers	ACM, AEM, EEA
chlorinated polyethylene	CM
chloroprene	CR
chlorosulfonated polyethylene	CSM
epichlorohydrin	ECO
fluoroelastomers	FKM, FPM, FFKM
fluorosilicone	FVMQ
polysulfides	T
silicone	MQ
urethanes	AU, EU
vinyl acetate copolymers	EVA

Chloroprene elastomers have a balance of good properties. Essentially an isoprene elastomer with chlorine present in the backbone, the polymer exhibits excellent tensile strength and low hysteresis, much like natural rubber (see ELASTOMERS, SYNTHETIC–POLYCHLOROPRENE).

Chlorinated polyethylene (CPE) has excellent ozone, oil, and heat resistance. In addition chlorinated polyethylene has replaced chloroprene elastomers. CPE has a lower specific gravity than chloroprene compounds and produces compounds that are similar to CR in properties but with lower costs.

Chlorosulfonated polyethylene is made by the simultaneous chlorination and chlorosulfonation of polyethylene in an inert solvent. The resulting polymer is an odorless, colorless chip that is mixed and processed on conventional rubber equipment. The polymer requires compounding with normal fillers to produce useful compounds. Chlorosulfonated polyethylene (CSM) excels in resistance to attack by oxygen, ozone, corrosive chemicals, and oil, and in addition has very good electrical properties (see ELASTOMERS, SYNTHETIC–CHLOROSULFONATED POLYETHYLENE).

Acrylic elastomers possess good oil and heat resistance. They are made by polymerizing monomeric acid esters of ethyl or butyl acrylate and methoxyethyl acrylate or ethoxyethyl acrylate. They can be polymerized in emulsion, suspension, or solution systems. The polyacrylates are inherently tacky and soft. They require compounding with carbon black or mineral fillers for achieving useful properties.

Commercial polyester elastomers include both the homopolymer and the copolymer of epichlorohydrin with ethylene oxide. The very polar chloromethyl groups create basic resistance to oil for these polymers, and they have been extensively used in fuel lines.

The fluoroelastomers are made by modification of Teflon polymers. They are designed to have excellent heat and chemical resistance, but remain elastomeric in nature (see ELASTOMERS, SYNTHETIC–FLUOROCARBON ELASTOMERS).

Thiokol elastomers (polysulfides) possess fairly low tensile and tear properties. However, they have excellent resistance to both aliphatic and aromatic solvents at room and slightly elevated temperatures (see POLYMERS CONTAINING SULFUR, POLYSULFIDES).

Silicone elastomers have a siloxy backbone with methyl, vinyl, and phenyl groups attached. They possess outstanding resistance to heat aging. The Si–O–Si backbone imparts resistance to oxygen, ozone, uv, and to some polar fluids. By fluorinating the silicone polymer molecule it is possible to improve the solvent, fuel, and oil resistance of this already heat-resistant class of elastomers. Fluorosilicones (FVMQ) have excellent low temperature flexibility properties coupled with good oil, fuel, and solvent resistance and excellent aging properties.

Urethane elastomers are prepared by the reaction of an isocyanate molecule with a high molecular weight ester or ether molecule. The result is either an elastomeric rubber form or a liquid prepolymer that can be vulcanized with an amine or a hydroxyl molecule. Urethanes are processed as rubber-like elastomers, cast systems, or thermoplastic elastomers (see URETHANE POLYMERS).

In vinyl acetate–ethylene copolymers, the ratio of ethylene to vinyl acetate (EVA) is varied from 30–60%. The main properties of these elastomers include heat resistance, moderate oil and solvent resistance, low compression set, good weather resistance, high damping, excellent ozone resistance, and they can be easily colored (see VINYL POLYMERS, VINYL ACETATE POLYMERS).

Reclaim Rubber. The process of reclaiming rubber by chemical digestion has been in use since the late 1800s. The principal benefit of using reclaim is its lower cost compared to virgin rubbers. Reclaimed rubber also imparts some desirable improvements in processing; it has much lower nerve than virgin polymers. As a result, compounds containing reclaim have much lower die swell and extrusion rates. It also increases calender rates and improves flow and mold filing (see RECYCLING, RUBBER).

Vulcanization

Vulcanization is a chemical process for improving an elastomer compound's performance. However, in most cases not all of the desired properties reach their optimum levels simultaneously. One of the rubber compounder's key responsibilities is to achieve a balance of the most important property requirements by the proper selection of cure system (chemical) and time–temperature cure cycle (physical).

Measuring Vulcanization. The formation of a three-dimensional structure during vulcanization increases the stiffness (modulus) of the compound. Therefore, following the modulus increase versus cure time provides a continuous picture of the vulcanization process. Oscillating disk rheometers provide a useful method to do this. As vulcanization proceeds, the compound's resistance to rotor movement increases and this resistance is followed as a function of time, thereby generating a continuous profile of cure behavior (Fig. 1).

Effect on Elastomer Physical Properties. Sulfur vulcanization is a complex reaction during which both the number (density) and type (structure) of cross-links are continuously changing as the reaction proceeds. The chemical structures formed at any given time during cure may favor one set of properties such as tear strength while not being optimum for others such as hysteresis and set. A cure system designed to minimize these trade-offs is always a goal of the compounder.

Vulcanizing Agents

Sulfur and Sulfur Donors. Sulfur is the most common vulcanizing agent for the widely used diene-containing elastomers, such as natural rubber, SBR, and polybutadiene. Rubbermakers' sulfur is a rhombic form existing as a cyclic or eight-member ring structure. Sulfur-containing chemicals such as dimorpholinyl disulfide (DTDM) and tetraethylthiuram disulfide (TMTD) are not only effective accelerators, but they can also be used as sulfur donors. As such, they are effective in controlling sulfur cross-link length to form primarily mono- and disulfide cross-links.

Nonsulfur Vulcanizing Agents. Many high performance specialty elastomers do not contain diene moieties in their molecular structure and therefore cannot be sulfur-cured. These elastomers require cross-linking agents capable of reacting with the specific functional group(s) contained by the specific elastomer. Some common nonsulfur curatives include peroxides, difunctional resins, and metal oxides.

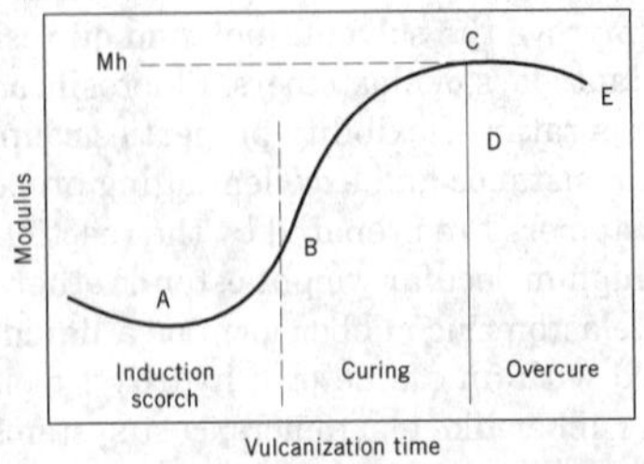

Figure 1. Cure curve from oscillating disk rheometer where A represents scorch safety; B, cure rate; C, state of cure; D, optimum cure time; and E, reversion.

Table 2. Accelerated Sulfur Vulcanization

Accelerator type	Scorch safety	Cure rate
none		very slow
guanidines	moderate	moderate
mercaptobenzothiazoles	moderate	moderate
sulfenamides	long	fast
thiurams	short	very fast
dithiocarbamates	least	very fast

Rubber Vulcanization Chemicals

Accelerators. During sulfur vulcanization of rubber, accelerators serve to control time to onset of vulcanization, rate of vulcanization, and number and type of sulfur cross-links that form. These factors in turn play a significant role in determining the performance properties of the vulcanizate. Choosing the best cure system is a responsibility of the rubber chemist and requires extensive knowledge of each accelerator type and its applicability in each elastomer. Table 2 shows a rule of thumb comparison of the scorch/cure rate attributes for the five most widely used classes of accelerators for processing the high volume diene-based elastomers (see RUBBER CHEMICALS).

Activators. Activators are chemicals that increase the rate of vulcanization by reacting first with the accelerators to form rubber soluble complexes. These complexes then react with the sulfur to achieve vulcanization. The most common activators are combinations of zinc oxide and stearic acid. Other metal oxides have been used for specific purposes. Natural rubber usually contains sufficient levels of fatty acids to solubilize the zinc salt. Synthetic rubbers, especially the solution polymers, do not contain fatty acids and require their addition to the cure system.

The role of activators in the mechanism of vulcanization is as follows. The soluble zinc salt forms a complex with the accelerator and sulfur. This complex then reacts with a diene elastomer to form a rubber–sulfur–accelerator cross-link cursor while also liberating the zinc ion. The final step involves completion of the sulfur cross-link to another rubber diene segment.

Retarders. The purpose of vulcanization retarders is to delay the initial onset of cure in order to guarantee sufficient time to process the unvulcanized rubber. Three main classes of materials are used commercially, including organic acids and anhydrides, cyclohexylthiophthalimide (Santogard PVI or CTP), and a sulfenamide material (Vulkalent E).

Cure System Design

Curatives and accelerators are combined to achieve desired performance properties through cure system design.

There are three generally recognized classifications for sulfur vulcanization: conventional, efficient (EV) cures, and semiefficient (semi-EV) cures. These differ primarily in the type of sulfur cross-links that form, which in turn significantly influences the vulcanizate properties. The term *efficient* refers to the number of sulfur atoms per cross-link.

Conventional cure systems use relatively high levels (2.5 + phr, 100 parts by weight of the rubber) of sulfur combined with lower levels of accelerator(s). These typically provide high initial physical properties, tensile and tear strengths, and good initial fatigue life, but with a greater tendency to lose these properties after heat aging.

In contrast, the EV cure systems employ much lower levels of free sulfur (0.1–1.0 phr) or they use sulfur donors such as TMTD or DTDM combined with higher accelerator levels. The short mono- and disulfide cross-links that form often do not exhibit the excellent physical properties of the conventional systems but they do retain their properties much better after aging.

Semi-EV cures represent a compromise between conventional and EV cures. Although semi-EV cures do yield polysulfide cross-links, they tend to minimize formation of inefficient moieties such as sulfur bridging with itself, accelerator-terminated sulfur linkages, etc. This cleaner usage of sulfur is the reason for their compromise properties between conventional and EV cures.

General Strategy (Statistical Techniques). Achieving the proper balance of properties by cure system design can be a tedious process. For this reason, compounders often use statistically designed experiments to optimize curative levels. By using properly designed experiments, ie, balanced or orthogonal arrays of curative levels, regression analyses can be performed which unequivocally quantify the effect of each ingredient, and combination of ingredients, on the properties of interest.

Effect of Other Compounding Ingredients on Vulcanization

Other ingredients besides the elastomer and the cure system itself influence cure and scorch behavior. Usually the effect of a material on cure is pH-dependent. Ingredients which are basic in nature tend to accelerate the rate of both scorch and cure, whereas acidic materials exhibit the opposite effect.

Antidegradants. Amine-type antioxidants (qv) or antiozonants (qv) such as the phenylenediamines (ppd) can significantly decrease scorch time. This is particularly true in metal oxide curing of polychloroprene or in cases where the ppd had suffered premature degradation prior to cure. In contrast, antioxidants can have an opposite effect when peroxide curing. Because peroxide cross-linking involves a free-radical mechanism, and antioxidants are designed to scavenge free radicals, peroxide efficiency can be compromised by the addition of antioxidants.

Fillers. Materials used as fillers (qv) in rubber can also be classified as acidic, basic, or neutral. Furnace blacks are somewhat basic. As such, they can have an activating effect on sulfur cure rates. Furthermore, carbon blacks have been found to promote formation of mono/disulfide cross-links thereby helping minimize reversion and enhance aging properties. Nonblack fillers such as the precipitated silicas can reduce both rate and state of cure.

Process Oils, Plasticizers. Petroleum-based rubber process oils generally contain a mixture of paraffinic, naphthenic, and aromatic components. These oils vary in composition from grade to grade, but most contain some unsaturated moieties and this unsaturation can compete with the polymer for curatives. Plasticizers (qv) can vary in composition such as the ester types, phosphates, amides, etc. The effect of any one of these on curing is usually pH-dependent.

Health and Safety Factors

Environmental Issues. The rubber industry has responded to concerns about the environment by developing new product forms for accelerators and other chemicals which improve industrial hygiene and minimize worker exposure to these materials by eliminating dust exposure and improving handling ease.

Nitrosamines. Findings that secondary amines, so common in rubber accelerators, can react with NO_x species to form the suspected human carcinogens, nitrosamines, have prompted active programs to develop alternative accelerators. Neither primary nor tertiary amines form stable nitrosamines, and they are generally considered to be safe materials.

The NO_x nitrosating agents present in the atmosphere are often caused by air pollution. High surface area fillers such as carbon black absorb NO_x and liberate it during the vulcanization process. Consequently, unacceptably high levels of nitrosamines have been detected near curing presses, near continuous curing lines, and in tire storage warehouses. Nitrosamines from vulcanizing agents can be controlled or eliminated by using only primary amine-based accelerators or accelerators which form stable fragments during cure. Another way to minimize nitrosamine exposure may be to use scavengers, which are reported to function by reacting rapidly with nitrosamines to render them harmless, but the relative effectiveness of each scavenger remains to be better quantified.

Fillers for Rubber

The rubber compound usually requires an inert inorganic filler and small particle size carbon particle for reinforcement. The rubber polymers vary in inherent tensile strength from very high in the case of natural rubber to almost nonexistent for some synthetic polymers. The fillers most commonly used for rubber compounds include carbon black, clay, calcium carbonate, silica, talc (qv), and several other inorganic fillers.

Carbon Black. This is the principal reinforcing filler used in rubber. Carbon black is made by three processes: the furnace process, the thermal process, and the channel process. Over 97% of black is made by the furnace process. The furnace process involves injecting low end fraction of crude oil into a heated chamber. The temperature, shape of the injectors of the oil, rate of injection, and other factors are controlled to produce black fillers of different particle size and structure. There are ~30 common grades of carbon black used in the rubber industry.

Silica. The main uses of silica are in the treads of off-the-road tires for improved chunking and tear resistance and as a component of the bonding system for brass and zinc-plated steel cord. In addition the body plies of steel radial truck tires, hoses and belts, and footwear use significant volumes of silica as a reinforcing filler.

Clay. Clays are a principal filler for many rubber goods. The natural air-floated and water washed clays are considered semireinforcing fillers for most rubber compounds. Silane-modified clays have found broad acceptance in a wide variety of rubber goods. The silane serves as a bridge between the organic rubber molecule and the inorganic filler particle. The result is a filler with higher reinforcing properties than regular clay (see CLAYS).

Processing Agents

To improve processing and to plasticize the rubber compound, numerous processing agents have been used over the years. Petroleum plasticizers are the most universally used plasticizers for all rubber compounds. They improve flow and processing characteristics and also reduce the cost of the final compound. Ester plasticizers are used mainly in very polar elastomers to improve low or high temperature performance or impart particular oil or solvent resistance to a compound (see PLASTICIZERS). Resins and tars are added to impart tack, soften the compound, improve flow, and in some cases improve filler wetting out. Resinous substances are also used as processing agents for homogenizing elastomer blends.

Stearic acid is used in many compounds as a cure activator; in addition the stearic acid acts as a processing agent to improve mill and processing equipment release properties. Thus stearic acid may be the most widely used processing agent in the rubber industry.

Processing agents can be prepared by blending fatty acids, zinc, and calcium soaps into waxes, such as myricyl palmitate, to overcome solubility limitations of individual products. The result is a blend of organic fatty acids, usually in paste form which serves to improve processing.

Antidegradants

Good aging properties of rubber compounds are essential for providing acceptable service life of rubber products. The type of elastomer used is the principal factor in considering aging properties. In general, the more saturated the main chain of elastomer, the better are the aging properties. Unsaturated sites are susceptible to oxidation. Antioxidants and antiozonants extend the service life of products.

Factors Affecting Aging of Rubber. Addition of only 1–2% combined oxygen is sufficient to cause significant deterioration of properties for most general-purpose elastomers. Oxidation proceeds by free-radical mechanisms and leads to chain scission and cross-linking.

Heat accelerates oxidation. Because oxidation is a chemical reaction, an increase of 10°C in temperature almost doubles the rate of oxidation.

Flex cracking is a common mode of fatigue failure in rubber compounds. Flex cracking involves not only a mechanical fatigue but also an oxidation that is accelerated by heat generated during flexing.

Although ozone concentration in the atmosphere is only a few pphm, it reacts with rubber double bonds rapidly and causes cracking of rubber products, especially under stress (stretching and bending). Antiozonants (qv) also protect rubber surfaces by the formation of a protection layer on the surface of rubber by reaction of the antiozonant with ozone.

Ultraviolet (uv) light promotes free-radical oxidation at the rubber surface which produces discoloration and a brittle film of oxidized rubber.

Low sulfur stocks and EV sulfur-accelerated systems have better aging resistance. Normally, the oxidation rate increases with the amount of sulfur used in the cure. Saturated sulfides are more inert to oxidation than allylic sulfides.

Transition-metal ions, when present even in small amounts, catalyze rubber oxidative reactions by affecting the breakdown of peroxides in such a way as to accelerate further attack by oxygen. Although some antioxidants (qv) are active against catalyzed oxidation of rubber, in general the standard antioxidants do not give protection against the metal ions.

Selection of Proper Antidegradant. Because the various antioxidants function by different mechanisms, an antioxidant under one condition may become an oxidation promoter in a different condition. Therefore, an antioxidant should be carefully selected depending on service requirements. Most antioxidants are either amines, phenols, or phosphates e.g., hindered phenols, hindered bisphenols, hindered thiobisphenols, hydroquinones, phosphites, diphenylamines, naphthylamines, quinolines, carbonylamines condensation products, and *para*-phenylene, diamines.

Antioxidant Types. Commercially available antioxidants may be divided into three general classes: secondary amines, phenolics, and phosphites. In general, the amines are more active than the phenolics which are in turn more active than the phosphites. Amine antioxidants, however, often cause staining problems and are therefore used mainly in black stocks. The phenolics and phosphites are relatively nonstaining and are normally used in light-colored rubbers.

Tire Compounding

The pneumatic rubber tire is a torodial air container that supports a vehicle and translates the forces of driving and braking. The forces that tires are subjected to are both complex and severe. Within a single component the tire rubber compound undergoes compression, extension, and torque forces simultaneously, or nearly simultaneously, in a rather small volume, such as a tread block segment in aggressive all-weather designs.

The pneumatic tire consists of two basic areas: the tread ares which is responsible for ground contact, and the casing which is responsible for supporting load and transmitting power to the tread area.

The role of the rubber compounds which are used in these basic components is threefold: (*1*) To provide the contact area between the vehicle and the surface; (*2*) to provide the cohesive material that holds the tire together such that it acts as an integral unit; and (*3*) to provide protection for the ultimate strength-bearing components, ie, the textiles, steel beads, and steel breakers in steel belted radial tires.

Component Parts. Desired properties in the components of a radial steel belted passenger care tire (PCT) are as follows. The *tread* is designed and compounded for abrasion resistance, traction, low rolling resistance, and protection of the carcass. The *steel belt,* which provides strength and protection for the ply or plies, is encased in a compound that must possess adhesion to the steel, which provides stress transfer from the very rigid steel to the many times more flexible tread, sidewall, and textile carcass components. The *carcass ply/plies coat compound* functions are basically the same as the steel breaker compound. Normally in the steel belted PCT the ply is textile cord of polyester or rayon fabrics which are soft and flexible (see TIRE CORD). The *sidewall compound* is compounded to protect the ply and must possess resistance to weathering, ozone, abrasion, and tearing while providing excellent flex fatigue resistance. The *apex* (often referred to as bead filler) compound must be formulated for excellent dynamic stiffness to facilitate stress distribution and provide good car handling properties. The *chafer/rim strip compound* protects the plies from rim abrasion and seals the tire to the rim. Table 3 summarizes desired properties of three tire components in generalized terms.

Table 3. Desirable Properties of Tire Components

Property[a]	Tread	Component	Sidewall
wear resistance	+	na	0
traction	+	na	na
rolling resistance	−	−	−
cut resistance	+	na	0
adhesion (to other components)	+	+	0
cut growth resistance	+	+	+
resistance	+	na	+
hysteresis (tan δ)	−	−	−
resilience (rebound)	+	+	+
heat resistance	+	+	+
flex fatigue	+	+	+
tear resistance	+	+	+
hardness	0	+	0
modules (stretch 100% elongation)	0	+	0
modulus (compression)			
static	0	+	0
dynamic	+	+	0
processing properties	+	+	+
Mooney viscosity	−	−	−
rheometer (rate/state of vulcanization)	v	v	v
swell	−	−	−
tack	+	+	+

[a] 0 = nominal, + = maximized, − = minimized; na = not applicable, and v = varied.

Materials

Tire compounds contain the following generalized ingredients in the approximate proportions noted. Industry practice is to formulate starting with 100 parts by weight of the rubber (phr).

Ingredient	Parts/hundred rubber (phr)
rubber (elastomers)	100
reinforcing fillers	30–90
softeners extenders	5–40
processing aids	0.1–15
protection agents	3–7
vulcanization agents	3–10

There may be as many as 12 different compounds in a single tire. Each compound uses 8–12 ingredients. There are approximately 750 different chemicals or materials used in the tire compounding field. Thus, the potential combinations/permutations are extremely high. In addition, interactions between materials are encountered and must be controlled.

Interactions are not limited to the 8–12 ingredients within a single formula, but may also occur between the compounds of the different components, through the processes of mixing, extruding, calendering (coating of steel and textiles), in prebuilding compilation of components for the tire building equipment, in storage of green (uncured) tires, in vulcanization of the tire, and in the warehousing and during operation of the tire on the vehicle.

Tire Elastomers. The rubbers used in various tire components are natural rubber (NR), styrene–butadiene (SBR), butadiene (BR), halogenated butyl (HIIR), and ethylene–propylene–diene monomer (EPDM). It is common to use more than one type of rubber within a given compound. In passenger car tires as many as four different polymers may be used for the tread compound totaling 100 phr. The use of different materials within a single grouping is commonly encountered, ie, two or more different grades of carbon black, oils, etc, may be used within one compound, as well as the use of two or more rubbers making up the 100 parts total.

Reinforcement for Tires. Carbon black is by far the most heavily used reinforcing filler for tire compounds. In tires, carbon black is important because of the high flex fatigue and tear strength requirements of this product.

Silica as a tire reinforcing filler was first introduced through the earthmover tire line, owing to its ability to improve tear resistance and lower tire running temperatures. These properties are of utmost importance in off-the-road tires which run at high operating temperatures and encounter rocks that may cut and tear out large chunks under high traction/torque conditions.

With demand for improved rolling resistance, silica usage has expanded into other tire lines. A silane coupling agent has been introduced that forms a chemical bond between the silica and the rubber, resulting in improved treadwear.

Clay, which is generally considered a mild reinforcing filler, is used sparingly in tires. It is most often used in white sidewalls or in low performance tires. Reinforcement and stiffness of a compound can also be achieved with the use of reactive resins. Resins consisting of two-component systems of resorcinol or resorcinol condensation products and a methylene donor are the most popular in tires.

Softeners, Extenders, and Plasticizers. In tires the most popular softeners/extenders/plasticizers are petroleum oils. They are utilized in considerable quantities for controlling green compound viscosity and vulcanized compound hardness. The term *extender* is used to denote their use to allow for extended use of reinforcing fillers. These materials also act as processing aids. The term *plasticizer* is used to denote their ability to act as an internal lubricant for processing purposes. *Bonding agents* are generally used only in wire cable coat compounds. They are basically organic complexes of cobalt and cobalt–boron.

Compound Formulating

Tire compounders generally use two basic approaches to formulate for achieving desired tire performance: the use of known relationships of a given compound to a specific performance parameter, and establishing the relationships of physical properties to a given performance parameter. The two main categories of measurement criteria are quasi static and dynamic mechanical properties.

Latex

The compounding technique for latex differs from that of dry rubber and is fundamentally simpler. Synthetic latices are mainly prepared by polymerizing low molecular weight monomers in a free-radical emulsion polymerization system. Artificial latices are water dispersions of reclaimed rubber, butyl rubber, *cis*-polyisoprene, *cis*-polybutadiene, or ethylene–propylene terpolymers. Compounding of latex is done in the liquid state. The use of latices avoids the hazards and additional expense associated with rubber solutions requiring toxic and inflammable solvents.

For dry-rubber compounding, the latex is coagulated, dried, and used in a solid state after mechanically inducing plasticity. Latex compounding, on the other hand, is done in the liquid state and the compounded latex used directly to form articles without being first converted to a separate solid.

The general principles of latex compounding are similar to those of dry rubber. The absence of mastication in the processing of latex saves time and power, and the resulting vulcanizate can be made reasonably resistant to aging. A disadvantage of latex is the difficulty in obtaining any measure of reinforcement through use of small particle-size inorganic fillers.

In order to obtain a homogenous and stable latex compound, it is necessary that insoluble additives be reduced in particle size to an optimum of ca 5 μm and dispersed or emulsified in water. The particle size of typical natural rubber latex ranges from slightly higher than 1 μm to as small as 20 nm, and can be destabilized by mechanical or chemical action. Dispersions to be added to latex must have good storage stability and be compatible with the latex; the pH of each should be similar to that of the latex. Latex particles are negatively charged. Because all the charges are in the same category in varying magnitude, the charges repel particles from each other and prevent coagulation. Therefore, materials in water that produce Zn^{2+}, Mg^{2+}, or Ca^{2+} ions must be added with care to maintain latex compound stability. Mixing of latex compounds is accomplished by stirring ingredients into the latex in the form of water solutions, dispersions, or emulsions. Although the rubber softeners needed to process dry rubber are not necessary for latex, use of emulsified softeners or polymeric plasticizers in natural or synthetic latex compounds provides lower modulus in the finished products.

Latex dipping compounds have generally higher tensile strength than those prepared from dry rubber dissolved in mineral solvent; the latter lose some strength during mastication and milling of the rubber. Without addition of reinforcing filler, latex rubber compounds can exhibit tensile strength values in excess of 35 MPa (5000 psi).

Sulfur is the almost universal, primary vulcanizing agent for all elastomers, whether in latex or dry form. Less sulfur is normally required for latex than for dry rubber compounds. The addition of zinc oxide to the latex compound increases the rate of vulcanization and generally improves the physical and aging properties. Fillers, eg, clays and whiting, are used to reduce cost or provide special properties. They are also used to increase viscosity for latex compound spreading suitability. The more active accelerators, particularly dithiocarbamates, are used more widely in latex than dry rubber compounding, because scorching during processing is not a problem. The use of fatty acids in latex compounds follows the same principles as in dry rubber; and although natural rubber latex contains some fatty acids or their soaps, it is sometimes necessary to add more. Antioxidants are used in latex compounding as in dry rubber. Reclaimed rubber, which is widely used in dry rubber, has little use in latex compounding.

Synthetic. The main types of elastomeric polymers commercially available in latex form from emulsion polymerization are butadiene–styrene, butadiene–acrylonitrile, and chloroprene (neoprene). There are also a number of specialty latices that contain polymers that are basically variations of the above polymers, eg, those to which a third monomer has been added to provide a polymer that performs a specific function. The most important of these are products that contain either a basic, eg, vinylpyridine, or an acidic monomer, eg, methacrylic acid. These latices are specifically designed for tire cord solutioning, papercoating, and carpet back-sizing.

The basic constituents of all commercial emulsion polymerization recipes are monomers, emulsifiers, and polymerization initiators. Other common components are modifiers, inorganic salts and free alkali, and shortstops.

Most synthetic latices contain 5–10 wt % of nonelastomeric components, of which more than half is an emulsifier or mixture of emulsifiers. One reason for this relatively high emulsifier concentration as compared with natural latex is that emulsifier micelles containing solubilized monomer play a principal role in the polymerization process. Secondly, a considerable fraction of the surface of the polymer particles must be covered by adsorbed soap or equivalent stabilizer to prevent flocculation of the latex during manufacture or subsequent use.

Neutral or alkaline salts, are often present in synthetic latices in quantities of $\sim\leq 1\%$, based on the weight of the rubber. During emulsion polymerization the salts help control viscosity of the latex and, in the case of alkaline salts, the pH of the system.

Styrene–Butadiene and Related Latices. The SBR latices can be classified into hot and cold types, as determined by polymerization temperatures, and subdivided into low and medium solids and high solids. There is a wide variety of high styrene resin latices for many diverse applications from carpet backing to paint and paper coating.

Butadiene–Acrylonitrile Latices. Nitrile latices are copolymers of butadiene and acrylonitrile in which those copolymerized monomers are the main constituents (see ELASTOMERS, SYNTHETIC–NITRILE RUBBER). Nitrile latices are used in a wide variety of applications, including production of dipped nitrile rubber products. Films deposited from compounded nitrile latices can be vulcanized with sulfur and accelerators, assisted by relatively high levels (ca 4.0–5.0 parts/100 DRC) of zinc oxide.

Neoprenes. Of the synthetic latices, a type that can be processed similarly to natural rubber latex and is adaptable to dipped product manufacture, is neoprene (polychloroprene). As in dry rubber compounding, neoprenes require sulfur and an accelerator for vulcanization, but zinc oxide can also be used to accelerate the cure and to function as an acid acceptor. Organic accelerators are used to improve the physical properties of neoprene latex films, but their effect is not generally as great as when used with other polymers. Use of a good antioxidant is recommended for almost all neoprene compounds; where color is not of importance, a staining antioxidant can be used.

Vinylpyridines. The vinylpyridine latices were developed specifically for adhering rubber stocks to fibers, particularly nylon. In general, the polymers are high diene types containing 10–15 wt % copolymerized 2-vinylpyridine and an approximately equal amount of styrene.

Carboxylated Latices. One advance in latex technology has been the introduction of latices in which the polymer phase contains functional groups. The functional groups are derived from the use of unsaturated monomers containing carboxy groups in the polymerization system. Carboxylated styrene–butadiene latices have been used increasingly in carpet-backing applications because of their self-curing feature, ie, the use of sulfur and accelerators is obviated, resulting in lower cost but more particularly very low odor compounds.

Test Methods

Because of the long-range elasticity of soft rubber vulcanizates and the various special conditions under which they are degraded in use, special test methods have been developed which in many respects are unlike those used for wood, metals, and hard plastics. The American Society for Testing Materials (ASTM) through its Committee D11 on Rubber and Rubber-like Materials is constantly developing and promulgating new and improved methods for testing rubber and rubber products.

WILLIAM KLINGENSMITH
FRANK JENKINS
Akron Consulting Company
WALTER KLAMP
Consultant
MICHAEL FATH
Consultant
KRIS BARANWAL
ARDL
ROBERT OHM
R. T. Vanderbilt Company
RAY RUSSEL FELL
Consultant
BRENDON RODGERS
Consultant

Worldwide Rubber Statistics, International Institute of Synthetic Rubber Producers, Inc., 1994.

A. K. Bhowmick, M. M. Hall, and H. A. Benarey, *Rubber Products Manufacturing Technology,* Marcel Dekker, Inc., New York, 1994.

Glossary of Terms Relating to Rubber and Rubber Technology, American Society for Testing and Materials, Philadelphia, Pa., 1972.

J. E. Mark, B. Erman, and F. R. Eirich, *Science and Technology of Rubber,* 2nd ed., Academic Press, Inc., New York, 1994.

RUBBER, NATURAL

Virtually all the natural rubber is obtained from the latex of the *Hevea brasiliensis* tree.

Rubber Production

Latex and Its Collection. The latex from the *Hevea brasiliensis* tree is a colloidal dispersion consisting of nonrubber substances and rubber particles in an aqueous serum phase. The rubber hydrocarbon constitutes 30–45% of the whole latex; the nonrubber substances account for 3–5%. Rubber particles vary in size from 0.15–3 μm and contain 90–95% natural rubber, ie, *cis*-1,4-polyisoprene, with a molecular weight distribution of 10^5–10^7 g/mol.

Rubber latex is formed and stored in rings of latex vessels found in the soft bark region of the tree, which lies between the inner cambium tissue and the outer hard bark layers. The physiological role of latex in the rubber tree is undetermined; however, the mechanism of production and methods of tapping are well understood. Latex is collected from the rubber tree by tapping once every two or three days. A tree is tapped by carefully cutting into the trunk with a special knife that removes a thin slice of bark and severs the latex vessels without affecting the normal sap circulatory system. The cut is made at an angle of 25–30° halfway around the circumference of the trunk, and at the lowest point a short metal spout is inserted, from which the latex flows into cups and is collected four to five hours later. Unless the latex is to be processed immediately, a small amount of preservative, eg, ammonia, is added to prevent premature coagulation before the latex is brought to a factory or processing center. Rubber collected in this manner is known as field latex. Latex that coagulates naturally is known as field coagulum.

Grade Production of Dry Natural Rubber

Field latex and field coagulum are the source materials for all varieties and grades of dry natural rubber that include the conventional International grades as well as the Technically Specified Rubbers (TSR).

Conventional Grades. Classification of traditional grades such as ribbed smoked sheet (RSS) is based on the source of raw materials and processing methods. Visual grading and comparison with international samples depend on factors such as color, cleanliness, and uniformity of appearance. Of all the 35 traditional grades, only ribbed smoked sheet and pale crepes are traded in any substantial quantity.

In large-scale operations, field latex is first preserved with either ammonia, sodium sulfite, or formalin and then diluted to ~15% dry rubber content (DRC) and strained into coagulating tanks. Formic acid (5 wt %) is added and the acidified latex left to stand for several hours in tanks to coagulate. After coagulation the spongy coagulum is in the form of a continuous sheet. Water is squeezed from the coagulum by passing it through a series of rollers to produce a thin sheet. Grooves on the last pair of rollers introduce the characteristic crisscross rib markings on the sheet to aid drying by increasing the rubber surface area. The resulting ribbed smoked sheets (RSS) are pressed into 112-kg bales.

Pale crepes are divided into one of three classes, depending on the manufacturing process: fractionated and bleached rubber (FB); unfractionated, but bleached rubber (UFB); or yellow fraction rubber (YF). Several grades of brown and blanket crepe are produced, and the source materials are field coagulum, unsmoked sheets, and smoked sheet cuttings.

Technically Specified Grades. The Standard Malaysian Rubber (SMR) scheme was introduced in 1965. It was so successful that it led to the adoption of an international scheme for Technically Specified Rubber (TSR). These schemes led to fewer grades, all with guaranteed specifications relating to quality, packed in small, polyethylene-wrapped 33.3-kg bales for easy transport, storage, and factory handling. The SMR scheme was revised in 1991.

The conversion of liquid latex to the SMR latex grades is a large-scale factory operation, involving the use of heavy-duty processing equipment. The incoming latex is bulked and blended and formic acid is added to reduce the pH to 5.0–5.2 and to produce rapid coagulation. The coagulum is covered with water and allowed to mature for about 16 h. Dewatering, cracking, and comminution of the coagulum is obtained by hammer mills, granulators, shredders, and pelletizers and produces small, uniformly sized crumbs. The final wet crumbs are dried in deep bed or conveyor driers at 100–120°C for four to five hours, followed by cooling, baling, and wrapping. Field coagulum is also converted to crumb form by similar size-reducing processes, but first it is important to wash the field coagulum thoroughly to remove bark and other contamination.

Properties of Natural Rubber

Crystallization. Raw natural rubber may freeze or crystallize during transit or prolonged storage, particularly at subzero temperatures. The rubber then becomes hard, inelastic, and usually much paler in color. This phenomenon is reversible and must be differentiated from storage hardening. The rate of crystallization is temperature-dependent and is most rapid at −26°C. Once at this temperature, natural rubber attains its maximum crystallinity within hours, and this maximum is no more than 30% of the total rubber.

Frozen rubber is normally thawed by storing in a hot room at 40–50°C. Because rubber is a relatively poor thermal conductor, melting is governed by the slow conduction of heat through the bale. Palletized frozen rubber can take a considerable time to thaw.

Storage Hardening. Storage hardening is a slow, irreversible increase in the Mooney viscosity of natural rubber that occurs during storage and transport. Hardening is thought to result from the interaction between the nonisoprenic groups, such as aldehydes, present on the rubber molecule to form cross-links; as the rubber hardens, the extent of long-chain branching and gel increases. Storage hardening is also accelerated by high temperature and low humidity.

Storage hardening can be effectively inhibited by the addition of a reagent such as hydroxylamine, which produces an oxime via a condensation reaction and thus blocks the cross-linking process. For viscosity-stabilized grades (CV) produced in this way, the viscosity increase during transit and storage is negligible.

Cure Characteristics. Methods of natural rubber production and raw material properties vary from factory to factory and area to area. Consequently, the cure characteristics of natural rubber can vary, even within a particular grade. Factors such as maturation, method and pH of coagulation, preservatives, dry rubber content and viscosity-stabilizing agents, eg, hydroxylamine-neutral sulfate, influence the cure characteristics of natural rubber.

Plasticity Retention Index. The oxidation behavior of natural rubber may affect both the processing characteristics and final vulcanizate performance, and the plasticity retention index (PRI) test can be used to give an indication of both. Natural antioxidants present in natural rubber give some protection. A measure of the efficacy of protection is given by PRI: $\text{PRI\%} = P_{30}/P_0 \times 100$, where P_0 is the initial Wallace plasticity and P_{30} is the plasticity after aging in an oven at 140°C for 30 min.

Chemistry and Technology

Natural rubber as obtained from *Hevea brasiliensis* is *cis*-1,4-polyisoprene with small amounts of nonrubber produced by the tree. Natural rubber is relatively unstable compared to the more modern almost fully saturated elastomers, such as EPDM rubber. In terms of physical properties, the strain crystallizing nature of natural rubber gives it high tensile and tear strengths, and it exhibits both low heat buildup and low rolling resistance. These latter two features make it especially attractive for tire manufacture. However, unless modified, it shows extensive swelling in oils, is relatively permeable to gases, and is not generally suitable for damping applications. In the majority of applications, natural rubber is used in blends and, until fairly recently, remarkably little was known about the distribution of cross-links in such blends. Modern techniques have been adapted to look into this subject, with a consequent improvement in properties brought about by achieving a better balance of cross-links in the blend.

Vulcanization. Natural rubber is generally vulcanized with a sulfur-based system, the mechanism of which is extremely complex. Basically, a high sulfur, low accelerator (2.5 parts phr of sulfur and 0.5 part of sulfenamide; phr = 100 parts by weight of the rubber) system results in a mainly polysulfidic cross-linked rubber that has a high tensile strength and exhibits good dynamic performance, but has relatively poor aging characteristics, even with an antidegradant present. On the other hand, an efficient vulcanizing (EV) system based on low sulfur and high accelerator levels leads to a mainly monosulfidic cross-linked network with consequent improvement in compression set and aging characteristics, but less good tensile and tear properties and inferior dynamic performance. Between them there are a multitude of semi-EV systems that fall between the high sulfur–low accelerator and low sulfur–high accelerator recipes and give a compromise in properties.

Protective Systems. The principal category of protective agents comprises the staining antidegradants, which protect against oxygen, ozone, and flex-cracking (see ANTIOXIDANTS; ANTIOZONANTS; AMINES, AROMATIC; RUBBER CHEMICALS). By far the most effective are the *p*-phenylenediamines. The second category is staining antioxidants with antiflex cracking capability only; examples are diphenylamine–acetone condensates and naphthylamine derivatives. They are good antiflex cracking agents, but not as effective as the *p*-phenylenediamines. The third type is made up of the low staining amine antioxidants. These impart good protection against heat and oxygen. Among nonstaining antioxidants, the least staining and discoloration is given by the phenolic antioxidants. These give reasonably good antioxidant protection, but some are susceptible to pinking on exposure to light. For nonstaining protection against ozone attack under static conditions, waxes (qv) are used. The type of wax is determined by the service condition involved and the temperature. Traces of certain metals such as copper, manganese, and iron act as catalysts in the oxidation of natural rubber vulcanizates. Many amines and some phenolics are effective inhibitors of metal-catalyzed oxidation.

Modified Forms of Natural Rubber. Poly(methyl methacrylate)-grafted natural rubber, superior processing natural rubber, and epoxidized natural rubber are some more common modified forms of natural rubber. Poly(methyl methacrylate)-grafted rubber (MG rubber), is made by the polymerization of vinyl monomers with the rubber to produce a material of increased modulus or hardness according to the percentage of grafting. The overall result is a modified natural rubber, which is a hard flexible material combining high rigidity with complete recovery from impact. The second modified form of natural rubber, superior processing rubber, comprises intimate mixtures of cross-linked and uncross-linked materials prepared by blending vulcanized latex with diluted field latex in proportion to percentage of vulcanized phase required. Superior processing (SP) rubbers provide significant manufacturing advantages. Their two-phase structure increases the stiffness and Mooney viscosity of rubber mixes and at the same time improves flow behavior. The result is faster production rates and greater throughput. The other advantage of these SP rubbers is that, unlike some other processing aids, these do not affect the final vulcanizate properties, being modified natural rubber themselves.

The third and newest modified natural rubber available is epoxidized natural rubber (ENR). It is also a strain crystallizing rubber and therefore retains the high tensile strength of natural rubber; however, in other respects they have very little in common. The epoxidation renders a much higher damping rubber, a much-improved resistance to oil swelling, and much-reduced air permeability.

Thermoplastic Natural Rubber. Natural rubber thermoplastic materials are based on blends of natural rubber and polypropylene. Basically, there are two types: those with a low natural rubber content, which are really only rubber toughened forms of polypropylene, and the softer grades with high natural rubber content, which are truly classed as thermoplastic elastomers. The latter have high strength and good recovery properties, better aging than conventional natural rubber vulcanizates, and excellent ozone resistance.

The more important grades of thermoplastic natural rubber, which fall into the olefinic class of thermoplastic elastomers, are prepared with the natural rubber phase partially cross-linked during blending, a process known as dynamic vulcanization. The hardness of the soft blends is controlled by the natural rubber content.

Natural Rubber–Rubber Blends. The majority of natural rubber is used in blends with other elastomers. Using a swollen-state nmr spectroscopy technique, it has been found possible not only to measure cross-link density in natural rubber itself, but also in the individual phases of vulcanized blends. This technique has enabled studies to be made of a range of natural rubber–synthetic rubber blends and has also shown a total imbalance of cross-link density in incompatible blends such as natural rubber and nitrile rubber. By careful selection of suitable curatives, which are likely to be more soluble in the natural rubber phase, an even distribution of cross-links between the two polymers has been achieved, with the resulting improvement in physical properties such as tensile strength.

Applications

In the same way that natural rubber is predominantly used in blends, it is also predominantly used in tire manufacture. Its excellent building tack, low heat buildup, low rolling resistance, and good low temperature performance make it the polymer of choice for many parts of tire construction, for both passenger and truck vehicles. The effects of radialization and demand for low rolling resistance and good low temperature performance have all tended to benefit natural rubber, especially in truck tire construction. Significantly large application areas for natural rubber other than tire use, are: tires and related products, latex products, footwear, nonautomotive engineering, belting and hose, automotive (nontire), and wire and cable.

Natural Rubber Latex

Preparation. At the processing factory, the field latex is bulked into batches of at least several tons and ammoniated to a level of about 0.5%, usually with ammonia gas. Bulking has the effect of ironing out variations in rubber content and other properties of latex from various sources. Ammonia has the role of preservative, but also raises the pH of the latex which promotes hydrolysis of some natural lipids to release stabilizing fatty acid soaps.

Concentration. The most common method of concentrating field latex is centrifugation. The two widely used types of centrifuge are the de Laval and Westfalia. Both operate by spinning the latex at high speed between a set of closely spaced conical disks. Small quantities of latex concentrate are produced by creaming and evaporation. Creaming takes advantage of the natural tendency of the rubber particles to rise to the surface of static latex. Commercially, creaming agents such as ammonium alginate are added to increase the rate of creaming. Evaporated latex is produced in continuous evaporators, which use a combination of heat and reduced pressure to remove water from the latex; a process in which 100% rubber recovery is achieved.

Latex Properties. Almost all latex concentrates are produced to meet an international standard, ISO 2004. This standard gives minimum requirements for rubber content and sets limits on some other latex properties. One of the most important properties is mechanical stability, which gives an indication of the resistance to flocculation or coagulation during processing of the latex. Copper and manganese concentrations are limited because of the severe degrading effects of these metals on finished products. The volatile fatty acid (VFA) number gives an indication of bacterial activity and thus shows how well the latex has been preserved. Color and odor specification are as follows: on visual inspection there should be no pronounced blue or gray; and after neutralization with boric acid, there should be no pronounced odor of putrefaction.

Processing and Products. The main distinguishing feature of rubber products made from latex rather than dry rubber is the rubber thickness, which is limited to a few millimeters. In producing latex products, the chemicals required for vulcanization, stiffening, coloring, antioxidant protection, or other purposes are added as solutions, emulsions, or fine dispersions to the latex before forming the product. Because no heat is generated during this mixing, it is possible to use ultrafast accelerators that would cause scorch problems in dry rubber compounds.

The most important group of latex products are the dipped goods. These are produced by dipping a shaped former into a suitably formulated latex compound, and then withdrawing it. The latex deposit is dried and vulcanized in hot air to give the product, which is then stripped from the former. Aside from dipping, the other main products produced from natural rubber latex are elastic thread and foam products.

Natural rubber latex also finds application in adhesives for tape, packaging, envelopes; in the footwear industry; and in the carpet industry.

C. S. L. BAKER
W. S. FULTON
Malaysian Rubber Producers' Research Association

"The Green Book," *International Standards of Quality and Packing for Natural Rubber Grades,* Rubber Manufacturers Association, Inc., Washington, D.C., 1975.

P. S. Brown, M. J. R. Loadman, and A. J. Tinker, *Rubb. Chem. Technol.* **65,** 744 (1992).

R. W. Keller, *Rubb. Chem. Technol.* **58,** 637 (1985).

A. D. T. Gorton and T. D. Pendle, *J. Nat. Rubber Res.* **1,** 122 (1986).

RUBBER, SYNTHETIC. See ELASTOMERS, SYNTHETIC.

RUBIDIUM AND RUBIDIUM COMPOUNDS

Rubidium, Rb is an alkali metal, in Group 1 (IA) of the Periodic Table. Its chemical and physical properties generally lie between those of potassium (qv) and cesium (see CESIUM AND CESIUM COMPOUNDS; POTASSIUM COMPOUNDS). Rubidium is the sixteenth most prevalent element in the earth's crust. Despite its abundance, it is usually widely dispersed and not found as a principal constituent in any mineral. Rather it is usually associated with cesium. Most rubidium is obtained from lepidolite, an ore containing 2–4% rubidium oxide.

After cesium, rubidium is the second most electropositive and alkaline element. The two isotopes of natural rubidium are ^{85}Rb (72.15%) and ^{87}Rb (27.85%). The latter is a beta-emitter having a half-life of 4.9×10^{10} yr. Twenty-four isotopes of rubidium are known.

Properties

Physical Properties. Rubidium, a soft, ductile, silvery-white metal, is the fourth lightest metallic element. Having a melting point of 39°C, it can be a liquid at ambient temperatures. Table 1 lists selected physical properties.

Chemical Properties. The reactions of rubidium are very similar to those of other alkali metals, especially potassium and cesium. Rubidium burns with a violet flame in the presence of air and reacts violently with water, liberating hydrogen which spontaneously explodes if oxygen or air are present. Like other alkali metals, rubidium reacts vigorously with lower alcohols, and violently with halogens, oxidizing agents, and chlorinated hydrocarbons. It also reacts with Teflon. Four oxides of rubidium are known: the yellow rubidium monoxide, Rb_2O; the dark brown rubidium peroxide, Rb_2O_2; the black rubidium trioxide, Rb_2O_3; and the dark orange rubidium superoxide, RbO_2. Commercial rubidium oxide is usually a mixture of these four oxides. Next to

Table 1. Properties of Rubidium

Property	Value
atomic weight	85.47
melting point, °C	39.0
boiling point, °C	689
density, g/cm^3	
solid, 18°C	1.522
liquid, 39°C	1.472
viscosity, at 39°C, mPa·s(= cP)	0.6713
surface tension at 39°C, mN/m(= dyn/cm)	75
heat of fusion, J/g[a]	25.69
heat of vaporization, J/g[a]	887
specific heat, J/(kg·K)[a]	
solid	331.37
liquid	368.19
vapor	241.83
thermal conductivity, liquid, W/(m·K)	29.3
critical temperature, T_c, K	1935 ± 87

[a] To convert J to cal, divide by 4.184.

cesium, rubidium is the second strongest Lewis base. Rubidium metal alloys with the other alkali metals, the alkaline-earth metals, antimony, bismuth, gold, and mercury.

Manufacture and Processing

Rubidium is found widely dispersed in potassium minerals and salt brines. Lepidolite, a lithium mica having the composition $KRbLi(OH,F)Al_2Si_3O_{10}$, contains up to 3.5% Rb_2O and is the principal source of the element. An ore that is basically pollucite, $Cs_2O{\cdot}Al_2O_3{\cdot}4\ SiO_2$, contains up to 1.5% RbO_2.

Pure rubidium metal is obtained by reducing pollucite or lepidolite ore using an active metal, followed by vacuum distillation. Another method is to reduce pure rubidium compounds thermochemically according to the following reactions:

$$2\ RbCl + Ca^0 \rightarrow CaCl_2 + 2\ Rb^0$$

$$2\ RbOH + Mg^0 \rightarrow Mg(OH)_2 + 2\ Rb^0$$

$$Rb_2CO_3 + 3\ Mg^0 \rightarrow 3\ MgO + C^0 + 2\ Rb^0$$

Although rubidium is more electropositive than either calcium or magnesium, the equilibrium is driven to the right because the rubidium is continuously distilled away from the reaction mixture. Rubidium metal can be purified by vacuum distillation.

Packaging, Shipping, and Storage. Owing to the reactivity of rubidium metal to oxygen and moisture, the material is considered hazardous and requires special packaging and shipping procedures.

Health and Safety Factors

Physiological experiments indicate exchangeability of rubidium for potassium in blood, plasma, and tissue. Medical and toxicological literature generally indicate a very low degree of toxicity. In many cases, the health risks of rubidium compounds are associated with the anion, eg, hydroxide, or fluoride, rather than the rubidium. Localized ventilation and the use of approved dust respirators are recommended when handling dry rubidium salts, especially hydroxides, oxides, and fluorides.

Table 2. Properties of Rubidium Compounds

	Solubility[a], g/100 mL		
Compound	Hot water	Cold water	Mp, °C
acetate	$86^{44.7}$		246
alum	43^{80}	1.3^{0}	99
bromide	$205^{113.5}$	98^{5}	682
chloride	139^{100}	77^{0}	715[b]
fluoride		130.6^{18}	760[c]
iodide	163^{25}	152^{17}	642[d]
sulfate	82^{100}	36^{0}	1073
chromate	95.7^{60}	62^{0}	
nitrate	452^{100}	19.5^{0}	
carbonate	450^{20}		837[e]
hydroxide		180^{15}	301
perchlorate	100^{18}	0.5^{0}	

[a] Superscripted number is temp in °C. [b] Bp = 1390°C. [c] Bp = 1410°C [d] Bp = 1300°C. [e] Bp = 740 dec.

Rubidium Compounds

Table 2 lists some properties of common rubidium compounds. Owing to the strong electropositive character of rubidium, its chemistry is normally restricted to ionic compounds. The element does not readily form metal–carbon bonds; therefore, alkyl and aryl compounds of rubidium are generally unstable. The rubidium cation forms a variety of chelated complexes with cyclic ethers, thiols, and amines. It also forms crown ether complexes, an example of which is shown. This chemistry has potential utility in the separation of alkali metals, from each other and from other metals.

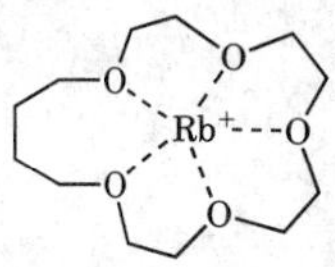

FRANK S. WAGNER
Strem Chemicals, Inc.

C. A. Hampel, *Rare Metals Handbook,* 2nd ed., Reinhold Publishing Corp., New York, 1961, pp. 434–440.

"Rubidium: Analysis, Biological Studies and Occurrence," *Chemical Abstracts-Chemical Substances Indexes,* American Chemical Society, Washington, D.C.

"Rubidium," in *Gmelins Handbuch der Anorganischen Chemie,* 8th ed., Vol. 24, Verlag Chemie, Berlin, 1955.

F. M. Perelman, *Rubidium and Cesium,* 1st English ed., The MacMillan Co., New York, 1965.

RUTHENIUM. See PLATINUM-GROUP METALS.

RUTHERFORDIUM. See ACTINIDES AND TRANSACTINIDES.

S

SALICYLIC ACID AND RELATED COMPOUNDS

The extract of roots, bark, leaves, and fruit of many common, widely distributed plants and trees contains the glucosides of methyl salicylate and salicyl alcohol. Salicylic acid, *o*-hydroxybenzoic acid, is easily derived from the glucosides by extraction and mild oxidation.

Related to salicylic acid as isomers are *m*-hydroxybenzoic acid and *p*-hydroxybenzoic acid. The three are commonly known as the monohydroxybenzoic acids and have the general structure shown below. They have a broad range of applications.

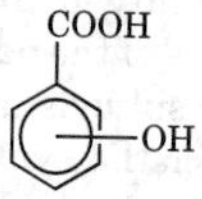

Salicylic acid and its derivatives exhibit activity as analgesics, antipyretics, and antiinflammatory agents, whereas the para-substituted acids and esters known as parabens have activity as preservatives for food.

Physical Properties

o-Hydroxybenzoic acid is obtained as white crystals, fine needles, or fluffy white crystalline powder. It is stable in air and may discolor gradually in sunlight. *m*-Hydroxybenzoic acid crystallizes from water in the form of white needles and from alcohol as platelets or rhombic prisms. *p*-Hydroxybenzoic acid crystallizes in the form of monoclinic prisms. Selected physical properties of hydroxybenzoic acids are listed in Table 1.

Reactions

The hydroxybenzoic acids have both hydroxyl and the carboxyl groups and, therefore, participate in chemical reactions characteristic of each of these moieties. In addition, these acids can undergo electrophilic ring substitution. The following reactions are discussed in terms of salicylic acid, but are characteristic of all the hydroxybenzoic acids.

Carboxylic Acid Group. Reactions of the carboxyl group include decarboxylation, reduction to alcohols, and the formation of salts, acyl halides, amides, and esters. Generally, the carboxyl group is not readily reduced. Lithium aluminum hydride is one of the few reagents that can reduce these organic acids to alcohols. The acid dissolves in aqueous sodium carbonate or sodium bicarbonate to form sodium salicylate; however, if salicylic acid is dissolved in the presence of alkali metals or excess sodium hydroxide, the disodium compound is formed.

Table 1. Physical Properties of Hydroxybenzoic Acids

Property	Isomer value		
	Ortho	Meta	Para
molecular weight	138.12	138.12	138.12
melting point, °C	159	201.5–203	214.5–215.5
boiling point, °C	211 sub		
density	$1.443^{20}{}_{4}$	$1.473^{25}{}_{25}$	$1.497^{20}{}_{20}$
refractive index	1.565		
flash point (Tag closed-cup), °C	157		
acid dissociation, K_a, at 25°C	1.05×10^{-3}	8.3×10^{-5}	2.6×10^{-5}
heat of combustion, mJ/mol[a]	3.026	3.038	3.035
heat of sublimation, kJ/mol[a]	95.14		116.1

[a] To convert J to cal, divide by 4.184.

Hydroxyl Group. Reactions of the phenolic hydroxyl group include the formation of salts, esters, and ethers. Acylation is achieved by reaction with an acid anhydride. The single most important commercial reaction of this type is the acetylation of salicylic acid with acetic anhydride to produce acetylsalicylic acid (aspirin).

Ring-Substitution Reactions. In the introduction of a third group into a disubstituted benzene, the position the group takes depends on the other groups present.

Manufacture of Salicylic Acid

Modern methods for the synthesis of salicylic acid use the general outline of the Kolbe-Schmitt reaction; that is, the reaction of dry sodium phenate with carbon dioxide under pressure at elevated temperature (180–200°C).

An interesting biochemical method of manufacture is the utilization of bioengineered *Pseudomonad* plasmid or *Pseudomonas stutzeri* in a culture medium to oxidize naphthalene or alkyl-substituted naphthalene. The metabolic oxidation products, unsubstituted or substituted salicylic acid, respectively, are recovered from the medium.

Health and Safety Factors. Because salicylic acid is a moderate respiratory irritant, skin and eye contact should be avoided and dust mask protection should be used. General good hygiene and good housekeeping should be incorporated into procedures. If large quantities are to be handled, protective clothing with long sleeves and gauntlets as well as approved dust respirators should be used. A further consideration is that dust concentrations as low as 9 g/m^2 can ignite. For safe practice, ignition and arcing sources should be recognized and eliminated. The single-dose oral toxicity of salicylic acid is moderate. The LD_{50} in rats is 400–800 mg/kg. Chemical goggles should be worn when handling salicylic acid because eye contact with the chemical can produce irritation and marked pain.

Uses of Salicylic Acid

Approximately 60% of the salicylic acid produced in the United States has been consumed in the manufacture of aspirin. However, several applications of the acid are growing. Salicylic acid USP, EP, and other pharmacopeial grades are used medically as antiseptic, disinfectant, antifungal, and keratolytic agents. Carbonless copypaper using salicylic acid and alkyl salicylic acid derivatives has become an active application area. The salicylic acid is incorporated into a resin or is encapsulated as one of the agents for pressure or thermally activated imaging.

Salicylic acid derivatives form surfactants that can influence the rheologic properties of solutions. They impart controllable and useful viscous and elastic properties to aqueous liquids.

Salts of Salicylic Acid

A large number of salts of salicylic acid have been prepared and evaluated for therapeutic or other commercial use. Sodium salicylate has analgesic, antiinflammatory, and antipyretic activities. Magnesium salicylate, an analgesic and antiinflammatory agent, appears to have exceptional ability to relieve backaches. It is also used for the symptomatic relief of arthritis. Bismuth subsalicylate is taken orally in combination with other ingredients for protective, antacid action as well as antidiarrheal and antiseptic effects.

Esters of Salicylic Acid

The esters of salicylic acid are commercially produced by esterification of salicylic acid with the appropriate alcohol using a strong mineral acid such as sulfuric as a catalyst. To complete the esterification, the excess alcohol and water are distilled away and recovered. The main commercial applications for salicylate esters are as uv sunscreen agents and as flavor and fragrance agents. Several have application as topical analgesics.

Other Derivatives of Salicylic Acid

p-Aminosalicylic acid and its salts have been used in the treatment of tuberculosis. *p*-Aminosalicylic acid can be prepared by the carboxylation of *m*-aminophenol. Methylene-5,5-disalicylic acid, produced by heating two parts salicylic acid with 1–1.5 parts of 30–40 wt % formaldehyde in the presence of an acid catalyst, is used as an intermediate in the production of bacitracin methylenedisalicylate.

Salicylamide, prepared by the reaction of methyl salicylate with ammonia, has mild analgesic, antiinflammatory, and antipyretic properties. Salicylanilide, prepared by heating salicylic acid and aniline in the presence of phosphorus trichloride, is used as an intermediate in the production of other chemicals and as a slimicide, fungicide, and medicament.

Acetylsalicylic Acid (Aspirin)

Aspirin is a registered trademark of Bayer in many nations, but in the United States and the United Kingdom, aspirin is accepted as the generic name for acetylsalicylic acid.

Physical Properties. Aspirin normally occurs in the form of white, flat platelets or needle-like crystals, or as a crystalline powder. It melts at 135–137°C and decomposes at 140°C. The solubility of aspirin is about 1 g/300 mL of water at 25°C and about 1 g/5 mL of ethanol at 25°C.

Manufacture. Aspirin (**1**) is manufactured by the acetylation of salicylic acid with acetic anhydride.

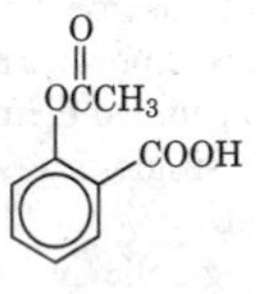

(**1**)

Production. Aspirin is produced in the United States by Rhodia, a division of Rhône-Poulenc. Globally, Rhône-Poulenc has additional production facilities in France and in Thailand.

Uses. Aspirin has analgesic, antiinflammatory, and antipyretic activity. It is used for the relief of less severe types of pain, such as headache, neuritis, acute and chronic rheumatoid arthritis, and toothache. Evidence has accumulated from a number of clinical trials that aspirin ingestion lowers the incidence of myocardial infarction, unstable angina, and stroke. These actions of aspirin are thought to result from its ability to reduce the production of prostaglandin formed by platelet cells without appreciably affecting the other important functions in these blood factors.

m-Hydroxybenzoic Acid

Of the three hydroxybenzoic acids, the metaisomer is of least commercial importance. It offers no outstanding points of chemical interest and is used industrially in small quantities in a limited number of applications.

p-Hydroxybenzoic Acid

p-Hydroxybenzoic acid is of significant commercial importance. The most familiar application is the use of several of its esters as preservatives, known as parabens. Also of interest is the use in liquid crystal polymer applications.

Reactions. *p*-Hydroxybenzoic acid undergoes the typical reactions of the carboxyl and hydroxyl moieties.

Manufacture. The commercial technique of preparation of *p*-hydroxybenzoic acid is similar to that of salicylic acid, ie, Kolbe-Schmitt carboxylation of phenol.

Uses. There are many polymer and plastic applications for *p*-hydroxybenzoic acid. An application of interest is in liquid crystal preparation. *p*-Hydroxybenzoic acid has been studied in liquid crystal systems either as part of a rod- or disk-shaped surfactant molecule in a solvent, or as a monomer in a polymer, such as a polyester, that exhibits a degree of order lower than the crystalline solid, but higher than the normal isotropic liquid (see LIQUID CRYSTALLINE MATERIALS). *p*-Hydroxy benzoic acid is used in the manufacture of the methyl, ethyl, *n*-propyl, *n*-butyl, and benzyl esters called parabens. These esters have been used as preservatives for food, pharmaceuticals, and cosmetics for many years; they are effective bacteriostatic and fungistatic agents against a wide variety of microorganisms.

Salicyl Alcohol

Salicyl alcohol (saligenin, *o*-hydroxybenzyl alcohol) crystallizes from water in the form of needles or white rhombic crystals. It occurs in nature as the bitter glycoside, salicin, which is isolated from the bark of *Salix helix, S. pentandra, S. praecos,* some other species of willow trees, and the bark of a number of species of poplar trees such as *Polpulus balsamifera, P. candicans,* and *P. nigra.*

Physical Properties. The alcohol, which sublimes readily and is very soluble in alcohol and ether, has the following properties: melting point, 86°C; density 13°/25°, 1.161 g/cm^3; heat of combustion, 3.542 mJ/mol (846.6 kcal/mol); and solubility in 100 mL water at 22°C, 6.7 g.

Manufacture. The hydrolysis of the naturally occurring β-glycoside (salicin) with hydrochloric or sulfuric acid affords saligenin (**2**) and glucose. Numerous methods for the synthesis of salicyl alcohol exist. These involve the reduction of salicylaldehyde or of salicylic acid and its derivatives.

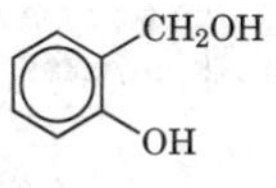

(**2**)

Uses. Saligenin has been used medically as an antipyretic and appears to possess marked topical analgesic powers in concentrations of 4–10%.

Thiosalicylic Acid

Thiosalicylic acid (*o*-mercaptobenzoic acid), a sulfur-yellow solid that softens at 158°C, has a melting point of 164°C. It sublimes, is slightly soluble in hot water but freely soluble in glacial acetic acid and alcohol, and yields dithiosalicylic acid upon exposure to air.

Uses. Thiosalicylic acid has been used as an anthelmintic, bactericide, and fungicide. It has also been used as a rust remover, a corrosion inhibitor for steel, and a polymerization inhibitor. In photography, it has application in print-out emulsions and as an activator for phtographic emulsions.

MARY R. THOMAS
The Dow Chemical Company

M. J. H. Smith and P. Smith, *The Salicylates: A Critical Bibliographic Review,* Wiley-Interscience, New York, 1966.

K. D. Rainsford, *Aspirin and the Salicylates,* Butterworths, London, 1984.

C. C. Mann and M. L. Plummer, *The Aspirin Wars,* Harvard Business School Press, Boston, Mass., 1991.

E. M. De Simone II, Ninth Edition, *Handbook of Nonprescription Drugs,* American Pharmaceutical Association, Washington, D.C., 1990, or most recent edition.

SABADILLA, SABADINE, SABALINE. See INSECT CONTROL TECHNOLOGY.

SACCHARIN. See SWEETENERS.

SAFETY. See INDUSTRIAL HYGIENE; MATERIALS RELIABILITY; PLANT SAFETY.

SAMARIUM. See LANTHANIDES.

SAMPLING

The chemical industry produces material in the gaseous, liquid, and solid phases that range from basic chemicals to functional specialties. In addition, many processes require the use of intermediates which may or not be in the physical form of the products. These various materials are sampled for the purpose of process control (qv), product quality control (see QUALITY ASSURANCE), environmental control, and occupational health control. Samplings of some of these materials can be hazardous, particularly those involving toxic, unstable, or pressurized substances, which require special safety precautions. However, excluding these special circumstances, the main problems encountered in sampling materials in the chemical industry are those of selecting an appropriate sampling procedure and device to obtain a representative sample. Experience involving a large variety of materials has produced many methods of sampling. Extensive coverage of sampling methods is available from the American Society for Testing Materials (ASTM).

Definitions and Problems

Sampling is the operation of removing a portion from a bulk material for analysis in such a way that the portion removed has representative physical and chemical properties of that bulk material. From a statistical point of view, sampling is expected to provide analytical data from which some property of the material may be determined. For pure liquids and gases, sampling is relatively easy. Sampling of these media becomes difficult, however, when particulates are involved. Almost all samples taken in the chemical industry contain solids in some form. Some materials exist in natural deposits, eg, strata, in a heap on the ground; in storage tanks, bins, pipes, or ducts; in railcars, drums, bottles, and bales; or in other containers that may or may not be subdivided easily into representative units. Many systems containing particulates tend to segregate during handling or storage, and this may introduce a sampling error in the form of bias. Because the distribution of chemical components in the material may be size-dependent, even a chemical assay from a sample obtained may be in error. Generally, little is known in advance concerning the degree of homogeneity of most sampled systems. In any bulk container, the product may be stratified into zones of variable properties. In gas and liquid systems, particulates segregate and concentrate in specific locations in the container as the result of sedimentation (qv) or flotation (qv) processes.

Although fluid systems containing particulates introduce sampling difficulties, these systems do conform to one rule of good sampling. They are in motion. For example, powders should be sampled from a moving stream rather than when at rest. Also the whole of the stream should be sampled, not just a part of it. For gas systems, whole stream sampling is usually not possible; for liquids, it can only be done from the outfall of a pipe. A third rule in sampling is that small quantities should be taken frequently rather than large quantities taken infrequently. The ideal place to sample is where the sample is well-mixed.

Chemical plants are rarely well designed in terms of sampling capabilities; therefore, most sampling involves some compromise. Decisions such as the sampling procedure to be employed, the quantity of material to be taken, and the permitted tolerance in the representativeness of the sample must be made on the basis of the use to which the subsequent analysis is to be put. Finally, chemical and physical changes during sampling and subsequent handling need to be minimized.

The quantity of sample required comprises two parts: the volume and the statistical sample size. The sample volume is selected to permit completion of all required analytical procedures. The sample size is the necessary number of samples taken from a stream to characterize the lot. In most sampling procedures, samples are taken at different levels and locations to form a composite sample. The need for skill and experience on the part of sample designers and personnel cannot be overemphasized in chemical plant sampling. Necessary steps must be taken to document the hazards involved in an operation and to ensure that the staff are well-trained, informed, protected, and capable.

Gases

By far the largest proportion of gas sampling operations in industry is carried out for environmental reasons. The preparation, precautions and equipment requirements involved in the sampling of air pollution sources are applicable to most other gaseous environments (see AIR POLLUTION CONTROL METHODS).

Before a source analysis program is undertaken, it is important to decide which information is really required. Sampling sites must be selected with care. Care must also be taken in the selection of sampling points at the site. Measurement usually involves the determination of temperature, concentration, and characterization of the gas contaminants. It also requires the mass rates of emission of each contaminant, therefore concentration and volumetric flow data are required.

Sampling Site. The location of a sampling site and the number of sampling points are based on the need to obtain representative data and whether the points are restricted by access problems. Sampling sites should be at least eight stack or duct diameters downstream and two diameters upstream from any disturbance. A disturbance is interpreted as a bend, expansion, contraction, valve, baffle, or visible flame.

Measurement of Gas Velocity and Temperature. Stack-gas velocity is determined at each traverse point, based on the gas density and a measurement of the average velocity head, using a pitot tube. The measured velocity pressure is the difference between the total pressure as measured against the gas flow and the static pressure measured perpendicular to the gas flow.

During sample and velocity traverses, the S-pitot tube is rarely used in isolation. It is necessary to measure stack or duct temperature profiles to determine variation in gas distribution, and this is usually done at the same time as the velocity profiles. For most purposes, the pitot tube is combined with a thermocouple and sampling nozzle in a sampling assembly.

Sample Extraction. Once the velocity and temperature profiles have been taken, gas samples can be withdrawn. In the sampling of noncondensable gases which are free of particulates, the gases are extracted from the duct by one of the following methods: a single-point grab sample, a single-point integrated sample, or a multipoint integrated sample.

Sample Extraction When Particulates Are Present. Different designs of probe and train may be required depending on the reason for sampling particulates. The simplest train is for the determination of mass loading only. For mass loadings, particulate matter is withdrawn isokinetically and collected on a glass fiber filter maintained at $120 \pm 4°C$. Particulate matter, present as solid or liquid at the sampling temperature, is collected on a preweighed filter and determined by weighing. Particulates are sometimes present in liquid rather than solid form, eg, as acid mists. For acid mists, the Brink impactor is often used. The mist is first drawn through a cyclone to remove particles larger than 3 μm. A five-stage impactor is used to classify mist particles of diameter 0.3–3.0 μm.

Liquids

In the chemical industry, liquids are sampled from process vessels, tanks, tanker trucks, tank cars, ships, barges, pipelines, transfer lines, drums, carboys, cans, bottles, open lagoons, settling ponds, sewers, and open flowing streams and rivers. For simplicity, the procedures for sampling liquids are divided into three categories: tanks and similar containers, pipelines (qv), and open streams and lagoons. The problems that arise in sampling liquids containing particulates are not as critical as those that arise in gaseous systems. There are two reasons. Viscosity tends to dampen the effects of sudden changes in flow direction, and particles do not separate as readily from streamlines unless the particle masses and velocities are large.

Tanks. Sampling methods for tanks, trucks, tank cars, barges, etc, usually involve the use of fixed sample taps, thief samplers, or bottle samplers. When particulates are present, as in slurries or suspensions, representative samples can only be obtained from agitated tanks (see TANKS AND PRESSURE VESSELS).

Pipes and Pipelines. Samples may be withdrawn both from closed pipe cross sections and from the outfall of open pipes. In the chemical industry, many pipe samples are taken from organic liquids, which may be toxic, highly reactive, and flammable. The amount of sample taken should be minimized in order to reduce worker exposure and sample disposal problems.

Open Streams and Lagoons. Open-stream discharges are encountered in wastewater plants in the chemical industry (see WASTES, INDUSTRIAL; WATER, POLLUTION). Settling ponds and lagoons are part of wastewater treatment plants. Details on the monitoring of industrial wastewater are spelled out in EPA regulations. Surface samples may be withdrawn using a dipper sampler of nonreactive material. Either grab or composite sampling may be used for lagoons.

Samplers must be designed and constructed to withstand the chemical composition extremes present in the individual discharges. Corrosion-resistant fabrication must be used in the equipment that comes in contact with many chemical industry discharges (see CORROSION AND CORROSION CONTROL).

Solids

Solids occur in several forms in the chemical industry. Raw material from natural deposits are compacted in the ground and sampling is performed during the exploration stages. This type of material is typified by minerals and fossil fuels. Before use, these must be crushed, ground into particulate form, mixed, cleaned, and stored. Sampling may be required before, during, or immediately after any one of these operations, and different sampling methods are used for most of them.

Flowing Streams. All free-flowing powders are transported at some time during manufacture as flowing streams. Hoppers are emptied by screw conveyors. Solids are transported to bagging operations by pneumatic conveyors, and most solids pass through transfer points in gravity-flow pipes and chutes. Even small samples in boxes, bags, cans, bottles, etc, can be made to flow by emptying them into volumetric feeders. Sampling is carried out on the resultant stream.

Efficiency. Sampling of bulk solids from grinding circuits or chemical plants represents tons of material per day. A primary sampler generally removes 10–100 kg as a gross sample, which is then subdivided into 1–10 kg laboratory samples by a secondary device. The samples may be examined as taken or may be crushed prior to examination. Measurement samples are needed by the laboratory in gram or milligram quantities and on the basis of subsequent measurement decisions are made as to the quality of tonnages of bulk material. It is therefore essential that the measurement sample be representative of the bulk. Bias at any stage of the sampling affects the final result. Rotary sampling is by far the best analytical sampling method to use for solids. This method follows all the rules for good sampling.

REG DAVIES
E. I. du Pont de Nemours & Co., Inc.

T. Allen, *Particle Size Measurement,* 5th ed., Chapman & Hall/Methuen, London/New York, 1997.

ASTM E122-5, *Standards Designation,* American Society for Testing and Materials, Philadelphia, Pa., 1993.

U.S. EPA, *Regulations on Standards of Performance for New Stationary Sources,* 40 CFR 60, Appendix A, Reference Methods, Washington, D.C., 1993.

ASTM D3685-92, *Standard Test Method for Sampling and Determination of Particulate Matter in Stack Gases,* American Society for Testing and Materials, Philadelphia, Pa., 1992.

SAPONINS. See STEROIDS.

SCANDIUM. See LANTHANIDES.

SCALEUP. See PILOT PLANTS; PROCESS CONTROL.

SCREEN PRINTING. See DYES, APPLICATION AND EVALUATION; ELECTROPHOTOGRAPHY; PRINTING PROCESSES; THERMOGRAPHY.

SCREENS, SCREENING. See SIZE ENLARGEMENT; SIZE MEASUREMENT OF PARTICLES; SIZE REDUCTION.

SCRUBBERS. See ABSORPTION; ADSORPTION; COAL CONVERSION PROCESSES, CLEANING AND DESULFURIZATION; GAS, NATURAL.

SEALANTS

A sealant is a material that is installed into a gap or joint to prevent water, wind, dirt, or other contaminants from passing through the joint or gap. Sealants, which can also be defined by how they are tested, are rated by their ability to stretch, twist, bend, and be compressed while maintaining their bulk properties so that they do not tear apart under stress. A most important rating of a sealant in many applications is the movement ability of the sealant. The adhesion required of a sealant is simply the strength to hold the sealant in position as it is stressed and strained. Some sealants are used as adhesives (qv) and some adhesives as sealants, and thus arises the occasional blurring of their roles. If the material's primary function is the exclusion of wind, water, dirt, etc, it is a sealant.

Performance Characteristics

Movement Capabilities. The movement capability of a sealant is the amount of displacement the sealant can endure in extension or compression without failing. In general, sealants with lower modulus of elasticity can handle higher movements.

Sealants can be broadly divided into classes according to the amount of movement they can successfully handle. High performance sealants such as silicones, urethanes, and polysulfides can typically handle movements of 25% or higher. Medium performance sealants such as some acrylics can handle movements of 10–25%. Low performance sealants such as butyls, putties, and caulks accommodate movements under 10%.

Modulus. The measure of the stress of a sealant at a specific strain is referred to as the modulus of elasticity, sometimes called the secant modulus. This important sealant property describes the force exerted by a sealant as it is stressed. Because a primary function of a sealant is to adhere to the substrate with which it is in contact, the forces generated by a joint opening or closing are transmitted by the sealant to the substrate–sealant bond line. For this reason, it is important to know the modulus of the sealant and also the strength of the substrate.

Durability. A primary factor in sealant durability is its ability to resist decay from environmental elements. For most typical applications this includes extremes of high and low temperature, water, oxidation, and sunlight.

Sealant Types and Formulations

Silicones. Commercially available silicone sealants are typically one of three curing types: moisture-reactive (curing) sealants, moisture-releasing (latex) sealants, and addition-curing sealants. Of these three types, moisture-curing silicones make up the vast majority of silicone sealants sold.

The formulation of moisture-curing silicones includes a silicone polymer, filler, a moisture-reactive cross-linker, and sometimes a catalyst. A newer class of silicone sealants are the silicone latex sealants. These sealants are silicone-in-water emulsions that cure by evaporation of the emulsifying water. Addition-curing silicones in general are two-part systems that cure by the platinum-catalyzed reaction of a silicone hydride with typically a vinyl group attached to silicon. Because no by-products are generated by the cure, there are few volatiles and no shrink in thick sections.

Urethanes. The basis for urethane chemistry is the reaction of an isocyanate group with a component containing an active hydrogen. The first step in formulating a urethane sealant is to prepare what is commonly called the prepolymer, typically by reaction of a hydroxy-terminated polyether with a stoichiometric amount of a diisocyanate.

Although polyethers are the main building blocks of the prepolymer, other materials such as polyesters, polythioethers, and polybutadienes are also used. Most urethanes use blends of polymers to achieve desired properties. Urethane sealants have good inherent adhesion to most substrates, but silane adhesion promoters are often used to improve this adhesion.

Polysulfides. Polysulfide sealants were the first high performance synthetic elastomeric sealants produced in the United States. The basic polymers are mercaptan-terminated (HS–R–SH), with molecular weights ranging from 1000 to ca 8000. Curing occurs through the mercaptan groups by oxidation and results in an S–S linkage. In addition to the terminal SH groups, pendent SH groups are located along the polymer chain and contribute to cross-linking. Polysulfide polymers provide inherent resistance to fuel and quite good resistance to alkali. Different curing agents have different effects on the properties of the polysulfide sealant. Therefore, the choice of curing agent depends on the sealant application. Adhesion-promoting silanes are often added to improve adhesion to various substrates. As is the case with urethane sealants, silanes with a dual-reactive nature are typically used.

Acrylics. There are two principal classes of acrylic sealants: latex acrylics and solvent-release acrylics. High molecular weight latex acrylic polymers are prepared by emulsion polymerization of alkyl esters of acrylic acid. The emulsion polymers are compounded into sealants by adding fillers, plasticizers, freeze–thaw stabilizers, thickeners, and adhesion promoters. As is true of the silicone latex sealants, the acrylic latex sealants are easy to apply and clean with water.

Another class of acrylic sealants are the solvent-releasing acrylics. Acrylic monomers are polymerized in a solvent. The natural adhesion of most of the solvent-releasing acrylics produces some of the best unprimed adhesion in the sealant industry. However, slow, continual cure generally produces large compression sets and limits their use to low movement applications. The relatively high amounts of solvent and traces of acrylic monomer in these formulations limits their use to outdoor applications, usually in construction.

Butyls. Butyl-based materials are sold in the form of preformed tapes, thermoplastic hot melts, and one-part solvent-releasing sealants. Formulations of butyl-based sealants also include plasticizer, filler, and tackifier resins. Solvents, such as mineral spirits, are used for the one-part solvent-releasing formulations. As the solvent leaves the typical one-part butyl, the sealant hardens and loses its elastomeric ability. This limits the use of solvents to low movement applications where durability is not of high concern.

Manufacture and Processing

Almost all sealants contain a mixture of a powdered filler incorporated into a viscous liquid, which results in a viscous liquid sealant having a paste-like consistency. Processing conditions can have a dramatic effect on sealant rheology, cure time, and physical properties. Typical processing variables are mixer speed (rpm), time, temperature, and vacuum. Order of ingredient addition is also important.

Health and Safety Factors

Sealant Manufacturing. Most sealants use mineral-based fillers which may contain small amounts of crystalline silica. This may require some manufacturing safeguards but after compounding no dust hazards exist. Crystalline silica is a known cause of silicosis, a debilitating disease of the lung. Some but not all sealants use flammable ingredients, but for those that do, proper inerting and grounding are needed to prevent potential explosions. For silicone manufacturing, handling of cross-linkers and catalysts in 100% pure form requires knowledge of any unusual hazard. Urethane sealant manufacturing requires precautions during the production of the prepolymers.

Sealant Application. In general, sealants should be used in areas that have good ventilation. A careful review of the material safety data sheet is fundamental before use of any sealant.

Uses

Each class of sealants has certain attributes inherent to the polymer on which it is based. These attributes often define the sealant's applications and limitations.

RICHARD A. PALMER
JEROME M. KLOSOWSKI
Dow Corning Corporation

J. M. Klosowski, ed., *Science and Technology of Building Seals, Sealants, Glazing, and Waterproofing,* Vol. 2, ASTM STP 1200, ASTM, Philadelphia, Pa., 1992.

J. M. Klosowski and D. Beers, in *Engineered Materials Handbook,* Vol. 3, ASM International, Materials Park, Ohio, 1990.

J. F. Regan, "Urethane Formulations," *Proceedings of Caulks and Sealants Short Course,* The Adhesive and Sealant Council, Dallas, Tex., 1990.

J. M. Klosowski, *Sealants in Construction,* Marcel Dekker, New York, 1989.

SEASONINGS. See FLAVORS AND SPICES.

SEAWEED COLLOIDS. See GUMS.

SEDIMENTATION

Sedimentation has been defined as "the separation of a suspension into a supernatant clear fluid and a rather dense slurry containing a higher concentration of solid". This definition is too broad. It does not specify the external field of acceleration, gravitational, centrifugal, magnetic, or electrostatic, that causes the separation. There is also a possible ambiguity whether the suspension is gaseous or liquid. Herein the term *sedimentation* is restricted to the most common definition, ie, the gravitational settling of solids in liquids (see also MINERAL RECOVERY AND PROCESSING; SEPARATION, CENTRIFUGAL SEPARATION; SEPARATION, MAGNETIC SEPARATION).

The uses of sedimentation in industry fall into the following categories: solid–liquid separation; solid–solid separation; particle-size measurement by sedimentation; and other operations such as mass transfer, washing, etc. In solid–liquid separation, the solids are removed from the liquid either because the solids or the liquid are valuable or because these have to be separated before disposal. If the primary purpose is to produce the solids in a highly concentrated slurry, the process is called thickening. If the purpose is to clarify the liquid, the process is called clarification. Usually, the feed concentration to a thickener is higher than that to a clarifier. Some types of equipment, if correctly designed and operated, can accomplish both clarification and thickening in one stage (see also EXTRACTION, LIQUID–SOLID).

In solid–solid separation, the solids are separated into fractions according to size, density, shape, or other particle property (see SIZE REDUCTION). Sedimentation is also used for size separation, ie, classification of solids (see SEPARATION, SIZE SEPARATION). One of the simplest ways to remove the coarse or dense solids from a feed suspension is by sedimentation. Successive decantation in a batch system produces closely controlled size fractions of the product. Generally, however, particle classification by sedimentation does not give sharp separation (see SIZE MEASUREMENT OF PARTICLES).

In particle-size measurement, gravity sedimentation at low solids concentrations (< 0.5% by vol) is used to determine particle-size distributions of equivalent Stokes' diameters in the range from 2 to 80 μm. Particle size is deduced from the height and time of fall using Stokes' law, whereas the corresponding fractions are measured gravimetrically, by light, or by x-rays. Some commercial instruments measure particles coarser than 80 μm by sedimentation when Stokes' law cannot be applied.

Sedimentation is also used for other purposes. For example, relative motion of particles and liquid increases the mass-transfer coefficient. This motion is particularly useful in solvent extraction in immiscible liquid–liquid systems (see EXTRACTION, LIQUID–LIQUID). An important commercial use of sedimentation is in continuous countercurrent washing, where a series of continuous thickeners is used in a countercurrent mode in conjunction with reslurrying to remove mother liquor or to wash soluble substances from the solids. Most applications of sedimentation are, however, in straight solid–liquid separation.

Coagulation and Flocculation. Both coagulation and flocculation are classical pretreatment methods used to increase the effective particle size, thereby improving sedimentation settling rates. Although these two terms are often used interchangeably, coagulation is sometimes defined as agglomeration of the primary particles into particles up to 1 mm in diameter. Flocculation, on the other hand, not only agglomerates particles but also interconnects them by means of long-chain molecules of the flocculating agent into giant loose flocs up to 1 cm in size (see FLOCCULATING AGENTS). The term *flocculation* is used here to include coagulation as defined. Chemical agents create favorable conditions for flocculation by neutralization of surface charges and thus reduce interparticle repulsion. Mineral coagulants are in the form of electrolytes, such as alum or lime, whereas flocculation agents are mostly synthetic polyelectrolytes of high molecular weight. Development of the latter group since the 1970s has resulted in a remarkable improvement in sedimentation equipment.

Equipment

Sedimentation equipment can be divided into batch-operated settling tanks and continuously operated thickeners or clarifiers. Most sedimentation processes are operated in continuous units.

Clarifiers. The largest user of clarifiers is probably the water-treatment industry. The conventional one-pass clarifier uses horizontal flow in circular or rectangular vessels (Fig. 1) with feed at one end and overflow at the other. The feed is preflocculated in an orthokinetic (paddle) flocculator which often forms an integral part of the clarifier. Settled solids are pushed to a discharge trench by paddles or blades on a chain mechanism or suspended from a traveling bridge (see WATER, INDUSTRIAL WATER TREATMENT).

Circular basin clarifiers are most commonly fed through a centrally located feed well. The overflow is led into a trough around the periphery of the basin. The bottom gently slopes to the center and the settled solids are pushed down the slope by a number of motor-driven scraper blades that revolve slowly around a vertical center shaft. This design closely resembles a conventional thickener. The conventional one-pass clarifier is designed for the lowest specific overflow rate

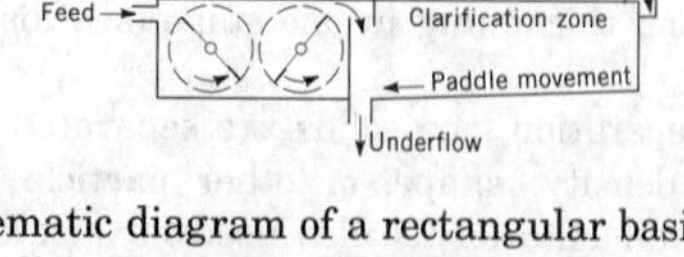

Figure 1. Schematic diagram of a rectangular basin clarifier having an orthokinetic flocculator where the feed is mixed with a flocculant.

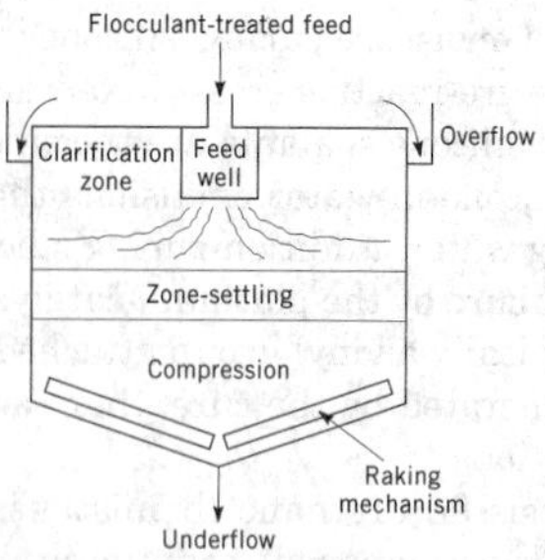

Figure 2. The circular basin continuous thickener.

(flow per unit area of liquid surface), which is usually 1–3 m/h depending on the degree of flocculation. These clarifiers can be started and stopped without difficulty.

Newer designs incorporate some vertical flow and combine flocculation, gravity and inertial clarification, and solids recirculation. Because such units achieve higher overflow rates, they are referred to as rapid settling or high rate clarifiers.

Flocculation is accelerated and higher overflow rates are achieved by external or internal recirculation of settled solids into the feed which leads to the collection of fine particles by interception. Addition of conditioned fine sand to the feed induces separation by differential sedimentation, and sometimes increases overflow rates to 6–8 m/h.

Thickeners

The most common thickener is the circular basin type shown in Figure 2. After treatment with flocculant, the feed stream enters the central feed well which dissipates the stream's kinetic energy and disperses it gently into the thickener. A typical thickener has three operating layers: clarification, zone-settling, and compression. Frequently, the feed is contained in the zone-settling layer which theoretically eliminates the need for the clarification zone because the particles would not escape through the interface. In practice, however, the clarification zone provides a buffer for fluctuations in the feed and the sludge levels.

LADISLAV SVAROVSKY
Consultant, Engineers and Fine Particle Software

L. Svarovsky, ed., *Solid–Liquid Separation,* 3rd ed., Butterworths, London, 1990.

D. B. Purchas, ed., *Solid–Liquid Separation Equipment Scale-Up,* Upland Press Limited, London, 1977.

L. Svarovsky, *Hydrocyclones,* H. H. Rinehart & Winston, London, 1984.

L. Svarovsky, in L. R. Weatherley, ed., *Engineering Processes for Bioseparations,* Butterworth-Heineman, Oxford, U.K., 1994, Chapt. 5.

SELENIUM AND SELENIUM COMPOUNDS

Selenium

Selenium, Se, atomic no. 34, atomic wt 78.96, lies between sulfur and tellurium in Group 16 (VIA) and between arsenic and bromine in Period 4 of the Periodic Table. Selenium exists in various allotropic modifications and forms many inorganic and organic compounds, often analogous and isomorphous with their sulfur equivalents.

Physical Properties. The six stable isotopes of selenium, ^{74}Se, ^{76}Se, ^{77}Se, ^{78}Se, ^{80}Se, and ^{82}Se, occur naturally. A number of artificial radioactive isotopes have been prepared by neutron activation. Solid selenium has several allotropic forms, including an amorphous one resembling plastic sulfur. The stable form at ordinary temperatures, ie, the gray or hexagonal selenium, is the most dense and is semimetallic in appearance. Some physical constants for selenium are given in Table 1.

Table 1. Physical Constants of Selenium

Property	Value
melting point, °C	217
boiling point, °C	ca 685
heat of fusion, trigonal liquid, kJ/mol[a]	5.2–5.4
heat of vaporization, kJ/mol[a]	59.7
heat of combustion, at 298 K, kJ/mol[a]	−225.1
heat capacity, J/(g·K)[a]	
trigonal	24.52
vitreous	25.627
liquid	29.288
thermal conductivity, W/(m·K)	248.1
thermal expansion coefficient[b], $°C^{-1}$	$3.24 \times 10^{-5} - 7.5 \times 10^{-5}$
viscosity, mPa·s(= Cp)	
at 220°C	221
360°C	70
density, g/cm^3	
trigonal at 298 K	4.819
monoclinic	4.4
liquid at 490 K	4.05
vitreous	4.285
standard reduction potential, V	
$Se + 2\,e^- \rightarrow Se^{2-}$	−0.78
$Se + 2\,H^+ + 2\,e^- \rightarrow H_2Se$ (aq)	−0.36
surface tension, liquid, mN/m(= dyn/cm)	
at 220°C	105.5
310°C	95.2
electronegativity, Pauling scale	2.4

[a] To convert J to cal, divide by 4.184. [b] Value depends on form.

Chemical Properties. The chemical properties of selenium are intermediate between those of sulfur and tellurium. Selenium reacts with active metals and gains electrons to form ionic compounds containing the selenide ion, Se^{2-}. With most other substances selenium forms covalent compounds.

Selenium combines with metals and many nonmetals directly or in aqueous solution. Selenium combines with oxygen yielding a number of oxides, the most stable being selenium dioxide, SeO_2. Under proper conditions, selenium forms selenides with hydrogen, carbon, nitrogen, phosphorus, and sulfur. Crystalline selenium does not react with water.

Selenium remains unaffected by dilute sulfuric acid or hydrochloric acid, but dissolves in a nitric–hydrochloric acid mixture, concentrated nitric acid (qv), and concentrated sulfuric acid. It is oxidized by ozone and solutions of alkali metal dichromates, permanganates, and chlorates and calcium hypochlorite. Selenium dissolves in strong alkaline solutions yielding selenides and selenites. It forms selenocyanates, MSeCN, with alkali metal cyanides, MCN, as well as many inorganic and organic derivatives of the corresponding acid, HSeCN. Selenium mixes in all proportions with sulfur and tellurium forming a continuous series of solid solutions and alloys.

Selenium can act both as an oxidant or reductant in chemical reactions. Selenium salts resemble the corresponding sulfur and tellurium salts in behavior. Selenium also forms a large number of organic compounds.

Occurrence. At 0.68 atom per 10,000 atoms Si, selenium is the 30th most abundant element. Selenium is widely dispersed in igneous rocks probably as selenide minerals; in volcanic deposits where it substitutes for some of the sulfur; in hydrothermal deposits, where it is associated isothermally with silver, gold, antimony, and mercury; and in massive sulfide and porphyry copper deposits, where it appears in large quantities but only in small concentrations. In sedimentary rocks such as sandstones, carbonaceous siltstones, phosphorite rocks, and limestones, selenium is syngenetic. The estimated selenium content of the oceans is ca 0.5 ppb, and only a small fraction of the selenium is transported into the sea by weathering and erosion. The principal species in the seas is SeO_4^{2-}. Adsorption by some marine organisms contributes to the removal of selenium from seawater.

Selenium occurs in alkaline soils chiefly as selenates which, when water soluble, are readily available to plants. In acid soils, selenides and to some degree elemental selenium are prevalent but little is available to plants. The concentration of selenium in most soils is between 0.01 and 2.0 mg/kg.

Selenium forms natural compounds with 16 other elements. It is a main constituent of 39 mineral species and a minor component of 37 others, chiefly sulfides. The minerals are finely disseminated and do not form a selenium ore.

Manufacture and Recovery. The main source of selenium is electrolytic copper refinery slimes. Selenium is an impurity present in the copper anode which is not solubilized during the refining process and ultimately accumulates in the bottom of the electrorefining tank. Slimes generated by the refining of primary copper, ie, copper produced from ores or concentrates, generally contain from 5–25% selenium as a constituent of various intermetallic compounds. Copper refinery slimes are always treated for the recovery of high value metals, such as gold, silver, platinum, and palladium, and generally most of these processes also include some method of recovering selenium as an additional by-product.

Purifying Selenium and Tellurium. Selenium recovery processes generally yield a metal product which contains some tellurium, and, correspondingly, recovered tellurium generally contains some small amount of selenium. Classically, the purification of elemental selenium from tellurium is achieved by distillation because of the much greater volatility of selenium compared to tellurium.

Selenium Compounds

Selenium is incorporated into a large variety of commercially used compounds, including barium selenate, nickel–selenium, and selenourea.

Inorganic Compounds. Inorganic selenium compounds are similar to those of sulfur and tellurium. The most important inorganic compounds are the selenides, halides, oxides, and oxyacids. Selenium oxidation states are −2, 0, +1, +2, +4, and +6.

Selenium forms compounds with most elements. The only important hyride of selenium is hydrogen selenide, H_2Se. Hydrogen selenide is a strong reductant. The aqueous solution is weakly acidic, although it is stronger than aqueous acetic acid.

Most binary selenides are formed by heating selenium in the presence of the element; reduction of selenites or selenates with carbon or hydrogen; and double decomposition of heavy-metal salts in aqueous solution or suspension with a soluble selenide salt.

Sulfur and selenium form homogeneous solutions in the liquid state, and these form a series of solid solutions on crystallization. Selenium and tellurium form a continuous series of solid solutions or alloys, which probably contain different simple and mixed chains of various lengths. Selenium combines directly with fluorine, chlorine, bromine, and iodine, and forms the monohalides, Se_2X_2; the dihalides, SeX_2; the tetrahalides, SeX_4; and the hexafluoride, SeF_6. The compounds are covalent and volatile. Selenium oxyhalides, $SeOX_2$, are prepared by the addition of a halogen to a dry mixture or a carbon tetrachloride suspension of selenium and selenium dioxide.

Organic Compounds. The chemical properties of organosulfur, organoselenium, and organotellurium compounds are markedly similar. Because bond stability decreases with increasing atomic number of the element, thermal stability and stability on exposure to light of all wavelengths decrease and oxidation susceptibility increases. They range from the simple COSe, CSSe, and CSe_2 to complex heterocyclic compounds, selenium-containing coordination compounds, and selenium-containing polymers.

Selenium dioxide, SeO_2, is prepared by burning selenium in a current of air or oxygen and, optionally, by passing it over a catalyst or by oxidation with nitric acid to selenous acid followed by evaporation to dryness by heating. The compound is white and crystalline with a tetragonal structure and is yellowish green as the vapor. Selenium trioxide, SeO_3, is white, crystalline, and hygroscopic. It can be prepared by the action of sulfur trioxide on potassium selenate or of phosphorous pentoxide on selenic acid. It forms selenic acid when dissolved in water.

The important oxyacids are selenous acid, H_2SeO_3, and selenic acid, H_2SeO_4. Selenous acid is an oxidant, and it is also readily oxidized to selenic acid. The selenate and hydrogen selenate salts are similar in their properties to the sulfates and hydrogen sulfates. Selenic acid and selenates are stronger oxidants than sulfuric acid or sulfates.

Specifications and Standards

There are no official specifications for selenium, apart from those of the U.S. Government for purchases under the national stockpile plan. Some producers publish standards, and some users specify a screen analysis and maximum content of certain impurities.

Health and Safety Factors

Commercial elemental selenium along with the stable metallic selenides are considered relatively nontoxic. However, other selenium compounds, particularly hydrogen selenide, the halides, oxides, and the organics, are highly toxic and must be handled with care. Selenium can enter the body through inhalation, ingestion, or absorption through the skin, where it accumulates primarily in the liver and kidney. Symptoms of selenium poisoning include bronchial irritation, gastrointestinal distress, nasopharyngeal irritation, and garlic odor on the breath.

Selenium plays a dual role in a living organism, depending on the compound and the amount adsorbed. Controlled small doses of some compounds are used in medicine and as diet supplements, for example, ca 0.1 ppm of diet dry matter for livestock (see FEED ADDITIVES; MINERAL NUTRIENTS). Larger amounts can be toxic.

Uses

Electrical and Optical. The electrical and optical uses are based on the semiconducting and photoresponsive properties of selenium in its amorphous and crystalline trigonal forms. In the metastable amorphous form, the electrical conductivity is very low, making selenium almost an insulator, but when exposed to light the conductivity is greatly increased. This effect is exploited in xerography and in vidicons.

Metallurgical. The metallurgical applications of selenium normally involve its use as a minor alloying additive to enhance the properties of both ferrous and nonferrous metals and alloys (see IRON; STEEL).

Glass and Ceramics. Selenium is used to decolorize the greenish tinge in glass caused by the presence of iron as an impurity.

Pigments. Selenium combines with cadmium sulfide to produce cadmium sulfoselenide pigments with colors in the range of orange to maroon depending on the Cd–S–Se ratio.

Rubber. The rubber industry uses both finely ground metallic selenium and selenium diethyl dithiocarbamate (Selenac), with natural rubber and styrene–butadiene rubber (SBR) to increase the rate of vulcanization and improve the aging and mechanical properties of sulfurless and low sulfur stocks (see RUBBER CHEMICALS).

Lubricants. Selenium and its compounds are added to lubricating oils and greases for normal and extreme pressure service (see LUBRICATION AND LUBRICANTS).

Organic Chemistry and Pharmaceuticals. Selenium dioxide is an important oxidizing agent and catalyst in the synthesis of organic chemical and drug products (see HORMONES; VITAMINS).

Medicine and Nutrition. A stabilized buffered suspension of selenium sulfide is marketed as Selsun Blue for control of seborrheic dermatitis of the scalp. Topical application of selenium sulfide controls dermatitis, pruritis, and mange in dogs (see COSMETICS; VETERINARY DRUGS). Selenium has been identified as an essential micronutrient in nutrition. In conjunction with vitamin E, selenium is effective in the prevention of muscular dystrophy in animals.

J. E. HOFFMANN
Jan Reimers and Associates USA Inc.
M. G. KING
Selenium–Tellurium Development Association

R. A. Zingaro and W. C. Cooper, *Selenium,* Van Nostrand Rheinhold Co., New York, 1974.

Proceedings of the Sixth International Symposium on Industrial Uses of Selenium and Tellurium, Phoenix, Ariz., 1998, Selenium–Tellurium Development Association, Grimbergen, Belgium.

J. E. Hoffmann, *J. Metal.,* 32–38 (July 1989).

C. Reilly, *Selenium in Food and Health,* Chapman Hall, London, 1996.

SEMICONDUCTORS

SILICON-BASED SEMICONDUCTORS

In the semiconductor industry, silicon is the basis for the technology sector, which includes computers, computer peripherals, and telecommunications. At present, the semiconductor market is predominantly metal oxide semiconductor (MOS) digital integrated circuits. To a great extent, silicon-based semiconductor technology has been driven by advances in MOS dynamic access memory (MOS DRAM) technology. Newer developments have been the result of steady advances in MOS manufacturing technology rather than the invention of new semiconductor devices.

Semiconductor Materials Theory

Silicon is a Group 14 (IV) element of the Periodic Table. This column includes C, Si, Ge, Sn, and Pb and displays a remarkable transition from insulating to metallic behavior with increasing atomic weight. Silicon and germanium are semiconductors; they look metallic, but they conduct poorly. Traditionally, semiconductors have been defined as materials whose resistance rises with decreasing temperature, unlike metals whose resistance falls. Table 1 indicates some of the key properties of Group 14 semiconductors.

The electrical behavior of many materials can be explained elegantly by band theory. When atoms are packed together their energy levels split into bands of closely spaced energy levels. Some of these bands overlap and some are separated by energy gaps. Whether a material is an insulator or conductor depends on whether a band is completely or partially filled. In the case of silicon, the atom contains two electrons which fill the 3*s* level and two electrons in the six states of the 3*p* level. As the atoms are brought together in the diamond (C) configuration, these eight states per atom hybridize and split, forming two bands of energy levels separated by an energy gap. At 0 K, the lower valence band with four bonding states per atom is filled and the upper conduction band, which also has four antibonding states per atom, is empty. Consequently, diamond, silicon, and germanium

Table 1. Important Properties[a] of Group 14 Semiconductors

Material	Cubic lattice constant, pm	Band gap, eV	Intrinsic carrier conc cm^{-3}	Relative dielectric constant, ϵ_r	Mobility, $cm^2/(V{\cdot}s)$ Electrons	Holes
C (diamond)	356.683	5.47	7.6×10^{-26}	5.7	1800	1200
6H–SiC	[b]	3.0	1.6×10^{-6}	9.66		
SiC					380	75
Si	543.095	1.12	1.45×10^{10}	11.9	1500	450
Ge	564.613	0.66	2.4×10^{13}	16.0	3900	1900

[a] At 300 K. [b] α-Sic crystallizes in the wurtzite lattice with $a = 308.6$ pm and $c = 1511.7$ pm.

are essentially insulators with an empty conduction band separated from a filled valence band by an energy gap, E_g. At higher temperatures, some electrons are excited to the conduction band, leaving behind empty states in the valence band. Applying pressure decreases the lattice spacing and increases E_g; thermal expansion decreases E_g. These effects can be exploited in pressure- and temperature-sensitive transducers.

Near a conduction band minimum, the energy of electrons depends on the momentum in the crystal. Thus, carriers behave like free electrons whose effective mass differs from the free electron mass. Their energy is given by equation 1, where E_c is the energy of the conduction band minimum, p is the

$$E - E_c = p^2/2m^* \tag{1}$$

effective momentum of the electrons in the crystal, and m^* is the effective mass. Because the crystal momentum p is related to the wave vector $k = 2\pi/\lambda$, the energy bands are often drawn as E vs k diagrams. The nearly filled states at the top of the valence band behave in conduction as positive charge carriers (holes) with a different effective mass. The density of states is the number of distinct states or k-values available for occupation within a given energy interval. There are two states for each unique k-value, corresponding to the two possibilities for electron spin. A low effective mass means that the E vs k diagram has a steep slope and a correspondingly low density of states. Figure **1a** shows the E vs k diagram for an idealized semiconductor in which the electrons and holes have nearly equal mass. The solid lines in Figure **1a** correspond to available states for electrons with quantized k-values. If the available states are counted as a function of energy, the density (per unit volume) of states (DOS) for electrons is given, as shown in Figure **1b**. Diamond, silicon, and germanium have an indirect band gap, ie, the conduction band minimum has a different momentum than the valence band maximum.

Semiconductor Statistics

Intrinsic Semiconductors. For semiconductors in thermal equilibrium, $\langle N(E)\rangle$, the average number of electrons occupying a state with energy E is governed by the Fermi-Dirac distribution. Because, by the Pauli exclusion principle, at most one electron (fermion) can occupy a state, this average number is also the probability, $P(E)$, that this state is occupied (see Fig. **1c**). In equation 2, K

$$\langle N(E)\rangle = P(E) = \frac{1}{1 + \exp(E - E_F/KT)} \tag{2}$$

is the Boltzmann constant and T is the absolute temperature in Kelvin. E_F, the Fermi energy, normalizes the distribution $P(E)$ at all temperatures. In intrinsic and extrinsic semiconductors the equilibrium concentrations of electrons and holes are described by a lever rule. In an intrinsic semiconductor, charge conservation gives $n = p = n_i$, where n_i is the intrinsic carrier concentration as shown

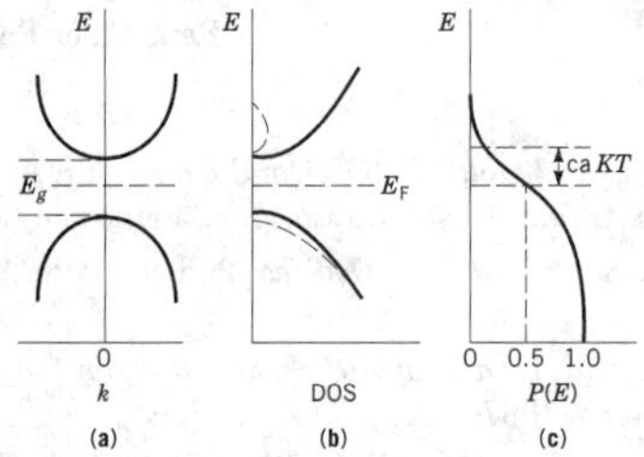

Figure 1. (**a**) Energy, E, versus wave vector, k, for free particle-like conduction band and valence band electrons; (**b**) the corresponding density of available electron states, DOS, where E_F is Fermi energy; (**c**) the Fermi-Dirac distribution, ie, the probability $P(E)$ that a state is occupied, where K is the Boltzmann constant and T is absolute temperature in Kelvin. The tails of this distribution are exponential. The product of $P(E)$ and DOS yields the energy distribution of electrons shown by (– – – –) in (**b**).

in Table 1. N_c and N_v are the effective densities of states per unit volume for the conduction and valence bands.

Extrinsic Semiconductors. The most common impurities or dopants introduced into the silicon crystal are Periodic Table Group 15 (V) and Group 13 (III) elements. These dopants are substitutional, ie, they replace silicon atoms in the crystal lattice. Group 15 elements contain five valence electrons, four of which bond covalently with the surrounding silicon atoms. These impurities are called donors because their fifth electron is weakly bound and can be donated to the conduction band by thermal excitation.

The carrier concentrations in doped or extrinsic semiconductors to which donor or acceptor atoms have been added can be determined by considering the chemical kinetics or mass action of reactions between electrons and donor ions or between holes and acceptor ions. The condition for electrical neutrality is given by equation 3. When the predominant dopants are donors, the semiconductor is

$$n + N_{a-} = N_{d+} + p \tag{3}$$

n-type. In n-type semiconductors the majority carriers are electrons and the minority carriers are holes, so that $N_{d+} > N_{a-}$ and $n \approx N_{d+}$. For p-type semiconductors, $N_{a-} > N_{d+}$ and $p \approx N_{a-}$.

For lightly doped n-type semiconductors at normal operating temperatures there is complete donor dissociation (donor saturation). $N_{d+} \approx N_d$, corresponding to E_F well below E_d. In the presence of acceptors, $n = N_{d+} - N_{a-} \approx N_d - N_a$. This frequently occurs because semiconductor devices are customarily made by the diffusion or implantation of excess donors into a p-type substrate or vice versa.

Semiconductor Transport

Charge carriers in a semiconductor are always in random thermal motion with an average thermal speed, v_{th}, given by the equipartion relation of classical thermodynamics as $m^* v_{th}^2/2 = 3KT/2$. As a result of this random thermal motion, carriers diffuse from regions of higher concentration. Applying an electric field superposes a drift of carriers on this random thermal motion. Carriers are accelerated by the electric field but lose momentum to collisions with impurities or phonons, ie, quantized lattice vibrations.

MOS Capacitance

So far, the fundamental factors which affect the current a device can carry have been considered. In practice, the speed of digital (and analog) devices is determined by the time it takes to charge or discharge a capacitor, C, where $\Delta Q = C\Delta V = I\Delta t$. Thus, from this general relationship, low values of Δt require some combination of low C, low ΔV, or large I. Different device technologies and circuit techniques achieve this in different ways.

Figure 2 shows the charge distributions and band bending in a MOS capacitor as the applied voltage is changed. As the voltage is increased, mobile electrons are drawn to the surface. Strong inversion is said to occur when the density of electrons at the surface equals the original density of holes. Strong inversion is generally used to define the threshold voltage for conduction in MOSFETs.

Thus, this MOS capacitor has three regions of operation: accumulation, for negative voltages when mobile holes are drawn to the surface; depletion, for positive voltages which repel mobile holes from the surface; and inversion, for positive voltages greater than the threshold voltage when mobile electrons are drawn to the surface. The foregoing description is of a NMOS capacitor where the semiconductor is p-type; a similar account can be given for a PMOS capacitor where the semiconductor is n-type, except that the metal plate voltages then have the reverse sign and the roles of electrons and holes are interchanged.

The excellence of a properly formed SiO_2–Si interface and the difficulty of passivating other semiconductor surfaces has been one of the most important factors in the development of the worldwide market for silicon-based semiconductors. MOSFETs are typically produced on $\langle 100\rangle$ silicon surfaces.

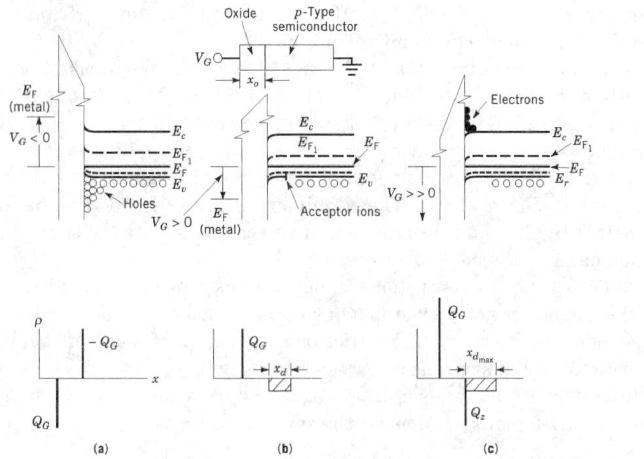

Figure 2. Charge distributions in the voltage-biased MOS capacitor: (**a**) accumulation of majority carriers near surface; (**b**) depletion of majority carriers from surface; (**c**) inversion, accumulation of minority carriers near surface. V_G = gate voltage; Q_G = gate charge; x_o, x_d, and $x_{d\max}$ = depletion widths; and ρ = charge density.

Device Physics

The ideal rectifier or diode is a two-terminal device that allows current flow in only one direction. The transistor is a three-terminal device in which current flow through two terminals is controlled by the third. Transistors can be used as analogue amplifiers or digital switches.

In addition to its use as a rectifier, the p–n junction is the fundamental building block for bipolar, junction FET (JFET), and MOSFET transistors. At an abrupt interface between n-type and p-type semiconductors there is an enormous difference between electron and hole concentrations on the two sides of the interface. This corresponds to a difference in Fermi energies on the two sides. Thermal equilibrium with no applied voltage and no current flow requires a constant Fermi energy throughout the material. This equilibrium is achieved if electrons diffuse into the p-region and holes diffuse into the n-region where they become minority carriers. This leaves behind a depletion region at the interface with ionized acceptors on the p-side and ionized donors on the n-side. This charge separation leads to a built-in or diffusion voltage difference, V_{bi}, across the interface. This built-in voltage corresponds to a difference in band energies.

In general, in a planar process, p–n junctions are formed just below the surface of a silicon wafer by the implantation of donor ions into a p-type region or acceptor ions into an n-type region. Thus, the general concern is with n^+–p or p^+–n junctions. As the initial wafer concentration of acceptors or donors in silicon increases from 10^{14} to 10^{18} cm^{-3}, V_{bi} increases from about 0.81 to 1.04 V for a p^+–n junction and is about 10 mV higher for an n^+–p junction.

When a positive voltage difference from p to n is applied across the p–n junction, the voltage drop is mostly across this depletion region and reduces the built-in voltage. When electrons are injected as minority carriers into a p-type semiconductor they may diffuse, drift, or disappear. That is, their electrical behavior is determined by diffusion in concentration gradients, drift in electric fields (potential gradients), or disappearance through recombination with majority carrier holes.

For positive voltages the current, I, can become exponentially large and is limited by junction heating and burnout. The depletion region width and minority carrier distributions near the junction must be altered in order to change the voltage across a p–n diode. This corresponds to two capacitive components, the junction capacitance which dominates in reverse bias and the diffusion or charge-storage capacitance which dominates in forward bias. The charge separation caused by the concentrations of ionized donors and acceptors in the depletion region forms a capacitor.

Schottky Diodes and Ohmic Contacts. A metal–semiconductor junction may be rectifying or ohmic. Such a rectifying junction (Schottky diode) has a nonlinear, asymmetric dependence of current on voltage like the p–n junction, which could be called a Shockley diode. An ohmic contact has a linear, symmetric dependence of current on voltage.

Both ohmic and rectifying behavior are possible. Because rectifying barriers are more common than ohmic barriers, dopant concentrations of about 10^{19} cm^{-3} are typically used to form ohmic contacts on silicon. Tunneling through the depletion region is possible if it is $<\sim 10$ nm, leading to ohmic behavior with a rectifying barrier.

The n–p–n bipolar junction transistor (BJT) may be regarded as two back-to-back p–n junctions separated by a thin base region. If external voltages are applied so that the base-emitter (BE) junction is forward biased and the base-collector (BC) junction is reverse biased, electrons injected into the base from the emitter can travel to the base-collector junction within their lifetime. If the time for minority carrier electrons to transit the base (base transit time) is short compared with their lifetime, the fraction of electrons which reaches the collector junction is nearly one. When they reach the collector they are accelerated across the depletion region, giving rise to a current.

Unlike bipolar transistors, field-effect transistors (FETs) are unipolar devices whose characteristics are determined by majority carrier transport through a channel. Current flow in this channel is altered by forming or constricting current flow by the application of an electric field. In the junction FET (JFET), a conducting channel between two ohmic contacts is constricted by the depletion regions of two opposing p–n junctions. In the MOSFET, a conducting channel between the two p–n junctions is formed by application of a field at the silicon surface by a MOS capacitor. MOSFETs dominate silicon device technology.

Several features should be noted about MOSFET operation. One is that MOSFETs are horizontal devices whose currents are limited by the depth of the channel. As a result, MOSFETs do not have as much current drive as bipolar transistors which, as vertical devices, have more area through which to pass current. This is less true for JFETs where current flow is not restricted to a thin surface channel. The MOSFET is a symmetrical device which can conduct equally well in either direction. This allows it to be used in switch or pass-transistor logic as well as static and dynamic logic families.

Newer Applications

Although the predominant applications for silicon-based semiconductors are transistors and integrated circuits (ICs) for computation and communication, several other applications have become significant. These include flat-panel displays, flash memories, power semiconductors, micromechanics, and cryoelectronics. Silicon devices correspond to front-end processing for fabricating integrated circuits. As the minimum feature sizes of devices shrink to submicrometer dimensions and chips increase in size, back-end processing of multilayer interconnect structures has become much more important in determining integrated circuit performance.

KENNETH ROSE
Rensselaer Polytechnic Institute

W. Shockley, *Electrons and Holes in Semiconductors with Applications to Transistor Electronics,* D. Van Nostrand Co., Inc., Princeton, N.J., 1950.

S. M. Sze, *Physics of Semiconductor Devices,* 2nd ed., John Wiley & Sons, Inc., New York, 1981.

A. S. Grove, *Physics and Technology of Semiconductor Devices,* John Wiley & Sons, Inc., New York, 1967.

N. H. E. Weste and K. Eshraghian, *Principles of CMOS VLSI Design,* Addison-Wesley Publishing Co., Reading, Mass., 1993.

C. W. Koburger and co-workers, *IBM J. Res. Devel.* **39,** 215 (1995).

AMORPHOUS SEMICONDUCTORS

Classification

Amorphous semiconductors are solids in which the atoms retain well-defined local, or nearest neighbor, order but lack long-range periodic order of crystalline solids, and the local bonding between atoms is characterized predominantly by covalent forces. These types of semiconductors are normally synthesized as metastable thin solid films; they are to be distinguished from glass, which is formed by continuous hardening of a cooled liquid. Amorphous material is a solid in internal equilibrium in which there is, just as in crystalline solids, a definite set of equilibrium positions about which the atoms oscillate. However, in contrast to crystalline solids, the amorphous materials do not exhibit translational symmetry, ie, there is no long-range order.

The tetrahedrally bonded materials, such as Si and Ge, possess only positional disorder; however, materials of this type exhibit high density of defect states (DOS). It is only with the addition of elements such as hydrogen and/or a halogen, typically fluorine, that the DOS is reduced to a point such that electronic device applications emerge. These materials contain up to ~10 atomic % hydrogen, commonly called hydrogenated amorphous silicon (*a*-Si:H).

Effect of Disorder

Inherent disorder has an effect on electronic and optical properties of amorphous semiconductors providing for distinct differences between them and the crystalline semiconductors. The inherent disorder provides for localized as well as nonlocalized states within the same band such that a critical energy, E_c, can be defined by distinguishing the two types of states.

Growth of Amorphous Silicon

By far the most widely studied and used deposition technique for preparing amorphous silicon is the plasma-enhanced chemical vapor deposition (PECVD) technique, which typically utilizes a capacitively coupled multichamber system, as shown in Figure 1. A gas, such as SiH_4, is allowed to pass at a controlled rate between the two plates and the plasma is generated using d-c, audio, radio-frequency (r-f), or microwave frequencies. A variety of species are produced, such as atoms, free radicals, and stable and unstable ions. The mean energy of the electron is on the order of a few eV and electron temperature can be up to 100 times higher than that of the gas. Hence the electrons possess enough energy to break down the molecular bonds.

Various plasma diagnostic techniques have been used to study the SiH_4 discharges and results have helped in the understanding of the growth kinetics. These processes can be categorized as r-f discharge electron kinetics, plasma chemistry including transport, and surface deposition kinetics.

The state-of-the-art *a*-Si:H films are deposited at the rate of 1–3 A/s with the gas utilization rate on the order of 15%. The use of higher excitation frequency can lead to deposition rates in excess of 15 A/s and still give relatively good film properties.

Properties

An inclusion of an element such as H plays a crucial role in determining the properties of the *a*-Si:H material. The optoelectronic properties of the *a*-Si:H films depend on many deposition

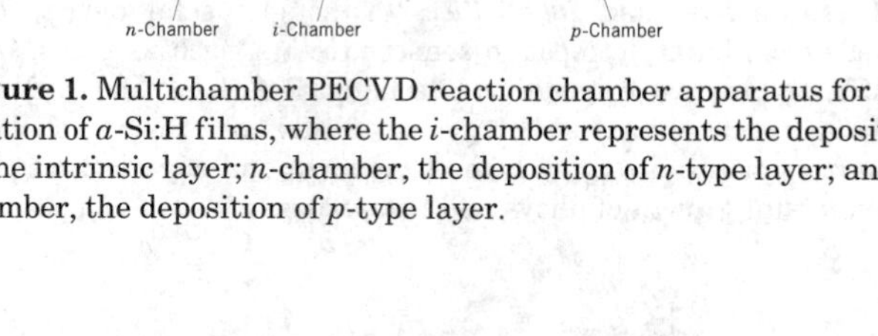

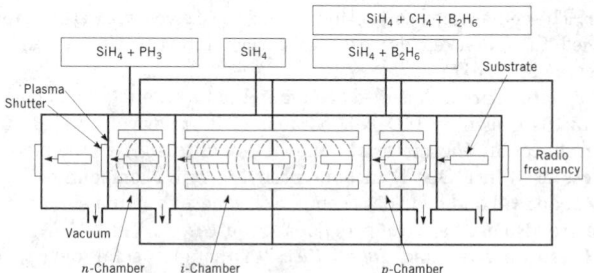

Figure 1. Multichamber PECVD reaction chamber apparatus for fabrication of *a*-Si:H films, where the *i*-chamber represents the deposition of the intrinsic layer; *n*-chamber, the deposition of *n*-type layer; and *p*-chamber, the deposition of *p*-type layer.

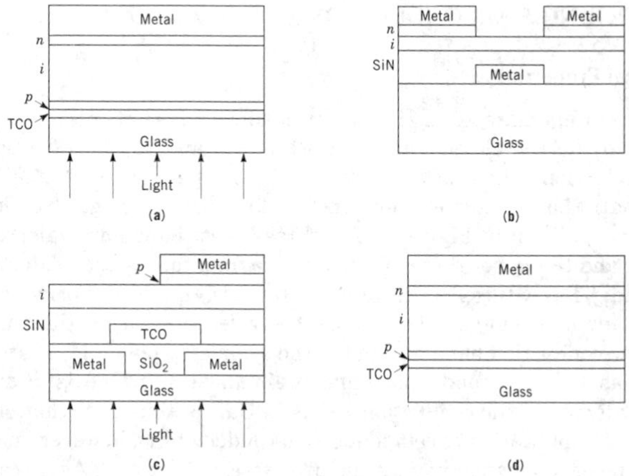

Figure 2. Some electronic device applications using amorphous silicon: (**a**) solar cell, (**b**) thin-film transistor, (**c**) image sensor, and (**d**) nuclear particle detector. TCO = transparent conducting oxide.

parameters such as the pressure of the gas, flow rate, substrate temperature, power dissipation in the plasma, excitation frequency, anode–cathode distance, gas composition, and electrode configuration. Deposition conditions that are generally employed to produce device-quality hydrogenated amorphous Si (*a*-SiH) are as follows: gas composition = 100%SiH_4; flow rate is high, ~40 cm^2; pressure is low, 26–80 Pa (200–600 mtorr); deposition temperature = 250°C; radio-frequency power is low, <25 mW/cm^2; and the anode-cathode distance is ~1–4 cm.

Another parameter of relevance to some device applications is the absorption characteristics of the films. Because the k quantum is no longer valid for amorphous semiconductors, *a*-Si:H exhibits a direct band gap (~1.70 eV) in contrast to the indirect band gap nature in crystalline Si.

Applications of *a*-Si-Based Semiconductors

The inherent disorder possessed by *a*-Si:H alloy limits the mobility of the free carriers (electrons and holes) to about 10 $cm^2/(s{\cdot}V)$; this is compared with crystalline Si, in which the electron mobility is 1500 $cm^2/(s{\cdot}V)$. However, crystalline Si is expensive to manufacture and its size is limited to about 20 cm in diameter. Many applications have either emerged or been identified which preclude the use of crystalline Si because of cost, size, or both. The basic commonality in these applications is the ability to fabricate devices on areas much larger than can be addressed by crystalline Si. Furthermore, these applications are not demanding in terms of speed, which then provides *a*-Si:H alloy with a distinct competitive advantage.

Figure 2 shows the basic construction of the devices used in different applications, involving the deposition of multilayers of *a*-Si:H of intrinsic (i), doped (n^+, p^+), and closely allied films, such as amorphous silicon nitride, SiN, and transparent conducting oxide (TCO). As in crystalline semiconductors, the basic building block is a junction (p–n) or thin-film transistor.

ARUN MADAN
MVSystems, Inc.

A. Madan and M. Shaw, *The Physics and Applications of Amorphous Semiconductors,* Academic Press, Inc., San Diego, Calif., 1988.

W. E. Spear and P. G. LeComber, *J. Noncryst Solids* **727,** 8–10 (1972); A. Madan, Ph.D dissertation, University of Dundee, U.K., 1973; W. E. Spear and P. G. LeComeber, *Phil. Mag.* **33,** 935 (1976).

R. Zallen, *Physics of Amorphous Solids,* John Wiley & Sons, Inc., New York, 1983.

G. Bruno, P. Capezzuto, and A. Madan, eds., *Plasma Deposition of Amorphous Silicon Based Materials,* Academic Press, Inc., New York, 1995.

COMPOUND SEMICONDUCTORS

Physical Properties

The defining characteristic of a semiconductor is that there is a gap between the completely filled valence states and the lowest excited or conduction states of the crystal, which is typically <2.5 eV. A schematic band structure for zincblende cubic direct gap semiconductors is shown in Figure 1. The three lowest bands are valence, v, bands and the upper three minima derive from the lowest conduction, c, band. The features of relevance to the design and performance of electronic and photonic devices are the valence band maximum, Γ_v, and the conduction band minima at the symmetry points Γ_c, L, and X. GaP has its lowest conduction band minimum at X; in GaAS it is at Γ_c. The ordering of the conduction bands in GaP is similar to that of Si. The wide application of compound semiconductors is, however, due to the type of conduction band structure associated with GaAs. Because the minimum band gap is $\Gamma_c - \Gamma_v$, and noting that the momentum of a photon is negligible, light emission via the direct, momentum conserving recombination of electrons (in the conduction band) with holes (in the valence band) is possible. In indirect semiconductors such as Si and GaP light emission is not readily obtained from recombination across the minimum band gap because momentum is not conserved in a simple photon emission process. For this reason virtually all light-emitting diodes (LEDs) and diode lasers are fabricated from compound semiconductors.

The primary advantage of a direct gap semiconductor is its ability to strongly absorb and emit light at the fundamental band gap. The secondary advantage is that a small isotropic electron-effective mass is typically associated with the Γ_c minimum.

The tertiary advantage of compound semiconductors is the ability to fabricate multilayer structures epitaxially, widely varying the material properties such as band gap and refractive index from layer to layer. Within the constraint that the lattice constants are well matched it is usually possible to grow heterostructures of different 13 and 15 and 2, 12, and 16 semiconductors with atomically abrupt interfaces and monolayer thickness control.

In the design of heteroepitaxial layers, which are not lattice matched, the elastic constants of the layers are essential to the calculation of the thermodynamically stable maximum layer thickness

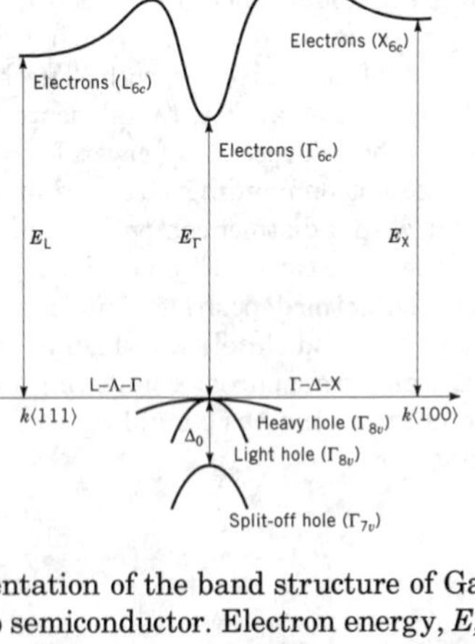

Figure 1. Representation of the band structure of GaAs, a prototypical direct band gap semiconductor. Electron energy, E, is usually measured in electron volts relative to the valence, v, band maximum which is used as the zero reference. Crystal momentum, k, is in the first Brillouin zone in units of $2\pi/a$ where a is the lattice constant. In semiconductors with an indirect band gap the ordering of the valence (hole) states is unchanged, whereas the ordering of the conduction, c, states is inverted to $X < L < \Gamma$. In the semimetal HgTe the conduction band lies between the Γ_{8v} and Γ_{7v} valence bands and the curvatures of the Γ_{8v} light hole band and the Γ_{6c} electron band are inverted with respect to those shown for GaAs. Δ represents the line of symmetry.

for elastic deformation as well as the shape of the deformed unit cell. In most cases the ⟨001⟩ surface is the preferred surface for crystal growth.

Metal Organic Chemical Vapor Deposition

Metal organic chemical vapor deposition (MOCVD) is a technique that has evolved into one of the two prime growth methods for compound semiconductor devices. This is because of the capability to grow high quality, very thin layers of most semiconductors of Groups 13 and 15 and 16, making it the most versatile of the growth methods in use. MOCVD is also referred to as organometallic vapor-phase epitaxy (OMVPE) as well as MOVPE and OMCVD. *Epitaxy* is the term used to describe the growth by chemical reaction of a thin layer of semiconductor on the surface of a bulk crystal (wafer) of similar or different material, where the layer has a well-defined crystalline orientation with respect to the bulk crystal. MOCVD growth typically involves decomposition in a cold wall reactor of chemical precursors, usually an organometallic Group 13 compound and a Group 15 hydride. For the growth of gallium arsenide from trimethylgallium and arsine a typical reactor is $Ga(CH_3)_3 + AsH_3 \rightarrow GaAs + 3\ CH_4$. This technique has been extended to all of the important 13 and 15 semiconductors such as AlGaAs, InP, InGaAs, and InGaAsP. Semiconductors grown by MOCVD for use as lasers, solar cells, detectors, and high speed transistors are in commercial production. MOCVD is a state-of-the-art technique for preparing epitaxial layers of compound semiconductors.

CVD. Chemical vapor deposition (CVD) is usually defined as a technique that involves a chemical reaction which results in the deposition of a solid on a specific surface. Growth techniques such as MOCVD, chloride or hydride vapor-phase epitaxy, and gas source molecular beam epitaxy are considered to be CVD techniques. Growth methods such as liquid-phase epitaxy, molecular beam epitaxy, and sputtering are not considered CVD techniques because they lack one or more of the criteria in the definition. CVD offers the advantages of near equilibrium growth which allows for the precise control of growth rates and improved purity. The use of CVD permits utilization of a growth pressure from atmospheric down to very low pressures ($<1.3 \times 10^{-3}$ Pa) as well as a large variety of reactants with widely varying chemical reactivities. The capability for using materials that would decompose in the liquid or high temperature gaseous state is also an advantage of CVD. MOCVD has demonstrated very precise and reproducible control of alloy composition, layer thickness, and dopant levels.

Transport Phenomena. One significant though often overlooked kinetic aspect of MOCVD is the influence that gas flow or hydrodynamics has on mass transport and uniformity of deposition. Because of the complexity of these effects, they are not intuitive and are often overlooked in the design of an MOCVD reactor.

Materials. Because of the increased demand for improved materials and increased understanding of the MOCVD process, new chemicals have been designed which take into consideration the decomposition mechanisms of the sources and how this impacts on the incorporation of impurities, particularly carbon, into the epitaxial layer. The replacement of methyl Group 13 sources with ethyl groups has led to a marked decrease in carbon incorporation in AlGaAs grown by MOCVD.

Dopants. Carbon has also been found to be an effective dopant for the 13:15s grown by MOCVD because of its relatively high solubility in Al-containing materials and its low diffusivity. Other dopant sources used in MOCVD include methyl or ethyl compounds Se, Te, Be, Zn, and Cd and hydrides S and Se. Tetraethyltin and silane or disilane are also used as Group 14 dopant sources.

Heterostructures and Superlattices. Although useful devices can be made from binary compound semiconductors, such as GaAs, InP, or InSb, the explosive interest in techniques such as MOCVD and molecular beam epitaxy (MBE) came about from their growth of ternary or quaternary alloy heterostructures and superlattices. For the successful growth of alloys and heterostructures the composition and interfaces must be accurately controlled. The composition of alloys can be predicted from thermodynamics if the flow in the reactor is optimized. Otherwise, composition and growth rate variations are observed across the surface of the substrate. The surface reactions can be complex and lead to deviations from the predicted compositions. The temperature profiles and hydrodynamics of the reactor also affect the resulting composition from the effects of the transport of the reactants to the surface and the incomplete decomposition of the reactants.

Electronic Devices

The main advantages that compound semiconductor electronic devices hold over their silicon counterparts lie in the properties of electron transport, excellent heterojunction capabilities, and semi-insulating substrates, which can help minimize parasitic capacitances that can negatively impact device performance. The ability to integrate materials with different band gaps and electronic properties by epitaxy has made it possible to develop advanced devices in compound semiconductors. The hole transport in compound semiconductors is poorer and more similar to silicon. For this reason the majority of products and research has been in *n*-type or electron-based devices.

Photonic Devices

A technology for which compound semiconductors are uniquely suited is that of photonics, devices that generate, amplify, detect, propagate, transmit, or modulate light. This field therefore covers a wide range of systems, including light-emitting diodes (LEDs), lasers and optical amplifiers, detectors, waveguides such as optical fibers, lenses, and other optical components, and optical modulators. Although silicon has contributed much to photonics technology with its applicability to visible and near-infrared (ir) light detection systems, the compound semiconductors, with their wide range of primarily direct bandages, enable the realization of efficient light-emitting devices, namely laser diodes and LEDs, and extension of detector technology into the far ir region of the spectrum.

The relevance of photonics technology is best measured by its omnipresence. Semiconductor lasers, for example, are found in compact disk players, CD-ROM drives, and bar code scanners, as well as in data communication systems such as telephone systems. Compound semiconductor-based LEDs are utilized in multicolor displays, automobile indicators, and in traffic lights. The trend to faster and smaller systems with lower power requirements and lower loss has led toward the development of optical communication and computing systems.

Device Fabrication Technology

Technology for fabricating compound semiconductor devices includes wet etching, dry etching, ion implantation, metallization, dielectric overlayers, and lithography.

Supported by the U.S. Dept. of Energy under contract #DE-AC04-94AL85000.

TIMOTHY J. DRUMMOND
ROBERT M. BIEFELD
MARC E. SHERWIN
MARY H. CRAWFORD
Sandia National Laboratories

S. Strite and H. Morkoç, *J. Vac. Sci. Technol. B* **10,** 1237 (1992).

H. Morkoç, S. Strite, G. B. Gao, M. E. Lin, B. Sverdlov, and M. Burns, *J. Appl. Phys.* **76,** 1363 (1994).

H. Luo and J. K. Furdyna, *Semicond. Sci. Technol.* **10,** 1041 (1995).

J. Brice and P. Capper, eds., *Properties of Mercury Cadmium Telluride,* Institution of Electrical Engineers (INSPEC), London, 1987.

ORGANIC

Organic materials are used in transistors, photochromic devices, and commercially viable light-emitting diodes. The study of organic semiconductors and conductors is highly interdisciplinary, involving the fields of chemistry, solid-state physics, engineering, and biology.

The theory of conduction in organic semiconductors conveniently begins with a discussion of bonding in extended solids, since the nature of conduction is intimately related to the extent of delocalization of orbitals in the material to be studied. For inorganic conducting solids, the orbital overlap of the atoms in the crystal lattice results in the atomic energy levels spreading out to form energy bands rather than discrete levels. These energy bands allow for motion of charge carriers over many lattice sites without interruption by trapping events. The interactions between molecules in a molecular organic solid similarly leads to the formation of energy bands, with the extent of interaction between molecules (the extent of overlap of interacting orbitals) determining the width of the bands.

The distinctions between metals, semiconductors, and insulators are based on the band structure of the materials as well as on the electron occupancy of these bands. Figure 1 is a diagram of energy bands and occupancies for various classes of solids. As can be seen in Figure 1, insulators have a filled band formed from the valence orbitals (a *valence* band) with a higher-lying unfilled band formed from higher-energy orbitals (the *conduction* band). The region between these has no allowed states, and the energy difference required to promote an electron from the valence to the conduction band is termed the band gap of the material. For an insulator, the band gap is large, typically >4 eV. Since all of the levels of the valence band are filled, no electrons are able to carry current.

Metallic behavior is observed for those solids that have partially filled bands (Fig. **1b**), that is, for materials that have their Fermi level within a band. Since the energy bands are delocalized throughout the crystal, electrons in partially filled bands are free to move in the presence of an electric field, and large conductivity results.

Semiconductors show thermally activated conductivity. The reason can be seen from Figure **1c.** The band gap of a semiconductor is small relative to that of an insulator. Similar to insulators, pure intrinsic semiconductors have a filled valence band and unfilled conduction band. However, this is strictly true only at absolute zero. At higher temperatures, the energy of the surroundings can become large enough to thermally excite electrons from the valence to the conduction band, with the result that semiconductors have partially unfilled orbitals, giving rise to conduction.

Certain one-dimensional, chain-like organic semiconductors are found to display metallic properties (decreasing conductivity with increasing temperature) above a certain temperature. The phase transition that occurs in these stacked materials is termed a metal-to-insulator transition and is the result of the highly anisotropic nature of these materials. A stack of equally spaced atoms or molecules is subject to a reorganization to a lower symmetry configuration. This reorganization, termed a Peierls distortion, is the condensed matter

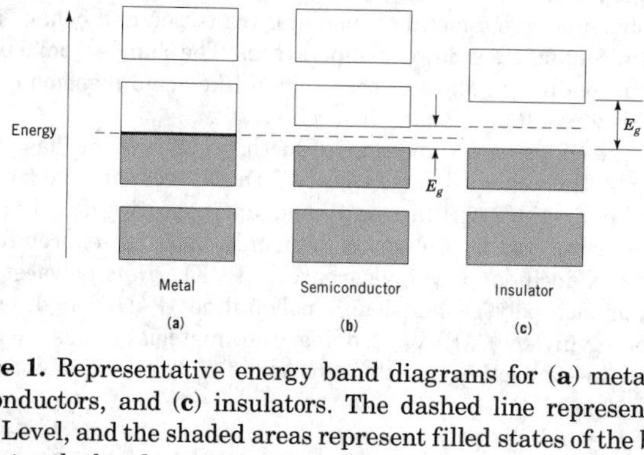

Figure 1. Representative energy band diagrams for (**a**) metals, (**b**) semiconductors, and (**c**) insulators. The dashed line represents the Fermi Level, and the shaded areas represent filled states of the bands. E_g denotes the band gap of the material.

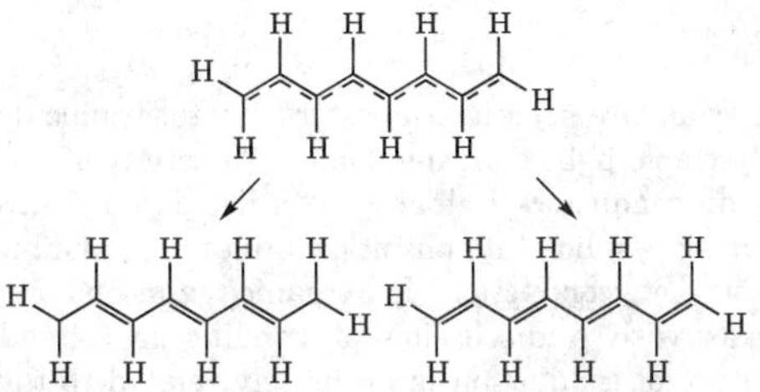

Figure 2. Representation of the Peierls distortion in *trans*-polyacetylene.

counterpart of the molecular Jahn-Teller distortion. The Peierls deformation results in pairing of molecules along the chain with the formation of short and long intermolecular spacings. This nuclear deformation causes changes in the electronic distribution of the lattice. The result of this redistribution is the opening of an energy gap at the Fermi level, with resulting semiconductivity. A Peierls distortion leads to alternating single and double bonds (Fig. 2) and the opening up of a band gap. As a result, undoped polyacetylene is a semiconductor.

Interaction of molecules to form charge-transfer complexes occurs via transfer of electron density from one molecule (an electron donor, D) to another (an electron acceptor, A), resulting in a partially ionic ground state. The amount of charge transfer is large when the ionization potential of the donor is low and the electron affinity of the acceptor is high. In a charge-transfer salt, the intermolecular interactions give rise to periodic arrays of donors and acceptors. Typical charge-transfer salts form as stacks of planar D and A molecules, though the ratio of D:A need not be 1:1, as the interaction can be spread over more than two molecules.

Some common single-carrier semiconducting salts are tetracyanoquinodimethane (TCNQ) radicals with alkali or other cations. The study of semiconductivity in these materials led to synthesis of more highly conductive charge-transfer salts based on the strong donor tetrathiafulvalene (TTF) with halogen or other acceptors.

Properties

Electrical properties have been the main focus of study of organic semiconductors, and conductivity studies on organic materials have led to the development of materials with extremely low resistivities and large anisotropies.

Most organic molecular crystals have only weak interactions between molecular units and have resulting low conductivities. Charge-transfer solids tend to have higher conductivities than the molecular compounds from which they are composed. The simple metal-ion radical salts of the composition $M^+(TCNQ)^-$ have conductivities intermediate between conductivity values for insulators and typical metals. The highly conductive class of solids based on TTF–TCNQ have less than complete charge transfer ($\sim$0.6 electrons/unit for TTF–TCNQ) and display metallic behavior above a certain temperature. However, these solids undergo a metal-to-insulator transition and behave as organic semiconductors at lower temperatures. The change from a metallic to semiconducting state in these chain-like one-dimensional (1-D) systems is a result of a Peierls instability.

The use of polymers in semiconducting applications has grown rapidly. In the undoped form, most of these π-conjugated systems are semiconducting, with a small bandgap separating filled valence and conduction bands. Polymers commonly used in semiconducting applications include polyvinylcarbazole (PVK), *trans*-polyacetylene, polythiophene, poly(*p*-phenylene), polyaniline (PANI), and poly(*p*-phenylene vinylene) (PPV). Doping the material leads to large increases in conductivity, with increases of several orders of magnitude possible.

One-dimensional polymers such as polyacetylene which have degenerate ground states are special cases. The nature of conduction in pure, *trans*-polyacetylene is a result of the alternating single bond–double bond structure that results from the Peierls instability. The energy of the system is the same if the alternating pattern of the bonds is reversed. Conductivity has been described for this system in terms of solitons (mobile kinks in the polymer chain that link alternating bond patterns), excitations that are unique to systems with degenerate ground states.

Optical Properties. New strong absorbance bands (charge-transfer bands) are often observed in spectra of donor–acceptor solids. These bands are the result of excitation from a ground state that is predominantly donor in character to an excited state that resides mainly on the acceptor. For a given acceptor, the frequency of this band varies linearly (over a small range) with the ionization potential of the donor molecule. By a suitable choice of donor and acceptor, it is possible to tune the frequency of this absorption band (and the associated fluorescence) throughout the visible region.

Polymers used in semiconducting applications such as light emitting diode displays are π-conjugated systems with small bandgaps of 1–3.5 eV. In the undoped state they give strong $\pi-\pi^*$ absorption bands in the visible or ultraviolet region, corresponding to transitions from the valence band to the conduction band levels. By tailoring the nature of the electron-donating and withdrawing substituents on the polymer backbone, absorbance and the resulting emission spectra can be shifted throughout the visible region.

Since charge-transfer and mobility of carriers are important features for organic semiconductors and devices, the energies of the highest occupied molecular orbital (HOMO) and lowest unoccupied molecular orbital (LUMO) are of interest. The energies of these levels are related to the efficiency of charge injection into a material as well as the degree of charge-transfer between donors and acceptors. Electrochemical measurements provide a convenient means for determining HOMO and LUMO levels by measuring the potentials required for oxidation and reduction of the material, respectively. Reversible oxidation and reduction of polymers is commonly used to increase conductivity in these systems (see ELECTRICALLY CONDUCTING POLYMERS).

Stability. The materials used in organic semiconducting applications are thermally labile upon exposure to high temperatures. For example, many of the compounds used in fabrication of organic light-emitting diodes (LEDs) are vapor deposited by resistive heating at relatively low temperatures in vacuum.

The presence of oxygen and water vapor has a profound effect upon the photostability of many organic semiconducting materials. It is necessary to control impurity and humidity levels during device production.

Structure

The structure of individual molecules can be tailored by synthetic methods, with molecules designed for specific applications. For example, rigid, planar structures are often used for dyes in photovoltaics, since this increases exciton mobility. Molecular packing and the structure of thin films can be dependent upon the method of fabrication and the structure of individual units. The nature of molecular and polymeric thin films for semiconducting applications ranges from highly crystalline to amorphous.

Single-Stack Acceptor. Simple charge-transfer salts formed from the planar acceptor TCNQ have a stacked arrangement with the $TCNQ^-$ units facing each other (intermolecular distances of ca 0.3 nm ($\sim$3 Å). Complex salts of TCNQ such as $TEA(TCNQ)_2$ consist of stacks of parallel TCNQ molecules, with cation sites between the stacks.

Single-Stack Donor. Ion-radical salts can also be formed from electron donors such as tetrathiafulvalene (TTF) or TMPD (*N,N,N′,N′*-tetramethyl-*p*-phenylene diamine) with inorganic acceptors such as halogens. The resulting structure of compounds such as $TTF(A)_x$ (A = acceptor) is a linear chain of parallel stacked TTF molecules. Unlike the situation found for the TCNQ salts, the TTF molecules in these solids are equally spaced along the chain. Substantial π-overlap exists between the TTF molecules in the chain, providing for conduction along the stacking direction.

Polymers. The individual units in a polymer are well-defined, but bulk polymer usually lacks the periodic structure characteristic of inorganic or small molecule crystalline solids. The structure of semiconducting and conducting polymers is determined by the units of the polymer backbone as well as pendant side-groups. Typical conducting polymers are conjugated systems that provide for motion of the electrons through the π-system, and both linear chain (polyacetylenes) and aryl chain poly(*p*-phenylene) polymers are commonly used.

Synthesis and Manufacture

Donors. Common selenium and sulfur-containing donors such as tetramethyltetraselenafulvalene, tetrathiafulvalene, and bis-(ethylenedithio)tetrathiafulvalene are commercially available, as are other common donors such as tetramethyl-*p*-phenylenediamine.

Acceptors. Most common acceptor molecules such as tetracyanoethylene or tetracyanoquinodimethane are commercially available.

Charge-Transfer Salts. Most charge-transfer salts can be prepared by direct mixing of donors and acceptors in solution. Semiconducting salts of TCNQ have been prepared with a variety of both organic and inorganic counterions.

Polymers. The π-conjugated polymers used in semiconducting applications are usually insulating, with semiconducting or metallic properties induced by doping (see ELECTRICALLY CONDUCTIVE POLYMERS). Most of the polymers of this type can be prepared by standard methods.

Uses

Organic semiconductors offer many advantages over inorganic-based materials for use in electronic and optoelectronic devices. Among the advantages are the possibility of tailoring properties through synthetic chemical modifications, the relative ease of generating large-area films, and low cost device manufacture. Applications include light-emitting diodes, photovoltaic devices, transistors, reproduction and resist materials, electrochromics, and switches and sensors.

SCOTT P. SIBLEY
Goucher College

F. Gutman, H. Keyzer, L. Lyons, and R. B. Samoano, *Organic Semiconductors,* part B, Robert E. Kreiger Publishing Co., Malabar, Fla., 1983, pp. 399–412.

M. H. Whangbo, in J. S. Miller, ed., *Extended Linear Chain Compounds,* Plenum Press, New York, 1982, p. 127.

R. S. Mulliken and W. B. Person, *Molecular Complexes,* John Wiley & Sons, Inc., New York, 1969.

SENSORS

Sensor technology has been revolutionized by advances in microelectronics and optoelectronics. The newer sensors have demonstrated improvements in the operation, reliability, safety, and efficiency of many engineering systems. Perhaps the most familiar consumer example of this technology is found in automobiles where silicon-micromachined pressure sensors, coupled to an electrochemical oxygen sensor, allow the engine management computer to provide the correct fuel and spark parameters for smooth, low pollution operation.

Increasingly, the word *sensor* is used interchangeably with or in place of the word *transducer* to represent the conversion of one type of energy into another. The word actuator, however, refers to a distinctly different set of devices capable of activating a device upon receiving an appropriate signal. Sensors and actuators may be found as part of a system on a single piece of silicon.

Various sensors are expected to respond in a predictable and controlled manner to such diverse parameters as temperature, pressure, velocity or acceleration of an object, intensity or wavelength of light or sound, rate of flow, density, viscosity, elasticity, and, perhaps most problematic, the concentration of any of millions of different chemical species.

Decision-Making Tool for Sensor Technology

The rapid advances in sensor technology have increased the difficulty of the task of finding the best sensor system for a given application. The sensor customer or engineer needs to be specific about defining the sensor problem to be solved, using language and descriptors that are commonly shared between the customer and a sensor technologist.

Sensors Based on Silicon Processing Technology

The tools for the development of new sensors include micromachining, batch processing, integrated sensors, readout and networking electronics, light-emitting and detecting structures, fiber optics (qv), etc. This has led to new applications of old fields of science from mechanics to physical chemistry, fluid mechanics, electrokinetics, capillary action, etc. These principles are being applied in a micro/nano-scale world, where batch processing means inexpensive, rugged systems (see NANOTECHNOLOGY).

Photodetectors. The most commercially successful silicon-based sensors are the photodetectors (qv). These are found in video cameras, security systems, ionizing radiation detectors, and many other applications. Photodetectors, closely related to solar cells in the principles of operation, all depend on light and ionizing radiation creating electron–hole pairs in the semiconducting material.

Micromachines. A fast-growing area of sensors based on silicon involves the use of micromachining to fabricate the sensor (see SEMICONDUCTORS). Chemical etching of the silicon is the key technology for the fabrication of these sensors. Isotropic etching refers to a process in which the etch rate of the materials is uniform in all directions. Anisotropic etching refers to etch rates that depend on the crystallographic orientation of the silicon wafer. Anisotropy ratios of 400:1 are possible for special directions. For pressure sensors, thin diaphragms of Si or related materials are etched into the wafer (see PRESSURE MEASUREMENTS).

Smart Sensors. A smart sensor is loosely defined as a sensing device with built-in intelligence that usually involves some kind of digital signal processing. Because the sensor is already being batch-fabricated on a silicon wafer, the signal-processing electronics can be created on the same chip as the sensor.

The integration of standard integrated circuitry and a sensor is not always a trivial extension of the IC process, because the processing steps involved in fabricating the sensor may be incompatible with the correct functioning of the IC. An alternative is the hybrid package where the sensor chip and the signal-processing chip are fabricated separately and then glued together in the same package.

Chemical Sensors. Silicon microelectronics are often cited as giving the most sensitive and stable measurements of physical phenomena, eg, charged particle detectors that can record a single x-ray photon or beta particle. In the realm of chemical detection, biological systems unequivocally outperform all of the microelectronically based chemical sensors in terms of sensitivity and selectivity. One of the best-studied examples of biological chemical detection is the sex attractant receptor of moths. The moth's microsensor is a receptor on a membrane protein that is part of the ion channel in the nerve cell wall.

A silicon-based sensor that operates on a similar principle is the chemically sensitive field-effect transistor (ChemFET) (8–10). The channel through which the electrical current flows is a very thin (about 10-nm) layer in the silicon crystal near a planar interface with an oxide of silicon that is thermally grown on the crystal. This interface forms the basis for most active devices in integrated circuits. The very high sensitivity of the ChemFET is in part a result of it acting as a chemical amplifier. In these small devices, as few as 10^6 hydrogen atoms occupying interfacial sites can cause an order of magnitude increase in conductivity.

Technologies Other Than Silicon

Acoustic Wave Sensors. Another physical transduction technique involves the use of acoustic waves to detect the accumulation of species in or on a chemically sensitive film. The device is operated in an oscillator configuration. Changes in resonant frequency are simply related to the areal mass density accumulated on the crystal face. These sensors, often referred to as quartz crystal microbalances (QCMs), have been coated with chemically sensitive films to produce gas and vapor detectors, and have been operated in solution as liquid-phase microbalances.

Another class of acoustic wave sensors uses surface acoustic waves (SAWs) to probe the accumulation of species in a surface film. These devices use interdigital electrodes that are photolithographically formed on the surface of a piezo-electric crystal to excite and detect the surface wave. The stress-free boundary condition imposed by the surface results in a mode of propagation that is confined to the surface and has acoustic energy distributed within one wavelength of the surface. This surface confinement of acoustic energy makes the SAW extremely sensitive to the properties of a surface film, particularly changes in its mass. The acoustic wave gravimetric technique can be extended to liquid-phase sensing. A number of gas and vapor sensors have been developed by placing chemically sensitive films on SAW devices and relying on their extreme mass sensitivity.

Microsensors Based on Optical Fibers. Optical fiber technology offers long-distance telemetry, freedom from electromagnetic interference, and small size, ie, approximately that of a human hair. The basic components necessary for sensor applications are being developed by the optical communications industry. Optical communications systems and optical sensors have the same functional design. Both need a light source, a detector, and a method of transmitting the light from the source and to the detector, namely, the optical fiber. For optical communications, a modulator is used to impose the information to be transmitted on the light beam. For optical sensing, the sensing element replaces the modulator and again imposes information on the light beam.

The sensing element can modify either the intensity, phase, or polarization state of the light; and physical sensors have been made that make use of all three techniques. For chemical sensing, most transduction mechanisms have focused on intensity modulation. Spectroscopy and fluorescence-based fiber optic sensors are in this group.

Electrochemical Microsensors. An example of a chemical microsensor is the oxygen sensor found in the exhaust system of almost all modern automobiles (see EXHAUST CONTROL, AUTOMOTIVE). It is an electrochemical sensor that uses a solid electrolyte, often doped ZrO_2, as an oxygen ion conductor. The sensor exemplifies many of the properties considered desirable for all chemical microsensors. It works in a process-control situation and has very fast (~100 ms) response time for feedback control. It is relatively inexpensive because it is designed specifically for one task and is mass-produced.

Semiconducting Metal Oxide Sensors. Another kind of conductometric sensor is formed from ceramic semiconducting metal oxides. The most common type of sensor consists of a thin porous film of tin oxide, SnO_2, deposited with a heater and electrodes to monitor the metal oxide conductivity. These are known as Taguchi gas sensors, after the inventor. They are used primarily in small stand-alone sensor packages for detecting leaks of hazardous gases such as carbon monoxide and natural gas. The mechanism for the change of conductivity in the presence of these reducing gases involves oxidative and reductive chemical reactions at the exposed surface and near-surface regions of the metal oxide. The dopant that makes these wide band gap materials semiconducting is actually a missing oxygen in the lattice, ie, a vacancy.

Written at Sandia National Laboratories with support provided by the U.S. Dept. of Energy under Contract No. DE-AC04-94AL85000.

R. C. HUGHES
A. J. RICCO
M. A. BUTLER
S. J. MARTIN
Sandia National Laboratories

Sensors 1995 Buyers Guide, Helmers Publishing Inc., St. Peterborough, N.H.

W. Göpel, J. Hesse, and J. N. Zemel, eds., *Sensors: A Comprehensive Survey,* VCH Publishers, New York, 1989.

J. Fraden, *AIP Handbook of Modern Sensors,* American Institute of Physics, New York, 1993.

J. Janata, *Principles of Chemical Sensors,* Plenum Publishing Corp., New York, 1989.

SEPARATION

CENTRIFUGAL SEPARATION

Centrifugal separation is a mechanical means of separating the components of a mixture of liquids or of liquids and solid particles. The material is accelerated in a centrifugal field which acts upon the mixture in the same manner as a gravitational field. The centrifugal field can, however, be varied by changes in rotational speed and equipment dimensions, whereas gravity is essentially constant.

Most centrifugation equipment is intended to separate immiscible or insoluble components from a liquid medium. The ultracentrifuge and gas centrifuge represent special cases that establish separation gradients on a molecular scale. The usual gravitational operations, such as sedimentation (qv) or flotation (qv) of solids in liquids, drainage or squeezing of liquids from solid particles, and stratification of liquids according to density, are accomplished more effectively in a centrifugal field (see SEPARATION, SIZE SEPARATION).

Separation by Density Difference

A single solid particle or discrete liquid drop settling under the acceleration of gravity in a continuous liquid phase accelerates until a constant terminal velocity is reached. At this point the force resulting from gravitational acceleration and the opposing force resulting from frictional drag of the surrounding medium are equal in magnitude. The terminal velocity largely determines what is commonly known as the settling velocity of the particle, or drop under free-fall, or unhindered conditions. For a small spherical particle, it is given by Stokes' law:

$$v_g = \frac{\Delta\delta d^2 g}{18\mu} \tag{1}$$

where v_g = the settling velocity of a particle or drop in a gravitational field; $\Delta\delta = \delta_S - \delta_L$ = the difference between true mass density of the solid particle or liquid drop, and that of the surrounding liquid medium; d = the diameter of the solid particle or liquid drop; g = the acceleration of gravity; and μ = the absolute viscosity of the surrounding medium.

Stokes' law can be readily extended to a centrifugal field:

$$v_{ss} = \frac{\Delta\delta d^2 \omega^2 r}{18\mu} = v_g\left(\frac{\omega^2 r}{g}\right) \tag{2}$$

where v_s = the settling velocity of a particle or drop in a centrifugal field; ω = the angular velocity of the particle in the settling zone; and r = the radius at which settling velocity is determined.

Separation by Drainage. The theory covering drainage in a packed bed or particles is incomplete, and requires more development for a centrifugal field. Liquid is held within the bed by various forces. Removal involves several flow mechanisms. In addition, the centrifugal acceleration changes with radius in the bed, causing changes in packing tendencies of particles and accelerating forces on the residual liquid.

Liquid–Liquid-Phase Behavior

Liquid drops, suspended in a continuous liquid medium, separate according to the same laws as solid particles. After reaching a boundary, these drops coalesce to form a second continuous phase separated from the medium by an interface that may be well- or ill-defined. The discharge of these separated layers is controlled by the presence of dams in the flow paths of the phases. The relative radii of these dams determine the radius of the interface between the two separated layers.

Centrifuge Components

Centrifuges accomplish their function by subjecting fluids and solids to centrifugal fields produced by rotation. Electric motors are the drive device most frequently used; however, hydraulic motors, internal combustion engines, and steam or air turbines are also used. One power equation applies to all types of centrifuges and drive devices.

The total power, P_T, needed to run a centrifuge, ie, delivered by the drive device, is equal to sum of all losses:

$$P_T = P_P + P_S + P_F + P_W + P_{BD} + P_{CP} \tag{3}$$

where P_P is the process power. Power for each liquid and the solid phase must be added to get P_P. P_S is the solids process power. P_F is the friction power, ie, loss in bearings, seals, gears, belts, and fluid couplings. P_W is the windage power. Increased density owing to gas pressure increases the windage power, and this may be very significant for high pressure applications. Windage power is a very important loss for large machines and must be determined. P_{CP} is the power absorbed by the centripetal pump. For scroll centrifuges having back-drives, P_{BD} is the back-drive power.

Materials of Construction and Operational Stress. Before a centrifugal separation device is chosen, the corrosive characteristics of the liquid and solids as well as the cleaning and sanitizing solutions must be determined. A wide variety of materials may be used. Once the material choice based on corrosion is made, a careful analysis of the stresses produced by rotation for the particular type of centrifuge is required, so that for the given liquid and solids specific gravities a maximum operating speed can be determined. The geometry of the bowl parts is important; and for intermittently discharging centrifuges, fatigue strength must be considered.

Equipment. Centrifugation equipment that separates by density difference is available in a variety of sizes and types and can be categorized by capacity range and the theoretical settling velocities of the particles normally handled. Centrifuges that separate by filtration produce drained solids and can be categorized by final moisture, drainage time, and physical characteristics of the system, such as particle size and liquid viscosity. For optimum results, a combination of several types of equipment may be used.

Centrifuges

Sedimentation Equipment. Centrifugal sedimentation equipment is usually characterized by limiting flow rates and theoretical settling capabilities. Feed rates in industrial applications may be dictated by liquid handling capacities, separating capacities, or physical characteristics of the solids.

In general, solids-retaining batch and batch automatic machines are limited to low feed concentrations to minimize the time required to unload the solids. Continuous disk centrifuges can have higher feed concentration. The limit is the underflow concentration. Conveyor discharge centrifuges can handle high feed concentration and are limited only by the volume of solids displacement, or torque capacity.

Centrifugal Sedimentation Equipment. Commercial sedimentation centrifuges are characterized principally by how solids are discharged, and the general dryness of these solids. There are batch and automatic batch solid bowl machines which collect the solids at the bowl wall. Solids are removed very dry. Almost any solid is collectable, even those that are very soft and compressible.

Disk-type solid bowl machines are batch, batch automatic, and continuous. The solids are removed in many different ways, but are usually wet. Scroll centrifuges discharge solids continuously and usually drier than disk and imperforate batch types. Generally disk centrifuges are superior for collecting fine particles at a high rate.

A bottle centrifuge is designed to handle small batches of material for laboratory separations, testing, and control. The basic structure is usually a motor-driven vertical spindle supporting various heads or rotors. A surrounding cover reduces windage, facilitates temperature control, and provides a safety shield. Accessories include timer, tachometer, and manual or automatic braking. Bench-top bottle centrifuges operate at 500–5000 rpm, producing centrifugal fields up to 3000 G in the lower speed range, and operate up to 20,000 rpm with 34,000 G in the high speed units. Larger models operate up to 6000 rpm and develop 8000 G, using special attachments that permit 40,000 G. These models may also be equipped with automatic temperature control down to -10°C and other programmable controls to manage the cycle.

Specialty rotors permit ordinary bottle centrifuges to achieve some of the results previously considered possible only in ultracentrifuges. Preparation ultracentrifuges are suitable for a range of applications, such as processing quantities of subcellular particles, viruses, and proteins (qv). Preparation ultracentrifuges range in operating speed from 20,000 rpm, generating about 40,000 G, to 75,000 rpm and about 500,000 G. The rotor is surrounded by a high strength cylindrical casing and underdriven by an electric motor. To avoid overheating of the rotor by air friction at these speeds, the pressure in the casing is reduced to about 0.13 Pa (1 μm Hg). Sensors (qv) monitor the temperature and a cooling system controls the temperature in the range of -15 to 30°C within ± 1°C. Electronic controls maintain the rotor speed within a required narrow range and may be automatically programmed for sequential changes in speed, including control of the acceleration and deceleration.

The use of density gradients in centrifuge rotors greatly increases the sharpness of separations and the quantities of material that can be handled. In principle, the density gradient is established normal to the axis of rotation of the rotor and the highest density is located at the outer radius of the rotor. A natural gradient may be formed by introducing a homogeneous solution and centrifuging for long periods of time. Continuous or step gradients may also be formed by introducing successive layers of solution, the composition of which varies continuously or stepwise from low to high density, where the latter displaces the former toward the center of the rotor.

Tubular centrifuges separate liquid–liquid mixtures or clarify liquid–solid mixtures having less than 1% solids content and fine particles. Liquid is discharged continuously, whereas solids are removed manually when sufficient bowl cake has accumulated. For industrial use, the cylindrical bowls are 100–180 mm in diameter with length-to-diameter ratios ranging from 4–8. Bowl speeds up to 17,000 rpm generate centrifugal accelerations up to 20,000 G at the bowl wall. The tubular centrifuge handles low to medium flows and theoretical particle settling velocities in the range of 5×10^{-6} to 5×10^{-5} cm/s. The tubular centrifuge was long used for the purification of contaminated lubricating oils because of the high centrifugal force developed and the simplicity of its operation.

Centrifuges that channel feed through a large number of conical disks to facilitate separation combine high flow rates with high theoretical capacity factors. For industrial units flow rates up to 250 m^3/h (1100 gpm) can be obtained on easy separations, and theoretical settling velocities may range from 8×10^{-6} to about 5×10^{-5} cm/s. Both liquid–liquid and liquid–solid separations are performed using feed solids concentration below 15% and small particle sizes.

The outstanding feature of the disk bowl design is a stack of thin cones, commonly referred to as disks, which are separated by thin spacers. These are so arranged that the mixture to be clarified must pass through the disk stack before discharge. The resulting stratification of the liquid medium greatly reduces the sedimenting distance required before a particle reaches a solid surface and can be considered removed from the process stream. The angle of the cones to the axis of rotation is great enough to ensure that solid particles deposited on the

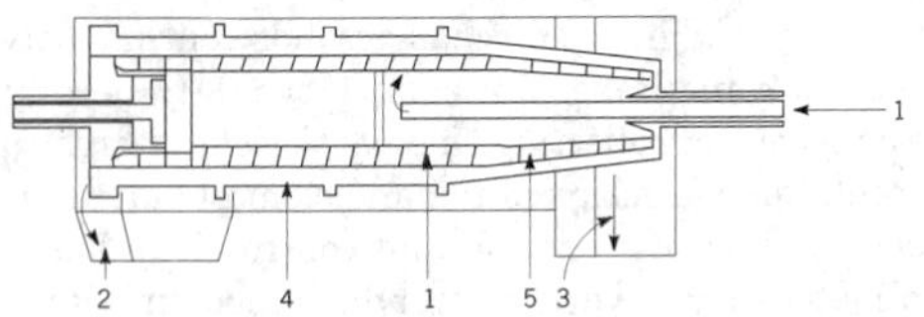

Figure 1. An imperforate bowl conveyor-discharge centrifuge, where 1 corresponds to feed suspension; 2, to liquid phase; 3, to solid phase; 4, to liquid pool; and 5, to dry beach.

surfaces slide, either individually, or as a concentrated phase according to the difference between their density and that of the medium.

Imperforate bowl conveyor-discharge centrifuges collect solids by sedimentation and continuously discharge both liquid and solid material. These centrifuges have bowl diameters of 150–1400 mm and are essentially tubular shells with a length-to-diameter ratio of 1.5–5.2, as shown in Figure 1. Deposited solids are moved by a helical screw conveyor operating at a differential speed of 0.5–100 rpm with respect to the bowl. Centrifugal fields are lower than in disk or tubular centrifuges because of the conveyor and its associated mechanism. Maximum speeds are in the range 300–9000 rpm. Particles of intermediate settling velocities, such as 1.5 to about 15×10^{-4} cm/s, are handled at medium to large flow rates. For clarification, this type of centrifuge recovers medium and coarse particles from feeds at high or low solids concentration. Particle sizes less than about 2 μm are normally not collected without the addition of flocculating agents.

Centrifugal Filtration Equipment. The important parameters of centrifugal filtration equipment are screen area, level of centrifugal acceleration in the final drainage zone, and cake thickness. The latter affects both residence time and volumetric throughput rate. The particle size of the solids and the kinematic viscosity of the mother liquor also strongly affect the final moisture content.

The simplest and most common form of centrifugal filter is a perforate-wall basket centrifuge, consisting of a cylindrical bowl having a diameter ranging from about 100–2400 mm and a diameter-to-height ratio ranging from 1–3. The wall is perforated with a large number of holes, more than adequate for the drainage of most liquid loads, and is lined with a filter medium. In the simplest case, the medium is a single layer of fabric or metal cloth or screen. In high speed basket centrifuges, one or more backup screens of relatively large mesh support a finer mesh filter surface. The method of discharging accumulated solids distinguishes three types of basket centrifuge: those that are stopped for discharge, those that are decelerated to a very low speed for discharge, and those that discharge at full speed.

Another batch automatic horizontal perforated bowl centrifuge inverts the flexible filter to discharge the solids. Feed slurry may be deposited on the inside surface of a cloth filter, with the bucket end completely closed. When the interior is full of dewatered solids, the bowl is decelerated to a slow speed and piston and closure plates move axially, inverting the filter cloth so that the solids reside on the outside diameter of the cloth. Very little residual material remains on the filter cloth surface for the next cycle.

Continuous filtering centrifuges are used for very fast draining that do not require extremely dry final products. Continuous centrifugal filters are equipped with either a cylindrical or a conical screen. Both types are made without a retaining lip on the solids discharge end of the bowl and employ various methods to move the solids through the bowl. The cylindrical screen centrifuge deposits solids at one end of the bowl in a layer 6–80 mm thick and pushes the annular ring of cake axially through the bowl by means of a reciprocating piston.

In conical screen centrifuges the angle of the bowl causes or assists the cake to move axially and redistributes it in an increasingly thin layer which improves drainage characteristics. The feed slurry is deposited at the small end of the cone, where most drainage occurs. The drained solids are discharged from the large end, which has no retaining ring.

Gas Centrifugal Separation

Highly developed centrifuges are used to enrich uranium for nuclear application (see NUCLEAR REACTORS; URANIUM AND URANIUM COMPOUNDS). Gaseous uranium hexafluoride, UF_6, is introduced into a very high speed tubular rotor, causing the lighter ^{235}U-fraction to separate from that of the heavier ^{285}U. The enrichment that is achieved using one centrifuge is not large enough, so in a commercial plant, the centrifuges are arranged in cascades, where the enriched fraction passes up the cascade for additional enrichment stages, while the depleted fraction passes down the cascade. Each stage of enrichment requires many centrifuges to multiply the total capacity, and many stages to result in an adequately enriched product. This ultimately results in thousands of individual centrifuges, interconnected to each other in the cascade.

ALAN LETKI
R. T. MOLL
Alfa Laval Separation Inc.
LEONARD SHAPIRO
Consultant

H. Axelsson, *Centrifugal Separations—Principles and Techniques,* Alfa Laval Separation, Tumba, Sweden, presented at the Bioprocess Technology Program, University of Virginia, Charlottesville, Va., Oct. 17–25 1991.

P. A. Vesilind, *Treatment and Disposal of Waste Water Sludges,* Ann Arbor Science Publishers, Ann Arbor, Mich., 1974.

O. M. Griffith, *Techniques of Preparative, Zonal and Continuous Flow Ultracentrifugation,* Spinco Division of Beckman Instruments, Inc., Palo Alto, Calif., 1975.

MAGNETIC SEPARATION

The application of magnetic separators relies on the behavior of individual particles under the influence of magnetic forces. Magnetic separation methods are used either to separate valuable minerals from nonmagnetic waste, or magnetic impurities or other valuable magnetic minerals from bulk nonmagnetic values. Magnetically susceptible mineral particles occur as individually discrete particles; as partially liberated particles consisting of two or more minerals, one of which is magnetically susceptible; or as particles stained or coated by a magnetically susceptible mineral such as geothite staining quartz or kaolin. Minerals normally considered nonmagnetic may be rendered magnetic by elemental substitution of a small amount of a magnetic element in the crystal lattice. Magnetic properties may also be affected by partial alteration in weathering effects.

Equipment. From the applications standpoint, magnetic equipment falls into one of four broad categories: tramp iron removal, magnetic particle separation and concentration, product cleaning, or eddy current separation on nonmagnetic metallics.

Tramp Iron Magnetic Separation

Tramp iron magnetic separators are used to protect handling and processing equipment such as crushers, pulverizers, and material handling equipment. Separators are usually applied to dry material or material that contains only surface moisture. Iron coarser than 13 mm is usually defined as tramp iron. Magnetic equipment developed for tramp iron removal may involve magnetic head pulleys, suspended magnets, magnetic drums, plate, or grate magnets.

Magnetic Pulleys. An easy and simple way to achieve tramp iron removal from material handled on a conveyor belt is by means of a magnetic head pulley (see CONVEYING). Magnetic pulleys accomplish both continuous and automatic tramp iron removal. A typical installation is shown in Figure 1.

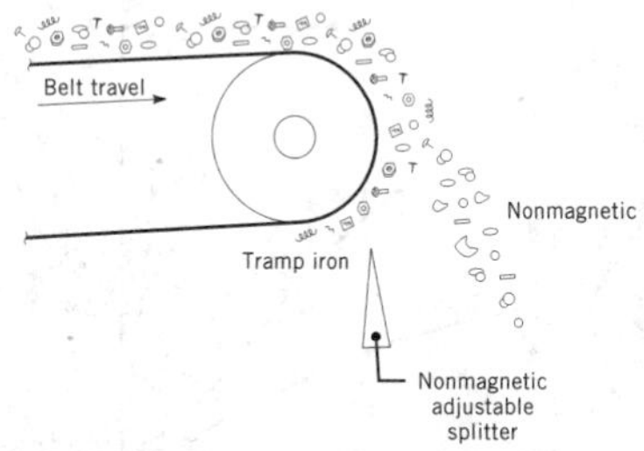

Figure 1. Principle of operation for magnetic pulleys.

Suspended Magnets. Suspended magnets can be installed at many points in a material handling system, at any point along a conveyor belt, at the discharge end of feeders or screens, and above chutes or launders. The preferred location for a suspended magnet is at an angle over the discharge of a conveyor. At this point the material moves into the magnet face and the load breaks open, making tramp iron removal easier. A suspended magnet can be made continuous in the discharge of tramp iron by placing a belt over the magnet face and driving the belt across it. This discharging feature is particularly effective where long pieces of tramp iron are encountered.

Tramp Iron Magnetic Drums. Magnetic drum separators are used for tramp iron removal where magnetic pulleys and suspended magnets are not feasible. In effect, these are individual pieces of process equipment inserted in the process line. The magnet assembly is held in a fixed position inside the drum shell and the drum shell is driven around this magnet assembly. The magnet assembly develops a field that typically covers 120–180° of the drum section. These drums can be fed at the top vertical centerline or near the bottom of the drum.

Plate Magnets. Plate magnets are simple devices usually mounted in the bottom of a chute or duct. The plate magnet traps the tramp iron against the magnet face so that the trapped material must be periodically.

Grate Magnets. The grate magnet consists of a series of magnetized tubes, round or square, mounted in a frame or housing. The collected tramp iron must be removed periodically. The grate magnet, largely restricted to use on free-flowing material finer than 13 mm in size, is typically mounted at the discharge of a hopper.

Magnetic Concentration and Purification

The magnetic responsiveness of mineral particles provides a positive means of concentrating and/or purifying ores. The type of magnetic separator used is influenced by feed condition, ie, wet or dry; the mineral to be concentrated or purified; the relative magnetic responsiveness of the mineral; the feed size; the purity to be obtained, in either the magnetic concentrate or the nonmagnetic product; the capacity to be handled; equipment operating and maintenance costs; and temperature of magnet applications (see MINERAL RECOVERY AND PROCESSING).

Magnetic Separator Classification. A fundamental equipment classification can be made on the basis of feed condition. A wet condition involves the treatment of a slurry or slip; a dry condition involves treatment where the particles are essentially free to move as independent particles. Both conditions depend on liberation of the magnetic particles. If liberation does not occur, the magnet traps the middling particles and reduces the concentration of magnetics. Wet or dry magnetic separators can be further divided based on the field intensity developed by the individual separator. Broadly speaking, classification by field strength is high, intermediate, or low intensity.

Low Intensity Wet Drum Magnetic Separators. Wet drum separators are used to recover ferromagnetic solids from a slurry feed. The principal areas of usage are in media recovery in heavy-media separation plants and in magnetite ore concentration. Physical construction of wet drum separators is slightly different for the two applications. The ore concentrator, which is more rugged in construction, is subject to more detailed specifications than the media recovery units. This is particularly true for feed and collection tank design, bearing construction, wear covers on the drums, magnet assembly design, and magnetic field strength ratings.

Wet High Intensity Magnetic Separators. There are several types of wet high intensity magnetic separators (WHIMS) commercially available. The first unit, developed in England for clay purification (the Jones' separator) was a cycling-type machine that later evolved into the continuous carousel unit. The Jones' units are capable of developing magnetic field intensities of 2.0 T. Other wet high intensity units provide configurations that have rotating matrixes similar to wet drum units having cooled electrocoils. Still others fall into the category of filters using cryogenically cooled coils and stationary matrixes.

Magnetic Filters. Small magnetic filters are simple devices in which an electrically energized coil or permanent magnets are used to magnetize a magnetic steel grid. The grids develop high points of magnetic strength on their edges. The material to be cleaned is passed through this series of magnetized grids and magnetic particles are attracted to and held onto the grid edges. Periodic cleaning of the magnetic filter is required. Magnetic filters have been used to clean paint slips and liquids.

Dry Magnetic Separators. Commercial types of dry magnetic separators fall into two broad areas. Low intensity magnetic pulley separators (see Fig. 1) and several types of magnetic drums have been used to concentrate iron ore, sponge iron, and to recover iron values from steel mill slags. These separators have effective working magnetic field strength of under 0.1 T and are largely limited to the recovery of ferromagnetic materials. High intensity dry magnetic separators include two basic types: the induced-roll and the cross-belt magnet. A high intensity inductively magnetized disk has been used in some areas of the world, but this is not well known in the United States.

Magnetic Pulleys. Magnetic pulleys of special design are used in the concentration of magnetite and other ferromagnetic minerals. For best results, the feed should be screened into various-sized fractions and each fraction treated on a separate pulley separator unit.

Magnetic Drums. Two types of magnetic drums, which vary in the construction of the magnet assembly inside the drum, are used in concentration service. The first incorporates a magnet assembly in which the poles vary in polarity across the drum width. This type of drum develops a strong holding force and is used in treatment of ore up to 200 mm in diameter. The second magnetic drum is used for the concentration of finer material, usually −25 mm or smaller in size, and incorporates a magnet assembly of from six to as many as 55 poles that vary in polarity around the circumference of the drum.

High Intensity Induced-Roll Magnetic Separators. Induced-roll magnetic separators are typically used for cleaning such materials as silica sand and feldspar. Both overfeed and underfeed roll designs have been built. The latter is used to obtain a higher degree of cleaning at a relatively high capacity. Owing to the narrow magnetic operating gap, the feed size to induced-roll separators is limited to material finer than 3-mm size. Furthermore, because of the surface activity of very fine material, there should be little 74-μm (−200 − mesh) material present, unless the loss of this material in the magnetic concentrate can be tolerated. For selective removal of minerals that vary in magnetic response, or for maximum removal of contaminants in silica sand or feldspars, multistage induced-roll separators are applied.

High Intensity Cross-Belt Magnetic Separators. For very selective concentration of weakly magnetic minerals, the high intensity cross-belt separator has been utilized. This separator utilizes a feed belt on which a thin layer of the feed material is introduced to a high intensity magnetic field. The magnetic materials are lifted to the upper pole of this field and transferred by a cross-belt to a collective hopper.

To obtain the high field gradient required for separation of the weakly magnetic minerals, the upper pole is shaped to a point or series of points for improved capacity. The bottom pole is flat. In cases where a variety of weakly magnetic minerals are to be concentrated, a series of high intensity poles is utilized, which have increasing coil strength at each succeeding pole, or by using a variation in the air gap.

Eddy-Current Separation of Nonferrous Metallics. The advent of material recovery facilities (MRFs) has increased the need for automatic removal of metallic values tenfold. Although many small facilities rely on hand sorting of commingled refuse, ie, glass, aluminum, tin, and plastic, as these plants grow in size, ways of reducing labor costs become more important. Suspended magnets are used to remove ferrous material. The removal of aluminum is relegated to a form of eddy-current separation.

The eddy-current concept was the basis for a separator patented by Thomas Edison in 1889. Only since the mid-1980s has this type of separator been utilized, however, because of the growth in the recycling (qv) and scrap industries. Although the recycling area is of main consideration, there are other areas of usage. These separators are used both in foundries and in autoshredder operations for the removal of nonferrous metallics from foundry sand and from shredder fluff, respectively.

Electrical current flows are induced in all conductors when exposed to an a-c field. These currents generate a magnetic field surrounding the conductors which oppose the field being produced by the a-c field with a force sufficient to repel the conductor.

The force exerted by a machine is proportional to the field intensity, and the frequency. In a rotating device, the frequency is proportional to the number of poles and the rotor rpm. With the advent of improved rare-earth magnet material, the field intensity for this type of rotor configuration can be greatly increased compared to other types of magnet material, such as ceramic, Alnico, and the early rare earth. Energy products that are nearly seven times greater than those employing a ceramic rotor can be attained.

DON MORGAN
O. S. Walker Company

W. J. Bronkala, "Magnetic Separation" in *Encyclopedia of Chemical Technology,* 2nd ed., Vol. 12, John Wiley & Sons, Inc., New York, 1967.

I. S. Wells, *Chem. Eng.* (1982).

J. E. Forciea, L. G. Hendrickson, and O. E. Palasvirta, *Mining Eng.* **10,** 339–349 (Dec. 1958).

J. Iannicelli, *Clays Clay Mineral,* **24,** 64–68 (1976).

SIZE SEPARATION

Size separation, the parceling of particulate material on the basis of size, is an important industrial unit operation that produces, on a continuous basis, coarser and finer streams from a feed stream in a single stage or multiple stages. Devices for size separation fall into two general categories: those basically involving the probability of passing through an aperture and those that separate by forces of fluid dynamics. The former, probability size separation, is based on the repeated presentations of particles to uniformly sized apertures in screen decks. The latter, fluid-dynamic size separation, takes advantage of the interplay between gravity and drag forces. If the particles are moving in a liquid, the separation is termed *wet* sizing; if moving in a gas, the separation is termed *dry* sizing.

Screening devices are used to make coarser separations; that is, fine products having 95% passing ca 100-mm–50-μm size. Dry-screening devices have a lower recommended size of a ca 500 μm. Wet-screening devices that produce 95% passing ca 500–50-μm size are continually being improved. Fluid-dynamic separating devices, called classifiers, are used to make finer separations; that is, fine products having 95% passing ca 1000–5-μm size.

Evaluation of Size Separations

Common separating devices are evaluated as follows. Consider a flat screen plate having square openings of dimension b and centers of dimension c. Ideally, the chance of a spherical particle having diameter d passing through an opening would be zero for all particles of relative size $d/b > 1$ and one for all particles of relative size $d/b < 1$. A plot of the probaility-of-passing vs size (Fig. 1, curve D) is a step function, and the separation size, so-called cut size, is $d/b = 1$. A perfect separation is one where all particles of size less than the cut size pass and all particles of size greater than the cut size are retained.

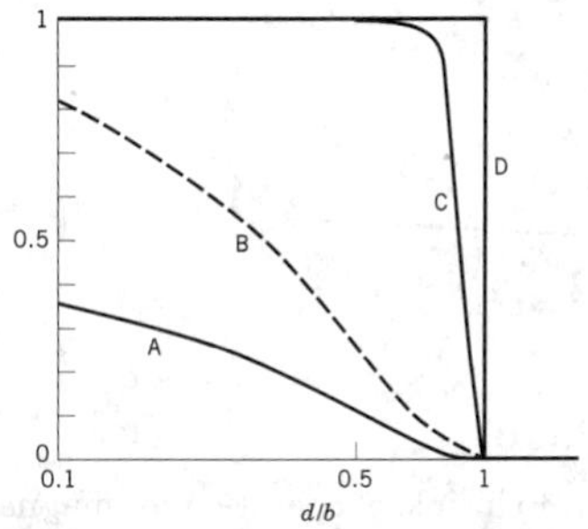

Figure 1. Probability of a particle size d passing through a square aperture of size b. See text for discussion of the various curves.

However, the probability of such a particle passing, when approaching the screen plate normally, is as shown in equation 1:

$$p = \left(\frac{b - d}{c}\right)^2 \tag{1}$$

If the c/b ratio is 1.5, a typical industrial value, then the plot of the probability-of-passing equation vs size (Fig. 1, curve A) is curvilinear. If each particle that did not pass is given another opportunity, then the probability of a particle passing after n opportunities is $1 - q^n$, where q, the probability of a particle not passing, is equal to $1 - p$. The plot of probability-of-passing vs size (Fig. 1, curve C) for a very good industrial screen ($n = 100$) is an improved curve approaching the perfect separation step function.

The characteristic curve is used to evaluate the separation because the deviation from the perfect step function represents a measure of the efficiency of the separation. A commonly used measure to characterize the shape of the curve for a size separation is the sharpness index, given the symbol κ. It is the ratio of the size of a particle having a 75% chance of passing or reporting to the finer stream, ie, d_{75}, to the size of a particle having a 25% chance of passing, ie, d_{25}. A perfect separation would give a value of 1, whereas merely a splitting action (no size separation) would give a value of 0. The sharpness index of curve C (Fig. 1) is 0.84/0.92 or 0.9. For size separations, the equiprobable size, ie, the size of a particle that has a 50/50 chance of passing, d_{50}, is used to define the cut size. The cut size for curve C is 0.9, which is less than the aperture size.

Curve A exhibits another characteristic: it does not go to one at d/b equal to zero (see eq. 1). This can be viewed as an apparent bypassing, or short-circuiting, of the feed material to the coarse stream, ie, a fraction of all of the particles in the feed did not pass through the screen. All three parameters, the cut size, sharpness index, and apparent bypass, are used to evaluate a size separation device, because these are assumed to be independent of the feed size distribution.

Screening

Screening is a process whereby particles are presented in an appropriate manner to a series of apertures of fixed dimensions that allow finer particles to pass through into the undersize, and coarser particles to be retained in the oversize.

Equipment. Screening devices can have stationary or moving screen decks. Moving screen decks can be designed as rotating cylinders (trommels) or vibrating surfaces. By far, the most popular screen deck design for sizing is the inclined vibrating screen. Components of vibrating screens are the vibrating frame, deck support frame, screening deck, motor/drive assembly, and feed box/distributor. Vibration is produced on inclined screen decks by circular motion in a vertical plane. The vibration lifts the material, producing stratification. Stratification places the larger particles on top, and the smaller particles on the bottom, next to the apertures. When the screen deck is used on an incline, the material cascades down the slope, introducing the probability that the particle either passes through the opening or over the screen surface. For horizontal screen decks, the motion must be capable of conveying the material without the assistance of gravity. Straight-line motion at an angle of approximately 45° to the horizontal produces a lifting component for stratification and a conveying component for probability of separation as the material passes across the horizontal screen surface.

Design Loading. The design loading, or capacity, is estimated by starting from a basic loading for square screen cloth having the desired aperture size. The base loading decreases with decreasing aperture size, partially because the open area decreases. The base value assumes an optimal bed depth, deck slope (flow velocity), quantity of near-size material, free-flowing material, and screen open area. Thus, the base loading must be modified for material properties, equipment design, and operating conditions that differ from the standard ones.

Deck Area. After the design loading has been estimated, the design feed rate is divided by the design loading to get the effective deck area. In order to convert the effective deck area into length and width dimensions, the deck width is calculated based on a standard bed depth. Then the length is calculated by dividing the effective area by the width. The top size of the feed to a screen deck should not be greater than two to four times the aperture size of the deck. Double- or triple-deck screen arrangements are used, requiring a separate sizing for each deck. Then the final screen size is set by the largest deck.

Slurries. There is a family of stationary screens known as crossflow screens, used mainly for dewatering or desliming, that have been applied to slurry sizing. These screens are characterized by a slotted deck, usually made of stainless steel profile wire and set at an angle. The slurry flow is at right angles to the slots and, depending on the angle of inclination, the cut size value is one-half to two-thirds of the slot dimension. Hence, it is possible to size at 0.5 mm using a 1-mm slot, which should reduce blinding problems.

Other approaches to reducing the blinding effect have included the development of so-called probability screens. The projected area of a rectangular aperture, normal to the horizontal, decreases as the angle of inclination of the screen deck containing the aperture increases. Thus, the probability of passage for a particle falling normal to the horizontal, which would normally pass through the aperture, decreases with an increasing angle of inclination. The probability of a particle lodging in the opening is minimal.

Wet Classification

Settling-Pool Classifiers. The settling-pool group of fluid-dynamic separating devices (classifiers) consists of a rectangular tank having an inclined floor that creates a pool. Feed slurry is introduced at the side of the tank and the overflow of fine particles and water exits through an overflow weir and box arrangement. The coarse particles settle to the bottom and are discharged over the upper edge of the tank after being dragged up the inclined floor by some mechanism, such as intermittently operating rakes, continuously operating drag conveyors, or continuously operating spirals.

Hydrocyclones. Increasing the gravitational force by developing centrifugal force decreases the settling time of smaller particles. A centrifugal force can be imposed on particles by rotating the slurry of particles (see SEPARATION, CENTRIFUGAL SEPARATION). The classifying hydrocyclone (Fig. 2) is designed to rotate the slurry of particles by introducing it tangentially into a cylinder, thereby achieving essentially free vortex motion. A second, smaller-diameter (20–40% of the cylinder diameter) pipe, called the vortex finder, is placed through the exact center of the solid top of the cylinder, extending substantially below the feed inlet but above the bottom of the cylinder. The slurry that exits through the vortex finder is termed the overflow and contains the finer particles. Attached to the bottom of the cylinder

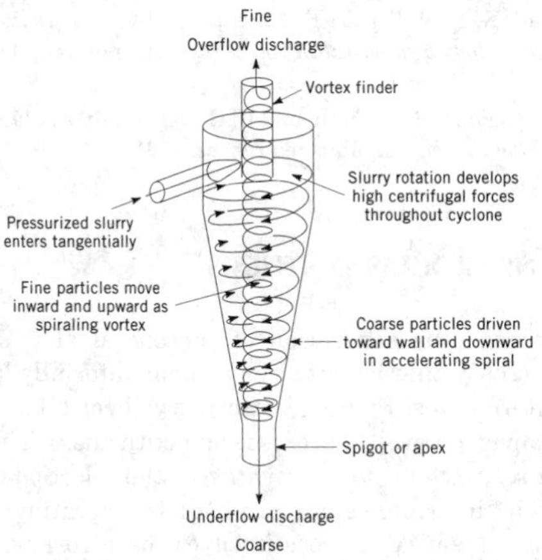

Figure 2. Hydrocyclone classifier.

is an inverted truncated cone section, the included angle of which ranges from 10–30° and is typically 15–17°. The slurry that exits through the cone section (apex) is termed the underflow and contains the coarser particles. The hydrocyclone is the most popular industrial wet classifier.

Dry Classification

Pneumatic classification can be partitioned conveniently into coarse, ie, fine products above 95% <100 μm; intermediate, ie, fine products ranging between 95% <100 and 30 μm; and fine, ie, fine products below 95% <30 μm. Pneumatic classification, like hydraulic classification, balances the force of gravity with drag forces (counter flow) in order to bring about a separation.

Health and Safety Factors

The pneumatic classification system should be designed to handle hazardous dust. A hazardous dust is one which, when finely divided and suspended in air in the proper concentration, burns, produces violent explosions, or is sufficiently toxic to be injurious to personnel health (see AIR POLLUTION CONTROL METHODS; POWDERS, HANDLING). At the least, almost any dust can be irritating to personnel because of inhalation or skin or eye contact. Fully oxidized and hydrated materials are generally considered safe. Dust explosions usually occur in pairs. The first explosion involves dust already in suspension. This jars dust from beams, ledges, etc, creating a second cloud to which the explosion propagates, resulting in a secondary explosion.

Prediction of Size Separation

Most corrected characteristic separation curves fit the following logistic function:

$$c(x_i; \kappa, d_{50}) = 1/(1 + e^{-Y_i}) \qquad (2)$$

where Y_i equals 1.0986 $(1 - z_i)(1 + \kappa)/(1 - \kappa)$ or 2.1972 ln $[z_i]$/ln $[\kappa]$ and z_i equals x_i/d_{50}, where x_i is the upper size of the narrow size interval, typically $x_i/x_{i+1} \leq \sqrt{2}$, hence x_1 is the maximum size. Thus, if the operating sharpness index and cut size are known for a separating device, the size distribution of the fine stream can be predicted.

PETER LUCKIE
Penn State University

L. G. Austin, C. H. Lee, F. Concha, and P. T. Luckie, *Min. Met. Process.*, **9**, 161–168 (1992).

Equipment Testing Procedure Committee, *Particle Size Classifiers, A Guide To Performance Evaluation,* 2nd ed., AIChE, New York, 1993.

R. J. Batterham, K. R. Weller, T. E. Norgate, and C. J. Birkett, *International Particle Technology Symposium Proceedings,* Amsterdam, the Netherlands, 1980.

J. Slechta, I. J. Taggart, B. A. Firth, and E. Gallagher, *An Evaluation of a Novel Vibrating Fine Screen,* unpublished report, 1981.

SEPARATIONS PROCESS SYNTHESIS

Process synthesis, the generation of conceptual flow sheets, is an open-ended activity characterized by a combinatorially large number of feasible alternatives. Finding better flow sheet alternatives has a significant impact on overall process competitiveness. This is particularly true for *separations process synthesis,* the selection of separation methods, their interconnection, and their operating parameters. Virtually every chemical process involves the recovery, isolation, or purification of products, by-products, intermediates, wastes, or raw materials. The separation systems to accomplish these tasks often dominate the capital and operating costs of chemical manufacturing processes.

Several approaches to the separations process synthesis problem have been formulated including *superstructure optimization, evolutionary modification,* and *systematic generation.* From a known feed composition, desired product compositions, and a well-defined set of separation methods, superstructure optimization approaches construct a hypothetical flow sheet which includes all applicable separation methods interconnected in every possible order so as to include all possible separations scheme alternatives. Separations synthesis then becomes a problem of systematically stripping away the less desirable parts of this superstructure while simultaneously optimizing the design and operating parameters of the remaining separation methods using mixed-integer nonlinear programming. Evolutionary modification starts with an existing flow sheet for a similar separation to which adaptations are made as necessary to meet the objectives of the specific case at hand. In the systematic generation approach, the separations flow sheet is constructed from a portfolio of basic components in a directed fashion so that a given feed stream is progressively transformed into one or more target (product) compositions.

Heuristic Distillation Sequencing for Nonazeotropic Mixtures

Every separation method is based on differences in one or more physical or chemical properties of the components in a mixture. For a large multicomponent mixture and a moderate number of potential separation methods, the number of possible separations precludes an exhaustive detailed analysis of all options. As a compromise between thoroughness, efficiency of evaluation, and guarantee of optimality, selection and sequencing methods often are reduced to design heuristics and simple ranked lists of physical and chemical properties characteristic of specific separation methods.

One practical systematic generation problem is the separation of nearly ideal liquid mixtures by simple distillation. If the components of the mixture do not form azeotropes, then all possible distillation sequences can be exhaustively enumerated from a list of the components ordered by boiling point. A feasible distillation operation can be made between any two components with adjacent boiling points. The distillate consists of the lower boiling group of components, the bottoms is the higher boiling group. The generated sequences can then be simulated in detail. Well-established distillation sequencing heuristics can be used to reduce the number of potential alternative sequences, often leading to near-optimal separation system designs. This is an example of an *opportunistic* systematic generation strategy. Any point in the partially completed flow sheet is a feasible consequence of the initial feed composition and distillations thus far specified. No attempt is made to anticipate any potential difficulties in separation problem resolution. Because distillation is assumed to be feasible for all necessary separations, this opportunistic strategy will not lead to intermediate streams from which it is impossible to resolve the remaining composition differences.

Separations Synthesis for Nonideal Liquid Mixtures

Typical liquid mixtures encountered in organic chemicals manufacturing often exhibit a wide range of melting and boiling points, reactivity, temperature sensitivity, and strong thermodynamic nonidealities resulting in azeotropism and liquid–liquid phase formation. Such mixtures show a diverse range of behaviors which tend to complicate separation operations. In spite of these added complications, simple distillation is the most widely used separation method and flow sheet generation can be thought of as a problem of finding applications for distillation and the identification and resolution of situations where distillation cannot be used.

An important aspect of separations process synthesis for nonideal liquid mixtures is effective problem representation and visualization. Commonly used methods, including variants of distillation and extraction, are equilibrium-based processes for which the pertinent thermodynamics can be represented conveniently with graphical methods. Triangular phase diagrams (3-component systems) and tetrahedral diagrams (4 components) are versatile tools for visualizing separation method behavior including material balances. Useful thermodynamic representations are *residue curve maps* (RCM) and *distillation region diagrams* (DRD) for vapor–liquid equilibria (VLE), miscibility diagrams for liquid–liquid equilibria (LLE), and solubility diagrams for visualizing solid–liquid equilibria (SLE). RCM and DRD are system-dependent collections of curves tracing the liquid phase compositions in batch still pots and continuous distillation columns, respectively. All residue curves orginate at low boiling pure components or azeotropic compositions (often referred to as nodes) and end at high boiling compositions. RCM and DRD with more than one origin or terminus have more than one distillation region. Intermediate boiling pure components which are not nodes are termed *saddles.* The pattern of boundaries, nodes, and saddles of a given multicomponent mixture is related to the boiling points of azetropes and pure components and is readily definable mathematically.

A single-feed distillation column can be designed with sufficient stages, reflux, and material balance control to produce a variety of different separations ranging from the *direct* mode of operation (pure low boiling node taken as distillate) to the *indirect* mode (pure high boiling node taken as bottoms). This range of operability results in a bow-tie shaped set of *reachable compositions* roughly bounded by the material balance lines corresponding to the sharpest direct and indirect separations. The exact shape of the reachable composition space is further limited by the requirement that the distillate and bottoms lie on the same residue curve (ie, must be in the same distillation region), and by material balance must be colinear with the feed. Except when highly curved, distillation boundaries act as barriers to single-feed distillations. Since residue curves are deflected by saddles, it is generally not possible to obtain a saddle product (pure component or azeotrope) from a simple single-feed column.

Relatively few points or composition regions in the phase diagram are of particular significance in separations process synthesis. These *compositions of interest* include the feed and desired product compositions, azeotropes, eutectics, and selected points on liquid–liquid binodal curves, especially reachable bottoms and distillate products as indicated by an RCM or DRD. The choice of composition for a mass separation agent, if required, is critical and usually is a composition which can be conveniently regenerated in the process (ie, some reachable composition).

Although distillation is favored, many situations prevent or interfere with the use of simple single-feed distillation schemes. The designer often is faced with avoiding, overcoming, or exploiting a limited set of *critical features* in order to accomplish the overall objective of the separation system. These critical features include distillation boundaries, ie, if a product (or MSA) composition is in a different distillation region from the feed mixture, the product cannot be obtained

Table 1. Some Strategic Separations

Strategy	Critical features: Distillation boundary	Pinched region	Saddle product	Vapor–solid–liquid equilibrium	Temp sensitivity
exploit SLE differences	+	+	+	+	+
change system *P* to alter critical feature	+	+			
exploit kinetic phenomena	+	+	+	+	+
exploit existing or new LLE	+	+	+		+
alter VLE with new component	+	+		+	

directly by simple single-feed distillation; saddle products, ie, if a product (or MSA) is a saddle in a particular distillation region, that product cannot be obtained at high purity directly by simple single-feed distillation; pinched or close boiling regions, ie, if a feed and product are separated by a region of low relative volatility, simple single-feed distillation tends to require a large number of stages and high reflux ratio; overlapping melting–boiling points, ie, a mixture, contain components with melting points higher that the boiling points of other components, resulting in solidification within the column; temperature-sensitivity, ie, a mixture contains components that degrade, decompose, polymerize, or react at the temperature conditions of distillation.

A purely opportunistic approach to separation synthesis for nonideal mixtures may be considered. Guided by the general sequencing heuristics, any feasible separation method applicable to the state under consideration may be picked, and progress toward the desired products may proceed. Distillation to the highest or lowest boiling node in a distillation region, and decantation of an existing liquid–liquid mixture are examples of opportunistic operations applicable at many points in the synthesis process. However, in systems exhibiting critical features, a purely opportunistic approach may lead to a partially completed solution from which it is impossible to reduce remaining composition differences by distillation or any other separation method. Further progress requires backtracking to an earlier intermediate composition, discarding part or all of the current flow sheet.

An alternative *strategic* approach makes use of known thermodynamic or physical property information to look ahead to potential difficulties, ie, critical features. Contingencies can be developed early before running into dead ends. Once a critical feature has been identified, it is useful to examine methods for crossing, breaking, by-passing, reaching, or exploiting the critical feature. The strategies and resulting separation methods for handling a given critical feature may differ considerably and the same strategy can often be implemented in several different fashions. For example, both decantation and extraction are methods for implementing the strategy of exploiting LLE tie-lines to cross a VLE distillation boundary. Table 1 lists strategies associated with several different critical features.

Early in the synthesis process it may not be known exactly where in the flow sheet a strategic separation will be used, only that it will be required some place in some form to overcome or exploit a particular critical feature. Thus, strategic separations often initially do not have well-defined feeds and products. The region where the feed must be located may be known, as well as a general idea of the types of products expected, but no definite compositions for either. A balanced synthesis method makes use of both opportunistic and strategic approaches. When critical features are present, opportunistic operations can be thought of as "links" between feeds, strategic separations, and products. A opportunistic operation often is used to reach a composition where a strategic separation is applicable (eg, opportunistically distill into a two-phase liquid region, then strategically decant the mixture to cross a VLE boundary).

It is often beneficial to re-examine a completed flow sheet and look for opportunities for simplification and consolidation of unit operations. A complicated series of unit operations can sometimes be replaced by a simpler structure that has equivalent material balance lines. When two or more sections of the flow sheet perform similar functions (ie, both produce the same product using the same or similar unit operations) one section often can be eliminated by recycling the stream to the input of the remaining section. An MSA contaminated by other components in the mixture will often function as effectively as a pure MSA, without the need for additional purification operations.

SCOTT D. BARNICKI
JEFFREY J. SIIROLA
Eastman Chemical Company

J. D. Seader and E. J. Henley, *Separation Process Principles,* 2nd ed., John Wiley & Sons, Inc., New York, 1998.

L. T. Biegler, I. E. Grossmann, and A. W. Westerberg, *Systematic Methods of Chemical Process Design,* Prentice Hall PTR, Upper Saddle River, N.J., 1997.

S. D. Barnicki and J. J. Siirola, "Enhanced Distillation", Section 13, in R. H. Perry and D. W. Greenwood, eds., *Perry's Chemical Engineers' Handbook,* 7th ed., McGraw-Hill Publishing Co., New York, 1997.

J. M. Douglas, *Conceptual Design of Chemical Processes,* McGraw-Hill Publishing Co., New York, 1988.

SEROLOGY, HUMAN. See FORENSIC CHEMISTRY.

SHALE OIL. See OIL SHALE.

SHAPE-MEMORY ALLOYS

Metallurgical Basis for the Shape-Memory Effect

Shape memory or the shape-memory effect (SME) is a phenomenon associated with the martensite transformation, a first-order displacive transformation usually related to the hardening of steel (qv). In steel, an alloy heated to the temperature where the elevated temperature face-centered cubic (fcc) austenite phase is stable is rapidly cooled to produce the hard martensite phase. In certain alloys, however, the martensite transformation is thermoelastic, ie, the martensite forms and disappears on heating and cooling over a relatively small temperature range. The intermetallic phase in these alloys undergoes a displacive, shear-like transformation when cooled below a critical temperature designated as M_S. Upon further cooling, to a temperature designated as M_F, the transformation is complete and the alloy is said to be in its martensitic state. When this martensite is deformed, it undergoes a strain that is completely recovered when the alloy is heated. The recovery process starts at the temperature A_S and is completed at a higher temperature, A_F. Because this is a first-order phase transformation, there is hysteresis associated with the formation of martensite and its reverse transformation to the elevated temperature parent phase, which is usually also referred to as austenite. The temperatures M_S, M_F, A_S, and A_F depend on the particular alloy, alloy composition, and processing. The hysteresis loop associated with the transformation in a typical shape-memory alloy is illustrated in Figure 1.

The martensite in shape-memory alloys (SMAs) may also be isothermally generated by applying stress at a temperature above M_S. Because the martensite is unstable at this temperature, when the

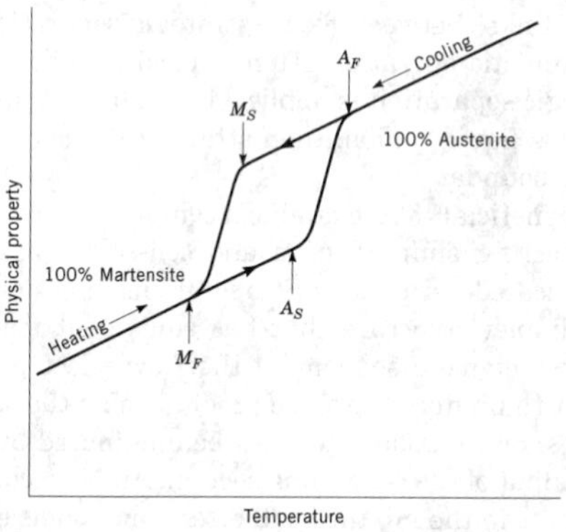

Figure 1. Schematic of the hysteresis loop associated with a shape-memory alloy transformation, where M_S and M_F correspond to the martensite start and finish temperatures, respectively, and A_S and A_F correspond to the start and finish of the reverse transformation of martensite, respectively. The physical property can be volume, length, electrical resistance, etc. On cooling the body-centered cubic (bcc) austenite (parent) transforms to an ordered B_2 or DO_3 phase and then to one of the various martensite structures.

stress is removed the martensite disappears. Above a temperature designated as M_D, martensite cannot be generated, no matter how high the stress. The stress-induced martensite (SIM) gives rise to a mechanical type of shape memory called pseudoelasticity. These alloys exhibit, in addition to SME and SIM, another unique property, that of very high damping. Damping is the property which causes a vibration, once induced in a material, to decay. This characteristic can be exploited in smart or adaptive materials (see SMART MATERIALS).

SME involves a martensitic transformation from an ordered parent phase, yielding an ordered martensite that is crystallographically reversible and thermoelastic. Simple shape-memory behavior, that is, one-way memory, may be extended to two-way memory. When a deformed martensite is heated to recover its original shape, it is a one-time event; to repeat the sequence the martensite must again be deformed. In two-way memory, the part spontaneously changes from one shape to another upon cooling or heating. This behavior, requiring a special conditioning of the martensite, is effected by limiting the number of martensite variants that form upon cooling through the application of an external stress during the transformation.

Another property peculiar to SMAs is the ability under certain conditions to exhibit superelastic behavior, also given the name *linear superelasticity*. This is distinguished from the pseudoelastic behavior, SIM. Many of the martensitic alloys, when deformed well beyond the point where the initial single coalesced martensite has formed, exhibit a stress-induced martensite-to-martensite transformation. In this mode of deformation, strain recovery occurs through the release of stress, not by a temperature-induced phase change.

Applications of the SME

Applications for nonferrous shape-memory alloys fall into the following categories: one-way SMA devices, virtual two-way devices using one-way SMAs, two-way SMA applications, and SIM and superelastic devices. The design of shape-memory devices is quite different from that of conventional alloys. These materials are nonlinear, have properties that are very temperature-dependent, including an elastic modulus that not only increases with increasing temperature, but can change by a large factor over a small temperature span. This difficulty in design has been addressed as a result of the demands made in the design of complicated smart and adaptive structures. Informative references on all aspects of SMAs are available.

One-Way SMA Applications. The first successful application for nitinol (a Ni–Ti alloy) was as a tube coupling for aircraft hydraulic control lines. Using the wide hysteresis alloy, seals of various types have been developed based on the closing force of a wire ring of suitable diameter. Such seals are used for a wide variety of electronic device closures (see PACKAGING, ELECTRONIC MATERIALS).

Virtual Two-Way Devices Using One-Way SMAs. Although alloys such as NiTi are considered to be one-way shape-memory alloys, by using a bias spring to provide the reverse motion (a recocking), a two-way device can be designed. A typical linear actuator consists of a shape-memory spring that produces motion and force in one direction and a steel spring that opposes this motion. When heated, the SMA spring has sufficient force to overcome the bias, but when the system cools, the bias spring has greater force and causes a reverse motion. Virtual two-way devices have found many applications, for example, as electrical connectors (qv) in computer systems and in circuit breakers. In addition to the circuit breaker, there have been a number of other SMA applications for various functions in electric power generation (qv), distribution, and transmission systems.

Successful actuators have been developed for various domestic safety devices. One such device is a shape-memory-actuated valve that fits in line with a bathroom sink, tub, or shower, and in the event that the water discharge approaches the temperature that would cause scalding, the valve automatically shuts off the flow and does not allow water flow to resume until the temperature is safe.

Superelastic and Pseudoelastic Devices. Medical instruments and devices have become an important application for shape-memory alloys (see PROSTHETICS AND BIOMEDICAL DEVICES). Both shape recovery and superelastic behavior are exploited. The first application that has achieved broad acceptance is the use of pseudoelastic SMA wire for orthodontic arch wires (see DENTAL MATERIALS). Tooth implant procedures involve the attachment of a metal peg in the jaw bone and when the two are stabilized, the new tooth is attached to the peg. Using shape-memory implant posts that lock into the bone immediately, the time for the procedure is reduced from six months to days. Shape-memory bone fasteners are proving superior to the usual screwed metal plates in bone fractures.

In addition to numerous medical applications, superelastic wires in very large quantities have been used for eyeglass frames. In the cellular telephone, the short antenna of superelastic shape-memory wire is very resistant to kinking and bending damage. Devices required to produce exceptionally high forces have also been developed. Industrial force generators, using the recovery of a prestrained shape-memory alloy shape having the capability of producing forces from 10 to 100 t, have been introduced.

Ferrous Shape-Memory Alloys. Ferrous shape-memory alloys have potential as low cost, readily fabricated one-way devices. The alloy family of greatest interest, Fe–Mn–Si, forms martensite on cooling. However, if this martensite is deformed, it does not exhibit shape memory on heating to the A_F temperature. On the other hand, if the alloy is subjected to sufficient stress, an epsilon martensite forms by the SIM reaction. This deformation is recovered on heating, yielding a one-way shape-memory behavior.

L. MCDONALD SCHETKY
Memory Corporation

J. D. Busch, *Proceedings of the First International Conference on Shape Memory and Superelastic Technoloiges,* Asilomar, Calif., 1994, p. 259.

T. Duerig and co-workers, eds., *Engineering Aspects of Shape-Memory Alloys,* Butterworth-Heinemann, London, 1990.

H. Funukubo, ed., *Shape Memory Alloys,* Gordon and Breach Science Publishers, New York, 1987.

Duerig, ed., *Proceedings of the 2nd International Conference on Shape-Memory and Superelastic Technologies,* Asilomar, Calif., 1997.

SHORTENINGS AND OTHER FOOD FATS. See FATS AND FATTY OILS.

SHRIMP MEAL. See AQUACULTURE CHEMICALS; FEEDS AND FEED ADDITIVES.

SIDERITE. See PIGMENTS.

SIENNA, BURNT. See PIGMENTS.

SIEVES. See SIZE MEASUREMENT OF PARTICLES.

SIGNALING SMOKES. See CHEMICALS IN WAR.

SILANES; SILANOLS. See SILICON COMPOUNDS.

SILICA

INTRODUCTION

The term *silica* denotes the compound silicon dioxide, SiO_2, and encompasses its various forms, including crystalline silicas, eg, quartz; microcrystalline silicas, eg, diatomaceous earth (see DIATOMITE); vitreous silicas, which are essentially supercooled liquid glasses; and amorphous silicas.

Silicon dioxide is the most common binary compound of silicon and oxygen, the two elements of greatest terrestrial abundance. Silica constitutes about 60 wt % of the earth's crust, occurring either alone or combined with other oxides in the silicates (see SILICON COMPOUNDS, SYNTHETIC INORGANIC SILICATES). Commercially, silica is the source of elemental silicon and is used in large quantities as a constituent of building materials (see BUILDING MATERIALS, SURVEY; SILICON AND SILICON ALLOYS). In its various amorphous forms it is used as a desiccant, adsorbent, reinforcing agent, filler, and catalyst component (see DESICCANTS; FILLERS). Silica also has numerous specialized applications, eg, as piezoelectric crystals, in vitreous silica optical elements, and as glassware. Silica is a basic material of the glass (qv), ceramics (qv), and refractories (qv) industries, and an important raw material for the production of soluble silicates, silicon and its alloys, silicon carbide, silicon-based chemicals, and the silicones (see ADVANCED CERAMICS; CARBIDES; INTEGRATED CIRCUITS; SILICON COMPOUNDS, SILICONES).

Structure and Bonding

Silicon shares with the other elements of Group 14 (IVA) of the Periodic Table the property of forming an oxide of formula MO_2. This oxide shows acidic properties. Like the dioxides of germanium, tin, and lead, silica is a solid of high melting point. Where silica differs is in having a three-dimensional lattice based on four-coordinate silicon.

The basic structural unit of most of the forms of silica and of the silicate minerals is a tetrahedral arrangement of four oxygen atoms surrounding a central silicon atom. Structurally, silica represents a limiting case in which an infinite three-dimensional network is formed by the sharing of all oxygen atoms of a given tetrahedron with neighboring groupings.

The structures in which SiO_4tetrahedra share all four oxygen atoms lead to the principal forms of silica. Replacement of some silicon atoms by aluminum gives a negatively charged framework of composition, $Al_xSi_yO_{2(x+y)}$, in which positive ions are accommodated in holes in the structure. Examples of these framework silicates include the feldspars, zeolites, and ultramarines (see CLAYS; MOLECULAR SIEVES). Because the bonding in the silicates is not dissimilar to that in other silicon compounds, the various silicate structures have analogues in molecular silicon–oxygen chemistry, where oxygen bridging between silicon atoms is commonly encountered. Thus, the silicon–oxygen ring of the cyclosilicate anions is found in the organocyclosiloxanes, whereas the single-chain silicates are formally similar to linear silicone polymers, in which the unshared oxygen atoms are replaced by univalent organic groups.

Properties

Reactions. At ordinary temperatures silica is chemically resistant to many common reagents. Reactivity is strongly dependent on the form, pretreatment, and state of subdivision of the particular sample investigated. Finely divided amorphous silica is in many circumstances considerably more reactive than bulk crystalline silica. Common aqueous acids do not attack silica, except for hydrofluoric acid, which forms fluorosilicate anions, eg, SiF_6^{2-}. Phosphoric acid attacks vitreous silica at elevated temperatures, forming a crystalline silicophosphate. Quartz and vitreous silica are affected only slightly by aqueous alkali at room temperature. Precipitated amorphous silica is more reactive than vitreous silica, which in turn is more reactive than quartz. Silica is reduced to silicon at 1300–1400°C by hydrogen, carbon, and a variety of metallic elements. Gaseous silicon monoxide is also formed.

Of the halogens, only fluorine attacks silica readily, forming SiF_4 and O_2. The acidic character of silica is shown by its reaction with a large number of basic oxides to form silicates. The reactions of silica with organic and organometallic compounds result in compounds containing Si–C and Si–O–C bonds.

Solubility. An important aspect of silica chemistry concerns the silica–water system. The interaction of the various forms of silica with water has geological significance and is applied in steam-power engineering where the volatilization of silica and its deposition on turbine blades may occur (see POWER GENERATION), in the production of synthetic quartz crystals by hydrothermal processes (qv), and in the preparation of commercially important soluble silicates, colloidal silica, and silica gel.

Polymerization. When a solution of $Si(OH)_4$ is formed, eg, by acidification of a solution of a soluble silicate, at a concentration greater than the solubility of amorphous silica (100–200 ppm), the monomer polymerizes to form dimers and higher molecular weight species. The polymerization occurs in such a way as to maximize formation of siloxane linkages, Si–O–Si, forming particles with internal siloxane linkages and external SiOH groups.

Vaporization. Silica vaporizes principally by dissociation. Gaseous SiO and O_2 are the predominant vapor species.

Forms of Silica

Crystalline Silica. Silica exists in a variety of polymorphic crystalline forms, in amorphous modifications, and as a liquid. There are three main forms of silica at atmospheric pressure: quartz, stable below about 870°C; tridymite, stable from about 870–1470°C; and cristobalite, stable from about 1470°C to the melting point at about 1723°C. In all of these forms, the structures are based on SiO_4 tetrahedra linked in such a way that every oxygen atom is shared between two silicon atoms. The structures, however, are quite different in detail. In addition, there are other forms of silica that are not stable at atmospheric pressure, including that of stishovite.

Microcrystalline Silicas. Various microcrystalline (cryptocrystalline) materials such as flint, chert, and diatomaceous earth are found in nature (see DIATOMITE). These may arise from amorphous silica, often of biogenic origin, which undergoes compaction and microcrystallization over geologic time.

Noncrystalline Silicas. The noncrystalline forms of silica include bulk vitreous silica and a variety of other amorphous types (see SILICA, AMORPHOUS SILICA; SILICA, VITREOUS SILICA). Vitreous silica (silica glass) is essentially a supercooled liquid formed by fusion and subsequent cooling of crystalline silica. Liquid silica is highly viscous, and supercooling to the glassy form occurs readily. Amorphous silica exists also in a variety of forms that are composed of small particles, possibly aggregated. Amorphous silicas are characterized by small ultimate particle size and high specific surface area. These silicas are frequently viewed as condensation polymers of silicic acid, $Si(OH)_4$. Silica sols are often called colloidal silicas, although other amorphous forms also exhibit colloidal properties owing to high surface areas. Sols are stable dispersions of amorphous silica particles in a liquid, almost always water.

Health and Safety Factors

Silica is chemically inert, and in bulk it is relatively biologically inert. It is listed in the *U.S. Food Chemicals Codex* and *National Formulary,* and is permitted as an additive in food for human and animal consumption subject to regulatory restrictions.

Inhalation of crystalline or fused vitreous silica dust, usually over long periods, causes a disabling, progressive pulmonary disease known as silicosis. Amorphous silicas can cause respiratory irritation. Silica has also come under scrutiny as a possible carcinogen.

ROBERT E. PATTERSON
The PQ Corporation

L. Pauling, *The Nature of the Chemical Bond,* 3rd ed., Cornell University Press, Ithaca, N.Y., 1960.

A. F. Wells, *Crystal Chemistry,* 4th ed., Oxford University Press, London, 1975, Chapt. 23.

R. K. Iler, *The Chemistry of Silica,* John Wiley & Sons, Inc., New York, 1979.

R. B. Sosman, *The Phases of Silica,* Rutgers University Press, New Brunswick, N.J., 1965.

AMORPHOUS SILICA

Amorphous silica, ie, silicon dioxide, SiO_2, does not have a crystalline structure as defined by x-ray diffraction measurements. Amorphous silica, which can be naturally occurring or synthetic, can be either surface-hydrated or anhydrous. Synthetic amorphous silica can be broadly divided into two categories of stable materials: vitreous silica or glass (qv), which is made by fusing quartz at temperatures greater than approximately 1700°C (see SILICA, VITREOUS SILICA), and microamorphous silica.

Microamorphous silica includes silica sols, gels, powders, and porous glasses. These consist of ultimate particles of the inorganic polymer $(SiO_2)_n$, where a silicon atom is covalently bonded in a tetrahedral arrangement to four oxygen atoms. Each of the four oxygen atoms is covalently bonded to at least one silicon atom to form either a siloxane, –Si–O–Si–, or a silanol, –Si–O–H–, functionality. Initially formed low molecular weight species condense to form ring structures so as to maximize siloxane and minimize silanol bonds. A random arrangement of rings leads to the formation of complex structures of generally spherical particles less than approximately 100 nm in diameter.

Chemical methods to determine the crystalline content in silica are based on the solubility of amorphous silica in a variety of solvents, acids or bases, with respect to relatively inert crystalline silica. Crystalline content is also determined using x-ray diffraction spectroscopy.

Microamorphous silica can be divided into microparticulate silica, ie, microscopic sheets and fibers, and highly hydrated silica. The microparticulate silicas are the most important group commercially. These include silicas precipitated from aqueous solution and silicas formed in the vapor phase, called pyrogenic or fumed silica. Several synthetic routes exist to prepare any form of microporous silica, where microporous often refers to the ability of the silica to adsorb gases or liquids.

Amorphous silica formed in aqueous solution can occur as sols, gels, or particles. A silica sol can either consist of polysilicic acid having a molecular weight of SiO_2 up to about 100,000, or be a stable dispersion of fine colloidal particles of diameters > 5 nm. A silica gel has a three-dimensional, continuous structure. Macroscopic particles formed by aggregation of a sol or very small particles of a gel to afford bunched aggregates of silica up to approximately 1 μm in diameter are called precipitated silica. This is collected as a powder by physically agglomerating these aggregates and evaporating the water.

Characterization

Amorphous silica is distinguished by chemical composition, primarily amount and type of non-SiO_2 content; by physical properties, such as surface area and ability to absorb liquids or gases; and by characteristics of the particle.

Commercial Production

Approximately 1,000,000 t of synthetic amorphous silica is produced each year with 40% of production in Europe, followed by North America at 30%, and Japan at 12%. Although deposits of naturally occurring amorphous silicas are found in all areas of the world, the most significant commercial exploitation is of diatomaceous earth in industrialized countries (see DIATOMITE). This is because of the high cost of transportation relative to the cost of the material.

Heterogeneous Reactions

Amorphous silica dissolves or depolymerizes in water according to equation 1:

$$SiO_2(s) + 2\ H_2O(l) \rightarrow H_4SiO_4(aq) \qquad (1)$$

The solubility of amorphous silica in water at 25°C ranges from 70 to 150 ppm SiO_2 (1.2–2.2 mmol/kg). The variation results from differences in particle size, degree of hydration, and level of trace impurities. The solubility of amorphous silica in water increases with increasing temperature and pressure. Hydrated amorphous silica dissolves more rapidly than does the anhydrous amorphous silica. Amorphous silica is essentially insoluble in methanol, thus solubility in methanol–water mixtures decreases with increasing percentages of methanol. Natural biogenic silica often has an organic coating that inhibits dissolution.

Polymerization. Dissolved silica undergoes polymerization to give discrete particles that can further react. The three stages of reaction are polymerization of monomers to form particles, growth of particles, and linking of particles to give chains and then networks. In basic solution, the particles grow in size and decrease in number. In acid solution or in the presence of flocculating salts, particles aggregate into three-dimensional networks and form gels.

Silica Sols and Colloidal Silica

Properties. Colloidal silica is a stable aqueous dispersion or sol of discrete amorphous silica particles having diameters of 1 to 100 nm. Silica sols do not gel or settle out of solution for at least several years of storage. The stability of a silica sol depends on several factors and can be controlled by the surface properties of the silica. The pH value should be above 7 to maintain sufficient negative charge on the silica particle surface to prevent aggregation. Silica sols can be destabilized by aggregation, gelation, crystallization, or particle growth plus settling. Aggregation occurs by coagulation in which particles collide or by flocculation in which particles become linked by bridges of flocculating agent (see FLOCCULATION AGENTS).

Preparation. To produce sols that are stable at relatively high silica concentrations, particles must be grown to a certain size in weakly alkaline aqueous dispersion. Silica sols can be purified with the aid of an ion-exchange resin, dialysis, electrodialysis, or washing. Sols having particle size >30 nm in diameter are concentrated by centrifugation.

Modifications. Coagulation, flocculation, or gelation of silica sols and their subsequent drying produce amorphous silica powders. Sol particles are extremely small, and particulate silica obtained from sols is composed of aggregates or porous particles that have a much higher specific surface area than estimated from apparent size. Aggregates are also called secondary particles. Once silica particles have been formed as a sol, the surface can be modified by the attachment of different atoms or groups to obtain specific properties.

Applications. Colloidal silicas are typically used for making silica gels having uniform pore sizes and pore volumes. There are numerous industrial applications.

Silica Gel

Properties. Silica gel is a coherent, rigid, continuous three-dimensional network of spherical particles of colloidal silica. Both siloxane, –Si–O–Si–, and silanol, –Si–O–H, bonds are present in the gel structure. The pores are interconnected and filled with water and/or alcohol from the hydrolysis and condensation reactions. A hydrogel is a gel in which the pores are filled with water. A xerogel is a gel from which the liquid medium has been removed, causing the structure to collapse, thus decreasing porosity. If the liquid medium is removed in a manner such that shrinkage and collapse of the gel structure do not occur, an aerogel is formed. Silica powder can be made from xerogels by grinding or micronizing, which decreases the size of the gel fragments but leaves the ultimate gel structure unchanged. Silica gels are classified into three types: regular, intermediate, and low density gels.

Preparation. Silica gels can be prepared by several methods. Most commonly, a sodium silicate solution is acidified to a pH value of $<$10–11. Syntheses include the bulk-set, slurry, and hydrolysis processes. The properties of a finished gel are determined by the size of the primary particles at the moment they aggregate into the gel network, called the gelation point (or gelation time). It is usually defined as the point at which aggregation in the sol forms a three-dimensional network that can support a stress elastically.

Characterization. When silica gel is used as an adsorbent, the pore structure determines the gel adsorption capacity. Pores are characterized by specific surface area, specific pore volume (total volume of pores per gram of solid), average pore diameter, pore size distribution, and the degree to which entrance to larger pores is restricted by smaller pores. Surfaces can be categorized as fully hydroxylated, in which the surface consists solely of silanol, –Si–O–H, groups, a siloxane, –Si–O–Si–, or an organic surface.

Modifications. Once a gel structure is formed, it can be modified in the wet state to strengthen the structure or enlarge the pore size and reduce surface area, through a process called aging. In this aging process, silica is dissolved from smaller particles and deposited at the points of contact between larger particles. Condensation reactions in the gels continue to occur because of the presence of the silanol, –Si–O–H, groups. Because of the formation of new siloxane bonds that bridge the particles, the continued polymerization reactions strengthen and stiffen the network. Sintering (heating to convert a powder into a continuous mass) a dried silica surface in air or in a vacuum causes shrinkage that decreases the surface area, whereas sintering in steam increases pore size.

Applications. The largest application areas for silica gels include health care products as thixotropic agents in cosmetics (qv) and dentrifices (qv); selective absorbents to maintain clarity during the brewing of beer; desiccants; thixotrope and flatting agents in coatings; and supports for polymerization catalysts.

Precipitated Silica

Particulate silica is composed of aggregates (or secondary particles) of ultimate (or primary) particles of collidal-size silica that have not become linked in a massive gel network during the preparation process. Particulate silicas are made either from the vapor phase to form pyrogenic (or fumed) silicas or by precipitation from solution, generally aqueous, to form precipitated silicas. Particulate silica powders have a more open structure with higher pore volume than do dried pulverized gels.

Properties. The physical and chemical properties of precipitated silicas can vary according to the manufacturing process. Ultimate particle and aggregate sizes of silicas precipitated from solution can be varied by reinforcement and control of suspension pH, temperature, and salt content value.

Preparation. Precipitated silica is formed from an alkaline metal silicate solution. Silica is precipitated by adding acid to sodium silicate to reduce the pH to a value of 9–10. Precipitation proceeds in three basic steps: forming colloidal particles through nucleation and particle growth to the desired ultimate particle size; coagulating colloids into aggregates to form a suspended precipitate, where the control of pH and sodium ion concentration is critical; and reinforcement of the aggregate particle to the desired degree without further nucleation so that the structure is not altered.

Modifications. Precipitated silicas are hydrophilic and may contain up to approximately 10% water as surface silanol, –Si–O–H, groups that remain after drying at 150°C. The interaction of the surface of amorphous silicas with organic molecules is strongly affected by the degree of hydroxylation of the silica surface. Thus, modification to hydrophobize the surface of a precipitated silica can be an important step toward increasing its compatibility with substrates such as nonpolar polymers.

Applications. Because of the small particle size and complex aggregate structure, precipitated silica imparts the highest degree of reinforcement to elastomer compounds, among all of the mineral fillers (see RUBBER, NATURAL). This superior reinforcement is employed in a variety of rubber compounds. Precipitated silica is used in a wide variety of applications including dentrifices, adsorbents, for food applications, and detergent extenders.

Pyrogenic Silica

Properties. Amorphous pyrogenic (fumed) silicas are generally less dense and of higher purity than are silicas precipitated from solution. Pyrogenic silicas have a much lower hydrated surface and are sometimes completely anhydrous.

Preparation. Pyrogenic silica can be prepared in several ways, including vaporizing silica and oxidizing organic or inorganic silicon compounds. Vaporizing silica at high temperatures of approximately 2000°C forms anhydrous amorphous silica particles upon cooling.

Applications. Pyrogenic silica is used in rubber applications that require a low level of surface water per unit surface area of the silica, primarily as a reinforcement in adhesives, sealants, and elastomer compounds based on silicone polymers. Pyrogenic silica also has a variety of other uses.

Fused Silica

Transparent fused silica can be formed at a temperature of 1200°C and a pressure of 13.8 MPa (2000 psi) from silica powder consisting of 15 nm ultimate particles or by electric arc fusion of pure silica sand having low iron and alkali metal contents. The cooled product is ground to the desired particle size. Fused silica is used as a sacrificial component or investment casting in the manufacture of metals and as a component in refractory materials.

Naturally Occurring Amorphous Silica

Formation. Most naturally occurring amorphous silica deposits are biogenic forms deposited from the exoskeleton, plates, or spines of aquatic organisms such as diatoms, radiolarians, silicoflagellates, and certain sponges. Large deposits of these materials are found as loosely coherent chalk-like sediments in the equatorial Pacific area, ie, radiolarians, and in high latitude areas of all oceans, ie, diatoms. All biogenic silicas are noncrystalline as formed. They are converted over times of 10^6 + years to crystalline forms. Amorphous silica can be formed by the alteration of sand to a colloidally subdivided high surface area, amorphous silica volcanic ash. Amorphous silica is sometimes precipitated from the hot supersaturated waters of hot springs (siliceous sinter) and geysers (geyserite) where it is often found along with calcareous sinter.

Applications. The most significant commercial material from the biogenic silicas is diatomaceous earth, also called kieselguhr or diatomite (qv). These deposits are the sediment of fragments and shells of one-celled algae, which form very fine particles having high surface area and as high as 94% silica content. Diatomaceous earth finds applications as absorbents, fillers, insulating materials, and polishing agents (see ABRASIVES; FILLERS).

Health and Safety Factors

Amorphous silica is classified as nontoxic by ingestion. Soluble silica is present in most drinking water. It has been observed that silica passes through the body in soluble form being excreted with the water. Unlike crystalline silica, amorphous silica has not been shown to cause silicosis, even in workers having long exposure to the product. Amorphous silica is classified by the Occupational Safety and Health Administration as a nuisance dust. The principal reported health reaction is contact dermatitis resulting from the absorption of protective oils from the skin.

WALTER H. WADDELL
Exxon Chemical Company
LARRY R. EVANS
J. M. Huber Corporation

R. K. Iler, *The Chemistry of Silica: Solubility, Polymerization, Colloid and Surface Properties, and Biochemistry,* John Wiley & Sons, New York, 1979.

P. Kleinschmit, *Special. Inorg. Chem.* **40,** 196–225 (1981).

H. D. Gesser and P. C. Goswami, *Chem. Rev.* **89,** 765–788 (1989).

L. L. Hench and W. Vasconcelos, *Ann. Rev. Mater. Sci.* **20,** 269–298 (1990).

VITREOUS SILICA

Vitreous silica is an amorphous phase of silicon dioxide, traditionally formed by heating crystalline SiO_2 above its melting point, 1730 ± 5°C, and then cooling it rapidly enough to avoid recrystallization. Because molten SiO_2 is extremely viscous, glass formation is possible at a relatively slow cooling rate. Vitreous silica is available in a variety of forms, including powders, coatings, fibers, porous bodies, and bulk (dense) glass.

Vitreous silica has a wide range of commercial and scientific applications. Its unique combination of physical properties includes good chemical resistance, minimal thermal expansion, high refractoriness, and excellent optical transmission from the ultraviolet to the near-infrared.

Although vitreous silica is a simple, single-component glass, its properties can vary significantly, depending on thermal history, the type and concentration of defects, and impurities.

Vitreous silica is, for the most part, a synthetic material, but there are instances of the material occurring in nature. Vitreous tubes called fulgurites are produced when lightning fuses quartz sand. Large deposits of fulgurite exist in the Libyan desert. Vitreous silica can also be produced by meteor impact.

The optical quality of vitreous silica can range from transparent to opaque. The degree of transparency depends on the amount of bubbles present. Opaque glass contains a large number of isolated bubbles ranging in size from 5 to 200 μm. The bubbles scatter light, giving the glass a milky appearance.

Structure

Vitreous silica is considered the model glass-forming material and as a result has been the subject of a large number of x-ray, neutron, and electron diffraction studies.

The basic structural element in both vitreous and crystalline silica is the SiO_4^{4-} tetrahedron, which arises from the sp^3 hybrid orbitals of the silicon. Each silicon atom sits in the center of the tetrahedron surrounded by four oxygen atoms that hold the corner positions. Tetrahedrons bond together by corner sharing. In a properly developed structure, each oxygen is shared by only two tetrahedrons. This bonding scheme can produce a large variety of three-dimensional structures and is the reason that silica has a number of crystalline phases.

In the crystalline form of silica, the SiO_4^{4-} tetrahedrons link together in a series of interconnecting, six-membered rings. In the glassy form, the tetrahedral network can be modeled using a distribution of ring sizes normally ranging from three- to eight-membered units. The six-membered ring is the most common size but it can exist in distorted configurations which are not present in the crystalline structures.

The tetrahedral network can be considered the idealized structure of vitreous silica. Disorder is present but the basic bonding scheme is still intact. An additional level of disorder occurs because the atomic arrangement can deviate from the fully bonded, stoichiometric form through the introduction of intrinsic (structural) defects and impurities. The intrinsic defects include paramagnetic and diamagnetic species. The impurity (extrinsic) defects include nonmetallic species (such as chlorine and hydroxyl ions) and metallic contaminants. The level of metallic contaminants can vary by 3 to 5 orders of magnitude, depending on the purity of the source material.

Manufacturing

Vitreous silica is a hard glass with a limited working range and cannot be produced using conventional glass-melting techniques. Its high melting point and high melt viscosity have forced the development of a number of unique forming methods which utilize sintering or some type of deposition process. Many of the key physical properties depend on the specific forming process used because the forming process defines the material's thermal history and impurity profile.

Glass Properties

Vitreous silica has many exceptional properties. Most are the expected result of vitreous silica being an extremely pure and strongly bonded glass. It is inert to most common chemical agents. It has a high softening point, low thermal expansion, excellent thermal shock resistance, and an excellent optical transmission over a wide spectrum. Compared to other technical glasses, vitreous silica is one of the best thermal and electrical insulators and has one of the lowest indexes of refraction.

Vitreous silica, however, also exhibits a number of abnormal behaviors for a glass. These are a consequence of its inherent atomic structure. For example, the expansion coefficient is negative below approximately −80 to 100°C and positive and very small above these temperatures. In contrast to most glasses, the equilibrium density decreases with heat treatment in the transformation range.

Chemical Properties. Stoichiometric vitreous silica contains two atoms of oxygen for every one of silicon, but it is extremely doubtful if such a material really exists. In general, small amounts of impurities derived from the starting materials are present and various structural defects can be introduced, depending on the forming conditions. Water is incorporated into the glass structure as hydroxyls. Vitreous silica does not react significantly with water under ambient conditions. It is susceptible to attack by alkaline solutions, especially at higher concentrations and temperatures. It is relatively inert to attack from most acids for temperatures up to 100°C. The main exceptions are phosphoric acid, which causes some corrosion above approximately 150°C, and hydrofluoric acid, which reacts readily at room temperature.

Metals do not generally react with vitreous silica below 1000°C or their melting point, whichever is lower. Exceptions are aluminum, magnesium, and alkali metals. Aluminum readily reduces silica at 700–800°C. Fused basic salts and basic oxides react with vitreous silica at elevated temperatures. Reaction with alkaline-earth oxides takes place at approximately 900°C. Halides tend to dissolve vitreous silica at high temperatures; fluorides are the most reactive. Dry halogen gases do not react with vitreous silica below 300°C. Hydrogen fluoride, however, readily attacks vitreous silica.

Vitreous silica is a highly refractory glass. Volatilization is significant only at elevated temperatures, occurring under a neutral atmosphere by dissociation to SiO gas and oxygen.

At atmospheric pressure, devitrification of vitreous silica can take place at temperatures from 1000 to 1723°C, ie, the cristobalite liquidus. The maximum growth rate occurs at approximately 1600–1675°C. Crystals form and grow from nuclei found predominantly at the glass surface. The devitrification rate is extremely sensitive to both surface and bulk impurities, especially alkali. Increased alkali levels tend to increase the devitrification rate and lower the temperature at which the maximum rate occurs.

The gaseous species for which diffusion parameters have been studied extensively in vitreous silica include both noble gases and molecular gases. In general, the activation energies for diffusion of the gaseous species scale with molecular size. The noble gases do not interact chemically with the glass structure. Helium is the fastest diffusing species in vitreous silica. Diffusion of the molecular gases can be complicated by reactions with the glass network, especially at the sites of structural defects.

Physical Properties. The density of transparent vitreous silica is approximately 2.20 g/cm^3. Translucent and opaque glasses have lower densities due to the entrapped bubbles.

The viscosity of vitreous silica in the transformation range depends on impurities and thermal history. Any contaminant that can break up the Si—O—Si, network, including OH, chlorine, fluorine, and metal ions, lowers the viscosity. The characteristic temperatures, ie, softening, anneal, and strain points, of vitreous silica are significantly higher than those of most commercial glasses. Vitreous silica starts to deform slowly at temperatures above 1200°C and, although it can be readily sagged at temperatures of 1400–1550°C, is a fairly viscous material even above the melting point of cristobalite.

Thermal Properties. Most manufacturers' literature quotes a linear expansion coefficient within the 0–300°C range of 5.4×10^{-7} to 5.6×10^{-7}/°C. The expansion coefficient of vitreous silica can be controlled by doping the glass with titania. At 7.4 wt % TiO_2, the room temperature expansion coefficient is effectively zero ($<10^{-8}$/°C). Thermal conductivity at 25°C is 1.38 W/(m · K). The thermal conductivity of opaque silica is 20% lower than that of clear vitreous silica. Phonon transport is the main conduction mechanism below 300°C.

Mechanical Properties. The Young's modulus of vitreous silica at 25°C is 73 GPa ($<1.06 \times 10^7$ psi), the shear modulus is 31 GPa ($<4.5 \times 10^6$ psi), and the Poisson's ratio is 0.17. Minor differences in values can arise owing to density variations. The elastic modulus decreases and Poisson's ratio increases with increasing density. The fracture strength of vitreous silica depends on its surface quality, which can be affected by thermal treatment and handling conditions. Microcracks, surface contamination, and crystallization can reduce the strength from the value of pristine vitreous silica by several orders of magnitude. The Knoop indentation hardness of vitreous silica is in the range of 473–593 kg/mm^2 and the diamond pyramidal (Vickers) hardness is in the range of 600–750 kg/mm^2.

Electrical Properties. Pure vitreous silica is an excellent electrical insulator. The synthetic glasses can have bulk resistivities as high as 10^{18} ohm·cm at room temperature. The surface conductivity of vitreous silica is also low compared to other silicate glasses. Because vitreous silica is not hydroscopic, water films containing exuded alkalies do not readily form on its surfaces. The surface conductivity, however, can increase significantly with increasing relative humidity.

Vitreous silica is also a very stable dielectric material. The dielectric constant is 3.8–4.0, depending on the grade of glass. The dielectric breakdown strength in vitreous silica depends on its impurity content, its surface texture, and the concentration of structural defects, such as cord and bubbles.

Optical Properties. The optical transmission of vitreous silica is influenced by impurities and the forming process. Ultrapure vitreous silica has the ability to transmit from the deep ultraviolet, through the visible, and to the near-infrared spectral range.

Radiation Effects. Significant structural changes are generally a result of high energy particle radiation. For example, irradiation by fast neutrons causes a densification of vitreous silica that reaches a maximum value of 2.26 g/cm^3, ie, an increase of approximately 3%, after a dose of 1×10^{20} neutrons per square centimeter. Electronic changes are usually the result of ionizing radiation, such as electron beams, x-rays, γ-rays, or protons. Damage by ionizing radiation manifests itself in the formation of optical absorption centers, also called color centers or defect centers. These absorption centers show up primarily in the visible and ultraviolet spectral regions. If the ionizing radiation is of very high intensity, structural defects may also occur.

Uses

Chemical Applications. Because of its excellent chemical durability, high purity, thermal shock resistance, and usefulness at high temperature, vitreous silica has a wide range of applications in chemical analysis and preparations. Tubing, rods, crucibles, dishes, boats, and other containers and special apparatus are available in both transparent and nontransparent varieties. Because of its inertness, vitreous silica is used as a chromatographic substrate in the form of microparticles, capillary tubing, and open columns for high resolution gas chromatography (see ANALYTICAL METHODS; CHROMATOGRAPHY).

Thermal Applications. The protection of precious-metal thermocouples in high temperature pyrometry is an important application of vitreous silica. Vitreous silica is used for gas-heated or electrically heated devices in various shapes. In its simplest form, an electric-resistance furnace consists of a vitreous silica tube or pipe on which the resistance element is wound (see FURNACES, ELECTRIC).

Optical Applications. Vitreous silica is ideal for many optical applications because of its excellent ultraviolet transmission, resistance to radiation darkening, optical polishing properties, and physical and chemical stability. It is used for prisms, lenses, cells, windows, and other optical components where ultraviolet transmission is critical.

Mechanical Applications. The volume of vitreous silica used for fibers is a very small part of the total consumption. However, some interesting and significant applications have been developed in the laboratory, particularly in the area of measurements. Because of its low and regular thermal expansion, vitreous silica is employed in apparatus used to measure the thermal expansion of solids.

Lighting. An important application of clear fused quartz is as envelop material for mercury vapor lamps. In addition to resistance to deformation at operating temperatures and pressures, fused quartz offers ultraviolet transmission to permit color correction. The color is corrected by coating the inside of the outer envelope of the mercury vapor lamp with phosphor.

Electronic Applications. In electronic systems, such as radar and computers, signal delay is sometimes necessary. A transducer converts electrical signals to ultrasonic elastic waves, which pass through a connecting medium to another transducer where the waves are reconverted to electrical signals. Vitreous silica is an ideal connecting medium because it has excellent physical stability and low ultrasonic transmission losses. Thin films (qv) of vitreous silica have been used extensively in semiconductor technology. They serve as insulating layers between conductor stripes and a semiconductor surface in integrated circuits, and as a surface passivation material in planar diodes, transistors, and injection lasers.

Space and Astronomy. Vitreous silica is used in several space-based applications because of its reduced static fatigue (slow crack growth), thermal stability, and radiation resistance. Every U.S. space vehicle having service personnel, including Mercury, Gemini, Apollo, and space shuttle vehicles, has been equipped with windows made of high optical-quality vitreous silica (Corning Code 7940) in order to have the clarity needed for visual, photographic, and television-based observations.

Optical Fibers. Pure and doped fused silica fibers have replaced copper lines in the telecommunication area (see FIBER OPTICS).

Other Uses. Vitreous silica also has applications in the form of powders, fibers, wool, and chips. The powder may be used as an inert filler or as a coating material in the investment-casting industry for the lost-wax process. Chips are used as an inert substrate in certain chemical process applications. Vitreous silica wool or fiber is an excellent insulation or packing material.

DANIEL R. SEMPOLINSKI
PAUL M. SCHERMERHORN
Corning, Inc.

R. B. Sosman, *The Phases of Silica,* Rutgers University Press, New Brunswick, N.J., 1965.

I. Fanderlik, ed., *Silica Glass and its Application,* Elsevier Science Publishers, New York, 1991.

E. B. Shand, *Glass Engineering Handbook,* McGraw-Hill Book Co., Inc., New York, 1958.

H. A. Miska, *Engineering Materials Handbook,* Vol. 4, *Ceramics and Glasses,* ASM International, Metals Park, Ohio, 1991.

SYNTHETIC QUARTZ CRYSTALS

Silicon dioxide, SiO_2, exists in both crystalline and glassy forms. In the former, the most common polymorph is α-quartz (low quartz). All commercial applications of crystalline quartz use α-quartz, which is stable only below a 573°C at atmospheric pressure. Some of the properties of α-quartz are listed in Table 1.

Quartz is used mainly in electronic applications, for which it must be free of electrical and optical twinning, voids, foreign minerals, and liquids. Moreover, it must be large enough for convenient processing. The principal source of electronic-grade natural quartz is Brazil, but manufacturers generally use synthetic quartz for electronic devices. The size, perfection, and properties of natural quartz vary. In addition, because natural crystals are generally irregular in shape, automated cutting is cumbersome and the yield is lower.

Synthesis

α-Quartz cannot be crystallized from its pure melt, because viscous SiO_2 melts almost always form silica glass upon cooling. When crystals are formed, these are high temperature polymorphs of SiO_2, eg, cristobalite and tridymite, that do not easily transform to untwinned α-quartz. The first known successful attempt to grow quartz crystals hydrothermally was in 1905. Small crystals were formed in a thermal gradient from a sodium silicate solution. All successful quartz growth processes depend on the supersaturation produced by dissolving small particles of quartz nutrient in a hot region of the high pressure system and crystallizing it onto α-quartz seeds in a cooler part of the system. Thus, it is necessary to employ a solvent in which quartz is the stable solid phase having reasonable solubility and in which the dependence of solubility upon temperature produces an appropriate supersaturation, ΔS, and an appropriate temperature differential, ΔT, between the dissolving and growth zones.

Table 1. Properties of α-Quartz[a]

Property	Value
crystal class, space group	32[b]
lattice constant, nm	
a	0.491
b	0.540[c]
indexes of refraction, Na D line	
n_o	1.5442
n_E	1.5533[d]
optically active, Na D line α, °/mm	27.71
transmission range, nm	150–3000
resistivity, Ω·cm	10^{15}
dielectric constant	
ϵ^T_1	4.58
ϵ_3	4.70
piezoelectric coupling coefficient, %	10
piezoelectric constant, FC/N[e]	
d_{11}	−23.12
d_{14}	727
hardness, Mohs'	7
thermal conductivity, W/(m·K)	6.69–12.13
acoustic Q	$(0.1-3) \times 10^6$

[a] Many properties are directionally dependent, therefore values listed are indicative only. [b] Trigonal trapezohedral class of the rhombohedral subsystem. [c] Variable, depending on purity. [d] Birefringent; $n_E - n_o = 0.0091$. [e] To convert FC/N to stat C/dyn, divide by 333×10^8.

All commercial processes use either NaOH or Na_2CO_3 as solvent systems. The dissolving mechanism is similar in both solvents because CO_3^{2-} hydrolyzes to OH^-. Sodium salts are required because insoluble sodium iron silicates form on the steel walls of the high pressure vessels when Na^+ and silicates are present (see SILICON COMPOUNDS). These compounds are relatively insoluble under hydrothermal conditions and form a protective coat on the walls, allowing the use on unlined, relatively inexpensive vessels.

Equipment. A typical commercial quartz-growing autoclave is constructed of material for use at 17 MPa (25,000 psi) and 400°C such as a low carbon steel. The pressure in the vessel is transmitted through a plunger to the steel surfaces which initially are nearly line contacts. Thus, the pressure in the seal surface greatly exceeds the pressure in the vessel because most of the area of the plunger is unsupported. Autoclaves up to a meter in diameter have been used to produce over 450 kg of synthetic crystal in a single run.

Modern process facilities are computer controlled. Temperature and ΔT are programmed upward during the start of a growth cycle, pressure and temperature are monitored and controlled, and pressure and temperature overshoot alarms and overrides are provided. Such systems also store data from previous runs for correlations with properties or for identical replication of past conditions.

Health and Safety Factors

The principal consideration in quartz synthesis is the safe management of the high pressures required for growth (see HIGH PRESSURE TECHNOLOGY). The autoclave must be designed on the basis of proper stress analysis and materials selection.

Uses

The principal uses of α-quartz are mostly related to the piezoelectric effect (see FERROELECTRICS). Crystals are used in electrical filters and oscillators for frequency and timing and frequency-division multiplexing and demultiplexing. They have also been used as electromechanical transducers in ultrasonic generators and electronic weighing microbalances. Crystal quartz is generally the material used wherever precise timing control is required. The advent of satellites for timing and communication requires the use of quartz that is not only radiation resistant but of greater mechanical strength and accuracy than material used for other purposes. Quartz also has modest but important uses in optical applications, primarily as prisms.

ALTON F. ARMINGTON
Consultant

A. A. Ballman and R. A. Laudise, in J. J. Gilman, ed., *The Art and Science of Growing Crystals,* John Wiley & Sons, Inc., New York, 1963, pp. 231–251.

D. M. Dodd and D. B. Fraser, *J. Phys. Chem. Solids,* **26,** 673 (1965).

SILICA, BRICK. See REFRACTORIES.

SILICATES. See SILICON COMPOUNDS.

SILICIDES. See SILICON AND SILICON ALLOYS.

SILICON AND SILICON ALLOYS

PURE SILICON

Silicon, Si, from the Latin *silex, silicis* for flint, has atomic number 14, atomic wt 28.083, and a room temperature density of 2.3 gm/cm^3. Silicon is brittle, has a gray, metallic luster, and melts at 1412°C. Elemental silicon does not occur in nature. As a constituent of various minerals, eg, silica and silicates such as the feldspars and kaolins, however, silicon comprises about 28% of the earth's crust. There are three stable isotopes that occur naturally and several that can be prepared artificially and are radioactive. This quality silicon is used primarily in the manufacture of semiconductor devices such as integrated circuits (qv).

Crystal Structure

At atmospheric pressure, silicon has a diamond cubic structure, ie, two interpenetrating face-centered cubes displaced 1/4, 1/4, 1/4 from each other. When subjected to ca 11 GPa (~110, 000 atm) hydrostatic pressure, the diamond structure is converted to a body-centered tetragonal lattice. Silicon produced at temperatures below about 500°C by vapor deposition is generally amorphous. In the early literature a hexagonal form stable at atmospheric pressure and thought to be analogous to the graphitic form of carbon was reported, but its existence has never been verified.

Chemical Properties

Silicon resembles metals in its chemical behavior. Silicon, carbon, germanium, tin, and lead comprise the Group 14 (IVA) elements. Silicon and carbon form the carbide, SiC (see CARBIDES). Silicon and germanium are isomorphous and thus mutually soluble in all proportions. Neither tin nor lead reacts with silicon. Molten silicon is immiscible in both molten tin and molten lead.

Silicon forms many compounds that are analogous to those of carbon and thus sometimes appears to give rise to an inorganic equivalent of carbon-based organic chemistry. However, the known silicon compounds (qv) are far fewer and much less complex than those of carbon. One example is the silane series, eg, SiH_4, Si_2H_6, Si_3H_8, and Si_4H_{10}.

For the manufacture of silicon semiconductor devices, oxide thicknesses of from <10 to >1000 nm are required on slices of single-crystal silicon. These oxide layers are formed at elevated temperatures, generally at about 1000°C, in an atmosphere of either oxygen or steam.

Silicon is virtually insoluble in any single acid but can be dissolved in a two-component, two-stage operation in which one acid component oxidizes the surface and the other etches away the oxide.

High Purity Silicon Preparation

Attempts to produce high purity silicon from metallurgical grade silicon by metallurgical methods have been underway since at least 1919. Such material was used for diodes during World War II and more recently has been used with limited success in the manufacture of solar cells. However, to produce the purity of silicon required for most semiconductor applications in the necessary quantities, a different approach is used. Metallurgical grade silicon is first converted to a silicon compound that can be easily purified (usually a halide such as $SiCl_4$ or $SiHCl_3$). After extensive purification by distillation, that compound is reduced to silicon using a high purity reducing agent (ordinarily hydrogen). A flow diagram of the common Siemens process is shown in Figure 1.

To produce the purity of silicon required for most semiconductor applications in the necessary quantities, a different approach is used.

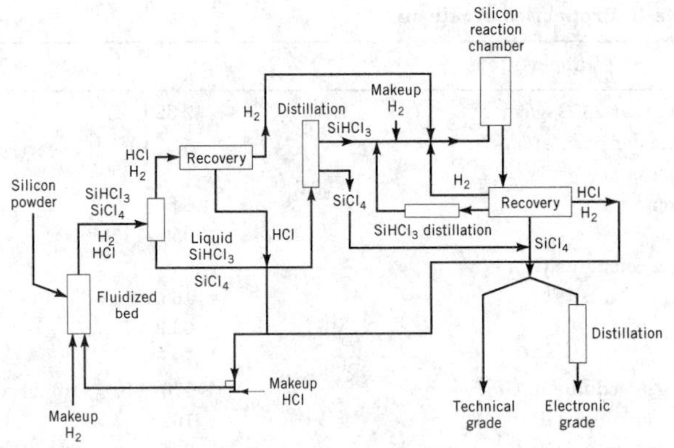

Figure 1. Flow sheet for high purity silicon manufacturing.

Metallurgical-grade silicon is first converted to a silicon compound that can be easily purified. After extensive purification, that compound is reduced to silicon by using a high purity reducing agent. Silicon tetrachloride, $SiCl_4$; trichlorosilane, $SiHCl_3$; and silane, SiH_4, are the silicon compounds most often used.

Crystal Growth. Most high purity silicon is used in the manufacture of semiconductor devices, and most such devices are made from single-crystal silicon. A number of processes for growing single crystals of silicon are available, but the most common is the pulling technique first described by Teal and Little in 1950. This is an adaptation of a much earlier crystal growing method for studying the speed of crystallization of metals, called the Czochralski method. Silicon, surrounded by an inert gas atmosphere, is first melted in a fused silica container. Then, the bottom end of a single-crystal silicon seed, capable of being both rotated and pulled up from the melt, is dipped into the melt and slowly withdrawn as silicon freezes on it. Crystals as small as a few mm in diameter and as large as 300 mm in diameter have been grown by this process.

Physical Properties

Unlike structural materials where properties are generally given for polycrystalline samples, many of the properties of the high purity silicon required for the manufacture of semiconductor devices are measured on single-crystal samples. For this reason, even though silicon was isolated in the early 1800s, most of the values for pure silicon have been measured since the introduction of single-crystal silicon growing techniques in the 1950s. Table 1 gives values for the more common physical properties and for some of the thermodynamic properties.

Electrical Properties

Silicon is a semiconductor and thus the electrical conductivity, ς, is determined by contributions from both electrons and holes, equation 1,

$$\sigma = q(\mu_n n + \mu_p p) \quad (1)$$

where q = charge on an electron, μ = carrier mobility in cm^2/(V·s), n = density of free electrons (electrons/cm^3), and p = density of free holes. Both hole and electron mobilities decrease as the number of carriers increase, but near room temperature and for concentrations less than about 10^{16} there is little change, and the values are ca 1400 cm^2/(V·s) for electrons and ca 475 cm^2/(V·s) for holes.

Radiation Effects

Gamma radiation produces free carriers much as does visible light. High energy protons and electrons produce defects that reduce minority carrier lifetime according to equation 2:

$$1/\tau_f = 1/\tau_0 + k\phi \quad (2)$$

Table 1. Properties of Silicon

Property	Value
density, at 25°C, g/cm^3	2.329
atomic density[a], atoms/cm^3	5×10^{22}
hardness	
Mohs'	6.5
Knoop	950–1150
elastic constants,GPa[b]	
C_{11}	165.7
C_{12}	63.9
C_{44}	79.57
Young's modulus in GPa[b]	170 ⟨110⟩ (110) direction
bulk modulus[a], GPa[b]	100
fracture stress, GPa[b]	2.8[c]
melting point, °C	1414
volume expansion on freezing, %	9.5
boiling point, °C	2355
debye temperature, K	645
critical temperature, °C	4886
critical volume, cm^3/mol	232.6
critical pressure, MPa[d]	53.6
heat of fusion, kJ/mol[e]	50.660
heat of vaporization, kJ/g[e]	16
vapor pressure, Pa[f]	
800°C	1.33×10^{-8}
1000°C	1.33×10^{-5}
1500°C	2.66
2000°C	80

[a] Calculated value. [b] To convert GPa to dyn/cm^2, multiply by 10^{10}. [c] Average value. [d] To convert MPa to atm, divide by 0.101. [e] To convert J to cal, divide by 4.184. [f] To convert Pa to mm Hg, multiply by 0.0075.

where τ_f is the lifetime after irradiation with a fluence ϕ, τ_0 is the original lifetime, and k is a radiation damage constant. Neutrons produce deep-level defects that not only degrade lifetime as in equation 2, but also increase resistivity through the removal of carriers, as described by equation 3,

$$N = N_0 - K\phi \tag{3}$$

where N_0 is the initial carrier density, N the density after irradiation, and K a different radiation damage constant.

Optical Properties

In general, as with other semiconductors, at wavelengths shorter than some critical value defined by the band gap energy, optical absorption generates free carriers and the transmissivity is very low. The absorption coefficient changes several orders of magnitude over the wavelength range 0.5–1 μm. This change is not nearly as abrupt as it is in direct band gap materials such as gallium arsenide, GaAs. Also, because Si is an indirect band gap material, it cannot be used for conventional light-emitting diodes (see LIGHT GENERATION, LIGHT-EMITTING DIODES). For wavelengths having photon energies less than that corresponding to the band gap, the transmissivity is substantial, but as wavelengths continue to increase, both impurity absorption bands and the gradually increasing free-carrier absorption reduce transmissivity. There are also a number of lattice absorption bands, the most pronounced of which is at 16.13 nm. Silicon is often used as an infrared optical element because of the good infrared transmissivity in the 1–8-μm region.

Health and Safety Factors

Elemental silicon is quite inert. In air, silicon is classified only as a nuisance particulate. However, SiO_2 dust, particularly that of crystalline quartz, is considered hazardous. Many silicon compounds are poisonous, and some, such as the silanes, are highly explosive as well.

W. R. Runyan
Texas Instruments Incorporated

A. S. Berezhnoi, *Silicon and its Binary Systems,* Consultants Bureau, New York, 1960; J. W. Mellor, *A Comprehensive Treatise on Inorganic and Theoretical Chemistry,* Vol. IV, Longmans, Green & Co., Inc., New York, 1957; M. C. Sneed and R. C. Brasted, eds., *Comprehensive Inorganic Chemistry,* Vol. VII; *The Elements and Compounds of Group IVA,* D. Van Nostrand Co., Inc., Princeton, N.J., 1958.

W. R. Runyan and K. E. Bean, *Semiconductor Integrated Circuit Processing Technology,* Addison-Wesley, Reading, Mass., 1990.

L. P. Hunt, in W. C. O'Mara, R. B. Herring, and L. P. Hunt, eds., *Handbook of Semiconductor Silicon Technology,* Noyes Publications, Park Ridge, N.J., 1990.

Properties of Silicon, EMIS Datareview Series No. 4, Inspec, London, 1988.

CHEMICAL AND METALLURGICAL

Silicon

Production. Silicon is typically produced in a three-electrode, a–c submerged electric arc furnace by the carbothermic reduction of silicon dioxide (quartz) with carbonaceous reducing agents. The overall reaction for the production of silicon is expressed as $SiO_2 + 2\,C \rightarrow Si^0 + 2\,CO$.

Uses. The market for silicon can be broken down into chemical, dominated by silicones; secondary aluminum; primary aluminum, recovery; and electronics. Silicones are remarkably versatile and can be produced in numerous forms. Examples range from fluids thinner than water to rigid plastics which can be clear or pigmented.

The secondary aluminum market is the largest consumer of silicon. The majority of the applications are in castings for the automotive industry. The most prominently used compositions in all casting processes are those of the aluminum–silicon family. The outstanding effect of silicon in aluminum alloys is the improvement of casting characteristics. The third largest silicon consumer is the primary aluminum market. A small (1%) percentage of silicon is added to the virgin aluminum ingot to remove oxygen during casting. Relatively higher purity silicon is required for this application. The electronics market uses silicon as trichlorosilane, which is decomposed with hydrogen at high temperatures to produce semiconductor-grade silicon (see SILICON COMPOUNDS).

Silicon is also used in the copper (qv) industry for production of silicon bronzes. The addition of silicon improves fluidity, minimizes dross formation, and enhances corrosion resistance and strength. Besides the chemical industry, silicon is used as a powder in the ceramics (qv) industry for the production of silicon carbide and silicon nitride parts (see ADVANCED CERAMICS). Silicon powder is also used as an explosive for defense applications and in the refractory industry for plasma spraying with other oxide mixtures (see REFRACTORY COATINGS).

Silicon Alloys and Silicides

Silicides are compounds of silicon and a metal. The silicides of Groups 4–6 (IVA, VA, VIA), the refractory metal silicides, and of Groups 8–10 (VIII), the near-noble metal silicides, are of importance for very large-scale integration (VLSI) electronic applications (see INTEGRATED CIRCUITS). Metal silicides form well-defined crystals with a bright metallic luster. These are usually hard and high melting.

Ferrosilicon and Alloys. Ferrosilicon is a grayish ferroalloy consisting mainly of the elements silicon and iron. Ferrosilicons for steelmaking and foundry uses have a silicon content of 14–95 wt %. Four compounds exist: Fe_2Si, Fe_5Si_3, Fe–Si, and $FeSi_2$. Silicon is an essential element in cast iron. The silicon level is often the controlling factor in whether cast iron solidifies as gray iron having its accompanying mechanical properties or as white iron or iron carbide, which has little engineering significance because of its brittle nature.

Ferrosilicon is produced in a three-phase submerged arc furnace by the carbothermic reduction of quartz in the presence of high quality iron ore or scrap steel. The smelting process for ferrosilicon is quite similar to silicon, with the exception of the amount of silicon carbide that is formed as an intermediate. As the iron content of ferrosilicon increases, the stability of the silicon carbide phase is known to decrease.

Specialty Silicon Alloy Uses. In the iron and steel industry, silicon alloys or silicides are used for alloying, deoxidizing, and reducing other elements such as manganese, chromium, tungsten, and molybdenum. Silicon in the form of 75% ferrosilicon is used as a reducing agent in magnesium manufacture.

Regular Ferrosilicon. Regular 50% ferrosilicon is a widely used, economical source of silicon for both iron and steel. This ferrosilicon is used as a deoxidizer and alloying agent in the production of killed and semikilled steels. Killed steels are those that have been made free from bubbling, while molten, by the addition of a deoxidizer. This process minimizes reaction of carbon and oxygen during solidification. Regular 75% ferrosilicon is widely used for deoxidizing and alloying additions to iron and steel that do not require tight control of residuals. This ferrosilicon is ideal for ladle additions because its higher silicon content permits smaller additions in order to reach desired silicon levels.

Specialty Silicon Alloys

Magnesium Ferrosilicons. Of the miscellaneous silicon alloys consumed in the United States, magnesium ferrosilicon is the most significant. Magnesium ferrosilicon alloys have a light gray to silvery crystalline appearance and are primarily used in the production of ductile iron. Magnesium ferrosilicon alloys are produced by post-taphole ladle additions of magnesium and at times rare earths to the ferrosilicon production process.

Rare-Earth Silicides. Rare-earth silicides, in the form of a ferroalloy that contain up to 33% rare earths, are used by the iron and steel industries. For nodular iron, addition of rare earths gives spheroidal rather than flaky graphite. Rare earths desulfurize and, combined with silicon, deoxidize and alloy steels. They may be substituted for mischmetal.

Manganese–Silicon Alloys. Manganese–silicon alloys usually contain various amounts of iron and carbon. They are made by carbon reduction of manganese ore or manganese slag in the presence of silica in an electric furnace. The lower silicon grades are referred to as silicomanganese. Silicomanganese is widely used in steelmaking furnaces and ladles for alloying and deoxidization. The use of silicomanganese is often preferred to separate additions of manganese and silicon, because it introduces less aluminum, carbon, nitrogen, and phosphorus. It also provides stronger deoxidization and produces higher purity steel. Barium- and calcium-bearing manganese silicon is used as an inoculant in gray and ductile iron. The zirconium-bearing manganese silicon is also used as an inoculant for gray and ductile iron castings. Manganese in the alloy enhances its dissolution rate, strengthens the matrix, and improves tensile strength without impairing machinability.

Calcium–Silicon. Calcium–silicon and calcium–barium–silicon are made in the submerged-arc electric furnace by carbon reduction of lime, silica rock, and barites. These alloys are used to deoxidize and degasify steel.

Barium–Silicon. Barium–silicon alloy is used as an inoculant in both gray and ductile iron.

Strontium–Silicon. The strontium–silicon alloy is made in an electric furnace by carbon reduction of silica- and strontium-bearing ore. In cast irons, this alloy effectively reduces chill.

Titanium–Silicon. The titanium–silicon alloy is made by adding titanium scrap to molten metallurgical silicon or silicon alloys. It is also made by carbon reduction of titanium ore, limestone, and quartz in a submerged arc furnace. Titanium–silicon alloy is an efficient graphitizing inoculant for chill reduction in gray cast iron. It is also a supplementary deoxidizer for wrought and cast steels.

Vanadium–Silicon. Vanadium–silicon alloy is made by the reduction of vanadium oxides with silicon in an electric furnace. Application is essentially the same as that of the titanium alloys.

Zirconium–Silicon. Zirconium–silicon alloy is made in an electric furnace by carbon reduction of the oxides. It is used by the iron and steel industries as a deoxidizer and scavenger.

Health, Safety, and Environmental Factors. The off-gas from the production of silicon and ferrosilicon alloys contains large amounts of dust having approximately 90% silica fume. This nonhazardous silica fume is amorphous and more of a nuisance than hazardous. The principal health hazard that may be associated with silicon and silicon alloys is caused by the crystalline form of the oxide, ie, quartz, used as a raw material. Silica in its crystalline form is the chief cause of disabling pulmonary fibrosis, such as silicosis. Silicon and ferrosilicons by themselves are nontoxic. Some ferrosilicons may produce phosphorus hydride on coming in contact with water. Ferrosilicon is classified by the U.S. Department of Transportation as "Dangerous When Wet."

VISHU DOSAJ
Dow Corning Corporation

H. Moissan, *The Electric Furnace,* 2nd ed., trans. by V. Lenker, Chemical Publishing Co., Easton, Pa., 1920.

Product data literature, Elkem Metals Co., Pittsburgh, Pa., 1996; Globe Metallurgical Sales Inc., Cleveland, Ohio, 1996; Hickman, Williams and Co., Lakewood, Ohio, 1996.

SILICON COMPOUNDS

SYNTHETIC INORGANIC SILICATES

Commercial soluble silicates have the general formula:

$$M_2O{\cdot}mSiO_2{\cdot}nH_2O$$

where M is an alkali metal and m, the modulus, and n are the number of moles of SiO_2 and H_2O, respectively, per mole of M_2O. The composition of commercial alkali silicates is typically described by the weight ratio of SiO_2 to M_2O; sodium silicates are the most common. Potassium silicate and lithium silicates are manufactured to a limited extent for special applications. These materials are usually manufactured as glasses that dissolve in water to form viscous, alkaline solutions. The modulus or ratio, m, in commercial sodium silicate products typically varies from 0.5 to 4.0. The most common form of soluble silicate has an m value of 3.3.

These soluble silicates have many uses based on their physical and chemical properties; however, their largest and most rapidly growing use classification derives from their ability to serve as a source of reactive primary silica (qv) species.

Structure

Silicate Glasses. Synthetic silicates and silica are composed of polymers of the orthosilicate ion, SiO_4^{4-}. Orthosilicate monomers have a

Table 1. Silicate Structural Categories

Silicate type	Unit structure	Name
oligosilicates, discreet		
orthosilicate	SiO_4^{4-}	zircon
pyrosilicate	$Si_2O_7^{6-}$	thortveitite
cyclosilicates, discrete cyclic		
trimer	$Si_3O_9^{6-}$	benitoite
tetramer	$Si_4O_{12}^{8-}$	papagoite
hexamer	$Si_6O_{18}^{12-}$	dioptase
polysilicates, chains		
pyroxenes	$(SiO_3^{2-})_n$	wollastonite
amphiboles	$(Si_8O_{22}^{12-})_n$	tremolite
phyllosilicates, sheets	$(Si_4O_{10}^{2-})_n$	talc
tectosilicates, frameworks	$(SiO_2)_n$	silica, quartz

tetrahedral structure. As shown in Table 1, structural categories are possible. The physical and chemical properties of silicate glasses depend on the composition of the material, ion size, and cation coordination number.

Crystalline Alkali Silicates. The most common crystalline soluble silicates belong to the metasilicate family, $Na_2O{\cdot}SiO_2{\cdot}nH_2O$. Anhydrous sodium monopolysilicate, Na_2SiO_3, contains SiO_2 chains, whereas the hydrates $Na_2H_2SiO_4{\cdot}xH_2O$, where $x = n - 1$, contain only the dihydrogen monosilicate ion. The structures of the series sodium dihydrogen monosilicate tetrahydrate, sodium dihydrogen monosilicate pentahydrate, sodium dihydrogen monosilicate heptahydrate, and sodium dihydrogen monosilicate octahydrate vary primarily by the order and coordination of the hydrated sodium ion.

Dissolution

The dissolution of soluble silicates is of considerable commercial importance. Its rate depends on the glass ratio, solids concentration, temperature, pressure, and glass particle size. Dissolution of sodium silicate glass proceeds through a two-step mechanism that involves ion exchange (qv) and network breakdown.

Ion exchange $$\equiv Si{-}ONa + H_2O \rightleftharpoons \equiv Si{-}OH + Na^+ + OH^-$$

Network breakdown $$\equiv Si{-}O{-}Si\equiv + HO^- \rightleftharpoons \equiv Si{-}O^- + HO{-}Si\equiv$$

Thus, silica is removed from the glass following leaching of the alkali cations.

Silicates in Solutions. The distribution of silicate species in aqueous sodium silicate solutions has long been of interest because of the wide variations in properties that these solutions exhibit with different moduli. Polymers in solutions of a variety of alkali silicates have been studied by gas chromatography of chemical derivatives and laser Raman, infrared (ftir), and ^{29}Si Fourier transform nuclear magnetic resonance (^{29}Si ft-nmr) spectroscopy. There is strong evidence for the presence of a variety of Si_1–Si_8 silicate structures, even in highly alkaline, dilute silicate solutions. Among the techniques for species determination in soluble silicates, ^{29}Si nmr spectroscopy gives the most information about equilibrium silicate solutions, but derivatization provides the best means for studying the dynamics of nonequilibrium systems.

Polymerization in Solution

Polymerization and depolymerization of silicate anions and their interactions with other ions and complexing agents are of great interest in sol–gel and catalyst manufacture, detergency, oil and gas production, waste management, and limnology. The complex silanol condensation process may be represented empirically by:

$$\equiv Si{-}OH \overset{K_a}{\rightleftharpoons} \equiv Si{-}O^- + H^+$$

$$\equiv Si{-}O^- + HO{-}Si\equiv \underset{k_{-2}}{\overset{k_2}{\rightleftharpoons}} \equiv Si{-}O{-}Si\equiv + HO^-$$

Condensation occurs most readily at a pH value equal to the pK_a of the participating silanol group. This representation becomes less valid at pH values above 10, where the rate constant of the depolymerization reaction (k_{-2}) becomes significant and at very low pH values where acids exert a catalytic influence on polymerization. The acidity of silanol functionalities increases as the degree of polymerization of the anion increases.

The state of ionization of the silica particle surface controls the rate of polymerization following homogeneous nucleation. The rate of reaction of dissolved silicate at the surface of amorphous SiO_2 is proportional to the density of ionized silanol groups. The degree of surface ionization also influences the value of the surface tension and hence the rate of homogeneous nucleation.

Silicate polymerization in dilute solutions at pH values up to ca 10 is sensitive to pH and other factors that generally influence colloidal systems, eg, ionic strength, dielectric constant, and temperature. Larger particles grow at the expense of smaller particles (Ostwald ripening), especially in strongly alkaline solutions where the latter dissolve more readily.

Silicate solutions of equivalent composition may exhibit different physical properties and chemical reactivities because of differences in the distributions of polymer silicate species. This effect is keenly observed in commercial alkali silicate solutions with compositions that lie in the metastable region near the solubility limit of amorphous silica.

Chemical Activity

Silica Polymer–Metal Ion Interactions in Solution. The reaction of metal ions with polymeric silicate species in solution may be viewed as an ion-exchange process. Consequently, it might be expected that silicate species acting as ligands would exhibit a range of reactivities toward cations in solution. Silica gel forms complexes with multivalent metal ions in a manner that indicates a correlation between the ligand properties of the surface Si—OH groups and metal ion hydrolysis. Metal ion adsorption on silica gel may be initiated at a pH value corresponding to surface nucleation. This seems to relate to a reduction of cation–solvent interactions leading to conditions favorable for adsorption of hydrated metal ions from solution.

At a given pH value, the solution activities of Ca^{2+}, Mg^{2+}, and Cu^{2+} decrease to a greater extent in the presence of SiO_2 derived from 2.0- and 3.8-ratio silicates than they do in solutions prepared from sodium orthosilicate. Thus, highly polymerized silicate anions appear to interact with metal ions in solution in a manner analogous to silica gel, and the interaction decreases as the degree of silicate polymerization decreases. This is consistent with the observation that silica suspended in solutions of polyvalent metal salts begins to adsorb metal ions when the pH value is raised to within 1–2 pH units of the OH^- concentration at which the corresponding metal hydroxides precipitate. The products of the reactions of soluble silicates with metal salts in concentrated solutions at ambient temperature are considered to be complex mixtures of metal ions and/or metal hydroxides, coagulated colloidal size silica species, and silica gels.

Effect on Oxide–Water Interfaces. The adsorption (qv) of ions at clay mineral and rock surfaces is an important step in natural and industrial processes. Silicates are adsorbed on oxides to a far greater extent than would be predicted from their concentrations. This adsorption maximum at a given pH value is independent of ionic strength, and maximum adsorption occurs at a pH value near the pK_a of orthosilicate. The pH values of maximum adsorption of weak acid anions and the pK_a values of their conjugate acids are correlated. This indicates that the presence of both the acid and its conjugate base is required for adsorption. The adsorption of silicate species is far greater at lower pH than simple acid–base equilibria would predict. Soluble silicates adsorb specifically to oxide surfaces and play a significant role in maintaining a negative surface charge on oxide particles in the presence of cations that could otherwise reverse the surface charge.

Characteristics

The characteristics of soluble silicates relevant to various uses include the pH behavior of solutions, the rate of water loss from films, and dried film strength. The pH values of silicate solutions are a function of composition and concentration. These solutions are alkaline, being composed of a salt of a strong base and a weak acid. The solutions exhibit up to twice the buffering action of other alkaline chemicals, eg, phosphate.

Table 2. Commercial Sodium and Potassium Silicates

Commercial silicates	SiO_2, wt %	Wt ratio[a], SiO_2:M_2O	Modulus[a], SiO_2:M_2O	H_2O, wt %	°Baumé[b], at 20°C	d^{20}_{20}, g/mc^3	Viscosity, at 20°C, Pa·s[c]	pH
anhydrous glasses								
sodium silicates	75.7	3.22	3.33					
	66.0	2.00	2.06					
potassium silicates	70.7	2.50	3.92					
hydrated amorphous powders								
sodium silicates	61.8	3.22	3.33	18.5				
	64.0	2.00	2.06	18.5				
solutions								
sodium silicates	31.5	1.60	1.65		58.5	1.68	7.00	12.8
	36.0	2.00	2.06		59.3	1.69	70.00	12.2
	26.5	2.50	2.58		42.0	1.41	0.06	11.7
	31.7	2.88	2.97		47.0	1.49	0.96	11.5
	28.7	3.22	3.32		41.0	1.39	0.18	11.3
	25.3	3.75	3.86		35.0	1.32	0.22	10.8
potassium silicates	20.8	2.50	3.93		29.8	1.259	0.04	11.30
	19.9	2.20	3.45		30.0	1.261	0.01	11.55
	26.3	2.10	3.30		40.0	1.381	1.05	11.70
	29.5	1.80	2.83		47.7	1.490	1.30	12.15
crystalline solids								
sodium orthosilicate	28.8	60.8[d]	0.50	9.5				
anhydrous sodium metasilicate	47.1	51.0[d]	1.00	2.0				
sodium metasilicate pentahydrate	26.4	29.3[d]	1.00	42.0				
sodium sesquisilicate	24.1	36.7[d]	0.67	38.1				

[a] M represents Na or K.
[b] To convert Be° to sp gr, divide 145 by (145 − Be°).
[c] To convert Pa·s to P, divide by 10.
[d] Value is wt % of M_2O.

Manufacture and Processing

Soluble silicate glasses are usually manufactured in oil- or gas-fired, open-hearth regenerative furnaces. Recent continuous-flow glass melters are equipped with high intensity gas burners or plasmas. These latter technologies offer significant advantages over conventional batch-melting processes. Glass composition and shutdown and start-up procedures can be changed rapidly since no large molten glass reservoir is maintained. In conventional processes, the glass is made by the reaction of sand and sodium carbonate (soda ash) at 1100–1300°C, a temperature sufficiently high to provide a reasonable rate of quartz dissolution in the molten batch and manageable melt viscosity. The rate of reaction of quartz with Na_2CO_3 is controlled by silica diffusion and varies inversely with the square of the radius of the quartz particles. As the Na_2CO_3 melt envelops the sand grains, the silica network breaks down and diffuses slowly into the melt. The glass product can be drawn and formed into solid lumps or drawn directly into a rotary dissolver. The glass lumps can also be dissolved in a pressure apparatus at pressures up to 690 MPa (100 psi).

Crystalline sodium metasilicates are manufactured by processing highly concentrated solutions of sodium silicate (1.0 ratio) or by direct fusion of sand and soda ash, followed by grinding and sizing.

Potassium silicates are manufactured in a manner similar to sodium silicates by the reaction of K_2CO_3 and sand. However, crystalline products are not manufactured and the glass is supplied as a flake. Lithium silicate solutions are usually prepared by dissolving silica gel in a LiOH solution or mixing silica sol with LiOH.

Commercial Products

The average composition and pertinent properties of commercial soluble silicates are given in Table 2.

Regulatory Status

Additives to Food and Potable Water. Sodium silicate is generally recognized as safe (GRAS) by the FDA in fabrics and when exposed to food. It is also recognized as a secondary direct food additive when used in boiler water for food-contact steam. In addition, it meets AWWA/ANSI standards as a corrosion inhibitor in potable water.

Transportation and Disposal. Only highly alkaline forms of soluble silicates are regulated by the U.S. Department of Transportation (DOT) as hazardous materials for transportation. When discarded, they are classified as hazardous waste under the Resource Conservation and Recovery Act (RCRA).

Consumer Products. In the absence of specific data, the Consumer Product Safety Commission (CPSC) requires specific cautionary statements for consumer products containing certain types and amounts of sodium silicates.

Occupational Safety and Health. OSHA has set no specific limits for sodium and potassium silicates.

Health, Safety, and Environmental Aspects

The primary hazard of commercial soluble silicates is their moderate-to-strong alkalinity. Contact–exposure effects can range from irritation to corrosion, depending on the concentration of the silicate solution, the silica-to-alkali ratio, the sensitivity of the tissue exposed, and the duration of exposure. Soluble silicates are rapidly absorbed and eliminated if ingested or inhaled. Trace quantities of silicon are essential in nutrition, possibly as a metal ion bioavailability attenuator, but siliceous urinary calculi may result if normal dietary amounts are greatly exceeded.

Uses

Alkali silicates are generally used as components, rather than reactants. In many cases they only contribute partially to the overall performance. Their benefit usually depends on the surface and solution chemical properties of the wide range of highly hydrophilic polymeric silicate ions deliverable from soluble silicate products or their proprietary modifications.

The largest single use for soluble silicates is in soaps and detergents (see DETERGENCY). The soluble silicates can enhance the effectiveness of surfactants, help provide a constant pH values in an effective range, and aid in the saponification of oils and fats by means of their alkalinity and buffering capacity.

Silicates are used in water treatment to prepare activated silica sol, a stable, acid-polymerized silicate suspension that functions as an aid in the alum coagulation of matter suspended in raw and wastewater streams; to inhibit corrosion of metal surfaces in contact with water; to stabilize reduced iron and manganese in water supplies; and in boiler water.

Silica in water exposed to various metals leads to the formation of a surface less susceptible to corrosion.

Sodium silicates are used in froth flotation as strong and selective settling agents, where they increase the hydrophilicity of the mineral particles.

Silicates are often used to bind sand in foundries or to bind other minerals. Silicates are used extensively as adhesives in spiral tube winding, fiber drums, end sealing, laminating metal foil to paper, and in corrugated boxes. They are also employed in the manufacture of refractory and acid-resistant mortars and cements.

Silicates are employed in enhanced oil recovery for reasons that are related to their function in detergency and ore beneficiation. The presence of silicate enhances surfactant effectiveness, especially in hard water reservoir brines.

Silicates are utilized in combination with hydrogen peroxide to improve pulp and textile bleaching efficiency.

Silicates are used as deflocculants, ie, agents that maintain high solids slurry viscosities at increased solids concentrations. Soluble silicates suppress the formation of ordered structures within clay slurries that creates resistance to viscous flow within the various systems.

Sodium silicates can be used for water control and soil stabilization in tunneling and excavation projects. These grouts are strong, reliable, and environmentally safe.

Derivatives

In the chemical processing industry (CPI), alkali silicates are valued as a reactive source of $(SiO_2)_n$ structural units. They may be viewed as silica dissolved and/or dispersed in an hydroxide ion-rich aqueous system. Soluble silica as an intermediate can be reacted with acids and bases to form a wide range of final products, from seemingly simple condensed forms of relatively pure noncrystalline silica, precipitates, gels, and sols, to highly complex crystalline metallosilicates like those found in the broad class of aluminosilicates, zeolites.

Silica Sols, Precipitated Silicas, and Silica Gels. Silica sols are manufactured from diluted 3.2-ratio sodium silicate solutions by H^+/Na^+ ion exchange and concentration. This process yields a dispersion of colloidal silica particles used in antislip agents, castings, binders, and polishing solutions for silicon wafers. Precipitated silicas are made by counteracting the interparticle forces that hold large polymeric silicate anions in solution by the addition of an organic compound or a sodium salt and mineral acid. Silica gels are prepared by acidifying concentrated 3.2-ratio sodium silicate solutions. Acidification rapidly produces a gel network, called a hydrogel, after it passes through the sol stage. Xerogels and aerogels (qv) are manufactured by gelation, followed by milling, washing, and drying.

Synthetic Insoluble Silicates. Insoluble crystalline silicates, ie, mineral-type compounds, are synthesized from soluble silicates by precipitation, gelation, ion exchange, and hydrothermal techniques. Hydrothermal treatment of partially neutralized, high mole ratio ($m = 12-50$), sodium silicate solutions yields neutral alkali polysilicates that exhibit a layered structure and high ion-exchange capacity.

The zeolites, $M_{x/n}[(AlO_2)_x(SiO_2)_y]\cdot zH_2O$, where M is usually an alkali or alkaline-earth metal ion, are of great commercial importance today. These materials, which serve as a model system for the advanced inorganic materials revolution, are made by hydrothermal methods and their architectural variations seem limitless. Synthetic zeolites are less expensive than their natural analogues and almost all commercial applications utilize the former.

Another insoluble silicate of commercial interest is the magnesium silicate clay, hectorite, which may be prepared from 3.2-ratio sodium silicate, LiF, $MgSO_4$, and Na_2CO_3. This inorganic thickening agent produces translucent, thixotropic gels that are not tacky, gummy, or stringy. The gels are used in antiperspirants, gel toothpastes, shampoos, cosmetics (qv), paint (qv), and cleaning products.

JAMES S. FALCONE, JR.
West Chester University

H. Bergna, ed., *Colloid Chemistry of Silica*, ACS Advances in Chemistry Series Number 234, ACS Books, Washington, D.C., 1994.

F. Liebau, *Structural Chemistry of Silicates*, Springer-Verlag, Berlin, 1985.

J. S. Falcone, Jr., ed., *Soluble Silicates, ACS Symposium Series, No. 194*, American Chemical Society, Washington, D.C., 1982.

D. Barby and co-workers, in R. Thomson, ed., *The Modern Inorganic Chemical Industry*, Chemical Society, London, 1977, p. 320.

SILICON HALIDES

Despite extensive research in silicon halides, only two of these chemicals are produced on a large industrial scale (excluding organohalosilanes). These are tetrachlorosilane, $SiCl_4$, and trichlorosilane, $HSiCl_3$.

Physical Properties

The physical properties of silicon tetrahalides are listed in Table 1; those of the halohydrides are listed in Table 2.

Chemical Properties

Silicon halides are typically tetrahedral compounds. The silicone–halogen bond is very polar; thus the silicon is susceptible to nucleophilic attack, which in part accounts for the broad range of reactivity with various chemicals. Furthermore, reactivity generally increases with the atomic weight of the halogen atom.

Halosilanes are very reactive toward protic chemicals. They generally react violently with water.

Table 1. Properties of Silicon Tetrahalides

Compound	Mp, °C	Bp, °C	Density[a], g/cm^3	Bond energy, kJ/mol[b]
SiF_4	−95.0	−90.3	1.66_{-95}	146
$SiCl_4$	−68.8	56.8	1.48_{20}	381
$SiBr_4$	5	155.0	2.81_{29}	310
SiI_4	124	290.0		234

[a] Subscripted values are temperature in °C. [b] To convert J to cal, divide by 4.184.

Table 2. Properties of Silicon Halohydrides

Compound	Mp, °C	Bp, °C	Density[a], g/cm^3
H_3SiF		−99.0	
H_2SiF_2	−122.0	−77.8	
$HSiF_3$	−131.2	−97.5	
H_3SiCl	−118.0	−30.4	1.145_{-113}
H_2SiCl_2	−122.0	8.3	1.42_{-122}
$HSiCl_3$	−128.2	31.8	1.3313_{25}
H_3SiBr	−94.0	1.9	1.531_{20}
H_2SiBr_2	−70.1	66.0	2.17_0
$HSiBr_3$	−73.0	111.8	2.7_{17}
H_3SiI	−57.0	45.4	$2.035_{14.8}$
H_2SiI_2	−1.0	149.5	$2.724_{20.5}$
$HSiI_3$	8.0	111.0[b]	3.314_{20}

[a] Subscripted values are temperature in °C. [b] At 2.9 kPa (21.8 mm Hg).

Manufacturing Processes

Silicon halides can be easily prepared by the reaction of silicon or silicon alloys and the respective halogens.

Hydrogen halides also react freely with elemental silicon at moderate temperatures to yield halosilanes.

$$2\,Si + 7\,HCl \rightarrow HSiCl_3 + SiCl_4 + 3\,H_2$$

Silicon Tetrachloride. Most commercially available silicon tetrachloride is made as a by-product of the production of alkylchlorosilanes and trichlorosilane and from the production of semiconductor-grade silicon by thermal reduction of trichlorosilane.

$$2\,HSiCl_3 \xrightarrow{>1000\,°C} Si + SiCl_4 + 2\,HCl$$

Trichlorosilane. The primary production process for trichlorosilane is the direct reaction of hydrogen chloride gas and silicon metal in a fluid-bed reactor. Although this process produces both trichlorosilane and silicon tetrachloride, production of the latter can be minimized by proper control of the reaction temperature. A significant amount of trichlorosilane is also produced by thermal rearrangement of silicon tetrachloride in the presence of hydrogen gas and silicon.

Health and Safety Factors

Halosilane vapors, except for fluorosilanes, react with moist air to produce the respective hydrohalogen acid mists.

Halosilanes should be handled only in areas that are equipped with adequate ventilation, eye-wash facilities, and safety showers. It is recommended that personnel handling halosilanes wear rubber aprons and gloves and chemical safety goggles. Furthermore, all personnel handling halosilanes should be thoroughly trained in safe handling procedures, the hazardous characteristics of halosilanes, and emergency procedures for all foreseeable emergencies.

Uses

Silicon Tetrachloride. The vast majority (ca 95%) of $SiCl_4$ use is in the manufacture of fumed silica. Silicon tetrachloride is also used to prepare silicate esters, which are used in the production of coating and refractories and in some semiconductor manufacturing operations.

Trichlorosilane. There are essentially only two large industrial applications for trichlorosilane. These are the synthesis of organotrichlorosilanes and the production of semiconductor-grade silicon metal (see SILICON AND SILICON ALLOYS, PURE SILICON).

Dichlorosilane. Dichlorosilane is produced in relatively modest commercial quantities compared to the above chlorosilanes. This silane is generally recovered as a by-product of the production of other silanes.

WARD COLLINS
Dow Corning Corporation

E. A. Ebsworth, *Volatile Silicon Compounds*, Pergamon Press Ltd., Oxford, U.K., 1963.

W. Noll, *Chemistry and Technology of Silicones*, Academic Press, Inc., New York, 1968.

A. G. MacDiarmid, *Organometallic Compounds of the Group IV Elements*, Vol. 2, *The Bond to Halogens and Halogenoids*, Marcel Dekker, Inc., New York, 1972.

V. Bazant, V. Choalovsky, and J. Rathousky, *Organosilicon Compounds*, Vol. 1, *Chemistry of Organosilicon Compounds*, Academic Press, Inc., New York, 1965.

SILANES

Silanes are compounds having a Si–H bond; halosilanes, a Si–X bond; and organosilanes, a Si–C bond. Compounds having Si–OSi bonds are called siloxanes or silicones. Those having a Si–OR bond are called silicon esters (see SILICON COMPOUNDS, SILICON ESTERS; SILICON COMPOUNDS, SILICONES).

Silane, SiH_4, is the simplest silicon compound and provides the basis of nomenclature for all silicon chemistry. Compounds are named as derivatives of silane. The substituents, whether inorganic or organic, are prefixed. Examples are trichlorosilane, $HSiCl_3$; disilane, H_3SiSiH_3; methyldichlorosilane, CH_3SiHCl_2; methylsilane, CH_3SiH_3; diethylsilane, $(C_2H_5)_2SiH_2$; and triethylsilane, $(C_2H_5)_3SiH$. Two or more substituents are listed alphabetically, adhering to the following rules: substituted organic moieties are named first, followed by simple organic fragments; alkoxy substituents are named next, followed by acyloxy, halogen, and pseudohalogen groups. For example, ethylmethylethoxysilane, $C_2H_5(CH_3)SiH(OC_2H_5)$, and 3-chloropropylmethylchlorosilane, $ClCH_2CH_2CH_2SiH(CH_3)Cl$, are correct.

Inorganic Hydride Functional Silanes

Hydride functional silanes are sometimes simply referred to as silanes. Only a few of the thousands of hydride functional silanes reported have any commercial significance. These include inorganic silanes, organic silanes, and polymeric siloxanes. Despite the small number, a wide range of applications has developed for such compounds, eg, in the manufacture of high purity and electronic-grade silicon metal (see SILICON AND SILICON ALLOYS, PURE SILICON) and in epitaxial silicon deposition (see ELECTRONIC MATERIALS; INTEGRATED CIRCUITS; SEMICONDUCTORS); as selective reducing agents; as monomers; and as elastomer intermediates (see ELASTOMERS, SYNTHETIC). Not least is the use of these materials as intermediates for production of other silanes and silicones.

The inorganic silanes of commercial importance include silane, dichlorosilane, and trichlorosilane. The latter is not only the preferred intermediate for the first two, but it is also used in the production of high purity silicon metal and as an intermediate for silane adhesion promoters, coupling agents, silicone resin intermediates, and surface treatments.

Physical Properties. Silanes and chlorosilanes have boiling points, melting points, and dipole moments comparable to those of simple hydrocarbons (qv) and chlorinated hydrocarbons. Moreover, both silanes and hydrocarbons are colorless gases or liquids at room temperature. The similarity ends, however, with these simple physical characteristics. Table 1 contains selected physical properties of inorganic silanes.

Thermal Properties. Silanes have less thermal stability than hydrocarbon analogues. Silane, however, is one of the most thermally stable inorganic silanes. Disilanes and other members of the binary series are less stable. Halogen-substituted silanes are subject to disproportionation reactions at higher temperatures.

Chemical Properties. All inorganic silicon hydrides are readily oxidized. Silane and disilane are pyrophoric in air and form silicon dioxide and water as combustion products.

Silanes do not react with pure water or slightly acidified water under normal conditions. A rapid reaction occurs, however, in basic solution with quantitative evolution of hydrogen. Alkali leached from glass is sufficient to lead to the hydrolysis of silanes.

Silane reacts with methanol at room temperature to produce methoxymonosilanes such as $Si(OCH_3)_4$, $HSi(OCH_3)_3$, and $H_2Si(OCH_3)_2$, but not H_3SiOCH_3.

Most compounds containing Si–H bonds react very rapidly with the free halogens.

Silane reacts with alkali metals dissolved in various solvents, forming as the chief product the silyl derivative of the metal, eg, $KSiH_3$.

Silanes react with alkyllithium compounds, forming various alkylsilanes.

Under certain conditions reasonably large quantities of higher silanes up to *n*-Si_4H_{10} and *iso*-Si_4H_{10} plus smaller amounts of various isomers of higher silanes up to Si_8H_{18} can be produced by

Table 1. Properties of Inorganic Silanes

Parameter	SiH_4	H_3SiCl	H_2SiCl_2	$HSiCl_3$	H_3SiSiH_3	$H_3SiOSiH_3$	$(H_3Si)_3N$
mp, °C	−185	−118	−122	−126.5	−132.5	−144	−105.6
bp, °C	−111.9	−30.4	8.2	31.9	−14.5	−15.2	52
vapor pressure,[a] Pa[b]	71_{-118}		13_{-34}	$53_{14.5}$			14.5_0
ΔH_{vap}, kJ/mol[c]	12.5	20	25.2	26.6	21.2	21.6	
ΔH_{fus}, kJ/mol[c]	0.67						
critical temperature, °C	−3.5		176	234	109		
critical pressure, MPa[d]	472		455	365			
ΔH_f, kJ/mol[c]	32.6		−314	−482			
dipole moment, C·m[e] $\times 10^{-30}$		4.347	3.913	3.24		0.8	0
density,[a] g/cm^3	0.68_{-185}	1.145_{-113}	1.22	1.34	0.69_{-15}	0.881_{-15}	0.895_{-106}
autoignition temperature, °C	f	f	55	215	f	<50	f

[a] Subscripted numbers are temperature in °C. [b] To convert Pa to mm Hg, multiply by 7.5. [c] To convert J to cal, divide by 4.184. [d] To convert MPa to psi, multiply by 145. [e] To convert C·m to debye, divide by 3.336×10^{-30}. [f] Pyrophoric.

decomposition of SiH_4 in an electrical discharge. Glow discharge also provides a method for a high deposition rate of silicon from disilane.

Mercury-sensitized photolysis of SiH_4 leads to the formation of H_2, Si_2H_6, Si_3H_8, and polymeric solid silanes. The photolysis of SiH_4 in the presence of GeH_4 or CH_3I produces SiH_3GeH_3 or CH_3SiH_3, respectively.

Manufacture and Processing. There are four methods of production of compounds containing a Si–H bond that are noteworthy. Silicides of magnesium, aluminum, lithium, iron, and other metals react with acids or their ammonium salts to produce silane and higher binary silanes. This method was generally abandoned in favor of methods involving reduction of silicon halides. Treatment of calcium silicide with HCl-ethanol or glacial acetic acid yields the complex polymer called siloxene.

The reductions of chlorosilanes by lithium aluminum hydride, lithium hydride, and other metal hydrides, MH, offers the advantages of higher yield and purity as well as flexibility in producing a range of silicon hydrides comparable to the range of silicon halides.

Direct synthesis is the preparative method that ultimately accounts for most of the commercial silicon hydride production. This is the synthesis of halosilanes by the direct reaction of a halogen or halide with silicon metal, silicon dioxide, silicon carbide, or metal silicide without an intervening chemical step or reagent. Trichlorosilane is produced by the reaction of hydrogen chloride and silicon, ferrosilicon, or calcium silicide with or without a copper catalyst.

Health and Safety Factors, Toxicology. The acute hazards of silicon hydrides are overwhelmingly important in considering worker safety. Silane is pyrophoric. Chlorine-containing compounds generate hydrogen chloride on contact with water and other protic materials. At low concentrations, chlorosilanes affect nasal and pulmonary membranes. The LD_{50} for trichlorosilane is 1050 mg/kg.

Organic Hydride Functional Silanes

The organohydrosilane of greatest commercial importance is methyldichlorosilane. Careful hydrolysis of this material with water affords polymethylhydrosiloxanes, primarily used in the textile industry to waterproof and improve the wear resistance of fabrics (see WATER–WATERPROOFING AND WATER/OIL REPELLANCY). This polymer is also used as a waterproofing agent in the leather (qv) industry, the paper (qv) industry for sizing, electronic applications, and in construction for enhancing the water resistance of gypsum board. Methyldichlorosilane is also used captively by silane and silicone producers in thermal condensation reactions to produce vinyl, phenyl, and cyanoalkyl precursors to silicone fluids. The addition reaction of methyldichlorosilane to fluorocarbon alkenes has enabled production of methyltrifluoropropyl silicone fluids, gums (qv), and rubbers. Trialkoxysilanes, eg, triethoxysilane and trimethoxysilane, are used to prepare a number of organic coupling agents utilized by the plastics industry as adhesion promoters (see ADHESIVES). Organosilanes containing one or more Si–H bonds have excellent reducing capabilities.

Physical Properties. The physical properties of organosilanes are determined largely by the properties of the silicon atom. Because silicon is larger and less electronegative than either carbon or hydrogen, the polarity of the Si–H bond is opposite to that of the C–H bond (Table 2). This difference in polarity imparts hydride character to the Si–H bonds of organosilanes. The size of the silicon atom is greater than the carbon atom, and this increase in atomic volume enables nucleophilic attack on the silicon to occur more readily than on carbon.

Chemical Properties. Organohydrosilanes undergo a wide variety of chemical conversions. The Si–H bond of organohydrosilanes reacts with elements of most groups of the Periodic System, especially Groups 16(VIA) and 17(VIIA). There are no known reactions if the Si–H bond is replaced by stable bonds of silicon with elements of Groups 2(IIA), 13(IIIA), and 8–10(VIII).

The oxidizability of the Si–H bonds is much greater than that of C–H bonds. This difference is manifested by the ease of oxidation of organohydrosilanes with metal oxides.

As for inorganic silanes, no reaction occurs between organohydrosilanes and water. The presence of acidic or alkaline catalysts, however, brings about the reaction according to the following scheme:

$$\equiv Si{-}H + H_2O \xrightarrow{\text{catalyst}} \equiv Si{-}OH + H_2$$

The ease of hydrolysis depends on the pH and is more rapid under alkaline than acidic conditions.

The catalyzed reaction of organosilanes with hydroxyl-containing organic compounds affords organoalkoxy and organoaryloxysilanes, usually in high yields. Alkali–metal oxides, hydrogen halides, and metal halides are most often used as catalysts.

As in reactions of alcohols and acids with organosilanes, reaction of the Si–H bond with amines and phosphines proceeds only under catalysis. Alkali metal amides or phosphines are the catalysts of choice and effect replacement of the Si–H bond with Si–N or Si–P bonds, respectively. Catalytic activity of the alkali metals for these reactions is $K > Na > Li$.

Table 2. Properties of Halosilanes

Property	SiF_4	$SiCl_4$	$SiBr_4$	SiI_4
mp, °C	−90.3	−70.4	5.4	120.5
bp, °C	−95.7 sub	57	155	287.5
vapor pressure,[a] kPa[b]	68.6_{-100}	25.9_{20}	0.24_0	
H_{vap}, kJ/mol[c]	18.7	28.7	37.9	26.6
bond energy, kJ/mol[c]	565	381	310	244
critical temperature, °	−14.15	233.6	383	
critical pressure, kPa[d]	37.3	36.8		
density,[a] g/cm^3	1.598_{-80}	1.4707	2.771	4.2

[a] The subscripted numbers are temperature in °C.
[b] To convert kPa to mm Hg, multiply by 7.5.
[c] To convert J to cal, divide by 4.184.
[d] To convert kPa to psi, multiply by 0.145.

The reaction of organosilanes with halogens and halogen compounds usually proceeds in good yield through cleavage of the Si–H bond and formation of the Si–X bond. This reaction can be achieved by direct action of halogen on the organosilane or by interaction with halogen-containing organic and inorganic compounds.

There are no reports of the direct reaction of the Si–H bond in organosilanes with magnesium, zinc, mercury, aluminum, and other elements of the Groups 2(IIA), 12(IIB), and 13(IIIA) metals. The alkali metals, ie, sodium, potassium, and their alloys, react with arylsilanes in amines or ammonia to produce the arylsilyl derivatives of these metals. However, these reactions should be regarded as indirect examples of hydrogen replacement on silicon because they probably go through an amide intermediate. Direct reaction between arylsilanes and alkali metals occurs when alloys of potassium and sodium are used.

The two principal categories of reductive chemistry of hydridosilanes are hydrosilylation and ionic reduction. Hydrosilylation is the catalyzed addition of a hydridosilane to a multiply bonded system. This chemistry is a principal technology in silicon–carbon bond formation. Ionic reduction by silanes is a class of chemistry more properly considered within the context of organic synthesis.

The hybridic nature of the Si–H bond is utilized to generate C–H bonds by ionic hydrogenation according to the following general mechanism, in which a hydride is transferred to a carbocation.

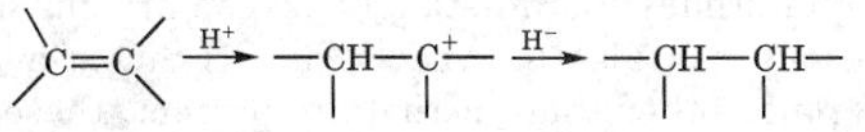

A catalyst, usually acid, is required to promote chemoselective and regioselective reduction under mild conditions. A variety of organosilanes can be used, but triethylsilane in the presence of trifluoroacetic acid is the most frequently reported.

Manufacture and Processing. The preparation of organosilanes by the direct process is the primary method used commercially. The synthesis involves the reaction of alkyl halides, eg, methyl and ethyl chloride, with silicon metal or silicon alloys in a fluidized bed at 250–450°C.

Organosilanes can be synthesized most conveniently on pilot, bench, and laboratory scale by reduction of organic-substituted halo- and alkoxysilanes, using metal hydrides. As for inorganic silanes, the most effective reducing agent is lithium aluminum hydride.

Disproportionation reactions have also been used to prepare organosilanes. These reactions involve interaction of organosilanes and other silicon compounds containing organic, alkoxy, and halogen groups bound to silicon. Reactions are catalyzed by a variety of materials.

Grignard reagents are utilized to transfer organic groups to silicon. In general, Grignard reagents are useful in the synthesis of mixed hydridochlorosilanes because these reagents can effect stepwise substitution of the halogen.

Organohydrosilanes can also be prepared by addition of halosilanes and organosilanes containing multiple Si–H bonds to olefins.

Halosilanes

Only three inorganic halosilanes are produced on a large industrial scale, ie, tetrachlorosilane, tetrafluorosilane, and trichlorosilane.

Physical Properties. All halosilanes fume in air from the presence of moisture, liberating the hydrogen halide. These compounds are thus extremely corrosive materials in open environments. Physical properties for halosilanes of commercial significance are given in Table 2.

Chemical Properties. Silicon halides are stable to oxygen at room temperature, but react at elevated temperatures to form, in the case of chlorides, oxychlorosilanes; tetrachlorosilane reacts form hexachlorodisiloxane, and tetrabromosilane reacts to form polybromosiloxanes.

At elevated temperature (1000°C) tetrachlorosilane attacks pure crystalline silicon to form a mixture of higher chlorosilanes including Si_2Cl_6, Si_3Cl_8, and Si_4Cl_{10}.

Tetrafluorosilane reacts with hydrogen only above 2000°C. Tetrachlorosilane can be reduced by hydrogen at 1200°C. Tetraiodosilane can be reduced to silicon at 1000°C. Reduction of tetrafluorosilane with potassium metal to silicon was the first method used to prepare silicon (see SILICON AND SILICON ALLOYS).

Manufacturing. Industrial tetrachlorosilane derives from two processes associated with trichlorosilane, the direct reaction of hydrogen chloride on silicon primarily produced as an intermediate for fumed silica production, and as a by-product in the disproportionation reaction of trichlorosilane to silane, utilized in microelectronics.

Tetrabromosilane and tetraiodosilane are produced by the direct reaction between silicon and bromine at 500°C and silicon and iodine at 600°C. There was no commercial production of these materials as of the late 1990s.

Health and Safety. Halosilane vapors react with moist air to produce the respective hydrohalogen acid mist. Federal standards in the United States have not set exposure to halosilanes, but it is generally believed that there is no serious risk if vapor concentrations are maintained below a level that produces an irritating concentration of acid mist.

Uses. The overwhelming use for tetrachlorosilane is in the production of fumed silica. Tetrafluorosilane is primarily an intermediate for the production of fluorosilicic acid, used as a sterilant for glass bottles, and in electroplating applications (see FLUORINE COMPOUNDS, INORGANIC).

BARRY ARKLES
Gelest, Inc.

V. Bazant, V. Chvalovsky, and J. Rathovsky, *Organosilicon Compounds*, Vol. 1, Academic Press, Inc., New York, 1965.

W. R. Runyan, *Silicon Semiconductor Technology*, McGraw-Hill Book Co., Inc., New York, 1965.

B. Arkles, *Silicon, Germanium, Tin, and Lead Compounds: A Survey of Properties and Chemistry*, Gelest Inc., Tullytown, Pa., 1995.

C. Eaborn, *Organosilicon Compounds*, Buttersworth Scientific Publications, Lander, U.K., 1960.

SILICON ESTERS

Silicon esters are silicon compounds that contain an oxygen bridge from silicon to an organic group, ie, Si–OR. The most conspicuous material is tetraethyl orthosilicate, $Si(OC_2H_5)_4$. The advent of organosilanes that contain silicon–carbon bonds, Si–C, initiated an organic nomenclature by which compounds are named as alkoxy derivatives. For example $Si(OC_2H_5)_4$ becomes tetraethoxysilane.

Applications for tetraalkoxysilanes cover a broad range. These compounds are classified roughly according to whether the Si–OR bond is expected to remain intact or to be hydrolyzed in the final application.

Properties

The tetraalkoxysilanes possess excellent thermal stability and liquid behavior over a broad temperature range that widens with length and branching of the substituents. The physical properties of the silane esters, particularly the polymeric esters containing siloxane bonds, ie, Si–O–Si, are often likened to the silicone oils. These have low pourpoints and similar temperature—viscosity relationships. The alkoxysilanes generally have sweet, fruity odors that become less apparent as molecular weight increases. With the exception of tetramethoxysilane, trimethoxysilane, triethoxysilane, and a few closely related compounds that can be absorbed into corneal tissue, causing eye damage, the alkoxysilanes generally exhibit low levels of toxicity.

Aryloxy- and acyloxysilanes are often solids. The aryloxysilanes have excellent thermal stability. Acyloxy and mixed acyloxyalkoxysilanes have poor thermal stability. The most significant difference between the alkoxysilanes and silicones is the susceptibility of the Si–OR bond to hydrolysis (see SILICON COMPOUNDS, SILICONES). The simple alkoxysilanes are often operationally viewed as liquid sources of silicon dioxide (see SILICA).

Sol–Gel Process Technology and Chemistry

The complete hydrolysis of tetraalkoxysilanes under highly controlled conditions, usually without the presence of fillers, is associated with sol–gel technology (qv). Sol–gel is a method for preparing specialty metal oxide glasses and ceramics by hydrolyzing a chemical precursor or mixture of chemical precursors that pass sequentially through a solution state and a gel state before being dehydrated to a glass or ceramic. The use of sol–gel technology has increased dramatically since 1980. The flexibility of sol–gel technology allows unique access to multicomponent oxide systems and low temperature process regimes.

Preparation

The principal method of silicon ester production is described by Von Ebelman's 1846 synthesis:

$$SiCl_4 + 4\ C_2H_5OH \rightarrow Si(OC_2H_5)_4 + 4\ HCl$$

The reaction is generalized to

$$R_{(4-n)}SiCl_n + n\ R'OH \rightarrow R_{(4-n)}Si(OR')_n + n\ HCl$$

Process considerations must take into account not only characteristics of the particular alcohol or phenol to be esterified, but also the self-propagating by-product reaction, which results in polymer formation.

Toxicity

The alkoxysilanes generally have a low order of toxicity. Notable exceptions are tetramethoxysilane and two hydridosilanes, trimethoxysilane and triethoxysilane. Vapors of these materials may be absorbed directly into corneal tissue, causing blindness.

Uses

The ethoxysilanes are used as binders in precision casting, deposition of silicon dioxide is used to impart a translucent coating on glass (qv), ethoxysilanes are used in high temperature, zinc-rich paints (see PAINT), protective and consolidating coatings for masonry and other applications are produced from methyl-, propyl-, isobutyl-, and octyltrialkoxysilanes, films of silicon dioxide are deposited on silicon substrates by the application of a partially hydrolyzed solution of tetraethoxysilane or methyltriethoxysilane, and chemical vapor deposition of silicon dioxide from tetraethoxysilane assisted by the presence of oxygen and a plasma is an important technology for the deposition of pure and modified dielectrics for microelectronics.

BARRY ARKLES
Gelest, Inc.

B. Arkles, *Silicon, Germanium, Tin, and Lead Compounds: A Survey of Properties and Chemistry*, Gelest, Inc., Tullytown, Pa., 1995.

N. Sax and R. Lewis, *Dangerous Properties of Industrial Materials*, Van Nostrand Reinhold Co., Inc., New York, 1989.

M. G. Voronkov, in G. Bendz and I. Lindqvist, eds., *Biochemistry of Silicon and Related Problems*, Plenum Publishing Corp., New York, 1977.

M. G. Voronkov, V. P. Mileshevich, and Yu. A Yuzhelevski, *The Siloxane Bond*, Plenum Publishing Corp., New York, 1978.

SILICONES

Silicones are a class of polymers having the formula $(R_mSi(O)_{4-m/2}))_n$, where $m = 1–3$ and $n \geq 2$. The most common are the polydimethylsiloxanes (PDMS).

Table 1. Silicone Products and Their Uses

Commercial product	Use
fluids: heat-stable liquids	lubricants, water repellents, defoamers, release agents, surfactants
filled fluids and gums	valve lubricants, moistureproof sealants for electrical connectors, pressure-sensitive adhesives, personal care products
grease: fluid and carbon black or soap	nonflow lubricants, polishes
resins: cross-linked materials	electrical insulation, lubricant and paint additives, release formulations, water repellents
rubbers: fluids or gums and surface-treated fillers; elastic with good tensile strength	electrical insulation, medical devices, seals, textile coatings, foams

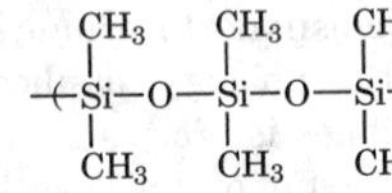

Various forms of silicones and examples of applications are listed in Table 1.

The designations M, D, T, and Q are used, respectively, for mono-, di-, tri-, and quaternary coordination of oxygen around silicon in silicones. A T group can also be written as $CH_3SiO_{3/2}$; a Q group as SiO_2. When groups other than methyl are present, these groups are indicated with a superscript; eg, D^{Vi} represents a methyl vinyl siloxy group, $(CH_3)(CH_2{=}CH)SiO$. Resins are often composed as M_xQ, M_xT, $M_xD_yT_zQ$, etc. This common shorthand notation for silicones is shown in Figure 1.

In 1940 Rochow discovered the direct process, also called the methylchlorosilane (MCS) process, in which methyl chloride is passed over a bed of silicon and copper to produce a variety of methylchlorosilanes, including dimethyldichlorosilane, $(CH_3)_2SiCl_2$. Working independently, Müller made a similar discovery in Germany. Consequently, the process is frequently called the Rochow process and sometimes the Rochow-Müller reaction. These discoveries were followed by two key publications describing the work that marked the beginning of the commercial silicone industry.

Chemistry

Direct Process. Passing methyl chloride through a fluidized bed of copper and silicon yields a mixture of chlorosilanes. The rate of methylchlorosilane (MCS) production and chemical selectivity, as determined by the ratio of dimethydichlorosilane to the other compounds formed, are significantly affected by trace elements in the catalyst bed; very pure copper and silicon gives poor yield and selectivity.

Processes to make phenyl and ethyl silicones have employed direct-process chemistry. Phenyl chloride has been used in place of methyl chloride to make phenylchlorosilanes. In addition, phenylchlorosilanes

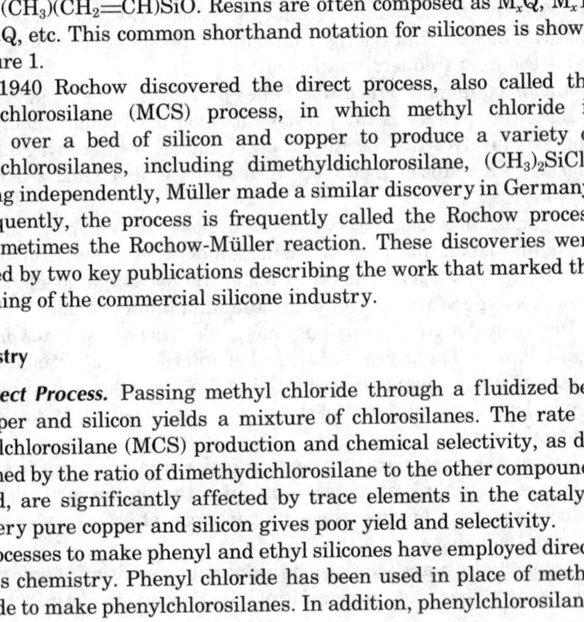

Figure 1. Widely accepted abbreviations used for silicone groups.

are produced by the reaction of benzene, $HSiCl_3$, and BCl_3. Ethylsilicones have been made primarily in the CIS, where the direct process is carried out with ethyl chloride in place of methyl chloride. Vinyl chloride can also be used in the direct process to produce vinylchlorosilanes.

Synthesis of Silicone Monomers and Intermediates. Another important reaction for the formation of Si—C bonds, in addition to the direct process and the Grignard reaction, is hydrosilylation (eq. 1), which is used for the formation of monomers for producing a wide range of organomodified silicones and for cross-linking silicone polymers. Formation of ether and ester bonds at silicon is important for the manufacture of curable silicone materials. Alcoholysis of the Si—Cl bond (eq. 2) is a method for forming silyl ethers. HCl removal is typically accomplished by the addition of tertiary amines or by using NaOR′ in place of R′OH to form NaCl.

$$R_3SiH + R'CH{=}CH_2 \xrightarrow{catalyst} R_3SiCH_2R' \qquad (1)$$

$$RSiCl_3 + 3\ R'OH \rightarrow RSi(OR')_3 \cdot 3HCl \qquad (2)$$

An important end group in silicone chemistry is the acetoxy group; the familiar silicone sealants release acetic acid during moisture cure of these acetoxy-stopped polymers. Acetoxysilanes hydrolyze more readily than alkoxy groups. Acylation of a chlorosilane can be accomplished by the addition of sodium acetate or by reaction with acetic anhydride.

Polymerization

The manufacture of polydimethylsiloxane polymers is a multistep process. The hydrolysis of the chlorosilanes obtained from the direction process yields a mixture of cyclic and linear silanol-stopped oligomers, called hydrolysate (eq. 3).

$$Cl-\underset{\underset{CH_3}{|}}{\overset{\overset{CH_3}{|}}{Si}}-Cl \xrightarrow[-HCl]{+H_2O} \left[HO-\underset{\underset{CH_3}{|}}{\overset{\overset{CH_3}{|}}{Si}}-OH \right] \xrightarrow{-H_2O} \left(\underset{\underset{CH_3}{|}}{\overset{\overset{CH_3}{|}}{Si}}-O \right)_m$$

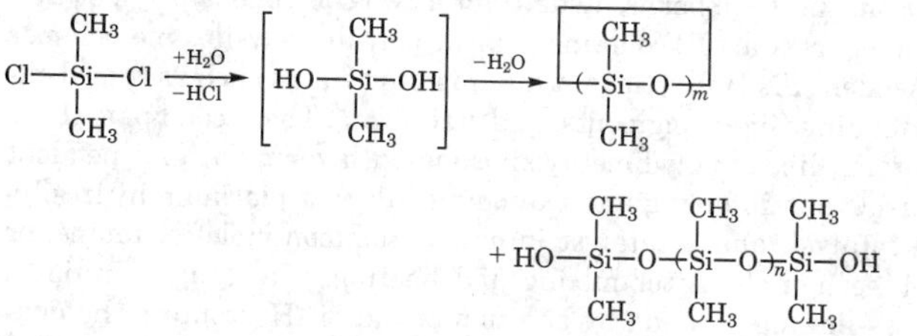

$$+\ HO-\underset{\underset{CH_3}{|}}{\overset{\overset{CH_3}{|}}{Si}}-O \left(\underset{\underset{CH_3}{|}}{\overset{\overset{CH_3}{|}}{Si}}-O \right)_n \underset{\underset{CH_3}{|}}{\overset{\overset{CH_3}{|}}{Si}}-OH \qquad (3)$$

In contrast to the hydrolysis technology, the methanolysis process allows for the one-step synthesis of organosiloxane oligomers and methyl chloride without formation of hydrochloric acid.

Polycondensation. The linear fraction of hydrolysate, ie, oligosiloxane-α,ω-diols whose viscosity is from 10 to 100 mPa (= cP), is converted further to silicone fluids and high molecular weight gums by polycondensation of the silanol end groups. Polycondensation is an equilibrium process.

Although the low molecular weight silanols such as trimethylsilanol or dimethylsilanediol undergo condensation thermally, the higher molecular weight oligomers are much more stable and their polycondensation must be catalyzed.

Ring-Opening Polymerization. Ring-opening polymerization of cyclic oligosiloxanes is an alternative to the polycondensation method of manufacturing siloxane polymers. Commercially, the polymerization of unstrained octamethyltetracyclosiloxane (D_4) is the most important. In the presence of catalysts such as strong acids or bases, D_4 undergoes equilibrium polymerization, which results in a mixture of high molecular weight polymer and low molecular weight cyclic oligomers.

Emulsion Polymerization. Even though siloxane bond formation is an equilibrium process, it is possible to form siloxane polymers by polycondensation or ring-opening polymerization in aqueous emulsions. D_4 can be converted into high molecular weight polymer by emulsion polymerization in the presence of dodecylbenzenesulfonic acid. (DBSA), which acts as both emulsifying surfactant and catalyst.

Radiation-Induced Polymerization. In 1956 it was discovered that D_3 can be polymerized in the solid state by γ-irradiation. The first successful polymerization of cyclic siloxanes in the liquid state and later work showed that the polymerization of cyclic siloxanes induced by γ-irradiation has a cationic nature. The polymerization is initiated by a cleavage of Si—C bond and formation of silylenium cation.

Plasma Polymerization. A need for well-defined, thin polymer films for applications in optics, electronics, or biomedicine stimulated the development of plasma-induced polymerization. Plasma-polymerized organosilicone films have the natural chemical affinity of a single-crystalline silicone, and the properties of these films can be varied widely by the choice of monomer and polymerization parameters. The mechanism of plasma polymerization is still not well understood as of this writing and differs substantially from the conventional ring-opening polymerization of cyclic monomers.

Silicone Network Formation

Silicone rubber has a three-dimensional network structure caused by cross-linking of polydimethylsiloxane chains. Three reaction types are predominantly employed for the formation of silicone networks: peroxide-induced free-radical processes, hydrosilylation addition cure, and condensation cure. Silicones have also been cross-linked using radiation to produce free radicals or to induce photoinitiated reactions.

Characterization of Silicone Networks.

The cross-linking of silicones as a function of time can be monitored using a variety of techniques such as infrared spectroscopy, dynamic mechanical analysis, dielectric spectroscopy, ultrasound, differential scanning calorimetry, and thermomechanical analysis.

Model Networks. Construction of model networks allows development of quantitative structure property relationships and provide the ability to test the accuracy of the theories of rubber elasticity. By definition, model networks have controlled molecular weight between cross-links, controlled cross-link functionality, and controlled molecular weight distribution of cross-linked chains. Silicones cross-linked by either condensation or addition reactions are ideally suited for these studies because all of the above parameters can be controlled.

Using both condensation-cured and addition-cured model systems, it has been shown that the modulus depends on the molecular weight of the polymer and that the modulus at rupture increases with increased junction functionality. However, if a bimodal distribution of chain lengths is employed, an anomalously high modulus at high extensions is observed.

Filled Silicone Networks. Few applications use silicone elastomers in the unfilled state. The addition of fillers (qv) results in a several-fold improvement in properties, and fillers can be broadly categorized as reinforcing and nonreinforcing (or semireinforcing). Reinforcing fillers increase tensile strength, tear strength, and abrasion resistance, whereas nonreinforcing fillers are used as additives for reducing cost, improving heat stability, imparting color, and increasing electrical conductivity.

The final mechanical properties of the compound are a function of the concentration of the reinforcing filler in the formulation. A good dispersion of the filler particles is essential for ensuring satisfactory ultimate properties.

Properties and Uses

Silicone Fluids. Silicone fluids are used in a wide variety of applications, including damping fluids, dielectric fluids, polishes, cosmetic and personal care additives, textile finishes, hydraulic fluids, paint additives, and heat-transfer oils. Polydimethylsiloxane oils are manufactured by the equilibrium polymerization of cyclic or linear dimethylsilicone precursors. Trifunctional organosilane end groups, typically trimethylsilyl (M), are used, and the ratio of end group to chain units (D), ie, M/D, controls the ultimate average molecular weight and viscosity. Low viscosity fluids, $<10^5$ mm^2/s($=$ cSt), are generally prepared by acid-catalyzed equilibration.

High molecular weight (>10^6 mm^2/s(= cSt)) silicone oils and gums are prepared by base-catalyzed, ring-opening polymerization of D_3 or D_4, or by condensation polymerization of silanol-terminated polydimethylsiloxane (PDMS). Both methods are practiced commercially.

Many of the applications for silicone oils are derived from the wide temperature range over which they can be used. Dimethylsilicone fluids decompose via two principal mechanisms: retrocyclization to low volatile cyclic siloxanes such as D_3 and D_4, and thermal oxidation of the alkyl side chains to give formaldehyde, CO_2, water, and T groups. The retrocyclization process is catalyzed by acids or bases and can occur at temperatures above 140°C. Catalytic acidic or basic sites on glassware and metallic containers are often the source of degradation of PDMS fluids.

Compared with petroleum-based fluids, silicone oils show relatively small changes in viscosity as a result of temperature change. A common measure of the viscosity change with temperature is the viscosity–temperature coefficient (VTC). Typical dimethylsilicone VTC is 0.6 or less. Phenylsilicones are slightly higher.

Many of the unique properties of silicone oils are associated with the surface effects of dimethylsiloxanes, eg, imparting water repellency to fabrics, antifoaming agents, release liners for adhesive labels, and a variety of polishes and waxes. Dimethylsilicone oils can spread onto many solid and liquid surfaces to form films of molecular dimensions. This phenomenon is greatly affected by even small changes in the chemical structure of siloxane in the siloxane polymer. Increasing the size of the alkyl substituent from methyl to ethyl dramatically reduces the film-forming ability of the polymer. The phenyl-substituted silicones are spread onto water or solid surfaces more slowly than PDMS.

Dimethylsilicone polymers are often described as having a combination of silicate and paraffin structures, and the orientation of the polymer chains onto surfaces, physically, by chemical affinity, or bonding, can contribute to the observed surface properties. Gases are soluble in dimethylsilicone polymers and PDMS is permeable to water vapor.

Silicone fluids have good dielectric properties, loss factor, specific resistance, and dielectric strength at normal operating conditions, and the properties vary only slightly with temperature. The properties in combination with relatively low flammability have led to the use of silicones in transformers and other large electrical applications.

Silicone oils are good hydrodynamic lubricants but have generally poor frictional lubricating properties. The latter can be improved by incorporating chlorophenyl groups into the polymer side chains.

Liquid silicone oils are highly compressible and remain liquid over pressure ranges where normal paraffin oils have already solidified. This property, combined with a wide temperature use range, is the reason for silicone use in a large number of hydraulic applications.

Silicone Heat-Cured Rubber. Silicone elastomers are made by vulcanizing high molecular weight (>5×10^5 mol wt) linear polydimethylsiloxane polymer, often called gum. Fillers are used in these formulations to increase strength through reinforcement. Extending fillers and various additives, eg, antioxidants, adhesion promoters, and pigments, can be used to obtain certain properties.

Unlike natural rubber, silicone rubber does not stress-crystallize when elongated, which leads to relatively poor physical properties. Unfilled silicone rubber has only a 0.35-MPa (50-psi) tensile stress at break. To overcome this, silicone rubber is compounded with 10 to 25 wt % reinforcing fillers, typically fumed silica, to improve the final rubber product properties.

The processing methods for silicone rubber are similar to those used in the natural rubber industry. Heat-cured silicone rubber is commercially available as gum stock, reinforced gum, partially filled gum, uncatalyzed compounds, dispersions, and catalyzed compounds. The latter is ready for use without additional processing.

It is common practice in the silicone rubber industry to prepare specific or custom mixtures of polymer, fillers, and cure catalysts for particular applications. The number of potential combinations is enormous.

Silicone rubber is most commonly fabricated by compression-molding-catalyzed gum stock at 100–180°C under 5.5–10.3 MPa (800–1500 psi) pressure. Mold release compounds are usually employed. Under these conditions the rubber is cured in a few minutes. Extrusion processing is used in the manufacture of tubes, rods, wire and cable insulation, and continuous profiles.

Vulcanized silicone rubber is characterized by its wide temperature use range (−50 to >200°C), excellent electrical properties, and resistance to air oxidation and weathering conditions. Silicone rubber is also extremely permeable to gases and water vapor. The mechanical properties of silicone rubber are generally inferior to most organic (butyl) rubbers at room temperature.

The electrical properties of silicone rubber are generally superior to organic rubbers and are retained over a temperature range from −50 to 250°C. Silicone rubber film is 10 to 20 times more permeable to gases and water vapor than organic rubber.

Solvent-resistant rubber based on either trifluoropropylmethylsiloxane or β-cyanoethylmethylsiloxane has been developed for applications, eg, as fuel tank sealants, where the material will be exposed to aggressive solvents. Pure water has little effect on silicone; however, long exposures in the presence of acid or base catalysts causes degradation and reversion of the rubber to a sticky gum. Silicone rubber burns with a high char yield, and the residual material is nonconducting silicon dioxide.

Silicone Liquid-Injection-Molding Rubber. An increasingly important processing technique for silicone rubber is liquid injection molding. Unlike heat-cured rubber, which is typically compression-molded from high viscosity gum stock, liquid-injection-molded (LIM) rubber is made from low viscosity starting materials, 1000−2000 mPa·s(= cP), and is cured in molds similar to those used for plastic injection molding. LIM processing is used for applications such as electrical connectors, O-ring seals, valves, electrical components, health care products, and sporting equipment such as goggles and scuba masks.

Silicone LIM rubber is made from a two-component polymer system. One part (Part B) contains a linear polydimethysiloxane polymer with pendent Si–H functionality, reinforcing fillers such as fumed silica, extending fillers, pigments, and stabilizers. The second part (Part A) contains linear polydimethylsiloxane with terminal and pendent vinyl groups; reinforcing and extending fillers; a platinum hydrosilylation catalyst; and a catalyst inhibitor, commonly olefin, amine, or phosphine ligands. After mixing and heating, the catalyst initiates the cross-linking reaction by addition of the Si–H group to the double bond.

Foam Rubber. Flexible foamed silicone rubber can be fabricated with a flame retardancy greatly superior to that of the urethane-type foam. A self-blowing, low to medium density (80–240-kg/cm^3(5–15-lb/ft^3)) silicone foamed rubber can be prepared using polymers similar to those used in LIM products. Because of its excellent flammability characteristics, silicone foam is used in building and construction fire-stop systems and as pipe insulation in power plants.

Silicone Resins. Silicone resins are an unusual class of organosiloxane polymers. Unlike linear poly(siloxanes), the typical silicone resin has a highly branched molecular structure. The most unique, and perhaps most useful, characteristics of these materials are their solubility in organic solvents and apparent miscibility in other polymers, including silicones. The incongruity between solubility and three-dimensional structure is caused by low molecular weight ($M_n < 10,000$ g/mol) and broad polydispersivity of most silicone resins.

A wide variety of organosilicone resins containing a combination of M, D, T, and/or Q groups have been prepared and many are commercially manufactured. In addition, resins containing hydrosilation-reactive SiH and SiVi groups or other functionalities, including OH and phenyl groups, are known. Two classes of silicone resins are most widely used in the silicone industry: MQ and TD resins.

MQ resins are composed of clusters of quadrafunctional silicate Q groups end-capped with monofunctional trimethylsiloxy M groups. The structure of an MQ resin molecule is defined by three characterization parameters: M/Q ratio, molecular weight, and % OH. The most prominent use of MQ resins is as the tackifying agent for silicone pressure-sensitive adhesives (PSA). The other main component of silicone PSA is a silicone gum.

TD resins are simply prepared by cohydrolyzing mixtures of chlorosilanes in organic solvents. TD resins are used as protective coatings, electrical coatings, saturants, laminates, and water repellents. TD resins are also useful in high performance paints. Compositions high in silanol content are utilized in reactive formulations. Low silanol TD resins are used as a nonreactive additive to alkyd paint formulations. Attractive features of TD resin-based protective coatings include superior, uv-resistant weatherability and excellent high and low temperature properties. Silicone electrical coatings are preferred when good dielectric insulation is required over a broad temperature range.

Organosilicone Coating Products. Silicone products are used in a large variety of coatings applications; most prominent among these are silicone pressure-sensitive adhesives (PSA), plastic hardcoats, and paper release coatings.

Silicone PSAs are used primarily in specialty tape applications that require the superior properties of silicones, including resistance to harsh chemical environments and temperature extremes.

Silicone hardcoat technology evolved from the need to develop thin film coatings to impart abrasion and chemical resistance to plastic substrates. The basic chemistry involves first hydrolyzing a trialkoxysilane in the presence of an aqueous colloidal silica solution. The resulting solution is then diluted with alcohols.

Advances in silicone hardcoat technology include the development of weathereable and uv-curable hardcoats. Weatherable hardcoats are used in exterior applications such as polycarbonate windows and automotive lighting.

Paper release coatings are used in label systems in which the silicone coating is part of the disposable paper liner. The role of the silicone is to provide a low surface energy interface between the paper liner and the adhesive label. Thermally cured paper release coatings are usually solvent-less mixtures of an SiVi-terminated PDMS, an SiH-containing cross-linker, a Pt hydrosilation catalyst, and a cure inhibitor. Typical substrates for silicone release coatings are supercalendered kraft paper, glassines, and thermally sensitive films such as polyethylene and polypropylene. Key properties for release coatings are cure speed, integrity of cure, and stable release values.

Room-Temperature Vulcanizable Silicones. Moisture-curable, room-temperature vulcanizable (RTV) silicones represent one of the largest-volume and commercially most successful silicone technologies. When exposed to atmospheric moisture, RTV silicones undergo hydrolysis and condensation reactions and cure into high strength elastomers. The cured elastomers have excellent primerless adhesion to substrates as varied as glass, metals, wood, masonry, and plastics. Additional benefits include excellent weatherability, durability, electrical insulation, chemical resistance, stability at high temperatures, and flexibility at low temperatures.

Health and Environmental Aspects

Few materials of commercial importance possess a range of applications as diverse as silicones. As a result of their widespread use and remarkable inertness toward light, heat, and chemical agents, an understanding of the environmental fate and distribution of silicones is important.

Organosilicones have been detected in terrestrial, aquatic, and atmospheric environments. The highest concentrations of environmental silicones are found in the sludges of wastewater treatment plants and upstream of process plants. Much lower levels are found in water or sediment samples except at certain point sources, such as in effluents of dyeing factories and other industries that employ silicones in their processes. In general, levels of organosilicones in sludges and sediments have been found to correlate with the total organic content of these media as determined by pyrolysis weight losses. This implies a migratory aptitude similar to that of other organic pollutants.

Because of their hydrophobic nature, silicones entering the aquatic environment should be significantly absorbed by sediment or migrate to the air-water interface. Volatile surface siloxanes become airborne by evaporation, and higher molecular weight species are dispersed as aerosols.

Samples of particulates taken at two New Jersey office buildings revealed silicone levels that were considerably higher indoors than outdoors. In these cases, indoor silicone aerosols are believed to be generated primarily by photocopiers, which use silicone fuser oils.

Knowledge of the transformation of silicones under various environmental conditions is key to understanding the fate of these materials. Model studies predict that a large fraction of silicones entering the environment through wastewater treatment systems, ie, municipal treatment plants or septic systems, will ultimately be deposited on soils as a result of absorption or of sludge amendment. The chemistry of silicones on soils is thus an important factor in assessing the overall environmental impact of these materials. Silicone fluids degrade when exposed to soils. The rate of degradation depends on soil type and moisture content. The products of soil-induced abiotic degradation are typically silanol-terminated monomers and oligomers. Dimethylsilane-1,1-diol, the simplest monomeric unit derived from silicone hydrolysis, is an important degradation product of abiotic silicone degradation. As a polar, water-soluble species, it exhibits environmental distribution properties drastically different from the silicones from which it is derived; it is expected to be the principal waterborne silicone species found in most environmental samples.

JONATHAN RICH
JAMES CELLA
LARRY LEWIS
JUDITH STEIN
NAVJOT SINGH
GE Corporate Research and Development
SLAWOMIR RUBINSZTAJN
JEFF WENGROVIUS
GE Silicones

D. Scott, *J. Am. Chem. Soc.* **68**, 2294 (1946).

E. G. Rochow, *Silicon & Silicones*, Springer-Verlag, Berlin, 1987.

F. O. Stark, J. R. Falender, and A. P. Wright, in G. Wilkinson, F. G. A. Stone, and E. W. Abel, eds., *Comprehensive Organometallic Chemistry*, Pergamon, Oxford, U.K., 1982.

R. R. McGregor, *Silicones & Their Uses*, McGraw-Hill Book Co., Inc., New York, 1954.

SILYLATING AGENTS

Silylation of Organic Compounds

Silylation is the replacement of one or more active hydrogens from an organic molecule by a trisubstituted silyl, R_3Si—, group. The active hydrogen is usually an alcohol, carboxylic acid, or phenol, ie, —OH; an amine, amide, or urea, —NH; or a thiol, —SH, and the silylating agent is usually a trimethylsilyl halide, dimethylsilyl dihalide, or a trimethylsilyl nitrogen-functional compound.

Derivatizing an organic compound for analysis may require only a few drops of reagent selected from silylating kits supplied by laboratory supply houses. Commercial synthesis of penicillins requires silylating agents purchased in tank cars from the manufacturer (see ANTIBIOTICS, β-LACTAMS–PENICILLINS AND OTHERS).

Typical commercial silylating agents are listed in Table 1. The first three silylating agents in the table are available in bulk quantities and are most suitable for large-scale commercial silylation. The chlorosilanes are generally used in combination with an acid acceptor, eg, triethylamine. The nitrogen-functional silanes each have certain advantages for particular applications. Fluorinated silylating agents give enhanced rates of reaction and more volatile by-products.

Table 1. Methyl Silylating Agents

Chemical name	Formula
trimethylchlorosilane (TMCS)	$(CH_3)_3SiCl$
dimethyldichlorosilane (DMCS)	$(CH_3)_2SiCl_2$
hexamethyldisilazane (HMDZ)	$(CH_3)_3SiNHSi(CH_3)_3$
chloromethyldimethyl-chlorosilane (CMDMS)	$ClCH_2(CH_3)_2SiCl$
N,N'-bis(trimethylsilyl)-urea (BSU)	$[(CH_3)_3SiNH]_2CO$
N-trimethylsilyldiethylamine (TMSDEA)	$(CH_3)_3SiNH(C_2H_5)_2$
N-trimethylsilylimidazole (TSIM)	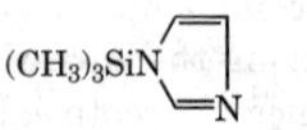
N,O-bis(trimethylsilyl)-acetamide (BSA)	$(CH_3)_3SiN{=}C(CH_3)OSi(CH_3)_3$
N,O-bis(trimethylsilyl)-trifluoroacetamide (BSTFA)	$(CH_3)_3SiN{=}C(CF_3)OSi(CH_3)_3$
N-methyl-*N*-trimethyl-silyltrifluoroacetamide (MSTFA)	$(CH_3)_3SiN(CH_3)COCF_3$
t-butyldimethylsilylimidazole (TBDMIM)	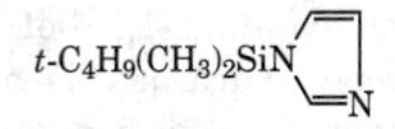
N-trimethylsilylacetamide (MTSA)	$(CH_3)_3SiNHCOCH_3$
trimethylsilyl trifluoromethanesulfonate (TMS triflate)	$(CH_3)_3SiOSO_2CF_3$
trimethylsilyl iodide (TMSI)	$(CH_3)_3SiI$

The chlorosilanes are clear liquids that should be treated as strong acids. They react readily with water to form corrosive HCl gas and liquid. Liquid chlorosilanes and their vapors are corrosive to the skin and extremely irritating to the mucous membranes of the eyes, nose, and throat. The nitrogen-functional silanes react with water to form ammonia, amines, or amides. Because ammonia and amines are moderately corrosive to the skin and very irritating to the eyes, nose, and throat, silylamines should be handled like organic amines. Trimethylsilyl trifluoromethanesulfonate and trimethylsilyl iodide form very corrosive acidic products.

Derivatization for Analysis. Silylation of organic materials has been an invaluable tool in analytical chemistry to allow ready analysis by gas–liquid chromatography (glc), mass spectrometry (qv), and thin-layer chromatography (tlc) (see CHROMATOGRAPHY). There are four main reasons to derivatize a compound for analysis: to increase volatility, to increase thermal stability, to enhance detectability, and to improve separation.

Silylation in Organic Synthesis. Silyl blocking agents are used in organic synthesis to protect sensitive functional groups, to alter reactivity and solubility, and to increase stability of intermediates. Silylation applications in pharmaceutical synthesis have been used to protect a wide range of OH groups, eg, alcohols in prostaglandins (qv) and steroid synthesis, enols in the synthesis of nucleosides and steroids (qv), and carboxylic acids and sulfenic acids in the synthesis of penicillins and cephalosporins (see ANTIBIOTICS).

Silylation of Inorganic Compounds

Silicate Modifications. Silicate minerals can be simultaneously acid-leached and trimethylsilyl end-blocked to yield specific trimethylsilyl silicates having the same silicate structure as the mineral from which these were derived. Certain anionic siliconates stabilize solutions of alkali silicates to give stable solutions in water or alcohols at any pH. Such silicate—siliconate mixtures are used as corrosion inhibitors in glycol antifreeze (see ANTIFREEZES AND DEICING FLUIDS; CORROSION AND CORROSION CONTROL).

Ziegler-Natta Polymerization. The polymerization of propylene with Ziegler-Natta catalysts, ie, complexes of $TiCl_3$–$(C_2H_5)_3Al$ on $MgCl_2$ supports, is significantly affected by external addition to the reactor of organo(alkoxy)silanes with the propylene feed. The nature of the organic group(s) and alkoxy group(s) affects the catalyst activity and the microstructure of the polymer.

Silylation of Inorganic Surfaces

Alkyl Silylating Agents. Alkyl silylating agents convert mineral surfaces to water-repellent, low energy surfaces useful in water-resistant treatments for masonry, electrical insulators, packings for chromatography, and noncaking fire extinguishers. Methylchlorosilanes react with water or hydroxyl groups at the surface to liberate HCl and deposit a thin film of methylpolysiloxane, which has a low critical surface tension and is therefore not wetted by water. Ceramic insulators can be treated with methylchlorosilane vapors or solutions in inert solvents to maintain high electrical resistivity under humid conditions. Hydrolyzed methylchlorosilanes also dissolve in aqueous alkali and are then applied as aqueous solutions of sodium methylsiliconates. The siliconates are neutralized by carbon dioxide in the air to form an insoluble, water-resistant methylpolysiloxane film within 24 hours. Treatment of brick, mortar, sandstone, concrete, and other masonry protects the surface from spalling, cracking, efflorescence, and other types of damage caused by water (see WATER–WATERPROOFING AND WATER/OIL REPELLENCY).

Organofunctional Silylating Agents

Whereas alkylsilylating agents provide low energy surfaces designed for release, a series of organofunctional silylating agents is offered commercially as adhesion promoters. Principal applications have been as coupling agents in glass- or mineral-reinforced organic resin composites and as adhesion promoters for paints, inks, coatings, and adhesives.

Liquid Crystals. In liquid crystal displays, clarity and permanence of image is enhanced if the display can be oriented parallel or perpendicular to the substrate. Oxide surfaces treated with dimethyloctadecyl-3-trimethoxysilylpropylammonium chloride, $C_{18}H_{37}N^+(CH_3)_2CH_2CH_2CH_2Si(OCH_3)_3Cl^-$, tend to orient liquid crystals perpendicular to the surface (see LIQUID CRYSTALLINE MATERIALS); parallel orientation is obtained on surfaces treated with *N*-methylaminopropyltrimethoxysilane, CH_3NH-$CH_2CH_2CH_2Si(OCH_3)_3$.

Ion Removal and Metal Oxide Electrodes. The ethylenediamine (*en*)-functional silane has been studied extensively as a silylating agent on silica gel to preconcentrate polyvalent anions and cations from dilute aqueous solutions. Numerous other chelate-functional silanes have been immobilized on silica gel, controlled-pore glass, and fiber glass for removal of metal ions from solution.

Metal oxide electrodes have been coated with a monolayer of this same diaminosilane by contacting the electrodes with a benzene solution of the silane at room temperature. Electroactive moieties attached to such silane-treated electrodes undergo electron-transfer reactions with the underlying metal oxide.

Antimicrobials. Surface-bonded organosilicon quaternary ammonium chlorides have enhanced antimicrobial and algicidal activity.

Polypeptide Synthesis and Analysis. Silica or controlled-pore glass supports treated with (chloromethyl)phenylethyltrimethoxysilane or its derivatives are replacing chloromethylated styrene—divinylbenzene (Merrifield resin) as supports in polypeptide synthesis. The silylated support reacts with the triethylammonium salt of a protected amino acid. Once the initial amino acid residue has been coupled to the support, a variety of peptide synthesis methods can be used (see PROTEIN ENGINEERING; PROTEINS).

Immobilized Enzymes and Metal-Complex Catalysts. The most frequently used technique for immobilizing enzymes on a solid support involves reducing *N*-(3-triethoxysilylpropyl)-*p*-nitrobenzamide after attachment to silica or controlled-pore glass to give aniline derivatives, then converting them to diazonium salt, and effecting coupling through azo linkage to the tyrosine of the proteins (see ENZYMES IN ORGANIC SYNTHESIS; ENZYME APPLICATIONS, INDUSTRIAL).

Reinforced Composites. Silane coupling agents modify the interface between inorganic surfaces and organic resins to improve the adhesion between resin and surface, thus improving physical properties and water resistance of reinforced plastics. Suitable coupling agents are available for any of the common plastics and metal, glass, or many other inorganic reinforcements. Principal applications for these coupling agents are in reinforced plastics for boats, storage tanks, pipes, automobiles, and architectural structures (see LAMINATED MATERIALS, PLASTIC). Other applications are in the treatment of mineral fillers and pigments for paint and rubber, in primers to improve the adhesion of paints, inks, coatings, and adhesives to metals and other inorganic surfaces, and in tarnish and corrosion inhibitors for silver, copper, aluminum, and steel (see FILLERS; CORROSION AND CORROSION CONTROL).

PETER G. PAPE
Dow Corning Corporation

A. E. Pierce, *Silylation of Organic Compounds*, Pierce Chemical Co., Rockford, Ill., 1968.

C. A. Roth, *Ind. Eng. Chem. Prod. Res. Develop.* **11**, 134 (1972).

E. P. Plueddemann, *Silane Coupling Agents*, 2nd ed., Plenum Press, New York, 1991, Chapt. 3.

B. Arkles, *Chemtech*, 768 (Dec. 1977).

SILK

Silks can be defined as externally spun fibrous protein secretions. Of all the natural fibers, silks represent the only ones that are spun. These fibers are remarkable materials displaying unusual mechanical properties. Strong, extensible, and compressible, silks display interesting electromagnetic responses, particularly in the uv range for insect entrapment; form liquid crystalline phases related to processing; and exhibit piezoelectric properties.

Types of Silk

Silks are synthesized by a variety of organisms, including silkworms, spiders, scorpions, mites, and flies. Few of these silks have been characterized. Silks differ in properties, composition, and morphology, depending on the source. Silkworm cocoon silk from *Bombyx mori* is the most well characterized owing to the extensive use of these fibers in the textile industry for over 5000 years in a practice originating in China. The dragline silk from the orb-weaving spider, *Nephila clavipes*, is the most well characterized of the different spider silks.

Silkworm Cocoon Silk. The cocoon silk from *B. mori* contains two structural fibroin filaments coated with a family of glue-like sericin proteins, resulting in a single thread having a diameter of 10 to 25 μm. The life cycle of *B. mori* runs for 55 to 60 days and the organism passes through a series of developmental stages or molts. Silk production occurs during cocoon formation around day 26 in the cycle during the fifth larval instar just before molt to the pupa.

Spider Silk. Spider silks function in prey capture, reproduction, and as vibration receptors, safety lines, and dispersion tools. Spider silks are synthesized in glands located in the abdomen and spun through a series of orifices (spinnerets). The types and nature of the various silks are diverse and dependent on the type of spider.

Structure

Composition. The silkworm cocoon silk contains two structural proteins, the fibroin heavy chain (mol wt ca 325,000) and fibroin light chain (mol wt ca 25,000), plus the family of sericin proteins (mol wt 20,000–310,000) to hold the fibroin chains together in the final cocoon fiber. Other silks, such as the caddis fly and aquatic midge, which spin silks underwater to form sheltered tubes, have also been characterized and consist of a family of proteins having high cysteine content and running from low to very high $>10^6$) molecular weights (see PROTEINS). The consensus crystalline amino acid repeat in the *B. mori* silkworm cocoon silk fibroin heavy chain is the 59mer: GAGAGSGAAG[SGAGAG]$_8$Y (see AMINO ACIDS). The spider dragline silk from the principal ampullate gland contains at least one protein, called MaSp1, for major ampullate silk protein, previously termed spidroin 1; mol wt is around 275,000.

Secondary Structure. The silkworm cocoon and spider dragline silks are characterized as an antiparallel β-pleated sheet wherein the polymer chain axis is parallel to the fiber axis. Other silks are known to form α-helical (bees, wasps, ants) or cross-β-sheet (many insects) structures. The cross-β-sheets are characterized by a polymer chain axis perpendicular to the fiber axis and a higher serine content. Most silks assume a range of different secondary structures during processing from soluble protein in the glands to insoluble spun fibers.

Two crystalline forms for silk have been characterized. The random coil or silk I, ie, the prespun pseudocrystalline form of silk present in the gland in a water-soluble state, is predominant in the gland; silk II, ie, the spun form of silk which is insoluble in water, becomes predominant once the protein is spun into fiber.

Crystallinity. Generally, spider dragline and silkworm cocoon silks are considered semicrystalline materials having amorphous flexible chains reinforced by strong stiff crystals. The orb web fibers are composite materials (qv) in the sense that they are composed of crystalline regions immersed in less crystalline regions, which have estimates of 30–50% crystallinity.

Structure of the Spider Orb Web. The construction of the orb web is a feat of engineering involving material tailoring, optimization of material interfaces, and conservation of resources to promote survival of the spider. In addition, the web absorbs water from the atmosphere, and ingestion by the spider may provide a significant contribution to water intake needs. Some orb webs appear to be at least in part recycled by ingestion as a conservation tool, and some of the amino acids are reused in new webs.

Processing

In Vivo Processing. Silks are synthesized in specialized glands within the organism. Initially, some degree of self-organization or assembly occurs as a result of protein–protein interactions among the crystalline repeats in the protein chains.

Rheological experiments indicate that crystallinity in the fiber correlates positively with shear and draw rates, and an extrusion rate of around 50 cm/min was found to be a minimum threshold for the appearance of birefringence and the conversion of the soluble silk solution in the gland to the β-sheet found in the spun fiber. In the posterior region of the gland, 0.4–0.8 mm in diameter, the silk solution is optically featureless, a range of secondary structures are present, including random coil and silk I, and the shear rate is low. In the middle region of the gland, the diameter is 1.2–2.5 mm, streaming birefringence is observed, and the shear rate is also low. In the anterior region of the gland, the diameter is narrow, 0.05–0.3 mm, the shear rate is high, water appears to be actively transported out of the gland, the pH decreases, and active ion exchange occurs. Viscosity also increases but presumably decreases prior to spinning as a result of the liquid crystalline phase. At this point the characteristic silk II structure forms. In the pair of major ampullate glands in the spider, which are the location of dragline protein synthesis, a similar process occurs as summarized for the silkworm. This gland is smaller, however, and there is no sericin contribution in the middle region of the gland. A lyotropic nematic liquid crystalline phase of the protein forms prior to spinning in both the spider and the silkworm, as well as in many of the different glands of the spider responsible for the different silks.

Commercial and Artificial Processing. Commercially, silkworm cocoons are extracted in hot soapy water to remove the sticky sericin protein. The remaining fibroin or structural silk is reeled onto spools, yielding approximately 300–1200 meters of usable thread per cocoon. These threads can be dyed or modified for textile applications.

Most solvents used to solubilize globular proteins do not suffice for silks. Silks are insoluble in water, dilute acids and alkali, and most organic solvents; they are resistant to most proteolytic enzymes. Silkworm fibroin can be solubilized by first degumming or removing the sericin using boiling soap solution or boiling dilute sodium bicarbonate solution, followed by immersion of the fibroin in high concentration salt solutions such as lithium bromide, lithium thiocyanate, or calcium chloride. These salt solutions can also be used to solubilize spider silk, as can high concentrations of propionic acid–hydrochloric acid mixtures and formic acid. After solubilization in these aggressive solvents, dialysis into water or buffers can be used to remove the salts or acids, although premature reprecipitation is a common problem.

Films or membranes of silkworm silk have been produced by air-drying aqueous solutions prepared from the concentrated salts, followed by dialysis. The films, which are water soluble, generally contain silk in the silk I conformation with a significant content of random coil. Many different treatments have been used to modify these films to decrease their water solubility by converting silk I to silk II in a process found useful for enzyme entrapment. Silk membranes have also been cast from fibroin solutions and characterized for permeation properties.

Properties

Mechanical Properties. The mechanical properties of silks are an intriguing combination of high strength, extensibility, and compressibility (Table 1).

Resistance to axial compressive deformation is another interesting property of the silk fibers. Based on microscopic evaluations of knotted single fibers, no evidence of kink-band failure on the compressive side of a knot curve has been observed. Synthetic high performance fibers fail by this mode even at relatively low strain levels.

Fibers. *B. mori* cocoon silk ranges from 10 to 25 μm in diameter; dragline silk from *N. clavipes* from 2.5 to 4.5 μm in diameter. Web fibers from some spiders have diameters as low as 0.01 μm.

Thermal Properties. Spider dragline silk was thermally stable to about 230°C based on thermal gravimetric analysis.

Genetic Engineering

An understanding of the genetics of silk production in silkworms and spiders should help in developing processes for higher levels of silk expression generated by recombinant deoxyribonucleic acid (DNA) methods. Genetically engineered or recombinant DNA silkworm and spider silks have been produced using either native genes or synthetic genes.

Applications

The ability to tailor polymer structure to a precise degree leads to interesting possibilities in the control of macroscopic functional properties of fibers, membranes, and coatings, as well as improved control of processing windows. Biotechnology offers the tools with which to solve limitations in spider silk production that have not been overcome with traditional domestication and breeding approaches, such as those used successfully with the silkworm. This is of interest because of the variety of silk structures available and the higher modulus and strength as compared to silkworm silk.

Hybrid silk fibers containing synthetic fiber cores having silk coextruded or grafted have been synthesized. Cosmetics (qv) and consumer products such as hair replacements and shampoos containing silk have also been marketed.

Table 1. Mechanical Properties of Silks and Other Fibers

Fiber	Elongation, %	Modulus, GPa[a]	Strength, GPa[a]	Energy to break, J/kg[b]
		Fibroins		
B. mori	15–35	5	0.6	7×10^4
other silkworms	12–50	2–4	0.1–0.6	$(3-6) \times 10^4$
		Draglines		
N. clavipes				
quasistatic[c]	9–11	22–60	1.1–2.9	$(3.7-12) \times 10^4$
high strain[d]	10	20		
other spiders	10–39	2–24	0.2–1.8	$(1-10) \times 10^4$
		Other fibers		
nylon	18–26	3	0.5	8×10^4
cotton	5–7	6–11	0.3–0.7	$(5-15) \times 10^3$
Kevlar	4	100	4	3×10^4
steel	8	200	2	2×10^3

[a] 1 GPa = 10^9 N/m^2. To convert GPa to psi, multiply by 145,000. [b] To convert J to cal, divide by 4.184. [c] Instron tensile test rates of 10%/s. [d] Rates of >500,000%/s.

DAVID L. KAPLAN
CHARLENE MELLO
STEPHEN FOSSEY
STEVEN ARCIDAICONO
U.S. Army Natick Research,
Development, & Engineering Center

D. L. Kaplan and co-workers, eds., *Silks: Materials Science and Biotechnology*, ACS, Washington, D.C., 1994.

R. D. B. Fraser and T. P. MacRae, *Conformation in Fibrous Proteins*, Academic Press, Inc., New York, 1973.

A. Simmons, C. Michal, and L. W. Jelinski, *Science* **271**, 84 (1996).

SILLEMANITE. See REFRACTORIES.

SILVER AND SILVER ALLOYS

Silver, Ag (at no. 47), is a white, lustrous, soft, malleable metal having the highest known electrical and thermal conductivities. It is the most highly reflective of all the metals in the visible spectrum and second only to gold in the long-wave infrared. Along with its colorful neighbors copper and gold in the Periodic Table Group 11 (IB) metals, silver occurs naturally in metallic form.

Properties

Selected properties of silver are summarized in Table 1. In the electromotive force series of the elements, silver is less noble than only Pd, Hg, Pt, and Au. All provide high corrosion resistance.

Oxidation States. The common oxidation state of silver is +1, ie, Ag^+, as found in AgCl, which is used with Mg in sea- or freshwater-activated batteries (qv); $AgNO_3$, the initial material for photographic materials, medical compounds, catalysts, etc; and silver oxide, Ag_2O, an electrode in batteries (see SILVER COMPOUNDS). Few Ag^{2+} compounds are known. Silver in the +3 oxidation state, including silver peroxide is obtained by the action of the vigorous oxidizing agent $S_2O_8^{2-}$ on Ag_2O or other Ag compounds.

Tetrasilver tetroxide Ag_4O_4 is a powerful oxidizer for sanitizing swimming pools, hot tubs, and industrial cooling system waters (see WATER, TREATMENT OF SWIMMING POOLS, SPAS, AND HOT TUBS). Bivalent and trivalent silver disinfectants have been shown to be from 50 to 200 times more effective as sanitizers than monovalent silver compounds.

Oxygen Reactivity. Silver is second only to gold as the element having the weakest interaction with oxygen, providing silver with its superior sparking and combustion resistance.

Solid silver is more permeable by oxygen than any other metal. Oxygen moves freely within the metallic silver lattice, not leaving the surface until two oxygen atoms connect to form O_2. This occurs at ~300°C. Below this temperature silver is an efficient catalyst for gaseous oxidative chemical reactions. Silver is also an extremely efficient catalyst for aqueous oxidative sanitation.

During casting some oxygen may be introduced to convert base metal impurities into oxides. Because these oxides do not enter into the solid solution, they have no effect on the annealing and recrystallization temperature of the silver critical to the silversmith.

Table 1. Selected Properties of Silver

Parameter	Value
atomic mass, amu	107.8682
melting point[a], °C	961.93
isotopic abundance, %	
106.9051	51.84
108.9048	48.16
electrochemical potential[b], V	0.798
density, at 20°C, g/cm^3	
annealed	10.492
hard drawn	10.43
at 0 K	10.63
boiling point, °C	2187
tensile strength, 5-mm dia wire, MPa[c]	
annealed at 600°C	125–186
50% cold worked	290
elongation, 5-mm dia wire, % in 5.08 cm	
at 20°C, annealed at 600°C	43–50
50% cold worked	3–5
electrical resistivity[d], ρ, Ω/m	
20 K	0.00422×10^{-8}
273.15 K (= 0°C)	1.467×10^{-8}
500 K	2.875×10^{-8}
hardness, Brinell, kg/mm^2[e]	25–30
elastic properties[f]	
Young's modulus at 20°C, GPa[g]	91.3
modulus of rigidity at 20°C, GPa[g]	26.9–29.7

[a] A partial pressure of 20 kPa (2.9 psi) O_2 results in a freezing point of ca 950°C. [b] For the equation, $Ag^+ + e^- \rightarrow Ag$. [c] To convert MPa to psi, multiply by 145. [d] Total electrical resistivity of 99.999% pure or purer bulk silver. Impurities increase resistivity. [e] At 20°C, annealed at 600°C. [f] 1 Pascal = 10 dyne/cm^2 = 1.45×10^{-4} lbf/in.^{2}. [g] To convert GPa to psi, multiply by 145,000.

Dissolution of Silver. Silver is dissolved by oxidizing acids and alkali metal cyanide solutions in the presence of oxygen. The latter method is the principal technique for dissolving silver from ore. Silver has extensive solubility in mercury (qv) and low melting metals such as sodium, potassium, and their mixtures. Cyanide solutions of silver are used for electroplating and electroforming.

Tarnish. No passivation treatment to prevent silver from tarnishing exists. Sulfides in the atmosphere react to form Ag_2S.

Occurrence

The American cordillera extending from Alaska to Bolivia has been the most productive source of silver wherever it is associated with Tertiary age intrusive volcanic rocks, mostly concentrated by hydrothermal action. The largest producing mine in the cordillera is at Potosi, Bolivia.

Some 60 silver minerals are known. The most important economically are argentite, Ag_2S; cerargyrite, AgCl; polybasite, $Ag_{16}Sb_2S_{11}$; proustite, Ag_3AsS_3; pyrargyrite, Ag_3SbS_3; stephanite, Ag_5SbS_4; tetrahedrite, $Cu_3(AsSb)S_3$; and the tellurides. Silver is commonly associated with gold (see GOLD AND GOLD COMPOUNDS), copper (qv), lead (qv), and zinc (see ZINC AND ZINC ALLOYS) ores.

Mining and Processing

In 1887 it was discovered that gold and silver can be recovered by sodium cyanide. By 1907 the cyanide process, where a cyanide solution is mixed with zinc dust to precipitate the silver, was universally in use.

In the 1980s, zinc precipitation was replaced by a method involving the passing of the solution over activated carbon to adsorb the precious metals, which are then stripped from the charcoal by a hot caustic solution. Electrowinning removes the precious metals from this solution, depositing them on the cathode.

Heap leaching, ie, spraying of sodium cyanide solution over roughly crushed ores heaped on an impervious pad, has become the most economical way of recovering precious metal values from very low grade ores. Recoveries are lower, however, ca 70%, for heap leaching; ca 85% for milling and cyanidization.

Secondary Silver Recovery. The consumption of silver normally exceeds its mine production; therefore recovery from scrapped products, such as electrical gear, coins, and photographic film and solutions, is critical to its supply (see RECYCLING, NONFERROUS METALS).

Standards and Specifications

Commodity exchanges require good delivery silver bullion to be 999 parts per 1000 fine silver. Specifications for silver bullion, brazing alloys, electrical contact alloys, etc, are published by ASTM, the American Welding Society, Japanese Industrial Standards, SAE (Aerospace Materials Specifications), and the U.S. Dept. of Defense.

Assaying and Analysis

The fire assay is the most reliable method for the accurate quantitative determination of precious metals in any mixture for concentrations from 5 ppm to 100%.

Chemical analysis methods may be used for assay of silver alloys containing no interfering base metals. Nitric acid dissolution of the silver and precipitation as AgCl, or the Gay-Lussac-Volhard titration methods are used interchangeably for the higher concentrations of silver.

Instrumental methods for quantitative determination of silver purity include (*1*) atomic absorption, (*2*) emission spectroscopy (including inductively coupled plasma), (*3*) mass spectrometry (qv), and (*5*) x-ray fluorescence (see PLASMA TECHNOLOGY).

Health and Safety Factors

Silver metal dissolves in water to the extent of 5 parts per billion (ppb), making the water sufficiently toxic to kill such organisms as *E. coli* and *B. typhosus*. The U.S. Environmental Protection Agency, under the Safe Drinking Water Act, set the secondary contaminant level for silver in drinking water at 0.1 mg/L.

Silver Alloys

The atomic radius of silver (144 pm) is within about 15% of many elements, permitting solid solutions with Al, Au, Be, Bi, Cu, Cd, Ge, In, Mn, Pb, Pd, Pt, Sb, Sn, Th, and Zn. These metals form useful brazing, jewelry, and soldering alloys.

Silver's advantageous physical, chemical, electrical, and thermal conductive properties are used in a variety of alloys. For example, a 3.5% Ag, 0.5% Zr copper alloy provides a high strength, thermally conductive, fatigue-resistant liner of the hottest portion of the space shuttle engine main combustion chamber (see ABLATIVE MATERIALS).

Brazes and Solders. Silver imparts high tensile strength, ductility, thermal conductivity, bactericidal properties, and unusual wettability to most metals for soldering and brazing alloys.

Uses

Coinage. Because silver is soft, the coin silver alloy (90 Ag, 10 Cu) was adopted in the United States in 1837 and also by other national governments. Following World War II, burgeoning industrial demand consumed much of mine production, and with the inflation of national currencies, the decreasing monetary value of silver coins to below that of their bullion value brought an end to circulating silver coinage.

Photography. The introduction into photographic techniques of tabular silver halide crystals in 1983 markedly improved the efficiency of silver halides for capturing light, decreasing light scattering, and allowing greater adsorption of sensitizing dyes, extending the film's sensitivity to 1400 nm in the infrared (see PHOTOGRAPHY).

Photochromic glass used for sunglasses contains silver, copper, and halogen salts. Sunlight reduces the silver ion to a silver atom, darkening the glass. The released halogen atoms are trapped by the copper, and on removal of sunlight are recovered by the silver, restoring the glass to colorlessness, a readily reversible process. The silver crystallites block up to 97% of harmful uv rays.

Electrical Contacts. Silver combines the highest electrical and thermal conductivity and freedom from corrosion with low price. Its high resistance to arcing and mechanical wear may be increased by combining it with other materials such as graphite and cadmium oxide, which have superior nonsticking and nonwelding properties, and tin and indium oxides.

Silver Thick Films. Silver is used for the preparation of thick-film pastes in circuit paths and capacitors for the electronics industry. These are silk-screened onto ceramic or plastic circuit boards for multilayer circuit sandwich components.

Electroplating. Silver is normally plated using potassium cyanide solutions, which exhibit the highest plating current densities. The advent of computer-monitored electroplating has allowed consistent production line plating of gold–silver alloys.

Electroless Plating. A two-solution silvering system of NaOH and ammoniacal $AgNO_3$, mixed with a glucose solution containing an amine group allows electroless plating of mirrors at relatively low temperatures with good adhesion. This system lends itself to automated production of video compact disks, ornaments, thermos bottles, bottle caps, etc.

Magnetron Sputtered Reflective Coatings. Silver atoms sputtered *in vacuo* onto glass and polyester films entered the domestic window market in 1979 and are primary contenders for energy savings.

Sputtered silver mirrors are used for solar energy (qv) collectors and astronomical telescope mirrors.

Dental Amalgam. Silver–mercury dental amalgams have proved to be the most successful tooth filling materials in terms of both bactericidal and physical properties. The silver alloys incorporated in the amalgam have varied widely in composition (see DENTAL MATERIALS).

Other Uses. Silver is plated in all main-shaft bearings of jet engines because the silver provides a low coefficient of friction and superior fatigue and corrosion resistance, and has sufficient lubricity to serve as an emergency lubricant in case of oil failure. Silver and cobalt, although mutually insoluble, may be ground together and fused to exhibit giant magnetoresistance when a magnetic field is applied (see MAGNETIC MATERIALS).

SAMUEL F. ETRIS
The Silver Institute

World Silver Survey, The Silver Institute, Washington, D.C., 1992, 1995.

T. P. Mohide, *Silver*, Ontario Ministry of Mineral Resources, Toronto, Canada, 1985.

Mineral Facts and Problems, U.S. Bureau of Mines, Washington, D.C., 1965.

I. E. Wachs and R. J. Madix, *Surface Sci.* **76**, 531 (1978).

SILVER COMPOUNDS

Silver, a white, lustrous metal, slightly less malleable and ductile than gold (see GOLD AND GOLD COMPOUNDS), has high thermal and electrical conductivity (see SILVER AND SILVER ALLOYS). Most silver compounds are made using silver nitrate, $AgNO_3$, which is prepared from silver metal.

Some silver metal is found in nature, frequently alloyed with other metals such as copper, lead, or gold. Naturally occurring silver compounds, however, are the primary sources of silver. The most abundant naturally occurring silver compound is silver sulfide (argentite), Ag_2S, found alone and combined with iron, copper, and lead sulfides.

Silver belongs to Group 11 (IB) of the Periodic Table. Silver has been shown to have three possible positive oxidation states, but only silver(I) is stable in aqueous solution.

Silver(I) Compounds

Silver acetate, $H_3CCOOAg$, is prepared from aqueous silver nitrate and acetate ion. Colorless silver acetate crystals and solutions made from this salt are unstable to light.

Silver azide, AgN_3, is prepared by treating an aqueous solution of silver nitrate with hydrazine (see HYDRAZINE AND ITS DERIVATIVES), or hydrazoic acid. It is shock-sensitive and decomposes violently when heated.

Silver acetylide (silver carbide), Ag_2C_2, is prepared by bubbling acetylene through an ammoniacal solution of silver nitrate. Silver acetylide is sensitive to the point of undergoing detonation on contact.

Silver bromide, AgBr, is formed by the addition of bromide ions to an aqueous solution of silver nitrate. Silver bromide is significantly more photosensitive than silver chloride, resulting in the extensive use of silver bromide in photographic products.

Silver carbonate, Ag_2CO_3, is produced by the addition of an alkaline carbonate solution to a concentrated solution of silver nitrate.

Silver chloride, AgCl, is a white precipitate that forms when chloride ion is added to a silver nitrate solution. The order of solubility of the three silver halides is $Cl^- > Br^- > I^-$. Because of the formation of complexes, silver chloride is soluble in solutions containing excess chloride and in solutions of cyanide, thiosulfate, and ammonia. Silver chloride is insoluble in nitric and dilute sulfuric acid. Treatment with concentrated sulfuric acid gives silver sulfate. The silver ion in silver chloride can be readily reduced by light, and is used to a great extent in photographic print papers. Sufficient light intensity and time leads to silver chloride decomposing completely into silver and chlorine.

Silver chromate, Ag_2CrO_4, is prepared by treating silver nitrate with a solution of chromate salt or by heating a suspension of silver dichromate.

Silver cyanide, AgCN, forms as a precipitate when stoichiometric quantities of silver nitrate and a soluble cyanide are mixed. Silver(I) ion readily forms soluble complexes, ie, $Ag(CN)_2^-$ or $Ag(CN)_3^{2-}$, in the presence of excess cyanide ion.

Silver fluoride, AgF, is prepared by treating a basic silver salt such as silver oxide or silver carbonate, with hydrogen fluoride. Ultraviolet light or electrolysis decomposes silver fluoride to silver subfluoride, Ag_2F, and fluorine.

Other Silver Halogen-Containing Salts. All silver halides are reduced to silver by treating an aqueous suspension with more active metals, such as magnesium, zinc, aluminum, copper, iron, or lead. Photolyzed silver halides are also reduced by organic reducing agents or developers, eg, hydroquinone, *p*-aminophenol, and *p*-phenylenediamine, during photographic processing (see PHOTOGRAPHY).

Silver chlorate, $AgClO_3$, silver bromate, $AgBrO_3$, and silver iodate, $AgIO_3$, have been prepared. The halates may decompose explosively if heated.

Silver iodide, AgI, precipitates as a yellow solid when iodide ion is added to a solution of silver nitrate. It dissolves in the presence of excess iodide ion, forming an AgI_2 complex; however, silver iodide is only slightly soluble in ammonia and dissolves slowly in thiosulfate and cyanide solutions. Although silver iodide is the least photosensitive of the three halides, it has the broadest wavelength sensitivity in the visible spectrum. This feature makes silver iodide particularly useful in the photographic industry.

Silver nitrate, $AgNO_3$, is the most important commercial silver salt because it serves as the starting material for all other silver compounds. It is prepared by the oxidation of silver metal with hot nitric acid. The by-products are nitrogen oxides, NO and NO_2, which are vented to the atmosphere or scrubbed out of the fumes with an alkaline solution. The manufacture of silver nitrate for the preparation of photographic emulsions requires silver of very high purity. In the absence of organic matter, silver nitrate is not photosensitive. It is easily reduced to silver metal by glucose, tartaric acid, formaldehyde, hydrazine, and sodium borohydride.

Silver nitrite, $AgNO_2$, is prepared from silver nitrate and a soluble nitrite, or silver sulfate and barium nitrite.

Slightly soluble or insoluble silver salts are precipitated when mono- and dicarboxylic aliphatic acids or their anions are treated with silver nitrate solutions.

Silver oxide, Ag_2O, a dark brown-to-black material, is formed when an excess of hydroxide ion is added to a silver nitrate solution. Silver oxide can also be prepared by heating finely divided silver metal in the presence of oxygen. When heated to 100°C, silver oxide decomposes into its elements, and is completely decomposed above 300°C.

Silver permanganate, $AgMnO_4$, is a violet solid formed when a potassium permanganate solution is added to a silver nitrate solution.

Perhalates. Whereas silver perchlorate, $AgClO_4$, and silver periodate, $AgIO_4$, are well known, silver perbromate, $AgBrO_4$, has more recently been described. Silver perchlorate is prepared from silver oxide and perchloric acid, or by treating silver sulfate with barium perchlorate.

Silver phosphate, or silver orthophosphate, Ag_3PO_4, is a bright yellow material formed by treating silver nitrate with a soluble phosphate salt or phosphoric acid. Silver pyrophosphate, $Ag_4P_2O_7$, is a white salt prepared by the addition of a soluble pyrophosphate to silver nitrate.

Silver selenate, Ag_2SeO_4, is prepared from silver carbonate and sodium selenate (see SELENIUM AND SELENIUM COMPOUNDS).

Silver sulfate, Ag_2SO_4, is prepared by treating metallic silver with hot sulfuric acid. Alternatively, a solution of silver nitrate is acidified with sulfuric acid and the nitric acid is evaporated, leaving a solution of silver sulfate.

Silver sulfide, Ag_2S, forms as a finely divided black precipitate when solutions or suspensions of most silver salts are treated with an alkaline sulfide solution or hydrogen sulfide. Silver sulfide is one of the most insoluble salts known. Silver and sulfur combine even in the cold to form silver sulfide. The tendency of silver to tarnish is an example of the ease with which silver and sulfur compounds react.

Silver sulfite, Ag_2SO_3, is obtained as a white precipitate when sulfur dioxide is bubbled through a solution of silver nitrate.

Silver tetrafluoroborate, $AgBF_4$, is formed from silver borate and sodium borofluoride or bromine trifluoride.

Silver thiocyanate, AgSCN, is formed by the reaction of stoichiometric amounts of silver ion and a soluble thiocyanate.

Silver thiosulfate, $Ag_2S_2O_3$, is an insoluble precipitate formed when a soluble thiosulfate reacts with an excess of silver nitrate. In order to minimize the formation of silver sulfide, the silver ion can be complexed by halides before the addition of the thiosulfate solution. In the presence of excess thiosulfate, the very soluble $Ag_2(S_2O_3)_3^{4-}$ and $Ag_2(S_2O_3)_5^{6}$ complexes form. These soluble thiosulfate complexes, which are very stable, are the basis of photographic fixers. Silver thiosulfate complexes are oxidized to form silver sulfide, sulfate, and elemental sulfur (see THIOSULFATES).

Silver(I) Complexes

Silver ions form a number of complexes with both π-bonding and non-π-bonding ligands.

Ammonia and Amine Complexes. In the presence of excess ammonia (qv), silver ion forms the complex ions $Ag(NH_3)_2^+$ and $Ag(NH_3)_3^+$.

Cyanide Complexes. Insoluble silver cyanide, AgCN, is readily dissolved in an excess of alkali cyanide. The predominant silver species present in such solutions is $Ag(CN)_2^-$, with some $Ag(CN)_3^{2-}$ and $Ag(CN)_4^{3-}$ (see CYANIDES).

Halide Complexes. Silver halides form soluble complex ions, AgX_2^- and AgX_3^{2-}, with excess chloride, bromide, and iodide. The relative stability of these complexes is $I^- > Br^- > Cl^-$.

Olefin Complexes. Silver ion forms complexes with olefins and many aromatic compounds.

Sulfur Complexes. Silver compounds other than sulfide dissolve in excess thiosulfate. Stable silver complexes are also formed with thiourea. Except for the cyanide complexes, these sulfur complexes of silver are the most stable.

Other Oxidation States

Silver(II) Compounds. Silver(II) is stabilized by coordination with nitrogen heterocyclic bases, such as pyridine and dipyridyl. These cationic complexes are prepared by the peroxysulfate oxidation of silver(I) solutions in the presence of an excess of the ligand.

Silver(II) fluoride, AgF_2, is a brown-to-black hygroscopic material obtained by the treatment of silver chloride with fluorine gas. Silver(II) oxide, AgO, is prepared by persulfate oxidation of Ag_2O in basic medium. Silver(II) oxide is a strong oxidant.

Silver(III) Compounds. No simple silver(III) compounds exist. When mixtures of potassium or cesium halides are heated with silver halides in a stream of fluorine gas, yellow $KAgF_4$ or $CsAgF_4$, respectively, are obtained.

Analytical Test Methods

The classic method for the *qualitative* determination of silver in solution is precipitation as silver chloride with dilute nitric acid and chloride ion.

Classically, silver concentration in solution has been determined by titration with a standard solution of thiocyanate. Gravimetrically, silver is determined by precipitation with chloride, sulfide, or 1,2,3-benzotriazole. Highly sensitive instrumental techniques, such as x-ray fluorescence, atomic absorption spectrometry, and inductively coupled plasma optical emission spectrometry, have wide application for the analysis of silver in a multitude of materials.

Health and Safety Factors

Silver compounds that generate significant quantities of free silver ion in solution, eg, silver nitrate, can be toxic to bacteria and freshwater aquatic organisms. In 1989, the medical profession determined that the deposition of silver in internal organs and skin (argyria and argyrosis) did not impair the functions of the affected organs. Thus argyria and argyrosis are considered to be cosmetic effects, not adverse health effects.

Environmental Impact

The impact that a silver compound has in water is a function of the free or weakly complexed silver ion concentration generated by that compound, not the total silver concentration.

Free ionic silver readily forms soluble complexes or insoluble materials with dissolved and suspended material present in natural waters, such as sediments and sulfide ions. The hardness of water is sometimes used as an indicator of its complex-forming capacity. In the manufacture of photographic materials, silver is originally present as a halide. Before disposing of exhausted fixing baths, most of the silver is recovered, frequently by metallic exchange or by electrolytic reduction.

In secondary wastewater treatment plants receiving silver thiosulfate complexes, microorganisms convert this complex predominately to silver sulfide and some metallic silver (see WASTES, INDUSTRIAL). These silver species are substantially removed from the treatment plant effluent at the settling step. Any silver entering municipal secondary treatment plants tends to bind quickly to sulfide ions present in the system and precipitate into the treatment plant sludge. Thus, silver discharged to secondary wastewater treatment plants or into natural waters is not present as the free silver ion but rather as a complexed or insoluble species.

Uses

Analysis. The ability of silver ion to form sparingly soluble precipitates with many anions has been applied to their quantitative determination. Bromide, chloride, iodide, thiocyanate, and borate are determined by the titration of solutions containing these anions using standardized silver nitrate solutions in the presence of a suitable indicator. Silver diethyldithiocarbamate is a reagent commonly used for the spectrophotometric measurement of arsenic in aqueous samples and for the analysis of antimony. Silver iodate is used in the determination of chloride in biological samples such as blood. Combination silver–silver salt electrodes have been used in electrochemistry.

Batteries. Primary, ie, nonrechargeable, batteries containing silver compounds have gained in popularity through use in miniaturized electronic devices. The silver oxide–zinc cell has a cathode of Ag_2O or AgO (see BATTERIES, PRIMARY CELLS).

Catalysts. Silver and silver compounds are widely used in research and industry as catalysts for oxidation, reduction, and polymerization reactions.

Cloud Seeding. Cloud seeding with silver iodide has been used in weather modifications attempts such as increases and decreases in precipitation (rain or snow) and the dissipation of fog.

Electroplating. Most silver-plating baths employ alkaline solutions of silver cyanide. The silver is added to the solution either directly as silver cyanide or by oxidation of a silver-rod electrode (see ELECTROPLATING).

Medicinal Preparations. Silver nitrate is used in medicine in the form of a stick, usually containing 1–3% silver chloride, or in solutions of varying concentrations. Uses of silver in medicine are much reduced because of the availability of a broad spectrum of other remedies. However, silver preparations are not likely to cause sensitization.

Mirrors. The use of silver for the production of mirrors results in a highly reflective coating.

Photography. The largest single use of silver and its compounds is in the photographic industry. Silver nitrate and a halide salt of an alkali metal or an ammonium halide give a light-sensitive silver halide. The silver halide can account for up to 30–40% of the total emulsion weight. Gelatin is the other primary constituent (see EMULSIONS; PHOTOGRAPHY). Newer methods of image recording seek to avoid the high cost of silver. However, continued research has not led to systems that are able to offer the same combination of high sensitivity, high image density, exceptional resolution, permanence, and tricolor recording.

Other Uses. Photochromic glass contains silver chloride and silver molybdate (see CHROMOGENIC MATERIALS).

C. ROBERT CAPPEL
Eastman Kodak Company

A. Andren, ed., *Proceedings from the 1st International Conference on Silver: Fate, Transport & Toxicity in the Environment*, University of Wisconsin, Madison, 1994.

A. Andren, ed., *Proceedings from the 2nd International Conference on Silver: Fate, Transport & Toxicity in the Environment*, University of Wisconsin, Madison, 1995.

A. Andren, ed., *Proceedings from the 3rd International Conference on Silver: Fate, Transport & Toxicity in the Environment*, University of Wisconsin, Madison, 1996.

A. Andren, ed., *Proceedings from the 4th International Conference on Silver: Fate, Transport & Toxicity in the Environment*, University of Wisconsin, Madison, 1997.

SIMULTANEOUS HEAT AND MASS TRANSFER

Heat transfer and mass transfer occur simultaneously whenever a transfer operation involves a change in phase or a chemical reaction. Of these two situations, only the first is considered herein, because in reacting systems the complications of chemical reaction mechanisms and pathways are usually primary (see HEAT-EXCHANGE TECHNOLOGY). Even in processes involving phase changes, design is frequently based on the heat-transfer process alone; mass transfer is presumed to add no complications. But in fact mass transfer effects do influence and can even limit the process rate.

In processes where a condensing vapor or vapor from a liquid phase moves through an inert gas, eg, condensation in the presence of air, drying, humidification, crystallization (qv), and boiling of a multicomponent liquid, mass-transfer as well as heat-transfer effects are important (see AIR CONDITIONING; DISTILLATION; EVAPORATION).

Condensation and Vaporization as Effected by Simultaneous Heat and Mass Transfer

An interphase transfer occurs when one or more components change phase in the presence of inert or less active components. The transferring component must be transported through its original phase to the boundary, and must then escape into the second phase. The phase change involves a heat effect. Energy is transported to or from the boundary to balance the phase-change heat effect. The boundary temperature is influenced by the rate of heat transfer, and this determines the fugacity of the diffusing component at the boundary. Thus, to describe the process, rate equations for heat transfer and mass transfer must be written along with material balances for the components present and an energy balance. Appropriate boundary conditions must be applied, and the resulting set of differential and algebraic equations solved. The rate equations for heat and mass transfer express the rate of transport in terms of the driving force divided by the resistance across the transfer path. For heat transfer the driving force is expressed as a temperature difference, whereas the resistance is the reciprocal of the transport area times a coefficient. Here the concentration driving force should be a fugacity or activity difference, with the coefficient in consistent units. However, these properties are not directly measurable, so the driving force is expressed in terms of mole fractions, partial pressures, or mole ratios. The use of these terms requires the use of consistent coefficient values and limits the usefulness of the equations to systems that obey Raoult's law, or requires the use of empirical nonideality coefficients. Herein it is assumed that Raoult's law holds.

This process has been used for various situations. For the condensation of a single component from a binary gas mixture, the gas-stream sensible heat and mass-transfer equations for a differential condenser section take the following forms:

$$G \cdot C_p \frac{dT_G}{dA} = -h_c \cdot (T_g - T_s) \frac{\epsilon}{e^\epsilon - 1} \quad (1)$$

$$\frac{dV}{dA} = -kg \cdot (P_g - P_s) \quad (2)$$

No condensation is taking place here in the bulk gas phase. If condensation does take place so that fogging occurs, these equations become

$$G \cdot C_p \frac{dT_G}{dA} = -h_c \cdot (T_g - T_s) \frac{\epsilon}{e^\epsilon - 1} + \lambda \frac{dF}{dA} \quad (3)$$

$$\frac{dV}{dA} = -k_g \cdot (P_g - P_s) - \frac{dF}{dA} \quad (4)$$

The term $\epsilon/(e^\epsilon - 1)$ in equations 1 and 3 was developed to account for the sensible heat transferred by the diffusing vapor. The quantity ϵ represents the group $M_i \cdot C_{pi}/h_c$, the ratio of total transported energy to convective heat transfer. Thus it may be thought of as the fractional influence of mass transfer on the heat-transfer process. The last term of equation 3 is the latent heat contributed to the gas phase by the fog formation. The vapor loss from the gas phase through both surface and gas-phase condensation can be related to the partial pressure of the condensing vapor by using Dalton's law and a differential material balance.

The effect on the coolant temperature of latent and sensible heat transferred to the surface from the condensing vapor is as shown in equation 5:

$$L \cdot M_L \cdot C_w \frac{dT_w}{dA} = \pm h_o \cdot (T_s - T_w) \quad (5)$$

where the $\pm$ sign is negative for countercurrent flow.

Assuming a linear relation between surface temperature and corresponding vapor pressure of the condensable component allows a heat balance to be written from gas phase to the surface:

$$h_o \cdot (T_s - T_w) = h_c(T_g - T_s) \frac{\epsilon}{1 - e^\epsilon} + k_g \cdot \lambda (P_g - P_s) \quad (6)$$

Combining equation 6 with the heat- and mass-transfer rate expressions gives

$$w \cdot C_w \frac{dT_w}{dA} = e^{\epsilon} \cdot G \cdot C_p \frac{dT_g}{dA} + \frac{V' \cdot \lambda P}{(P_t - P_{go})(P_t - P_{gf})} \cdot \frac{dP_g}{dA} \qquad (7)$$

Equations 6 and 7 are not affected by fogging because the latent heat thus obtained is retained as sensible heat in the gas phase.

These basic relations have been solved for a wide range of cooler–condenser conditions and for different complexities of systems (see also HEAT-EXCHANGE TECHNOLOGY, HEAT TRANSFER).

Description of Gas–Vapor Systems

In engineering applications, the transport processes involving heat and mass transfer usually occur in process equipment involving vapor–gas mixtures where the vapor undergoes a phase transformation, such as condensation to or evaporation from a liquid phase. In the simplest case, the liquid phase is pure, consisting of the vapor component alone.

The system of primary interest, then, is that of a condensable vapor moving between a liquid phase, usually pure, and a vapor phase in which other components are present. Some of the gas-phase components may be noncondensable. A simple example would be water vapor moving through air to condense on a cold surface. Here the condensed phase, characterized by T and P, exists pure. The vapor-phase description requires y, the mole fraction, as well as T and P. The nomenclature used in the description of vapor-inert gas systems is given in Table 1.

The humidity term and such derivatives as relative humidity and molal humid volume were developed for the air–water system. Use is generally restricted to that system. These terms have also been used for other vapor–noncondensable gas phases.

For the air–water system, the humidity is easily measured by using a wet-bulb thermometer. Air passing the wet wick surrounding the thermometer bulb causes evaporation of moisture from the wick. The balance between heat transfer to the wick and energy required by the latent heat of the mass transfer from the wick gives, at steady state,

$$-k_Y \cdot A \cdot (Y_1 - Y_w) \cdot \lambda_w = (h_c + h_r) \cdot A \cdot (T_1 - T_w) \qquad (8)$$

$$T_1 - T_w = \frac{k_Y \lambda_w}{(h_c + h_r)}(Y_w - Y_1) \qquad (9)$$

If radiant energy transfer can be prevented, the following equation is used:

$$T_1 - T_w = \frac{k_Y \cdot \lambda_w}{h_c}(Y_w - Y_1) \qquad (10)$$

Thus, a measurement of the wet-bulb temperature, T_w, and the temperature T_1, allows the molal humidity, Y_1, to be calculated because Y_w is known. The use of molal humidity as the mass-transfer driving force is conventional and convenient because of the development of humidity data for, especially, the air–water system. The mass-transfer coefficient must be expressed in consistent units.

Calculations for Humidification and Dehumidification Processes

Figure 1 shows the general arrangement and nomenclature for a humidification or dehumidification process, where the subscript 1 refers to the bottom of the column, and subscript 2 to the top. Steady state is assumed. Flow rates and compositions are given in molar terms because this simplifies the results.

Total material, condensable component, and energy balances can be written for the entire column:

$$L_1 - L_2 = V_1 - V_2 \qquad (11)$$

$$V' \cdot (Y_2 - Y_1) = L_2 - L_1 \qquad (12)$$

$$L_2 \cdot H_{L_2} + V' \cdot H_{V_1} + q = L_1 \cdot H_{L_1} + V' \cdot H_{V_2} \qquad (13)$$

Table 1. Definitions of Humidity Terms

Term	Meaning	Units	Symbol[a]
humidity	vapor content of a gas	mass vapor/mass noncondensable gas	$Y' = Y\frac{M_a}{M_b}$
molal humidity	vapor content of a gas	mol vapor/mol noncondensable gas	Y
relative sat, rh	ratio of partial pressure of vapor to partial pressure of vapor (sat)	kPa/kPa, or mol fraction/mol fraction, %	$\frac{y}{y_s} \times 100$
% sat, % humidity	ratio of vapor conc to vapor conc at sat; conc expressed as mol ratios	mole ratio/mole ratio %	$\frac{Y}{Y_s} \times 100$
molal humid volume	vol of 1 mol of dry gas plus associated vapor	$m^3/mol^{a,b}$	$V_h = (1 + Y) \times 0.0224\frac{T}{273} \times 1.013P^{-1}$
molal humid heat	heat required to raise the temp of 1 mol of dry gas + associated vapor 1°C	J/(mol·°C)[a]	$c_h = c_b + Y_{ca}$
adiabatic sat temp	temp attained if gas were saturated adiabatically	°C or K	T_{sa}
wet-bulb temp	steady-state temp attained by wet-bulb thermometer (standard cond)	°C or K	T_w
dew-point temp	temp when vapor condenses if gas phase is cooled at constant pressure	°C or K	T_d

[a] Concentration (conc) is on the basis of dry gas. [b] When T is in K and P is in Pa.

Generally, q is small because the outside area is not large in comparison to the amount of heat being transferred, and the energy balance can be simplified. In these conditions it is also convenient to write balances over a differential section of the column. These balances yield the following:

$$V' \cdot dY = dL \qquad (14)$$

$$V' \cdot dH_V = d(L \cdot H_L) \qquad (15)$$

If the amount of evaporation is small, the change in enthalpy in the liquid phase can be taken as a result of temperature change alone.

Determination of the Gas-Phase Temperature. Often the temperature of the vapor phase is important to the designer, either as one of the variables specified or as an important indicator of fogging conditions in the column. Such a condition would occur if the gas temperature equaled the saturation temperature, that is, the interface temperature. When fogging does occur, the column can no longer be expected to operate according to the relations presented herein but is basically out of control.

The temperature and enthalpy values of the bulk gas phase continuously approach the interface condition at the same point in the column as that for which the gas-phase conditions apply. This method allows the development of a conceptual understanding of the limits of operation of a humidification column. For actual design, the simplifications may be avoided by handling the fundamental equations numerically by computer.

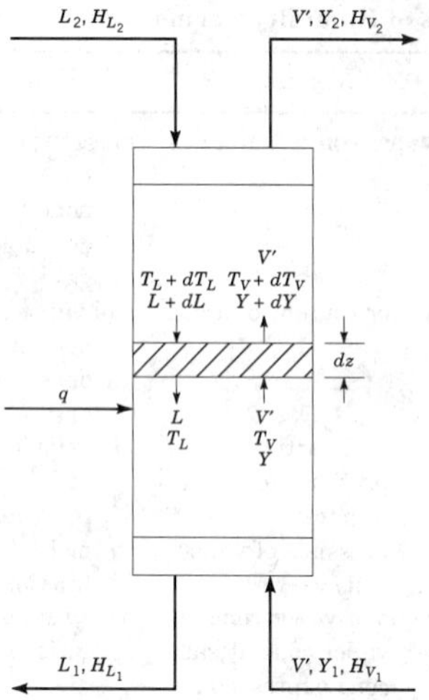

Figure 1. Arrangement and nomenclature for general humidification–dehumidification process.

Transfer Coefficient. The design method described depends for its utility on the availability of mass- and heat-transfer coefficients. Typically, $k_Y \cdot a$ and $h_L \cdot a$ are needed. These must be obtained from the standard correlations for mass and heat transfer, from data reported in the literature, or from data presented by equipment makers for particular packing. When this type of information is not available, it is possible to determine heat- and mass-transfer coefficients by a single test using the packing material of interest in a pilot-sized tower. If a steady state is obtained, measurement of air- and water-inlet and -outlet temperatures, and air-inlet and -outlet wet- and dry-bulb temperatures comprises all the necessary information.

Humidification and Dehumidification Equipment

The addition or removal of a condensable component to or from a noncondensable gas can be accomplished by direct contact between the vapor and the gas. This may be done in a countercurrent tower (see ADSORPTION; DISTILLATION). The direction of transfer depends on the temperatures of the two streams. Such operations can also be done using spray ponds in which a grid of nozzles sprays liquid, usually water, into the gas phase, usually air. If the air is relatively dry, liquid evaporates into it, both humidifying the air and cooling the liquid. If a large surface of water is available, the same process may be carried out through evaporation from the surface of lake or pond. Usually the purpose is the cooling of process water.

Water-Cooling Towers. By far the most common and large-scale mode of humidification processing is in water-cooling towers. As supplies of cooling water become more strained, and as discharge water temperatures are more closely controlled, water cooling and recirculation rather than once-through water use become more common. Two general types of direct-contact cooling towers are in use: forced-draft and natural-draft. The forced-draft tower depends on fans to move the air through the tower. Typically, the tower consists of a set of louvres and baffles over which the water falls, breaking into films and droplets. Air flow may be across this cascading liquid or countercurrent of it.

In a natural-draft cooling tower, the driving force for the air is provided by the buoyancy of the air column in a very tall stack. Stack heights of 100 m are common, and as power-plant sizes increase, the size of single towers is likely to increase also. In the absence of a fan, the air flow rate, G, is no longer an independent variable, but is dependent on the design and operating conditions of the tower. The governing equation for air flow becomes

$$z_t \cdot \Delta\rho = N \frac{G^2}{\rho g_c} \qquad (16)$$

where $\Delta\rho$ is the average density difference between the outside air and the air in the stack; z_t is the height of the tower, and N, the resistance to air flow through the tower in velocity heads, is specific for a given tower and can usually be expressed as a constant.

Trends in Cooling-Tower Use and Development. Natural-draft cooling towers had been rare in the United States because ample water supplies were available for power-plant cooling, and natural-draft towers are best suited for large heat loads. Mechanical-draft towers were used in large numbers for industrial applications and occasionally for power plants. However, almost all large post-1980 power plants require cooling towers. Limitations on use of cooling water and the return of warm water to rivers has forced the use of all cooling towers. Towers allow the heat load to be dissipated to the air rather than into natural water. The treating of recycled cooling water may pose hazards to aquatic animals, crops, etc, near the tower.

Cooling-Tower Plumes. An important consideration in the acceptability of either a mechanical-draft or a natural-draft tower cooling system is the effect on the environment. The plume emitted by a cooling tower is seen by the surrounding community and can lead to trouble if it is a source of severe ground fog under some atmospheric conditions. The natural-draft tower is much less likely to produce fogging than is the mechanical-draft tower. Nonetheless, it is desirable to devise techniques for predicting plume trajectory and attenuation.

Not only may the cooling-tower plume be a source of fog, which in some weather conditions can ice roadways, but the plume also carries salts from the cooling water itself. These salts may come from salinity in the water, or may be added by the cooling-tower operator to prevent corrosion and biological attack in the column.

Efforts to combat bacteria and corrosion in cooling towers have gone through a long development and are both complex and specific to different waters. Systems may include chromates, nitrites, orthophosphates, and ferrocyanides as cathodic inhibitors; zinc, nickel, lead, tin, copper, and silicates as anodic inhibitors; and possibly added materials as biocides and pH controllers. These chemicals can also be carried onto fields surrounding the cooling tower, seriously affecting crop yield or ornamental plantings (see WATER, INDUSTRIAL WATER TREATMENT).

Trends

Cooling water is a necessity for temperature control. Most industrially generated heat must be dissipated; water is an obvious receptor because heat transfer is relatively rapid and compact. When the heat load is large it is usually necessary to cool and reuse the water. Thus cooling towers are integral parts of power plants, chemical processing operations, and compression steps, and proper design and operation are critical to the entire process.

Simultaneous heat and mass transfer also occurs in drying processes, chemical reaction steps, evaporation, crystallization, and distillation. In all of these operations transfer rates are usually fixed empirically. The process can be evaluated using either the heat- or mass-transfer equations. However, if the process mechanism is to be fully understood, both the heat and mass transfer must be described (see PROCESS ENERGY CONSERVATION).

NOMENCLATURE

Symbol	Definition
A	interfacial area
a	interfacial area/unit column vol
C	heat capacity
F	rate of fog formation
G	mass flow rate of gas phase
g_c	force–mass conversion constant
H	enthalpy
h_c	convective heat-transfer coefficient
h_r	coefficient for heat transsfer by radiative mechanism

k_g	gas-phase mass-transfer coefficient in partial pressure driving force units
k_L	liquid-phase thermal conductivity
k_Y	mass-transfer coefficient in gas-phase mole ratio units
L	liquid stream molar flow rate
L_e	Lewis number
M	mol wt
N	resistance to air flow in velocity heads
NTU	number of transfer units
P	total pressure
q	heat flux
r	psychrometric ratio
T	temperature
V	specific volume
V	gas-phase molar flow rate
V'	molar flow rate of noncondensable component
w	total flow rate
Y	mole ratio
Y'	mass ratio
y	mole fraction in gas phase
z	height of column
ϵ	Ackerman correction term, $= m_i \cdot c_{pi}/h_g$
λ	latent heat of vaporization
	Subscripts
BM	mean value for noncondensing component
e	effective
g	gas phase
h	humid value, including gas and vapor
i	interface condition
L	liquid phase
o	at reference condition
p	at constant pressure
s	at saturation
V	in the vapor phase
w	for water, or at wet-bulb temperature
Y	the mole ratio driving force
$1,2$	end points in the process
	Superscript
$'$	mass rather than mole basis

LEONARD A. WENZEL
Lehigh University

H. S. Mickley, *Chem. Eng. Prog.* **45**, 739 (1949).

Cooling Towers, CEP Technical Manual, American Institute of Chemical Engineering, New York, 1972.

R. E. Treybal, *Mass Transfer Operations*, 2nd ed., McGraw-Hill Book Co., Inc., New York, 1968.

A. S. Foust and co-workers, *Principles of Unit Operations*, 2nd ed., John Wiley & Sons, Inc., New York, 1980, Chapt. 17.

SINGLE CRYSTALS, GROWTH. See ZONE REFINING.

SIZE ENLARGEMENT

Size enlargement processes bring together fine powders into larger masses in order to improve the powder properties. Usually, they produce relatively permanent entities in which the original particles can still be identified.

Particle-Bonding Mechanisms

The mechanisms by which particles bond together and grow into agglomerates are affected by the specific size enlargement method. Nevertheless there are certain aspects of the bonding process that are essentially independent of the equipment and method (see COATING PROCESSES, POWDER TECHNOLOGY).

A classification of bonding mechanisms based on the fundamental nature of the interparticle bonds has been widely adopted in the literature. This classification into five categories of particle–particle bridging is summarized in Table 1.

Table 1. Classification of Binding Mechanisms[a]

Class	Mechanism	Representative examples
solid bridges	sintering, heat hardening	induration of iron-ore pellets
	chemical reaction, hardening binders, "curing"	cement binder for flue-dust pellets
	incipient melting owing to pressure, friction	briquetting of metals, plastics
	deposition through drying	crystallization of salts in fertilizer granulation
immobile liquids	viscous binders, adhesives	sugars, glues, gums in pharmaceutical tablets
	adsorption layers	instantizing food powders by steam condensation
mobile liquids	liquid bridges (pendular state)	flocculation of fine particles in liquid suspension by immiscible liquid wetting
	void space filled or partly filled with liquid (capillary and funicular states)	balling (wet pelletization of ores)
intermolecular and long-range forces	van der Waals, electrostatic, and magnetic	adhesion of fine powders during storage, flow, and handling
mechanical interlocking	shape-related bonding	fracturing and deformation of particles under pressure

[a] According to H. Rumpf.

Theoretical Strength of Agglomerates. Based on statistical-geometrical considerations, H. Rumpf developed the following equation for the mean tensile strength of an agglomerate in which bonds are localized at the points of particle contact:

$$\sigma_T \approx \frac{9}{8}\left(\frac{1-\epsilon}{\epsilon}\right)\frac{H}{d^2} \tag{1}$$

where σT is the mean tensile strength per unit section area, Pa; ϵ is the void fraction of the agglomerate; d is the diameter of the (assumed) monosized spherical particles in m; and H is the tensile strength in N, of a single particle–particle bond. To convert Pa to psi, multiply by 0.145×10^{-3}.

In a second main class of agglomerate bonding, particles are embedded in or surrounded by an essentially continuous matrix of binding material rather than having bonding localized at points of particle contact. An important example of this second type of binding is the case in which particles are held together by mobile liquids where adhesion occurs as a result of interfacial tension at the liquid surface and the pressure deficiency (capillary suction) created within the liquid phase by curvature at the liquid surface.

It has been demonstrated that the viscosity of the binder liquid plays a key role during agglomerate formation, where it provides strength under dynamic strain, such as would occur with intensive mixing. It has been found that under industrially relevant conditions, the strength of the dynamic bridge exceeds the static strength by more than an order of magnitude.

Measuring and Correlating Agglomerate Strength. In practice, simple and quick test methods are most often used to assess the quality of bonding and other desirable properties of product agglomerates.

Compression, impact, abrasion, and other types of tests are widely accepted in industry to characterize agglomerate quality in relation to subsequent handling and processing.

Compression tests, in which agglomerates are crushed between parallel platens, are probably most universal.

For approximately spherical agglomerates, compression strength (σ_C) is calculated as follows:

$$\sigma_C = L/(\pi D^2/4) \qquad (2)$$

where L is the compression force at failure in N, and D is the agglomerate diameter in m. Because agglomerates often fail undertension, correlations between tensile and compressive strength are important.

Types of Size Enlargement

Classification of size enlargement methods reveals two distinct categories. The first is forming-type processes in which the shape, dimensions, composition, and density of the individual larger pieces formed from finely divided materials are of importance. The second is those processes in which creation of a coarse granular material from fines is the objective, and the characteristics of the individual agglomerates are important only in their effect on the properties of the bulk granular product.

Four principal mechanisms are used to bring fine particles together into larger agglomerates: agglomeration by tumbling and other agitation methods; pressure compaction and extrusion methods; heat reaction, fusion, and drying methods; and agglomeration from liquid suspensions.

Agglomeration by Tumbling and Other Agitation Methods

When fine particles, usually in a moist state, are brought into intimate contact through agitation, binding forces come into action to hold the particles together as an agglomerate. Capillary binding forces caused by wetting with water or aqueous solutions is the most common binding mechanism, but others such as intermolecular forces developed in extremely fine dry powders may also be used to form the agglomerates. Several different forms of agitation may be used. The rolling cascading action of disk and drum devices produces rounded or roughly spherical agglomerates; other types of mixers generally yield more irregular shapes.

Agglomerate growth can occur through a number of mechanisms, such as coalescence, crushing and layering, layering of fines, and abrasion transfer. More than one mechanism may occur simultaneously in a given process. To survive and grow in an agitated system, agglomerates must be able to withstand the destructive forces generated by the moving charge of powder.

Drum and Inclined-Disk Agglomerators. Although a wide variety of agitation equipment is used industrially to produce agglomerates, rotary drums or cylinders and inclined disks or pans are the most important. Drum agglomerators consist of an inclined rotary cylinder powered by a fixed- or variable-speed drive. Feed material, containing the correct amount of liquid phase, agglomerates under the rolling, tumbling action of the rotating drum. The pitch of the drum (up to 10° from the horizontal) assists material transport down the length of the cylinder.

An inclined-disk agglomerator consists of a tilted rotating plate equipped with a rim to contain the agglomerating charge. Solids are fed continuously from above or from the front onto the central part of the disk, and product agglomerates discharge over the rim. Moisture or other binding agents can be sprayed on at various locations on the plate surface. Adjustable scrapers and plows maintain a uniform protective layer of product over the disk surface and also control the flow pattern of material on the disk.

Operation and control of tumbling agglomerators is affected primarily by the character of the feed powder (size distribution, solubility), by its optimum liquid content for agglomeration, and by retention time in the device. An important parameter for drum agglomeration is the amount of undersize product that must be recycled. An important feature of the inclined-disk agglomerator is its size separating ability. Feed particles and the smaller agglomerates sift down to the bottom of the tumbling load where, because of their high coefficient of friction, they are carried to the highest part of the disk before rolling downward in an even stream. Larger agglomerates remain closer to the top of the bed, where they travel shorter paths.

The inherent classifying action of inclined disks offers an advantage in applications that require accurate agglomerate sizing. Advantages claimed for the drum compared with the disk agglomerator are greater capacity, longer residence time for difficult materials, and less sensitivity to upsets owing to the damping effect of a larger recirculating load.

Mixer Agglomerators. Mixers are used by various industries in which, by contrast with drum or disk equipment, internal agitators of several designs provide a positive rubbing and shearing action to accomplish both mixing and size enlargement. Horizontal pan mixers, pugmills, and other types of intensive agitation devices are used. The positive cutting-out action of such equipment can handle plastic and sticky powder feeds and its kneading action is claimed to produce denser and stronger granules than the tumbling methods although agglomerates of more irregular shape usually result (see MIXING AND BLENDING).

Pug mixers (blungers, pugmills, paddle mixers) consist of a horizontal trough with a rotating shaft to which mixing blades or paddles of various designs may be attached. Twin shafts rotate in opposite directions throwing the materials forward and to the center as the pitched blades on the shaft pass through the charge.

Shaft mixers operating at very high rotational speeds are also used to granulate extreme fines, such as clays and carbon black, which may be highly aerated when dry, and plastic or sticky when wet. These machines are generally single-shaft devices in which the paddles are replaced by a series of pins, pegs, or blades.

Powder Clustering. Many applications of size enlargement require only relatively weak, small, cluster-type agglomerates to improve behavior of the original powder in flow, wetting, dispersion, or dissolution. Tableting feeds in pharmaceutical manufacture, detergent powders, and "instant" food products are examples. In these cases, agglomeration is accomplished by superficially wetting the feed powder. Continuous-flow mixing systems are commonly used in the agglomeration of powdered food products (see FOOD PROCESSING).

Pressure Compaction and Extrusion Methods

The compression techniques of size enlargement produce agglomeration by application of suitable forces to particulates held in a confined space. In roll-pressing equipment, the particulate material is compacted by squeezing as it is carried into the gap between two rotating rolls. In extrusion systems the particulates undergo shearing and mixing as they are consolidated while being forced through a die or orifice under the action of a screw or roller.

Tableting, pressing, molding, and extrusion operations are commonly used to produce agglomerates of well-defined shape, dimensions, and uniformity (see CERAMICS, CERAMICS PROCESSING; PHARMACEUTICALS; METALLURGY, POWDER METALLURGY; PLASTICS PROCESSING).

The compaction process of void reduction may be considered to occur by two essentially independent mechanisms. The first is the filling of the holes of the same order of size as the original particles. The second process consists of the filling of voids that are substantially smaller than the original particles by plastic flow or by fragmentation. The success of the compaction operation depends partly on the effective utilization and transmission of applied forces and partly on the physical properties and condition of the mixture being compressed.

Compacting Presses. In the automotive industry and other metalworking industries, coarse scrap-metal particulates are compressed and recycled to melting operations through piston-type briquetting presses. Feed materials are typically cast-iron and steel borings or turnings, which tend to bond under pressure at least partially by mechanical interlocking. Briquettes of cylindrical shape 7.6 cm in length are produced at ca 3–4 t/h.

Roll Briquetting and Compacting Machines. In roll presses, particulate material is compacted by squeezing as it is carried into the gap between two rolls rotating at equal speed. This is probably the most versatile method of size enlargement, because most materials can be agglomerated by this technique with the aid of binders, heat, and/or very high pressures if needed. The method generally requires less binder and therefore there is little or no requirement for drying the agglomerates.

Roll presses consist of the frame, the two rolls that do the pressing, and the associated bearings, reduction gear, and fixed or variable-speed drive. For fine powders that tend to bridge or stick and are of low bulk density, some form of forced feed must be used.

Pellet Mills. Pellet mills differ from roll briquetting and compacting machines in that the particulates are compressed and formed into agglomerates by extrusion through a die rather than by squeezing as they are carried into the nip between two rolls. The action of the roller and die assembly to produce a shearing and mixing action yields a plastic mix to be pushed through the die. Binders, plasticizing agents, and lubricants may be used to facilitate the process. Scores of materials can be pelleted, from catalysts and carbon materials to rubber crumb, wood pulp and bark, compound animal feeds, and many chemicals.

Heat Reaction, Fusion, and Drying Methods

These methods of size enlargement depend on heat transfer to accomplish particle bonding. Heat may be transferred to the particle agglomerates, as in the drying of a concentrated slurry or paste, the fusion of a mass of fines, or by chemical reaction into agglomerates suitable for drying, firing, or chilling.

Sintering and Pelletizing. In extractive metallurgy (see METALLURGY–EXTRACTIVE METALLURGY), sintering and pelletizing processes have been developed to allow processing of fine ores, concentrates, and recyclable dusts. In this connection, sintering refers to a process in which fuel (5–10% coke fines) is mixed with an ore and burned on a grate. A cake of hardened porous material is the resulting agglomerate. Four separate sequential processes take place during these high temperature operations: drying, preheating, firing or high temperature reaction, and cooling. Simultaneous useful processes may also occur, such as the elimination of sulfur and the decomposition of carbonates and sulfates.

Drying and Solidification on Surfaces. In this type of equipment, granular products are formed directly from fluid pastes and melts, without intermediate preforms, by drying or solidification on solid surfaces. Drum dryers consist of one or more heated metal rolls on which solutions, slurries, or pastes are dried in a thin film. Drying takes place in less than one revolution of the slowly revolving rolls, and a doctor blade scrapes the product off in flake, chip, or granular form. Molten materials can also be cooled to solid products on endless-belt systems.

Suspended Particle Techniques. In these methods, granular solids are produced directly from a liquid or semiliquid phase by dispersion in a gas to allow solidification through heat and/or mass transfer. The feed liquid, which may be a solution, gel, paste, emulsion, slurry, or melt, must be pumpable and dispersible. Equipment used includes spray dryers, prilling towers, spouted and fluidized beds, and pneumatic conveying dryers. In the suspension methods, agglomerate formation occurs by hardening of feed droplets into solid particles, by layering of solids deposited from the feed onto existing nuclei, and by adhesion of small particles into aggregates as binding solids from the dispersed feed are deposited (see DRYING).

In spray drying, the largest particles produced are normally ca 1 mm diameter; however, larger agglomerated particles can be produced, and this technique is used to produce granular dried products in the pharmaceuticals and ceramics industries as well as coarse food powders with instant properties.

Spray cooling or solidification, more commonly known as prilling, is similar to spray drying in that liquid feed is dispersed into droplets at the top of a chamber. These congeal into a solid granular product as they travel down the chamber. Unlike in spray drying, the liquid droplets are produced from a melt that solidifies primarily by cooling in the chamber with little, if any, drying.

Agglomeration from Liquid Suspensions

Size enlargement of particles contained in liquids is a frequent aid to other operations such as filtration, dewatering, settling, etc. Flocculation procedures are the traditional means used to promote such size enlargement (see FLOCCULATING AGENTS); the product is usually in the form of loose aggregates of an open network structure.

Agglomeration by Competitive Wetting. Fine particles in liquid suspension can readily be formed into large dense agglomerates of considerable integrity by adding a second or bridging liquid under suitable agitation conditions. This second liquid should be effectively immiscible with the suspending liquid and must preferentially wet the solid particles to be agglomerated. A simple example is the addition of oil to an aqueous suspension of fine coal. The oil readily adsorbs preferentially on the carbon particles and forms liquid bridges between these particles by coalescence during the collisions produced by agitation. Inorganic impurity (ash) particles are not wetted by the oil and remain in unagglomerated form in the aqueous slurry (see COAL CONVERSION PROCESSES, CLEANING AND DESULFURIZATION).

As in the flotation process, selective agglomeration by immiscible liquids depends on the relative wettability of surfaces, and the same fundamentals of surface chemistry apply to the conditioning of particles to yield the required affinity for the wetting liquid (see FLOTATION).

A most useful feature of the agglomeration technique is its ability to work with extreme fines. Even particles of less than nanometer size (ca 10^{-10} m) can be treated, if appropriate, so that ultrafine grinding can be applied to materials with extreme impurity dissemination to allow recovery of agglomerates of higher purity.

Liquid Waste Treatment. Persistent crude petroleum–water emulsions are produced in the recovery and extraction of heavier oils, eg, in the hot water processing of surface-mined oil sands and during *in situ* methods such as steam or water flooding. Oily sludges, formed from water contaminated by fine solids or soil and crude petroleum or bitumens, are an undesirable by-product of most oil recovery or refinery operations.

Liquid-phase agglomeration can play a significant role in oily waste treatment. In liquid waste treatment, the waste stream provides the immiscible liquids, and it is then necessary to provide the solid adsorbent for the oil. Dispersed, emulsified, and dissolved oils are all susceptible to removal in various degrees by this method. This approach to liquid waste treatment requires synergy between different process stages or between various industries (industrial ecology). Suitably matched solid adsorbents and waste liquids must be available so that selective wetting and agglomeration take place.

Spherical Oxide Fuel Particles. The sol–gel process is a related technique which has been actively developed for the preparation of spherical oxide fuel particles, up to ca 1 mm diameter, for nuclear reactors. In agglomeration by immiscible liquid wetting, small amounts of a bridging phase adsorbed on the particles coalesce to draw the particles into larger entities. In the sol–gel process, fine particles are initially suspended in an excess of a bridging phase, the suspension is formed into spherical droplets and the excess bridging phase is removed to solidify the droplets into a particulate product.

Soil Remediation. There are many sources of oil-contaminated soils as a result of petrochemical spills and industrial activity in general. For high levels of oil loading (eg, greater than a few weight percent), solvent extraction methods provide important cleaning process options, but have difficulty in handling fine, clay-containing soils because of difficulties in separating extreme fines from the extraction solvent.

Liquid-phase agglomeration can overcome the fines separation problem by forming aggregates of controlled size tailored to the chosen downstream separation method. A size range of 0.5–2 mm has been recommended for soil remediation. This provides optimal drainage and aeration for subsequent bioremediation, which may be needed as a final polishing step. In addition, this size mimics the natural soil size distribution which is important in returning the clean aggregates to the remediated site for fertile agricultural use.

C. EDWARD CAPES
K. DARCOVICH
National Research Council of Canada

W. Pietsch, *Size Enlargement by Agglomeration*, John Wiley & Sons, Inc., New York, 1991.

C. E. Capes, *Particle Size Enlargement*, Elsevier, Amsterdam, the Netherlands, 1980.

SIZE MEASUREMENT OF PARTICLES

The size distribution of particles in a wide variety of particulate systems is of paramount importance in the chemical processing industries.

A particle is a single unit of material having discrete physical boundaries which define its size, usually in micrometers, μm (1 μm(10^4 Å) = 1×10^{-4} cm = 1×10^{-6} m). The size of a particle is usually expressed by the dimension of its diameter. Typically, particle science is limited to particulate systems within a size range from 10^4 to 10^{-2} μm.

A limitation of the linear dimensional size descriptor is that only particles having simple or defined shapes, such as spheres or cubes, can be uniquely defined by a linear dimension. The common solution to this problem is to describe a nonspherical particle to be equivalent in diameter to a sphere having the same mass, volume, surface area, or settling speed (uniquely defined parameters) as the particle in question. Therefore, a particle can be described as behaving as a sphere of diameter d.

Data Representation

Particulate systems composed of identical particles are extremely rare. It is therefore useful to represent a polydispersion of particles as sets of successive size intervals, containing information on the number of particle, length, surface area, or mass. The entire size range, which can span up to several orders of magnitude, can be covered with a relatively small number of intervals. This data set is usually tabulated and transformed into a graphical representation. Size distributions can also be reduced to a single average diameter, such as the mean, median, or mode.

Distribution Averages. The most commonly used quantities for describing the average diameter of a particle population are the mean, mode, median, and geometric mean. The mean diameter, d, is statistically calculated and in one form or another represents the size of a particle population. It is useful for comparing various populations of particles.

The simplest calculation of the mean, referred to as arithmetic mean (count mean diameter) for data grouped in intervals, consists of the summation of all diameters forming a population, divided by the total number of particles. It can be expressed mathematically by equation 1:

$$\bar{d} = \frac{\sum n_i d_i}{\sum n_i} \qquad (1)$$

where n_i is the number of particles in group i having midpoint diameter d_i. The most appropriate definition of the mean diameter to be used in any specific application should be the one corresponding most closely to the relevant property of the particle system under study.

The mode of distribution is simply the value of the most frequent size present. The median particle diameter is the diameter which divides half of the measured quantity (mass, surface area, number), or divides the area under a frequency curve in half. The median, a useful measure of central tendency, can be easily estimated, especially when the data are presented in cumulative form. Another frequently used average is the geometric mean. The geometric mean diameter, d_g, is calculated using the logarithm values of the measured diameters:

$$\ln d_g = \frac{\sum n_i \ln d_i}{N} \qquad (2)$$

Tabular. A typical distribution as measured by modern instrumentation can include size information on tens of thousands and even millions of individual particles. These data can be listed in a computer and then sorted into a series of successive size intervals, keeping track of the measured quantity, such as number, surface area, or mass, within each group. For narrow size distributions it may be sufficient to group the data in linear intervals, such as 0–1, 1–2, 2–3 μm, etc, and then list the intervals as a percent value of the whole.

Classifying data on a geometric scale, eg, 1–2, 2–4, 4–8, 8–16, has the virtue of maintaining a consistent band resolution over the entire distribution. Typical particle size data are given in Table 1, along with the percent represented by each interval. This interval can be based on the total number of particles measured, the total sample weight, total volume, or any other basis upon which data might be acquired.

Graphical. A tabular presentation offers the ultimate in precision because all data are included. It is, however, inconvenient to compare tables. A graph usually offers advantages. The two common types of plots are the frequency histogram, sometimes referred to as differential plots, and the cumulative frequency plots. A histogram provides a simple means for graphically presenting size distributions. Normally, the percentage of particles in a given size interval, called the frequency, is plotted against size as shown in Figure 1. It is standard practice to plot data that have been grouped in geometric progression on a logarithmic scale (Fig. **1b**). Histograms are particularly useful for comparisons among similar distributions.

The second type of graphical representation, the cumulative frequency, shown in Figure 2 for the data in Table 1, is most useful when several distributions need to be compared. Cumulative distributions by number are generated by summing the contributions of all the particles less than a certain diameter range and plotting this total contribution versus the lower boundary of the diameter range. From

Table 1. Particle Size Distribution

Diameter range, μm	Number	%	Cumulative finer %
1–2	12	1.2	1.2
2–4	62	6.2	7.4
4–8	185	18.5	25.9
8–16	250	25.0	50.9
16–32	295	29.5	80.4
32–64	172	17.2	97.6
64–128	22	2.2	99.8
128–256	2	0.2	100.0

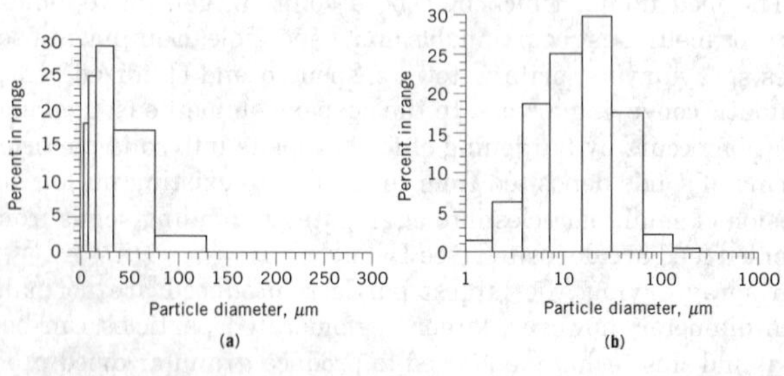

Figure 1. Histograms of the data from Table 1, plotted on (**a**) a linear and (**b**) a logarithmic scale.

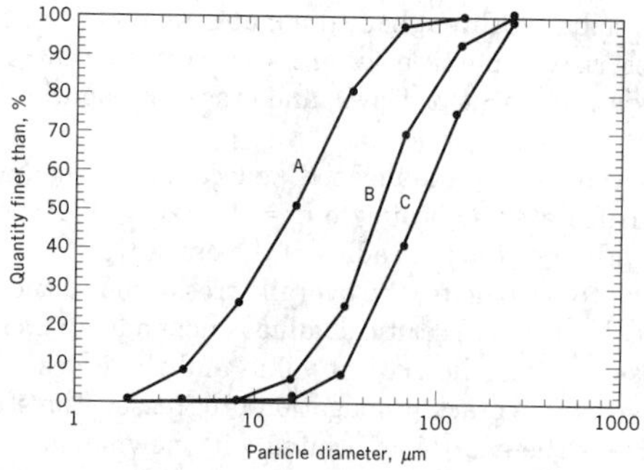

Figure 2. Cumulative frequency distribution plotted by A, number; B, surface area; and C, volume, for the data in Table 1.

Table 1, because the 1.2% in the 1–2 μm interval is finer than 2 μm, the amount of finer material is 1.2%, the amount finer than 4 μm is 7.4%, and so on.

Sampling

Sampling (qv) of powders is carried out at two different levels. First there is the taking of a sample from a gross supply of powder such as a rail car or a large heap. Sampling from larger (ie, tons) supplies of powders can be achieved using a device known as a thief sampler. When a sample of ca 100 g has been obtained, a representative sample for use in size characterization equipment must then be taken. Some of the more modern methods of size characterization require as little as one milligram of powder; thus obtaining a representative sample can be quite difficult.

A device which enables a small sample to be taken efficiently is the free-fall tumbling mixer equipped with sampling scoops. The powder to be sampled is placed in a sample jar so that it fills no more than half the jar. The lid of the jar is equipped with a stirrup that holds a sampling scoop which is covered by the powder when the jar is upright. The jar of powder to be sampled is assembled and placed in a carrier cube which is then tumbled chaotically inside a rotating drum.

Measurement Methods

A wide variety of particle size measurement methods has evolved to meet the almost endless variability of industrial needs. Every method, with the exception of imaging technologies, provides the measurement of an equivalent spherical diameter in one form or another. The spherical diameter information can be deduced indirectly from the behavior of the particles passing through restricted volumes or channels under the influence of gravity or centrifugal force fields, and from interaction with many forms of radiation.

Sieving. The oldest and still one of the most widely employed sizing methods determines particle size by the degree to which a powder is retained on a series of sieves having different opening dimensions. This technique is straightforward and requires simple equipment. A typical sieve is a shallow pan having a wire-mesh bottom or an electroformed grid.

Dry-sieving is typically performed using a stack of sieves having openings diminishing in size from the top downward. The lowest pan has a solid bottom to retain the final undersize. Powders are segregated according to size by placing the powder on the uppermost sieve and then shaking the stack manually, using a mechanical vibrator, or with air pulses of sonic frequency until all particles fall onto sieves through which they are unable to pass or into the bottom pan.

Wet-sieving is performed using a stack of sieves in a similar manner except that water or another liquid which does not dissolve the material is continually applied to facilitate particle passage. A detergent is frequently added to promote particle dispersion.

Computer-Automated Image Analysis. Particle characterization by image analysis consists of examining and measuring the size or shape of a relatively small number of particles that have been magnified. The ever-increasing power of data processing capability coupled with the high performance and falling costs of television cameras and scanners has led to the development of highly sophisticated and powerful image processing and analysis systems. Image analyzers can extract information from negatives, photomicrographs, or directly from microscopic (both optical and electron) images by scanning or digitization techniques. Optical microscopy is normally used for particles >1.0 μm in diameter that have been deposited on a microscope slide. Electron microscopy is applicable to particles having diameters from 0.002 to 15 μm.

Sedimentation. Measurement of the settling rate for particles under gravitational or centrifugal acceleration in a quiescent liquid provides the basis of a variety of techniques for determining particle sizes. In liquid-phase sedimentation, the particles initially may be distributed uniformly throughout a liquid (homogeneous start) or concentrated in a narrow band or layer at the liquid's surface (line start). The particle movement is monitored using light and/or x-ray beams.

The particle size determined by sedimentation techniques is an equivalent spherical diameter, also known as the equivalent settling diameter, defined as the diameter of a sphere of the same density as the irregularly shaped particle that exhibits an identical free-fall velocity. Gravitational sedimentation is intended for larger and higher density particles, which exhibit a relatively high settling rate, in order to obtain a distribution in a reasonable amount of time. The use of centrifugal force to accelerate the settling rate of particles is essential to monitor the movement of smaller, lower density particles. Centrifugal sedimentation permits evaluation of smaller diameters but adds mechanical complexity.

Field-Flow Fractionation. Field-flow fractionation is a general name for a class of separation techniques that fractionate a particle population into groups according to size. Particulate species to be fractionated are placed in a tube or other similar device in which a liquid is flowing. The tube is subjected to either a gravitational or centrifugal force confining the particles to be separated to regions of the tube. As the liquid flows along the tube, the combined effect of the field and the flow is to separate the profiles.

Photon Correlation Spectroscopy. Photon correlation spectroscopy (pcs), also commonly referred to as quasi-elastic light scattering (qels) or dynamic light scattering (dls), is a technique in which the size of submicrometer particles dispersed in a liquid medium is deduced from the random movement caused by Brownian diffusion motion. This technique has been used for a wide variety of materials.

Ultrasonic Spectroscopy. Information on size distribution may be obtained from the attenuation of sound waves traveling through a particle dispersion. Two distinct approaches are being used to extract particle size data from the attenuation spectrum: an empirical approach based on the Bouguer-Lambert-Beer law and a more fundamental or first-principle approach. The first-principle approach implies that no calibration is required, but certain physical constants of both phases, ie, speed of sound, density, thermal coefficient of expansion, heat capacity, thermal conductivity, attenuation of sound, viscosity for fluid phase, and shear rigidity for solid phase, are required for accurate measurements.

Time-of-Flight Instrumentation

In the late 1980s and early 1990s an instrument was developed which characterizes the size of particles by confining the particles to be characterized in an inspection zone scanned by a laser beam. The size of a particle is deduced from the time required for the laser beam to traverse a particle in the inspection zone.

Resistazone Counters. In a resistazone counter the particles to be characterized are suspended in a conducting electrolyte and drawn through an orifice situated between two electrodes. The presence of a particle within the zone alters the electrical resistance of the electrolyte in the inspection zone. The change in the resistance of this electrolyte plus particle system can be used to measure the size of the particle.

Laser Diffraction Equipment. Another type of size analysis equipment is one in which the powder to be characterized is presented as a random array to a laser beam. The diffraction pattern of this randomized array is then deconvoluted using optical theories to generate the size distribution of the powders in random array.

REMI TROTTIER
Aluminum Company of America
BRIAN KAYE
Laurentian University

K. Sommer, *Sampling of Powders and Bulk Materials*, Springer-Verlag, New York, 1986.

T. Allen, *Particle Size Analysis*, 4th ed., Chapman and Hall, London, 1992.

J. K. Beddow, *Particulate Science and Technology*, Chemical Publishing Co., Inc., New York, 1980.

K. D. Caldwell, in H. G. Bart, ed., *Modern Methods of Particle Size Analysis*, Vol. 73, *Monographs on Analytical Chemistry and its Applications*, John Wiley & Sons, Inc., New York, 1984.

SIZE REDUCTION

Size reduction is an extremely important unit operation in which materials are subjected to stress in order to reduce the size of individual pieces. The stress is applied by transmitting mechanical force to the solid. Size reduction typically involves preparation of naturally occurring raw materials for subsequent separation processes. Size reduction of solids is an extremely energy-intensive operation. Only a small fraction of the energy is used efficiently for size reduction, with the remainder being converted mainly into heat. The science of size reduction relies heavily on experience, and it is important that requirements be discussed extensively with equipment manufacturers and large-scale test work carried out prior to decisions being made on the most suitable methods to achieve a given requirement.

Particle Breakage and Fracture Mechanics

Size reduction causes particle breakage by subjecting the material to contact forces or stresses. The applied forces cause deformation which generates internal stress in the particles, and when this stress reaches a certain level, particle breakage occurs.

It is important to differentiate between brittle and plastic deformations within materials. With brittle materials, the behavior is predominantly elastic until the yield point is reached, at which breakage occurs. When fracture occurs as a result of a time-dependent strain, the material behaves in an inelastic manner. Most materials tend to be inelastic. Figure 1 shows a typical stress–strain diagram. The section A–B is the elastic region where the material obeys Hooke's law, and the slope of the line is Young's modulus. C is the yield point, where plastic deformation begins. The difference in strain between the yield point C and the ultimate yield point D gives a measure of the brittleness of the material, ie, the less difference in strain, the more brittle the material.

The total area under the curve A–D, shown as shaded in Figure 1, is the strain energy stored in a body. This energy is not uniformly distributed throughout the material, and it is this inequality that gives rise to particle failure. Stress is concentrated around the tips of existing cracks or flaws, and crack propagation is initiated therefrom.

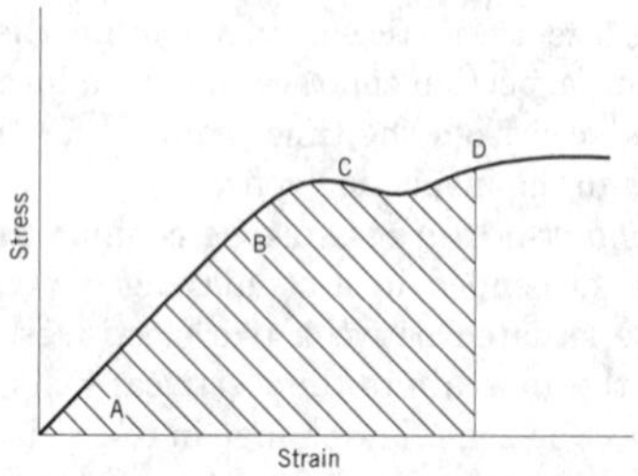

Figure 1. Typical stress–strain diagram. See text.

Stress concentration K is defined as local stress/mean stress in a particle and calculated according to $K = 1 + 2(LR)^{1/2}$, where L is half the crack length and R is the radius of the crack tip.

For a crack to propagate, the overall stress around the crack must reach a critical value. This critical value is dependent on crack length, so once stress reaches the critical value and the crack lengthens, it continues to grow. As crack propagation progresses, the strain energy released exceeds the energy associated with new surface and the excess energy concentrates around other cracks in the material, causing a multiple fracture. This is typical behavior for brittle materials.

With tough or plastic products the excess strain energy causes internal deformation. With decreasing particle size, materials exhibit increasing plastic behavior, and this explains why it is more difficult to break small particles than large particles. In small particles, the crack length is limited by the size of the particle. In practice this is seen where a limit of grindability is reached with many materials and with subjection to further grinding, no decrease in particle size can be observed.

Models Predicting Grindability and Energy Requirements

Many attempts have been made to develop models which predict the behavior of materials undergoing size reduction. One proposal is that the energy expended in size reduction is proportional to the new surface formed. Another theory is that the energy required to produce a given reduction ratio feed (size ÷ product size) is constant, regardless of initial feed particle size. Practical results show, however, that both these theories are limited in their usefulness.

A more practically useful work index based on empirical results from ball milling trials has been developed and is expressed by equation 1:

$$E = W_i = \left(\frac{10}{\sqrt{x_p}} - \frac{10}{\sqrt{x_f}} \right) \qquad (1)$$

where E is the energy required, W_i is the Bond work index, and x_f and x_p represent the particle size through which 80% of the feed and products, respectively, pass. A standard grindability test, using a specifically sized ball mill of a given design, is used to measure the work index.

The Hardgrove index is more usually used when predicting energy requirements for pendulum-type roller mills. The portion of a product passing a 75-μm sieve is measured after a specified test procedure. This is related to the Hardgrove index by reference to a calibration graph. Although neither the Bond index nor the Hardgrove index can be considered to give absolute values, both are useful in predicting comparative power consumption and output when scaling up from testwork or in estimating the performance of an existing mill if new products are processed.

Predicting Particle-Size Distribution

The breakage process can be modeled using two basic functions. The specific breakage rate, S_j, is the probability that particles in size class j are broken. This probability can be related to either time or energy input, ie, number of mill rotations, to give S_j. The breakage distribution function, b_{ij} is the mass fraction of particles falling into size class i that are formed by the breakage of particles in size class j. The breakage distribution can also be expressed in a cumulative form B_{ij}, which is the mass fraction of particles below size class i that are produced when breaking material of size class j.

Using the above concepts, models have been developed to predict size distribution from comminution devices. An assumption is that the rate of breakage of material of a particular size is proportional to the mass of that size present in the comminution zone of a machine.

Both the need to reduce experimental costs and increasing reliability of mathematical modeling have led to growing acceptance of

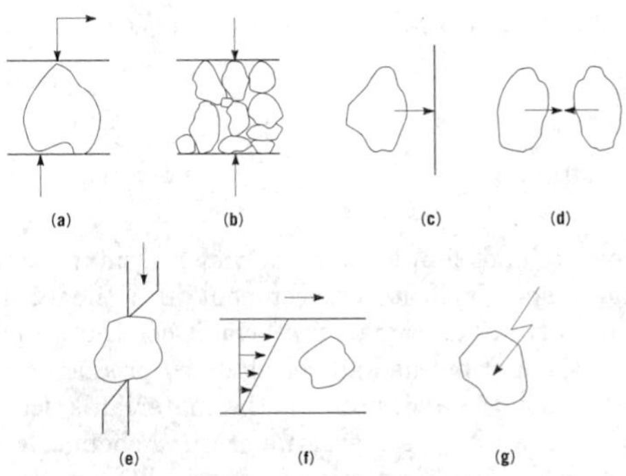

Figure 2. Stressing mechanisms: (**a**) single particles or (**b**) a bed of particles crushed between two solid surfaces; impact of a particle against (**c**) a solid surface or (**d**) another particle; (**e**) cutting; (**f**) shearing forces or pressure waves; and (**g**) plasma reaction, an example of size reduction by nonmechanical energy.

computer-aided process analysis and simulation, although modeling should not be considered a substitute for either practical experience or reliable experimental data.

Methods for Applying Stress

Equipment for size reduction can be categorized by the method in which the necessary stress is applied to the particles (Fig. 2).

Stressing Between Two Solid Surfaces: Crushing. Either single particles (Fig. 2**a**) or a bed of particles (Fig. 2**b**) are crushed between two solid surfaces.

Stressing by Impact. Size reduction is achieved by the impact of a particle against a solid surface (Fig. 2**c**) or another particle (Fig. 2**d**).

Stressing by Cutting. This method (Fig. 2**e**) is useful for materials which exhibit plastic behavior.

Stressing by the Surrounding Medium. Size reduction is effected by shearing forces or pressure waves (Fig. 2**f**). The amount of energy that can be transferred is very limited and this method is used mainly to break agglomerates.

Stressing by Nonmechanical Energy. Such processes are not fully developed but examples exist of a plasma reaction (Fig. 2**g**) being used for size reduction.

Selection Criteria for Size Reduction

Selection of the most suitable machine for a given requirement is an extremely complex process. Added to variations in the properties of the different materials, many of the machines involved have been specifically developed or adapted to perform only particular tasks. The principal factors which must be addressed are toughness/brittleness, hardness, abrasiveness, feed size, cohesity, particle shape and structure, heat sensitivity, toxicity, explodability, and specific surface.

Toughness/Brittleness. In tough materials the excess strain energy causes plastic deformation, whereas in brittle materials new cracks are propagated. Brittle materials are able to be reduced relatively easily, whereas tough materials present problems.

Hardness. There are several hardness (qv) scales. In selecting size reduction equipment generally hardness is expressed according to the Mohs' scale.

Moh's hardness	Material	Mohs' hardness	Material
1	talcum	6	feldspar
2	gypsum	7	quartz
3	calcite	8	topaz
4	fluorite	9	corundum
5	apatite	10	diamond

High speed machines such as impact mills begin to suffer high wear rates when processing materials above Mohs' hardness 3, unless very special wear-resisting measures can be taken.

Abrasiveness is closely related to hardness in homogenous materials. A small proportion, as low as 0.5%, of a hard impurity is enough to cause severe wear to many high speed machines.

The acceptable feed size for a given machine is governed by the type of feed device and physical characteristics of the machine.

Many materials stick together and adhere to machine parts (cohesity), depending on their condition, particle size, and temperature.

Some materials exhibit particular properties owing to their particle shape or form. It is often desired to maintain particle shape; in such cases, an impact-type mill is usually chosen rather than a ball mill.

Only around 1–2% of applied energy is effectively used for size reduction. The remainder is mainly converted to heat, which is absorbed by the grinding air, product, and equipment. Equipment with a high air throughput is often chosen, as it provides an economical method of dissipating heat and thus limiting the temperature rise of the end product.

Toxicity has no influence on the actual size reduction, but equipment is often selected for ease of product containment or safe cleaning.

Any material which is flammable in air can potentially support a dust explosion when it is finely divided and dispersed in air. Most organic materials, many metals, and other products fall into this category. Equipment has to be protected by inerting, explosion containment, explosion venting, or suppression.

If a defined specific surface area is required, this can affect the choice of equipment. Machines that apply stress by crushing generally create more ultrafines, and hence higher surface area, than impact mills.

Equipment Survey

Crushers and Roller Mills. In this equipment group, stress is applied by either crushing single particles or a bed of particles between two solid surfaces. In general, most machines are used for coarse and medium-size reduction, with the exception of the high pressure roller mill which can achieve extremely fine particle distributions.

Jaw Crushers. Both single-toggle and double-toggle designs are still widely used. In both, the principle of operation is the same, with feed material being stressed between a stationary jaw and a reciprocating jaw, driven by an eccentric shaft. Jaw crushers are used in primary size reduction of minerals.

Gyratory and Cone Crushers. Both of these designs utilize the principle of an eccentrically driven rotor crushing material against a stationary mantle. Depending on the particular duty, the rotor is supported by bearings at either one or both of its ends. The size of the end product can be varied by changing the clearance between the rotor and mantle.

Roll Crushers. Traditional roll crushers effect size reduction by crushing single particles between two counter-rotating rolls; hence they are crushed between the two surfaces. Roll mills are employed in a wide variety of applications for medium-coarse down to fine size reduction. By far the most prevalent use is in flour milling. In recent years high compression roll mills have become commercially important. In contrast to traditional roll mills, these machines apply stress to a bed of material rather than to single particles. The end fineness is dependent on the crushing pressure, and the output is determined by the roll speed.

Roller Mills: Pendulum and Table Type. This is a group of machines commonly applied for grinding of mineral powders down to approximately 97% below 75 μm, or even finer in some instances. The mills operate at medium speed, up to approximately 30 m/s, and can handle materials with up to Mohs' hardness 5 before wear rates become prohibitive. The two most commonly encountered variants are pendulum mills and the table roller mill.

Pendulum mills have a central, driven shaft. From this shaft several rollers are suspended on pivots. As the central shaft turns, the centrifugal force causes the pivoted rollers to press against the outer, stationary grinding ring. Material is stressed by compression between the roller and the outer ring. The grinding zone is swept by an airstream and the partially ground material is carried to the upper section of the unit, where an air classifier is located. The table roller mill employs a ring of pivoted rollers which are forced down toward a driven table by either springs or hydraulic pressure. Material is stressed between the rotating table and the rollers to achieve size reduction. Again, the unit is air-swept and the same choice of classifier units is available as for pendulum mills. Both of these roller mill types are widely employed for limestone, barytes, phosphates, dolomite, and many similar minerals.

Ball Mills and Rod Mills. Ball mills have been utilized since the late 1800s and the construction and principles remain essentially very simple. The machine consists of a cylindrical or conical tube into which loose grinding balls are filled up to a certain level. Size reduction is achieved by rotating the tube so that the balls either roll against each other or, if the speed is sufficient, they are lifted and fall. In general, the action is a combination of rolling and lifting.

The grinding action ensures that a very high stress can be applied to the particles so that a high portion of ultrafine product is produced in a ball mill, compared to impact grinding, for example.

Impact Mills. In this equipment group, stress is applied by transferring kinetic energy by either particle–particle contact or machine–particle contact. Impact mills can broadly be separated into mechanical types where high speed beaters impact the material to apply stress, and fluid energy mills, where particles are accelerated by the surrounding medium and impact against each other or a target.

Cutting Mills. The machines applying stress by cutting are described in Figure 2e. They are usually employed for size reduction of ductile materials such as plastics, vegetables, and animal products.

Wet Grinding. In certain circumstances it is advantageous to grind in a wet state: (*1*) where a suspension is required as an end product; (*2*) where the required particle size cannot be achieved in a dry state; (*3*) where toxic or flammable emissions must be avoided; and (*4*) where chemical or physical surface reactions are desired. Providing the material does not require subsequent drying after grinding, wet milling can give energy savings of $\sim$ 30%; however, wear rates are usually three to five times higher.

The most commonly applied wet grinding device is the stirred ball mill also referred to as sand mill, pearl mill, bead mill, or agitated ball mill. The units consist essentially of a grinding container which is partially filled with loose grinding media, usually in the size range 0.5–10 mm dia. The chamber is filled with the slurry to be ground, and some form of stirrer accelerates the grinding media. Size reduction takes place as particles are crushed between the media.

Uses range from dispersion of pigments and filler for paints, to ceramics and ultrafine minerals such as kaolin. These products usually demand a high percentage of submicrometer particles, which wet grinding can achieve.

MICHAEL PRIOR
Hosokawa Micron Ltd.

H. Rumpf, *Powder Technol.* **7**, 145–159 (1973).

F. C. Bond, *AIME Min. Eng. Trans.* **193**, 484–494 (1952).

R. M. Hardgrove, *The Relation Between Pulverising Capacity, Power and Grindability*, ASME, Chicago, Ill., 1993.

J. Liu and K. Schönert, *Int. J. Miner. Proc.* **44–45**, 101–115 (1996).

SLAGCERAM. See GLASS-CERAMICS.

SLIMICIDES. See INDUSTRIAL ANTIMICROBIAL AGENTS.

SKIN, PERCUTANEOUS ABSORPTION. See ANTIAGING AGENTS; CONTROLLED RELEASE TECHNOLOGY; DRUG DELIVERY SYSTEMS.

SMOKES, FUMES, AND SMOG. See AIR POLLUTION; AIR POLLUTION CONTROL METHODS.

SMART MATERIALS

From a technical and simple point of view, a smart material is a material that responds to its environment in a timely manner. To expand on this definition, a smart material is one that receives, transmits, or processes a stimulus and responds by producing a useful effect, which may include a signal that the material is acting upon it. Stimuli may include strain, stress, temperature, chemicals, an electric field, a magnetic field, hydrostatic pressures, different types of radiation, and other forms of stimuli. Transmission or processing of the stimulus may be in the form of an absorption of a photon, of a chemical reaction, of an integration of a series of events, of a translation or rotation of segments within the molecular structure, of a creation and motion of crystallographic defects or other localized conformations, of an alteration of localized stress and strain fields, and of others. The useful effects produced could be a change in color, index of refraction, stress or strain distribution, or volume. Also, incorporated within the definition of smart materials is the ability to be reversible.

Under the proper set of environments and circumstances all materials are smart and depict smart behavior at some point during their life cycle. Some examples of technically smart behaving materials are piezoelectric materials, electrostrictive materials, magnetostrictive materials, electrorheological materials, magnetorheological materials, thermoresponsive materials, pH-sensitive materials, uv-sensitive materials, smart polymers, smart gels (hydrogels), smart catalysts, and shape memory alloys.

Smart structures are structures that incorporate at least one smart material within itself and from the effort produced by the smart material causes an action.

Piezoelectric Materials

Piezoelectric materials are materials that exhibit a linear relationship between electric and mechanical variables. The direct piezoelectric effect can be described as the ability of materials to convert mechanical stress into an electric field; and the reverse, to convert an electric field into a mechanical stress. The use of the piezoelectric effect in sensors is based upon the latter property.

There are two principal types of materials that can function as piezoelectrics: the ceramics and polymers. The piezoelectric materials most widely used are the piezoceramics based upon the lead zirconate titanate, PZT. The advantages of these piezoceramics are that they have a high piezoelectric activity and they can be fabricated in many different shapes.

A newer class of materials called smart tagged composites has been developed for structural health monitoring applications. These composites consist of PZT-5A particles embedded into the matrix resin (unsaturated polyester) of the composite.

Electrostrictive Materials

Electrostrictive materials are materials that exhibit a quadratic relationship between mechanical stress and the square of the electric polarization. Electrostriction can occur in any material. Whenever an electric field is applied, the induced charges attract each other, thus, causing a compressive force. This attraction is independent of the sign of the electric field. Typical electrostrictive materials include such compounds as lead manganese niobate, lead titanate (PMN:PT), and lead lanthanium zirconate titanate (PLZT).

Magnetostrictive Materials

As materials show mechanical deformation induced by electric fields, the same type of material response can be observed when the stimulus is a magnetic field. Shape changes are the largest in ferromagnetic and ferrimagnetic solids. The repositioning of domain walls that occur when these solids are placed in a magnetic field leads to hysteresis between magnetization and an applied magnetic field.

Materials that have shown a response to magnetic stimuli have primarily been inorganic in chemical composition, alloys of iron, nickel, and cobalt doped with rare earths. However, there has been a great interest in the development of organic and organometallic magnets.

In comparing organic magnets with organometallic magnets there are several key differences between the two types. The first is that the organic-based magnets do not contain metal atoms. The second difference involves the fact that in organic-based magnets, the coupled spins residue entirely in the *p* orbitals; whereas in the organometallic-based magnets, they are either in the *p* or *d* orbitals, or a combination of the two.

Electrorheological Materials

Electrorheological materials are fluids whose viscous properties are modified by applying an electric field. There are many electrorheological fluids, which are usually a uniform dispersion or suspension of particles within a fluid. In an applied electric field the particles orient themselves in fiberlike structures (fibrils). When the electric field is off, the fibrils disorient themselves. The damping characteristics of the system can be changed (flexible to rigid). Electrorheological fluids are non-Newtonian fluids, that is, the relationship between shear stress and shear strain rate is nonlinear. The changes in viscous properties of electrorheological fluids are obtained only at relatively high electric fields in the order of 1 kV/mm.

Magnetorheological Materials

Magnetorheological materials (fluids) are the magnetic equivalent of electrorheological fluids. In this case, the particles are either ferromagnetic or ferrimagnetic solids that are either dispersed or suspended within a liquid and the applied field is magnetic.

An adaptation of magnetorheological fluids is a series of elastomeric matrix composites embedded with magnetic particles such as iron. During the thermal cure of the elastomer, a strong magnetic field was applied to align the iron particles into chains. These chains of iron particles were locked into place within the composite through the cross-linked structure of the cured elastomer. The resistance of the composite to changes in modulus or deformation was controlled by an external magnetic field. When stimulated by a compressive force, the composite was 60% more resistant to deformation in a magnetic field.

Thermoresponsive Materials

Polymeric materials are unique because of the presence of a glass-transition temperature. At the glass-transition temperatures, the specific volume of the material and its rate of change changes, thus, affecting a multitude of physical properties.

Materials that typify thermoresponsive behavior are polyethylene–poly(ethylene glycol) copolymers that are used to functionalize the surfaces of polyethylene films (smart surfaces). When the copolymer is immersed in water, the poly(ethylene glycol) functionalities at the surfaces have solvation behavior similar to poly(ethylene glycol) itself. The ability to design a smart surface in these cases is based on the observed behavior of inverse temperature-dependent solubility of poly(alkene oxide)s in water.

pH-Sensitive Materials

By far the most widely known classes of pH-sensitive materials are those classes of chemical compounds that include the acids, bases, and indicators. The most interesting of these are the indicators. These materials change colors as a function of pH and usually are totally reversible (see HYDROGEN-ION ACTIVITY).

In addition to acting as a means of observing changes in pH in titrations and in chemical reactions, indicators have been used in the development of novel chemical indicating devices.

Other examples of pH-sensitive materials are the smart hydrogels and smart polymers.

Light-Sensitive Materials

There are several different types of material families that exhibit different kinds of responses to a light stimuli. One type comprises materials that exhibit electrochromism. This is a change in color as a function of an electrical field. Other types of behaviors include thermochromism (color change with heat), photochromic materials (reversible light-sensitive materials), photographic materials (irreversible light-sensitive materials), and photostrictive materials (shape changes due to light usually caused by changes in electronic structure).

Smart Polymers

Even though smart polymers have been used in all types of applications and can exhibit all types of stimuli–response behaviors, the term, smart polymers, has been used as a separate category of smart materials. In medicine and biotechnology, smart polymer systems usually involve aqueous polymer solutions, interfaces, and hydrogels. These are polymeric systems that are capable of responding strongly to slight changes in the external medium; a first-order transition, accompanied by a sharp decrease in the specific volume of the system. The presence of a poor solvent is one of the main conditions for this phenomenon in swollen polymer networks or linear polymers to occur. A poor solvent causes the forces of attraction between the polymer chain segments to overcome the repulsion forces associated with the extended volume, thus, leading to the collapse of the polymer chain.

Smart polymers can respond to environmental stimuli such as temperature, pH, ions, solvents, reactants, light or uv radiation, stress, recognition, electric fields, and magnetic fields. These stimuli once acted upon, result in changes in phases, shape, optics, mechanics, electric fields, surface energies, recognition, reaction rates, and permeation rates. The polymers that fit into this category include the naturally occurring polymers, acrylic polymers and copolymers, and polymers based on combining acid monomers with basic monomers.

Smart Gels (Hydrogels)

Smart (intelligent) gels (or hydrogels) are not new. They are finally reaching commercialization after thirty years of research and development. The concept of smart gels is also more complex than the simple concept of solvent-swollen polymer networks. It is the behavior of the solvent-swollen polymer networks in conjunction with the material being able to respond to other types of stimuli; such as temperature, pH, and concentrations of solvents. An example of a smart gel chemical composition consists of an entangled network of two polymers; one is a poly(acrylic acid) (PAA), and the second, a tri-block copolymer containing poly(propylene oxide) (PPO) and poly(ethylene oxide) (PEO) in a PEO–PPO–PEO sequence. The PPA portion of the smart gel system is a bioadhesive and is pH responsive. The PPO segments are hydrophobic that help solubilize lipophilic substances in medical applications, and the PPO segments tend to aggregate, thus, resulting in gelation at body temperatures.

Smart Catalysts

One class of smart catalysts is based on homogeneous rhodium-based poly(alkene oxide)s, in particular those with a poly(ethylene oxide) backbone. Traditionally chemical catalyzed reactions proceed in a manner in which the catalysts become more soluble and active as the temperature is raised. This can lead to exothermal runaways, thus, posing both safety and yield problems. The behavior of these smart catalysts is different from that of traditional catalysts. As the temperature increases, they become less soluble, thus precipitating out of solution and becoming inactive. As the reaction mixture cools down, a smart catalyst redissolves and becomes active again.

Smart Memory Alloys

Shape-memory alloys undergo thermomechanical changes as they pass from one phase to another. The crystalline structure of such alloys based on nickel and titanium enters the martensitic phase as the alloy is cooled below a critical temperature. In this stage, the alloy is easily manipulated through large strains with a little change in stress. As the temperature of the alloy is increased above the critical (transformation) temperature it changes into the austentic phase. In the austentic phase, the alloy regains its high strength and high modulus. It behaves like a "normal" metal. The alloy shrinks during the transformation from the martensitic to austentic phase.

The use of shape-memory alloys as actuators depends on their use in the plastic martensitic phase that has been constrained within the structural device. Shape-memory alloys (SMAs) can be divided into three functional groups; one-way SMAs, two-way SMAs, and magnetically controlled SMAs. The magnetically controlled SMAs show great potential as actuator materials for smart structures because they could provide rapid strokes with large amplitudes under precise control. The most extensively used conventional shape-memory alloys are the nickel–titanium- and copper-based alloys (see SHAPE-MEMORY ALLOYS).

Elastorestrictive Materials

This class of smart materials is the mechanical equivalent of electrostrictive and magnetostrictive materials. Elastorestrictive materials exhibit high hysteresis between strain and stress. This hysteresis can be caused by motion of ferroelastic domain walls. This behavior is more complicated and complex near a martensitic phase transformation.

Materials with Unusual Behaviors or Unusual Materials

Only a few materials fit into this category; they seldom can be categorized into one of the above material classes. Water fits into the category of materials with unusual behavior. Water is one of the few materials that expands upon freezing. It changes volume by approximately 8% transiting from the liquid to the solid state.

Fullerene and its derivatives can be included in the unusual material category. One interesting application of fullerenes as smart materials has been in the area of embedding fullerenes into sol–gel matrices for the purpose of enhancing optical limiting properties. A semiconducting material with a magnetic ordering at 16.1 K was produced from the reaction of the fullerene C_{60} with tetra(dimethylamino)ethylene.

JAMES A. HARVEY
Hewlett-Packard Company
Oregon Graduate Institute of Science & Technology

Committee on New Sensor Technologies: Materials and Applications, National Materials Advisory Board, Commission on Engineering and Technical Systems, *National Research Council Report: Expanding the Vision of Sensor Materials*, National Academy Press, Washington, D.C. 1995.

C. A. Rogers, *Scientific American* **273**(3), 122–126 (Sept. 1995).

E. Udd, *Fiber Optic Smart Structures*, John Wiley & Sons, Inc., New York, 1995.

J. S. Miller and A. J. Epstein, *Chem. Eng. News*, 30–41 (Oct. 2, 1995).

SOAP

Soap is one example of a broader class of materials known as surface-active agents, or surfactants (qv). This class of materials is simultaneously soluble in both aqueous and organic phases or preferential aggregate at air–water interfaces. It is this special chemical structure that leads to the ability of surfactants to clean dirt and oil from surfaces and produce lather.

Although soaps have many physical properties in common with the broader class of surfactants, they also have several distinguishing factors. First, soaps are most often derived directly from natural sources of fats and oils (see FATS AND FATTY OILS).

Second, soaps form insoluble complexes, commonly referred to as curd, in the presence of calcium and magnesium ions in solution. Calcium and magnesium ions are the principal metal ions found in water and their level is commonly referred to as the hardness of the water. This curd reduces the effectiveness of soap as a surfactant and gives rise to other undesirable properties during use, eg, precipitation on surfaces. Many synthetic surfactants are considerably less susceptible to water hardness.

Physical Properties of Surfactants

Surfactants, including soap, possess a bipolar structure, comprised of both a hydrophobic tail and a hydrophilic head group. As a result of this bifunctional structure, surfactants possess many unique physical properties. In solution, surfactants preferentially concentrate as monolayers at the interfacial region between any two phases of dissimilar dielectric constants or polarity. Examples of interfacial regions are the interfaces between oil and water, or air and water. The hydrophilic portion preferentially solubilizes in the polar or higher polarity phase whereas the hydrophobic portion preferentially solubilizes in the nonpolar or lower polarity phase. The presence of surfactants at the interface provides stability to the interface by lowering the total energy associated with maintaining the boundary. Thus, surfactants facilitate stabilization of intermixed, normally immiscible phases, such as oil in water, by decreasing the energy necessary to maintain the large interfacial region associated with mixing.

Another property of surfactants is their ability to aggregate in solution to form various composite structures or phase states, such as micelles and liquid crystals, as a function of concentration and temperature. At very low surfactant levels, the surfactant exists as individual molecules in solution associating primarily with water molecules. It also concentrates or partitions at the interfacial regions as described above. However, as the concentration of surfactant in solution is increased, a point is reached where the molecules aggregate to form micelles. The micellar structure minimizes energy through surfactant self-association; the micelle in water is typically characterized with the hydrophobic tails pointing to the center and the head groups pointing out toward the water in spherical superstructures. As the concentration of surfactant in solution is further increased, the micelles elongate into long tubules which align with each other to form a hexagonal arrangement when viewed end-on. As the surfactant concentration is further increased, the tubules expand in a second direction to form large, stacked lamellar sheets of surfactants. These liquid crystals are very important in soap making.

Because the core of an aqueous micelle is extremely hydrophobic, it has the ability to solubilize oil within it, as well as to stabilize a dispersion. These solubilization and suspension properties of surfactants are the basis for the cleansing ability of soaps and other surfactants. Furthermore, the ability of surfactants to stabilize interfacial regions, particularly the air–water interface, is the basis for lathering, foaming, and sudsing.

Soap Raw Materials and Their Processing

Carboxylate soaps are most commonly formed through either direct or indirect reaction of aqueous caustic soda, ie, alkali earth metal hydroxides such as NaOH, with fats and oils from natural sources, ie, triglycerides. Fats and oils are typically comprised of both saturated and unsaturated fatty acid molecules containing between eight and 20 carbons randomly linked through ester bonds to a glycerol backbone. Overall, the reaction of caustic with triglyceride yields glycerol (qv) and soap in a reaction known as saponification.

Compositional differences in the fats and oils give rise to their significantly different physical properties and those of the resulting fatty acids and soaps. Fats and oils are triglycerides composed of glycerol, ester-linked to three fatty acids. The main compositional difference is the chain length distribution of the fatty acids associated with the fats or oils. High levels of unsaturated (containing double bonds) or short-chain length components produce fatty acids that are liquid and soaps that have high water solubilities at room temperature. Conversely, high levels of saturated, long-chain length components produce waxy and hard fatty acids, eg, candle wax, and soaps that are essentially insoluble at room temperature. A key to producing soaps with acceptable qualities is the proper blending of these fats and oils.

The quality, ie, level of impurities, of the fats and oils used in the manufacture of soap is important in the production of commercial products. For commercial soaps, it is desirable to keep impurities at the absolute minimum for both storage stability and finished product quality considerations.

There are a number of processing steps that can be utilized to improve the quality and stability of the fats and oils raw material. These include water washing, alkali refining, physical (steam) refining, deodorization, bleaching, and hydrogenation. Industry utilizes a number of analytical methods to characterize fats and oils, which include moisture, titre (solidification point), free fatty acid, unsaponifiable material, iodine value, peroxide value, and color.

Soap Solution-Phase Properties

Commercially, soap is most commonly produced through either the direct saponification of fats and oils with caustic or the hydrolysis of fats and oils to fatty acids followed by stoichiometric (equal molar) neutralization with caustic.

The Binary Soap–Water System. Mixtures of soap in water exhibit a rich variety of phase structures, depending on temperature and concentration of the mixture. Phase diagrams chart the phase structures, or simply phases, as a function of temperature (on the y-axis) and concentration (on the x-axis). Figure 1 shows a typical soap–water binary phase diagram, in this case for sodium palmitate–water. Sodium palmitate is a fully saturated, sixteen-carbon chain length soap. At lower temperatures, soap crystals coexist with a dilute isotropic soap solution. Upon heating, liquid crystalline phases may form, depending on the relative concentration of soap in water. In dilute solutions, the crystals disproportionate and form simple micellar isotropic solutions (nigre phase). As the soap concentration is increased, hexagonal (hex) liquid crystal (middle phase) is formed. At even higher soap levels, lamellar (lam) liquid crystal (neat phase) is formed.

At typical soap processing temperatures (80–95°C), three liquid soap phases are possible: nigre (sometimes called isotropic), middle, and neat soap. Nigre soap is observed in dilute soap solutions and is characterized as very fluid. However, because of the dilute nature of this phase, it is not of practical use in the manufacture of soap. Middle soap is a liquid crystalline phase that is extremely viscous and difficult to handle and work. Neat soap is considerably more fluid

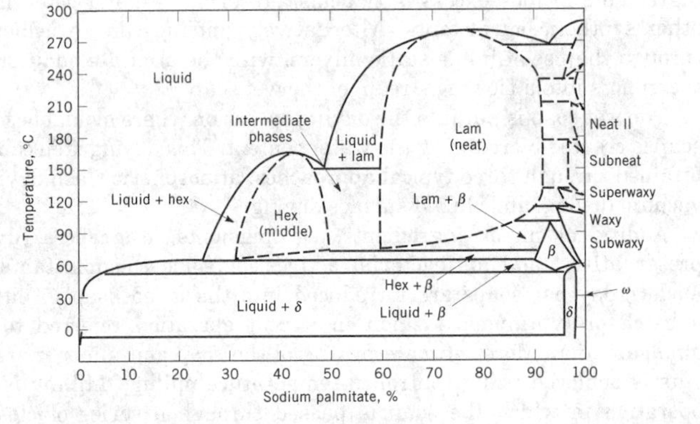

Figure 1. Binary soap–water phase diagram for sodium palmitate. Courtesy of Academic Press, Ltd.

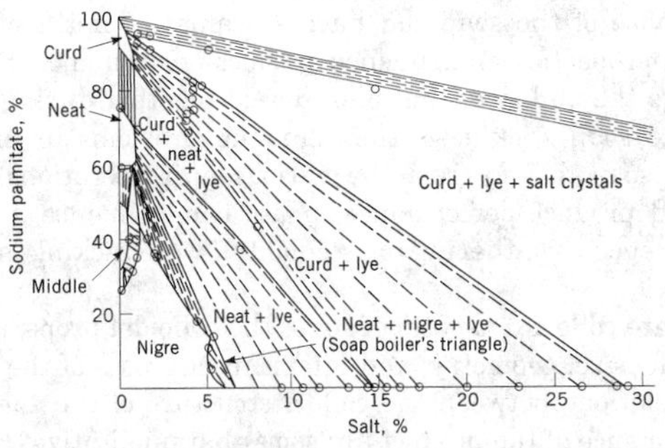

Figure 2. Ternary soap–NaCl–water phase diagram for sodium palmitate. Courtesy of Academic Press, Ltd.

than middle phase and is readily pumped and mixed. Neat phase is always approached from the more concentrated soap direction. This is the phase most commonly desired for soap making.

Ternary Soap–Water–Salt systems. A variety of components such as salt, fatty acid, and glycerol can alter the general phase characteristics of the soap–water system. Ternary phase diagrams are constructed to account for the presence of a third material.

The soap–water–salt diagram is typically shown graphically with the 90° vertices (Fig. 2). At 0% salt, the phases along the axis present a slice of the binary soap–water-phase diagram at 90°C (sodium palmitate in this case). The addition of salt to the system greatly reduces the concentration ranges for the liquid crystalline phases and increases the ranges for the isotropic phases: nigre and lye (a caustic rich aqueous phase). Further increase in the salt concentration drives the system into a biphasic region in which both a concentrated soap and a nigre (or lye) phase coexist. This ability of salt to drive the system into a biphasic, neat soap–nigre/lye phase structure is the basis for the direct saponification approach to soap making.

Soap Solid-Phase Properties and Crystallization

Soap crystallizes into bilayer structures comprised of stacks of soap layers arranged head-to-head and tail-to-tail, with water of hydration present in the head-to-head intralamellar region. It is generally accepted that only four common, distinct pure sodium soap crystalline phases exist. These phases are designated α, β, δ, and ω and were originally identified through powder x-ray diffraction. Only ω, β, and δ are observed in conventional soaps. Most soap in the solid state exists in one or a combination of these three phases. The phase diagram (Fig. 1) refers to equilibrium states. In practice, the drying routes and other mechanical manipulation utilized in the formation of solid soap can result in the formation of nonequilibrium phase structure. This point is important when dealing with the manufacturing of soap bars and their performance.

There are some general guidelines that allow the prediction of which crystal structure will form upon cooling of a hot soap mixture. The most pronounced is the fatty acid chain length distribution. Those mixtures which have more shorter or unsaturated chain lengths (the more soluble soaps) typically result in ω- or β-phase soap crystals. The cis-configuration unsaturated isomers have a greater impact than the trans-configuration isomers. Those hot soap mixtures enriched in the longer saturated chain length (palmitate and stearate) soaps typically result in the formation of δ-phase soap crystals.

Other factors also impact the type of crystals formed upon cooling of hot soap. Water activity or moisture content contribute to the final crystal state as a result of the different phases containing different levels of hydration. Any additive that changes the water activity changes the crystallization pathway.

Phase diagrams can be used to help understand the resulting crystal formation. Upon definition of the starting conditions (temperature and composition), the crystal phases can be estimated by following a line down in temperature at a constant composition.

The phase of the soap can have a dramatic impact on both the in-process properties and finished product performance. Soap phase can change the ability of the soap to weld together during finishing, and induce soap stickiness, thus creating problems in milling and conveying soap to and from the various unit operations. In terms of finished product performance, phase can influence a variety of attributes such as lathering (or solubility), wet cracking, smear, and firmness.

There are colloidal aspects of soap. The colloidal properties are defined by the size, geometry, and interconnectiviness of the soap crystals. Correlations between the colloid structure of the soap bar and the performance of the product are somewhat qualitative. However, it might be anticipated that smaller crystals would lead to a softer product. Furthermore, these smaller crystals might also be expected to dissolve more readily, leading to more lather. Translucent and transparent products rely on the formation of extremely small crystals to impart optical clarity.

Commercial Processing

Direct Saponification. Direct saponification of fats and oils is the traditional process utilized for the manufacturing of soap. Commercially this is done through either a kettle boiling batch process or a continuous process.

The kettle boiled batch process produces soap in large, open steel tanks (kettles), cylindrical tanks with conical bottoms, which contain open steam coils for heating and agitation. Fats and oils, caustic soda, salt, and water are simultaneously added to the kettle. The addition of steam to the system facilities mixing and the saponification reaction. To complete the saponification process, the soap batch is boiled for a period of time using steam sparging.

Upon completion of the saponification reaction, additional salt is added to the kettle while boiling with steam to convert the mixture from a pure neat-soap phase composition into the curd soap–lye seat biphasic composition (opening the grain of the soap). The mixture is allowed to separate for several hours, after which the aqueous solution or lye seat is removed from the bottom of the kettle. The curd soap remaining in the kettle is typically washed a few times by adding water, converting it back into neat soap and repeating the salt addition, boiling, and separation process. After the final wash, the water level in the curd soap remaining in the kettle is adjusted to achieve the proper physical properties for additional processing as (fitting), resulting in the formation of a neat soap–nigre-phase mixture, which facilitates further removal of impurities through the settling of the nigre phase. What remains in the kettle is pure neat soap at ~70% concentration with low levels of salt and glycerol. This process requires several days to complete.

Continuous saponification systems have led to improved manufacturing efficiency and considerably shorter processing times.

Blended fat and oil feedstocks are continuously and accurately metered into a pressurized, heated vessel, an autoclave, along with the appropriate amount of caustic, water, and salt. The concentrations of these ingredients are adjusted to yield a mixture of neat soap and a lye seat. At the temperatures (~120°C) and pressures (~200 kPa) utilized, the saponification reaction proceeds quickly (<30 min). After a relatively short resident time in the autoclave, the neat soap and lye seat reaction blend is pumped into a cooling mixer where the saponification reaction is completed, and the reaction product is cooled to below 100°C. The reaction product is pumped next into a static separator, where the lye phase containing a high level of glycerol (25–30%) is separated from the neat soap.

The neat soap is then washed with a lye and salt solution using a countercurrent flow process. As with the kettle process, it is important to have a proper level of electrolyte (salt and lye) for effective removal of the glycerol. Final separation of the lye seat from the neat soap is commonly achieved using centrifugation, after which the remaining caustic or residual alkalinity in the separated neat soap is neutralized. The soap is now ready for use in the manufacturing of soap bars.

Fatty Acid Neutralization. Another approach to produce soap is through the neutralization of fatty acids with caustic. In this approach, fatty acids are produced through the hydrolysis of fats and oils by water, followed by subsequent neutralization with appropriate caustics. This approach has a number of inherent benefits over the saponification process.

The hydrolysis of fats and oils by water requires intimate mixing of these two normally immiscible phases. The reaction is carried out under conditions where water possesses appreciable solubility (10–25%) in fats and oils. In practice, this is achieved under high pressure 4–5.5 MPa (580–800 psi) and with high temperatures (~240–270°C) in stainless steel columns of around 24–31 m in height and 50–130 cm in diameter. ZnO is sometimes added as a catalyst.

The fatty acids obtained from the process can be used directly or further manipulated for improved or modified performance and stability, as in hardening or speciation.

The formation of soap from fatty acids is achieved through the reaction of the fatty acid with the appropriate caustic. This reaction is extremely rapid for most common caustics, eg, NaOH or KOH, and requires proper stoichiometry and rigorous mixing to ensure processing effectiveness (avoidance of undesirable phases).

Bar Soap Manufacturing

The conversion of wet base soap into consumer-acceptable bar soaps can be achieved using one of three commonly utilized manufacturing processes: framing, milling, and hot extrusion, all of which use a variety of processing unit operations or finishing steps. These steps include wet mixing or crutching, drying, dry mixing or compounding, and bar forming.

Framing. In the framed bar process, the wet base soap is pumped into a heated, agitated vessel commonly referred to as a crutcher. The ingredients such as fragrance or preservative are added to the wet soap in the crutcher or injected in-line after reduction of product stream temperature. The hot mixture is then pumped into molds and allowed to cool.

These molds can be either finished bar-shape molds or large blocks. For bar-shaped molds, the solid bar is removed from the mold and packaged as desired upon cooling. For the large blocks, the mold is pulled apart, the block of solid soap is removed, and wire cutters are employed to cut the blocks, and the cut shapes go through final stamping to emboss the logo on the brick.

Milled Bar Process. Most bar soaps are produced by the milled bar process, which yields high quality soap bars. The process requires drying as well as milling and plodding (extrusion) of low moisture soap and is capable of high efficiency and throughput (~300–400 bars/min for a given packing line). However, it is also equipment-intensive due to the number of process unit operations required.

The wet soap is pumped into a mixing vessel (crutcher) where the addition and mixing of minor ingredients may be achieved. Minor ingredients include excess fatty acids, preservatives, and potentially other synthetic surfactants. Alternatively, mixing can be achieved through the use of inline static mixers, with the accurate addition of the minors into a flowing stream of the wet soap.

The wet soap is put into the drying operation where moldable solid soap is created by reducing the water content. This drying step can be attained through three typical approaches: atmospheric flash drying, vacuum drying, and chilled surface drying.

Additional minor ingredients, eg, pigments, fragrances, dyes, preservatives, and antibacterial actives, or some co-surfactants in modern-day bar soaps are introduced into the dried soap in either a batch or continuous fashion in a unit operation referred to as amalgamation. More intimate mixing of the soap and minor ingredients is achieved using controlled-temperature milling. Milling is an operation in which the soap is passed through a series of closely spaced, temperature-controlled steel rolls which dictates product temperature, inputs work into the soap mixture, and provides efficient micromixing. The formation of finished bar soaps is accomplished through the continuous extrusion of a shaped plug of soap. This extrusion is commonly referred to as plodding and is achieved using a two-stage single- or twin-worm-screw extruder.

The plug at the exit of the plodder is cut into the appropriate length and directed into the stamping and packaging operations. Product can be stamped into the desired shape on account of its intrinsic plasticity using either fixed capacity or box dies.

Hot Extrusion Process. The hot extrusion process utilizes a scraped-wall heat exchanger (SWHX) to provide controlled crystallization and plug forming in one step.

At the outlet of the SWHX there emerges a partially crystallized strip that is firm enough to maintain its shape once extruded. The extruded strip is taken away on a conveyor belt, cut to appropriate dimensions, and allowed to cool further under specific temperature and humidity conditions to ensure proper crystallization. The cut strips are then finished into bars through stamping process.

Formulation of Soaps

Soap Bars. In soap bars the primary surfactant is predominantly sodium salts of fatty acids. These products typically contain between 70 and 85% soap. Soap performance can be controlled through the proper blending of fats and oils to specific ratios, and the formation of the proper phase and colloidal structure.

Bar Soap Additives. Additives may be formulated into soap bars to provide additional consumer benefits or modify the performance of the products. These include free fatty acids; glycerol; colorants, dyes, and pigments; fragrance; chelants and antioxidants; mildness and skin additives; and antimicrobial agents.

Specialty soaps require certain additives to deliver the special consumer needs for which they were developed. Scouring soaps contain an abrasive agent homogeneously distributed throughout the soap to aid in the cleaning properties of the product. Transparent soap bars, often called glycerin soap, are formed through the quiescent cooling of a high solubility soap system containing a high level of solvent.

Much of the development of new soap bar technologies has been focused on products containing some level of synthetic surfactants. The primary benefits of synthetic surfactants over soaps are their intrinsic lower sensitivity to water hardness, which improves their rinsing profiles, their lathering ability, and their effects on skin feel and mildness. Anionic, nonionic, and amphoteric surfactants have all been formulated into bar soaps. These surfactants, in conjunction with soap, produce bars that may possess superior lathering and rinsing in hard water, greater lather stability, and improved skin effects. Beauty and skin care bars are becoming very complex formulations.

Liquid Soaps and Body Washes. Liquid soaps have been offered as a practical replacement of soap bars for use at sinks in the bathroom and kitchen. Manufacturers have taken two basic approaches to the formulation of these products: soap-based and synthetic-based formulations. Soap-based formulas use potassium or ammonium salts to yield soaps that are highly soluble at room temperature. More recently, soap-based formulations have been replaced by synthetic surfactant-based formulations. Synthetic surfactant formulations have the advantages of being milder-to-skin, cleaner rinsing, higher lathering, and less sensitive to water hardness. A typical synthetic surfactant formulation is around 80% water and may contain sodium alkyl sulfate and sodium alkylethoxy sulfate as the primary surfactants, a nonionic surfactant such as lauramide DEA, and potentially a lather-building amphoteric surfactant such as cocamidopropylbetaine. Most products now in the 1990s contain both antibacterial agents (usually TCS) and moisturizers.

Body washes are another more recent introduction into the marketplace. These can be simple formulas similar to those used for liquid handsoaps or complex 2-in-1 oil-in-water emulsion, moisturizing formulations. These products contain a wide range of synthetic surfactants not typically found in bar soaps or liquid handsoaps, such as sodium monoalkyl phosphate.

Analytical Characterization of Soap

A number of analytical methods are commonly utilized for the characterization of neat soap and bar soaps. Many of these methods have been published as official methods by the American Oil Chemists' Society. Instrumental techniques are frequently utilized.

Health, Safety, and Toxicology

Soap manufacture poses some material handling concerns because of the reaction of strong caustics with either neutral fats and oils or fatty acids at relatively high temperatures. The caustics represent the primary hazard. They are extremely corrosive and may cause serious body burns and eye injuries if not removed quickly through rinsing with copious amounts of water. Appropriate protective clothing is strongly urged when handling these materials.

Modern soaps have been specifically formulated to be compatible with skin and to be used on a daily basis with minimal side effects. Excessive use of soap for skin cleansing can disrupt the natural barrier function of skin through the removal of skin oils and disruption of the lipid bilayer in skin. This can result in imperfect desquamation or a dry appearance to skin and cause an irritation response or erythema, ie, reddening of the skin.

Additional Uses of Soap

The primary use of carboxylate soaps is in the manufacture of personal cleansing products, principally bar soaps. There are also other applications. Soluble soaps such as potassium and sodium are utilized in cleansing applications where their detersive and emulsification properties can be leveraged, for example, in the textile industry where the cleansing of various in-process fibers and leather (defatting) is desired. Soaps are also utilized in emulsion polymerization. Furthermore, soaps are used in a variety of cosmetic products.

Metal soaps, such as calcium, magnesium, and aluminum soaps, are commonly utilized for the thickening of hydrocarbon lubricating greases, mold release, and suspending agents in paints (see DRIERS AND METALLIC SOAPS). Magnesium stearate is frequently used as a filler material and binder in drug tablets and as either inert binder or emulsification agent in cleansing products and cosmetics, respectively.

ROBERT G. BARTOLO
MATTHEW L. LYNCH
The Procter & Gamble Company

D. Small, ed., *Handbook of Lipid Research 4, The Physical Chemistry of Lipids from Alkanes to Phospholipids*, Plenum Press, New York, 1986, Chapts. 3 and 9.

R. G. Laughlin, *The Aqueous Phase Behaviour of Surfactants*, Academic Press, Ltd., London, 1994.

E. Woollatt, *The Manufacture of Soaps, Other Detergents and Glycerine*, John Wiley & Sons, Inc., New York, 1985.

L. Spitz., ed., *Soaps and Detergents, A Theoretical and Practical Review*, AOCS Press, Champaign, Ill., 1996.

SODA. See ALKALI AND CHLORINE PRODUCTS.

SODIUM AND SODIUM ALLOYS

SODIUM

Sodium, Na, an alkali metal, is the second element of Group 1 (IA) of the Periodic Table, atomic wt 22.9898. Sodium is not found in the free state in nature because of its high chemical reactivity. It occurs naturally as a component of many complex minerals and of such simple ones as sodium chloride, sodium carbonate, sodium sulfate, sodium borate, and sodium nitrate. Soluble sodium salts are found in seawater, mineral springs, and salt lakes. (see CHEMICALS FROM BRINE). Sodium-23 is the only naturally occurring isotope.

Physical Properties

Sodium is a soft, malleable solid readily cut with a knife or extruded as wire. It is commonly coated with a layer of white sodium monoxide, carbonate, or hydroxide, depending on the degree and kind of atmospheric exposure. In a strictly anhydrous inert atmosphere, the freshly cut surface has a faintly pink, bright metallic luster. Liquid sodium in such an atmosphere looks much like mercury.

Only body-centered cubic crystals, lattice constant 428.2 pm at 20°C, are reported for sodium. The atomic radius is 185 pm, the ionic radius 97 pm, and electronic configuration is $1s^22s^22p^63s^1$. Selected properties of sodium are given in Table 1. Sodium is paramagnetic. The vapor is chiefly monatomic. Thin films are opaque in the visible range but transmit in the ultraviolet at ca 210 nm.

High Surface Sodium. Liquid sodium readily wets many solid surfaces. This property may be used to provide a highly reactive form of sodium without contamination by hydrocarbons. Powdered solids having a high surface area per unit volume, eg, completely dehydrated activated alumina powder, provide a suitable base for high surface sodium.

Chemical Properties

Sodium forms unstable solutions in liquid ammonia, where a slow reaction takes place to form sodamide and hydrogen, as follows:

$$Na + NH_3 \rightarrow NaNH_2 + 0.5\ H_2$$

Iron, cobalt, and nickel catalyze this reaction. At high temperature, sodium and its fused halides are mutually soluble.

The solubility–temperature relationships of sodium, sodium compounds, iron, chromium, nickel, helium, hydrogen, and some of the rare gases are important in the design of sodium heat exchangers, especially those used in liquid-metal fast-breeder reactors (LMFBR). The solubility of oxygen in sodium is particularly important because of its marked effect on the corrosion of containment metals and because of problems of plugging narrow passages.

Sodium Reactions. Sodium reacts with many elements and substances and forms well-defined compounds with a number of metals. Some of these alloys are liquid below 300°C. When heated in air, sodium ignites at about 120°C and burns with a yellow flame, evolving a dense white acrid smoke. In the presence of air or oxygen a monoxide or peroxide is formed (see PEROXIDES AND PEROXIDE COMPOUNDS, INORGANIC PEROXIDES). Sodium does not react with extremely dry oxygen or air, except for the possible formation of a surface film of transparent oxide.

Hydrogen and sodium do not react at room temperature, but at 200–350°C sodium hydride is formed. Sodium and graphite form lamellar intercalation compounds. Sodium reacts with carbon monoxide to give sodium carbide, and with acetylene to give sodium acetylide, $NaHC_2$, and sodium carbide (disodium acetylide), Na_2C_2 (see CARBIDES). In vapor phase, sodium forms halides with all halogens. At room temperature, chlorine and bromine react rapidly with thin films of sodium, whereas fluorine and sodium ignite (see SODIUM COMPOUNDS, SODIUM HALIDES).

Table 1. Properties of Sodium

Property	Value
specific heat, solid, at 20°C, kJ/(kg·K)[a]	2.01
melting point, °C	97.82
heat of fusion, kJ/kg[a]	113
boiling point, °C	881.4
heat of vaporization at bp, MJ/kg[a]	3.874
density at 20°C, g/cm^3	0.968
electrical resistivity at 20°C, $\mu\Omega$·cm	4.69
thermal conductivity, at 20°C, W/(m·K)	1323

[a] To convert J to cal, divide by 4.184.

Organosodium compounds are prepared from sodium and other organometallic compounds or active methylene compounds by reaction with organic halides, cleavage of ethers, or addition to unsaturated compounds. Some aromatic vinyl compounds and allylic compounds also give sodium derivatives.

Manufacture

Thermal Reduction. Metallic sodium is produced by thermal reduction of several of its compounds. The earliest commercial processes were based on the carbon reduction of sodium carbonate or sodium hydroxide.

Electrolysis of Fused Sodium Hydroxide. The first successful electrolytic production of sodium was achieved with the Castner cell:

Cathode $$4\ Na^+ + 4\ e^- \rightarrow 4\ Na^0$$

Anode $$4\ OH^- - 4\ e^- \rightarrow 2\ H_2O + O_2$$

Electrolysis of Fused Sodium Chloride. Although many cells have been developed for the electrolysis of fused sodium chloride, the Downs cell (Fig. 1) has been most successful (see ELECTROCHEMICAL PROCESSING). The cell consists of three chambers. The upper chamber is outside the chlorine dome and above the sodium-collecting ring. The other two chambers are the chlorine-collecting zone inside the dome and diaphragm, and the sodium-collecting zone outside the diaphragm, and under the sodium-collecting ring of the collector unit. This arrangement prevents recombination of the sodium and chlorine.

The chlorine emerges through the nickel dome, H, and is removed through the chlorine line, I, to a header (see ALKALI AND CHLORINE PRODUCTS). Sodium, J, is channeled to a riser pipe, K, which leads to a discharge point above the cell wall. The sodium, still containing some calcium, electrolyte, and oxide, overflows into a receiver, L. The calcium precipitated in the riser pipe tends to adhere to the wall from which location it is dislodged by the scraper, M, and returned to the base of the riser. The cell is fitted with a smoke-collection cover to collect particulate emissions and to protect the operators.

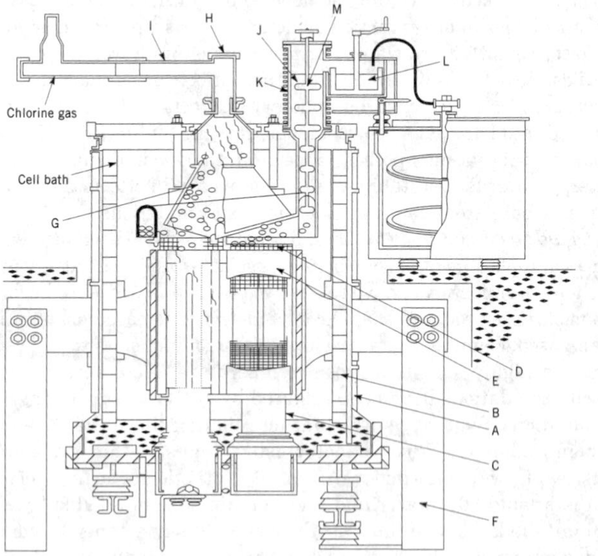

Figure 1. Downs cell: A, the steel shell, contains the fused bath; B is the fire-brick lining; C, four cylindrical graphite anodes project upward from the base of the cell, each surrounded by D, a diaphragm of iron gauze, and E, a steel cathode. The four cathode cylinders are joined to form a single unit supported on cathode arms projecting through the cell walls and connected to F, the cathode bus bar. The diaphragms are suspended from G, the collector assembly, which is supported from steel beams spanning the cell top. For descriptions of H–M, see text.

Electrolysis of Amalgam. Sodium in the form of amalgam as made by the electrolysis of sodium chloride brine in mercury cathode cells is much less expensive than any other form of the metal, but commercial use of amalgam is restricted largely to production of caustic soda (see ALKALI AND CHLORINE PRODUCTS).

Electrolysis Based on Cationically Conducting Ceramics. Searching for a method for using sodium and sulfur (qv) as reactants in a secondary battery, the Ford Motor Company developed a polycrystalline β-alumina ceramic material that selectively transports sodium cations when subjected to an electric field (see BATTERIES; CERAMICS AS ELECTRICAL MATERIALS). This ceramic, or any of its many variants, is useful as a diaphragm or divider in a two-compartment cell. In one compartment, the sodium is in contact with the ceramic; in the other, a suitable liquid electrolyte is in contact with the opposite side of the ceramic. Thus, the sodium is in electrochemical but not physical contact with the liquid electrolyte. Because sodium is not in contact with the liquid electrolyte, the various reactions that usually lower the current efficiency of commercial cells do not occur. Cells based on this principle generally operate at close to 100% current efficiency.

Analytical Methods

Sodium is identified by the intense yellow color that sodium compounds impart to a flame or spectroscopically by the characteristic sodium lines. The latter test is extremely sensitive, yet because many materials contain traces of sodium salts as impurities, it is not conclusive evidence of the presence of sodium in any considerable quantity. Metallic sodium is determined with fair accuracy by measuring the hydrogen liberated on the addition of ethyl alcohol. Sodium amalgam is analyzed by treating a sample with a measured volume of dilute standard acid. After the evolution of hydrogen stops, the excess acid is titrated with a standard base. Total alkalinity is calculated as sodium.

Health and Safety Factors

Using properly designed equipment and strict safe-handling procedures, sodium is used in large- and small-volume applications without incident. The hazards of handling sodium are no greater than those encountered using many other industrial chemicals. Direct contact of the skin with sodium can cause deep, serious burns from the action of sodium with the moisture present and the subsequent corrosive action of the caustic formed. Goggles, face shields, hard hat, hoods, long-gauntlet mittens, and multiple layers of flame-retardant protective clothing are recommended when working with molten sodium.

Perhaps the greatest hazard presented by metallic sodium stems from its extremely vigorous reaction with water to form sodium hydroxide and hydrogen with the evolution of heat. In the presence of air this combination usually results in explosion; in a closed system where an inert atmosphere is present, the hydrogen evolved can cause a rapid increase in pressure.

Another hazard arises from the oxidation of sodium in air. Liquid sodium can autoignite at 120°C, although under some conditions dispersed or high surface sodium may ignite at much lower temperatures. In the laboratory, sodium is best handled in a glove box filled with nitrogen or another inert gas, or in a water-free hood. When sodium is handled on the bench top, water and aqueous solutions must be excluded from the area. Contact of sodium with air should be kept to a minimum because moisture in the air reacts rapidly with sodium. Most reactions of sodium are heterogeneous, occurring on the surface of solid or liquid sodium. Such reactions are accelerated by extending the sodium surface exposed. Dispersions of sodium spilled on cloth or other absorbent material may ignite quickly.

Uses

As of the mid-1990s, the largest consumption of sodium worldwide was the production of tetraethyllead and tetramethyllead antiknock compounds for gasoline. This production was outside of North America.

Refractory metals such as titanium, zirconium, and hafnium are manufactured by sodium reduction of their halides. Calcium metal and calcium hydride are prepared by the reduction of granular calcium chloride with sodium or sodium and hydrogen, respectively, at temperatures below the fusion point of the resulting salt mixtures. Sodium peroxide is an excellent agent for liberating metal from complex ores, eg, silver tetrahedrites. Sodium hydride is employed as catalyst or reactant in numerous organic reactions and for the production of other hydrides, eg, sodium borohydride (see HYDRIDES). Sodium is used indirectly for the descaling of metals such as stainless steel and titanium.

Sodium is employed as a reducing agent in numerous preparations, including the manufacture of dyes (see DYES AND DYE INTERMEDIATES); herbicides (qv); pharmaceuticals (qv); high molecular weight alcohols; and perfume materials (see PERFUMES).

Sodium is a catalyst for many polymerizations. Sodium as an active electrode component of primary and secondary batteries offers the advantages of low atomic weight and high potential. Sodium is used as a heat-transfer medium in primary and secondary cooling loops of liquid-metal fast-breeder power reactors. Low neutron cross section, short half-life of the radioisotopes produced, low corrosiveness, low density, low viscosity, low melting point, high boiling point, high thermal conductivity, and low pressure make sodium systems attractive for this application.

Sodium has also been essential to new developments in heat transfer in advanced solar energy collectors for powering systems remote from electrical distribution systems, and aerospace. In metallurgical practice, sodium uses include preparation of powdered metals; removal of antimony, tin, and sulfur from lead; modification of the structure of silicon–aluminum alloys; application of diffusion alloy coatings to substrate metals; cleaning and desulfurizing alloy steels via NaH; nodularization of graphite in cast iron; deoxidation of molten metals; heat treatment; and the coating of steel using aluminum or zinc.

SODIUM ALLOYS

Sodium is miscible with many metals in liquid phase and forms alloys or compounds. Important examples are listed in Table 2. The brittleness of metals is frequently increased by the addition of sodium to form alloys. The metals vary in their ability to dilute the natural reactivity of sodium. Most binary alloys are unstable in air and react with water. Ternary and quaternary alloys are more stable. Sodium–potassium alloy is easily prepared by melting the clean metals in an inert atmosphere or under an inert hydrocarbon, or by the reaction of sodium with molten KCl, KOH, or solid K_2CO_3 powder. Alloys of lead and sodium containing up to 30 wt % sodium are obtained by heating the metals together in the desired ratio. The brittle alloys can be ground to a powder and should be stored under a hydrocarbon or in airtight containers.

Sodium–lead alloys that contain other metals, eg, the alkaline-earth metals, are hard even at high temperatures, and are thus suitable as bearing metals (see BEARING MATERIALS). Up to ca 0.6 wt %

Table 2. Metal-Sodium Systems

Metal	Alloy formation	Compound formation	Consolute temperature, °C
barium	+	+	miscible
calcium	+		ca 1200
lead	+	+	
lithium	+		306
magnesium	+		>800
mercury	+	+	
potassium	+	+	miscible
rubidium	+		miscible
tin	+	+	
zinc	+	+	>800

sodium dissolves readily in mercury to form amalgams that are liquid at room temperature. Amalgams are useful in many reactions in place of sodium because the reactions are easier to control. Sodium does not form alloys with aluminum but is used to modify the grain structure of aluminum–silicon alloys and aluminum–copper alloys for improved machinability. Sodium–gold alloy is photoelectrically sensitive and may be used in photoelectric cells.

CHARLES H. LEMKE
E. I. du Pont de Nemours & Co., Inc.
University of Delaware
VERNON H. MARKANT
E. I. du Pont de Nemours & Co., Inc.

O. J. Foust, ed., *Sodium–NaK Engineering Handbook*, Gordon and Breach, New York, 1972.

J. W. Mellor, *The Alkali Metals*, Vol. II, Suppl. II of *Comprehensive Treatise on Inorganic and Theoretical Chemistry*, John Wiley & Sons, Inc., New York, 1961, Pt. 1.

T. P. Whaley, in A. F. Trotman-Dickenson, ed., *Comprehensive Inorganic Chemistry*, Pergamon Press, Oxford, U.K., 1973, Chapt. 8.

"Natrium," *Gmelins Handbuch der Anorganischen Chemie*, 8th ed., Vol. 2, Verlag Chemie GmbH, Weinheim, Germany, 1965.

SODIUM CARBONATE. See ALKALI AND CHLORINE PRODUCTS.

SODIUM COMPOUNDS

SODIUM HALIDES

SODIUM CHLORIDE

Salt producers classify sodium chloride, NaCl, also known as common salt, by the three methods used for its production: mechanical evaporation of solution-mined brine, ie, evaporated–granulated salt; underground mining of halite deposits, ie, rock salt; and solar evaporation of seawater, natural brine, or solution-mined brine, ie, solar salt. Salt in brine is a fourth classification, for solution-mined brine that is typically used as a feedstock for chemical production. Salt is a readily available, inexpensive bulk commodity and a basic requirement for all life. It is found throughout the world in natural underground deposits as the mineral halite and, in some locations, as mixed evaporites in saline lakes. Sodium chloride, as a compound, is the largest component of dissolved solids found in seawater, where it averages 2.6% by weight.

Salt Deposition

Salt deposits are widely distributed both in location and in geologic time. There are two basic types of salt formations: bedded deposits and salt structures or diapirs, also known as salt domes. In bedded deposits, layers of halite are separated by layers of the mineral anhydrite (calcium sulfate). Salt structures or diapirs are formed from bedded salt deposits by isostatic salt movement. The salt dome or diapir is typically composed of relatively pure sodium chloride in a vertically elongated, roughly cylindrical, or inverted teardrop-shaped mass.

Properties

Sodium chloride precipitates in cubic, crystalline form. When pure, it is colorless and consists of 60.663 wt % Cl and 39.337 wt % Na. Table 1 shows selected properties of sodium chloride. Salt is soluble in polar solvents, insoluble in nonpolar. The aqueous solution has a pH of 7 in the absence of impurities.

Table 1. Properties of Pure Sodium Chloride

Property	Value
molecular wt	58.44
crystalline form	cubic
color	clear to white
index of refraction, n_D^{20}	1.5442
density or specific gravity, g/cm^3	2.165
mp, °C	801
bp, °C	1413
hardness, Mohs' scale	2.5
specific heat, J/(g·K)[a]	0.853
heat of fusion, J/g[a]	517.1
critical humidity at 20°C, %	75.3
heat of solution, 1 kg H_2O, 25°C, kJ/mol[a]	3.757

[a] To convert J to cal, divide by 4.184.

Processing

Salt production refers to the production of dry, crystalline salt from seawater, or solution-mined or natural brine, and the conventional mining of rock salt. Salt in brine produced by solution mining is an intermediate product that is processed into evaporated–granulated salt or other chemical products (see CHEMICALS FROM BRINE). In the strictest sense, all natural salt is oceanic in origin.

Solution Mining and Mechanical Evaporation. Bedded and domal salt deposits are solution-mined by drilling wells into halite deposits and injecting fresh and recycled water through the well casings, dissolving the salt.

Table salt is typical of the fine, evaporated–granulated salt produced in vacuum pan evaporators. Virtually all food-grade salt sold or used in the United States is produced by vacuum evaporation (qv) of brine. Prior to mechanical evaporation, the brine may be treated to remove minerals that can cause scaling in the evaporators and adversely affect salt purity. Chemical treatment of the brine, followed by settling, reduces levels of dissolved calcium, magnesium, and sulfate.

Evaporated (grainer) salt can also be produced with the addition of heat in open pans. The resulting grainer salt consists of flakes and is used for food applications where a coarser, flake salt product is desired. Three specialty types of salt are produced by variations of the evaporating process. Alberger salt is produced with a modified grainer method and results in a combination of flakes and cubic crystals of salt. It is typically used where coarser salt is required, as in koshering. Dendritic salt is made by adding yellow prussiate of soda (YPS) (sodium ferrocyanide) to the brine; the YPS modifies the crystal structure of sodium chloride to form a branch-like, low density dendritic salt crystal. The recrystallizer process is a means of producing purified salt. Brine made by dissolving rock salt or solar salt is purified and fed to vacuum pan evaporators. The result is granulated–evaporated salt of excellent purity, it has many uses where purity and consistency of grain size and shape are important.

Conventional Underground Mining. Rock salt is extracted from a conventional mine by drilling and blasting. Access is through a circular shaft, typically about 6 m dia and as deep as 700 m, depending on the depth of the salt deposit. Shafts are lined with concrete at least through the overburden and into the top of the salt deposit, and sometimes all the way to the shaft bottom.

Domal salt deposits, because of the potentially great thickness of the salt in them, are mined first by using horizontal entries or rooms, followed by excavating or benching the floor downward to a depth of 20 m or more. Benching improves the efficiency of mining salt domes because larger quantities of salt can be blasted in a single shot. Salt is mined by undercutting, drilling, blasting, loading, and transporting the broken salt for processing. Salt mining operations are highly mechanized, with extensive use of large, diesel-powered equipment.

Continuous miners or boring machines have been used since the late 1950s for salt mine development and, in some cases, for production. These machines have movable, rotating heads with carbide-tipped cutting bits. Continuous miners bore into the salt and eliminate undercutting, drilling, and blasting. After the salt is blasted or mined by a continuous miner, it is loaded into trucks or shuttle cars and hauled to a primary crusher. The salt is then transported by conveyor belt to second- and third-stage crushers and screening stations for further size reduction and screening into standard product grades for specific end uses. The salt, either as mine run grade or as screened grades, is then conveyed to the shaft bottom for hoisting to the surface.

Solar Evaporation. Solar salt is produced by natural evaporation of seawater or brine in large, diked, earthen ponds. Solar radiation and wind action concentrate the seawater or natural brine until sodium chloride crystallizes. Solar salt production requires a large area of flat, low cost land: Climatic conditions must ensure high evaporation rates and low rainfall. Although most solar salt production facilities use seawater as feedstock, natural brine and solution-mined brine are also used. Solar ponds are constructed to take advantage of natural ground contours to aid in brine movement. At some point in most pond systems, seawater or brine must be elevated by pumping. Solar salt production is a form of fractional crystallization. Brine reaching the crystallizers contains in solution calcium sulfate, magnesium sulfate, magnesium chloride, and small amounts of potassium chloride, plus minuscule amounts of other elements. The saturated brine is fed onto level, rectangular crystallizing ponds to maintain a brine depth of 30 cm or less. As evaporation proceeds, sodium chloride precipitates and forms a salt layer 10–25 cm thick. When the brine reaches 25–30° Bé, the bittern is discharged. Most of the magnesium sulfate and magnesium and potassium chlorides remains in the bittern.

The salt crop is harvested loaded into trucks, and transported to a washplant. The salt is washed with clean, nearly saturated brine. Uncontaminated brine is made and recycled in a settling pond by adding seawater to dissolve fine salt collected by the wash brine. After washing, the salt is stockpiled and allowed to drain. Limited rainfall is relied on to improve salt quality by rinsing action. Solar salt typically drains naturally to a moisture level of about 3.5%. Similar to processing mined rock salt, solar salt may be crushed, screened, and kiln dried or fluidized-bed dried. Coarse solar salt is a premium product because of high purity and relatively large crystal size (see ION EXCHANGE; WATER, INDUSTRIAL WATER TREATMENT).

Electrodialysis. Electrodialytic membrane process technology is used extensively in Japan to produce granulated–evaporated salt. Filtered seawater is concentrated by membrane electrodialysis and evaporated in multiple-effect evaporators.

Salt Standards and Specifications

Mechanically evaporated salt made using purified brine generally has the highest purity; rock salt generally has the lowest.

Additives. Sodium chloride is hygroscopic, at high relative humidity, and individual salt crystals can adsorb enough moisture during storage to result in formation of brine on crystal surfaces. Subsequent evaporation causes recrystallization, and the salt crystals bond firmly together. Free-flow or anticaking agents are sometimes added to salt to prevent caking.

Analytical Methods

The most common impurities, depending on type of salt, are calcium sulfate, calcium chloride, magnesium chloride or magnesium sulfate, sodium sulfate, and water-insoluble material. Surface moisture is determined by drying, water-insoluble material by weighing, calcium and magnesium by EDTA titration, and sulfate gravimetrically.

Health and Safety Factors

Salt is an essential nutrient without which life could not exist (see MINERAL NUTRIENTS). Sodium is the principal cation in blood plasma and body tissue fluids. Regulation of sodium in the body maintains osmotic pressure, acid–base balance, and volume of circulating body fluids. Extracellular fluid volume is maintained by the total body sodium content. Brain mechanisms control the elective intake of salt, and physiological control systems regulate the conservation and excretion of sodium. The sodium concentration of blood and other body fluids is maintained by a complex mechanism involving the kidneys and the adrenal glands.

Sodium Intake. The U.S. FDA's GRAS review puts the amount of naturally occurring sodium in the American diet at 1000–1500 mg/d, equivalent to the amount of sodium in approximately 2500–3800 mg NaCl. The requirement for salt in the diet has not been precisely established, but the safe and adequate intake for adults is reported as 1875–5625 mg.

Sodium Restriction. There is no consensus in the scientific community about the relationship between sodium or salt intake and hypertension. Whereas there is no evidence that high levels of salt intake can cause high blood pressure in healthy, normotensive people, there is evidence that severe salt restriction can lower blood pressure in one-third to one-half of individuals with hypertension.

Toxicity. The U.S. Food and Drug Administration regards common salt, ie, NaCl, as GRAS for its intended use as a food additive. Oral toxicity for mammals is reported in mg/kg: for humans, TD_{LO}: 12,357; 23D-C (daily-continuous) for mice, LD_{50}: 4000; for rats, LD_{50}: 3000; and for rabbits, LD_{LO}: 8000. TD_{LO} and LD_{LO} are lowest level toxic and lethal dosages, respectively.

Environment and Infrastructure. Both sodium and chloride are present in the environment naturally. Ecosystems supporting aquatic species and biota as well as vegetation, including roadside grasses, shrubs and trees, and food crops, can tolerate various concentrations of sodium and chloride. Environmental effects of elevated salinity levels resulting from uses of salt are highly site-specific, such as soil texture, permeability, drainage, amount of water applied, and the salt tolerance of the vegetation or crops.

Uses

Salt can be classified under five principal use categories, plus a catch-all classification that includes most industrial uses, as (*1*) food-grade, (*2*) agriculture, (*3*) highway, (*4*) water conditioning, (*5*) chemical, and (*6*) miscellaneous.

Foods and Food Processing. Sodium chloride performs several necessary functions in food processing and cooking. In addition to use as a flavor enhancer, salt is used in food processing as a preservative, a color developer, binder, texturizer, and as a fermentation-control agent.

Iodized Salt. Iodized table salt has been used to provide supplemental iodine to the U.S. population since 1924, when producers, in cooperation with the Michigan State Medical Society, began a voluntary program of salt iodization in Michigan that ultimately led to the elimination of iodine deficiency in the United States. Iodine deficiency in less developed countries is still a serious problem.

Agriculture. Salt supplementation is a critical part of a nutritionally balanced diet for animals. In addition, because animals have a definite appetite for salt, it can be used as a delivery mechanism to ensure adequate intake of less palatable nutrients and as a feed limiter (see FEEDS AND FEED ADDITIVES).

Highway. Rock salt, solar salt, and in some cases in Europe, evaporated salt are used to maintain traffic safety and mobility during snow and ice conditions in snowbelt regions throughout the world. Sodium chloride melts ice at temperatures down to its eutectic point of −21.12°C.

Water Conditioning. Well water and many public drinking water supplies contain elevated levels of calcium and magnesium. The resulting hard water reduces the sudsing action of soaps and detergents and causes a greasy, curd-like deposit when the water is used for laundering, cleaning, and bathing. Water is softened by removing calcium and magnesium ions from hard water in exchange sodium ions at sites on cation-exchange resin.

Chemical Uses. Salt in brine is produced and used by chemical manufacturers to produce chlorine and caustic soda. Chlorine and caustic soda are used by the pulp (qv) and paper (qv) industry in multistage bleaching; caustic soda is used in wood fiber processing. Reduced use of chlorine in the pulp and paper industry owing to environmental concern has contributed to the decline in chlor–alkali production. Salt is used to make sodium chlorate and metallic sodium by electrolysis and to a lesser degree is used as a reactant for sulfuric acid to produce sodium sulfate and hydrochloric acid.

Industrial Uses. Salt is reported to have more than 14,000 uses. Among these, salt is used by the textile and dyeing industry to fix dyes and to standardize dye batches (see DYE AND DYE INTERMEDIATES; TEXTILES); by metal processors, such as secondary aluminum refiners, to remove impurities (see ALUMINUM AND ALUMINUM ALLOYS); in rubber manufacturing to separate rubber from latex (see RUBBER, NATURAL); as a filler and grinding agent in pigment and dry-detergent processes; in ceramics (qv) manufacture for surface vitrification of heated clays; in soapmaking to separate soap (qv) from water and glycerol; in oil and gas well drilling muds to inhibit fermentation, increase density, and to stabilize rock salt strata; and in animal-hide processing and leather (qv) tanning to cure, preserve, and tan hides.

BRUCE M. BERTRAM
Salt Institute

R. P. Multhauf, *Neptune's Gift*, The Johns Hopkins University Press, Baltimore, Md., 1978.

D. Kaufman, ed., *Sodium Chloride*, ACS Monograph Series, No. 145, The American Chemical Society, Reinhold Publishing Corp., New York, 1960.

C. A. Baar, *Applied Salt-Rock Mechanics 1, The In Situ Behavior of Salt Rocks*, Elsevier Science, Inc., New York, 1977.

I. Lerche and J. J. O'Brien, eds., *Dynamical Geology of Salt and Related Structures*, Academic Press, Inc., Orlando, Fla., 1987.

SODIUM BROMIDE

Sodium bromide, NaBr, the most common and available alkali bromide, is a salt of hydrobromic acid (see BROMINE COMPOUNDS). Sodium bromide crystallizes from aqueous solution as a dihydrate, $NaBr{\cdot}2H_2O$, below 51 °C. Above 51 °C, it crystallizes as the anhydrous compound. Crystals of the dihydrate belong to the monoclinic system. The anhydrous crystal belongs to the cubic system, $a = 596$ pm. Other properties of the anhydrous salt are listed in Table 1. The anhydrous salt is hygroscopic but not deliquescent.

Preparation and Manufacture

Small quantities of pure sodium bromide can be prepared by neutralizing solutions of either sodium hydroxide or sodium carbonate using hydrobromic acid which is free of bromine, followed by evaporation and crystallization. Commercial quantities of sodium bromide are usually prepared by adding excess bromine to a solution of sodium hydroxide, evaporating to dryness, and treating with a reducing agent to reduce sodium bromate to sodium bromide.

Table 1. Properties of Sodium Bromide

Property	Value
molecular weight	102.89
melting point, °C	755
boiling point, °C	1390
density, g/cm^3	3.203
refractive index, n_D^{25}	1.6412
heat of fusion, J/g[a]	254.9
ΔH_f, kJ/mol[a]	−361.414
S, J/(mol·K)[a]	86.82
heat capacity, J/(mol·K)[a,b]	$47.92 + 1.331 \times 10^{-3}\,T$

[a] To convert J to cal, divide by 4.184.
[b] Temperature, T, is in kelvin.

Health and Safety

Sodium bromide is moderately toxic by ingestion and can affect the gastrointestinal and central nervous systems and the skin. Ingestion of large amounts in a single dose causes immediate abdominal pain. The LD_{50} for sodium bromide taken orally by rats is 3.5 g/kg body weight, and the TD_{LO} orally in rats is 720 mg/kg.

Uses

The oil and gas drilling industry is a principal consumer of sodium bromide. Because of its high solubility in water, clear brine fluids of densities up to 1.547 g/cm^3 (12.9 lb/gal) at 25°C can be obtained. These are used directly or in blends with other clear brine fluids for well completions or workovers. An increasingly important use of sodium bromide is as a biocide, particularly in industrial cooling water towers and in swimming pool water treatment, replacing the more hazardous chlorine (see WATER, TREATMENT OF SWIMMING POOLS, SPAS, AND HOT TUBS).

Other applications of sodium bromide include use in the photographic industry both to make light-sensitive silver bromide emulsions and to lower the solubility of silver bromides during the developing process; use as a wood (qv) preservative in conjunction with hydrogen peroxide; as a cocatalyst along with cobalt acetate for the partial oxidation of alkyl side chains on polystyrene polymers; and as a sedative, hypnotic, and anticonvulsant.

ROGER N. KUST
Tetra Technologies, Inc.

R. E. Gosselin, H. C. Hodge, R. P. Smith, and M. N. Gleason, *Clinical Toxicology of Commercial Products,* 4th ed., Williams and Wilkins, Baltimore, Md., 1976.

"Sodium Bromide" in R. J. Lewis, ed., *SAX's Dangerous Properties of Industrial Materials,* 8th ed., Van Nostrand Reinhold Co., New York, 1994.

SODIUM IODIDE

Sodium iodide NaI, occurs as colorless crystals or as a white crystalline solid. It has a salty and slightly bitter taste. In moist air, it gradually absorbs as much as 5% water, which causes caking or even liquefaction (deliquescence) (see IODINE AND IODINE COMPOUNDS).

Properties

Sodium iodide crystallizes in the cubic system. Physical properties are given in Table 1. Sodium iodide is soluble in water as well as in methanol, ethanol, acetone, glycerol, and several other organic solvents.

Manufacture

Bulk production of *United States Pharmacopeia* (USP) and reagent grades is based on the reaction of sodium carbonate or hydroxide with an acidic iodide solution, typically hydriodic acid or a metal iodide.

Economic Aspects and Uses

The principal use of sodium iodide is in scintillation crystals, which are used for gamma-ray counters, and in medicine as the detectors in computer-assisted tomography (CAT) scan and positron emission tomography (PET) equipment. A small amount is used in the wet extraction of silver, in iodized salt (see FOOD ADDITIVES), animal feeds to prevent hoofrot (see FEEDS AND FEED ADDITIVES), and in the manufacture of organic chemicals. It has also been used in cloud seeding and in halogen discharge lamps.

Table 1. Physical Properties of Sodium Iodide

Property	Value
mol wt	149.895
mp, °C	651
bp, °C	1304
d_4^{25}, g/cm^3	3.667
specific heat, J/(kg·K)[a] at 0°C	350
50°C	360

[a] To convert J to cal, divide by 4.184.

PHILIP H. MERRELL
ELIZABETH M. PETERS
Mallinckrodt, Inc.

J. C. Bailar, Jr., H. J. Emelius, R. Nyholm, and A. F. Trotman-Dickenson, eds., *Comprehensive Inorganic Chemistry,* Pergamon Press, Inc., Elmsford, N.Y., 1973, Vols. 1 and 2.

N. Gehrels and R. M. Candey, *AIP Conf. Proc.,* 211, 213–223, (1990) for information on gamma-ray spectroscopy.

M. J. Geagan, B. B. Chase, and G. Muehllehner, *Nucl. Instrum. Methods Phys. Res., Sect. A,* **353**(1–3), 379–383 (1994) for information on CAT scan and PET.

The United States Pharmacopeia XXII(USP XXII-NF XVII), The United States Pharmacopeial Convention, Inc., Rockville, Md., 1990, p. 1261.

SODIUM NITRATE

Sodium nitrate, $NaNO_3$, is found in naturally occurring deposits associated with sodium chloride, sodium sulfate, potassium chloride, potassium nitrate, magnesium chloride, and other salts. The only deposits being commercially exploited are in Chile, South America. Natural sodium nitrate is also referred to as Chilean saltpeter or Chilean nitrate.

Deposits

The nitrate ore, caliche, is a conglomerate of insoluble and barren material such as breccia, sands, and clays (qv), firmly cemented by soluble oxidized salts that are predominantly sulfates, nitrates, and chlorides of sodium, potassium, and magnesium.

Properties

Selected physical and chemical properties of sodium nitrate are listed in Table 1. At room temperature, sodium nitrate is an odorless and colorless solid, moderately hygroscopic, saline in taste, and very soluble in water, ammonia, and glycerol.

Table 1. Selected Properties of Sodium Nitrate

Property	Value
mol wt	84.99
crystal system	trigonal, rhombohedral
mp, °C	308
refractive index, n_D	
trigonal	1.587
rhombohedral	1.336
density, solid, g/cm^3	2.257
solubility in H_2O, molality (±2%)	
at 0°C	8.62
40°C	12.39
120°C	24.80
specific conductivity, at 300°C, S/cm	0.95
viscosity[a], η, mPa·s(= cP)	
at 590 K	2.85
730 K	1.53
heat of fusion, J/g[b]	189.5
heat capacity, J/g[b]	
solid at 0°C	1.035 ± 0.005
liquid at 350°C	1.80 ± 0.02

[a] Measurement method: capillary; $\eta = 25.0987 - 6.0544 \times 10^{-2}\,T + 3.8709 \times 10^{-5}\,T^2$. Precision = ca 0.6%; uncertainty = ca 3%. [b] To convert J to cal, divide by 4.184.

Manufacture and Processing

Natural Sodium Nitrate. The manufacture of natural sodium nitrate is carried out by its extraction from the ore by leaching with a brine, followed by fractional crystallization. Historically the Shanks process was utilized, but the last plant closed in 1977.

The Guggenheim process, introduced in the late 1920s, was developed to permit the treatment of low grade caliche ores, making it possible to mine by mechanical methods instead of by hand. Because the ore quality is variable, large open-pit mining areas are first identified by general exploration; specific mining strips are later identified by further exploration and testing. Surface mining methods are used. After removal of the overburden, the exposed caliche is drilled, blasted, and loaded into 80-metric ton trucks that deliver the ore to a transfer rail station for transportation to the plants.

The mineral brought from the mine varies in size from fine particles to chunks of 3–5 metric tons. Crushing is carried out in three stages by means of jaw and cone-type crushers. Before crushing, a selecting screen rejects material smaller than 0.42 mm. About 80% of the crushed material, ie, that having diameter >0.42 mm, is sent to the leaching plant. The remaining 20% is sent to the fines treatment ponds. These two fractions have to be leached separately because clay present in the fine fraction swells when in contact with rich brines, occurrence of which would retard the rate of leaching of the coarser fraction.

Once the coarser fraction is transferred to the leaching plant, leaching takes place in a series of 10,000-m^3 leaching vats built of reinforced concrete. The process consists of countercurrent leaching, with one cycle involving 10 vats, one of which is being loaded with crushed ore while another one is being unloaded. The leaching is carried out at a temperature of 40°C, using a mother liquor entering the process at a concentration of 320 g/L of sodium nitrate and ending at a concentration of 440–450 g/L. The leaching process terminates with a final washing with water, where Glauber's salt, $Na_2SO_4 \cdot 10H_2O$, is obtained by crystallization (qv). The depleted ores, containing ca 1% sodium nitrate, are removed by means of electromechanical dredgers into trucks that haul them to tailing disposal areas.

The strong solution from the leaching process is cooled to crystallized sodium nitrate in a series of shell-and-tube heat exchangers. The crystallized sodium nitrate can be dried and prilled to produce fertilizer-grade sodium nitrate, or recrystallized to remove impurities and obtain technical grades of sodium nitrate for industrial uses. Otherwise, crystallized sodium nitrate can be sent to the potassium nitrate plant, where potassium nitrate is produced through a direct reaction of sodium nitrate and potassium chloride (see POTASSIUM COMPOUNDS). Crystallized sodium nitrate can also be combined with existing potassium nitrate to obtain a potassium sodium nitrate.

Synthetic Sodium Nitrate. Sodium nitrate can be obtained synthetically by absorption of nitrous gases or by neutralization of nitric acid (qv). Whereas low NO_x content in waste gases from nitric acid plants makes these gases useless for producing nitric acid, one way to avoid emission of nitrous gases to the atmosphere consists of using an alkaline solution of NaOH or Na_2CO_3 to absorb them. Products are mainly sodium nitrite or sodium nitrate. Sodium nitrate can also be produced by neutralizing nitric acid with sodium hydroxide or sodium carbonate.

Health and Safety Factors

The acceptable daily intake by human adults for nitrates suggested by the World Health Organization (WHO) is 5 mg/kg body wt per day (expressed as sodium nitrate). High doses of nitrates are lethal. Victims of sodium nitrate or potassium nitrate poisoning contract severe gastroenteritis. Outbreaks of poisoning from the ingestion of meats containing sodium nitrate and sodium nitrite have occurred from the accidental incorporation of excessive amounts of nitrate–nitrite mixtures. The health hazards associated with nitrates result mainly from the bacterial conversion of ingested nitrates to nitrites.

Uses

Sodium nitrate is used as a fertilizer as well as in a number of industrial applications. As a fertilizer, it provides nitrogen, needed in large quantities by plants and commonly in shortage in soils. Sodium nitrate is particularly effective as a nitrogen source for sugar beet, vegetable crops, tobacco, and cotton (qv), and for any crop in acid soils.

Sodium nitrate is used in a number of industrial processes, in most of them acting primarily as an oxidizing agent. A primary use is in the manufacture of medium and high quality glass (qv), such as optical and artistic glass, television and computer screens, and fiber glass. In the manufacture of explosives, sodium nitrate is used mainly in blasting agents. In slurries and emulsions, sodium nitrate improves stability and sensitivity.

Another large application is as an ingredient in the production of charcoal briquettes. Typically charcoal briquettes contain up to almost 3% sodium nitrate. Sodium nitrate is also used in formulations of heat-transfer salts for heat-treatment baths for alloys and metals, rubber vulcanization, and petrochemical industries.

LUDWIK POKORNY
IGNACIO MATURANA
SQM Nitratos SA

Grossling and G. E. Ericksen, *Computer Studies of the Composition of Chilean Nitrates Ores,* U.S. Geological Survey, Washington, D.C., Dec. 1970.

G. E. Ericksen, *Geology and Origin of the Chilean Nitrate Deposits,* U.S. Geological Survey, Professional Paper 1188, U.S. Government Printing Office, Washington, D.C., 1981.

V. Sauchelli, ed., *Fertilizer Nitrogen, Its Chemistry and Technology,* Reinhold Publishing Corp., New York, 1964.

Sodium Nitrate, U.S. Military Technical Specifications, MIL-S-322C, U.S. Government Printing Office, Washington, D.C., Feb. 5, 1968.

SODIUM NITRITE

Sodium nitrite, $NaNO_2$, a stable, odorless, pale yellow or straw-colored compound of molecular weight 69.00, is the sodium salt of nitrous acid, HNO_2.

Properties

Pure anhydrous crystalline sodium nitrite has a specific gravity of 2.168 at 0°C/0°C. The crystal structure is body-centered orthorhombic. Sodium nitrite melts at ~284°C and decomposition begins above 320°C, yielding N_2, O_2, NO, and Na_2O. The heat of formation is −362.3 kJ/mol (−86.6 kcal/mol) at 25°C. Sodium nitrite has a transition point at 158–165°C and displays significant changes in physical properties within this temperature range.

Sodium nitrite is hygroscopic and very soluble in water. It has limited solubility in most organic solvents. Sodium nitrite serves as the primary industrial source for nitrous acid in organic syntheses. As an oxidizer, sodium nitrite can convert ammonium ion to nitrogen, urea to carbon dioxide and nitrogen, and sulfamate to sulfate and nitrogen. The oxidizing properties of sodium nitrite contribute to its application as a corrosion inhibitor (see CORROSION AND CORROSION CONTROL). Because it is a strong oxidizer, sodium nitrite is capable of supplying oxygen and thus accelerating the combustion of organic matter. It functions as a reducing agent to more powerful oxidizers such as dichromate, permanganate, chlorate, and chlorine.

Manufacturing

Sodium nitrite has been synthesized by a number of chemical reactions involving the reduction of sodium nitrate, $NaNO_3$. Industrial production of sodium nitrite is by absorption of nitrogen oxides (NO_x) into aqueous sodium carbonate or sodium hydroxide. NO_x gases originate from catalytic air oxidation of anhydrous ammonia, a practice common to nitric acid plants. Solutions of sodium nitrite thus produced are concentrated and a slurry of crystals obtained in conventional evaporation (qv) and crystallization (qv) equipment. The crystals are typically separated from the mother liquor by centrifugation and subsequently dried. Because of its tendency to lump and cake rapidly in storage, dry sodium nitrite products are frequently treated with an anticaking agent to keep them free-flowing.

Shipment and Storage

Dry products of sodium nitrite are most commonly packaged into multi-ply paper bags which contain a polyethylene moisture barrier. Fiber drums and semibulk sacks are also utilized. Liquid sodium nitrite products are typically 40–42% $NaNO_2$ and can be shipped in tank cars or tank trucks when volume and freight considerations allow. Solution products are often preferred because of more convenient, efficient, and cost-effective handling. Care must be exercised in using sodium nitrite near other chemicals. Sodium nitrite exhibits good shelf-life characteristics if stored in secure containers in a cool, dry place, segregated from combustible and incompatible materials.

Health and Safety Factors

Sodium nitrite is poisonous and prolonged contact with dry sodium nitrite or its solutions can cause irritation to the skin, eyes, and mucous membranes. For handling dry products, a hard hat, safety glasses, impervious gloves, and long sleeves should be worn as a minimum. Where dusty or misty conditions prevail or when handling solutions, a NIOSH-approved respirator, chemical goggles, and full impervious clothing may be required.

Uses

The many industrial uses for sodium nitrite are based primarily on its oxidizing properties or its liberation of nitrous acid in acidic solutions.

Dyes. Sodium nitrite is a convenient source of nitrous acid in the nitrosation and diatozation of aromatic amines. When primary aromatic amines react with nitrous acid, the intermediate diamine salts are produced which, on coupling to amines, phenols, naphthols, and other compounds, form the important azo dyes (qv).

Rubber Chemicals. Sodium nitrite is an important raw material in the manufacture of rubber processing chemicals. Accelerators, retarders, antioxidants (qv), and antiozonants (qv) are the types of compounds made using sodium nitrite. (see RUBBER CHEMICALS).

Heat Treatment and Heat-Transfer Salts. Mixtures of sodium nitrite, sodium nitrate, and potassium nitrate are used to prepare molten salt baths and heat-transfer media. The salts can be used for indirect heating or cooling or as quenching baths in the annealing of iron and steel.

Corrosion Inhibition. Sodium nitrite acts as an anodic inhibitor toward ferrous metals by forming a tightly adhering oxide film over the steel, preventing the dissolution of metal at anodic areas. Sodium nitrite is used in boiler water treatment, as a dip or spray for protection of metals in process and storage, and in concrete. Sodium nitrite should not be used as a corrosion inhibitor in amine-based metalworking fluids because of the formation of potentially carcinogenic nitrosamines.

Metal Finishing. In phosphating solutions, sodium nitrite performs as an accelerator and oxidizer, serving to reduce processing times and control buildup of ferrous ions in solution, respectively. In gold–sulfite-plating baths, sodium nitrite functions in the formation a gold–sulfite–nitrite complex, $Na_4Au(SO_3)_2NO_2$, from which the gold can be electrolytically deposited (see GOLD AND GOLD COMPOUNDS).

Meat Curing. Sodium nitrite is used extensively in curing meat and meat products (qv), particularly pork products such as ham, bacon, frankfurters, etc. As an ingredient in curing brines, sodium nitrite acts as a color fixative and inhibits bacteria growth, including *Clostridium botulinum,* the source of the botulism toxin. Certain fish and poultry products are also cured with brines containing sodium nitrite. In the United States all food uses of sodium nitrite are strictly regulated by the FDA and USDA.

Other Uses. Other applications for sodium nitrite include the syntheses of saccharin (see SWEETENERS), synthetic caffeine, fluoroaromatics, and other pharmaceuticals (qv), pesticides (qv), and organic substances; as an inhibitor of polymerization; in the production of foam blowing agents; in removing H_2S from natural gas; in textile dyeing (see TEXTILES); as an analytical reagent; and as an antidote for cyanide poisoning (see CYANIDES).

WALTER H. BORTLE
General Chemical Corporation

"Natrium," in *Gmelins Handbuch der Anorganischen Chemie,* System 21, Vol. 3, Verlag Chemie, Weinheim, Germany, 1966.

J. W. Mellor, *A Comprehensive Treatise on Inorganic and Theoretical Chemistry,* Vol. 8, Longmans, Green & Co., London, 1928; J. W. Mellor, *Supplement to Mellor's Treatise on Inorganic and Theoretical Chemistry,* Vol. VIII, Suppl. II, Part II, John Wiley & Sons, Inc., New York, 1967.

R. H. Perry, ed., *Chemical Engineer's Handbook,* 5th ed., McGraw-Hill Book Co., Inc., New York, 1973.

Sodium Nitrite, Product brochure, GC7767, General Chemical Corp., Parsippany, N.J., 1989.

SODIUM SULFATES

Sulfates of sodium are industrially important materials commonly sold in three forms (Table 1). About half of all sodium sulfate produced is a synthetic by-product of rayon, dichromate, phenol (qv), or potash (see CHROMIUM COMPOUNDS; FIBERS, REGENERATED CELLULOSICS; POTASSIUM COMPOUNDS). Sodium sulfate made as a by-product is referred to as synthetic. Sodium sulfate made from mirabilite, thenardite, or naturally occurring brine is called natural sodium sulfate. Sodium sulfate is mined commercially from three types of mineral evaporites: thenardite, mirabilite, and high sulfate brine deposits (see CHEMICALS FROM BRINE).

Physical and Chemical Properties

Physical and chemical properties of the three most important forms of sodium sulfate are summarized in Table 2. The aqueous solubility of sodium sulfate changes rapidly from 0 to 40°C, and addition of NaCl to a saturated solution of Na_2SO_4 dramatically suppresses this solubility. These two effects are exploited by all manufacturers of sodium sulfate. The reactivity of Na_2SO_4 is relatively low at room temperature.

Manufacture and Processing

There are only two significant producers of natural sodium sulfate: Ozark-Mahoning (Texas), and North American Chemical (California). In Texas, subterranean sulfate brines are pumped to the surface where the brines are first saturated with NaCl before they are cooled by mechanical refrigeration to form Glauber's salt. Processing at Searles Lake, California, is similar to that of Texas brines.

Nearly all manufacturers of sodium sulfate use Glauber's salt in an intermediate process step. Glauber's salt is then converted to anhydrous sodium sulfate. Glauber's salt can be converted to anhydrous sodium sulfate by simply drying it in rotary kilns. Direct drying forms a fine, undesirable powder. This process is not used in the United States but is used in other countries.

The Mannheim process produces sodium sulfate by reaction of sodium chloride and sulfuric acid.

$$NaCl + H_2SO_4 \rightarrow Na_2SO_4 + HCl$$

This reaction takes place in a fluidized-bed reactor or a specially made furnace (the Mannheim furnace). In another process, SO_2, O_2, and H_2O react with NaCl.

$$4\ NaCl + 2\ SO_2 + O_2 + 2\ H_2O \rightarrow 2\ Na_2SO_4 + 4\ HCl$$

This is called the Hargreaves process. Both the Hargreaves and the Mannheim processes are used widely in the rest of the world.

Table 1. Sulfates of Sodium

Chemical name	Mineral name	Common name	Formula
sodium sulfate	thenardite	salt cake[a]	Na_2SO_4
sodium sulfate decahydrate	mirabilite	Glauber's salt	$Na_2SO_4 \cdot 10H_2O$
sodium bisulfate		niter cake	$NaHSO_4$

[a] Increasingly salt cake is used as another name for both high and low grade Na_2SO_4.

Table 2. Properties of Sodium Sulfates

Property	Sodium sulfate, Anhydrous	Sodium sulfate, Decahydrate	Sodium hydrogen sulfate
mol wt	142.05	322.21	120.06
mp, °C	884	32.4	315
specific gravity	2.664	1.464	2.435
specific heat, J/(g·K)[a]	0.845		
heat of formation, kJ/mol[a]	−1385	−4322	−1125
heat of solution, kJ/mol[a]	1.17	−78.41	7.28
heat of crystallization, kJ/mol[a]	−8.8	78.2	
refractive index	1.464	1.394	1.459
crystalline form	rhombic, monoclinic, and hexagonal	monoclinic	triclinic

[a] To convert J to cal, divide by 4.184.

Health and Safety

In general, Na_2SO_4 is not considered an environmentally dangerous material, but the Mannheim and Hargreaves processes are practically nonexistent in the United States because it is difficult to keep emissions of particulate Na_2SO_4 low and keep HCl from escaping to the atmosphere. Sodium sulfate in moderation is used as a diuretic and cathartic for humans and animals (see GASTROINTESTINAL AGENTS). It is also used in consumer products such as laxatives, antacids, and as a natural filler it is used extensively in powdered laundry detergents (see DETERGENCY).

Uses

The principal uses of Na_2SO_4 are in the manufacture of paper, soaps, and detergents. The kraft paper process uses a mixture of sodium sulfide and sodium hydroxide to digest wood chips. Both the sulfide and hydroxide are generated, starting with sodium sulfate as the raw material. At low temperatures, Na_2SO_4 is nonreactive; because of this, it is used as a filler in household soaps and detergents. Properties of sodium sulfate help speed up the melting process in glassmaking. Sodium sulfate improves the working properties of high silica glasses.

Both Na_2SO_4 and $NaHSO_4$ are used to adjust pH and dilute dyes. Sodium sulfate is used in cattle feed (see FEEDS AND FEED ADDITIVES), in cellulose-sponge, as a cement and plaster hardener, and as an aid in metallurgy refining. Sodium bisulfate, $NaHSO_4$, is a convenient mild acid and is safe for uses as a household toilet-bowl cleaner, automobile-radiator cleaner, and for swimming pool pH adjustment. It is used for metal pickling, as a dye-reducing agent, for soil disinfecting, and as a promoter in hardening certain types of cement.

DAVID BUTTS
Great Salt Lake Minerals Corporation

Mellor's Comprehensive Treatise on Inorganic and Theoretical Chemistry, Vol. II, Suppl. II, John Wiley & Sons, Inc., New York, 1961.

Mines FaxBack, U.S. Geological Survey, Document on Demand, on-line 24 h/d, 7 d/wk, (703) 648 4999.

Chemical Engineer's Handbook, 7th ed., McGraw-Hill Book Co., Inc., New York, 1997, Sect. 2.

SODIUM SULFIDES

Sodium sulfide, Na_2S, mol wt 78.05; sodium hydrosulfide (sodium sulfhydrate, sodium bisulfide, sodium hydrogen sulfide), NaHS, mol wt 56.06; and sodium tetrasulfide, Na_2S_4, mol wt 174.24, are somewhat interchangeable in many applications. These compounds are used in the pulp (qv) and paper (qv) industries, in mining and leather (qv) tanning applications, as chemical intermediates, and in dye production (see DYES AND DYE INTERMEDIATES; MINERALS RECOVERY AND PROCESSING). Environmental applications of these sulfides, including heavy metal precipitation from wastewater and the removal of nitrogen oxides from emissions, are of particular interest to many industrial chemical consumers.

Sodium Hydrosulfide

Properties. Pure sodium hydrosulfide is a white, crystalline solid, mp 350°C, sp gr 1.79. The commercial product, available in flake form at approximately 73% strength, is yellow in color and highly deliquescent. The average water of hydration may be expressed as $NaHS \cdot 0.81\ H_2O$. The flake is highly soluble in water, alcohol, or ether. The heat of formation of NaHS is −237.6 kJ/mol (−56.79 kcal/mol); the heat of solution is 15.9 kJ/mol (10.7 kcal/mol) (2). In aqueous solution NaHS has an alkaline pH. When exposed to air, sodium hydrosulfide undergoes autoxidation and gradually forms polysulfur, thiosulfate, and sulfate. It also absorbs carbon dioxide, forming sodium carbonate.

Manufacture. Production is closely related to the supply of hydrogen sulfide, which reacts with sodium hydroxide to form NaHS. Hydrogen sulfide can also be obtained by the reaction of hydrogen and sulfur.

Economic Aspects and Uses. Use of NaHS in ore flotation (qv) has decreased over the years owing to the substitution of more environmentally sound methods, but this usage may fluctuate along with copper (qv) production. Use in dye production has remained constant, whereas the use of NaHS in pulp processing is increasing as the demand for paper products increases. Modest growth in the leather (qv) tanning sector has resulted from strengthening of demand in the automotive industry and increases in leather exports. The engineering plastic poly(phenylene sulfide) uses NaHS as a raw material (see ENGINEERING PLASTICS; POLYMERS CONTAINING SULFUR).

In many applications sodium hydrosulfide and sodium sulfide are interchangeable. Where either chemical may be used, 28% less sodium hydrosulfide is required by weight to achieve a given level of sulfidity and is therefore the more economical choice. If desired, the sodium hydrosulfide can be converted to sodium sulfide by the addition of sodium hydroxide. Sodium sulfides are very effective as heavy metal precipitants owing to the extremely low solubilities exhibited by metal sulfides.

Sodium Sulfide

Properties. Pure sodium sulfide is a white, crystalline solid, mp 1180°C, sp gr 1.856. The commercial product is available in flake form at approximately 60% strength, is a light tan-to-yellow color, and is deliquescent. The heat of formation for the crystalline state is −373 kJ/mol (−89.1 kcal/mol), and the heat of solution is −63.5 kJ/mol (−15.2 kcal/mol). In solution, Na_2S is strongly alkaline. The average water of hydration may be expressed as $Na_2S \cdot 2.71H_2O$. Sodium sulfide crystallizes from aqueous solutions as the nonahydrate, $Na_2S \cdot 9H_2O$. The flake is readily soluble in water, slightly soluble in alcohol, and insoluble in ether.

Manufacture. One commercial process for producing sodium sulfide is as a by-product of barium carbonate production (see BARIUM COMPOUNDS). Barite ore, $BaSO_4$, is reduced with carbon at 800°C to produce crude barium sulfide (black ash), which is then leached to dissolve the barium sulfide in solution. The solution is then reduced using sodium carbonate to produce barium carbonate, leaving a weak sodium sulfide solution as the by-product. The sodium sulfide solution may then be concentrated and flaked or crystallized. Another process involves two steps. Sodium hydrosulfide produced from hydrogen sulfide and caustic soda reacts with sodium hydroxide to yield sodium sulfide.

Economic Aspects and Uses. In the late 1990s, about 65% U.S. sodium sulfide usage was in the leather industry for dehairing hides before tanning. The production of miscellaneous chemicals included the production of polysulfide elastomers and plastics as well as a variety of organic chemicals. In the dye industry, Na_2S is used as a solvent for water-soluble dyes and as a reducing agent. In ore flotation, the mining industry uses Na_2S to form insoluble metal sulfides of copper, lead, molybdenum, nickel, and cobalt. Technology has been developed for the absorption of nitrogen oxides from gas streams via sodium sulfide scrubber systems.

Sodium Tetrasulfide. Sodium tetrasulfide is prepared by the reaction of sodium sulfide with sulfur. The 34 wt % solution is normally dark red, solidifies at −15 to −9°C, boils at 113°C, and has a specific gravity at 15.5°C of 1.268. The chemical formula is written Na_2S_4, but the product is better regarded not as a compound but as a mixture of sodium sulfide with free, elemental sulfur, ie, $Na_2S \cdot S_3$. Sodium tetrasulfide. It is used in leather processing, dye manufacturing, wastewater treatment, in metals finishing, ore manufacturing, and in lubricant manufacturing. No commercial polysulfide of significance is produced other than the tetrasulfide.

Analysis

A double end point, acid–base titration can be used to determine both sodium hydrosulfide and sodium sulfide content. Standardized hydrochloric acid is the titrant; thymolphthalein and bromophenol blue are the indicators.

Health and Safety

The combination of organic matter and the sodium sulfides can cause combustion to occur. Personnel handling sodium sulfides must be equipped with goggles, a full face shield, and rubber or plastic protective clothing. Chemical cartridge escape respirators should be available at all storage and use locations, owing to the potential for hydrogen sulfide formation. The sodium sulfides are similar to sodium hydroxide and other alkalies as corrosive substances on animal tissues. Contact with skin can be very irritating and is especially harmful to soft tissues.

DAVID R. BUSH
PPG Industries, Inc.

Sulfur Chemicals, PPG Industries, Inc., Pittsburgh, Pa., 1992.

Miscellaneous Sulfur Chemicals—United States, Chemical Economics Handbook, 780.4000M-X, Stanford Research Institute International, Menlo Park, Calif., May 1992.

Sodium Sulfide and Sodium Hydrosulfide as Heavy Metal Precipitants, Technical Service Bulletin, PPG Industries, New Martinsville, W. Va., 1995.

Sodium Sulfide: Wet Scrubber for Oxides of Nitrogen (NO_x) Absorption, Technical Service Bulletin, PPG Industries, New Martinsville, W. Va., 1995.

SODIUM HYDROXIDE. See ALKALI AND CHLORINE PRODUCTS.

SODIUM SILICATE. See SILICON COMPOUNDS.

SODIUM SULFITES. See SULFUR COMPOUNDS.

SODIUM TRIPOLYPHOSPHATE. See PHOSPHORIC ACIDS AND PHOSPHATES.

SOIL CHEMISTRY OF PESTICIDES

The detection of trace amounts of organic pesticides (qv) in surface and groundwater has been a significant environmental issue since the early 1980s. From a national perspective in the United States, particular concern was focused on the rural drinking water supplies for which groundwater is the principal source (see GROUNDWATER MONITORING). Soils play a significant role in modifying the amounts and kinds of pesticides ultimately detected in water. There is a strong relationship between the amount of pesticide applied and the amount detected in soil and water.

Pesticide Usage

Pesticide is a generic name for compounds used in pest control (see PESTICIDES). The three principal groups of pesticides, and the pests they control, are insecticides for insects, herbicides (qv) for weeds, and fungicides (see FUNGICIDES, AGRICULTURAL) for plant diseases (see INSECT CONTROL TECHNOLOGY).

Pesticides are further subdivided into classes of compounds. Historically, insecticides included the organochlorine, methyl carbamate, and organophosphate classes of pesticides. Herbicides comprise about 10–12 principal classes of compounds. Within each class of pesticide there may be several hundred active ingredients. Agriculture is the largest user of pesticides on a weight basis (77%), but significant amounts are also used by the industrial, commercial, and government sectors (16%) and for home and garden use (6%).

Pesticide Properties and Detection

One of the first problems encountered by scientists attempting to get a national perspective on the potential magnitude of the groundwater pollution problem was the large number of soil types and pesticides involved. The use of models to predict the potential movement of pesticides in soils under a variety of conditions began in earnest about 1980. An integral component of these models deals with chemical and physical properties of the pesticides.

By early 1995, the Agricultural Research Service (ARS) of the U.S. Department of Agriculture (USDA) had developed a computerized pesticide property database containing 17 physical properties for 330 pesticide compounds. The primary user of this data has been the USDA's Natural Resources Conservation Service (formerly the Soil Conservation Service) for leaching models to advise farmers on any combination of soil and pesticide properties that could potentially lead to substantial groundwater contamination.

Limits of Detection. One reason for the concern about pesticides in groundwater has been the ability to detect trace amounts of these compounds by more sophisticated analytical methodology. Limits of residue detection have increased progressively from parts per million (ppm), parts per billion (ppb), to parts per trillion (ppt).

Pesticide Metabolism and Chemical Degradation

Pesticides are susceptible to a variety of transformations in the environment, including both chemical degradation and microbial metabolism. Microbial transformations are catalyzed exclusively by enzymes, whereas chemical transformations are mediated by a variety of organic and inorganic compounds. Many pesticide transformations can occur either chemically or biologically.

Microbial Metabolism. Studies indicate that, for many pesticides, metabolism by microorganisms is the most important environmental fate. Pesticide-degrading microorganisms are found in soils, aquatic environments, and wastewater treatment plants, although the greatest number and variety are probably in agricultural soils. A wide variety of pesticide-degrading microorganisms have been identified, including over 100 genera of bacteria and fungi. The rate and extent of pesticide metabolism can vary dramatically, depending on chemical structure, the number of specific pesticide-degrading microorganisms present and their affinity for the pesticide, and environmental parameters.

Transformations/Metabolic Pathways. The initial enzymatic transformation of most pesticides can be characterized as oxidative, reductive, or hydrolytic. In general, oxidative and hydrolytic reactions are typical of both fungi and bacteria, whereas reductive reactions are most typical of bacteria. Oxidative reactions occur only under aerobic conditions; reductive reactions typically occur under anaerobic conditions; hydrolytic reactions occur under both. Many, if not most, pesticides are susceptible to several kinds of transformations. Some are susceptible to complete mineralization. Consequently, it is difficult to predict the fate of any given pesticide at any given site.

Oxidative Reactions. The majority of pesticides, or pesticide products, are susceptible to some form of attack by oxidative enzymes. For more persistent pesticides, oxidation is frequently the primary mode of metabolism, although there are important exceptions, eg, DDT. For less persistent pesticides, oxidation may play a relatively minor role, or be the first reaction in a metabolic pathway. Oxidation generally results in degradation of the parent molecule. However, attack by certain oxidative enzymes (phenol oxidases) can result in the condensation or polymerization of the parent molecules; this phenomenon is referred to as oxidative coupling.

Reductive Reactions. A number of pesticides are susceptible to reductive reactions under anaerobic conditions, depending on the substituents present on the molecule. Reductive reactions can be either chemically or enzymatically mediated. The only definitive means of distinguishing between chemical vs biological (enzymatic) reactions is to determine whether the reaction rate is consistent with enzyme kinetics.

Hydrolytic Reactions. Many pesticides possess bonds that are susceptible to hydrolytic attack. These reactions are most easily characterized according to the type of bond hydrolyzed: ester, carbamate, organophosphate, urea, or chlorine (hydrodechlorination). In many instances the specific hydrolytic enzymes have been purified and characterized and the genes encoding for the enzymes isolated and cloned. It is commonly observed that there are multiple forms of the enzymes catalyzing a particular hydrolytic reaction, which suggests that these catalytic functions have evolved independently in different bacteria.

Metabolic Pathways. Most pesticides are susceptible to complete degradation, ie, mineralization. This typically requires a sequence of enzymatic transformations, ie, metabolic pathway in which the product(s) are utilized as growth substrates by microorganisms or consortia of microorganisms. Most pesticides are susceptible to mineralization only under aerobic conditions, although a few can also be mineralized under anaerobic conditions. One of the first pesticides demonstrated to be mineralized by soil microorganisms was 2,4-D. The metabolic pathway of 2,4-D biodegradation has been elucidated and shown to consist of the steps shown in Figure 1.

Pesticides that are susceptible to mineralization are not typically found in, or considered to be a threat to, groundwater supplies because of their rapid degradation, ie, nonpersistence.

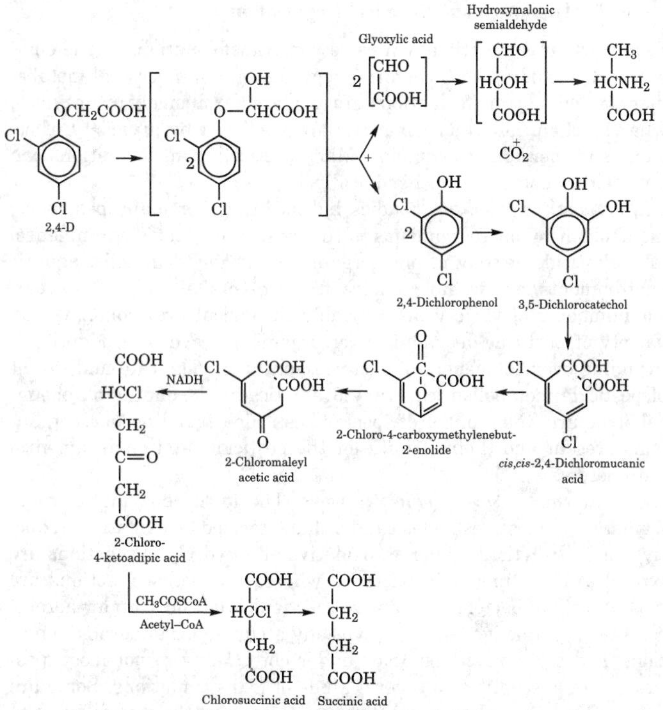

Figure 1. Metabolic pathway of 2,4-D biodegradation.

Kinetics of Pesticide Biodegradation. Rates of pesticide biodegradation are important because they dictate the potential for carryover between growing seasons, contamination of surface and groundwaters, bioaccumulation in macrobiota, and losses of efficacy. Pesticides are typically considered to be biodegraded via first-order kinetics, where the rate is proportional to the concentration. For those pesticides which are utilized as microbial growth substrates, sigmoidal rates of biodegradation are frequently observed. Sigmoidal data are more difficult to summarize than exponential (first-order) data because of their inherent nonlinearity.

Variability (spatial and temporal) in the rate of biodegradation of specific pesticides is frequently observed. Rates of biodegradation tend to be site-specific because of the differences in the numbers of specific pesticide degraders, pesticide bioavailability, and soil parameters such as temperature, moisture, and pH.

Chemical Degradation

Chemical, or abiotic, transformations are an important reaction of pesticides. Such transformations are ubiquitous, occurring in either aqueous solution or sorbed to surfaces. Rates can vary dramatically depending on the reaction mechanism, chemical structure, and relative concentrations of such catalysts as protons, hydroxyl ions, transition metals, and clay particles. Chemical transformations can be generically classified as hydrolytic, photolytic, or redox reactions (transfer of electrons).

Hydrolytic and Substitution Reactions. A variety of functional groups common to many pesticides is susceptible to hydrolysis. Hydrolysis reactions are catalyzed by acids (low pH), bases (high pH), and/or transition metals (Cu^{2+}, Fe^{3+}, Mn^{2+}). Consequently, environmental parameters such as pH, mineral composition and concentration, and clay content can have significant effects on rates of hydrolysis. In addition, the reaction mechanism in conjunction with chemical structure is of critical importance in dictating the rate of reaction. For instance, in the case of aromatic pesticides, if the reaction mechanism involves attack by a nucleophile (OH^-), then the presence of electron-withdrawing substituents such as NO_2^- and Cl^- causes the bond to be more electron-poor (more positive), resulting in faster rates of hydrolysis; whereas the presence of electron-donating substituents such as NH_2^- and CH_3^- causes the bond to be more electron-rich (more negative), resulting in slower rates of hydrolysis. If the reaction mechanism involves attack by an electrophile ($OH\cdot$), then electron-withdrawing substituents cause the rate of hydrolysis to be slower, whereas electron-donating substituents cause the rate of hydrolysis to be faster.

Photolytic Reactions. Extensive pesticide photodegradation in soil is problematic for many compounds, because light penetration into soils is extremely limited. The most likely candidate pesticides for soil photolysis are those that are water-soluble, weakly sorbed to soil surfaces, and have low vapor pressure; such compounds are most likely to rise with capillary water to the soil–atmospheric interface where photodegradation can occur (see PHOTOCHEMICAL TECHNOLOGY, PHOTOCATALYSIS).

Redox Reactions. Oxidative reactions typically occur as a consequence of the light-mediated production of singlet oxygen or hydroxyl radical, which are both potent oxidants. This process, termed indirect photolysis, involves the initial absorption of light energy by organic molecules, eg, humic substances, which either is directly transferred to oxygen (sensitization) or results in a chain reaction leading to the formation of oxidants. In contrast, soil organic matter has also been shown to quench photolysis of certain sorbed molecules. Chemical oxidative reactions in soil are generally less importance environmentally than biological oxidative reactions, because observed reaction rates are slower on account of competition for oxidants by organic matter.

Physical Processes Affecting Pesticides in Soil and Water

Persistence of pesticides in the environment is controlled by retention, degradation, and transport processes and their interaction. *Retention* refers to the ability of the soil to bind a pesticide, preventing its movement either within or outside of the soil matrix. *Transport* processes describe movement of the pesticide from one location to another or from one phase to another. Transport processes include both downward leaching, surface runoff, volatilization from the soil to the atmosphere, as well as upward movement by capillary water to the soil surface. Transport of pesticides is a function of both retention and transport processes.

Sorption and Desorption Processes. *Sorption* is a generalized term that refers to surface-induced removal of the pesticide from solution; it is the attraction and accumulation of pesticide at the soil–water or soil–air interface, resulting in molecular layers on the surface of soil particles. Sorption is generally considered a reversible equilibrium process. *Desorption* is the reverse of the sorption process. If the pesticide is removed from solution that is in equilibrium with the sorbed pesticide, pesticide desorbs from the soil surface to reestablish the initial equilibrium. Desorption replenishes pesticide in the soil solution as it dissipates by degradation or transport processes. Sorption/desorption therefore is the process that controls the overall fate of a pesticide in the environment. It accomplishes this by controlling the amount of pesticide in solution at any one time that is available for plant uptake, degradation or decomposition, volatilization, and leaching.

Pesticides are sorbed on both inorganic and organic soil constituents. The sorptive reactivity of soil organic and inorganic surfaces to pesticides is dependent on the number and type of functional groups at accessible surfaces. When a pesticide reacts with the surface functional groups, either an inner- or an outer-sphere surface complex is formed. Although functional groups account for much of the reactivity of soil to pesticide retention, accessibility of the functional groups to the pesticide is also an important factor. For instance, steric hinderance caused by a large neighboring substituent or chemical may preclude the pesticide from interacting with the functional group.

Inorganic solids are composed of crystalline and noncrystalline amorphous minerals. The principal functional groups on inorganic surfaces contributing to the sorptive capacity are siloxane ditrigonal cavities in phyllosilicate clays and inorganic hydroxyl groups generally associated with metal (hydrous) oxides. Organic components of the solid phase include polymeric organic solids, decomposing plant residues, and soil organisms. The variety of functional groups in soil organic matter and the steric interactions between functional groups lead to a continuous range of reactivities in soil organic matter. The relative importance of organic vs inorganic constituents on pesticide sorption depends on the amount, distribution, and properties of these constituents, and the chemical properties of the pesticide. Soil organic matter is the principal sorbent for many organic compounds.

For any given compound, there is likely a continuum of mechanisms with differing energy relationships that is responsible for sorption onto soil. London and van der Waals forces are short-range interactions resulting from a correlation in electron movement between two molecules to produce a small net electrostatic attraction. These interactions are particularly important for neutral high molecular weight compounds. Hydrogen bonds are dipole–dipole interactions involving an electrostatic attraction between an electropositive hydrogen nucleus on functional groups such as –OH and –NH and exposed electron pairs on electronegative atoms such as –O and –N. Hydrogen bonding is probably most prevalent in the bonding of pesticides to organic surfaces in the soil. Cation and water bridging involve complex formation between an exchangeable cation and an anionic or polar functional group on the pesticide. Protonation of a pesticide, or formation of charge-transfer complexes, at a mineral surface occurs when an organic functional group forms a complex with a surface proton. This retention mechanism is particularly important for basic functional groups at acidic mineral surfaces at low pH and low water content, particularly in the presence of aluminum or other metal cations.

Anion-exchange mechanisms involve a nonspecific electrostatic attraction of an anion to a positively charged site on the soil surface, involving the exchange of one anion for another at the binding site. Ligand exchange is a sorption mechanism that involves displacement of an inorganic hydroxyl or water molecule from a metal ion at a hydrous oxide surface by a carboxylate or hydroxyl on an organic molecule. Cation exchange is an electrostatic attraction that involves the exchange of a cation for a cation sorbed at a negatively charged site on the soil surface.

A variety of mechanisms or forces can attract organic chemicals to a soil surface and retain them there. For a given chemical, or family of chemicals, several of these mechanisms may operate in the bonding of the chemical to the soil. For any given chemical, an increase in polarity, number of functional groups, and ionic nature of the chemical can increase the number of potential sorption mechanisms for the chemical.

Sorption in the soil is generally controlled by the rate of molecular diffusion into soil aggregates and the rate of reaction (rate of sorption) at the soil–water interface. Diffusion has been found to be the rate-limiting step. Solute moves from mobile pore water to the sorbent surface surrounded by immobile pore water, limiting the initial rate of sorption as sorption slows down. The actual retention reactions tend to be relatively rapid, particularly the exchange-type reactions.

Pesticide Transport Mechanisms

Pesticides can be transported away from the site of application either in the atmosphere or in water. The process of volatilization transfers the pesticide from the site of application to the atmosphere. The off-site transport and deposition can be at scales ranging from local to global. Once the pesticide is in the atmosphere, it is subject to chemical and photochemical processes, wet deposition in rain or fog, and dry deposition.

Water leaves the field either as surface runoff, carrying pesticides dissolved in the water or sorbed to soil particles suspended in water, or as water draining through the soil profile, carrying dissolved pesticides to deeper depths. The distribution of water between drainage and runoff is dependent on the amount of water applied to the field, the physical and chemical properties of the soil, and the cultural practices imposed on the field. These factors also impact the retention and transformation processes affecting the pesticide.

Runoff is an important pathway for transport of pesticide away from the site of application. Field application of pesticides inevitably leads to pesticide contamination of surface runoff water unless runoff does not occur while pesticide residues remain on the surface of the soil.

Leaching. Pesticides in groundwater are present as the result of agricultural practices and may be the product of both point source and nonpoint source pollution. The movement of pesticides in soil water depends on rainfall or irrigation water, the macroscopic and microscopic structure of the soil, and the sorption–desorption characteristics of the pesticide on the soil. Water moves through the soil under both saturated and unsaturated conditions. When the soil is saturated with water, the pores are filled with water and transport occurs at the maximal rate. Movement of water and pesticide occurs at much slower rates under unsaturated conditions because only the smaller pores are filled and water moves in response to water potential gradients. Generally, coarse-textured soils have greater rates of water movement than fine-textured soils when saturated. However, under unsaturated conditions, fine-textured soils may have greater transport rates. Because many pesticides are applied to the soil surface, the transport of pesticide during water infiltration is important. Water infiltration is characterized by high initial infiltration rates which decrease rapidly to a nearly constant rate. Dry soils have greater rates of infiltration than wet soils during the initial application of water. Sorbed pesticides are not available for transport, but if water having lower pesticide concentration moves through the soil layer, pesticide is desorbed from the soil surface until a new equilibrium is reached. Thus, the kinetics of sorption and desorption relative to the water conductivity rates determine the actual rate of pesticide transport.

PHILIP C. KEARNEY
University of Maryland
DANIEL R. SHELTON
WILLIAM C. KOSKINEN
USDA-Agricultural Research Service

J.-M. Bollag and S.-Y. Liu, in H. H. Cheng, ed., *Pesticides in the Soil Environment: Processes, Impacts, and Modeling,* Soil Science Society of America, Madison, Wis., 1990.

B. Burgoa and R. D. Wauchope, in T. R. Roberts and P. C. Kearney, eds., *Progress in Pesticide Biochemistry and Toxicology,* Vol. 9, John Wiley & Sons, Ltd., Chichester, U.K., 1995.

A. G. Hornsby, R. D. Wauchope, and A. E. Herner, *Pesticide Properties in the Environment,* Springer Verlag, New York.

P. C. Kearney and T. R. Roberts, *Pesticide Remediation in Soils and Water,* John Wiley & Sons, Ltd., Chichester, U.K., 1998.

SOIL STABILIZATION

Chemical grouting is the practice of injecting liquid solutions of cement or organic materials into soil, rock, or concrete in order to form solid inorganic or organic masses that impart desirable permanent physical characteristics in the soil, rock, or concrete. The solutions that are injected, ie, chemical grouts, undergo either polymerization of monomers or cross-linking of soluble polymers to form insoluble polymer masses. Soil conditioners are materials that measurably improve the physical characteristics of the soil as a plant growth medium. Typical uses include erosion control, prevention of surface sealing, and improvement of water infiltration and drainage.

Chemical Grouting

The ideal chemical grout is a low viscosity solution capable of penetrating finely divided profiles as easily and to the same extent as water. The grout solution viscosity remains low for a predetermined time to allow the desired penetration of the profile. Rapid gelation then occurs to form a water-impermeable barrier filling all the voids in the formation, thereby waterproofing it. The barrier's durability should be adjustable. Long-lived barriers are needed in applications such as repair of sewers or tunnels, but short-lived grouts are useful for temporary stabilization of excavations such as in construction. Grout systems must be chemically compatible with the profiles to be grouted and capable of being applied with available equipment.

Silicate Grouts. Sodium silicate has been most commonly used in the United States. Its properties include specific gravity, 1.40; viscosity, 206 mPa·s(= cP) at 20°C; $SiO_2 : Na_2O = 3.22$. Reaction of sodium silicate solutions with acids, polyvalent cations, such organic compounds as formamide, or their mixtures, can lead to gel formation at rates which depend on the quantity of acid or other reagent(s) used. In the Joosten or two-shot method, successive injections are made of a concentrated solution of sodium silicate and a calcium chloride solution into a single pipe. In the Siroc or one-shot method, formamide is used to coagulate sodium silicate.

For waterproofing, sodium silicate concentrations below 30% are adequate; concentrations between 35 and 70% are used for strength improvement. Various additives can impart desired handling and performance properties. Portland cement is used as a reactant with sodium silicate grouts to obtain short gel times, useful to stop flowing water and seal grouting cavities. Addition of particulates to grouts generally reduces their ability to penetrate finely divided formations. The Siroc grouting system is considered nonhazardous and nonpolluting. Sodium silicate is essentially nontoxic. Formamide is toxic and corrosive, but does not present a serious hazard if normal safety precautions are followed.

Organic Polymer Grouts. There are several types of organic grouting systems. Aqueous acrylamide solution grouts were introduced in the United States in 1955 and rapidly became popular because of lower cost, better flexibility, and superior performance compared to other grouts then commercially available. Acrylamide grouts came closest to matching performance requirements for an ideal grout. Specialized equipment was designed for injection of acrylamide grouts. Acrylamide has been the most-often selected grout for use in sewer applications. Polymer gels resulting from acrylamide polymerization are generally regarded as nonhazardous. Acrylamide (qv) monomer, however, has been reported to have potential neurotoxic effects in higher animals. Citing the reported neurotoxicity of acrylamide and the availability of other grouting systems, the EPA proposed in 1991 to ban the use of grouts based on acrylamide and *N*-methylolacrylamide (NMA). Advantages of acrylamide grouts include low cost, quick and controllable gel times from a few seconds to several hours, low viscosity, and a long history of reliable performance. A low viscosity grout is advantageous for penetration of finely divided formations. Acrylamide grouts are among the lowest viscosity grouts available.

The equipment and processes necessary for sewer rehabilitation and manhole sealing, which are the main uses of NMA-based grouts, are the same as those for acrylamide-based grouts, although a different persulfate catalyst is typically used.

Acrylate grouts were developed specifically as acrylamide grout replacements intended for sewer rehabilitation and similar applications. The objective was to provide low viscosity, low toxicity grouting systems that could be used in the same equipment and with the same catalysts as acrylamide grouts to provide controllable gel times and strong, durable, and water-impermeable grouts. The viscosity of a 10% acrylate grout formulation is approximately the same as for acrylamide grouts. Grouts formed are reported to be roughly comparable to those obtained with acrylamide grouts. Field experience has given mixed results.

Polyacrylamide gels were designed to obtain the advantages of the performance of acrylamide-based gels while substantially avoiding exposure to acrylamide monomer. Low molecular weight polyacrylamides, which form low viscosity solutions in water, are cross-linked on demand to form insoluble gels (see ACRYLAMIDE POLYMERS). This system forms gels similar to those obtained with acrylamide grout and can thus be operated with equipment designed to handle the latter.

Urethane grouts comprise low molecular weight prepolymers of polyethylene or polypropylene glycol, end-terminated with toluene diisocyanate (TDI) or methylenebis(phenyl isocyanate). Gel-strengthening agents, such as aqueous polymeric latex, may be added to reduce shrinkage (see URETHANE POLYMERS). Polymerized forms of these grouts are considered nonhazardous. Isocyanate monomers are toxic but are converted to nonhazardous ureas on contact with the environment (see ISOCYANATES, ORGANIC).

Published comparisons indicate that properly used urethane grouts offer cost–performance equivalent or superior to that of acrylamide grouts in many applications.

Grouts based on epoxy resins (qv) are commercially available both as coatings and as gels. Two-part systems are mixed on-site and applied promptly. Application to wet surfaces is possible and sewer repair has been demonstrated.

Several other organic polymer systems have been used as grouts. These include urea–formaldehyde resins (aminoplasts), phenol–formaldehyde resins (phenoplasts), and lignosulfonates (see AMINO RESINS AND PLASTICS; PHENOLIC RESINS). When strengthening of soil to support the weight of roads, railroads, or building foundations is desired, combinations of grouts, admixtures with other materials, and geosynthetics may be used (see GEOTEXTILES). Such soil reinforcements to support foundations are widely used, but long-term durability is a concern.

Soil Conditioners

Agricultural Applications. The emphasis in soil conditioning for agriculture is on formation of soil aggregates that support seed germination, seedling emergence, efficient use of irrigation water, and erosion prevention; and stabilization of these aggregates against the impact of wind, rain, and irrigation water. Chemical treatments can be a useful supplement to other methods used to prepare fields for agriculture. Cost-effective commercial soil conditioners are emerging because improvements in the understanding of soil structure and in organic polymer science have led to better polymers and more efficient ways to apply them for soil conditioning. However, chemical treatment alone cannot be used to recover or prepare fields that are too wet or otherwise unsuitable for agriculture.

Slaking of weak soil aggregates leads to formation of finely divided material that is deposited on the surface of the soil, forming seals and blocking soil pores. Surface seals impede water infiltration, promote ponding, runoff, and erosion, and reduce water use efficiency, soil aeration, and root respiration. When surface seals dry, they form hard crusts that mechanically impede the emergence of seedlings, reduce stands, lower yields, and require expensive overplanting and thinning or even replanting of crops. Poor water infiltration and internal drainage are common problems on arable soils of the arid southwestern United States. Surface crusting and plugging of soil pores caused by fine clay particles and swelling in clay heavy soils result in poor infiltration and drainage. This impairs management of salty or sodic soils, which require adequate leaching and drainage to prevent accumulation of salt and sodium.

The serious levels of erosion associated with irrigation and especially with furrow irrigation have been recognized. Known erosion-control practices coupled with conservation tillage and selected crop sequences can substantially eliminate erosion. Furrow erosion can be reduced using settling ponds, minibasins and buried pipe to control runoff, straw placed in furrows, and sodded furrows.

Overland water flow applies shear forces to soil surfaces. When shear forces exceed the stress required to overcome cohesive forces between soil particles, the particles are detached and suspended in the flow. Suspended particles are carried into surface soil with infiltrating water where they block pores and initiate seal formation. Thus, erosion results in reduced water infiltration as well as loss of soil from the field and consequent downstream water pollution. If erosion is controlled, good water infiltration is maintained.

Both synthetic organic polymers and polysaccharides have been shown in laboratory tests to maintain structure and permeability of soils under artificial rainfall conditions. Several field studies have demonstrated that 1–10-ppm levels of polyacrylamides dissolved in irrigation water, approximately 1 lb/acre, can eliminate most erosion during furrow irrigation. Use of anionic polyacrylamides in erosion control is synergistic with nonchemical erosion control strategies and is a recommended erosion control practice of the Natural Resources Conservation Service (NRCS) in the United States. Different soils, terrains, and irrigation practices may require different application strategies for polyacrylamides, but their effective use to eliminate most silt and mineral, eg, nitrate and phosphate, losses from irrigated fields has been demonstrated at many test sites.

Acid-based fertilizers, acidic polymers, and anionic polyacrylamides have all been used successfully in at least some circumstances to prevent the formation of soil crusts.

Sodium acrylate–acrylamide copolymers cross-linked with methylenebisacrylamide, the so-called superabsorbent polymers, have been used to improve soil properties, specifically water distribution, availability, and drainage characteristics in situations where soil texture is coarse (sandy) or rainfall is marginal for agriculture. The cross-linked gels absorb large volumes of water, swelling and preventing gravity-induced downward flow in the soil. The absorbed water can later be lost to evaporation or extracted by plant roots. Proper placement of the polymers by spraying or other means can disrupt undesirable soil capillary action while providing a water reservoir in the right location to promote desirable root system growth.

Other Applications. Construction of highways creates many steep slopes which must be stabilized against erosion by water or wind. Excavations as part of construction projects and natural disasters also create severe erosion problems. Many types of inert structures are available for slope stabilization and erosion control, including retaining structures of various types, revetment systems, and ground covers such as artificial mulches (cellulose fibers, fiber glass); chemical systems such as tackifiers and emulsions; blankets, mats, and nettings to cover slopes; and cellular confinement systems. Although designed for long life, these inert systems, whether based on steel, concrete, or synthetic polymers, slowly degrade with time. Hence, reestablishment of vegetation is highly desirable and is possible with porous retaining structures, revetments, or ground covers. Hydroseeding is widely used in slope stabilization. A mixture comprising grass seeds, fertilizer, synthetic polymer, and water is sprayed onto banks.

DONALD VALENTINE, JR.
Cytec Industries Inc.

R. H. Borden, R. D. Holtz, and I. Juran, eds., *Proceedings of the 1992 ASCE Specialty Conference on Grouting, Soil Improvement, and Geosynthetics, New Orleans, La., Feb. 25–28, Geotechnical Special Publication* Vol. 1, no. 30, ASCE, New York, 1992.

M. F. De Boodt, *Soil Colloids and Their Association in Aggregates,* NATO ASI Ser. B 215, 1990.

F. Barvenik, *Soil Sci.* **158**(4), 235–243 (1994).

SOLAR ENERGY

Solar energy represents a potentially limitless source of energy. Roughly 10,000 times as much solar energy falls on the surface of the earth each year as is consumed in the form of fossil and nuclear fuel. The technologies for utilizing solar energy are perceived as being environmentally benign. These technologies use sunlight, rainwater, or wind as the energy resource and thus generally do not produce gaseous emissions or waste materials having adverse environmental impact (see AIR POLLUTION). The intensity of incident sunlight is diffuse, having a peak power density of only 1 kW/m^2 at the earth's surface at noon in the tropics. The efficiency of conversion of solar energy varies from a few percent for photosynthetic production of biomass to as much as 15–20% for production of electricity for some photovoltaic modules (see PHOTOCHEMICAL TECHNOLOGY; PHOTOVOLTAIC CELLS). The trend toward electrification worldwide militates in favor of a shift to solar or renewable energy, because many of the most promising solar conversion technologies naturally produce electricity.

Wind Energy Technology

The use of wind as a renewable energy source involves the conversion of power contained in moving air masses to rotating shaft power. These air masses represent the complex circulation of winds near the surface of the earth caused by the earth's rotation and by convective heating from the sun. The actual conversion process utilizes basic aerodynamic forces, ie, lift or drag, to produce a net positive torque on a rotating shaft, resulting in the production of mechanical power, which can then be used directly or converted to electrical power.

The scope of the wind resource is widespread and less dependent upon latitude than other solar technologies. The intermittency of the wind resource, however, makes it impractical to base more than 10–20% of electricity generation on this resource until a suitable storage technology is developed. Wind is a very complex resource, existing in three dimensions, rather than the two associated with other solar resources. It is intermittent and strongly influenced by terrain effects. Moreover, there is a nonlinear (cubic) relationship between wind speed and power or energy available. Wind machines can be classified as either horizontal-axis or vertical-axis designs and typically utilize either two or three airfoils. Recent commercial designs have evolved toward machines having capacities between 200–500 kW each. These are usually grouped into wind farms of total capacity of 20 MW or more. Wind turbines of much smaller (10 kW) capacity are finding increased application for rural electrification, particularly in developing countries. Wind technology provides economical energy to remote areas and for specialized applications. The modularity and wide size range of wind turbines available enable wind energy to serve many applications. By combining the intermittent wind energy with backup power sources such as diesel generators or storage devices, most loads can be reliably and competitively served. Wind energy has little or no impact on flora, fauna, climate, materials, or in terms of human health hazards. It does, however, have a potential negative impact on land use. On the negative side, there are three siting considerations: the visual impact of large, rotating structures; the nearby acoustic disturbance associated primarily with the generation of aerodynamic forces on the rotating airfoils. and concerns about the possibility of bird kills from the rotating blades. On the positive side, the three-dimensional nature of the resource provides it with a distinct advantage compared to other solar technologies. Specifically, because siting usually involves placing the individual turbines as high as possible, typically spacing turbines about two to three blade diameters apart crosswind and 10 diameters apart downwind, only a small fraction of a wind farm area is actually occupied. The rest of the land remains available for other applications, such as crop production or livestock grazing. Performance of wind turbines, as well as other sources of energy, must be judged by the cost of energy (COE), ie, the levelized cost per kilowatt hour of electricity produced. For wind turbines, this cost can be determined from only a few parameters: the capital cost, the annual energy capture, and the operation and maintenance/replacement costs.

Solar Thermal Electric Technology

Use of concentrated sunlight to generate electricity by thermodynamic processes is well documented. Reflective surfaces concentrate incident sunlight onto a receiver, where it is absorbed into a working fluid that powers a thermal conversion–generator device. Solar thermal systems, operating either with storage or in a hybrid mode with an

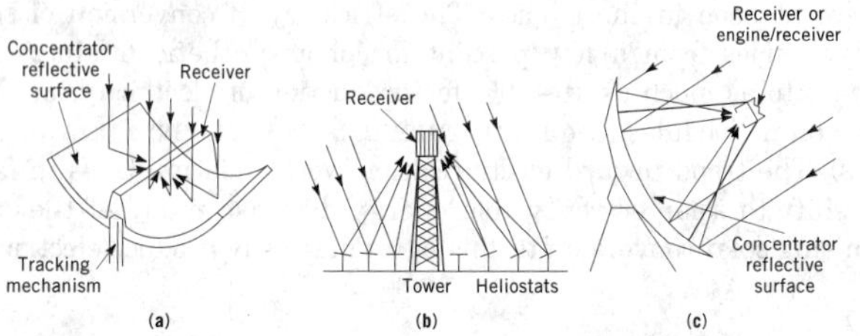

Figure 1. Solar thermal designs: (**a**) parabolic trough; (**b**) central receiver; and (**c**) parabolic dish.

auxiliary fuel, offer significant potential as capacity to meet utility peaking or intermediate electric power-generation needs.

Three main types of concentrating collectors have evolved for use in solar thermal systems: low concentration parabolic troughs, high concentration parabolic dishes, and central receivers (Fig. 1). Higher concentration produces higher temperatures in a working fluid and makes electrical generation more efficient.

Parabolic trough systems use surface reflectors to concentrate sunlight onto a fluid-filled receiver tube that is positioned along the line of focus. Concentration ratios of more than 100 times are typically used to generate temperatures of 400–500°C. Troughs are modular and many can be grouped together to produce large amounts of heated fluid. The fluid is then transported to a nearby facility to generate electricity.

The modular parabolic troughs and dishes (Fig. **1a** and **b**) are classified as distributed systems, whereas central receiver systems, in which heliostats are deployed in a central receiver configuration by placing large numbers of them around a tower-mounted receiver (Fig. **1b**), are more centralized. All concentrating systems have their best annual output in regions where direct insolation is highest. Prototype parabolic dish electric systems totaling about 5 MWe have been operated in utility settings in Georgia and in southern California. Development of a dish-mounted engine–generator has led to significant increases in system performance as compared to the earlier designs which collect the heat as thermal energy and transport it to a central location for electric generation. A dish/Stirling engine–generator model has achieved a 29% overall system conversion of sunlight to electricity.

Photovoltaics

Photovoltaic devices typically consist of a series of thin semiconductor layers that are designed to convert sunlight to direct-current electricity (see SEMICONDUCTORS). As long as the device is exposed to sunlight, a photovoltaic (PV) cell produces an electric current proportional to the amount of light it receives. Photovoltaic devices produce electricity from incident direct or diffuse sunlight. These devices have no moving parts and thus are extremely reliable. Moreover, their operation does not release any effluent to the atmosphere. Costs, however, have been relatively high compared to the operation of bulk electricity generation technologies, and the output of a photovoltaic cell is intermittent because of variations in sunlight.

The smallest unit of a PV system is called the PV cell. Cells are manufactured using crystalline and amorphous forms of silicon, copper indium diselenide (CIS), cadmium telluride, and gallium arsenide, as well as even more exotic materials. Photovoltaic systems generally consist of a flat layer of semiconductor material encapsuled by a glass or plastic cover, or of individual high efficiency PV cells incorporated in an optical arrangement to concentrate the sunlight. This latter arrangement often requires a solar tracking system, whereas the former, flat plate arrangement is normally installed at a fixed angle determined by the latitude of the site.

Modules, the building blocks of large PV systems, are aggregates of PV cells large enough to provide convenient levels of electrical power. These modules can be 0.1 m^2 to more than 2 m^2 and can be expected to produce from 0.5 to 2 W/m^2 of power during a clear midday, depending on the conversion efficiency of the cell material. The efficiency is defined as the ratio of electricity produced to the amount of sunlight incident on the PV device, and it is a critical figure-of-merit characterizing all PV cells, modules, and systems. The output power of a module at noon on a clear day is called its peak-watt power, because it represents a maximum typical output. A module characterized as 100 Wp produces 100 W of power during a clear midday. Unlike solar thermal systems or PV concentrator systems, the PV flat plate systems work well in cloudy locations, because these latter convert diffuse as well as direct sunlight to electricity. As a result, it is practical to use photovoltaic systems in normally cloudy locations such as Seattle or northern Maine.

The terrestrial PV market has three principal segments: consumer products, remote power, and utility generation. The consumer product market was one of the first economic applications of the technology and is characterized by millions of small, milliwatt-sized cells powering calculators and watches (see BATTERIES).

The largest use of PV is the remote power market. The self-contained and modular nature of PV systems has led to their adoption to meet power loads remote from the electric utility. These applications have come to be referred to as stand-alone because all of the energy needed by the load must come from on-site sources. Typical stand-alone uses are power for telecommunications, lighting, security systems, water supply, battery charging, cathodic protection, vaccine refrigeration, remote monitoring, rural housing, and small villages. The systems are economical because there is no reasonable alternative, such as for a microwave repeater on an inaccessible mountaintop, or because the alternative (often diesel generators) is too costly to install, operate, and refuel.

Biomass and Biofuels

Biomass is the term used to describe all plant-derived materials, whether wood (qv) or wood wastes, residue of wood-processing industries, food industry waste products, sewage or municipal solid waste (MSW), herbaceous or other biological materials cultivated as energy crops, or other biological materials. Biomass is both a principal and a prospective source of energy. Green plants use the sun's energy to convert CO_2 from the atmosphere to sugars during photosynthesis. Hence biomass is considered a form of solar or renewable energy. Unlike direct solar or wind, the solar energy in biomass is stored for later use. The conversion efficiency of photosynthesis is very low, however. The key feature of the biomass technology is the rapid (MATHEQ) recycling of the carbon fixed in the biological process. Unlike the burning of fossil fuels, earth's reserves of which were ultimately derived from solar energy because these materials consist of degraded residues of plants and animals, combustion of biomass merely recycles the carbon fixed by photosynthesis in the growth phase and typically has no net impact on global carbon dioxide levels.

There are four principal ways in which biomass is used as a renewable energy resource. The first, and most common, is as a fuel used directly for space and process heat and for cooking. The second is as a fuel for electric power generation. The third is by gasification into a fuel used on the site. The fourth is by conversion into a liquid fuel that provides the portability needed for transportation and other mobile applications of energy.

Thermal Combustion of Biomass. Direct combustion in air is the principal mechanism used to convert biomass into useful energy. The heat or steam (qv) produced is used to generate electricity or provide thermal requirements for industrial processes, building heating, cooking, or district heating in municipalities.

The residential sector uses biomass for direct applications such as cooking and space heating. Considerable progress has been made in the design of household cooking and heating appliances that use biomass, primarily wood, to improve efficiency and reduce CO and particulate emissions. The industrial sector uses biomass for both process and space heating, as well as power generation, often jointly in cogeneration projects. The technology available to these larger consumers is equivalent to that used in burning conventional fossil fuels and has been widely implemented. Somewhat lower efficiency is obtained compared to fossil fuels as the result of the high moisture content of biomass. The small amount of ash produced is usually suitable as a soil supplement.

Generation of Electric Power Using Biomass. By the end of the 20th century, electric utilities were making only limited use of biomass as a fuel for power generation, although the utilities often bought power from cogenerators who used biomass as fuel. Most power generation from biomass, whether generated by industry or utilities, was via steam turbines. Research is continuing to develop gas clean-up technologies that would permit use of gasified biomass as a fuel for gas turbines. Wood and wood wastes and by-products are the principal fuels. Greater use of biomass resources (exclusive of MSW) in electricity generation is constrained by delivered resource costs. At high levels of utilization, competition among energy and nonenergy uses may tend to bid up biomass resource prices. To stabilize price and availability, a potential long-term solution is the development of dedicated high productivity herbaceous or short-rotation woody crops for feedstock production.

Gasification of Biomass. The third energy conversion mechanism is the production of biogas, a mixture of methane and carbon dioxide, CO_2, which can be produced from either thermal conversion or the biological anaerobic digestion of biomass materials. The methane can be subsequently separated from the CO_2 using conventional technology and the resultant gas supplied to a natural gas system or other consumer. MSW may be processed anaerobically to produce methane from the digestible components. The volume of gasification residue, which includes materials such as burnable plastics, is greater than the residues from combustion processes. Combustible plastics and similar materials could be separated if required before feeding the remaining material to the digester. Landfills are also a source of methane produced from the decomposition of MSW although the economics of recovery of the naturally occurring methane are not universally favorable. A lower heat-content gas, syngas, consisting primarily of carbon monoxide and hydrogen (CO and H_2), can also be produced for use as a fuel or as an intermediate feedstock.

Biofuels. Biofuels are liquid fuels, primarily used in transportation (qv), produced from biomass feedstocks. Identified liquid fuels and blending components include ethanol (qv), methanol (qv), and the ethers-ethyl *t*-butyl ether (ETBE) and methyl *t*-butyl ether (MTBE), as well as synthetic gasoline, diesel, and jet fuels.

Ethanol can be produced from sugar (qv), starch (qv), or cellulosic feedstocks, ie, from wood, energy crops, and municipal and other wastes. In the United States, the primary pathway for conversion of biomass to alcohol fuels is the fermentation of corn to ethanol. In the biochemical conversion process, the biomass feedstock is first separated into its three main components, cellulose (qv), hemicellulose (qv), and lignin (qv). The cellulose is hydrolyzed to sugars, primarily glucose, which are then fermented easily to produce ethanol. The hemicellulose portion is more readily converted to sugars, primarily xylose; however, xylose is more difficult to ferment to ethanol. Finally, the lignin, although it cannot be fermented, can be converted to a high octane liquid fuel or, as is more common, burned to provide process energy.

Methanol is made from biomass by first gasifying the feedstock to form a syngas, a mixture of CO, H_2, CO_2, higher hydrocarbons, and tar. A gas shift reaction is employed to adjust the chemical structure of the components of the gas mixture to the requisite H_2-to-CO ratio. The syngas is then cleaned and conditioned before being converted, in the presence of standard commercial catalysts, to form methanol.

The basic approach used in converting biomass to traditional hydrocarbon fuels is to first pyrolyze the biomass feedstock to form an intermediate biocrude liquid product. The second step is to catalytically convert the biocrude to gasoline (see FUELS, SYNTHETIC). The technology uses a fast pyrolysis step that obtains higher yields of desired liquid components than those achieved in longer residence time processes. There are two potential routes for the second step: hydrogenation at high pressures and zeolite cracking at low pressures.

An alternative method of producing hydrocarbon fuels from biomass uses oils that are produced in certain plant seeds, such as rape seed, sunflowers, or oil palms, or from aquatic plants (see SOYBEANS AND OTHER OILSEEDS). Certain aquatic plants produce oils that can be extracted and upgraded to produce diesel fuel. The primary processing requirement is to isolate the hydrocarbon portion of the carbon chain that closely matches diesel fuel and modify its combustion characteristics by chemical processing.

ROBERT A. STOKES
Stokes Associates

K. W. Boer, *Advances in Solar Energy: An Annual Review of Research and Development,* Vol. 7, American Solar Energy Society, Boulder, Colo., 1992.

Renewing Our Energy Future, Office of Technology Assessment, U.S. Congress, Washington, D.C., Sept. 1995.

K. Ahmed, *Renewable Energy Technologies: A Review of the Status and Costs of Selected Technologies,* World Bank Technical Page Number 240, Washington, D.C., Jan. 1994.

T. B. Johansson and co-workers, eds., *Renewable Energy—Sources for Fuels and Electricity,* Island Press, Washington, D.C., and Covelo, Calif., 1993.

SOLDERS AND BRAZING FILLER METALS

The ultimate goal of the brazing and soldering processes is to join parts into an assembly through metallurgical bonding. A relatively low melting temperature alloy, a filler metal (FM), is placed in the clearance (gap) between the pieces of base materials (BM) to be joined, and the assembly is subsequently heated until the FM has melted and spread throughout the gap. The molten metal fills the gap and reacts with parts to be brazed, forming after solidification an integral solid whole. Assembly heating can be carried out by various means. These include electromagnetic induction, Joule heating, or use of an oven, flame, etc. Joining temperatures above 723 K are arbitrarily associated with brazing rather than with soldering. However, these processes are essentially similar. In order to distinguish joining materials, those used at temperatures below 723 K are called solders; those above, brazing filler metals.

There are three principal stages in any brazing or soldering process. The first occurs during the heating of an assembled workpiece. At this stage, the FM melts and flows, filling completely the gap normally existing between the parts. The second stage, which normally sets in at a given joining temperature, is characterized by an intensive solid—liquid interaction accompanied by a substantial mass transfer through the interface at strongly uneven rates. Indeed, BM immediately adjoining the liquid filler metal dissolves in this stage. At the same time, a small amount of material from the liquid phase penetrates into the solid BM. Such mass-transfer unbalance results from significantly different diffusion rates in the solid and liquid phases. The final stage of the brazing (soldering process) overlaps with the second and is characterized by the formation of the final joint microstructure, progressing vigorously during assembly cooling while the liquid phase is still present.

In any brazing/soldering process, a molten alloy comes in contact with a surface of solid, which may be an alloy, a ceramic, or a composite material (see CERAMICS; COMPOSITE MATERIALS). For a molten alloy to advance over the solid surface, a special relationship has to exist between surface energies of the liquid–gas, solid–gas, and liquid–solid interfaces. The same relationships should, in principle, hold in joining processes, where a molten alloy has to fill the gaps existing between surfaces of the parts to be joined. In general, the molten alloy should have a lower surface tension than that of the base material. It is of paramount importance to keep the BM surface clean from the so-called virgin oxides and oxides that may have formed during the brazing operation if the heat-treating atmosphere is oxidizing. Cleaning of the BM pieces to be joined and their subsequent protection from oxidation during heating are essential steps in joining operations. Chemically active substances (fluxes) are used to accomplish cleaning and to provide protective shielding when vacuum or protective atmospheres are not available (see METAL SURFACE TREATMENTS). If the brazing environment is not oxidizing, then boron, silicon, and phosphorus, which are used as inherent constituents of FM alloys, can play the role of a flux, ie, reducing original oxide films and making the FM self-fluxing.

Basic Forms of Filler Metal

Filler metal forms include solid preforms and powders, used mostly in the compound form of paste, plastic-bonded tape, and, in the case of soldering, rosin core wire. In special soldering applications, solders may be used as a liquid medium of the soldering baths in which electronic boards are immersed for a short time to solder multiple joints. Conventional paste forms of joining alloys, eg, FM powder plus fluxing agent plus binder/solvent, are applied at externally accessible locations of the clearance between BM pieces. Organic binders decompose when compound FM forms are used in the high vacuum brazing of parts intended for critical high temperature service. Such decomposition of binders can result in the formation of soot in the joint.

Joint Requirements

A basic requirement for any joint is that its strength and ductility be equal or at least close to that of the BM. In general, the strength of a material increases with decreasing grain size, whereas ductility is affected by the presence of brittle phases. Ideally, the maximum size that the grains in a joint may achieve is equal to half the clearance between the BM pieces. Thus, the use of a smaller clearance during BM joining necessarily limits maximum grain size, promotes higher cooling rates of the FM alloy, and thereby results in a refined joint microstructure. A smaller clearance in brazing also promotes improved retention of BM properties because of curtailed BM erosion by the use of a smaller volume of FM. For these reasons, a preplaced self-fluxing thin FM foil used as a preform is superior to the use of powder-containing paste. The grain size and brittle intermetallic, if prescribe, size and extent in the parent FM have a direct effect on the strength of the formed joint. The newer rapidly solidified microcrystalline and amorphous joining materials possess ultimate uniformity of their elemental components, which is beneficial for formation of a fine microcrystalline joint microstructure in both soldering and brazing.

Solder and Brazing Filler Metals

Solders. Most principal solders are lead- or tin-based alloys to which a small amount of silver, zinc, antimony, bismuth, and indium or a combination thereof are added. The principal criterion for choosing a certain solder is its melting characteristics, ie, solidus and liquidus temperatures and the temperature spread or pasty range between them. Other criteria are mechanical properties such as strength and creep resistance, physical properties such as electrical and thermal conductivity, and corrosion resistance.

The majority of solders can be divided into three categories: tin–lead alloys, tin–lead alloys with antimony, and lead–tin–silver alloys. The tin–lead eutectic and near-eutectic alloys are the most commonly used solders that find application in electronics and in general purposes such as joining pipelines (qv). Because of relatively high melting temperature, lead—silver solders are used in applications where joint strength at moderately high temperatures is needed. The addition of tin improves wetting and flow and reduces corrosion in a humid atmosphere. These solders are used in cryogenic apparatuses and for soldering of fine copper wires because of their lower tendency to dissolve copper. The presence of the so-called heavy metals, eg, lead, cadmium, and antimony, in traditional solders has become an important environmental issue owing to concerns for health and safety. As a result, solders containing no lead and antimony such as tin–silver are finding a growing number of applications.

Table 1. AWS Classes of Brazing Filler Metals

Class	Alloy type (and family)	AWS designation
1	Al–Si, eutectic (I)	BAlSi
2	Cu–X, solid solution (II)	BCu
3	Cu–Zn, peritectic (III)	RBCuZn
	Cu–Sn, peritectic (III)	
4	Cu–P, eutectic (I)	BCuP
5	Cu–Ag, eutectic (IV)	BAg
6	TM–Si–B[a], eutectic (I)	BNi
	(Ni/Fe + Cr)–Si–B (I)	
7	(Co, Cr)–Si–B	BCo
	(Ni, Pd)–Si–B	
8	Au–Ni, solid solution (II)	BAu
	Cu–(Ti, Zr)–Ni eutectic and peritectic (V)	

[a] This family includes alloys based on transition metals, such as nickel, iron, cobalt, and palladium.

Brazing Filler Metals. Ideally, the composition of an FM alloy must be such that the following four functions are achieved: (*1*) the FM melting temperature must be lower than that of the corresponding BM; (*2*) the FM surface tension for both solid and liquid states must be lower than that of the BM to provide a driving force for wetting; (*3*) the FM must be compatible with the BM in order to form good metallic bonding, ie, the FM and BM structure, composition, and properties should be similar; and (*4*) the FM-containing elements should be able to bring about chemical reduction/decomposition or physical removal of BM oxide film.

Five families of brazing FM alloy compositions have emerged with respect to metallurgical nature or type. Of these, the first four families of conventional brazing FM have been classified by the American Welding Society (AWS) into eight well-defined classes as indicated in Table 1. The preferred base metals with which each specific family is most compatible as well as the principal areas of FM applications are also given.

The first and largest family of FM alloys, Family I, contains eutectic-type alloys having aluminum, nickel, cobalt, and copper as a base to which silicon–boron (aluminum- and nickel-base alloys) and phosphorus (copper- and nickel-base alloys) are added. Family I encompasses AWS Classes 1, 4, 6, and 7 as well as the nonclassified nickel- and palladium-based alloys. Most of these alloys are brittle, because various intermetallic phases precipitate when processed by conventional technology. This brittle character has limited the available forms of conventionally produced brazing alloys to powder. The presence, however, of silicon, phosphorus, and boron in many conventional FM alloys having near-eutectic compositions facilitates the conversion of such alloys into ductile, thin amorphous alloy foil form when rapid solidification (RS) technology is used.

The second family of brazing FM, Family II, consists of solid solution alloys based on copper (Class 2) and gold–nickel (Class 8). These alloys are used mainly in vacuum-brazing applications and therefore require no alloying elements playing the role of fluxing agents (see VACUUM TECHNOLOGY). Family III consists of alloys having a phase diagram where a peritectic reaction exists, such as copper–zinc alloys (Class 3) and nonclassified Cu–Sn alloys. Alloys of Family IV (Class 5) are probably the most widely used. Family IV is based on the copper–silver binary eutectic system modified by substantial additions of zinc and cadmium, both of which provide fluxing activity, and minor additions of tin and nickel. The fifth family of brazing FM alloys, Family V, although so far unclassified, consists of purely metallic eutectic/peritectic titanium–zirconium-base alloys to which copper and/or nickel are added.

The majority of all these classes, even noneutectic alloys, have been processed successfully by rapid solidification technology. This technology provides a beneficial alternative in the form of an amorphous flexible ductile foil when materials that are inherently brittle are used.

Joining Process Technology

Joint design must ensure a variety of service criteria such as joint mechanical strength; resistance to service environment; electrical conductivity, which is of prime importance in soldering of electrical circuitries; ease of manufacturing; and economics. Compatibility of the parts to be joined (BM) with the forming braze (FM) is always considered from minimization of mechanical stresses, which may appear after brazing/soldering owing to the difference in coefficient of thermal expansion that always exists between the BM and the FM. These stresses may be very high.

Preparation and Protection of the Parts. One, if not the most, important technological step to guarantee success of a joining operation is the preparation of the parts to be joined. Cleaning part surfaces of oxide films, oil, grease, and dirt includes mechanical and/or chemical means, solvent usage, and, finally, rinsing and drying (see METAL SURFACE TREATMENTS). Parts must be protected from oxidation during heat treating in joining. Protection from oxidation may be accomplished by using self-fluxing brazing filler metals, protective fluxes, protective atmospheres, and simply by using vacuum furnaces that have no trace of oxygen.

Fluxes. Fluxes often play multiple roles. Not only do fluxes protect parts from oxidation during heating but they also clean up virgin surfaces from existing tarnishing oxide films by reducing or scaling the oxides. In soldering these fluxes are often used for joining electrical circuitries. Organic-acid or water-soluble fluxes are also frequently preferred in soldering operations in which a final cleaning of soldered boards can be achieved using noncorrosive water solutions.

Heat Treatments. Heating methods in joining can be divided into two principal categories: the local one, where heat is supplied predominantly to the joint area, and the overall one, where the brazed assembly is heated to a certain temperature.

Health and Safety Factors

The specifics of brazing, which should be considered in addition to the conventional safety requirements applied to manufacturing environments, mostly relate to the metal fumes and metal and oxide particulars evolved during processing. The exposure limits to these substances and to the various active fluxes and the solvents used are primary safety parameters regulating the workplace. Sufficient ventilation must be provided. Stringent OSHA composition limits exist for applications of brazing filler metals and solders.

ANATOL RABINKIN
AlliedSignal Inc.

Welding, Brazing and Soldering, American Society for Metals, Metals Park, Ohio, 1993.

Brazing Handbook, American Welding Society, Miami, Fla., 1991.

A. Rabinkin and H. H. Liebermann, in H. H. Liebermann, ed., *Rapidly Solidified Alloys,* Marcel Dekker, Inc., New York, 1993.

F. G. Yost, F. M. Hosking, and D. R. Frear, *The Mechanics of Solder Alloy Wetting and Spreading,* Van Nostrand Reinhold Co., Inc., New York, 1993.

SOL–GEL TECHNOLOGY

The goal of sol–gel technology is to use low temperature chemical processes to produce net-shape, net-surface objects, films, fibers (qv), particulates, or composites that can be used commercially after a minimum of additional processing steps (see COMPOSITE MATERIALS; THIN FILMS). Sol–gel processing can provide control of microstructures in the nanometer size range, ie, 1–100 nm (0.001–0.1 μm), which approaches the molecular level. These materials often have unique physical and chemical characteristics (see NANOTECHNOLOGY).

Sols are dispersions of colloidal particles in a liquid. *Colloids* are nanoscaled entities dispersed in a fluid. *Gels* are viscoelastic bodies that have interconnected pores of submicrometric dimensions. A gel typically consists of at least two phases, a solid network that entraps a liquid phase. *Sol–gel technology* is the preparation of ceramic, glass, or composite materials by the preparation of a sol, gelation of the sol, and removal of the solvent.

Compositions

Nucleation of particles in a very short time followed by growth without supersaturation yields monodispersed colloidal particles that resist agglomeration. A large range of colloidal powders having controlled size and morphologies have been produced using these concepts. Materials include oxides, hydroxides, carbonates, sulfides, as well as various mixed phases or composites and coated particles. Controlled hydrolysis of alkoxides has also been used to produce submicrometer TiO_2, doped TiO_2, ZrO_2, doped ZrO_2, doped SiO_2, $SrTiO_3$, and even cordierite powders. Emulsions have been employed to produce spherical powders of mixed cation oxides, such as yttrium aluminum garnets (YAG), and many other systems. Sol–gel powder processes have also been applied to fissile elements. Spray-formed sols of UO_2 and UO_2–PuO_2 were formed as rigid gel spheres during passage through a column of heated liquid. Abrasive grains based on sol–gel-derived mixed alumina are important commercial products. Powders for superconductors and magnetic ceramics were also developed using the sol–gel technology (see MAGNETIC MATERIALS).

Glass and polycrystalline ceramic fibers have been prepared using the sol–gel method.

Sol–Gel Process Steps

Overview. Three approaches are used to make most sol–gel products: method 1 involves gelation of a dispersion of colloidal particles; method 2 employs hydrolysis and polycondensation of alkoxide or metal salts precursors followed by supercritical drying of gels; and method 3 involves hydrolysis and polycondensation of alkoxide precursors followed by aging and drying under ambient atmospheres.

Production of net-shape silica (qv) components serves as an example of sol–gel processing methods. A silica gel may be formed by network growth from an array of discrete colloidal particles (method 1) or by formation of an interconnected three-dimensional network by the simultaneous hydrolysis and polycondensation of a chemical precursor (methods 2 and 3). When the pore liquid is removed as a gas phase from the interconnected solid gel network under supercritical conditions (critical-point drying, method 2), the solid network does not collapse and a low density aerogel is produced. Aerogels can have pore volumes as large as 98% and densities as low as 80 kg/m^3.

When the pore liquid is removed at or near ambient pressure by thermal evaporation, ie, by drying (methods 1 and 3), shrinkage occurs and the monolith is termed a xerogel. If the pore liquid is primarily alcohol-based, the monolith is often termed an alcogel. The generic term gel is usually applied to either xerogels or alcogels, whereas aerogels are usually specified as such. A gel is defined as dried when the physically adsorbed solvent is completely evacuated. This occurs between 100 and 180°C.

A dried gel still contains a very large concentration of chemisorbed hydroxyls on the surface of the pores. Thermal treatment in the range of 500–800°C desorbs the hydroxyls and thereby decreases the contact angle and the sensitivity of the gel to rehydration stresses, resulting in a stabilized gel. Heat treatment of a gel at elevated temperatures substantially reduces the number of pores and their connectivity owing to viscous phase sintering. This is termed *densification.* The density of the material increases and the volume fraction of porosity decreases during sintering. The porous gel is transformed to a dense glass when all pores are eliminated. Densification is complete at 1250–1500°C for silica gels made by method 1 and as low as 1000°C for gels made by method 3. The densification temperature decreases as the pore radius decreases and surface area of the gels increases. Silica glass made by densification of porous silica gel is amorphous and nearly equivalent in structure and density to vitreous silica made by fusing quartz crystals or sintering of SiO_2 powders made by chemical vapor deposition (CVD) of $SiCl_4$.

Seven processing steps are involved to various degrees in making sol–gel-derived silica monoliths by methods 1, 2, and 3. The emphasis herein is primarily on net-shape sol–gel-derived silica monoliths made by the alkoxide process (method 3) prepared under ambient pressures.

In method 1, a suspension of colloidal powders, or sol, is formed by mechanical *mixing* of colloidal particles in water at a pH that prevents precipitation. In method 2 or 3, a liquid alkoxide precursor such as $Si(OR)_4$, where R is CH_3 (TMOS), C_2H_5 (TEOS), or C_3H_7, is hydrolyzed by mixing with water (eq. 1).

Hydrolysis

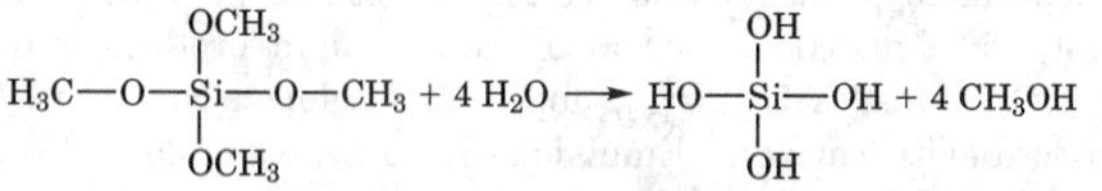

(1)

As soon as any hydrolyzed species is present, condensation proceeds. The hydrated silica tetrahedra interact in a condensation reaction (eq. 2), forming ≡Si—O—Si≡ bonds.

Condensation

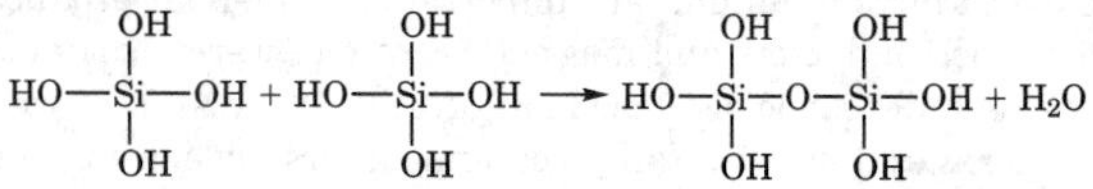

Linkage of additional ≡Si—OH tetrahedra occurs as a polycondensation reaction (eq. 3) and eventually results in a SiO_2 network. The H_2O and alcohol expelled from the reaction remain in the pores of the network.

Polycondensation

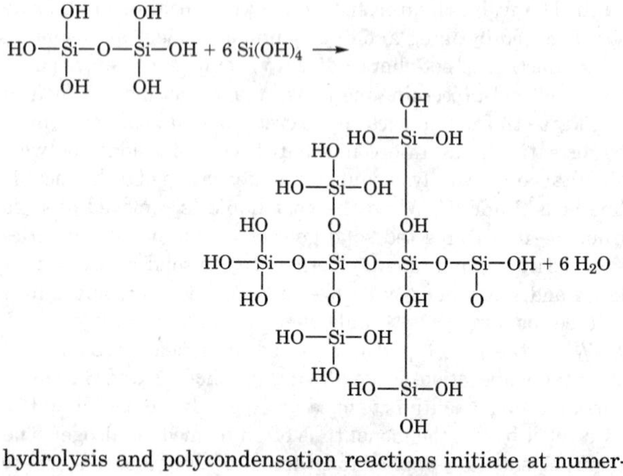

The hydrolysis and polycondensation reactions initiate at numerous sites within the TMOS/H_2O solution as mixing occurs. When sufficient interconnected Si–O–Si bonds are formed in a region, the material responds cooperatively as colloidal (submicrometer) particles or a sol. The size of the sol particles and the cross-linking within the particles, ie, the density, depends on the pH and R ratio, where $R = [H_2O]/[Si(OR)_4]$.

Because the sol is a low viscosity liquid, it can be *cast* into a mold.

After some time the colloidal particles and condensed silica species link together to become a three-dimensional network. The physical characteristics of the gel network depend greatly on the size of particles and extent of cross-linking prior to *gelation.* At gelation, the viscosity increases sharply, and a solid object results in the shape of the mold.

The process that involves a continuous change in structure and properties of a completely immersed gel in liquid after the gel point is called *aging.* The shrinkage of the gel and the resulting expulsion of liquid from the pores during aging is called syneresis. During aging, polycondensation continues along with localized solution and reprecipitation of the gel network, which increases the thickness of interparticle necks and decreases the porosity.

During *drying,* the liquid is removed from the interconnected pore network. Large capillary stresses can develop during drying when the pores are small (<20 nm). These stresses can cause gels to crack catastrophically unless the drying process is controlled by either decreasing the liquid surface energy by addition of surfactants, elimination of very small pores (method 1), supercritical evaporation which avoids the vapor–liquid interface (method 2), or producing a homogenous structure free of defects by controlling the rates of hydrolysis and condensation (method 3).

Dehydration or chemical stabilization, the removal of surface silanol (Si–OH) bonds from the pore network, results in a chemically stable ultraporous solid. Porous gel–silica made in this manner by method 3 is optically transparent, having both interconnected porosity and sufficient strength to be used as unique optical components when impregnated with optically active polymers, such as fluors, wavelength shifters, dyes, or nonlinear polymers.

Heating the porous gel at high temperatures causes *densification.* The pores are eliminated and the density ultimately becomes equivalent to quartz or fused silica.

Hydrolysis and Polycondensation. At gel time, events related to the growth of polymeric chains and interaction between colloids slow down considerably and the structure of the material is frozen. Post-gelation treatments (aging, drying, stabilization, and densification) alter the structure of the original gel, but the resultant structures all depend on the initial structure. Relative rates of hydrolysis (eq. 1) and condensation (eq. 2) determine the structure of the gel. Many factors influence the kinetics of hydrolysis and condensation, because both processes often occur simultaneously. The most important variables are temperature, nature and concentration of electrolyte, nature of solvents, and type of alkoxide precursor. Pressure also influences the gelation process. Condensation may result in a spectrum of structures ranging from molecular networks to colloidal particles. Under acidic conditions, more linear structures are formed prior to gelation. Under basic conditions, the distribution of polysilicate species is much broader and characteristic of branched polymers having a high degree of cross-linking, whereas for acidic conditions there is a lower degree of cross-linking. Thus, the shape and size of polymeric structural units are determined by the relative values of the rate constants for hydrolysis and polycondensation reactions.

Gelation. A sol becomes a gel when it can support a stress elastically, defined as the gelation point or gelation time, t_g. A sharp increase in viscosity accompanies gelation. A sol freezes in a particular polymer structure at the gelation point. This frozen-in structure may change appreciably with time, depending on the temperature, solvent, and pH conditions or on removal of solvent.

The time of gelation changes significantly with sol–gel chemistry. One method of measuring t_g determines the viscoelastic response of the gel as a function of shear rate.

The system evolves from a sol, where individual particles interact more or less weakly with each other, to a gel, which is a continuous network occupying the entire volume. The techniques available to follow structural evolution at the nanometer scale of sol–gel networks include small-angle x-ray scattering (saxs), neutron scattering, light scattering, and transmission electron microscopy. Scattering studies show that acid-catalyzed sols develop a more linear structure with less branching, whereas base-catalyzed systems have highly ramified structures.

The classical or mean field theory of polymerization is useful for visualizing the conditions for gelation. This model yields a degree of reaction, p_c, of one-third at the time of gelation for chemical species having functionality equal to four. Two-thirds of the possible connections are still available and therefore may play a role in subsequent processing. This value is lower than the experimental evidence but represents the minimum degree of reaction before gelation can occur. Experimental results indicate that $0.6 < p_c \leq 0.84$ for silica sol–gel systems. Percolation theory is also used to represent gelation.

Aging. When a gel is maintained in its pore liquid, the structure and properties continue to change long after the gel point. This process is called aging. Four aging mechanisms can occur, singly or simultaneously: polycondensation, syneresis, coarsening, and phase transformation. Polycondensation reactions (eqs. 2 and 3), continue to occur within the gel network as long as neighboring silanols are close enough to react. This increases the connectivity of the network and its fractal dimension. Syneresis is the spontaneous shrinkage of the gel and resulting expulsion of liquid from the pores. Coarsening is the irreversible decrease in surface area through dissolution and reprecipitation processes.

During aging, there are changes in most textural and physical properties of the gel. Inorganic gels are viscoelastic materials responding to a load with an instantaneous elastic strain and a continuous viscous deformation. Because the condensation reaction creates additional bridging bonds, the stiffness of the gel network increases, as does the elastic modulus, the viscosity, and the modulus of rupture.

Drying. For porous systems, there are three stages of drying. During the first stage of drying the decrease in volume of the gel is equal to the volume of liquid lost by evaporation. The compliant gel network is deformed by the large capillary forces that cause shrinkage of the object. Changes in pore size during drying as well as a shift in composition of pore liquid can affect the rate of drying in stage 1. For large- or small-pore gels the greatest changes in volume, weight, density, and structure occur during stage-1 drying. Stage 1 ends when shrinkage ceases. The second stage begins when the critical point is reached. The critical point occurs when the strength of the network has increased owing to the greater packing density of the solid phase, which is sufficient to resist further shrinkage. In stage 2, liquid transport occurs by flow through the surface films that cover partially empty pores. The liquid flows to the surface where evaporation takes place. The flow is driven by the gradient in capillary stress. Because the rate of evaporation decreases in stage 2, this is termed the first falling rate period. The third stage of drying is reached when the pores have substantially emptied and surface films along the pores cannot be sustained. The remaining liquid can escape only by evaporation from within the pores and diffusion of vapor to the surface. During this stage, called the second falling rate period, there are no further dimensional changes, only a slow progressive loss of weight until equilibrium is reached, which is determined by the ambient temperature and partial pressure of water.

When gels crack, they do so at distinct points within the drying sequence. Cracking during stage 1 is rare but can occur when the gel has had insufficient aging and strength, and therefore does not possess the dimensional stability to withstand the increasing compressive stress. Most failures occur during the early part of stage 2, the point at which the gel stops shrinking. Cracking during stage 3 seldom occurs.

Stabilization. A critical step in preparing sol–gel products and especially Type VI silica optical components is stabilization of the porous structure. Both thermal and chemical stabilization is required in order for the material to be used in an ambient environment. The reason for the stabilization treatment is the large concentration of hydroxyls on the surface of the pores of these high ($>400\ m^2/g$) surface area materials. *Chemical stabilization* involves removing the concentration of surface hydroxyls and surface defects, such as metastable three-membered rings, below a critical level so that the surface is not stressed by rehydroxylation in use. Thermal stabilization involves reducing the surface area sufficiently to enable the material to be used at a given temperature without reversible structural changes. The mechanisms of thermal and chemical stabilization are interrelated because of the extreme effects that surface hydroxyls and chemisorbed water have on structural changes. Full densification of gels, such as the transformation of gel–silica to a glass, is nearly impossible without dehydration of the surface prior to pore closure. *Dehydration* of a gel requires removal of two forms of water: free water within the ultraporous gel structure, ie, physisorbed water, and hydroxyl groups associated with the gel surface, ie, chemisorbed water. The amount of physisorbed water adsorbed to the silica particles is directly related to the number of hydroxyl groups existing on the surface. When a silica gel has been completely dehydrated, there are no surface hydroxyl groups to adsorb the free water, and the surface is hydrophobic. It is the realization of this critical point that is the focus for making stable gel products.

Densification. Densification, the final treatment process of gels, occurs between 1000–1700°C, depending on the radii of the pores and the surface area. Controlling the gel–glass or gel–ceramic transition to retain the initial shape of the starting material is difficult. It is essential to eliminate volatile species prior to pore closure and density gradients owing to nonuniform thermal or atmosphere gradients. Using appropriate successful stabilization treatments, it is possible to produce monolithic, dense, gel-derived glasses without pressure or heating to temperatures above the melting point. There are at least four mechanisms responsible for the shrinkage and densification of gels: capillary contraction, condensation, structural relaxation, and viscous sintering. It is likely that several mechanisms operate at the same time, eg, condensation and viscous sintering.

Alumina Derived from Sol–Gel

Aluminum oxide (alumina), Al_2O_3, has high technological value. Sol–gel processing of alumina has created novel applications and improved some of its properties. Products such as catalyst carriers, abrasives, fibers, films for electronic applications, aerogels, and membranes for molecular filtration have been developed based on sol–gel processing (see ALUMINUM COMPOUNDS–ALUMINUM OXIDE (ALUMINA)).

Hydrolysis and Condensation Reactions of Aluminum Alkoxides. Aluminum is less electronegative than silicon, causing it to be more electrophilic and thus less stable toward hydrolysis. These features are responsible for the greater rates of hydrolysis and condensation of aluminum alkoxides when compared to the rates of silicon alkoxide reactions. Hydrolysis and condensation reactions probably occur by nucleophilic addition, followed by proton transfer and elimination of either alcohol or water under neutral conditions. Both reactions are catalyzed by the addition of acid or base. When acids are added, they protonate organic or hydroxyl groups, creating reactive species and eliminating the requirement for proton transfer as an intermediate step. Bases deprotonate water or OH groups, leading to the formation of strong nucleophiles.

Peptization. Aggregation of small inorganic polymeric chains and clusters produced from hydrolysis and condensation reactions of aluminum alkoxides forms macroparticles and nonuniform aggregates. These aggregates that are broken during the peptization step effect the formation of a clear sol having a narrow distribution of particle size. Another mechanism associated with peptization is the production of surface charges on the colloids that eventually leads to either gelation or dispersion.

Gelation. Mechanisms of gelation of alumina sols derived from alkoxides differ from gelation of silicon alkoxide sols. Whereas the sol–gel transition for silica sols is basically a consequence of interactions between long inorganic chains, the gel transition in alumina sols results from colloidal growth by dissolution and reprecipitation processes (Ostwald ripening), followed by formation of linkages between particles. These linkages are initialized by physical–chemical interactions between surface-charged colloids that eventually produce a three-dimensional network, formed by interconnected colloids. The initial contact points between colloids are responsible for neck formation by a coarsening process, followed by particle reshaping and densification. Gelation can be induced either by eliminating excess of water added during hydrolysis or by adding electrolytes to the peptized sol prepared in temperatures above 60°C that leads to sol flocculation.

Drying of Gels. Drying of alumina gels prepared using high temperatures of hydrolysis consists of two steps: sol concentration and pore liquor removal. The rate of sol concentration can dictate some gel properties. High rates of sol concentration lead to less efficiency in packing of colloids. Transparent monolit's can be prepared by drying high density gels.

Applications

Sol–Gel Processing of Thin Films. The sol–gel method enables the production of ceramic films having thickness from 10–1000 nm (see

THIN FILMS). The rheological characteristics of the sol allow the deposition of a film by several procedures: dip coating, spin coating, electrophoresis, thermophoresis, and settling (see COATING PROCESSES). Dip and spin coating are the most frequently used procedures. Dip coating can be divided into five stages: immersion, startup, deposition, drainage, and evaporation. The fluid mechanical boundary layer, which is pulled with the substrate, splits into two. The inner layer moves upward with the substrate, while the outer layer is returned to the bath. The thickness where the split occurs is responsible for the thickness of the film. The spin coating process can be divided into four stages: deposition, spinup, spinoff, and evaporation. In the first stage, an excess of liquid is distributed along the surface that is to undergo deposition. The spinup stage is related to flow of liquid along all the surface, driven by centrifugal force. In the spinoff stage, the excess of liquid flows to the perimeter and is eliminated as droplets. The solvent is eliminated in the fourth stage by evaporation, which leads to the thinning of the film.

Sol–Gel Fibers. During sol-to-gel evolution, changes in the rheology of the sol can be used to allow fiber pulling. Formation of elongated polymers in a solution is a requirement for spinnability, ie, the ability to form fibers. Reduced viscosity for solutions of chain-like or spherical polymers is independent of concentration, whereas linear polymers give a direct relation between reduced viscosity and concentration. Acidic pHs and low values for the molar ratio between water and alkoxide result in the production of linear polymers that exhibit spinnability.

Organic–Inorganic Hybrids. Ceramics and polymers have been combined into high performance composites. The association between high modulus and high strength ceramic fibers, such as glass, carbon, and boron fibers, having the inherent ductility and toughness of some polymers, enables the fabrication of materials having special properties. The integration of different types of materials is restricted by the high temperature processing conditions usually employed in ceramic fabrication. The sol–gel method enables preparation of ceramic materials in a temperature range compatible with organic polymer stability, and involves mechanisms of network formation such as hydrolysis and polycondensation reactions, that are similar to the polymerization reactions of polymers. Thus, sol–gel can be used to produce new types of composites involving ceramics and polymers. These are called organic–inorganic hybrids or ceramers.

Several types of organic–inorganic hybrids have been prepared by using different polymers coupled with TEOS. The basic procedure involves dissolution of the polymer in THF (20 wt %) followed by addition of TEOS with an acidic water solution. Some of the polymers,ie, poly(methyl methacrylate), poly(vinyl acetate), poly(vinyl pyrrolidone), and poly(*N,N*-dimethylamide), yielded transparent films, this demonstrating the absence of macrophase separation. Polycarbonate, poly(acrylic acid), and Nylon trogamid lead to the production of opaque films. Extensive work has been done in terms of combining nanometric clay particles using either Nylon or polyimide. Montmorillonite has been modified by cation exchange using aminolauric acid, and the new groups attached on the surface of the clay bonded to the polymer by initiation of polymerization.

Sol–Gel Bioactive Glasses. Bioactive glasses and ceramics bond to both soft and hard tissue. The chemical bond formed between the implant and tissue can provide the desired adhesion required in many medical and dental applications. Two types of sol–gel processing yield bioactive materials in the SiO_2–CaO–P_2O_5 system having high bioactivity index (see DENTAL MATERIALS; PROSTHETIC AND BIOMEDICAL DEVICES). The bioactivity of the gel-derived materials is equivalent or greater than melt-derived glasses.

LARRY L. HENCH
RODRIGO OREFICE
University of Florida

J. D. Mackenzie, in L. L. Hench and D. R. Ulrich, eds., *Ultrastructure Processing of Ceramics, Glasses and Composites*, John Wiley & Sons, Inc., New York, 1984.

R. K. Iler, *The Chemistry of Silica*, John Wiley & Sons, Inc., New York, 1979.

C. J. Brinker and G. W. Scherer, *Sol–Gel Science*, Academic Press, New York, 1990.

C. Brinker and co-workers, in L. L. Hench and J. K. West, eds., *Chemical Processing of Advanced Materials*, John Wiley & Sons, Inc., New York, 1992.

SOLVENT RECOVERY, CONDENSATION

The removal of solvents by condensing them from an inert gas in a water-cooled condenser is inefficient, and may create a potentially hazardous condition at the point of discharge. Removing most of the solvent from a vent stream by condensation, can drastically reduce the size and cost of a downstream cleanup system.

For optimum solvent removal, the heat exchanger should be arranged so the exiting gas temperature approaches that of the incoming cooling medium. This is best accomplished in a reflux condenser (dephlegmator) wherein solvent laden vapor is introduced into the bottom head of a vertical tubular heat exchanger, and lean vapor leaves from the top. The condensed solvent will flow countercurrent to an upflowing gas and is withdrawn from the bottom head. Cooling medium is introduced at the top of the shell to assure that uncondensed gas leaves at the lowest possible temperature.

The principal concern in designing a tubeside reflux condenser is that an excessive velocity of the entering gas will prevent downflow of the condensed solvent. This is usually not a problem in atmospheric condensation.

The thermal design of a vent condenser is best accomplished by estimating an overall heat-transfer coefficient based on past experience in a similar service. If solvent recovery is maximized by minimizing the temperature approach, the overall heat-transfer coefficient in the condenser will be reduced. This is due to the fact that a large fraction of the heat transfer area is now utilized for cooling a gas rather than condensing a liquid. Depending on the desired temperature approach, the overall heat transfer coefficients in vent condensers usually range between 85 and 170 W/m^2K (ca 15 and 30 $Btu/h{\cdot}ft.^2{\cdot}°F$).

OTTO FRANK
Air Products and Chemicals, Inc.

Perry's Chemical Engineers Handbook, 6th ed., McGraw-Hill, New York, 1984, pp. 5–59.

SOLVENTS, INDUSTRIAL

The term *industrial solvents* is generally applied to liquid organic compounds used on a large scale to perform numerous functions in industry. Useful solvents are found in various chemical classes, the most widely used of which are aromatic and aliphatic hydrocarbons. Other classes include alcohols, ketones, esters, ethers, glycols, glycol ethers, chlorinated hydrocarbons, amines, and aldehydes. In most applications, industrial solvents serve a transitory function by facilitating a process or performing a task, then exiting the process. Even though the solvent may have a temporary function, the success of the industrial process and the quality of the product from the process depends on solvent selection. Solvents perform numerous functions, and the characteristics of the specific application determine which function or functions are most important (Table 1).

Regulations

Regulations have a significant impact on the selection and use of industrial solvents. Since 1966, new and more restrictive regulations have been adopted in the United States at both the federal and state level, which influence the kind of solvents used, the amount used, and the way they are used. Regulations have given impetus to the

Table 1. Functions of Industrial Solvents in End Uses

Solvent function	Purpose
dissolving	prepare solutions of polymers, resins, and other substances
softening	used as tackifiers, improve adhesion to substrate for better bonding
suspension/dispersion	pigments and other particulates
extraction	separate one material from another by selective dissolution
viscosity reduction	thin coatings to application viscosity
chemical intermediate	react with other compounds to form new substances
mfg/processing	improved workability during processing
heat-transfer fluid	remove heat of reaction in chemical manufacturing processes
reaction medium	an inert medium in which other compounds react

development of new technologies and products which decrease solvent emissions to minimize environmental impact.

National Ambient Air Quality Standards (NAAQS) have been established for ozone (qv), nitrogen oxides, lead (qv), carbon monoxide (qv), sulfur dioxide, and particulates. The standards have been set to safeguard human health and the environment. Many areas of the United States violate one or more of the NAAQS. To achieve healthful air for all citizens, the states establish an implementation plan for each area that violates the NAAQS for any of the six criteria pollutants. The plan is a strategy designed to achieve sufficient emission reductions to meet the NAAQS within the deadline. Based on the implementation plan, specific regulations are written which govern the operations emitting the pollutant.

Clean Air Act as Amended in 1990. The Clean Air Act Amendments of 1990 represent an effort by the U.S. Congress to address clean air concerns. The seven titles of the 1990 Act not only extended previous measures, but also broke new conceptual ground.

VOC Emissions Reduction/Ozone Attainment. Title I of the 1990 Amendments continued the process of diminishing emissions of volatile organic compounds (VOC) from all sources to reduce ozone concentrations. A compliance timetable by ozone nonattainment category was established.

Hazardous Air Pollutants. Title III of the Amendments extended the air toxics component of the Clean Air Act. It attacked the issue of air toxics by establishing a long list of toxic pollutants to be regulated by source category on a strict timetable.

Stratospheric Ozone Protection. Title VI of the 1990 Amendments dealt with stratospheric ozone protection. Certain chlorinated fluorocarbon (CFC) compounds, eg, 1,1,1-trichloroethane and carbon tetrachloride, would be phased out over a scheduled time period.

Proposition 65. The formal designation of Proposition 65 is the Safe Drinking Water and Toxic Enforcement Act of 1986. This act was passed overwhelmingly by California's voters in 1986. The principal objectives of Proposition 65 are to protect the State's drinking water sources from toxic contamination and to warn the public of possible carcinogenic and reproductive hazards associated with certain identified chemicals.

Reg Neg Process. Over the years, more than a dozen regulatory negotiations (reg negs) have been held to develop consensus for regulations for diverse, complex industries. In nearly all cases, the negotiations achieved consensus and led to more realistic, efficient regulations. Under charter from the EPA, a reg neg committee conducts negotiations to establish the basis for regulations for a particular industry. Parties to the negotiations include the EPA, state and local environmental authorities, public interest groups, and industry. The primary objective is to develop a regulation agreeable to all parties which recognizes the realities of the industry and provides an improved environment. The reg neg process may then help develop a meaningful, effective regulation more quickly.

Solvent Selection and Formulation

Practical Solubility Concepts. Solution theory can provide a convenient, effective framework for solvent selection and blend formulation. When a solute dissolves in a solvent, a change in free energy occurs as a result of solvent—solute interactions. The change in free energy of mixing must be negative for dissolution to occur. In equation 1,

$$\Delta G = \Delta H - T\Delta S \qquad (1)$$

ΔG is free energy; ΔH, enthalpy; and ΔS, entropy. One way to evaluate the enthalpy of mixing, ΔH, is through the concept of cohesive energy density (CED), the potential energy per cm^3 of solvent in equation 2, where ΔE is the energy of vaporization, V is the molar volume, and

$$\delta^2 = \text{CED} = -\Delta E/V \qquad (2)$$

the Hildebrand solubility parameter, δ, is the square root of the CED. It is a measure of all interactions that occur between molecules of the solvent.

The solubility parameter concept can provide some useful guidelines, but it does not by itself differentiate polar effects and hydrogen bonding forces. For many solutes, polar interactions and hydrogen bonding are important, and a solvent blend designed on the basis of just the solubility parameter may not be adequate. Three parameter approaches, in which the root mean square of dispersion, and polar and hydrogen bonding components of the solubility parameter equals the solubility parameter, have been devised. Polar interactions can be described based on dipole moment or on fractional polarity. A hydrogen bonding index based on infrared measurements, such as the Shell net hydrogen bonding index, can describe the hydrogen bonding capabilities of a solvent.

Computerized optimization using the three-parameter description of solvent interaction can facilitate the solvent blend formulation process, because numerous possibilities can be examined quickly and easily and other properties can also be considered. This approach is based on the premise that solvent blends with the same solvency and other properties have the same performance characteristics.

Solvent Characteristics. Solvents are conveniently classified as hydrocarbons or nonhydrocarbons. The latter are generally oxygenated compounds.

Hydrocarbon Solvents. Most hydrocarbon solvents are mixtures. Toluene is an exception. Hydrocarbon solvents are usually purchased and supplied on specification. The most important specification properties are distillation range, solvency as expressed by aniline cloud point and Kauri-Butanol (KB) value, specific gravity, and flash point. Composition requirements such as aromatic content and benzene concentration are also important in many applications. The key to selecting a hydrocarbon solvent is understanding the requirements of the application. The solubility parameter and evaporation time are useful in designing blends containing both oxygenated solvents and hydrocarbons, but the preferred properties for hydrocarbon solvent blends are aniline cloud point, KB value, and distillation range.

Nonhydrocarbon and Oxygenated Solvents. Most industrial solvents that are not hydrocarbons are pure chemical compounds. As such, they have sharp boiling points and well-defined properties. Specifications for these solvents focus mostly on impurities such as water and other contaminants.

Solvent Selection. A thorough knowledge of the requirements of each solvent application is necessary to formulate a solvent system successfully and meet all needs at the lowest possible cost. The most important properties are solvency, evaporation rate, flash point, and solvent balance. In terms of general solvency, solvents may be described as active solvents, latent solvents, or diluents. This differentiation is particularly popular in coatings applications, but the designations are useful for almost any solvent application. Active solvents are strong solvents for the particular solute in the application, and are most commonly ketones or esters. Latent solvents function as active solvents in the presence of a strong active solvent. Alcohols exhibit this effect in nitrocellulose and acrylic resin solutions. Diluents, most often hydrocarbons, are nonsolvents for the solute in the application.

Solvent Blend Design. Similar guidelines apply regardless of the specific application. The base of the solvent system should be a strong active solvent for the solute in question. If that is a hydrocarbon solvent, then specific hydrocarbon solvent components can be chosen on the basis of aniline cloud point and KB value. If the solute requires a stronger solvent, the most commonly used nonhydrocarbon strong solvents are ketones and esters, which function as true solvents for many solutes. Depending on the solute, a latent solvent, typically an alcohol, may be included, which contributes solvency while lowering blend cost. The remainder of the blend would be composed of diluent solvents, usually hydrocarbons. The evaporation profile of the solvent blend is also important, as well as the related property, flash point. The evaporation rate or boiling range selected should be appropriate for the needs of the application method. In coatings formulations, this would mean a fast evaporating blend for a coating applied by spray, and a slow evaporating blend for brush or roller application. As a safety-related property, flash point must be a consideration. Composition changes during evaporation must also be considered. This characteristic, referred to as solvent balance, is important because adequate solvency must be maintained throughout the entire processing cycle. In most solvent blends, the highest boiling, slowest-evaporating solvent should be a strong solvent for the solute.

Formulator's Dilemma. Emissions of solvents that are hazardous air pollutant (HAP) compounds must be significantly reduced. In a situation where the allowed VOC emission levels are also being reduced, the formulator would like to use the most effective solvents available, but some of the most cost-effective solvents are now designated as HAPs.

Reformulating to reduce HAP solvents frequently means that solvent blend costs increase. The newer blends are generally not as effective. For example, many coatings were usually formulated using ketones as the active solvents with aromatic hydrocarbons as diluents. Esters are the most common ketone replacements, and aliphatic diluents would replace the aromatic hydrocarbons. In this situation, more strong solvent is required compared to the ketone/aromatic formulation and costs increase. The combination of reduced VOC emissions and composition constraints in the form of HAP restrictions have complicated the formulator's task.

Solvent Uses

Although regulatory pressures are gradually reducing the amount of solvents used overall, solvents will continue to be used in significant amounts. A gradual shift away from HAP solvents, combined with a general decrease in solvent usage resulting from VOC emission reduction requirements is expected.

Coatings. Paints (see PAINT) and coatings (qv) comprise the largest single category of solvent consumption, accounting for nearly half the solvent used. Lower solvent emission technologies such as high solids solvent-based coatings, water-based coatings, and nonsolvent technologies such as powder coatings and uv/electron beam coatings will continue to increase in usage at the expense of the traditional low solids, higher solvent content coating.

Cleaning. Perchloroethylene remains the dominant dry-cleaning solvent, accounting for more than 80% of solvent usage. This solvent exhibits good solvency, easy recovery, and a lack of flammability that makes it very suitable for dry cleaning. Emissions have decreased significantly with the adoption of improved recovery techniques, modified operating procedures designed to reduce emissions, and improved equipment.

Cold solvent cleaning is used to degrease metal parts and other objects in many operations, including automotive repair facilities. Mineral spirits have been popular in cold cleaning, but are being supplanted by higher flash point hydrocarbon solvents on account of emissions and flammability concerns (see METAL TREATMENTS).

In vapor degreasing, the solvent is vaporized and the cold part is suspended in the vapor stream. The solvent condenses on the part, and the liquid dissolves and flushes dirt, grease, and other contaminants off the surface. The part remains in the vapor until it is heated to the vapor temperature. Drying is almost immediate when the part is removed and solvent residues are not a problem. The most common solvent used in vapor degreasing operations has been 1,1,1-trichloroethane. Because production and use of 1,1,1-trichloroethane is being phased out, vapor degreasing with that solvent can no longer be used. This has stimulated interest in returning to trichloroethylene, which can be an excellent solvent choice.

Printing Inks. Printing ink preparation is similar to many coating systems. The resin is dissolved in the solvent, followed by pigment dispersion to produce the ink. In most printing operations, the solvent must evaporate fast for best production speed. Alcohol–hydrocarbon solvent combinations are used with polyamide resins for some printing processes (see INKS).

Agricultural Products. Pesticides are frequently applied as emulsifiable concentrates. The active insecticide or herbicide is dissolved in a hydrocarbon solvent which also contains an emulsifier. Hydrocarbon solvent selection is critical for this application. It can seriously impact the efficacy of the formulation. The solvent should have adequate solvency for the pesticide, promote good dispersion when diluted with water, and have a flash point high enough to minimize flammability hazards. When used in herbicide formulas, low solvent phytotoxicity is important to avoid crop damage. Hydrocarbon solvents used in post-harvest application require special testing to ensure that polycyclic aromatics are absent.

Reaction and Heat-Transfer Solvents. Many industrial production processes use solvents as reaction media. Ethylene and propylene are polymerized in hydrocarbon solvents, which dissolves the gaseous reactant and also removes the heat of reaction. Because the polymer is not soluble in the hydrocarbon solvent, polymer recovery is a simple physical operation.

Process Raw Material. Industrial solvents are raw materials in some production processes. For example, only a small proportion of acetone is used as a solvent, most is used in producing methyl methacrylate and bisphenol A. Alcohols are used in the manufacture of esters and glycol ethers. Diethylenetriamine is also used in the manufacture of curing agents for epoxy resins. Traditionally, chlorinated hydrocarbon solvents have been the starting materials for fluorinated hydrocarbon production.

Solvent Extraction. Extraction processes, used for separating one substance from another, are commonly employed in the pharmaceutical and food processing industries. Extraction-grade hexane is the solvent used to extract soybeans, cottonseed, corn, peanuts, and other oilseeds to produce edible oils and meal used for animal feed supplements. Tight specifications require a narrow distillation range to minimize solvent losses as well as an extremely low benzene content. The specification also has a composition requirement, which is very unusual for a hydrocarbon, where the different components of the solvent must be present within certain ranges (see EXTRACTION).

DON A. SULLIVAN
Shell Chemical Company

A. F. M. Barton, *CRC Handbook of Solubility Parameters and Other Cohesion Parameters*, CRC Press, Inc., Boca Raton, Fla., 1983.

E. W. Flick, ed., *Industrial Solvents Handbook*, 4th ed., Noyes Data Corp., Park Ridge, N.J., 1991.

R. J. Lewis, Sr., *Sax's Dangerous Properties of Industrial Materials*, 9th ed., John Wiley & Sons, Inc., New York, 1995.

W. L. Archer, *Industrial Solvents Handbook*, Marcel Dekker, Inc., New York, 1996.

SOMATOTROPINS. See GENETIC ENGINEERING, ANIMALS; GROWTH REGULATORS; HORMONES, HUMAN GROWTH HORMONE.

SONOCHEMISTRY

Ultrasonic irradiation of liquids causes high energy chemical reactions to occur, often with the emission of light. The origin of sonochemistry and sonoluminescence is acoustic cavitation: the formation, growth, and implosive collapse of bubbles in liquids irradiated with high intensity sound. The collapse of bubbles caused by cavitation produces intense local heating and high pressures, with very short lifetimes. In clouds of cavitating bubbles, these hot-spots have equivalent temperatures of roughly 5000 K, pressures of about 1000 atmospheres, and heating and cooling rates above 10^{10} K/s. In single-bubble cavitation, conditions may be even more extreme. Thus, cavitation can create extraordinary physical and chemical conditions in otherwise cold liquids. Sonoluminescence in general may be considered a special case of homogeneous sonochemistry; however, recent discoveries in this field have heightened interest in the phenomenon in and by itself.

Acoustic Cavitation

The chemical effects of ultrasound do not arise from a direct interaction with molecular species. Ultrasound spans the frequencies of roughly 15 kHz to 1 GHz. With sound velocities in liquids typically about 1500 m/s, acoustic wavelengths range from roughly 10 to 10^{-4} cm. These are not molecular dimensions. Consequently, no direct coupling of the acoustic field with chemical species on a molecular level can account for sonochemistry or sonoluminescence. Instead, sonochemistry and sonoluminescence derive principally from acoustic cavitation, which serves as an effective means of concentrating the diffuse energy of sound. Compression of a gas generates heat. The compression of bubbles during cavitation is more rapid than thermal transport, which generates a short-lived, localized hot-spot.

If the acoustic pressure amplitude of a propagating acoustic wave is relatively large (greater than ≈ 0.5 MPa), local inhomogeneities in the liquid (eg, gas-filled crevices in particulates) can give rise to the explosive growth of a nucleation site into a cavity of macroscopic dimensions, primarily filled with vapor. Such a bubble is inherently unstable, and its subsequent collapse can result in an enormous concentration of energy. This violent cavitation event has been termed "transient cavitation". A normal consequence of this unstable growth and subsequent collapse is that the cavitation bubble itself is destroyed. Gas-filled remnants from the collapse, however, may give rise to reinitiation of the process. The generally accepted explanation for the origin of sonochemistry and sonoluminescence is the hot-spot theory, in which the potential energy given the bubble as it expands to maximum size is concentrated into a heated gas core as the bubble implodes.

Two-Site Model of Sonochemical Reactivity

The transient nature of the cavitation event precludes conventional measurement of the conditions generated during bubble collapse. Chemical reactions themselves, however, can be used to probe reaction conditions. The effective temperature realized by the collapse of clouds of cavitating bubbles can be determined by the use of competing unimolecular reactions whose rate dependencies on temperature have already been measured. The sonochemical ligand substitutions of volatile metal carbonyls were used as these comparative rate probes (eq. 1), where the symbol $\xrightarrow{)))}$ represents ultrasonic irradiation of a solution, and L represents a substituting ligand). These kinetic studies revealed that there were in fact

$$M(CO)_x \xrightarrow{)))} M(CO)_{x-n} + n\,CO \xrightarrow{L} M(CO)_{x-n}(L)_n$$

$$\text{where } M = \text{Fe, Cr, Mo, W} \qquad (1)$$

two sonochemical reaction sites: the first (and dominant site) is the bubble's interior gas-phase while the second is an *initially* liquid phase. The latter corresponds either to heating of a shell of liquid around the collapsing bubble or to droplets of liquid ejected into the hot-spot by surface wave distortions of the collapsing bubble.

Microjet Formation during Cavitation at Liquid–Solid Interfaces

A very different phenomenon arises when cavitation occurs near extended liquid–solid interfaces. There are two proposed mechanisms for the effects of cavitation near surfaces: microjet impact and shockwave damage. Whenever a cavitation bubble is produced near a boundary, the asymmetry of the liquid particle motion during cavity collapse can induce a strong deformation in the cavity. The potential energy of the expanded bubble is converted into kinetic energy of a liquid jet that extends through the bubble's interior and penetrates the opposite bubble wall. Because most of the available energy is transferred to the accelerating jet, rather than the bubble wall itself, this jet can reach velocities of hundreds of meters per second. Because of the induced asymmetry, the jet often impacts the solid boundary and can deposit enormous energy densities at the site of impact. Such energy concentration can result in severe damage to the boundary surface. The second mechanism of cavitation-induced surface damage invokes shockwaves created by cavity collapse in the liquid. The impingement of microjets and shockwaves on the surface creates the localized erosion responsible for much of ultrasonic cleaning and many of the sonochemical effects on heterogeneous reactions. In this process, the erosion of metals by cavitation generates newly exposed, highly heated surfaces that are highly reactive.

Sonoluminescence

In addition to driving chemical reactions, ultrasonic irradiation of liquids can also produce light. As with sonochemistry, sonoluminescence derives from acoustic cavitation. There are two separate forms of sonoluminescence: multiple-bubble sonoluminescence (MBSL) and single-bubble sonoluminescence (SBSL). Since cavitation is a nucleated process and liquids generally contain large numbers particulates that serve as nuclei, the cavitation field generated by a propagating or standing acoustic wave typically consists of very large numbers of interacting bubbles, distributed over an extended region of the liquid. If this cavitation is sufficiently intense to produce sonoluminescence, then this phenomenon is called multiple-bubble sonoluminescence (MBSL).

Under the appropriate conditions, the acoustic force on a bubble can be used to balance against its buoyancy, holding the single bubble isolated in the liquid by acoustic levitation. This permits examination of the dynamic characteristics of the bubble in considerable detail, from both a theoretical and an experimental perspective. Such a bubble is typically quite small, compared to an acoustic wavelength (eg, at 20 kHz, the resonance size is approximately 150 μm). For rather specialized but easily obtainable conditions, a single, stable, oscillating gas bubble can be forced into such large amplitude pulsations that it produces sonoluminescence emissions on each (and every) acoustic cycle. This phenomenon is called single-bubble sonoluminescence.

Sonochemistry

In a fundamental sense, chemistry is the interaction of energy and matter. In large part, the properties of a specific energy source determine the course of a chemical reaction. Ultrasonic irradiation differs from traditional energy sources (such as heat, light, or ionizing radiation) in duration, pressure, and energy per molecule. The immense local temperatures and pressures and the extraordinary heating and cooling rates generated by cavitation bubble collapse mean that ultrasound provides an unusual mechanism for generating high energy chemistry. Furthermore, sonochemistry has a high-pressure component, which suggests that one might be able to produce on a microscopic scale the same macroscopic conditions of high temperature–pressure "bomb" reactions or explosive shockwave synthesis in solids.

Experimental Design. A variety of devices have been used for ultrasonic irradiation of solutions. There are three general designs in use presently: the ultrasonic cleaning bath, the direct immersion ultrasonic horn, and flow reactors. The originating source of the ultrasound is generally a piezoelectric material, usually a lead zirconate titanate ceramic (PZT), which is subjected to a high a-c voltage with an ultrasonic frequency (typically 15 to 50 kHz).

The ultrasonic cleaning bath has been used successfully for a variety of liquid–solid heterogeneous sonochemical studies. Lower acoustic intensities can often be used in liquid–solid heterogeneous systems, because of the reduced liquid tensile strength at the liquid–solid interface. The low intensity available in these devices ($\approx$1 W/cm^2), however, can prove limiting. The most intense and reliable source of ultrasound generally used in the chemical laboratory is the direct immersion ultrasonic horn (50 to 500 W/cm^2) which can be used for work under either inert or reactive atmospheres or at moderate pressures (<10 atmospheres). Commercially available flow-through reaction chambers which will attach to these horns allow the processing of multiliter volumes. Homogeneous sonochemistry typically is not a very energy efficient process (although it can be more efficient than photochemistry), whereas heterogeneous sonochemistry is several orders of magnitude better. Unlike photochemistry, whose energy inefficiency is inherent in the production of photons, ultrasound can be produced with nearly perfect efficiency from electric power. A primary limitation of sonochemistry remains the small fraction of the acoustic power actually involved in the cavitation events. Sonochemistry is strongly affected by a variety of external variables, including acoustic frequency, acoustic intensity, bulk temperature, static pressure, ambient gas, and solvent. The frequency of the sound field is not a commonly altered variable in most sonochemistry. Changing sonic frequency alters the resonant size of the cavitation event and to some extent, the lifetime of the bubble collapse, but the overall process remains unchanged.

Acoustic intensity has a dramatic influence on the observed rates of sonochemical reactions. Below a threshold value, the amplitude of the sound field is too small to induce nucleation or bubble growth. Above the cavitation threshold, increased intensity of irradiation (from an immersion horn, for example) will increase the effective volume of the zone of liquid which will cavitate, and thus, increase the observed sonochemical rate.

The effect of the bulk solution temperature lies primarily in its influence on the bubble content before collapse. With increasing temperature, in general, sonochemical reaction rates are *slower*. This reflects the dramatic influence which solvent vapor pressure has on the cavitation event: the greater the solvent vapor pressure found within a bubble prior to collapse, the less effective the collapse. Increases in the applied static pressure increase the acoustic intensity necessary for cavitation, but if equal numbers of cavitation events occur, the collapse should be more intense. In contrast, as the ambient pressure is reduced, eventually the gas-filled crevices of particulate matter which serve as nucleation sites for the formation of cavitation in even "pure" liquids, will be deactivated, and therefore the observed sonochemistry will be diminished.

The choice of ambient gas will also have a major impact on sonochemical reactivity. Monatomic gases give much more heating than diatomic, which are much better than polyatomic gases (including solvent vapor). The choice of the solvent also has a profound influence on the observed sonochemistry. In addition to vapor pressure, other liquid properties, such as surface tension and viscosity, will alter the threshold of cavitation. The chemical reactivity of the solvent is often much more important. No solvent is inert under the high temperature conditions of cavitation.

Homogeneous Sonochemistry: Bond Breaking and Radical Formation. The primary products of ultrasound in aqueous solutions are H_2 and H_2O_2; there is strong evidence for various high-energy intermediates, including HO_2, H·, OH·, and perhaps $e^-_{(aq)}$. The sonolysis of water, which produces both strong reductants and oxidants, is capable of causing secondary oxidation and reduction reactions.

Suslick and co-workers established that virtually all organic liquids will generate free radicals upon ultrasonic irradiation, as long as the total vapor pressure is low enough to allow effective bubble collapse. The sonolysis of simple hydrocarbons (for example, *n*-alkanes) creates the same kinds of products associated with very high temperature pyrolysis.

Applications of Sonochemistry to Materials Synthesis. Of special interest is the development of sonochemistry as a synthetic tool for the creation of unusual inorganic materials. More generally, ultrasound has proved extremely useful in the synthesis of a wide range of nanostructured materials, including high surface area transition metals, alloys, carbides, oxides and colloids. Sonochemical decomposition of volatile organometallic precursors in high boiling solvents produces nanostructured materials in various forms with high catalytic activities. Nanometer colloids, nanoporous high surface area aggregates, and nanostructured oxide supported catalysts can all be prepared by this general route. Sonochemistry has important applications with polymeric materials. Substantial work has been accomplished in the sonochemical initiation of polymerization and in the modification of polymers after synthesis.

Another important application has been the sonochemical preparation of biomaterials, most notably protein microspheres. Using high intensity ultrasound and simple protein solutions, a remarkably easy method to make both air-filled microbubbles and nonaqueous liquid-filled microcapsules has been developed. These microspheres are stable for months, and being slightly smaller than erythrocytes, can be intravenously injected to pass unimpeded through the circulatory system. The mechanism responsible for microsphere formation is a combination of *two* acoustic phenomena: emulsification and cavitation. Ultrasonic emulsification creates the microscopic dispersion of the protein solution necessary to form the proteinaceous microspheres. The long life of these microspheres comes from a sonochemical cross-linking of the protein shell. These protein microspheres, have a wide range of biomedical applications.

Heterogeneous Sonochemistry: Reactions of Solids and Liquids. The use of ultrasound to accelerate chemical reactions in heterogeneous systems has become increasingly widespread. The use of high intensity ultrasound to enhance the reactivity of reactive metals as stoichiometric reagents has become an especially routine synthetic technique for many heterogeneous organic and organometallic reactions. Applications of ultrasound to electrochemistry have also seen substantial recent progress. Another important application for sonoelectrochemistry is the electroreductive synthesis of sub-micrometer powders of transition metals.

Sonocatalysis. Ultrasound has potentially important applications in both homogeneous and heterogeneous catalytic systems. The inherent advantages of sonocatalysis include *(1)* the use of low ambient temperatures to preserve thermally sensitive substrates and to enhance selectivity; *(2)* the ability to generate high energy species difficult to obtain from photolysis or simple pyrolysis; and *(3)* the mimicry of high temperature and pressure conditions on a microscopic scale. Homogeneous catalysis of various reactions often uses organometallic compounds that are often catalytically inactive until loss of metal-bonded ligands (such as carbon monoxide) from the metal. Ultrasound can induce ligand dissociation, permitting initiation of homogeneous catalysis by ultrasound. Heterogeneous catalysts often require rare and expensive metals. The use of ultrasound may permit activating less reactive, but also less costly, metals.

KENNETH S. SUSLICK
University of Illinois at Urbana-Champaign

T. J. Mason, ed., *Advances in Sonochemistry*, vols. 1–4, JAI Press, New York, 1990, 1991, 1993, 1996.

T. J. Mason and J. P. Lorimer, *Sonochemistry: Theory, Applications and Uses of Ultrasound in Chemistry*, Ellis Horword, Ltd., Chichester, U.K., 1988.

G. J. Price, ed., *Current Trends in Sonochemistry*, Royal Society of Chemistry, Cambridge, 1992.

K. S. Suslick, "Sonochemistry of Transition Metal Compounds," in R. B. King, ed., *Encyclopedia of Inorganic Chemistry*, John Wiley & Sons, Inc., New York, Vol. 7.

SORBIC ACID

Sorbic acid (*trans,trans*-2,4-hexadienoic acid), is a white crystalline solid. It is widely used as a preservative in foods having a pH of 6.5 or below, where control of bacteria, molds, and yeasts is essential for obtaining safe and economical storage life.

Physical Properties

The sorbic acid crystal has a well-ordered morphology as a result of its hydrogen bonding and trans,trans structure (**1**).

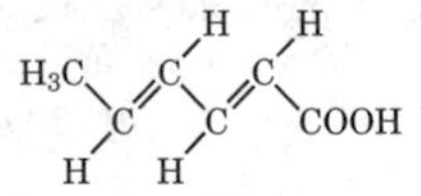

(**1**)

Physical properties are given in Table 1, along with those of the commercially most used salt, *E,E*-potassium sorbate. Sorbic acid dust, as well as any organic dust, can accumulate a static charge and become an explosion hazard, particularly when mixed with highly flammable solvents or oxidizing agents. Minimum explosive limits are 0.02 g/L of air for sorbic acid.

Chemical Properties

The chemical reactivity of sorbic acid is determined by the conjugated double bonds and the carboxyl group.

Conjugated Double Bonds. Sorbic acid is brominated faster than other olefinic acids. Reaction with hydrogen chloride gives predominately 5-chloro-3-hexenoic acid. Sorbic acid is oxidized rapidly in the presence of molecular oxygen or peroxide compounds. The decomposition products indicate that the double bond farthest from the carboxyl group is oxidized. More complete oxidation leads to acetaldehyde, acetic acid, fumaraldehyde, fumaric acid, and polymeric products. Polymerization catalyzed by free radicals occurs with sorbic acid. The ability of sorbic acid to polymerize, particularly on metallic surfaces, has been used to explain its corrosion inhibition for steel, iron, and nickel.

Table 1. Physical Properties of Sorbic Acid

Properties	Sorbic acid
mol wt	112.13
melting point, °C[a]	134.5
boiling point at 101.3 kPa (= 1 atm), °C	228
density, g/cm^3, at 19°C[b]	1.204
flash point, °C	126–130
dissociation constant at 25°C, mol/L	1.73×10^{-5}
pK_a^{25}	
H_2O	4.76
50 wt % ethanol	4.62
pK_a, 0.1 *M* NaCl	4.51
dissociation constant of dimer, K^{24} (CCl_4), mol/L	1.96×10^{-4}
specific heat, J/(g·K)[c]	1.84
latent heat of fusion, kJ/mol[c]	13.6
heat of combustion, kJ/mol[c]	3107
heat of neutralization, kJ/mol[c]	6.07
vapor pressure, kPa[d]	
130°C	1.3
150°C	3.7
170°C	9.3

[a] Potassium sorbate, mol wt 150.22, decomposes at 270°C. [b] Density for potassium sorbate at 20°C = 1.363 g/cm^3. [c] To convert J to cal, divide by 4.184. [d] To convert kPa to mm Hg, multiply by 7.5.

Carboxylic Acid Group. Sorbic acid undergoes the normal acid reactions forming salts, esters, amides, and acid chlorides. Industrially, the most important compound is the potassium salt because of stability and high water solubility.

Synthesis and Manufacture

One of the first commercial methods for synthesis of sorbic acid involved the reaction of ketene and crotonaldehyde in the presence of boron trifluoride in ether at 0°C.

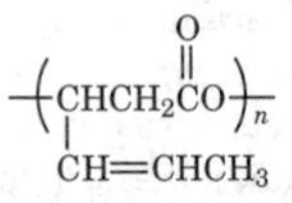

(**2**)

Most commercial sorbic acid is produced by a modification of this route. Catalysts composed of metals (zinc, cadmium, nickel, copper, manganese, and cobalt), metal oxides, or carboxylate salts of bivalent transition metals (zinc isovalerate) produce a condensation adduct with ketene and crotonaldehyde, which has been identified as (**2**). After removal of unreacted components and solvent, the adduct referred to as polyester is decomposed in acidic media or by pyrolysis. Proper operation of acidic decomposition can give high yields of pure *trans,trans*-2,4-hexadienoic acid, whereas the pyrolysis gives a mixture of isomers that must be converted to the pure trans,trans form. Food-grade specifications require further purification in the form of carbon treatments and recrystallization from aqueous or other solvent systems.

Purification Specifications

Sorbic acid and its salts are highly refined to obtain the necessary purity for use in foods. The quality requirements are defined in the United States by the *Food Chemicals Codex*. Codistillation or recrystallization from water, alcoholic solutions, or acetone is used to obtain sorbic acid and potassium sorbate of a purity that passes not only the *Codex* requirements but is sufficient for long-term storage. Measurement of the peroxide content and heat stability can further determine the presence of low amounts of impurities. The presence of isomers, other than the trans,trans form, causes instability and affects the melting point.

Analytical Techniques. Sorbic acid and potassium sorbate are assayed titrimetrically. The quantitative analysis of sorbic acid in food or beverages employs various techniques. The two classical methods are both spectrophotometric. In the ultraviolet method, the prepared sample is acidified and the sorbic acid is measured at ~250–260 nm. In the colorimetric method, the sorbic acid in the prepared sample is oxidized and then reacts with thiobarbituric acid; the complex is measured at ~530 nm. Chromatographic techniques are also used for the analysis of sorbic acid.

Uses

Sorbic acid and its potassium salt, collectively called sorbates, are used primarily in a wide range of food and feed products and to a lesser extent in certain cosmetics, pharmaceuticals, and tobacco products. Sorbates have been shown to inhibit a wide spectrum of yeasts, molds, and bacteria, including most foodborne pathogens. As bacterial inhibitors, sorbates are least effective against lactic acid bacteria. Although this can be a problem for foods that suffer from lactic spoilage, it has proven to be a positive point in cases of yeast and mold suppression during lactic fermentations. The effectiveness of sorbates can be influenced by a number of factors, including pH, microbial load, water activity, temperature, and atmosphere. The inhibitory activity of sorbates is attributed to the undissociated acid molecule. The activity, therefore, depends on the pH of the substrate. The upper limit for activity is approximately pH 6.5 in moist applications; the degree of activity increases as the pH decreases.

The activity of the sorbates at a higher pH is one distinct advantage over the two other most commonly used food preservatives, benzoic and propionic acids, because the upper pH limits for activity of these compounds are approximately pH 4.5 and 5.5, respectively. Although the effect of sorbates can be microbiocidal under certain conditions, activity is most often manifested as a microbial growth retardant.

Sorbates are classified as generally recognized as safe (GRAS) in the United States, with no upper limit set for foods that are not covered by Standards of Identity. They are also allowed in more than 70 food products having Standards of Identity. Examples of products that often contain sorbates are natural and processed cheeses, other cheese products, salad dressings, bakery products, prepared salads, fermented vegetable products, dried fruits, fruit juices, margarine, wine, fish products, jams, and jellies. Compared with other antimicrobial preservatives, sorbates can be used in higher concentrations without affecting the flavor of foods. Maximum shelf life extension with sorbates is achieved when products have low initial levels of microbial contamination and are properly handled and stored. Therefore, the preservative cannot be used to mask poor quality product or poor handling practices. Sorbates can be applied to food by any of several methods, including direct addition, dipping in or spraying with an aqueous sorbate solution, dusting with sorbate powder, or addition to food packaging materials. The potassium salt is used in applications where high water solubility is desired.

Margarine. Sorbates and benzoates are both used in margarine.

Wine. Sorbic acid is used in table wines to prevent secondary fermentation of residual sugar. Adding sorbic acid affords protection against recontamination by yeasts for wines that have been heated or filter-sterilized, but at those low levels it does not provide adequate protection against undesirable malolactic or acetic acid bacteria.

Dairy Products. The dairy industry is the largest commercial user of sorbates, with the largest portion used in processed cheeses. The most common application methods include dipping or spraying with potassium sorbate solutions for natural cheeses and direct addition of sorbic acid to processed cheeses and cold-pack cheeses. Sorbate-impregnated wrapping material can be used for packaged cheese slices and pieces.

Seafood. Sorbates are used to extend the shelf life of many seafood products, both fresh and processed. Sorbates inhibit the growth of psychrotrophic spoilage bacteria, but the treatment must be applied while the product is fresh.

Fruit and Vegetable Products. Sorbates are applied at 0.05–0.1 wt % as a fungistat for prunes, pickles, relishes, maraschino cherries, olives, and figs. The same levels extend shelf life of prepared salads such as potato salad, cole slaw, and tuna salad. Sorbates reduce post-harvest losses of fresh citrus fruit, particularly when the spoilage fungi are resistant to chemical treatments. Post-harvest treatment of apples and apple juice with potassium sorbate decreases spoilage and may prevent mycotoxin production.

Bakery Products. Sorbates are used in and/or on yeast-raised and chemically leavened bakery products. The internal use of sorbates in yeast-raised products at one-fourth the amount of calcium–sodium propionate that is normally added provides a shelf life equal to that of propionate without adversely affecting the yeast fermentation. This internal treatment combined with an external spray of potassium sorbate can provide the same or an increased shelf life of pan breads, hamburger and hot-dog buns, English muffins, brown-and-serve rolls, and tortillas.

Meat and Poultry. The only sorbate treatment of meat permitted by the United States Department of Agriculture (USDA) is a potassium sorbate dip for dry sausages to prevent mold growth. Numerous research studies support increased sorbate use in meat and poultry, but most of these applications have not been approved for use in the United States.

Pet Foods and Commercial Animal Feeds. In intermediate-moisture pet foods, the antimicrobial effectiveness of sorbates is enhanced by a combination of moderate heat treatment, pH adjustment, and reduced water activity via humectants such as propylene glycol, or by adjusting sugar and salt content. Because of the broad activity spectrum, sorbates are extremely effective in the preservation of wet by-products, eg, brewers' and distillers' grains, beet pulp, citrus pulp, and condensed whey.

Health and Safety Factors

The extremely low toxicity of sorbic acid enhances its desirability as a food preservative. The oral LD_{50} for sorbic acid in rats is 7–10 g/kg body weight compared to 5 g/kg for sodium chloride. Studies of the long-term toxicity of sorbic acid in mice and in rats indicate no carcinogenic effects at dietary levels up to 10%. As a result of the favorable toxicological and physiological aspects, the World Health Organization (WHO) has allowed sorbic acid at the highest acceptable daily intake of all food preservatives, 25 mg/kg body weight.

Food Chemicals Codex, 3rd ed. and 1st, 2nd, 3rd, and 4th Supplements to the 3rd ed., National Academy of Science, National Academy Press, Washington, D.C., 1981.

Sorbic Acid and Potassium Sorbate for Preserving Food Freshness, Eastman Chemical Co. Publication ZS-1C, Kingsport, Tenn., 1995.

J. N. Sofos and F. F. Busta, in P. M. Davidson and A. L. Branen, eds., *Antimicrobials in Foods*, 2nd ed., Marcel Dekker, New York, 1993.

J. N. Sofos, *Sorbate Food Preservatives*, CRC Press, Inc., Boca Raton, Fla., 1989.

SOYBEANS AND OTHER OILSEEDS

Soybeans, cottonseed, peanuts, and sunflowers are the four principal oilseed crops grown in the United States. Except for cottonseed, these are consumed directly as foods to varying extents, but all serve as sources of edible oils commonly referred to as vegetable oils. After removal of the oils, the resulting oilseed meals are rich in proteins (qv) and find widespread use as animal feeds (see FEEDS AND FEED ADDITIVES).

Physical Characteristics

Plants and seeds of the four oilseeds vary in growth habit, size, shape, and other features. A common feature of the four oilseeds is storage of the bulk of the protein and oil in distinct membrane-bound, subcellular organelles called protein bodies and lipid bodies, respectively.

Soybean. Soybeans grow on erect, bushy annual plants, 75–125 cm high, having hairy stems and trifoliate leaves. Seeds are produced in pods, usually containing three almost spherical-to-oval seeds weighing 0.1–0.2 g.

Cottonseed. Cotton (qv) grows as an annual or perennial herb or shrub, sometimes tree-like. The seeds are produced in leathery capsules (bolls), covered with lint and fuzz fibers. The seeds are ovoid, 8–12 mm long having brown to nearly black seed coats.

Peanut. Peanuts grow on annual herbaceous plants, bushy upright (erect) or spreading (runner) types, 25–50 cm high with bijugate leaves, hairy stems, and bright yellow flowers having peduncles that bend after fertilization and push the pods underground, where they develop and ripen. The seeds are produced in pods containing two or three seeds. The kernels are almost spherical to roughly cylindrical (0.4–1.1 g each).

Sunflower. Two types of sunflowers are grown in the United States. Varieties grown for oilseed production, ca 85% of crop, are generally black-seeded. These contain 40–50% oil and ca 20% protein. Nonoilseed varieties, ca 15% of crop, sometimes referred to as confectionery, have striped, relatively thick hulls that do not adhere to the kernels. The sunflower seed (achene) is four-sided and flattened, ca 9 mm long ×4–8 mm wide, having a black or striped gray and black seed coat (pericarp) enclosing a kernel.

Chemical Composition

Compositions of the four oilseeds are given in Table 1. All except soybeans have a high content of seed coat or hull. Because of the high

Table 1. Composition[a] of Oilseeds, Wt %

Oilseed	Hulls	Oil	Protein[b]	Protein in dehulled, defatted meal [c]
soybean	8	20	43	52
cottonseed				
acid delinted	36	21.6	21.5	
kernels		36.4	32.5	63
peanut	20–30			
kernels	2–3.5[d]	50.0	30.3	57
sunflower				
arrowhead variety, low oil type	47	29.8	18.1	~68[e]
armavirec variety				~60[e]
high oil type	30	48.0	16.9	
kernels		64.7	21.2	

[a] Approximate; moisture-free basis. [b] As nitrogen, N, ×6.25. [c] Data vary with efficiency of dehulling and oil extraction, variety of seed, and climatic conditions during growth. [d] Red skins or testa. [e] Calculated from kernels on oil-free basis.

hull content, the crude fiber content of the other oilseeds is also high. Soybeans differ from the other oilseeds in their high protein and low oil content. All these oilseeds, however, yield high protein meals when dehulled and defatted.

Proteins. The proteins found in the four oilseeds are complex mixtures consisting of four characteristic fractions having molecular weights of ca 9,000–700,000. The major fractions are considered to be storage proteins and are located in the protein bodies. Of the four oilseeds, the proteins of soybeans are the best characterized. In addition to the storage proteins, oilseeds contain a variety of minor proteins, including trypsin inhibitors, hemagglutinins, and enzymes.

Lipids. Cottonseed, peanut, and sunflower oils are classified as oleic–linoleic acid oils because of the high (>50%) content of these fatty acids (see FATS AND FATTY OILS). Soybean oil is called a linolenic acid oil because it contains 4–10% of this fatty acid. In addition to the triglycerides, the four oilseeds also contain phosphatides. Sterols are present in concentrations of 0.2–0.4% in the oils. The sterols exist in the seeds in four forms: free, esterified, nonacylated glucosides, and acylated glucosides. Soybeans contain a total of 0.16% of these sterol forms in the ratio of ca 3:1:2:2 (see STEROIDS).

Carbohydrates. Oilseeds contain two types of carbohydrates (qv): soluble mono- and oligosaccharides and largely insoluble polysaccharides. Sucrose, raffinose, and stachyose are the principal sugars present. Polysaccharides make up about 12–22% of defatted flours.

Minor Constituents. All four oilseeds contain minor constituents, eg, phytic acid, that affect the use of the defatted seeds, especially in feeds and foods. Soybeans contain the isoflavones genistein, daidzein, and glycetein in the form of malonyl glucosides, acetylated glucosides, glucosides, and aglycones.

Harvesting and Storage

Soybeans. Primary production of the U.S. soybean crop occurs in the Midwest. Harvest begins in September or October. Ideal moisture for harvesting is 13%, and the crop can be successfully stored at this moisture content until the following summer. Soybeans at ≤12% moisture can be stored for two years or more with no significant deterioration, although the entire crop is usually processed in little over a year after harvest.

Cottonseed. In the United States, cotton harvesting begins in late July and is usually completed by the end of December. After picking, the cotton is processed in gins to separate the lint from the seed. Cottonseed is usually stored in Muskogee-type warehouses, ie, low, flat metal buildings having roofs that slope close to the angle of repose of cottonseed, and that are equipped with aeration and temperature-monitoring systems. High temperatures cause rapid deterioration of the seed (see FOOD TOXICANTS, NATURALLY OCCURRING).

Peanuts. When the kernels are fully developed and taking on a mature color, the plants are dug mechanically, shaken to remove the soil, and inverted into windrows to dry (cure) and mature completely. Ideally, the peanuts are left to cure for several days until the moisture content drops to ca 10%. The cured peanuts are stored or shelled, and <10% of the crop is retailed in the shell. Shelled peanuts should have a moisture content of ca 7% (6.5% is optimum) for safe storage and are best stored under refrigeration. Moisture control is critical to maintain quality.

Sunflowers. In the United States, sunflower seeds are harvested after a killing frost in late September and October. The crop dries easily because air readily passes through beds of the large seed. A moisture content of 9.5% is safe for short-term storage, but ≤7% is recommended for long-term storage without aeration.

Processing

Soybeans. Virtually all soybeans processed in the United States are solvent-extracted with hexane to recover the oil. The beans are cleaned and may be dried and allowed to equilibrate at 10–11% moisture to facilitate loosening of the seed coat or hull. They are then cracked, dehulled by screening and aspiration, and conditioned by treatment with steam (qv) to facilitate flaking. The conditioned meats are flaked and extracted with hexane to remove the oil.

Cottonseed. In the United States, cottonseed is processed into oil and meal by screw-pressing or solvent extraction. In screw-pressing the seed is cleaned, delinted, dehulled, flaked, and cooked prior to pressing. Screw-pressing yields a cake containing 2.5–4.0% residual oil. The cake is ground into a meal, and ground cottonseed hulls are blended back to adjust protein content to trading standards. In the solvent extraction procedure, the flakes are often processed through an extruder to form collets, which are then extracted with hexane. Meal emerging from the solvent extractors is freed of hexane by heating. The resulting meal contains about 1% residual oil (see EXTRATION, LIQUID–SOLID).

Peanuts. Only 15–20% of the U.S. peanut crop is converted into oil and meal. Processing is carried out by screw-pressing or prepressing, followed by solvent extraction. In screw-pressing, the peanuts are shelled, cooked, and pressed to yield a crude oil plus a cake containing ca 5% residual oil. In prepressing–solvent extraction, the cooked meats are screw-pressed at low pressure to remove a portion of the oil and then extracted with hexane, which is later recovered by evaporation.

Sunflowers. Processing of sunflowers consists of screw-pressing, direct extraction with hexane, or prepress–solvent extraction. The first step is cleaning, followed by dehulling. The dehulled seed is conditioned by heating and then goes to screw presses or is flaked as in the case of direct solvent extraction with hexane.

Nutritional Properties and Antinutritional Factors

Oil. Because of their high linoleic acid contents, unhydrogenated and partially hydrogenated soybean, cottonseed, peanut, and sunflower oils are good sources of this essential fatty acid. Soybean oil is the principal vegetable oil consumed in the United States, and approximately three-fourths is partially hydrogenated to improve flavor stability, increase resistance to oxidation, and raise the melting point. Linoleic and linolenic acid contents of soybean oil are reduced by hydrogenation, but the process is also accompanied by migration of double bonds up and down the fatty acid chain and the conversion of cis to trans isomers. Interesterification of fats of different compositions is being used in Europe and Canada as an alternative to hydrogenation to modify the physical properties of oils. Heating of fats during the frying of foods results in hydrolysis and oxidation reactions that generate various compounds, including free fatty acids, alcohols, aldehydes, ketones, dimer acids, and polymeric fatty acids. Cyclopropenoid fatty acids found in cottonseed oil are biologically active in several animal species. For example, they act as synergists with aflatoxins and as liver carcinogens when fed to trout. The long history of cottonseed oil use in the human diet, however, has not revealed any adverse effects.

Proteins and Meals. Nutritional properties of the oilseed protein meals and their derived products are determined by the amino acid compositions, content of biologically active proteins, and various nonprotein constituents found in the defatted meals. Phytic acid, present as salts in all four meals, is believed to interfere with dietary absorption of minerals such as zinc, calcium, and iron (see FOOD TOXICANTS, NATURALLY OCCURRING; MINERAL NUTRIENTS).

Soybeans. Numerous studies have demonstrated that methionine is the first limiting amino acid in soybean proteins. That is, methionine is in greatest deficit for meeting the nutritional requirements of a given species. Although it is common practice to add synthetic methionine to broiler feeds to compensate for this deficiency, methionine supplementation is not necessary for humans except for infants. Soybean proteins meet or exceed the essential amino acid requirements of FAO/WHO/UNU for children from age two to adults. The presence of trypsin inhibitors in soybeans is well-documented, and when ingested by laboratory animals, these inhibit growth and affect organs such as the pancreas. The inhibitors are largely inactivated by moist heat.

Cottonseed. When compared with FAO/WHO/UNU essential amino acid requirements, cottonseed proteins are low in lysine, threonine, and leucine for two to five-year-old children, yet meet all requirements for adults. Cyclopropenoid fatty acids exist in hexane-defatted meal at levels of 21–76 ppm. In rainbow trout, the cyclopropenoid acids cause cancer of the liver. However, similar effects in mammals or humans have not been demonstrated. Contamination of cottonseed by aflatoxins is a perennial concern.

Peanuts. The proteins of peanuts are low in lysine, threonine, cystine plus methionine, and tryptophan when compared to the amino acid requirements for children but meet the requirements for adults. Peanut flour can be used to increase the nutritive value of cereals such as cornmeal, but further improvement is noted by the addition of lysine. Peanuts are prone to contamination by aflatoxin.

Sunflower Seed. Compared to the FAO/WHO/UNU recommendations for essential amino acids, sunflower proteins are low in lysine, leucine, and threonine for two to five-year-olds but meet all the requirements for adults. There are no major antinutritional factors known to exist in raw sunflower seed.

Oilseed Products and Uses

Oil. Most crude oil obtained from oilseeds is processed further and converted into edible products. Only a small fraction of the total oil is used for industrial (nonedible) purposes.

For edible uses, oilseed oils are processed into salad and cooking oils, shortenings, margarines, and confectionery fats such as for candy, toppings, icings, and coatings. Degumming removes the phosphatides and gums, which are refined into commercial lecithin (qv) or returned to the defatted flakes. Then, free fatty acids, color bodies, and metallic prooxidants are removed using aqueous alkali. High vacuum deodorization yields a salad oil. Partial hydrogenation, under conditions where linolenate is selectively hydrogenated, results in greater stability to oxidation and flavor deterioration. Hydrogenation is also used to produce hardened fats for shortenings and margarines.

Vegetable oils are utilized in a variety of nonedible applications, but only a few percent of the U.S. soybean oil production is used for such products. Soybean oil is converted into alkyd resins (qv) for protective coatings, plasticizers, dimer acids, surfactants (qv), and printing inks.

Protein Products. The bulk of the meal obtained in processing of oilseeds is used as protein supplements in animal feeds. Since the 1960s appreciable amounts have been also converted into products for human consumption, the majority of which have been derived from defatted soybean flakes.

The high protein content of oilseed meals has made them essential ingredients of poultry and livestock feeds. Soybean meal, especially the dehulled product, is low in crude fiber and high in lysine. Although limiting in methionine, soybean meal is a key ingredient for blending with corn in formulating feeds for nonruminants, eg, poultry and swine. Cottonseed meal is used primarily for beef and dairy feeds. Less than one-third of the U.S. cottonseed meal is used for poultry feeds. Peanut meal is used mainly for dairy and beef cattle feeds. Sunflower meal is also fed mainly to ruminants, ie, sheep, beef, and dairy cattle (see FEEDS AND FEED ADDITIVES).

As of the mid-1990s only peanuts and soybeans were being converted into protein ingredients for use in food products. Peanuts are hydraulically pressed to remove about 55% of the oil and the pressed peanuts are then ground into flours for use in baked products, snacks, and confections. Starting materials for soybean protein ingredients are defatted flakes that are processed into three classes of products differing in minimum protein content (expressed on a dry basis): flours and grits (50% protein); protein concentrates (65% protein); and protein isolates (90% protein). Flours and grits are made by grinding and sieving flakes. Concentrates are prepared by extracting and removing the soluble sugars from defatted flakes. Protein isolates are obtained by extracting the soluble proteins from defatted flakes. Flours, concentrates, and isolates are also processed, commonly by extrusion, to texturize them for use as meat extenders and substitutes.

Oilseed proteins are used as food ingredients at concentrations of 1–2% to nearly 100%. At low concentrations, the proteins are added primarily for their functional properties, eg, emulsification, fat absorption, water absorption, texture, dough formation, adhesion, cohesion, elasticity, film formation, and aeration (see FOOD PROCESSING). Use of some oilseed proteins in foods is limited by flavor, color, and flatus effects. Flatus production by defatted soy flours has been attributed to raffinose and stachyose, which are removed by processing the flours into concentrates and isolates.

Among the oilseeds, only soybean protein isolates are used for industrial applications in the United States. Industrial isolates are used primarily as adhesives for binding pigments to paper and paperboard to make the surfaces suitable for printing.

Food Products. Soybeans, peanuts, and sunflower seeds are consumed as such or are processed into edible products.

Soybeans. Small amounts of soybeans are roasted and salted for snacks. Nut substitutes for baked products and confections are also manufactured from soybeans. Larger amounts are used in Oriental foods, such as soymilk, tofu, miso, tempeh, and soy sauce.

Peanuts. About 65% of U.S. peanuts are consumed directly in the form of peanut butter, roasted peanuts, and confections (see NUTS).

Sunflower Seed. The large seed is dry-roasted, salted, and sold in the shell. The medium sized seeds are dehulled, roasted dry or in oil and used in cookies, salad toppings, ice cream toppings, trail and snack mixes, and breads and rolls. The small-sized, off-sized, and broken seed is sold as bird and pet feeds.

WALTER J. WOLF
U.S. Department of Agriculture

J. F. Carter, ed., *Sunflower Science and Technology*, American Society of Agronomy, Madison, Wis., 1978.

D. R. Erickson, ed., *Practical Handbook of Soybean Processing and Utilization*, American Oil Chemists' Society Press, Champaign, Ill. and United Soybean Board, St. Louis, Mo., 1995.

R. J. Kohel and C. F. Lewis, eds., *Cotton*, American Society of Agronomy, Madison, Wis., 1984.

J. G. Woodroof, ed., *Peanuts: Production, Processing, Products*, 3rd ed., AVI Publishing Co., Inc., Westport, Conn., 1983.

SPACE CHEMISTRY. See EXTRATERRESTRIAL MATERIALS; SPACE PROCESSING.

SPACE PROCESSING

Because the residual accelerations resulting from atmospheric drag and gravity gradient effects in low earth orbit are typically on the order of 10^{-6} times earth-gravity, the term microgravity generally is used to describe this acceleration environment. Because by the mid-1990s the space shuttle had been operational for more than a decade, a large number of microgravity experiments were conducted on various flights. Several microgravity-emphasis missions, in which the shuttle was flown in an attitude that minimized acceleration disturbances, were also flown to accommodate experiments that were exceptionally sensitive to accelerations. The more academically oriented experiments designed to address fundamental issues of materials processing are sponsored by the NASA Office of Microgravity Science and Applications; experiments designed to address issues of more direct interest to industrial research are sponsored by the NASA Office of Space Access and Technology (formerly the Office of Commercial Programs).

Materials Experiments in Space

Protein Crystal Growth. By the mid-1990s, the protein crystal growth experiments produced the most spectacular results of all the space processing experiments. The importance of x-ray crystallography as a mechanism for determining three-dimensional structure of complex macromolecules has placed new demands on the ability to grow large (ca 0.5 mm on a side), highly ordered crystals of a vast variety of biological macromolecules in order to obtain high resolution x-ray diffraction data. There are numerous difficulties encountered in attempts to grow macromolecular crystals of biological interest, and the ability to grow such crystals of sufficient size and quality has become an important step for advancement in this field. The difficulties encountered prompted some investigators to consider growing protein crystals in reduced gravity. The first protein growth experiment in reduced gravity was carried out in 1983 on Spacelab 1; crystals of lysozyme and β-galactosidase grew substantially larger in space than in ground control experiments.

The first U.S. protein crystal growth experiment under reasonably well-controlled conditions was carried out in September 1988 on a space transportation system (STS-26), the first shuttle flight after the Challenger accident. The γ-interferon sample and the porcine elastase sample grew much larger than the ground control samples. The isocitrate lyase sample grew as discrete prisms, whereas the ground control crystals always grew dendritically. Crystals of all three of these proteins exhibited significantly higher x-ray diffraction resolution than any produced on Earth. This was true even when some of the smaller space-grown crystals were compared with larger Earth-grown crystals.

Following STS-26, there have been many other attempts to grow a variety of protein crystals in space. Considered as a whole, these experiments have produced a mixed set of results. In some cases the space experiments yielded no crystals or produced crystals that were inferior to those grown on Earth. However, there have been a number of cases in which the space-grown crystals were larger and better ordered than the best ever grown on the earth.

Solution Growth of Small-Molecule Crystals. At least some of the advantages obtained from growth of macromolecular crystals in microgravity appear to carry over to the growth of small-molecule crystals. Triglycine sulfate (TGS) crystals were grown from solution on Spacelab-3 and again on IML-1 using a novel cooled sting method. Supersaturation was maintained by extracting heat through the seed, which was mounted on a small heat pipe, and in turn was attached to a thermoelectric device. By growing under diffusion-controlled transport conditions, it was hoped it would be possible to avoid liquid–vapor inclusions. These inclusions are the most common types of defect in solution-grown crystals and are believed to be caused by unsteady growth conditions resulting from convective flows.

Typically, TGS crystals are grown on ⟨001⟩ oriented seeds, because growth on the ⟨010⟩ face tends to be nonuniform and multifaceted. However, in the absence of convection, growth on the ⟨001⟩ seeds on Spacelab-3 was mostly around the periphery of the seed. Therefore, seeds with a natural ⟨010⟩ face were cut from a polyhedral TGS crystal for the experiments on IML-1. The crystal was grown in space with a 4°C undercooling, which produced a growth rate of 1.6 mm/d. Even though this is somewhat larger than typical growth rates (because of the limited time available) the quality of the space-grown crystal was extremely good. Growth was very uniform and the usual growth defects in the vicinity of the seed that form during the transition from dissolution to growth, known as ghost of the seed, were notably absent.

Vapor Crystal Growth. For materials that lend themselves to physical or chemical vapor transport, growth from the vapor offers some attractive alternatives to growth from the melt. Growth can take place at temperatures considerably lower than the melting point, thus avoiding some of the higher temperature problems associated with melt growth. Several crystal growth experiments on the shuttle have produced provocative results that are not at all understood, eg, growth of unseeded GeSe crystals by physical vapor transport using an inert noble gas as a buffer in a closed tube on STS-7. In the ground control experiment, many small crystallites formed a crust inside the growth ampul at the cold end. The flight experiment produced dramatically different results; the crystals apparently nucleated away from the walls and grew as thin platelets that eventually became entwined with one another, forming a web that was loosely contained by the ampul. Even more striking was the appearance of the surfaces of the space-grown crystals. These were mirror-like and almost featureless, exhibiting only a few widely spaced growth terraces. By contrast, the crystallites in the ground control experiments conducted under identical thermal conditions had many pits and irregular closely spaced growth terraces.

Test of Dendritic Growth Models. The microgravity environment provides an excellent opportunity to carry out critical tests of fundamental theories of solidification without the complicating effects introduced by buoyancy-driven flows. This advantage was used to carry out a series of experiments to elucidate dendrite growth kinetics under well-characterized diffusion-controlled conditions in pure succinonitrile (SCN). Dendrite tip velocities were measured as a function of undercooling over a range from 0.05–1.5 K. Comparing these measurements with ground-based measurements, it was possible to show that effects of convection are more significant at the smaller undercoolings and are still important up to undercoolings as large as 1.3 K. Even in microgravity, a slight departure in the data was noted at the smallest undercooling, which was attributed to the residual acceleration of the spacecraft. These data also allow the determination of the scaling constant important in the selection of the dynamic operating state, which the present theories have been unable to provide.

Electrodeposition. Electrodeposition experiments in reduced gravity have produced some intriguing results. An early experiment on the German TEXUS rocket, using higher current densities than can normally be used on Earth, reported the deposition of amorphous Ni on Au substrates. In a series of experiments on the Consort Rocket, it was possible to repeat this result.

ROBERT J. NAUMANN
University of Alabama in Huntsville

L. J. DeLucas and co-workers, *Science* **246**, 651 (1989).

L. J. DeLucas and co-workers, *J. Cryst. Growth* **135**, 172 (1994).

M. E. Glicksman, M. B. Koss, and E. A. Winsa, *Phys. Rev. Lett.* **73**, 573 (July 25, 1994).

C. Riley, H. Abi-Akar, B. Benson, and G. Maybee, *J. Spacecraft Rockets* **29**, 386 (1990).

SPANDEX AND OTHER ELASTOMERIC FIBERS. See FIBERS, ELASTOMERIC.

SPECTROSCOPY, OPTICAL

Spectroscopy, the study of electromagnetic radiation and its interaction with matter as a function of frequency or wavelength, is a versatile and powerful tool for investigating atomic and molecular structure, as well as for qualitative and quantitative analysis. Optical spectroscopy conventionally implies the ultraviolet, visible, and infrared spectral regions. Herein coverage is extended to include shorter and longer wavelengths that interact with matter by the same basic mechanism of coupling with the electric vector of the electromagnetic field; mass, acoustic, particle energy, and magnetic resonance spectroscopies are thus excluded (see MASS SPECTROSCOPY; MAGNETIC SPIN RESONANCE). The analytical applications of spectroscopy, range from bench analyses of chemical samples in the laboratory, to process monitoring in chemical plants, to the detection and monitoring of pollutants in the atmosphere.

The objective in any analytical procedure is to determine the composition of the sample (speciation) and the amounts of different species present (quantification). Spectroscopic techniques can both identify and quantify in a single measurement. A wide range of compounds can be detected with high specificity, even in multicomponent mixtures. Many spectroscopic methods are non-invasive, involving no sample collection, pretreatment, or contamination (see NONDESTRUCTIVE EVALUATION). Because only optical access to the sample is needed, instruments can be remotely situated for environmental and process monitoring (see ANALYTICAL METHODS; PROCESS CONTROL). Spectroscopy provides rapid real-time results, and it is easily adaptable to continuous long-term monitoring. Spectra also carry information on sample conditions such as temperature and pressure.

In spectroscopic analysis, species are identified by the frequencies and structures of absorption, emission, or scattering features, and quantified by the intensities of these features. The many applications of optical methods to chemical analysis rely on just a few basic mechanisms of light–matter interaction.

Absorption spectroscopy records depletion by the sample of radiant energy from a continuous or frequency-tunable source, at resonance frequencies that are characteristic of various energy levels in atoms or molecules. The basic law of absorption, credited to Bouguer-Lambert-Beer, states that in terms of the incident, I_0, and transmitted, I_t, light intensities, the absorbance, A (or transmittance, T), is given by equation 1:

$$A(\nu) = -\log T(\nu) = \log(I_0/I_t) = a(\nu)cl \tag{1}$$

where c is the concentration of the absorbing species, l is the path length through the sample, and $a(v)$ is an intensive property that specifies the absorption strength of the analyte at frequency v. Terminology for $a(v)$ varies depending on the units of c, on whether A is a decadic or Napierian logarithm, and on conventions employed in different spectral regions. The terms absorption coefficient, absorptivity, extinction coefficient, molar absorptivity, and molar extinction coefficient may each be encountered in different contexts for $a(v)$. In multicomponent systems, $A(v)$ is simply the sum of the contributions from the individual absorbers. The linear relationship between A and c in equation 1 usually holds over a wide range of c, making this law the basis for quantitative spectroscopy.

Emission spectroscopy is the analysis, usually for elemental composition, of the spectrum emitted by a sample at high temperature, or that has been excited by an electric spark or laser. The direct detection and spectroscopic analysis of ambient thermal emission, usually in the infrared or microwave regions, without active excitation, is often termed radiometry. In emission methods the signal intensity is directly proportional to the amount of analyte present.

Scattering techniques record the change of a usually monochromatic probe signal scattered by a sample. It can involve elastic (energy-conserving) interactions, such as Rayleigh scattering, where photons undergo only a change in momentum, or the inelastic (energy-changing) Raman effect, in which scattering is accompanied by discrete changes in frequency. Rayleigh scattering occurs for all species having dimensions much smaller than the wavelength of the probe light, Mie scattering occurs from larger dielectric particles, and Tyndall scattering from discontinuities such as interfaces. These elastic processes provide little chemical information, but in atmospheric applications can furnish a return signal for laser infrared radar (lidar) sounding. The Raman effect is weaker by factors of $\sim 10^3$, but spectroscopic analysis of scattered Raman light reveals spectral shifts characteristic of different chemical species (see INFRARED TECHNOLOGY AND RAMAN SPECTROSCOPY).

Fluorescence and phosphorescence are types of luminescence, ie, emission attributed to selective excitation by previously absorbed radiation, chemical reaction, etc, rather than to the temperature of the emitter. Laser-induced and x-ray fluorescence are important analytical techniques (see LUMINESCENT MATERIALS, CHEMILUMINESCENCE).

Background

The Electromagnetic Spectrum. Electromagnetic radiation is characterized by its wavelength, λ, frequency, ν, or wavenumber, ν ; which are related by equation 2:

$$\nu = c/\lambda \quad \tilde{\nu} = 1/\lambda = \nu/c \tag{2}$$

where c is the speed of lights. Units for wavelength are commonly nm or μm (1 nm = 10 Å = 10^{-3} μm = 10^{-9} m); for frequency, some multiple of cycles per second (hertz); and for wavenumber, cm^{-1}(1 $cm^{-1} \cong$ 30 GHz). The photon energy is $E = h\nu = hc\tilde{\nu}$, where h is Planck's constant, and so is proportional to frequency and wavenumber (1 eV $\cong$ 8066 cm^{-1}). The electromagnetic spectrum is conventionally divided into several energy regions characterized by the different experimental techniques employed and the various nuclear, atomic, and molecular processes that can be studied; these are summarized in Table 1.

Atomic and Molecular Energy Levels. Absorption and emission of electromagnetic radiation can occur by any of several mechanisms. Those important in spectroscopy are resonant interactions in which the photon energy matches the energy difference between discrete stationary energy states (eigenstates) of an atomic or molecular system: $\Delta E_{\text{system}} = E_{\text{photon}} = h\nu$. This is known as the Bohr frequency condition. Transitions between different types of eigenstates have characteristic energies (see Table 1), and so occur in different spectral regions. All of these regions have at least some applications to chemical analysis, but most useful are rotational, vibrational, and electronic transitions. Molecules and molecular ions exhibit all three types of spectra; atoms and atomic ions undergo only electronic transitions.

Rotational transitions in gaseous molecules occur in the far-infrared and microwave regions, generally $\lambda > 100$ μm. Very light species absorb at shorter wavelengths, and in fact the strong, dense rotational spectrum of water vapor for $\lambda > 15$ μm makes operation in the fir difficult. Microwave spectroscopy is an important discipline oriented more toward molecular structure research than chemical analysis (see MICROWAVE TECHNOLOGY). The radar region, which includes longer microwaves and shorter radio waves ($\lambda \cong 0.54-133$ cm), is used for the detection and ranging of extended objects, from raindrops to aircraft to large weather systems. Microwave and radio wave spectroscopy is useful for detecting molecules in astronomical sources (radio astronomy).

Molecules vibrate at fundamental frequencies that are usually in the mid-infrared. Their spectrum of a gas consists of rovibrational bands in which each vibrational transition is accompanied by numerous simultaneous rotational transitions. In condensed phases the rotational structure is suppressed, but the vibrational frequencies remain highly specific, and information on the molecular environment can often be deduced from linewidths, frequency shifts, and additional spectral structure owing to phonon (thermal acoustic mode) and lattice effects.

Shorter-wavelength radiation promotes transitions between electronic orbitals in atoms and molecules. Valence electrons are excited in the near-uv or visible. At higher energies, in the vacuum uv (vuv), inner-shell transitions begin to occur. Electronic transitions in molecules are accompanied by structure from vibrational and, in

Table 1. Regions of the Electromagnetic Spectrum

Region	Wavelength limits[a]	Frequency or photon energy[a]	Transitions observed or excited
radio waves	>30 cm	<1,000 MHz	hyperfine structure from nuclear spins and isotopic shifts
microwaves	1 mm–30 cm	300–1 GHz	rotation and inversion of molecules; cyclotron resonance of electrons in solids
far-infrared (fir) (sub-mm waves)	50–1000 μm	200–10 cm^{-1}	molecular rotations and certain low frequency bending, torsional, and skeletal vibrations; lattice modes in solids
mid-infrared	2.5–50 μm	4,000–200 cm^{-1}	fundamental molecular vibrations (rovibrational spectra)
near-infrared (nir)	0.8–2.5 μm	12,500–4,000 cm^{-1}	vibrational overtones and combinations
visible (vis)	400–800 nm (0.4–0.8 μm)		valence electrons
near-ultraviolet (uv)	200–400 nm		valence electrons (atomic and rovibronic molecular spectra)
vacuum ultraviolet (vuv)	10–200 nm	125–6 eV	inner-shell electrons; ionization
x-rays	0.01–10 nm (0.1–100 Å)	125–0.125 keV	inner-shell electrons; nuclear
gamma-rays	<0.01 nm	>0.125 MeV	nuclear

[a] Values are approximate.

gases, rotational transitions (vibronic and rovibronic bands). Deep inner-shell electronic transitions can be induced by x-ray excitation, useful for elemental analysis (see X-RAY TECHNOLOGY).

Transition Widths and Strengths. The widths and strengths of spectroscopic transitions determine the information that can be extracted from a spectrum, and are functions of the molecular parameters summarized in Table 2.

Individual transitions have natural linewidths $\Delta\tilde{\nu}_N$ that are broadened in the gas phase by molecular collisions and Doppler effects arising from random thermal motion. In condensed phases, broadening occurs from short-range intermolecular interactions, including solute–solvent effects in solutions, and matrix, lattice, and phonon effects in solids. Finally, instrumental broadening results from limitations to the spectral resolving power of the equipment used, ie, its ability to distinguish transitions of nearly equal wavelength.

The strength of a photon–molecule interaction is determined by the frequency-dependent cross section $\varsigma(\nu)$, expressed in cm^2 for absorption and related to $a(\nu)$ in equation 1. Both the transition strength

Table 2. Line-Shape Parameters

Parameter	Transitions: Rotational	Vibrational	Valence electronic
line frequency, $\nu_{\tilde{\nu}}$ cm^{-1}	1–100	100–4,000	<50,000(> 200 nm)
natural linewidth, $\Delta\tilde{\nu}_N$, cm^{-1}	$<10^{-11}$	$<10^{-7}$	$<3\times10^{-3}$
Doppler width at 300 K, $\Delta\tilde{\nu}_D$, cm^{-1}	<0.0005	<0.01	0.01–0.2 (~0.0005 nm)
peak Doppler-broadened absorption cross section, ς_A, cm^2	$<10^{-20}$	$\leq10^{-18}$	10^{-11} – 10^{-16}[a] $<10^{-17}$[b]

[a] Values are for atoms. [b] Values are for molecules.

and the effective spectral resolution determine the detectivity of the technique.

General Instrumental Considerations. A spectroscope disperses light for visual observation, using a slit to define the source, a collimating lens, a dispersive prism or grating, and an objective lens or telescope. Spectroscopes fitted with wavelength scales and cameras are termed spectrometers and spectrographs, respectively. A monochromator is a spectrometer having an exit slit in the focal plane to isolate a narrow wavelength region. If the output is focused on a detector for quantitative intensity measurements it becomes a spectrophotometer. The use of such instrumentation constitutes the field of spectroscopy (sometimes called spectrometry). Table 3 indicates components and materials typically used in the important near-uv, vis, and ir regions.

Microwave Spectroscopy

The longest wavelengths of the electromagnetic spectrum are sensitive probes of molecular rotation and hyperfine structure. Microwave spectroscopy is used for studying free radicals and in gas analysis. The technique is highly sensitive. At microwave resolution, frequencies are so specific that a single line can unambiguously identify a component of a gas mixture.

Microwave sources are electronic rather than thermal. Klystrons, used most frequently, induce radio-frequency (r-f) fields in a resonant cavity using a modulated electron beam, providing coherent output tunable over a 10% frequency range. Magnetrons, traveling-wave tubes, and backward-wave oscillators are also useful sources. No dispersion is required, because these sources are essentially monochromatic. Microwaves are transmitted through metal waveguides that also serve as gas sample cells. Thin mica (qv) windows are used when necessary. Detectors are crystal rectifiers and sometimes bolometers.

Infrared Spectroscopy

Infrared frequencies depend on the masses of the atoms involved in the various vibrational motions, and on the force constants and geometry of the bonds connecting them; band shapes are determined by the rotational structure and hence by the molecular symmetry and moments of inertia. The rovibrational spectrum of a gas thus provides direct molecular structural information, resulting in very high specificity.

Instrumentation. The ir region was developed using dispersive techniques adapted as appropriate from uv–vis spectroscopy. Greatly improved performance has been achieved using the Fourier-transform spectrometer (fts), essentially a Michelson interferometer in which a collimated light beam divided by a partially reflecting beam splitter is recombined after the optical delay (retardation) of one arm is changed by a scanning mirror. The resulting signal strength as a function of mirror travel is an interferogram, from which the desired spectrum (intensity vs wavenumber) can be obtained by performing a Fourier transform. The discovery of the Cooley-Tukey fast Fourier transform (fft) algorithm in 1962 and the availability of powerful and

Table 3. Components and Materials Used in Uv–Vis–Ir Spectroscopy[a]

Component	Far-infrared[b]	Mid-infrared[c]	Near-uv, visible, and near-ir[d]
broad-band thermal sources	plasma emission from high pressure mercury arc lamp	blackbodies: Nernst glower (zirconia), globar (silicon carbide), nichrome wire coil	deuterium lamp (160–370 nm)
			Xe arc lamp (300 nm–1.3 μm)
			quartz-envelope tungsten-halogen lamp (350 nm–2.5 μm)
continuously tunable laser sources	CO_2 laser frequency difference generation (65–1000 μm)	OPO (<16 μm)	organic dye lasers (320 nm–1.3 μm)
	microwave sideband mixing (>70 μm)	color-center lasers (1.5–4 μm)	OPO (>410 nm)
		nonlinear optical mixing (2–9 μm)	Ti:sapphire laser (700–1000 nm)
		semiconductor diodes (3–28 μm)	frequency conversion of other lasers (>190 nm)
		spin-flip Raman (5–6 μm)	diode lasers (>670 nm)
		waveguide CO_2 (9–11 μm)	
detectors	pyroelectric (DTGS, LT)	photoconductors (<40 μm) (InSb; doped germanium; MCT)	photomultipliers (120 nm–1.1 μm)
	doped-germanium and InSb bolometers	thermal (thermocouples, bolometers)	photodiodes (InAs, InGaAs, Si, Ge)
	Golay pneumatic cell	pyroelectric (DTGS, LT)	photoconductors (Si, Ge)
array detectors	none	PtSi (<5 μm)	silicon arrays (175 nm–1.1 μm)
		InSb (<5 μm)	photographic emulsions (<1.2 μm)
		MCT (<12 μm)	Ge arrays (700 nm–1.5 μm)
		doped Si (10–30 μm)	InSb, PbSe, PtSi
general optical materials	polyethylene, Mylar	alkali halide crystals	alkali halide crystals
	quartz	fluorite, CaF_2 (<9 μm)	glass (300 nm–2.2 μm)
	diamond (>6 μm)	ZnSe (800 nm–20 μm)	quartz (170 nm–3.5 μm)
		AgCl (420 nm–25 μm)	diamond (220 nm–4 μm)
		KRS-5 (550 nm–40 μm)	sapphire (150 nm–6 μm)
			fluorite (120 nm–9 μm)
fiber optics	none	fluoride glasses (900 nm–5 μm)	silica glasses
		chalcogenide glasses (2.2–12 μm)	quartz (250 nm–1.3 μm)
		hollow metal fibers (~10 μm)	

[a] DTGS = deuterated triglycine sulfate; KRS − 5 = mixed thallium bromide-iodide; LT = lithium tantalate; MCT = mercury cadmium telluride; and OPO = optical parametric oscillator. [b] 50–1000 μm. [c] 2.5–50 μm. [d] 200 nm–2.5 μm.

inexpensive computers, led to Fourier spectroscopy becoming a practical technique.

The development of nearly monochromatic lasers that can be continuously frequency-tuned throughout much of the ir (see Table 3) has revolutionized vibrational spectroscopy. By far the most used are lead salt semiconductor tunable diode lasers (TDLs), which can be tuned over 50−100 cm^{-1} intervals in continuous scans of >1 cm^{-1} with resolutions of ~0.003 cm^{-1}.

Sampling. Almost any sample can be prepared for ir analysis. Cells for gases and liquids, typically 10 cm and 0.1 mm paths, respectively, are available in many configurations. For trace gases, compact small-volume folded-path cells offer adjustable optical paths up to 200 m. Gaseous species can be isolated in crystal lattices by dilution in an inert gas that is condensed at cryogenic temperatures (matrix isolation). Solutions and gels can be coated directly on ir-transmitting salt plates or microporous plastic films. Immersion probes are used for *in situ* analysis in chemical reaction vessels and process streams.

Solids can be observed directly in thin sections, as mulls in mineral or halocarbon oils, as finely ground dispersions in pressed disks of an ir-transparent salt such as KBr, as films cast from solutions, or as solutions in solvents having few infrared absorptions, such as CCl_4 or CS_2.

Specialized sampling techniques include ir microscopes that focus down to 10 μm (approximately the ir diffraction limit) for spectra of less than nanogram samples. This is useful in fiber analysis and forensics. Attenuated total reflection (ATR), provides a useful sampling technique for gels, slurries, strongly absorbing liquids, and coatings. External reflection methods are also used for solids and films.

For process monitoring and analysis in chemical plants, transmission of the ir beam through fiber optics allows safe access to a small sample region in harsh environments far removed from the spectrometer.

Null-Background Techniques. In conventional absorption spectroscopy the difference between two large quantities, the incident and transmitted intensities, is measured, thus limiting the minimum detectable absorbance, A_{min}, to ~10^{-3}. This can be greatly improved using null-background techniques, where the detected signal is (within limits) proportional to the source intensity. An example is harmonic or derivative spectroscopy, in which a narrow-band light source is frequency-modulated and synchronously amplified at the modulation frequency, yielding the first derivative of the absorption profile. This eliminates the background, reduces effects of low frequency drifts in the source, and discriminates against spurious signals lacking a sharp wavelength dependence at the modulated frequency.

Another such technique is direct calorimetric measurement of the radiant energy absorbed as the latter is converted into kinetic motion (heat). In photoacoustic spectroscopy (pas) pressure modulation caused by absorption of a modulated source is synchronously detected by a sensitive microphone transducer (spetrophone).

Applications. Infrared spectroscopy is broadly applicable to analytical problems of molecular speciation and quantification, because most molecules have strong fundamental vibrational transitions in the mid-infrared. In the region 3–8 μm, many chemical functional moieties exhibit characteristic group frequencies that are relatively independent of the molecular environment, providing information on the chemical nature of the absorber. At longer wavelengths the frequencies are influenced more by the skeletal vibrations of the molecule. This is known as the fingerprint region where even similar species may have sufficiently different spectra to be readily distinguished. Single-component unknowns can be identified by simply comparing their spectra with reference spectra, of which many catalogues are available. Reference spectra are also available in digitized versions, and searches of databases (qv) can be made rapidly by computer.

The ir spectral region is particularly useful for organic compounds, which have sufficiently distinctive spectra to be easily identified and quantified. Areas of application include drug analysis, biological systems, indoor and outdoor atmospheres, surface analysis, isotopic analysis, polymers, and electrode processes in solutions (spectroelectrochemistry) (see AIR POLLUTION; INFRARED TECHNOLOGY AND RAMAN SPECTROSCOPY, INFRARED TECHNOLOGY).

Transmission and reflection nir spectroscopies have emerged as important probes for industrial and process analysis, such as monitoring moisture, saturation of oils, and protein and fat content in the food and agricultural industries. Nir radiation can probe, with minimal sample preparation, long-path, concentrated, and aqueous samples that would totally absorb longer wavelengths.

Radiometry. Radiometry is the measurement of radiant electromagnetic energy, considered herein to be the direct detection and spectroscopic analysis of ambient thermal emission, as distinguished from techniques in which the sample is actively probed. At any temperature above absolute zero, some molecules are in thermally populated excited levels, and transitions from these to the ground state radiate energy at characteristic frequencies. This radiation occurs at just the energies of molecular rovibrational transitions, so thermal emission carries much the same information as an ir absorption spectrum. Detection of the emissions of remote thermal sources is the ultimate passive and noninvasive technique, requiring not even an optical probe of the sampled volume.

Molecular Uv–Vis Absorption Spectroscopy

Spectroscopy in the uv–vis detects electronic transitions, and so is applicable to both atoms and molecules. This technique has important qualitative and quantitative applications.

Instrumentation and Sampling. Modern commercial recording instruments employ holographic gratings and photomultiplier tubes or (in the nir) avalanche photodiode detectors to cover ca 190 nm to 3 μm (see HOLOGRAPHY; PHOTODETECTORS). A typical benchtop spectrophotometer has 2 nm resolution in the uv–vis, but dual-grating research instruments can resolve 0.05 nm. In the vacuum ultraviolet (vuv), special techniques are required. Tunable uv–vis lasers are well developed. Optically pumped organic dye lasers provide especially useful continuously tunable high power sources, widely used in laboratory research on spectroscopy and photochemistry (see PHOTOCHEMICAL TECHNOLOGY). Other useful laser sources are available (see Table 3), including conversion of longer-wavelength tunable lasers by Raman shifting or by frequency doubling in nonlinear media. Most of the specialized ir sampling methods described have uv equivalents with appropriate modifications. Photoacoustic methods are employed for strongly absorbing samples. Glass and polymer fiber optics are available for process monitoring.

Applications. Uv–vis absorption results from transitions between outer electron shells. The specific moiety or structure responsible is termed the chromophore. Compounds having only single ς-bonds generally absorb only in the vuv. Saturated organic compounds containing heteroatoms exhibit uv transitions, and the π-electrons of an unsaturated bond or aromatic nucleus are strong chromophores, absorbing more strongly and at longer wavelengths (to the visible) in conjugated and fused-ring systems. Inorganic species having incomplete electron shells, notably transition-metal cations, also absorb in the uv–vis.

Uv–vis spectra do not offer the unique group frequencies and fingerprinting ability of the ir, but different chromophores exhibit absorptions at specific wavelengths, λ, and have characteristic intensities. These are tabulated in handbooks as λ_{max} and ϵ_{max}, where ϵ_{max} is a molar decadic absorption coefficient equivalent to the $a(v)$ of equation 1, but in units typically of L/(mol·cm). Photon energies sufficient to promote electronic transitions can also excite vibrational and rotational transitions, so the electronic spectra of gaseous molecules consist of highly structured rovibronic bands, having very high specificity. In condensed phases, rotational structure is suppressed, but the molecules still vibrate, resulting in vibronic bands with progressions of characteristic vibrational frequencies that accompany each electronic transition. The uv offers high sensitivity and excellent quantitative accuracy. The greater uv absorption cross sections (see Table 2) and more efficient uv sources and quantum detectors result in detection limits several orders of magnitude better than in the ir.

Atomic Uv/Vis Spectroscopy

Narrow-line uv–vis spectra of free atoms, corresponding to transitions in the outer electron shells, have long been employed for elemental analysis using both atomic absorption (AAS) and emission (AES) spectroscopy. Atomic spectroscopy is sensitive but destructive, requiring vaporization and decomposition of the sample into its constituent elements.

Atomic Absorption Spectroscopy. Samples for AAS can be prepared by either flame or electrothermal atomization. In the former, the sample is dissolved if necessary, nebulized into an aerosol by a high velocity gas jet, and sprayed into a burner where dissociation produces neutral free atoms. In electrothermal atomization the sample is usually placed in a graphite tube furnace that follows a programmed heating cycle to dry, ash, and finally atomize it. Free-atom production is most efficient in a furnace, resulting in greater sensitivity. For quantitative analysis, the resolution of the spectral analyzer must be significantly narrower than the absorption lines, which are ~0.002 nm at 400 nm for $M = 50$ amu at 2500°C.

Atomic Emission Spectroscopy. AES is the primary technique for metals analysis in ferrous and other alloys; geological, environmental, and biological samples; water analysis; and process streams. Traditionally, large and expensive concave- or plane-grating, high resolution instruments have been needed to resolve narrow and possibly interfering atomic lines. Echelle gratings, designed to be used in high orders with a prism cross-dispersing element, allow for more compact short focal length spectrographs. These produce a two-dimensional output of wavelength vs grating order in the focal plane. Photographic plates, read using microdensitometers, provide simultaneous recording of the whole spectrum and allow for long integration times for detection of weak lines. Photomultiplier tubes, quicker and more sensitive, are used in either a scanning mode, recording different lines sequentially, or a direct-reading configuration (quantometer) in which as many as 60 tubes are precisely positioned in the focal plane for simultaneous recording of as many lines. Uv-sensitive charge-coupled device (CCD) arrays are increasingly used.

Many techniques exist for volatilizing and exciting samples. Flame emission spectroscopy uses much the same source apparatus as AAS, but records the emission lines that in the latter method must be minimized. Line interference is more serious than for absorption, and chemiluminescence in the flame may be a problem. AES is most appropriate for alkali metals, for which it offers the best detectability, alkaline-earth metals, rare earths, and trace metals.

Noncombustion plasma sources offer improved accuracy, less background interference, better dynamic range, and easier sample handling (especially for solutions), and have replaced flame and arc/spark excitation in many applications (see PLASMA TECHNOLOGY).

Laser-generated plasmas are also useful sources for AES. In laser-induced breakdown spectroscopy (libs) the plasma emission is dispersed and analyzed directly, but laser ablation can be used simply as a sampling process followed by ICP analysis, cross-excitation with a spark discharge, or mass spectrometry.

Scattering Techniques

Spectroscopic examination of light scattered from a monochromatic probe beam reveals the expected Rayleigh, Mie, and/or Tyndall elastic scattering at unchanged frequency, and other weak frequencies arising from the Raman effect. Both types of scattering have applications to analysis.

Elastic Scattering. Elastic scattering is not, in its simplest form, a spectroscopic technique, and conveys no chemical information about the sample, though it can provide physical properties useful in analytical applications. Rayleigh scattering and polarization measurements on Mie-scattered light furnish particle size and distribution, absolute number density, and mean molecular weight.

Elastic scattering is also the basis for lidar, in which a laser pulse is propagated into a telescope's field of view, and the return signal is collected for detection and in some cases spectral analysis. An important modification of lidar is two-frequency differential absorption lidar (dial), which combines lidar and absorption spectroscopy, using two alternating frequencies from a tunable laser, one on and one off an absorption resonance. This generates a differential absorption signal from the Raleigh or Mie scattering returns, which are relatively insensitive to small changes in frequency, and yields range-resolved concentrations of specific molecules.

Spontaneous Raman Spectroscopy. The Raman effect is the inelastic scattering of an incident photon $\tilde{\upsilon}_i$ to a new frequency $\tilde{\nu}_R = \tilde{\nu}_i \pm \tilde{\nu}_{\text{mol}}$, where $\tilde{\upsilon}_{;\text{mol}}$ is the energy acquired (−) or lost (+) by the molecule during a vibrational, rotational, or (less commonly) electronic transition. It is an important technique for the investigation of molecular structure, providing information similar, and often complementary, to ir spectroscopy, and for chemical analysis, both in the laboratory and for remote atmospheric monitoring.

The spectrum of the scattered light contains a strong Rayleigh line at the exciting frequency, and much weaker (by factors of 10^3 or more) red-shifted Raman lines (Stokes lines) from interactions in which the molecules have gained rovibrational energy from the photons, and corresponding but even weaker blue-shifted anti-Stokes lines, where the molecules have lost energy to the photons. The resulting rovibrational spectra permit species identification and quantification (see INFRARED TECHNOLOGY AND RAMAN SPECTROSCOPY, RAMAN SPECTROSCOPY).

Raman Instrumentation and Sampling. Raman instrumentation must provide an intense monochromatic source, means for resolving the desired frequency-shifted signal from the strong elastically scattered background, and provision for recording the Stokes Raman intensity as a function of $\tilde{\upsilon}_{\text{mol}}$. Early Raman spectroscopy employed mercury-arc sources. The development of lasers revitalized Raman spectroscopy. In fact, lasers found perhaps their first practical use as Raman sources.

Modern Raman spectroscopy uses both high throughput grating spectrometers, often having double monochromators, and Fourier-transform (ft) interferometers. Common sources are He:Ne, Ar^+, K^+, and Nd:YAG lasers; the last, at 1.064 μm, is especially useful with fluorescent samples. Detectors are InGaAs photodiodes or multichannel charge-coupled (CCD) array detectors. Raman spectra can routinely be acquired in a few seconds using a dispersive spectrograph and CCD detector; ft instrumentation typically takes some minutes.

Sample containers can be of glass or quartz, which are weak Raman scatterers, and aqueous solutions pose no problems. Fiber-optic probes can be used in process monitoring.

Special Raman Spectroscopies. In resonance Raman (RR) scattering, the excitation frequency is tuned to within a few linewidths of an allowed electronic absorption. This increases the intensities of both fluorescence and Raman scattering by several orders of magnitude, allowing the highly selective enhancement of a particular component of a mixture, with sensitivity adequate for monitoring trace constituents. It requires an analyte with an accessible electronic transition. Applications include pigment analysis, biochemical systems, and atmospheric oxygen and trace gases. Surface-enhanced Raman scattering (sers) exploits certain imperfectly understood mechanisms that can enhance by more than 10^6 the Raman signal from molecules at colloidal metal surfaces. Detection limits in solutions can reach the ppm level.

Certain nonlinear spectroscopic phenomena that result from intense laser excitation can greatly improve the efficiency of the Raman scattering process. An example is coherent anti-Stokes Raman spectroscopy (cars), which is especially useful in combustion and flame diagnostics.

A coherent Raman process of great promise for sensitive analysis is degenerate four-wave mixing (DFWM), which uses a strong nonresonant pump and weaker probe beam from the same laser to induce a large back-scattered coherent signal. DFWM is simpler to align than cars, requires 10^3-fold less irradiance, and produces an intense and highly directional return even in inhomogeneous or turbulent media. DFWM has been used for high fidelity spatial mapping of analytes in flames, and has shown excellent detection sensitivities for molecular species; for example, OH at concentrations of 10^{10} cm^{-3}.

Applications. Raman spectroscopy has important biological applications because of its suitability for aqueous solutions. Low frequency vibrational modes important in inorganic and organometallic chemistry can often be studied more easily by Raman than by far-ir spectroscopy. Sampling ease and the ability to probe through packaging materials makes ft-Raman useful in industries such as pharmaceuticals.

Scattering techniques have important applications to atmospheric and environmental monitoring. Much of the instrumentation can be shared by the various techniques and by fluorescence methods, and they are often combined into one instrument.

Optical Activity

Although the usual absorption and scattering spectroscopies cannot distinguish enantiomers, certain techniques are sensitive to optical activity in chiral molecules. These include optical rotatory dispersion (ORD), the rotation by the sample of the plane of linearly polarized light, used in simple polarimeters; and circular dichroism (CD), the differential absorption of circularly polarized light.

Circular dichroism employs standard dispersive or interferometric instrumentation, but uses a thermal source that is rapidly modulated between circular polarization states using a photoelastic or electro-optic modulator. As the stronger effect, electronic circular dichroism is well established in stereochemistry, but is applicable only to easily accessible electronic transitions. Although weaker, vibrational optical activity can be observed on many bands, and helps in elucidating short-range structural features.

Luminescence Spectroscopies

Luminescence is spontaneous light emission during a transition from a nonthermally excited level to a lower state. Such processes are classified as short-lived (on a ns time scale) emission or fluorescence that continues only as long as the sample is being excited; and longer-lived phosphorescence, persisting after excitation ceases. For the important case of electronic excitation, fluorescence typically represents temperature-independent spin-allowed transitions from singlet excited states; phosphorescence is longer-wavelength spin-forbidden emission from lower lying triplet states.

Fluorometry and Phosphorimetry. As null-background techniques, fluorometry and phosphorimetry are orders of magnitude more sensitive than absorption spectroscopy, having achievable detectivities of ng/L. Modern spectrofluorometers can record both fluorescence and excitation spectra. Excitation is furnished by a broad-band xenon arc lamp followed by a grating monochromator. The selected excitation frequency, λ_{ex}, is focused on the sample; the emission is collected at usually 90° from the probe beam and passed through a second monochromator to a photomultiplier detector. Scan control of both monochromators yields either the fluorescence spectrum, ie, emission intensity as a function of wavelength λ_{em} for a fixed λ_{ex}, or the excitation spectrum, ie, emission intensity at a fixed λ_{em} as a function of λ_{ex}. Fluorescence and phosphorescence can be distinguished from the temporal decay of the emission.

Molecular fluorescence extends over a broad wavelength region, limiting its usefulness for analyzing multicomponent systems. By scanning both the excitation and fluorescence wavelengths synchronously with an empirically chosen fixed wavelength separation, $\lambda_{ex}-\lambda_{em}$, the fluorescence can be reduced to a narrow signal at the region of overlap between the excitation and fluorescence spectra.

This allows the components in complex mixtures to be distinguished. Samples are usually in solution, but solids (often frozen solutions) yield narrow-line spectra that are useful in distinguishing components of mixtures. In phosphorimetry solid sampling may be necessary to minimize quenching processes. Atomic fluorescence spectroscopy (afs) using either a laminar-flow flame similar to AAS or an inductively coupled plasma (ICP) source provides multi-element analysis having relative freedom from background and interferences, and detection limits similar to AAS.

Laser-Induced Fluorescence. Laser-induced fluorescence (lif) provides, much as does ir spectroscopy, fingerprints of different organic molecules, which can be quantified by measuring fluorescence intensities. Selectivity is excellent, as both pump and fluorescence frequencies can be individually chosen for optimum performance, and it can be improved with measurements of fluorescence lifetimes and polarization behavior. The enhanced null-background sensitivity can achieve single-atom or single-molecule detection. Lif has important applications in gas analysis and combustion and plasma diagnostics.

Chemiluminescence. Chemiluminescence is the emission of light during an exothermic chemical reaction, generally as fluorescence. It often occurs in oxidation processes, and enzyme-mediated bioluminescence has important analytical applications. Chemiluminescence analysis is highly specific and can reach ppb detection limits with relatively simple instrumentation.

X-Ray Spectroscopy

X-rays provide an important suite of methods for nondestructive quantitative spectrochemical analysis for elements of atomic number $Z > 12$ (see X-RAY TECHNOLOGY).

Instrumentation. An x-ray tube accelerates electrons from a heated cathode through a high voltage field to a target anode, where kinetic energy is converted to a continuum of x-rays having wavelengths $\lambda > hc/eV = 1240/V$ for λ in nm and exciting voltage V in volts. X-ray spectroscopy by wavelength dispersion employs a scanning crystal spectrometer conceptually not unlike visible and ir spectrometers. The beam is collimated using parallel slits or a bundle of fine-diameter tubes, and dispersed by Bragg's law diffraction from a goniometer-mounted crystal where the lattice spacing is chosen to act as a grating for the wavelengths of interest. An alternative spectroscopic technique is energy dispersion, which exploits the energy resolution of solid-state detectors. The beam is not dispersed, but strikes a high purity (intrinsic) germanium or silicon semiconductor, where generation of electron–hole pairs allows the energy of each photon to be recorded by a multichannel pulse-height analyzer.

X-Ray Absorption Spectroscopy. As the excitation energy incident on a sample is increased, sharp rises in absorption occur at the $K, L, \ldots,$ absorption edges where the energy just matches that required for ionization by ejection of an electron from the $K, L, \ldots,$ shells. These energies are characteristic for each element. The absorption follows Beer's law (eq. 1). X-ray absorption is especially useful in determining heavy elements in mixed materials of lower atomic number, such as lead in gasoline and uranium in aqueous solution.

X-Ray Emission and Fluorescence. X-ray analysis by direct emission following electron excitation is of limited usefulness because of inconveniences in making the sample the anode of an x-ray tube. An important exception is the x-ray microphobe, in which an electron beam focused to $\sim 1\ \mu m$ diameter excites characteristic x-rays from a small sample area. Surface corrosion, grain boundaries, and inclusions in alloys can be studied with detectability limits of $\sim 10^{-14}$ g (see SURFACE AND INTERFACE ANALYSIS).

X-ray fluorescence employs sample excitation using radiation from a standard x-ray tube, providing weaker (by factors of $\sim 10^3$) signals than electron excitation, but having greater flexibility. The characteristic fluorescence emission is analyzed using a crystal spectrometer. Examples of x-ray fluorescence applications are in specialty chemicals, hazardous waste analysis, thin films, and geochemistry. X-radiation can also be induced by high energy (several MeV) proton beams from ion accelerators. Such particle-induced x-ray emission (PIXE) is useful for thin samples and particulates, having detection limits of $\sim 10^{-12}$ g.

Gamma-Ray Spectrometry

A nuclear spectrum represents the energy distribution of particles and radiation emitted during nuclear processes such as natural radioactive decay, nuclear reaction, and fission. These emissions may include alpha-particles (helium nuclei), beta-particles (electrons or positrons), neutrinos, and gamma-rays (high energy photons). Gamma-ray energies correspond to transitions between energy levels in atomic nuclei, and provide a unique signature of nuclear processes. Gamma-ray spectrometry is a probe of nuclear rather than chemical processes, but its high specificity and sensitivity have applications in analysis of materials. It is especially suited for activation analysis. Unstable nuclides produced by nuclear bombardment can be identified by their characteristic gamma-ray decay emissions. An important example is the remote, nondestructive detection of possible hidden explosives (see EXPLOSIVES AND PROPELLENTS).

Mössbauer Spectroscopy. The low resolution of gamma-ray spectrometry can be dramatically improved, greatly increasing its usefulness, by exploiting the Mössbauer effect, also known as recoilless gamma-ray resonance absorption. Chemical applications of Mössbauer spectroscopy are broad, with important applications to materials science and metallurgy (see SURFACE AND INTERFACE ANALYSIS).

Ultrasensitive Techniques

Certain very high resolution or ultrasensitive spectroscopies have been used primarily for laboratory research, but eventual application to analytical or sensor applications is expected.

Beam Spectroscopy. Both specificity and sensitivity can be greatly enhanced by suppressing collisional and Doppler broadening. This is accomplished in supersonic atomic and molecular beams by probing the beam transversely to its direction of flow in a near-collisionless regime. When a gas adiabatically expands through a nozzle into a vacuum, its thermal energy is converted into kinetic energy of mass flow, cooling the gas to effective temperatures of <10 K, and collapsing any rovibrational structure into a few transitions originating from the lowest lying quantum states. Beam work has emphasized basic atomic and molecular structure and aggregation, but the methods may eventually be applied to analysis.

Multiphoton Absorption and Ionization. High laser powers can induce the simultaneous absorption of two or more photons that together provide the energy necessary to excite a transition; this transition may be one that is forbidden as a single-photon process. Such absorption can be made Doppler-free by propagating two laser beams of frequency v in opposite directions, so the Doppler shifts cancel and a two-photon transition occurs at $2v$ for any absorber velocity. Photoionization of atoms or molecules permits sensitive detection because the ionic signal is a clear null-background signature. Tunable dye lasers have made this a practical technique. Multiphoton ionization (mpi) has been applied to liquids and interfaces, but is most useful in probing low pressure gases and molecular beams, where the charged photofragments can be extracted with high efficiency and analyzed by electron or mass spectrometers.

Frequency-Modulation Spectroscopy. Frequency-modulation spectroscopy (fms) is a high sensitivity null-background infrared technique for measuring absorbances down to 10^{-8} with fast acquisition speeds. Fms involves frequency-modulating a laser source at ω_0 to produce a carrier frequency having sidebands at $\omega_0 \pm n\omega_m$ is an integral multiple of the modulation frequency. The frequency modulated (FM) laser light is passed through the sample, where near-resonant absorption (or dispersion) leads to differential absorption (scattering) of the sidebands. This produces an amplitude modulated (AM) beat frequency that is coherently detected by standard phase-sensitive mixing. If the sideband modulation ω_m is small compared to the spectral linewidth, the observed signature is a derivative of the line shape.

Other Techniques. Under proper conditions, a strong laser probe can interact with only a particular velocity subset of gaseous absorbers, allowing homogeneous line shapes to be resolved in static cells. This phenomenon, saturated absorption, offers excellent selectivity at higher sample densities than in molecular beams. Saturated absorption is used mainly as a sensitive null-background spectroscopy to measure precise line positions.

Photothermal spectroscopy includes two related high sensitivity approaches, thermal lensing and photothermal deflection. The former can detect absorbances of $<10^{-7}$.

A promising new technique is cavity ringdown laser absorption spectroscopy, in which the rate of decay of laser pulses injected into an optical cavity containing the sample is measured. Absorption sensitivities of 5×10^{-7} have been achieved on a μs time scale.

ROBIN S. MCDOWELL
JAMES F. KELLY
Pacific Northwest National Laboratory

H. H. Willard, L. J. Merritt, Jr., J. A. Dean, and F. A. Settle, Jr., *Instrumental Methods of Analysis*, 7th ed., Wadsworth Publishing Co., Belmont, Calif., 1988.

J. D. Ingle, Jr., and S. R. Crouch, *Spectrochemical Analysis*, Prentice-Hall, Englewood Cliffs, N.J., 1988.

J. M. Hollas, *Modern Spectroscopy*, 3rd ed., John Wiley & Sons, Inc., New York, 1996.

Analytical Chemistry, "Fundamental Reviews" (June 1994, June 1996, June 1998); analytical applications of infrared, ultraviolet, atomic absorption, emission, Raman, fluorescence, phosphorescence, chemiluminescence, and x-ray spectroscopy.

SPRAYS

A spray comprises a cloud of liquid droplets randomly dispersed in a gas phase. Depending on the application, sprays may be produced in many different ways. The purposes of most sprays are *(1)* creation of a spectrum of droplet sizes to increase the liquid surface-to-volume ratio, *(2)* metering or control of the liquid throughput, *(3)* dispersion of the liquid in a certain pattern, or *(4)* generation of droplet velocity and momentum. The mechanical devices designed to generate sprays are commonly called atomizers. In the past, the design of atomizers and spray processes was based on traditional fluid dynamic principles and empirical methods. The technology of spraying has advanced rapidly through computer-aided design, mathematical modeling, and sophisticated instrumentation.

Liquid Atomizers

The transformation of bulk liquid to sprays can be achieved in many different ways. Basic techniques include applying hydraulic pressure, electrical, acoustic, or mechanical energy to overcome the cohesive forces within the liquid. Atomizers can be classified according to the energy source used to achieve liquid breakup. Table 1 provides a summary of various atomizers.

Table 1. Summary of Atomizers Based on Source of Energy

Atomizer	Description
air-assisted	pneumatic atomizer in which pressurized air is utilized to enhance atomization produced by pressurized liquid
airblast	pneumatic atomizer that utilizes a relatively large volume of low pressure air
centrifugal	rotating solid surface is the primary source of energy utilized to produce spray
electrostatic	electric charge is the primary source of energy utilized to produce spray
piloted airblast	airblast atomizer combined with a lower capacity pressure atomizer
pneumatic (twin-fluid)	movement of gas/vapor is primary source of energy utilized to produce spray
pressure	pressurized liquid is primary source of energy utilized to produce spray
sonic	pneumatic or vibratory atomizer in which energy is imparted (frequencies >20 KHz) to liquid
ultrasonic	pneumatic or vibratory atomizer in which energy is imparted at high frequency to liquid
vibratory (piezoelectric)	oscillating solid surface is primary source of energy

Physics of Liquid Atomization

Liquid atomization involves a series of complicated physical processes. These processes can generally be divided into three different flow regimes: internal flow, breakup, and droplet dispersion. The internal flow regime extends from the atomizer inlets to the discharge orifice where liquid emerges. The liquid breakup regime starts at the atomizer exit plane and ends at a certain distance downstream where primary atomization is complete. The final process of atomization is the dispersion regime where spherical droplets gradually evolve into a particular spray pattern.

Droplet Dispersion. The primary feature of the dispersed flow regime is that the spray contains generally spherical droplets. In most practical sprays, the volume fraction of the liquid droplets in the dispersed region is relatively small compared with the continuous gas phase. Depending on the gas-phase conditions, liquid droplets can encounter acceleration, deceleration, collision, coalescence, evaporation, and secondary breakup during their evolution. Through droplet and gas-phase interaction, turbulence plays a significant role in the redistribution of droplets and spray characteristics.

After breakup, droplets continue to interact with the surrounding environment before reaching their final destination. In theory, each droplet group produced during primary breakup can be traced by using a Lagrangian calculation procedure. Droplet size and velocity can be determined as a function of spatial locations.

Spray Characteristics

Spray characteristics are those fluid dynamic parameters that can be observed or measured during liquid breakup and dispersal. They are used to identify and quantify the features of sprays for the purpose of evaluating atomizer and system performance, for establishing practical correlations, and for verifying computer model predictions. Spray characteristics provide information that is of value in understanding the fundamental physical laws that govern liquid atomization.

Spray Parameters. There are several common spray parameters.

Droplet Size Distribution. Most sprays comprise a wide range of droplet sizes. Some knowledge of the size distribution is usually required, particularly when evaluating the overall atomizer performance.

Mean Diameters. Several mean diameters are frequently used to represent the statistical properties of droplets produced by liquid atomizers. These mean diameters include volume mean and Sauter mean diameters.

Median Diameter. The median droplet diameter is the diameter that divides the spray into two equal portions by number, length, surface area, or volume. Median diameters may be easily determined from cumulative distribution curves.

Number Density and Volume Flux. The determination of number density and volume flux requires accurate information on the sample volume cross-sectional area, droplet size and velocity, as well as the number of droplets passing through the sample volume at any given instant of time. Volume flux is the volume contained by the droplets passing through a unit cross-sectional area per unit interval of time.

Cone Angle. The spray cone angle is one of the most important parameters in the specification of atomizers. A common method of defining the spray cone angle is to draw two tangent lines originating at the

Table 2. Summary of Atomizer Sprays for Specific Applications

Atomizer spray	Special application
cone spray, hollow or solid	aerating water, brine sprays, chemical processing, coil defrosting, dust control, evaporative condensers, evaporative coolers, industrial washers, roof cooling, spray ponds, spray coating, spray drying, gas scrubbing and washing, humidification, gas cooling, cooling towers, coal washing, degreasing, gravel washing, dish washing, foam control, suspensions and slurries for food and chemical products, pollution control, and oil heating
flat spray	asphalt or tar laying, bottle washing, coal and gravel washing, foam control, degreasing, metal cleaning–rinsing, spray coating, vehicle washing and water misting, descaling, roll cooling, quenching, and agricultural spraying
plain jet spray	rocket engines, diesel engines, agitation, mixing of liquids, cataphoresis plants, and cutting
air atomizing spray	chemical processing, continuous casting, cooling casting and molds, curing concrete products, evaporative coolers, foam control, incineration, quenching, spray coating, spray painting, spray drying, flue gas desulfurization, pollution control, gas turbine engines, and medical spray

orifice and extending to the outermost spray edges at a specified axial distance.

Patternation. The spray pattern provides important information for many spray applications. It is directly related to the atomizer performance. The pattern information must be able to reveal characteristics such as skewness, degree of pattern hollowness, and the uniformity of liquid flux over the entire cross-sectional area.

Spray Dynamic Structure. Detailed measurements of spray dynamic parameters are necessary to understand the process of droplet dispersion. Improvements in phase Doppler particle analyzers (PDPA) permit *in situ* measurements of droplet size, velocity, number density, and liquid flux, as well as detailed turbulence characteristics for very small regions within the spray.

Spray Correlations. One of the most important aspects of spray characterization is the development of meaningful correlations between spray parameters and atomizer performance. The parameters can be presented as mathematical expressions that involve liquid properties, physical dimensions of the atomizer, as well as operating and ambient conditions that are likely to affect the nature of the dispersion. Empirical correlations provide useful information for designing and assessing the performance of atomizers. Dimensional analysis has been widely used to determine nondimensional parameters that are useful in describing sprays.

Spray Instrumentation

An ideal droplet measurement instrument should (*1*) not interfere with the spray pattern or breakup process, (*2*) provide for large representative samples, (*3*) permit rapid sampling or counting, (*4*) have adequate resolution and accuracy over a wide range of droplet sizes, and (*5*) accommodate variations in the liquid and ambient gas properties. Significant advances have been made in the development of laser diagnostic techniques for measuring sprays. Prior to selecting such an instrument, users should have a thorough understanding of its capabilities and limitations. Existing droplet measurement techniques may be classified into three broad categories: (*1*) optical nonimaging techniques; (*2*) imaging techniques; and (*3*) nonoptical methods.

Industrial Applications

Although atomizers are usually small components in many industrial spray applications, they play an important role in determining the performance and efficiency of the entire process. It has long been recognized that atomizers must be properly selected to achieve optimum performance. More recently it has become necessary to comply with stringent environmental regulations to reduce waste and pollution. Though spray requirements differ from one application to another, the spray pattern or shape appears to be a sensible criterion for selecting liquid atomizers for certain processes. Table 2 lists a variety of applications that are based on the pattern of the spray.

CHIEN-PEI MAO
ROGER TATE
Delavan Inc.

A. H. Lefebvre, *Atomization and Sprays*, Hemisphere Publishing Corp., New York, 1989.

E. Giffen and A. Muraszew, *The Atomization of Liquid Fuels*, John Wiley & Sons Inc., New York, 1953.

A. H. Lefebvre, *Atomization and Sprays*, Hemisphere Publishing Corp., New York, 1989.

W. D. Bachalo and M. J. Houser, *Opt. Eng.* **23**(5), 583 (1984).

L. Bayvel and Z. Orzechowski, *Liquid Atomization*, Taylor & Francis Ltd., London, 1993.

SPUTTERING. See THIN FILMS, FILM FORMATION TECHNIQUES; METALLIC COATINGS.

STAINS, INDUSTRIAL

With few exceptions, stains used in wood finishing are formulated to improve the appearance of the substrate. Unlike paints, sealers, and topcoats, stains are utilized either to accentuate the natural beauty of the wood or to hide inherent defects found in most species of wood.

Dye Stain or Soluble Colors

A defining characteristic of dyes is the ability to dissolve in a given medium. Dissolution leaves no particles to refract or scatter light; thus a dye solution is transparent. A distinct advantage of a soluble-type stain is this transparency and the brightness afforded by use of various dye types.

Solvent Dyes. These dyes are soluble in a variety of industrial solvents. Color strength, brilliance, lightfastness, and the ability to formulate with solvents other than alcohol and water have helped to increase their use and popularity.

Acid Dyes. These dyes are sodium salts of color acids and are the most widely used dyes for nongrain-raising wood stains.

Theory and Practice of Wood Staining

Color Mixing. The various types of dye powders used to make dye stains are blended to achieve the desired color. In the wood-finishing industry, various shades of brown are the most common. These colors are usually blended from primary colors. By slightly altering the levels in which each of the colors is present, literally thousands of variations of brown can be achieved.

Types of Wood Stains. The term wood stain in this article applies only to those stains that are applied directly to the wood or over another wood stain. The types of stains used are dependent on the application, finished product, and regional terminology. Multiple wood stains are frequently used in the residential furniture industry to achieve a desired look. They include equalizers and nongrain-raising (NGR) stains.

Equalizers can be either pigmented or dye-type stains used to tone down or lighten dark areas of wood prior to finishing. Another method of equalizing stains is through the use of sap stain. Sap stains are usually alcohol-based dye stains that tie lighter areas of the wood into darker areas.

Nongrain-raising (NGR) stains are usually sprayed overall and contribute the greatest to the overall undertone color of a finished piece. These stains are also sometimes referred to as body stains or overall stains.

Pigmented Stains

Pigment colors are finely divided color particles which do not dissolve in any available solvent; they can be dispersed only by grinding them in a liquid. Pigmented wood stains are used in a variety of applications, and in many instances such use shortens the number of steps needed to achieve a desired appearance.

Three basic groups of pigments were used in the industry: *(1)* natural or earth colors consisting of umbers, siennas, yellow oxide, and red oxide; *(2)* organic pigments, ie, carbon compounds which are usually derived from coal-tar bases or petroleum (see PIGMENTS, ORGANIC); and *(3)* chemical pigments or synthesis, usually metal compounds.

Penetrating Stains. Penetrating (or no-wipe) stains are sprayed on and require little if any wiping. The solvent itself penetrates into the pore and allows the pigment and a small amount of binder to remain on the surface. These stains usually are composed of an oil-type vehicle and a combination of earth pigments reduced in a combination of aliphatic and aromatic hydrocarbons such as naphtha and toluene. The solvent system itself plays a big role in the appearance of the stain. Because of the uniform appearance provided by penetrating stains, many finishing applications utilize penetrating stains as the only color step.

Toners and Tinted Sealers. These materials are usually pigments dispersed in nonpenetrating lacquers. Toners can be sprayed either directly onto the wood or over other wood stains. Although the difference between toners and tinted sealers may not be clearly defined, it is usually the role of the tinted sealer to provide both color and sealing properties. Therefore the tinted sealer usually is higher in solids and provides the majority of color to the finish. Both toners and tinted sealers provide uniform pigment distribution without penetration.

Overtone Stains

Washcoats. Washcoats for wood finishing can be defined as thin coats of sealer applied to control the amount of penetration and subsequent staining from overtone stains and fillers. The solids content of a washcoat determines the amount of penetration of an overtone stain. Washcoats are usually sanded prior to application of a glaze, wiping stain, or filler.

Fillers. Wood fillers are applied directly over the washcoat. Fillers (qv) perform two significant functions: they fill pores and give color to the pores.

Glazes and Wiping Stains. Some applications utilize wiping stains direct-to-the-wood. In most fine furniture applications, wiping stains and glazes are applied over the washcoat or sealer step. These overtone stains are normally composed of pigments, oils, solvents, and driers. The important quality of glazes and wiping stains is the ability to apply a color coat which can be wiped on and then highlighted to add depth and contrast to the overall appearance of the finish.

Shade Stains. These stains are usually applied after the sealer or first topcoat and are typically sprayed on specific areas to compensate for uneven color distribution during the initial finishing process.

Pad Stains. More progressive or higher end furniture finishers add color or pad stains to enhance grain patterns, produce shadows, and create hues found only in exceptionally fine veneers and woods. These pads are applied at varying levels to create the illusion of the third dimension. Like the shade stains, they are usually applied over the sealer or first topcoat and subsequently are topcoated themselves.

Distressing Stains. Interest and charm can be added to furniture by the deliberate infliction of imperfections. These imperfections may be caused by physical distress or staining to simulate past abuse.

Environmental Considerations

Stains are solvent-borne and therefore possess inherent properties such as flammability, toxicity, and reactivity. Adequate ventilation and avoidance of any source of possible ignition are key in the use and storage of these materials. Personal protection when using stains should include safety glasses or goggles, respirators, aprons, and rubber gloves.

RON W. TUCKER
Lilly Industries, Inc.

J. A. Hager, Guardsman Products Inc., High Point, N.C.

Technical data, Sandoz Chemical Corp., Charlotte, N.C., 1996; Crompton and Knowles Corp., Charlotte, N.C., 1996.

STARCH

Starch, $(C_6H_{10}O_5)_n$, the main reserve food of plants, constitutes two-thirds of the carbohydrate caloric intake of most humans. Commercial starches are obtained from seeds, particularly corn, waxy corn, high amylose corn, wheat, and rice, and from tubers or roots particularly potato, sweet potato, and tapioca (cassava).

Physical Properties

Starch granules in plants vary in diameter from 1–150 μm. Among commercial starches, rice starch (3–9 μm) has among the smallest granules and potato starch (15–100 μm), the largest. Corn starch granules are 5–26 μm with an average diameter of 15 μm. Amaranth starch granules are 1–3 μm in diameter. Undamaged starch granules are insoluble in cold water but imbibe water reversibly accompanied by a slight swelling. In hot water a larger irreversible swelling occurs producing gelatinization, which takes place over a discrete temperature range that depends on starch type. At a specific temperature during heating (lower limit of gelatinization temperature), the kinetic energy of the molecules is sufficient to overcome intermolecular hydrogen bonding in the interior of the starch granule. The amorphous regions of the granule are initially solvated and the granule swells rapidly, eventually to many times its original size. When a cooked starch paste containing a mixture of linear amylose molecules, swollen granules, and granule fragments is cooled, the dispersion thickens, and if sufficiently concentrated may form a gel. This property of forming thick pastes or gels is the basis of most starch uses. Physical properties of starch can be altered by mechanical treatments or by chemical modification.

Chemical Properties

Most normal starches contain two distinct types of D-glucopyranose polymers. Amylose is an essentially linear polymer of α-D-glucopyranosyl units linked $(1 \rightarrow 4)$ (**1**). Although amylose molecules are generally thought of as being linear chains of α-D-glucopyranosyl units, most amylose preparations contain amylose molecules with two to five branches. These long branches allow

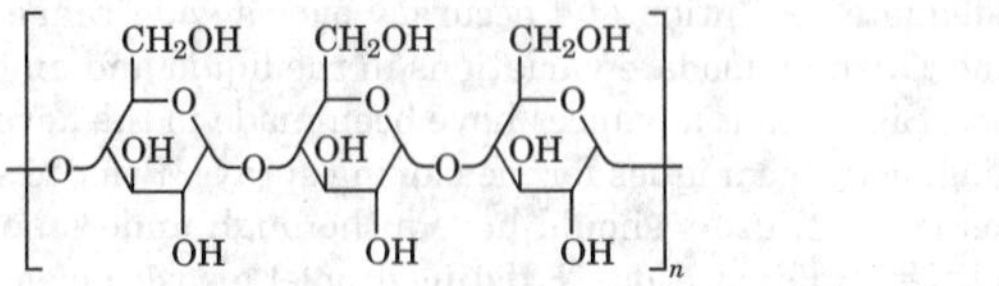

(1)

the molecules to possess nearly the same properties as truly linear molecules. Starch gives a characteristic blue color when stained with iodine that has been employed both as a qualitative and quantitative test for starch. Amylopectin is a highly branched polymer of α-D-glucopyranosyl units containing $1 \rightarrow 4$ links with $1 \rightarrow 6$ branch points (**2**). Amylopectin and its simple modifications do not easily retrograde and are usable in food systems requiring freeze–thaw stability. Starch not only varies in polymer structure but also in molecular weight distribution. Osmotic pressure measurements provide a weight range of 10,000–60,000 for amylose. The degree of polymerization (DP) of amylose usually falls in the range of 200–20,000 DP. Using anaerobic techniques measurements suggest that amylose has a molecular weight range of $(1.6-7.0) \times 10^5$. Amylopectin, a much larger molecule, has a molecular weight range of $4-5 \times 10^8$.

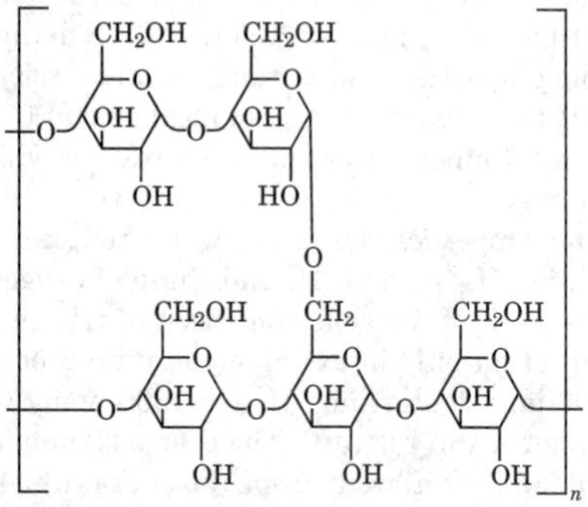

(2)

Starch hydrolysis is accomplished in industry by using acid, enzymes, or both sequentially. Acid treatment of starch causes random cleavage of $\alpha - 1 \rightarrow 4$ and $\alpha - 1 \rightarrow 6$ linkages. Other products are oligosaccharides and acid breakdown products of D-glucose.

α-Amylase, a common enzyme present in human and animal digestive tracts and in plants and microorganisms, hydrolyzes starch mainly to a mixture of D-glucose, maltose and limit dextrins derived from the amylopectin. β-Amylase occurs in many plants, where it is generally accompanied by some α-amylase. β-Amylase initiates hydrolysis at the nonreducing end of an amylose or amylopectin chain, and removes maltose units.

Oxidation of starch hydroxyl groups by hypochlorite gives aldehydes, ketones, or carboxylic acids. Starch is oxidized with other reagents including nitrogen dioxide, chlorine, permanganate, dichromate, and ozone to produce various aldehydes, ketones, and carboxylic acids and derivatives. Etherification and esterification of hydroxyl groups produce derivatives; they may also be obtained by graft polymerization wherein free radicals, initiated on the starch backbone, react with monomers such as vinyl or acrylyl derivatives.

Manufacture

Wet-Milling of Corn Starch. Milling of corn, *Zea mays*, provides a quality starch, used in food and nonfood applications. Other grains are milled for starch, but corn is the principal source. Corn wet-milling processes are fully automated to provide well-separated grain components.

Corn is first cleaned by screening to remove cob, sand, and other foreign material, followed by aspiration to remove lighter dust and chaff. The grain is then transferred with water containing ~0.1% sulfur dioxide at pH 3–4 to large vats (steeps) which softens the kernels for milling.

Steeped corn is coarsely ground in an attrition mill to free the germ. Germ is removed from the aqueous slurry in a hydroclone (cyclone separator), where suspended components are separated by density, the endosperm and fiber exiting in the hydroclone underflow and the germ from the center. The germ fraction is processed to produce corn oil.

The cyclone underflow is re-milled to complete the release of starch granules. After the second milling, the suspension contains starch, gluten, and fiber. Fiber is removed and later combined with the grain's gluten to a content of 21% for animal feed use. The resultant starch–gluten suspension, known as mill starch, is concentrated by centrifugation.

Chemical Modification. Acidic treatment below the gelatinization temperatures initially attacks amorphous regions of the granule but leaves the crystalline regions relatively unaffected. In corn starch modification, amylopectin is more extensively depolymerized than amylose. Properties of acid-treated starches, as compared to the unmodified starches, include decreased hot paste viscosity, decreased intrinsic viscosity, lower gel strength, and higher gelatinization temperature range. However, mild acid treatment of starch produces what is known as thin-boiling starch, which has excellent cooled gel strength.

To produce oxidized starch, a slurry of starch granules is treated with alkaline hypochlorite for an appropriate period, neutralized, washed to remove the inorganic salts, and finally dried to a moisture content of 10–12%. Temperature is usually in the range of 21–38°C. As with acid hydrolysis, most oxidation occurs primarily in the loosely organized, amorphous region of the starch granule. Oxidation results in lower gelatinization temperatures, decreased hot-paste viscosity, and lower paste setback.

In manufacture of pyrodextrins, dry starch is sprayed with dilute inorganic acid, usually hydrochloric, nitric, or acid salts, and redried to 1–5% water content. To produce a white dextrin, well-mixed acidified starch is hydrolyzed and heated to a final temperature of 110–150°C; at 135–160°C, a canary yellow dextrin is produced. When made with small amounts of acid, and a longer heating time to a temperature of 150–180°C, the product is known as a British gum.

Hydroxylalkyl Starch Ethers. The preparation of starch hydroxyethyl ethers begins at the end of the wet-milling process, utilizing a high solids–starch suspension.

Cationic Starches. The two general categories of commercial cationic starches are tertiary and quaternary aminoalkyl ethers. Tertiary aminoalkyl ethers are prepared by treating an alkaline starch dispersion with a tertiary amine containing a β-halogenated alkyl, 3-chloro-2-hydroxypropyl radical, or a 2,3-epoxypropyl group. Quaternary ammonium alkyl ethers are prepared by treating an alkaline starch with a quaternary ammonium salt containing a 3-chloro-2-hydroxypropyl or 2,3-epoxypropyl radical.

Starch Phosphates. Starch phosphate monoesters may be prepared by heating a dry mixture of starch and acid salts of ortho-, pyro-, or tripolyphosphoric acid at 50–60°C for one hour. Starch in aqueous suspension may react to form diesters with phosphorus oxychloride, phosphorus pentachloride, and thiophosphoryl chloride. Low DS starch acetates are manufactured by treatment of native starch with acetic acid or acetic anhydride, either alone or in pyridine or aqueous alkaline solution. Of particular importance for modifications of starch are the enzyme degradation products such as glucose syrups, cyclodextrins, maltodextrins, and high fructose corn syrups (HFCS) (see SYRUPS).

New Starches

Banana starch is a large-granule starch with many properties of potato starch. Another promising starch is that from amaranth seed. Amaranth starch granules (1–3 micrometers dia) have potential for numerous food applications, one of which is as a fat replacer (see FAT REPLACERS).

Uses

Nonfood Uses. Native corn starch is principally used in nonfood applications in mining, adhesives, and paper industries. Pregelatinized starches are used to decrease water losses in oil-well drilling muds, in cold water-dispersable wallpaper pastes, and in papermaking as an internal fiber adhesive.

Various organic chemicals, eg, ethanol, isopropyl alcohol, *n*-butanol, acetone, 2,3-butylene glycol, glycerol, fumaric acid, citric acid, and lactic acid, are derived from starch by fermentation. Other compounds derived from starch include D-glucose, sorbitol, methyl α-D-glucopyranoside, and glycerol or glycol D-glucopyranosides.

Food Uses. Unmodified starch is used in foods that require thickening or gelling. Pregelatinized starch is used in products where thickening is required but cooking must be avoided.

Acid-modified starches are used in the manufacture of gum candies. Thermalized starches are used in foods to bind and carry flavors and colors. Sweetening agents (corn syrup, HFCS) are made from starch by enzymatic or acid treatment.

Derivatives

Starches, as organic polyhydroxy compounds, undergo many reactions characteristic of alcohols, such as esterification and etherification. Commercial starch derivatives are generally very lightly derivatized (degree of substitution, DS < 0.1). Such modification produces distinct changes in colloidal properties and generally produces polymers with properties useful under a variety of conditions. Hydroxyethyl group introduction at low DS results in distinct modification of physical properties. Among these are decreased gelatinization temperature range, increased granule swelling rate, and decreased ability of starch pastes to gel and retrograde.

Cationic starches show decreased gelatinization temperature range and increased hot paste viscosity. Quaternary ammonium starches, like tertiary ammonium derivatives, show lower gelatinization temperature ranges, increased paste clarity and viscosity, and reduced retrogradation. Starch phosphates have increased paste viscosity and clarity and decreased retrogradation. Their properties are in many ways similar to those of potato starch. Starch monophosphates are useful in foods because of their superior freeze–thaw stability. Starch phosphate diesters show a significant increase in stability of the swollen granule. Starches with high DS have exceptional stability to high temperatures, low pH, and mechanical agitation.

Starch acetates may have low or high DS. The industrial importance of low DS acetates results from their ability to stabilize aqueous polymer sols. Highly derivatized starches (DS 2–3) are useful because of their solubility in organic solvents and ability to form films and fibers.

ROY L. WHISTLER
JAMES R. DANIEL
Purdue University

J. R. Daniel, R. L. Whistler, A. C. J. Voragen, and W. Pilnik, in B. Elvers, S. Hawkins, and W. Russey, eds., *Ullmann's Encyclopedia of Industrial Chemistry*, 5th ed., Vol. A25, VCH, Weinheim, Germany, 1994.

W. Banks and C. T. Greenwood, *Starch and Its Components*, Edinburgh University Press, Edinburgh, Scotland, 1975.

R. L. Whistler, J. N. BeMiller, and E. F. Paschall, eds., *Starch: Chemistry and Technology*, 2nd ed., Academic Press, Inc., New York, 1984.

O. B. Wurzburg, ed., *Modified Starches: Properties & Uses*, CRC Press, Boca Raton, Fla., 1986.

STEAM

Steam, gaseous H_2O, is the most important industrially used vapor and, after water (qv), the most common and important fluid used in chemical technology. Steam is generated from water by boiling, flash evaporation, and throttling from high to low pressure. The phase change occurs along the saturation line such that the specific volume of steam is larger than that of the boiling water. Thermal energy, ie, the heat of evaporation, is absorbed during the process. At the critical and supercritical pressures, the water–steam distinction disappears, and the fluid can go from water-like properties to steam-like properties without an abrupt change in density or enthalpy. Properties of steam can be divided into thermodynamic, transport, physical, and chemical properties.

Physical Properties

Official Properties. The International Association for Properties of Water and Steam (IAPWS), an association of national committees that maintains the official standard properties of steam and water for power cycle use, maintains two formulations of the properties of water and steam. The first is an industrial formulation, the official properties for the calculation of steam power plant cycles. This formulation is appropriate from 0.001 to 100 MPa (0.12–1450 psia) and from 0 to 800°C (32–1472°F) and also from 0.001 to 10 MPa (0.12–145 psia) between 800 and 2000°C (1472–3632°F). This formulation is used in the design of steam turbines and power cycles. IAPWS maintains a second formulation of the properties of water and steam for scientific and general use from 0.01 MPa (extrapolating to ideal gas) at 0°C (1.45 psia at 32°F) to the highest temperatures and pressures for which reliable information is available.

Thermodynamic Properties. Ordinary water contains three isotopes of hydrogen (qv), ie, ^{1}H, ^{2}H, and ^{3}H, and three of oxygen (qv), ie, ^{16}O, ^{17}O, and ^{18}O. The bulk of water is composed of ^{1}H and ^{16}O. Tritium, ^{3}H, and ^{17}O are present only in extremely minute concentrations, but there is about 200-ppm deuterium, ^{2}H, and 1000-ppm ^{18}O in water and steam (see DEUTERIUM AND TRITIUM). The thermodynamic properties of heavy water are subtly different from those of ordinary water. The properties given herein are for ordinary water having the usual mix of isotopes.

Vapor pressure is one of the most fundamental properties of steam. Figure 1 shows the vapor pressure as a function of temperature for temperatures between the melting point of water and the critical point. This line is called the saturation line. Liquid at the saturation line is called saturated liquid; liquid below the saturation line is called subcooled. Similarly, steam at the saturation line is saturated steam; steam at higher temperature is superheated. Properties of the liquid and vapor converge at the critical point, such that at temperatures above the critical point, there is only one fluid. Along the saturation line, the fraction of the fluid that is vapor is defined by its quality, which ranges from 0 to 100% steam.

The density of saturated water and steam is a function of temperature. As the temperature approaches the critical point, the densities of the liquid and vapor phase approach each other. This fact is crucial to boiler construction and steam purity, because the efficiency of separation of water from steam depends on the density difference.

The enthalpies and internal energies of steam and water also converge at the critical point. The heat capacity at constant pressure, C_p, is defined as the derivative of enthalpy with respect to temperature. The value of C_p becomes very large in the vicinity of the critical point. The variation is much smaller for the heat capacity at constant volume, C_v.

Transport Properties. Viscosity, thermal conductivity, the speed of sound, and various combinations of these with other properties are called steam transport properties, which are important in engineering calculations. The speed of sound is important to choking phenomena, where the flow of steam is no longer simply related to the difference

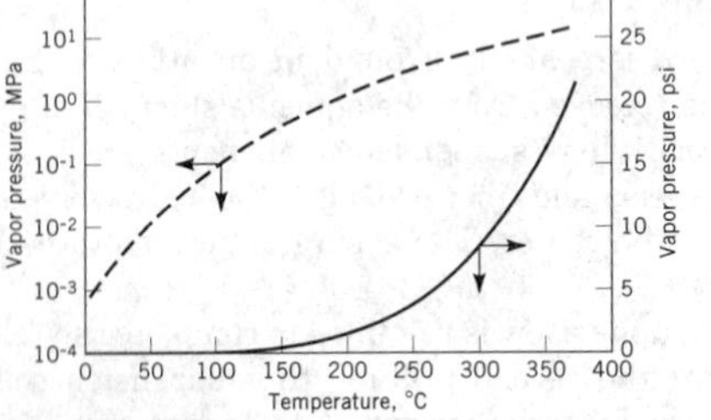

Figure 1. Vapor pressure of ordinary water, where (——) represents linear and (– – –) logarithmic scale. To convert MPa to psi, multiply by 145.

in pressure. Thermal conductivity is important to the design of heat-transfer apparatus (see HEAT-EXCHANGE TECHNOLOGY). Sharp declines in each of these properties occur at the transition from liquid to gas phase, ie, from water to steam.

Miscellaneous Properties. The dielectric constant is a physical property having great importance to the chemical properties of hot water and steam. Along the saturation line, the steam and water values converge at the critical point. The ability of water to dissolve salts results from the high dielectric constant. The precipitous drop in water dielectric constant in the region of the critical point is very important to the solubility of salts in water near the critical temperature. Many salts exhibit declining solubilities as the critical temperature is approached and then exceeded. The drop in dielectric constant is largely a result of the decline in density.

The ion product of water is the product of the molality of the hydrogen and hydroxide ions, $K_w = m_{H^+}m_{OH^-}$. The ion product increases with temperature to 250°C and then declines. The initial increase is the temperature effect, and the later decline is on account of the decline in the dielectric constant of water. This variation means that neutral pH, which is the square root of the ion product, varies with temperature.

Chemical Properties

Molecular Nature of Steam. The molecular structure of steam is not as well known as that of ice or water. There are indications that in the steam phase some H_2O molecules are associated in small clusters of two or more molecules.

Solvent. The solvent properties of water and steam are a consequence of the dielectric constant. At 25°C, the dielectric constant of water is 78.4, which enables ready dissolution of salts. As the temperature increases, the dielectric constant decreases. The solubility of many salts declines at high temperatures. As a consequence, steam is a poor solvent for salts. Although the solubility of salts in steam is small, it has great significance to corrosion of steam system components, particularly steam turbines. At the critical point and above, water is a good solvent for organic molecules.

Reactant. Steam can behave as an oxidant. Steam reacts with salts so that the salts dissociate into the respective hydroxide and acid. For sodium salts, the sodium hydroxide is largely in a liquid solution and the acid is volatile.

Generation

Simplified Cycle. The water accumulates in the bottom of the condenser, called the hotwell. It goes through a feed pump to pressurize it. The pressurized water passes through one or more feedwater heaters, which raise the temperature. The water then enters the boiler where heat from the fuel converts it to steam. The steam expands through the engine, usually a turbine, which extracts work. In the middle of the turbine some of the steam is extracted to supply heat to the feedwater heater. The remainder expands through the turbine and is condensed. The rejected heat is carried away by the condenser coolant, which is usually water, but sometimes air. The condensed steam then returns to the hotwell to repeat the cycle.

Steam Generators or Boilers. Steam is produced in a boiler or steam generator. The term *steam generator* is used when the heat source is nuclear power (see NUCLEAR REACTORS) and is often used for fossil-fired boilers, particularly supercritical once-through units, where the fluid changes gradually from liquid-like to vapor-like properties without really boiling. Boilers using hot gas as a heat source are generally called heat recovery steam generators (HRSGs).

There are two basic types of boilers: recirculating and once-through. The difference between the boiler types is the manner in which heat is absorbed from the fuel to generate steam. In a recirculating boiler, water enters the boiler at the bottom of the waterwall. Approximately 25 to 33% of the water boils by the time it reaches the top of the waterwall. In the steam drum, the steam is separated from the remaining water. The mixture of steam and water passes through steam-separating devices, where the direction of flow changes rapidly. Steam follows the change in direction but water hits the metal and is drained away. The water flow in a recirculating boiler may be either convective or forced.

In a once-through system the feedwater enters the steam generator at the bottom of the waterwall and passes out of the boiler through the superheater. Most once-through steam generators are supercritical (>23 MPa (3300 psia)).

There are three basic types of nuclear heat sources. Most common are pressurized water reactors having steam generators. In this system, the nuclear reactor heats water in a high pressure loop. The water is circulated through tubes in a steam generator. On the outside of the tubes, water is boiled to steam which goes to the steam turbine. In boiling-water reactors, the second type, the nuclear heat is used to boil the feedwater directly. In gas-cooled reactors, the third type, the gas takes the same role as the pressurized water in a pressurized-water reactor (PWR) and transfers the heat to the steam generator.

Heat recovery steam generators comprise a special class of boilers where essentially all heat transfer takes place convectively; they are used to extract energy from hot gas streams. One of the principal uses is extraction of heat from the exhaust gases of a combustion turbine.

Feedwater Heaters. Feedwater heaters use steam from the turbine to preheat the feedwater before it reaches the boiler. Feedwater heaters increase efficiency of steam cycles because heat comes from a source having a lower temperature than the fire in the boiler.

Condenser. Water-cooled condensers are constructed of tubes between tubesheets. Cooling water is pumped from the source into the inlet waterbox, through the tubes, and through the outlet waterbox. The cooling water may be drawn from lakes or rivers or may be recirculated from cooling towers (see WATER, INDUSTRIAL WATER TREATMENT). Air-cooled condensers are being used more frequently, because cooling water sources are being exhausted.

Water Chemistry in Steam-Generating Systems. The usual function of steam purity limits is to protect the turbine from deposition and subsequent corrosion. In systems where the steam is used for chemical processes, the specific process may create additional requirements for steam purity.

The steam purity limits define boiler-water limits, because the steam cannot be purified once it leaves the boiler. For a once-through boiler, the boiler water must have the same specifications as the steam. A recirculating boiler is a still, and there can be considerable purification of the steam as it boils and is separated from the water in the steam drum. In addition to the requirement to conform to steam purity needs, there are concerns that the boiler water not corrode the boiler tubes nor produce deposits, known as scale, on these tubes. Three important components of boiler tube scale are iron oxides, copper oxides, and calcium salts, particularly calcium carbonate (see also WATER, INDUSTRIAL WATER TREATMENT).

The feedwater for a steam cycle must be purified. The degree of purity depends on the pressure of the boiler. Higher pressure boilers require higher feedwater purity.

Makeup water is the water supplied to replenish the steam system for any losses. In most systems it is introduced into the condenser or the feed pump suction. In steam systems where the makeup is a small fraction of the total feedwater, its purity may be somewhat lower than the feedwater requirement because it is diluted by condensate. In systems where there is little condensate return the makeup purity must be essentially the same as the feedwater.

Water Treatment. Water and steam chemistry must be rigorously controlled to prevent deposition of impurities and corrosion of the steam cycle. Deposition on boiler tubing walls reduces heat transfer and can lead to overheating, creep, and eventual failure. Additionally, corrosion can develop under the deposits and lead to failure. If steam is used for chemical processes or as a heat-transfer medium for food and pharmaceutical preparation there are limitations on the additives that may be used. Steam purity requirements set the allowable impurity concentrations for the rest of most cycles. All purification must be carried out in the boiler or preboiler part of the cycle.

Most steam systems are maintained in a reducing state to avoid oxidation of the steel piping and other components. Oxidation is further suppressed by raising the pH, which is commonly controlled by using ammonia or organic amines. In plants where the steam may be in contact with food, sodium sulfite, ascorbic acid, or erythorbic acid is commonly used as an oxygen scavenger.

Oxidizing Chemistry. In high pressure boilers systems having the ability to maintain very pure feedwater, the feedwater may be treated with oxygen. Oxygen is added either as gaseous oxygen or as hydrogen peroxide. The pH may be neutral or elevated with ammonia. The goal of this treatment is to maintain all iron alloy surfaces in a passivated state.

Other Water Treatment Issues. Maintaining steam temperatures at correct values may require a process called attemperation, where water is sprayed into the inlet of the superheater or reheater to lower the temperature of the entering and, consequently, exiting steam.

In order to maintain the feedwater purity required for once-through boilers, but also as an aid to maintaining feedwater purity for drum boilers and nuclear steam generators, condensate polishers are used to reduce contaminants. They are large ion-exchange (qv) systems designed to pass high flows. These reduce contamination.

Makeup treatment depends extensively on the source water. Some steam systems use municipal water as a source. These systems may require dechlorination followed by reverse osmosis (qv) and ion exchange. Other systems use wellwater. In hard water areas, these systems include softening before further purification.

To maintain appropriate steam and water purity requires analysis at the concentrations of interest. Feedwater and steam purity for most boiler systems approaches the detection limit of on-line monitoring instruments. One of the most reliable monitoring devices is electrical conductivity. Sodium and chloride may be measured using ion-selective electrodes (see ELECTROANALYTICAL TECHNIQUES). Silica and phosphate may be monitored colorimetrically. Chloride, sulfate, phosphate, and other anions may be monitored by ion chromatography.

Uses

Power Production. Steam cycles for generation of electric power use various types of boilers, steam generators, and nuclear reactors; operate at subcritical or supercritical pressures; and use makeup and often also condensate water purification systems as well as chemical additives for feedwater and boiler-water treatment. These cycles are designed to maximize cycle efficiency and reliability.

Turbines. The structure of steam turbines varies with the size. Very large (>500 MW) utility turbines have several individual turbines, usually classified as high pressure (HP), intermediate pressure (IP), and low pressure (LP). On smaller turbines, a combined HP–IP turbine is joined with an LP turbine. Large steam turbines are generally designed as a rotor inside one or two cylinders or casings. The low pressure nuclear turbine having a partial integral rotor features many of the aspects of both older and newer turbines. Modern utility turbines are generally rather massive devices. Turbine cylinders are pressure vessels (see TANKS AND PRESSURE VESSELS). The high pressure cylinder of a fossil unit must withstand 16.5 to 24.1 MPa (2400–3500 psig) at 538°C (1000°F), at which temperatures and pressures creep becomes significant during the design life, even for the thick metal walls. To reduce the pressure and temperature differentials, either the exhaust or the extraction steam from the HP turbine is used to pressurize and cool the cavity between the inner and outer cylinders.

Rankine Cycle Thermodynamics. Carnot cycles provide the highest theoretical efficiency possible, but these are entirely gas phase. A drawback to a Carnot cycle is the need for gas compression. The Rankine cycle overcomes the problem of an efficient gas compressor by compressing the liquid. Efficient pumps (qv) are much easier to construct than efficient compressors.

In a regenerative cycle, some of the steam is removed from the expansion in the turbine and used to heat the water. In essence, regeneration uses some of the latent heat of the steam vaporization to heat feedwater, rather than rejecting this heat to the condenser. Regeneration improves cycle efficiency considerably.

Fossil Fuel Cycles. Modern fossil fuel cycles have turbine throttle pressures at 16.6 MPa (2415 psia) or supercritical 23.9 MPa (3515 psia). The more feedwater heaters, the more efficient the cycle, but eight is usually the economically practical limit. The boiler feed pump suction comes from the deaerating heater. Although electric motor-driven pumps are used for start-up and some cycles, the boiler feedpump is usually driven by steam extracted from the cycle just before the low pressure turbine.

Nuclear Steam Cycles. Most nuclear power plant cycles begin with saturated steam. This steam is expanded through the high pressure turbine. In the middle of this expansion, steam is extracted for the highest pressure feedwater heater. After heating the feedwater, the water in the shell side of the heater is drained to the shell side of the second highest pressure heater. Because the inlet steam has such low enthalpy, the steam becomes very moist in the high pressure turbine. Moisture reduces the efficiency of the turbine because the droplets do not follow the steam flow. Typically, 10% efficiency is lost in wet steam. After the high pressure turbine, the steam is dried and superheated in a moisture separator reheater (MSR). The steam then expands through the low pressure turbine to the condenser.

Relatively volatile components, such as amines and organic acids, can appear at rather high concentrations in the high pressure region of nuclear power steam cycles. Because of the variation in distribution between water and steam phases with temperature and the generation of significant moisture, the chemical relationships can be quite complex.

Combined Cycle. Combined cycles use a steam turbine system as a bottoming cycle for a combustion turbine. Exhaust gases leave the combustion turbine at 510–593°C (950–1100°F). This heat is wasted in the simple combustion turbine cycle. However, it can be used to boil water in a heat recovery steam generator (HRSG). This process lowers the temperature of much of the rejected heat to that leaving the steam turbine. However, to prevent HRSG corrosion, the stack gas temperature is nominally 150°C (302°F), causing additional heat rejection to the environment. Many older conventional steam plants have been converted to combined cycle.

Cogeneration. Cogeneration is another modification of the Rankine cycle used when steam is required to heat a process. It is common in pulp (qv) and paper (qv) mills, chemical plants, and municipal or district heating systems. The steam is produced at higher pressure and temperature than would be required for the process. The steam is expanded through a turbine and steam at the desired pressure is extracted. The turbine may be used to drive a generator or other machinery.

Mechanical Drives. Steam turbines are very efficient at high load ratings and, depending on the steam balance in a plant, are normally considered as drive units if more than 37 kW (50 hp) is required. If many small loads are to be handled separately in a plant, it may be preferable to generate electric power by passing the steam through a back-pressure (no condenser) turbine connected to an electric generator. The generated electricity in turn can be fed into the motors throughout the plant. Steam turbines operate very effectively at high speeds (3,000–10,000 rpm) and thus lend themselves well to large power output, high speed drives.

Steam Heating. Wet and saturated steam has a definite pressure for each fixed boiling or condensing temperature. Therefore, the control of the desired temperature for any process heating requirement may be fixed by choosing the steam pressure. It is customary for steam to be generated under slightly superheated conditions if the steam generator is to be located at a considerable distance from the various users (see HEAT-EXCHANGE TECHNOLOGY).

Evaporators. Steam heating systems have often been installed in a cascade system, as in multiple-effect evaporators (see EVAPORATION). This arrangement makes possible the recovery of heat at several successive levels merely by reducing the pressure at each of the stages.

Control. When close temperature control is required in order to prevent overheating of material being processed or to ensure a high heating density, steam is the medium normally used. A pressure regulator controlling the steam pressure of a heating unit maintains temperatures usually within degrees of the design conditions on the process side.

Steam Reforming Processes. In the steam reforming process, light hydrocarbon feedstocks (qv), such as natural gas, liquefied petroleum gas, and naphtha, or in some cases heavier distillate oils are purified of sulfur compounds (see SULFUR REMOVAL AND RECOVERY). These then react with steam in the presence of a nickel-containing catalyst to produce a mixture of hydrogen, methane, and carbon oxides. Essentially total decomposition of compounds containing more than one carbon atom per molecule is obtained (see AMMONIA; HYDROGEN; PETROLEUM).

Synthetic Ammonia. Steam-methane reforming is used to produce hydrogen for the production of ammonia (qv), which is synthesized by using nitrogen from the air at high pressure and temperature over a suitable catalyst. The exhaust steam is used for the gas reforming reaction. The high pressure steam thus generated from waste heat is the heart of the energy system for the process, which in turn is the key to the process economics (see PROCESS ENERGY CONSERVATION).

Coal Gasification. Coal gasification processes involve the reaction of coal at high temperature with steam and air or oxygen to produce a mixture of gases, typically carbon monoxide, carbon dioxide, hydrogen, and methane. Sulfur (qv) in the coal (qv) reacts to form hydrogen sulfide. By reaction with steam, the carbon monoxide in the gas is converted to hydrogen and carbon dioxide. Hydrogen sulfide and carbon dioxide can be removed in a purification system and the hydrogen converted to methane by reaction with carbon monoxide (see COAL CONVERSION PROCESSES–GASIFICATION; FUELS, SYNTHETIC–GASEOUS FUELS).

Coal Liquefaction. Steam is used to produce hydrogen for the liquefaction of coal. In the liquefaction process, coal is crushed, dried, pulverized, and then added to a solvent to produce a slurry. The slurry is heated, usually in the presence of hydrogen to dissolve the coal. The extract is cooled to remove hydrogen, hydrocarbon gases, and hydrogen sulfide. The liquid is then flashed at low pressure to separate condensable vapors from the extract. Mineral matter and organic solids are separated and used to produce hydrogen for the process (see COAL CONVERSION PROCESSES–LIQUEFACTION; FUELS, SYNTHETIC–LIQUID FUELS).

Petroleum Recovery. Steam is injected into oil wells for tertiary petroleum recovery. Steam pumped into the partly depleted oil reservoirs through input wells decreases the viscosity of crude oil trapped in the porous rock of a reservoir, displaces the crude, and maintains the pressure needed to push the oil toward the production well (see PETROLEUM, ENHANCED OIL RECOVERY).

Evaporation and Distillation. Steam is used to supply heat to most evaporation (qv) and distillation (qv) processes. In evaporation, pure solvent is removed and a low volatility solute is concentrated. Distillation transfers lower boiling components from the liquid to the vapor phase. The vapors are then condensed to recover the desired components. In steam distillation, the steam is admitted into direct contact with the solution to be evaporated and the flow of steam to the condenser is used to transport distillates of low volatility.

Desalination. A special case of distillation is water desalination. In places where energy is abundant but fresh water is not, eg, the Arabian Peninsula, water may be produced from seawater in flash evaporators. Low pressure turbine steam is extracted to provide heat for the evaporators.

Steam Cleaning. High pressure steam can be used to produce a high velocity jet with some superheating by expansion through a suitable nozzle to atmospheric pressure. The high velocity is effective in removing dirt and loose scale from solid surfaces.

Hydrothermal Treatment of Wastes. Hydrothermal processing (qv) of materials appears to be a promising method of disposal for many noxious materials and of conversion of some wastes to valuable byproducts. This method consists of mixing reactants and water; pressurizing, heating to reaction temperatures, and cooling the products; and then carrying out secondary processing of the products. For efficient oxidation, oxygen is added to the water.

Corrosion in Steam

Use of metals in hot steam is limited by oxidation rate, mechanical strength, and creep resistance (see CORROSION AND CORROSION CONTROL). General corrosion rates in pure steam are about the same as in high purity deoxygenated water, except for gray iron, nickel, lead, and zirconium, which corrode faster in steam. Iron-base alloys, including the austenitic and ferritic stainless steels, are used extensively in contact with steam. These oxidize to form a protective film of the spinel oxide, Fe_3O_4 (magnetite), or, in the case of stainless steels, M_3O_4, where M is iron, chromium, or nickel. Where corrosive impurities from steam or water concentrate on metal surfaces, corrosion can be severe.

JAMES BELLOWS
Westinghouse Electric Corporation

C. A. Meyer and co-workers, *Steam Tables*, 6th ed., ASME, New York, 1993.

H. J. White and co-workers, *Proceedings of the 12th International Conference on the Properties of Water and Steam, Orlando, Fla., 1994*, Begell House, New York, 1995.

P. Cohen, *ASME Handbook on Water Technology for Thermal Power Systems*, ASME, New York, 1989.

Steam: Its Generation and Use, 40th ed., Babcock and Wilcox Co., New York, 1992.

STEARIC ACID. See CARBOXYLIC ACIDS.

STEATITE. See TALC.

STEEL

Steel, the generic name for a group of ferrous metals composed principally of iron (qv), is the most useful metallic material known on account of its abundance, durability, versatility, and low cost. Most modern steels contain >98% iron. Steel also contains carbon, which, if present in sufficient amounts (up to ca 2%), gives the steel the ability to become extremely hard if cooled very quickly (quenched) from a high enough temperature.

The most widely used processes for making liquid steel are the oxygen steelmaking processes, in which commercially pure oxygen (99.5% pure) is used to refine molten pig iron. In the top-blown basic oxygen process, oxygen is blown down onto the surface of the molten pig iron and such coolants as scrap steel or iron ore. In the bottom-blown basic oxygen process, the oxygen is blown upward through the molten pig iron.

Pig iron and iron and steel scrap are the sources of iron for steelmaking in basic-oxygen furnaces. Electric furnaces have relied on iron and steel scrap, although newer iron sources such as direct-reduced iron (DRI), iron carbide, and even pig iron are becoming both desirable and available (see IRON BY DIRECT REDUCTION). In basic-oxygen furnaces, the pig iron is used in the molten state as obtained from the blast furnace; in this form, pig iron is referred to as hot metal. Pig iron consists of iron combined with numerous other elements, depending on the composition of the raw materials used in the blast furnace. Steelmaking processes were historically either acid or basic, but acid processes have virtually disappeared. Basic slags, rich in lime, are able to absorb much of the unwanted sulfur and phosphorus.

Oxidation is employed to convert a molten bath of pig iron and scrap, or scrap alone, into steel. Each steelmaking process has been devised primarily to provide some means by which controlled amounts of oxygen can be supplied to the molten metal undergoing refining. In general, steel having similar chemical compositions have similar mechanical and physical properties, no matter by which process they are made.

Electric Furnace Processes

The principal electric furnace steelmaking processes are the electric-arc furnace, induction furnace, consumable-electrode melting, and electroslag remelting. The main raw material for all these processes is solid steel. In the first two processes steel is made that is different in composition and shape from the starting material, which is usually steel scrap and/or DRI. The starting material used for the last two processes closely resembles the desired steel ingot subsequently rolled or forged, and yields very high quality steel for applications with extremely strict requirements.

Electric-Arc Furnace. The electric-arc furnace is by far the most popular electric steelmaking furnace. It offers the advantages of low construction costs, flexibility in the use of raw materials, and the ability to produce steels over a wide range of compositions (carbon, alloy, and stainless) and to operate below full capacity (see FURNACES, ELECTRIC).

The biggest change in steelmaking in the last quarter of the twentieth century was the fraction of steel made by remelting scrap or, increasingly, other iron units in an electric-arc furnace (EAF).

Modern processes mainly use a single d-c electrode with a conducting hearth. Scrap is charged into the furnace, which usually contains some of the last heat as a liquid heel to improve efficiency. Oxygen is blown to speed the reactions. Various techniques are applied to ensure adequate separation of melt and slag to permit effective ladle treatment. Final treatment takes place in a ladle furnace, which allows refining, temperature control, and alloying additions to be made without delaying the next heat.

Induction Furnace. The high frequency coreless induction furnace is used in the production of complex, high quality alloys such as tool steels. It is used also for remelting scrap from fine steels produced in arc furnaces, for melting chrome–nickel alloys and high manganese scrap, and more recently for vacuum steelmaking processes. Most commonly, the melting procedure is essentially a dead-melt process. Little if any refining is attempted in ordinary induction melting.

Vacuum and Atmosphere Melting. A coreless high frequency induction furnace is enclosed in a container or tank which can be either evacuated or filled with a gaseous atmosphere of any desired composition or pressure. Provision is made for additions to the melt, and tilting the furnace to pour its contents into an ingot mold also enclosed in the tank or container without disturbing the vacuum or atmosphere in the tank. Although vacuum melting has often been employed as a remelting operation for very pure materials or for making electrodes for the vacuum consumable-electrode furnace, it is generally more useful in applications that include refining.

Consumable-Electrode Melting. This refining process produces special-quality alloy and stainless steels by casting or forging the steel into an electrode that is remelted and cast into an ingot in a vacuum. A consumable-electrode furnace consists of a tank that encloses the electrode and a water-cooled copper mold. After the furnace has been evacuated, power is turned on and an arc is struck between the electrode and a starting block that is placed in the mold before operation begins. Heat from the arc progressively melts the end of the electrode. Melted metal is deposited in a shallow pool of molten metal on the top surface of the ingot being built up in the mold. The remelting operation removes gases from the steel, improves its cleanliness, and improves hot workability and the mechanical properties of the steel.

Electroslag Remelting. Electroslag remelting has the same general purpose as consumable-electrode melting, and a conventional air-melted ingot serves as a consumable electrode. No vacuum is employed. Melting takes place under a layer of slag that removes unwanted impurities.

Powder Techniques. A method for avoiding or reducing segregation of alloying elements involves preparing small spheres of material by the atomization of a liquid stream through a nozzle to produce a powder. This powder can be compacted, often hot and triaxially by gas pressure, to form a material where, on further heating, the residual pores close by diffusion to approach 100% density.

Oxygen Steelmaking Processes

In oxygen steelmaking, 99.5% pure oxygen gas is mixed with hot metal, causing the oxidation of the excess carbon and silicon in the hot metal and thereby producing steel. In the United States, this process is called the basic-oxygen process (BOP).

Top-Blown Basic Oxygen Process. The top-blown basic oxygen process is conducted in a cylindrical furnace. This furnace has a dished bottom without holes and a truncated cone-shaped top section in which the mouth of the vessel is located. The furnace shell is made of steel plates and lined with refractory.

A jet of gaseous oxygen is blown at high velocity onto the surface of a bath of molten pig iron and scrap at the bottom of the furnace by a vertical water-cooled retractable lance inserted through the mouth of the vessel. The furnace is mounted in a trunnion ring and can be tilted backward or forward.

When the furnace is tilted toward the charging floor, which is on a platform above ground level, solid scrap is dumped by an overhead crane into the mouth of the furnace. The crane then moves away from the furnace and another crane carries a transfer ladle of molten pig iron to the furnace and pours the molten pig iron on top of the scrap.

After charging, the furnace is immediately returned to the upright position, the lance is lowered into it to the desired height above the bath, and the flow of oxygen is started. Striking the surface of the liquid bath, the oxygen immediately forms iron oxide, part of which disperses rapidly through the bath.

The manganese residue of the blown metal before ladle additions is closely related to the amount of manganese in the furnace charge. High slag fluidity and good slag–metal contact promote transfer of phosphorus from the metal into the slag, even before the carbon reduction is complete.

Bottom-Blown Basic-Oxygen Process. The bottom-blown basic-oxygen process, called oxygen bottom metallurgy (OBM) in Europe and Quelle basic-oxygen process (Q-BOP) in the United States and Japan, is conducted in a furnace comprising two parts, the bottom and the barrel, which consist of an outer steel shell lined with refractory. The bottom contains 6–24 tuyeres or double pipes. Oxygen is blown into the furnace through the center pipes and natural gas or some other hydrocarbon is blown into the furnace through the annular space between the two pipes.

Both the OBM and the Q-BOP processes are operated in about the same way as the top-blown process. However, the method of adding lime to the furnace is different. In the bottom-blown processes, powdered lime is added to the oxygen before it is blown into the furnace. The chemistry of the bottom-blown processes is similar to that of the top-blown process. However, the slags in the bottom-blown processes contain significantly less iron oxide, causing a $\sim$ 2% increase in yield.

Top- and Bottom-Blown Basic-Oxygen Processes. In the lance bubbling equilibrium (LBE) process, nitrogen or argon is injected through a number of porous refractory plugs installed in the furnace bottom, while oxygen is top-blown into the furnace through a lance. The bottom injection of nitrogen or argon causes more intimate mixing of slag and metal, and hence most of the advantages of the Q-BOP bottom-blowing process are obtained in furnaces designed for top blowing.

Scrap as Raw Material

Scrap consists of the by-products of steel fabrication and worn-out, broken, or discarded articles containing iron or steel. It is a principal source of the iron for steelmaking; the other source is iron from blast furnaces, either molten as it comes from the furnace (hot metal) or in solid pig form. Purchased scrap comes in two basic categories. Industries that have a steady supply of discards from such operations as stamping, drawing, machining, and forging count on resale of what is often a premium product as part of their routine costs. Less economically desirable material comes from salvaging the values in products such as ships, automobiles, railroad equipment, and buildings (see RECYCLING, METALS).

Chemical Composition. A standard problem in purchased scrap is the accuracy of analysis of composition. For home scrap, this may not be much of an problem, but sampling (qv) difficulties and analytical costs can be troublesome in other cases. The difficulty of estimating the alloy content, is a continuing worry.

Addition Agents

In steelmaking, various elements are added to the molten metal to effect deoxidation, control of grain size, improvement of the mechanical, physical, thermal, and corrosion properties, and other specific results. The term *addition agent* is preferred when describing the materials added to molten steel for altering its composition or properties; ferroalloys are a special class of addition agents.

Included in the ferroalloy class are alloys of iron with aluminum, boron, calcium, chromium, niobium, manganese, molybdenum, nitrogen, phosphorus, selenium, silicon, tantalum, titanium, tungsten, vanadium, and zirconium.

Ladle Metallurgy

The finished steel from any furnace, whether basic-oxygen or electric, is tapped into ladles. Most ladles hold all the steel produced in one furnace heat. Some slag is allowed to float on the surface of the steel in the ladle to form a protective blanket. A ladle consists of a steel shell, lined with refractory brick, having an off-center opening in its bottom equipped with a nozzle that makes it possible to enlarge or close the opening to control the flow of steel. Ladle metallurgy is the treatment of liquid steel in the ladle.

Argon Treatment. In the argon–oxygen decarburization (AOD) process, argon or argon–oxygen mixtures are blown in through bottom tuyeres to lower the carbon content of the bath by making escape of carbon monoxide in the argon bubbles easier. The predominant use of AOD is in making stainless steels more weldable by lowering the carbon content to 0.03% max.

Vacuum Processes. More complete control over ladle treatment is achieved by the ability to seal a vessel well enough so that a good vacuum can exist over the steel. This is used for steels with the most difficult property requirements; for many purposes less elaborate treatments are adequate.

Casting

Continuous Casting. Although the continuous casting of steel appears deceptively simple in principle, many difficulties are inherent to the process. When molten steel comes into contact with a water-cooled mold, a thin solid skin forms on the wall. However, because of the physical characteristics of steel, and because thermal contraction causes the skin to separate from the mold wall shortly after solidification, the rate of heat abstraction from the casting is low enough that molten steel persists within the interior of the section for some distance below the bottom of the mold. The thickness of the skin increases because the action of the water sprays as the casting moves downward and, eventually, the whole section solidifies. The mass of the solidifying section is supported as it descends by driven pinch rolls that control the speed of descent. Continuous casting is almost universal. It gives a higher yield than ingot casting and avoids the cost of rolling ingots into slabs; the slabs are produced directly from liquid steel. Much progress has been made on casting directly to sheet of the order of 1 mm thick and this will in all probability soon be a new standard practice.

Plastic Working of Steel

Plastic working of a metal such as steel is the permanent deformation accomplished by applying mechanical forces to a metal surface. The primary objective is usually the production of a specific shape or size (mechanical shaping), although increasingly it also involves the improvement of certain physical and mechanical properties of the metal (mechanical treatment). These two objectives can be readily attained simultaneously. Plastic deformation of steel can be accomplished by hot working or cold working.

The principal hot-working techniques are hammering, pressing, extrusion, and rolling, the first two of which are called forging. Other methods include rotary swaging, hot spinning, hot deep-drawing, roll forging, and die forging. Although at high temperatures the material is not very strong when compared to its room temperature strength, the shape changes are made in several stages for reasons of mechanics. Cold working is generally applied to bars, wire, strip, sheet, and tubes. It reduces the cross-sectional area of the piece being worked on by cold rolling, cold drawing, or cold extrusion. Cold working imparts improved mechanical properties, better machinability, good dimensional control, bright surface, and production of thinner material than can be accomplished economically by hot working (see METAL TREATMENTS).

Metallography and Heat Treatment

The physical and mechanical properties of steel depend on its microstructure, that is, the nature, distribution, and amounts of its metallographic constituents as distinct from its chemical composition. The amount and distribution of iron and iron carbide determine most of the properties, although most plain carbon steels also contain manganese, silicon, phosphorus, sulfur, oxygen, and traces of nitrogen, hydrogen, and other chemical elements such as aluminum and copper. These elements may modify, to a certain extent, the main effects of iron and iron carbide, but the influence of iron carbide always predominates. This is true even of medium alloy steels, which may contain considerable amounts of nickel, chromium, and molybdenum.

There are two allotropic forms of iron: ferrite and austenite. *Cementite*, the term for iron carbide in steel, is the form in which carbon appears in steels. It has the formula Fe_3C, and thus consists of 6.67 wt % carbon and the balance iron. Cementite is very hard and brittle. As the hardest constituent of plain carbon steel, it scratches glass and feldspar, but not quartz. Most commercial steels contain at most a few tenths of a percent of carbon. The limiting temperatures for austenite are given by phase diagram (Fig. 1). By controlling the cooling rate from the austenite range, the carbon comes out of solution in a range of microstructures that have very specific properties.

A nonalloyed carbon steel having 0.76% carbon, the eutectoid composition, consists of austenite above its lowest stable temperature, 727°C (the eutectoid temperature). Above (hyper-) or below (hypo-) eutectoid, a primary precipitation of cementite or ferrite occurs until the austenite reaches the eutectoid composition when it transforms as before. The amount of proeutectoid constituents offers a rough guide to the composition of the steel.

If small specimens are prepared in which the austenite can be cooled to 250–500°C sufficiently rapidly to avoid the above microconstituents, and transformed at temperatures in this range, the formation of a completely different phase, a bcc α-phase supersaturated with carbon and containing small cementite particles (bainite), which is both strong and tough, occurs. If all of the above sets of phases can

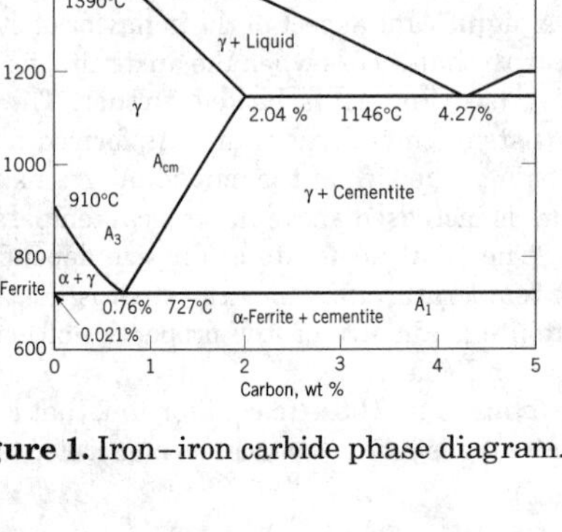

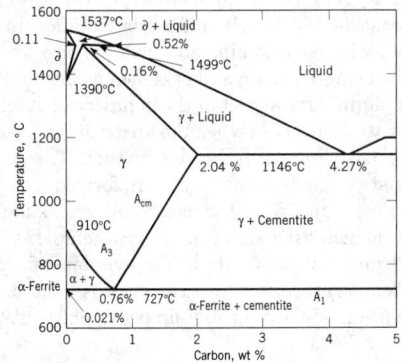

Figure 1. Iron–iron carbide phase diagram.

be suppressed by sufficiently rapid cooling, the austenite transforms to a phase known as martensite. This phase contains all of the carbon in the austenite. Martensite forms by a lattice shear reaction without diffusion of the carbon atoms, and usually has a plate or needle-like morphology.

Iron–Iron Carbide Phases

The iron–iron carbide phase diagram (see Fig. 1) shows the ranges of compositions and temperatures in which the various stable or metastable phases, such as austenite, ferrite, and cementite, are present in slow-cooled steels. In steels and cast irons, carbon can be present either as iron carbide (cementite) or as graphite. Under equilibrium conditions, only graphite is present, because iron carbide is metastable with respect to iron and graphite. However, in commercial steels, iron carbide is present instead of graphite. When a steel containing carbon solidifies, the carbon in the steel usually solidifies as iron carbide, which, under normal conditions, is quite stable for many years.

Changes on Heating and Cooling Pure Iron. The only changes occurring on heating or cooling pure iron are the reversible changes at ca 910°C from bcc α-iron to fcc γ-iron and from the fcc γ-iron to bcc δ-iron at ca 1390°C.

Changes on Heating and Cooling Eutectoid Steel. Eutectoid steels are those that contain 0.76% carbon. The phase change on heating a eutectoid carbon steel occurs at 727°C, which is designated as A_1, the eutectoid or lower critical temperature. On heating such a steel just above this temperature, all ferrite and cementite transform to austenite. When a eutectoid steel is slowly cooled from the austenite range, the ferrite and cementite form in alternate layers of microscopic thickness. Under the microscope at low magnification, the diffraction effects from this mixture of ferrite and cementite give an appearance similar to that of a pearl, hence the material is called pearlite.

Changes on Heating and Cooling Hypoeutectoid Steel. Hypoeutectoid steels are those that contain less carbon than the eutectoid steels. If the steel contains more than 0.02% carbon, the constituents present at and below 727°C are usually ferrite and pearlite. The first phase change on heating, if the steel contains more than 0.02% carbon, occurs at 727°C. On heating just above this temperature, the pearlite slowly changes to austenite. On slow cooling the reverse changes occur. Ferrite precipitates, generally at the grain boundaries of the austenite. Just above A_1, the austenite is substantially of eutectoid composition.

Changes on Heating and Cooling Hypereutectoid Steel. The behavior on heating and cooling hypereutectoid steels (steels containing >0.76% carbon) is similar to that of hypoeutectoid steels, except that the excess constituent is cementite rather than ferrite.

Effect of Alloys on the Equilibrium Diagram. The iron–carbon diagram may be significantly altered by alloying elements, and its quantitative application should be limited to plain carbon and low alloy steels.

Grain Size. Crystal structures exist as a regular array over a distance of 10,000 or more unit cells in a given direction to give a crystal commonly called a grain. Eventually, however, these run into a region where the orientation of the cells differs by rotation in three directions. The region where the grains abut is a surface where no atoms are at their equilibrium spacing and is known as a grain boundary.

Austenite. A significant aspect of the behavior of steels on heating is the grain growth that occurs when the austenite, formed on heating above A_3 or A_{cm} (see Fig. 1), is heated further. The austenite, like any metal, consists of polyhedral grains. As formed at a temperature just above A_3 or A_{cm}, the size of the individual grains is small, but as the temperature is increased above the critical temperature, the grain size increases. The final austenite grain size depends primarily on the maximum temperature to which the steel is heated. The general effects of austenite grain size on the properties of heat-treated steel are summarized in Table 1.

Phase Transformations. Close to equilibrium, that is with very slow cooling, austenite transforms to pearlite when cooled below the A_1 (see Fig. 1) temperature. When austenite is cooled more rapidly, this transformation occurs at a lower temperature. The faster the cooling rate, the lower the temperature at which transformation occurs. Thus, heat treatment involves a controlled supercooling of austenite.

Table 1. Trends in Heat-Treated Products

Property	Coarse-grained austenite	Fine-grained austenite
	Quenched and tempered products	
hardenability	increasing	decreasing
toughness	decreasing	increasing
distortion	more	less
quench cracking	more	less
internal stress	higher	lower
	Annealed or normalized products	
machinability		
rough finish	better	inferior
fine finish	inferior	better

Constituent Properties. Pearlites are less ductile than the lower temperature bainites and, for a given hardness, far less ductile than tempered martensite. Although as a class pearlite tends to be soft and not very ductile, its hardness and toughness both increase markedly with decreasing transformation temperatures.

In a given steel, bainite microstructures are generally found to be both harder and tougher than pearlite, although less hard than martensite. Bainite properties generally improve as the transformation temperature decreases.

Martensite is the hardest and most brittle microstructure obtainable in a given steel. The hardness of martensite increases with increasing carbon content up to the eutectoid composition. The hardness of martensite at a given carbon content varies only very slightly with the cooling rate.

Martensite is tempered by heating to a temperature ranging from 170–700°C for 30 min to several hours. This treatment causes the martensite to transform to ferrite interspersed with small particles of cementite. Tempered martensitic structures are, as a class, characterized by very desirable toughness at almost any strength.

Transformation Rates. The main factors affecting transformation rates of austenite are composition, grain size, and homogeneity. In general, increasing carbon and alloy content as well as increasing grain size tend to lower transformation rates. These effects are reflected in the isothermal transformation curve for a given steel. In practice, it is generally desirable to use as low a carbon content as possible for achieving the desired mechanical properties. Toughness, internal stress, distortion, and weldability are thus improved.

Hardenability

Hardenability refers to the depth of hardening or to the size of a piece that can be hardened under given cooling conditions, and not to the maximum hardness that can be obtained in a given steel. Although the critical cooling rate can be used to express hardenability, cooling rates ordinarily are not constant but vary during the cooling cycle. Hardenability is most conveniently measured by a test in which a steel sample is subjected to a continuous range of cooling rates.

Heat-Treating Processes

In almost all heat-treating processes, steel is heated above the A_3 point and then cooled at a rate that results in the microstructure that gives the desired properties. Process annealing and stress-relieving are exceptions.

Austenitization. The steel is first heated above the temperature at which austenite becomes stable. The actual austenitizing temperature should be high enough to dissolve the carbides completely and take advantage of the hardening effects of any alloying elements present. The temperature should not be high enough to produce pronounced grain growth.

Quenching. The primary purpose of quenching is to cool rapidly enough to suppress at least some, and perhaps all, transformation at temperatures above the M_s temperature, the temperature at which transformation to martensite starts on cooling.

Tempering. Quenching forms hard, brittle martensite having high residual stresses. Tempering relieves these stresses and precipitates excess carbon as carbides; it improves ductility, although at some expense of strength and hardness. The operation consists of heating at temperatures below the lower critical temperature, A_1.

A modified quenching procedure known as *martempering* minimizes the high stresses created by the transformation to martensite during the rapid cooling characteristic of ordinary quenching.

Lower bainite is generally as strong as and somewhat more ductile than tempered martensite. *Austempering,* which is an isothermal heat treatment that results in lower bainite, offers an alternative heat treatment for obtaining optimum strength and ductility.

Normalizing. In this operation, steel is heated above its upper critical temperature (A_3) and cooled in air. The purpose of this treatment is to refine the hot-rolled structure (often quite inhomogeneous), depending on the finishing temperature, and to obtain a carbide size and distribution that is favorable for carbide solution on subsequent heat treatment.

Annealing. Annealing has two different purposes: to relieve stresses induced by hot- or cold-working, and to soften the steel to improve its machinability or formability. The most favorable microstructure for machinability in the low or medium carbon steels is coarse pearlite. The customary heat treatment to develop this microstructure is a full annealing. It consists of austenitizing at a relatively high temperature to obtain full carbide solution, followed by slow cooling to give transformation exclusively in the high temperature end of the pearlite range. This simple heat treatment is reliable for most steels.

Isothermal Annealing. Annealing to coarse pearlite can be carried out isothermally by cooling to the proper temperature for transformation to coarse pearlite and holding until transformation is complete. This method is called isothermal annealing.

Spheroidization Annealing. Coarse pearlite microstructures are too hard for optimum machinability in the higher carbon steels. Such steels are customarily annealed to develop spheroidized microstructures.

Process Annealing. Process annealing is the term used for subcritical annealing of cold-worked materials. It involves heating at a temperature high enough to cause recrystallization of the cold-worked material and to soften the steel.

Carburizing. In carburizing, low carbon steel acquires a high carbon surface layer by heating in contact with carbonaceous materials. On quenching after carburizing, the high carbon skin forms martensite, whereas the low carbon core remains comparatively soft. The result is a highly wear-resistant exterior over a very tough interior.

Nitriding. The nitrogen case-hardening process, termed nitriding, consists of subjecting machined and, preferably, heat-treated Cr–Mo steel parts at about 500°C to the action of a nitrogenous medium, commonly ammonia gas, under conditions whereby surface hardness is imparted without requiring any further treatment.

Carbon Steels

Plain carbon steels, by far the largest volume of steel produced, have the most diverse applications of any metallic engineering materials. Carbon steels are made by all modern steelmaking processes and, depending on their carbon content and intended purpose, may be rimmed, semiskilled, or fully killed.

Properties. The properties of plain carbon steels are governed principally by carbon content and microstructure. These properties can be controlled by heat treatment.

Microstructure and Grain Size. The carbon steels having relatively low hardenability do not contain martensite or bainite in the cast, rolled, or forged state. The constituents of the hypoeutectoid steels are therefore ferrite and pearlite, and of the hypereutectoid steels, cementite and pearlite.

Microstructure of Cast Steels. Cast steel is generally coarse-grained because austenite forms at high temperature and the pearlite is usually coarse, in as much as cooling through the critical range is slow.

Hot Working. Many carbon steels are used in the form of as-rolled finished sections. The microstructure and properties of these sections are determined largely by composition, rolling procedures, and cooling conditions.

Cold Working. The manufacture of wire, sheet, strip, bar, and tubular products often includes cold working, with effects that may be eliminated by annealing. The most pronounced effects of cold working are increased strength and hardness and decreased ductility.

Heat Treatment. Although many wrought (rolled or forged) carbon steels are used without a final heat treatment, this may be employed to improve the microstructure and properties for specific applications.

Residual Elements. In addition to carbon, manganese, phosphorus, sulfur, and silicon which are always present, carbon steels may contain small amounts of hydrogen, oxygen, or nitrogen, introduced during the steelmaking process; nickel, copper, molybdenum, chromium, and tin, which may be present in the scrap; and aluminum, titanium, vanadium, or zirconium, which may have been introduced during deoxidation.

Alloy Steels

For slightly less than 10% of products, alloying elements are introduced to produce properties not available for carbon steels where the functional elements are usually considered to be carbon, silicon (to 0.6%), and manganese (to 1.65%). The principal classes of alloy steels in decreasing order of volume are high strength low alloy (HSLA) steels (usually hot- or cold-rolled), American Iron and Steel Institute (AISI) alloy steels (usually quenched and tempered), stainless steels (cast or wrought), electrical steels (largely iron–silicon), alloy tool steels, and nonstainless heat-resistant steels. One or more alloying elements may be critical to the resulting properties (see HIGH TEMPERATURE ALLOYS; TOOL MATERIALS).

Functions of Alloying Elements. Alloy steels may contain up to ca 50% of alloying elements which directly enhance properties, although 30% is a more common upper limit.

High Strength Low Alloy Steels. Several types of processing that contribute to desirable properties are as follows. (*1*) Controlled rolling of steels having small amounts of carbon and nitrogen containing V, Ti, Nb, and/or Zr in the austenite produces fully recrystallized fine austenite grains. (*2*) Accelerated cooling from the hot mill after controlled rolling in the range produces equiaxed ferrite while avoiding acicular ferrite. (*3*) Sufficiently rapid cooling of low (< 0.08%) carbon in steels having enough hardenability produces low carbon bainite. (*4*) Simple normalizing for some steels, eg, those containing vanadium, can give sufficient ferrite grain refinement. (*5*) Intercritical annealing, ie, in the $\alpha + \gamma$ range, of low carbon steels containing 1.5% manganese gives martensite islands in a ferrite matrix after rapid cooling. There are several hundred types of steels that use these principles separately or in combination.

Interstitial-Free Steels. In some ways, interstitial-free (IF) steels are primarily carbon steels having deliberately low yield strength.

AISI Alloy Steels. AISI defines alloy steels as follows: "steel is considered to be alloy steel when the maximum of the range given for the content of alloying elements exceeds one or more of the following limits: manganese, 1.65%; silicon, 0.60%; copper, 0.60%; or in which a definite range or a definite minimum quantity of any of the following elements is specified or required within the limits of the recognized field of constructional alloy steels: aluminum, boron, chromium up to 3.99%, cobalt, columbium (niobium), molybdenum, nickel, titanium, tungsten, vanadium, zirconium, or any other alloying element added to obtain a desired alloying effect" (*Steel Products Manual: Strip Steel*, 1978). Steels that fall within the AISI definition have been standardized and classified jointly by AISI and the Society of Automotive Engineers (SAE).

Quenched and Tempered Low Carbon Constructional Alloy Steels. A class of quenched and tempered low carbon constructional alloy steels has been very extensively used in a wide variety of applications such as pressure vessels, mining and earth-moving equipment, and in large steel structures (see TANKS AND PRESSURE VESSELS). As a general class, these steels are referred to as low carbon martensites to differentiate them from constructional alloy steels of higher carbon content.

Alloy Tool Steels. Alloy tool steels are classified roughly into three groups. The first consists of alloy tool steels to which alloying elements have been added to impart hardenability higher than that of plain carbon tool steels. The second group is that of intermediate alloy tool steels, which usually contain elements such as tungsten, molybdenum, or vanadium. The last are high speed tool steels that contain large amounts of carbide-forming elements.

Stainless Steels. Stainless steels are more resistant to rusting and staining than plain carbon and low alloy steels. This superior corrosion resistance results from the presence of chromium.

High Temperature Service, Heat-Resisting Steels. The term *high temperature service* covers many types of operations in many industries. Numerous steels are available. Where unusual conditions occur, modification of the chemical composition may adapt an existing steel grade to service conditions (see HIGH TEMPERATURE ALLOYS).

Miscellaneous High Strength Steels. Strengths above 1400–1700 MPa (203,000–246,500 psi) are not often used or required in steels because of difficulties in joining, ductility, and/or toughness. For specialized uses, however, it is possible to achieve such strength (see HIGH PRESSURE TECHNOLOGY). An important item is the class known as maraging steels. This group of high nickel martensitic steels contain so little carbon that they are often referred to as carbon-free iron–nickel martensites.

Silicon Steel Electrical Sheets. The silicon steels are characterized by relatively high permeability, high electrical resistance, and low hysteresis loss when used in magnetic circuits (see MAGNETIC MATERIALS).

Health and Safety Factors

The hazards and environmental problems associated with steelmaking have been sharply reduced by industrywide efforts. Since the early 1980s, increasingly stringent environmental regulations have been imposed by local, state, and federal government in the United States.

Health Hazards. There are many sources of industrial hygiene (qv) and occupational health hazards. The use of protective clothing, masks, eye shields, ear protection, etc, is routine and rigorously enforced. Hazards are contained by means of sensors monitoring potential dangers along with good work practices and engineering controls such as radiation shields, ventilation, air conditioning of control stations, mufflers, and soundproof enclosures.

HARRY PAXTON
Carnegie Mellon University

U.S. Steel Corp., *Making, Shaping and Treating of Steel*, 10th ed., Association of Iron and Steel Engineers, Pittsburgh, Pa., 1985.

B. Deo and R. Boom, *Fundamentals of Steelmaking Metallurgy*, Prentice Hall, N.J., 1993.

W. C. Leslie, *The Physical Metallurgy of Steels*, McGraw-Hill Book Co., Inc., New York, 1981.

Metals Handbook, Vol. 1, 10th ed., ASM International Materials Park, Ohio, 1990.

STERILIZATION TECHNIQUES

Sterilization is defined as rendering a substance incapable of reproduction. Whereas this is often taken to mean total absence of living organisms, a more accurate representation is that the substance is free from living microorganisms with a probability previously agreed to be acceptable.

Generally, sterilization technology is more readily associated with the health-care profession and industry or with the electronics industry.

A distinction must be made between sterilization and certain other processes often called sterilization as a result of popular misconception. Methods and procedures less rigorous than sterilization, such as disinfection, sanitization, and the use of antiseptics and bacteriostats (see DISINFECTANTS AND ANTISEPTICS; FOOD PROCESSING; INDUSTRIAL ANTIMICROBIAL AGENTS), are often applied to render the object safe for certain applications. In most instances, a judgment on the suitability of a sterilization or a substitute process can only be made by a microbiologist. In order to design acceptable test procedures, the kinetics and thermodynamics of the sterilization process must be understood.

Kinetics

The rate of destruction of microorganisms is logarithmic, ie, first order with respect to the concentration of microorganisms. The process can be described by the following expression:

$$\frac{N_o}{N_t} = e^{-kt} \tag{1}$$

in which N_t = the number of organisms alive at time t, N_0 = the initial number of organisms, and k = the kinetic rate constant. It can be seen that N_t approaches zero as t approaches infinity. Absolute sterility, accordingly, is impossible to attain.

It has been found convenient to express the rate of microbial kill in terms of a decimal reduction rate or *D*-value. The *D*-value represents the time of exposure (at given conditions for a given microorganism) required for a 10-fold decrease in the viable population. The practical significance of the *D*-value is that it simplifies the design of sterilization cycles. A sufficiently large D for any process results in a negative log N_t which, in a practical sense, represents the probability of survival of the last remaining microorganism. In the health sciences, a 10^{-6} residual concentration of microorganisms is generally regarded as an acceptable criterium for sterility.

Thermodynamics

The Eyring equation for the theory of absolute reaction rates can be applied to the sterilization process:

$$k = \frac{k_B T}{h} e^{(T\Delta S - \Delta H/RT)} \tag{2}$$

where k_B = Boltzman's constant, h = Plank's constant, and ΔS and ΔH are the standard entropy and enthalpy changes, respectively.

The relationship between the *D*-value and k can be derived by considering the meaning of D:

$$D = \frac{t_2 - t_1}{\log_{10} N_1 - \log_{10} N_2} \tag{3}$$

where $N_1 = 10\, N_2$.

Substituting into equation 1:

$$D = 2.3/k \tag{4}$$

Testing and Monitoring

Direct testing for sterility by culturing is a destructive test method, ie, the product is rendered useless for food or medical purposes. Indirect testing methods usually rely on a statistically valid sampling (qv) pattern for a product. Product monitoring for sterilization using a test sample is a limited value; it is utilized only if no other information is available for the particular process cycle for a given lot of products. Because sterilization is a highly reproducible and well-understood process, it has been found that process monitoring is far more suitable for purposes of sterility assurance. Process monitoring can be accomplished by measurement of individual parameters.

Table 1. Performance Characteristics of Some Biological Indicators

Culture spores	Sterilization process	Approximate *D*-value
Bacillus subtilis	ethylene oxide (50% rh, 54°C)	
	600 mg/L	3 min
	1200 mg/L	1.7 min
B. stearothermophilus	saturated steam (121°C)	1.5 min
B. pumilus	γ-radiation	
	wet preparations	2×10^{-6} Gy[a]
	dry preparations	1.5×10^{-6} Gy[a]
B. subtilis	dry heat (121–170°C)	60–1 min
Clostridium sporogenes	saturated steam (112°C)	3.5–0.7 min

[a] To convert Gy to rad, multiply by 100.

Biological Monitoring. Biological indicators are preparations of specific microorganisms particularly resistant to the sterilization process they are intended to monitor. When designing industrial sterilization cycles, the bioburden or bioload is determined first. The bioload is the average number of organisms present on or in an article that is to be sterilized.

The carriers containing the biological indicators are retrieved following exposure, transferred aseptically into sterile culture media, and incubated for the required length of time. Some unexposed indicators are also incubated to prove that the spores were viable. If no growth is observed while the viability control displays the required growth, the conclusion is made that the sterilization cycle was successful. Typical performance characteristics for some of the most widely used biological indicators are given in Table 1.

Monitoring by Electromechanical Instrumentation. All sterilizers are equipped with gauges, sensors (qv), and timers for the measurement of the various critical process parameters; many are equipped with computerized control. It is not possible to install the sensors inside the packages which are to be sterilized. Electromechanical instrumentation is capable of providing information only on the conditions to which the packages are exposed; it cannot detect failures as the result of inadequate sterilization conditions inside the packages.

Chemical Monitoring. Chemical indicators are devices employing chemical reactions or physical processes designed in such a way as to permit observation of changes in a physical condition and to monitor one or more process parameters. Chemical indicators can be located inside the packages, and the results are observable immediately when the package is opened for use. Chemical indicators have historically been used only to differentiate between sterilized and nonsterilized packages. Chemical dosimeters of sufficient accuracy have been developed that permit their application either as total monitors or as critical detectors of specific parameters.

Dry-Heat Sterilization

Dry-heat sterilization is generally conducted at 160–170°C for ≥2 h. Specific exposures are dictated by the bioburden concentration and the temperature tolerance of the products under sterilization. Appropriate conditions must be established throughout the material to be sterilized. Chemical indicators for dry-heat sterilization are available either in the form of pellets enclosed in glass ampuls, or in the form of paper strips containing a heat-sensitive ink (see CHROMOGENIC MATERIALS).

Steam Sterilization

Steam (qv) sterilization is a process that is invariably carried out under high pressure in autoclaves using saturated steam. The elimination of air from the chamber and complete steam penetration of the load is of critical importance. This may be accomplished by gravity displacement or prevacuum techniques.

The gravity-displacement-type autoclave relies on the relative nonmiscibility of steam and air to allow the steam that enters to rise to the top of the chamber and fill it. The air is pushed out through the steam-discharge line located at the bottom of the chamber. The prevacuum technique, eliminates air by creating a vacuum. This procedure facilitates and permits more rapid steam penetration (see VACUUM TECHNOLOGY).

The critical parameters of steam sterilization are temperature, time, air elimination, steam quality, and the absence of superheating. Temperature and time are interrelated. The success of steam sterilization is dependent on direct steam contact, which can be prevented by the presence of air in the chamber. Air elimination is regarded as an absolute parameter. If the required amount of air has not been eliminated from the chamber and the load, no combination of time and temperature results in complete sterilization. The term *steam quality* refers to the amount of dry steam present relative to liquid water in the form of droplets. Excessive amounts of liquid water can result in air entrapment, drying problems following exposure, and unacceptable steam levels. Superheated steam results when steam is heated to a temperature higher than that which would produce saturated steam.

The selection of an appropriate steam-sterilization cycle must be made after a careful study of the nature of the articles to be sterilized, the type and number of organisms present, type and size of each package, and type of packaging material used. Biological indicators for steam sterilization utilize *Bacillus stearothermophilus*. Electromechanical monitors for steam sterilization include pressure, temperature, time-recording charts, and pressure–vacuum gauges.

There are various chemical indicators for steam sterilization. Some indicators integrate the time–temperature of exposure. Some are capable of monitoring the safety factor in the exposure; others only minimal conditions. All are capable of indicating incomplete steam penetration. Some indicators can determine whether a specific temperature has been achieved. Indicators can also be utilized to distinguish packages that have been processed from those that have not been processed.

Gas Sterilization

When articles that cannot withstand the temperatures and moisture of steam sterilization or exposure to radiation require sterilization, gaseous sterilants that function at relatively low temperatures offer an attractive alternative. A practical requirement is that the gas selected should allow safe handling, and that any residue should volatilize relatively quickly if absorbed by components of the article sterilized. When properly applied, ethylene oxide satisfies most of these requirements, and is the most frequent choice. The critical parameters of ethylene oxide sterilization are temperature, time, gas concentration, and relative humidity. Biological monitoring of ethylene oxide sterilization is essential and is conducted using spores of *Bacillus subtilis* (see Table 1).

Other Sterilization Techniques

Ionizing Radiation. Radiation sterilization employs electron accelerators or radioisotopes (qv). Gamma-radiation sterilization usually employs ^{60}Co and occasionally ^{137}Cs as the radioisotope source.

Filtration. The filtration process depends on the physical retardation of microorganisms from a fluid by a filter membrane or similarly effective medium.

Liquid Sterilants. Formalin, a solution of methanal (formaldehyde (qv)) in water, has sterilizing properties, as do glutaraldehyde and hydrogen peroxide (qv). Liquid sterilants are known to corrode the metal parts of articles and instruments that are to be sterilized, although articles composed exclusively of glass or certain type of corrosion-resistant metal alloys can be safely processed. While liquid sterilization is an extremely useful method for articles that cannot withstand the conditions of steam sterilization, the problems associated with its use limit its application.

Other Sterilants. Ultraviolet light has sterilizing properties, but cannot penetrate many materials; however, some substances are destroyed by exposure to a sterilizing dose. The manufacture and sale of chemical sterilants and disinfectants is regulated in the United States.

Sterilization Packaging

Sterilization generally takes place at one location prior to use of an article at another location. The main purpose of packaging (qv) is to protect the sterility of the contents. When an article is placed in its protective container and subsequently sterilized, the process is called terminal sterilization. When it is sterilized first and then placed in a presterilized container, the process is called sterile filling. For any packaging method, provision must be made for opening of the package and retrieval of the sterilized article in a manner that does not compromise its sterility.

Related Techniques

Procedures less thorough than sterilization may be used for the preparation of foods and medical supplies. Some of these processes are capable of rendering an object microbiologically safe for a given purpose when employed using proper safeguards.

Pasteurization, the heating of certain fluids, frequently milk or dairy products (see MILK AND MILK PRODUCTS), destroys potentially harmful organisms such as mycobacteria, *M. tuberculosis, M. bovis*, or *M. avium*. However, pasteurization, carried out at 62°C for 30 min or at 72°C for 15 s, is not a sterilization procedure. Disinfection destroys pathogenic organisms. This procedure can render an object safe for use. Disinfectants include solutions of hypochlorites, tinctures of iodine or iodophores, phenolic derivatives, quaternary ammonium salts, ethyl alcohol, formaldehyde, glutaraldehyde, and hydrogen peroxide (see DISINFECTANTS AND ANTISEPTICS). Bacteriostasis is the process of preventing the growth and reproduction of microorganisms. When the bacteriostat is removed or its power is exhausted, however, the organisms can resume growth.

Sanitization is a cleaning procedure that reduces microbial contaminants on certain surfaces to safe or relatively safe levels, as defined by the EPA or public health authorities. Decontamination is a procedure to render safe for handling, disposal, or the subsequent processing of an article that may contain a large amount of potentially infectious organisms. Decontamination and sterilization are similar procedures, except that in the former case the bioburden is higher. Decontamination is not expected to result reliably in the 10^{-6} probability of microbial survival, as in sterilization, because of the higher bioburden. Germicides are agents capable of killing some specific forms of microorganisms.

Sterilization in the Food-Processing Industry

The problem of microbial contamination of foods is twofold: foods may act as nutrients for, and carriers of, pathogenic organisms; additionally, foods may be spoiled by the action of certain organisms (see FOOD PROCESSING). The most widely used sterilization method in the food industry is moist heat. The cooking and sterilization processes can frequently be combined into one.

Acidic foods such as fruits tend to retard microbial growth and resist certain types of contamination. For this reason, the standards adopted industry-wide have been based on the processing of foods of high acidity (low pH). In the U.S., the FDA has regulatory responsibility over the preparation, sterilization, and distribution of foods.

There are four types of food sterilization processes: terminal sterilization in prefilled containers in a batchwise process; terminal sterilization in prefilled containers heated to the required temperatures in a continuous process; aseptic filling following batchwise cooking in an appropriate retort; and aseptic filling in a continuous cooking system.

THOMAS A. AUGURT
Propper Manufacturing Co., Inc.
J. A. A. M. VAN ASTEN
National Institute for Public Health and the Environment

S. S. Block, *Disinfection, Sterilization, and Preservation*, 2nd ed., Lea and Febiger, Philadelphia, Pa., 1977.

G. Sykes, *Disinfection and Sterilization*, 2nd ed., Spon, London, 1965.

S. Turco and R. E. King, *Sterile Dosage Forms, Their Preparation and Clinical Application*, Lea and Febinger, Philadelphia, Pa., 1974.

N. A. M. Eskin, H. M. Henderson, and R. S. Townsend, *Biochemistry of Foods*, Academic Press, Inc., New York, 1971.

STEROIDS

Steroids are members of a large class of lipid compounds called terpenes that are biogenically derived from the same parent compound, isoprene, C_5H_8. Steroids contain or are derived from the perhydro-1,2-cyclopentenophenanthrene ring system (**1**) and are found in a variety of different marine, terrestrial, and synthetic sources. The vast diversity of the natural and synthetic members of this class depends on variations in side-chain substitution (primarily at C17), degree of unsaturation, degree and nature of oxidation, and the stereochemical relationships at the ring junctions.

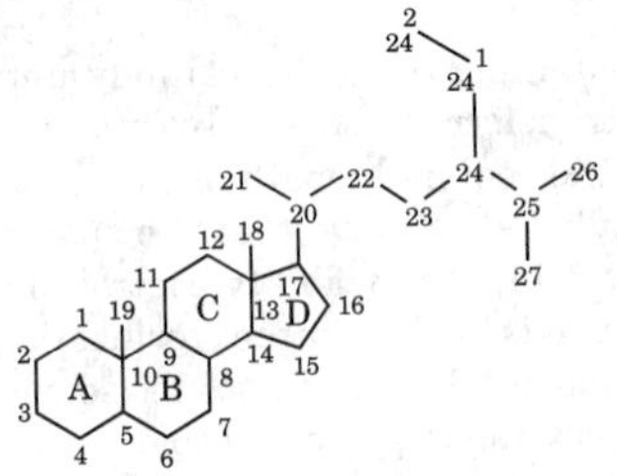

(**1**)

There are many classes of natural and synthetic steroids best known for their wide array of biological activity. The naturally occurring steroids can be subdivided into several categories that include (*1*) nonhormonal, mammalian steroids; (*2*) vitamin D; (*3*) hormonal steroids; and (*4*) other naturally occurring steroids (see HORMONES; VITAMINS, VITAMIN D).

Structure and Nomenclature

The position-numbering and ring-lettering conventions for steroids are shown in (**1**). Positions 18 and 19 are often angular methyl groups; in addition, position 19 can be a hydrogen and is not substituted when the A-ring is aromatic. Position 17 can be substituted, unsubstituted, and/or oxygenated. Compounds are systematically named as derivatives of the parent hydrocarbons shown in Figure 1. Substituents that extend below the plane of the steroid are referred to as α and are designated by a broken line; those attached to the plane of the steroid from above are called β and are shown by a bold or solid line. Substituents of unknown configuration are indicated by a wavy line. Generally, the ring junctions have an all-trans relationship with the hydrogen attached to C9 on the α-face, unless otherwise indicated. Changes in steroid nomenclature that have been introduced since 1972 include a wider use of the (*R*),(*S*)-system for designating the stereochemistry in the side chain. Although the systematic nomenclature for steroids has been firmly established, the most common and most important steroids are often designated by trivial names.

Classification of Biologically Active, Natural Steroids

Nonhormonal Mammalian Steroids. *Sterols and Cholesterol.* Natural sterols are crystalline C_{26}–C_{30} steroid alcohols containing an aliphatic side chain at C17. Sterols were first isolated as nonsaponifiable fractions of lipids from various plant and animal sources and have been identified in almost all types of living organisms. By far, the most common sterol in vertebrates is cholesterol (**8**). Cholesterol serves two principal functions in mammals. First, cholesterol plays a role in the structure and function of biological membranes. Secondly, cholesterol

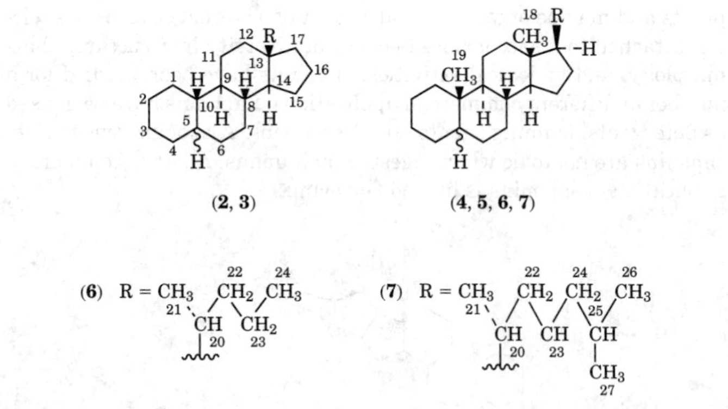

Figure 1. Nomenclature of the parent hydrocarbon ring skeletons. Gonane (**2**) R = H; estrane (**3**) R = CH_3; androstane (**4**) R = H; pregnane (**5**) R = C_2H_5; cholane (**6**) R as shown; and cholestane (**7**) R as shown.

serves as a central intermediate in the biosynthesis of many biologically active steroids, including bile acids, corticosteroids, and sex hormones.

Bile Acids and Alcohols. Bile acids have been detected in all vertebrates that have been examined and are a result of cholesterol metabolism. The C_{24} acid, 5β-cholanic acid (**9**) is the structural derivative of the majority of bile acids in vertebrates. Most mammalian bile acids have a cis-fused A–B ring junction resulting in a nonplanar steroid nucleus. Bile acids, like sterols, typically contain a C3α-hydroxyl group (lithocholic acid: 3α-hyroxycholanic acid.

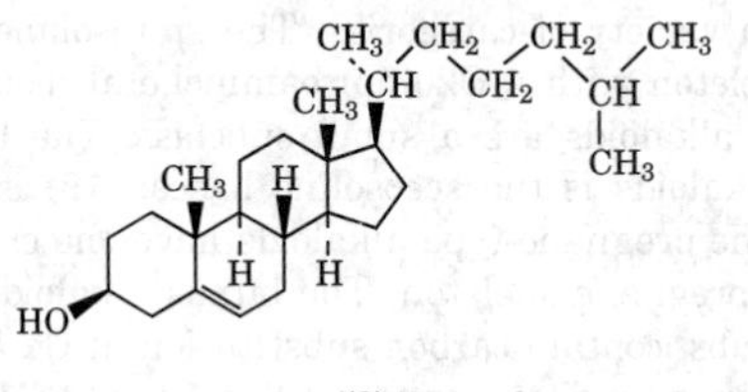

(**8**)

Along with the C3α-hydroxyl group, bile acids may contain a hydroxyl at C7α, at C12α, and at other positions. Bile salts, cholesterol, phospholipids, and other minor components are secreted by the liver.

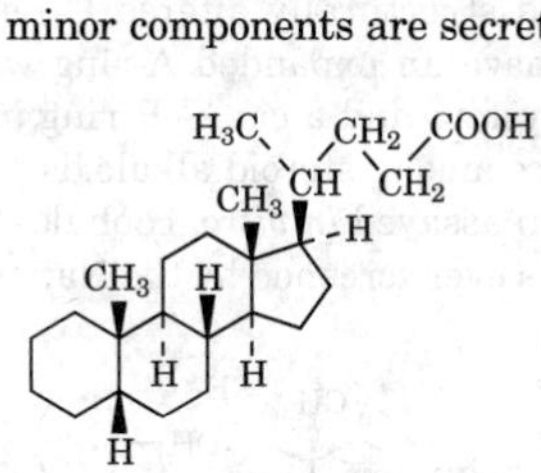

(**9**)

Vitamin D. The term vitamin D refers to a group of seco-steroids that possess a common conjugated triene system of double bonds. Vitamin D_3 (**10a**) and vitamin D_2 (**10b**) are the best-known examples (Fig. 2). Vitamin D_3 (**10a**) is found primarily in vertebrates, whereas vitamin D_2 (**10b**) is found primarily in plants. The term *vitamin* is a misnomer. Vitamin D_3 is a prohormone that is converted into physiologically active form, primarily 1,25-dihydroxyvitamin D_3 (**11**), by successive hydroxylations in the liver and kidney. This active form is part of a hormonal system that regulates calcium and phosphate metabolism in the target tissues.

Steroid Hormones. Generally, steroid hormones are metabolically short-lived steroids produced in small amounts by various endocrine glands. They serve as chemical messengers that regulate a variety of physiological and metabolic activities in vertebrates. Steroid hormones bind to soluble, intracellular receptor molecules. In the nucleus of the cell, dimeric steroid receptors bind to DNA and together with a heteromeric complex of proteins, regulate gene transcription. Molecules that interfere with steroid hormone gene regulation are called antagonists or antihormones. Steroid hormones can be subdivided into sex hormones (androgens, estrogens, and progestins) and corticosteroids (glucocorticoids and mineralocorticoids).

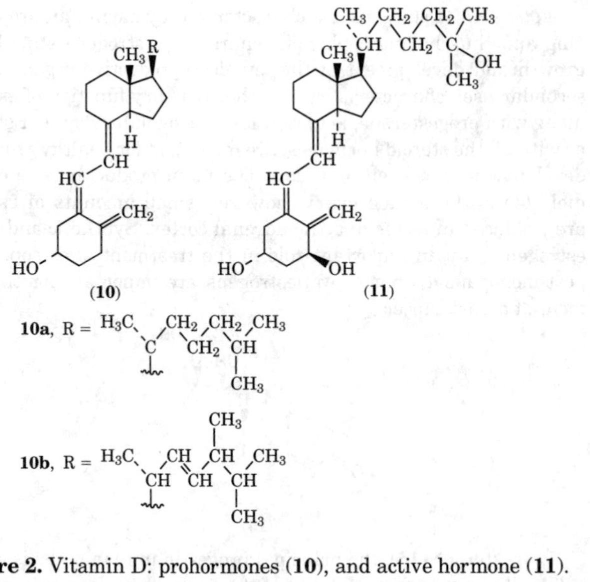

Figure 2. Vitamin D: prohormones (**10**), and active hormone (**11**).

Sex Hormones. Androgens, estrogens, and progestins are steroids that are secreted primarily by the genital glands. From a chemical point of view, the division of the sex hormones into these three groups is convenient; however, they may possess common physiological properties. Therefore, the sex hormones are organ-specific rather than sex-specific.

Androgens are C_{19} steroids that contain the basic perhydro-1,2-cyclopentenophenanthrene ring system with the C18 and C19 angular methyl group. A primary function of androgens is to maintain the male sex organs and secondary sex characteristics. Examples of androgens are testosterone (**12**) and dehydroepiandrosterone (DHEA) (**13**). DHEA is one of the most abundant steroids in human males; however, it is not a potent androgen.

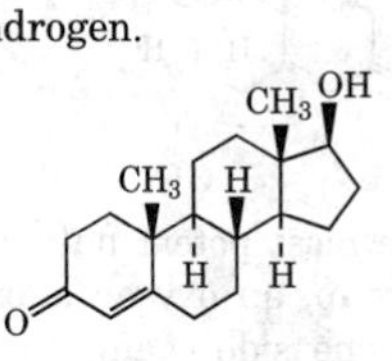

(**12**)

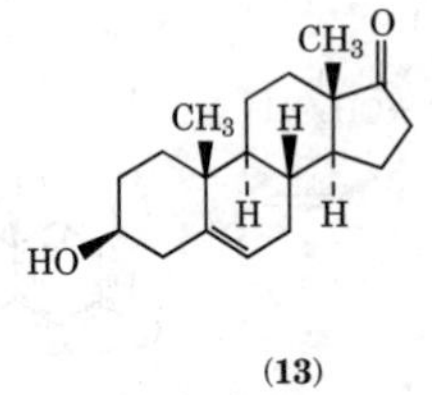

(**13**)

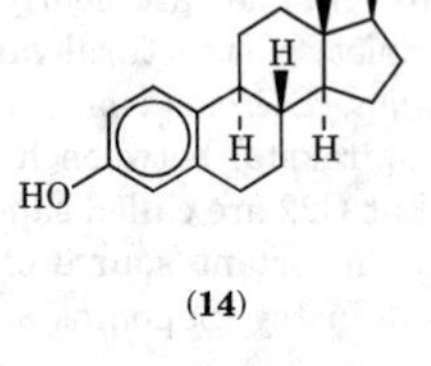

(**14**)

Estrogens. Estrogens are characterized by having an aromatic A-ring and thus having a phenolic character. Estrogens stimulate the growth and development of the female reproductive organs and the secondary sex characteristics. Another primary function of estrogens along with progesterone, is to regulate the ovulatory cycle. Estrogens, as with all the steroid hormones, are important for healthy growth and development in women and men. The main production site of estradiol (**14**) is the female ovary; however, small amounts of estrogens are produced in testes and the adrenal cortex. Synthetic and natural estrogens play an important role in the treatment of osteoporosis in post-menopausal women. Antiestrogens are important for the treatment of breast cancer.

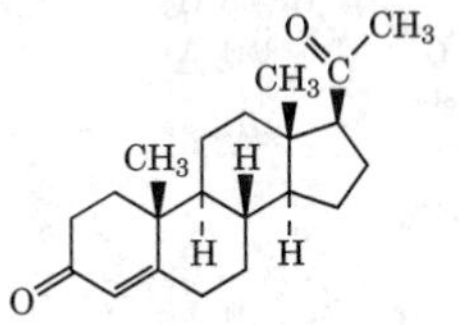

(**15**)

Progesterone (**15**), the principal progestin in mammals, is secreted primarily by the corpus luteum of the ovary. A main responsibility of progesterone, together with estrogen, is to prepare the endometrium for pregnancy.

Corticosteroids. Although the adrenal cortex secretes small amounts of androgens and estrogens, the major secretory steroids from this gland are called corticosteroids. Corticosteroids have several biological activities, including the regulation of electrolyte balance by mineralocorticoids and carbohydrate and protein metabolism by glucocorticoids.

Natural, potent glucocorticoids possess a Δ^4-3-one group, an oxygen substituent at C11β (necessary for agonism), and a C17β-2-hydroxyethan-1-one sidechain. Atypical example is cortisol (**16**). The principal effects of glucocorticoids are to mobilize fat and protein from tissues, utilize these nutrients to supply energy for the body, and decrease the rate of carbohydrate utilization. Thus, they are diabetogenic and act as functional insulin antagonists. Glucocorticoids are also potent inhibitors of inflammation, and they are used as therapeutics.

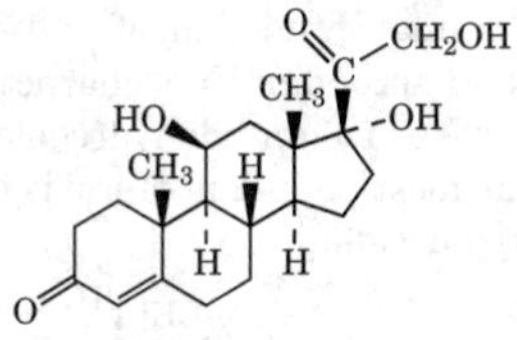

(**16**)

Aldosterone (**17**), the most potent natural mineralocorticoid, also possesses a Δ^4-3-one group, an oxygen substituent at C11β, and a C17β-2-hydroxyethan-1-one side chain. In addition, the C18 of aldosterone is oxidized to an aldehyde. Mineralocorticoids act to retain sodium and to prevent the retention of excess potassium.

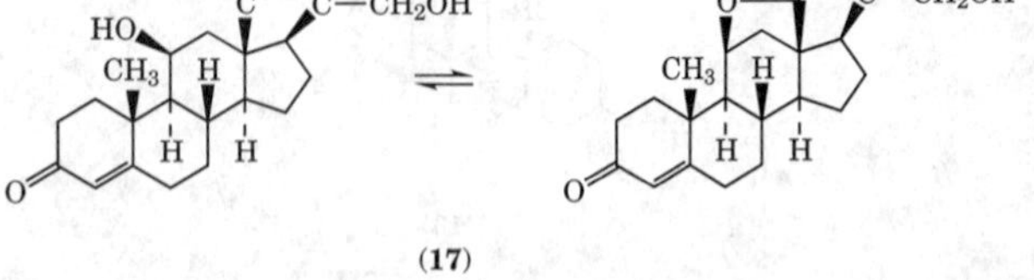

(**17**)

Other Natural Steroids. Steroids are nearly ubiquitous to all living organisms and have a variety of structural variations.

Sapogenins and Saponins. Steroids isolated from a variety of plant sources that contain a spiroketal between hydroxyl moieties at C16 and C26 and a carbonyl at C22 are called sapogenins (**18**). Sapogenin aglycones have been an important source of starting materials for the commercial steroid industry. Saponins are widely distributed in plants and marine organisms and consist of a steroid or terpene skeleton attached to a saccharide. Because of diversity in structure, pharmacology, and biological activities, saponins have been studied for a number of different commercial applications. Saponins have been used as detergents, foaming agents, and fish toxins. Although toxic to fish, saponins are nontoxic when ingested by humans. Another commercial application of saponins is in food flavoring.

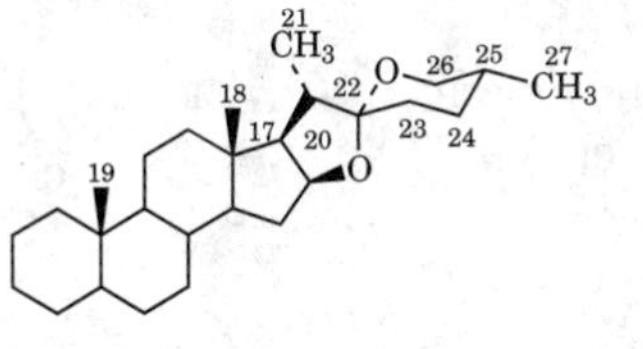

(**18**)

Plant Sterols. Sterols have been identified in almost all types of living organisms and can be isolated, in varying quantities, from many different plants. Similar to cholesterol, plant sterols have a structural and functional role in biological systems and serve as intermediates in the biosynthesis of an assortment of biologically active steroids.

Steroid Alkaloids. Steroid alkaloids are compounds isolated from plants and some higher animals that possess the basic steroidal skeleton with nitrogen(s) incorporated as an integral part of the molecule. The nitrogen can be located within the perhydro-1,2-cyclopentenophenanthrene ring system or in a side chain.

Steroid alkaloids have been isolated from four families of terrestrial plant sources (*Solanaceae, Liliaceae, Apocynaceae,* and *Buxaceae*), two animal sources (*Salamandra* and *Phyllobates*), and several marine sources. Steroid alkaloids can be classified based on structure and fall into a variety of categories. The spirosolanes contain a C_{27} cholestane skeleton with a C20 spiroaminoketal moiety. Solanidine-type steroidal alkaloids are a small subclass. The largest subclass of steroidal alkaloids is the secosoline bases; (**19**) is a general secosolanidine. The pregnane-type alkaloids have one or more nitrogens attached to a pregnane skeleton. The buxus alkaloids, isolated from evergreen shrubs, contain carbon substitution at C4 and C14 and either a cyclopropane moiety between C9, C10, and C19 or the B-ring expanded diene. Buxus alkaloids have been used as folk remedies for a variety of disorders, including venereal disease, tuberculosis, cancer, and malaria. The samanine, jerveratrum, and ceveratrum-type compounds all have a structurally altered C_{27} steroid skeleton. The samanine alkaloids have an expanded A-ring with the formation of an isoxazoline ring system and a cis-A–B ring junction. Ritterazines and cephalostatins are among steroid alkaloids recently isolated from marine sources. When assayed *in vitro,* cephalostatins are among the most potent cytotoxins ever screened by the National Cancer Institute.

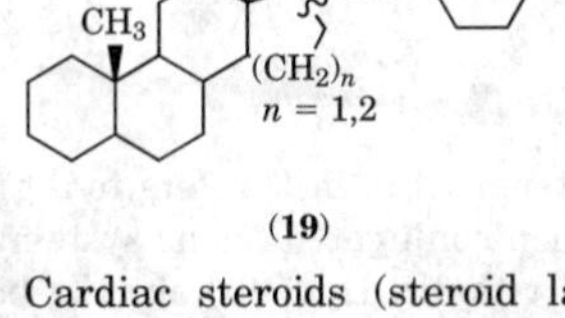

(**19**)

Cardiac Steroids. Cardiac steroids (steroid lactones) and corresponding glycosides are characterized by their ability to exert a powerful inotropic (increasing the force of cardiac contraction) effect, and are used both for their inotropic and antiarrhythmic properties. The two most prevalent cardiac aglycones are the cardenolides and bufadienolides. The cardenolides are C_{23} steroids that have a C17β-substituted five-membered lactone that is generally α,β-unsaturated, an unusual β-faced oxygen on C14, and a bile acid-like cis-A–B ring junction. Cardenolides are exemplified by digitoxigenin (**20**) which is an active ingredient in digitalis. The bufadienolides differ in that they are C_{24}-steroids that possess a C17β-substituted six-membered lactone ring that generally has two degrees of unsaturation. Other structural variations in both series are the stereochemistry at C3 and the degree of oxidation on the nucleus and side chains.

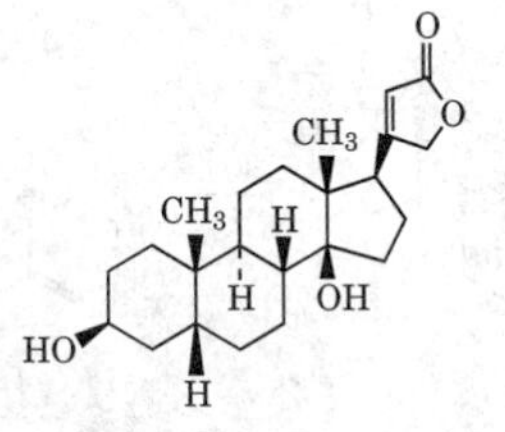

(20)

Withanolides. Withanolides are C_{28}-steroidal lactones that are isolated from the Solanaceae plant family. Withanolides are characterized by an ergostane-type skeleton, the C17-side chain of which is transformed into a six-member lactone ring. The withanolides and the related ergostanes are the only known natural steroids obtained from the same family that have representatives with both α- and β-orientations of the C17 side chain.

Ecdysteroids. Ecdysteroids can be isolated from many species of the animal kingdom that belong to the phyla Protomia, eg, insects, worms, and arthropods, as well as a variety of different plant species. Ecdysteroids include the molting hormones; however, not all the over 60 ecdysteroids that have been isolated are active hormones. Ecdysteroids from animals are referred to as zooecdysteroids and from plants are referred to as phytoecdysteroids.

Marine Sterols. Several hundred unique sterol structures have been elucidated from a variety of marine invertebrates. A single nucleus can be used to describe most terrestrial sterols, but no single template suffices for marine sterols. Similar to cholesterol, marine sterols play a critical role in both the physiology and biochemistry of biological systems.

Steroid Antibiotics. The steroid antibiotics are a structurally diverse class of steroids that have a common biological function, ie, antibacterial, antifungal, antiviral, or antitumor activities. This group of compounds can overlap with other steroid classes listed above. Fusidic acid, helvolic acid, and cephalosporin P_1 (**21**) exemplify a set of antibacterial steroids that contain a prolanostane skeleton with an unique trans–syn–trans–antitrans stereochemistry. These compounds inhibit the growth of gram-positive bacteria by inhibiting protein synthesis, but have little activity against gram-negative bacteria. An antibiotic isolated from the tissues of the dogfish shark is the steroid alkaloid squalamine, a broad-spectrum antibiotic that exhibits potent antimicrobial activity against fungi, protozoa, viruses, and both gram-negative and gram-positive bacteria.

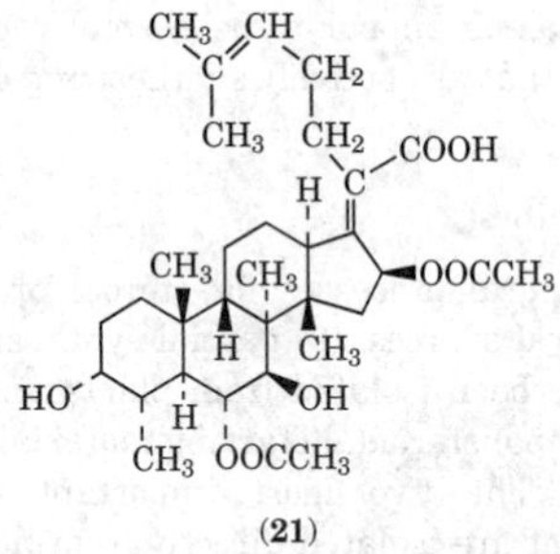

(21)

Biosynthesis

Steroids are members of a large class of lipid compounds called terpenes. Using acetate as a starting material, a variety of organisms produce terpenes by essentially the same biosynthetic scheme (Fig. 3). The self-condensation of two molecules of acetyl coenzyme A (CoA) forms acetoacetyl CoA. Condensation of acetoacetyl CoA with a third molecule of acetyl CoA, then followed by an NADPH-mediated reduction of the thioester moiety produces mevalonic acid (**22**). Phosphorylation of (**22**) followed by concomitant decarboxylation and dehydration processes produce isopentenyl pyrophosphate. Isopentenyl pyrophosphate isomerase establishes an equilibrium between isopentenyl pyrophosphate and 3,3-dimethylallyl pyrophosphate (**23**). The head-to-tail addition of these isoprene units forms geranyl pyrophosphate. The addition of another isopentenyl pyrophosphate unit results in the sesquiterpene (C_{15}) farnesyl pyrophosphate (**24**). Both of these

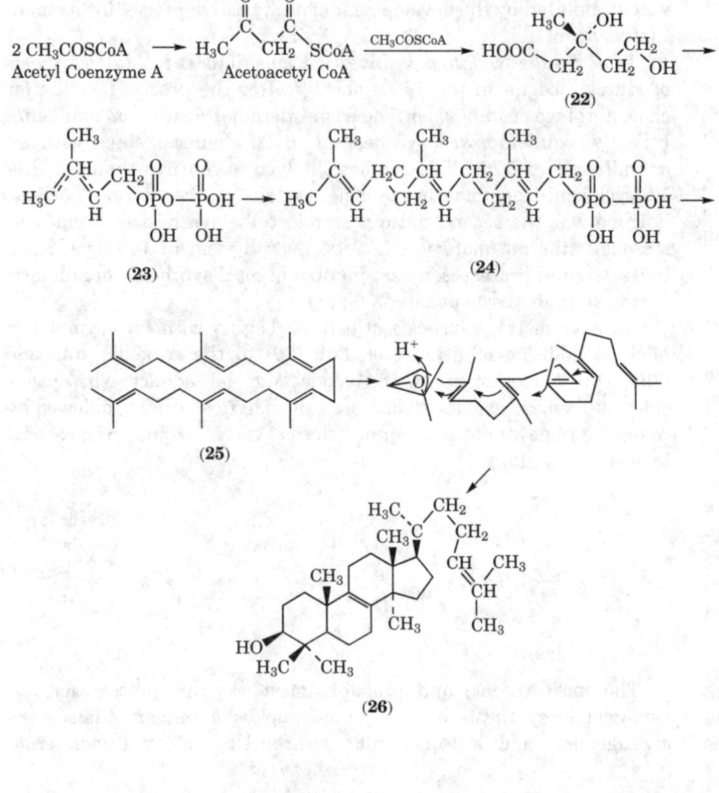

Figure 3. Abbreviated terpene biosynthesis.

head-to-tail additions are catalyzed by prenyl transferase. Squalene synthetase catalyzes the head-to-head addition of two achiral molecules of farnesyl pyrophosphate, through a chiral cyclopropane intermediate, to form the achiral triterpene, squalene (**25**).

Stereospecific 2,3-epoxidation of squalene, followed by a nonconcerted carbocationic cyclization and a series of carbocationic rearrangements, forms lanosterol (**26**) in the first steps dedicated solely toward steroid synthesis. Cholesterol is the principal starting material for steroid hormone biosynthesis in animals. The cholesterol biosynthetic pathway is composed of at least 30 enzymatic reactions. Lanosterol and squalene appear to be normal constituents, in trace amounts, in tissues that are actively synthesizing cholesterol.

Manufacture and Synthesis

There are three general processes for steroid production: (*1*) direct isolation from natural sources, (*2*) partial synthesis from steroid raw materials that have been isolated from plants and animals, and (*3*) total synthesis from nonsteroidal starting materials.

Direct Isolation. The two most important classes of steroid pharmaceuticals that are isolated directly from natural products are some estrogens and most cardiac steroids. Compounds with estrogenic activity have been isolated from different sources, including urine from pregnant women and from pregnant mares. Cardiac steroids occur in small amounts in various plants with a wide geographical distribution.

Partial Syntheses.

Raw Materials and Extraction. The variety of natural sources of steroid raw materials is vast, and the exact details of manufacturing processes are ambiguous closely held industrial secrets. However, the most widely utilized raw materials for the partial synthesis of steroids appear to be the following: (*1*) the sapogenins, for example, diosgenin (**27**), (*2*) the structurally related steroid alkaloids, (*3*) sterols, such as cholesterol (**8**), and (*4*) bile acids.

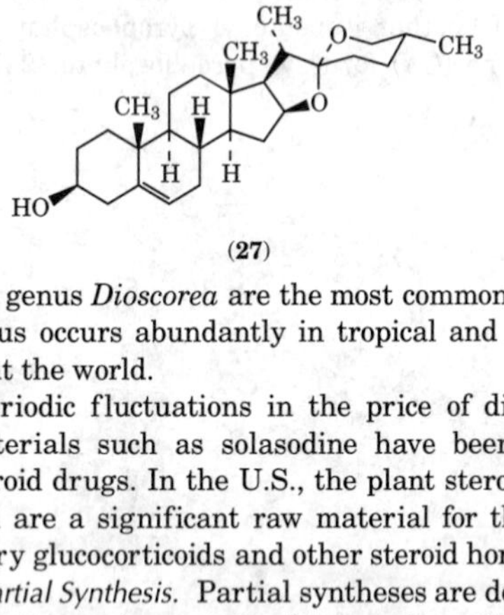

(**27**)

Plants of the genus *Dioscorea* are the most common source of diosgenin. This genus occurs abundantly in tropical and subtropical regions throughout the world.

Owing to periodic fluctuations in the price of diosgenin, alternative raw materials such as solasodine have been used for the synthesis of steroid drugs. In the U.S., the plant sterols stigmasterol and β-sitosterol are a significant raw material for the synthesis of antiinflammatory glucocorticoids and other steroid hormones.

Methods of Partial Synthesis. Partial syntheses are done typically by chemical degradation or fermentation/biotransformation.

An important commercial method for the commercial synthesis of steroids is the chemical degradation of diosgenin. The Marker degradation became the principal method for commercial steroid synthesis in the 1940s and 1950s, and modifications of this process are still in use. When diosgenin is heated to approximately 200°C in acetic anhydride, elimination and acetylation of the oxygen in the F-ring produce the bis-acetylated enol ether. Oxidative cleavage of the enol ether with chromium trioxide followed by elimination of the C16-acyl-oxygen results in steroid. Selective hydrogenation of the α,β-unsaturated ketone in the D-ring from the sterically less hindered α-face forms pregnenolone (**28**). Pregnenolone is readily converted into progesterone (**15**) under oxidative conditions.

This process was improved and expanded to provide starting materials for the C19-sex hormones that include estrogens and androgens. Another commercial method that has been used for the production of progesterone is the chemical degradation of the side chain of stigmasterol.

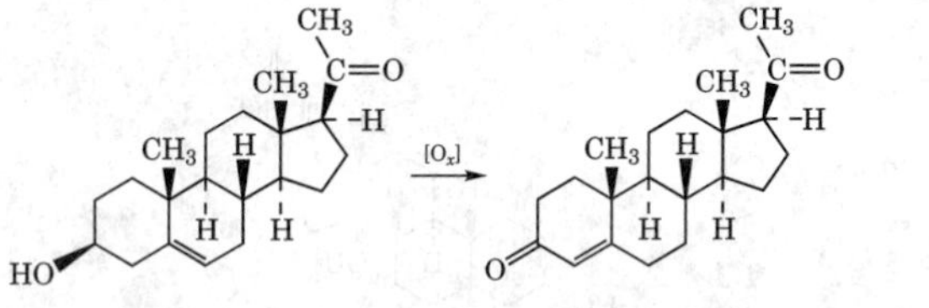

(**28**) (**15**)

Fermentation/Biotransformation. Commercial biotechnology operations have focused on microbial agents for specific transformations of individual steroid substrates. The regio- and stereoselective hydroxylation of every site on virtually every steroid nucleus is possible. Many of these hydroxylation steps are of commercial importance. For example, the 9α-, 11α-, 11β-, and 16α-hydroxylations are key steps in the industrial synthesis of synthetic corticosteroid antiinflammatory drugs. These steps are accomplished almost exclusively by microbial transformations. In addition to hydroxylations, other useful microbial oxidations of steroids include alcohol oxidations, epoxidations, oxidative cleavage of carbon–carbon bonds, introduction of double bonds, peroxidations, and heteroatom oxidations. Other invaluable microbial steroid transformations include reductions, degradations, A-ring aromatization, resolutions, isomerizations, conjugations, hydrolyses, heteroatom introduction, and sequential reactions.

There are two principal biotechnological applications dealing with steroids. Microbial agents are used for processing raw materials into useful intermediates for general steroid production and for specific transformations of steroids to advanced intermediates or finished products (Table 1).

Processing Raw Materials. Along with the aforementioned chemical methods of processing steroid raw materials, microbial transformations have been and are used in a number of commercial degradation processes. The microbial degradation of the C17 side chain of the two most common sterols, cholesterol (**8**) and β-sitosterol, is a principal commercial method for the preparation of starting materials in Japan and the U.S.

Representative Partial Syntheses. The synthesis of 19-nor-steroids was stimulated by the development of orally active progestins as birth control agents.

Total Synthesis. *Estranes.* Investigations into the total synthesis of steroids began in the 1930s shortly after the precise formula for cholesterol was established. The earliest studies focused on equilenin. Initially, equilenin was synthesized in 20 chemical steps with an overall yield of 2.7%. This synthesis helped to confirm the perhydro-1,2-cyclopentenophenanthrene ring system of the steroid nucleus. Estrone was the second natural steroid to be synthesized from nonsteroidal starting materials in 0.1% overall yield in 18 steps. Since these original processes, a vast number of total syntheses of aromatic A-ring steroids have appeared.

An asymmetric synthesis of estrone begins with an asymmetric Michael addition of lithium enolate (**29**) to the scalemic sulfoxide (**30**). Direct treatment of the crude Michael adduct with *meta*-chloroperbenzoic acid to oxidize the sulfoxide to a sulfone, followed by reductive removal of the bromine affords (**31**) X = α and βH; R = H) in over 90% yield.

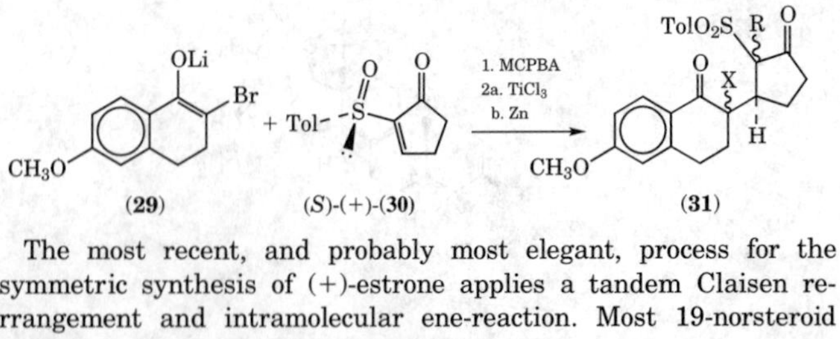

(**29**) (*S*)-(+)-(**30**) (**31**)

The most recent, and probably most elegant, process for the asymmetric synthesis of (+)-estrone applies a tandem Claisen rearrangement and intramolecular ene-reaction. Most 19-norsteroid

Table 1. Commercial Microbial Transformations Used To Produce Advanced Intermediates or Finished Products

Substrate[a]	Transformation	Product	Organism
progesterone	11α-hydroxylation oxidation/ lactonization		*Rhizopus nigricans Cylindrocarpon radicicola*
11-deoxycortisol (17α-derivatives)	11β-hydroxylation	cortisol/ derivatives	*Curvularia lunata*
6α-fluoro-16α-methyl-21-hydroxypregn-4-ene-3,20-dione	11β-hydroxylation	Paramethasone	*Curvularia lunata*
11-deoxy-16-methylenecortisol	11β-hydroxylation	Prednylidene	*Curvularia lunata*
9α-fluorohydrocortisone	1-dehydrogenation 16α-hydroxylation	Triamcinolone	*Arthrobacter simplex*
hydrocortisone	1-dehydrogenation	Prednisolone	*Arthrobacter simplex* or
6α-fluoro-16α-methyl corticosterone	1-dehydrogenation	Fluocortolone	*Bacillus lentus*
11β,21-dihydroxypregna-4,17(20)-dien-3-one	1-dehydrogenation		*Septomyxa affinis*
rac-3-methoxy-8,14-secoestra-1,3,5-(10),9(11)-tetra-ene-14,17-dione (Secosteroid)[b]	17-ketone reduction		*Saccharomyces uvarum*
androst-4-ene-3,17-dione[c]	17-ketone reduction		*Saccharomyces* sp.
21-acetoxy-17α-hydroxypregnenolone	Δ^5-3β-alcohol dehydrogenase		*Flavobacterium dehydrogenans*
6α-fluoro-21-hydroxy-16α-methyl-pregn-4-ene-3,20-one	9α-hydroxylation		*Curvularia lunata*

[a] Class is corticosteroid unless otherwise noted.
[b] Class is estrogen–progestin.
[c] Class is androgen.

contraceptive agents are produced by total synthesis from nonsteroidal starting materials.

Androstanes and Pregnanes. The first total syntheses of nonaromatic steroids that contain the C19-angular methyl substituent were accomplished in the early 1950s. These syntheses all began with starting materials containing a two-ring system. A more recent ring annulation strategy for the total synthesis of steroids begins with the formation of the C–D-ring system as a suitably functionalized indane. Condensation of the pyrrolidine enamine of cyclopentanone with ene-one results in the bicyclic keto-ester in 60–70% yield. Treatment of the latter with isopropenyl acetate and sulfuric acid produces a dienol acetate. Treatment of this dienol acetate with acetic anhydride and boron trifluoride etherate forms (**32**) as the major product.

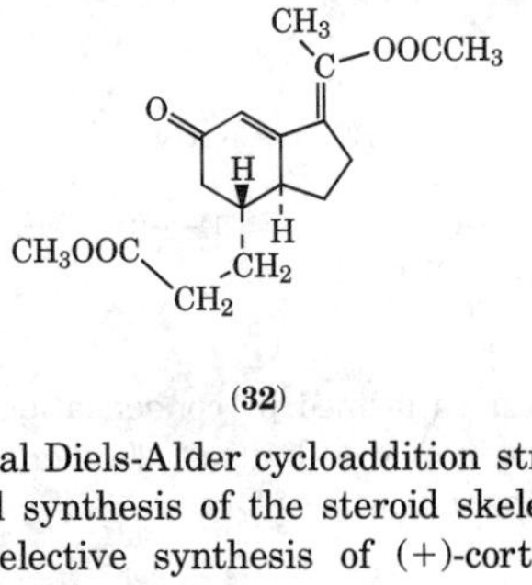

(**32**)

Several additional Diels-Alder cycloaddition strategies have been applied to the total synthesis of the steroid skeleton. For example, the first enantio-selective synthesis of (+)-cortisone was accomplished by the intramolecular [4 + 2] cycloaddition of an olefinic *o*-quinodimethane that contained an optically active stereodirecting group as the key chemical step.

Other approaches to the stereoselective total synthesis of nonaromatic steroids include the carbocationic, biomimetic cyclization reactions. Generally, these cyclizations begin with the synthesis of an appropriately functionalized cyclopentenol. Acid-catalyzed cyclization forms the B–C–D rings of the steroid nucleus with the natural relative stereochemistry in a single step.

Removable cation-stabilizing auxiliaries have been investigated for polyene cyclizations. For example, a silyl-assisted carbocation cyclization has been used in an efficient total synthesis of lanosterol. Other conditions for the cyclization of polyenes and of ene-ynes to steroids have been investigated. Oxidative free-radical cyclizations of polyenes produce steroid nuclei with exquisite stereocontrol. Besides the aforementioned A-ring aromatic steroids and contraceptive agents, partial synthesis from steroid raw materials has also accounted for the vast majority of industrial-scale steroid synthesis.

An interesting breakthrough in steroid endocrinology occurred with the discovery of a novel class of steroid antihormones. Several 11β-substituted 19-norsteroids display potent antiprogestinal activity. For example, RU-486 (**33**) is marketed in Europe as a contragestive agent. The synthesis of RU-486 demonstrates a unique method for functionalization of the 11β-position of a steroid nucleus.

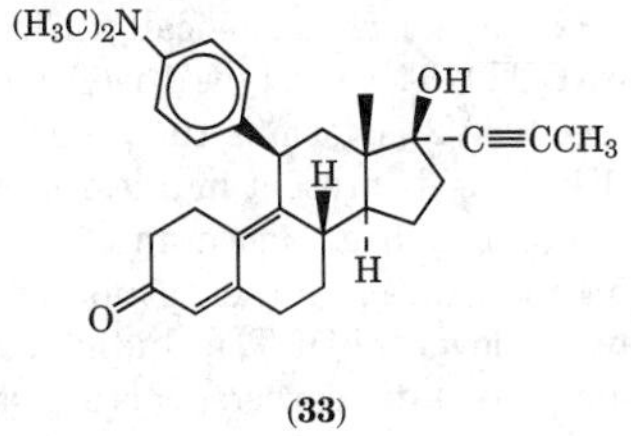

(**33**)

Uses: Therapeutics and Toxicology

Steroid Hormones. *Sex Hormones.* The largest economic impact of synthetic estrogen and progestin production has been for use as contraceptive agents and for treatment and prevention of osteoporosis. Mixtures of estrogens and progestins have been used as contraceptive agents since the early 1960s. The principal mode of steroid contraceptive action is exerted at the hypothalamic–pituitary–ovarian and uterine sites. Thus, contraceptive steroid mixtures have been used to treat a variety of related abnormal states including endometriosis, dysmenorrhea, hirsutism, polycystic ovarian disease, dysfunctional uterine bleeding, benign breast disease, and ovarian cyst suppression.

Estrogens are routinely prescribed to post-menopausal women to prevent the development and exacerbation of osteoporosis, because it can increase bone density and reduce fractures. Estradiol (**14**) or conjugated estrogens are typical agents used for the prevention and treatment of osteoporosis.

Antiprogestins, such as RU-486 (17β-hydroxy-11β-(4-dimethylaminophenyl-1)-17α-(prop-1-ynyl)-estra-4,9-diene-3-one) (**33**) and ZK98299 (11β-(4-dimethylaminophenyl)-17α-hydroxy-17β-(3-hydroxypropyl-13α-methyl-4,9-gonadien-3-one) represent a new class of drugs for fertility regulation. Also, these drugs have potential applications in the treatment of uterine cancer.

During the 1960s and 1970s a wide range of estrogens and antiestrogens were synthesized primarily to study reproductive endocrinology. The focus of clinical applications of many of these antiestrogens has shifted to breast cancer therapy. Although structurally different, these antiestrogens bind to the estrogen receptor in the breast cancer cell and exert a profound influence on cell replication.

Testosterone, alkylated testosterone, or testosterone esters are the primary anabolic–androgenic steroid drugs. The medicinal uses for these drugs include treatment of certain types of anemias, hereditary angioedema, certain gynecological conditions, protein anabolism, certain allergic reactions, and use in replacement therapy in gonadal failure states. However, anabolic–androgenic steroids are best known for their nonmedical, and illegal, use to aid in body-building or to increase skeletal muscle size, strength, and endurance.

Corticosteroids. The greatest portion of steroid drug production is aimed at the synthesis of glucocorticoids, which are highly effective agents for the treatment of chronic inflammation. Glucocorticoids exert their effects by binding to the cytoplasmic glucocorticoid receptor within the target cell and thus either increase or decrease transcription of a number of genes involved in the inflammatory process. Glucocorticoids are used to treat a variety of different diseases that are exacerbated by inflammation, such as arthritis, asthma, rhinitis, and skin irritations.

Corticosteroids are the most efficacious treatment available for the long-term treatment of asthma, and inhaled corticosteroids are considered to be a first-line therapy for asthma. Rhinitis is characterized by nasal stuffiness with partial or full obstruction, and itching of the nose, eyes, palate, or pharynx, sneezing, and rhinorrhoea. If left untreated it can lead to more serious respiratory diseases such as sinusitis or asthma. Nasal spray topical corticosteroids are widely regarded as the reference standard in rhinitis therapy.

Other Therapeutics Steroids. *Saponins.* Synthetic steroids that are structurally related to saponins have been shown to lower plasma cholesterol in a variety of different species.

Heterocyclic Steroids. Steroid 5α-reductase (types 1 and 2) converts testosterone (**12**) to the physiologically more potent androgen dihydrotestosterone (DHT) (**34**). The type 1 isoform occurs in nongenital skin, whereas the type 2 isoform is the predominant form in the prostate (the type 1 isoform is present in a lesser extent) and genital skin fibroblasts. There has been much interest in developing inhibitors of steroid 5α-reductase as a therapy for a variety of disorders associated with elevated levels of DHT, including benign prostatic hyperplasia (BPH), some prostatic cancers, certain skin disorders, and male pattern baldness.

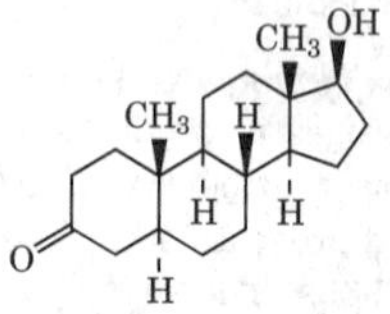

(**34**)

Analytical Methods

The field of steroid analysis includes identification of steroids in biological samples, analysis of pharmaceutical formulations, and elucidation of steroid structures. Many different analytical methods, such as ultraviolet (uv) spectroscopy, infrared (ir) spectroscopy, nuclear magnetic resonance (nmr) spectroscopy, x-ray crystallography, and mass spectroscopy, are used for steroid analysis.

Generally, the most powerful method for structural elucidation of steroids is nuclear magnetic resonance (nmr) spectroscopy. A definitive method for structural determination is x-ray crystallography. Extensive x-ray crystal structure determinations have been done on a wide variety of steroids. In addition, other analytical methods for steroid quantification or structure determination include, mass spectrometry, polarography, fluorimetry, radioimmunoassay, and various chromatographic techniques.

BRADLEY P. MORGAN
MELINDA S. MOYNIHAN
Pfizer, Inc.

L. F. Fieser and M. Fieser, *Steroids,* Reinhold Publishing Corp., New York, 1959.

E. Heftmann, *Steroid Biochemistry,* Academic Press, Inc., New York, 1970.

M. H. Briggs and J. Brotherton, *Steroid Biochemistry and Pharmacology,* Academic Press, London, 1970.

W. L. Duax and D. A. Norton, *Atlas of Steroid Structure,* IFI/Plenum Data, New York, 1975.

STILBENE DYES

Stilbene dyes of importance are mostly direct yellow dyes for cellulosic fibers, especially paper. Stilbene, $C_6H_5CH{=}CHC_6H_5$, is a crystalline hydrocarbon used in the manufacture of dyes. However, there is not a single commercial dye derived directly from stilbene itself. In most cases, the starting material for stilbene dyes is 4-nitrotoluene-2-sulfonic acid, which is oxidized under alkaline conditions to 4,4′-dinitro-2,2′-dinitrostilbenedisulfonic acid (**1**) as the first descriptive substance.

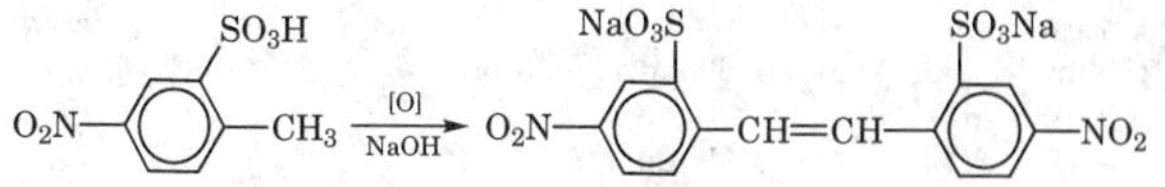

(**1**)

Stilbene dyes are classed as a subgroup of azo dyes having excellent colorfastness and typical direct dye wash fastness on cotton and are arranged into six categories by the Society of Dyers and Colourists, as in the following descriptions.

(*1*) Self-condensation products of 4-nitrotoluene-2-sulfonic acid or its derivative 4,4′-dinitro-2,2′-stilbenedisulfonic acid or 4,4′-dinitro-2,2′-dibenzyldisulfonic acid and products of their treatment with reducing or oxidizing agents. An example is Direct Yellow 11 (**2**).

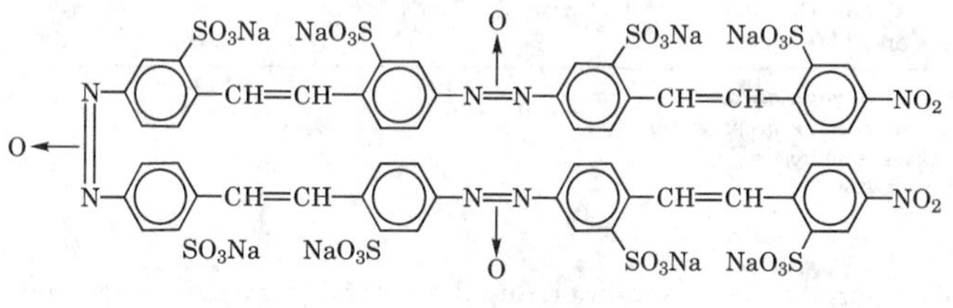

(**2**)

(*2*) Condensation products of 4-nitrotoluene-2-sulfonic acid or its derivatives together with phenols, naphthols, or aminophenols. An example here is Direct Yellow 19 (**3**):

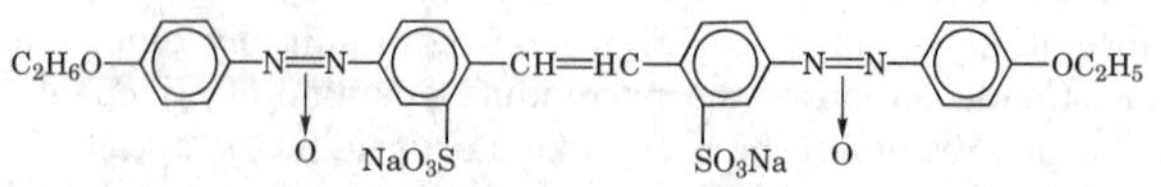

(**3**)

(*3*) Condensation products of 4-nitrotoluene-2-sulfonic acid or its derivatives together with aromatic amines. Direct Orange 28 (**4**) is an example.

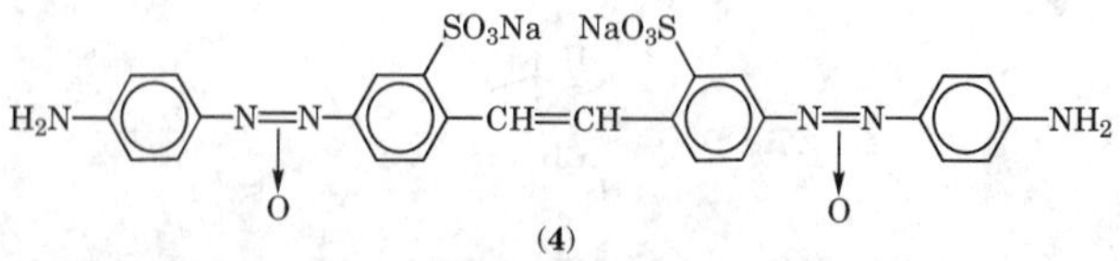

(**4**)

(*4*) Azo-stilbene dyes formed by condensation of 4,4′-dinitro-2,2′-stilbene-disulfonic acid or 4,4′-dinitro-2,2′-dibenzyldisulfonic acid (**1**) with aminoazo compounds. Direct Orange 34 (**5**) is a representative:

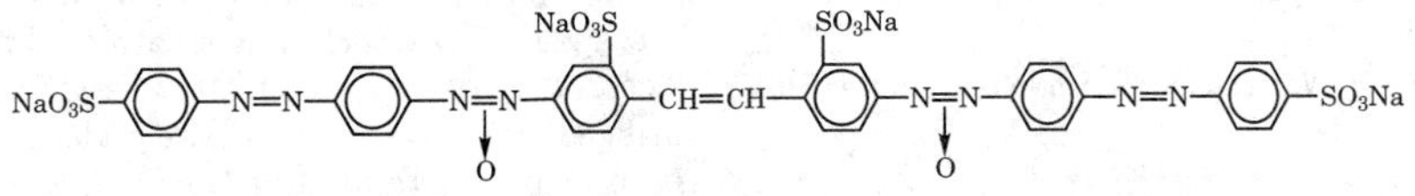

(5)

(5) Azo-stilbene dyes formed by diazotization of a condensation product containing primary amino groups and coupling with azo dye coupling components, eg, Direct Brown 29 (**6**):

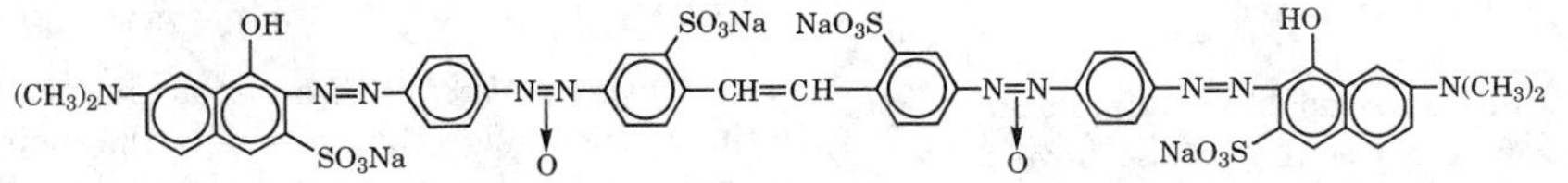

(6)

(6) Stilbene-azo dyes of more precise constitution prepared in the usual way by tetrazotization and coupling of 4,4′-diamino-2,2′-stilbenedisulfonic acid (**7**). The product in this example is Direct Yellow 4 (**8**):

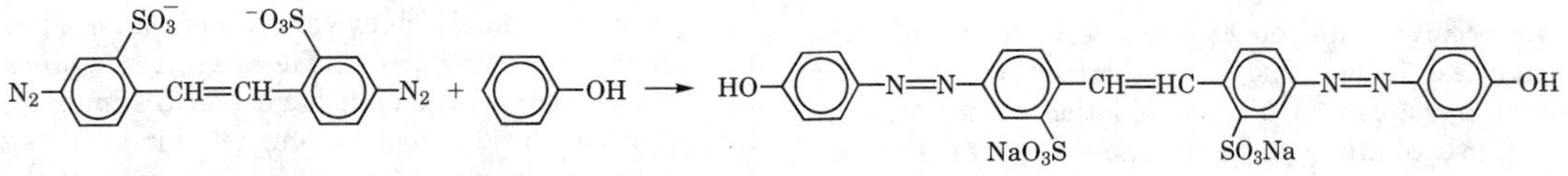

(7) (8)

(6a) Stilbene-azo dyes from 4-nitro-4′-amino-2,2′-stilbenedisulfonic acid include Direct Yellow 106 (**9**). Most stilbene dyes derived from 4-nitrotoluene-2-sulfonic acid are of nondefinitive structure even though structures are proposed which are descriptive of the major components.

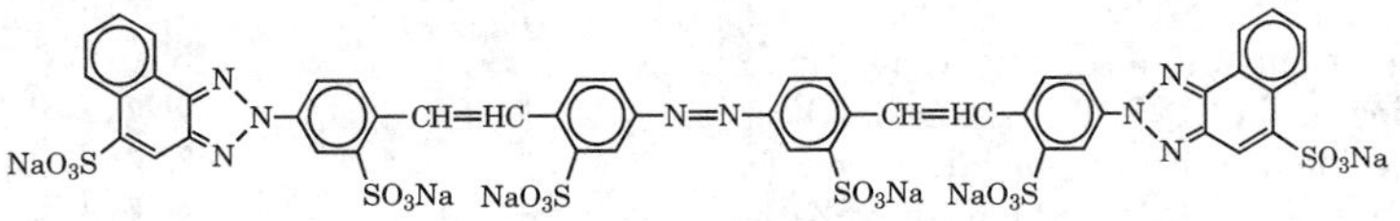

(9)

Stilbene dyes have generally been important as direct dyes and fluorescent brighteners for cellulosic fibers. Most stilbene dyes are yellow and orange, with some examples of reds and browns and even a few blues. Brown stilbene dyes have commercial value as leather dyes.

Direct Yellow 4 (CI 24890) (**10**) is the most familiar dye made from tetrazotized 4,4′-diamino-2,2′-stilbenedisulfonic acid (**7**) and phenol.

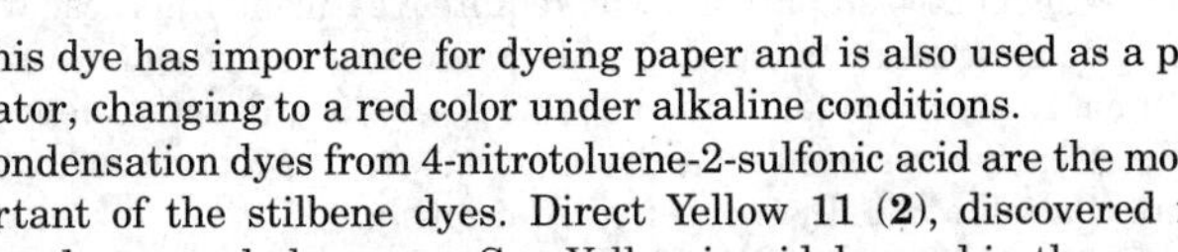

(10)

This dye has importance for dyeing paper and is also used as a pH indicator, changing to a red color under alkaline conditions.

Condensation dyes from 4-nitrotoluene-2-sulfonic acid are the most important of the stilbene dyes. Direct Yellow 11 (**2**), discovered in 1883 and commonly known as Sun Yellow is widely used in the paper industry.

Direct Orange 15 (**11**) is an important paper dye used in dyeing brown paper for bags. Activity in this class of compounds since the early 1980s has been mostly with stilbene fluorescent brighteners (see FLUORESCENT WHITENING AGENTS). There has, however, been some other activity in most of the six listed categories of dyes as well.

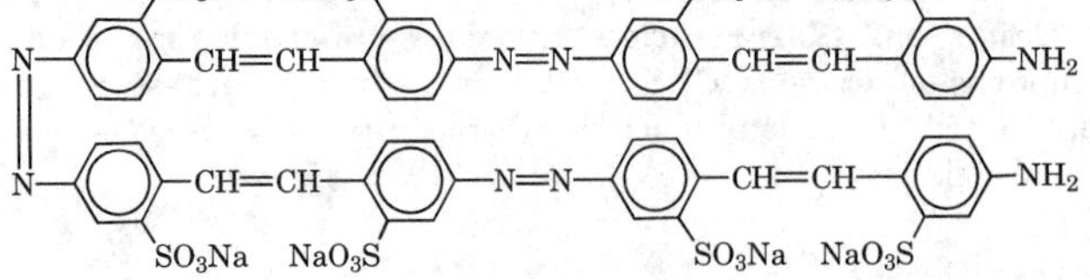

(11)

Health and Safety Factors

Stilbene dyes are similar to azo dyes in their resistance to biological degradation. Typically the BOD is only a small percentage of the COD.

ROY E. SMITH
Consultant

The Colour Index, 3rd ed., Vol. 3, Society of Dyers and Colorists, Bradford, Yorkshire, U.K., 1971, pp. 4212–4214, 4365–4371, and rev. 3rd ed., Vol. 6, 1st Suppl., p. 6400.

K. Venkataraman, *The Chemistry of Synthetic Dyes and Pigments,* Vol. 1, Academic Press, Inc., New York, 1952.

East Ger. Pat. 236,338 (June 4, 1986), W. Hepp and co-workers (to Bitterfeld).

STIMULANTS

A variety of chemical agents have the capacity to stimulate the central nervous system (CNS) of mammals. Some have therapeutic uses; others are primarily of toxicological importance. Herein stimulants are separated into three more or less distinct pharmacological categories: analeptics, psychomotor stimulants, and antidepressants (see PSYCHOPHARMACOLOGICAL AGENTS).

Analeptics

Analeptics are respiratory stimulants capable of stimulating respiratory and vasomotor centers in the medulla. These have been used to revive individuals poisoned by central nervous depressants such as barbiturates, alcohol, and general anesthetics (see ANESTHETICS; HYPNOTICS, SEDATIVES, ANTICONVULSANTS, AND ANXIOLYTICS). The action is not confined only to the medulla; at doses only slightly higher than those that stimulate the medulla, analeptics can stimulate the motor cortex and produce seizures. Although the clinical usefulness is limited, analeptics continue to be valuable tools in the study of CNS neurotransmitters (see NEUROREGULATORS).

Some naturally occurring analeptics have been known for centuries. Two of the best known and most thoroughly studied are strychnine (**1**) and picrotoxin, a 1:1 combination of picrotoxinin (**2**) and picrotin. These continue to be of interest in the study of mammalian neurotransmission.

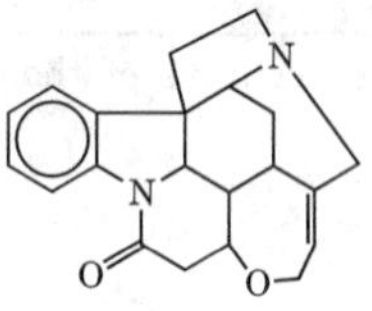

(**1**)

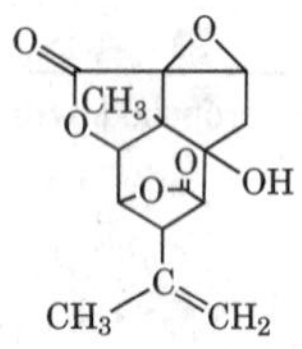

(**2**)

There is good evidence that strychnine is a specific, competitive, post-synaptic antagonist of glycine in the CNS. Picrotoxin has been instrumental in establishing an inhibitory neurotransmitter role for the amino acid, gamma-aminobutyric acid (GABA), quantitatively the most important inhibitory neurotransmitter in the mammalian CNS. Benzodiazepines and similar agents have largely replaced barbiturates and barbiturate-like agents for use as anxiolytics and sedative–hypnotics. Because benzodiazepines rarely produce levels of CNS depression that require therapeutic intervention, the need for analeptics has decreased considerably.

Health and Safety Factors. Overdoses of analeptics produce symptoms of extreme CNS excitation, including restlessness, hyperexcitability, skeletal muscle hyperactivity, and in some cases convulsions.

Psychostimulants

Compounds having relatively specific cerebral stimulant properties are classified as psychostimulants or psychoanaleptics. Caffeine (**3**), a mild psychostimulant, has been called the most widely used psychoactive substance on earth. Caffeine, theophylline (**4**), and theobromine (**5**) are three closely related alkaloids known as methylxanthines that occur in plants widely distributed throughout the world (see ALKALOIDS). The first two have CNS stimulant properties; the last is virtually inactive as a stimulant. The word caffeine is used exclusively herein even though some of the effects of caffeinated beverages may result from the theophylline content.

The effects of low to moderate amounts of caffeine ingestion are generally salutary. At higher levels, however, more serious signs of CNS stimulation may be elicited. Methylxanthines have a few valid therapeutic uses, including treatment of asthma and relief of dyspnea (see ANTIASTHMATIC AGENTS). The CNS stimulatory effects are also utilized for the treatment of the prolonged apnea that may be observed in premature infants.

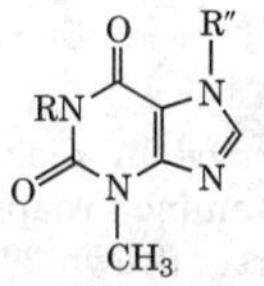

(**3**, R = R′ = CH_3)
(**4**, R = CH_3; R′ = H)
(**5**, R = H; R′ = CH_3)

Sympathomimetics. Sympathomimetics are a group of mostly synthetic compounds that resemble the neurotransmitters epinephrine and norepinephrine pharmacologically and to some extent chemically (see EPINEPHRINE AND NOREPINEPHRINE). These agents have wide-ranging pharmacological effects, including, in some cases, profound CNS excitatory actions. Sympathomimetics that have selective central effects have been used in the treatment of narcolepsy, as an aid in weight reduction (see ANTIOBESITY DRUGS), and in the treatment of attention-deficit hyperactivity disorder (ADHD).

The oldest of the centrally acting sympathomimetics is ephedrine (**6**). Ephedrine occurs in many varieties of plants of the genus *Ephedra* and may also be synthesized. There are three other indirectly acting adrenomimetic compounds having CNS stimulant properties that have been employed clinically: amphetamine (**7**), methamphetamine (**8**), and methylphenidate. D-Amphetamine is three to four times more potent in producing CNS stimulation than is L-amphetamine. The development of tolerance, the abuse potential, and the insomnia and nervousness it causes have led to a marked reduction in amphetamine usage. Methylphenidate is reported to have less abuse potential and is the drug of choice for ADHD.

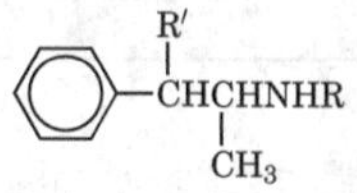

(**6**, R = CH_3; R′ = OH)
(**7**, R = R′ = H)
(**8**, R = CH_3; R′ = H)

Pemoline structurally dissimilar to amphetamine or methylphenidate, appears to share the CNS-stimulating properties.

Side Effects and Abuse Potential. Sympathomimetics are one of the most abused classes of drugs marketed in the United States. The abuse characteristics of amphetamines and related drugs are similar to that of cocaine, also a sympathomimetic. Because of the potential for abuse, the manufacture, distribution, and use of sympathomimetics are strictly controlled in the United States by the Drug Enforcement Agency (DEA).

Antidepressants

Depression. Disorders of mood or affect may be either a pathological state or a normal human emotion. The American Psychiatric Association has established diagnostic criteria that allow clinicians to distinguish between patients who require treatment and those who do not. The most common mood disorder, known as reactive depression, is commonly observed following adverse life events. Reactive depression is frequently expressed by depression, anxiety, or feelings of stress or guilt. Patients usually recover spontaneously.

The most common type of depression, known as unipolar disorder (major depression), accounts for about 25% of all depression. Signs include weight loss, loss of libido, alterations in sleep pattern, symptoms of negative self-image, suicidal thoughts, and overwhelming grief. The other type is known as bipolar disorder (manic–depressive disorder), and includes about 10–15% of depressions. Typically, the patient having bipolar disorder alternates between depression as seen in unipolar depression, and periods in which the symptoms are exactly the opposite. The patient having bipolar illness cycles between depression and mania, in which the duration of each cycle is commonly measured in months. The patient having unipolar disorder, on the other hand, is usually in a state of constant depression.

Mechanism. The monoamine oxidase inhibitors were the first effective drugs for the treatment of depression. Their mechanism of action is to elevate levels of those endogenous agents, eg, monoamines such as norepinephrine, dopamine, and serotonin, that are substrates for the enzyme monoamine oxidase (MAO). Many studies monitoring regulation of various monoamine receptors in the central nervous system have led to two important experimental findings. One is a down-reulation of β-adrenoceptors following chronic administration of many drugs effective in depression. The other is an enhancement of transmission through a particular receptor, HT_{1A}, after chronic administration of all clinically effective antidepressants and after electroconvulsive treatment.

Treatment. Most, although not all, of the drugs effective in the treatment of depression are CNS stimulants. Until the middle of the twentieth century, pharmacological treatment was symptomatic, supportive, and frequently ineffective. The discovery in the 1950s of agents known as monoamine oxidase inhibitors (MAOI) and later of the tricyclic antidepressants (TCA) has led to more effective and safer preparations.

Monoamine Oxidase Inhibitors. MAOIs inactivate the enzyme MAO, which is responsible for the oxidative deamination of a variety of endogenous and exogenous substances. Among the endogenous substances are the neurotransmitters, norepinephrine, dopamine, and serotonin. Because of toxicity only three agents are available in the United States: isocarboxazid, phenelzine, and tranylcypromine. Nialamide and mebanazine are two MAO inhibitors marketed in Europe. Use of MAOIs for the treatment of depression is severely restricted because of potential side effects, the most serious of which is hypertensive crisis.

Tricyclic Antidepressants. Imipramine (**9**), which was the first tricyclic antidepressant to be developed, is one of many useful psychoactive compounds derived from systematic molecular modifications of the antihistamine promethazine (see HISTAMINE AND HISTAMINE ANTAGONISTS).

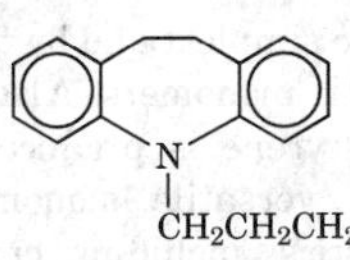

(**9**)

Following the successful introduction of imipramine (**9**), many related compounds were prepared and clinically evaluated for antidepressant effects, including amitriptyline, trimipramine, desipramine, nortriptyline, and protriptyline.

Side Effects and Toxicity. Adverse effects to the tricyclic antidepressants are primarily the result of the actions of these compounds on either the autonomic, cardiovascular, or central nervous systems. The most serious side effects of the tricyclics concern the cardiovascular system. Arrhythmias, which are dose-dependent and rarely occur at therapeutic plasma levels, can be life-threatening.

Selective Serotonin Reuptake Inhibitors. In 1987, the FDA approved fluoxetine (**10**) for use in the treatment of major depression. Fluoxetine (Prozac) and related compounds, sertraline, and paroxetine appear to inhibit selectively the reuptake of serotonin while having virtually no effect on the uptake of norepinephrine or dopamine. These selective serotonin reuptake inhibitors (SSRIs) do not appear to be more effective than the tricyclics for the treatment of depression. However, the SSRIs do appear to lack many of the side effects associated with the tricyclics and other antidepressants, and are therefore both safer for and more readily accepted by the patient.

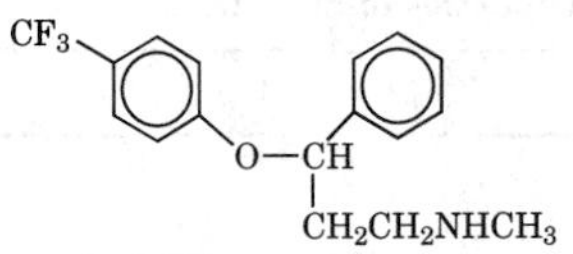

(**10**)

Miscellaneous Antidepressants. There are a few agents that either chemically or pharmacologically do not fit neatly into any of the categorized antidepressant agents. Trazodone was introduced as a safer, less toxic, and faster-acting antidepressant. It is effective in some patients, virtually ineffective in others. Nefazodone is similar in structure to trazodone and appears to share most of its clinical and pharmacological effects. Buprion is devoid of inhibitory actions on both serotonin and norepinephrine uptake systems. However, it is a potent inhibitor of the uptake system for dopamine and is also not a monoamine oxidase inhibitor.

Other Drugs. Agents not considered to be CNS stimulants yet employed for the treatment of certain types of depression includes lithium carbonate for the treatment of bipolar disorder. In most patients, lithium is the sole agent used to control manic behavior and is very effective (see PSYCHOPHARMACOLOGICAL AGENTS).

CHARLES R. CRAIG
West Virginia University

C. R. Craig, in C. R. Craig and R. E. Stitzel, eds., *Modern Pharmacology,* 5th ed., Little, Brown & Co., Boston, Mass., 1997, p. 294.

G. L. Gessa, W. Fratta, L. Pani, and G. Serra, *Depression and Mania From Neurobiology to Treatment,* Lippincott-Raven, Hagerstown, Md., 1995.

D. A. Taylor, in C. R. Craig and R. E. Stitzel, eds., *Modern Pharmacology,* 5th ed., Little, Brown Co., Boston, Mass., 1997, p. 365.

STREPTOMYCIN AND RELATED ANTIBIOTICS. See ANTIBIOTICS.

STRESS CRACKING. See FRACTURE MECHANICS; MATERIALS RELIABILITY.

STRONTIUM AND STRONTIUM COMPOUNDS

Strontium

Strontium, Sr, is in Group 2 (IIA) of the Periodic Table, between calcium and barium. These three elements are called alkaline-earth metals. Strontium forms 0.02–0.03% of the earth's crust and its minerals are usually found in or near sedimentary rocks.

Properties. Strontium is a hard white metal having physical properties shown in Table 1. It has four stable isotopes, atomic weights 84, 86, 87, and 88; and one radioactive isotope, strontium-90, which is a product of nuclear fission. The most abundant isotope is strontium-88. The chemical properties of strontium are intermediate between those of calcium and barium. The metal and its salts impart a brilliant red color to flames.

Occurrence. The principal strontium mineral is celestite, naturally occurring strontium sulfate. Strontianite is the naturally occurring form of strontium carbonate; no economically workable deposits are known.

Production. Strontium oxide is reduced thermally with aluminum in a vacuum according to the following reaction:

$$3\ SrO + 2\ Al \rightarrow 3\ Sr + Al_2O_3$$

All strontium metal is produced commercially by this process.

Uses. The main application for strontium is in the form of strontium compounds. The carbonate, used in cathode ray tubes (CRTs)

Table 1. Physical Properties of Strontium

Property	Value
atomic weight	87.62
melting range, °C	768–791
boiling range, °C	1350–1387
density, g/cm^3	2.6
crystal system	face-centered cubic
lattice constant, pm	605
latent heat of fusion, kJ/kg[a]	104.7
electrical resistivity, $\mu\Omega$/cm	22.76

[a] To convert J to cal, divide by 4.184.

for color televisions and color computer monitors, is used both in the manufacturing of the glass envelope of the CRT and in the phosphors which give the color.

Strontium metal is used as a eutectic modifier in aluminum–silicon casting alloys. The addition of strontium changes the microstructure of the alloy, imparting improved ductility and strength in cast aluminum automotive parts. A second use for strontium metal is as an inoculant in ductile iron castings.

Strontium Compounds

Strontium has a valence of +2 and forms compounds that resemble the compounds of the other alkaline-earth metals (see BARIUM COMPOUNDS; CALCIUM COMPOUNDS). Although many strontium compounds are known, there are only a few that have commercial importance and, of these, strontium carbonate, $SrCO_3$, and strontium nitrate, $Sr(NO_3)_2$, are made in the largest quantities. The mineral celestite, $SrSO_4$, is the raw material from which the carbonate or the nitrate is made.

Production. Mexico is the principal producer of strontium minerals.

Uses. The primary application for strontium carbonate is in the manufacture of cathode ray tubes (CRTs). The other principal application is in permanent ceramic magnets. Other applications for strontium compounds include pyrotechnics (qv), pigments (qv), and electrolytic zinc refining.

Health and Safety Factors. The strontium ion has a low order of toxicity, and strontium compounds are remarkably free of toxic hazards. Strontium nitrate is an oxidizer and should not be stored in areas of potential fire hazards.

Strontium Acetate. Strontium acetate, $Sr(CH_3\ CO_2)_2$, is a white crystalline salt. Its solubility in water is 36.4 g/100 mL at 97°C. When heated, strontium acetate decomposes.

Strontium Carbonate. Strontium carbonate, $SrCO_3$, is a colorless or white crystalline solid. It is insoluble in water but reacts with acids, and is soluble in solutions of ammonium salts.

Ground celestite ore is mixed with ground coke and is reduced in kilns to strontium sulfide, known as black ash. This product is treated with soda ash, $CaCO_3$, or carbon dioxide to cause precipitation of strontium carbonate crystals.

Strontium Chromate. Strontium chromate, $SrCrO_4$, is made by precipitation of a water-soluble chromate solution using a strontium salt of chromic acid. It is used as a yellow pigment and as an anticorrosive primer (see CORROSION AND CORROSION CONTROL).

Strontium Hexaferrite. Strontium hexaferrite, $SrO{\cdot}6Fe_2O_3$, is made by combining powdered ferric oxide, Fe_2O_3, and strontium carbonate, $SrCO_3$, and calcining the mixture at ca 1000°C in a rotary kiln. It finds application as magnets in small electric motors and as flexible magnets.

Strontium Halides. Strontium halides are made by the reactions of strontium carbonate with the appropriate mineral acids. They are used primarily in medicines as replacements for other bromides and iodides.

Strontium Nitrate. Strontium nitrate, $Sr(NO_3)_2$, in the anhydrous form is a colorless crystalline powder. Strontium nitrate is made by the reaction of milled strontium carbonate with nitric acid.

Strontium Oxide, Hydroxide, and Peroxide. Strontium oxide, SrO, is a white powder. It is made by heating strontium carbonate with carbon in an electric furnace. It reacts with water to form strontium hydroxide and is used as the source of strontium peroxide. Strontium hydroxide, $Sr(OH)_2$, resembles slaked lime but is more soluble in water (21.83 g per 100 g of water at 100°C). It is a white deliquescent solid. Strontium soaps, made by combining strontium hydroxide with soap stocks, are used to make strontium greases.

Strontium peroxide, SrO_2, is a white powder that decomposes in water. It is made by the reaction of hydrogen peroxide with strontium oxide.

Strontium Sulfate. Strontium sulfate, $SrSO_4$, occurs as celestite deposits in beds or veins in sediments or sedimentary rocks. It forms colorless or white rhombic crystals.

Strontium Titanate. Strontium titanate, $SrTiO_3$, is a ceramic dielectric material that is insoluble in water. It is used in the form of 0.5-mm thick disks as electrical capacitors in television sets, radios, and computers.

STEPHEN G. HIBBINS
Timminco Metals

D. L. Stein, in T. J. Gray, ed., *Conference on Strontium Containing Compounds*, Nova Scotia Technical College, Halifax, Nova Scotia, Canada, 1973.

G. D. Parkes, ed., *Mellor's Modern Inorganic Chemistry*, John Wiley & Sons, Inc., New York, 1967.

STRUCTURAL FOAMS. See FOAMED PLASTICS.

STRUCTURE-ACTIVITY RELATIONSHIPS. See PHARMACODYNAMICS.

STYRENE

Styrene, $C_6H_5CH{=}CH_2$, is the simplest and by far the most important member of a series of aromatic monomers. Also known commercially as styrene monomer (SM), styrene is produced in large quantities for polymerization. It is a versatile monomer extensively used for the manufacture of plastics, including crystalline polystyrene, rubber-modified impact polystyrene, expandable polystyrene, acrylonitrile–butadiene–styrene copolymer (ABS), styrene–acrylonitrile resins (SAN), styrene–butadiene latex, styrene–butadiene rubber (SBR), and unsaturated polyester resins (see ACRYLONITRILE POLYMERS; STYRENE PLASTIC).

Properties

Styrene is a colorless liquid with an aromatic odor. Important physical properties of styrene are shown in Table 1. Styrene is infinitely soluble in acetone, carbon tetrachloride, benzene, ether, *n*-heptane, and ethanol. Polymerization generally takes place by free-radical reactions initiated thermally or catalytically. Styrene undergoes many reactions of an unsaturated compound, such as addition, and of an aromatic compound, such as substitution.

Ethylbenzene Manufacture

Styrene is manufactured from ethylbenzene. Ethylbenzene is produced by alkylation of benzene with ethylene. The reaction takes place on acidic catalysts and can be carried out either in the liquid or vapor phase.

Table 1. Physical Properties of Styrene Monomer

Property			Value		
boiling point (at 101.3 kPa = 1 atm, °C)			145.0		
freezing point, °C			−30.6		
flash point (fire point), °C					
Tag open-cup			34.4 (34.4)		
Cleveland open-cup			31.4 (34.4)		
autoignition temperature, °C			490.0		
explosive limits in air, %			1.1–6.1		
refractive index, n_D^{20}			1.5467		
	at 0°C	20°C	60°C	100°C	140°C
viscosity, mPa·s(= cP)	1.040	0.763	0.470	0.326	0.243
surface tension, mN/m (= dyn/cm)	31.80	30.86	29.01	27.15	25.30
density, g/cm^3 [a]	0.9237	0.9059	0.8702	0.8346	
heat of formation (liquid) at 25°C, ΔH_f, kJ/mol[b]			147.36		
heat of polymerization, kJ/mol[b]			74.48		

[a] Density at 150°C is 0.7900 g/cm^3.
[b] To convert J to cal, divide by 4.184.

Commercial ethylbenzene is manufactured almost exclusively for captive use to produce styrene. Most of the ethylbenzene plants built before 1980 are based on use of aluminum chloride catalysts. Aluminum chloride is an effective alkylation catalyst but is corrosive. The newer plants are based on zeolite catalysts.

Zeolite-Based Alkylation. Zeolites have the advantage of being noncorrosive and environmentally benign. The Mobil-Badger vapor-phase ethylbenzene process was the first zeolite-based process to achieve commercial success. It is based on a synthetic zeolite catalyst, ZSM-5, and has the desirable characteristics of high activity, low oligomerization, and low coke formation (see MOLECULAR SIEVES).

In the Mobil-Badger vapor-phase process, fresh and recycled benzene are vaporized and preheated to the desired temperature and fed to a multistage fixed-bed reactor. Ethylene is distributed to the individual stages. Alkylation takes place in the vapor phase. Separately, the polyethylbenzene stream from the distillation section is mixed with benzene, vaporized and heated, and fed to the transalkylator, where polyethylbenzenes react with benzene to form additional ethylbenzene. The combined reactor effluent is distilled in the benzene column. Benzene is condensed in the overhead for recycle to the reactors. The bottoms from the benzene column are distilled in the ethylbenzene column to recover the ethylbenzene product in the overhead. The bottoms stream from the ethylbenzene column is further distilled in the polyethylbenzene column to remove a small quantity of residue. The overhead polyethylbenzene stream is recycled to the reactor section for transalkylation to ethylbenzene.

A liquid-phase process based on an ultraselective Y (USY)-type zeolite catalyst, called the Lummus-UOP process, is similar to the Mobil-Badger vapor-phase process. The differences are primarily in the catalysts, reaction conditions, reactor sizes, yields, and product specifications. The zeolite-based processes require more benzene recycle than the aluminum chloride-based processes. The EBMax technology, based on a Mobil zeolite catalyst called MCM-22, overcomes the oligomerization problem that plagues other liquid-phase alkylation processes. The catalyst is highly active for alkylation but inactive for oligomerization and cracking.

Aluminum Chloride-Based Alkylation. An improved aluminum chloride-based process was developed by Monsanto in the 1970s. Using a presynthesized aluminum chloride complex and operating the reactor at higher temperature and pressure, the catalyst inventory is reduced to below its solubility in the reaction mixture. The reactants and the catalyst complex are mixed in the reactor to form a homogeneous liquid. The transalkylation of polyethylbenzenes is carried out separately. These improvements result in a higher yield.

Other Technologies. Ethylbenzene can be recovered from mixed C_8 aromatics by superfractionation. The Alkar process, commercialized in 1960, uses boron trifluoride on alumina support as the catalyst. It has been used for polymer-grade as well as dilute ethylene feeds.

Styrene Manufacture

Styrene manufacture by dehydrogenation of ethylbenzene is used for nearly 90% of the worldwide styrene production. The rest is obtained from the coproduction of propylene oxide (PO) and styrene (SM).

Dehydrogenation. The dehydrogenation of ethylbenzene to styrene takes

$$C_6H_5CH_2CH_3 \rightleftharpoons C_6H_5CH{=}CH_2 + H_2$$

place on a promoted iron oxide–potassium oxide catalyst in a fixed-bed reactor at the 550–680°C temperature range in the presence of steam. The reaction is limited by thermodynamic equilibrium. Low pressure favors the forward reaction. Dehydrogenation is an endothermic reaction. High temperature favors dehydrogenation both kinetically and thermodynamically but also increases by-products from side reactions and decreases the styrene selectivity.

The main by-products in the dehydrogenation reactor are toluene and benzene. The formation of toluene accounts for the biggest yield loss. Other by-products include carbon dioxide and various hydrocarbons.

Dehydrogenation catalysts usually contain 40–90% Fe_2O_3, 5–30% K_2O, and promoters such as chromium, cerium, molybdenum, calcium, and magnesium oxides. Dehydrogenation is carried out either isothermally or adiabatically. In principle, isothermal dehydrogenation has the dual advantage of avoiding a very high temperature at the reactor inlet and maintaining a sufficiently high temperature at the reactor outlet. In practice, these advantages are negated by formidable heat-transfer problems. In an adiabatic reactor, the endothermic heat of reaction is supplied by the preheated steam that is mixed with ethylbenzene upstream of the reactor. As the reaction progresses, the temperature decreases. To obtain a high conversion of ethylbenzene to styrene, usually two, and occasionally three, reactors are used in series with a reheater between the reactors to raise the temperature of the reaction mixture.

Other than the reactor system, the distillation column that separates the unconverted ethylbenzene from the crude styrene is the most important and expensive equipment in a styrene plant. To minimize yield losses and to prevent equipment fouling by polymer formation, polymerization inhibitors are used in the distillation train, product storage, and in vent gas compressors.

The qualities of the styrene product and toluene by-product depend primarily on three factors: the impurities in the ethylbenzene feedstock, the catalyst used, and the design and operation of the dehydrogenation and distillation units. Other than benzene and toluene, the presence of which is usually inconsequential, possible impurities in ethylbenzene are C_7–C_{10} nonaromatics and C_8–C_{10} aromatics. The condensed reactor effluent is separated in the settling drum into vent gas (mostly hydrogen), process water, and organic phase. The organic phase with polymerization inhibitor added is pumped to the distillation train.

Benzene and toluene by-products are recovered in the overhead of the benzene–toluene distillation column. The bottoms from the benzene–toluene column are distilled in the ethylbenzene recycle column, where the separation of ethylbenzene and styrene is effected. The bottoms, are pumped to the styrene finishing column. The overhead product from this column is purified styrene. The bottoms are further processed in a residue-finishing system to recover additional styrene from the residue.

PO–SM Coproduction. The coproduction of propylene oxide and styrene includes three reaction steps: (*1*) oxidation of ethylbenzene to ethylbenzene hydroperoxide, (*2*) epoxidation of ethylbenzene hydroperoxide with propylene to form α-phenylethanol and propylene oxide, and (*3*) dehydration of α-phenylethanol to styrene.

$$C_6H_5CH_2CH_3 + O_2 \rightarrow C_6H_5CH(CH_3)OOH$$

$$C_6H_5CH(CH_3)OOH + CH_2{=}CHCH_3 \longrightarrow C_6H_5CH(CH_3)OH + \underset{\diagdown O \diagup}{CH_2CHCH_3}$$

$$C_6H_5CH(CH_3)OH \rightarrow C_6H_5CH{=}CH_2 + H_2O$$

The recovery and purification facilities in such a process are complex. One reason is that oxygenated by-products are made in the reactors. Oxygenates hinder polymerization of styrene and cause color instability. Elaborate purification is required to remove the oxygenates.

Specifications and Analysis

The freezing point measurement, standard method for the determination of styrene assay until the 1970s, has been largely replaced by gas chromatography. Color is measured spectrophotometrically and registered on the APHA or the platinum–cobalt scale.

Health and Safety Factors

Styrene is mildly toxic, flammable, and can be made to polymerize violently under certain conditions. However, handled according to proper procedures, it is a relatively safe organic chemical.

While styrene is not confirmed as a carcinogen, it is considered a suspect carcinogen. Styrene liquid is inflammable and has sufficient vapor pressure at slightly elevated temperatures to form explosive mixtures with air. Properly inhibited and attended, styrene can be stored for an extended period of time.

Uses

Commercial styrene is used almost entirely for the manufacture of polymers.

Common applications for polystyrene include packaging, food containers, and disposable tableware; toys; furniture, appliances, television cabinets, and sports goods; and audio and video cassettes. Expandable polystyrene is widely used in construction for thermal insulation.

Uses for ABS are in sewer pipes, vehicle parts, appliance parts, business machine casings, sports goods, luggage, and toys.

SB latex is used in coatings, carpet backing, paper adhesives, cement additives, and latex paint.

SBR is used primarily in tires, vehicle parts, and electrical components.

The principal uses for UPR are in putty, coatings, and adhesives. Glass-reinforced UPR is used for marine, construction, and vehicle materials, as well as for electrical parts.

Derivatives

A large number of compounds related to styrene have been reported in the literature. Those having the vinyl group $CH_2{=}CH$— attached to the aromatic ring are referred to as styrenic monomers. Several of them have been used for manufacturing small-volume specialty polymers. The specialty styrenic monomers that are manufactured in commercial quantities are vinyltoluene, *para*-methylstyrene, α-methylstyrene, and divinylbenzene. In addition, 4-*tert*-butylstyrene (TBS) is a specialty monomer that is superior to vinyltoluene and *para*-methylstyrene in many applications. Other styrenic monomers produced in small quantities include chlorostyrene and vinylbenzene chloride. With the exception of α-methylstyrene, which is a by-product of the phenol–acetone process, these specialty monomers are more difficult and expensive to manufacture than styrene.

Vinyltoluene. Vinyltoluene (VT) is a mixture of *meta*- and *para*-vinyltoluenes, typically in the ratio of 60:40. This isomer ratio results from the ratio of the corresponding ethyltoluenes in thermodynamic equilibrium. Vinyltoluene is produced for special applications. Its copolymers are more heat-resistant than the corresponding styrene copolymers, and it is used as a specialty monomer for paint, varnish, and polyester applications.

para-Methylstyrene. PMS is the para isomer of vinyltoluene in high purity. PMS is made by alkylation of toluene with ethylene to *p*-ethyltoluene, followed by dehydrogenation of *p*-ethyltoluene.

Divinylbenzene. This is a specialty monomer used primarily to make cross-linked polystyrene resins. The largest use of divinylbenzene (DVB) is in ion-exchange resins for domestic and industrial water softening. Ion-exchange resins are also used as solid acid catalysts for certain reactions, such as esterification. Divinylbenzene is manufactured by dehydrogenation of diethylbenzene, which is an internal product in the alkylation plant for ethylbenzene production.

α-Methylstyrene. This compound is not a styrenic monomer in the strict sense. The methyl substitution on the side chain, rather than the aromatic ring, moderates its reactivity in polymerization. It is used as a specialty monomer in ABS resins, coatings, polyester resins, and hot-melt adhesives. As a copolymer in ABS and polystyrene, it increases the heat-distortion resistance of the product. In coatings and resins, it moderates reaction rates and improves clarity. α-Methylstyrene (AMS) is produced as a by-product in the production of phenol and acetone from cumene.

SHIOU-SHAN CHEN
Raytheon Engineers & Constructors

B. Maerz, S. S. Chen, C. R. Venkat, and D. Mazzone, EBMax: *Leading Edge Ethylbenzene Technology from Mobil/Badger,* 1996 DeWitt Petrochemical Review, Houston, Tex., Mar. 19–21, 1996.

U.S. Pat. 4,066,706 (Jan. 3, 1978), J. P. Schmidt (to Halcon International, Inc.).

Styrene/Ethylbenzene, PERP report 94/95-8, Chem Systems, Tarrytown, N.Y., Mar. 1996.

Chem. Mark. Rep. **248**(5), 41 (July 24, 1995).

STYRENE–BUTADIENE RUBBER

Styrene–butadiene rubber (SBR), an elastomer, is a copolymer of three parts 1,3-butadiene and one part styrene. It is a synthetic rubber used mainly in the manufacture of automobile tires.

In the late 1920s Bayer & Company began studies of the emulsion polymerization process of polybutadiene for producing synthetic rubber. Incorporation of styrene (qv) as a comonomer produced a superior polymer compared to polybutadiene. The product, Buna S, was the precursor of the single largest-volume polymer produced in the 1990s, emulsion styrene–butadiene rubber (ESBR).

In the mid-1950s, the Nobel Prize-winning work of K. Ziegler and G. Natta introduced anionic initiators which allowed the stereospecific polymerization of isoprene to yield high cis-1,4 structure, much like natural rubber. At almost the same time, another route to stereospecific polymer architecture by organometallic compounds was announced.

In the 1960s, anionic polymerized solution SBR (SSBR) began to challenge emulsion SBR in the automotive tire market. Organolithium compounds allow control of the butadiene microstructure, not possible with ESBR. Because this type of chain polymerization takes place without a termination step, an easy synthesis of block polymers is available, whereby glassy (polystyrene) and rubbery (polybutadiene) segments can be combined in the same molecule. These thermoplastic elastomers (TPE) have found use in nontire applications.

Physical Properties

Desirable properties of elastomers include elasticity, abrasion resistance, tensile strength, elongation, modulus, and processibility. These properties are related to and dependent on the average molecular weight and mol wt distribution, polymer macro- and microstructure, branching, gel (cross-linking), and glass-transition temperature (T_g) (see ELASTOMERS, SYNTHETIC).

Emulsion polymerization gives SBR polymer of high molecular weight. Because it is a free-radical-initiated process, the composition of the resultant chains is heterogeneous, with units of styrene and butadiene randomly spaced throughout. Unlike natural rubber, which is polyisoprene of essentially all cis-1,4 configuration, giving an ordered structure and hence crystallinity, ESBR is amorphous. Unlike SSBR, the microstructure of which can be modified to change the polymer's T_g, the T_g of ESBR can be changed only by a change in ratio of the monomers. Glass-transition temperature is that temperature where a polymer experiences the onset of segmental motion.

The glass-transition temperatures for solution-polymerized SBR as well as ESBR are routinely determined by nuclear magnetic resonance (nmr), differential thermal analysis (dta), or differential scanning calorimetry (dsc).

For routine analysis of SBR polymers, gpc is widely accepted.

Advantages of natural rubber and isoprene rubber (NR/IR) are high resilience and strength, and abrasion-resistance. BR shows low heat buildup in flexing, good resilience, and abrasion-resistance. Random SBR is low in price, wears well, and bonds easily. Block SBR is easily injection-molded, and is not cross-linked. Applications of NR/IR include tires, tubes, belts, bumpers, tubing, gaskets, seals, foamed mattresses, and padding. BR is used in tire treads and mechanical goods, as is random SBR. Block SBR is used in toys, rubber bands, and mechanical goods.

Raw Materials

The monomers butadiene (qv) and styrene (qv), are the most important ingredients in the manufacture of SBR polymers. For ESBR, the largest single material is water; for solution SBR, the solvent.

The quality of the water used in emulsion polymerization affects the manufacture of ESBR. Water hardness and other ionic content can directly affect the chemical and mechanical stability of the polymer emulsion (latex). Solution polymerization can use various solvents, primarily aliphatic and aromatic hydrocarbons. SSBR polymerization depends on recovery and reuse of the solvent for economical operation as well as operation under the air-quality permitting of the local, state, and federal mandates involved.

Styrene. Commercial manufacture of this commodity monomer depends on ethylbenzene, which is converted by several means to a low purity styrene, subsequently distilled to the pure form. A small percentage of styrene is made from the oxidative process, whereby ethylbenzene is oxidized to a hydroperoxide or alcohol and then dehydrated to styrene. A popular commercial route has been the alkylation of benzene to ethylbenzene, with ethylene, after which the crude ethylbenzene is distilled to give high purity ethylbenzene (see STYRENE).

Butadiene. Economic considerations favor recovering butadiene from by-products in the manufacture of ethylene. Butadiene is a by-product in the C4 streams from the cracking process. For use in polymerization, the butadiene must be purified to 99 + %. Crude butadiene is separated from C_3 and C_5 components by distillation. Separation of butadiene from other C_4 constituents is accomplished by salt complexing/solvent extraction (see BUTADIENE).

Soap. A critical ingredient for emulsion polymerization is the soap (qv), which performs a number of key roles, including production of oil (monomer) in water emulsion, provision of the loci for polymerization (micelle), stabilization of the latex particle, and impartation of characteristics to the finished polymer. Both fatty acid and rosin acid soaps, mainly derived from tall oil, are used in ESBR.

Polymerization

ESBR and SSBR are made from two different addition polymerization techniques: one radical and one ionic. ESBR polymerization is based on free radicals that attack the unsaturation of the monomers, causing addition of monomer units to the end of the polymer chain, whereas the basis for SSBR is by use of ionic initiators (qv).

Free-radical initiation of emulsion copolymers produces a random polymerization in which the trans/cis ratio cannot be controlled. The nature of ESBR free-radical polymerization results in the polymer being heterogeneous, with a broad molecular weight distribution and random copolymer composition. The microstructure is not amenable to manipulation, although the temperature of the polymerization affects the ratio of trans to cis somewhat.

In solution-based polymerization, use of the initiating anionic species allows control over the trans/cis microstructure of the diene portion of the copolymer. In solution SBR, the alkyllithium catalyst allows the 1,2 content to be changed with certain modifying agents such as ethers or amines. Anionic initiators are used to control the molecular weight, molecular weight distribution, and the microstructure of the copolymer.

SBR Compounding and Processing

The art of compounding requires extensive experience and knowledge of the many compound ingredients. A typical rubber compound in addition to polymer contains one or more ingredients from the following general classes: vulcanizing agents, accelerators, accelerator activators, antioxidants, pigments,and softeners (see RUBBER CHEMICALS).

The vulcanizing agent, which supplies the bridge between the polymer chains, is furnished predominantly by the sulfur molecule in commercial formulations. Peroxide vulcanizers that produce carbon-to-carbon cross-links are also important.

Accelerators are chemical compounds that increase the rate of cure and improve the physical properties of the compound. Accelerator activators are chemicals required to initiate the acceleration of the curing process.

Antioxidants (qv) are routinely added to the compounds over and above those contained in the polymer at manufacture. Antiozonants (qv) prevent or reduce polymer degradation by the active ozone molecule. Some antioxidant compounds, such as the *para*-phenylenediamines, are excellent as antiozonants.

Pigments (qv) improve or change polymer properties as well as lower product costs. Reinforcement of SBR by carbon blacks allows this family of polymers to compete with natural rubber (see CARBON, CARBON BLACK). It is the most important attribute of the pigment in SBR processing. Softeners, ie, plasticizers, reinforcing agents, extenders, lubricants, tackifiers, and dispersing aids, are used as processing aids to enhance mixing of uncured stocks and soften cured compounds.

Economic Aspects and Uses

Styrene–butadiene elastomers, emulsion and solution types combined, are reported to be the largest-volume synthetic rubber. The actual percentage has decreased steadily since 1973. The decline has been attributed to the switch to radial tires (longer milage) and the growth of other synthetic polymers. SBR is forecast to remain the dominant elastomer of all synthetic polymers. In the late 1990s, use of SBR has encompassed the following: tires and tire-related products, including tread rubber, 80%; mechanical goods, 11%; other automotive uses, 6%; and adhesives, chewing gum base, shoe products, flooring, etc, for the remaining 3%.

Health and Safety Factors

Air quality and plant effluent have been monitored and more or less regulated from the inception of SBR manufacture. Most local and state governments have strict discharge permits that limit what kind of chemicals and how much of it can be emitted into the environment. Both styrene and butadiene are considered suspect carcinogens.

There is an industry trend to supply SBR certifiably free of volatile nitrosamines or nitrosatable compounds. Of primary concern to local, state, and federal governments is the growing stockpile of scrap tires. The threat of huge piles of scrap tires catching fire is cited as a principal concern. Such fires pollute the air and threaten groundwater as the large quantities of oil released in the incomplete burning become a serious runoff problem. Although use of scrap tires is projected to increase rapidly, the only economically feasible use has been as a fuel or fuel supplement in utility and industrial applications (see RECYCLING, RUBBER).

RICHARD R. LATTIME
The Goodyear Tire & Rubber Company

G. S. Whitby, ed., *Synthetic Rubber,* John Wiley & Sons, Inc. New York, 1954.

The Vanderbilt Rubber Handbook, 13th ed., R. T. Vanderbilt Co., Inc., Norwalk, Conn., 1990.

STYRENE–BUTADIENE SOLUTION COPOLYMERS.

See ELASTOMERS, SYNTHETIC.

STYRENE PLASTICS

Polystyrene (PS), the parent of the styrene plastics family, is a high molecular weight linear polymer which, for commercial uses, consists of ~1000 styrene units. Its chemical formula (**1**), where n =~ 1000, tells little of its properties. The main commercial form of PS is amorphous and hence PS is highly transparent. Addition of butadiene-based rubbers increases impact resistance, and copolymerization of styrene with acrylonitrile produces heat- and solvent-resistant plastics (see ACRYLONITRILE POLYMERS).

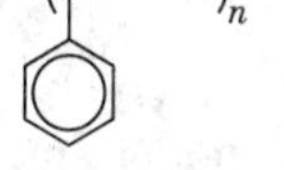

(**1**)

Properties

The general mechanical properties of styrene polymers are given in Table 1. Considerable differences in performance can be achieved by using the various styrene plastics. In choosing an appropriate resin for a given application, other properties and polymer behavior during fabrication must be considered. Consideration must be given to such factors as the surface appearance of the part and the development of anisotropy, and the effect of anisotropy on mechanical strength, ie, long-term resistance of the molding to external strain.

Physical. In general, a polymer must have a weight-average molecular weight (M_w) about 10 times higher than its chain entanglement molecular weight (M_e) to have optimal strength.

Stress–Strain Properties. The strain energy, derived from the area under the stress–strain curve, is considered to indicate the level of toughness of a polymer. High impact PS (HIPS) has a higher strain energy than an acrylonitrile–butadiene–styrene (ABS) plastic. ABS materials are generally tougher than HIPS materials. Tensile strengths of styrene polymers vary with temperature. Increased temperature lowers the strength. The molecular orientation of the polymer in a fabricated specimen can significantly alter the stress–strain data.

Creep, Stress Relaxation, and Fatigue. Creep tests involve the measurement of deformation as a function of time at a constant stress or load. Creep curves for styrene and its copolymers at room temperature show low elongation having only small variation with stress.

Stress-relaxation measurements, where stress decay is measured as a function of time at a constant strain, have also been used extensively to predict the long-term behavior of styrene-based plastics. Fatigue is another property of considerable interest to the design engineer. Cyclic deflections of a predetermined amplitude, short of giving immediate failure, are applied to the specimen, and the number of cycles to failure is recorded. In addition to mechanically induced periodic stresses, fatigue failure can be studied when developing cyclic stresses by fluctuating the temperature.

Melt Properties. The melt properties of PS at temperatures between 120 and 260°C are important, because it is in this temperature range that PS is extruded to make sheets, foams, and films, or molded into parts. Generally it is desirable to make parts having high strength from materials having low melt viscosity for easy melt processing. In applications where heat resistance is important, melt processibility can be influenced, without a significant effect on heat resistance, by control of the polydispersity, by branching, or by the introduction of pendent ionic groups.

Impact Strength. PS and styrene copolymers are brittle polymers under normal use conditions. Rubber-modified styrene polymers, however, are significantly more impact-resistant. Embrittlement of otherwise tough rubber-modified styrene polymers occurs through aging of these polymers.

Surface Appearance. HIPS materials can have a glossy or a dull surface appearance. This surface appearance is a function of a surface roughness caused by rubber particles disrupting the surface regularity. The irregularities are caused not only by the nature of the rubber particles, eg, size and shape, but also by processing conditions.

Material Types

General-Purpose Polystyrene. Polystyrene is a high molecular weight ($M_w = 2 - 3 \times 10^5$), crystal-clear thermoplastic that is hard, rigid, and free of odor and taste. In addition, PS materials have excellent thermal and electrical properties that make them useful as low cost insulating materials (see INSULATION, ELECTRIC; INSULATION, THERMAL). When additional lubricants, eg, mineral oil and butyl stearate, are added to PS, easy-flow materials are produced.

Specialty Polystyrenes. These include ionomers and PS of specified tacticity, as well as stabilized PS.

PS ionomers are typically prepared by copolymerizing styrene with an acid functional monomer or by sulfonation of PS followed by neutralization of the pendent acid groups with monovalent or divalent alkali metals.

Isotactic (IPS) and syndiotactic (SPS) polystyrenes can be obtained by the polymerization of styrene with stereospecific catalysts of the Ziegler-Natta type. As a result of the regular tactic structure, both IPS (phenyl groups cis) and SPS (phenyl groups alternating trans) can be crystallized. In the crystalline state, both IPS and SPS are opaque and insoluble in most common PS solvents.

Stabilized polystyrenes are materials with added stabilizers, eg, uv-screening agents, antioxidants (qv), and synergistic agents. Polymers containing flame retardants (qv) have been developed. The addition of flame retardants does not make a polymer noncombustible, but rather increases the polymer's resistance to ignition and reduces the rate of burning with minor fire sources.

Styrene Copolymers. Acrylonitrile, butadiene, α-methylstyrene, acrylic acid, and maleic anhydride have been copolymerized with styrene to yield commercially significant copolymers. Acrylonitrile copolymer with styrene (SAN), the largest volume styrenic copolymer, is used in applications requiring increased strength and chemical resistance over PS. (see ACRYLONITRILE POLYMERS; COPOLYMERS).

SAN is extremely incompatible with PS. Copolymers with over 30 wt % acrylonitrile have good barrier properties. If the acrylonitrile content of the copolymer is increased to > 40 wt%, the copolymer becomes ductile. These copolymers also constitute the rigid matrix phase of the ABS engineering plastics.

Butadiene copolymers are mainly prepared to yield rubbers (see STYRENE–BUTADIENE RUBBER). Most of the block copolymers prepared by anionic catalysts, eg, butyllithium, are also elastomers; however, some are thermoplastic rubbers. Diblock (styrene–butadiene (SB)) and triblock (styrene–butadiene–styrene (SBS)) copolymers represent a class of new and interesting polymeric materials (see POLYMER BLENDS).

Maleic anhydride readily copolymerizes with styrene to form an alternating structure. Depending on their molecular weights, these copolymers can be used as chemically reactive resins or as high heat-deformation molding materials.

Table 1. Mechanical Properties of Injection-Molded Specimens of Main Classes of Styrene-Based Plastics

						Acrylonitrile–butadiene–styrene terpolymer (ABS)[a]			
Property	Polystyrene (PS)	Poly(styrene-*co*-acrylonitrile) (SAN)[b]	Glass-filled PS[c]	High impact PS[a]	HIPS	Type 1	Type 2	Standard ABS	Super ABS
specific gravity	1.05	1.08	1.20	1.05	1.05	1.05	1.05	1.04	1.04
Vicat softening point, °C	96	107	103	103	95	99	108	103	108
tensile yield, MPa[d]	42.0	68.9	131	39.6	29.6	31.0	53.8	41.4	34.5
elongation, rupture, %	1.8	3.5	1.5	15	58	55	10	20	60
modulus, MPa[d]	3170	3790	7580	2690	2140	2620	2620	2070	1790
impact strength (notched Izod), J/m[e]	21	21	80	96	134	193	187	267	428
dart-drop impact strength	very low	very low	medium high	low	medium high	medium high	high	very high	very high
relative ease of fabrication	excellent	excellent	poor	excellent	excellent	excellent	good	good	medium good

[a] Medium mol wt. [b] 24 wt % acrylonitrile. [c] 20% glass fibers. [d] To convert MPa to psi, multiply by 145. [e] To convert J/m to ft · lbf/in., divide by 53.38.

Polymers of Styrene Derivatives. Many styrene derivatives have been synthesized and the corresponding polymers and copolymers prepared. The highest T_g is that of poly(α-methylstyrene), which can be prepared by anionic polymerization. The polymer, which is difficult to fabricate because of its high melt viscosity, is more brittle than PS but can be toughened with rubber.

Poly(sodium styrenesulfonate), a versatile water-soluble polymer, is used in water-pollution control and as a general flocculant (see WATER, INDUSTRIAL WATER TREATMENT; FLOCCULATING AGENTS). Poly(vinylbenzyl ammonium chloride) has been useful as an electroconductive resin (see ELECTRICALLY CONDUCTIVE POLYMERS).

Rubber-Modified Polystyrene. Rubber is incorporated into PS primarily to impart toughness. The resulting materials, commonly called high impact polystyrene (HIPS), are available in many different varieties. In rubber-modified PS, the rubber is dispersed in the PS matrix in the form of discrete particles.

Acrylonitrile–butadiene–styrene (ABS) polymers, like HIPS, are two-phase systems in which the elastomer component is dispersed in the rigid SAN copolymer matrix (see ACRYLONITRILE POLYMERS).

The combination of stiffness, impact strength, and solvent resistance makes ABS polymers particularly suitable for demanding applications. Another important attribute of several ABS polymers is their minimum tendency to orient or develop mechanical anisotropy during molding. ABS can be blended with bisphenol A polycarbonate resins to make a material having excellent low temperature toughness. Not only are ABS polymers useful engineering plastics, but some of the high rubber compositions are excellent impact modifiers for poly(vinyl chloride) (PVC).

Rubber modification of styrene copolymers other than HIPS and ABS has been useful for specialty purposes. Transparency has been achieved with the use of methyl methacrylate as a comonomer.

Glass-Reinforced Styrene Polymers. Glass reinforcement of PS and SAN markedly improves mechanical properties. The strength, stiffness, and fracture toughness are generally doubled at least. Creep and relaxation rates are significantly reduced and creep rupture times are increased. The coefficient of thermal expansion is reduced by more than one half, and generally response to temperature changes is minimized. Normally glass fibers can be used to achieve these improvements.

Degradation

Styrene plastics are susceptible to degradation by heat, oxidation, uv radiation, high energy radiation, and shear, although in normal use only uv radiation imposes any real limit on the general usefulness of these plastics. It is generally recommended that the use of styrene plastics in outdoor applications be avoided. A clear difference in thermal stability has been shown between PS produced using free-radical (FRPS) and anionic (APS) polymerization. This difference is due mainly to the initiator-derived fragments that remain in the polymer after isolation. Thermal oxidative degradation of PS occurs rapidly, leading to additional volatile components consisting of aldehydes and ketones, yellowing of the polymer with a very dramatic drop in molecular weight, and some cross-linking. Rates and yields are highly oxygen- and temperature-sensitive.

Environmental Considerations

Environmental Degradation. Polystyrene is quite photodegradable and must be stabilized by the addition of uv-absorbing additives if it is to be used in outdoor applications where durability is important. Even though PS is naturally quite photodegradable, there have been considerable efforts to accelerate the process to produce so-called photodegradable PS. The approach is to add photosensitizers, typically ketone containing molecules, that absorb sunlight. The absorbed light energy is then transferred to the polymer to cause backbone scission via an oxidation mechanism. Photodegradable PS is useful in litter-prone applications, eg, fast food packaging.

A concern in the use of photodegradable PS is the environmental impact of the products of photooxidation. However, photodegraded PS is expected to be more susceptible to biodegradation because the molecular weight has been reduced, the PS chains have oxidized end groups, the incorporation of oxygen as alcohol and ketone groups has increased the hydrophilicity of the PS fragments, and the surface area has increased.

Biodegradation. Efforts have been made to enhance the biodegradability of PS by inserting hydrolyzable linkages, eg, ester and amide, into its backbone.

Polymerization

Styrene and most of its derivatives are among the few monomers that can be polymerized by all four distinct mechanisms, ie, anionic, cationic, free-radical, and Ziegler-Natta. Each of the mechanisms used to polymerize styrene has its own unique advantages and disadvantages. Styrene–butadiene block copolymers are made with anionic chain carriers, and low molecular weight PS is made by a cationic mechanism.

Free-Radical Polymerization. The styrene monomers are almost unique in their ability to undergo spontaneous or thermal polymerization merely by heating to >100°C. Styrene in essence acts as its own initiator.

The mechanism for spontaneous polymerization proposed by F. R. Mayo involves the Diels-Alder reaction of two styrene molecules to form a reactive dimer (DH) followed by a molecular assisted homolysis between DH and another styrene molecule.

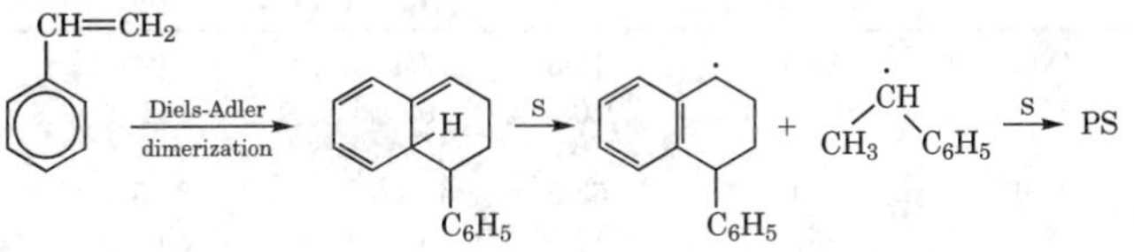

S DH D· SH·

Initiators (qv) that have been utilized to initiate styrene polymerization can be generally categorized into three types: peroxides, azo, and carbon–carbon. Below 80°C, radical combination is the primary termination mechanism; above it, both disproportionation and chain transfer with the Diels-Alder dimer are increasingly important.

Chain-transfer (CT) agents are occasionally added to styrene to reduce the molecular weight of the polymer; however polymerization temperature alone is generally sufficient to achieve molecular weight control. High levels of CT agents can be used to control the termination process. If the CT agent has a functional group attached to it, the functional group ends up becoming attached to the end of the PS chain. Another approach to control end-group structure is the use of CT agents that operate by an addition-fragmentation mechanism. This approach can lead to the formation of PS having functional groups at both ends if both the initiating and terminating fragments contain functional groups.

Ionic Polymerization. Instead of a neutral unpaired electron, styrene polymerization can proceed with great facility through a positively charged species (cationic polymerization) or a negatively charged species (anionic polymerization). *n*-Butyllithium (NBL) is the most widely used initiator for anionic polymerization of styrene. In solution, it exists as six-membered aggregates, and a key step in the initiation sequence is dissociation yielding at least one isolated molecule. Anionic polymerization, if carried out properly, can be truly a living polymerization. Addition of a second monomer to polystyryl anion results in the formation of a block polymer with no detectable free PS. This technique is of considerable importance in the commercial preparation of styrene–butadiene block copolymers, which are used either alone or blended with PS as thermoplastics.

Cationic polymerization of styrene can be initiated either by strong acids or by Friedel-Crafts reagents with a proton-donating activator. The solvent plays an important role, and chain-transfer reactions are very common. As a result, high molecular weights are more difficult to achieve. Commercial use of cationic styrene polymerization is reported only where low molecular weight polymers are desired.

Living Styrene Polymerization. The requirements for a polymerization to be truly living are that the propagating chain ends must not terminate during polymerization. If the initiation, propagation, and termination steps are sequential, ie, all of the chains are initiated and then propagate at the same time without any termination, then monodisperse (ie, $M_w/M_n = 1.0$) polymer is produced. In general, anionic polymerization is the only mechanism that yields truly living styrene polymerization and thus monodisperse PS.

Copolymerization. Styrene readily copolymerizes with many other monomers spontaneously. Copolymerization makes possible dramatic improvements in one or more properties. Styrene–acrylonitrile (SAN) copolymers are large-volume thermoplastics having improved mechanical properties and heat and solvent resistance with some loss of color stability (see ACRYLONITRILE POLYMERS). Copolymers with butadiene, ie, those containing at least 60 wt % butadiene, are an important family of rubbers.

Styrene–maleic anhydride (SMA) copolymers are used where improved resistance to heat is required. Polymerization of styrene in the presence of polybutadiene rubber yields a much tougher material than PS. Although the material is usually opaque and has a somewhat lower modulus, the gain in impact resistance has placed this plastic in wide commercial usage.

Commercial Processes

There are two problems in the manufacture of PS: removal of the heat of polymerization (ca 700 kJ/kg (300 Btu/lb)) of styrene polymerized and the simultaneous handling of a partially converted polymer syrup with a viscosity of ca 10^5 mPa(= cP). For the four mechanisms free radical, anionic, cationic, and Ziegler, several processes can be used.

Two types of reactors are used for continuous solution polymerization: the linear-flow reactor (LFR), approximating in the ideal case a plug-flow reactor, and the continuous-stirred tank reactor (CSTR), which ideally is isotropic in composition and temperature (see REACTOR TECHNOLOGY). LFRs usually involve conductive heat transfer to many tubes through which a heat-transfer fluid flows. Reactors of this type operate for long periods and handle very high viscosity partial polymers with reliable heat-transfer coefficients. CSTRs are either of the recirculated coil configuration and rely on conduction to the cooled jacket for heat removal, or ebullient and makes use primarily of evaporative cooling to achieve temperature control. Most general-purpose, ie, unmodified PS is manufactured in the United States in such facilities.

Rubber-modified PS manufacture places several additional demands: dissolving ca 5–10 wt % polybutadiene in styrene, often a 10–20-h process; shear conditions to achieve phase inversion and the desired particle size for a given product; control of phase compositions to produce the desired particle morphology; and sufficient time and temperature at the end of the process to achieve the necessary cross-linking, gel formation, and swelling index. In the case of LFR system, only addition of a rubber dissolver is needed to meet the above requirements. For a CSTR-based system, multiple reactors in series are required to avoid rapid shifts in phase compositions on mixing of one reactor effluent with the next reaction contents with loss of desirable morphology.

Polybutadiene rubbers are used almost exclusively as reinforcing agents for styrenic plastics. The very low glass-transition temperature (−80°C or lower) gives good low temperature properties, and the allylic hydrogen atom and weakly active double bonds can provide the desired degree of grafting and cross-linking.

Fabrication

Injection Molding. There are two basic types of injection-molding machines in use: the reciprocating screw and the screw preplasticator. The injection-molding process is basically the forcing of melted polymer into a relatively cool mold where it freezes and is removed in a minimum time. The shape of a molding is defined by the cavity of the mold. Injection molding of styrene-based plastics is usually carried out at 200–300°C. For ABS polymers, the upper limit may be somewhat less. To obtain satisfactory moldings with good surface appearance, contamination, including that by moisture, must be avoided.

Extrusion. Extrusion of styrene polymers is one of the most convenient and least expensive fabrication methods, particularly for obtaining sheet, pipe, irregular profiles, and films. Extrusion is also the method for plasticizing the polymer in screw injection-molding machines and is used to develop the parison for blow molding. Extrusion is a continuous method involving relatively inexpensive equipment (see FILM AND SHEETING MATERIALS). Many rubber-modified styrene plastics are fabricated into sheet by extrusion primarily for subsequent thermoforming operations. With the advent of low cost computers, closed-loop control of the extrusion system has become commonplace. Lamination of polymer films to styrene-based materials can be carried out during the extrusion process.

Thermoforming and Orientation. Thermoforming of HIPS and ABS extruded sheet is of considerable importance in several industries. Thermoforming is usually accomplished by heating a plastic sheet above its softening point and forcing it against a mold by applying vacuum, air, or mechanical pressure. On cooling, the contour of the mold is reproduced in detail.

Blow Molding. Blow molding is a multistep fabrication process for manufacturing hollow symmetrical objects. The granules are melted and a parison is obtained by extrusion or by injection molding. The parison is then enclosed by the mold, and pressure or vacuum is applied to force the material to assume the contour of the mold. After sufficient cooling, the object is ejected. PS or copolymers are used extensively in injection blow molding. Tough and craze-resistant PS containers have been made by multiaxially oriented injection-molded parisons. This process permits the design of blow-molded objects with a high degree of controlled orientation, independent of blow ratio or shape.

Additives. Processing aids, eg, plasticizers and mold-release agents, are often added to PS (see RELEASE AGENTS). For food contact applications, the additives must be FDA-approved. Important additives used in styrene plastics are listed in Table 2.

Characterization

Four modes of characterization are of interest: chemical analyses, ie, qualitative and quantitative analyses of all components; mechanical characterization, ie, tensile and impact testing; morphology of the rubber phase; and rheology at a range of shear rates. Other properties measured are stress crack resistance, heat distortion temperatures, flammability, creep, etc, depending on the particular application.

There are three components of chemical analysis: the high molecular weight portion, the additives, and the residuals remaining from the polymerization process. The high molecular weight portion is best characterized using gel permeation chromatography. The additives in the polymer are best characterized by extracting them from the polymer and analyzing the extract using high performance liquid chromatography. Mechanical testing is carried out according to standard ASTM methods.

Rubber particle size distribution is usually measured with a Coulter counter or directly from electron photomicrographs. Rheological studies are made using a modified tensile tester with capillary rheometer (ASTM D1238-79) or with the powerful Rheometrics mechanical spectrometer. Molecular weight distribution and residuals analyses are used for product specifications.

Health and Safety Factors

In pellet form, styrene-based plastics have a very low degree of toxicity. Under normal conditions of handling and use, they should pose no unusual problems from ingestion, inhalation, or eye and skin contact. Heating of these polymers usually results in the release of some vapors. Adequate ventilation should be provided. Styrene polymers burn under the right conditions of heat and oxygen supply. Combustion products from any burning organic material should be considered toxic. Fires can be extinguished by conventional means, ie, water and water fog.

Table 2. Additives Used in Styrene Plastics

Type	Compounds	Amount wt %
plasticizers (qv)	mineral oil, phthalate esters, adipate esters	<4
mold-release agents	steric acid, metal sterates, sterate esters, silicones, amide waxes	0.1
		3
antioxidants (qv)	alkylated phenols, phosphite esters, thioesters	1
		1
		1
antistatic agents	quaternary ammonium compounds (qv)	2
uv stabilizers	benzotriazoles, benzophenones	0.25
ignition suppression agents	halogenated compounds, antimony oxide, aluminum oxide, phosphate esters	

Uses

The following are commercially significant foamed PSs. Extruded planks and boards that are largely flame-retardant and in the density range of 29 kg/m^3 are used for low temperature thermal insulation, buoyancy, floral display, novelty, packaging (qv), and construction purposes. Foamed boards and shapes from foaming-in-place (FIP) beads (density 16–32 kg/m^3) are used for packaging, buoyancy, insulation, and numerous other applications. Extruded foamed PS sheet 17-mm thick with densities of 64–160 kg/m^3 are used largely in packaging applications. High density extruded planks (density 35–64 kg/m^3), are used for heavy-duty structural applications. Foamable ABS systems, eg, laminates with ABS skins and a heat-foamable core, are used for structural applications, as in car body parts. Of growing commercial significance is coextruded, foam-core ABS pipe (see FOAMED PLASTICS).

Foamed PSs are used for insulation against ambient temperatures in the form of perimeter insulation and insulation under floors and in walls and roofs (see ROOFING MATERIALS).

Expandable PS (EPS) is in wide use in packaging. Applications range from pallet shipping containers for computer terminals, material-handling stacking trays, to packages for fresh fruits and vegetables.

Use of injection-molded structural foams, based on HIPS and ABS and containing flame-retardant additives, has grown rapidly. Because of environmental concerns over the use of halogenated flame-retardant chemicals, research is focusing on the development of halogen-free systems (see FLAME RETARDANTS).

Foamed Sheet. PS foamed sheet is used for foamed trays, egg cartons, disposable dinnerware, and packaging.

Oriented PS Film and Sheet. Oriented PS film, in addition to being heat-shrinkable and having the lowest cost of any of the rigid plastic materials, offers a high degree of optical clarity, high surface gloss, and excellent dimensional stability, particularly with regard to relative humidity. It is not a barrier film; in fact, one of its largest uses, that of packaging field-fresh produce, depends on its being highly permeable to oxygen and water vapor. Although PS is normally considered a rather brittle material, biaxial orientation imparts some extremely desirable properties, particularly in regard to an increase in elongation.

PS film contains no plasticizers, absorbs negligible moisture, and exhibits exceptionally good dimensional stability. It does not become brittle with age nor distort when exposed to low or high humidity. Excellent clarity, stability, and machinability, ie, ability to pass through packaging machinery at high speeds and be cut and sealed, etc, are central factors in the use of oriented PS in various applications. PS sheet possesses excellent dimensional stability and resistance to moisture.

DUANE B. PRIDDY
The Dow Chemical Company

R. F. Boyer, H. Keskkula, and A. E. Platt, in N. M. Bikales, ed., *Encyclopedia of Polymer Science and Technology,* Vol. 13, John Wiley & Sons, New York, 1970, pp. 128–447; E. R. Moore and co-workers, *ibid.*, Vol. 13, 1989, pp. 1–246.

J. F. Rudd, in J. Brandrup and E. H. Immergut, eds., *Polymer Handbook,* 3rd ed., John Wiley & Sons, Inc., 1989, p. V-81–86.

C. B. Bucknall, *Toughened Plastics,* Applied Science Publishers, London, 1977, Chapts. 7 and 10.

C. E. Rogers, in E. Baer, ed., *Engineering Design for Plastics,* Reinhold Publishing Corp., New York, 1964, pp. 609–688.

SUBERIC ACID. See DICARBOXYLIC ACIDS.

SUCCINIC ACID AND SUCCINIC ANHYDRIDE

Succinic acid, (**1**) (butanedioic acid; 1,2-ethanedicarboxylic acid; amber acid), $C_4H_6O_4$, occurs frequently in nature as such or in the form of its esters. It can be found in animal tissues, in vegetables and fruit, or in spring water, and has also been identified in meteorites.

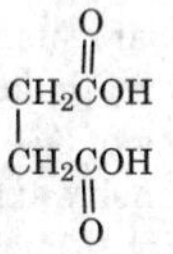

(**1**)

Succinic anhydride (3,4-dihydro-2,5-furandione; butanedioic anhydride; tetrahydro-2,5-dioxofuran; 2,5-diketotetrahydrofuran; succinyl oxide), $C_4H_4O_3$, was first obtained by dehydration of succinic acid. From the chemical point of view, succinic acid and its anhydride are characterized by the reactivity of the two carboxylic functions and of the two methylene groups.

Physical Properties

The acid occurs both as colorless triclinic prisms (α-form) and as monoclinic prisms (β-form). The β-form is triboluminescent and is stable up to 137°C; the α-form is stable above this temperature. Both forms dissolve in water, alcohol, diethyl ether, glacial acetic acid, anhydrous glycerol, acetone, and various aqueous mixtures of the last two solvents. Succinic acid sublimes with partial dehydration to the anhydride when heated near its melting point.

Succinic anhydride forms rhombic pyramidal or bipyramidal crystals. It is relatively insoluble in ether, but soluble in boiling chloroform and ethyl acetate. Succinic anhydride reacts with water and alcohols, giving the acid and monoesters, respectively. Physical properties of the acid and its anhydride are summarized in Table 1.

Table 1. Physical Properties of Succinic Acid and Succinic Anhydride

Property	Succinic acid	Succinic anhydride
molecular weight	118.09	100.08
melting point, °C	188	119.6
boiling point at 101.3 kPa (= 1 atm), °C	dehydration, 235	261
density, g/cm^3		
solid, 20°C		1.2
solid, 25°C	1.572	
solubility, g/100 g solvent		
in water at 0°C	2.8	
in water at 100°C	121	
96% ethanol at 15°C	10	
ethyl ether at 15°C	1.2	
methylene chloride at bp	insoluble	6.6
chloroform at bp	insoluble	3.7
enthalpy of combustion, kJ/mol[a]	−1491	−1537.9
enthalpy of formation at 298.15 K, kJ/mol[a]	−940.5	−607.8
heat capacity at 298.15 K, J/(mol·K)[a]	153	
enthalpy of solution, kJ/mol[a]	−27.3	
enthalpy of sublimation, kJ/mol[a]		80.7
enthalpy of fusion, kJ/mol[a]		20.41
dielectric constant at 3–97°C, 5 kHz	2.29–2.90	
flammability point, °C		147

[a] To convert kJ/mol to kcal/mol, divide by 4.184.

Chemical Properties

Succinic acid and anhydride undergo most of the reactions characteristic of dicarboxylic acids and cyclic acid anhydrides, respectively. Other interesting reactions take place at the active methylene groups.

Heat. When heated, succinic acid loses water and forms an internal anhydride (**2**) with a stable ring structure. Further heating of succinic anhydride causes decarboxylation and the formation of the dilactone of gamma ketopimelic acid (eq. 1). The same reaction takes place at lower temperatures in the presence of alkali.

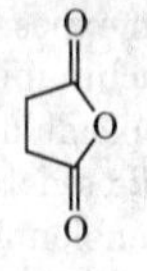

(**2**)

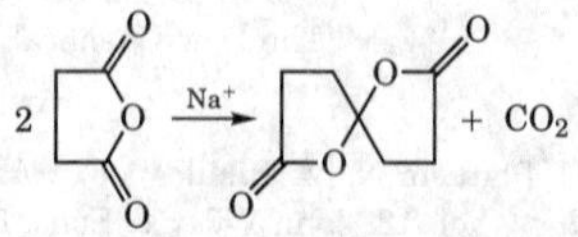

(1)

At higher temperatures the presence of alkali causes an explosive reaction. Precautions must therefore be taken to exclude traces of alkali when handling succinic anhydride.

Succinic anhydride is stabilized against the deteriorative effects of heat by the addition of small amounts (0.5 wt %) of boric acid.

Hydration and Dehydration. Succinic anhydride reacts slowly with cold water and rapidly with hot water to give the acid. Succinic acid can be dehydrated to the anhydride by heating at 200°C.

Esterification. Succinic anhydride reacts readily with alcohols to give monoesters of succinic acid, which are readily esterified to diesters by the usual methods. Dimethyl succinate (mp 19°C, bp 196°C at atmospheric pressure) can be produced from methanol and the anhydride or the acid, or by hydrogenation of dimethyl maleate. The same methods can be used to prepare diethyl succinate (mp −18°C, bp 216.5°C at atmospheric pressure) and diisopropyl succinate.

An important use of dialkyl succinates is in the preparation of dialkyl succinyl succinates, which are intermediates in the manufacture of quinacridone pigments. The reaction is carried out in the presence of alkali metal alkoxides (eq. 2).

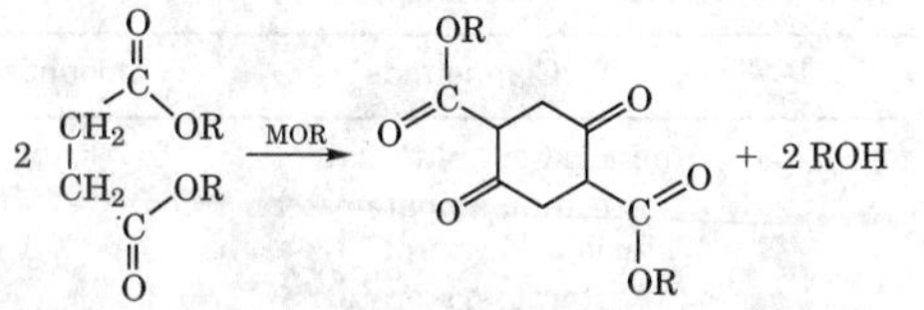

(2)

Oxidation. Succinic acid reacts with hydrogen peroxide, giving different products that depend on the experimental conditions: peroxysuccinic acid $(CH_2COOOH)_2$, oxosuccinic acid (oxaloacetic acid); malonic acid, or a mixture of acetaldehyde, malonic acid, and malic acid. Succinic anhydride in dimethylformamide (DMF) with H_2O_2 gives monoperoxysuccinic acid, $HOOCCH_2CH_2COOOH$, mp 107°C.

Hydrogenation. Gas-phase catalytic hydrogenation of succinic anhydride yields γ-butyrolactone (GBL), tetrahydrofuran (THF), 1,4-butanediol (BDO), or a mixture of these products, depending on the experimental conditions.

Halogenation. Succinic acid and succinic anhydride react with halogens through the active methylene groups. Succinic acid heated in a closed vessel at 100°C with bromine yields 2,3-dibromosuccinic acid almost quantitatively. The anhydride gives the mono- or dibromo derivative, depending on the equivalents of bromine used.

Condensation with Aldehydes and Ketones. Succinic esters condense with aldehydes and ketones in the presence of bases, eg, sodium alkoxide or piperidine, to form monoesters of alkylidenesuccinic acids, eg, condensation of diethyl succinate with acetone yields ethyl 2-isopropylidenesuccinate (eq. 3). This reaction, known as Stobbe condensation, is specific for succinic esters and substituted succinic esters.

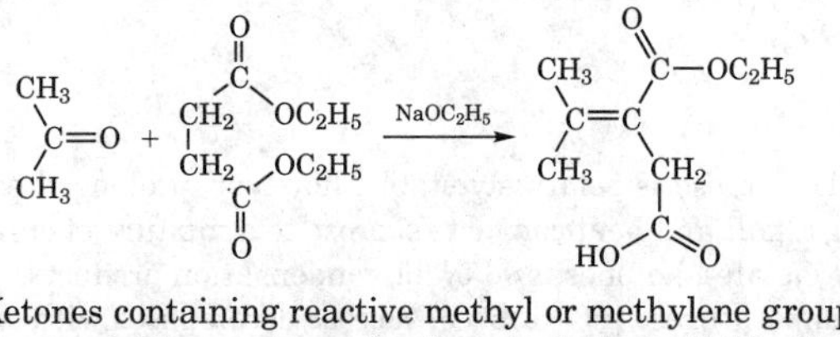
(3)

Ketones containing reactive methyl or methylene groups give with succinates, in the presence of sodium hydride, both the Stobbe condensation and the formation of diketones by a Claisen mechanism (eq. 4).

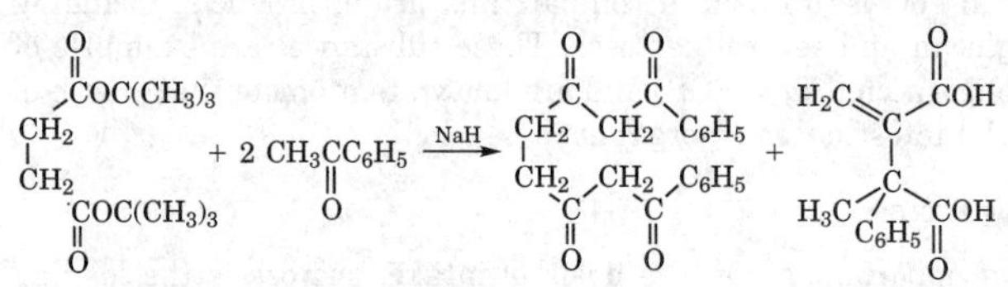
(4)

Friedel-Crafts Reactions. In the presence of Friedel-Crafts catalysts, succinic anhydride reacts with alkyl benzenes to form alkylbenzoylpropionic acids, eg, the reaction with indane gives a 97% yield of 4-oxo-(4,5-indanyl)butyric acid (eq. 5).

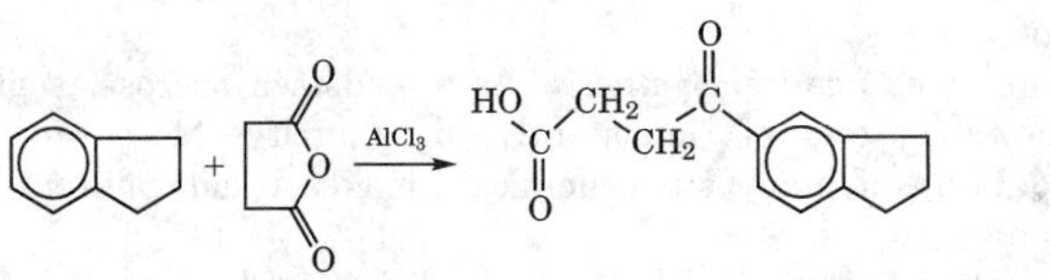
(5)

Reactions with Nitrogen Compounds. Succinimide, mp 126°C, can be prepared by reaction of aqueous solutions of the acid with ammonia or urea. The solution is heated until water and ammonia are no longer evolved and the molten crude succinimide is purified by fractionation.

Reactions with Sulfur Compounds. Thiosuccinic anhydride is obtained by reaction of diethyl or diphenyl succinate with potassium hydrogen sulfide followed by acidification. Sulfur trioxide reacts with both methylene groups to yield 2,3-disulfosuccinic acid (see SULFUR COMPOUNDS).

Miscellaneous Reactions. Radiolysis at room temperature of diluted aqueous solutions of succinic acid produces 1,2,3,4-butane tetracarboxylic acid, which has numerous industrial and agricultural applications.

Manufacture and Processing

Succinic anhydride is manufactured by catalytic hydrogenation of maleic anhydride. In the most widely used commercial process this reaction is performed in the liquid phase, at temperatures of 120–180°C and at moderate pressures. Catalysts include nickel, Raney nickel, palladium on different carriers, and palladium complexes.

The simplest route to succinic acid is by hydration of its anhydride. Pure succinic anhydride is dissolved in hot water, succinic acid is formed, separated as crystals upon cooling, filtered, and dried. Succinic acid can also be produced by catalytic hydrogenation of aqueous solutions of maleic or fumaric acid in the presence of noble metal catalysts.

A mixture of succinic (15–25 wt %), glutaric (45–55 wt %), and adipic acid (25–35 wt %) is obtained as a by-product in the oxidation of cyclohexane to adipic acid. Various techniques have been proposed for the recovery of pure succinic acid, including extraction, selective crystallization, heating to dehydrate the acid and subsequent recovery of succinic anhydride by distillation, esterification followed by fractionation of the mixture of the esters, and separation as urea adduct. Succinic anhydride can be prepared from succinic acid by dehydration; it operates in high boiling solvent, in the presence of clays as a catalyst, or at room temperature with triphosgene.

Specifications and Analysis

Methods used to analyze succinic acid and succinic anhydride include acidimetric titration for total acidity or purity; comparison with Pt–Co standard calibrated solutions for color; oxidation with potassium permanganate for unsaturated compounds; atomic absorption or plasma spectroscopy for metals; titration with $AgNO_3$ or $BaCl_2$ for chlorides and sulfates, respectively; and comparison of the color of the sulfide solution of the metals with that of a solution with a known Pb content for heavy metals. Techniques used for the determination of small concentrations of succinic acid or anhydride in various substances include gc or capillary gc, ion chromatography, gc/ms hplc, and polarography.

Health and Safety Factors

Succinic acid is Generally Recognized As Safe (GRAS) by the U.S. FDA and is approved as a flavor enhancer, as a pH control agent, and for use in meat products. It causes irritation to the eyes, skin, mucous membranes, and upper respiratory tract. Succinic acid, like most materials in powder form, can cause dust explosion. Succinic anhydride is extremely irritating to the eyes. It may be a sensitizer. There is no evidence of carcinogenic activity in male or female rats given 50 or 100 mg/kg succinic anhydride; the Ames test is negative. Succinic acid and anhydride should be handled with rubber or plastic gloves safety goggles and a dust filter are recommended when handling the products in powder form. A full-face gas mask with a type A (brown) filter cartridge should be worn when handling molten products.

Uses

Succinic acid has many applications. The main use of succinic acid in Japan is for bath preparations. An important emerging use of succinic acid and anhydride is the production of inherently degradable polymers (see POLYMERS, ENVIRONMENTALLY DEGRADABLE). Other important applications are in the fields of food additives (qv), detergents, cosmetics (qv), pigments (qv), toners, cement additives, soldering fluxes, as well as in the synthesis of pharmaceutical products.

CARLO FUMAGALLI
LONZA SpA

SUGAR

PROPERTIES OF SUCROSE

Sucrose (β-D-fructofuranosyl-α-D-glucopyranoside), $C_{12}H_{22}O_{11}$, formula weight 342.3, is a disaccharide composed of glucose and fructose residues joined by an α, β-glycosidic bond (Fig. 1).

The most common sugar in plants, sucrose is formed as a result of photosynthesis and occurs in abundance in sugarcane (*Saccharum officinarum*) and sugarbeets (*Beta vulgaris*). Commercial quantities are provided only from these two sources.

Sugarcane is cultivated in tropical and semitropical regions, eg, Central and South America, Cuba, India, Australia, Africa, and the Far East. Sugarbeets are grown in more temperate climates such as North America, Europe, and the former Soviet Union. In some nations, eg, the United States, China, and Japan, sucrose is produced from both sources.

Physical Properties of Sucrose

Physical properties of sucrose are summarized in Table 1. Sucrose is one of the purest substances available in bulk quantities, with purities

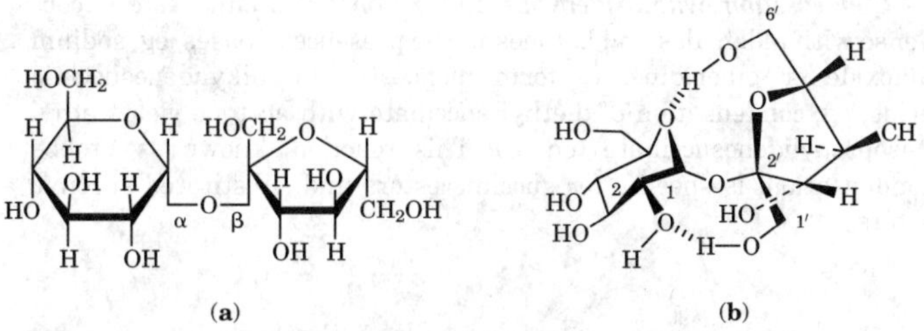

Figure 1. Structural representations of sucrose: (**a**) Haworth perspective formula, and (**b**) conformational structure of sucrose in solid crystals.

Table 1. Physical Properties of Sucrose

Property	Value
density, $d^{15}{}_4$, kg/m^3	1587.9
melting point, °C	160–186
specific rotation, degrees	+66.5
solubility in water at 20°C, g/g	2.00
apparent molar volume at 20°C, cm^3/mol	209.5
specific heat, J/mol^a	
crystalline, at 20°C	415.8
amorphous, at 22°C	90.2
heat of solution, kJ/mol^a	4.75 ± 0.26
dipole moment, C·m^b	3.1×10^{-18}
enthalpy of crystallization at 30°C, kJ/mol^a	10.5
bulk density, kg/m^3	
crystalline	930
powdered	600
normal entropy, J/(mol·K)	360.5
angle of repose, degree	34

[a] To convert J to cal, divide by 4.184.
[b] To convert C·m to debye, multiply by 3×10^{29}.

averaging ~99.96%. Water accounts for about half of the nonsugar impurities.

Sucrose crystals are triboluminescent and emit light when they are fractured. Aqueous sucrose solutions rotate polarized light in direct proportion to sugar concentration. The sweet taste of sucrose is its most notable and important physical property and is regarded as the standard against which other sweeteners (qv) are rated.

Sucrose has eight hydroxyl groups capable of hydrogen-bond formation. The glucose and fructose residues in crystalline sucrose are nearly perpendicular to each other and are held in this conformation by hydrogen bonds between the C_1-OH of fructose and the C_2-oxygen of glucose, and C_6-OH of fructose and the ring oxygen of glucose (Fig. **1b**). The hydroxyl groups of sucrose contribute to its very high water solubility. Sucrose forms sphenoidic monoclinic crystals, the shapes of which are affected by syrup impurities like raffinose and dextran. Impurities included within sucrose crystals impart color, raise ash levels, and reduce sugar quality. The high dielectric constant of sucrose (3.50–3.85) gives it substantial microwave-absorbing capacity and makes it a valuable ingredient in formulating microwavable foods.

Chemical Properties of Sucrose

The carbonyl groups of fructose and glucose partake in the glycosidic bond of sucrose, making the latter nonreducing. The hydroxyl groups of sucrose are very weakly acidic. The hydroxyls dissociate in alkali to form alcoholates called saccharates, and can be derivatized to produce valuable sucrochemicals.

Sucrose can be oxidized by HNO_3, $KMnO_4$, and peroxide.

Sucrose can be enzymatically hydrolyzed to glucose and fructose by invertase. During this reaction, the optical rotation falls to a negative value owing to the large negative specific rotation of fructose. The reversal is called inversion and the resulting glucose-fructose mixture is called invert.

At high pH, sucrose is relatively stable; however, prolonged exposure to strong alkali and heat converts sucrose to a mixture of organic acids (mainly lactate), ketones, and cyclic condensation products.

At high temperatures (160–186°C), sucrose decomposes with charring, emitting an odor of caramel. Acid-catalyzed thermolysis causes decomposition to glucose and fructofuranosyl cation. The latter reacts with sucrose to form a complex mixture of products, including fructoglucan and several kestoses. These substances are examples of fructooligosaccharides (FOS) and are known to promote the growth of beneficial intestinal microorganisms.

Uses for Sucrose

Food Applications. On the basis of intake, sucrose is the leading food additive. Its principal contribution to food is sweetness. However, it provides many other functionalities, eg, body, mouthfeel, texture, and moisture retention.

Subthreshold levels of sucrose enhance meat flavor; in cured meats, sucrose is a preservative and improves flavor by reducing the salty taste.

Feedstock for Chemical Synthesis. As a feedstock sucrose is plentiful, renewable, and of consistently high purity. Moreover, the biodegradability of many sucrochemicals makes them environmentally friendly.

Fermentation Feedstock. Sucrose, in the form of beet or cane molasses, is a fermentation feedstock for production of a variety of organic compounds. An abundantly produced substance is ethanol (qv) for use in alcoholic beverages, and as a fuel, solvent, and feedstock for organic syntheses. A valuable by-product of ethanol fermentation is industrial CO_2 (see CARBON DIOXIDE).

Pharmaceutical Applications. Sucrose imparts body to syrups and medicinal liquids and masks unpleasant tastes. Sucrose also functions as a diluent to control drug concentrations in medicines, as an ingredient binder for tablets, and to impart chewiness to the latter.

Sucralfate, an aluminum salt of sucrose octasulfate, is used as an antacid and antiulcer medication.

WILLIAM J. COLONNA
UPASIRI SAMARAWEERA
American Crystal Sugar Company

M. Mathlouthi and P. Reiser, eds., *Sucrose: Properties and Applications*, Chapman and Hall, New York, 1995.
N. L. Pennington and C. W. Baker, eds., *Sugar, A User's Guide to Sucrose*, Van Nostrand Reinhold Publishing Co., New York, 1990.
F. W. Lichtenthaler, ed., *Carbohydrates as Organic Raw Materials*, VCH Press, New York, 1990.
J. C. P. Chen and C.-C. Chou, eds., *Cane Sugar Handbook*, John Wiley & Sons, Inc., New York, 1993.
R. A. McGinnis, ed., *Beet Sugar Technology, 3rd ed.*, Beet Sugar Development Foundation, Fort Collins, Colo., 1982.

SUGAR ANALYSIS

Standards and Definitions

The International Commission for Uniform Methods of Sugar Analysis (ICUMSA) promulgates official methods of sugar analysis for the cane and beet sugar industry (see SUGAR, CANE SUGAR; SUGAR, BEET SUGAR).

Physical Methods of Sugar Analysis

The concentration of a pure sugar solution is determined by measurements of polarization (optical rotation), refractive index, and density.

Polarimetry. Polarimetry, or polarization, is defined as the measure of the optical rotation of the plane of polarized light as it passes through a solution. Saccharimeters are polarimeters in which the scales have been modified to read directly in percent sucrose based on the normal sugar solution reading 100%. The International Sugar Scale is calibrated in °Z. The 100°Z point is the optical rotation of the normal solution of pure sucrose at specified standard conditions.

The Clerget double polarization method is a procedure that attempts to account for the presence of interfering optically active compounds. Two polarizations are obtained: a direct polarization, followed by acid hydrolysis and a second polarization. The rotation of substances other than sucrose remains constant, and the change in polarization is the result of inversion (hydrolysis) of the sucrose.

Refractive Index. The refractometric value of sugar solutions is used as a rapid method for the approximate determination of the solids content (also known as dry substance), because it is assumed that the nonsugars present have a similar influence on the refractive index as sucrose. Measurement is usually carried out on a Brix refractometer, which is graduated in percentage of sucrose on a wt/wt basis (g sucrose/100 g solution) according to ICUMSA tables.

Density. Measurement of density to determine sugar concentration is made with an instrument called a hydrometer or a spindle. When it is graduated in sucrose concentration (percent sucrose by weight), it is called a Brix hydrometer or a Brix spindle. Brix is defined as the percent of dry substance by hydrometry, using an instrument or table calibrated in terms of percent sucrose by weight in water solution. Hydrometers are also graduated in °Baumé, still in use in some industries. The relationship between °Baumé and density, d, in g/cm^3, is °Baumé $= 145(1 - 1/d)$.

Purity. This is a widely used expression in the industry and represents, as a percentage, the proportion between polarization (considered a measure of sucrose) and dry solids (usually obtained by refractometry), ie, purity $= 100 \times$ sucrose/dry substance.

The Determination of Reducing Sugars

The most common methods for determining reducing sugars are based on the reduction of the copper(II) complex with tartaric acid in alkaline solutions.

Lane and Eynon Constant Volume Procedure. This method is based on the reduction of Fehling's solution, Soxhlet's modification. The constant volume modification determines reducing sugars in the presence of sucrose.

Berlin Institute Method. This is a copper reduction method that utilizes Müller's solution, which contains sodium carbonate.

Emmerich Method. This method carried out in a nitrogen atmosphere and is based on the reduction of 3,6-dinitrophthalic acid.

Knight and Allen. This method utilizes EDTA to determine excess unreacted copper.

Luff Schoorl. This method is for the determination of total reducing sugars in molasses and refined syrups after hydrolysis.

Ofner Method. This is a copper-reduction method that uses Ofner's solution instead of Fehling's. The reduced cuprous oxide is treated with excess standardized iodine, which is black-titrated with thiosulfate using starch indicator.

Other Methods

Colorimetric Methods. Numerous colorimetric methods exist for the quantitative determination of carbohydrates as a group. Among the most popular of these is the phenol–sulfuric acid method of Dubois, which relies on the color formed when a carbohydrate reacts with phenol in the presence of hot sulfuric acid. The Somogyi method relies on the reduction of cupric sulfate to cuprous oxide and is applicable to reducing sugars.

Enzymatic Methods. Commercial enzyme analyzers are based on immobilized enzymes embedded in membranes. When the membrane or biosensor contacts a solution of the material to be analyzed, glucose is oxidized by glucose oxidase, releasing hydrogen peroxide, which is then measured electronically, giving an estimation of the amount of glucose present.

Chromatographic Methods. Chromatographic methods have their widest application in research and in commercial laboratories dealing with food analysis, where both gas liquid chromatography (glc) and high performance liquid chromatography (hplc) are in use.

The glc analysis of sugars requires chemical derivatization to produce a volatile molecule. Simplest and most rapid derivatization method for routine analysis is silylation to produce trimethylsilyl derivatives.

Sugar analysis by hplc has advanced greatly as a result of the development of columns specifically designed for carbohydrate separation.

Near-Infrared Spectroscopy. Near-infrared spectroscopy (nir) is applied in the sugar industry for several types of analyses. The technique has the advantage of requiring little or no sample preparation. It is a secondary technique, and it can only be as precise and accurate as the primary method used for calibration.

Determination of Other Components

In the sugar industry, the analysis of other components is essential to determine purity. The most important of these, besides reducing sugars, are moisture, ash, and color. Also relevant are methods used to determine particle-size distribution and insoluble matter.

Moisture. In relatively pure sugar solutions, moisture is determined as the difference between 100 and Brix. In crystalline products, it is usually determined by loss-on-drying under specified conditions by commercial moisture analyzers. Moisture in molasses and heavy syrups is determined by a special loss-on-drying technique.

Ash and Inorganic Constituents. Ash may be measured gravimetrically by incineration in the presence of sulfuric acid or by conductivity measurement.

Tests for elements such as arsenic, lead, and copper are usually of the colorimetric or atomic absorption types.

Color. The visual color, from white to dark brown, of sugar and sugar products is used as a general indication of quality and degree of refinement.

Particle-Size Distribution. Grain-size distribution is determined by using a series of sieves, either hand-sieved or machine-sieved.

Insoluble Matter. Insoluble matter in sugar is determined as the dry weight of material left on a filter or membrane after passage of a sugar solution.

MARY AN GODSHALL
Sugar Processing Research Institute, Inc.

ICUMSA Methods Book, International Commission for Uniform Methods of Sugar Analysis, Norwich, U.K., 1994; supplement, 1998

Standard Analytical Methods, Corn Industries Research Foundation, Washington, D.C., binder is constantly updated with new and revised methods.

K. Helrich, ed., *Official Methods of Analysis of the Association of Official Analytical Chemists*, 15th ed., Arlington, Va., 1990.

Quality Specifications and Test Procedures for Bottlers' Granulated and Liquid Sugar, National Soft Drink Association, Washington, D.C., 1975.

CANE SUGAR

The term sugar describes the chemical class of carbohydrates (qv) of the general formula $C_n(H_2O)_{n-1}$ or $(CH_2O)_n$ for monosaccharides.

Colloquially, sugar is the common name for sucrose. Cane sugar is the sugar extracted from sugarcane.

Sugarcane is a large perennial tropical grass belonging to the tribe *Andropogoneae* of the family *Gramineae* and the genus *Saccharum*. The sucrose in cane sugar is identical to that in beet sugar; both white refined products are 99.9% sucrose, with water as the principal nonsucrose component. Trace components from the plant indicate the origin of the sugar.

Physical and Chemical Properties

Cane sugar is generally available in one of two forms: crystalline solid or aqueous solution, and occasionally in an amorphous or microcrystalline glassy form. Microcrystalline is here defined as crystals too small to show structure on x-ray diffraction. Colligative properties vary with concentration of sucrose in solution. The strong effect of cane sugar on freezing point depression is widely used in frozen desserts; the reduction in vapor pressure and increase in boiling point are essential for manufacture of hard candy and other confectionery. The high osmotic pressure generated by sucrose in solution reduces the water activity and therefore the equilibrium relative humidity, so that insufficient moisture remains to sustain microorganisms, as in jams and preserves. Among chemical properties of cane sugar that affect daily use are color, flavor, sweetness, antioxidant properties, and reactions in aqueous solution.

Color of cane sugar depends on its nonsucrose content; sucrose, glucose, and fructose are white crystalline materials. Colorant compounds are in two classes: one from the cane plant, including flavonoid and polyphenolic compounds, and one from process-developed colorant, based on sucrose degradation products.

Cultivation, Harvesting, and Processing of Sugarcane

Cane sugar production is accomplished in one or two stages. At sugarcane factories, located in cane-growing areas, harvested sugarcane is brought in, sugar-containing juice is extracted, and sugar crystallized from the concentrated juice. In the single-stage process, the juice is purified and bleached for the manufacture of plantation white (mill white, direct white) sugar, usually for local consumption. In the two-stage process, partially purified, unbleached juice is crystallized into yellow to brown-colored raw sugar; this is shipped in bulk to the countries of principal cane sugar consumption in North America and northern Europe, where it is refined into white and colored products for industrial and home use. Sugarcane, once cut (harvested), immediately begins to lose sucrose to deterioration by enzyme, or chemical inversion. The two-stage production system arose because sugarcane cannot be stored for long periods (ie, shipping times) because it contains more water and invert than does refined sugar, and discolors and becomes hardened and lumpy.

Cultivation. Sugarcane requires at least 60 cm moisture each year, whether from rainfall or irrigation. It is propagated vegetatively, from cuttings. Most of the world's cane is planted by hand, and some 60% is still harvested by hand in the tropics.

Diseases and Pests. Sugarcane is subject to a number of bacterial, fungal, and viral diseases. Pests include rats, wild animals, nematodes, and a number of insects. The most severe insect pests are the various types of borers, ie, the sugarcane borer, *Diatrea saccharalis* (F.) and the eldana borer, *Eldana saccharina*.

Weeds cause problems in sugarcane culture by competing for nutrients and crowding or overgrowing the young plants.

Harvesting. Harvest season is during the cooler, drier part of the year. Generally, replanting is not necessary after each harvest; buds on the plant base and roots remaining sprout again to produce another crop, called ratoon or stubble.

In hand cutting practice, cane knives range from long machetes to shorter-handled Australian and Brazilian knives with hand guards. Cane leaves and tops (known as trash), which contain little sugar, are removed first by burning the cane field and then by hand or mechanical harvesters. Cane stalks are sufficiently high in moisture so that controlled and rapid burns incinerate only the leaves, tops, and trash. Most common are combine harvesters, or chopper harvesters, which cut cane stalks at the base, cut the stalk into billets, 28–38 cm long, blow excess leaves and trash off the billets, and drop the billets into a cane cart pulled alongside the combine harvester.

Transportation. Cane loading in the field is accomplished by hand, grab loaders, or continuous belt loaders, into small bins or wagons, which collect at transloader stations for transfer to larger transport containers.

Processing. After weighing, sugarcane is washed, then cut into chips, and often further broken up by a shredder. Shredded cane then moves through a series of mills, usually four to seven mills with four rolls each. Imbibition water, or water of maceration, is run countercurrent to the cane. Juices are pumped to the heaters and to the clarification station. Bagasse comes off the mills and goes directly to factory boilers as fuel. To heated (98–105°C) juice is added lime (milk of lime, usually in sugar solution) to pH 7, and flocculation aids, usually polyacrylamides. Solids are allowed to settle out of juice in juice clarifiers, large settling tanks, with various arrangements of baffles. Heat and lime stop enzyme action in juice and raise pH to minimize inversion. Clear juice flows off the upper part of the clarifier; muds are withdrawn below. Evaporator syrup is sent to vacuum pans, where syrup is heated, under vacuum, to supersaturation: fine seed crystals are added, and the sugar mother liquor yields about 50% by weight crystalline sugar. This is a serial process. The first crystallization of A-sugar or A-strike yields a residual mother liquor (A-molasses) that is concentrated to yield a B-strike. Open crystallizers stir lowest grade massecuite (a mixture of crystals and mother liquor) to yield C-sugar and final molasses (blackstrap) from which no more sugar can economically be removed.

After crystallization, crystals and mother liquor are separated in basket-type centrifuges. Mother liquor is spun off the crystals, and a fine jet of water is sprayed on the wall of sugar against the centrifugal basket to reduce the syrup coating on each crystal. Raw sugar is dumped onto moving belts, on which it dries as it is moved to storage. Composition of raw cane sugar is shown in Table 1.

A cane factory generates its own requirements for energy from burning bagasse to produce electricity.

Diffusion. The alternative to extraction by milling is extraction by diffusion, a process developed from sugarbeet diffusion. Here, cane from the shredder must be prepared further in a fiberizer, or extended shredder, for best extraction. Finely prepared cane enters a multicell, countercurrent diffuser. In the diffusers, shredded cane moves countercurrent to hot water (75°C).

Direct Consumption Sugar. This sugar (plantation white, mill white, crystal, superior) is the regular table and industrial product in most cane-growing countries outside the United States. This white (but not sparkling white) crystalline cane sugar product is produced from sugarcane juice by the raw sugar production process, with the addition of sulfur dioxide gas, SO_2, that is injected into juice where it

Table 1. Composition of Sugars

Component	Raw cane sugar	White refined cane sugar	Mill white	Blanco Directo	Brown cane sugars
sucrose, %	96–99	99.3	99.6	99.9	92.96
glucose, %	0.2–0.3	0.007	0.07	0.02	1–2
fructose, %	0.2–0.3	0.006	0.06	0.03	2–3
color, ICU	900–8,000	35	100–200	40–80	2000–9000
ash, %	0.3–0.6	0.012	0.15	0.05	1–2
moisture, %	0.3–0.7	0.015	0.15	0.03	1–2
organic non-sugars, %	0.3–0.8	0.014	0.40	0.03	1–2
SO_2, mg/kg			20–50	1–5	

bleaches colorant and is itself oxidized to sulfate. Sulfate is a major anion in sulfitation sugars. Nonsugar components are not removed in process.

Sulfitation sugar, the most common type of white sugar in the world, is therefore not suitable for industrial use or food and beverage manufacture; it contains high ash, turbidity, and reducing sugars, and generally has a high sediment content. As demand for higher quality (refined quality) sugars increases in the cane-growing countries, there is increasing production of "improved" plantation white sugar. The best direct production sugars are made by the Blanco Directo process, where color precipitating reagents are used to remove nonsugars rather than bleach them.

Cane Sugar Refining

Refining cane sugar processes raw cane sugar into very high purity white and brown sugars and liquid products, including edible molasses. Content of water, ash, and reducing sugars is controlled.

Refinery input (melt) is raw cane sugar at 96° to 98°Z polarization (% sucrose read by rotation of polarized light). The brown products have characteristic palatable cane and cane molasses flavors, not available from sugarbeet.

Raw sugar is weighed into the refinery from rail car, ship, or raw sugar warehouse, and conveyed to the affination station, where it is mingled with a heavy syrup (80% solids content, or 80° Bx where Bx = Brix, wt %), then spun in basket centrifugals and washed with a spray of water to remove the added and the integral syrup coatings. The washed raw sugar is dissolved (melted) to give a washed sugar liquor of ca 70% solids content, which is pumped to clarification.

Phosphoric acid to give a concentration up to 400 mg/kg as P_2O_5 and calcium hydroxide as milk of lime or sugar solution of lime, up to pH 7.5–8.3, are combined with the sugar liquor in an aerated flotation clarifier. Calcium phosphate forms, occluding suspended solids and inorganics in its mass, and floats to the surface where it is scraped off by rotating blades. Clarified liquor (syrups are called liquors in refineries) is pumped out from the bottom of the clarifier.

Talo phosphatation is performed as described above with the addition of color-precipitating chemicals and a series of mud-desweetening steps. It has almost replaced traditional phosphatation.

In this process, called carbonation in Europe, lime and carbon dioxide are mixed in liquor in a two-stage process similar to that for beet sugar processing.

Any type of clarification is followed by filtration through leaf-type vertical or horizontal pressure filters. Carbonatated liquors and phosphatated liquors are generally filtered with the addition of diatomaceous earth as precoat.

Filtration is followed by decolorization with bone char (traditional), granular activated carbon (now most common), ion-exchange resins, or any combination of these.

Decolorized liquor, or fine liquor of very pale yellow color, is evaporated further to 72–74% solids and sent to crystallization in a series of vacuum pans, as with raw cane sugar. Refinery strikes are designated 1, 2, 3, etc.

Refined brown sugars are made by crystallizing sugar from a mixture of third and fourth runoff syrups and affination syrup (boiled brown sugars), or by coating white sugar crystals with a brown sugar liquor–caramel syrup. Compositions of raw cane sugar, refined granulated, direct mill white, and Blanco Directo sugar are shown in Table 1.

After storage, sugar can become moist from water that has been trapped under the outside syrup coating of the crystal by the very high rate of crystallization and drying. The moisture is removed by a process known as conditioning, in which the sugar is stored for about four days with a current of air passing through it to carry away the moisture.

Newest among cane sugar manufacturing systems are processes using membrane filtration to remove nonsucrose solids from juices and syrups. There are two basic classes of membrane: plastic types with metal ions in the matrix, and ceramic types with a porous layer (stainless steel is a variation on ceramic), all with controlled porosity.

Health and Safety Factors

Sugar is one of the purest foods made, from natural sources, and has never been known to contain any toxic or harmful components. Intensive investigations by the U.S. Food and Drug Administration resulted in the conclusion that sugar has no deleterious effect on health in regard to heart disease, diabetes, or other metabolic disorder. Dental cavities appear to be the only disease for which sucrose could be a cause.

Cane Sugar Products

Refined granulated sugar is the principal output of a cane sugar refinery. Large-grain specialty sugars are used for candy and cookies. White large-grain sugar can be made only from the very purest of liquors. Fine-grain sugar, or fruit sugar, consists of small crystals obtained by screening. Powdered sugar is made by grinding granulated sugar and adding 3% corn starch (in the United States) to help prevent caking. In other countries, calcium phosphate, or maltodextrins are used as hygroscopic additives.

Cubes are made by mixing a syrup with granulated sugar to the right consistency to form cubes. These are then dried.

Liquid sucrose and liquid invert are refinery products in Europe and outside the United States. In the United States they have been almost completely replaced by cheaper corn syrups made by enzymatic hydrolysis of starch and isomerization of glucose.

Brown sugar is not raw sugar, but rather it is refined. The difference between raw sugar and brown sugar is not so much the sucrose content, the color, or taste, but rather the absence of field soil, cane fiber, bacteria, yeasts, molds, and insect parts which may be present in raw sugars. Composition is outlined in Table 1.

Other Products. Other products from sugarcane, in addition to cane sugar, are cane fiber (known as bagasse) and molasses, the final thick syrup from which no more sugar may be economically removed by crystallization. In some cane-growing countries, cane tops and leaves, separated during harvest, are used for cattle feed.

Sucrochemistry. A wide range of fermentation and chemical products can be made from sucrose either *per se* or in juice or molasses. Among the classes of products chemically derived from sucrose are the following. (*1*) ethers; (*2*) esters of fatty acids; (*3*) other esters; (*4*) acetals, thioacetals, and ketals; (*5*) oxidation; products, (*6*) halogen and sulfur derivatives and metal complexes; (*7*) polymers and resins. Because sugarcane is the most efficient plant at converting photosynthesis into chemical bonds, it can be the basis of a renewable resource economy.

MARGARET A. CLARKE
Sugar Processing Research Institute, Inc.

M. Mathlouthi and P. Reiser, eds., *Sucrose: Properties and Applications*, Blackie and Son, Ltd., London, 1995.

J. P. C. Chen and C-C. Chou, eds., *Cane Sugar Handbook*, 12th ed., John Wiley & Sons, Inc., New York, 1993.

Sugar and Sweetener Reports, Economic Research Service, U.S. Dept. of Agriculture, Washington, D.C.

BEET SUGAR

Agricultural Practices

The sugar beet, *Beta vulgaris*, is a hearty biennial which produces crops of commercial impact in a wide range of climates. A successful crop depends on seed quality and varietal characteristics, weed and pest control, timely irrigations or timely rains (not all beet crops are grown on irrigated land), disease control, crop rotations of at least three years, and a nitrogen management program designed to limit the amount of leaf growth to the minimum necessary to cover the rows and take full advantage of available sunlight.

Beet Receiving, Storage, and Handling Before Processing

Beets are loaded into side-dump or end-dump trucks in the field and taken to a receiving station. At the receiving station the beets are unloaded from the truck and passed over a series of rotating grab-rolls arranged to allow trash, dirt, and small pieces of beets to fall out of the main stream. This first separation is especially important if the beets are destined for storage of more than a day or two.

Commercial strategies for maintaining effective storage are (*1*) careful monitoring for unremoved leaves, trash, dirt, and early signs of rot or frost damage as the crop is received; (*2*) initially piling roots with temperatures between 0 and 5°C (never >10°C); (*3*) building large piles to stabilize temperatures and minimize surface area exposed to the elements; (*4*) carefully monitoring the condition of the piles to detect hot spots that can be processed immediately or discarded; (*5*) protecting the piles by covering with plastic sheets or straw; (*6*) mechanical ventilation; (*7*) using mechanical ventilation to deep-freeze the beets and stop respiration altogether.

Processing of Beets to Sugar

Whether beets are processed on the day of harvest or several months afterward, they arrive at the factory and are dropped into a cement trough of moving water which flumes them past a series of weed rakes and trash collectors.

Beets are either elevated or pumped to an agitated beet washer, after which they are rinsed with clean water. The beets are fed from bins to slicing machines. The slices are known as cossettes. The knives can be adjusted in blocks to vary the thickness and length of the cossettes. Long cossettes are always desirable and thickness is kept to the minimum which still allows the beet pieces to maintain integrity in the extraction step.

Continuous Countercurrent Extraction of Sucrose

Extraction processes applied to sound, unfrozen beets are designed to allow the sucrose to diffuse from the beet cell wall mass and leave the intercellular material within the cells. Partial denaturing of the cell walls is accomplished by pre-scalding or heating cossettes after they leave the slicers. There is a variety of diffuser configurations. Residence time within all the diffusers is typically 45 to 60 minutes.

Spent cossettes (pulp) exit the diffuser with a moisture content of ca 92% and a sucrose content of ca 1%. To maximize the amount of sucrose returned the wet pulp is pressed in tapered twin-screw presses fitted with perforated side screens. On exiting the presses the moisture content of the pulp is ca 75% and the sucrose content ca 1%, which means that ca 2% of the sucrose that enters the factory with the cossettes leaves with the wet pulp (pulp loss).

The diffusion process has not been designed to ensure sterility, although temperatures above 65°C significantly retard microbial activity. Sulfur dioxide, thiocarbamates, glutaraldehyde, sodium bisulfite, and chlorine dioxide are all used to knock down or control infections.

The raw juice exiting the diffuser is a murky dark gray solution occluded by colloidal materials from the ruptured beet tissue, small pieces of cossettes, and fine soil that escaped the fluming and washing processes. It is microbiologically and chemically unstable and unsuitable for concentration and crystallization.

Juice Purification

Raw juice is heated, treated sequentially with lime (CaO) and carbon dioxide, and filtered. This accomplishes three objectives: (*1*) microbial activity is terminated; (*2*) the thin juice produced is clear and only lightly colored; and (*3*) the juice is chemically stabilized.

Unit Operations. The chemistries involved in purification are described by seven unit operations.

Lime slurry, 0.25% lime on juice (0.250 g of CaO/100 g juice), is added to bring the pH of the mixture into the alkaline range. Insoluble calcium salts are precipitated as finely dispersed colloids. Calcium carbonate in the form of recycled first carbonation sludge is added to provide colloid absorption and stabilization.

A further 1.50% CaO on juice is added, and the juice is brought to its maximum alkalinity and pH. The invert sugars (glucose and fructose) are converted to organic acids, which do not form insoluble calcium salts.

The process stream *p*OH is raised to 3.0 with carbon dioxide. Juice is recycled either internally or in a separate vessel to provide seed for calcium carbonate growth.

Clarification is also referred to as sludge separation. First carbonation effluent is passed through a continuous clarifier where the precipitated calcium carbonate is allowed to settle while the clear juice overflows to second carbonation.

Calcium is reduced to the practical minimum by the addition of carbon dioxide at a *p*OH of 4.5 at a temperature of as near to 100°C as possible. The sludge from this unit operation is easily removed by in-line filters.

Sulfur dioxide is added to a level of about 150 ppm on juice to discourage further color-forming Malliard reactions.

Juice Purification Chemistry. Lime in juice purification serves as a source of calcium, a source of alkalinity, and a source of calcium carbonate which serves as the clarification–filtration medium.

As a source of calcium, lime reacts with nonsucrose components to form a precipitate.

The stoichiometric ratio of lime to removable nonsucrose components is nearly five or six to one. This excess calcium is removed by the addition of carbon dioxide to form calcium carbonate, which is the primary clarification agent. Suspended or colloidal materials are adsorbed onto freshly precipitated calcium carbonate.

High alkalinities of limed juice serve several functions. Foremost is to retard sucrose hydrolysis.

High alkalinity also helps produce a stable thin juice by acting on specific nonsucroses. Invert sugars are oxidized to acids and glutamine, the most abundant amino acid in sugar beets and raw juice, is converted to 2-pyrollidinone-5-carboxylic acid and ammonia, a reaction promoted by heat and the high pH of the process streams.

Nonsucrose Components from Storage or Damage of Beets. Some nonsucrose components are associated with the conditions under which the beets have been stored prior to processing. They directly and indirectly reduce sucrose yield and may cause other processing problems. In addition to glucose and fructose, these components include raffinose, a nonreducing galactose–glucose–fructose trisaccharide; betaine, zwitterionic trimethyl glycine; and levans and dextrans, the slimy products of microbial attack. The presence of these nonsucroses is usually associated with degraded beets; although they can be a sign of factory process infections.

Nitrite is usually one indicator of the infection level in the diffuser.

Removal of Calcium Prior to Evaporation and Crystallization. The second carbonation step is designed to minimize the amount of calcium in thin juice by removing it as $CaCO_3$, the solubility of which is minimized by high temperature, high carbonate concentrations, and low nonsucrose concentrations. Counterbalancing these factors is the high solubility of calcium bicarbonate, which predominates as the pH drops below 8.4. The addition of CO_2 provides carbonate ion, but also lowers the pH.

Crystallization and Recovery of Sucrose

The three-boiling scheme is typical of U.S. beet sugar production. Incoming thick juice is combined with recycled lower grade sugars, filtered, and crystallized under vacuum to yield about half the sucrose separated from the mother liquor and washed in batch centrifugals. The process is repeated two more times using continuous centrifugals for the separation of syrups and crystals; only the first crop of crystals is used for finished product. The third crop is washed with

Table 1. Material Balance and Purity Effects of Thick Juice Purity[a]

Material	Low purity juice		High purity juice	
	Sugar, kg	Purity, %	Sugar, kg	Purity, %
thick juice	100.0	88.0	100.0	92.0
white pan	208.2	93.5	143.8	93.9
intermediate pan	72.8	83.4	50.3	84.5
low pan	32.8	69.1	22.6	70.8
molasses	16.2	54.3	11.2	56.3
product	83.8		88.8	

[a] Purity = sugar content as percent of total dissolved solids content.

a slightly higher purity syrup (affined) to reduce the load of nonsugars and color. Both the second and third crops are returned to the first stage. After the liquor has been concentrated to about 10% past supersaturation, crystal growth is initiated by fully seeding each pan with enough crystals, 1–5 μm in diameter, to account for the finished pan.

All of these processes are carried out under kinetically controlled conditions, and vacuum pan operations may be followed by a crystallizer in which the massecuite (crystals/mother liquor) is allowed more time for crystal growth.

Balancing these crystallization steps to maximize the yield and maintain high product quality is a significant operational challenge, especially when confronted with changing thick juice quality, ie, color, purity, and pH stability (Table 1). These values reveal the critical dependence of performance on the quality of the thick juice and, by inference, on the nature of the beets.

Conditioning and Storage of Sucrose

Washed, wet sugar contains about 1% moisture which must be reduced to about 0.03% without glazing over the crystals. This is accomplished in a rotating drum granulator–cooler in which warm air is passed over the crystals as they roll down the length of the drum. This is followed by cooler air to stabilize the crystals at <35°C before conditioning.

A factory may store as much as half of the production sugar in order to provide continuous distribution to customers. This is accomplished either in sets of tall, vertical cement silos, each capable of holding 6800 t, or single large-diameter Weibul-type curing silos of 23,000 t capacity. A Weibul silo has a central distribution column with a rotating arm which scatters the sugar, allowing it to fall through dehumidified air in even layers about the bin. Withdrawing sugar is accomplished by using the same arm to rake sugar from the top (LIFO) toward the center, where it falls to the bottom into an annular opening around the central column.

Molasses Desugarization

Chromatographic separation of diluted molasses streams into a high purity fraction suitable for concentration and crystallization and a second low purity by-product, which can be concentrated and sold as an animal feed product, is employed originally in Finland and now in the United States and Japan.

The separation uses either of two modes of operation: (*1*) a pulse method in which batches of feed are sequentially placed on the column and eluted with water, taking product and by-product fractions as desired, and (*2*) a simulated moving bed (SMB) configuration in which the feed is continually placed on the column at the point where its purity matches the separation.

The product has purities typically in the 90–92% range and can be combined with thin juice, concentrated and crystallized, or concentrated and stored for later use. The desugarization by-product is normally sold as a low value molasses.

Product Quality and Requirements

The most common parameters used to measure product quality are moisture, color, granulation, sediment, and ash.

Moisture is usually determined by a vacuum oven-dry method at 80°C.

Color is usually specified as white and measured as a solution color using the specific absorbance at 420 nm.

Crystal size distribution is determined on a stack of three to eight sieves of decreasing sizes using U.S. Sieve values for reference.

Sediment. Sediment measurement is normally done by passing the 50% solution used for the color determination through a half black–half white filter pad and visually counting the white and black specks.

Ash is usually determined conductometrically referenced to a gravimetric method using sulfuric acid to digest the sugar followed by burning in a muffle oven at 650°C.

MICHAEL CLEARY
Imperial Holly Corporation

R. A. McGinnis, ed., *Beet Sugar Technology*, 3rd ed., Beet Sugar Development Foundation, Denver, Colo., 1983, for an overview of the processes involved in all but molasses desugarization.

M. A. Clarke and M. A. Godshall, *Chemistry and Processing of Sugarbeet and Sugarcane*, Elsevier Science Publishers BV, Amsterdam, the Netherlands, 1988.

P. van der Poel, H. Schiweck, and T. Schwartz, eds., *Sugar Technology* Verlag Dr. Albert Bartens AG, Berlin, Jan. 1999. A compendium of European and North American Practices.

SUGAR DERIVATIVES

Sucrochemistry has seen rapid advances since the 1960s. Over 300 well-identified sucrose compounds have been described.

Ethers

Trityl Ethers. Treatment of sucrose with four molar equivalents of chlorotriphenylmethyl chloride (trityl chloride) in pyridine gives, after acetylation and chromatography, 6,1′,6′-tri-*O*-tritylsucrose and 6,6′-di-*O*-tritylsucrose. Conventional acetylation of 6,1′,6′-tri-*O*-tritylsucrose, followed by detritylation and concomitant C-4 to C-6 acetyl migration using aqueous acetic acid, yields a pentaacetate, which on chlorination 4,1′,6′-trichloro-4,1′,6′-trideoxygalactosucrose (sucralose), a low calorie sweetener.

Methyl Ethers. Methylation of sucrose is generally conducted under basic conditions. Etherification occurs initially at the most acidic hydroxyl groups, HO-2, HO-1′, and HO-3′, followed by the least hindered groups, HO-6 and HO-6′. Several reagents have found use in the methylation of sucrose, including dimethyl sulfate-sodium hydroxide; methyl iodide–silver oxide–acetone, methyl iodide–sodium hydride in *N,N*-dimethylformamide (DMF), and diazomethane–boron trifluoride etherate (20).

Other Alkyl Ethers. Sucrose has been selectively etherified by electrochemical means to generate a sucrose anion followed by reaction with an alkyl halide. The benzylation of sucrose using this technique gives 2-*O*-benzyl- (49%), 1′-*O*-benzyl- (41%), and 3′-*O*-benzyl- (10%) sucrose. The benzylation of sucrose with benzyl bromide and silver oxide in DMF also produces the 2-*O*-benzyl ether as the principal product.

Silyl Ethers. The preparation of per-*O*-trimethylsilyl ethers of sucrose is generally achieved by reaction with chlorotrimethylsilane and/or hexamethyldisilazane in pyridine. However, this reaction is not selective and in general per-trimethylsilyl ethers are only used as derivatives for gas chromatographic studies.

Cyclic Acetals. One of the most significant developments in the chemistry of sucrose was the synthesis of cyclic acetals. The first synthesis of 4,6-*O*-benzylidenesucrose was achieved from the reaction of sucrose with α,α-dibromotoluene in pyridine. Since then, many new acetalating reagents have been used to give a variety of sucrose acetals, generally by transacetalation reactions.

Esters

Acetates. The synthesis of partially acetylated sucroses has generally been achieved either by way of selectively protected derivatives such as trityl ethers and cyclic acetals or by direct selective acetylation and deacetylation reactions.

Benzoates. The selective debenzoylation of sucrose octabenzoate using isopropylamine in the absence of solvents causes deacylation in the furanose ring to give 2,3,4,6,1′,3′,6′-hepta- and 2,3,4,6,1′,6′-hexa-*O*-benzoyl-sucroses, in 24.1 and 25.4% after 21 and 80 hours, respectively. Identification of any benzoylated sucrose derivative can be achieved by comparison of its ^{13}C-nmr carbonyl carbon resonances with those of the assigned octabenzoate derivative after benzoylation with 10 atom % benzoyl-carbonyl ^{13}C chloride in pyridine.

Pivalates. The reactivity of sucrose toward pivaloylation has been shown to be significantly different from other sulfonic or carboxylic acid chlorides. For example, reaction of sucrose with four molar equivalent of toluene-*p*-sulfonyl chloride in pyridine revealed, based on product isolation, the reactivity order of $O-6 \cong O-6' > O-1' > O-2$. In contrast, a reactivity order for the pivaloylation reaction, under similar reaction conditions, was observed to be $O-6 \cong O-6' > O-1' > O-4$.

Fatty Acid Esters. These derivatives can be produced on an industrial scale by solventless or melt reactions. The transesterification reaction of sucrose with fatty acid methyl esters or triglycerides in the presence of a base gives the corresponding fatty acid esters.

The monofatty acid esters are surfactants and emulsifiers and have the advantage that they are biodegradable, nontoxic, edible, and can inhibit the growth of microorganisms in some cases. Sucrose monofatty acid esters are used in food formulations and, because of their excellent skin compatibility, find application in shampoos and cosmetics.

A commercially interesting low calorie fat has been produced from sucrose. Proctor & Gamble has patented a mixture of penta- to octafatty acid ester derivatives of sucrose under the brand name Olestra.

It is prepared by a solventless transesterification process in which sucrose is treated with methyl ester of fatty acids in the presence of sodium methoxide.

Other Carboxylic Esters. Selective 2-*O*-acylation of sucrose has been achieved by way of the 2-oxyanion compound. Treatment of sucrose in DMF with 3-lauryl-, 3-stearyl-, 3-hydrocinnamoyl-, and 3-(4-phenylbutyryl)-thiazolidine-2-thione derivatives and sodium hydride produced the corresponding 2-*O*-acyl derivatives.

Orthoesters. The value of cyclic orthoesters as intermediates for selective acylation of carbohydrates has been demonstrated. Treatment of sucrose with trimethylorthoacetate and DMF in the presence of toluene-*p*-sulfonic acid followed by acid hydrolysis yields the 6-*O*-acetylsucrose as the major and the 4-*O*-acetylsucrose as the minor component.

Phosphate Esters. The phosphorylation of sucrose using sodium metaphosphate has been reported.

Sulfonate Esters. Sucrose sulfonates are valuable intermediates for the synthesis of epoxides and derivatives containing halogens, nitrogen, and sulfur. In addition, the sulfonation reaction has been used to determine the relative reactivity of the hydroxyl groups in sucrose. The general order of reactivity in sucrose toward the esterification reaction is $OH-6 \cong OH-6' > OH-1' > HO-2$.

Deoxyhalogeno Derivatives

The application of bimolecular, nucleophilic substitution (S_N2) reactions to sucrose sulfonates has led to a number of deoxhalogeno derivatives. Selective displacement reactions of tosyl, mesyl, and tripsyl derivatives of sucrose with different nucleophiles have been reported. The order of reactivity of the sulfonate groups in sucrose toward S_N2 reaction has been found to be $6 > 6' > 4 > 1'$. 4,1′,6′-Trichloro-4,1′,6′-trideoxygalactosucrose (sucralose) has 650 times the sweetness of sucrose. It is poorly absorbed by the intestines and passes unchanged through the body.

Anhydrides and Epoxides

Anhydride derivatives of sucrose are generally synthesized by intramolecular nucleophilic displacement reactions of the respective sulfonate or deoxyhalogeno derivatives.

Nitrogen-Containing Compounds

The aminodeoxy derivatives of carbohydrates are of interest because they are components of such biologically active materials as antibiotics, glycoproteins, and bacterial polysaccharides. They are usually synthesized by catalytic reduction of the corresponding azido derivatives.

Sulfur-Containing Compounds

The reaction of sucrose 2,3-manno-epoxide with potassium thioacetate and ammonium chloride in aqueous ethanol yields expected 3-*S*-acetyl-3-thioaltropyranoside. Treatment of 6,6′-dibromo-6,6′-dideoxysucrose hexaacetate with potassium thioacetate and *N,N*-dimethylthiocarbamate yields the corresponding derivatives of 6,6′-dithiosucrose.

Oxidation Products

Sucrose can undergo two distinct types of oxidation reaction: (*1*) conventional oxidation of primary hydroxyl groups to aldehydes or carboxylic acid residues and a secondary hydroxyl to a ketone, and (*2*) oxidative cleavage of vicinal diols to produce dialdehyde species. The catalytic oxidation of sucrose with platinum and oxygen at pH 7 and at 100°C yields selectively the 6- and 6′-mono- and 6,6′-dicarboxylic acid derivatives. When the reaction is performed at pH 9 and 100°C, some oxidation occurs at the C-1′ position to produce sucrose 6,6′-dicarboxylate as the major and sucrose 6,1′,6′-tricarboxylate as the minor product. The carboxylate derivatives of sucrose are of interest as chelators, detergent builders, and for application in food and drink formulations.

Compounds from Enzymic Isomerization

The synthesis of some commercially important bulk sweeteners such as isomaltulose (Palatinose), isomaltitol (Palatinit), and Actilight (formerly Neosugar) has been achieved by enzymatic transformations of sucrose.

Leucrose, 6-*O*-(α-D-glucopyranosyl)-β-D-fructopyranose, is synthesized from sucrose using a dextranase enzyme from *Leuconostoc mesenteriodes* and a small proportion of fructose (2%).

RIAZ KHAN
PAUL A. KONOWICZ
POLYtech

H. Schiweck and co-workers, in W. Lichtenthaler, ed., *Carbohydrates as Organic Raw Materials*, VCH Verlagsgesselschaft, Wienheim, Germany, 1991, pp. 57–94.

SPECIAL SUGARS

Although sucrose is commercially the most important sugar, there are also special sugars with special applications, among which fructose is the most important.

Fructose

D-Fructose (levulose, fruit sugar) is a monosaccharide constituting one-half of the sucrose molecule. Fructose comprises 4–8 wt % of many fruits, where it primarily occurs with glucose (dextrose) and sucrose (see CARBOHYDRATES; SWEETENERS).

Pure D-fructose is a white, hygroscopic, crystalline substance, that is highly soluble in water.

The sweetness of fructose is 1.3–1.8 times that of sucrose. This property makes fructose attractive as an alternative for sucrose and other commercially available sweeteners. Fructose is probably sweetest in comparison with sucrose when cold and freshly made up in low concentrations at a slightly acidic pH. This relative sweetness difference is commonly attributed to changes in fructose structure when cold (β-D-fructopyranose, sweet) as compared to the structure when the sweetener is warm (β-D-fructofuranose, less sweet).

Fructose possesses colligative properties that distinguish it from sucrose, glucose, and other nutritive sweeteners. It is one of the more effective monosaccharide humectants, binding moisture and lowering water activity, A_w, in food applications, thereby rendering the food products less susceptible to microbial growth and more stable to moisture loss.

Fructose is a highly reactive molecule. When stored in solution at high temperatures, fructose not only browns rapidly but also polymerizes to dianhydrides. Fructose also reacts rapidly with amines and proteins in the nonenzymatic or Maillard browning reaction.

Modern technologies for production involve ion-exchange separation of fructose from glucose in a mixture obtained through the isomerization of glucose by means of immobilized glucose isomerase or microbial cells containing the enzyme. In another procedure for making crystalline fructose, glucose (dextrose) is oxidized by glucose-2-oxidase to glucosone, which is then selectively hydrogenated to fructose.

Fructose has in the 1990s been successfully incorporated into formulas for the preparation of light and reduced calorie beverages and sports beverages; table syrup and table top sweeteners; baked goods; dairy products, including yogurt and chocolate milk; jams, jellies, and preserves; dry mix beverages, puddings, gelatins, and cake mixes; confectionery caramel fillings and starch-based jelly candies; and frozen dairy products and novelties (see FOOD PROCESSING).

Maltose

Maltose (malt sugar) is a disaccharide, 4-*O*-α-D-glucopyranosyl-D-glucose, comprising two molecules of glucose (dextrose). Although occurring in some plants and fruits, it is more frequently recognized as a structural component of starch. Pure maltose is isolated with difficulty from a directed starch hydrolysate, ie, high maltose corn syrup, by precipitation with ethanol.

Important physical and functional properties of maltose and maltose syrups include sweetness, viscosity, color stability, humectancy, freezing point depression, and promotion of beneficial human intestinal microflora growth. Maltose possesses ca 30–40% of the sweetness of sucrose in the pure state.

Lactose

Lactose (milk sugar) is the only commercially available sugar that is derived from animal rather than plant sources. It is a disaccharide consisting of one galactose and one glucose moiety, 4-*O*-β-D-galactopyranosyl-D-glucose. Lactose is isolated commercially as the crystalline α-monohydrate from the whey by-products of cheese or caseinate production. It is available in varying degrees of purity. The sugar is not very soluble in water, nor is it very sweet (ca one-fifth the sweetening power of sucrose). Lactose is a reducing sugar that reacts with amines and amino compounds with resultant browning.

Uses of lactose production by application include baby and infant formulations, human food (30%), pharmaceuticals (25%), and fermentation and animal feed (15%) (39).

Lactose (and the lactose in substances such as milk and whey) has been hydrolyzed commercially by enzymes to yield products that can be tolerated physiologically much more easily by people who have a lactose intolerance.

J. S. WHITE
White Technical Research Group

L. M. Hanover, in F. W. Schenck and R. E. Hebeda, eds., *Starch Hydrolysis Products: Worldwide Technology, Production and Applications*, VCH Publishers, Inc., New York, 1992.

C. A. M. Hough, in C. A. M. Hough, K. J. Parker, and A. J. Vlitos, eds., *Developments in Sweeteners*, Vol. 1, Applied Science Publishers, Ltd., London, 1979.

W. A. Roelfsema and co-workers, in B. Elvers, S. Hawkins, and G. Schulz, eds., *Ullman's Encyclopedia of Industrial Chemistry*, 5th ed., VCH Publishers, Inc., Germany, 1990.

SUGAR ALCOHOLS

The sugar alcohols bear a close relationship to the simple sugars from which they are formed by reduction and from which their names are often derived (see CARBOHYDRATES). The polyols discussed herein contain straight carbon chains, each carbon atom usually bearing a hydroxyl group. Also included are polyols derived from disaccharides. Most of the sugar alcohols have the general formula $HOCH_2(CHOH)_n—CH_2OH$, where $n = 2–5$. They are classified according to the number of hydroxyl groups as tetritols, pentitols, hexitols, and heptitols. Polyols from aldoses are sometimes called alditols. Each class contains stereoisomers. Counting meso and optically active forms, there are three tetritols, four pentitols, ten hexitols, and sixteen heptitols, all of which are known either from natural occurrence or through synthesis.

Physical Properties

In general, these polyols are water-soluble, crystalline compounds with small optical rotations in water and a slightly sweet to very sweet taste. Selected physical properties of many of the sugar alcohols are listed in Table 1.

Polymorphism has been observed for both D-mannitol and sorbitol. Three different forms exist for each hexitol. Bond lengths of crystalline pentitols and hexitols are all similar. The average C–C distance is 152 pm; the average C–O distance is 143 pm. Conformations in the crystal structures of sugar alcohols are rationalized by Jeffrey's rule that "the carbon chain adopts the extended, planar zigzag form when the configurations at alternate carbon centers are different, and is bent and nonplanar when they are the same". Conformations are adopted which avoid parallel C–O bonds on alternate carbon atoms. Very little, if any, intramolecular hydrogen bonding exists in the crystalline sugar alcohols, but an extensive network of intermolecular hydrogen bonds has been found. Usually each hydroxyl group is involved in two hydrogen bonds, one as a donor and one as an acceptor.

The small optical rotations of the alditols arise from the low energy barrier for rotation about C–C bonds, permitting easy interconversion and the existence of mixtures of rotational isomers (rotamers) in solution.

Occurrence

D-Arabinitol (lyxitol) is found in lichens and in a variety of fungi. Xylitol is found in the primrose and in minor quantity in mushrooms.

Sorbitol (D-glycitol) is found in many fruits in the exudate of fruit blossoms; and in the leaves and bark of some fruit trees.

D-Mannitol is widespread in nature. It is found to a significant extent in the exudates of trees and shrubs. Mannitol also occurs in the fruit, leaves, and other parts of various plants. Dulcitol (galactitol) is found in red seaweed, in various shrubs, and in the mannas from a wide variety of plants.

Chemical Properties

Anhydrization. The sugar alcohols can lose one or more molecules of water internally, usually under the influence of acids, to form

Table 1. Physical Properties of the Sugar Alcohols

Sugar alcohol	Melting point, °C	Optical activity in H_2O, $[\alpha]^{20-25}_D$	Solubility, g/100 g H_2O[a]	Heat of solution, J/g[b]	Heat of combustion, constant vol, kJ/mol[b]
			Tetritols		
erythritol	120	meso	61.5	23.3[c]	−2091.6
threitol					
D-threitol	88.5–90	+4.3	very sol		
L-threitol	88.5–90	−4.3			
D,L-threitol	69–70				
			Pentitols		
ribitol	102	meso	very sol		
arabinitol					
D-arabinitol	103	+131[c]	very sol		
L-arabinitol	102–103	−130[c]			−2559.4
D,L-arabinitol	105				
xylitol	61–61.5 (metastable) 93–94.5 (stable)	meso	179	−153.1	−2584.5
			Hexitols		
allitol	155	meso	very sol		
dulcitol (galactitol)	189	meso	3.2[d]		−3013.7
glucitol					
sorbitol (D-glucitol)	93 (metastable) 97.7 (stable)	−1.985	235	−111.5	−3025.5
L-glucitol	89–91	+1.7			
D,L-glucitol	135–137				
D-mannitol	166	−0.4	22	−120.9	−3017.1
L-mannitol	162–163				
D,L-mannitol	168				
altritol					
D-altritol	88–89	+3.2	very sol		
L-altritol	87–88	−2.9			
D,L-altritol	95–96				
iditol					
D-iditol	73.5–75.0	+3.5			
L-iditol	75.7–76.7	−3.5	449		
			Disaccharide alcohols		
maltitol	147–150	+90	175	−78.2	
lactitol	146	+14		−52.7	
isomalt	145–150		24.5	−38.5	

[a] At 25°C unless otherwise indicated. [b] To convert J to cal, divide by 4.184. [c] In aqueous molybdic acid. [d] At 15°C.

cyclic ethers. Nomenclature is illustrated by the hexitol derivatives. Monoanhydro internal ethers are called hexitans, and the dianhydro derivatives are called hexides. The main dehydration involves loss of water from the primary hydroxyl groups.

Esterification. The most important method for the preparation of partial fatty esters involves the interaction of polyols and fatty acids at 180–250°C. During direct esterification of the sugar alcohols, anhydrization occurs to varying degrees depending upon the conditions. Thus, esterification of sorbitol with stearic acid leads to a mixture of stearates of sorbitan and isosorbide as well as of sorbitol. Unanhydrized esters may be prepared by reaction with acid anhydrides or acid chlorides or by ester interchange reactions. In general, use of an excess of these reagents leads to esterification of all hydroxyl groups. Cyclic carbonates result from polyols by transesterification using organic carbonates. Mannitol hexanitrate is obtained by nitration of mannitol with mixed nitric and sulfuric acids.

Etherification. The reaction of alkyl halides with sugar polyols in the presence of aqueous alkaline reagents generally results in partial etherification.

Reaction of olefin oxides (epoxides) to produce poly(oxyalkylene) ether derivatives is the etherification of polyols of greatest commercial importance. The products of oxyalkylation have the same number of hydroxyl groups per mole as the starting polyol.

Acetal Formation. The sugar alcohols react with aldehydes and ketones to yield cyclic acetals and ketals. Five-membered rings are formed from adjacent hydroxyls and six-membered rings result from 1,3-hydroxyls.

Oxidation. Sorbitol is oxidized by fermentation with *Acetobacter suboxydans* to L-sorbose, an intermediate in the synthesis of ascorbic acid (see VITAMINS). The same organism, *Acetobacter xylinium* and related bacteria convert erythritol to L -erythrulose, D-mannitol to D-fructose, and allitol to L-ribohexulose (see MICROBIAL TRANSFORMATIONS). Careful oxidation using aqueous bromine produces mixtures of aldoses and ketoses.

Reduction. Sorbitol and mannitol are each converted by the action of concentrated hydriodic acid to secondary iodides (2- and 3-iodohexanes).

Metal Complexes. The sugar alcohols form complexes in solution with most metal ions.

Isomerization

Isomerization of sorbitol, D-mannitol, L-iditol, and dulcitol occurs in aqueous solution in the presence of hydrogen under pressure and a nickel-kieselguhr catalyst at 130–190°C.

Manufacture of Sorbitol, Mannitol, and Xylitol

Sorbitol is manufactured by catalytic hydrogenation of glucose using either batch or continuous hydrogenation procedures. Corn sugar (qv) is the most important raw material, but other sources of glucose, such as hydrolyzed starch (qv), also may be used. Both supported nickel and Raney nickel are used as catalysts. When invert sugar is used as a starting material, sorbitol and mannitol are produced simultaneously. Mannitol crystallizes from solution after the hydrogenation owing to its lower solubility in water.

Extraction of mannitol from seaweed is a method of lesser importance commercially. Xylose is obtained from sulfite liquors, particularly from hardwoods, by methanol extraction of concentrates or dried sulfite lyes, ultrafiltration and reverse osmosis, ion exchange, ion exclusion, or combinations of these treatments.

Biological Properties

Sugar alcohols are classified as relatively harmless.

All sugar alcohols have the potential to cause diarrhea or flatulence owing to their slow absorption from the small intestine.

In humans, ingestion of sugar alcohols has shown a significantly reduced rise in blood glucose and insulin response, owing to slow absorption by the body. As a result, many foods based on sugar alcohols have been used safely in the diets of diabetics.

Sugar alcohols are not fermented to release acids that may cause tooth decay by the oral bacteria which metabolize sugars and starches.

Uses

The hexitols and their derivatives are used in many fields, including foods, pharmaceuticals, cosmetics, textiles, and polymers.

Aqueous sorbitol solutions are hygroscopic and are used as humectants, softeners, and plasticizers in many different types of formulation. Mannitol is considerably less hygroscopic in its crystalline form. Many applications of mannitol take advantage of its low hygroscopicity and its resistance to occlusion of water.

Sweetness is often an important characteristic of sugar alcohols in food and pharmaceutical applications. Erythritol and xylitol are similar to or sweeter than sucrose. Sorbitol is about 60% as sweet as sucrose, and mannitol, D-arabinitol, ribitol, maltitol, isomalt, and lactitol are generally comparable to sorbitol (see SWEETENERS).

The partial fatty acid esters of the hexitols, usually anhydrized, find extensive use in surface-active applications, such as emulsification, wetting, detergency, and solubilization.

Foods. Sugar alcohols can replace sugar as a bulking agent in foods. As a result, sugar alcohols have increased the number and variety of sugar-free foods which utilize their functional advantages of anticariogenicity, insulin-independent metabolism, and lower caloric values. Mannitol is a food additive permitted in food on an interim basis.

Sugar alcohols have also found application in foods containing sugars. In candy manufacture, sorbitol is used in conjunction with sugars to increase shelf life.

In artificially sweetened canned fruit, addition of sorbitol syrup provides body. Sorbitol has the property of reducing the undesirable aftertaste of saccharin.

Sorbitan fatty esters and their poly(oxyethylene) derivatives are used as shortening emulsifiers.

Pharmaceuticals. Mannitol finds its principal use in pharmaceutical applications (see PHARMACEUTICALS). It is used as a base in chewable, multilayer, and press-coated tablets of vitamins, antacids, aspirin, and other pharmaceuticals, sometimes in combination with sucrose or lactose. Sorbitol solution finds use as a bodying agent in pharmaceutical syrups and elixirs.

A major pharmaceutical use of poly(oxyethylene) sorbitan fatty acid esters is in the solubilization of the oil-soluble vitamins A and D.

Sorbitan sesquioleate emulsions of petrolatum and wax are used as ointment vehicles in skin treatment.

Manufacture of vitamin C starts with the conversion of sorbitol to L-sorbose. Sorbitol and xylitol have been used for parenteral nutrition following severe injury, burns, or surgery. Mannitol hexanitrate and isosorbide dinitrate are antianginal drugs (see CARDIOVASCULAR AGENTS).

Cosmetics. Sorbitol is widely used in cosmetic applications, both as a humectant, and as an emollient (see COSMETICS). Xylitol has been used as a humectant in toothpaste, both alone and in combination with sorbitol and glycerol (see DENTIFRICES).

Poly(oxyethylene(20)) sorbitan monolaurate is useful in shampoo formulations (see HAIR PREPARATIONS).

Textiles. Sorbitol sequesters iron and copper ions in strongly alkaline textile bleaching or scouring solutions (see TEXTILES).

Sorbitan fatty acid esters and their poly(oxyethylene) derivatives are used both to emulsify textile-treating chemicals and, by themselves, as finishes for textile processing. Poly(oxyethylene(20)) sorbitan monolaurate and its homologues are used as antistatic agents on textiles.

Polymers. In combination with various metal salts, sorbitol is used as a stabilizer against heat and light in poly(vinyl chloride) resins and, with a phenolic antioxidant, as a stabilizer in uncured styrene–butadiene rubber compositions and in polyolefins (see HEAT STABILIZERS; OLEFIN POLYMERS; RUBBER COMPOUNDING). Heat-sealable films are prepared from a dispersion of sorbitol and starch in water.

Sorbitol, together with other polyhydric alcohols such as glycerol or pentaerythritol, can serve as the polyol component of alkyd resins and rosin esters for use in protective coatings and core binders (see ALKYD RESINS).

Miscellaneous Uses. Sorbitol is used in flexible glues, cork binders, and printers' rollers, frequently in combination with glycerol, to confer strength and flexibility, as well as stability to humidity change. Glue-type products in which sorbitol is used include bookbinding and magazine and paper tape adhesives.

Mannitol and dulcitol are used as components of bacteriological media. Poly(oxyethylene) derivatives of sorbitol and sorbitan fatty acid esters, usually in blends with anionic surfactants, are used as emulsifiers for insecticides, herbicides, and other pesticides. Oil slicks on sea water are dispersed when sprayed with mixtures of sorbitan monooleate and poly(oxyethylene(20)) sorbitan monooleate (see SURFACTANTS).

In the explosives industry, mannitol hexanitrate is used as an initiator in blasting caps. Nitration of glycerin–ethylene glycol solutions of sorbitol yields low freezing, liquid, high explosive mixtures of value for dynamite formulas (see EXPLOSIVES AND PROPELLANTS).

MARY E. LAWSON
SPI Polyols, Inc.

F. T. Jones and K. S. Lee, *Microscope* **18**, 279 (1970).
G. A. Jeffrey, *Carbohydr. Res.* **28**, 233 (1973).

SULFAMIC ACID AND SULFAMATES

Sulfamic acid (amidosulfuric acid), HSO_3NH_2, molecular weight 97.09, is a monobasic, inorganic, dry acid and the monoamide of sulfuric acid.

Properties

Sulfamic Acid. Sulfamic acid is a dry acid having orthorhombic crystals. The pure crystals are nonvolatile, nonhygroscopic, colorless, odorless, and soluble in water. The acid is highly stable up to its melting point and may be kept for years without change in properties. Selected physical properties of sulfamic acid are listed in Table 1.

Selected chemical properties of sulfamic acid are listed in Table 2.

Whereas sulfamic acid is a relatively strong acid, corrosion rates are low in comparison to other acids. The low corrosion rate can be further reduced by addition of corrosion inhibitors (see CORROSION AND CORROSION CONTROL).

Table 1. Physical Properties of Sulfamic Acid

Property	Value
mol wt	97.09
mp, °C	205
decomposition temperature, °C	209
density at 25°C, g/cm^3	2.126
refractive indexes, 25 ± 3°C	
α	1.533
β	1.563
γ	1.568
solubility, wt %	
aqueous	
at 0°C	12.08
20°C	17.57
80°C	32.01
nonaqueous, at 25°C	
formamide	16.67
methanol	4.12
ethanol (2% benzene)	1.67
acetone	0.40
ether	0.01
71.8% sulfuric acid	0.00

Table 2. Chemical Properties of Sulfamic Acid

Property	Value
dissociation constant, at 25°C	0.101
heat of formation, kJ/mol[a]	−685.9
heat of solution, kJ/mol[a]	19.10
pH of aqueous solutions, at 25°C	
1.00 *N*	0.41
0.75 *N*	0.50
0.50 *N*	0.63
0.25 *N*	0.87
0.10 *N*	1.18
0.05 *N*	1.41
0.01 *N*	2.02

[a] To convert J to cal, divide by 4.184.

Thermal decomposition of liquid sulfamic acid begins at 209°C. At 260°C, sulfur dioxide, sulfur trioxide, nitrogen, water, and traces of other products, chiefly nitrogen compounds, result.

Aqueous sulfamic acid solutions are quite stable at room temperature. At higher temperatures, however, acidic solutions and the ammonium salt hydrolyze to sulfates. Rates increase rapidly with temperature elevation, lower pH, and increased concentrations. These hydrolysis reactions are exothermic. Concentrated solutions heated in closed containers or in vessels having adequate venting can generate sufficient internal pressure to cause container rupture. An ammonium sulfamate, 60 wt % aqueous solution exhibits runaway hydrolysis when heated to 200°C at pH 5 or to 130°C at pH 2. Alkali metal sulfamates are stable in neutral or alkaline solutions even at boiling temperatures.

Sulfamic acid readily forms various metal sulfamates by reaction with the metal or the respective carbonates, oxides, or hydroxides. The ammonium salt is formed by neutralizing the acid with ammonium hydroxide.

Nitrous acid reacts very rapidly and quantitatively with sulfamic acid: $HSO_3NH_2 + HNO_2 \rightarrow H_2SO_4 + H_2O + N_2$. This reaction can be used for the quantitative analysis of nitrites.

Chlorine, bromine, and chlorates oxidize sulfamic acid to sulfuric acid and nitrogen. Chromic acid, permanganic acid, and ferric chloride do not oxidize sulfamic acid.

Sodium sulfate and sulfamic acid form the complex $6\ HSO_3NH_2 \cdot 5\ Na_2SO_4 \cdot 15H_2O$.

Sulfamic acid and its salts retard the precipitation of barium sulfate and prevent precipitation of silver and mercury salts by alkali.

Primary alcohols react with sulfamic acid to form alkyl ammonium sulfate salts. Sulfation by sulfamic acid has been used in the preparation of detergents from dodecyl, oleyl, and other higher alcohols. It is also used in sulfating phenols and phenol–ethylene oxide condensation products. Amides react in certain cases to form ammonium salts of sulfonated amides. Primary, secondary, and tertiary amines react with sulfamic acid to form ammonium salts. Aldehydes form addition products with sulfamic acid salts. These are stable in neutral or slightly alkaline solutions but are hydrolyzed in acid and strongly alkaline solutions.

The *N*-alkyl and *N*-cyclohexyl derivatives of sulfamic acid are comparatively stable. The *N*-aryl derivatives are very unstable and can only be isolated in the salt form. A series of thiazolylsulfamic acids has been prepared.

Sulfamates. Sulfamates are formed readily by the reaction of sulfamic acid and the appropriate metal or its oxide, hydroxide, or carbonate. Sulfamates prepared from weak bases form acidic solutions, whereas those prepared from strong bases produce neutral solutions. Crystals of ammonium sulfamate deliquesce at relative humidity of 70% and higher. Both ammonium sulfamate and potassium sulfamate liberate ammonia at elevated temperatures and form the corresponding imidodisulfonate. Inorganic sulfamates are quite water-soluble, except for the basic mercury salt.

Manufacture

Sulfamic Acid. Sulfamic acid is manufactured by the reaction of urea and fuming sulfuric acid. The reaction between urea and fuming sulfuric acid is rapid and exothermic. It may proceed with violent boiling unless the reaction temperature is controlled.

The reaction takes place at atmospheric pressure. After completion of reaction, the slurry is diluted to about 70% sulfuric acid solution, and crude sulfamic acid crystals are separated by centrifuge.

By-Products. A by-product of sulfamic acid manufacturing is fuming sulfuric acid or dilute sulfuric acid. This by-product also contains ammonium salts and is therefore normally used as raw material for fertilizer (see FERTILIZERS).

Analysis

The assay determination of sulfamic acid is made by titration using sodium nitrite solution and an external potassium iodide starch-paste indicator. For sulfamate assay determination, the same procedure is used as for sulfamic acid.

Health and Safety Factors

Contact with sulfamic acid and its solutions can cause eye burns and irritate the nose, throat, and skin. Workers should wear cup-type, rubber, or soft plastic-framed goggles, equipped with approved impact-resistant glass or plastic lenses. Exposure to the skin can be minimized by wearing rubber gloves and hands should be washed thoroughly after handling. Breathing of the dust should be avoided and adequate ventilation should be provided.

Uses

Sulfamic Acid. Properties of sulfamic acid make it particularly well suited for scale-removal operations and chemical cleaning in a large variety of applications.

Use of sulfamic acid in the manufacture of dyes and pigments involves removal of excess nitrite from diazotization reactions (see AZO DYES). Sulfamic acid is also used in some dyeing operations for pH adjustment. It also is useful in lowering pH levels in a variety of other systems.

Sulfamic acid additions to chlorination bleaching stages are effective in reducing pulp-strength degradation associated with high temperatures (see BLEACHING AGENTS).

Sulfamic acid reacts with hypochlorous acid to produce *N*-chlorosulfamic acids, compounds in which the chlorine is still active but more stable than in hypochlorite form. The commercial interest in this area is for chlorinated water systems in paper mills, ie, for slimicides, cooling towers, and similar applications (see INDUSTRIAL ANTIMICROBIAL AGENTS).

Sulfamic acid has been recommended as a reference standard in acidimetry.

Sulfation of mono-, ie, primary and secondary, alcohols; polyhydric alcohols; unsaturated alcohols; phenols; and phenolethylene oxide condensation products has been performed with sulfamic acid (see SULFONATION AND SULFATION). The best-known application of sulfamic acid for sulfamation is the preparation of sodium cyclohexylsulfamate, which is a synthetic sweetener (see SWEETENERS).

Sulfamates. A number of flame retardants used for cellulosic materials, including fabrics and paper products, are based on ammonium sulfamate.

Ammonium sulfamate is highly effective in nonselective herbicides to control weeds, brush, stumps, and trees (see HERBICIDES).

Nickel sulfamate is made by the combination of nickel carbonate and sulfamic acid. It is almost exclusively used in the plating industry, with its solutions used for both plating and electroforming.

Calcium sulfamate and magnesium sulfamate are used effectively as a stiffening promoter of concrete and hydraulic cement (see CEMENT).

KATSUMASA YOSHIKUBO
MICHIO SUZUKI
Nissan Chemical Industries, Ltd.

K. Andersen, *Comprehensive Organic Chemistry*, Vol. 3, Pergamon Press, Oxford, U.K., 1979.

J. Donnay and H. Ondik, *Crystal Data Determinative Tables*, Vol. 2, 3rd ed., U.S. Dept. of Commerce, National Bureau of Standards, Joint Committee on Powder Standards, Washington, D.C., 1975, pp. 1–202.

SULFOLANE AND SULFONES

Sulfolane

Properties. Sulfolane, $C_4H_8SO_2$ (**1**), also known as tetrahydrothiophene-1,1-dioxide and tetramethylene sulfone, is a colorless, highly polar, water-soluble compound. Physical properties are given in Table 1.

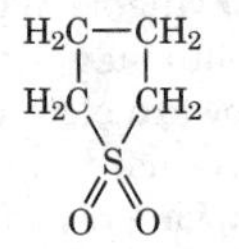

(**1**)

Whereas sulfolane is relatively stable to about 220°C, above that temperature it starts to break down. Sulfolane, also stable in the presence of various chemical substances, is relatively inert except toward sulfur and aluminum chloride. Despite this relative chemical inertness, sulfolane does undergo certain reactions, for example, halogenations, ring cleavage by alkali metals, ring additions catalyzed by alkali metals, reaction with Grignard reagents, and formation of weak chemical complexes.

Production. Industrially, sulfolane is synthesized by hydrogenating 3-sulfolene (2,5-dihydrothiophene-1,1-dioxide), the reaction product of butadiene and sulfur dioxide.

Toxicity. Sulfolane causes minimal and transient eye and skin irritation. Inhalation of sulfolane vapors in a saturated atmosphere is not considered biologically significant. However, when aerosol dispersions have been used to elevate atmospheric concentration, blood changes and convulsions have been observed in laboratory animals.

Table 1. Physical Properties of Sulfolane

Property	Value
molecular weight	120.17
boiling point, °C	287.3
melting point, °C	28.5
specific gravity	
30/30	1.266
100/4	1.201
density, 15°C, g/cm^3	1.276
flash point, °C	165–178
viscosity, mPa·s(= cP)	
at 30°C	10.5
100°C	2.5
200°C	1.0
refractive index, n_D, at 30°C	1.48
heat of fusion, kJ/kg[a]	11.44
dielectric constant	43.3
surface tension, at 30°C, mN/m(= dyn/cm)	35.5

[a] To convert J to cal, divide by 4.184.

Uses. Sulfolane is used principally as a solvent for extraction of benzene, toluene, and xylene from mixtures containing aliphatic hydrocarbons.

The urea-adduction method for separating normal and branched aliphatic hydrocarbons can be carried out in sulfolane.

Sulfolane exhibits selective solvency for fatty acids and fatty acid esters which depends on the molecular weight and degree of fatty acid unsaturation. Applications for this process are enriching the unsaturation level in animal and vegetable fatty oils to provide products with better properties for use in paint, synthetic resins, food products, plastics, and soaps.

Extractive distillation is a technique for separating components in narrow boiling range mixtures which are difficult to separate by ordinary fractionation. Sulfolane is a suitable extractive-distillation solvent for carrying out the separation of such mixtures.

Another large commercial use for sulfolane is the removal of acidic components and mercaptans, from sour gas streams.

Sulfolane is a solvent for a variety of polymers. Sulfolane can be used alone or in combination with a cosolvent as a polymerization solvent.

Sulfolane has been tested quite extensively as the solvent in batteries, particularly for lithium batteries. Sulfolane is also used in a wide variety of other electronic and electrical applications.

Textile applications for sulfolane include preparation of concentrated, storage-stable basic dyes; fabric treating prior to dyeing to improve the dye adsorption; and fiber treating to improve the tensile strength, pilling resistance, and drawing properties.

Sulfones

3-Sulfolene is the next most commercially important sulfone after sulfolane. Besides its precursor role in sulfolane manufacture, 3-sulfolene is an intermediate in the synthesis of sulfolanyl ethers, which are used as hydraulic fluid additives (see HYDRAULIC FLUIDS). 3-Sulfolene or its derivatives also have been used in cosmetics and slimicides.

Other sulfones with commercial potential include dimethyl sulfone (sulfonylbismethane), diiodomethyl *p*-tolylsulfone, 4,4′-dihydroxydiphenyl sulfone (Bisphenol S), bis(*p*-chlorophenyl) sulfone, and 4,4′-bis(*p*-chlorophenylsulfonyl) biphenyl.

EARL CLARK
Phillips Research Center

J. A. Reddick and W. E. Bunger, *Organic Solvents*, 3rd ed., Vol. 2, Wiley-Interscience, New York, 1970, pp. 467–468.

Technical Information on Sulfolane, Bulletin 524, Phillips Chemical Co., Bartlesville, Okla.

SULFONAMIDES.

See ANTIBACTERIAL AGENTS, SYNTHETIC–SULFONAMIDES.

SULFONATION AND SULFATION

Sulfonation and sulfation, chemical methods for introducing the SO_3 group into organic entities, are related and usually treated jointly.

In sulfonation, an SO_3 group is introduced into an organic molecule to give a product having a sulfonate, CSO_3, moiety. The compound may be a sulfonic acid, a salt, or a sulfonyl halide requiring subsequent alkaline hydrolysis. Aromatic hydrocarbons are generally directly sulfonated using sulfur trioxide, oleum, or sulfuric acid. Sulfonation of unsaturated hydrocarbons may utilize sulfur trioxide, metal sulfites, or bisulfites. The latter two reagents produce the corresponding hydrocarbon metal sulfonate salts in processes referred to as sulfitation and bisulfitation, respectively. Organic halides react with aqueous sodium sulfite to produce the corresponding organic sodium sulfonate. In instances where the sulfur atom at a lower valance is attached to a carbon atom, the sulfonation process entails oxidation. Thus the reaction of a paraffin hydrocarbon with sulfur dioxide and oxygen is referred to as sulfoxidation; the reaction of sulfur dioxide and chlorine is called chlorosulfonation. The sulfonate group may also be introduced into an organic molecule by indirect methods through a primary reaction, eg, esterification, with another organic molecule already having an attached sulfonate group.

Sulfation is defined as any process of introducing an SO_3 group into an organic compound to produce the characteristic $C–OSO_3$ configuration. Typically, sulfation of alcohols utilizes chlorosulfuric acid or sulfur trioxide reagents. Unlike the sulfonates, which show remarkable stability even after prolonged heat, sulfated products are unstable toward acid hydrolysis. Hence, alcohol sulfuric esters are immediately neutralized after sulfation in order to preserve a high sulfation yield.

In sulfamation, also termed *N*-sulfonation, compounds of the general structure R_2NSO_3H are formed as well as their corresponding salts, acid halides, and esters. The reagents are sulfamic acid (amido-sulfuric acid), SO_3–pyridine complex, SO_3–tertiary amine complexes, aliphatic amine–SO_3 adducts, and chlorine isocyanate–SO_3 complexes.

Uses for Derived Products and Sulfonation Technology

Sulfonation and sulfation processes are utilized in the production of water-soluble anionic surfactants (qv) as principal ingredients in formulated light-duty and heavy-duty detergents, liquid hand cleansers, general household and personal care products (see COSMETICS), and dental care products (see DENTIFRICES). Other commercially significant product applications include emulsifiers, lube additives, sweeteners, pesticides, medicinals, ion-exchange resins, dyes and pigments.

Sulfonation and sulfation processes are important tools for organic synthesis of specific molecules and positional isomers.

Application chemists are most interested in physical and functional properties contributed by the sulfonate moiety, such as solubility, emulsification, wetting, foaming, and detersive properties. Products can be designed to meet various criteria including water solubility, water dispersibility, and oil solubility. The polar SO_3 moiety contributes detersive properties to lube oil sulfonates and dry-cleaning sulfonates.

Process Selection and Options. Because of the diversity of feedstocks, no one process fits all needs. An acceptable sulfonation/sulfation process requires (*1*) the proper reagent for the chemistry involved and the ability to obtain high product yields; (*2*) consistency with environmental regulations such that minimal and disposable by-products are formed; (*3*) an adequate cooling system to control the reaction and to remove significant heat of reaction; (*4*) intimate mixing or agitation of often highly viscous reactants to provide adequate contact time; (*5*) products of satisfactory yields and marketable quality; and (*6*) acceptable economics. Viscosity constraints may play a significant role not only dictating agitation/mixing requirements but also seriously affecting heat-exchange efficiency.

Reagents

Reagents for direct sulfonation and sulfation reactions are listed in Table 1. Unlike sulfuric acid reactions which usually require 3–4 moles per mole of organic feedstock resulting in substantial "spent acid" requiring disposal, SO_3 generally reacts essentially stoichiometrically thus producing high-purity products directly.

By 1987, sulfur trioxide reagent use in the United States exceeded that of oleum for sulfonation. Sulfur trioxide source is divided between liquid SO_3 and *in situ* sulfur burning. The latter is integrated into sulfonation production facilitates.

Liquid SO_3 is commercially available as both unstabilized and stabilized liquids. Unstabilized liquid SO_3 can be utilized without problem as long as moisture is excluded, and it is maintained at ca 27–32°C. Stabilized liquid SO_3 has an average in that should the liquid freeze (16.8°C), in the absence of moisture pickup, the SO_3 remains in the gamma-isomer form and is readily remeltable. Gaseous SO_3 can also be obtained by stripping 70% oleum (70%SO_3:30%H_2SO_4) or by utilizing SO_3 converter gas (6–8% SO_3) from H_2SO_4 production, or by vaporizing liquid SO_3 which is then generally diluted with moisture free air.

Sulfur trioxide is an extremely strong electrophile that rapidly seeks to enter into transient or permanent relationships or reactions with organics containing electron donor elements, such as oxygen, nitrogen, halogen, and phosphorus. In some instances, SO_3 may first form a transient intermediate adduct at some moderate temperature, which at some higher temperature becomes unstable, liberating SO_3. This subsequently may react to produce a stable sulfonated product, often accompanied by a difficult to control strong or violent exotherm. The reactivity of SO_3 can be moderated by the use of solvents (such as liquid SO_2, or halogenated hydrocarbons), or by the use of SO_3 adducts, (such as SO_3–Trimethylamine or SO_3–Pyridine).

Sulfonation

All sulfonation is concerned with generating a carbon sulfur(VI) bond in the most controlled manner possible using some form of the sulfur trioxide moiety. Sulfonation can be carried out in a number of ways using the reagents listed in Table 1. Sulfur trioxide is a much more reactive sulfonating reagent than any of its derivatives. Care should be taken with all sulfonating reagents owing to the general exothermic nature of the reaction.

The variety of reagents available makes possible the conversion of a wide range of aromatics into sulfonic acids. The reactivity of compounds that are activated toward electrophilic attack are so high that often alternative reagents are used in order to minimize undesirable by-products largely formed owing to excessive heating.

Aromatic Compounds. The accepted general mechanism for the reaction of an aromatic compound with sulfur trioxide involves an activated intermediate as shown in equation 1.

$$R—C_6H_5 + SO_3 \rightarrow [R—C_6H_5SO_3]^* \rightarrow R—C_6H_4SO_3H \quad (1)$$

The reaction of sulfur trioxide and benzene in an inert solvent is very fast at low temperatures. Yields of 90% benzenesulfonic acid can be expected. Increased yields of about 95% can be realized when the solvent is sulfur dioxide.

Several thousand different synthetic dyes are known, having a total worldwide consumption of 298 million kg/yr. Many dyes contain some form of sulfonate as $–SO_3H$, $–SO_3Na$, or $–SO_2NH_2$.

The world's largest volume synthetic surfactant is linear alkylbenzene sulfonate (LAS), which was developed as a biodegradable

Table 1. Reagents for Direct Sulfonation and Sulfation Reactions[a]

Reagent	Formula	Physical form	Advantages	Disadvantages	Applications
sulfur trioxide liquid	SO_3	liquid	low cost, concentrated reagent	extremely reactive; charring	very few
gas	SO_3	gas, 3–8% SO_3	low cost, stoichiometric reactions; preferred reagent	requires significant dry diluent gas; mole ratio sensitive; liquid storage	most every sulfonation and sulfation reaction
sulfur burning	SO_3	gas *in situ*, 3–8% SO_3	lowest cost SO_3 produced *in situ*; preferred reagent	catalyst requires startup time; higher investment cost	most every sulfonation and sulfation reaction
chlorosulfuric acid	$ClSO_3H$	liquid	stoichiometric reactions	expensive; produces HCl gas, disposal problem	alcohol sulfation, dyes, etc
oleum	$H_2SO_4{\cdot}SO_3$	liquid	low cost	reactions not stoichiometric; 3–4 mol generally required	dyes, alkylated aromatic sulfonation; continuous sulfation of alcohols
sulfuric acid	H_2SO_4[b]	liquid	low cost, easily handable	reactions not stoichiometric; generally requires 3–4 mol	hydrotrope sulfonation of aromatics using azeotropic water removal, etc
sodium bisulfite	$NaHSO_3$	solid, 38% liquid	simple processing	higher cost, except for sulfur burning	sulfosuccinates, lignin, olefins, Streker reaction
sodium sulfite	Na_2SO_3	solid, 38% liquid	simple processing	higher cost	Streker reaction, etc
sodium bisulfite, hydroperoxide catalyst	$NaHSO_3$, O_2	solid, 38% liquid	sulfonation of olefins	requires hydroperoxide catalyst; costly	sulfonation of olefinic hydrocarbons producing primary paraffin sulfonation
sulfamic acid	H_2NSO_3H	solid	stoichiometric reaction, mild, simple	high cost; limited to NH_4 salt; heating to ca 150°C	small specialties, sulfations
sulfuryl chloride	SO_2Cl_2	liquid	few	expensive, usually required catalyst	chlorosulfonation reactions; mostly research
sulfur dioxide and chlorine	SO_2, Cl_2	gases	few, relatively inexpensive	not generally stoichiometric; need catalyst	chlorosulfonation of paraffins, produces HCl
sulfur dioxide and oxygen	SO_2, O_2	gases	few, inexpensive	not stoichiometric; requires catalyst	sulfoxidation of paraffins

[a] In order of descending reactivity.
[b] 93–100%.

replacement for nonlinear alkylbenzene sulfonates (BAB). LAS is derived from the sulfonation of linear alkylbenzene (LAB). Detergent sulfonates use LAB in the 236 to 262 molecular weight range, having a C_{11}–C_{13} alkyl group. The simplest sulfonation route uses 100% sulfuric acid. Continuous falling film SO_3 sulfonation systems utilizing either vaporized and dry air diluted gaseous SO_3 (3–8% SO_3) or *in-situ* sulfur burning integrated air diluted SO_3 generating systems (3–8% SO_3) have become the method of choice for the sulfonation of most aromatics, as well as for the sulfation of alcohols.

Sulfonated toluene, xylene, and cumene, neutralized to the corresponding ammonium or sodium salts, are important industrially as hydrotropes or coupling agents in the manufacture of liquid cleaners and other surfactant compositions.

Sulfitation and Bisulfitation of Unsaturated Hydrocarbons. Sulfites and bisulfites react with compounds such as olefins, epoxides, aldehydes, ketones, alkynes, aziridines, and episulfides to give aliphatic sulfonates or hydroxysulfonates.

Sulfosuccinates and Sulfosuccinamates. The principal sulfonating reagent in these cases is the bisulfite molecule which readily attacks electron-deficient carbon centers. Variations in the choice of starting material can give a broad spectrum of products of widely varying chemical and physical properties.

Unsaturated Hydrocarbons. The reaction of long-chain, ie, C_{12}–C_{18}, α-olefins with strong sulfonating agents leads to surface-active materials (see SURFACTANTS). The overall product of continuous falling film SO_3 sulfonation of α-olefins, termed α-olefin sulfonate (AOS), is really a mixture containing both alkenesulfonates (65–70%) and hydroxyalkanesulfonates (20–25%), along with small amounts of disulfonated products (7–10%). The composition of the final product varies as a result of manufacturing conditions.

AOS prepared from α-olefins in the C_{12}–C_{18} range are most suitable for detergent applications.

Fatty Acid Esters. Fatty acid ester sulfonates are manufactured by reaction of the corresponding hydrogenated (usually methyl) ester and a strong sulfonating agent, such as sulfur trioxide, in order to sulfonate on the alpha-position of the ester. The procedure for the reaction and equipment requirements are very similar to those for the production of LAS. Sodium fatty acid ester sulfonates are known to be highly attractive as surfactants, because they are produced from renewable natural resources and their biodegradability is almost as good as alkyl sulfates.

Petroleum and Related Feedstocks. Petroleum sulfonate by-products were the first petrochemical product. Since that time, By-product petroleum sulfonates have gradually found utilization in a great many applications, including as lubricant additives for high performance engines; as emulsifiers, flotation agents, and corrosion inhibitors; and for enhanced oil recovery. The importance of petroleum sulfonates has grown to the point where these compounds are produced as coproducts, or even as primary petrochemicals.

Factors impacting petroleum sulfonation operations since the late 1970s include the many significant changes and modernizations petroleum refineries have undergone leading to the closing of many refineries practicing oil sulfonation processes; white oil manufacturing technology has eliminated sulfonation and thus sludge disposal to utilize the more cost-efficient hydrogenation process; a principal shift has developed in the use of first-intent oil-soluble synthetic alkylated aromatic sulfonates in place of the traditional petroleum sulfonates for lube additives, and the synthetic sulfonates are made by continuous SO_3 sulfonation processes; and the large projected need for petroleum sulfonates for enhanced oil recovery processes has ceased owing to a significant and prolonged drop in crude oil market prices. Hence there has been a significant drop in the production of natural petroleum sulfonates.

Lignin. Lignosulfonates are complex polymeric materials obtained as by-products of wood pulping where lignin (qv) is treated with sulfite reagents under various conditions (see PULP). Lignin polymers contain substantial amounts of guaiacyl units, followed by *p*-hydroxyphenyl and syringyl units. Two principal wood pulping processes are utilized: the sulfite process and the kraft process. Sulfonation of lignin mainly occurs on the substituted phenyl–propene precursors at the alpha-carbon next to the aromatic ring.

Styrene and Vinyl Monomer, Polymer, and Copolymer Sulfonates. The incorporation of sulfonates into polymeric material can occur either after polymerization or at the monomer stage. The sulfonic acid group is strongly acidic and can therefore be used to functionalize the polymer backbone to the desired degree. The ability of sulfonic acids to exchange counterions has made these polymers prominent in industrial water treatment applications, separators in electrochemical cells, and selective membranes of many types.

The simplest monomer, ethylenesulfonic acid, is made by elimination from sodium hydroxyethyl sulfonate and polyphosphoric acid. Ethylenesulfonic acid is readily polymerized alone or can be incorporated as a copolymer using such monomers as acrylamide, allyl acrylamide, sodium acrylate, acrylonitrile, methylacrylic acid, and vinyl acetate. Styrene and isobutene fail to copolymerize with ethylene sulfonic acid.

Sulfation

Sulfation is the generation of an oxygen sulfur(IV) bond, where the oxygen is attached to the carbon backbone, in the most controlled manner possible, using some form of sulfur trioxide moiety. When sulfating alcohols, the reaction is strongly exothermic. Examples of feedstocks for such a process include alkenes, alcohols, or phenols. Unlike the sulfonates, which exhibit excellent stability to hydrolysis, the alcohol sulfates are readily susceptible to hydrolysis in acidic media. The sulfation of fatty alcohols and fatty polyalkoxylates has produced a substantial body of commercial detergents and emulsifiers.

Linear ethoxylates are the preferred raw materials for production of ether sulfates used in detergent formulations because of uniformity, high purity, and biodegradability. The alkyl chain is usually in the C_{12} to C_{13} range having a molar ethylene oxide: alcohol ratio of anywhere from 1:1 to 7:1. Propoxylates, ethoxylates, and mixed alkoxylates of aliphatic alcohols or alkyl phenols are sulfated for use in specialty applications.

Alcohols and Alkoxylates. The preferred method of sulfation uses some form of a continuous thin-film SO_3 reactor.

Sulfamation

Sulfamation is the formation of a nitrogen sulfur(VI) bond by the reaction of an amine and sulfur trioxide, or one of the many adduct forms of SO_3. Heating an amine with sulfamic acid is an alternative method. A practical example of sulfamation is the artificial sweetener sodium cyclohexylsulfamate, produced from the reaction of cyclohexylamine and sulfur trioxide (see SWEETENERS). Sulfamic acid is prepared from urea and oleum. Whereas sulfamation is not greatly used commercially, sulfamic acid has various applications (see SULFAMIC ACID AND SULFAMATES).

Industrial Processes

A wide array of industrial processes is suitable for the manufacture of sulfated and sulfonated products. Process selection is dependent on the specific chemistry involved, choice and cost of reagents, physical properties of feedstocks and derived products, product volume requirements, operational mode (batch, continuous), quality of derived products, and possible generation and disposal of by-products as well as operating and equipment investment costs. Another important consideration is the location of the sulfonation plant relative to raw material suppliers, particularly for the more limited liquid SO_3 supplier's plants. On the other hand, molten sulfur used for *in situ* sulfur burning and gaseous SO_3 generation is readily available throughout the United States and worldwide. Another consideration for process selection is plant versatility in sulfonating a variety of feedstocks.

The handling of highly acidic sulfonation reagents and the actual sulfonation processing conditions for the production of acidic reaction products and by-products present a number of corrosion problems which must be carefully addressed. Special stainless steel alloys or glass-lined equipment are often used, although the latter generally has poorer heat-exchange properties. All environment regulations or restrictions must also be met. For example, in utilizing $ClSO_3H$ reagent, HCl gaseous by-product is generated requiring its recovery by adsorption or neutralization.

The viscosity of sulfonation and sulfation reaction mixtures increases with conversion, often producing extremely high viscosities. Sulfonation process design must accommodate such viscosities.

Batch processes are currently used for the manufacture of small volume specialty sulfonates based on H_2SO_4, oleum, $ClSO_3H$, sulfite, or SO_3 reagents. Production of large volume sulfonates or alcohol sulfates generally utilize continuous SO_3 falling-film processes based on multitubular or concentric designed reactor systems.

EDWARD A. KNAGGS
Consultant
MARSHALL J. NEPRAS
Stepan Company

W. H. deGroot, *Sulphonation Technology in the Detergent Industry*, Kluwer Academic Publishers, Dorrecht, the Netherlands, 1991.

E. E. Gilbert, *Sulfonation and Related Reactions*, Interscience Publishers, New York, 1965; reprinted by R. E. Kreiger Publ. Co., Melbourne, FL.

S. Patai and Z. Rappoport, eds., *The Chemistry of Sulphonic Acids, Esters and Their Derivatives*, John Wiley & Sons, Ltd., Chichester, U.K., 1991.

K. K. Andersen, in D. N. Jones, ed., *Sulphonic Acids and Their Derivatives*, Vol. 3, Pergamon Press, Oxford, U.K., 1991.

SULFONIC ACIDS

Sulfonic acids are classically defined as a group of organic acids which contain one or more sulfonic, $-SO_3H$, groups, The general formula of organic sulfonic acids RSO_3H, where the R-group may be derived from many different sources. Typical R-groups are alkane, alkene, alkyne, and arene. The R-group may contain a wide variety of secondary functionalities such as amine, amide, carboxylic acid, ester, ether, ketone, nitrile, phenol, etc. Sulfonic acid derivatives, where the R-group is derived from an inorganic source such as a halide, oxygen (ie, sulfate), or amine (ie, sulfamic acid), are often referred to as sulfuric acid derivatives (see CHLOROSULFURIC ACID).

Physical Properties

The physical properties of sulfonic acids vary greatly depending on the nature of the R-group. Sulfonic acids can be described as having similar acidity characteristics to sulfuric acid. Sulfonic acids are prone to thermal decomposition, ie, desulfonation, at elevated temperatures. However, several of the alkane-derived sulfonic acids show excellent thermal stability, as shown in Table 1. Arene-based sulfonic acids are thermally unstable.

Sulfonic acids are such strong acids that in general they can be considered greater than 99% ionized.

Chemical Properties

Sulfonic acids are prepared on a commercial scale by the sulfonation of organic substrates using a variety of sulfonating agents, including

Table 1. Physical Properties of Sulfonic Acids

Acid	Mp, °C	Bp,[a] °C	Density d_4^{25}, g/cm^3
methanesulfonic acid	20	122	1.48
ethanesulfonic acid	−17	123	1.33
propanesulfonic acid	−37	159	1.19
butanesulfonic acid	−15	149	1.19
pentanesulfonic acid	−16	163	1.12
hexanesulfonic acid	16	174	1.10
benzenesulfonic acid	44	172[b]	
p-toluenesulfonic acid	106	182[b]	
1-naphthalenesulfonic acid	78	dec	
2-naphthalenesulfonic acid	91	dec	1.44
trifluoromethanesulfonic acid	none	162[c]	1.70

[a] At 133 Pa (1 mm Hg) unless otherwise noted.
[b] At 13.3 Pa (0.1 mm Hg).
[c] At 101.3 kPa = 760 mm Hg.

sulfur trioxide (diluted in air), sulfur trioxide (in sulfur dioxide), sulfuric acid, oleum (fuming sulfuric acid), chlorosulfuric acid, sulfamic acid, trialkylamine–sulfur trioxide complexes, and sulfite ions. Other methods of sulfonic acid production, practiced on an industrial scale, include the oxidation of thiols, sulfide, disulfides, sulfoxides, sulfones, and sulfinic acids (see SULFONATION AND SULFATION).

General Reaction Chemistry of Sulfonic Acids. Sulfonic acids may be used to produce sulfonic acid esters, which are derived from epoxides, olefins, alkynes, allenes, and ketenes, as shown in Figure 1. Phosphorus pentachloride and phosphorus pentabromide can be used to convert sulfonic acids to the corresponding sulfonyl halides.

Halogenation of sulfonic acids, which avoids production of a sulfonyl halide, can be achieved under oxidative halogenation conditions.

Sulfonic acids may be subjected to a variety of transformation conditions. Sulfonic acids may be hydrolytically cleaved, using high temperatures and pressures, to drive the reaction to completion.

Aromatic sulfonic acid derivatives can be nitrated using nitric acid, in H_2SO_4. Sulfones may be treated with hydrazine derivatives to give the corresponding ring-opened sulfonic acid.

Production

At the end of the 1990s, there were four primary methods of sulfonic acid production in the United States: falling film sulfonation; oleum sulfonation; chlorosulfuric acid sulfonation; and SO_3 solvent-based sulfonation.

The vast majority of sulfonic acids were produced using continuous falling film sulfonation technology, which utilizes vaporized SO_3 mixed with air. This technology dominates the sulfonation industry owing to the capability of high product throughput and low by-product waste streams.

Analytical and Test Methods

Modern analytical techniques have been developed for complete characterization and evaluation of a wide variety of sulfonic acids and sulfonates. Titration is the most straightforward method of evaluating sulfonic acids. Spectroscopic methods for sulfonic acid analysis include ultraviolet spectroscopy, infrared spectroscopy, and ^{1}H and ^{13}C nmr spectroscopy. Modern separation techniques of sulfonates include liquid chromatography and ion chromatography (see CHROMATOGRAPHY).

Health and Safety Factors

In general, unneutralized sulfonic acids are regarded as moderate to highly toxic substances. However, slight detoxification, via the introduction of a sulfonic acid moiety, is observed for nitrobenzene and aminobenzene. Sulfonic acids emit toxic SO_x fumes upon heating to decomposition. Halogenated sulfonic acids, such as trifluoromethane sulfonic acid, also release toxic halogen-containing fumes when heated to decomposition.

Sulfonic acids have essentially the same corrosive characteristics as does concentrated sulfuric acid. Detergent-based sulfonic acids pose a contact hazard, as they are very corrosive to the skin.

When sulfonic acids are neutralized to sulfonic acid salts, the materials become relatively innocuous and low in toxicity, as compared to the parent sulfonic acid.

Environmental Issues

Linear alkylbenzenesulfonic acid is the largest intermediate used for surfactant production in the world. Owing to the large volumes of production and consumption of linear alkylbenzenesulfonate, much attention has been paid to its biodegradation and a series of evaluations have been performed to thoroughly study its behavior in the environment. Much less attention has been paid to the environmental impact of other sulfonic acid-based materials.

Linear alkylbenzenesulfonate showed no deleterious effect on agricultural crops exposed to this material. Kinetics of biodegradation have been studied in both wastewater treatment systems and natural degradation systems. Studies have concluded that linear alkylbenzenesulfonate does not pose a risk to the environment. Linear alkylbenzenesulfonate has a half-life of approximately one day in sewage sludge and natural water sources and a half-life of one to three weeks in soils. Aquatic environmental safety assessment has also shown that the material does not pose a hazard to the aquatic environment.

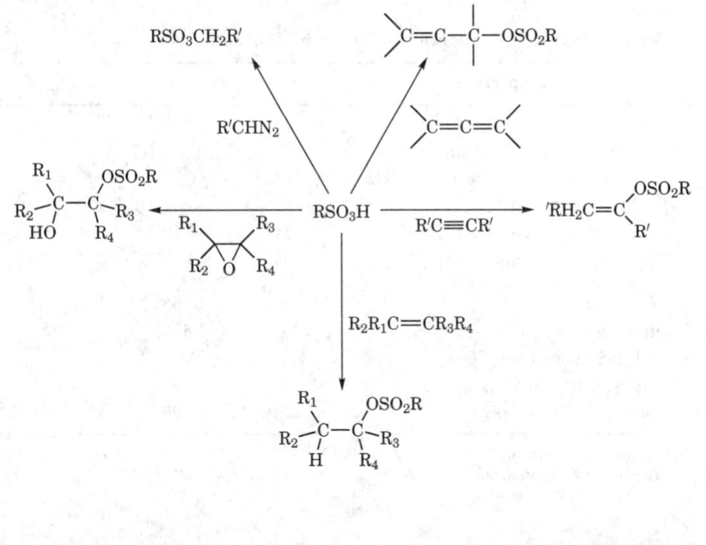

Figure 1. Reaction chemistry of sulfonic acids.

Uses

Surfactants and Detergents Uses. Perhaps the largest use of sulfonic acids is the manufacture of surfactants and surfactant formulations. In almost all cases, the parent sulfonic acid is an intermediate which is converted to a sulfonate prior to use. The largest volume uses for sulfonic acid intermediates are the manufacture of heavy-duty liquid and powder detergents, light-duty liquid detergents, hand soaps (see SOAP), and shampoos (see HAIR PREPARATIONS).

Lignosulfates, a complex mixture containing sulfonated lignin, are used as dispersing agents, wetting agents, binding agents, and sequestering agents (see LIGNIN). Dry forms of the materials are used as road binders, concrete additives, animal feed additives, and in vanillin production.

Naphthalenic, lignin, and melamine-based sulfonic acids are used as dispersion and wetting agents in industry. The sulfonate (**1**) is also widely used as a dispersing agent in dyestuff manufacture and high temperature dyeing of polyester fibers. A derivative of (**1**) based on 4-aminobenzene sulfonic acid has also been produced.

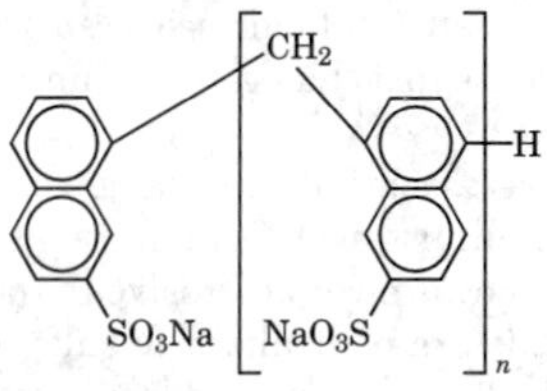

(**1**)

Other commercial naphthalene-based sulfonic acids, such as dinonylnaphthalene sulfonic acid, are used as phase-transfer catalysts and acid reaction catalysts in organic solvents.

Sulfonic Acid-Based Dyestuffs. Sulfonic acid-derived dyes are utilized industrially in the areas of textiles, paper, cosmetics, foods, detergents, soaps, leather, and inks, both as reactive and disperse dyes. Of the principal classes of dyes, sulfonic acid derivatives find utility in the areas of acid, azoic, direct, disperse, and fiber-reactive dyes. Sulfonic acid-based azo dyes (qv) and intermediates are characterized by the presence of one or more azo, RN=NR, groups.

Amide-Based Sulfonic Acids. The most important amide-based sulfonic acids are the alkenylamidoalkanesulfonic acids. These include 2-acrylamidopropanesulfonic acid, 2-acrylamido-2-methylpropanesulfonic acid, 3-acrylamido-2,4,4-trimethylpentanesulfonic acid, 2-acrylamido-2-(*p*-tolyl)ethanesulfonic acid, and 2-acrylamido-2-pyridylethanesulfonic acid.

Biological Uses

Taurine (2-aminoethanesulfonic acid), is the only known naturally occurring sulfonic acid. The material is an essential amino acid for cats and is used extensively by Ralston Purina Company as a food supplement in cat food manufacture.

Sulfonic acids have found greatly expanded usage in biological applications. Whereas the toxicity of sulfonic acids is in general rather high, several sulfonic acids are beneficially utilized *in vivo*. Taurocholic acid is an important bile component, aiding in the digestion of fat.

Potent inhibition of the herpes simplex virus has been observed using biphenyl disulfonic acid urea copolymers. Sulfonic acid derivatives have been shown to be potent antihuman immunodeficiency virus (anti-HIV) agents (see ANTIVIRAL AGENTS).

Other Applications. Hydroxylamine-*O*-sulfonic acid has many applications in the area of organic synthesis. The acid has found application in the preparation of hydrazines from amines, aliphatic amines from activated methylene compounds, aromatic amines from activated aromatic compounds, amides from esters, and oximes.

Petroleum sulfonates have found wide usage in enhanced oil recovery technology.

A variety of barium sulfonates have found use in antifriction lubricants for high speed bearing applications. Calcium and sodium salts of sulfonated olefins, esters, or oils are used for the enhancement of extreme pressure properties of grease and gear lubricants.

PAUL S. TULLY
Stepan Company

S. R. Sandler and W. Karo, *Organic Functional Group Preparation*, Vol. I, Academic Press, Inc., New York, 1983.

United States International Trade Commission, *Synthetic Organic Chemicals, United States Production and Sales, 1991*, USITC Publication 2607, Washington, D.C., Feb. 1993, pp. 12-3, 12-11–12-14.

E. A. Knaggs, *CHEMTECH*, 436–445 (July 1992), for a review of major surfactant sulfonic acids.

R. E. Bank and R. N. Hazeldine, *The Chemistry of Organic Sulfur Compounds*, Vol. 2, Pergamon Press, Inc., New York, 1966.

SULFOXIDES

Sulfoxides are compounds that contain a sulfinyl group covalently bonded at the sulfur atom to two carbon atoms. They have the general formula RS(O)R′, ArS(O)Ar′, and ArS(O)R, where Ar and Ar′ = aryl. Sulfoxides represent an intermediate oxidation level between sulfides and sulfones. The naturally occurring sulfoxides often are accompanied by the corresponding sulfides or sulfones. The only commercially important sulfoxide is the simplest member, dimethyl sulfoxide (DMSO) or sulfinylbismethane.

Sulfoxides occur widely in small concentrations in plant and animal tissues.

Properties

For the most part, sulfoxides are crystalline, colorless substances, although the lower aliphatic sulfoxides melt at relatively low temperatures. The lower aliphatic sulfoxides are water soluble; but as a class the sulfoxides are not soluble in water. They are soluble in dilute acids and a few are soluble in alkaline solution. DMSO is a colorless liquid; selected properties are listed in Table 1. Dimethyl sulfoxide generally undergoes typical sulfoxide reactions. It is used herein as an illustrative example.

Thermal Stability. Dimethyl sulfoxide decomposes slowly at 189°C to a mixture of products that includes methanethiol, formaldehyde, water, bis(methylthio)methane, dimethyl disulfide, dimethyl sulfone, and dimethyl sulfide. The decomposition is accelerated by acids, glycols, or amides. Sulfoxides undergo oxidation, reduction, carbon-sulfide cleavage, and Pummerer reactions.

Methylsulfinyl Carbanion. Strong bases, eg, sodium hydride or sodium amide, react with DMSO producing solutions of methylsulfinyl carbanion, known as the dimsyl ion, which are synthetically

Table 1. Selected Properties of Dimethyl Sulfoxide

Property		Value	
boiling point, °C		189.0	
conductivity, at 20°C, S/cm		3×10^{-8}	
dielectric constant, at 25°C, 10 MHz		46.7	
dipole moment, C·m[a]		1.4×10^{-29}	
entropy of fusion, J/(mol·K)[b]		45.12	
free energy of formation gas, C_{graph}, S_2(g), at 25°C, kJ/mol[b]		115.7	
freezing point, °C		18.55	
refractive index, $n^{25}{}_{D}$		1.4768	
flash point, open cup, °C		95	
density, g/cm^3, at 25°C		1.0955	
viscosity, mPa·s(= cP)	1.996^{25}	1.396^{45}	0.68^{100}

[a] To convert C·m to debye, divide by 3.336×10^{-30}.
[b] To convert J to cal, divide by 4.184.

useful. The solutions also provide a strongly basic reagent for generating other carbanions.

Methoxydimethylsulfonium and Trimethylsulfoxonium Salts. Alkylating agents react with DMSO at the oxygen. For example, methyl iodide gives methoxydimethylsulfonium iodide as the initial product. The alkoxysulfonium salts are quite reactive and, upon continued heating, either decompose to give carbonyl compounds or rearrange to the more stable trimethylsulfoxonium salts.

Complexes. The sulfoxides have a high (ca 4) dipole moment, which is characteristic of the sulfinyl group, and a basicity about the same as that of alcohols. They are strong hydrogen-bond acceptors. They would be expected, therefore, to solvate ions with electrophilic character, and a large number of DMSO complexes of metal ions have been reported.

Synthesis and Manufacture

The sulfoxides are most frequently synthesized by oxidation of the sulfides.

Dimethyl Sulfoxide. Dimethyl sulfoxide is manufactured from dimethyl sulfide (DMS), which is obtained either by processing spent liquors from the kraft pulping process or by the reaction of methanol or dimethyl ether with hydrogen sulfide.

Health and Safety Factors

Dimethyl sulfoxide is a relatively stable solvent of low toxicity. However, DMSO can penetrate the skin and may carry with it certain chemicals with which it is combined under certain conditions. Dimethyl sulfoxide has received considerable attention as a useful agent in medicine. In veterinary medicine, DMSO is used for horses and dogs as a topical application to reduce swelling resulting from injury or trauma (see VETERINARY DRUGS).

Uses of Dimethyl Sulfoxide

Polymerization and Spinning Solvent. Dimethyl sulfoxide is used as a solvent for the polymerization of acrylonitrile and other vinyl monomers, and as a reaction solvent for other polymerizations. It is also used as a solvent for displacement reactions, solvent for base-catalyzed reactions, extraction solvent, solvent for electrolytic reactions, cellulose solvent, pesticide solvent, and clean-up solvent.

W. W. Epstein and F. W. Sweat, *Chem. Rev.* **67**(3), 247 (1967).

D. Martin and H. G. Hauthal, *Dimethyl Sulfoxide*, Halsted Press, a division of John Wiley & Sons, Inc., New York, 1975.

B. S. Thyagarajan and N. Kharasch, *Intrascience Sulfur Reports*, Vol. 1, The Chemistry of DMSO, Intrascience Research Foundation, Santa Monica, Calif., 1966.

SULFUR

Sulfur S, a nonmetallic element, is the second element of Group 16 (VIA) of the Periodic Table, coming below oxygen and above selenium. In massive elemental form, sulfur is often referred to as brimstone. Sulfur is one of the most important raw materials of the chemical industry. It is of prime importance to the fertilizer industry (see FERTILIZERS) and its consumption is generally regarded as one of the best measures of a nation's industrial development and economic activity (see SULFUR COMPOUNDS; SULFUR REMOVAL AND RECOVERY; SULFURIC ACID AND SULFUR TRIOXIDE).

As of the 1990s, sulfur recovered as a by-product, involuntary sulfur, accounts for a larger portion of world supply than does mined or voluntary material. Sulfur is obtained from hydrogen sulfide, which evolves when natural gas, crude petroleum, tar sands, oil shales, coal, and geothermal brines are desulfurized

Sulfur constitutes about 0.052 wt % of the earth's crust. The forms in which it is ordinarily found include elemental or native sulfur in unconsolidated volcanic rocks, in anhydrite over salt-dome structures, and in bedded anhydrite or gypsum evaporate basin formations; combined sulfur in metal sulfide ores and mineral sulfates; hydrogen sulfide in natural gas; organic sulfur compounds in petroleum and tar sands; and a combination of both pyritic and organic sulfur compounds in coal.

Properties

Allotropy. Sulfur occurs in a number of different allotropic modifications, that is, in various molecular aggregations which differ in solubility, specific gravity, crystalline form, etc. Like many other substances, sulfur also exhibits dynamic allotropy, ie, the various allotropes exist together in equilibrium in definite proportions, depending on the temperature and pressure. The molecular formulas for the various allotropes are $S{-}S_n$, where n is a large but unidentified number, such as $n \geq 10^6$. Sulfur crystallizes in at least two distinct systems: the rhombic and the monoclinic forms.

The molecular constitution of liquid sulfur undergoes significant and reversible changes with temperature variations. These changes are evidenced by the characteristic temperature dependence of the physical properties of sulfur. In most studies of liquid sulfur, some striking changes in its physical properties are observed at about 160°C.

Constants and Chemical Properties. The constants of sulfur are presented in Table 1.

Sulfur falls between oxygen and selenium in Group 16 and resembles oxygen in its chemical reactions with most of the elements. The normal orbital electron structure is of the arrangement $1s^2\ 2s^2\ 2p^1\ 3s^2\ 3p^4$. Sulfur has valences of −2, +2, +3, +4, and +6. Selenium is a closely related element having a similar group of valences and analogous allotropy. Sulfur is insoluble in water but soluble to varying degrees in many organic solvents. Sulfur combines directly and usually energetically with almost all of the elements. Exceptions include gold, platinum, iridium, and the helium-group gases. In the presence of oxygen or dry air, sulfur is very slowly oxidized to sulfur dioxide. When burned in air, it forms predominantly sulfur dioxide with small amounts of sulfur trioxide. When burned in the presence of moist air, sulfurous acid and sulfuric acids are slowly generated.

Elemental Sulfur

Occurrence. The sulfur deposits associated with salt domes in the Gulf Coast regions of the southern United States and Mexico are an important segment of both U.S. and world sulfur supply.

Elemental sulfur occurs in another type of subsurface deposit similar to the salt-dome structures in that the sulfur is associated with anhydrite or gypsum. The deposits are sedimentary and occur in huge evaporite basins.

Elemental sulfur occurs in other types of surface or underground deposits throughout the world, but seldom in sufficient concentration to be commercially important. Volcanic deposits of sulfur usually occur in tufas, lava flows, and similar volcanic rocks but also in sedimentary and intrusive formations.

Extraction. Extraction processes include the Frasch process, the hydrodynamic process, distillation, flotation, autoclaving, fitration, solvent extraction, or a combination of these processes.

Sulfide Ores

Occurrence. Metal sulfides have been an important source of elemental sulfur. These sources are less attractive economically and technologically than other sources of sulfur; nevertheless sulfide ores are an important source of sulfur in other forms, such as sulfur dioxide and sulfuric acid.

Some of the most important metal sulfides are pyrite FeS_2 which is the most abundant form; chalcopyrite, $CuFeS_2$; pyrrhotite, $Fe_{n-1}S_n$; sphalerite, ZnS; galena, PbS; arsenopyrite, $FeS_2{\cdot}FeAs_2$; and pentlandite, $(Fe,Ni)_9S_8$.

Pyrometallurgical Processes. Recovery of sulfur from cuprous pyrite deposits is accomplished by using various smelting processes, which

Table 1. Physical Constants of Sulfur

Property	Value: Ideal	Value: Natural
freezing point of solid phase, °C		
rhombic	112.8	110.2
monoclinic	119.3	114.5
boiling point, °C		444.6
density of solid phase, 20°C, g/cm^3		
rhombic		2.07
monoclinic		1.96
amorphous		1.92
density of liquid, g/cm^3		
125°C		1.7988
130°C		1.7947
140°C		1.7865
150°C		1.7784
density of vapor, 444.6°C and 101.3 kPa(= 1 atm), g/L		3.64
refractive index, $n^{110}{}_D$		1.929
vapor pressure[a], P in Pa, T in K		
rhombic, 20–80°C	$\log P = 16.557 - 5166/T$	
monoclinic, 96–116°C	$\log P = 16.257 - 5082/T$	
liquid		
120–325°C	$\log P = 19.6 - 0.0062238T - 5405.1/T$	
325–550°C	$\log P = 12.3256 - 3268.2/T$	
surface tension, mN/m (= dyn/cm)		
120°C		60.83
150°C		57.67
critical temperature, °C		1040
specific heat, C_p, J/(kg·K)[b]		
rhombic, 24.9–95.5°C	$C_p = 468 + 0.814T$	
monoclinic, −4.5 to 118.9°C	$C_p = 465 - 0.908T$	
liquid, Sλ, 118.9–444.6°C	$C_p = 706 - 0.65T$	
gas, S, 25–1727°C	$C_p = 709 - 0.034T - 3.5 \times 10^6 T^{-2}$	
gas, S_2, 25–1727°C	$C_p = 558 + 0.018T - 5.2 \times 10^6 T^{-2}$	
linear thermal expansion of rhombic sulfur		
0–13°C		4.567×10^{-5}
50–78°C		8.633×10^{-5}
98–110°C		103.2×10^{-5}
latent heat of vaporization, L, J/g[c]	L^c	L^d
200°C	308.6	
400°C	286.4	278.0
440°C	290.1	274.6
electrical resistivity, ohm·cm		
20°C		1.9×10^{17}

[a] To convert log P_{Pa} to log P_{psi}, subtract 3.8384 from the constant.
[b] Includes heat of dissociation to S_2 present in vapor.
[c] Includes heat of dissociation to S_2 present in vapor.
[d] Minus heat of dissociation to S_2 present in vapor.

include the Orkla process, the Noranda process, and the Autokumpu process.

Hydrometallurgical Processes. Recovery of sulfur in the processing of nonferrous metal sulfides has been in the form of SO_2 and/or H_2SO_4 when smelter pyrometallurgical) operations are employed. However, there have been accounts of processes, mainly hydrometallurgical, in which sulfur is recovered in the elemental form (see METALLURGY, EXTRACTIVE). These include the CLEAR process, the Cymet process, the Electroslurry process, and the Sherritt-Cominco (SC) copper process.

Sulfates

Occurrence. The largest untapped source of sulfur occurs in the ocean as dissolved sulfates of calcium, magnesium, and potassium. The average sulfur concentration in seawater is 880 ppm. Thus, 1 km^3 of seawater contains about 0.86×10^6 t of elemental sulfur in the form of sulfates. Natural and by-product gypsum, $CaSO_4{\cdot}2H_2O$, and anhydrite, $CaSO_4$, rank second only to the oceans as potential sources of sulfur. Mineral deposits of gypsum and anhydrite are widely distributed in extremely large quantities. Gypsum is a by-product waste material from several manufacturing processes; most notable is the waste gypsum produced in manufacturing phosphoric acid from phosphate rock and sulfuric acid.

Extraction. Although many processes have been developed to recover elemental sulfur from gypsum or anhydrite, high capital and operating costs have precluded widespread use of these processes and are expected to continue to do so while less expensive sources remain available.

Phosphogypsum. Phosphogypsum is produced in tremendous quantities in the manufacture of phosphate fertilizers. As of the late 1990s, no commercial process existed for economical sulfur recovery from phosphogypsum.

Bacteriological Sulfur. Anaerobic, sulfate-reducing bacteria burn hydrocarbons as a source of energy, but combine sulfur instead of oxygen with the hydrogen to form hydrogen sulfide. Several experimenters have tried to utilize this knowledge in a controlled process for producing sulfur from gypsum or anhydrite. This process requires a strain of sulfate-reducing bacteria, an organic substrate whose hydrocarbons provide food for the bacteria, and close control of environmental conditions in order to obtain maximum sulfur yields.

Operating Factors. Increasing environmental concerns and subsequent governmental regulations have had a large impact on the sulfur industry. At first, recovered sulfur was considered a waste material, not a commercial by-product. As time went on, the importance of recovered sulfur increased as sulfur demand increased faster than the supply of Frasch and other native sulfur. Recovered sulfur became the primary domestic source of elemental sulfur in 1982.

The U.S. Clean Air Act set limits on the quantity of pollutants that could be released into the atmosphere. Sulfur dioxide was identified as one of the most common pollutants and also one of the principal contributors of acid rain, known to damage both natural and artificial environments.

Analysis

Elemental sulfur in either its ore or its refined state can generally be recognized by its characteristic yellow color or by the generation of sulfur dioxide when it is burned in air. Quantitatively, sulfur in a free or combined state is generally determined by oxidizing it to a soluble sulfate, by fusion with an alkali carbonate if necessary, and precipitating it as insoluble barium sulfate.

The National Safety Council, National Fire Protection Association, and other similar organizations publish technical information that describes general safety practices for use during the testing, handling, storage, and transport of sulfur. Each of these publications include a list of references for additional health and safety information.

Uses

Sulfur is used mainly as a chemical reagent rather than as a component of a finished product. Its predominant use as a process chemical generally requires that it first be converted to an intermediate chemical product prior to use in industry. In most of the ensuing chemical reactions between these sulfur-containing intermediate products and other minerals and chemicals, the sulfur values are not retained.

Rather, the sulfur values are most often discarded as a component of the waste product.

Sulfuric acid is the most important sulfur-containing intermediate product. More than 85% of the sulfur consumed in the world is either converted to sulfuric acid or produced directly as such (see SULFURIC ACID AND SULFUR TRIOXIDE). Worldwide, well over half of the sulfuric acid is used in the manufacture of phosphatic fertilizers and ammonium sulfate for fertilizers. The sulfur source may be voluntary elemental, such as from the Frasch process; recovered elemental from natural gas or petroleum; or sulfur dioxide from smelter operations.

In recent years, the largest demand for sulfur has been for agricultural purposes. The principal use is for phosphatic fertilizer processing. Other uses have been in petroleum refining; leaching of copper and uranium ores; and production of organic and inorganic chemicals, paints and pigments, pulp and paper, and synthetic materials; as well as numerous agricultural and industrial uses.

The Sulphur Institute

B. Meyer, *Sulfur, Energy, and Environment*, Elsevier Science Publishing Co., Inc., New York, 1977.

W. N. Tuller, ed., *The Sulphur Data Book*, McGraw-Hill Book Co., Inc., New York, 1954.

U.S. Bureau of Mines Annual Report, U.S. Bureau of Mines, Washington, D.C., 1994.

Sulphur Outlook, The Sulphur Institute, Washington, D.C., 1995 and 1996.

SULFUR COMPOUNDS

Carbon Sulfides

The only commercial carbon sulfide is carbon disulfide, CS_2. There are several unstable carbon sulfides. Carbon subsulfide, C_3S_2, is a red liquid (mp −0.5°C, bp 60–70°C at 1.6 kPa (12 mm Hg)) produced by the action of an electric arc on carbon disulfide.

Carbon monosulfide, CS, is an unstable gas produced by the decomposition of carbon disulfide at low pressure in a silent electrical discharge or photolytically.

Incompletely characterized carbon sulfides include a poorly characterized black solid, known as carsul. It occurs as a residue in sulfur distillation or as a precipitate in molten Frasch sulfur.

Carbonyl Sulfide. *Physical Properties.* Carbonyl sulfide (carbon oxysulfide), COS, is a colorless gas that is odorless when pure; however, it has been described as having a foul odor. Physical constants and thermodynamic properties are listed in Table 1.

Table 1. Physical and Thermodynamic Properties of Carbonyl Sulfide

Property	Value
mol wt	60.074
mp, °C	−138.8
bp, °C	−50.2
ΔH fusion, at 134.3 K, kJ/mol[a]	4.727
ΔH vaporization, at 222.87 K, kJ/mol[a]	18.57
density at 220 K, 101.3 kPa(= 1 atm), g/cm^3	1.19
sp gr, gas at 298 K (air = 1)	2.10
critical temperature, °C	105
critical pressure, kPa[b]	5946
critical volume, cm^3/mol	138
triple point, K	134.3
autoignition temperature in air, °C	ca 250
flammability limits in H_2O-saturated air at 17.9°C, vol %	
upper limit	9.6
lower limit	33.2
solubility in water at 101.3 kPa(= 1 atm), vol %	
0°C	0.356
20°C	0.149

[a] To convert J to cal, divide by 4.184.
[b] To convert kPa to psi, multiply by 0.145.

Chemical Properties. Carbonyl sulfide is a stable compound and can be stored under pressure in steel cylinders as compressed gas in equilibrium with liquid. At ca 600°C carbonyl sulfide disproportionates to carbon dioxide and carbon disulfide; at ca 900°C it dissociates to carbon monoxide and sulfur. It burns with a blue flame to carbon dioxide and sulfur dioxide. Carbonyl sulfide reacts only slowly with water to form carbon dioxide and hydrogen sulfide. Much technology has been developed for hydrolysis of carbonyl sulfide in gas streams to permit the removal of the sulfur content as hydrogen sulfide.

Occurrence and Preparation. Carbonyl sulfide is formed by many high temperature reactions of carbon compounds with donors of oxygen and sulfur. A principal route is the following reaction: $CO + S^0 \rightleftarrows COS$.

Carbonyl sulfide occurs as a by-product in the manufacture of carbon disulfide and is an impurity in some natural gases, in many manufactured fuel gases and refinery gases, and in combustion products of sulfur-containing fuels.

Carbonyl sulfide is overall the most abundant sulfur-bearing compound in the earth's atmosphere: 430–570 parts per trillion (10^{12}), although it is exceeded by H_2S and SO_2 in some industrial urban atmospheres.

Health and Safety Factors. Carbonyl sulfide is dangerously poisonous, especially because it is practically odorless when pure. It is lethal to rats at 2900 ppm. Studies show an LD_{50} (rat, ip) of 22.5 mg/kg. It acts principally on the central nervous system with death resulting mainly from respiratory paralysis.

Uses. There may be some captive use of carbonyl sulfide for production of certain thiocarbamate herbicides.

Thiophosgene

Physical Properties. Thiophosgene (thiocarbonyl chloride), $CSCl_2$, is a malodorous, red-yellow liquid (bp 73.5°C, d_{20}^{15} 1.509, n_D^{20} 1.5442). It is only slightly soluble with decomposition in water, but it is soluble in ether and various organic solvents.

Chemical Properties. Thiophosgene is more resistant to hydrolysis than its oxygen analogue, phosgene, but it is slowly hydrolyzed to carbon dioxide, hydrogen sulfide, and hydrochloric acid.

Most of the reactions of thiophosgene involve the expected chemistry of an acid chloride, in which the chlorine atoms are replaceable by various nucleophiles.

Preparation. Thiophosgene forms from the reaction of carbon tetrachloride with hydrogen sulfide, sulfur, or various sulfides at elevated temperatures. Of more preparative value is the reduction of trichloromethanesulfenyl chloride by various reducing agents.

Health and Safety Factors. Thiophosgene has an LD_{50} (rat, oral) of 929 mg/kg and an LC_{50} (inhalation, rat) of 370 mg/m^3. It has both irritant and systemic toxic properties.

Trichloromethanesulfenyl Chloride

Physical Properties. Trichloromethanesulfenyl chloride (perchloromethyl mercaptan, a misnomer but used as the common commercial name), CCl_3SCl, is a strongly acrid, pale yellow liquid, boiling at 149°C with some decomposition at atmospheric pressure, 68°C at 6.93 kPa (52 mm Hg), and 25°C at 0.8 kPa (6 mm Hg); sp gr (20°C/4°C) 1.6996; n_D^{23} 1.541. It slowly hydrolyzes and is soluble in most organic solvents.

Chemical Properties. Trichloromethanesulfenyl chloride is stable for prolonged periods at ambient temperature but decomposes slowly at its atmospheric boiling point forming sulfur monochloride, carbon tetrachloride, carbon disulfide, and polymeric oils. It is hydrolyzed very slowly by water at room temperature. At 160°C, hydrolysis is rapid and leads ultimately to the formation of carbon dioxide, hydrochloric acid, and sulfur. Trichloromethanesulfenyl chloride reacts rapidly with sodium hydroxide forming sodium dichloromethanesulfinate, $CHCl_2SO_2Na$.

The oxidation of trichloromethanesulfenyl chloride by nitric acid or oxidative chlorination in the presence of water yields trichloromethanesulfonyl chloride, CCl_3SO_2Cl, a lacrimatory solid.

Manufacture. Trichloromethanesulfenyl chloride is made commercially by chlorination of carbon disulfide with the careful exclusion of iron or other metals, which catalyze the chlorinolysis of the C–S bond to produce carbon tetrachloride.

Health and Safety Factors. Trichloromethanesulfenyl chloride is extremely toxic and mutagenic. It has an LD_{50} in rabbits (percutaneous) of 1410 mg/kg. Severe local irritation can result from contact of the liquid or vapor with the skin, eyes, mucous membranes, and upper respiratory tract.

Uses. The principal commercial application for trichloromethanesulfenyl chloride is as an intermediate for the manufacture of fungicides, the most important being captan, *N*-(trichloromethylthio)-4-cyclohexene-1,2-dicarboximide, and folpet, *N*-(trichloromethylthio)phthalimide, (see FUNGICIDES, AGRICULTURAL).

Hydrogen Sulfide

Hydrogen sulfide is present in the gases from many volcanoes, sulfur springs, undersea vents, swamps, and stagnant bodies of water. Bacterial reduction of sulfates and bacterial decomposition of proteins forms hydrogen sulfide. Of greater importance as a source of sulfur are sour gases, which occur in large amounts in several locations.

Hydrogen sulfide is a by-product of many industrial operations, such as the hydrodesulfurization of crude oil and of coal. (see PETROLEUM, REFINERY PROCESSES). A large source of hydrogen sulfide may result if coal liquefaction attains commercial importance (see COAL CONVERSION PROCESSES).

Physical Properties. Hydrogen sulfide, H_2S, is a colorless gas having a characteristic rotten-egg odor. The physical properties of hydrogen sulfide are given in Table 2.

Chemical Properties. Although hydrogen sulfide is thermodynamically stable, it can dissociate at very high temperatures.

Hydrogen sulfide is oxidized by a number of oxidizing agents.

Table 2. Selected Physical and Thermodynamic Properties of Hydrogen Sulfide

Property	Value
mol wt	34.08
mp, °C	−85.53
bp, °C	−60.31
ΔH fusion, kJ/mol[a]	2.375
ΔH vaporization, kJ/mol[a]	18.67
density at −60.31°C, kg/m^3	949.6
sp gr, gas (air = 1)	1.182[b]
critical temperature, °C	100.38
critical pressure, kPa[c]	9006
critical density, kg/m^3	346.0
C_p, at 27°C, J/(mol·K)[a]	34.2
autoignition temperature in air, °C	ca 260
vapor pressure, kPa[c]	
0°C	1049
40°C	2937
solubility in water[d], g/100 g soln	
0°C	0.710
20°C	0.398

[a] To convert J to cal, divide by 4.184.
[b] Based on air = 79 mol % nitrogen plus 21 mol % oxygen.
[c] To convert kPa to psi, multiply by 0.145; to convert kPa to bars, divide by 100.
[d] At 101.3 kPa(= 1 atm) total pressure.

Some of these reactions are of practical importance. The oxidation of hydrogen sulfide in a flame is one means for producing the sulfur dioxide required for a sulfuric acid plant. Oxidation of hydrogen sulfide by sulfur dioxide is the basis of the Claus process for sulfur recovery.

Anhydrous gaseous or liquid hydrogen sulfide is practically nonacidic, but aqueous solutions are weakly acid. Anhydrous hydrogen sulfide does not react at ordinary temperatures with metals, eg, mercury, silver, or copper. However, in the presence of air and moisture, the reaction is rapid, leading to tarnishing in the case of silver and copper. Hydrogen sulfide causes the precipitation of sulfides from many heavy-metal salts.

Manufacture. Small cylinders of hydrogen sulfide are readily available for laboratory purposes, but the gas can also be easily synthesized by action of dilute sulfuric or hydrochloric acid on iron sulfide, calcium sulfide, zinc sulfide, or sodium hydrosulfide. Small laboratory quantities of hydrogen sulfide can be easily formed by heating at 280–320°C a mixture of sulfur and a hydrogen-rich, nonvolatile aliphatic substance, eg, paraffin. Gas evolution proceeds more smoothly if asbestos or diatomaceous earth is also present.

Commercial-scale processes have been developed for the production of hydrogen sulfide from heavy fuel oils and sulfur as well as from methane, water vapor, and sulfur.

It has also been produced in commercial quantities by the direct combination of the elements.

Recovery from Gas Streams. The crude oil refined in the United States contains varying amounts of sulfur, eg, 0.04 wt % in Pennsylvania crude to ca 5 wt % in heavy Mississippi crude. Hydrodesulfurization is becoming increasingly important as a refinery operation. More than 90% of the sulfur in crude oils is accounted for in the gas–oil and coke-distillate fractions. The sulfur compounds are removed by passing the sulfur-rich fractions through a fixed-bed catalyst with hydrogen, which is generally by-product hydrogen from catalytic reforming. Besides conversion of 80–90% of the sulfur compounds to hydrogen sulfide, the hydrocarbon saturation is increased. (see SULFUR REMOVAL AND RECOVERY).

Corrosivity. Anhydrous hydrogen sulfide has a low general corrosivity toward carbon steel, aluminum, Inconel, Stellite, and 300-series stainless steels at moderate temperatures. Temperatures greater than ca 260°C can produce severe sulfidation of carbon steel.

Wet hydrogen sulfide can be quite corrosive to carbon steel; corrosion rates can exceed 2.5 mm/yr. In addition to general corrosion, wet hydrogen sulfide service can cause sulfide stress cracking. An important factor in reducing the likelihood of sulfide stress cracking is to limit the hardness of the base metal, weld metal, and base metal's heat-affected zone.

Health and Safety Factors. Hydrogen sulfide has an extremely high acute toxicity and has caused many deaths both in the workplace and in areas of natural accumulation, eg, cisterns and sewers. Brief exposure to hydrogen sulfide at a concentration of 140 mg/m^3 causes conjunctivitis and keratitis (eye damage), and exposures at above ca 280 mg/m^3 cause unconsciousness, respiratory paralysis, and death.

Hydrogen sulfide is especially dangerous when it occurs in low lying areas or confined workspaces or when it exists in high concentrations under pressure. Protective measures involve prompt detection and adequate ventilation.

Uses.

Most of the hydrogen sulfide recovered as a by-product is converted to elemental sulfur by the Claus process or to sulfuric acid where there is a market for the acid near the source of the hydrogen sulfide. Hydrogen sulfide is also used to prepare various inorganic sulfides, notably sodium sulfide and sodium hydrosulfide, which are used in the manufacture of dyes, rubber chemicals, pesticides, polymers, plastics additives, leather, and pharmaceuticals. A large amount of sodium hydrosulfide or sodium sulfide is used and largely recycled in kraft pulping; hydrogen sulfide can be used for replenishing the sulfide content (see PULP). An important industrial application is the reaction of hydrogen sulfide with alcohol or olefins (alkenes) to produce thiols or mercaptans.

Hydrogen Polysulfides

Individual hydrogen polysulfides (sulfanes) have been characterized from H_2S_2 up to at least H_2S_8. These are of no commercial utility by themselves, although sodium and calcium polysulfides, which are made by addition of sulfur to the corresponding monosulfides, are used commercially. The atmospheric boiling point of H_2S_2 is 70.7°C and the boiling point of H_2S_3 is 69°C at 0.3 kPa (2 mm Hg).

Except for hydrogen sulfide the H_2S_x sulfanes have no practical utility.

Sulfur Halides and Oxyhalides

Sulfur forms several series of halides with all of the halogens except iodine. The fluorides include the commercially important sulfur hexafluoride (see FLUORINE COMPOUNDS, INORGANIC).

Sulfur Monochloride. *Properties.* Sulfur monochloride, S_2Cl_2, is a yellow-orange liquid with a characteristic pungent odor and boiling point of 137.8°C.

Sulfur monochloride is stable at ambient temperature but undergoes exchange with dissolved sulfur at 100°C, indicating reversible dissociation. When distilled at its atmospheric boiling point, it undergoes some decomposition to the dichloride.

Manufacture. Sulfur monochloride is made commercially by direct chlorination of sulfur, usually in a heel of sulfur chloride from a previous batch.

The principal commercial uses of sulfur monochloride are in the manufacture of lubricant additives and vulcanizing agents for rubber (see LUBRICATION AND LUBRICANTS; RUBBER CHEMICALS).

Health and Safety Factors. Sulfur monochloride is highly toxic and irritating by inhalation, and is corrosive to skin and eyes. Pulmonary edema may result from inhalation.

Sulfur Dichloride. *Properties.* Sulfur dichloride SCl_2, is a reddish or yellow fuming liquid which decomposes in moist air with the evolution of hydrogen chloride. Pure sulfur dichloride is unstable and is supplied commercially as a 72–82 wt % SCl_2 mixture, with sulfur monochloride comprising the remaining percentage. The melting point is reported in the range −121.5 to −61°C, and the boiling point with decomposition is 59°C.

Sulfur dichloride in the liquid state at ambient temperature is in equilibrium with sulfur monochloride and dissolved chlorine. Sulfur dichloride is oxidized by sulfur trioxide or chlorosulfuric acid to form thionyl chloride, $SOCl_2$.

Sulfur dichloride reacts with an excess of sulfur trioxide forming pyrosulfuryl chloride, $S_2O_5Cl_2$.

Manufacture. The manufacture of sulfur dichloride is similar to that of sulfur monochloride, except that the last stage of chlorination proceeds slowly and must be conducted at temperatures below 40°C.

Uses. Sulfur dichloride is used as a chlorinating agent in the manufacture of parathion insecticide intermediates (see INSECT CONTROL TECHNOLOGY). It is also useful in the rapid vulcanization of rubber, eg, in the preparation of thin rubber goods by coating molds or fabrics with rubber latex.

Oxyhalides

Thionyl Chloride. *Properties.* Thionyl chloride, $SOCl_2$, is a colorless fuming liquid with a choking odor. Selected physical and thermodynamic properties are listed in Table 3.

Significant inorganic reactions of thionyl chloride include its reactions with sulfur trioxide to form pyrosulfuryl chloride and with hydrogen bromide to form thionyl bromide. With many metal oxides it forms the corresponding metal chloride plus sulfur dioxide.

Alkyl chlorides, alkyl sulfites, or alkyl chlorosulfites form from the reactions of thionyl chloride with aliphatic alcohols.

Table 3. Selected Physical and Thermodynamic Properties of Thionyl Chloride

Property	Value
mol wt	118.98
mp, °C	−104.5
bp, °C	76
specific gravity, 25°C	1.63
latent heat of vaporization, kJ/mol[a]	31.
viscosity, mPa·(= cP)	
0°C	0.81
refractive index, n^{20}_D	1.517
vapor pressure, kPa[b]	
−20°C	1.5
0°C	4.5
20°C	11.6

[a] To convert J to cal, divide by 4.184.
[b] To convert kPa to mm Hg, multiply by 7.5.

Manufacture. Thionyl chloride may be made by any of the following reactions:

$$SCl_2 + SO_3 \rightarrow SOCl_2 + SO_2$$

$$SCl_2 + SO_2 + Cl_2 \rightarrow 2\ SOCl_2$$

$$SCl_2 + SO_2Cl_2 \rightarrow 2\ SOCl_2$$

The sulfur dichloride can be fed as such or produced directly in the reactor by reaction of chlorine with sulfur monochloride.

Health and Safety Factors. Thionyl chloride is a reactive acid chloride which can cause severe burns to the skin and eyes and acute respiratory tract injury upon vapor inhalation.

Uses. A principal use of thionyl chloride is in the conversion of acids to acid chlorides, which are employed in many syntheses of herbicides, surfactants, drugs, vitamins (qv), and dyestuffs. Possible larger-scale applications are in the preparation of engineering thermoplastics of the polyarylate type made from iso- and terephthaloyl chlorides, which can be made from the corresponding acids plus thionyl chloride (see ENGINEERING PLASTICS).

Sulfuryl Chloride. *Properties.* Sulfuryl chloride, SO_2Cl_2, is a colorless to light yellow liquid with a pungent odor. Physical and thermodynamic properties are listed in Table 4.

Sulfuryl chloride is stable at room temperature but readily dissociates to sulfur dioxide and chlorine when heated.

The decomposition of sulfuryl chloride is accelerated by light and catalyzed by aluminum chloride and charcoal. Sulfuryl chloride reacts with sulfur at 200°C or at ambient temperature in the presence of aluminum chloride producing sulfur monochloride. It liberates bromine or iodine from bromides or iodides.

Manufacture. The preparation of sulfuryl chloride is carried out by feeding dry sulfur dioxide and chlorine into a water-cooled glass-lined steel vessel containing a catalyst, eg, activated charcoal. Alternatively, chlorine is passed into liquefied sulfur dioxide at ca 0°C in the presence of a dissolved catalyst.

Health and Safety Factors. Sulfuryl chloride is both corrosive to the skin and toxic upon inhalation. The vapors irritate the eyes and upper respiratory tract, causing prompt symptoms ranging from coughing to extreme bronchial irritation and pulmonary edema.

Uses. Uses of sulfuryl chloride include the manufacture of chlorophenols, eg, chlorothymol for use as disinfectants. It is also used in the manufacture of alpha-chlorinated acetoacetic derivatives, eg, $CH_3COCHClCOOC_2H_5$, which are precursors for important substituted imidazole drugs, phosphate insecticides, and fungicides. Sulfuryl chloride is used captively by DuPont on a large scale in the

Table 4. Selected Physical and Thermodynamic Properties of Sulfuryl Chloride

Property	Value
mol wt	134.968
mp, last crystal point, °C	−54
bp, °C	69.1
density, at 25°C, g/cm^3	1.6570
latent heat of vaporization, kJ/mol[a]	27.95
surface tension at 23.5°C, mN/m(= dyn/cm)	35.26
viscosity, mPa·s(= cP)	
0°C	0.918
refractive index, n^{20}_D	1.443
vapor pressure, kPa[b]	
0°C	5.45
coefficient of expansion, 0–38°C, °C^{-1}	0.0012
electrical conductivity, S/cm	3×10^{-8}

[a] To convert J to cal, divide by 4.184.
[b] To convert kPa to mm Hg, multiply by 7.5.

manufacture of chlorosulfonated polyethylene (see ELASTOMERS, SYNTHETIC–CHLOROSULFONATED POLYETHYLENE).

Sulfur Nitrides

Although no commercial applications have as yet been developed for these compounds, some interest was stimulated by the discovery that polythiazyl, a polymeric sulfur nitride, $(SN)_x$, with metallic luster, is electroconductive. Other sulfur nitrides are unstable. Tetrasulfur nitride is explosive and shock-sensitive.

Sulfur Oxides

Numerous oxides of sulfur have been reported; those that have been characterized are SO, S_2O, S_nO (n = 6–10), SO_2, SO_3, and SO_4. Among these, SO_2 and SO_3 are of principal importance (see SULFURIC ACID AND SULFUR TRIOXIDE).

Sulfur Dioxide. *Properties.* Sulfur dioxide, SO_2, is a colorless gas with a characteristic pungent, choking odor. Its physical and thermodynamic properties are listed in Table 5.

Sulfur dioxide is extremely stable to heat, even up to 2000°C. It is not explosive or flammable in admixture with air. The oxidation of sulfur dioxide by air or pure oxygen is a reaction of great commercial importance and is commonly conducted at 400–700°C in the presence of a catalyst, eg, vanadium oxide.

Reduction of sulfur dioxide to sulfur includes an industrially important group of reactions.

The Claus process, which involves the reaction of sulfur dioxide with hydrogen sulfide to produce sulfur in a furnace, is important in the production of sulfur from sour natural gas or by-product sulfur-containing gases (see SULFUR REMOVAL AND RECOVERY).

Manufacture. For most chemical process applications requiring sulfur dioxide gas or sulfurous acid, sulfur dioxide is prepared by the burning of sulfur or pyrite, FeS_2.

Corrosivity. Almost all common materials of construction are resistant to commercial dry liquid sulfur dioxide, dry sulfur dioxide gas, and hot sulfur dioxide gas containing water at above the dew point.

Health, Safety, and Environmental Factors. Sulfur dioxide has only a moderate acute toxicity. Sulfur dioxide shows some mutagenic effects in microorganisms and fruit flies. Human lymphocyte DNA damage has been observed. Sulfur dioxide is a strong irritant; concentrations even as low as 2 ppm can have a respiratory irritant, choking, and sneeze/cough inducing effect. The symptoms at 50 ppm are sufficiently disagreeable that most persons would not tolerate them for more than a few minutes. Acutely toxic levels can cause suppurative bronchitis and asthma-like or influenza-like symptoms. At very high levels, asphyxia leading to death, or chemical bronchopneumonia may develop which can be fatal after several days. Overexposure to sulfur dioxide has formerly been widespread in the smelting and paper industries. However, animal and human studies suggest that sulfur dioxide has only a low degree of chronic toxicity at subacute levels.

Plants and animals have a natural tolerance to low levels of sulfur dioxide. Natural sources include volcanoes and volcanic vents, decaying organic matter, and solar action on seawater. Sulfur dioxide is believed to be the main sulfur species produced by oxidation of dimethyl sulfide that is emitted from the ocean.

At low levels, sulfur dioxide in the atmosphere is not harmful to crops, but damage can occur at excessive levels. Sulfur dioxide itself has been found useful in drip irrigation systems and in calcareous soils.

Sulfur Dioxide Emissions and Control. A substantial part of the sulfur dioxide in the atmosphere is the result of burning sulfur-containing fuel, notably coal, and smelting sulfide ores. (see also AIR POLLUTION CONTROL METHODS; COAL CONVERSION PROCESSES, CLEANING AND DESULFURIZATION; EXHAUST CONTROL, INDUSTRIAL; SULFUR REMOVAL AND RECOVERY).

Uses. The dominant use of sulfur dioxide is as a captive intermediate for production of sulfuric acid. There is also substantial captive production in the pulp and paper industry for sulfite pulping, and it is used as an intermediate for on-site production of bleaches, eg, chlorine dioxide or sodium hydrosulfite (see BLEACHING AGENTS). In food processing, sulfur dioxide has a wide range of applications as a fumigant, preservative, bleach, and steeping agent for grain and dried fruit. Because of the sensitivity of some persons to sulfur dioxide, it has been banned for use on fresh produce by the U.S. FDA.

In water treatment, sulfur dioxide is often used to reduce residual chlorine from disinfection and oxidation. This technology is used in potable water treatment, in sewage treatment, and especially in industrial wastewater treatment.

In petroleum technology, sulfur dioxide, or sodium sulfite, is used as an oxygen scavenger. This use is particularly important in secondary and tertiary oil recovery processes involving flooding of underground oil formations using water or aqueous solutions (see PETROLEUM, ENHANCED OIL RECOVERY).

In mineral technology, sulfur dioxide and sulfites are used as flotation depressants for sulfide ores. A newer use for sulfur dioxide is in cyanide detoxification in connection with cyanide leaching of precious metals from mine dumps.

Table 5. Selected Physical and Thermodynamic Properties of Sulfur Dioxide

Property	Value
mol wt	64.06
mp, °C	−72.7
bp, °C	−10.02
ΔH fusion, kJ/mol[a]	7.40
ΔH vaporization, at −10.0°C, kJ/mol[a]	24.92
vapor density, at 0°C, 101.3 kPa(= 1 atm), air = 1	2.263
liquid density, at −20°C, g/cm^3	1.50
critical temperature, °C	157.6
critical pressure, kPa[b]	7911
dielectric constant at −16.5°C	17.27
dipole moment at 25°C, C · m[c]	3.87×10^{-30}
vapor pressure, kPa[b]	
10°C	230
30°C	462
solubility in water, at 101.3 kPa, g/100 g H_2O	
0°C	22.971
20°C	11.577
40°C	5.881

[a] To convert J to cal, divide by 4.184.
[b] To convert kPa to psi, multiply by 0.145.
[c] To convert C · m to D, divide by 3.336×10^{-30}.

Sulfur Oxygen Acids and Their Salts

Sulfuric acid, H_2SO_4, the most important commercial sulfur compound (see SULFURIC ACID AND SULFUR TRIOXIDE), and peroxymonosulfuric acid (Caro's acid), H_2SO_5, are discussed elsewhere (see PEROXIDES AND PEROXIDE COMPOUNDS, INORGANIC). The lower valent sulfur acids are not stable species at ordinary temperatures. Dithionous acid, $H_2S_2O_4$, sulfoxylic acid, H_2SO_2, and thiosulfuric acid, $H_2S_2O_3$ are unstable species.

Sodium Sulfite. *Properties.* Anhydrous sodium sulfite, Na_2SO_3, is an odorless, crystalline solid and most commercial grades other than by-product materials are colorless or off-white. It melts only with decomposition. The specific gravity of the pure solid is 2.633 (15.4°C). Sodium sulfite is quite soluble in water; it is soluble in glycerol but insoluble in alcohol, acetone, and most other organic solvents.

Anhydrous sodium sulfite is stable in dry air at ambient temperatures or at 100°C, but in moist air it undergoes rapid oxidation to sodium sulfate. On heating to 600°C, sodium sulfite disproportionates to sodium sulfate and sodium sulfide.

Aqueous solutions of sodium sulfite are alkaline, and they are oxidized readily by air.

Manufacture. In a typical process, a solution of sodium carbonate is allowed to percolate downward through a series of absorption towers through which sulfur dioxide is passed countercurrently. The solution leaving the towers is chiefly sodium bisulfite of typically 27 wt % combined sulfur dioxide content. The solution is then run into a stirred vessel where aqueous sodium carbonate or sodium hydroxide is added to the point where the bisulfite is fully converted to sulfite. The solution may be filtered if necessary to attain the required product grade. A pure grade of anhydrous sodium sulfite can then be crystallized above 40°C, because the solubility decreases with increasing temperature.

Analytical Methods. A classical and still widely employed analytical method is iodimetric titration.

Health and Safety Factors. Although sodium sulfite has no detectible odor, its dust and solutions are irritating to the skin, eyes, and mucous membranes. The ingestion of sodium sulfite causes gastric irritation resulting from the liberation of sulfurous acid.

Uses. Sodium sulfite is utilized in neutral semichemical pulping, acid sulfite pulping, high yield sulfite cooling, and some kraft pulping processes. Sodium sulfite is useful as a reducing agent in certain photographic fixing baths, developers, hardeners, and intensifiers. However, the principal use is as a film preservative and discoloration preventative (see PHOTOGRAPHY).

Sodium Bisulfite. Sodium bisulfite, $NaHSO_3$, exists in solution but is not a stable compound in the solid state. The anhydrous sodium bisulfite of commerce consists of sodium metabisulfite, $Na_2S_2O_5$. Aqueous sodium bisulfite solution, having specific gravity 1.36 and containing the equivalent of 26–27 wt % SO_2, is a commercial product.

Sodium Metabisulfite. *Properties.* Sodium metabisulfite (sodium pyrosulfite, sodium bisulfite (a misnomer)), $Na_2S_2O_5$, is a white granular or powdered salt (specific gravity 1.48) and is storable when kept dry and protected from air. The chemistry of sodium metabisulfite is essentially that of the sulfite–bisulfite–metabisulfite–sulfurous acid system.

Manufacture. Aqueous sodium hydroxide, sodium bicarbonate, sodium carbonate, or sodium sulfite solution are treated with sulfur dioxide to produce sodium metabisulfite solution.

Health and Safety Factors. Sodium metabisulfite is nonflammable, but when strongly heated it releases sulfur dioxide. The oral acute toxicity is slight and the LD_{50} (rat, oral) is 2 g/kg. The solid product and its aqueous solutions are mildly acidic and irritate the skin, eyes, and mucous membranes. Food-grade sodium metabisulfite is permitted in those foods that are not recognized as sources of vitamin B_1, with which sulfur dioxide reacts.

Uses. Sodium metabisulfite is extensively used as a food preservative and bleach in the same applications as sulfur dioxide. In tanneries, sodium bisulfite is used to accelerate the unhairing action of lime. It is also used as a chemical reagent in the synthesis of surfactants. The reversible addition of sodium bisulfite to carbonyl groups is used in the purification of aldehydes. Sodium bisulfite also is employed in polymer and synthetic fiber manufacture.

Sodium Dithionite. *Properties.* Sodium dithionite (sodium hydrosulfite, sodium sulfoxylate), $Na_2S_2O_4$, is a colorless solid and is soluble in water to the extent of 22 g/100 g of water at 20°C.

Anhydrous sodium dithionite is combustible and can decompose exothermically if subjected to moisture.

Sodium dithionite is most stable and effective as a reducing agent in alkaline solutions. Dithionite is a stronger reducing agent than sulfite.

The most important organic chemistry of sodium dithionite involves its use in reducing dyes, eg, anthraquinone vat dyes, sulfur dyes, and indigo, to their soluble leuco forms (see DYES, ANTHRAQUINONE).

Manufacture. Commercial processes for production of sodium dithionite are based on reduction of sulfite or bisulfite. The electrochemical process, commercialized in the late 1980s, is the newest available technology and utilizes only caustic and sulfur dioxide as raw materials.

Sodium dithionite solution can be produced on-site utilizing a mixed sodium borohydride–sodium hydroxide solution to reduce sodium bisulfite. Electrochemical technology is also being offered for on-site production of sodium hydrosulfite solution.

Analytical Methods. Various analytical methods involve titration with oxidants, eg, hexacyanoferrate (ferricyanide), which oxidize dithionites to sulfite.

Health and Safety Factors. Dry sodium dithionite, when exposed to moist air, heats and can ignite spontaneously. Sodium dithionite is considered only moderately toxic. As a food additive, sodium dithionite is generally recognized as safe (GRAS).

Uses. Textile applications have historically been primary uses for dithionite, including dye reduction, dye stripping from fabric, bleaching, and equipment cleaning.

Pulp and paper bleaching applications mainly involve brightening mechanical pulps.

Miscellaneous uses include reductive bleaching of glue, gelatin, soap, oils, food products, and oxygen scavenging in water (used for high pressure boilers or for synthetic rubber polymerization).

Zinc Dithionite. Zinc dithionite, ZnS_2O_4, is a white, water-soluble powder. Although it exhibits somewhat greater stability in aqueous solution compared to sodium dithionite at a given temperature and pH, it is no longer used in the United States because of regulatory constraints on pollution of water by zinc.

Sodium and Zinc Formaldehyde Sulfoxylates. Although free sulfoxylic acid, H_2SO_2, has not been isolated and its salts are in doubt, organic derivatives, which may be viewed as adducts of sulfoxylic acid, are commercially made. The latter are mainly sodium formaldehyde sulfoxylate, $HOCH_2SO_2Na$ (commercially sold as the dihydrate) and zinc formaldehyde sulfoxylate. These compounds are water-soluble reducing agents with uses similar to the dithionites but are more stable. They can be used in reducing and bleaching applications at lower pH values and at somewhat higher temperatures than the dithionites.

In addition to applications in dyeing, sodium formaldehyde sulfoxylate is used as a component of the redox system in emulsion polymerization of styrene–butadiene rubber recipes.

Thiocyanic Acid and Its Salts

Free thiocyanic acid, HSCN, can be isolated from its salts, but is not an article of commerce because of its instability, although dilute solutions can be stored briefly. Commercial derivatives of thiocyanic acid are principally ammonium, sodium, and potassium thiocyanates, as well as several organic thiocyanates.

Ammonium Thiocyanate. *Properties.* Ammonium thiocyanate, NH_4SCN, is a hygroscopic crystalline solid which deliquesces at high humidities. It melts at 149°C with partial isomerization to thiourea. It is soluble in water to the extent of 65 wt % at 25°C and 77 wt % at 60°C. It is also soluble in methanol and ethanol. It is highly soluble in liquid ammonia and liquid sulfur dioxide, and moderately soluble in acetonitrile.

Ammonium thiocyanate rearranges upon heating to an equilibrium mixture with thiourea. Aqueous solutions of ammonium thiocyanate are weakly acidic.

Manufacture. The principal route used in the United States is the reaction of carbon disulfide with aqueous ammonia, which proceeds by way of ammonium dithiocarbamate. Upon heating, the ammonium dithiocarbamate decomposes to ammonium thiocyanate and hydrogen sulfide.

Analytical Methods. Thiocyanate is quantitatively precipitated as silver thiocyanate, and thus can be conveniently titrated with silver nitrate.

Health and Safety Factors. The lowest published human oral toxic dose is 430 mg/kg, causing nervous system disturbances and gastrointestinal symptoms. The LD_{50} (rat, oral) is 750 mg/kg. Thiocyanates are destroyed readily by soil bacteria and by biological treatment systems in which the organisms become acclimatized to thiocyanate.

Uses. Ammonium thiocyanate is a chemical intermediate for the synthesis of several proprietary agricultural chemicals, mainly herbicides. There are many smaller specialized uses for ammonium thiocyanate, including stabilization of glue formulations, as an ingredient in antibiotic fermentations, and as an adjuvant in textile dyeing and printing. A newer use is as a tracer in oil fields.

Sodium and Potassium thiocyanates. *Properties.* Sodium thiocyanate, NaSCN, is a colorless deliquescent crystalline solid (mp 323°C). It is soluble in water to the extent of 58 wt % NaSCN at 25°C and 69 wt % at 100°C. It is also highly soluble in methanol and ethanol, and moderately soluble in acetone. Potassium thiocyanate, KSCN, is also a colorless crystalline solid (mp 172°C) and is soluble in water to the extent of 217 g/100 g of water at 20°C and in acetone and alcohols. Much of the chemistry of sodium and potassium thiocyanates is that of the thiocyanate anion.

Manufacture, Shipment, and Analysis. In the United States, sodium and potassium thiocyanates are made by adding caustic soda or potash to ammonium thiocyanate, followed by evaporation of the ammonia and water. Analysis, and safety factors are similar to those of ammonium thiocyanate, except that the alkali thiocyanates are more thermally stable. The acute LD_{50} (rat, oral) of sodium thiocyanate is 764 mg/kg, accompanied by convulsions and respiratory failure; LD_{50} (mouse, oral) is 362 mg/kg. The lowest published toxic dose for potassium thiocyanate is 80–428 mg/kg, with hallucinations, convulsions, or muscular weakness.

Uses. The largest use for sodium thiocyanate is as the 50–60 wt % aqueous solution, as a component of the spinning solvent for acrylic fibers (see FIBERS, ACRYLIC; ACRYLONITRILE POLYMERS). Other textile applications are as a fiber swelling agent and as a dyeing and printing assist. A newer commercial use for sodium thiocyanate is as an additive to cement in order to impart early strength to concrete.

Sodium thiocyanate and other thiocyanate salts are used to prepare organic thiocyanates.

Lesser amounts of sodium thiocyanate are used in color toning photographic paper, as a stabilizer in rapid film development, and as a sensitizing agent in color negative-film emulsions.

Methanesulfonyl Chloride

Properties. Methanesulfonyl chloride (MSC), CH_3SO_2Cl, is a clear liquid, and is soluble in a wide variety of organic solvents, eg, methanol and acetone.

Methanesulfonyl chloride (MSC) is a reactive chemical which allows introduction of the mesyl group, $CH_3SO_2^-$, into a wide range of substrates. MSC undergoes free-radical-initiated addition to olefins to produce chloro-substituted sulfones. With strong bases, MSC can undergo dehydrochlorination to a transitory reactive intermediate, $CH_2{=}SO_2$, which can dimerize or undergo various addition reactions.

Manufacture. Methanesulfonyl chloride is made commercially either by the chlorination of methyl mercaptan or by the sulfochlorination of methane.

Health and Safety Factors. MSC is a lachrymator and in order to prevent contact with eyes, goggles should be worn. It is also corrosive to skin and therefore chemically resistant gloves and protective clothing should be worn to prevent contact with skin.

Uses. Most applications of MSC are for intermediates in the pharmaceutical, photographic, fiber, dye, and agricultural industries. There also are miscellaneous uses as a stabilizer, catalyst, curing agent, and chlorination agent.

Methanesulfonic Acid

Properties. Methanesulfonic acid (MSA), CH_3SO_3H, is a clear, colorless, strong organic acid. MSA is soluble in water and in many organic solvents.

MSA combines high acid strength with low molecular weight. MSA finds use as catalyst for esterification, alkylation, and in the polymerization and curing of coatings. The anhydrous acid is also useful as a solvent.

The metal salts of MSA are highly soluble in water as well as in some organic solvents, making MSA useful in electroplating operations.

MSA also finds use in preparing biological and agricultural chemicals, textile treatment chemicals, and for plastics and polymers.

Health and Safety Factors. MSA is a strong toxic acid and is corrosive to skin. Certain reaction products of MSA are suspected of mutagenic, teratogenic, and carcinogenic activity.

Manufacture. Methanesulfonic acid is made commercially by oxidation of methyl mercaptan by chlorine in aqueous hydrochloric acid to give methanesulfonyl chloride which is then hydrolyzed to MSA.

EDWARD D. WEIL
Polytechnic University
STANLEY R. SANDLER
Elf Atochem North America, Inc.

Sulfur Products Handbook on Sulfur Monochloride and Sulfur Chloride, Bulletin SPE-SUL-HB 10/9, Oxychem Basic Chemicals Group, Occidental Chemical Corp., Dallas, Tex., 1993, p. 3.

F. J. Dinan and J. F. Bieron, *A Survey of Reactions of Thionyl Chloride, Sulfuryl Chloride and Sulfur Chlorides*, Occidental Chemical Corp., Niagara Falls, N.Y., 1990.

J. O. Nriagu, ed., *Sulfur in the Environment*, John Wiley & Sons, Inc., New York, 1978.

A. Senning, ed., *Sulfur in Organic and Inorganic Chemistry*, Marcel Dekker, New York, 1972.

SULFUR DYES

Sulfur dyes are used mainly for dyeing textile cellulosic materials or blends of cellulosic fibers with synthetic fibers such as acrylic fibers, polyamides (nylons), and polyesters. They are also used for silk and paper in limited quantities for specific applications. Solubilized sulfur dyes are used on certain types of leathers.

From an applications point of view, the sulfur dyes are between vat, direct, and fiber-reactive dyes. They give good to moderate lightfastness and good wetfastness at low cost and rapid processing (see DYES, APPLICATION AND EVALUATION).

Traditionally, these dyes are applied from a dyebath containing sodium sulfide. However, development in dyeing techniques and manufacture has led to the use of sodium sulfhydrate, sodium polysulfide, sodium dithionite, thiourea dioxide, and glucose as reducing agents. In the reduced state, the dyes have affinity for cellulose and are subsequently exhausted on the substrate with common salt or sodium sulfate and fixed by oxidation.

The range of colors covers all hue classification groups except a true red. As a rule, the hues are dull compared with other dye classes. Black is the most important, followed by blues, olives, and browns (see DYES AND DYE INTERMEDIATES).

Table 1. Sulfur-Bake Dyes

Intermediates	Shade	CI designation	
		Name	Number
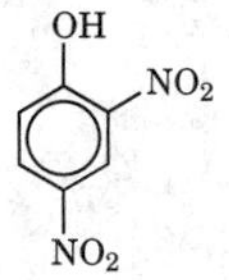	orange	Sulfur Orange 1	53050
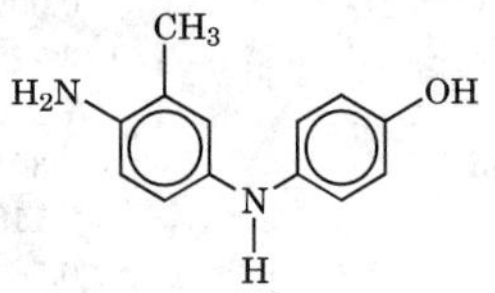	yellowish brown	Sulfur Brown 12	53065
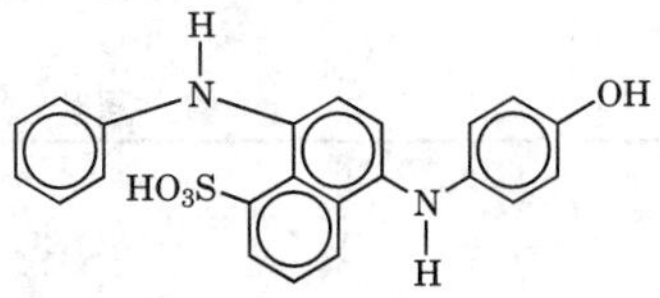	reddish yellow	Sulfur Yellow 1	53040

Sulfur dyes have good storage stability and fastness properties, and are generally easily adaptable to modern dyeing methods.

Chemical Properties

Classification and Structure. Little is known about the structure of sulfur dyes, and therefore, they are classified according to the chemical structure of the starting materials.

The process of sulfurization is usually carried out by a sulfur bake, in which the dry organic starting material is heated with sulfur between 160 and 320°C; a polysulfide bake, which includes sodium sulfide; a polysulfide melt, in which aqueous sodium polysulfide and the organic starting material are heated under reflux or under pressure in a closed vessel; or a solvent melt, in which butanol, Cellosolve, or dioxitol are used alone or together with water. In the last two methods, hydrotropes may be added to enhance the solubility of the starting material. The hydrotropes improve yield and quality of the final dyestuff.

Sulfur Bake. The yellow, orange, and brown sulfur dyes belong to this group (Table 1). The dyes are usually made from aromatic amines, diamines, and their acyl and nuclear alkyl derivatives. These may be used in admixture with nitroanilines and nitrophenols or aminophenols to give the desired shade.

Polysulfide Bake. Although most dyes made formerly by this method are made by a polysulfide melt, some dyes where certain nitro and phenolic bodies prevent color formation in the sulfur bake are still made by this method (Table 2). Pre-reduction of nitro groups substantially reduces the explosion hazard associated with heating these compounds with sulfur alone. Included in this class are also the exceptionally lightfast sulfur dyes derived from the polynitrodecacyclenes.

Polysulfide Melt. CI Sulfur Black 1 (CI 53185), derived from 2,4-dinitrophenol, is the most important dye in this group (Table 3), which also includes the indophenol-type intermediates. The latter are applied in the stable leuco form. The derived dyes are usually confined to violet, blue, and green shades. Other members of this group are intermediates capable of forming quinoneimine or phenazone structures that produce red-brown or Bordeaux shades.

Table 2. Polysulfide-Bake Dyes

Intermediate(s)	Shade	CI designation	
		Name	Number
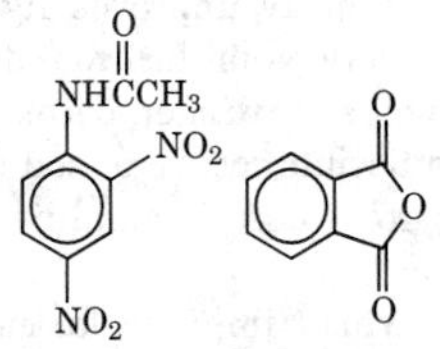	yellow	Sulfur Yellow 9	53010
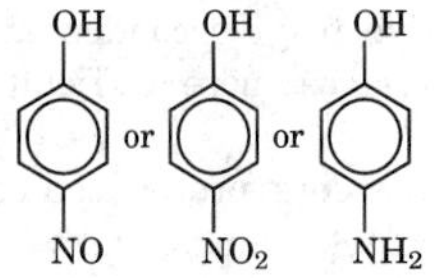	olive	Sulfur Green 11	53165
	dull green	Sulfur Green 1	53166
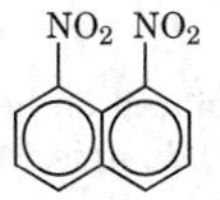	dull reddish brown	Sulfur Brown 7	53275
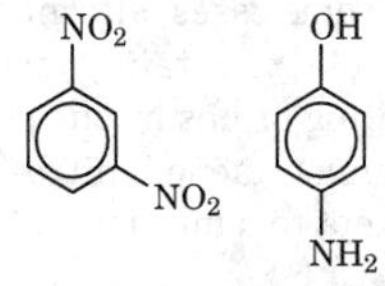	dull green	Sulfur Green 9	53005

Table 3. Polysulfide-Melt Dyes

Intermediates	Shade	CI designation	
		Name	Number
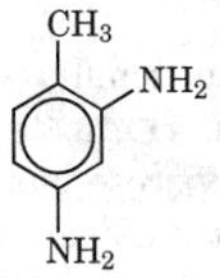	greenish black	Sulfur Black 1	53185
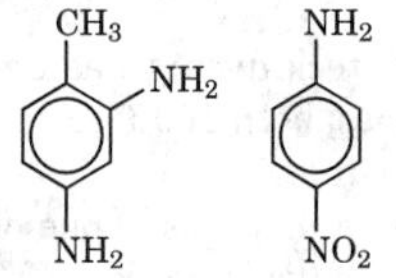	reddish blue-bluish violet	Sulfur Blue 7	53440
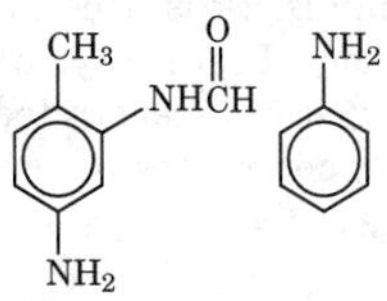	green	Sulfur Green 3	53570

Application

Sulfur dyes are applied to leuco form. In this form, the dye has affinity for the fiber. After the dye is completely absorbed by the fiber, it is reoxidized *in situ*. In dyes, such as the bright blues which contain quinonimine groups, further reduction takes place in a manner similar to the reduction of the keto group in vat dyes.

The reducing agent traditionally employed with sulfur dyes is sodium sulfide, but sodium sulfhydrate, NaHS, together with a small quantity of alkali such as sodium carbonate or sodium hydroxide is also widely used. Effluent control has resulted in a search for alternative reducing agents (see DYES, ENVIRONMENTAL CHEMISTRY).

Aftertreatment. Because the dye is applied by reduction and oxidation, many methods are available to obtain the correct hue. Air oxidation takes place gradually after the residual reducing agent has been rinsed away, but in general chemical oxidation is faster. The traditional oxidizing agents include sodium or potassium bichromate mixed with acetic acid; addition of copper sulfate slightly improves lightfastness. However, because of ecological restrictions on bichromates, other oxidizing agents have come into use, such as hydrogen peroxide, sodium perborate, and products based on potassium iodate or sodium bromate mixed with acetic acid. Sodium chlorite is used in alkaline solution together with detergent.

Aftertreatments include resin finishes, which improve fastness properties, and dye-fixing agents of the epichlorhydrin–organic amine type. These agents react with the dye to give condensation products that are not water soluble and hence more difficult to remove.

Commercial forms of sulfur dyes include powders, prereduced powders, grains, dispersed powders, dispersed pastes, liquids, and water-soluble brands.

Health and Safety

Since the 1980s, much more emphasis has been placed on health and safety aspects within the chemical and dyestuff industries. As a consequence, benzidine and β-naphthylamine, which are known carcinogens, have been banned in many countries. In some cases, alternatives to these intermediates have been found in order to retain dyestuffs with similar shades and properties. The handling of hazardous chemicals such as nitroanilines, dinitro- and diaminotoluenes, nitro- and dinitrophenols, and chlorodinitrobenzenes is kept to a minimum. In all cases, suitable protective gear is employed.

The effluent from the dye manufacture and textile dyeing industry is usually treated by aeration in large tank farms. By these means, ca 10–12% of the oxygen in the air can be effectively utilized in the degradation of the sulfur dye waste, rendering it acceptable to municipal sewage treatment farms. Small dye houses utilize spent flue gases or hydrogen peroxide as well as ferrous salts to treat plant effluent.

Uses

The sulfur dyes are widely used in piece dyeing of traditionally woven cotton goods such as drill and corduroy fabrics (see TEXTILES). The cellulosic portion of polyester–cotton and polyester–viscose blends is dyed with sulfur dyes. Their fastness matches that of the disperse dyes on the polyester portion, especially when it is taken into account that these fabrics are generally given a resin finish.

Yarn is dyed with sulfur dyes, although raw stock dyeing has declined in recent years. The dyeing of knitted fabrics, both 100% cotton or blends of cotton with synthetic fibers, is increasing.

Continuous dyeing of piece goods by pad–steam methods is one of the principal outlets for sulfur dyes, mostly in prereduced liquid form. Nontextile uses for sulfur dyes are limited. Sulfur Black is used for dyeing paper, particularly for lamination applications. Leather dyeing is done with the solubilized sulfur Episol dyes that give well-penetrated dyeings with far better wetfastness than is usually obtained with acid or direct dyes.

J. SENIOR
R. A. GUEST
W. E. WOOD
James Robinson, Ltd.

H. A. Lubs, *The Chemistry of Synthetic Dyes and Pigments*, Reinhold Publishing Corp., New York, 1955, Chapt. 6.

J. F. Thorpe and M. A. Whiteley, eds., *Thorpe's Dictionary of Applied Chemistry*, 4th ed., Vol. 11, Longmans, Green & Co., London, 1954.

K. Venkataraman, *The Chemistry of Synthetic Dyes and Pigments*, Vol. 2, 1952, Chapt. 35; Vol. 3, 1970, Chapt. 1; Vol. 7, 1974, Chapt. 1, Academic Press, London and New York.

Colour Index, 3rd ed., The Society of Dyers and Colourists, Bradford, U.K., and The American Association of Textile Chemists and Colorists, Research Triangle Park, N.C., 1975, rev. ed., 1982.

SULFURIC ACID AND SULFUR TRIOXIDE

Sulfuric acid, H_2SO_4, is a colorless, viscous liquid having a specific gravity of 1.8357 and a normal boiling point of approximately 274°C. Its anhydride, sulfur trioxide, SO_3, is also a liquid, having a specific gravity of 1.857 and a normal boiling point of 44.8°C. Sulfuric acid is by far the largest-volume chemical commodity produced. It is sold or used commercially in a number of different concentrations, including 78 wt % (60° Bé), 93 wt % (66° Bé), 96 wt %, 98–99 wt %, 100%, and as various oleums, ie, fuming sulfuric acid, $H_2SO_4 + SO_3$. Stabilized and unstabilized liquid SO_3 are items of commerce.

Sulfuric acid has many desirable properties that lead to its use in a wide variety of applications, including production of basic chemicals, steel (qv), copper, fertilizers, fibers, plastics, gasoline, explosives, electronic chips, batteries, and pharmaceuticals. It typically is less costly than other acids; it can be readily handled in steel or common alloys at normal commercial concentrations. It is available and readily handled at concentrations >100 wt % (oleum). Sulfuric acid is a strong acid; it reacts readily with many organic compounds to produce useful products. Sulfuric acid forms a slightly soluble salt or precipitate with calcium oxide or hydroxide, the least expensive and most readily available base. This is a useful property when it comes to disposing of sulfuric acid. Concentrated sulfuric acid is also a good dehydrating agent and under some circumstances it functions as an oxidizing agent.

Physical Properties

Sulfur Trioxide. Pure sulfur trioxide at room temperature and atmospheric pressure is a colorless liquid that fumes in air. This material can exist in both monomeric and polymeric forms. In the gaseous and liquid state pure SO_3 is an equilibrium mixture of monomeric SO_3 and trimeric S_3O_9 also called γ-SO_3.

If the SO_3 is pure, it freezes to γ-SO_3, also called ice-like SO_3, at 16.86°C. Table 1 presents a summary of the properties of pure sulfur trioxide.

Table 1. Properties of Sulfur Trioxide

Property	Value
critical temperature, °C	217.8
critical pressure, kPa[a]	8208
critical density, g/cm^3	0.630
normal boiling point, °C	44.8
melting point, γ-phase, °C	16.8
density, γ-phase, g/cm^3	
liquid at 20°C	1.9224
solid at −10°C	2.29
coefficient of thermal expansion at 18°C, °C^{-1}	0.002005
liquid heat capacity, at 30°C, kJ/(kg·°K)[b]	3.222
heat of formation of gas at 25°C, kJ/mol[b]	−395.76
free energy of formation of gas at 25°C, kJ/mol[b]	−371.07
heat of dilution, kJ/kg[b]	2.110
heat of vaporization, γ-liquid, MJ/kg	0.5843
diffusion in air, at 80°C, m/s	0.000013
liquid dielectric constant, at 18°C	3.11
electric conductivity	negligible

[a] To convert kPa to psi, multiply by 0.145.
[b] To convert J to cal, divide by 4.184.

Sulfuric Acid. Sulfuric acid is a dense, colorless liquid at room temperature. Historically, the concentration of sulfuric acid has been reported as specific gravity (sp gr) in degrees Baumé. In the United

States, the Baumé scale is calculated by the following formula:

$$\text{Beacute;} = 145 - \left(\frac{145}{\text{sp gr}}\right)$$

In Germany and France the Baumé scale is calculated using 144.3 as the constant. The Baumé scale includes only the sulfuric acid concentration range of 0–93.19% H_2SO_4. Higher concentrations are not included in the Baumé scale because density is not a unique function of concentration between 93% and 100%.

At atmospheric pressure, sulfuric acid has a maximum boiling azeotrope at approximately 98.48%. The vapor pressure exerted by sulfuric acid solutions below the azeotrope is primarily from water vapor; above the azeotropic concentration SO_3 is the primary component of the vapor phase. At the boiling point, sulfuric acid solutions containing <85% H_2SO_4 evaporate water exclusively; those containing >35% free SO_3 (oleum) evaporate exclusively sulfur trioxide.

Oleum. Oleum strength is usually reported as weight percent free SO_3 or percent equivalent sulfuric acid. The formula for converting percent oleum to equivalent sulfuric acid is $\%H_2SO_4 = 100 + \%\text{oleum}/4.444$.

Oleum is thought of as a mixture of sulfuric acid and free sulfur trioxide. In various strength oleums the "free" SO_3 forms disulfuric acid ($H_2S_2O_7$) and trisulfuric acid (HS_3O_{10}).

Manufacture

Sulfuric acid may be produced by the contact process from a wide range of sulfur-bearing raw materials by several different process variants, depending largely on the raw material used. In some cases sulfuric acid is made as a by-product of other operations, primarily as an economical or convenient means of minimizing air pollution or disposing of unwanted by-products.

The principal direct raw materials used to make sulfuric acid are elemental sulfur, spent (contaminated and diluted) sulfuric acid, and hydrogen sulfide. Elemental sulfur is by far the most widely used.

Generation of Sulfur Dioxide Gas. *Sulfur Burning.* There is a trend toward very large single-train plants. Because of this the usual practice is to use horizontal, brick-lined combustion chambers with dried air and atomized molten sulfur introduced at one end. Atomization typically is accomplished either by pressure spray nozzles or by mechanically driven spinning cups. Because the degree of atomization is a key factor in producing efficient combustion, sulfur nozzle pressures are typically 2.76 MPa (150 psi) or higher. Sulfur burners are typically designed as proprietary items by companies specializing in acid plant design and construction.

Spent Acid or H_2S Burning. Burners for spent acid or hydrogen sulfide are generally similar to those used for elemental sulfur, with a few critical differences. Special types of nozzles are required both for H_2S, a gaseous fuel, and for the corrosive and viscous spent acids. In a few cases, spent acids may be so viscous that only a spinning cup can satisfactorily atomize them. Because combustion of H_2S is highly exothermic, careful design is necessary to avoid excessive temperatures.

Ore Roasting, Sintering, or Smelting. Generation of SO_2 at nonferrous metal smelters is determined primarily by the needs of the various metallurgical processes and only incidentally by requirements of the sulfuric acid process. Traditionally, sulfur recovery from copper (qv), nickel, lead (qv), and zinc (qv), smelters has been limited to treatment of gases from roasters, sintering machines, and converters. Roasters and sintering machines operate continuously and produce a fairly uniform but low concentration off-gas. Traditional converters, which operate as batch reactors, produce gases having varying concentrations of SO_2. Treatment of these cyclical and low grade gases is costly and involves technical problems that make efficient acid production difficult.

The most modern smelters use continuous smelting and converting processes which utilize high levels of oxygen-enriched air to produce a uniform flow of high strength process gas. This allows efficient acid plant design including high levels of energy recovery (see METALLURGY, EXTRACTIVE METALLURGY).

Process Details and Flow Sheets. The stoichiometric relation between reactants and products for the contact process may be represented as follows:

$$SO_2 + 1/2\ O_2 \rightleftharpoons SO_3, \quad SO_3 + H_2O \rightarrow H_2SO_4$$

There are three important characteristics of the first of these equations. It is exothermic, reversible, and shows a decrease in molar volume on the right-hand side, ie, in the direction of the desired product. To improve equilibrium or driving force for the reaction, the sulfuric acid industry has attempted one or a combination of the following process design modifications: increasing concentration of SO_2 in the process gas stream; increasing concentration of O_2 in the process gas stream by air dilution or oxygen enrichment; increasing the number of catalyst beds; removing the SO_3 product by interpass absorption, known as the double absorption process; lowering catalytic converter inlet operating temperatures, ie, using better catalysts; and increasing the catalytic converter operating pressure (pressure plants).

Single Absorption Sulfur-Burning Plants. Single absorption sulfuric acid plants were standard in the industry for many years. These used either relatively low strength (approximately 8 vol %) SO_2 gas without air dilution, or air dilution designs and higher (approximately 10 vol %) inlet gas strength.

Double-Absorption Plants. In the United States, newer sulfuric acid plants are required to limit SO_2 stack emissions to 2 kg of SO_2 per metric ton of 100% acid produced, equivalent to a sulfur dioxide conversion efficiency of 99.7%. This high conversion efficiency is not economically achievable by single absorption plants using available catalysts, but it can be attained in double absorption plants when the catalyst is not seriously degraded.

A typical double absorption plant design uses intermediate SO_3 absorption after the second or, more commonly, the third converter pass.

Approximately 90–95% of total sulfur trioxide produced by the double absorption process is absorbed in the interpass absorption tower. The sulfur trioxide produced in subsequent converter passes is absorbed in the final absorbing tower.

Another large difference between single and double absorption processes is that, after interpass absorption, the process gas has to be reheated before reentering the converter. Reheating the process gas is accomplished in gas-to-gas heat exchangers, using some of the heat from the initial converter passes. All other plant operations are very similar to the corresponding single absorption processes.

Oleum Manufacture. To produce fuming sulfuric acid (oleum), SO_3 is absorbed in one or more special absorption towers irrigated by recirculated oleum. Because of oleum vapor pressure limitations the amount of SO_3 absorbed from the process gas is typically limited to less than 70%. Because absorption of SO_3 is incomplete, gas leaving the oleum tower must be processed in a nonfuming absorption tower.

Sulfur Trioxide. The anhydride of sulfuric acid, SO_3, is a strong organic sulfonating and dehydrating agent which has some specialized uses (see SULFONATION AND SULFATION). Its principal applications are in production of detergents and as a raw material for chlorosulfuric acid and 65% oleum. More recently, SO_3 gas has been added to cooled combustion gases at many coal-burning power plants to improve dust removal in electrostatic precipitators (see AIR POLLUTION CONTROL METHODS).

Liquid SO_3 is a difficult material to handle because of its relatively low (44.8°C) boiling point, its tendency to form solid polymers below 30°C, plus high reactivity with almost all organic substances and water. Liquid SO_3 is usually produced by distilling SO_3 vapor from oleum and condensing it.

Special Plant Designs. A dramatic increase in energy cost, plus governmental regulation regarding cogeneration of electric power, have led to significant changes in plant design. The need for additional electric generating capacity in many regions of the United States has become a force in the development of energy recovery (see ENERGY MANAGEMENT; PROCESS ENERGY CONSERVATION).

Design changes have included increased (up to 12% SO_2) gas strength; increased (up to 6.2 MPa (900 psi) and 480°C) steam pressure and superheat; low temperature economizers; suction side dry towers, ie, dry towers placed on the suction side of the main compressor; reduced plant pressure drop via new catalyst shapes and low pressure equipment design; and installation of turbogenerators to convert steam to electricity. These changes have allowed sulfur-burning plants to recover about 70% of the available energy.

Analysis.

Concentrations of 78 wt % and 93 wt % H_2SO_4 are commonly measured indirectly by determining specific gravity. Higher acid concentrations are normally determined by titration with a base.

Health and Safety

Shipping and Handling. Sulfuric acid is injurious to the skin, mucosa, and eyes. Moreover, dangerous amounts of hydrogen may be evolved in reactions between weakened acid and metals. Sulfuric acid at high concentrations reacts vigorously with water, organic compounds, and reducing agents. Oleums and liquid SO_3 frequently react with explosive violence, particularly with water.

General handling precautions should be observed. Sulfuric acid must not come in contact with eyes, skin, or clothing. When handling containers or operating equipment containing sulfuric acid, equipment appropriate for exposure conditions should be worn. These include chemical splash goggles, face shield and chemical splash goggle combination (not face shield alone), rubber acid proof gloves, and a full acid-proof suit, hood, and boots. Personnel must avoid breathing mist or vapors. The acid should be handled only in areas having sufficient ventilation to prevent irritation. Alternatively, an appropriate NIOSH/MSHA approved respirator should be worn.

Sulfuric Acid Toxicity. Sulfur trioxide does not exist in the atmosphere except in trace amounts; it rapidly combines with moisture in air to form sulfuric acid mist. Sulfuric acid aerosol or mist is a significantly more powerful pulmonary irritant than sulfur dioxide. Studies of prolonged exposure to sulfuric acid fumes have been performed on workers in plants manufacturing lead acid batteries. Prolonged exposure to mineral acid fumes causes the teeth of the exposed subject to deteriorate. Overexposure to sulfuric acid aerosols results in pulmonary edema, chronic pulmonary fibrosis, residual bronchiectasis, and pulmonary emphysema.

The International Agency for Research on Cancer (IARC) has classified "occupational exposure to strong inorganic acid mists, containing sulfuric acid," as a Category 1 carcinogen, ie, a substance that is carcinogenic to humans.

The Threshold Limit Value (TLV) of sulfuric acid mist for humans agreed upon by the ACGIH, OSHA, and NIOSH is 1 mg/m^3 of air. Sulfuric acid aerosols below the TLV are commonly not detected by odor, taste, or irritation. A TLV of 1 mg/m^3 is recommended by the ACGIH to prevent pulmonary irritation and injury to the teeth at particle sizes likely to occur in industrial situations.

THOMAS L. MULLER,, P.E.
E. I. du Pont de Nemours & Co., Inc.

W. W. Duecker and J. R. West, eds., *The Manufacture of Sulfuric Acid*, Reinhold Publishing Co., New York, 1959; reprinted by Robert E. Krieger Publishing Co., Huntington, N.Y., 1971.

S. K. Brubaker, *Materials of Construction for Sulfuric Acid*, Process Industries Corrosion, National Association of Corrosion Engineers meeting, Houston, Tex., 1986.

O. T. Fasullo, *Sulfuric Acid Use and Handling*, McGraw-Hill Book Co., Inc., New York, 1965.

SULFUR DIOXIDE. See SULFUR COMPOUNDS.

SULFURIC AND SULFUROUS ESTERS

Sulfuric and sulfurous acids form a series of esters analogous to those from other acidic materials. The hydrogen of the acid is replaced by a carbon-containing group. Because two hydrogens are present in the sulfur-based acids, there are two series of esters. Replacement of one hydrogen results in an acid ester. If both hydrogens are replaced, whether with the same, with different, or with bifunctional substituents, symmetrical and unsymmetrical diesters form. The two series are represented by the following general formulas, where R is a carbon

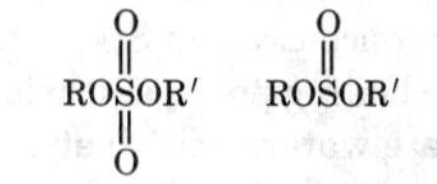

group and R′ is a carbon group, hydrogen, or metal cation. In the acid ester series a chlorine or amine group may be present in place of the hydroxy group. These compounds can be used as intermediates in making diesters. The carbon groups most commonly present are those from short- and long-chain alcohols and from hydroxyaromatic and heterocyclic compounds. Table 1 illustrates the variety of known compounds and some of their properties.

Dimethyl sulfate, diethyl sulfate, and long-chain monoalkyl alkali metal sulfates are the compounds of practical interest.

Physical Properties

The physical properties of sulfuric and sulfurous esters are best understood as resulting from the blending of the polar contribution of an acid or neutral sulfate group with the contribution of the usually nonpolar carbon group. With lower alkyl groups, the polar effects dominate, whereas with higher alkyl groups, the nonpolar interactions of the alkyl dominate.

The lower alkyl hydrogen sulfates are moderately viscous liquids. Higher alkyl hydrogen sulfates are hygroscopic, low melting solids. They do not have boiling points, but decompose on heating.

Chemical Properties

Sulfates. The chemistry of alkyl sulfates is dominated by two fundamental process types: reaction with nucleophiles and reaction as acids. Reaction with nucleophiles results in alkylation.

Alkylation. In alkylation, the dialkyl sulfates react much faster than do the alkyl halides, because the monoalkyl sulfate anion ($ROSO_3^-$) is more effective as a leaving group than a halide ion.

Although the first alkyl group is the most reactive, the second alkyl group on the intermediate anion can also alkylate.

Manufacture

Monoester Hydrogen Sulfates. The hydrogen sulfates are prepared by the action of a sulfating agent on the corresponding alcohol or phenol. The following reagents are used for sulfation: sulfur trioxide, sulfuric acid, chlorosulfuric acid, sulfur trioxide–amine complex, and sulfamic acid.

Diorgano Sulfates. Dialkyl sulfates up to octadecyl can be made from the alcohols by a general method involving the following reactions:

$$ROH + SO_2Cl_2 \rightarrow ROSO_2Cl + HCl$$

$$2\ ROH + SOCl_2 \rightarrow (RO)_2SO + 2\ HCl$$

$$ROSO_2Cl + (RO)_2SO \rightarrow (RO)_2SO_2 + RCl + SO_2$$

Mixed esters are synthesized by the reaction of one alkyl chlorosulfate with a different sodium alkoxide or a different dialkyl sulfite as follows:

$$ROSO_2Cl + NaOR' \rightarrow ROSO_2OR' + NaCl$$

Table 1. Some Sulfuric and Sulfurous Acid Esters

Ester	Bp_{kPa}[a], °C	Mp, °C
	Sulfates, $(RO)_2SO_2$	
R		
open-chain		
methyl	$188.8_{101.3}$, $69.70_{1.33}$	
ethyl	208, $89_{1.20}$	
n-propyl	$95_{0.670}$	
n-butyl	$103_{0.200}$	
phenyl	$194.6_{2.66}$	
cyclic		
ethylene		99
1,3-propylene		63
methylene (dimer)		155
	Sulfites, $(RO)_2SO$	
R		
open-chain		
methyl	$126–127_{101.3}$	
ethyl	$159–160_{101.3}$	
n-propyl	$82_{2.00}$	
phenyl	$185_{2.00}$	13–16
cyclic		
ethylene	$80_{3.72}$	
1,2-propylene	$84_{3.92}$	
pentaerythritol,di		154
	Organo hydrogen sulfates, $(RO)SO_3H$	
R		
methyl	$130–140_{101.3}$ dec	
n-decyl	liquid	
n-dodecyl		25–27
phenyl		
	Organo halosulfates, $(RO)SO_2X$	
R, X		
methyl,chloro	$134–135_{101.3}$, $42_{2.13}$	
methyl,fluoro	$92_{101.3}$, $45_{21.4}$	
ethyl,chloro	$75_{15.6}$, $58_{2.66}$	
ethyl,fluoro	113_{100}, $21_{1.60}$	
	Organo halosulfites, (RO)SOX	
R, X		
methyl,chloro	$35_{8.00}$	
ethyl,chloro	$50–53_{8.00}$	

[a] To convert kPa to mm Hg, multiply by 7.5; 101.3 kPa = 1 atm.

$$ROSO_2Cl + (R'O)_2SO \rightarrow ROSO_2OR' + R'Cl + SO_2$$

Diorgano Sulfites. Symmetrical or mixed dialkyl sulfites are prepared by the stepwise reaction of thionyl chloride either with two molecules of an alcohol or with stoichiometric quantities of two alcohols in pyridine.

$$ROH + SOCl_2 + C_5H_5N \rightarrow ROSOCl + C_5H_5N{\cdot}HCl$$

$$R'OH + ROSOCl + C_6H_5N \rightarrow R'OSOOR + C_5H_5N{\cdot}HCl$$

Halosulfates and Halosulfites. A general method for the preparation of alkyl halosulfates and halosulfites is the treatment of the alcohol with sulfuryl or thionyl chloride at low temperatures while passing an inert gas through the mixture to remove hydrogen chloride.

Health and Safety Factors

The most commonly used dialkyl sulfate is dimethyl sulfate. This is also the most hazardous in liquid and vapor forms. The hazard arises from its toxicity, high reactivity, and to some extent, combustibility. Dimethyl sulfate is corrosive and poisonous, and its effects may be either acute or chronic. Because it has an analgesic effect on many body tissues, even severe exposures may not be immediately painful. Dimethyl sulfate is particularly dangerous to the eyes and respiratory system. It causes severe burns, but symptoms may be delayed. Exposed workers must be immediately and properly treated to avoid either permanent injury to eyes and lungs or even death. Skin burns can also be severe. Ingestion causes convulsions and paralysis, with later damage to the kidneys, liver, and heart. Genetic effects have been extensively reviewed, and its mutagenicity is correlated with carcinogenicity.

According to the U.S. Department of Labor (OSHA), exposure to dimethyl sulfate shall not exceed an eight-hour time-weighted average of 1 ppm in air. Because both liquid and vapor can penetrate the skin and mucous membranes, control of vapor inhalation alone may not be sufficient to prevent absorption of an excessive dose. Dimethyl sulfate is listed as an industrial substance with suspected carcinogenic potential in humans. The ACGIH recommends a time-weighted average threshold limit value of 0.1 ppm.

Uses

The sulfuric acid esters as compared to the sulfurous esters are the most widely used. In nature they appear as solubilizing groups in detoxification–excretory mechanisms and as sulfated carbohydrate groups in modified proteins. The significant uses are alkylation, formation of long-chain alcohol monosulfates as surfactants, and formation of intermediates in preparation of some lower alcohols. Alkylation (qv) involves primarily dimethyl and diethyl sulfates in the preparation of a wide variety of intermediates and products, especially in the fields of dyes, agricultural chemicals, drugs, and other specialties. In particular, dimethyl sulfate is a powerful reagent yielding quaternary salts in the form of the methosulfates (see QUATERNARY AMMONIUM COMPOUNDS).

Organometallic usage is shown in the preparation of titanium- or vanadium-containing catalysts for the polymerization of styrene or butadiene by the reaction of dimethyl sulfate with the metal chloride.

A number of mixed alkyl and aryl sulfites have been patented as insecticides and biocides (see INSECT CONTROL TECHNOLOGY).

W. B. McCORMACK
B. C. LAWES
E. I. du Pont de Nemours & Co., Inc.

E. E. Gilbert, *Sulfonation and Related Reaction*, Wiley-Interscience, New York, 1965, Chapt. 6.

C. M. Suter, *Organic Chemistry of Sulfur*, John Wiley & Sons, Inc., New York, 1944, Chapt. 1.

SULFURIZATION AND SULFURCHLORINATION

Sulfur reacts with alkanes to dehydrate; oxidize, forming carbon disulfide and hydrogen sulfide; or cyclize, forming thiophenes. The products of alkane sulfurization depend on the temperature, the time at the temperature, and the structure of the hydrocarbon.

Generally, unsaturated compounds, eg, alkenes and natural fats and their derivatives, are much more reactive toward sulfur than alkanes. Sulfur reacts with unsaturated compounds at temperatures of 120–215°C, forming products that are usually dark and often viscous cross-linked mixtures of dithiole-3-thiones and sulfides.

The mechanisms for the reaction of sulfur with alkanes and unsaturated compounds are strongly influenced by the specific structure of the substrate and by the conditions (particularly temperature) of reaction. Sulfurization of unsaturated compounds in the presence of hydrogen sulfide also affords polysulfides (see SULFUR COMPOUNDS).

Sulfur monochloride S_2Cl_2, and sulfur dichloride, SCl_2, react with unsaturated materials, forming products that are cross-linked by sulfur but which also contain chlorine.

Table 1. Chemical and Physical Properties of Sulfurized and Sulfurchlorinated Unsaturated Compounds and Mercaptans

Product[a] designation	Sulfur, wt %	Chlorine, wt %	Active sulfur, wt %	Viscosity, mm^2/s (= cSt) 40°C	100°C	Density, at 25°C, g/cm^3	Total acid number, mg KOH/g	Pour point, °C
Base 10-L[b]	10.0	0	0	1100	80	0.98	25	18
Base 10-SE[c]	10.0	0	0	20	4	0.94	5	10
Base 14-L[b]	14.0	0	4	1700	100	0.99	25	27
Base 401[d]	40.0	0	25	80	10	1.07	7	−50
Base L-66[e]	5.5	5.5	0	1800	100	0.99	5	10
Sulperm 18[f]	18.0	0	8	600	55	1.02	12	16
Sulperm 110[f]	10.0	0	0	500	40	0.98	8	10
DTNPS[g]	37.0	0	25	35		1.04	5	−50
DTBTS[h]	44.0	0	0	3		1.00	0	<−60

[a] Products and properties submitted by Keil Chemical Division of Ferro Corp. [b] Based on animal fat. [c] Based on methyl ester of vegetable oil. [d] Based on olefin. [e] Sulfurchlorinated animal fat. [f] Based on a mixture of animal and vegetable oils plus synthetic esters. [g] Di-*t*-nonyl pentasulfide is based on mercaptan. [h] Di-*t*-butyl trisulfide is based on mercaptan.

Properties

Properties of typical commercial sulfurized unsaturated compounds and mercaptans are listed in Table 1.

Uses

Sulfurized and sulfurchlorinated unsaturated compounds and mercaptans are used as lubricant additives (antiwear, friction modification, load-carrying, extreme pressure and temperature, corrosion inhibition, and antioxidants), refinery catalyst regeneration compounds, steel processing (annealing) aids, and vulcanization catalysts (see LUBRICATION AND LUBRICANTS).

Manufacture

Sulfurization of unsaturated compounds and mercaptans is normally carried out at atmospheric pressure, in a mild or stainless steel, batch-reaction vessel equipped with an overhead condenser, nitrogen atmosphere, an agitator, heating media capable of 120–215°C temperatures and a scrubber capable of handling hydrogen sulfide.

Sulfurchlorination of unsaturated compounds or mercaptans is normally carried out at atmospheric pressure in a glass-lined reaction vessel equipped with a cooling jacket or coils, a nitrogen or dry air sparging system, an overhead condenser, and a caustic or bleach scrubber.

Health and Safety Factors

Sulfurized and sulfurchlorinated unsaturated compounds or mercaptans are normally considered nonhazardous. These materials may, however, liberate H_2S and HCl at elevated (>200C) temperatures and during combustion.

MICHAEL P. DUNCAN
Keil Chemical Division
Ferro Corporation

S. Oae, *Organic Chemistry of Sulfur*, Plenum Press, New York, 1977.

W. A. Pryor, *Mechanisms of Sulfur Reactions*, McGraw-Hill Book Co., Inc., New York, 1962.

SULFUR REMOVAL AND RECOVERY

Sulfur can be produced directly via Frasch mining or conventional mining methods, or it can be recovered as a by-product from sulfur removal and recovery processes. Production of recovered sulfur has become more significant as increasingly sour feedstocks are utilized and environmental regulations concerning emissions and waste streams have continued to tighten worldwide.

Sulfur recovery processes have historically been focused primarily on the removal and conversion to elemental sulfur of two sulfur species: hydrogen sulfide and sulfur dioxide.

Hydrogen Sulfide Removal and Recovery

Hydrogen sulfide, H_2S, represents the largest source of recovered sulfur. Hydrogen sulfide is often present in natural gas or refinery streams, occurring naturally or as the by-product of processing operations such as hydrotreating. A number of processes have been developed to remove and recover hydrogen sulfide. These processes are generally categorized according to the primary mechanism, as adsorption, absorption, or conversion.

Adsorption Processes. The processes based on adsorption of hydrogen sulfide onto a fixed bed of solid material are among the oldest types of gas treating applications. Two common sorbent materials for low concentration gas streams are iron oxide and zinc oxide.

Hydrogen sulfide reacts with iron oxide, Fe_2O_3, to form iron sulfide, Fe_2S_3.

The sulfur is thus removed from the gas stream and trapped in the sorbent as iron sulfide. Over time all of the iron oxide becomes sulfided and the adsorptive capacity of the sorbent becomes exhausted. The bed can be partially regenerated by oxidation.

The zinc oxide process is similar to the iron oxide process, except that the zinc sulfide, ZnS, formed cannot be oxidized back to zinc oxide, and therefore the sorbent bed must be replaced once the capacity is fully utilized.

The sulfur removed via these fixed-bed metal oxide processes is generally not recovered. Rather the sulfur and sorbent material both undergo disposal.

Absorption Processes. Absorption-based processes are the most widely practiced hydrogen sulfide removal techniques. These processes use a liquid solvent to absorb hydrogen sulfide and render a sweet gas stream. The hydrogen sulfide-rich solvent can be regenerated for reuse.

Absorption processes are categorized based on the mechanism of absorption, as either chemical or physical. In addition, a number of hybrid absorption processes featuring both chemical and physical solvents have been developed.

Chemical Absorption. In chemical absorption processes, the solvent reacts with the hydrogen sulfide and other species to form new complex compounds which are held in the solvent. These reactions occur in an absorber tower where the solvent and the sour streams are contacted countercurrently across trays or packing. A regenerator is employed to release the hydrogen sulfide and other absorbed compounds from the solvent. Two common chemical solvents are aqueous solutions of alkanolamines or alkali carbonate salts.

Physical Absorption. Whereas chemical absorption relies on solvent reactions to hold acid gas components in solution, physical absorption exploits gas–liquid solubilities. These processes are most applicable in situations involving high pressure feed streams containing significant concentrations of acid gas components.

The process flow sheet for a physical absorption unit is similar to that of the chemical absorption processes, featuring an absorber and regenerator; however, the solvent is regenerated primarily through reduction in pressure, although heating or stripping may also be required. Generally, the regeneration energy requirements for physical absorption solvents are lower than those for chemical solvents.

Hybrid Processes. These processes use both chemical and physical absorption solvents to offer high purity treat gas and low energy solvent regeneration. The operation of these processes is usually similar to that of the individual chemical or physical absorption processes and solvent composition is typically customized to meet the requirements of individual applications.

Conversion Processes. Most of the adsorption and absorption processes remove hydrogen sulfide from sour gas streams thus producing both a sweetened product stream and an enriched hydrogen sulfide stream. Conversion processes treat the hydrogen sulfide stream to recover the sulfur as a salable product.

The Claus process is the most widely used to convert hydrogen sulfide to sulfur. While the process was developed in 1883, it did not become widely used until the 1950s. A Claus sulfur recovery unit consists of a combustion furnace, waste heat boiler, sulfur condenser, and a series of catalytic stages each of which employs reheat, catalyst bed, and sulfur condenser. Typically, two or three catalytic stages are employed.

The Claus process converts hydrogen sulfide to elemental sulfur via a two-step reaction. The first step involves controlled combustion of the feed gas to convert approximately one-third of the hydrogen sulfide to sulfur dioxide and noncatalytic reaction of unburned hydrogen sulfide with sulfur dioxide. In the second step, the Claus reaction, the hydrogen sulfide and sulfur dioxide react over a catalyst to produce sulfur and water.

Oxygen enrichment of the combustion air can be used either to increase the processing capacity of an existing unit or to extend operation to low concentration hydrogen sulfide feeds.

Sulfur recovery offered by a Claus plant is limited to about 97%. A number of processes have been developed to recover the residual sulfur present in the Claus plant tail gas. Comprimo's Superclaus process and Parson's Hi-Activity process are two examples of direct oxidation tail gas treating processes. Both consist of the replacement of the final Claus stage by, or the addition of, a reaction stage featuring proprietary catalyst to promote the direct oxidation of hydrogen sulfide in the Claus tail gas to sulfur selectively. Another variety of tail gas treatment process is an extension of the Claus reaction by operating the final reactor at temperatures below the sulfur dew point. Elf Aquitaine's Sulfreen process, Amoco's cold bed adsorption (CBA) process, and the Mineral and Chemical Resource Company (MCRC) process licensed by Delta Hudson are all variations on the cold bed sub-dew point process.

IFP has developed a process suitable for application to Claus unit tail gas treatment which also extends the Claus reaction in the liquid phase.

Hydrogenation amine based processes are the most common type of tailgas cleanup. The Claus tail gas reacts with a reducing gas to convert all sulfur species to hydrogen sulfide. Absorption of the hydrogen sulfide into a selective amine, usually MDEA, solution follows. The hydrogen sulfide is recycled to the claus plant.

Sulfur Dioxide Removal and Recovery

Sulfur can be recovered from sulfur dioxide as liquid sulfur dioxide, sulfuric acid, or elemental sulfur. As with hydrogen sulfide, sulfur dioxide removal processes are categorized as adsorption, absorption, or conversion processes.

Adsorption Processes. Sulfur dioxide can be adsorbed on a solids bed, in a fixed-, fluidized-, or moving-bed configuration. Copper oxide is an effective sorbent. Sulfur dioxide is adsorbed onto the sorbent via chemical reaction to form copper sulfate. The sulfated sorbent material can be regenerated by reduction using hydrogen, carbon monoxide, or methane to produce sulfur dioxide and hydrogen sulfide, which must be further processed to recover the sulfur. Removal efficiencies of greater than 95% are possible.

An additional benefit of adsorption-based sulfur dioxide removal processes is that nitrogen oxides, NO_x, are also removed by the sorbent. Nitrogen oxides desorb when the sorbent is heated using hot air.

Absorption Processes. Most flue gas desulfurization (FGD) systems are based on absorption of the sulfur dioxide into a nonregenerable alkali-salt solvent. Sulfur absorbed using nonregenerable solvents is not recovered and the alkali sulfite–sulfate produced presents a disposal problem.

In regenerable absorption processes, the solvent releases the sulfur dioxide in a regenerator and then is reused in the absorber. The Wellman-Lord process is typical of a regenerable process. Sulfur dioxide removal efficiency is from 95–98%. The gas is prescrubbed with water, then contacts a sodium sulfite solution in an absorber. The sulfur dioxide is absorbed into solution.

Conversion Processes. A number of options exist for handling concentrated sulfur dioxide streams. One option is the sale of a liquid sulfur dioxide product. Alternatively, the sulfur dioxide can be converted to elemental sulfur or to sulfuric acid.

MICHAEL CAPONE
EXXON Engineering

S. A. Newman, ed., *Acid and Sour Gas Treating Processes*, Gulf Publishing, Houston, Tex., 1985.

A. L. Kohl and F. C. Riesenfeld, *Gas Purification*, 3rd ed., Gulf Publishing, Houston, Tex., 1979.

J. B. Pfeiffer, ed., *Sulfur Removal and Recovery from Industrial Processes*, American Chemical Society, Washington, D.C., 1975.

SUPERCOMPUTERS. See COMPUTER TECHNOLOGY.

SUPERCRITICAL FLUIDS

The supercritical region of a pure fluid, which may be defined as the area above both the critical pressure and critical temperature, is shown in Figure 1. A unique feature of supercritical fluids (SCFs) may be demonstrated by beginning with a subcritical liquid at point A on Figure 1. If the liquid is depressurized isothermally in a view cell from point A to point E, the presence of a meniscus is observed as the vapor pressure line is crossed. However, if the liquid takes the path of A–B–C–D–E, the fluid then passes from a liquid phase to a gas and no meniscus is seen. On this path, if one were looking only inside the view cell, one could not tell whether the component was in the gas, liquid, or fluid state. This A–B–C–D–E pathway is used in supercritical drying to avoid collapse of delicate microstructures by the strong surface tension forces that arise at liquid–vapor interfaces.

Frequently, the term compressed fluid, a more general expression than supercritical fluid, is used. A compressed fluid can be either a supercritical fluid, a near-critical fluid, an expanded liquid, or a highly compressed gas, depending on temperature, pressure, and composition.

The supercritical fluid carbon dioxide, CO_2, is of particular interest. This compound has a mild (31°C) critical temperature (Table 1); it is nonflammable, nontoxic, and, especially when used to replace freons and certain organic solvents, environmentally friendly. Moreover, it can be obtained from existing industrial processes without further contribution to the greenhouse effect (see AIR POLLUTION). Carbon dioxide is fairly miscible with a variety of organic solvents, and is readily recovered after processing owing to its high volatility.

Water has an unusually high (374°C) critical temperature owing to its polarity. At supercritical conditions water can dissolve gases

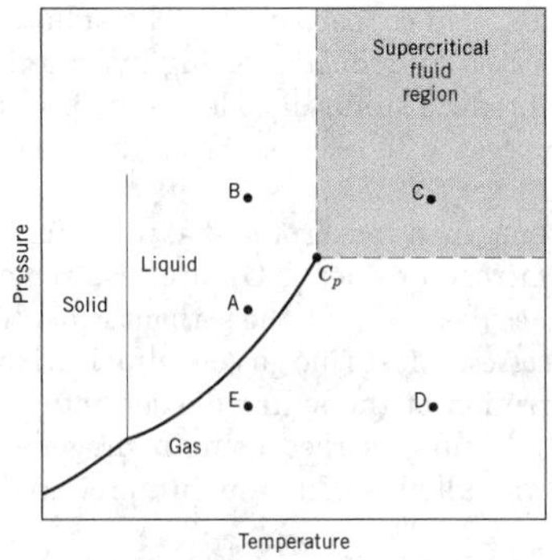

Figure 1. Schematic pressure–temperature diagram for a pure material showing the supercritical fluid region, where C_p is the pure component critical point and dots A to E are points on the diagram. See text.

Table 1. Critical Properties for Common Supercritical Fluids[a]

Solvent	T_c, °C	P_c, MPa[b]	ρ_c, g/cm^3
ethylene	9.3	5.04	0.22
xenon	16.6	5.84	0.12
carbon dioxide	31.1	7.38	0.47
ethane	32.2	4.88	0.20
nitrous oxide	36.5	7.17	0.45
propane	96.7	4.25	0.22
ammonia	132.5	11.28	0.24
n-butane	152.1	3.80	0.23
n-pentane	196.5	3.37	0.24
isopropanol	235.2	4.76	0.27
methanol	239.5	8.10	0.27
toluene	318.6	4.11	0.29
water	374.2	22.05	0.32

[a] T_c = critical temperature; P_c = critical pressure; ρ_c = critical density.
[b] To convert MPa to psi, multiply by 145.

such as O_2 and nonpolar organic compounds as well as salts. This phenomenon is of interest for oxidation of toxic wastewater (see WASTE TREATMENTS, HAZARDOUS WASTE).

Properties of Supercritical Fluids and Their Mixtures

Solvent Strength of Pure Fluids. The density of a pure fluid is extremely sensitive to pressure and temperature near the critical point.

The solvation strength of a given supercritical compressed fluid is related directly to the fluid density. Thus solvent strength may be manipulated over a wide range by making small changes in temperature and pressure. In general, the greater the density, the greater the ability of a given compressed fluid to solvate a component.

A particularly attractive and useful feature of supercritical fluids is that these materials can have properties somewhere between those of a gas and a liquid (Table 2). A supercritical fluid has more liquid-like densities, and subsequent solvation strengths, while possessing transport properties, ie, viscosities and diffusivities, that are more like gases. Thus, an SCF may diffuse into a matrix more quickly than a liquid solvent, yet still possess a liquid-like solvent strength for extracting a component from the matrix.

Phase Behavior. In terms of the solubilities of solutes in a supercritical phase, the following generalizations can be made. Solute solubilities in supercritical fluids approach and sometimes exceed those of liquid solvents as the SCF density increases. Solubilities typically increase as the pressure is increased. Increasing the temperature can cause increases, decreases, or no change in solute solubilities, depending on the temperature effect on solvent density and/or the solute vapor pressure. Also, at constant SCF density, a temperature increase increases the solute solubility.

Table 2. Comparison of Properties of Gases, Supercritical Fluids, and Liquids

Physical property	Gases	Supercritical fluids	Liquids
density, g/cm^3	0.001	0.2–1.0	0.6–1.6
diffusivity, cm^2/s	0.1	0.001	0.00001
viscosity, g/(cm·s)	0.0001	0.001	0.01

Compressed gases and fluids have the ability to dissolve in and expand organic liquid solvents at pressures typically between 5–10 MPa (50–100 bar). This expansion nearly always decreases the liquid's solvent strength. If enough compressed fluid is added, eventually the mixture solvent strength is comparable to that of the pure compressed fluid. Knowledge of when a solute begins to precipitate can be important. Such information, for instance, helps to determine whether heavy hydrocarbons precipitate when using CO_2 injection in an oil reservoir.

Classification of Phase Boundaries for Binary Systems. Classifications are typically based on pressure–temperature (P–T) projections of mixture critical curves and three-phase equilibria lines. Experimental data are usually obtained by a simple synthetic method in which the pressure and temperature of a homogeneous solution of known concentration are manipulated to precipitate a visually observed phase.

Polymers and Supercritical Fluids. A supercritical or compressed gas can be sorbed into a polymer to act as a plasticizing agent. As the concentration of the compressed fluid is increased in the polymer phase, the sorption and subsequent swelling of an amorphous polymer can cause a glass-to-liquid phase transition. The glass-transition temperature (T_g) of a polymer can be depressed to below the normal T_g by 100°C or more. Certain polymers can exhibit an isobaric liquid-to-glass transition with a temperature increase, defined as retrograde vitrification.

Molecular Modeling of Phase Behavior

Although modeling of supercritical phase behavior can sometimes be done using relatively simple thermodynamics, this is not the norm. Especially in the region of the critical point, extreme nonidealities occur and high compressibilities must be addressed.

Thermodynamic models are based on the principle that the fugacities (escaping tendencies) of component i, f_i, are equal for all phases at equilibrium under constant temperature and pressure.

Specific chemical interactions, eg, associations resulting from hydrogen bonding or donor-acceptor interactions, can have a pronounced effect on SCF solution phase behavior.

Various equations of state have been developed to treat association in supercritical fluids. Two of the most often used are the statistical association fluid theory (SAFT) and the lattice fluid hydrogen bonding model (LFHB). These models include parameters that describe the enthalpy and entropy of association. The most detailed description of association in supercritical water has been obtained using molecular dynamics and Monte Carlo computer simulations but this requires much larger amounts of computer time.

Experimental Techniques

Physical confirmation of the actual phase behavior is important. A number of different experimental techniques are available for determining phase behavior. These include dynamic flow-through cells, static systems using visual observations in a variable-volume view cell, static systems with sampling for analysis, and the use of static or dynamic optical transmission cells for uv—visible, Fourier transform infrared spectroscopy (ftir), and Raman spectroscopy for analysis.

Processes and Applications

A variety of SCF processes have been commercialized. These include, among others, the residuum oil supercritical extraction (ROSE) process, which uses supercritical alkanes as the extracting solvent to separate heavy components in crude oil; the high pressure polyethylene production process; processes for coffee decaffeination using CO_2; and the supercritical water oxidation process, which is used for purifying aqueous wastes containing organics.

Supercritical Fluid Chromatography and Analytical Extraction.

As an analytical tool, the use of supercritical fluids has gained broad acceptance. As in other SCF processes, density is used as the controlling feature. Many SCF chromatographic separations use a programmed density profile; the mobile phase is typically the supercritical fluid and the stationary phase can be in a packed or capillary column. Because the supercritical fluid density can be controlled accurately by manipulation of pressure and temperature conditions, fractionation of an oligomeric mixture or other mixtures can be tuned to give a desired separation.

Extractions. SCFs are used as extraction solvents in commercial food, pharmaceutical, environmental, and petroleum applications.

Reactions. Supercritical fluids are attractive as media for chemical reactions such as; hydrothermal oxidations, bioreactions, and polymerizations. Solvent properties such as solvent strength, viscosity, diffusivity, and dielectric constant may be adjusted over the continuum of gas-like to liquid-like densities by varying pressure and temperature. Subsequently, these changes can be used to affect reaction conditions.

Materials. Supercritical fluids offer many opportunities in materials processing, such as crystallization, recrystallization, comminution, fiber formation, blend formation, and microcellular (foam) formation.

DAVID J. DIXON
South Dakota School of Mines & Technology
KEITH P. JOHNSTON
University of Texas at Austin

M. A. McHugh and V. J. Krukonis, *Supercritical Fluid Extraction Principles and Practice*, 2nd ed., Butterworth-Heinemann, Boston, Mass., 1994.

T. J. Bruno and J. F. Ely, *Supercritical Fluid Technology: Reviews in Modern Theory and Applications*, CRC Press, Inc., Boca Raton, Fla., 1991.

M. E. Paulaitis and co-workers, eds., *Chemical Engineering at Supercritical-Fluid Conditions*, Ann Arbor Science Publishers, Mich., 1983.

SURFACE AND INTERFACE ANALYSIS

Surfaces are formed in the transition from one state of matter to another, whether the two phases are chemically distinct or not. Thus, surfaces exist at interphases or interfaces between two phases of either the same or different materials.

Many important chemical processes occur at solid–solid, solid–liquid, solid–gas, liquid–liquid, and liquid–gas interfaces. For example, many catalytic processes such as the conversion of petroleum to high-octane fuels are routinely performed at the surface of metal oxides. These surfaces can act either to provide sites enhancing proximity of reactants or mechanistically be involved in the chemical transformation itself. Environmental concerns often center on the extent to which organic pollutants are transported through solid soil, clay or mineral surfaces; organic pollutants that strongly interact with these surfaces will be transported less efficiently than those which do not experience such interactions. Surfactant molecules are essential ingredients in soaps, foams and emulsions; the unique chemistry of these systems arises from the formation of interfaces between two liquids or between a liquid and a gas. Thus, the ability to characterize surfaces and interfaces is essential to better understanding a variety of technologically and societally important chemical processes.

Overview of Surface and Interface Analysis Methods

A plethora of characteristics of surfaces and interfaces can be analyzed using a staggering variety of techniques. The focus of this article will be on those methods that are deemed by the author to be of the greatest importance.

The chemical, structural, and electronic characteristics of surfaces and interfaces are usually different from those of the bulk phase(s). Thus, methods to be used for the analysis of surfaces must be selective in response to the surface or interfacial region relative to the bulk. Surfaces and interfaces are most commonly explored using techniques based on the interaction of photons, electrons, or ions with the surface or using a force such as electric field or Van der Waals attraction. These excitations generate a response involving the production of photons, electrons, ions or the alteration of a force that is then sensed in the analysis.

The choice of which of these exciting species or forces, and hence technique, to use depends on the nature of the information sought about the surface. The analysis methods chosen seek to answer the following important fundamental questions that are typically asked about surfaces and interfaces: What does the surface or interface look like? What is the elemental composition of the surface or interface? What molecules are at the surface or in the interface?

Table 1 provides an overview of many of the techniques available for the characterization of surfaces and interfaces. These techniques are categorized on the basis of the nature of the exciting and detected species (or force). Many approaches are available for the study of surfaces.

Several methods utilizing photons as the exciting and detected species are commonly employed. Ellipsometry provides information about thin film thickness. Information about the vibrational modes of molecules at surfaces or within interfaces can be obtained using one of several vibrational spectroscopies. These approaches are based on visible or infrared wavelength photons that are either absorbed, scattered, or generated. X-ray diffraction (xrd) is a surface structural technique that relies on the diffraction of x-ray wavelength photons. Extended x-ray absorption fine structure (exafs) is a technique in which the x-ray absorption cross-section of a material is modulated by the local surface structure and/or coordination environment of an atom.

Other techniques in which incident photons excite the surface to produce detected electrons are also listed in Table 1.

Electrons are used for excitation and detection in the various electron microscopy methods.

Ions can also be used as both the excitation and detection species in several surface analysis techniques.

In choosing appropriate techniques with which to interrogate surfaces in search of answers to the above three questions, one must remain cognizant of the lateral resolution and depth at which the chemistry of interest occurs as well as the lateral resolution and depth sensitivity of each technique. Photon beams cannot be focused as tightly as electron or ion beams, for example, implying that techniques using photon beams will have poorer lateral resolution than techniques using electron beams. This characteristic may or may not be a limitation, however, depending on the scale on which the surface chemistry of interest is occurring.

Depth sensitivity is an equally important consideration in the analysis of surfaces. Techniques based on the detection of electrons or ions derive their surface sensitivity from the fact that these species cannot travel long distances in solids without undergoing interactions which cause energy loss. The depth sensitivity of a particular method will depend on the nature and energy of both the probe and detected species, and may or may not be sensitive to the entire depth to which the chemistry of interest occurs.

Techniques

Imaging of Surfaces–Analysis of Surface Morphology. *Electron Microscopy.* The techniques of primary importance in surface imaging are the electron microscopies. These techniques take advantage of the fact that accelerated beams of electrons typically have higher energies than commonly available photon beams, and therefore, can be focused more tightly.

Electron microscopy is done in one of two ways: scanning electron microscopy (sem) or transmission electron microscopy (tem). Scanning

Table 1. Overview of Common Surface Analysis Techniques

	Detection			
Excitation	Photons	Electrons	Ions	Force
photons	ellipsometry	xps[a] (esca[b])	psd[c]	
	ftir[d], irras[e], drift[f]	ups[g]		
	Raman, sers[h]			
	shg[i], sfg[j]			
	xrd[k]			
	exafs[l], nexafs[m],			
electrons	ip[o] sexafs[n]	sem[p], fe-sem[q], tem[r], stem[s]	esdiad[t]	
		aes[u], aps[v]		
		leed[w], rheed[x]		
		eels[y], hreels[z], ceels[aa]		
ions		ins[bb]	sims[cc]	
			iss[dd], rbs[ee]	
			leis[ff]	
force		fem[gg]	fim[hh]	afm[ii]
		stm[jj]		

[a] xps = x-ray photoelectron spectroscopy.
[b] esca = electron spectroscopy for chemical analysis.
[c] psd = photon stimulated ion angular distribution.
[d] ftir = Fourier transform infrared spectroscopy.
[e] irras = infrared reflection-absorption spectroscopy.
[f] drift = diffuse reflectance infrared Fourier transform spectroscopy.
[g] ups = ultraviolet photoelectron spectroscopy.
[h] sers = surface enhanced Raman scattering.
[i] shg = second harmonic generation.
[j] sfg = sum-frequency generation.
[k] xrd = x-ray diffraction.
[l] exafs = extended x-ray absorption fine structure.
[m] nexafs = near-edge x-ray absorption fine structure.
[n] sexafs = surface extended x-ray absorption fine structure.
[o] ip = inverse photoemission.
[p] sem = scanning or secondary electron microscopy.
[q] fe-sem = field emission scanning electron microscopy.
[r] tem = transmission electron microscopy.
[s] stem = scanning transmission electron microscopy.
[t] esdiad = electron-stimulated ion angular distribution.
[u] aes = Auger electron spectroscopy.
[v] aps = appearance potential spectroscopy.
[w] leed = low energy electron diffraction.
[x] rheed = reflected high energy electron diffraction.
[y] eels = electron energy loss spectroscopy.
[z] hreels = high resolution electron energy loss spectroscopy.
[aa] ceels = core level electron energy loss spectroscopy.
[bb] ins = ion neutralization spectroscopy.
[cc] sims = secondary ion mass spectroscopy.
[dd] iss = ion scattering spectroscopy.
[ee] rbs = Rutherford backscattering spectroscopy.
[ff] leis = low energy ion scattering.
[gg] fem = field emission microscopy.
[hh] fim = field ion microscopy.
[ii] afm = atomic force microscopy.
[jj] stm = scanning tunneling microscopy.

transmission electron microscopy (stem) is simply tem carried out in a scanning mode. Sem, the most common and well-known electron microscopy method for the physical imaging of surfaces, is based on the interaction with a surface of a primary beam of electrons with energy typically in the range of 0.5–40 keV. Sem must be done in vacuum so that the electrons can travel unimpeded for adequate distances. The typical surface magnification realized is on the order of 10–10,000 X depending on the energy of the primary electron beam.

In most sem analyses, secondary electrons created near the position of the impinging primary beam are detected. However, surface images can also be obtained through collection of backscattered primary electrons. Sem is a highly surface sensitive tool even though the primary beam electrons may penetrate into the sample to depths of 0.5–5 μm.

The detectors used in sem are of two types depending on whether secondary or backscattered electrons are being detected. Backscattered electrons are detected with an annular detector placed above the sample such that electrons backscattered along the surface normal can be detected. Secondary electrons are detected with a detector that is positively charged to attract these electrons. The absolute efficiencies of both detectors are relatively low; therefore, the signals must be amplified before being sent to an image intensifier (CRT) for viewing.

A second important mode of electron microscopy is transmission electron microscopy (tem). An image of a sample in tem is obtained using the transmission of electrons by a sample in a method analogous to optical microscopy using photons. This method provides a magnified image of a transparent sample with an electron beam using objective and projector lenses.

Instrumentation for tem is somewhat similar to that for sem; however, because of the need to keep the sample surface as clean as possible throughout the analysis to avoid imaging surface contamination as opposed to the sample surface itself, ultrahigh vacuum conditions (ca10^{-7} – 10^{-8} Pa) are needed in the sample area of the microscope.

Several lenses are used in a transmission electron microscope. The final magnification of the image is performed by one or more projector lenses. The final image is typically recorded on a fluorescent or phosphorescent screen where it can be captured by a video camera for viewing. Tem is a very powerful surface imaging tool with atomic resolution in some cases, providing sample magnifications between 100–500,000 X.

Scanning Probe Microscopies. The most common of these tools are scanning tunneling microscopy (stm) and atomic force microscopy (afm).

Stm is based on the tunneling of electrons between a very sharp, electrically conductive tip (hopefully terminating in a single atom) and a conductive sample when the tip is brought very close (<1 nm) to the sample. The magnitude of the tunneling current will vary in a well-defined exponential manner with tip-to-sample distance. Thus, the magnitude of the tunneling current can be used to generate an image of the surface. Stm requires that both the tip and the sample be electrically conductive.

A second scanning probe microscopy is afm. This technique is related to stm, but provides a direct topographical image of a surface without the complications arising from the electronic surface structure that cloud image interpretation in stm.

Analysis of Surface Elemental Composition. The most common techniques used for determination of elemental composition are the electron spectroscopies in which electrons or x-rays are used to stimulate either electron or x-ray emission from the atoms in the surface (or near-surface region) of the sample. These electrons or x-rays are emitted with energies characteristic of the energy levels of the atoms from which they came, and therefore, contain elemental information about the surface.

X-Ray Photoelectron Spectroscopy. X-ray photoelectron spectroscopy (xps) and Auger electron spectroscopy (aes) are related techniques that are initiated with the same fundamental event, the stimulated ejection of an electron from a surface.

Xps is based on the photoelectric effect when an incident x-ray causes ejection of an electron from a surface atom. In this process, an incident x-ray photon of energy hv impinges on the surface atom causing ejection of an electron, usually from a core electron energy level. This primary photoelectron is detected in xps.

Auger Electron Spectroscopy. Auger electron spectroscopy (aes) is also based on an electron ejection process like xps, but the electrons that are monitored in aes are secondary electrons. These secondary or Auger electrons arise from a process that occurs after primary electron emission such that a core level hole exists. Incident electrons are conventionally used in aes to stimulate primary electron emission, although incident x-rays can also be used as in xps.

Quantitation of Auger spectra follows the same lines of reasoning as for quantitation of xps data.

Xps and Aes Instrumentation. Beyond the requirement for UHV conditions (and the hardware that accompanies it), the most important components of an electron spectrometer system are the source, the electron energy analyzer, and the electron detector.

Xps requires a source that can provide a single x-ray line reasonably narrow in energy. The absolute energy requirement for this x-ray line is that it must be energetic enough to generate photoelectrons from core levels of a majority of the elements with reasonable resolution.

The most common detector for electron spectroscopy is the channel electron multiplier. The raw currents encountered for surface-ejected electrons in a typical xps or aes experiment are on the order of 10^{-14} to 10^{-16} A, far too low to be measured directly. When electrons emerging from the analyzer impinge on the inside surfaces of the electron multiplier, multiple electrons are ejected. Electron amplification on the order of 10^7 to 10^8 is achieved making the currents produced by the electron multiplier easily measurable.

Electron Microprobe Analysis. Electron microprobe analysis (ema) is a technique based on x-ray fluorescence from atoms in the near-surface region of a material stimulated by a focused beam of high energy electrons. Essentially, this method is based on electron-induced x-ray emission as opposed to x-ray-induced x-ray emission, which forms the basis of conventional x-ray fluorescence (xrf) spectroscopy. Ema usually refers to a stand-alone instrument for electron-induced x-ray analysis in either the edx or wavelength dispersive x-ray spectroscopy mode (wds). Ema is used for the determination of elemental constituent identification due to the dependence of x-ray energy on atomic number (Z) of the atom from which the x-ray originates. Ema is also commonly employed for compositional mapping of elements across a surface. Ema can provide elemental maps of both major and minor constituents at detectabilities of a few hundred parts per million in the best cases. This elemental mapping can be accomplished with a lateral resolution of ca 1 μm.

Analysis of Surface Molecular Composition. There is a variety of methods available for elucidating the nature of the molecules that exist on a surface or within an interface.

The most common and readily accessible methods are those based on ftir spectroscopy.

Transmission Fourier Transform Infrared Spectroscopy. Transmission ftir spectroscopy is most often used to study surface species on metal oxides. These solids leave reasonably large spectral windows within which the spectral behavior of the surface species can be viewed. The advantage of this approach for characterization of surface species is the ease of sample preparation and straightforward data interpretation. Spectra of such samples resemble normal transmission ir spectra and can be quantitated using conventional methods.

Diffuse Reflectance Infrared Fourier Transform Spectroscopy. An alternative approach to the acquisition of surface ir spectra for particulate or powdered samples is diffuse reflectance infrared Fourier transform spectroscopy (drifts). The technique is based on the diffuse reflectance of radiation that occurs when it is directed onto a surface with a matte finish or a sample comprised of a powder. This diffuse reflectance is different than specularly reflected radiation in that it penetrates and interacts with a sample before emerging.

The amount of diffusely reflected light cannot be measured directly; instead, it is typically measured relative to a nonabsorbing reference material to allow adequate correction for scattering characteristics of the powdered sample. The signal that is measured is the ratio of the diffuse reflectance of the sample to that of a nonabsorbing reference material.

Attenuated Total Reflectance Fourier Transform Infrared Spectroscopy. Attenuated total reflectance (atr) ftir spectroscopy is based on the principle of total internal reflection. The implementation of internal reflection in the ir region of the spectrum provides a means of obtaining ir spectra of surfaces or interfaces, thus providing molecularly-specific vibrational information.

Infrared Reflection-Absorption Spectroscopy. For adsorbed surface species or thin films on ir reflective surfaces such as metals, an alternative method is infrared reflection-absorption spectroscopy (irras). This technique is based on the external reflection on an infrared beam of light at such surfaces, the characteristics of which are highly polarization dependent. Upon reflection of a polarized beam of light at a surface, phase shifts occur that can be quite significant in magnitude. Given that the electric field experienced by a molecule at a reflective surface equals the sum of the fields from the incident and reflected beams, the magnitude of this phase shift is critical in determining the resulting light intensity available with which a surface molecule can couple.

The ir spectra acquired in this way are extremely sensitive to the orientation of the surface molecules. Molecules must have a significant component of a molecular vibration perpendicular to the surface to be sensed by coupling with the highly directional electric field. Molecules whose dipole moments are perfectly parallel to the surface cannot couple to the existing electric fields, and therefore, are ir transparent by this method. This selectivity of the approach for molecule dipole moments perpendicular as opposed to parallel to the surface is known as the surface selection rule of irras.

This approach works quite well for species at metal surfaces. It has been used extensively to ascertain information about organic thin films on metal surfaces. Of particular interest in many of these studies has been the determination of molecular orientation on surfaces from such studies. Few other techniques are quite so useful for unambiguously ascertaining molecular orientation.

JEANNE E. PEMBERTON
University of Arizona

B. W. Rossiter and R. C. Baetzold, eds., *Investigations of Surfaces and Interfaces*, Part A. *Physical Methods of Chemistry*, Vol. 9A, 2nd ed., John Wiley & Sons, Inc., New York, 1993.

C. R. Brundle and A. D. Baker, eds., *Electron Spectroscopy: Theory, Techniques and Applications*, Vol. 4, Academic Press, New York, 1981.

D. J. Connor, B. A. Sexton, and R. St. C. Smart, eds., *Surface Analysis Methods in Materials Science, Springer Series in Surface Sciences*, Vol. 23, Springer-Verlag, Berlin, 1992.

J. C. Riviere, *Surface Analytical Techniques. Monographs on the Physics and Chemistry of Materials*, Oxford University Press, New York, 1990.

SURFACTANTS

The term surfactant, contraction of surface-active agent, is used to describe organic substances having certain characteristics in structure and properties. The term detergent is often used interchangeably with surfactant. As a designation for a substance capable of cleaning, detergent can also encompass inorganic substances when these do in fact perform a cleaning function. More often, however, detergent refers to a combination of surfactants and other substances, organic or inorganic, formulated to enhance functional performance, specifically cleaning, over that of the surfactant alone. It is so used herein.

Surfactants are characterized by the following features. Amphipathic structure: surfactant molecules are composed of groups of opposing solubility tendencies, typically an oil-soluble hydrocarbon chain and a water-soluble ionic group; solubility: a surfactant is soluble in at least one phase of a liquid system; adsorption at interfaces: at equilibrium, the concentration of a surfactant solute at a phase interface is greater than its concentration in the bulk of the solution; orientation at interfaces: surfactant molecules and ions form oriented monolayers at phase interfaces; micelle formation: surfactants form aggregates of molecules or ions called micelles when the concentration of the surfactant solute in the bulk of the solution exceeds a limiting value, the so-called critical micelle concentration (CMC), which is a fundamental characteristic of each solute–solvent system; and functional properties: surfactant solutions exhibit combinations of cleaning (detergency), foaming, wetting, emulsifying, solubilizing, and dispersing properties.

The presence of two structurally dissimilar groups within a single molecule is the most fundamental characteristic of surfactants. The surface behavior (surface activity) of the surfactant molecule is determined by the makeup of the individual groups, solubility properties, relative size, and location within the surfactant molecule.

Different designations describe the opposing groups within the surfactant molecules, eg, hydrophobic (water hating) and hydrophilic (water liking), lipophobic (fat hating) and lipophilic (fat liking), oleophobic (fat (oil) hating) and oleophilic (fat (oil) liking), and lyophobic (solvent hating) and lyophilic (solvent liking). The terms polar and nonpolar are also used to designate water-soluble and water-insoluble groups, respectively.

Surface activity is not limited to aqueous systems; however, because water is present as the solvent phase in the overwhelming proportion of commercially important surfactant systems, its presence is assumed in much of the common terminology of industry. Thus, the water-soluble amphipathic groups are often referred to as solubilizing groups.

Surfactants are classified depending on the charge of the surface-active moiety. In anionic surfactants, this moiety carries a negative charge. In cationic surfactants, the charge is positive. In nonionic surfactants, there is no charge on the molecule, the solubilizing contribution can be supplied by side groups. Finally, in amphoteric surfactants, solubilization is provided by the presence of positive and negative charges in the molecule. In general, the hydrophobic group consists of a hydrocarbon chain containing ca 10–20 carbon atoms. The chain may be interrupted by oxygen atoms, a benzene ring, amides, esters, other functional groups, and/or double bonds. A propylene oxide hydrophobe can be considered a hydrocarbon chain in which every third methylene group is replaced by an oxygen atom. In some cases, the chain may carry substituents, most often halogens. Siloxane chains have also served as the hydrophobe in some surfactants.

Hydrophilic, solubilizing groups for anionic surfactants include carboxylates, sulfonates, sulfates, and phosphates. Cationics are solubilized by amine and ammonium groups. Ethylene oxide chains and hydroxyl groups are the solubilizing groups in nonionic surfactants. Amphoteric surfactants are solubilized by combinations of anionic and cationic solubilizing groups.

The molecular weight of surfactants may be as low as ca 200 up to the thousands for polymeric structures. A surfactant with a straight-chain C_{12}-hydrophobe and a solubilizing group is generally an effective structure. The optimum can be higher by several carbon atoms or even slightly lower than 12 depending on the nature of the polar group and the desired function of the surfactant.

In the application of surfactants, physical and use properties, precisely specified, are of primary concern. Chemical homogeneity is of little significance in practice. In fact, surfactants are generally polydisperse mixtures, such as the natural fats as precursors of fatty acid-derived surfactant structures; eg, coconut oil contains glycerol esters of C_6–C_{18} fatty acids. Nonionic surfactants of the alcohol ethoxylate type are polydisperse not only with respect to the hydrophobe but also in the number of ethylene oxide units attached.

Commercial surfactants are complicated mixtures exceedingly difficult to separate into pure molecular species.

Physical Chemistry of Interfaces

The usefulness of surfactants stems from the effects that they exert on the surface, interfacial, and bulk properties of their solutions and the materials their solutions come in contact with.

Phenomena at Liquid Interfaces. The area of contact between two phases is called the interface; three phases can have only a line of contact, and only a point of mutual contact is possible between four or more phases. Combinations of phases encountered in surfactant systems are L–G, L–L–G, L–S–G, L–S–S–G, L–L, L–L–L, L–S–S, L–L–S–S–G, L–S, L–L–S, and L–L–S–G, where G = gas, L = liquid, and S = solid. An example of an L–L–S–G system is an aqueous surfactant solution containing an emulsified oil, suspended solid, and entrained air (see EMULSIONS; FOAMS). This embodies several conditions common to practical surfactant systems. First, because the

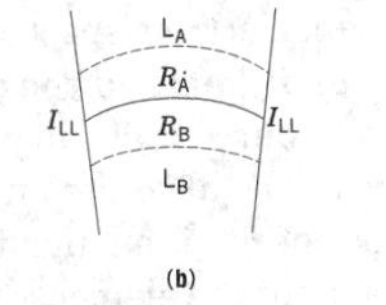

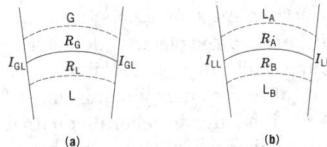

Figure 1. (**a**) Gas–liquid (GL) interface; (**b**) liquid–liquid (LL) interface.

surface area of a phase increases as particle size decreases, the emulsion, suspension, and entrained gas each have large areas of contact with the surfactant solution. Next, because interfaces can exist only between two phases, analysis of phenomena in the L–L–S–G system breaks down into a series of analyses, ie, surfactant solution to the emulsion, solid, and gas. It is also apparent that the surfactant must be stabilizing the system by preventing contact between the emulsified oil and dispersed solid. Finally, the dispersed phases are in equilibrium with each other through their common equilibrium with the surfactant solution.

Figures **1a** and **1b** represent typical gas–liquid and liquid–liquid interfaces at equilibrium. Assuming that gas, G, consists of air and vapor of the liquid, L, at equilibrium, there is continuous movement of liquid molecules through the gaseous interfacial region R_G because rates of evaporation and condensation at the interface I_G are equal (Fig. 1). Liquid molecules are also moving continuousy into and out of I_G through the liquid interfacial region, R_L. R_G and R_L represent nonhomogeneous transitional regions between the homogeneous phases, G and L. Systems are known in which R_G and R_L have thicknesses equivalent to two or more layers of molecules, but for most analyses the interface I_G can be considered as consisting of a single layer of molecules.

For thermodynamic treatment of surface phenomena, the thickness of the boundary regions can often be ignored or their effect eliminated by selection of a convenient location for the interface I_{GL}. The liquid–liquid interface, I_{LL} (Fig. **1b**) is similarly associated with interfacial regions, R_A and R_B, which can be treated like the gas–liquid interface in most analyses. Because few liquids are completely immiscible, mutual saturation is taken as the equilibrium condition.

Energy of Adhesion. The interfacial energy between two mutually insoluble saturated liquids, A and B, is equal to the difference in the separately measured surface energies of each phase: $\gamma_{AB} = \gamma_A - \gamma_B$, where γ is free-surface or interfacial energy. The term γ_{AB} represents the energy that must be added to the system to separate the liquids.

Contact Angle. The line of contact between the three phases of a G–L–S system is the locus of all points from which the angle of contact between the liquid and the solid can be measured. The drop of liquid, L, is resting on the solid, S, and both phases are exposed to the gas, G, at equilibrium saturation of the liquid in air (gas). The drop is assumed to be small enough for the flattening pressure of gravity to be negligible. The vector X_G is tangent to the liquid at its contact with the solid. The angle between the tangent and the surface of the solid is called the contact angle, θ. The equilibrium value of θ is an indicator of the energy relationships between liquid–liquid and liquid–solid interfaces.

Effects of Surfactants on Solutions. A surfactant changes the properties of a solvent in which it is dissolved to a much greater extent than is expected from its concentration effects. This marked effect is the result of adsorption at the solution's interfaces, orientation of the adsorbed surfactant ions or molecules, micelle formation in the bulk of the solution, and orientation of the surfactant ions or molecules in the micelles, which are caused by the amphipathic structure of a surfactant molecule. The magnitude of these effects depends to a large extent on the solubility balance of the molecule. An efficient surfactant is usually relatively insoluble as individual ions or molecules in the bulk of a solution.

Positive adsorption, the concentration of one component of a solution at a phase boundary, results in a lowering of the free-surface energy of the solution. Accumulation of a surfactant at a solution interface means that the attractive forces between surfactant and solvent are less than the attraction between solvent molecules. As thermal diffusion brings surfactant molecules into the surface, accumulation occurs because the solute molecules cannot re-enter the solution against the stronger mutual attraction of the solvent molecules. Negative adsorption occurs when the attraction between solute and solvent molecules is greater than that between solvent molecules, and exists in concentrated aqueous solutions of inorganic compounds such as NaOH. It is associated with a surface tension slightly higher than for pure water.

Practical applications of surfactants usually involve some manner of surfactant adsorption on a solid surface. This adsorption is always associated with a decrease in free-surface energy, the magnitude of which must be determined indirectly. The force with which the adsorbate is held on the adsorbent may be roughly classified as physical, ionic, or chemical. Physical adsorption is a weak attraction caused primarily by van der Waals forces. Ionic adsorption occurs between charged sites on the substrate and oppositely charged surfactant ions, and is usually a strong attractive force. The term chemisorption is applied when the adsorbate is joined to the adsorbent by covalent bonds or forces of comparable strength.

Physical and ionic adsorption may be either monolayer or multilayer. Capillary structures in which the diameters of the capillaries are small, ie, one to two molecular diameters, exhibit a marked hysteresis effect on desorption. Sorbed surfactant solutes do not necessarily cover all of a solid interface and their presence does not preclude adsorption of solvent molecules. The strength of surfactant sorption generally follows the order cationic > anionic > nonionic.

Micelles. Surfactant molecules or ions at concentrations above a minimum value characteristic of each solvent-solute system associate into aggregates called micelles. The formation, structure, and behavior of micelles have been extensively investigated. The term critical micelle concentration (CMC) denotes the concentration at which micelles start to form in a system comprising solvent, surfactant, possibly other solutes, and a defined physical environment.

Micelle size is expressed as the micellar molecular weight or, more generally, the aggregation number, ie, the number of monomers making up the micelles. Micellar aggregation numbers generally lie between 20 and 100, for single-chain anionic and cationic surfactants. Large aggregation numbers (> 1000) have been reported for nonionic micelles, especially as the cloud point is approached.

Small micelles in dilute solution close to the CMC are generally believed to be spherical. Under other conditions, micellar materials can assume structures such as oblate and prolate spheroids, vesicles (double layers), rods, and lamellae.

Micellar properties are affected by changes in the environment, eg, temperature, solvents, electrolytes, and solubilized components. These changes include complicated phase changes, viscosity effects, gel formation, and liquefication of liquid crystals.

Measurement of Surface Activity. Each surface-active property can be measured in a variety of ways and the method of choice depends on the characteristics of the substance to be tested. The most frequently determined properties are surface tension (γ_{SG}, γ_{LG}), interfacial tension (γ_{LL}, γ_{LG}), contact angle (θ), and CMC.

Anionic Surfactants

Carboxylate, sulfonate, sulfate, and phosphate are the polar, solubilizing groups found in most anionic surfactants. In dilute solutions of soft water, these groups are combined with a 12–15 carbon chain hydrophobe for best surfactant properties. In neutral or acidic media, or in the presence of heavy-metal salts, eg, Ca, the carboxylate group loses most of its solubilizing power.

Of the cations (counterions) associated with polar groups, sodium and potassium impart water solubility, whereas calcium, barium, and magnesium promote oil solubility. Ammonium and substituted ammonium ions provide both water and oil solubility. Triethanolammonium is a commercially important example. Salts (anionic surfactants) of these ions are often used in emulsification. Higher ionic strength of the medium depresses surfactant solubility. To compensate for the loss of solubility, shorter hydrophobes are used for application in high ionic-strength media.

Carboxylates. Soaps represent most of the commercial carboxylates. The general structure of soap is $RCOO^-M^+$, where R is a straight hydrocarbon chain in the C_9–C_{21} range and M^+ is a metal or ammonium ion. Interruption of the chain by amino or amido linkages leads to other structures which account for the small volumes of the remaining commercial carboxylates.

Large volumes of soap are used in industrial applications as gelling agents for kerosene, paint driers, and as surfactants in emulsion polymerization (see DRIERS AND METALLIC SOAPS; EMULSIONS; SOAP). Concern over water eutrophication resulted in a ban of phosphorus in laundry detergents. Phosphates have been effectively replaced by combinations of zeolite, citrate, and polymers, coupled with rebalanced synthetic active systems. Soap itself is generally present only as a minor component of surfactants.

Polyalkoxycarboxylates surfactants are produced either by the reaction of sodium chloroacetate with an alcohol ethoxylate or from an acrylic ester and an alcohol alkoxylate. Because of the presence of the ethylene oxide linkages, these products possess a higher aqueous solubility which manifests itself in greater compatibility with cationic surfactants and polyvalent cations.

N-Acylsarcosinates. Sodium *N*-lauroylsarcosinate is a good soap-like surfactant. The amido group in the hydrophobe chain lessens the interaction with hardness ions. *N*-Acylosarcosinates are prepared from a fatty acid chloride and sarcosine.

Acylated Protein Hydrolysates. These surfactants are prepared by acylation of protein hydrolysates with fatty acids or acid chlorides. Acylated protein hydrolysates are mild surfactants recommended for personal-care products (see COSMETICS).

Sulfonates. The sulfonate group, $-SO_3M$, attached to an alkyl, aryl, or alkylaryl hydrophobe, is a highly effective solubilizing group. Sulfonic acid surfactants are strong, their salts are relatively unaffected by pH, they are stable to oxidation, and because of the strength of the C–S bond also stable to hydrolysis. Sulfonates interact moderately with the hardness, Ca^{2+} and Mg^{2+}, but significantly less so than carboxylates. Sulfates can be tailored for specific applications by introduction of double bonds or ester or amide groups, either into the hydrocarbon chain or as substituents. Because the introduction of the SO_3H function is inherently inexpensive, eg, by oleum, SO_3, SO_2Cl_2, or $NaHSO_3$, sulfonates are heavily represented among high volume surfactants (see SULFONATION AND SULFATION).

Sulfonates include alkylbenzenesufonates (ABS), the most widely used of the non-soap surfactants; short-chain alkylarenesulfonates; lignosulfonates; napthalenesulfonates; α-olefinsulfonates; petroleum sulfonates; sulfonates with ester, amide, and ether linkages; and fatty acid ester sulfonates.

Sulfates and Sulfated Products. The sulfate group, $-OSO_3M$, where M is a cation, represents the sulfuric acid half-ester of an alcohol and is more hydrophilic than the sulfonate group because of the presence of an additional oxygen atom. Attachment of the sulfate group to a carbon atom of the hydrophobe through the C–O–S linkage limits its hydrolytic stability, particularly under acidic conditions. Usage of sulfated alcohols and sulfated alcohol ethoxylates has expanded dramatically since the 1970s as the detergent industry reformulates consumer products to improve biodegradability, lower phosphate content, and move from powder to liquid.

Sulfates and sulfated products include alcohol sulfates, ethoxylated and sulfated alcohols; ethoxylated and sulfated alkylphenols; and sulfated natural oils and fats.

Phosphate Esters. Mono and diesters of orthophosphoric acid:

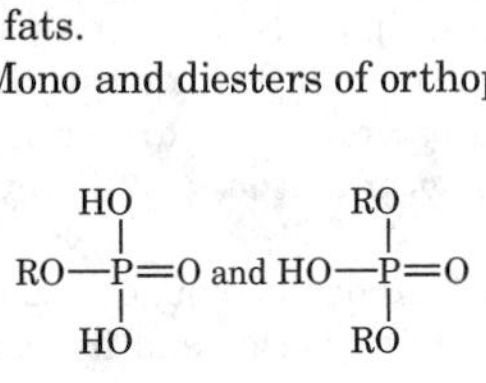

and their salts are useful surfactants. In contrast to sulfonates and sulfates, the resistance of alkyl phosphate esters to acids and hard water is poor. Calcium and magnesium salts are insoluble. In the acid form, the esters show limited water solubility, although their alkali metal salts are more soluble. The surface activity of phosphate esters is good, although in general it is somewhat lower than that of the corresponding phosphate-free precursors. Thus, a phosphated nonylphenol ethoxylated with 9 mol of ethylene oxide is less effective as a detergent in hard water than its nonionic precursor. At higher temperatures, however, the phosphate surfactant is significantly more effective.

Because of high costs and the limitations noted above, phosphate surfactants find application in specialty situations where such limitations are of no concern. As specialty surfactants, phosphate esters and their salts are remarkably versatile. Applications include emulsion polymerization of vinyl acetate and acrylates; dry-cleaning compositions where solubility in hydrocarbon solvents is a particular advantage; textile mill processing where stability and emulsifying power for oil and wax under highly alkaline conditions is necessary; and industrial cleaning compositions where tolerance for high concentrations of electrolyte and alkalinity is required. In addition, phosphate surfactants are used as corrosion inhibitors, in pesticide formulations, in papermaking, and as wetting and dispersing agents in drilling mud fluids.

Nonionic Surfactants

Unlike anionic or cationic surfactants, nonionic surfactants carry no discrete charge when dissolved in aqueous media. Hydrophilicity in nonionic surfactants is provided by hydrogen bonding with water molecules. Oxygen atoms and hydroxyl groups readily form strong hydrogen bonds, whereas ester and amide groups form hydrogen bonds less readily. Hydrogen bonding provides solubilization in neutral and alkaline media. In a strongly acid environment, oxygen atoms are protonated, providing a quasi-cationic character. Each oxygen atom makes a small contribution to water solubility; more than a single oxygen atom is therefore needed to solubilize a nonionic surfactant in water. Nonionic surfactants are compatible with ionic and amphoteric surfactants. Because a polyoxyethylene group can easily be introduced by reaction of ethylene oxide with any organic molecule containing an active hydrogen atom, a wide variety of hydrophobic structures can be solubilized by ethoxylation.

Polyoxyethylene Surfactants. Polyoxyethylene-solubilized nonionics (ethoxylates) are moderate foamers and do not respond to conventional foam boosters. Foaming shows a maximum as a function of ethylene oxide content. Low foaming nonionic surfactants are prepared by terminating the polyoxyethylene chain with less soluble groups such as polyoxypropylene and methyl groups. Ethoxylates can be prepared to attain almost any hydrophilic–hydrophobic balance. For incorporation into powdered products, they suffer from the disadvantage of being liquids or low melting waxes, which complicates the manufacture of free-flowing, crisp powders. Solid products are manufactured with ethoxylates of high ethylene oxide content. The latter, however, are too water soluble to provide good surface activity.

Base-catalyzed ethoxylation of aliphatic alcohols, alkylphenols, and fatty acids can be broken down into two stages: formation of a monoethoxy adduct and addition of ethylene oxide to the monoadduct to form the polyoxyethylene chain. Polyoxyethylene surfactants include alcohol ethoxylates and akylphenol ethoxylates.

Carboxylic Acid Esters. In the carboxylic acid ester series of surfactants, the hydrophobe, a naturally occurring fatty acid, is solubilized with the hydroxyl groups of polyols or the ether and terminal hydroxyl groups of ethylene oxide chains. Included in this group are glycerol esters; polyoxyethylene esters; and hydrosorbitol esters; ethoxylated anhydrosorbitol esters; natural ethoxylated fats, oils, and waxes; and glycol esters of fatty acids.

Carboxylic Amides. Carboxylic amide nonionic surfactants are condensation products of fatty acids and hydroxyalkyl amines. They include diethanolamine condensates, monoalkanolamine condensates, and polyoxyethylene fatty acid amides.

Fatty Acid Glucamides. Fatty acyl glucamides (FAGA) or polyhydroxyamides (PHA) have been adopted by detergent manufacturers in the United States and Europe. FAGA is produced via reaction of fatty acid methyl ester with *N*-methyl glucamine and attendant elimination of methanol. The methyl ester would be produced via the standard route of transesterification with fatty triglycerides; the glucamine, via reaction between glucose and methylamine with attendant hydrogenation and elimination of water.

Fatty acid glucamides are used in dishwashing liquids and heavy-duty liquids. Benefits include improved mildness for dishwashing liquids and improved enzyme stability in fabric washing detergents.

Polyalkylene Oxide Block Copolymers. The higher alkylene oxides derived from propylene, butylene, styrene (qv), and cyclohexene react with active oxygens in a manner analogous to the reaction of ethylene oxide. Because the hydrophilic oxygen constitutes a smaller proportion of these molecules, the net effect is that the oxides, unlike ethylene oxide, are hydrophobic. The higher oxides are not used commercially as surfactant raw materials except for minor quantities that are employed as chain terminators in polyoxyethylene surfactants to lower the foaming tendency. The hydrophobic nature of propylene oxide units, $-CH(CH_3)CH_2O-$, has been utilized in several ways in the manufacture of surfactants.

Block polymer nonionic surfactants are not strongly surface-active but exhibit commercially useful surfactant properties. Aqueous solutions characteristically foam less than those of other surfactant types. They act as detergents, wetting and rinsing agents, demulsifiers and emulsifiers, dispersants, and solubilizers. They are used in automatic dishwashing detergent compositions, cosmetic preparations, spin finishing compositions for textile processing, metal-cleaning formulations, papermaking, and other technologies.

Cationic Surfactants

The hydrophobic moiety of a cationic surfactant carries a positive charge when dissolved in aqueous media, which resides on an amino or quaternary nitrogen. A single amino nitrogen is sufficiently hydrophilic to solubilize a detergent-range hydrophobe when protonated in dilute acidic solution; eg, laurylamine is soluble in dilute hydrochloric acid. For increased water solubility, additional primary, secondary, or tertiary amino groups can be introduced or the amino nitrogen can be quaternized with low molecular weight alkyl groups such as methyl or hydroxyethyl. Quaternary nitrogen compounds are strong bases that form essentially neutral salts with hydrochloric and sulfuric acids. Most quaternary nitrogen surfactants are soluble even in alkaline aqueous solutions. Polyoxyethylated amino surfactants behave like nonionic surfactants in alkaline solutions and like cationic surfactants in acid solutions.

Cationic surfactants are widely used in acidic aqueous and nonaqueous systems as textile softeners, conditioning agents, dispersants, emulsifiers, wetting agents, sanitizers, dye-fixing agents, foam stabilizers, and corrosion inhibitors. To some extent, the usage pattern mirrors that of the anionic surfactants in neutral and alkaline solutions. The positively charged cationic surfactants are more strongly adsorbed than anionic or nonionic surfactants on a variety of substrates including textiles, metal, glass (qv), plastics, minerals, and animal and human tissue, which can often carry a negative surface charge. Substantivity of cationic surfactants is the key property in many applications. In general, they are incompatible with anionic surfactants. Reaction of the two large, oppositely charged ions gives a salt insoluble in water. Ethoxylation moderates the tendency to form insoluble products with anionic surfactants.

Many benzenoid quaternary cationic surfactants possess germicidal, fungicidal, or algicidal activity. Solutions of such compounds, alone or in combination with nonionic surfactants, are used as detergent sanitizers in hospital maintenance. Classified as biocidal products, their labeling is regulated by the U.S. EPA.

Amines. Aliphatic mono-, di-, and polyamines derived from fatty and rosin acids make up this class of surfactants. Primary, secondary, and tertiary monoamines with C_{18} alkyl or alkenyl chains constitute the bulk of this class. The products are sold as acetates, naphthenates, or oleates. Principal uses are as ore-flotation agents, corrosion inhibitors, dispersing agents, wetting agents for asphalt, and as intermediates for the production of more highly substituted derivatives.

In addition to the mono- and dialkylamines, representative structures of this class of surfactants include *N*-alkyltrimethylene diamine, $RNH(CH_2)_3NH_2$, where the alkyl group is derived from coconut, tallow, and soybean oils; or is 9-octadecenyl, 2-alkyl-2-imidazoline, where R is heptadecyl, heptadecenyl, or mixed alkyl, and 1-(2-aminoethyl)-2-alkyl-2-imidazoline, where R is heptadecyl, 8-heptadecenyl, or mixed alkyl.

This group includes amine oxides, ethoxylated alkylamines, 1-(2-hydroxyethyl)-2-imidazolines, and alkoxylates of ethylenediamine.

Amine oxides have attracted widespread interest as replacements for alkanolamides as foam builders in liquid hand-dishwashing compositions.

2-Alkyl-1-(2-hydroxyethyl)-2-imidazolines are used in hydrocarbon and aqueous systems as antistatic agents, corrosion inhibitors, detergents, emulsifiers, softeners, and viscosity builders. They are prepared by heating the salt of a carboxylic acid with (2-hydroxyethyl)ethylenediamine at 150–160°C to form a substituted amide; 1 mol water is eliminated to form the substituted imidazoline with further heating at 180–200°C. Substituted imidazolines yield three series of cationic surfactants: by ethoxylation to form more hydrophilic products; quaternization with benzyl chloride, dimethyl sulfate,and other alkyl halides; and oxidation with hydrogen peroxide to amine oxides.

Quaternary Ammonium Salts. The quaternary ammonium ion is a much stronger hydrophile than primary, secondary, or tertiary amino groups, strong enough to carry a hydrophobe into solution in the surfactant molecular weight range, even in alkaline media. The discrete positive charge on the quaternary ammonium ion promotes strong adsorption on negatively charged substrates, such as fabrics, and is the basis for the widespread use of these surfactants in domestic fabric-softening compositions (see QUATERNARY AMMONIUM COMPOUNDS).

Amphoteric Surfactants

Amphoteric surfactants contain both an acidic and basic hydrophilic group. Ether or hydroxyl groups may also be present to enhance the hydrophilicity of the surfactant molecule. Examples of amphoteric surfactants include amino acids and their derivatives in which the nitrogen atom tends to become protonated with decreasing pH of the solution. Amino acid salts, under these conditions, contain both a positive and a negative charge on the same molecule.

Amphoteric surfactants are generally considered specialty surfactants, however, usage has expanded significantly. They do not irritate skin and eyes, exhibit good surfactant properties over a wide pH range, and are compatible with anionic and cationic surfactants. A basic nitrogen and an acidic carboxylate group are the predominant functional groups.

Imidazolinium Derivatives. Amphoteric imidazolinium derivatives are prepared from the 2-alkyl-1-(2-hydroxyethyl)-2-imidazolines and from sodium chloroacetate.

Imidazolinium derivatives are recommended as detergents, emulsifiers, wetting and hair conditioning agents, foaming agents, fabric softeners, and antistatic agents. There is some evidence that in cosmetic formulations certain imidazolinium derivatives reduce eye irritation caused by sulfate and sulfonate surfactants present in these products.

Uses

Detergency, ie, cleaning, is the primary function of household and personal products. More recently, a secondary function, such as softening in combination with detergency in laundry detergents or conditioning in combination with detergency in shampoos, has been offered as an additional product benefit. In general, products have tended toward functional specialization.

Surfactants are widely used outside the household for a variety of cleaning and other purposes. Often the volume or cost of the surfactant consumed in industrial applications is small compared to benefit.

JESSE L. LYNN, JR.
BARBARA H. BORY
Lever Company

L. Spitz, ed., *Soaps and Detergents*, AOCS Press, Champaign, Ill., 1996.

R. D. Swisher, ed., *Surfactant Biodegradation, Surfactant Science Series*, Vol. 3, Marcel Dekker, Inc., New York, 1970.

R. D. Swisher, ed., *Surfactant Biodegradation, Surfactant Science Series*, 2nd ed., Vol. 18, Marcel Dekker, Inc., New York, 1987.

T. F. Tadros, *Surfactants*, Academic Press, London, 1984.

R. Zana, ed., *Surfactant Solutions: New Methods of Investigation, Surfactant Science Series*, Vol. 22, Marcel Dekker, Inc., New York, 1986.

SUTURES

Surgical sutures are sterile, flexible strands used to close wounds or to tie off tubular structures such as blood vessels. Made of natural or synthetic fiber and usually attached to a needle, they are available in monofilament or multifilament forms. Sutures are classified as either absorbable or nonabsorbable.

Sutures are required to hold tissues together until the tissues can heal adequately to support the tensions exerted on the wound during normal activity. Sutures can be used in skin, muscle, fat, organs, and vessels. Nonabsorbable sutures are designed to remain in the body for the life of the patient, and are indicated where permanent wound support is required. Absorbable sutures are designed to lose strength gradually over time by chemical reactions such as hydrolysis. These sutures are ultimately converted to soluble components that are then metabolized and excreted in urine or feces, or as carbon dioxide in expired air. Absorbable sutures are indicated only where temporary wound support is needed.

Monofilament and multifilament sutures behave very differently in surgery. Monofilaments, which pass easily through living tissue and generate little frictional resistance, contain no pores or interstices that might harbor infectious organisms. Multifilament sutures, on the other hand, are frequently coated to reduce frictional drag and damage to tissue, to fill the interstices between fibers, and to ease the repositioning of already-tied knots.

Sutures must be knotted in use, but the knot raises the local stress, so that the suture is most likely to fail at the knot. For this reason, there are minimum knot-pull tensile strength requirements for sutures.

Similar to any implanted foreign body, sutures produce an initial inflammatory response at the site of the injury. Granulation tissue consisting of fibrous connective tissue and new blood vessels gradually encapsulates the suture as the wound begins to heal. Using absorbable sutures, this tissue is ultimately resorbed as the suture disappears. Natural products such as silk and gut tend to provoke a more severe response than do the synthetic materials. Before marketing a new suture, manufacturers conduct preclinical studies to evaluate the safety and effectiveness of a suture.

Absorbable Sutures

Absorbable sutures are classified into collagen and synthetic sutures. Synthetic absorbable sutures are available as braids or monofilaments. Absorbable sutures are intended only for indications where temporary wound support is needed.

Natural Absorbable Sutures. Natural absorbable sutures (collagen) are composed primarily of collagen and are sold as surgical gut sutures. Contrary to popular belief, catgut sutures are not derived from cats. Collagen sutures are derived from the serosa layer of beef intestine (the gut) or the submucosa layer of sheep intestine, which are mechanically processed and bleached. The resulting tissue, called goldbeater, is sliced into thin ribbons and may be treated with chromium salts to cross-link the collagen protein chains. The ribbons are either left undyed or dyed black using pyrogallol and ferric ammonium citrate. They are then sorted by size and twisted together under tension to make multiple-plied strands in a range of sizes.

All surgical gut sutures lose strength gradually over a period of days, but retain some residual strength for several weeks. Absorption involves enzymatic digestion and is usually complete after several months, depending on the type of gut. Animal collagen is a foreign protein; surgical gut frequently provokes an intense inflammatory response.

Braided Synthetic Absorbable Sutures. Polyglycolic acid sutures, introduced commercially in 1970, were a milestone in suture development. This only the first synthetic alternative to surgical gut, but also the first synthetic fiber specifically designed for suture use.

Polyglycolic acid and poly(glycolide—lactide) copolymers are made by bulk polymerization in the presence of a Lewis acid catalyst. The polymer is melt-spun into fine filaments (yarns) which are then braided to form sutures in a range of different diameters. The larger sizes may contain a core of twisted or untwisted yarns in addition to the braided sleeve yarns. Most polyglycolic acid and poly(glycolide—lactide) sutures are coated with proprietary formulations to reduce drag as the suture is passed through tissue.

Monofilament Synthetic Absorbable Sutures. The first monofilament synthetic absorbable suture was introduced in 1984. The polymer for Polydioxanone sutures is produced by the bulk polymerization of 2,5-*p*-dioxanone.

The suture material polyglyconate is made by the bulk copolymerization of a mixture of 67% glycolide and 33% trimethylene carbonate.

Poliglecaprone-25 is a copolymer of 75% glycolide and 25% caprolactone.

Glycomer-631 is a terpolymer produced by the bulk polymerization of a mixture of 60% glycolide, 14% 2,5-*p*-dioxanone, and 26% trimethylene carbonate.

There are significant differences between absorbable sutures in rate of loss of strength. Polydioxanone and polyglyconate sutures retain some strength up to six weeks in the body. Polyglycolic acid, poly(glycolide–lactide) copolymer, and glycomer-631 sutures retain some strength up to three weeks, and poliglecaprone-25 up to two weeks.

Nonabsorbable Sutures

Nonabsorbable Natural Sutures. Cotton and silk are the only nonabsorbable sutures made from natural fibers that are still available in the United States. Cotton suture is made from fibers harvested from various species of plants belonging to the genus *Gossipium*. The fiber is composed principally of cellulose. The cotton yarns are braided or twisted to form sutures in a range of sizes. The suture is bleached and coated. It may be white or dyed blue with D&C Blue No. 9.

Silk suture is made from the threads spun by the silkworm *Bombyx mori*. The fiber is composed principally of the protein fibroin and has a natural coating composed of sericin gum. The gum is usually removed before braiding the silk yarns to make sutures in a range of sizes. Fine silk sutures may be made by simply twisting the gum-coated silk yarns to produce the desired diameter. The sutures are undyed or dyed black with logwood extract or blue with D&C Blue No. 9.

Although silk and cotton are classified as nonabsorbable sutures, these do lose strength gradually in living tissue and slowly break up after long periods of implantation.

Braided Synthetic Nonabsorbable Sutures. Braided synthetic nonabsorbable sutures are made by melt-spinning thermoplastic polymers into fine filaments (yarns), and braiding them, with or without a core, to form multifilament sutures in a range of sizes. Braided nylon-6,6 synthetic nonabsorbable sutures may be coated using either high molecular weight polydimethylsiloxane or wax. The sutures are available undyed (clear monofilaments and white braids), or post-dyed black (with logwood extract), blue (FD&C Blue No. 2), or green (D&C Green No. 5). Although Nylon-6,6 is classified as a nonabsorbable suture, it does slowly hydrolyze in tissue and is not indicated for attachment of synthetic prostheses such as heart valves.

Polyethylene terephthalate is a polyester produced by the condensation polymerization of dimethyl terephthalate and ethylene glycol. Polyethylene terephthalate sutures are available white (undyed), or dyed green with D&C Green No. 6, or blue with D&C Blue No. 6.

Monofilament Synthetic Nonabsorbable Sutures. Monofilament synthetic nonabsorbable sutures are made from thermoplastic resins melt-spun to form monofilaments. Spinnerets of different capillary diameter are used to make a range of suture sizes.

Nylon-6 is made by the bulk addition polymerization of caprolactam. Monofilament Nylon-6 sutures are available undyed (clear), or in post-dyed black (with logwood extract), blue (FD&C Blue No. 2), or green (D&C Green No. 5).

Polybutester is a polyether-ester produced by the condensation polymerization of dimethyl terephthalate, polytetramethylene ether glycol, and 1,4-butanediol. Polybutester sutures are available in clear, ie, undyed, or blue, ie, melt-pigmented with (phthalocyaninato(2-)) copper.

Polypropylene is made by the polymerization of propylene gas. Polypropylene (qv) sutures are available in clear (undyed) or blue (melt-pigmented with [phthalocyaninato(2-)] copper).

Polytetrafluoroethylene suture is composed of expanded polytetrafluoroethylene (ePTFE), resulting in a porous microstructure having longitudinally oriented nodes and fibrils.

Steel suture is made from 316-L stainless steel wire. The suture may be monofilament, known as fixation wire, or multifilament twisted wires. The multifilament strands are either uncoated or coated.

Needles

Most sutures are equipped with needles made from heat-treatable stainless steel. They can be straight or bent. The points of the needles may be ground to a taper (round shaft) for use in tissues that are easy to penetrate, or ground to a triangular or polyhedral geometry to form cutting edges for use in tough tissue such as skin, sclera, or even bone. Not illustrated is the blunt point, a taper needle having a rounded point sharp enough to penetrate skin, but not sharp enough to penetrate a rubber glove.

Needles are commonly attached to sutures by mechanical crimping. Needles may also be attached with adhesives, or by means of a short length of shrink-fit tubing slipped over the ends of the needle and strand. Needles are often treated with a curable silicone compound to reduce friction, making penetration easier.

Packaging

A suture package must maintain sterility as well as the physical and chemical integrity of the suture–needle combination. The suture must be sealed in packaging that presents a barrier to infectious agents. In the case of absorbable sutures, the package must also present a barrier to water vapor.

Sterilization

Commercially available sutures are subjected to a validated sterilization process, and carefully sealed.

Regulatory Requirements

In the United States sutures are regulated by the Food and Drug Administration (FDA) as medical devices intended for human use. The FDA classifies medical devices (sutures) in one of three categories as follows.

Class I: General Controls. This category regulates devices for which performance standards or premarket approvals are not required.

Class II: Performance Standards. This category regulates devices for which General Controls are not sufficient to ensure safety and effectiveness.

Class III: Premarket Approval. Similar to a new drug approval, a premarket approval grants the applicant a license to market a specific well-characterized device.

USP Standards

Voluntary standards for sutures are published by the *United States Pharmacopeia* (USP). If a suture complies with all the requirements of the *Pharmacopeia* it may be labeled "USP".

O. GRIFFIN LEWIS
Consultant
WALTER FABISIAK
Sherwood-Davis & Geck

The United States Pharmacopeia, **23**, 1475–1477, 1835–1836, The United States Pharmacopoeial Convention, Inc., Rockville, Md., 1995.

D. J. Casey and O. G. Lewis, in A. F. von Recum, ed., *Handbook of Biomaterials Evaluation*, Macmillan Publishing Co., New York, 1986, pp. 86–94.

C. C. Chu, J. A. von Fraunhofer, and H. P. Greisler, eds., *Wound Closure Biomaterials and Devices*, CRC Press, New York, 1997.

SWEETENERS

Sugar (sucrose) imparts a sweet taste that is quick, clean, and short-lived. These desirable qualities render sugar the gold standard for sweet taste. Sugar is also an important functional ingredient for preparing attractive foods. It provides the support for bulkiness, texture, preservation, flavor, and color. However, sugar is a nutritive sweetener. It is easily metabolized, yielding an energy of ca 4 kcal/g (16.7 kJ/g). Furthermore, metabolism of sugar and other fermentable carbohydrates by the microorganisms in the oral cavity contributes to tooth decay. Alternative sweeteners can be classified into two groups: nutritive and nonnutritive. Alternative nutritive sweeteners are less caloric than sugar, but retain many of sugar's desirable chemical and physical properties. Hence these are useful as bulking agents in sugar-free products. Examples of alternative nutritive sweeteners are the sugar alcohols hydrogenated starch hydrolysate; and isomalt, a mixture of glucosyl sorbitol and glucosylmannitol. These alcohols are reduced saccharides resulting from catalytic hydrogenation and, for the most part, are less sweet and less caloric than sugar and mostly noncariogenic. Erythritol reportedly yields only 0.4 kcal/g (1.67 kJ/g). Sorbitol, mannitol, and xylitol are approved food additives in the United States.

Another example of an alternative nutritive sweetener is fructose. Although naturally occurring and yielding ca 4 kcal/g (16.7 kJ/g), the same as sugar, fructose does not cause a fluctuation in blood sugar, ie, glucose levels after ingestion, making fructose a better choice for diabetics. The most popular form of fructose used in beverage products is a 55:45 mixture with glucose, usually referred to as high fructose corn syrup (HFCS) (see SYRUPS).

Nonnutritive sweeteners are potently sweet in general, and only minute quantities are required for sweetening foods. As such, foods containing nonnutritive sweeteners generate no or negligible calories from the sweeteners themselves, regardless of whether or not these sweeteners are caloric.

Sweetness potency denotes how many times a given compound is more potent than sugar on the same weight basis.

In addition to being both sweet and safe, a good alternative sweetener should have other qualities similar to those of sucrose. These include stability as a function of temperature and pH, clean sweet taste, quick onset, no lingering aftertaste, compatibility with other food ingredients, high water solubility, high dissolution rate, and ease of handling.

Nonnutritive Sweeteners

By the late 1990s, there were six principal nonnutritive sweeteners being used throughout the world. In descending order of usage by weight, these were aspartame, saccharin, cyclamate, acesulfame-K, stevioside, and glycyrrhizin.

Aspartame. Aspartame (APM, L- aspartyl-L-phenylalanine methyl ester) (**1**), also known under the trade names of Nutra-Sweet and EQUAL, is the most widely used nonnutritive sweetener worldwide.

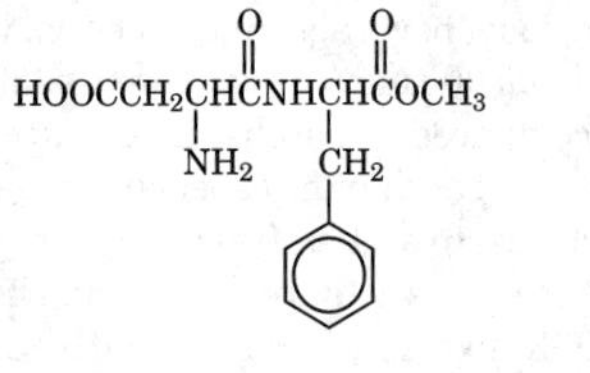

(**1**)

Aspartame can be used legally in the United States in just about all food categories.

In soft drinks, aspartame is 180 times more potent than sugar. In water, it is ca 133X. Aspartame has a good sweet taste but its time–intensity profile is different from that of sugar. Aspartame has a slower onset and a lingering sweet taste.

Aspartame is caloric. As a peptide, it yields ca 4 kcal/g (16.7 kJ/g). However, because of its high sweetness potency, only a minute quantity is consumed, resulting in a negligible calorie contribution.

The methyl ester group of aspartame is very susceptible to bond cleavage. When this cleavage takes place, the sweetness is lost. Avoidance of excessive heat exposure to aspartame is therefore desirable.

The safety of aspartame for human consumption has been studied extensively. However, because phenylalanine is a metabolite of aspartame, people who lack the ability to metabolize this amino acid should refrain from using aspartame.

Acesulfame-K. Acesulfame-K (**2**), the potassium salt of acesulfame (6-methyl-1,2,3-oxathiazin-4(3*H*)-one, 2,2-dioxide), is a sweetener that resembles saccharin in structure and taste profile. Acesulfame-K has been approved for dry product use and for food categories such as yogurts, frozen and refrigerated desserts, and baked goods. Acesulfame-K has been approved for many food categories including non-alcoholic beverages.

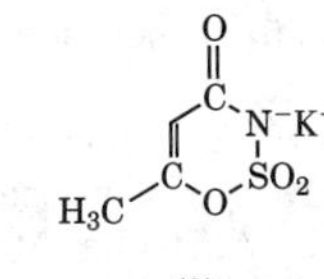

(**2**)

Acesulfame-K is a white crystalline powder having a long (six years or more) shelf life. It readily dissolves in water (270 g/L at 20°C). Like saccharin, acesulfame-K is stable to heat over a wide range of pH. At higher concentrations, there is a detectable bitter and metallic off-taste similar to saccharin. The sweetness potency of acesulfame-K (100 to 200X) is considered to be about half that of saccharin, which is about the same as that of aspartame.

Acesulfame-K–aspartame blends exhibit a significant synergistic effect. Each sweetener apparently masks the off-taste associated with the other.

Since that time, saccharin has been used in many parts of the world.

In 1969, a chronic toxicity study on a cyclamate:saccharin (10:1) blend indicated bladder cancer problems in rats. Cyclamate was soon banned by the FDA, but saccharin remained an approved sweetener. In 1977, the FDA proposed a ban on saccharin because of the discovery of bladder tumors in some male rats fed with high doses of saccharin. In December, 1991, the FDA withdrew its proposed ban. All saccharin-containing packaged products are required to carry a warning label indicating that saccharin has been determined to cause cancer in laboratory animals.

Saccharin imparts a sweetness that is pleasant at the onset but is followed by a lingering, bitter aftertaste. Saccharin is synergistic with other sweeteners of different chemical classes. The blends, as a rule, exhibit less aftertaste than each of the component sweeteners by themselves.

Cyclamate. Sodium cyclamate the sodium salt of cyclamic acid, was so widely used that it was often just called cyclamate. The other common salt, calcium cyclamate, is useful in low sodium diets. Although banned in the United States cyclamate is allowed for use in any or all three categories, ie, food, beverage, and tabletop, in about 50 countries.

Cyclamate is about 30 times more potent than sugar. Its bitter aftertaste is minor compared to saccharin and acesulfame-K. The mixture of cyclamate and saccharin, especially in a 10:1 ratio, imparts both a more rounded taste and a 10–20% synergy.

Stevioside. Stevioside is a naturally occurring sweetener (sweetness potency ca 300X) extracted from a South American plant, *Stevia rebaudiana* Bertoni. The dried leaves, the water extract of leaves, and the refined chemical ingredients, eg, stevioside and rebaudioside A, can all be used as sweetening agents. These are collectively referred to as stevia. Stevia is cultivated primarily in southern China, Taiwan, Thailand, and Malaysia.

Stevioside and rebaudioside A are diterpene glycosides. The sweetness is tainted with a bitter and undesirable aftertaste. The time–intensity profile is characteristic of naturally occurring sweeteners: slow onset but lingering. The aglycone moiety, steviol, which is the principal metabolite, has been reported to be mutagenic. Wide use of stevia in Japan for over 20 years did not produce any known deleterious side effects. However, because no food additive petition has been presented to the FDA, stevioside and related materials cannot be used in the United States as sweeteners. Stevia is allowed for use as diet supplements.

Glycyrrhizin. Glycyrrhizin, also known as glycyrrhizic acid, is a glycoside isolated from the roots of licorice, *Glycyrrhiza glabra* L. For improved water solubility, an ammoniated salt is commonly used.

The sweetness potency of glycyrrhizin is about 33X. Its taste, however, is accompanied by a characteristic licorice flavor, making it incompatible with many other food ingredients. The time–intensity profile is similar to that of other naturally occurring high potency sweeteners: slow onset followed by lingering aftertaste. It is claimed to be heat-stable. Ammonium glycyrrhizinate, which tends to precipitate below pH 4.5, is affirmed in the United States as a GRAS flavoring agent (FEMA no. 2528), but not approved for use as a sweetener.

Sucralose. Sucralose is a trichlorodisaccharide sweetener developed in England during the 1970s. Sucralose has been approved for use as a sweetener by Canada, Australia, Mexico, Russia, Romania, New Zealand, and United States.

The disaccharide structure (**3**) is emphasized by the manufacturer as responsible for a taste quality and time–intensity profile closer to that of sucrose than any other high potency sweetener. The sweetness potency is between 450 and 500X, or about two and one-half times that of aspartame.

Sucralose is stable to heat over a wide range of pH.

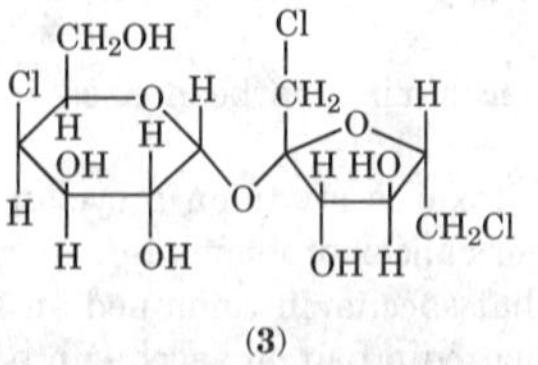

(**3**)

Alitame. A new group of aspartyl-dipeptide sweeteners became known to the public in 1983. Alitame, L-aspartyl-D-alanine *N*-(2,2,4,-tetramethylthietan-3-yl)amide, was selected for commercial development. Alitame has been approved for use as a sweetener by Australia, China, Mexico, New Zealand, Indonesia, and Colombia.

Alitame (trade name: Aclame) is a water-soluble, crystalline powder of high sweetness potency (2000X). The sweet taste is clean, and the time–intensity profile is similar to that of aspartame.

Neohesperidin Dihydrochalcone. In the 1960s, there was a strong effort by the U.S. Department of Agriculture (USDA) to study the structure–activity relationship of citrus-derived chemicals. The goal was to reduce the bitter taste of citrus juices derived from bitter principles, such as, neohesperidin, a glycoside composed of a flavanone and a disaccharide (glucose and L-rhamnose). Upon treatment with potassium hydroxide, the flavanone ring opens up to yield a chalcone. Catalytic hydrogenation of this chalcone produces neohesperidin dihydrochalcone (NHDC), which tastes sweet.

NHDC imparts a sweetness that has a much slower onset and much greater lingering than sucrose. There is a slight aftertaste. The sweetness potency is overshadowed by inferior taste qualities.

Thaumatin. Thaumatin is a mixture of proteins extracted from the fruit of a West African plant, *Thaumatococcus daniellii* (Bennett) Benth. Thaumatin (trade name: Talin) is a very potent sweetener (sweetness potency 2000X). However, its potency is overshadowed by inferior taste qualities. Primarily owing to poor taste quality, thaumatin is not considered a practically useful sweetener. It is, however, used as a flavor enhancer, especially in products such as chewing gum.

New Sweetness

Inhibitors. Sugar is used in large quantities in fruit jams as a preservative. The strong sweetness, however, prevents fruity flavors from being noticed. For these and other foods that must use a large amount of sugar for purposes other than sweet taste, there is need for a sweet-taste inhibitor.

Lactisole, the sodium salt of racemic 2(4-methoxyphenoxy)-propionic acid, is a sweet-taste inhibitor marketed by Domino Sugar. It was affirmed as a GRAS flavor (FEMA no. 3773). At a concentration of 100 to 150 ppm, lactisole strongly reduces or eliminates the sweet taste of a 10% sugar solution.

THOMAS D. LEE
Kraft Foods

1. A. Bar, in L. O. Nabors and R. C. Gelardi, eds., *Alternative Sweeteners*, 2nd ed., Marcel Dekker, Inc., New York, 1991.
2. J. M. Janusz, in T. H. Grenby, ed., *Progress in Sweeteners*, Elsevier Science Publishing Co., Inc., New York, 1989, p. 42.
3. R. Rohse and H.-D. Belitz, in D. E. Walters, F. T. Orthoefer, and G. E. DuBois, eds., *Sweeteners: Discovery, Molecular Design and Chemoreception*, American Chemical Society, Washington, D.C., 1991.

SYNTHETIC LUBRICANTS.

See LUBRICATION AND LUBRICANTS.

SYRUPS

Corn sweeteners, maple syrup, and molasses, all commercially available syrups, are concentrated solutions of carbohydrate. These products are produced for a variety of food and nonfood applications. Corn sweeteners are prepared from hydrolyzed starch and include dextrose (D-glucose), high fructose corn syrup (HFS), regular corn syrup, and maltodextrin (see SWEETENERS), which all have in common the raw material source, general methods of preparation, and many properties and applications. Dextrose, the common or commercial name for D-glucose, is available as a syrup or as a pure crystalline solid. HFS is produced by the partial enzymatic isomerization of dextrose. Corn syrups and maltodextrins are clear, colorless, viscous liquids prepared by hydrolysis of starch to solutions of dextrose, maltose, and higher

Table 1. Physical Properties of D-Glucose

Property	α-D-Glucose	α-D-Glucose hydrate	β-D-Glucose
molecular formula	$C_6H_{12}O_6$	$C_6H_{12}O_6 \cdot H_2O$	$C_6H_{12}O_6$
mp, °C	146	83	150
solubility at 25°C, g/100 g solution	$62 \rightarrow 30.2 \rightarrow 51.2^a$	$30.2 \rightarrow 51.2^{a,b}$	$72 \rightarrow 51.2^a$
$[\alpha]_D$	$112.2 \rightarrow 52.7^a$	$112.2 \rightarrow 52.7^{a,b}$	$18.7 \rightarrow 52.7^a$
heat of solution at 25°C, J/g[c]	−59.4	−105.4	−25.9

[a] Initial value through solution equilibrium value. See text.
[b] Anhydrous basis.
[c] To convert J to cal, divide by 4.184.

molecular weight saccharides. Maple syrup, like corn syrup, is a nutritive sweetener produced as a concentrated carbohydrate (sucrose) solution. Molasses is a syrup produced as a by-product of sugar (qv) manufacture.

Dextrose

Dextrose (D-glucose) is by far the most abundant sugar in nature. It occurs either in the monosaccharide form (free state) or in a polymeric form of anhydrodextrose units. As a monosaccharide, dextrose is present in substantial quantities in honey, fruits, and berries. As a polymer, dextrose occurs in starch, cellulose (qv), and glycogen. Sucrose is a disaccharide of dextrose and fructose.

Properties. Physical properties of the three crystalline forms of dextrose are listed in Table 1. In solution, dextrose exists in both the α- and β-forms. When α-dextrose dissolves in water, its optical rotation, $[\alpha]_D$, diminishes gradually as a result of mutarotation until, after a prolonged time, an equilibrium value is reached (see Table 1).

Dextrose shows the reactions of an aldehyde, a primary alcohol, a secondary alcohol, and a polyhydric alcohol. Catalytic hydrogenation of dextrose is practiced commercially to manufacture sorbitol.

Dextrose is the fundamental intermediary metabolite in carbohydrate metabolism. Other utilizable monosaccharides are largely converted to dextrose before being further metabolized. Starch, glycogen, and the common disaccharides are hydrolyzed enzymatically in the alimentary canal. The resulting monosaccharides are absorbed into the portal vein blood, by which they are transported first to the liver and then to all other parts of the body. In the liver, monosaccharides other than dextrose, eg, galactose and fructose, are largely converted to dextrose before they reach other tissues. When dextrose is taken up by body tissues, it is phosphorylated to glucose-6-phosphate, which can enter the glycolytic pathway or be stored as glycogen.

Manufacture. Dextrose is manufactured almost exclusively from corn (maize) starch in the United States. In other countries, starch from sorghum (milo), wheat, rice, potato, tapioca (yucca, cassava), arrowroot, and sago are used to varying degrees along with corn starch.

Health Factors. Dextrose products are substances that are presumed to be GRAS by the FDA. A study of the health aspects of dextrose, fructose, and corn syrups has indicated that these sweeteners are not hazardous at levels of normal human consumption with the exception of a small contribution to the formation of dental caries.

Uses. The main use of dextrose is in food processing, where it is of value for its physical, chemical, and nutritive properties. Dextrose is also used in nonfood applications in the chemical, drug, and pharmaceutical industries. In the baking industry, dextrose is used as a fermentable sugar to provide crust color and, also, to supply strength, develop flavor, and optimize texture. In the beverage industry, it is used as a source of fermentables in low calorie beer and as a sweetener in beverage powders. In the canning industry, it supplies sweetness, body, and osmotic pressure. Dextrose is also used in the pharmaceutical industry for intravenous feeding as well as for tableting and other formulations. In fermentation, dextrose is a raw material for biochemical synthesis of organic acids, vitamins, antibiotics, enzymes, amino acids, and polysaccharides. Reaction of dextrose with sorbitol and citric acid produces a low calorie dextrose polymer, which is used as a bulking agent in reduced-calorie foods.

High Fructose Corn Syrups

High fructose corn syrups (HFS, HFCS, isosyrup, isoglucose) are concentrated carbohydrate solutions containing primarily fructose and dextrose as well as lesser quantities of higher molecular weight saccharides. A 42 wt % fructose syrup is produced by partial enzymatic isomerization of dextrose hydrolyzate. A 55 wt % fructose syrup is produced by a combination of enrichment and blending. Liquid products containing 80–95 wt % fructose are also manufactured. Pure crystalline fructose is produced in a dry form.

In nature, fructose (levulose, fruit sugar) is the main sugar in many fruits and vegetables. Fructose exists in polymeric form as inulin in plants such as Jerusalem artichokes, chicory, dahlias, and dandelions, and is liberated by treatment with acid or enzyme.

Properties. Fructose, a ketohexose monosaccharide, crystallizes as β-D-fructopyranose and has a molecular weight of 180 and a melting point of 102–104°C. The crystalline form of fructose is 10–80% sweeter than sucrose and 50–100% sweeter than dextrose, depending on temperature, pH, concentration, and the presence of other additives.

Manufacture. HFS containing 42% fructose is produced commercially by column isomerization of clarified and refined dextrose hydrolyzate using an immobilized glucose isomerase. Enriched syrup containing 90% fructose is prepared by chromatographic separation and blended with 42% HFS to obtain 55% HFS.

Glucose isomerase is produced for commercial use from a variety of bacterial organisms. HFS is produced from 93–96% dextrose hydrolyzate that has been clarified, carbon-treated, ion-exchanged, and evaporated to 40–50% dry basis.

Health Factors. Health aspects are the same as those for dextrose. HFS is presumed to be GRAS by the FDA.

Uses. High fructose syrup is used as a partial or complete replacement for sucrose or invert sugar in food applications to provide sweetness, flavor enhancement, fermentables, or humectant properties. The primary application of HFS is in soft drinks.

Corn Syrups

Corn syrups (glucose syrup, starch syrup) are concentrated solutions of partially hydrolyzed starch containing dextrose, maltose, and higher molecular weight saccharides. In the United States, corn syrups are produced from corn starch by acid and enzyme processes. Other starch sources such as wheat, rice, potato, and tapioca are used elsewhere depending on availability. Syrups are generally sold in the form of viscous liquid products and vary in physical properties, eg, viscosity, humectancy, hygroscopicity, sweetness, and fermentability.

Properties. Corn syrups are defined as those starch hydrolysis products exhibiting a dextrose equivalent (DE) of 20–99.4.

Fermentability of corn syrups by yeast is important in certain food applications, eg, baking and brewing. The fermentable sugars present in corn syrup are dextrose, maltose, and maltotriose.

Viscosity of corn syrup is a function of DE value, temperature, and solids concentration.

The hygroscopic and humectant properties of corn syrups are of great importance in many applications. Depending on the type of syrup and on the specific conditions of temperature and humidity, the products may either resist or facilitate moisture loss or moisture absorption. The ability to attract moisture or retard its loss increases with increasing DE value.

Sweetness is primarily a function of the levels of dextrose and maltose present.

Manufacture. Corn syrups are manufactured by acid, acid–enzyme, or enzyme–enzyme hydrolysis processes. The standard acid-converted syrup is typically about 42 DE.

Health Factors. Maltodextrin and corn syrup products are substances that are presumed to be GRAS by the FDA. Health and safety aspects are the same as those for dextrose.

Uses. Principal uses of corn syrups are in the confectionery, beverage, processed foods, dairy products, baked goods, and cereal industries. The specific type of syrup employed depends on the properties desired in the final product.

Maple Syrup

Maple syrup is prepared by concentrating (evaporation or reverse osmosis) sap from the maple tree to a concentrated solution containing predominantly sucrose. Its characteristic flavor and color are formed during evaporation. Maple syrup is produced from the sap of several varieties of mature maple trees, eg, the sugar maple (*Acer saccharum*) and black maple (*Acer nigrum*).

Syrup is clarified, graded as to color, flavor, and density, and finally packaged in small containers for retail sale as table syrup. Typically the product contains 88–99 wt % sucrose and 0–12 wt % invert sugar. Maple syrup is also used in candy manufacture by blending with sucrose. Other applications include addition to cookies, ice cream, baked beans, baked ham, and baked apples.

Molasses

Molasses, another type of syrup, is a by-product of the sugar industry. It is the mother liquor remaining after crystallization and removal of sucrose from the juices of sugar cane or sugar beet and is used in a variety of food and nonfood applications.

Molasses from other sources include citrus and corn sugar (hydrol) molasses. Citrus molasses is produced from citrus waste. Corn sugar molasses is the mother liquor remaining after dextrose crystallization.

Composition. Molasses composition depends on several factors, eg, locality, variety, soil, climate, and processing. Cane molasses is generally at pH 5.5–6.5 and contains 30–40 wt % sucrose and 15–20 wt % reducing sugars. Beet molasses is ca 7.5–8.6 pH, and contains ca 50–60 wt % sucrose, a trace of reducing sugars, and 0.5–2.0 wt % raffinose.

Uses. The primary use of molasses is in animal feed. Molasses, which provides a carbohydrate source, salts, protein, vitamins, and palatability, may be used directly or mixed with other feeds.

Molasses is also used as an inexpensive source of carbohydrate in various fermentations for the production lactic acid, citric acid, monosodium glutamate, lysine, and yeast. Blackstrap molasses is used for the production of rum and other distilled spirits.

Food applications utilize molasses in baking (bread, cakes, cookies) for the molasses flavor. Molasses is also used in curing of tobacco and meats, in confections such as toffees and caramels, and in baked beans and glazes.

Ronald E Hereda
Corn Products International

P. J. Mulvihill, in F. W. Schenck and R. E. Hebeda, eds., *Starch Hydrolysis Products*, VCH Publishers, Inc., New York, 1992.

R. E. Hebeda, in T. Nagodawithana and G. Reed, eds., *Enzymes in Food Processing*, 3rd ed., Academic Press, Inc., San Diego, Calif., 1993.

R. E. Hebeda, in G. Reed and T. W. Nagodawithana, eds., *Enzyme, Biomass, Food and Feed: Biotechnology*, 2nd ed., VCH Verlagsgesellschaft mbH, Weinheim, Germany, 1995.

F. W. Schenck, in *Ullmann's Encyclopedia of Industrial Chemistry*, Vol. A 12, VCH Publishers, Inc., New York, 1989.

T

TACK. See RUBBER COMPOUNDING.

TALC

Talc, a naturally occurring mineral of the general chemical composition $Mg_3Si_4O_{10}(OH)_2$, is a crystalline hydrous magnesium silicate belonging to the general mineral family of the layered silicates. Other layered silicates are kaolin, mica, and pyrophyllite.

Geology and Occurrence

Talc deposits are of four types and each has a different group of accessory minerals. The most common type is of ultramafic origin, where talc is formed by alteration of serpentinite, $Mg_3Si_2O_5(OH)_4$, to talc—carbonate rock. This type of talc deposit has magnesite, $MgCO_3$, and chlorite, $Mg_3Al_3Si_4O_{10}(OH)_8$, as accessory minerals.

Talc of metasedimentary origin is formed by hydrothermal alteration of a dolomitic host rock by a silica-containing fluid. It is usually quite pure with talc content of 90 to 98% and often very white as well. Dolomite, $CaMg(CO_3)_2$, is the most common accessory mineral. The fourth type is of metamorphic origin, where a silicaceous dolostone is first converted to tremolite or actinolite and then partially converted to talc. Tremolite, dolomite, and serpentine are common accessory minerals.

Mining and Processing

The commercial value of a talc ore is based on its color, purity, accessibility, proximity to the market, and accessory minerals. Of these the most critical is color. Most talc is mined by open-pit methods, but there are also underground mines in the United States, Canada, Italy, India, and China. All the mining is highly selective, using much smaller shovels and trucks than those for conventional base-metal mining. A typical mine can produce multiple grades based on color, purity, and accessory minerals.

Properties

The crystal structure of talc, consists of repeating layers of a sandwich of brucite, $Mg(OH)_2$, between sheets of silica, SiO_2. When fractured, talc has a natural balance of hydrophilic and hydrophobic surfaces, giving it surfactant properties.

The mineral talc is extremely soft (Mohs' hardness = 1), has good slip, a density of 2.7 to 2.8 g/cm^3, and a refractive index of 1.58. It is relatively inert and nonreactive with conventional acids and bases. It is soluble in hydrofluoric acid. Although it has a pH in water of 9.0 to 9.5, talc has Lewis acid sites on its surface and at elevated temperatures is a mild catalyst for oxidation, depolymerization, and cross-linking of polymers.

Pure talc is thermally stable up to 930°C. Talc is an insulator for both heat and electricity.

Uses

Talc is used in a wide variety of applications, including paper, ceramics, roofing, paint, plastics, rubber, cosmetics, pharmaceuticals, adhesives, sealants, and animal feedstuffs. In all of these applications it is a functional ingredient with specific beneficial properties. Talc is rarely used as a filler, because it is much more expensive than alternative minerals such as limestone and clay.

Health and Safety

Talc is considered a nuisance dust and subject to regulation in the workplace in the United States by both the Occupational Health and Safety Administration and the Mine Safety and Health Administration. Eight-hour exposure limits for talc dust are two milligrams of talc per cubic meter. Used for decades in a wide variety of cosmetic and other applications, talc has proven to be among the safest of all consumer products.

EDWARD F. MCCARTHY
Luzenac America

Industrial Minerals and Rocks, Society of Mining Engineers, Littleton, Colo., 1994, pp. 1049–1069.

Mineral Industry Surveys: Talc and Pyrophyllite, U.S. Geological Survey, Washington, D.C., July 1996.

E. F. McCarthy, *Talc/Polyolefin Composites for Exceptional Cost Performance in Automotive Applications*, Schotland Business Research, Princeton, N.J., 1992.

TALL OIL

Tall oil is a by-product of kraft pulping of pine wood. Crude tall oil (CTO), formed by acidifying black liquor soap skimmings with sulfuric acid, is a dark oily liquid with 26–42% resin acids or rosin, 36–48% fatty acids, and 10–38% neutrals. CTO is an excellent source of oleic/linoleic fatty acids and rosin. Fractional distillation is required, not only to separate these desirable products, but also to remove the neutrals. The bulk of these neutrals, largely esters of fatty acids, sterols, resin and wax alcohols, and hydrocarbons, boil at either lower or higher temperatures than the boiling range of the fatty and resin acids.

Variations in CTO compositions result primarily from the species of wood pulped and the location and climate where the trees are grown. Pulping process variations further affect CTO composition. The best CTO is produced from pine wood. However, many U.S. mills mix hardwood with pine to reduce fiber costs, or mix hardwood black liquor with pine black liquor. This lowers the rosin content.

Tall Oil Processing

The first step in CTO fractionation is depitching. A relatively small distillation column is used as a *pitch stripper*. The vapor from the pitch stripper is fed directly into the *rosin column*, where rosin and fatty acids are separated. Rosin is taken from the bottoms of the column and fatty acids as a sidestream near the top. Palmitic acid and light neutrals are removed in the rosin column as heads.

The crude tall oil fatty acids obtained from the rosin column usually contain about 5% rosin because the boiling points of the heavier fatty acids and the lighter resin acids overlap. By adding the intermediate fraction to the fatty acid, rosin does not have to be redistilled.

The *rosin column* split is controlled by the fatty acid content specified for rosin. This is usually set at 2% fatty acids.

Higher grade fatty acids with less than 2% rosin are obtained by further distillation.

Environmental Considerations. All CTO processing plants have environmental safeguards. CTO contains small amounts of volatile odor compounds that are exhausted by the vacuum systems of the distillation columns. They are captured, along with the more volatile fatty acids, in the cooling water of the barometric condensers. This oily water is recycled from the receivers or hot wells to the barometric condensers through heat exchangers, where it is cooled *indirectly* with water from the cooling towers. The oily water purge stream is skimmed and passed to the wastewater treatment system.

All vapors, including hotwell odors, are captured in a header system linked with the incineration air of a steam boiler or hot oil vaporizer. Drain seals avoid escape of odors from the sewer lines.

Uses of Tall Oil Products

Tall Oil Rosin. Tall oil rosin (TOR) was introduced in the 1950s as a low cost substitute for gum and wood rosin, particularly for paper size.

TOR is used in the United States for oligomeric resins in printing inks and in adhesives. Another use of rosin is as an emulsifier in the manufacture of synthetic rubber, eg, styrene–butadiene rubber (SBR), by emulsion polymerization.

Tall Oil Fatty Acids. TOFA can replace fatty acid mixtures from vegetable oil sources in industrial applications, such as the manufacture of drying alkyd resins (qv). At least one-third of TOFA is turned into dimer acids. They, in turn, are converted to noncrystalline polyamides. These polyamides find application in hot-melt adhesives, printing ink resins, and epoxy curing agents.

Distillation By-Products. Of the CTO distillation by-products, ie, pitch, heads, and distilled tall oil (DTO), only the last, a unique mixture of rosin and fatty acids, has significant commercial value. Pitch and heads are used as fuel. Tall oil heads are outstanding solvent properties, but hard-to-remove bad odor.

DERK T. A. HUIBERS
Union Camp Corporation

E. Fritz and R. W. Johnson, in R. W. Johnson and E. Fritz, eds., *Fatty Acids in Industry*, Marcel Dekker, Inc., New York, 1989.

J. Drew and M. Probst, *Tall Oil*, Pulp Chemicals Association, New York, 1981.

D. F. Zinkel and J. Russel *Naval Stores Production, Chemistry, Utilization* Pulp Chemicals Association, New York.

TANKS AND PRESSURE VESSELS

Tanks are used in innumerable ways in the chemical process industry, not only to store every conceivable liquid, vapor, or solid, but also in a number of processing applications. For example, as well as reactors, tanks have served as the vessels for various unit operations such as settling, mixing, crystallization, phase separation, and heat exchange. The main focus here is on the use of tanks as liquid storage vessels.

The most fundamental classification of storage tanks is based on whether they are above or below ground. For flat bottom tanks which are the most common, the underside of aboveground tanks is usually placed directly on an earthen or a concrete foundation. Sometimes these tanks are placed on grillage, structural members, or heavy screen so that the bottom of the tank can be inspected on the underside and leaks can be easily detected.

The other important type of tanks is the underground tank. These are usually limited to between 500–20,000 gal (2–75 m^3), although most are under 12,000 gal (45 m^3). Used to store fuels as well as a variety of chemicals, these tanks require special consideration for the earth pressure and settlement loads to which they are subjected. Because buoyancy must also be taken into account, underground tanks are often anchored into the ground so that they do not pop out when surrounded by groundwater. In addition, these tanks are subject to severe conditions of corrosion. Placement of special backfill, cathodic protection, and coatings and liners are some of the corrosion-prevention measures necessary to ensure good installations.

Basic Concepts

Most tanks store liquid rather than gases or solids. Characteristics and properties such as corrosiveness, internal pressures of multicomponent solutions, tendency to scale or sublime, and formation of deposits and sludges are vital for the tank designer and the operator of the tank.

Density and Specific Gravity. All things being equal, liquids having greater densities require thicker tank shells.

Specific gravity (sp gr) is a ratio of the density of a liquid divided by the density of liquid water at 16°C (60°F).

In the petroleum industry, a common indicator of specific gravities, known as the API gravity or °API, is usually applied to crude oils. The formula for the API gravity is °API = 141.5/sp gr − 131.5. Thus, the higher the specific gravity, the lower the API gravity. Water, having a specific gravity of 1, has an API gravity of 10.

Another common indicator of specific gravities used in the chemical industry is degrees Baumé (°Bé). For liquids heavier than water, °Bé = 140/sp gr − 145. For liquids lighter than water, °Bé = 140/sp gr − 130.

Temperature. Temperature may be measured on an absolute or relative scale. The two most common relative scales are the Celsius and the Fahrenheit scales.

The absolute temperature scale that corresponds to the Celsius scale is the Kelvin scale; for the Fahrenheit scale, the absolute scale is called the Rankine scale.

Tanks are used to store liquids over a wide temperature range. Cryogenic liquids, such as liquefied hydrocarbon gases, can be as low as −201°C (−330°F). Some hot liquids, such as asphalt (qv) tanks, can have a normal storage temperature as high as 260–316°C (500–600°F). However, most storage temperatures are either at or a little above or below ambient temperatures.

At very high and very low temperatures, material selection becomes an important design issue. At low temperatures, the material must have sufficient toughness to preclude transition of the tank material to a brittle state. At high temperatures, corrosion is accelerated, and thermal expansion and thermal stresses of the material occur.

Vapor Pressure and Boiling Point. Vapor pressure is important in liquid storage tank considerations. It affects the design and selection of the tank evaporation losses and is crucial for characterizing fire hazards of flammable and combustible liquids. The boiling point is also important because liquids should usually be stored at temperatures well below the boiling point. Flammable and combustible liquids are expressly prohibited by the fire codes for storage at temperatures above the boiling points.

The vapor pressure of a pure liquid is the pressure of the vapor space of a closed container. It is a specific function of temperature and always increases with increasing temperature. If the temperature of a liquid in an open container is increased until its vapor pressure reaches atmospheric pressure, boiling occurs. The temperature of a pure liquid does not increase beyond its boiling point as heat is supplied. Rather, all the liquid evaporates at the boiling point. Because standard atmospheric pressure at sea level is 14.7 psia (101.3 kPa), this also presents the vapor pressure of a boiling liquid. Atmospheric pressure varies, however, with altitude. Changes in barometric pressure slightly alter the boiling point of liquids in tanks as well.

In the fire codes, the atmospheric boiling point is an important physical property used to classify the degree of hazardousness of a liquid.

In the United States vapor pressure has also become a means of regulating storage tank design by the Environmental Protection Agency (EPA).

Flash Point. As a liquid is heated, its vapor pressure and, consequently, its evaporation rate increase. Although a liquid does not really burn, its vapor mixed with atmospheric oxygen does. The minimum temperature at which there is sufficient vapor generated to allow ignition of the air–vapor mixture near the surface of the liquid is called the flash point.

For flammable and combustible liquids, flash point is the primary basis for classifying the degree of fire hazardousness.

Pressure. Pressure, defined as force per unit area, can be expressed as an absolute or relative value. Although atmospheric pressure constantly fluctuates, a standard value of 101.3 kPa (14.7 psia) has been assigned as the accepted value at sea level. The "a" in the psia stands for absolute, ie, the pressure is 14.7 psi (101.3 kPa) above zero pressure or a vacuum. Most ordinary pressure-measuring instruments do not measure true pressure, but rather a pressure relative to the barometric or atmospheric pressure known as gauge pressure. The atmospheric pressure is defined to be 1 psig, in which the "g" indicates that it is relative to atmospheric pressure. Vacuum is the pressure below atmospheric pressure and is, therefore, a relative pressure measurement as well. The re-

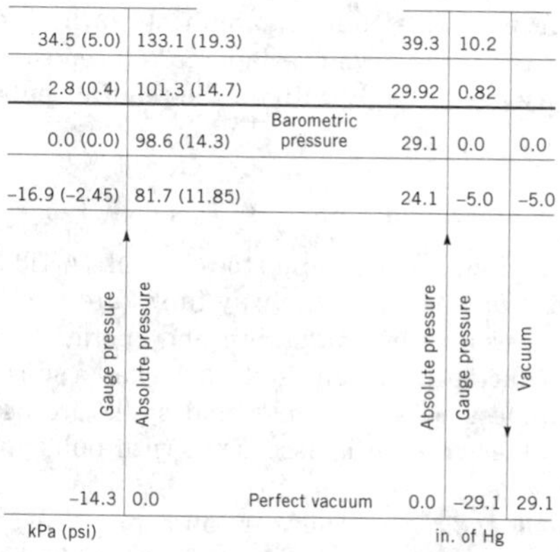

Figure 1. Standard system of pressure measurement, where the bold line represents standard atmospheric pressure at sea level.

lationship between absolute and relative pressure is shown in Figure 1 (see PRESSURE MEASUREMENT; VACUUM TECHNOLOGY).

For tank work, inches water column (in. wc) or ounces per square inch (osi) are commonly used to express the value of pressure or vacuum in the vapor space of a tank. These pressures are usually very low relative to atmospheric pressure. The common measures of pressure are compared as follows:

psi	in. wc	osi	Pa
1	27.68068	16	6894.7
0.03612628	1	0.5780205	249
0.0625	1.730042	1	431

The difference in pressure between the inside of a tank or its vapor space and local barometric or atmospheric pressure is called internal pressure. When the internal pressure is negative it is simply called a vacuum. The pressure is measured at the top of the liquid in the tank because the liquid itself exerts hydrostatic pressure, thus increasing to a maximum value at the base of the tank.

When the vapor space of a tank is open to the atmosphere or is freely vented, then the internal pressure is always zero or atmospheric. No pressure buildup can occur, except in the case of dynamic conditions that occur in explosions or deflagrations. Most tanks, however, are not open to the atmosphere, but are provided with some form of venting device usually called a pressure-vacuum (PV) valve. Such a valve is designed to open when the internal pressure builds up to some level in excess of atmospheric pressure, and keep the internal pressure from rising high enough to damage the tank. Internal pressure may be caused by several potential sources. One source is the vapor pressure of the liquid itself.

The most fundamental limitation on pressure is at 15 psig (101.4 kPa). Containers built to pressures exceeding this value are usually called pressure vessels. For all practical purposes, tanks are defined to have internal pressures below this value.

Other properties such as viscosities, solidification temperature, pour point, and cubical rate of thermal expansion are all important for the tank designer or operator to consider and understand.

Tank Classification

There are many ways to classify a tank, although there is no universal method. A classification commonly employed is based on the tank's internal pressure.

By far, the most common type of tank is the atmospheric tank. These tanks are usually operated at internal pressures slightly above atmospheric pressure. Fire codes define an atmospheric tank as operating from atmospheric up to 0.5 psi (3448 Pa) above atmospheric pressure.

Low pressure in the context of tanks means tanks designed for a higher pressure than atmospheric tanks. In other words, these are relatively high pressure tanks, designed to operate from atmospheric pressure up to 15 psig (101.4 kPa).

High pressure tanks are vessels operating above 15 psig (101.3 kPa). These are really pressure vessels, and the term high pressure tank is basically never used. Pressure vessels are a specialized form of container treated separately from tanks by all codes, standards, and regulations.

Tank Components

To a large extent, the vapor pressure of the substance stored determines the shape and, consequently, the type of tank used. The roof shape of a tank may be used to classify the type of tank. Also important is the tank bottom.

Fixed-Roof Tanks. The effect of internal pressure on plate structures, including tanks and pressure vessels, is important to tank design. If a flat plate is subjected to pressure on one side, it must be made quite thick to resist bending or deformation. A shallow cone-roof deck on a tank approximates a flat surface and is typically built of 3/16-in. (4.76-mm) thick steel. The larger the tank, the more severe the effect of pressure on the structure. As pressure increases, the practicality of fabrication practice and costs force the tank builder to use shapes more suitable for internal pressure. The cylinder is an economic and easily fabricated shape for pressure containment. Indeed, almost all large tanks are cylindrical. The problem, however, is that the ends must be closed. The relatively flat roofs and bottoms or closures of tanks do not lend themselves to much internal pressure. As internal pressure increases, tank builders use roof domes or spheres. The spherical tank is the most economic shape for internal pressure storage in terms of required thickness.

Floating-Roof Tanks. All floating-roof tanks have vertical, cylindrical shells just like a fixed-cone-roof tank. These common tanks have a cover that floats on the surface of the liquid. The floating cover or roof is a disk structure that has sufficient buoyancy to ensure that the roof floats during all expected conditions, even if leaks in the roof develop.

If the tank is open on top, it is called an external floating-roof (EFR) tank. If the floating roof is covered by a fixed roof on top of the tank, it is called an internal floating-roof (IFR) tank. The function of the cover is to reduce evaporation losses and air pollution by reducing the surface area of liquid that is exposed to the atmosphere.

Tank Bottoms. The shape of cylindrical tank closures, both top and bottom, is a strong function of the internal pressure. Because of the varying conditions to which a tank bottom may be subjected, several types of tank bottoms have evolved. These may be broadly classified as flat bottom, conical, or domed or spherical. Flat-bottom tanks only appear flat. These usually have designed slope and shape and are subclassified according to the following: flat, cone up, cone down, or single slope.

Small Tanks

Numerous types of small tanks have been developed as a result of increasingly stringent regulations regarding leaks, spills, and containment.

Single-wall tanks are usually cylindrical and may have either vertical or horizontal orientation.

Double-wall tanks are usually cylindrical tanks and may have either vertical or horizontal orientation.

Small tanks can have a secondary-containment dam built integrally into the tank. These tanks may be either vertically or horizontally oriented in both cylindrical and rectangular shapes.

Vaulted tanks are installed inside a concrete vault, which itself is a liquid-tight compartment, these tanks are classed as fire-resistant aboveground storage tanks.

Engineering Considerations

Required Component Thicknesses. The tank design codes consider all strength calculations to be independent of temperature from ambient up to some upper limit. At high design temperatures, the various codes provide derating factors for steel, aluminum, and stainless steel.

When the design temperatures are significantly below ambient temperature, the primary threat to tank integrity is failure of the material by brittle fracture.

In the fabrication of large steel structures, the minimum thickness of the tank bottom is often governed by the minimum necessary for weldability and fabricability and not necessarily by strength requirements.

Another example of where thickness is set by minimums for fabricability but not for strength is in small-diameter tanks.

In the large-diameter vertical cylindrical tanks, because hoop stress is proportional to diameter, the thickness is set by the hydrostatic hoop stresses. Although the hydrostatic forces increase proportionally with the depth of liquid in the tank, the thickness must be based on the hydrostatic pressure at the point of greatest depth in the tank.

Materials of Construction. Tanks are constructed from a number of materials based on cost and availability of the material, ease of fabrication, resistance to corrosion, and compatibility with stored fluid.

Carbon steel, or mild steel, is by far the most common material for tank construction. Austenitic 300 series stainless steel is another important material used for storage of corrosive chemicals and liquids. Fiber glass-reinforced plastic (FRP) tanks are noted for resistance to chemicals.

Tank Selection Criteria. The selection of tanks is a complex process. Once the specific liquid(s) to be stored is established, the liquid's physical properties determine the range of possible tank types. Material selection, corrosion prevention systems, as well as environmental requirements and considerations may also influence the selection.

Special Engineering Considerations. Because tanks are used in so many different ways, some specialized applications have been developed that have become fairly commonplace.

Cryogenic Tanks. These are low temperature tanks used for liquefied hydrocarbon gases (LHG); liquefied natural gas (LNG); various liquefied gases such as air, nitrogen, or oxygen, and ammonia; and other refrigerated liquids (see CRYOGENICS; LIQUEFIED PETROLEUM GAS). As a general rule, larger quantities of stored liquids that have high vapor pressures favor low temperature or cryogenic storage.

Heated Tanks. Many compounds either freeze, solidify, or thicken to the point where they cannot be transferred through piping and equipment unless maintained at some minimum temperature. Examples are heavy oils, asphalts, sulfur, highly concentrated salt solutions, caustic soda solutions, or even molasses and foodstuffs. The storage tanks for these fluids must be heated and maintained to some minimum temperature.

Design Considerations. Most of the design codes and standards for tanks provide checklists and concepts to prevent the designer from making gross mistakes. In particular, tank standards are issued by the American Petroleum Institute (API), which remove most of the risks of the catastrophic results that can occur without considering material selection, brittle fracture, insufficient welding or joining methods, fabrications methods, etc. These standards are recognized worldwide. As a result, they have been used in industries such as chemicals, pulp and paper, food, and a host of others.

Regulations

Regulations and laws are mandatory requirements with which a tank owner or operator must comply. Most regulatory requirements are channeled through an agency whose general responsibility is the safety, well-being, and protection of the public or the environment. The authority having jurisdiction may be a federal, state, local, or regional agency, an individual such as a fire chief or marshal, a labor or health department, or a building official or inspector.

Most tank facilities are subject to multiple authorities. When this is the case and the rules have overlapping or even conflicting provisions, the facility must comply with all the requirements of the multiple authorities.

Spills, Leaks, and Prevention

Leaks and spills from aboveground storage tank (AST) facilities have had more impact on change in the way tanks are regulated and will be regulated, as well as on the design and operation of tanks, than any other single factor. Leaks and spills have had a substantial impact on public awareness as well. Leaks and spills are associated with groundwater, which in turn is associated with public water supplies and irrigation.

Causes of Spills and Leaks. There are numerous causes of tank leaks and spills. They include corrosion, over fill of tanks, tank breakage, insufficient maintenance, vandalism, piping, and design deficiencies. The best prevention is to use established codes, such as those provided by the API.

Spill Prevention and Detection. The fundamental rule of leak and spill prevention is to reduce the possibility for contamination by directing resources as close to the source as possible.

Leak and spill prevention comprises a system of management or a program embodying many facets which, when all working together, virtually eliminate the possibility of leaks and spills.

PHILIP MYERS
Chevron Products Company

API Standard 2610, "API Design, Construction, Operation, Maintenance and Inspection of Terminal and Tank Facilities," American Petroleum Institute.

R. P. Benedetti, *Flammable and Combustible Liquids Code Handbook*, Quincy, Mass., 5th ed., 1994.

TANTALUM AND TANTALUM COMPOUNDS

Tantalum, atomic number 73, is the heaviest element in Group 5 (VA) of the Periodic Table. This tough, ductile, silvery gray metal has an atomic weight of 180.948 amu.

Physical and Chemical Properties

The physical properties of tantalum are presented in Table 1.

Tantalum metal is easily oxidized to the +5 valence state although this reactivity is usually partly masked by the presence of a stable, adherent, passivating oxide layer on the surface of the metal.

Occurrence

In nature, tantalum occurs in several complex oxidic minerals, often in solid solution with a variety of other elements. The main source of tantalum is an isomorphous series of minerals containing oxides of tantalum, niobium, iron, and manganese, the tantalite–columbite series, occurs as an accessory mineral distributed in granitic rocks or pegmatites. A considerable amount of tantalum slags from the tin smelting processes in Thailand, Malaysia, and Brazil is also an important source.

Processing

Ore Dressing. The mining of pegmatite deposits, either open-pit or underground is done by conventional techniques like blasting and crushing (see MINERALS RECOVERY AND PROCESSING). The materials are then dressed mainly by gravity concentration.

Upgrading of Tin Slags. The 0.2–17% Ta_2O_5-containing tin slags are upgraded in a sequence of three pyrometallurgical processing steps. In the first step the slags are intensively mixed with additives, and the mixture is continuously fed into a three-stage electric arc furnace. The

Table 1. Physical Properties of Tantalum

Property	Value
at no.	73
at wt	180.9479
at vol, cm^3/mol	10.9
atomic radius, pm	147
crystal structure	bcc
lattice constant, nm	0.33
coordination number	8
density, at 20°C, g/cm^3	16.62
mp, °C	2996
bp, °C	5427 ± 100
enthalpy of fusion, J/mol[a]	31,400
enthalpy of vaporization, J/mol[a]	7.53×10^5
heat capacity, J/(K·mol)[a], t from 25 to 2000°C	$24.2 + 3.0\,t + 0.2 \times 10^{-6}\,t^2$
vapor pressure, log P_{kPa}[b], T from 290 K–mp	$-40,800/T + 9.41$
thermal conductivity, J/(cm·°C)[a]	
20°C	0.540
568°C	0.680
828°C	0.720
coefficient of linear thermal expansion, $°C^{-1}$	
at 20°C	6.5×10^{-6}
for t = 20–500°C	$L_0(1 + (6.59\,t + 0.00008\,t^2) \times 10^6$

[a] To convert J to cal, divide by 4.184.
[b] $P_{mmHg} = -40,800/T + 10.29$ log.

second step involves oxidation of the comminuted and crushed alloy. In the last processing step, the oxidized material is reduced in an electric arc furnace. Under strictly controlled conditions.

Separation of Tantalum and Niobium. Solvent extraction is used to separate tantalum from niobium.

Production

Tantalum Compounds. Potassium heptafluorotantalate, K_2TaF_7, the most important tantalum compound produced at plant scale, is used in large quantities for tantalum metal production. The fluorotantalate is prepared by adding potassium salts such as KCl and KF to the hot aqueous tantalum solution produced by the solvent extraction process.

Tantalum pentoxide, Ta_2O_5, is prepared by calcining tantalic acid or hydrated tantalum oxide, $Ta_2O_5 \cdot nH_2O$, at temperatures between 800 and 1100°C. This oxide hydrate is produced by adding gaseous or aqueous ammonia to aqueous tantalum solution.

Tantalum. Numerous methods have been developed to extract tantalum metal from compounds. The only processes that ever achieved commercial significance are the electrochemical reduction of tantalum pentoxide in molten K_2TaF_7/KF/KCl mixtures and the reduction of K_2TaF_7 with sodium.

Post-Reduction Processing. The primary tantalum powder produced by the sodium reduction process is treated to convert the metal to a form suitable for use as capacitor-grade powder and feedstock for wire and sheet.

Uses of Tantalum Metal

Tantalum is used in chemical industry equipment for reaction vessels and heat exchangers in corrosive environments. It is usually the metal of choice for heating elements and shields in high temperature vacuum sintering furnaces. Much of the tantalum produced in the world goes to the manufacturing of solid tantalum capacitors for use in the most demanding electronic applications.

Solid tantalum capacitors are extremely reliable and, therefore, are often the capacitor of choice in critical applications like spacecraft electronics, pacemakers, and safety equipment.

Tantalum wire is used primarily as the anode lead wire in solid tantalum capacitors.

The starting point for manufacturing tantalum fabricated parts is mainly high grade tantalum scrap that is converted to ingots by a combination of electron beam and arc melting.

The biocompatibility of a metal is related to its corrosion resistance and toxicity of the metal ion. The very low corrosion rate and inertness of tantalum make it attractive for many biomedical applications, such as sutures, bone screws, and plates. Other uses include cartilage wire, nets to hold bone grafts in place, braided suture wire for skin closures and tendon repair, clips for ligature of vessels and bile ducts, mesh for abdominal wall reconstruction after hernias, plates for cranioplasty, and dental implants (see DENTAL MATERIALS).

Corrosion of Tantalum

Technically, the excellent corrosion resistance of tantalum reflects the chemical properties of the thermal oxide always present on the surface of the metal. This very adherent oxide layer makes tantalum one of the most corrosion-resistant metals to many chemicals at temperatures below 150°C. Tantalum is not attacked by most mineral acids, and it is inert to most organic compounds.

Tantalum is not resistant to substances that can react with the protective oxide layer. The most aggressive chemicals are hydrofluoric acid and acidic solutions containing fluoride.

The excellent corrosion resistance means that tantalum is often the metal of choice for processes carried out in oxidizing environments or when freedom from reactor contamination of the product or side reactions are necessary, as in food and pharmaceutical processing.

Health and Safety

Tantalum powder can be ignited by the localized heating associated with a static electrical discharge or contact with a hot surface.

The only effective way to control a tantalum fire is to smother it with a nonoxidizing chemical like sodium chloride or argon. (Under no circumstances should burning tantalum be allowed to come in contact with water.)

Exposure to tantalum metal dust may cause eye injury and mucous-membrane irritation.

Anodic Oxide Films on Tantalum

Anodic Oxidation. The ability of tantalum to support a stable, insulating anodic oxide film accounts for the majority of tantalum powder usage (see THIN FILMS). The film is produced or formed by making the metal, usually as a sintered porous pellet, the anode in an electrochemical cell.

Tantalum Compounds

Potassium Heptafluorotantalate. Potassium heptafluorotantalate, K_2TaF_7, crystallizes in colorless, rhombic needles. It hydrolyzes in boiling water containing no excess of hydrofluoric acid.

Tantalum Oxides. Tantalum pentoxide, Ta_2O_5, is a white powder. It reacts slowly with hot hydrofluoric acid but is insoluble in water.

Tantalum(II) oxide, TaO, is the only other oxide the existence of which has been confirmed.

Tantalum Halides. Tantalum pentafluoride, TaF_5, is used in petrochemistry as an isomerization and alkalation catalyst. In addition, the fluoride can be utilized as a fluorination catalyst for the production of fluorinated hydrocarbons. The pentafluoride is produced by the direct fluorination of tantalum metal or by reacting anhydrous hydrogen fluoride with the corresponding pentoxide or oxychloride.

Tantalum pentachloride, $TaCl_5$, forms strongly hydroscopic, needle-shaped white crystals. It is produced by the chlorination of tantalum scrap or ferrotantalum with $NaFeCl_4$ or by the reductive chlorination of natural ores or synthetic raw materials.

$TaBr_5$ and TaI_5 are well known but do not find industrial application.

Tantalum Carbide. Tantalum monocarbide, TaC, is a gold-colored powder produced industrially by direct reaction of carbon with either tantalum scrap or tantalum pentoxide at temperatures up to 1900°C. It is added in small amounts in the form of TaC or mixed carbides like TaNbC and WTiTaC to tungsten carbide–cobalt-based cutting tools in order to reduce grain growth.

Tantalum Nitrides. Tantalum nitride, TaN, is produced by direct synthesis of the elements at 1100°C. Ta_3N_5 is used as a red pigment in plastics and paints.

Tantalic Acid and Tantalates. Tantalic acid, $Ta_2O_5 \cdot nH_2O$, is a white insoluble precipitate formed by hydrolysis of alkali hydroxide or alkali carbonate fusions containing tantalum, or by adding ammonia to an acidic solution containing tantalum ions. Tantalic acid is characterized by a high surface acidity, affording it potential use as a catalyst.

Lithium tantalate, $LiTaO_3$, is the most important tantalate. The crystal structure is related to that of perovskite. Applications include electronic devices such as surface acoustic wave (SAW) filters, second harmonic generators (SHG), and wave guides. Other tantalates like $Ba_3MgTa_2O_9$ and $Ba_3ZnTa_2O_9$ also have the perovskite structure and are used in high frequency resonators in satellite communication systems.

Hazards of Tantalum Compounds. The toxicity of tantalum compounds depends on their solubility. Tantalum pentoxide is poorly absorbed and nontoxic perorally. The pentachloride, on the other hand, shows an LD_{50} of 985 mg/Kg administered perorally.

TERRANCE B. TRIPP
J. ECKERT
H. C. Starck Inc.

D. Brown, *Comprehension Inorganic Chemistry*, Vol. 5, Pergamon Press, Oxford, U.K., 1973.

F. Fairbrother, *The Chemistry of Niobium and Tantalum*, Elsevier Publishing Company, New York, 1967.

G. L. Miller, *Tantalum and Niobium*, Academic Press, Inc., New York, 1959.

B. Elvers, S. Hawkins, and W. Russey, eds., *Ullmann's Encyclopedia of Industrial Chemistry*, Vol. A26, VCH Verlagsgesellschaft GmbH, Weinheim, Germany, 1995.

TAR AND PITCH

Most organic substances, other than those of simple structure and low boiling point, when pyrolyzed, ie, heated in the absence of air, yield dark-colored, generally viscous liquids termed tar or pitch. The differentiation between these terms is not precise. When the by-product is a liquid of fairly low viscosity at ordinary temperature, it is regarded as a tar; if of very viscous, semisolid, or solid consistency, it is designated as a pitch.

Large amounts of tar or pitch by-products are produced by industrial processes. The distillation of crude petroleum yields a pitch-like residue termed bitumen or asphalt (see ASPHALT).

Wood Tar

The pyrolysis or carbonization of hardwoods, eg, beech, birch, or ash, in the manufacture of charcoal yields, in addition to gaseous and lighter liquid products, a by-product tar in ca 10 wt % yield. Dry distillation of softwoods, eg, pine species, for the production of the so-called DD (destructively distilled) turpentine yields pine tar as a by-product in about the same amount.

Composition, Processing, and Uses. There are no statistics available for the amount of wood tar processed, but almost all of it is burned. The commercial by-products from wood carbonization are limited to methanol, denatured methanol, methyl acetate, and acetic acid.

Small amounts of the sedimentation tar, ie, the separated organic layer from the condensed wood-carbonization vapors, are distilled, first at atmospheric pressure to give wood spirit, crude acetic acid, and light wood oils. Further distillation under reduced pressure yields wood creosote.

Chemically, wood tar is a complex mixture that contains at least 200 individual compounds, among which the following have been isolated: 2-methoxyphenol, 2-methoxy-4-ethylphenol, 5-methyl-2-methoxyphenol, 2,6-xylenol, butyric acid, crotonic acid, 1-hydroxy-2-propanone, butyrolactone, 2-methyl-3-hydroxy-4*H*-pyran-4-one, 2-methyl-2-propenal, methyl ethyl ketone, methyl isopropyl ketone, methyl furyl ketone, and 2-hydroxy-3-methyl-2-cyclopenten-1-one.

Coal Tar

By far the largest source of tar and pitch is the pyrolysis or carbonization of coal (qv). Generally, the terms tar and pitch are synonymous with coal tar and the residue obtained by its distillation (see COAL-CONVERSION PROCESSES, CARBONIZATION).

Until the end of World War II, coal tar was the main source of aromatic chemicals. However, the enormously increased demands by the rapidly expanding plastics and synthetic-fiber industries have greatly outstripped the potential supply from coal carbonization. Over 90% of the world production of aromatic chemicals at present is derived from the petrochemical industry, whereas coal tar is chiefly a source of anticorrosion coatings, wood preservatives, feedstocks for carbon-black manufacture, and binders for road surfacings and electrodes.

Coal-tar consists essentially of two parts. The first, which at atmospheric pressure distills up to about 400°C, is primarily a complex mixture of mono- and polycyclic aromatic hydrocarbons, a proportion of which are substituted with alkyl, hydroxyl as well as amine and/or hydro sulfide groups, and to a lesser extent their sulfur-, nitrogen-, and oxygen-containing analogues. The second part is the residue from the distillation, amounting to at least 50% of the coal-tar products by high temperature carbonization and consisting of a continuation of the sequence of polynuclear aromatic, aromatic, and heterocyclic compounds.

Manufacture and Processing. *Primary Distillation.* As produced, crude coal tar is of value only as a fuel. Although formerly large amounts were burned, the practice has largely been abandoned. Today, 99% of the tar produced in the United Kingdom and Germany and 75% of U.S. production is distilled. Most of the crude tar regarded as being burned in the United States is first topped in simple continuous stills to recover a chemical oil, ie, a fraction distilling to 235°C that contains the bulk of naphthalene and phenols.

Continuous stills that have daily capacities of 100–700 t are used in most of the world.

Secondary Processing of Tar Distillate Oils. The only processing that light oils might receive at the refinery is a fractional distillation into crude benzene (formerly called benzol or benzole) distilling up to 150°C, a naphtha fraction distilling from 150 to 190°C, and a creosote residue (see BTX PROCESSING; BENZENE). Crude benzene from coke-oven tar is normally refined together with crude benzene separated from coke-oven gas. The naphtha is washed with alkali and acid to remove tar acids and bases, then treated with a small amount of sulfuric acid to remove sulfur compounds and olefins, and finally redistilled to give refined solvent naphtha.

In the refining of the combined crude benzene, a defronting steam-stripping operation removes the lower boiling components. The defronted benzene may be fractionated in batch or continuous stills to yield a mixture of crude benzene, toluene, and mixed xylenes, and a naphtha residue. Each fraction is purified to meet grade specifications.

Coumarone–indene resins should be called polyindene resins (see HYDROCARBON RESINS). They are derived from a close-cut fraction of a coke-oven naphtha free of tar acids and bases. This feedstock, distilling between 178 and 190°C and containing a minimum of 30% indene, is warmed to 35°C and polymerized by adding 0.7–0.8% of the phenol or acetic acid complex of boron trifluoride as catalyst.

Formerly, pyridine bases were recovered from coal-tar light oils, but in more recent years synthetic pyridine and methylpyridine have mostly replaced the coal-tar products.

Carbolic Oils and Low Temperature Tar Middle Oil, Tar Acids are in the fractions of some coke-oven tars, distilling in the range of 180–240°C, and the middle oil fraction (180–310°C) from low temperature tars are treated for the recovery of tar acids.

In the recovery and refining of tar acids, water and pitch are removed from the crude tar acids in a continuous-vacuum still heated by superheated steam or circulating hot oil. The aqueous phenol overhead distillate is recycled, the stream of once-run tar acids is refined, and the phenolic pitch bottoms are burned.

The once-run tar acids are fractionated in three continuous-vacuum stills heated by superheated steam or circulating hot oil. The overhead product from the first column is 90–95% phenol; from the second, 90% *o*-cresol; and from the third, a 40:60 *m*-cresol-*p*-cresol mixture. Further fractionation gives the pure products.

Naphthalene is the principal component of coke-oven tars and the only component that can be concentrated to a reasonably high content on primary distillation. Naphthalene oils from coke-oven tars distilled in a modern pipe still generally contain 60–65% of naphthalene. They are further upgraded by a number of methods.

The modern processes adopted in the United Kingdom and some European plants are based on crystallization of the primary naphthalene oil, which is diluted with lower crystallizing material to give a feedstock crystallizing point at 55°C. This material is cooled in closed, stirred tanks to 30–35°C; and the resultant slurry of naphthalene crystals and mother liquor is centrifuged, washed, and spun-dried. These operations are automatically timed and controlled.

No tar chemicals are extracted commercially from tar oils distilling in the range of 250–300°C. Although the wash-oil fraction of coke-oven tars, distilling mainly in the range of 250–280°C, is employed at coking installations to scrub benzene from coal gas, most oils in this boiling point range are used in creosote blends.

In Europe but not in the United States, crude anthracene is isolated from coke-oven anthracene oils.

Physical Properties. The physical properties of crude tars vary over a wide range. Investigation has been mainly concerned with establishing correlations between the more readily determined chemical and physical properties of the distillate oils and residual pitch, and other properties. Based on the correlations, other properties can be predicted with an accuracy sufficient for such purposes as plant design (Table 1).

Viscosity of Coal-Tar Pitch and Change with Temperature. Because pitch is mainly used as a hot-applied binder or adhesive, the viscosity and its change with temperature are important in industrial practice. Some useful correlations, by which the viscosity of pitch at any temperature can be predicted, have been developed. The data on which such correlations are based may be from one of the fixed equiviscous points that characterize a pitch (Table 2).

The viscosity of a straight-run or fluxed-back pitch can be calculated from the R-and-B (ring-and-ball) softening point: $\log \eta_t = -4.175 + 711.8/86.1 - t_S + t$ where η_t is the viscosity at temperature t °C in Pa · s × 10^{-1}(= P) and t_S is the R-and-B softening point in °C.

Chemical Composition. The tars recovered from commercial carbonization plants are not the primary products of the thermal decomposition of coal. The initial products undergo a complex series of secondary reactions.

The nature of the secondary reactions is uncertain. Some believe that the primary tar components are broken down to small free radicals that recombine as they travel toward the retort exit; others suggest that some components remain relatively intact except for the removal of peripheral substituent groups and that the higher molecular weight components of coal tar are, in effect, slightly altered fragments of the original coal structure. However, it is clear that even tars produced at the lowest commercial carbonization temperatures are very different from primary tars.

Coke-oven tar is an extremely complex mixture, the main components of which are aromatic hydrocarbons ranging from the monocyclics benzene and alkylbenzenes to polycyclic compounds containing as many as twenty or more rings. Small amounts of paraffinic, olefinic, and partly saturated aromatic compounds also occur.

Table 1. Correlations for Predicting the Physical Properties of Tars and Tar Products

Property	Applicable to	Correlation expression
density, d at 20°C, g/cm^3	coke-oven dry tars and tar oils	$d_{20} = 1.877 \times 10^{-3}\ M^a + 0.808$
		$d_{20} = 7.337 \times 10^{-4}\ t_b{}^b + 0.890$
variation of sp gr with temperature, t, g/cm^3	dry tars	$\frac{\text{sp gr}}{t} = 0.001778 - 0.00098$ sp gr at 15.6°C
variation of density with temperature, t, g/cm^3	dry tars, tar oils, and pitches	$d_{t_1} = d_{t_2} - b(t_1 - t_2)^c$
	dry coke-oven tars and tar oils	$\beta = \frac{d_0 - d_t}{d_0 t}$
		$\beta = (-10.5433 + \frac{0.0122}{t_b}) \times 10^{-4}$, $\beta = (-12.6114 + 0.0346M)^a \times 10^{-4}$
viscosity, η, mPa·s(= cP)	tar oils	$\log \eta_{20} = 0.0078t_b - 1.123^d$
specific heat capacity, C, kJ/(kg·K)^e	tar oils	$C_t = ((0.7360 + 0.8951d_{20} + 0.00360t)/d_{20}) + (0.00904 - 0.0000221t_b)T_a{}^f$
	pitches	$C_t = \frac{3.665}{d_{20}} - 1.729 + 0.00389t$
thermal conductivity, K, W/(m·K)	tar oils	$K = (1.34 - 0.00084t \pm 0.084) \times 10^{-5}$
	pitches	$K = (1.423 \pm 0.084) \times 10^{-5}$
surface tension, S, mN/m (= dyn/cm)	dry tars, tar oils, and pitches	$S = 93.8d_{20} - 0.0496t_b, -47.5$, $S_t = 18.4d_t^4$, $S_t = \frac{S_{20}d_t}{d_{20}^4}$
latent heat of vaporization, L, kJ/kg^b	tar oils	$L = d_{20}(486.1 - 0.599t_b)^g$

[a] Average molecular weight.
[b] Average boiling point defined as the mean of the temperatures in °C at which 10%, 20%, ... 90% by volume distills in a standard flask distillation.
[c] $b = 0.00068 \pm 0.00005$ for dry tars and $(162.7 - 86.2d_{20} \pm 8) \times 10^{-5}$ for tar oils and pitches.
[d] At other temperatures, log η varies linearly with the absolute temperature. At t_b, the viscosity of any tar oil is approximately 0.25 mPa·s(= cP).
[e] To convert J to cal, divide by 4.184.
[f] T_a, % tar acids, is defined as the percentage by volume extracted by 10% aqueous caustic soda.
[g] At the average boiling point.

Although gas chromatography and high pressure liquid chromatography have assisted in the elucidation of the structure of tar distillate oils, has restricted applicability to pitch. Application of high pressure liquid chromatography, thin-layer chromatography, and low voltage mass spectrometer examinations of pitches and pitch solvent fractions indicate that pitch contains many high molecular weight constituents: aromatic hydrocarbons having four rings, five rings, and six rings.

Health and Safety Factors. The volatile components of coal tar, ie, mononuclear aromatic hydrocarbons, phenols, and pyridine bases, are toxic when ingested, inhaled, or absorbed through the skin and the usual precautions must be taken when crude benzene or tar light oils are handled. Most polynuclear aromatic compounds are primary skin and eye irritants but are tolerated internally.

The main health hazard usually associated with coal tar and its products is carcinogenicity. There is no evidence that the use of tar products in road surfacing, preservation of telegraph and transmission poles, pitch fiber or coal-tar enamels for water pipes constitutes any danger to the general public.

Cancerous skin lesions of workers exposed to pitch dust support the belief that these lesions are caused by polynuclear aromatic hydro-

Table 2. Viscosities of Fixed Points for Coal-Tar Pitch

Fixed point	Viscosity, Pa·s(= 10^3 cP)	Difference between fixed point and EVT, °C
equiviscous temperature, EVT, °C	25	
R-and-B softening point[a], °C	800	−19
K-and-S softening point, °C	5500	−27
penetration of 200[b]	2×10^4	−30
penetration of 10[b]	10^7	−51
ductility point, °C[c]	10^7	−51
Fraas brittle point, °C	4×10^7	−65
transition to glassy state	10^{12}	−90

[a] Ring-and-ball method (ASTM D36-26). In the United States, two other softening point methods are employed: cube-in-air and cube-in-water. Cube-in-air softening point = R-and-B softening point +14°C. Cube-in-water softening point = R-and-B softening point +10°C.
[b] Penetration of 20 or 1 mm using a 100-g weight for 5 s at 25°C.
[c] When a pitch is tested for ductility, the sample either suffers brittle fracture without elongation or elongates to the maximum distance without breaking. When tested at increased temperatures at a particular point, ie, the ductility point, the behaviour changes from the first type to the second.

carbons, although it had not been possible to demonstrate their carcinogenic action in animals more closely related to humans, such as monkeys.

The main risk is long-term, continual exposure of the skin to finely divided solid pitch (dust). In a relatively small, but statistically significant proportion of persons exposed in this manner, premalignant pitch warts appear, usually around the scrotum, hands, and face, particularly around the nostrils. These lesions, if untreated, develop into malignant epitheliomas.

A more serious hazard might be thought to exist in the pitch-roofing and road-tar industries where some personnel are continually exposed to the inhalation of pitch fumes, which contain benzo[a]pyrene (BaP) and other carcinogens.

Uses. Coumarone-indene resins have outlets in paints, as tackifiers in rubber compounding, and as adhesives in the manufacturing of flooring tiles (see HYDROCARBON RESINS).

Cresylic Acids. The higher boiling cresylic acids are mixtures of cresols or xylenols with higher boiling phenols. Their main uses are in phenol–formaldehyde resins, solvents for wire-coating enamels, as metal-degreasing agents, froth-flotation agents, and synthetic tanning agents.

Naphthalene. Naphthalene is used to produce phthalic anhydride and for the manufacture of β-naphthol and for dye stuff intermediates (see DYES AND DYE INTERMEDIATES; PHTHALIC ACIDS).

Naphthalene has been used in condensation products from naphthalene sulfonic acids, utilizing formaldehyde as additives to improve the flow properties of concrete; these are referred to as superplasticizers. Another newer application is the production of diisopropylnaphthalenes, used for dyes in the production of carbonless copy paper.

Creosote. In coal-tar refining, the recovery of tar chemicals leaves residual oils, including heavy naphtha, dephenolated carbolic oil, naphthalene drained oil, wash oil, strained anthracene oil, and heavy oil. These are blended to give creosotes conforming to particular specifications.

Timber-preservation creosotes are mainly blends of wash oil, strained anthracene oil, and heavy oil. Coal-tar creosote is also a feedstock for carbon black manufacture (see CARBON, CARBON BLACK).

Pitch. The principal outlet for coal-tar pitch is as the binder for the electrodes used in aluminum smelting. More recent uses are in the manufacture of carbon fibers and premium coke.

In North America, coal-tar pitch is used as an adhesive in membrane roofs (see ADHESIVES; ROOFING MATERIALS).

Other uses for coal-tar pitch include production as a binder for foundry cores, as a sealant for dry batteries, and in the manufacture of clay pigeons.

Fluxed Pitches and Refined Tars. *Road Tars.* In the United States, which has a large supply of bitumen, tar is little used in road construction or maintenance, but in Europe road binders still constitute an important, though declining, market for tar bulk products.

Surface Coatings. Tar-based surface coatings range from the so-called black varnishes, which consist of a soft pitch fluxed back to brushing or spraying consistency using coal-tar naphtha, to pipe-coating enamels and pitch-polymer coatings.

W. D. BETTS
Tar Industries Services

Standard Methods for Testing Tar and Its Products, 7th ed., Serial No. COI 3-79, Standardization of Tar Products Tests Committee, Chesterfield, U.K., 1979.

Council Regulation (EEC) No. 793/93, *Official Journal of the EC*, L84 5.4.93, p. 1–75, 1993.

S. A. Henry, *Burt. Med. Bull. H*, 389–401 (1947).

H. G. Franck and J. W. Stadelhofer, *Industrial Aromatic Chemistry*, Springer Verlag, New York, 1988, pp. 368 and 382.

TAR SANDS

In addition to conventional petroleum and heavy crude oil, there remains another subclass of petroleum, the bitumen found in tar sand deposits. Tar sands, also known as oil sands and bituminous sands, are sand deposits impregnated with dense, viscous petroleum. Tar sands are found throughout the world, often in the same geographical areas as conventional petroleum.

Tar sand is a mixture of sand, water, and bitumen, but many of the tar sand deposits in the United States lack the water layer that is believed to cover the Athabasca sand in Alberta, Canada, thereby facilitating the hot-water recovery process from the latter deposit. The heavy asphaltic organic material has a high viscosity under reservoir conditions and cannot be retrieved through a well by conventional production techniques.

It is incorrect to refer to bitumen as tar or pitch. Although the word tar is somewhat descriptive of the black bituminous material, it is best to avoid its use in referring to natural materials (see TAR AND PITCH).

Physical methods of fractionation of tar sand bitumen usually indicate high proportions of nonvolatile asphaltenes and resins. In addition, the presence of ash-forming metallic constituents, including such organometallic compounds as those of vanadium and nickel, is also a distinguishing feature of bitumen.

Occurrence

Many of the reserves of bitumen in tar sand formations are available only with some difficulty; optional refinery methods are necessary for future conversion of these materials to liquid products because of the substantial differences in character between conventional petroleum and bitumen (Table 1).

Tar sand deposits are widely distributed throughout the world, and the various deposits have been described as belonging to two types: stratigraphic traps and structural traps. In general terms, the entrapment character of the very large tar sand deposits involves a combination of both stratigraphic and structural traps.

The largest tar sand deposits are in Alberta, Canada, and in Venezuela. Smaller tar sand deposits occur in the United States (mainly in Utah), Peru, Trinidad, Madagascar, the former Soviet Union, Balkan states, and the Philippines.

The Alberta (Athabasca) tar sand deposits are located in the northeast part of that Canadian province. These are the only mineable tar sand deposits undergoing large-scale commercial exploitation at present.

Table 1. Bitumen vs Conventional Petroleum Properties

Property		Bitumen	Conventional
gravity, °API		8.6	25–37
distillation	Vol %	IBP[a], °C	
	5	221	
	10	293	
	30	437	
	50	543	
viscosity, suspension			
at 38°C		35,000	<30
at 99°C		513	
pour point, °C		10	≤0
elemental analysis, wt %			
carbon		83.1	86
hydrogen		10.6	13.5
sulfur		4.8	0.1–2.0
nitrogen		0.4	0.2
oxygen		1.1	
hydrocarbon type, wt %			
asphaltenes		19	≤5
resins		32	
oils		49	
metals, ppm			
vanadium		250	
nickel		100	
iron		75	≤75
copper		5	
ash, wt %		0.75	0
net heating value, kJ/g[b]		40.68	ca 45.33

[a] IBP = initial boiling point.
[b] To convert kJg tp btu/lb, multiply by 430.2.

Properties

Tar sand has been defined as sand saturated with a highly viscous crude hydrocarbon material not recoverable in its natural state through a well by ordinary production methods. Technically the material should perhaps be called bituminous sand rather than tar sand because the hydrocarbon is bitumen, ie, a carbon disulfide-soluble oil.

Bulk properties of samples from several locations (Table 2) show that there is a wide range of properties. Substantial differences exist between the tar sands in Canada and those in the United States; a difference often cited is that the former is water-wet and the latter, oil-wet.

The sand component is predominantly quartz in the form of rounded or angular particles, each of which is wet with a film of water. Surrounding the wetted sand grains and somewhat filling the void among them is a film of bitumen. The balance of the void volume in the Canadian sands is filled with connate water plus, sometimes, a small volume of gas. The sand grains are packed to a void volume of ca 35%, corresponding to a mixture of ca 83 wt % sand; the remainder is bitumen and water, which constitute ca 17 wt % of the tar sands.

Bitumen. There are wide variations both in the bitumen saturation of tar sand (0–18 wt % bitumen), even within a particular deposit, and the viscosity.

The API gravity of tar sand bitumen varies from 5 to ca 10°API, depending on the deposit, and the viscosity is very high. This offers a formidable obstacle to bitumen recovery and, as a result of the high viscosity, bitumen is relatively nonvolatile under conditions of standard distillation.

Minerals. Usually >99% of the tar sand mineral is composed of quartz sand and clays (qv). In the remaining 1%, more than 30 minerals have been identified, mostly calciferous or iron-based. Particle sizes range from large grains (99.9% finer than 1000 μm) to 44 μm (325 mesh), the smallest size that can be determined by dry screening.

Clays are aluminosilicate minerals, some of which have definite chemical compositions. In regard to tar sands, however, clay is only a size classification and is usually determined by a sedimentation method.

Recovery

Oil prices and operating costs are the key to economic development of tar sand deposits. However, two technical conditions of vital concern for economic development are the concentration of the resource (percent bitumen saturation) and its accessibility, usually measured by the overburden thickness.

Recovery methods are based either on mining combined with some further processing or operation on the oil sands *in situ*. The mining methods are applicable to shallow deposits, characterized by an overburden ratio (ie, overburden depth-to-thickness of tar sand deposit) of ca 1.0.

Nonmining Methods. Nonmining (*in situ*) processes depend on injecting a heating-and-driver substance into the ground through injection wells and recovering bitumen through production wells. Such processes need a relatively thick layer of overburden to contain the driver substance within the formation between injection and production wells.

In principle, the nonmining recovery of bitumen from tar sand deposits is an enhanced oil recovery technique and requires the injection of a fluid into the formation through an injection well. This leads to the *in situ* displacement of the bitumen from the reservoir and bitumen production at the surface through an egress (production) well. There are, however, several serious constraints that are particularly important and relate to the bulk properties of the tar sand and the bitumen. In fact, both recovery by fluid injection and the serious constraints on it must be considered *in toto* in the context of bitumen recovery by nonmining techniques (see PETROLEUM, ENHANCED OIL RECOVERY).

Another general constraint to bitumen recovery by nonmining methods is the relatively low injectivity of tar sand formations. It is usually necessary to inject displacement/recovery fluids at a pressure such that fracturing (parting) is achieved. Such a technique, therefore, changes the reservoir profile and introduces a series of channels through which fluids can flow from the injection well to the production well. On the other hand, the technique may be disadvantageous insofar as the fracture occurs along the path of least resistance, giving undesirable or inefficient flow characteristics within the reservoir between the injection and production wells, which leave a part of the reservoir relatively untouched by the displacement or recovery fluids.

Mining Methods. The alternative to *in situ* processing is to mine the tar sands, transport them to a processing plant, extract the bitumen value, and dispose of the waste sand. Such a procedure is often referred to as oil mining. This is the term applied to the surface or subsurface excavation of petroleum-bearing formations for subsequent removal of the oil by washing, flotation, or retorting treatments. Oil mining also includes recovery of oil by drainage from reservoir beds to mine shafts or other openings driven into the oil rock, or by drainage from the reservoir rock into mine openings driven outside the oil sand but connected with it by bore holes or mine wells.

There are two approaches to open-pit mining of tar sands. The first uses a few mining units of custom design, which are necessarily expensive, eg, bucket-wheel excavators and large drag lines in conjunction with belt conveyors. In the second approach, a multiplicity of smaller mining units of conventional design is employed at relatively much lower unit costs.

Processing

Hot-Water Process. The hot-water process is the only successful commercial process to be applied to bitumen recovery from mined tar sands in North America. The process utilizes linear and nonlinear

Table 2. Bulk Properties of Tar Sands

Property	Alberta	Asphalt Ridge[a]	P.R. Springs[a]	Sunnyside[a]	Tar Sand Triangle[a]	Texas	Alabama
bulk density, g/cm^3	1.75–2.19		1.83–2.50				
porosity, vol %	27–56	16–27	6–33	16–28	9–32	32	6–25
permeability, $m^2 \times 10^{-16}$[b]	99–5,900	4,905–5,950	553–14,902	5,265–7,402	2,043–7,777	3158	9.9–6,316
specific heat, J/(g·°C)[c]	1.46–2.09						
thermal conductivity, J/(s·°C·cm)[c]	0.0071–0.0015						

[a] Deposit in Utah.
[b] To convert m^2 to millidarcies, multiply by 1.013×10^{12}.
[c] To convert J to cal, divide by 4.184.

variations of bitumen density and water density, respectively, with temperature so that the bitumen that is heavier than water at room temperature becomes lighter than water at 80°C. Surface-active materials in tar sand also contribute to the process. The essentials of the hot-water process involve conditioning, separation, and scavenging.

The hot-water separation process involves extremely complicated surface chemistry with interfaces among various combinations of solids (including both silica sand and aluminosilicate clays), water, bitumen, and air.

One problem resulting from the hot-water process is disposal and control of the tailings.

Environmental regulations in Canada and the United States do not allow the discharge of tailings streams into the river, onto the surface, or onto any area where contamination of groundwater domains or the river may occur; hence the need for tailings ponds, where some settling of the clay occurs.

Cold-Water Process. The cold-water bitumen separation process has been developed to the point of small-scale continuous pilot plants. The process uses a combination of cold water and solvent. The first step usually involves disintegration of the tar sand charge, which is mixed with water, diluent, and reagents. The diluent may be a petroleum distillate fraction such as kerosene and is added in a ca 1:1 weight ratio to the bitumen in the feed. The effluent is mixed with more water, and in a raked classifier the sand is settled from the bulk of the remaining mixture. The water and oil overflow the classifier and are passed to thickeners, where the oil is concentrated.

The sand reduction process is a cold-water process without solvent. The objective is removal of sand to provide a feed suitable for a fluid coking process.

Solvent Extraction. An anhydrous solvent extraction process for bitumen recovery has been attempted and usually involves the use of a low boiling hydrocarbon. Although solvent extraction processes have been attempted and demonstrated for the Athabasca, Utah, and Kentucky tar sands, solvent losses influence economics of such processes and they have not yet been reduced to commercial practice.

Bitumen Conversion

Bitumen is a hydrogen-deficient oil that is upgraded by carbon removal (coking) or hydrogen addition (hydrocracking). There are two methods by which bitumen conversion can be achieved: by direct heating of mined tar sand and by thermal decomposition of separated bitumen. The latter is the method used commercially, but the former has potential for commercialization (see FUELS, SYNTHETIC)

Conversion of Separated Bitumen. The overall upgrading process by which bitumen is converted to liquid products is accomplished in two steps. The first step is the primary upgrading process, which improves the hydrogen-to-carbon (H/C) ratio by either carbon removal or hydrogen addition, cracking the bitumen to lighter products which are more easily processed downstream. The secondary upgrading process involves hydrogenation of the primary products and is the means by which sulfur and nitrogen are removed from the primary products. The upgraded or synthetic crude can then be refined to consumer goods such as gasoline, jet fuel, and home heating oil by conventional means.

Health and Safety Factors

Health and safety factors associated with tar sand processing depend on the nature of the process and products. Issues arising from tar sand mining are similar to other large-scale surface mining operations involving large equipment and the movement of huge quantities of material. The principal environmental consideration relating to the mining process is land reclamation following the completion of the mining and, in particular, those areas affected by the deposition of tailings. Both air and liquid effluents are subject to controls.

Health and safety factors in *in situ* operations are associated with high temperature, high pressure steam, or high pressure air. Environmental considerations relate to air and water quality and surface reclamation.

JAMES G. SPEIGHT
Western Research Institute

N. Berkowitz and J. G. Speight, *Fuel* **54**, 138 (1975).

J. G. Speight, in J. G. Speight, ed., *Fuel Science and Technology Handbook*, Part II, Marcel Dekker, Inc., New York, 1990.

J. G. Speight, *The Chemistry and Technology of Petroleum*, 3rd ed., Marcel Dekker, Inc., New York, 1998.

M. R. Gray, *Upgrading Petroleum Residues and Heavy Oils*, Marcel Dekker, Inc., New York, 1994.

TEA

Tea is prepared from the leaves of the plant *Camellia sinensis*. Traditionally, *Camellia sinensis* was propagated and bred through seeds. Vegetative propagation has become a common practice. This helps to maintain genetic purity and aids in more rapid establishment of new productive stands of tea.

The flush of a tea shoot is defined as the apical bud and two new leaves below it. This is the ideal target for harvesting fresh tea of optimum quality.

Composition of Fresh Tea and Biosynthesis of Tea Polyphenols

The leaves of *Camellia sinensis* are similar to most plants in general morphology and contain all the standard enzymes and structures associated with plant cell growth and photosynthesis. Unique to tea plants are large quantities of flavonoids and methylxanthines, compounds which impart the unique flavor and functional properties of tea. The general composition of fresh tea leaves is presented in Table 1.

Flavonoids. Green tea leaves contain many types of flavonoids, the most important of which are the flavanols (catechins) (**1**), the flavonols, and flavanol glycosides. Tea catechins are water-soluble,

Table 1. General Composition of Fresh Tea Leaves

Components	Quantity, wt %[a]
flavanols	25.0
flavonols and flavonol glycosides	3.0
polyphenolic acids and depsides	5.0
other polyphenols	3.0
caffeine	3.0
theobromine	0.2
amino acids	4.0
organic acids	0.5
monosaccharides	4.0
polysaccharides	13.0
cellulose	7.0
protein	15.0
lignin	6.0
lipids	3.0
chlorophyll and other pigments	0.5
ash	5.0
volatiles	0.1

[a] On a dry weight basis.

colorless substances which impart the bitter and astringent taste characteristic of green teas. Localized within the cytoplasmic region of leaf cells, the flavanols (catechins) generally make up 25–40% of the water-soluble solids of tea. Catechins are easily oxidized. The oxidation products of catechins are the red-brown pigments found in brewed and instant teas. They also form complexes with many other substances such as proteins and caffeine. These polyphenolic constituents are the key reactants involved with the enzymatic fermentation of green tea to black tea. The quality of tea infusions correlates with the flavanol content of fresh green leaves. A number of flavonols including quercetin, kaempferol, myricetin, and their glycosides are also found in tea leaves. Flavonol glycosides generally make up 2–3% of the water-soluble solids of tea.

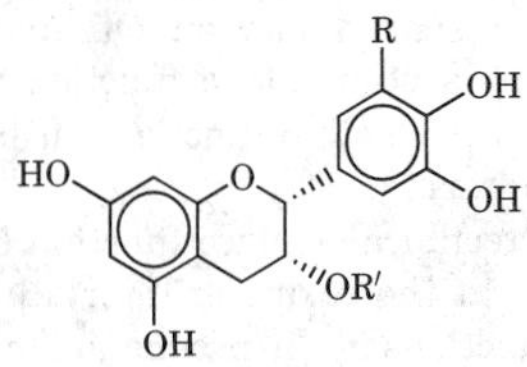

(1)

epicatechin; R = H, R′ = H	H
epicatechin gallate; R = H, R′ = gallate	H
epigallocatechin; R = OH, R′ = H	
epigallocatechin gallate; R = OH, R′ = gallate	

Other Phenolic Compounds. There are several phenolic acids important to tea chemistry. Gallic acid (**2**) and its quinic acid ester, theogallin (**3**), have been identified in tea.

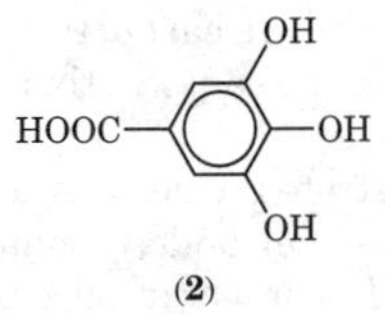

(2)

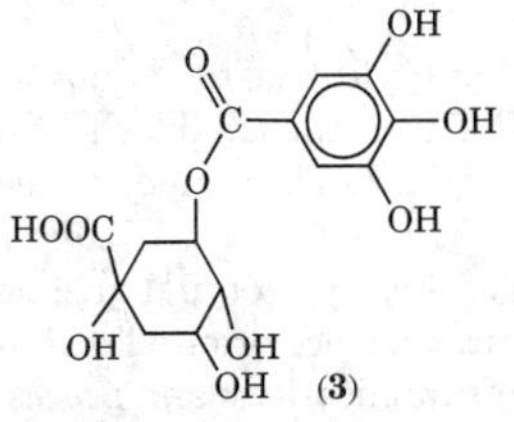

(3)

Caffeine and Other Xanthines. Tea flush contains 2.5–4.0% caffeine (**4**) on a dry weight basis and much smaller quantities of the related methylxanthine theobromine (**5**). On average, a 6-ounce (180–cm^3) cup of tea contains 20–70 mg of caffeine, compared to 40–155 mg of caffeine in a 6-ounce (180-cm^3) cup of freshly brewed coffee (qv). Infusions of black, green, and oolong teas all contain about the same amounts of caffeine when prepared using similar amounts of leaves. Caffeine forms complexes with the polyphenolic constituents in tea, and these complexes have poor solubility and often precipitate under cold storage. This precipitate is called cream because of its milky appearance. This physical and chemical property of tea affects the behavior of iced tea beverages as well as the technology of instant-tea manufacture.

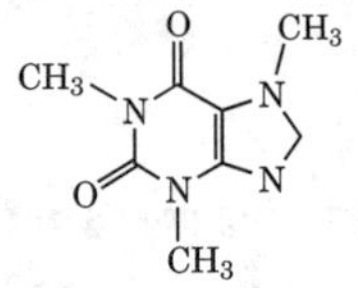

(4)

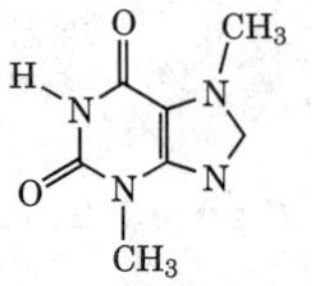

(5)

Theanine and Other Amino Acids. Amino acids (qv) make up 4–8% of the soluble solids found in brewed tea. There is an amino acid unique to tea, γ-*N*-ethyl glutamine, called theanine.

Minerals and Ash. The water-soluble extract solids which infuse from tea leaves contain 10–15% ash. The tea plant has been found to be rich in potassium, calcium, magnesium, and aluminum. Tea beverages are also a significant source of fluoride.

Volatiles or Aroma. Black tea aroma contains over 300 characterizing compounds, the most important of which are terpenes, terpene alcohols, lactones, ketones, esters, and spiro compounds.

Enzymes. The enzymes most important to the chemistry and manufacturing of tea are those responsible for the biosynthesis of tea flavonoids and those involved in the conversion of fresh leaf into manufactured commercial teas.

Alcohol dehydrogenase and leucine α-ketoglutarate transaminase contribute to the development of aroma during black tea manufacturing. Polyphenol oxidase and peroxidase are essential to the formation of polyphenols unique to fermented teas.

Polyphenol oxidase (PPO) (EC 1.14.18.1; monophenol monooxygenase [tyrosinase] or EC 1.10.3.2; *o*-diphenol: O_2-oxidoreductase) is one of the more important enzymes involved in the formation of black tea polyphenols. The enzyme is a metallo-protein thought to contain a binuclear copper active site.

Peroxidase (POD) (EC 1.11.1.7) is thought to play in integral role in the fermentation process and is found in fresh green leaf. It is a heme-based enzyme which catalyzes the reductive decomposition of hydrogen peroxide to water, and organic peroxide species to the corresponding alcohol. Along with PPO, POD plays a role in the oxidation processes involved with the formation of the black tea components.

Manufacturing

Freshly harvested tea leaves require manufacturing to be converted into green, oolong, and black teas (Fig. 1). Black tea, the dominantly manufactured tea product worldwide, is made through a polyphenol oxidase-catalyzed oxidation of fresh leaf catechins. Green tea is processed in a manner designed to prevent the enzymatic oxidation of catechins before drying. Oolong is a partially oxidized tea. Instant tea, usually a powder, is generally prepared by the aqueous extraction of tea leaves, followed by concentration and drying.

Chemistry of Tea Fermentation Oxidation. The chemical changes which take place during the manufacture of green, oolong, and black teas are responsible for the unique color and flavors characteristic of each tea type. The most significant and well-understood of these reactions occurs during the manufacture of black tea. During the process called tea fermentation the colorless catechins (**1**) found in green tea proceed through a series of oxidative condensation reactions leading to the formation of a range of products of orange-yellow to red-brown color, plus the development of a large number of unique volatile con-

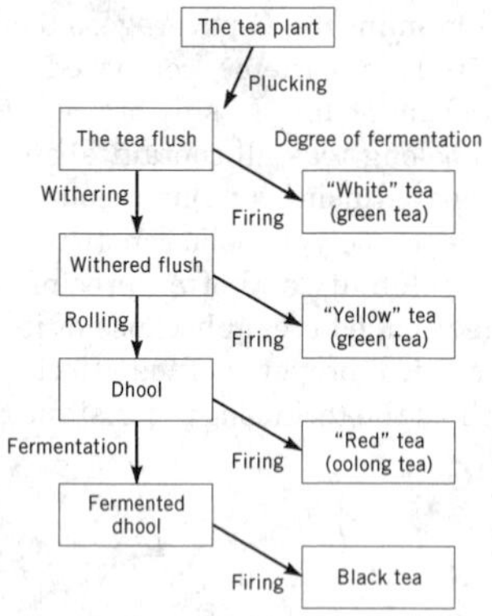

Figure 1. Tea manufacturing processes.

stituents. These changes are reflected in the red amber color, reduced astringency, and more complex flavor found in black tea beverages.

Black Tea. The black tea manufacturing process consists of the unit operations of withering (reduction of moisture), rolling, fermentation, firing, and sorting.

Green Tea. Green tea is made by rolling and firing without enzymic oxidation. The green tea beverage is pale yellow to green, slightly more astringent than black tea, and has a brothy characteristic imparted by theanine and other amino acids.

Other Types. For oolong tea, withered leaf is lightly rolled and only partially fermented. Enzymatic action is terminated by a heating process. In some respects, the beverage characteristics are between those of green tea and black tea, but oolong contains a unique array of volatile and polyphenolic compounds not found in either green or black tea.

Blending, Packaging, and Forms of Consumption

Blending. The tea taster plays an important role in purchasing and blending tea. The goal is usually the establishment and maintenance of a chosen standard of tea under constantly changing conditions. It is necessary to include teas from many countries and many gardens within a country in a blend to ensure constancy in flavor, color, and price over a long period of time.

Packaging. In most countries, tea is sold in packets. In the United States, more than 90% of leaf tea is sold in tea bags. An important consideration is tea bag paper quality, which must be selected so as to avoid flavor contamination, allow efficient infusion, and retain tea fines.

The outer packaging must protect the tea from light and moisture absorption. Polypropylene or coated cellophane outer wraps for paper board tea packages provide a barrier to loss of tea aroma and retard permeation of oxygen and foreign flavors.

Instant Teas

The basic process for manufacture of instant tea as a soluble powder from dry tea leaf includes extraction, concentration, and drying. In practice, the process is considerably more complicated because of the need to preserve the volatile aroma fraction, and produce a product which provides color yet is soluble in cold water, all of which are attributes important to iced tea products.

Decaffeination

Caffeine can be removed from tea leaves or instant tea by a variety of processes. Solvent extraction uses ethyl acetate or methylene chloride. Supercritical carbon dioxide is also used to decaffeinate teas (see SUPERCRITICAL FLUIDS).

Physiological and Health Effects of Tea

There are numerous synthetic and natural compounds called antioxidants which regulate or block oxidative reactions by quenching free radicals or by preventing free-radical formation. The antioxidant properties of tea flavonoids have been characterized using models of chemical and biological oxidation reactions.

Animal studies have shown that teas are effective in blocking or slowing carcinogenesis.

Studies to determine the physiological effects of tea consumption associated with antioxidant activity and other relevant biomarkers of cancer risk have been conducted with human volunteers. While more research is required, these studies demonstrated that tea polyphenols are absorbed into the body and appear to have physiological effects that are consistent with antioxidant activity.

DOUGLAS A. BALENTINE
Lipton

B. Banerjee, in K. C. Wilson and M. N. Clifford, eds., *Tea: Cultivation to Consumption*, Chapman and Hall, London, 1992, p. 25.

W. H. Ukers, *All About Tea*, The Tea and Coffee Trade & Journal Co., New York, 1935.

TECHNICAL SERVICE

The principal objective of technical service in the chemical industry is to provide timely and professional information and support to downstream customers regarding chemical products and their uses.

The general focus of technical service in the chemical industry was, at the outset, largely tied to a firm's direct sales and marketing efforts. As applications became more complex, however, and customer requirements became more and more specific, a need evolved in many areas of the chemical industry to provide in-depth technical support having direct ties to the research and manufacturing arms.

The Spectrum of Technical Service

Some firms have seen fit to blur the distinction between technical service *per se* and the research function. Others maintain a technical service organization as a stand-alone function while maintaining a high level of integration with other functions such as sales, marketing, research, and manufacturing.

A well-integrated technical service function having strong ties to the sales organization at the customer interface and to the research and manufacturing functions at the production interface allows a company to provide rapid and accurate responses to customer needs as a singular function.

Technical Service Functions

The largest number of technical service inquiries from customers involve questions regarding the performance of an existing product already in use by the customer. A typical question is whether product X would work in application Y. The answer may be quite straightforward, or it may require a substantial applications research effort. Test results can be relatively quick and easy to obtain or lengthy and considerably more difficult. Once the results have been obtained and discussed with the supplier's technical service personnel, the customer can make an informed decision regarding the use of the production question.

Another typical customer question is "why did product X do Y in my process?" This is the troubleshooting, consultative part of the technical service function. The range of effort required to answer this question is broad.

Both of these prototypal questions illustrate the need for a successful technical service professional to have a strong understanding of the customer's applications and processes, within proper intellectual property considerations.

Simulation of a customer process at the laboratory scale is sometimes requested, usually to allow the ready surveying of a variety of raw materials or the evaluation of the impact of process changes at a scale more economical than full-plant capability. This is simple for certain requests, ie, the evaluation of a series of antioxidants (qv) in a given polymer formulation, yet nearly impossible for others, ie, studies of flocculation/dispersion phenomena at plant-scale shear rates in multicomponent systems. Generally, laboratory-scale efforts are quite valuable, as they provide a means to identify any gross incompatibilities or other system problems prior to carrying out plant-scale changes.

A common requirement of the technical service professional is the support of either semiworks or full plant-scale trials of a material in a new application at a customer's site. The technical service person must be familiar with the customer's processing equipment, operating practices, and raw materials if a smooth evaluation is to be ensured. As for any complex manufacturing process, experimental design is a crucial aspect of a successful line evaluation, both for the execution of the test and for the evaluation of data from the test.

In-house training at a customer site is another valuable technical service function.

Technical Service Organization. There are many structures for technical service organizations. Requisite elements of a successful technical service organization generally include the following: (*1*) highly trained technical professionals having a high level of expertise; (*2*) appropriate support staffing, ie, laboratory technicians, clerical staff, etc; (*3*) sufficient physical facilities, such as a laboratory, instrumentation, computational support, etc; (*4*) strong integration with other company functions, such as manufacturing, marketing, sales, engineering, and research; and (*5*) the ability to leverage support from others.

Training Requirements for Personnel. Specialized training is an absolute requirement for technical service personnel. Both internal and external resources are used to provide specialized training in appropriate areas.

The area of regulatory requirements is of particular importance. The extreme complexity and constant changes of regulations applying to chemicals used in commerce dictate that personnel providing technical service support be well-versed in regulations specific to the product(s) they support.

Technical Service and the Development Process

The technical service function acts as the interface between the user and the research and manufacturing arms of the supplier for many firms. A consequence of this is that the technical service function can act as a conduit of information between the customer and personnel involved in the development process.

The troubleshooting aspect of technical service *de facto* results in the technical service professional having the best perspective on any faults that may be identified by a customer for a given application. Taking timely action on such observations enhances a supplier's credibility with a given customer and quite often provides an improvement to the product of general value to other members of the supplier's customer base.

A critical but often overlooked aspect of the technical service function is the value of relationships that develop between more senior members of the technical service staff and their colleagues at customer sites. Professional relationships provide value both to supplier and customer.

Technology

Computers. The increase in availability of low cost computing power has greatly enhanced the problem-solving capability of the technical service professional. The availability of highly portable notebook computers allows great freedom to transport substantial computational support anywhere in the world. Some remote control aspects of computers include the ability to access, via telephone line connections, databases in the suppliers' computers, using powerful in-house computers to remotely execute molecular mechanics simulations of reaction pathways, or carrying out gas chromatographic analyses while away from the laboratory.

The Internet. One of the most important changes in the manner in which technical service is provided to customers of the manufacturing and service industries is the advent of the Internet and its use in providing direct access to a wide range of technical service and related information. It is possible, either through the use of an in-house computer acting as a server or by renting space on a server at an internet service provider (ISP) site, to provide almost any sort of information that a customer might request.

AUSTIN H. REID, JR.
E. I. du Pont de Nemours & Co.

P. G. Smith, *Developing Products in Half the Time*, Van Nostrand Rheinhold, New York, 1991.

Science: The Endless Frontier, National Science Foundation Report, NSF 90-8, The National Science Foundation, Washington, D.C., 1990.

J. F. Rabek, *Photostabilization of Polymers*, Elsevier Science Publishers Ltd., Essex, U.K., 1990.

A. Roussel, K. N. Saad, and T. J. Erikson, *Third-Generation R&D: Managing the Link to Corporate Strategy*, Harvard Business School Press, Boston, Mass., 1991.

TELLURIUM AND TELLURIUM COMPOUNDS

Tellurium, Te, at no. 52, at wt 127.61, is a member of the sixth main group, Group 16 (VIA) of the Periodic Table, located between selenium and polonium. Tellurium is in the fifth row of the Table, between antimony and iodine, and has an outer electron configuration of $5s^2 5p^4$. The four inner principal shells are completely filled. Tellurium is more metallic than oxygen, sulfur, and selenium, yet it resembles them closely in most of its chemical properties. Whereas oxygen and sulfur are nonmetals and electrical insulators, selenium and tellurium are semiconductors, and polonium is a metal. Tellurium forms inorganic and organic compounds superficially similar to the corresponding sulfur and selenium compounds, yet dissimilar in properties and behavior. The valence states assigned to the central atom in tellurium compounds are −2, 0, +2, +4, +6.

Although widely disseminated, tellurium minerals do not form ore bodies. Hence, there are no deposits that can be mined for tellurium alone, and there are no formally stated reserves. Large resources however, are present in the base-metal sulfide deposits mined for copper, nickel, gold, silver, and lead, where the recovery of tellurium, like that of selenium, is incidental.

Physical Properties

At least 21 tellurium isotopes are known, with mass numbers from 114 to 134. Of these, eight are stable, ie, 120, 122–126, 128, 130. The others are radioactive and have lifetimes from 2 min to 154 d; the heaviest six, 131m, 131, 132, 133m, 133, and 134, are fission products (see RADIOISOTOPES). Tellurium illustrates the rule that elements having even atomic numbers have more isotopes than elements having odd atomic numbers.

The physical properties of tellurium are given in Table 1. At ordinary temperature and pressure, solid tellurium, unlike sulfur and selenium, has only one structural form. Tellurium crystallizes in a trigonal lattice. The crystal structure may be considered as a set of helical chains parallel to the *c*-axis, held together by relatively weak atomic forces.

The physical properties of tellurium are generally anistropic. Owing to its weak lateral atomic bonds, crystal imperfections readily occur in single crystals as dislocations and point defects.

Table 1. Physical Properties of Tellurium

Property	Value
specific gravity[a] at 18°C	
crystalline	6.24
amorphous	6.0–6.2
hardness, Mohs[b]	2.0–2.5
modulus of elasticity, MPa[c]	4140
Poisson ratio at 30°C	0.33
heat capacity at 25°C, kJ/mol[d]	25.70
entropy at 25°C, J/(K·mol)[d]	49.70
heat of fusion, kJ/mol[d]	17.87
mp, °C	450
viscosity at mp, mPa·s(= cP)	1.8–1.95
bp, °C[e]	990
heat of formation[f], kJ/mol[d]	171.5
heat of vaporization, kJ/g[d]	46.0
thermal conductivity at 20°C[g], W/(m·K)	0.060

[a] Increases under pressure.
[b] Anisotropic.
[c] To convert MPa to psi, multiply by 145.
[d] To convert J to cal, divide by 4.184.
[e] Extrapolated.
[f] Te (g) atom to Te_2 (g) molecule.
[g] Polycrystalline material; in single crystals, it is anisotropic and affected by impurities and lattice imperfections.

Chemical Properties

Although tellurium resembles sulfur and selenium chemically, it is more basic, more metallic, and more strongly amphoteric. Its behavior as an anion or a cation depends on the medium. Tellurium forms ionic tellurides with active metals, and covalent compounds with other elements. The valence states are −2, +4, and +6. Molten tellurium is readily oxidized to tellurium dioxide. Tellurium reacts with halogens and halogenating agents, and mixes in all proportions with sulfur and selenium. Oxidation with nitric acid, and ignition of the resulting $2\ TeO_2 \cdot HNO_3$ yields very pure TeO_2.

Elemental tellurium liberates chlorine from compounds such as $AsCl_3$ and $AuCl_3$; it reduces $FeCl_3$ partially to $FeCl_2$, and SO_2Cl_2 to SO_2 and Cl_2 gases. Oxidation of metals by tellurium gives metallic tellurides. Tellurium itself is oxidized by strong reagents such as $Na_2Cr_2O_7$, $KMnO_4$, $Ca(OCl_2)_2$, H_2O_2, and $HClO_3$. Tellurium dioxide and tellurous acid and its salts are readily reduced to the element, Te^0 with $SnCl_2$, H_2S, and $Na_2S_2O_4$. Solid tellurium oxides can be reduced by heating with hydrogen, carbon, and carbon monoxide.

The stability of organic chalcogen compounds decreases mostly in the order sulfur > selenium > tellurium.

Manufacture

Recovery. Most commercial tellurium is recovered from electrolytic copper refinery slimes. The tellurium content of slimes can range from a trace up to 10% (see SELENIUM AND SELENIUM COMPOUNDS).

Today, the removal of copper in slimes is normally accomplished by the autoclaving of slimes at elevated temperature with sulfuric acid and oxygen. Use of temperatures of 120°C and oxygen pressures of 345 kPa (50 psig) allows almost complete copper extraction and this is accompanied by frequently tellurium extractions ranges from 50 to 80%. The tellurium solubilized by autoclaving may be present in both the tetravalent and hexavalent forms.

"Copper slimes are first treated for the removal of the copper values. Normally, this is accomplished by autoclaving the slimes at elevated temperature with sulfuric acid and oxygen. Use of temperatures of 120° C and oxygen pressures of 345 kPa (50 psig) allows almost complete copper extraction and, additionally, this is accompanied by tellurium extraction ranging from 50–80%"

Tellurium is recovered from solution by cementation with copper at elevated (>90°C) temperature.

$$H_2TeO_3 + 4\ Cu^0 + 2\ H_2SO_4 \rightarrow Cu_2Te + 2\ CuSO_4 + 3\ H_2O$$

$$H_2TeO_4 + 5\ Cu^0 + 3\ H_2SO_4 \rightarrow Cu_2Te + 3\ CuSO_4 + 4\ H_2O$$

Although this procedure yields tellurium as the same compound found in the original feedstock, the copper telluride is recovered in a comparatively pure state which is readily amenable to processing to commercial elemental tellurium or tellurium dioxide.

Purification. Tellurium can be purified by distillation at ambient pressure in a hydrogen atmosphere. However, because of its high boiling point, tellurium is also distilled at low pressures.

Ultrahigh (99.999 + %) purity tellurium is prepared by zone refining in a hydrogen or inert-gas atmosphere.

Commercial Products. These include tellurium dioxide, TeO_2; sodium tellurate, Na_2TeO_4; ferrotellurium or iron telluride; FeTe, and tellurium diethyldithiocarbamate, $[(C_2H_5)_2NC(S)S]_4Te$.

Metal tellurides for semiconductors are made by direct melting, melting with excess tellurium and volatilizing the excess under reduced pressure, passing tellurium vapor in an inert gas carrier over a heated metal, and high temperature reduction of oxy compounds with hydrogen or ammonia.

Health and Safety Factors

Elemental tellurium and the stable tellurides of heavy nonferrous metals are relatively inert and do not represent a significant health hazard. Other, more reactive tellurides, including soluble and volatile tellurium compounds such as hydrogen telluride, tellurium hexafluoride, and alkyl tellurides, should be handled with caution. Some of these materials can enter the body by absorption through the skin or by inhalation and ingestion of dust or fumes.

The soluble tellurites are more toxic than the selenites and arsenites.

The unusual physical complaints and findings in workers overexposed to tellurium include somnolence, anorexia, nausea, perspiration, a metallic taste in the mouth and garlic-like odor on the breath.

Industrial precautions for handling tellurium include the commonsense measures of good housekeeping, adequate ventilation, personal cleanliness, and frequent changes of clothing. Gloves and safety glasses should be worn at all times, and dust masks and chemical goggles should be used where needed.

The U.S. Occupational Safety and Health Administration (OSHA) has established its permissible exposure limit (PEL) for tellurium and its compounds at 0.1 mg/m^3.

Inorganic Compounds

Tellurium forms inorganic compounds very similar to those of sulfur and selenium. The most important tellurium compounds are the tellurides, halides, oxides, and oxyacids.

Tellurides. Most elements form compounds with tellurium. Most tellurides are prepared by direct reaction, varying from very vigorous with alkali and some alkaline-earth metals to sluggish and requiring a high temperature with hydrogen. The alkali and alkaline-earth tellurides are colorless ionic solids rapidly decomposed by air, especially in the presence of atmospheric moisture. The alkali metal tellurides are strong reductants. Hydrotellurides, such as sodium hydrotelluride, NaHTe, are also known. Some metals form more than one telluride, and some metal tellurides show nonstoichiometry; many of them exhibit semiconductor properties.

Tellurium Halides. Tellurium forms the dihalides $TeCl_2$ and $TeBr_2$, but not TeI_2. However, it forms tetrahalides with all four halogens. Tellurium decafluoride and hexafluoride can also be prepared. No monohalide, Te_2X_2, is believed to exist. Tellurium does not form well-defined oxyhalides unlike sulfur and selenium. The tellurium halides show varying tendencies to form complexes and addition compounds

Table 2. Isolated Organic Tellurium Compounds

Compounds	Formula	Mp, °C	Bp, °C
tellurols or tallanes	RTeH		>90
ethanetellurol	C_2H_5TeH		
benzenetellurol	C_6H_5TeH		
alkyl, aryl, and cyclic tellurides			
dimethyltelluride (methyl telluride)	$(CH_3)_2Te$		82
diphenyl telluride	$(C_6H_5)_2Te$	53.4	182
tetraphenyl telluride	$(C_6H_5)_4Te$	104–106	
tetrahydrotellurophene	$TeCH_2(CH_2)_2CH_2$ (ring)		
ditellurides	R_2Te		
diphenyl ditelluride	$C_6H_5Te{-}TeC_6H_5$	53–54	
alkyltellerium dihalides	$RTeX_3$		
methyltellurium tribromide	CH_3TeBr_3	dec 140 C	
dialkyltellurium dihalides	R_2TeX_3		
dimethyltellurium dichloride	$(CH_3)_2TeCl_2$	92(α), 134(β)	
tellurium salts	R_3TeX		
trimethyltellurium iodide	$(CH_3)_3TeI$		
telluroxides	RTeO		
diphenyl telluroxide	$(C_6H_5)_2TeO$	185	
tellurones	R_2TeO_2		
dimethyl tellurone	$(CH_3)_2TeO_2$		
telluroketones	R_2CTe		
dimethyl telluroketone	CH_3CTeCH_3	63–66	
tellurinic acid	RTeO·OH		
phenyltellurinic acid	$C_6H_5Te{\cdot}OH$	211	
heterocyclic compounds			
1,4-oxatellurane	$OCH_2CH_2TeCH_2CH_2$ (ring)		
3,5-telluranedione	$CH_2COCH_2TeCH_2CO$ (ring)		

with nitrogen compounds such as ammonia, pyridine, simple and substituted thioureas and anilines, and ethylenediamine, as well as sulfur trioxide and the chlorides of other elements.

Tellurium Oxides, Oxyacids and Salts. These include tellurium oxide, TeO_2; tellurium trioxide, TeO_3; tellurous acid, H_2TeO_3, an unstable white powder that dehydrates to TeO_2; orthotelluric acid, H_6TeO_6; polymetatelluric acid, H_2TeO_{4n} ($n \simeq 10$); and tellurates, for example, potassium tellurate, K_2TeO_3.

Organic Compounds

Organotellurium compounds range from the simple carbon sulfotelluride to complex heterocyclic compounds and organotellurium ligands. Tellurium analogues of alcohols and mercaptans are prepared by reacting their vapors with aluminum telluride and protecting the products in an atmosphere of hydrogen. Various types of tellurium compounds and specific examples are listed in Table 2.

Uses

Tellurium has been shown to be the best additive for improving machinability in several types of ferritic steel. It is used to control the chill depth of iron castings (see IRON), and as a component in alloys of copper, lead, and other metals. Tellurium finds application as a pigment for ceramics and glass, in lubricant compositions, as a secondary vulcanizing agent in hard rubber compositions, and as a jelling component in explosives.

A very small, yet very important application of tellurium is in organic derivatives and radioactive isotopes for use as biological tracers, x-ray-contrast agents, and diagnostic aids, and for the treatment of thyroid diseases (see MEDICAL IMAGING TECHNOLOGY; RADIOACTIVE TRACERS).

Most metal tellurides are semiconductors with a large range of energy gaps and can be used in a variety of electrical and optoelectronic devices.

JAMES E. HOFFMANN
Jan Reimers and Associates USA Inc.
MICHAEL G. KING
Selenium–Tellurium Development Association
S. C. CARAPELLA
Consultant
J. E. OLDFIELD
Oregon State University
R. D. PUTNAM
Putnam Environmental Services

W. C. Cooper, ed., *Tellurium*, Van Nostrand Reinhold Co., New York, 1971.

Proceedings of the 6th Symposium on the Industrial Uses of Selenium and Tellurium, Phoenix, Arizona, May 1998, Selenium–Tellurium Development Association, Grimbergen, Belgium, 1998.

J. E. Hoffmann, *J. Metals*, 50–54 (1990).

A. A. Kudryavtsev, *The Chemistry and Technology of Selenium and Tellurium*, Collets Publishers Ltd., London and Wellingborough, U.K., 1974.

TEMPERATURE MEASUREMENT

The Kelvin Thermodynamic Temperature Scale

Temperature is a measure of the hotness of something. For a measure to be rational, there must be agreement on a scale of numerical values defining hotness and on devices for realizing and displaying these values. The single temperature scale having an absolute basis in nature is the Kelvin thermodynamic temperature scale (KTTS), or absolute scale, which is based on principles that can be deduced from the first and second laws of thermodynamics (qv). The most commonly used practical scale, however, is the International Temperature Scale (ITS), on which temperatures are designated as Celsius degrees (°C). Because the lower limit of the KTTS is absolute zero and the scale extends indefinitely upward and is by definition linear, only one nonzero reference point is required to stipulate its slope. The first such reference point was the equilibrium temperature of pure water in its liquid and solid phases at 101.3 kPa (1 atm) pressure, which was assigned the value 0°C, or 273.15 K. In 1954 the present KTTS was established by changing the reference point to the more reproducible equilibrium temperature of water in liquid–solid equilibrium under its own vapor pressure, ie, the triple point of water, assigned the value 273.16 K, or 0.01°C. (The unit of temperature on the KTTS is the kelvin, abbreviated K. The interval 1 K is identical to the interval 1°C. Thus these symbols may be used interchangeably to indicate an interval, but not to indicate a temperature. The symbol for a KTTS temperature is T; that for a temperature on any other scale is t).

The zero of the Celsius scale is therefore 273.15 K, and $T = t + 273.15$. The zero and the interval of the KTTS are defined without reference to properties of any specific substance.

The Fixed Points of Practical Temperature Scales

Accurate temperature measurements in real-life situations are difficult to make using the KTTS. Most easily used thermometers are not thermodynamic; that is, they do not operate on principles of the first and second laws. Most practicable thermometers depend upon some principle that is a repeatable and single-valued analogue of temperature, and they are used as interpolation devices of practical and utilitarian temperature scales which are themselves artifacts. Such principles include the expansion and contraction of liquids and solids, changes in the electrical properties of conductors and semiconductors, and the color and brilliance of light emitted from a very hot source.

Any such principles may be used to make a thermometer. Because they are nonthermodynamic, they require construction of a consensus scale to relate the properties of a prescribed interpolation device to the KTTS.

The KTTS depends upon an absolute zero and one fixed point through which a straight line is projected. Because they are not ideally linear, practicable interpolation thermometers require additional fixed points to describe their individual characteristics. Thus a suitable number of fixed points, ie, temperatures at which pure substances in nature can exist in two- or three-phase equilibrium, together with specification of an interpolation instrument and appropriate algorithms, define a practical temperature scale. The temperature values of the fixed points are assigned values based on adjustments of data obtained by thermodynamic measurements such as gas thermometry.

Two-phase equilibria may be solid–liquid, liquid–vapor, or solid–vapor. As is evident from the phase rule of Gibbs, two-phase equilibria are pressure-dependent: $P + V = C + 2$ where P is an integer equal to the number of phases present, C is the number of components (for an ideally pure material $C = 1$), and V is an integer giving the number of degrees of freedom. Phase equilibria involving a vapor phase are much more pressure-dependent than liquid–solid equilibria. Triple points, ie, equilibria of all three phases, are independent of external pressure.

The triple point of water is the most important of the fixed points, because it is the one point that the KTTS and the ITS have in common, and because it can be realized with great accuracy.

An important class of fixed points comprises the freezing points of high purity metals. As solid–liquid equilibria these are pressure-dependent, but this dependence is small (for tin, eg, 3.0×10^{-8} K/Pa (0.003 K/atm)) and can be corrected for pressure. The metals used in fixed-point cells are better than 99.9999% (6 nines) pure.

The ITS and Interpolation Instruments

The ITS is an artifact scale, designed to relate temperature measurements made with practicable instruments as closely as possible to the thermodynamic scale. The scale is established and controlled by the International Committee of Weights and Measures (BIPM) through its Consultative Committee on Thermometry.

Through the years the ITS has been changed a number of times, the better to fit its values to new information regarding the thermodynamic values of the fixed points. These changes have included revisions of (*1*) the range of the scale, (*2*) the interpolation instruments and their algorithms, and (*3*) values of the defining fixed points. The changes have been, for most purposes, nontrivial, and it is essential in reading a temperature from the literature or in citing a temperature to specify which version of the IPTS or ITS is meant.

The ITS-90 (the scale effective as of January 1, 1990) has its lowest point at 0.65 K and extends upward without specified limit. The standard platinum resistance thermometer (SPRT) is specified as the interpolation standard from 13.8033 K to 961.78°C, and the interpolation standard above 961.78°C is a radiation thermometer based on Planck's radiation law. Between 0.65 and 13.8033 K interpolation of the scale relies upon vapor pressure and constant-volume gas thermometry.

Platinum Resistance Thermometer Range: 13,8033–1234.93 K

Temperatures on the ITS-90 are expressed in terms of the ratio of the resistance of the SPRT at temperature to the resistance at the triple point of water. The resistance ratio $W(T_{90})$ is calculated as

$$W(T_{90}) = \frac{R(T_{90})}{R(273.16\ \text{K})}$$

In all previous scales the denominator of the ratio W was the resistance at 0°C = 273.15 K.

The temperature T_{90} is calculated from the resistance-ratio relationship, $W(T_{90}) - W_r(T_{90}) = \Delta W(T_{90})$, where $W(T_{90})$ is the observed value, $W_r(T_{90})$ is a value calculated from a reference function, and $\Delta W(T_{90})$ is the deviation of the observed $W(T_{90})$ of the individual SPRT from the reference function.

The ITS-90 provides for 11 SPRT ranges. The specification of so many ranges benefits the user of the SPRT over limited ranges, in that (*1*) accuracy of interpolation may be higher as the range is limited and (*2*) limited ranges may require fewer fixed-point determinations and consequently substantially lower calibration cost.

Resistance Thermometers

The standard platinum resistance thermometer (SPRT) must be made from almost ideally pure platinum wire mounted in a physical construction which will keep it in a strain-free condition.

SPRTs are devices of superb accuracy and resolution, but they are fragile and can be easily broken. They can also be put out of calibration by strain, induced by even slight mechanical shock or vibration. The principal use of SPRTs in science and industry is to maintain the calibrations of working thermometers.

Working-grade metallic resistance thermometers are made of metal wire, usually high-purity platinum, but also, more commonly in past years, copper, nickel, and alloys. The sensitive element is usually protected by a covering of ceramic tubing or ceramic cement. Such thermometers, conventionally called industrial resistance thermometers, are generally smaller than the SPRT element and may be as small as 2.5 mm in diameter and 10 mm in length. These are available in various 0°C resistances, eg, 100, 200, and 500 Ω. They are available as unsheathed elements or in a wide variety of sheaths and enclosures, both standard and custom. They are relatively inexpensive. They are usually made to be interchangeable, without relying on individual calibration, within limits of 0.25 K, or closer upon special order.

Among nonmetallic resistance thermometers, an important class is that of thermistors, or temperature-sensitive semiconducting ceramics. The variety of available sizes, shapes, and performance characteristics is very large.

The thermistor material is usually a metal oxide, eg, manganese oxide. Dopants, eg, nickel oxide or copper oxide, may be added to obtain a variety of resistance and slope characteristics. The material is usually sintered into a disk or bead with integral or attached connecting wires.

Two characteristics of thermistors are distinctly different from those of metallic resistance thermometers. For all but special types, the temperature-resistance characteristic is negative; that is, resistance decreases with increasing temperature (all metal resistance thermometers exhibit a positive characteristic). Also, the temperature-resistance characteristic is very nonlinear. Over favorable portions of the nonlinear characteristic, the change in resistance can be more than an order of magnitude larger than that of metallic resistance thermometers. This high dR/dT is a distinct advantage where high sensitivity is required over narrow ranges, as it is in many biological, medical, and environmental situations.

Seebeck Effect Thermometers (Thermocouples)

Thermocouples are composed of two dissimilar materials, usually in the form of wires, that accomplish a net conversion of thermal energy into electrical energy (E) with the occurrence of an electrical current. Unlike resistance thermometers, where the response is proportional to temperature (T), the response of thermocouples is proportional to the temperature difference between two junctions. Figure 1 illustrates such a circuit.

Thermocouple junctions can be made very small and in almost infinite variety. Properties of various types of working thermocouple are shown in Table 1.

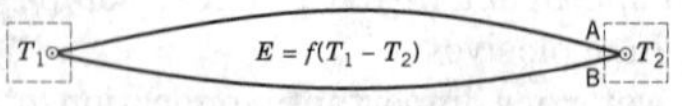

Figure 1. Basic thermocouple circuit. A and B are wires of different materials.

Table 1. Properties of Various Thermocouple Pairs

ASTM Type	Materials	Range °C	$\mu V/°C^a$
J	Iron-constantan	0–760	64.3
K	Chromel-alumel	0–1260	36.5
R	Pt/87Pt-13Rh[b]	0–1450	13.8
S	Pt/90Pt-10Rh[b]	0–1450	11.8
T	Copper-constantan	−183−375	53.0
E	Chromel-constantan	0–875	78.5
B	70Pt30Rh/94Pt6Rh[b]	870–1700	11.6

[a] Called the Seebeck coefficient. It is given here at the high-temperature end of the range; many thermocouples are nonlinear.
[b] Indicates, for example, one leg of pure platinum, the other of an alloy of platinum and rhodium. Alloy percentages are shown.

The thermal emf of the thermocouple is a function of the difference between the hot end and the cold end, the latter usually located at the readout instrument; thus the measurement can be no more accurate than the isothermality of the leads at the cold end and the accuracy with which this temperature is known.

Measurement of Emitted Radiation

Above 962°C, the freezing point of silver, temperatures on the ITS-90 are defined by a thermodynamic function and an interpolation instrument is not specified. The interpolation instrument universally used is an optical pyrometer, manual or automatic, which is itself a thermodynamic device.

An optical pyrometer contains an optical system for viewing the radiating target, a lamp filament, appropriate filters, an eyepiece, and a calibrated means for varying the lamp current and consequently the brightness of the filament. In the optical path, the filament is visible against the radiating source. The lamp filament current is varied until the image of the glowing filament is of such brilliance that its separate image is indistinguishable from the brilliance of the radiating source. The lamp current is then noted. If a second radiating source also causes the lamp filament to be indistinguishable at that lamp current, the temperature of the second source is said to be the same as that of the first.

Absolute calibrations on the ITS-90 are performed by sighting on a target which is a blackbody at the freezing temperature of silver, gold, or copper.

A strip lamp is a convenient means for the calibration of secondary pyrometers. The notched portion of a tungsten strip is the target. A pyrometer which has been calibrated against a radiating blackbody is sighted on the target, and the strip lamp current is adjusted to radiate at the intensity of the blackbody, as transferred by the primary pyrometer. The secondary pyrometer is then substituted for the primary, and the current required to raise the lamp filament to the brilliance of the target of the strip lamp is noted.

Measurement of Radiation From Real Objects

Real objects, when they are in an environment generally hotter or cooler than themselves, radiate, absorb, and reflect energy. The portion radiated is called the emissivity ϵ. If the portion reflected is r, then $\epsilon + r = 1$. No object can radiate more energy than can a blackbody at the same temperature, because a blackbody in equilibrium with a radiation field at temperature T radiates exactly as much energy as it absorbs. Any object exhibiting surface reflection must have emissivity of less than 1. Pyrometers are usually calibrated.

Other Electrical Thermometers

Many special-purpose electrical thermometers have been developed, either for use in practical temperature measurement, or as research devices for the study of temperature and temperature scales. Among the latter are thermometers which respond to thermal noise (Johnson noise) and thermometers based on the temperature dependence of the speed of sound.

A novel and useful thermometer, based on the change in resonant frequency of a quartz crystal, was produced by Hewlett-Packard. The sensor probe contain was a precisely cut quartz crystal about 8 mm in diameter. Each quartz sensor had a unique temperature-vs-frequency characteristic, and each sensor is individually calibrated with its operating electronics. Sensitivities of 100 μK are obtained at −50 to 150°C, the effective limits of measurement, and digital readout is provided. This system is a useful and practicable device in many environmental and energy situations, and it is particularly favored by oceanographers.

Nonelectrical Thermometers (Liquid in Glass)

The thermal expansions of liquids are reliable analogues of temperature. Liquids are suitable for use between their freezing and boiling points; mercury and colored alcohols are common materials. The typical liquid-in-glass thermometer includes a thin-walled glass bulb attached to a capillary stem partially filled with a visible liquid and sealed against the environment. The portion of the capillary above the liquid is usually filled with a dry a gas under some pressure to avoid column separation. A scale is etched or fused onto the stem.

Other Nonelectrical Thermometers

Other nonelectrical thermometers are bimetal, filled-system, and pyrometric cone thermometers.

HENRY E. SOSTMANN
Consultant

H. Preston-Thomas, *Metrologia* **27**, 3–10 (1990).
T. J. Quinn, *Temperature*, Academic Press, Inc., New York, 1983.
J. F. Schooley, *Thermometry*, CRC Press, Boca Raton, Fla., 1986.
J. V. Nicholas and D. R. White, *Traceable Temperatures*, John Wiley & Sons, Inc., New York, 1994.

TEREPHTHALIC ACID.

See PHTHALIC ACID AND OTHER BENZENEPOLYCARBOXYLIC ACIDS.

TERPENES AND TERPENOIDS.

See TERPENOIDS.

TERPENOIDS

Terpenes are found as constituents of essential oils and oleoresins of plants. Many important constituents of the essential oils have been identified and syntheses for them developed (see OILS, ESSENTIAL).

Terpenes are characterized as being made up of units of isoprene (**1**) in a head-to-tail orientation. This isoprene concept, invented to aid in the structure determination of terpenes found in natural products, was especially useful for elucidation of structures of more complex sesquiterpenes, diterpenes, and polyterpenes. The hydrocarbon, myrcene, and the terpene alcohol, α-terpineol, can be considered as being made up of two isoprene units in such a head-to-tail orientation.

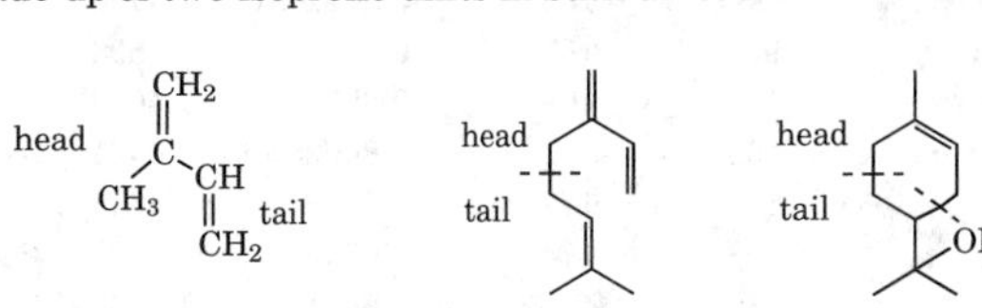

isoprene myrcene α-terpineol

(**1**)

The terpenes are further classified by the number of isoprene units in their carbon skeletons, as shown in Table 1. Many natural products contain an isolated isoprene unit derived from mevalonic acid and

Table 1. Classification of Terpenes

Isoprene units	Carbon atoms	Classification
1	5	hemiterpene
2	10	monoterpene
3	15	sesquiterpene
4	20	diterpene
5	25	sesterterpene
6	30	triterpene
8	40	tetraterpene
>8	>40	polyterpene

these compounds are thus considered hemiterpenoid derivatives. Turpentine is the largest-volume natural terpene source in commercial use in. The term terpenoids designates the various derivatives of the terpene hydrocarbons; and many of the oxygenated derivatives such as alcohols, aldehydes, esters, and ketones are very important flavor and fragrance chemicals.

Terpene chemists use trivial names for most of the compounds because the systematic names are much more complex. Common or trivial names and properties of selected terpenes and terpenoids are listed in Table 2. For commercial products, a material safety data sheet (MSDS), which is required by OSHA, frequently lists multiple names such as a product name, trivial name, IUPAC name and the TSCA name. The MSDS is a good source of information about physical properties, potential health hazards, and other useful information for the safe handling of the materials.

Terpene chemists use mainly gas chromatography in dealing with terpene mixtures in research and development as well as in quality control. Capillary gas chromatography with stable bonded-phase columns, the primary analytical method, is also used in product quality control, because its greater resolution is helpful in producing consistent products.

Fractional vacuum distillation is the method used to separate terpene mixtures into their components.

Capillary gc/ms, hplc, nmr, ir, and uv are all analytical methods used by the terpene chemist; with a good library of reference spectra, capillary gc/ms is probably the most important method used in dealing with the more volatile terpenes used in the flavor and fragrance industry (see FLAVORS AND SPICES). The physical properties of density, refractive index, boiling point, melting point of derivatives, and specific rotation are used less frequently but are important in defining product specifications.

Table 2. Properties of Selected Monoterpene [c]Hydrocarbons

	Bp, kPa[a], °C				
Common name	13.33	101.3	d^{20}, g/cm^3	n_D^{20}	$[\alpha]_D$
α-pinene	89	156	0.8595	1.4658	±51
α-fenchene	91	157	0.8697	1.4740	±44
camphene	91	158			±108
β-pinene	98	165	0.8722	1.4790	±22
myrcene (4)	93[b]	167	0.7880	1.4692	0
cis-pinane	101	168	0.8575	1.4629	±23
cis-*p*-menthane	105	172	0.8002	1.4431	0
1,4-cineole	105	172	0.8986	1.4446	0
1,8-cineole	108	174	0.9245	1.4574	0
limonene	110	176.5	0.8411	1.4730	±124
p-cymene	110	177	0.8570	1.4905	0
terpinolene	120	186	1.4861	1.4861	0

[a] To convert kPa to mm Hg, multiply by 7.5.
[b] At 9.33 kPa.
[c] At 2.0 kPa.

Monoterpenes

Production of Hydrocarbons from Turpentine. The majority of the turpentine comes from the southeastern United States, which consists of 60–70% α-pinene, 20–25% β-pinene, and 6–12% other components. Some of the other components consist of *p*-menthadienes, alcohols, ethers, and the sesquiterpene hydrocarbon, β-caryophyllene.

The crude sulfate turpentine coming from the mills of the Kraft sulfate process for paper is a commercially important raw material for obtaining α- and β-pinene for synthesizing aroma chemicals identical to natural products.

Synthesis of β-Methylheptenone from Petrochemical Sources. β-Methylheptenone (**1**) is an important intermediate in the total synthesis of terpenes. Continuous hydrochlorination of isoprene produces prenyl chloride, which then reacts with acetone with a quaternary ammonium catalyst and sodium hydroxide to give β-methylheptenone (6-methylhept-5-en-2-one).

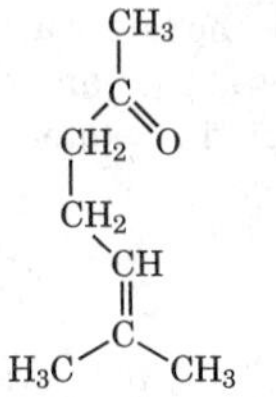

(**1**)

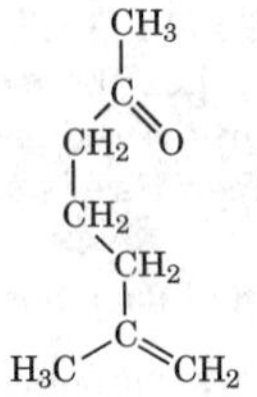

(**2**)

Another process involves a one-step reaction of isobutylene with formaldehyde and acetone under high temperature and pressure. α-Methylheptenone (**2**) (6-methylhept-6-en-2-one) is the product, but it is easily catalytically isomerized to β-methylheptenone.

The methylheptenones are intermediates for the synthesis of linalool (**3**).

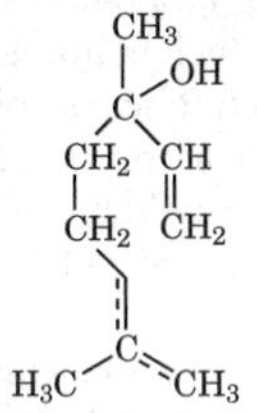

α-,β-linalool

(**3**)

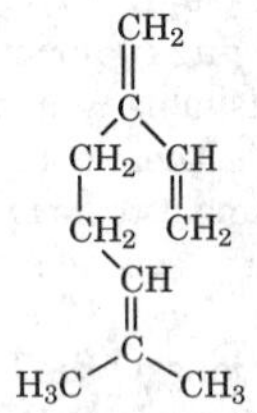

(**4**)

Dimerization of Isoprene. Isoprene is an important raw material for the production of terpenes. For example, myrcene (**4**) can be produced by the dimerization of isoprene (2-methyl-1,3-butadiene); and myrcene is very useful for synthesizing a number of oxygenated terpenes important in the flavor and fragrance industry.

α-Pinene Manufacture. Industrially, α-pinene produced from the fractionation of sulfate turpentine can be used directly for most of its applications.

Uses and Reactions. α-Pinene is useful for synthesizing a wide variety of terpenoids. Hydration to pine oil, acid-catalyzed isomerization to camphene, thermal isomerization to ocimene and alloocimene, and polymerization to terpene resins are some of its direct uses.

Another important use of α-pinene is the hydrogenation to *cis*-pinane.

Pyrolysis of the *cis*-pinane produces dihydromyrcene (citronellene) as the major product. Fractionation of the crude product then gives an 87 wt % dihydromyrcene. Dihydromyrcenol produced from the dihydromyrcene is becoming increasingly important as a fragrance material. It has excellent stability and has a powerful, fresh, lime-like aroma.

β-Pinene Manufacture. β-Pinene is obtained by fractionation of turpentine. It is used in flavor and perfumery applications. Most of the β-pinene produced by the turpentine fractionators is used captively for producing fragrance chemicals or for β-pinene resins.

Uses and Reactions. Some of the principal uses for β-pinene are for manufacturing terpene resins and for thermal isomerization (pyrolysis) to myrcene. β-Pinene polymerizes much easier than α-pinene, and the resins are useful in pressure-sensitive adhesives, hot-melt adhesives and coatings, and elastomeric sealants.

The production of myrcene (**4**) from β-pinene is important commercially for the synthesis of a wide variety of flavor and fragrance materials. Some of those include nerol and geraniol, and citral (**5**), and citronellol (**6**).

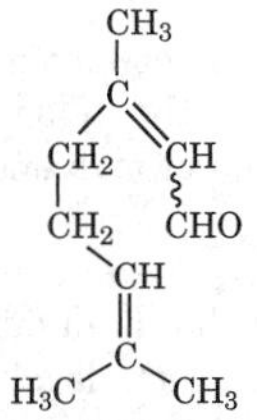

(**5**)

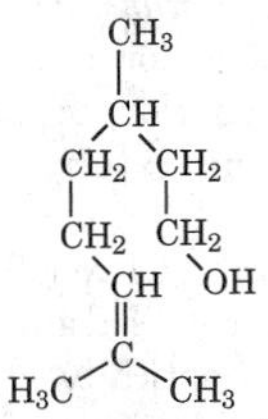

(**6**)

3-Carene Manufacture. 3-Carene is obtained by fractional distillation of turpentine. Turpentine from the western United States and Canada averages about 25% 3-carene; much of it is unutilized although it is obtained in high optical purity.

p-Menthadienes and p-Cymene Manufacture. The *p*-menthadienes are mainly produced as by-products from the manufacture of synthetic pine oil. The mixtures of the *p*-menthadienes are also commonly referred to as dipentene.

Uses and Reactions. Dipentene is a good solvent for paints, varnishes, and enamels that contain synthetic resins, particularly phenolic resins. It is also used as an antiskinning agent and as a wetting agent in the dispersion of pigments. The solvency of dipentene for rubber and its swelling and softening properties make it useful in rubber reclaiming and in the processing of natural and synthetic rubbers. Dipentene is also formulated into a variety of cleaners similar to pine oil cleaners.

Camphene Manufacture. Camphene is produced by the reaction of α-pinene with a TiO_2 catalyst. Tricyclene is formed as a coproduct, but it undergoes the same reactions as camphene; thus the product is generally used as a mixture.

Uses and Reactions. Camphene is used for preparing a number of fragrance compounds. Condensation with acids such as acetic, propionic, isobutyric, and isovaleric produce useful isobornyl esters. Isobornyl acetate has the greatest usage as a pine fragrance. The isobornyl esters of acrylic and methacrylic acids are also useful in preparing acrylic polymers.

Myrcene Manufacture. An important commercial source for myrcene is its manufacture by pyrolysis of β-pinene at 550–600°C.

Purified myrcene has minimal use in flavor and fragrance applications.

Uses and Reactions. The largest use of myrcene is for the production of the terpene alcohols nerol (**7**), geraniol (**8**), and linalool (**3**). The nerol and geraniol are further used as intermediates for the production of other large-volume flavor and fragrance chemicals such as citronellol, dimethyloctanol, citronellal, hydroxycitronellal, racemic menthol, citral, and the ionones and methylionones.

Nerol, geraniol, and linalool, known as the rose alcohols, are found widely in nature. Nerol and geraniol have mild, sweet odors reminiscent of rose flowers. They are manufactured by the hydrochlorination of myrcene at the conjugated double bonds when a copper catalyst is used.

Alloocimene Manufacture. α-Pinene is converted thermally first to *cis*-ocimene, which rearranges to give the two alloocimene isomers, ie, 4-*trans*-6-*cis*-alloocimene and 4-*trans*-6-*trans*-alloocimene, along with other terpenoids.

Oxidation of alloocimene in the presence of a catalyst produces a polymeric peroxide, which can be thermally isomerized to produce alloocimene diepoxide, which has been used in the manufacture of resins and as an acid scavenger for halogenated solvents.

Dihydromyrcene Manufacture. 2,6-Dimethyl-2,7-octadiene, commonly known as dihydromyrcene or citronellene, is produced by the pyrolysis of pinane, which can be made by hydrogenation of α- or β-pinene.

Uses and Reactions. Dihydromyrcene is used primarily for manufacture of dihydromyrcenol, an important fragrance material. Dihydromyrcene can be catalytically hydrated to dihydromyrcenol by a variety of methods.

Monoterpene Alcohols

Pine Oil Manufacture. Synthetic pine oil manufacture is one of the principal uses of turpentine. Natural pine oil is a product derived from the extraction of aged pine stumps, and sulfate pine oil is a product separated from crude sulfate turpentine in about 5% yield.

Synthetic pine oil is produced by the acid-catalyzed hydration of mainly α-pinene derived from sulfate turpentine, followed by distillation of the crude mixture of hydrocarbons and alcohols. The predominant alcohol obtained is α-terpineol.

Pine oils can be fractionally distilled to produce a higher α-terpineol product, but usually contain borneol and γ-terpineol, along with small amounts of other components. High grade perfumery α-terpineol can be made by the partial dehydration of *p*-menthane-1,8-diol (terpin hydrate) under mildly acidic conditions.

Uses and Reactions. The largest use of pine oil is in the manufacture of cleaners and disinfectants. It is effective against gram-negative enterobacteria but not against gram-positive organisms. Lower grades such as sulfate pine oil are used for the flotation of metallic sulfide ores. In textiles, the most important property of pine oil is its ability to reduce surface tension and interfacial tension between fiber and solution. Also, because of its bacteriocidal activity, it is used in almost all wet-processing of cotton, silk, rayon, and woolen goods.

Nerol and Geraniol Manufacture. Citronella oil is the principal natural source of geraniol. It occurs as a mixture with citronellol and citronellal and can be separated by fractional distillation. Nerol and geraniol are produced synthetically from turpentine sources, either from α-pinene or β-pinene in far greater quantities than are produced from natural sources. A considerable amount of nerol and geraniol is used captively by the producers as intermediates for manufacture of other terpene products, namely citral, ionones, methyl ionones, citronellol, citronellal, hydroxycitronellal, (±)-menthol, vitamins A and E, and carotenoids.

Uses and Reactions. Nerol (**7**) and geraniol (**8**) can be converted to citronellol by hydrogenation over a copper chromite catalyst. If a nickel catalyst is used, a mixture of nerol, geraniol, and citronellol is obtained, and such a mixture is useful in perfumery.

Linalool Manufacture. The most important natural source of linalool is bois de rose oil from which it is separated by distillation. By far, more linalool (**3**) is produced synthetically than is obtained from the natural sources.

Isoprene (2-methyl-1,3-butadiene) can be telomerized in diethylamine with *n*-butyllithium as the catalyst to a mixture of *N,N*-diethylneryl- and geranyl-amines. Oxidation of the amines with hydrogen peroxide gives the amine oxides, which, by the Meisenheimer rearrangement and subsequent pyrolysis, produce linalool in an overall yield of about 70%.

Uses and Reactions. Linalool can be esterified to linalyl acetate by reaction with acetic anhydride. Linalyl acetate has a floral-fruity odor, reminiscent of bergamot and lavender.

Linalool can be hydrogenated to dihydrolinalool and tetrahydrolinalool, both of which are used in perfumery.

Linalool has been used to prepare a mixture of terpenes useful for enhancing the aroma or taste of foodstuffs, chewing gums, and perfume compositions.

Citronellol Manufacture. Citronellol is found widely in nature and in both optically isomeric forms. Prior to the development of synthetic citronellol, this alcohol was obtained from certain oils of the *Rosaceae* family or by hydrogenation of citronellal isolated from citronella oil. Citronellol has a floral odor resembling that of roses.

Citronellol is manufactured on a commercial scale by the hydrogenation of nerol and geraniol made either from α- or β-pinene. Hydrogenation of nerol and geraniol over a copper chromite catalyst gives a high yield of citronellol. Fractional distillation of the crude product produces a perfumery-quality citronellol.

Uses and Reactions. The main application for citronellol is for use in soaps, detergents, and other household products. It is also important as an intermediate in the synthesis of other important fragrance compounds, such as citronellyl acetate and other esters, citronellal, hydroxycitronellal, and menthol.

Menthol Manufacture. Of the menthol isomers, only (−)-menthol and (±)-menthol are of commercial importance. The most important natural sources of (−)-menthol are the oils of *Mentha arvensis* and *Mentha piperita*. (−)- Menthol is known for its refreshing, diffusive odor characteristic of peppermint. It also is known for its strong physiological cooling effect, which is useful in cigarettes, dentrifices, cosmetics, and pharmaceuticals.

Natural menthol is obtained by freezing the essential oil, eg, *Mentha arvensis*, and the menthol crystals are separated by centrifuging the supernatant liquid away from the crystals.

All (±)-menthol is made by synthetic methods. One method involves the cyclization of (+) -citronellal. Using a mild acid catalyst, (+) -citronellal undergoes an ene-reaction to produce a mixture of isopulegols. Catalytic hydrogenation of the isopulegol mixture gives a mixture of menthol and its isomers. The (±)-menthol is obtained after efficient fractional distillation.

(±)-Menthol can also be made synthetically by hydrogenation of thymol, which can be produced by isopropylation of *m*-cresol with propylene.

Monoterpene Aldehydes and Ketones

Campholenic Aldehyde Manufacture. Campholenic aldehyde is readily obtained by the Lewis-acid-catalyzed rearrangement of α-pinene oxide. It has become an important intermediate for the synthesis of a wide range of sandalwood fragrance compounds. Epoxidation of (+)-α-pinene also gives the (+)-*cis*-α-pinene epoxide, and rearrangement with zinc bromide is highly stereospecific and gives (−)-campholenic aldehyde.

Aldol condensation of compholenic aldehyde with propionic aldehyde or n-butyraldehyde yields the intermediate conjugated aldehyde, which can be selectively reduced to the saturated alcohol with a sandalwood odor.

Citral Manufacture. Natural sources of citral are lemongrass oil and *Litsea cubeba*. Both oils contain 70–80 wt % citral. Synthetic citral is made from terpene sources such as nerol and geraniol and in multitonnage quantities from petrochemical sources. Most terpene-based citral (**5**) produced is based on the catalytic oxidative dehydrogenation of nerol (**7**) and geraniol (**8**), or by the Oppenauer oxidation of nerol and geraniol.

Petrochemical-based methods of citral manufacture are very important for the large-scale manufacture of Vitamin A and carotenoids.

Reactions. Besides the large quantity of citral used in the Vitamin A and carotenoid industry, a large amount is used to produce perfume ionones and methyl ionones.

Ionones and Methyl Ionones Manufacture. The discovery of ionones and methylionones was an early example of the need to develop synthetic fragrance materials because of the high cost of natural materials. The aroma of violet flowers was important to perfumery and led to the development of ionones and methylionones at the end of the nineteenth century.

Citral reacts in an aldol condensation using excess acetone and a basic catalyst. The resulting intermediate pseudoionone after cyclization with phosphoric acid gives predominantly α-ionone, which is the isomer commercially important in flavors and fragrances.

Camphor Manufacture. Camphor is obtained both naturally and synthetically. Natural camphor is obtained from the wood of the camphor tree, *Cinnamormum camphora*, which grows in China and Japan. The camphor is isolated by combination of steam distillation, filtration, distillation, and sublimation. Natural camphor is the (+)-camphor, whereas synthetic camphor is racemic; both products are recognized by the USP. The largest single use of camphor is for religious purposes in Asian countries.

Most synthetic camphor is produced from camphene made from α-pinene. The conversion to isobornyl acetate followed by saponification produces isoborneol.

Citronellal Manufacture. Natural sources of citronellal are citronella oil and *Eucalyptus citridora*.

Synthetic methods for the production of citronellal include the catalytic dehydrogenation of citronellol, the telomerization of isoprene, and the lithium-catalyzed reaction of myrcene with secondary alkylamines.

Uses and Reactions. Citronellal undergoes an ene-reaction to form isopulegols, which are used mainly for the manufacture of menthol. If the citronellal is optically active, the isopulegol is also optically active. This is very important for the synthesis of (−)-menthol from (+)-citronellal. Citronellal can also be converted to the *cis*- and *trans*-*p*-menthane-3,8-diol by reaction with dilute acids. The glycol mixture can be readily purified by distillation and the two isomers easily separated. The glycols are useful as insect repellents (qv) and are especially effective against mosquitos.

Sesquiterpenes

Sesquiterpenes are formed by the head-to-tail arrangement of three isoprene units (15 carbon atoms); there are, however, many exceptions to the rule. Because of the complexity and diversity of the substances produced in nature, it is not surprising that there are many examples of skeletal rearrangements, migrations of methyl groups, and even loss of carbon atoms to produce norsesquiterpenoids.

Important commercial sesquiterpenes mostly come from essential oils, for example, cedrene and cedrol from cedarwood oil. Many sesquiterpene hydrocarbons and alcohols are important in perfumery as well as being raw materials for synthesis of new fragrance materials. There are probably over 3000 sesquiterpenes that have been isolated and identified in nature.

The sesquiterpenes found in essential oils have low volatilities compared with monoterpenes and so are isolated mainly by steam distillation or extraction, but some are also isolated by distillation or crystallization. Most of the sesquiterpene alcohols are heavy viscous liquids and many crystallize when they are of high enough purity. Sesquiterpene alcohols are important in perfume bases for their odor value and their fixative properties as well. They are valuable as carriers of woody, balsamic, or heavy oriental perfume notes.

Diterpenes. Diterpenes contain 20 carbon atoms. The resin acids and Vitamin A are the most commercially important group of diterpenes. Gibberellic acid (**9**), produced commercially by fermentation processes, is used as a growth promoter for plants, especially seedlings.

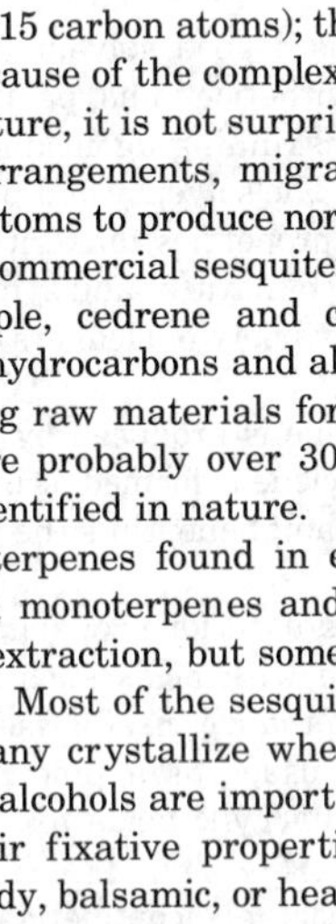

nerol(cis-) (**7**)
geraniol(trans-) (**8**)

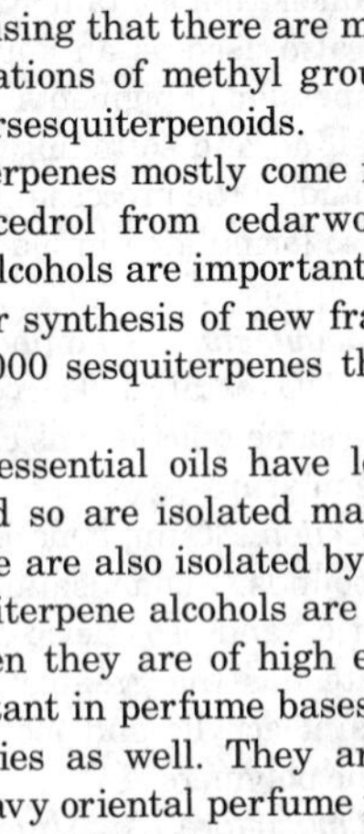

(**9**)

Phytol and isophytol are important intermediates used in commercial synthesis of Vitamins E and K.

The tricyclic diterpenoid, paclitaxel (Taxol), is obtained by extraction from the bark of the Pacific Yew tree, *Taxus brevifolia*, which grows in forests of the western United States and Canada. Taxol (**10**) has shown promising results in fighting advanced stages of ovarian, breast, and other cancers. The yew trees, however, are slow growing and isolation of the Taxol is difficult and expensive. The patent and chemical literature abound with efforts to synthesize Taxol and analogues.

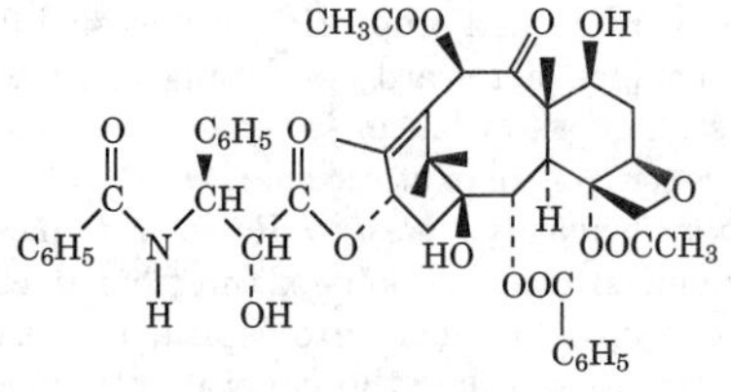

(10)

Triterpenes. The triterpenes (30 carbon atoms) are widely found in nature, especially plants, both in the free state and as esters or glycosides. A smaller but important group, including lanosterol, occurs in animals. The triterpene hydrocarbon, squalene, occurs in the liver oils of certain fish, especially those of sharks. Squalene is also an intermediate in the synthesis of cholesterol. Structurally, chemically, and biogenetically, many of the triterpenes have much in common with steroids. It has been verified experimentally that *trans*-squalene is the precursor in the biosynthesis of all triterpenes through a series of cyclization and rearrangement reactions.

Squalane (fully saturated squalene) is produced synthetically by the coupling of two molecules of geranyl acetone with diacetylene, followed by dehydration and complete hydrogenation.

Tetraterpenes. Carotenoids make up the most important group of C_{40} terpenes and terpenoids, although not all carotenoids contain 40 carbon atoms. They are widely distributed in plant, marine, and animal life.

Carotenoids may be acyclic, monocyclic, or bicyclic. The respective parent compounds for these categories are lycopene, γ-carotene, and β-carotene. The geometrical configuration of the double bonds is usually trans. The prefix neo is often used to designate isomers containing at least one cis-configuration. The prefix apo indicates carotenoids that are oxidative degradation products retaining more than half of the carotene structure. There are about 400 naturally occurring carotenoids of known structure.

An important function of certain carotenoids is their provitamin A activity. Provitamin A compounds are converted to Vitamin A by an oxidative enzyme system present in the intestinal mucosa of animals and humans. This conversion apparently does not occur in plants (see VITAMINS, VITAMIN A).

Carotenoids are also used as pigments and dietary supplements in animals and poultry feedstuffs.

β-Carotene is prescribed in the treatment of the inherited skin disorder erythropoietic protoporphyria (EPP) to reduce the severity of photosensitivity reactions in such patients.

JAMES O. BLEDSOE, JR.
Bush Boake Allen, Inc.

W. F. Erman, *Studies in Organic Chemistry*, Vol. II, *Chemistry of the Monoterpenes: An Encyclopedic Handbook*, Marcel Decker, New York, 1985.

G. Ohloff, *Scent and Fragrances, The Fascination of Odors and Their Chemical Perspectives*, transl. by W. Pickenhagen and B. M. Lawrence, Springer-Verlag, Berlin, 1994.

E. T. Theimer, ed., *Fragrance Chemistry*, Academic Press, Inc., New York, 1982.

D. F. Zinkel and J. Russell, eds., *Naval Stores*, Pulp Chemicals Association, Inc., New York, 1989.

TEXTILES

SURVEY

Textiles are manufactured from staple fibers (qv), which have finite lengths, and filaments, which have continuous lengths, by a variety of processes to form woven, knitted, and nonwoven or feltlike fabrics. In woven and knitted fabrics, the fibers and filaments are formed into continuous-length yarns, which are then either interlaced by weaving or interlooped by knitting into planar, flexible, sheet-like structures known as fabrics. Nonwoven fabrics (qv) are formed directly from fibers and filaments by chemically or physically bonding or interlocking fibers that have been arranged in a planar configuration.

Classification

Textile fibers may be classified according to their origin, as follows.

Naturally occurring fibers: Vegetable, (based on cellulose) cotton, linen, hemp, jute, ramie; Animal, (based on proteins) wool, mohair, vicuna, other animal hairs, silk; Mineral, asbestos.

Manufactured fibers based on natural organic polymers: rayon, regenerated cellulose (viscose and cuprammonium); lyocell, regenerated cellulose (solvent process); acetate, partially acetylated cellulose derivative; triacetate, fully acetylated cellulose derivative; and azlon, regenerated protein

Manufactured fibers based on synthetic organic polymers: acrylic, polyacrylonitrile (also modacrylic); aramid, aromatic polyamides; nylon, aliphatic polyamides; olefin, polyolefins (polyethylene and polypropylene); polyester, polyesters of aromatic dicarboxylic acids and dihydric alcohols; spandex, segmented polyurethane; vinyon, poly(vinyl chloride); vinal (or vinylon), poly(vinyl alcohol); carbon/graphite, derived from polyacrylonitrile, rayon or pitch; and specialty fibers such as those based on poly(phenylene sulfide), polyetheretherketone, polyimides, and others.

Manufactured fibers based on inorganic substances: glass, metallic, and ceramic.

With the exception of silk (qv), which is extruded by the silkworm as a continuous filament, natural fibers are of finite length and are used directly in textile manufacturing operations after preliminary cleaning. These fibers are known as staple fibers, and an estimate of their average length is referred to as their staple length.

The other major category of textile fibers is that of the manufactured fibers, which are produced from natural organic polymers, synthetic organic polymers, and inorganic substances. Glass fiber is the only inorganic manufactured fiber in common use, although other ceramic and metallic fibers are being developed.

Manufactured Fibers

Manufactured fibers produced from natural organic polymers are either regenerated or derivative. A regenerated fiber is one which is formed when a natural polymer or its chemical derivative is dissolved and extruded as a continuous filament, and the chemical nature of the natural polymer is either retained or regenerated after the fiber-formation process. A derivative fiber is one which is formed when a chemical derivative of the natural polymer is prepared, dissolved, and extruded as a continuous filament, and the chemical nature of the derivative is retained after the fiber-formation process.

Manufactured fibers based on synthetic organic polymers are generally referred to as the synthetic fibers. The production of manufactured synthetic fibers is based on three methods of fiber formation, or extrusion spinning. In this context, spinning refers to the overall process of polymer liquefaction (dissolution or melting), extrusion, and fiber formation. The three principal methods are melt spinning, dry spinning, and wet spinning, although there are many variations and combinations of these basic processes. (The term spinning is otherwise customarily reserved for that textile manufacturing operation wherein staple fibers are formed into continuous textile yarns by several consecutive attenuating and twisting steps. A yarn so formed from natural or manufactured staple fibers is referred to as a staple or spun yarn.)

In melt spinning, the polymer is heated above its melting point and the molten polymer is forced through a spinneret, a die with many small orifices. The jet of molten polymer emerging from each orifice in the spinneret is guided to a cooling zone where the polymer solidifies to complete the fiber-formation process. In dry spinning, the polymer is dissolved in a suitable solvent and the resultant solution is extruded under pressure through a spinneret. The jet of polymer solution is guided to a heating zone where the solvent evaporates and the filament solidifies. In a wet spinning, the polymer is also dissolved in a suitable solvent and the solution is forced through a spinneret which is submerged in a coagulation bath. As the polymer solution emerges from the spinneret orifices in the coagulating bath, the polymer is either precipitated or chemically regenerated.

In most instances the filaments formed by melt, dry, or wet spinning are not suitable textile fibers until they have been subjected to one or more successive drawing operations. Drawing is the hot or cold stretching and attenuation of manufactured filaments to induce molecular orientation with respect to the fiber axis, and to develop a fiber fine structure.

In the production of manufactured fibers, the filaments are obtained in continuous form. When several such filaments are combined together and slightly twisted to maintain unity, the product is called a multifilament yarn. Individual filaments, may also be used in certain applications, and these are referred to as monofilaments.

Types of Textile Materials

Textile yarns are produced from staple (finite-length) fibers by a combination of processing steps, referred to collectively as yarn spinning. After preliminary fiber alignment, the fibers are locked together by twisting the structure to form the spun yarn, which is continuous in length and remarkably strong and uniform. The staple fibers may be either natural fibers, such as cotton or wool, or any of a number of manufactured fibers.

Yarns are used principally in the formation of textile fabrics either by weaving or knitting processes. In a woven fabric, two systems of yarns, known as the warp and the filling, are interlaced at right angles to each other in various patterns. The woven fabric can be viewed as a planar, sheetlike material with pores or holes created by the yarn interlacing pattern.

Knitted fabrics are produced from one set of yarns by looping and interlocking processes to form a planar structure. The pores in knitted fabrics are usually not uniform in size and shape.

In the production of nonwovens, the fibers are processed directly into a planar, sheet-like fabric structure, bypassing the intermediate one-dimensional yarn state, and then are either bonded chemically or interlocked mechanically, or both, to achieve a cohesive fabric.

Dyeing and Finishing

After the fabric formation process, textiles are generally subjected to either dyeing or printing and to a variety of mechanical and chemical finishing operations. The specific nature of the dyeing and finishing operations depends on the fiber type and on the intended use of the fabric.

LUDWIG REBENFELD
TRI/Princeton

B. C. Goswami, J. G. Martindale, and F. L. Scardino, *Textile Yarns: Technology, Structure and Applications*, John Wiley & Sons, Inc., New York, 1977.

J. Lunenschloss and W. Albrecht, *Nonwoven Bonded Fabrics*, Halsted Press, a division of John Wiley & Sons, Inc., New York, 1985.

H. Zollinger, *Color Chemistry: Syntheses, Properties and Applications of Organic Dyes and Pigments*, VCH Publishers, New York, 1987.

G. C. Tesoro, in M. Lewin and S. B. Sello, eds., *Functional Finishes Part A, Handbook of Fiber Science and Technology*, Vol. II, Marcel Dekker, Inc., New York, 1983, pp.

FINISHING

Textile finishing includes various efforts to improve the properties of textile fabrics. In particular, these processes are directed toward modifying either the fiber characteristics themselves or the gross textile end properties. Such modifications may be chemical or mechanical in nature.

Textile finishing encompasses a broad range of approaches and may be directed toward needed properties such as shrinkage control or smooth-dry performance or toward developing properties for specific end uses such as flame retardance, soil release, smolder resistance, weather resistance, or control of static charges.

Treatments with Chemicals or Resins. Resin treatments are divided into topical or chemical modifications of the fiber itself. Most chemical treatments of synthetic fibers are topical because of the inert character of the fiber itself and the general resistance of the fiber to penetration by reagents. By contrast, cellulosics and wool possess chemical functionality that makes them reactive with reagents containing groups designed for such purchases. Natural fibers also provide a better substrate for nonreactive topical treatments because they permit better penetration of the reagents.

Chemical Treatments

Chemical treatments of textiles encompass a variety of approaches. In one, the finisher may be attempting to form a chemical bond between the reactive group of the fiber (–OH group of cellulose) and the applied reagent. If the bond is resistant to hydrolysis, the property conferred to the garment, for example, smooth-dry performance, will be durable to laundering. If the bond is not durable, the change accomplished by finishing will gradually disappear.

Another approach in chemical finishing is to use reagent systems that are reactive with themselves but only to a limited extent or not at all with the fiber substrate. An example of such approaches are *in situ* polymer systems that form a condensed fiber system within the fiber matrix. A third type of approach may be the deposition of a polymer system on the fiber substrate. Once deposited, such systems may show a strong affinity to the fiber and may be quite durable to laundering.

Methods of Application. The predominant system used for finishing cotton and other cellulosics is the so-called pad, dry, and cure process. The padding is done normally by immersing the fabric in an aqueous solution, followed by squeezing it between two rollers, and finally drying and curing.

Following fabric padding, the fabric in conventional processing is dried open-width in ranges at high temperatures.

To many textile chemists, finishing refers primarily to chemical or resin finishing. The principal chemicals used in early finishing processes were aminoplasts capable of cross-linking cellulose as well as of homopolymerization. These agents not only have the ability to improve smooth-dry performance and dimensional stability of fabrics, but also to improve other useful properties.

Several other processes should also be mentioned. One fabric treatment is mercerization. This process is done to improve fabric strength, luster, and dye yield for those fabrics that are to be subsequently dyed.

A second treatment that has gained a certain amount of use in modern textile practice is that of ammonia mercerization. Fabrics treated with liquid ammonia are generally softer and exhibit better smooth-dry performance than fabrics without such treatment.

Cross-Linking of Cellulosics Fabrics. Essentially, any compound containing two reactive groups can be used to cross-link cotton. Exceptions are those that are too large to penetrate the fiber and perhaps those in which the reactive groups are widely separated. In the cross-linking reaction of equation 1, the fabric takes on a memory for its state at the moment of cross-linking:

$$2\ \text{Cell—OH} + \text{HO—X—OH} \rightleftharpoons \text{Cell—O—X—O—Cell} + 2\ H_2O \quad (1)$$

Thus, if the fabric is flat and smooth, it will tumble dry in that configuration. On the other hand, if the fabric is cross-linked in a creased condition, as in a pleated skirt, the original pleated skirt configuration should return on laundering and tumble drying.

Early Cross-Linking Agents. Formaldehyde, urea–formaldehyde, and melamine–formaldehyde were among the earliest agents utilized for resin finishes. Concerns about the safety of formaldehyde, the need for lower formaldehyde release values, and the safety of exposure to melamine have reduced the use of these early cross-linking agents by industry substantially.

Chemistry of N-Methylol Agents. The reaction of dimethylolurea and cellulose is illustrated in equation 2:

$$HOCH_2NH\overset{O}{\overset{\|}{C}}NHCH_2OH + 2\ Cell\text{—}OH \rightleftharpoons Cell\text{—}OCH_2NH\overset{O}{\overset{\|}{C}}NHCH_2O\text{—}Cell + 2\ H_2O \quad (2)$$

First, it should be noted that the N-methylol group (CH_2OH) is activated by the carbonyl group. This reactive group is present in almost all N-methylol systems. The reaction is an equilibrium reaction, so that both forward and reverse reactions can occur. Third, the agent is not simply a dimethylol agent, but is predominantly a mixture of mono- and di-substituted ureas.

Mechanisms for Formation and Hydrolysis of Finishes. The general mechanism for acid-catalyzed formation and hydrolysis of N-methylol cellulose cross-links has been shown to pass through a carbonium ion intermediate as in equations 3 and 4:

$$\text{—}\overset{O}{\overset{\|}{C}}NHCH_2OH + H^+ \rightleftharpoons \text{—}\overset{O}{\overset{\|}{C}}NHCH_2\overset{H^+}{\overset{|}{O}}H \rightleftharpoons \text{—}\overset{O}{\overset{\|}{C}}NHCH_2^+ + H_2O \quad (3)$$

$$\text{—}\overset{O}{\overset{\|}{C}}NHCH_2^+ + Cell\text{—}OH \rightleftharpoons \text{—}\overset{O}{\overset{\|}{C}}NHCH_2O\text{—}Cell + H^+ \quad (4)$$

Based on the principle of microscopic reversibility, it has been reasoned that a highly reactive agent would form an easily hydrolyzed product. This proved true in practice, and the industry has adopted the phrase "easy on means easy off".

Curing Catalysts for N-Methylol Agents. Catalyst selection must take into consideration not only achievement of the desired chemical reaction, but also such secondary effects as influence on dyes, effluent standards, formaldehyde release, discoloration of fabric, chlorine retention, and formation of odors. In much of the industry, the chemical supplier specifies a catalyst for the agent so the exact content of the catalyst may not be known by the finisher. Types of catalysts used include mineral or organic acids and latent acids such as ammonurea salts, amine salts, and metal salts.

Melamine and Other Amino-S-Triazines. In the case of melamines, hypochlorite does not cause fabric degradation but does lead to yellow fabric. A partially methylated di- or trimethylol melamine is an excellent polymer-former as well as an effective cross-linking agent. However, this polymer-forming characteristic can lead to poor shelf life and complications in finishing because the degree of self-polymerization before fabric treatment is a hard-to-control variable.

Melamines have found utility as rotproofing and weatherproofing cellulosics, as binders for pigments and for transfer printing of cotton and cotton blends, as well as in numerous other applications.

Triazones, Urons, and Alkylene Ureas. Triazones, urons, and alkylene ureas were the first commercial N-methylol cross-linking agents for reducing or eliminating the hypochlorite bleach problem. These agents are still commercially available; however, the release of formaldehyde by these agents has contributed to their decrease in popularity.

Delayed Cure and Permanent Press. Delayed cure systems are designed so that the chemical agent is applied in the textile mill, but final curing is delayed until after the garment is fabricated and pressed into the final desired configuration. This approach has led to garments having a new, higher level of performance. Thus garments not only can be smooth-drying, but also have desired shapes fixed into them, eg, the permanent crease of trousers.

Incorporation of a level of nylon or polyester in the fabric substantially increases the garments' abrasion resistance. The 50% cotton–50% polyester fabric seems to contain sufficient cellulosic to benefit from a chemical finish and sufficient synthetic to provide strength and abrasion resistance.

Two types of approaches are available. In one, the fabric is padded with the cross-linker finish, dried, then sent to the garment cutter. The garments are then pressed and cured. In the second, the fabric is cured in fabric form, then fabricated into garments. It is then pressed and recured in hot-head presses.

Sources of Formaldehydes in Textiles and Formaldehyde Analysis. Modern textile mill finishes need to be concerned with formaldehyde in the finishing garment plants and in the textile product itself. Formaldehyde can be present in the air, in cross-linking agent pad baths, and in the finished fabrics. The maximum concentration of formaldehyde in the air of workplaces should be dictated; in work areas, concentrations in air above 1 ppm are unusual. The free formaldehyde in finishing formulations has been substantially reduced and mill conditions in finishing areas and garment plants have been greatly improved.

The American Association of Textile Chemists and Colorists (AATCC) Test Method 112, or the sealed-jar test, developed in the United States and used extensively for 25 years, measures the formaldehyde release as a vapor from fabric stored over water in a sealed jar. Results from this test have been used to eliminate less stable finishes.

Control of Formaldehyde Release. Once the sealed-jar test became a factor in measuring the formaldehyde release of fabrics supplied to garment cutters, limitations were placed on the allowable limits acceptable to the garment producers. These limits brought to the fore two classes of reagents: those based on 1,3-dimethylol-4,5 dihydroxyethyleneurea (DMDHEU), and those based on the N,N-dimethylolcarbamates.

Several factors were utilized in bringing formaldehyde release down. In particular, resin manufacturers executed more careful control of variables such as pH, formaldehyde content, and control of methylolation. There has also been a progressive decrease in the resin content of pad baths.

Carbamates. Carbamates as finishing agents were an alternative to DMDHEU for a number of years and have gained wide commercial usage. Examples of carbamates used for commercial purposes are methyl, ethyl, isopropyl, isobutyl, hydroxyethyl, and methoxyethyl carbamates. However, their usage has declined substantially because it is much easier to control formaldehyde levels by using the DMDHEU system.

Carboxylic Acids and Cross-Linkers. It has been shown that cotton fabric can be cross-linked with an agent such as 1,2,3,4-butanedicarboxylic (BTCA) to yield fabric quite comparable to DMDHEU in terms of smooth-dry performance. Several factors emerged to slow commercial acceptance of this finish. First, regarding DMDHEU, the polycarboxylic acid finish utilized much more expensive chemicals in both the agent and catalyst. Second, curing conditions tend to be more demanding in either time or temperature. Considerable effort has been expended for an acceptable product using a cheaper chemical such as citric acid.

Researchers are exploring combinations of acids, additives, and catalysts to achieve a suitable economic finish.

Special Finishing

Stone and Enzyme Treatments. In the 1980s, stone washing of heavy-weight denim fabrics burst on the scene. The initial treatments consisted of various methods of tumbling garments using high ratios of pumic stones in garment-dyeing equipment.

The difficulties of maintaining a textile plant in which rocks were strewn through the plumbing system have led to the use of cellulose enzymes. These enzymes, which hydrolyze the 1,4-glycoside bond of the cellulose molecules, can also lead to a soft fabric having a different surface effect, eg, a denim fabric. The other effect produced by such enzymes has been termed biopolishing, which arises because the enzyme tends to hydrolyze cellulose in protruding fibers, thus giving the fabric a smooth or glossy appearance.

Modifying Dyeing. *Characteristics of Cotton.* The long-term approach in producing colored, smooth-dried garments has been to dye the fabrics first in a textile mill, then to apply the cross-linking finish, and, finally, to prepare the garments.

However, it was considered better for just-in-time merchandising if the garment could be dyed just before product shipment. Therefore, an intense effort was made to produce a dyeable smooth-dye fabric. Although several functional moieties were grafted to cellulose and evaluated, the moiety that seemed to hold the most promise was the quaternary group. The addition of this group to cotton led to a cationic cotton, which was dyeable using anionic dyes.

Outdoor Fabrics. There has been an increasing use of synthetics for outdoor fabrics, particularly in tents and awnings. Cotton in outdoor usage needs protection for mildew and algae, protection against sunlight, and a degree of waterproofing and water oil repellency (qv). Another property that is sometimes required of outdoor fabric is flame retardancy.

Flame Retardants. A number of flame-retardant (FR) finishes have been developed for outdoor cotton fabrics. Various experimental and commercial finishes have been compared. While THPOH-NH_3 and THPOH-NH_3 precondensate finishes do not perform well outdoors. Antimony oxide–halogen finishes perform exceptionally well.

In the case of polyester, acetate, or triacetate fabrics, a tris(2,3-dibro-mopropyl)phosphate, generally referred to as Tris, was padded onto the fabric and forced into the fiber using a thermal treatment. However, in 1977, the Consumer Product Safety Commission in the United States labeled Tris as carcinogenic. Although Tris was not used on cotton, the ban seemed to affect all sleepwear finishing, so the total volume of treated goods was substantially reduced.

Another fire-related problem that has seen some research effort is that of smolder resistance of upholstery and bedding fabrics. Finishing techniques have been developed to make cotton smolder-resistant, but the use of synthetic barrier fabrics appears to provide a degree of protection.

In addition to FR treatments that are durable to laundering and weathering, work has also been done on a variety of treatments for the production of FR fabrics using inorganic salt mixtures (see FLAME RETARDANTS).

Miscellaneous Finishing

Finishing of Wool. Wool is desirable where warmth, wrinkle recovery, and ability to set in creases are important. Wool problems relate to shrinkage, particularly to its tendency to felt. This is caused by scaly structure, which tends toward fiber entanglement when wet and subjected to mechanical action. In order to compensate for this tendency, wool needs to be set and also made shrinkproof if it is to be laundered.

The mechanism of setting involves bond fusion and rebuilding of disulfide linkage. Chemicals used for this purpose are thiols, bisulfates, and thioglycolates. The problem of felting has been solved by altering the scaly structure of wool. Processors have used chlorination or some oxidation treatments. The chlorination step is followed by the application of a polymer to achieve desired shrinkage control.

Finishing of Synthetics. Although finishing is not as important to synthetics as it is to cotton and other cellulosics, there are still many opportunities for its use in synthetics. However, the finishing of synthetics, particularly polyester fabric, is not directed toward improving resiliency and smooth-dry performance by chemical treatments as in the case of cellulosics. In the case of polyester fabric, simple heat-setting treatment suffices to set the fabric for desired smoothness and resiliency.

Synthetic fabrics can also be finished to achieve a number of specific characteristics. For example, increased electrical conductivity can improve the antistatic character of polyester. Similarly, finishes that improve hydrophilic character also improve properties related to soil release and soil redeposition.

TIMOTHY A. CALAMARI, JR.
ROBERT J. HARPER
United States Department of Agriculture

B. A. Andrews, *Colourage Annual*, 77–83 (1991).

M. Lewin and S. B. Sello, *Handbook of Fiber Science and Technology*, Vol. II, Part A, Marcel Dekker Inc., New York, 1983, p. 47.

C. S. Whewell, *Colourage*, **26**, (22), (1979).

H. M. Elder, *Textile Finishing* (1978).

TESTING

Knowledge of fiber and yarn properties, including mechanical, physical, and chemical behaviors, is fundamental in understanding textile structures, whether the fabric structure is woven, knitted, or nonwoven. Because fabric performance is a function of the application of chemical, thermal, or mechanical finishes, the effects of these treatments must be studied. Textile testing is the use of engineering principles in the measurement of properties of textile fibers, yarns, and fabrics. Tests performed on textile structures relate to, but do not necessarily define, textile's use performance. These tests can be categorized as either objective or subjective. Objective testing relates to physical performances, including strength, elastic behavior, shrinkage, color fastness, as well as tear, abrasion, pilling, and degradation resistance. Subjective testing relates to aesthetics and includes fabric hand, appearance following laundering or dry cleaning, luster, and comfort.

Properties of finished textile structures or fabrics are an accumulation of properties of the fiber (or fibers if it is a blend), yarn configuration, construction, and selected finish. Fiber properties include length, crimp, transverse dimensions, density, cross-sectional shape, shrinkage, friction, moisture absorption, electrostatic and thermal properties, as well as optical, tensile, and elastic properties. Yarn properties include size and number, twist, strength, evenness, friction, and texture. Fabric properties include construction, thickness, air permeability, strength and elongation, snag, pilling and abrasion resistance, shrinkage, thermal and moisture transmission, color and wash fastness, flammability, hand, drape, wrinkle resistance, luster, and comfort. The properties of textile structures are dictated by interactions of all of these parameters.

In general, textile materials are moisture- and temperature-sensitive. As a result, all tests should be performed at ambient conditions of 21°C and 65% relative humidity unless otherwise stated. The textile industry, the American Society for Testing Materials (ASTM), and the American Association of Textile Chemists and Colorists (AATCC) have developed standards and testing procedures that can be used as guidelines for predicting satisfactory performance for textile materials.

Geotextile Testing

Geotextiles (qv) are defined as any permeable technical material used with foundation, soil, rock, earth, or any geotechnical-related material as an integral part of an artificial project, structure, or system. Geotextiles function mainly in separation, reinforcement, filtration, drainage and/or moisture barrier applications. In separation, geotextiles are used between different layers of earth. In reinforcing, geotextiles have the ability to distribute a concentrated load over a large area of subgrade, thus avoiding local overloading. Filtration in fabric-to-soil systems means water can move across the plane of the geotextile, whereas soil is retained indefinitely on the upstream side without clogging the fabric. This is especially important under road and railroad beds and airfields. Moisture barriers use geotextiles called geomembranes for the purpose of preventing or slowing down water passage in either liquid or vapor state. Geomembranes are used as pond liners and covers, underneath landfills, etc.

ASTM has test methods specifically for testing of geotextiles.

PAMELA BANKS-LEE
JAN PEGRAM
North Carolina State University

N. R. S. Hollies and L. Fourt, *Clothing: Comfort and Function*, Marcel Dekker Inc., New York, 1970.

Fabric Assurance by Simple Testing: Instruction Manual, CSIRO Division of Wool Technology, Sydney, Australia, 1989.

R. V. van Zanten, ed., *Geotextiles and Geomembranes in Civil Engineering*, A. A. Balkema Publishing, Rotterdam, The Netherlands, 1986.

THALLIUM AND THALLIUM COMPOUNDS

Thallium Metal

Occurrence. Thallium, Tl, belongs to Group 13 (IIIA) of the Periodic Table along with boron, aluminum, gallium, and indium. ^{203}Tl (29.5%) and ^{205}Tl (70.5%) are the two stable naturally occurring isotopes, whereas $^{191-202}$Tl, and $^{206-210}$Tl are the decomposition products of natural radioactive series (see RADIOISOTOPES).

Thallium is not a particularly rare metal and its abundance in the earth's crust is ca 0.3 ppm. It occurs not only in oxide minerals but also as a chalcophilic element. The metal cation commonly occurs in potash minerals and in a number of thallium-containing minerals, eg, crookesite, $(Cu,Tl,Ag)_2Se$; lorandite, $TlAgS_2$ (59% Tl), hutchinsonite, $(Tl,Pb)S{\cdot}(Ag,Cu)_2S{\cdot}5\ As_2S_3$, in Switzerland; vrbaite, $TlSbAs_2S_5$ (30% Tl); and avicennite, $7\ Tl_2O_3{\cdot}Fe_2O_3$. Of these, crookesite and lorandite are the most important.

Properties. Thallium is grayish white, heavy, and soft. When freshly cut, it has a metallic luster that quickly dulls to a bluish gray tinge like that of lead. The physical properties of thallium are summarized in Table 1.

Table 1. Physical Properties of Thallium

Property	Value
atomic weight	204.37
melting point, °C	303
boiling point, °C	1457
density, g/cm^3	11.85
thermal conductance, W/(cm·K)[a]	0.39
specific heat, 20°C, J/g[b]	0.13
heat of fusion, J/g[b]	21.1
heat of vaporization, J/g[b]	795
Brinell hardness	2
linear coefficient of expansion	28×10^{-6}
electrical resistivity, $\mu\Omega$·cm	18
tensile strength, MPa[c]	9.0
crystal structure	
below 230°C	close-packed hexagonal
above 230°C	body-centered cubic[d]

[a] To convert W/(cm·K) to (cal·cm/(s·cm^2·°C), divide by 4.184.
[b] To convert J to cal, divide by 4.184.
[c] To convert MPa to psi, multiply by 145.
[d] To convert J/T to Bohr magneton, divide by 9.274×10^{-24}.

Production and Economic Aspects. Thallium is obtained commercially as a by-product in the roasting of zinc, copper, and lead ores. The thallium is collected in the flue dust in the form of oxide or sulfate with other by-product metals, eg, cadmium, indium, germanium, selenium, and tellurium.

Uses. Thallium has limited commercial applications because of its toxic nature. Thallium forms alloys readily with many metals and some of these alloys have unique properties. A number of binary, ternary, and quaternary eutectic alloys are known and have very low coefficients of friction and good resistance to acids. These alloys can be used in bearings, eg, Ag–Tl and Au–Al–Tl. Alloys of silver and thallium are used in contact points, and an alloy of lead, tin, and thallium is used in the production of anodes. The most important alloy of thallium is the mercury–thallium alloy, which forms a eutectic at 8.7 wt % Tl and has a melting point of 60°C. It can be used as a substitute for mercury in switches and seals for equipment used in the polar region, stratosphere, or space program.

Thallium Compounds

Unlike boron, aluminum, gallium, and indium, thallium exists in both stable univalent (thallous) and trivalent (thallic) forms. There are numerous thallous compounds, which are usually more stable than the corresponding thallic compounds. The thallium(I) ion resembles the alkali metal ions and the silver ion in properties. The properties of the more important inorganic thallium compounds are listed in Table 2.

Organometallics. Organothallium compounds have attracted intense interest mostly because of the applications of these compounds in organic synthesis. Stable compounds occur in both Tl(I) and Tl(III) oxidation states.

Organothallium (III) derivatives can be classified into three types: R_3Tl, R_2TlX, and $RTlX_2$. The trialkyl derivatives are reactive, unstable compounds, whereas the dialkyl derivatives are among the most stable and least reactive organometallic compounds. The monoalkyl compounds are rather unstable and often cannot be isolated. They are important intermediates in some Tl(III)-promoted organic reactions.

Uses. Thallium compounds have limited use in industrial applications. Thallium sulfide has been used in photoelectric cells (see PHOTOVOLTAIC CELLS). Many thallium compounds have been used as reagents in organic synthesis in research laboratories.

Toxicology

The relative toxicities of thallium compounds depend on their solubilities and valence states. Soluble univalent thallium compounds, are especially toxic. They are rapidly and completely absorbed from the gastrointestinal tract, skin peritoneal cavity, and sites of subcutaneous and intramuscular injection. Thallium is also rapidly absorbed from the mucous membranes of the respiratory tract, mouth, and lungs following inhalation of soluble thallium salts. Insoluble compounds, eg, thallous sulfide and iodide, are poorly absorbed by any route and are less toxic.

The usual symptoms in human thallotoxicosis resulting from acute, subacute, or chronic intoxication are generally the same. Common symptoms include nausea, vomiting, abdominal colic, pain in legs, nervousness and irritability, chest pain, gingivitis or stomatitis, and anorexia.

Table 2. Properties of Selected Inorganic Thallium Compounds

Compound	Formula	Color and crystal form	Mp, °C	Density, g/cm^3	Solubility, g/100 g solvent[a]	
					Water	Other
thallous carbonate	Tl_2CO_3	colorless, monoclinic	272	7.16	5.2 (18), 22.4 (100)	
thallous formate	TlO_2CH	colorless, needles	101	4.967	500 (10)	methanol
thallous acetate	TlO_2CCH_3	silky white, needles	131	3.765	very soluble	alcohols
thallous fluoride	TlF	colorless, cubic	327	8.23	78.6 (15)	alcohols, HF
thallic fluoride	TlF_3	olive green, orthorhombic	550 dec	8.65	insoluble	
thallous chloride	TlCl	colorless, cubic	430	7.0	0.32 (20), 2.38 (100)	
thallic chloride	$TlCl_3$	hexagonal plate	155		very soluble	alcohols, ether
thallous bromide	TlBr	green–yellow, cubic	456	7.5	0.05 (20), 0.25 (60)	
thallic bromide	$TlBr_3$	yellow			soluble	alcohols
thallous iodide	TlI	yellow, rhombic	440	7.29	0.0006 (20)	
thallous oxide	Tl_2O	black	300	9.52	dec	
thallic oxide	Tl_2O_3	black	7.7	10.11	insoluble	
thallous hydroxide	TlOH	yellow, needles	139 dec	7.44	25.9 (0), 52 (40)	alcohols
thallous nitrate	$TlNO_3$	colorless	206	5.55	8 (15), 594 (104.5)	
thallic nitrate	$Tl(NO_3)_3$	colorless			dec	
thallous sulfate	Tl_2SO_4	colorless, prisms	632	6.77	4.87 (20), 18.45 (100)	

[a] Value in parentheses is temperature in °C of solubility.

BOB BLUMENTHAL
Noah Technologies Corporation

N. V. Sidgwick, *The Chemical Elements and Their Compounds*, Vol. I, Oxford University Press, London, 1950.

A. G. Lee, *The Chemistry of Thallium*, Elsevier, Amsterdam, the Netherlands, 1971.

K. Wade and A. J. Banister, in J. C. Bailar, Jr., and co-workers, eds., *Comprehensive Inorganic Chemistry*, Vol. 1, Pergamon Press, Oxford, 1973.

A. N. Nesmeyanov and R. A. Sokolik, *The Organic Compounds of Boron, Aluminum, Gallium, Indium and Thallium*, North-Holland Publishing Co., Amsterdam, the Netherlands, 1967.

THERMAL POLLUTION

An important by-product of most energy technologies is heat. Few energy conversion processes are carried out without heat being rejected at some point in the process stream. Historically, it has been more convenient as well as less costly to reject waste heat to the environment rather than to attempt significant recovery. The low temperatures of waste heat in relation to process requirements often make reuse impractical and disposal the only attractive alternative.

Concern over heat rejection arose when quantities at localized sites rose dramatically as the electric utility industry shifted to water-cooled, thermal-electric generating stations of high unit capacity in the 1950s. The term thermal pollution took on fearsome portents among aquatic scientists, fishery managers, and eventually water pollution control agencies. Directly lethal effects of high temperatures on aquatic life were predicted and, where sublethal temperatures were maintained, effects on reproductive cycles, growth rates, migration patterns, and interspecies competition were hypothesized.

Research involving monitoring of thermal effects at power stations resulted in increased knowledge; and more stringent regulations led to approaches that use biological requirements of aquatic organisms plus local environmental characteristics of the rivers, lakes, and estuaries used for cooling to design nondamaging, site-specific cooling systems. Because the rate of increase in demand for electricity and development of new generating stations also diminished, new power plants were evaluated more thoroughly and located in less susceptible environments. Approaches for regulatory thermal power stations that evolved by the mid-1980s were still in practice (see POWER GENERATION).

Cooling Techniques

Power station cooling is fairly straightforward. Generation of electricity by the steam cycle, the most common method regardless of fuel type, entails production of waste heat. Although there are some atmospheric losses in the steam cycle, most of the heat is rejected to flowing water through the heat of condensing steam. For a modern lightwater nuclear power, this release is approximately twice the thermal equivalent of the generated electric energy. The amount of reject heat is proportionately less in fossil-fueled plants. The flowing water for the steam condenser was traditionally pumped from a nearby lake or river, then returned to this water body at an elevated temperature in the once-through or open-cycle system.

Risks

A risk is considered herein to be the biological or ecological damage that could be done by a human alteration of the environment with the likelihood or probability that the specific damaging alteration will actually occur. For purposes of clarifying risks to aquatic life, a clear distinction must be made between heat, a quantitative measure of energy that depends on the mass of an object (in this case the volume of cooling water), and temperature, a measure of energy intensity. An amount of heat distributed in two unit volumes of water yields half the elevation of temperature as the same amount of heat distributed in one unit. Although heat is the waste product of electricity generation, temperature is the environmental characteristic to which organisms respond. Thus, the quantity of water used to carry away the load of reject heat is crucial to determining the temperatures created in the environment.

Risk Minimization

Selection of the best cooling system in terms of minimal environmental damages involves matching engineering options to the local aquatic system potentially at risk. General principles of aquatic ecology and of the life histories and environmental requirements of species represented locally can be adapted to local water resource goals using detailed understanding of the local aquatic setting to achieve site-specific risk prevention.

Thermal Effects. Temperature is the most all-pervasive environmental factor that influences aquatic organisms. There is always an environmental temperature, whereas other factors may or may not be present. Nearly all aquatic organisms, with the exception of marine mammals, are for all practical purposes thermal conformers. As such, they are not able to exert significant influence on maintaining a certain body temperature by physiological means. Their body temperatures fluctuate in close accord with the temperature of the immediate surrounding water. Especially large, active-moving fish, such as tunas, do maintain deep-muscle temperatures slightly higher than the surrounding water. Intimate contact between body fluids and water at the gills and the high specific heat of water assure a near-identity of internal and external temperatures. Behavioral thermoregulation or the control of body temperature by selection of water temperature in natural gradients is, however, a common feature of many fish. Behavioral thermoregulation serves an important ecological role in partitioning aquatic habitats among species. Its understanding is a powerful feature for estimating and ameliorating the impacts of thermal discharges.

Organisms evolving under annual temperature cycles and in environments with varying temperatures spatially have incorporated thermal cues in reproductive behavior, habitat selection, and certain other features which act at the population level. Thus, the balance of births and mortalities, which determines whether a species survives, is akin to the metabolic balance at the physiological level in being dependent upon the match, within certain limits, to prescribed temperatures at different times of year. At the ecosystem level, relationships among species, eg, predators, competitors, prey animals, and plant foods, are related to environmental temperatures in complex ways. Many of these interactions are poorly understood.

Preventing Mortality. Upper and lower temperature tolerances of aquatic organisms have been well conceptualized and standardized methods are available for determining a species' tolerance ranges under different conditions of thermal history. On a short time scale, mortality is highly dependent on duration of exposure, so that brief exposures to potentially lethal temperatures are not actually lethal. Temperature elevations and duration of exposure can be tailored in a plant's piping or in the effluent mixing zone to maximize organism survival. This is usually accomplished using detailed mathematical models of cooling-water-effluent dispersion and heat dissipation in the near field where temperatures are highest.

Preventing Stressful High Temperatures Over Long Periods. For the long term ($>1-2$ d), simply preventing mortality is insufficient for protecting aquatic species. All of the physiological functions normally performed must be carried out to maintain healthy individuals that are capable of competing in the natural ecosystem. An aggregate measure, growth rate, has proved useful as an integrator of all physiological functions and some behavioral ones, eg, feeding rate. Growth occurs only if all other metabolic demands are being met and when sufficient food energy is left over for adding biomass. Typically, many physiological functions of well-fed, cold-blooded organisms proceed optimally over a temperature range in which growth rate is maximal. Above the temperatures of maximum growth rate, the rate typically declines steeply to a temperature of zero growth, which often occurs 1–2°C below the temperature at which direct mortalities begin.

Standards for upper limits on water temperatures for particular water bodies over periods of about one week or more can be based on species-specific growth rates. Using an inventory of important species and life stages in the area during the warm season, the upper temperature limit which does not stress those in the desired aquatic assemblage can be ascertained. Hydrothermal models of heat dissipation in the far field beyond the zone of effluent mixing are important for estimating the zones that may present a long-term risk from elevated temperatures.

The preferred and avoided temperatures in a gradient have been used as surrogates for optimum growth and upper danger levels for fish. Long-term abundance of species at power plant sites appears to be generally correlated with preferred and avoided temperatures.

Preserving Reproduction Cycles. Reproduction success depends in part on the preservation of an annual temperature pattern, although the precise timing is usually not critical. Thermal discharges can be designed, usually with the help of far-field mathematical models, to assure the necessary thermal periodicity. Thus, annual thermal cycles are generally maintained despite anthropogenic heat rejection.

Maintaining Ecosystem Structure and Function. Thermal heterogeneity of water bodies is an important structural feature of the environment and plays a large role in determining the composition and functioning of most aquatic systems.

Careful planning of thermal additions, including creation of new cooling reservoirs, can yield thermal structures which enhance rather than damage desirable aquatic species. Knowledge of the thermal niches of these species and their potential competitors or predators can permit special provisions for thermal refuges, eg, cool summer zones in a heavily loaded cooling pond. Cooling-water circulation can be designed so that the thermal stratification patterns that are essential for some desired species are maintained. From a different perspective, aquatic species introduced to waters used for power-station cooling can be selected so that their thermal niche matches the thermal structure that the facility creates.

Impingement. Repellents, eg, sound, electricity, and light, were used to keep susceptible fish away from intakes with only moderate success. More useful were orientations of the inlet structure such that screens flush with the shoreline allowed lateral escape. Where this screen orientation was impossible, guidance systems were developed for intake bays to direct fish from the incoming water flows to alternative escape routes.

In general, minimization of impingement risks focuses on site-specific analyses of potentially vulnerable species and selection of engineering designs which, within acceptable cost limits, keep impingement deaths low.

Biocides. Chlorine and other biocides are used occasionally in cooling water to kill and dispose of organic growths on heat-exchange surfaces and on piping where water flow could be hampered by such growth. Of necessity, organisms passing through the cooling circuit or residing in the effluent area during periodic chlorine injections experience the potentially lethal exposures. The objective is to maximize the intended kill and minimize extraneous damages, particularly in the receiving water.

Chlorine is a toxicant with a typical dose-response pattern for biota. There is a time-dependent mortality at high concentrations and a low concentration above which long-term chronic effects are shown. Early methods for using the time-dependent effects of high temperature for predicting safe temperatures and exposures have times during cooling-water exposures have led to a similar approach for chlorine.

Entrainment. Stresses to small nonscreenable organisms, eg, fish larvae, during passage through the cooling circuit come from a combination of thermal shock, physical abuses, and periodic injections of biocide.

A principal frustration in attempts to minimize entrainment damages has been the contradictory demands of thermal and physical stresses. Thermal stresses can be quantitatively predicted based on dose-response data and minimized by increasing water-flow volumes, which dilute the fairly constant supply of rejected heat. The added volume of cooling water, however, includes proportionately more planktonic organisms, which are subjected to physical stresses.

The assumption of high percentage mortality resulting from physical stresses has been criticized. It appears that ameliorations of thermal stress with flow increases generally are not cancelled by additional physical mortalities except in cases of exceedingly high flows.

An optimization procedure, such as that suggested by the Committee on Entrainment, appears fruitful for identifying on a site-specific and seasonally varying basis the most appropriate cooling-water flow regime.

Gas Balance. When the solution of oxygen in the intake water is undersaturated, the agitation of the cooling circuit generally yields an increase. Polluted receiving water does, however, have an accelerated microbial oxygen demand because of raised temperature, and this microbial deoxygenation can yield exaggerated dissolved oxygen sag zones downstream. Engineering models for predicting oxygen sag zones are generally capable of incorporating temperature changes.

Damaging supersaturation of dissolved gases has occurred in some cooling-water discharges. The practice of winter increases in temperature rise across condensers by cutting back on pumping capacity has either ceased in general practice or the immediate discharge areas have been engineered to prevent long-term residence by susceptible biota. These remedial measures can be completely effective.

Cooling-Tower Chemicals. The risk posed by changing power-station systems from traditional once-through or open-cycle cooling to cooling towers is from chemicals added to the recirculating water. Blowdown to aquatic systems and drift to the terrestrial landscape carry these chemicals to locations where natural biota can be damaged through direct poisoning or where toxicants can accumulate to potentially detrimental levels by food-chain transfer. Such risks can be minimized using fairly straightforward engineering approaches. Airborne drift has been reduced significantly by installation of physical baffles at the air outlets. Blowdown can be treated for removal of chemical constituents. Chemical-laden sludges become a more long-term disposal problem. Practices include ponding and landfills. Chemical-recovery processes are also available for sludge treatment.

Human Pathogens. Proper designs and good maintenance practices can reduce the potential risks to humans from pathogens stimulated by environmental conditions in cooling systems, eg, amoebae *Legionella*, to very low levels. The Legionnaires' disease microorganism, identified as a common inhabitant of many cooling-tower systems, especially small ones used for building-temperature control, can be held in check by systematic use of biocides in recirculating cooling water. Contact with the infectious organisms can be minimized if care is taken to isolate inlets for ventilating air from the drift aerosols which are emitted from cooling towers.

Maximizing the Benefits

Increasing attention has been given to finding productive uses for power plant waste heat. Potential physical applications of power plant rejected heat include industrial heating and biological applications, such as fish culture, soil warming, heating greenhouses, and livestock shelters.

The most important factor determining the possible uses of such heat is the temperature of the heat source. There is a threshold between low and high grade heat for engineering uses at ca 100°C. Most waste heat from electric power stations is in the form of low grade heat. Discharges from once-through cooling systems of thermal power plants have outlet temperatures of 10–40°C, depending on the season. Circulating water in closed-cycle cooling systems is only slightly warmer (20–50°C). These low temperatures are the result of careful engineering of power stations to extract the maximum amount of electrical energy from the fuel before the rejected heat is dumped to circulating water in steam condensers.

A more useful, higher grade heat is in a form that has not been degraded to the low temperatures of most power plant thermal discharges but which remains at 40–200°C following production of initial amounts of electricity.

Ideally, a power station could be a potential supplier of electricity, high grade heat, and low grade heat, even though electricity is the main product, and most new power stations are optimized accordingly. Depending on the desired uses at particular sites, future power stations could be designed to alter the ratios of electricity and different grades of heat to produce the most efficient total energy use.

Aquaculture. Culture of some aquatic species in essentially unmodified thermal effluents of power stations has been attempted both experimentally and commercially. Use of rejected heat to prolong optimal growth temperatures in cool months can significantly increase the sizes attained in a year.

Open-Field Agriculture. Use of warmed water for field crops and orchards has been tested in several studies. Buried pipes can convey thermal-discharge heat to soils where warming aids plant growth and extends the growing season.

Greenhouse Agriculture. Greenhouse agriculture is well known for its many advantages over open-field agriculture for certain crops. One drawback is high expense for heating in winter, which can be the most costly part of greenhouse operation. The use of waste heat from steam-electric power plants therefore appears promising as a source of low cost heat for use in greenhouses.

Animal Shelters. The advantages of temperature control for maximizing weight gain and avoiding animal losses in livestock and poultry are well known. Low grade heat from power station cooling offers the possibility of low cost heating of animal shelters.

Space Heating. A large percentage of the energy requirements of most countries in temperate zones is for heating and cooling of living and working spaces and for hot water. The historical use of dual-purpose power plants for electricity generation and central district heating in the United States and their extensive use in such countries as the Russia, Sweden, and Germany suggests that expansion of this form of waste-heat utilization can contribute significantly to energy conservation and control of concentrated thermal discharges worldwide.

Industrial Process Heat. Many industries use process heat at 77–110°C. Much of this heat is supplied by combustion of oil and natural gas (qv). Equipment manufacturers are developing industrial heat pumps to capture free industrial plant waste heat and regenerate it to the desired process heat temperature, thereby greatly reducing energy costs associated with direct heating (see PROCESS ENERGY CONSERVATION).

Cooling Reservoirs. The most extensively developed productive use for power plant cooling is in multiple-purpose cooling reservoirs. Small impoundments built especially for heat dissipation have been managed for extensive recreational uses as well. Highly productive fisheries for warm-water species have made cooling reservoirs highly popular.

CHARLES C. COUTANT
Oak Ridge National Laboratory

Environmental Effects of Cooling Systems, Technical Report Series No. 202, International Atomic Energy Agency, Vienna, 1980.

H. Precht, J. Christophensen, H. Hensel, and W. Larcher, *Temperature and Life*, Springer-Verlag, New York, 1973.

L. B. Goss and L. Scott, *Factors Affecting Power Plant Waste Heat Utilization*, Pergamon Press, New York, 1980.

M. C. Bell, *Fisheries Handbook of Engineering Requirements and Biological Criteria*, U.S. Army Corps of Engineers, North Pacific Division, Portland, Ore., 1991.

THERMODYNAMICS

Thermodynamics is a deductive science built on the foundation of two fundamental laws that circumscribe the behavior of macroscopic systems. The first law of thermodynamics affirms the principle of energy conservation; the second law states the principle of entropy increase.

In the formal application of these laws, attention is focused on a specific object, a particular quantity of matter, or a bounded region of space, specified as the system. All else then constitutes the surroundings. The coordinates which characterize the systems of interest in chemical technology are temperature, T, pressure, P, molar volume, V, and composition. Such a PVT system may be open or closed to the exchange of matter with its surroundings, or it may be isolated from its surroundings, in which case it can exchange neither matter nor energy. Once isolated, a system is independent of its surroundings and can only progress toward an equilibrium state whose thermodynamic coordinates have no further tendency to change. Systems not in equilibrium states undergo processes by which their coordinates alter; during such processes the system and its surroundings may exchange energy in the two forms of heat, Q, and work, W.

The coordinates of thermodynamics do not include time; thermodynamics does not predict rates at which processes take place. It is concerned with equilibrium states and with the effects of temperature, pressure, and composition changes on such states. For example, the equilibrium yield of a chemical reaction can be calculated for given T and P, but not the time required to approach the equilibrium state. It is however true that the rate at which a system approaches equilibrium depends directly on its displacement from equilibrium. One can therefore imagine a limiting kind of process that occurs at an infinitesimal rate by virtue of never being displaced more than differentially from its equilibrium state. Such a process may be reversed in direction at any time by an infinitesimal change in external conditions, and is therefore said to be reversible. A system undergoing a reversible process traverses equilibrium states characterized by the thermodynamic coordinates.

The Laws of Thermodynamics

First Law of Thermodynamics. The energy change of any system together with its surroundings is zero. Implicit in this declaration is the affirmation that there exists a form of energy, known as internal energy, which for systems in equilibrium states is an intrinsic property of the system. Internal energy is separate from the external-energy forms, ie, the kinetic and potential energies of macroscopic bodies. Although a macroscopic property, internal energy originates in the kinetic and potential energies of molecules and submolecular particles. In applications of the first law of thermodynamics, energy in all its forms, both internal and external, must be considered.

Applied to a closed system which undergoes only an internal-energy change, the first law of thermodynamics is given be equation 1:

$$dU^t = dQ + dW \tag{1}$$

where U^t is the total internal energy of the system and dU^t represents a differential change in this property. On the other hand, dQ and dW are differential quantities of heat and work representing energy in transit across the boundary of the system, and serving to account for the energy change of the surroundings. Integration of equation 1 gives for a finite process (eq. 2):

$$\Delta U^t = Q + W \tag{2}$$

where ΔU^t is the finite change from the initial to the final value of U^t. The heat, Q, and work, W, are finite quantities, neither properties of the system nor functions of the thermodynamic coordinates that characterize the system. By convention, the numerical values of Q and W are positive for the transfer of heat and work to the system and negative for transfer from the system.

Second Law of Thermodynamics. The entropy change of any system together with its surroundings is positive for a real process, approaching zero as the process approaches reversibility:

$$\Delta S_{\text{total}} \geq 0 \tag{3}$$

where S_{total} is the entropy of the system and its surroundings. Implicit in this declaration is the affirmation that there exists a thermodynamic property, known as entropy, which for systems in equilibrium states is an intrinsic property of the system. For reversible processes, changes in the entropy of the system are given by equation 4:

$$dS^t = dQ_{\text{rev}}/T \tag{4}$$

where S^t is the total entropy of the system and T is the absolute temperature of the system.

Each of the two laws of thermodynamics asserts the existence of a primitive thermodynamic property, and each provides an equation connecting the property with measurable quantities. These are not defining equations; they merely provide a means to calculate changes in each property.

Heat Engines and Heat Pumps

A heat engine is a device operating in cycles that takes in heat, Q_H, from a heat reservoir at temperature T_H, discards heat, Q_C, to another heat reservoir at a lower temperature T_C, and produces work. A heat reservoir is a body that can absorb or reject unlimited amounts of heat without change in temperature. Entropy changes of a heat reservoir depend only on the absolute temperature and on the quantity of heat transferred, and are always given by the integrated form of equation 4:

$$\Delta S^t = \frac{Q}{T} \tag{5}$$

Here, Q_{rev} is replaced by Q, because the effect of heat transfer on a heat reservoir does not depend on its reversibility. Thus the entropy changes of the two heat reservoirs associated with a heat engine are given by equations 6 and 7:

$$\Delta S_H^t = \frac{Q_H}{T_H} \tag{6}$$

$$\Delta S_C^t = \frac{Q_C}{T_C} \tag{7}$$

In these equations Q_H and Q_C refer to the respective heat reservoirs, and numerical values are positive when heat flows into the reservoir and negative when heat flows out.

Since the engine operates in cycles, it experiences no change in its own properties; therefore the total entropy change of the engine and its associated heat reservoirs is given by equation 8:

$$\Delta S_{\text{total}} = \frac{Q_H}{T_H} + \frac{Q_C}{T_C} \tag{8}$$

The first law (eq. 2) is applied to the engine taken as the system:

$$\Delta U_{\text{engine}} = -Q_H - Q_C + W \tag{9}$$

As in equation 8, the heat quantities Q_H and Q_C are here written with respect to the heat reservoirs.

For a reversible process, $\Delta S_{\text{total}} = 0$, and for a cyclic engine, $\Delta U_{\text{engine}} = 0$. Under these conditions equations 8 and 9 can be combined to yield equations 10 and 11:

$$\frac{|Q_C|}{|Q_H|} = \frac{T_C}{T_H} \tag{10}$$

$$\frac{|W|}{Q_H} = 1 - \frac{T_C}{T_H} \tag{11}$$

These are Carnot's equations, and they apply to all reversible heat engines operating between a heat source and a heat sink at fixed temperature levels, ie, to Carnot engines.

Equation 11 expresses the thermal efficiency of a Carnot heat engine, ie, the fraction of the heat taken in that is converted into work. Since a Carnot engine is reversible, it can be run as a heat pump or refrigerator, as indicated in Figure 1. Equations 10 and 11 apply to the Carnot heat pump or refrigerator as well as to the Carnot engine, because all quantities are the same for the two cases.

Fundamental Property Relations

The systems of interest in chemical technology are usually comprised of fluids not appreciably influenced by surface, gravitational, electrical, or magnetic effects. For such homogeneous fluids, molar or specific volume, V, is observed to be a function of temperature, T, pressure, P, and composition. This observation leads to a basic postulate that macroscopic properties of homogeneous PVT systems at internal

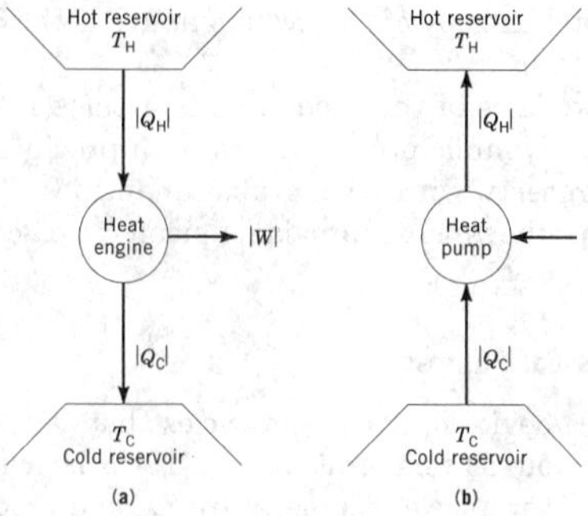

Figure 1. Schematic representation of (**a**) Carnot heat engine and (**b**) Carnot refrigerator used as a heat pump.

equilibrium can be expressed as functions of temperature, pressure, and composition only. Thus the internal energy and the entropy are functions of temperature, pressure, and composition. These molar or unit-mass properties, represented by the plain symbols V, U, and S, are independent of system size, and are intensive. Total-system properties (V^t, U^t, S^t) do depend on system size, and are extensive. Thus, if the system contains n moles of fluid, $M^t = nM$, where M is a molar property. Temperature and pressure are also intensive, but have no extensive counterparts.

Consider a closed, nonreacting PVT system containing 1 mole of a homogeneous fluid mixture of constant composition. Equation 1, applied to this system as it undergoes a differential change of state in a reversible (rev) process, may be written as follows:

$$dU = dQ_{\text{rev}} + dW_{\text{rev}} \tag{12}$$

According to equation 4,

$$dQ_{\text{rev}} = TdS \tag{13}$$

and by the definition of work,

$$dW_{\text{rev}} = -PdV \tag{14}$$

These three equations combine to give

$$dU = TdS - PdV \tag{15}$$

Although derived for a reversible process, equation 15 relates properties only, irrespective of the process, and therefore applies to any change in the equilibrium state of a homogeneous, closed, nonreacting system. This fundamental property relation is the basis for development of all other equations relating the properties of PVT systems.

Equation 15 implies that U is a function of S and V, a choice of variables that is not always convenient. Alternative fundamental property relations may be formulated in which other pairs of variables appear. They are found from the following definitions for the enthalpy, H, the Helmholtz energy, A, and the Gibbs energy, G:

$$H \equiv U + PV \tag{16}$$

$$A \equiv U - TS \tag{17}$$

$$G \equiv H - TS \tag{18}$$

The total differentials of these three equations in combination with equation 15 yield

$$dU = TdS - PdV \tag{19}$$

$$dH = TdS + VdP \tag{20}$$

$$dA = -SdT - PdV \tag{21}$$

$$dG = -SdT + VdP \tag{22}$$

Since these are exact differential expressions, Maxwell equations can be written by inspection; the two most useful ones are derived from equations 21 and 22:

$$\left(\frac{\partial S}{\partial V}\right)_T = \left(\frac{\partial P}{\partial T}\right)_V \tag{23}$$

$$\left(\frac{\partial S}{\partial P}\right)_T = -\left(\frac{\partial V}{\partial T}\right)_P \tag{24}$$

Enthalpy and Entropy as Functions of T and P

For a homogeneous fluid of constant composition,

$$H = H(T,P) \tag{25}$$

$$S = S(T,P) \tag{26}$$

Therefore,

$$dH = \left(\frac{\partial H}{\partial T}\right)_P dT + \left(\frac{\partial H}{\partial P}\right)_T dP \tag{27}$$

$$dS = \left(\frac{\partial S}{\partial T}\right)_P dT + \left(\frac{\partial S}{\partial P}\right)_T dP \tag{28}$$

By definition,

$$C_P \equiv \left(\frac{\partial H}{\partial T}\right)_P \tag{29}$$

is the heat capacity at constant pressure. If equation 20 is divided by dT and restricted to constant P,

$$\left(\frac{\partial H}{\partial T}\right)_P = T\left(\frac{\partial S}{\partial T}\right)_P \tag{30}$$

With equation 29 this becomes

$$\left(\frac{\partial S}{\partial T}\right)_P = \frac{C_P}{T} \tag{31}$$

When equation 20 is divided by dP and restricted to constant T,

$$\left(\frac{\partial H}{\partial P}\right)_T = T\left(\frac{\partial S}{\partial P}\right)_T + V \tag{32}$$

In view of equation 24, this becomes

$$\left(\frac{\partial H}{\partial P}\right)_T = V - T\left(\frac{\partial V}{\partial T}\right)_P \tag{33}$$

Combination of equations 27, 29, and 33 gives

$$dH = C_P dT + \left[V - T\left(\frac{\partial V}{\partial T}\right)_P\right]dP \tag{34}$$

Similarly, combination of equations 28, 31, and 24 gives

$$dS = \frac{C_P}{T}dT - \left(\frac{\partial V}{\partial T}\right)_P dP \tag{35}$$

For an ideal gas,

$$PV^{ig} = RT \tag{36}$$

and

$$\left(\frac{\partial V^{ig}}{\partial T}\right)_P = \frac{R}{P} = \frac{V^{ig}}{T} \tag{37}$$

Thus, for an ideal gas, equations 34 and 35 reduce to

$$dH^{ig} = C_P^{ig} dT \tag{38}$$

$$dS^{ig} = \frac{C_P^{ig}}{T}dT - \frac{R}{P}dP \tag{39}$$

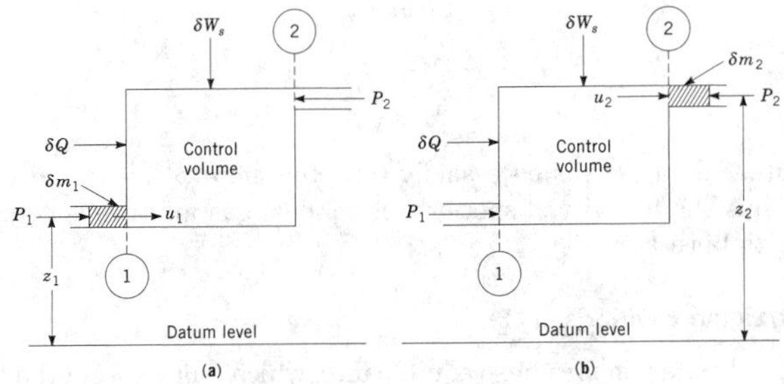

Figure 2. Schematic diagram of a simple steady-flow process. See text.

Energy Equations for Steady-State Steady-Flow Processes

Industrial production of chemicals is largely by continuous processes in which rates of inflow and outflow of mass are constant. Moreover, conditions at all points in the process are maintained constant with time. A simple steady-state steady-flow process is represented in Figure 2. A process occurs within the fixed control volume between points 1 and 2 that changes the properties of a fluid element as it flows from point 1 to point 2. The system is the fluid in the control volume plus the fluid element of mass δm_1 (Fig. 2**a**) that enters the control volume at point 1 during time $\delta\tau$. After time interval $\delta\tau$, the system appears as in Figure 2**b** and consists of fluid in the control volume plus the fluid element of mass δm_2 that has left the control volume. Since the process is one of steady-state steady flow, $\delta m_2 = \delta m_1 = \delta m$, and the properties of the fluid in the control volume are unchanged. The fluid elements δm_1 and δm_2 have properties as measured at points 1 and 2, including a velocity u and an elevation above a datum level z. Each mass element therefore has kinetic energy and potential energy as well as internal energy, and the general energy balance as given by equation 2, must be expanded to include terms for changes in these forms of energy. Moreover, the work is of two kinds: the shaft work δW_s shown in Figure 2, and the work done by pressures P_1 and P_2 moving through the volumes occupied by the fluid masses δm_1 and δm_2. When all of these forms of energy are taken into account, and the energy balance is reduced to its simplest form, it becomes

$$\Delta\left(H + \frac{u^2}{2} + zg\right)\delta m = \delta Q + \delta W_s \qquad (40)$$

where Δ denotes the change from entrance to exit of the control volume. In terms of rates, equation 40 is written:

$$\Delta\left(H + \frac{u^2}{2} + zg\right)\dot{m} = \dot{Q} + \dot{W}_s \qquad (41)$$

where $\dot{m}$; is mass flow rate, $\dot{Q}$; is heat-transfer rate, and $\dot{W}_s$ is power. Multiplication by the time required for a unit mass of fluid to enter and leave the control volume gives:

$$\Delta H + \frac{\Delta u^2}{2} + g\Delta z = Q + W_s \qquad (42)$$

where each term is for a unit mass of fluid. These equations are readily generalized to apply for any number of entrance and exit streams.

Partial Molar Properties

Because the macroscopic intensive properties of homogeneous fluids in equilibrium states are functions of T, P, and composition, the total property of a phase nM can be expressed functionally by equation 43:

$$nM = m(T, P, n_1, n_2, n_3, \cdots) \qquad (43)$$

The total differential of nM is therefore

$$d(nM) = \left[\frac{\partial(nM)}{\partial T}\right]_{P,n} dT + \left[\frac{\partial(nM)}{\partial P}\right]_{T,n} dP + \sum_i \left[\frac{\partial(nM)}{\partial n_i}\right]_{T,P,n_j} dn_i \qquad (44)$$

or

$$d(nM) = n\left(\frac{\partial M}{\partial T}\right)_{P,x} dT + n\left(\frac{\partial M}{\partial P}\right)_{T,x} dP + \sum_i \left[\frac{\partial(nM)}{\partial n_i}\right]_{T,P,n_j} dn_i \qquad (45)$$

The derivatives in the summation are partial molar properties, denoted by $\bar{M}_i$; thus,

$$\bar{M}_i \equiv \left[\frac{\partial(nM)}{\partial n_i}\right]_{T,P,n_j} \qquad (46)$$

This definition is the means by which partial properties are calculated from solution properties. Equation 45 can now be written as equation 47:

$$d(nM) = n\left(\frac{\partial M}{\partial T}\right)_{P,x} dT + n\left(\frac{\partial M}{\partial P}\right)_{T,x} dP + \sum_i \bar{M}_i dn_i \qquad (47)$$

Important equations follow from this result through the identities:

$$d(nM) \equiv ndM + Mdn \qquad (48)$$

$$dn_i \equiv d(nx_i) = ndx_i + x_i dn \qquad (49)$$

Combining these equations with equation 47 yields equation 50:

$$\left[dM - \left(\frac{\partial M}{\partial T}\right)_{P,x} dT - \left(\frac{\partial M}{\partial P}\right)_{T,x} dP - \sum_i \bar{M}_i dx_i\right] + \left[M - \sum_i x_i \bar{M}_i\right] dn = 0 \qquad (50)$$

Because n and dn are independent and arbitrary, the terms in brackets must each be zero; whence

$$dM = \left(\frac{\partial M}{\partial T}\right)_{P,x} dT + \left(\frac{\partial M}{\partial P}\right)_{T,x} dP + \sum_i \bar{M}_i dx_i \qquad (51)$$

and

$$M = \sum_i x_i \bar{M}_i \qquad (52)$$

This summability equation, the counterpart of equation 46, provides for the calculation of solution properties from partial properties.

Differentiation of equation 52 yields equation 53:

$$dM = \sum_i x_i d\bar{M}_i + \sum_i \bar{M}_i dx_i \qquad (53)$$

Because equations 51 and 53 are both valid in general, their right-hand sides can be equated, yielding:

$$\left(\frac{\partial M}{\partial T}\right)_{P,x} dT + \left(\frac{\partial M}{\partial P}\right)_{T,x} dP - \sum_i x_i d\bar{M}_i = 0 \qquad (54)$$

This result, known as the Gibbs/Duhem equation, imposes a constraint on how the partial molar properties of any phase may vary with temperature, pressure and composition.

Residual Properties

A class of thermodynamic functions called residual properties is given generic definition by equation 55:

$$M^R \equiv M - M^{ig} \qquad (55)$$

where M is the molar value of any extensive thermodynamic property of a fluid in its actual state and M^{ig} and is the corresponding value for the fluid in its ideal-gas state, ie, as an ideal gas at the same temperature, pressure, and composition. Residual property M^R reflects the contributions made to property M by intermolecular forces. For H^R and S^R,

$$li_{P \to 0} M^R = 0 \qquad (56)$$

It follows from equations 55 and 56 (constant T,x) that

$$M^R = \int_0^P \left[\left(\frac{\partial M}{\partial P}\right)_T - \left(\frac{\partial M^{ig}}{\partial P}\right)_T\right] dP \qquad (57)$$

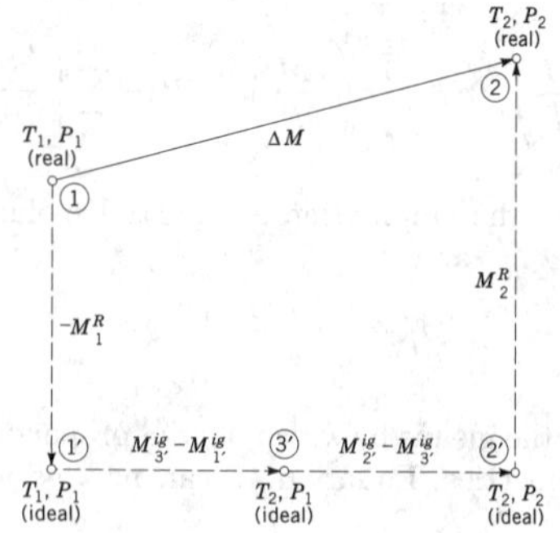

Figure 3. Calculation path for determination of property change ΔM by residual properties (eq. 46).

Values and expressions for H^R and S^R are found from equation 57 through the use of *PVT* data.

Property changes are readily determined for fluids in the ideal-gas state, and these in combination with residual properties are used to compute property changes of real fluids. The computational scheme is suggested in Figure 3, and is based on the identity:

$$\Delta M \equiv -M_1^R + \left(M_{3'}^{ig} - M_{1'}^{ig}\right) + \left(M_{2'}^{ig} - M_{3'}^{ig}\right) + M_2^R \tag{58}$$

Thus, for $M = H$ (eq. 38),

$$\Delta H = -H_1^R + \int_{T_1}^{T_2} C_P^{ig}\, dT + H_2^R \tag{59}$$

and, for $M = S$ (eq. 39),

$$\Delta S = -S_1^R + \int_{T_1}^{T_2} \frac{C_P^{ig}}{T}\, dT - R \ln \frac{P_2}{P_1} + S_2^R \tag{60}$$

The residual Gibbs energy is related to H^R and S^R by equation 61:

$$G^R = H^R - TS^R \tag{61}$$

Fugacity and Fugacity Coefficient

The Gibbs energy plays a vital role in both phase and chemical-reaction equilibria. However, it exhibits certain unfortunate characteristics which discourage its use in the solution of practical problems. The Gibbs energy is defined in relation to the internal energy and entropy, both primitive quantities for which absolute values are unknown. The application of equilibrium criteria is facilitated by introduction of a new quantity, the fugacity. For a pure fluid, the fugacity, f_i, is given partial definition by equation 62:

$$G_i \equiv \Gamma_i(T) + RT \ln f_i \tag{62}$$

This definition leads to the following relation for the residual Gibbs energy of a pure fluid:

$$G_i^R = RT \ln \phi_i \tag{63}$$

where ϕ_i is another new quantity, the fugacity coefficient of a pure species, defind by equation 64:

$$\phi_i \equiv \frac{f_i}{P} \tag{64}$$

The definition of fugacity is completed by setting the ideal-gas-state fugacity of pure species i equal to its pressure:

$$f_i^{ig} = P$$

Thus for the special case of an ideal-gas, $\phi_i^{ig} = 1$, and $G_i^R = 0$.

The definition of the fugacity of a species in solution, $\hat{f}_i$, is parallel to the definition of the pure-species fugacity. Thus, the analogues of equations 62, 63, and 64 are as follows:

$$\bar{G}_i \equiv \Gamma_i(T) + RT \ln \hat{f}_i \tag{66}$$

$$\bar{G}_i^R = RT \ln \hat{\phi}_i \tag{67}$$

$$\hat{\phi}_i \equiv \frac{\hat{f}_i}{x_i P} \tag{68}$$

For an ideal gas, $\bar{G}^R{}_i$ is necessarily zero; therefore $\hat{\phi}_i^{ig} = 1$, and $\hat{f}_i^{ig} = x_i P$. Thus the fugacity of species i in an ideal-gas mixture is equal to its partial pressure.

Vapor/Liquid Equilibria (VLE)

A general criterion for phase equilibrium, which follows from the first and second laws of thermodynamics, is given by equation 69:

$$\bar{G}_i^\alpha = \bar{G}_i^\beta = \dots = \bar{G}_i^\pi \tag{69}$$

where superscripts α, β, and π denote phases. The criterion for vapor/liquid equilibrium then follows directly from equation 66:

$$\hat{f}_i^l = \hat{f}_i^v \tag{70}$$

Effective use of this general equation requires explicit introduction of the compositions of the phases. This is done either through the fugacity coefficient, $\phi\hat{}_i$, or through the activity coefficient, $\gamma_i \equiv \hat{f}_i / x_i f_i$. Two procedures are in common use. By the "gamma/phi" approach, activity coefficients are substituted for the liquid phase and fugacity coefficients for the vapor phase (eq. 68); equation 70 then becomes:

$$x_i \gamma_i f_i = y_i \hat{\phi}_i P \tag{71}$$

The second common procedure for VLE calculations is the "phi/phi" approach. Here, fugacity coefficients replace the fugacities for both the liquid and vapor phases, and equation 70 becomes

$$x_i \hat{\phi}_i^l = y_i \hat{\phi}_i^v \tag{72}$$

Chemical-Reaction Stoichiometry

In a system containing N chemical species, any or all of which can participate in r chemical reactions, the r reactions can be represented schematically by the algebraic equation, where $j = I, \mathrm{II}, \dots, r$:

$$0 = \sum_i \nu_{i,j} A_i \tag{73}$$

The A_i represent formulas for the chemical species and $\upsilon_{i,j}$ is the stoichiometric number for species i in reaction j. Each $\upsilon^{i,j}$ has a magnitude and a sign:

$$\mathrm{sign}(\nu_{i,j}) = \begin{cases} -\text{for a reactant species} \\ +\text{for a product species} \end{cases}$$

If species i does not participate in reaction j, then $\upsilon_{i,j} = 0$.

The stoichiometric numbers provide relations among the changes in mole numbers of chemical species which occur as the result of chemical reactions. Thus, for reaction j:

$$\frac{\Delta n_{1,j}}{\nu_{1,j}} = \frac{\Delta n_{2,j}}{\nu_{2,j}} = \dots = \frac{\Delta n_{N,j}}{\nu_{N,j}} \tag{74}$$

Because all of these terms are equal, they can be equated to the change in a single quantity ϵ_j, called the reaction coordinate for reaction j, thereby giving:

$$\Delta n_{i,j} = \nu_{i,j} \Delta \epsilon_j \begin{cases} i = 1, 2, \cdots, N \\ j = I, II, \cdots, r \end{cases} \tag{75}$$

Now the total change in mole number n_i is just the sum of the changes $\Delta n_{i,j}$ resulting from the various reactions. Thus, by equation 75:

$$\Delta n_i = \sum_j \Delta n_{i,j} = \sum_j \nu_{i,j} \Delta \epsilon_j \tag{76}$$

where $i = 1, 2, \dots, N$. If the initial number of moles of species i is n_{i0} and if the conventions is adopted that $\epsilon_j = 0$ for each reaction in this initial state, then, for $i = 1, 2, \dots, N$,

$$n_i = n_{i_0} + \sum_j \nu_{i,j} \epsilon_j \tag{77}$$

Equation 77 is the basic expression of material balance for a closed system in which r chemical reactions occur. It asserts the that in such a system there are at most r mole-number-related quantities, ϵ_j, capable of independent variation. Note the absence of implied restrictions with respect to chemical-reaction equilibria; the reaction-coordinate formalism is merely an accounting scheme, valid for tracking the progress of each reaction to any arbitrary level of conversion.

Criteria for Chemical-Reaction Equilibria

A general criterion of chemical-reaction equilibria is that the total Gibbs energy of a closed system be a minimum at constant, uniform T and P. If the T and P of a single-phase, chemically reactive system are constant, then the quantities capable of change are the mole numbers n_i. The independently variable quantities are just the r reaction coordinates, and thus the equilibrium state is characterized by the r necessary derivative conditions of equation 78:

$$\left(\frac{\partial G^t}{\partial \epsilon_j}\right)_{T,P,\epsilon_k} = 0 \tag{78}$$

where $j = I, II, \ldots, r$. The material-balance constraints of equation 77 also apply.

In the case of a single-phase, multicomponent system undergoing just a single reaction, the total Gibbs energy is

$$G^t = n^t G = \sum_i n_i \bar{G}_i \tag{79}$$

and is minimized subject to the constraints

$$n_i = n_{i0} + \nu_i \epsilon \tag{80}$$

Equation 78 requires that

$$\left(\frac{\partial G^t}{\partial \epsilon}\right)_{T,P} = 0 \tag{81}$$

If equation 81 and the appropriate form of equation 54 are applied to equation 79, with $\partial n_i/\partial \epsilon = \upsilon_i$, then

$$\sum_i \nu_i \bar{G}_i = 0 \tag{82}$$

which is the familiar algebraic criterion for single-reaction equilibria. This is easily extended to an arbitrary number of independent reactions r and produces the expected generalization of equation 82:

$$\sum_i \nu_{i,j} \bar{G}_i = 0 \tag{83}$$

where $j = I, II, \ldots, j = I, II, \ldots, r$.

NOMENCLATURE

Symbol	Definition	Units
A	Helmholtz energy	J
C_P	heat capacity at constant pressure	J/(mol·k)
C_V	heat capacity at constant volume	J/(mol·k)
E_K	kinetic energy	J
E_P	gravitational potential energy	J
f_i	fugacity of pure species i	kPa
$\hat{f}_i$	fugacity of species i in solution	kPa
G	molar or unit-mass Gibbs energy	J/mol or J/kg
H	molar or unit-mass enthalpy	J/mol or J/kg
M	molar or unit-mass extensive thermodynamic property	
$\overline{M}_i$	partial molar property of species i in solution	
m	mass	kg
$\dot{m}$;	mass flow rate	kg/s
N	number of chemical species	
n	number of moles	
n_i	number of moles of species i	
P	absolute pressure	kPa
Q	heat	J
$\dot{Q}$;	rate of heat transfer	J/s
R	universal gas constant	J/(mol·k)
r	number of independent chemical reactions	
S	molar or unit-mass entropy	J/(mol·k) or
T	absolute temperature	K J/(kg·K)
U	molar or unit-mass internal energy	J/mol or J/kg
u	velocity	m/s
V	molar or unit-mass volume	m^3/mol or m^3/kg
W	work	J
W_s	shaft work for flow process	J
$\dot{W}_s$	shaft power for flow process	J/s
x_i	liquid-phase mole fraction of species i in solution	
y_i	vapor-phase mole fraction of species i in solution	
z	elevation above a datum level	m
Superscripts		
ig	value for an ideal gas	
l	liquid phase	
R	residual thermodynamic property	
t	total value of a thermodynamic property	
v	vapor phase	
Subscripts		
C	value for a colder heat reservoir	
H	value for a hotter heat reservoir	
Greek letters		
α, β	as superscripts, identify phases	
γ_i	activity coefficient of species i in solution	
ϵ_j	reaction coordinate for reaction j	
$\upsilon_{i,j}$	stoichiometric number of species i in reaction j	
π	number of phases	
ϕ_i	fugacity coefficient of pure species i	
$\hat{\phi}_i$	fugacity coefficient of species i in solution	

HENDRICK C. VAN NESS
MICHAEL M. ABBOTT
Rensselaer Polytechnic Institute

M. M. Abbott and H. C. Van Ness, *Schaum's Outline of Theory and Problems of Thermodynamics*, 2d ed., McGraw-Hill, New York, 1989.

S. I. Sandler, *Chemical and Engineering Thermodynamics*, 3rd eds., Wiley, New York, 1998.

J. M. Smith, H. C. Van Ness, and M. M. Abbott, *Introduction to Chemical Engineering Thermodynamics*, 5th ed., McGraw-Hill, New York, 1996.

R. C. Reid, J. M. Prausnitz, and B. E. Poling. *The Properties of Gases and Liquids*, 4th ed., McGraw-Hill, New York, 1987.

THERMOGRAPHY

Thermography, also known as thermal imaging, is the technique of detecting spatial and time variations of radiated heat energy and transforming them to a visible display. The equipment, called the sensor, consists of an (ir) camera connected to a television and video tape recorder. The heart of the ir camera is an array of infrared detectors sensitive from visible to at least 5 μm and preferably to 14 μm wavelength (see PHOTODETECTORS). The detector array and a set of optics provide for two-dimensional spatial resolution of the thermal scene. The high speed response of the sensor results in "real time" television displays. Since thermal radiation is a function of emissivity and temperature, a calibrated thermographic sensor can remotely measure the

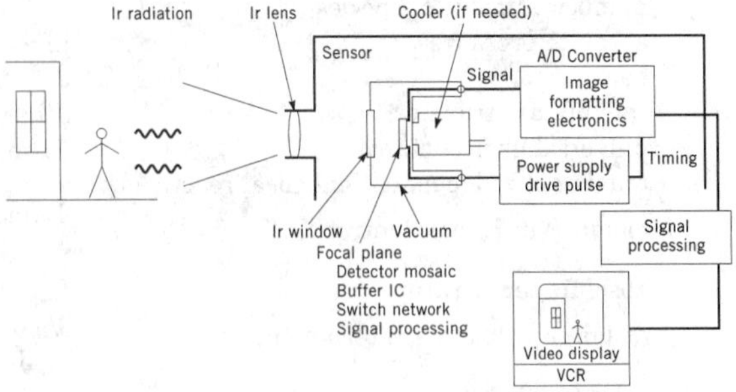

Figure 1. Schematic for thermographic imaging. The ambient thermal radiation is imaged on the focal plane which converts the infrared to an electrical signal for display on a video monitor.

surface temperature of objects. When the camera is coupled to a spectrometer or narrow band filters, the sensor is capable of spectral discrimination. Thus thermography can become both spatial radiometry and spectral imaging.

A schematic of a thermographic sensor for thermal imaging is illustrated in Figure 1. Typically, the focal plane signal is an analog of the photon intensity variations. The analog signals are converted to a digital format (A/D conversion) for signal processing.

Infrared focal plane array technology is making possible spectrographic imaging in the infrared portion of the electromagnetic spectrum which is rich in molecular absorption lines. This will have a major impact on industrial processing where control of the chemistry is essential for cost reduction and safety.

Principles of Thermography

Noise Equivalent Power. The total system electronic noise V_T may be combined with the responsivity, R_V to give the noise equivalent power equation (NEP):

$$NEP = \frac{V_T}{R_V} \tag{1}$$

The NEP may be written in terms of the detector element active area, A_p, the number of detector pixels elements connected for additive output n_p, the electronic noise bandwidth B and the detector element detectivity, D^*. Typically $n_p = 1$, but may be increased for improved sensitivity with an attendant loss in resolution.

$$NEP = \frac{1}{D^*}\sqrt{\frac{BA_p}{n_p}} \tag{2}$$

Thermographic Sensitivity. The noise equivalent temperature difference (sensitivity to scene temperature variations in degrees C) may be expressed in terms of the NEP:

$$NE\Delta T = \frac{\pi}{\Omega A_p n_p \varepsilon_0}\left[\int_{\lambda 1}^{\lambda 2} \frac{dJ_s}{d} d\lambda\right]^{-1} (NEP) \tag{3}$$

where ϵ_0 = the optical efficiency of the sensor and is typically 85%, A_p = the detector element (pixel) active area, typically 1.5 E-5 cm^2, Ω = the collection solid angle of the optical system ($= \pi \sin^2(\tan^{-1} 1/2f_n)$), typically 0.15 str, f_n = optical f number, typically 2.2 or less. For fast optics $f_n = 1.0$, J_s = spectral power density in watts/cm^{2}cmK emitted by an element of the scene.

In all cases, the imaging is conducted in real-time with a near standard television read-out display. This results in a 40,000 Hz bandwidth for scanning an a 100 Hz bandwidth for staring. It is possible to do frame addition and effectively reduce the bandwidth to achieve more sensitivity. The display can be repeated at the normal 60 Hz rate to avoid the appearance of flicker.

The thermographic sensor is used as a remote sensing radiometer when a reference target is imaged. It is usually necessary to correct for emissivity and atmospheric transmission to determine surface temperature with a reasonable degree of accuracy.

Thermographic Spectroscopy

The broad-band spectral capability of most thermographic sensors is generating a lot of interest for developing a low-cost imaging spectrometer. The focal plane itself is sensitive from visible well into the infrared. The cooled HgCdTe array responds to 12 μm wavelength and the uncooled pyroelectric and bolometer arrays can efficiently detect from the uv to 50 μm. When the entire array detects the same portion of the spectrum the display shows the spatial features in that spectral band.

Although thermal imaging can be achieved by observing the irradiation of the ambient environment, spectral imaging is a different matter. Bodies, including liquids and gases emit radiation according to their temperature and emissivity and therefore act as emitting gray bodies. In addition to gray body emission substances have specific spectral emissions depending on excitation energy somehow transferred to the substance. When there is no external excitation spectral discrimination by a remote spectrometer is difficult relying on second order effects. In general, detection of small amounts of a specific material such as a toxic gas will depend on a powerful excitation source or absorption of excess ir radiation. If only internal radiation (by the material's own temperature) is available the material will be spectrally gray and will not be detected. However an external ir radiation source such as a heated body can provide enough excess radiation for spectral absorption by the material and therefore spectral discrimination and identification.

Applications

Thermal Imaging. Thermography as a night vision tool in the nonmilitary sector is being used by law enforcement agencies to detect the movement of suspects in the complete dark. Sensors are finding use by rescue services looking for survivors floating in the water at night. All mammals are easy to detect in relatively clear weather making it easier to count livestock at night than in daylight.

Industrial use of thermal imaging typically is the detection of thermal anomalies such as leaking pipes and valves, overheating boilers, transformers and power lines, and friction generated heat in bearings. Defects in materials and circuit boards can be detected by the associated discontinuities in thermal conduction.

Chemical Gas Detection. Spectral identification of gases in industrial processing and atmospheric contamination is becoming an important tool for process control and monitoring of air quality.

Spectral monitoring for chemical process control has been demonstrated for semiconductor integrated circuit manufacture and detection of contamination of processing gases and liquids.

SEBASTIAN BORRELLO
Texas Instruments

M. C. Dudzik, ed., *Electro-Optical Systems Design, Analysis and Testing*, Vol. 4 of J. Accetta and D. Shumaker, eds., *The Infrared and Electro-Optical Systems Handbook*, Environmental Research Institute of Michigan, 1993.

F. G. Smith, ed., *Atmospheric Propagation of Radiation*, Vol. 2 of the above reference.

THERMOELECTRIC ENERGY CONVERSION

Thermoelectric energy conversion is the science of the interchange of thermal and electrical energy in simple solid-state devices, generally heavily doped semiconductors. If thermal energy is applied at the top of the device from some heat source, and removed at the bottom by a heat sink, an electrical potential appears across the device as indicated by the polarity signs. The size of the Δ change in temperature, T, and the basic material properties determine the voltage, V, across the couple. Potential per couple is usually on the order of tenths of a volt. Practical working voltages are obtained by connecting a large number of these thermocouples in series. The amount of current is dependent on the cross-sectional area and length of the legs.

If an electric current were to be passed through the device from an outside source, heat would be absorbed at one junction and released at the other. The device then operates as a solid-state heat pump.

Thermoelectric devices can be used to generate electrical power, to refrigerate, or to transfer large amounts of heat without the use of freons or compressors or any rotating machinery. However, the overall efficiency of these devices is somewhat lower than other systems, and the initial cost is often higher. Mass production, improved design and manufacturing techniques, and an extremely long (decades or more) operating life for many of these solid-state devices should both decrease initial cost and serve to aid in the recovery of the costs over time.

Segmenting of legs to take advantage of the higher performance of certain materials in a given temperature range has been used to increase conversion efficiency. However, this technique is limited because all of the thermal and electrical currents must flow through each segment.

In thermoelectric cooling applications, extensive use has been made of cascaded systems to attain very low temperatures, but because the final stage is so small compared to the others, the thermal flux is limited. The relative sizes of the stages are adjusted to obtain the maximum ΔT. Thus, for higher cooling capacity, the size of each stage is increased while the area ratios are maintained.

Thermoelectric Effects

The primary thermoelectric phenomena considered in practical devices are the reversible Seebeck, Peltier, and, to a lesser extent, Thomson effects, and the irreversible Fourier conduction and Joule heating. The Seebeck effect causes a voltage to appear between the ends of a conductor in a temperature gradient. The Seebeck coefficient, S, is given by

$$S = \frac{\delta V}{\delta T}$$

The Seebeck voltage is often referred to as being generated only at the junction of dissimilar conductors in an electrical circuit. Consideration of the Thomson effect, however, leads to the conclusion that the Seebeck voltage is generated along the entire thermoelectric element in a temperature gradient. The Thomson effect is the generation or absorption of heat, q_{τ};, other than I^2R heat, in a current-carrying conductor subjected to a thermal gradient: $q_{\tau} = \tau I$. The Thomson coefficient, τ, is given by

$$\tau = -T\frac{\delta S}{\delta T}$$

Thermoelectric properties are expressed in a term known as the figure of merit, Z. The important factors in determining the efficiency, η, of a device are Z and ΔT.

For insulators, Z is very small for metals, Z is very small. Z peaks for semiconductors at $\sim 10^{19}$ cm^{-3} charge carrier concentration, which is about three orders of magnitude less than for free electrons in metals. Thus for electrical power production or heat pump operation the optimum materials are heavily doped semiconductors.

The basic thermoelectric parameters are all functions of carrier concentration. Thus adjusting the dopant level to increase the output voltage generally also increases the electrical resistance. In addition, it affects the electronic component of the thermal conductivity. However, there are limitations on what can be accomplished by simply varying the carrier concentration in any given material.

Technology

Because the Seebeck coefficient, the electrical resistivity, and the thermal conductivity of thermoelectric materials all vary with temperature, no single material can function well over the entire temperature range in which thermoelectric devices are used. Table 1 indicates the generally accepted operating range categories and the primary thermoelectric materials used in each range. Many other materials have been or are being developed to obtain either higher conversion efficiencies or less expensive manufacturing.

Table 1. Operating Modes and Temperatures of Thermoelectric Materials

Temperature, °C			
Hot	Cold	Predominant materials	Operating mode
200	−130	bismuth telluride; bismuth antimony telluride	cooling or power
600	100	lead telluride; silver antimony germanium telluride	power
1000	300	silicon germanium	power

The physical form of the thermocouples varies significantly according to applications. Most spacecraft power supplies utilize separate thermocouples that can be checked for performance at successive stages of manufacturing and be replaced if necessary. This approach fits in very well with the extremely high reliability requirements imposed on such systems. In terrestrial systems where such individualized attention is not economically feasible, modular assemblies are generally used, which can contain tens to hundreds of couples in a single unit.

Thermoelectric Power Generation

Thermoelectric devices represent niche markets, but as economic and environmental conditions continue to change, they appear poised to advance into more common use. Thermoelectric power generators are in use in many areas, including satellites, deep-space probes, remote-area weather stations, undersea navigational devices, military and remote-area communications, and cathodic protection.

Thermoelectric generators have made extended spacecraft missions possible for exploration of the giant outer planets and the regions beyond the far reaches of the solar system, eg, the two Pioneer and the two Voyager missions.

For terrestrial applications, 77 radioisotope-powered thermoelectric generators were put into operation between the years 1961 and 1995. Their electrical power outputs range from one to 500 watts. They have been used in a wide range of applications, mostly involving remote locations where supply and servicing is difficult and long-term unattended operation is of primary importance.

More mundane but, in the long run, much more important economically and environmentally are the terrestrial systems for power generation and cooling. The primary focus of the power-generating devices is to recapture some of the waste heat dumped by commercial processes and vehicle engines, or to make use of the heat from geothermal sources (see GEOTHERMAL ENERGY). In cooling devices the emphasis is on eliminating the use of refrigerant gases that could be environmentally harmful, and in providing very small systems for specialized uses.

E. A. SKRABEK
Orbital Sciences Corporation

A. F. Joffe, *Semiconductor Thermoelements and Thermoelectric Cooling*, Infosearch Limited, London, 1957.

D. M. Rowe, ed., *CRC Handbook of Thermoelectrics*, CRC Press, London, 1993.

R. R. Heikes and R. W. Ure, Jr., *Thermoelectricity: Science and Engineering*, Interscience Publishers, New York, 1961.

S. W. Angrist, *Direct Energy Conversion*, Allyn and Bacon, Inc., Boston, Mass., 1965.

THIN FILMS

FILM FORMATION TECHNIQUES

A deposited thin film is a layer on a surface having properties that differ from those of the bulk material (substrate) that has been formed by the addition of solid material to the surface. Generally, the substrate material cannot be detected in the film, which can be an organic or inorganic material. This surface layer differs from surface conversion where the surface is chemically converted to another material, eg, anodization of aluminum (see METAL SURFACE TREATMENTS). The term thin film is generally applied to layers that have thicknesses on the order of several micrometers or less. These films may be as thin as a few atomic layers. In many cases, thin films are formed by adding atoms or molecules to a substrate surface one at a time. Thicker layers are generally called coatings.

The properties of thin films generally differ from the values for the material in bulk form. In many cases, the growth and properties of thin films are affected by the properties of the underlying substrate material. The properties of the film can also be affected by the high surface-to-volume ratio of the film.

Physical Deposition Techniques

Physical vapor deposition (PVD) processes are film deposition processes in which atoms or molecules of a material are vaporized from a solid or liquid source, transported in the form of a vapor through a vacuum or low pressure gaseous environment, and condense on the substrate. PVD processes, by far the largest-volume techniques for forming thin films, use vacuum to control the gaseous contamination in the deposition system and to provide a long mean-free path for collision for the vaporized material as it passes from the source to the substrate. PVD processes can be used to deposit films of elemental, alloy, and compound materials as well as atomically dispersed mixtures and some polymeric materials. The films can be of single materials or multilayers of different materials, have a graded composition, or be very thick deposits (coatings). The deposited material can be amorphous, fine or coarse grained, or single-crystal, depending on the material and deposition conditions. Compounds can be deposited by vaporizing the compound material directly, or by allowing the depositing material to react with an ambient gaseous environment or a codeposited species to form films of compound materials. Deposition with reaction is called reactive deposition.

Vacuum Deposition. Vacuum deposition, sometimes called vacuum evaporation, is a PVD process in which the material is thermally vaporized from a source and reaches the substrate without collision with gas molecules in the space between the source and substrate. The trajectory of the vaporized material is therefore line-of-sight.

The vapor pressure is an important property of the material to be thermally vaporized. In a closed container at equilibrium, the number of atoms returning to the surface is the same as those leaving the surface, and the pressure above the surface is the equilibrium vapor pressure. Vapor pressures are strongly dependent on the material and the temperature.

Materials having a higher vapor pressure at low temperatures are typically vaporized from resistively heated sources. Refractory materials require a high temperature to be vaporized; a focused high energy electron-beam heating is necessary.

Vacuum deposition is used to form optical interference coatings, reflecting coatings, decorative coatings, permeation barrier films on flexible packaging materials, electrically conducting films, and corrosion protective coatings. The sophisticated vacuum deposition technique of molecular beam epitaxy (MBE) is used in the semiconductor industry to form single-crystal films of elemental and compound semiconductors.

Sputter Deposition. Sputter deposition, often called sputtering, is the deposition of atoms vaporized from a surface (sputtering target) by physical sputtering. Physical sputtering is a nonthermal vaporization process where surface atoms are physically ejected by momentum transfer from an atomic-sized energetic bombarding particle. This particle is usually a gaseous ion accelerated from a plasma or an ion gun (see PLASMA TECHNOLOGY). In the collision of atomic-sized particles, energy is transferred from the incident particle, to the target particle. The maximum energy is transferred when the masses are equal and the collision is along a line joining their centers.

The surface atom that is struck can strike other atoms in the near-surface region, resulting in a collision cascade. Multiple collisions in the near-surface region can result in some momentum being directed back toward the surface. If the energy attained by a surface atom that is struck from below is sufficient, this atom can be physically ejected from the surface, ie, sputtered.

The sputtering yield is the number of surface atoms that are sputtered for each incident energetic bombarding particle. The sputtering yield depends on the bombarding particle energy, relative masses of the bombarding and target species, the angle of incidence of the bombarding species, and the chemical bonding of the surface atoms. The most common inert gas used for sputter deposition is argon.

The most simple plasma configuration for sputtering is d-c diode sputtering, where a high negative d-c voltage is applied to the surface of an electrically conductive material to be deposited in a low pressure inert gas. A plasma is formed, which fills the container, and positive ions are accelerated to the surface. In the d-c diode glow discharge, the electric field strength is high near the cathode and most of the applied voltage is dropped across a region called the cathode dark space near the cathode surface. This region is the primary region of ionization by electron atom collision. In the cathode dark space region, ions are accelerated from the plasma to impinge on the target surface with a high kinetic energy. This ion bombardment causes the ejection of secondary electrons, which are accelerated away from the cathode and cause ionization and atomic excitation by electron-atom collision. At equilibrium, enough electrons are created to cause sufficient ionization to sustain the discharge. The rest of the space between the cathode and the anode is filled with a plasma where there is little potential gradient. This is called the plasma region.

Typically, d-c diode sputtering discharges are controlled by regulating the sputtering gas pressure, target voltage, and the target power (W/cm^2). Because most of the bombarding energy is given up as heat, the sputtering target must be actively cooled.

The advent of the use of magnetic fields to confine the plasma near the target surface, ie, magnetron configuration, to increase the plasma density as well as the sputtering rate has allowed d-c diode magnetron sputtering to be used to deposit elements and compound materials at high rates.

In reactive sputter deposition where reactive gases or gas mixtures are used, the gas pressure and composition are important. During film growth, new film material is continuously being deposited and the deposited film material must react rapidly with the ambient to form the desired compound. Too high a reactive gas pressure can result in poisoning the target surface with an associated reduction in sputtering rate. Too low a gas pressure can result in not enough reaction with the depositing film, thus resulting in an undesirable film composition. Unbalanced magnetron sputtering is particularly useful in reactive sputter deposition because it activates the reactive species in the vicinity of the substrate. However, the plasma generated in various regions near the target by the unbalanced configurations is generally nonuniform in plasma properties because of the nonuniform escape of the electrons from the target region. Often, when using reactive sputter deposition, mixtures of inert and reactive gases are employed because the mass of the reactive species is rather low, eg, MN = 14, thus giving low sputtering yields.

When coating three-dimensional parts, the magnetron and unbalanced magnetron targets are often arranged so that the parts to be coated are passed between opposing targets. This allows deposition on all sides of the part even though the average angle of incidence of the depositing material varies over the surface of the part.

Principal sputter deposition processing variables for a specific material for nonreactive sputter deposition include sputtering gas; sputtering pressure; substrate temperature; deposition rate; target voltage; target power; deposition geometry, ie, angle of incidence of depositing flux; and level of gaseous contamination in the system. For reactive sputter deposition, the variables include those listed, plus partial pressures of the gases and vapors, availability of reactive species over the surface of the depositing material, and chemical reactivity (activation) of the reactive species.

Sputter deposition is widely used to deposit thin-film metallization on semiconductor material, energy-conserving coatings on architectural glass, transparent conductive coatings on glass, reflective coatings on compact disks, magnetic films, dry film lubricants, wear-resistant coatings, and decorative coatings.

Ion Plating. Controlled concurrent energetic bombardment of the depositing film material by particles of atomic or molecular size (atomic peening) can be used to modify and tailor the properties of the deposited film material. This form of PVD is called ion plating. In ion plating, the source of material to be deposited can be evaporation, sputtering, arc erosion, laser ablation, or other vaporization source. The energetic particles used for bombardment are usually ions of an inert or reactive gas. However, when using an arc erosion source, a high percentage of the vaporized materials is ionized and ions of the film material (film-ions) can be accelerated and used to bombard the growing film. Ion plating can be produced in a plasma environment where ions for bombardment are extracted from the plasma, or it may be produced in a vacuum environment where ions for bombardment are formed in a separate ion gun. The latter ion plating configuration is often called ion-beam-assisted deposition (IBAD).

The most common form of ion plating is the plasma-based process, in which the substrate and/or its fixture is an electrode used to generate a d-c or an r-f plasma in contact with the surface being coated. If an elemental or alloy material is being deposited, the plasma can be of an inert gas, usually argon. In reactive ion plating, the plasma, generally a mixture of inert and reactive gases, provides ions of reactive species such as nitrogen or oxygen that are accelerated to the surface to form compounds such as oxides, nitrides, carbides, or carbonitrides. Typically, in plasma-based ion plating, the substrate fixture is the cathode of the d-c circuit. However, the plasma can also be formed independent of the substrate and ions accelerated from the plasma to the surface of the growing film.

Concurrent bombardment during film growth affects film properties such as the film—substrate adhesion, density, surface area, porosity, surface coverage, residual film stress, index of refraction, and electrical resistivity. In reactive ion plating, the use of concurrent bombardment allows the deposition of stoichiometric, high density films of compounds such as TiN, ZrN, and ZrO_2 at low substrate temperatures.

The principal processing variables for ion plating for nonreactive deposition include substrate temperature; deposition rate; species, flux, and energy distribution of bombarding species; bombardment uniformity over the substrate surface; and gaseous contamination level. For reactive ion plating, the following are also necessary: partial pressures of the gases and vapors, availability of reactive species, chemical reactivity of the reactive species, and adsorption of reactive species on the surface.

Ion plating is used to deposit hard coatings of compound materials, adherent metal coatings, optical coatings having high densities, and conformal coatings on complex surfaces.

Thin Films from Vapors

Chemical Vapor Deposition. Chemical vapor deposition (CVD) or thermal CVD is the deposition of atoms or molecules from a chemical vapor precursor, which contains the film material to be deposited. Decomposition of the vapor is by chemical reduction or thermal decomposition. The reduction is normally accomplished by hydrogen at an elevated temperature. Some vapors, such as the carbonyls, can be thermally decomposed at relatively low temperatures. The deposited material may react with gaseous species such as oxygen, with a hydrocarbon gas such as methane or ammonia, or with a codeposited species to give compounds such as oxides, nitrides, carbides, and borides. These reactions are called synthesis reactions. CVD has numerous other names and adjectives. Examples include vapor–phase epitaxy (VPE), which takes place when CVD is used to deposit single-crystal films; metalorganic CVD (MOCVD), when the precursor gas is a metalorganic species; plasma-enhanced CVD (PECVD), when a plasma is used to induce or enhance decomposition and reaction; and low pressure CVD (LPCVD), when the reaction chamber pressure is less than ambient.

CVD reactions are most often produced at ambient pressure in a freely flowing system. The gas flow, mixing, and stratification in the reactor chamber can be important to the deposition process. CVD can also be performed at low pressures (LPCVD) and in ultrahigh vacuum (UHVCVD) where the gas flow is molecular. The gas flow in a CVD reactor is very sensitive to reactor design, fixturing, substrate geometry, and the number of substrates in the reactor, ie, reactor loading. Flow uniformity is a particularly important deposition parameter in VPE and MOCVD.

The CVD process is accomplished using either a hot-wall or a cold-wall reactor. In the former, the whole chamber is heated and thus a large volume of processing gases is heated as well as the substrate. In the latter, the substrate or substrate fixture is heated, often by inductive heating. This heats the gas locally.

The gas flow over the substrate surface establishes a boundary layer across which precursor species must diffuse in order to reach the surface and deposit. In the cold-wall reactor configuration, the boundary layer defines the temperature gradient in the vapor in the vicinity of the substrate. This boundary layer can vary in thickness and turbulence, depending on the direction of gas flow. Direct impingement of the gas on the surface reduces the boundary layer thickness and increases the temperature gradient, whereas stagnant flow regions give much thicker boundary layers.

The gases used in the CVD reactor may be either commercially available gases in tanks, for example, Ar; liquids such as chlorides and carbonyls; or solids such as Mo carbonyl. Vapor may also come from reactive-bed sources where a flowing halide, such as chlorine, reacts with a hot-bed material, such as chromium or tantalum, to give a gaseous species.

Vapors from liquids can be put into the gas stream by bubbling the hydrogen or a carrier gas through the liquid or by using a hot surface to vaporize the liquid into the gas stream. Liquid precursors are generally metered onto a hot surface using a peristaltic pump and the gas handling system is kept hot to keep the material vaporized.

Reactive-bed sources use heated solid materials, eg, chips and shavings, over which a reactive gas flows. The reaction produces a volatile gaseous species that can then be used as the precursor gas. By controlling the reaction-bed parameters, the stoichiometry of the resulting gas can be controlled.

The morphology, composition, crystalline structure, defect concentration, and properties of CVD-deposited material depends on a number of factors. An important variable in the CVD reaction is the effect of vapor supersaturation over the substrate surface and the substrate temperature. At low supersaturations, which also give a low deposition rate, nuclei initiate on isolated sites and grow over the surface, giving a high density film. At high temperatures and low supersaturations of the vapor, epitaxial growth (oriented overgrowth) can be obtained on appropriate substrates. This vapor-phase epitaxial (VPE) growth is used to grow doped layers of semiconductors. At intermediate concentrations, a nodular growth structure may form. At high supersaturations, the decomposition gives whiskers and dendritic structures having a low film density. High temperatures often lead to significant reaction between the deposited material and the substrate, which can also introduce stresses.

Plasmas can be used in CVD reactors to activate and partially decompose the precursor species and perhaps form new chemical species. This allows deposition at a temperature lower than thermal CVD. The process is called plasma-enhanced CVD (PECVD). The plasmas are generated by direct-current, radio-frequency (r-f), or electron-cyclotron-resonance (ECR) techniques.

The CVD mechanism under plasma conditions is complicated, being a combination of plasma processes and surface processes. The radicals, unique species, and excited species formed in the plasma can play an important role in adsorption and deposition from a gaseous precursor.

Processing variables that affect the properties of the thermal CVD material include the precursor vapors being used, substrate temperature, precursor vapor temperature gradient above substrate, gas flow pattern and velocity, gas composition and pressure, vapor saturation above substrate, diffusion rate through the boundary layer, substrate material, and impurities in the gases. For PECVD, plasma uniformity, plasma properties such as ion and electron temperature and densities, and concurrent energetic particle bombardment during deposition are also important.

The density of CVD deposits is generally high (>99%), but dendritic or columnar growth can decrease the density. The thickness of a CVD deposit is determined by the processing parameters and can range from very thin films to thick coatings to free-standing shapes.

Applications of CVD thin films exist in the semiconductor industry for semiconductor materials, eg, Si and Ge epitaxy, 3–5 compounds (GaAs), doped (As, P, B) epitaxial silicon (epi-silicon), polycrystalline silicon (polysilicon), and amorphous silicon for solar cells.

A unique CVD technique is used to deposit polycrystalline diamond films by passing a precursor gas mixture of H_2 plus ~5% CH_4 over a hot (~2200°C) tungsten or tantalum filament. The hot filament is carburized and then dissociates the CH_4 to carbon plus other species and the H_2 to hydrogen radicals. The activated hydrogen preferentially reacts with and etches the deposited carbon that has the sp^2 bonding (graphitic-type bonding), as opposed to the sp^3-bonded carbon (diamond-type bonding), leaving predominately sp^3-bonded carbon in the film. At substrate temperatures above 600°C, the atoms arrange into the tetrahedral diamond structure, giving a polycrystalline diamond film.

Plasma Polymerization. Many organic and inorganic monomers can be cross-linked in a plasma environment. Plasmas can be used to polymerize organic monomers to form thin polymer films. For example, a plasma is used to polymerize organosilicone thin films for protective coatings on aluminum reflector films for the automotive headlight industry.

Thin Films from Chemical Solutions

When a metal is dipped into a solution containing its own ions, some of the surface atoms dissolve and some of the ions in solution deposit. The difference in rates establishes a potential specific to the material. To measure this potential, a second electrode is needed. All electrode potentials, reported with respect to hydrogen, on platinum give the relative tendency of the material to gain or lose electrons.

Displacement Plating. Displacement or immersion plating results from the differences arising in electromotive potential between a surface and the ions in solution. In displacement plating, ions from the surface go into solution to be replaced by ion from the solution.

Electroplating. When ionically bonded molecules are dissolved in a solvent, some of the molecules dissociate into ions, whether the solvent is water, organic solvent, or a fused salt.

If the cations in solution are condensable as a solid,such as copper, they can plate out on the cathode of the cell. Thus the anode is a consumable electrode in the process.

Electroplating (qv), the deposition of metallic ions on a cathode in an electrolysis cell, is a way of depositing a limited number of materials on electrically conductive surfaces. Electroplating is often used to form coatings many micrometers thick. A thin film formed by electroplating is called a flash or strike.

Faraday's Law of electrolysis states that the amount of chemical change, ie, amount dissolved or deposited, produced by an electric current is proportional to the quantity of electricity passed, as measured in coulombs; and that the amounts of different materials deposited or dissolved by the same quantity of electricity are proportional to their gram-equivalent weights (GEW) defined as the atomic weight divided by the valence.

The deposition of ions at the cathode creates a depletion layer across which the ions must migrate in order to deposit. This layer can vary in thickness according to surface morphology. The depletion layer is more or less defined as the region where the ion concentration differs from that of the bulk solution by >1%. The layer thickness can be decreased by agitation.

Electroless Electrolytic Plating. In electroless or autocatalytic plating, no external voltage/current source is required. The voltage/current is supplied by the chemical reduction of an agent at the deposit surface. In order to initiate the electroless deposition process, a catalyst must be present on the surface. Often an accelerator is needed to remove the protective coat on the catalyst and start the reaction.

Important bath constituents in electroless plating are metal ion concentration, catalyst, reducing agent(s), complexing agent(s), and bath stabilizer(s), along with pH adjusters. Important deposition parameters are temperature, pH, metal ion and reducer concentrations, stabilizer concentration, and trace impurities that can catalyze the decomposition of the solution.

Reduction Reactions. Some thin films can be deposited from chemical solutions at low temperatures by immersion in a two-part solution that gives a reduction reaction. Chemical silvering of mirrors and dewar flasks is a common example.

Properties of Thin Films

Thin films formed by atomistic deposition techniques are unique materials that seldom have handbook properties. Properties of these thin films depend on several factors, including substrate surface condition, the deposition process used, details of the deposition process and system geometry, details of film growth on the substrate surface, and post-deposition processing and reactions. For some applications, such as wear resistance, the mechanical properties of the substrate is important to the functionality of the thin film. In order to have reproducible film properties, each of these factors must be controlled.

DONALD M. MATTOX
Management Plus, Inc.

D. M. Mattox *Handbook of Physical Vapor Deposition (PUD) Processing* Noyes Publications, 1998.

Surface Engineering, ASM Handbook, Vol. 5, ASM International, 1994.

H. K. Pulker, *Coatings on Glass*, Elsevier Science Publishing Co., Inc., New York, 1984.

D. A. Glocker and S. Ismatshah, eds., *Handbook of Thin Film Process Technology*. Institute of Physics Publishing, 1995.

MONOMOLECULAR LAYERS

Molecularly engineered materials can be fabricated from the molecular level up, and their physical properties can be both predicted and designed. Surface analytical tools enable investigations of monomolecular layers in previously unprecedented detail, leading to understanding of molecular packing and ordering. These tools also provide information to aid in understanding the relationships between the structure and properties of the individual molecule as well as of the material it forms (see SURFACE AND INTERFACE ANALYSIS).

Supramolecular assemblies are fabricated by assembling molecules that interlock in a planned, hierarchical manner, forming structures having specific desired functions. There are two principal methods. Using the Langmuir-Blodgett (LB) technique, molecular layers are formed at air–water interfaces under programmed external influence. The different kinds of monolayers are superimposed in an intelligently planned sequence, forming increasingly more complex supramolecular structures. In the self-assembly technique, layers are formed spontaneously by molecules self-organizing at a solid–liquid interface, and multilayer structures are formed only after the monolayer surface has been chemically modified.

Langmuir-Blodgett Films

Langmuir-Blodgett was the first technique to provide a practical route for the construction of ordered molecular assemblies. These monolayers, which provide design flexibility both at the individual molecular and at the material levels, are prepared at the water–air interface using a fully computerized trough.

Monolayers at the Air–Water Interface. Molecules that form monolayers at the water–air interface are called amphiphiles or surfactants (qv). Such molecules are insoluble in water. One end is hydrophilic, and therefore is preferentially immersed in the water; the other end is hydrophobic, and preferentially resides in the air, or in a nonpolar solvent. Complex organic amphiphiles containing chromophores, various donor or acceptor groups, etc, can be designed and synthesized. Understanding the structure of such monolayers can be assisted by computer modeling (see COMPUTER TECHNOLOGY; MOLECULAR MODELLING).

The monolayer resulting when amphiphilic molecules are introduced to the water–air interface was traditionally called a two-dimensional gas owing to what were the expected large distances between the molecules. However, it has become quite clear that amphiphiles self-organize at the air–water interface even at relatively low surface pressures.

The Transfer of Monolayers to a Solid Substrate. Two methods of transfer of monolayers from the water–air interface onto a solid substrate are important. The first, and more conventional, method is the vertical deposition. A monolayer of amphiphiles at the water–air interface can be deposited by the displacement of a vertical plate. When such a plate is moved through the monolayer at the water–air interface, the monolayer can be transferred during immersion (retraction or upstroke) or immersion (dipping or downstroke). A monolayer is usually transferred during retraction when the substrate surface is hydrophilic, and the hydrophilic head groups interact with the surface. On the other hand, if the substrate surface is hydrophobic, the monolayer is transferred in the immersion, and the hydrophobic alkyl chains interact with the surface. If the deposition process starts with a hydrophilic substrate, the surface becomes hydrophobic after the first monolayer transfer. Thus the second monolayer is transferred in the immersion. This is the most usual mode of multilayer formation for amphiphilic molecules in which the head group is very hydrophilic and the tail is an alkyl chain. This mode is called the Y-type deposition (Fig. **1a**). For very hydrophilic head groups, eg, –COOH, $-PO_3H_2$, etc, this is the most stable deposition mode, because the interactions between adjacent monolayers are then either hydrophobic–hydrophobic, or hydrophilic–hydrophilic. This mode produces centrosymmetric films comprised of bilayers.

Films may be formed only in downstroke (X-type, Fig. **1b**). The deposition speed may affect the deposition mode. If deposition occurs only when films are formed in upstroke, Z-type films result (Fig. **1c**).

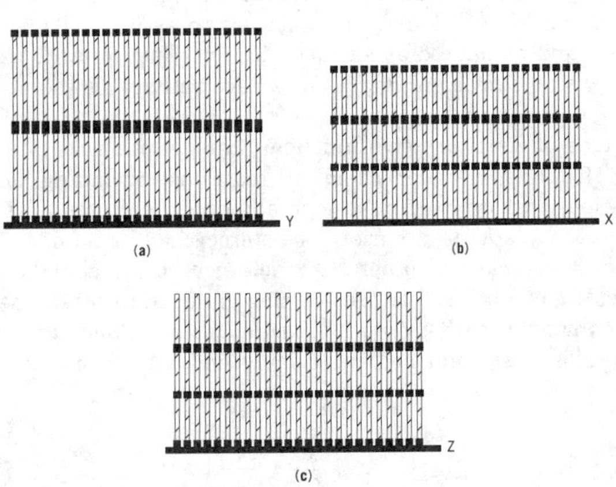

Figure 1. Multilayer films where (▯) represent a hydrophobic group and (■) a hydrophilic one: (**a**) Y-type, (**b**) X-type, and (**c**) Z-type.

These are cases where the head group is not as hydrophilic, eg, $COOCH_3$, or where the alkyl chain is terminated by a weak polar group, eg, NO_2. In both cases the interactions between adjacent monolayers are hydrophilic—hydrophobic. These multilayers are therefore less stable than the Y-type systems. Both X- and Z-type depositions are noncentrosymmetric.

Another way of building LB multilayer structures is the horizontal lifting or Schaefer's method, which is useful for the deposition of very rigid films. In this method, a compressed monolayer is formed at the water–air interface and then a flat substrate is placed horizontally on the monolayer film. When this substrate is lifted and separated from the water surface, the monolayer is transferred onto the substrate, in theory, keeping the same molecular direction (X-type, Fig. **2b**).

LB Films of Long-Chain Fatty Acids. The most stable films of long-chain fatty acids are formed by cadmium arachidate deposited from a buffered $CdCl_2$ subphase. These films, considered to be standards, have been widely used as spacer layers and for examining new analytical techniques.

LB Films of Liquid-Crystalline Amphiphiles. Liquid-crystal (LC) phases are materials that have inherently ordered-layer structures, formed by self-organization of mesogenic compounds (see LIQUID CRYSTALLINE MATERIALS). Therefore, by having a liquid-crystalline group in an amphiphile, enhanced order, thermal stability, and interesting physical properties can result. Furthermore, the study of liquid crystals at the water–air interface in a systematic way should add to the understanding of the two-dimensional organization, and the effect of the director on the relative orientation of molecules in the layers of a multilayer film. There have been a large number of studies on LB films of LCs.

LB Films of Porphyrins and Phthalocyanines. The porphyrin is one of the most important among biomolecules. Many porphyrin and phthalocyanine (PC) derivatives form good LB films. Both these molecules are important for applications such as hole-burning that may allow information storage using multiple frequency devices.

The first synthesis of amphiphilic porphyrin molecules involved replacement of the phenyl rings in 5,10,15,20-tetraphenylporphyrin (TPP) with pyridine rings, quaternized with $C_{20}H_{41}Br$ to produce tetra(3-eicosylpyridinium)porphyrin bromide **(1)**. The pyridinium nitrogen is highly hydrophilic: the long C_{20} hydrocarbon serves as the hydrophobic part.

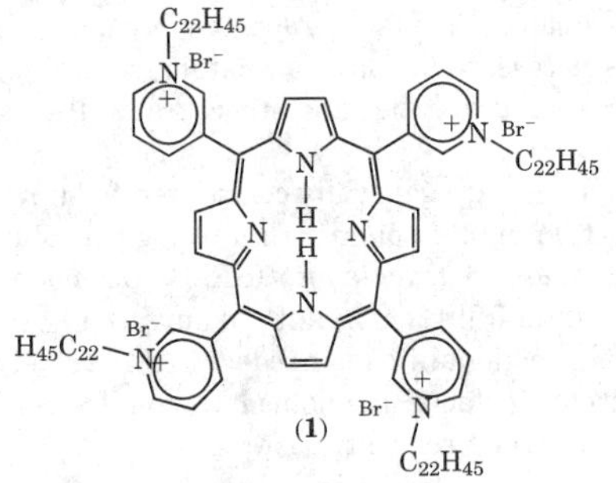

(1)

LB Films of Polymerizable Amphiphiles. Studies of LB films of polymerizable amphiphiles include simple olefinic amphiphiles, conjugated double bonds, dienes, and diacetylenes. In general, a monomeric amphiphile can be spread and polymerization can be induced either at the air–water interface or after transfer to a solid substrate. The former polymerization results in a rigid layer that is difficult to transfer.

Self-Assembled Monolayers

The formation of monolayers by self-assembly of surfactant molecules at surfaces is one example of this general phenomenon. In nature, self-assembly results in supermolecular hierarchical organizations of interlocking components providing very complex systems. In the 1980s it was shown that self-assembled monolayers (SAMs) of alkanethiolates on gold can be prepared by adsorption of di-*n*-alkyl disulfides from dilute solutions. Many self-assembly systems have since been investigated, but monolayers of alkanethiolates on gold are probably the most studied.

The ability to tailor both head and tail groups of the constituent molecules makes SAMs excellent systems for a more fundamental understanding of phenomena affected by competing intermolecular, molecular—substrate and molecule—solvent interactions, such as ordering and growth, wetting, adhesion, lubrication, and corrosion. Because SAMs are well-defined and accessible, they are good model systems for studies of physical chemistry and statistical physics in two dimensions, and the crossover to three dimensions.

SAMs are ordered molecular assemblies formed by the adsorption of an active surfactant on a solid surface. This simple process makes SAMs inherently manufacturable and thus technologically attractive for building superlattices and for surface engineering. The order in these two-dimensional systems is produced by a spontaneous chemical synthesis at the interface, as the system approaches equilibrium. Although the area is not limited to long-chain molecules, SAMs of functionalized long-chain hydrocarbons are most frequently used as building blocks of supermolecular structures.

Monolayers of Organosilicon Derivatives. SAMs of alkylchlorosilanes, alkylalkoxysilanes, and alkylaminosilanes require hydroxylated surfaces as substrates for their formation. The driving force for this self-assembly is the *in situ* formation of polysiloxane, which is connected to surface silanol groups (–Si–OH) via Si–O–Si bonds. Substrates on which these monolayers have been successfully prepared include silicon oxide, aluminum oxide, quartz, glass, mica, zinc selenide, germanium oxide, and gold.

Patterns of ordered molecular islands surrounded by disordered molecules are common in Langmuir layers, where even in zero surface pressure molecules self-organize at the air–water interface. The difference between the two systems is that in SAMs of trichlorosilanes the island is comprised of polymerized surfactants, and therefore the mobility of individual molecules is restricted.

Surface Engineering Using SAMs. Independent control of surface structure and chemical properties and the resulting structure property relationships are scientifically interesting and technologically important. For many applications, controlling the properties of interfaces is very important. However, in real-life circumstances, interfaces that contain at least one polymer surface are typically irregular. Surface properties of polymers depend critically upon the chemical and physical details of molecular structure at the surface of the polymer. To control surface properties by manipulating surface structure, it is necessary to have an extensive database of detailed correlations between properties and structure for the polymer surface of interest.

Surface properties are generally considered to be controlled by the outermost 0.5–1.0 nm at a polymer film. A logical solution, therefore, is to use self-assembled monolayers (SAMs) as model polymer surfaces. To understand fully the breadth of surface interactions, a portfolio of chemical functionalities is needed. SAMs are especially suited for the studies of interfacial phenomena owing to the fine control of surface functional group concentration.

ABRAHAM ULMAN
Polytechnic University

G. L. Gaines, *Insoluble Monolayers Liquid-Gas Interfaces*; Wiley-Interscience, New York, 1966.

A. Ulman, *An Introduction to Ultrathin Organic Films From Langmuir-Blodgett to Self-Assembly*, Academic Press, Boston, 1991.

For an introduction to liquid crystals see P. J. Collings, *Liquid Crystals Nature's Delicate Phase of Matter*, Princeton Science Library, Princeton, N.J., 1990.

THIOGLYCOLIC ACID

Thioglycolic acid (2-mercaptoacetic acid, $HSCH_2COOH$, molecular weight 92.11, is the first member of the mercaptocarboxylic acids series. The advent of the PVC industry in the 1950s brought thioglycolic

acid a significant application as a raw material in the manufacture of organotin stabilizers (see HEAT STABILIZERS; VINYL POLYMERS). These stabilizers improved thermal stability and prevented discoloration during polymer processing. As large-scale commercial chemical production grew, thioglycolic acid began to be used more and more as a raw material in the manufacture of fine and specialty chemicals in the pharmaceutical and agricultural sectors.

Properties

Pure thioglycolic acid is a water-white liquid that freezes at $-16°C$ and distills under reduced pressure. Reported constants are bp at 3.9 kPa (29 mm Hg), 123°C, bp at 1.33 kPa (10 mm Hg), 96°C; d^{20}_4, 1.325 g/cm^3; viscosity at 20°C, 6.55 mPa·s(= cP); refractive index n^{20}_D, 1.5030; heat of vaporization, 627.2 J/g (149.9 cal/g); heat of combustion, 1446 kJ/mol (345.6 kcal/mol); electrical conductivity at 20°C, $2 \times 10^6 (\Omega \cdot cm)^{-1}$; dielectric constant, 7.4×10^{-30} C·m (2.25 debye); dipole moment, 7.6×10^{-30} C·m (2.28 debye); flash point in closed cup, 132°C.

Both the carboxyl and the mercapto moieties of thioglycolic acid are acidic. Dissociation constants at 25°C are for pK_1, 3.6; pK_2, 10.5. Thioglycolic acid is miscible in water, ether, chloroform, dichloroethane and esters. It is weakly soluble in aliphatic hydrocarbons such as heptane, hexane. Solvents such as alcohols and ketones can also react with thioglycolic acid.

Reactions

Thioglycolic acid is altered by self-esterification. This alteration depends on temperature and concentration of aqueous solutions. Under an air atmosphere oxidation to disulfide occurs. Many self-esterification products, called thioglycolides, have been detected by nmr. The self-esterification products can be reversed in the presence of dilute acids or alkalies, and aged thioglycolic acid can be almost completely recovered by hydrolysis.

Thioglycolic acid undergoes reactions typical of carboxylic acids, forming salts, esters, amides, and reactions typical of mercaptans, and forming thioethers with olefins or halogenated compounds, disulfides by oxidation, and metal mercaptides by reaction with metal oxides or metal chlorides. Thioglycolic acid is a powerful reducing agent in neutral or alkaline solutions. One of the most widely used reactions of thioglycolic acid is the thiol disulfide interchange reaction, particularly with the disulfide bond of cystine (1) in protein material, eg, wool and hair, to form the amino acid cysteine (2). The rate at which equilibrium is established is a function of pH; it is low at pH <6 and rapid at pH 8–10.

Long-chain thioacetic acids are obtained by reaction of primary alkenes with thioglycolic acid, by using uv lamps or radical initiators. Other products of commercial value, such as laurylthiopropionic acid, $C_{12}H_{24}SCH_2CH_2COOH$, are produced starting from 3-mercaptopropionic acid, $HSCH_2CH_2COOH$, and unsaturated products. S-Alkyl-thiocarboxylic acids and their potassium salts have been described and evaluated as surfactants (qv). They provide excellent thermally stable behavior and good surface activity for their alkaline salts.

Because of its two active functions, thioglycolic acid is an ideal reagent for a variety of chemical reactions, including addition, elimination, and cyclization. Under alkaline conditions, interesting routes are available to a variety of heterocyclic compounds, such as thiophene, thiazole and other N–S heterocyclic compounds. Methyl thioglycolate can be used as the starting material to obtain methyl 3-amino-2-thiophenecarboxylate by reaction with 2-chloroacrylonitrile. Compound is a key intermediate to drugs and agrochemicals.

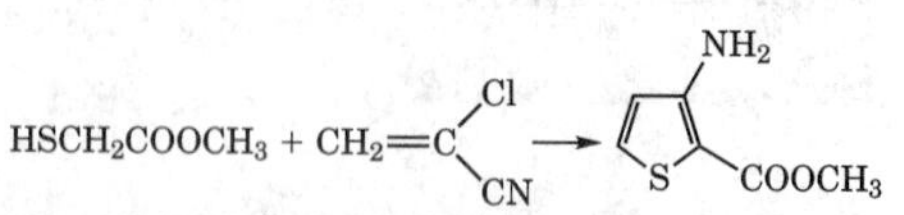

$$HSCH_2COOCH_3 + CH_2=C \begin{smallmatrix} Cl \\ CN \end{smallmatrix} \longrightarrow \quad (1)$$

(1)

Table 1. Specifications for Thioglycolic Acid and Some Derivatives

Property	Thioglycolic acid	Thioglycolic acid	Ammonium thioglycolate	Glyceryl monothioglycolate	Monoethanolamine thioglycolate	Calcium thioglycolate
EINECS[a]	2006774	2006774	2265409	2502648	2048154	
assay as TGA, %	98–99	80.0–80.3	60.0–60.3	75–76	50.0–50.4	50
specific gravity, 20°C	1.32	1.26–1.27	1.205	1.285	1.25	
iron[b], ppm	0.4	0.3	0.4	2	0.4	
pH			7.0–7.2		6.9–7.3	11–12
refractive index, 20°C	1.5030			1.5020–1.5030		
appearance	water-white	water-white	clear	clear, colorless	clear	white crystalline powder

[a] EINECS = European Inventory of Existing Commercial Chemical Substances.
[b] Value is maximum.

Several kinds of products can be obtained by reaction of thioglycolic acid and its esters with aldehydes to form mercaptals, $RCH(SCH_2COOH)_2$, or with ketones to form thiolketals, $RR'C(SCH_2COOH)_2$. Reaction with formaldehyde (qv) yields di-*n*-butylmethylene-bisthioglycolate (MBT ester): Thioglycolic acid forms a multiplicity of stable complexes with metal ions.

Manufacturing, Processing, and Storage

Thioglycolic acid is manufactured by the reaction of monochloracetic acid or its salts with alkali hydrosulfides, eg, NaSH or NH_4SH, in aqueous medium, under controlled conditions of pressure, temperature, pH, concentration, to give a higher yield of thioglycolate salt and minimize the formation of such by-products as thiodiglycolic and dithiodiglycolic acids. The reaction mixture is acidified to liberate thioglycolic acid, which is extracted from the aqueous solution into an organic solvent, such as an ether, and then purified by vacuum distillation.

Commercial monochloroacetic acid contains many other organic acids, particularly dichloroacetic acid which has to be completely converted into sulfur derivatives to avoid residual chlorine compounds which are harmful for cosmetic applications. Many other routes to produce thioglycolic acid have been investigated. These alternative methods, which require reduction of the disulfides or hydrolysis of carboxymethylthio derivatives, seem less competitive than those using alkali sulfhydrates. Manufacture of thioglycolic acid is associated with the production of aqueous salt waste, so problems of waste disposal have to be resolved in each plant.

Esters of thioglycolic acid are to a large extent manufactured by conventional esterification processes. Manufacture at a larger scale of 2-ethylhexyl thioglycolate and isooctyl thioglycolate by a continuous process gives esters of higher consistency. Thioglycolic acid is marketed as pure product or at 80–85 wt % aqueous solution. The ammonium salts are available in aqueous solutions containing 50–60 wt % thioglycolic acid; the monoethanolamine salts are available as solutions containing 40–50 wt % thioglycolic acid. Glycerol monothioglycolate is supplied in anhydrous form. Calcium thioglycolate is supplied as a crystalline powder. Potassium thioglycolate and sodium thioglycolate are also available as aqueous solutions. Thioglycolic acid is stored in reinforced polyethylene or polypropylene tanks or containers. It is advisable to keep thioglycolic acid at low (<10°C) temperature to slow down self-esterification.

Among other mercaptocarboxylic acids, the mercaptopropionic acids have undergone a promising development. Thiolactic acid or 2-mercaptopropionic acid is manufactured by using 2-chloropropionic acid. 3-Mercaptopropionic acid, competes with thioglycolic acid in plastic additives or as modifiers in various polymers. Mercaptopropionic acid is produced by using acrylic monomers, eg, acrylic acid, acrylonitrile, methyl acrylate, as raw materials.

Economic Aspects and Specifications

Since its development in cosmetics in the 1940s, thioglycolic acid has become a widespread thiochemical, used all over the world as the acid or in the form of its salts or esters. Because of the several derivatives of thioglycolic acid used, the market size for this chemical is often expressed in terms of thioglycolic acid equivalents. In 1994 the total world market was estimated at around 15,000 to 20,000 metric tons of thioglycolic acid equivalents. Table 1 lists some properties and commercial specifications of thioglycolic and its salts.

Health and Safety Factors

A safety assessment of ammonium and glyceryl thioglycolates and thioglycolic acid has been published. Hair products containing ammonium thioglycolate and glyceryl thioglycolate may be used safely, at infrequent intervals, at concentrations of ammonium thioglycolate and glyceryl thioglycolate up to 14.5% (as thioglycolic acid). Hairdressers should avoid skin contact and minimize consumer skin exposure. Handling of thioglycolic acid and mercaptopropionic acids requires the usual precautions observed for strong acids and corrosive chemicals.

Thioglycolic acid is harmful to fish. 2-Ethylhexyl thioglycolate is toxic to fish. The LC_{50}, 48 h to Leuciscus idus was found to be 9 mg/L.

Uses

The thioglycolates continue to be the most commonly used active ingredient in permanent waving, in hair straightening lotions, and in depilatory creams.

Properties of polymers obtained by free-radical polymerization depend on their molecular weight and molecular weight distribution, and substances capable of regulating these are called chain-transfer agents. Depending on the purpose for which a thioglycolate additive is used in polymerization, it is called a catalyst, a promoter, an accelerator, or a chain transfer agent.

The most important development of thioglycolic acid, and especially its isooctyl and 2-ethylhexyl esters, concerns its use as raw material for tin stabilizers, to prevent discoloration during thermal processing of PVC, and also to assure good compatibility and diffusion of the stabilizers through the resins.

In the leather industry, dehairing is a specific step in the treatment of animal skins, where the reducing properties of thioglycolic salts for the disulfide links of keratins are exploited along side those of other reducing agents such as sodium sulfide and sodium hydrosulfide. Wool treatment by thioglycolic acid improves elongation, elasticity, and strength properties of fibers; prevents shrinkage; and makes dyeing easier. In wood industries, addition of thioglycolic acid to pulping liquor during alkaline cooking of pinewood led to an increase in the degree of delignification of wood. Thioglycolic acid is recommended as a cocatalyst with strong mineral acid in the manufacture of bisphenol A by the condensation of phenol and acetone. Thioglycolic acid is described in descaling compositions for iron oxide removal. It can also be used as ammonium or ethanolamine salts to remove rust without attacking the metal substrate (see METAL SURFACE TREATMENTS). Thioglycolic acid can modify the fluidity of concretes and cements in a new class of superplasticizers, where the polymeric melamine structures are grafted with a sulfur atom instead of nitrogen in the case of the well-known sulfanilic acid. This newer way of linking using an S atom modifies the structure of the resin and is responsible to a great extent for high fluidity values, a water-reducing content in concrete, and an improvement in mechanical properties (see CEMENT).

In the oil field industry, in drilling activities the sequestering properties of thioglycoic acid for iron are reflected in its use as an iron-controlling agent, for acidizing in well stimulation, and also as corrosion inhibitors in high density completion fluids. In the lightening of petroleum hydrocarbon oil, esters of mercaptocarboxylic acids can modify radical behavior during the distillation step. Thioesters of dialkanol and trialkanolamine have been found to be effective multifunctional antiwear additives for lubricants and fuels.

Sulfur compounds are traditionally used as rubber and plastic additives. Esters of thioglycolic acid and various glycols, eg, ethylene glycol, propylene glycol, pentaerythritol, and particularly the butyl ester of methylene bisthioglycolic acid are used as a polar plasticizer and softener for synthetic rubber, especially for nitrile rubber and chloroprene. Dodecylthioacetic acid is a viscosity modifier. Esters of thiodipropionic acid and long-chain alcohols, in the C_{12}–C_{18} range, are largely used as thioester antioxidant additives in polyolefins and styrene–butadiene latex resins. The pentaerythritol ester of dodecylthiopropionic acid is marketed for the same purpose. Thioglycolic acid and esters are used in the manufacture of sulfur dyes (qv), thioindigo pigments, and as additives for dyeing baths.

Thioglycolic acid and its esters are useful as a raw material to obtain biologically active molecules. Various cyclic compounds can be built using thioglycoic acid, eg, thiazolidinone, thiazole, isothiazole, and thiazine-type structures, leading to intermediates for the agricultural and pharmaceutical industries. Fungicidal organotin mercaptocarboxylates have also been claimed.

Y. LABAT
Elf Atochem

U.S. Pat. 3,927,085 (Dec. 10, 1975), H. G. Zenkel and co-workers (to Akzo); Jpn. Pat. 55,145,663 (Nov. 14, 1980), K. Tamashima and S. Nagasaki (to Denki Kagaku Kogyo); Jpn. Pat. 56,097,264 (Aug. 5, 1981), K. Tamashima and S. Nagasaki (to Denki Kagaku Kogyo); U.S. Pat. 5,023,371 (June 11, 1991), M. E. Tsui and M. B. Sherwin (to W. R. Grace).

G. T. Walker, *Seifen Ole Fette Wachse* **13**, 402–404 (1962); **14**, 431 (1962).

J. Am. Coll. Toxicol. **10**(1), 135–192 (1991).

H. Andreas, *Plastics Additives Handbook*; 3rd ed., Hanser Publishers, New York, 1990, pp. 271–325.

THIOLS

Thiols, or mercaptans as they were originally called, are essential as feedstocks in the manufacture of many types of rubber (qv) and plastics (qv). They are utilized as intermediates in agricultural chemicals, pharmaceuticals (qv), in flavors and fragrances, and as animal feed supplements.

Nomenclature

Thiols are still commonly named as mercaptans, although the proper nomenclature is that established by the International Union of Pure and Applied Chemists (IUPAC).

Occurrence

Cysteine is a thiol-bearing amino acid which is readily isolated from the hydrolysis of protein. There are only small amounts of cysteine and its disulfide, cystine, in living tissue. Glutathione contains a mercaptomethyl group, $HSCH_2$–, and is a commonly found tripeptide in plants and animals. Coenzyme A is another naturally occurring thiol that plays a central role in the synthesis and degradation of fatty acids.

Table 1. Properties of Thiols

Compound	Mol wt	Mp, K	Bp, K
methanethiol	48.11	150.18	279.11
ethanethiol	62.14	125.26	308.15
2-mercaptoethanol	78.14	200.00	430.90
1,2-ethanedithiol	94.20	231.95	419.20
mercaptoacetic acid	92.12	256.65	493.00
1-propanethiol	76.16	159.95	340.87
2-propanethiol	76.16	142.61	325.71
3-mercaptopropionic acid	106.15	290.65	501.00
1-butanethiol	90.19	157.46	371.61
2-butanethiol	90.19	133.02	358.13
2-methyl-1-propanethiol	90.19	128.31	361.64
2-methyl-2-propanethiol	90.19	274.26	337.37
2,2′-oxybisethanethiol	138.26	193.15	490.15
1-pentanethiol	104.22	197.45	399.79
cyclohexanethiol	116.23	189.64	431.95
1-hexanethiol	118.24	192.62	425.81
benzenethiol	110.18	258.26	442.29
1-heptanethiol	132.27	229.92	450.09
α-toluenethiol	124.21	243.95	472.03
1-octanethiol	146.30	223.95	472.19
2,4,4-trimethyl-2-pentanethiol	146.30	199.00	428.65
1-nonanethiol	160.32	253.05	492.95
1-decanethiol	174.35	247.56	512.35
1-undecanethiol	188.38	270.15	530.55
1-dodecanethiol	202.40	265.15	547.75
tert-dodecanethiol[a]	202.40		515.65
1-hexadecanethiol	258	291–293	$396–401_{0.07}$[b]
1-octadecanethiol	286	301	$461_{0.1}$[b]

[a] *tert*-Dodecanethiol is a mixture of isomers.
[b] Subscripted value represents pressure in kPa at which boiling point was taken. To convert kPa to mm Hg, multiply by 7.5.

Physical Properties

The physical characteristic of thiols that most distinguishes them is their odor. Thiols range over the gamut of physical properties. Most of the important thiols are liquids, however methanethiol is a gas, and 1-hexadecanethiol and 1-octadecanethiol are waxy solids. Table 1 lists a variety of physical properties for the more important thiols.

Preparation

Thiols can be prepared by a variety of methods. The most-utilized of these synthetic methods for tertiary and secondary thiols is acid-catalyzed synthesis; for normal and secondary thiols, the most-utilized methods are free-radical-initiated, alcohol substitution, or halide substitution; for mercaptoalcohols, the most-utilized methods is oxirane addition; and for mercaptoacids and mercaptonitriles, the most-utilized methods are Michael-type additions.

Reactions of Thiols

Disulfides are prepared commercially by two types of reactions. The first is an oxidation reaction utilizing the thiol and a suitable oxidant. The most common oxidants are chlorine, oxygen, elemental sulfur, or hydrogen peroxide. The second type of reaction is the reaction of a sulfenyl halide with a thiol. This process is used to prepare unsymmetric disulfides, RSSR′ such as 4,4-dimethyl-2,3-dithiahexane.

Thiols react readily with alkenes under the same types of conditions used to manufacture thiols. In this way, dialkyl sulfides and mixed alkyl sulfides can be produced. Sulfides are a principal by-product of thiol production.

Thiols react with carbonyl compounds to form dithioacetals and thioacetals. The dithioacetals have been utilized as protecting groups

Table 2. Odor Threshold Levels and Threshold Limit Values (TLV)

Thiol	Odor threshold, ppb	TLV, ppm
methanethiol	1.05	0.5
ethanethiol	1.07	0.5
1,2-ethanedithiol	19.50	
1-propanethiol	1.26	
2-propanethiol	0.35	
2-propene-1-thiol	0.40	
1-butanethiol	1.41	0.5
2-butanethiol	0.18	0.5
2-butene-1-thiol	0.13	
2-methyl-1-propanethiol	1.12	0.5
2-methyl-2-propanethiol	0.33	0.5
1-pentanethiol	0.12	0.5
2-methyl-1-butanethiol	0.26	
2-methyl-2-butanethiol	0.72	
benzenethiol	0.31	0.1
4-methylbenzenethiol	1.70	
α-toluenethiol	1.58	
1-dodecanethiol	0.25	
2-methyl-1-undecanethiol	190.55	

for carbonyl compounds. They generally require stringent conditions to remove the blocking groups. The thioacetals are not overly stable.

Thiols decompose by two principal paths. These are the carbon–sulfur bond homolysis and the unimolecular decomposition to alkene and hydrogen sulfide.

Thiols undergo photolytic reactions.

The interchange between thiols and disulfides is base-catalyzed. It involves the nucleophilic attack of a thiolate ion on a disulfide. In rubber production, the thiol acts as a chain-transfer agent, in which it functions as a hydrogen atom donor to one rubber chain, effectively finishing chain growth for that polymer chain. The sulfur-based radical then either terminates with another radical species or initiates another chain. The thiol is used up in this process.

In general, rubber manufacturers balance thiol reactivity and odor. The odor of light thiols is generally too strong for most rubber manufacturers, as it is generally hard to remove residual odors from polymers.

Another area in which sulfur compounds have long found use is in the area of agricultural chemicals. Many of these materials had been produced by the manufacturer of the agricultural chemicals, but difficulties in containing odor and the use of hydrogen sulfide in heavily populated areas again pushed toward specialization by several companies. Over the years, a diverse group of products made use of the rather unique chemistry of organosulfur chemicals. These include the following:

Thiol	Application
methanethiol	DL-methionine
ethanethiol	propane odorant
2-methyl-2-propanethiol	natural gas odorant components
1-octanethiol	water repellants
cyclohexanethiol	prevulcanization inhibitor
2-propanethiol	flocculent natural gas odorant component
tert-nonanethiol	polysulfides for oil additives
tert-dodecanethiol	polysulfides for oil additives surfactants
2-mercaptoethanol	PVC stabilizers solvent for acrylic fibers

Toxicology and Handling

The toxicology and procedures for handling thiols used in gas odorants have been reviewed, as have thiols as a whole. A listing of known TLV values and odor thresholds for a variety of thiols is given in Table 2. Thiol spills are handled in the same manner that all chemical spills are handled, with the added requirement that the odor be eliminated as rapidly as possible.

JOHN S. ROBERTS
Phillips Petroleum Company
CH&A Corporation

S. Oae, ed., *Organic Chemistry of Sulfur*, Plenum Press, New York, 1977.

S. Patai, ed., *The Chemistry of the Thiol Group*, John Wiley & Sons, Ltd., London, 1974.

S. Patai, ed., *The Chemistry of Functional Groups, Supplement E*, John Wiley & Sons, Inc., New York, 1980.

S. Patai and Z. Rappoport, eds., *The Chemistry of Sulphur-Containing Functional Groups, Supplement S*, John Wiley & Sons, Inc., New York, 1993.

THIOPHENE AND THIOPHENE DERIVATIVES

Thiophene and a number of its derivatives are significant in fine chemical industries as intermediates to many products for pharmaceutical, agro-chemical, dyestuffs, and electronic applications.

The basic nomenclature of the thiophene ring system and its derivatives is indicated by the following: the sulfur atom is number 1, positions 2 and 5 are equivalent in the parent ring, as are the 3 and 4 positions.

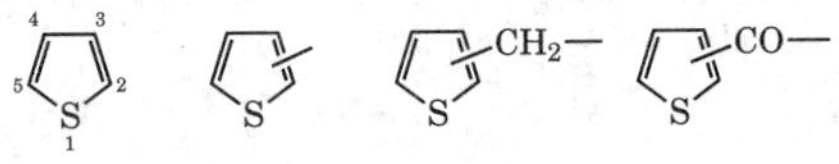

thiophene thienyl thenyl thenoyl

Table 1 indicates the significant physical properties of thiophene and 2- and 3-methylthiophene.

Reactions

Electrophilic substitution of thiophene occurs largely at the 2-position and the reactivity of the ring is greater than that of benzene. 3-Substituted derivatives are generally prepared by indirect means or through ring cyclization reactions.

Thiophenes can be alkylated in the 2-position using alkyl halides, alcohols, and olefins. Choice of catalyst is important; the weaker Friedel-Crafts catalysts, eg, $ZnCl_2$ and $SnCl_4$, are preferred.

To achieve acylation of thiophenes, acid anhydrides with phosphoric acid, iodine, or other catalysts have been widely used. Acid chlorides with $AlCl_3$, $SnCl_4$, $ZnCl_2$, and BF_3 also give 2-thienylketones. All reactions give between 0.5 and 2.0% of the 3-isomer.

Many different halogenating reagents have been used to accomplish halogenation of the thiophene ring. Excess of reagent gives di-, tri-, and even tetrahalogenation of thiophene itself. The bromothiophenes are often the preferred target. Bromine in acetic acid or chloroform has been the traditional route, but utilizes just one of the atoms of bromine. Addition products are often observed, particularly when proceeding through to di- and tribromothiophenes.

It is difficult to control nitration of thiophene, which yields 2-nitrothiophene. The strongly electrophilic nitronium ion leads to significant yields (12–15%) of 3-isomer. A preferred procedure is the slow addition of thiophene to an anhydrous mixture of nitric acid, acetic acid, and acetic anhydride.

Direct reaction of thiophene and butyllithium in diethyl ether gives 2-thienyllithium, a valuable intermediate and source of many further derivatives.

Table 1. Physical Properties of Thiophene and Methylthiophenes[a]

Property	Thiophene	2-Methylthiophene	3-Methylthiophene
freezing point, °C	−38.3	−63.4	−68.9
boiling point at 101.3 kPa[a], °C	84.16	113	115
flash point, °C	−7	16	15.5
density, d_4^0, kg/m^3	1087.3	1025.0	1025.0
density, d_4^{25}, kg/m^3	1057.3	1014.0	1016.2
refractive index, n_D^{25}	1.52572	1.5174	1.5172
viscosity at 25°C, mPa·s(= cP)	0.621	0.669	0.642
surface tension at 20°C, mN/m (= dyn/cm)	31.34	30.95	32.37
vapor pressure, kPa[a], at			
0°C	2.86		1.33 at 11°C
20°C	8.36	2.66 at 21°C	3.99 at 31°C
50.1°C	31.16	10.66 at 49.1°C	10.66 at 51.5°C
95.9°C	143.3	133.3 at 122.4°C	133.3 at 125.4°C
heat of vaporization at bp, kJ/mol[b]	31.47	33.90	34.25
heat of formation at 298.16 K, kJ/mol[b]			
liquid	81.67	45.44	43.89
gaseous	115.44	84.35	83.43
heat of combustion at 101.3 kPa[a] and 25°C, kJ/mol[b]	−2435.2	−3471.3	−3469.0
dielectric constant at 20°C	2.74		

[a] To convert kPa to mm Hg, multiply by 7.5.
[b] To convert kJ/mol to kcal/mol, divide by 4.184.

Strong oxidizing agents can rupture the thiophene ring structure, eg, nitric acid gives maleic acid. In the vapor phase, oxidation can lead to loss of aromaticity, producing thiomaleic anhydride. Oxidation of alkylthiophenes can lead to carboxylic acids.

Reduction of thiophene to 2,3- and 2,5-dihydrothiophene and ultimately tetrahydrothiophene can be achieved by treatment with sodium metal–alcohol or ammonia. Hydrogen with Pd, Co, Mo, and Rh catalysts also reduces thiophene to tetrahydrothiophene, a malodorous material used as a gas odorant.

Reaction of thiophene with aqueous formaldehyde solution in concentrated hydrochloric acid gives 2-chloromethylthiophene. This relatively unstable, lachrymatory material has been used as a commercial source of further derivatives such as 2-thiopheneacetonitrile and 2-thiopheneacetic acid.

Useful nucleophilic substitutions of halothiophenes are readily achieved in copper-mediated reactions. Of particular note is the ready conversion of 3-bromoderivatives to the corresponding 3-chloroderivatives with copper(I)chloride in hot *N,N*-dimethylformamide.

Manufacture of Thiophenes

The thiophene ring system has been synthesized through numerous reaction types. A systematic review of these has been made and is based on the number of components utilized in the construction of the ring. Some 19 combinations are possible, utilizing five method types. Not all combinations have been reported and some would be of only minor benefit.

Manufacture of thiophene on the commercial scale involves reactions of the two component method type wherein a 4-carbon chain molecule reacts with a source of sulfur over a catalyst which also effects cyclization and aromatization. A range of suitable feedstocks has included butane, *n*-butanol, *n*-butyraldehyde, crotonaldehyde, and furan; the source of sulfur has included sulfur itself, hydrogen sulfide, and carbon disulfide.

Manufacturing Processes for Thiophene Derivatives

Halothiophenes. The bromothiophenes, commercially the most important of the halothiophenes, are readily made and can be further derivatized. Manufacture of 2-bromothiophene involves the reaction of thiophene with a solution of sodium bromide/sodium bromate in acid solution. Such a reaction is controlled by the rate of addition of the acid. The two-phase system is stirred throughout the reaction; the heavy product layer is separated and washed thoroughly with water and alkali before distillation. The alkali treatment is particularly important and serves not just to remove residual acidity but, more importantly, to remove chemically any addition compounds that may have formed. Distillation of material containing residual addition compounds is hazardous, because traces of acid become self-catalytic, causing decomposition of the still contents and much acid gas evolution. Bromination of alkylthiophenes follows a similar pattern. The route to 3-bromothiophene utilizes a variation of the halogen dance technology.

Acylthiophenes. Manufacturing methods introducing the carboxaldehyde group into the 2- or 5-positions of thiophene and alkylthiophenes utilize the Vilsmeier-Haack reaction. To synthesize 2-thiophenecarboxaldehyde, a controlled addition of phosphorus oxychloride to thiophene in *N,N*-dimethylformamide is carried out, causing the temperature to rise. Completion of the reaction is followed by an aqueous quench, neutralization, and solvent extraction to isolate the product. 3-Thiophenecarboxaldehyde has been commercially available via carbonylation of 2,5-dimethoxy-2,5-dihydrofuran, followed by treatment with hydrogen sulfide, which introduces the sulfur atom with loss of methanol, inducing aromaticity and producing 3-thiophenecarboxaldehyde directly. Manufacture of 2-acetylthiophenes involves direct reaction of thiophene or alkylthiophene with acetic anhydride or acetyl chloride. The need for low levels of 3-isomer in 2-thiophenecarboxylic acid, which is produced by oxidation of 2-acetylthiophene and used in drug applications, has been the driving force to find improved acylation catalysts. The most widely used oxidant is sodium hypochlorite, which produces a quantity of chloroform as by-product, a consequence that detracts from its simplicity. Alternative oxidants have included sodium nitrite in acid solution, which has some advantages, but, like the hypochlorite method, also involves very dilute solutions and low throughput volumes. The long-standing manufacturing route to 2-thiopheneacetic acid has also involved 2-acetylthiophene. Oxidation with potassium permanganate under controlled conditions leads to 2-thiopheneglyoxylic acid, which may be isolated as ammonium salt. The salt is then carried through a reduction stage involving the Wolff-Kishner reaction in aqueous solution and utilizing hydrazine hydrate. Workup via acidification gives the unpleasant smelling 2-thiopheneacetic acid.

Methyl 3-aminothiophene-2-carboxylate. Synthesis of this amino ester has been variously described in the literature; it is a key intermediate to both pharmaceutical and agrochemical products. The main synthetic schemes use thioglycollate esters as starting materials.

Toxicity

Thiophene is harmful by inhalation, in contact with skin, or if swallowed; it is also a skin-irritant. Thiophene is harmful in the aqueous environment and is nonbiodegradable. Thiophene and 3-methylthiophene are listed on the TSCA chemical substances inventory. Thiophene is regulated as a hazardous material under OSHA and also regulated under the Clean Air Act, Section 110, 40 CFR 60.489, but there are no exposure limits or controls set for 3-methylthiophene. Bromothiophenes are toxic materials by all routes.

Uses of Thiophene and Derivatives

Pharmaceuticals. Thiophene and its derivatives find applications in the pharmaceutical area over a wide range of drug types, which can be divided into four main groups: (*1*) Nonsteroidal antiinflammatory rheumatoid and osteoarthritis drugs; (*2*) Hypertension and heart drugs; (*3*) Antibiotics; (*4*) Other pharmaceuticals (see ANTIASTHMATICS; ANTIFUNGAL; ANTIGLAUCOMA DRUGS). Other pharmaceutical products incorporating the thiophene ring include the antiasthmatic drug Ketotifen (Sandoz), which is particularly marketed in Japan. The antifungal drug Tioconazole (Pfizer) is based on 2-chloro-3-methylthiophene. The antiglaucoma drug Dorzolamide is made from a range of fused thiophene derivatives developed by Merck.

Veterinary Products. Principal users of thiophene are anthelmintics and fungicidal products.

Other. Thiophenes are also used in agrichemical products, dyestuffs, and in conjugated polythiophenes (used because of their potential electrical conductivity) and as nonlinear optical devices.

LANCE S. FULLER
Synthetic Chemicals Limited

S. Gronowitz, ed., *Thiophene and Thiophene Derivatives*, Vols. 1–4, Wiley-Interscience, New York, 1985.

O. Meth-Cohn, *Comp. Org. Chem.* **4**, 789 (1979).

R. M. Kellog, *Comp. Heterocyclic Chem.* **4**, 713 (1984); S. J. Rajappa, *Comp. Heterocyclic Chem.* **4**, 741 (1984); E. Campaigne, *Comp. Heterocyclic Chem.* **4**, 863 (1984).

M. Szajda, *Comp. Heterocyclic Chem. II*, **2**, 437 (1996); S. J. Rajappa, *Comp. Heterocyclic Chem. II*, **2**, 491 (1996); J. Nakayama, *Comp. Heterocyclic Chem. II*, **2**, 607 (1996); R. K. Russell and J. B. Press, *Comp. Heterocyclic Chem. II*, **2**, 679 (1996).

THIOSULFATES

The thiosulfate ion, $S_2O_3^{2-}$, is a structural analogue of the sulfate ion where one oxygen atom is replaced by one sulfur atom. The two sulfur atoms of thiosulfate thus are not equivalent. Indeed, the unique chemistry of the thiosulfate ion is dominated by the sulfide-like sulfur atom which is responsible for both the reducing properties and complexing abilities. The ability of thiosulfates to dissolve silver halides through complex formation is the basis for their commercial application in photography.

Physical Properties

Thermodynamic Properties. The heat of formation of the thiosulfate ion, −5.75 kJ/g (−1.37 kcal/g). The standard free energy of formation is −4.58 kJ/g (−1.09 kcal/g). The partial molal entropy is 62.8 ± 25.1 J/K (15 ± 6 cal/K).

The oxidation potential for the reaction $2\ S_2O_3^{2-} \leftrightarrows S_4O_6^{2-} + 2\ e^-$ ranges from 0.2 to 0.4 V in neutral solution, depending on the method of measurement. Electrolytic reduction with a mercury or platinum electrode produces equimolar amounts of sulfide and sulfite: $S_2O_3^{2-} + 2e \rightarrow S^2 + SO_3^{2-}$.

Chemical Properties

Thiosulfuric Acid. Thiosulfuric acid is relatively unstable and thus cannot be recovered from aqueous solutions. Pure thiosulfuric acid has been prepared in liquid CO_2 at −50°C or in diethyl ether at −78°C. It decomposes at −30°C to $H_2S_3O_6$ and H_2S, and rapidly at higher temperatures to H_2O, SO_2, and sulfur.

Thiosulfates. The ammonium, alkali metal, and alkaline-earth thiosulfates are soluble in water. Neutral or slightly alkaline solutions containing excess base or the corresponding sulfite are more stable than acid solutions. Acidification of thiosulfate with strong acid invariably leads to decomposition with the formation of colloidal sulfur and sulfur dioxide. In dilute aqueous solution, the following equilibrium is established:

$$S_2O_3^{2-} + H^+ \leftrightarrows HSO_3^- + S \quad K = 0.013 \text{ at } 11°C$$

Reactions. Catalytic amounts of arsenic, antimony, or tin salts promote the formation of pentathionate. Mild oxidizing agents such as hydrogen peroxide in acid solutions produce tetrathionates and trithionates. The presence of Fe^{2+} promotes oxidation to the sulfate. The reaction with iodine in neutral or slightly acid solution is the basis of a volumetric analytical procedure. Stronger oxidizing agents such as chlorine, bromine, permanganate, chromate, or alkaline hydrogen peroxide oxidize thiosulfate quantitatively to sulfate. Hypochlorite, hypobromite, and hypoiodite are also strong enough to oxidize thiosulfate to sulfate. Thiosulfates are reduced to sulfides by metallic copper, zinc, or aluminum. The thiosulfate reaction with cyanide to give thiocyanate is the basis for the use of thiosulfate as an antidote in cyanide poisoning. Thiosulfates form complex ions with a number of metal ions by the coordination of more than one thiosulfate ion.

Corrosion. The preferred material of construction for pumps, piping, reactors, and storage tanks is austenitic stainless steels such as 304, 316, or Alloy 20. The corrosion rate for stainless steels is <440 g/(m^2·yr) at 100°C (see also CORROSION AND CORROSION CONTROL).

Preparation

Thiosulfates are normally prepared by the reaction of sulfur and sulfite in neutral or alkaline solution. Polysulfides react similarly. Sulfides react with sulfur dioxide, sulfite, or bisulfite. These three methods are employed commercially.

Sodium Thiosulfate

Sodium thiosulfate, either the anhydrous salt, $Na_2S_2O_3$, or the crystalline pentahydrate, is commonly referred to as hypo or crystal hypo.

Selected physical properties of sodium thiosulfate pentahydrate are shown in Table 1. Sodium thiosulfate has been produced commercially by the air oxidation of sulfides, hydrosulfides, and polysulfides.

Health and Safety Factors. The LD_{50} of anhydrous sodium thiosulfate for mice is 7.5 ± 0.752 g/kg. Because of low toxicity, it can be safely used in veterinary medicine. Sodium thiosulfate pentahydrate is affirmed as a GRAS indirect and direct human food ingredient under the Federal Food, Drug, and Cosmetic Act (see FOOD ADDITIVES).

Uses. The principal use for sodium thiosulfate continues to be as fixative in photography (qv) to dissolve undeveloped silver halide from negatives or prints. In applications where rapid processing is required, such as the processing of x-ray film, sodium thiosulfate has been largely replaced by ammonium thiosulfate.

Sodium thiosulfate is also used in leather tanning, paper and textile manufacture, flue-gas desulfurization, silver-cleaning formulations, thermal-energy storage, cyanide antidote, cement additive,

Table 1. Physical Properties of Sodium Thiosulfate Pentahydrate

Property	Value
refractive index, n_D^{20}	1.4886
density, d_4^{25}, g/cm^3	1.750
heat of solution in water at 25°C, J/g[a]	−187
heat of formation, kJ/g[a]	−10.48
heat of fusion, J/g[a]	200
specific heat J/(g·K)[a]	
solid	1.84
molten salt	2.38
dissociation pressure, kPa[b], 20°C	0.796
vapor pressure of saturated solutions, kPa[b], 33°C	1.33
density of aqueous solutions, d_{20}^{20}, g/cm^3 at wt %, $Na_2S_2O_3$, 1.00	1.0083

[a] To convert J to cal, divide by 4.184.
[b] To convert kPa to mm Hg, multiply by 7.5.

aluminum-etching solutions, removal of nitrogen dioxide from flue gas, concrete-set accelerator, stabilizer for acrylamide polymers, extreme pressure additives for lubricants, multiple-use heating pads, in soap and shampoo compositions, and as a flame retardant in polycarbonate compositions. Moreover, precious metals can be recovered from difficult ores using thiosulfates. Use of thiosulfates avoids the environmentally hazardous cyanides.

Ammonium Thiosulfate

Ammonium thiosulfate, $(NH_4)_2S_2O_3$, commonly referred to as ammo hypo, has displaced sodium thiosulfate in photography. It is normally sold in the United States only as the aqueous solution. In addition, a crystal slurry and anhydrous crystal are available in Europe. The anhydrous monoclinic crystalline form has a density of 1.679 g/cm^3, no hydrates are known.

Manufacture. Ammonium thiosulfate has been produced by the reaction of ammonium sulfite with sulfur, sulfides, or polysulfides. This reaction series continues until the last polysulfide is ammonium sulfide and the process is completed by reaction with sulfur dioxide. Other commercial processes are based on the direct reaction of ammonium sulfite and sulfur.

Health and Safety Factors (Toxicology). The toxicological properties of ammonium thiosulfate are generally considered to be the same as those of sodium thiosulfate and thiosulfates in general.

Uses. The use distribution of ammonium thiosulfate in 1995 was estimated to be photography, 48%; agricultural applications, 50%; and others, including dechlorination, 2%. The principal use of photochemical-grade ammonium thiosulfate continues to be in photography, where is dissolves undeveloped silver halides from negatives and prints.

Other Thiosulfates

Many other metal thiosulfates, eg, magnesium thiosulfate and its hexahydrate have been prepared on a laboratory scale, but with the exception of the calcium, barium, and lead compounds, these are of little commercial or technical interest.

Complexes and Organic Thiosulfates

Gold thiosulfate complexes of the form $Na_3[Au(S_2O_3)_2]{\cdot}2H_2O$ are prepared by addition of gold trichloride to concentrated sodium thiosulfate solution. This complex has been used in the treatment of rheumatoid arthritis. Sodium ethyl thiosulfate is also known as Bunte's salt after the name of its discoverer. Bunte salts may be thought of as esters of thiosulfuric acid. Bunte salts have bacterial, insecticidal, and fungicidal properties, and are also used as chelating agents or surfactants (qv). Bunte salts have been tested for preirradiation protection for mammals exposed to lethal radiation doses (see RADIOPROTECTIVE AGENTS).

S. L. BEAN
General Chemical Corporation

Gmelins Handbuch der Anorganischen Chemie, 8th ed., Schwefel, Part B, No. 2, Verlag Chemie, G.m.b.H., Weinheim/Bergstrasse, 1960, p. 868; 1966, pp. 1162–1174; 1965, p. 614; Part B, No. 3; 1961, pp. 785–793; Natrium; 1964, pp. 247–252.

THORIUM AND THORIUM COMPOUNDS

Thorium, a naturally occurring radioactive element, atomic number 90, atomic mass 232.0381, is the second element of the actinide (5*f*) series (see ACTINIDES AND TRANSACTINIDES; RADIOISOTOPES). Thorium is rather unique, generally having an oxidation state of +4, and thus no 5*f* valence electrons in any of its compounds.

The chemistry of thorium, dominated by the +4 oxidation state, is similar to that of the lanthanides (qv) and the Group 4 (IVB) elements, Ti, Zr, and Hf.

Twenty-five isotopes of thorium have been observed having masses ranging from 212 to 236. Radioactive half-lives range from 0.1 μs for ^{218}Th to 1.405×10^{10} yr for ^{232}Th. The latter is the predominant isotope in nature. Thorium has a wide distribution in nature and is present as a tetravalent oxide in a large number of minerals in minor or trace amounts. Thorium is significantly more common in nature than uranium, having an average content in the earth's crust of approximately 10 ppm. There are only a few minerals where thorium occurs as a significant constituent. The commercially important ore is the golden-brown, lanthanide phosphate, monazite, $LnPO_4$, where Ln = Ce, La, or Nd, in which thorium is generally present in a 1–15% elemental composition. Monazite is widely distributed around the world. Some deposits are quite large.

Recovery from Ores

There are a number of minerals in which thorium is found. Thus a number of basic process flow sheets exist for the recovery of thorium from ores. The extraction of monazite from sands is accomplished via the digestion of sand using hot base, which converts the oxide to the hydroxide form. The hydroxide is then dissolved in hydrochloric acid and the pH adjusted to between 5 and 6, affording the separation of thorium from the less acidic lanthanides. Thorium hydroxide is dissolved in nitric acid and extracted using methyl isobutyl ketone or tributyl phosphate in kerosene to yield $Th(NO_3)_4$, which can then be removed from the organic solvent.

Uses

Thorium is mainly used in the production of commercial lantern mantles, refractory materials (see REFRACTORIES), electronic components, alloys utilized for components of jet engines, and as a catalyst in the chemical industry. The isotope ^{232}Th is used in nuclear reactor fuels. The oxide finds application in electrodes for arc welding, in the manufacturing of ceramics, and as a minor component in the catalytic production of hydrocarbon mixtures for use as liquid motor fuel. Over a ton of thorium is produced annually, approximately half of which is devoted to the production of gas mantles. Tetravalent thorium is used to replace trivalent lanthanides in *n*-type doped superconductors, $R_{2-x}Th_x\ CuO_{4-\delta}$;, where R = Pr, Nd, or Sm, producing a higher T_c superconductor. Thorium also forms alloys with a wide variety of metals.

The most important thorium compound is ThO_2 owing to its high chemical and thermal stability. This oxide has a high melting point of nearly 3400°C, the highest for any metal oxide. The inherent radioactivity of ^{232}Th (the most important isotope) and the formation of radioactive daughter products are important limiting factors in the uses of thorium.

Thorium Metal

Pure thorium metal is a dense, bright silvery metal having a very high melting point. Thorium is a reactive, soft, and ductile metal which tarnishes slowly on exposure to air. A survey of the physical properties of thorium is summarized in Table 1.

Pure thorium metal is very difficult to prepare owing to high reactivity with H_2, O_2, N_2, and C at the high temperatures necessary for production. Thorium metal can be produced by a variety of reduction techniques, all of which have unique difficulties. The most advantageous and common method of preparation is the molten salt reduction of thorium fluoride, ThF_4, in a blend of Ca and $ZnCl_2$ in a dolomite-lined reactor at 660°C.

Thorium Compounds

Oxo Ion Salts. Salts of oxo ions, eg, nitrate, sulfate, perchlorate, hydroxide, iodate, phosphate, and oxalate, are readily obtained from aqueous solution, eg, thorium nitrate is readily formed by dissolution of thorium hydroxide in nitric acid from which, depending on the pH of solution, crystalline $Th(NO_3)_4{\cdot}5H_2O$ or $Th(NO_3)_4{\cdot}4H_2O$ can be obtained. Thorium nitrate is very soluble in water and in a host of oxygen-containing organic solvents, including alcohols, ethers, esters, and ketones.

Table 1. Physical Properties of Thorium Metals

Property	Value
crystal structure	
fcc to 1360°C, a_0, pm	508.42
density, g/cm^3	11.724
atoms per unit cell, Z	4
bcc, 1360–1750°C, a_0, pm	411
density, g/cm^3	11.10
atoms per unit cell, Z	2
mp, °C	1750
bp, °C	~3800
enthalpy of vaporization, 25°C, kJ/mol^a	598
enthalpy of fusion, kJ/mol^a	14
vapor pressure, 1757–1956 K	$\log(p/\text{atm}) =$
thermal conductivity, 25°C, W/(cm·K)	0.6 $-28,780(T/\text{K})^{-1} + 5.991$
work function, eV	3.49
Hall coefficient, 24°C, cm^3/C	-11.2×10^{-5}
elastic constants	
Young's modulus, kPa^b	7.2×10^7
shear modulus, kPa^b	2.8×10^7
Poisson's ratio	0.265
compressibility, cm^2/dyn	17.3×10^{-13}

^{a} To convert kJ to kcal, divide by 4.184.
^{b} To convert kPa to psi, multiply by 0.145.

Coordination Complexes. The coordination and organometallic chemistry of thorium is dominated by the extremely stable tetravalent ion. Except in a few cases where large and sterically demanding ligands are used, lower thorium oxidation states are generally unstable. The chemistry of Th(IV) has expanded greatly since the mid-1980s. Being a hard metal ion, Th(IV) has the greatest affinity for hard donors such as N, O, and light halides such as F^- and Cl^-. Coordination complexes that are common for the *d*-block elements have been studied for thorium. These complexes exhibit coordination numbers ranging from 4 to 11.

Oxides. Owing to the importance as nuclear fuel material, actinide oxides have been intensively investigated. These are very complicated compounds because of the formation of nonstoichiometric or polymorphic materials. Actinide oxides are very heat-resistant and ThO_2 is the highest (3390°C) melting of any metal oxide.

Hydroxides. Thorium(IV) is generally less resistant to hydrolysis than similarly sized lanthanides, and more resistant to hydrolysis than tetravalent ions of other early actinides, eg, U, Np, and Pu. Many of the thorium(IV) hydrolysis studies indicate stepwise hydrolysis to yield monomeric products of formula $Th(OH)_n^{(4-n)+}$, where n is integral between 1 and 4, in addition to a number of polymeric species.

Carbonates. There has been a great deal of interest in carbonate complexes of thorium owing to their environmental relevance. Solution studies for thorium have been reported. For example, the solubility of microcrystalline ThO_2, examined as a function of pH and CO_2 partial pressure, gave results consistent with the presence of the mixed hydroxocarbonatothorium complex, $Th(OH)_3(CO_3)^-$, and pentacarbonatothorium(IV), $Th(CO_3)_5^{6-}$ (**1**).

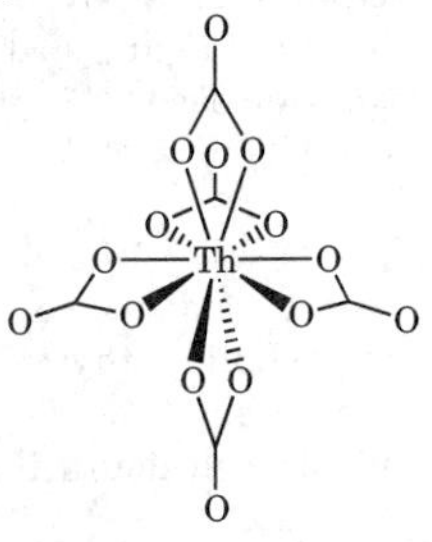

(**1**)

Phosphates. Thorium phosphates are of considerable interest because of their potential as radioactive waste forms and as xerogel thin films for light waveguides. Binary and ternary thorium phosphates have been synthesized having varying ratios of nonthorium metal, thorium, and phosphate. Binary compounds having ThO_2:P_2O_5 ratios of 1:2, 1:1, 3:2, and 3:1 have been reported.

Oxygen-Containing Organics. Neutral and anionic oxygen-containing organic molecules form complexes with thorium. Recent work has focused on alkoxides, aryloxides, and carboxylates; however, complexes with alcohols, ethers, esters, ketones, aldehydes, ketoenolates, and carbamates are also well known.

Halides. Thorium(IV) fluoride is widely used as a source of elemental thorium through electrochemical reduction. Anhydrous ThF_4 is insoluble in water and has been isolated a number of ways. Anhydrous $ThCl_4$ has usually been prepared by direct interaction of thorium metal, hydride, or carbide with chlorine. An alternative to this approach is the reaction of anhydrous HCl with the metal or the hydride at elevated temperatures (700–900°C). Anhydrous $ThBr_4$ and ThI_4 have been prepared in a similar fashion as the chloride, ie, interaction of thorium metal or the hydride with the elemental halide of choice, or at high (700–900°C) temperatures with HX, where X is Br or I. Organic solvent-soluble tetrahalides $ThX_4(THF)_4$, where X = Br or I, have been obtained through the interaction of thorium metal turnings with elemental halides in THF at 0°C.

Organometallic Complexes. The organometallic chemistry of thorium has been widely studied owing to potential utility in homogeneous and heterogeneous catalysis. Activities range from the hydrogenation and polymerization of olefins to the selective activation of alkanes. Although there are no examples of thorium carbonyl complexes, hydrocarbyls, allyls, arenes, cyclooctatetraenyl, and a host of cyclopentadienyl-based ligand complexes have been reported.

Health and Safety Factors

Thorium is potentially hazardous. Finely divided thorium metal and hydrides can be explosive or inflammatory hazards with respect to oxygen and halogens. Finely divided ThO_2 and other inorganic salts also present an inhalation and irritation hazard. The long half-life of ^{232}Th makes it a minimal radiation hazard. When inhaled, ingested, or adsorbed through the skin, thorium isotopes are potentially harmful because of ionizing radiation and chemical toxicity. Thorium compounds are legally classified as source materials for nuclear energy and thus are regulated by various government agencies, eg, the Nuclear Regulatory Commission. The relevant regulations cover licensing and safety aspects.

DAVID L. CLARK
D. WEBSTER KEOGH
MARY P. NEU
WOLFGANG RUNDE
Glenn T. Seaborg Institute for Transactinium Science
Los Alamos National Laboratory

S. Ahrland and co-workers, eds., *Gmelin Handbook of Inorganic Chemistry, Thorium, Suppl. Vol. D1, Properties of Thorium Ions in Solutions*, 8th ed., Springer-Verlag, Berlin, 1988.

R. Ditz and co-workers, eds., *Gmelin Handbook of Inorganic Chemistry, Thorium, Suppl. Vol. A1a, Natural Occurrence, Minerals (Excluding Silicates)*, Springer-Verlag, Berlin, 1990.

L. I. Katzin, in J. J. Katz, G. T. Seaborg, and L. R. Morss, eds., *The Chemistry of the Actinide Elements*, Chapman and Hall, London, 1986.

N. N. Greenwood and A. Earnshaw, *Chemistry of the Elements*, Pergamon Press, Oxford, 1988.

D. L. Clark and A. P. Sattelberger, in R. B. King, ed., *Encyclopedia of Inorganic Chemistry*, Vol. 1, Wiley-Interscience, New York, 1994, p. 24.

M. Ephritikhine, *New J. Chem.* **16**, 451 (1992).

J. Fuger and co-workers, *The Actinide Aqueous Inorganic Complexes*, Part 12, International Atomic Energy Agency, Vienna, 1992.

THROMBOLYTIC AGENTS. See BLOOD COAGULANTS AND ANTICOAGULANTS.

THULIUM. See LANTHANIDES.

THYROID AND ANTITHYROID PREPARATIONS

The main role of the human thyroid gland is production of thyroid hormones (iodinated amino acids), essential for adequate growth, development, and energy metabolism. Thyroid underfunction is an occurrence that can be treated successfully with thyroid preparations. In addition, the thyroid secretes calcitonin (also known as thyrocalcitonin), a polypeptide that lowers excessively high calcium blood levels. Thyroid hyperfunction, another important clinical entity, can be corrected by treatment with a variety of substances known as antithyroid drugs.

Thyroid Function and Malfunction

Human life without thyroid hormones is possible but of minimal quality. In the fetus, thyroid hormones affect growth and differentiation; in the mature human, they regulate metabolism. The two principal thyroid hormones, L-thyroxine (T_4) (**1**) and L-triiodothyronine (T_3) (**2**) are produced by the thyroid gland and secreted into the blood stream. The minute amounts secreted are regulated by a complex system that originates in the central nervous system (CNS).

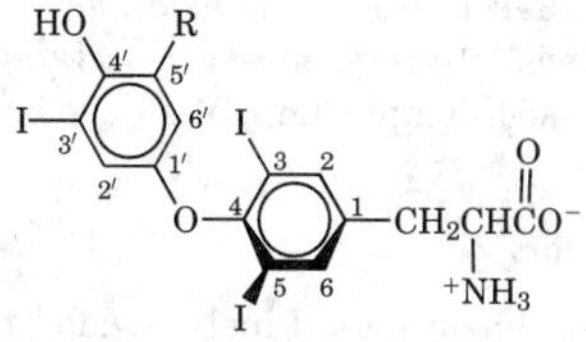

(**1**) R = I
(**2**) R = H

Thyroid hormones affect growth and development by stimulating protein synthesis. In mature animals, the main action of the thyroid hormones is their calorigenic effect which is caused by an increase in the basal metabolic rate (BMR). Thyroid underfunction results in a series of hypothyroid states clinically known as cretinism if present in a fetus or an infant, and myxedema in an adult. If the hypothyroidism is owing to insufficient iodine intake, it is known as simple goiter, a state characterized by an enlarged but functionally underactive thyroid gland. Thyroid hyperfunction occurs as diffuse toxic goiter, also known as Graves' disease. Other forms of thyroid hyperfunction are thyrotoxicosis and toxic nodular goiter (Plummer's disease).

Thyromimetic Compounds

The main thyroid hormone. It is the 5′-desiodo analogue of thyroxine, T_3 (**2**).

The main structural requirements for thyromimetic activity can be summarized as follows: two aromatic rings insulated electronically from each other by connecting oxygen, sulfur, or carbon bridges, forming a central lipophilic core; substitution at the 3 and 5 positions with alkyl groups or with halogens; at position 1, an acidic side chain two or three carbons long should be present; the presence of a small substituent capable of forming hydrogen bonds in the 4′-position; and one lipophilic substituent ortho to the 4′-OH group.

Biosynthesis, Distribution, and Metabolism. Iodine is a trace element in the environment and the diet. The thyroid gland extracts it from the blood as iodide ion via an active transport system. It is converted to di-, tri-, and tetraiodothyronines.

Synthesis. The most widely employed route for T_3 and T_4 is the so-called Glaxo method. This versatile synthetic route has been used extensively with a great variety of phenols and thiophenols to establish structure–activity relationships for thyromimetic activity; L-tyrosine is the starting material.

Chemical Assay. The main problem in the chemical analysis of the thyroid hormones is their separation. A USP procedure gives the details of a paper chromatographic separation in which T_3 is examined for contamination by T_4 and 3,5-diiodothyronine. Body fluids are analyzed for T_3 and T_4 by a variety of radioimmunoassay procedures (see IMMUNOASSAYS).

Bioassay. Although the chemical assays described above have replaced bioassays for the determination of T_3 and T_4, several *in vivo* and *in vitro* bioassays are used to determine the potency of thyroglobulin preparations and to establish the thyromimetic or antithyroid potency of new compounds.

Antithyroid Substances

In principle, antithyroid effects can be produced by destroying excess thyroid gland tissue surgically or by treatment with radioiodine; blocking synthesis of thyroid hormones with goitrogens such as certain thionamides; inhibition of thyroid-hormone release with lithium; inhibition of the peripheral deiodination of T_4 to the more active T_3 with thiouracils; increasing excretion of thyroid hormone (*n*-butyl-3,5-diiodo-4-hydroxybenzoate as a result of displacing T_3 and T_4 from serum proteins; and competitive antagonism at the receptor level (r-T_3).

When large doses of iodide ion are administered, a transient inhibition of synthesis and release of the thyroid hormones is brought about by the so-called Wolff-Chaikoff effect. The selective uptake of iodide ion by the thyroid gland is the basis of radioiodine treatment in hyperthyroidism, mainly with ^{131}I, although various other radioactive isotopes are also used. Thiocyanate ion, inhibits formation of thyroid hormones by inhibiting the iodination of tyrosine residues in thyroglobulin by thyroid peroxidase.

Although several hundred compounds incorporating thionamide are known, only four are used clinically and only two are accepted by the USP XX. In the lithium carbonate treatment of certain psychotic states, a low incidence (3.6%) of hypothyroidism and goiter production have been observed as side effects. Lithium salts have not found general acceptance in the treatment of hyperthyroidism.

Peripheral Antagonists. The relatively long duration of action of the thyroid hormones makes it desirable to have compounds capable of blocking them competitively at their site of action. This is desirable in the treatment of thyroid storm where the reduction of circulating hormone levels brought about by the inhibition of their synthesis is too slow. A large number of thyroid hormone analogues have been tested for this effect, but no potent or clinically useful peripheral antagonists have been found. The level of circulating hormones is lowered indirectly by *n*-butyl 3,5-diiodo-4-hydroxybenzoate which displaces them from their carriers (TBG and TBPA) and thus accelerates their metabolism and excretion.

Calcitonin

The thyroid gland is also a source of a hypocalcemic hormone having effects in general opposition to those of the parathyroid hormone. Originally called thyrocalcitonin, it is now referred to as calcitonin (CT). Calcitonin is used clinically in various diseases in which hypercalcemia is present, eg, Paget's disease.

Commercial Preparations

Basic preparations for hypothyroidism are sodium levothyroxine, sodium liothyronine, thyroglobulin, thyroid extracts (*Glandulae Thyroideae siccatae*) and calcitonin.

Antithyroid drugs include propylthiouracil and methimazae.

S. C. Werner and S. H. Ingbar, eds., *The Thyroid*, 4th ed., Harper & Row Publishers Inc., New York, 1978.

R. C. Haynes, Jr. and F. Murad, in A. Goodman Gilman and co-workers, eds., *The Pharmacological Basis of Therapeutics*, 6th ed., Macmillan Publishing Co., New York, 1980, pp. 1397–1419.

E. C. Jorgensen, in M. E. Wolff, ed., *Burger's Medicinal Chemistry*, Pt. III, John Wiley & Sons, Inc., New York, 1981, pp. 103–145.

R. Pitt-Rivers and J. R. Tata, *The Thyroid Hormones*, Pergamon Press, New York, 1959.

TIN AND TIN ALLOYS

Tin is one of the world's most ancient metals. Of the nine different tin-bearing minerals found in the earth's crust, only cassiterite, SnO_2, is of importance. Over 80% of the world's tin ore occurs in low grade alluvial or alluvial placer deposits where the tin content of the ore can be as low as 0.015%. Complex tin sulfide minerals are found in the lode deposits of Bolivia and Cornwall associated with cassiterite and granitic rock. No workable tin deposits have been found in the US. Tin-mining methods depend on the character of the deposit. Gravel-pump mining is widely used in southeast Asia and probably accounts for 40% of the world's tin production. Hydraulicking and open-pit mining methods also involve gravity separation with water in palongs (see MINERALS RECOVERY AND PROCESSING).

Dredging is mining with a floating dredge on an artificial pond in a placer. Access to the lode deposits is by shaft sinking or by adits, ie, passages driven into the side of a mountain, depending on the terrain. Tin concentrates from the lode deposits are 40–60 wt % tin and must be further upgraded before smelting.

Properties

Some physical, mechanical, and thermal constants of tin are shown in Table 1. Although the pure metal has a silvery-white color, in the cast condition it may have a yellowish tinge caused by a thin film of protective oxide on the surface. When highly polished, it has high light reflectivity. It retains its brightness well during exposure, both outdoors and indoors. Tin exists in two allotropic forms: white tin (β) and gray tin (α).

Tin, at wt 118.69, falls between germanium and lead in Group IV A of the Periodic Table. It has 10 naturally occurring isotopes. A reversal of potential of the tin–iron couple occurs when tin-coated steel (tin-plate) is in contact with acid solutions in the absence of air. The tin coating acts as an anode; it is the tin that is slowly attacked and not the steel. This unique property is the keystone of the canning industry because dissolved iron affects the flavor and appearance of the product. Thus, the presence of tin protects the appearance and flavor of the product.

Processing

The crude tin obtained from slags and by smelting ore concentrates is refined by further heat treatment or sometimes electrolytic processes. The conventional heat-treatment refining includes liquidation or sweating and boiling, or tossing. Iron, copper, arsenic, and antimony can be readily removed by pyrometallurgical processes. For the removal of large quantities of lead or bismuth, either separately or together, conventional electrolysis or a newly developed vacuum-refining process is used. Electrolytic refining is more efficient in regard to both the purity of the product and the ratio of tin to impurities in by-products.

Secondary Tin. In 1990, >7700 metric tons of tin were recovered in the United States from scrap. Sources include bronze rejects and used parts, solder in the form of dross or sweepings, dross from tinning pots, sludges from tinning lines, babbitt from discarded bearings, type-metal scrap, and clean tinplate clippings from container manufacturers. High purity tin is recovered by detinning clean tinplate (see RECYCLING). Alloy scrap containing tin is handled by secondary smelters as part of their production of primary metals and alloys; lead refineries accept solder, tin drosses, babbitt, and type metal.

Table 1. Physical Properties of Tin

Property	Value
mp, °C	231.9
bp, °C	2625
sp gr	
α-form (gray tin)	5.77
β-form (white tin)	7.29
liquid at mp	6.97
transformation temp $\beta \leftrightarrows \alpha$, °C	13.2
vapor pressure, Pa[a], at 1000 K	986×10^{-6}
surface tension, at mp, mN/M (= dyn/cm)	544
viscosity, at mp, mPa·s(= cP)	1.85
specific heat, at 20°C, J/(kg·K)[b]	222
latent heat of fusion, kJ/(g·atom)	7.08
thermal conductivity, at 20°C, W/(m·K)	65
coefficient of linear expansion, $\times 10^{-6}$, at 0°C	19.9
100°C	23.8
shrinkage on solidification, %	2.8
volume conductivity, % IACS	15
Brinell hardness, 10 kg, 5 mm, 180 s	15
at 20°C	3.9
tensile strength, as cast, MPa[c], at 15°C	14.5
latent heat of vaporization, kJ/mol[b]	296.4

[a] To convert Pa to mm Hg, multiply by 0.0075.
[b] To convert J to cal, divide by 4.184.
[c] To convert MPa to psi, multiply by 145.

Economic Aspects

Tin has long been regarded as a strategic metal because of its importance in canning, electrical, and transportation applications. Accordingly, it is stockpiled by the General Services Administration (GSA) at various locations in the country. Since the mid-1950s, the price of tin has been subject to an international agreement between producing and consuming nations. Under the International Tin Agreement (ITA), the ITC seeks to deal effectively with situations where a shortage or surplus of tin arises and prevent excessive price fluctuations. World trading in tin occurs mostly at Penang, London, and New York. The U.S. is by far the largest consumer of tin, followed by Japan and Germany. Tinplate provides an outlet for over one-third of the primary tin used in the U.S.

Specifications and Analytical Methods

The ASTM Classification of Pig Tin B 339 lists three grades, as shown in Table 2.

The tin content of ores, concentrates, ingot metal, and other products is determined by fire assay, fusion method, and volumetric wet analysis.

The purity of commercial tin is under strict control at the smelters. Photometric, chemical, atomic absorption, fluorimetric, and spectrographic methods are available for the determination of impurities.

Table 2. Classification of Pig Tin

ASTM designation	Tin, %[a]	General applications
ultrapure	99.95	analytical standards and research, pharmaceuticals, fine chemicals
A standard	99.85	collapsible tubes, unalloyed (block) tin products, electrotinning, tin-alloyed cast iron, high grade solders
grade A tinplate	99.85	food containers, foil

[a] Percent shown is minimum. A more complete description of these grades is given in full ASTM Standard B339-93.

Health and Safety Factors

Tests have shown that considerable quantities of tin can be consumed without any effect on the human system.

Uses

Tin is used in various industrial applications as cast and wrought forms obtained by rolling, drawing, extrusion, atomizing, and casting; tinplate, ie, low carbon steel sheet or strip rolled to 0.15–0.25 mm thick and thinly coated with pure tin; tin coatings and tin alloy coatings applied to fabricated articles (as opposed to sheet or strip) of steel, cast iron, copper, copper-base alloys, and aluminum; tin alloys; and tin compounds.

Cast and Wrought Forms. Thousands of tons of tin ingots are cast into anodes for plating processes. Tin foil is used for electrical condensers, bottle-cap liners, gun charges, and wrappings for food. Tin wire is used for fuses and safety plugs. Extruded tin pipe and tin-lined brass pipe are the first choice for conveying distilled water and carbonated beverages. Tin powder is used in powder metallurgy, for coating paper, and for solder pastes.

Tinplate. The development of tinplate was associated with the need for a reliable packaging material for preserving foods. It comprises in one inexpensive material the strength and formability of steel and the corrosion resistance, solderability, absence of toxicity, and good appearance of tin.

Tin Coatings. The coating may be applied by hot-dipping the fabricated article in liquid tin, by electroplating using either acid or alkaline electrolytes, and by immersion tinning. The hot-tinned coating is bright. The coating thickness may range from 0.0025 to 0.05 mm, depending on the type of protection required. Pure tin coatings are used on food-processing equipment, milk cans, kitchen implements, electronic and electrical components, fasteners, steel and copper wire, pins, automotive bearings, and pistons.

Tin Alloys. Tin-alloy coatings provide harder, brighter, and more corrosion-resistant coatings than tin alone. Tin–copper electrodeposited coatings (12 wt % tin) have the appearance of 24-carat gold and provide a bronze finish for furniture hardware, trophies, and ornaments (see COATINGS). They also provide a stop-off coating (resist) for nitriding.

Tin–lead coatings (10–60 wt % tin) can be applied by hot-dipping or electrode position to steel and copper fabricated articles and sheet. A special product is terne plate used for roofing and flashings, automobile fuel tanks and fittings, air filters, mufflers, and general uses such as covers, lids, drawers, cabinets, consoles for instruments, and for radio and television equipment. Terne plate is low carbon steel, coated by a hot-dip process with an alloy of tin and lead, commonly about 7–25 wt % tin, remainder lead. Electroplating is another possibility. Tin–cadmium coatings are particularly resistant to marine atmospheres and have applications in the aviation industry.

Tin and lead combine easily to form a group of alloys known generally as soft solders. The joining of metals with tin-containing solders can be attributed to several properties. Their low melting point allows simple equipment to be used for melting and joining, the alloys are unsurpassed in wetting and adhering to clean metal surfaces and flowing into small spaces, and they are relatively cheap. The tin–lead solders have no serious competitors in the field of low temperature joining (see SOLDERS AND BRAZING ALLOYS).

Tin is the important constituent in solders because it is the element that wets the base metal, such as copper and steel, by alloying with it. Solders are used mainly in auto radiators, air conditioners, heat exchangers, plumbing and sheet-metal joining, container seaming, electrical connections in radio and television, generating equipment, telephone wiring, electronic equipment and computers, and aerospace equipment.

Bronze. Copper–tin alloys, with or without other modifying elements, are classed under the general name of bronzes. They can be wrought, sand-cast, or continuously cast into shapes. Bronzes are especially applicable to marine and railway engineering pumps, valves and pipe fittings, bearings and bushings, gears and springs, and ship propellers. Included in special bronze alloys is bell metal, known for its excellent tonal quality, containing 20 wt % tin, and statuary bronze (see COPPER ALLOYS).

Bearing Metals. Metals used for casting or lining bearing shells are classed as white bearing alloys, but are known commercially as babbitt (see BEARING MATERIALS). The term white metal was used by Isaac Babbitt in 1839 in his description of tin-base bearing metals supported by a stronger shell. The term *babbitt* includes high tin alloys (substantially lead-free) containing $>$80 wt % tin, and high lead alloys containing $\geq$70 wt % lead and $\leq$12 wt % tin. Both have the characteristic structure of hard compounds in a soft matrix, and although they contain the same or similar types of compounds, they differ in composition and properties of the matrix. Babbitt alloys are suitable for hundreds of types of installations involving the movement of machinery, eg, the main, crankshaft, connecting rod big end, camshaft, and journal bearings associated with marine propulsion, railroad and automotive transportation, compressors, motors, generators, blowers, fans, rolling-mill equipment, etc.

Pewter. Modern pewter may have a composition of 90–95 wt % tin, 1–8 wt % antimony, and 0.5–3 wt % copper. Lead should be avoided by contemporary craftsman because it causes the metal surface to blacken with age. Pewter metal can be compressed, bent, spun, and formed into any shape, as well as being easily cast. A wide variety of consumer articles are available. The annual U.S. production of pewter exceeds 1100 t.

Type Metals. The printing trade still requires some amounts of lead-based alloys containing 10–25 wt % antimony and 3–13 wt % tin. By varying the tin and antimony content, type suitable for each printing process can be obtained.

Alloyed Iron. Tin-alloyed flake and nodular cast irons are widely used. Tin-inoculated iron has a uniformity of hardness, improved machinability, wear resistance, and better retention of shape on heating. Where pearlitic and heat-resistant cast irons are required, such as for engine blocks, transmissions, and automotive parts, tin additions may provide a suitable material.

Special Alloys. Alloys of tin with the rarer metals, such as niobium, titanium, and zirconium, have been developed. The single-phase alloy Nb_3Sn has the highest transition temperature of any known superconductor (18 K) and appears to keep its superconductivity in magnetic fields up to at least 17 T (170 kG). Niobium–tin ribbon, therefore, is of practical importance for the construction of high field superconducting solenoid magnets.

Tin is an important addition to titanium. Zirconium is an attractive structural material and fuel cladding for nuclear power reactors, but it has low strength and highly variable corrosion behavior. However, Zircalloy-2, with a nominal composition of 1.5 wt % tin, 0.12 wt % iron, 0.05 wt % nickel, 0.10 wt % chromium, and the remainder zirconium, can be used in all nuclear power reactors that employ pressurized water as coolant and moderator (see NUCLEAR REACTORS). Dental amalgams, mainly silver–tin–mercury alloys, have been used as fillings for many years (see DENTAL MATERIALS). The most common alloy contains 12 wt % tin.

Other Uses. The electronics and aerospace industries have for a number of years used gold-plated printed circuit boards and component leads where highest reliability is desired. Problems in the use of gold coatings have plagued the industry and the trend is toward the substitution of tin–lead or tin coatings for the gold coatings. Tin–nickel coatings with a thin flash of tin or gold are also used as a substitute for heavy gold coatings (see ELECTRICAL CONNECTORS).

CHARLES C. GAVER, JR.
Consultant

C. L. Mantell, *Tin: Its Mining, Production, Technology, and Applications*, 2nd ed., Reinhold Publishing Corp., New York, 1949.

P. M. Dinsdale, *A Guide to Tin*, Publication No. 540, International Tin Research Institute, London.

Nonferrous Metals; Book of ASTM Standards, Sect. 2, Vol. 2.04, American Society for Testing Materials, Philadelphia, Pa., 1985.

TIN COMPOUNDS

Tin has valences of +2 and +4 and forms stannous, ie, tin(II) compounds and stannic, ie, tin(IV) compounds. Tin compounds are present in natural waters, in soil, in marine organisms, and in meteorites.

Inorganic Tin Compounds

Tin reacts with strong acids and bases. A thin oxide film forms on tin exposed to oxygen. Sulfides are formed by a vigorous reaction when tin and sulfur are heated.

Halides. Properties of tin chlorides are given in Table 1. They are prepared by the reaction of chlorine with tin metal. Anhydrous stannous chloride, a water-soluble white solid, is used in redox and plating reactions. Solutions are widely employed as reducing agents. Stannous chloride is also used as a food additive. Stannous fluoride (opaque white water-soluble crystals) is used in toothpaste and other dental preparations (see DENTIFRICES).

Oxides. Stannous oxide (dec > 385°C) reacts readily with acids, which accounts for its primary use in the manufacture of other tin compounds. It is used in binary catalyst systems and in the ceramics and glass industries as an opacifier.

Metal Stannates. Many stannates of the formula $M_nSn(OH)_6$ are known. Both potassium and sodium stannates are colorless, water-soluble crystals used for alkaline tin electroplating. Insoluble metal stannates are used as additives for ceramic dielectrics.

Salts. Stannous sulfate is a white crystalline powder prepared from sulfuric acid and granulated tin at 100°C. It is used in tin plating, as is stannous fluoroborate. Solutions of the latter have good throwing and covering power. Stannous pyrophosphate is used in toothpaste and in x-ray technology.

Health and Safety Factors. Tin compounds are generally low in toxicity because of poor absorption and rapid excretion. Tin is present in all animals and in the human body in small amounts in all organs.

Organotin Compounds

Mono-, di-, tri-, tetra-, and hexaorganotin compounds are known. Most important are those where the organic radical is methyl, butyl, octyl, chclohexyl, phenyl, or β,β-dimethylphenethyl (neophyl).

Tetraorganotin Compounds. These compounds are insoluble in water but soluble in organic solvents. Their most important reaction is the Kocheshkov redistribution reaction, by which organotin halides are prepared.

$$R_4Sn + SnCl_4 \rightarrow 2\ R_2SnCl_2$$

$$R_2SnCl_2 + R_4Sn \rightarrow 2\ R_3SnCl$$

$$3\ R_4Sn + SnCl_4 \rightarrow 4\ R_3SnCl$$

$$R_4Sn + 3\ SnCl_4 \rightarrow 4\ RSnCl_3$$

Table 1. Physical Properties of Tin Chlorides

Property	$SnCl_2$	$SnCl_2.2H_2O$	$SnCl_4$	$SnCl_4.5H_2O$
mol wt	189.60	225.63	260.50	350.58
mp, °C	246.8	37.7	−33	ca 56 dec
bp, °C	623		114	
density (at 25°C), g/cm^3	3.95	2.63	2.23[a]	2.04

[a] At 20°C.

Table 2. Physical Properties of Commercially Important Triorganotin Compounds

Compound	Mp, °C	Bp, °C	n_D^{20}	d^{20}, g/cm^3
$[(C_4H_9)_3Sn]_2O$	<−45	$210–214_{1.3\ kPa}$[a]	1.488	1.17
$(C_4H_9)_3SnF$	218–219 dec			1.27[b]
$(C_4H_9)_3SnOCOC_6H_5$		$166–168_{0.13\ kPa}$[a]	1.5157	1.1926
$(C_4H_9)_3SnOCOCH_3$	80–85			1.27
$(C_6H_5)_3SnOH$	118–120 dec			1.552[b]
$(C_6H_5)_3SnF$	357 dec			1.53

[a] To convert kPa to mm Hg, multiply by 7.5.
[b] At 25°C.

The main use for tetraorganotin compounds is as intermediates for the tri-, di-, and monocompounds. Application as Ziegler-Natta-type catalysts has been reported (see OLEFIN POLYMERS).

Triorganotin Compounds. Triorganotin halides are soluble in organic solvents; only $(CH_3)_3SnCl$ is soluble in water. The utility of triorganotin chlorides results from the ease of nucleophilic displacement. They are generally prepared by Kocheshkov redistribution. Tribenzyltin chloride, however, is prepared directly from the halide and tin. Physical properties of triorganotin compounds are given in Table 2.

Uses. Triorganotin compounds are widely used as biocides, agricultural chemicals, wood preservatives, and marine antifoulants. The most useful biological control agents are the tributyl-, triphenyl-, and tricyclohexyltin compounds. Triorganotin compounds are used as marine antifouling agents, alone and in combination with cuprous oxide. Preferred for this application are tributyltin fluoride, triphenyltin hydroxide, and triphenyltin fluoride. Eroding antifouling paints are based on tributyltin acrylate or methacrylate copolymers (see COATINGS, MARINE). Triorganotin compounds have been used experimentally to control the snail vector in schistosomiasis and to control mosquitos.

Diorganotin Compounds. Diorganotin dichlorides are the intermediates for the preparation of all commercial diorganotin compounds (Table 3). Dibutyltin dichloride is manufactured by Kocheshkov redistribution from tetrabutyltin and stannic chloride.

$$(C_4H_9)_4Sn + SnCl_4 \rightarrow 2\ (C_4H_9)_2SnCl_2$$

Many organic halides, especially alkyl bromides and iodides, react with tin metal at elevated temperatures. Treatment of molten tin with methyl chloride gives good yields of dimethyl tin dichloride, an important intermediate in the manufacture of dimethyltin-based PVC stabilizers. Tin reacts directly with activated organic halides, eg, allyl bromide and benzyl chloride.

Uses. The largest industrial application for organotin compounds is in the stabilization of PVC; dialkyltin compounds are the best general-purpose PVC stabilizers. In the building industry, rigid PVC is stabilized with diorganotin carboxylates. In addition, dibutyltin

Table 3. Physical Properties of Diorganotin Compounds

Compound	Mp, °C	Bp, °C	n_D^{20}	d^{20}, g/cm^3
$(CH_3)_2SnCl_2$	107–108	185–190		
$(C_4H_9)_2SnCl_2$	41–42	$140–143_{1.3\ kPa}$[a]		
$(C_4H_9)_2SnBr_2$	21–22	$90–92_{0.04\ kPa}$[a]	1.5400	1.3913[b]
$(C_4H_9)_2SnI_2$		$145_{0.8\ kPa}$[b]	1.6042	1.996[b]
$(C_6H_5)_2SnCl_2$	42–44	$180–185_{0.7\ kPa}$[a]		
$(CH_3OC(O)CH_2CH_2)_2SnCl_2$	132			
$(CH_3)_2Sn(SC_4H_9)_2$		$81_{0.013\ kPa}$[a]	1.5400	1.280
$(C_4H_9)_2Sn(OCH_3)_2$		$126–128_{7\ Pa}$[a]	1.4880	

[a] To convert kPa to mm Hg, multiply by 7.5.
[b] At 25°C.

compounds are used as catalysts in the preparation of rigid foams (see URETHANE POLYMERS). Dibutyltin as well as monobutyltin compounds are used as catalysts in the manufacture of organic esters for plasticizers, lubricants, and heat-transfer fluids. Other uses include the application of dibutyltin dilaurate as a coccidiostat in fowl. Dimethyltin dichloride forms a thin coating of stannic oxide on glass to improve abrasion resistance and bursting strength of bottles.

Monoorganotins. Monoorganotin trihalides are strong Lewis acids and resemble acid chlorides. The halides are easily replaced. Monoorganotin halides are raw materials for triorganotin compounds and are generally prepared by Kocheshkov redistribution. Butylthiostannoic acid anhydride is used in Germany as sole stabilizer for certain grades of PVC.

Compounds with Tin–Tin Bonds. Ditin compounds are prepared by reductive coupling of a triorganotin halide with sodium in liquid ammonia:

$$R_3SnCl + 2\ Na \rightarrow R_3SnNa + NaCl$$

$$R_3SnNa + R_3SnCl \rightarrow R_3SnSnR_3 + NaCl$$

Hexaorganoditin compounds with short-chain aliphatic groups are colorless liquids of little commercial importance.

Health and Safety Factors. Most toxic are the lower trialkyltin compounds. The toxicity is strongly dependent on the organic groups. Most triorganotin compounds are eye and skin irritants. The diorganotin compounds are less toxic, and the monoorganotin compounds present no special problems. They show the familiar trend of decreasing toxicity with increasing alkyl length but lower than the diorganotin compounds. The current OSHA TLV for exposure to organotin compounds is 0.1 mg (as tin)/m^3 air over an 8-h work shift.

MELVIN H. GITLITZ
MARGUERITE K. MORAN
M & T Chemicals, Inc.

W. Neumann, *The Organic Chemistry of Tin,* John Wiley & Sons, Inc., New York, 1970.

A. G. Davies and P. J. Smith, *Adv. Inorg. Chem. Radiochem.* **23,** 1 (1980).

A. F. Trotman-Dickenson, ed., *Comprehensive Inorganic Chemistry,* Vol. 2, Pergamon Press, Oxford, U.K., 1973, pp. 43–104.

TIRE CORD

Since the introduction of the pneumatic tire the reinforcing strength or tensile member has been some form of textile or steel fiber, usually in the form of a cord. The development of textile reinforcements for tire cords has been driven by the tire need for better mechanical properties and therefore better tire performance. Tire cord development has been paralleled by the emergence of other markets for new fibers (qv) in the apparel and industrial textile sectors. Thus, tire cords have progressed from natural fibers (cotton), through man-made (rayon) to the present totally synthetic suite of reinforcement candidates (nylons and polyesters).

Cotton (qv) was the first material to be used in mass production of tires. Its relative abundance and well-established handling make the fiber naturally amenable to industrial usage. However, as tire usage and performance demands rose, specifically in the areas of strength and fatigue resistance, the need for better reinforcement arose.

Rayon was introduced commercially in 1938 and developed through generations of product to the present Super III rayon, rayon is a man-made regenerated cellulose fiber derived from wood pulp (qv). Rayon fibers do not possess the intrinsic ability to bond with rubber and thus require development of suitable adhesive systems in the form of resorcinol–formaldehyde–latex (RFL) dip systems (*vide infra*). Although in declining use due to the increased cost associated with environmental aspects of the fiber production, rayon continues to be useful in specialty high performance tires.

Nylon-reinforced tires use nylon-6 polymer (polycaprolactam) fibers as well as nylon-6,6 (poly(hexamethylenediamine adipamide)) fibers. Nylon tire cords are characterized by extremely good fatigue resistance in compression and good adhesion to most rubber compounds with simple RFL adhesives.

Polyester is the newest of the tire-reinforcement fibers in the form of poly(ethylene terephthalate) (PET) fiber introduced in 1962. It has developed through several product generations up to the generation of high modulus low shrink (HMLS) or dimensionally stable polyester (DSP) forms of the 1990s. With respect to nylon, polyester fibers display lower thermal shrinkage and higher modulus. More complex adhesive systems tend to be required for polyesters to bond to rubber.

As of the late 1990s, only specialized applications have emerged for aramid tire cord that draw on their high strength-to-weight ratio to produce tires with lower weight.

For several years fiber glass was used extensively in bias-belted and radial tires, but was ultimately replaced by steel belts in radial tires.

Few materials have emerged to displace the current two major fibers, nylon and polyester. Nonetheless, many examples of fibers offering attractive properties for tire cords have been reported in the literature, eg, polyethylene ketone, poly(paraphenylene benzobisoxazole), acrylics, and high strength poly(vinyl alcohol) (see VINYL POLYMERS).

Fiber Development for Tire Cord Use

The fundamental requirements for a useful tire cord fiber are high strength and modulus coupled with good dimensional stability (ie, resistance to deformation under temperature and load), and durability (fatigue and chemical stability) at favorable economics. The aspects of processing must also be considered when forming a fiber with suitable properties that will take full advantage of the polymer's potential and maintain viable economics. Thus, the market for organic tire cord remains divided between nylon and polyester fibers and will likely remain so for the medium-term future.

Fiber Properties

A tire reinforcement's use is dependent on several physical properties. Some of the most important are tabulated in Table 1. These properties effectively screen candidates for use in tires. The secondary features define a fiber's potential for tire use. A key feature implicitly included in the use of organic fibers for tire reinforcement is the ability to retain strength and modulus characteristics at elevated temperatures (80–120°C for most tires) sufficient to sustain service demands. Thus, a tire which may appear to be overdesigned at room temperature is, in many cases, reflecting the changes in properties experienced at operating temperatures.

Processing

The basis for reinforcement of a requires placing the strength or tensile member in a preferred direction, depending on the location and cord function in the tire. An overview of the tire production process, including essential elements of transforming a continuous yarn into a useful embodiment for tire reinforcement, is shown in Figure 1.

Steel Tire Cord

Steel tire cord provides a unique combination of strength, ductility, dimensional stability, resistance to fatigue, rubber adhesion, and consumer value that led to a dramatic increase in steel cord consumption in the 1990s. The steel chosen for tire cord is a eutectoid carbon steel containing 0.7% carbon, 0.5% manganese, 0.2% silicon, and a very low amount of sulfur and phosphorus. Filaments or wires produced are twisted into strands and then combined into cords. The lay length is the axial distance required to make a 360° revolution of any component in a strand or cord and is expressed in millimeters. Direction of lay is the helical disposition of components of a strand or cord. The

Table 1. Mechanical Properties for Reinforcing Fibers[a]

Name	Strength, cN/Tex[b]	Modulus, cN/Tex[b]	Thermal shrinkage	Density	T_g, °C	Mp, °C	Compression fatigue resistance
rayon(III)	52	1,200	0.2	1.53		210	good
nylon-6	84	300	7.5	1.13	20	220	excellent
nylon-6,6	86.5	400	7.4	1.14	50	254	excellent
polyester							
standard modulus	90	1,000	13.0	1.38			good
high modulus	70	1,000	3.9	1.38	75	260	good
aramid	200	4,900	0.1	1.44		454[d]	poor
carbon fiber	300–370	33,500–15,250	0.0	1.83		3700[d]	poor
high tensile steel	43.4	2,680	0.0	7.85		1600	good
fiber glass (E-glass)	100	2,200	0.0	2.55		1180	poor

[a] Strength and modulus normalized to linear density (Tex = (g·wt)/km), an appropriate basis for materials of similar density; however, this breaks down for dense materials (eg, steel) because the basis for tire use is better described by properties per unit cross section (MPa) rather than weight.
[b] To convert cN/Tex to MPa (N·mm^{-2}), multiply by density (g·cm^{-3}) by 10.
[c] To convert cN/Tex to MPa ((N·mm^{-2}), multiply by density (g·cm^{-3}) by 10.
[d] Decomposition temperature.

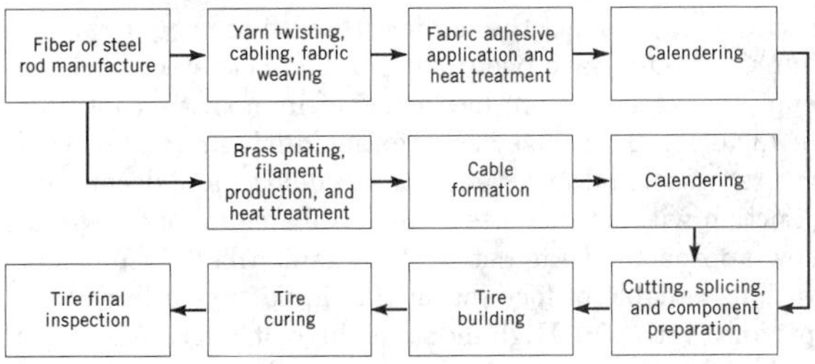

Figure 1. Tire production process.

way the filament, strand, and spiral wrap are assembled, and the filament diameters used (usually in millimeters), determine the cord construction. The tire industry has adopted an ASTM wire nomenclature. The description of the construction follows the sequence of manufacturing of the cord, ie, starting with the innermost strand or filament and moving outward. A plus (+) sign separates each layer. If the filament diameters are the same in two or more components, the diameter is omitted for all but the last component. If a strand is a single filament, the numerical designation "1" is omitted. The full description of the construction is given by the following formula:

$$(N \times F) \times D + (N \times F) \times D + F \times D \qquad (1)$$

where N = number of strands, F = number of filaments, and D = nominal diameter of filaments, in mm. The cords can be classified into (*1*) regular, (*2*) Lang's lay, (*3*) open, (*4*) compact, and (*5*) high elongation. The most common cord is the regular cord in which the direction of lay in strands is opposite to direction of lay in closing the cord. Lang's lay cord is formed such that the direction of lay in the strands is the same as the direction of lay in closing the cord.

A thin layer of brass coating on steel cord facilitates adhesion between the metal and rubber compound.

Twisting two or more filaments together to form a cord reduces the overall modulus and strength of the construction. The degree of the reduction depends on the individual filaments.

Cord Impact on Tire Performance

The primary load-bearing member of cord–rubber composites is the cord, which provides strength and many other critical properties essential for tire performance. Cords in plies form the structural backbone of the tire. The rubber plays the important but secondary role of transmitting load to the cords via shearing stresses at the cord–rubber interface. Other expected performance characteristics of the tire are due to design and manufacturing processes.

Burst Strength. The burst strength of the tire is derived from the cord strength and is related by equation 2

$$\text{burst strength} = Nt_uK = \frac{\Pi P_b(r_c^2 - r_{\max}^2)}{\sin \alpha} \qquad (2)$$

where N is total number of cords in a tire, t_u is average ultimate tensile strength of cord, P_b is burst pressure, r_c is radius from the center of rotation to the crown of tire, $r_{\max}$ is radius from the center of rotation to the maximum section width of a tire, α is the crown angle between cord path and circumferential plane through the crown of a tire, and K is the efficiency factor which depends on the distribution of ultimate cord strength and is always less than one. Tires are designed with a very high factor of safety, about six for radial medium truck tires and ten or higher for radial passenger tires.

Bruise Resistance. Bruise resistance of a rolling tire describes its ability to resist impact failure. Bruise resistance is tested by measuring the energy required to break a passenger tire under inflation at room temperature when a 19-mm diameter plunger is pushed through the crown at 51 mm/min crosshead speed.

Tire Endurance. The interlaminar shear stress and deformation in the composite laminate of tire can produce delamination-induced failures, especially at the belt edges. The tire composite can be designed to minimize interlaminar shear stress through proper selection of cord properties, cord orientation, cord end count, and rubber properties. Another separation mechanism can involve cord–rubber adhesion. Another failure phenomenon is fatigue which is mostly observed on sidewall near the shoulder and bead regions due to high temperature and stresses. To some extent fatigue performance can be controlled through proper selection of cord material and twist. High speed performance tires may generate enough heat to cause tread separation.

Cornering and Ride. The stiffness of the belt package primarily determines the cornering and ride characteristics of a radial passenger tire. The belt package in contact with the road is a fairly complex composite consisting of tire components including the innerliner, carcass, belts, and tread. The in-plane flexural rigidity of the belt package is the most important parameter controlling cornering, whereas out-of-plane flexural rigidity controls ride. Flexural rigidity, in turn, depends on the stiffness of the belt package.

Treadwear. Treadwear is a complex physical–chemical process driven by the frictional energy developed at the interface between tread pattern elements and the pavement. The rate of wear is influenced by the loss modulus property of the tread compound, the microstructure (or abrasiveness) of the pavement, environmental conditions, and vehicle operation. The wear of passenger tires occurs mainly during cornering maneuvers, where the tread center line distortion in the footprint impacts slippage of the tread rubber relative to road surface mainly in the region at the rear of the footprint. Belted radial tires experience less slippage than nonbelted tires in cornering.

Flatspotting. Flatspotting in tires is based on the viscoelastic behavior of tire cords. Tire cords such as nylon and polyester tend to shrink when heated above their glass-transition temperature, T_g, as in the case of a running tire. When the tire stops rotating, the cord elements in the footprint cool to ambient temperature shrink more than the cords in the remainder of tire. When the tire starts to rotate again, persistence of this difference causes flatspot, which remains until the tire is reheated to a temperature at which the flatspot was introduced.

Power Loss on Tire Rolling Resistance. About 95% of rolling resistance or power loss is due to the viscoelastic loss of the tire. Tires having carcass constructed with aramid cords have a lower rolling resistance than those constructed with steel cords. The challenge is to understand the exact contribution of cord to tire power loss in radial tires and to develop cord material with properties that significantly reduce power loss.

Test Methods

Tire cords are characterized for their physical, adhesion, and fatigue properties for use in tires. These characterizations are conducted under normal and varying test conditions to predict their performance during tire operation.

ASTM standard D2969-92 includes test methods for steel cords that are specifically designed for use in the reinforcement of pneumatic tires. It describes test methods determining steel cord construction, break strength, elongation at break, modulus, flare, linear density, straightness, residual torsion, brass coating composition, and mass of steel cords.

Steel cords are vulcanized into a block of rubber and the force necessary to pull the cords linearly out of the rubber is measured as adhesive force. ASTM method D2229-93a can be used for evaluating rubber compound performance with respect to adhesion to steel cord.

A transverse impact test method for steel cords has been designed to determine the resistance to cutting (puncture resistance) when used as a tire belt reinforcement. The test is a modified charpy test.

Rotating beam fatigue test for steel cords evaluates steel cord for pure bending fatigue.

Rotoflex test for steel cords determines the bending fatigue limit as a function of pretension in the carcass cords of truck tires.

ASTM standard D885M-94 includes test methods for characterizing tire cord twist, break strength, elongation at break, modulus, tenacity, work-to-break, toughness, stiffness, growth, and dip pickup for industrial filament yarns made from organic base fibers, cords twisted from such yarns, and fabrics woven from these cords that are produced specifically for use in the manufacture of pneumatic tires. These test methods apply to nylon, polyester, rayon, and aramid yarns, tire cords, and woven fabrics.

ASTM test method D4974-89 is used for measuring thermal shrinkage of yarn and cords with linear density ranging from $20-700 \times 10^{-6}$ kg/m (20–700 tex) using the Testrite thermal shrinkage oven. The most commonly used static adhesion tests are the H-test (ASTM D4776-88), U-test (ASTM D4777-88), T-test, I-test, and strip–peel test (ASTM D4393-85). These tests derive their name from the shape of the test specimen.

Adhesion gradually deteriorates with repeated deformation. Dynamic adhesion evaluation is characterized by number of cycles of deformation to reach limiting value. Various dynamic adhesion test methods have been developed based on the type of deformation. The two most commonly used methods are the Goodrich Disk Fatigue test and the Dynamic Flex Strip Adhesion (ASTM D430-59).

Tire cords experience large number of stress–strain cycles which lead to their degradation. There are a number of laboratory tests used to estimate fatigue life of tire cords (eg, ASTM D885-64). Some of the most commonly used test methods include the following: the Firestone Compression Flex test (ASTM D885-64); Goodrich Disk Fatigue test (ASTM D885-64); The Mallory Tube Fatigue test (ASTM D885-64); and the Dunlop Fatigue test.

Tire Cord Status

As a load-carrying member, the type of tire cord used for a specific tire depends on the use requirements. To some extent the preference for a tire cord is dictated by tire designs, ie, bias vs radial. The belt material for radial tires requires high stiffness/modulus vs carcass material requirements of flexibility and fatigue resistance for comfort and durability. For bias applications, nylon tire cord remains dominant, particularly in heavy tires with high load-carrying capacity. The use of nylon-6 vs nylon-6,6 follows the basic criteria: value to customer in that market. Polyester has also been used in bias tires (passenger and light trucks). For radial tires the choice of tire cord for belt application varies among glass, rayon, and steel.

Requirements for tire cord material will to some extent be driven by new vehicle trends. For example, the clean air emphasis in North America places lightweight vehicles and materials at a premium. For tire cord the fuel economy or rolling resistance provided by the cord–rubber composite may shift the pattern of usage. A common requirement for all types of tire cord surfaces is a high strength-to-weight ratio.

For steel cord, to meet the higher strength-to-weight ratio, continuous improvements have been made to increase steel cord strength, including steel composition, low levels of impurities, controlled nonmetallic inclusion, minimized segregation during metallurgical solidification processes, and minimum surface defects and decarbonization, in conjunction with an increase in the total amount of drawing.

A new approach for high strength organic fibers includes polymer compositions capable of forming in the liquid crystalline state (thermotropic and lyotropic). High modulus, high strength fibers from aromatic polyamides and aromatic copolyesters have been manufactured utilizing conventional dry-jet wet spinning or melt spinning technology.

ROOP S. BHAKUNI
SURENDRA K. CHAWLA
D. K. KIM
D. SHUTTLEWORTH
The Goodyear Tire & Rubber Company

F. J. Kovac, *Tire Technology,* The Goodyear Tire & Rubber Co., Akron, Ohio, 1978.

R. S. Bhakuni, S. K. Mowdood, W. H. Waddell, I. S. Rai, and D. L. Knight, in J. I. Kroschwitz, ed., *Encyclopedia of Polymer Science and Engineering,* Vol. 16, John Wiley & Sons, Inc., New York, 1989, p. 834.

S. K. Clark, ed., *Mechanics of Pneumatic Tires,* DOT HS No. 805,952, U.S. Dept. of Transportation, Washington, D.C., 1981.

TITANIUM AND TITANIUM ALLOYS

Titanium, a metal element of Group IVB, has a melting point of 1675°C and an atomic weight of 47.90. Titanium metal has become known as a space-age metal because of its high strength-to-weight ratio and inertness to many corrosive environments. Its principal use, however, is as TiO_2 as paint filler (see PAINT; PIGMENTS). The whiteness and high refractive index of TiO_2 are unequaled for whitening paints, paper, rubber, plastics, and other materials. A small amount of mineral-grade TiO_2 is used in fluxes and ceramics.

Titanium mineral occurs in nature as ilmenite, $FeTiO_3$; rutile, tetragonal TiO_2; anatase, tetragonal TiO_2; brookite, rhombic TiO_2; perovskite, $CaTiO_3$; sphene, $CaTiSiO_5$; and geikielite, $MgTiO_3$. Ilmenite is by far the most common, although rutile has been an important source of raw material.

The principal titanium mineral, ilmenite, is found in either alluvial sands or hard-rock deposits. In metallic form, titanium has a dull silver luster and an appearance similar to stainless steel.

Titanium ore bodies are uniformly distributed throughout the continents of the world. They occur either as hard-rock deposits, magnetic in origin, or as secondary placer deposits. The titanium oxide contained in known deposits of ilmenite is close to 1 billion metric tons, whereas only about 50 million metric tons of titanium oxide in rutile are known to exist. The largest known reserves of titanium are in Canada and China.

Manufacture

Ore-Concentrate Refining. The TiO_2 content of ore concentrates determines further processing steps. High grade ore such as rutile, synthetic rutile, or slag from Richard's Bay is refined to pigment-grade TiO_2 via chlorination. Lower grade ore is processed via the sulfate route. The chlorination process produces a better quality pigment, requires less processing energy than the sulfate process (1800 kWh/t compared to 2500 kWh/t), and has less waste discharge. The sulfate process produces approximately 6 t of waste per ton of TiO_2, whereas only one ton of waste is produced through the chloride process. However, high grade ore is required for the latter process, ie, TiO_2 content >70%, with <1% MgO and 0.2% CaO, because ores that have high MgO and CaO can clog the chlorinator. Environmental problems have forced the industry either to shut down sulfate plants or to install expensive pollution-control equipment. Because of the shortage of high grade TiO_2 reserves, the pigment industry must adapt the ore to the chloride process. As of 1997, the trend was toward ore beneficiation. Ore containing 50–60% TiO_2 content is beneficiated by partial reduction, then leached with sulfuric or hydrochloric acid to yield a concentrate containing >90% TiO_2, the so-called synthetic rutile.

Sulfate Process. In the sulfate process, ilmenite ore is treated with sulfuric acid at 150–180°C. The undissolved solids are removed and the liquid is evaporated under vacuum and cooled. The precipitated $FeSO_4 \cdot 7H_2O$ is filtered and the filtrate concentrated to ca 230 g/L. Heating to 90°C hydrolyzes titanyl sulfate to insoluble titanyl hydroxide.

Chloride Process. In the chloride process, a high grade titanium oxide ore is chlorinated in a fluidized-bed reactor in the presence of coke at 925–1010°C:

$$TiO_2 + 2\,C + 2\,Cl_2 \rightarrow 2\,CO + TiCl_4 \quad \Delta G_{1300°C} = -125\ \text{kJ}(30\ \text{kcal})$$

The volatile chlorides are collected and the unreacted solids and nonvolatile chlorides are discarded. Titanium tetrachloride is separated from the other chlorides by double distillation.

Tetrachloride-Reduction Process. Titanium tetrachloride for metal production must be of very high purity. Titanium tetrachloride for metal production is prepared by the same process as described above, except that a greater effort is made to remove impurities, especially oxygen- and carbon-containing compounds.

Magnesium-Reduction (Kroll) Process. In the 1990s, nearly all sponge is produced by the magnesium reduction process.

$$TiCl_4(g) + 2\,Mg(l) \rightarrow Ti(s) + 2\,MgCl_2(l)\ \Delta G_{900°C} = -301\ \text{kJ}(-72\ \text{kcal})$$

The production is in batches up to 10 metric tons of titanium. The product, the so-called sponge, is further processed to remove the unreacted titanium chlorides, magnesium, and residual magnesium chlorides.

Sodium-Reduction Process. The sodium-reduction process was employed as an alternative to magnesium reduction. The last large production plant was closed in the early 1990s. Although the process was more costly than magnesium reduction, the product contained less metallic impurities, ie, Fe, Cr, and Ni. This product is desirable for a growing titanium market in the electronics industry.

Other Reduction Processes. Other methods of producing titanium have been studied in the hope of finding another reduction route, collecting the metal as an ingot instead of as a sponge, and designing a continuous process. The most successful and widely studied noncommercialized process is the electrolytic reduction of $TiCl_4$. Primary electrical energy reduces $TiCl_4$ to titanium metal at the cathode and chlorine gas at the anode.

Sponge Consolidation. The next step is the consolidation of the sponge into ingot. The crushed sponge is blended with alloying elements or other sponge. Consumable electrodes are produced by welding 45–90-kg sponge compactions (electrode compacts) in an inert atmosphere, and then double-vacuum-arc-remelted (VAR). A portion of the elemental sponge compacts are often replaced with bulk scrap. The ingots are ca 71–91-cm dia and long enough to weigh 4.5 to 9.0 t.

Alloys

Titanium alloy systems have been extensively studied. Alloy development has been aimed at elevated-temperature aerospace applications, strength for structural applications, biocompatibility, and corrosion resistance. The original effort has been in aerospace applications to replace nickel- and cobalt-base alloys in the 250–600°C range. The useful strength and corrosion-resistance temperature limit is ca 550°C.

The important α-phase stabilizing alloying elements include aluminum, tin, zirconium, and the interstitial alloying elements, ie, elements that do not occupy lattice positions, oxygen, nitrogen, and carbon. Small quantities of interstitial alloying elements, generally considered to be impurities, are always present, and have a great effect on strength. In sufficient amounts they can embrittle the titanium at room temperature. Oxygen is often used as an alloying element, ranging from as low as 500 ppm to as high as 3000 ppm, whereas carbon and nitrogen are maintained at their residual level. Nitrogen has the greatest effect and commercial alloys specify its limit to be less than 0.05 wt %. It may also be present concentrated as high melting point nitride inclusions (TiN), referred to as Type 1 defects, which are detrimental to critical aerospace structural applications. Carbon does not affect strength at concentration above 0.25 wt % because carbides (TiC) are formed.

The most important α-stabilizing alloying element is aluminum, which is inexpensive and has an atomic weight less than that of titanium. Aluminum additions slightly lower the density. The mechanical strength of titanium can be increased considerably by aluminum additions. Even though the solubility range of aluminum extends to 27 wt %, above 7.5 wt % the alloy becomes too difficult to fabricate and is embrittled. The embrittlement is caused by a coherently ordered phase based on Ti_3Al. The important β-stabilizing alloying elements are the bcc elements vanadium (most important), molybdenum, tantalum, and niobium of the β-isomorphous type and manganese, iron, chromium, cobalt, nickel, copper, and silicon of the β-eutectoid type. Alloys of the β-type respond to heat treatment, are characterized by higher density than pure titanium, and are more easily fabricated.

Physical Properties

The physical properties of titanium are given in Table 1. The most important physical property of titanium from a commercial viewpoint is the ratio of its strength (ultimate strength >690 MPa (100,000 psi)) at a density of 4.507 g/cm^3. Titanium alloys have a higher yield strength to density rating, between −200 and 540°C, than either aluminum alloys or steel. Titanium alloys can be made to have strength equivalent to high strength steel, yet having density ca 60% that of iron alloys. Because of its high melting point, titanium can be alloyed to maintain strength well above the useful limits of magnesium and aluminum alloys. This property gives titanium a unique position in applications between 150–550°C when the strength-to-weight ratio is the sole criterion.

Corrosion Resistance. Titanium is immune to corrosion in all naturally occurring environments. It resists decomposition because of a tenacious protective oxide film. However, when this oxide film is broken, the corrosion rate is very rapid. Usually the presence of a small amount of water is sufficient to repair the damaged oxide film. Titanium is resistant to corrosion attack in oxidizing, neutral, and inhibited reducing conditions. Titanium's resistance to aqueous chloride solutions and chlorine accounts for most of its use in corrosion-resistance applications.

Table 1. Physical Properties of Titanium

Property	Value
melting point, °C	1668 ± 5
boiling point, °C	3260
density, g/cm^3	4.54
thermal conductivity, at 25°C, W/(m·K)	21.9
emissivity	9.43
electrical resistivity, at 20°C, nΩ·m	420
magnetic susceptibility, mks	180×10^{-6}
modulus of elasticity, GPa[a]	
tension	ca 101
compression	103
shear	44
Poisson's ratio	~0.41
vapor pressure, kPa[b]	$\log P_{kPa} = 5.7904 - 24644/T - 0.000227T$
specific heat, J/(kg·K)[c]	$C_p = 669.0 - 0.037188T - 1.080 \times 10^7/T^2$

[a] To convert GPa to psi, multiply by 145,000.
[b] To convert log P_{kPa} to logP_{atm}, add 2.0056 to the constant.
[c] $T > 298$ K.

Titanium corrodes very rapidly in acid fluoride environments, and is also attacked by hot caustic solutions. Titanium is susceptible to pitting and crevice corrosion in aqueous chloride environments. Titanium does not stress-crack in environments that cause stress-cracking in other metal alloys, eg, boiling 42% $MgCl_2$, NaOH, sulfides, etc. Titanium is susceptible to failure by hydrogen embrittlement. Titanium resists erosion–corrosion by fast-moving sand-laden water. In galvanic coupling, titanium is usually the cathode metal and consequently not attacked.

Casting. Consolidated titanium is cast either by precision castings or investment casting. The metal is melted using a consumable titanium electrode in a protected atmosphere, usually in a water-cooled copper crucible. In precision casting, rammed graphite molds are used; in investment casting, ceramic molds. Reliable, low cost powder-production techniques have not yet been fully developed.

Metal Working. The ingots are further processed by the conventional methods of forging, hot-rolling, cold-rolling, extrusion, etc. The mill product forms include billet, bar, plate, sheet, strip, foil, extrusion, wire, pipe, and welded tubing. Mill practices differ somewhat from those of other metal products.

Sheet, thin plate, welded tubing, and small-diameter bar of commercially pure titanium are manufactured into parts by conventional cold-working techniques.

Fabrication. Fabrication of titanium into useful parts, such as tanks, heat exchangers, and pressure vessels, is comparable to the fabrication of austenitic steel in method, degree of difficulty, and cost. Commercial-grade titanium can be bent 105° without cracking around a radius of 2–2.5 times the sheet thickness. Heat is required to form most titanium alloy parts. Super plastic forming (SPF) is used to form complex shapes in $\alpha-\beta$-type alloys such as Ti–6Al–4V. The forming is conducted at ca 900°C where the alloy becomes super plastic, ie, elongates without necking. The process is sometimes combined with diffusion bonding (SPF/DB) to form complex structures.

Welding (qv) of titanium requires a protected atmosphere of inert gas. Titanium cannot be fusion-welded to other metals because of formation of brittle intermetallic phases in the weld zone.

Aerospace Alloys. The alloys of titanium for aerospace use can be divided into three categories: an all-α structure, a mixed $\alpha-\beta$structure, and an all-βstructure. The $\alpha-\beta$-structure alloys are further divided into near-α alloys (<2% β-stabilizers). Most of the ca 100 commercially available alloys are of the $\alpha-\beta$-structure type. The most important commercial alloy is Ti–6Al–4V, an $\alpha-\beta$-alloy having a good combination of strength and ductility. This alloy represents over 65% of all the titanium alloys produced. It can be age-hardened in thin sections and has both moderate ductility and an excellent record of successful applications. It is mostly used for compressor blades, fan blades, and rotating disks in aircraft gas-turbine engines, but also used for rocket motor cases, structural airframe forgings, steam-turbine blades, and cryogenic parts for which ELI grades are usually specified.

Other commercially important $\alpha-\beta$-alloys used in aircraft applications include Ti–3Al–2.5V, Ti–6Al–6V–2Sn, Ti–6Al–2Sn–4Zr–6Mo, Ti–5Al–2Sn–2Zr–4Mo–4Cr (Ti-17), Ti–6Al–2Sn–2Zr–2Mo–2Cr–0.15Si, and Ti–10V–2Fe–3Al. As a group, these alloys have good combination of strength and ductility. Weldability becomes more difficult as β-constituents increase.

The only α-alloy of commercial importance is Ti–5Al–2.5Sn. This alloy is weldable, has good elevated temperature stability and good oxidation resistance to about 600°C, and is used for forgings and sheet-metal parts, such as aircraft engine compressor cases because of its weldability. The commercially important near-α-alloys used in the United States are Ti–8Al–1Mo–1V and Ti–6Al–2Sn–4Zr–2Mo–0.1Si. Both of these alloys exhibit good creep resistance and the excellent weldability and high temperature strength of α-alloys; temperature limit is ca 470°C. β-Titanium alloys represent only 1% of the commercial titanium production.

Nonaerospace Alloys. In order to optimize corrosion resistance, many new alloys and modification of existing alloys have been developed. The four basic grades of pure titanium, ie, ASTM Grades 1 through 4, differ primarily in oxygen and iron content and represent about 24% of the titanium production. ASTM Grade 1 has the highest purity and the lowest strength. The α-alloys in this group are distinguished by excellent weldability, formability, and corrosion resistance. The strength, however, is not maintained at elevated temperatures. The primary use of alloys in this group is in industrial processing equipment, eg, tanks, heat exchangers, pumps, and electrodes, though there is also some use in airframes and aircraft engines.

Other Alloys. Other alloying ranges include the aluminides, TiAl and Ti_3Al; the superconducting alloys, Ti–Nb type; the shape-memory alloys, Ni–Ti type; and the hydrogen storage alloys, Fe–Ti (see SHAPE-MEMORY ALLOYS). The aluminides are intended for both static and rotating parts in the turbine section of newer gas-turbine aircraft engines.

Titanium alloyed with niobium exhibits superconductivity, and a lack of electrical resistance below 10 K. Their use is of interest for power generation, propulsion devices, fusion research, and electronic devices. Titanium alloyed with nickel exhibits a memory effect, ie, the metal form switches from one specific shape to another in response to temperature changes. Titanium alloyed with iron is a candidate for solid-hydride energy storage material for automotive fuel.

Health and Safety Factors

Titanium and its corrosion products are nontoxic. A safety problem does exist with titanium powders, grindings, turnings, and some corrosion products that are pyrophoric. Powders can ignite at about 250°C and should be handled in small quantities at room temperature.

Uses

Titanium is primarily used in the form of high purity titanium oxide as a pigment in surface coatings. Other uses reflect its special properties, which include high refractory index that inputs good hiding power; high reflectivity that inputs great brightness and brilliant whiteness; chemical inertness that contributes to excellent color retention; and thermal stability over a wide range of temperatures. Although the principal application of high purity (pigment-grade) TiO_2 is in paint pigments, other important uses are in plastics for color in floor-covering products and to help protect plastic products and foodstuffs contained in plastic bags from uv radiation deterioration; in paper as a filler and whitener; and in rubber. Titanium metal was first used as a material for aerospace. Aerospace applications have shaped and controlled the titanium metal industry. The use of titanium in aircraft is divided about equally between engines and airframes. Another outstanding property of titanium metal is its corrosion resistance. The largest single application is related to heat-exchanger pipes and tubing for the power industry, as well as marine and desalination applications, where titanium provides protection against corrosion by seawater, brackish water, and other estuary waters containing high concentrations of chlorides and industrial wastes (see HEAT-EXCHANGE TECHNOLOGY).

Titanium metal is especially utilized in environments of wet chlorine gas and bleaching solutions. Titanium is used in heat-exchanger tubing for salt production, in the production of ethylene glycol, ethylene oxide, propylene oxide, and terephthalic acid, and in industrial wastewater treatment. Titanium is used for ore-leaching solutions and as racks for metal plating. In oil and gas refinery applications, titanium is used as protection in environments of H_2S, SO_2, CO_2, NH_3, caustic solutions, steam, and cooling water. It is used in heat-exchanger condensers for the fractional condensation of crude hydrocarbons, NH_3, propane, and desulfurization products using seawater or brackish water for cooling.

Titanium alloys are being used in deep-water hydrocarbon and geothermal wells for risers. In consumer applications, titanium is used in golf club heads, jewelry, eyeglass frames, and watches. Other application areas include nuclear-waste storage canisters, pacemaker castings, medical implants, high performance automotive applications, and ordnance armor.

STAN R. SEAGLE
Consultant

Titanium Annual Review-1994, Mineral Industry Surveys, U.S. Department of the Interior, Bureau of Mines, Washington, D.C., 1995.

L. C. Covington and R. W. Schutz, in E. W. Kleefisch, ed., *Industrial Applications of Titanium and Zirconium,* STP 728, American Society for Testing and Materials, Philadelphia, Pa., 1981, p. 163.

M. J. Donachie, *Titanium, A Technical Guide,* ASM International, Metals Park, Ohio, 1988.

P. A. Blenkinsop, W. J. Evans, and H. M. Flower, *Titanium '95: Science and Technology 8th World Conference on Titanium,* Institute of Materials University Press, Cambridge, 1966.

TITANIUM COMPOUNDS

INORGANIC

The titanium compounds of greatest technological importance are titanium dioxide, the predominant white pigment, which sold ca 4×10^6 metric ton worldwide in 1996 (see PIGMENTS, INORGANIC); titanium esters, eg, titanium isopropoxide, and derived compounds, which are used for applications such as structuring agents in paint (qv); titanium metal, which has excellent corrosion resistance and a high strength–weight ratio (see TITANIUM AND TITANIUM ALLOYS); and barium titanate, an important electroceramic (see ADVANCED CERAMICS; CERAMICS AS ELECTRICAL MATERIALS).

Titanium–Hydrogen System

Titanium metal readily absorbs hydrogen. Absorption rates above 400°C are normally high, but at lower temperatures the rates depend critically on the cleanliness of the surface. The hydrogen dissociates prior to absorption. Because the absorption is a reversible process, there is thus an equilibrium pressure of hydrogen at each temperature and composition. The limiting stoichiometry of the system is normally accepted as TiH_2, although higher hydrides have been reported under special conditions. For example, TiH_4 has been reported in the products formed when mixtures of titanium tetrachloride and hydrogen were irradiated with uv radiation at 254 nm.

Titanium hydrides are grey powders. Their density decreases with increasing hydrogen content from 4410 kg/m^3 for pure titanium metal to ca 3800 kg/m^3 for TiH_2. The applications of titanium hydrides are related to their ability to store hydrogen reversibly. Their possible use as a hydrogen storage medium in hydrogen-fueled vehicles has received particular attention (see HYDROGEN ENERGY). Titanium hydrides may also be used to purify cylinder hydrogen or as high purity hydrogen sources for, eg, plasma guns.

Titanium Borides

Five phases of titanium boride have been reported. TiB_2, Ti_2B, TiB, Ti_2B_5, and TiB_{12}. The most important of these is the diboride, TiB_2, which has a hexagonal structure and lattice parameters of $a = 302.8$ pm and $c = 322.8$ pm. Titanium diboride is a gray crystalline solid. It is not attacked by cold concentrated hydrochloric or sulfuric acids, but dissolves slowly at boiling temperatures. It dissolves more readily in nitric acid/hydrogen peroxide or nitric acid/sulfuric acid mixtures. It also decomposes upon fusion with alkali hydroxides, carbonates, or bisulfates.

Research-grade material may be prepared by reaction of pelleted mixtures of titanium dioxide and boron at 1700°C in a vacuum furnace. Technical-grade (purity >98%) material may be made by the carbothermal reduction of titanium dioxide in the presence of boron or boron carbide.

The applications of titanium diboride depend on such properties as its hardness, electrical conductivity, and ability to be wetted by metals. In monolithic form, sintered titanium diboride can be used as a strong but lightweight armor. It is an important constituent of cermets, eg, composites based on TiB_2–TiC–Fe mixtures show excellent performance as cutting tools for aluminum alloys. Titanium diboride also shows excellent wettability and stability in liquid metals. The use of titanium diboride as an inert solid cathode, which would replace the molten aluminum cathode in aluminum reduction refining cells, has been proposed.

Titanium Carbide

Titanium carbide, TiC, has the fcc, NaCl structure in which the carbon atoms can be regarded as occupying the octahedral interstices in a slightly expanded cubic close-packed arrangement of titaniums. Titanium carbide forms solid solutions with TiO and TiN. Sintered titanium carbide is light gray when fractured but can be polished to a silver gray. The maximum melting point in the Ti–C system is 3067°C for $TiC_{0.8}$, that of $TiC_{1.0}$ is slightly lower. The boiling point of titanium carbide is 4800°C.

Titanium carbide is resistant to aqueous alkali except in the presence of oxidizing agents. It is resistant to acids except nitric acid, aqua regia, and mixtures of nitric acid with sulfuric or hydrofluoric acid. In oxygen at 450°C, a nonprotecting anatase coating forms. Titanium carbide is manufactured mainly in-house by cutting-tool manufacturers by the reduction of titanium dioxide with carbon.

A number of high temperature processes for the production of titanium carbide from ores have been reported. The aim is to manufacture a titanium carbide that can subsequently be chlorinated to yield titanium tetrachloride. Titanium carbide may also be made by the reaction at high temperature of titanium with carbon; titanium tetrachloride with organic compounds such as methane, chloroform, or poly(vinyl chloride); titanium disulfide with carbon; organotitanates with carbon precursor polymers and titanium tetrachloride with hydrogen and carbon monoxide. The primary commercial applications of titanium carbide are in wear-resistant components and cutting tools.

Titanium–Nitrogen Compounds

Titanium Nitride. Titanium nitride has the cubic NaCl structure, but the structure is stable over a wide range of either anion or cation deficiency. The nitride is a better conductor of electricity than titanium metal. It becomes superconductive at 1.2–1.6 K. Titanium nitride has a density of 5213 kg/m^3 and melts at 2950°C.

Table 1. Properties of the Lower Oxides of Titanium

Property	TiO	Ti_2O_3	Ti_3O_5
color	golden yellow	violet	blue-black
density, kg/m^3	4888	4486	421(0)
melting point, °C	1737	2127	
structure	fcc	hexagonal	monoclinic
lattice parameters, pm			
a	417	515.5	975.2
b		515.5	380.2
c		1316.2	944.2
solubility			
HF[a]	dissolves rapidly	dissolves	
HCl[b]	slow attack	no action	
H_2SO_4[b]	slow attack	slow attack	
HNO_3[b]	surface attack	no action	
NaOH[b]	slow attack	no action	

[a] Hot 40 wt %.
[b] Hot concentrated.

The powder, usually described as chocolate-brown, is actually bluish black when sufficiently fine. The ease of sintering depends on particle size. It is resistant to attack by acids except boiling aqua regia, but is decomposed by alkalies with the evolution of ammonia.

Direct synthesis from nitrogen and finely divided titanium metal can be achieved at temperatures of >ca1200°C. Typically, titanium sponge or powder is heated in an ammonia- or nitrogen-filled furnace and the product is subsequently milled and classified.

$$2\ Ti + N_2(\text{or } NH_3) \rightarrow 2\ TiN(+1.5\ H_2)$$

TiN is used in cutting tools and wear parts, often in conjunction with TiC. Because of its excellent electrical conductivity, TiN can be used as an additive to confer electrical conductivity on silicon nitride and sialon ceramic components.

Titanium Nitrate. Titanium nitrate, $Ti(NO_3)_4$, is a white powder, mp 58°C. It is stable in a sealed tube at room temperature, is less sensitive to moisture than titanium tetrachloride, and does not fume in air but reacts vigorously with water, liberating nitrogen oxides. It may be prepared by prolonged reaction of N_2O_5 on titanium tetrachloride at −60 to −20°C. This produces an intermediate compound that decomposes on warming in the presence of excess nitrogen dioxide to form NO_2Cl and titanium nitrate.

Titanium Oxides

Metallic α-titanium can dissolve oxygen up to a composition of $TiO_{0.42}$, retaining the hexagonal structure but showing an increase in the lattice parameters. One consequence of the absorption of oxygen is that the transition temperature from α-titanium to the high temperature β-phase increases steeply from ca 900°C at 0 atom % to ca 1750°C at ca 15 atom %. As the oxygen content is increased above $TiO_{0.42}$, titanium oxides of increasing oxygen content are formed.

Lower Oxides of Titanium. The properties of lower oxides of titanium are summarized in Table 1.

Titanium monoxide, TiO, has a rock-salt structure but can exist with both oxygen and titanium vacancies. Titanium monoxide may be made by heating a stoichiometric mixture of titanium metal and titanium dioxide powders at 1600°C.

Ti_2O_3 has the corundum structure. At room temperature it behaves as a semiconductor having a small (0.2 eV) band gap. At higher temperatures, however, it becomes metallic. Titanium sesquioxide, Ti_2O_3, may be made by heating a stoichiometric mixture of titanium metal and titanium dioxide powders at 1600°C under vacuum in an aluminum or molybdenum capsule.

Trititanium pentoxide, Ti_3O_5, may be made by the reduction of titanium dioxide by hydrogen at 1300°C.

Hydrated Titanium Oxides. Hydroxides of Ti(II) (black) and Ti(III) (brown) are precipitated when an alkali metal hydroxide is added to a solution of the corresponding salt. These precipitates are powerful reducing agents and readily oxidize in air to form a hydrated titanium dioxide.

Hydrolysis of solutions of Ti(IV) salts leads to precipitation of a hydrated titanium dioxide. The composition and properties of this product depend critically on the precipitation conditions, including the reactant concentration, temperature, pH, and choice of the salt.

Titanium Dioxide. Titanium dioxide occurs in nature in three crystalline forms: anatase, brookite, and rutile. Anatase and rutile are produced commercially, whereas brookite has been produced by heating amorphous titanium dioxide, which is prepared from an alkyl titanate or sodium titanate with sodium or potassium hydroxide in an autoclave at 200–600°C for several days. Only rutile has been synthesized from melts in the form of large single crystals. It is accepted that, at normal pressures, rutile is the thermodynamically stable form of titanium dioxide at all temperatures. Representative physical properties are collected in Table 2.

The reactivity of titanium dioxide toward acid is dependent on the temperature to which it has been heated. Titanium dioxide that has been calcined at 900°C is almost insoluble in aqueous alkalies but dissolves in molten sodium or potassium hydroxide, carbonates, or borates.

Normally, the first stage in the preparation of pure titanium dioxide is repeated distillation of titanium tetrachloride. A number of different routes may then be followed, the choice depending on the use for which the titanium dioxide is required. Hydrolysis in aqueous solution precipitates hydrated titanium dioxide which, after washing and drying, can be calcined at 800°C to remove water and residual Cl. This method has been the basis of producing titanium dioxide of 99.999% purity.

A high purity titanium dioxide of poorly defined crystal form (ca 80% anatase, 20% rutile) is made commercially by flame hydrolysis of titanium tetrachloride.

Sales of titanium oxide for various nonpigmentary applications are of the order of 100,000 t/yr. The traditional ceramic applications of titanium dioxide are in vitreous enamels, thread guides for the fibers industry, and electroceramics.

Titanium dioxide is also used as an inorganic sunblock. Titanium dioxide finds limited, specialist application as a catalyst or catalyst support. Mineral rutile is used as an ingredient of welding-rod coatings, for which impurities such as iron are acceptable.

The most important commercial use of titanium dioxide is as a white pigment in a wide range of products, including paint (qv), plas-

Table 2. Physical Properties of Anatase and Rutile

Property	Anatase	Rutile
refractive index, 550 nm	2.54[a]	2.75[a]
dielectric constant		
static	48[a]	114[a]
parallel to *c* axis		170
perpendicular to *c* axis		86
high frequency		7.37[a]
parallel to *c* axis		8.43
perpendicular to *c* axis		6.84
band gap, eV	3.25	3.05
melting point, °C	converts to rutile	1830–1850
electrical conductivity, S/cm		
parallel to *c* axis		
30°C		10^{-13}
227°C		10^{-6}
perpendicular to *c* axis		
30°C		10^{-10}
227°C		10^{-7}
breakdown voltage, mV/m		15.2/17.8
hardness, Mohs' scale	5.5–6	7–7.5

[a] Weighted mean values.

tics, paper (qv), and inks (qv). TiO_2 is used not just in whites but to opacify colored systems also. Titanium dioxide is the predominant white pigment both because of its high refractive index and because technology has been developed to allow the right size range and the necessary chemical purity.

Two pigment production routes are in commercial use. In the sulfate process, the ore is dissolved in sulfuric acid, the solution is hydrolyzed to precipitate a microcrystalline titanium dioxide, which in turn is grown by a process of calcination at temperatures of ca 900–1000°C. In the chloride process, titanium tetrachloride, formed by chlorinating the ore, is purified by distillation and is then oxidized at ca 1400–1600°C to form crystals of the required size. In both cases, the raw products are finished by coating with a layer of hydrous oxides, typically a mixture of silica, alumina, etc.

The titanium-bearing ores ilmenite, natural rutile, and leucoxene used in the production of titanium dioxide pigment occur as mineral sands and massive hard rock in many parts of the world. The choice of ore depends on the production process used. Ilmenites can be attacked by sulfuric acid, the first step in the sulfate process.

There are two routes for the manufacture of raw pigmentary titanium dioxide. Both processes provide high quality products having good color (chemical purity) and opacity (mean particle size and narrow size distribution). Almost all pigments are then finished by coating with inorganic oxides and phosphates in order to control such properties as dispersion, dispersion stability, gloss, and durability.

Since 1970, there has been virtually no change in the global sulfate-route pigment production capacity. One reason for this is that wastes from the TiO_2 industry have received special and unfavorable attention from environmental and governmental agencies. The main manufacturers of TiO_2 by the sulfate route have, in the late 1980s and early 1990s, made significant and successful efforts to introduce a variety of processes to meet the resulting environmental requirements. As a result of these developments, there is no longer an inherent difference between the environmental acceptability of the chloride and the sulfate processes.

The first stage in the process, carbothermal chlorination of the ore to produce titanium tetrachloride, is carried out in a fluid-bed chlorinator at ca 950°C. If mineral rutile is used as the feedstock, the dominant reaction is chlorination of titanium dioxide.

$$2\ TiO_2 + 4\ Cl_2 + 3\ C^0 \rightarrow 2\ TiCl_4 + 2\ CO + CO_2$$

The exothermic oxidation reaction is carried out in the gas phase at temperatures of 1200°C or higher.

$$TiCl_4(g) + O_2(g) \rightarrow TiO_2(s) + 2\ Cl_2(g)$$

Aluminum chloride is added to ensure that the product is rutile.

Since 1970, the global production capacity by the chloride route has increased approximately eightfold, largely because of the lower capital costs associated with this process. Chlorination of low Ti ores leads to iron chloride wastes that must be disposed of. In the United States, deep wells are often used but this may not always be allowed.

Rutile pigments produced by the chloride and sulfate routes are basically similar and require coating for the same reasons, ie, to optimize dispersibility, dispersion stability, opacity, gloss, and durability. The coating techniques are essentially common to sulfate and chloride-route base pigments. Because the treatments are tailored to the requirements of the final application, the details are specific to the needs of the different market sectors, eg, paints, plastics, and paper.

Colored pigments may be derived from titanium dioxide by substituting some of the titanium in the rutile lattice by small amounts of transition-metal ions. The most important are a yellow pigment formed by introducing nickel ions, and a buff pigment formed by introducing chromium.

Peroxidic Compounds. When hydrogen peroxide is added to a solution of titanium(IV) compounds, an intense, stable, yellow solution is obtained, which forms the basis of a sensitive method for determining small amounts of titanium.

Inorganic Titanates. Titanium forms a series of mixed oxide compounds with other metals. Only in one of these, Ba_2TiO_4, are there discrete $[TiO_4]^{2-}$ ions.

Alkali metatitanates may be prepared by fusion of titanium oxide with the appropriate alkali metal carbonate or hydroxide. The alkali metal titanates tend to be more reactive and less stable than the other titanates, eg, they dissolve relatively easily in dilute acids. The polytitanates, $K_2Ti_4O_9$ and $K_2Ti_6O_{13}$, are of considerable interest because they can be manufactured in fibrous form. They are chemically stable and melt at 1370°C. Production methods include hydrothermal synthesis in which titanium dioxide reacts with aqueous potassium hydroxide at high (ca 20 MPa (200 atm) and 600–700°C) pressure and temperature. Preparation from alkali halide melts and, by slow cooling after calcination, has also been described. Potassium titanate has a high refractive index and a low thermal conductivity. Moreover, its size is in the right range to scatter infrared radiation. Thus it has potential use as an insulating and ir-reflective material. Other potential applications of potassium titanate include its use as a filtration medium, a reinforcement material for organic polymers, and an asbestos replacement in friction brakes.

The most important applications of these titanates are in the manufacture of electronic components. The most important member of the class is barium titanate, $BaTiO_3$, which owes its significance to its exceptionally high dielectric constant and its piezoelectric and ferroelectric properties.

Other alkaline-earth titanates may be synthesized by heating together stoichiometric amounts of the oxide or by decomposing double salts such as strontium titanium oxalate. The primary use of magnesium, calcium, and strontium titanates is as additives to modify the properties of electroceramic components.

Other Titanates. Aluminum titanate, Al_2TiO_5, a white solid, density 3700 kg/m^3, mp ca 1860°C, has an orthorhombic, pseudobrookite, structure. Al_2TiO_5 may be made by calcination of the stoichiometric amount of oxides at temperatures above 1280°C. The uses of aluminum titanate are as a low thermal expansion ceramic which has a good thermal shock resistance. The material finds applications in thermally insulating exhaust port liners in engine cylinder heads and in burner nozzles and thermocouple sleeves.

Ferrous metatitanate, $FeTiO_3$, mp ca 1470°C, density 472(0), an opaque black solid having a metallic luster, occurs in nature as the mineral ilmenite. This ore is used extensively as a feedstock for the manufacture of titanium dioxide pigments. Artificial ilmenite may be made by heating a mixture of ferrous oxide and titanium oxide for several hours at 1200°C or by reducing a titanium dioxide/ferric oxide mixture at 450°C.

Ferrous orthotitanate, Fe_2TiO_4, is orthorhombic and opaque. It has been prepared by heating a mixture of ferrous oxide and titanium dioxide. Ferrous dititanate, $FeTi_2O_5$, is orthorhombic and has been prepared by reducing ilmenite with carbon at 1000°C.

Lead titanate, $PbTiO_3$, is a yellow solid having a density of 73(00). It can be made by heating the calculated amounts of the two oxides together at 400°C. Lead titanate is a ferroelectric and the static dielectric constant shows a strong maximum, at ca 500°C, associated with the phase change from ferroelectric to paraelectric behavior. Lead titanate zirconate (PZT) is widely used as a piezoelectric ceramic.

Nickel titanate, $NiTiO_3$, is a canary-yellow solid having a density of 73(00). When a mixture of antimony oxide, nickel carbonate, and titanium dioxide is heated at 980°C, nickel antimony titanate forms, which is used as a yellow pigment.

Zinc orthotitanate, Zn_2TiO_4, a white solid having a density of 512(0) and a spinel structure, is obtained by heating the calculated amounts of the two oxides at 1000°C.

Titanium Halides

The most important halides and oxyhalides are shown in Table 3.

Titanium Fluorides. Unlike other titanium dihalides, titanium difluoride is known only from mass spectra of gases.

Table 3. Titanium Halides

Titanium oxidation state	Fluoride	Chloride	Bromide	Iodide
Ti(II)	TiF_2	$TiCl_2$	$TiBr_2$	TiI_2
Ti(III)	TiF_3, TiOF	$TiCl_3$, TiOCl	$TiBr_3$	TiI_3
Ti(IV)	TiF_4	$TiCl_4$, $TiOCl_2$	$TiBr_4$, $TiOBr_2$	TiI_4, $TiOI_2$

The trifluoride is a blue crystalline solid, density 2980 kg/m^3. Titanium trifluoride is stable in air at room temperature but decomposes to titanium dioxide when heated to 100°C. It is insoluble in water, dilute acid, and alkalies but decomposes in hot concentrated acids. The compound sublimes under vacuum at ca 900°C but disproportionates to titanium and titanium tetrafluoride at higher temperatures.

Titanium trifluoride may be prepared in 90% yield by the reaction of gaseous hydrogen fluoride, with either titanium metal or titanium hydride at 900°C.

Titanium tetrafluoride is a white hygroscopic solid, density 2798 kg/m^3, that sublimes at 284°C. The preferred method of preparation is by direct fluorination of titanium sponge at 200°C in a flow system. The reaction of titanium tetrachloride with cooled, anhydrous, liquid hydrogen fluoride may be used if pure hydrogen fluoride is available.

$$TiCl_4 + 4\ HF \rightarrow TiF_4 + 4\ HCl$$

Anhydrous potassium fluorotitanate, K_2TiF_6, may be prepared by dissolving titanium dioxide in dilute hydrofluoric acid to form a clear solution of H_2TiF_6. The principal use of potassium fluorotitanate is as a grain-refining agent for aluminum and its alloys, ie, to promote the production of a small grain size as the molten metal cools and solidifies. There are minor uses in the preparation of dental fillings and in abrasive grinding wheels.

Titanium Chlorides. Titanium dichloride is a black crystalline solid (mp > 1035 at 10°C, bp > 1500 at 40°C, density 31(40) kg/m^3). $TiCl_2$ reacts vigorously with water to form a solution of titanium trichloride and liberate hydrogen. The dichloride is difficult to obtain pure because it slowly disproportionates.

Titanium trichloride exists in four different solid polymorphs that have been much studied because of the importance of $TiCl_3$ as a catalyst for the stereospecific polymerization of olefins. The α-, γ-, and δ-forms are all violet and have close-packed layers of chlorines. Titanium trichloride is almost always prepared by the reduction of $TiCl_4$, most commonly by hydrogen. The primary use of $TiCl_3$ is as a catalyst for the polymerization of hydrocarbons. The Ziegler-Natta catalysts used to produce stereoregular polymers of several olefins and dienes, eg, polypropylene, are based on α-$TiCl_3$ and $Al(C_2H_5)_3$.

Titanium trichloride hexahydrate can be prepared by dissolving anhydrous titanium trichloride in water or by reducing a solution of titanium tetrachloride. The hydrated salt has had some commercial application as a stripping or bleaching agent in the dyeing industry, particularly where chlorine must be avoided.

Titanium tetrachloride is a clear, colorless liquid, normally made by the chlorination of titanium dioxide at ca 1000°C in the presence of a reducing agent. Its main uses are in the manufacture of titanium dioxide pigments and titanium metal, as a starting material in the manufacture of a wide range of commercially important titanium organic compounds, which are mostly alkoxides rather than true organometallic compounds, and as a starting material in the production of Ziegler-Natta catalysts.

Physical properties of titanium tetrachloride are given in Table 4.

Titanium tetrachloride's affinity for water is so high that it acts as a desiccating agent. It is readily hydrolyzed by water and fumes strongly when in contact with moist air. The dense white fumes, consisting of finely divided oxychlorides, are the basis of the use of titanium tetrachloride as white smoke. Hence $TiCl_4$ is used by the military for smoke screen purposes (see CHEMICALS IN WAR).

Titanium tetrachloride is also miscible with other common liquids, including organic solvents such as hydrocarbons, carbon tetrachloride,

Table 4. Physical Properties of Titanium Tetrachloride

Property	Value
color	none
density, 20°C, g/cm^3	1.7
freezing point, °C	−24.1
heat of fusion, kJ/mol[a]	9.966
boiling point, °C	136.5
vapor pressure, kPa[b], 20°C	1.33
heat of vaporization, kJ/mol[a], 25°C	41.087
specific heat, 20°C, J/(k·mol)[a]	145.21
critical temperature, °C	370
heat of liquid formation, 25°C, kJ/mol[a]	−804.2 ± 4.2
viscosity, mPa·s(= cP)	0.079
refractive index, n	1.6985
magnetic susceptibility	-0.287×10^{-6}
dielectric constant, 20°C	2.79

[a] To convert kJ to kcal, divide by 4.184

[b] To convert kPa to psi, multiply by 0.145.

and chlorinated hydrocarbons. With those containing hydroxyl, carboxyl, or diketone (in the enol form) groups, reaction occurs. Substitution products are formed with the elimination of hydrogen chloride. Thus, with alcohols, alkoxides, also called esters of titanic acid, are formed (see ALKOXIDES, METAL).

Addition compounds form with those organics that contain a donor atom, eg, ketonic oxygen, nitrogen, and sulfur. Thus, adducts form with amides, amines, and N-heterocycles, as well as acid chlorides and ethers. Addition compounds also form with a number of inorganic compounds, eg, $POCl_3$. Titanium tetrachloride is readily reduced.

Titanium chloride is manufactured by the chlorination of titanium compounds. The feedstocks usually used are mineral or synthetic rutile, beneficiated ilmenite, and leucoxenes. The reaction is normally carried out as a continuous process in a fluid-bed reactor.

H_2TiCl_6 may be made by dissolving anhydrous hydrogen chlorine in titanium tetrachloride (C13). Ammonium hexachlorotitanate, potassium hexachlorotitanate, rubidium hexachlorotitanate, and cesium hexachlorotitanate are light-green to yellow crystalline solids. They may be prepared either by direct interaction of the alkali-metal chloride with titanium tetrachloride or by reaction in fuming HCl. Both the acids and its salts are more susceptible to hydrolysis than the corresponding fluorotitanates. They are also thermally unstable and decompose to the alkali metal chloride and titanium tetrachloride.

Hydrolysis of $TiCl_4$ yields a number of products, the composition of which depends on the hydrolysis conditions. Titanium oxide dichloride, $TiOCl_2$, is a yellow hygroscopic solid that may be prepared by bubbling ozone or chlorine monoxide through titanium tetrachloride. It is insoluble in nonpolar solvents but forms a large number of adducts with oxygen donors, eg, ether. It decomposes to titanium tetrachloride and titanium dioxide at temperatures of ca 180°C.

Titanium Bromides. Titanium dibromide, a black crystalline solid, density 4310 kg/m^3, mp 1025°C, has a cadmium iodide-type structure and is readily oxidized to trivalent titanium by water. Spontaneously flammable in air, it can be prepared by direct synthesis from the elements, by reaction of the tetrabromide with titanium, or by thermal decomposition of titanium tribromide. This last reaction must be carried out either at or below 400°C, because at higher temperatures the dibromide itself disproportionates.

Titanium tribromide crystallizes in two different habits: hexagonal plates or blue-black needles. It can be prepared by the reaction of $TiBr_4$ with either titanium or hydrogen.

Titanium tetrabromide is an amber-yellow, easily hydrolyzed, crystalline solid, having a density of 3250 kg/m^3. The crystal structure depends on temperature. Titanium tetrabromide is soluble in dry chloroform, carbon tetrachloride, ether, and alcohol. Like titanium tetrachloride, $TiBr_4$ forms a range of adducts with molecules such as ammonia, amines, nitrogen heterocycles, esters, and ethers. Analogous with titanium tetrachloride, the tetrabromide may be made by the carbothermal bromination of titanium dioxide at ca 700°C, and also by direct bromination of titanium at 300–600°C in a flow system.

Titanium Iodides. Titanium diiodide is a black solid (ρ = 499(0) kg/m^3) that has the cadmium iodide structure. Titanium diiodide may be prepared by direct combination of the elements, the reaction mixture being heated to 440°C to remove the tri- and tetraiodides. It can also be made by either reaction of solid potassium iodide with titanium tetrachloride or reduction of TiI_4 with silver or mercury.

Titanium triiodide is a violet crystalline solid having a hexagonal unit cell. The crystals oxidize rapidly in air but are stable under vacuum up to 300°C; above that temperature, disproportionation to the diiodide and tetraiodide begins. Titanium triiodide can be made by direct combination of the elements or by reducing the tetraiodide with aluminum at 280°C in a sealed tube. TiI_3 reacts with nitrogen, oxygen, and sulfur donor ligands to give the corresponding adducts.

Titanium tetraiodide forms reddish-brown crystals. TiI_4 melts at 150°C, boils at 377°C, and has a density of 440(0) kg/m^3. It forms adducts with a number of donor molecules and undergoes substitution reactions. It also hydrolyzes in water and is readily soluble in nonpolar organic solvents. Titanium tetraiodide can be prepared by direct combination of the elements at 150–200°C; it can be made by reaction of gaseous hydrogen iodide with a solution of titanium tetrachloride in a suitable solvent.

Titanium Silicon Compounds

Titanium Silicides. The titanium–silicon system includes Ti_3Si, Ti_5Si_3, TiSi, and $TiSi_2$. Direct synthesis by heating the elements *in vacuo* or in a protective atmosphere is possible. Other preparative methods include high temperature electrolysis of molten salt baths containing titanium dioxide and alkalifluorosilicate; reaction of $TiCl_4$, $SiCl_4$, and H_2 at ca 1150°C, using appropriate reactant quantities for both TiSi and $TiSi_2$ (156); and, for Ti_5Si_3, reaction between titanium dioxide and calcium silicide at ca 1200°C, followed by dissolution of excess lime and calcium silicate in acetic acid.

Titanium disilicide is a silvery-gray, crystalline material that oxidizes slowly in air when heated to 700–800°C. It is resistant both to mineral acids (except hydrofluoric) and to aqueous solutions of alkalies, but reacts with fused borax, sodium hydroxide, and potassium hydroxide. It reacts explosively with chlorine at high temperatures.

Titanium silicides are used in the preparation of abrasion- and heat-resistant refractories. Compositions based on mixtures of Ti_5Si_3, TiC, and diamond have been claimed to make wear-resistant cutting-tool tips. Titanium silicide can be used as an electric–resistant material, in electrically conducting ceramics, and in pressure-sensitive elastic resistors, the electric resistance of which varies with pressure.

Titanium Silicates. A number of titanium silicate minerals are known. In most cases, it is convenient to classify these on the basis of the connectivity of the SiO_4 building blocks, eg, isolated tetrahedra, chains, and rings, that are typical of silicates in general.

Titanium Phosphorus Compounds

Titanium Phosphides. The titanium phosphides include Ti_3P, Ti_5P_3, and TiP. Titanium monophosphide, TiP, can be prepared by heating phosphine with titanium tetrachloride or titanium sponge. Alternatively, titanium metal may be heated with phosphorus in a sealed tube. The gray metallic TiP is slightly phosphorus-deficient ($TiP_{0.95}$), has a density of 408(0) kg/m^3, and displays considerable mechanical hardness (700 kg/mm^2). It is oxidized on heating in air but is stable when heated to 1100°C in either vacuum or a protective atmosphere; it is resistant to concentrated acid (except aqua regia) and alkalies; and it is reported to act as a catalyst in polycondensation reactions.

Titanium Phosphates. Titanium(III) phosphate (titanous phosphate) is a purple solid, soluble in dilute acid, giving relatively stable solutions. It can be prepared by adding a soluble phosphate to titanous chloride or sulfate solution and raising the pH until precipitation occurs.

Titanium(IV) phosphate gel may be prepared by adding an alkali phosphate to titanium(IV) sulfate or chloride solution, followed by filtering, leaching, and drying the derived gel. Titanium phosphate prepared in this way has been used in the dyeing and leather (qv) tanning industries. Titanium pyrophosphate, TiP_2O_7, a possible uv reflecting pigment, is a white powder that crystallizes in the cubic system and has a theoretical density of 3106 kg/m^3. It is insoluble in water and can be prepared by heating a stoichiometric mixture of hydrous titania and phosphoric acid at 900°C.

Titanium Sulfur Compounds

Titanium Sulfides. Titanium subsulfide, Ti_2S, forms as a gray solid of density 4600 kg/m^3 when titanium monosulfide, TiS, is heated at 1000°C with titanium in a sealed tube. It can also be formed by heating a mixture of the two elements at 800–1000°C. The sulfide, although soluble in concentrated hydrochloric and sulfuric acids, is insoluble in alkalies.

Titanium disulfide can also be made by pyrolysis of titanium trisulfide at 550°C. Titanium trisulfide, TiS_3, a black crystalline solid having a monoclinic structure and a theoretical density of 3230 kg/m^3, can be prepared by reaction between titanium tetrachloride vapor and H_2S at 480–540°C.

The principal use of titanium sulfides is as a cathode material in high efficiency batteries. In these applications, the titanium disulfide acts as a host material for various alkali or alkaline-earth elements. The titanium sulfide is able to act as a lithium reservoir. Small button cells have been developed, incorporating lithium perchlorate in propylene carbonate electrolyte, for use in watches and pocket calculators (see BATTERIES). Titanium disulfide has been proposed as a solid lubricant.

Titanium Sulfates. Solutions of titanous sulfate are readily made by reduction of titanium(IV) sulfate in sulfuric acid solution by electrolytic or chemical means, eg, by reduction with zinc, zinc amalgam, or chromium(II) chloride. The reaction is the basis of the most used titrimetric procedure for the determination of titanium. Titanous sulfate solutions are violet and, unless protected, can slowly oxidize in contact with the atmosphere.

Titanium(IV) sulfate can be prepared by the reaction of titanium tetrachloride with sulfur trioxide dissolved in sulfuryl chloride.

In addition to its role as an intermediate in the production of titanium dioxide, titanyl sulfate is used for treatment of metals and in the dyeing industry for the preparation of titanous sulfate. Titanyl sulfate may also be used as a tanning agent for the production of leather. In this application, it is often complexed, ie, masked, with, eg, sodium gluconate to prevent unwanted precipitation of hydrous titania.

Health and Safety Aspects

Apart from very few exceptions, the inorganic compounds of titanium are generally regarded as having low toxicity.

Titanium metal is frequently used as a surgical implant, and several titanium compounds have been used in medical, food, food-contact, and cosmetic products. Titanium dioxide, because of its insolubility and low toxicity, has been allocated occupational exposure limits by the ACGIH of 10-mg/m^3 total dust, or 3-mg/m^3 respirable dust averaged over an eight-hour exposure period. Because titanium dioxide has the ability to dry and defat the skin by adsorption, prolonged skin contact should be avoided.

TERRY A. EGERTON
Tioxide Group Services Limited

J. Barksdale, *Titanium, Its Occurrence, Chemistry and Technology,* 2nd ed., Ronald Press Co., New York, 1966.

J. B. Goodenough and A. Hamnett, in O. Madelung, ed., *Landolt-Bornstein Semiconductors,* Group 3, Vol. 17, Springer-Verlag, Berlin, Germany, pp. 133–166, 1984.

T. A. Egerton and A. Tetlow, in R. Thompson, ed., *Industrial Inorganic Chemicals: Production and Uses,* The Royal Society of Chemistry, Cambridge, U.K., 1995, Chapt. 13.

R. J. H. Clark, in A. F. Trotman Dickenson, ed., *Comprehensive Inorganic Chemistry,* Pergamon, London, U.K., 1973, Chap. 32.

ORGANIC

Organic titanium compounds contain a covalent bond between titanium and another atom that is also bonded to a carbon-based group. Titanium tetrachloride, $TiCl_4$, the basic raw material from which organic titanate compounds are made (see TITANIUM COMPOUNDS, INORGANIC), is readily converted to tetraisopropyl titanate, TYZOR TPT, by the Nelles process. This ester can be converted by alkoxy exchange (transesterification) to a wide variety of tetraalkyl titanates, sold commercially worldwide. The tetraalkyl titanates react with other ligands and chelating agents (qv), such as glycols (qv), β-diketones and ketoesters, α-hydroxycarboxylic acids, and alkanolamines (qv), to give complexes having properties significantly different from the starting materials. These complexes are also important items of commerce.

True organometallic compounds having a titanium–carbon bond are prepared from $TiCl_4$ by reaction with main-group organometallics such as organomagnesium, sodium, or lithium reagents. Most simple C–Ti bonds are very unstable.

Titanium alkoxides (titanate esters) are superb catalysts for esterification, transesterification, and cross-linking of ester-containing resins and epoxides (see ALKOXIDES, METALS; CATALYSIS). Water-soluble titanium chelates are widely used in oilfield fracturing fluid applications (see PETROLEUM). Titanium's great affinity for oxygen atoms is reflected in its bonding to oxide surfaces, such as glass (qv) and plastic, to yield a scratch-resistant oxide coating (see COATINGS). Titanates are also used to cross-link silicone resins and as curing agents for wire coatings (see SILICONES). Titanates disperse pigments (qv) and often help bond resins to fillers and reinforcement agents such as fiber glass.

Alkoxides

The standard manufacturing method for tetraalkyl titanates, such as TYZOR TPT, or tetra-*n*-butyl titanate, TYZOR TBT, involves the addition of $TiCl_4$ to an alcohol. In a series of reversible displacement reactions, the alkoxy substitution products and hydrochloric acid form. The reaction can be driven to the tetraalkoxide stage by addition of an amine or ammonia to scavenge the liberated hydrochloric acid. The amine or ammonium hydrochloride that forms can be filtered from the reaction mass and the tetraalkyl titanate purified by distillation. Higher alkoxides can be prepared by alcohol interchange (transesterification) in a solvent, such as benzene or cyclohexane, to form a volatile azeotrope with the displaced alcohol, or by a solvent-free process involving vacuum removal of the more volatile displaced alcohol. The affinity of an alcohol for titanium decreases in the order: primary > secondary > tertiary, and branched > unbranched.

Tetrahexafluoroisopropyl titanate can be prepared by the reaction of $TiCl_4$ and hexafluoroisopropyl alcohol, in a process similar to that used for TYZOR TPT. Alternatively, it can be prepared by the reaction of sodium hexafluoroisopropoxide and $TiCl_4$ in excess hexafluoroisopropyl alcohol.

The reaction of $TiCl_4$ with epoxides, such as ethylene or propylene oxide (qv), gives β-chloroalkyl titanates. One example is $Ti(OCH_2CH_2Cl)_4$. The β-chloroalkoxy titanates can be used to bind refractory powders and in admixture with diethanolamine to impart thixotropy to emulsion paints.

Tetraallyl titanate can be prepared by reaction of TYZOR TPT with allyl alcohol, followed by removal of the by-product isopropyl alcohol. A vinyloxy titanate derivative can be formed by reaction of TYZOR TPT with vinyl alcohol formed by enolization of acetaldehyde

$$4\ CH_3CH{=}O \rightarrow 4\ CH_2{=}CHOH + Ti(OR)_4 \rightarrow Ti(OCH{=}CH_2)_4 + 4\ ROH$$

Properties and Reactions. Organic titanates tend to associate. Although a titanium(IV) atom strives to achieve a coordination number of 6 by sharing electron pairs from nearby ester molecules, this tendency may be opposed by stearic crowding. X-ray diffraction (xrd) experiments performed on tetramethyl and tetraethyl titanate single crystals show that, in the solid state, these are tetramers. In benzene solution, however, cryoscopic measurements suggest that tetraethyl and tetra-*n*-butyl titanate are trimeric; tetraisopropyl titanate is monomeric in nature.

Increased molecular association increases viscosity. Tetra-*t*-butyl titanate and tetraisopropyl titanate are mobile liquids at room temperature; tetra-*n*-butyl titanate and tetra-*n*-propyl titanate, TYZOR NPT, are thick and syrupy. The boiling points of these materials also reflect association.

The rate of hydrolysis of the tetraalkyl titanates is governed by the nature of the alkoxy groups. The lower titanium alkoxides, with the exception of tetramethyl titanate, are rapidly hydrolyzed by moist air or water, giving a series of condensed titanoxanes, $(Ti{-}O{-}Ti{-}O{-})_x$. As the chain length of the alkyl group increases, the rate of hydrolysis decreases. Titanium methoxides, aryloxides, and C-10 and higher alkyl titanates are hydrolyzed much more slowly. For example, Tyzor BTP, is believed to be a linear C-8 oligomer.

Complete hydrolysis to TiO_2 is difficult to achieve without heating the reaction mixture.

Higher aliphatic alcohol and phenolic group-containing polytitanates may be prepared by transesterification of TYZOR BTP.

Titanoxanes can also be prepared by reaction of a tetraalkyl titanate and carboxylic acids.

Titanoxanes can also be prepared by pyrolysis of tetraalkyl titanates at 200–250°C. Higher temperatures, however, can lead to thermal decomposition. Alternatively, titanoxanes can be prepared by reaction of tetraalkyl titanates with carboxylic acid anhydrides. The polymeric acyl titanate esters are viscous liquids or waxes that are soluble in hydrocarbon solvents and can be used as TiO_2-dispersing agents, water-repellent agents for textile fabrics, and rust inhibitors for steel.

Organic-solvent-soluble, higher molecular weight polytitanoxanes, having a proposed rudder-shaped structure, can be prepared by careful addition of an alcohol solution of 1.0–1.7 moles of water per mole of tetraalkyl titanate, followed by distillation of the low boiling alcohol components. Polytitanoxanes having molecular weights up to 20,000 have been prepared by this method.

The tendency of titanium(IV) to reach coordination number six accounts for the rapid exchange of alkoxy groups with alcohols. Departure of an alkoxy group with the proton is the first step in the ultimate exchange of all four alkoxyls. For preparative purposes, the equilibrium must be shifted either by using an excess of the exchanging alcohol or by distilling the more volatile lower alcohol.

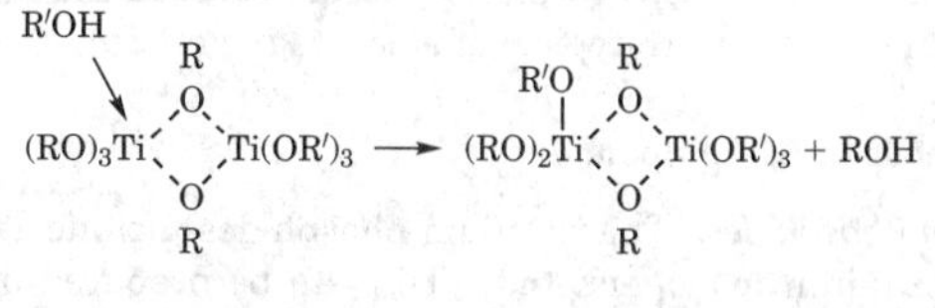

Ester interchange catalyzed by titanates is an important industrial reaction. One commercial example is the formation of basic methacrylates from dialkyl-aminoethanols and methyl methacrylate. The same reaction can be used to convert one alkoxide to another by distillation of a lower boiling ester.

Reaction with Lactones. Hydroxycarboxylic acid ester complexes of titanium are formed by reaction of a tetraalkyl titanate with a lactone, such as β-propiolactone, γ-butyrolactone, or valerolactone.

Organic acids form acylates when heated with tetraalkyl titanates. Best results are obtained using only one or two moles of acid, as attempts to force the reaction with three or four moles of acid can yield polymers.

On heating two moles of an α-amino acid, such as alanine, in the presence of a tetraalkyl titanate and an alcohol, reaction that gives a 2,5-piperazinedione and an oxytitanate occurs.

Tetraalkyl titanates react in benzene with salicylaldehyde in a 1:1 or 1:2 molar ratio to give salicylaldehydotrialkoxy and dialkoxy products, which when heated at reflux seem to undergo a Meewein-Ponndorf reaction to give an aldehyde derived from the alcohol group on the titanate and a reduced titanate complex.

TYZOR TPT catalyzes the trimerization of isocyanates and polyisocyanates to isocyanurates and polyisocyanurates.

Thermolysis. Lower tetraalkyl titanates are reasonably stable and can be distilled quickly at atmospheric pressure. Protracted heating forms condensation polymers plus, usually, alcohol and alkene. Longer or more branched chains are less stable. Thermolysis is used in the coating of glass and other surfaces with a film of titanium dioxide. When a lower alkoxide, eg, TYZOR TPT, vaporizes in a stream of dry air and is blown onto hot glass bottles above ca 500°C, a thin, transparent protective coating of TiO_2 is deposited.

The alkoxides are manufactured by a modification of the classical Nelles process, in which $TiCl_4$ in an inert solvent, eg, heptane, is treated with a monohydric lower alcohol. If the hydrogen chloride by-product is expelled by heating or by sweeping with dry nitrogen, the product is the dialkoxydichlorotitanate. Expulsion of hydrogen chloride is facilitated by adding the required alcohol gradually with continuous sweep. An acid acceptor is required to obtain the tetraalkoxide; ammonia is preferred because of its low cost. Higher alkoxides are readily prepared from TYZOR TPT or TYZOR TBT by alcohol interchange (transesterification), with removal of the by-product isopropyl alcohol or *n*-butanol by distillation.

Chelates are made by mixing the chelating agent with TYZOR TPT or another alkoxide. The liberated alcohol is usually left in the product to maintain the products fluidity. It may, however, be removed by distillation if desirable. Most titanates are moisture-sensitive and must be handled with care, preferably under dry nitrogen.

Alkoxy Halides

Titanium alkoxy halides have the formula $Ti(OR)_nX_{4-n}$, where R may be alkyl, alkenyl, or aryl, and X is F, Cl, or Br, but not I.

Properties. Alkoxytitanium fluorides and chlorides are colorless or pale-yellow solids or viscous liquids that darken on standing, especially in the light. Bromides are yellow crystalline solids. Aryloxytitanium halides are orange-to-red solids. The alkoxy halides are hygroscopic. Most dissolve in water without immediate decomposition, but hydrolyze slowly to hydrogen halide, alcohol, alkyl halide, and hydrous titanium dioxide. Using less than one mole of water, poly(alkoxytitanium) compounds can be prepared if the liberated hydrogen halide is neutralized by ammonia. Primary alkoxy halides are thermally stable, although they disproportionate on heating. Secondary and tertiary alkoxy halides decompose gradually on standing (more rapidly on heating) yielding alkyl halides, polymers, and titanium oxychloride. Physical properties of some alkoxytitanium halides are shown in Table 1.

Synthesis. Titanium alkoxy halides are intermediates in the preparation of alkoxides from a titanium tetrahalide (except the fluoride) and an alcohol or phenol. If $TiCl_4$ is heated with excess primary alcohol, only two chlorine atoms can be replaced and the product is dialkoxydichlorotitanium alcoholate. The yields are poor. Using excess $TiCl_4$ at 0°C, the trichloride $ROTiCl_3$ is obtained nearly quantitatively, even from *sec-* and *tert*-alcohols.

Table 1. Titanium(IV) Alkoxyhalides and Aryloxyhalides

Compound	Formula	Mp, °C	Bp[a], °C	Other properties
propoxytrichloride	$TiOC_3H_7Cl_3$	65–66	$83–85_{147}$	
isopropoxytrichloride	$TiOC_3H_7\text{-}i\text{-}Cl_3$	78–79	65_{13}	
dipropoxydichloride	$Ti(OC_3H_7)_2Cl_2$	53–57	159_{240}	
tripropoxychloride	$Ti(OC_3H_7)_3Cl$		168_{160}	d_4^{25} 1.1348 g/cm^3

[a] Subscripted values are pressure in Pa. To convert Pa to mm Hg, divide by 133.

Tetraalkoxides can be cleaved by hydrogen chloride or bromide in an inert solvent. Dialkoxytitanium dichloride is obtained as an alcoholate.

More useful is cleavage of alkoxides by acetyl chloride or bromide. One, two, three, or four alkoxyls can be replaced by chloride or bromide. The tri- and tetrachlorides, coordinate with the alkyl acetate formed and yield distillable complexes.

A principal use for the alkoxy halides is their reaction with organolithium or -magnesium compounds, R′M, in a Wurtz-type reaction to form compounds having carbon–titanium bonds.

In a given $(RO)_nTiCl_{4-n}$, the alkoxy group can be exchanged by a higher alcohol if the resulting lower alcohol is removed by distillation. In the intermediate, HOR departs much more easily than HCl.

Chelates

Titanium chelates are formed from tetraalkyl titanates or halides and bi- or polydentate ligands. One of the functional groups is usually alcoholic or enolic hydroxyl, which interchanges with an alkoxy group, RO, on titanium to liberate ROH. If the second function is hydroxyl or carboxyl, it may react similarly. Diols and polyols, α-hydroxycarboxylic acids and oxalic acid are all examples of this type. β-Keto esters, β-diketones, and alkanolamines are also excellent chelating ligands for titanium.

Glycol Titanates. Primary diols (HOGOH), such as ethylene glycol and 1,3-propanediol, react by alkoxide interchange at both ends, yielding insoluble, white solids that are polymeric in nature. The 1:1 molar addition products of a primary diol and a tetraalkyl titanate, $Ti(OGO)(OR)_2$ may react with water to give either $Ti(OGO)(OH)_2$ or condensed products $(Ti(OGO)O)_n$, which can be used as esterification catalysts.

Silanediols, eg, $(C_6H_5)_2Si(OH)_2$, yield four- and six-membered rings with titanium alkoxides. Pinacols and 1,2-diols form chelates rather than polymers. The more branched the diol molecule, the more likely are its titanium derivatives to be soluble and even monomeric.

Reaction of 2,4-diorgano-1,3-diols, such as 2-ethylhexane-1,3-diol, with TYZOR TPT in a 2:1 molar ratio gives the solvent soluble titanate complex, TYZOR OGT.

α-Hydroxycarboxylic Acid Complexes. Titanate α-hydroxy acid complexes are prepared in aqueous solutions and are stable in water over a wide pH range. Their structures are uncertain and probably depend on pH and concentration. The alkoxytitanate solution in acetone or THF is added to a solution of the α-hydroxy acid in the same solvent, or an aqueous titanium(IV) chloride, sulfate, or nitrate is treated with an α-hydroxy acid. Oxalic acid behaves as an α-hydroxy acid and yields crystalline ammonium or potassium salts from aqueous titanium(IV) solution or tetraalkoxytitanium compounds. Succinic and adipic acids, however, do not dissolve titanic acid.

β-Diketone Chelates. β-Diketones, reacting as enols, readily form orange-red chelates with titanium alkoxides which are soluble in common solvents. Since they are coordinately saturated (coordination number 6), they are more resistant to hydrolysis than the parent alkoxides (coordination number 4). Similar chelates can be prepared from $TiCl_4$ and β-diketones.

β-Ketoester Chelates. β-Ketoesters react in a fashion similar to the β-diketones. TYZOR DC is the light-yellow liquid from TYZOR TPT and two moles of ethyl acetoacetate (eaa) after removal of the isopropyl alcohol.

Alkanolamine Chelates. Alkanolamine chelates, which are prepared by reaction of tetraalkyl titanates with one or more alkanolamines, are used primarily in cross-linking water-soluble polymers (see ALKANOLAMINES). The products are used in thixotropic paint emulsion paints, in hydraulic fracturing and drilling of oil and gas wells, and in many other fields. The structure of the product indicates that titanium has received electron pairs from both nitrogens to complete the coordination shell. No role is, however, ascribed to the four free hydroxyls.

The reactions of simpler alkanolamines with tetraalkyl titanates are not completely understood.

Acylates. Titanium acylates are prepared either from $TiCl_4$ or tetraalkyl titanates. Because it is difficult to obtain titanium tetraacylates, most compounds reported are either chloro- or alkoxyacylates. Under most conditions, $TiCl_4$ and acetic acid give dichlorotitanium diacetate. The best method involves passing preheated (136–170°C) $TiCl_4$ and acetic acid simultaneously into a heated chamber. The product separates as an HCl-free white powder.

$$TiCl_4 + 2\ HOOCCH_3 \rightarrow Cl_2Ti(OOCCH_3)_2 + 2\ HCl$$

Tetraacylates can be prepared from titanium tetrabromide and excess carboxylic acid in an inert solvent. Tetraacylates have been prepared in this way from stearic, benzoic, cinnamic, and other acids, as well as from diacids such as succinic and adipic acids. In some cases, $TiCl_4$ may also be used. The usual products from alkoxides and acids are dialkoxytitanium diacylates. The third acyl group, but not the fourth, can often be introduced by azeotroping the lower alcohol with benzene.

Titanium(IV) Complexes with Other Ligands

The d^0-titanium(IV) atom is hard, ie, not very polarizable, and can be expected to form its most stable complexes with hard ligands, eg, fluoride, chloride, oxygen, and nitrogen. Soft or relatively polarizable ligands containing second- and third-row elements or multiple bonds should give less stable complexes. The stability depends on the coordination number of titanium, on whether the ligand is mono- or polydentate, and on the mechanism of the reaction used to measure stability.

Peroxide Titanate Complexes. Titanates may influence reactions of organic peroxides (see ORGANIC PEROXIDES). For example, *t*-butyl hydroperoxide epoxidizes olefins:

Titanates trigger peroxide-initiated curing of unsaturated polyesters to give products of superior color. Hydrogen peroxide (qv) produces an intense yellow color with Ti(IV) in aqueous solution and has long been used as a qualitative test for Ti.

Fluorocarbon Group Containing Titanium Complexes. Fluorocarbon groups containing carboxylic acids and alcohols, such as perfluorooctanoic acid or *1H,1H,5H*-octafluoropentanol, react with tetraalkyl titanates to give complexes that are useful either in the treatment of fabrics to render them water-repellent, or as gasoline additives to minimize deposits and improve performance.

Chiral Titanium Complexes. Chiral titanium complexes are useful for the enantioselective addition of nucleophiles to carbonyl groups.

Composite Oxyalkoxides. Composite oxyalkoxides can be prepared by reaction of tetraalkyl titanates and alkaline-earth metal hydroxides. These oxyalkoxides and their derivatives can be hydrolyzed and thermally decomposed to give alkaline-earth metal titanates such as barium titanate. Reaction of a tetraalkyl titanate with potassium hydroxide, gives oxyalkoxide derivatives $(KTi_xO(OR)_y)_n$, which can be further processed to give alkali metal titanate powders, films, and fibers.

$$M(OH)_2 + n\ Ti(OR)_4 \rightarrow (MTiO_2(OR)_2)_n + 2n\ ROH$$

Titanium–Vanadium Mixed Metal Alkoxides. Titanium–vanadium mixed metal alkoxides, $VO(OTi(OR)_3)_2$, are prepared by reaction of titanates, eg, TYZOR TBT, with vanadium acetate in a high boiling hydrocarbon solvent. The by-product butyl acetate is distilled off to yield a product useful as a catalyst for polymerizing olefins, dienes, styrenics, vinyl chloride, acrylate esters, and epoxides.

Titanium Amides. The reaction of lithium amides, $LiNR_2$, with $TiCl_4$ gives tetrakisdialkylaminotitanates, $(R_2N)_4Ti$, which can react with alcohols or other ligands to produce tetraalkyl titanates and chelated derivatives.

Schiff Bases. The nitrogen of a Schiff base unit in a polydentate ligand coordinates readily.

Polymetallocarbosilanes. Polymetallocarbosilanes having a number-average molecular weight of 700–100,000 can be prepared by reaction of polycarbosilane, $\text{-}(Si(R)_2\text{–}CH_2)_x\text{-}$, where R is H, or lower alkyl, with a tetraalkyl titanate, to give a mono-, di-, tri-, or tetrafunctional polymer containing at least one Si–O–Ti bond. By firing polytitanocarbosilanes in a vacuum, an inert gas, or a nonoxidizing atmosphere, they can be converted to a molded article consisting mainly of SiC and TiC and having a higher mechanical strength and better oxidation resistance at higher temperatures than SiC itself.

Titanium Silicates

The synthesis of titanium silicate catalysts, such as TS-1 can be formed by controlled hydrolysis of a mixture of tetraethylorthosilicate (TEOS) and TYZOR TBT was first reported in the mid-1980s. Since that time, several alternative methods of synthesis have been developed. The addition of sodium hydroxide during formation of the TS-1 precursor gel gives a titanium silicate having fewer acidic sites, which modifies its catalytic activity. Titanium silicates are excellent catalysts for the selective oxidation of alkanes to alkenes and for the hydroxylation of benzene and phenols.

Lower Valent Titanates

Titanium(III) alkoxides can be produced by photoreduction of the tetraalkyl titanates in the presence of a base, such as pyridine, and by reduction of tetraalkyl titanates by organosilicon compounds containing Si–H groups. Table 2 is a list of a few of the organotitanium compounds of valency lower than four.

A family of Ti(III) derivatives roughly parallels those of Ti(IV). Titanium(III) chelates are known, eg, titanium trisacetylacetonate prepared in benzene from titanium trichloride, acetylacetone, and ammonia. This deep-blue compound is soluble in benzene but insoluble in water.

A group of violet titanium(III) acylates has been prepared from $TiCl_3$ and alkali carboxylates. All of the acylates are strong reducing agents similar to $TiCl_3$. A broad selection of Ti(III) compounds coordinated to α-hydroxy acids, diboric acids, and 8-hydroxyquinoline has been prepared.

Organometallics

Titanium(IV) Organometallics. In classical organometallic chemistry, Grignard reagents (qv) or organolithium compounds react with halides of less active metals to form new carbon–metal bonds. This type of reaction with titanium halides invariably failed, until it was

Table 2. Some Organotitanium Compounds of Lower Valence

Compound	Type[a]	Appearance	Mp, °C	Other properties
$(C_5H_5)_2TiCl$	Ti(III) Cp_2 halide	green crystals	279–281	
$C_5H_5TiCl_2$	Ti(III) Cp halide	violet	sublimes *in vacuo* at 150	insoluble in hydrocarbons; very sensitive to oxygen; blue solution in acetonitrile
$(C_5H_5)_3Ti$	Ti(III) Cp_3	green	sublimes at 125	extremely air-sensitive; gives $(C_5H_5)_2Ti(CO)_2$ with CO under pressure
$(C_5H_5)_2Ti$	Ti(II) Cp_2	dark green	200	pyrophoric; catalyst for polymerization of olefins and acetylenes

[a] Cp = cyclopentadienyl.

realized that many simple titanium alkyls are extraordinarily unstable thermally as well as to moisture and air. This thermal instability is derived from the presence of unfilled, low lying $3d$ orbitals. In titanium metal, the electron configuration is $3s^2 3p^6 3d^2 4s^2$; in simple tetralkyltitaniums, it is $3s^2 3p^6 3d^6 4s^2$, and may also include hybrid $3d$–$4s$ orbitals. A source of instability is the availability of facile decomposition mechanisms, eg, β-elimination:

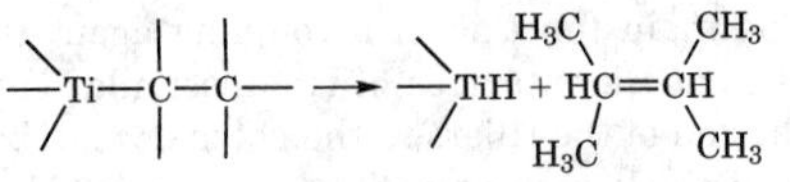

This can be circumvented by choosing alkyl groups with no β-H, eg, methyl, neopentyl, trimethylsilylmethyl, phenyl and other aryl groups, and benzyl. The linear transition state for β-elimination can also be made sterically impossible. The pentahaptocyclopentadienyl ring anion $\eta^5 - C_5H_5^-$ (Cp) has six π-electrons available to share with titanium. Biscyclopentadienyltitanium dichloride (titanocene dichloride), Cp_2TiCl_2, melts at 289°C, can be sublimed at 190°C at 267 Pa (2 mm Hg), and can be recovered almost quantitatively from its solution in boiling dilute hydrochloric acid (pH <1). The Cp ligand and its substitution products, abbreviated as Cp′, can also stabilize the otherwise labile Ti–R compounds. Thus, $Cp_2Ti(C_6H_5)_2$ is unchanged for several days at room temperature.

Covalent Noncyclopentadienyl Compounds. The general synthesis of covalent non-Cp compounds, R_nTiX_{4-n}, where R = alkyl or aryl and X = halogen, alkoxyl, or amido, involves a lithium, sodium, or magnesium organometallic with a titanium–halogen compound in an inert atmosphere. Solvents are usually either ethers, eg, $(C_2H_5)_2O$, THF, or glyme, or hydrocarbons, eg, hexane or benzene. In addition, a low temperature is required to compensate for the low thermal stability.

There are numerous alkyltitaniums, and many of their reactions resemble those of alkyllithiums and alkylmagnesium halides. They are protolyzed by water and alcohols, $R{-}Ti(R')_3 + HA \rightarrow RH + A{-}Ti(R')_3$; they insert oxygen, $R{-}TiR' + O_2 \rightarrow ROTiR'$; and they add to a carbonyl group:

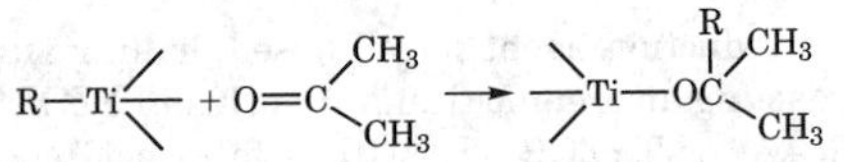

Grignard reagents and lithium alkyls add to ester groups, but the CH_3Ti reagents do not. This selectivity has synthetic value. Titanium alkyls and aryls discriminate between aldehydes and ketones. Titanium alkyls minimize side reactions, eg, elimination, rearrangement, and enolization, which often occur with aluminum alkyls and other active organometallics. The solvent is important.

Titanium alkyls, known as tamed Grignard reagents, do not add to esters, nitriles, epoxides, or nitroalkanes at low temperatures. Rather, they add exclusively in a 1,2 fashion to unsaturated aldehydes.

Carbometalation, an important reaction of RTi(IV) compounds in which RTi adds to a C=C or C≡C multiple bond and results in a net R–H addition, is involved in Ziegler-Natta polymerization as follows:

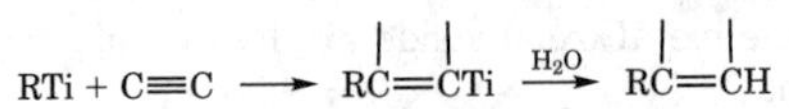

Olefin isomerization is often catalyzed by titanium. An example is the conversion of vinylnorbornene to the comonomer ethylidenenorbornene.

Cyclopentadienyltitanium Compounds. The structure of Cp_2TiCl_2 has been shown by x-ray diffraction to be a distorted tetrahedron. Changes in the structure are imposed by bridging the Cp rings with $-(CH_2)_n-$ or other groups. Titanium tetrachloride is a Lewis acid having many useful synthetic properties. Replacing Cl by OR weakens the acid. Acid–base complexes having dimethyl sulfoxide (DMSO), dimethylformamide (DMF), dimethylacetamide (DMAc), which are all O-bonded, as well as pyridine and ethylenediamine, have been reported. $CpTiCl_3$ forms stable, sublimable complexes having ditertiary amines or arsines. The less-basic O, S, or P analogues do not form stable complexes.

The basic laboratory synthesis involves a salt of cyclopentadiene, eg, lithium, sodium, potassium, or magnesium, and a titanium(IV) halide, usually in an ether solvent. However, this is probably too expensive for industrial use. Different cyclopentadienyl groups can be introduced in two separate steps.

Monocyclopentadienyl compounds can be prepared by the same techniques with appropriate control of the stoichiometric proportion of reagents or by use of reagents such as $ClTi(OR)_3$. Biscyclopentadienyltitanium dichloride can be carefully chlorinolyzed with Cl_2 or with SO_2Cl_2 in refluxing $SOCl_2$. A wide variety of ring-substituted cyclopentadienes have been converted to Cp′–Ti compounds. Methyl and other alkyl and aryl groups and $(CH_3)_3Si-$ are most common.

Titanium-containing polyethers have been prepared by the reaction of dicyclopentadienyltitanium dichloride with aromatic and aliphatic diols via an interfacial and/or aqueous solution polycondensation technique.

Cyclopentadienyltitanium trichloride and, particularly, Cp_2TiCl_2 react with RLi or with RA1 compounds to form one or more R–Ti bonds. Methyl and aryl groups are most commonly used. Higher alkyltitaniums tend to decompose by β-elimination. Both vinylic and ethynylic groups can be attached to the Cp_2Ti framework. These tend to be stable thermally and to air and moisture. Alkyl and aryl groups are cleaved by iodine, but Cp groups are not affected.

Photolysis has been intensively studied. For example, $Cp_2Ti(C_6H_5)_2$ yields a green polymer $(Cp_2TiH)_x$. At low temperature in benzene or THF, a dark-green transient Cp_2Ti–solvent species forms and quickly dimerizes. The phenyl groups appear as benzene and biphenyl.

Photolysis of Cp_2TiAr_2 in benzene solution yields titanocene and a variety of aryl products derived both intra- and intermolecularly. Dimethyltitanocene photolyzed in hydrocarbons yields methane. Pyrolysis of solid $Cp_2Ti(CD_3)_2$ yields CD_3H but not CD_4. Pyrolysis of $(C_5D_5)_2Ti(CH_3)_2$ yields CH_3D. These results show that the radical attacks the Cp rings.

Cyclopentadienyltitanium halides undergo displacements with a wide variety of nucleophiles. Hydroxylic reagents cleave Ti–R bonds. Amides are formed with amines, often with strong base assistance. Occasionally Cp groups are lost.

In titanium acylates, the carboxylate ligands are unidentate, not bidentate, as shown by ir studies. The ligands are generally prepared from the halide and silver acylate.

Sulfur dioxide yields sulfones and ultimately sulfinates. The latter are available also from RSO_2Na, where R is CH_3, C_2H_5, C_4H_9, or C_6H_5, and X is F or Cl. Such titanium sulfinates are reported to increase crop yields. Isocyanides insert to yield imines as follows:

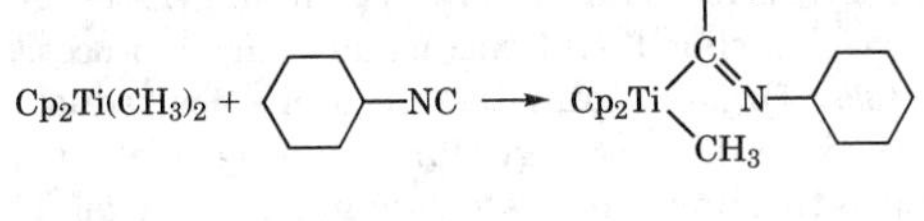

Health and Safety Factors

Commercial titanates should be handled according to good industrial practice. The tetraalkyl titanates have a low acute oral toxicity, LD_{50}, of 7,500–11,000 mg/kg in rats. Because of their rapid hydrolysis, these titanates can cause severe eye damage. The chelates containing isopropyl alcohol have flash points of 12–27°C.

Uses

Organic titanates perform three important functions for a variety of industrial applications. These are (*1*) catalysis, especially polyesterification and olefin polymerization; (*2*) polymer cross-linking to enhance performance properties; and (*3*) surface modification for adhesion, lubricity, or pigment dispersion.

Glass-Surface Coating. A thin (<100 nm) film of $(TiO_2)_x$ is virtually transparent and imparts considerable scratch resistance to glass and consequently greatly reduce its fragility. The lower alkoxides, particularly TYZOR TPT, are preferred for glass treatment. They can be applied undiluted, ie, in a hot process, or in a solvent.

The $(TiO_2)_x$ films are also applied to glass or vitreous enamel for decorative purposes.

Nonemulsion Paints. Pyrolysis of oligomeric titanates, obtained by controlled hydrolysis of TYZOR TBT, furnishes adherent films of nearly inorganic $(TiO_2)_x$. These oligomers suspend pigments. Paints were formulated from these treated pigments, oligomers, ethylcellulose (to prevent pigment settling), and mineral-spirit solvent. This application may be used in heat-resistant paints.

Adhesives. Tetrafunctional titanates react with hydroxyl, ester, amide, imide, and other functions, and with oxide groups on metals and nonmetal oxides. Titanates bond such materials together. Titanates bond to polyethylene and fluorinated polymers. Packaging films, such as Mylar polyester or aluminized films, are coated with a titanate.

Water Repellents. Titanate–wax compositions have been used for the reproofing of textiles that have been dry-cleaned. Leather waterproofed with silicones possesses improved properties when the resin is bonded with TYZOR TBT (see WATERPROOFING AND WATER/OIL REPELLENCY).

Catalysts. Titanates accelerate many organic reactions and frequently provide significant advantages in product purity and yield over conventional catalysts (see CATALYSIS). Their polyfunctionality permits assembling oxygen-containing reactants at one location in a geometrical, usually octahedral, arrangement, which permits facile shuffling of groups to yield products.

Olefin Polymerization. Titanates having a carbon–titanium bond are extensively involved in Ziegler-Natta and metallocene polymerization of olefins.

Esterification. Esterification of an acid and an alcohol can be catalyzed by small quantities of tetraalkyl titanates (see ESTERIFICATION). Although the water that forms can hydrolyze and inactivate the titanate, titanoxane oligomers are cleaved by carboxylic acids to bicoordinated monomeric acylates.

Thixotropic Paints. Water-based latex emulsion paints may be made thixotropic or nondrip by the addition of alkanolamine-based titanium chelates. Thixotropic paints are very viscous, yet thin out enough when applied to a surface. They remain thin long enough to allow leveling to occur. Such easy applicability is the result of shear forces.

Incorporation of an α-hydroxycarboxylic acid in the formulation slows the rate of metal gelation of the paint, making it possible to fill more containers during the packing operation before the paint becomes too viscous to flow, and improves the ability to blend the chelated titanate into the paint without localized areas of gelation. These paints are also less likely to lose potential gel strength resulting from excessive shearing during the manufacturing process.

Printing Inks. Organic titanates are useful for reducing the drying time of flexographic and gravure printing inks. In the case of uv-curable printing inks, tetraalkyl titanates such as TYZOR TPT or TYZOR TBT are believed to catalyze the polymerization of vinyl monomers, such as styrene, with unsaturated polyesters comprising the ink vehicle to form a hard surface in a matter of seconds (see INKS).

The alkanolamine titanates, such as TYZOR TE, when mixed with a coloring agent used to print fibrous materials such as cotton, wool, or silk, promote adhesion of the dye molecule to the fiber, thus minimizing bleeding of the printed design. The titanium triethanolamine chelates, such as TYZOR TE, are excellent adhesion promoters for use in water-based laminating inks typically used to bond two similar or dissimilar plastic films together. The use of α-hydroxycarboxylic acid titanates, such as TYZOR LA, in nonaqueous or aqueous ink-jet printing ink minimizes premature destruction of bubble-jet printer nozzles by forming a protective coating on the printing head.

Polysiloxane Resin Coatings. Various organopolysiloxane waterproofing compositions have been proposed, which use organic titanates as catalysts, curing agents, and adhesion-promoting materials.

Room-Temperature Vulcanizable Silicone Rubber. Organic titanates are incorporated into room-temperature vulcanizable (RTV) silicone rubber formulations to provide a one-component system that is stable in the absence of moisture, but that also cures spontaneously at room temperature on exposure to moisture.

Coupling Agents for Polymer Composites. Organic titanate esters and their chelates or functionalized derivatives are finding increasing utility as coupling agents for reinforcing fillers in polymeric resin composites. The main function of a coupling agent is to serve as a molecular bridge at the interface of two dissimilar surfaces, thereby promoting adhesion of the fillers to the polymer matrix.

Other Uses. Polyols such as natural polysaccharides, eg, cellulose, starch, guar gum and their derivatives, and poly(vinyl alcohol) and its derivatives can be cross-linked by organic titanates. Aqueous solutions of water-dispersible, nonionic, natural hydroxylated polymers, such as galactomannans and their derivatives, can be cross-linked with organic titanates, such as TYZOR TE, TYZOR AA, and TYZOR LA to give water-bearing gels, which can be used to form gelled explosives.

Cellulose or starch xanthate cross-linked by titanates can adsorb uranium from seawater. Carboxymethylcellulose cross-linked with TYZOR ISTT is the bonding agent for clay, talc, wax, and pigments to make colored pencil leads of unusual strength.

Various materials are added to paper to improve its wet strength and ink acceptance, to make possible clay coating, etc (see PAPERMAKING ADDITIVES). Titanic acid precipitation of paper fibers can be controlled.

Poly(vinyl alcohol) (PVA) is used extensively as a paper size, alone or in combination with dyes or pigments; it is rendered insoluble by cross-linking with titanates. A rayon-based paper has been made resistant to boiling water.

Hydraulic fracturing can be used to stimulate production of oil and gas from subterranean formations. A typical fluid consists of an aqueous liquid containing a polysaccharide gelling agent; a borate, an organic titanate, or a zirconate cross-linking agent; pH buffers; thermal stabilizers (sodium thiosulfate); and, if necessary, an enzyme or peroxygen gel breaker. Water-compatible chelated titanates, such as TYZOR TE, TYZOR LA, and TYZOR AA, are excellent cross-linking agents.

A fluid-loss reduction agent for oil-based drilling fluids has been prepared by reacting an organic titanate, such as TYZOR TPT, TYZOR TBT, TYZOR AA, or TYZOR TE, with a fatty acid such as oleic or stearic acid and a metal oxide. The reaction product of an organic titanate, such as TYZOR TOT, TYZOR AA, or TYZOR TE, with an anionic emulsifying agent, such as calcium dodecylbenzene sulfonate, can be used to stabilize oil-based drilling fluids against inorganic salt contamination.

A hydrocarbon solution of TYZOR TPT, TYZOR TBT, or TYZOR TOT can be pumped into the porous zones of an oil-bearing formation; upon contact with water, an amorphous, gelatinous TiO_2 plug is formed, which allows water to be diverted to less porous zones.

The sol–gel process involves conversion of a metal alkoxide or mixture of metal alkoxides, dissolved in an organic solvent (generally the parent alcohol) into a hydroxooxyalkoxide sol, followed by gelation and sintering to give the desired ceramic material. The steps involved in formation of the metal oxoalkoxide are hydrolysis, dehydration, and dealkoxylation.

Spherical, fine-particle titanium dioxide that has no agglomeration and of monodispersion can be manufactured by carrying out a gas-phase reaction between a tetraalkyl titanate vapor and methanol vapor in a carrier gas to form an initial fine particle, which can then be hydrolyzed with water or steam.

Hollow fibers of titanium dioxide can be manufactured by preparing a solution of a tetraalkyl titanate, an acid such as HCl, and an alcohol such as isopropyl alcohol, followed by spinning and drying the resultant fiber.

DONALD E. PUTZIG
THOMAS W. DEL PESCO
E. I. du Pont de Nemours & Company, Inc.

J. H. Clark, *The Chemistry of Titanium and Vanadium,* Elsevier, Amsterdam, the Netherlands, 1968.

R. J. H. Clark, in J. C. Bailar and co-workers, eds., *Comprehensive Inorganic Chemistry,* Vol. 3, Pergamon Press, London, 1973, pp. 355–417.

P. C. Wailes, R. S. P. Coutts, and H. Weigold, *Organometallic Chemistry of Titanium, Zirconium, and Hafnium,* Academic Press, Inc., New York, 1974.

R. J. H. Clark, D. C. Bradley, and P. Thornton, *Chemistry of Titanium, Zirconium, and Hafnium,* Pergamon Press, New York, 1975.

D. C. Bradley, in F. G. A. Stone and W. A. Graham, eds., *Inorganic Polymers,* Academic Press, Inc., New York, 1962, pp. 410–446.

TOCOPHEROLS. See VITAMINS, VITAMIN E.

TOLUENE

Toluene, C_7H_8, is a colorless, mobile liquid with a distinctive aromatic odor somewhat milder than that of benzene. Prior to World War I, the main source of toluene was coke ovens. Petroleum became the source for toluene with the advent of catalytic reforming and the need for large quantities of toluene for use in aviation fuel during World War II. Since then, manufacture of toluene from petroleum sources has continued to increase, and manufacture from coke ovens and coal-tar products has continued to decrease.

Toluene is generally produced along with benzene, xylenes, and C_9-aromatics by the catalytic reforming of C_6–C_9 naphthas. There have been, ca 1997, recent technological developments to produce benzene, toluene, and xylenes from pyrolysis of light hydrocarbons C_2–C_5, LPG, and naphthas (see XYLENES AND ETHYLBENZENE). About 85–90% of the toluene produced annually in the U.S. is not isolated, but is blended directly into the gasoline pool as a component of reformate and of pyrolysis gasoline.

Derivatives are formed by substitution of the hydrogen atoms of the methyl group, by substitution of the hydrogen atoms of the ring, and by addition to the double bonds. Substitutions on the methyl group are generally high-temperature, free-radical reactions. Thus, chlorination at ca 100°C, or in the presence of uv or other free-radical initiators, successively gives benzyl chloride, benzal chloride, and benzotrichloride (see BENZALDEHYDE; BENZOIC ACID). With oxygen in the liquid phase, particularly in the presence of a catalyst, good yields of benzoic acid are obtained. In the presence of alkali metals, toluene is alkylated. With a lithium catalyst and a chelating compound, telomers are obtained with ethylene.

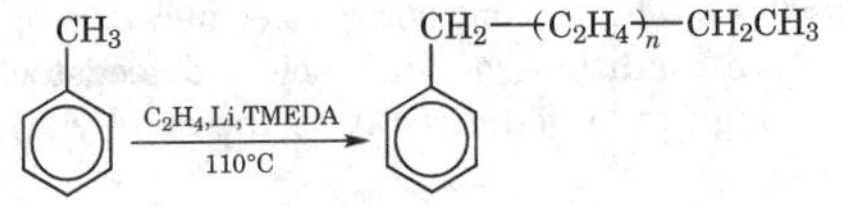

Additions to the double bonds results from both free-radical and catalytic reactions, eg, chlorination at 0°C and hydrogenation. Usually, all three double bonds react. Substitution of the ring hydrogen atoms by electrophilic attack takes place with the same reagents that react with benzene. Some of the common groups with which toluene can be substituted directly include

$$—Cl,\ —Br,\ —\overset{O}{\overset{\|}{C}}CH_3,\ —SO_3H,\ —NO_2,\ —(CnH_{2n+1}),\ \text{and}\ —CH_2Cl.$$

Under the same conditions, toluene reacts more rapidly than benzene. These reactivities and the related selectivity to the ortho and para positions can be explained in terms of the inductive effect of the methyl group. Toluene requires substitution by strongly negative groups, such as NO_2, to react with anions.

Substitution Reactions on the Methyl Group. The reactions that give substitution on the methyl group are generally high temperature and free-radical reactions. Thus, chlorination at ca 100°C, or in the presence of ultraviolet light and other free-radical initiators, successively gives benzyl chloride, benzal chloride, and benzotrichloride.

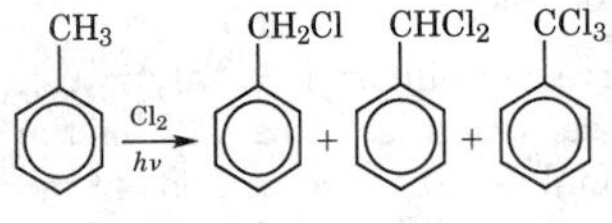

This oxidation reaction which yields benzoic acid is another example of this type of reaction.

In the presence of alkali metals such as potassium and sodium, toluene is alkylated with ethylene on the methyl group to yield, successively, normal propylbenzene, 3-phenylpentane, and 3-ethyl-3-phenylpentane.

In the formation of π-complexes with electrophiles such as silver ion, hydrogen chloride, and tetracyanoethylene, toluene differs from either benzene or the xylenes by a factor of less than two in relative basicity.

Properties

Physical and thermodynamic properties are given in Table 1. Toluene forms azeotropes with many hydrocarbons and most alcohols that boil in a similar range; all are minimum-boiling azeotropes. Toluene, water, and alcohols frequently form ternary azeotropes.

Because of the high electron density in the aromatic ring, toluene behaves as a base both in the formation of charge-transfer π-complexes and in the formation of sigma complexes. When only π-electrons are involved, toluene behaves much like benzene and

Table 1. Physical Properties of Toluene

Property	Value
molecular weight	92.14
melting point, K	178.15
normal bp, K	383.75
critical temperature, K	591.80
critical pressure, MPa[a]	4.108
critical volume, L/(g·mol)	0.316
critical compressibility factor	0.264
acentric factor	0.262
flash point, K	278
autoignition temperature, K	809
Gas properties, 298.15 K	
H_f, kJ/mol[b]	50.17
G_f, kJ/mol[b]	122.2
C_p, J/(mol·K)[b]	104.7
H_{vap}, kJ/mol[b]	38.26
H_{comb}, kJ/mol[b]	−3734.
viscosity, mPa·s(= cP)	0.00698
flammability limits, in air[c], vol%	
lower limit at 1 atm	1.2
upper limit at 1 atm	7.1
Liquid properties, 298.15 K	
density, L/mol	9.38
C_p, J/(mol·K)[b]	156.5
viscosity, mPa·s(= cP)	0.548
thermal conductivity, W/(m·K)	0.133
surface tension, mN·m(= dyn/cm)	27.9
Solid properties	
density at 93.15 K, L/mol	11.18
C_p at 178.1 K, J/(mol·K)[b]	90.0
heat of fusion at 178.15 K, kJ/mol[b]	6.62

[a] To convert MPa to psi, multiply by 145.
[b] To convert J to cal, divide by 4.184.
[c] At 101.3 kPa (1 atm).

xylene. When σ-bonds and complexes are involved, toluene reacts much faster than benzene and much slower than xylenes.

Manufacture and Processing

The principal source of toluene is catalytic reforming of refinery streams. This source accounts for ca 79% of the total toluene produced. An additional 16% is separated from pyrolysis gasoline produced in steam crackers during the manufacture of ethylene (qv) and propylene (qv). The reactions taking place in catalytic reforming to yield aromatics are dehydrogenation or aromatization of cyclohexanes, dehydroisomerization of substituted cyclopentanes, and the cyclodehydrogenation of paraffins. The formation of toluene by these reactions is shown.

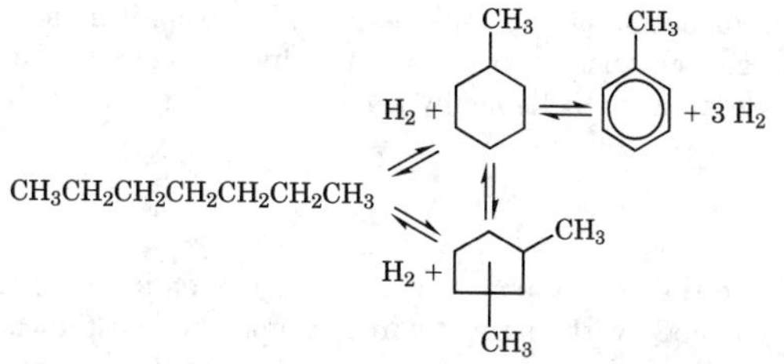

Of the main reactions, aromatization takes place most readily and proceeds ca 7 times as fast as the dehydroisomerization reaction and ca 20 times as fast as the dehydrocyclization. Hence, feeds richest in cycloparaffins are most easily reformed.

Because catalytic reforming is an endothermic reaction, most reforming units comprise about three reactors with reheat furnaces in between to minimize kinetic and thermodynamic limitations caused by decreasing temperature. There are three basic types of operations, ie, semiregenerative, cyclic, and continuous. In the semiregenerative operation, feedstocks and operating conditions are controlled so that the unit can be maintained on-stream from 6 mo–2 yr before shutdown and catalyst regeneration. In cyclic operation, a swing reactor is employed so that one reactor can be regenerated while the other three are in operation. Regeneration, which may be as frequent as every 24 h, permits continuous operation at high severity. Since ca 1970, continuous units have been used commercially. In this type of operation, the catalyst is continuously withdrawn, regenerated, and fed back to the system.

Toluene, Benzene, and BTX Recovery. The composition of aromatics centers on the C_7- and C_8-fraction, depending somewhat on the boiling range of the feedstock used. Most catalytic reformate is used directly in gasoline. That part which is converted to benzene, toluene, and xylenes for commercial sale is separated from the unreacted paraffins and cycloparaffins or naphthenes by liquid–liquid extraction or by extractive distillation.

Proper choice of feedstocks and use of relatively severe operating conditions in the reformers produce streams high enough in toluene to be directly usable for hydrodemethylation to benzene without the need for extraction. Toluene is recovered from pyrolysis gasoline, usually by mixing the pyrolysis gasoline with reformate and processing the mixture in a typical aromatics extraction unit. Yields of pyrolysis gasoline and the toluene content depend on the feedstock to the steam-cracking unit. Pyrolysis gasoline is hydrotreated to eliminate dienes and styrene before processing to recover aromatics.

Emerging Technologies for Production of BTX from Light Hydrocarbons. Recent (ca 1997) technological developments have centered on high temperature pyrolysis of light hydrocarbons C_2 to C_5, LPG, and naphtha to form aromatics in higher yields. Conversions were traditionally low because they were accompanied by a high degree of degradation to carbon and hydrogen. Recent improvements include modification of the thermal cracking process to produce higher yields of liquid products rich in aromatics and the extension of the catalytic hydroforming process to promote oligomerization and dehydrocyclization of the lower olefins. The common core of these developments is the use of shape-selective zeolite catalysts to promote the various reactions.

Specifications, Standards, and Quality Control

Toluene is marketed mostly as nitration and industrial grades. The generally accepted quality standards for the grades are given by ASTM D841 and D362, respectively.

Purity of toluene samples as well as the number, concentration, and identity of other components can be readily determined using standard gas chromatography techniques.

Health and Safety Factors

Permissible exposure limits established by the U.S. Department of Health and Human Services and the U.S. Department of Labor are summarized below, with the more restrictive levels proposed by NIOSH.

	OSHA, mg/m^3 (ppm)	NIOSH, mg/m^3 (ppm)
average during 8-h shift (TWA)	752 (200)	376 (100)
not to exceed	1129 (300)	
except for 10-min average (TLV)	1181 (500)	752 (200)

Toluene generally resembles benzene closely in its toxicological properties; however, it is devoid of benzene's chronic negative effects on blood formation.

Uses

About 90% of the toluene generated by catalytic reforming is blended into gasoline as a component of $>C_5$ reformate. The octane number $(R + M/2)$ of such reformates is typically in the range of 88.9–94.5, depending on severity of the reforming operation. Toluene itself has a blending octane number of 103–106, is exceeded only by oxygenated compounds such as methyl *tert*-butyl ether, ethanol, and methanol.

Toluene is a valuable blending component, particularly in unleaded premium gasolines. Although reformates are not extracted solely for the purpose of generating a high octane blending stock, the toluene that is co-produced when xylenes and benzene are extracted for use in chemicals, and that exceeds demands for use in chemicals, has a ready market as a blending component for gasoline.

Toluene is converted to benzene by hydrodemethylation either under thermal or catalytic conditions. Benzene produced from this source generally supplies 25–30% of the total benzene demand. The feedstock is usually extracted toluene, but some reformers are operated under sufficiently severe conditions or with selected feedstocks to provide toluene pure enough to be fed directly to the dealkylation unit without extraction.

Toluene is more important as a solvent than either benzene or xylene. Solvent use accounts for ca 14% of the total U.S. toluene demand for chemicals. About two-thirds of the solvent use is in paints and coatings; the remainder is in adhesives, inks, pharmaceuticals, and other formulated products utilizing a solvent carrier. Use of toluene as solvent in surface coatings has been declining, primarily because of various environmental and health regulations. It is being replaced by other solvents.

Potential Uses. Because much toluene is demethylated for use as benzene, considerable effort has been expended on developing processes in which toluene can be used in place of benzene to make directly from toluene the same products that are derived from benzene. Such processes both save the cost of demethylation and utilize the methyl group already on toluene. Most of this effort has been directed toward manufacture of styrene. An alternative approach is the manufacture of *para*-methylstyrene by selective ethylation of toluene, followed by dehydrogenation. Resins from this monomer are expected to displace polystyrene because of price and performance advantages.

Derivatives

Toluene Diisocyanate. Toluene diisocyanate is the basic raw material for production of flexible polyurethane foams. It is produced by the reaction in which toluene is dinitrated, the dinitrotoluene is hydrogenated to yield 2,4-diaminotoluene, and this diamine in turn is treated with phosgene to yield toluene 2,4-diisocyanate.

Benzoic Acid. Benzoic acid is manufactured from toluene by oxidation in the liquid phase using air and a cobalt catalyst. Typical conditions are 308–790 kPa (30–100 psi) and 130–160°C. The crude product is purified by distillation, crystallization, or both. Yields are generally >90 mol%, and product purity is generally >99%.

Benzyl Chloride. Benzyl chloride is manufactured by high temperature free-radical chlorination of toluene. The yield of benzyl chloride is maximized by use of excess toluene in the feed. More than half of the benzyl chloride produced is converted by butyl benzyl phthalate by reaction with monosodium butyl phthalate. The remainder is hydrolyzed to benzyl alcohol, which is converted to aliphatic esters for use in soaps, perfume, and flavors. Benzyl salicylate is used as a sunscreen in lotions and creams.

Disproportionation to Benzene and Xylenes. With acidic catalysts, toluene can transfer a methyl group to a second molecule of toluene to yield one molecule of benzene and one molecule of mixed isomers of xylene. This disproportionation is an equilibrium reaction. Disproportionation generates benzene from toluene and at the same time takes full advantage of the methyl group to generate a valuable product, ie, xylene. Economic utility of the process is strongly dependent on the relative values of toluene, benzene, and the xylenes.

Vinyltoluene. Vinyltoluene is used as a resin modifier in unsaturated polyester resins. Its manufacture is similar to that of styrene; toluene is alkylated with ethylene, and the resulting ethyltoluene is dehydrogenated to yield vinyltoluene.

Toluenesulfonic Acid. Toluene reacts readily with fuming sulfuric acid to yield toluene–sulfonic acid. By proper control of conditions, *p*-toluenesulfonic acid is obtained. The primary use is for conversion, by fusion with NaOH, to *p*-cresol. The resulting high purity *p*-cresol is then alkylated with isobutylene to produce 2,6-di-*tert*-butyl-*p*-cresol (BHT), which is used as an antioxidant in foods, gasoline, and rubber. Mixed cresols can be obtained by alkylation of phenol and by isolation from certain petroleum and coal–tar process streams.

Benzaldehyde. Annual production of benzaldehyde requires ca 6,500–10,000 t ($2-3 \times 10^6$ gal) of toluene. It is produced mainly as by-product during oxidation of toluene to benzoic acid, but some is produced by hydrolysis of benzal chloride. The main use of benzaldehyde is as a chemical intermediate for production of fine chemicals used for food flavoring, pharmaceuticals, herbicides, and dyestuffs.

Toluenesulfonyl Chloride. Toluene reacts with chlorosulfonic acid to yield both *o*- and *p*-toluenesulfonyl chlorides. The ortho isomer is converted to saccharin. The para isomer is used for preparation of specialty chemicals. Annual toluene requirements are ca 6500 t (2×10^6 gal).

Miscellaneous Derivatives. Other derivatives of toluene, none of which is estimated to consume more than ca 3000 t (10^6 gal) of toluene annually, are mono- and dinitrotoluene hydrogenated to amines; benzotrichloride and chlorotoluene, both used as dye intermediates; *tert*-butylbenzoic acid from *tert*-butyltoluene, used as a resin modifier; dodecyltoluene converted to a benzyl quaternary ammonium salt for use as a germicide; and biphenyl, obtained as by-product during demethylation, used in specialty chemicals. Toluene is also used as a denaturant in specially denatured alcohol (SDA) formulas 2-B and 12-A.

ACKNOWLEDGMENT

To R. A. Wilsak and M. E. Carrera (Amoco Chemical Co.) and O. C. Okoroafor (Cooper Union for the Advancement of Science and Arts).

E. DICKSON OZOKWELU
Amoco Chemical Company

J. Wisniak and A. Tamir, *Liquid–Liquid Equilibrium and Extraction,* Elsevier Scientific Publishing Co., Amsterdam, Pt. A, 1980; Pt. B, 1981; Suppl. 1, 1985; Suppl. 2, 1987.

K. Weissermel and H. Arpe, *Industrial Organic Chemistry,* Verlag Chemie, New York, 1978, pp. 288–289.

D. R. Stull and co-workers, *Chemical Thermodynamics of Hydrocarbon Compounds,* John Wiley & Sons, Inc., New York, 1969, p. 368.

1996–97 Toluene–Xylenes Annuals, Dewitt & Co., Inc., Houston, Tex., Jan. 1997.

TOLUENEDIAMINES. See AMINES, AROMATIC–DIAMINOTOLUENES.

TOOL MATERIALS

Machining of materials with a cutting tool is a common operation in which the unwanted material is removed in the form of chips. A successful cutting tool must resist severe conditions of high temperature, high pressure, and chemically reactive surfaces, and provide a sufficiently long tool life (see also CERAMICS; CARBON, DIAMOND; BORON COMPOUNDS, REFRACTORY BORON COMPOUNDS; HIGH TEMPERATURE ALLOYS).

A tool material must meet stringent requirements. Both deformation energy and frictional energy are converted into heat, and consequently, tool temperatures are very high (ca 1000°C).

A wide range of materials is available with a variety of properties, performance capabilities, and costs. High speed steels (HSS) and cemented carbides (coated and uncoated) are currently the most extensively used materials. Ceramics, diamond, and cubic boron nitride (cBN) are used for special applications. Guidelines for selection for different cutting operations are given in Table 1.

Carbon Steels and Low–Medium Alloy Steels. Low–medium carbon steels have since then been largely superseded by other tool materials, except for some low speed applications.

High Speed Steels. High speed steels (HSS) contain significant amounts of W, Mo, Co, V, and Cr in addition to Fe and C. The presence of these alloying elements strengthens the matrix beyond the tempering temperature, increasing the hot hardness and wear resistance. Tool steels are broadly classified as T-type or M-type depending on whether W or Mo is the principal alloying element. The two types T- and M- can be used interchangeably because they possess more or less the same properties and have comparable cutting performance. In general, M-type tool steels are more popular, representing ca 85% of all tool steels, because they are less expensive by ca 30% than the corresponding T-type steels.

High speed steel tools are available in cast, wrought, and sintered forms.

A processing technique introduced in the late 1960s involves atomization of the prealloyed molten tool steel alloy into fine powder, followed by consolidation under hot isostatic pressure (HIP). This technique, termed *consolidation by powder metallurgy* (CPM), when combined with suitable hardening and tempering, provides a microstructure consisting of a uniform and fine dispersion of carbides in a fine-grained, tempered, martensite matrix. Tool steels made in this manner grind more easily, especially the highly alloyed tool steels, with grinding ratios two to three times better; exhibit more uniform properties; and perform more consistently.

Shortages and escalating costs of Co in the 1970s prompted tool-steel producers to seek an appropriate substitute. Hot hardness can be maintained without Co by appropriate increases of Mo–W or V content, or both.

HSS tools are used mostly for low speed, heavy-duty applications. Thus built-up edge, adhesion of the chips to the tool, and high friction are the primary concerns for these tools rather than high tool temperatures.

A newer tool steel material having a fine grain size of TiC (40–55%) in a steel matrix (45–60%) with several unique characteristics was developed. Additional Cr (3–17.5%), Mo (0.5–4%), Ni (0.5–12%),

Table 1. Guidelines for Tool Materials

Tool materials[a]	Work materials	Modes of tool wear or failure[b]
carbon steels	low strength, softer materials, nonferrous alloys, plastics	buildup, plastic deformation, abrasive wear, microchipping
low–medium alloy steels	low strength–soft materials, nonferrous alloys, plastics	buildup, plastic deformation, abrasive wear, microchipping
HSS and TiN-coated HSS	all materials of low–medium strength and hardness	flank wear, crater wear
cemented carbide	all materials up to medium strength and hardness	flank wear, crater wear, nose wear thermal, cracks, deformation, fracture
coated carbides	cast iron, alloy steels, stainless steels, superalloys	flank wear, crater wear nose wear thermal, cracks, deformation, fracture
ceramics	cast iron, Ni-base superalloys, nonferrous alloys, plastics	DCL notching, microchipping, gross fracture
cBN	hardened alloy steels, HSS, Ni-base superalloys, hardened chill-cast iron, commercially pure Ni	DCL notching, chipping, oxidation, graphitization
diamond	pure, Cu, pure Al, Al–Si alloys, cold-pressed cemented carbides, rock, cement, plastics, glass–epoxy composites, nonferrous alloys, hardened high carbon alloy steels, fibrous composites	chipping, oxidation, graphitization

[a] HSS = high speed steel; cBN = cubic boron nitride.
[b] DCL = depth of cut line.

Co (5–5.7%), Ti (0.5–0.7%), and C (0.4–0.85%) are made to provide solid solution strengthening as well as hot hardness of the matrix material. This material, which combines the hardness (consequently, the wear resistance) of cemented carbides with the heat treatability of HSS, responds to heat treatment, such as annealing and quench hardening, and can be machined in the annealed condition.

Cast-Cobalt Alloys. Cast-cobalt alloys were introduced about the same time as HSS for cutting tool applications. Popularly known as Stellite tools, these materials are Co-rich Cr–W–C cast alloys having properties and applications in the intermediate range between HSS and cemented carbides. Although comparable in room-temperature hardness to HSS tools, cast-cobalt alloy tools retain their hardness to a much higher temperature and hence can be used at higher (25%) cutting speeds than HSS tools. Cast-cobalt alloys have, however, been phased out owing to the high cost of Co, safety in handling Co-base alloys, and availability difficulties.

Cemented Carbides. Cemented carbides contain a large-volume fraction (≥90%) of fine-grain, refractory carbides in a metal binder. By varying the amount of the binder phase, cemented carbides tool materials of different toughness values can be obtained.

There are at least four different classification systems for cemented carbides. The U.S. system is based on relative performance; the U.K. system is based on properties, and the former USSR system on composition; the fourth system, widely used in Europe and supported by the ISO, is based on application and chip form.

In the U.S., the C-classification (C-1 to C-8) for cemented carbide tools, used unofficially for machining applications, was originally developed by the automobile industry to obtain a relative performance index of tools made by different tool producers. This is by far the simplest system. The grades are broadly divided into two classes (C-1 to C-4 and C-5 to C-8), according to the type of work material to be machined. Grades C-1 to C-4 are recommended for machining nonsteels, ie, cast iron, nonferrous alloys, and nonmetallics (nonsteel workmaterials), whereas C-5 to C-8 are recommended for machining carbon steels and alloy steels.

Within each class, ie, C-1 to C-4 and C-5 to C-8, each grade is distinguished by the type of machining operation: C-1 and C-5 for roughing, C-2 and C-6 for general purpose, C-3 and C-7 for semifinishing, and C-4 and C-8 for precision-finishing operations. In general, from grades C-1 to C-4 or C-5 to C-8 within each class, the shock resistance decreases, hardness increases, high temperature deformation resistance and wear resistance increase, and the Co content and carbide grain size decrease.

The ISO classification for cemented-carbide cutting tools broadly divided into three categories and, for convenience as well as easy identification, color coded when used on the shop floor. P-grades (blue) are highly alloyed multicarbides used mainly for machining hard steels and steel castings; M-grades (yellow) are low alloy multicarbide alloys which are multipurpose nonsteel grades used for machining high temperature alloys, low strength steels, gray cast iron, free machining steels, and nonferrous metals and their alloys; and K-grades (red) are straight WC grades for machining very hard gray cast iron, chilled castings, nonferrous metals and their alloys, and nonmetallics such as plastics, glass, glass–epoxy composites, hard rubber, and cardboard.

The original objective of the ISO classification was to issue detailed standards for cemented carbides in terms of microstructure, composition, and properties for quality control and performance reliability. This objective, however, is yet to be realized.

In selecting a carbide grade for a given application, the following general guidelines should be followed: the grade with the lowest Co content and the finest grain size consistent with adequate strength to eliminate chipping should be chosen; straight WC grades can be employed if cratering, seizure, or galling is not experienced and for work materials other than steels; to reduce cratering and abrasive wear when machining steels, TiC grades are preferred; and for heavy cuts in steel where high temperature and high pressure deform the cutting edge plastically, a multicarbide grade containing W–Ti–Ta(Nb) with low binder content should be used.

Composition, microstructure, and performance of cemented carbides depend on Co binder content, carbide grain size, and type and composition of various carbides.

A breakthrough occurred in the development of cermets when in 1956 additions of Mo to TiC–Ni cermet were shown to improve wetting of the carbide by forming a mixed carbide shell (Ti,Mo)C around the TiC grains, thereby inhibiting carbide coalescence and grain growth. The resulting microstructure gives improved hardness and impact resistance. However, this material was not as hard and tough as the cemented multicarbide counterpart. Table 2 shows a comparison between the original TiC–Ni–Mo cermet and the cermet containing TiN, WC, TaC, and Co.

Although cemented-carbide tools can be brazed, most of the carbide tools are available in insert form such as squares, triangles, diamonds, and rounds. These can be easily clamped on to the tool shank, thereby avoiding the problems and complexities associated with brazing. It is this feature that widely extended the applications of cemented carbide tools.

Cemented carbides are not generally recommended for low speed cutting operations because the chips tend to weld to the tool face and

Table 2. A Comparison of Properties of Cermets

Composition of cermet	HV,[a] at 1000°C	TRS, 900°C, N/mm^2	Oxidation resistance wt gain after 1 h at 1000°C, mg/cm^2	Thermal conductivity at 1000°C, W/(K · m)
TiC–16.5% Ni–9% Mo	500	1050	11.8	24.7
TiC–20% TiN–15% WC–10% TaC–5.5% Ni–11% Co–9% Mo	650	1360	1.66	42.3

[a] HV = Vicker's hardness.

cause microchipping and there is no economic incentive to use them at lower speeds. However, for applications requiring higher stiffness, and higher wear resistance, such as broaches and shaper cutters, they are used extensively at lower speeds. Thin coatings of TiN made this application even more attractive. To conserve the strategic materials (W, Co, and Ta) and reduce costs, recycling of used cemented-carbide inserts (so-called disposable or throwaway inserts) is growing steadily.

Coated Tools

Although rapid advances in coated cemented-carbide technology have taken place, coating technology for HSS is still limited to coating of TiN by PVD. An analysis of the cutting process indicates that the material requirements at or near the surface of the tool are different from those of the tool body. A thin, chemically stable, hard, refractory binderless coating often satisfies these requirements.

An effective coating should be hard, refractory, chemically stable, chemically inert to shield the constituents of the tool and the work-material from interacting chemically under the conditions of cutting, binder free, of fine grain size with no porosity, metallurgically bonded to the substrate with a graded interface to match the properties of the coating and the substrate, thick enough to prolong tool life but thin enough to prevent brittleness, free of the tendency of metal chips to adhere to or seize to the tool face, able to provide residual compressive stress, easy to deposit in bulk quantities, and inexpensive. In addition, coatings should have low friction and exhibit no detrimental effects on the substrate or bulk properties of the tool.

Several refractory coatings (qv) have been developed including single coatings of TiC, TiN, Al_2O_3, HfN, or HfC, and multiple coatings of Al_2O_3 or TiN on top of TiC, generally deposited by CVD. TiC is used as a hard wear-resistant coating at low speeds. Multiple coatings prolong tool life, as the thickness of the coating can be increased from ca 5 to 10 μm without inducing brittleness, provide a strong metallurgical bond between the coating and the substrate by choosing appropriate coatings that would provide graded interface(s), and provide protection for machining a range of work materials.

Coated tools, are used for a wide range of materials, including various types of steels, cast iron, stainless steel, nickel and Co-base superalloys, and titanium alloys. The tool manufacturers often encounter a challenge between developing a general-purpose grade that covers a wide range of work materials, cutting conditions, and machining operations to reduce the number of grades the user can stock vs the coatings targeted for niche areas. The selection of an appropriate tool material and proper tool geometry, chip groove geometry, coating, and substrate properties for machining a given material is not a simple selection based on rule of thumb, but rather is a more sophisticated decision-making process based on detailed knowledge and experimentation. Of the carbide tools used as of the mid-1990s, some 65% are coated.

Multiple Nanolayered Coatings

The refractory hard materials used for coatings on cutting tools are generally brittle and hence not tough. The fracture mechanism consists of crack initiation at stress concentrations and its rapid propagation to failure (see FRACTURE MECHANICS). By arresting the propagation of the cracks, it is possible to increase the toughness of these hard coatings significantly without compromising on hardness. This is accomplished by applying multiple nanolayer coatings of alternating hard and tough materials (see NANOTECHNOLOGY). The number of nanolayers can be several hundred in contrast to the few layers used on cutting tools prepared by CVD techniques. Nanolayer coatings are generally expected to be harder, tougher, and chemically more stable than coatings of several micrometers or of bulk materials.

The number of material systems that can be used for nanolayer coatings is virtually unlimited. Any refractory hard material can be used as the hard material; compatible metal can be used as the tough material. Multiple nanolayer coatings are deposited by PVD; chiefly magnetron sputtering using multiple targets is employed. As of 1997, no commercial nanocoatings for cutting tool applications are available in the marketplace. This technology is expected to become widely used in the twenty-first century.

Ceramics

Ceramics (qv), one of the newest classes of advanced tool materials, are used on the one hand for high speed finishing operations involving light feeds and on the other for high removal-rate machining involving low speeds and large depths of cut of some difficult-to-machine steels and cast irons. The ceramics used initially were predominantly alumina based, although silicon nitride-based materials (also called nitrogen ceramics) have been found to be very attractive for high speed machining of gray cast iron (1500 m/min) and nickel-base superalloys (200 m/min or higher) (see ADVANCED CERAMICS). Ceramics, in general, are harder, more wear-resistant, more highly refractory, and chemically more stable than cemented carbides and HSS.

Several factors have rejuvenated interest in the development and application of ceramic cutting tools. Applications of advanced ceramics for structural applications, advances in the ceramic-processing technology; progress in the understanding of the toughening mechanisms in ceramics; rapidly rising manufacturing costs; the need to use materials that are increasingly more difficult to machine; rapidly increasing costs and decreasing availability of W, Ta, and Co, which are the principal and strategic raw materials in the manufacture of cemented-carbide tools; and advances in machining science and technology have all played a role.

A comparison of the physical properties of ceramic tools and carbide tools is given in Table 3. The next advancement in alumina-based ceramics is the development of pure alumina and alumina–TiC dispersion-strengthened ceramics. Alumina–TiC-based ceramics contain ca 30 wt % TiC and small amounts of yttria as a sintering agent, resulting in a density close to 99.50% theoretical.

An Al_2O_3–ZrO_2 ceramic (Cer Max 460) performs exceptionally well in the grinding industry as a tough abrasive in heavy-stock grinding operations, such as cut-off and snagging.

The second interesting class of ceramic tool material under development is based on Si_3N_4, either nearly pure Si_3N_4 (except for some minor additions of sintering aids) or having various additions of aluminum oxide, yttrium oxide, and TiC. It is a spin-off of the high temperature–structural ceramics technology developed in the 1970s for automotive gas turbines and other high temperature applications. This material provides a number of favorable properties, including higher elevated temperature strength, thermal stability, low thermal expansion coefficient, higher thermal conductivity, and higher fracture toughness than alumina.

Si_3N_4 is marketed by most tool manufacturers and used extensively in high speed machining of gray cast iron. In the late 1970s a ceramic tool of Si–Al–O–N with additions of yttria was marketed under the trademark SYALON.

Even though Si_3N_4 and SiAlON tool materials are used extensively for high speed machining of cast iron and machining of nickel-base superalloys, respectively, they could not be used for machining steels. To take advantage of their improved fracture toughness as well as their ability to machine ductile C-1, ceramic coatings on Si_3N_4, SiAlON, and modified compositions of the two were developed.

Table 3. Physical Properties of Ceramic and Cemented-Carbide Cutting Tools

Property[a]	Ceramics	Cemented carbide[b]
hardness, HRA[c]	91–95	90–93
TRS[d] for alumina-based ceramics, MPa[e]	690–930	1590–2760
melting range, °C	ca 2000	ca 1350
density, g/cm^3	3.9–4.5	12.0–15.3
modulus of elasticity, *E*, GPa[f]	410	70–648
grain size, μm	1–3	0.1–6
compressive strength, MPa[e]	2760	3720–5860
tensile strength, MPa[e]	240	1100–1860
thermal conductivity, W/(m·K)		41.8–125.5
thermal expansion coefficient, 10^{-6}/°C	7.8	4–6.5

[a] Exact properties depend on materials used, grain size, binder content, vol. fraction of each constituent, and processing method.
[b] Coated carbides not included.
[c] Rockwell hardness A scale.
[d] Transverse rupture strength.
[e] To convert MPa to psi, multiply by 145.
[f] To convert GPa to psi, multiply by 145,000.

A newer whisker-reinforced ceramic composite (SiC whisker-reinforced alumina) material possessing improved fracture toughness has been introduced as a cutting tool for machining of nickel–iron-base superalloys used in aircraft engines. Ceramic tools are inherently more brittle than cemented carbides. Certain ceramic tools, especially those based on alumina, are not suitable for machining aluminum, titanium, and similar materials because of a strong tendency to react chemically.

Diamond

Diamond is the hardest (Knoop hardness ca 78.5 GPa (ca 8000 kgf/mm^2)) of all known materials. Both the natural (single-crystal) and synthetic (polycrystalline sintered body) forms can be used for cutting-tool applications. Diamond tools exhibit high hardness, good thermal conductivity, ability to form a sharp edge by cleavage (especially the single-crystal natural diamond), low friction, nonadherence to most work materials, ability to maintain a sharp edge for a long period of time, especially when machining soft materials like copper and aluminum; and high wear resistance. Disadvantages of diamond tools include extensive chemical interaction with metallic elements of Groups (4–10) (IVB–VIII) of the Periodic Table (diamond wears rapidly when machining or grinding mild steel; it wear less rapidly with high carbon alloy steels than with low carbon steel and is occasionally employed to machine gray cast iron (high carbon content) with long life); a tendency to revert at higher (ca 700°C) temperatures to graphite and oxidize in air; extreme brittleness (single-crystal diamond cleaves easily); difficulty in shaping and reshaping after use; and high cost.

Limited supply, increasing demand, and high cost have led to an intense search for an alternative, dependable source of diamond. This search led to the high pressure (ca 5 GPa (0.5×10^6 psi)), high temperature (ca 1500°C) (HP–HT) synthesis of diamond from graphite in the mid-1950s in the presence of a catalyst–solvent material, eg, Ni or Fe, and the subsequent development of polycrystalline sintered diamond tools in the late 1960s.

Sintered polycrystalline diamond tools of various grain sizes are fabricated in an assortment of shapes (squares, rounds, triangles, and sectors of a circle of different included angles) and sizes from round blanks. The main advantages of sintered polycrystalline tools over natural single-crystal tools are better control over inclusions and imperfections, higher quality, and greater toughness and wear resistance (resulting from the random orientation of the diamond grains and the corresponding lack of simple cleavage planes). In addition, the availability of sintered diamond tools is not dictated by nature or some artificial control. Sintered diamond tools are used for applications similar to the lower quality industrial diamonds.

Cubic Boron Nitride (cBN)

Cubic boron nitride (cBN) is next only to diamond in hardness (Knoop hardness 46.1 GPa (ca 4700 kgf/mm^2)). It is a remarkable material in that it does not exist in nature and is produced by HP–HT synthesis in a process similar to that used to produce diamond from graphite. Sintered cBN tools are fabricated in the same manner as sintered diamond tools and are available in the same sizes and shapes.

Polycrystalline cBN is used extensively for machining of high hardness steels (Rc > 45), nickel-base superalloys, and alloyed cast iron. However, the development of other, less expensive tool materials, chiefly ceramics (SiC whisker-reinforced alumina and SiAlON) for machining of nickel-base alloys is challenging the use of this material. The two predominant wear modes of cBN tools are DCL (depth of cutline) notching and microchipping. These tools have been used successfully for heavy interrupted cutting and for milling white cast iron and hardened steels. Diamond and cBN tools provide significantly higher performance capability, and demands are being placed on the machine tools and manufacturing practice in order to take full advantage of the potential of these materials.

Health and Safety Factors

Threshold limit values for the components of cemented carbides and tool steels range from 0.1 mg/m^3 for cobalt to 5 mg/m^3 for tungsten. There is generally no fire or explosion hazard involved with tool steels, cemented carbides, or other tool materials. During machining operations, eye protection is recommended; during grinding operations, NIOSH-approved respirators for metal fumes and dust are recommended. Safety is of particular concern in metal-cutting and metal-forming operations. Some cutting fluids, and work materials, may present a fire hazard. Toxic vapors, unpleasant odors, smoke fumes, skin irritations (dermatitis), or effects from bacteria cultures from the cutting fluid are all factors for the operator.

Future Outlook

The implementation of coated tool technology on the shop floor is proceeding at a significant pace. New coating combinations, tailored substrates, CVD/PVD technologies for different types of engineered coatings for different workmaterials, cutting conditions, machining operations are being addressed effectively. This effort is expected to continue. Multiple nanolayer coatings which have more recently been developed have improved hardness, strength, and chemical stability. The practically unlimited choice of coating combinations, ie, alternative hard and tough materials, or alternative hard material and solid lubricant, etc, should lead to numerous multiple nanolayer coating applications. Nanolayer coatings may be ideal for hard refractory coatings which have difficulty in bonding with the substrate or other coatings.

SiC whisker-reinforced alumina is a major advance in tool material development, as it provides a means to increase the fracture toughness of the material via the composite material approach. Another interesting concept, one used in the development of superalloys, is the strengthening of the matrix by orderly precipitation of the second-phase materials and strengthening by dispersoids, such as Al_2O_3.

Cemented carbides are fairly expensive owing to the use of hard, refractory materials. This is expected to become even more the case as some of the strategic materials used in these tools become more expensive or newer but more expensive materials such as HfC or HfN come into more common use. It may be economical, therefore, to use these materials at or near the cutting edge instead of as the whole insert. The development of tools of TiC (40–55%) or TiN (30–60%) in a steel matrix on a steel core using powder metallurgy technology suggests a similar approach for cemented carbides as the need arises.

RANGA KOMANDURI
Oklahoma State University

Machinability Data Handbook, 3rd ed., Vols. 1 and 2, Machinability Data Center, Metcut Research Associates, Cincinnati, Ohio, 1980.

E. Dow Whitney, ed., *Ceramic Cutting Tools,* Noyes Publications, Park Ridge, N.J., 1994.

R. Komanduri, *ASME Appl. Mechan. Rev.* **46**(3), 80–132 (Mar. 1993).

V. A. Tipnis, in M. B. Peterson and W. O. Winer, eds., *Wear Control Handbook,* ASME, New York, 1980.

TOOTHPASTE. See DENTIFRICES.

TOXAPHENE. See INSECT CONTROL TECHNOLOGY.

TOXICOLOGY

Natural and synthetic chemicals affect every phase of our daily lives in both good and noxious manners. There are about as many definitions of toxicology as there exist textbooks on the subject. Although they differ in detail, all good definitions embrace the concept that toxicology is concerned with the potential of chemicals, or mixtures of them, to produce harmful effects in living organisms. Toxicology is a study of the interactions between chemicals and biological systems in order to determine quantitatively the potential for such chemicals to produce injury that results in adverse health effects in intact living organisms, and to investigate the nature, incidence, mechanism of production, and reversibility of such adverse effects.

Although studies are carried out by *in vitro* and *in vivo* procedures, it is the primary aim of toxicology to determine the potential for harmful effects in the intact living organism, usually with an ultimate goal of assessing the significance of the findings with respect to humans. There are four components to a complete risk assessment:

(*1*) Hazard identification involves gathering and evaluating data on the types of health injury or disease that may be produced by a chemical and on the conditions of exposure under which injury or disease is produced. It may also involve characterization of the behavior of a chemical within the body and the interactions it undergoes with organs, cells, or even parts of cells. Hazard identification is not risk assessment. It is a scientific determination of whether observed toxic effects in one setting will occur in other settings.

(*2*) Dose–response evaluation is used in describing the quantitative relationship between the amount of exposure to a substance and the extent of toxic injury or disease. Data may be derived from animal studies or from studies in exposed human populations. Dose–response toxicity relationship for a substance varies under different exposure conditions. The risk of a substance can not be ascertained with any degree of confidence unless dose–response relations are described.

(*3*) Human exposure evaluation is used in describing the nature and size of the population exposed to a substance and the magnitude and duration of their exposure. The evaluation could concern past or current exposures, or exposure anticipated in the future.

(*4*) Risk characterization is defined as the integration of the data and analysis of the above three components to determine the likelihood that humans will experience any of the various forms of toxicity associated with a substance. When the exposure data are not available, hypothetical risk is characterized by the integration of hazard identification and dose–response evaluation data.

Toxicity, the potential to produce harmful effects, is to be clearly differentiated from *hazard,* which is the likelihood that a particular material will exhibit its known toxicity under specific conditions of use.

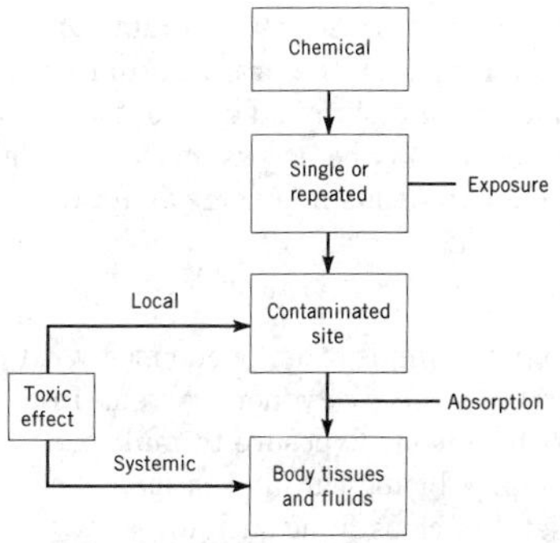

Figure 1. Schematic representation showing the basis for classification of toxic effects into local and systemic by single or repeated exposures.

Classification of Toxic Effects

The diagram shown in Figure 1 gives a basis for the classification of toxic effects according to site and degree of exposure. The nature of a toxic effect and the probability of its occurring are often related to the number of exposures. *Latent effects* occur either only after there has been a significant period free of toxic signs following exposure, or after resolution of acutely toxic effects which appeared immediately following exposure. *Persistent effects* do not resolve, and may even become more severe after removal from the source of exposure. They can occur as a consequence of acute or repeated-exposure conditions. *Cumulative effects* are those where there is progressive injury and worsening of the toxic effect as a result of repeated-exposure conditions. *Transient effects* are those where there is repair of toxic physical injury or the reversal of induced biochemical aberrations. *Acute exposures* involve a single exposure to the test chemical in order to determine if this is effective in producing immediate, delayed, or persistent effects. *Short-term repeated exposures* involve consecutive daily exposures to the test chemical which are continued over a period of a few days to a few weeks but usually not more than 5% of the lifespan of the animal. *Subchronic exposures* involve consecutive daily exposures to the test material for a period amounting to usually no more than 10–15% of the lifespan of the test species. *Chronic exposures* involve consecutive daily exposure to the test material over the lifespan of the test species, or a great portion of it.

The Nature of Toxic Effects

The biological response to chemical insult may take numerous forms, depending on the physicochemical properties of the material and the conditions of exposure. The following are some of the more significant and frequently encountered types of injury or toxic response. They may be defined in terms of tissue pathology, altered or aberrant biochemical processes, or extreme physiological responses: inflammation; degeneration; necrosis; immune-mediated hypersensitivity reaction; immunosuppression; neoplasia; mutagenesis; enzyme inhibition; biochemical uncoupling; lethal synthesis; teratogenesis; sensory irritation; and endocrine system disruption.

Factors Influencing Toxicity

During the design, conducting, and evaluation of toxicology studies, there is a constant need to be aware of the numerous factors that may influence the nature, severity, and probability of induction of toxic injury. Some of the more important are number of exposures; magnitude of exposure; species tested; route of exposure; time of dosing; formulation; and impurities.

Routes of Exposure

In order to induce a toxic effect, local or systemic, the causative material must first come into contact with an exposed body surface; these are the routes of exposure. In normal circumstances, and depending on the nature of the material, the practical routes of exposure are by swallowing, inhalation, and skin and eye contact. In addition, and for therapeutic purposes, it may be necessary to consider intramuscular, intravenous, and subcutaneous injections as routes of administration.

Multiple Exposures

Although toxicology testing is often performed with only a single material or a material in a relatively inert solvent, in most practical situations there is simultaneous exposure to multiple chemicals and thus a potential for complex biological interactions. The following descriptive terms are useful in classifying such effects.

Independent is an effect in which each material exerts its own effect irrespective of the presence of another. *Additive effects* involve materials producing similar toxic effects where the magnitude of the response is numerically equal to the sum of the effect produced by each individual material. *Antagonism* is applied to a situation where two chemicals, given together, interfere with each other's action or where one interferes with the action of the other. *Potentiation* is applied to a condition where one material, of relatively low toxicity, enhances the expression of toxicity by another chemical. *Synergism* is applied to a situation where the effect of two or more chemicals that have common mechanism of toxicity, given together, is significantly greater than that expected from considerations on the toxicity of each material alone. This differs from potentiation in that both materials contribute to the toxic injury, and the net effect is always greater than additive.

Dose–Response Relationships and Their Toxicological Significance

The importance of determining a relationship between the magnitude of the exposure and the frequency of occurrence of a toxic effect is considered in detail below.

An observation which is fundamental to the interpretation of toxicology information is the variation in susceptibility to potentially harmful chemicals of individual members within a given population.

Dose–response relationships are useful for many purposes; in particular, the following: if a positive dose–response relationship exists, then this is good evidence that exposure to the material under test is causally related to the response; the quantitative information obtained gives an indication of the spread of sensitivity of the population at risk, and hence influences hazard evaluation; the data may allow assessments of no effects and minimum effects doses, and hence may be valuable in assessing hazard; and by appropriate considerations of the dose–response data, it is possible to make quantitative comparisons and contrasts between materials or between species.

Testing Procedures

For descriptive purposes, toxicology testing procedures can be conveniently subdivided into general and specific forms. General toxicology studies are those in which animals are exposed to a test material under appropriate conditions and then examined for all types of toxic effects that the monitoring procedures employed allow. Specific toxicological studies are those in which exposed animals are monitored specifically for a defined toxic end point or effect.

There are many guidelines that need to be followed and which are common to all types of toxicity testing, the most important of which are as follows:

(*1*) There should be sufficiently large numbers of animals to allow a quantitative determination of the average response and the range of responses, including the demonstration of hypersensitive populations. When objective procedures are undertaken, these should be sufficient to allow valid statistical comparison to be made between treated and control groups.

(*2*) Sufficient numbers of control animals should be employed. The use of such controls allows a determination of normal values for features monitored in the study and background incidence of pathology in the population studied; detection of the onset of adverse conditions, eg, infection, which are unrelated to, and detrimental to, the conduct of the study; and deviation of monitored features between controls and exposed animals, which may indicate a treatment-related effect.

(*3*) Vehicle control animals may be necessary to allow an assessment of the possible contribution of the vehicle to any effects observed in exposed animals.

(*4*) Exposure should be by the practical route. Other conditions, such as number and magnitude of exposures, should include at least one level representative of the practical situation; monitoring should be appropriate to the needs for conducting the study; and when practically and economically possible, pharmacokinetic observations should be undertaken in order to better define the relationship of dose to metabolic thresholds.

General Toxicology Studies. Studies may be conducted in live specimens (*in vivo*), or in test tubes (*in vitro*). For reasons inherent in both the toxicity assessment procedure and the design of studies, it is usual to proceed in sequence from acute to the various stages of multiple-exposure studies. The types of monitoring employed to assess the functional status of the living animal and for the detection of injury in dead animals may include the following: general observations; body weight; food and water consumption; hematology; chemical pathology; urinalysis; pathology; organ weight determinations; and special investigations.

Types of Studies. Studies may be conducted in live specimens (*in vivo*), or in test tubes (*in vitro*). Studies may be carried out by single exposure or by repeated exposure over variable periods of time. The design of any one study, including the monitoring procedures, is determined by a large number of factors, including the nature of the test material, route of exposure, known or suspected toxicity, practical use of the material, and the reason for conducting the study.

Acute toxicity studies are often dominated by consideration of lethality, including calculation of the median lethal dose.

Short-term repeated studies should give information about the potential for cumulative toxicity and allow the detection of toxicity, other than neoplasia, not detected in acute studies.

Although short-term repeated exposure studies provide valuable information about toxicity over this time span, they may not be relevant for assessment of hazard over a longer time period. For example, the minimum and no-effects levels determined by short-term exposure may be significantly lower if exposure to the test material is extended over several months. Also, certain toxic effects may have a latency which does not allow their expression or detection over a short-term repeated-exposure period; for example, kidney dysfunction or disturbances of the blood-forming tissues may not become apparent until subchronic exposure studies are undertaken.

With the exception of tumorigenesis, most types of repeated exposure toxicity are detected by subchronic exposure conditions. Therefore, chronic exposure conditions are usually conducted for the following reasons: if there is a need to investigate the tumorigenic potential of a material; if it is necessary to determine a no-effects or threshold level of toxicity for lifetime exposure to a material; and if there is reason to suspect that particular forms of toxicity are exhibited only under chronic exposure conditions.

Specific Toxicology Studies. Many procedures, both *in vivo* and *in vitro,* are available to detect specific organ toxicity or quantitatively monitor for particular end points or effects. Although many of these studies are directed at measuring a particular toxic effect for hazard-evaluation purposes, some are employed as screening or short-term tests to determine the potential of a material to induce chronic toxic effects or those with a long latency period. In this context, screening means an experimental approach that allows the rapid and cost-effective prediction of the likelihood that a material exerts a particular type of adverse biological activity. Such approaches should be based on studies showing the method gives a high degree of correlation with conventional and credible methods for detecting the particular toxic end point. Some of the most commonly employed special toxicology methods and approaches are **1.** Primary irritancy studies. **2.** Studies for immune-mediated hypersensitivity. **3.** Neurological and behavioral toxicology. **4.** Sensory irritation. **5.** Teratology. **6.** Reproductive toxicology tests. **7.** Metabolism and pharmacokinetics. **8.** Mutagenicity.

Review of Toxicology Studies

The review and interpretation of toxicology studies is a professional matter, requiring experience in both the laboratory conduct of such studies and the practice of applied toxicology. Some general considerations to be kept in mind during the review process, described below.

The reviewer should establish that the laboratory reporting the study has the necessary professional reputation, scientific experience, and expertise in the area investigated. The objectives of the study should be precisely stated and the work presented in a clear and coherent matter, with all the detail necessary to allow the reviewer to make his or her own assessment of the study. The material tested should be specified, including nature, relative proportions of any impurities, and stability over the test period. All details of the conduct of the study should be presented. Attention should be paid to the sufficiency of the study with respect to determining significance and assessing hazard. There should be sufficient dose–response information to allow decisions on causal relationships and relevance. The results of the study should allow decisions on whether injury is a direct result of toxicity or secondary to other events. In evaluating numerical information, it is important to remember that, although an effect may be statistically significant, this does not necessarily imply that the effect is of adverse biological significance. Conversely, a change or trend which is determined not to be statistically significant may be of biological consequence. Quantitative information, particularly when this involves does–response considerations, should be reviewed against the background of the study as a whole and the perspective of normal biological variations.

Hazard Evaluation Procedures and the Role of Toxicology

Hazard is the likelihood that the known toxicity of a material will be exhibited under specific conditions of use. It follows that the toxicity of a material, ie, its potential to produce injury, is but one of many considerations to be taken into account in assessment procedures with respect to defining hazard. The following are equally important factors that need to be considered: physicochemical properties of the material; use pattern of the material and characteristics of the environment where the material is handled; source of exposure, normal and accidental; control measures used to regulate exposure; the duration, magnitude, and frequency of exposure; route of exposure and physical nature of exposure conditions, eg, gas, aerosol, or liquid; population exposed and variability in exposure conditions; and experience with exposed human populations. Consideration of the above information allows the exposure conditions to be defined and reviewed in the light of the known toxicity of the material being examined.

Relevance of Toxicology in Hazard Evaluation

Ideally, available information on the toxicology of a material should allow the following to be determined as part of a hazard evaluation procedure: nature of potential adverse effects; relevance of the conditions of the toxicology studies to the practical in-use situation; the average response, range of responses, the presence of a hypersensitive group, and an indication of minimal or no-effects levels; identification of factors likely to modify the toxic response; effects of acute gross overexposure, ie, accident situations; effects of repeated exposures; recognition of adverse effects; assistance in the definition of allowable and nonallowable exposure conditions; assistance in the definition of monitoring requirements; guidance on the need for personal and collective protection measures; guidance on first-aid, antidotal, and medical support needs; relevance of toxicity to coincidental disease; and definition of "at risk" individuals, eg, pregnant and fertile females; genetically susceptible individuals. Information of this type can only be obtained from carefully designed studies.

The cost of toxicology studies varies, for example, from tens of thousands of dollars for simple acute toxicity tests to perhaps several million dollars for an extensive study. Those needing toxicology testing of their materials should understand the requirements of the end user and arrange for a careful and critical independent opinion on the nature of the testing required, and then obtain estimates for the timing and costing of these from several laboratories. Advice should be sought about the reputation of these laboratories and their ability to conduct particular studies. Sponsors should arrange for an independent audit of the conduct and reporting of their studies.

BRYAN BALLANTYNE
Union Carbide Corporation
WILLIAM GEORGE FONG
Florida Department of Agriculture and Consumer Services

P. A. Fenner-Crisp, "Risk Assessment Methods for Pesticides in Food and Drinking Water," Office of Pesticide Programs, U.S. Environmental Protection Agency, presented at the Florida Pesticide Review Council Meeting, July 7, 1989.

V. W. Mayer and W. G. Flamm in A. L. Reeves, ed., *Toxicology: Principles and Practice,* Vol. 1, John Wiley & Sons, Inc., New York, 1981.

C. L. Galli, S. D. Murphy, and R. Paoletti, eds., *The Principles and Methods in Modern Toxicology,* Elsevier/North Holland, Amsterdam, 1980.

Principles and Methods for Evaluating the Toxicity of Chemicals, Part 1, Environmental Health Criteria No. 6, World Health Organization, Geneva, Switzerland, 1978, Chapt. 5.

TRACE AND RESIDUE ANALYSIS

Trace analysis is the detection of minute quantities of organic and inorganic materials. As of the mid-1990s, trace analysis is generally recognized as those determinations that represent around 0.0001%, ie, at the parts per million (ppm) level, where 1 ppm is equivalent to 1 μg/g. Ultratrace analysis, ie, determination below trace analysis, corresponds to levels below ppm or $<\mu$g/g. Residue analysis is the analysis of material left from an operation, ie, residual.

There are numerous applications for trace or ultratrace analyses in the chemical process industry. Environmental toxicology, in particular, is an area where determination of residues and traces of pesticides and other toxic substances is frequently employed.

Frontiers of Low Level Detection

Extremely low level detection work is being performed in analytical chemistry laboratories. Detection of rhodamine 6G at 50 yoctomole (50×10^{-24} mol) has been reported using a sheath flow cuvette for fluorescence detection following capillary electrophoresis. This represents 30 molecules of rhodamine, a highly fluorescent molecule (see ELECTROSEPARATIONS, ELECTROPHORESIS; SPECTROSCOPY, OPTICAL).

Claims of single molecule detection in liquid samples have been made by combining the high sensitivity of laser-induced fluorescence (lif) and the spatial localization and imaging capabilities of optical microscopy. This technique combines confocal microscopy, diffraction-limited laser excitation, and a high efficiency detector. The probe volume is defined latitudinally by optical diffraction and longitudinally by spherical aberration. Using an unlimited excitation throughout and a low background level, this technique allows fluorescence detection of single rhodamine molecules at a signal-to-noise (S/N) ratio of approximately 10 in 1 ms. The use of confocal fluorescence microscopy can be extended to individual, fluorescently tagged biomolecules, including deoxynucleotides, whether single-stranded primers or double-stranded deoxyribonucleic acid (DNA).

Analysis of single mammalian cells by capillary electrophoresis has been reported using on-column derivatization and laser-induced fluorescence detection. Radioactive tracers (qv) are powerful tools for trace detection. A method of labeling proteins using ^{99m}Tc has been described. Radiotracer imaging agents have been used for mapping sympathetic nerves of the heart. The radioiodination of analogues of a calichemicin constituent have been employed as a possible brain-imaging agent (see MEDICAL IMAGING TECHNOLOGY).

Samples

Sampling. A sample used for trace or ultratrace analysis should always be representative of the bulk material. The principal considerations are determination of population or the whole from which the sample is to be drawn, procurement of a valid gross sample, and reduction of the gross sample to a suitable sample for analysis (see SAMPLING).

Sample Preparation. Sample contamination must be prevented throughout the sampling procedures. No significant changes should occur in the sample when it is being held for analysis. Sample stabilization generally includes storage at low temperatures; however, any stabilization generally includes storage at low temperatures; however, any stabilization step should be validated. A review of sample composition and properties is advised. This would include number of compounds present, chemical structures (functionality) of compounds, molecular weights of compounds, pK_a values of compounds, uv spectra of compounds, nature of sample matrix (solvent, fillers, etc), concentration range of compounds in samples of interest, and sample solubility. Significant sample losses can occur during this step because of very small volume losses to glass walls of the recovery containers, pipets, and other glassware.

Solid-phase microextraction (SPME), used as a sample introduction technique for high speed gc, utilizes small-diameter fused-silica fibers coated with polymeric stationary phase for sample extraction and concentration. SPME has been utilized for determination of pollutants in aqueous solution by the adsorption of analyte onto stationary-phase coated fused-silica fibers, followed by thermal desorption in the injection system of a capillary gas chromatograph. Full automation can be achieved using an autosampler.

Polycyclic aromatic hydrocarbons (PAHs) have been extracted from contaminated land samples by *supercritical fluid extraction* (SFE) with both pure and modified carbon dioxide. Removing an analyte from a matrix using SFE requires knowledge about the solubility of the solute, the rate of transfer of the solute from the solid to the solvent phase, and interaction of the solvent phase with the matrix. These factors collectively control the effectiveness of the SFE process, if not of the extraction process in general. The range of samples for which SFE has been applied continues to broaden. Applications have been in the environment, food, and polymers.

Sample preparation techniques that prevent or minimize pollution in analytical laboratories, improve target analyte recoveries, and reduce sample preparation costs were evaluated with regard to the *microwave-assisted extraction* (MAE) procedure for 187 compounds and four Aroclors listed in EPA Methods 8250, 8081, and 8141A (see MICROWAVE TECHNOLOGY). The results indicate that most of these compounds can be recovered in good yields from the matrices investigated. Comparative studies were performed to evaluate microwave digestion with conventional sample destruction procedures. These included the analysis of shellfish, meats, rocks, and soils. Generally, comparable accuracy at much shorter digestion time was found for the MAE vs the classical digestion method.

Sample Cleanup. The recoveries from a quick cleanup method for waste solvents based on sample filtration through a Florisil and sodium sulfate column are available. This method offers an alternative for analysts who need to confirm the presence or absence of pesticides or PCBs.

Method Validation

Statistically designed studies should be performed to determine accuracy, precision, and selectivity of the methodology used for trace or ultratrace analyses. The reliability requirements for these studies are that the data generated withstand interlaboratory comparisons.

The following principles should be used to establish a valid analytical method: A specific detailed description and protocol should be written (standard operating procedure (SOP)). Each step in the method should be investigated to determine the extent to which environmental, matrix, material, or procedural variables, from time of collection of material until the time of analysis and including the time of analysis, may affect the estimation of analyte in the matrix. A method should be validated for its intended use with an acceptable protocol. Wherever possible, the same matrix should be used for validation purposes. The concentration range over which the analyte will be determined must be defined in the method, on the basis of actual standard samples over the range (standard curve). It is necessary to use a sufficient number of standards to adequately define the relationship between concentration and response. Determination of accuracy and precision should be made by analysis of replicate sets of analyte samples of known concentration from equivalent matrix.

Methodologies

The commonly used methods for ultratrace analyses together with the accepted detection limits are given in Table 1.

Atomic absorption or emission spectrometric methods are commonly used for inorganic elements in a variety of matrices. A radiochemical neutron activation analysis technique for determination of 26 elements, including the emitting elements Th and U and Cu, Fe, K, Na, Ni, and Zn, has been developed. The radiochemical separation was performed by anion exchange on Dowex 1 × 8 column from HF and HF–NH_4F medium, leading to selective removal of the matrix-produced radionuclides ^{46}Sc, ^{47}Sc, ^{48}Sc, and nearly selective isolation of ^{239}Np and ^{233}Pa, the indicator radionuclides of U and Th, respectively. For K, Na, Th, and U, a limit of detection of 30, 0.05, 0.03, and 0.07 ng/g, respectively, was achieved.

The most commonly used approach in thin-layer chromatography (tlc) entails separations on a silica (qv) gel plate where the silica gel is coated as a thin layer on a glass plate. The plate is developed using the mobile phase of choice after a sample has been applied to the starting line of the plate. Quantification is achieved directly by scanning the plate or indirectly by scraping and eluting the sample.

A rapid tlc immunoaffinity chromatographic method has been reported for quantitation in serum of an acute phase reactant, C-reactive protein (CRP), which can differentiate between viral and bacterial infections.

A number of compounds have been quantified by tlc or high performance thin-layer chromatography (hptlc) using absorption or fluorescence scanning densitometry.

Gas chromatography is a technique utilized for separating volatile substances (or those that can be made volatile) between two phases,

Table 1. Ultratrace Analyses Methods

Method	Minimum amount detected, g
mass spectrometry	
electron impact	10^{-12}
spark source	10^{-13}
ion scattering	10^{-15}
flame emission spectrometry	10^{-12}
liquid chromatography	
ultraviolet detection	10^{-11}
fluorescence detection	10^{-12}
gas chromatography	
flame ionization	10^{-12} to 10^{-14}
electron capture	10^{-13}
combination techniques	
liquid chromatography/mass spectrometry	10^{-12}
gas chromatography/mass spectrometry	10^{-12}
electron capture (negative)/ionization mass spectrometry	10^{-15}

one of which is a gas. Purge-and-trap methods are frequently used for trace analysis. Various detectors have been employed in trace analysis, the most commonly used being flame ionization and electron capture detectors.

High pressure liquid chromatography (hplc), frequently referred to as simply lc or as high performance liquid chromatography, is used in virtually all fields of chemistry. Nonvolatile or thermally labile compounds are best separated by hplc. Although techniques such as adsorption and ion-exchange chromatography have been used, the technique of choice is reversed-phase liquid chromatography (rplc). In rplc the stationary phase is nonpolar and the mobile phase is polar and its polarity can be suitably changed.

In hplc, detection and quantitation have been limited by availability of detectors. Using a uv detector set at 254 nm, the lower limit of detection is 3.5×10^{-11} g/mL for a compound such as phenanthrene. A fluorescence detector can increase the detectability to 8×10^{-12} g/mL. The same order of detectability can be achieved using amperometric, electron-capture, or photoionization detectors. Hplc is capable of routine determination at the nanogram range.

Increased use of liquid chromatography/mass spectrometry (lc/ms) for structural identification and trace analysis has become apparent. Thermospray lc/ms has been used to identify by-products in phenyl isocyanate precolumn derivatization reactions. Liquid chromatography/thermospray mass spectrometric characterization of chemical adducts of DNA formed during *in vitro* reaction has been proposed as an analytical technique to detect and identify those contaminants in aqueous environmental samples which have a propensity to be genotoxic, ie, to covalently bond to DNA.

Using capillary hplc, femtomole amounts of recombinant DNA-derived human growth hormone (rhgh) have been successfully detected from solutions at nanomolar concentrations. A sample of rhgh that was recovered from rat serum was analyzed by capillary reversed-phase hplc, using both acidic- and neutral-pH mobile phases, as well as by capillary ion-exchange chromatography. Submicrogram amounts of rhgh were also analyzed by tryptic mapping, using capillary hplc, and the resulting peptides were identified by capillary lc/ms.

Capillary electrophoresis (ce) is an analytical technique that can achieve rapid high resolution separation of water-soluble components present in small sample volumes. The separations are generally based on the principle of electrically driven ions in solution. Selectivity can be varied by the alteration of pH, ionic strength, electrolyte composition, or by incorporation of additives. Typical examples of additives include organic solvents, surfactants (qv), and complexation agents (see CHELATING AGENTS).

Supercritical fluid chromatography (sfc) combines the advantages of gc and hplc in that it allows the use of gc-type detectors when supercritical fluids are used instead of the solvents normally used in hplc. Carbon dioxide, *n*-petane, and ammonia are common supercritical fluids (qv). For example, carbon dioxide (qv) employed at 7.38 MPa (72.9 atm) and 31.3°C has a density of 448 g/mL. Derivatization of primary and secondary amines using 9-fluorenylmethyl chloroformate to form a nonpolar, uv-absorbing derivative has been reported.

Immunoassays (qv) may be simply defined as analytical techniques that use antibodies or antibody-related reagents for selective determination of sample components. These make up some of the most powerful and widespread techniques used in clinical chemistry. The main advantages of immunoassays are high selectivity, low limits of detection, and adaptibility for use in detecting most compounds of clinical interest. Because of their high selectivity, immunoassays can often be used even for complex samples such as urine or blood, with little or no sample preparation.

A two-site immunometric assay of undecapeptide substance P (SP) has been developed. A number of solid-phase automated immunoassay analyzers have been used for performing immunoassays. A number of immunoassay methods have been found useful for environmental analysis (see AUTOMATED INSTRUMENTATION).

Applications

Trace or ultratrace and residue analyses are widely used throughout chemical technology. Areas of environmental investigations, explosives, food, pharmaceuticals, and biotechnology rely particularly on these methodologies.

Environment. Detection of environmental degradation products of nerve agents directly from the surface of plant leaves using static secondary ion mass spectrometry (sims) has been demonstrated.

Some of the methods used for determination of organic pollutants in the environment follow. The most notable are polyaromatic hydrocarbons (PAHs) and volatile organic compounds (VOCs).

Analytes	Methods
n-alkanes	gc/fid
VOCs	prefraction and gc
PAHs	sample clean up and gc
bromoform and other bromine compounds	gc for flame-retardant compounds in air
nitrodibenzopyranone	isomeric separations based on gc
PCDDs and PCDFs	gc/ms
PAH	on-line lc/gc for PAH in air
chlorinated PAH	lc/gc/ms
nitrated PAH	hplc with electrochemical detection
azaarenes	column chromatography, cleanup, and tlc
toluene diisocyanate	micro lc
dimethyl sulfoxide or sulfone	atmospheric pressure chemical ionization (APCI) ms
VOCs	ion trap monitoring
volatile chlorine compounds	fiber-optic emission sensor-based on AA of chlorine
formaldehyde	monitoring tape: hydroxylamine sulfate, methyl yellow
methyl nitrite	nitrogen oxide-indicating tubes with ir detection
VOCs	ir-based methods for on-site analysis
isocyanate species	chemiluminescent techniques
benzene	photoionization detector
hydrazine	coulometric methods
nitrobenzene	piezoelectric sensors
fluorocarbon	metal oxide sensors
perchloroethylene	quartz balance and calorimetric transducer

A variety of organic and inorganic analytes have been analyzed in air and in water.

Explosives. Explosives can be detected using either radiation- or vapor-based detection. The aim of both methods is to respond specifically to the properties of the energetic material that distinguish it from harmless material of similar composition. These techniques are useful for detecting organic as well as inorganic explosives (see EXPLOSIVES AND PROPELLANTS).

Food. Laws and regulations controlling contamination of food were once the province of religious organizations. As far back as the Dark Ages, government set standards. Significant changes have occurred only relatively recently, since the time analytical chemists could characterize most of the substances that comprise food and thus more effectively control contaminants in it. The reliability of data regarding medicated feeds (see FEEDS AND FEED ADDITIVES), dairy products (see MILK AND MILK PRODUCTS), seafood, meat products (qv), fruits and vegetables, and beverages has been reviewed.

The development of analytical strategies for the regulatory control of drug residues in food-producing animals has also been reviewed. Because of the complexity of biological matrices such as eggs (qv), milk, meat, and drug feeds, well-designed off-line or on-line sample treatment procedures are essential. For example, methylmercury in fish was extracted and cleaned using column chromatography and then determined using flameless atomic absorption. A rapid and sensitive electrothermal AA spectrophotometric method using a combination of microwave digestion and palladium as a stabilizer was used for detecting Cd and Pb at sub ppm levels in vegetables and protein in foodstuffs.

Inorganic elements in food can be determined by atomic absorption (AA) methods. These methods have been extensively reviewed.

Pharmaceuticals. Examples of trace and ultratrace analyses of various drugs and pharmaceuticals (qv) have been provided throughout. The purity of the active ingredient, its content and availability in dosage form, therapeutic blood levels, delivery to target areas, elimination (urine, feces, and metabolites), and toxicity are always of importance.

A safety profile of a generic drug can differ from that of the brand-name product because different impurities may be present in each of the drugs. Impurities can arise out of the manufacturing processes and may be responsible for adverse interactions that can occur.

The subject of impurity analysis of pharmaceutical compounds has been insufficiently addressed in the scientific literature. Many monographs in the *U.S. Pharmacopeia* have nonspecific assay methods. An attempt has been made to address this problem by focusing on specific methodologies and delineating origination and concentration of impurities found in pharmaceutical compounds.

Chiral separations (qv) have become of significant importance because the optical isomer of an active component can be considered an impurity. Optical isomers can have potentially different therapeutic or toxicological activities. The pharmaceutical literature is trying to address the issues pertaining to these compounds. Frequently separations can be accomplished by glc, hplc, or ce.

Biotechnology. Particular attention must be paid to the detection of DNA in all finished biotechnology products because of the possibility that such DNA could be incorporated into the human genome and thus become a potential oncogene. The absence of DNA at the picogram-per-dose level should be demonstrated in order to assure the safety of biotechnology products.

The isolation and purification of DNA and ribonucleic acid (RNA) restriction fragments are of great importance in the area of molecular biology. These fragments are the product of site-specific digestion of large pieces of DNA and RNA with enzymes called restriction endonucleases. The fragments may range in size from a few base pairs to tens of thousands of base pairs. An ion-exchange column can provide DNA and RNA separations within one hour, giving resolution equivalent to that obtained with gel electrophoresis. Nucleic acid fragments are then visualized, using on-line uv detection and sample loading from 500 ng to 50 mg.

Molecular biologists are utilizing hplc for characterization and purification of proteins (qv), peptides, and antibodies.

Miscellaneous. Trace analyses have been performed for a variety of other materials.

Thermal neutron activation analysis has been used for archeological samples, such as amber, coins, ceramics, and glass; biological samples; and forensic samples (see FORENSIC CHEMISTRY); as well as human tissues, including bile, blood, bone, teeth, and urine; laboratory animals; geological samples, such as meteorites and ores; and a variety of industrial products.

SATINDER AHUJA
Ahuja Consulting

S. Ahuja, *Impurities Evaluation of Pharmaceuticals,* Marcel Dekker, New York, 1998.

S. Ahuja, *Trace and Ultratrace Analysis by HPLC,* John Wiley & Sons, Inc., New York, 1992.

P. Kolla, *Anal. Chem.* **67,** 184A (1995).

S. Ahuja, *Chiral Separations Applications & Technology,* American Chemical Society, Washington, D.C., 1997.

TRACERS. See IMAGING TECHNOLOGY; RADIOACTIVE TRACERS.

TRANQUILIZERS. See PSYCHOPHARMACOLOGICAL AGENTS.

TRANSISTORS. See SEMICONDUCTORS.

TRANSPORTATION

The transportation of chemicals and related products is unusual in that substantial quantities are moved in packages as well as in bulk. Other materials, such as coal (qv), grain, and ore, are transported in bulk but seldom in packaged form. Moreover, most other bulk commodities, including petroleum and its products, are limited in the diversity of their chemical and physical characteristics and, therefore, do not require as wide a variety of packaging and bulk conveyances as is necessary for the movement of chemicals. Virtually all railroad tank cars are supplied by chemical producers rather than railroad companies, which furnish at least a portion of most other types of equipment used in rail transportation.

Since the late nineteenth century, the U.S. federal government and almost all states have regulated both the supply and pricing of transportation service, and such regulation has had a profound effect on the chemical industry. In 1993, railroads carried 24.3% of all intercity tonnage; trucks, 45.1%; oil pipelines, 15.7%; water, 14.7%; and air transportation, 0.1%.

Transportation Modes

Railroads. Traditionally, railroads have furnished boxcars of varying sizes and capacities for general-purpose rail movement of packaged freight, including chemicals and related products, in drums, barrels, bags, and other containers. More recently, the introduction of energy-absorbing underframes and other technical innovations has enhanced the ability of boxcars to carry greatly increased loads without excessive damage to the lading.

Railroads generally do not supply tank cars and other special-purpose rail cars such as covered hopper cars for the movement of bulk plastic materials. Rather, shippers or receivers must furnish such equipment, usually through a purchase or lease arrangement with car manufacturers or lessors. Car manufacturers and intermediaries offer various forms of rail-car leases, ranging from short-term, full-maintenance rentals to long-term leases requiring outside financing.

At many chemical plants, as well as other manufacturing or receiving facilities dependent on rail transportation, railroad tracks are constructed within the plant to permit the shipment or receipt of rail cars. Such tracks, usually called industry or private tracks or sidetracks, connect directly with the tracks of the railroad(s) serving the plant. Because the sidetrack must be compatible with the railroad track to permit railroad switch engines and crews to enter the plant, the industry and railroad enter into a written sidetrack agreement, which defines their respective rights and obligations with regard to track construction, maintenance, and operation. Included in such agreements are provisions pertaining to required lateral and overhead track clearances and maintenance of hoppers, pits, or other loading or unloading devices.

When private tracks have insufficient capacity for the number of freight cars required to be stored, shippers or receivers of freight may lease additional trackage from a railroad in the vicinity of the plant. Frequently, tracks located at strategic places remote from a plant facility may be leased for storage of loaded cars in order to have them available for prompt delivery to customers or distributors in the vicinity of the track. When a railroad provides transportation services under a continuing contract, many of these matters are addressed in the contract.

Motor Carriage. Since the 1930s, motor carriage has been an essential part of the U.S. transportation system. Improved public highways, including the interstate system, helped to develop a network of transportation competitive with railroads in both rates and service.

Less capital-intensive and, therefore, more numerous than railroads, and unconfined by the rigidity of tracks, motor carriers demonstrated a flexibility that broke historic patterns of industrial concentration in transportation centers, thereby contributing to the dispersion of manufacturing and other commercial enterprises to suburban and rural areas. Motor carriage may fall within one of three different categories. In 1994, U.S. Congress eliminated most tariff filing requirements for common carriers, allowing shippers and common carriers to negotiate individually determined rates, terms, and conditions. Tariffs are required for chemicals traffic only if the shipment moves in noncontiguous domestic trade, ie, transportation from or to Alaska, Hawaii, or a territory or possession of the United States. Some limited volume of traffic also may be subject to certain rates, terms, and conditions collectively established by a group of carriers pursuant to an agreement between such carriers, if approved and exempted from the antitrust laws by the Surface Transportation Board (STB). In addition to common carriers, motor contract carriers have been widely used in the chemical industry, in part because relaxation of federal regulations governing contract carriage resulted in a proliferation of such service in recent years.

A third type of motor carriage of considerable importance to many industries, including the chemicals industry, is proprietary or private carriage. Such transportation is conducted in furtherance of a primary business other than transportation. Thus, manufacturers transporting goods that they have manufactured or processed or that they will use in such manufacturing or processing, or for purposes of bona fide sale or purchase, are engaged in private carriage. It is generally not required that a company use only vehicles that it owns rather than leases, or that it directly employ the drivers of such vehicles, provided that such company actually controls the transport operation and bears its characteristic burdens and financial risks.

Corporate members of a single group of corporations may lawfully perform such transportation for a parent or subsidiary, or for a sister subsidiary, provided that the one corporation wholly owns the other or that both are wholly owned by a common parent.

Motor carriers use a wide variety of highway vehicles, including trucks, tractors, trailers, tank vehicles, hopper vehicles, low-boys, vans, and others. Unlike railroads, commercial motor carriers of bulk liquids or solids in tank or hopper trucks usually offer shippers both power equipment (tractor) and freight-carrying trailers, although shippers frequently supply such trailers under special arrangements. The development of the interstate highway system and more permissive federal and state legislation have allowed the use of vehicular equipment of increased length and other dimensions, as well as higher weight-carrying capacity, thereby contributing to more economical motor transportation. Such legislation, however, has in turn given rise to disputes concerning highway tolls, fuel taxes, registration fees, and similar assessments, against both for-hire and proprietary truck operators, to permit adequate maintenance of highways.

Waterborne Transport. Despite natural limitations, the transportation of chemicals by water has enjoyed substantial growth, especially since the end of World War II. Although water carriers are sometimes classified as common or contract carriers, such distinctions are frequently insignificant, because water carriage of bulk chemicals in the United States is essentially unregulated. In conformity with longstanding practice in the maritime field, such transportation is often provided under various forms of agreement, such as bareboat charters, time charters, or voyage charters. Barges, like other transportation vehicles, are available in a variety of types, sizes, and capacities. As in the case of highways, considerable contention results from public maintenance of the inland waterways for recreation, flood control, and other purposes, as well as for the transportation of barges and other freight-carrying vessels. Because barge transportation of chemicals is considered essential to economical distribution, governmental tolls assessed for such maintenance are of critical interest to the chemicals industry.

Most oceangoing vessels, particularly those used between North America and other continents, are self-propelled. For the movement of packaged freight in international commerce, ocean transportation in recent years has been dominated by container ships designed to load and carry large, trailer-sized containers. Such ships can be loaded and unloaded more quickly than traditional freight-carrying vessels. Other types of ocean vessels include tankers and dry-bulk ships for the transportation of a wide variety of liquid hydrocarbons, chemicals, and materials such as coal, coke, and ores in large quantities. Chemical tankers tend to be smaller in size than petroleum tankers and usually have several compartments, each designed to carry one or more products.

Pipelines. The feasibility of pipeline transportation depends on the availability of very large quantities of compatible materials between locations with sufficient storage facilities. Thus, pipeline transportation is predominantly, but not exclusively, limited to the movement of hydrocarbons, many of which are raw materials in the production of petrochemicals.

Air Transport. Relatively small quantities of chemicals are transported by air, although availability of such service for the movement of samples, emergency shipments, and radioactive chemicals with a short half-life is important. Both economic and safety considerations impede the development of air carriage as a significant means of transporting a substantial volume of chemicals.

Other Services. Domestic freight forwarders, although sometimes treated as common carriers, do not provide any physical transportation service. Instead, they arrange transportation services for their customers, usually the underlying shipper of goods, and perform related functions such as the booking of space with a carrier and preparing necessary documentation. One important function commonly performed by freight forwarders is the consolidation of multiple small shipments into carload or truckload lots, which are forwarded to a central location for subsequent distribution to individual destinations. In the export and import trade, where transportation is provided in whole or in part by a water carrier, similar services are provided by commercial operators known as nonvessel operating common carriers (NVOCC).

The diversity and flexibility of a highly developed transportation structure is demonstrated by intermodal transportation, ie, the combination of two or more transportation modes. Traditional combinations, such as rail and water or truck and water, are essentially end-to-end arrangements. There has been substantial growth in combinations of the various transportation modes, such as the piggyback transportation of trucks or trailers on railroad flat cars, and similar loadings of trucks or containers on ships or barges. These methods of transportation are largely deregulated and have led to substantial economies for both carriers and shippers.

Warehouses and Terminals. Warehousing constitutes an integral part of the distribution system of the United States. Although employed primarily to store inventory, warehouses are also used to assure timely deliveries to customers remote from a production facility. Additionally, warehousing may facilitate the aggregation of large shipments, thus reducing transportation costs. Warehouses may be owned and operated by individual companies for their own purposes or they may be available to the public for storage of goods. The chemicals industry makes extensive use of bulk terminals for storage of liquid- and dry-bulk materials in a wide variety of sizes and types of tanks, silos, bins, and other facilities. Warehouse and terminal operators, who offer their facilities and services for compensation, are liable for goods in their custody if they are negligent.

Shipping

Shipping Terms. Although frequently referred to as shipping terms, *fob, fas,* and *cif* are actually terms of sale because they pertain to the relationship between vendor and vendee, rather than between shipper and carrier. Fob means free on board and usually indicates that delivery of the goods to the vendee will occur when the goods, packaged in accordance with the terms of the sales agreement, are delivered aboard a vehicle of the type agreed upon at the fob point named.

The selection of shipping terms has a material effect on the sales contract. The party with the risk of loss must decide whether or not to insure against such risk and must prepare and file a claim against the transportation carrier when goods are lost or damaged in transit. Unless otherwise agreed, that party must also pay transportation charges and file any claims for freight overcharges. In export or import transactions, shipping terms such as fas (free alongside ship) or cif (cost, insurance, freight) may also determine the party responsible for preparing required documents, obtaining customs clearances, and similar matters.

Shipping Documents. The bill of lading serves as a receipt for goods delivered to a carrier as well as a contract of carriage. Bills of lading may be negotiable documents and as such constitute evidence of title or the right to possession of the goods described in the document.

In international trade, the ocean bill of lading serves essentially the same purposes, although it may differ in form and content and is frequently negotiated in such a manner that payment by the foreign consignee is required before delivery of goods by the carrier.

Interstate and Intrastate Commerce

The applicability of various federal and state transportation laws and regulations depends on whether transportation constitutes interstate or intrastate commerce. The transportation laws and regulations that may apply are both economic (ie, rates and routes) and safety related. Beginning in January 1995, however, Congress preempted all state economic regulation of intrastate motor carriage, but did not substitute federal regulation for the preempted state regulations. As a result, although limited economic regulation of interstate commerce by motor carrier remains, there is no economic regulation of intrastate commerce by motor carrier.

Except in rare instances, it can be assumed that transportation requiring physical movement across state boundaries is interstate commerce. On the other hand, transportation that takes place wholly within the confines of a single state is not necessarily intrastate commerce, because such transportation may be a portion of a continuous movement in interstate commerce.

Economic Regulation

In the U.S., transportation has long been subjected to regulation by both federal and state governments and has been directed at operational safety or toward economic concerns such as discrimination in rates and services or excessive competition. In addition, regulatory statutes have provided for control of entry into the transportation business, regulation of freight rates and charges, and various finance, accounting, and insurance requirements, although there are numerous exceptions. Among the exceptions of particular importance to the chemical industry is that afforded to water carriage of liquid and dry-bulk commodities.

At the federal level, the STB, the Federal Maritime Commission (FMC) the Department of Transportation (DOT), and the Federal Energy Regulatory Commission (FERC) are all concerned with economic regulation of various modes of transportation. The STB regulates interstate railroads, motor and water carriers, and pipelines (other than water, gas, and oil). The DOT regulates motor carriers and international airlines, and the FMC regulates water transportation in foreign commerce. The FERC regulates pipeline transportation of oil, natural gas, water, and other energy resources.

There has been a significant trend toward relaxation of many economic regulatory controls since the 1970s which has resulted in substantial change in the transportation industry. Regulation, however, has not been entirely abandoned and in many respects at least the form of railroad regulation remains essentially intact. With respect to freight rates, historic rules requiring that rates be reasonable and prohibiting discrimination or preference as between particular shippers or geographic areas have been phased out to varying degrees among different modes of transportation. To the extent these historic rules remain at all, their impact has been largely dissipated by provisions placing increased reliance on competitive forces.

A significant result of regulatory relaxation is an increase in the authority of the STB to grant administrative exemptions from railroad and motor-carrier regulation. Such authority has been exercised in a variety of ways, including the virtually complete deregulation of most piggyback transportation. Thus, railroads are not treated as regulated carriers when performing piggyback service. Regulated railroads, when not operating under contracts with shippers, are required to provide a shipper, upon request, its rates and terms for transportation as a common carrier. A rail common carrier cannot increase a rate or change the terms of service for a shipper who has requested this information within the previous 12 months unless the carrier provides 20 days prior notice. Shippers must bring a lawsuit to recover any overcharges by a railroad and carriers must bring a lawsuit to recover any undercharges within three years after the claim accrues.

STB credit regulations require railroads to collect all freight charges within five days after issuance of a freight bill, and motor carriers must collect charges within seven days after billing. Motor carriers must also maintain certain minimal liability insurance coverage for the protection of the public, as well as insurance covering carrier liability for loss of or damage to freight. In the past, many state governments regulated economic activity in intrastate transportation in a manner similar to federal regulation. However, beginning in 1995, the federal government preempted all state economic regulation of intrastate motor carriage, although the federal government did not supplant state regulation. Instead, the market for intrastate motor carriage was left open to competitive market forces. With respect to railroads, federal legislation has effectively compelled the states to adopt federal regulatory requirements or to abandon railroad regulation entirely.

Freight Rates and Allowances. The establishment of freight rates, ie, a transportation price structure, embraces virtually all articles of commerce in a multitude of packages and quantities, via numerous routes and between innumerable locations. A variety of intermediate services are also often included, such as storage or reconsignment in transit and stop-offs to partially load or unload. The classification of freight into various categories or classes is the result of an effort to systemize the various factors considered in fixing a particular rate. Freight classifications are generally based on freight density, susceptibility to damage or theft, value of the goods, etc.

Freight classification has established a more or less standard nomenclature to identify the numerous products shipped in commerce. This standardization facilitates preparation of shipping documents, determination of freight rates, and free interchange of freight between connecting or competing carriers. Chemical and freight nomenclatures, however, are frequently different.

In the 1990s, the variety of possible transportation arrangements is virtually without limit, but it may be useful to generally describe the most common types of freight rates and charges. (*1*) Less-than-carload or less than-truckload rates are applicable to quantities of particular commodities less than a specified volume considered to constitute a carload or truckload quantity of such commodities. (*2*) Carload or truckload rates are applicable to quantities of a commodity sufficient to constitute a specified minimum carload or truckload volume. (*3*) Multiple car rates are applicable only when a specified number of carloads is tendered to a railroad for transportation in a single shipment. For commodities such as coal, which move in large volumes, trainload rates may be provided. (*4*) Annual (or periodic) volume rates are applicable to individual shipments that are part of an aggregate tonnage of a particular commodity or commodities that a shipper has agreed to ship between specified points in a specified period. (*5*) Accessorial charges are for services that are ancillary to line-haul transportation, such as switching, demurrage, storage or stopping in transit, reconsignment, and similar services.

Empty return movements are usually made without charge, provided that aggregate loaded and empty distances on each railroad are maintained in equilibrium. Such allowances, paid for loaded miles of car movement, represent large revenues for the industry and are an important consideration in calculating the actual (net) cost of transporting a given shipment. The amount of the per-mile allowance varies with the fair market value and age of a car, and may be eliminated altogether when freight rates are contractually established "net" of such allowances.

Freight Loss and Damage. In the early 1900s, liability was codified in the Interstate Commerce Act for railroads, motor carriers, and freight forwarders and provided that such carriers were liable for the full, actual loss of or damage to the goods. This codification is commonly referred to as the Carmack Amendment. Freight rates applicable to shipments subject to limited carrier liability are known as released rates.

Recent amendments to the Carmack Amendment, arguably allow rail and motor carriers to unilaterally limit their liability by the establishment of released value rates, without the shipper's knowledge. The Act also provides that claims for loss or damage of goods transported via joint through routes may be filed against either the originating or destination carrier or against an intermediate carrier on whose line(s) the goods are known to have been damaged. Additionally, the Act provides a minimum time limit for filing a claim (nine months) and commencing suit (two years). Carriers generally are liable only for full actual loss.

When goods consigned to a shipper's warehouse or terminal are damaged, disputes frequently arise as to their value. Such disputes are sometimes resolved by payment of the sales price less costs not incurred, such as the cost of delivery from the warehouse to the consignee.

Contract carriers generally are not held to the same standard of liability as common carriers because they are considered ordinary for-hire bailees and, therefore, are liable only for their failure to exercise a reasonable degree of care for goods in their custody or possession, although such liability may be varied by the contract. However, motor carriers providing service under contract are held to the same liability standard applicable to common carriers unless the statutory provisions imposing the standard are waived in the contract.

The liability of water carriers is established under principles of traditional admiralty law, which generally reflect the fundamental concept of liability for negligence, modified to accommodate risks peculiar to the long and dangerous voyages in ancient times. Carriers are not liable for the fault or negligence of the shipper as, for example, in using faulty or defective packaging or in improper loading. Nevertheless, most transit loss or damage is recoverable from the carrier and, as a result, many shippers find it unnecessary to insure freight transported by land carriers, unless carrier liability has been limited in accordance with the legal principles discussed above. On the other hand, it is common practice for shippers or receivers to insure cargo transported by water, unless the carrier has contractually agreed to purchase such insurance.

Safety Regulation

Upon the establishment of DOT in 1967, the Federal Aviation Administration (FAA) and Coast Guard were transferred to that department and the safety functions formerly administered by ICC were assumed by DOT.

DOT safety regulations fall into two categories: The first pertains to the qualifications and hours of service of carrier employees and the safety of transport operations and equipment. The second, of special concern to the chemical industry, pertains to the transportation of hazardous materials and related commodities. The National Transportation Safety Board is responsible for investigating serious accidents by all modes of transportation. The Hazardous Materials Transportation Act of 1974 consolidated the authority of DOT with respect to safety.

Many private industrial or professional organizations, eg, ASME, the Compressed Gas Association, the CMA, and the Bureau of Explosives, publish standards for containers, materials of construction, and tests. Such standards are frequently incorporated by reference in the DOT regulations. In 1990 and 1991, however, the DOT issued comprehensive regulations adopting international performance-oriented packaging standards for nonbulk packaging (drums, boxes, etc). Contrary to prior packaging requirements, which included the prescription of package designs in considerable detail, performance standards rely primarily on package tests intended to simulate the transportation environment, such as drop, stacking, vibration, or pressure testing to demonstrate package integrity. Simultaneously, the DOT adopted the international system of identifying hazard classes by class or division numbers. Although the hazardous materials regulations are frequently characterized as highly complex, the application is greatly facilitated through the use of the Hazardous Materials Table at 49 CFR § 172.101, which alphabetically lists hundreds of hazardous material descriptions by proper shipping names as prescribed by DOT.

The RPSA adoption of the four-digit United Nations' numbering system for identification of hazardous materials reflects a worldwide effort to improve response to transportation emergencies. Although accidental release of hazardous materials in transit is relatively rare, the potential for significant harm is of constant concern to the public and industry and is magnified by the fact that many public emergency response agencies have had little, if any, training or experience in dealing with chemical emergencies. In an effort to provide immediate and reliable information to carriers and public officials at the scene of an emergency, the Chemical Manufacturers' Association established the Chemical Transportation Emergency Center (CHEMTREC) in Washington, D.C. Since its formation in 1971, CHEMTREC has responded to hundreds of thousands of emergency calls, providing information from its files containing data on nearly 1.5 million chemical products.

Despite its traditional emphasis on product-based regulation, various DOT rules adopted in the early 1990s were more concerned with regulated persons than with regulated products. Thus, as mandated by statute, DOT issued new rules in 1992 requiring shippers and carriers of certain materials or types of shipment to register with DOT and pay an annual registration fee. Similarly, DOT adopted regulations requiring all hazmat employees to be periodically trained and tested and to maintain certified records of such training.

Outlook

Transportation and distribution costs constitute a substantial portion of the total cost of the chemical industry. Rail carriers, armed with the freedom to price their services, to abandon unprofitable lines, and to merge with other railroads, have already demonstrated a tendency to increase rates on captive chemicals traffic. At the same time, however, the removal of regulatory restraints on contracts between shippers and railroads has generated a revolution in transport pricing.

In the motor-carrier field, increased competition resulting from the virtual elimination of economic regulatory controls has given motor carriers a degree of efficiency enabling them to challenge proprietary transportation in both cost and service. The availability of energy resources and the adequacy of the highway infrastructure may impose the most substantial constraints on the continued growth of motor transportation. Without a breakthrough in energy usage, the technology of transportation is not expected to change dramatically in the foreseeable future. Transportation safety will continue to be of concern to government at all levels but such concern may be directed less at new regulations and restraints and more on the application of existing computer and communications technology to more effective emergency response.

STANLEY HOFFMAN
Consultant
JEFFREY O. MORENO
KARYN A. BOOTH
ANTOINE P. COBB
RICHARD D. FORTIN
Donelan, Cleary, Wood & Maser PC

W. J. Augello, *Freight Claims in Plain English,* Shippers National Freight Claim Council, Inc., Huntington, N.Y., 1979.

Traffic World, The Journal of Commerce, Inc., Washington, D.C., published weekly. Widely read news magazine for the transportation industry.

W. J. Augello and S. Hoffman, *Transportation Contracts In Plain English,* Transportation Claims and Prevention Council, Inc., Huntington, N.Y., 1991. Review of laws and regulations pertaining to transportation contracts, with examination of numerous specific contract provisions.

Hazmat Transport News, biweekly newsletter, Business Publishers, Inc., Silver Spring, Md.

TRANSURANIUM ELEMENTS. See ACTINIDES AND TRANSACTINIDES.

TRIPHENYLMETHANE AND RELATED DYES

Triphenylmethane dyes are of brilliant hue, exhibit high tinctorial strength, are relatively inexpensive, and may be applied to a wide range of substrates. They are seriously deficient in fastness properties, especially fastness to light and washing. Consequently, the use of triphenylmethane dyes on textiles such as wool, silk, and cotton has decreased as dyes from other classes with superior lightfastness and washfastness properties have become available (DYES AND DYE INTERMEDIATES). Interest in this class of dyes was revived with the introduction of polyacrylonitrile fibers (see ACRYLONITRILE POLYMERS; FIBERS, ACRYLIC). Triphenylmethane dyes are readily adsorbed on this fiber and show surprisingly high lightfastness and washfastness properties, compared with the same dyes on natural fibers. However, the durability of acrylic fibers created an even greater demand for fastness properties.

The triarylmethane dyes are broadly classified into the triphenylmethanes (CI 42000–43875), diphenylnaphthylmethanes (CI 44000–44100), and miscellaneous triphenylmethane derivatives (CI 44500–44535). The triphenylmethanes are classified further on the basis of substitution in the aromatic nuclei, as follows: (*1*) diamino derivatives of triphenylmethane, ie, dyes of the malachite green series (CI 42000–42175); (*2*) triamino derivatives of triphenylmethane, ie, dyes of the fuchsine, rosaniline, or magenta series (CI 42500–42800); (*3*) aminohydroxy derivatives of triphenylmethane (CI 43500–43570); and (*4*) hydroxy derivatives of triphenylmethane, ie, dyes of the rosolic acid series (CI 43800–43875). Monoaminotriphenylmethanes are known but they are not included in the classification because they have little value as dyes.

Chemically, the triarylmethane dyes are monomethine dyes with three terminal aryl systems of which one or more are substituted with primary, secondary, or tertiary amino groups or hydroxyl groups in the para position to the methine carbon atom. Additional substituents such as carboxyl, sulfonic acid, halogen, alkyl, and alkoxy groups may be present on the aromatic rings. The number, nature, and position of these substituents determine both the hue or color of the dye and the application class to which the dye belongs.

Structure

The triarylmethane dyes are generally considered to be resonance hybrids. However, for convenience, usually only one hybrid is indicated, as shown for crystal violet, CI Basic Violet 3 (**1**), for which λ_{max} = 589 nm.

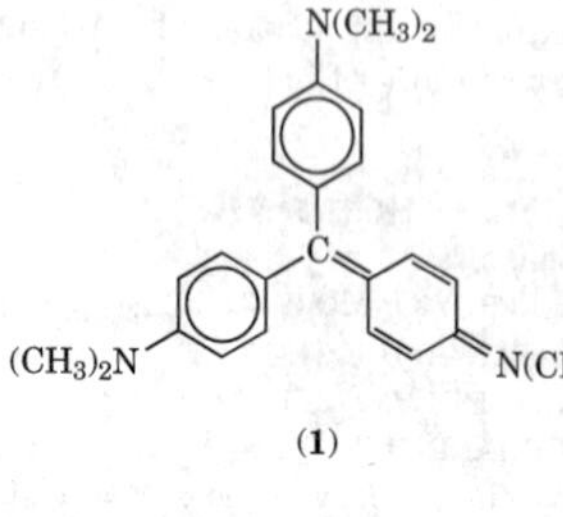

(**1**)

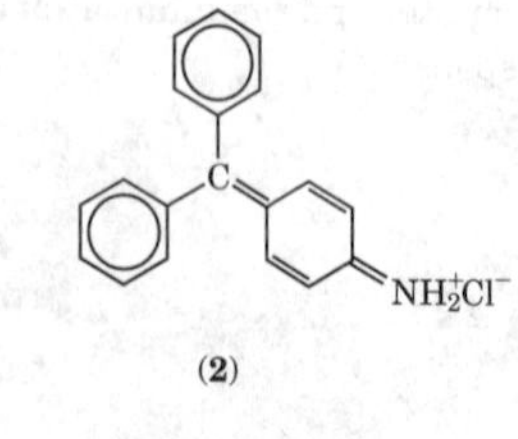

(**2**)

The ortho hydrogen atoms surrounding the central carbon atom show considerable steric overlap. Therefore, it can be assumed that the three aryl groups in the dye are not coplanar, but are twisted in such a fashion that the shape of the dye resembles that of a three-bladed propeller. Substitution in the para position of the three aryl groups determines the hue of the dye. When only one amino group is present, as in fuchsonimine hydrochloride, λ_{max} = 440 nm (**2**), the shade is a weak orange-yellow.

When at least two or more amino groups are present in different rings, the resonance possibility is greatly increased, resulting in a much greater intensity of absorption and in a strong bathochromic shift to longer wavelengths, eg, Doebner's violet (**3**), λ_{max} = 562 nm, which is a reddish violet, and pararosaniline (**4**) λ_{max} = 538 nm, which is a bluish violet. The amino derivatives of commercial value contain two or three amino groups.

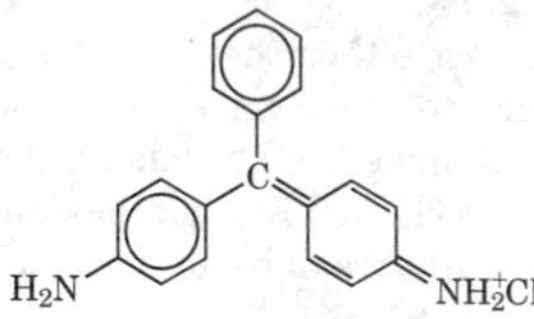

(**3**)

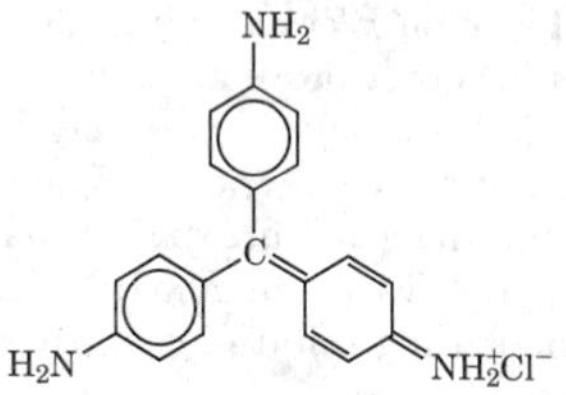

(**4**)

A further strong bathochromic shift is observed as the basicity of the primary amines is increased by *N*-alkylation, eg, malachite green, CI Basic Green 4, λ_{max} = 621 nm.

Chemical Properties

Although many triarylmethane dyes are prepared by the oxidation of leuco bases, they are usually destroyed by strong oxidizing agents. The triarylmethane dyes are extremely sensitive to photochemical oxidation, a fact which accounts for their poor lightfastness on natural fibers. There are many factors which affect the rate of fading (degradation) of the triarylmethane dyes on natural and synthetic fibers. Several studies have revealed that *N*-dealkylation occurs simultaneously with the cleavage and contributes to the photodegradation. *N*-Dealkylation is a general phenomenon in dye photochemistry.

Triarylmethane dyes are reduced readily to leuco bases with a variety of reagents. Reduction with titanium trichloride (Knecht method) is used for rapidly assaying triarylmethane dyes.

The direct sulfonation of alkylaminotriphenylmethane dyes gives mixtures of substituted products. Although dyes containing anilino or benzylamino groups give more selective substitution, a sulfonated intermediate such as 3[(*N*-ethyl-*N*-phenylamino)methyl]benzenesulfonic acid (ethylbenzylanilinesulfonic acid) is the preferred starting material.

Dyes containing highly alkylated amino groups are prepared from highly alkylated intermediates and not by direct alkylation of dyes carrying primary amino groups. 4,4′,4″-Triaminotriphenylmethane (pararosaniline) may, however, be *N*-phenylated with excess aniline and benzoic acid to give the greenish blue, *N,N′,N″*-triphenylaminotriphenylmethane hydrochloride, CI Solvent Blue 23, λ_{max} = 586 nm.

Triarylmethane dyes can be converted into two types of insoluble compounds, which are used industrially as pigments (qv). Both are salts of triarylmethane dyes. Known as pigment lakes, these complexes provide clean, brilliant red and violet shades. These pigments are used in printing inks, especially packaging and special printing inks. The second type of pigments derived from triarylmethane dyes are known as alkali blues. The main use of the alkali blues is as shading pigments in inks based on carbon black, where an inexpensive blue component is needed to correct the natural brown tone of the base pigment. The main area of application is in printing inks, particularly offset, letterpress, and to a lesser extent in aqueous flexographic inks.

They are used to color ribbons for typewriters and also to blue copy paper.

Manufacture

The preparation of triarylmethane dyes proceeds through several stages: formation of the colorless leuco base in acid media, conversion to the colorless carbinol base by using an oxidizing agent, eg, lead dioxide, manganese dioxide, or alkali dichromates, and formation of the dye by treatment with acid. The oxidation of the leuco base can also be accomplished with atmospheric oxygen in the presence of catalysts.

Aldehyde Method. This method is generally used for the preparation of diaminotriphenylmethane dyes or hydroxytriphenylmethane dyes. The central carbon atom is derived from an aromatic aldehyde or a substance capable of generating an aldehyde during the course of the condensation. For example, Malachite green is prepared by heating benzaldehyde under reflux with a slight excess of dimethylaniline in aqueous acid.

Ketone Method. In the ketone method, the central carbon atom is derived from phosgene (qv). A diarylketone is prepared from phosgene and a tertiary arylamine and then condenses with another mole of a tertiary arylamine (same or different) in the presence of phosphorus oxychloride or zinc chloride. The dye is produced directly without an oxidation step. Thus, ethyl violet, CI Basic Violet 4, is prepared from 4,4′-bis(diethylamino)benzophenone with diethylaniline in the presence of phosphorus oxychloride. This reaction is very useful for the preparation of unsymmetrical dyes.

Diphenylmethane Base Method. In this method, the central carbon atom is derived from formaldehyde, which condenses with two moles of an arylamine to give a substituted diphenylmethane derivative. The methane base is oxidized with lead dioxide or manganese dioxide to the benzhydrol derivative. The reactive hydrols condense fairly easily with arylamines, sulfonated arylamines, and sulfonated naphthalenes. The resulting leuco base is oxidized in the presence of acid.

Benzotrichloride Method. The central carbon atom of the dye is supplied by the trichloromethyl group from *p*-chlorobenzotrichloride. Both symmetrical and unsymmetrical triphenylmethane dyes suitable for acrylic fibers are prepared by this method.

Health, Safety, and Environmental Information

In the 1960s, problems were encountered with the interpretation of toxicological studies on animals given triarylmethane dyes used as food colorants. Conflicting test data have been obtained on various triarylmethane dyes. Positive and negative results for the same dye in different assays have only led to more genotoxic studies. In the U. S. CI Food Violet 2 was delisted for use in food, drugs, and cosmetics in 1973. There are triarylmethane dyes which have also been delisted worldwide for use in food, eg, Guinea Green B, CI Food Green 1 (CI 42085), and Violet BNP, CI Food Violet 3.

The triarylmethane dyes of the rosaniline family, eg, fuchsine and crystal violet, show similar toxic responses in assays. They are moderately toxic after acute exposure, but the effects usually pass within a couple of days. These effects are no cause for alarm as long as the dyes are not permitted for food use or contact. There is evidence that the toxicological effects might be the result of impurities, eg, aromatic amines, or of certain functional groups, notably amino substituents found in the dyes.

The toxicity of dyes to aquatic organisms has also been investigated, with most of the work done on fish. In these studies, over 3000 dyes in common use were tested, of which 27 dyes had a LC_{50} around 0.05 mg/L. Ten of these cases had triarylmethane structures.

Environmental Concerns

The main route by which dyes enter the environment is via wastewater, both from their manufacture and their use. Accurate data on dyes released into the environment are not available, although lists of materials released to the environment from the processes operated for the production of some triarylmethane dyes have been reported.

Uses

Present usage of triarylmethane dyes is confined mainly to nontextile applications. Substantial quantities are used in the preparation of organic pigments for printing inks, pastes, and for the paper printing trade, where cost and brilliance of shade are more important than lightfastness. Triarylmethane dyes and their colorless precursors, eg, carbinols and lactones, are used extensively in heat-, light-, and pressure-sensitive recording materials for high speed photoduplicating and photoimaging systems and for the production of printing plates and integrated circuits. They are also used for specialty applications such as tinting automobile antifreeze solutions and toilet sanitary preparations, in the manufacture of carbon paper, in ink for typewriter ribbons, and ink jet printing for high speed computer printers.

In addition to the dyeing and printing of natural and acrylic fibers, triarylmethane dyes are suitable for the coloration of other substrates such as paper, ceramics, leather, fur, anodized aluminium, waxes, polishes, soaps, plastics, drugs, and cosmetics. Several triarylmethane dyes are used as food colorants and are manufactured under stringent processing controls (see COLORANTS FOR FOODS, DRUGS, COSMETICS, AND MEDICAL DEVICES).

Triarylmethane dyes can be used for the coloration of glass. Other high technology applications using triarylmethane dyes include electrophotography and optical data storage.

Related Dyes

Diphenylmethane Dyes. The diphenylmethane dyes are usually classed with the triarylmethane dyes. The dyes of this subclass are ketoimine derivatives, and only three such dyes are registered in the *Colour Index*. They are Auramine O CI Basic Yellow 2 (CI 41000), Auramine G, Basic Yellow 3 (CI 41005), and CI Basic Yellow 37 (CI 41001). These dyes are still used extensively for the coloration of paper and in the preparation of pigment lakes.

Phthaleins. Dyes of this class are usually considered to be triarylmethane derivatives. Phenolphthalein and phenol red are used extensively as indicators in colorimetric and titrimetric determinations (see HYDROGEN-ION ACTIVITY).

Heteroarylmethane Dyes. Dyes of this class usually have either one or two heteroaryl groups attached to the methane carbon atom. Trihetarylmethane dyes are known and have been investigated for their pharmacological activity as well as their color characteristics. These types of triarylmethane dyes and their derivatives are used as color formers in thermoreactive and pressure-sensitive recording materials.

Triarylmethane Dyes with Near-Infrared Absorption. The long wavelength absorption bands of triarylmethane dyes can be shifted into the near-infrared region, but the dyes still remain colored because other absorption bands are shifted to or stay in the visible region. These types of triarylmethane dyes and their derivatives have been claimed as infrared absorbers for optical information recording media and security devices, and as organic photoconductors for use in lithographic plate production.

DEAN THETFORD
Zeneca Specialties

D. R. Waring and G. Hallas, eds., *The Chemistry and Applications of Dyes,* Plenum Publishing Corp., New York, 1990.

P. Gregory, *High-Technology Applications of Organic Colorants,* Plenum Publishing Corp., New York, 1990.

E. N. Abrahart, *Dyes and their Intermediates,* 2nd ed., Chemical Publishing, New York, 1977.

P. Rys and H. Zollinger, *Fundamentals of the Chemistry and Applications of Dyes,* Wiley-Interscience, New York, 1972.

TRYPTOPHAN. See AMINO ACIDS.

TUNGSTEN AND TUNGSTEN ALLOYS

Tungsten (wolfram), atomic number 74, atomic weight 183.85, is a silver-gray metallic element that appears in Group VIB of the Periodic Table, below chromium and molybdenum. There are 31 isotopes of this element, ranging from 160 to 190. Tungsten has a very low vapor pressure, the highest melting point (3695 K) of any metal, as well as the highest tensile strength of any metal above 1650°C.

Tungsten is the eighteenth most abundant metal, having an estimated concentration in the earth's crust of 1–1.3 ppm. Of the more than 20 tungsten-bearing minerals, only four are of commercial importance: ferberite (iron tungstate), huebnerite (manganese tungstate), wolframite (iron–manganese tungstate containing ca 20–80% of each of the pure components), and scheelite (calcium tungstate). Tungsten deposits occur in association with metamorphic rocks and granitic igneous rocks throughout the world. Deposits in China constitute over half of the world reserves and over five times the reserves of the second largest source, Canada.

Physical Properties

Some of the physical properties of tungsten are given in Table 1.

Chemical Properties

The oxidation states of tungsten range from +2 to +6; compounds that have zero oxidation state also exist. Above 400°C, tungsten is very susceptible to oxidation. Tungsten is stable in nitrogen to over 2300°C. Carbon monoxide and hydrocarbons react with tungsten to give tungsten carbide at 900°C. Carbon dioxide oxidizes tungsten at 1200°C. Fluorine is the most reactive halogen gas and attacks tungsten at room temperature. Chlorine reacts at 250°C, whereas bromine and iodine require higher temperatures. Tungsten is resistant to many chemicals. At room temperature, it is rapidly attacked only by a mixture of hydrofluoric and nitric acids. Tungsten resists attack by many molten metals.

Manufacture

Tungsten mines are generally small, producing less than 200 metric tons of raw ore per day. Worldwide, there are only about 20 mines producing over 300 t/d. Mining is almost exclusively by underground methods. Because tungsten minerals have a high specific gravity, they can be beneficiated by gravity separation, usually by tabling. Flotation is used for many scheelites having a fine liberation size, but not for wolframite ores. Magnetic separators can be used for concentrating wolframite ores or cleaning scheelites.

Table 1. Physical Properties of Tungsten

Property	Value
crystal structure	bcc
lattice constant at 298 K, nm	0.316524
shortest interatomic distance at 298 K, nm	0.2741
density[a] at 298 K, g/cm^3	19.254
mp, K	3660
bp, K	5936
thermal conductivity at K, W/(cm·K)	
0	0.0
10	97.1
100	2.08
500	1.46
1000	1.18
3400	0.90

[a] Determined by x-ray.

In extractive metallurgy, a relatively impure ore concentrate is converted into a high purity tungsten compound that can subsequently be reduced to metal powder. The two most common intermediate tungsten compounds are tungstic acid, H_2WO_4, and ammonium paratungstate (APT), $(NH_4)_{10}W_{12}O_{41}\cdot 5H_2O$.

Metal powder is obtained from APT by stepwise reduction with carbon or hydrogen. Because of its high melting point, tungsten is usually processed by powder metallurgy techniques (see POWDER METALLURGY). Small quantities of rod are produced by arc or electron-beam melting. Tungsten is unusual in that its ductility increases with working. As-sintered or after a full recrystallization anneal, it is as brittle as glass at room temperature. For this reason, tungsten is initially worked at very high temperatures, and large reductions are required to achieve ductility. Furthermore, the low specific heat of tungsten causes it to cool very rapidly. A working operation also requires rapid transfer from furnace to working equipment and frequent reheating during the working operation.

Swaging, the oldest process used for the metalworking of tungsten, is the method used first for the manufacture of lamp wire (Coolidge process). Swaging temperatures start at 1500–1600°C and decrease to ca 1200°C as the bar is worked. Reductions per pass start as low as 5% but then increase to as high as 40% as the size decreases. Total reductions of ca 60–80% are typical between anneals. For rod and wire production, rod rolling or, more recently, Kocks rolling is also applied, at least in the initial breakdown stages. For larger-diameter rod or plate, rolling is also employed, with temperatures starting at 1600°C. Forging is also used on tungsten. Hammer forging is generally preferred to press forging because the temperature is better maintained for the higher rate of deformation.

Health and Safety Factors

There are no documented cases of tungsten poisoning in humans. However, numerous cases of pneumoconiosis have been reported in the cemented-carbide industry, but its cause, ie, WC or cobalt, has not been determined.

Occupational exposure to insoluble tungsten needs to be controlled so that employees are not exposed to insoluble tungsten at a concentration greater than 5 mg tungsten/m^3 air, determined as a TWA concentration for up to a 10-h workshift in a 40-h workweek. An STEL value of 10 mg/m^3 has been set by ACGIH in 1983.

Uses

Tungsten is used in four forms: as tungsten carbide, as an alloy additive, as essentially pure tungsten, and as tungsten chemicals. Tungsten carbide, because of its high hardness at high temperatures, is used for cutting tools, abrasion-resistant surfaces, and forming tools. Cemented carbides are used for cutting tools, mining and drilling tools, forming and drawing dies, bearings, and numerous other wear-resistant applications.

About 16% of tungsten usage is as an alloy additive. Tungsten added to steels forms a dispersed tungsten carbide phase that imparts a finer-grain structure and increases the high temperature hardness (see HIGH TEMPERATURE ALLOYS).

Metallic tungsten accounts for 16% of tungsten consumption. Frequently, tungsten is used because of its high melting point and low vapor pressure. The best known use is the manufacture of lamp filaments, where potassium, silicon, and aluminum dopants are added to the oxide. Tungsten is widely employed as an electron emitter because it can be used at very high temperatures. Tungsten is used as the target in high intensity x-ray tubes. Other high temperature applications include furnace elements, heat shields, vacuum metallizing coils and boats, glass-melting equipment, and arc-lamp electrodes. Alloys of tungsten with various combinations of iron, nickel, and copper are called *heavy alloys,* which have the high density of tungsten and are easier to machine. These are used as counterweights, armor-piercing penetrator cores, x-ray shielding, gyroscope rotors, dart bodies, and other high density applications. A similar material is made by infiltrating porous tungsten with copper or silver. Such alloys are used as electrical contact materials and rocket nozzles. Composite materials with barium and strontium compounds are used in electron-emitting devices.

Nonmetallurgical uses include brilliant organic tungsten dyes and pigments that can be used in a variety of materials. Tungstates are used as phosphors in fluorescent lights, cathode-ray tubes, and x-ray screens, whereas tungsten compounds are used as catalysts in petroleum refining.

Recycling. In more recent years, processes that can convert used carbide cutting tools and used tungsten alloy penetrators back into powdered form that can be used directly into new products have been developed. It is estimated that in 1996 ca 25% of cutting inserts used in the United States were recycled in this way. A viable method for recovery of used carbide tools is by immersing them in molten zinc followed by vacuum distillation. High density tungsten alloy machine chips are recovered by oxidation at about 850°C, followed by reduction in hydrogen at 700–900°C.

THOMAS W. PENRICE
Consultant

M. Hoch. *High Temp. High Pressures,* **1,** 531 (1969).

Mineral Commodity Summaries 1981, U.S. Bureau of Mines, Washington, D.C., 1981.

C. C. Clark and J. B. Sutliff, *Am. Metal Market* (Jan. 23, 1981).

Handbook of Toxic and Hazardous Chemicals and Carcinogens, 2nd ed., M. Sittig Marshall Publishers, Park Ridge, N.J.

TUNGSTEN COMPOUNDS

Tungsten is a Group VIB transition element having an atomic number of 74 and a valence state of 0, +2, +3, +4, +5, or +6 in compounds. However, tungsten alone has not been observed as a cation. Its most stable, and therefore most common, valence state is +6. Tungsten complexes vary widely in stereochemistry and oxidation states. Complex formation is exemplified by the large number of polytungstates. Simple tungsten compounds, such as the halides, are also known.

The chemical uses of tungsten have increased substantially in more recent years. Catalysis (qv) of photochemical reactions and newer types of soluble organometallic complexes for industrially important organic reactions are among the areas of these new applications.

Tungsten Hexacarbonyl. Tungsten hexacarbonyl, $W(CO)_6$, may be prepared in yields >90% by the aluminum reduction of tungsten hexachloride in anhydrous ether under a pressure of 0.1 MPa (ca 1 atm) of carbon monoxide at 70°C. It is fairly stable in air, water, or acid, but is decomposed by strong bases and attacked by halogens. Tungsten carbonyl is slightly soluble in organic solvents but insoluble in water (see CARBONYLS). Various applications such as lubricant additives, dyes, pigments, and catalysts are under investigation.

Tungsten Halides and Oxyhalides. Tungsten forms binary halides for all oxidation states between +2 and +6; oxyhalides are only known for oxidation states +5 and +6. In general, tungsten halogen compounds are reactive toward water and oxygen in the air and must therefore be handled in an inert atmosphere. These are all solid, colored compounds at room temperature, except the fluorides, and many decompose on heating before melting.

Tungsten hexafluoride, WF_6, is a colorless gas at room temperature, sp gr 12.9 with respect to air. At 17.5°C, it condenses into a pale yellow liquid, and at 2.5°C, a white solid is formed. It may be prepared by treating hydrogen fluoride, arsenic trifluoride, or antimony pentafluoride with tungsten hexachloride or by direct fluoridation of tungsten. Tungsten hexafluoride dissolves in benzene or cyclohexane to give a bright red color, in dioxane a pale red, and in ether a violet-brown. Tungsten pentafluoride, WF_5, is prepared by the reduction of the hexafluoride on a hot tungsten filament in almost quantitative yield. Tungsten tetrafluoride, WF_4, is a nonvolatile, hygroscopic, reddish brown solid. It has been prepared in low yields by the reduction of the hexafluoride with phosphorus trifluoride in the presence of liquid anhydrous hydrogen fluoride at room temperature. Tungsten oxytetrafluoride, WOF_4, mp 110°C, bp 187.5°C, forms colorless plates. It is prepared by the action of an oxygen–fluorine mixture on the metal at elevated temperatures. The compound is extremely hygroscopic and decomposes to tungstic acid in the presence of water. Tungsten oxydifluoride, WO_2F_2, is a white solid prepared by the hydrolysis of WOF_4. Its chemistry has not been investigated.

Tungsten hexachloride, WCl_6, mp 275°C, bp 346.7°C, is a blue-black crystalline solid. It is prepared by the direct chlorination of pure tungsten in a flow system at atmospheric pressure at 600°C. Solidification usually occurs without incident, but further cooling may result in a violent, explosion-like expansion of the solid mass at 168–170°C. However, tungsten hexachloride may be safely cooled if it occupies not more than one-half of the containing vessel. In the presence of moisture or oxygen, some $WOCl_4$ is formed as an impurity. Tungsten hexachloride is very soluble in carbon disulfide but decomposes in water to form tungstic acid. The hexachloride is easily reduced by hydrogen to the lower halides and finally to the metal itself. Tungsten pentachloride, WCl_5, mp 243°C, bp 275.6°C, is a black, crystalline, deliquescent solid. It is only slightly soluble in carbon disulfide and decomposes in water to the blue oxide, $W_{20}O_{58}$. Tungsten pentachloride may be prepared by the reduction of the hexachloride with red phosphorus. Tungsten tetrachloride, WCl_4, is obtained as a coarse, crystalline, deliquescent solid that decomposes upon heating. It is diamagnetic and may be prepared by the thermal-gradient reduction of WCl_6 with aluminum. Tungsten dichloride, WCl_2, is an amorphous powder. It is a cluster compound and may be prepared by the reduction of the hexachloride with aluminum in a sodium tetrachloroaluminate melt. Tungsten oxytetrachloride, $WOCl_4$, mp 211°C, bp 327°C, is a red crystalline solid. It is soluble in carbon disulfide and benzene and is decomposed to tungstic acid by water. It may be prepared by refluxing sulfurous oxychloride, $SOCl_2$, on tungsten trioxide and purified after evaporation by sublimation. Tungsten oxydichloride, WO_2Cl_2, a pale yellow crystalline solid having an mp of 266°C, is soluble in cold water and in alkaline solution, although partly decomposed by hot water. It is prepared by the action of carbon tetrachloride on tungsten dioxide at 250°C in a bomb. Tungsten oxytrichloride, $WOCl_3$, a green solid, is prepared by the aluminum reduction of $WOCl_4$ in a sealed tube at 100–140°C.

Tungsten hexabromide, WBr_6, bluish black crystals having an mp of 232°C, is formed by metathetical exchange reaction of BBr_3 with tungsten hexachloride. Tungsten pentabromide, WBr_5, violet-brown crystals having an mp of 276°C and a bp of 333°C, is extremely sensitive to moisture. It is prepared by the action of bromine vapor on tungsten at 450–500°C. Tungsten tetrabromide, WBr_4, black orthorhombic crystals, is formed by the thermal-gradient reduction of WBr_5 with aluminum, similar to the reduction of WCl_4. Tungsten tribromide, WBr_3, prepared by the action of bromine on WBr_2, in a sealed tube at 50°C, is a thermally unstable black powder that is insoluble in water. Tungsten dibromide, WBr_2, formed by the partial reduction of the pentabromide with hydrogen, is a black powder that decomposes at 400°C. Tungsten oxytetrabromide, $WOBr_4$, black, deliquescent needles having an mp of 277°C and a bp of 327°C, is formed by the action of carbon tetrabromide on tungsten dioxide at 250°C. Tungsten oxydibromide, WO_2Br_2, light-red crystals, is formed by passing a mixture of oxygen and bromine over tungsten at 300°C.

Tungsten tetraiodide, WI_4, is a black powder that is decomposed by air. It is prepared by the action of concentrated hydriodic acid on tungsten hexachloride at 100°C. Tungsten triiodide, WI_3, is prepared by the action of iodine on tungsten hexacarbonyl in a sealed tube at 120°C. Tungsten diiodide, W_6I_2, sp gr 6.79, is a brownish crystalline substance. It can be prepared by the reaction between iodine and $W(CO)_6$ in a nitrogen atmosphere. Tungsten oxydiiodide, WO_2I_2, is prepared by heating a mixture of tungsten and tungsten trioxide with excess iodine in a 500–700°C temperature gradient for 36 h.

Tungsten oxides form a series of well-defined ordered phases to which precise stoichiometric formulas can be assigned. The composition of the tungsten oxides may vary over a fixed range without change in crystalline structure.

Tungsten trioxide, WO_3, is a yellow powder. However, the smallest diminution of oxygen brings about a change in color. Tungsten trioxide, which is pseudorhombic at room temperature but tetragonal above 700°C, is usually prepared from tungstic acid or tungstates. It is the most important tungsten oxide and is the starting material for the production of tungsten powder.

Tungsten dioxide, WO_2, is a brown powder formed by the reduction of WO_3 with hydrogen at 575–600°C. The oxide W_3O is regarded as both an oxide and a metal phase. It is gray and has a density of 14.4 g/cm^3, and is prepared by the electrolysis of fused mixtures of WO_3 and alkali-metal phosphates. At ca 700°C, it decomposes into W and WO_2; β-tungsten is W_3O.

Tungsten bronzes constitute a series of well-defined nonstoichiometric compounds of the general formula $M_{1-x}WO_3$, where x is a variable between 0 and 1, and M is some other metal, generally an alkali metal, although many other metals can also be substituted. The systems most extensively investigated are the sodium tungsten bronzes. These compounds are intensely colored, ranging from golden-yellow to bluish black, depending on the value of x, and, in crystalline form, exhibit a metallic sheen. The compounds have a positive temperature coefficient of resistance for Na:WO_3 ratios of <0.3, and a negative temperature coefficient of resistance at lower ratios. Sodium tungsten bronzes are inert to chemical attack by most acids, but may be dissolved by basic reagents. Sodium tungsten bronzes serve as promoters for the catalytic oxidation of carbon monoxide and reformer gas in fuel cells (see BATTERIES, SECONDARY CELLS).

The mild reduction, eg, by Sn, of acidified solutions of tungstates, tungsten trioxide, or tungstic acid in solutions gives intense blue products, which are referred to by the general name of tungsten blues, and which resemble molybdenum blues in many respects. Blue hydrogen tungsten bronzes, $H_{1-x}WO_3$, are prepared by the wet reduction of tungstic acid and are structurally related to the alkali tungsten bronzes. Tungsten blues have a strong tendency to form colloids.

Tungstic acid, H_2WO_4 or $WO_3 \cdot H_2O$, is an amorphous yellow powder that is practically insoluble in water or acid solution, but dissolves readily in a strongly alkaline medium. It may be precipitated from hot tungstate solutions with strong acids. If the tungstate solution is acidified in the cold, a white voluminous precipitate of hydrated tungstic acid forms. This is converted to the yellow form by boiling in an acid medium. Some properties of tungstates are given in Table 1.

Ammonium tungstate, $(NH_4)_2WO_4$, cannot be obtained from an aqueous solution because it decomposes when such a solution is concentrated. It is prepared by the addition of hydrated tungstic acid to liquid ammonia.

Anhydrous sodium tungstate, Na_2WO_4, is prepared by fusing tungsten trioxide in the proper proportion with sodium hydroxide or sodium carbonate.

An important and characteristic feature of the tungstate ion is its ability to form condensed complex ions of isopolytungstates in acid solution. As the acidity increases, the molecular weight of the isopolyanions increases until tungstic acid precipitates. If polytungstates are considered as formed by the addition of acid to WO_4^{2-}, then a series of isopolytungstates appears, in which the degree of aggregation in solution increases with decreasing pH. Metatungstates of the alkali, alkaline-earth, rare-earth, and transition metals have been reported. However, classical synthesis rarely gives high yields of the pure compounds.

Commercially, heteropolytungstates, particularly the heteropolytungstates, are produced in large quantities as precipitants for basic dyes, with which they form colored lakes or toners (see also DYES AND DYE INTERMEDIATES). They are also used in catalysis, passivation of steel, etc.

Table 1. Properties of Normal Tungstates

Compound	Properties	Sp gr
$BaWO_4$	colorless, tetragonal, $a = 0.564$ nm, $c = 1.270$ nm	5.04
$CdWO_4$	yellow, rhombic	
$CaWO_4$	white, tetragonal, $a = 0.524$ nm, $c = 1.138$ nm, $n_D^{20} = 1.9263$	6.06
$Ce_2(WO_4)_3$	yellow, monoclinic, $a = 1.151$ nm, $b = 1.172$ nm, $c = 0.782$ nm, $\beta = 109° 48'$, mp 1089°C	6.77
$PbWO_4$	colorless, monoclinic, mp 1123°C	8.46
Ag_2WO_4	pale yellow	
Na_2WO_4	white, rhombic, mp 689°C	4.179
$Na_2WO_4 \cdot 2H_2O$ white, rhombic, loses $2H_2O$ at 100°C	3.245	
$SrWO_4$	white, tetragonal, $a = 0540$ nm, $c = 1.190$ nm	6.187

Sulfides. Tungsten disulfide, WS_2, although found in nature, is usually prepared by heating tungsten powder with sulfur at 900°C. It is a soft, grayish black powder, relatively inert and unreactive, with sp gr of 7.5. This disulfide is insoluble in water, hydrochloric acid, alkali, and organic solvents or oils, and decomposes in hot, strong oxidizing agents, eg, aqua regia, concentrated sulfuric acid, and nitric acid. Heating in air or in the presence of oxygen yields WO_3. However, its thermal stability in air is ca 90°C higher than that of MoS_2. Tungsten disulfide forms adherent, soft, continuous films on a variety of surfaces and exhibits good lubricating properties similar to molybdenum disulfide and graphite (see also LUBRICATION AND LUBRICANTS). It is also reported to be a semiconductor (qv). Tungsten trisulfide, WS_3, is a chocolate-brown powder, slightly soluble in cold water, but readily forming a colloidal solution in hot water. It is prepared by treating an alkali-metal thiotungstate with HCl. Tungsten trisulfide is soluble in alkali carbonates and hydroxides.

Potassium tetrathiotungstate, K_2WS_4, forms yellow rhombic crystals that are soluble in water. Ammonium tetrathiotungstate, $(NH_4)_2WS_4$, forms bright orange crystals that exhibit a metallic iridescence. These crystals are stable in dry air and soluble in water. Ammonium tetrathiotungstate is generally prepared by treating a solution of tungstic acid with excess ammonia and saturating with hydrogen sulfide. It is readily decomposed in a nonoxidizing atmosphere to WS_2, for which it is a convenient source.

Interstitial Compounds. Tungsten forms hard, refractory, and chemically stable interstitial compounds with nonmetals, particularly C, N, B, and Si. These compounds are used in cutting tools, structural elements of kilns, gas turbines, jet engines, sandblast nozzles, protective coatings, etc (see also REFRACTORIES; REFRACTORY COATINGS).

Toxicity

A considerable difference in the toxicity of soluble and insoluble compounds of tungsten has been reported. In view of the degree of systemic toxicity of soluble compounds of tungsten, a threshold limit of 1 mg of tungsten per m^3 of air is recommended. A threshold limit of 5 mg of tungsten per m^3 of air is recommended for insoluble compounds.

Uses

Tungsten compounds, especially the oxides, sulfides, and heteropoly complexes, form stable catalysts for a variety of commercial chemical processes, eg, petroleum processing. Tungsten hexachloride is used for preparing tungsten metathesis catalysts, which are very interesting because they form double and triple bonds with carbon. Tungsten disulfide, applied as a dry powder, suspension, bonded film, or aerosol, can be an effective lubricant in wire drawing, metal forming, valves, gears, bearings, packing materials, etc. Oil-soluble tungsten compounds, such as the ammonium salts of tungstate or tetrathiotungstate, are reported to be effective lubricating-oil additives.

Sodium tungstate is used in the manufacture of heteropolyacid color lakes, which are used in printing inks, plants, waxes, glasses, and textiles. It is also used as a fuel-cell electrode material and in cigarette filters. Other uses include the manufacture of tungsten-based catalysts, for fireproofing of textiles, and as an analytical reagent for the determination of uric acid. Calcium tungstate is fluorescent when exposed to ultraviolet radiation and is therefore widely used in the manufacture of phosphors. It is also used in lasers, fluorescent lamps, high voltage sign tubes, and oscilloscopes for high speed photographic processes. Ammonium paratungstate is commercially significant because it is the precursor of high purity tungsten oxides, tungsten, and tungsten carbide powders. Tungsten trioxide is a principal source of tungsten metal and tungsten carbide powders. It is used as a pigment in oil and water colors (see PIGMENTS). It is also used in a wide variety of catalysts and, more recently, in the control of air pollution and industrial hygiene. Tungsten carbides, on the other hand, are widely used in the manufacture of hard carbides for high speed machining tools, wire-drawing dies, wear surfacing, drills, etc.

Heteropoly tungstic acids are useful in analytical chemistry and biochemistry as reagents; in atomic-energy work as precipitants and inorganic ion exchangers; in photographic processes as fixing agents and oxidizing agents; in plating processes as additives; in plastics, adhesives, and cements for imparting water resistance; and in plastics and plastic films as curing or drying agents.

In the textile industry, the salts are useful as antistatic agents. The acids are used in diverse applications, eg, printing inks, paper coloring, nontoxic paints, and wax pigmentation. The tungstates and molybdates are good corrosion inhibitors and have been used for some time in antifreeze solutions. In addition, they are used as laser-host materials, phosphors, and for the flameproofing of textiles.

THOMAS W. PENRICE
Consultant

S. W. H. Yih and C. T. Wang, *Tungsten,* Plenum Press, New York, 1979.

G. D. Rieck, *Tungsten and Its Compounds,* Pergamon Press, London, 1967.

K. C. Li and C. Y. Wang, *Tungsten,* Reinhold Publishing Corp., New York, 1955.

J. H. Canterford and R. Colton, *Halides of the Transition Elements,* John Wiley & Sons, Inc., New York, 1978.

TURBIDITY AND NEPHELOMETRY. See ANALYTICAL METHODS.

TYROCIDINE. See ANTIBIOTICS, PEPTIDES.

TYROTHRICIN. See ANTIBIOTICS, PEPTIDES.

U

ULTRAFILTRATION

Ultrafiltration is a pressure-driven filtration separation occurring on a molecular scale (see DIALYSIS; FILTRATION; HOLLOW-FIBER MEMBRANES; MEMBRANE TECHNOLOGY; REVERSE OSMOSIS). Typically, a liquid including small dissolved molecules is forced through a porous membrane. Large dissolved molecules, colloids, and suspended solids that cannot pass through the pores are retained.

Ultrafiltration separations range from ca 1 to 100 nm. Above ca 50 nm, the process is often known as microfiltration. Below ca 2 nm, interactions between the membrane material and the solute and solvent become significant. That process, called reverse osmosis or hyperfiltration, is best described by solution–diffusion mechanisms.

Membrane-retained components are collectively called concentrate or retentate. Materials permeating the membrane are called filtrate, ultrafiltrate, or permeate.

Media

Most ultrafiltration membranes are porous, asymmetric, polymeric structures produced by phase inversion, ie, the gelation or precipitation of a species from a soluble phase (see MEMBRANE TECHNOLOGY). Membrane structure is a function of the materials used (polymer composition, molecular weight distribution, solvent system, etc) and the mode of preparation (solution viscosity, evaporation time, humidity, etc). Commonly used polymers include cellulose acetates, polyamides, polysulfones, dynels (vinyl chloride–acrylonitrile copolymers) and poly(vinylidene fluoride).

Modification of the membranes affects the properties. Cross-linking improves mechanical properties and chemical resistivity. Fixed-charge membranes are formed by incorporating polyelectrolytes into polymer solution and cross-linking after the membrane is precipitated, or by substituting ionic species onto the polymer chain (eg, sulfonation). Polymer grafting alters surface properties. Enzymes are added to react with permeable species and reduce fouling.

Polyelectrolyte complex membranes are phase-inversion membranes where polymeric anions and cations react during the gelation. Inorganic ultrafiltration membranes are formed by depositing particles on a porous substrate. Dynamic membranes are concentration–polarization layers formed *in situ* from the ultrafiltration of colloidal material analogous to a precoat in conventional filter operations. Track-etched membranes are made by exposing thin films (mica, polycarbonate, etc) to fission fragments from a radiation source.

Process

Pore-flow models most accurately describe ultrafiltration processes. Other membrane transport mechanisms, which may occur simultaneously although generally at a much lower rate, include dialysis (diffusion), osmosis (solvent by osmotic gradient), anomalous osmosis (osmosis with a charged membrane), reverse osmosis (solvent by pressure gradient larger and opposite to osmotic gradient), electrodialysis (solute ions by electric field), piezodialysis (solute by pressure gradient), electroosmosis (solvent in electric field), Donnan effects, Knudsen flow, thermal effects, chemical reactions (including facilitated diffusion), and active transport.

When pure water is forced through a porous ultrafiltration membrane, Darcy's law states that the flow rate is directly proportional to the pressure gradient:

$$J = \frac{V}{A \cdot t} = \frac{K_m \Delta P}{\mu} \tag{1}$$

where J is permeate flux in units of volume V per membrane area A, at time t, K_m is the membrane hydraulic permeability, μ is the fluid viscosity, and ΔP is the membrane pressure drop between the retentate and permeate. These parameters change with the membrane pressure history.

The addition of small membrane-permeable solutes to the water affects permeate transport in the following ways. (*1*) Solute–solvent interactions change the permeating fluid viscosity. (*2*) Solute adsorption reduces the apparent membrane-pore diameter. Because of high interfacial tension between water and certain materials, the water phase in the pores can be replaced. Surfactants suppress hydrophobic adsorption. Adsorption of permeate species is characterized by a lag in permeate concentration as a function of time. (*3*) The interfacial charge between the membrane-pore wall and the liquid affects permeate transport when the Debye screening length approaches (ca 10%) the membrane-pore size. (*4*) High surface tension on hydrophobic membranes forces water molecules to form large clusters in the pores. (*5*) Solvents, swelling agents, and plasticizers that diffuse into the polymer structure can change the apparent pore size (K_m in eq. 1) or increase the rate of long-term compaction. The rejection R of a solute is defined as:

$$R = 1 - \frac{C_{pi}}{C_{bi}} \tag{2}$$

where C_p is the permeate concentration of species i and C_b is the concentration of that species in the retentate. There are two components of rejection. Observed rejection, R_o, is based on the concentration of the solute in the bulk solution, C_b. The intrinsic rejection, R_i, is based on the concentration of the solute on the surface of the membrane, C_w.

$$R_o = 1 - \frac{C_p}{C_b}$$

$$R_i = 1 - \frac{C_p}{C_w}$$

If the solute size is approximately the (apparent) membrane-pore size, it interferes with the pore dimensions. The solute concentration in the permeate first increases, then decreases with time. The point of maximum interference is further characterized as a minimum flux. If the solute size is greater than the pore dimensions, the solute is retained by mechanical sieving, forming a gel-polarization layer.

The gel usually has a much lower hydraulic permeability and smaller apparent pore size than the underlying membrane. (The gel layer and the concentration gradient between the gel layer and the bulk concentration are called the gel-polarization layer.)

The gel-layer thickness is limited by mass transport back into the solution bulk at the rate. At steady state,

$$JC_b = K\frac{dC}{dX} \tag{3}$$

where C_b is the bulk concentration of all retained species. Integration gives

$$J = K \cdot \ln \frac{C_g}{C_b} \tag{4}$$

In a static system, the gel-layer thickness rapidly increases and flux drops to uneconomically low values.

The gel-polarization layer has an hydraulic permeability of K_g. Equation 4 states that flux is independent of pressure, and K_g must therefore decrease with increasing pressure. Equation 1 becomes

$$J = \frac{\Delta P}{\mu\left(\frac{1}{K_m} + \frac{1}{K_g}\right)} = \frac{\Delta P}{\mu(R_m + R_g)} \tag{5}$$

where R_m and R_g are the hydraulic resistances of the membrane gel.

Flux is independent of pressure when the process flux is much less than the wafer flux ($K_g \ll K_m$). If $K_g > K_m$, the process is limited by the membrane water flux and flux would flatten out at low concentrations of solids.

Fouling. If the gel-polarization layer is not in hydrodynamic equilibrium with the fluid bulk, the membrane may be fouled. Fouling is caused either by adsorption of species on the membrane or on the surface of the pores, or by deposition of particles on the membrane or within the pores. Fouled systems are characterized as follows: flux is a function of total permeate production when hydrodynamic conditions are constant; if hydrodynamic conditions are changed, hydraulic permeability response of the gel layer is not reversible; and theoretical permeate flux (TPF) changes with time. A sensitive test for predicting fouling or process instability is to measure change in TPF after subjecting the system to process extremes.

Fouling is controlled by selection of proper membrane materials, pretreatment of feed and membrane, and operating conditions. Control and removal of fouling films is essential for industrial ultrafiltration processes.

When fouling is present or possible, ultrafiltration is usually operated at high liquid shear rates and low pressure to minimize the thickness of the gel polarization layer.

Cleaning. Fouling films are removed from the membrane surface by chemical and mechanical methods. Dissolved fouling material may pass into the membrane pores. Reprecipitation upon rinsing must be avoided. Membrane-swelling agents, such as hypochlorites, flushout material which may be lodged in the pores.

Certain applications require that the equipment meet FDA and USDA sanitary requirements. These requirements ensure that the products are not contaminated by extractables or microorganisms from the equipment. Special considerations are given to the design of such equipment (see STERILIZATION TECHNIQUES).

Practical Aspects

The theoretical models cannot predict flux rates. Plant-design parameters must be obtained from laboratory testing, pilot-plant data, or in the case of established applications, performance of operating plants.

Flux is maximized when the upstream concentration is minimized. For any specific task, therefore, the most efficient (minimum membrane area) configuration is an open-loop system where retentate is returned to the feed tank. When the objective is concentration (eg, enzyme), a batch system is employed. If the object is to produce a constant stream of uniform-quality permeate, the system may be operated continuously (eg, electrocoating).

Open-loop systems have inherently long residence times which may be detrimental if the retentate is susceptible to degradation by shear or microbiological contamination. A feed-bleed or closed-loop configuration is a one-stage continuous membrane system. At steady state, the upstream concentration is constant at C_f. For concentration, a single-stage continuous system is the least efficient (maximum membrane area).

The single-pass system and the staged cascade have high flux at low residence time. Both trade the concentration dependence of the batch system on time for concentration dependence on position in the system. Thus, a uniform flux is maintained (assuming no fouling) allowing continuous process integration. In practice, the single-pass system is difficult to implement, and therefore most commercial systems are multistaged cascade. The more stages used, the closer the average flux approaches the batch flux.

Electroultrafiltration

Electroultrafiltration (EUF) combines forced-flow electrophoresis (see ELECTROSEPARATIONS, ELECTROPHORESIS) with ultrafiltration to control or eliminate the gel-polarization layer. Placing an electric field across an ultrafiltration membrane facilitates transport of retained species away from the membrane surface. Thus, the retention of partially rejected solutes can be dramatically improved (see ELECTROSEPARATIONS–ELECTRODIALYSIS).

Electroultrafiltration has been demonstrated on clay suspensions, electrophoretic paints, protein solutions, oil–water emulsions, and a variety of other materials.

Diafiltration

Diafiltration is an ultrafiltration process where water or an aqueous buffer is added to the concentrate and permeate is removed. The two steps may be sequential or simultaneous. Diafiltration improves the degree of separation between retained and permeable species.

Constant-volume batch diafiltration is the most efficient process mode. Sequential batch diafiltration is a series of dilution–concentration steps. Continuous diafiltration practiced in one or more stages of a cascade system has the same volume turnover relationship for overall recoveries as sequential batch diafiltration. The residence time however is dramatically reduced. If recovery of permeable solids is of primary importance, the permeate from the last stage may be used as diafiltration fluid for the previous stage. This countercurrent diafiltration arrangement results in higher permeate solids at the expense of increased membrane area.

Membrane Equipment

Commercial industrial ultrafiltration equipment first became available in the late 1960s. Since that time, the industry has focused on five different configurations.

Parallel-Leaf Cartridge. A parallel-leaf cartridge consists of several flat plates, each having membrane sealed to both sides. Cartridges are inserted in series into plastic or stainless-steel tubular pressure housings of square cross section. Feed flows parallel to the leaf surface. A permeate fitting secures each cartridge to the housing wall, which allows permeate egress and facilitates sealing between concentrate, atmosphere, and permeate channels.

Plate and Frame. Plate-and-frame systems consist of plates each with a membrane on both sides.

At least one hole near the perimeter of each plate connects the flow channels from one side of the plate to the other. The membrane is sealed around the hole to isolate the permeate from the concentrate. Permeate collects in a drain grid behind the membrane and exits from a withdrawal port on the frame perimeter.

Spiral Wound. A spiral-wound cartridge has two flat membrane sheets (skin side out) separated by a flexible, porous permeate drainage material.

Supported Tube. There are three types of supported tubular membranes: cast in place (integral with the support tube), cast externally and inserted into the tube (disposable linings), and dynamically formed membranes. The most common supported tubes are those with membranes cast in place.

Self-Supporting Tubes. Depending on the membrane material and operating pressure, self-supporting tubes are less than 2-mm ID; inside diameters as small as 0.04 mm are commercially available.

A large number of fibers are cut to length, and potted in epoxy resin at each end (see EMBEDDING). The fiber bundle is shrouded in a cylinder which aids in permeate collection, reduces airborne contamination,

Table 1. Ultrafiltration Applications

Application	Process
electrophoretic paint	control of properties, recovery of solids from rinse systems
dairy wheys	protein recovery, concentration, purification, diafiltration
milk	cheese and yogurt mfg, 15–20% yield improvement, standardization
oil–water emulsions	concentration
effluents of wool, yarn scouring	lanolin recovery, pollution abatement
enzymes	concentration, purification
biological reactors	antibiotic mfg, alcohol fermentation, sewage treatment
vegetable proteins	
latex concentration	
production of pure[a] water	
pulp and paper	lignosulfonate sprn from spent liquor
blood and blood products	fractionation, purification
vaccines	conc, purification
biotechnology products	conc, purification

[a] Virus-free.

and allows back pressing of the membrane. Hollow-fiber membranes (qv) have also found use in ultrafiltration.

Each of the membrane devices may be assembled by connecting the modules into combinations of series, parallel-flow paths, or both. These assemblies are connected to pumps, valves, tanks, heat exchangers, instrumentation, and controls to provide complete systems.

Because of the broad differences between ultrafiltration equipment, the performance of one device cannot be used to predict the performance of another. Comparisons can only be made on an economic basis and only when the performance of each is known.

Uses

Applications of ultrafiltration are summarized in Table 1.

RALF KURIYEL
Millipore Corporation

S. Hwang and K. Kammermeyer, *Membranes in Separations,* John Wiley & Sons, Inc., New York, 1975; good study of membrane transport phenomenon.

R. E. Kesting, *Synthetic Polymeric Membranes,* McGraw-Hill, New York, 1971; good bibliographies.

A. R. Cooper, ed., *Ultrafiltration Membranes and Applications, Proceedings of 178th National ACS Meeting,* Washington, D.C., 1979, Plenum Press, New York, 1980.

L. J. Zeman and A. L. Sydney, *Microfiltration and Ultrafiltration,* Marcel Dekker, Inc., New York, 1996.

UNDERGROUND STORAGE TANKS. See TANKS AND PRESSURE VESSELS.

UNITS AND CONVERSION FACTORS

In 1790, the French National Assembly requested of the French Academy of Sciences that it work out a system of units suitable for adoption by the whole world. This system was based on the meter as a unit of length and the gram as a unit of mass. Industry, commerce, and especially the scientific community benefited greatly. In 1893, the United States actually adopted the meter and the kilogram as the fundamental standards of length and mass. In 1954, the 10th General Conference on Weights and Measures (CGPM) added the degree Kelvin as the unit of temperature and the candela as the unit of luminous intensity. In 1960, this new system with six base units was formalized with the title International System of Units. Its abbreviation in all languages is SI, from the French *Le Système International d'Unités.* Since 1960, various refinements to the system have been made.

In 1995 the 20th CGPM approved eliminating the class of supplementary units as a separate class in SI. Thus the new SI consists of only two classes of units: base units and derived units, with the radian and steradian subsumed into the class of derived units of the SI.

Advantages of SI

SI is a decimal system. Fractions have been eliminated, and multiples and submultiples are formed by a system of prefixes ranging from yotta, for 10^{24}, to yocto, for 10^{-24}. Calculations, therefore, are greatly simplified. Each physical quantity is expressed in one and only one unit, eg, the meter for length, the kilogram for mass, and the second for time. Derived units are defined by simple equations relating two or more base units. Some are given special names, such as newton for force and joule for work and energy.

The system is coherent. There is no duplication of units for a quantity, and all derived units are obtained by a direct one-to-one relation of base units or derived units.

The International System of Units

SI rests on seven base units and a number of derived units, some of which have special names. A list of these units is given in the introduction to this volume.

The base units are meter, kilogram, second, ampere, kelvin, mole, and candela.

The largest class of SI units, the derived units, consists of a combination of base and derived units according to the algebraic relations linking the corresponding quantities.

In SI, 20 prefixes are used and are directly attached to form decimal multiples and submultiples of the units. Prefixes indicate the order of magnitude, thus eliminating nonsignificant digits and providing an alternative to powers of 10; eg, 45 300 kPa becomes 45.3 MPa and 0.0043 m becomes 4.3 mm.

It is usually recommended that only one prefix be used in forming a multiple of a compound unit, and that it should be attached to the numerator. An exception is the base unit kilogram, where it appears in the denominator.

A number of non-SI units are used in SI (Table 1).

Units Used Temporarily with SI. Additional non-SI units are used with SI units until the CIPM considers their use no longer necessary. They are nautical mile, knot, hectare, kilowatt-hour, barn, bar, curie, roentgen, rad, and rem.

Except for the non-SI units referred to in the two preceding sections, a great many other metric units should be avoided in order to maintain the advantages of using one common coherent system of units, eg, units of the cgs system with special names such as the erg, dyne, poise, stokes, gauss, oersted, maxwell, stilb, phot, and angstrom. Other unit names to be deprecated are the kilogram-force, calorie, torr, millimeter of mercury, and the mho.

Weight is a force: the weight of a body is the product of its mass and the acceleration due to gravity.

The use of the same term for units of force and mass causes confusion. When the non-SI units are used, a distinction should be made between force and mass.

The term load means either mass or force, depending on its use. A load that produces a vertically downward force because of the influence of gravity acting on a mass may be expressed in mass units. Any other load is expressed in force units.

The kelvin is the SI unit of thermodynamic temperature, and is generally used in scientific calculations. Wide use is made of the degree Celsius (°C) for both temperature and temperature interval.

Pressure is usually designated as gauge pressure, absolute pressure, or, if below ambient, vacuum. Pressures are expressed in pascals with appropriate prefixes. When the term vacuum is used, it should be made clear whether negative gauge pressure or absolute pressure is meant.

Impact energy absorption, often incorrectly called impact resistance or impact strength, is measured in terms of the work required to break a standard specimen; the proper unit is joule.

Some dimensions do not have an SI equivalent because their values are nominal, that is, a value is assigned for the purpose of convenient designation. For example, a 1-in. pipe has no dimension that is 25.4 mm.

Table 1. Units in Use with SI

Unit	Symbol	Value in SI units
minute	min	1 min = 60 s
hour	h	1 h = 60 min = 3600 s
day	d	1 d = 24 h = 86400 s
degree	°	$1° = (\pi/180)$ rad
minute	′	$1' = (1/60)° = (\pi/10800)$ rad
second	″	$1'' = (1/60)' = (\pi/648000)$ rad
liter	L	$1\ L = 1\ dm^3 = 10^{-3}\ m^3$
metric ton	t	$1\ t = 10^3$ kg

Certain quantities, eg, refractive index and relative density (formerly specific gravity), are expressed by pure numbers.

Density is mass per unit volume and in SI is normally expressed as kilograms per cubic meter (density of water = 1000 kg/m^3 or 1 g/cm^3).

Style and Usage. If the advantages of SI are to be realized, everyone must use the system in the same manner. A number of editorial rules that must be followed are SI symbols are always in roman type, not italics; a space is required between the number and the unit; a period is not placed after a symbol; the plural form of a symbol is the same as the singular; Y, Z, E, P, T, G, and M, the prefixes for 10^6 and above, are capitalized, as are the symbols whose unit names have been derived from proper names (an exception is the use of L for liter); the product of two or more symbols is indicated by a centered dot and the product of unit names preferably by just a space; a solidus indicates the quotient of two unit symbols and the word per the division of two unit names; an exponent attached to a symbol containing a prefix indicates that the multiple of the unit is raised to the power expressed by the exponent; compound prefixes are not used; a comma should not be used to separate groups of digits; the term billion must be avoided; the prefix giga is unambiguous; when using powers with a unit name, the modifier squared or cubed is used after the unit name, except for areas and volumes.

Conversion and Rounding. Conversion of quantities should be handled with careful regard to the implied correspondence between the accuracy of the data and the number of digits. In all soft conversions (a soft conversion being defined as the conversion of an existing non-SI measurements to acceptable SI units without a significant change in size or magnitude), the number of significant digits retained should be such that accuracy is neither sacrificed nor exaggerated.

Conversion Factors. Excellent tables of conversion factors are available from ASTM, IEEE, and NIST.

ROBERT P. LUKENS
American Society for Testing and Materials
Committee E-43 on SI Practice

Standard for Use of the International System of Units (SI): The Modern Metric System, IEEE/ASTM SI 10-1997, Institute of Electrical and Electronics Engineers, New York, and American Society for Testing and Materials, West Conshohocken, Pa.

B. N. Taylor, *Guide for the Use of the International System of Units (SI),* NIST Special Publication 811, Superintendent of Documents, U.S. Government Printing Office, Washington, D.C., 1995.

UNSATURATED POLYESTERS. See POLYESTERS, UNSATURATED.

URANIUM AND URANIUM COMPOUNDS

Uranium is a naturally occurring radioactive element with atomic number 92 and atomic mass 238.03.

In 1939, Hahn and Strassman reported their discovery of nuclear fission which announced the dawn of the nuclear age. Uranium gained importance as fuel for nuclear reactors and as starting material for the synthesis of plutonium. There are 19 isotopes of uranium with masses 218, 222, 225–240, and 242 and radioactive half-lives ranging from 1 μs (^{222}U) to 4.468×10^9 yr, the latter for the main naturally occurring (99.27%) uranium isotope, ^{238}U.

Uranium is the fourth element of the actinide (5*f*) series. Of its four oxidation states (III, IV, V, and VI), only the IV and VI states are stable enough to be of general importance.

Isotopes

Natural uranium is a mixture of three α-emitting isotopes: ^{238}U (99.274%, half-life = 4.47×10^9 yr, 4.15 MeV α), ^{235}U (0.7202%, half-life = 7.08×10^8 yr, 4.29 MeV α), and ^{234}U (0.0057%, half-life of 2.45×10^5 yr, 4.78 MeV α). Uranium is the progenitor of two naturally occurring decay series, ^{238}U ($4n + 2$), and ^{235}U ($4n + 3$) which terminates at stable ^{207}Pb. The manmade Np series, which ends in ^{209}Bi, includes ^{233}U.

Uranium isotopes and their radioactive decay products, from thorium to lead, are used extensively in determining the geochronology and geochemistry of a wide variety of minerals, rocks, and geologic formations. A mineral can be dated once the concentration ratios of ^{238}U and He are known.

Occurrence in Nature

Uranium is widely distributed in nature. It is found in significant concentrations in rocks, oceans, lunar rocks, and meteorites. Uranium is present at about 2 ppm in the earth's crust. In general, igneous rocks with a high silicate content, such as granite, contain an above average uranium concentration; whereas basic rocks, such as basalts contain a below average uranium content. Sedimentary rocks also generally contain below average uranium concentrations. Despite this low uranium content, sedimentary rocks like sandstones and conglomerates contain approximately 90% of the world's uranium resources.

Uranium resources can be assigned on the basis of their geological setting to fifteen main categories of uranium ore deposit types arranged according to their approximate economic significance: (*1*) unconformity-related deposits; (*2*) sandstone deposits; (*3*) quartz pebble conglomerate deposits; (*4*) vein deposits; (*5*) breccia complex deposits; (*6*) intrusive deposits; (*7*) phosphorite deposits; (*8*) collapse breccia pipe deposits; (*9*) volcanic deposits; (*10*) surficial deposits; (*11*) metasomatite deposits; (*12*) metamorphite deposits; (*13*) lignite; (*14*) black shale deposits; and (*15*) other deposits.

Approximately 155 minerals are known that contain uranium as an important, or major constituent, and another 60 that contain minor amounts of uranium, or contain uranium as an impurity. Uranium minerals can be divided into two mineral classes, primary and secondary. Uraninite and pitchblende are important uranium minerals with a composition that varies from UO_2 to $UO_{2.67}$ and are found in veins, pegmatites, and unweathered portions of conglomerate and sandstone ores which contain the bulk of the world's economic uranium deposits. Secondary uranium minerals are produced by hydration, metathesis, oxidation or possibly transport and redeposition. Primary minerals are generally black and contain uranium in an average oxidation state less than VI, while secondary minerals are generally yellow, green, or orange, and contain uranium in the hexavalent state. Uraninite can be considered both a primary and secondary mineral, and there are a wide variety of theories regarding the mechanism of formation of uraninite veins.

Resources

Reasonably Assured Resources (RAR) refers to uranium in known mineral deposits of size, grade, and configuration such that recovery is within the given production cost ranges with currently proven mining and processing technology. The majority of these resources are found in Australia, Brazil, Canada, Namibia, Niger, South Africa, and the United States.

Estimated additional resources (EAR) is a term that applies to resources that are inferred to occur as extensions of well-explored deposits, little-explored deposits, or undiscovered deposits believed to exist along a well-defined geological continuity with known deposits.

In January 1993, RAR recoverable at costs of $\leq$\$130/kg U, for selected countries, were estimated at 2.093×10^6 t of uranium. Estimates of total RAR recoverable at costs between \$80–\$130/kg U accounted for 660,000 t. Total RAR recoverable at costs of $\leq$\$80/kg U were estimated at 1.424×10^6 t uranium, and RAR at costs <\$80/kg were estimated at 670,000 t. This represents a decrease of ~2.4% from the 1991 values and is related to mine closures in traditional supplier countries.

Uranium exploration has decreased significantly since 1990, primarily due to decreased expenditures in France and the U.S. However, exploration programs are still being conducted in Australia, Canada, France, India, and the U.S.

The demand for uranium in the commercial sector is primarily determined by the consumption and inventory requirements of nuclear power reactors. During the past few years, governments and governmental agencies have become involved in regulating international trade in uranium involving supply from the CIS republics.

The extraction of uranium from ores varies widely, and depends on the nature of the ore involved. The ore may vary from hard, igneous rock to soft, weakly cemented sedimentary rock. The principal gangue mineral may be quartz, which is chemically inactive, or an acid-consuming mineral, such as calcite. Some ores are highly refractory and require intensive processing, whereas others break down between the mine and the mill. In order to recover the uranium from ores, a series of steps is often required including crushing and concentrating by conventional physical means; roasting and leaching the ore with acid in the presence of an oxidant to ensure conversion to UO_2^{2+}; recovery of the uranium from the leach solution, and refining to a high purity product.

Uranium Metal

Properties. Uranium metal is a dense, bright silvery, ductile, and malleable metal. Uranium is highly electropositive, resembling magnesium, and tarnishes rapidly on exposure to air. Even a polished surface becomes coated with a dark-colored oxide layer in a short time upon exposure to air. At elevated temperatures, uranium metal reacts with most common metals and refractories. Powdered uranium is usually pyrophoric, an important safety consideration in the machining of uranium parts.

In the solid state, uranium metal exists in three allotropic modifications. Uranium metal is weakly paramagnetic, with a magnetic susceptibility of 1.740×10^{-5} A/g at 20°C, and 1.804×10^{-5} A/g (A = 10 emu) at 350°C. Uranium is a relatively poor electrical conductor. Uranium metal exhibits three crystalline forms before finally melting at 1132.4°C. The α-phase exists at room temperature and consists of corrugated sheets of atoms. The β-phase exists between 668 and 775°C, and the γ-phase is formed at temperatures above 775°C.

Preparation of Uranium Metal. Uranium is a highly electropositive element, and extremely difficult to reduce. As such, elemental uranium cannot be prepared by reduction with hydrogen. Instead, uranium metal must be prepared using a number of rather forcing conditions. Uranium metal can be prepared by reduction of uranium oxides (UO_2 or UO_3 with strongly electropositive elements (Ca, Mg, Na), reduction of uranium halides (UCl_3, UCl_4, UF_4 with electropositive metals (Li, Na, Mg, Ca, Ba), electrodeposition from molten salt baths, and decomposition of uranium halides (the van Arkel-de Boer method).

A combination of technical considerations makes the reduction of UF_4 by Mg or Ca the preferred method for the preparation of uranium metal. Most important is that the reaction mixture must be fluid for the molten uranium metal to collect into an ingot at the bottom of the reaction vessel. This is an important safety consideration because finely divided uranium metal is pyrophoric.

In practice, uranium ore concentrates are first purified by solvent extraction with tributyl phosphate in kerosene to give uranyl nitrate hexahydrate. The purified uranyl nitrate is then decomposed thermally to UO_3, which is reduced with H_2 to UO_2, which in turn is converted o UF_4 by high temperature hydrofluorination. The UF_4 is then converted to uranium metal with Mg; $UF_4 + 2\ Mg \rightarrow U(0) + 2\ MgF_2$. Reduction of uranium tetrafluoride by magnesium metal is often referred to as the Ames process. The reaction is very exothermic and the reduction process is carried out in a sealed bomb due to volatility at the temperatures reached in the reaction (Fig. 1).

Isotope Enrichment

The enrichment of uranium is expressed as the weight percent of ^{235}U in uranium. For natural uranium the enrichment level is 0.72%.

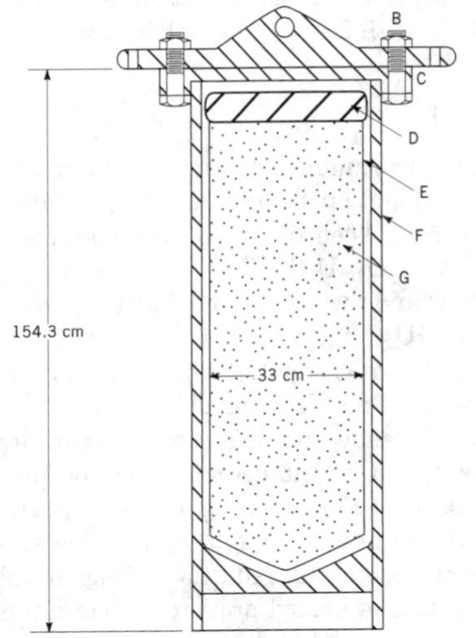

Figure 1. Bomb reactor for the reduction of UF_4 with Mg by the Ames process (capacity 144.2 kg uranium metal): A, steel cover flange with lifting eye; B, bolt and nut; C, top flange of bomb; D, graphite cover; E, liner of fused dolomitic oxide; F, steel bomb, and G, charge, where ▧ represents steel, □ liner, ▨ graphite, and ■ charge.

Many applications of uranium require enrichment levels above 0.72%, such as nuclear reactor fuel. Normally for lightwater nuclear reactors (LWR), the 0.72% natural abundance of ^{235}U is enriched to 2–5%. There are special cases such as materials-testing reactors, high flux isotope reactors, compact naval reactors, or nuclear weapons where ^{235}U enrichment of 96–97% is used.

Uranium isotope enrichment can be achieved in a number of ways. At the time of writing, the methods that have been, or are currently in use include gaseous diffusion, gaseous centrifugation, electromagnetic separation, chemical exchange, laser photoionization and photodissociation, separation nozzle, and cyclotronic resonance isotope separation. Most of these processes are of historical significance, and have been described. The gaseous diffusion and centrifugation processes (GDP and GDC) are the only methods employed on an industrial scale in the United States and Europe. Vigorous research programs are under way in the United States, France, and Japan, for a new industrial technique known as atomic-vapor laser isotope separation (AVLIS, or SILVA in France). The AVLIS process is expected to have a lower cost per separative work unit (SWU) than the diffusion process for uranium enrichment.

Uses and Economic Aspects

Uranium is a synthetic precursor of transuranium elements and the source of the light isotope, ^{235}U. The primary use of ^{235}U, is as a source of nuclear energy for nuclear power generators and nuclear weapons. Uranium carbide (UC) has been utilized in sodium or lead cooled reactors, whereas uranium silicides have been proposed as a fuel source in lightwater reactors. Uranium–zirconium and uranium–aluminum alloys are used in materials and research testing reactors. Uranium–zirconium alloys are also widely used in marine reactors, while the hydrogenated U–Zr alloys ($UZrH_x$) are fuels for spacecraft reactors.

Depleted uranium (^{238}U), which is about 0.2% ^{235}U, has a density more than twice that of steel. This property has been utilized for military purposes in the production of armor and armor-piercing projectiles, also known as kinetic energy penetrators. The high density of uranium makes it attractive for flywheels, and its density and effectiveness at absorbing gamma-rays also suggests a possible use for shielding of spent nuclear fuels. One of the difficulties in designing metallic uranium shields is the tendency of uranium to corrode in air, forming oxide surfaces. A corrosion-resistant material can be obtained by alloying uranium with 2–8 wt % molybdenum.

Uranium Compounds

Oxides. Oxides of uranium are some of the most prevalent and technologically important binary uranium compounds known. Numerous oxide phases have been observed and characterized, including uranium oxide, UO, UO_2; U_4O_9; U_3O_7; U_3O_8; UO_3; hydrated species such as $UO_3 \cdot xH_2O$ and the peroxo complex, $UO_4 \cdot xH_2O$; and anionic uranates including $[U_2O]^{2-}7$ and $[U_4O]^{2-}13$. Of these oxide phases, UO_2, U_3O_8, and UO_3 are extremely important both industrially and in the nuclear energy cycle. Uranium dioxide, UO_2, is found in nature as the mineral pitchblende and as a component in uraninite. Industrially, UO_2 is prepared by the decomposition of ammonium uranyl carbonate on the scale of 10 kg/d, using a fluidized-bed furnace. In addition to the industrial process, pure UO_2 has been synthesized by (*1*) oxidation of uranium metal, (*2*) reduction of higher valent oxides, (*3*) thermal decomposition of uranyl uranates, (*4*) oxidation or reduction of uranium halides, (*5*) decarboxylation of uranium compounds of carbonic acids, (*6*) hydrometallurgical preparation, and (*7*) electrolysis of uranium halides.

The main technological uses for UO_2 are found in the nuclear fuel cycle as the principal component for light and heavy water reactor fuels. Uranium dioxide is also a starting material for the synthesis of UF_4, UF_6 (both critical for the production of pure uranium metal and isotopic enrichment), UCl_4, and $UO_2(NO_3)_2{\cdot}6H_2O$. Uranium dioxide has been found to exhibit a majority of desirable properties for nuclear fuels, with an average thermal coefficient of expansion of 10.8×10^{-6} (20–946°C), specific heat from 0.237 to 0.338 J/gK (300–1773°C), and a thermal conductivity of 8.281 to 2.353 W/mK (300–1773°C) at 0 atomic % burnup UO_2. The sintered complex has also been found to be chemically stable toward air and H_2O up to 300°C.

For most nuclear applications, UO_2 must be produced as uniform spheres and pellets. Three techniques utilized for microsphere fabrication are sol-gel, gel-precipitation, and plasma spheroidization. Details on the sol-gel and gel-purification processes, the two most popular, can be obtained from the *Gmelin Handbook*. The common method for producing UO_2 pellets consists of pressing granules in the presence of binding agents and lubricants with a subsequent sintering, after the organics have been removed.

Triuranium octaoxide, U_3O_8, is a greenish black material which is also a constituent of pitchblende. This complex has been identified with a number of different oxygen deficiencies, depending mostly on the temperature and partial pressure of O_2 used in the preparation.

Industrially, U_3O_8 has been shown to be active in the decomposition of organics, including benzene and butanes and as supports for methane steam reforming catalysts. In the nuclear fuel industry, U_3O_8 is an oxidation product of UO_2 (SIMFUEL), and thus, a large component of spent fuel rods. U_3O_8 is less dense than UO_2 and as a result, the production of U_3O_8 in nuclear fuel can lead to the destruction of the UO_2 pellet by pulverization. Triuranium octaoxide is not always a destructive force in the fuel cycle, it is actually quite useful in the initial production of UO_2 pellets for fuel, in the manufacturing of mixed oxide (MOX) pellets, as well as being a dispersive nuclear fuel itself.

Uranium trioxide, UO_3, is a versatile solid that also has important applications in the nuclear fuel cycle. The trioxide has been isolated in six well-defined stoichiometric modifications as well as a hypostoichiometric modification, $UO_{2.9}$. Similar to U_3O_8, the trioxide decomposes into lower oxides prior to melting or subliming. The preparation of UO_3 has been accomplished by a variety of means. Industrially, the complex is prepared by three main routes, thermal decomposition of $UO_4{\cdot}xH_2O$, $(NH_4)_2U_4O_{13}$, or $UO_2(NO_3)_2{\cdot}6H_2O$ under O_2. For the latter complex, the techniques utilized to acomplish the decomposition include batch decomposition, continuous stirred-bed, fluidized bed, and spray decomposition. The trioxide can also be synthesized by the oxidation of lower oxides, UI_3 (lower temperature), UI_4 (low temperature), UC, or UN with O_2, and by the calcination of $(NH_4)_4UO_2(CO_3)_3$.

As mentioned above, uranium trioxide exists in six well-defined modifications with colors ranging from yellow to brick-red. Of these phases, the γ-phase has been found to be the most stable.

The most important role of UO_3 is in the production of UF_4 and UF_6, which are used in the isotopic enrichment of uranium for use in nuclear fuels. The trioxide also plays a part in the production of UO_2 for fuel pellets. Microspheres of UO_3 can themselves be used as nuclear fuel. Fabrication of UO_3 microspheres has been accomplished using sol-gel or internal gelation processes. UO_3 is also a support for destructive oxidation catalysts of organics.

Nitrides. Uranium nitrides are well known and are used in the nuclear fuel cycle. There are three nitrides of exact stoichiometry, uranium nitride, UN; U_2N_3; and U_4N_7. The brown mononitride, which is the only nitride complex stable above 1300°C, melts at 2600°C. Uranium mononitride is the most dense of the nitrides with a density of 14.31 g/cm^3. The magnetic properties of the nitrides are extremely dependent on the phase and stoichiometry of the complex. Classically, the different nitrides have been prepared from direct interaction of the elements under the appropriate conditions. A number of alternatives to this preparation have been investigated, including uranium metal under static NH_3 at 300–350°C to yield U_2N_3, uranium metal or uranium carbides with NH_3 or N_2 at 600–900°C to produce U_2N_{3+x}, uranium carbide fuels reacted with N_2/H_2 to form UN, and a self-propagating metathetical reaction, thermolysis at 500°C of UCl_4 with Li_3N, yielding UN and U_2N_3.

Uranium and mixed uranium–plutonium nitrides have a potential use as nuclear fuels for lead cooled fast reactors. Reactors of this type have been proposed for use in deep-sea research vehicles. The nitrides must have an appropriate size and shape, ie, spheres. Microspheres of uranium nitrides have been fabricated by internal gelation and carbothermic reduction. Uranium nitrides are used as a catalyst for the cracking of NH_3.

Carbides. Uranium carbides, UC, U_2C_3, and UC_2 are all dark gray solids with a metallic luster. The melting points of UC and U_2C_3 are 2400°C and 2417°C, respectively, and the dicarbide melts at 2475°C and boils at 4370°C (760 mm Hg). The monocarbide is the most dense of the carbide series with a room temperature density of 13.60 g/cm^3, whereas U_2C_3 and UC_2 have densities of 12.85 g/cm^3 and 11.69 g/cm^3, respectively. All three materials are paramagnetic at room temperature, with U_2C_3 becoming antiferromagnetic at low temperatures. The typical techniques involved in the synthesis of the carbides include the reaction of carbon or hydrocarbons with uranium metal or UH_3 at elevated temperature, precipitation from metal melts, and reduction of uranium halides. Techniques for the synthesis include the carbothermic reduction of UO_2 and the direct interaction of uranium and carbon under highly exothermic conditions.

Oxo Ion Salts. Salts of oxo anions, such as nitrate, sulfate, perchlorate, iodate, hydroxide, carbonate, phosphate, oxalate, etc, are important for the separation and reprocessing of uranium, hydroxide, carbonate, and phosphate ions are important for the chemical behavior of uranium in the environment.

Nitrate complexes are very weak, and the determination of the formation constants for aqueous nitrate solution species is extremely difficult. Solid uranyl nitrate, $UO_2(NO_3)_2{\cdot}xH_2O$, is obtained as the orthorhombic hexahydrate from dilute nitric acid solutions, and as the trihydrate from concentrated acid. The melting point of the hexahydrate is at 118°C. Uranyl nitrate plays an important role in the reprocessing of uranium in spent fuel and in uranium extraction from aqueous solutions. The preparation of the anhydrous uranyl nitrate by dehydration is extremely difficult.

Hydroxides. The hydrolysis of uranium has been recently reviewed, yet as noted in these compilations, studies are ongoing to continue identifying all of the numerous solution species and solid phases.

Carbonates. Actinide carbonate complexes are of interest not only because of their fundamental chemistry and environmental behavior, but also because of extensive industrial applications, primarily in uranium recovery from ores and nuclear fuel reprocessing.

The aqueous U(VI) carbonate system has been very thoroughly studied, and there is little doubt about the compositions of the three monomeric complexes $UO_2(CO_3)$, $UO_2(CO_3)_2^{2-}$, and $UO_2(CO_3)_3^{4-}$ present under the appropriate conditions. There is also a great deal of evidence from emf, solubility, and spectroscopic data supporting the existence of polymeric solution species of formulas $(UO_2)_3(CO_3)_6^{6-}$, $(UO_2)_2(CO_3)(OH)_3^-$, $(UO_2)_3O(OH)_2(HCO_3)^+$, and $(UO_2)_{11}(CO_3)_6\text{-}(OH)_{12}^{2-}$ which form only under conditions of high metal ion concentration or high ionic strength.

The known uranium(VI) carbonate solids have empirical formulas, $UO_2(CO_3)$, $M_2UO_2(CO_3)_2$, and $M_4UO_2(CO_3)_3$. The solid of composition $UO_2(CO_3)$ is a well-known mineral, rutherfordine, and its structure has been determined from crystals of both the natural mineral and synthetic samples.

Biscarbonato complexes of uranium(VI) are well-established in solution and there are many reports dating from the late 1940s through the 1960s of solid phases with the general stoichiometry $M_2UO_2(CO_3)_2$, where M is a monovalent cation (Na^+, K^+, Rb^+, Cs^+, NH_4^+, etc). A summary of the preparative details is available, as is a listing of the compounds in the *Gmelin Handbook*.

The triscarbonato solids, $M_4UO_2(CO_3)_3$ (M = monovalent cation) are the most thoroughly studied uranium(VI) carbonate solids. These solid phases are generally prepared by evaporation of an aqueous solution of the components, or by precipitation of the UO_2^{2+} ion with an excess of carbonate.

Although there is a great deal of qualitative information regarding anionic carbonato complexes of the tetravalent actinides, reliable quantitative data are rare. All of the uranium(IV) complexes are readily air-oxidized to uranium(VI) complexes, and therefore there is no structural information for the uranium(IV) analogues.

Phosphates. Inorganic phosphate ligands are important with respect to the behavior of uranium in the environment and as potential waste forms. There have been a number of experimental studies to determine the equilibrium constants in the uranium–phosphoric acid system, but they have been complicated by the formation of relatively insoluble solid phases and the formation of ternary uranium complexes in solution. In acidic solution (hydrogen-ion concentration range 0.25–2.00 *M*) H_3PO_4 and $H_2PO_4^-$ are potential ligands, whereas in neutral to basic solution, HPO_4^{2-} and PO_4^{3-} ligands are predominant. Numerous U(VI) phosphate complexes have been identified and their formation constants determined. However, relatively little thermodynamic data have been recommended with confidence.

Coordination Complexes

Considered "hard" metal ions, U(III to VI) have the greatest affinity for hard donor atoms such as N, O, and the light halides. Tetravalent and hexavalent uranium coordination complexes are the most common, however trivalent and pentavalent complexes have been identified with increasing frequency. The ionic radius of any uranium ion is significantly larger compared to a transition metal ion in an identical oxidation state. The result of this increased ionic radius is an expansion of the possible coordination environments (3- to 14-coordinate) and electron counts (up to 24 electrons).

Nitrogen Donors. There are numerous *N*-donating ligands which have been complexed with uranium. Classic examples range from neutral mono-, bi-, and polydentate ligands, ie, ammonia, primary, secondary, and tertiary amines, alkyl–aryldiamines (en = ethylenediamine, 1,4-diaminobenzene), *N*-heterocycles (py = pyridine, bipy = bipyridine, terp = terpyridyl), nitriles (CH_3CN), to anionic amides [$N(C_2H_5)_2$, $N(Si(CH_3)_3)_2^-$], thiocyanates, and polypyrazolyl-borates. A complete listing of ligands can be found in the general references, *Gmelin Handbook* and *Comprehensive Coordination Chemistry*.

Very few U(III) coordination complexes with neutral *N*-donor ligands have been identified due in part to the ease of oxidation.

N-Donor coordination complexes of U(IV) are numerous and have been well characterized. Adducts of UX_4 (X = halogen, alkoxide) have been isolated with all of the ligand types described above. The most common coordination environments for U(IV) are 8–12, as exemplified by UX_4L_n [X = Cl, L = NH_3 (n = 1–10), en (n = 4), bipy (n = 2); X = Br, I L = NH_3 (n = 4–6), 1,4-diaminobenzene (n = 4); X = alkyl, L = CH_3CN (x = 4)].

As in the case of U(III), coordination chemistry of U(V) *N*-donor complexes are relatively unexplored owing to disproportionation which yields U(IV) and U(VI) complexes. Typically, ammonia, secondary amines, pyridines, pyrazines, and nitrile adducts of $U(OR)_5$ and UX_5 have been isolated with coordination numbers ranging from 6–8.

The majority of U(VI) coordination chemistry has been explored with the *trans*-dioxo uranyl cation, UO_2^{2+}. The simplest complexes are ammonia adducts, of importance because of the ease of their synthesis and their versatility as starting materials for other complexes.

Phosphorus Donors. Phosphine coordination complexes of uranium are rare owing to the preference of uranium for hard donor atoms. However, complexes with monodentate phosphines, ie, $P(CH_3)_3$, have been identified. The benefit of these complexes is their versatility in synthetic uranium organometallic chemistry. Uranium(V) phosphine complexes have been synthesized, using amino ligands with a phosphine appendage, such as $UCl_2[N(CH_2CH_2PPri_2)_2]_3$. The phosphido complex, $U(PPP)_4$ (PPP = P $(CH_2CH_2P\text{-}(CH_3)_2)$, was prepared and fully characterized. This complex was one of the first actinide complexes containing exclusively metal-phosphorus bonds.

Oxygen Donors. A wide variety of *O*-donors have been used to complex uranium. The predominate oxidation states are IV and VI; however, complexes with U(III) and U(V) are also known. The majority of the complexes have coordination numbers of 6 to 12, depending mostly on the steric bulk of the ancillary ligands. Owing to the prevalence of *O*-donating ligands in natural systems, ie, aquo, hydroxide, carbonate, phosphate, carboxylate, and catecholate, understanding the complexation of uranium and other radioactive nuclides is important to environmental, waste processing and storage, and bioinorganic chemistry. Some of the other *O*-donating ligands which have been studied are crown ethers, Schiff bases, polyglycols, and cryptands. A complete listing of the *O*-donating ligands complexed with uranium can be found in the *Gmelin Handbook* and in *Comprehensive Coordination Chemistry*.

Halides. Uranium halide complexes can be found in all four of the available metal oxidation states, III, IV, V, and VI. In general, fluoride ligands tend to favor higher oxidation states, and iodide ligands tend to favor the lower oxidation states. As a result of the important industrial applications of binary fluorides and chlorides (*vide infra*), the majority of the halide discussion focuses on the binary systems. A selected listing of physical constants for the binary uranium halides is provided in Table 1.

Fluorides. Uranium fluorides play an important role in the nuclear fuel cycle as well as in the production of uranium metal. The dark purple UF_3 has been prepared by two different methods neither of which neither have been improved. The first involves a direct reaction of UF_4 and uranium metal under elevated temperatures, while the second consists of the reduction of UF_4 by UH_3. The local coordination environment of uranium in the trifluoride is pentacapped trigonal

Table 1. Physical Constants for Selected Uranium Halides

Compound	Density, g/mL	Mp, °C	Bp, °C
UF_6	4.68	64.5–64.8	56.2^{765}
UF_4	6.70	960	
UF_3		>1000 dec	
UCl_5	3.81	>300 dec	
UCl_4	4.87	590	792^{760}
UCl_3	5.44	842	
UBr_4	5.35	516	792^{760}
UBr_3	6.53	730	volatile
UI_4	5.6	506	759^{760}

prismatic with an 11-coordinate uranium atom. The trifluoride is insoluble in H_2O but is soluble in strong acids, ie, nitric, hot sulfuric and perchloric.

The tetrafluoride, UF_4, is a green solid, which can be isolated with high purity and has industrially important properties, ie, high stability and low volatility. As a result of these properties UF_4 is widely used as a starting material in uranium production processes. The preparation of UF_4 has been accomplished by reaction of HF with UO_2 at elevated temperatures or by electrolytic reduction of uranyl fluoride in aqueous HF.

Uranium pentafluoride, UF_5, has been isolated under different conditions, leading to two different modifications, α and β. The former is a grayish white solid, which is synthesized from the interaction of UF_6 and HBr or by heating UF_4 and UF_6 to 80–100°C. The yellowish white β-modification is also obtained by reacting UF_4 and UF_6, but at higher temperatures (150–200°C).

Uranium hexafluoride, UF_6, is an extremely corrosive, colorless, crystalline solid, which sublimes with ease at room temperature and atmospheric pressure. The complex can be obtained by multiple routes, ie, fluorination of UF_4 with F_2, oxidation of UF_4 with O_2, or fluorination of UO_3 by F_2. UF_6 is soluble in H_2O, CCl_4 and other chlorinated hydrocarbons, is insoluble in CS_2, and decomposes in alcohols and ethers. The importance of UF_6 in isotopic enrichment and the subsequent applications of uranium metal cannot be overstated.

Chlorides. The olive-green trichloride, UCl_3, has been synthesized by chlorination of UH_3 with HCl. This reaction is driven by the formation of gaseous H_2 as a reaction by-product. The structure of the trichloride has been determined and the central uranium atom possesses a nine-coordinate tricapped trigonal prismatic coordination geometry. The solubility properties of UCl_3 are as follows: soluble in H_2O, methanol, glacial acetic acid; insoluble in ethers.

Uranium tetrachloride, UCl_4, has been prepared by several methods. The first method, which is probably the best, involves the reduction/chlorination of UO_3 with boiling hexachloropropene. The second consists of heating UO_2 under flowing CCl_4 or $SOCl_2$. The structure of the dark green tetrachloride is identical to that of Th, Pa, and Np, which all show a dodecahedral geometry of the chlorine atoms about a central actinide metal atom. The tetrachloride is soluble in H_2O, alcohol, and acetic acid, but insoluble in ether, and chloroform.

The reddish brown pentachloride, uranium pentachloride, UCl_5, has been prepared in a similar fashion to UCl_4 by reduction–chlorination of UO_3 under flowing CCl_4, but at a lower temperature. Another synthetic approach which has been used is the oxidation of UCl_4 by Cl_2. The pentachloride decomposes in H_2O and acid, is soluble in anhydrous alcohols, and insoluble in benzene and ethers.

The hexachloride, uranium hexachloride, UCl_6, is best prepared by chlorination of UCl_4 with $SbCl_5$. An alternative preparative approach is the disproportionation UCl_5 to UCl_4 and UCl_6 under reduced pressure.

Bromides and Iodides. The red-brown tribromide, UBr_3, and the black triiodide, UI_3, may both be prepared by direct interaction of the elements, ie, uranium metal with X_2 (X = Br, I). The tribromide has also been prepared by interaction of UH_3 and HBr, producing H_2 as a reaction product. The tribromide and triiodide complexes are both polymeric solids with a local bicapped trigonal prismatic coordination geometry. The tribromide is soluble in H_2O and decomposes in alcohols.

The best synthetic approach to isolate UBr_4 and uranium tetraiodide, UI_4, is by direct interaction of the elements. This is typically accomplished by heating uranium turnings under flowing nitrogen–halogen gas. The tetrabromide is dark brown and hygroscopic. The black tetraiodide is unstable, undergoing reduction to uranium triiodide, UI_3, and I_2. Structural details of the tetrabromide and the tetraiodide are not available. The tetrabromide is soluble in H_2O and liquid NH_3, but decomposes in alcohols, whereas the tetraiodide is soluble in cold H_2O and acetonitrile, and decomposes in hot H_2O.

Uranium pentabromide, UBr_5, is unstable toward reduction and the pentaiodide is unknown. Two synthetic methods utilized for the production of UBr_5 involve the oxidation of uranium tetrabromide, UBr_4, by Br_2 or by bromination of uranium turnings with Br_2 in acetonitrile. The metastable pentabromide is isostructural with the pentachloride, being dimeric with edge-sharing octahedra U_2Br_{10}.

Organometallic Complexes

The organometallic chemistry of uranium has grown rapidly since the 1970s. The majority of the organouranium complexes are found with U(IV) centers; however, there are some examples of higher and lower valent species being isolated. Uranium organometallic compounds have potential uses in homogeneous and heterogeneous catalysis with activities ranging from the hydrogenation and polymerization of olefins to the selective activation of alkanes. In addition to these potentially important industrial uses, uranium complexes are also used as innocuous models for other more radioactive actinides. A wide range of organic molecules have been complexed with uranium including: hydrocarbyl, allyl, arene, cyclooctatetraenyl, and a host of cyclopentadienyl-based ligands.

Health and Safety Factors

Exposure and Health Effects. Uranium is a general cellular poison which can potentially affect any organ or tissue. Uranium and its compounds can be damaging due to chemical toxicity and by the injury caused by ionizing radiation. The chemical toxicity of uranium compounds depends on their solubility in biological media. Highly soluble and therefore highly transportable and toxic compounds include fluorides, chlorides, nitrates, and carbonates of uranium(VI); moderately transportable compounds include corresponding uranium(IV) compounds; slightly transportable compounds include oxides, hydrides, and carbides.

Occupational Protection and Radiation Considerations. The main adverse factor during the mining and processing of uranium and uranium-containing minerals is airborne dust. Personal protection should be used. Finely divided uranium metal, some alloys, and uranium hydride are pyrophoric.

The toxicity of uranium caused by its radiation depends on the isotopes present. Natural uranium does not constitute an external radiation hazard since it emits mainly low energy α-radiation. It does, however present an internal radiation hazard if it enters the body by inhalation or ingestion. The concentration of 1 mg U/g biological tissue corresponds to an absorbed dose of 0.006 Sv per year. Large quantities of fissile isotopes, ^{233}U and ^{235}U, should be handled and stored appropriately to avoid a criticality hazard. Clear and relatively simple precautions, such as dividing quantities so that the minimum critical mass is avoided, following administrative controls, using neutron poisons, and avoiding critical configurations (or shapes), must be followed to prevent an extremely treacherous explosion.

DAVID L. CLARK
D. WEBSTER KEOGH
MARY P. NEU
WOLFGANG RUNDE
Glenn T. Seaborg Institute for Transactinium Science
Los Alamos National Laboratory

Uranium, Resources, Production and Demand: a joint report by the OECD Nuclear Energy Agency and the International Atomic Energy Agency, Organisation for Economic Cooperation and Development, Nuclear Energy Agency, Paris, France, 1993.

K. W. Bagnal, in Sir G. Wilkinson, R. D. Gillard, and J. A. McCleverty, eds., *Comprehensive Coordination Chemistry: The Synthesis, Reactions, Properties, and Applications of Coordination Compounds,* 1st ed., Pergamon Press, New York, 1987, pp. 1120–1130.

F. T. Edelman, in E. W. Abel, F. G. A. Stone, Sir G. Wilkinson, eds., *Comprehensive Organometallic Chemistry II: A Review of the Literature 1982–1994,* 1st ed., Pergamon Press, New York, 1995, pp. 12–192.

M. Peehs, T. Walter, and S. Walter, in B. Elvers and S. Hawkins, eds., *Ullmann's Encyclopedia of Industrial Chemistry,* VCH, Weinheim, Germany, 1996, p. 281.

UREA

Properties

Urea (**1**) can be considered the amide of carbamic acid, NH_2COOH, or the diamide of carbonic acid, $CO(OH)_2$.

$$NH_2\overset{\overset{O}{\parallel}}{C}NH_2$$

(**1**)

At room temperature, urea is colorless, odorless, and tasteless. Selected properties are shown in Table 1. Dissolved in water, it hydrolyzes very slowly to ammonium carbamate and eventually decomposes to ammonia and carbon dioxide. This reaction is the basis for the use of urea as fertilizer (see FERTILIZERS).

Urea is classified as a nontoxic compound.

At atmospheric pressure and at its melting point, urea decomposes to ammonia, biuret (**2**), cyanuric acid (qv), ammelide, and triuret. Biuret is the main and least desirable by-product present in commercial urea. An excessive amount (>2 wt%) of biuret in fertilizer-grade urea is detrimental to plant growth.

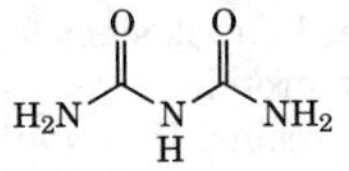

(**2**)
biuret

Urea acts as a monobasic substance and forms salts with acids. With nitric acid, it forms urea nitrate, $CO(NH_2)_2{\cdot}HNO_3$, which decomposes explosively when heated. Solid urea is stable at room temperature and atmospheric pressure. Heated under vacuum at its melting point, it sublimes without change. At 180–190°C under vacuum at its melting point, it sublimes without change. At 180–190°C under vacuum, urea sublimes and is converted to ammonium cyanate, NH_4OCN. When solid urea is rapidly heated in a stream of gaseous ammonia at elevated temperature and at a pressure of several hundred kPa (several atm), it sublimes completely and decomposes partially to cyanic acid, HNCO, and ammonium cyanate. Solid urea dissolves in liquid ammonia and forms the unstable compound urea–ammonia, $CO(NH_2)_2NH_3$. Urea–ammonia forms salts with alkali metals, eg, NH_2CONHM or $CO(NHM)_2$. The conversion of urea to biuret is promoted by low pressure, high temperature, and prolonged heating. At 10–20 MPa (100–200 atm), biuret gives urea when heated with ammonia.

The reaction of urea with alcohols yields carbamic acid esters, commonly called urethanes (see URETHANE POLYMERS):

$$NH_2\overset{\overset{O}{\parallel}}{C}NH_2 + ROH \longrightarrow NH_2\overset{\overset{O}{\parallel}}{C}OR + NH_3$$

Urea reacts with formaldehyde and forms compounds such as monomethylolurea, $NH_2CONHCH_2OH$, dimethylolurea, $HOCH_2NHCONHCH_2OH$, and others, depending upon the mol ratio of formaldehyde, to urea and upon the pH of the solution. Hydrogen peroxide and urea give a white crystalline powder, urea peroxide, $CO(NH_2)_2{\cdot}H_2O_2$, known under the trade name of Hypersol, an oxidizing agent.

Urea and malonic acid give barbituric acid (**3**), a key compound in medicinal chemistry (see also HYPNOTICS, SEDATIVES, AND ANTICONVULSANTS):

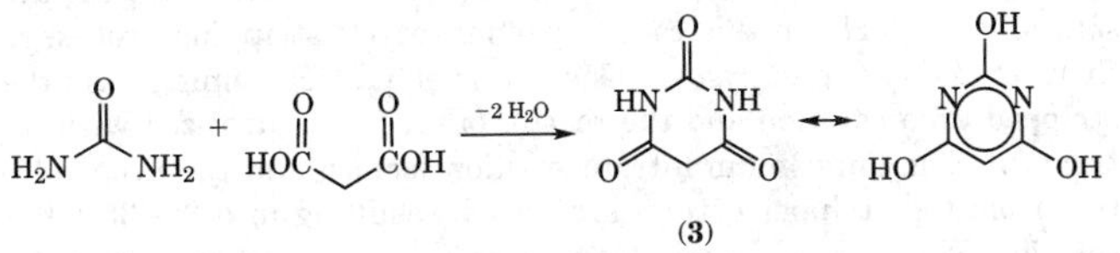

(**3**)
malonyl urea or
barbituric acid

Manufacture

Urea is produced from liquid NH_3 and gaseous CO_2 at high pressure and temperature; both reactants are obtained from an ammonia-synthesis plant. The latter is a by-product stream, vented from the CO_2 removal section of the ammonia-synthesis plant. The two feed components are delivered to the high pressure urea reactor, usually at a mol ratio >2.5 : 1.

The formation of ammonium carbamate and the dehydration to urea take place simultaneously, for all practical purposes:

$$2\,NH_3 + CO_2 \rightleftharpoons \underset{\text{ammonium carbamate}}{NH_2\overset{\overset{O}{\parallel}}{C}ONH_4} \qquad (1)$$

$$NH_2\overset{\overset{O}{\parallel}}{C}ONH_4 \rightleftharpoons \underset{\text{urea}}{NH_2\overset{\overset{O}{\parallel}}{C}NH_2} + H_2O \qquad (2)$$

Reaction 1 is highly exothermic. The excess heat must be removed from the reaction. The rate and the equilibrium of reaction 1 depend greatly on pressure and temperature, because large volume changes take place. This reaction may occur only at a pressure that is below the pressure of ammonium carbamate at which dissociation begins or, conversely, the operating pressure of the reactor must be maintained above the vapor pressure of ammonium carbamate. Reaction 2 is endothermic. It takes place mainly in the liquid phase.

Ammonium Carbamate. Ammonium carbamate is a white crystalline solid which is soluble in water. It forms at room temperature by passing ammonia gas over dry ice. In an aqueous solution at room temperature, it is slowly converted to ammonium carbonate, $(NH_4)_2CO_3$, by the addition of one mol of water. Above 60°C, the ammonium carbonate solution reverts to carbamate solution, and at 100°C, only carbamate is present in the solution. Above 150°C, ammonium carbamate loses a mol of water and forms urea. Ammonium

Table 1. Properties of Urea

Property	Value
mp, °C	135
index of refraction, n_D^{20}	1.484, 1.602
density, d_4^{20}, g/cm^3	1.3230
crystalline form and habit	tetragonal, needles, or prisms
free energy of formation, at 25°C, J/mol[a]	−197.150
heat of fusion, J/g[a]	251[b]
heat of solution in water, J/g[a]	243[b]
heat of crystallization, 70% aq urea soln, J/g[a]	460[b]
bulk density, g/cm^3	0.74
specific heat, J/(kg·K)[a]	
at 0°C	1.439
50	1.661
100	1.887
150	2.109[c]

[a] To convert J to cal, divide by 4.184.
[b] Endothermic.
[c] Exothermic.

carbamate melts at ca 150°C, and has a heat of fusion of ca 16.74 kJ/mol (4.0 kcal/mol).

Conversion at Equilibrium. The maximum urea conversion at equilibrium attainable at 185°C is ca 53% at infinite heating time. The conversion at equilibrium can be increased either by raising the reactor temperature or by dehydrating ammonium carbamate in the presence of excess ammonia. Excess ammonia shifts the reaction to the right side of the overall equation:

$$2\ NH_3 + CO_2 \leftrightarrows NH_2CONH_2 + H_2O$$

Water, however, has the opposite effect.

Processing

At this time over 95% of all new urea plants are licensed by Snamprogetti (SNAM), Stamicarbon (STAC), or Toyo Engineering. SNAM utilizes thermal stripping while STAC and Toyo use CO_2 stripping.

STAC, with their current new design (pool reactor), feel the only size limitation will be vessel size. Unless both the vessel fabricator and the intended plant site are "on water," a 4-m-diameter is the maximum that can be transported.

The urea produced is normally either prilled or granulated. In some countries there is a market for liquid urea-ammonium nitrate solutions (32% N). In this case, a partial-recycle stripping process is the best and cheapest system. The unconverted NH_3 coming from the stripped urea solution and the reactor off-gas is neutralized with nitric acid. The ammonium nitrate solution formed and the urea solution from the stripper bottom are mixed, resulting in a 32–35 wt % solution. This system drastically reduces investment costs as evaporation, finishing, (prill or granulation), and wastewater treatment are not required.

Finishing Processes

Urea processes provide an aqueous solution containing 70–87% urea. This solution can be used directly for nitrogen-fertilizer suspensions or solutions such as urea–ammonium nitrate solution. Urea solution can be concentrated by evaporation or crystallization for the preparation of granular compound fertilizers and other products. Concentrated urea is solidified in essentially pure form as prills, granules, flakes, or crystals. Solid urea can be shipped, stored, distributed, and used more economically than in solution. Furthermore, in the solid form, urea is more stable and biuret formation less likely.

Prilling. The manufacture of prills is rapidly decreasing owing to both environmental problems and product quality as compared to granules. In a prilling plant the urea solution from the recovery section is evaporated in two stages to +99.8% strength. It is then pumped to the top of a 50–60 m cylindrical concrete tower where it is fed into a spinning bucket containing many (+2000) small holes. The emerging small liquid droplets solidify as they fall and are cooled by a forced or induced draft air flow. The very fine dust that is formed and exits at the top of the tower with the air flow is an environmental problem. Prill size must necessarily be small in order to obtain proper solidification and cooling in the fall height that is practical. Generally, both the crushing and impact strength of the prill is much less than for a granule. If it were not for environmental considerations, prilling would still be a cheaper option than granulation in a small-scale marketing area (ie, not on a global scale).

Granulation. Almost all new plants produce granules, and the Hydro-Agri process by NSM of Holland is used in the majority of plants.

Wastewater Treatment

Under the pressure of progressively more stringent government regulations with regard to permissible levels of residual NH_3 and urea content in wastewaters, the fertilizer industry has made an effort to improve wastewater treatment (see also WATER, SEWAGE).

For each mol of urea produced in a total-recycle urea process, one mol of water is formed. It is usually discharged from the urea concentration and evaporation section of the plant. Small amounts of urea are usually found in wastewaters because of entrainment carry-over.

The problem in reducing the NH_3 and urea content in the wastewaters to below 100 ppm is that it is difficult to remove one in the presence of the other. The wastewater can be treated with caustic soda to volatilize NH_3. However, in a more efficient method, the urea is hydrolyzed to ammonium carbamate, which is decomposed to NH_3 and CO_2; the gases are then stripped from the wastewater.

All process licensors also feature wastewater treatment systems. Stamicarbon guarantees the lowest NH_3–urea content and has plants in operation confirming the low NH_3–urea (1 ppm NH_3–1 ppm urea). This water is very satisfactory to use as boiler feed water.

Uses

Solid urea containing 0.8–2.0 wt % biuret is primarily used for direct application to the soil as a nitrogen-release fertilizer. Weak aqueous solutions of low biuret urea (0.3 wt% biuret max) are used as plant food applied to foliage spray.

Mixed with additives, urea is used in solid fertilizers of various formulations, eg, urea–ammonium phosphate (UAP), urea–ammonium sulfate (UAS), and urea–phosphate (urea + phosphoric acid). Concentrated solutions of urea and ammonium nitrate (UAN) solutions (80–85 wt %) have a high nitrogen content but low crystallization point, suitable for easy transportation, pipeline distribution, and direct spray application.

Urea is also used as feed supplement for ruminants, where it assists in the utilization of protein. Urea is one of the raw materials for urea-formaldehyde resins. Urea (with ammonia) pyrolyzes at high temperature and pressure to form melamine plastics (see also CYANAMIDES). Urea is used in the preparation of lysine, an amino acid widely used in poultry feed (see AMINO ACIDS; FEEDS AND FEED ADDITIVES). It also is used in some pesticides.

Partially polymerized resins of urea are used by the textile industry to impart permanent-press properties to fabrics (see also TEXTILES, FINISHING).

Reagent-grade urea as used in some pharmaceutical preparations must meet the purity specifications issued by the ACS.

Clathrates

Urea has the remarkable property of forming crystalline complexes or adducts with straight-chain organic compounds. These crystalline complexes consist of a hollow channel, formed by the crystallized urea molecules, in which the hydrocarbon is completely occluded. Such compounds are known as clathrates. This property of urea clathrates is widely used in the petroleum-refining industry for the production of jet aviation fuels (see AVIATION AND OTHER GAS-TURBINE FUELS) and for dewaxing of lubricant oils (see also PETROLEUM, REFINERY PROCESSES SURVEY).

IVO MAVROVIC
Consultant
A. RAY, JR. SHIRLEY
Applied Chemical Technology
G. R. "BUCK" COLEMAN
Consultant

J. Meesen, "Urea" in *Ullmann's Encyclopedia of Industrial Chemistry,* 5th ed., vol. A27, VCH Publishers, 1996, Chapts., 1–7.

K. Jonckers, V. Mennen, J. Meesen, and W. Lemmen, *Stamicarbon New Process Urea 2000* and *Stamicarbon, Wastewater Process,* Stamicarbon AB Geleen, The Netherlands, Feb. 1997.

World Fertilizer Market Information Services, TVA, National Fertilizer Development Center, Muscle Shoals, Ala.

Knop and Hufner in A. E. Werner, ed., *Chemistry of Urea,* Longmans, Green and Co., London, 1923, p. 161.

UREA–FORMALDEHYDE RESINS. See AMINO RESINS AND PLASTICS.

URETHANE POLYMERS

The rapid formation of high molecular weight urethane polymers from liquid monomers, which occurs even at ambient temperature, is a unique feature of the polyaddition process, yielding products that range from cross-linked networks to linear fibers and elastomers. The enormous versatility of the polyaddition process allowed the manufacture of a myriad of products for a wide variety of applications.

Polyurethanes contain carbamate groups, —NHCOO—, also referred to as urethane groups, in their backbone structure. They are formed in the reaction of a diisocyanate with a macroglycol, a so-called polyol, or with a combination of a macroglycol and a short-chain diol extender. In the latter case, segmented block copolymers are generally produced. The macroglycols are based on polyethers (qv), polyesters, or a combination of both. A linear polyurethane polymer has the structure of (**1**), whereas a linear segmented copolymer obtained from a diisocyanate, a macroglycol, and a diol extender, $HO(CH_2)_xOH$, has the structure of (**2**).

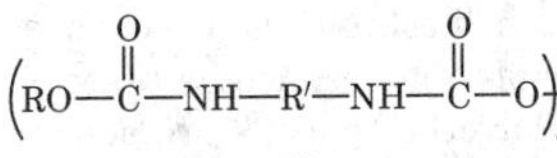

(**1**)

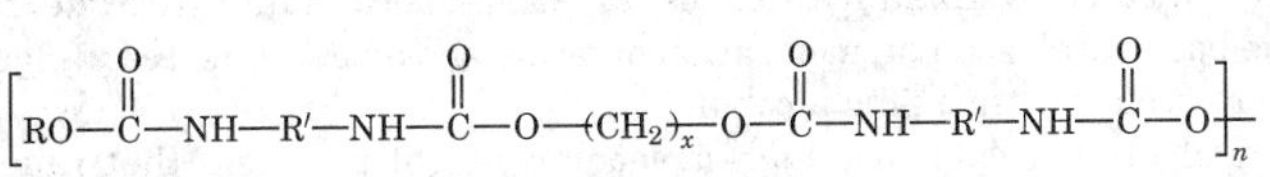

(**2**)

In addition to the linear thermoplastic polyurethanes obtained from difunctional monomers, branched or cross-linked thermoset polymers are made with higher functional monomers. Linear polymers have good impact strength, good physical properties, and excellent processibility, but, owing to their thermoplasticity, limited thermal stability. Thermoset polymers, on the other hand, have higher thermal stability but sometimes lower impact strength (rigid foams). The higher functionality is obtained with higher functional isocyanates (polymeric isocyanates), or with higher functional polyols. Cross-linking is also achieved by secondary reactions. Urea-modified segmented polyurethanes are manufactured from diisocyanates, macroglycols, and diamine extenders. Urethane network polymers are also formed by trimerization of part of the isocyanate groups. This approach is used in the formation of rigid polyurethane-modified isocyanurate (PUIR) foams (**3**).

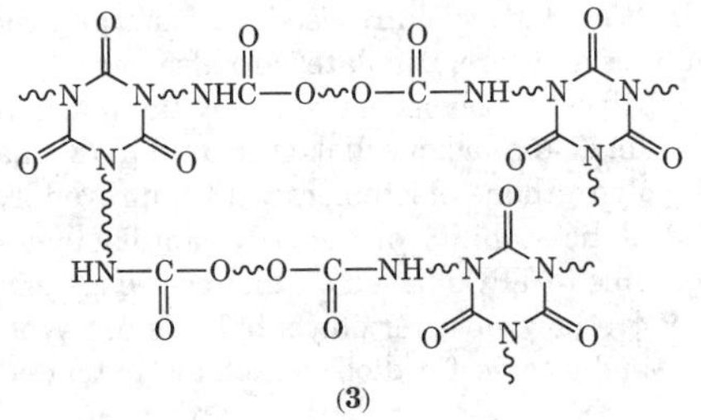

(**3**)

Formation and Properties

Polyurethane Formation. The polarization of the isocyanate group enhances the addition across the carbon-nitrogen double bond, which allows rapid formation of addition polymers from diisocyanates and macroglycols.

The liquid monomers are suitable for bulk polymerization processes. The reaction can be conducted in a mold (casting, reaction injection molding), continuously on a conveyor (block and panel foam production), or in an extruder (thermoplastic polyurethane elastomers and engineering thermoplastics). Also, spraying of the monomers onto the surface of suitable substrates provides insulation barriers or cross-linked coatings.

The polyaddition reaction is influenced by the structure and functionality of the monomers, including the location of substituents in proximity to the reactive isocyanate group (steric hindrance) and the nature of the hydroxyl group (primary or secondary). Impurities also influence the reactivity of the system.

The steric effects in isocyanates are best demonstrated by the formation of flexible foams from TDI. In the 2,4-isomer (**4**), the initial reaction occurs at the nonhindered isocyanate group in the 4-position. The unsymmetrically substituted ureas formed in the subsequent reaction with water are more soluble in the developing polymer matrix. Low density flexible foams are not readily produced from MDI or PMDI; enrichment of PMDI with the 2,4′-isomer of MDI (**5**) affords a steric environment similar to the one in TDI, which allows the production of low density flexible foams that have good physical properties. The use of high performance polyols based on a copolymer polyol allows production of high resiliency (HR) slabstock foam from either TDI or MDI.

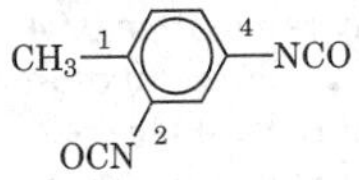

(**4**)

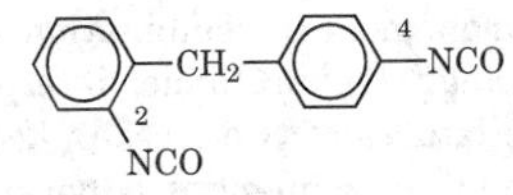

(**5**)

Tailoring of performance characteristics to improve processing and properties of polyurethane products requires the selection of efficient catalysts. In flexible foam manufacturing a combination of tin and tertiary amine catalysts are used in order to balance the gelation reaction (urethane formation) and the blowing reaction (urea formation). The tin catalysts used include dibutyltin dilaurate, dibutylbis(laurylthio)stannate, dibutyltinbis(isooctylmercapto acetate), and dibutyltinbis(isooctylmaleate). Strong bases, such as potassium acetate, potassium 2-ethylhexoate, or amine–epoxide combinations are the most useful trimerization catalysts.

The formation of cellular products also requires surfactants to facilitate the formation of small bubbles necessary for a fine-cell structure. The most effective surfactants are polyoxyalkylene–polysiloxane copolymers. The physical properties of polyurethanes are derived from their molecular structure and determined by the choice of building blocks as well as the supramolecular structures caused by atomic interaction between chains. The ability to crystallize, the flexibility of the chains, and spacing of polar groups are of considerable importance, especially in linear thermoplastic materials. In rigid cross-linked systems, eg, polyurethane foams, other factors such as density determine the final properties.

Thermoplastic Polyurethanes. The unique properties of polyurethanes are attributed to their long-chain structure. In segmented polyether- and polyesterurethane elastomers, hydrogen bonds form between —NH— groups (proton donor) and the urethane carbonyl, polyether oxygen, or polyester carbonyl groups. The symmetrical MDI is more suitable for the preparation of segmented polyurethane elastomers having excellent physical properties. Segmented polyurethanes are also obtained from 2,6-TDI, but an economically attractive separation process for the TDI isomers has yet to be developed.

The melt viscosity of a thermoplastic polyurethane (TPU) depends on the weight-average molecular weight and is influenced by chain length and branching. TPUs are viscoelastic materials, which behave like a glassy, brittle solid, an elastic rubber, or a viscous liquid, depending on temperature and time scale of measurement. With increasing temperature, the material becomes rubbery because of the onset of molecular motion. At higher temperatures a free-flowing liquid forms. The melt temperature of a polyurethane is important for processibility. Melting should occur well below the decomposition temperature.

Thermoset Polyurethanes. The physical properties of rigid urethane foams are usually a function of foam density. A change in strength properties requires a change in density. Rigid polyurethane foams that have densities of <0.064 g/cm^5, used primarily for thermal insulation, are expanded with HCFCs, HFCs, or hydrocarbons (see INSULATION, THERMAL). Often water or a carbodiimide catalyst is added to the formulation to generate carbon dioxide as a coblowing agent. In addition to density, the strength of a rigid foam is influenced by the catalyst, surfactant, polyol, isocyanate, and the type of mixing. By changing the ingredients, foams can be made that have high modulus, low elongation, and some brittleness (friability), or relative flexibility and low modulus (see FOAMED PLASTICS).

The properties of thermoset flexible polyurethane foams are also related to density; load-bearing properties are likewise important. Under normal service temperatures, flexible foams exhibit rubber-like elasticity to deformations of short duration, but creep under long-term stress. Maximum tensile strength is obtained at densities of ca 0.024–0.030 g/cm^3. g/cm^3. The densities are controlled by the amount of water in the formulation and may range from 0.045 to 0.020 g/cm^3 by raising the amount of water from 2 to 5%. Auxiliary blowing agents are also used to reduce density and control hardness. The size and uniformity of the cells are controlled by the efficiency of mixing and the nucleation of the foam mix.

Hyperbranched polyurethanes are constructed using phenol-blocked trifunctional monomers in combination with 4-methylbenzyl alcohol for end capping. Polyurethane interpenetrating polymer networks (IPNs) are mixtures of two cross-linked polymer networks, prepared by latex blending, sequential polymerization, or simultaneous polymerization. IPNs have improved mechanical properties, as well as thermal stabilities, compared to the single cross-linked polymers. In pseudo-IPNs, only one of the involved polymers is cross-linked. Numerous polymers are involved in the formation of polyurethane-derived IPNs.

Raw Materials

Isocyanates. The commodity isocyanates TDI and PMDI are most widely used in the manufacture of urethane polymers (see also ISOCYANATES, ORGANIC). The former is an 80:20 mixture of 2,4- and 2,6-isomers, respectively; the latter a polymeric isocyanate obtained by phosgenation of aniline–formaldehyde-derived polyamines. A coproduct in the manufacture of PMDI is 4,4′-methylenebis(phenyl isocyanate) (MDI). The manufacture of TDI involves the dinitration of toluene, catalytic hydrogenation to the diamines, and phosgenation. Separation of the undesired 2,3-isomer is necessary because its presence interferes with polymerization. Polymeric isocyanates (PMDI) are crude products that vary in exact composition. The basic raw materials for the manufacture of PMDI and its coproduct MDI is benzene. Nitration and hydrogenation affords aniline (see AMINES, AROMATIC). Reaction of aniline with formaldehyde in the presence of hydrochloric acid gives rise to the formation of a mixture of oligomeric amines, which are phosgenated to yield PMDI. The coproduct, MDI, is obtained by continuous thin-film vacuum distillation.

Urethanes obtained from aromatic diisocyanates undergo slow oxidation in the presence of air and light, causing discoloration, which is unacceptable in some applications. Polyurethanes obtained from aliphatic diisocyanates are color-stable, although it is necessary to add antioxidants (qv) and uv-stabilizers to the formulation to maintain the physical properties with time. The least costly aliphatic diisocyanate is hexamethylene diisocyanate (HDI). Isophorone diisocyanate (IPDI) and its derivatives are also used in the formulation of rigid coatings; hydrogenated MDI (HMDI) and cyclohexane diisocyanate (CHDI) are used in the formulation of flexible coatings and polyurethane elastomers.

Masked or blocked diisocyanates are used in coatings applications. The blocked diisocyanates are storage-stable, nonvolatile, and easy to use in powder coatings. Blocked isocyanates are produced by reaction of the diisocyanate with blocking agents such as caprolactam, 3,5-dimethylpyrazole, phenols, oximes, acetoacetates, or malonates.

Table 1. Commercial Polyether Polyols

Product	Nominal functionality	Initiator	Cyclic ether[a]
poly(ethylene glycol) (PEG)	2	water or EG	EO
poly(propylene glycol) (PPG)	2	water or PG	PO
PPG/PEG[b]	2	water or PG	PO/EO
poly(tetramethylene glycol)	2	water	THF
glycerol adduct	3	glycerol	PO
trimethylolpropane adduct	3	TMP	PO
pentaerythritol adduct	4	pentaerythritol	PO
ethylenediamine adduct	4	ethylenediamine	PO
phenolic resin adduct	4	phenolic resin	PO
diethylenetriamine adduct	5	diethylenetriamine	PO
sorbitol adduct	6	sorbitol	PO/EO
sucrose adduct	8	sucrose	PO

[a] EO = ethylene oxide; PO = propylene oxide; THF = tetrahydrofuran.
[b] Random or block copolymer.

Polyether Polyols. Polyether polyols are addition products derived from cyclic ethers (Table 1). The alkylene oxide polymerization is usually initiated by alkali hydroxides, especially potassium hydroxide. In the base-catalyzed polymerization of propylene oxide, some rearrangement occurs to give allyl alcohol.

Polyether polyols are high molecular weight polymers that range from viscous liquids to waxy solids, depending on structure and molecular weight. Most commercial polyether polyols are based on the less expensive ethylene or propylene oxide or on a combination of the two. Block copolymers are manufactured first by the reaction of propylene glycol with propylene oxide to form a homopolymer.

With amine initiators the so-called self-catalyzed polyols are obtained, which are used in the formulation of rigid spray foam systems. The rigidity or stiffness of a foam is increased by aromatic initiators, such as Mannich bases derived from phenol, phenolic resins, toluenediamine, or methylenedianiline (MDA). In the manufacture of highly resilient flexible foams and thermoset RIM elastomers, graft or polymer polyols are used.

Polyester Polyols. Polyester polyols are based on saturated aliphatic or aromatic carboxylic acids and diols or mixtures of diols. The carboxylic acid of choice is adipic acid (qv) because of its favorable cost/performance ratio. For elastomers, linear polyester polyols of ca 2000 mol wt are preferred. Branched polyester polyols, formulated from higher functional glycols, are used for foam and coatings applications. Phthalates and terephthalates are also used.

In addition, polyester polyols are made by the reaction of caprolactone with diols. Poly(caprolactone diols) are used in the manufacture of thermoplastic polyurethane elastomers with improved hydrolytic stability. The hydrolytic stability of the poly(caprolactone diol)-derived TPUs is comparable to TPUs based on the more expensive long-chain diol adipates. Polyether/polyester polyol hybrids are synthesized from low molecular weight polyester diols, which are extended with propylene oxide.

Uses

Flexible Foam. Flexible slab or bun foam is poured by multicomponent machines at rates of >45 kg/min. One-shot pouring from traversing mixing heads is generally used. A typical formulation for furniture-grade foam having a density of 0.024 g/cm^3 includes a polyether triol, mol wt 3000; TDI; water; catalysts, ie, stannous octoate in combination with a tertiary amine; and surfactant. Coblowing agents are often used to lower the density of the foam and to achieve a softer hand. Coblowing agents are methylene chloride, methyl chloroform, acetone, and CFC 11, but the last has been eliminated because of its ozone-depletion potential. Additive systems and new polyols are being developed to achieve softer low density foams. Higher density (0.045 g/cm^3) slab or bun foam, also called high resiliency (HR) foam, is similarly produced, using polyether triols having molecular weight of 6000. The use of polymer polyols improves the load-bearing properties.

Flame retardants (qv) are incorporated into the formulations in amounts necessary to satisfy existing requirements. There are four main types of flexible slabstock foam: conventional, high resiliency, filled, and high load-bearing foam.

Most flexible foams produced are based on polyether polyols; ca 8–10% (15–20% in Europe) of the total production is based on polyester polyols. Flexible polyether foams have excellent cushioning properties, are flexible over a wide range of temperatures, and can resist fatigue, aging, chemicals, and mold growth. Polyester-based foams are superior in resistance to dry cleaning and can be flame-bonded to textiles.

Molded flexible foam products are becoming more popular. The bulk of the molded flexible urethane foam is employed in the transportation industry, where it is highly suitable for the manufacture of seat cushions, back cushions, and bucket-seat padding. TDI prepolymers were used in flexible foam molding in conjunction with polyether polyols. The need for heat curing has been eliminated by the development of cold-molded or high resiliency foams.

Semiflexible molded polyurethane foams are used in other automotive applications, such as instrument panels, dashboards, arm rests, head rests, door liners, and vibrational control devices. An important property of semiflexible foam is low resiliency and low elasticity, which results in a slow rate of recovery after deflection. The isocyanate used in the manufacture of semiflexible foams is PMDI, sometimes used in combination with TDI or TDI prepolymers. Both polyester as well as polyether polyols are used in the production of these water-blown foams.

Rigid Foams. Rigid polyurethane foam is mainly used for insulation (see INSULATION–THERMAL). The configuration of the product determines the method of production. Rigid polyurethane foam is produced in slab or bun form on continuous lines, or it is continuously laminated between either asphalt or tar paper, or aluminum, steel, and fiberboard, or gypsum facings. Rigid polyurethane products, for the most part, are self-supporting, which makes them useful as construction insulation panels and as structural elements in construction applications. Polyurethane can also be poured or frothed into suitable cavities, ie, pour-in-place applications, or be sprayed on suitable surfaces.

Some formulations, particularly those for refrigerator and freezer insulation, are based on modified TDI (golden TDI) or TDI prepolymers, but these are being replaced by PMDI formulations. The polyols used include propylene oxide adducts of polyfunctional hydroxy compounds or amines (Table 1). The amine-derived polyols are used in spray foam formulations where high reaction rates are required. Crude aromatic polyester diols are often used in combination with the multifunctional polyether polyols. Blending of polyols of different functionality, Polyether–polyester polyol hybrids are also synthesized from low mol wt polyesters, which are subsequently propoxylated. Reactive or nonreactive fire retardants, containing halogen and phosphorous, are often added to meet the existing building code requirements. The most commonly used reactive fire retardants are Fyrol 6, chlorendic anhydride-derived diols, and tetrabromophthalate ester diols (PHT 4-Diol). Because the reactive fire retardants are combined with the polyol component, storage stability is important. Nonreactive fire retardants include halogenated phosphate esters, such as tris(chloroisopropyl) phosphate (TMCP) and tris(chloroethyl) phosphate (TCEP), and phosphonates, such as dimethyl methylphosphonate (DMMP). Also used are borax and melamine.

Because of the mandatory phaseout of CFCs by Jan. 1, 1996, it had become necessary to develop blowing agents that have a minimal effect on the ozone layer. As a short-term solution, two classes of blowing agents are considered: hydrochlorofluorocarbons (HCFCs) and hydrofluorocarbons (HFCs). For example, HCFC 141b, CH_3CCl_2F (bp 32°C), is a drop-in replacement for CFC-11, and HFC 134a, CF_3CH_2F (bp −26.5°C), was developed to replace CFC-12. HCFC 142b, CH_3CClF_2 (bp −9.2°C), is the blowing agent used in the 1990s. Addition of water or carbodiimide catalysts to the formulation generates carbon dioxide as a coblowing agent. Longer-range environmental considerations have prompted the use of hydrocarbons such as pentanes and cyclopentane as blowing agents.

From the onset of creaming to the end of the rise during the expansion process, the gas must be retained completely in the form of bubbles, which ultimately result in the closed-cell structure. Addition of surfactants facilitates the production of very small uniform bubbles necessary for a fine-cell structure. The catalysts used in the manufacture of rigid polyurethane foams include tin and tertiary amine catalysts. Many of the rigid insulation foams produced in the 1990s are urethane-modified isocyanurate (PUIR) foams. In the formulation of poly(urethane isocyanurate) foams an excess of PMDI is used. The isocyanate index can range from 105 to 300 and higher. PUIR foams have a better thermal stability than polyurethane foams.

The formation of isocyanurates in the presence of polyols occurs via intermediate allophanate formation, ie, the urethane group acts as a cocatalyst in the trimerization reaction. By combining cyclotrimerization with polyurethane formation, processibility is improved, and the friability of the derived foams is reduced. Modification of cellular polymers by incorporating amide, imide, oxazolidinone, or carbodiimide groups has been attempted but only the urethane-modified isocyanurate foams are produced in the 1990s. PUIR foams often do not require added fire retardants to meet most regulatory requirements. A typical PUIR foam formulation is shown in Table 2.

CASE Polyurethanes. CASE is the acronym for coatings, adhesives, sealants, and elastomers. Polyurethane coatings are mainly based on aliphatic isocyanates and acrylic or polyester polyols because of their outstanding weather-ability. For flexible elastomeric coatings, HMDI and IPDI are used with polyester polyols, whereas higher functional derivatives of HDI and IPDI with acrylic polyols are mainly used in the formulation of rigid coatings. Plastics coatings, textile coatings, and artificial leather are based on either aliphatic or aromatic isocyanates. For light-stable textile coatings, combinations of IPDI and IPDA (as chain extender) are used. The poly(urethane urea) coatings are applied either directly to the fabric or using transfer coating techniques. Microporous polyurethane sheets (poromerics) are used for shoe and textile applications. Polyurethane binder resins are also used to upgrade natural leather.

Blocked aliphatic isocyanates or their derivatives are used for one-component coating systems. Masked polyols are also used for this application. Water-borne polyurethane coatings are formulated by incorporating ionic groups into the polymer backbone. These ionomers are dispersed in water through neutralization. Ionic polymers are also formulated from TDI and MDI. Poly(urethane urea) and polyurea ionomers are obtained from divalent metal salts of *p*-aminobenzoic acid, MDA, dialkylene glycol, and 2,4-TDI. Polyurethane adhesives are known for excellent adhesion, flexibility, toughness, high cohesive strength, and fast cure rates. Two-component adhesives consist of an isocyanate prepolymer, which is cured with low equivalent weight diols, polyols, diamines, or polyamines. Such systems can be used

Table 2. Typical PUIR Foam Formulation

Ingredients	Parts
PMDI (250 index)	208.7
Terate 203[a]	100.0
Dabco K-15	5.2
Dabco TMR 30	1.2
surfactant	2.0
HCFC 141b	35.0

[a] Crude aromatic polyester diol.

neat or as solution. The two components are kept separately before application. Two-component polyurethane systems are also used as hot-melt adhesives.

Water-borne adhesives are preferred because of restrictions on the use of solvents. Low viscosity prepolymers are emulsified in water, followed by chain extension with water-soluble glycols or diamines. As cross-linker PMDI can be used. Water-borne polyurethane coatings are used for vacuum forming of PVC sheeting to ABS shells in automotive interior door panels, for the lamination of ABS/PVC film to treated polypropylene foam for use in automotive instrument panels, as metal primers for steering wheels, in flexible packaging lamination, as shoe sole adhesive, and as tie coats for polyurethane-coated fabrics. PMDI is also used as a binder for reconstituted wood products and as a foundry core binder.

Polyurethane sealant formulations use TDI or MDI prepolymers made from polyether polyols. The sealants contain 30–50% of the prepolymer; the remainder consists of pigments, fillers, plasticizers, adhesion promoters, and other additives.

Polyurethane elastomers are either thermoplastic or thermoset polymers, depending on the functionality of the monomers used. Thermoplastic polyurethane elastomers are segmented block copolymers, comprising of hard- and soft-segment blocks. The soft-segment blocks are formed from long-chain polyester or polyether polyols and MDI; the hard segments are formed from short-chain diols, mainly 1,4-butanediol, and MDI. Thermoset polyurethanes are cross-linked polymers, which are produced by casting or reaction injection molding (RIM). For cast elastomers, TDI in combination with 3,3′-dichloro-4,4′-diphenylmethanediamine (MOCA) are often used.

Polyurethane engineering thermoplastics are also manufactured from MDI and short-chain glycols.

Segmented elastomeric polyurethane fibers (Spandex fibers) based on MDI have also been developed.

Recycling. The methods proposed for the recycling of polyurethanes include pyrolysis, hydrolysis, and glycolysis. Regrind from polyurethane RIM elastomers is used as filler in some RIM as well as compression molding applications. The RIM chips are also used in combination with rubber chips in the construction of athletic fields, tennis courts, and pavement of working roads of golf courses.

The use of rebound flexible foam for carpet underlay and for high load-bearing padding for furniture or for gymnasium mats is already a reality. Rebound flexible foam can also be used for sound dampening in cars. Rebounding of rigid foam particles with PMDI produces polyurethane particle boards. These boards are unaffected by water and are therefore used in furniture aboard ships. Rigid foam scrap is also used as filler in the manufacture of building products. The most convenient chemical recycling process, consists of glycolysis of solid polyurethane products. Heating of polyurethane scrap in a mixture of glycols and diethanolamine converts the cross-linked polymers into linear soluble oligomers via a transesterification process.

Health and Safety Factors

Fully cured polyurethanes present no health hazard; they are chemically inert and insoluble in water and most organic solvents. Dust can be generated in fabrication, and inhalation of the dust should be avoided. Polyether-based polyurethanes are not degraded in the human body, and are therefore used in biomedical applications. Some of the chemicals used in the production of polyurethanes, such as the highly reactive isocyanates and tertiary amine catalysts, must be handled with caution. The other polyurethane ingredients, polyols and surfactants, are relatively inert materials having low toxicity.

Isocyanates in general are toxic chemicals and require great care in handling. Respiratory effects are the primary toxicological manifestations of repeated overexposure to diisocyanates.

There are a multitude of governmental requirements for the manufacture and handling of isocyanates. The U.S. EPA mandates testing and risk management for TDI and MDI under Toxic Substance Control Administration (TSCA). Annual reports on emissions of both isocyanates are required by the EPA under SARA 313.

The liquid tertiary aliphatic amines used as catalysts in the manufacture of polyurethanes can cause contact dermatitis and severe damage to the eye. Inhalation can produce moderate to severe irritation.

Polyurethanes can be considered safe for human use. However, exposure to dust, generated in finishing operations, should be avoided. Polyurethanes are combustible. An approved fire-resistive thermal barrier must be applied over foam insulation on interior walls and ceilings. Under no circumstances should direct flame or excessive heat be allowed to contact polyurethane or polyisocyanurate foam.

Commercial Applications

The largest markets for flexible polyurethane foam are in the furniture, transportation, and bedding industries.

The bulk of the rigid polyurethane and polyisocyanurate foam is used in insulation (see INSULATION, THERMAL). More than half (60%) of the rigid foam consumed in 1994 was in the form of board or laminate; the remainder was used in pour-in-place and spray foam applications.

Polyurethane surface coatings are used wherever applications require abrasion resistance, skin flexibility, fast curing, good adhesion, and chemical resistance (see COATINGS). Synthetic leather products are also produced using a urethane binder. These poromeric materials are produced from textile-length fiber mats impregnated with DMF solutions of polyurethanes. Permeability to moisture vapor is the key property needed in synthetic leather. In addition to shoe applications, poromerics are used for handbags, luggage, and apparel (see LEATHER-LIKE MATERIALS). Polyurethane films having oxygen and water permeability are applied in bandages and wound dressings and as artificial skin for burn victims.

Polyurethane elastomers are used in applications where toughness, flexibility, strength, abrasion resistance, and shock-absorbing qualities are required. Thermoplastic polyurethane elastomers and polyurethane engineering thermoplastics are molded or extruded to produce elastomeric products used as automobile parts, shoe soles, ski boots, roller skate and skateboard wheels, pond liners, cable jackets, and mechanical goods. Cast and RIM elastomers are used in auto fascia, bumper and fender extensions, printing and industrial rolls, industrial tires, and industrial and agricultural parts, such as oil well plugs and grain buckets. Elastomeric spandex fibers are used in hosiery and sock tops, girdles, brassieres, support hose, and swim wear. The use of spandex fibers in sport clothing is increasing.

HENRI ULRICH
Consultant

G. Oertel, *Polyurethane Handbook,* 2nd ed., Carl Hanser Publishers, Munich, Germany, 1993.

G. Woods, *The ICI Polyurethanes Book,* John Wiley & Sons, Inc., New York, 1987.

R. Herrington and K. Hook, eds., *Flexible Polyurethane Foams,* Dow Chemical Company, Midland, Mich., 1991.

H. Ulrich, *The Chemistry and Technology of Isocyanates,* John Wiley & Sons, Inc., New York, 1996.

V

VACCINE TECHNOLOGY

A vaccine is a preparation used to prevent a specific infectious disease by inducing immunity in the host against the pathogenic microorganism. The practice is also called immunization.

With the discovery and widespread use of antibiotics, beginning in the 1950s, the interest in vaccine research disappeared. It was anticipated that infectious diseases would no longer be a threat to human health. This expectation turned out to be too optimistic. Today (ca 1998), there are still numerous infection diseases, for which antibiotic has not been effective. The development of biotechnology and modern immunology created new opportunities for producing new antigens and vaccine research has become a primary focus in recent years. As a result, several vaccines such as Hepatitis B, Hepatitis A, *H. influenza,* and Varicella have been approved. A new vaccine against pertussis has been recently approved in the U.S.

Commercial Vaccines

Vaccines can be roughly categorized into killed vaccines and live vaccines. A killed vaccine can be (*1*) an inactivated, whole microorganism such as pertussis, (*2*) an inactivated toxin, called toxoid, such as diphtheria toxoid, or (*3*) one or more components of the microorganism commonly referred to as subunit vaccines.

Vaccines for human use are regulated by the FDA in the U.S. and Boards of Health in other countries. The manufacturing of vaccines requires adherence to strict current good manufacturing practices (cGMPs) and in the U.S. licenses for both the process and the facility where the vaccine is produced are required. The Center for Biologics Evaluation and Research is the branch of the FDA that regulates vaccines. Basic requirements are described in the *Code of Federal Regulations* (CFR).

Vaccines for the General Population

Vaccines in this category protect children and adults from polio, diphtheria, tetanus, pertussis (whooping cough), measles (rubeola), mumps, rubella (German measles), hepatitis B, and haemophilus disease (meningitis, epiglotitis). The basic schedule is given in Table 1.

Poliomyelitis. Two vaccines are licensed for the control of poliomyelitis in the United States. The live, attenuated oral polio virus (OPV) vaccine can be used for the immunization of normal children. The killed or inactivated vaccine is recommended for immunization of adults at increased risk of exposure to poliomyelitis and of immunodeficient patients and their household contacts. Both vaccines protect against the three serotypes of poliomyelitis that cause disease.

Diphtheria, Tetanus, and Pertussis. These vaccines in combination (DTP) have been routinely used for active immunization of infants and young children since the 1940s. The recommended schedule calls for immunizations at 2, 4, and 6 months of age with boosters at 18 months and 4–5 years of age. Since 1993 these vaccines have been available in combination with a vaccine that protects against *Haemophilus* disease, thus providing protection against four bacterial diseases in one preparation. A booster immunization with diphtheria and tetanus only is recommended once every 10 years after the fifth dose.

Measles, Mumps, Rubella. Live, attenuated vaccines are used for simultaneous or separate immunization against measles, mumps, and rubella in children from around 15 months of age to puberty. Two doses, one at 12–15 months of age and the second at 4–6 or 11–12 years are recommended in the U.S.

Haemophilus influenza serotype b. Three vaccines are available for immunizing infants. Two of these vaccines are administered at 2, 4, and 6 months of age with a booster given at 12–15 months of age, and the third vaccine is administered at 2 and 4 months of age with a booster at 12–15 months of age.

Hepatitis B. Although Hepatitis B (Hep B) is not an infant disease, it is recommended for infant immunization to better control spread, because compliance with vaccine immunization programs is easier to achieve in an infant population. Infants receive immunizations at birth, 1–2 months, and a third dose at 6 months. Other schedules are available for immunization of adolescents and adults who have not previously received the vaccine.

Varicella. The varicella (chicken pox) vaccine was approved in April 1995 for immunization of children. A single dose at one year of age is recommended. In the future it may be combined with measles, mumps, and rubella.

Vaccines for Special Populations

Vaccines for special populations are listed in Table 2. Two vaccines that are in fairly widespread use in the adult population are vaccines that prevent viral influenza and pneumococcal pneumonia.

Influenza. The ACIP recommends annual influenza vaccination for all persons who are at risk from infections of the lower respiratory tract and for all older persons. Influenza viruses types A and B are responsible for periodic outbreaks of febrile respiratory disease.

Pneumococcal Polysaccharide. Pneumococcal polysaccharide vaccine may be used for immunization of persons two years of age or older who are at increased risk of pneumococcal disease.

Vaccines Being Developed

Despite the tremendous advances since the 1960s in the biomedical fields, there remains a large number of diseases that are endemic in many parts of the world. The Third World or developing countries bear the brunt of several of these, eg, malaria, trypanosomiasis, and schistosomiasis. In developed countries, diseases such as herpes and

Table 1. Recommended U.S. Childhood Immunization Schedule, Jan. 1995

Vaccine	Birth	2 Months	4 Months	6 Months	12 Months	15 Months	18 Months	4–6 yr	11–12 yr	14–16 yr
Hepatitis B		Hep 1 →	Hep 2 →		Hep 2 →					
diphtheria–tetanus–pertussis (DTP)		DTP[a]	DTP[a]	DTP[a]		→		DTP or DTaP	Td	→
Haemophilus influenzae type b (Hib)		Hib	Hib	Hib	→					
poliovirus	OPV[b]	OPV[b]			→		OPV			
measles–mumps–rubella (MMR)					→			MMR →	MMR	

[a] As of mid-1996, DTaP can be used.
[b] As of 1997, IPV can be used.

Table 2. Selectively Used Vaccines

Type	Composition
	Viral
rabies	inactivated rabies grown in cultures of human diploid cells
yellow fever	live, attenuated virus grown in embryonated chicken eggs
adenovirus	live virus, types 4 and 7
hepatitis A	inactivated virus
	Bacterial
meningitis	purified capsular polysacharides of *Neisseria meningitidis* serogroups A, C, Y, W_{135}
cholera	inactivated *Vibrio* cholera strains Inaba and Ogawa
typhoid	inactivated whole cells, capsular polysaccharide, or live attenuated bacteria
plague	inactivated Yersinia pestis
anthrax	inactivated cell culture filtrate of *Bacillus anthracis*
tuberculosis	mixture of live and killed bacteria, Bacille Calmette-Guerin

gonorrhea are becoming increasingly prevalent. Some of these vaccines have been developed and licensed, whereas good progress is advancing in other areas. In the meantime, emerging exotic viruses such as HIV and drug-resistant pathogens continue to appear. There is an urgent need to expand vaccine R&D.

Meningitis. *Haemophilus influenze,* type b (Hib), *Streptococcus pneumoniae,* and *Neisseria meningitidis* are the major cause of meningitis in infants. Vaccines against Hib disease prepared using conjugate technology have been in use worldwide, and have been efficacious in eliminating the disease from the population. This same technology is being applied to the development of vaccines for *S. pneumoniae* and *N. meningitidis.*

S. pneumoniae has more than 80 sero-types. The current polysaccharide vaccine consists of 23 serotypes and covers about 87% of all pneumococcal diseases in the United States. Current vaccine development is based on conjugate technology and concentrates on the most prevalent 7–9 serotypes. Three multivalent vaccine candidates are in clinical trials. All are based on conjugating the polysaccharide to a T-dependent protein carrier. The results of phase I and II trials in infants have demonstrated the safety and immunogenicity of these vaccines. Phase III trials to demonstrate efficacy are in progress and final approval of this vaccine for infant immunization will be by the year 2000.

N. meningititidis also has several groups and serotypes. Most of the diseases are caused by groups A, B, and C. A multivalent polysaccharide vaccine consisting of types A, C, Y, and W_{135} is available. However, like other polysaccharide vaccines, it is not immunogenic in infants. Conjugate vaccines against groups A and C are being developed, using different protein carries and conjugate chemistries. Clinical trials of these vaccines are in progress.

The capsular polysaccharide of group B meningococcus is not immunogenic in humans. Thus, a conjugate vaccine of the group B polysaccharide will not improve its efficacy, and this remains a major challenge in developing the vaccine against group B organisms.

Rotavirus. Rotavirus causes infant diarrhea, a disease which has major socio-economic impact. In developing countries it is the major cause of death in infants worldwide. In the U. S., diarrhea is still a primary cause of physician visits and hospitalization, although the mortality rate is relatively low. Two membrane proteins (VP4 and VP7) of the virus have been identified as protective epitopes and most vaccine development programs are based on these two proteins as antigens. Both live attenuated vaccines and subunit vaccines are being developed. By using the technique of viral gene re-assortant, a multivalent live attenuated vaccine for rotavirus has been developed. The vaccine candidates are generated by transferring the VP7 gene from a human rotavirus to Rhesus monkey (or bovine) rotavirus. On August 31, 1998 the FDA licensed a tetravalent rhesus rotavirus vaccine for use in infant. It will be well into the twenty-first century before a vaccine will be available.

Respiratory Syncytial Virus. RSV causes severe lower respiratory tract disease in infants. It is the major cause of hospitalization in the U. S. and it has a high mortality rate in neonates and other high risk populations, such as the geriatric population.

Both subunit and live, attenuated vaccine approaches are being developed for RSV. A candidate subunit vaccine based on the surface (F-) protein is being tested. Live attenuated vaccines for RSV are also being developed.

Parainfluenza. Parainfluenza viruses (PIV) also causes viral pneumonia in infants. It is similar to RSV, therefore similar approaches are being used for developing a vaccine. A live attenuated PIV-3 vaccine has been in clinical trial.

Otitis Media. Otitis media is thought to be caused by several bacteria, significantly *S. pneumoniae,* nontypable *H. influenza,* and *Moraxella catarrhalis.* Viruses such as influenza, RSV, and PIV may also play a role in the disease. The use of a pneumococcal vaccine is the first step in the development of an otitis media vaccine. Vaccines against nontypable *H. influenza* and *Moraxella* are at the development stages. Both vaccine candidates are derived from the exposed proteins of the bacteria.

Herpes Simplex. There are two types of herpes simplex virus (HSV) that infect humans. Type I causes orofacial lesions and 30% of the U.S. population suffers from recurrent episodes. Type II is responsible for genital disease and anywhere from $3 \times 10^4 - 3 \times 10^7$ cases per year (including recurrent infections) occur. The primary source of neonatal herpes infections, which are severe and often fatal, is the mother infected with type II. In addition, there is evidence to suggest that cervical carcinoma may be associated with HSV-II infection. Vaccine development is hampered by the fact that recurrent disease is common. Thus, natural infection does not provide immunity and the best method to induce immunity artificially is not clear. A much better understanding of the pathogenesis of the virus and virus-host interactions are required for the efficient development of the vaccine.

Influenza. Although current influenza vaccine (subunit split vaccine) has been in use yearly for the elderly, it is not recommended for the general population or infants. Improvements to increase or prolong the immunogenicity, reduce the side-effects (due to egg production procedure), and provide mass protection are still being pursued. One approach is to use a live, attenuated virus though cold adaptation. Subunit vaccines based on the surface proteins of virus are also being explored. It has been demonstrated that the two major protective antigens are haemagglutinin (HA) and neuraminidase (NA). The genes for these antigens have been cloned and expressed in baculovirus in insect cell culture.

Malaria. Malaria infection occurs in over 30% of the world's population and almost exclusively in developing countries. The majority of the disease in humans is caused by four different species of the malarial parasite. Vaccine development is problematic for several reasons. The parasites have a complex life cycle; malaria is difficult to grow in large quantities outside the natural host. Despite these difficulties, vaccine development has been pursued for many years. An overview of the state of the art is available.

Gonorrhea. Gonorrhea, caused by *Neisseria gonorrheae,* is the most commonly reported communicable disease in the United States. An increasing number of strains are becoming resistant to penicillin, the antibiotic that is usually used to treat this disease. Development of a vaccine is problematic because natural infection does not necessarily provide immunity. Studies are being carried out on various structural components of the gonococcal bacterium, including pili, outer membrane proteins, lipopolysaccharide, and the outer capsule, in an effort to develop a vaccine. One of the more promising approaches involves a vaccine made with pili. Human studies indicate that a pili vaccine stimulates antibody formation that is 50–100 times the prevaccination level and is effective in preventing disease after challenge.

Human Immunodeficiency Virus. HIV causes Acquired Immunodeficiency Syndrome (AIDS). HIV infects the cells of the human immune system, such as T-lymphocytes, monocytes, and macrophages. After a long period of latency and persistent infection, it results in the progressive decline of the immune system, and leads to full-blown AIDS, resulting in death.

The development of vaccine against HIV-1 has been a top priority of the national public health agencies and medical research institutes. The effort in developing a vaccine has not been as successful as expected. The main problem is the tremendous antigenic variability of the virus. An antigen derived from the cultured strain might not be the same as the clinical strain. Another problem is the fact that the virus infects the cells of the human immune system, making the design of the vaccine more complex. It will require certain combinations of immune responses to provide long-term protection or eliminate the virus from the host. So far, the proper immune mechanism for achieving this goal has not been identified, although it is generally agreed that a cell-mediated immune response (CMI) is essential. Up to the 1990s, most of the vaccine candidates have been derived from the surface proteins of the virus. Although these candidates all show immunogenicity and are protective in animal models, clinical studies of these proteins have not been able to demonstrate protection against disease. Efforts in development of the vaccine are being continued in the public and private research institutes.

Other Vaccines

There are many other diseases which do not have effective vaccines. These diseases are mostly regional in nature, epidemic in the developing world. Vaccines against parasites are also becoming critical to public health. Vaccines are being developed for Lyme disease, dengue, *Helicobacter pylori,* Japanese encephalitis, Equine encephalitis, Tick-borne encephalitis, cholera, shigellas, schistosomiasis, group B streptococcus, and other sexually transmitted diseases.

Future Technology

Vaccines for many diseases are unavailable because of an inability to determine the appropriate method for vaccination or difficulty in obtaining large quantities of antigens. Advances in medical science and immunology have substantially improved the understanding of the design and delivery of antigens. Genetic engineering offers further advances in providing the techniques for construction and production of large quantities of antigens. Development of these fields has been responsible for the rapid advances of vaccinology. Development of new vaccines also requires different process technology for the production of antigens and preparation of delivery system for vaccines.

Genetic Engineering. Genetic engineering involves preparation of DNA fragments (passengers) coding for the substance of interest, inserting the DNA fragments into vectors (cloning vehicles), and introducing the recombinant vectors into living host cells where the passenger DNA fragments replicate and are expressed, ie, transcribed and translated, to yield the desired substance.

Since the 1970s, genetic engineering (qv) has evolved to become the most powerful and routine tool in the study of immunology and the development of new vaccines. It offers new, and in some instances safer and more effective methods for production of vaccines of higher quality. It has allowed an efficient way for the study of construction of new attenuated live viral or bacterial vaccines. It can also be used to study the pathogenicity and immunology of viruses or bacteria. Recombinant hepatitis B vaccine is the first approved human vaccine based on a genetic engineering technique. The genetic engineering techniques can also be used to reduce the virulence of a pathogen which can then be used to produce vaccines. Live vectors are another application of genetic engineering. In this case, the genes from a pathogen are inserted into a vaccine vector, such as salmonella or vaccinia.

The use of naked DNA as a vaccine is the most recent development in this field. Since the demonstration of the possibility of genetic immune response by direct injection of DNA into muscle cells, the field is developing rapidly. Clinical trials for influenza, hepatitis, HIV-1, and herpes simplex are being initiated.

Adjuvants. Adjuvants are substances which can modify the immune response of an antigen. With better understanding of the functions of different arms of the immune system, it is possible to explore the effects of an adjuvant, such that the protective efficacy of a vaccine can be improved. At present, aluminum salt is the only adjuvant approved for use in human vaccines.

Peptide Vaccines. Development of a peptide vaccine is derived from the identification of the immunodominant epitope of an antigen. A polypeptide based on the amino acid sequence of the epitope can then be synthesized. Preparation of a peptide vaccine has the advantage of allowing for large-scale production of a vaccine at relatively low cost. It also allows for selecting the appropriate T- or B-cell epitopes to be included in the vaccine, which may be advantageous in some cases.

Process Technology

In the preparation of classical killed or toxoid vaccines, simple process technology was used. With the advance of new vaccines, far more sophisticated process technologies are needed. The desire to reduce side effects of vaccination requires processes which will yield antigens of extreme purity. The new regulation in cGMP requires consistent production procedures, and global competition also demands that the most efficient process technology be applied.

The basic process technology in vaccine production consists of fermentation for the production of antigen, purification of antigen, and formulation of the final vaccine. In bacterial fermentation, technology is well established. For viral vaccines, cell culture is the standard procedure. Different variations of cell line and process system are in use. For most of the live viral vaccine and other subunit vaccines, production is by direct infection of a cell substrate with the virus. Alternatively, some subunit viral vaccines can be generated by rDNA techniques and expressed in a continuous cell line or insect cells.

Development of conjugate and peptide vaccines requires the typical organic synthesis process and purification. This is a new area for vaccine technologists. Again, the main concern is to maintain the immunogenicity of the vaccine candidate during the chemical reaction and purification steps. Most of these procedures are proprietary.

Economic Aspects

Costs of vaccine manufacture vary according to the type of vaccine produced and how it is supplied. Live virus vaccines are generally less expensive because the quantitative mass to be given to the recipient is less than an inactivated or subunit vaccine. The purification process and yield and the number of strains or components in any given vaccine also affect the cost of manufacture. New vaccines often have a royalty cost, in addition to manufacturing and testing costs. Filling and packaging is often the most expensive part of the manufacturing process and the cost varies by how many doses are filled and packed into one unit.

Another important aspect of vaccine technology is the cost–benefit relationship between prevention vaccination and disease treatment. Generally the cost savings are high.

Liability for adverse reaction events associated in time with immunization have also played a principal role in vaccine economics. Prior to 1988, compensation for any adverse reaction associated in time with vaccination required that the vaccine recipient bring suit against the manufacturer or the health care provider that administered the vaccine. The uncertainty of numbers and costs associated with lawsuits contributed to the decline in the number of providers of routine childhood vaccines. The enactment in 1988 of the National Vaccine Injury Compensation Program was provided as a nonfault alternative to the tort system for resolving claims resulting from adverse reactions to mandated childhood vaccines, and has achieved its goal of providing compensation to those injured by rare adverse events associated with vaccination and providing some stability for the vaccine market.

CHIA-LUNG HSIEH
MARY B. RITCHEY
Wyeth-Lederle Vaccine and Pediatrics

S. A. Plotkin and E. A. Mortimer, *Vaccines,* 2nd ed., W. B. Saunders Co., Philadelphia, Pa., 1994.

R. Eby, in M. Powell and M. Newman, eds., *Vaccine Design: The Subunit and Adjuvant Approach,* Plenum Press, New York, 1995, Chapt. 31.

R. Ellis, "The Application of rDNA Technology to Vaccines," in S. A. Plotkin and B. Fantini, eds., *Vaccinia, Vaccination and Vaccinology,* Elsevier, Paris, 1996.

J. Perez Tirse and P. A. Gross, *Pharmaco. Economics,* **2**(3), 198 (1992).

VACUUM TECHNOLOGY

Vacuum technology concerns the means to predict, effect, and control subatmospheric environments (vacuum). Each vacuum environment must be safe and cost-, energy-, and materials-effective and also tailored to serve each use profitably. Vacuum environments can be grouped into the following operational levels: crude (CR), rough (R), controlled (C), highly controlled (HC), and ultracontrolled (UC). In some instances, these correspond to the traditional categories where vacuum is referred to as low, medium, high, ultrahigh, and beyond ultrahigh. The traditional categorization focuses on the magnitude of pressure rather than on the parameters and their magnitudes that are essential to a given use.

Within vessels, vacuum environments comprise gaseous molecular phase(s) in contact but not necessarily in equilibrium with condensed molecular phases. Nonmolecular species, including radiant quanta, electrons, holes, and phonons, may interact with the molecular environment. Molecular concentrations and species that are anathema in one application may be tolerable or even desirable in another. Toxic and other types of dangerous gases are handled or generated in vacuum systems. Safety procedures have been discussed.

The American Vacuum Society has produced a nearly complete bibliography (to 1996), a dictionary of terms, a monograph series, and a number of other useful publications.

Vacuum Dynamics

Units and Concentration. In the gaseous as well as the condensed phases, molecular concentration by molecular species is of prime importance. By convention, total pressure in a Maxwellian gas is used as though it indicates the quality of the vacuum and as though Maxwellian gases were the rule rather than the exception. In general, in dynamic systems, gas pressure (or its partial pressure components) is neither isotropic nor an adequate indicator of molecular significance.

Interaction between Gaseous and Condensed Phases. In a closed vessel of volume V containing a nonionized, unexcited molecular gas having total number of molecules N, the change in the pressure P in the gas can often be predicted if the steady-state absolute temperature T is changed to another steady, constant level:

$$PV = NkT \tag{1}$$

where k is the Boltzmann constant, relating the steady-state absolute temperature T and the equilibrium pressure P in the gas.

Kinetics Modified by Dynamic Interaction. The kinetic theory of gases is a valuable tool for vacuum technology. The unmodified kinetic theory must not be applied when the gas interacts significantly with itself or with the molecular phases that bound it. When interaction occurs, as it does for many molecular species in the systems considered here, the kinetic predictions must be modified by dynamic considerations. The condensed phase dominates the behavior of the gaseous phase in almost every respect under free molecular conditions. In general, measuring vacuum is not equivalent to measuring any single parameter.

All gaseous components, at the same partial pressure or absolute pressure or ratios thereof, are not likely to have the same significance to any or all vacuum applications.

Essential Parameters. Traditionally, all vacuum environments are characterized in terms of one parameter, ie, pressure in the gaseous phase. When costs, energy, safety, hazardous wastes, and other requirements are taken into account, each system must be characterized by a host of parameters, such as electrical breakdown, zinc coating of capacitors, electron phenomena, film contamination from bulk phase, oil contamination of helium gas, and field emission of electrons.

For service beyond the atmosphere, the vacuum environment allows materials to evaporate or decompose under the action of various forces encountered. These forces include the photons from the sun, charged particles from solar wind, and dust. The action of space environment on materials and spacecraft can be simulated by a source–sink relationship in a vacuum environment. Thus, for example, the lifetime of a solar panel in space operation may be tested (see PHOTOVOLTAIC CELLS). In general, the test object cannot be heated above its operating temperature in space.

In special electronic vacuum diode tubes, with spacing between the cathode and anode of 10 μm, high gas concentrations of some types are beneficial to the operation of the tube under proper control.

Pump Down

Initially, the vessel is filled with ambient air. Any given macrosample of air may contain at least 1600 substances.

Gross cracks and voids are usually lined with microstructure that holds water. Such structures can be removed effectively by chemical and other methods.

Turbulent Gas Flow (Rough Pumping). An oil-sealed mechanical pump in good condition, having vented or trapped exhaust, is gas purged by running for several hours. A liquid-nitrogen (LN) trap is between pump and vessel. The trap can be filled automatically from a local reservoir, from a built-in LN supply line, or by hand. As the pressure falls, the composition of the gas in the vessel begins to change. At a pressure of ca 13 Pa (0.097 mm Hg), water is the dominant species in the gas phase. The surface phases then change appreciably, although initially water was the dominant species on the surfaces. The bulk phase is unlikely to contain any water molecules as such, except in voids and gross defects. Water is desorbed from glass as a result of OH radicals changing to H_2O at the surface.

Diffusion Pump System. After the pump line and trap have been shut off, a large valve is opened slowly enough that the mass flow of gas from the chamber through the valve into the oil-diffusion pump system does not disrupt the top jet of the diffusion pump (DP).

Pumping-speed efficiency depends on trap, valve, and system design. For gases having velocities close to the molecular velocity of the DP top jet, system-area utilization factors of 0.24 are the maximum that can be anticipate. The system speed factor can be quoted together with the rate of contamination from the pump set. Utilization factors of <0.1 for N_2 are common.

The rate of contamination from the pump set is $<10^9$ molecule/(m^2·s) for molecular weights >44. This is the maximum contamination rate for routine service for a well-designed system that is used constantly and subject to automatic liquid-nitrogen filling and routine maintenance.

Leaks

A vacuum system can be stalled by gas leaks. Traditionally, leaks are categorized as real or virtual. A real leak refers to permeation processes or cracks or holes that allow external gas (air) to seep into the vacuum environment. Virtual leaks refer to gases that originate from within, eg, from trapped volumes, the gauges, pumps and the bulk and surface-phase species. Proper instruments readily distinguish real leaks from virtual leaks. The question of virtual leak vs real leak may be important in every system.

Molecular Transport

Molecular transport concerns the mass motion of molecules in condensed and gaseous phases. The mass motions are driven primarily by temperature. As time progresses, the initial mass motion results in concentration gradients. In the condensed phase, flow along concentration gradients is described by Fick's law.

No noble gas permeates a metal. Metals are, however, permeated readily by hydrogen. The least permeable material for hydrogen is carbon. Glasses are permeable, especially by the light noble gases at elevated temperatures.

Gas Transport. Initially, in a vessel containing air at atmospheric pressure, mass motion takes place when temperature differences exist and especially when a valve is opened to a gas pump.

Low velocity viscous laminar flow in gas pipes is commonplace. Practical gas flow can be based on pressure drops of <50% for low velocity laminar flow in pipes whose length-to-diameter ratio may be as high as several thousand. Under laminar flow, bends and fittings add to the frictional loss, as do abrupt transitions.

Wall Geometries. Rougher-than-rough wall geometries can reduce transmission probabilities in Knudsen flow by as much as 25% compared to so-called rough walls. Therefore, conductance calculations that claim accuracy beyond a few percent may not be realistic.

The probability of a random gas entering a duct is not a random function but is proportional to the cosine of the angle between the molecular trajectory and the normal to the entrance plane of the duct. The latter assumption is consistent with the second law of thermodynamics, whereas assuming a random distribution entry is not.

The probability of passage is independent of the entrance velocity of free molecules and the subsequent velocity ($V = 0$) of these molecules within the tube. It depends upon the entering angular and wall-refraction distributions of the molecules. For engineering surfaces and gases at room temperature, reasonable results within $\pm 10\%$ are obtained by assuming that a statistical number of molecules impinging on a surface exhibits a cosine distribution upon reflection from the surface. In free molecular flow, all tubes of similar shape have the same probability of free molecular passage for the same entering gas distribution.

Combining Conductances. Combining short conductances may be difficult because, if a free molecular gas that is Maxwellian in steady state enters conductance 1 (length = 0), the gaseous distribution is no longer Maxwellian at exit 1. This corresponds to the so-called beaming effect. The overall conductance can be estimated if the probabilities of passage of the individual components are known and if the juxtaposed components do not vary more than about a factor of two in cross-sectional areas.

Pumping Speed. If the standard formulas for gas flow in vacuum are applied, it is assumed that a Maxwellian free molecular gas is entering the pump. The operational speed of the pump is a systems effect. All current pumps perform more than one function. Pumps are sinks and at the same time sources for molecules.

Instrumental Measurement

Especially important under dynamic conditions, the role of a system and of each component can be disclosed by appropriate measurements. Thus, it can be established when the system environment is ready for a dynamic use, eg, if the pump is likely to perform as a molecular sink, a source, or some combination of these.

Gauges. Because there is no way to measure and/or distinguish molecular vacuum environment except in terms of its use, readings related to gas-phase concentration are provided by diaphragm, McCleod, thermocouple, Pirani gauges, and hot and cold cathode ionization gauges (manometers).

Residual Gas Analyzers. A gaseous molecular phase is analyzed using a mass spectrometer (see ANALYTICAL METHODS). If heat is delivered to the condensed phase or if electrons are caused to strike its surface, molecular desorption provides a signal for the mass spectrometer analysis. Ultrasound frequencies can be introduced into the walls of the vacuum system. If a source of ultrasound is placed on the wall of an ultrahigh vacuum system, a large hydrogen peak is observed.

Vacuum Systems and Equipment

Glass (qv) is often the material of choice for small laboratory systems and sealed systems in commercial practice. Glass has a wide range of useful properties, including high compressive strength but relatively low tensile strength, requiring careful selection of the glass and design. Evacuated glass tubes, such as photomultiplier tubes, are exposed to temperatures as high as 720 K. In some applications, glass has been displaced by high alumina ceramic. Demountable joints are commercially available in great variety in stainless steel, but less so in aluminum alloy or related materials.

The cross sections of a vacuum brazing furnace are of the bell-jar type. Contamination is very low even when rapidly cycled from one work load to the next. The bell jar can serve on one hearth while another hearth is being loaded for the next operation.

Liquid-Nitrogen Traps. The principal reason that cold traps are frequently ineffective in preventing the passage of oil or mercury is the warming of the trap and its internal filling lines when LN is added and/or as LN depletes. However, some designs have eliminated this problem. A liquid-nitrogen trap need not have an active chemical surface, but in some cases, it is advantageous to cool a chemically active surface to liquid-nitrogen temperature in order to obtain effective trapping of methane and other low molecular weight species. This method, however, has a predictable finite life determined by the formation of monolayers on the chemically active surface. A plain metal surface cooled to LN temperature in a trap has two primary functions, namely, to act as a cryopump for water vapor and to prevent contamination from DP working fluid from reaching a given vacuum environment. A well-designed LN trap can provide a pumping speed of at least 10^2 m^3/s per m^2 pf of system entrance area for water vapor (at room temperature, under free molecular conditions) and confine oil contamination to negligible levels. In general, uncracked oil from a DP is completely inhibited from creeping by a surface temperature of <223 K. The effectiveness of a LN trap can be observed by the absence of pressure pips on an ionization gauge when LN is replenished in the reservoir.

Process Equipment

Rough vacuum ranges from 101 kPa to ca 100 Pa (760 to 0.75 torr); medium vacuum ranges from 100 Pa to ca 0.1 Pa (0.75 to 0.00075 torr). The chemical engineer is likely to work in the rough vacuum region in which distillation, evaporation, drying, and filtration are normally conducted. The medium vacuum range is employed in molten-metal degassing, molecular distillation, and freeze drying. Vacuum equipment requires strength to withstand the pressure of the surrounding atmosphere. The full load is ca 101.3 kPa when the internal gas pressure in the system is sufficiently reduced (see PUMPS).

Steam Ejectors. Ejectors are simple vacuum pumps. They have no moving parts but can accomplish compression through fluid-momentum transfer.

Liquid-Ring Pumps. In a liquid-ring pump, the rotor is the only moving part. The liquid ring performs all the functions normally done by mechanical pistons or vanes.

Rotary-Piston Pumps. These are positive-displacement, oil-sealed machines that compress a specific volume of gas with each revolution, and compress and exhaust it to the atmosphere. The oil-sealed piston traps the aspirated gas ahead of it by closing the inlet port. The gas is compressed, the discharge valve opens, and the gas is exhausted to the atmosphere. Compression ratios can be as high as more than 10^6:1 for a single-stage pump.

Rotary-Vane Pumps. Rotary-vane pumps are positive-displacement machines having spring-loaded vanes that contact the inside of the pump casing. Gas entering the pump is trapped between adjacent blades, compressed, and forced out to the atmosphere through the discharge point.

Rotary-Blower Pumps. This type of pump employs two interlocking rotors to trap and compress gases. The rotors are prevented from touching one another, and there is no sealing liquid in the pump. Because there is no positive seal between the rotors, the rotary blower is limited to small compression ratios, though it can also be designed for higher throughput than any other mechanical pump.

NORMAN MILLERON
SEN Vac Services

P. A. Redhead, *Vacuum Science and Technology: Pioneers of the 20th Century, History of Vacuum Science and Technology,* Vol. 2, AIP Press for the American Vacuum Society, New York, 1994.

G. L. Saksaganskii, *Molecular Flow in Complex Vacuum Systems,* Gordon & Breach, New York, 1988.

R. V. Latham, ed., *High Voltage Vacuum Insulation: Basic Concepts and Technological Practice,* Academic Press, Inc., New York, 1995.

A. Berman, *Vacuum Engineering Calculations, Formulas, and Solved Exercises,* Academic Press, Inc., New York, 1992.

VANADIUM AND VANADIUM ALLOYS

Vanadium, a member of Group VB of the periodic system and of the first transition series, is a gray bcc metal. It occurs in the uranium-bearing minerals of Colorado, in copper, lead and zinc vanadates of Africa, and with certain phosphatic shales and phosphate rocks in the western U.S. It is a constituent of titaniferous magnetites with large deposits in the former Soviet Union, South Africa, Finland, the People's Republic of China, and Australia. There are more than 65 known vanadium-bearing minerals.

The U.S. dominated world production of vanadium until the late 1960s when several other countries, notably the Soviet Union, expanded production significantly. At about the same time, the U.S. shifted from being a net exporter to a net importer.

Properties

In pure form, vanadium is a soft, ductile metal, which is hardened and embrittled by oxygen, nitrogen, carbon, and hydrogen. Addition of selected metals results in alloys of higher strength. Physical properties are given in Table 1.

Vanadium exists in the oxidation states +2, +3, +4, and +5. When heated in air, it is oxidized to a brownish black trioxide, a blue-black tetroxide, or a reddish orange pentoxide, depending on the temperature. With chlorine at 180°C, it readily forms VCl_4, and with carbon and nitrogen at high temperatures, it forms VC and VN, respectively. Vanadium exhibits corrosion resistance in alkali solutions and in acids and liquid metals.

Manufacture

The vanadium-bearing ores are generally crushed, ground, screened, and mixed with sodium chloride or carbonate. The mixture is roasted at ca 850°C and the oxides are converted to the water-soluble sodium metavanadate, $NaVO_3$. The vanadium is extracted by leaching with water and precipitates at pH 2–3 as sodium hexavanadate, $Na_4V_6O_{17}$, a red cake, by addition of sulfuric acid. The hexavanadate is fused at 700°C to yield a dense, black product, which is sold as technical-grade vanadium pentoxide.

Ferrovanadium. Ferrovanadium, an important additive to steel, is produced by the reduction of vanadium ore, slag, or technical-grade oxide with carbon, ferrosilicon, or aluminum. Another steel additive is vanadium carbide, which is produced by the solid-state reduction of vanadium oxide with carbon in a vacuum furnace. A silicon reduction process has been developed by the Foote Mineral Co.

In the aluminothermic process, technical-grade vanadium oxide, aluminum, iron scrap, and a flux are charged into an electric furnace. The temperature of the highly exothermic reaction is controlled by the particle size and the feed rate. In the thermite reaction, vanadium and iron oxides are reduced together by aluminum granules in a magnesite-lined steel vessel or in a water-cooled copper crucible.

Pure Vanadium. Vanadium metal is prepared either by the reduction of vanadium chloride with hydrogen or magnesium, or by the reduction of vanadium oxide with calcium, aluminum, or carbon. In the calcium reduction, the exothermic reaction is carried out adiabatically in a sealed vessel or bomb using $CaCl_2$ as a flux for the CaO slag. The vanadium metal is recovered in the form of droplets or beads. A massive ingot or regulus is obtained by using iodine as both a flux and a thermal booster.

In the aluminothermic process, vanadium pentoxide is heated with high purity aluminum in a bomb; a massive regulus of vanadium–aluminum alloy forms. Proprietary additives are used to increase the temperature, decrease the melting point of the slag or metal, or increase the fluidity of the two phases. A metallic regulus forms, which can be used as fuel-element cladding material following purification to 99% purity using electron-beam melting (see NUCLEAR REACTORS). In addition to electron-beam purification, vanadium can be refined by iodine refining (van Arkel-deBoer process), electrolytically in a fused salt, or by electrotransport.

Alloys. Consolidation by the consumable-electrode electric-arc melting technique is used extensively for the preparation of ingots. An electrode consisting of carefully weighed portions of each constituent is prepared, which is welded in vacuum or under an inert gas. Multiple-arc melting for a minimum of two melts ensures a homogenous ingot.

Forging or extrusion is used for primary or initial fabrication. To avoid oxidation, the machined ingot is clad and sealed in a steel container. Intermediate and final recrystallization is performed at 650–1000°C in vacuum or under an inert gas. Fabrication of most vanadium alloys is more difficult than the pure metal because of increased strength and decreased ductility, especially at low temperatures.

Vanadium alloys are used for fuel cladding and other structural components in liquid metal-cooled fast-breeder reactors. Selection is based on neutron considerations, corrosion resistance, ductility, and strength. Some alloys with room-temperature strength of 1.2 GPa (175,000 psi) exhibit strengths of up to 1 GPa (145,000 psi) at 600°C. Beyond this temperature, most alloys lose tensile strength rapidly. Weld ductility of the alloys is usually not as good as that of the pure metal.

Table 1. Physical Properties of Vanadium Metal

Property	Value
mp, °C	1890 ± 10
bp, °C	3380
density, g/cm^3	6.11
specific heat (at 20–100°C), J/g[a]	0.50
latent heat of fusion, kJ/mol[a]	16.02
latent heat of vaporization, kJ/mol[a]	458.6
thermal conductivity (at 100°C), W/(cm·K)	0.31
electrical resistance (at 20°C), $\mu\Omega$·cm	24.8–26.0

[a] To convert J to cal, divide by 4.184.

Health and Safety Factors

Vanadium metal and its alloys pose no particular health or safety hazard. Dust or fine powder present a moderate fire hazard. Vanadium compounds may irritate the conjunctivae and respiratory tract.

Uses

Vanadium as an alloying element in steel increases grain refinement and hardenability. Vanadium alloys are used in dies or taps and for cutting tools. The titanium 6–4 alloy (6% A1, 4% V) is used in aircraft where strength-to-weight ratio is important. Because of its low capture cross section for fast neutrons as well as its resistance to corrosion by liquid sodium and its high temperature creep strength, vanadium is being developed as a fuel-cladding element for fast-breeder reactors.

A potential use is in the field of superconductivity. The compound V_3Ga exhibits a critical current of 20 T (20×10^4 G) at 20°C, which is one of the highest of any known material (see SUPERCONDUCTING MATERIALS).

Metals Handbook, 9th ed., Vol. 2, American Society for Metals, Metals Park, Ohio, 1979, p. 822.

R. W. Buchman, Jr., *International Metals Reviews,* 158 (1980).

G. A. Morgan, "Vanadium" in *Mineral Facts and Problems,* Bureau of Mines Bulletin 671, U.S. Bureau of Mines, Washington, D.C., 1980.

R. Rostoker, *The Metallurgy of Vanadium,* John Wiley & Sons, Inc., New York, 1965.

VANADIUM COMPOUNDS

Vanadium is widely dispersed in the earth's crust at an average concentration of ca 150 ppm. Deposits of ore-grade minable vanadium are rare. Vanadium is ordinarily recovered from its raw materials in the form of the pentoxide, but sometimes as sodium and ammonium vanadates. For metallurgical uses, which represent ca 90% of vanadium consumption, some oxides are prepared for conversion to master alloys by fusion and flaking to form glassy chips. Only about a dozen vanadium compounds are commercially significant; of these, vanadium pentoxide is dominant.

Physical Properties

Some properties of selected vanadium compounds are listed in Table 1. Vanadium, a typical transition element, displays well-characterized valence states of 2–5 in solid compounds and in solutions. Valence states of −1 and 0 may occur in solid compounds. All compounds of vanadium having unpaired electrons are colored, but because the absorption spectra may be complex, a specific color does not necessarily correspond to a particular oxidation state.

Chemical Properties

The chemistry of vanadium compounds is related to the oxidation state of the vanadium. Thus, V_2O_5 is acidic and weakly basic, VO_2 is basic and weakly acidic, and V_2O_3 and VO are basic. Coordination compounds (qv) of vanadium are mainly based on six coordination, in which vanadium has a pseudooctahedral structure.

Interstitial and Intermetallic Compounds. In common with certain other metals, eg, Hf, Nb, Ti, Zr, Mo, W, and Ta, vanadium is capable of taking atoms of nonmetals into its lattice. Such uptake is accompanied by a change in the packing pattern to a cubic close-packed structure. Carbides, hydrides, and nitrides so formed are called interstitial compounds. Their composition is determined by geometrical packing arrangements rather than by valence bonding. Large-atomed nonmetals, eg, Si, Ge, P, As, Se, and Te, form compounds with vanadium that are intermediate between being interstitial and intermetallic.

Vanadium Oxides. Vanadium pentoxide (V_2O_5) is intermediate in behavior and stability between the highest oxides of titanium, ie, TiO_2, and of chromium, ie, CrO_3. It is thus less stable to heat than TiO_2 and more heat-stable than CrO_3. Also, V_2O_5 is more acidic and a stronger oxidant than TiO_2, but less so than CrO_3.

Vanadium(IV) oxide (vanadium tetroxide, V_2O_4) is a blue-black solid, having a distorted rutile (TiO_2) structure. It can be prepared from the reaction of V_2O_5 at the melting point with sulfur or carbonaceous reductants such as sugar or oxalic acid.

Vanadium(III) oxide (vanadium sesquioxide, V_2O_3) is a black solid, having the corundum (Al_2O_3) structure. It can be prepared by reduction of the pentoxide by industry most often using NH_3 or CH_4. Air oxidation proceeds slowly at ambient temperatures, but oxidation by chlorine at elevated temperatures to give $VOCl_3$ and V_2O_5 is rapid.

Vanadium(II) oxide is a nonstoichiometric material with a gray-black color, metallic luster, and metallic-type electrical conductivity. Metal–metal bonding increases as the oxygen content decreases, until an essentially metal phase containing dissolved oxygen is obtained.

Vanadates. Ammonium metavanadate and, to a lesser extent, potassium and sodium metavanadates, are the main vanadates of commercial interest. The pure compounds are colorless crystals.

Vanadium Halides and Oxyhalides. Only vanadium(V) oxytrichloride ($VOCl_3$) and the tetrachloride (VCl_4) have appreciable commercial importance. The halides and oxyhalides have been well-characterized.

Vanadium(V) oxytrichloride ($VOCl_3$) is readily hydrolyzed and forms coordination compounds with simple donor molecules, eg, ethers, but is reduced by reaction with sulfur-containing ligands and molecules. It is completely miscible with many hydrocarbons and nonpolar metal halides, eg, $TiCl_4$, and it dissolves sulfur.

Vanadium(IV) chloride (vanadium tetrachloride, VCl_4) is a red-brown liquid, is readily hydrolyzed, forms addition compounds with donor solvents such as pyridine, and is reduced by such molecules to trivalent vanadium compounds. Vanadium tetrachloride dissociates slowly at room temperature and rapidly at higher temperatures, yielding VCl_3 and Cl_2.

Vanadium(III) chloride (vanadium trichloride, VCl_3) is a pink-violet solid, is readily hydrolyzed, and is insoluble in nonpolar solvents but dissolves in donor solvents, eg, acetonitrile, to form coordination compounds. Chemical behavior of the tribromide (VBr_3) is similar to that of VCl_3.

Vanadium Sulfates. Sulfate solutions derived from sulfuric acid leaching of vanadium ores are industrially important in the recovery of vanadium from its raw materials. Vanadium in tetravalent form may be solvent-extracted from leach solutions as the vanadyl ion $(VO)^{2+}$ Pentavalent vanadium does not form simple sulfate salts.

Vanadium(IV) oxysulfate pentahydrate (vanadyl sulfate), $VOSO_4 \cdot 5H_2O$) is an ethereal blue solid and is readily soluble in water. It forms from the reduction of V_2O_5 by SO_2 in sulfuric acid solution. Vanadium(III) sulfate ($V_2(SO_4)_3$) is a powerful reducing agent and has been prepared in both hydrated and anhydrous forms. The anhydrous form is insoluble in either water or sulfuric acid. Vanadium(II) sulfate heptahydrate ($VSO_4 \cdot 7H_2O$) is a light red-violet crystalline powder that can be prepared by electrolytic reduction of $VOSO_4$.

Manufacture

Primary industrial compounds produced directly from vanadium raw materials are principally 98 wt % fused pentoxide, air-dried (technical-grade) pentoxide, and technical-grade ammonium metavanadate (NH_4VO_3). Much of the fused and air-dried pentoxides produced at the millsite is made by thermal decomposition of ammonium vanadates.

Vanadium raw materials are processed to produce vanadium chemicals, eg, the pentoxide and ammonium metavanadate (AMV) primary kkcompounds, by salt roasting or alkaline leaching. Interlocking circuits, in which unfinished or scavenged material from one process is diverted to the other, are sometimes used. Such interlocking to enhance vanadium recovery and product grade became more feasible in the late 1950s with the advent of solvent extraction.

Ore is ordinarily ground to pass through a ca 1.2-mm (14-mesh) screen, mixed with 8–10 wt % Na_2CO_3 and other reactants that may be needed, and roasted under oxidizing conditions in a multiple-hearth furnace or rotary kiln at 800–850°C for 1–2 h.

Hot calcine from the kiln is water-quenched or cooled in air before being lightly ground and leached. Leaching and washing of the residue is by percolation in vats or by agitation and filtration. Extraction of vanadium is 65–85%. Vanadium solution from water leaching of the calcine has a pH of 7–8 and a vanadium content of ca 30–50 g V_2O_5/L. The process routes favored for recovering vanadium from the leach solution are solvent extraction and precipitation of ammonium vanadates or vanadic acid.

Table 1. Physical Properties of Some Vanadium Compounds

Compound	Formula	Appearance	Mol wt	Density, g/cm^3	Mp, °C	Soly
vanadic acid, meta	HVO_3	yellow scales	99.95			sol in acid, alkali
ammonium metavanadate	NH_4VO_3	white–yellowish or colorless crystals	116.98	2.326	200 dec	slightly sol in H_2O
potassium metavanadate	KVO_3	colorless crystals	134.04			sol in hot H_2O
sodium metavanadate	$NaVO_3$	colorless, monoclinic prisms	121.93		630	sol in H_2O
vanadium carbide	VC	black cubic	62.95	5.77	2810[a]	insol in H_2O; sol in HNO_3 with decomposition
vanadium nitride	VN	black cubic	64.95	6.13	2320	sol in *aqua regia*
vanadium(IV) tetrachloride	VCl_4	red-brown liquid	192.75	1.816	28 ± 2[b]	sol (deliquescent) in H_2O, methanol, ether, chloroform
vanadium(V) oxytrichloride	$VOCl_3$	yellow liquid	173.30	1.829	−77 ± 2[c] sol (deliquescent) in H_2O, methanol, ether, acetone, acid	
vanadium(II) oxide	VO	light green crystals	66.95	5.758	ignites	sol in acid
vanadium(III) oxide	V_2O_3	blue crystals	149.88	4.87	1970	sol in HNO_3, HF, and alkali in presence of oxide
vanadium(V) oxide	V_2O_5	yellow-red rhombohedra	181.88	3.357	690[d]	sl sol in H_2O; sol in acid and alkali
vanadium(IV) disilicide	VSi_2	metallic prisms	107.11	4.42		sol in HF
vanadium(III) acetylacetonate	$V(C_5H_7O_2)_3$	brown crystals	348.27	0.9–1.2	178–190	sol in methanol acetone, benzene, chloroform
biscyclopentadienyl-vanadium chloride	$(C_5H_5)_2VCl_2$	pale green crystals	252.04		<250 dec	sol in methanol, chloroform

[a] Bp = 3900°C.
[b] Bp = 148.5°C.
[c] Bp = 126.7°C.
[d] Bp = 1750°C dec.

A calcareous carnotite ore at Yeelirrie, Australia, is ill-suited for salt roasting and acid leaching. Dissolution of vanadium and uranium by leaching in sodium carbonate solution at elevated temperature and pressure has been tested on a pilot-plant scale.

Halides and Oxyhalides. Vanadium(V) oxytrichloride is prepared by chlorination of V_2O_5 mixed with charcoal at red heat. The tetrachloride (VCl_4) is prepared by chlorinating crude metal at 300°C and freeing the liquid from dissolved chlorine by repeated freezing and evacuation. It now is made by chlorinating V_2O_5 or $VOCl_3$ in the presence of carbon at ca 800°C. Vanadium trichloride (VCl_3) can be prepared by heating VCl_4 in a stream of CO_2 or by reaction of vanadium metal with HCl.

Production

The bulk of world vanadium production is derived as a by-product or coproduct in processing iron, titanium, and uranium ores, and, to a lesser extent, from phosphate, bauxite, and chromium ores and the ash, fume, or coke from burning or refining petroleum. Most foreign vanadium is obtained as a coproduct of iron and titanium. South Africa, Norway, and Finland are suppliers.

Analytical and Test Methods

A delicate qualitative test for the presence of vanadium is the formation of brownish red pervanadic acid upon addition of hydrogen peroxide to a solution of a vanadate. Quantitative determinations over a wide range of vanadium content are readily performed by spectroscopy. Volumetric, colorimetric, and spectrographic methods for vanadium are well-developed. Conversely, gravimetric methods are seldom used. X-ray absorption spectroscopy is a convenient means for identifying traces of vanadium in coal.

Health, Safety, and Environmental Considerations

In humans, toxic effects have been observed from occupational exposure to airborne concentrations of vanadium compounds that were probably several milligrams or more per cubic meter of air. Oral vanadium toxicity in humans is minimal.

Uses

Conversion of fused pentoxide to alloy additives is by far the largest use of vanadium compounds. Air-dried pentoxide, ammonium vanadate, and some fused pentoxide, representing ca 10% of primary vanadium production, are used as such, purified, or converted to other forms for catalytic, chemical, ceramic, or specialty applications. The dominant single use of vanadium chemicals is in catalysts (see CATALYSIS). Much less is consumed in ceramics and electronic gear, which are the other significant uses (see BATTERIES). Minor uses of vanadium chemicals are preparation of vanadium metal from refined pentoxide or vanadium tetrachloride; liquid-phase organic oxidation reactions, eg, production of aniline black dyes for textile use and printing inks; color modifiers in mercury-vapor lamps; vanadyl fatty acids as driers in paints and varnish; and ammonium or sodium vanadates as corrosion inhibitors in flue-gas scrubbers.

MIKE WOOLERY
U.S. Vanadium

R. J. H. Clark, *The Chemistry of Titanium and Vanadium,* Elsevier Publishing Co., Amsterdam, the Netherlands, 1968.

P. M. Busch, *Vanadium,* IC-8060, U.S. Bureau of Mines, Washington, D.C., 1961.

Medical and Biological Effects of Environmental Pollutants—Vanadium, National Academy of Sciences, Washington, D.C., 1974.

VANILLIN

Vanillin, a natural product, can be found as a glucoside (glucovanillin) in vanilla beans, at concentrations of about 2%. It can be extracted with water, alcohol, or other organic solvents. Approximately 250 by-products have been identified in natural vanilla, out of which 26 are present at levels in excess of 1 ppm. The balance of all these products contributes to the subtle taste of vanilla beans. The vanilla bean contains about 2% vanillin, but the 10% extract prepared from beans has several times the strength of a solution of 2% vanillin. The best known natural source of vanillin is the vanilla plant, *Vanilla planifolia* A., which belongs to the orchid family. It is cultivated mainly in Mexico, Madagascar, Reunion, Java, and Tahiti.

The long and expensive process of extracting vanillin from vanilla beans yields a product that has an inconsistent quality. The demand for this universally popular flavoring cannot be satisfied by vanilla beans alone. The consumption of naturally occurring vanilla has gradually given way to synthetic vanillin. Synthetic vanillin is identical to that contained in the pod, but differs in smell and flavor from natural vanillin as a result of the various compounds in the natural extract that do not exist in artificial vanillin. These other compounds represent only 2% of the extract; the remaining 98% is vanillin. Vanillin is the common name for 3-methoxy-4-hydroxybenzaldehyde (**1**).

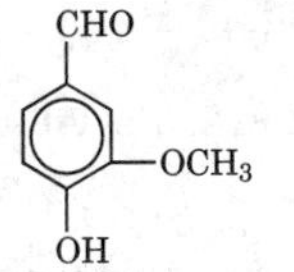

(**1**)

Production

The manufacture of vanillin shows the progress made in the chemistry and chemical engineering of the substance. Most commercial vanillin is synthesized from guaiacol; the remainder is obtained by processing waste sulfite liquors. Preparation by oxidation of isoeugenol is of historical interest only.

Preparation from Guaiacol and Glyoxylic Acid. Several methods can be used to introduce an aldehyde group into an aromatic ring. Condensation of guaiacol (**2**) with glyoxylic acid (**3**), followed by oxidation of the resulting mandelic acid (**4**) to the corresponding phenylglyoxylic acid (**5**) and decarboxylation continues to be a competitive industrial process for vanillin synthesis.

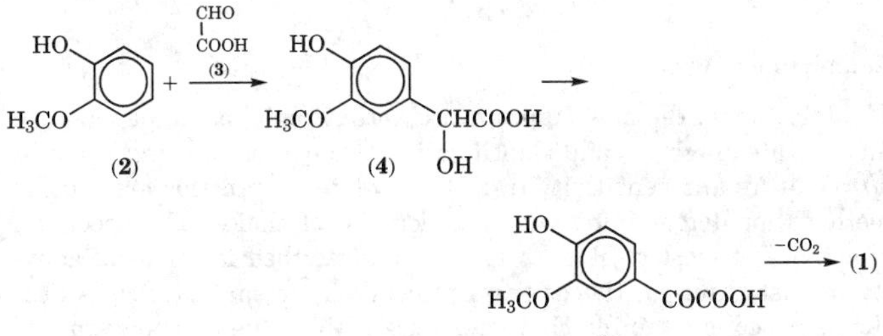

(**2**) (**4**) (**5**)

Preparation from Waste Sulfite Liquors. The starting material for vanillin production can also be the lignin (qv) present in sulfite wastes from the cellulose industry. The concentrated mother liquors are treated with alkali at elevated temperature and pressure in the presence of oxidants. The vanillin formed is separated from the by-products, particularly acetovanillone, 4-hydroxy-3-methoxyacetophenone, by extraction, distillation, and crystallization. In contrast to vanillin from lignin, the principal impurity found in vanillin from guaiacol is 5-methyl vanillin, typically present at levels of about 100 ppm, although levels as high as 3000 ppm have been found. This impurity is completely odorless.

No residual guaiacol can be found in vanillin produced by the guaiacol process. In contrast to vanillin from lignin, vanillin from guaiacol is extremely consistent in quality owing to the consistency of the supply source, and shows no variation in taste, odor, or color.

Table 1. Physical Properties of Rhovanil Extra Pure Vanillin

Property	Value
white to off-white nonhygroscopic crystalline powder	
mp, capillary, °C	81–83
assay, %	99.96 min
bulk density	~0.6
fp, °C	153
bp, °C	
at 101.3 kPa[a]	284–285
10 mm Hg	154
sublimation[b], °C	70

[a] To convert kPa to mm Hg, multiply by 7.5.
[b] At normal pressure.

Specifications

The physical properties of Rhovanil Extra Pure vanillin of Rhône-Poulenc, the leading company in this area, are shown in Table 1.

Solubility. Solubility in water is less than 2%; the solubility in ethanol is given by the ratio one part vanillin to two parts alcohol. Certain manufacturing processes require that the product be in liquid form.

Particle-Size Distribution. Particle size, crystal shape, and distribution of vanillin are important and greatly affect parameters such as taste, flavor, solubility, ease of dispersion in solvent, flowability of the powder, caking effect, and production of dust.

Taste and Flavor. The taste effect is generally sweet, but depends strongly on the base of preparation. For tasting purposes, vanillin is often evaluated in ice-cold milk with about 12% sugar. A concentration of 50 ppm in this medium is clearly perceptible. Vanilla is undoubtedly one of the most popular flavors; its consumption in the form of either vanilla extracts or vanillin is almost universal. The food flavor industry is the largest user of vanillin, an indispensable ingredient in chocolate, candy, bakery products, and ice cream. It is easy to smell a difference in the quality of vanillins from different origins, but it is normally difficult to taste the same difference, provided the various samples are of good quality.

Available Grades. Rhovanil Extra Pure is the trade name of the food-grade vanillin of Rhodia, worldwide leader in the diphenols area. The following grades are commercially available: Rhovanil Extra Pure crystallized, Rhovanil Fine Mesh, Rhovanil Free Flow, and Rhovanil Liquid.

Rhovanil Extra Pure is the standard mesh, multipurpose quality of food-grade extra pure vanillin. Rhovanil Fine Mesh, a specially calibrated extra pure vanillin that avoids demixing with other very fine dry ingredients such as sucrose, flour, and dextrose, provides a faster dissolution rate at lower stirring, at lower temperature, in low acidity medium, or in viscous liquids.

Rhovanil Free Flow is obtained by adding an anticaking agent (0.5% max) to the extra pure vanillin. The flowability is increased, making it particularly suitable for self-dispensing equipment (instant beverage), while both mixability and dispersion/dissolution ratios remain as good as the standard Rhovanil Extra Pure vanillin.

Chemical Properties

Vanillin is a compound that possesses both a phenolic and an aldehydic group. It is capable of undergoing a number of different types of chemical reactions. Addition reactions are possible owing to the reactivity of the aromatic nucleus. On distillation at atmospheric pressure, vanillin undergoes partial decomposition with the formation of pyrocatechol. Exposure to air causes vanillin to oxidize slowly to vanillic acid. Reduction of vanillin by means of platinum black in the presence of ferric chloride gives vanillin alcohol in excellent yields. Because vanillin is a phenol aldehyde, it is stable to autooxidation and does not undergo the Cannizzarro reaction. All three functional groups in vanillin are highly reactive.

Applications

In flavor formulations, vanillin is used widely either as a sweetener or as a flavor enhancer, not only in imitation vanilla flavor, but also in butter, chocolate, and all types of fruit flavors, root beer, cream soda, etc. It is widely acceptable at different concentrations; 50–1000 ppm is quite normal in these types of finished products. Concentrations up to 20,000 ppm, ie, one part in fifty parts of finished goods, are also used for direct consumption such as toppings and icings. Ice cream and chocolate are among the largest outlets for vanillin in the food and confectionery industries, and their consumption is many times greater than that of the perfume and fragrance industry.

When vanillin is not used as a single flavoring ingredient, it is a key part of flavor compounding. At least 30% of food-grade vanillin consumed in the world is through flavoring compounds.

In the industrial production of dry cookies, cakes, and pastries, the vanillin content ranges between 20 and 50 g per 100 kg of dough. Often, vanillin is added at the dry stage of dough preparation as the flour and sugar are being mixed. In fat-free recipes, it is possible to add and mix vanillin powder with eggs.

Vanillin is added, in powder form during the manufacturing process of chocolate, in average amounts of 20 g per 100 kg of the finished product. However, this amount varies according to the quality of the chocolate being made.

Main applications in confections are sugared almonds, caramel, nougat, and sweets. For sugared almonds and caramel, vanillin is mixed into the sugar in the dry phase of the recipe. For nougat, Vanillin is added during the liquid phase of manufacturing. In sweets, vanillin is added in the form of a 10% ethanol solution. Vanillin is used in flavored milk, desserts, yogurts, sorbets, and ice cream.

Vanillin sugar is prepared by dry mixing or impregnating the sugar with a vanillin alcohol solution and evaporating the alcohol. Vanillin confers a pleasant note to liqueur flavoring and improves the flavor of fortified wines by giving them a greatly enhanced bouquet. Vanillin is used as a palatability enhancer to make animal feed more appetizing by flavor-masking minerals with off-taste. Vanillin, a crystal, is the main constituent of the vanilla bean. Its importance can be illustrated by the fact that human preferences in fragrances and in flavors, as determined by various studies, comprise three main smells or tastes: rose, vanilla, and strawberry. Vanillin is used in perfumes and cosmetics as a fragrance and as a washing deodorant.

A flaked technical-grade vanillin, Vaniltek, to be used in pharmaceutical applications. The single largest use for vanillin is as a starting material for the manufacture of an antihypertensive drug having the chemical name of Methyldopa or L-3-(3,4-dihydroxyphenyl)-2-methylalanine. Other drugs made from vanillin include L-Dopa, Trimethoprim Mebeverine, and Verazide for example.

Vanillin itself has some bacteriostatic properties and therefore has been used in formulations to treat dermatitis. Hydrazones of vanillin have been shown to have a herbicidal action similar to that of 2,4-D, and the zinc salts of dithiovanillic acid. A new potential use for vanillin is as a ripening agent to increase the yield of sucrose in sugarcane by the treatment of the cane crop a few weeks before harvest.

The antiultraviolet protection properties of vanillin have been patented and look promising for the plastics and cosmetics (suncreams) industries.

Health and Safety Factors

Vanillin is listed in the *Code of Federal Regulations* by the FDA as a Generally Recognized As Safe (GRAS) substance. The Council of Europe and the FAO/WHO Joint Expert Committee on Food Additives have both given vanillin an unconditional Acceptable Daily Intake (ADI) of 10 mg/kg.

Vanillin has a low potential for acute and chronic toxicity. Vanillin is known to cause allergic reactions in people previously sensitized to balsam of Peru, benzoic acid, orange peel, cinnamon, and clove, but vanillin itself is not an allergic sensitizer. Vanillin has been reported to be a bioantimutagen, demonstrating the ability to protect against mutagenic effects by enhancement of an error-free post-replication repair pathway.

LAWRENCE J. ESPOSITO
K. FORMANEK
G. KIENTZ
F. MAUGER
V. MAUREAUX
G. ROBERT
F. TRUCHET
Rhône-Poulenc

K. Bauer, in D. Garbe, ed., *Common Fragrance and Flavor Materials: Preparation, Properties and Uses,* VCH, Weinheim, Germany, 1985.

P. Z. Bedoukian, ed., *Perfumery and Flavoring Synthetics,* Allured Publishing Corp.

S. Arctander, ed., *Perfume and Flavor Chemicals (Aroma Chemicals),* Montclair, N.J., 1969.

Ullmann's Encyclopedia of Industrial Chemistry, 5th ed., Vol. A 11, VCH, Weinheim, Germany, 1988, pp. 199–200.

VAPOR–LIQUID EQUILIBRIA. See ABSORPTION; DISTILLATION.

VARNISH. See INSULATION, ELECTRIC; RESINS, NATURAL.

VETERINARY DRUGS

The use of pharmaceuticals (qv) in the treatment and prevention of animal diseases has expanded greatly since the 1960s. The modern veterinarian, whether in companion-animal practice, equine service, feedlot medicine, swine and poultry dairy work, zoo management, or any animal medical or surgical specialty, has a wide range of products from which to choose.

All drugs used for veterinary purposes are subject to governmental regulations. In the U.S., pharmaceuticals are regulated by the Center for Veterinary Medicine of the FDA. Vaccines and many immunotherapeutics are classified as biologicals, and thus, are regulated by the USDA. Topical insecticides and growth regulators are regulated by the EPA.

Antimicrobial Agents

The selection of the most appropriate antimicrobial agent depends on an accurate diagnosis and identification of the offending organism. *In vitro* culture and sensitivity testing of isolated organisms are routine methods for determining the antimicrobial of choice. The spectrum of activity of most antibiotics is broadly described in terms of activity against gram-positive or gram-negative organisms. This classification is based on staining characteristics with a blue primary stain of crystal violet with iodine and a red counterstain, usually safranin. Microbes are constantly undergoing genetic change. Some variant strains have the ability to deactivate certain antimicrobials or grow in the presence of antimicrobials to which earlier generations were sensitive. Under antimicrobial therapy, the resistant strains may survive and render the agent less effective.

In addition to the various families of drugs discussed herein, antimicrobial agents include carbadox, the cephalosporins, nitrofurans, oxytetracycline, cefluofor, tilmecosen, trimethoprim sulfa, florfenicol, and tylosin.

Sulfonamides. The sulfonamides (sulfas) are derivatives of *para*-aminobenzenesulfonamide. These agents are active against a broad spectrum of gram-positive and gram-negative organisms. Their mode of action is by competitive antagonism of *para*-aminobenzoic acid (PABA), a folic acid precursor.

Administration is typically oral or by injection. Toxic reactions or untoward side effects have been characterized as blood dyscrasias;

crystal deposition in the kidneys, especially with insufficient urinary output; and allergic sensitization.

Penicillins. Since the discovery of penicillin in 1928 as an antibacterial elaborated by a mold, *Penicillium notatum,* the global search for better antibiotic-producing organism species, radiation-induced mutation, and culture-media modifications have been used to maximize production of the compound. A variety of natural penicillins differing in side chains from the basic molecule, 6-aminopenicillanic acid have been developed. These chemical variations have produced an assortment of drugs having diverse pharmacokinetic and antibacterial characteristics (see ANTIBIOTICS, β-LACTAMS).

The mammalian toxicity of the penicillins is low. Allergic phenomena in patients following sensitization may occur.

The penicillins as natural and semisynthetic agents are used primarily against susceptible *Pasteurella* sp., staphylococci, streptococci, clostridia, and *Corynebacterium* sp. Penicillin is widely used for therapeutic purposes against these organisms and in animal feeds as a growth promoter.

Aminoglycosides. The aminoglycosides, such as streptomycin, neomycin, kanamycin, and gentamycin, have a hexose nucleus joined to two or more amino sugars (see ANTIBIOTICS, AMINOGLYCOSIDES). They are rapidly bactericidal by inhibiting intracellular protein synthesis. Toxicity following exaggerated or prolonged dosage schedules is characterized by renal failure or damage to the eighth cranial nerve (auditory) with auditory- or vestibular-balance dysfunction.

Tetracyclines. The tetracyclines, including chlortetracycline and oxytetracycline, are produced as fermentation products (see ANTIBIOTICS, TETRACYCLINES). These have a broad antibacterial spectrum including gram-positive and gram-negative organisms, rickettsiae, *Chlamydia* sp., and *Mycoplasma* sp. The tetracyclines are commonly employed at low dosages as growth promoters in the main food-producing species. Toxicity is rare.

Growth Promoters. The tetracyclines and penicillin, when administered to food-producing species (poultry, swine, and cattle) during active growth, improve the rate of weight gain and efficiency of feed utilization significantly (see FEEDS AND FEED ADDITIVES). Other antibacterials are also used for this purpose, including avoparcin, monensin, bacitracins, virginiamycin, lincomycin, tylosin, and flavomycin. These effects are related to subtle shifts in enteric processes.

Antifungal Agents. Favorable responses of the superficial infections have been observed following exposure to sunlight or administration of vitamin A. More often, topical application of an antifungal such as nystatin or cuprimyxin or systemic griseofulvin over six to 12 weeks is justified. The systemic mycoses are sensitive to very few therapeutic agents.

Parasiticides

Parasiticides can be roughly divided according to parasites, host species, or chemical classification.

The main pharmacologic action of organophosphates and carbamates is the inhibition of the cholinesterase enzymes, primarily acetylcholinesterase (AChE) (see ENZYME INHIBITORS).

They are typically lipid-soluble and are, as a consequence, rapidly absorbed following inhalation or oral, parenteral, or topical administration. Once absorbed, metabolism is primarily by hepatic hydrolysis or oxidation. Various organophosphates (O–Ps) and carbamates are used against virtually all animal parasites. Concurrent exposure to more than one agent results in cumulative physiologic effect, and therefore the incidence of toxic effect is relatively high. Atropine is an excellent antidote. Pralidoxime chloride is another antidote frequently used as an adjunct to atropine specifically for O–P toxicity.

The avermectins are fermentation products derived from *Streptomyces avermitilis* (see ANTIPARASITIC AGENTS, AVERMECTINS). They are macrocyclic lactones having a very broad spectrum of insecticidal and anthelmintic activity.

The racemic mixture of the *d* and *l* isomers of tetramisole is used as an anthelmintic against a wide variety of nematodes, including lungworms, of ruminants, swine, horses, dogs, and poultry.

The benzimidazoles include a large family of anthelmintics, eg, thiabendazole, albendazole, cambendazole, fenbendazole, mebendazole, oxfendazole, and oxibendazole. Administration is oral, and the spectrum of activity is broad against nematode parasites of the intestinal tract. Benzimidazoles have the advantage of a low mammalian toxicity, ca 10–30 times the recommended dosage.

Other parasiticides have a relatively limited usage owing to a narrow spectrum of antiparasitic activity or because of the introduction of inherently safer or more effective products. Immiticide (melarsonine hydrochloride) is now the drug of choice in dogs against the adult stage of heartworm infection. Carbon disulfide (qv) is used, in combination with other orally administered anthelmintics, by stomach tube for bots (*Gastrophilus sp.* larvae) and ascarids (roundworms) of horses.

Antiinflammatory Agents

Aspirin, acetylsalicylic acid, well tolerated by the dog and the horse, but is relatively toxic to cats. Under the proper clinical circumstances, it can be used for prolonged therapy in chronic inflammatory diseases such as arthritis. Rimadyl is presently used. Ketoprofen is a new drug of choice for equine practitioners.

The most widely used group of antiinflammatory drugs are the corticosteroids and their synthetic analogues. This group of compounds has several physiologic actions, including effects on sodium retention and liver glycogen deposition as well as inhibitory effects on wound healing and, more recently recognized, proliferation of cancer cells.

The natural compounds cortisol, cortisone, and corticosterone vary only slightly in structures and pharmacologic properties (see STEROIDS). The synthetic analogues in more modern practice, prednisolone, dexamethasone, triamcinolone, and betamethasone have greater antiinflammatory potency, and their effects on sodium retention tend to be less severe.

Hormones

Hormones (qv) as naturally occurring, semisynthetic, or synthetic compounds are used to regulate reproductive cycles, gestation, and parturition. They are also used as therapeutics for hormonal imbalances and responsive physical or physiological abnormalities, or as growth promoters in ruminants (see GROWTH REGULATORS, ANIMAL). The application of other than sex-related hormones is not as complex as in human therapy because of the relatively short life spans of animals and high cost.

Estrogens, testosterone, or compounds such as zeranol or trenbolone which can mimic their effects, have shown utility in accelerating the rate of weight gains and decreasing the amount of feed required to produce these gains in food-producing animals. The potential for human consumption of these compounds via the food supply has come under severe regulatory scrutiny.

Tranquilizers and Anesthetics

Tranquilizers find their niche in veterinary medicine in the management of excitement in individual animals (see PSYCHOPHARMACOLOGICAL AGENTS). This group of compounds allows the practitioner to examine the frightened or injured patient with less chance of further damage or injury to the animal, the owner, and the veterinarian. Tranquilizers are also useful in the management of stress to avoid injury during the shipping of animals (see also ANESTHETICS; HYPNOTICS, SEDATIVES, ANTICONVULSANTS AND ANXIOLYTICS).

Acepromazine, a phenothiazine, is used in most animal species in both oral and injectable forms. It can be used at varying dosages to provide the state of tranquilization desired by the veterinarian. The product has a good margin of safety and has been used successfully by the veterinary profession for many years. Xylazine hydrochloride is another product used for both large and small animals.

Tranquilizers are employed for restraint in minor surgical procedures. The cardiovascular system, respiratory system, and blood chemistry are all greatly altered by general anesthesia. Lidocaine is preferred in veterinary medicine.

A drug combination in popular use in dogs is a mixture of fentanyl, a narcotic analgesic, and droperidol, a butyrophenone tranquilizer. This combination produces a state of neuroleptanalgesia in which sedation and analgesia are achieved. The mixture is sold commercially and can be administered by both subcutaneous and intramuscular injection. Because the combination contains a narcotic, it has the advantage of being rapidly reversible with narcotic antagonists such as naloxone and nalorphine once the effects are no longer needed.

Another injectable anesthetic widely used in feline and primate practice is ketamine hydrochloride. Ketamine, a derivative of phencyclidine, can be chemically classified as a cyclohexamine and pharmacologically as a dissociative agent. Analgesia is produced along with a state that resembles anesthesia but in humans has been associated with hallucinations and confusion. For these reasons, ketamine is often combined with a tranquilizer. The product is safe when used in accordance with label directions, but the recovery period may be as long as 12–24 h. Another group of anesthetics is comprised of barbiturates. In veterinary medicine, the list of inhalation anesthetics generally includes only two agents, halothane and methoxyflurane.

Cancer Chemotherapy

In the veterinary as in the human patient, neoplasms are often metastatic and widely disseminated throughout the body. Surgery and irradiation are limited in use to well-defined neoplastic areas and, therefore, chemotherapy is becoming more prevalent in the management of the veterinary cancer victim (see CHEMOTHERAPEUTICS, ANTICANCER). Because of the expense and time involved, such management must be restricted to individual animals for which a favorable risk–benefit evaluation can be made and treatment seems appropriate to the practitioner and the owner. In general, treatment must be viewed not as curative, but as palliative.

Chemotherapeutic agents are grouped by cytotoxic mechanism. The alkylating agents, such as cyclophosphamide and melphalan, interfere with normal cellular activity by alkylation deoxyribonucleic acid (DNA). Antimetabolites, interfering with complex metabolic pathways in the cell, include methotrexate, 5-fluorouracil, and cytosine arabinoside hydrochloride. Antibiotics such as bleomycin and doxorubicin have been used, as have the plant alkaloids vincristine and vinblastine. These compounds vary in their specific mechanism of action and often have different effects on the individual patients.

Immunostimulation

The body's immune mechanism, both humoral and cell-mediated, affords a primary defense against invasion by foreign substances, ie, exogenous entities that the body may encounter, including viruses, bacteria, chemicals, drugs, grafts, and transplants. The reaction by the immune system kills, neutralizes, or rejects the entity. The mechanisms involved in this complex system are under intense investigation, and a better understanding of the immune system will, in the future, permit the control of disease by means only speculated about today.

In 1971, levamisole, an anthelmintic compound widely used in cattle and swine, was shown to improve the effects of an experimental *Brucella abortus* vaccine in mice. New immunostimulants include *Staph Lysate acemannon, MAB-31.*

Discovered in 1957, a group of natural substances called interferons has been the subject of therapeutic interest. Interferon seems to be an integral part of the body's basic defense mechanism, but more recent work indicates broad therapeutic activity and the possibility of cross-species efficacy. Human interferons are used in cats with FIV and FetV. The emergence of recombinant DNA technology might allow sufficient quantities of interferon to be produced at a reasonable cost and thus may make interferon therapy a practical reality (see GENETIC ENGINEERING, ANIMALS).

O. H. Siegmund, *The Merck Veterinary Manual,* 5th ed., Merck and Co., Inc., Rahway, N.J., 1979.

R. W. Kirk, *Current Veterinary Therapy, Small Animal Practice,* 7th ed., W. B. Saunders Co., Philadelphia, Pa., 1980.

J. L. Howard, *Current Veterinary Therapy, Food Animal Practice,* W. B. Saunders Co., Philadelphia, Pa., 1981.

VETIVER. See OILS, ESSENTIAL.

VINEGAR

Vinegar is the liquid condiment or food flavoring used to give a sharp or sour taste to foods. It is also used as a preservative in pickling and as the sour component in many different sauces, dressings, and gravies.

Vinegar results from the action of the enzymes of bacteria of the genus *Acetobacter* and some others on dilute solutions of ethyl alcohol such as cider, wine, beer, or diluted distilled alcohol (see ETHANOL). Most vinegars for table use, eg, in the dressing of salads, derive from the acetic acid–bacterial fermentation of wine or cider (see ACETIC ACID AND DERIVATIVES–ACETIC ACID). These latter, in turn, are produced by alcoholic fermentation (see FERMENTATION) of dilute sugar solutions such as grape juice, apple juice, or malt. *Saccharomyces cerevisiae* is the yeast involved most frequently in alcoholic fermentation, ie, in the enzymatic conversion of fermentable sugars to dilute alcoholic solutions (see YEASTS).

Some raw materials used for vinegar production are listed in Table 1.

A number of factors govern the composition of vinegar: the nature of the raw material, the substances added to promote alcoholic fermentation and the growth and activity of *Acetobacter,* the procedure used for the acetification, and finally the aging, stabilization, and bottling operations.

Table 1. Raw Materials Used to Make Vinegar

Raw material
Mainly sugary
jujube
sweet potato
dates
citrus
persimmon
pear
sugar cane
plum
tomato
kiwi fruit
pineapple
molasses
honey
palm sap
muscavado (brown sugar)
Mainly starchy
potato, corn flour
soybean
seaweed
rice
grain starch
Various
onions
bamboo grass
wood
whey
coconut water
vinasse (distillation residue)

Manufacture

Only since the mid-1800s has it been known that yeast and bacteria are the cause of fermentation and vinegar formation.

Starch Hydrolysis and Alcoholic Fermentation. In general, because yeasts cannot utilize starch directly as a carbon source, the starch must first be hydrolyzed to sugar. Malt vinegars are made from malted barley or a mixture of malted barley with other starchy grains. Malt enzymes convert starch to sugars readily fermentable by *Saccharomyces* yeasts. In Japan, where vinegars are made from rice, a mixture of hydrolyzing enzymes produced by the fungus *Aspergillus oryzae* converts rice starches to sugars. Alcoholic fermentation is frequently conducted in two phases, although in a modern vinegar plant it can be conducted in one. The first phase is a vigorous fermentation during which the rapid evolution of carbon dioxide protects the alcoholic solution from air. The second or slower phase is fermentation of the residual sugar at a lower rate, during which, again, protection from air is required. In the first phase, 50–100 mg SO_2/L of sugar-containing mash is added, followed after approximately 1 h by 1–3% of an actively fermenting, pure-culture starter of *Saccharomyces cerevisiae.* The fermentation process is monitored for disappearance of sugar and increase in temperature. Rates of alcoholic fermentation are highest at ca 25–30°C; higher temperatures tend to damage enzyme systems.

Both the fermentation of hexose sugars to ethanol and carbon dioxide and the oxidation of ethanol to acetic acid are exothermic (heat yielding) processes (see SUGAR). Depending on the size of the fermenter and the rates of fermentation and aeration, loss of waste heat is apportioned among radiation, conduction, and vaporization of water and ethanol plus carbon dioxide. Although small fermenters may require no cooling, large ones require more cooling than occurs through natural radiation and conductance.

Acetic Acid. Ethyl alcohol is converted to acetic acid by air oxidation catalyzed by the enzymes within bacteria of the genus *Acetobacter:*

$$46\ \mathrm{g}\ C_2H_5OH + 32\ \mathrm{g}\ O_2 \rightarrow 60\ \mathrm{g}\ CH_3COOH + 18\ \mathrm{g}\ H_2O$$
$$+ 487.2\ \mathrm{kJ}(116.4\ \mathrm{kcal})$$

One gram of ethanol should yield 1.304 g acetic acid. Practical yields are 77–85%. In contrast with the well-known Embden-Meyerhof-Parnass glycolysis pathway for the conversion of hexose sugars to alcohol, the steps in conversion of ethanol to acetic acid remain in some doubt.

Orleans Process. In the Orleans process, wine oxidizes slowly in a barrel where it is covered with a film of *Acetobacter.* Holes that are covered with screens to exclude insects are bored in each barrel head to permit access to air. Wine is added through the bung hole with a long-stemmed funnel below the surface of the bacterial film and without disturbing the film. Wines with 10–12% ethanol give vinegars of 8–10% acetic acid concentration. Orleans vinegars are characterized by a relatively high concentration of ethyl acetate, detected by its pleasantly strong fruity odor.

Modena-Style or Balsamic Vinegar Process. Balsamic vinegars are made in and around the city of Modena in central Italy and consist of two general types. The traditional product is made from juice of the Trebbiano grape concentrated to about 40% sugar by direct flame heating of the container. *Zygosaccharomyces, Saccharomyces cerevisiae,* and *Gluconobacter* in barrels convert the sugars to ethyl alcohol and gluconic acid as principal products. The ethyl alcohol is converted to acetic acid by the *Acetobacter* and *Gluconobacter.* Subsequent reactions form many taste and odor substances, among which the ester ethyl acetate is of key importance. Years of age in barrels result in a condiment having a harmonious balance of sweetness and tartness. Aceto Balsamico di Modena differs from the Tradizionale in that it is a blend of ordinary wine vinegar with concentrate. Lacking the complex character of Tradizionale and being younger, it is much less expensive.

Generator Process. Generators are packed with shavings of beech wood, which tend to curl and thus provide packing that does not consolidate but allows open spaces for the free flow of liquid and air. Beech wood does not contribute undesirable flavors or impurities to the vinegar. In the modern generator, a recirculating pump transfers the partially acetified alcoholic mixture from the bottom section of the generator to a distributing system at the top of the packed section. Cooling coils may be located in the packed section, but more frequently are placed at the bottom of the receiver section or are incorporated in the line for recirculating the liquid.

Various species and many strains of *Acetobacter* are used in vinegar production. Aeration rates, optimum temperatures and nutrient requirements vary with individual strains. In general, fermentation alcohol substrates require minimal nutrient supplementation while their addition is necessary for distilled alcohol substrates.

Adaptation of the surface-film growth procedure for producing antibiotics to an aerated submerged-culture process has been successful in making vinegar.

The most widely used submerged-culture oxidizer is the Frings acetator. It uses a bottom-driven hollow rotor turning in a field of stationary vanes arranged in such a way that the air which is drawn in is intimately mixed with the liquid throughout the whole bottom area of the tank.

Submerged-culture oxidizers are usually operated on a semicontinuous basis. In most cases, ca half the liquid in the tank is removed every 1–2 d, when the alcohol concentration has dropped to 0.1–0.2 vol %. The removed vinegar is replaced with wine or mash of richer ethanol and lower acetic acid concentration, giving a mixture in the tank of 5–6 vol % ethanol and 6–8 vol % of acetic acid. These are the optimum conditions for *Acetobacter* growth.

Foam production is most troublesome under conditions adverse to bacterial growth and thus can be minimized by keeping nutrient, ethanol, and acetic acid concentrations in the optimum ranges (see DEFOAMERS). Temperature and aeration rate are also critical.

Submerged culture oxidizers can also be operated on a continuous basis. Optimum production, however, is achieved by semicontinuous operation because the composition of vinegar desired in the withdrawal stream is so low in ethanol that vigorous bacterial growth is impeded. Bacterial concentrations up to 100×10^6 cells/cm^3 have been reported in generators making about 20% vinegars.

Fluidized-Bed Vinegar Reactors. Intimate contact of air with *Acetobacter* cells is achieved in fluidized-bed or tower-type systems. Air introduced through perforations in the bottom of each unit suspends the mixture of liquid and microorganisms within the unit. Air bubbles penetrating the bottom plate keep *Acetobacter* in suspension and active for the ethanol oxidation in the liquid phase.

Vinegars with High Concentrations of Acetic Acid. The U.S. regulations require at least 4 g acetic acid/100 cm^3 vinegar. Commercial vinegar and many quality table vinegars are significantly more concentrated. Submerged-culture oxidizers easily give acetic acid concentrations of 10–13 g/cm^3. Submerged-culture oxidizer techniques that produce vinegars with acetic acid concentrations ranging from 15–20% are now in commercial use. Continuous aeration, careful stepwise addition of ethanol as it is oxidized and careful control of temperature seem to be the keys to successful operation.

Vinegar Eels and Mother of Vinegar. Although esthetically undesirable, nematodes, known as vinegar eels, may actually be of some assistance in consuming dead bacteria from the surface of the packing material in the vinegar tank, and thus may aid in prolonging the operation of the system. Vinegar eels are removed from the raw vinegar by filtration and pasteurization before the vinegar is sold or used further in pickling or other processes.

Processing

Raw vinegars vary widely in stability and may contain materials that form cloudiness or deposits. Vinegars that carry a high and cloud suspension of bacterial cells are clarified and stabilized with bentonite and similar agents or activated carbon. Membrane filtration can be combined with aseptic bottling to provide a product free of all microorganisms. Wine, cider, and malt vinegars benefit from aging, which improves the character of the product. Vinegars bottled for table use and pickling usually are pasteurized at 77–78°C.

A. DINSMOOR WEBB
University of California, Davis

J. A. Ekundayo, *Brit. Mycol. Soc. Symp. Ser. 3 (Fungal Biotechnol.)*, 243–271, 1980.

M. A. Amerine and co-workers, *Technology of Wine Making,* 4th ed., AVI, Westport, Conn., 1980.

M. Ameyama and S. Otsuka, eds., *Science of Vinegar,* Asakura Publishing Co., Ltd., Tokyo, 1990.

H. Ebner and H. Follmann, in G. Reed, ed., *Biotechnology,* Vol. 5, Verlag Chemie, Weinheim, Germany, 1983, pp. 425–446.

VINYL CHLORIDE

Vinyl chloride, $CH_2{=}CHCl$, by virtue of the wide range of application for its polymers in both flexible and rigid forms, is a major commodity chemical in the U.S. and an important item of international commerce. Growth in vinyl chloride production is directly related to demand for its polymers and, on an energy-equivalent basis, rigid poly(vinyl chloride) (PVC) is one of the most energy-efficient construction materials available.

Vinyl chloride (also known as chloroethylene or chloroethene) is a colorless gas at normal temperature and pressure, but is typically handled as the liquid (bp −13.4°C). However, no human contact with the liquid is permissible. Vinyl chloride is an OSHA-regulated material.

Physical Properties

The physical properties of vinyl chloride are listed in Table 1. Vinyl chloride and water are nearly immiscible. Vinyl chloride is soluble in hydrocarbons, oil, alcohol, chlorinated solvents, and most common organic liquids.

Table 1. Physical Properties of Vinyl Chloride

Property	Value
mol wt	62.4985
melting point (1 atm), K	119.36
boiling point (1 atm), K	259.25
heat capacity at constant pressure, J/(mol·K)[a]	
vapor at 20°C	53.1
liquid at 20°C	84.3
critical temperature, K	432
critical pressure, MPa[b]	5.67
critical volume, cm^3/mol	179
critical compressibility	0.283
acentric factor	0.100107
dipole moment, C · m	4.84×10^{-30}
enthalpy of fusion (mp), kJ/mol[a]	4.744
enthalpy of vaporization (298.15 K), kJ/mol[a]	20.11
enthalpy of formation (298.15 K), kJ/mol[a]	28.45
Gibbs energy of formation (298.15 K), kJ/mol[a]	41.95
vapor pressure, kPa[b]	
−30°C	49.3
−20°C	78.4
−10°C	119
0°C	175
viscosity, mPa·s	
−40°C	0.345
−30°C	0.305
−20°C	0.272
−10°C	0.244
explosive limits in air, vol %	
lower limit	3.6
upper limit	33
autoignition temp, K	745

[a] To convert J to cal, divide by 4.184.
[b] To convert MPa, to psi, multiply by 145.

Reactions

Polymerization. The most important reaction of vinyl chloride is its polymerization and copolymerization in the presence of a radical-generating initiator.

Substitution at the Carbon–Chlorine Bond. Vinyl chloride is generally considered inert to nucleophilic replacement compared to other alkyl halides. However, the chlorine atom can be exchanged under nucleophilic conditions in the presence of palladium and certain other metal chlorides and salts. Vinyl alcoholates, esters, and ethers can be readily produced from these reactions. Vinylmagnesium chloride (Grignard reagent) can be prepared from vinyl chloride and then used to make a variety of useful end products or intermediates by adding a vinyl anion to organic functional groups. Vinyl chloride similarly undergoes Grignard reactions with other organomagnesium halide compounds. Vinyllithium, another reactive intermediate, can be formed directly from vinyl chloride by means of a lithium dispersion. Vinyl chloride reacts with sulfides, thiols, alcohols, and oximes in basic media. Reaction of vinyl chloride with hydrogen fluoride over a chromia on alumina catalyst yields vinyl fluoride. The carbon–chlorine bond can also be activated at high temperatures.

Oxidation. The chlorine atom-initiated, gas phase oxidation of vinyl chloride yields 74% formyl chloride and 25% CO at high oxygen to Cl_2 ratios. It is unique among chloro olefin oxidations because CO is a major initial product and because the reaction proceeds by a nonchain path. The oxidation of vinyl chloride with oxygen in the gas phase proceeds by a nonradical path which, again, is unique among the chloro olefins. No C_2 carbonyl compounds are made; the major products are formyl chloride, CO, HCl, and formic acid. Oxidation of vinyl chloride with ozone in either the liquid or the gas phase gives formic acid and formyl chloride. Vinyl chloride can be completely oxidized to CO_2 and HCl using potassium permanganate in an aqueous solution at pH 10. The combustion of vinyl chloride in air produces mainly CO_2 and HCl, along with CO and a trace of phosgene.

Addition. Vinyl chloride undergoes a wide variety of addition reactions. Chlorine adds to vinyl chloride to form 1,1,2-trichloroethane by either an ionic or a radical path. Hydrogen halides add to vinyl chloride, usually to yield the 1,1-adduct. Many other vinyl chloride adducts can be formed under acid-catalyzed Friedel-Crafts conditions. Vinyl chloride can be hydrogenated to ethyl chloride and ethane over a platinum on alumina catalyst.

Photochemistry. Vinyl chloride is subject to photodissociation. Photoexcitation at 193 nm results in the elimination of HCl fragments and Cl atoms in an approximately 1.1:1 ratio. Both vinylidene and acetylene have been observed as photolysis products, as have H_2 molecules and H atoms.

Pyrolysis. Vinyl chloride is more stable than saturated chloroalkanes to thermal pyrolysis. That is why nearly all vinyl chloride made commercially comes from thermal dehydrochlorination of ethylene dichloride (EDC). When vinyl chloride is heated to 450°C, only small amounts of acetylene form. Decomposition of vinyl chloride via a free-radical chain process begins at approximately 550°C, and increases with increasing temperature. Acetylene, HCl, chloroprene, and vinylacetylene are formed in about 35% total yield at 680°C. At higher temperatures, tar and soot formation becomes increasingly important. When dry and in contact with metals, vinyl chloride does not decompose below 450°C. However, if water is present, vinyl chloride can corrode iron, steel, and aluminum because of the presence of trace amounts of HCl. This HCl may result from the hydrolysis of the peroxide formed between oxygen and vinyl chloride.

Manufacture

Vinyl chloride monomer was first produced commercially in the 1930s from the reaction of HCl with acetylene derived from calcium carbide. After ethylene became plentiful in the early 1950s, commercial processes were developed to produce vinyl chloride from ethylene and chlorine. These processes included direct chlorination of ethylene to form EDC, followed by pyrolysis of EDC to make vinyl chloride. However, because the EDC cracking process also produced HCl as a coproduct, the industry did not expand immediately, except in conjunction with acetylene-based technology. The development of ethylene oxychlorination technology in the late 1950s encouraged new growth in the vinyl chloride industry. In this process, ethylene reacts with HCl and oxygen to form EDC. Combining the component processes of direct chlorination, EDC pyrolysis, and oxychlorination provided the so-called balanced process for production of vinyl chloride from ethylene and chlorine with no net consumption or production of HCl.

Although a small fraction of the world's vinyl chloride capacity is still based on acetylene or mixed acetylene–ethylene feedstocks, nearly all production is conducted by the balanced process based on ethylene and chlorine. The reactions for each of the component processes are shown in equations 1–3 and the overall reaction is given by equation 4:

$$\textit{Direct chlorination}\ CH_2{=}CH_2 + Cl_2 \rightarrow ClCH_2CH_2Cl \qquad (1)$$

$$\textit{EDC pyrolysis}\ 2\ ClCH_2CH_2Cl \rightarrow 2\ CH_2{=}CHCl + 2\ HCl \qquad (2)$$

$$\textit{Oxychlorination}\ CH_2{=}CH_2 + 2\ HCl + 1/2\ O_2 \rightarrow ClCH_2CH_2Cl + H_2O \qquad (3)$$

$$\textit{Overall reaction}\ 2\ CH_2{=}CH_2 + Cl_2 + 1/2\ O_2 \rightarrow 2\ CH_2{=}CHCl + H_2O \qquad (4)$$

Direct chlorination of ethylene is usually conducted in liquid EDC in a bubble column reactor. Under typical process conditions, the reaction rate is controlled by mass transfer, with absorption of ethylene as the limiting factor. Ferric chloride is a highly selective and efficient catalyst for this reaction, and is widely used commercially. The direct chlorination process may be run with a slight excess of either ethylene or chlorine, depending on how effluent gases from the reactor are subsequently processed. Conversion of the limiting component is essentially 100%, and selectivity to EDC is greater than 99%. The direct chlorination reaction is exothermic ($\Delta H = -180$ kJ/mol for eq. 1) and requires heat removal for temperature control. One widely used method involves operating the reactor at the boiling point of EDC, allowing the pure product to vaporize, and then either recovering heat from the condensing vapor or replacing one or more EDC fractionation column reboilers with the reactor itself.

In oxychlorination, ethylene reacts with dry HCl and either air or pure oxygen to produce EDC and water. While commercial oxychlorination processes may differ from one another to some extent because they were developed independently by many different vinyl chloride producers, in each case the reaction is carried out in the vapor phase in either a fixed- or fluidized-bed reactor containing a modified Deacon catalyst. Cupric chloride is usually the primary active ingredient of the catalyst, supported on a porous substrate such as alumina, silica–alumina, or diatomaceous earth. The oxychlorination reaction is highly exothermic ($\Delta H = -239$ kJ/mol for eq. 3) and requires heat removal for temperature control.

Fluidized bed reactors typically are vertical cylindrical vessels equipped with a support grid and feed sparger system for adequate fluidization and feed distribution, internal cooling coils for heat removal, and either external or internal cyclones to minimize catalyst carryover. Fluidization of the catalyst assures intimate contact between feed and product vapors, catalyst, and heat-transfer surfaces, and results in a uniform temperature within the reactor. Reaction heat can be removed by generating steam within the cooling coils or by some other heat-transfer medium.

Fixed-bed reactors resemble multitube heat exchangers, with the catalyst packed in vertical tubes held in a tubesheet at top and bottom. Reaction heat can be removed by generating steam on the shell side of the reactor or by some other heat-transfer fluid. However, temperature control is more difficult in a fixed-bed than in a fluidized-bed reactor because localized hot spots tend to develop in the tubes.

In the air-based oxychlorination process with either reactor type, ethylene and air are fed in slight excess of stoichiometric requirements to ensure high conversion of HCl and to minimize losses of excess ethylene that remains in the vent gas after product condensation. Under these conditions, typical feedstock conversions are 94–99% for ethylene and 98–99.5% for HCl with EDC selectivities of 94–97%. The use of oxygen instead of air in the oxychlorination process with either reactor type allows operation with excess ethylene and at lower temperatures. This enables recycling of unconverted ethylene, resulting in improved operating efficiency and product yield. An important advantage of oxygen-based oxychlorination technology over air-based operation is the drastic reduction in volume of the vent gas discharged. Since nitrogen is no longer present in the reactor feed streams, only a small amount of purge gas (about 2–5% of the vent gas volume for air-based operation) is vented.

Direct chlorination usually produces EDC with a purity greater than 99.5 wt %, so that, except for removal of the $FeCl_3$, little further purification is needed before it can be cracked. EDC from the oxychlorination process, however, is less pure and requires purification by distillation.

Thermal pyrolysis or cracking of EDC to vinyl chloride and HCl occurs as a vapor-phase, homogeneous, first-order, free-radical chain reaction. The endothermic cracking of EDC ($\Delta H = 71$ kJ/mol EDC reacted for eq. 2) is relatively clean at atmospheric pressure and at temperatures of 425–550°C. Commercial pyrolysis units, however, generally operate at gauge pressures of 1.4–3.0 MPa (200–435 psig) and at temperatures of 475–525°C to provide for better heat transfer and reduced equipment size, and to allow separation of HCl from vinyl chloride by fractional distillation at noncryogenic temperatures. EDC conversion per pass through the pyrolysis reactor is normally maintained at 53–63%. Cracking reaction selectivity to vinyl chloride of >99% can be achieved at these conditions. Increasing cracking severity beyond this level gives progressively smaller increases in EDC conversion, with progressively lower selectivity to vinyl chloride. Higher conversion also increases pyrolysis tube coking rates and causes problems with downstream product purification. To minimize coke formation, it is necessary to quench or cool the pyrolysis reactor effluent quickly. Therefore, the hot effluent gases are normally quenched and partially condensed by direct contact with cold EDC in a quench tower. Alternatively, the pyrolysis effluent gases can first be cooled by heat exchange with cold liquid EDC furnace feed in a transfer line exchanger (TLE) prior to quenching in the quench tower.

Quenched pyrolysis product is typically distilled to remove first HCl and then vinyl chloride. The vinyl chloride is usually further treated to produce specification product, recovered HCl is sent to the oxychlorination process, and unconverted EDC is purified for removal of by-products before it is recycled to the cracking furnace.

By-product disposal from vinyl chloride manufacturing plants is complicated by the need to process a variety of gaseous, organic liquid, aqueous, and solid streams, while ensuring that no chlorinated organic compounds are inadvertently released. Each class of by-product streams requires its own treatment and disposal system.

Vent streams from the different unit operations may contain traces (or more) of HCl, CO, methane, ethylene, chlorine, and vinyl chloride. The common treatment method is either incineration or catalytic combustion, followed by removal of HCl from the effluent gas. Organic liquid streams contain a variety of chlorinated compounds. When there is economic justification, these streams can be fractionated to recover specific, useful components, and the remainder subsequently incinerated and scrubbed to remove HCl. Alternative methods include catalytic oxidation or combustion, and may involve recycle of HCl to oxychlorination and recovery of the heat of combustion to make high pressure steam. Process water streams from vinyl chloride manufacture are typically steam-stripped to remove volatile organics, neutralized, and then treated in a conventional wastewater treatment process. Solid by-products include sludge from wastewater treatment, spent catalyst, and coke from the EDC pyrolysis process. These need to be disposed of in an environmentally sound manner, eg, by sludge digestion, incineration, landfill, etc.

Environmental Considerations

Since the early 1980s, there has been much debate among environmental activist organizations, industry, and government about the impact of chlorine chemistry on the environment. One aspect of this debate involves the incidental manufacture and release of trace amounts of hazardous compounds such as polychlorinated dibenzodioxins, dibenzofurans, and biphenyls (PCDDs, PCDFs, and PCBs, respectively, but often referred to collectively as dioxins) during the production of chlorinated compounds like vinyl chloride. In 1994, the EPA released a review draft of its reassessment of the impact of dioxins in the environment on human health, which prompted speculation as to the amount of dioxins that might be attributed to chlorine-based industrial processes. The U.S. vinyl industry responded by committing to a voluntary characterization of dioxin levels in its products and in emissions from its facilities to the environment. The results of this study to date support the vinyl industry's position that it is a minor source of dioxins in the environment. In addition, a global benchmark study released recently by the American Society of Mechanical Engineers found no relationship between the chlorine content of waste and dioxin emissions from combustion processes.

Because of the toxicity of vinyl chloride, the EPA in 1975 proposed emission standards for vinyl chloride manufacture. This proposal was subsequently enacted as EPA Regulation 40 CFR 61, Subpart F. Compliance testing began in 1978. Environmental concerns and government regulations have prompted a major increase in the amount of add-on technology used in U.S. vinyl chloride production plants.

Technology Trends

The ethylene-based, balanced vinyl chloride process, which accounts for nearly all capacity worldwide, has been practiced by a variety of vinyl chloride producers since the mid-1950s. The technology is mature, so that the probability of significant changes is low. New developments in production technology will likely be based on incremental improvements in raw material and energy efficiency, environmental impact, safety, and process reliability.

More recent trends include widespread implementation of oxygen-based oxychlorination, further development of new catalyst formulations, a broader range of energy recovery applications, a continuing search for ways to improve conversion and minimize by-product formation during EDC pyrolysis, and chlorine source flexibility. The application of computer model-based process control and optimization is growing as a way to achieve even higher levels of feedstock and energy efficiency and plant process reliability.

Health and Safety Factors

Vinyl chloride is an OSHA-regulated substance. Current OSHA regulations impose a permissible exposure limit (PEL) to vinyl chloride vapors of no more than 1.0 ppm averaged over any 8-h period. Short-term exposure is limited to 5.0 ppm averaged over any 15-min period. Wherever exposure is above the OSHA limit, respirators are required. Contact with liquid vinyl chloride is prohibited.

Chronic exposure to vinyl chloride at concentrations of 100 ppm or more is reported to have produced Raynaud's syndrome, lysis of the distal bones of the fingers, and a fibrosing dermatitis. However, these effects are probably related to continuous intimate contact with the skin. Chronic exposure to large amounts of vinyl chloride gas over a period of many years is also reported to have produced a rare cancer of the liver (angiosarcoma) in a small number of workers.

Vinyl chloride also poses a significant fire and explosion hazard. It has a wide flammability range, from 3.6% to 33.0% by volume in air. Large fires of the compound are very difficult to extinguish, while vapors represent a severe explosion hazard.

Vinyl chloride is generally transported via pipeline, and in railroad tank cars and tanker ships. Because hazardous peroxides can form on standing in air, especially in the presence of iron impurities, vinyl chloride should always be handled and transported under an inert atmosphere.

Uses

Vinyl chloride has gained worldwide importance because of its industrial use as the precursor to PVC. It is also used in a wide variety of copolymers. The inherent flame-retardant properties, wide range of plasticized compounds, and low cost of polymers from vinyl chloride have made it a significant industrial chemical. About 95% of current vinyl chloride production worldwide ends up in polymer or copolymer applications. Vinyl chloride also serves as a starting material for the synthesis of a variety of industrial compounds. The primary nonpolymeric uses of vinyl chloride are in the manufacture of vinylidene chloride and tri- and tetrachloroethylene.

JOSEPH A. COWFER
MAXIMILIAN B. GORENSEK
The Geon Company

L. G. Shelton, D. E. Hamilton, and R. H. Fisackerly, in E. C. Leonard, ed., *Vinyl and Diene Monomers, P. 3,* Wiley-Interscience, New York, 1971, pp. 1205–1289.

R. W. McPherson, C. M. Starks, and G. J. Fryar, *Hydrocarbon Process.* **58**(3), 75 (1979).

K. Weissermel and H.-J. Arpe, *Industrial Organic Chemistry,* 2nd ed., VCH Publishers, Inc., New York, 1993, pp. 215–218.

1997/98 World Vinyls Analysis, Chemical Market Associates, Inc., Houston, Tex., May 1998.

VINYL ETHER. See ANESTHETICS; VINYL POLYMERS.

VINYL FIBERS. See FIBERS, POLY(VINYL ALCOHOL).

VINYLIDENE CHLORIDE MONOMER AND POLYMERS

Vinylidene chloride copolymers' most valuable property is low permeability to a wide range of gases and vapors. From the beginning in 1939, the word Saran has been used for polymers with high vinylidene chloride content, and it is still a trademark of The Dow Chemical Company in some countries. Sometimes Saran and poly(vinylidene chloride) are used interchangeably in the literature.

Three types of comonomers are commercially important: vinyl chloride; acrylates, including alkyl acrylates and alkylmethacrylates; and acrylonitrile. When extrusion is the method of fabrication, the formulation includes plasticizers, stabilizers, and extrusion aids.

Monomer

Properties. Pure vinylidene chloride (1,1-dichloroethylene) is a colorless, mobile liquid with a characteristic sweet odor. Its properties are summarized in Table 1. Vinylidene chloride is soluble in most polar and nonpolar organic solvents. Its solubility in water (0.25 wt %) is nearly independent of temperature at 16–90°C.

Manufacture. Vinylidene chloride monomer can be conveniently prepared in the laboratory by the reaction of 1,1,2-trichloroethane with aqueous alkali:

$$2\ CH_2ClCHCl_2 + Ca(OH)_2 \rightarrow CH_2{=}CCl_2 + CaCl_2 + 2\ H_2O$$

Vinylidene chloride (VDC) is prepared commercially by the dehydrochlorination of 1,1,2-trichloroethane with lime or caustic in slight excess (2–10%). A continuous liquid-phase reaction at 98–99°C yields ~90% VDC. Commercial grades contain 200 ppm of the monomethyl ether of hydroquinone (MEHQ).

For many polymerizations, MEHQ need not be removed; instead, polymerization initiators are added. Vinylidene chloride from which the inhibitor has been removed should be refrigerated in the dark at −10°C, under a nitrogen atmosphere, and in a nickel-lined or baked phenolic-lined storage tank. If not used within one day, it should be reinhibited.

Health and Safety Factors. Vinylidene chloride is highly volatile and, when free of decomposition products, has a mild, sweet odor. A single , brief exposure to a high concentration of vinylidene chloride vapor, eg, 2000 ppm, rapidly causes intoxication, which may progress to unconsciousness on prolonged exposure. Vinylidene chloride is hepatotoxic, but does not appear to be a carcinogen. The liquid is irritating to the skin after only a few minutes of contact. In the presence of air or oxygen, uninhibited vinylidene chloride forms a violently explosive complex peroxide at temperatures as low as 40°C. Vinylidene chloride containing peroxides may be purified by being washed several times, either with 10 wt % sodium hydroxide at 25°C or with a fresh 5 wt % sodium bisulfite solution.

Table 1. Properties of Vinylidene Chloride Monomer

Property	Value
odor	pleasant, sweet
color (APHA)	0–10
sol of monomer in H_2O at 25°C, wt %	0.25
sol of H_2O in monomer at 25°C, wt %	0.035
normal bp, °C	31.56
fp, °C	−122.56
flash point, °C	
Tag closed cup	−28
Tag open cup	−16
flammable limits in air (ambient conditions), vol %	6.5–15.5
autoignition temp, °C	513[a]
latent ΔH°_v, kJ/mol[b]	
at 25°C	26.48 ± 0.08
at normal bp	26.14 ± 0.08
latent ΔH_m fp, J/mol[b]	6514 ± 8
at 25°C, ΔH_p, kJ/mol[b]	−75.3 ± 3.8
ΔH_c, liquid monomer at 25°C, kJ/mol[b]	1095.9
ΔH_f, at 25°C, kJ/mol[b]	
liquid monomer	−25.1 ± 1.3
gaseous monomer	1.26 ± 1.26
C_p, at 25°C, J/(mol·K)[b]	
liquid monomer	111.27
gaseous monomer	67.03
T_c, °C	220.8
P_c, MPa[c]	5.21
V_c, cm^3/mol	218
liquid density, at 20°C, g/cm^3	1.2137
index of refraction at 20°C, n_D	1.42468
absolute viscosity at 20°C, mPa·s(= cP)	0.3302[d]

[a] Inhibited with methyl ether of hydroquinone.
[b] To convert J to cal, divide by 4.184.
[c] To convert MPa to atm, divide by 0.101.
[d] P measured from 6.7–104.7 kPa. To convert kPa to mm Hg, multiply by 7.5 (add 0.875 to the constant to convert $\log_{kPa}$ to $\log_{mm\ Hg}$).

Polymerization

Homopolymerization. The free-radical polymerization of VDC has been carried out by solution, slurry, suspension, and emulsion methods. Slurry polymerizations are usually used only in the laboratory. The heterogeneity of the reaction makes stirring and heat transfer difficult; consequently, these reactions cannot be easily controlled on a large scale. Aqueous emulsion or suspension reactions are preferred for large-scale operations. The spontaneous polymerization of VDC, so often observed when the monomer is stored at room temperature, is caused by peroxides formed from the reaction of VDC with oxygen. Very pure monomer does not polymerize under these conditions. Heterogeneous polymerization is characteristic of a number of monomers, including vinyl chloride and acrylonitrile.

Emulsion and suspension reactions are doubly heterogeneous; the polymer is insoluble in the monomer and both are insoluble in water.

The instability of PVDC is one of the reasons why ionic initiation of VDC polymerization has not been used extensively. Many of the common catalysts either react with the polymer or catalyze its degradation.

Copolymerization. The importance of VDC as a monomer results from its ability to copolymerize with other vinyl monomers. Bulk copolymerizations yielding high VDC-content copolymers are normally heterogeneous. During copolymerization, one monomer may add to the copolymer more rapidly than the other. Batch reactions carried to completion usually yield polymers of broad composition distribution. More often than not, this is an undesirable result.

Polymer Structure and Properties

The chemical composition of poly(vinylidene chloride) has been confirmed by various techniques, including elemental analysis, x-ray diffraction analysis, degradation studies, and ir, Raman, and nmr spectroscopy. The polymer chain is made up of vinylidene chloride units added head-to-tail:

$$—CH_2CCl_2—CH_2CCl_2—CH_2CCl_2—$$

Molecular weights of PVDC can be determined directly by dilute solution measurements in good solvents. Viscosity studies indicate that polymers having degrees of polymerization from 100 to more than 10,000 are easily obtained. Dimers and polymers having DP < 100 can be prepared by special procedures.

The crystal structure of PVDC is fairly well established. Several unit cells have been proposed. The unit cell contains four monomer units with two monomer units per repeat distance. The calculated density, 1.96 g/cm^3, is higher than the experimental values, which are 1.80–1.94 g/cm^3 at 25°C, depending on the sample. The melting temperature, T_m, of PVDC is independent of molecular weight above DP = 100. The properties of PVDC (Table 2) are usually modified by copolymerization.

The highly crystalline particles of PVDC precipitated during polymerization are aggregates of thin lamellar crystals.

Melting temperatures of as-polymerized powders are high, ie, 198–205°C. As-polymerized PVDC does not have a well-defined glass-transition temperature because of its high crystallinity. The amorphous polymer has a glass-transition temperature of −17°C. Once melted, PVDC does not regain its as-polymerized morphology when subsequently crystallized.

Poly(vinylidene chloride) does not dissolve in most common solvents at ambient temperatures. Copolymers, particularly those of low crystallinity, are much more soluble. However, one of the outstanding characteristics of vinylidene chloride polymers is resistance to a wide range of solvents and chemical reagents. The insolubility of PVDC results less from its polarity than from its high melting temperature. It dissolves readily in a wide variety of solvents above 130°C.

Poly(vinylidene chloride) also dissolves readily in certain solvent mixtures. One component must be a sulfoxide or *N,N*-dialkylamide. Effective cosolvents are less polar and have cyclic structures.

Mechanical Properties. Because PVDC is difficult to fabricate into suitable test specimens, very few direct measurements of its mechanical properties have been made. Some characteristic properties of high

Table 2. Properties of Poly(vinylidene chloride)

Property	Best value	Reported values
T_m, °C	202	198–205
T_g, °C	−17	−19 to −11
density at 25°C, g/cm^3		
amorphous	1.775	1.67–1.775
unit cell	1.96	1.949–1.96
crystalline		1.80–1.97
refractive index (crystalline), n_D	1.63	
ΔH_m, J/mol[a]	6275	4600–7950

[a] To convert J to cal, divide by 4.184.

VDC content, unplasticized copolymers are listed in Table 3. The performance of a given specimen is sensitive to morphology, including the amount and kind of crystallinity, as well as orientation. Tensile strength increases with crystallinity, whereas toughness and elongation decrease. Orientation, however, improves all three properties.

In cases where the copolymers have substantially lower glass-transition temperatures, the modulus decreases with increasing comonomer content.

The long side chains of the acrylate ester group can apparently act as internal plasticizers. Substitution of a carboxyl group on the polymer chain increases brittleness. Copolymers of VDC with *N*-alkylacrylamides are more brittle than the corresponding acrylates even when the side chains are long.

Vinylidene chloride polymers are more impermeable to a wider variety of gases and liquids than other polymers. For example, commercial copolymers are available with oxygen permeabilities of 0.05 nmol/m·s·GPa. This is a consequence of the combination of high density and high crystallinity in the polymer. An increase in either tends to reduce permeability. Permeability is affected by the kind and amounts of comonomer as well as crystallinity. A more polar comonomer, eg, an AN comonomer, increases the water-vapor transmission more than VC when other factors are constant. All VDC copolymers, are very impermeable to aliphatic hydrocarbons. Plasticizers increase permeability. However, water does not alter the permeability.

Degradation Chemistry

Vinylidene chloride polymers are highly resistant to oxidation, permeation of small molecules, and biodegradation, which makes them extremely durable under most use conditions. However, these materials are thermally unstable and, when heated above about 120°C, undergo degradative dehydrochlorination.

The principal steps in the thermal degradation of VDC polymer are formation of a conjugated polyene sequence followed by carbonization.

On being heated, the polymer gradually changes color from yellow to brown and finally to black. In general, the stability of the polymer reflects the method of preparation, with bulk > solution > suspension ≫ emulsion.

Table 3. Mechanical Properties of High Vinylidene Chloride Copolymers

Property	Range
tensile strength, MPa[a]	
unoriented	34.5–69.0
oriented	207–414
elongation, %	
unoriented	10–20
oriented	15–40
softening range (heat distortion), °C	100–150
flow temp, °C	>185
brittle temp, °C	−10 to 10
impact strength, J/m[b]	26.7–53.4

[a] To convert MPa to psi, multiply by 145.
[b] To convert J/m to ft·lbf/in., divide by 53.38 (see ASTM D256).

To some extent, the stability of VDC polymers is dependent on the nature of the comonomer present. Copolymers with acrylates degrade slowly. Copolymers with acrylonitrile or methacrylate undergo degradation more readily.

The degradation of VDC polymers in nonpolar solvents is comparable to degradation in the solid state. However, these polymers are unstable in many polar solvent. The rate of dehydrochlorination increases markedly with solvent polarity. This reaction is clearly unlike thermal degradation and may well involve the generation of ionic species as intermediates.

Stabilization. The ideal stabilizer system should (*1*) absorb or combine with evolved hydrogen chloride irreversibly under conditions of use, but not strip hydrogen chloride from the polymer chain; (*2*) act as a selective uv absorber; (*3*) contain a reactive dienophilic moiety capable of preventing discoloration by reacting with and disrupting the color-producing conjugated polymer sequences; (*4*) possess nucleophilicity sufficient for reaction with allylic dichloromethylene units; (*5*) possess antioxidant activity so as to prevent the formation of carbonyl groups and other chlorine-labilizing structures; (*6*) be able to scavenge chlorine atoms and other free radicals efficiently; and (*7*) chelate metals, eg, iron, to prevent chlorine coordination and the formation of metal chlorides.

Commercial Methods of Polymerization and Processing

Emulsion polymerization and suspension polymerization are the preferred industrial processes. Either process is carried out in a closed, stirred reactor, which should be glass-lined and jacketed for heating and cooling. The reactor must be purged of oxygen, and the water and monomer must be free of metallic impurities to prevent an adverse effect on the thermal stability of the polymer.

Emulsion polymerization is used commercially to make vinylidene chloride copolymers. The principal advantages are high molecular weight polymers can be produced in reasonable reaction times, especially copolymers with vinyl chloride and monomer can be added during the polymerization to maintain copolymer composition control. The disadvantages of emulsion polymerization result from the relatively high concentration of additives in the recipe. The water-soluble initiators, activators, and surface-active agents generally cause the polymer to have greater water sensitivity, poorer electrical properties, and poorer heat and light stability.

Suspension polymerization of vinylidene chloride is used commercially to make molding and extrusion resins. The principal advantage is the use of fewer ingredients that might detract from the polymer properties. Stability is improved and water sensitivity is decreased. Extended reaction times and the difficult preparation of higher molecular weight polymers are disadvantages of the suspension process compared to the emulsion process, particularly for copolymers containing vinyl chloride.

The batch-suspension process does not compensate for composition drift, whereas constant-composition processes have been designed for emulsion or suspension reactions. It is more difficult to design controlled-composition processes by suspension methods.

Applications

Vinylidene chloride–vinyl chloride copolymers were originally developed for thermoplastic molding applications, and small amounts are still used for this purpose. Extrusion of VDC–VC copolymers is the main fabrication technique for filaments, films, rods, and tubing or pipe, and involves the same concerns for thermal degradation, streamlined flow, and noncatalytic materials of construction as described for injection-molding resins. A significant application for vinylidene chloride copolymer resins is in the construction of multilayer film and sheet. This permits the design of a packaging material with a combination of properties not obtainable in any single material. Rigid containers for food packaging can be made from coextruded sheet that contains a layer of a barrier polymer.

Vinylidene chloride polymers have several properties that are valuable in the coatings industry: excellent resistance to gas and moisture vapor transmission, good resistance to attack by solvents and by fats and oils, high strength, and the ability to be heat-sealed.

Vinylidene chloride polymers are often made in emulsion, but usually are isolated, dried, and used as conventional resins. Stable latices have been prepared and can be used directly for coatings. The principal applications for these materials are as barrier coatings on paper products and, more recently, on plastic films.

Vinylidene chloride emulsion copolymers are used in a variety of ignition-resistant binding applications.

R. A. WESSLING
D. S. GIBBS
P. T. DELASSUS
B. E. OBI
The Dow Chemical Company
B. A. HOWELL
Central Michigan University

R. A. Wessling, *Polyvinylidene Chloride,* Gordon & Breach, New York, 1977.

P. T. DeLassus, *J. Vinyl Technol.* **1,** 14 (1979).

J. D. Danforth, in P. O. Klemchuk, ed., *Polymer Stabilization and Degradation,* American Chemical Society, Washington, D.C., 1985, Chapt. 20, and references cited therein.

VINYLIDENE POLYMERS, POLY(VINYLIDENE FLUORIDE) ELASTOMERS.

See FLUORINE COMPOUNDS, ORGANIC.

VINYL POLYMERS

VINYL ACETAL POLYMERS

Vinyl acetal polymers are made by the acid-catalyzed acetalization of poly(vinyl alcohol) with aldehydes.

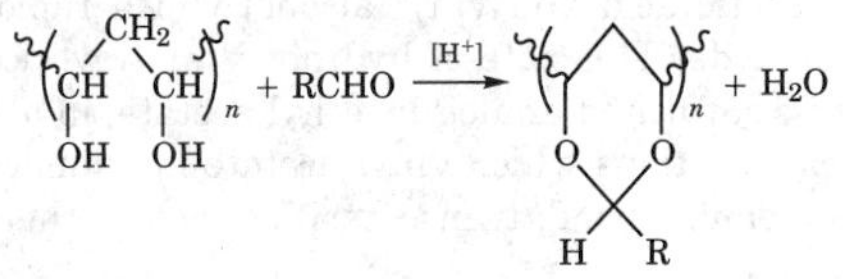

Although many members of this class of resins have been made, only poly(vinyl formal) (PVF) and poly(vinyl butyral) (PVB) are made in significant commercial quantities.

Synthesis and Structure

Poly(vinyl acetals) are made from poly(vinyl alcohol) and aldehydes by acid-catalyzed addition–dehydration. The degree of acetalization and the conditions used during the reaction significantly affect product properties. Batch and continuous processes in both aqueous and organic media are used during manufacturing. In single-stage batch processes, hydrolysis of poly(vinyl acetate) and acetalization of the poly(vinyl alcohol) hydrolysis product are carried out in the same kettle at the same time. In two-stage batch processes, hydrolysis and acetalization take place in separate kettles.

Physical Properties

Unformulated poly(vinyl acetal) resins form hard, unpliable materials which are difficult to process without using solvents or plasticizers. Plasticizers aid resin processing, lower the glass-transition temperature, T_g, and can profoundly change other physical properties of the resins.

Table 1. Physical Properties of Butvar Resins

Property	Method	B-72	B-74	B-76	B-90	B-98
mol mass × 10^3 (avg)	[a]	170–250	120–250	90–120	70–100	40–70
viscosity 15 wt %, Pa·s[b]	[c]	7–14	3–7	0.5–1	0.6–1.2	0.2–0.4
viscosity 10 wt %, Pa·s[b]	[d]	1.6–2.5	0.8–1.3	0.2–0.45	0.2–0.4	0.07–0.2
Ostwald soln viscosity, mPa·s[b] (= cP)	[e]	170–260	40–50	18–28	13–17	6–9
specific gravity, 23°C/23°C	ASTM D792-50	1.100	1.100	1.083	1.100	1.100
refractive index	ASTM D542-50	1.490	1.490	1.485	1.490	1.490
vinyl alcohol content, wt %		17–20	17–20	11–13	18–20	18–20
vinyl acetate content, wt %		0–2.5	0–2.5	0–1.5	0–1.5	0–2.5

[a] Determined by size exclusion chromatography in tetrahydrofuran with low angle light scattering.
[b] To convert Pa·s to P, multiply by 10.
[c] Measured in 60:40 toluene:ethanol at 25°C using a Brookfield viscometer.
[d] Measured in 95% ethanol at 25°C using an Ostwald-Cannon-Fenske viscometer.
[e] B-72 in 7.5 wt % anhydrous methanol at 20°C; B-76 and B-79 in 5.0 wt % SD 29 ethanol at 25°C; B-74, B-90, and B-98 in 6.0 wt % anhydrous methanol at 20°C, all using an Ostwald-Cannon-Fenske viscometer.

Poly(vinyl acetal)s can be formulated with other thermoplastic resins and with a variety of multifunctional cross-linkers. When cross-linking takes place the resin becomes thermoset. Thermosetting generally increases thermal stability, rigidity, and abrasion resistance, and improves resistance to solvents and to acids and bases. It also severely limits processibility by making the resin insoluble and impossible to extrude.

Health and Safety Factors

Representative unformulated PVB and PVF resins are practically nontoxic orally (rats) and no more than slightly toxic after skin application (rabbits).

Some forms of these products may contain sufficient fines to be considered nuisance dust and present dust explosion potential if sufficient quantities are dispersed in air. Unformulated PVB and PVF resins have flash points above 370°C. The lower explosive limit (lel) for PVB dust in air is about 20 g/m^2.

Poly(vinyl butyral)

Several grades are available that differ primarily in residual vinyl alcohol content and molecular weight. Both variables strongly affect solution viscosity, melt flow characteristics, and other physical properties. Some physical, properties of various grades of Monsanto's Butvar resins are listed in Table 1. In general, resin melt and solution viscosity increase with increasing molecular weight and vinyl alcohol content, whereas the tensile strength of materials made from PVB increases with vinyl alcohol content for a given molecular weight.

Commercially available PVB resins are generally soluble in lower molecular weight alcohols, glycol ethers, and certain mixtures of polar and nonpolar solvents. A common solvent for all of the Butvar resins is a combination of 60 parts of toluene and 40 parts of ethanol (95%) by weight.

PVB resins are also compatible with a limited number of plasticizers and resins. Plasticizers (qv) improve processibility, lower T_g, and increase flexibility and resiliency over a broad temperature range.

PVB combinations with the thermoplastic resins nitrocellulose or shellac have been used as sealers for wood finishing. In these applications the PVB component adds flexibility and adhesion. Tough, optically clear blends have been made with aliphatic polyurethanes. Thermosets are prepared with cross-linkers that form covalent bonds with hydroxyl groups.

Manufacture. PVBs are manufactured by a variety of two-stage heterogeneous processes. In one of these an alcohol solution of poly(vinyl acetate) and an acid catalyst are heated to 60–80°C with strong agitation. As the poly(vinyl alcohol) forms, it precipitates from solution. As the reaction approaches completion the reactants go into solution. When the reaction is complete, the catalyst is neutralized and the PVB is precipitated from solution with water, washed, and centrifuged and dried. Resin from this process has very low residual vinyl acetate and very low levels of gel from intermolecular acetalization.

In the second stage of a representative aqueous process, an aqueous solution of poly(vinyl alcohol) is heated with butyraldehyde and an acid catalyst. PVB precipitates from solution as it forms. Because PVB resin precipitates early in the reaction there is a tendency toward high levels of intermolecular acetalization. Cross-linking can be minimized by adding emulsifiers to control particle size or substances like ammonium thiocyanate or urea to improve the solubility of PVB in the aqueous phase.

Applications. During 1994, about 68,000 t of unplasticized PVB was manufactured worldwide. Of this, the overwhelming majority, about 66,000 t, was plasticized and extruded into sheet for use in laminated safety glass. Only about 2,300 t of unplasticized PVB was used for noninterlayer applications.

Plasticized PVB is uniquely suited for safety and security glazing applications. It is easily extruded into sheet. The laminated sheet exhibits high adhesion to glass, optical clarity, stability to sunlight, and high tear strength and impact-absorbing characteristics, all of which are demanded for safety glazing use.

Most laminated safety glazings are glass-PVB-glass trilayer composites. Adhesion of plasticized PVB to clean glass is very high.

Some categories of nonglazing uses for PVB are as follows: phenolic/adhesives; metal/glass binders; hard copy printing; and coatings/additives.

Poly(vinyl formal)

Estimated worldwide production of poly(vinyl formal) resin was about 2700 t in 1994. PVF resins are currently manufactured by Wacker Chemie (Pioloform F) in Germany and by Chisso (Vinylec) in Japan. Poly(vinyl formal) resins are free-flowing white powders with a poly(vinyl formal) content of about 81 wt %. Chemical resistance of poly(vinyl formal) to acids, bases, and aliphatic hydrocarbons is excellent, and chemical resistance to alcohols, aromatic hydrocarbons, esters, and ketones is good; however, chemical resistance to chlorinated solvents is rated poor. In general, PVF resins are soluble in a limited number of solvents and in certain mixtures of alcohols and aromatic hydrocarbons. The solubility of PVF resins in polar solvents increases with increasing proportions of residual vinyl acetate content. Increasing vinyl acetate content also reduces resin stiffness and tensile and impact strength. Solution viscosity increases with increasing vinyl alcohol content and with the average molecular weight of the resin.

PVF resins are generally compatible with phthalate, phosphate, adipate, and dibenzoate plasticizers, and with phenolic, melamine–formaldehyde, urea–formaldehyde, unsaturated polyester, epoxy, polyurethane, and cellulose acetate butylate resins. They are incompatible with polyamide, ethyl cellulose, and poly(vinyl chloride) resins.

Commercial PVF is manufactured by a single-stage batch process in acetic acid. The ratio of vinyl acetate and vinyl alcohol components in the acetal product is controlled by the ratio of acetic acid, water, and formaldehyde used.

Applications. PVF resins are used almost exclusively to make electric and magnetic wire insulation. In these applications the PVF resin component helps provide toughness, as well as abrasion and thermal resistance.

PVF resins have also been used in a variety of other applications, including conductive films, electrophotographic binders, as a component for inks and in membranes, photoimaging, solder masks, and reprographic toners.

JEROME W. KNAPCZYK
Solutia

Butvar, Polyvinyl Butyral Resin, Technical Bulletin No. 8084A, Monsanto Chemical Co., St. Louis, Mo., 1991.

Vinylec, Polyvinyl Formal Resins, Technical Bulletin, Chisso America Corp., New York, 1994.

T. P. Blomstrom in J. I. Kroschwitz, ed., *Encyclopedia of Polymer Science and Engineering,* Vol. 17, Wiley-Interscience, New York, 1989, pp. 136–167.

P. H. Farmer and B. A. Jemmott, in I. Skeist, ed., *Handbook of Adhesives,* 3rd ed., Van Nostrand Reinhold, New York, 1990, pp. 423–436.

VINYL ACETATE POLYMERS

Vinyl acetate is a colorless, flammable liquid having an initially pleasant odor which quickly becomes sharp and irritating. Table 1 lists the physical properties of the monomer.

The most important chemical reaction of vinyl acetate is free-radical polymerization. The reaction is summarized as follows:

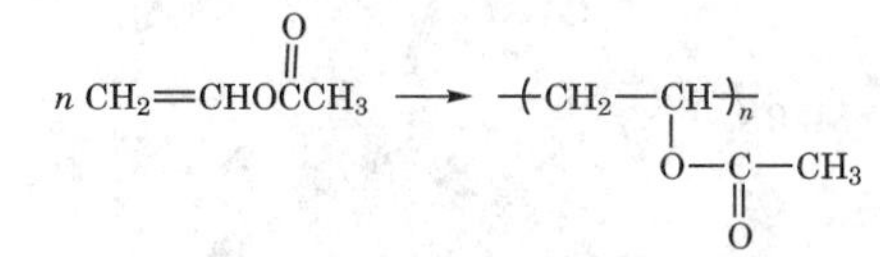

Vinyl acetate has been polymerized by bulk, suspension, solution, and emulsion methods. It copolymerizes readily with some monomers but not with others.

Hydrolysis of vinyl acetate is catalyzed by acidic and basic catalysts to form acetic acid and vinyl alcohol which rapidly tautomerizes to acetaldehyde. This rate of hydrolysis of vinyl acetate is 1000 times that of its saturated analogue, ethyl acetate, in alkaline media. Other chemical reactions which vinyl acetate may undergo are addition across the double bond, transesterification to other vinyl esters, and oxidation.

Vinyl acetate monomer is supplied in three grades, which differ in the amount of inhibitor they contain but otherwise have identical specifications. In storage vinyl acetate should be kept away from ignition sources. It should be stored in a cool environment away from heat, direct sunlight, oxidizing materials, and free-radical generating chemicals to avoid rapid uncontrolled polymerization. Bulk storage should be blanketed with dry nitrogen.

Health and Safety Aspects

NIOSH recommends a ceiling limit of 4 ppm for 15 minutes of exposure to vinyl acetate vapor. The ACGIH recommends an 8-hour TLV Time Weighted Average of 10 ppm and a 15 minute short-term exposure limit of 15 ppm to its vapor. Vinyl acetate is a severe eye and skin irritant, forming blisters on the skin, and redness, swelling, or corneal burns on the eyes. The vapor is irritating to the nose and throat, and high levels of exposure may result in pulmonary edema. Vinyl acetate has moderate acute toxicity if ingested.

Vinyl Acetate Polymers

Properties. Poly(vinyl acetate) (PVAc) polymer resins are manufactured in a variety of molecular weights. Some physical properties of the polymer are listed in Table 2. With increasing molecular weight,

Table 1. Some Properties and Characteristics of Vinyl Acetate

Property	Value
formula weight	86.09
physical state	liquid
flammable limits in air (101.3 kPa[a]), vol %	LEL 2.6, UEL 13.4
flash point, °C	
Tag closed cup (ASTM D56)	−8
Tag open cup (ASTM D1310)	−4
autoignition temperature, °C	426.9
bp at 101.3 kpa[a], °C	72.7
relative evaporation rate (n − butyl acetate = 1)	8.9
vapor pressure at 20°C, kPa[a]	11.8
critical temp, °C	246
critical pressure, kPa[b]	3950
color	clear and colorless
sp gr, 20/20°C	0.934
vapor density (air = 1.00)	2.97
viscosity at 20°C	0.43 cps
fp, °C	−92.8
heat of combustion at 25°C	−495.0 kcal/mol
heat of vaporization (1 atm)	87.6 cal/g
heat of formation (liquid at 25°C), kJ/mol[c]	−349.4
heat of polarization, kJ/mol[c]	89.1
specific heat at 20°C (liquid)	0.46 cal/g °C
odor	sweetish smell in small quantities
reactivity	reactive with self and variety of other chemicals; stable when properly stored and inhibited
light sensitivity	light promotes polymerization
electrical conductivity at 23°C	2.6 × 104 pS/m(1 S = 1 mho)
refractive index, n_D^{20}	1.3953
surface tension at 20°C, mN/m(= dyn/cm)	23.6
coefficient of cubical expansion	0.00137/°C at 20°C

[a] To convert kPa to mm Hg, multiply by 7.5.
[b] To convert kPa to atm, divide by 101.3.
[c] To convert kJ/mol to kcal/mol, divide by 4.184.

properties vary from viscous liquids to low melting solids to tough, horny materials. They are neutral, water-white to stray-colored, tasteless, odorless, and nontoxic. The resins have no sharply defined melting points but become softer with increasing temperature. Due to their solubility parameter, they are soluble in organic solvents, but are insoluble in the lower alcohols (excluding methanol), glycols, water, and nonpolar liquids.

The chemical properties of PVAc are those of an aliphatic ester. Thus, acidic or basic hydrolysis produces poly(vinyl alcohol) and acetic acid or the acetate of the basic cation.

Poly(vinyl acetate) emulsion films adhere well to most surfaces that have relatively high surface energy, (eg, wood, paper, glass, and metal), and have good binding capacity for pigments (qv) and fillers (qv). PVAc film can be laid down on a damp surface with trapped moisture gradually passing through the film without lifting or blistering it. Poly(vinyl acetate) polymers are environmentally friendly because they easily biodegrade.

Polymerization Processes. Vinyl acetate has been polymerized industrially by bulk, solution, suspension, and emulsion processes. Perhaps 90% of the material identified as poly(vinyl acetate) or copolymers that are predominantly vinyl acetate are made by emulsion techniques.

Emulsion Polymerization. Poly(vinyl acetate)-based emulsion polymers are produced by the polymerization of an emulsified monomer through free-radicals generated by an initiator system. An emulsion recipe, in general, contains monomer, water, protective colloid or surfactant, initiator, buffer, and perhaps a molecular weight regulator.

Many different combinations of surfactant and protective colloid are used in emulsion polymerizations of vinyl acetate as stabilizers. The properties of the emulsion and the polymeric film depend to a large extent on the identity and quantity of the stabilizers. Poly(vinyl acetate) emulsions can be made with a surfactant alone or with a protective colloid alone, but the usual practice is to use a combination of the two. In general, the greater the quantity of stabilizers in a recipe, the smaller the particle size of the emulsion.

The initiators used in vinyl acetate polymerizations are the familiar free-radical types. Buffers are frequently added to emulsion recipes. Vinyl acetate emulsion polymerization recipes are usually buffered to pH 4–5. The pH of most commercially available emulsions is 4–6.

A chain-transfer agent is added to vinyl acetate polymerizations to control the polymer molecular weight.

A polymerization process may consist of simply charging all ingredients to the reactor, heating to reflux, and stirring until the reaction is over while controlling the heat removal at the reaction temperature using cooling systems. However, this simple procedure is seldom followed.

Industrially, polymerizations are carried out to over 99% conversion and thus there is no need to reduce the unreacted monomer unless very low levels are required to meet regulatory, product, or workplace requirements. Most poly(vinyl acetate) emulsions contain less than 0.5 wt % unreacted vinyl acetate. All of the processes are operated in conventional glass-lined or stainless steel kettles or reac-

Table 2. Physical Constants for Poly(vinyl acetate)

Property	Value
absorption of water at 20°C for 24–144 h, %	3–6
coefficient of thermal expansion, K^{-1}	
cubic	6.7×10^{-4}
linear, below T_g	7×10^{-5}
above T_g	22×10^{-5}
compressibility, cm^3/(g·kPa)[a]	17.8×10^{-6}
decomposition temperature, °C	150
density at 20°C, g/cm^3	1.191
dielectric constant at 50°C, at 2 MHz	3.5
dielectric dissipation factor at 50°C, at 2 MHz, tan δ	150
dielectric strength at 30°C, V/L	0.394
dipole moment at 20°C, C·m[b] per monomer unit	2.30
elongation at break, at 20°C and 0% rh, %	10–20
glass-transition temperature, T_g, °C	28–31
pressure dependence, °C/100 MPa[c]	0.22
hardness, at 20°C, Shore units	80–85
heat capacity, at 30°C, J/g[d]	1.465
heat distortion point, °C	50
heat of polymerization, kJ/mol[d]	87.5
refraction index at 20.7°C, n_D	1.4669
interfacial tension, mN/m (= dyn/cm)	
at 20°C with polyethylene	14.5
20°C with polystyrene	4.2
modulus of elasticity, GPa[c]	1.275–2.256
notched impact strength, J/m[e]	102.4
softening temperature, °C	35–50
surface resistance (ohm/cm)	5×10^{11}
surface tension, mN/m(= dyn/cm)	
at 20°C	36.5
180°C	25.9
tensile strength, MPa[c]	29.4–49.0
thermal conductivity, mW/(m · K)	159
Young's modulus, MPa[c]	600

[a] To convert kPa to atm, divide by 101.3.
[b] To convert C · m to debye, divide by 3.336×10^{-30}.
[c] To convert MPa to psi, multiply by 145; GPa to psi, multiply by 145,000.
[d] To convert J to cal, divide by 4.184.
[e] To convert J/m to lbf/in., divide by 53.38.

tors. Control of the process is important to ensure reproducibility of the product.

Bulk Polymerizations. In the bulk polymerization of vinyl acetate the viscosity increases significantly as the polymer forms making it difficult to remove heat from the process. Low molecular weight polymers have been made in this fashion. Continuous processes are known to be used for bulk polymerizations.

Suspension Polymerization. The suspension or pearl polymerization process has been used to prepare polymers for adhesive and coating applications and for conversion to poly(vinyl alcohol). Suspension polymerizations are carried out with monomer-soluble initiators predominantly, with low levels of stabilizers. Continuous tubular polymerization of vinyl acetate in suspension yields stable dispersions of beads with narrow particle size distributions at high yields.

Solution Polymerization. Solution polymerization of vinyl acetate is carried out mainly as an intermediate step to the manufacture of poly(vinyl alcohol). A small amount of solution-polymerized vinyl acetate is prepared for the merchant market. When solution polymerization is carried out, the solvent acts as a chain-transfer agent, and depending on its transfer constant, has an effect on the molecular weight of the product. Continuous solution polymers of poly(vinyl acetate) in tubular reactors have been prepared at high yield and throughput.

Propagation. The rate of emulsion polymerization has been found to depend on initiator, monomer, and emulsifier concentrations. Vinyl acetate polymerizes chiefly in the usual head-to-tail fashion, but some of the monomers orient head-to-head and tail-to-tail as the chain grows. In vinyl acetate polymerizations, the molecular weights of the products increase with the extent of conversion: the ratio of weight-to-number-average-degree-of-polymerization also changes, becoming larger at higher conversions.

Chain Transfer. At the molecular scale, vinyl acetate polymerizations generally are understood as free-radical polymerizations, but are characterized in particular by a relatively large amount of chain transfer.

Grafting and Stabilizers. The degree of grafting of poly(vinyl acetate) (PVAc) on poly(vinyl alcohol) (PVA) and other stabilizers during emulsion polymerization strongly affects latex properties such as viscosity, rheology, and polymer solubility.

Copolymers. Vinyl acetate copolymerizes easily with a few monomers, eg, ethylene, vinyl chloride, and vinyl neodecanoate, which have reactivity ratios close to its own. Block copolymers of vinyl acetate with methyl methacrylate, acrylic acid, acrylonitrile, and vinyl pyrrolidinone have been prepared by copolymerization in viscous conditions, with solvents that are poor solvents for the vinyl acetate macroradical.

Blends. Latex film properties are commonly modified through the blending of latexes, eg, a "soft" polymer is made slightly harder by blending with a "hard" latex.

Specifications and Standards. Borax stability is an important property in adhesives, paper (qv), and textile applications. Other emulsion properties tabulated by manufacturers include tolerance to specific solvents, surface tension, minimum film-forming temperature, dilution stability, freeze–thaw stability, percent soluble polymer, and molecular weight.

Poly(vinyl acetate) and its copolymers with ethylene are available as spray-dried emulsion solids with average particle sizes of 2–20 μm; the product can be reconstituted to an emulsion by addition of water or it can be added directly to formulations, eg, concrete.

Uses. The uses of poly(vinyl acetate) adhesive are packaging (qv) and wood gluing. PVAc copolymer adhesives are finding application in more diverse areas such as construction and adhesion to more difficult to bond surfaces because of the range of adhesion and the flexibility that may be built into the polymer.

Poly(vinyl acetate) emulsions can be used in high speed gluing equipment. Poly(vinyl acetate) homopolymers adhere well to porous or cellulosic surfaces, eg, wood, paper, cloth, leather (qv), and ceramics (qv). Homopolymer films tend to creep less than copolymer or terpolymer films. They are especially suitable in adhesives for high speed packaging operations.

Poly(vinyl acetate) dry resins and ethylene–vinyl acetate (EVA) copolymers are used in solvent adhesives, which can be applied by total industrial techniques, eg, brushing, knife-coating, roller-coating, spraying, or dipping.

Paints (see PAINT) prepared from poly(vinyl acetate) and its copolymers form flexible, durable films with good adhesion to clean surfaces, including wood, plaster, concrete, stone, brick, cinder blocks, asbestos board, asphalt, tar paper, wallboards, aluminum, and galvanized iron.

Poly(vinyl acetate) emulsions and resins have been used as the binder in coatings for paper and paperboard since 1955. The coatings may be clear, colored, or pigmented, and are glossy, odorless, tasteless, grease-proof, nonyellowing, and heat-sealable. Conventional paper-coating equipment is used.

The use of vinyl acetate copolymers as binding agents for nonwoven fabrics has grown rapidly. Poly(vinyl acetate) was first used in concrete in the 1940s as a thermoplastic polymer to strengthen the concrete matrix. Vinyl acetate resins are useful as antishrinking agents for glass fiber-reinforced polyester molding resins and as binders for numerous materials. Emulsions containing added poly(vinyl alcohol) and dichromate are used to make light-sensitive stencil screens for textile printing and ceramic decoration. Vinyl acetate polymers have long been used as chewing gum bases.

CAJETAN F. CORDEIRO
Air Products and Chemicals, Inc.

Chemical Economics Handbook Marketing Research Report, "Vinyl Acetate," (July 1993).

M. K. Lindemann, in G. E. Ham, ed., *Vinyl Polymerization,* Vol. 1, Marcel Dekker, Inc., New York, 1967, Part 1, Chapt. 4.

J. Brandrup and E. H. Immergut, eds., *Polymer Handbook,* Interscience Publishers, a division of John Wiley & Sons, Inc., New York, 1966.

M. K. Lindemann, in N. M. Bikales, ed., *Encyclopedia of Polymer Science and Technology,* Vol. 15, John Wiley & Sons, Inc., New York, 1971, p. 636.

VINYL ALCOHOL POLYMERS

Poly(vinyl alcohol) (PVA), a polyhydroxy polymer, is the largest-volume synthetic, water-soluble resin produced in the world. It is commercially manufactured by the hydrolysis of poly(vinyl acetate), because monomeric vinyl alcohol cannot be obtained in quantities and purity that makes polymerization to poly(vinyl alcohol) feasible.

The main uses of PVA are in textile sizing, adhesives, protective colloids for emulsion polymerization, fibers, production of poly(vinyl butyral), and paper sizing. Significant volumes are also used in the production of concrete additives and joint cements for building construction and water-soluble films for containment bags for hospital laundry, pesticides, herbicides, and fertilizers. Smaller volumes are consumed as emulsifiers for cosmetics, temporary protective film coatings, soil binding to control erosion, and photoprinting plates.

Physical Properties

The physical properties of poly(vinyl alcohol) are highly correlated with the method of preparation. The final properties are affected by the polymerization conditions of the parent poly(vinyl acetate), the hydrolysis conditions, drying, and grinding. Further, the term poly(vinyl alcohol) refers to an array of products that can be considered copolymers of vinyl acetate and vinyl alcohol. Representative properties are shown in Table 1.

Table 1. Physical Properties of Poly(Vinyl Alcohol)

Property	Value
appearance	white to ivory-white granular powder
specific gravity	1.27–1.31
tensile strength, MPa[a]	67–110[b]
elongation, %	0–300
specific heat, J/(g·K)[c]	1.67
thermal conductivity, W/(m·K)	0.2
T_g, K	358[b]
mp, K	503[b]
electrical resistivity, Ω·cm	$(3.1-3.8) \times 10^7$
refractive index, n_D^{20}	1.55
degree of crystallinity	0–0.54
storage stability (solid)	indefinite when protected from moisture
flammability	similar to paper
stability in sunlight	excellent

[a] To convert MPa to psi, multiply by 145.
[b] 98–99% hydrolyzed.
[c] To convert J to cal, divide by 4.184.

The ability of PVA to crystallize is the single most important physical property of PVA as it controls water solubility, water sensitivity, tensile strength, oxygen barrier properties, and thermoplastic properties.

The glass-transition temperature, T_g, of fully hydrolyzed PVA has been determined to be 85°C for high molecular weight material. Poly(vinyl alcohol) is only soluble in highly polar solvents, such as water, dimethyl sulfoxide, acetamide, glycols, and dimethylformamide. The solubility in water is a function of degree of polymerization (DP) and hydrolysis.

The viscosities of PVA solutions are mainly dependent on molecular weight and solution concentration. The viscosity increases with increasing degree of hydrolysis and decreases with increasing temperature.

The tensile strength of unplasticized PVA depends on degree of hydrolysis, molecular weight, and relative humidity. Tensile elongation of PVA is extremely sensitive to humidity and ranges from <10% when completely dry to 300–400% at 80% rh. Addition of plasticizer can double these values. Poly(vinyl alcohol) is virtually unaffected by hydrocarbons, chlorinated hydrocarbons, carboxylic acid esters, greases, and animal or vegetable oils. Resistance to organic solvents increases with increasing hydrolysis.

The oxygen-barrier properties of PVA at low humidity are the best of any synthetic resin. However, barrier performance deteriorates above 60% rh.

The surface tension of aqueous solutions of PVA varies with concentration, temperature, degree of hydrolysis, and acetate distribution on the PVA backbone. Surface tension decreases slightly as the molecular weight is reduced. The relationship between the intrinsic viscosity and molecular weight changes with degree of hydrolysis of the polymer.

Chemical Properties

Poly(vinyl alcohol) participates in chemical reactions in a manner similar to other secondary polyhydric alcohols. Of greatest commercial importance are reactions with aldehydes to form acetals, such as poly(vinyl butyral) and poly(vinyl formal).

Boric acid and borax form cyclic esters with poly(vinyl alcohol).

An unlimited number of organic esters can be prepared by reactions of poly(vinyl alcohol) employing standard synthesis. Ethers of poly(vinyl alcohol) are easily formed. Insoluble internal ethers are formed by the elimination of water, a reaction catalyzed by mineral acids and alkali.

Poly(vinyl alcohol) and aldehydes form products which find use in the manufacture of safety glass and as adhesives for hydrophilic surfaces.

Poly(vinyl alcohol) can be readily cross-linked using a multifunctional compound that reacts with hydroxyl groups. These types of reactions are of significant industrial importance as they provide ways to obtain improved water resistance of the poly(vinyl alcohol) or to increase the viscosity rapidly. The thermal decomposition of poly(vinyl alcohol) in the absence of oxygen occurs in two stages. The first stage begins at about 200°C and is mainly dehydration, accompanied by the formation of volatile products. Further heating to 400–500°C yields carbon and hydrocarbons.

Poly(vinyl alcohol) is one of the few truly biodegradable synthetic polymers; the degradation products are water and carbon dioxide. At least 55 species or varieties of microorganisms have been shown to degrade or participate in the degradation of PVA.

Manufacture

Poly(vinyl alcohol) can be derived from the hydrolysis of a variety of poly(vinyl esters), such as poly(vinyl acetate), poly(vinyl formate), and poly(vinyl benzoate), and of poly(vinyl ethers). However, all commercially produced poly(vinyl alcohol) is manufactured by the hydrolysis of poly(vinyl acetate). The manufacturing process can be viewed as one segment that deals with the polymerization of vinyl acetate and another that handles the hydrolysis of poly(vinyl acetate) to poly(vinyl alcohol).

Vinyl acetate is polymerized commercially using free-radical polymerization in either methanol or, in some circumstances, ethanol.

Poly(vinyl acetate) can be converted to poly(vinyl alcohol) by transesterification, hydrolysis, or aminolysis. Industrially, the most important reaction is that of transesterification, where a small amount of acid or base is added in catalytic amounts to promote the ester exchange.

Copolymers

Numerous vinyl alcohol copolymers have been prepared. Copolymers with ethylene and methacrylate are the only copolymers that have found sizable commercial utility. Ethylene–vinyl alcohol (EVOH) copolymers containing 20–30 mol % ethylene are used as an oxygen barrier in food packaging. Vinyl alcohol–methyl methacrylate copolymers are used as sizing agents in the textile industry. The presence of the methacrylate unit disrupts the crystallinity, making the product easier to remove during the desizing operation. The product is especially useful as an alkaline-resistant textile size.

Specifications and Standards

Poly(vinyl alcohol) is produced mainly in five molecular weight ranges. Industry practice expresses the molecular weight of a particular grade in terms of the viscosity of a 4% aqueous solution. An unlimited number of viscosities can be generated by blending the available molecular weights. Products having different degree of hydrolysis can also be blended to obtain a particular performance characteristic. Poly(vinyl alcohol) is an innocuous material having unlimited storage stability.

Health and Safety Factors

Poly(vinyl alcohol) is a nonhazardous material according to the American Standard for Precautionary Labeling of Hazardous Industrial Chemicals (ANSI 2129.1-1976). Extensive tests indicate a very low order of oral toxicity. Short-term inhalation of PVA dust has no known health significance, but can cause discomfort and should be avoided in accordance with industry standards for exposure to nuisance dust. During transport and handling, granular PVA may form an explosive mixture with air, which shows a low severity rating (Bureau of Mines Rating) of 0.1 on a scale in which coal dust has a rating of 1.0.

Processing

Poly(vinyl alcohol) is not considered a thermoplastic polymer because the degradation temperature is below that of the melting point. Thus, industrial applications of poly(vinyl alcohol) are based on and limited by the use of water solutions.

Uses

The main applications for PVA are in textile sizing, adhesives, polymerization stabilizers, paper coating, poly(vinyl butyral), and PVA fibers. In terms of percentage, and omitting the production of PVA not isolated prior to conversion into poly(vinyl butyral), the principal applications are textile sizes, at 30%; adhesives, including use as a protective colloid, at 25%; fibers, at 15%; paper sizes, at 15%, poly(vinyl butyral), at 10%; and others, at 5%, which include water-soluble films, nonwoven fabric binders, thickeners, slow-release binders for fertilizer, photoprinting plates, sponges for cosmetic, and health care applications.

F. LENNART MARTEN
Air Products and Chemicals, Inc.

C. A. Finch, ed., *Polyvinyl Alcohol Developments,* John Wiley & Sons, Inc., New York, 1992.

I. Sakurada, *Poly(Vinyl Alcohol) Fibers,* Marcel Dekker, Inc., New York, 1985.

C. A. Finch, ed., *Polyvinyl Alcohol,* John Wiley & Sons, Inc., New York, 1973.

VINYL CHLORIDE POLYMERS

Poly(vinyl chloride) (PVC), commanding large and broad uses in commerce, is second in volume only to polyethylene, having a volume sales in North America in 1995 of 6.2×10^9 kg (13.7×10^9 lb). Vinyl compounds usually contain close to 50% chlorine, which not only provides no fuel, but acts to inhibit combustion in the gas phase, thus supplying the vinyl with a high level of combustion resistance, useful in many building as well as electrical housings and electrical insulation applications.

PVC has a unique ability to be compounded with a wide variety of additives, making it possible to produce materials that range from flexible elastomers to rigid compounds, that are virtually unbreakable. Compounds are also made that have stiff melts for profile extrusion or low viscosity melts for thin-walled injection molding.

Produced by free radical polymerization, PVC has the structure of $\text{-(}CH_2CHCl\text{)-}_n$, where the degree of polymerization, *n,* ranges from 500 to 3500.

PVC Morphology

The principal type of polymerization of PVC is the suspension polymerization route. The morphology formed during polymerization strongly influences the processibility and physical properties.

In the suspension polymerization of PVC, droplets of monomer 30–150 μm in diameter are dispersed in water by agitation. A thin membrane of a graft copolymer of poly(vinyl chloride) and poly(vinyl alcohol) is formed at the H_2O–monomer interface. Primary particles, 1 μm in diameter, deposit onto the membrane from the monomer side.

Mass-polymerized PVC also has a skin of compacted PVC primary particles very similar in thickness and appearance to the suspension-polymerized PVC skin. However, mass PVC does not contain the thin-block copolymer membrane.

In suspension PVC polymerization, droplets of polymerizing PVC, 30–150-μm dia agglomerate to form grains at 100–200-μm dia. With one droplet per grain, the shape is quite spherical. With several droplets making up the grain, the shape can be quite irregular and knobby. The grain shape plays an important role in determining grain packing and bulk density of a powder.

For both suspension and mass polymerizations at less than 2% conversion, PVC precipitates from its monomer as stable primary particles, slightly below 1-μm dia.

On an even smaller scale is the microdomain structure at 0.01-μm spacing. This is interpreted as a structure where the crystallites of about 0.01-μm spacing are tied together by molecules in the amorphous regions. Plasticizer swells the amorphous regions without dissolving the crystallites.

PVC has structure that is built upon structure which is, in turn, built upon even more structure. These many layers of structure are all important to performance and are interrelated. A summary of these structures is listed in Table 1.

The first step in processing is usually powder mixing in a high speed, intensive mixer. PVC resin, stabilizers, plasticizers, lubricants, processing aids, fillers, and pigments are added to the powder blend for distributive mixing.

In plasticized PVC, liquid plasticizers first fill the voids or pores in the PVC grains fairly rapidly during powder mixing. If a large amount of plasticizer is added, the excess plasticizer beyond the capacity of the pores initially remains on the surface of the grains, making the powder somewhat wet and sticky. Continued heating increases the diffusion rate of plasticizer into the PVC mass where the excess liquid is eventually absorbed and the powder dries.

PVC powder compounds are heated, sheared, and deformed during melt processing. During this process, the grains of PVC are broken down. A processing window of stable primary particles exists even

Table 1. Summary of Poly(vinyl chloride) Morphology

Feature	Size	Description
droplets	30–150 μm dia	dispersed monomer during suspension polymerization
membranes	0.01–0.02 μm thick	membrane at monomer–water interface in suspension PVC (usually graft copolymer of PVC and dispersant, such as poly(vinyl alcohol))
grains	100–200 μm dia	after polymerization, free-flowing powder usually made up of agglomerated droplets; in mass polymerization, is free-flowing powder
skins	0.5–5 μm thick	shell on grains made up of PVC deposited onto membrane during suspension polymerization; in mass polymerization, is PVC compacted on grain surface
primary particles	1 μm dia	formed as single polymerization site in both suspension and mass polymerization by precipitation of polymer from monomer; made up of over a billion molecules, it is often melt flow unit established during melt processing (in emulsion polymerization, it is emulsion particle)
agglomerates of primary particles	3–10 μm dia	formed during polymerization by merging of primary particles
domains	0.1 μm dia	formed under special conditions such as high temperature melting (205°C) followed by lower temperature mechanical work (140–150°C); water-phase polymerization also produces domain-sized structure
microdomains	0.01 μm spacing	crystallite spacing
secondary crystallinity	0.01 μm spacing	crystallinity reformed from amorphous melt and responsible for fusion (gelation)

with continued melt processing. The primary particle is about a billion molecules of PVC held together by a structure of crystallites and tie molecules.

The PVC crystallites are small, average 0.7 nm (3 monomer units), in the PVC chain direction, and are packed laterally to a somewhat greater extent.

PVC Fusion (Gelation). The PVC primary particle flow units (billion molecule bundles) can partially melt, freeing some molecules of PVC that can entangle at the flow unit boundary. These entangled molecules can recrystallize upon cooling, forming secondary crystallites, and tie the flow units together into a large three-dimensional structure. This process is known as fusion or gelation.

The strength created by the fusion process is strongly dependent on the previous processing temperature and the molecular weight of the PVC. PVC normally improves in properties with increasing fusion (or increasing melt temperature).

Plasticized PVC has the same structures as rigid PVC, except that plasticizer enters the amorphous phase of PVC and makes the tie molecules elastomeric. The grains break down to 1-μm primary particles which become the melt flow units. The crystallites are not destroyed by plasticizer. Table 2 provides a list of the PVC physical parameters.

Chemical Properties

The addition of vinyl monomer to a growing PVC chain can be considered to add in a head-to-tail fashion, resulting in a chlorine atom on every other carbon atom, ie,

$$-\!\!\left(CH_2CHClCH_2CHClCH_2CHCl\right)\!\!-_n$$

or in a head-to-head, tail-to-tail fashion, resulting in chlorine atoms on adjacent carbon atoms, ie,

$$-\!\!\left(CH_2CHClCHClCH_2CH_2CHClCHClCH_2\right)\!\!-_n$$

Table 2. PVC Physical Parameters

PVC property	Value		
	orthorhombic, two monomer units/cell		
crystallographic data	*a*	*b*	*c*
commercial PVC, nm	1.06	0.54	0.51
single crystal, nm	1.024	0.524	0.508
crystallinity, %			
as polymerized		19	
from melt		4.9	
density (uncompounded), g/cc			
whole		1.39	
crystallites		1.53	
oxygen permeability, cc/(cm·s)cm^2 cm Hg		$238e^{-13.3/RT}$	
Poisson ratio (rigid PVC)		0.41	
refractive index		1.54	
glass-transition temperature, °C		83	
coefficient of linear thermal expansion (unplasticized), °C		7×10^{-5}	
specific heat	temp, °C		value, J/g °C[a]
rigid PVC	23		0.92
	80		1.45
plasticized PVC (50 phr DOP)	23		1.54
	80		1.75
thermal conductivity (unplasticized), J/(cm·s)°C		17.5×10^{-4}	
dielectric strength			
kV/mil		0.5	
kV/mm		20	
solubility parameter, (J/cm^3)$^{0.5}$		40.7 (av)	

[a] To convert J to cal, divide by 4.184.

Dechlorination studies show that the head-to-tail structure is predominate.

Both saturated and unsaturated end groups can be formed during polymerization by chain transfer to monomer or polymer and by disproportionation.

PVC polymerization has a high chain-transfer activity to monomer; about 60% of the chains have unsaturated chain ends and the percentage of chain ends containing initiator fragments is low. Chain transfer to polymer leads to branching.

The addition of monomer fixes the tacticity of the previous monomer unit. The tacticity of PVC is nearly random, with syndiotacticity slightly favored at lower polymerization temperature.

Polymerization Kinetics of Mass and Suspension PVC. The polymerization kinetics of mass and suspension PVC are considered together because a droplet of monomer in suspension polymerization can be considered to be a mass polymerization in a very tiny reactor. During polymerization, the polymer precipitates from the monomer when the chain size reaches 10–20 monomer units. The precipitated polymer remains swollen with monomer, but has a reduced radical termination rate. This leads to a higher concentration of radicals in the polymer gel and an increased polymerization rate at higher polymerization conversion.

Polymerization in two phases, the liquid monomer phase and the swollen polymer gel phase, forms the basis for kinetic descriptions of PVC polymerization.

Chain transfer to monomer is the main reaction controlling molecular weight and molecular weight distribution. PVC molecular weights are usually determined in the United States using inherent viscosity or relative viscosity measured according to ASTM D1243.

PVC Resin Manufacturing Processes

Mass Polymerization. Mass or bulk polymerization of PVC is normally difficult. At high conversions the mixture becomes extremely viscous, impeding agitation and heat removal, causing a high polymerization temperature and broad molecular weight distribution. A two-stage process overcomes these problems. The first stage of the process, which forms a skeleton seed grain for polymerization in a second stage, is carried out in a prepolymerizer with flat blade agitator and baffles to about 7–10% conversion. The number of grains remain constant throughout this polymerization.

Suspension Polymerization. Suspension polymerization is carried out in small droplets of monomer suspended in water. The monomer is first finely dispersed in water by vigorous agitation. Suspension stabilizers act to minimize coalescence of droplets by forming a coating at the monomer-water interface. The hydrophobic-hydrophilic properties of the suspension stabilizers are key to resin properties and grain agglomeration.

Kinetics of suspension PVC are identical to the kinetics of mass PVC.

Emulsion Polymerization. Emulsion polymerization takes place in a soap micelle where a small amount of monomer dissolves in the micelle. The initiator is water-soluble. Polymerization takes place when the radical enters the monomer-swollen micelle. Termination takes place in the growing micelle by the usual radical–radical interactions. The high solubility of vinyl chloride in water, 0.6 wt %, accounts for a strong deviation from true emulsion behavior. Also, PVC's insolubility in its own monomer accounts for such behavior as a rate dependence on conversion. Emulsions of up to 0.2-μm dia are sold in liquid form for water-based paints, printing inks, and finishes for paper and fabric. Other versions, 0.3–10-μm dia and dried by spray-drying or coagulation, are used as plastisol resins.

Microsuspension Polymerization. Microsuspension polymerization uses a monomer-soluble initiator. The monomer is homogenized in water along with emulsifiers or suspending agents to control the particle sizes. Microsuspension paste resins at 0.3–1-μm dia are used to make plastisols for flooring, seals, barriers, etc. These plastisols are also dispersions of PVC in liquid plasticizer and are cured by heating.

Microsuspension blending resins at 10–100-μm dia are used as extenders to paste resins in plastisols.

Solution Polymerization. In solution polymerization, a solvent for the monomer is often used to obtain very uniform copolymers. Polymerization rates are normally slower than those for suspension or emulsion PVC.

Copolymerization. Vinyl chloride can be copolymerized with a variety of monomers. Vinyl acetate, the most important commercial comonomer, is used to reduce crystallinity, which aids fusion and allows lower processing temperatures. Copolymers are used in flooring and coatings. This copolymer sometimes contains maleic acid or vinyl alcohol (hydrolyzed from the poly(vinyl acetate)) to improve the coating's adhesion to other materials, including metals. Copolymers with vinylidene chloride are used as barrier films and coatings. Copolymers of vinyl chloride with acrylic esters in latex form are used as film formers in paint, nonwoven fabric binders, adhesives, and coatings. Copolymers with olefins improve thermal stability and melt flow, but at some loss of heat-deflection temperature.

Compounding

The additives found in PVC help make it one of the most versatile, cost-efficient materials in the world. Without additives, literally hundreds of commonly used PVC products would not exist. Many materials are useless until they undergo a similar modification process.

Stabilizers. Lead stabilizers, particularly tribasic lead sulfate, is commonly used in plasticized wire and cable compounds because of its good nonconducting electrical properties. Organotin stabilizers are commonly used for rigid PVC, including pipe, fittings, windows, siding profiles, packaging, and injection-molded parts.

Antimony tris(isooctylthioglycolate) has found use in pipe formulations at low levels. Barium–zinc stabilizers have found use in plasticized compounds, replacing barium–cadmium stabilizers. These are used in moldings, profiles, and wire coatings. Cadmium use has decreased because of environmental concerns surrounding certain heavy metals.

Calcium–zinc stabilizers are used in both plasticized PVC and rigid PVC for food contact where it is desired to minimize taste and odor characteristics. Applications include meat wrap, water bottles, and medical uses. Many stabilizers require costabilizers.

Impact Modifiers. Rubbery polymers are added to PVC to improve toughness. Rubbery particles are added to act as stress concentrators or multiple weak points, leading to crazing or shear-banding under impact load. This can result in cavitation and/or cold drawing, thus allowing the PVC to absorb large amounts of energy.

Processing Aids. PVC often flows in the form of billions of molecule primary particles. Processing aids glue these particles together before the PVC melts, thus acting as a fusion promoter. Processing aids also modify melt rheology by increasing melt elasticity and die swell; some by reducing melt viscosity and melt fracture.

Lubricants. A model for the lubrication mechanism has been developed that explains synergy between certain lubricants. This model treats lubricants as surface-active agents. Some lubricants have polar ends that are attracted to other polar ends and to polar PVC flow units and to polar metal surfaces. These also have nonpolar ends that are repelled by the polar groups. Synergy happens when nonpolar lubricants are added, which are attracted to the nonpolar ends and act as a slip layer.

Plasticizers. Solutions of PVC, prepared at elevated temperatures with high boiling solvents, possess unusual elastic properties when cooled to room temperature. Such solutions are flexible, elastic, and exhibit a high degree of chemical inertness and solvent resistance.

This unusual behavior results from unsolvated crystalline regions in the PVC that act as physical cross-links. These allow the PVC to accept large amounts of solvent (plasticizers) in the amorphous regions, lowering its T_g to well below room temperature, thus making it rubbery. PVC was, as a result, the first thermoplastic elastomer (TPE). This rubber-like material has stable properties over a wide temperature range.

Fillers. Fillers are used to improve strength and stiffness, to lower cost, and to control gloss. The most common filler is calcium carbonate.

Pigments. A variety of pigments are added to PVC to give color, including titanium dioxide and carbon black.

Ultraviolet Light Stabilizers. Both titanium dioxide and carbon black are strong ultraviolet light absorbers and effective in protecting the PVC. For ultraviolet light absorption in transparent PVC or for improvement of pigmented systems, various derivatives of benzotriazole are used.

Biocides. Although PVC itself and rigid PVC compounds are resistant to attack by microorganisms, plasticized PVC, in specific applications such as flashing and sealing boots on roofs, shower curtains, and swimming pools, may need protection. Many biocides, often containing arsenic compounds, are available for a balance of stability, compatibility, weatherability, and biocidal effectiveness.

Flame Retardants. Because PVC contains nearly half its weight of chlorine, it is inherently flame-retardant.

Foaming or Blowing Agents. Cellular PVC can be made by a variety of techniques, such as whipping air into a plastisol, incorporating a gas under pressure, incorporating a physical blowing agent into the melt, or using a chemical blowing agent which releases a gas when it decomposes with heat. The most common chemical blowing agent is 1,1′-azobisdicarbonamide, which decomposes with heat to release nitrogen gas.

Uses

PVC is so versatile that it can be compounded for a wide range of properties and used in a wide variety of markets. Most of the products are durable goods and have long life spans. Its use in short-term, one-time-use products is limited.

Pipe and fittings, a principal market for PVC, are a prime example of PVC as an engineering thermoplastic.

PVC is accepted commercially as an excellent weathering material. PVC's chemical response to weathering is well understood so that compounds and products can be designed for satisfactory outdoor performance. Products include siding, windows, and doors.

Complex profiles require specialty manufacturing skills to build, maintain, and operate extrusion dies as well as cooling and sizing equipment to deliver the exact dimensions required. Cubed compound, where the PVC grains are already broken down, can be run faster on simple single-screw extruders on account of the typical low melt temperatures.

PVC has been used in wire and cable applications since World War II. The compounds are optimized for the requirements, including low temperature flexibility, high use temperature, especially low combustibility, weatherability, and high resistance to cutthrough.

Numerous housings, electrical enclosures, and cabinets are injection-molded from rigid PVC. These take advantage of PVC's outstanding UL flammability ratings and easy molding into thin-walled parts.

Health and Safety Factors

There are no significant health hazards arising from exposure to poly(vinyl chloride) at ambient temperature. At processing temperatures, most polymers emit fumes and vapors that may be irritating to the respiratory tract. Decomposition of plastics, eg, through greatly elevated temperatures above normal operating temperatures, can result in personnel exposure to decomposition or combustion products. In the case of PVC compounds, such decomposition involves hydrogen chloride, which causes irritation of the respiratory tract, eyes, and skin.

Poly(vinyl chloride) resin has a flash point of approximately 391°C (735°F) and a self-ignition temperature of approximately 454°C (850°F) (ASTM D1929). In general, PVC burns with difficulty because a substantial amount of energy is required to break down the polymer. Poly(vinyl chloride) powder has a very low tendency to explode. In firefighting where PVC is involved, water, ABC dry chemical, or protein-type air foams should be used as extinguishing media.

In the U.S., poly(vinyl chloride) is an EPA hazardous air pollutant under the Clean Air Act Section 112 (40 CFR 61) and is covered under the New Jersey Community Right-to-Know Survey: N.J. Environmental Hazardous Substances (EHS) List as "chloroethylene, polymer" with a reporting threshold of 225 kg (500 lb).

Environmental Considerations and Recycling

Over 30% of the chlorine produced on a global basis goes to make PVC. Chlorine makes PVC inherently flame-retardant. PVC is over 50% chlorine and, as a result, one of the most energy-efficient polymers. Chlorine makes PVC far more environmentally acceptable than other materials that are totally dependent on petrochemical feedstocks. In addition, recycling PVC is easier because the chlorine in PVC acts as a marker, enabling automated equipment to sort PVC containers from other plastics in the waste stream.

Although vinyl is the world's second most widely used plastic, less than one-half percent by weight is found in the municipal solid waste stream. Most of that consists of vinyl packaging, bottles, blister packaging, and flexible film. This is because most vinyl applications are long-term uses, such as pipe and house siding, and are not disposed of quickly. Vinyl wastes are handled by all conventional disposal methods, ie, recycling, landfilling, and incineration (including waste-to-energy).

A study sponsored by the American Society of Mechanical Engineers (ASME), involving the analysis of over 1700 test results from 155 large-scale, commercial incinerator facilities throughout the world, found no relationship between the chlorine content of waste and dioxin emissions from combustion processes. Instead, the study stated, the scientific literature is clear that the operating conditions of combustors are the critical factor in dioxin generation.

Incinerator scrubbing systems can remove about 99% of the hydrogen chloride generated by incinerating vinyl plastics and other chlorine-containing compounds and materials. Municipal incinerators are often targeted as a primary cause of acid rain. In fact, power plants burning fossil fuels, which produce sulfur dioxide and nitrogen oxide, are actually the leading cause of acid rain, along with automotive exhaust. In Europe and Japan, studies show that only about 0.02% of all acid rain can be traced to incineration of PVC.

Industrial scrap vinyl has been recycled for years, but in more recent years, post-consumer vinyl recycling is growing. In 1991, there were an estimated 1100 municipal recycling programs in place or planned in the United States that include vinyl.

In municipal recycling contamination occurs whether or not vinyl is present. Other resins are just as much a contamination problem as vinyl. Except for commingled plastics applications, different plastic materials cannot be mixed successfully in most recycled products applications. This is why it is crucial to separate efficiently one plastic from another.

It is not true that poly(ethylene terephthalate) (PET) and high density polyethylene (HDPE) packaging are listed as 1 and 2 in the Society of the Plastics Industry (SPI) recycling coding system because they are the most recyclable. The numbers assigned to each plastic in the SPI coding system are purely arbitrary and do not reflect the material's recyclability.

Vinyl plastics do not decompose in landfills and give off vinyl chloride monomer, because like all plastics, vinyl is an extremely stable landfill material. It resists chemical attack and degradation, and is so resistant to the conditions present in landfills that it is often used to make landfill liners. A recent study compared vinyl to a number of other packaging materials and found that vinyl consumed the least amount of energy, used the lowest level of fossil fuels, consumed the least amount of raw materials, and produced the lowest levels of carbon dioxide of any of the plastics studied.

It is not true that in a fire, vinyl is unusually hazardous and damaging. The real hazards in a fire are carbon monoxide and heat; these are especially a problem with other materials that readily burn. Because vinyl products contain chlorine, they are inherently flame-retardant and resist ignition. When it does burn, however, vinyl produces carbon monoxide, carbon dioxide, and hydrogen chloride. Of these, the most hazardous is carbon monoxide. Hydrogen chloride is an irritant gas that can be lethal at extremely high levels. However, research indicates that those levels are never reached or even approached in real fires.

JAMES W. SUMMERS
The Geon Company

J. A. Davidson and D. E. Witenhafer, *J. Polym. Sci.: Polym. Phys. Ed.* **18**, 51 (1980).

J. T. Lutz, Jr. and D. L. Dunkelberger, *Impact Modifiers for PVC; The History and Practice,* John Wiley & Sons, Inc., New York, 1992.

H. G. Rigo, A. J. Chandler, and W. S. Lanier, *The Relationship Between Chlorine In Waste Streams and Dioxin Emissions From Waste Combustor Stacks,* CRTD, Vol. 36, The American Society of Mechanical Engineers, United Engineering Center, New York, 1995.

J. W. Summers, *J. Vinyl Additive Technol.* **3**(2), 130 (1997).

VINYL ETHER MONOMERS AND POLYMERS

Because of the strong electron-donating oxygen, the polymerization of vinyl ethers (VE) can be readily accomplished using cationic initiators, resulting in polymers and copolymers that have the potential for significant variety. However, only poly(methyl vinyl ether) (PMVE)) achieved commercial success among the homopolymers, and its commercial importance has faded. Divinyl ethers are emerging as important ingredients in radiation-cured coatings, whereas copolymers of methyl vinyl ether (MVE) and maleic anhydride, easily prepared by free-radical initiation, continue to be valued as ingredients in personal care and pharmaceutical products.

Monomers

The most general commercial process for the manufacture of mono- and divinyl ethers, developed by Reppe in the 1930s at BASF, is by treating alcohols with acetylene under pressure of ≥6.8 atm (100 psi) at temperatures of 120–180°C in the presence of catalytic amounts of the corresponding metal alcoholate. The danger of handling acetylene under pressure in concentrated form requires sophisticated equipment and should only be attempted experimentally in an appropriately barricaded high pressure autoclave.

$$HC{\equiv}CH \overset{ROM}{\rightarrow} (ROCH = CHM) \overset{ROH}{\rightarrow} ROCH = CH_2 + ROM$$

Alternatively, thermal cracking of acetals or metal-catalyzed transvinylation can be employed. Some physical properties of the lower homologues of vinyl ether are presented in Table 1.

Reactions of Vinyl Ethers. Vinyl ethers undergo the typical reactions of activated carbon-carbon double bonds. A key reaction of VEs is acid-catalyzed hydrolysis to the corresponding alcohol and acetaldehyde, ie, addition of water followed by decomposition of the hemiacetal. MVE is a reactive flammable gas and must be handled safely.

Homopolymerization

VEs such as MVE polymerize slowly in the presence of free-radical initiators to form low mol wt products of no commercial importance. Examples of anionic polymerization are unknown, whereas cationic initiation promotes rapid polymerization to high mol wt polymers in excellent yield and has been extensively studied.

A typical cationic polymerization is conducted with highly purified monomer free of moisture and residual alcohol, both of which act as inhibitors, in a suitably dry unreactive solvent such as toluene with a Friedel-Crafts catalyst, eg, boron trifluoride, aluminum trichloride, and stannic chloride. Usually low temperatures (−40 to −70 °C) are favored in order to prevent chain-transfer or side reactions.

Table 1. Physical Properties of the Lower Vinyl Ethers

Property	Methyl	Ethyl	Isopropyl	*n*-Butyl	Isobutyl
odor	sweet, pleasant	pleasant	pleasant	pleasant	pleasant
bp, °C	5.5	35.6	55–56	94.3	83
fp, °C	−122	−115.3	−140	−112.7	−132.3
sp gr at 20/4°C	0.7511	0.753	0.753	0.778	0.767
refractive index, n_D^{25}	1.3947	1.3734	1.3829	1.3997	1.3946
sol in water at 20°C, wt %	0.97	0.039	0.6	0.1	0.1
flash point, °C	−56[a]	−18[b]		0.55	−9.4
heat of vaporization at 101.3 kPa		367		316	323

[a] Cleveland open cup.
[b] Tag open cup (ASTM D1310).

The nature of the side chain R group exerts considerable influence on the reactivity of vinyl ethers toward cationic polymerization. The rate is fastest when the alkyl substituent is branched and electron-donating. Aromatic vinyl ethers are inherently less reactive and susceptible to side reactions.

Stereoregular Polymerization. In order to generate stereoregular (usually isotactic) polymers, the polymerization is conducted at low temperatures in nonpolar solvents. A variety of soluble initiators can produce isotactic polymers, but there are some initiators, eg, $SnCl_4$, that produce atactic polymers under isotactic conditions. The nature of the pendant group can influence tacticity.

The low temperature limitation of homogeneous catalysis has been overcome with heterogeneous catalysts such as modified Ziegler-Natta solid-supported protonic acids and metal oxides.

It has been suggested that the mechanism of stereoregular vinyl ether polymerization heavily depends on the degree of association of the counterion with the growing terminal carbocation. An incoming monomer can approach the carbocation terminus from either the front- or back-side attack, which attack is prevalent depends on the tightness of the growing ion pair and the steric requirements of the particular vinyl ether monomer. Front-side attack is favored by a loose ion pair associated with polar solvents, whereas a back-side attack is favored by nonpolar solvents where a tight ion pair prevails.

Living Polymerization

Living polymerization is characterized by an increasing number-average molecular weight as the monomer is consumed. The rate of M_n increase is inversely proportional to the initial concentration of hydrogen iodide, not iodine, and the molecular weight distribution (MWD) of the polymer is very narrow throughout the course of the polymerization ($M_w/M_n < 1.1$). Thus, this type of polymerization can be stopped and started by consumption or addition of fresh monomer. It is similar to ethylene oxide/propylene oxide (EO/PO) anionic polymerization in this regard, but the initiation system is longer lived.

Living VE polymerization is usually terminated by addition of alcohols, phenols, amines, etc, that can replace iodide. Without some base present to neutralize generated HI, an aldehyde end group forms if moisture is present because of acid-catalyzed hydrolysis.

The living polymerization process offers enormous flexibility in the design of polymers. It is possible to control terminal functional groups, pendant groups, monomer sequencing along the main chain (including the order of addition and blockiness), steric structure, and spatial shape.

Homopolymer Properties

Physical properties, which depend on molecular weight, the nature of the alkyl group, the nature of the initiator, stereospecificity, and crystallinity, range from viscous liquids, through sticky liquids and rubbery solids, to brittle solids. Polyethers with long alkyl side chains are waxy, however, as the alkyl group in such cases dominates physical properties.

The glass-transition temperatures of the amorphous straight-chain alkyl vinyl ether homopolymers decrease with increasing length of the side chain. Also, the melting points of the semicrystalline poly(alkyl vinyl ether)s increase with increasing side-chain branching.

Commercial Aspects

No crystalline polymers are known to have been commercialized. This lack of commercial success results from the economically competitive situation concerning vinyl ether polymers versus other, more readily available polymers such as those based on acrylic and vinyl ester monomers.

Copolymerization

VEs do not readily enter into copolymerization by simple cationic polymerization techniques; instead, they can be mixed randomly or in blocks with the aid of living polymerization methods. Reactivity ratios must be taken into account if random copolymers, instead of mixtures of homopolymers, are to be obtained by standard cationic polymerization. VEs can also copolymerize by free-radical initiation with a variety of comonomers.

MVE/MAN Copolymers. Various mol wt grades of poly(methyl vinyl ether-*co*-maleic anhydride) (PMVEMA) are available. PMVEMA, supplied as a white, fluffy powder, is soluble in ketones, esters, pyridine, lactams, and aldehydes, and insoluble in aliphatic, aromatic, or halogenated hydrocarbons, as well as in ethyl ether and nitroparaffins. When the copolymer dissolves in water or alcohols, the anhydride group is cleaved, forming the polymers in free acid form or the half-esters of the corresponding alcohol, respectively.

International Specialty Products (ISP) supplies ethyl, isopropyl, and *n*-butyl half-esters of PMVEMA as 50% solutions in ethanol or 2-propanol. These half-esters do not dissolve in water but are soluble in dilute aqueous alkali and in aqueous alcoholic amine solutions. The main application for the half-esters is in hairsprays where they combine excellent hair-holding properties at high humidity without making the hair stiff or harsh. These half-esters are easily removed during shampooing, have a very low order of toxicity, and form tack-free films that exhibit good gloss, luster, and sheen (see HAIR PREPARATIONS).

Health and Safety Factors

Poly(methyl vinyl ether-*co*-maleic anhydride) and their monoalkyl ester derivatives have been shown on rabbits to be neither primary irritants nor primary sensitizers to skin and eyes. The acute oral toxicities on white rats of the two copolymers are, respectively, 29 g/kg and 25 g/kg body weight.

Applications

Radiation-Curable Coatings. A wide variety of monovinyl and divinyl ethers are commercially available for this application, which allows the formulator greater latitude. For example, triethylene glycol divinyl ether (DVE-3) and 1,4-cyclohexanedimethanol divinyl ether (CHVE) can be combined as reactive diluents, with each contributing quite different properties to the subsequently cured coating. CHVE offers hard brittle films, whereas DVE-3 produces films that have greater flexibility.

Vinyl ethers can also be formulated with acrylic and unsaturated polyesters containing maleate or fumarate functionality. Because of their ability to form alternating copolymers by a free-radical polymerization mechanism, such formulations can be cured using free-radical photoinitiators. With acrylic monomers and oligomers, a hybrid approach has been taken using both simultaneous cationic and free-radical initiation.

Polymer–Polymer Compatibility. Frequently when polymers are mixed together they are immiscible because the combinatorial entropy of mixing is too small to overcome the enthalpy changes, which are usually positive. This small entropy of mixing is a result of the high mol wt nature of the component polymers. If the component polymers exhibit a specific interaction such as hydrogen bonding, Van der Waals, or electrostatic, etc, then miscibility can occur. In the case of PMVE–polystyrene, the blend presents a lower critical solution temperature (LCST) and the miscibility region depends on the molecular weight of the polymers. The interaction in this case is between the electrons of the ether groups and the aromatic polystyrene ring. In fact, PMVE can function as a diluent for isotactic polystyrene enhancing spherulite formation. Depending on the molecular weight and tacticity of the PMVE employed, separated regions of crystallized PMVE can function as reinforcement for polystyrene blends and offer improved plastic properties.

ROBERT B. LOGIN
Sybron Chemicals Inc.

N. D. Field and D. H. Lorenz, in E. C. Leonard, ed., *Vinyl and Diene Monomer,* Part I, John Wiley & Sons, Inc., 1970, p. 365.

T. Higashimura and M. Sawamoto, in G. Allen and J. Bevington, eds., *Comprehensive Polymer Science* Pergamon, Oxford, U.K., 1989, p. 673.

J. M. G. Cowie, *Alternating Copolymers,* Plenum Publishing Corp., New York, 1985.

J. A. Dougherty and F. J. Vara, *Proceedings of Radtech 88-North Americal Conference,* New Orleans, La., 1988.

N-VINYLAMIDE POLYMERS

N-Vinylamide-based polymers, especially the *N*-vinyllactams, such as poly(*N*-vinyl-2-pyrrolidinone) or simply polyvinylpyrrolidinone (PVP), continue to be of major importance to formulators of personal-care, pharmaceutical, agricultural, and industrial products because of desirable performance attributes and very low toxicity profiles. Because of hydrogen bonding of water to the amide group, many of the *N*-vinylamide homopolymers are water-soluble or dispersible. Like proteins, they contain repeating (but pendant) amide (lactam) linkages and share several protein-like characteristics. Many studies have actually employed PVP as a substitute for proteins, eg, in simplifying the chemistry of the effects of radiation on polymers. Proteins are extremely complicated molecules with not only sequence distribution but tertiary bonding and structural complexity and it is an oversimplification to compare them to PVP, but the effects of radiation on PVP can be more readily studied. PVP can even be considered as a uniform synthetic protein-like analogue. By itself it does not enter into intermolecular hydrogen bonding, thus affording low viscosity concentrates, and also, unlike the proteins, PVP is soluble in polar solvents like alcohol. But even given these differences, the chemistry of PVP, the most commercially successful polymer of the class, is in many respects similar to that of proteins because of amide linkages sharing with them complexation to large anions such as polyphenols, anionic dyes, and surfactants. In addition to the ability to complex, PVP and its analogues along with a large assortment of copolymers are excellent film-formers. They exhibit the ability to interact with a variety of surfaces by hydrogen or electrostatic bonding, resulting in protective coatings and adhesive applications of commercial significance such as hair-spray fixatives, tablet binders, disintegrants, iodophors, antidye redeposition agents in detergents, protective colloids, dispersants, and solubilizers, among many others.

Monomers

N-Vinylamides and *N*-vinylimides can be prepared by reaction of amides and imides with acetylene, by dehydration of hydroxyethyl derivatives, by pyrolysis of ethylidenebisamides, or by vinyl exchange, among other methods; the monomers are stable when properly stored. Only *N*-vinyl-2-pyrrolidinone (VP) is of significant commercial importance. Vinylcaprolactam is available and is growing in importance, and vinyl formamide is available as a developmental monomer.

Table 1. Properties of *N*-Vinyl-2-Pyrrolidinone (Commercial Production)

Property	Value
mol wt	111
assay, %	98.5 (min)
moisture content, %	0.2 (max)
color (APHA)	100 (max)
vapor pressure, Pa[a]	
at 17°C	6.7
45°C	67
64°C	266
77°C	667
bp at 400 mm Hg, °C	193
fp, °C	13.5
flash point (open cup), °C	98.4
fire point, °C	100.5
viscosity at 25°C, mPa·s(= cP)	2.07
sp gr (25/4°C)	1.04
refractive index, n_D^{25}	1.511
solubility	miscible in H_2O and most organic solvents
uv spectrum	no significant absorption at wavelengths >220 nm

[a] To convert Pa to mm Hg, multiply by 0.0075.

***N*-Vinyl-2-Pyrrolidinone.** Commonly called vinylpyrrolidinone or VP, *N*-vinyl-2-pyrrolidinone is a clear, colorless liquid that is miscible in all proportions with water and most organic solvents. It can polymerize slowly by itself but can be easily inhibited by small amounts of ammonia, sodium hydroxide (caustic pellets), or antioxidants such as *N,N'*-di-*sec*-butyl-*p*-phenylenediamine. It is stable in neutral or basic aqueous solution but readily hydrolyzed in the presence of acid to form 2-pyrrolidinone and acetaldehyde. Properties are given in Table 1.

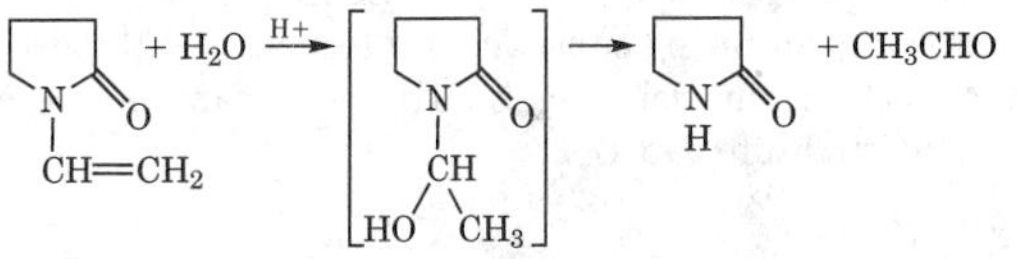

Commercially available VP is usually over 99% pure but does contain several methyl-substituted homologues and 2-pyrrolidinone. The vinylation of 2-pyrrolidinone is carried out under alkaline catalysis analogous to the vinylation of alcohols. 2-Pyrrolidinone is treated with ca 5% potassium hydroxide, then water and some pyrrolidinone are distilled at reduced pressure. A ca 1:1 mixture (by vol) of acetylene and nitrogen is heated at 150–160°C and ca 2 MPa (22 atm). Fresh 2-pyrrolidinone and catalyst are added continuously while product is withdrawn. Conversion is limited to ca 60% to avoid excessive formation of by-products. The *N*-vinyl-2-pyrrolidinone is distilled at 70–85°C at 670 Pa (5 mm Hg) and the yield is 70–80%.

One of the manufacturers, ISP, recommends that an appropriate workplace exposure limit be set at 0.1 ppm (vapor). In case of accidental eye contact, immediately flush with water for at least 15 minutes and seek medical attention.

Homopolymerization of *N*-Vinyl-2-Pyrrolidinone

Ammonia H_2O_2 Initiation. The lower molecular weight grades (K-15 and K-30) of PVP are prepared industrially with an ammonia/H_2O_2 initiation system. Such products are the standards for the pharmaceutical industry and conform to the various national pharmacopeias.

Organic Peroxides and Azo Initiation. The H_2O_2/ammonia initiation system is not employed commercially in the manufacture of higher molecular weight homologues; they are prepared with organic initiators. Such polymerizations follow simple chain theory and are usually performed in water commercially. The rate of polymerization is at a maximum in aqueous media at pH 8–10 and at 75 wt % monomer. Polymerization rates follow the polarity and hydrogen bonding capability of the solvent.

Cationic Polymerization. VP polymerizes to low molecular weight (oligomers) with typical cationic initiators, such as boron trifloride etherate. This reaction requires high concentrations, if not neat, of monomer and scrupulously anhydrous conditions for high yields; VP will readily hydrolyze to 2-pyrrolidinone and acetaldehyde even in the presence of trace moisture when catalyzed by strongly acidic reagents.

Proliferous Polymerization. Early attempts to polymerize VP anionically resulted in proliferous or "popcorn" polymerization. This was found to be a special form of free-radical addition polymerization, and not an example of anionic polymerization, as originally thought. VP contains a relatively acidic proton alpha to the pyrrolidinone carbonyl. In the presence of strong base such as sodium hydroxide, VP forms cross-linkers *in situ.* Both ethylidene vinyl pyrrolidinone (EVP) and ethylidene-bis-vinylpyrrolidinone (EBVP) are generated in about a 10:1 ratio, respectively. At the temperature required to generate these cross-linkers and when their concentration reaches some minimum level, usually a few percent, proliferous polymerization begins.

Crospovidones are produced commercially by two processes, ie, *in situ* generation of cross-linker or addition of divinylimidazoline, and they are indistinguishable by ir. Both types exhibit a T_g of 190–195°C, which is not that much above the 175°C of high molecular weight, soluble PVP. Proliferous polymers prepared with easily hydrolyzed cross-linker containing an imine linkage do not further swell even when the cross-links are hydrolyzed. The crospovidones are unusually high molecular weight, highly chain-entangled polymers having covalent cross-links that most likely retard the termination reaction during polymerization and are not entirely responsible for the resulting mechanical properties, such as swell ratio.

The crospovidones are easily compressed when anhydrous but readily regain their form upon exposure to moisture. This is an ideal situation for use in pharmaceutical tablet disintegration and they have found commercial application in this technology. PVP strongly interacts with polyphenols, the crospovidones can readily remove them from beer, preventing subsequent interaction with beer proteins and the resulting formation of haze. The resin can be recovered and regenerated with dilute caustic.

PVP Hydrogels

Cross-linked versions of water-soluble polymers swollen in aqueous media are broadly referred to as hydrogels and have a growing commercial utility in such applications as oxygen-permeable soft contact lenses (qv) and controlled-release pharmaceutical drug delivery devices. Cross-linked PVP and selected copolymers fit this definition and are of interest because of the following structure/performance characteristics:

Structure	Performance	Benefit
nonionic	compatibility with other ingredients	stable formulation
pyrrolidinone	low toxicity	nonirritating/ nonthrombogenic
	complexation–actives/ O_2	controlled release-transport
	high T_g	mechanical stability
	hydrolytic stability	storage-stable
ethylene backbone	nonbiodegradable, hydrolytic stability	resists biocontamination storage-stable
cross-links	swell volume/viscosity	mechanical stability/ diffusion control

Cross-linked PVP can be prepared by several routes other than proliferous polymerization PPVP (crospovidones). Although a hydrogel, the swell volume of this type of polymer cannot be controlled over a large increment because the granular particles cannot be formed into larger uniform assemblies. These limitations can be overcome by the polymerization of VP in the presence of a few percent of suitable cross-linker utilizing standard free-radical initiation. The solution to this problem is to balance the reactivity ratios of the cross-linker and other comonomers with those of VP to obtain uniform copolymerization and cross-linking.

Cross-linked PVP can also be obtained by cross-linking the preformed polymer chemically (with persulfates, hydrazine, or peroxides) or with actinic radiation. If the starting PVP homopolymer is too low in molecular weight or too dilute, cyclization or cleavage is preferred.

Poly(*N*-Vinyl-2-Pyrrolidinone)

Poly(*N*-vinyl-2-pyrrolidinone) (PVP) is undoubtedly the best-characterized and most widely studied *N*-vinyl polymer. It derives its commercial success from its biological compatibility, low toxicity, film-forming and adhesive characteristics, unusual complexing ability, relatively inert behavior toward salts and acids, and thermal and hydrolytic stability.

Poly(*N*-vinyl-2-pyrrolidinone) is described in the *U.S. Pharmacopeia* as consisting of linear *N*-vinyl-2-pyrrolidinone groups of varying degrees of polymerization. The molecular weights of PVP samples are determined by size exclusion chromatography (sec), osmometry, ultracentrifugation, light-scattering, and solution viscosity techniques. The most frequently employed method of determining and reporting the molecular weight of PVP samples utilizes the sec/low angle light scattering (lalls) technique. A frequently used and commonly recognized method of distinguishing between different molecular weight grades of PVP is the *K* value. The relative viscosity is obtained with an Ostwald-Fenske or Cannon-Fenske capillary viscometer, and the *K* value is derived from Fikentscher's equation.

$$\log \frac{\eta_{\mathrm{rel}}}{c} = \frac{75K_0^2}{1 + 1.5K_0c} + K_0 \tag{1}$$

where $K = 1000\ \mathrm{K}_0$, η_{rel} = relative viscosity, and c = concentration of the solution in g/100 mL. Solving directly for *K*, the Fikentscher equation is converted to:

$$K = [300c \log Z + (c + 1.5c \log Z)^2 + 1.5c \log Z - c]/(0.15c + 0.0003c^2)$$

where $Z = \eta_{\mathrm{rel}}$.

Specifications for Pharmaceutical grade is given in Table 2.

The T_g of PVP is sensitive to residual moisture and unreacted monomer. It is even sensitive to how the polymer was prepared, suggesting that MWD, branching, and cross-linking may play a part. Polymers presumably with the same molecular weight prepared by bulk polymerization exhibit lower T_gs compared to samples prepared

Table 2. Specifications of Pharmaceutical PVP Grades (Povidone)

Assay	Value (max)
K value	
10–15	85–115%[a]
16–90	90–107%[a]
moisture, %	5
pH[b]	3.0–7.0
residue on ignition, %	0.02
aldehydes, %[c]	0.02
N-vinyl-2-pyrrolidinone, %	0.20
lead, ppm	10
arsenic, ppm	1
nitrogen, %	11.5–12.8

[a] Of stated supplier's value.
[b] Of a 5% solution in distilled water.
[c] Calculated as acetaldehyde.

by aqueous solution polymerization, lending credence to an example of branching caused by chain-transfer to monomer.

Molecular weight also plays a significant role in T_g, which increases to a limiting value of 180°C for high purity samples above K-90 in molecular weight. The following equation applies:

$$T_g(°\text{C}) = 175 - \frac{9685}{K^2}$$

One of PVP's more outstanding attributes is its solubility in both water and a variety of organic solvents. PVP is soluble in alcohols, acids, ethyl lactate, chlorinated hydrocarbons, amines, glycols, lactams, and nitroparaffins. PVP is insoluble in hydrocarbons, ethers, ethyl acetate, *sec*-butyl-4-acetate, 2-butanone, acetone, cyclohexanone, and chlorobenzene.

Complexation

The combination of electrostatic interaction (induced dipole–dipole interaction) with an increase in entropy resulting from the discharge of bound water is fundamental to PVP's ability to complex with a variety of large anions.

Other factors that can stabilize such a forming complex are hydrophobic bonding by a variety of mechanisms (Van der Waals, Debye, ion–dipole, charge-transfer, etc). Such forces complement the stronger hydrogen-bonding and electrostatic interactions.

Approximately a minimum $\overline{M}_n$ of 1 to 5,000 is required before complexation is no longer dependent on molecular weight for small anions such as KI_3 and 1-anilinonaphthaline-8-sulfonate (ANS).

Equilibrium dialysis studies indicate around 10 repeat VP units (base moles) are required to form favorable complexes. This figure can rise to several hundred for methyl orange and other anions depending on structure.

Iodine Complexes. The small molecule/PVP complex between iodine and PVP is probably the best-known example and can be represented as follows:

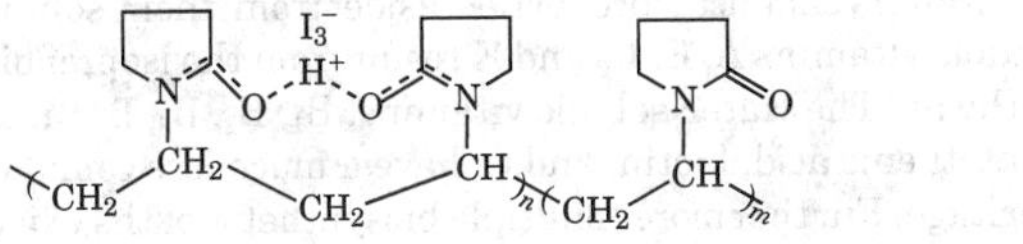

It is widely employed as a disinfectant in medicine (Povidone-iodine) because of its mildness, low toxicity, and water solubility. According to the *U.S. Pharmacopeia,* Povidone-iodine is a free-flowing, brown powder that contains from 9–12% available iodine. It is soluble in water and lower alcohols. When dissolved in water, the uncomplexed free iodine level is very low; however, the complexed iodine acts as a reservoir and by equilibrium replenishes the free iodine to the equilibrium level. This prevents free iodine from being deactivated because the free form is continually available at effective biocidal levels from this large reservoir. PVP will interact with other small anions and resembles serum albumin and other proteins in this regard. It can be "salted in" with anions such as NaSCN or "out" with Na_2SO_4 much like water-soluble proteins.

Phenolics. PVP readily complexes phenolics of all types to some degree, the actual extent depending on structural features such as number and orientation of hydroxyls and electron density of the associated aromatic system. A model has been proposed. Complexation with phenolics can result in reduced PVP viscosity and even polymer-complex precipitation.

One practical result of this strong interaction is the employment of PVP to remove unwanted phenolics such as bitter tanins from beer and wine.

Dyes. PVP is currently (ca 1997) employed in a variety of antidye redeposition detergents as a result of its strong interaction with fugitive anionic dyes. This interaction depends on the structure of the dye. Cationic dyes complex only if they also contain hydrogen-bonding functionality. Anionic dyes complex more easily, depending on the number of anionic groups, size of the aromatic nucleolus, and number and orientation of phenolic hydroxyl groups, etc.

Anionic Surfactants. PVP also interacts with anionic detergents, another class of large anions. The addition of PVP results in the formation of micelles at lower concentration than the critical micelle concentration (CMC) of the free surfactant the mechanism is described as a "necklace" of hemimicelles along the polymer chain, the hemimicelles being surrounded to some extent with PVP. The effective lowering of the CMC increases the surfactant's apparent activity at interfaces. PVP will increase foaming of anionic surfactants for this reason. Because of this interaction, PVP has found application in surfactant formulations.

Polymer/Polymer Complexes. PVP complexes with other polymers capable of interacting by hydrogen-bonding, ion-dipole, or dispersion forces. The interest in compatibility on a molecular level, an interesting phenomenon rarely found to exist between dissimilar polymers, is favored by the ability of PVP to form polymer/polymer complexes. Practical applications have been reported for PVP/cellulosics and PVP/polysulfones in membrane separation technology. Electrically conductive polymers of polyaniline are rendered more soluble and hence easier to process by complexation with PVP. Addition of small amounts of PVP to nylon 66 and 610 causes significant morphological changes, resulting in fewer but more regular spherulites.

Copolymerization

Copolymerizations can be conveniently carried out in aqueous solution or in a variety of solvents, depending on monomer/polymer solubilities. Various strategies have been employed to compensate for the divergence in reactivity ratios in order to form uniform (statistical) copolymers such as semibatch or mixed monomer feeds, the goal being to add the more reactive monomer at the rate at which it is being consumed. Clearly, if the difference in reactivity is too great, then the amount of more reactive monomer that can be uniformly incorporated is significantly reduced.

Table 3. Properties and Applications of Commercial PVPs

Polymer	Properties/applications
	Homopolymers
PVP	film former, adhesive, binder, complexant, stabilizer, crystallization inhibitor, dye scavenger, detoxicant, viscosity modifier
	Cross-linked
proliferous polymerization	pharmaceutical tablet disintegrant, adsorbent for polyphenols (tanins), beverage clarification
	Copolymers
PVP/VA	film-forming adhesives for hair preparations, bio-adhesives, water-remoistenable or removable adhesives
PVP/DMAEMA	mildly cationic, hair styling aids and conditioners with strong hold; substantive, lustrous film-formers
PVP/DMAEMA DES quaternary	strongly cationic, substantive, hair fixative ingredients
PVP/imidazolinum quaternary	
PVP/styrene[a]	opacifier for personal care products; stable styrene emulsion
PVP/alpha-olefins[a]	surface-active film formers; waterproofing of sunscreens
	Terpolymers
VP/VCl/DMAEMA	cationic water-soluble hair styling aid
VP/tBMA/MA	hair fixatives

[a] Graft copolymers.

Poly(Vinylpyrrolidinone-co-Vinyl Acetate). The first commercially successful class of VP copolymers, poly(vinylpyrrolidinone-co-vinyl acetate) is currently manufactured in sizeable quantities. A wide variety of compositions and molecular weights are available as powders or as solutions in ethanol, isopropanol, or water (if soluble).

An important reason for the ongoing interest in these copolymers is that vinyl acetate reduces hydrophilicity so that applications that require less moisture-sensitive films such as those employed to set hair are less prone to plasticize and become tacky under high humidity conditions.

Desirable fixative properties superior to PVP homopolymer can be specified by judicious selection of the amount of vinyl acetate. Hair sprays are limited in the molecular weight of the resin because if they are too high the resulting viscosity of the formulation will result in a poor (coarse) spray pattern. Increasing the VP/VA ratio causes properties to increase in the direction shown by the arrows. Other applications for VP/VA copolymers are uses as water-soluble or remoistenable hot melt adhesives, pharmaceutical tablet coatings, binders, and controlled-release substrates.

Tertiary Amine-Containing Copolymers. Copolymers based on DMAEMA (dimethylaminoethyl methacrylate) in either free amine form or quaternized with diethyl sulfate or methyl chloride have achieved commercial significance as fixatives in hair-styling formulations, especially in the well-publicized "mousses" or as hair-conditioning shampoo additives.

The most successful of these products contain high ratios of VP to DMAEMA and are partially quaternized with diethyl sulfate (Polyquaternium 11). They afford very hard, clear, lustrous, non-flaking films on the hair that are easily removed by shampooing. More recently, copolymers with methylvinylimidazolium chloride (Polyquaternium 16) or MAPTAC (methacrylamidopropyltrimethyl ammonium chloride) (Polyquaternium 28) have been introduced. Unquaternized DMAEMA copolymers afford resins that are mildly cationic and less hydroscopic.

Copolymers Containing Carboxylic Groups. A new line of VP/acrylic acid copolymers in powdered form prepared by precipitation polymerization from heptane have been introduced commercially. A wide variety of compositions and molecular weights are available, from 75/25 to 25/75 wt % VP/AA and from 20×10^3 to 250×10^3 molecular weights.

The copolymers are insoluble in water unless they are neutralized to some extent with base. They are soluble, however, in various ratios of alcohol and water, suggesting applications where delivery from hydroalcoholic solutions but subsequent insolubility in water is desired, such as in low volatile organic compound (VOC) hair-fixative formulations or tablet coatings. Unneutralized, their T_gs are higher than expected, indicating interchain hydrogen bonding.

Miscellaneous Copolymers. VP has been employed as a termonomer with various acrylic monomer–monomer combinations, especially to afford resins useful as hair fixatives.

Applications

An overview of the various product categories is given in Table 3.

ROBERT B. LOGIN
Sybron Chemicals Inc.

B. V. Robinson and co-workers, *PVP, A Critical Review of the Kinetics and Toxicology of Polyvinylpyrrolidone (Povidone)*, Lewis, Chelsea, Mich., 1990.

F. Haaf, A. Sanner, F. Straub, *Polym. J.* **17**(1), 143 (1985).

Y. E. Kirsh, *Prog. Polym. Sci.* **18**, 519–542 (1993).

P. Molyneuz and S. Vekavakaynondha, *J. Chem. Soc., Faraday Trans. 1* **82,** 291 (1986).

VINYL POLYMERS, POLY(VINYL FLUORIDE).

See FLUORINE COMPOUNDS, ORGANIC.

VINYLTOLUENE.

See STYRENE.

VIRAL INFECTIONS, CHEMOTHERAPY.

See ANTIVIRAL AGENTS.

VITAMINS

SURVEY

Vitamins are specific organic compounds that are essential for normal metabolism. These micronutrients are not synthesized by humans, either at all or in sufficient quantity, and must be obtained from the diet or as synthetic supplements. The principal vitamins are given in Table 1.

Structure

Common names, chemical structure, and synonyms for the vitamins are given in Table 1. The names given to the vitamins and/or their letter designations do not follow a logical pattern and are of historical significance only. Despite this, the nomenclature is in common use.

Because of their diverse structures, there are few common threads to vitamin chemical properties aside from their solubility. The fat-soluble vitamins A, E, D_3 and K result from the isoprenoid biosynthetic pathway. The water-soluble vitamins, B_1, B_2, B_6, B_{12}, niacin, folic acid, pantothenic acid, biotin, and C, have a much less common biosynthetic heritage. Furthermore, multiple biosynthetic paths exist for several of the vitamins.

The first commercial synthesis of a vitamin occurred in 1933 when the Reichstein approach was employed to manufacture vitamin C. Preparation of the vitamins in commercial quantities can involve isolation, chemical synthesis, fermentation, and mixed processes. The choice is process is economic.

Contrary to popular assumption, ongoing studies indicate that the majority of the U.S. population is not receiving the Recommended Daily Allowance (RDA) of vitamins through diet. Supplementary vitamins are thus needed.

Along with increasing evidence of health benefits from consumption of vitamins at levels much higher than RDA recommendations comes concern over potential toxicity. Very high doses of some water-soluble vitamins, especially niacin and vitamin B_6, are associated with adverse effects. In contrast, some fat-soluble micronutrients, especially vitamin E, are safe at doses many times higher than recommended levels of intake. Chronic intakes above the RDA for vitamins A and D especially are to be avoided, however.

Humans consume ca 40% of vitamins made worldwide. The majority of the vitamins, particularly in countries outside the United States, are used in animal husbandry.

Approximately $3,000,000,000 of vitamins were sold in 1996.

Table 1. Vitamin Names and Function[a]

Vitamin	Common name and synonyms	Major commercial forms	Function	RDA
A	retinol, axerophthol	vitamin A acetate, vitamin A propionate, vitamin A palmitate, vitamin A acid, β-carotene, β-apo-8′-carotenal	needed for normal vision, reproduction, and maintenance of healthy skin, mucous membranes, bones, red blood cells, cell differentiation, and immune function	800–1200 μg
D	vitamin D_3, cholecalciferol, antirachitic vitamin	cholecalciferol (vitamin D_3), ergocalciferol (vitamin D_2)	enhances calcium absorption, regulates calcium and phosphorus metabolism, promotes mineralization of bones, plays role in cell differentiation	10 μg
E	α-tocopherol, (*RRR*)-α-tocopherol, antisterility vitamin	(*RRR*)-α-tocopherol, (*RRR*)-α-tocopheryl acetate, all-*rac*-α-tocopherol, all-*rac*-α-tocopheryl acetate, all-*rac*-α-tocopheryl succinate	lipid-soluble antioxidant, prevents lipid oxidation of membranes, needed for healthy blood cells and tissues, blocks nitrosamine formation, protects PUFAs from autoxidation, important for normal immune function	10–12 mg
K	vitamin K_1, phylloquinome, coagulation vitamin, antihemorrhagic vitamin, phytonadione	2-(*R,S*)-vitamin K_1 (cis, trans-mixture), menadione sodium bisulfite, menadione dimethylpyrimidinol bisulfite	essential for normal blood clotting, needed for formation and maintenance of healthy bones, essential for formation of osteocalcin (bone protein)	65 μg
B_1	thiamine, aneurin	thiamine chloride, thiamine mononitrate, thiamine diphosphate, thiamine disulfide	helps convert carbohydrates from food into energy, required by muscle, nervous system, and brain	1.6 mg
B_2	riboflavin, vitamin G, lactoflavin, hepatoflavin, ovoflavin, verdoflavin	riboflavin, riboflavin sodium phosphate	involved in carbohydrate, protein and fat metabolism	1.8 mg
niacin	vitamin B_3, nicotinic acid	niacin, niacinamide	involved in carbohydrate, protein, and fat metabolism; at pharmacologic levels used to lower blood cholesterol	17–20 mg
B_6	pyridoxine, pyridoxol, antiacrodynia factor	pyridoxine hydrochloride, pyridoxal-5′-phosphate	active in protein metabolism, required for bone health, involved in homocysteine metabolism	2.2 mg
B_{12}	cobalamin, cyano-cobalamin	cyanocobalamin, hydroxocobalamin		2.2–2.6 μg
folic acid	folacin, pteroylglutamic acid, antianemia factor, fermentation *L. casei* factor	folic acid	involved in amino acid interconversions, needed for red blood cell formation and DNA and RNA synthesis, protects against neural tube birth defects, involved in homocysteine metabolism	400 μg
pantothenic acid	chick antidermatitis factor	calcium pantothenate, *d*-panthenol, *d,l*-panthenol	active in carbohydrate, protein, and fat metabolism	
biotin	*d*-biotin, vitamin H, anti-egg-white-injury factor, bios II, coenzyme R	*d*-biotin	involved in protein, fat, and carbohydrate metabolism	
C	ascorbic acid, antiscorbutic vitamin	ascorbic acid, ascorbyl palmitate, sodium ascorbate, calcium ascorbate, ascorbyl-2-polyphosphate	helps form and maintain collagen; important for healthy cartilage, bones, teeth, and gums; needed for wound healing; enhances nonheme iron absorption; water-soluble antioxidant protects cells from free-radical damage; blocks nitrosamine formation; increases body's resistance to infection	70–95 mg

[a] Since a general consensus is lacking that the following ought to be called vitamins, they are not included here: choline, myoinositol, vitamin F (essential fatty acids), α-lipoic acid, ubiquinones, plastoquinone, vitamin P (bioflavonoids), vitamin L (*o*-aminobenzoic acid), vitamin T (mixture including folic acid and vitamin B_{12}), vitamin U (L-methioninylmethylsulfonium chloride), orotic acid (vitamin B_{13}), and pyrroloquinoline quinone.

JOHN W. SCOTT
Hoffmann-La Roche Inc.

O. Isler and G. Brubacher, *Vitamine I: Fettlösliche Vitamine,* Georg Thieme Verlag, Stuttgart, Germany, 1982.

L. J. Machlin, ed., *Handbook of Vitamins,* 2nd ed., Marcel Dekker, New York, 1991.

S. K. Gaby, A. Bendich, V. N. Singh, and L. J. Machlin, *Vitamin Intake and Health: A Scientific Review,* Marcel Dekker, New York, 1991.

O. Isler, G. Brubacher, S. Ghisla, and B. Kräutler, *Vitamine II: Wasserlösliche Vitamine,* Georg Thiem Verlag, Stuttgart, Germany, 1988.

ASCORBIC ACID

Ascorbic acid (**1**) is the name recognized by the IUPAC-IUB Commission on Biochemical Nomenclature for Vitamin C. Other names are (L-ascorbic acid, L-xyloascorbic acid, and L-*threo*-hex-2-enoic acid γ-lactone). The name

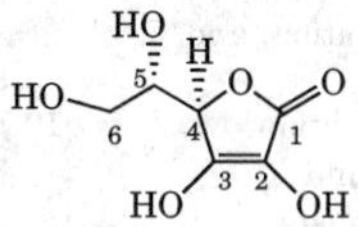

(**1**) L-Ascorbic acid

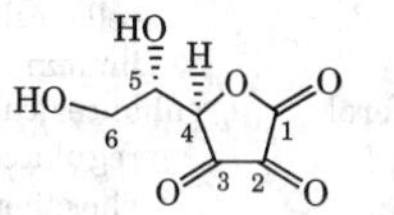

(**2**) Dehydro-L-ascorbic acid

implies the vitamin's antiscorbutic properties, eg, the prevention and treatment of scurvy, one of the oldest diseases known. L-Ascorbic acid is widely distributed in plants and animals. Some mammals (including humans), birds, and fish lack a liver enzyme (L-gulono-γ-lactone oxidase) necessary for synthesizing ascorbic acid and must ingest the vitamin. The pure vitamin, $C_6H_8O_6$, mol wt 176.13, is a water-soluble, strongly reducing, optically active (chiral) white crystalline substance that is synthesized both biologically and chemically from D-glucose. The vitamin and its main derivatives, sodium ascorbate, calcium ascorbate, and ascorbyl palmitate, are officially recognized the *United States Pharmacopeia/National Formulary* (*USP/NF*) and the *Food Chemicals Codex* (*FCC*).

L-Ascorbic acid's reversible oxidation to dehydro-L-ascorbic acid (**2**) (L-*threo*-2,3-hexodiulosonic acid γ-lactone), is the basis for its known physiological activities, stabilities, and technical applications.

Albert Szent-Györgyi first isolated vitamin C. Ultraviolet absorption studies, followed by x-ray crystallographic techniques elucidated its structure.

Properties

Table 1 contains a summary of the physical properties of L-ascorbic acid.

The most significant chemical property of L-ascorbic acid is its reversible oxidation to dehydro-L-ascorbic acid (**2**). Dehydro-L-ascorbic acid has been prepared by uv irradiation and by oxidation with air and charcoal, halogens, ferric chloride, hydrogen peroxide, 2,6-dichlorophenolindophenol, neutral potassium permanganate, selenium oxide, and many other compounds. It has been reduced to L-ascorbic acid by hydrogen iodide, hydrogen sulfide, 1,4-dithiothreitol and the like. Stable to air when dry, ascorbic acid in solution oxidizes in response to air, light, and elevated temperatures.

Degradation may result also from enzymes or various oxidation agents such as metals, L-ascorbic acid can be stabilized with oxalic acid, metaphosphoric acid, amino acids, 8-hydroxyquinoline, glycols, sugars, and trichloracetic acid.

Table 1. Physical Properties of L-Ascorbic Acid

Property	Characteristic(s)
appearance	white, odorless, crystalline solid with a sharp acidic taste
formula; mol. wt.	$C_6H_8O_6$; 176.13
crystalline form	monoclinic; usually plates, sometimes needles
mp °C	190–192 (dec)
optical rotation	$[\alpha]^{25}$ + 20.5° to +21.5°(c = 1 in water)
pK_1	4.17
pK_2	11.57
redox potential	first stage: E + 0.166 V (pH 4)
solubility, g/mL water	0.33
absolute ethanol	0.02
ether	insoluble (also insoluble in chloroform, benzene, oils, fat solvents)
	Spectral Properties
uv	pH 2: E_{max} (1%, 1 cm) 695 at 245 nm (nondissociated form)
ir (KBr)	characteristic wavelengths, cm^{-1} 3455, 3405, 3155 ν OH groups 2570 associated OH groups 1770, 1670 carbonyl lactone
nmr[a]	^1H nmr (D_2O) δ 4.97 (d, 1 H, $J_{4.5}$ = 2 Hz, H-4), 4.10 (ddd, 1 H, $J_{5.6}$ = 5 and 7 Hz, $J_{4.5}$ = 2 Hz, H-5), 3.78 (m, 2 H, C-6)

[a] d = Doublet; ddd = doublet of doublet of doublet; m = multiplet; and t = triplet.

Synthesis

Biosynthesis. In all plants and most animals, L-ascorbic acid is produced from D-glucose and D-galactose (which can be converted to D-glucose). Amphibians and reptiles carry the enzyme L-gulono-γ-lactone oxidase, which can transform a sugar-like glucose or galactose into ascorbic acid in the kidneys. In mammals and birds this enzyme system has been transferred from the kidneys to the liver. Humans, other primates, guinea pigs, fruit bats, and monkeys from the top of the evolutionary tree, as well as insects and invertebrates from the bottom end of the evolutionary tree, cannot synthesize L-ascorbic acid. Thus, they must consume vitamin C from exogenous sources to survive.

The biosynthesis of L-ascorbic acid is inhibited by deficiencies of certain vitamins, eg, vitamin A, vitamin E, and biotin, but is stimulated by certain drugs, eg, barbiturates, chlorobutanol, aminopyrine, and antipyrine, and by carcinogens, eg, 3-methylcholanthrene and 3,4-benzpyrene. It has been proposed that the excretion of D-glucaric acid can be used to diagnose both the exposure to the body for foreign substances and the drug metabolic capacity of the liver.

As in animals, L-ascorbic acid is also the product of hexose phosphate metabolism in plants, but plants have two biosynthetic pathways for the conversion of D-glucose or D-galactose to L-ascorbic acid. The main pathway is postulated as involving retention of configuration by oxidation at C-1 to yield D-gluconic-acid. The other biosynthetic pathway is similar to that in animals, in which the C- and C-6 of D-glucose become C-6 and C-1, respectively of L-ascorbic acid. Little is known about the functions of ascorbic acid in plants, except that it is involved in cellular respiration and may contribute to plant growth.

Chemical Synthesis. In 1933, the structure of ascorbic acid was confirmed by synthesis from L-xylose. The following year, ascorbic acid was synthesized from D-glucose, a pathway made possible by the chiral centers at C-2 and C-3 of D-glucose being in the correct configuration to become C-4 and C-5, respectively of L-ascorbic acid.

Most current industrial vitamin C production is based on the second synthesis developed by Reichstein and Grüssner.

Fermentation A mixed-culture fermentation, now practiced on an industrial scale produces an intermediate, which is then converted chemically to ascorbic acid. Because of the instability of ascorbic acid in water in the presence of oxygen, it seems highly unlikely that direct fermentation to ascorbic acid will be economically viable, although it is biologically possible.

Manufacture

The first vitamin to be manufactured by chemical synthesis on an industrial scale; L-ascorbic acid is produced in large, integrated, automated facilities, involving both continuous and batch operations.

Environmental issues.

The environmental concerns of an ascorbic acid manufacturing facility are those typical of a chemical processing plant. Its operating design must conform to environmental protection regulations. Measures must be taken to contain solvents and to keep emissions within official guidelines. Special condensers, continuous instrumental monitoring, and emergency containment and cleanup systems are required. Wastewater-treatment facilities have to be provided to remove by-product organics and inorganics from effluent streams before disposal.

Analytical Methods

Many different methods have been developed to determine L-ascorbic acid in feed, biological, and pharmaceutical samples, including uv absorption, redox and derivatization reactions, electrochemical and enzymatic oxidation reactions, chromatographic (eg, hplc methods), and biological methods with animals. The official assay is the iodimetric titration with 0.1 N iodine solution and starch as the indicator.

Uses

L-Ascorbic acid is used in pharmaceutical, food, feed, and beverage products, as well as in cosmetic applications.

Industrial uses of L-ascorbic acid relate to its antioxidant and reducing properties. It is used as an antioxidant in the commercial preparation of beer, fruit juices, cereals, and canned and frozen foods, etc.

Ascorbic acid is also used in agriculture as an abscission agent for fruit, in photography as a developing agent, in metallurgy as a reducing agent, in the polymer industry as a catalyst, in the manufacture of inks, in explosives, and in a variety of other applications.

Derivatives

Ascorbic acid has a variety of reactive positions that can be used to synthesize derivatives. Only L-ascorbic acid and its salts and C-6-substituted esters have full vitamin activity; sodium L-ascorbate, calcium L-ascorbate, and L-ascorbyl palmitate are commercially significant. L-Ascorbic acid 2-sulfate is bioavailable to fish. The activity of 6-chloro-6-deoxy-L-ascorbic acid compared to L-ascorbic acid is 4/5. Derivatives, eg, 6-deoxy-L-ascorbic acid, L-ascorbic acid 3-*O*-methylether, and 2-amino-2-deoxy-L-ascorbic acid have been prepared; their respective activities compared to L-ascorbic acid are 1/3, 1/25–1/50, and 0. The highest vitamin C activity correlates with the enediol lactone group, D-configuration for the C-4 hydrogen group, at least a two-carbon substituent on C-4, and L-configuration for the C-5 hydroxyl group. The primary C-6 hydroxyl group has minor impact on the biological activity.

Physiology and Biochemistry

Ascorbic acid has various biochemical functions, involving, for example, collagen synthesis, immune function, drug metabolism, folate metabolism, cholesterol catabolism, iron metabolism, and carnitine biosynthesis. In addition, ascorbic acid can act as a reducing agent and as an antioxidant. Ascorbic acid also interferes with nitrosamine formation by reacting directly with nitrites, and consequently may potentially reduce cancer risk.

Enzymatic Reactions. Vitamin C's role in collagen formation (which involves enzymatic hydroxylations of proline and of lysine) is of importance in wound healing. Daily intake of 8–50 times the RDA level of 60 mg ascorbic acid before and after surgery increases the rate of wound healing considerably. Many of the clinical signs of scurvy are attributed to defects in collagen synthesis.

Ascorbic acid's role in carnitine biosynthesis is important because carnitine—a component of heart muscle, skeletal tissue, and liver is involved in the transport of fatty acids into mitochondria, where they are oxidized to provide energy for the cell and animal.

Ascorbic acid has important biochemical functions with various hydroxylase enzymes in steroid, drug, and lipid metabolism. The cytochrome P-450 oxidase catalyzes the conversion of cholesterol to bile acids and the detoxification process of aromatic drugs and other xenobiotics, eg, carcinogens, pollutants, and pesticides, in the body. L-Ascorbic acid affects histamine metabolism related to scurvy and anaphylactic shock and is involved in the conversion of folate to tetrahydrofolate.

Antioxidant Activity. Ascorbic acid acts as an antioxidant by scavenging free radicals and forming the less reactive ascorbyl radical. The ascorbyl radical can be either reduced to ascorbic acid or oxidized to dehydroascorbic acid. Ascorbic acid protects membrane and other hydrophobic compartments by regenerating the antioxidant form of vitamin E. It thus protects the lungs by quenching free radicals generated by smoke, ozone, and singlet oxygen. The possibility that free-radical damage is also involved in the pathogenesis of HIV and that antioxidants may reduce infection is an area of intense interest.

Iron Absorption. Ascorbic acid enhances the reduction of ferric iron to ferrous iron, which is important both in increasing iron absorption and in its function in many hydroxylation reactions. In addition, ascorbic acid is involved in iron metabolism. Ascorbic acid in the presence of iron can exhibit either prooxidant or antioxidant effects, depending on the concentration used. The combination of citric acid and ascorbic acid may enhance the iron load in aging populations. Iron overload may be the most important common etiologic factor in the development of heart disease, cancer, diabetes, osteoporosis, arthritis, and possibly other disorders.

Absorption, Transport, and Excretion. The vitamin is absorbed into the bloodstream through the small intestine and is excreted in urine. Ascorbic acid is widely distributed to the cells of the body and is mainly present in the white blood cells (leukocytes). The adrenal and pituitary glands have the highest tissue concentration of ascorbic acid. The brain, liver, and spleen, however, represent the largest contribution to the body pool.

Mobilization and Metabolism. Approximately 3–4% of the body pool's ~1.5 g of ascorbic acid turns over daily, representing 40–60 mg/d of metabolized vitamin C. Smokers have a higher metabolic turnover rate of vitamin C (approximately 100 mg/d) and a lower body pool than nonsmokers. The metabolism of ascorbic acid varies among different species.

In rats and guinea pigs, respiratory carbon dioxide is the major oxidation product of vitamin C whereas in humans, urinary oxalic acid is the predominant metabolite. Excess vitamin is excreted largely unmetabolized.

Sources of Vitamin C

The vitamin C content of some representative foods is listed in Table 2. Potatoes and cabbage have traditionally been the most important sources of vitamin C for the majority of people in the Western World during the winter season. During storage after harvesting, the vitamin C content of fruit and vegetables will decrease depending on time and temperature of storage. Ascorbic acid readily oxidizes both enzymatically and chemically on exposure to oxygen. Losses

Table 2. Content of L-Ascorbic Acid in Representative Foods

Food substances	L-Ascorbic acid, mg/100 g
Meat, fish, and milk	
beef, pork, fish, cow's milk	≤2
liver, kidney	10–40
Vegetables	
asparagus, leek, onion, peas, beans, potatoes	10–30
brussel sprouts, broccoli	90–150
cabbage	30–60
cauliflower	60–80
kale	120–180
parsley	170
peppers	125–200
spinach	50–90
tomatoes	20–33
Fruit	
apples	10–30
guava	300
hawthorne berries	160–800
oranges, lemons	50
rose hips	1000
strawberries	40–90

are caused also through leaching during washing and blanching. Heat sterilization and freezing are good methods for preserving the vitamin C content of foodstuffs since they inactivate the enzyme (ascorbic acid oxidase) that destroys vitamin C.

Requirements. An extensive lack of knowledge exists about the biochemical and physiological functions of vitamin C. Although as little as 10 mg/d of ascorbic acid can prevent clinical scurvy, this intake is insufficient to maintain an adequate body pool of the vitamin for peak physical and mental health. The RDA for vitamin C in the United States is 60 mg, except for pregnant and lactating women (70–90 mg/day) and cigarette smokers (100 mg/day). The current RDA for vitamin C may not be adequate for elderly individuals. A recent study recommended that the current RDA be increased to 200 mg. Humans have ingested 10 g vitamin C daily without toxicity, although the value of doses above 400 mg daily has been questioned.

VOLKER KUELLMER
Hoffmann-La Roche Inc.

Ascorbic Acid: Chemistry, Metabolism, and Uses, P. A. Seib and B. M. Tolbert, eds., American Chemical Society, Washington, D.C., 1982.

O. Isler, G. Brubacher, S. Ghisla, and B. Kräutler, *Vitamin II,* Georg Thieme Verlag, Stuttgart, Germany, 1988, pp. 396–444.

J. N. Counsell and D. H. Hornig, eds., *Vitamin C (Ascorbic Acid),* Applied Science Publishers, London, 1981.

L. Packer and J. Fuchs, eds., *Vitamin C in Health and Disease,* J. Marcel Dekker Inc., New York, 1997.

BIOTIN

Biotin (vitamin H, vitamin B_8, bios IIB, and coenzyme R) (**1**) is a water-soluble B complex vitamin. A complex molecule having three stereocenters, biotin has eight stereoisomers, but only the naturally occurring one is active in metabolism. The richest sources of biotin are yeast, liver, kidney, egg yolks, pancreas, and milk. Plant materials (nuts, seeds, oats, bulgar wheat, pollen, molasses, rice, soybeans, mushrooms, fresh vegetables—eg, cauliflower and legumes—and some fruits) are also good sources. In addition, small amounts of biotin are found in most fish. Biotin-producing microorganisms exist in the large bowel but the extent and significance of this internal synthesis is unknown.

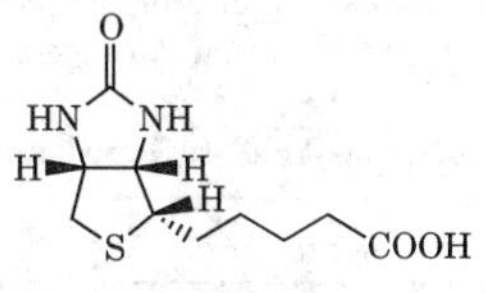

(**1**)

Biotin is listed GRAS for use as a nutrient or supplement. The biotin market is divided between agricultural and human use, with ~90% of biotin used in the animal health care market and ~10% for the human nutritional market.

Biochemical Function

Biotin is necessary for normal growth and body function. It acts as a cofactor for enzymes involved in carbon dioxide fixation and transfer. These reactions are important in the metabolism of carbohydrates, fats, and proteins, as well as formation of nicotinic acid, fatty acids, glycogen, and amino acids. Highest concentrations of biotin are found in the liver and kidneys.

Nutritional Requirements

Dietary biotin deficiencies are extremely rare, perhaps because intestinal microorganisms synthesize it and only minute amounts support body functions. Most susceptible to biotin deficiency are infants under six months of age and people who lack biotinidase, the enzyme catalyzing the cleavage of the bound form of biotin. Low circulation biotin levels have been observed also in smokers, alcoholics, and patients treated long-term with anticonvulsant drugs. Biotin deficiency can be induced by ingesting large amounts of raw egg white, because the avidin in egg white binds with the ureido group to form a complex that is resistant to digestive enzymes. The first clinical sign of biotin deficiency is an erythematous exfoliative dermatitis. Other symptoms include anorexia, fatigue, nausea, alopecia, abnormal sensitivity to stimulation, an inflamed tongue, pallor, and depression. Biotin-responsive disease conditions not caused by deficiency (eg, fatty liver and kidney syndrome) have been observed in animals.

Since an exact requirement for biotin is uncertain, no RDA exists. Adequate dietary intake for adults is 30–100 μg, according to the U.S. National Research Council. Therapeutic dose for deficiency is three times as much. Biotin supplementation reduces the incidence and severity of claw lesions in pigs and weak hoof horn in horses. No toxicity for biotin has been found.

Table 1. Physical Properties of *d*-Biotin

Property	Characteristic
appearance	fine long needles, white
molecular weight	244.31
molecular formula	$C_{10}H_{16}N_2O_3S$
melting point, °C	232–233
α^{21}, specific optical rotation, degrees	+89–91[a]
dissociation constant, pK_A	6.3×10^{-6}
isoelectric point, pH	3.5
solubility, mg/mL	
H_2O, RT	~0.22[b]
95% alcohol, RT	~0.80
common organic solvents	insoluble

[a] $c = 1$ in 0.1 *N* NaOH.
[b] Higher in dil alkali.

Chemical and Physical Properties

The physical properties of *d*-biotin are found in Table 1. Chemically pure biotin $C_{10}H_{16}N_2O_3S$, is stable to air and heat. Although biotin is not affected by reducing agents, it is incompatible with formaldehyde, chloramine T, oxidizing agents such as hydrogen peroxide and potassium permanganate, strong acids, and alkali above pH 9. In most foodstuffs, biotin is bound to proteins, from which it is released in the intestine by protein hydrolysis and the enzyme biotinidase. biotin is a highly stable, water-soluble vitamin that is resistant to most processing procedures, long-term storage, and normal cooking heat.

Chemical Synthesis

The first attempted synthesis of *d*-biotin in 1945 afforded racemic biotin. In this synthetic pathway, L-cysteine (**2**) was converted to the methyl ester (**3**). The nonselective catalytic hydrogenation and the epimerization that occurred during the conversion of (**3**) to (**4**) were the major factors leading to formation of racemic biotin.

Original Asymmetric Synthesis. The efficient introduction of the three stereocenters in the all-cis configuration was first accomplished by Sternbach in 1949. This process, the Hoffmann-La Roche industrial synthesis of biotin, is still the basis of industrial preparations.

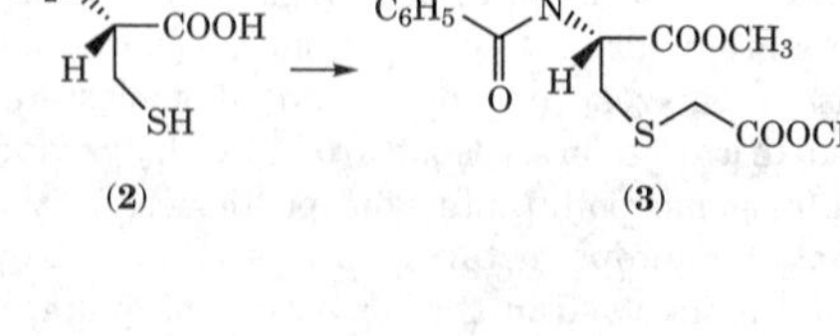

(**2**) (**3**)

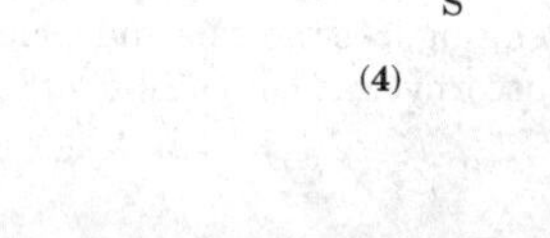

(**4**)

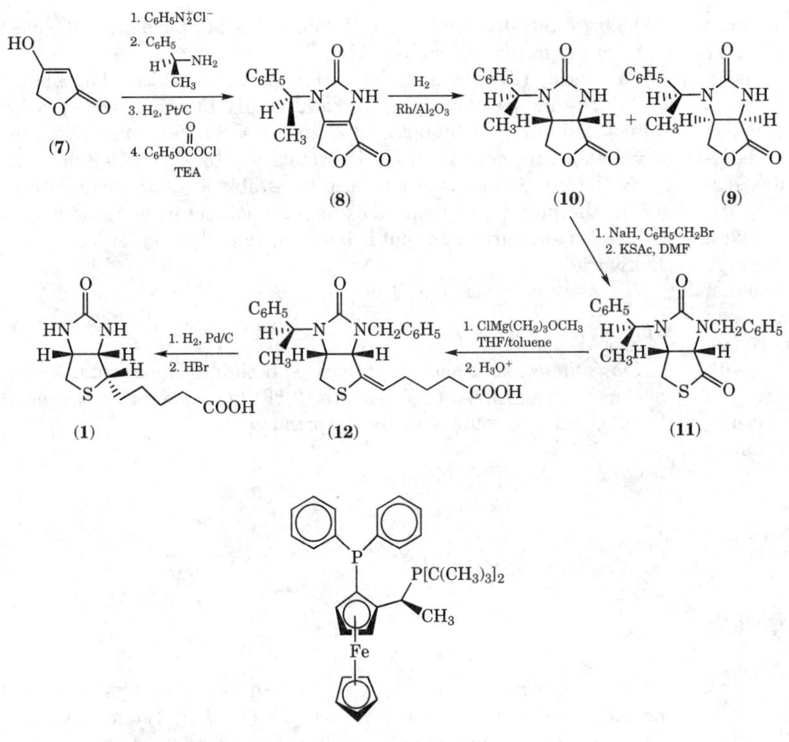

Figure 1. Synthetic pathway for *d*-biotin (Lonza synthesis). An improved process uses the chiral ferrocenyldisphosphine (**13**) to introduce stereospecificity during the hydrogenation of lactone (**8**).

The major drawbacks in the original Sternbach synthesis are the resolution/recycling of the intermediate that leads to *d*-lactone and the multiple manipulations required to add the five-carbon side chain. This sequence is inefficient, bringing with it a net loss of methanol, hydrogen bromide, carbon dioxide, and water that were once part of the molecule. In the resolution of the intermediate that leads to *d*-lactone, 50% of the product is the undesirable isomer, although this material is converted to the cycloacid for recycling. This is inherently inefficient and limits single-run production capacity by at least 50%. Recycling the undesired isomer also requires additional labor.

Industrial Synthetic Improvements. One significant modification of the Sternbach process is the result of work by Sumitomo chemists in 1975, in which the optical resolution—reduction sequence is replaced with a more efficient asymmetric conversion of the *meso*-cycloacid (**5**) to the optically pure *d*-lactone (**6**).

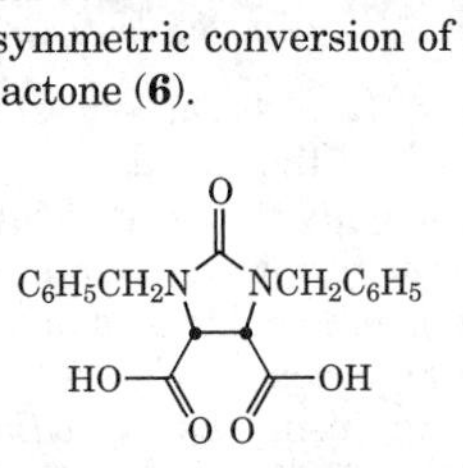

(**5**)

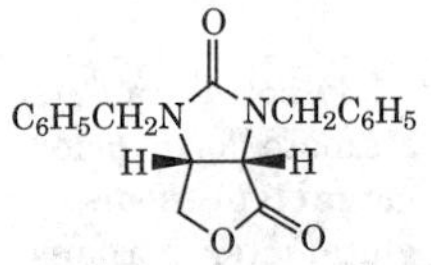

(**6**)

A final variation of the Sternbach method was reported by Lonza. This method (Fig. 1) not only involves direct attachment of the C-5 side chain to the lactone (**10**), it also introduces the key stereocenters early in the synthetic pathway via a chiral mono-protected imidazole intermediate (**8**). A recent modification of this Lonza process uses a chiral-substituted ferrocenyldisphosphine with a rhodium catalyst to stereoselectively reduce the enamine (**8**) to the *d*-lactone (**10**) in 99% yield. This improvement eliminates the need to separate the undesired *l*-lactone by chromatography or crystallization and increases the yield of the desired lactone from 54 to 99%.

Biosynthesis.

Biotin is produced by a multistep pathway in a variety of fungi, bacteria, and plants. Pimelic acid is believed to be the natural precursor of biotin for some microorganisms.

On the other hand, *E. coli* does not seem to rely on pimelic acid as a starting material for biotin synthesis. *E. coli* seems to form pimelyl-CoA by a pathway similar to that of fatty acid and polyketide synthesis. Desthiobiotin is converted to biotin, supposedly by the enzyme biotin synthetase, which is encoded by the Bio B gene.

To date, no commercial scale total synthesis of biotin by fermentation is known. However, several microorganisms and mutant strains have been evaluated. One of the problems associated with biotin biosynthesis via fermentation is that total biotin production decreases with increasing biotin concentrations in the fermentation broth. In fact, biotin strongly inhibits all steps of the biosynthesis except the synthesis of pimelyl-CoA.

Analytical Methods

Various analytical methods, including microbiological, biological, chemical, enzymatic and chromatographic assays, have been used to determine biotin levels in food, feed, and body fluids.

For the microbiological assay, capable of detecting as little as 0.05 ng of biotin/mL, *Lactobacillus plantarum* ATCC 8014 is the test organism usually employed. Since biotin occurs in both the free and bound forms in nature and *L. plantarum* ATCC 8014 responds only to free biotin, all available biotin must be converted to free biotin prior to the microbiological assay. Bound biotin is usually converted to free biotin by acid hydrolysis with sulfuric acid or by digestion with papain. Hydrochloric acid, should not be used in place of sulfuric acid for the acid hydrolysis because it may inactivate biotin. Lipids stimulating the growth of *L. plantarum* interfere with the assay and have to be removed. After the cell growth stops, the biotin content of the test material is determined by measuring the growth response of the organism (eg, colorimetrically, spectrophotometrically, or titrimetrically with sodium hydroxide) as compared to cell growth with known biotin concentration standards.

An isotope dilution assay for biotin, based on the high affinity of avidin for the ureido group of biotin, compares the binding of radioactive biotin and nonradioactive biotin with avidin. This method is sensitive to a level of 1–10 ng biotin.

Biotin has been analyzed in B-complex tablets, vitamin premixes, and multivitamin–multimineral preparations by reverse-phase, high performance liquid chromatographic methods (hplc) using a C^{18} column and uv detection at either 230 nm or 200 nm. This method is not sensitive enough to determine typical biotin levels in food or feed. A reverse-phase ion-interaction reagent hplc method has greater sensitivity, with a biotin detection limit of 4 μg. An hplc method with a sensitivity of 5 ng for fluorescent derivatives involves deriving biotin prior to chromatographic separation. From a sensitivity standpoint (approximately 0.3 μg), a comparable technique is gas chromatography.

ROBERT A. OUTTEN
Hoffmann-La Roche Inc.

O. Isler, G. Bracher, S. Ghisla, and B. Kräutler, *Vitamine II. Wasserlösliche Vitamine,* Georg Thieme Verlag, Stuttgart, Germany, 1988, p. 231.

J. P. Bonjour, in L. Machlin, ed., *Handbook of Vitamins,* 2nd ed., Marcel Dekker Inc., New York, 1991, p. 393.

FOLIC ACID

Folic acid (**1**) belongs to the group of B-vitamins. The term folate is used to designate all members of the family of compounds based on the *N*-[(6-pteridinyl)methyl]-*p*-aminobenzoic acid skeleton conjugated with one or more L-glutamic acid units. Folic acid is *N*-[4-[(2-amino-1,4-dihydro-4-oxo-6-pteridinyl)methylamino]benzoyl]-L-glutamic acid. The metabolically active forms of folic acid have a reduced pteridine ring and several glutamic acids residues.

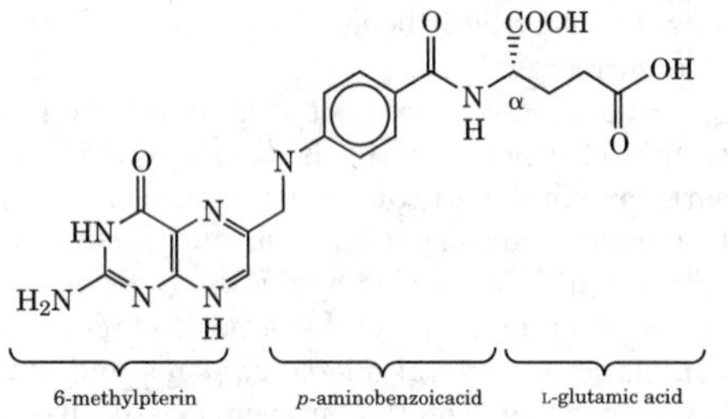

(1)

Sources and Bioavailability

Good food sources of folate are liver; fresh, dark green, leafy vegetables; beans; wheat germ; and yeasts. Folic acid is synthesized only by microorganisms and plants. Most dietary folates exist in the polyglutamate form, which are converted to the more readily bioavailable monoglutamate form in the small intestine by the jejunal brush border folate conjugase. Certain foods such as cabbage and legumes contain conjugase inhibitors, which can decrease folate absorption.

Different forms of folates occur in nature and the stability or bioavailability of each form varies. Most folates in food are susceptible to oxidation under aerobic conditions during storage and processing. Folic acid (commercial form) has superior bioavailability because it is more readily absorbed than the tri- or heptaconjugates. Deficiencies of iron, vitamin C, and zinc impairs utilization of dietary folate.

Chemical and Physical Properties

Enzymatic reduction of folic acid leads to the 7,8-dihydrofolic acid (H_2 folate), a key substance in biosynthesis. Further reduction provides (6*S*)-5,6,7,8-tetrahydrofolic acid (H_4 folate), the key biological intermediate for the formation of other folates.

Folic acid is found as yellow, thin platelets which char above 250°C. The uv spectrum of L-folic acid at pH 13 shows absorptions at $\lambda = 256$ nm ($\epsilon = 30{,}000$), 282 nm ($\epsilon = 26{,}000$), and 365 nm ($\epsilon = 9800$). Folic acid has a specific rotation of $[\alpha]_D^{27} = +19.9°$ ($c = 1$, 0.1 *N* NaOH). Solutions of folic acid are stable at room temperature and in the absence of light. It is slightly soluble in aqueous alkali hydroxides and carbonates but is insoluble in cold water, acetone, and chloroform.

Synthesis

Biosynthesis. Folic acid is synthesized both in microorganisms and in plants. Guanosine-5-triphosphate (GTP) (**2**), *p*-aminobenzoic acid (PABA), and L-glutamic acid are the precursors.

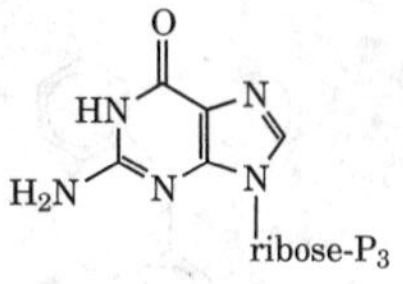

(2)

Chemical Synthesis. The first L-folic acid synthesis was based on the concept of a three-component, one-pot reaction. Triamino-4(3*H*)-pyrimidinone (**3**) was reacted simultaneously with C_3-dibromo aldehyde (**4**) and *p*-amino-benzoyl-L-glutamic acid (**5**) to yield folic acid (**1**).

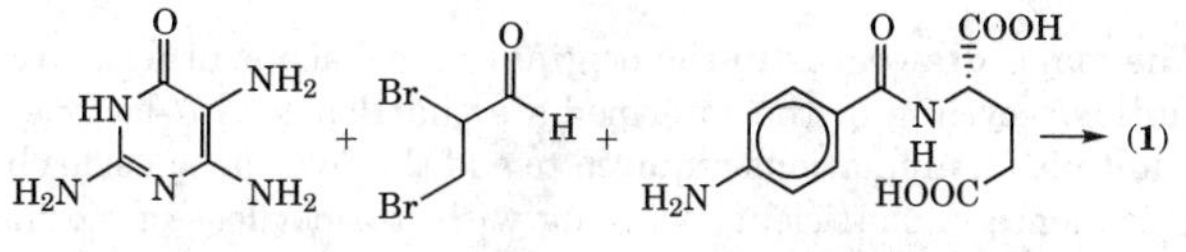

(3) **(4)** **(5)**

All known commercial syntheses are based on this approach. Alternatively, L-folic acid has been prepared in two steps by condensing 6-bromomethylpterin with *p*-aminobenzoyl-L-glutamic acid in 80% yield. However, the starting material is not easily synthesized. Another viable method is via Schiff base formation, but is economically feasible only if 6-formylpterin is readily available.

Analytical Methods

Analysis of folic acid is difficult because most natural folates exist in the polyglutamate form and the oxidation state of the single-carbon substituent varies. To simplify the assay, folate in the monoglutamyl or diglutamyl forms is determined after enzymatic deconjugation. Methods for determining folic acid in food and feed include biological, microbiological, chemical, chromatographic, and radiometric assays. The microbiological assay using *Lactobacillus casei* is the official method of the Association of Official Analytical Chemists (AOAC).

Biochemistry

Metabolism. Dietary folylpolyglutamate is enzymatically hydrolyzed to the monoglutamate form prior to transport across the intestinal mucosa. A single protein seems to be responsible for transportation of the monoglutamate, perhaps by an anion-exchange mechanism. The metabolism of folic acid occurs in the liver and mitochondria and involves reduction of the pterin ring to different forms of tetrahydrofolylglutamate. The metabolic roles of the folate coenzymes are to serve as acceptors or donors of one-carbon units in a variety of reactions. These one-carbon units exist in different oxidation states and include methanol, formaldehyde, and formate. The resulting tetrahydrofolylglutamate is an enzyme cofactor in amino acid metabolism and in the biosynthesis of purine and pyrimidines.

Folic acid is a precursor of several important enzyme cofactors required for the synthesis of nucleic acids and the metabolism of certain amino acids. Folic acid deficiency results in an inability to produce DNA, RNA, and certain proteins.

Requirements. The RDA for folic acid is 400 μg for adults and 800 μg for pregnant women.

Deficiency. Megaloblastic anemia is a common symptom of both folate and vitamin B_{12} deficiency. One of the clinical signs of acute folate deficiency includes a red and painful tongue. One study indicates that the substantial consumption of alcohol, when combined with an inadequate intake of folate and methionine, may increase the risk of colon cancer.

Incomplete closure of neural tube during the embryo development in humans can lead to spina bifida. The condition is characterized by an opening in the spinal cord and results in physical disability in a child. Incomplete closure of the skull produces anencephaly. These and similar conditions are collectively called neutral tube defects (NTD). It has been shown that folic acid given at 400 μg/day prevents the recurrence of NTD and that doses of 800 μg/day prevent both the occurrence and recurrence of NTD in the majority of cases. Folic acid supplementation taken six weeks before conception may reduce the risk of neural tube defects by at least 50%.

Incidence of cleft lip/cleft palate has been reported in animal studies due to folic acid deficiency. Poor folic acid status has been associated with megaloblastic changes in the cells of the uterine, cervix, and intestinal epithelium. For optimal performance of farm animals, folic acid supplementation is required.

Homocysteine arises from dietary methionine. High levels of homocysteine (hyperhomocysteinemia) are a risk factor for occlusive vascular diseases including atherosclerosis and thrombosis. Clinically, homocysteine levels can be lowered by administration of vitamin B_6, vitamin B_B, and folic acid.

Toxicity. Folic acid is safe, even at levels of daily oral supplementation up to 5–10 mg. A high intake of folic acid can mask the clinical signs of pernicious anemia and recurrence of epilepsy in patients using drugs with antifolate activity. LD_{50} is approximately 500–600 mg per kg body weight for rats and mice, respectively.

Uses

L-Folic acid is available as a crystalline dihydrate containing 8% water. Approximately 80% of the commercial production is consumed for feed enrichment in animal nutrition. Folic acid is being offered by the pharmaceutical industry for therapeutic and prophylactic use (see PHARMACEUTICALS). Pharmacological doses of folic acid are commonly used as a rescue dose during cancer chemotherapy, in women using oral contraceptives, and alcoholics. Several studies have provided evidence that multivitamins or folic acid (0.8–4 mg/day) supplementation prevent the majority of neural tube defects.

CONCLUSION

All known commercial syntheses of folic acid are based on the three-component process developed in the late 1940s. Industrial production of folic acid by genetically engineered microorganisms or extraction from natural sources is not yet economically viable. The mechanism governing folate turnover and excretion is still poorly understood. Folic acid is associated with a decreased occurrence of both neural tube defects and cardiovascular disease, possibly by reducing homocysteine levels. Folic acid analogues are used in antibody production, chemotherapy, and treatment of malaria and protozoal diseases.

THIMMA R. RAWALPALLY
Hoffmann-La Roche Inc.

O. Isler and G. Brubacher, in O. Isler, G. Brubacher, S. Ghisla, and B. Krätler, eds., *Vitamine II,* Georg Thieme Verlag, New York, 1982, p. 264.

S. K. Gaby and A. Bendich, in S. K. Gaby, A. Bendich, V. N. Singh, and L. J. Machlin, eds., *Vitamin Intake and Health,* Marcel Dekker Inc., New York, 1991, p. 175.

T. Brody, in L. J. Machlin, ed., *Handbook of Vitamins,* Marcel Dekker Inc., New York, 1991, p. 453.

J. F. Gregory, in L. B. Bailey, ed., *Folate in Health and Disease,* Marcel Dekker Inc., 1995, p. 195.

NIACIN, NICOTINAMIDE, AND NICOTINIC ACID

3-Pyridine carboxamide is also known as nicotinamide (**1**) or vitamin B_3. 3-Pyridine carboxylic acid is commonly called nicotinic acid (**2**), which is also known as niacin or vitamin PP (pellagra preventing).

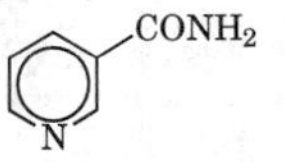

(**1**)

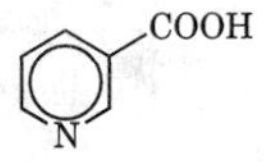

(**2**)

The biological importance of these compounds stems from their use as cofactors. Both nicotinamide and nicotinic acid are building blocks for coenzyme I (Co I), nicotinamide–adenine dinucleotide (NAD) and coenzyme II (Co II), nicotinamide–adenine dinucleotide phosphate (NADP).

Chemical and Physical Properties

Physical properties are listed in Tables 1 and 2. Nicotinamide is a colorless, crystalline solid that is very soluble in water and in 95% ethanol. The compound is soluble in butanol, amyl alcohol, ethylene glycol, acetone, and chloroform, but is only slightly soluble in ether or benzene.

Nicotinic acid is an amphoteric solid with needle-shape crystals. It is less soluble than nicotinamide and its poor solubility in diethyl ether can be used as a basis to separate these compounds.

The ring nitrogen of both the amide and the carboxylic acid can be quaternized and oxidized. Acid chlorides, esters, amides, and anhydrides have been prepared from nicotinic acid. Both the corresponding aldehyde and alcohol are available from the acid with a variety of reducing agents. Nicotinamide can be converted to nicotinic acid esters, the nitrile and acylamidines by routine methods.

Synthesis

Biosynthesis. A significant pathway for the formation of nicotinic acid mononucleotides begins with tryptophan. This first step

Table 1. Physical Properties of Nicotinamide

Property	Value
molecular weight	122.12
melting point, °C	
stable modification	129–132
boiling point, °C (0.067 Pa)	150–160
sublimation range, °C	80–100
true dissociation constants in water, at 20°C	
K_{b1}	2.24×10^{-11}
K_{b2}	3.16×10^{-14}
specific heat, kJ/(kg·K)[a]	
solid, 55°C	1.30
65°C	1.34
75°C	1.39
liquid, 135°C	2.18
heat of solution in water, kJ/kg[a]	−148
heat of fusion, kJ/kg[a]	381
density of melt, at 150°C, g/cm^3	1.19

[a] To convert J to cal, multiply by 4.184.

Table 2. Physical Properties of Nicotinic Acid

Property	Value
molecular weight	123.11
melting point, °C	236–237
sublimation range	≥150
density of crystals, g/cm^3	1.473
dissociation constants in water, at 25°C	
K_a	1.50×10^{-5}
K_b	1.04×10^{-12}
isoelectric point in water, at 25°C, pH	3.42
pH of saturated aqueous solution	2.7
solubility, g/L	
water	
38°C	24.7
100°C	97.6
ethanol, 96%, 78°C	76.0
methanol, 62°C	345.0

is rate determining unless it is hormonally induced by glucocortocoids and even tryptophan itself. The resulting *N'*-formylkynurenine is deformylated to the free aniline derivative, kynurenine. Hydroxylation to yield is followed by oxidative removal of alanine to form 3-hydroxyanthranilic acid. Oxidative cleavage of the aromatic ring to the semialdehyde is followed by dehydration to the important metabolite, quinolinic acid. Quinolinic acid is decarboxylated with concomitant alkylation at nitrogen to yield nicotinic acid mononucleotide.

The result of this biosynthesis is that the product is nicotinic acid mononucleotide rather than free nicotinic acid. Ingested nicotinic acid is converted to nicotinic acid mononucleotide which, in turn, is converted to nicotinic acid adenine dinucleotide. Nicotinic acid adenine dinucleotide is then converted to nicotinamide adenine dinucleotide. If excess nicotinic acid is ingested, it is metabolized into a series of detoxification products. Physiological metabolites include *N*-methylnicotinamide and *N*-methyl-6-pyridone-2-carboxamide.

Nicotinamide is incorporated into NAD and nicotinamide is the primary circulating form of the vitamin. NAD has two degradative routes: by pyrophosphatase to form AMP and nicotinamide mononucleotide and by hydrolysis to yield nicotinamide adenosine diphosphate ribose.

Chemical Synthesis. Key intermediates in the industrial preparation of both nicotinamide and nicotinic acid are alkyl pyridines such as 2-methyl-5-ethylpyridine.

In the case of nicotinic acid, the transformations can occur by either chemical or biological means. From an industrial standpoint, the majority of nicotinic acid is produced by the nitric acid oxidation of 2-methyl-5-ethylpyridine. Isocinchomeronic acid (**4**) (Fig. 1) is formed as an intermediate.

Although an inherently more efficient process, the direct chemical oxidation of 3-methylpyridine does not have the same commercial significance. Liquid-phase oxidation procedures are typically used. A Japanese patent describes a procedure that uses no solvent and avoids the use of acetic acid. In this procedure, 3-methylpyridine is combined with cobalt acetate, manganese acetate and aqueous hydrobromic acid in an autoclave. The mixture is pressurized to 101.3 kPa (100 atm) with air and allowed to react at 210°C. At a 32% conversion of the picoline, 19% of the acid was obtained. Electrochemical methods have also been described.

Fermentation. Several groups have reported on fermentative approaches to nicotinic acids from 3-picoline. *Rhodococcus, Acinetobactorr,* and *Pseudomonas* have found utility in this application.

Nicotinonitrile is produced by ammoxidation of alkylpyridines. A wide variety of different catalysts have been developed for this application.

Conversion of the nitrile to the amide has been achieved by both chemical and biological means. Several patents have described the use of modified Raney nickel catalysts in this application. Alkali metal perborates have demonstrated their utility. Typically, the hydrolysis is conducted in the presence of sodium hydroxide. Other catalysts such as magnesium oxide, ammonia, and manganese dioxide have also been employed.

Organisms such as *Achromobacter, Agrobacterium, Streptomyces Rhodococcus,* and *Cornebacterium* have been used to prepare the amide from the nitrile. Purified enzymes in either free or immobilized form have also been used in this application.

Nicotinic acid has also been produced by microbial (usually *Rhodococcus*) means from the nitrile. Irradiation of a *Corynebacterium* suspension during the fermentation led to higher yields of nicotinic acid.

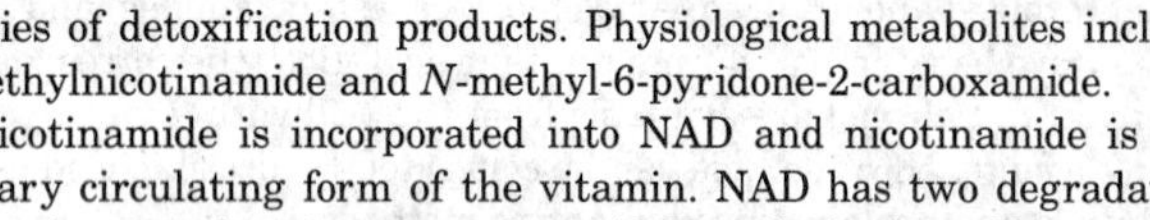

(**3**) (**4**) (**2**)

Figure 1. Formation of isocinchomeronic acid (**4**).

Analytical Methods

Both nicotinic acid and nicotinamide have been assayed by chemical and biological methods. In the König reaction quaternization of the pyridine nucleus by cyanogen bromide is followed by ring opening to generate the putative dialdehyde intermediate. Reaction of this compound with an appropriate base (eg, *p*-methylaminophenol sulfate or sulfanilic acid) generates a dye whose concentration is determined colorimetrically. Because in the case of nicotinamide, the color yield is often low, the amide is either hydrolyzed to nicotinic acid or converted to a fluorescent compound.

Using reverse-phase columns and uv detection, hplc methods have been applied to the analysis of nicotinic acid and nicotinamide in biological fluids such as blood and urine and in foods such as coffee and meat. Derivatization techniques have also been employed to improve sensitivity.

As with many of the vitamins, biological assays (eg, using microbes or chicks) have an important historical role and are widely used. Selective detection of nictonic acid is possible if *Leuconostoc mesenteroides* ATCC No. 9135 is used as the test organism.

Sources and Bioavailability

Nicotinamide and nicotinic acid occur in nature almost exclusively in the bound form. In plants, nicotinic acid is prevalent (associated with peptides and polysaccharides), whereas in animals nicotinamide is the predominant form. This nicotinamide is exclusively in the form of NAD and NADP.

From a bioavailability standpoint, the fact that a significant amount of nicotinic acid is in a bound form has important biological consequences since the ester linkage is resistance to digestive enzymes. Pretreatment with alkali is helpful, as in the preparation of tortillas from corn.

Nicotinamide and nicotinic acid are prevalent in many common foodstuffs and are especially concentrated in brewer's yeast, wheat germ and liver. In this regard, tryptophan is considered a provitamin and is assigned a niacin equivalent of 1/60. Coffee, a fertile source of nicotinic acid, contains 1–2 mg per average cup. The amount of nicotinic acid depends on the roasting conditions, during which trigonelline is demethylated to nictonic acid.

Biochemistry

NAD and NADP are required as redox coenzymes by a large number of enzymes and in particular dehydrogenases. NAD$^+$ is utilized in the catabolic oxidations of carbohydrates, proteins, and fats, whereas $NADPH_2$ is the coenzyme for anabolic reactions and is used in fats and steroid biosynthesis. $NADP^+$ is also used in the catabolism of carbohydrates.

Deficiency. A deficiency of niacin manifests itself in the disease pellagra. The symptoms include dermatosis (expressed as lesions) dementia, and diarrhea. A deficiency of niacin also affects the nervous system. Symptoms progress from paralysis to a spastic gait and severe thought disorder.

Requirements. The RDA for niacin (20 mg daily for adults) is based on the concept that niacin coenzymes participate in respiratory enzyme function and 6.6 niacin equivalents (NE) are needed per intake of 239 kJ (1000 kcal). One NE is equivalent to 1 mg of niacin.

Safety

Despite structural similarities, the pharmacological consequences of excesses of these substances are quite different. Due to the interest in the effects of nicotinic acid on atherosclerosis, and in particular its use based on its ability to lower serum cholesterol, the toxicity of large doses of nicotinic acid has been evaluated. For example, in a study designed to assess its ability to lower serum cholesterol, only 28% of the patients remained in the study after receiving a large initial dose of 4 g of nicotinic acid due to intolerance at these large doses.

Nicotinamide can also be toxic to cells at concentrations that increase the NAD levels above normal. At levels of 3 g/d for extended periods (3–36 months) individuals have experienced various side effects such as heartburn, nausea, headaches, hives, fatigue, sore throat, dry hair, and tautness of the face.

Uses

Both nicotinic acid and nicotinamide have been used in the enrichment of bread, flour, and other grain-derived products. Animal feed is routinely supplemented with nicotinic acid and nicotinamide. Nicotinamide is also used in multivitamin preparations. Nicotinic acid is rarely used in this application. The amide and carboxylic acid have been used as a brightener in electroplating baths and as stabilizer for pigmentation in cured meats.

Nicotinic acid and its derivatives have widespread use as antihyperlipidemic agents and peripheral vasodilators.

Derivatives

Nicotinyl alcohol (3-pyridinylcarbinol, 3-pyridinemethanol) has use as an antilipemic and peripheral vasodilator. It is available from either the reductions of nicotinic acid esters or, preferably, the reduction of the nitrile to the amine followed by diazotation and nucleophilic displacement.

SUSAN D. VAN ARNUM
Hoffmann-La Roche Inc.

O. Isler, G. Brubacher, S. Ghisla, and B. Kräutler, "Die Niacin-Gruppe (Vitamin PP)" in *Vitamine II,* Georg Thieme Verlag Stuttgart, New York, 1988, pp. 160–192.

J. van Eys, in L. Machlin, ed., *Handbook of Vitamins,* Marcel Dekker, New York, 1991, pp. 311–340.

PANTOTHENIC ACID

(*R*)-Pantothenic acid (**1**) is a member of the B-complex vitamins. Also known as vitamin B_5, it is a water-soluble vitamin.

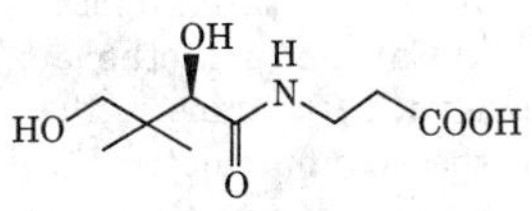

(**1**)

Pantothenic acid is one of the components of coenzyme A, an acyl group carrier that is a cofactor for various enzymatic reactions and serves as either a hydrogen donor or acceptor. Pantothenic acid is also a structural component of acyl carrier protein (ACP), which is required for fatty acid synthesis.

Sources

Pantothenic acid is synthesized naturally only by plants and microorganisms. Good food sources of pantothenic acid include yeast, chicken, beef, potatoes, vegetables, legumes, and whole grain. Estimation of the dietary intake of pantothenic acid is difficult because the acid exists in the free form and is also incorporated in coenzyme A and fatty acid synthetase.

Only approximately 50% of pantothenic acid present in the diet is actually absorbed.

Physical and Chemical Properties

Table 1 lists some physical properties of pantothenic acid and its derivatives. Free pantothenic acid is isolated from liver, and is a pale yellow, viscous, and hygroscopic oil. (*R*)-Pantothenic acid contains two subunits, (*R*)-pantoic acid and β-alanine. The chemical abstract name is *N*-(2,4-dihydroxy-3,3-dimethyl-1-oxobutyl)-β-alanine. Only (*R*)-pantothenic acid is biologically active. Pantothenic acid is unstable under alkaline or acidic conditions. It is extremely hygroscopic and the sodium salt is unstable. The major commercial source of this vitamin is thus the stable calcium salt (calcium pantothenate).

Panthenol is the reduced form of pantothenic acid and is the pure form most commonly used. The alcohol is more easily absorbed and is converted into the acid *in vivo*.

The biologically active *R*- or *d*-pantothenic acid can be obtained upon hydrolysis of coenzyme A with alkaline phosphatase and pantotheinase. The phosphatase catalyzes the selective cleavage of the phosphate bond in coenzyme A to afford adenosin-3′5′-diphosphate and 4-phosphopantetheine. The latter substance is dephosphorylated enzymatically to yield pantetheine, which is rapidly converted by pantotheinase to pantothenic acid.

Synthesis

Biosynthesis. The metabolically active form of pantothenic acid is coenzyme A. Coenzyme A is produced only by microorganisms, which enzymatically couple (*R*)-pantoate with β-alanine.

Table 1. Physical Properties of Pantothenic Acid and Derivatives

Compound	α_D^{20a}	bp/mp, °C	Molecular formula	Molecular weight
(*R*)-pantothenic acid			$C_9H_{17}NO_5$	219.24
(*S*)-pantothenic acid			$C_9H_{17}NO_5$	219.24
(*R*)-pantothenic acid calcium salt	+28.2°	195°C	$(C_9H_{16}NO_5)_2Ca$	476.5
(*R*)-pantothenic acid sodium salt	+27.7°	171–178°C	$C_9H_{16}NO_5Na$	241.2
(*R*)-panthenol	+29.5°	120°C[b]	$C_9H_{19}NO_4$	205.25
(*S*)-panthenol	−29.5°	120°C[b]	$C_9H_{19}NO_4$	205.25
(*R,S*)-panthenol	0	66–69°C	$C_9H_{19}NO_4$	205.25
(*R,S*)-ethylpanthenol	0	210°C	$C_{11}H_{23}NO_4$	233.1
(*R*)-ethylpanthenol			$C_{11}H_{23}NO_4$	233.1

[a] C = 5 in H_2O for all values other than 0.
[b] At 0.02 mm Hg.

Chemical Synthesis. Currently, pantothenic acid is produced mainly by chemical methods. (*R*)-Calcium pantothenate is prepared by condensing (*R*)-pantolactone with β-alanine in the presence of base, followed by treatment of the sodium salt with calcium hydroxide. An alternative procedure is to condense (*R*)-pantolactone with the preformed calcium salt of β-alanine. Despite the progress made in the stereoselective synthesis of (*R*)-pantothenic acid since the mid-1980s, the commercial chemical synthesis still involves resolution of racemic pantolactone. Recent synthetic efforts have been directed toward developing a method for enantioselective synthesis of (*R*)-pantolactone by either chemical or microbial reduction of ketopantolactone.

A variety of organisms have been evaluated in the microbial reduction of ketopantoate to (*R*)-pantoate and of ketopantolactone to (*R*)-pantolactone. Results have been favorable. In a novel approach, enantiomerically enriched (*R*)-pantolactone (**2**) is obtained in a enzymatic two-step process starting from racemic pantolactone (**3**).

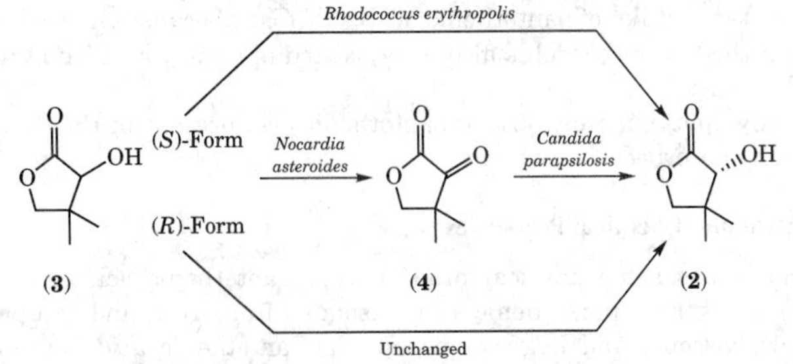

First, *Nocardia asteroides* selectively oxidizes only (*S*)-pantolactone to ketopantolactone (**4**), whereas the (*R*)-pantolactone remains unaffected. Then, the accumulated ketopantolactone is stereospecifically reduced to (*R*)-pantolactone with *Candida parapsilosis.*

Derivatives and Analogues

Methyl and diacetyl derivatives of pantothenic acid are prepared via methylation and acetylation of pantothenic acid. Pantethein is one of the biochemical degradation products derived from coenzyme A. Synthetically, it can be prepared in several different ways, starting from either methyl pantothenate or (*R*)-pantolactone. Phosphopantetheine analogue was prepared starting from (*R*)-pantothenic acid. Enzymatic condensation of this compound with adenosin-3,5-diphosphate gave CoA analogue.

Analytical Methods

Chemical, physical, animal, microbiological, and biochemical assays have been used to determine pantothenic acid. Pantothenic acid in foods is found in both the free form and as the vitamin moiety of coenzyme A and phosphopantetheine. For most assays, the incorporated pantothenic acid has to be liberated enzymatically. Usually, a combination of pantotheinase and alkaline phosphatase is used. The official method for pantothenic acid of the Association of Official Analytical Chemists (AOAC) is the microbiological assay that uses *L. Plantarium* (ATCC 8014) as the test organism.

A radioimmunoassay method is a sensitive method for the determination of small amounts of pantothenic acid in biological fluids. The assay is based on the binding of an enzyme specific for pantothenic acid. In the case of pharmaceutical preparations, vitamin concentrations are determined by using spectrophotometric and fluorometric methods. Food pantothenate assays involve hydrolysis of the food by 25% hydrochloric acid, extraction of the pantoyl lactone into dichloromethane and analysis by gas chromatography with flame ionization detection. Recently, clinical separation and simultaneous determination of (*R*)- and (*S*)-pantothenic acids in rat plasma using gc–ms has been described.

Biochemistry

The most important functions of pantothenic acid are its incorporation in coenzyme A and acyl carrier protein (ACP), which function metabolically as carriers of acyl groups. Coenzyme A forms high-energy thioester bonds with carboxylic acids. The active acetate group of acetyl CoA can enter the Krebs cycle and is used in the synthesis of fatty acids or cholesterol. ACP is a component of the fatty acid synthase multienzyme complex. This complex catalyzes several reactions of fatty acid synthesis (condensation and reduction).

The utilization of the vitamin depends upon release of the free vitamin by the hydrolytic digestion of the CoA and acyl-carrier protein found in food. Coenzyme A is hydrolyzed in the intestinal lumen before absorption into the cell by a sodium ion-dependent, passive mechanism. Pantothenic acid utilization in the formation of acetyl CoA is impaired in alcoholics because the ethanol metabolite acetaldehyde inhibits the conversion. Absorption is inhibited by oral contraceptives and such antagonists as ω-methylpantothenic acid, and to some extent (*S*)-pantothenic acid. A high fat diet lowers the utilization of pantothenic acid in the formation of coenzyme A in the liver because of changes in lipid metabolism, but high levels of dietary protein may promote its utilization.

Deficiency A deficiency of pantothenic acid in humans has not been detected (except in severe cases of malnutrition) because it is widely available in food. Experimentally induced pantothenic acid deficiency in humans results in fatigue, headache, sleep disturbance, nausea, vomiting, and cardiovascular instability. Impaired responses to insulin, histamine, and ACTH (stress hormone) have also been observed.

Nutritional Requirements The RDA for pantothenic acid has not been established. In 1989, the Food and Nutrition Board of the United States National Research Council suggested a safe intake of 4–7 mg/d for adults. The provisional allowance for infants is 2–3 mg daily (90).

Toxicity Pantothenic acid toxicity has not been reported in humans. Massive doses (10 g/d) in humans have produced mild intestinal distress and diarrhea. Acute toxicity was observed in case of mice and rats by using calcium pantothenate at fairly large doses.

Uses The bulk of the industrial supply of the calcium salt of (*R*)-pantothenic acid is used in food and feed enrichment (breakfast cereals, beverages, dietetic, and baby foods). Animal feed is fortified with calcium-(*R*)-pantothenate which functions as a growth factor.

Panthenol is used in skin care products to alleviate itching and to keep the skin moist and supple, stimulates cell growth, and accelerates wound healing by increasing the fibroblast content of scar tissue (93). It is also used as a conditioner in hair care products.

THIMMA R. RAWALPALLY
Hoffmann-La Roche Inc.

O. Isler, G. Brubacher, S. Ghisla, and B. Kräutler, *Vitamine II,* Georg Thieme Verlag, New York, 1988, p. 309.

S. Shimizu and H. Yamada, in T. O. Baldwin, F. M. Raushel, and A. I. Scott, eds., *Chemical Aspects of Enzyme Biotechnology,* Plenum Press, New York, 1990, p. 151.

S. Shimizu, S. Hattori, H. Hata, and H. Yamada, *Appl. Environ. Microbiol.* **53,** 519 (1987).

PYRIDOXINE (B_6)

Pyridoxine is the official name of 5-hydroxy-6-methyl-3,4-bis(hydroxymethyl)pyridine (**1**), which usually comprises only a fraction of the total amount of vitamin B_6 in natural sources. Closely related, biologically interconvertible 3-hydroxy-2-methylpyridines that exhibit vitamin B_6 activities include pyridoxal (**2**) and pyridoxamine (**3**) and their respective 5′-phosphates (**4–6**), as well as to pyridoxine hydrochloride (**7**). The term vitamin B_6 most often refers to the latter, which is the most important commercial form (Fig. 1).

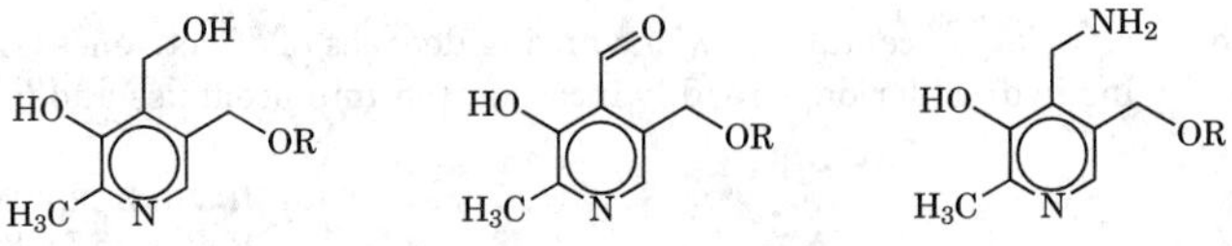

R = H (1)
R = OPO_3H_2 (4)
R = H HCl salt (7)

(2)
(5)

(3)
(6)

Figure 1. Forms (vitamers) of vitamin B_6.

Table 1. Properties of Pyridoxine, Pyridoxamine, Pyridoxal, and Derivatives

Substance	Formula	Mol wt	Form	Mp in water, °C	Solubility, g/100 mL	*pK* values
pyridoxine	$C_8H_{11}NO_3$	169.18	needles	160	soluble	5.0, 9.0
pyridoxine HCl	$C_8H_{12}ClNO_3$	205.64	platelets	204–206 dec	22	
pyridoxine-5′-PO_4	$C_8H_{12}NO_6P$	249.16	needles	212–213 dec	soluble	5.0, 9.4
pyridoxamine	$C_8H_{12}N_2O_2$	168.20	crystals	193	soluble	3.4, 8.1, 10.5
pyridoxamine di-HCl	$C_8H_{14}Cl_2N_2O_2$	241.12	platelets	226–227 dec	50	
pyridoxamine-5′-PO_4	$C_8H_{13}N_2O_5P$	248.18			soluble	3.5, 5.8, 8.6, 10.8
pyridoxal	$C_8H_9NO_3$	167.16			50	4.2, 8.7, 13
pyridoxal HCl	$C_8H_9NO_3$	203.63	rhombic	173 dec	50	
pyridoxal-5′-PO_4	$C_8H_{10}NO_6P$	247.14			soluble	3.5, 8.4

Vitamin B_6 is widely distributed in small amounts in the tissues and fluids of nearly all living substances. Bound to enzymes as the active form, pyridoxal phosphate, it acts as a coenzyme for a number of biochemical conversions of amino acids important to cell life organ functions. No natural sources are rich enough, nor is biosynthetic understanding sufficiently advanced, to allow cost-effective production from biological sources. All commercially available pyridoxine hydrochloride is manufactured by chemical processes.

Chemical and Physical Properties

The B_6 vitamers are high melting crystalline solids that are very soluble in water and insoluble in most other solvents. Their properties are listed in Table 1. The commercially important form of vitamin B_6, pyridoxine hydrochloride, is an odorless crystalline solid composed of colorless platelets melting at 204–206°C (with decomposition). It has a density of ~0.4 kg/L. It is very soluble in water (ca 0.22 kg/L at 20°C), soluble in propylene glycol, slightly soluble in acetone and alcohol (ca 0.014 kg/L), and insoluble in most lipophilic solvents. A 10% water solution shows a pH of 3.2.

Physical Properties. Ultraviolet absorptions of the B_6 vitamers vary with pH and the substituent at position 4. As hydroxypyridines, the B_6 vitamers show strong fluorescence. Ir and ^1H, ^{13}C, and ^{15}N nmr show the expected effects of N-protonation, hydrogen bonding, and exchange of the acidic phenolic proton. In mass spectrometry (ms), the molecular ion and an intense peak for a quinone–methide fragment from loss of water at the 4-hydroxymethyl position are evident.

Chemical Properties. The B_6 vitamers show weakly basic and acidic properties and undergo reactions expected of their functionalities. Pyridoxine gives the characteristic reactions of phenols such as color tests, ether formation, diazo coupling, etc. The B_6 vitamers are labile. Pyridoxine is the most stable and pyridoxal the least. The reactions of the substituents at the 4-carbon dominate. On mild oxidation, pyridoxine is converted to pyridoxal (**2**). Air oxidation gives pyridoxic acid, an intensely blue fluorescent substance of value for quantitation. Heating pyridoxine with ammonia converts it to pyridoxamine (**3**) and with alcohols to the 4-ethers.

Pyridoxal and its 5′-phosphate undergo typical aldehyde reactions, such as aldol condensations, bisulfite addition, etc. Both form imines. The epsilon-amino group of a lysine unit is usually involved in the active sites of enzymes but other groups can bind pyridoxal during thermal processing and storage, reducing bioavailability. Reactions of imines and their tautomers are the basis of the important biochemistry of pyridoxal 5′-phosphate, the coenzyme form for over 100 enzymatic reactions. Known reactions involving the alpha-, beta-, and gamma-carbons of the amino acid and their substituents include decarboxylations, transaminations, racemizations, aldol and retroaldol reactions, eliminations, and hydration reactions. Covalent bonding of a substrate to other reactive residues in the active site can occur, resulting in irreversible "suicide inactivation" of the enzyme. This is a mechanism of action of some naturally occurring toxins and a protocol in rational drug design.

Pyridoxine hydrochloride is stable if kept dry or in acidic solution. Decomposition in foods on processing and storage depends mainly on pH, moisture, and temperature. All B_6 vitamers are susceptible to decomposition by near-uv light and gamma irradiation.

Occurrence

Pyridoxal phosphate bound to protein is the main form in animals. Pyridoxine, pyridoxamine, and their phosphates are the usual forms in plants. Significant percentages in plants occur as pyridoxine glycosides, which are substantially less bioavailable. Relatively rich sources of vitamin B_6 are edible yeast, meats (especially salmon and calf liver), whole grain cereals, legumes, and nuts.

Biochemistry and Physiology

Vitamin B_6 is essential in human and animal nutrition. As enzyme cofactors, pyridoxal 5′-phosphate and pyridoxamine 5′-phosphate are required for important biotransformations of amino acids. Decarboxylations of amino acids give rise to a number of important amines, among them the vasodilator histamine, the vasoconstrictor and neurotransmitter serotonin, and the major neurotransmitters gamma-aminobutyric acid, epinephrine, norepinephrine, tyramine, and dopamine. Transamination interconverts specific pairs of amino acids and their corresponding ketoacids and amino groups and aldehydes, key functions in amino acid biosynthesis and catabolism. Racemization provides D-amino acids, constituents of the cell walls of bacteria. Many other important transformations require pyridoxal. Consequently, vitamin B_6 levels affect gluconeogenesis, nervous system activity, red cell formation and metabolism, immune function, and hormone functions.

Deficiency. Deficiency in animals leads to a variety of symptoms ranging from mild (dermatitis, loss of appetite, irritability, and muscular weakness) to serious (anemia, weight loss, and nerve degradation leading to convulsive seizures) and even death. Depressed antibody, nucleic acid, and protein synthesis are characteristic.

Requirements. The RDA ranges from 0.7 mg daily for infants to 2.5 mg for pregnant women. Humans cannot biosynthesize vitamin

B_6 and must obtain it from dietary sources. Ruminant animals derive some benefit from microfloral supply in the stomach. All animals interconvert the six major forms. Those on oral contraceptives, alcoholics, and users of drugs that interfere with vitamin B_6 require more. Oral doses 150 times the RDA have been used in studies without adverse effects. More recent studies show that long-term consumption of very large doses can lead to sensory nerve damage.

Synthesis

Biosynthesis. Almost all microorganisms synthesize B_6 vitamers intracellularly. Biosynthesis in *E. coli* is known to be controlled by both feedback inhibition and repression. The mechanisms of vitamin B_6 biosynthesis are partly understood. In *E. coli,* all carbons are derived from glucose and the nitrogen from glycine. Union of the intermediates 1-deoxy-D-xylulose and 4-hydroxy-L-threonine gives pyridoxine in several steps, only two of which require enzymatic catalysis. In aerobic bacteria, pyridoxal phosphate is made as needed by the action of oxidase or dehydrogenase enzymes on pyridoxine phosphate or pyridoxamine phosphate.

Chemical Synthesis. Pyridoxal can be prepared by oxidation of pyridoxine. Pyridoxamine can be made by reduction of pyridoxal oxime. Only pyridoxine, the most stable of the three, is manufactured. The first syntheses to be applied on a large scale relied on classical condensation reactions, but had low overall yields, in part due to difficulties of obtaining the correct oxidation levels of the 4- and 5-substituents.

The current paradigm for B_6 syntheses is an adaptation of a 1957 synthesis of pyridines by cycloaddition reactions of oxazoles. Preferred oxazoles bear 5-ethoxy- or 5-cyanosubstituents.

For the alkene partners, current manufacturers generally use the cyclic acetal of butenediol with isobutyraldehyde, a liquid of conveniently high boiling point that readily allows the product to be deprotected with no waste.

All routes for deriving oxazoles require dehydration of amides as a critical step. In some cases, the oxazoles are generated *in situ* in the cycloaddition reactions from *N*-formyl compounds by the isonitriles under dehydrating conditions.

Analytical Methods

Determining vitamin B_6 in foods and tissues is complicated by the number of forms, typical low levels, varying degrees of ionization, lability to heat, light, and base, interconversion, and binding to proteins and glucose. Colorimetric, fluorometric, microbiological, animal, and chromatographic methods are used. Few are fully suitable for simultaneous analysis of all six bioactive forms at the typical low levels. Colorimetric assays involve derivatization and are relatively insensitive. Microbiological assays (the official AOAC method) require prior hydrolysis, overestimate bioavailability, and do not respond to all forms. The comprehensive method of choice generally is high pressure liquid chromatography (hplc). Immunoassay methods are being developed.

Analogues and Antagonists

Over 250 analogues of the B_6 vitamers have been reported. Nearly all have low vitamin B_6 activity and some show antagonism. Among these are the 4-deshydroxy analogue, pyridoxine 4-ethers, and 4-amino-5-hydroxymethyl-2-methylpyrimidine, a biosynthetic precursor to thiamine. Structurally unrelated antagonists include drugs such as isoniazid, cycloserine, and penicillamine, which are known to bind to pyridoxal enzyme active sites.

Uses

The primary uses of pyridoxine hydrochloride are in multivitamin supplement tablets and for fortification of human food and animal feed, especially for poultry and pigs. Most breakfast cereals and infant formulas in the United States are supplemented. Lesser amounts are used therapeutically to correct deficiencies or to treat specific disorders. Pyridoxine hydrochloride has been used experimentally to treat a variety of conditions with varying degrees of effectiveness. Pyridoxine hydrochloride is readily incorporated into premixes and foods.

DAVID BURDICK
Hoffmann-La Roche Inc.

D. Dolphin, R. Poulson, and O. Avramovic, eds., *Coenzymes and Cofactors,* Vols. 1A & 1B, John Wiley & Sons, Inc., New York, 1986.

O. Isler, G. Brubacher, S. Ghisla, and B. Kraeutler, eds., *Vitamine II,* Verlag, New York, 1981, pp. 193–227.

J. M. Osbond, *Vitamins and Hormones* **22,** 387 (1964).

RIBOFLAVIN (B_2)

Riboflavin (vitamin B_2, vitamin G, lactoflavin, ovoflavin, lyochrome, hepatoflavin, uroflavin) has the chemical name 1-deoxy-1-(3,4-dihydro-7,8-dimethyl-2,4-dioxobenzo[*g*]pteridin-10(2*H*)-yl)-D-ribitol, 7,8-dimethyl-10-D-ribitylisoalloxazine, $C_{17}H_{20}N_4O_6$ (**1**), mol wt 376.37.

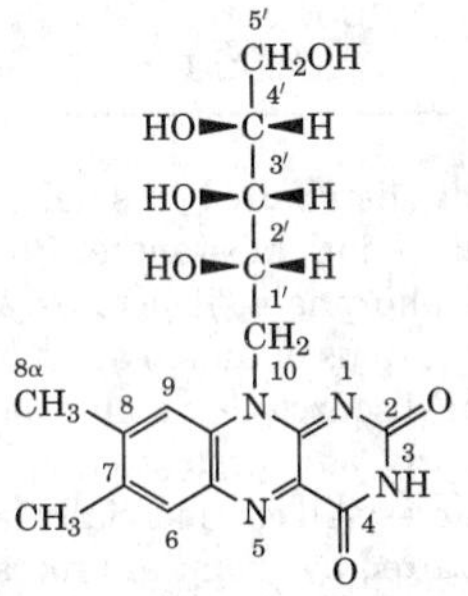

(**1**)

In the free form, riboflavin occurs in the retina of the eye, in whey, and in urine. Principally, however, riboflavin fulfills its metabolic function in a complex form. In general, riboflavin is converted into flavin mononucleotide (FMN, riboflavin-5′-phosphate) and flavin–adenine dinucleotide (FAD), which serve as the prosthetic groups (coenzymes), ie, they combine with specific proteins (apoenzymes) to form flavoenzymes, in a series of oxidation–reduction catalysts widely distributed in nature.

As a coenzyme component in tissue oxidation–reduction and respiration, riboflavin is distributed in some degree in virtually all naturally occurring foods. Liver, heart, kidney, milk, eggs, lean meats, malted barley, and fresh leafy vegetables are particularly good sources of riboflavin. It does not seem to have long stability in food products.

Properties

Riboflavin forms fine yellow to orange-yellow needles with a bitter taste from 2 *N* acetic acid, alcohol, water, or pyridine. It melts with decomposition at 278–279°C. The solubility of riboflavin in water is 10–13 mg/100 mL at 25–27.5°C, and in absolute ethanol 4.5 mg/100 mL at 27.5°C; it is slightly soluble in amyl alcohol, cyclohexanol, benzyl alcohol, amyl acetate, and phenol, but insoluble in ether, chloroform, acetone, and benzene. It is very soluble in dilute alkali, but these solutions are unstable.

Neutral aqueous solutions of riboflavin have a greenish-yellow color and an intense yellowish green fluorescence with a maximum at ca 530 nm. The optical activity of riboflavin in neutral and acid solutions is $[\alpha]_D^{20} = +56.5 - 59.5°$ (0.5%, dil HCl). Borate-containing solutions are strongly dextrorotatory, because borate complexes with the ribityl side chain of riboflavin; $[\alpha]_D^{20} = +340°$ (pH 12).

Photochemical decomposition of riboflavin in neutral or acid solution gives lumichrome (7,8-dimethylalloxazine). In alkaline solution, the irradiation product is lumiflavin (7,8,10-trimethylisoalloxazine).

Riboflavin is stable against acids, air, and common oxidizing agents (except chromic acid, $KMnO_4$, and potassium persulfate). Upon reduction, riboflavin readily takes up two hydrogen atoms to form the almost colorless 1,5-dihydroriboflavin, which is reoxidized by shaking with air. This oxidation–reduction system has considerable stability and is probably responsible for the physiological functions of riboflavin. The flavins are reduced to dihydroflavins through intermediate semiquinone radicals. Riboflavin forms a deep-red silver salt.

Synthesis

For therapeutic use, riboflavin is produced by chemical synthesis, whereas concentrates for poultry and livestock feeds are manufactured by microbial fermentation.

Chemical Synthesis. In 1935, Karrer and Kuhn independently synthesized riboflavin by a condensation of 6-D-ribitylamino-3,4-xylidine with alloxan in acid solution.

Although not suitable for large-scale manufacture, the synthesis of riboflavin from lumazine derivatives is interesting in connection with the biosynthesis of riboflavin. Thus, 5-amino-6-D-ribitylaminouracil was condensed with a dimeric or trimeric aldol of biacetyl to give riboflavin through the formation of intermediary 6,7-dimethyl-8-D-ribityllumazine.

A later, more convenient synthesis of riboflavin and analogues consists of the nitrosative cyclization of 6-(*N*-D-ribityl-3,4-xylidino)uracil with excess sodium nitrite in acetic acid or with potassium nitrate in acetic in the presence of sulfuric acid to give riboflavin-5-oxide in high yield. Reduction with sodium dithionite gives riboflavin.

Microbial Synthesis. Riboflavin is produced by many microorganisms, including *Ashbya gossypii, Asperigillus* sp, *Eremothecium ashbyii, Candida* yeasts, *Debaryomyces* yeasts, *Hansenula* yeasts, *Pichia* yeasts, *Azotobactor* sp, *Clostridium* sp, and *Bacillus* sp. The biosynthetic pathways deduced include conversion of a purine such as guanosine triphosphate (GTP) to 6,7-dimethyl-8-D-ribityllumazine which is then converted to riboflavin.

Fermentative Manufacture. A suitable carbohydrate-containing mash is prepared and sterilized, and the pH adjusted to 6–7. The mash is buffered with calcium carbonate, inoculated with *Clostridium acetobutylicum,* and incubated at 37–40°C for 2–3 d. Most varieties of *Candida* yeasts produce substantial amounts of riboflavin when glucose is the carbon source. However, most of the commercial riboflavin production by aerobic fermentation is obtained by biosynthesis with the yeastlike fungus *Eremothecium ashbyii.*

In recent (ca 1997) years, fermentative manufacturing methods using recombinant microorganisms have been developed. The recombinant mutant clone GA18Y8-6#2 of *C. flaveri* produced riboflavin at 7.0–7.5 g/L/6 d. Riboflavin overproducing bacteria prepared by expression of the cloned *rib* operon of *Bacillus subtilis* showed increases in riboflavin manufacture of up to a hundredfold. Culturing recombinant *Corynebacterium ammoniagenes* KY13313 harboring a gene for guanidine triphosphate cyclohydrolase and riboflavin synthase produced riboflavin ~30 – fold higher than that with the controlled bacteria.

Analytical Methods

Riboflavin can be assayed by chemical, enzymatic, and microbiological methods. The most commonly used chemical method is fluorometry (maximum fluorescence at 565 nm for neutral aqueous solutions). A laser–fluorescence technique has extended the limits of detection for riboflavin by two orders of magnitude.

The microbial assay is based on titrating the lactic acid formed by the microorganism (e.g., *Lactobacillus casei, Leuconostoc mesenteroides*) photometrically measuring its turbidity. A very useful method for measuring total riboflavin in body fluids and tissues is based on the riboflavin requirement of the protozoan cliate *Tetrahymena pyriformis.* Nutritional studies depend upon a growth response, usually in rats or chicks. This method is particularly useful for assaying riboflavin derivatives, since the substituents frequently reduce or eliminate the biological activity.

An enzymatic method for assessing riboflavin deficiency in humans has been developed. It is based on the fact that NADPH-dependent glutathione reductase of red cells reflects riboflavin fluctuations.

High pressure liquid chromatography (hplc) has been used in combination with fluorometric detection for the riboflavin assay in foods, as well as in a simple assay. A rapid and efficient reversed-phase hplc method is described for the quantitative separation of flavin coenzymes and their structural analogues.

Deficiency, Requirements, and Toxicity

Riboflavin is essential for mammalian cells. Deficiency is characterized by sore throat, hyperemia, cheilosis, angular stomatitis, glossitis (magenta tongue), a generalized seborrheic dermatitis, scrotal and vulval skin changes, and a normocytic anemia. Because riboflavin is essential to the functioning of vitamins B_6 and niacin, some symptoms may be due to the failure of these other nutrients to operate effectively.

The RDA stipulates 0.6 mg riboflavin per 239 kJ (1000 kcal) from 0.4 mg/day for early infants to 1.8 mg/day for young males. Riboflavin is essentially nontoxic.

Derivatives

Derivatives include riboflavin-5′-phosphate (FMN), riboflavin-5′-adenosine diphosphate (FAD), covalently bound flavins, 6-hydroxyriboflavin, 8-nor-8-hydroxyriboflavin, roseoflavin, and 5-deazariboflavin.

FUMIO YONEDA
Fujimoto Pharmaceutical Corporation

H. Baker and O. Frank in R. S. Rivilin, ed., *Riboflavin,* Plenum Press, New York, 1975.

F. Müller, ed., *Chemistry and Biochemistry of Flavoenzymes,* Vol. 1, 1991; Vol. 2, 1991; and Vol. 3, 1992 CRC Press, Boca Raton.

D. B. McCormick in M. E. Sils and V. R. Young, eds., *Modern Nutrition in Health and Disease,* Lea and Febiger, Philadelphia, Pa., 1988, p. 362.

THIAMINE (B_1)

Thiamine is the official IUPAC-IUB name for 3-(4-amino-2-methyl-5-pyrimidinyl) methyl-5-(-2-hydroxyethyl)-4-methylthiazolium chloride, $C_{12}H_{17}$-N_4OSCl (**1**). The chloride hydrochloride (**2**) is the common biological and commercial form.

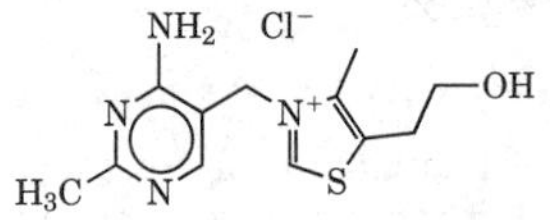

(**1**)

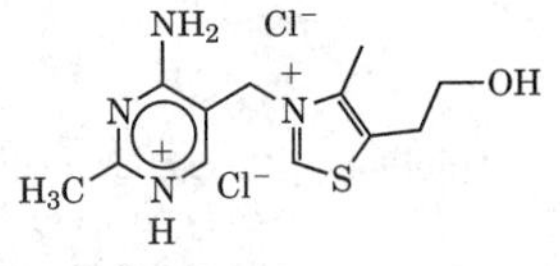

(**2**)

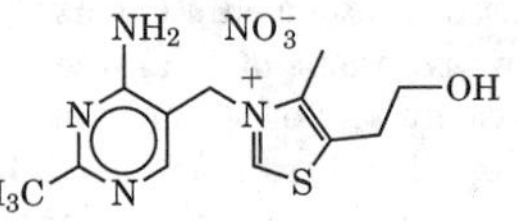

(**3**)

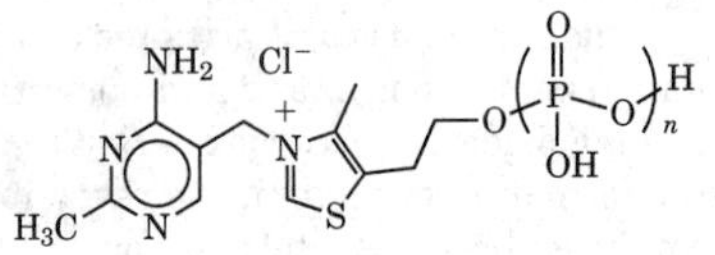

(**4**, n = 1)
(**5**, n = 2)
(**6**, n = 3)

Physical and Chemical Properties

Salt Formation. As a weakly basic pyrimidine and a thiazolium cation, thiamine forms both mono- and dipositive salts, eg, the two commercial forms.

Thiamine chloride hydrochloride (**2**), crystallizes as colorless monoclinic needles, mp 248–250°C, density ~0.4 kg/L. The salt has a characteristic thiazole meat-like odor and a slightly bitter taste. It can adsorb up to one mole of water from air. It is very soluble in water (over 1 kg/L at 25°C), soluble in glycerol (0.056 kg/L), propylene glycol, and methanol, sparingly soluble in 95% ethanol (0.01 kg/L), and practically insoluble in less polar organic solvents. In 1–5% solutions in water it shows a pH of 3–3.5.

Thiamine mononitrate (**3**) is a colorless crystalline solid with a typical odor, melting point of ca 196–200°C and a density of ~0.5 kg/L. It is much less soluble in water than the hydrochloride (0.027 kg/L at 25°C, 0.030 kg/L at 100°C) and practically nonhygroscopic. Dilute solutions in water show a pH of 6.5–7.

At pH 7 aqueous thiamine is destroyed at room temperature. Concentrated solutions of thiamine in alcohol when neutralized degrade rapidly at room temperature to liberate 5-(2-hydroxyethyl)-4-methylthiazole and form oligomers of the pyrimidine where the 5-methylene is linked to N-1 of another pyrimidine unit.

Basification allows intramolecular attack of the 4-amino function on the thiazole ring to generate dihydrothiochrome (**7**), which eliminates thiolate to generate the salt of the yellow form, 5,6-dihydropyrimido(4,5*d*)pyrimidine (**8**), as a kinetic product (Fig. 1). More slowly, thiamine thiol, or the thiol form (**9**). Also present in the system is a small amount of thiamine ylid (**10**), formed by deprotonation of C-2.

The yellow form (**8**) on acidification is converted to the more stable thiol form (**9**). On oxidation, typically with alkaline ferricyanide, yellow form is irreversibly converted to thiochrome (**12**) which exhibits an intense blue fluorescence, a property used for the quantitative determination of thiamine.

The thiol form (**9**) undergoes reactions mainly via its *N*-formyl or ene–thiol groups. Acylation of the thiolate occurs on both sulfur and oxygen to give mono- or diacyl thiamines, some of which are interesting fat-soluble depot forms of thiamine.

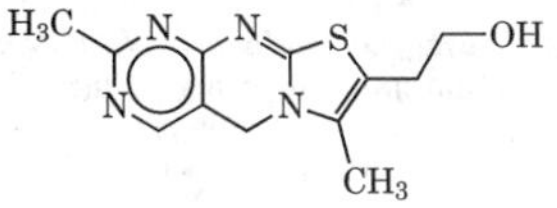

(**12**)

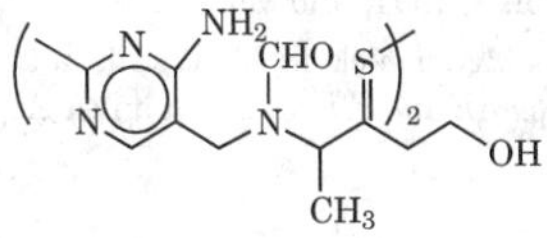

(**13**)

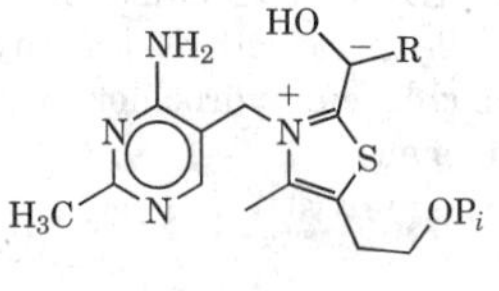

(**14**)

The thiol form (**9**) is susceptible to oxidation. Iodine treatment regenerates thiamine in good yield. Heating an aqueous solution at pH 8 in air gives rise to thiamine disulfide (**13**), thiochrome (**12**), and other. The disulfide is readily reduced to thiamine *in vivo* and is as biologically active.

Thiamine ylid (**10**) is accepted as an unstable intermediate in explanations of the enzymatic and nonenzymatic reactions of thiamine; including oxidative decarboxylation of pyruvic and 2-ketoglutaric acids, the formation of acetoin, the reversible alpha-ketol transfer reactions catalyzed by transketolase, and the nonenzymatic acyloin condensation. Apparently, ylid (**10**) reacts reversibly with carbonyl reagents, facilitating aldol–retroaldol and decarboxylation reactions, via the acyl anion equivalent thiazolium ylids (**14**), the so-called active aldehydes, which are the key intermediates.

Thiamine is susceptible to reaction with various nucleophilic species and is readily ruptured by sulfite treatment. This occurs slowly at pH 3 and rapidly at pH 5 and above. Similar reactions are observed in the degradation of thiamine by thiaminases. Thiaminase I, found in shellfish, ferns, some vegetables and some bacteria, promotes the replacement of the thiazole by other organic bases, including purines. Thiaminase II, found mostly in bacteria, catalyzes the cleavage of thiamine into a pyrimidine that is an antagonist of pyridoxine (vitamin B_6). These enzymes can have strong effects on animals as they promote thiamine deficiency. Fortunately, they are thermolabile, and hence a problem mostly in uncooked foods or feeds.

Thermostable substances which inactivate thiamine have been found in a large number of plants and in some animal tissues. Among these are polyphenols such as caffeic acid, tannic acid, hydroxylated derivatives of tyrosine, and some flavonoids. Thiamine is susceptible to destruction by x-rays, gamma rays, and ultraviolet light to generate a diamine by cleavage of the thiazole ring. Photochemical and thermal degradation of thiamine gives rise to numerous sulfur-containing heterocycles, some with meaty or bread aromas of interest to the flavor industry.

Thiamine forms the expected derivatives of the thiazole alcohol function, such as carboxylic and phosphate esters.

Natural Occurrence

Thiamine is widespread in nature, although generally in only relatively minute quantities. In microorganisms it is found mainly intracellularly. In higher plants the most abundant form is free thiamine along with lesser amounts of the phosphate esters (**4–6**). In the tissues of animals, most thiamine is found as its phosphorylated esters and is predominantly bound to enzymes as the pyrophosphate (**5**), the active coenzyme form. As expected for a factor involved in carbohydrate metabolism, the highest concentrations are generally found in organs with high activity, such as the heart, kidney, liver, and brain. Almost no excess is stored.

Good natural human dietary sources of thiamine are unrefined cereal grains (most of the thiamin is in the germ layer), organ meats, pork, legumes, and nuts.

Biochemical and Physiological Functions

Thiamine deficiency causes particularly deleterious effects on an organisms's energy status and, in higher organisms, its nerve functions. In living systems, the only established biochemically active form is the pyrophosphate (cocarboxylase) (**5**), which plays a vital role in intermediate metabolism as a cofactor for some important enzymatic reactions. Dehydrogenase enzymes require cocarboxylase for oxidative decarboxylation of 2-ketoacids, notably pyruvate in glycolysis, 2-oxoglutarate in the citric acid cycle, and other ketoacids from amino acid decarboxylation. Transketolase enzymes require cocarboxylase for the reversible transfer of alpha-ketols in ketose–aldose transformations important in the production of pentoses for RNA and DNA synthesis. Thiamine triphosphate (**6**) occurs in unusually higher concentrations in nerve tissues and the brain and may play an essential role in the stimulation of peripheral nerves.

Deficiency. Thiamine deficiency is characterized by anorexia and mental disturbances, such as irritability, inattention, memory defects, depression, and insomnia. If left untreated, beriberi develops, the symptoms of which include mental changes, peripheral neuritis, paresthesias, muscle cramps, edema, muscular atrophy, and cardiac failure. The most commonly encountered type of thiamine deficiency in Western countries is associated with alcohol abuse (high intake of empty calories and low intake of nutritionally adequate foods).

Requirement. Because a constant intake of thiamine is needed, deficiency can occur relatively quickly. The RDA is based on calorie intake at the level of 0.50 mg/4184 kJ (1000 kcal), or from 0.7 mg/d for infants to 1.5 mg/d for adults.

There is no evidence of toxicity from thiamine taken orally, even at doses over 300 times the RDA.

Synthesis

Biosynthesis. Higher plants, most bacteria, and some fungi biosynthesize thiamine. Many microbial species are self-sufficient, others can synthesize thiamine if one or both of immediate precursors are available, and still others require the complete substance. Few produce much

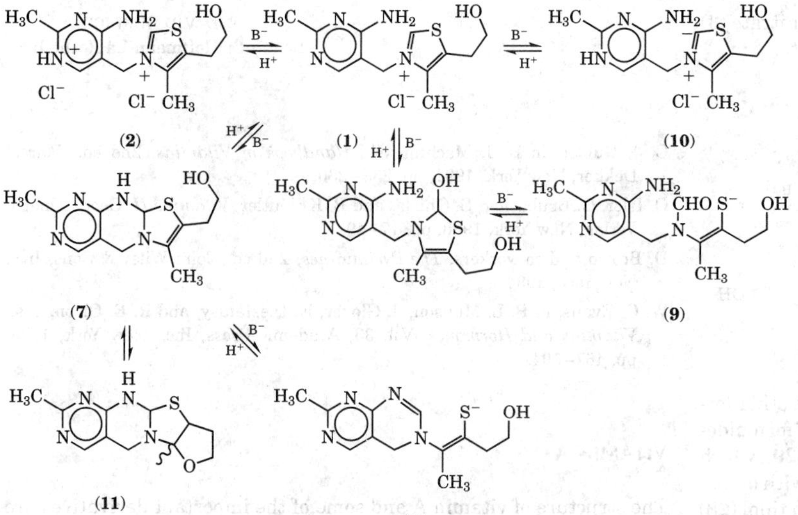

Figure 1. Structural changes of thiamine with pH.

of an excess over their own requirements because thiamine biosynthesis is tightly controlled by feedback mechanisms.

The pathways for thiamine biosynthesis have been elucidated only partly. Thiamine pyrophosphate is made universally from the precursors 4-amino-5-hydroxymethyl-2-methylpyrimidine pyrophosphate and 4-methyl-5-(2-hydroxyethyl)thiazole phosphate, but there appear to be different pathways in the earlier steps. In bacteria, the early steps of the pyrimidine biosynthesis are same as those of purine nucleotide biosynthesis.

Chemical Synthesis. All of the thiamine produced worldwide is manufactured chemically. Two major synthetic routes have been used: alkylation of a preformed thiazole, or construction of the thiazolium salt from a pyrimidine carrying the ultimate thiazole nitrogen. The latter approach is now generally preferred for manufacturing.

The first approach parallels the known biosynthetic pathway where the alkylating agent is the pyrophosphate ester of alcohol (**15**).

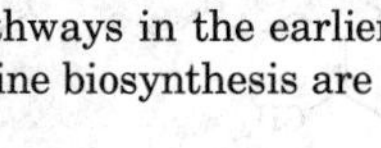

(15)

Synthetic 4-amino-5-bromomethyl-2-methylpyrimidine (**16**) and thiazole (**17**), or its *O*-acetate, is condensed and then the bromide is replaced by use of silver salts.

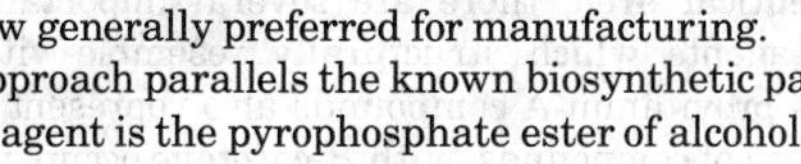

(16) **(17)**

In the second general method, the amine (**18**), known as Grewe diamine, is the paradigm intermediate onto which the thiazolium ring is constructed. Differences occur only in the raw materials and methods used for the manufacture of the diamine.

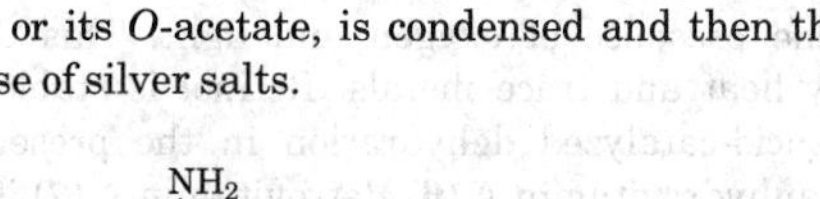

(18) **(19)**

The synthesis which forms the basis of production at Hoffmann-La Roche proceeds via the pyrimidinenitrile (**19**) made from malononitrile, trimethylorthoformate, ammonia, and acetonitrile. High pressure catalytic reduction of the nitrile furnishes diamine (**18**).

Other syntheses produce Grewe diamine using inexpensive acrylonitrile and alkyl formates as raw materials. Such routes deliver the pyrimidine bridge carbon at the correct aminomethyl oxidation level without a need for reduction.

A new variation for diamine, also based on acrylonitrile and alkyl formate, delivers the pyrimidine bridge carbon at the carbonyl level. Unlike other methods in which the intermediate enolate is alkylated, in this process the enolate is acetalized and the acetal thermally converted to the acetal enol ether (**20**). Condensation with acetamidine provides the remaining carbons. After hydrolysis of the acetal (**21**), reductive amination at high pressure with a specialized catalyst provides diamine (**18**) in very high overall yield.

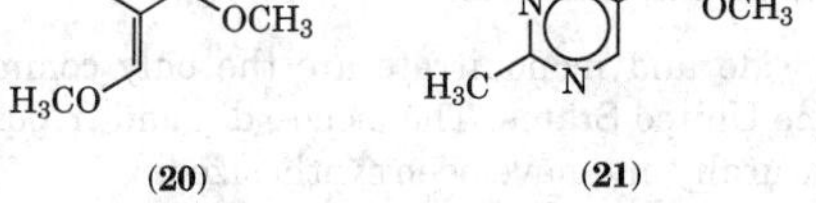

(20) **(21)**

The process is operated in a new, large-scale, automated, continuous, technically complex plant in Japan. Advantages include use of inexpensive acrylonitrile and carbon monoxide, avoidance of the costs of an alkylating agent, and consumption of only one equivalent of acetamidine.

In a much different approach based on cyanamide, acrylonitrile, and acetonitrile, cyanoacetamidine (**22**) is cyanoethylated and the condensation product (**23**) is dehydrogenated and hydrogenated directly to Grewe diamine in the presence of Raney cobalt and ammonia.

(22) **(23)**

Later it was shown that the 2-carbon and the sulfur atom can be supplied very efficiently via carbon disulfide, the extra sulfur atom being readily removed by oxidation in nearly quantitative yield. Typically, an aqueous solution of diamine and alkali is treated in succession with carbon disulfide to form the dithiocarbamate (**27**), then with chloroketone (**25**), then acid to form the relatively insoluble, thio-thiamine (**26**). Oxidation with hydrogen peroxide forms thiamine sulfate, which is converted by ion exchange to a solution of the hydrochloride (**2**) which is concentrated, crystallized, and dried. The much less soluble nitrate (**3**) is precipitated from aqueous solution with alkali metal nitrate.

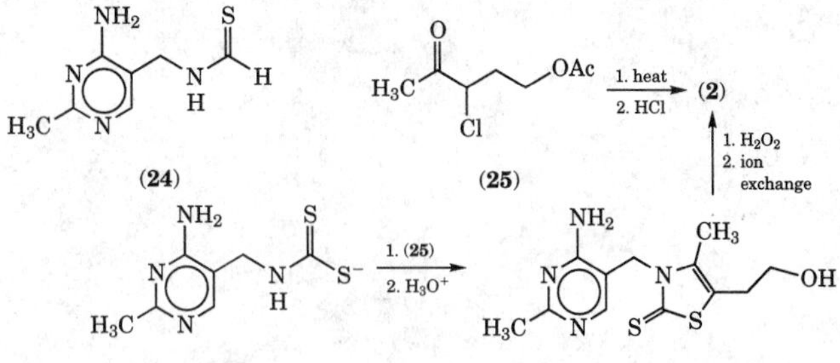

(**24**) (**25**) (**27**) (**26**)

In two unique, convergent approaches to the thiazole ring, other formate synthons are used for the 2-carbon. In one, reaction of formaldehyde with diamine (**18**) and thiol (**28**) gives dihydrothiamine (**29**), which is oxidized to thiamine. In another, reaction of diamine (**18**) with orthoformate gives intermediate dihydropyrimidine (**30**), to which thiol (**28**) is added. Acidic rearrangement gives thiamine in high yield.

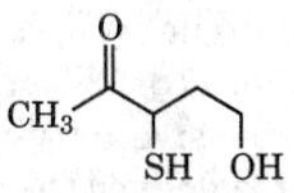

(**28**)

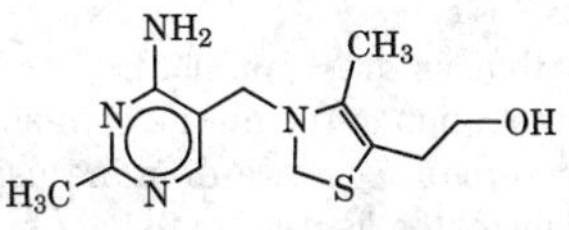

(**29**)

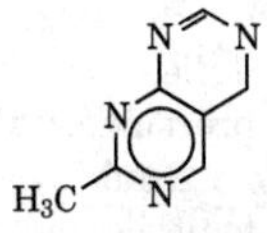

(**30**)

Analytical Methods

Fluorometric, chromatographic, microbiological, and animal assays have been used for thiamine and its derivatives. The most widely used and officially sanctioned method has been the fluorometric assay (excitation 365 nm/emission 435 nm), although high performance liquid chromatography has been increasingly employed. In natural materials, thiamine is often present as its phosphate esters and is protein-bound, therefore procedures to free it are necessary steps in most assays.

Forms, Derivatives, and Analogues

The hydrochloride and mononitrate are the only commercial forms approved in the United States. The mono-, di-, and triposphate esters (**4–6**) occur naturally and have been synthesized.

Treatment of thiamine with extracts of garlic or other *Allium* species converts it to a lipid-soluble disulfide derivative which is a very physiologically active depot form of thiamine. Because of their greater lipid solubility, such alkylated thiamine disulfides are absorbed and retained more strongly in the body.

Numerous analogues of thiamine have been synthesized by structural modifications of the pyrimidine or thiazole rings or the bridging atoms. Substitution at the 2-methyl, 4-amino, or 6-position of carbon for ring nitrogens, or of the methylene bridge, results in loss of activity. In the thiazolium ring, the sulfur atom, hydrogen at the 2-position, and the 2-hydroxyethyl group are essential for activity. The pyridine analogue, pyrithiamine (**31**) is the most potent thiamine antagonist known.

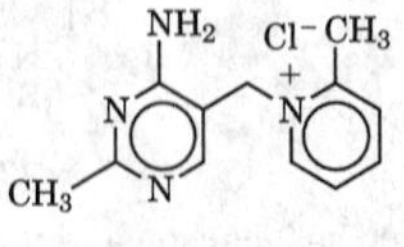

(**31**)

Uses

Most of the thiamine sold worldwide is used for dietary supplements. Small amounts are used in medicine to treat deficiency diseases and other conditions, in agriculture as an additive to fertilizers, and in foods as flavorings.

DAVID BURDICK
Hoffmann-La Roche Inc.

C. J. Gubler, in L. J. Machlin, ed., *Handbook of Vitamins,* 2nd ed., Marcel Dekker, New York, 1991, pp. 233–280.

O. Isler, G. Brubacher, S. Ghisla, and B. Kraeutler, *Vitamine II,* Georg Thieme Verlag, New York, 1988, pp. 17–49.

D. Brown and co-workers, *The Pyrimidines,* 2nd ed., John Wiley & Sons, Inc., New York, 1994.

W. C. Evans, in P. L. Munson, J. Glover, E. Diezfalusy, and R. E. Olson, eds., *Vitamins and Hormones,* Vol. 33, Academic Press, Inc., New York, 1975, pp. 467–504.

VITAMIN A

The structure of vitamin A and some of the important derivatives are shown in Figure 1. The parent structure is all-*trans*-retinol and its IUPAC name is (all-*E*)-3,7-dimethyl-9-(2,6,6-trimethyl-1-cyclohexen-1-yl)-2,4,6,8-nonatetraen-1-ol (**1**). The numbering system for vitamin A derivatives parallels the system used for the carotenoids. In older literature, vitamin A compounds are named as derivatives of trimethyl cyclohexene and the side chain is named as a substituent.

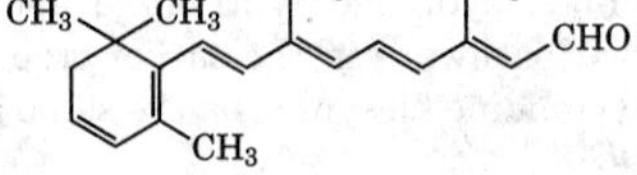

(**5**)

Vitamin A constitutes the most significant sector of the commercial retinoid market and is used primarily in the feed area. In the pharmaceutical area, there are several important therapeutic dermatologic agents which structurally resemble vitamin A. The carotenoids as provitamin A compounds also represent an important commercial class of compounds, with β-carotene occupying the central role.

Chemical and Physical Properties

The chemical and physical properties of the retinoids and carotenoids are dominated by the presence of an extended polyene chain. Vitamin A and related substances are yellow compounds which are unstable in the presence of oxygen and light. This decay can be accelerated by heat and trace metals. Retinol is stable to base but is subject to acid-catalyzed dehydration in the presence of dilute acids to yield anhydrovitamin A (**6**) Retrovitamin A (**7**) is obtained by treatment of retinol in the presence of concentrated hydrobromic acid. In the case of retinoic acid and retinal, reisomerization is possible after conversion to appropriate derivatives such as the acid chloride or the hydroquinone adduct. Table 1 lists the physical properties of β-carotene and vitamin A.

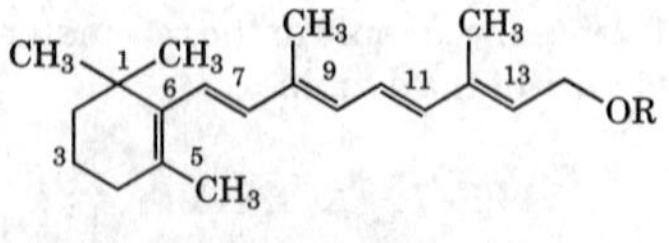

(**1**, R = H) retinol
(**2**, R = CH_3CO) retinyl acetate
(**3**, R = $CH_3(CH_2)_{14}CO$) retinyl palmitate
(**4**, R = CH_3CH_2CO) retinyl propionate

Figure 1. Vitamin A and derivatives, and vitamin A_2: retinol (**1**), retinyl acetate (**2**), retinyl palmitate (**3**), and retinyl propionate (**4**).

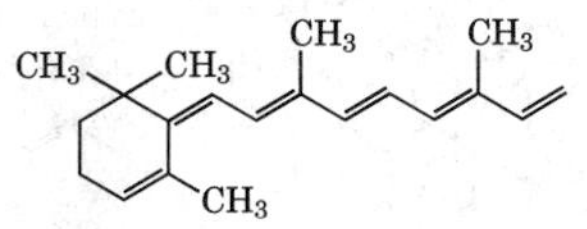

(6)

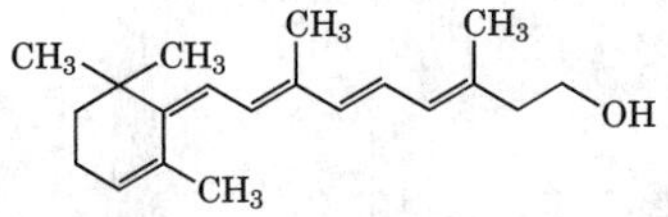

(7)

Of the 16 possible geometric isomers of vitamin A, only the all-trans form has full vitamin A activity. From a biological standpoint, 11-*cis*-retinal plays a critical role in vision (*vide infra*); 13-*cis*-retinoic acid is important as a dermatological agent; and 9-*cis*-retinoic acid has been identified as a novel endogenous hormone in mammalian tissues. The other 13 isomers have no biological or commercial significance.

The most conspicuous physical feature of the retinoids and of the carotenoids is the uv spectrum. Factors influencing the spectra include the length of the polyene chain, the number of cis double bonds, and end group functionality. Both hypsochromic shifts and hypochromic effects are observed when the stereochemistry of the double bond is changed from trans to cis.

Table 1. Properties of β-Carotene and Vitamin A Derivatives

Property	Retinol	Retinyl acetate	Retinyl palmitate	Retinyl propionate	β-carotene
appearance	crystalline	crystalline	crystalline or amorphous	oil	crystalline
color	yellow	yellow	yellow	light yellow	dark red-dark purple
odor	faint, hay-like	faint, hay-like	faint, hay-like	slight	faint, hay-like
mol wt	286.46	328.50	534.88	342.50	536.85
mol formula	$C_{20}H_{30}O$	$C_{22}H_{32}O_2$	$C_{36}H_{60}O_2$	$C_{23}H_{34}O_2$	$C_{40}H_{56}$
mp, °C	63–64	57–59	28–29		180–182
solubility, g/100 mL					
water	insoluble	insoluble	insoluble	insoluble	insoluble
ethanol	soluble	soluble	soluble	soluble	slightly soluble
isopropanol	soluble	soluble	soluble		slightly soluble
chloroform	soluble	soluble	soluble		soluble
acetone	soluble	soluble	soluble		slightly soluble
fats, oils	750	750	750		0.05–0.08
spectrophotometric properties, nm, max	375[a]	326[a]	325[a]		497,466[b]
fluorescence					
excitation, max, nm	325	325	325		
emission, max, nm	470	470	470		

[a] Isopropanol.
[b] Chloroform.

Synthesis

Biosynthesis. In nature, vitamin A aldehyde is produced by the oxidative cleavage of β-carotene by 15,15′-β-carotene dioxygenase. Alternatively, retinal is produced by oxidative cleavage of β-carotene to β-apo-8′-carotenal followed by cleavage at the 15,15′-double bond to vitamin A aldehyde. The carotenoid skeleton is assembled in a primary step by the coupling of geranylgeranyl pyrophosphate by an enzyme which is encoded by the crt B (carotenogenic) gene. The resulting prephytoene pyrophosphate is further transformed to phytoene possibly by products also derived from the crt B gene. The phytoene is converted to lycopene in sequential dehydration steps. By further biosynthetic transformations, β-carotene is produced from lycopene and zeaxanthin from β-carotene.

Chemical Synthesis. Vitamin A acetate (**2**) is the commercially significant form of the vitamin (Fig. 2). All of the commercial processes have β-ionone (**8**) as their key intermediate. Their differences lie in methodology to this key intermediate as well as in the methodology to elaborate the side chain.

When a $C_{13} + C_1 + C_6$ strategy is employed, β-ionone is subjected to a Darzen's condensation to yield the C_{14} aldehyde (**9**). Construction of the side chain is completed by a metal acetylide coupling reaction. Acetylation, partial reduction of the triple bond, and acid-catalyzed

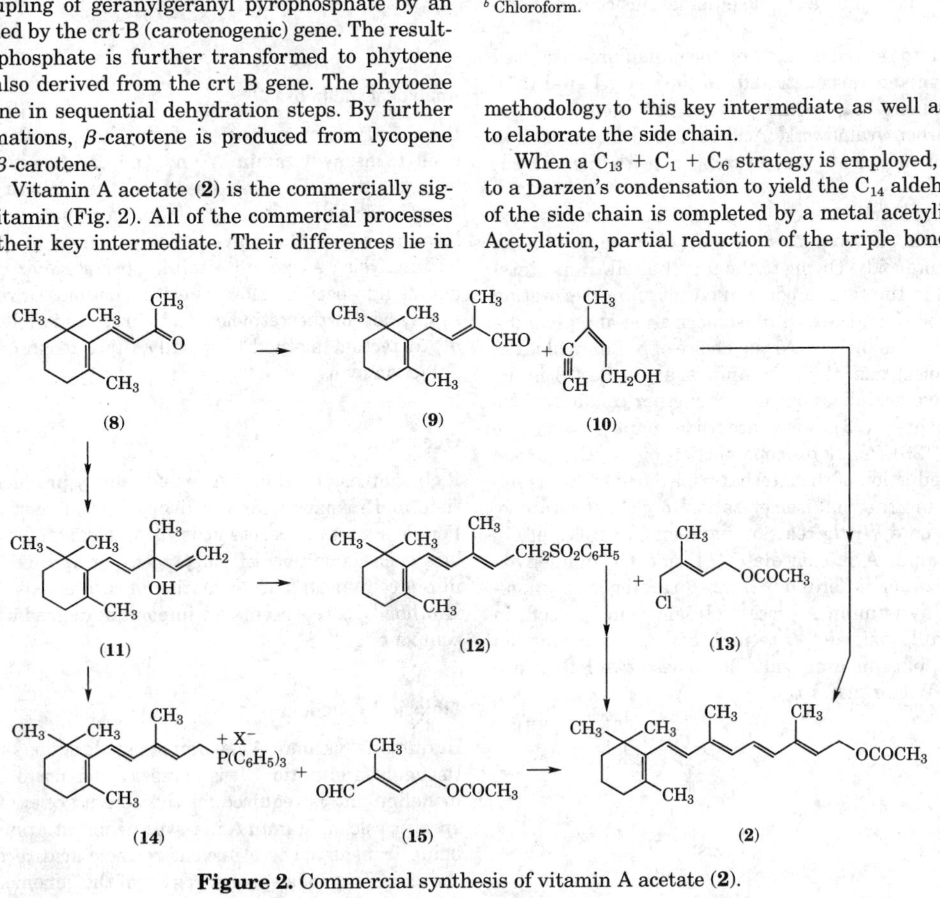

Figure 2. Commercial synthesis of vitamin A acetate (**2**).

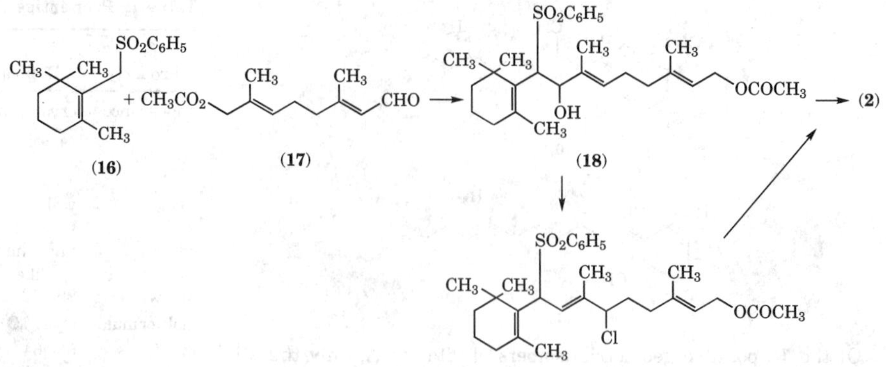

(19)

Figure 3. Kuraray synthesis.

elimination of water completes the synthesis. An alternative scheme extends the side chain of β-ionone in an initial step via a Grignard reaction with a metal acetylide. Semihydrogenation yields vinyl β-ionol (**11**). The carbon terminus of vinyl β-ionol is activated by conversion to the sulfone (**12**) followed by reaction with with C_5-chloroacetate (**13**) to yield vitamin A acetate; or a Wittig olefination may be used to prepare the phosphonium salt of the C_{15} unit. Reaction of the salt (**14**) with the C_5 aldehyde (**15**) leads to vitamin A acetate.

Vitamin A palmitate (**3**), a commercially important form of the vitamin, is produced from vitamin A acetate (**2**) via a transesterification reaction with methyl palmitate. Enzymatic preparation of the palmitate from the acetate has also been described.

In other work, sulfone chemistry plays an integral part of the syntheses of both β-carotene and vitamin A. In this approach, the anion of C_{10} β-cyclogeranyl sulfone (**16**) is condensed with the C_{10} aldehyde (**17**). The resulting β-hydroxy sulfone (**18**) is treated with dihydropyran followed by a double elimination to yield vitamin A acetate. Alternatively, the β-hydroxy sulfone (**18**) can be converted to the δ-halo sulfone (**19**) followed by a double elimination scheme is employed (Fig. 3).

A different approach to retinal is based on the palladium-catalyzed rearrangement of the mixed carbonate (**20**) to the allenyl enal (**21**). Isomerization of the allene (**21**) to the polyene (**22**) completes the construction of the carbon framework. Acid-catalyzed isomerization yields retinal, but in insufficient yield to compete with the current commercial syntheses.

In contrast to the similarities seen in the majority of the industrial syntheses of vitamin A, substantial diversity is observed in the preparations of the carotenoids. Owing to the fact that all-trans stereochemistry is required in the final product and that many olefination reactions yield the cis product, the ease of isomerization at a given locus on the polyene chain has influenced the choice of building blocks.

A five-carbon homologation of C_{14} vitamin A aldehyde (**9**) is accomplished by successive acetalizations and enol ether condensations to prepare the C_{19} aldehyde (**23**). Metal acetylide coupling with two molecules of aldehyde (**23**) completes construction of the C_{40} carbon framework. Selective reduction of the internal triple bond of (**24**) is followed by dehydration and thermal isomerization to yield β-carotene. Another method relies on a Wittig reaction between two moles of C_{15} phosphonium salt (vitamin A intermediate (**14**) and C_{10} dialdehyde (**25**). Thermal isomerization affords all *trans-β*-carotene. In an alternative preparation, by vitamin A process streams can be used, in which, retinol is carefully oxidized to retinal, and a second portion is converted to the C_{20} phosphonium salt (**26**). These two halves are united using standard Wittig chemistry.

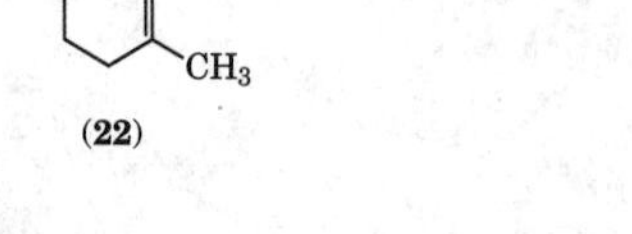

(**22**)

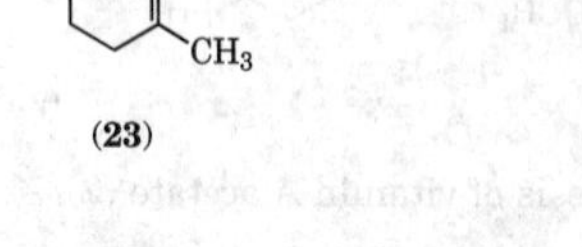

(**23**)

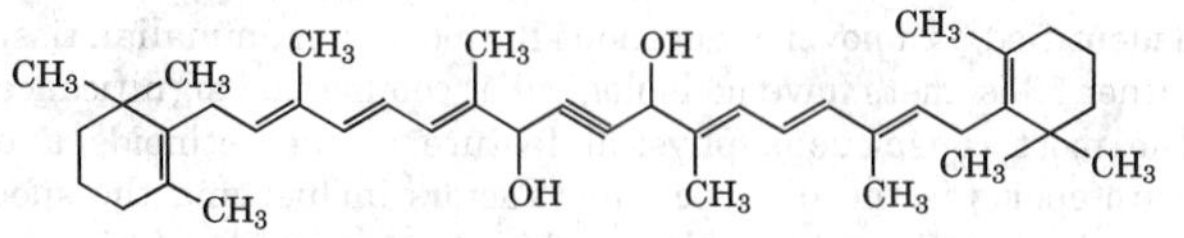

(**24**)

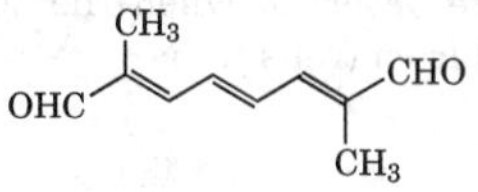

(**25**)

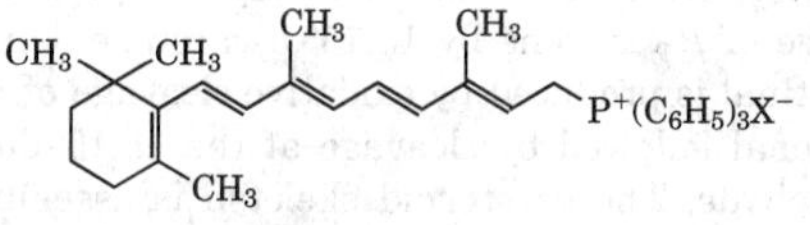

(**26**, X = Cl^-, 1/2 SO_4^{2-})

Analytical Methods

Biological, spectroscopic, and chromatographic methods have been used to assay vitamin A and the carotenoids. Biological methods have traditionally been based on the growth response of vitamin A-deficient rats. Spectroscopic methods such as uv and fluorescence have relied on the polyene chromophore of vitamin A as a basis for analysis. Although useful, spectroscopic methods suffer from not being specific. More specific methods involve chromatographic separation of the retinoids and carotenoids followed by an appropriate detection method. Typically, hplc techniques are coupled with detection by uv.

Occurrence

Rich sources of vitamin A include dairy products, eggs, organ, meat, fish, and in particular the liver oil from certain marine organisms. Fertile sources of carotenoids include carrots and leafy green vegetables such as spinach. Food processing and oxidation affect the vitamin A content in food. Oxidation of carotenoids to biologically inactive xanthophylls represents an important degradation pathway for these compounds.

Biological Functions

In humans, vitamin A has important functions in vision, growth, and tissue differentiation. It is necessary for normal bone growth and remodeling and is required for the activity of epiphyseal cartilage. The specific role of vitamin A in tissue differentiation involves retinoic acid being transferred to a nuclear retinoic acid receptor which enhances the expression of a specific region of the genome. Transcription occurs

Table 2. Provitamin A Activity of Selected Carotenoids

Provitamin A activity	No provitamin A activity
β-carotene	astaxanthin
α-carotene	canthaxanthin
γ-carotene	lutein
β-crypotoxanthin	lycopene
β-zeacarotene	zeaxanthin

and new proteins appear during the retinoic acid-induced differentiation of cells.

In contrast to vitamin A, the carotenoids are important in both plant and animal kingdoms. In plants, carotenoids are associated with photosynthetic structures where the pigments (*1*) serve as energy-transfer agents and (*2*) protect the organism from light-induced photooxidative damage by quenching of the excited triplet state of chlorophyll (itself a producer of singlet oxygen) and quenching of singlet oxygen.

Many carotenoids function in humans as vitamin A precursors; however, not all carotenoids have provitamin A activity (Table 2). Of the biologically active carotenoids, β-carotene has the greatest activity, with 6 μg of β-carotene being equivalent to 1 μg of vitamin A. Owing to the presence of an extended polyene chain, all carotenoids are effective single-oxygen quenchers and antioxidants. In addition, they can stimulate the immune response and, as a result, may protect against certain forms of cancer.

Requirements. Animals cannot synthesize vitamin A-active compounds and must ingest vitamin A or consume appropriate provitamin A compounds such as β-carotene. The RDA of vitamin A ranges from 1500 IU for infants to 3000 IU for pregnant women.

Deficiency. In humans, vitamin A deficiency manifests itself in night blindness, xerophthalmia, Bitot's spots, and corneal ulceration. Changes in the skin have also been observed. Vitamin A deficiency is the principal cause of blindness in the very young.

Toxicity. Acute vitamin A toxicity results from extremely high doses ($\geq$500,000 IU of vitamin A) and is characterized by headache, blurred vision, loss of coordination, nausea, and peeling and itchy skin. Chronic vitamin A toxicity occurs in adults with long-term intakes of $\geq$50,000 IU/d. Symptoms include hair loss, weakness, headache, bone thickening, enlarged liver and spleen, anemia, abnormal menstrual periods, and joint pain. Most of these symptoms are reversible. In animals, extremely high doses of vitamin are teratogenic. On the other hand, the carotenoids are generally nontoxic.

Uses

Vitamin A is generally used in feeds, foods, and pharmaceutical applications; however, the principal use of carotenoids is as a colorant.

SUSAN D. VAN ARNUM
Hoffmann-La Roche Inc.

M. A. Livrea and G. Vidali, eds., *Retinoids: From Basic Science to Clinical Applications,* Birkhauser Verlag, Germany, 1994.

L. J. Machlin, ed., *The Handbook of Vitamins,* Marcel Dekker, New York, 1991.

J. Ganguly, *Biochemistry of Vitamin A,* CRC Press, Boca Raton, Fla., 1989.

VITAMIN B_{12}

Vitamin B_{12} (**1**) is the generic name for a closely related group of substances of microbial origin. Although it is the generic descriptor for all corrinoids exhibiting the biological activity of cyanocobalamin (**1a**). Here vitamin B_{12} and cyanocobalamin are used interchangeably.

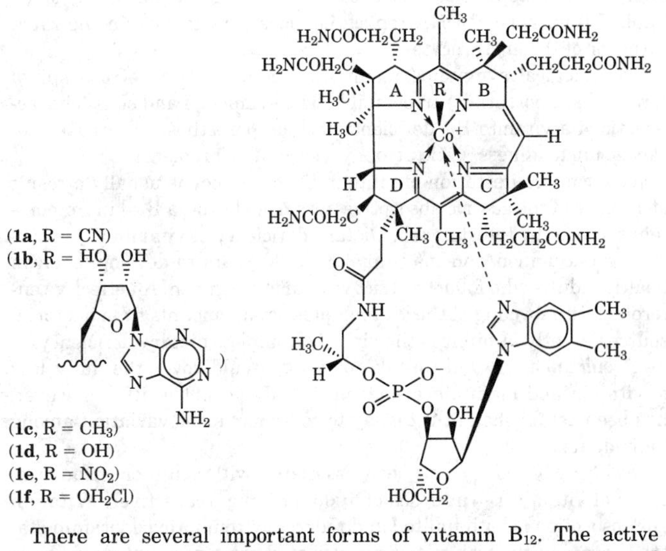

There are several important forms of vitamin B_{12}. The active coenzyme forms are adenosylcobalamin (coenzyme B_{12} (**1b**)) and methylcobalamin (**1c**). These, along with hydroxocobalamin (vitamin B_{12a} (**1d**)), are the forms found in humans and other animals. Other forms of interest are nitrocobalamin (vitamin B_{12c} (**1e**)) and aquacobalamine chloride (vitamin B_{12b} (**1f**)). The primary commercial form is cyanocobalamin, due to its ease of isolation and purification as well as its stability.

Vitamin B_{12} belongs to the class of molecules known as corrins, which consist of four linked, partially saturated pyrrole rings. Corrin is a truncated form (no CH_2 between rings A and D) of the more common porphyrin skeleton. The corrin ring in vitamin B_{12} is substituted in a highly regio- and stereospecific manner with eight methyl groups, three acetic acid chains, and four propionic acid chains. The six conjugated double bonds of the corrin system give octahedral complexes with a number of metals. However, only complexes with cobalt exhibit vitamin B_{12} activity.

Occurrence

For many years, it was thought that the occurrence of vitamin B_{12} was limited to animal tissues and bacteria, with all of the material originating from bacterial sources. More recently, the presence of vitamin B_{12} and/or vitamin B_{12}-like activity at low levels in plants has been recognized. Herbivorous animals satisfy their vitamin B_{12} needs by absorption of material produced by rumenal or intestinal flora. In humans and carnivorous/omnivorous animals, intestinal production of vitamin B_{12} is insufficient and consumption of foods rich in vitamin B_{12} (eg, organ meat, egg yolk, and some fish and shellfish) is needed.

Biochemical Functions

Methylcobalamin and adenosylcobalamin are the two coenzyme forms of vitamin B_{12} in animals and humans. Adenosylcobalamin (coenzyme B_{12}) is required in a number of rearrangement reactions; that occurring in humans is the methylmalonyl-CoA mutase-mediated conversion of (*R*)-methylmalonyl-CoA to succinyl-CoA. The reaction is involved in the catabolism of valine and isoleucine. Methylcobalamin is involved in a critically important physiological transformation, namely the methylation of homocysteine to methionine. The reaction involves transfer of a methyl group first from N^5-methyltetrahydrofolate to cobalamin (yielding methylcobalamin) and thence to homocysteine. Demethylation of tetrahydrofolate to tetrahydrofolic acid is a step in the formation of thymidine phosphate, in turn required for DNA synthesis. Homocysteine has been identified as an independent risk factor for atherosclerosis and thus metabolic control over homocysteine levels has major health implications.

Deficiency. Macrocytic anemia, megaloblastic (pernicious) anemia, and neurological symptoms characterize vitamin B_{12} deficiency. Alterations in hematopoiesis occur because of the high requirement for vitamin B_{12} for normal DNA replication necessary to sustain the rapid turnover of the erythrocytes.

Neurological symptoms result from demyelination of the spinal cord and are potentially irreversible. The symptoms and signs characteristic of a vitamin B_{12} deficiency include paresthesia, decreased reflexes, unsteadiness, optic atrophy, psychiatric problems.

Clinical manifestation of vitamin B_{12} deficiency is usually a result of absence of the gastric absorptive (intrinsic) factor, a B_{12}-binding protein secreted by the stomach. Dietary deficiency of vitamin B_{12} is otherwise uncommon and may take 20 to 30 years to develop, even in healthy adults who follow a strict vegetarian regimen. An effective enterohepatic recycling of the vitamin plus small amounts from bacterial sources greatly minimizes the risk of a complete dietary deficiency.

Requirement. A daily intake of 1 μg should cover the daily loss of vitamin and maintain an adequate body pool. The RDA however, has been established at 2 μg/day to cover metabolic variation among individuals.

Safety. No toxicity has been associated with acute or chronic intakes of vitamin B_{12} in doses of 100 and 1 mg, respectively. Vitamin B_{12} absorption is both limited and reduced with improved vitamin status, lessening the risk of toxicity. Isolated cases of anaphylaxis have been reported with intravenous administration.

Properties

Table 1 lists some of the physical and chemical properties of vitamin B_{12} Crystalline vitamin B_{12} is stable in air and is not affected by moisture. The anhydrous compound, however, is very hygroscopic and may absorb about 12% of water from air. Aqueous solutions of vitamin B_{12} at pH 4.0 to 7.0 show no decomposition during extended storage at 25°C.

Redox Reactions. Critical to the function of cobalamins as enzyme cofactors is the ability of the cobalt atom to exist in the Co(III) and Co(I) oxidation states. Each oxidation state has different ligand-accepting abilities. The chemical (**1a, 1d–1f**) and coenzyme (**1b,1c**) forms of cobalamin are trivalent. These compounds are readily reduced to Co(II) and Co(I)cobalamins. Co(I)cobalamin is a powerful reducing agent and is therefore unstable in aqueous acid solution, less so in neutral or basic aqueous solution. It is frequently used for the preparation of organocobalamins, eg, adenosylcobalamin and methylcobalamin.

Exchange of Axial Ligands. Many ligand-exchange reactions involve groups in which the coordination to the metal is through nitrogen (NH_3,N_3^-), oxygen (H_2O,OH^-), sulfur (SH^-,SO_3^{-2}), halogen, or carbon (CN^-,CH_3^-. Important reactions involve displacement of the heterocyclic base from the alpha-coordination position by a solvent, usually water. This displacement occurs in acidic solution and results from the protonation (associated with a characteristic change in the spectrum) of the heterocyclic base. The pK_a for the base-on/base-off equilibrium depends on the nature of the beta-ligand. In the presence of cyanide ion, aquacobalamin and adenosylcobalamin are converted to cyanocobalamin. In contrast, methylcobalamin and other alkyl corrinoids are stable in the presence of 0.1 *M* cyanide in the dark.

Chemical Reactions. Vitamin B_{12} is slowly decomposed by ultraviolet or strong visible light is also inactivated by treatment with strong acids or bases.

Vitamin B_{12} derivatives can catalyze chemical as well as biochemical processes, eg, the electrochemical reduction of alkyl halides and formation of C—C bonds, as well as the zinc–acetic acid-promoted reduction of nitriles, alpha, beta-unsaturated nitriles, alpha, beta-unsaturated carbonyl derivatives and esters, and olefins. It is assumed that these reactions proceed through intermediates containing a Co–C bond which is then reductively cleaved.

Analytical Methods. Vitamin B_{12} can be determined by microbiological, radioisotope dilution, spectrophotometric, chemical, or biological methods employing animals. Microbiological assays involve the extraction and stabilization of vitamin B_{12} from the food or pharmaceuticals prior to assay. The official method of the AOAC accomplishes this by autoclaving samples in a phosphate-citric acid buffer containing metabisulfite.

The assay is based on the graded growth of the bacterium *Lactobacillus leichmannii* ATCC 7830 in a medium lacking only vitamin B_{12}. Results are obtained by determining the transmittance of the sample and standard tubes after 16–24 h incubation at 30–40°C or by titrating the acid produced after 72 h incubation. Derivatives of deoxynucleic acid that stimulate the growth of *L. leichmannii* in the absence of vitamin B_{12} can be measured after destroying the vitamin B_{12} by autoclaving the sample at pH 11–12 and determining the residual activity. Radioisotope dilution assays are based on the principle of competition between radioactive labeled (^{57}Co) vitamin B_{12} and cobalamins extracted from matrices for binding sites on the intrinsic factor (a glycoprotein). Free cobalamins are separated from those bound on the intrinsic factor by absorption onto treated charcoal and the amount of free-labeled vitamin B_{12} is determined. Vitamin B_{12} content of the sample is determined from a standard curve.

Spectrophotometric determination at 550 nm is relatively insensitive and is useful for the determination of vitamin B_{12} in high potency products such as premixes. A high performance liquid chromatographic (hplc) method is reported to require a sample containing 20–100 μg cyanocobalamin and is suitable for premixes, raw material, and pharmaceutical products.

Synthesis

Chemical Synthesis. All commercially available vitamin B_{12} is produced by microbial fermentation. The complexity of the molecule makes it extremely unlikely that commercial quantities will ever be prepared by total chemical synthesis.

The core of the first synthesis of vitamin B_{12} involved condensation of the A–D ring fragment (**2**) with the B–C fragment (**3**).

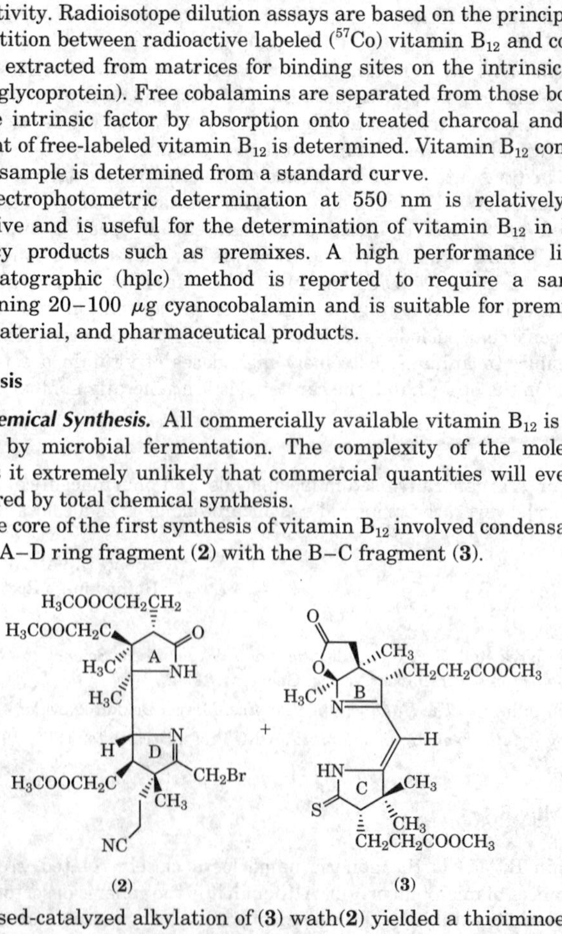

(**2**) (**3**)

Based-catalyzed alkylation of (**3**) wath(**2**) yielded a thioiminoether which underwent sulfide contraction upon treatment with acid to

Table 1. Physical and Chemical Properties of Vitamin B_{12} (Cyanocobalamin)[a]

Property	Characteristic
appearance	crystalline
color	dark red
mol wt	1355.42
empirical formula	$C_{63}H_{88}CoN_{14}O_{14}P$
mp, °C	darkens 210–220; does not melt <300
specific rotation	$\alpha_{656}^{23} = -59 \pm 9°$ (diluted aqueous)
uv absorption	278,361,550 nm (water)
pH (aq)[b]	neutral
solubility %	
H_2O	1.25
alcohol	2.03
acetone	insoluble
ether	insoluble

[a] Properties given for the anhydrous compound.
[b] Vitamin B_{12} exhibits maximum stability between pH 4.5 and 5.0

give the tetrapyrrole (**4**). The cobalt atom was then added. The stereo-organizing template effect of the complex allowed base-catalyzed cyclization to complete the corrin ring system. Functional group exchange and addition of the final methyl group then yielded cobyric acid, a compound that can be converted to vitamin B_{12}.

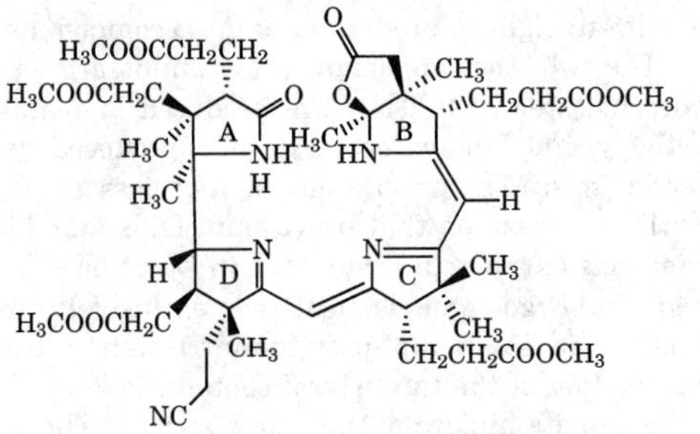

(**4**)

A second synthesis of cobyric acid involves photochemical ring closure of an A–D secocorrinoid. Thus, the Diels-Alder reaction between butadiene and *trans*-3-methyl-4-oxopentenoic acid was used as starting point for all four ring A–D synthons (**5–8**). These were combined in the order B + C → BC + D → BCD + A → ABCD. The resultant cadmium complex (**9**) was photocyclized in buffered acetic acid to give the metal-free corrinoid (**10**), which was converted to cobyric acid (**11**).

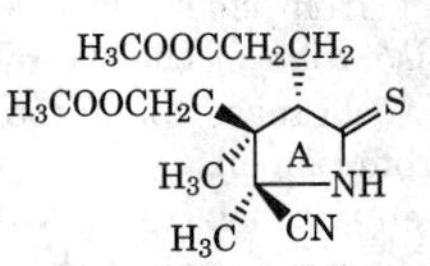

(**5**)

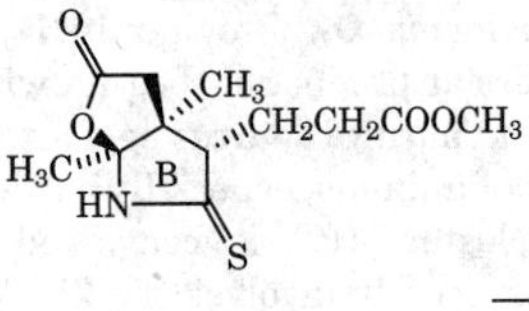

(**6**)

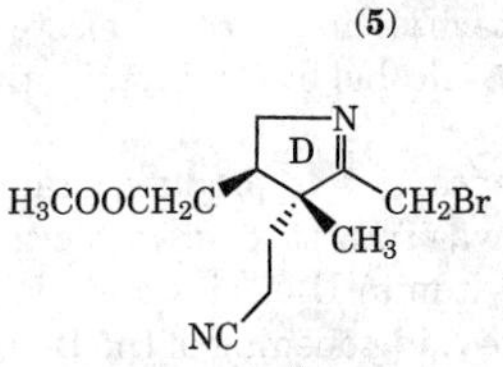

(**8**)

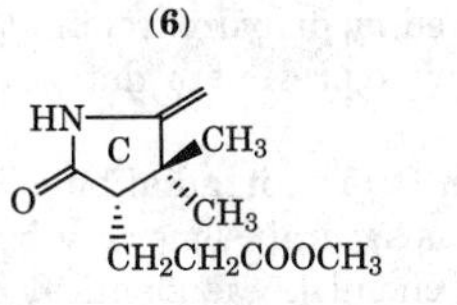

(**7**)

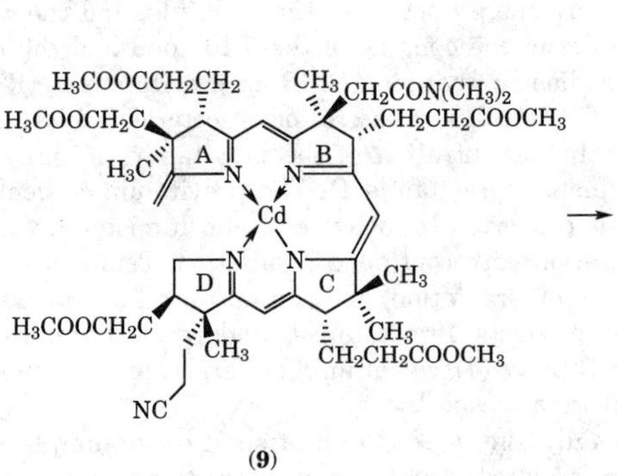

(**9**)

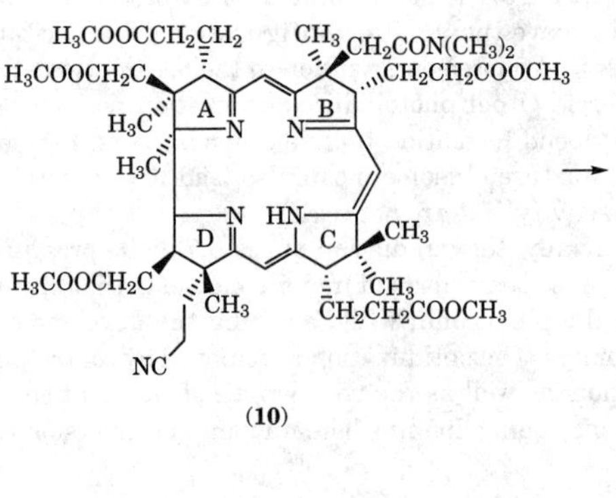

(**10**)

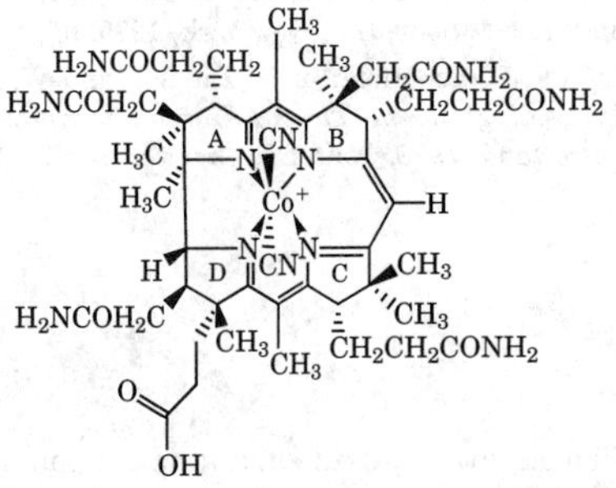

(**11**)

One approach to synthesizing other cobalamins, eg, methylcobalamin and adenosylcobalamin, involves the oxidative addition of the appropriate alkyl halide (eg, CH_3I to give methylcobalamin) or tosylate (eg, 5′-*p*-tosyladenosine to yield adenosylcobalamine) to cobalt(I)alamine.

Biosynthesis. The biosynthesis of vitamin B_{12} in *Pseudomonas dentrificans* is important commercially and is representative of the process in aerobic bacteria. The 22 genes involved have been identified and the sequence of the resulting proteins described. The sequence is initiated from 5-aminolevulinic acid, which is biosynthesized from succinyl-CoA and glycine. Condensation yields porphobilinogen, which is tetramerized as a linear, head-to-tail arrangement to give hydroxymethylbilane. Ring closure and rearrangement (formally on exchange of acetate and propionate substituents in ring C) gives uroporphyrinogen III (uro'gen III). Uro'gen III is a key biosynthetic branchpoint, leading not only to vitamin B_{12} but also to chlorophyll and heme.

The sequence to vitamin B_{12} proceeds by the introduction of the eight methyl groups, punctuated by ring contraction to the corrin, an oxidation and reduction, and a methyl migration to give hydrogenobyrinic acid.

Introduction of the cobalt atom into the corrin ring is preceeded by conversion of hydrogenobyrinic acid to the diamide. The resultant cobalt(II) complex is reduced to the cobalt(I) complex prior to adenosylation. Four of the six remaining carboxylic acids are converted to primary amides (adenosylcobyric acid) and the other amidated to provide adenosylcobinamide. Completion of the nucleotide loop involves conversion to the monophosphate followed by reaction with guanosyl triphosphate to give diphosphate. Reaction with α-ribazole 5′-phosphate, derived biosynthetically in several steps from riboflavin, and dephosphorylation complete the synthesis.

In microaerophilic bacteria such as *Propionibacterium shermanii,* a fundamental difference occurs in that the cobalt atom is introduced at a much earlier stage. A fermentation such as that of *Pseudomonas dentrificans* typically requires 3–6 days. A submerged culture is employed with glucose, cornsteep liquor and/or yeast extract, and a cobalt source (nitrate or chloride) plus other minerals. The required pH of 6–7 is achieved by ammonium or calcium salts.

The fermentation product, which is primarily adenosylcobalamin, is retained to the largest extent within the cell. Centrifugation of the broth yields a sludge which, when dispersed in a minimum of water, water–alcohol, or water–acetone and heated, releases the vitamin into the solution. Addition of cyanide converts the cobalamins into cyanocobalamin which is then extracted from the filtered solution. Chromatography on aluminum oxide and crystallization from methanol–acetone or water–acetone complete the process.

JOHN W. SCOTT
Hoffmann-La Roche Inc.

J. Marks, *A Guide to the Vitamins, Their Role in Health and Disease,* Medical and Technical Publishing Co., Lancaster, U.K., 1975, p. 118.

J. P. Glusker, in G. Litwack, ed., *Vitamins and Hormones,* Vol. 50, Academic Press, Inc., New York, 1995, pp. 1–76.

H. P. C. Hogenkamp, in B. M. Babior, ed., *Cobalamin-Biochemistry and Pathophysiology*, John Wiley & Sons, Inc., New York, 1975, p. 21.

G. Rytz, L. Walder, and R. Scheffold, in B. Zagalak and W. Friedrich, eds., *Vitamin B_{12}, Proceedings of the Third European Symposium on Vitamin B_{12}, Zurich, Switzerland, 1979,* Walter de Gruyter, Berlin, 1979, p. 173.

VITAMIN D

Vitamin D is formed in the skin of animals upon irradiation by sunlight and serves as a precursor for metabolites which control the animal's calcium homeostasis and act in other hormonal functions. A deficiency of vitamin D can cause rickets, as well as other disease states.

The vitamins D are 9,10-secosteroids, that is, steroid molecules with an opened 9,10 bond of the B-ring. Vitamin D_2 and vitamin D_3 are the two economically important forms. The other D vitamins have relatively little biological activity. Vitamin D_2 (ergocalciferol; ercalciol), (5*Z*,7*E*,22*E*)-(3*S*)-9,10-*seco*-5,7,10(19)-22-ergostatraene-3-ol **(2)**, is active in humans and other mammals, although recently (ca 1997) it has been shown to be less active than vitamin D_3 in cattle, swine, and horses. It is relatively inactive in poultry. It is prepared by the uv irradiation of ergosterol (provitamin D_2), (24-methylcholesta-5,7,22-triene-3B-ol (**1**), a plant sterol.

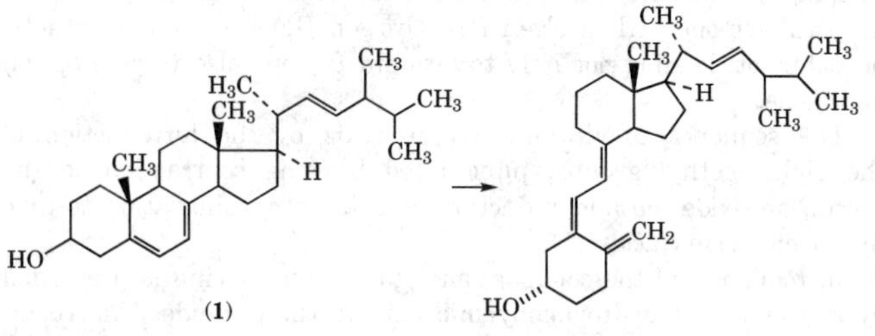

(1) (2)

Vitamin D_3 (cholecalciferol; calciol), (5*Z*,7*E*)-(3*S*)-9,10-seco-5,7,10(19) cholestatri-ene-3-ol (**4**), is the naturally occurring active material found in all animals. It is produced in the skin by the irradiation of stored 7-dehydrocholesterol (provitamin D_3), cholesta-5,7-diene-3B-ol (**3**).

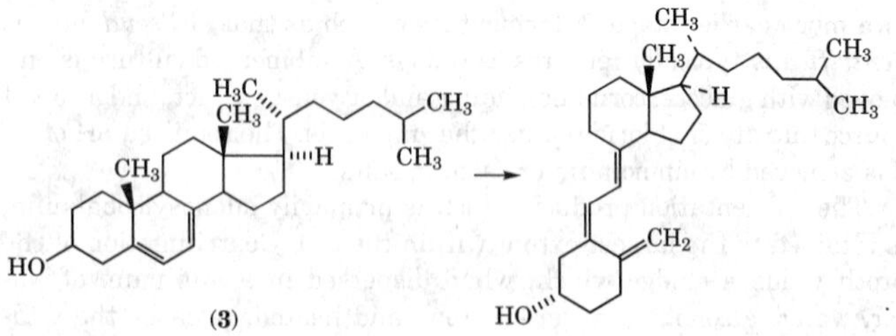

(3) (4)

Occurrence

The provitamins, precursors of the vitamins D, are distributed widely in nature, whereas the vitamins themselves are less prevalent. Provitamin D_2 is found in plants and D_3 almost exclusively in animals. Fish-liver oil, liver, milk, and eggs are good natural sources of the D_3 vitamin. Fish oil is the only commercial source of natural vitamin D_3, and the content of the vitamin varies according to species as well as geographically.

Chemical and Physical Properties

The chemistry of the D vitamins is intimately involved with that of their precursors, the provitamins. The manufacture of the vitamins and their derivatives usually involves the synthesis of the provitamins, from which the vitamin is then generated by uv irradiation.

3β-Hydroxy steroids which contain the 5,7-diene system and can be activated with uv light to produce vitamin D compounds are called provitamins. The two most important provitamins are ergosterol (**1**) and 7-dehydrocholesterol (**3**). They are produced in plants and animals, respectively, and 7-dehydrocholesterol is produced synthetically on a commercial scale. The provitamins do not possess physiological activities, with the exception that provitamin D_3 is found in the skin of animals and acts as a precursor to vitamin D_3.

Provitamin D_2. Ergosterol is isolated exclusively from plant sources. Usually, the isolation of provitamin D_2 from natural sources involves the isolation of the total sterol content, followed by the separation of the provitamin from the other sterols. The isolation of the sterol fraction involves extraction of the total fat component, its saponification, and then reextraction of the unsaponifiable portion with an ether. The sterols are in the unsaponifiable portion. Another method is the saponification of the total material, followed by isolation of the nonsaponifiable fraction. Separation of the sterols from the unsaponifiable fraction is done by crystallization from a suitable solvent, eg, acetone or alcohol. In the case of yeasts, it is particularly difficult to remove the ergosterol by simple extraction. Industrially, therefore, the ergosterol is obtained by preliminary digestion with hot alkalies or with amines.

Provitamin D_3. Provitamin D_3 is made from cholesterol, and its commercial production begins with the isolation of cholesterol from one of its natural sources eg, eluted from wool grease spinal cords and brains of animals, especially cattle, and from fish oils.

Cholesterol (**7**) is converted to 7-dehydrocholesterol (**3**). This process usually involves the Ziegler allylic bromination of the 7 position followed by dehydrobromination. Esterification of the cholesterol is necessary to prevent oxidation of the 3β-alcohol by the brominating agent.

Vitamin D. The irradiation of the provitamins to produce vitamin D as well as several isomeric substances was first studied with ergosterol. The chemistry is identical for the vitamin D_3 series and yields analogous isomers. The initial step involves ring opening of the B-ring of the sterol by ultraviolet activation of the conjugated diene. The absorbance of uv energy activates the molecule, and the $\pi \rightarrow \pi^*$ excitation, results in the opening of the 9,10 bond and the formation of the (*Z*)-hexadiene, previtamin D_3 (R) or previtamin D_2 (R′). The uv irradiation of 7-dehydrocholesterol or ergosterol results in the steady diminution in concentration of the provitamin, initially giving rise to predominantly previtamin D. The previtamin concentration then falls as it is converted to tachysterol and lumisterol, which increase in concentration with continued irradiation. Temperature, frequency of light, time of irradiation, and concentration of substrate all affect the ratio of products. Previtamin D undergoes thermal equilibration to vitamin D (**2**) or (**4**), *cis*-vitamin D_2 (ergocalciferol) and *cis*-vitamin D_3 cholecalciferol, respectively.

Additionally, the $\pi \rightarrow \pi^*$ excitation of previtamin D can result in ring closure back to the provitamin (**1**) or (**3**) or to lumisterol$_2$ or lumisterol$_3$, which have the 9β,10α configuration. This excitation can also exhibit $(Z) \rightleftharpoons (E)$ photoisomerization to the 6,7-(*E*)-isomer, tachysterol$_2$ or tachysterol$_3$. Other photoinduced cyclization reactions can occur by conrotatory bond formation to give other isomers. Prolonged irradiation of the mixture of isomers can also lead to toxisterols (which lack biological activity, but are not toxic).

Commercially, the irradiation of the 5,7-diene provitamin to make vitamin D must be performed under conditions that optimize the production of the previtamin while avoiding the development of the unwanted isomers. The optimization is achieved by controlling the extent of irradiation, as well as the wavelength of the light source, with the best frequency being 295 nm. The unwanted conversion of previtamin

Table 1. Physical Properties of Provitamins and Vitamins D_2 and D_3

	Substance			
Properties	7-Dehydrocholesterol	Ergosterol	Vitamin D_2 (ergocalciferol)	Vitamin D_3 (cholecalciferol)
melting point, °C	150–151	165	115–118	84–85
color and form	solvated plates from ether-methanol	hydrated plates from alcohol; needles from acetone	colorless prisms from acetone	fine colorless needles from dilute acetone
optical rotation (α_D^{20}), °				
acetone			82.6	83.3
ethanol			103	105–112
chloroform	−113.6	−135	52	51.9
ether			91.2	
petroleum ether			33.3	
benzene	−127.1			
uv max, nm	282	281.5	264.5	264.5
specific absorption, E_{max}, at 1% conc	308		458.9 ± 7.5	473.2 ± 7.8
potency[a], IU/g			40×10^6	40×10^6
biological activity			in mammals	n mammals and birds
chicken efficacy, %			8–10[b]	100
solubility, g/100 mL				
acetone at 7°C			7	
acetone at 26°C			25	
absolute ethanol	sl sol		28	
ethyl acetate at 26°C			31	
water	insol	insol	insol	insol

[a] The international standard for vitamin D is an oil solution of activated 7-dehydrocholesterol. The IU is the biological activity of 0.025 μg of pure cholecalciferol.
[b] Studies have claimed an efficacy as high as 10%.

to tachysterol is favored when 254 nm light is used. Sensitized irradiation, eg, with fluorenone, has been used to favor the reverse, triplet-state conversion of tachysterol to previtamin D.

Physical Properties. The physical properties of the provitamins and vitamins D_2 and D_3 are listed in Table 1. The D vitamins are fat-soluble and, as such, are hydrophobic. Vitamin D and its products are sensitive to uv light, heat, air, and mineral acids. Its sensitivity to these conditions is exaggerated by the presence of heavy-metal ions, eg, iron.

Shipping vitamin D in crystalline or resin form should be done in containers marked appropriately to indicate the material is toxic by DOT standards.

Analytical and Test Methods

The development of reliable uv analysis permitted the dependable detection and assay of the provitamins and vitamins.

A radioimmunodiffusion assay allows assay of the material for vitamin D activity down to 1–200 IU.

The standard chemical and biological methods of analysis are those accepted by the *United States Pharmacopeia XXIII* as well as the ones accepted by the AOAC in 1995. The USP method involves saponification of the sample; solvent extraction; solvent removal; chromatographic separation of vitamin D from extraneous ingredients; and colormetric determination with antimony trichloride.

The AOAC procedure, requires removal of the trans-isomer before colorimetric assay. However, certain inactive isomers, eg, isotachysterol, if present, give rise to a falsely high analysis.

The USP and AOAC recognize a biological method for the determination of vitamin D; but the rat line test is slow, expensive and not as accurate as the chemical or chromatographic methods.

Vitamins D_2 and D_3 exhibit uv absorption curves that have a maximum at 264 nm and an E_{max} (absorbance) of 450–490 at 1% concentration. The various isomers of vitamin D exhibit characteristically different uv absorption curves, but require prior separation by hplc.

Synthesis

Manufacture. Most of the vitamin D produced in the world is made by the photochemical conversion of 7-dehydrocholesterol. Irradiation of 7-dehydrocholesterol or ergosterol is carried out by dissolving the steroid in an appropriate solvent, eg, peroxide-free diethyl ether. Best results are achieved at 275–300 nm.

The ether solution is recirculated through a quartz uv reactor (275–300 nm), and the ether is distilled off. Methanol is added to the 7-dehydrocholesterol-vitamin D_3 mixture, and the remaining ether is azeotroped. The resulting solution is transferred to a crystallizer, and the 7-dehydrocholesterol is crystallized and recovered by filtration. The methanol is distilled, and the vitamin D_3 resin is heated to isomerize the pre-vitamin D to the cis-vitamin D isomer.

Total Synthesis. In 1959 Inhoffen synthesized vitamin D_3 from 3-methyl-2-(2-carboxyethyl)-2-cyclohexenone, using the Wittig reaction extensively. 5,6-*trans*-Vitamin D_3 was prepared, followed by photochemical isomerization into vitamin D_3.

The most useful synthetic routes include the following pathways to the vitamin D structures and their derivatives:

Photochemical ring opening of 7-dehydrocholesterol derivatives which have ring A or the side chain modified.

A phosphine oxide of type (**5**) can be coupled with Grundman's ketone (**6**) to produce the D_3 skeleton.

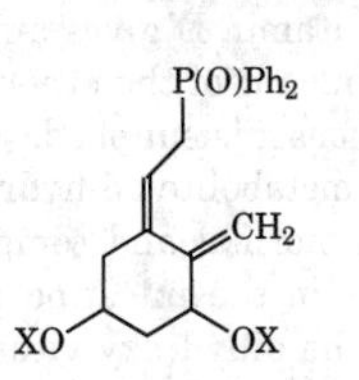

(**5**, X = a protective group)

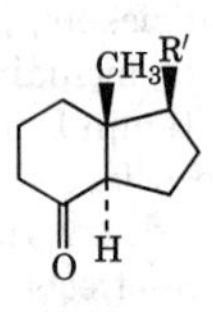

(**6**, R′ = cholesterol side chain or derivative)

Synthetic dienynes like (**7**) are semihydrogenated to form previtamin I and are then rearranged to the D structure.

Vinyl allenes (**8**) are rearranged with heat or metal catalysis and photo sensitized isomerization to produce the vitamin D triene.

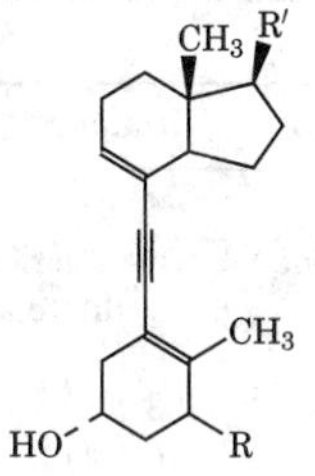

(7, R = H, OH)

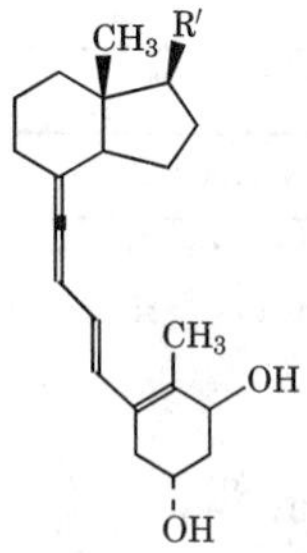

(8)

R′ = cholesterol side chain or derivative

Direct modification of vitamin D_2 or D_3 or its metabolites to give derivatives by a variety of synthetic methods.

Biochemistry

Vitamin D metabolites behave as hormones. As such they play an active role in the endocrine system. Although vitamin D metabolites influence many different biological functions, metabolism primarily occurs to maintain the calcium homeostasis of the body. When calcium serum levels fall below the normal range, 1α,25-dihydroxyvitamin D_3 is made; when calcium levels are at or above this level, 24,25-dihydroxycholecalciferol is made, and 1α-hydroxylase activity is discontinued. The calcium homeostasis mechanism involves a hypocalcemic stimulus, which induces the secretion of parathyroid hormone. This causes phosphate diuresis in the kidney, which stimulates the 1α-hydroxylase activity and causes the hydroxylation of 25-hydroxy-vitamin D to 1α,25-dihydroxycholecalciferol. Parathyroid hormone and 1,25-dihydroxycholecalciferol act at the bone site cooperatively to stimulate calcium mobilization from the bone. Calcium blood levels are also influenced by the effects of the metabolite on intestinal absorption and renal resorption.

Vitamin D_3 Deficiency. Vitamin D_3 deficiency is uncommon in normal adults. It can occur because of a large reduction of fat intake, which decreases D_3 absorption; minimal exposure to sunlight; or clinical stresses that interfere with vitamin D_3 metabolism, including malabsorption, stomach bypass surgery, obstructive jaundice, alcoholism, liver or kidney failure, inborn error of metabolism, and use of anticonverdiants.

Rickets is the most common disease associated with vitamin D deficiency. Other disease states may involve a lack of the vitamin, deficient synthesis of the metabolites from the vitamin, deficient control mechanisms, or defective organ receptors. Vitamin D's control of calcium and phosphorus homeostasis is essential in the maintenance of normal cellular biochemistry, eg, muscle contraction, nerve conduction, and enzyme function. The vitamin D metabolites also affect gene transcription and cell proliferation.

Vitamin D_3 metabolites have been successfully used for treatment of milk fever in cattle, turkey leg weakness, plaque psoriasis, and osteoporosis and renal osteodystrophy in humans.

Dietary Requirements. According to the National Research Council (NRC), the vitamin D requirement for optimum health is ca 400 IU/d in humans, regardless of age. This amount of vitamin D gives ample protection from rickets, provided a sufficient amount of the other essential nutrients, including calcium and phosphorus, is supplied.

Toxicity. Vitamin D intoxication causes the metabolite 25-hydroxy vitamin D_3 blood levels to rise 80-fold above a normal and compete with 1α-25-dihydroxy vitamin D_3 for receptors in the intestine and bone and induce effects usually attributed to the dihydroxy vitamin D_3. Vitamin D_2 is metabolized slower than vitamin D_3 and thus appears to be less toxic. The overall effect in most animals is to stimulate intestinal absorption of calcium with a concomitant increase in serum calcium, reduction in parathyroid hormone (PTH), and the subsequent fall in urinary calcium. Polyuria, along with vomiting can cause extracellular fluid to be reduced, contributing to further renal function disruption. The metabolites of vitamin D are usually more toxic than the vitamin because the feedback mechanisms that regulate vitamin D concentrations are circumvented.

Uses

Most of the vitamin D sold is synthetic. Vitamin D_2 as a concentrate or in microcrystalline forms is used in many pharmaceutical preparations, although vitamin D_3 is preferred by many manufacturers and consumers. Vitamin D_2 in the form of irradiated yeast has been used as a feed supplement for cattle, swine, and dogs, but its use has declined in favor of Vitamin D_3. Essentially all milk produced in the United States is fortified with vitamin D_3. Cereals and margarine are also fortified with vitamin D_3.

Approximately 200 kg/yr of Vitamin D_3 formulations are also marketed as rat poisons. The metabolites of vitamin D_3 and synthetic derivatives are being used or developed for treatment of osteoporosis, skin psoriasis, and other diseases in humans. 1α-Hydroxy vitamin D_3 is being used for milk fever in cows, and 25-hydroxy vitamin D_3 has been proposed for eggshell thickness in poultry and is being marketed as an animal dietary nutritional supplement.

ARNOLD L. HIRSCH
Alpharma Inc.

G. F. Combs and H. F. DeLuca, *The Vitamins: Fundamental Aspects in Nutrition and Health,* Academic Press, Inc., 1992.

E. D. Collins and A. W. Norman, in L. J. Macklin, eds., *Handbook of Vitamins,* 2nd ed., Marcel Dekker, New York, 1991, Chapt. 2.

H. F. DeLuca in R. B. Olfin-Slater and D. Kritchevsky, eds., *Nutrition and the Adult: Micronutrients,* Plenum Press, New York, 1986.

W. Friedrich, *Vitamins,* Walter de Gruyter, New York, 1988.

VITAMIN E

Vitamin E was the name originally applied to a material found in vegetable oils that was essential for fertility in rats. Subsequently a group of substances (Fig. 1), which fall into either the family of tocopherols or tocotrienols, were found to act like vitamin E. In label claims for vitamin E, only the *RRR* or *all-rac-α*-tocopherol and its esters can be claimed.

Chemical, Biological, and Physical Properties

The structures of all vitamin E compounds are characterized by a 6-chromanol ring with a C_{16} side chain. As a result of three asymmetric carbons, there are eight possible enantiomers. However, it is the configuration around the 2-position on the ring which determines, to the largest extent, biological activity. Natural vitamin E is characterized by all R stereochemistry. The tocopherols are distinguished by a saturated phytol side chain and the tocotrienols by an unsaturated side chain. The isomers within the two families differ by the extent and position of the methylation on the ring. Physical properties of α-tocopherols are given in Table 1. Although they are generally stable to heat and alkali, and acids in the absence of oxygen, free tocopherols and tocotrienols can be oxidized by atmospheric oxygen in the presence of light, unsaturated fatty acids, heat, or metal ions. As a result of the ease of oxidation, significant amounts of vitamin E (tocopherols) may be lost during food processing. The stability of α-tocopherol is significantly enhanced by esterification.

Sources

Unesterified tocopherols are found in a variety of foods. Almost all of the tocopherol in meat, fish, and dairy is α-tocopherol. Vegetable oils contain significant levels of γ-, β-, and δ-tocopherol, along with α-tocopherol.

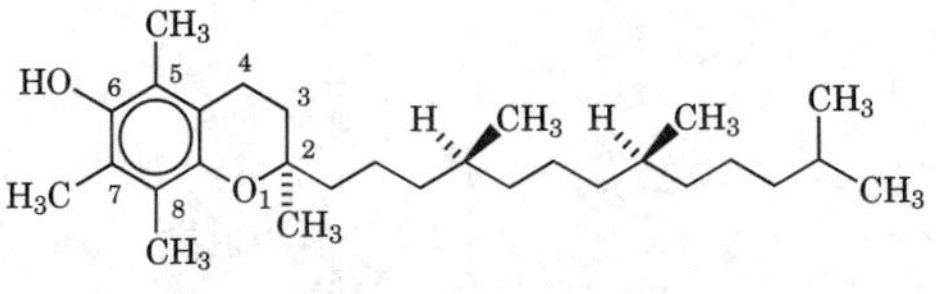

(1)

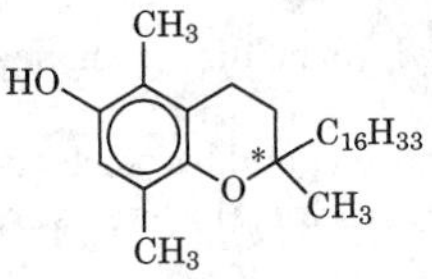

(2)

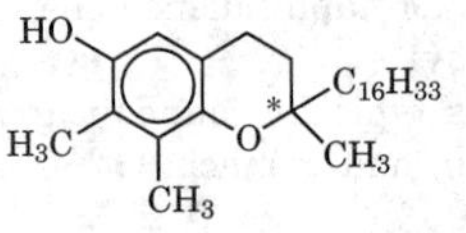

(3)

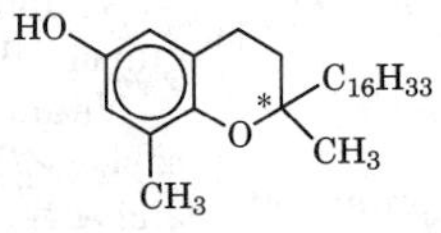

(4)

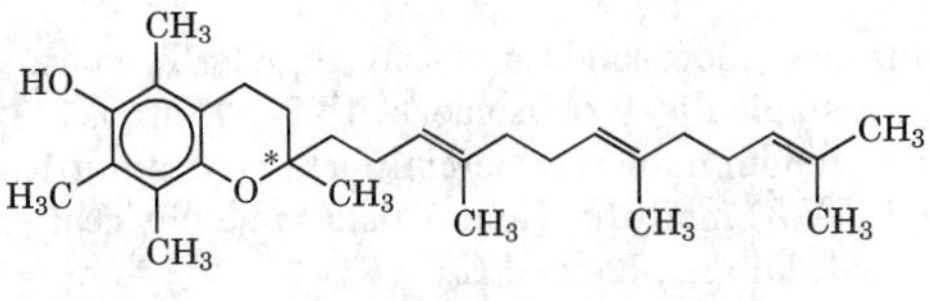

(5)

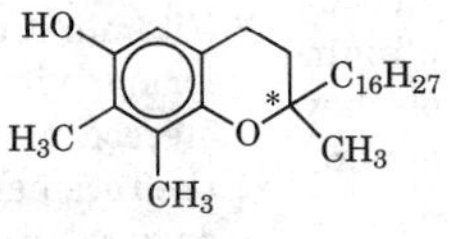

(6)

Figure 1. The four naturally occurring tocopherols (α-tocopherol, *RRR*/*all-rac* (**1**); β-tocopherol (**2**); γ-tocopherol (**3**); δ-tocopherol (**4**)), α-tocotrienol (**5**), and β-tocotrienol (**6**) where asterisks denote asymmetric centers and the absolute configuration of *RRR*-α-tocopherol is shown.

Table 1. Physical Properties of Tocopherols

Property	*all-rac*-α-Tocopherol	(*RRR*)-α-Tocopherol	*all-rac*-α-Tocopheryl Acetate	*RRR*α-Tocopheryl-Acetate
color	←colorless to pale yellow→			
form	←viscous oil→			
bp, °C	200–220 (0.1 mm)		224 (0.3 mm)	
sp gr$^{25}_{25}$	0.947–0.958	0.950–0.964	0.950–0.964	
n_D^{20}	1.5030–1.5070	1.4940–1.4985	1.4940–1.4985	
mol wt	430.69	430.69	472.73	472.73
uv absorption				
maxima, nm	292–294	292–294	285.5	285.5
$E^{1\%}_{1\,cm}$ (ethanol)	71–76	71–76	40–44	40–44

Synthesis

Although all four tocopherols have been synthesized as their *all-rac* forms, the commercially significant form of tocopherol is *all-rac*-α-tocopheryl acetate. The commercial processes use a Friedel-Crafts-type condensation of 2,3,5-trimethylhydroquinone with either phytol, a phytyl halide, or phytadiene. The principal synthesis in current commercial use involves condensation of 2,3,5-trimethylhydroquinone with synthetic isophytol in an inert solvent, such as benzene or hexane, with an acid catalyst, such as zinc chloride, boron trifluoride, or orthoboric acid/oxalic acid to give the *all-rac*-α-tocopherol. Free tocopherol is protected as its acetate ester by reaction with acetic anhydride. Purification of tocopheryl acetate is readily accomplished by high vacuum molecular distillation and rectification. *RRR*-α-tocopherol and the other natural forms of vitamin E can be isolated from deodorizer distillates produced as by-products of vegetable oil processing. Typically alkali-treated soybean oil is distilled at high vacuum in a continuous molecular still. This minimizes thermal degradation. The distillate, which contains α- γ-, and δ-tocopherols, is then cooled (∼−10°C) to remove impurities such as sterols. The other impurities, such as fatty acids, can be removed by saponification. Further molecular distillation produces a fraction of (≥60%) tocopherols. The sterols and fatty acids can be sold. The yield of the more active *RRR*-α-tocopherol can be improved by selective methylation of the other tocopherol isomers or by hydrogenation of α-tocotrienol.

Although apparently not commercially important, fermentation processes have been reported for the production of tocopherols.

Physiological Effects and Requirements

The symptoms of vitamin E deficiency in animals are numerous and vary by species. Although the deficiency of the vitamin can affect different tissue types (reproductive, gastrointestinal, vascular, neural, hepatic, and optic) necrotizing myopathy is common to most species. In humans, vitamin E deficiency can result from poor fat absorption and leads to hemolytic anemia, reduction in red blood cell lifetimes, retinopathy, and neuromuscular disorders.

Vitamin E can also act as an antioxidant alone or in combination with vitamin C. Vitamin C acts to preserve the levels of vitamin E, which in turn can preserve the levels of vitamin A. Vitamin E reduces the incidence of cardiovascular disease. Most likely because it inhibits the oxidation of low density lipoproteins (LDLs).

The recommended daily allowance for vitamin E ranges from 10 International Units (1 IU = 1 mg *all-rac*-α-tocopheryl acetate) for children under 4 years of age to 30 IU for adults. Vitamin E is considered nontoxic at levels up to 3200 mg/day, but can heighten the effect of vitamin K deficiency (coagulation defect) or anticoagulation therapy.

Analytical Methods

Analysis of vitamin E can be done by a variety of methods depending on the form and level present and the preparation in which the form is found. For pure or highly concentrated forms, this is accomplished by reaction with Emmerie-Engel reagent (2,2′-bipyridine (α,α′-dipyridyl) and ferric chloride) to give a red color. This color results from the combination of bipyridine with ferrous ions formed from the reduction of the ferric ions by the tocopherol and is directly proportional to the amount of tocopherol present. The colorimetric method is nonspecific. The AOAC Official Methods of Analysis describes a packed column gas chromatographic method for the analysis of tocopherol isomers in mixed concentrates. This method separates α- and δ-tocopherols as discrete peaks, but β- and γ-tocopherols elute as a combined peak.

Some high pressure liquid chromatographic techniques require a saponification step to remove fats, to release tocopherols from cells, and/or to free tocopherols from their esters. All require an extraction step. The methods include both normal and reverse-phase hplc with either uv absorbance or fluorescence detection.

Bioassay Method. Amodification of the Evans resorption–gestation method in rats can be used to determine α-tocopherol activity in supplements. Unless a vitamin E supplement containing more than 0.3–1.0 mg (depends on methodology) of α-tocopherol is administered in the first 10–12 days of pregnancy, the embryos die and are reabsorbed without apparent harm to the mother.

Applications and Product Forms

Both α-tocopherol and its esters are constituents of multivitamin and single-dose nutrient capsules or liquid dietary supplements. Supplements for human use range from a few milligrams in multivitamin preparations to 500–1000 mg in single-dose supplements. Specialty items, such as ointments, creams, salves, and suppositories containing vitamin E provide other outlets for α-tocopherol. Tocopherols have significant application in cosmetics and dermatology. Tocopherols can be used in topical cream or oil forms to treat chronic skin diseases, reduce scarring in wound healing, reduce inflammation, and protect against uv radiation. Tocopherols are also incorporated into cosmetic formulations to reduce nitrosamine and nitrosamide formation from amines and amides also present in the formulation.

Animal feeds consume approximately 40% of the commercial production of α-tocopherol acetate. Tocopherols are finding additional applications as antioxidants in foods as the less expensive BHT and BHA addition are being prohibited in more and more countries. The meat from pigs fed vitamin E supplements shows improved color and/or storage stability. Tocopherol can be used in bacon and similar foods to prevent the formation of nitrosamines.

ROBERT CASANI
Hoffman-La Roche Inc.

L. J. Machlin, ed., *Vitamin E: A Comprehensive Treatise,* Marcel Dekker, Inc., New York, 1980.

G. W. Burton and K. U. Ingold, in L. Packer and G. Fuchs, eds., *Vitamin E in Health and Disease,* Marcel Dekker, Inc., New York, 1993, pp. 329–344.

L. J. Machlin, in L. Machlin, ed., *Handbook of Vitamins,* Marcel Dekker, Inc., New York, 1990, pp. 99–144.

VITAMIN K

Vitamin K represents a class of substances which contain the 2-methyl-1,4-naphthoquinone moiety and are characterized by their antihemorrhagic properties. The chemical name and common names and structures for some important forms of the vitamin follow.

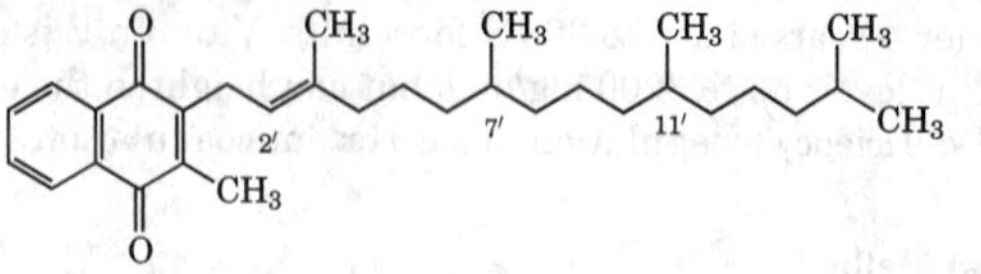

(1)

Vitamin K_1 1,4-naphthalenedione, 2-methyl-3-(3,7,11,15-tetramethyl-2-hexadecenyl)-, [*R*-[*R**,*R**-(*E*)]]-(phylloquinone, phytomenadione, phytonadione)

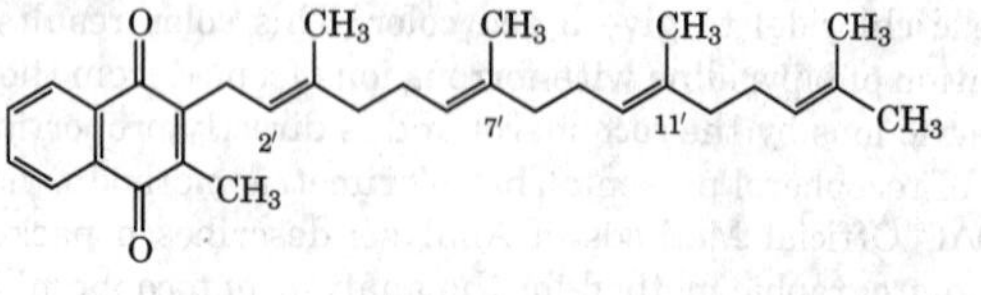

(2)

Vitamin $K_{2(20)}$ 1,4-naphthalenedione, 2-methyl-3-(3,7,11,15-tetramethyl-2,6,10,14-hexadecatetraenyl)-, (*E,E,E,*)-menaquinone 4, menaquinone K4, MK-4

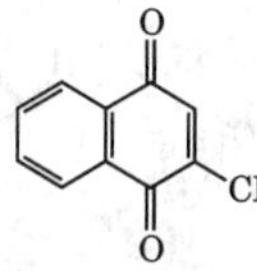

(3)

Vitamin K_3 1,4-naphthalenedione, 2-methyl (menadione; menaquinone-0; synkay)

Vitamin K is typically found in green, leafy vegetables such as cabbage, broccoli, and spinach. Other sources are cauliflower, liver, and eggs.

Chemical and Physical Properties

Vitamin K compounds are yellow solids or viscous liquids. The natural form of vitamin K_1 is a single diastereoisomer with 2′(*E*), 7′(*R*), 11′(*R*) stereochemistry. The predominant commercial form of vitamin K_1 is the racemate and a 2′(*E*)/(*Z*) mixture. Table 1 lists some physical and spectral properties of vitamin K_1, K_2, and K_3.

Vitamin K_1 is insoluble in water and is soluble in 70% alcohol, chloroform, petroleum ether, benzene, and hexane. Vitamin K_1 is stable in air but should be protected from light. Although unstable in alkali, the vitamin is stable in acidic medium. Its facile decomposition in basic solution forms the basis of the Dam-Karrer color test. It reacts with both reducing and oxidizing agents.

Analytical Methods

The classical method for the determination of vitamin K is based on the clotting time of a vitamin K-deficient chick. This method has been supplanted by modern chromatographic techniques.

The majority of analytical methods use the presence of a naphthoquinone nucleus, as a basis for analysis (eg, as its reaction with sodium bisulfite or its uv spectrum). Although not specific, titrimetric, polarographic, and potentiometric methods have also been used.

Chromatographic methods include thin-layer, hplc, and gc methods.

Synthesis

Biosynthesis. Animals cannot synthesize the naphthoquinone ring of vitamin K, but necessary quantities are obtained by ingestion and from manufacture by intestinal flora. In plants and bacteria, the desired naphthoquinone ring is synthesized from 2-oxoglutaric acid and shikimic acid. Chorismic acid reacts with a putative succinic semialdehyde TPP anion to form *o*-succinyl benzoic acid. In a second step, *ortho*-succinyl benzoic acid is converted to the key intermediate, 1,4-dihydroxy-2-naphthoic acid. Prenylation with phytyl pyrophosphate is followed by decarboxylation and methylation to complete the biosynthesis.

Table 1. Physical and Spectral Properties of Vitamin K_1, K_2, and K_3

Item	Vitamin K_1	Vitamin $K_{2(30)}$[a]	Vitamin $K_{2(35)}$[a]	Vitamin K_3
color	yellow, viscous oil	yellow crystals	yellow crystals	bright yellow crystals
melting point, °C	−20	50	54	105–107
molecular weight	450.68	580.9	649.02	172.17
molecular formula	$C_{31}H_{46}O_2$	$C_{41}H_{56}O_2$	$C_{46}H_{64}O_2$	$C_{11}H_8O_2$
spectrophotometric data, λ_{max}, nm	242, 248, 260, 269, 325	243, 248, 261, 270, 325	243, 248, 261, 270, 325	244
ϵ, petroleum ether	396, 419, 383, 387, 68	304, 320, 290, 292, 53	278, 195, 266, 267, 48	1150 (hexane)

[a] The subscript in parentheses denotes the number of carbon atoms in the side chain.

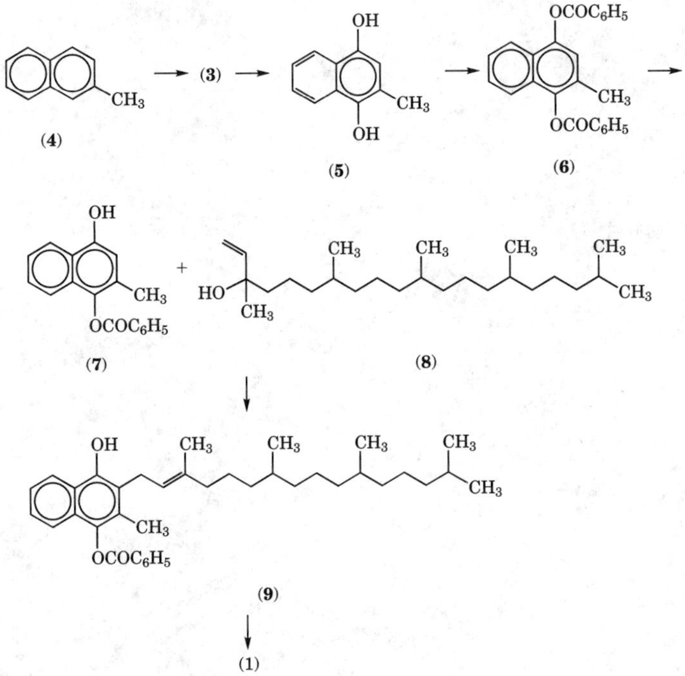

Figure 1. Synthesis of vitamin K_1.

Chemical Synthesis. *Vitamin K_1.* The earliest synthesis of vitamin K_1 involved the reaction of menadione with phytyl bromide in the presence of zinc. Vitamin K_1 has been synthesized also from the condensation of the monosodium salt of menadiol with phytyl bromide.

The commercial synthesis of vitamin K_1, is outlined in Figure 1. Oxidation of 2-methylnaphthalene (**4**) yields menadione (**3**). Catalytic reduction to the naphthohydroquinone (**5**) is followed by reaction with a benzoating reagent to yield the bis-benzoate (**6**). Selective deprotection yields the less hindered benzoate (**7**). Condensation of isophytol (**8**) with (**7**) under acid-catalyzed conditions yields the coupled product (**9**). Saponification followed by an air oxidation yields vitamin K_1 (**1**).

In a novel approach to vitamin K_1, Hoffmann-La Roche has exploited the potential acidity at C-3 as a means to attach the side chain of vitamin K_1. Menadione is reacted with cyclopentadiene to yield the Diels-Alder adduct. The adduct is treated with base and alkylated at C-3 with phytyl chloride. A retro Diels-Alder reaction yields vitamin K_1.

Aside from chemical methods, several patents have appeared on the biochemical production of natural vitamin K_1 from callus tissue cultures.

Vitamin K_2. As compared to vitamin K_1, vitamin K_2 is relatively unimportant industrially. The industrial synthesis parallels that of vitamin K_1 and involves as a key step alkylation of monosubstituted menadione with an appropriate (all-*E*) reagent.

In contrast to vitamin K_1, there has been considerably more activity on fermentative approaches to vitamin K_2. The biosynthetic pathway to vitamin K_2 is analogous to that of vitamin K_1 except that poly(prenylpyrophosphates) are the reactive alkylating agent.

Vitamin K_3. Industrially, vitamin K_3 is prepared by the chromic acid oxidation of 2-methylnaphthalene. The process is complicated by the formation of isomeric 6-methyl-1,4-naphthoquinone.

A process has been disclosed in which the mixture of naphthoquinones is reacted with a diene such as butadiene. Owing to the fact that the undesired product is an unsubstituted naphthoquinone, this dieneophile readily reacts to form a Diels-Alder adduct. By appropriate control of reaction parameters, little reaction is observed with the substituted naphthoquinone. Differential solubility of the adduct and vitamin K_3 allows for a facile separation.

For a less environmentally toxic and more selective oxidizing agent than chromium, organorhenium compounds, ceric sulfate, and electrochemistry have been used.

In a biotechnology-based approach, Japanese workers have reported the microbial conversion of 2-methylnaphthalene to both 2-methyl-1-naphthol and menadione by *Rhodococcus.* The intermediate 2-methyl-1-naphthol can readily be converted to menadione by a variety of oxidizing agents.

In addition to its industrial importance as an intermediate in the synthesis of vitamin K_1, menadione, or more specifically, salts of its bisulfite adduct, are important commodities in the feed industry.

Biochemistry

The important function of vitamin K in the clotting mechanism is to act as a cofactor in the biosynthesis of the active form of prothrombin. The vitamin K-dependent step in prothrombin synthesis is the conversion of glutamyl residues to γ-carboxyglutamyl residues. This carboxylation reaction is essentially a two-step process which first involves generation of a carbanion at the γ-position of the glutamyl (Gla) residue. This event is coupled with the epoxidation of the reduced form of vitamin K and in a subsequent step, the carbanion is carboxylated. Proteins containing Gla residues have been found in other tissues, eg, chick bones.

In addition, vitamin K-dependent carboxylase activity has been observed in cell cultures, eg, a vitamin K-dependent protein, Gas6, has been identified as a ligand for tyrosine kinase. This suggests that vitamin K may have a more general metabolic role.

Requirements. Owing to the ubiquitous natural occurrence of vitamin K and its production by intestinal bacteria, vitamin K deficiencies are rare. However, they can be caused by certain antibiotics (v) coupled with a reduced dietary intake. Newborn infants who do not possess the necessary intestinal bacterial population are at danger for vitamin K deficiency. RDA for vitamin K ranges from 5 μg for infants to 60 μg for adults.

SUSAN D. VAN ARNUM
Hoffmann-La Roche Inc.

H. Dam and E. Sondegarrd, in P. György and W. N. Pearson, eds., *The Vitamins,* 2nd ed., Vol. VI, Academic Press, Inc., New York, 1967, pp. 245–260.

O. Isler and G. Brubacher, *Vitamine I,* Georg Thieme Verlag, New York, 1982, pp. 152–176.

S. Patai and Z. Rappoport, eds., *The Chemistry of Quinonoid Compounds,* Vol. 2, Parts 1 and 2, John Wiley and Sons, Ltd., Chichester, U.K., 1988.

J. W. Suttie, in L. J. Machlin, ed., *The Handbook of Vitamins,* Marcel Dekker, Inc., New York, 1991.

W

WASTE TREATMENT, HAZARDOUS WASTE

Hazardous waste treatment systems incorporate many different unit processes. Recycling technology, an important part of pollution prevention, incorporates both old and new unit processes.

Table 1 provides a list of some of the technologies used to treat hazardous waste.

Physical–Chemical Treatment

Air Stripping. Compounds that have relatively high volatilities and low water solubilities can be transferred (stripped) from aqueous streams into the air or an air stream. Compounds easily air-stripped include gasoline and jet fuels (benzene, toluene, ethylbenzene, and xylenes), solvents such as tetra- and trichloroethylene, and ammonia.

The air stream exiting a stripper may require some type of emissions control, such as carbon adsorption a catalytic oxidation.

Carbon Adsorption. This is a well established and widely used technology for the removal of organics from wastewaters and gaseous streams.

Contaminants are physically attracted or adsorbed on the surface of activated carbon.

Carbon adsorption is most effective at removing organic compounds that have low polarity, high molecular weight, low water solubility, and high boiling point.

New areas in adsorption technology include carbonaceous and polymeric resins. Based on synthetic organic polymer materials, these resins may find special uses where compound selectivity is important, low effluent concentrations are required, carbon regeneration is impractical, or the waste to be treated contains high levels of inorganic dissolved solids.

Dissolved Air Flotation. Dissolved air flotation (DAF) is used to separate suspended solids and oil and grease from aqueous streams and to concentrate or thicken sludges. Air bubbles (formed by pressurizing either the influent wastewater or the effluent) carry or float these materials to the surface where they can be removed.

Table 1. Summary of Hazardous Waste Treatment Technologies

Physical-Chemical Treatment	
air stripping	neutralization
carbon adsorption	oil–water separation
dissolved air flotation	oxidation/reduction
distillation	precipitation
evaporation	sedimentation-clarification
ion exchange	solvent extraction
membrane filtration	stabilization–solidification
	steam stripping
	supercritical fluid extraction
	supercritical water oxidation
	wet air oxidation
Biological Treatment: Activated Sludge	
Thermal Treatment	
catalytic oxidation	multiple hearth furnaces
fluidized beds	rotary kilns
liquid injection	thermal desorption
Soil and Ground Water Treatment	
in situ air stripping	soil flushing
in situ and *ex situ* bioremediation	soil leaching
electrokinetics	soil vapor extraction
pump and treat	soil washing
	vitrification

Distillation. Distillation separates volatile components from a waste stream by taking advantage of differences in vapor pressures or boiling points among volatile fractions and water. There are two general types of distillation, batch or differential distillation and continuous fractional or multistage distillation (see also DISTILLATION).

Evaporation. Evaporation can be used to separate volatile compounds from nonvolatile components and often is used to remove residual moisture or solvents from solids or semisolids. Thin-film evaporators and dryers are examples of evaporation equipment used for this type of application. Some evaporators are also appropriate for aqueous solutions.

Ion Exchange. Ion exchange is an adsorption process where ionic species are adsorbed from solution by exchanging places with a similarly charged ion on the exchange media. Ion exchange is used primarily to remove metals, although nonmetallic inorganic and organic ions can also be removed. The adsorptive materials used for ion exchange are called zeolites. Naturally occurring zeolites belong to a group of hydrous aluminum silicate minerals. Synthetic zeolites or resins are based on organic polymers and are in more common usage today.

Organics that can foul resins can be removed by carbon adsorption. Iron and manganese, commonly present in ground waters, should be removed because they precipitate on the resin.

Membrane Filtration. Membrane filtration describes a number of well-known processes including reverse osmosis, ultrafiltration, nanofiltration, microfiltration, and electrodialysis which uses an electric current. The basic principle behind this technology is the use of a driving force (electricity or pressure) to filter particles, ions, and organic molecules through a membrane, producing a clean stream on one side and a concentrated stream on the other. Applications in waste treatment include metals removal–recovery, oil–water separation, and removal of toxic organic compounds.

Membranes are subject to fouling which can be caused by metal oxides, precipitating salts, colloids, and biological growth. Pretreatment prior to membrane filtration is required to reduce heavy solids and remove free oil.

Neutralization. pH adjustment of waste to a neutral range is a very common treatment step for wastewaters and gases; pH is less of a problem with solid wastes. Neutralization of acid gases with liquid caustic solutions (wet scrubbing) is very common. For example, the flue gas from an incinerator may be scrubbed with a soda ash solution. There are also dry scrubbing systems for acid gases where the neutralizing agent, for example, lime, is sprayed into the gas stream.

Oil–Water Separation. Gravity separation of oil and water is a very simple process, often used as a pretreatment step in an overall wastewater treatment system. Free oil floats to the surface where it can be skimmed off. Water emulsions may be broken for separation by using coagulants, acids, pH adjustment, heat, centrifuging, or high-potential alternating current.

Oxidation and Reduction. Oxidation and reduction (redox) reactions are used for both partial and complete degradation of many organic and inorganic compounds. A substance is oxidized when its oxidation state is increased, likewise it is reduced when its oxidation state is reduced. Oxidizers do not discriminate among compounds. Common chemical oxidants for waste treatment include chlorine, ozone, and hydrogen peroxide.

Chlorine is a well known disinfectant for water and wastewater treatment, however, it can react with organics to form toxic chlorinated compounds. Chlorine dioxide may be used instead to avoid chlorinated by-products but it is not suitable for waste streams containing cyanide. Oxidation with ozone avoids chlorinated by-products but the cost of producing ozone is high. Hydrogen peroxide is typically used as an oxidizer of organic compounds in combination with uv light, ozone, and/or metal catalysts.

Precipitation. Precipitation processes have been used for many years to remove metals from aqueous streams. Hydroxide precipitation, sulfide precipitation, and carbonate precipitation are commonly used.

The typical precipitation process takes place in a series of tanks beginning with chemical addition, followed by flocculation and coagulation of precipitated solids, sludge thickening, and finally, sludge dewatering. Sand filtration of the wastewater effluent to remove residual solids may be used as a final polishing step.

Sedimentation–Clarification. Sedimentation–clarification is a process by which hazardous and nonhazardous grits, fines, and other suspended solids are removed from the waste stream through gravity settling. The process typically is used as a pretreatment step or in conjunction with another treatment process such as chemical or biological treatment.

Solvent Extraction. With solvent extraction, organics are separated from a waste by mixing the waste with a solvent capable of dissolving or extracting the organics. The extraction process can have multiple stages and be operated where the solvent and waste pass concurrently or countercurrently.

Stabilization–Solidification. Stabilization and solidification are technologies that are often used together to improve or strengthen the physical nature of a waste and to bind, immobilize, or otherwise prevent the migration of toxic constituents contained in the waste. Stabilization generally refers to processes that reduce the toxicity, leaching, or mobility of toxic constituents, perhaps through chemical reactions, but the physical form of the waste is not necessarily changed or improved. Solidification generally refers to processes that change the physical nature of the waste to make it more manageable, increase its structural strength and load-bearing capacity, or reduce its permeability, but not necessarily by chemical reaction.

Common stabilization–solidification processes include the following. With sufficient water for hydration, cement forms a calcium alumino–silicate crystalline structure that can physically and/or chemically bind with toxic constituents such as metal hydroxides and carbonates. Use of lime (calcium hydroxide) involves similar reactions where alumino–silicates are supplied by the waste or treatment additives. Wastes containing heavy metals and/or radioactive materials may be mixed with molten thermoplastics, such as bitumen, asphalt, polyethylene or polypropylene. The process is not suitable for organic wastes (without air emissions controls), hygroscopic wastes, or strongly oxidizing constituents, anhydrous inorganic salts, and aluminum salts.

Radioactive and acid wastes may be combined with materials, eg, urea–formaldehyde, phenolics, polyesters, epoxides, and vinyls) that form thermosetting reactive polymerized material when mixed with a catalyst. The treated waste forms a sponge-like material which traps the solid particles, but not the liquid fraction. Also, spills of chemicals that are monomers or low-order polymers can be polymerized by adding a catalyst.

Steam Stripping. Steam stripping is used to separate volatile components from wastewater. The process differs from distillation in that there is no rectifying section and the waste steam is introduced at the top of the column (packed or tray). Steam stripping is amenable to volatile components that phase separate from water.

Supercritical Fluid Extraction. Supercritical fluid extraction takes advantage of the enhanced solvent power of compounds at supercritical temperature and pressure conditions to extract organics from oily and/or organic liquids and sludges. Carbon dioxide is a particularly attractive solvent at supercritical conditions (greater than 7400 kPa and 31°C) because it is nontoxic and leaves no residue when it is reverted to gaseous form during the recovery step. However, it is not suitable for wastes with high alkalinity.

Supercritical Water Oxidation. Supercritical water oxidation is similar to wet air oxidation except that high temperature and pressure (greater than 22,000 kPa and 374°C) form water that has properties between those of a liquid and a gas. Organics are easily solubilized in supercritical water and once solubilized, they are completely oxidized at the high temperature and pressure conditions. Supercritical water oxidation has the capability of treating many different kinds of organics including polycyclic aromatic hydrocarbons, chlorinated hydrocarbons, PCBs, paint, oil, dyes, pulp and paper wastes, chemical warfare agents, and missile propellants.

Wet Air Oxidation. With wet air oxidation, increased temperature and pressure are used to oxidize dilute concentrations of organics and some inorganics, such as cyanide, in aqueous wastes which contain too much water to be incinerated, but are too toxic to be treated biologically. In general, wet air oxidation provides primary treatment for wastewaters that are subsequently treated by conventional methods.

Biological Treatment

Biodegradability of Compounds. Biological processes are effective in treating a wide variety of organic and inorganic compounds. Most biological processes are aerobic and use the oxygen available in supplied air or pure oxygen. The chief products of aerobic biodegradation of carbonaceous and nitrogenous compounds are carbon dioxide, ammonia, water, and biomass (cell growth of the microorganisms). The ease with which a compound is degraded is related to its chemical structure. Some compounds are not degraded completely, but are instead broken down into simpler compounds.

Activated Sludge. Activated sludge is widely used to treat both municipal and industrial wastewater. The process contains three basic elements: an aeration basin, clarification, and sludge recycle. The aeration basin contains a suspension of *activated* microorganisms, primarily bacteria and protozoa, that biodegrade the wastewater under aerobic conditions. Aerators supply oxygen for the microorganisms and mixing energy to keep them in suspension and in contact with the waste. The microorganisms are separated from the treated wastewater by clarification and a portion of this biological sludge is recycled back to the aeration basin to maintain the microbial population and begin a new treatment cycle. The process must be protected from shock loadings that are toxic or excessively high in concentration.

Thermal Treatment

Thermal treatment is used to destroy, break down, or aid in the desorption of contaminants in gases, vapors, liquids, sludges, and solids. Most thermal processes are classified as incineration.

There are many factors that affect both the choice of a particular thermal treatment and its performance. Chief among these are waste characteristics, temperature, residence time, mixing or turbulence, and air supply.

Catalytic oxidation is used only for gaseous streams because combustion reactions take place on the surface of the catalyst which otherwise would be covered by solid material. Common catalysts are palladium and platinum. In a fluidized bed the waste and an inert bed material, sand or aluminum oxide, are fluidized by blowing heated air through a distributor plate at the bottom of the bed. Fluidization results in good mixing and uniform distribution of materials within the bed which results in low operating temperatures (450–710°C) and low excess air requirements.

Liquid injection units are the most common type of incinerator used today for the destruction of liquid hazardous wastes. Rotary kilns can be used to incinerate gases, liquids, sludges, and solids. A rotary kiln is a long, cylindrical incinerator that is sloped a few degrees from horizontal. Waste is introduced at the upper end. The gentle slope and slow rotation of the kiln continually mix and re-expose the waste to the hot refractory walls, moving the waste toward the exit point. Thermal desorption has been applied primarily to soils. Wastes are heated to temperatures of 200 to 600°C to increase the volatilization of organic contaminants. Volatilized organics in the gas stream are removed by a variety of methods including incineration, carbon adsorption, and chemical reduction.

Soil and Ground Water Treatment

Contaminated soil and ground water present special challenges because the contamination is often diffuse and difficult to access. *In situ* processes, those that leave both soil and ground water in place during treatment, are often favored because they are less disruptive; they take advantage of natural, existing conditions; and they preserve the existing subsurface environment. Both *in situ* and *ex situ* treatment for soil and ground water rely on a combination of processes, which often include biological degradation of organics.

Plume Containment. Wells can be placed at a contaminated site to prevent the contamination from spreading further and offer cost and flexibility advantage over older methods such as bentonite slurry trenches, grout curtains, sheet pilings, and fixative injections. Two other methods of plume containment are biofilters and a funnel-and-gate system.

Bioremediation. It is a natural process that degrades hazardous organic chemicals into innocuous carbon dioxide and water or nonhazardous byproducts and it is often less expensive and more effective than pump and treat methods.

Biodegradation is based on the ability of aerobic and anaerobic microorganisms to degrade organic compounds to obtain energy to grow and reproduce. Bacteria and fungi catabolize (degrade) compounds to produce chemicals with lower free energy, and in the case of aerobic biodegradation, this often results in complete decomposition of a hydrocarbon to carbon dioxide and water.

In Situ Bioremediation. *In situ* bioremediation can be an aerobic or anaerobic process, or a combination of the two. The vast majority of *in situ* bioremediations are conducted under aerobic conditions. Besides being slower, anaerobic treatment is more difficult to manage and can generate by-products that are more mobile or toxic than the original compound, for example, the daughter products of TCE, ie, dichloroethenes and vinyl chloride.

However, some contaminants can be degraded only anaerobically. Anaerobic respiration can degrade organics by using nitrate or sulfate as oxygen sources.

Combined aerobic–anaerobic systems use sequenced aerobic and anaerobic conditions to degrade compounds that are resistant to aerobic biodegradation.

Soils with hydraulic conductivities greater than 10^{-4} centimeters per second (cm/s) are good candidates for *in situ* bioremediation because they are permeable enough to allow the transport of oxygen and nutrients through the aquifer. Because biological growth and biodegradation of hydrocarbons is oxygen-limited, oxygen may need to be distributed to ground water and soil for bioremediation. Instead of pure oxygen, hydrogen peroxide can be used and it has the advantages of stimulating microbial activity and chemically degrading contaminants partially or fully.

Solid phase oxygen in the form of peroxides (calcium peroxide, magnesium peroxide) can be used as a slow, continuous release system that avoids problems associated with the transient nature of molecular oxygen and hydrogen peroxide. Other nutrients such as nitrogen and phosphorus may also need to be added.

An additional bioremediation technique is air sparging. Air is injected in the saturated zone to volatilize organics and to deliver oxygen for biodegradation. Biosparging is a form of air sparging, but the difference is that the primary purpose of biosparging is to deliver just enough air (or ozone) to meet oxygen requirements for bioremediation. Biofilters, also known as biobarriers or microbial fences, are used to hinder migration of a contaminant plume. A biofilter is essentially a zone of biological activity that treats the contaminant as the ground water flows through the area.

Bioventing is soil venting that enhances biodegradation while extracting volatile compounds from the unsaturated zone. In the funnel-and-gate bioremediation method ground water is funneled through openings or gates in an impermeable sheet piling or slurry wall. Zones of biological activity are created at the gates through air sparging. Contaminants are biodegraded as ground water is forced to pass through the gates. Finally, phytoremediation is a developing technology that uses plants, trees, and grasses to biodegrade, extract, or stabilize organic and metal contaminants in soil and ground water (eg, engineered wetlands).

Ex Situ Bioremediation. Ground water and leachate can be removed and treated in a biological wastewater treatment system such as rotating biological contactors (RBCs), trickling filters, sequencing batch reactors, fluidized bed reactors, activated sludge, or aerated impoundments.

Contaminated soil can be treated biologically in slurry form, not unlike biological treatment of wastewater. However, one important difference with contaminated soil is that the contaminants preferentially adsorb unto the soil particles and must be desorbed before the microorganisms can effectively degrade the compounds.

Alternatively, contaminated soil can be excavated and placed in heaps, piles, beds, or windrows. Other organic or bulking agents such as wood chips or straw, may be added to aid in composting, mixing, and aerating the soil. Volatilized contaminants that are removed by vacuum may be treated by incineration, carbon adsorption, or vapor-phase biodegradation. Leachate may be treated biologically in a wastewater treatment system.

Physical–Chemical–Thermal *In Situ* Treatment Related Specifically to Soils and Ground Water

Electrokinetics. Primarily used for metals removal, electrokinetics utilizes an electrical field to generate a flow and concentration gradient in porous and semiporous soils. An innovative technology called the "lasagna" process is based on the electrokinetic phenomenon called electroosmosis. Created to treat difficult wastes in low permeability, silt- and clay-laden soils, the lasagna process consists of a number of (horizontally or vertically) layered subsurface electrodes and treatment zones. A low voltage electric current is applied by subsurface electrodes which forces the migration of contaminants from low permeable soils into high permeable treatment zones. Treatment in these zones may include physical, chemical, or biological processes.

In Situ Air Stripping. In an innovation to conventional pump and treat air stripping, two horizontal wells are installed, one below the water table and one in the vadose zone. Air is injected in the lower well while contaminated soil vapor is extracted by vacuum through the upper well.

Soil Flushing. Soil flushing is similar to soil washing of excavated soils except flushing is done *in situ* below the ground surface. With soil flushing, the contaminated area is flooded with water. Surfactants or detergents may be added to the water to enhance removal of organics.

Soil Vapor Extraction. Volatile compounds can be extracted from subsurface soils by applying a vacuum. An added benefit is that the vacuum pulls in air from the ground surface that stimulates biodegradation through increased oxygen transport.

An innovative companion technology to soil vapor extraction is radio frequency heating of the soil to 100 to 150°C, which increases the volatility of the contaminants.

Vitrification. Vitrification turns contaminated soils into a glasslike monolithic mass. Heat is applied through electrodes placed in the ground, the soil reaching temperatures of 1600 to 2000°C. A layer of graphite and glass frit is first placed on the surface of the ground between the electrodes to act as the initial conductive starter path. The conductive layer and adjacent soils become a molten mass that moves both outward and downward. Organic contaminants in the soil vaporize and eventually pyrolize and inorganic contaminants are immobilized in the molten material. Off-gases, are captured above the site and treated to meet air emissions standards. Once cooled, the vitrified mass is very stable with low leaching potential.

Pump and Treat

In the early years of ground water and soil remediation, pump and treat was the conventional technology. Contaminated ground water is pumped to the surface where it is treated and reinjected or discharged to surface waters or wastewater treatment plants.

The general consensus is that pump and treat can reduce contamination or keep it from spreading, but it has failed in many cases to remediate aquifers to stringent cleanup goals.

Ex Situ Soil Nonbiological Treatment

Soil Leaching. Soil leaching or acid extraction uses acid to solubilize metals for removal from soils, a technique akin to that in the mining industry. After treatment with an acid, the soil is separated from the acid, rinsed with water to remove excess acid and metals, dewatered, and neutralized. The extracted metals can be precipitated and the acid recycled.

Soil Washing. Soil is excavated, physically broken up, and washed with a water-based solution to which extractants, or detergents have been added.

DIANNA S. KOCUREK
Tischler/Kocurek
GAYLE WOODSIDE
IBM Corporation

U.S. EPA, *Manual: Ground Water and Leachate Treatment Systems,* EPA/625/R-94/005, Washington, D.C., 1995.

R. Noyes, *Unit Operations in Environmental Engineering,* Noyes Publications, Park Ridge, N.J., 1994.

G. R. Chaudhry, ed., *Biological Degradation and Bioremediation of Toxic Chemicals,* Dioscorides Press, Portland, Oreg., 1994.

M. D. LaGrega, P. L. Buckingham, and J. C. Evans, *Hazardous Waste Management,* McGraw-Hill, Inc., New York, 1994.

WASTES, INDUSTRIAL

Recent legislation in the United States (ca 1997) has added several new parameters to the requirements for effluent permits, namely, biochemical oxygen demand (BOD), total suspended solids (TSS), chemical oxygen demand (COD), volatile organic compounds (VOC), priority pollutants, aquatic toxicity, heavy metals, nitrogen, and phosphorus.

Waste Minimization

Generally, waste minimization techniques can be grouped into four major categories: inventory management and improved operations, modification of equipment, production process changes, and recycling and reuse. Such techniques can have applications across a range of industries and manufacturing processes, and can apply to hazardous as well as nonhazardous wastes. Waste minimization approaches as developed by the EPA are shown in Table 1.

The six major ways of reducing pollution are as follows: *(1)* Recirculation of materials. *(2)*) Segregation of clean streams for direct discharge. *(3)* Disposal in a semidry or a sludge state rather than flushing to the sewer. *(4)* Reduction of wastewater volume. *(5)*) The use of drip pans to catch products instead of flushing material to the sewer. *(6)*) Substitution of chemical additives with less polluting ones in processing operations.

Table 1. Waste Minimization Approaches and Techniques

Approach	Related techniques
inventory management and improved operations	inventory and trace all raw materials
	purchase fewer toxic and more nontoxic production materials
	implement employee training and management feedback
	improve material receiving, storage, and handling practices
modification of equipment	install equipment that produces minimal or no waste
	modify equipment to enhance recovery or recycling options
	redesign equipment or production lines to produce less waste
	improve operating efficiency of equipment
	maintain strict preventive maintenance program
production process changes	substitute nonhazardous for hazardous raw materials
	segregate wastes by type for recovery
	eliminate sources of leaks and spills
	separate hazardous from nonhazardous wastes
	redesign or reformulate end products to be less hazardous
	optimize reactions and raw material use
recycling and reuse	install closed-loop systems
	recycle on-site for reuse
	recycle off-site for reuse
	exchange wastes

Definition of Wastewater Constituents

Parameters used to characterize wastewaters can be classified as organic and inorganic analyses. The organic content of wastewater is estimated in terms of oxygen demand using biochemical oxygen demand (BOD), chemical oxygen demand (COD), or total oxygen demand (TOD). Additionally, the organic fraction can be expressed in terms of carbon, using total organic carbon (TOC). It is important to identify volatile organic carbon (VOC) and the presence of specific priority pollutants, in addition to the total organic content. The inorganic characterization schedule for wastewaters to be treated using biological systems should provide information concerning (*1*) potential toxicity, such as heavy metal, ammonia, etc; (*2*) potential inhibitors, such as total dissolved solids (TDS) and chlorides; (*3*) contaminants requiring specific pretreatment such as pH, alkalinity, acidity, suspended solids, etc; and (*4*) nutrient availability.

Aquatic toxicity is becoming (ca 1997) a permit requirement on all discharges. Aquatic toxicity is generally reported as an LC_{50} (the percentage of wastewater which causes the death of 50% of the test organisms in a specified period) or as a No Observed Effect Level (NOEL), in which the NOEL is the highest effluent concentration at which no unacceptable effect will occur, even at continuous exposure. Toxicity is also frequently expressed as toxicity units (TU), which is 100 divided by the toxicity measured. Therefore, an effluent having an LC_{50} of 10% contains 10 toxic units.

Wastewater Treatment

A summary of available technologies is given in Tables 2 and 3. (See also WASTE TREATMENT, HAZARDOUS WASTE.)

Pre- or Primary Treatment

The principal objectives of pretreatment are to remove heavy metals prior to subsequent treatment, to neutralize the wastewater to a suitable pH for discharge or subsequent treatment; to remove high concentrations of suspended solids, to eliminate or reduce toxicity, and to eliminate or reduce volatiles.

Neutralization. Wastewater discharge usually requires a pH between 6 and 9. Exceptions are a biological process in which microbial respiration degrades acidity (acetic acid is oxidized to CO_2 and H_2O), or one in which the CO_2 generated by microbial respiration neutralizes caustic alkalinity (OH^-) to bicarbonate HCO_3. Acidic wastewaters can be neutralized with lime, magnesium hydroxide, caustic, or limestone. Alkaline wastewaters can be neutralized with H_2SO_4 or HCl, or by using flue gas (CO_2).

Removal of Oil and Grease. High concentrations of oil and grease can be removed in a gravity separator where the lighter oils and greases float to the surface, where they are skimmed off.

Low concentrations of oil can be removed by dissolved air flotation (DAF) or, induced air flotation (IAF), processes that rely on air bubbles to capture suspended solids and oil globules.

Suspended Solids Removal. Depending on the concentration and characteristics of the suspended solids, they can be removed by filtration, flotation, or sedimentation. Coarse solids are removed by screening. Settleable suspended solids are removed in a clarifier.

Table 2. Chemical Waste Treatment

Treatment method	Type of waste	Mode of operation	Degree of treatment	Remarks
ion exchange	plating, nuclear	continuous filtration with resin regeneration	demineralized water recovery; product recovery	may require neutralization and solids removal from spent regenerant
reduction and precipitation	plating, heavy metals	batch or continuous treatment	complete removal of chromium and heavy metals	one day's capacity for batch treatment; 3-h retention for continuous treatment; sludge disposal or dewatering required
coagulation	paperboard, refinery, rubber, paint, textile	batch or continuous treatment	complete removal of suspended and colloidal matter	flocculation and settling tank or sludge blanket unit; pH control required
adsorption	toxic or organics, refractory	granular columns of powdered carbon	complete removal of most organics	powdered carbon (PAC) used with activated sludge process
chemical oxidation	toxic and organics, refractory	batch or continuous ozone or catalyzed hydrogen peroxide	partial or complete oxidation	partial oxidation to render organics more biodegradable

Inorganic and organic colloidal suspensions in wastewater can be removed by chemical coagulation, defined as the addition of a positively charged ion or cationic polyelectrolyte that results in particle destabilization and charge neutralization. Coagulation involves the formation of complex oxides that form flocculent suspensions which subsequently are separated from the liquid by sedimentation. Filtration is employed when the suspended solids concentration is less than 100 mg/L and high effluent clarity is required.

Heavy Metals Removal. Heavy metals should be removed prior to biological treatment or use of other technologies which generate sludges. The technologies available for metals removal are summarized in Table 4.

Removal of Volatile Organics. Volatile organics, eg, benzene and toluene, should usually be removed prior to biological treatment. Air stripping, activated carbon and chemical oxidation are the methods commonly used.

Biological Treatment

Aerobic treatment is generally applied to lower strength wastewaters, whereas anaerobic treatment is employed as a pretreatment for high strength wastewaters. The choice of process depends both on the concentration of organics and the volume of wastewater to be treated. The objective of biological treatment is to remove biodegradable organics. In an aerobic biological treatment process, organic removal can occur through biodegradation, stripping, or sorption on the biological flocs.

Approximately half of the organics removed are oxidized to CO_2 and H_2O, and half synthesized to biomass. Three to 10 percent of the organics removed result in soluble microbial products (SMP). The SMP is significant because it causes aquatic toxicity.

Nitrogen (from ammonia or nitrates) and phosphorus are required in the reaction at an approximate ratio of BOD:N:P of 100:5:1. Nitrogen and phosphorus are amply available in municipal wastewaters, but frequently are deficient in industrial wastewaters.

Biological sludges generally fall into one of three classifications—filamentous bulking, nonbulking, and pinpoint. A flocculent sludge is one in which the major part of the biomass consists of flocculent organisms, with some filaments growing within the floc. Filamentous bulking occurs when the filaments grow out from the floc in the bulk of the liquid. This condition hinders sludge settling. The pinpoint case occurs at very low loadings, causing floc dispersion.

Nutrient Removal

In many locations, nitrogen and phosphorus must be removed in order to meet effluent limitations.

Nitrogen. Nitrogen is most commonly removed by biological nitrification and denitrification. In this process, organic nitrogen is hydrolyzed to ammonia. Ammonia, in turn, is oxidized to nitrite by *Nitrosommonas.* The nitrite, in turn, is oxidized to nitrate by *Nitrobacter.* This is followed by denitrification, a process in which facultative organisms reduce nitrate to nitrogen gas in the absence of molecular oxygen.

Phosphorus Removal. Phosphorus can be removed from wastewater either chemically or biologically. It can be precipitated with lime to form $Ca_3(PO_4)_2$. Alum or iron will precipitate phosphorus as $AlPO_4$ or $FePO_4$. Certain organisms normally present in activated sludge have the ability to store phosphorus. The process involves an anaerobic step in which phosphorus is released and acetate taken up by the bio-P organisms, followed by an aerobic step in which phosphorus is rapidly taken up by the bio-P.

Alternative Biological Treatment Technologies

Lagoons. Where large land areas are available, lagooning provides a simple and economical treatment for nontoxic or nonhazardous wastewaters. There are several lagoon alternatives. The impounding and absorption lagoon has no overflow and is particularly suitable to short seasonal operations in arid regions.

Anaerobic ponds require the addition of sodium nitrate to control odors. An alternative is the use of a stratified facultative lagoon, in which aerators are suspended three meters below the liquid surface in order to maintain aerobic surface conditions, with anaerobic digestion occurring at the lower depths. Aerobic lagoons depend on algae to produce oxygen by photosynthesis.

Aerated Lagoons. An aerated lagoon system is a two- or three-basin system designed to remove degradable organics (BOD). The first basin is fully mixed, maximizing the organic removal rate. A second basin permits solids to deposit on the bottom for anaerobic degradation and stabilization. A third basin is frequently employed for further removal of suspended solids and enhanced clarification. In order to avoid groundwater pollution, these basins must usually be lined.

Activated Sludge. There are several generic activated sludge processes presently available. Complete Mix (CMAS) is applicable to refractory-type wastewaters in which filamentous bulking is not a problem. Plug flow is applicable for readily degradable wastewaters subject to filamentous bulking. The selector process is applicable also for readily degradable wastewaters; but because degradable organics are removed by the floc formers by biosorption, they are not available as a food source for the filaments. The sequencing batch reactor

Table 3. Biological Waste Treatment

Treatment method	Mode of operation	Degree of treatment	Land requirements	Equipment	Remarks
lagoons	intermittent or continuous discharge; faculative or anaerobic	intermediate	earth dug; 10–60 days' retention (may require lining)		odor control frequently required groundwater considerations
aerated lagoons	completely mixed or faculative continuous basins	high in summer; less in winter	lined earth basin, 2.44–4.88 m deep; 8.55–17.1 $m^3/(m^3 \cdot d)$	pier-mounted or floating surface aerators or subsurface diffusers	solids separation in lagoon; periodic dewatering and sludge removal groundwater considerations
activated sludge	completely mixed or plug flow; sludge recycle	>90% removal of organics	earth or concrete basin; 3.66–6.10 m deep; 0.561–2.62 $m^3/(m^3 \cdot d)$	diffused or mechanical aerators; clarifier for sludge separation and recycle	excess sludge dewatered and disposed of
trickling filter	continuous application; may employ effluent recycle	intermediate or high, depending on loading	5.52–34.4 $m^3/(10^3\ m^3 \cdot d)$	plastic packing 6.10–12.19 m deep	pretreatment before POTW or activated sludge plant
RBC	multistage continuous	intermediate or high		plastic disks	solids separation required
anaerobic	complete mix with recycle; upflow or downflow filter, fluidized bed; upflow sludge blanket	intermediate		gas collection required; pretreatment before POTW or activated sludge plant	
spray irrigation	intermittent application of waste	complete; water percolation into groundwater and runoff to stream	6.24×10^{-7}–; $4.68 \times 10^{-6} m^3/(s \cdot m^2)$)	aluminum irrigation pipe and spray nozzles; movable for relocation	solids separation required; salt content in waste limited

Table 4. Heavy Metals Removal Technologies

Process type	Examples
conventional precipitation	hydroxide
	sulfide
	carbonate
	coprecipitation
enhanced precipitation	dimethyl thiocarbamate
	diethyl thiocarbamate
trimercapto-*s*-triazine, trisodium salt	
other methods	ion exchange
	adsorption
recovery opportunities	ion exchange
	membranes
	electrolytic techniques

(SBR) or intermittent process is a combination of complete mix and plug flow, and usually controls filamentous bulking. The nature of the process eliminates the need for an external clarifier. The oxidation ditch process is usually considered when nitrogen removal is required. Other processes include deep tank aeration, the use of high purity oxygen, and the Deep Shaft process.

Fixed-Film Processes

Trickling Filter. Used as a pretreatment process, a trickling filter is a packed bed, usually plastic on which a biofilm grows. As a wastewater passes over the film, organics and oxygen diffuse into the film, where they undergo biodegradation.

Rotating Biological Contactor (RBC). An RBC is a fixed-film process in which a biofilm is developed on a rotating plastic cylinder that passes through the wastewater. As the cylinder passes through, the wastewater organics diffuse into the film. As the cylinder passes through the air, oxygen diffuses into the biofilm, causing degradation of the organics. Increased treatment is achieved by increasing the number of stages.

Anaerobic Treatment

Anaerobic treatment is usually employed for high strength wastewaters (eg, when the BOD exceeds 1000 mg/L). In anaerobic treatment, complex organics are broken down through a sequence of reactions to end products of methane gas, CH_4, and carbon dioxide, CO_2. Because anaerobic treatment will not reach usual permit discharge levels, it is employed as a pretreatment process prior to discharge to a publicly owned treatment works or to a subsequent aerobic process.

There are five principal process variants which are proprietary in nature: **1.** *Anaerobic Filter.* The anaerobic filter is similar to a trickling filter in that a biofilm is generated on media. **2.** *Anaerobic Contact.* Sludge is recycled from a clarifier or separator to the reactor. Since the material leaving the reactor is a gas–liquid–solid mixture, a vacuum degasifier is required. **3.** *Fluidized Bed.* This reactor consists of a sand bed on which the biomass is grown. **4.** *Upflow Anaerobic Sludge Blanket (UASB).* Under proper conditions anaerobic sludge will develop as high density granules that form a sludge blanket in the reactor. The wastewater is passed upward through the blanket. **5.** *ADI Process.* The ADI is a low rate anaerobic process which is operated in a reactor resembling a covered football field.

With the exception of the ADI process, anaerobic processes usually operate at a temperature of 35°C. In order to maintain this temperature, the methane gas generated in the process is used to heat the reactor.

Advanced Wastewater Treatment

New regulations for toxics and priority pollutants frequently cannot be met by conventional technology. Other physical–chemical technologies must therefore be applied. These include chemical oxidation and carbon adsorption.

Chemical Oxidation. Chemical oxidation can be applied in industrial wastewater pretreatment for reduction of toxicity, to oxidize metal complexes to enhance heavy metals removal from wastewaters, or as a posttreatment for toxicity reduction or priority pollutant removal. Complex organics and toxics are chemically oxidized to end products of CO_2 and H_2O or to intermediate products which are nontoxic and biodegradable.

Carbon Adsorption. Carbon can be employed either as granular carbon in columns (GAC) or as powdered carbon added to an activated sludge plant (PACT). Carbon removes most organics except low molecular weight soluble organics such as sugars and alcohols. In general, those organics which adsorb the poorest biodegrade the best, whereas those which biodegrade poorly adsorb well on carbon.

Sludge Handling and Disposal

Municipal primary sludge consists of organic and inorganic particulates. The sludge must be stabilized before land disposal. Biological sludge consists of organisms and other particulates not degraded in the biological process. Chemical sludges consist of chemical precipitates, heavy metals, and other contaminants such as color precipitated from industrial wastewaters.

Land disposal of wet sludges can be accomplished in a number of ways: Lagooning or the application of liquid sludge to land by truck or spray system. Biological sludges can be incorporated into the soil. The heavy metal content of the sludge will dictate the total number of years sludge can be applied. Dewatered sludges can be employed as a landfill. Incineration can be accomplished in multiple-hearth furnaces.

Storm-Water Control

In most industrial plants, it is now necessary to contain and control pollutional discharges from storm water. Pollutional discharges can be minimized by providing adequate diking around process areas, storage tanks, and liquid transfer points, with drainage into the process sewer.

W. WESLEY ECKENFELDER, JR.
Eckenfelder Inc.

R. A. Conway and R. D. Ross, *Handbook of Industrial Waste Disposal,* Van Nostrand Reinhold Co., Inc., New York, 1980, p. 125.

WATER

SOURCES AND QUALITY ISSUES

Factors Affecting Stream Water Quality

Chemical Reactions. Weathering is a general term for mechanical and chemical alteration of rock minerals that are exposed to the atmosphere and circulating water at and near the land surface. Chemical reactions that occur during weathering produce both water-soluble and nonwater-soluble products; those that are water soluble are transported from the reaction site in surface runoff or in moving groundwater.

Mineral dissolution reactions of importance generally require a continuous supply of hydrogen ions in the incoming solution. To some degree, reacting hydrogen ions are supplied from the water itself, which always dissociates to some extent, into hydrogen and hydroxide ions. However, in natural systems the most effective sources of hydrogen ions generally are chemical reactions involving dissolved constituents. An important source is carbon dioxide gas, which is present in weathering solutions as a result of contact with air; it is produced in larger quantities by plant root respiration and decay of soil organic matter and by the metabolic processes of various organisms in water and sediment. Equation 1 shows that some of the carbon dioxide that dissolves forms carbonic acid:

$$CO_2 + H_2O \rightleftharpoons H_2CO_3 \qquad (1)$$

The acid dissociates to form bicarbonate and hydrogen ions:

$$H_2CO_3 \rightleftharpoons HCO_3^- + H^- \qquad (2)$$

and carbonate anions can be formed in a second dissociation step:

$$HCO_3^- \rightleftharpoons CO_3^{2-} + H^+ \qquad (3)$$

The H^+ that is supplied by reactions in Eqs. 2 and 3 can react with silicate minerals such as the sodium-bearing form of feldspar:

$$\underset{\text{albite}}{2\,NaAlSi_3O_8} + \underset{\text{hydrogen}}{2\,H^+}\ \underset{\text{water}}{9\,H_2O} \rightarrow \underset{\text{kaolinite}}{Al_2Si_2O_5(OH)_4} + \underset{\text{silicic acid}}{4\,H_4SiO_4(aq)} + \underset{\text{sodium}}{2\,Na^-} \qquad (4)$$

to produce the clay mineral kaolinite, undissociated silicic acid (H_4SiO_4 which also can be written as SiO_2), and sodium ions.

The hydrogen ion flux that is provided by carbonic acid dissociation also can attack calcite ($CaCO_3$):

$$CaCo_3 + H^+ \rightleftharpoons HCO_3^- + Ca^{2+} \qquad (5)$$

Effects of Human Activities. Human activities that alter stream flow characteristics and thus cause water quality changes include the building of structures that impound or regulate rates of stream flow, diversion of water from one drainage basin to another, irrigation of land adjacent to streams, and lowering of tributary groundwater tables by pumping from wells. Waste disposal, directly or indirectly, into streams also influences water quality by adding chemicals and suspended matter. Disposal of untreated organic waste into streams was common in urban and rural areas of the nation until the early twentieth century. Besides pathogenic bacteria in the waste, large amounts of organic chemicals and suspended material depleted the dissolved oxygen of receiving waters and killed much of the aquatic biota in some streams. Thus, the concentration of dissolved oxygen in stream water is considered a contamination index. Additionally, phosphate (PO_4^{-3}) concentrations are indicative of contamination from waste sources. Phosphate is a constituent of domestic and industrial waste, in part, because of the widespread use of phosphate compounds as detergent additives.

Land use changes and related developments also can affect stream water quality. Examples include urbanization, clearing of forests, various agricultural practices, such as use of fertilizers and pesticides and return flow of drainage water from irrigated fields, and industrialization.

Circulation Rates of Elements. The concept of cycling of individual elements, in part coupled to the hydrologic cycle, has been developed and quantified over the past half century.

Carbon. Most of the Earth's supply of carbon is stored in carbonate rocks in the lithosphere. Normally the circulation rate for lithospheric carbon is slow compared with that of carbon between the atmosphere and biosphere.

The efficiency of the weathering of rocks in using carbonic acid produced in the carbon cycle is affected by various hydrologic, environmental, and cultural controls. The fact that the principal anion in fresh surface water worldwide almost always is bicarbonate attests to the overriding importance of this process.

Sulfur. The cycle of sulfur in weathering environments is affected by a more diverse set of reactions than for carbon. As is the case with carbon, most of the Earth's supply of sulfur is stored in the lithosphere. In igneous and metamorphic rocks, sulfur generally is present in the chemically reduced sulfide form (S^{2-}) and commonly is associated with metals. In weathering environments where oxygen is continuously available, negatively charged reduced forms of sulfur are converted to positively charged oxidized forms such as sulfate (SO_4^{2-}). Oxidation and reduction reactions of sulfur commonly are bacterially mediated. The calcium sulfate minerals gypsum ($CaSO_4 \cdot 2\,H_2O$) and anhydrite ($CaSO_4$) are common constituents of evaporite rocks, and in semiarid regions where such rocks are near the land surface, stream water can contain substantial concentrations of sulfate. However, streams in humid regions generally carry relatively low concentrations of sulfate unless human activities have intervened. For instance, any reduced species of sulfur may be converted to sulfate by oxidation when wetlands are drained or soils are converted from their natural state to agricultural cropland.

Some of the compounds produced in the sulfur cycle are gases. For example, combustion of fossil fuels, especially coal, produces sulfur dioxide gas (SO_2), which is further oxidized in the atmosphere to sulfur trioxide (SO_3) that combines with water to form sulfuric acid (H_2SO_4). Reduction of sulfate in anaerobic soils and sediment produces hydrogen sulfide (H_2S) gas that also is reoxidized to sulfate in the presence of air. As a result, precipitation from the atmosphere is a major source of sulfate in streams in parts of North America and Europe.

Another source of sulfur in the global hydrologic cycle that is not related to human activities is gaseous emission of hydrogen sulfide and sulfur dioxide from volcanoes and other geothermal sources. Although effects of these emissions can be locally intense, they generally are thought to be much less significant on a global scale than the human sources.

Chlorine. Nearly all chlorine compounds are readily soluble in water. Widespread increases in chloride concentration in runoff in much of the United States can be attributed to the extensive use of sodium chloride and calcium chloride for deicing of streets and highways. Increases in chloride concentration also can occur as a result of disposal of sewage, oil field brines, and various kinds of industrial waste. Thus, chloride concentration trends also can be considered as an index of the alternation of streamwater chemistry by human development in the industrialized sections of the world.

Nitrogen. About three-fourths of the Earth's nitrogen is present in the atmosphere as nitrogen gas. Because of its importance as an essential element in plant and animal nutrition, nitrogen, in its various oxidation states and its yield and concentration, is of considerable interest in studies of human influences on stream water composition. Certain small- and medium-sized streams in the intensively developed agricultural areas of the United States have been strongly affected by nitrogen-bearing runoff from fertilized soil. However, nitrogen is more difficult to use than sulfate or chloride as an index of human effects on water composition.

Stream Water Quality Trends

Those streams that have been subjected to contaminant additions through disposal of industrial and other forms of waste generally show substantial decreases in the diversity and abundance of instream biota. These effects are often first observed in decreases of native game fish such as brook or rainbow trout, but more extensive research has shown many more subtle changes.

Summary

Under pristine conditions, that is, in the absence of human civilization and development, the chemical composition of inland stream and lake waters is, ideally, controlled by the alteration of rock minerals through chemical weathering processes, which liberate soluble products. These processes in turn are controlled or influenced by climatic factors such as rainfall, air temperature, and evaporation and by associated biological or biochemical processes, such as photosynthesis and transpiration by plants, decay of vegetative debris, and the effects of aquatic life processes. Circulation of essential nutrient elements, including carbon, sulfur, chlorine, and nitrogen, generally is bound to elemental oxygen from the atmosphere and provides most anionic species occurring in natural water, such as bicarbonate, sulfate, chloride, and nitrate. Other constituents of natural surface waters, including calcium, magnesium, sodium, potassium and silica, can be correlated in general with the chemical composition of rocks and soils in a given drainage basin; the first four of these are found as principal cationic species and are in electrochemical balance with anions in these waters.

Examples In the Great Lakes–Upper St. Lawrence River basin, the yield of sulfate in tons per square mile per year in the St. Lawrence River nearly doubled between 1905 and 1956 and continued to increase, but at a lesser rate, until about 1970, when the yield leveled off or perhaps even declined slightly. The continuing yield of between about 19 and 25 tons/mi^2 indicates that the basin may have reached a steady state between the natural and human-induced loading of sulfur to the basin and its removal by the St. Lawrence River. In contrast, sulfate concentrations in the upper Columbia River basin at Northport, Washington, show less clearly defined trends in sulfate concentrations and yields during the century. These data indicate that human-induced effects are largely masked by the large amount of runoff available in the Columbia River basin and by the effects of storage and mixing in lakes and reservoirs.

Coal mining was extensive in the Allegheny River drainage basin in Pennsylvania in this century, and sulfate concentrations in the river near Pittsburgh increased substantially between the early 1900s and 1962 as a result of drainage from many active and abandoned coal mines. The operation of flood control and multipurpose reservoirs in the basin has caused a nearly 50% decline in sulfate concentration in the Allegheny River near Pittsburgh at low flow, as shown in analysis of samples collected in 1947 and compared to samples collected in 1975 and 1989. This flow augmentation by timed reservoir releases has had a beneficial effect on the quality of the Allegheny River.

The Mississippi River drains more than 1,125,000 mi^2 of the conterminous United States and integrates the effect on stream water quality of a large range of human activities across a large continental area. The calculation of sulfate yield in the basin has many uncertainties because water quantity and water quality data are incomplete, and the effects of flow control and water diversions are difficult to measure. However, from the available data it can be estimated that sulfate concentrations in the lower Mississippi River at and upstream from New Orleans, Louisiana, probably have doubled between 1905–1906 and 1989; most of this increase seems to have occurred before 1980. Estimates of the increase in annual sulfate yield due to human activities since 1905 ranging from about 9 to about 14 tons/mi^2 are consistent with the increase in sulfate yields in the Great Lakes–St. Lawrence River system. In both instances, yields seem to have leveled off around 1970 or 1980 and to have remained fairly stable since. Possibly, both drainage systems have reached a steady state, and the natural and human-induced sulfate loading to the basins, much of it from atmospheric deposition, is now stable. Data for total SO_2 emissions for each state from 1965 to 1980 show that in most states adjacent to the Great Lakes the emission rates decreased significantly after 1970.

JOHN D. HEM
U. S. Geological Survey
Water Resources Division

American Water Works Association, American Public Health Association, and Water Environment Foundation, *Standard Methods for the Examination of Water and Wastewater,* 18th ed., American Public Health Assc., Washington, D.C., 1992.

J. D. Hem, *Study and Interpretation of the Chemical Characteristics of Natural Water,* 3rd ed., U.S. Geological Survey Water-Supply Paper 2254, U.S. Geological Survey, Reston, Va., 1985.

N. E. Peters , *Evaluation of Environmental Factors Affecting Yields of Major Dissolved Ions of Streams in the United States,* U.S. Geological Survey Water-Supply Paper 2228, U.S. Geological Survey, Reston, Va., 1984.

W. Stumm and J. J. Morgan, *Aquatic Chemistry,* 2nd ed., Wiley-Interscience, New York, 1981.

PROPERTIES

Water is the solvent and transport medium, participant, and catalyst in nearly all chemical reactions occurring in the environment. It is a necessary condition for life and represents a necessary resource for humans. It is an extraordinarily complex substance (Tables 1 and 2). Structural models of liquid water depend on concepts of the electronic structure of the water molecule and the structure of ice. Hydrogen bonding between H_2O molecules has an effect on almost every physical property of liquid water.

Water has a very high boiling point and high heat of vaporization; ice has a very high melting point. The maximum density of liquid water is near 4°C, not the freezing point, and water thus expands upon freezing. It has a very high surface tension. It is an excellent

Table 1. Thermodynamic Constants for Phase Changes of H_2O[a]

Property	Fusion[b]	Vaporization[b]	Sublimation[c]
temperature, K	273.15	373.15	273.16
isopiestic heat capacity change Δc_p, J/(mol°C)[d]	37.28	−41.93	
enthalpy change ΔH, kJ/mol[d]	6.01	40.66	51.06
entropy change ΔS, J/(mol°C)[d]	22.00	108.95	186.92
volume change ΔV, cm^3/mol	−1.621	3.01×10^4	
internal energy change ΔE, kJ/mol[d]	6.01	37.61	48.79

[a] Eisenberg and Kansmann.
[b] At 101.3 kPa (1 atm).
[c] At ice–liquid–vapor triple point.
[d] To convert J to cal, divide by 4.184.

Table 2. Physical and Chemical Properties of Liquid Water[a]

Property	Comparison with other substances	Importance to environment
density	maximum density at 4°C, not at freezing point; expands upon freezing (both properties unusual)	in lakes prevents freezing up and causes seasonal stratification
melting and boiling points	abnormally high	permits water to exist as a liquid at earth's surface
heat capacity	highest of any liquid except ammonia	moderates temperature by preventing extremes
heat of vaporization	one of the highest known	important to heat transfer in atmosphere and oceans; moderates temperature extremes
surface tension	very high	regulates drop formation in clouds and rain
absorption of radiation	large in infrared and ultraviolet regions; less in visible regions	important control on biological activity (photosynthesis) in water bodies and on atmospheric temperature
solvent properties	excellent solvent for ionic salts and polar molecules because of dipolar nature	important in transfer of dissolved substances in hydrological cycle and in biological systems

[a] Berner and Berner.

solvent for salts and polar molecules. It has the greatest dielectric coefficient of any liquid. These unusual properties are a consequence of the dipolar character of the H_2O molecule.

Dipole–dipole interaction between two water molecules forms a hydrogen bond, which is electrostatic in nature. The energy of each H bond is estimated to be about 20 kJ mol^{-1} (about 5% of the strength of a covalent bond).

The persistence, or the reformation, of hydrogen bonding in liquid water is a key to understanding the physical properties of water as well as its poorer solvent properties for nonpolar, hydrophobic solutes. The highly structured water linked by H bonds must be disrupted by any solute. When the solute is ionic, the attractive interactions between ion and water molecule favor dissolution. When the solute is nonpolar molecule, the structural cost to the hydrogen-bonded water makes the probability of dissolution unfavorable. Water's extremely high heat of vaporization, relatively low heat of fusion, and the unusual values of the other thermodynamic properties, including melting point, boiling point, and heat capacity, can be explained also by the presence of hydrogen bonding.

Acquisition of Solutes and Circulation of Water

In hydrological studies, the transfer of water between reservoirs is of primary interest. The oceans hold ca 76% of all the earth's water. Most of the remainder, ie, ca 21%, is contained in pores of sediments and in sedimentary rocks. A little more than 1% (or 73% of freshwater) is locked up in ice. The other freshwater reservoir of significant size is groundwater. The exchange of water between the various reservoirs includes ocean mixing, evaporation, precipitation, runoff, and percolation.

Hydrogeochemical cycles couple atmosphere, land, and water. Natural waters acquire their chemical characteristics by dissolution and by chemical reactions with solids, liquids, and gases with which they have come into contact during the various parts of the hydrological cycle. As a first approximation, seawater may be interpreted as the result of a gigantic acid–base titration, ie, acid of volcanoes vs bases of rocks (oxides, carbonates, silicates). The composition of freshwater may similarly be represented as resulting from the interaction of the CO_2 of the atmosphere with mineral rocks. Freshwater is now scarce in many regions of the world, resulting in severe ecological degradation, limits on agricultural and industrial production, threats to human health, and increased potential for international conflict. Only freshwater flowing through the solar-powered hydrological cycle is renewable. Nonreplenishable (fossil) groundwater can be tapped, but such extraction depletes reserves.

Atmosphere–Water Interaction. Although water is a very minor component of the atmosphere, less than 10^{-6} vol %, many important reactions occur in the water droplets of cloud, fog, and rain. The atmosphere is an oxic environment; in its water phase, gigantic quantities of reductants, such as organic substances, Fe(II), SO_2, CH_3SCH_3 (dimethyl sulfide), and nitrogen oxides, are oxidized by oxidants such as O_2, $OH^{\cdot}$ radicals, H_2O_2, and Fe(III).

The atmosphere is an important conveyor belt for many pollutants. Water and atmosphere are interdependent systems. Many pollutants, especially precursors of acids and photooxidants, originate directly or indirectly from the combustion of fossil fuels, as well as such natural activities as volcanic eruptions and the metabolic activity of organisms.

The following reactions are of particular importance in the formation of acid precipitation: oxidative reactions, either in the gaseous phase or in the aqueous phase, leading to the formation of oxides of C, S, and N; absorption of gases into water (cloud droplets, falling raindrops, or fog) and interaction of the resulting acids with ammonia and the carbonates of airborne dust; and the scavenging and partial dissolution of aerosols into water.

The products of the various chemical and physical reactions in the atmosphere are eventually returned to the earth's surface, either by wet deposition (ie, rain or snow) or dry deposition.

Interactions with Rocks.

A general relationship between the composition of the water and that of the solid minerals with which the water has come into contact can be expected. Chemical weathering is one of the major processes controlling the global hydrogeochemical cycle of elements. The atmosphere provides a reservoir for carbon dioxide and for oxidants required in the weathering reactions. The biota assists the weathering processes by providing organic ligands and acids and by supplying locally, upon decomposition, increased CO_2 concentrations.

During chemical weathering, rocks and primary minerals become transformed to solutes and soils, and eventually to sediments and sedimentary rocks.

Calcareous minerals and evaporite minerals (halides, gypsum) are very soluble and dissolve rapidly. Estimates indicate that evaporites and carbonates contribute approximately 17% and 38%, respectively, of the total dissolved load in the world's rivers. The remaining 45% is the result of the weathering of silicates.

In all cases, water and carbonic acid, the latter of which is the source of protons, are the main reactants. The net result of the reaction is the release of cations (Ca^{2+}, Mg^{2+}, K^+, Na^+) and the production of alkalinity via HCO_3^-.

The HCO_3^- and Ca^{2+} ions produced by weathering of $CaCO_3$ precipitate in the ocean as $CaCO_3$ through incorporation by marine organisms and the CO_2 consumed in carbonate weathering is released again upon formation of $CaCO_3$ in the ocean. Thus, globally, carbonate weathering results in no net loss of atmospheric CO_2. The weathering of calcium silicates also produces Ca^{2+} and HCO_3^-, which form $CaCO_3$ in the sea, but only half of the CO_2 consumed in the weathering is released and returned to the atmosphere upon $CaCO_3$ formation. Thus, silicate weathering results in a net loss of atmospheric CO_2.

Biochemical Cycle and Oxidation–Reduction Processes

The ecological system may be considered as a unit of the environment that contains a biological organization made up of all the organisms interacting reciprocally with the chemical and physical environment. The maintenance of life in aquatic ecosystems results directly or indirectly from the steady impact of solar energy. Photosynthesis may be conceived as a disproportionation of water into an oxygen reservoir and hydrogen, which forms high energy bonds with C, N, S, and P compounds that are incorporated as organic matter in the biomass. Various organisms catalytically decompose the unstable products of photosynthesis through energy-yielding electron-transfer reactions, eg, reduction–oxidation processes and respiration.

The composition of natural waters is strongly influenced by the growth, distribution, and decay of algae and other organisms. Organisms regulate the oceanic and lacustrine composition and its variation with depth. Dissolved constituents are taken up by organisms. Their remains sink under the influence of gravity and are gradually destroyed by oxidation. The superposition of this particular cycle upon the ordinary mixing cycle in the ocean or in lakes accounts for the variation in depth and distribution of chemical properties.

Chemical Composition

The concentrations of biologically regulated components, ie, C, N, P, and Si, vary with depth and are markedly influenced by the growth, distribution, and decay of phytoplankton and other organisms. The concentrations of other constituents, especially the salts, ie, Cl^-, SO_4^{2-}, Mg^{2+}, and Na^+, are remarkably constant and are different from those in fresh waters. For most elements, a mass balance appears to exist between input into the sea, mostly by rivers, and its removal, mostly by sedimentation. Thus, the oceans are often assumed to be at steady state.

The flow hydrothermal vents, ie, springs coming from the sea floor in areas of active volcanism, and the chemical reactions occurring there by high temperature alteration of basalts are of significance in the mass balance of Mg^{2+} and K^+. Furthermore, during the cycling of seawater through the earth's crust along the mid-ocean ridge system, geothermal energy is transferred into chemical energy in the form of reduced inorganic compounds, which are used by chemolithotrophic bacteria as sources of energy for the reduction of carbon dioxide (assimilation) to organic carbon.

The transition between rivers and oceans occurs in estuaries, which are semienclosed coastal bodies of water within which seawater is diluted with freshwater from land drainage. Estuaries are of great environmental concern because many marine organisms require the estuarine environment for part of their life cycle.

Estuaries with their salinity gradients and tidal movements represent gigantic natural coagulation tanks. Clay minerals, organic matter, colloidal hydroxides of iron(III), and heavy metals tend to coagulate and accumulate in the estuary.

Chemical Variety. The term species refers to the actual form in which a molecule or ion is present in solution. Dissolved inorganic species present in seawater are principally electrolytes, uncharged species, and dissolved gases.

Metals and metalloids that form alkyl compounds, eg, methylmercury and methylarsenic acid, tributyltin, deserve special concern because these compounds are volatile and accumulate in cells; they are poisonous to the central nervous system of higher organisms.

Pollution

The products of human activities find their way into the environment and disturb ecosystems. Pollution has altered the surroundings to the detriment of humanity. In the last several decades, the pollutional load has increased, and its character has changed (see WATER-POLLUTION).

JAMES J. MORGAN
California Institute of Technology
WERNER STUMM
Swiss Federal Institute of Technology

F. M. M. Morel and J. G. Hering, *Principles and Applications of Aquatic Chemistry,* Wiley-Interscience, New York, 1993, pp. 374.

D. Eisenberg and W. Kauzmann, *The Structure and Properties of Water,* Oxford University Press, London, 1969.

W. Stumm and J. J. Morgan, *Aquatic Chemistry,* 3rd ed., Wiley-Interscience, New York, 1996.

E. K. Berner and R. A. Berner, *Global Environment: Water, Air, and Geochemical Cycles,* Prentice Hall, Inc., Englewood Cliffs, N.J., 1996.

POLLUTION

Water is omnipresent on the earth. Constant circulation of water from the ocean to the atmosphere (evaporation) and from the atmosphere to land and the oceans (precipitation, runoff, etc) is generally known as the hydrologic cycle. Within the hydrologic cyclic, there are several subcycles where water is used and returned to the environment.

Freshwater is withdrawn from various sources (rivers, lakes, groundwater, etc) and used many times before its discharge to the ocean. Water uses can generally be classified as follows: public water supply (domestic); industrial; commercial and institutional, eg, restaurants, schools; agricultural; and livestock.

The largest consumers of water in the United States are thermal power plants (eg, steam and nuclear power plants) and the iron and steel, pulp and paper, petroleum refining, and food-processing industries. They consume >60% of the total industrial water requirements (see also POWER GENERATION; WASTES, INDUSTRIAL).

Water Quality Management

Over the past decade, increased emphasis has been placed on the removal of secondary pollutants, such as nutrients and refractory organics, and on water reuse for industrial and agricultural purposes. This in turn has improved both the design and operation of wastewater treatment facilities.

Wastewaters emanate from four primary sources: municipal sewage, industrial wastewaters, agricultural runoff, and stormwater and urban runoff. In the future, water quality management in highly urbanized areas will have to consider stormwater as a primary pollutant.

As municipal and industrial wastewaters receive treatment, increasing emphasis is being placed on the pollutional effects of urban and agricultural runoff (Table 1).

Table 1. Pollution from Urban and Agricultural Runoff

Constituent	Urban runoff (stormwater)	Agricultural runoff
suspended solids, mg/L	5–1200	
chemical oxygen demand (COD), mg/L	20–610	
biological oxygen demand (BOD), mg/L	1–173	
total phosphorus, mg/L	0.02–7.3	0.10–0.65
nitrate nitrogen, mg/L		0.03–5.00
total nitrogen, mg/L	0.3–7.5	0.50–6.50
chlorides, mg/L	3–35	

Agricultural runoff is a large contributor to etrophication in lakes and other natural bodies of water. Effective control measures have yet to be developed for this problem. Runoff of pesticides is also receiving increasing attention.

Water Quality Standards. Water quality standards are usually based on one of two primary criteria, stream standards or effluent standards. Stream standards are based on the intended use of the water and have dilution requirements based on a threshold value of specific pollutants. Effluent standards are based on the concentration of pollutants that can be discharged or on the degree of treatment required.

Although stream standards are the more realistic, they are difficult to administer and control. The equitable allocation of pollutional loads for many industrial and municipal complexes also poses political and economic difficulties. One variation of stream standards is the specification of a maximum concentration of a pollutant (ie, the BOD) in the stream after mixing at a specified low flow condition.

Note that the maintenance of water quality and hence stream standards are not static, but subject to change with the municipal and industrial environment. For example, as the carbonaceous organic load is removed by treatment, the detrimental effect of nitrification in the receiving water increases.

In 1972 the U.S. Legislature passed Public Law 92-500, which requires certain levels of treatment for industrial wastewater discharges. Effluent guideline criteria (expressed as kilograms pollutant per unit of production) have been developed for each industrial category to be met by specified time periods.

The BPT is defined as the level of treatment that has been proven to be successful for a specific industrial category and that is currently in full-scale operation. The BAT is defined as the level of treatment beyond BPCTCA that has been proven feasible in laboratory and pilot studies and that is, in some cases, in full-scale operation.

The U.S. Environmental Protection Agency (EPA) has also developed pretreatment guidelines for those industrial plants which discharge into municipal sewer systems. In general, compatible pollutants such as BOD, suspended solids, and coliform organisms can be discharged providing the municipal plant has the capability of treating these wastewaters to a satisfactory level. Noncompatible pollutants, such as grease and oil, heavy metals, etc, must be pretreated to specified levels. Rigid limitations have been developed for the discharge of toxic substances to the nation's waterways.

Recent air pollution regulations limit the amount of volatile organic carbon (VOC) that can be discharged from wastewater treatment plants. Benzene is a particular case in which air emission controls are required if the concentration of benzene in the influent wastewater exceeds 10 mg/L.

Pollution Control

In past years, municipal wastewaters were treated to improve their appearance and bacteriological safety. Treatment included reduction in biochemical oxygen demand (BOD), suspended solids, pathogens, and inorganic dissolved solids.

The post-World War II growth in industrial activity has significantly altered the composition of wastewaters in urban treatment facilities. Pollutants that are resistant to biological oxidation have become predominant (eg, synthetic detergents, petrochemicals, synthetic rubber, etc), requiring the development of new nonbiological processes and approaches to water-pollution control. Today, the industrial-wastewater engineer must be familiar with the manufacturing process and the chemistry of the raw materials, products, and by-products.

A wastewater management plan comprises the following steps: segregation of clean water for potential recovery and reuse; isolation and segregation of noncompatible waste streams for separate pretreatment (eg, ion-exchange-regeneration wastes, inorganic and organic waste streams, waste streams that contain potentially toxic compounds, etc); isolation and segregation of concentrated waste streams and streams containing solvents for possible recovery or separate treatment and disposal (ie, thermal decomposition with heat recovery); and identification and characterization of batch discharges and spills in order to incorporate protective systems (eg, equalization) for the treatment facility.

In-Plant Waste Control. Pollution can be reduced or eliminated by process modification, chemical and raw materials substitution, or recovery of by-products. In addition, process modification generally increases product yield by incorporating control devices.

Toxic Organic Materials. The term toxic organics includes synthetic organic compounds such as pesticides, herbicides, PCBs, and chlorinated hydrocarbons. Because these compounds persist over a long period of time in a natural environment, the most effective treatment technology at present is incineration (see INCINERATORS).

Membrane Separation. If the concentration of organic material is too low for incineration, reverse osmosis or ultrafiltration can be used to concentrate toxic organic substances, depending on the type of compounds and the stability of the membrane against chemical attack (see also MEMBRANE TECHNOLOGY).

Chemical Treatment. Some organic compounds are attacked by chemical reagents such as potassium permanganate, sodium hydroxide, calcium hypochlorite, and ozone.

Adsorption. Many organic compounds can be adsorbed on activated carbon and synthetic resins. This technique depends on the properties of the compound being removed and the regenerative capability of the adsorbent.

Heavy Metals. Heavy metals of particular concern in the treatment of wastewaters include copper, chromium, zinc, cadmium, mercury, lead, and nickel. They are usually present in the form of organic complexes, especially in wastewaters generated from textiles finishing and dye chemicals manufacture.

Inorganic heavy metals are usually removed from aqueous waste streams by chemical precipitation in various forms (carbonates, hydroxides, sulfide) at different pH values.

Other methods, including activated carbon, ion exchange, and reverse osmosis, can be used to concentrate waste streams and remove the heavy metals.

See WASTE TREATMENT, HAZARDOUS WASTE and WASTES, INDUSTRIAL for more information on wastewater treatment.

W. WESLEY ECKENTELDER, JR.
Eckenfelder Inc

W. W. Eckenfelder, Jr., "Water Quality Management," in A. Bisio and S. Boots, eds., *Encyclopedia of Energy Technology and the Environment,* Vol. 4, John Wiley & Sons, Inc., New York, 1995, pp. 2823–2859.

Roy F. Weston, Inc., *Process Design Manual for Upgrading Existing Wastewater Treatment Plants,* prepared for U.S. EPA, technology transfer, program no. 17090 GNQ, 1971.

ANALYSIS

Since 1970, new analytical techniques, eg, ion chromatography, have been developed, and others, eg, atomic absorption and emission, have been improved. Detection limits for many chemicals have been dramatically lowered. Many wet chemical methods have been automated and are controlled by microprocessors which allow greater data output in a shorter time. Perhaps the best known continuous-flow analyzer for water analysis is the Autoanalyzer system manufactured by Technicon Instruments Corp. (Tarrytown, N.Y.) Recently, flow-injection analysis has also become popular.

Although simple analytical tests often provide the needed information regarding a water sample, such as the formation and presence of chloroform and other organohalides in drinking water, require some very specialized methods of analysis. The separation of trace metals into total and uncomplexed species also requires special sample handling and analysis.

A list of all water analyses would be extremely long since, under some conditions and with enough time, water can solubilize everything to some extent. Fortunately, a great deal can be learned about a water supply by carrying out a few physical and chemical tests. These simple tests might be all that are needed to characterize a water supply for many purposes, and it is usually the purpose for which the water is to be used that determines the type and extent of testing. The methods described in this review are intended primarily for freshwater analysis and may not be suitable for the analysis of saline water.

Physical Properties

An analysis of physical properties would include temperature, specific conductance, color, turbidity, taste and color, dissolved solids, suspended solids, and pH.

Principal Mineral Constituents and Gases

An analysis of the principal mineral constituents and gases in water would include measures of alkalinity, acidity, hardness (calcium and magnesium), sodium and potassium, chloride, sulfate, nitrate and nitrite, fluoride, phosphate, boron and borates, silica, and oxygen.

Minor Mineral Constituents and Gases

An analysis of minor mineral constituents and gases would include measurements of metals nonmetals (arsenic, selenium, cyanide, bromide and iodide), and hydrogen sulfide and ammonia.

Organic Materials

The following organic factors and materials would be measured biochemical oxygen demand (BOD), chemical oxygen demand (COD), total organic carbon (TOC), detergents, oil and grease, pesticides, and trihalomethanes (THN).

Radioactive Materials

Radioactivity in environmental waters can originate from both natural and artificial sources. The natural or background radioactivity usually amounts to $\leq$100 mBq/L. The development of the nuclear power industry as well as other industrial and medical uses of radioisotopes (qv) necessitates the determination of gross alpha and beta activity of some water samples.

Bacteria

A bacteriological examination of water is primarily carried out to determine the possible presence of harmful microorganisms. Testing is actually done to detect relatively harmless bacteria called *colon bacilli,* commonly called the coliform group, which are present in the intestinal tract of humans and animals. If these organisms are present in a water in sufficient number, then this is taken to be evidence that other harmful pathogenic bacteria may also be present.

J. D. Mulik and E. Sawicki, eds., *Ion Chromatographic Analysis of Environmental Pollutants,* Vol. 2, Ann Arbor Science Publishers, Inc., Ann Arbor, Mich., 1979.

J. K. Foreman and P. B. Stockwell, eds., *Topics in Automatic Chemical Analysis,* Vol. 1, Ellis Horwood Ltd., W. Sussex, UK, 1979.

Standard Methods for the Examination of Water and Wastewater, 15th ed., American Public Health Association, Washington, D.C., 1981.

SUPPLY AND DESALINATION

Water desalination is a process that separates the water from a saline water solution. The natural water cycle is the prevalent, and best, example of water desalination. Ocean waters evaporate as a result of solar heating and atmospheric influences, the vapor, consisting mostly of freshwater because of the negligible volatility of the salts at these temperatures, rises buoyantly and condenses into clouds in the cooler atmospheric regions, is transported across the sky by cloud motion, and is eventually deposited back on the earth's surface as freshwater rain, snow, and hail. The global freshwater supply (about 0.06% of the total water of the planet) from this natural cycle is ample (1.3×10^{12} m^3/d), but many regions on earth do not receive an adequate share.

Wells produce groundwater, stored from previous rains. However, the fact that in recent years wells have had to be made deeper and deeper to reach water shows that groundwater is being used faster than it is being replenished. Water lying in deep strata for millions of years is being mined like other minerals, never to be replaced.

Even rain is not pure water. Reports from the U.S. Geological Survey show that it contains 2.3–4.6 ppm of solids, or a yearly precipitation of 2.5–5 t/km^2. Recently (ca 1997), work conducted in the United States and Europe has underscored the rather dangerous results of increased use of fossil fuels, where the SO_x and NO_x emissions that end up in the rain lower its pH from 5.6 (slightly acidic) for uncontaminated rain, to <pH 4 for acid rains. Such acid rain has serious effects on surface waters.

Population growth, rapidly increasing demand for freshwater, and increasing contamination of the available natural freshwater resources render water desalination increasingly attractive. Improved energy economy can be obtained when desalination plants are coupled with power generation plants.

Of the surface of the earth, 71% (3.60×10^8 km^2) is covered by oceans, which have volume is 8.54×10^8 km^3. Unfortunately, this water with over 1000 ppm (parts per million by weight, or mg/L) salt is usually considered unfit for human consumption, and water with over 500 ppm is considered undesirable, but in some parts of the world, people and land animals are forced to survive with much higher concentrations of salts, sometimes of over 2500 ppm.

Transport of Freshwater

For centuries, containers have been used to carry freshwater, usually for longer distances than would be practical for conduits. Systems of dams, canals, and aqueducts were developed to carry freshwater considerable distances to growing cities, and to irrigate agricultural lands. Records of ancient Rome indicate that 14 aqueducts, averaging over 144 km in length, carried 1.175×10^6 m^3/d of water from the surrounding highlands by gravity. The aqueducts of Istanbul are even more dramatic. The engineers' competence developed with increasing needs for water, and conduits became longer. In more modern times, New York City's water system, initiated ca 135 yr ago, stands comparison as a true engineering marvel. Farsighted action in the late nineteenth century also gave the city extensive upstate watershed rights. This system, with a storage capacity of 2.07×10^9 m^3, safely furnishes about 8.5 million people with an average of 5.3×10^6 m^3/d of water. The system is not adequate today, but it has served well except in years of serious drought.

The 1974 Safe Drinking Water Act put all public water supplies in the United States under Federal supervision. For example, the 94th Congress authorized a six-state water study, the High Plains Study Council, to develop a set of plans for increasing water supply in a region which supplies about 15% of the nation's wheat, corn, sorghum, and cotton, and 38% of its livestock, and which depends on its rapidly dwindling groundwater (the Ogallala aquifer). In 1990 it was estimated that the aquifer would be practically depleted by the year 2020. It was proposed that huge amounts of water from the Missouri and Arkansas Rivers be diverted to the High Plains, with an estimate by the Army Corps of Engineers that it would take as long as 9 yr to design, 25 yr to build, and would cost $\$6-25 \times 10^9$ in 1982 dollars. Rising energy costs, costs of construction, local objections to Federal intervention, environmental and navigation considerations, and various other problems have prevented this and many other large water supply programs from materializing.

The problem of bringing water to southern California continues to be one producing controversy. The region is essentially a desert, devoid of any significant water resources, yet its population is growing at a fast pace and is expected to reach 49 million in 2020. Water is transported from Northern California and from the Colorado River, at distances of up to about 1000 km, expected to increase to about 1600 km by the year 2000, often against the objections by those living around these sources. The project to build the so-called Peripheral Canal, a 67-km long, 120-m wide channel to carry water from the Sacramento River delta to an existing aqueduct and then to the agriculturally productive San Joaquin Valley and thence to Southern California, was passed by California's legislature in 1980, and then defeated in 1982 owing to political opposition from a coalition of environmental and local urban and agricultural interests. Elevated salinity in the California Aqueduct and some other water canals has recently revived some interest in this project.

In some places and under certain conditions, freshwater can be obtained more cheaply by desalination of seawater than by transporting water.

Because construction costs have remained approximately constant since the mid-1980s, despite inflation, desalination has a high potential of becoming less costly with improved technology, whereas the cost of transporting water is not likely to decrease substantially. Desalination increases the total amount of freshwater; transportation never does. However, desalination is an energy-intensive process, and no presently conceivable alternatives promise any reduction of energy costs.

The Water Problem

The use of ground and surface freshwater in the United States from 1950–1990 is shown in Figure 1. Whereas the overall increase in use had been slightly reduced and then arrested after 1980, this was primarily because of a reduction in use for irrigation and industry, owing in large part to a decline in agricultural and industrial production. Public supply continues to increase, approximately keeping pace with population growth.

An analysis of freshwater use in the United States predicted that by the year 2040, barring major changes in use patterns and natural water cycles, withdrawals and consumption will increase by about 40%. Most of this increase must be met by reuse. For example, in manufacturing, recycling currently provides a reuse ratio of 1.3, which must be increased by the year 2020 to above 6.3.

In some areas of the United States, twice as much water is pumped from the ground as soaks into it. Not only is the groundwater being depleted, but often this withdrawal causes sinking of the ground level, as it has near Houston, Texas; Mexico City; and in Florida.

Water is far from evenly distributed in the United States, with major shortages in some very populous areas. California and the Southwest have always been water-short, but in the first half of the 1960s and the late 1970s, the northeast, too, experienced water shortages.

Increasing pollution of water sources worldwide is a major contributor to the climbing shortage of usable water. A recent report of the U.S. Environmental Protection Agency points out that about 40% of the U.S. rivers, lakes, and estuaries surveyed are too polluted for even basic uses such as fishing or swimming. Especially grim is the condition of the Great Lakes, which contain one-fifth of the world's fresh surface water. About 97% of the Great Lakes waters were found to be substandard for designated uses, and particularly worrisome is the fact that the lakes are continuously polluted by toxic chemicals. In this context it is important to note the vital importance of water-quality legislation. Along with the Clean Air Act, the Clean Water Act of 1972 is of unprecedented value in U.S. environmental protection history. Launched to "restore and maintain the chemical, physical, and biological integrity of the Nation's waters," it has resulted in a significant slowdown of the pollution processes and in many cases reversed the trend. The activities in support of the Clean Water Act incur, however, an economic cost, and the Act is continuously threatened by interest groups which object to this cost and inconvenience. Many countries in the world, especially the developing ones, and those currently or formerly under communist regimes, still do not have such legislation or enforcement and have severely polluted waters. A World Bank report issued in 1995 stated that 80 countries with 40% of the world's population already have water shortages which could cripple agriculture and industry, and that about 95% of the world's urban areas dump raw sewage into rivers and other water bodies, with dirty water killing 10 million or more people every year and causing untold economical damage.

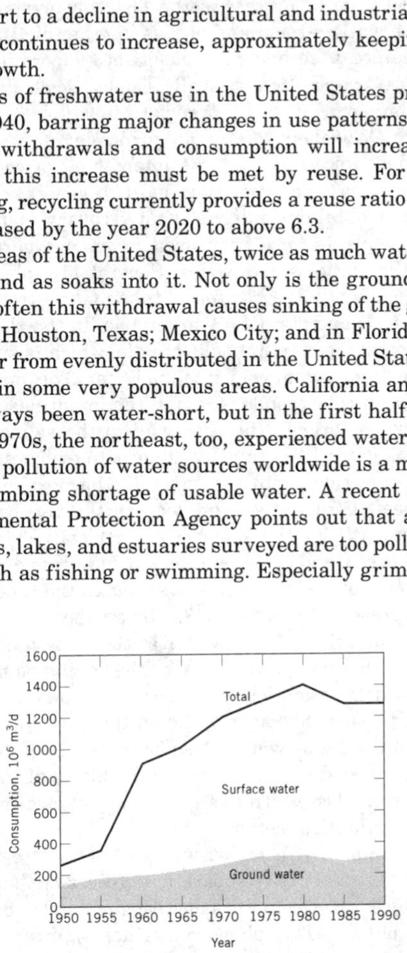

Figure 1. Trends of estimated ground and surface freshwater use in the United States, 1950–1990.

Some of the most attractive areas of the world, particularly islands and beaches, are almost devoid of freshwater. The biggest and one of the fastest growing of the industries of the world is tourism. It is a particularly attractive industry to developing countries, and in some of these it may be almost the only nonagricultural industry. Tourism accounts for a substantial use of the available freshwater, and it may be stifled or entirely prevented in otherwise attractive places if there is insufficient freshwater. To conserve available water supplies, some hotels might use double water systems, ie, seawater for flushing, but they still must provide on the order of 400–600 L/d per tourist to assure a comfort to guests unaccustomed to water shortages. Production of these quantities of water by desalination techniques has become an important expense to the hotels. Today, the cost of production of desalinated water can be as high as \$4.00–5.25/m^3, and in some locations even higher.

The newly acquired wealth of the countries in the Middle East has given an unexpected boost in desalination for the region. Starting with nearly zero, Saudi Arabia, Kuwait, and the United Arab Emirates have over the past three decades added plants producing about 9×10^6 m^3/d of freshwater, representing nearly one-half the worldwide desalination capacity. Following the Arab Peninsula in order of desalination capacity are the United States, North Africa, the rest of the Middle East, the European Mediterranean area, Southeast Asia, the rest of Europe, and the Caribbean islands. Water desalination is increasingly used for the treatment of effluent waters for reuse, and of river and city waters to improve purity for various industrial applications such as boiler feed and ultrapure water for the electronics industry.

In many of the developing nations, lack of water hampers the profitable exploitation of material resources. For example, proven mines on Egypt's Red Sea Coast cannot be operated; others on the same coast are operable only with desalination water for processing phosphate ores and for mining lead and zinc. Fishing industries on South America's arid Pacific Coast cannot be expanded for lack of water to process the haul. These represent great losses in the world's supplies of minerals and foods; there are many others.

Saline Water for Municipal Distribution. Only a very small amount of potable water is actually taken by people or animals internally, and it is quite uneconomical to desalinate all municipally piped water, although all distributed water must be clear and free of harmful bacteria. Most of the water piped to cities and industry is used for little more than to carry off small amounts of waste materials or waste heat. In many locations, seawater can be used for most of this service. If chlorination is required, it can be accomplished by direct electrolysis of the dissolved salt. Arrayed against the obvious advantage of economy, there are several disadvantages: use of seawater requires different detergents; sewage treatment plants must be modified; the usual metal pipes, pumps, condensers, coolers, meters, and other equipment corrode more readily; chlorination could cause environmental pollution; and dual water systems must be built and maintained.

Water in Industry. Freshwater for industry can often be replaced by saline or brackish water, usually after sedimentation, filtration, and chlorination (electrical or chemical), or other treatments. Such treatment is not necessary for the largest user of water, the electric power industry, which in the United States passed through its heat exchangers in 1990 about 40% of the total supply of surface water, a quantity similar to that used for agriculture, and it was 48% of the combined fresh and saline water withdrawals.

Water for Agriculture. Two liters of water in some form is the daily requirement of the average human, depending on many personal and external conditions. However, at least several hundred liter per day are required for the growing of vegetables, fruits, and grain that make up the absolute minimal daily food ration for a vegetarian.

(A loaf of bread alone contains little of the more than a ton of water necessary to grow the wheat therein.)

The one-seventh of the world's crop lands that are irrigated produce one-quarter of the world's crops. Irrigation's main losses result directly from seepage and evaporation from the open water-carrying channels and the soil. Only a small fraction of the water withdrawn from the irrigation ditch or pipe is absorbed by the plants. Plastic films, as ground covers through which the plants protrude, prevent some losses but at great expense for film and labor. Cheaper systems are necessary to assure better water utilization by plants. Other possible goals would be food plants with membranes capable of separating freshwater from brackish water, to give a non-salty crop. Progress has been made in both of these directions, and some plants have been developed that accumulate salt from the ground.

Desalination: Manufactured Freshwater

The possibility of producing freshwater from seawater or brackish water by separation of the salts opens a new dimension in the supply of freshwater. Areas bordering the sea would have an available raw material without limit or cost of transportation to the water facility. The successful realization of desalination by the combined effort of chemists, and chemical, mechanical, and materials engineers, as opposed to the search for and transport of existing freshwater, gives new hope for adequate water in many, but not all, cases.

Many ways are available for separating water from a saline water solution. The oldest and still prevalent desalination process is distillation. The evaporation of the solution is effected by the addition of heat or by lowering of its vapor pressure, and condensation of these vapors on a cold surface produces freshwater. The three dominant distillation processes are multistage flash (MSF), multieffect (ME), and vapor compression (VC). Until the early 1980s, the MSF process was prevalent for desalination. Now membrane processes, especially reverse osmosis (RO), are economical enough to have gained about one-third of the market. In all membrane processes, separation occurs as a result of the selective nature of a membrane's permeability, which, under the influence of an external driving force, permits the preferential passage of either water or salt ions, but not both. The force driving the process may be pressure (as in RO), electric potential (as in electrodialysis, ED), or heat (as in membrane distillation, MD). A process used for low salinity solutions is the well-known Ion Exchange (IE), in which salt ions are preferentially adsorbed onto a material that has the required selective adsorption property, and thus reduce the salinity of the water in the solution.

Two aspects of the basically simple desalination process require special attention. One is the high corrosivity of seawater, especially pronounced in the higher temperature distillation processes, which requires the use of corrosion-resistant, and therefore expensive, materials (eg, copper–nickel alloys, stainless steel, titanium, and, at lower temperatures, fiber-reinforced polymers and special concrete compositions. The other aspect is scale formation. The term scale describes deposits of calcium carbonate, magnesium hydroxide, or calcium sulfate which can form in the brine heater and the heat-recovery condensers. To reduce corrosion, scaling, and other problems, the water to be desalted is pretreated (by filtration, deaeration, decarbonation, and softening).

Ingestion into the plant of saline water feed which has been contaminated by undesirable components may not only impair product water quality, but also impair or incapacitate the plant for future operation. Must be kept out of the plant intakes. Screens, filters, and oil booms are commonly used for protecting the plant intake from spilled oil, heavy metals, detergents and other undesirable components which increasingly pollute the seas and oceans.

Distillation Processes

Multistage Flash Evaporation (MSF). Almost all of the large desalination plants use the MSF process. The seawater feed is preheated by internal heat recovery from condensing water vapor during passage through a series of stages, and then heated to its top temperature by steam generated by an external heat source. The hot seawater then flows as a horizontal free-surface stream through a series of stages, created by vertical walls which separate the vapor space of each stage from the others. These walls allow the vapor space of each stage to be maintained at a different pressure, which is gradually decreased along the flow path as a result of the gradually decreasing temperature in the condenser/seawater-preheater installed above the free stream. The seawater is superheated by a few °C relative to the vapor pressure in each stage it enters, and consequently evaporates in each stage along its flow path. The evaporation is vigorous, causing intensive bubble generation and growth with accompanying stream turbulence, a process known as flash evaporation. One of the primary advantages of the MSF process is the fact that evaporation occurs from the saline water stream and not, as in other distillation process such as submerged tube and multiple-effect evaporation, on heated surfaces, where evaporation typically causes scale deposition and thus gradual impairment of heat-transfer rates.

Multi-Effect Distillation (ME). The principle of the multi-effect (ME) distillation process is that the latent heat of condensation of the vapor generated in one effect is used to generate vapor in the next effect, thus obtaining internal heat recovery and good energy efficiency. Such plants have been used for many years in the salt, sugar, and other process industries. Several ME plant configurations, most prominently the horizontal tube multi-effect (HTME) and the vertical tube evaporator (VTE) are in use. In the HTME, vapor is circulated through a horizontal tube bundle, which is subjected to an external spray of somewhat colder saline water. The vapor flowing in these spray-cooled tubes condenses, and the latent heat of condensation is transferred through the tube wall to the saline water spray striking the exterior of the tube, causing it to evaporate. The vapor generated thereby flows into the tubes in the next effect, and the process is repeated from effect to effect. In the vertical tube multi-effect evaporator (VTE), the saline water typically flows downward inside vertical tubes, and evaporates as a result of condensation of vapor coming from a higher temperature effect, on the tube exterior.

Vapor Compression Distillation (VC). The vapor pressure of saline water is lower than that of pure water at the same temperature, the pressure difference being proportional to the boiling point elevation of the saline water. Desalination is attained by evaporating the saline water and condensing the vapor on top of the pure water. Therefore, the pressure of the saline water vapor must be raised by the magnitude of that pressure difference, plus some additional amount to compensate for various losses. This is the principle governing the vapor compression desalination method.

Freeze-Desalination. It is rather well known that freezing of saline water solutions is an effective separation process in that it generates ice crystals which are essentially salt-free, surrounded by a more salt-concentrated solution. This phenomenon has significant underlying appeal for its use as a water desalination process: (*1*) compared with distillation, it is more energy efficient; (*2*) operation at freezing temperatures reduces the problems of corrosion and scaling significantly;

and (3) the low temperature also allows the use of less expensive construction materials.

Freeze-desalination has not yet (ca 1997) reached commercial introduction for several reasons, such as the difficulty in developing efficient and economical compressors for vapor having the extremely high specific volume at the low process pressure, and difficulties in maintaining the vacuum system leak-free and in effecting reliable washing of the ice crystals.

Membrane Separation Processes

Reverse Osmosis (RO). If a pressure higher than the osmotic pressure is applied to the concentrated solution side of a semipermeable membrane, the water will move across the membrane in the reverse direction, from the saline solution side to the pure water one. This process is called reverse osmosis (and sometimes hyperfiltration), and is the basic principle underlying reverse-osmosis desalination.

Unlike filtration of particulates, the selective filtration of the water in reverse osmosis is not the result of the relationship of the membrane pore size to the relative sizes of the salt and water molecules. One way to view the process is that the very thin active surface layer of the membrane forms hydrogen bonds with water molecules and thus makes them unavailable for dissolving salt. Salt thus cannot penetrate through that layer. Water molecules approaching that pure water layer are, however, transported through it by forming such hydrogen bonds with it and in that process displacing water molecules that were previously hydrogen-bonded at these sites. The displaced water molecules then move by the forces of the hydraulic pressure difference through the pores of the remainder of the membrane, emerging at its other side. It is noteworthy that RO also separates small-molecule organic solutes from the water.

Electrodialysis. In electrodialysis (ED), the saline solution is placed between two membranes, one permeable to cations only and the other to anions only. A direct electrical current is passed across this system by means of two electrodes, causing the cations in the saline solution to move toward the cathode, and the anions to the anode. The anions can only leave one compartment in their travel to the anode, because a membrane separating them from the anode is permeable to them. Cations are both excluded from one compartment and concentrated in the compartment toward the cathode. This reduces the salt concentration in some compartments, and increases it in others. Tens to hundreds of such compartments are stacked together in practical ED plants, leading to the creation of alternating compartments of fresh and salt-concentrated water. ED is a continuous-flow process, where saline feed is continuously fed into all compartments and the product water and concentrated brine flow out of alternate compartments.

Solar Desalination

The benefits of using the nonpolluting and inexhaustible energy of the sun for water desalination are obvious. Furthermore, many water-poor regions also have a relatively high solar flux for a large fraction of the time. The major impediment to the use of solar energy is economical: the diffuse nature of solar energy, even at its highest, dictates the need for constructing a large solar energy collection area. The still area required for producing 1 m^3 of fresh water per day is about 500 m^2.

A typical solar still consists of a saline water container in which the water is exposed to the sun and heated by it. The temperature rise to above ambient causes net evaporation of the saline water, thus separating pure water vapor from the solution. The transparent cover of the still serves several important functions: it prevents vapors from escaping the still; exposed to air at temperatures lower than the heated saline water in the basin, it serves as a condenser for the vapor; given proper inclination and geometric configuration, it channels the condensed fresh water to product collection troughs; it reduces heat loss from the warm saline water to the outside; and it prevents dirt from entering the still. However, it also must be as transparent as possible to solar radiation so that maximal heat gain by the still can be accomplished.

Solar stills integrate the desalination and solar energy collection processes. Another approach to solar desalination is to use separately a conventional desalination process and a solar energy supply system suitable for it. Any compatible desalination and solar energy collection processes could be used. Distillation, such as MSF or ME, can be used with heat input from solar collectors or concentrators or from solar ponds.

Still in the R&D stage, open-cycle ocean–thermal energy conversion (OTEC) plants using surface condensers produce desalted water. The warmer surface ocean water is pumped into a flash evaporation vessel from which the air has been evacuated. The vapor generated by flashing passes through a turbine and thus produces power, and is then condensed in a deaerated surface condenser cooled by colder (typically by about 20°C) seawater pumped to it from the ocean depths. The condensed vapor is freshwater, which can then be pumped out and used. In addition to requiring no fuel or other form of nonrenewable energy, OTEC and solar pond desalination have an advantage in that thermal energy storage is naturally included in the large thermal mass of the ocean or pond.

Solar or wind energy can also be used for desalination processes which are driven by mechanical or electrical power, such as VC, RO, and ED.

NOAM LIOR
University of Pennsylvania
ROBERT BAKISH
Bakish Materials Corporation

N. Lior, ed., *Measurements and Control in Water Desalination,* Elsevier, Amsterdam, the Netherlands, 1986.

K. S. Spiegler and Y. M. El Sayed, *A Desalination Primer,* Balaban Desalination Publications, Mario Negri Sud Research Institute, Santa Maria Imbaro (Ch), Italy, 1994.

M. W. Kellogg Co., *Saline Water Data Conversion Engineering Data Book,* 3rd ed., Office of Saline Water Contract No. 14-30-2639, U.S. Department of the Interior, Washington, D.C., 1975.

Z. Amjad, ed., *Reverse Osmosis: Membrane Technology, Water Chemistry and Industrial Applications,* Van Nostrand Reinhold, New York, 1993.

INDUSTRIAL WATER TREATMENT

Abundant supplies of fresh water are essential to the development of industry. Enormous quantities are required for the cooling of products and equipment, for process needs, for boiler feed, and for sanitary and potable water supply. Water impurities that may pose problems include dissolved and suspended solids.

Clarification

The suspended matter in raw water can consist of large solids, settleable by gravity alone and nonsettleable material, often colloidal in nature, that is removed by coagulation, flocculation, and sedimentation. The combination of these three processes is referred to as conventional clarification.

Coagulation involves neutralizing charged particles to destabilize suspended solids. In most clarification processes, a flocculation step then follows. Flocculation starts when neutralized or entrapped particles begin to collide and fuse to form larger particles. This process can occur naturally or can be enhanced by the addition of polymeric flocculant aids.

Typical inorganic coagulants are iron and aluminum acid salts that lower the pH of the treated water by hydrolysis.

Polyelectrolytes refers to all water-soluble organic polymers used for clarification, whether they function as coagulants or flocculants. Water-soluble polymers may be classified as anionic, cationic, or nonionic.

The use of organic polymers offers several advantages over the use of inorganic coagulants:

The amount of sludge produced during clarification can be reduced by 50–90%.

The resulting sludge contains less chemically bound water and can be more easily dewatered.

Polymeric coagulants do not affect pH; therefore, the need for supplemental alkalinity, such as lime, caustic, or soda ash, is reduced or eliminated.

Polymeric coagulants do not add to the total dissolved solids concentration.

Polymeric coagulants do not add soluble iron or aluminum carryover in the clarifier effluent and thereby minimize later deposition of these metals in filters, ion-exchange units, and cooling systems.

In certain instances, even high doses of primary coagulant will not produce the desired effluent clarity and a polymeric coagulant aid maybe.

Generally, very high molecular weight anionic polyacrylamides are the most effective coagulant aids. Essentially, the polymer bridges the small floc particles and causes them to agglomerate rapidly into larger, more cohesive flocs that settle quickly.

Precipitation Softening

Precipitation softening processes are used to reduce raw water hardness, alkalinity, silica, and other constituents. The water is treated with lime or a combination of lime and soda ash (carbonate ion). These chemicals react with the hardness and natural alkalinity in the water to form insoluble compounds. The compounds precipitate and are removed by sedimentation and, usually, filtration. Waters with moderate to high hardness and alkalinity concentrations (150–500 ppm as $CaCO_3$) are often treated in this fashion. The methods used include the following: **cold lime softening,** which is precipitation softening accomplished at ambient temperatures; **warm lime softening,** which operates at 49–60°C; and **hot process softening,** which is usually carried out under pressure at temperatures of 108–116°C.

Filtration

Filtration is used in addition to regular coagulation and sedimentation or precipitation softening for removal of solids from surface water or wastewater. Clarifier effluents of 2–10 NTU may be improved to 0.1–1.0 NTU by conventional sand filtration. Filtration ensures acceptable suspended solids concentrations in the finished water even when upsets occur in the clarification processes.

Ion Exchange

Ion exchangers exchange one ion for another, hold it temporarily, and then release it to a regenerant solution. Ionizable groups attached to the resin bead determine the functional capability of the resin. Industrial water treatment resins are classified into four basic categories: strong acid cation (SAC); weak acid cation (WAC); strong base anion (SBA); and weak base anion (WBA).

Sodium Zeolite Softening. Sodium zeolite softening is the most widely applied use of ion exchange. Water containing scale-forming ions, such as calcium and magnesium, passes through a resin bed containing SAC resin where the hardness ions are exchanged with the sodium, and the sodium diffuses into the bulk water solution.

Demineralization. Softening alone is insufficient for most high-pressure boiler feed waters and for many process streams, especially those used in the manufacture of electronics equipment. Demineralization of water is the removal of essentially all inorganic salts by ion exchange. In this process, strong acid cation resin in the hydrogen form converts dissolved salts into their corresponding acids, and strong base anion resin in the hydroxide form removes these acids. Demineralization produces water similar in quality to distillation at a lower cost for most fresh waters.

Condensate Polishing. Ion exchange can be used to purify or polish returned condensate, removing corrosion products that could cause harmful deposits in boilers. Typically, the contaminants in the condensate system are particulate iron and copper.

Membrane Processes

Common membrane processes include ultrafiltration (UF), reverse osmosis (RO), electrodialysis (ED), and electrodialysis reversal (EDR). These processes (with the exception of UF) remove most ions; RO and UF systems also provide efficient removal of nonionized organics and particulates. Because UF membrane porosity is too large for ion rejection, the UF process is used to remove contaminants, such as oil and grease, and suspended solids.

Boilers

Boiler System Corrosion. The dissolved gases normally present in water cause many corrosion problems. For instance, oxygen in water produces pitting that is particularly severe because of its localized nature. Carbon dioxide corrosion is frequently encountered in condensate systems and less commonly in water distribution systems. Water containing ammonia, particularly in the presence of oxygen, readily attacks copper and copper-bearing alloys. The resulting corrosion leads to deposits on boiler heat transfer surfaces and reduces efficiency and reliability. In addition to the gases, corrosion can also be caused by the concentration of caustic or acidic species in the feed water. If combined with mechanical or thermal stresses, stress corrosion cracking may occur.

Boiler Deposits. Deposition is a principal problem in the operation of steam generating equipment. The accumulation of material on boiler surfaces can cause overheating and/or corrosion. Common feedwater contaminants that can form boiler deposits include calcium, magnesium, iron, copper, aluminum, silica, and (to a lesser extent) silt and oil. Most deposits can be classified as one of two types: scale that crystallized directly onto tube surfaces or sludge deposits that precipitated elsewhere and were transported to the metal surface by the flowing water.

Boiler Water Treatment. To meet industrial standards for both oxygen content and the allowable metal oxide levels in feed water, nearly complete oxygen removal is required. This can be accomplished only by efficient mechanical deaeration supplemented by an effective and properly controlled chemical oxygen scavenger.

Scale and deposits are controlled through the use of phosphates, chelants, and polymers. Phosphates are precipitating treatments, and chelants are solubilizing treatments. Polymers are most widely used to disperse particulates but they are also used to solubilize contaminants under certain conditions.

Boiler water solids can be carried over with steam to form deposits in nonreturn valves, superheaters, and turbine stop and control valves. Carry-over can contaminate process streams and affect product quality. Deposition in superheaters can lead to failure due to overheating and corrosion. Effective boiler water blowdown control and the application of a boiler water antifoam may be required to mitigate these problems.

Chemical Treatment of Condensate Systems

Condensate systems can be chemically treated to reduce metal corrosion. Treatment chemicals include neutralizing amines, filming amines, and oxygen scavenger-metal passivators. Neutralizing amines hydrolyze when added to water, generating hydroxide ions required to neutralize the acid (H^+) formed by the dissolution of carbon dioxide or other acidic process contaminants in the condensate.

Filming amines protect against oxygen and carbon dioxide corrosion by replacing the loose oxide scale on metal surfaces with a thin amine film barrier.

Where oxygen invades the condensate system, corrosion of iron and copper-bearing components can be overcome through proper pH control and the injection of an oxygen scavenger. Hydroxylamines have been shown to be particularly effective for most systems. The use of neutralizing amines in conjunction with an oxygen scavenger–metal passivator improves corrosion control.

Cooling Systems

Corrosion. Corrosion can be defined as the destruction of a metal by chemical or electrochemical reaction with its environment. In cooling systems, corrosion occurs at the anode, where metal dissolves. Often, this is separated by a physical distance from the cathode, where a reduction reaction takes place. An electrical potential difference exists between these sites, and current flows through the solution from the anode to the cathode. This is accompanied by the flow of electrons from the anode to the cathode through the metal.

Types of Corrosion. The formation of anodic and cathodic sites, necessary to produce corrosion, can occur for any of a number of reasons: impurities in the metal, localized stresses, metal grain size or composition differences, discontinuities on the surface, and differences in the local environment (eg, temperature, oxygen, or salt concentration). When these local differences are not large and the anodic and cathodic sites can shift from place to place on the metal surface, corrosion is uniform. With uniform corrosion, fouling is usually a more serious problem than equipment failure.

Localized corrosion, which occurs when the anodic sites remain stationary, is a more serious industrial problem. Forms of localized corrosion include the following.

Pitting. Pitting is one of the most destructive forms of corrosion and occurs when anodic and cathodic sites become stationary due to large differences in surface conditions. It is generally promoted by low velocity or stagnant conditions (eg, shellside cooling) and by the presence of chloride ions. Once a pit is formed, the solution inside it is isolated from the bulk environment and becomes increasingly corrosive.

Selective Leaching. Selective leaching is the corrosion of one element of an alloy. The most common example in cooling systems is dezincification, which is the selective removal of zinc from copper–zinc alloys. Low pH conditions (<6.0) and high free chlorine residuals (<1.0 ppm) are particularly aggressive in producing dezincification.

Galvanic Corrosion. Galvanic corrosion occurs when two dissimilar metals are in contact in a solution. The contact must be good enough to conduct electricity, and both metals must be exposed to the solution. The driving force for galvanic corrosion is the electric potential difference that develops between two metals.

The most serious form of galvanic corrosion occurs in cooling systems that contain both copper and steel alloys. Galvanic corrosion can be controlled by the use of sacrificial anodes.

Crevice Corrosion. Crevice corrosion is intense localized corrosion that occurs within any area that is shielded from the bulk environment. Solutions within a crevice are similar to solutions within a pit in that they are highly concentrated and acidic.

Stress Corrosion Cracking. Stress corrosion cracking (SCC) is the brittle failure of a metal by cracking under tensile stress in a corrosive environment. Commonly used cooling system alloys that may crack due to stress include austenitic stainless steels (300 series) and brasses. Chloride is the main contributor to SCC of stainless steels. For brasses, the ammonium ion is the principal cause of SCC. The most likely places for SCC to be initiated are crevices or areas where the flow of water is restricted. This is due to the buildup of corrodent concentrations in these areas.

Microbiologically Influenced Corrosion (MIC). Microorganisms in cooling water form "biofilms" on cooling system surfaces. Biofilms consist of populations of sessile organisms and their hydrated polymeric secretions. The presence of a biofilm can contribute to corrosion in three ways: physical deposition, production of corrosive byproducts, and depolarization of the corrosion cell caused by chemical reactions.

Erosion Corrosion. Erosion corrosion is the increase in the rate of metal deterioration from abrasive effects. It is increased by high water velocities and suspended solids and is often localized at areas where water changes direction. Cavitation (damage due to the formation and collapse of bubbles in high velocity turbines, propellers, etc) is a form of erosion corrosion.

Control of Cooling System Corrosion. Corrosion control requires a change in either the metal or the environment. The first approach, changing the metal, is expensive. Also, highly alloyed materials, which are resistant to general corrosion, are more prone to such failure as stress corrosion cracking.

The second approach, changing the environment, is a widely used, practical method of preventing corrosion. In aqueous systems, there are three ways to effect a change in environment to inhibit corrosion: (*1*) form a protective film of calcium carbonate on the metal surface using the natural calcium and alkalinity in the water, (*2*) remove the corrosive oxygen from the water, either by mechanical (vacuum) or chemical (sodium sulfite) deaeration, and (*3*) add corrosion inhibitors.

A corrosion inhibitor is any substance that effectively decreases the corrosion rate when added to an environment. An inhibitor can be identified most accurately in relation to its function:

Deaeration (mechanical or chemical) removes the corrosive substance—oxygen.

Passivating (anodic) inhibitors (chromate, nitrite, molybdate, orthophosphate) form a protective oxide film on the metal surface; they are the best inhibitors because they can be used in economical concentrations and their protective films are tenacious and tend to be rapidly repaired if damaged.

Precipitating (cathodic) inhibitors are simply chemicals (zinc, calcium carbonate, calcium orthophosphate) that form insoluble precipitates that can coat and protect the surface.

Copper corrosion inhibitors are aromatic triazoles, which bond with cuprous oxide to form a "chemisorbed" film at the metal surface.

Adsorption inhibitors have polar properties that cause them to be adsorbed on the surface of the metal; they are usually organic materials containing nitrogen sulfur, hydroxyl groups.

Silicates inhibit acqueous corrosion, but their mechanism of inhibition is not yet firmly established, although it may involve adsorption.

Cooling System Deposits. Deposit accumulations in cooling water systems reduce the efficiency of heat transfer and the carrying capacity of the water distribution system. In addition, the deposits cause oxygen differential cells to form, which accelerate corrosion and lead to process equipment failure. Deposits range from thin, tightly adherent films to thick, gelatinous masses, depending on the depositing species and the mechanism responsible for deposition.

Deposit formation is influenced strongly by system parameters, such as water and skin temperatures, water velocity, residence time, and system metallurgy. The most severe deposition is encountered in process equipment operating with high surface temperatures and/or low water velocities. With the introduction of high efficiency film fill deposit accumulation in the cooling tower packing has become an area of concern. Deposits are broadly categorized as scale or foulants.

Scale. Scale deposits are formed by precipitation and crystal growth at a surface in contact with water. Precipitation occurs when solubilities are exceeded either in the bulk water or at the surface. The most common scale-forming salts that deposit on heat transfer surfaces are those that exhibit retrograde solubility with temperature (eg, calcium carbonate, calcium phosphate, and magnesium silicate).

Deposit control agents that inhibit precipitation at dosages far below the stoichiometric level required for sequestration or chelation are called *threshold inhibitors.* They work by adsorbing on the newly emerging crystal, blocking active growth sites. The precipitate dissolves and releases the inhibitor, which is then free to repeat the process.

The most commonly used scale inhibitors are low molecular weight acrylate polymers and organophosphorus compounds (phosphonates).

Fouling. Fouling occurs when insoluble particulates suspended in recirculating water form deposits on a surface. Fouling mechanisms are dominated by particle–particle interactions that lead to the formation of agglomerates. Most potential foulants enter with makeup water as particulate matter, such as clay, silt, and iron oxides.

Both iron and aluminum are particularly troublesome because of their ability to act as coagulants. Also, their soluble and insoluble hydroxide forms can each cause precipitation of some water treatment chemicals, such as orthophosphate. Airborne contaminants usually consist of clay and dirt particles but can include gases such as hydrogen sulfide, which forms insoluble precipitates with many metal ions. Process leaks introduce a variety of contaminants that accelerate deposition and corrosion.

Methods that minimize fouling include removal of particulate matter by filtration and/or sedimentation processes and the use of high water velocities, dispersants (which suspend particulate matter by adsorbing onto the surface of particles and imparting a high charge), and surfactants (wetting agents that emulsify insoluble hydrocarbons).

Biofouling. Microbiological fouling in cooling systems is the result of abundant growth of algae, fungi, and bacteria on surfaces. Once-through and open or closed recirculating water systems may support microbial growth, but fouling problems usually develop more quickly and are more extensive in open recirculating systems. Process leaks may contribute further to the nutrient load of the cooling water. Reuse of waste water for cooling adds nutrients and also contributes large amounts of microbes to the cooling system.

The outcome of uncontrolled microbial growth on surfaces is "slime" formation. Slimes typically are aggregates of biological and nonbiological materials. The biological component, known as the biofilm, consists of microbial cells and their by-products. The predominant by-product is extracellular polymeric substance (EPS), a mixture of hydrated polymers. These polymers form a gel-like network around the cells and appear to aid attachment to surfaces. The nonbiological components can be organic or inorganic debris from many sources that have become adsorbed to or embedded in the biofilm polymer. Microbial activity under deposits or within slimes can accelerate corrosion rates and even perforate heat exchanger surfaces.

Both oxidizing and nonoxidizing antimicrobials are used to rectify biofouling. The oxidizing antimicrobials commonly used in industrial cooling systems are the halogens, chlorine and bromine, in liquid and gaseous form; organic halogen donors; chlorine dioxide; and to a limited extent, ozone. Oxidizing antimicrobials oxidize or accept electrons from other chemical compounds. Their mode of antimicrobial activity can be direct chemical degradation of cellular material or deactivation of critical enzyme systems within the bacterial cell. An important aspect of antimicrobial efficiency is the ability of the oxidizing agent to penetrate the cell wall and disrupt metabolic pathways.

Nonoxidizing antimicrobials usually control growths by one of two mechanisms. In one, microbes are inhibited or killed as a result of damage to the cell membrane by quaternary ammonium compounds (quats). In the other, microbial death results from damage to the biochemical machinery involved in energy production or energy utilization. Antimicrobials known to inhibit energy metabolism include organotins, bis(trichloromethyl) sulfone, methylenebis-(thiocyanate) (MBT), β-bromo-β-nitrostyrene (BNS), dodecyl-guanidine salts, and bromonitropropanediol (BNPD).

Macrofouling Organisms. Fouling caused by large organisms, such as oysters, mussels, clams, and barnacles, is referred to as macrofouling. Typically, organisms are a problem only in large once-through cooling systems or low cycle cooling systems that draw cooling water directly from natural water sources. Water that has been processed by an influent clarification and disinfection system is usually free of the larvae of macrofouling organisms. In the last 15 years, the incidence of problems in the United States caused by macrofouling has increased dramatically due primarily to the "invasion" of two organisms that were accidentally introduced to this country: the Asiatic clam and the zebra mussel.

The organisms that cause macrofouling can be killed by heated water (thermal backwash) and by nonoxidizing antimicrobials. Although oxidizing antimicrobial are toxic to all living organisms from bacteria to humans, exposure is not easily accomplished. Some mollusks (eg, oysters, blue mussels, Asiatic clams, and zebra mussels) and crustaceans (eg, barnacles) have sensitive chemoreceptors that detect the presence of oxidizing chemicals such as chlorine (hypochlorite), bromine (hypobromite), ozone, and hydrogen peroxide. When oxidizers are detected at life-threatening levels, the animal withdraws into its shell and closes up tightly (for days to weeks, if necessary) to exclude the hostile environment. If treatment is interrupted long enough (1 h or possibly less), the animals have time to reoxygenate their tissues and survive even longer.

JAMES ROBINSON
Betz Dearborn

C. E. Hamilton, ed., *Manual on Water,* ASTM Special Publication 442A, American Society for Testing and Materials, Philadelphia, Pa., 1978, 472 pp.

The Nalco Water Handbook, McGraw-Hill, Inc. Publishers, New York, NY, 1979.

Betz Handbook of Industrial Water Conditioning, Ninth Edition, Trevose, PA, 1991.

Metals Handbook, Ninth Edition, Volume 13, ASM International, Metals Park, Ohio, 1987. "A Practical Guide to Avoiding Steam Purity Problems in The Industrial Plant," *CRTD* **35,** (1995).

MUNICIPAL WATER TREATMENT

Sedimentation and Filtration

Most surface waters contain varying amounts of suspended solids, including silt, clay, bacteria, viruses, etc, and it is necessary to remove these prior to distribution to the domestic or industrial consumer. Suspended solids not only affect the acceptability of the water, they also interfere with disinfection. The principal treatment processes are sedimentation (qv) and filtration (qv). Sedimentation alone is rarely adequate for the clarification of turbid waters and is of little or no value for the removal of such very fine particles as clay, bacteria, etc.

In many plants that treat surface waters, there is a presedimentation reservoir ahead of the treatment units. The reservoir allows the larger particles to settle and provides a volume buffer against changes in water quality. Further treatment, which is necessary to produce potable water, may involve only filtration through sand or multiple-medium filters or may require considerable pretreatment, eg, coagulation and flocculation before filtration.

The two steps in the removal of a particle from the liquid phase by the filter medium are the transport of the suspended particle to the surface of the medium and interaction with the surface to form a bond strong enough to withstand the hydraulic stresses imposed on it by the passage of water over the surface. The transport step is influenced by such physical factors as concentration of the suspension, medium particle size, medium particle-size distribution, temperature, flow rate, and flow time.

Coagulation and Flocculation. The removal of colloidal particles, eg, turbidity, color, and bacteria, requires agglomeration of these particles prior to filtration. Agglomeration can be carried out chemically, that is by interaction of the colloidal particles with materials added as coagulants, or physicochemically by neutralizing the charge on the particles or by interparticle bridging. The term coagulation is applied to the addition of any material that causes agglomeration of the colloids; flocculation refers to the process of gentle agitation which builds floc particles large enough to settle rapidly. A physicochemical mechanism has been proposed for coagulation and is based on charge neutralization by double-layer compression, which allows the particles to approach closely enough for short-range van der Waals forces to cause agglomeration. Flocculation is the growth of a three-dimensional structure by interparticle bridging.

There are two classes of coagulants used in municipal water treatment: inorganic metal salts, containing iron and aluminum, and organic polymers. Many metal ions react with water to produce hydrolysis products that are multiply charged inorganic polymers. These may react specifically with negative sites on the colloidal particles to form relatively strong chemical bonds, or they may be adsorbed at the interface. In either case, the charge on the particle is reduced. The metal salts reduce the alkalinity in the water; therefore, it may be necessary to add base in the form of lime or soda ash.

The organic coagulants in use may be either natural or synthetic. Among the useful natural polymers are starches, gums, and gelatin. There are four classes of organic polymeric coagulants: cationic, anionic, ampholytic (frequently referred to as polyelectrolytes), and nonionic. The cationic class comprises quaternary ammonium compounds. Poly(sodium methacrylate) is an example of an anionic polyelectrolyte. Poly(lysine–glutamic acid), having both $-NH_3^+$ and $-COO^-$ groups, is an example of an ampholytic polyelectrolyte.

Color can be removed effectively and economically with either alum or ferric sulfate at pH values of 5–6 and 3–4, respectively. Raw-water colors may be as high as 450–500 units on the APHA color scale. The secondary MCL (maximum contaminant level) for color in the finished water is 15 units, although most municipal treatment plants produce water that seldom exceeds 5 units.

The treatment units used for color removal are the same as those used for turbidity removal. However, the pH must be increased prior to filtration so that the metal hydroxides are removed by the filters.

Softening. A water is classified as hard if it contains more than 120 mg of divalent ions per liter (454 mg/L), usually calcium and magnesium, expressed as $CaCO_3$. Hard water is problematic as cleaning in it requires a greater amount of soap than is needed in soft water to produce lather. Also, the calcium and magnesium salts have negative solubility coefficients and precipitate with an increase in temperature. A deposit of scale forms on heat-transfer surfaces, leading to localized overheating or clogging of service lines.

The American Water Works Association (AWWA) Water Quality Goals recommend a maximum total hardness of 80 ppm for municipal purposes. Municipal softening plants, however, distribute waters containing 70–150 ppm; the final quality is established based on such factors as public demand and economics. More than 1000 municipalities soften water. Most are in the Midwest and in Florida. However, concern for the adverse effect of soft water on cardiovascular disease (CVD) may limit the number of plants that introduce softening.

The two principal methods of softening water for municipal purposes are addition of lime or lime-soda and ion exchange. The choice method depends upon such factors as the raw-water quality, the local cost of the softening chemicals, and means of disposing of waste streams.

The lime or lime–soda process results in the precipitation of calcium as calcium carbonate and magnesium as magnesium hydroxide. One of the main problems associated with this method is the disposal of the sludge. The principal methods of disposal have been lagooning, discharge into the nearest water course, or discharge to the sanitary sewer system. The latter two are no longer acceptable methods of disposal because of the resultant pollution load. A method of disposal for large plants is recalcination, ie, regeneration of lime from the $CaCO_3$ by heating the $CaCO_3$ in a kiln to remove the carbon dioxide. The lime can be reused in the plant, the CO_2 used for recarbonation, and the excess lime sold. In cases where $Mg(OH)_2$ has precipitated with $CaCO_3$, it is necessary to remove the magnesium prior to recalcination. In some plants, the sludge is vacuum-filtered to ca 40–50% solids and then used for agricultural purposes or disposed of in a land fill.

For waters with high noncarbonate hardness or high magnesium content, or both, the use of lime or lime-soda may be very expensive. In such cases, ion-exchange softening may be more economical, particularly if brines are locally available at little cost for regeneration of the resins. Ion-exchange softening involves exchanging sodium or hydrogen ions for the calcium or magnesium ions in the raw water. The water passing through the exchanger is reduced to 0–2 mg/L (0–7.6 mg/gal) hardness. Since this is usually much lower than the final acceptable hardness, only enough of the water is passed through the exchanger to give the desired final hardness.

The two types of installations used for ion exchange are fixed-bed and continuous-regeneration units. The continuous units consist of a closed circuit containing two sections, one for softening and one for regeneration; the resin is circulated countercurrent to the rawwater flow through the softening tank. The resin then is pulsed periodically into the regeneration tank, where it is regenerated and rinsed; then it is pulsed back to the softening tank. Thus, there is continuous softening without the downtime customary with fixed-bed units.

Taste and Odor Control. Tastes and odors in surface waters result from the action of biological organisms, eg, algae, or from various minerals, pollution by industry, domestic seepage, or agriculture. Groundwaters may have taste and odor if they are polluted or if they contain gases, eg, H_2S or CH_4; the latter always contains associated impurities that have taste and odor. Removal of these gases can be accomplished by adsorption (qv) with activated carbon (qv); oxidation with chlorine, potassium permanganate, or ozone; or aeration.

Organic materials are generally removed by addition of powdered activated carbon. Oxidation with chlorine, potassium permanganate, or ozone may destroy tastes and odors or it may intensify them, depending upon the particular compounds involved. For example, chlorination of phenolic compounds leads to greatly increased tastes and odors. Hydrogen sulfide and methane can be removed by aeration, although the largest reduction in hydrogen sulfide may result from oxidation by the dissolved oxygen introduced during the aeration.

Control of Organic Compounds. There are two groups of organic compounds of concern: the trihalomethanes, which result from the use of chlorine as a disinfectant, and volatile organic solvents, which percolate through the soil thereby contaminating groundwaters. Trihalomethanes (THMs) result from the reaction of chlorine with natural organic precursors. The THMs of concern include chloroform, dichlorobromomethane, chlorodibromomethane, and bromoform. Control is based on the use of an alternative disinfectant, removal of precursors prior to chlorination, or removal of the THMs after formation. Use of alternative disinfectants has had varying degrees of success. The removal of precursors has been accomplished by optimizing coagulation, using activated carbon, or oxidizing with ozone, $KMnO_4$, or ClO_2. The removal of THMs after formation can be accomplished by air stripping or adsorption, or by a combination of these processes.

Volatile organic contaminants occur primarily in groundwaters as a result of the disposal of industrial solvents on the ground or in soakage pits. The removal of these compounds has best been accomplished by the use of air stripping or adsorption on activated carbon.

Iron and Manganese Removal. Groundwaters or water withdrawn from the depths of reservoirs may contain soluble iron and manganese in the +2 oxidation state. If the reduced metal ions are allowed to remain in the finished water and then come into contact with the atmosphere, the oxidized forms ($Fe(OH)_3$ and MnO_2) precipitate upon domestic fixtures or clothes, yielding a reddish-brown stain from the iron and a dark-brown-to-black stain from the manganese. Both ions can be removed by oxidation and subsequent filtration. Aeration is adequate for iron(II) oxidation at above pH 6, but the oxidation of manganese(II) requires potassium permanganate or chlorine dioxide for effective removal; however, their use must be followed by coagulation prior to filtration because of the formation of colloidal MnO_2.

J. E. SINGLEY
Environmental Science & Engineering, Inc.

J. L. Cleasby, *Proceedings, American Water Works Association Seminar*, No. 20153, American Water Works Association, Denver, Colo., 1980.

SEWAGE

Sewage is the spent water supply of a community. Because of infiltration of groundwater into loose-jointed sewer pipes, the total amount of water treated may exceed the amount consumed. Sewage contains about 99.95% water and ca 0.05% waste material.

Some materials, such as gasoline and heavy metals, may damage the collection system, cause explosions, or upset the treatment processes. Therefore, they cannot be discharged directly into the system. These prohibitions are governed by sewer ordinances.

Strength of sewage is expressed in terms of BOD, total solids, suspended solids, fixed solids, volatile solids and filterable solids. The BOD is a measure of the load placed on the oxygen resources of the receiving waters. Treatment efficiency is usually evaluated on the basis of BOD removal by the plant. Unless otherwise stated, BOD signifies the biochemical oxygen demand for five days at 20°C (BOD_5). The significant parameters of wastewater are evaluated according to the analytical methods set forth by the U.S. Public Health Association (3).

Sewage Treatment and Disposal Systems

Sewage works were originally constructed for reasons based primarily on public health concepts. However, prevention of disease is not the sole purpose of modern water sanitation practice; it is also a means to protect the oxygen resources of the receiving water. An effluent that does not significantly reduce the oxygen concentration of the receiving water into which it is discharged can be expected to have a low concentration of food for microorganisms that deplete the dissolved oxygen. Thus, protection of the oxygen resources of the receiving water prevents the spreading of disease and satisfies aesthetic considerations.

Private and Rural Disposal Systems. In areas not served by sewers, human and other water-carried wastes are disposed of in primitive privies, cesspools, or septic tanks.

Small Communities. Small communities and recent subdivision additions to larger communities, which have not yet been connected to municipal collection systems, must have a means of waste disposal. Septic tanks are a possibility, but require periodic servicing and cleaning. Furthermore, the soil is not always suitable for accepting the effluent. An alternative is the package plant. These units furnish primary treatment and some secondary treatment, and require only minimal operating supervision. Capacity can be varied as needs dictate. In general, public health authorities prefer such installations instead of septic tanks.

City Disposal Systems. For large population centers, the large particles are removed by screening and sedimentation; 15 to 60% of the BOD can be removed in this manner. Colloidal and dissolved substances are removed by secondary, or biological treatment where the microorganisms of the process utilize the waste material for food. A commonly accepted definition of tertiary treatment is the use of any process or operation for further removals.

Sludge Digestion and Disposal

Digestion. Digestion reduces the volume and the pathogenic organisms. Digested sludge is black and granular and has a tarry color. Sludge is withdrawn periodically from the settling tanks and allowed to flow by gravity to a collection well. From there it is pumped to a digester where it is thoroughly mixed. Sludge digestion can be aerobic or anaerobic. Generally, the gases generated in sludge digestion are sufficient to provide for the immediate fuel needs of the plant in order to maintain the digester temperature, provide hot water, and fuel for incineration of sludge (if practiced) and generators.

Disposal. Digested sludge is reasonably inert but has a high water content. It can be dewatered by filtration or by drying on open or covered sand beds. The dried sludge may be incinerated or used for landfill or fertilizer. The ultimate disposal site must be chosen with great care and contained in order to prevent release of toxic substances to the surrounding environment.

Ocean Disposal. Disposal of raw or treated sludge by barging to sea was practiced for many years by some coastal cities, but today is highly controversial, and it appears that this method will be no longer economically feasible.

Chlorination of Effluent

It is common practice to chlorinate wastewater effluents for bacterial control. Regulations vary from state to state, but all require chlorination to specified residual concentrations. Because some organic industrial chemicals pass unchanged through conventional treatment plants and upon chlorination maybe carcinogenic, there is renewed interest in ozone, other halogens, and ultraviolet irradiation as means of bacterial control.

Other Disposal Problems

Other disposal problems include watercraft-waste disposal, surfactants, and phosphates.

Health and Safety Factors

Wastewater-treatment plants have numerous hazards to be expected in a chemical-process plant, but several hazards require special notice. The potential for infection by pathogenic organisms is always present, and plant workers require inoculation against the common waterborne diseases. In addition, since wastewater treatment plants utilize deep water-filled tanks, provision must be made against drowning. A special hazard is the biomass concentration in aeration tanks; ingestion of this floc has caused a number of fatalities.

G. M. Fair, J. C. Geyer, and D. A. Okun, *Water and Wastewater Engineering*, Vols. 1 and 2, John Wiley & Sons, Inc., New York, 1968.

C. A. MacInnis, "Municipal Wastewater," in *The Encyclopedia of Environmental Science and Engineering*, Vol. 1, Gordon and Breach Science Publishers, New York, 1976, pp. 587–600.

W. W. Eckenfelder, Jr. in A. Bisio and S. Boots, eds., *Encyclopedia of Energy Technology and Environment*, Vol. 4, John Wiley & Sons, Inc., New York, 1995, pp. 2835–2837.

REUSE

Use dictates the quality required. Potable water must be bacteriologically safe, toxic substances must not be present above safe levels, and the water must be aesthetically acceptable. Water that is suitable for drinking may be inadequate for many industrial processes. On the other hand, many industrial processes can use water that is not pure enough to drink.

Today, used water is treated in such a manner that it can be used again before ultimate disposal. Furthermore, a distinction can be made between direct reuse, where the water is reclaimed without dilution or natural purification, and indirect use, where treated used water is returned to the environment for subsequent utilization as a raw water supply.

Water can seldom be reused directly. Conventional secondary wastewater treatment does not produce an effluent suitable for direct reuse, and a tertiary treatment step is required. Contaminants that cannot be removed by conventional means are termed refractory, and they range from simple inorganic salts to complex organic substances such as pesticides, herbicides, and surfactants.

Tertiary Treatment

Tertiary treatments include chlorination, chemical precipitation, filtration, microstrainers, effluent polishing, foam separation, activated-carbon adsorption, ion exchange, oxidation ponds, reverse osmosis, electrodialysis, and vapor-compression evaporation and waste heat evaporation (see EVAPORATION).

Water: Conservation and Reclamation, The Global Cities Project, San Francisco, Calif., 1991.

TREATMENT OF SWIMMING POOLS, SPAS, AND HOT TUBS

Swimming Pools

Most swimming pools are of the recirculating type. Through filtration, chemical treatment, and dilution with rain and makeup water, the water can be reused without draining and refilling, although some health codes require partial replacement of pool water (based on bather load). Sanitizing chemicals must be added regularly to oxidize pool contaminants and kill and control disease-carrying bacteria and other organisms introduced by swimmers and dirt entering the water. It is also necessary to destroy algae whose spores are carried into the water by wind and rain. Unchecked algal growth results in discolored water, unsightly growth on pool surfaces, clogging of filters, and provides breeding ground for bacteria. The pH of pool water must be maintained within a desirable range for swimmer comfort and optimal effectiveness of chlorine sanitizers. In order to control the corrosive or scaling tendencies of pool water, it is also necessary to maintain a proper balance between pH, alkalinity, and hardness. Undesirable trace metals such as iron, manganese, or copper are sometimes found in source water or formed by corrosion of pool equipment. Unless removed, these metals discolor pool water and cause stains, especially damaging to plaster pool surfaces. Filtration of pool water is necessary for removal of suspended solids which otherwise cloud water and interfere with the disinfection process.

Design. The basic recirculation loop of a swimming pool consists of strainers (for removing hair and lint), skimmers (in outdoor pools for removing large floating debris such as leaves, twigs, and insects), a pump, a filter, and in some cases, a heater (typically gas). Large municipal pools typically have balancing tanks to hold water displaced by bathers and are equipped with instrumentation for automatic measurement and control of available chlorine and pH via sanitizer and chemical feeders.

Chlorine-Based Sanitizers. Chlorine compounds are the most commonly used swimming-pool sanitizers and include: chlorine gas, sodium hypochlorite solution, calcium hypochlorite, trichloroisocyanuric acid (Trichlor), and sodium dichloroisocyanurate (Dichlor) (Table 1) (see also DISINFECTANTS). Chlorine gas is used as a sanitizer mainly in large commercial pools requiring a high feed rate to maintain the desired residual. Because of the acidity formed by chlorine gas, addition of soda ash (Na_2CO_3) or sodium sesquicarbonate ($Na_2CO_3 \cdot NaHCO_3$) is necessary to maintain the proper pH and to replenish alkalinity. Because of the gas's toxicity, elaborate precautions against leakage are needed. All chlorine sanitizers provide free available chlorine (FAC), ie, $HOCl^{+}ClO^{-}$.

Small electrolytic chlorine generators are available and are based on electrolysis of brine (ie, aqueous NaCl). The gaseous chlorine formed in the anode compartment is dissolved in the recirculating pool water and the sodium hydroxide formed in the cathode compartment can be used for adjusting the pH of the pool.

Sodium hypochlorite (liquid bleach) is a popular sanitizer with the home-pool owner because of low cost per unit package. Disadvantages of sodium hypochlorite solution are low storage stability and bulkiness. Calcium hypochlorite has superior storage stability and much higher available Cl_2 concentration than liquid bleach, which reduces storage requirements and purchasing frequency. Lithium hypochlorite is similar in action.

The cyanuric acid-based sanitizers (chlorinated isocyanurates) are stable crystalline compounds with moderate-to-high available Cl_2.

Other sanitizers and swimming pool treatment systems used to a limited extent are bromine, quats, ozone, ionizers, electrolyzers, uv, uv/H_2O_2, and Ag/H_2O_2. Various problems are associated with their use, including limited effectiveness.

Sanitizer and Chemical Feeders. Feeders dispense the chemicals in gaseous, liquid, and solid (both granular and compacted) forms. Many health departments require that public pools have approved feeding devices for the daily application of all chemicals, including sanitizers. A slurry feeder for diatomaceous earth (DE) on diatomite filter installations may also be required.

Water Quality Maintenance. In addition to controlling algae and microorganisms such as bacteria, proper swimming pool maintenance requires control of free and combined available chlorine, pH, alkalinity, hardness, and saturation index.

Disinfection. The biocidal properties of chlorine and bromine are due principally to the formation of hypochlorous and hypobromous acids. Disinfection with bromine is less sensitive to pH than with chlorine. Although, bromamines are better disinfectants than chloramines, they are less stable. The presence of interfering substances such as turbidity (ie, suspended solids), ammonia, and organic matter can adversely affect disinfection. Whereas halogens kill bacteria by diffusion of hypohalous acid through the cell membrane, ozone inactivates bacteria by rupturing the cell wall (ie, lysing).

Table 1. Chlorine Sanitizers Used in Swimming Pools

Compound	Formula	Typical form	Cl, % av
chlorine	Cl_2	liquefied gas	100
trichloroisocyanuric acid	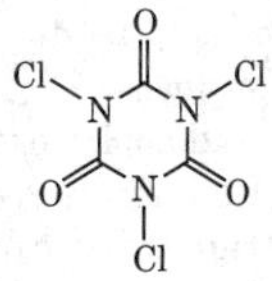	tablets, sticks	89–91
calcium hypochlorite	$Ca(OCl)_2$	granules, tablets, briquettes	65–75
sodium dichloroisocyanurate	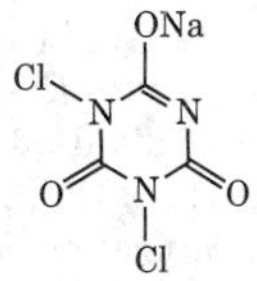	granules	62–63
sodium dichloroisocyanurate dihydrate	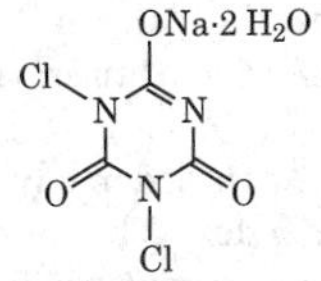	granules	55–56
lithium hypochlorite	LiOCl	granules	35
sodium hypochlorite	NaOCl	solution	10–15

Superchlorination–Shock Treatment. Superchlorination or shock treatment of pool water is necessary since accumulation of organic matter, nitrogen compounds, and algae consumes free available chlorine and impedes disinfection. During superchlorination or shock treatment, ammonium ion is oxidized to nitrogen by breakpoint chlorination. Urea, amino acids, and creatinine are also decomposed during superchlorination or shock treatment, with formation of N_2 and other oxidation products (see CHLORAMINES AND BROMAMINES).

Superchlorination typically refers to adding FAC equal to 10 × ppm CAC (combined available chlorine, ie, chloramines), whereas shock treatment generally involves addition of 10 ppm FAC. Calcium hypochlorite, because of its convenience, is widely used for superchlorination and shock treatment. Chloroisocyanurates are not recommended since their use would result in excessive cyanuric acid concentrations.

pH Control. The optimal range for bather comfort and efficiency of disinfection of chlorine sanitizers is pH 7.2–7.8 where the biocidal agent HOCl represents 69–35% of FAC. The pH of pool water is readily controlled with inexpensive chemicals. Hydrochloric acid solution or sodium bisulfate lower it, whereas sodium carbonate raises it.

Alkalinity. In swimming-pool water at its normal pH range, the so-called carbonate alkalinity is due primarily to bicarbonate with a very small contribution from carbonate. Because of its buffering intensity, alkalinity resists changes in pH when sanitizing chemicals are added to pool water. Since cyanuric acid is largely neutralized at pool pH, cyanurate ion also contributes to alkalinity (and buffering). Therefore, in stabilized pools the alkalinity determination is corrected for 1/3 × ppm cyanuric acid in the normal pH range (7.2–7.8). Sodium bicarbonate is generally added to increase alkalinity and muriatic acid (HCl) or sodium bisulfate ($NaHSO_4$) to reduce it.

Hardness. Water hardness is caused by certain polyvalent metals and is expressed in terms of ppm of equivalent $CaCO_3$. In swimming-pool water, the hardness is caused primarily by calcium ions and to a lesser extent magnesium ions. To ensure proper water balance the calcium hardness should be at or near the $CaCO_3$ saturation value. Hardness is raised with $CaCl_2$ and is lowered by draining some of the pool water and adding water of lower hardness.

Saturation Index. Materials used in pool construction are subject to the corrosive effects of water, eg, iron and copper equipment can corrode whereas concrete and plaster can undergo dissolution, ie, etching. The corrosion rate of metallic surfaces has been shown to be a function of the concentrations of Cl^-, SO_4^{2-}, dissolved O_2, alkalinity, and Ca hardness as well as buffer intensity, time, and the calcium carbonate saturation index. Deposition of a protective layer of crystalline $CaCO_3$ has been proposed for protection of metallic surfaces against corrosion by using the natural calcium and alkalinity in water. Prevention of etching and scaling (ie, precipitation of $CaCO_3$) can be controlled by maintenance of a proper degree of saturation with respect to calcium carbonate. The tendency of a water to precipitate or dissolve calcium carbonate can be determined by the calcium carbonate saturation index (SI), which is the common logarithm of the degree of $CaCO_3$ saturation.

The saturation index is rooted in thermodynamics. It is a scaling index and not a corrosion index. The numerical value of SI for a given water determines its departure from equilibrium, ie, $> 0 =$ oversaturated, $0 =$ saturated, and $< 0 =$ undersaturated. The higher the positive value of the index ($SI > 0$), the greater the tendency for $CaCO_3$ precipitation or scaling and the lower the negative value ($SI < 0$), the greater the tendency for etching. Water with a sufficiently negative index will not only etch concrete, plaster, and tile grout but it can also dissolve an existing protective film of $CaCO_3$. To avoid problems such as etching or scaling, pool water should have a small positive saturation index. Water can be brought to a state of approximate equilibrium (balance) by suitable adjustments in pH, alkalinity, and hardness. However, the SI will drift, primarily because the pH changes due to CO_2 loss and addition of sanitizer. Many swimming-pool chemical dealers and some pool service companies have computer programs for calculating necessary water balance adjustments.

Ancillary Chemicals. In addition to sanitizers, various ancillary chemicals are employed in swimming pools for stabilizing chlorine sanitizers, water balance, control of algae, scale and stain prevention, water clarification, filter degreasing, and cleaning of pool surfaces. Compatibility with chlorine must be considered in selecting ancillary chemicals since many of these additives are organic materials which have chlorine demands and therefore may interfere with pool disinfection. Cyanuric acid is used to stabilize available chlorine derived from chlorine gas, hypochlorites or chloroisocyanurates against decomposition by sunlight.

Water balance chemicals include muriatic acid, sodium bisulfate, and soda ash for pH control, sodium bicarbonate for alkalinity adjustment, and calcium chloride for hardness adjustment. A recent development is use of buffering agents for pH control.

Algal growth in pools is unsightly, a potential safety hazard to swimmers, and usually a result of poor pool maintenance. It can cause slipperiness, development of odors, cloudy and discolored water, chloramine formation, increased chlorine demand, bacterial growth, and stubborn stains. Low FAC, high temperatures, sunlight, and certain mineral nutrients promote algae growth. Such growth can be prevented by maintaining the proper pH range, free available chlorine content, and cyanuric acid concentration supplemented by periodic shock treatment. The most widely used algicides other than chlorine (or bromine) are the quaternary ammonium compounds (quats). Quats are absorbed on filter media and consequently require high initial doses and frequent replenishments. They are surfactants and therefore can cause foaming at sufficiently high concentrations. Other algicides used to a smaller extent contain silver compounds, copper compounds, and $La_2(CO_3)_3$. Zinc compounds can also be used as algicides. Some algicides have chlorine demands which must be taken into account in maintaining the proper FAC.

Scale and stain controllers are polyacrylates (low molecular weight) and organic phosphonates which prevent or control precipitation of $CaCO_3$ by acting as chelating agents or dispersants.

To clarify poolwater, water-soluble, high molecular weight polymers can be used as coagulating agents for removal of precipitated trace metals or other undesirable turbidity. These polymers can also serve as flocculating agents to increase the settling rate or strength of a chemical floc. Inorganic compounds such as aluminum sulfate and polyaluminum chloride are sometimes also used as flocculating agents for clarification or improving filtration.

Grease and oils from bathers can affect filtration. Surfactant or enzyme based degreasers are employed to clean diatomaceous earth (DE) and sand filters.

Test Kits. Proper pool management requires routine analysis for free and combined chlorine and pH, and less frequently, alkalinity, hardness, and cyanuric acid. These analyses can conveniently be carried out at poolside with simple test kits which are available at moderate cost.

Filtration. Efficient filtration for removal of suspended particles is essential for sparkling pool water. Filters may be of the fixed-bed, precoat, or cartridge type and operate under vacuum or pressure. The most common filter media are sand, anthracite, diatomaceous earth (DE), and paper or cloth cartridges.

Spas and Hot Tubs

The basic principles of swimming-pool water treatment also apply to spas and hot tubs. However, spas and tubs are not miniature swimming pools but are unique in treatment requirements because of use patterns and a high ratio of bather to water. For example, four people in a 1.9-m^3 (500-gal) spa or tub have a sanitizer demand equal to 160 people in a 75.7-m^3 (20,000-gal) swimming pool.

Spa and hot-tub sanitation is dominated by chlorine- and bromine-based disinfectants. In addition to replacing spa or tub water because of cyanuric acid build-up from chloroisocyanurate sanitizers, the NSPI recommends that the water be replaced at least monthly or more frequently when often used because bathers contribute microorganisms, perspiration, body oils, lotions, etc, which affect water chemistry, total dissolved-solids concentration, water surface tension (resulting in foaming), and sanitizer demand.

Various chemicals are used in spa and hot tub maintenance including aqueous silicone solutions (polydimethylsiloxane) for defoaming, sequestrants (organic phosphonates) for scale and stain prevention, flocculating agents (polyelectrolytes) for clarification, cleaners, tints, and fragrances. Since these chemicals have significant chlorine demands, they might reduce the bacterial kill efficiency by lowering the FAC or by interfering in the disinfection process. The sanitizer demand of these chemicals must be compensated for to ensure the maintenance of the residual concentrations stated on the product's label.

Health Factors

Improperly maintained swimming pools, spas and hot tubs can result in a variety of illnesses ranging from minor to life threatening.

Reported illnesses and outbreaks have been attributed to bacteria, protozoa, and viruses (adenovirus, enterovirus, and Hepatitis A virus). The protozoa *Giardia* and *cryptosporidium* are more difficult to inactivate than enteric bacteria such as *E. coli.* Diseases can be transmitted by contact with and ingestion of the water, inhalation of microorganisms contained in aerosols produced by the aerated water, and infection through close personal contact.

The water temperature of spas and tubs is normally maintained at 36.7–40.0°C. Long exposure at these temperatures may result in nausea, dizziness, or fainting. Spas and tubs should never be used while under the influence of alcohol. Elderly persons and those suffering from heart disease, diabetes, or high or low blood pressure should not use a spa or hot tube without prior consultation with a physician.

Skin irritations are known to occur from *Pseudomonas aeruginosa* and sometimes bromine sanitizers. Chlorine-based swimming-pool and spa and hot-tub sanitizers irritate eyes, skin, and mucous membranes and must be handled with extreme care.

JOHN A. WOJTOWICZ
Consultant

C. White, *Handbook of Chlorination,* Van Nostrand Reinhold Co., New York, 1986.

J. A. Wojtowicz, *J. Sw. Pool and Spa Ind.* **1**(2), 7, 9 (1995).

J. A. Wojtowicz, *J. Sw. Pool and Spa Ind.,* **3**(1), 26, 35 (1997).

WATERPROOFING AND WATER/OIL REPELLENCY

Principles of Repellency

When a drop of liquid is placed on a solid surface, the shape of the drop depends on the equilibrium among three forces, as shown in Figure 1. On an ideal surface, the contact angle θ depends on three surface tensions: the liquid–air interfacial tension, γ_{LA}; the solid–air interfacial tension, γ_{SA}; and the solid–liquid interfacial tension, γ_{SL}. Wetting of the solid by the liquid results from the reduction of the contact angle of the advancing liquid so that the liquid spreads easily. A surface is made repellent by raising the advancing contact angle, so that spreading and migration into capillaries do not occur.

Textiles

Early waterproofing of textiles was generally accomplished by the use of impermeable coatings, such as natural fats, oils, waxes, pitch, and asphalt. Later, vulcanized natural rubber became an important waterproofing material. By the 1930s, repellent finishes durable to washing and dry cleaning were introduced. The following definitions apply to fabrics with water-repellent properties.

Water-repellent fabrics resist wetting or repel waterborne stains; they pass AATCC Test Method 22 (Spray Test).

Water-resistant fabrics protect against water penetration during a light or brief shower and pass AATCC Test Methods 22 and 42 (Impact Penetration Test).

Rain-resistant fabrics protect against water penetration during a rain of moderate intensity and pass AATCC Test Methods 22 and 35 (Rain Test).

Waterproof is normally used to describe plastic, plastic-coated, or nonbreathable fabrics. However, the term has been used for some chemically coated fabrics that allow the penetration of air.

Durable finishes maintain a high level of performance after laundering, dry cleaning, or both. Durable has also been used to describe resistance to abrasion.

The levels of water repellency required to pass the preceding tests depend upon the specifications for particular fabrics.

Oil-repellent fabrics resist wetting by oily liquids and repel oilborne stains. The level of performance of such fabrics is judged by AATCC Test Method 118.

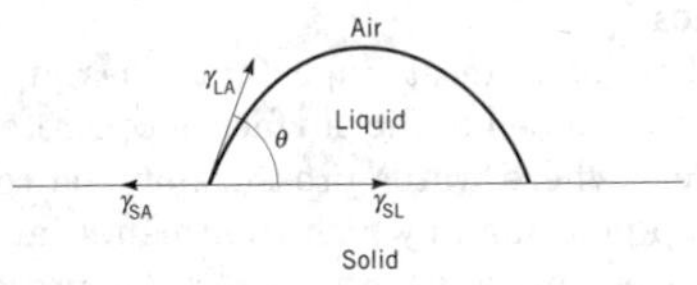

Figure 1. Diagram of forces determining shape of a drop of liquid placed on a solid surface. See text.

Waterproof Finishes. Waterproofing results from coating a fabric and filling the pores with film-forming material such as varnish, rubber, nitrocellulose, wax, tar, or plastic. Fabrics are also commonly laminated to films, such that the total structure is waterproof, or in some cases water-resistant but breathable.

Repellent Finishes. The following are classes of repellent textile finishes. Fluorochemicals are the only repellents that provide repellency to water, waterborne stains, oil, oilborne stains, and oily particulates. Silicones provide water-based stain resistance; good durability to washing; improved tear strength; a soft, slick hand; and improved fabric sewability. Resin-based water repellents are durable finishes that are modified melamine resins, blended with waxes. They are widely used as extenders for fluorochemical repellents.

Waxes and wax–metal emulsions are the lowest priced, widely used water repellents and fluorochemical extenders.

Organometallic complexes consist of Werner-type complexes of chromium and long-chain carboxylic acids, eg, stearic acid. The complexes have a small market in the textile industry.

Modification of Textile Fibers. The reaction of hydrophobic chemicals with textile fibers offers the possibility of permanent repellency without alteration of the other physical properties of fibers. However, the disadvantages caused by complex processing, and resultant higher costs of carrying out chemical reactions on fiber in commercial textile plant operations, have limited the commercial applications. The etherification and esterification of cellulose have been most effective in terms of achieving durable water repellency. Radiation grafting of reactive repellents onto fibers has been studied as a potential commercial process, as has modification by plasma polymerization of gas monomers or plasma initiated polymerization of liquid monomers.

Fabric Construction for Water Repellency. Fabric construction, including twist, ply, and coarseness of yarns, affects the performance of water repellents. Waterproof films can more easily be formed on close weaves than on open-weave fabrics. Hydrophobic finishes, which make individual fibers repellent without altering fabric porosity, are generally applied to fabrics whose pores are small. Some reports indicate that fabric roughness reduces repellency. Mechanical action on fabrics, even after treatment, can reduce repellency if the action increases fiber roughness or exposes fibers that have little repellent treatment.

Carpet. Carpet, an important textile, may also be treated to provide water and oil repellency; however, the principal functions of the current carpet treatments are to provide soil and stain resistance. High quality carpets, especially those made from nylon, polyester, or wool, have a significant proportion of the surface coated with fluorochemical materials. The treatments can be spray-applied to a finished carpet or applied directly to the fiber during the spinning or dyeing operations. Suitable fluorinated resin materials are readily available from 3M or DuPont.

Health and Safety Factors. The Material Safety Data Sheets provided by the suppliers should be consulted for each product. In general, products are aqueous emulsions with low levels of toxicity. Products with high solvent content have mostly been eliminated. Personnel handling the chemicals should always avoid contact of the products with skin and eyes, and avoid exposure to vapors if the product contains volatile components.

Test Methods.

The most widely used tests for water and oil repellency and for the durability of finishes to washing and dry cleaning are described in the *Technical Manual* of the American Association of Textile Chemists and Colorists. Some widely used repellency tests are as follows. The spray test is one of the most commonly used tests for fabrics and nonwoven products. The rain test simulates the effects of rainfall. Oil repellency is measured by observing a fabric's resistance to wetting by a selected series of numbered test liquid hydrocarbons with a range of surface tensions. Alcohol holdout tests, which are also used to measure aqueous fluid repellency, involve placing drops of aqueous isopropyl alcohol solutions of concentrations 10, 20, ... 100 wt % on a fabric surface. The rating for the fabric is based on the most concentrated solution that does not penetrate the fabric in the specified time frame. The nonwovens industry also uses a saline repellency test, especially for medical fabrics. In the hydrostatic-pressure test, is the fabric is subjected to increasing water pressure at a constant rate until leakage occurs at three points on the fabric's undersurface. The dynamic absorption test measures the resistance of fabrics to wetting by water, not the repellency of the total fabric surface.

Leather

Water-resistant treatments are useful to reduce the tendency of leather to become stiff and uncomfortable after wetting and drying.

Commonly used repellents for leather are silicones, chrome complexes of long chain fatty acids, and fluorochemicals. Fluorochemical repellents also provide repellency to oils and greases.

Paper and Paperboard

Paper or paperboard that repels water may be designated waterproof or water-resistant. In this context, waterproof means resistant to the penetration of both water and water vapor. A chemical additive, called a sizing agent, can provide water resistance by lowering the surface energy of the paper without greatly affecting the porosity. A water-vapor barrier or moisture-vapor barrier (MVB) is provided by a film or film-forming coating. Acrylic-based overprint varnishes are often used to provide water resistance, as well as abrasion resistance, to a printed surface. Paper or paperboard can also be made resistant to oil-based fluids. Food oil resistance is the most common application for this property. To lower the surface energy enough to provide oil repellency, a fluorochemical additive is required (for pigment or clay-coated paper or paperboard, see PAPER; PAPERMAKING ADDITIVES).

Waterproof. Waterproof paper or paperboard is commonly made by lamination of a polyethylene film or wax that is applied by extrusion coating. The MVB properties derived from wax emulsions are generally inferior to those derived from extrusion coatings, but both forms provide excellent waterproof properties. Polyester film is used in food board and other applications where the composite material must withstand cooking temperatures. Currently (ca 1997), there are efforts to replace the standard film laminants with aqueous coatings, but the wax residues in aqueous wax-based coatings cause problems in the recycling process. Polyvinylidene chloride-based aqueous coatings provide excellent MVB properties, but paper and paperboard with chlorine-containing coatings are becoming less common because of concerns regarding cost, aging characteristics, and residual chlorine-containing products in the environment. To provide protection to exposed cut edges or to protect from failure at pinholes, it may be beneficial to use paper or board that is sized or has an oil repellent treatment.

Water-Resistant. Paper or paperboard that is resistant to water can be produced by chemical addition of a sizing agent at several points in the manufacturing process or during subsequent converting operations. There are two types of sizing: surface sizing and internal sizing. If the sizing is added to the aqueous suspension of pulp fibers before sheet formation, it is called internal or beater sizing. This gives chemical modification of the fiber surface throughout the sheet or ply.

Surface sizing done on a paper machine can actually be a deep treatment from a pond size press, or a shallow treatment applied by a premetered size press, blade, or rod coater. A surface treatment can also be applied at a printing station in a converting operation.

The most widely used internal sizes are alkyl ketene dimers (AKD), alkenylsuccinic anhydrides (ASA), and rosin-based sizes that are used with papermaker's alum (aluminum sulfate with 14 waters of hydration), polyaluminum chloride (PAC), or polyaluminum silicosulfate (PAS). Anionic starch is perhaps the most common additive or co-additive used for surface sizing. Most of the surface sizes used in North America are modified styrene maleic anhydride (SMA) copolymers. Other materials used as surface sizes include acrylonitrile acrylate copolymer, stearylated melamine resin, polyurethane, and diisobutylene maleic anhydride copolymers. Chromium complexes of long-chain fatty acids are excellent water repellents which are also used for their food-release properties in certain packaging applications. The presence of chromium has raised environmental concerns, despite the fact that the metal is in the trivalent rather than in the highly toxic hexavalent state. This material is available as Quilon (DuPont).

Oil Repellent. Fluorochemicals are the only class of material that can provide oil repellency without altering the porosity of the paper or paperboard. Physical barriers to oil penetration are used primarily for their moisture- or gas-barrier properties, with retarded oil penetration as a secondary benefit. The most common oil-repellent additives are long-chain perfluoroalkyl phosphate salts of ammonia or diethanol amine. The packaging of many items such as snack food, french fries, convenience food, and pet food is facilitated by the oily-stain protection provided by fluorochemicals.

FDA Regulations. Most of the sizing agents, waterproof coatings, and oil repellents mentioned are regulated by the U.S. Food and Drug Administration (FDA) for use in paper or paperboard which comes in direct contact with food. To avoid raising contamination issues, paper and paperboard mills often require all chemicals to be FDA-regulated if any of their products are used in food contact applications. The regulations often include limitations on the intended use, method, or point of application; upper limits on the amount that can be present on the paper or paperboard; and guidelines related to the filling, storage, and final use of the paper or paperboard.

Concrete and Masonry

The term waterproof describes concrete and masonry that is completely impervious to water and its vapor, whether or not the water is under pressure. Waterproof construction involves the use of some type of barrier that covers all surface pores or capillaries. Water repellent describes concrete or masonry that repels water without significantly reduced permeability to water vapor.

Waterproof. Waterproofing barrier systems may be either hot- or cold-applied. The hot-applied generally involve a bituminous material such as asphalt used in conjunction with a reinforcing fabric such as roofing felt, cotton, or glass cloth. Cold-applied can be bituminous or elastomeric materials either in liquid or sheet form, with or without fabric reinforcement. Liquid elastomeric treatments include neoprene, polyurethanes, and blends of these or epoxies with bituminous materials. Among the commonly used precured elastomeric sheet materials are neoprene, polyisobutylene, EPDM rubber, and plasticized PVC. Polyethylene and PVC films and nonwoven plastic or glass fabric coated with bituminous materials also find use. Because these treatments can trap moisture in a wall, the proper selection and application of treatment materials depend on the nature of the concrete structure, such as whether it is made up of exterior or interior walls above or below grade, floors, or roofs.

Water-Repellent. Three techniques used for water repellency are modification of cement by the addition of waterproofers, use of repellent additives to the concrete mix, and surface treatment of concrete structures with repellents. The modification of portland cement by intergrinding with stearate salts or other water-repellent material can reduce the water permeability of mortar, although considerable controversy exists on this point.

Admixtures are sometimes used to reduce permeability of concrete. These include pore-filling materials such as chalk, Fuller's earth, or talc; water repellents such as mineral oil, asphalt, or wax emulsions; organic polymers (acrylic latexes, epoxies); and salts of fatty acids, especially stearates.

The third and perhaps most important class of water repellents consists of materials applied to the surface of concrete for above-grade structures or others where water pressure on the concrete is small. This includes damp-proofing in which treatments cannot be subjected to continuous or even intermittent hydrostatic pressure. Repellents that may be used are oils, waxes, soaps, resins, and silicon-based systems.

Wood

Wood is subject to water infiltration by both liquid and vapor. As the moisture content increases, the wood will swell until it reaches its maximum dimension at its fiber-saturation point (about 30% moisture). Rapid dimensional changes resulting from changes in the level of bound water cause the wood to shrink or swell and hence to crack and split. These cracks will then allow moisture to absorb easily and quickly into the wood. Above the fiber-saturation point, moisture will be present as free water, which in turn promotes the rate of wood decay.

After-market water repellents, ie, those applied to wood structures after they are built, are widely sold in the consumer market. Materials used in formulating them can be organic resins, waxes, metal stearates, and acrylics. Three components are generally used for a robust water repellent: a water-repellent material that penetrates into the wood pores, a material that remains on the surface to give the visual effect of water beading, and a material that reduces the rate of water-vapor transmission into and out of the wood to improve dimensional stability. To penetrate the wood with a solvent or water-based material, a molecular weight of about 1000 or less is generally needed. Water-based short-chain siloxanes which penetrate and bond with the surface are especially effective. Paraffin waxes are typically used to provide surface-water beading. Because it is very difficult to ensure total moisture vapor impermeability, it can be beneficial to use blends of water vapor-permeable materials and water vapor-occlusive materials in order to allow moisture that has entered the wood to exit.

RICHARD D. NOWELLS
3M

J. L. Moillet, ed., *Waterproofing and Water Repellency*, Elsevier Publishing Co., Amsterdam, the Netherlands, 1963.

R. E. Johnson and R. H. Dettre, in J. C. Berg, ed., *Wettability*, Marcel Dekker, Inc., New York, 1993, Chapt. 1.

AATCC Technical Manual, American Association of Textile Chemists and Colorists, Research Triangle Park, N.C., 1996.

WATER-SOLUBLE POLYMERS

Water-soluble polymers find application in a wide variety of areas that include polymers as food sources, plasma substitutes, and as diluents in medical prescriptions. Other areas of importance for water-soluble polymers include detergents, cosmetics, sewage treatment, stabilizing agents in the production of commodity plastics, rheology modifiers in the various processes for petroleum, textile, paper, and latex coatings production. The water-soluble polymers discussed in this article have significant commercial impact.

Hydrophilic Groups. Water solubility can be achieved through hydrophilic units in the backbone of a polymer, such as O and N atoms that supply lone-pair electrons for hydrogen bonding to water. Solubility in water is also achieved with hydrophilic side groups (eg, OH, NH_2, CO_2^-, SO_3^-). Truly unique in its ability to interact and promote water solubility is the —O—CH_2—CH_2— group. The interactions of these groups with water and their placement in the polymer structure influence the water solubility of the polymer and its hydrodynamic volume.

Viscosity Efficiency. A majority of the applications of water-soluble polymers revolve around their role in increasing the viscosity of water solutions. The hydrodynamic volume of the polymer influences its viscosity efficiency, and thereby its ability to modify the rheology of an application formulation. The primary parameter influencing the hydrodynamic volume is the polymer's molecular weight. The second is in the conformational rigidity or extension of the polymer chain in solution. For example, carbohydrate polymers contain repeating ring structures that facilitate a more rigid structure than observed in nonionic synthetic polymers. At a given molecular weight this effects a greater viscosity efficiency. The rigidity also can be increased by inclusion of charged groups in both synthetic and carbohydrate polymers.

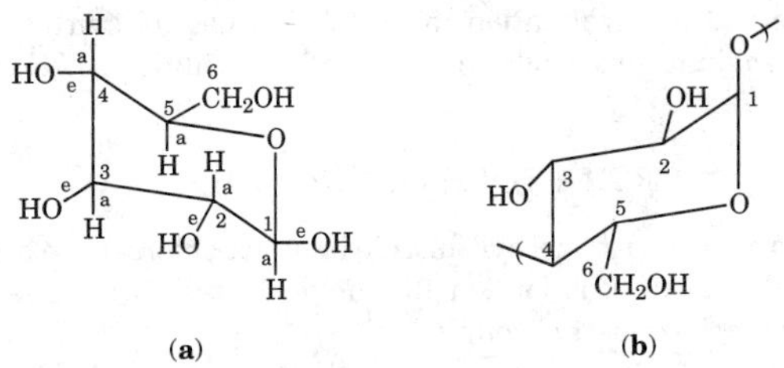

Figure 1. (**a**) The glucopyranosyl basic unit, where e = equatorial and a = axial bond; (**b**) amylose with equatorial–axial interunit bonding (the C—H axial bonds have been omitted).

Such groups lead to electrostatic repulsions and extension of the chain in deionized aqueous solutions and an increase in the hydrodynamic volume of the polymer. Greater rigidity also can be achieved in carbohydrate polymers if helical conformations are realized.

Carbohydrate Polymers

An anhydroglucose ring or glucopyranosyl unit (Fig. **1a**) is the structural unit on which human metabolism is dependent. The anhydroglucose unit provides, through the four hydroxyl units, a diversity of polymer structures that can be formed through variations in positional bonding between the rings (ie, $1 \rightarrow 2$, $1 \rightarrow 3$, $1 \rightarrow 4$, and $1 \rightarrow 6$). Examples of different positional bonding in carbohydrate polymers are illustrated in Figure 2.

Symmetrical $1 \rightarrow 4$ bonding provides the world's most abundant polymer, cellulose (Fig. **2a**). The bonds linking hydroxyl units in the plane of the puckered ring are referred to as equatorial bonds. The glucopyranosyl unit is also present in amylose (Fig. **1b**). The variance between cellulose and amylose (the latter is a component of starch (qv)) is in the interunit bonding between anhydroglucose rings. In cellulose, the rings are connected through equatorial–equatorial bonding (beta-linkage); in amylose the $1 \rightarrow 4$ inter-ring bonding is equatorial–axial (alpha-linkage; axial denotes a bond perpendicular to the general plane of the ring). The beta-linkage in cellulose (qv), complemented by the intrahydrogen bonding among rings facilitates a linear projection of the polymer. This, complemented by interhydrogen bonding among polymer chains, provides crystallinity and a rigid structure. To transform the world's most abundant polymer, cellulose, into a water-soluble species requires replacement of some of the hydroxyls to disrupt the extensive hydrogen bonding.

Amylose (Fig. **1b**), with alpha-interconnecting linkages, is soluble in hot water, (unlike cellulose), but it retrogrades in low-temperature aqueous solutions into a helical conformation and precipitates. Glucopyranose branching from the C-6 hydroxyl provides solubility at low temperatures. Thus this anomeric bonding difference (ie, alpha-($1 \rightarrow 4$) instead of the beta ($1 \rightarrow 4$) linkage in cellulose) provides a readily accessible source of energy and is designated as a storage source in nature.

The anomeric difference (ie, the alpha- and beta-linkages) between cellulose and amylose is important. Of greater importance in determining the aqueous solution properties of water-soluble polymers is the interunit bonding patterns between rings. As with amylose, branches from the C-6 position promote solubility in water. Two naturally occurring carbohydrate polymers with structural main chain similarity to cellulose (ie, in the beta-($1 \rightarrow 4$) linkage), but with water solubility without derivatization due to side branch units, are the nonionic fraction of flaxseed gum and guaran (Fig. **2b** and **2c**, respectively).

Cellulose

Commercial Derivatization of Cellulose. Cellulose, the world's most abundant polymer, is derivatized for use in a variety of markets. Cellulose ethers are an important segment of water-soluble polymers.

Commercial derivatization of cellulose begins with the addition of sodium hydroxide to form alkali cellulose (AC) (eq. 1, R = carbohydrate).

$$ROH + NaOH \rightleftharpoons RO^-Na^+ + H_2O \qquad (1)$$

The AC may react with methyl chloride or alpha-chloroacetic acid via a direct displacement reaction (eq. 2). The derivatives would be methyl cellulose or carboxymethyl cellulose.

$$RO^-Na^+ + R'X \rightarrow ROR' + NaX \qquad (2)$$

Alternatively, the AC may react with oxiranes (eg, ethylene oxide ($R'' = H$) or propylene oxide ($R'' = CH_3$) (eq. 3).

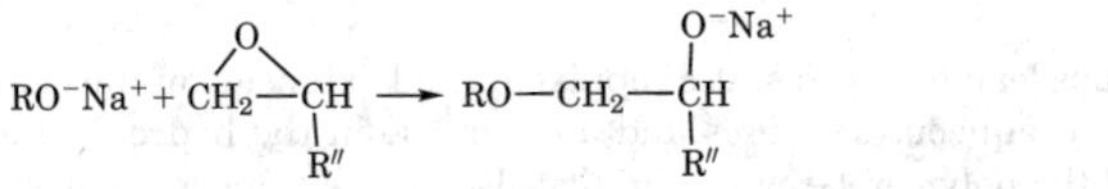

(3)

The derivatives are hydroxyethyl and hydroxypropyl cellulose. All four derivatives find numerous applications, and there are other reactants that can be added to cellulose, including the mixed addition of reactants leading to adducts of commercial significance (see CELLULOSE ETHERS).

Biosynthesis. Although cellulose can be derivatized, such materials do not provide the optimum properties desired in many applications, such as retention of viscosity at higher solution temperatures, greater mechanical stability, and greater thickening efficiency. These properties can be approached with carbohydrate polymers in helical conformations, which can be achieved in some carbohydrate polymers prepared by fermentation processes (eg, SGPS; see Fig. 2**e**). Yeasts have an advantage over fermentation processes using bacteria, because the fermentation can be conducted in low pH media. Yeast are also larger and easier to remove by filtration. However, the most successful commercial fermentation polymer is XCPS, synthesized by a bacteria, *Xanthomonas campestris.* In this polymer the main chain is simply cellulose, but with three pyranosyl rings branched from the C-3 position of every other repeating ring. This arrangement promotes a helical conformation (see YEASTS).

Recently, two fermentation polymers have produced optimum properties through variations in positional and interunit bonding patterns: gellan and wellan.

Other Considerations. With multiple hydroxyl units on every repeating ring, most commercial carbohydrate particles are surface treated with glyoxal. After adequate dispersion of the particles in water, a small quantity of base solution is added to remove the acetal cross-linked structure. The individual particles then readily hydrate without the agglomeration of partially hydrated particles before complete hydration is achieved. The segmentally rigid and conformationally rigid (in helical polymers) ring structures provide a more viscous aqueous shear viscosity, but lower extensional solution viscosity for a given molecular weight. The rigidity provides greater mechanical stability relative to synthetic polymers discussed below.

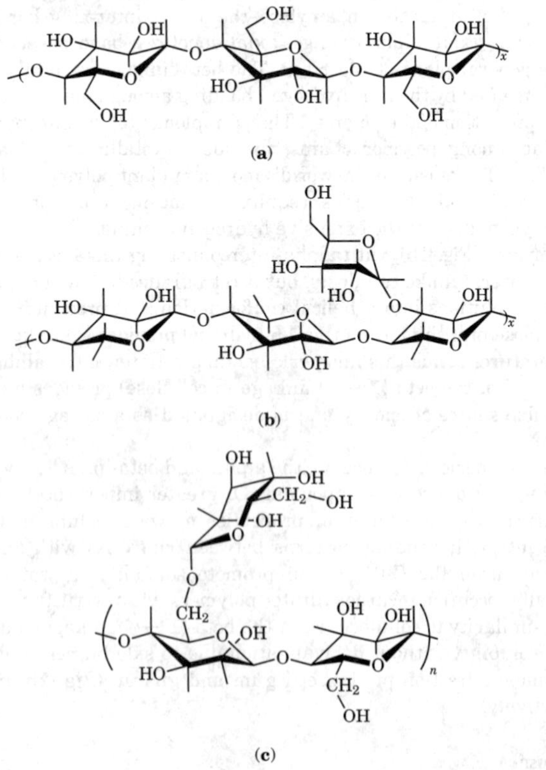

Figure 2. Examples of carbohydrate polymers with interunit and branch position differences: (**a**) cellulose; (**b**) flaxseed gum; (**c**) guaran.

Synthetic Water-Soluble Polymers

Polyoxyethylene. Synthetic polymers with a variety of compositionally similar chemical structures are as follows. Based on polarity, poly(oxymethylene) (**1**) would be expected to be water soluble. It is a highly crystalline polymer used in engineering plastics but it is not water-soluble (see ACETAL RESINS). Polyoxypropylene (PPO) (**2**) and poly(methyl vinyl ether) (PMVE) (**3**) have

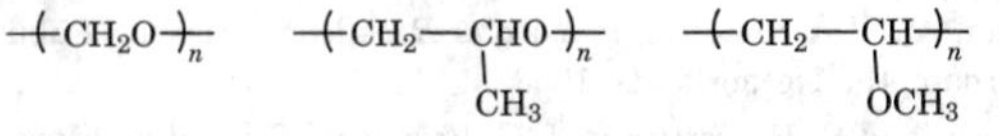

(**1**) (**2**) (**3**)

identical chemical compositions; PMVE is water soluble up to a modest temperature (~50°C), but PPO is not soluble in water. PPO is soluble only in the oligomeric form with less than 10 PO units. PO defines oxypropyl or propylene oxide monomer overcome the terminal hydroxyl groups facilitating water solubility when there are more than 10 repeating units. In view of the lack of water solubility of two compositionally similar polymers, the water solubility of polyoxyethylene POE (**4**) is notable. It occurs because of the unique interaction of water with the

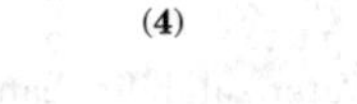

(**4**)

oxyethylene chain.

Polyoxyethylene (POE) is synthesized employing different catalyst systems depending on the desired mol wt. For molecular weight below 20,000 (generally referred to as poly(ethylene glycol)s, PEGs), base or the Na^+ or K^+ alkoxides of methanol or butanol are used. Without the numerous hydroxyls present in carbohydrate polymers, POE particle structures cannot be treated to minimize their hydration prior to dissolution. The particle surfaces are covered with silica to minimize hydration and blocking of POE particles on storage. POE and PEGs of various molecular weights have found numerous applications.

Commodity Chain-Growth Polymers. Two of the largest commodity water-soluble polymers are poly(vinyl alcohol) (PVA) and polyacrylamide (PAM). They are prepared by the free-radical initiation of vinyl monomers, a chain-growth polymerization technique.

Poly(vinyl alcohol). The vinyl alcohol monomer is unstable and isomerizes to acetaldehyde. The polymer is obtained by the hydrolysis of poly(vinyl acetate). The vinyl acetate monomer produces a very high energy radical during chain-growth propagation. This accounts for the high chain transfer to monomer, a higher than normal head to head addition of propagating species, and grafting of some of the propagating species to polymer that is formed during an earlier stage of the polymerization. This leads to a more complex variation in structure than observed in PAM polymers, and this, with the differences that are realized with different methods of hydrolysis, can result in different Poly(vinyl alcohol) (PVA). These factors, on hydrolysis, lead to PVA below 50,000 molecular weights (see VINYL POLYMERS, POLY(VINYL ACETATE); POLY(VINYL ALCOHOL).

PVA is isomeric with POE; however, it can enter into hydrogen bonds with both water and with the other hydroxyl units of the repeating polymer chain, forming both inter- and intrahydrogen bonds. The extensive hydrogen bonding can lead to crystallinity, an occurrence that complicates its water solubility. Commercial PVA is essentially atactic. With most chain-growth polymers, crystallinity is associated with stereoregularity, but the small size of the hydroxyl substituent group promotes crystallinity even in the atactic polymer. For this reason poly(vinyl acetate) is seldom hydrolyzed completely; it is manufactured retaining three acetate levels: 25, 12, and 2 mole percents (these numbers represent averages). With higher acetate percentages the polymer is more surface active, which is important in its role as a suspending agent in poly(vinyl chloride) commodity resin production. Also, it is readily soluble in water at ambient temperatures. With low acetate percentages, PVA is difficult to dissolve unless the water temperature is high, and on cooling the low acetate PVA may precipitate.

Hydrolyzed Polyacrylamide. HPAM can be prepared by a free-radical process in which acrylamide is copolymerized with incremental amounts of acrylic acid or through homopolymerization of acrylamide followed by hydrolysis of some of the amide groups to carboxylate units. The carboxylated units, ionized, decrease adsorption on subterranean substrates, in proportion to the number of units, an important parameter in petroleum recovery processes. In waste treatment processes cationic acrylamide comonomer units are often used to increase adsorption and thereby flocculation of solids in wastewater (see ACRYLAMIDE POLYMERS FLOCCULATING AGENTS). Because the synthesis of HPAM can be conducted in water, the problem of dissolution in many applications is addressed by polymerizing the monomer in water-in-oil emulsions.

Poly(vinyl pyrrolidone). Another commercial polymer with significant usage is PVP. It was developed in World War II as a plasma substitute for blood. This monomer polymerizes faster in 50% water than it does in bulk, an abnormality inconsistent with general polymerization kinetics. This may be due to a complex with water that activates the monomer; it may also be related to the impurities in the monomer that are difficult to remove (see VINYL POLYMERS, *N*-VINYLAMIDE POLYMERS).

Poly(acrylic acid) and Poly(methacrylic acid). Poly(acrylic acid) (PAA) may be prepared by polymerization of the monomer with conventional free-radical initiators using the monomer either undiluted (with cross-linker for superadsorber applications) or in aqueous solution. Photochemical polymerization (sensitized by benzoin) of methyl acrylate in ethanol solution at −78°C provides a syndiotactic form that can be hydrolyzed to syndiotactic PAA.

Cationic Water-Soluble Polymers

Cationic monomers are used to enhance adsorption on waste solids and facilitate flocculation. One of the first used in water treatment processes (**5**) is obtained by the cyclization of dimethyldiallylammonium chloride in 60–70 wt % aqueous solution (see WATER). Another cationic water-soluble polymer, poly(dimethylamine-*co*-epichlorohydrin), prepared by the step-growth

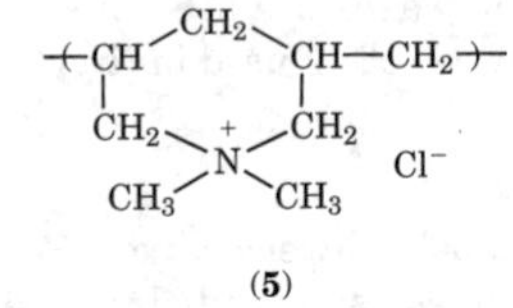

(**5**)

polymerization of dimethyl amine and epichlorohydrin, is also used for adsorption on clays.

Inorganic Water-Soluble Polymers

Two inorganic water-soluble polymers, both polyelectrolytes in their sodium salt forms, have been known for some time: poly(phosphoric acid) and poly(silicic acid). A more exciting inorganic water-soluble polymer with nonionic characteristics has been reported. This family of phosphazene polymers is prepared by the ring-opening polymerization of a heterocyclic monomer (**6**) followed by replacement of the chlorine atoms in the resultant polymer.

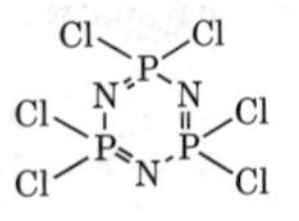

(**6**)

The hydrophobic inorganic backbone provides an easy route to variable water-solubilizing side nonionic or ionizing groups. These are stable to hydrolysis at room temperature.

One of the main advantages of water-soluble polyphosphazenes is the ease with which water-solubilizing side groups can sufficiently cross-link in a stable matrix with high energy radiation such as x-rays, gamma-rays, electron beams, or ultraviolet light.

The versatility of water-soluble polyphosphazenes is in the variations in the structures that can be prepared. Structures with a low glass-transition temperature backbone can be modified with a variety of versatile side units.

New Commercial Water-Soluble Polymers

Two recently developed water-soluble polymers have achieved limited market acceptance.

One product is poly(2-ethyl-2-oxazoline) (PEOX). It is prepared by the ring-opening polymerization of 2-ethyl-2-oxazoline with a cationic initiator. Most of the polymer's characteristics stem from its molecular structure, which like POE, promotes solubility in a variety of solvents in addition to water. It exhibits Newtonian rheology and is mechanically stable relative to other thermoplastics. It also forms miscible blends with a variety of other polymers. Another product is prepared from *N*-ethenylformamide formed from the reaction of acetaldehyde and formamide. The vinyl amide is polymerized with a free-radical initiator, then hydrolyzed.

The protonated form of poly(vinyl amine) (PVAm–HCl) has two advantages over many cationic polymers: high cationic charge densities are possible and the pendent primary amines have high reactivity. It has been applied in water treatment, paper making, and textiles.

Hydrophobe-Modification of Water-Soluble Polymers

Although many of the new water-soluble polymers discussed above have not achieved large-scale commercial acceptance, there is a class that has achieved outstanding success since the early 1980s: hydrophobically modified water-soluble polymers (HM-WSPs). They have filled certain voids in a number of applications that include cosmetic, paper, architectural, and original equipment manufacturing (OEM) coating areas and have found unsuspected application in the airplane de-icers market. The driving force for the development of HMWSPs is threefold in most application areas: **1.** The achievement of high viscosities at low shear rates without high molecular weights. **2.** Minimization of the elastic behavior of the fluid at high deformation rates that are present when high molecular weight water-soluble polymers are used. **3.** Providing colloidal stability to disperse phases in aqueous media, not achievable with traditional water-soluble polymers.

Preparation of hydrophobically modified, water-soluble polymer in aqueous media by a chain-growth mechanism presents a unique challenge in that the hydrophobically modified monomers are surface active and form micelles.

The hydrophobe modification of acrylic acid represents an important class of hydrophobe-modified thickeners prepared by a chain-growth free-radical process. They differ slightly from other examples in that these products are generally cross-linked.

Hydrophobe-modification of hydroxyethylcellulose produces what should be considered model associative thickeners, for the distribution of hydroxyethyl units has been characterized. The commercial material, a Hercules product, contains three hydrophobes per chain.

HEUR associative thickeners are in effect poly(oxyethylene) polymers that contain terminal hydrophobe units. They can be synthesized via esterification with monoacids, tosylation reactions, or direct reaction with monoisocyanates.

J. EDWARD GLASS
North Dakota State University

P. Molyneux, *Water-Soluble Synthetic Polymers: Properties and Behavior,* Vols. I and II, CRC Press, Boca Raton, Fla., 1982.

J. E. Glass, ed., *Water-Soluble Polymers: Beauty with Performance,* Advances in Chemistry Series 213, American Chemical Society, Washington, D.C., 1986.

J. E. Glass, ed., *Polymers in Aqueous Media, Performance through Association,* Advances in Chemistry 223, American Chemical Society, Washington, D.C., 1989.

J. E. Glass, ed., *Hydrophilic Polymers: Performance with Environmental Acceptance,* Advances in Chemistry Series 248, American Society, Washington, D.C., 1995.

WAXES

Wax usually refers to a substance that is a plastic solid at ambient temperature and that, on being subjected to moderately elevated temperatures, becomes a low viscosity liquid. Because it is plastic, wax usually deforms under pressure without the application of heat. The chemical composition of waxes is complex; all of the products have relatively wide molecular weight profiles, with the functionality ranging from products which contain mainly normal alkanes to those which are mixtures of hydrocarbons and reactive functional species.

Insect and Animal Waxes

Insect and animal waxes include beeswax and spermaceti. White and yellow beeswax has been known for over 2000 years, especially through its use in the fine arts. Beeswax is secreted by bees and is used to construct the combs in which bees store their honey. The wax is harvested by removing the honey and melting the comb in boiling water; the melted product is then filtered and cast into cakes. The yellow beeswax cakes can be bleached with oxidizing agents to white beeswax, a product much favored in the cosmetic industry. Beeswax typically has a melting point of 64°C, a penetration (hardness) of 20 dmm at 25°C and 76 dmm at 43.3°C (ASTM D1321), a viscosity of 1470 mm^2/s at 98.9°C, an acid number of 20, and a saponification number of 84.

Beeswax has GRAS status from the FDA. The major use of beeswax is in the cosmetic industry, with smaller amounts used in pharmaceuticals and candle production.

Spermaceti was derived from the head oil of the sperm whale. Owing to the present status of the sperm whale as an endangered species, however, the material is no longer an item of commerce and has been replaced by other natural and synthetic waxes.

Vegetable Waxes

Vegetable waxes include carnauba, candelilla, Japan wax, ouricury wax, rice bran wax, jojoba, castor wax, and bayberry wax. The source of carnauba wax is the palm tree, whose wax-producing stands grow almost exclusively in the semiarid northeast section of Brazil. Its hardness and high melting point, when combined with its ability to disperse pigments such as carbon black, allows Carnauba wax increasing use in the thermal printing inks. Carnauba is also widely used to gel organic solvents and oils, making the wax a valuable component of solvent and oil paste formulations. Carnauba polishes to a high gloss and thus is widely used as a polishing agent for items such as leather, candies, and pills. Other uses include cosmetics and investment casting applications (see COPPER ALLOYS, CAST COPPER ALLOYS).

Candelilla wax is harvested from shrubs in Mexico and Texas. Principal markets for candelilla include cosmetics, foods, and pharmaceuticals.

Japan wax is a fat and is derived from the berries of a small tree native to Japan and China. Principal markets include the formulation of candles, polishes, lubricants, and as an additive to thermoplastic resins.

Ouricury wax is a brown wax obtained from the fronds of a Brazilian palm tree and is sometimes used as a replacement for carnauba.

Rice-bran wax is extracted from crude rice-bran oil and may be used in some food applications.

Jojoba oil is obtained from the seeds of the jojoba plant grown in semiarid regions of Costa Rica, Israel, Mexico, and the United States. Hydrogenated jojoba oil is a wax used in candles and other low volume specialty applications.

Castor wax is catalytically hydrogenated castor bean oil. It is used primarily in the formulation of cosmetics. Derivatives of castor wax are used as surfactants and plastics additives. Bayberry wax is removed from the surface of the berry of the bayberry (myrtle) shrub by boiling the berries in water and skimming the wax from the surface of the water. The wax has an aromatic odor and is used primarily in the manufacture of candles and other products where the distinctive odor is desirable.

Mineral Waxes

Mineral waxes include montan, peat waxes, ozokerite and ceresin, and petroleum waxes. Montan wax is derived by solvent extraction of lignite. The largest traditional use for montan waxes was as a component in one-time hot-melt carbon-paper inks. The refined grades have found use in the formulation of polishes and as plastics lubricants. Peat waxes are much like montan waxes in that they contain three main components: a wax fraction, a resin fraction, and an asphalt fraction. Ozokerite wax and its refined derivative, ceresin, were products of Poland, Austria, and in the former USSR where ozokerite was mined. They have been replaced with blends of petroleum-derived paraffin and microcrystalline waxes.

Waxes derived from petroleum are hydrocarbons of three types: paraffin (clay-treated); semimicrocrystalline or intermediate; and microcrystalline (clay-treated). Semimicrocrystalline waxes are not generally marketed as such.

A paraffin wax is a petroleum wax consisting principally of normal alkanes. Microcrystalline wax is a petroleum wax containing substantial proportions of branched and cyclic saturated hydrocarbons, in addition to normal alkanes. Semimicrocrystalline wax contains more branched and cyclic compounds than paraffin wax, but less than microcrystalline. A classification system based on the refractive index of the wax and its congealing point as determined by ASTM D938. Typical properties of petroleum waxes are listed in Table 1.

Petroleum wax is widely used in chewing gum. It can also be used as protective coatings for fruits, vegetables, and cheeses. Petroleum wax is outstanding as a cost-effective moisture and gas barrier, and food packaging applications are a major market for refined food-grade petroleum wax. Other industrial applications are as an additive to protect rubber from ozone degradation, lubricate plastics, help control the properties of hot-melt adhesives, and improve slip and rub properties in inks. Petroleum waxes are used in cosmetics, polishes, and candles. Unrefined, they are often used in fireplace logs.

Synthetic Waxes

Synthetic waxes include polyethylene waxes, Fischer-Tropsch waxes, chemically modified waxes, substituted amide waxes, and polymerized α-olefius. Low molecular weight (less than ca 10,000 Mn) polyethylenes having waxlike properties are made either by high

Table 1. Typical Properties of Petroleum Waxes

Property	Wax	
	Paraffin	Microcrystalline
flash point, closed cup, °C	204[a]	260[a]
viscosity at 98.9°C, mm^2/s	4.2–7.4	10.2–25
melting range, °C	46–68	60–93
refractive index at 98.9°C	1.430–1.433	1.435–1.445
number average molecular weight	350–420	600–800
carbon atoms per molecule	20–36	30–75
ductility/crystallinity of solid wax	friable to crystalline	ductile-plastic to tough-brittle

[a] Value is minimum.

pressure polymerization or low pressure (Zeigler-type catalysts) polymerization. All the products have the same basic structure, but the processes yield products with distinctly different properties. Some polyethylenes have fairly low densities, owing to branching that occurs during the polymerization. Molecular weight distributions, expressed as the weight average molecular weight divided by the number average molecular weight, or polydispersity, also varies widely among the different processes, as does the range of molecular weights available.

Major uses include hot-melt adhesives for applications requiring high temperature performance, additives to improve the processing of plastics, slip and rub additives for inks and paints, and cosmetic applications.

Products used in food applications require regulatory approvals. Some by-product polyethylene waxes have been recently introduced. Uses include additives for inks and coatings, pigment dispersions, plastics, cosmetics, toners, and adhesives.

Polymethylene wax production is based on the Fischer-Tropsch synthesis, which is basically the polymerization of carbon monoxide under high pressure and over special catalysts to produce hydrocarbons. Uses are similar to those for polyethylene waxes, including hot-melt adhesives and additives for inks and coatings.

Hydrocarbon waxes of the microcrystalline, polyethylene, and polymethylene classes are chemically modified to meet specific market needs. In the vast majority of cases, the first step is air oxidation of the wax with or without catalysts. The product has an acid number usually no higher than 30 and a saponification number usually no lower than 25. Oxidized wax is easily emulsified in water through the use of surfactants or simple soaps, and is widely used in many coating and polish applications.

The product of fatty acid amidation has unique waxlike properties. Probably the most widely produced material is *N,N′*-distearylethylenediamine, which has a melting point of ca 140°C, an acid number of ca 7, and a low melt viscosity. It is used in additive quantities to raise the apparent melting point of thermoplastic resins and asphalts, as an internal–external lubricant in the compounding of a variety of thermoplastic resins, and as a processing aid for elastomers.

Some polymers of higher α-olefins, eg, $C_{>20}$, have waxlike properties and are sold as synthetic waxes. The polymerization process yields highly branched materials, with broad molecular weight distributions. The unique structure makes these products very effective when used in additive amounts to modify the properties of paraffin wax, primarily for use in candles. The products can increase the hardness and opacity of the paraffin, without increasing the cloud point or viscosity. Other uses include mold release for polyurethane foams, additives for casting wax, and additives for leather treating.

Analytical Techniques

Most waxes are complex mixtures of molecules with different carbon lengths, structures, and functionality. Attempts to measure the exact chemical composition are extremely difficult, even for the vegetable waxes, which are based on a relatively few number of basic molecules. Products such as oxidized microcrystalline wax not only have a mixture of hydrocarbon lengths and types as starting materials, but also add complexity through the introduction of various types of carboxylic functionality onto those hydrocarbons during the oxidation process.

Because of the difficulty in analysis of chemical composition, most of the routine test procedures on waxes are for the measurement of the physical properties and are used to compare the properties of waxes within a class. Some properties, such as acid number or saponification number, give insight into the chemical functionality of the product, and are widely used for products which contain carboxyl groups such as vegetable, montan, and oxidized waxes. Increasingly, instrumental methods such as gas chromatography (gc), gel permeation (also known as size exclusion) chromatography (gpc), refractive index (ri), differential scanning calorimetry (dsc), infrared spectroscopy (ir), and nuclear magnetic resonance (nmr) are being used to further characterize the products. Properties such as molecular weight distribution, degree of branching, degree of crystallinity, and functionality can be readily measured with these techniques.

WILLIAM P. COTTOM
Baker Petrolite Corporation

P. E. Kolattukudy, *Chemistry and Biochemistry of Natural Waxes,* Elsevier Scientific Publishing Co., Amsterdam, the Netherlands, 1976.

A. H. Warth, *The Chemistry and Technology of Waxes,* Reinhold Publishing Corporation, New York, 1947, p. 49.

Petroleum Waxes—Characterization, Performance, and Additives, STAP No. 2, Technical Association of the Pulp and Paper Industry, Atlanta, Ga., 1963, p. 16.

F. J. Ludwig, *Anal. Chem.* **37,** 1732 (Dec. 1965).

WEED KILLERS. See HERBICIDES.

WEIGHING AND PROPORTIONING

Weighing is the operation of determining the mass of any material as represented by one or more objects or by a quantity of bulk material. Proportioning is the control, by weighing, of relative quantities of two or more ingredients according to a specific recipe in order to make a mixed product, or to prepare the ingredients for use in a chemical process.

Mass is a fundamental physical property of matter; the mass of an object does not change with its location on the earth. Mass is what is measured in the act of weighing, and scales are calibrated to read in units of mass, eg, kilograms. The terms weight and weighing are used somewhat loosely in this connection, in that weight is a measure of the gravitational force of the earth acting on a mass. Thus, weight depends on the location of the scale relative to the equator and on its altitude. It is common practice to calibrate scales *in situ* using mass standards (test weights); hence, for practical purposes scales actually determine mass.

The terms balance and scale are currently used to describe weighing machines of various forms. Over the centuries *bilanx* evolved into balance. Balance refers to sensitive weighing devices with two pans although it is also used in a more general sense to refer to any very sensitive laboratory weighing device. Scale is used to describe general-purpose weighing machines found in industrial and retail applications.

Weighing Principles

There are many types of scales using many different principles of operation. There are, however, three distinct elements which can be identified regardless of the principle of operation (Fig. 1).

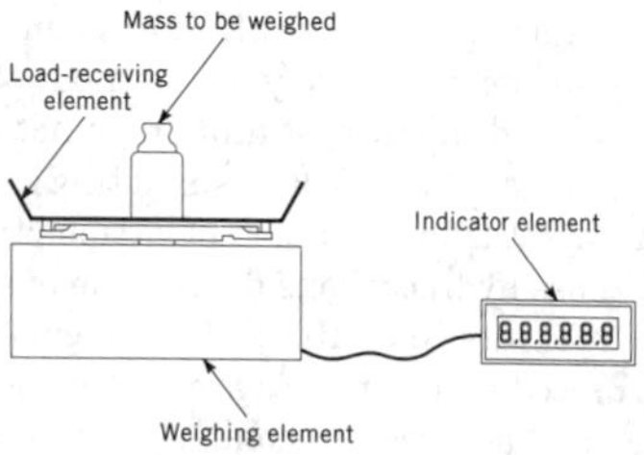

Figure 1. Basic elements of a scale.

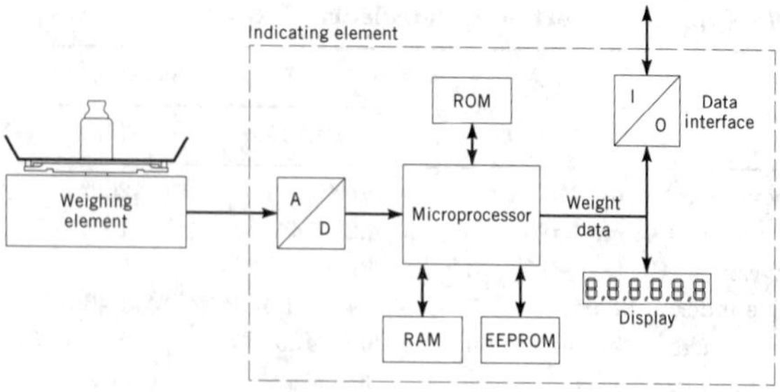

Figure 2. Basic components of an electronic scale.

The load-receiving element supports the load during the weighing operation. It may take any configuration appropriate to the material being weighed.

The weighing element supports the load-receiving element. It produces a signal proportional to the load and sends this signal to the indicator element.

The indicator element measures the signal from the weighing element, and converts it into a readable form. It may be any of several different types, eg, a graduated beam, a Bourdon tube pressure gauge, or a numeric display device. The indicator element may not display the weight but may instead transmit it electronically to a controller.

The principal weighing technologies in use currently are mechanical, hydraulic, strain-gauge, electromagnetic force compensation, and nuclear.

Mechanical. Mechanical scales, measure a load either by comparing it with a known weight or by measuring the distortion of a spring caused by the load. With the even-balance type, the load is measured by direct comparison with known weights. With beam scales, the force resulting from the load on the platform is reduced to a more manageable magnitude by the multiple of the lever system. It is then measured by the position of a known weight, or poise, on a graduated beam.

Automatic indicating scales eliminate the manual operation of placing tip weights and positioning a poise on a graduated beam. In spring-type automatic indicating scales, the steelyard rod is connected to a spring whose deflection is indicated by a pointer against a graduated scale.

Hydraulic. A hydraulic scale measures weight by supporting the load receiver on the piston of a hydraulic load cell and measuring the resulting hydraulic pressure with a Bourdon tube or similar device. Many weighing operations require that the load be supported at three or more points, requiring the use of three or more hydraulic load cells. The individual hydraulic pressures developed by these cells must be converted back to forces for summing and measurement; this is often accomplished using strain-gauge load cells. The hydraulic equipment performs the same function as a mechanical lever system in reducing the load to an easily measured force.

Strain-Gauge Load Cells. The majority of modern industrial scales today use strain-gauge load cells as the weighing element. The strain-gauge load cell is a device which, when a force is applied to it, gives an electrical output proportional to the applied load.

A typical metallic foil strain gauge consists of an etched grid of very thin foil attached to a thin insulating backing material. The gauges are bonded firmly to a surface that prevents buckling. When stretched or compressed along the grid lines, the resistance measured between the ends of the grid (solder pads) increases in the case of the former and decreases in the case of the latter.

The central mechanism of a load cell is the spring element that supports the load, and it is designed to have areas of both tensile and compressive strain suitable for application of strain gauges. Typically, four gauges are applied in such a way that two are in compression and two in tension as the spring element is loaded.

The four gauges are wired together to form a Wheatstone bridge. An input voltage (typically 10 V dc) is applied, and the output voltage resulting from applied load varies from omV at no load to 20 mV at full load.

Figure 2 shows the basic components of an electronic scale based on strain-gauge load cells. A power supply (not shown, but usually housed in the indicator) provides operating power for the indicator and the input voltage for the load cell. The analogue signal produced by the load cell is sensed by the analogue-to-digital (A/D) converter, which sends digital weight information to the digital computer. The digital computer comprises the microprocessor, the working storage (RAM), a permanent memory (EEPROM) for storing calibration data, and the program memory (ROM). The computer filters the raw digital signal and processes it into calibrated weight for the display. The weight is also available at the data interface, from which it can be transmitted to any external device. Also, the indicator can be remotely controlled via the data interface.

Electromagnetic Force Compensation. Traditionally, precision balances were very finely made, with agate pivots and bearings and were often housed in an enclosure of glass and hardwood. Today, the principle of electromagnetic force compensation (EMFC), also referred to as magnetic force restoration (MFR), is used where extremely high accuracy is required. The fundamental principle of an EMFC balance is that the load added to the pan of the balance is maintained at the same vertical height by varying the current supplied to an electrodynamic converter that supports the pan; the current required is proportional to the load in the pan.

The electronics for a balance based on EMFC technology is very similar to that shown in Figure 2 for a strain-gauge-based scale, with the exception that the weighing element consists of the EMFC cell, together with the circuitry to control the current flow to the electrodynamic converter. Also, the signal sent to the A/D converter is not a voltage, but rather only the current necessary to return the pan to its null position. For balances, the weighing and indicating elements are usually integrated into the same housing.

Nuclear. Mass can be determined directly by measuring changes in the absorption, reflection, or transmission of alpha or beta rays, which changes in proportion to the amount of material present. This method is primarily used to approximate the mass of bulk material moving on a conveyor.

Commercial Regulatory Aspects and Performance Characteristics

Scale Performance. The ideal scale would have a perfectly linear relationship, as represented by the straight line, and the results from increasing and decreasing load tests would fall on this line. The two broken lines represent the typical calibration curve (greatly exaggerated); the arrows indicate increasing and decreasing load directions.

U.S. Regulations. Weighing equipment used in commercial transactions is subject to regulation and inspection by state or local weights and measures agencies. It must comply with certain design and performance requirements, and must have a certificate of conformance issued prior to installation. *Handbook 44* issued by the National Institute of Standards and Technology is the basis for weights and measures enforcement activities in the United States.

Regulation Outside the United States. Each country establishes its own weights and measures requirements. The majority of these are based on the recommendations of the Organisation Internationale de Métrologie Légale (OIML), in Paris. *R76-1* is the OIML equivalent of *Handbook 44.*

Factors Affecting Weighing Accuracy

These include variations in the force due to gravity; moisture content; temperature effects; the buoyant effect of air; and electrostatic and magnetic effects.

Types of Scales

Scales are available in a variety of designs and configurations to facilitate different weighing operations. The two principal categories are industrial and retail scales, and precision scales and balances.

Industrial and Retail Scales. Scales using strain-gauge load cells predominate in this market segment, although mechanical and hydraulic scales are also used to some extent.

Precision Scales and Balances. These products today rely predominantly on electromagnetic force compensation (EMFC) technology.

Weighing Methodologies

Scales can be portable so that they can be moved about and temporarily located where the weighing operation is to take place; they may be light enough to be carried by hand, or they may be on wheels or be transported using a lift truck.

Some scales are fixed in position and the material to be weighed is brought to the scale. The object of a weighing operation may be to determine the gross weight (the weight of the material plus the tare weight of its container) or, more commonly, the net weight (the weight of material exclusive of its container).

Weighing Predefined Quantities. In this form of weighing, the amount of material has been defined prior to weighing, eg, by filling a truck to a certain level, and the weighing operation is carried out simply to determine what the weight is. An object may be placed on a scale and its net weight will be displayed directly (assuming the object is not packaged). Bulk materials may be handled in anything from small containers to railroad cars; the weighing operation consists of measuring the gross weight and subtracting the container's tare weight to arrive at the material's net weight.

Controlling Weight. Filling containers to predetermined weights is the most common example of controlling weight. One method is the net-weighing process, in which material is fed into a hopper or tank scale until the desired weight is reached; it is then discharged into the container. In the gross-weighing method, the container is placed directly on the scale and filled to the required weight. Either filling operation may be controlled manually by an operator observing the scale display; more typically, the filling is completely automated for better consistency and greater speed.

Continuous Weighing and Controlling. Scales can be used to determine the weight per unit length of material in sheet or strip form as it is being manufactured or transported. The material can pass over a roll that is weighed, or over a belt-conveyor scale. Direct mass measurement with a nuclear scale can also be used for this application.

Process industries frequently need to weigh and control the flow rate of bulk material for optimum performance of such devices as grinders or pulverizers, or for controlling additives, eg, to water supplies. A scale can be installed in a belt conveyor, or a short belt feeder can be mounted on a platform scale. Either can be equipped with controls to maintain the feed rate within limits by controlling the operation of the device feeding the material to the conveyor.

The most accurate flow-rate control can be achieved by using the loss-in-weight method. The total amount of material required for a downstream process is first added to a tank or hopper scale. As the material is discharged, the loss-in-weight is monitored and used to modulate the discharge valve or gate to achieve the desired flow rate.

Material Proportioning. In proportioning operations, two or more materials are weighed and mixed according to a recipe for use in a chemical process or to make a mixed product, such as animal feed. Proportioning can be performed manually on a bench scale or automatically in tank and hopper scales employed with elaborate material handling systems and controls. Proportioning can be done on a continuous basis or, more typically, in batches, which yield greater accuracy.

The three principal methods of batch proportioning are accumulative, sequential, and simultaneous; the best method for a given application depends on the processing equipment available and the accuracy required.

In the case of accumulative proportioning, a single-weigh hopper having adequate capacity for the entire batch is used. Each ingredient is weighed in turn and accumulated in the hopper scale until completion of the batch.

In sequential proportioning, a single-weigh hopper is used and each material is weighed and discharged before weighing of the next ingredient begins.

In the case of simultaneous proportioning, a weigh hopper is provided for each material.

Scale Functionality Mechanical scales indicate the weight applied to the scale through the manipulation of a poise weight on a beam, or automatically using a pointer and dial.

Electronic scales are microprocessor-based devices that provide a range of features for various applications. At its most basic, an electronic scale provides a digital display of the applied weight.

Some indicators are designed for truck in–out operations. The indicator can store tare weights for several hundred vehicles, linking them to the trucks' registration numbers.

Counting scales display the number of parts on the scale as well as the total weight. They establish the average piece weight (APW) of the parts to be counted, which can be stored in memory along with the part number and an associated tare weight.

Over/under scales are often used for manually filling containers to a desired weight. Scales can be connected with printers, remote displays, bar code scanners, keyboards, relays, computers, and various controllers such as programmable logic controllers (PLCs).

TOM LEAHY
Mettler-Toledo, Inc.

Handbook 44, Specifications, Tolerances, and Other Technical Requirements For Weighing and Measuring Devices, National Institute of Standards and Technology, Gaithersburg, Md., 1996.

Glossary of Weighing Terms, A Practical Guide to the Terminology of Weighing, Mettler-Toledo Inc., Worthington, Ohio, 1994.

OIML R76-1, Non Automatic Weighing Instruments, Organisation Internationale De Metrologie Legale, Paris, 1992.

Do-It-Yourself Guide to Building Tank Scales, Mettler-Toledo Inc., Worthington, Ohio, pp. 34–35, 1997.

WELDING

Welding comprises a group of processes whereby the localized coalescence of materials is achieved through application of heat and/or pressure. The American Welding Society recognizes nearly seventy methods of welding and more than twenty allied processes employing thermal cutting, thermal spraying, and adhesive bonding, some of which are identified in Table 1.

Arc-Welding Processes

In arc welding, the coalescence of metals is achieved through the intense heat of an electric arc, which is established between the base metal and an electrode. The processes listed in Table 1 are differentiated by various means of shielding the arc from the atmosphere.

Shielded-Metal Arc Welding. The arc acts between the consumable electrode wire and the base metal, and droplets of molten filler metal are transferred to the weld pool. The unique feature of this widely used

Table 1. Welding Processes

Process	Abbreviation	Process	Abbreviation
arc welding		resistance welding	RW
shielded-metal arc welding	SMAW	resistance spot welding	RSW
		resistance seam welding	RSEW
gas–tungsten arc welding	GTAW	projection welding	RPW
		flash welding	FW
plasma arc welding	PAW	solid-state welding	SSW
gas–metal arc welding	GMAW	diffusion welding	DFW
flux-cored arc welding	FCAW	explosion welding	EXW
stud arc welding	SW	forge welding	FOW
submerged arc welding	SAW	friction welding	FRW
oxyfuel–gas welding	OFW	ultrasonic welding	USW
oxyacetylene welding	OAW	other welding processes	
oxyhydrogen welding	OHW	electron beam welding	EBW
pressure-gas welding	PGW	laser beam welding	LBW
brazing	B	electroslag welding	ESW
resistance brazing	RB	thermit welding	TW
furnace brazing	FB	allied processes	
induction brazing	IB	thermal spraying	THS
soldering	S	adhesive bonding	ABD
dip soldering	DS	thermal cutting	TC
wave soldering	WS		
torch soldering	TS		

process is the role of the electrode coating. This coating is decomposed by the heat of the arc and provides both a necessary shielding atmosphere for the arc as well as a slag coating over the weld metal, thus affording still further protection from the atmosphere.

In addition to these functions, the coating introduces fluxing agents to the weld pool, assists in establishing the electrical characteristics of the arc, and can be used to provide additional filler metal to the weld.

Gas–Tungsten Arc Welding. In the GTAW process, often called the tungsten–inert gas or TIG process, the electrode is nonconsumable and shielding is provided by the flow of inert gas through the welding-torch nozzle.

Plasma Arc Welding. In the transferred-arc mode of the PAW process, the arc is between a nonconsumable electrode and the base metal, in a manner similar to the GTAW process. The unique feature is the flow of inert gas around the electrode and through a restricted orifice, which constricts the arc to form a plasma jet. A second, outer stream of shielding gas protects the molten metal from atmospheric contamination. In the nontransferred arc mode, the arc is between the electrode and the constricting orifice.

Gas–Metal Arc Welding. In the GMAW process, also called the metal–inert gas or MIG process, a consumable bare-wire electrode is fed continuously through the welding torch, and a flow of inert gas through the nozzle provides shielding.

Flux-Cored Arc Welding. The characteristic feature of the FCAW process is the hollow, flux-filled, consumable electrode, which is fed continuously through the welding torch. The decomposition of the flux provides both arc shielding and a slag blanket over the weld.

Submerged-Arc Welding. In the SAW process, a loose flux is blanketed over the region of the arc. A consumable, bare-wire electrode is fed continuously through the torch into the weld. The molten- and granular-flux blankets provide the necessary arc shielding and slag cover for the solidified weld.

Arc-Welding Systems

The various welding processes result in systems of varying complexity. They include at least the electrode and a device for holding or feeding it, the work piece, the power source, and heavy-duty cabling to provide a complete electrical circuit. Provisions for supply and control of gas and control of wire feed and movement of the electrode assembly are required, depending on process type and degree of automation.

Welding systems are generally classified as manual, semiautomatic, mechanized, and automatic. In manual welding, the operator must maintain the arc, feed in filler metal, and provide travel and guidance along the joint. In semiautomatic welding, the welding machine maintains the arc and feeds filler metal, and the operator controls joint travel and provides guidance. In mechanized welding, the welding machine maintains the arc, feeds filler metal, and moves a mounted torch along the joint; the operator manually adjusts the welding parameters based on observation of the process. In automatic welding, the machine assumes all of the preceding functions. Automatic processes may be further classified, depending on the degree of feedback control used in controlling the welding variables. Automated systems may be dedicated to a specific type of production, or they may be flexible, programmable robot systems.

Other Important Fusion Welding Processes

Oxyfuel–Gas Welding. This process, commonly called gas welding, uses the heat of combusting gases to melt and coalesce base metals. Although several different fuel gases, eg, propylene, hydrogen, MAPP, or methane, can be added to the oxygen, the oxyacetylene flame is the most widely used because its high (3100°C) flame temperature is needed to weld steel.

Resistance Welding. As noted in Table 1, resistance welding comprises several processes; the most widely used is resistance spot welding (RSW). The principles of RSW are quite different from those underlying the processes previously described. The workpieces are firmly clamped between copper electrodes, and an electric current is passed through the assembly. Heat is generated by the electrical resistance of the components, the maximum heat occurring at the interface between the workpieces. A nugget of metal at the interface region is melted, at which time the current is shut off and the clamping force on the electrodes released.

Electroslag Welding. In this process, the heat of a molten slag coalesces the base and filler metals. Electric current flows from a consumable electrode through a molten metallurgical slag into the molten weld metal. The electric resistance of the slag provides the heat to maintain the slag in a molten state and to melt the weld and base metal.

Electron–Beam Welding. This welding process achieves the heat necessary for coalescence by bombarding the base metals with a concentrated stream of electrons. Electrons emitted by a heated filament are accelerated between cathode and anode and then focused into a narrow beam onto the base metal.

Laser-Beam Welding. The heat of coalescence is produced by focusing the beam of light (photons) from a high power laser on the base metal. Laser welding is considered a high energy–density (ca 10^6 W/cm^2) welding process.

Nonfusion Welding Processes

Solid-State Welding. Solid-state welding comprises a group of welding processes wherein a bond is made between two base materials upon the application of pressure at a temperature below the solidus of the base materials (Table 1). Interlayers are sometimes used. By joining materials in the solid state, many of the difficulties of the fusion processes are avoided.

Brazing and Soldering. In brazing and soldering processes, a molten filler metal flows by capillary action into the closely fit joint between the base metals at a temperature below the melting point of the base metal. Bonding is accomplished by metal-to-metal adhesion, which may involve the formation of intermetallic compounds at the interface between the base and filler metals. Although these characteristics are common features of brazing and soldering, the two processes are differentiated by the melting temperature of the filler metal. In brazing processes, the filler metal melts at a temperature above 450°C, whereas in soldering, the melting point of the filler metal is below 450°C.

Joining of Polymers and Adhesive Bonding

Polymers are characterized as thermosetting and thermoplastic with respect to the methods by which they are joined. Thermosetting polymers are permanently hard and do not soften upon the application of heat; they are joined by mechanical fasteners and adhesives. Several methods have been devised to join thermoplastic polymers, as well as thermoplastic composite materials, which soften upon heating.

Hot Plate, Infrared, and Hot Gas Welding. These processes involve external means to heat thermoplastic polymers to a viscous state in which the interdiffusion of polymer chain molecules can occur with the application of pressure. In hot plate welding, the two surfaces to be joined are forced against a platen heated to a desired temperature based on the composition of the polymer. In hot gas welding, a stream of heated gas or air is directed at both the joint surfaces and a polymer filler material.

Friction and Ultrasonic Welding. Both rotational and linear friction can be used to melt the interface between two thermoplastics. The parts are then aligned and a weld is formed as the interface solidifies. Ultrasonic welding of polymers involves oscillations of 10–50 kHz, which are dissipated at the bond line to produce heat through both friction and hysteresis. The surfaces to be joined are held together as the sound energy generated by the welding machine is transferred through the parts at right angles to the contacting surfaces.

Adhesive Bonding. As one of the processes allied to welding, adhesive bonding is used particularly in applications where welding or mechanical fastening is either impossible or undesirable (Table 1). Adhesives, which are derived from polymers, are not structural materials and act only to distribute stress over the bond area.

Physics and Metallurgy

In most welding processes, the local regions of the adjoining base metals and any added filler metal are melted and resolidified. Welding differs from conventional casting by the speed of the solidification process. Only a local region of material is melted in welding, and the surrounding, low temperature base metal acts as a large heat sink, producing rapid heat flow from, and the solidification of, the weld zone. This produces a complex metallurgical microstructure and physical properties not typical of casting.

Heat Flow. An important aspect of welding heat flow is the thermal cycle at a given location in the material. The nature of the cycle depends on the intensity of the heat source, the speed of welding, the thermal characteristics of the material, and the location in the material. The weld thermal cycle, involving peak temperatures achieved and the speed of heating and cooling, and possibly preheating and postheating, accounts for many of the subsequent complexities of welding metallurgy.

Solidification. The heat of the electric are melts a portion of the base metal and any added filler metal. The force of the arc produces localized flows within the weld pools, thus providing a stirring effect, which mixes the filler metal and that portion of the melted base metal into a fairly homogeneous weld metal. There is a very rapid transfer of heat away from the weld to the adjacent, low temperature base metal, and solidification begins nearly instantaneously as the welding heat source moves past a given location.

Metallurgy. Welding metallurgy deals with the interactions of the base and filler metals and the interactions of these materials with various chemicals injected into the weld via gases, electrode coverings, fluxing and slagging agents, and surface contaminants. Oxygen, nitrogen, water vapor, and carbon dioxide are gases that react with ferrous metals to yield products harmful to the metallurgical properties of a weld. The nature of the slag–metal reactions that occur in the molten state strongly depends on the composition of the flux or the electrode coating. Flux chemistry may be altered to control removal of specific weld-metal impurities, such as the addition of manganese or silicon to provide strong deoxidizing action, or additions to enhance slag removal, with all of these additions influencing the final metallurgical characteristics of the weld.

Material Properties. The properties of materials are ultimately determined by the physics of their microstructure. For engineering applications, however, materials are characterized by various macroscopic physical and mechanical properties. Among the former, the thermal properties of materials, including melting temperature, thermal conductivity, specific heat, and coefficient of thermal expansion, are particularly important in welding. The last property named greatly influences structural distortion that can occur in welding. The electrical conductivity of a material is important in any welding process where base or filler metal is part of the welding electrical circuit.

The response of materials to force is characterized by mechanical properties, eg, elastic modulus, yield stress, tensile strength, ductility, hardness, and impact or fracture strength. Fatigue strength, which is the ability of a material to withstand cyclic loading, is of particular importance to welded structures.

Base and Filler Metals

The term weldability specifies the capacity of a metal, or combination of metals, to be welded under fabrication conditions into a suitable structure that provides satisfactory service. It encompasses a range of conditions, eg, base- and filler-metal combinations, type of process, procedures, surface conditions, and joint geometries of the base metals.

Base Metals. Base metals include carbon steels, low alloy steels, stainless steels, aluminum and aluminum alloys, magnesium alloys, nickel and nickel alloys, titanium and titanium alloys, copper and copper alloys, and reactive and refractory metals.

Filler metals are added to a weld by melting a consumable electrode or a separate wire fed into the weld pool. In the first category, the filler metal is part of the welding electrical circuit and may be in the form of short lengths of covered wire, as in shielded-metal arc welding, or in the form of continuous reels of wire used in semiautomatic, mechanized, and automatic welding processes. Solid wire is used in the gas–metal and submerged-arc welding processes, whereas a hollow, flux-filled wire is used in flux-cored arc welding. More filler metal in the form of iron powder is sometimes added to the electrode coating or flux. In the second category, the filler metal may be in the form of short lengths of bare solid wire, as used in gas welding or manual gas–tungsten arc welding, or in continuous reel form, used in automatic gas–tungsten arc welding.

Design

Welded Joints. The weld joint is the geometric arrangement between two pieces of base metal brought together for purposes of welding. There are only five recognized weld-joint configurations: corner, butt, tee, lap, and edge joints.

Weld Types. The fillet weld joins corners and tees. In the plug weld, the two pieces are joined by weld metal deposited in a prepared hole in the overlying piece. For a spot weld, heat is applied to the overlying plate, creating fusion at the interface. Resistance welding is generally used for the spot weld and the seam weld. The groove weld joins base metals in a butt joint. For the backing weld, the root of the original weld is first removed by chipping or gouging. Surface welds are used to build up the surface of parts with special materials or to replace worn material. Flange welds are applied to edge joints, particularly to thin section materials.

Stress and Distortion. The forces acting on a structure are transmitted through the welded joints; that is, the joint is subjected to simple tension (or compression), bending, shear, or torsional stresses, or to combinations of these stresses owing to combined loading situations. Weldments must be of a proper size, length, and location to withstand the loads imposed during service.

Health and Safety Factors

Welders are subject to the same hazards as all other workers in the metalworking trades. Specific additional hazards of welding include electrical shock, arc radiation, fumes and gases, fires and explosions, compressed gases, cutting and chipping operations, and high noise levels.

RANDY J. BOWERS
KARL F. GRAFF
Edison Welding Institute

H. B. Cary, *Modern Welding Technology,* Prentice-Hall, Inc., Englewood Cliffs, N.J., 1994.

Welding Handbook, 8th ed., Vol. 1, American Welding Society, Miami, Fla., 1987; see also other volumes in this series.

Welding and Brazing, Vol. 6 of *Metals Handbook,* 10th ed., ASM International, Materials Park, Ohio, 1993.

D. Simonson, *The History of Welding,* Monticello Books, Inc., 1969.

WETTING AGENTS. See SURFACTANTS; DETERGENCY.

WHEAT AND OTHER CEREAL GRAINS

Origins and History

Botanically, cereals are simple, one-seed fruits. They include wheat, rice, corn (also called maize in some parts of the world), rye, barley, oats, sorghum, and millet. All cereals contain large amounts of starch and little fat; the fat is associated primarily with the germ and scutellum (single cotyledon or the first leaf). In most cases, the lipids in cereals contain a high concentration of unsaturated fatty acids, which are protected from oxidation by the presence of tocopherols. Actually, the oil from wheat germ was the starting material for the isolation of α-tocopherol (see VITAMINS, VITAMIN E). As long as the antioxidants are in proximity to the lipids, oxidation of the lipids and the accompanying rancidity is minimized. This is one reason why wheat is stored in the kernel stage until it is milled.

Cereals grow in a wide variety of climatic and soil conditions, and they successfully compete with weeds for the limited amounts of nutrients and water where these plant factors are in short supply.

Nutritional Value of Cereals

Deficiency Diseases. Not only did cereals make an important contribution to improving the general status of humankind, but they also were important dietary components of some groups of people who showed certain nutritional deficiencies. This observation led to the discovery of some of the vitamins. These deficiency diseases have been most prominently associated with use of rice, corn, and wheat. They include beriberi (thiamine deficiency), pellegra (niacin deficiency), zinc deficiency, poor absorption of calcium, and insufficiency of lysine.

Health Advantages of High Cereal Consumption

Health advantages include a reduction in the level of blood urea and improved kidney function, prevention of osteoporosis, prevention of obesity, and reduction in dental cares.

Health Problems Associated with the Consumption of Cereals

Celiac Disease. A disturbance of the lower gastrointestinal tract, celiac disease is a chronic disease characterized by loss of appetite and weight, depression and irritability, and diarrhea frequently followed by constipation.

This disturbance has been recognized as being related to the ingestion of wheat. Although considerable work has been done attempting to identify the mechanism whereby gluten causes the disturbance characterizing celiac disease, it appears to be an allergic reaction.

Disturbances Associated with Flour-Aging Agents. Some health problems are associated with the various agents used to age flour in order to toughen the gluten content.

Ergot. Ergotism is only indirectly related to health hazards associated with the consumption of cereals because the alkaloid which is responsible for ergotism is produced by a fungus (*Claviceps purpurea*) that grows primarily on rye. The fungus may also grow on other cereals (see ALKALOIDS).

Colonic Cancer. Another area in which cereals may play an important role in maintaining the health of human beings is as a source of dietary fiber. Evidence has revealed that certain diseases, to which people in the more-developed countries are prone, owe their origin to the almost complete removal on the fibrous parts of cereals consumed in the diet.

Formerly, dietary fiber, of which cellulose is one of the more important constituents, was considered important primarily as a means of preventing or overcoming constipation. Otherwise, dietary fiber was considered to be a metabolically inert substance. Many diseases, such as appendicitis, hiatus hernia, gallstones, ischemic heart disease, diabetes, obesity, dental caries, and duodenal ulcers, are now suspected to be associated with the consumption of a highly refined diet.

Diabetes. Fiber has also been shown to be an aid in treating diabetes in patients with adult onset diabetes. Practically all evidence indicates that dietary fiber decreases the maximum blood glucose level following a dose of glucose, as in the typical glucose tolerance test. The lower level of blood glucose presumably occurs because its absorption from the intestinal tract is slowed.

Cardiovascular Diseases. For some time it was believed that dietary fiber protected the individual from various heart disturbances, especially ischemic heart disease. That presumably occurred as a result of the action of dietary fiber in lowering serum cholesterol levels. Inclusion of large amounts of high fiber foods in the diet may decrease the intake of cholesterol-containing foods and those high in saturated fats. The intake of these two dietary components is probably closely related to the level and nature of blood cholesterol.

Wheat

Production, Trade, and Uses. Wheat is cultivated in most countries on all continents. Wheat is grown in most of the 50 states of the United States. The kind of wheat grown and quantity vary widely from one region to another. Winter wheats are planted in the fall. After the grasslike seedlings emerge from the ground, they lie dormant during the winter. They come up again in the spring, ripen, and are harvested in early summer. Spring wheats are planted in the spring and harvested in late summer.

Winter wheats grow best in areas of the country where the winters are not too harsh for the young plants.

Milling of Wheat. Before wheat can be used in the production of most foods, it must undergo several mechanical and chemical changes. The first change involves milling of wheat into flour.

The mill flow diagram begins with a separator, where the wheat first passes through a vibrating screen that removes straw and other coarse materials, and then over a second screen through which drop small foreign materials like seeds. An aspirator lifts off lighter impurities in the wheat. After the aspirator, wheat moves into a disk separator, consisting of disks revolving on a horizontal axis. The revolving disks discharge the wheat into a hopper or into the continuing stream. The wheat then moves into a scourer—a machine in which beaters attached to a central shaft throw the wheat against a surrounding drum. Air currents carry off the dust and loosened particles of bran coating.

The stream of wheat next passes over a magnetic separator that pulls out iron and steel particles. A washer–stoner may be the next piece of equipment. High-speed rotors spin the wheat in a water bath.

Excess water is thrown out by centrifugal force. Stones drop to the bottom and are removed. Lighter materials float off, leaving only the clean wheat. In the production lines of some mills, a dry stoner is also used. The wheat passes over an inclined, vibrating table that pushes stones and heavy material away from the lighter wheat, discharged separately.

The clean wheat is then tempered before the start of grinding; in the process moisture is added. Tempering aids the separation of the bran from the endosperm and helps provide a constant, controlled amount of moisture and temperature throughout the milling process. The sound wheat flows to a grinding bin or hopper from which it is fed in a continuous metered stream into the mill itself.

The first-break rolls of a mill are corrugated rather than smooth like the reduction rolls that reduce the particles of endosperm further along in the process.

The next important step introduces the broken-ground particles of wheat and bran into a sifter where they are shaken through a series of bolting cloths or screens to separate the larger from the smaller particles. The reduction rolls reduce the purified, granular middlings or farina to flour.

The process is repeated until the maximum amount of flour is separated, consisting of at least 72% of the wheat. This flour is made up of various grades.

Toward the end of the millstream, the finished flour flows through a device that releases a bleaching–maturing agent in measured amounts. If the flour is self-rising, a leavening agent and salt are also added.

In milling of durum wheat for the macaroni trade, special equipment is required, especially additional purifiers to separate the bran from the semolina, a coarse granulation of the endosperm. Semolina is prepared by grinding and bolting durum wheat, separating the bran and germ, to produce a granular product with no more than 3% flour. Durum millers also make granulars, or a coarse product with greater amounts of flour; or they grind the wheat into flour for special use in macaroni products, particularly in noodles.

Flour is often bleached. Bleaching improves the color of the flour, and some bleaching agents mature the flour and condition the gluten, improving baking quality. In aged, unbleached flour or in treated flour, the proteins are slightly oxidized by oxygen from air. The oxidation of flour makes gluten stronger or more elastic and produces better baking results.

Flour Types. Hard-wheat flours are usually higher in protein than are soft-wheat flours. They may be milled from either winter or spring wheat varieties. Those with highest protein content, characterized by their capacity to develop the strongest gluten, are used in commercial bread production where doughs must withstand the rigors of machine handling.

Soft-wheat flours are sold for general family use, as biscuit or cake flours, and for the commercial production of crackers, pretzels, cakes, cookies, and pastry. The protein in soft wheat flour runs from 7 to 10%.

Whole wheat flour, according to FDA specifications, is a coarse-textured flour ground from the entire wheat kernel. It contains the bran, germ, and endosperm. All-purpose flours are designed for home baking of a wide range of products—yeast breads, quick breads, cakes, cookies, and pastries. Self-rising flours are all-purpose flours to which leavening agents and salt have been added.

Cake flours are milled from low-protein soft wheat especially suitable for baking cakes and pastries or from low-protein fractions derived in the milling process. Semolina is the coarsely ground endosperm of durum wheat. Additional basic wheat products are wheat berry (kernel), bulgar, cracked wheat, wheat germ, bran, and commercial cereals.

Rice

Production and Consumption. Rice is grown in more than 100 countries and on every continent except Antarctica. In the world economy rice is an extremely important food, second only to wheat in total world production, and its yield per hectare exceeds that of wheat. Rice is the main staple food for more than half of the world's population.

The major rice-exporting countries are Thailand, United States, Pakistan, the European community (EC-12), China, Burma, and Australia.

Processing of Rice. Rice processing involves harvesting, drying, storage, and milling. These operations vary considerably throughout the world. The most important consideration during rice processing is prevention of breakage of the endosperm.

The purpose of rice milling is to remove the hulls and bran from harvested, dried rough rice. Shelling refers to the removal of the outer hull or shell from rough rice. The operation is conducted in machines that are known by different names such as shellers, hullers, huskers, dehuskers, and decorticators. The term hulling in most parts of the United States has the same meaning as shelling; however, in some areas hulling also refers to the removal of both hulls and bran. After removal of the hulls, the rice is called brown rice.

In some rice growing areas rice milling is accomplished by very primitive methods such as pounding the rough rice in a wooden mortar and pestle followed by winnowing. At the other extreme are very modern methods where milling is accomplished in large, highly automated plants. Thus there is no typical rice mill. However, the modern processing of rice consists of essentially these steps: (1) cleaning the incoming rough rice, (2) shelling the rough rice, (3) milling to remove the bran from the brown rice, (4) grading the milled rice by length into whole grain and different sizes of brokens, (5) mixing milled whole grain and brokens to meet specifications of buyers, and (6) packaging.

A hydrothermic process, parboiling, will greatly improve the milling quality of rice such that head yields will approach total yields (ie, zero breakage). When properly dried, the rice kernels are resistant to mechanical breakage.

Corn

Production and Economics. Every continent, except Antarctica, grows corn; 40% of the present world crop is produced in the United States. Yield is influenced by many factors, including climate, pest control, planting density, and fertilization.

A crop of corn is always maturing somewhere in the world. It grows from north latitude 58° to south latitude 40° and from below sea level to altitudes of 4,000 meters. Early varieties, that have been adapted to cold climates, mature in as little as 60 days. Late varieties grown in the tropics need nearly a year to reach maturity. After harvesting, kernels are dried to less than 15% moisture content to maintain grain quality and prevent long term storage spoilage.

Use in Animal Feed. Although corn was originally grown as food, the single largest use today is as feed for farm animals. Corn is an important feed ingredient, because it supplies the energy component and a large portion of the protein input to the animal's diet.

Also, the co-products of the various industries that use corn to produce beverage alcohol, starch, corn sweeteners, corn oil, and dry milled products provide the concentrated protein and vitamin content of corn, making them valuable feed ingredients as well.

Wet Milling. Shelled corn is shipped to wet millers by truck, rail, or barge. After cleaning to remove coarse material, ie, cobs, and fines (broken corn, dust, etc), the corn is steeped in a sulfurous acid solution to soften the corn and render the starch granules separable from the protein matrix that envelopes them.

The softened kernels are coarsely ground (first grind) to release the germs. Because of their high oil content, the germs are lighter than the starch, protein, and fiber fractions and can be separated in hydrocyclones. The germs are washed free of remaining starch, dried, and the valuable corn oil is removed by expelling, solvent extraction, or a combination of both.

The starch-protein-fiber slurry is subjected to an intense milling to release additional starch from the fiber. The fiber is then wet-screened from the starch-protein slurry, washed free of starch, and dried to form the major component of corn gluten feed. The starch-protein slurry is separated into its component parts. Separation is usually done with combinations of disk-nozzle centrifuges and banks of hydrocyclones.

The protein fraction is filtered and dried to become high (60%) protein content corn gluten meal. The starch slurry can be dewatered and dried to produce regular corn starch.

Starch. Corn starch is a principal ingredient in many food products, providing texture and consistency, as well as energy. However, more than half of the corn starch sold is used in industrial applications. Starch is a polymer consisting of α-linked anhydroglucopyranose units. Two forms exist: amylose is an essentially linear molecule in which the anhydroglucopyranose units are linked almost exclusively via α-1,4 bonds. Amylopectin is a much larger, branched molecule (the mol wt is ca 1,000 times greater than amylose); α-1,4 linkages predominate, but there is a significant number of α-1,6 linkages that result in the branched structure.

When heated in the presence of water to 62–72°C, normal starch granules swell, forming high viscosity pastes or gels. This process is called gelatinization.

Corn Starch-Based Sweeteners. Acid or enzyme catalysts can be used to break the linkages between the anhydroglucopyranose units in the starch molecule with the addition of a molecule of water at the break site. This process, called hydrolysis, produces a variety of corn-based sweeteners. Enzymatic isomerization of glucose to fructose produces high fructose corn syrups (HFCS) that are as sweet as sucrose syrups, thus allowing corn sweeteners to replace sugar in liquid applications (such as soft drinks).

Corn Oil. The crude corn oil recovered from the germs consists of a mixture of triglycerides, free fatty acids, phospholipids, sterols, tocopherols, waxes, and pigments. Refining removes the substances that detract from the quality. The first refining step is degumming; 1–3% water is added and the dense, hydrated gums are removed by centrifugation. The degummed oil is then refined.

Corn oil's flavor, color, stability, retained clarity at refrigerator temperatures, polyunsaturated fatty acid composition, and vitamin E content make it a premium vegetable oil.

Drying Milling of Corn. Dry corn mills tend to be smaller than wet mills. About one-fifth of the corn ground is white corn, the remainder being yellow dent. Dry millers use three processes; tempering-degerming, stone-grinding or nondegerming, and alkali-cooking.

Food Uses for Corn. Corn as corn flakes, sweet corn, corn as various types of flour and meal, popcorn, other snacks foods such as chips, and corn juice as sweeteners, corn used in fermentation for beer and in the production of alcohol, and corncobs and stalks used as carriers for various chemicals and medications, as fiber sources, and for the improvement of soil condition by plowing under stalks, are some of the uses for this versatile crop.

W. J. Darby, P. Ghalioungiu, and L. Grivetti, *Food: The Gift of Osiris,* Vol. 2, Academic Press, New York, 1977, pp. 492, 493.

D. K. Mecham in Y. Pomeranz, ed., *Wheat, Chemistry and Technology,* American Association of Cereal Chemists, Inc., St. Paul, Minn., 1971, p. 395.

J. I. Wadsworth, "Rice" in Y. H. Hui, ed., *Encyclopedia of Food Science and Technology,* Vol. 4, John Wiley & Sons, Inc., New York, 1991, pp. 2264–2279.

F. W. Schenck, "Corn and Corn Products," in Y. H. Hui, ed., *Encyclopedia of Food Science and Technology,* Vol. 1, John Wiley & Sons, Inc., 1991, pp. 482–490.

WHISKEY. See BEVERAGE SPIRITS, DISTILLED.

WINE

The unmodified word wine signifies the juice of grapes, particularly wine grapes, fermented by yeast and appropriately (and legally) finished into an alcoholic beverage. Wines are made from other fruits in much smaller amounts. Rice, being starchy, when fermented has more in common with beer, and its product should be considered sake instead of rice wine.

Magnitude

Production of grapes is a very large endeavor in a number of countries and, notwithstanding the importance of table fruit and raisins, commercial grapes mostly go into wine. More than 8 million hectares of grapes produce annually about 60 million tons of grapes and 30 billion liters of wine.

Grape and wine production, marketing, and serving worldwide are estimated to employ directly at least tens of millions of persons (there are 2 million grape growers in France alone), and grapes occupy nearly 1% of the total agricultural land, worldwide. Wines are important economically, not only as both everyday and premium beverages, but also in their contribution to feelings of contentment and perceptions of material well-being and sophistication.

Classification of Types and Styles of Wine

In classifying wines, many parameters might be used: fruit (species, variety, or condition), composition (color, alcohol, sugar, acid, or carbon dioxide), fermentation procedure (carbonic maceration, alcoholic, aerobic yeast film, malolactic), regulations (taxes, etc), geography, climate, weather (vintage). The last three are considered important in determining prices and reputations for individual wines, but are not very helpful in wine group classification. All of the above, and legally permitted variations, as well as different producer and advertising descriptions contribute to an almost infinite number of kinds of wines.

Most wines with <14% alcohol are classed as table wines because they are usually consumed with meals. Note that as used here, premium wines are included. In the countries of the European Union (EU), table wine means only ordinary or everyday wine. Sparkling wines are included in this group because producing the sparkle and retaining it during consumption of a bottle by few people necessitates a modest alcohol level. The "generous" group of wines have had distilled grape spirits added in order to reach, usually, about 18% ethanol. Such dessert and aperitif wines are intended to be consumed after or before a meal. Sweet and otherwise strongly flavored wines have gravitated here because they are more stable and thus can be consumed over several sittings. The significant classes of wine can be summarized as follows:

"Natural" wines, <14% alcohol. Their nature and keeping qualities have traditionally depended heavily on complete fermentation and protection from air.

Wines without obvious carbon dioxide

1. Wines containing anthocyanin (red) pigments. **2.** White wines not containing anthocyanin (red) pigments. **3.** Specialty products.

Wines with obvious carbon dioxide

1. From fermentation of added sugar; usual gauge pressure about $2.026-6.078 \times 10^5$ Pa (2–6 atm). **2.** Wines with excess carbon dioxide, not from refermentation of added sugar; Usual gauge pressure about $0.2026-2.026 \times 10^5$ Pa (0.2–2.0 atm). **3.** Carbonated wines: relatively rare; may be white or red.

Wines with 14–17% alcohol

Miscellaneous, white and red, usually sweet types, mainly with proprietary names

Special types

"Generous" or Fortified Wines, 17–21% alcohol. Their nature and keeping qualities depend heavily on the addition of distilled wine spirits.

Without added food flavoring materials

Table 1. Estimates of Typical Gross Composition of Wines, wt %

Component	Table wines		Dessert wines	
	White	Red	White	Red
water	87	87	76	74
ethanol	10	10	14	14
other volatiles	0.04	0.04	0.05	0.05
extract	2.6	2.7	10.1	12.2
sugar	0.05	0.05	8	10
pectin and related substances	0.3	0.3	0.25	0.25
glycerol and related substances	1.1	1.1	0.9	0.9
acids	0.7	0.6	0.5	0.5
ash	0.2	0.2	0.2	0.2
phenols	0.01	0.2	0.01	0.1
amino acids and related substances	0.25	0.25	0.2	0.2
fats, terpenoids	0.01	0.02	0.01	0.02
miscellaneous and vitamins	0.01	0.01	0.01	0.01
Total	*100*	*100*	*100*	*100*

1. Containing anthocyanins and related (red) pigments. **2.** Not containing red pigments.

With added (natural) food flavors or plant extracts

1. With a red color. **2.** Without a red color.

Composition

With so many types (eg, dry red table wine) and styles (eg, oaky or not) within a type of wine, composition is quite variable; but table wines have certain consistencies in constituents, and sweet wines overall tend to be similar, with some allowance given for their extra sugar and perhaps alcohol in certain cases. Components tend to be in two categories: those produced by fermentation (ethanol is a prime example) and those from the grapes (eg, red anthocyanin pigments).

A rough comparison of typical compositions of wines without unusual additions is given in Table 1. The apparent simplicity is, however, deceptive. Extensive listings of several hundreds of identified wine constituents are available, of which the data in Table 1, are only a useful summary. Each group of chemical constituents usually includes several, sometimes many, individual substances. As groups, sugars contribute sweetness and "body" and acids tartness. These are important sensory properties, but are relatively obvious and qualitatively if not quantitatively similar among compounds of the group. Together they account for much of the extract (nonvolatile solids) of wines. Glycerol is produced by fermentation, along with ethanol and a number of the volatile substances, especially higher alcohols and certain esters. These contribute to the typical vinous flavors of wine, but are rather similar among wines. A number of the other constituents, eg, pectins, ash, vitamins, fats, amino acids, and proteins, may be useful in comparing wines analytically, but do not ordinarily contribute to recognized sensory distinctions among wines.

Most of the sensory differences among wines, other than sweetness and acidity, are attributable to certain volatiles, terpenoids, and phenols. Rarely is a single volatile compound responsible for differences characteristic of a grape variety or wine type.

Red wines are notably higher in phenol content (Table 1). This is partly a result of the fact that the red pigments themselves, anthocyanins, are flavonoids. In making such wines, the skins (and seeds) of the grape are more thoroughly extracted so that not only the anthocyanins of red grapes are extracted, but also other phenols, including astringent tannins, bitter catechins, and oxidation-browning substrates. Volatiles account for the aromas and bouquets of wines. Monoethyl tartrate forms slowly as part of aging owing to high content of both tartaric acid and ethanol. Whereas this ester is not volatile or odorous, a sufficient amount of it forms to lower the free acid and the tartness of the aging wine, thus mellowing it.

Sensory Analysis, Wine Appreciation, and Spoilage

There is no combination of chemical or physical analyses which can, or is ever likely to, replace human sensory evaluation completely. Sensory examination of wines employs two major approaches: detecting differences and evaluating quality or, more briefly, analytical and hedonic. The former can be objective and the latter is inevitably somewhat subjective regardless of the expertise of the judges.

Few commercial wines today are frankly spoiled. It must be noted that spoilage of wine does not pose a health risk to consumers. Because some of the spoilage odors and flavors are not easily analyzed chemically, sensory evaluation is the final judge.

Healthfulness

Clinical, experimental, epidemiological, and historical evidence indicates that moderate consumption of wine is beneficial. The proven benefit is in lowered incidence of cardiovascular complications in wine consumers.

Recent studies suggest that wine, particularly red table wine, has an additional favorable effect. This is attributed to the antioxidant, free-radical chain-breaking effect of the wine's natural phenolic compounds.

Technology

Certain operations are required, unique, and irreversible in order to produce certain types of wine, eg, oxidation for sherries. Other operations such as clarification and tartrate stabilization are similar for all wines.

Variety Selection, Fruit Production, and Harvest. The wine grape is *Vitis vinifera,* and there are estimated to be a few hundred varieties of some note commercially worldwide. The list of important, widely planted varieties is much smaller. The grape variety involved is one of the most important factors in the final wine's characteristics.

With most vinifera varieties the flavor is subtle and more intensity is often sought. With varieties and hybrids from other species, such as Concord from *Vitis labrusca* or *labruscana,* the flavor may be too strong.

Varietal labeling is an important quality factor in the United States, and indirectly elsewhere because only certain specific varieties are planted in each prestigious foreign area. U.S. law currently requires that 75% of the wine must come from the *V. vinifera* variety named on the label.

Vineyard site is important to wine quality and character and interacts with variety. The general climate must not be too cold, too hot, or too humid. Proper management of the vines and vineyard is important for yield of high quality wine grapes. The proper stage of grape ripeness is different for different wines. The decision to harvest is based on sugar content, acidity, and close observation of the fruit's manner of changing in other parameters, including taste.

Must Processing. Ideally the grapes are picked quickly at the chosen stage and transported to the winery. As soon as possible after harvest, the grapes are ordinarily passed through destemmer–crusher machinery to remove the stems and break each berry open. If white table wine is to be made, rapid juice separation is recommended, whereas for red, fermentation of the whole pomace (pulp, skins, seeds) is usual. There are several types and capacities of equipment for destemming, crushing, pumping, pressing, juice clarification, etc. Much of it is specially designed for winery use.

The must comprises the whole crushed mass (pomace, ie, seeds and skins, plus juice) of red grapes, if the wine is to be red. Frequently, but not invariably, a small amount, 60 mg/L or so, of sulfur dioxide in one form or another is added to minimize any effects of oxidation and inhibit any undesirable microorganisms.

Fermentation. Today it is almost universal to inoculate the must with a selected yeast strain.

Among the reasons for choosing a particular yeast strain are its ability to ferment at cold temperatures or under pressure, separate for clarification easily, or avoid production of hydrogen sulfide during typical fermentations. Heat is released as well as a considerable volume of carbon dioxide (2 moles per mole of sugar fermented, about 50 times the must volume). Heat liberated during fermentation must be removed to prevent ill effects. The temperature rise can become inhibitory, even fatal at about 35°C, to the wine yeasts. This is prevented by circulating refrigerant through jackets on stainless steel tanks to control the fermentation temperature at the desired level. A large winery needs large refrigeration capacity. White table and sparkling wines are ordinarily better (more fruity, etc) if fermented cool, at about 10–16°C. Of course, the fermentation is slower the lower the temperature.

Red table wines should be fermented at about 21–29°C for proper character.

In addition to alcoholic fermentation, a malolactic fermentation by certain desirable strains of lactic acid bacteria is desirable in many red table wines for increased stability, more complex flavor, and sometimes for decreased acidity.

Sparkling Wines. In making sparkling wines, a light, dry, moderate alcohol, tart, stable, white or nearly so table wine is the preferred base wine. Sugar is added in the precise amount needed to give the desired pressure (about 24 g/L for 6.078×10^5 Pa (6 atm) at the end of refermentation).

Wines can be made effervescent by carbonation rather than refermentation in a closed system, but that must be stated on the label.

Sweet Wines. Slightly sweet table wines can be made by adding a small amount of reserved grape juice or concentrate, but they are often made today by refrigerating, centrifuging, or filtering out the yeasts before fermentation is complete. In any case, such wines are unstable and the residual sugar will ferment if not prevented from doing so.

Table wines of higher sweetness can be made by extension of the same treatments, but Sauternes (French), Trockenbeerenauslesen (German), and specialty wines elsewhere are made from shriveled grapes. Such grapes have been colonized by the mold *Botrytis cinerea,* which makes the skin porous. In dry weather the grapes then lose water and the must becomes concentrated. Not only are special, concentrated flavors produced, but also the sugar becomes so high the yeasts cannot ferment it all and the wine remains sweet.

A third method of making sweet wines is by arresting the fermentation through use of alcohol so that the level reached prevents further fermentation and leaves the wine sweet. Most of the traditional sweet wines of the world, port and other types, are made by such fortification.

Oxidized Wines. Oxidized color and flavor are defects in most wines, but certain ones capitalize upon oxidation. Sherries fall in this group and include wines that are at least moderately high in alcohol and may or may not be sweet. Access to air is required during processing or maturation and may take the form of chemical autoxidation (as in oloroso sherries) or aerobic microbiological metabolism (fino, Manzanilla, etc).

Post-Fermentation Processing. When the wine has been fermented and matured to the desired stage, it is clarified and stabilized so that it will remain clear and not be undesirably changed when bottled and marketed. Processing is held to the essential minimum to avoid flavor change and loss.

Chemical stabilization is considered necessary in the United States to prevent formation of crystals of potassium bitartrate in the bottled wine. This is commonly done by cooling the wine to barely above its freezing point, holding it to allow crystallization, and filtering it cold.

Waste Disposal. Table-wine wineries and other large-volume food processors generate proportionally little waste. It can create a nuisance if allowed to develop off-odors, insect attraction, or oxygen demand in runoff, but generally it is innocuous. Wastes consist of stems, pomace, lees, and stillage from the production of beverage or fortifying brandy. In many instances, potential problems are avoided by frequent removal to and scattering in the vineyards.

Storage (Maturation, Aging) and Blending. One of the prime requisites of an interesting premium wine is complexity of flavor. Maturation (bulk storage and associated final processing), aging (properly speaking, the storage of packaged wine ready for the consumer), and blending are the principal ways of achieving that complexity.

In oaken barrels, slow evaporation of water and alcohol through the staves is ordinarily compensated for by refilling and topping each barrel by addition of the required amount of the same wine every week or so. This special process and the transfers necessary in the course of normal processing inevitably slowly permit a little oxidation, but wet wood and cork essentially do not pass oxygen.

When they have suitably matured and been fully processed and blended, wines are bottled. Ideally, all the changes which should result during the bulk stage have been completed and the bottling itself should preserve all the desirable characteristics.

Conditions for proper storage of bottled wine include a fairly low and constant (about 13°C) temperature, restricted light and agitation, and bottles stored with corks wetted by the contents.

Regulations

In the United States a special agency, the Bureau of Alcohol, Tobacco, and Firearms (BATF) within the U.S. Treasury Department, regulates the alcoholic beverage industries. The regulations and their application are voluminous and detailed. They specify not only label compliance and matters relating to taxation that are of direct interest to consumers, but contain all the details of permitted processes for and additions to wines.

Codification to protect regional names for wines was perfected in France and, by treaty, extended for their wines internationally under *appelation d'origine contrôlée* nomenclature (AOC, or just AC). In general these regulations delimit the region protected, limit the varieties of grapes that can be used, restrict the wine production per hectare, and require approved enological practices. Several other countries, notably Italy, Germany, and South Africa, have adopted somewhat similar regulations.

VERNON L. SINGLETON
CHRISTIAN E. BUTZKE
University of California, Davis

R. B. Boulton, V. L. Singleton, L. F. Bisson, and R. E. Kunkee, *Principles and Practices of Winemaking,* Aspen Publishers, Gaithersburg, Md, 1995.

WOOD

Wood is an important natural resource, one of the few that are renewable. It is prevalent in our everyday lives and the economy; in wood-frame houses and furniture; newspapers, books, and magazines; bridges and railroad ties; fence posts and utility poles; fuelwood; textile fabrics; and organic chemicals. Wood and wood products are also a store for carbon, thus, helping to minimize carbon dioxide in the atmosphere.

Wood supplies the solid raw material for products, such as lumber, plywood, and wood pallets, and the fiber for paper, paperboard, fiberboard panels, rayon, and acetate (see also WOOD-BASED COMPOSITES AND LAMINATES; PAPER; PULP). Many wood products can be recovered for reuse or recycling, thus extending our wood supply into the future.

Structure

The anatomical structure of wood affects strength properties, appearance, resistance to penetration by water and chemicals, resistance to decay, pulp quality, and the chemical reactivity of wood. To use wood most effectively requires a knowledge of not only the amounts of various substances that make up wood, but also how those substances are distributed in the cell walls.

Woods are either hardwoods or softwoods. Hardwood trees (angiosperms, ie, plants with covered seeds) generally have broad leaves, are deciduous in the temperate regions of the world, and are porous, ie, they contain vessel elements. Softwood trees (conifers or gymnosperms, ie, plants with naked seeds) are cone bearing, generally have scalelike or needlelike leaves, and are nonporous, ie, they do not contain vessel elements. The terms hardwood and softwood have no direct relation to the hardness or softness of the wood.

Many mechanical properties of wood, such as bending and crushing strength and hardness, depend upon the density of wood; denser woods are generally stronger. Wood density is determined largely by the relative thickness of the cell wall and by the proportions of thick-walled and thin-walled cells present.

Just under the bark of a tree is a thin layer of cells, not visible to the naked eye, called the cambium. Here, cells divide and eventually differentiate to form bark tissue outside of the cambium and wood or xylem tissue inside of the cambium. This newly formed wood on the inside contains many living cells and conducts sap upward in the tree, and hence, is called sapwood. Eventually, the inner sapwood cells become inactive and are transformed into heartwood. This transformation is often accompanied by the formation of extractives that darken the wood, make it less porous, and sometimes provide more resistance to decay.

Composition

Wood is a complex polymeric structure consisting of lignin and carbohydrates [cellulose and hemicelluloses], which form the visible lignocellulosic structure of wood. Also present, but not contributing to wood structure, are minor amounts of other organic chemicals and minerals. The organic chemicals are diverse and can be removed from the wood with various solvents. The minerals constitute the ash residue remaining after ignition at a high temperature.

Wood–Liquid Relationship

Adsorption. Wood is highly hygroscopic. The amount of moisture adsorbed depends mainly on the relative humidity and temperature (Fig. 1). Exceptions occur with species with high extractive contents (eg, redwood, cedar, and teak). The equilibrium moisture contents of such woods are generally somewhat lower than those given in Figure 1.

Shrinking and Swelling. The adsorption and desorption of water in wood is accompanied by external volume changes. At moisture contents below the fiber saturation point, the relationship may be a simple one, merely because the adsorbed water adds its volume to that of the wood, or the desorbed water subtracts its volume from the wood. Total volumetric shrinkage or swelling between fully saturated and completely dry wood ranges from 10 to 20%, depending on species. The relationship may be complicated by the development of stresses, causing changes in volume or shape to occur.

Permeability. Although wood is a porous material (60–70% void volume), its permeability (ie, flow of liquids under pressure) is extremely variable. This is due to the highly anisotropic shape and arrangement of the component cells and to the variable condition of the microscopic channels between cells. In the longitudinal direction, the

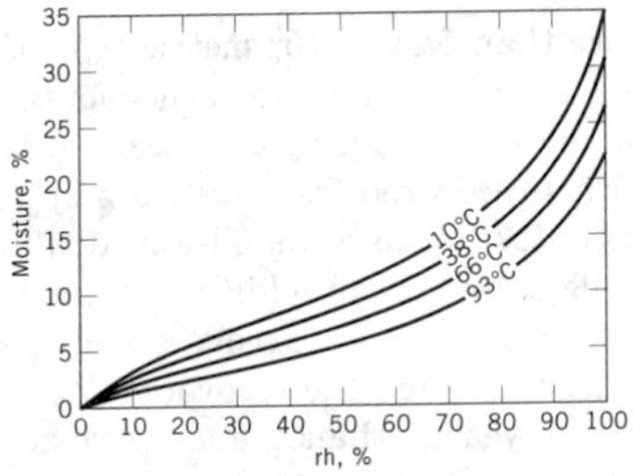

Figure 1. Relationship between the moisture content of wood (% of dry wood) and relative humidity at different temperatures.

permeability is 50 to 100 times greater than in the transverse direction. Sapwood is considerably more permeable than heartwood. In many instances, the permeability of the heartwood is practically zero.

Transport. Wood is composed of a complex capillary network through which transport occurs by capillarity, pressure permeability, and diffusion.

Drying. There are a number of important reasons for drying: it reduces the likelihood of stain, mildew, or decay developing in transit, storage, or use; the shrinkage that accompanies drying can take place before the wood is put to use; wood increases in most of its strength properties as it dries below the fiber saturation point (30% moisture content); the strength of joints made with fasteners, such as nails and screws, is greater in dry wood than in wet wood dried after assembly; the electrical resistance of wood increases greatly as it dries; dry wood is a better thermal insulating material than wet wood; and the appreciable reduction in weight that accompanies drying reduces shipping costs (see DRYING).

Ideally, the temperature and relative humidity during drying should be controlled; if wood dries too rapidly, it is likely to split, check, warp, or honeycomb because of stresses.

Air drying is a process of stacking lumber outdoors to dry. Kiln drying is a controlled drying process widely used for drying both hardwoods and softwoods.

Structural Material

Strength and Related Properties. In the framing of a building or the construction of an industrial unit, where wood is used because of its unique physical properties, strength and stiffness are primary requirements. Different species of wood have different mechanical properties that relate to the amount of wood substance per unit volume, ie, its specific gravity. The strength of a piece of lumber depends also upon its grade or quality. The strength values of a lumber grade depend upon the size and number of such characteristics as knots, cross grain, shakes, splits, and wane. Wood free from these defects is known as clear wood.

Most strength properties of clear wood improve markedly as moisture is reduced below ~30%, based on the ovendry weight. However, some properties (eg, tensile strength parallel and perpendicular to the grain) may be reduced if the wood is overly dried.

The mechanical properties of wood tend to increase when it is cooled and to decrease when it is heated. Temperatures ≥70°C result in permanent degradation. The amount of this irreversible loss in mechanical properties depends upon moisture content, heating medium, temperature, exposure period, and, to some extent, species.

The effect of absorption of various liquids upon the strength properties of wood largely depends on the chemical nature and reactivity of the absorbed liquid. In general, neutral, nonswelling liquids have little if any effect upon the strength properties. Any liquid that causes wood to swell causes a reduction in strength.

Wood preservatives are applied either from an oil system, such as creosote, petroleum solutions of pentachlorophenol, or copper naphthanate, or a water system. Oil treatments are relatively inert with wood material, and thus, have little effect on mechanical properties. However, most oil treatments require simultaneous thermal treatments, which are specifically limited in treating standards to preclude strength losses.

The mechanical properties of wood can be damaged by preservative systems. In North America, the primary which may contain chromic acid and arsenic, ammoniacal or amine-based copper systems with supplemental zinc, arsenate, or quaternary ammonium chloride. Waterborne preservatives are copper-based systems. The effects of waterborne preservative treatments are directly related to treatment processing factors.

Fire retardant treatments generally reduce allowable design stresses by 10–25%. The magnitude of these reductions varies depending upon the fire retardant chemical, the severity of treatment and processing conditions, and the property being considered.

Reaction to Heat and Fire. In general, the thermal degradation of wood and other cellulosic substances proceeds along one of two competing reaction pathways. At temperatures up to ~200°C, carbon dioxide and traces of organic compounds are formed, in addition to the release of water vapor. These gases are not readily ignitable. Exothermic reactions may occur near 200°C and, in situations where heat is conserved, self-ignition at temperatures as low as 100°C has been observed. Times and temperatures that might result in smoldering initiation can be determined. To provide a margin of safety, 77°C should be the upper limit in prolonged exposure near heating devices. Wood in its untreated form has good resistance or endurance to fire penetration when used in thick sections for walls, doors, floors, ceilings, beams, and roofs. This endurance is due to low thermal conductivity, which reduces the rate at which heat is transmitted to the interior.

The surface burning characteristics (flame spread index and smoke developed index) for wood and wood products can be reduced with fire retardant treatments, either chemical impregnation or coatings. Fire retardant treatments also reduce the heat release rate of a burning piece of wood. Fire retardant chemicals, such as ammonium phosphate, ammonium sulfate, zinc chloride, guanylurea phosphate, dicyandiamide phosphate, borax, and boric acid, are often used in combinations.

Resistance to Chemicals. Different species of wood vary in their resistance to chemical attack. The significant properties are believed to be inherent to the wood structure, which governs the rate of ingress of the chemical and the composition of the cell wall, which affects the rate of action at the point of contact.

Wood is widely used as a structural material in the chemical industry because it is resistant to a large variety of chemicals. Its resistance to mild acids is far superior to that of steel but not as good as some of the more expensive acid-resistant alloys.

Alkaline solutions attack wood more rapidly than acids of equivalent concentrations, whereas strong oxidizing chemicals are harmful. Wood is seldom used where resistance to chlorine and hypochlorite solutions is required.

Because traces of iron reduce the brilliance of many dyes, wood tanks have long been preferred to steel in the manufacture of dyes. Similarly, vinegar and sour foodstuffs are processed in wood tanks because common metals impart a metallic taste.

Resistance to chemical attack is generally improved by resin impregnation, which protects the underlying wood and reduces movement of liquid into the wood. Resistance to acids can be obtained by impregnating with phenolic resin and to alkalies by impregnating with furfural resin (see FURAN DERIVATIVES; PHENOLIC RESINS).

Biodeterioration. The principal organisms that degrade wood are fungi, bacteria, insects, and marine borers. Decay, molds, and stain are caused by fungi. Decay is the most serious kind of damage because it causes structural failure and consequently, tremendous economic losses.

Termites are the most destructive insects that attack wood. Their attack can be prevented or lessened by using naturally resistant wood or by treating wood with preservatives. Despite great differences between fungi and termites, chemicals that inhibit fungi usually also inhibit termites.

For practical purposes, the sapwood of all species may be considered to be susceptible to biodeterioration. The heartwood of some species, however, contains toxic extractives that protect it against biological attack. Among the native species that have decay-resistant or highly decay-resistant heartwood are bald cypress, redwood, cedars, white oak, black locust, and black walnut.

Modified Wood

In addition to preservation or fire protection, wood is modified to reduce the rate that moisture is sorbed by the wood (water repellency) and/or to reduce the shrinking and swelling that occur at equilibrium (dimensional stability) under conditions of fluctuating relative humidity. Certain species with high extractives content, especially in the cell walls, have greater water repellency and, in some cases, greater dimensional stability than species with low extractives content. This suggests a means of obtaining still greater reductions in the rate and extent of swelling and shrinking, that is, by filling the voids in the wood to reduce rate or deliberately adding large amounts of bulking agents to the cell walls to improve dimensional stability.

Low molecular weight, nonswelling vinyl-type monomers can be impregnated into woods that polymerize *in situ* into the void structure by radiation or heat and a catalyst. The wood cell wall can be bulked with leachable polyethylene glycol (PEG) to reduce the swelling and shrinking tendency by about 80%. Also, it is possible to react an organic moiety to the hydroxyl groups on cell wall components. This type of treatment also bulks the cell with a permanently bonded chemical. Many compounds modify wood chemically. The best results are obtained by the hydroxyl groups of wood reacting under neutral or mildly alkaline conditions below 120°C.

Thermosetting phenolic resins have been used successfully to penetrate and polymerize in the cell wall. The mechanical properties of resin-impregnated wood are improved or not affected except for toughness, which is reduced by as much as 60%. Treatment with phenol–formaldehyde resins increases the decay resistance. Heat resistance is improved markedly by resin impregnation. The largest industrial application of impreg is in die models for automobile body parts and other model dies.

Bending is another treatment process. Above 80°C, green wood becomes readily deformable. The new shape persists after cooling to room temperature and drying under restraint. This is the basis for commercial bending of wood to various shapes.

Chemical Raw Material

Wood is one of our most important renewable biomass resources. Unlike most biomass sources, wood is available year round and is more stable on storage than other agricultural residues. Increasingly, wood residues from industrial by-products are incorporated into manufactured wood products and are used as a fuel, replacing petroleum; some is converted to charcoal but most is used in the pulp and paper industry. Residues are also available for manufacturing chemicals, generally at a cost equivalent to their fuel value (see FUELS FROM BIOMASS; FUELS FROM WASTE).

Wood can be pyrolyzed (heated to 400°C or higher in the absence of oxygen) to produce a variety of chemical compounds; for example, charcoal, acetic acid, methanol, tar, and gases (predominately hydrogen, carbon monoxide, carbon dioxide, and methane).

Wood is about 65–75% carbohydrate and has been considered as a potential source of ethanol for fuel. The carbohydrate material can be hydrolyzed to monomer sugars, which in turn can be fermented to produce ethanol. However, wood carbohydrates are expensive to hydrolyze. Hydrolysis with acids and enzymes is impeded by the crystalline structure of cellulose. Lignin interferes with processing, and hydrolytic by-products such as furfural, acetic acid, and derivatives of lignin and extractives can inhibit fermentation.

The principal chemical industry based on wood is pulp and paper. Pure cellulose is the raw material for a number of products, eg, rayon, cellulose acetate film base, cellulose nitrate explosives, cellophane, celluloid, carboxymethylcellulose, and chemically modified cellulosic material.

Most of the more than 54.5×10^6 metric tons of organic material removed from wood during pulping are burned for their energy content and to recover the inorganic pulping chemicals. A large proportion of the organic chemicals recovered from sulfite pulp mills are lignosulfonates. Lignosulfonates are used as oil-well drilling fluids, binders for animal food pellets, and as an additive to improve the structural properties of concrete. Some of the recovered lignin-derived material is used to produce vanillin, and the recovered sugars are used to produce small amounts of yeast, ethanol, and acetic acid. The most valuable chemical by-products are isolated at kraft (sulfate) pulp mills. These chemicals are sulfate turpentine and tall oil (a mixture of fatty acids and rosin).

Hydrolysis

In the acid hydrolysis process, wood is treated with concentrated or dilute acid solution to produce a lignin-rich residue and a liquor containing sugars, organic acids, furfural, and other chemicals. The process is adaptable to all species and all forms of wood waste. The liquor can be concentrated to a molasses for animal feed, used as a substrate for fermentation to ethanol or yeast, or dehydrated to furfural and levulinic acid. Attempts have been made to obtain marketable products from the lignin residue rather than using it as a fuel, but currently only carbohydrate-derived products appear practical.

Fuel Properties

The fuel properties of wood can be summarized by ultimate and proximate analyses and determination of heating value. The analytical procedures are the same as those for coal, but with some modifications. Analytical results generally vary about as much within a species as they do between species, except that softwood species generally have a higher carbon content and higher heating values than hardwood species because of the presence of more lignin and resinous materials in softwood species (see FUELS FROM WASTE).

Charcoal Production

Charcoal is produced by heating wood under limited access of oxygen. When wood is heated slowly to ~280°C, an exothermic reaction occurs. In the usual carbonization procedure, heating is prolonged to 400 to 500°C in the absence of air. The term charcoal also includes charcoal made from bark. Charcoal is produced commercially from primary wood-processing residues and low quality roundwood in either kilns or continuous furnaces. To alleviate the associated air pollution problem, the gases from the kilns or furnaces can be burned with additional fossil fuel to recover heat and steam, or in afterburners to nearly eliminate visible air pollution and odors (see AIR POLLUTION CONTROL METHODS).

IRENE DURBAK
DAVID W. GREEN
TERRY L. HIGHLEY
JAMES L. HOWARD
DAVID B. MCKEEVER
REGIS B. MILLER
ROGER C. PETTERSEN
ROGER M. ROWELL
WILLIAM T. SIMPSON
KENNETH E. SKOG
ROBERT H. WHITE
JERROLD E. WINANDY
JOHN I. ZERBE
USDA Forest Service

A. J. Panshin and C. deZeeuw, *Textbook of Wood Technology: Structure, Identification, Uses, and Properties of the Commercial Woods of the United States*, 4th ed., McGraw-Hill Book Co., Inc., New York, 1980.

Wood Handbook: Wood as an Engineering Material, Agriculture Handbook 72, U.S. Department of Agriculture, Forest Service, Forest Products Laboratory, Madison, Wisc., revised, 1987.

D. Fengel and G. Wegener, *Wood: Chemistry, Ultrastructure, Reactions,* W. de Gruyter, Berlin and New York, 1984.

F. F. P. Kollmann and W. A. Côté, Jr., *Principles of Wood Science and Technology,* Vol. 1, Springer-Verlag New York, Inc., New York, 1968.

WOOD-BASED COMPOSITES AND LAMINATES

Virtually the entire woody stem of the tree is converted to a family of wood and wood-based products. These developments have literally doubled, and perhaps with recycling, more than doubled the yield of usable products from a given volume of wood. Most of these products are bonded together with adhesives.

Adhesives for Wood

Most of the adhesives which can be or have been used in wood and wood composite bonding applications include natural adhesives, such as starch, soy flour, blood, casein, hide/bone; and synthetic adhesives, such as urea–formaldehyde, melamine –formaldehyde, phenol–formaldehyde, resorcinol–formaldehyde, isocyanate/urethanes, poly(vinyl acetate) (PVA), contact adhesives, epoxy resins, and hot-melt adhesives.

Only limited quantities of natural adhesives are still used, largely in conjunction with a synthetic adhesive.

Synthetic adhesives have resulted from developments in the field of organic chemistry. Virtually all are obtained from petroleum, natural gas, or coal by-products.

Plywood

Plywood is a panel made from wood veneers (thin slices or sheets) bonded to one another. Generally each ply is oriented at right angles to the adjacent ply, and the two face plies should have the grain direction parallel to each other.

Hardwood Plywood Processing and Products. Hardwood is generally considered to be that coming from a tree having distinct leaves as opposed to needles, and usually but not always deciduous. Production of hardwood plywood follows one of two general processes: one using sliced decorative veneers and the other using rotary-cut veneers.

Hardwood plywood logs are almost always soaked or steamed to soften the wood, making it more amenable to slicing or peeling.

Decorative veneers from the slicer are passed through a long tunnel dryer. As they emerge from the dryer at 3–5% moisture content (MC), they are restacked in the order they emerge. They proceed to a trimmer, which trims the stack to obtain parallel edges, at the maximum obtainable width. A coating of adhesive, usually a contact adhesive, is applied to the edges of the stack. Then the pieces are passed edge-to-edge through an edge-gluer, from which sheets of increasing width result with each pass through the machine. When the desired width, usually about 1270 mm, is achieved, the sheets are stacked and moved to the panel lay-up area.

Veneers from the rotary lathe pass through a high speed clipper or knife which cuts out the usable widths, the objective being to obtain as many full-width sheets as possible. Veneers are again sorted after emerging from the dryer.

The stacks of veneer are moved to the panel lay-up area, where adhesive is applied and veneers assembled into lay-ups ready to be pressed.

The primary adhesive used in hardwood plywood is urea–formaldehyde (UF) mixed with wheat flour as an extender to improve spreadability, reduce penetration, and provide dry-out resistance.

For exterior applications, where water exposure is expected, phenol–formaldehyde (PF) or phenol–resorcinol–formaldehyde (PRF) adhesives are used.

Lay-up proceeds by laying down the veneer which is to be the back surface of the panel. Then a sufficient number of pieces of core veneer are passed through the glue spreader to form the next layer of cross-oriented veneer. Adhesive is applied only to the cross-plies and in sufficient quantity to provide a continuous layer on both opposing faces of veneer. Then the top surface veneer, which is normally the decorative surface, is placed on the assembly.

As succeeding panels are laid-up, a stack of panels is formed. The stacks are moved to a large one-opening pre-press where they are cold-pressed as a unit for several minutes.

After prepressing, the stacks move to the press loader. The press loader raises the stack vertically beside the press. As the top panel reaches each opening, the panel is manually pushed into the press. The press presses the entire group of panels at 1035–1205 kPa (ca

150–175 psi) for several minutes at a temperature of about 140°C. Heating greatly accelerates the cure of the adhesive.

Uses and Treatments of Hardwood Plywood. Most early applications of hardwood plywood were those where the hardwood plywood was better adapted to the use than solid wood. One of the most important early uses was in curved or formed parts. As furniture manufacturers realized the inherently superior stability of plywood compared to solid wood, lumber-core or plywood panels began to be used for most flat-panel constructions in furniture. Both lumber core and plywood core have been almost totally displaced in recent years by particleboard or medium-density fiberboard.

Specifications, Standards, Quality Control, and Health and Safety. Specifications and standards for hardwood plywood products are found in the American National Standard Institute (ANSI) standard for *Hardwood and Decorative Plywood.* At present, there is no emission regulation for products used in buildings, other than homes, but the majority of products are now manufactured to pass a stringent emissions test. The flamespread test is designed to measure the potential of a product to ignite and contribute to the spread of a flamefront in a fire.

In addition, many hardwoods also may now be used in softwood plywood as core veneers.

Softwood Plywood Processing and Products. Softwood is generally considered to be that coming from a coniferous tree. In the traditional softwood plywood process, veneer logs are kept wet until cut to peeler block lengths, about 2500 mm (100 in.). The block is placed in a charger, which loads and positions the block into the lathe after the previous block is finished.

The lathe turns the block against the knife and peels the veneer in a continuous sheet as the knife moves toward the center of the block. When the knife cannot advance further without moving into the metal chucks, the lathe is stopped, the core of the block is dropped, the lathe is recharged, and the cycle repeated.

As the veneer leaves the lathe, it moves into a series of trays, long storage conveyors. The veneer sheets move along another long conveyor table where they are pulled and sorted into stacks by grade and width. The stacked veneer is moved to the dryer area and moved through the long conveyor dryer. Hot air is blown onto the surfaces of the veneer to hasten the drying process. The dry veneers are again sorted by grade and width and moved to the splicing and patching area. Here the narrow veneers are spliced together into full-size sheets. Then the stacks of veneer are moved to the gluing and lay-up area.

The adhesive used in virtually all softwood plywood has a phenol—formaldehyde (PF) base to provide an exterior-grade, durable, waterproof bond. There are two types of gluing and lay-up areas, manual and automated.

The assembled veneers for each panel are placed in stacks equivalent to the number of panels to be pressed in each pressload. Each stack is then placed in a cold single-opening prepress and pressed for a short time to assure transfer of adhesive from the spread faces to the unspread faces. The stacks of panels are moved to a lift on the side of the press. The lift raises the stack and the top panel is pushed into succeeding panel holders on the press loader as the stack passes each holder. The press opens after completion of pressing and curing the adhesive on the current pressload.

Secondary Treatments and Uses. Depending on grade, panels may receive from no processing to significant processing beyond this point. Structural sheathing panels, a nonappearance grade used as structural wall sheathing, roof sheathing or flooring, may be simply trimmed to size, grade-stamped, and packaged for shipment. Appearance-grade panels will have surface defects shimmed, patched, or filled. Then the panels will be sanded to a uniform thickness, trimmed to final dimension, regraded, and packaged for shipment.

Specifications, Standards, Quality Control, and Health and Safety Factors. APA–The Engineered Wood Association represents the softwood plywood and oriented strandboard industries in the areas of specification, standards, and quality control (QC). An APA product standard, PS1-95, discusses the above areas in detail covered.

Insulation Board and Structural Fiberboards

Production, Processing and Shipment. *Insulation Board.* The panel products known as insulation board are fiber-base products with a density less than 500 kg/m^3.

In the process using wood residues, the raw material (usually wood chips) is passed through a chip washer to remove extraneous materials. The washed chips move to a refiner where the chips pass between rotating steel disks with molded patterns in the faces of the disks. The chips are converted into fiber and small fiber bundles as they pass between the disks. The fiber drops into a stock chest, where it is mixed with a large quantity of water, forming a uniform suspension. This fiber slurry is then moved to a smaller stock chest, where additives such as wax emulsion, starch or PF resin as binder, and alum are metered into the slurry.

The fiber slurry is pumped to the mat-forming box of a wet-process forming machine. A wide endless wire screen belt, usually about 2540 mm (100 in.) wide, moves under the flow of fiber and water. The water drains through the fiber and through the belt to be recirculated back to the stock chest. The fibers are deposited on the belt. The belt then passes over a number of suction boxes, in which a partial vacuum is maintained. The air drawn through the fiber mat draws more water from the mat through the screen and through a series of holes in the top of the suction boxes, further consolidating the mat as well as removing excess moisture.

The formed mat is carried along on the wire screen and is trimmed to the desired width and length by high pressure water jets. The mats then pass into a long heated drying chamber. Bonding occurs at points within the mat where individual fibers or fiber bundles touch one another. As the dry (5–7% mc) mat emerges from the dryer, it is already a useful product, having its own inherent properties and characteristics.

Structural Insulation Boards. Structural insulation boards are made by a process similar to that used for insulation board, with the exception of another additive which provides additional weight, strength, and water resistance. The additive is normally asphalt, which is added in a pulverized form in the additive stock chest.

Secondary Treatments and Uses. Insulation boards normally have few secondary treatments. Insulation boards are used for economical, insulative wall paneling, ceiling tiles, bulletin boards, and similar uses. Structural insulation boards are used primarily for wall sheathing in constructions where wall diaphragm racking resistance is provided by other means.

Specifications, Standards, Quality Control, and Health and Safety Factors. Specifications and standards are found in American National Standards Institute (ANSI) standard for *Cellulosic Fiberboard.*

Hardboard and Hardboard Siding

Production, Processing, and Shipment. Hardboards and hardboard siding are fiber-base panel products having densities in the 500–1000-kg/m^3 range. Hardboards are generally thin products, 2.5–9.5 mm in thickness, whereas the siding products are usually 11.1–12.7 mm in thickness.

The first industrial hardboard was developed by W. Mason in the mid-1920s. The product was patented in 1928, trademarked as Masonite. Over time several other processes for producing hardboards have been developed from modifications of the original process.

The Mason (Masonite) Process. In this system, wood chips are sealed in a small steel digestor and steamed at high pressure. The fibers move in a slurry from a stock chest into a Fourdrinier-type mat-forming machine. The mats are transferred from the Fourdrinier wire screen to another wire screen that carries the mat through the pressing process. The mats and screen move into a press loader.

The press closes and the initial pressure and heat force much of the water held in the fiber out of the mat and press as water or steam. The press continues to close until the target thickness and density of the product are reached. The press then maintains this position as the mats continue to absorb heat from the press. The heat converts the remaining water to steam as the temperature in the pressed mat rises above the boiling point of water, 100°C. The steam pressure forces steam through the wire screen under the mat and toward the edges of the press. In this manner the mat continues to dry, and two natural bonding processes take place. When the pressed mats are dry and the press is opened, strong, stable, and durable panels are removed from the press without the requirement of added binder. The Masonite process is an example of a wet–wet hardboard process; that is, the mat is formed wet and pressed wet.

Wet-Process Hardboard From Refined Fiber. This hardboard is another form of wet–wet hardboard. The principal difference between this process and the Masonite process lies in the method of preparing fiber and the additives used in the boards.

Dry-Process Hardboard. Dry-process hardboard is produced by a dry-dry system where dry fiber is formed into mats, which are then pressed in a dry condition. In this process, wood chips, sawdust, or other residues are refined to fiber in pressurized refiners. Wax and PF resin may be added in the refiner or immediately outside of the refiner, in the fiber-ejection tube or "blowline."

As the fiber exits the refiner, it moves into and through the blowline. From the blowline, the fiber is blown into a tube dryer. From the tube dryer, fiber drops into a cyclone which separates the fiber from the moist, warm air.

Mat forming is usually accomplished by 2–5 similar machines in sequence, each depositing a portion of the fiber. A plastic or wire screen moves under each forming head, which results in a cloud of fiber falling onto the screen. Under the screen are suction boxes which aid in pulling the loose fiber toward the screen and providing a small amount of mat densification. As the mat emerges from the last forming machine and scalper, it should contain enough fiber per unit area to produce the desired thickness and density of the final product. The mats are moved along the line to the press loader. Because no screens are used, the products are called smooth-two-sides (S-2-S).

Wet–Dry Hardboard. Wet–dry hardboards are a special class of boards in which the fiber is processed through the mat-forming stage in the same manner as in the case of wet-process board. However, an emulsion of a drying oil such as linseed oil is used as the binder and is applied in the stock chest preceding the former.

Dry–Wet Hardboard. Dry–wet hardboards are the other possible manufacturing alternative. Fiber is processed as in the dry process up through the mat-forming stage. PF binders are used in this system. Then, as the mats are ready to enter the press loader, a large quantity of water is sprayed onto the surface of the mat. This water saturates the top surface of the mat and is sufficient to raise the total moisture content of the mat to 20–40% mc, depending on the thickness and density of the product.

Hardboard Siding. Hardboard siding is intended for use as an exterior siding material for buildings. These products have been made using all of the previously outlined hardboard processes, using PF or drying oil binders.

Regardless of which of the various types of production processes hardboard panels follow, processing and shipment are generally quite similar for each kind of hardboard panel. Depending on the projected end-use of the panels, some will be tempered or heat-treated immediately after pressing. These treatments provide an added measure of resistance to water, or retard the rate at which the products will absorb water during exposure.

Upon completion of the various treatments, panels or strips are stacked, wrapped, labelled, and shipped.

Secondary Treatments and Uses. Because hardboard products are utilized in many different ways, the variety of secondary treatments used by customers are practically unlimited. Post-treatments may include cutting-to-size, finishing treatments with roll-applied patterns, melamine overlays, printed paper overlays, paints, and even some extremely durable and water-resistant coatings used in tub and shower linings or other uses where water contact is frequent and extreme.

Specifications, Standards, Quality Control, and Health and Safety Factors. Specifications and standards are contained in several ANSI standards. These standards define the various hardboard product categories as well as specific product qualities required for each group. In addition to the previously noted safety factors associated with these processes, there are additional needs for dust control and ventilation for dissipation of various vapors from pressing, tempering/heat treatment, and machining and finishing operations.

Particleboard

Production, Processing, and Shipment. Particleboards are composites made from particles or small pieces of wood or other lignocellulosic residues, in contrast to the fibers used in the various types of hardboard; and the former are bonded together with an adhesive under heat and pressure. Particleboards are made in thicknesses of 3.2–44.4 mm (1/8–1 3/4 in.). Particleboards are made across a wide range of densities, 415–1000 kg/m^3.

To begin the process, the particulate raw materials are rough-screened to remove oversize materials. The material then moves to the milling/drying/screening area (material preparation area). From this point to the mat-formers, materials usually flow in two streams, one of smaller particles for the surfaces of the board and one of larger particles for the core or middle of the board. Milling generally precedes drying.

Drying is almost always done in large, rotating drum dryers.

A measurement of particle moisture content will normally be taken at the exit of the dryer. At this point, the dry particles and flakes may pass through a system designed to remove grit. Grit and metal removal equipment is a necessity for those mills reusing recycled urban waste woods.

The dry material then passes to the screening area where separations are made, based primarily on suitability as surface (fine fractions) or core (coarse fractions) materials. After screening, the materials are stored in large bins or silos.

From the storage units, the raw materials are metered at desired and uniform flow rates to blending area. Additives are metered into the blender and are mixed by the paddles as the material proceeds through the blender.

A small amount of water may also be added to improve blending or to assure proper furnish (blended materials) mc. The scavenger, normally urea, is added to react with excess formaldehyde from the resin and prevent excessive formaldehyde emissions from the product. Catalyst is used to hasten the cure of the core adhesive, along with the possible use of small amounts in the surface.

The blended materials, now called furnish, move from the blenders to the formers. After the mat is formed, it is trimmed to width and length.

The pressing operation in a composites facility is a critical one. The press cycle should be as short as possible while producing high quality board and maintaining a consistently uniform operation. A recent development is the continuous press, which has two opposing heated steel belts which transport, press, and cure the adhesives as the mat is moved through the press. The press may operate over a wide range of temperatures, a range of 154–177°C being most common.

As the panels leave the press area, they may pass through two sensing units. The first is a blow detector, which can locate delaminations that may or may not be seen by visual inspection. The second unit is an automatic thickness sensor which, by means of several sensing heads across the board, measures and averages the thickness across and along the board.

Panels then move into a cooling device, where they are held individually and air is circulated between them. From these stacks the boards are sanded to final thickness.

Boards then proceed to the final processing steps, including trimming to the exact desired sizes, stacking, strapping, and shipping.

Secondary Treatments and Uses. Particleboards are the jack-of-all-trades products for interior use panels. They have been and are being used for virtually every conceivable interior use for which panels or strips of material can be used. The principal desirable features of particleboards are flat, smooth, warp-free, stable, and cost-efficient panels. A small amount of particleboard is made with a fire-retardant treatment for use in locations where codes require this material. A small amount of particleboard is also made in the form of shaped, molded articles such as furniture parts, paper roll plugs, brush bases, and even toilet seats.

Specifications, Standards, Quality Control, and Health and Safety Issues. Specifications, standards, and quality control (QC) procedures are outlined in *Particleboard,* ANSI A208.1-1993.

In addition to the always-present concerns for worker safety around hot or moving machinery, additional measures are warranted by the dusty conditions that generally prevail around a particleboard mill. Constant clean-up is necessary to prevent excessive dust buildup, which left unchecked can become a fire and explosion hazard. Some workers in dusty areas wear dust masks. Most workers wear safety glasses and often use hearing protection devices, safety shoes, and hardhats.

Medium-Density Fiberboards

Production, Processing, and Shipment. Medium-density fiberboards (MDF) are panels made of fibrous raw material and used in most of the same applications as particleboard. MDF products generally have more smooth surfaces and edges than particleboards and are thus preferred for some uses. The manufacture of MDF, with a few exceptions, duplicates the manufacture methods for dry-process hardboard.

The major difference from dry-process hardboard manufacture is in the pressing area. Because MDF is usually a lower density product, lower pressing pressures are needed. The use of radio-frequency (RF) heating allows the center of the board to heat almost as quickly as the surface, and this occurs during closure of the press. Thus, the heated core, compresses almost as easily as the faces, resulting in a quite uniform density profile through the thickness of the board. The other major benefit of RF heating is in reduced presstimes. After pressing, the MDF process basically duplicates the particleboard process with the steps of cooling, sanding, trimming, cut-to-size, stacking, strapping, and shipping.

Secondary Treatments and Uses. MDF competes with particleboard in virtually every application, especially in furniture and cabinetry, but not as floor underlayment or decking.

MDF is an extremely expensive product and thus is used only for special applications requiring its special properties.

Specifications, Standards, Quality Control, and Health and Safety Aspects. Specifications and standards are found in *Medium-Density Fiberboard (MDF),* ANSI A208.2-1994.

The health and safety issues outlined herein for particleboard also apply to MDF. However, there exists in MDF a large component of very small, broken, dust-like wood fibers.

Waferboards and Oriented Strand Boards

Production, Processing, and Shipment. The waferboard and oriented strand board (OSB) industries are based upon the use of special forms of wood flakes generated from small logs. A flake is a long, flat section of wood which may be 25–100 mm (1–4 in.) in length and in the grain (longitudinal) directions of the wood. Thickness may be 0.25–1.00 mm (0.010–0.040 in.), and width is usually variable. Normally, a good flake has a length-to-thickness ratio of at least 100.

Wafers as used in the industry are large, flat flakes of about 0.6–1.0 mm (0.025–0.040 in.) in thickness, 38–50 mm (1 1/2–2 in.) in length, and 13–50 mm (1/2–2 in.) in width. Strands are long, narrow flakes of about 0.6–1.0 mm (0.025–0.040 in.) in thickness, 75–100 mm (3–4 in.) in length, and 6–25 mm (1/4–1 in.) in width.

Excellent strength and stability can be achieved with boards made from flakes. However, the requirement of solid wood as a starting material and uneven swelling properties of the boards prevented them from being used in many particleboard applications.

In the early 1980s the concept of OSB was realized. OSB is a panel product made from wood strands and somewhat like plywood in that the strands on the two faces are oriented in the long direction of the panel and the core strands are oriented in the cross-panel direction. The use of orientation yields panels having excellent directional properties, much like plywood, and thus an excellent and economical structural sheathing material is created.

The manufacture of waferboard and OSB has many of the same process steps as particleboard, but adapted to the special needs of producing an exterior quality panel with large wafers or strands. Waferboard has been almost entirely replaced by OSB, and most of the early waferboard mills have now been converted to production of OSB.

The process begins with small logs of almost any species. The logs to be processed are debarked and the bark used for fuel. The debarked logs are placed in soaking tanks. As the logs reach the end of the soaking tanks, they are removed and prepared for the flakers. Flakers are of two types, the first using huge rotating disks containing a number of long knives extending outward from the axis of the disk. Log sections are fed into the disk and each knife removes a slice from the log. Spur knives cut the slice into many separate lengths, and these pass through an opening in the disk. Another type of flaker is a revolving ring flaker. The ring is about 600 mm (24 in.) in depth, with many flaker knives and spur knives protruding inward on the inner periphery of the ring. Slices of wood are removed and are again broken into strands as they pass through the ring and fall into a transport conveyor. It is possible to flake whole logs one section at a time in both of these flaker types.

The strands move through large drum dryers which reduce the moisture content. From the dryer, the strands are screened to remove fines and small particles, which would detract from board quality and economy. The strands are metered from the bins into the blenders, which are large rotating drums which tumble the flakes as they move from the infeed to the exit end. Resin adhesive and wax are applied. Other liquid additives are normally applied by centrifugal disk atomizers rotating at high speed. Wax is applied to provide water repellancy in the finished product. Another binder being used in some locations is isocyanate resin. Isocyanates have several desirable features, such as excellent bonding abilities, high moisture tolerance, faster curing than PF, and lower adhesive usage than PF.

Strands move from the blenders to the formers. Mats are formed on wire-screen cauls. As the screen-cauls pass under the formers, the lower surface strands are placed on the screen-caul such that they are generally aligned with the long direction of the screen. The core strands are then laid down so that their lengthwise dimension is generally perpendicular to the surface strands. Finally, the top layer of strands is applied to the mat, also in the long direction of the screen and mat.

Mats then move to the press loader and into the press. After the press, the panels are immediately cut into ordered sizes. Panels are then printed with appropriate use information, stacked, and usually edge coated with a moisture-proof barrier coating prior to banding and shipping.

Secondary Treatment and Uses. Most OSB panels are used "as is," without further processing or treatment. Primary uses are as wall and roof sheathing, floor decking, and other construction panel uses in home and commercial construction.

The major secondary use of OSB has been as exterior wall-paneling or siding.

Specifications, Standards, Quality Control and Health and Safety Aspects. The OSB industry is represented by the APA—The Engineered Wood Association in areas of specifications and standards. These are outlined in APA *Performance Standard for Wood-Based Structural-Use Panels,* PS2-92.

Health and safety factors in the OSB industry are similar to those in other composite mills. An area of special concern in some OSB mils, those using isocyanate adhesives, is awareness of the toxic nature of this adhesive in the uncured state and the requirements of personal care, housekeeping, and ventilation and air handling in the process areas from blending through pressing.

Structural Composite Lumber

Production, Processing, and Shipment. Structural composite lumber is a group of composite or veneer products which can be used in place of structural lumber in many applications. While composites in general may not be as strong as clear, straight-grain solid wood, they may be significantly stronger than knotty or angled-grain lower grades. Also, the variation in properties is much less than in run-of-the-mill solid lumber. There are three distinct types of structural composite lumber.

Laminated Veneer Lumber. Laminated veneer lumber was introduced in the late 1960s. The product was made of many layers of 2.5–3.2 mm (1/10–1/8 in.) veneer bonded with PRF adhesive in a long continuous press. Since that time, other manufacturers have entered the laminated veneer business. Properly made, these materials approach the load carrying abilities of high grade solid lumber.

Veneer-Composite I-Beams. One of the earliest applications of laminated veneer lumber was as flange stock in a veneer-base I-beam. Narrow strips, 50–100 mm (2–4 in.) were cut from the 1220 mm (4 ft.) wide panel and a groove about 9.5 mm (3/8 in.) wide and deep was cut in the surface of two of these strips, and the two strips were placed in a machine with a set distance separation between them. As the strips moved through the machine, a PRF adhesive was metered into the grooves. Then cross-cut sections of 9.5 mm (3/8 in.) plywood were placed in the grooves end-to-end to make web-stock for the I-beam. Beams of almost any grade and size can be made by changing the width of the flange and with width of the web material. In recent years, OSB has replaced much of the plywood as web material in these I-beams.

Laminated Strand Products. The most recent developments in the family of wood-based composites comprises a group of laminated strand products, made with strands oriented in the long direction of the product and marketed as structural composite lumber. One product is made with long, narrow strips of softwood veneer, coated with a PRF adhesive, and pressed under heat and pressure into large blocks.

Another product is made with strands produced by flakers. The strands are dried and then coated with isocyanate adhesives.

Secondary Treatments and Uses. The structural composites lumber group requires little secondary processing. The products are being used in most of the applications formerly and almost solely filled by large structural timbers and large-dimension lumber.

Specifications, Standards, Quality Control, and Health and Safety Aspects. In general, structural composites lumber products are tested and rated in the same manner as structural lumber. Terminology, procedures for determining allowable design stresses, and quality assurance requirements are set forth in the ASTM *Standard Specification for Evaluation of Structural Lumber Composite Products* (D5456-93).

Health and safety concerns in the structural composite lumber industry are similar to those in the other composite industries.

WILLIAM F. LEHMANN
Consultant

For more information, see publications of the following associations.

American Hardboard Association (AHA), 887-B Wilmette Road, Palatine, IL 60067, Tel. (708) 934-8800.

APA—The Engineered Wood Association, 7011 So. 19th St., Tacoma, WA 98411, Tel. (253) 565-6600.

National Particleboard Association, 18928 Premiere Court, Gaithersburg, MD 20879, Tel. (301) 670-0604.

The Hardwood Plywood and Veneer Association (HPVA), 1825 Michael Faraday Drive, P.O. Box 2789, Reston, VA 22090, (703) 435-2900.

WOOD PULP. See PULP.

WOOL

Wool is the fibrous covering from sheep and is by far the most important animal fiber used in textiles. Wool belongs to a family of proteins, the keratins, that also includes hair and other types of animal protective tissues such as horn, nails, feathers, and the outer skin layers. It is also an extremely important export for several nations, notably Australia, New Zealand, South Africa, and Argentina and commands a price premium over most other fibers because of its outstanding natural properties of soft handle (the feel of the fabric), moisture absorption abilities (and hence comfort), and superior drape (the way the fabric hangs) (see FIBERS; TEXTILES).

The principal characteristics of clean wool types are average diameter, measured in micrometers, and average length, measured in millimeters (Table 1).Raw wool from sheep contains other constituents considered contaminants by wool processors. These can vary in content according to breed, nutrition, environment, and position of the wool on the sheep. The main contaminants are a solvent-soluble fraction called wool grease; protein material; a water-soluble fraction (largely perspiration salts collectively termed suint); dirt; and vegetable matter, eg, burrs and seeds from pastures.

Fiber Growth

Wool follicles are produced from multicellular tube-like structures known as follicles. These follicles are located in the skin layers (dermis and epidermis) of sheep, and are of two types, primary and secondary. The primary follicles develop first, in the unborn lamb. Secondary follicles develop later, and in finer wooled sheep derived secondary follicles subsequently form by branching from the original secondaries with which they share a common orifice. Each primary follicle has a sebaceous gland and a sweat gland together with an arrector muscle, whereas secondary follicles usually have only an associated sebaceous gland.

Fiber Morphology

Wool fibers consist of flattened overlapping cuticle cells that form a protective sheath around cortical cells. In some coarser fibers, a central vacuolated medullary cell type may be present. Sections of cuticle cells show an internal series of laminations comprising outer sulfur-rich bands known as the exocuticle and inner regions of lower sulfur content called the endocuticle. On the exposed surface of cuticle cells, a membrane-like proteinaceous band (epicuticle) and a unique lipid component form a hydrophobic resistant barrier. These lipid and protein components are the functional moieties of the fiber surface and are important in fiber protection and textile processing.

The cortex comprises the main bulk and determines many mechanical properties of wool fibers. Cortical cells consist of a class of biological filaments known as intermediate filaments embedded in a

Table 1. Main Types of Wool

Wool type	Breed examples	Fiber diameter, μm	Fiber length, mm
fine	merino	10–30	35–100
medium	cheviot	20–40	50–100
	dorset		
	hampshire		
	southdown		
crossbred	columbia	20–30	75–150
	corriedale		
	polwarth		
	targhee		
coarse (long)	cotswold	25–50	125–350
	leicester		
	lincoln		

sulfur-rich matrix. The intermediate filaments (originally termed microfibrils) together with the matrix are organized into large macrofibrillar units. The main cortices are designated as ortho and para, according to their intermediate filament/matrix packing arrangements. The arrangement of ortho- and paracortical cells differs among wool types.

Between cuticle cells and cortical cells a continuous intercellular material is present which, despite being a relatively minor fraction of the total fiber weight, is of increasing interest owing to its presumed role in water and reagent-penetration of fibers. This region of the fiber is known as the cell membrane complex and comprises the intercellular material together with the apposing cellular membranes of cuticle and cortical cells.

Physical Properties

Moisture Sorption (Regain). One of wool's major attributes as a textile fiber is its ability to absorb and desorb large amounts of moisture as the relative humidity surrounding the fiber changes. The absorption of water by wool results in improved comfort during wear, with the heat of absorption buffering the wearer against sudden environmental changes; during active sports, perspiration transport away from the skin is enhanced. As can be seen from Table 2, most physical properties of the fiber are affected by the moisture level.

Diameter Distribution. Wool fibers exhibit a range of diameters from coarse fibers (25 μm–70 μm), used in carpets, to fine Merino fibers (10 μm–25 μm), used in apparel. Coarse fibers rather than buckling, indent the skin and activate nerve receptors. This gives rise to a sensation of prickle and itch that has been incorrectly assumed, by some consumers, to be an allergic reaction. The irritation is mechanical and not immunological.

Table 2. The Effect of Moisture Content on the Physical Properties of Wool Fibers at 25°C

	Moisture content, %							
Property	0	5	10	15	20	25	30	33
relative humidity, % (absorption)	0	15	42	68	85	94	99	100
relative humidity, % (desorption)	0	8	32	58	79	92	98	100
specific gravity, kg/m^3	1.304	1.314	1.315	1.313	1.304	1.292	1.277	1.268
volume swelling, %	0	4.24	9.07	14.25	20.0	26.2	32.8	36.8
length swelling, %	0	0.55	0.93	1.08	1.15	1.17	1.18	1.19
radial swelling, %	0	1.82	4.00	6.32	8.88	11.69	14.57	16.26
heat of wetting^a, kJ/kg wool	101	64.4	38.1	20.5	10.0	4.2	1.13	0
Young's modulus (relative)	1.00	0.96	0.87	0.76	0.66	0.56	0.44	0.38
torsional rigidity modulus, GPa^b	1.76	1.60	1.26	0.90	0.50	0.28	0.16	0.11
electrical resistivity, MΩ·m		3×10^4	400	8	0.40	0.06		
dielectric constant at 10^4 Hz	4.6	5.1	6.2	8.3	12.8			

^{a} Heat evolves when wool, dry mass of 1 kg, at a particular regain is immersed in water.
^{b} To convert GPa to psi, multiply by 145,000.

Felting. Because wool fiber consists of overlapping cuticle cells, the protruding scale edges result in differential friction between the with-scale and against-scale direction. As a consequence, wool can be readily felted to produce a dense matting of fibers for use as hats, polishing pads, table covers, and piano hammers and dampers.

Thermal Properties. The regular packing of α-helical polypeptide chains within the intermediate filaments forms a crystalline phase that occupies about 70% of the dry volume of the fiber. This phase melts irreversibly at a temperature that is both time- and regain-dependent. The amorphous matrix phase contains a high concentration of the amino acid cystine and is therefore highly cross-linked. Similarly to other amorphous materials, a glass transition, T_g, has been detected in the wool fiber. Water acts as a plasticizer, lowering the glass-transition temperature of the dry fiber from 165°C to below −10°C when wet. The glass-transition temperature is an important parameter, as the properties and performance of wool are influenced by the environmental conditions (temperature and humidity) relative to the glass transition. The viscoelastic properties, physical aging and the water absorption isotherm of wool are also influenced by the glass transition.

Mechanical Properties. The mechanical properties of wool are largely understood in terms of a two-phase composite model. Water-impenetrable crystalline regions (generally associated with the intermediate filaments) oriented parallel to the fiber axis are embedded in a water-sensitive matrix to form a semicrystalline biopolymer. The parallel arrangement of these filaments produces a fiber that is highly anisotropic. Whereas the longitudinal modulus of the fiber decreases by a factor of 3 from dry to wet, the torsional modulus, a measure of the matrix stiffness, decreases by a factor of 10.

Chemical Structure

Wool belongs to the family of proteins (qv) called keratins. However, morphologically the fiber is a composite and each of the components differs in chemical composition. Principally the components are proteinaceous, although wool cleaned of wax, suint, and other extraneous materials acquired during growth contains small amounts of lipids (structural and free), trace elements, and, in colored fibers, pigments called melanin.

Protein Components. Complete acid hydrolysis of wool yields 18 amino acids, the relative amounts of which vary considerably from one wool to another. The side groups of the amino acids vary markedly and play an important role in the chemical reactions of the fiber. For example, the sulfur containing cysteine provides a disulfide bridge (–S–S–) between different polypeptide molecules or between segments of the same molecules. The disulfide bridges prevent movement of chains and chain segments, and thus are responsible for the high form stability of wool fabrics. The disulfide cross-links are readily rearranged under the influence of heat and moisture, or alkaline reducing agents, a property employed in the permanent setting of wool fabrics.

Lipids. The lipids of the fiber are believed to be constituents of the cell membranes (see FIBERS). They constitute about 1% of the fiber but play an important role in many properties, such as the intercellular diffusion of reagents. The free lipids consist of fatty acids, sterols, and trace amounts of glycerides, sphinolipids, and glycolipids. In addition to the free lipids, wool contains some lipids that are assumed to be covalently bound to proteins. Covalently bound surface lipids account for the hydrophobicity of the fiber surface.

Wool Processing

The conversion of raw wool into a textile fabric or garment involves a long series of separate processes. There are two main processing

Worsted

Sorting → Blending → Scouring → Drying → Carding → Gilling → Combing → Gilling → Drawing → Spinning → Twisting → Winding → Weaving/knitting → Fabric finishing

Woolen

Sorting → Blending → Scouring → Drying → Carding → Spinning → Winding → Weaving/knitting → Fabric finishing

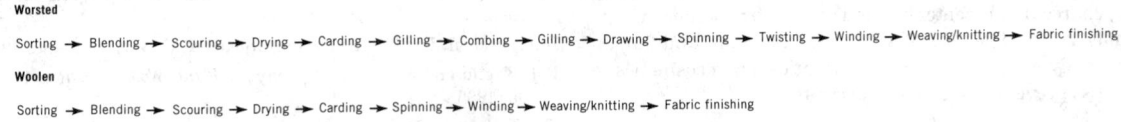

Figure 1. The principal stages in the worsted and woolen systems.

systems, ie, worsted and woolen, although an appreciable volume of wool is also processed on the short-staple (cotton) system or on the semiworsted system for carpet use. The principal stages in the woolen and worsted systems shown in Figure 1.

Dyeing

Wool is dyed from aqueous solutions. The majority of dyes used on wool are sodium salts of aromatic anions. Water solubility is usually provided by sulfonic acid groups, but in a few cases carboxyl or hydrophilic, nonionic substituents are used. The affinity and wetfastness properties of dyes on wool are largely determined by van der Waals forces and interactions between hydrophobic regions in the fiber and hydrophobic parts of the dye molecules.

Insectproofing

Wool, as a keratin, is a highly cross-linked, insoluble proteinaceous fiber, and few animals have developed the specialized digestive systems that allow them to derive nutrition from it. In nature, these few keratin-digesting animals, principally the larvae of clothes moths and carpet beetles, perform a useful function in scavenging the keratinous parts of dead animals and animal debris (fur, skin, beak, claw, feathers) that are inaccessible to other animals. It is only when these keratin digesting animals attack processed wool goods that they are classified as pests. The principal insects that attack wool are the common clothes moth, the case-bearing clothes moths, the brown house moth, and the variegated carpet beetle. Usually, an insecticide is incorporated into the wool from the dyebath.

The use of the dyebath as an application medium is attractive, not only because it avoids the needs for an additional wet processing step, but also because the pesticide has a high probability of being taken inside the highly swollen fiber. This ensures that the pesticide will be relatively fast on the wool, as diffusion, especially from the dry fiber, will be slow. As an example, permethrin, the main active agent currently used for insect-resist treatments of wool goods, has a half-life of only 19 d in storage at 60°C when sprayed onto wool, but its lifetime exceeds the usual life of wool goods when applied from boiling dyebath. With the pesticide located inside the fiber, the active agent is released primarily in the insect gut when the fiber is completely degraded. Unfortunately, it is impossible to ensure 100% exhaustion (transfer of pesticide from dyebath to fiber) and as a result, there is inevitably some environmental contamination upon disposal.

Shrink-Resist Treatment

When a wool fabric or yarn is immersed for the first time in water, it contracts owing to a combination of reversible hygral expansion and irreversible relaxation shrinkage. Relaxation shrinkage results from the release of strains imparted during previous processing operations such as spinning, knitting, and fabric finishing. Felting shrinkage, characterized by the irreversible migration of individual fibers, occurs subsequently when the relaxed fabric is subject to more severe mechanical action as in laundering, but it can also occur in other operations which involve mechanical action, such as tumble drying or dry cleaning. Shrink-resist treatments are essentially directed at the prevention of felting shrinkage, whereas control of relaxation shrinkage requires attention to processing, particularly to fabric-finishing routines.

Table 3. Some Physical and Chemical Data for Wool Wax

Property	Value
density, at 15°C	0.94–0.97
refractive index, at 40°C	1.48
mp, °C	35–40
free acid content, %	4–10
free alcohol content, %	1–3
iodine value (Wijs)	13–30
saponification value	95–120
mol wt[a]	790–880
fatty acids, %	50–55
alcohols, %	50–45
acids	
mp, °C	40–45
iodine value (Wijs)	10–20
mean mol wt	330
alcohols	
mp, °C	55–65
iodine value, Wijs	40–50
mean mol wt	370

[a] Rast method, in phenyl salicylate.

Mechanistically, shrink-resist treatments are often divided into degradative and additive types. Degradative treatments aim to remove the differential friction effect of the surface scales by the use of an oxidizing agent to modify or remove the fiber scales, and in some cases a polymer is also applied to the surface to mask the scales. The additive approach prevents the migration of fibers by bonding fibers together with elastomeric polymers.

Wool Grease

In wool scouring, the contaminants on the wool, mainly grease, dirt, suint, and protein material, are washed off the fiber and remain in the wastewaters either in emulsions or suspension (grease, dirt, protein) or in solution (suint). Centrifugal extraction of the wastewaters produces a grease contaminated with detergent and suint. This product is called wool grease.

Lanolin is wool grease that has been refined to lighten its color and reduce its odor and free fatty acid content. Wool wax is the pure lipid material of the fleece, yellow to pale brown in color and extractable with the usual fat solvents such as diethyl ether and chloroform. Wool grease is a mixture of compounds that are classed as waxes but lack the physical characteristics usually displayed by waxes. Table 3 gives some physical and chemical data for wool wax.

Chemical Composition. Wool wax is a complex mixture of esters of water-soluble alcohols and higher fatty acids with a small proportion (ca 0.5%) of hydrocarbons.

Wool-Grease Recovery. The principal recovery process in use involves centrifuging in a cream-separator type of centrifuge modified by the addition of peripheral jets or other mechanical devices to remove dirt.

Grease Refining and Fractionation. The refining process most commonly used involves treatment with hot aqueous alkali to convert free fatty acids to soaps, followed by bleaching, usually with hydrogen peroxide, although sodium chlorite, sodium hypochlorite, and ozone have also been used. Other techniques include distillation, steam stripping, neutralization by alkali, liquid thermal diffusion, and the use of active adsorbents, eg, charcoal and bentonite, and solvent fractionation.

Uses of Wool Grease. The uses of wool grease, lanolin, and lanolin derivatives are many, ranging from pharmaceuticals and cosmetics to printing inks, rust preventatives, and lubricants.

Acknowledgment

Much of the work supporting the findings in this article was carried out at CSIRO Wool Technology in Geelong, Australia. Funding for this research came from IWS and the Australian Government.

PATRICK T. NAUGHTIN
BRETT O. BATEUP
JOHN R. CHRISTOE
ALLAN G. DE BOOS
RON J. DENNING
DAVID J. EVANS
G. BRUCE GUISE
BARRY V. HARROWFIELD
BILL HUMPHRIES
MICKEY G. HUSON
FRANCIS W. JONES
LES N. JONES
KEITH R. MILLINGTON
DAVID G. PHILLIPS
TONY J. PIERLOT
JOHN A. RIPPON
ROB A. ROTTENBURY
IAN M. RUSSELL
CSIRO Wool Technology

Textile Terms and Definitions, 10th ed., Textile Institute, Manchester, England, 1996.

International Wool Textile Organization (IWTO) Wool Statistics 1994–1995.

I. J. Kaplin and K. J. Whiteley, *Proc. 7th Int. Wool Text. Res. Conf. (Tokyo)* **1,** 95 (1985).

Top-Tech '96 Papers, Seminar Proceedings, CSIRO, Geelong, Australia, 1996.

X

XANTHATES

The salts of the *O*-esters of carbonodithioic acids and the corresponding *O,S*-diesters are xanthates. The free acids decompose on standing.

Properties

The free xanthic acids are unstable, colorless, or yellow oils, and may decompose with explosive violence. They are soluble in the common organic solvents and are slightly soluble in water: methyl xanthic acid at 0°C, 0.05 mol/L; ethyl xanthic acid at 0°C, 0.02 mol/L; and *n*-butyl xanthic acid at 0°C, 0.0008 mol/L. Values for the dissociation constant for ethyl xanthic acid are $(2.0–3.0) \times 10^{-2}$. Potentiometric determinations for $C_1–C_8$ xanthic acids show a decreasing acid strength with increasing molecular weight. The alkali metal salts, in contrast to the free acids, are relatively stable solids, are pale yellow when pure, and sometimes have a disagreeable odor.

When exposed to air, the sodium salts tend to take up moisture and form dihydrates. The alkali metal xanthates are soluble in water, alcohols, the lower ketones, pyridine, and acetonitrile (Table 1). They are not particularly soluble in nonpolar solvents.

The heavy metal salts, in contrast to the alkali metal salts, have lower melting points and are more soluble in organic solvents. They are slightly soluble in water, alcohol, aliphatic hydrocarbons, and ethyl ether. Alkalies stabilize xanthate solutions somewhat and the solutions readily decompose at acidic pHs.

Reactions. The chemistry of the xanthates is essentially that of the dithio acids.

Table 1. Solubilities of Some Alkali-Metal Xanthates

		Solubility, g/100 g soln			
Xanthate	Solvent	0°C	10°C	20°C	35°C
sodium ethyl	water	40.8	46.0	52.0	
potassium *n*-propyl	water	43.0			58.0
	n-propyl alcohol	1.9			8.9
sodium *n*-propyl	water	17.6			43.3
	n-propyl alcohol	10.2			22.5
potassium isopropyl	water	16.6			37.2
	isopropyl alcohol				2.0
	IPA-H_2O azeotrope			6.9	
sodium isopropyl	water	12.1			37.9
		24.5	27.3	30.5	
		24.0	27.5	31.0	37.5
	isopropyl alcohol				19.0
potassium *n*-butyl	water	32.4			47.9
	n-butyl alcohol				36.5
sodium *n*-butyl	water	20.0			76.2
	n-butyl alcohol				39.2
potassium isobutyl	water	10.7			47.7
	isobutyl alcohol	1.6			6.2
sodium isobutyl	water	11.2			33.4
		46.2	48.2	50.5	
		44.0	49.0	51.0	57.3
	isobutyl alcohol	1.2			20.5
sodium *sec*-butyl	water	29.4	34.0	38.8	
potassium isoamyl	water	28.4			53.5
		16.9	26.0	35.0	
		28.5	39.0	45.6	52.5
	isoamyl alcohol	10.9			15.5

Peroxyxanthates. A new factor in the theory and practice of flotation was found in the Mount Isa, Australia, copper flotation solution. Secondary butyl perxanthate was formed by the reaction of the xanthate with hydrogen peroxide in dilute alkaline aqueous solution and was found to be identical to a substance from the flotation solution. The perxanthate was isolated as the ammonium salt.

Preparation and Manufacture

The alkali metal xanthates are generally prepared from the reaction of sodium or potassium hydroxide with an alcohol and carbon disulfide.

Many of the heavy metal xanthates have been prepared from aqueous solutions of the alkali metal xanthates and the water-soluble compound of the heavy metal desired.

Alkali Metal Xanthates. The commercially available xanthates are prepared from various primary or secondary alcohols. The alkyl group varies from C_2 to C_5 and the alkali metal may be sodium or potassium.

The alkali metal xanthates are fairly safe to handle. The standard precautions of rubber gloves, dust mask, and goggles are sufficient when handling the solid or the solution.

Under regulations for the enforcement of the Federal Insecticide, Fungicide, and Rodenticide Act, products containing over 50 wt % sodium isopropyl xanthate must bear the label "Caution. Irritating dust. Avoid breathing dust, avoid contact with skin and eyes" (104). Rubber goods in repeated contact with food may contain diethyl xanthogen disulfide not to exceed 5 wt % of the rubber products.

Xanthate drums should be kept as cool and dry as possible. Protection from moisture is the most important factor. A combination of moisture and hot weather causes sodium ethyl xanthate to ignite spontaneously.

Environmental Concerns. Concern for the well-being of the environment has resulted in studies on the effects of mining chemicals, including the xanthates, on various aquatic organisms worldwide. In a thorough and detailed study of three typical mill operations, it was concluded that residual organic flotation reagents do not seem to present widespread problems in effluent disposal.

GUY H. HARRIS
University of California at Berkeley

G. Gattow and W. R. Bahrendt, *Topics in Sulfur Chemistry, Vol. 2: Carbon Sulfides and Their Inorganic and Complex Chemistry,* Georg Thieme, Stuttgart, Germany, 1977.

G. H. Harris, *Reagents in Mineral Technology,* Surfactant Science Series, Vol. 27, Marcel Dekker, Inc., New York, 1988, pp. 371–383.

J. Leja, *Surface Chemistry of Froth Flotation,* Plenum Press, New York, 1982.

S. R. Rao, *Xanthates and Related Compounds,* Marcel Dekker, Inc., New York, 1971.

XANTHAN GUM. See GUMS; MICROBIAL POLYSACCHARIDES.

XANTHENE DYES

Xanthene dyes are those containing the xanthylium (**1a**) or dibenzo-γ-pryan nucleus (xanthene) (**1b**) as the chromophore with amino or hydroxy groups meta to the oxygen as the usual auxochromes. They are important because of their brilliant hues; shades between greenish yellows to dark violets and blues are obtainable, the most important being reds and pinks. Xanthenes are often fluorescent, which adds to their strength and brightness; but, as is often the case with fluorescent dyes, they have lower light-fastness compared to other chromophores. Their use is concentrated on those areas in which lightfastness is relatively unimportant compared to economy (eg, paper dyes) or where lightfastness can be achieved by modification. They are used for the direct dyeing of wool and silk and mordant dyeing of cotton. Paper, leather, woods, food, drugs, and cosmetics are dyed with xanthene dyes (see DYES, APPLICATION AND EVALUATION–APPLICATION).

Brilliant insoluble lakes are used in paints and varnishes. Recent applications for xanthene dyes include use in ink-jet printers, as markers in biological and medical research, and even as insecticides.

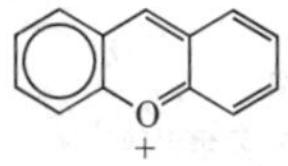

(1a)

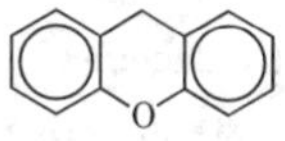

(1b)

Xanthenes date from 1871 when von Bayer synthesized fluorescein (**2**).

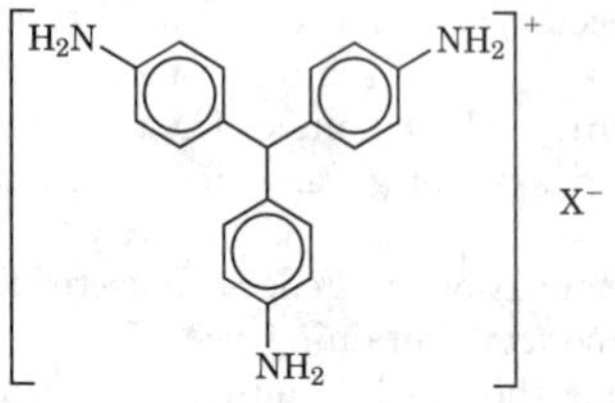

(3)

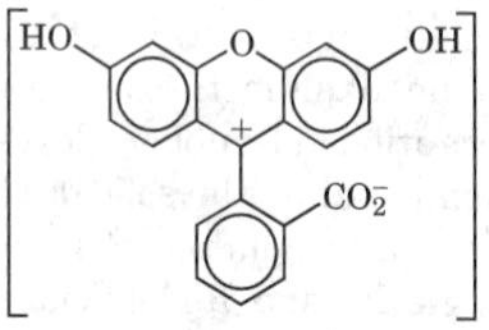

(2)

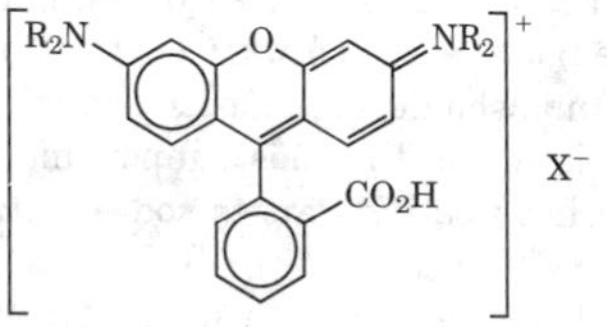

(4)

The xanthene dyes may be classified into two main groups: diphenylmethane derivatives, called pyronines, and triphenylmethane derivatives (eg, (**3**)), which are mainly phthaleins made from phthalic anhydride condensations. A third much smaller group of rosamines (9-phenylxanthenes) is prepared from substituted benzaldehydes. The phthaleins may be further subdivided into the following: fluoresceins (hydroxy substituted); Rhodamines (amino substituted), eg, (**4**); and mixed hydroxy/amino substituted.

Diphenylmethane Derivatives

Pyronines. Pyronines are diphenylmethane derivatives synthesized by the condensation of *m*-dialkylaminophenols with formaldehyde, followed by oxidation of the xanthene derivative to the corresponding xanthydrol, which in the presence of acid forms the dye (**5**). If R is methyl, the dye is pyronine G (CI 45005); if R is ethyl, pyronine B (CI 45010) is obtained.

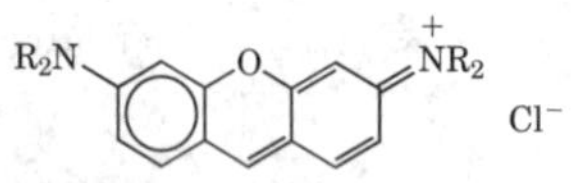

(5)

Succineins. Succineins are carboxyethyl-substituted pyronines made by substituting succinic anhydride for formaldehyde in the basic synthesis.

Triphenylmethane Derivatives

Amino-Derivatives. *Rhodamines.* Rhodamines are commercially the most important aminoxanthenes. If phthalic anhydride is used in place of formaldehyde in the above condensation reaction with *m*-dialkylaminophenol, a triphenylmethane analogue, 9-phenylxanthene, is produced. Historically, these have been called rhodamines. Rhodamine B (Basic Violet 10, CI 45170) (**6**) is usually manufactured by the condensation of two moles of *m*-diethylaminophenol with phthalic anhydride.

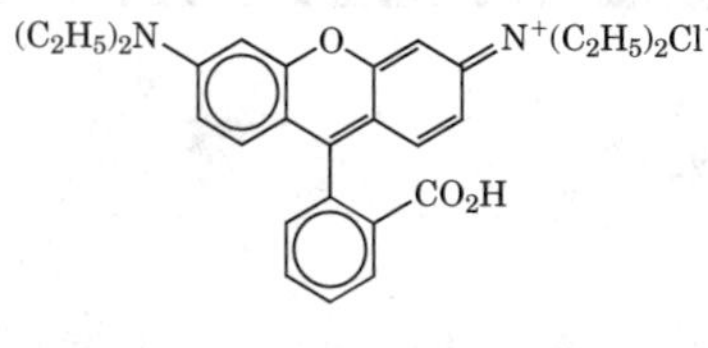

(6)

The rhodamines described thus far are basic rhodamines. They are used primarily for the dyeing of paper and the preparation of lakes for use as pigments. They are also used in the dyeing of silk and wool where brilliant shades with fluorescent effects are required, but where lightfastness is unimportant. Many new uses for rhodamine dyes have been reported. For example, when vacuum-sublimed onto a video disk, Rhodamine B loses its color to form a clear stable film which becomes permanently colored on exposure to uv light. This can be used in optical recording for computer storage or video recording. Acid rhodamines are made by the introduction of the sulfonic acid group to the aminoxanthene base.

Some acid rhodamines are used for silk and wool. Highly substituted acid rhodamines have been reported for fiber-reactive dye applications.

Rosamines. Rosamines are 9-phenylxanthene derivatives prepared from substituted benzaldehydes instead of phthalic anhydride. Sulforhodamine B (Acid Red 52; CI 45100) is the most important rosamine.

Hydroxyl Derivatives. The building block of most hydroxyl-substituted xanthenes, or fluorones, is fluorescein (**2**). The sodium or potassium salt of fluorescein, commonly called uranine (CI 45350), is used for dyeing wool and silk brilliant yellow shades. However, the principal use of fluorescein is as an intermediate for more highly substituted hydroxyxanthenes.

Aminohydroxy Derivatives

Aminohydroxy-substituted xanthenes are of little commercial importance. They are synthesized by condensing one mole of *m*-dialkylaminophenol with phthalic anhydride, and then condensing that product with an appropriately substituted phenol.

Miscellaneous Derivatives

Two additional xanthene analogues are termed fluorescent brighteners. Fluorescent Brightener 74 (CI 45550) and Fluorescent Brightener 155 (CI 45555) are used in the formulation of solid dielectric compositions for application in high voltage cables to prevent conductive treeing.

Health and Safety Factors, Toxicology

Xanthene dyes have not exhibited health or safety properties warranting special precautions; however, standard chemical labeling instructions are required.

PAUL WIGHT
Zeneca Specialties

P. Gregory, *High-Technology Applications of Organic Colorants*, Plenum Publishing Corp., New York, 1990.

K. Ventkataraman, ed., *The Chemistry of Synthetic Dyes*, Vol. 2, Academic Press Inc., New York, 1952.

H. A. Lubs, ed., *The Chemistry of Synthetic Dyes and Pigments*, American Chemical Society Monograph Series, Reinhold Publishing Corp., New York, 1955.

XENON. See HELIUM GROUP, GASES.

XEROGRAPHY. See ELECTROPHOTOGRAPHY.

X-RAY TECHNOLOGY

As with medical x-ray instruments there are analytical x-ray instruments that can produce images of internal structures of objects that are opaque to visible light. There are instruments that can determine the chemical elemental composition of an object, that can identify the crystalline phases of a mixture of solids, and others that determine the complete atomic and molecular structure of a single crystal. The determination of particle size and structural information for fibers and polymers, and the study of stress, texture, and thin films are x-ray applications that are growing in importance.

Characterization and Generation of X-Rays

X-Ray Electromagnetic Spectrum. X-rays are a form of electromagnetic radiation and have a wavelength, λ, much shorter than visible light. The center of the visible light spectrum has a wavelength of about 0.56×10^{-6} m. The range of wavelengths for x-rays used in the applications discussed here are from about 0.01×10^{-9} m to about 7.0×10^{-9} m. The most commonly used methods for generating x-rays are the synchrotron and x-ray tubes.

Synchrotron Radiation. X-rays are produced when very energetic electrons traveling close to the speed of light are decelerated. In synchrotrons, Electrons are accelerated with electromagnets while traveling along a linear path long. Then they are inserted into a nearly circular path which is maintained by bending magnets. In this circular path, the electrons lose energy by producing x-ray photons whose paths are tangential to the circle. X-ray instruments designed for many different applications are placed around the synchrotron ring.

X-Ray Tubes. X-ray tubes are the most widely used source for the generation of x-rays. In these tubes, electrons are accelerated by a high electric potential (20–120 kilovolts). These electrons strike the target (anode) of the tube and decelerate as they pass through the electron clouds of the atoms. This phenomenon produces a continuous spectrum similar, but much less intense, to that of the synchrotron. In addition, some high energy electrons knock electrons out of the atomic orbitals of the atoms of the target material. When these orbitals are refilled by electrons, x-ray photons are generated. The resulting x-ray spectrum of intensity vs wavelength has a series of peaks known as characteristic lines. The wavelengths and intensities of these lines are dependent on the elemental composition of the target material (synchrotron radiation has no characteristic lines). The most intense peaks are the two lines known as $K_{\alpha 1}$ and $K_{\alpha 2}$ (or together known as K_{α}). The next most intense lines are the K_{β} lines. The materials that are used as targets in x-ray tubes depend on the application.

Properties of X-Rays

An x-ray photon can interact with an object in the following ways: (*1*) The x-ray photon is transmitted through the object without any interaction neither the energy nor the direction of the photon changes. (*2*) The x-ray photon is completely absorbed. All of the energy of the x-ray photon is transferred to the electrons within the object. (*3*) The x-ray photon is absorbed and another x-ray photon of longer wavelength (lower energy) is produced. An electron (or electrons) gains the lost energy. The new x-ray photon may travel in any direction from the site of the event. This phenomenon is known as incoherent, inelastic, or Compton scattering. Because there is a transfer of energy at a localized position within the object the incident x-ray photon can be considered to behave as a particle. (*4*) The x-ray photon produces an oscillating electric field in the object; thereby generating a photon of the same wavelength. The resulting photon may leave the object at many different angles with respect to the incident photon. No energy is transferred to the object. This phenomenon is known as coherent, elastic, or Rayleigh scattering. The incident x-ray photon can be considered to behave as a wave and the phenomenon of diffraction takes place.

X-ray applications can be placed into three categories based on which of the above phenomena are exploited x-ray radiography permits the imaging of the internal structure of an object (eg, bones of a hand). It is based on the comparative observance of photons (phenomena 1 and 2 above) as they travel through the different materials making up the object.

X-ray fluorescence spectrometry consists of the measurement of the incoherent scattering of x-rays (phenomenon 3). It is used primarily to determine the elemental composition of a sample. X-ray diffraction consists of the measurement of the coherent scattering of x-rays (phenomenon 4). X-ray diffraction is used to determine the identity of crystalline phases in a multiphase powder sample and the atomic and molecular structures of single crystals. It can also be used to determine structural details of polymers, fibers, thin films, and amorphous solids and to study stress, texture, and particle size.

X-Ray Diffraction Principles

Interference of Waves. The coherent scattering property of x-rays is used in x-ray diffraction applications. Two waves traveling in the same direction with identical wavelengths, λ, and equal amplitudes (the intensity of a wave is equal to the square of its amplitude) can interfere with each other so that the resultant wave can have anywhere from zero amplitude to two times the amplitude of one of the initial waves. The resultant amplitude is a function of the phase difference between the two initial waves.

Diffraction Patterns. A classical experiment in physics is the diffraction of light through slits or holes. If the size of the slits or holes are about the same size as the wavelength of the incident light, a diffraction pattern results. The analog of the slit or hole in an x-ray diffraction experiment is an electron which is the entity that produces the scatter of an x-ray photon. An electron in an atom has a size on the same order of magnitude as the wavelength of the x-rays used in a diffraction experiment. Diffraction of x-rays from many atoms with many electrons results in a two-dimensional continuous pattern of peaks and valleys.

The amount of information that is determined from a crystal structure experiment is much greater and more precise than for any other analytical tool for structural chemistry or structural molecular biology. Indeed, almost all of the structural information that has been determined for these two fields has been derived from x-ray single crystal diffraction experiments.

Bragg's Law. In 1913 W. L. Bragg showed that the positions of the discrete x-ray spots in the diffraction pattern can be explained by assuming that the diffracted x-ray photons behave as if they were "reflected" from certain families of equally spaced parallel planes passing through the crystal lattice.

Instruments for X-Ray Single-Crystal Diffraction

An x-ray single-crystal diffraction experiment consists of mounting a single crystal on a diffractometer, finding a few reflections, assigning indices to these reflections, determining the unit cell parameters and the orientation of the crystal on the diffractometer and then systematically measuring the intensities of all of the reflections. An x-ray single-crystal diffractometer consists of: (*1*) *An x-ray source.* Either synchrotron radiation or x-radiation from an x-ray tube is used. The target materials for x-ray tubes for single-crystal diffraction experiments are usually copper or molybdenum. (*2*) *An x-ray monochromator.* A monocrhomator is a large single crystal (usually graphite) that is oriented so that a very intense reflection is directed toward the sample. All wavelengths are absorbed by the monochromator except a small range of wavelengths used for the diffraction experiment. (*3*) *A goniometer.* The crystal is mounted in the center of a goniometer, which with computer control, can orient the single crystal in many different directions. (*4*) *An x-ray detector.* Two different types of detectors are commonly used. A single-reflection detector measures the intensity of one reflection at a time. It is necessary to orient the crystal and the detector (with the goniometer) with respect to the monochromatic incident x-ray beam for each reflection so that the Bragg condition is satisfied. An x-ray area detector can be used to collect the intensities of many reflections at a time. The crystal must be oriented in many different settings with respect to the incident beam but the detector needs to be positioned at only a few positions to collect all the data.

Instruments for Powder Diffraction

Bragg-Brentano Powder Diffractometer. A powder diffraction experiment differs in several ways from a single-crystal diffraction experiment. The sample, instead of being a single crystal, usually consists of many small single crystals that have many different orientations. It may consist of one or more crystalline phases (components).

The instruments are designed so that a divergent beam of x-rays impinges on the sample. A convergent beam of x-rays is diffracted from the sample and passes through a narrow slit. The normal to the flat surface of the sample bisects the angle between the incident and diffracted beams. Usually a single-crystal monochromator is placed after the diffraction slit to remove x-ray photons of unwanted wavelengths. A conventional scintillation detector is usually used to measure the intensity of the diffracted beam. This kind of powder diffractometer is called a Bragg-Brentano diffractometer and gives best results if the sample is very flat.

Graded Multilayer Device. Recently, a new diffractometer design takes advantage of the graded multilayer device. With this device, parallel incident and diffracted beams are produced. No diffracted beam monochromator is necessary. The advantage over the Bragg-Brentano design is that the sample does not have to be flat.

Applications. The most common application for powder diffractometers is to measure the 2Θ-values (2Θ is the angle between the incident and diffracted x-ray beams) and intensities for the peaks in a powder diffraction pattern.

The computer identifies which crystalline phases (components) match the unknown pattern by using a file of known powder patterns maintained by the International Center for Diffraction Data (ICDD). Current search-match programs can successfully identify up to seven components in an unknown pattern.

Once the identity of the components in a sample are known, it is possible to determine the relative amounts of each component.

From a powder pattern of a single component it is possible to determine the indices of many reflections. From this information and the 2Θ-values for the reflections, it is possible to determine the unit cell parameters. As with single crystals this information can then be used to identify the material by searching the NIST Crystal Data File.

In many situations where it is impossible to obtain a suitable single crystal, the Rietveld method can produce adequate atomic and molecular structures from a powder pattern.

From the width of the peaks the computer can determine the size of the crystallites in the sample. The smaller the crystallite size, the broader are the diffraction peaks. This kind of analysis is important where crystallite size may be a health hazard if inhaled into the lungs.

The displacement of the 2Θ-value of a particular line in a diffraction pattern from its nominal, nonstressed position gives a measure of the amount of stress retained in the crystallites during the crystallization process. Thus, metals prepared in certain ways (eg, cold rolling) have stress in their polycrystalline form. Strain is a function of peak width, but the peak shape is different than that due to crystallite size.

For samples that consist of a mixture of crystalline and amorphous material, it is possible to determine the percent of crystallinity by measuring the integrated intensity of sharp Bragg reflections and the integrated intensity of the very broad regions due to the amorphous scattering.

For many rigid polycrystalline materials (eg, metals) an analysis of the orientation of the crystallites is important to understanding the mechanical properties of the sample.

Instruments for Special Applications

Because so many new and exotic materials are being manufactured, special x-ray instruments have been designed to measure properties of these materials. These include x-ray reflectometer, position sensitive detectors, and area detectors. The reflectometer is useful in measuring the thickness of thin layers (films). Position sensitive detectors speed data collection by measuring many degrees of powder pattern simultaneously. Area detectors also greatly decrease collection time and are used for determining texture, characterizing large-scale structure (eg, polymers, fibers) from small angle scattering, collecting two-dimensional images to determine polymer and fiber diffraction patterns, and characterizing very small regions of a sample (microdiffraction).

X-Ray Fluorescence Spectrometry

X-ray fluorescence spectrometry is a technique for measuring the elemental composition of samples. The basis of the technique is the relationship between the wavelength or energy of the emitted incoherently scattered x-ray photons and the atomic number of the element. When an atom is bombarded with x-ray photons of sufficient energy, an inner-orbital electron may be displaced, leaving the atom in an excited state. The atom can return to the ground state by transference of a higher orbital electron into the vacancy (the resulting higher level vacancy is filled by an electron from a still higher level and so on). In so doing, the difference in energy between the electron ousted from the lower shell and the energy of the higher orbital electron is emitted as an x-ray photon. Each element produces a fluorescence spectrum of intensity vs wavelength which is characteristic of that element.

X-Ray Spectrometers. An x-ray spectrometer is an instrument that measures the fluorescence spectra of samples. The associated computer software then determines the qualitative and quantitative elemental composition of the samples from the resulting spectra. The primary x-ray source is usually a sealed tube with a Rh anode.

Wavelength Dispersive Spectrometer. A wavelength dispersive spectrometer uses an analyzing crystal to measure the spectrum. The analyzing crystal is a single crystal positioned so that a specific Bragg reflection can be used to measure the intensity of the various wavelengths of the fluorescence spectrum. The analyzing crystal is continuously rotated. At each position of the crystal, x-ray photons with only a narrow band of wavelengths are diffracted in to a detector. Because such a wide range of wavelengths must be measured, no single analyzing crystal is best for all elements.

Energy Dispersive Spectrometer. An energy dispersive spectrometer uses a detector that can measure the energy of each detected x-ray photon. The detector is usually a lithium-drifted silicon [Si(Li)] detector which is a proportional detector of high intrinsic resolution. A multichannel analyzer is used to sort the arriving pulses into memory locations corresponding to the energy of the x-ray photons. No analyzing crystal is needed for the energy dispersive systems.

X-Ray Radiography X-ray imaging tests are widely used to examine interior regions of metal castings, fusion welds, composite structures, and brazed components. Radiographic tests are made on pipeline welds, pressure vessels, nuclear fuel rods, and other critical materials and components that may contain three-dimensional voids, inclusions, gaps or cracks. Since penetrating radiation tests depend upon the absorption properties of materials on x-ray photons, the tests can reveal changes in thickness and density and the presence of inclusions in the material.

X-ray fluoroscopy is used for direct on-line examination. A fluorescent screen is used to convert x-ray photons into visible light photons. A television camera receives the visible image and displays it on a television screen. This type of system is used for security screening of carry-on luggage at airports.

As in medical x-ray imaging, computerized tomography (CT) can reveal the details of the internal structure of complex objects. Many detectors are used to measure the transmittance of x-rays along many lines through the object. A computer uses this information to produce an image of the internal structure of a slice of the object.

Applications Using Synchrotron Radiation

Because of the unique features of the x-ray radiation available at synchrotrons, many novel experiments are being conducted at these sources. Some of these unique features are the very high intensity and the brightness (number of photons per unit area per second), the nearly parallel incident beam, the ability to choose a narrow band of wavelengths from a broad spectrum, the pulsed nature of the radiation (the electrons or positrons travel in bunches), and the coherence of the beam (the x-ray photons in a pulse are in phase with one another). The applications are very diverse. Among them are the following

For samples which are not single crystals it is possible to obtain structural information around the site of certain heavy atoms. The procedure is called Extended Absorption Fine Structure (EXAFS). Much information about the structures of catalysts have been obtained using EXAFS.

A new technique, the Laue method, uses a wide spectrum of incident x-rays instead of monochromated x-rays. All of the reflections that are diffracted onto an area detector are recorded at just one setting of the detector and the crystal. By collecting many complete data sets over a short period of time, the Laue method can be used to follow the reaction of an enzyme with its substrate.

ROBERT A. SPARKS
Siemens Analytical X-Ray Systems, Inc.

J. P. Glusker and K. N. Trueblood, *Crystal Structure Analysis,* 2nd ed., Oxford University Press, New York, 1985.

R. Jenkins and R. L. Snyder, *Introduction to X-ray Powder Diffractometry,* John Wiley & Sons, Inc., New York, 1996.

R. E. Van Grieken and A. A. Markowicz, *Handbook of X-ray Spectrometry, Methods and Techniques,* Marcel Dekker, Inc., New York, 1993.

H. Winick and J. Doniach, eds., *Synchrotron Radiation Research,* Plenum Press, New York, 1980.

XYLENES AND ETHYLBENZENE

Xylenes and ethylbenzene (EB) are C_8 aromatic isomers having the molecular formula C_8H_{10}. The xylenes consist of three isomers: *o*-xylene (OX), *m*-xylene (MX), and *p*-xylene (PX). These differ in the positions of the two methyl groups on the benzene ring. The molecular structures are shown below.

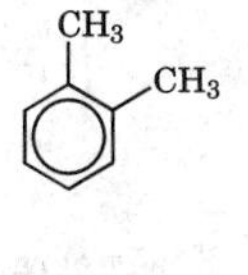
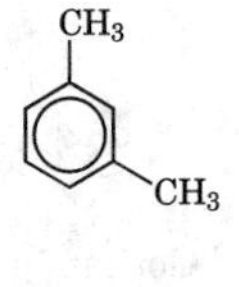

o-xylene (OX) *m*-xylene (MX)

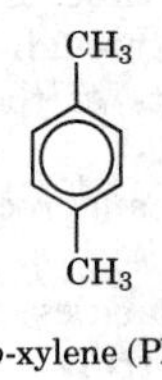
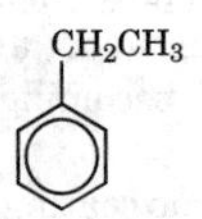

p-xylene (PX) ethylbenzene (EB)

Sources and Uses

The term mixed xylenes describes a mixture containing the three xylene isomers and usually EB. Commercial sources of mixed xylenes include catalytic reformate, pyrolysis gasoline, toluene disproportionation product, and coke-oven light oil. Ethylbenzene is present in all of these sources except toluene disproportionation product. Catalytic reformate is the product obtained from catalytic reforming processes.

Table 1. Physical Properties for C_8 Aromatic Compounds

Property	*p*-Xylene	*m*-Xylene	*o*-Xylene	Ethylbenzene
molecular weight	106.167	106.167	106.167	106.167
density at 25°C, g/cm^3	0.8610	0.8642	0.8802	0.8671
boiling point, °C	138.37	139.12	144.41	136.19
freezing point, °C	13.263	−47.872	−25.182	−94.975
refractive index at 25°C	1.4958	1.4971	1.5054	1.4959
surface tension, mN/m (= dyn/cm)	28.27	31.23	32.5	31.50
dielectric constant at 25°C	2.27	2.367	2.568	2.412
dipole moment of liquid ^{a}Cm·	0	0.30	0.51	0.36
critical properties:				
critical density, mmol/cm^3	2.64	2.66	2.71	2.67
critical volume, cm^3/mol	379.0	376.0	369.0	374.0
critical pressure, MPa^b	3.511	3.535	3.730	3.701
critical temperature, °C	343.05	343.90	357.15	343.05
thermodynamic properties:				
C_s at 25°C, J/(mol·K)^c	181.66	183.44	188.07	185.96
S_s at 25°C, J/(mol·K)^c	247.36	253.25	246.61	255.19
H_o–H_o at 25°C, J/mol^c	44.641	40.616	42.382	40.219
−(G_s–H_o/T) at 25°C, J/(mol·K)^c	97.633	117.03	104.46	120.29
heats of transition, J/(mol·K)^c				
vaporization at 25°C	42.036	42.036	43.413	42.226
formation at 25°C	−24.43	−25.418	−24.439	−12.456

^{a} To convert Cm· to D, divide by 3.336×10^{-30}.
^{b} To convert MPa to psi, multiply by 145.
^{c} To convert J to cal, divide by 4.184.

Pyrolysis gasoline is a by-product of steam cracking of hydrocarbon feeds in ethylene. Coke over light oil is a by product of the manufacture of coke for the steel industry.

Properties

Because of their similar molecular structures, the three xylenes and EB exhibit many similar properties (see Table 1).

Chemical reactions that the xylenes participate in include (*1*) migration of the methyl groups, (*2*) reaction of the methyl groups, (*3*) reaction of the aromatic ring, and (*4*) complex formation.

Migration of the Methyl Groups. Reactions that involve migration of the methyl groups include isomerization, disproportionation, and dealkylation. The interconversion of the three xylene isomers via isomerization is catalyzed by acids.

Reactions of the Aromatic Ring. The reactions of the aromatic ring of the C_8 aromatic isomers are generally electrophilic substitution reactions. All of the classical electrophilic substitution reactions are possible (see FRIEDEL-CRAFTS REACTIONS), but in most instances they are of little practical significance.

Complex Formation. All four C_8 aromatic isomers have a strong tendency to form several different types of complexes. Complexes with electrophilic agents are utilized in xylene separation.

Manufacture of Xylenes

The initial manufacture of mixed xylenes and the subsequent production of high purity PX and OX consists of a series of stages in which (*1*) the mixed xylenes are initially produced; (*2*) PX and/or OX are separated from the mixed xylenes stream; and (*3*) the PX- (and perhaps OX-) depleted xylene stream is isomerized back to an equilibrium mixture of xylenes and then recycled back to the separation step. These steps are discussed below.

Mixed Xylenes Production Via Reforming. Two principal methods for producing xylenes are catalytic reforming and toluene disproportionation. (see BTX PROCESSING). In reforming, a light fraction from a straight run petroleum fraction or from an isocracker is fed to a catalytic reformer. This is followed by heart-cutting and extraction. The mixed xylenes stream must then be processed further to produce high purity PX and/or OX. However, because of the close boiling points of PX and MX, using distillation to produce high purity PX is impractical. Instead, other separation methods such as crystallization and adsorption are used.

Xylenes Production Via Toluene Transalkylation and Disproportionation. The toluene that is produced from processes such as catalytic reforming can be converted into xylenes via transalkylation and disproportionation. Toluene disproportionation is defined as the reaction of 2 mol of toluene to produce 1 mol of xylene and 1 mol of benzene. Toluene transalkylation is defined as the reaction of toluene with C_9 or higher aromatics to produce xylenes.

Separation Processes for PX. There are essentially two methods that are currently used commercially to separate and produce high purity PX: (*1*) crystallization and (*2*) adsorption. A third method, a hybrid crystallization/adsorption process, has been successfully field-demonstrated.

Low temperature fractional crystallization was the first and for many years the only commercial technique for separating PX from mixed xylenes. PX has a much higher freezing point than the other xylene isomers. Thus, upon cooling, a pure solid phase of PX crystallizes first. Eventually, upon further cooling, a temperature is reached where solid crystals of another isomer also form. This is called the eutectic point. The solid PX crystals are typically separated from the mother liquor by filtration or centrifugation.

Adsorption represents the second and newer method for separating and producing high purity PX. In this process, adsorbents such as molecular sieves are used to produce high purity PX by preferentially removing PX from mixed xylene streams. Separation is accomplished by exploiting the differences in affinity of the adsorbent for PX, relative to the other C_8 isomers. The adsorbed PX is subsequently removed from the adsorbent by displacement with a desorbent.

In 1994, IFP and Chevron announced the development of a hybrid process that reportedly combines the best features of adsorption and crystallization. In this process, the adsorbent bed is used to initially produce PX of 90–95% purity. The PX product from the adsorption section is then further purified in a small single-stage crystallizer and the filtrate is recycled back to the adsorption section. It is reported that ultra-high (99.9 + %) purity PX can be produced easily and economically with this scheme.

MX Separation Process. The Mitsubishi Gas–Chemical Company (MGCC) has commercialized a process for separating and producing high purity MX. This process is based on the formation of a complex between MX and $HF-BF_3$. MX is the most basic xylene, and its complex with $HF-BF_3$ is the most stable.

MX of >99% purity can be obtained with the MGCC process with <1% MX left in the raffinate by phase separation of hydrocarbon layer from the complex-HF layer. The latter undergoes thermal decomposition, which liberates the components of the complex.

Xylene Isomerization. After separation of the preferred xylenes, ie, PX or OX, using the adsorption or crystallization processes discussed herein, the remaining raffinate stream, which tends to be rich in MX, is typically fed to a xylenes isomerization unit in order to further produce the preferred xylenes. Isomerization units are fixed-bed catalytic processes that are used to produce a close-to-equilibrium mixture of the xylenes. The catalysts are also designed to convert EB to either xylenes, benzene and lights, or benzene and diethylbenzene.

Health and Safety Factors

The xylene isomers are flammable liquids and should be stored in approved closed containers with appropriate labels and away from heat and open flames. The vapor can travel along the ground to an ignition source. In the event of fire, foam, carbon dioxide, and dry chemical are preferred extinguishers. The xylenes are mildly toxic. They are mild skin irritants, and skin protection and the cannister-type masks are recommended.

Uses

The majority of xylenes, which are mostly produced by catalytic reforming or petroleum fractions, are used in motor gasoline (see GASOLINE AND OTHER MOTOR FUELS). The majority of the xylenes that are recovered for petrochemicals use are used to produce PX and OX. PX is the most important commercial isomer.

Almost all of the OX that is recovered is used to produce phthalic anhydride, the phthalic anhydride is a basic building block for plasticizers used in flexible PVC resins, for polyester resins used in glass-reinforced plastics, and for alkyd resins used for surface coatings. Some of the mixed xylenes that are produced are used as solvents in the paints and coatings industry (see SOLVENTS, INDUSTRIAL). However, this use has declined.

WILLIAM J. CANNELLA
Chevron Research & Technology Co.

Y. Igarashi and T. Ueno, *A New Xylene Separation Process*, ACS Meeting, Atlantic City, N.J., 1968.

Chemical Economic Handbook, SRI International, Menlo Park, Calif., 1996.

J. J. Jeanneret, in R. A. Meyers, ed., *Handbook of Petroleum Refining Processes*, 2nd ed., McGraw-Hill Book Co., Inc., New York, 1997.

XYLYLENE POLYMERS

In a process capable of producing pinhole-free coatings of outstanding conformality and thickness uniformity through the unique chemistry of *p*-xylylene (PX) (**1**), a substrate is exposed to a controlled atmosphere of pure gaseous monomer. The coating process is best described as a vapor deposition polymerization (VDP). The monomer molecule is thermally stable, but kinetically very reactive toward polymerization with other molecules of its kind. Although it is stable as a rarified gas, upon condensation it polymerizes spontaneously to produce a coating of high molecular weight, linear poly(*p*-xylylene) PPX) (**2**).

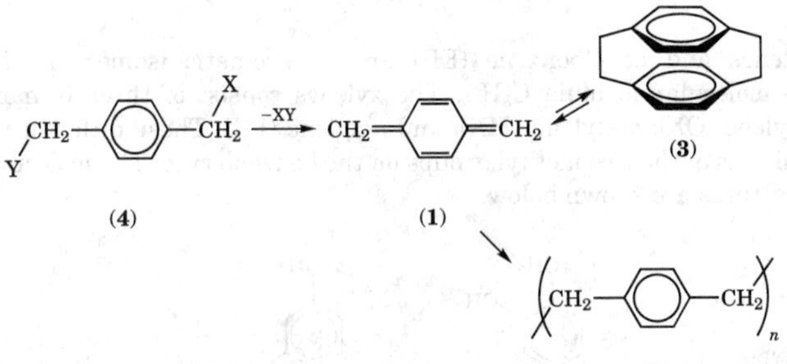

In the commercial Gorham process, PX is generated by the thermal cleavage of its stable dimer, *cyclo*-di-*p*-xylylene (DPX), a [2.2]paracyclophane (**3**). In many instances, substituents attached to the paracyclophane framework are carried through the process unchanged, ultimately becoming substituents of the polymer in the coating.

The PPXs formed as coatings in the Gorham process are referred to generically as the parylenes.

The parylene process has certain similarities with vacuum metallizing. The principal distinction is that truly conformal parylene coatings are deposited even on complex, three-dimensional substrates, such as on sharp points and in hidden or recessed areas. Vacuum metallizing, on the other hand, is a line-of-sight coating technology. VACUUM TECHNOLOGY).

The *p*-xylylene species plays a central role in the coating process itself as well as in the making of the dimers which are used as feedstocks for the coating process. Polymers and dimers have both been made from precursor *p*-xylene compounds (**4**) featuring a variety of X and Y leaving groups.

Gorham Process Monomers

The eight-carbon monomer PX is generated in the first stage of the parylene process by heating gaseous dimer as it passes through a high temperature zone.

Chemical Evidence for PX Monomer. Establishing early on that PX is indeed the pyrolysis product, rather than the molecule formed by breaking only one of the original dibenzyl bonds, the dimer diradical (**5**), proved to be an important development.

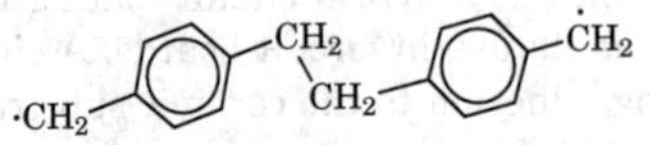

(**5**)

When the pyrolysis gases are quenched with a molar excess of iodine vapor, a yield of greater than 50% *p*-xylylene diiodide is recovered. The observation of this effect offered the first direct chemical support for the idea that DPX pyrolysis results in PX (**1**).

Monomer Properties. Despite difficulties involved in studying it owing to its great reactivity, a great deal is known about the structure of the parylene. The eight-carbon framework of the monomer PX is planar. The molecule is diamagnetic, ie, all electron spins are paired in the ground state (spectroscopically, a singlet). The PX molecule is a conjugated tetra olefin whose particular arrangement gives it extreme reactivity at its end carbons (**6**).

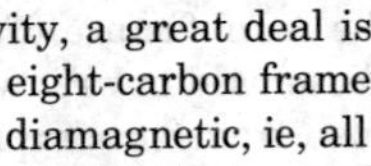

(**6**)

A particularly useful property of the PX monomer is its enthalpy of formation. Using a semiempirical molecular orbital technique, the heat of formation of *p*-xylylene has been computed to be 234.8 kJ/mol (56.1 kcal/mol).

Successful p-Xylylene VDP Monomers. Within the limits mentioned above, it is frequently possible, and often desirable, to modify the *p*-xylylene monomer by attaching to it certain substituents. Limitations on such modifications lie in the three areas: reactivity, performance in the coater (deposition equipment) and cost.

Other, Related Processes

VDP processes using means other than the pyrolytic cleavage of DPX (Gorham process) to generate the reactive monomer are also known, although none are practiced commercially at present.

Dimer

In contrast to the extreme reactivity of the monomeric PX (**1**) generated from it, the dimer DPX (**3**) feedstock for the parylene process is an exceptionally stable compound. At present only three dimers are commercially available: DPXN, DPXC, and DPXD, which give rise to Parylene N, Parylene C, and Parylene D, respectively.

The unsubstituted C-16 hydrocarbon, [2.2]paracyclophane (**3**), is DPXN. Both DPXC and DPXD are prepared from DPXN by aromatic chlorination and differ only in the extent of chlorination; DPXC has an average of one chlorine atom per aromatic ring and DPXD has an average of two.

Manufacture. For the commercial production of DPXN (di-*p*-xylylene) (**3**), two principal synthetic routes have been used: the direct pyrolysis of *p*-xylene (**4**, X = Y = H) and the 1,6-Hofmann elimination of ammonium (HNR_3^+) from a quaternary ammonium hydroxide (**4**, X = H, Y = NR_3^+). Most of the routes to DPX share a common strategy: PX is generated at a controlled rate in a dilute medium, so that its conversion to dimer is favored over the conversion to polymer.

Purification. Unsubstituted di-*p*-xylylene (DPXN) is readily purified by recrystallization from xylene. It is a colorless, highly crystalline solid. The principal impurity is polymer, which is insoluble in the recrystallization solvent and easily removed by hot filtration. In purifying DPXC and DPXD, care is taken not to disturb the homologue composition, so that product uniformity is maintained.

Properties. The DPXs are all crystalline solids; melting points and densities are given in Table 1. Their solubility in aromatic hydrocarbons is limited.

The structure of DPXN was determined from x-ray diffraction studies. There is considerable strain energy in the buckled aromatic rings and distorted bond angles. The release of this strain energy is the principal reason for success in the preparation of monomer in the parylene process.

Polymer

The linear polymer of PX, poly(*p*-xylylene) (PPX) (**2**), is formed as a VDP coating in the parylene process. The energetics of the polymerization set it apart from all other known polymerizations and enable it to proceed as a vapor deposition polymerization.

Thermodynamic Considerations. On the basis of the value for the enthalpy of formation of *p*-xylylene, the enthalpy of polymerization, can be estimated.

Polymerization Mechanism. The physical processes of condensation and diffusion must be considered along with the *p*-xylylene polymerization chemistry for a proper understanding of what happens microscopically during vapor deposition polymerization. These processes point to an important distinction between VDP and vacuum metallization, ie, that in the latter, adsorption is followed by a surface reorganization of the existing deposited material, and diffusion of incoming species through the bulk is nonexistant. In most parylene depositions, a coating forms from gaseous monomer under steady-state conditions.

The monomer is consumed by two chemical reactions: initiation, in which new polymer molecules are generated, and propagation, in which existing polymer molecules are extended to higher molecular weight. In steady-state VDP, both reactions proceed continuously inside polymeric coating, in the reaction zone just behind the growth interface.

The concentration of monomer within the coating decreases approximately exponentially with distance from the growth interface. With this decrease in monomer concentration, the rates of initiation and propagation reactions also decrease. Moving back into the polymer from the growth interface, through the reaction zone where polymer is being manufactured, a region in which the polymer formation is essentially complete is gradually entered. Under conditions prevailing during a typical deposition, the characteristic depth of the reaction zone is a few hundred nanometers, and the maximum concentration of monomer, ie, the concentration at the growth interface, is of the order of a few tenths percent by weight. Thus the parylene polymerization takes place just behind the growth interface in a medium that is best described as a slightly swollen, solid polymer.

During the vapor deposition process, the polymer chain ends remain truly alive, ceasing to grow only when they are so far from the growth interface that fresh monomer can no longer reach them. No specific termination chemistry is needed.

Polymer Properties. The single most important feature of the parylenes, that feature which dominates the decision for their use in any specific situation, is the vapor deposition polymerization (VDP) process by which they are applied. VDP provides the room temperature coating process and produces the films of uniform thickness, having excellent thickness control, conformality, and purity. The

Table 1. Properties of Parylene Dimers

Dimer	Melting point, °C	Density, g/cm^3
DPXN	284[a]	1.22
DPXC	140–160[b]	1.30
DPXD	170–195[b]	1.45

[a] Decomposes.
[b] Mixture of homologues and their isomers.

Table 2. Typical Engineering Properties of Commercial Parylenes

Property	Parylene N	Parylene C	Parylene D
density, g/cm^3	1.110	1.289	1.418
refractive index, n_D^{23}	1.661	1.639	1.669
tensile modulus, GPa[a]	2.4	3.2	2.8
tensile strength, MPa[b]	45	70	75
yield strength, MPa[b]	42	55	60
elongation to break, %	30	200	10
yield elongation, %	2.5	2.9	3
Rockwell hardness	R85	R80	
Thermal			
melting point, °C	420	290	380
heat capacity at 25°C, J/(g·K)[c]	1.3	1.0	
thermal conductivity at 25°C, W/(m·K)	0.12	0.082	
dielectric constant			
1 kHz	2.65	3.10	2.82
1 MHz	2.65	2.95	2.80
dissipation factor			
1 kHz	0.0002	0.019	0.003
1 MHz	0.0006	0.013	0.002

[a] To convert GPa to psi, multiply by 145,000.
[b] To convert MPa to psi, multiply by 145.
[c] To convert J to cal, divide by 4.184.

engineering properties of commercial parylenes once they have been formed are given in Table 2.

The most important mode of degradation for parylenes is oxidative chain scission. Oxidative degradation limits the use of parylenes at elevated temperatures in many common applications. Conventional antioxidants, incorporated during or after VDP, can extend the life of the parylenes at elevated temperatures.

The oxidation of parylene appears to be enhanced by ultraviolet radiation.

The bulk barrier properties of parylenes are among the best of organic polymeric coatings.

The parylenes do not absorb visible light, and absorb only at the shorter wavelength, high energy end of the ultraviolet range. Films and coatings are colorless in the visible, becoming opaque to sufficiently short wavelength uv light.

The crystallinity of the parylenes determines two of their most important practical characteristics: mechanical strength at elevated temperatures and solvent resistance. The crystallinity of parylenes is confined to small-submicrometer domains that are randomly dispersed throughout an amorphous continuum. Because the crystalline domains are much more resistant to permeation than the amorphous phase, they retain their reinforcing structural role even in the presence of permeants in the amorphous phase, thus giving the parylenes their resistance to solvent attack.

At temperatures below the melting of the crystallites, the parylenes resist all attempts to dissolve them.

Applications

Because the parylenes are generally insoluble in most solvents, even at elevated temperatures, they cannot be used as solvent-based coatings; neither can they be cast as films nor spun as fibers from solution. Because of their high crystalline melting points, melt-working (molding, extrusion, calendering, etc) is also difficult. Yet it is often just these features of solvent resistance and high temperature mechanical strength that constitute the advantages of PPX materials.

The most important application of parylenes is as a conformal coating for printed wiring assemblies. These coatings provide excellent chemical resistance, and resistance to fungal attack. In addition, they exhibit stable dielectric properties over a wide range of temperatures.

The use of parylenes as a hybrid circuit coating is based on much the same rationale as its use in circuit boards. A significant distinction lies in obtaining adhesion to the ceramic substrate material, the success of which determines the eventual performance of the coated part. Adhesion to the which must be achieved using adhesion promoters, such as the organosilanes.

Parylenes are superior candidates for dielectrics in high quality capacitors. Their dielectric constant and loss remain constant over a wide temperature range. The thermistor sensing probe of a disposable bathythermograph is coated with parylene. This instrument is used to chart the ocean water temperature as a function of depth.

Parylene is used in the manufacture of high quality miniature stepping motors, such as those used in wristwatches, and as a coating for the ferrite cores of pulse transformers, magnetic tape-recording heads, and miniature inductors.

Parylene's use in the medical field is linked to electronics. For example, as a protective conformal coating on pacemaker circuitry.

As books age, the paper of their pages becomes brittle. A relatively thin coating of parylene can make these embrittled pages stronger.

By separating the coating from the substrate after deposition, the unique coating features of parylenes, especially continuity and thickness control and uniformity, can be imparted to a freestanding film. Applications include optical beam splitters, a window for a micrometeoroid detector, a detector cathode for an x-ray streak camera, and windows for x-ray proportional counters.

Parylenes can be used for contamination control, that is, securing small particles to prevent them from damaging a surface in a sealed unit; barrier coating; coating for corrosion control; and as dry lubricants.

Health and Safety

Provided the vacuum pump exhaust is appropriately vented and suitable caution is observed in cleaning out the cold trap, the VDP parylene process has an inherently low potential for operator contact with hazardous chemicals. Before using the process chemicals, operators must read and understand the current Material Safety Data Sheets, which are available from the manufacturers.

W. F. BEACH
Alpha Metals

L. A. Errede and M. Szwarc, *Q. Rev. Chem. Soc.* **12**, 301 (1958); M. Szwarc, *Polym. Eng. Sci.* **16**(7), 473 (1976); W. F. Gorham and W. D. Niegisch "Xylylene Polymers" in N. M. Bikales, ed., *Encyclopedia of Polymer Science and Technology*, Vol. 15, Interscience Publishers, a Division of John Wiley & Sons, Inc., New York, 1989, pp. 98–124.

Y

YEASTS

Yeasts are probably the oldest cultivated plants: their use dates to 3000 BC or earlier. They are eukaryotic organisms, ie, they have a defined nucleus surrounded by a nuclear membrane and organized cytoplasmic organelles (mitochondria, peroxisomes, vacuoles, etc). They are fungi that exist as single cells during at least some part of their life cycle, they have no photosynthetic ability, and are not motile. The genus *Saccharomyces* is of greatest practical and economic importance for the baking, beer, and wine industries, as well as for the production of biomass. Other yeasts participate in alcoholic fermentations and occur as food-spoilage organisms or pathogens.

Morphology, Reproduction, and Life Cycle

A single yeast cell is usually spherical to ellipsoidal in shape but may be cylindrical, ogival, pyramidal, or apiculate. Often the shape reflects the specific site of bud formation. Yeasts may form a pseudomycelium consisting of single or branched chains of cells if daughter cells do not separate from their mothers; starvation for nitrogen induces such a morphology in *S. cerevisiae*.

The yeast cell is surrounded by a mechanically refractory cell wall. It surrounds the plasma membrane, which regulates the transport of chemical compounds into and out of the cell either by simple diffusion or by active transport. The cell wall consists of alkali-soluble β-glucan, alkali-insoluble glucan, and glucan- and mannan-containing glycoproteins. The small nucleus of the yeast cell is surrounded by a membrane or tonoplast, which has many pores. In haploid cells of *S. cerevisiae*, the nucleus contains 90% of the DNA of the cell (the other 10% residing in the mitochondria), organized into 16 linear chromosomes ranging in size from about 150 to about 1600 kbp. The total is about 1.5×10^7 bp, or three times the size of the *E. coli* genome. The small average chromosome size makes it possible to separate intact chromosomes elecrophoretically using a variety of pulsed-field electrophoresis systems, which have led to simple strain identification procedures and also allow rapid assignment of cloned genes to specific chromosomes.

Vegetative reproduction in yeast occurs mostly by budding or, in the instance of *Schizosaccharomyces* and *Endomycopsis*, by fission. Yeasts belonging to *Basidiomycetes* form external spores. In *Ascomycetes*, sexual reproduction occurs with the formation of spores in a cell that serves as a spore sac or ascus. In principle, sexual reproduction in yeasts consists of an alternation between the haploid phase (one set of chromosomes) and the diploid phase (two sets of chromosomes). However, production strains of bakers' yeast (*S. cerevisiae*) are generally polyploid and sporulate with low efficiency. Other yeasts may exist predominantly in the haploid state, eg, *S. rouxii*; in the diploid state as for *S. cerevisiae*; or in both states, eg, for *S. lipolytica*. An alternation between the haploid and the diploid state may also occur without production of sexual spores (parasexuality).

Strain Improvement and Development

A variety of transformation techniques using *E. coli*-yeast shuttle vectors and yeast selectable markers, as well as efficient yeast promoters and signal sequences, is generally available. Public disinclination to purchase genetically engineered strains of yeast for food use, eg, in baking, brewing, and wine making, has limited the use of such strains in most of the industry. Generally, work on strain improvement is, thus, carried out using the classical genetic approaches of hybridization and mutagenesis.

Fermentative and Respiratory Metabolism

Although most yeast species are strict aerobes, strongly fermenting yeasts such as *S. cerevisiae* grow well under anaerobic conditions, but with a lower yield. The yield of yeast based on weight of sugar consumed is greater in the presence of oxygen, and oxygen decreases the rate of sugar consumption. Under aerobic conditions, alcohol is produced as the end product of metabolism if the sugar concentration exceeds 0.1–0.2% by weight. This is the reverse Pasteur or Crabtree effect and is also known as glucose inhibition or catabolite repression. In the presence of higher sugar concentrations, synthesis of respiratory enzymes such as cytochromes is inhibited.

The anaerobic pathways for glucose utilization by yeasts and other microbes lead to the formation of various alcohols and organic acids. Ethanol is the principal product of yeasts used in the beverage industry, formed from pyruvate via acetaldehyde by the enzyme alcohol dehydrogenase.

Composition, Nutrients, and Growth Rate

The elemental and vitamin compositions of some representative yeasts are listed in Table 1. The principal carbon and energy sources for all yeasts are carbohydrates (usually sugars), alcohols, and organic acids, as well as a few other specific hydrocarbons. Nitrogen is usually supplied as ammonia, urea, amino acids or oligopeptides. The main essential mineral elements are phosphorus (supplied as phosphoric acid), and potassium, with smaller amounts of magnesium and trace amounts of copper, zinc, and iron. The vitamin requirements differ among species.

The specific growth constant for exponential growth, μ, is defined as $\mu \times dt = dM/M$, where M is the yeast mass and t is time.

Yeast-Fermented Foods and Beverages

Table 2 shows the production of alcoholic beverages, baked goods, and yeast biomass and the production of fuel alcohol by fermentation.

Table 1. Composition of Yeast

Component	Bakers' yeast	Brewers' yeast	*Candida* sp.
C, wt %	47.0		45.9
H, wt %	6.0		6.7
N, wt %	8.5		7.3
O, wt %	32.5		32.1
ash, wt %	6.0	6.4	7.8
Ca, wt %	0.06	0.13	0.57
Fe, wt %	0.003	0.01	0.01
Mg, wt %	0.13	0.23	0.13
P, wt %	1.0	1.4	1.7
K, wt %	2.0	1.7	1.9
Na, wt %	0.03	0.07	0.01
Co, mg/kg		0.2	
Cu, mg/kg	8.0	33.0	12.4
Mn, mg/kg	5.9	5.7	38.7
Zn, mg/kg	197	38.7	99.2
dry matter, wt %	94.0	93.0	93.0
crude fiber, wt %		3.0	2.0
ether extract, wt %		1.1	2.5
protein (N × 6.25, wt %	45	44.6	48.3
digestible protein, wt %		38.4	41.5
thiamine, mg/kg	90	91.7	6.2
riboflavin, mg/kg	45	35	44
nicotinic acid, mg/kg		448	500
pyridoxine, mg/kg	40	43	30
biotin, mg/kg	1.3		1.1
pantothenate, mg/kg	65	110	83
folic acid, mg/kg	15	10	23
choline, mg/kg	4000	3885	2911

Table 2. Production of Yeast and Yeast-Fermented Foods, Beverages, and Alcohol

Product	Production, 10^6 m^3	Yeast solids produced[a], 10^3 t
beer	22.9	44
wine	1.68	6
distilled beverages (50% ethanol)	0.36	4
baked goods (yeast raised)	11.3	90
fuel and industrial alcohol (100% ethanol)	3.0	67.5
yeast biomass (inactive)		10

[a] Estimates based on production of 2.0–2.5 kg yeast solids per cubic meter of beverage at 8–12% ethanol.

Yeasts that have been grown in clear media, eg, clarified molasses or brewers' worts, can be recovered by centrifuging or filtration followed by pressing. Yeasts that have been grown in distillers' mashes or grape musts, that contain insoluble particles, cannot be separated economically. While yeast produced in wine fermentations is either discarded or used as fertilizer, the excess yeast produced during the fermentation of beer, distilled beverages, and in fuel-ethanol production is recovered with the spent grains and sold as feed.

Bakers' Yeast Production. Bakers' yeast is grown aerobically in fed-batch fermentors under conditions of carbohydrate limitation. This maximizes the yield of yeast biomass and minimizes the production of ethanol. Yeasts grown under these conditions have excellent dough leavening capability and perform much better in the bakery than yeast grown under anaerobic conditions. All bakers' yeast strains are *Saccharomyces cerevisiae*.

Harvesting and Packaging of Yeast. Yeast cream (so called because of its off-white color) is harvested by centrifugation and washed. Stored in refrigerated tanks, it may be sold directly in this form, since in large baking facilities, it is piped to the desired location. Compressed yeast, a moist press cake, is obtained by pressing yeast cream in plate-and-frame filter presses or filtration through a rotary vacuum filter. Compressed yeast is a perishable commodity and must be refrigerated. Compressed yeast sold in supermarkets contains approximately 10% added starch to increase its shelf life and so has a lower protein content and fermentative activity than the compressed yeast sold to bakeries.

The production of active dry yeast (ADY) is very similar to the production of compressed yeast. However, a different strain of yeast is used and the nitrogen content is reduced to 7% of solids compared with 8–9% for compressed yeast. Where storage stability and convenience are important considerations, ADY has largely replaced compressed yeast.

Instant ADY (IADY) production is similar to ADY production but requires a different strain of yeast. Unlike ADY, instant active dry yeast does not require separate rehydration. On an equivalent solids basis, the activity of IADY is greater than that of regular ADY, but still less than that of compressed yeast.

Yeast in Baked Goods. Baked goods can be leavened with yeast, chemicals, the foam of egg white, or steam. There is a preference for yeast-leavened baked goods because of their desirable flavor. Therefore, only those products are chemically leavened in which yeast does not perform well, generally as a result of high osmotic pressure. The osmotic pressure of doughs depends inversely on moisture content and directly on the concentration of salt and sugars. Cookies and highly sweetened cake doughs have an osmotic pressure too high for adequate fermentation by yeasts.

Another requirement for proper leavening is the presence of a protein matrix sufficiently elastic to trap small carbon dioxide bubbles. Wheat gluten fulfills this requirement. Rye protein is less suitable and the proteins of other cereals, eg, rice, oats, or corn, are practically useless.

Brewing. The basic raw materials for the production of beer are sweet worts formed by enzymatic hydrolysis of cereal starches. The principal cereal is barley which, after malting, is also the source of enzymes that hydrolyze starches, glucans, and proteins. In some countries, eg, Germany, the mash bill consists entirely of malted barley. In other countries, adjuncts such as corn, rice, corn syrup, or glucose are common.

Two pure-culture yeasts, top- and bottom-fermenting, are used in brewing. Although both are biologically members of the species *S. cerevisiae*, bottom-fermenting yeasts are referred to as *S. uvarum* or *S. carlsbergensis*. Top-fermenting *S. cerevisiae* is used for the production of ale and *S. uvarum* for the production of lagers. The rate-limiting step in brewing appears to be transport of maltose across the membrane.

The flavor compounds, fermentation by-products, are higher alcohols (known as fusel oils), esters, diketones, aldehydes, organic acids, and sulfur compounds. The particular yeast strain used affects the formation of higher alcohols and esters. Although the flavor compounds can be identified analytically, it is not always possible to specify their effect on flavor. Beer taste can be spoiled by contaminating bacteria or yeasts.

Wine. Grapes (about 21–23% sugar) are pressed; the liquid must is either separated and allowed to settle (for white wines) before inoculation with yeast, or the whole mass is directly inoculated with yeast (for red wines). In either case, while the initial fermentation takes place, the carbon dioxide formed by fermentation excludes air and prevents oxidation. White wines are transferred to a second fermentor (racked) near the end of fermentation and kept isolated from the air while solids, including yeast, settle out. Additional rackings, filtration or centrifugation may be required to clarify the wine and remove tartrates prior to bottling.

Red wines are fermented in the presence of the grape skins to extract pigments and tannins, until 70–90% of the sugar has been fermented, and then pressed. The juice undergoes a residual fermentation to consume the remaining sugar. At this point the wine is racked to separate the yeast; a second, bacterial malolactic fermentation, which transforms malic acid to lactic acid, occurs at this stage and reduces the acidity of the wine. Red wines may be aged in wooden casks to improve flavor and precipitate tannins and tartrates. Additional filtrations or centrifugations may be employed before bottling to clarify the wine.

In addition to strains of *S. cerevisiae*, the term wine yeast refers to genera that occur naturally in grape musts and participate in spontaneous fermentation in wine. Most yeasts that take part in spontaneous wine fermentations derive from cells adhering to wine-making equipment such as crushers and presses.

Spontaneous fermentations are used for wine production in France, some other European countries, South America, and in recent years, some California wineries. They generally start more slowly than fermentations inoculated with commercial dried yeast, are more difficult to control, and may suffer from growth of undesirable contaminants. However, it is claimed that the resulting wines possess better organoleptic properties, particularly more complex flavor and aroma.

Distilled Beverages. Distilled alcoholic beverages are made by fermenting sugars from grains or fruits followed by distillation. The common cereal grains used for whiskey are barley, wheat, corn, or rye. Vodka is made from potatoes, rum from molasses, and brandy from grape or other fruits. Grain or potato starch must be converted to a fermentable form before use. Except for some Scotch or Irish whiskeys, *whole mashes* of grains are fermented. That is, fermentation and conversion of starch to fermentable sugar proceed simultaneously.

Distillers' yeasts are specialized strains. Many distillers maintain their own proprietary yeasts or have them grown by commercial yeast producers. Commercial bakers' yeasts may be used for the production of neutral grain spirits.

Lactic acid bacteria are common contaminants of distillers' fermentations, producing excessive amounts of volatile acids and an acrid odor.

Sake. Production of Japanese rice wine begins with the preparation of a culture of *Aspergillis oryzae* by inoculating steamed polished rice. The mold produces extracellular amylases and proteases that act on the rice, liberating glucose and amino acids. This *koji* is inoculated into a mixture of water and additional steamed rice that is allowed to ferment for a period of about 10 days (the moto). At this time, it may be inoculated with a culture of sake yeast, *Saccharomyces sake*, a specialized subspecies of *S. cerevisiae*. Further inoculation and fermentation ensue.

Sake yeasts are unusual. They will grow at relatively high level of lactic acid (over 2%), and excrete succinic acid while fermenting. They continue to ferment at alcohol levels much higher than tolerated by other yeasts and will grow, albeit very slowly, at the very high osmotic pressures found early in the *moto* stage; some strains will continue to grow at sugar concentrations of 45%.

Soy Sauce (Shoyu). Production of soy sauce is a two-step process in which a mixture of boiled soybeans and crushed roasted wheat is first inoculated with *Aspergillis oryzae* or *A. soyae*. Several days later brine is added and the subsequent fermentation, which traditionally lasts 2–3 years, is first dominated by lactic acid-producing bacteria. As the pH of the fermenting mash (*moroni*) drops, a yeast fermentation begins, dominated by strains of *S. rouxii* tolerant of high salt and osmotic pressure. Proper flavor development, however, requires the presence of species of *Torulopsis* as well.

Microbial Biomass and Single-Cell Protein

In all fermented foods, microbes contribute as preservatives, ie, by lowering the pH and producing ethanol, or by making the food more palatable. The deliberate use of yeasts as food in themselves is less common. Small beer, the sediment from beer, has been traditionally used as a vitamin supplement for infants. Beginning in 1910, dried, spent brewers' yeast was developed as a food, and *Candida utilis* was used as a food supplement in Germany during World War II.

In the 1960s and 1970s, the world protein shortage stimulated the development of additional processes for producing microbial biomass both from traditional substrates, eg, carbohydrates, and from alternative materials such as *n*-paraffins, ethanol, and methanol. Generally, industrial processes for biomass production use yeast cultures. The most widely available yeast biomass is a by-product of the brewing industry. Historically the carbon source for biomass production is fermentable sugars. Because of their low cost and contribution of some nitrogen, minerals and vitamins, beet, and cane molasses are preferred feedstocks. One of the most promising substrates for future production of microbial biomass is the cellulose contained in agricultural residues such as wood pulp, sawdust, feed-lot waste, corn stover, rice hulls, nut shells, and bagasse. In addition to these carbon sources, yeasts require large supplements of nitrogen, phosphorus, potassium, calcium, and magnesium, as well as trace minerals.

Like bakers' and brewers' yeasts, yeast for biomass need not be produced under conditions of complete sterility. Most fermentations are carried out at a pH less than 4.5, which limits bacterial growth. The most common contaminants are lactic and acetic acid-producing bacteria. Processes designed for yeast growth are highly aerobic and require heat removal. The supply of oxygen is also costly, since oxygen transfer rates from air to liquid are low.

The main importance of microbial biomass is in its contribution as a protein supplement. Yeast solids contain about 7.5–9% nitrogen. The essential amino acid contents of dry yeasts are quite similar regardless of the species. Lysine content is high, and sulfur-containing amino acids such as methionine are low.

Brewers' and bakers' dried yeasts are used as dietary supplements. They contribute some protein and trace minerals, and some B vitamins, but no vitamin C, vitamin B_{12} or fat-soluble vitamins. The glucose tolerance factor (GTF) of yeast, chromium nicotinate, mediates the effect of insulin. It seems to be important for older persons who cannot synthesize GTF from inorganic dietary chromium. The cell wall fraction of bakers' yeast reduces cholesterol levels in rats fed a hypercholesteremic diet. Bakers' inactive dry yeast is also widely used in the food industry. This yeast may be grown specifically as a food supplement and consequently there is a choice in its composition by varying growth conditions and feedstock makeup. It can possibly produce high levels of nicotinic acid and thiamin, the crude protein content can be raised to 50–55% and it can be used as a vehicle for the incorporation of micronutrients such as selenium or chromium into the diet.

Recovery of biomass from the fermentor requires centrifugation or filtering, washing, concentration, and drying. Extraction of protein requires breaking the cell wall to release the cytoplasmic contents. This can be achieved by high speed ball or colloid mills or by high pressure (50–60 Mpa) extrusion. Protein is extracted by alkaline treatment followed by precipitation after enzymatic hydrolysis of nucleic acids.

The presence of nucleic acids in yeast is one of the main problems with their use in human foods. Purines ingested by humans and some other primates are metabolized to uric acid, which may precipitate out in tissue to cause gout. The daily human diet should contain no more than about 2 g of nucleic acid, which limits yeast intake to a maximum of 20 g.

Uses. Inactive dried yeasts are used as ingredients in many formulated foods: baby foods, soups, gravies, and meat extenders; as carriers of spice and smoke flavors; and in baked goods. Yeasts used in the health food industry are generally fortified with minerals and contain higher concentrations of the B vitamins, especially thiamin, riboflavin, and niacin (see VITAMINS).

Dried yeasts are used extensively in animal feeds for monogastric animals. Most distillers' and brewers' yeasts are incorporated by codrying with the spent grains. Active (ie, live) dry yeast is incorporated into feed for ruminants and is believed to aid digestion, increases in weight gain, and productivity. Appreciable amounts of brewers' yeast and torula (*C. utilis*) are also used in pet foods, principally for dogs and cats but also in feeds for birds, fish, mink, and bees.

Yeast-Derived Products

Yeast derived products include enzymes, and extracts. The enzyme invertase (β-fructofuranosidase) is used in sucrose hydrolysis in high-test molasses and in the production of cream-centered candies. Lactase (β-galactosidase) is used in the hydrolysis of lactose in milk or skim milk.

Extracts are used in fermentation media for production of antibiotics, in cheese starter cultures, and in the production of vinegar. They are also extensively used in the food industry as condiments to provide savory flavors for soups, gravies and bouillon cubes, and as flavor intensifiers in cheese products.

Food Preservation and Food Spoilage

Many foods such as alcoholic beverages, pickles, cheese, and fish sauce are preserved by fermentation. Preservation results from a lowering of pH or the formation of ethanol. Yeasts do not produce antibiotics, although isolates of a number of species produce a toxin ("killer factor") lethal to other yeasts.

Yeasts also act as spoilage organisms. Jams, jellies, and honey can be fermented by osmophilic yeasts such as *S. mellis* or *S. rouxii*. Wild yeasts may also spoil wines or beers. Film-forming yeasts, such as *P. membranefaciens*, may grow on the surface of sauerkraut, pickles, or other fermented vegetables. *K. fragilis* or other lactose-fermenting yeasts occur in milk products. *C. lipolytica* may occasionally spoil butter or margarine.

Pathogenic Yeasts

Few yeasts are pathogenic to healthy individuals, but immunocompromised persons can suffer from infections from a number of normally innocuous yeasts. *Candida albicans* is the best-known of the pathogenic species. *Cryptococcus neoformans* may cause serious infections in a number of organs, particularly the meninges, with sometimes fatal results. *Histoplasma capsulatum* infection may result in histoplasmosis, which can be fatal.

Outlook

Yeast is of great importance for the study of eukaryotic cell biology: for example, elucidation of the mechanism of cell cycle control and of proteins involved in regulating cell division has been highly dependent on studies undertaken in yeast. The results could be directly applicable to understanding the mechanism of carcinogenesis in humans. With the complete sequencing of the yeast genome, reported in late 1995, molecular studies will undoubtedly increase. Although genetically engineered yeasts are currently not available for food use, such yeasts have been developed. Continuous fermentations and fermentations using immobilized cells are under investigation.

DANIEL MALONEY
Universal Foods Technical Center

A. H. Rose and J. E. Harrison, eds., *The Yeasts*, 3rd ed., Academic Press, New York, 6 vols., 1987–1996.

J. Broach, E. Jones, and J. Pringle, eds., *The Molecular and Cellular Biology of the Yeast Saccharomyces*, Cold Spring Harbor Press, New York, 3 vols., 1991–1997.

J. White, *Yeast Technology*, Chapman and Hall, London, 1954.

G. Reed and T. W Nagodawithana, *Yeast Technology*, 2nd ed., Van Nostrand Reinhold, New York, 1991.

YTTERBIUM. See LANTHANIDES.

Z

ZEOLITES. See MOLECULAR SIEVES.

ZIEGLER-NATTA CATALYSTS. See CATALYSIS; OLEFIN POLYMERS; ORGANOMETALLICS.

ZINC AND ZINC ALLOYS

Zinc is a relatively active metal and its compounds are stable. Since it is not found free in nature, it was discovered much later than less-reactive metals such as copper, gold, silver, iron, and lead.

The main application of zinc is to protect iron and other metals from corrosion. Zinc in contact with iron and other metals, as a coating or attached anode, corrodes sacrificially and protects the iron. Another important use is in alloys for die casting. These alloys are used extensively because of their high quality and low cost. Brass and bronze products account for the third largest usage.

Occurrence

Zinc ores are widely distributed throughout the world; 55 zinc minerals are known. Sphalerite provides ca 90% of the zinc produced today.

In the United States, the richest zinc district is in Alaska; however, Australia leads the world by every measure of reserves followed by Canada and the United States. Zinc minerals tend to be associated with those of other metals; the most common are zinc–lead or lead–zinc, depending upon the dominant metal, zinc–copper or copper–zinc, and base metal such as silver.

Physical Properties

Zinc is a lustrous, blue-white metal, which can be formed into virtually any shape by the common metal-forming techniques such as rolling, drawing, extruding, etc. The hexagonal close-packed crystal structure governs the behavior of zinc during fabrication. Physical properties are given in Table 1.

Table 1. Some Physical Properties of Zinc

Property	Value
ionic radius, Zn^{2+}, nm	0.074
covalent radius, nm	0.131
metallic radius, nm	0.138
ionization potential, eV	
first	9.39
second	17.87
third	40.0
density solid, g/cm^3	
at 25°C	7.133
liquid, g/mL at 419.5°C	6.620
melting point, °C	419.5
boiling point, °C	907
heat of fusion at 419.5°C, kJ/mol[a]	7.387
heat of vaporization at 907°C, kJ/mol[a]	114.8
thermal conductivity, W/(m·K) solid	
at 419.5°C	96.0
liquid at 419.5°C	60.7
heat capacity, J/(mol·K)[a]	
solid	$22.39 + 10^{-2}\ T$[b]
liquid	31.39
gas	20.80

[a] To convert J to cal, divide by 4.184.
[b] T = 298–692.7 K.

Chemical Properties

The most significant chemical property of zinc is its high reduction potential. Zinc, which is above iron in the electromotive series, displaces iron ions from solution and prevents dissolution of the iron.

In batteries, a zinc anode undergoes the oxidation reaction, $Zn \rightarrow Zn^{2+} + 2e(+0.763\ V)$ to provide a flow of electrons to the external circuit.

The capability of zinc to reduce the ions of many metals to their metallic state is the basis of important applications.

Zinc hydrosulfite (zinc dithionite) is a powerful reducing agent used in bleaching paper and textiles. Another hydrosulfite reducing agent is zinc formaldehyde sulfoxylate, $Zn(HSO_2{\cdot}CH_2O)_2$. Surface amalgamation with mercury is useful in alkaline-battery applications because it increases the hydrogen overvoltage and reduces hydrogen gassing.

Processing

Processing includes roasting and reduction (electrolytic and pyrometallurgical processes). Zinc ores are too low in zinc content for direct reduction and must be concentrated.

Roasting. Copper and lead sulfides are directly smelted but not zinc sulfide. The zinc sulfide in the concentrate is always converted to oxide by roasting. An exception is the direct leach process.

For environmental and economic reasons, the early practice of roasting zinc sulfide and discharging the sulfur dioxide to the atmosphere gave way to plants where the sulfur dioxide is converted to sulfuric acid. Desulfurization takes place while the ore particles are suspended in hot gases. These processes are called flash- and fluid-bed roasters. Some plants use combinations of roasters and sintering for desulfurization.

Sintering completes the roast, eliminates volatile material, and aggregates fine calcine.

Reduction. The electrolytic process, where zinc is deposited from an aqueous solution onto the cathode, treats complex ores that do not lend themselves to pyrometallurgical recovery (see also ELECTROCHEMICAL PROCESSING). The sulfide concentrate is oxidized to crude oxide and leached with return acid from the cells. The zinc sulfate solution is purified and electrolyzed. Alternatively, in an adaptation that avoids roasting, the concentrate can be leached directly with return and from the cells and the sulfide converted to free sulfur.

Zinc pyrometallurgy is based upon reduction of zinc oxide. The lowest temperature for the reaction is 857°C but 1100–1300°C is required for acceptable rates. Carbon monoxide and dioxide oxidize zinc vapor below 1100–1300°C although only the carbon dioxide reaction is significant. Rapid condensation of the zinc vapor avoids the formation of zinc-oxide-coated droplets, so-called blue powder.

Secondary Recovery. Zinc is recovered as metal, dust, and chemicals (including oxide) from secondary sources, mostly scrap (see also RECYCLING). Recovery practices vary greatly, but zinc products are usually made from metallic scrap by melting and distillation. Dust is made by rapid condensation of zinc vapor in oxygen-free atmospheres. In a novel process, pyrite cinders are given a chloridizing roast and leached; solvent extraction in two stages recovers 98% of the zinc.

Analytical Methods

Zinc in ores at the concentrating mill is often determined polarographically and mill slurries are commonly monitored continuously for zinc by x-ray fluorescence to control addition of flotation reagent. Low zinc concentrations in solution are analyzed polarographically or by atomic absorption spectroscopy (AAS). In concentrates where zinc is high and great precision is required, wet methods are often used, such as the titration with potassium ferrocyanide. Finished zinc and zinc alloys are usually analyzed for metals other than zinc by emission spectroscopy and the zinc determined by difference.

Health and Safety Factors; Environmental Aspects

The level of natural versus man-made emissions to the environment are of similar magnitude. Soil erosion is the major contributor of natural emissions with zinc mining, zinc production facilities, iron and steel production, corrosion of galvanized structures, coal and fuel combustion, waste disposal and incineration, and the use of zinc fertilizers and pesticides being the principal anthropogenic contributors. Zinc is an essential element and thus can exist in both the deficient or toxic state in plants and animals. Like plants, marine and fresh water organisms vary significantly in their response to zinc. The range of concentration among species is relatively narrow (3–30 ppm) except for oysters, which tend to accumulate zinc to very high levels (100–2000 ppm).

Likewise, too little zinc in the human diet can lead to poor health, reproductive problems and a lowered ability to resist disease. The levels of zinc that produce adverse effects in humans are higher than the Recommended Daily Allowances, which range from 15 mg/day for men to 5 mg/day for infants.

In most zinc mines, zinc is present as the sulfide and coexists with other minerals, especially lead, copper, and cadmium. Therefore, the escape of zinc from mines and mills is accompanied by these other often more toxic materials. Mining and concentrating, usually by flotations, does not present any unusual hazards to personnel. Atmospheric pollution is of little consequence at mine sites, but considerable effort is required to flocculate and settle fine ore particles, which would find their way into receiving waters. The most significant occupational exposures to zinc would occur during the smelting and refining of zinc ore. The standards for occupational exposure have been established at a level to prevent the onset of metal fume fever. This temporary condition is caused by excessive exposure to freshly formed fumes of zinc oxide and results in flulike symptoms of fever, chills, headache, muscle pain, nausea and vomiting.

Uses

Zinc is used in metallic coatings, die-casting alloys, foundry alloys (eg, high strength alloys, see Table 2; slush alloys; forming-die alloys), rolled zinc, and zinc dust and powder.

By-Product Metals

Ores that are exploited primarily for zinc invariably contain one or more other valuable metals. Various ores contain different combinations of such other metals. Cadmium and mercury are usually recovered in separate processes at the zinc plant. The others are shipped as enriched residues to plants that specialize in their recovery.

Table 2. Properties of High Strength Zinc Foundry Alloys

		Alloy No. 12		Alloy No. 27	
Property	Alloy No. 8, Permanent-Mold Cast	Sand Cast	Permanent-Mold Cast	Sand Cast[a]	Sand Cast, H.T.[b]
physical					
density, g/cm^3	6.37	6.03	6.03	5.01	5.01
melting or solidification range, °C	375–404	377–432	377–432	375–487	375–487
mechanical					
tensile strength, MPa[c]	221–225	276–310	310–345	400–441	310–324
yield strength, 0.2% MPa[d]	207	207	214	365	255
elongation, %	1–2	1–3	4–7	3–6	8–11
Brinell hardness[d]	85–90	92–96	105–125	110–120	90–100

[a] Primary purpose. [b] Heat treated. [c] To convert MPa to psi, multiply by 145. [d] 500 kg for 30 s.

FRANK E. GOODWIN
International Lead and Zinc Research Organization, Inc.

C. H. Mathewson, ed., *Zinc*, Reinhold Publishing Co., New York, 1959.

J. M. Cigan, T. S. Mackey, and T. J. O'Keefe, eds., *Lead-Zinc-Tin '80*, American Institute of Mining Engineers, Warrendale, Pa., 1979.

IPCS Task Group on Zinc, *International Programme on Chemical Safety, Environmental Health Criteria for Zinc*, September 20, 1996.

Engineering Properties of Zinc Alloys, International Lead Zinc Research Organization, New York, 1980.

ZINC COMPOUNDS

Zinc usually occurs as the sulfide but significant quantities of the oxide, carbonate, silicate, and basic compounds of the latter two are also mined (see ZINC AND ZINC ALLOYS). Table 1 lists properties and uses of zinc compounds.

Table 1. Properties and Uses of Zinc Compounds

			Solubility[a], g/100 g solvent			
Zinc compound	Sp gr	Mp, °C	Water		Other	Uses
acetate	1.735	237	$40^{25°C}$	$67^{100°C}$	$3^{25°C}$ alcohol	wood preservative, mordant antiseptics, catalyst, waterproofing
ammonium chloride	1.88	150 dec	$66^{0°C}$	$69^{30°C}$		galvanizing, solder flux, adhesives
diborate	3.64		$0.007^{25°C}$		sl sol HCl	fireproofing, ceramics, fungicide
dodecaborate	4.22	980	insol			fire retardant
bromide	4.21	394	$471^{25°C}$	$675^{100°C}$	sol alcohol, ether	photographic paper, catalyst, batteries
carbonate	4.40	$-CO_2$ at 300	$0.001^{15°C}$		insol alcohol	ceramics, rubber, astringent (lotions)
chloride	2.91	275	$432^{25°C}$	$614^{100°C}$	sol ether	textiles, adhesives, flux, wood preservative, antiseptic, astringent
cyanide	1.85	800 dec	$0.005^{20°C}$		sol alkali, CN^-	electroplating, gold extraction
dithiocarbamates						
dibutyl	1.21	106	insol		sol C_6H_6, CS_2, $CHCl_3$	vulcanization accelerator, lube oil
diethyl	1.48	176	insol		sol C_6H_6, CS_2, $CHCl_3$	vulcanization accelerator
ethylenebis			insol		sol C_6H_6, CS_2, $CHCl_3$	fungicide, insecticide
dimethyl	1.71	249	$0.0065^{25°C}$		sol CS_2, acetone, alkali	vulcanization accelerator, fungicide

Table 1. Properties and Uses of Zinc Compounds *(continued)*

Zinc compound	Sp gr	Mp, °C	Solubility[a], g/100 g solvent			Uses
			Water	Other		
2-ethylhexanoate	0.90		insol		sol hydrocarbon	paint drier, silicone rubber cure
fluoroborate		$-H_2O$ at 60°C	$>100^{25°C}$		sol alcohol	plating, bonderizing, textile resin cure
fluoride	4.95	872	$1.6^{18°C}$		sol hot acid, NH_4OH	ceramics, impregnating wood, galvanizing
formaldehyde sulfoxylate		90 dec	$60^{25°C}$		insol alcohol	reducing agent, drying, polymerization
hydrosulfite		200 dec	$40^{20°C}$			bleach, especially textile, paper, reducing agent
iodide	4.70	446	$432^{18°C}$	$510^{100°C}$	sol alcohol	medicine, photography
2-mercaptobenzothiazole	1.70	300 dec	insol	insol	sol C_6H_6, CS_2, $CHCl_3$	vulcanization accelerator for latex
naphthenate			insol	insol	sol hydrocarbon, acid	paint film improver, rot proofer
nitrate	2.065	−18	93	$900^{70°C}$	sol alcohol	textiles as resin catalyst, mordant latex coagulant
oxide	5.47, 5.61	1800 sublimes	$0.00042^{18°C}$		sol acid, alkali, NH_4OH	vulcanization accelerator, mildewstat, pigment, supplement in feed and fertilizer, catalyst, ceramics, intermediate
peroxide	1.57	212 explodes	insol		sol acid	cosmetic powders as antiseptic
phosphate	4.00	900	$2.6^{25°C}$	insol	sol NH_4OH, insol alcohol	metal coatings, dental cement
potassium chromate	3.36–3.46		$0.24^{25°C}$			rust-inhibiting pigment
resinate	1.24	205	insol	insol	sol hydrocarbon	inks, paint drier
selenide	5.42	>1100	insol		sol acid	phosphor
silicofluoride	2.10	100 dec	$77^{10°C}$	$93^{60°C}$		laundry sour, wood preservative, plaster additive
stearate	1.09	120	insol	insol	sol hydrocarbon	lubricant, mold release, vinyl stabilizer, anti-cake, water repellant
sulfate	3.54	680 dec	$41.9^{0°C}$	$91^{70°C}$	sol glycerol	rayon bath, agriculture, zinc plating, intermediate, flotation, mordant
sulfate	3.28	238 dec	$101^{70°C}$	$87^{105°C}$		rayon bath, agriculture, zinc plating, intermediate, flotation, mordant
sulfide	3.98, 4.10	1185 sublimes	$0.0007^{18°C}$	insol	sol acid	phosphor, white pigment, dental materials
tetroxychromate	3.87–3.97		$0.01^{25°C}$			wash primer
undecylenate		115	insol	insol	sol hydrocarbon	dermal fungicide

[a] Insol = insoluble, v sol = very soluble, dec = decompose.

FRANK E. GOODWIN
International Lead and Zinc Research Organization, Inc.

H. E. Brown, *Zinc Oxide, Properties and Applications*, International Lead Zinc Organization, Inc., New York, 1976, pp. 56–59.

A. P. Thompson in C. H. Mathewson, ed., *Zinc*, Reinhold Publishing Co., New York, 1959.

ZIRCONIUM AND ZIRCONIUM COMPOUNDS

Zirconium is classified in subgroup IVB of the periodic table with its sister metallic elements titanium and hafnium. Zirconium forms a very stable oxide. The principal valence state of zirconium is +4, its only stable valence in aqueous solutions. The naturally occurring isotopes are given in Table 1. Zirconium compounds commonly exhibit coordinations of 6, 7, and 8. The aqueous chemistry of zirconium is characterized by the high degree of hydrolysis, the formation of polymeric species, and the multitude of complex ions that can be formed.

Zirconium occurs naturally as a silicate in zircon, as an oxide in baddeleyite, and in other oxide compounds. Zircon is an almost ubiquitous mineral, occurring in granular limestone, gneiss, syenite, granite, sandstone, and many other minerals, albeit in small proportion, so that zircon is widely distributed in the earth's crust. The average concentration of zirconium in the earth's crust is estimated at 220 ppm, about the same abundance as barium (250 ppm) and chromium (200 ppm).

Table 1. Naturally Occurring Zirconium Isotopes

Isotope	Occurrence, %	Thermal-neutron capture cross section, 10^{-28} m^{2a}
^{90}Zr	51.45	0.03
^{91}Zr	11.32	1.14
^{92}Zr	17.49	0.21
^{94}Zr	17.28	0.055
^{96}Zr	2.76	0.020

[a] To convert m^2 to barns, multiply by 10^{28}.

Zirconium is used as a containment material for the uranium oxide fuel pellets in nuclear power reactors. Zirconium is particularly useful for this application because of its ready availability, good ductility, resistance to radiation damage, low thermal-neutron absorption cross section of 18×10^{-30} m^2 (0.18 barns), and excellent corrosion resistance in pressurized hot water up to 350°C. Zirconium is used as an alloy strengthening agent in aluminum and magnesium, and as the burning component in flash bulbs. It is employed as a corrosion-resistant metal in the chemical process industry, and is accepted as a pressure-vessel material of construction in the ASME Boiler and Pressure Vessel Codes.

Occurrence and Mining

Zircon occurs worldwide as an accessory mineral in igneous, metamorphic, and sedimentary rocks. Weathering has resulted in segregation and concentration of the heavy mineral sands in layers or lenses of placer deposits in river beds and ocean beaches. All commercial sources of zircon are derived from the mining of these ancient, unconsolidated beach deposits, the largest of which are in Kerala State in India, Sri Lanka, the east and west Coasts of Australia, on the Trail Ridge in Florida, and at Richards Bay in the Republic of South Africa.

The deposits, usually ca 4% heavy minerals, are mined with front-end loaders or sand dredges. Typically, the overburden is bulldozed away, the excavation is flooded, and the raw sand is handled by a floating sand dredge capable of dredging to a depth of 18 m. Initial wet concentration using screens, Reichert cones, spirals, and cyclones removes the coarse sand, slimes, and light-density sands to produce a 40 wt % heavy-mineral concentrate. The tailings are returned to the back end of the excavation and used for rehabilitation of worked-out areas. The concentrate is dried and iron oxide and other surface coatings are removed; various combinations of gravity separation, magnetic separation, and electrostatic separation yield individual concentrates of rutile, ilmenite, leucoxene, zircon, monazite, and xenotime.

Physical Properties

Zirconium is a hard, shiny, ductile metal, similar to stainless steel in appearance. Physical properties are given in Table 2.

Table 2. Some Physical Properties of Zirconium

Property	Value
atomic weight	91.22
density at 298.15 K, g/cm^3	6.5107
crystal structure	
αZr	
close-packed hexagonal space group	P6$_3$/mmc
a, nm	0.3231
c, nm	0.5146
c/a	1.5927
βZr	
body-centered cubic space group	Im3m
$\alpha-\beta$ transition temperature, K	1136 ± 5
melting temperature, K	2125 ± 10
boiling temperature, K	4577 ± 100
vapor pressure, $T < 2125$ K, $\log_{10} P_{kPa}$[a]	
βZr	$8.956 \pm 0.080 - (30810 \pm 240)T^{-1}$
liquid Zr	$8.547 \pm 0.080 - (29940 \pm 240)T^{-1}$
heat of transition, kJ/mol[b]	3.89 ± 0.08
heat of melting, kJ/mol[b]	18.8 ± 2.1
heat of boiling, kJ/mol[b]	573.2 ± 4.6
heat of sublimation at 298 K, kJ/mol[b]	600.8
entropy at 298.15 K,J/(mol·K)[b]	38.99 ± 0.21
thermal expansion	
single crystal	
perpendicular to c-axis, $L_{TC} = L_{0°C}$	$1 + 5.145 \times 10^{-6}T$
parallel to c-axis, $L_{TC} = L_{0°C}$	$1 + 9.213 \times 10^{-6}T - 6.385 \times 10^{-9}T^2 + 18.491 \times 10^{-12}T^3 - 9.856 \times 10^{-15}T^4$
volumetric, $V_{TC} = V_{0°C}$	$1 + 19.756 \times 10^{-6}T - 7.023 \times 10^{-9}T^2 + 19.146 \times 10^{-12}T^3 - 9.890 \times 10^{-15}T^4$
polycrystalline linear, for random orientation, $L_{TC} = L_{0°C}$	$1 + 6.499 \times 10^{-6}T - 2.096 \times 10^{-9}T^2 + 6.108 \times 10^{-12}T^3 - 3.259 \times 10^{-15}T^4$
electrical resistivity, Zr, at 5°C, Ω·cm	$43.74 \pm 0.08 \times 10^{-6}$
temperature coefficient, 0–200°C, per °C	42.5×10^{-4}
Poisson's ratio	0.33
Brinell hardness number, HB[c]	90–130

[a] To convert KPa to mm Hg, multiply by 7.5.
[b] To convert J to cal, divide by 4.184.
[c] At good Kroll-process purity, lower for iodide zirconium.

Chemical Properties

Zirconium forms anhydrous compounds in which its valence may be 1, 2, 3, or 4, but the chemistry of zirconium is characterized by the difficulty of reduction to oxidation states less than four. In aqueous systems, zirconium is always quadrivalent. It has high coordination numbers, and exhibits hydrolysis which is slow to come to equilibrium, and as a consequence, zirconium compounds in aqueous system are polymerized.

Zirconium is a highly active metal that, like aluminum, seems quite passive because of its stable, cohesive, protective oxide film which is always present in air or water.

Corrosion resistance. Zirconium is resistant to corrosion by water and steam, mineral acids, strong alkalies, organic acids, salt solutions, and molten salts.

Processing

Decomposition of Zircon. Zircon is a highly refractory mineral as shown by its geological stability; the ore is cracked only with strong reagents and high temperatures. Methods include the carbothermic reduction, caustic fusion, fluorosilicate fusion, chlorination, and thermal dissociation.

Separation of Hafnium. Zirconium and hafnium always occur together in natural minerals and, therefore, all zirconium compounds contain hafnium, usually about 2 wt% Hf/(Hf + Zr). However, the only applications that require hafnium-free material are zirconium components of water-cooled nuclear reactors. Today, the separation of zirconium and hafnium by multistage counter-current liquid-liquid extraction is routine.

Reduction. The Kroll Process is the most common method of reduction, in which zirconium tetrachloride vapor is reduced by molten magnesium in a protective argon atmosphere.

Refining. Zirconium sponge produced by the Kroll process has adequate purity and ductility for most uses. For applications requiring extremely soft metal and for research studies on the properties of the pure metal, it can be further purified by the van Arkel-de Boer (iodide-bar) process using a selective vapor transport.

Health and Safety Factors

Zirconium is generally nontoxic as an element or in compounds.

The oral toxicity is low; OSHA standards for pulmonary exposure specify a TLV of 5 mg zirconium per m^3.

Finely divided zirconium is classified as a flammable solid and shipping regulations are prescribed accordingly. Metal powder finer than 74μm (270 mesh) is limited to 2.26 kg per individual container.

Uses

The largest use is foundry sand, where zircon is used as the basic mold material, as facing material on mold cores, and in ram mixes.

Zirconium oxide is used in the production of ceramic colors or stains for ceramic tile and sanitary wares.

Zirconium oxide increases the refractive index of some optical glasses, and is used for dispersion hardening of platinum and ruthenium.

Yttria or calcia-stabilized cubic zirconias are used extensively as the solid electrolyte in oxygen sensors in automotive and boiler exhausts, and oxygen-content probes for molten copper or iron i smelters.

Zirconate compounds exhibit several interesting properties. Lead zirconate–titanate compositions display piezoelectric properties which are utilized in the production of FM-coupled mode filters, resonators in microprocessor clocks, photoflash actuators, phonograph cartridges, gas ignitors, audio tweeters and beepers, and ultrasonic transducers (see Ferrites). Lanthanum-modified lead zirconate-titanate ceramics have been studied for photoferroelectric image storage. Alkaline-earth zirconate dielectrics are used in ceramic capacitors.

The uses of other zirconium compounds result primarily from the ability of zirconium to complex with carboxyl groups to form an insoluble organic compound.

Zirconium metal is marketed in three forms: zirconium-containing silicon-manganese, iron, ferrosilicon, or magnesium master alloys; commercially-pure zirconium metal; and hafnium-free pure zirconium metal.

Silicon-manganese-zirconium, ferrozirconium, and ferrosilicon–zirconium (and some pure zirconium) are used in the steel industry for deoxidizing. Magnesium–zirconium is added to magnesium and aluminum alloys for grain refining and strengthening. Most magnesium alloys used at elevated temperatures contain zirconium.

Pure zirconium is being increasingly used as a corrosion-resistant material of construction for chemical process industry equipment.

Hafnium-free zirconium alloys containing tin or niobium are used for tubing to hold uranium oxide fuel pellets inside water-cooled nuclear reactors.

Compounds include zirconium hydride, ZrH_2, zirconium carbide, ZrC, zirconium nitride, ZrN, zirconium diboride, ZrB_2, zirconium dodecaboride, ZrB_{12}, phosphides, ZrP_3, ZrP_2, and $ZrP_{0.6}$, chalcogenides, ZrS_3, $ZrSe_3$, and $ZrTe_3$, zirconium dioxide, zirconium silicate, $ZrSiO_4$, zirconium tetrafluoride, ZrF_4, potassium hexafluorozirconate, K_2ZrF_6, zirconium tetrachloride, $ZrCl_4$, zirconium tetrabromide, $ZrBr_4$, and zirconium tetraiodide, ZrI_4, zirconium trichloride, $ZrCl_3$, tribromide, $ZrBr_3$, triiodide, ZrI_3, zirconium monochloride, ZrCl, hydrous zirconium oxide hydrate, $ZrO_2{\cdot}nH_2O$, zirconium oxide dichloride, $ZrOCl_2{\cdot}8H_2O$, anhydrous zirconium oxide chloride, $ZrOCl_2$, anhydrous zirconium tetranitrate, $Zr(NO_3)_4$, zirconium carbonate, $2\,ZrO_2{\cdot}CO_2{\cdot}xH_2O$, zirconium sulfate, $Zr(SO_4)_2{\cdot}4\,H_2O$, zirconium bis(monohydrogen phosphate), $Zr(HPO_4)_2{\cdot}H_2O$, zirconium alkoxides, $ZrX_{4-n}(OR)_n$, zirconium hydroxy carboxylates, zirconium dichlorobis(dimethylamide), $ZrCl_2[N(CH_3)_2]_2$, and zirconium arylamines.

Organometallic Compounds. Certain zirconium organometallic compounds are highly reactive toward low molecular weight unsaturated molecules. Some of these compounds are useful in various syntheses; others function effectively as catalysts for polymerization, hydrogenation, or isomerization. Compounds include hydrides, carbonyl complexes, dinitrogen complexes, alkyl and aryl complexes, and mixed-metal systems.

Catalysts. Several types of zirconium organometallic compounds are useful catalysts. In addition to the catalytic properties of the molecules, the fact that they can be bound to a relatively inert substrate increases their utility.

O. Kubaschewski, ed., *Zirconium: Physico-Chemical Properties of its Compounds and Alloys*, International Atomic Energy Agency, Vienna, 1976.

A. M. Alper, ed., *High Temperature Oxides*, Academic Press, New York, 1970.

A. Clearfield, *Inorganic Ion Exchange Materials*, CRC Press, Boca Raton, Fla., 1982.

ZONE REFINING

Zone refining is one of a class of techniques known as fractional solidification, in which a separation is brought about by crystallization of a melt without solvent being added (see also CRYSTALLIZATION). A massive solid is formed slowly and a sizable temperature gradient is imposed at the solid–liquid interface.

Zone refining can be applied to the purification of almost every type of substance that can be melted and solidified, eg, elements, organic compounds, and inorganic compounds. Because the solid-liquid phase equilibria are not favorable for all impurities, zone refining often is combined with other techniques to achieve ultrahigh purity.

The high cost of zone refining has thus far limited its application to laboratory reagents and valuable chemicals such as electronic materials. The cost arises primarily from the low processing rates, handling, and high energy consumption owing to the large temperature gradients needed.

Equipment and Techniques

Containers. The ideal container for zone melting should not contaminate the melt nor be damaged by the melt or subsequent contraction of the solid. For organic materials, borosilicate glasses are especially suitable, although metals and fluorocarbon and other polymers have also been successfully employed. For many metals and semiconductors, fused silica (often erroneously called quartz) is ideal. High melting inorganic salts and oxides are conveniently, albeit expensively, treated in noble-metal containers, platinum, rhodium, and iridium. The container may either be in the form of a tube or a horizontal boat.

Drive Mechanisms. Either the heaters or the sample may be moved. The optimal zone travel rates are typically rather slow, ie, ca 1 cm/h. Thus, electric-motor drives require gearing systems, resulting in an undesirably jerky stick-slip movement. This is avoided with double-rod supports with linear ball bearings and with low backlash gears, where tension on the moving piece pushes in the direction of motion.

Heating and Cooling. Heat must be applied to form the molten zones, and this heat much be removed from the adjacent solid material. The most common method is to place electrical resistance heaters around the container.

Heat is often removed by simply allowing it to escape by convection, radiation, and conduction. However, such uncontrolled escape can lead to very large temperature fluctuations. It is better to surround the entire container, heaters and all, with a controlled-temperature cooled chamber.

Floating-Zone Melting. No completely satisfactory container material exists for many high melting materials. In such cases, floating-zone melting may be employed. The primary application for floating-zone melting is crystal growth rather than purification.

Stirring. As noted earlier, stirring increases the optimal zone-travel rate by lowering the film thickness and aids significantly in removal of foreign particles.

WILLIAM R. WILCOX
Clarkson College of Technology

W. G. Pfann, *Zone Melting*, 2nd ed., John Wiley & Sons, Inc., New York, 1966.

J. C. Brice, *The Growth of Crystals from Liquids*, North-Holland Publishing Co., Amsterdam, the Netherlands, 1973.

J. S. Shah in B. R. Pamplin, ed., *Crystal Growth*, Pergamon, Oxford, U.K., 1975, Chapt. 4.

ZYMURGY. See BEER; FERMENTATION; WINE.

INDEX